国际材料前沿丛书
International Materials Frontier Series

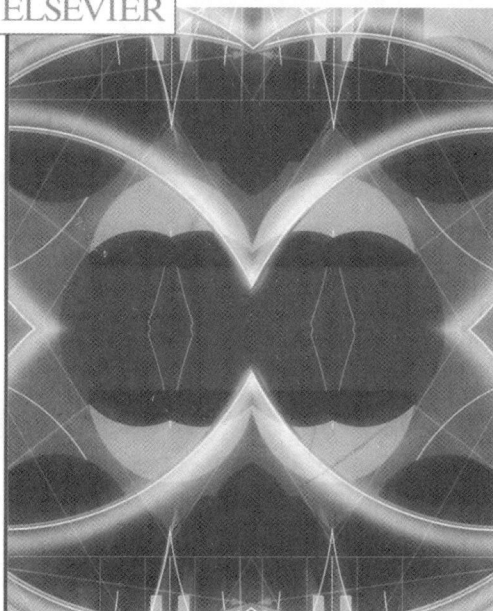

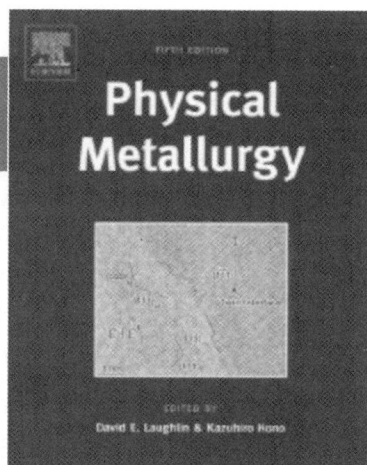

Physical Metallurgy

FIFTH EDITION

EDITED BY
David E. Laughlin & Kazuhiro Hono

David E. Laughlin

Kazuhiro Hono

物理冶金学（第5版）
Physical Metallurgy (Fifth Edition)

（上）
影印版

中南大学出版社
www.csupress.com.cn
·长沙·

图字:18 - 2017 - 166 号

Physical Metallurgy(Fifth Edition)

David E. Laughlin, Kazuhiro Hono

ISBN: 9780444537706

Elsevier (Singapore) Pte Ltd.

3 Killiney Road

#08 - 01 Winsland House I

Singapore 239519

Tel: (65) 6349 - 0200

Fax: (65) 6733 - 1817

First Published <2017>

<2017>年初版

内容简介

　　该书全面系统地涵盖了材料科学领域中的相关知识,深刻地描述和解释了物理冶金学中的大多数方法,其中轻合金物理冶金、钛合金物理冶金、原子探针场离子显微镜、计算冶金和取向成像显微镜等都是当前科技发展的前沿。全书分为 3 册(上、中、下):上册包含第 1 章 ~ 第 9 章的内容,中册包含第 10 章 ~ 第 19 章的内容,下册包含第 20 章 ~ 第 27 章的内容。

　　该书为材料类的经典书籍,第 5 版在第 4 版(1996 年)的基础上做了全面修改和扩充,且新增了几个主题以反映过去 18 年来物理冶金学的新进展。

　　本书可供相关学科领域(冶金、材料、物理、化学、生物医学等)的科研人员、工程技术人员使用,也可作为本科生、研究生等的参考用书。

作者简介

David E. Laughlin　美国匹兹堡卡内基梅隆大学教授，TMS 和 ASM International 的会士。他的主要研究方向为材料结构的电子显微镜观察、相变和磁性材料。1969 年获得美国德雷塞尔大学冶金工程专业学士学位，1973 年获得美国麻省理工学院冶金和材料科学专业博士学位。他从 1974 年起在美国卡内基梅隆大学任教。他发表了超过 450 篇同行评审的研究论文，并且共有 11 项美国专利。

Kazuhiro Hono　日本国立材料研究所研究员，磁性材料部主任，筑波大学教授。主要研究方向是金属材料，特别是磁性材料微观结构与性能的关系。1982 年获得日本东北大学学士学位，1988 年获得美国宾夕法尼亚州立大学材料科学与工程博士学位。

序

    这套 3 卷本的《物理冶金学》，是剑桥大学 Robert W. Cahn 教授和哥廷根大学 Peter Haasen 教授的享有崇高声望的同类著作的第 5 版。《物理冶金学》的第 1 版于 1965 年以单卷本的形式出版。关于本系列出版物的历史请参见第 4 版《序》。本系列权威参考工具书提供了物理冶金学——材料科学与工程领域的最大学科的全面知识。本系列著作广泛而深刻地描述和解释了物理冶金学的大多数方法。书中的每篇文章或由新作者重写，或由第 4 版作者单独或联合新的合作者全面修改和扩充。

    在《物理冶金学》第 1 版的《序》中，主编之一 R. W. Cahn 教授说："物理冶金学是现代材料科学赖以蓬勃发展的根源。"（R. W. Cahn，1965，《物理冶金学》，北荷兰出版公司，阿姆斯特丹 – 伦敦）。50 年过去了，这一说法仍然正确。本版的两位主编均是作为物理冶金学家培养的，但我们通常称自己为材料科学家。事实上，我们各自所在的部门并不使用"冶金学"一词。但材料科学的核心概念（有时称为"理论框架"），即材料的性能取决于其加工工艺和由此产生的微观结构，直接来源于物理冶金学和加工冶金学。若想了解材料科学的详尽历史，请查看 R. W. Cahn 主编的《材料科学的未来》，并与 R. F. Mehl 主编的《金属科学简史》进行对比。

    在本系列著作的第 1 版和第 2 版中，R. F. Mehl 在《物理冶金学的发展历史》一文中写道："物理冶金学和提取冶金学交织在一起。从历史的角度看，在很长一段时间内，冶金学的两个分枝构成的一门统一的艺术，并由同样的艺术家实现。"早在 2014 年面世的 3 卷本《冶金学论丛》，由 Seshadri Seetharaman 主编，同样由 ELSEVIER 出版，是本系列著作的同类出版物。这两套共 6 卷的著作无疑地全面覆盖了冶金学。

    《物理冶金学》第 5 版是继第 4 版（1996 年）大约 18 年之后出版的。2007 年，创始主编不幸辞世，延缓了第 5 版的面世。最后，由我们负责第 5 版的出版工作。

    本版作者具有更多的国际元素。21 世纪通信的便捷确实使编辑工作变得容易。然而，这似乎并没有加快所有作者写作的进程！

    本版增加了几个主题，以反映过去 18 年来物理冶金学的最新进展。部分章节由第 4 版的相同作者撰写，但均进行了更新，涵盖了新的论题和方法。我们感谢 45 名作者，他们各自勤奋地完成了有关章节的撰写和校对工作。这套著作获得的声誉当然应主要归功于作者而不是现任主编。当今，在我们的研究机构中，"精打细算的人"并不总是认可撰写类

似套书中的某一章节这样的工作。因此我们对每一位作者心存感激,他们将个人利益置之度外,详细报道他们精通的研究工作,而不在意其是否被 Web of Science 收录。

1970 年,本套著作主编之一(DEL)参加博士生入学资格考试,使用的教材就是《物理冶金学》的第 1 版。他从来不曾想过,近 50 年后他会亲自主编这套深受好评的《物理冶金学》著作。

我们愿将这 3 卷的著作献给我们的前任主编 Robert W. Cahn 教授和 Peter Haasen 教授。我们相信,我们在续写这套专著中付出的努力,可以达到他们曾经制定的高标准。

David E. Laughlin
Kazuhiro Hono

目　录

PHYSICAL METALLURGY

VOLUME I

FIFTH EDITION

PHYSICAL METALLURGY

VOLUME I

FIFTH EDITION

EDITED BY

DAVID E. LAUGHLIN

ALCOA Professor of Physical Metallurgy
Materials Science and Engineering
Carnegie Mellon University
Pittsburgh, PA, USA

KAZUHIRO HONO

Magnetic Materials Unit
National Institute for Materials Science
Tsukuba-city Ibaraki, Japan

ELSEVIER

AMSTERDAM • WALTHAM • HEIDELBERG • LONDON • NEW YORK
OXFORD • PARIS • SAN DIEGO • SAN FRANCISCO • SYDNEY • TOKYO

Elsevier
Radarweg 29, PO Box 211, 1000 AE Amsterdam, Netherlands
The Boulevard, Langford Lane, Kidlington, Oxford OX5 1GB, UK
225 Wyman Street, Waltham, MA 02451, USA

Fifth edition

Notice
No responsibility is assumed by the publisher for any injury and/or damage to persons or property as a matter of products liability, negligence or otherwise, or from any use or operation of any methods, products, instructions or ideas contained in the material herein. Because of rapid advances in the medical sciences, in particular, independent verification of diagnoses and drug dosages should be made.

British Library Cataloguing in Publication Data
A catalogue record for this book is available from the British Library

Library of Congress Cataloging in Publication Data
A catalog record for this book is available from the Library of Congress

Volume I ISBN: 978-0-444-59598-0
SET ISBN: 978-0-444-53770-6

For information on all Elsevier publications
visit our website at www.store.elsevier.com

Printed and bound in the UK

Working together
to grow libraries in
developing countries

www.elsevier.com • www.bookaid.org

CONTENTS

LIST OF CONTRIBUTORS TO VOLUME I

Zoltan Balogh
Institute of Materials Physics, University of Münster, Münster, Germany

Dilip K. Banerjee
Materials Science and Engineering Division, Materials Measurement Laboratory, NIST, Gaithersburg, MD, USA

H.K.D.H. Bhadeshia
University of Cambridge, UK

William J. Boettinger
Materials Science and Engineering Division, Materials Measurement Laboratory, NIST, Gaithersburg, MD, USA

A.L. Greer
Department of Materials Science & Metallurgy, University of Cambridge, Cambridge, UK

NUMAKURA Hiroshi
Department of Materials Science, Osaka Prefecture University, Naka-ku, Sakai, Japan

M. Inukai
Toyota Technological Institute, Nagoya, Japan

David E. Laughlin
Department of Materials Science and Engineering, Carnegie Mellon University, Pittsburgh, PA, USA

U. Mizutani
Nagoya Industrial Science Research Institute, Nagoya, Japan

Arthur D. Pelton
Centre de Recherche en Calcul Thermochimique (CRCT), École Polytechnique de Montréal, Montréal, Canada

H. Sato
Aichi University of Education, Kariya-shi, Aichi, Japan

Guido Schmitz
Institute of Materials Physics, University of Münster, Münster, Germany

W.A. Soffa
Department of Materials Science and Engineering, University of Virginia, USA

Walter Steurer
Laboratory of Crystallography, ETH Zurich, Zurich, Switzerland

C.M. Wayman
Deceased but originally from University of Illinois at Urbana-Champaign, USA

E.S. Zijlstra
Theoretical Physics, University of Kassel, Kassel, Germany

PREFACE TO THE FIFTH EDITION

These three volumes represent the fifth edition of *Physical Metallurgy*, a prestigious and famous family formerly edited by Robert Cahn (University of Cambridge) and Peter Haasen (Universität Göttingen). Physical Metallurgy was first published as a single volume in 1965. See the preface to the fourth edition for a history of this series. It is an authoritative reference tool, providing a complete knowledge set in Physical Metallurgy, the largest discipline in the fields of Materials Science and Engineering. This series describes and explains most aspects of physical metallurgy across the full breadth and in considerable depth. Each article has been either rewritten by new authors, or thoroughly revised and expanded, either by the 4th edition authors alone or jointly with new co-authors.

In the preface to the first edition of Physical Metallurgy, the founding editor of this series stated that "Physical metallurgy is the root from which the modern science of materials has principally sprung." (R. W. Cahn (1965), Physical Metallurgy North Holland Publishing Company, Amsterdam-London). Over the next five decades this has continued to ring true. While both of the editors of this edition were educated as physical metallurgists, nowadays it is more common to call ourselves Materials Scientists, and indeed our respective departments do not utilize the word "metallurgy." But the core concept (or sometimes called its paradigm) of Materials Science, that the properties and performance of materials have their origin in their processing and resulting microstructure, is derived directly from Physical and Process Metallurgy. For an exhaustive history of Materials Science see "The Coming of Materials Science" by R. W. Cahn and compare this to "A Brief History of the Science of Metals," by R. F. Mehl.

In the article "The Historical Development of Physical Metallurgy" by R. F. Mehl, which appeared in the first two editions of this series, Mehl wrote: "*Physical Metallurgy* has been ... interwoven with *Extractive Metallurgy*, and for a long time, historically, these two branches constituted a common art, practiced by the same artisans..." Early in 2014 there appeared the three volume *Treatise on Process Metallurgy*, edited by Seshadri Seetharaman and also published by ELSEVIER, which may be said to be the companion to this set. These six volumes certainly cover Metallurgy comprehensively.

This fifth edition of Physical Metallurgy is published some eighteen years after the 4th edition of this series, which was published in 1996. The lamented death of the founding editor in 2007 slowed down the appearance of this 5th edition. Finally we are ready to present the 5th edition.

This edition has a more international flavor to the listing of authors. Indeed in this 21st Century the ease with which correspondence can be sent makes the task of editing easier. It does not seem to speed up the writing and response of all authors however!

Several new subjects were added in this edition to update the progress in *physical metallurgy* in the last 18 years. Several of the chapters are written by the same authors as those in the fourth edition; but they have all been updated to include new topics and approaches.

We do thank the 45 authors for their hard work and diligence to get their chapters and proofs in. It is of course to authors, more than the current editors, that a series such as this gets its reputation. In a day when "bean counters" in our institutions do not always appreciate the work that goes into writing a chapter in a series such as this, we are grateful that each of the authors put that aside and wrote up work about which they are experts, whether or not it gets indexed in the Web of Science!

In 1970 one of the editors of these volumes (DEL) studied for his qualifying examinations for entrance into the Ph.D. program from the first edition of the series of *Physical Metallurgy*. Little did he suspect that nearly five decades later he would be editing the 5th edition to this well received series on *Physical Metallurgy*.

We wish to dedicate these volumes to our predecessor editors: Prof. Robert W. Cahn and Prof. Peter Haasen. We trust that our efforts in the continuation of this series will be up to the high standards which they have set.

David E. Laughlin
ALCOA Professor of Physical, Metallurgy,
Department of Materials Science and Engineering,
Carnegie Mellon University, Pittsburgh, PA USA

Kazuhiro Hono
ZNIMS Fellow, Naitonal Institute for
Materials Science, Tsukuba, 305-0047, Japan

PREFACE TO THE FOURTH EDITION

The first, single-volume edition of this Work was published in 1965 and the second in 1970; continued demand prompted a third edition in two volumes which appeared in 1983. The first two editions were edited by myself alone, but in preparing the third, which was much longer and more complex, I had the crucial help of Peter Haasen as co-editor. The third edition came out in 1983, and sold steadily, so that the publishers were motivated to propose the preparation of yet another version of the Work; we began the joint planning for this in early 1992. We agreed on the changes and additions we wished to make: the responsibility for commissioning chapters was divided equally between us, but the many policy decisions, made during a series of face-to-face discussions, were very much a joint enterprise. Peter Haasen was able to commission all the chapters which he had agreed to handle, and this task (which involved detailed discussions with a number of authors) was completed in early 1993. Thereupon, in May 1993, my friend of many years was suddenly taken ill; the illness worsened rapidly, and in October of the same year he died, at the early age of 66. When he was already suffering the ravages of his fatal illness, he yet found the resolve and energy to revise his own chapter and to send it to me for comments, and to modify it further in the light of those comments. He was also able to examine, edit and approve the revised chapter on dislocations, which came in early. These were the very last professional tasks he performed. Peter Haasen was in every sense co-editor of this new edition, even though fate decreed that I had to complete the editing and approval of most of the Chapter I am proud to share the title-page with such an eminent physicist.

The first edition had 22 chapters and the second, 23. There were 31 chapters in the third edition and the present edition has 32. The first two editions were single volumes, the third had to be divided into two volumes, and now the further expansion of the text has made it necessary to go to three volumes. This fourth edition is nearly three times the size of the first edition 30 years ago; this is due not only to the addition of new topics, but also to the fact that the treatment of existing topics has become much more substantial than it was in 1965. There are those who express the conviction that physical metallurgy has passed its apogee and is in steady decline; the experience of editing this edition, and the problems I have encountered in holding enthusiastic authors back from even more lengthy treatments (to avoid exceeding the agreed page limits by a wholly unacceptable margin), have shown me how mistaken this pessimistic assessment is! Physical metallurgy, the parent discipline of materials science, has maintained its central status undiminished.

The first three editions each opened with a historical overview. We decided to omit this in the fourth edition, for two main reasons: the original author had died and it would have fallen to others to revise his work, never an entirely satisfactory proceeding; it had also become plain (especially from the reaction of the translators of the earlier editions into Russian) that the overview was not well balanced

between different parts of the world. I am engaged in writing a history of materials science, as a separate venture, and this will incorporate proper attention to the history of physical metallurgy as a principal constituent. — It also proved necessary to leave out the chapter on superconducting alloys: the ceramic superconductor revolution has virtually removed this whole field from the purview of physical metallurgy. — Three entirely new topics are treated in this edition: one is oxidation, hot (dry) corrosion and protection of metallic materials, another is the dislocation theory of the mechanical behavior of intermetallic compounds. The third new topic is a leap into very unfamiliar territory: it is entitled "A Metallurgist's Guide to Polymers". Many metallurgists — including Alan Windle, the author of this chapter — have converted in the course of their careers to the study of the more physical aspects of polymers (regarded by many materials scientists as *the* "materials of the future"), and have had to come to terms with novel concepts (such as "semicrystallinity") which they had not encountered in metals: Windle's chapter is devoted to analysing in some depth the conceptual differences between metallurgy and polymer science, for instance, the quite different principles which govern alloy formation in the two classes of materials. I believe that this is the first treatment of this kind.

Six of the existing chapters (now numbered 1, 4, 21, 22, 27, 30) have been entrusted to new authors, while another five chapters have been revised by the previous authors with the collaboration of additional authors (8,13,16, 17, 19). Chapter 19, originally entitled "Alloys rapidly quenched from the melt" has been broadened and retitled "Metastable states of alloys". A treatment of quasicrystals has been introduced in the form of an appendix to Chapter 4, which is devoted to the solid-state chemistry of intermetallic compounds; this seemed appropriate since quasicrystallinity is generally found in such compounds. — Only three chapters still have the same authors they had in the first edition, written some 32 years ago.

27 of the 29 new versions of existing chapters have been substantially revised, and many have been entirely recast. Two Chapters (11 and 25) have been reprinted as they were in the third edition, except for corrected cross-references to other chapters, but revision has been incorporated in the form of an Addendum to each of these chapters; this procedure was necessary on grounds of timing.

This edition has been written by a total of 44 authors, working in nine countries. It is a truly international effort.

I have prepared the subject index and am thus responsible for any inadequacies that may be found in it. I have also inserted some cross-references between chapters (internal cross- references within chapters are the responsibility of the various authors), but the function of such cross-references is better achieved by liberal use of the subject index.

As always, the editors have been well served by the exceedingly competent staff of North—Holland Physics Publishing (which is now an imprint of Elsevier Science B.V. in Amsterdam; at the time of the first two editions, North—Holland was still an independent company). My particular thanks go to Nanning van der Hoop and Michiel Bom on the administrative side, to Ruud de Boer who is responsible for production and to Chris Ryan and Maurine Alma who are charged with marketing. Mr. de Boer's care and devotion in getting the proofs just right have been extremely impressive. My special thanks also go to Professor Colin Humphreys, head of the department of materials science and metallurgy in Cambridge University, whose warm welcome and support for me in my retirement made the creation of this edition feasible. Finally, my thanks go to all the authors, who put up with good grace with the numerous forceful, sometimes impatient, messages which I was obliged to send in order to "get the show on the road", and produced such outstanding chapters under pressure of time.

I am grateful to Dr. W.J. Boettinger, one of the authors, and his colleague Dr. James A. Warren, for kindly providing the computer-generated dendrite microstructure that features on the dust-cover.

The third edition was dedicated to the memory of Robert Franklin Mehl, the author of the historical chapter and a famed innovator in the early days of physical metallurgy in America. I would like to dedicate this fourth edition to the memory of two people: my late father-in- law, *Daniel Hanson* (1892–1953), professor of metallurgy at Birmingham University for many years, who did more than any other academic in Britain to foster the development and teaching of modern physical metallurgy; and the physical metallurgist and scientific publisher — and effective founder of Pergamon Press — *Paul Rosbaud* (1896–1963), who was retained by the then proprietor of the North–Holland Publishing Company as an adviser and in 1960, in the presence of the proprietor, eloquently urged upon me the need for a new, advanced, multiauthor text on physical metallurgy.

Robert W. Cahn
Cambridge
November 1995.

PREFACE TO THE THIRD EDITION

The first edition of this book was published in 1965 and the second in 1970. The book continued to sell well during the 1970s and, once it was out of print, pressure developed for a new edition to be prepared. The subject had grown greatly during the 1970s and R. W. C. hesitated to undertake the task alone. He is immensely grateful to P. H. for converting into a pleasure what would otherwise have been an intolerable burden!

The second edition contained 22 chapters. In the present edition, 8 of these 22 have been thoroughly revised by the same authors as before, while the others have been entrusted to new contributors, some being divided into pairs of chapters. In addition, seven chapters have been commissioned on new themes. The difficult decision was taken to leave out the chapter on superpure metals and to replace it by one focused on solute segregation to interfaces and surfaces—a topic that has made major strides during the past decade and which is of great practical significance. A name index has also been added.

Research in physical metallurgy has become worldwide and this is reflected in the fact that the contributors to this edition live in no fewer than seven countries. We are proud to have been able to edit a truly international text, both of us having worked in several countries ourselves. We would like here to express our thanks to all our contributors for their hard and effective work, their promptness and their angelic patience with editorial pressures!

The length of the book has inevitably increased, by 50% over the second edition, which was itself 20% longer than the first edition. Even to contain the increase within these numbers has entailed draconian limitations and difficult choices; these were unavoidable if the book was not to be priced out of its market. Everything possible has been done by the editors and the publisher to keep the price to a minimum (to enable readers to take the advice of G. Chr. Lichtenberg (1775): "He who has two pairs of trousers should pawn one and buy this book".).

Two kinds of chapters have been allowed priority in allocating space: those covering very active fields and those concerned with the most basic topics such as phase transformations, including solidification (a central theme of physical metallurgy), defects, and diffusion. Also, this time we have devoted more space to experimental methods and their underlying principles, microscopy in particular. Since there is a plethora of texts available on the standard aspects of X-ray diffraction, the chapter on X-ray and neutron scattering has been designed to emphasize less familiar aspects. Because of space limitations, we regretfully decided that we could not include a chapter on corrosion.

This revised and enlarged edition can properly be regarded as to all intents and purposes a new book.

Sometimes it was difficult to draw a sharp dividing line between physical metallurgy and process metallurgy, but we have done our best to observe the distinction and to restrict the book to its intended

theme. Again, reference is inevitably made occasionally to nonmetallics, especially when they serve as model materials for metallic systems.

As before, the book is designed primarily for graduate students beginning research or undertaking advanced courses, and as a basis for more experienced research workers who require an overview of fields comparatively new to them, or with which they wish to renew contact after a gap of some years.

We should like to thank Ir J. Soutberg and Dr A. P. de Ruiter of the North-Holland Publishing Company for their major editorial and administrative contributions to the production of this edition, and in particular we acknowledge the good-humored resolve of Dr W. H. Wimmers, former managing director of the Company, to bring this third edition to fruition. We are grateful to Dr Bormann for preparing the subject index. We thank the hundreds of research workers who kindly gave permission for reproduction of their published illustrations: all are acknowledged in the figure captions.

Of the authors who contributed to the first edition, one is no longer alive: Robert Franklin Mehl, who wrote the introductory historical chapter. What he wrote has been left untouched in the present edition, but one of us has written a short supplement to bring the treatment up to date, and has updated the bibliography. Robert Mehl was one of the founders of the modern science of physical metallurgy, both through his direct scientific contributions and through his leadership and encouragement of many eminent metallurgists who at one time worked with him. We dedicate this third edition to his memory.

Robert W. Cahn, Paris
Peter Haasen, Göttingen.
April 1983.

PREFACE TO THE FIRST AND SECOND EDITIONS

This book sets forth in detail the present state of physical metallurgy, which is the root from which the modern science of materials has principally sprung. That science has burgeoned to such a degree that no author can do justice to it at an advanced level; accordingly, a number of well-known specialists have consented to write on the various principal branches, and the editor has been responsible for preserving a basic unity among the expert contributions. This book is the first general text, as distinct from research symposium, which has been conceived in this manner. While principally directed at senior under-graduates at universities and colleges of technology, the book is therefore also appropriate for post-graduates and particularly as a base for experienced research workers entering fields of physical metallurgy new to them.

Certain topics have been left to one side or treated at modest length, so as to limit the size of the book, but special stress has been placed on others, which have rarely been accorded much space. For instance, a good deal of space is devoted to the history of physical metallurgy, and to point defects, structure and mechanical properties of solid solutions, theory of phase transformations, recrystalliza-tion, superpure metals, ferromagnetic properties, and mechanical properties of two-phase alloys. These are all active fields of research. Experimental techniques, in particular diffraction methods, have been omitted for lack of space; these have been ably surveyed in a number of recent texts. An exception has however been made in favor of metallographic techniques since, electron microscopy apart, recent innovations have not been sufficiently treated in texts.

Each chapter is provided with a select list of books and reviews, which will enable readers to delve further into a particular subject. Internal cross-references and the general index will help to tie the various contributions together.

I should like here to acknowledge the sustained helpfulness and courtesy of the publisher's staff, and in particular of Mr A. T. G. van der Leij, and also the help provided by Prof P. Haasen and Dr T. B. Massalski in harmonizing several contributions.

Brighton, June 1965 (and again 1970) R. W. Cahn

ABOUT THE EDITORS

David E. Laughlin is the ALCOA Professor of Physical Metallurgy in the Department of Materials Science and Engineering at Carnegie Mellon University, Pittsburgh, PA. He obtained his B.S. in Metallurgical Engineering from Drexel University in 1969 and his Ph.D. in Metallurgy and Materials Science from MIT in 1973. He has taught at CMU since 1974. He is Principal Editor of *Metallurgical and Materials Transactions* and has co-edited eight books. His research has centered on the structure of materials as observed by electron microscopy, phase transformations and magnetic materials. He has published more than 450 peer reviewed research papers and is co-inventor on eleven US patents. Laughlin is a Fellow of TMS and ASM International.

Kazuhiro Hono is NIMS Fellow and Director of Magnetic Materials Unit at National Institute for Materials Science and Professor of Materials Science at the University of Tsukuba. He obtained his B.S. from Tohoku University in 1982, and Ph.D. in Materials Science and Engineering from the Pennsylvania State University in 1988. His research interests are mirostructure-property relationships of metallic materials, in particular of magnetic materials.

1 Crystal Structures of Metallic Elements and Compounds

Walter Steurer, Laboratory of Crystallography, ETH Zurich, Zurich, Switzerland

Glossary

Allotrope Modification of an element at a given temperature and pressure range.

Alloy A homogenous mixture of two or more (inter)metallic phases that may form a solid solution or remain phase separated.

Aperiodic crystal The signature of ideal aperiodic crystals is that they show a pure-point Fourier spectrum (Bragg diffraction pattern) as periodic crystals do but have no translationally periodic structures. Examples are incommensurately modulated structures, composite (host/guest) structures and quasicrystals.

Atomic environment type (AET) Local coordination of an atom by the surrounding atoms (first coordination polyhedron).

Cluster Polyhedral arrangement of atoms in several cluster shells.

Crystal structure It is defined by its chemical composition, metrics (lattice parameters), space group symmetry, equipoint (Wyckoff) positions occupied by the different types of atoms (crystallographic orbits) and their specific coordinates in one asymmetric unit. The metrics may differ for all chemical compounds or phases adopting one particular structure type.

Intermetallic compound Special case of an intermetallic phase with sharply defined composition ("*line compound*").

Intermetallic phase A phase, constituted by two or more metallic elements, which may have a narrow or an extended compositional stability range.

Intermetallics Short form for intermetallic phase.

Pearson notation Symbol for the shorthand characterization of crystal structures in combination with the structure type. It consists of a lower-case letter denoting the crystal system, an upper-case letter giving Bravais lattice type and the number of atoms per unit cell.

Quasicrystal Intermetallic phase with noncrystallographic symmetry and quasiperiodic structure. The Fourier module of a quasiperiodic structure is of rank $n > d$, with d, the dimension of the quasiperiodic structure in physical space, and n, the number of basis vectors spanning the Fourier module.

Solid solution Solution of one or more elements in the structure of another element or of an intermetallic phase by substituting atoms without changing the structure type.

Sphere packing Infinite set of noninterpenetrating hard spheres with the property that any pair of spheres is connected by a chain of spheres with mutual contact. A sphere packing is called *homogenous* for all spheres being symmetrically equivalent, otherwise it is called *heterogenous*.

Structure type Crystal structure representing a whole set of similar crystal structures, i.e. structures with the same space group symmetry and atomic environment types (AET).

Superspace group The symmetry of modulated structures can be described by superspace groups. The symbol describing a $(3 + 1)$-dimensional superspace group, for instance, consists of the space group of the basic structure and the components of the modulation wave vector. For instance, the

superspace group $I4/\mathrm{mcm}(00\gamma)$ refers to a tetragonal basic structure that is periodically distorted by a modulation wave along the [001] direction, i.e. along the fourfold axis.

Tiling A tessellation of the plane or the space where the unit tiles (copies of prototiles) fill the plane or space without gaps or overlaps.

$\mathbb{Z}$-**module** A vector module of rank n is an infinite set of vectors resulting from all possible linear combinations of its n basis vectors. It is also called $\mathbb{Z}$-module to emphasize that the coefficients of all linear vector combinations are integers.

1.1   Introduction

Intelligent materials design depends on a deep understanding of the relationships between chemical composition, crystal structure and physical properties. Particularly, in case of multiphase materials, the microstructure, which strongly depends on processing parameters, can drastically modify physical properties provided by individual components. One also has to keep in mind that structure and properties of materials on the nanoscale can strongly differ from those on the macroscale due to size effects. However, in the present chapter, we will exclusively discuss ideal structures of macroscopic single-crystalline elements and compounds in thermodynamic equilibrium and, consequently, will not have to deal with any size- or processing-related topics.

Although only a very few technologically and commercially important materials consist of metallic elements in their (almost) pure form such as Cu, Au, Ag, Pd, Pt, etc., their crystal structures are of more than academic interest. Just to give an example, the crystal structure of a pure metal remains unchanged if it forms a solid solution (single-phase alloy) when one or several other elements are added for tuning or modifying its properties. This technique has been used since time immemorial by alloying gold with copper or silver, for instance, to make jewelry or coins more mechanically resistant. Especially, closest packed structures and their derivatives, typical for many metallic elements, are also characteristic for numerous materials consisting of multicomponent solid solutions.

Even though elements are chemically simpler model systems than intermetallic phases, their structures can be quite complex, mainly originating from intricate electronic interactions. Most elements are polymorphous, i.e. they adopt several different crystal structures as a function of ambient conditions (temperature, pressure, ...). The understanding of the phase transformations in these homoatomic cases is not only of interest on its own but also very helpful for understanding in general why a phase with given chemical composition is adopting a particular crystal structure under given conditions.

The structural chemistry and crystallography of intermetallic phases is incredibly rich. More than 2000 different structure types are known, with the number of atoms per unit cell ranging between one and more than 20,000; in the case of aperiodic crystals such as incommensurate structures, composite crystals or quasicrystals (QCs), three-dimensional (3D) unit cells even do not exist at all. However, only a comparably small number of intermetallics have found so far important applications since their physical properties have rarely been studied in the past. Well known is the usage of some intermetallics as strengthening phases in precipitation hardening systems such as $LiAl_3$ in Al–Li alloys or Ni_3Al in superalloys (both of the $cP4$-$AuCu_3$ type). There are also several intermetallic phases known that have important applications as functional materials such as $cP8$-Cr_3Si (A15)-type structures as superconductors (Nb_3Sn), $cF24$-Cu_2Mg-type Laves phases as magnetostrictive materials (Terfenol D) or $hP6$-$CaCu_5$-type structures as powerful permanent magnets ($SmCo_5$).

Neither is it possible nor would it make sense to discuss all structure types of intermetallic phases in a similar way as we do for the metallic elements; their crystallographic data can be found in *Pearson's Handbook of Crystallographic Data for Intermetallic Phases* (Villars and Calvert, 1991) or in the database *Pearson's Crystal Data* (Villars and Cenzual, 2011), anyway. Besides reviewing the most frequent structure types as well as those of technologically and commercially important materials, we will illustrate the major structural building principles on typical examples and underline interesting structural interrelationships wherever existing.

The visualization of a crystal structure usually aims at elucidating the structural building principles in as general as possible way. In case of intermetallic phases, contrary to ionic or molecular crystals, the chosen visualization is not always reflecting the sometimes very complex underlying crystal-chemical reality. However, it provides a very useful construction kit that can make clear complex packing principles or structural relationships, particularly, if different visualization tools are applied to one and the same structure:

- Ball-and-stick model for highlighting atomic connectivity.
- Space-filling atomic representation for illustrating the packing density.
- Polyhedra model for depicting atomic environment types (AET) and their connectivity.
- Cluster model for visualizing the packing of larger structural subunits.
- Layer decomposition for demonstrating the modularity.
- Framework representation to emphasize the role of substructures.

Some comments to our terminology: by *intermetallic compound*, we mean a phase, constituted by two or more metallic elements, with sharply defined composition ("*line compound*"), while the compositional stability range of an *intermetallic phase* in general may be extended; a *solid solution* is formed if an element or an intermetallic phase can dissolve one or more elements that are not constituents of the base element or intermetallic phase; by *alloy*, we denote a mixture of two or more (inter)metallic phases that remain phase separated; the short form "*intermetallics*" will be used for both intermetallic compounds and intermetallic phases. The arrangement of elements in the chemical formula of a compound, $A_xB_yC_z...$, follows increasing Mendeleev numbers (**Table 1**) $M_A < M_B < M_C < ...$ These numbers are related to a chemical scale χ introduced by Pettifor (1988 and references therein) mainly for the derivation of structure maps.

For the shorthand characterization of crystal structures, the *Pearson notation* in combination with the prototype formula defining the structure type is used throughout this chapter. In accordance with the IUPAC recommendations (Leigh, 1990), the old *Strukturbericht* designation (A3 for $hP2$-Mg, for instance) should not be used any longer. A comparison of the Pearson notation, prototype formula, space group and Strukturbericht designation for a large number of crystal structure types is given in Massalski (1990), for instance.

The Pearson symbol consists of two italic letters and a number. The first (lower case) letter denotes the crystal family and the second (upper case) letter, the Bravais lattice type (**Table 2**). The symbol is completed by the number of atoms in the unit cell. The symbol $cF4$, for instance, classifies a structure type to be cubic (c), all-face centered (F) with four atoms per unit cell. In the case of rhombohedral structures such as the $hR3$-Sm type, for instance, the number of atoms per unit cell in the rhombohedral setting ($a = b = c$, $\alpha = \beta = \gamma \neq 90°$) is given. This notation is used by Villars and Calvert (1991), Villars

Table 1   Periodic table of the elements. The structures of the elements below the thick line and with $Z \leq 103$ are discussed in this chapter. Ge, Sb and Po are included, although usually considered as metalloids or semimetals, because they can form compounds with metallic character. Above each element the atomic number Z is given, below the Mendeleev number M (Pettifor, 1988) is listed

1	2	3	4	5	6	7	8	9	10	11	12	13	14	15	16	17	18
1 H 103																	2 He 1
3 Li 12	4 Be 77											5 B 86	6 C 95	7 N 100	8 O 101	9 F 102	10 Ne 2
11 Na 11	12 Mg 73											13 Al 80	14 Si 85	15 P 90	16 S 94	17 Cl 99	18 Ar 3
19 K 10	20 Ca 16	21 Sc 19	22 Ti 51	23 V 54	24 Cr 57	25 Mn 60	26 Fe 61	27 Co 64	28 Ni 67	29 Cu 72	30 Zn 76	31 Ga 81	32 Ge 84	33 As 89	34 Se 93	35 Br 98	36 Kr 4
37 Rb 9	38 Sr 15	39 Y 25	40 Zr 49	41 Nb 53	42 Mo 56	43 Tc 59	44 Ru 62	45 Rh 65	46 Pd 69	47 Ag 71	48 Cd 75	49 In 79	50 Sn 83	51 Sb 88	52 Te 92	53 I 97	54 Xe 5
55 Cs 8	56 Ba 14	57* La 33	72 Hf 50	73 Ta 52	74 W 55	75 Re 58	76 Os 63	77 Ir 66	78 Pt 68	79 Au 70	80 Hg 74	81 Tl 78	82 Pb 82	83 Bi 87	84 Po 91	85 At 96	86 Rn 6
87 Fr 7	88 Ra 13	89+ Ac 48	104 Rf	105 Db	106 Sg	107 Bh	108 Hs	109 Mt	110 Ds	111 Rg							

* Lanthanoids	58 Ce 32	59 Pr 31	60 Nd 30	61 Pm 29	62 Sm 28	63 Eu 18	64 Gd 27	65 Tb 26	66 Dy 24	67 Ho 23	68 Er 22	69 Tm 21	70 Yb 17	71 Lu 20
+ Actinoids	90 Th 47	91 Pa 46	92 U 45	93 Np 44	94 Pu 43	95 Am 42	96 Cm 41	97 Bk 40	98 Cf 39	99 Es 38	100 Fm 37	101 Md 36	102 No 35	103 Lr 34

Table 2   Meaning of the letters included in the Pearson Symbol. In case of symbols of the type *hRn*, we use the symbol in the same way as it is done by Villars and Calvert (1991): *n* means the number of atoms in the rhombohedral unit cell, i.e., 1/3 of that in the hexagonal setting

	Crystal family		Bravais lattice type
a	triclinic (anorthic)	P	primitive
m	monoclinic	I	body centered
o	orthorhombic	F	all-face centered
t	tetragonal	S, C	side- or base-face centered
h	hexagonal, trigonal (rhombohedral)	R	rhombohedral
c	cubic		

and Cenzual (2011) and Daams et al. (1991) in all their large databases. The number of atoms in the corresponding hexagonal setting ($a = b \neq c$, $\alpha = \beta = 90°$, $\gamma = 120°$) would be three times as large and *hR*3-Sm would be given as *hR*9-Sm. Unfortunately, in the literature, both notations are used without further explanation.

We will use the Pearson symbol for characterizing a structure type, for instance, *cI*2-W. We use it as well if we want to denote a particular modification of an element such as *cI*2-Li, for example, which has the *cI*2-W structure type. If several different structure types have the same Pearson symbol, then we explicitly name the structure type in addition. To give an example, *tI*2-Hg has the *tI*2-Pa structure and not that of *tI*2-In.

This chapter contains four main sections. First, some factors governing the formation of crystal structures, in general, are introduced. This is followed by a comprehensive discussion of the structures of the metallic elements and, subsequently, by a presentation of the most relevant (frequency, properties, didactical content) structure types of intermetallic phases. Finally, a few representative examples of the structures of QCs will be given.

1.2   Factors Governing Formation and Stability of Crystal Structures

Crystalline order is the outstanding characteristics of condensed matter in thermodynamic equilibrium. It is usually realized in the form of a 3D translationally periodic repetition of a particular atomic configuration (recurrent structural subunit). However, crystalline order is also present in aperiodic crystals with incommensurately modulated structures, composite or host/guest structures as well as in quasiperiodic structures. The term *aperiodic crystals* refers to the fact that these structures can be described as irrational 3D sections of virtual *n*D ($n > 3$) periodic hypercrystals. The signature of ideal aperiodic crystals is that they show a pure-point Fourier spectrum (Bragg diffraction pattern) as periodic crystals do.

The kind of atomic interactions (chemical bonding), the electronic band structure and geometrical factors, such as atomic size ratios and packing density, determine which kind of crystal structure has the lowest Gibbs free energy, $G = H - TS$, for a given chemical composition and temperature/pressure. At finite temperature, entropic contributions from atomic vibrations or intrinsic structural disorder can also be crucial for the stability of a structure. This particularly applies to the so-called high-entropy alloys (HEA), which are equiatomic solid solutions of four or more elements. Their usually simple body centered cubic (*bcc*) or face centered cubic (*fcc*) structures are stabilized by mixing entropy, preventing the formation of intermetallic phases at sufficiently high temperatures. There have been many attempts to predict the stability of crystal structures based on either first-principles calculations (Oganov et al., 2011; and references therein) or (semi)empirical parameters. Quite successful has been the prediction of stability fields of binary and some groups of ternary phases based on the Mendeleev numbers (Pettifor, 1986, 1988) and, additionally, some other empirical parameters (Villars et al., 2001).

For the study of chemical bonding and the electronic structure, first-principles calculations are indispensable. Since it is not possible to solve analytically the Schrödinger equation for a crystal, numerous approximations have been developed. In most approximations, the many electron problem is reduced to a one electron problem by the assumption that the electrons, surrounded by a mutual exclusion zone, are moving independently from each other in the average field of all other ones (*local-density-functional approximation, LDA*). This works quite well for modeling known structures up to a few hundred atoms per unit cell, depending on the kind of atoms involved. On the contrary, structure prediction based on first-principles calculations is much more computer time-consuming and still limited to a few tens of atoms per unit cell. Beside this powerful quantum-mechanical approach for understanding and predicting crystal structures, there are a number of useful methods and rules in use that cannot be considered here.

In the following, we want to discuss just a few aspects of two important factors crucial for structure formation: the *chemical bond factor*, which is particularly taking into account the directionality of chemical bonds, and the *geometrical factor*, which is considering optimum space filling, symmetry and connectivity. Under these aspects, a crystal structure is considered to result from packing optimization under the constraints imposed by chemical bonding. In case of metallic bonding, the stability of particular structures can be greatly enhanced by nonlocal interactions such as Fermi-surface/Brillouin-zone nesting, which can lead to the opening of (pseudo)gaps in the electronic density of states at the Fermi-energy (Hume-Rothery-type stabilization). Such structure-stabilizing pseudogaps can also be caused by hybridization effects in particular atomic configurations (clusters).

1.2.1 Chemical Bonding

The concept of chemical bonding interprets and explains formation and stability of molecules and/or crystals based on local interactions. These atomic interactions are usually described in terms of idealized limiting types of chemical bonding:

- *Ionic (heteropolar)*: isotropic electrostatic interaction between ions that can also consist of larger structural units (complexions, polyions).
- *Covalent (homeopolar)*: directed bonds based on electrons localized between the bonding partners, which have contributed all of them.
- *Metallic (delocalized)*: isotropic interactions based on delocalized electrons in which the ionized bonding partners, which have contributed all of them, are embedded.
- *van der Waals*: (in case of atoms isotropic) dipole–dipole interactions, where the dipoles may be permanent or induced.
- *Hydrogen bonds*: directed bonds based on a kind of resonance state of the H^+ proton between two strongly electronegative atoms.

In the first approximation, the collective interaction of all atoms in a crystal may be replaced by the sum of nearest neighbor interactions. A further simplification comes in by the fact that only electrons of the outer electron shells (valence electrons) contribute to chemical bonding.

The types of bonding in crystals of the metallic elements at ambient conditions range from typical metallic in alkali metals to increasingly covalent for zinc or cadmium, for instance. The structural implications of these two bond types, which are just two contrary limiting manifestations of electronic interactions with a continuously changing degree of electron localization, shall be characterized in the following a bit more detailed.

The covalent bond may be described and illustrated in terms of the valence bond (VB) theory by overlapping atomic orbitals occupied by unpaired valence electrons. Its strength depends on the degree of overlapping and is given by the exchange integral. In terms of the linear combination of atomic orbitals–molecular orbitals (LCAO–MO) theory, molecular orbitals are constructed by linear combination of atomic orbitals. The resulting bonding, nonbonding and antibonding molecular orbitals, filled up with all valence electrons of the bonding partners according to the Pauli principle, are localized between the bonding atoms with well-defined geometry. Generally, covalent bonds can be characterized as strong, directed bonds.

Increasing the number of atoms contributing to the bonds increases the number of molecular orbitals and their energy differences become smaller and smaller. Finally, the discrete energy levels of the molecular orbitals condense to quasicontinuous bands separated by energy gaps. Since, in a covalent bond, each atom reaches its particular stable noble gas configuration (filled shell), the energy bands are either completely filled or empty. Due to the localization of the electrons, it needs much energy (several electron volt) to lift them from the last filled valence band into the empty conduction band. The classic example of a crystal built from only covalently bonded atoms is diamond: all carbon atoms are bonded via tetrahedrally directed sp^3 hybrid orbitals. Thus, the crystal structure of diamond results as a framework of tetrahedrally coordinated carbon atoms.

The metallic bond can be described in a similar way as the covalent bond. The main difference between these two bonding types is that the ionization energy for electrons occupying the outer orbitals of the metallic elements is much smaller. In typical metals, such as the alkali metals, these outer orbitals are spherical s orbitals, allowing overlaps with up to 12 further s orbitals of the surrounding atoms. Thus, the well-defined electron localization in bonds, connecting pairs of atoms with each other, loses its meaning. Quantum mechanical calculations show that in large aggregates of metal atoms, the delocalized bonding electrons occupy lower energy levels than in the free atoms; this would not be true for isolated "metal molecules." Thus, contrary to the other bond types that also can occur in isolated small molecules, metallic bonding can take place in larger arrays of atoms, only. The metallic bond in typical metals is nondirected, favoring structures based on closest sphere packings. With increasing localization of valence electrons, covalent interactions cause deviations from spherically symmetric bonding leading to more complex structures.

Since the interaction of electron orbitals of different atoms depends on their distance and mutual orientation, the bond type may change with the structure during phase transformations. An example is the transition from metallic white tin to nonmetallic gray tin ("tin pest") that takes place at 286 K deteriorating organ pipes in churches that are not heated properly during winter time; another example is the transformation from molecular to metallic hydrogen under very high pressure.

1.2.2 Geometrical Factors

A *crystal structure* is fully defined by its chemical composition, metrics (lattice parameters), space-group symmetry, equipoint (Wyckoff) positions occupied by the different types of atoms (crystallographic orbits) and their specific coordinates in one asymmetric unit (**Figure 1**). The metrics may differ for all chemical compounds or phases adopting one particular *structure type*. Also, for general Wyckoff positions, the numerical values of the coordinates may vary in a range not destroying the characteristics of this crystal structure such as coordination polyhedra (AET) or clusters and their connectivity. In our

(a)

$F\bar{4}3m$ T_d^2 $\bar{4}3m$ Cubic

No. 216 $F\bar{4}3m$

Patterson symmetry $Fm\bar{3}m$

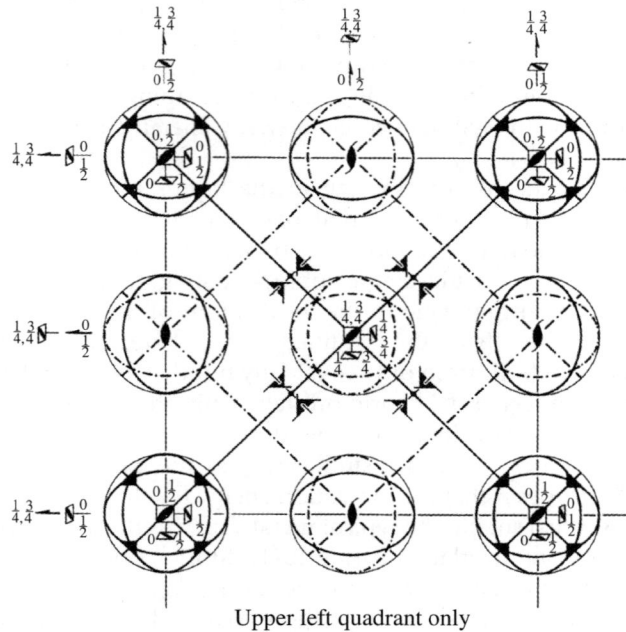

Upper left quadrant only

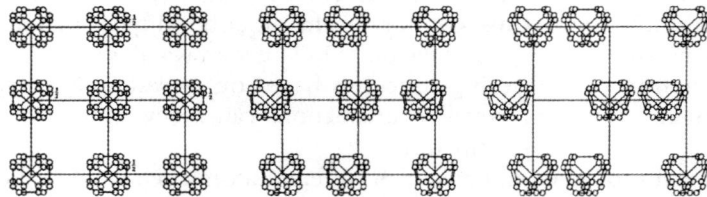

Origin at $\bar{4}3m$

Asymmetric unit $0 \le x \le \frac{1}{2}$; $0 \le y \le \frac{1}{4}$; $-\frac{1}{4} \le z \le \frac{1}{4}$; $y \le \min(x, \frac{1}{2} - x)$; $-y \le z \le y$

Vertices $0,0,0$ $\frac{1}{2},0,0$ $\frac{1}{4},\frac{1}{4},\frac{1}{4}$ $\frac{1}{4},\frac{1}{4},-\frac{1}{4}$

Figure 1   (a) Information given in the *International Tables for Crystallography* (Hahn, 2002) on the example of the space group $F\bar{4}3m$. Left side, top line: space group symbol in short Hermann–Mauguin and in Schoenflies notation, point group (crystal class), crystal system. Second line: consecutive space group number, full Hermann–Mauguin space group symbol, Patterson symmetry and its short Hermann–Mauguin space group symbol. Upper drawing: projection along [001] of the framework of symmetry elements in a unique part of one unit cell (upper left quadrant in this case). Inclined reflection and glide planes around high-symmetry points are shown in their orthographic projection. Lower drawings: point complexes generated by the action of symmetry operations in stereoscopic projection along [001] and rotated by 6° and 12°, respectively. Below: unit-cell origin and its site symmetry, definition of the asymmetric unit. (b) Sets of symmetry operators and their orientation as well as generating symmetry operators. Below: information about the different equipoint positions. The Wyckoff letters *a, b, c, …, i* denote the equipoint positions with multiplicities 4, 4, … 96, and their respective site symmetries. The reflection conditions in the right column refer to systematically extinct Bragg reflections in diffraction experiments.

(b)

CONTINUED   No. 216 $F\bar{4}3m$

Symmetry operations

For $(0,0,0)+$ set

(1) 1	(2) 2 $0,0,z$	(3) 2 $0,y,0$	(4) 2 $x,0,0$
(5) 3^+ x,x,x	(6) 3^+ $\bar{x},x,\bar{x}$	(7) 3^+ $x,\bar{x},\bar{x}$	(8) 3^+ $\bar{x},\bar{x},x$
(9) 3^- x,x,x	(10) 3^- $x,\bar{x},\bar{x}$	(11) 3^- $\bar{x},\bar{x},x$	(12) 3^- $\bar{x},x,\bar{x}$
(13) m x,x,z	(14) m $x,\bar{x},z$	(15) $\bar{4}^+$ $0,0,z;\ 0,0,0$	(16) $\bar{4}^-$ $0,0,z;\ 0,0,0$
(17) m x,y,y	(18) $\bar{4}^+$ $x,0,0;\ 0,0,0$	(19) $\bar{4}^-$ $x,0,0;\ 0,0,0$	(20) m $x,y,\bar{y}$
(21) m x,y,x	(22) $\bar{4}^-$ $0,y,0;\ 0,0,0$	(23) m $\bar{x},y,x$	(24) $\bar{4}^+$ $0,y,0;\ 0,0,0$

For $(0,\tfrac{1}{2},\tfrac{1}{2})+$ set

(1) $t(0,\tfrac{1}{2},\tfrac{1}{2})$	(2) $2(0,0,\tfrac{1}{2})$ $0,\tfrac{1}{4},z$	(3) $2(0,\tfrac{1}{2},0)$ $0,y,\tfrac{1}{4}$	(4) 2 $x,\tfrac{1}{4},\tfrac{1}{4}$
(5) $3^+(\tfrac{1}{3},\tfrac{1}{3},\tfrac{1}{3})$ $x-\tfrac{1}{3},x-\tfrac{1}{6},x$	(6) 3^+ $\bar{x},x+\tfrac{1}{2},\bar{x}$	(7) $3^+(-\tfrac{1}{3},\tfrac{1}{3},\tfrac{1}{3})$ $x+\tfrac{1}{3},\bar{x}-\tfrac{1}{6},\bar{x}$	(8) 3^+ $\bar{x},\bar{x}+\tfrac{1}{2},x$
(9) $3^-(\tfrac{1}{3},\tfrac{1}{3},\tfrac{1}{3})$ $x-\tfrac{1}{6},x+\tfrac{1}{6},x$	(10) $3^-(-\tfrac{1}{3},\tfrac{1}{3},\tfrac{1}{3})$ $x+\tfrac{1}{6},\bar{x}+\tfrac{1}{6},\bar{x}$	(11) 3^- $\bar{x}+\tfrac{1}{2},\bar{x}+\tfrac{1}{2},x$	(12) 3^- $\bar{x}-\tfrac{1}{2},x+\tfrac{1}{2},\bar{x}$
(13) $g(\tfrac{1}{4},\tfrac{1}{4},\tfrac{1}{2})$ $x-\tfrac{1}{4},x,z$	(14) $g(-\tfrac{1}{4},\tfrac{1}{4},\tfrac{1}{2})$ $x+\tfrac{1}{4},\bar{x},z$	(15) $\bar{4}^+$ $\tfrac{1}{4},\tfrac{1}{4},z;\ \tfrac{1}{4},\tfrac{1}{4},\tfrac{1}{4}$	(16) $\bar{4}^-$ $-\tfrac{1}{4},\tfrac{1}{4},z;\ -\tfrac{1}{4},\tfrac{1}{4},\tfrac{1}{4}$
(17) $g(0,\tfrac{1}{2},\tfrac{1}{2})$ x,y,y	(18) $\bar{4}^+$ $x,\tfrac{1}{2},0;\ 0,\tfrac{1}{2},0$	(19) $\bar{4}^-$ $x,0,\tfrac{1}{2};\ 0,0,\tfrac{1}{2}$	(20) m $x,y+\tfrac{1}{2},\bar{y}$
(21) $g(\tfrac{1}{4},\tfrac{1}{4},\tfrac{1}{2})$ $x-\tfrac{1}{4},y,x$	(22) $\bar{4}^-$ $\tfrac{1}{4},y,\tfrac{1}{4};\ \tfrac{1}{4},\tfrac{1}{4},\tfrac{1}{4}$	(23) $g(-\tfrac{1}{4},\tfrac{1}{2},\tfrac{1}{4})$ $\bar{x}+\tfrac{1}{4},y,x$	(24) $\bar{4}^+$ $-\tfrac{1}{4},y,\tfrac{1}{4};\ -\tfrac{1}{4},\tfrac{1}{4},\tfrac{1}{4}$

For $(\tfrac{1}{2},0,\tfrac{1}{2})+$ set

(1) $t(\tfrac{1}{2},0,\tfrac{1}{2})$	(2) $2(0,0,\tfrac{1}{2})$ $\tfrac{1}{4},0,z$	(3) 2 $\tfrac{1}{4},y,\tfrac{1}{4}$	(4) $2(\tfrac{1}{2},0,0)$ $x,0,\tfrac{1}{4}$
(5) $3^+(\tfrac{1}{3},\tfrac{1}{3},\tfrac{1}{3})$ $x+\tfrac{1}{6},x-\tfrac{1}{6},x$	(6) $3^+(\tfrac{1}{3},-\tfrac{1}{3},\tfrac{1}{3})$ $\bar{x}+\tfrac{1}{6},x+\tfrac{1}{6},\bar{x}$	(7) 3^+ $x+\tfrac{1}{2},\bar{x}-\tfrac{1}{2},\bar{x}$	(8) 3^+ $\bar{x}+\tfrac{1}{2},\bar{x}+\tfrac{1}{2},x$
(9) $3^-(\tfrac{1}{3},\tfrac{1}{3},\tfrac{1}{3})$ $x-\tfrac{1}{6},x-\tfrac{1}{3},x$	(10) 3^- $x+\tfrac{1}{2},\bar{x},\bar{x}$	(11) 3^- $\bar{x}+\tfrac{1}{2},\bar{x},x$	(12) $3^-(\tfrac{1}{3},-\tfrac{1}{3},\tfrac{1}{3})$ $\bar{x}-\tfrac{1}{6},x+\tfrac{1}{3},\bar{x}$
(13) $g(\tfrac{1}{4},\tfrac{1}{4},\tfrac{1}{2})$ $x+\tfrac{1}{4},x,z$	(14) $g(\tfrac{1}{4},-\tfrac{1}{4},\tfrac{1}{2})$ $x+\tfrac{1}{4},\bar{x},z$	(15) $\bar{4}^+$ $\tfrac{1}{4},-\tfrac{1}{4},z;\ \tfrac{1}{4},-\tfrac{1}{4},\tfrac{1}{4}$	(16) $\bar{4}^-$ $\tfrac{1}{4},\tfrac{1}{4},z;\ \tfrac{1}{4},\tfrac{1}{4},\tfrac{1}{4}$
(17) $g(\tfrac{1}{2},\tfrac{1}{4},\tfrac{1}{4})$ $x,y-\tfrac{1}{4},y$	(18) $\bar{4}^+$ $x,\tfrac{1}{4},\tfrac{1}{4};\ \tfrac{1}{4},\tfrac{1}{4},\tfrac{1}{4}$	(19) $\bar{4}^-$ $x,-\tfrac{1}{4},\tfrac{1}{4};\ \tfrac{1}{4},-\tfrac{1}{4},\tfrac{1}{4}$	(20) $g(\tfrac{1}{2},-\tfrac{1}{4},\tfrac{1}{4})$ $x,y+\tfrac{1}{4},\bar{y}$
(21) $g(\tfrac{1}{2},0,\tfrac{1}{2})$ x,y,x	(22) $\bar{4}^-$ $\tfrac{1}{2},y,0;\ \tfrac{1}{2},0,0$	(23) m $\bar{x}+\tfrac{1}{2},y,x$	(24) $\bar{4}^+$ $0,y,\tfrac{1}{2};\ 0,0,\tfrac{1}{2}$

For $(\tfrac{1}{2},\tfrac{1}{2},0)+$ set

(1) $t(\tfrac{1}{2},\tfrac{1}{2},0)$	(2) 2 $\tfrac{1}{4},\tfrac{1}{4},z$	(3) $2(0,\tfrac{1}{2},0)$ $\tfrac{1}{4},y,0$	(4) $2(\tfrac{1}{2},0,0)$ $x,\tfrac{1}{4},0$
(5) $3^+(\tfrac{1}{3},\tfrac{1}{3},\tfrac{1}{3})$ $x+\tfrac{1}{6},x+\tfrac{1}{3},x$	(6) 3^+ $\bar{x}+\tfrac{1}{2},x,\bar{x}$	(7) 3^+ $x+\tfrac{1}{2},\bar{x},\bar{x}$	(8) $3^+(\tfrac{1}{3},\tfrac{1}{3},-\tfrac{1}{3})$ $\bar{x}+\tfrac{1}{6},\bar{x}+\tfrac{1}{3},x$
(9) $3^-(\tfrac{1}{3},\tfrac{1}{3},\tfrac{1}{3})$ $x+\tfrac{1}{3},x+\tfrac{1}{6},x$	(10) 3^- $x,\bar{x}+\tfrac{1}{2},\bar{x}$	(11) $3^-(\tfrac{1}{3},\tfrac{1}{3},-\tfrac{1}{3})$ $\bar{x}+\tfrac{1}{3},\bar{x}+\tfrac{1}{6},x$	(12) 3^- $\bar{x},x+\tfrac{1}{2},\bar{x}$
(13) $g(\tfrac{1}{2},\tfrac{1}{2},0)$ x,x,z	(14) m $x+\tfrac{1}{2},\bar{x},z$	(15) $\bar{4}^+$ $\tfrac{1}{4},0,z;\ \tfrac{1}{4},0,0$	(16) $\bar{4}^-$ $0,\tfrac{1}{4},z;\ 0,\tfrac{1}{4},0$
(17) $g(\tfrac{1}{4},\tfrac{1}{4},\tfrac{1}{4})$ $x,y+\tfrac{1}{4},y$	(18) $\bar{4}^+$ $x,\tfrac{1}{4},-\tfrac{1}{4};\ \tfrac{1}{4},\tfrac{1}{4},-\tfrac{1}{4}$	(19) $\bar{4}^-$ $x,\tfrac{1}{4},\tfrac{1}{4};\ \tfrac{1}{4},\tfrac{1}{4},\tfrac{1}{4}$	(20) $g(\tfrac{1}{4},\tfrac{1}{4},-\tfrac{1}{4})$ $x,y+\tfrac{1}{4},\bar{y}$
(21) $g(\tfrac{1}{4},\tfrac{1}{4},\tfrac{1}{4})$ $x+\tfrac{1}{4},y,x$	(22) $\bar{4}^-$ $\tfrac{1}{4},y,-\tfrac{1}{4};\ \tfrac{1}{4},\tfrac{1}{4},-\tfrac{1}{4}$	(23) $g(\tfrac{1}{4},\tfrac{1}{4},-\tfrac{1}{4})$ $\bar{x}+\tfrac{1}{4},y,x$	(24) $\bar{4}^+$ $\tfrac{1}{4},y,\tfrac{1}{4};\ \tfrac{1}{4},\tfrac{1}{4},\tfrac{1}{4}$

Generators selected   (1); $t(1,0,0)$; $t(0,1,0)$; $t(0,0,1)$; $t(0,\tfrac{1}{2},\tfrac{1}{2})$; $t(\tfrac{1}{2},0,\tfrac{1}{2})$; (2); (3); (5); (13)

Positions

Multiplicity, Wyckoff letter, Site symmetry	Coordinates $(0,0,0)+\quad (0,\tfrac{1}{2},\tfrac{1}{2})+\quad (\tfrac{1}{2},0,\tfrac{1}{2})+\quad (\tfrac{1}{2},\tfrac{1}{2},0)+$	Reflection conditions

96 i 1	(1) x,y,z (2) $\bar{x},\bar{y},z$ (3) $\bar{x},y,\bar{z}$ (4) $x,\bar{y},\bar{z}$ (5) z,x,y (6) $z,\bar{x},\bar{y}$ (7) $\bar{z},\bar{x},y$ (8) $\bar{z},x,\bar{y}$ (9) y,z,x (10) $\bar{y},\bar{z},\bar{x}$ (11) $y,\bar{z},\bar{x}$ (12) $\bar{y},\bar{z},x$ (13) y,x,z (14) $\bar{y},\bar{x},z$ (15) $y,\bar{x},\bar{z}$ (16) $\bar{y},x,\bar{z}$ (17) x,z,y (18) $\bar{x},z,\bar{y}$ (19) $\bar{x},\bar{z},y$ (20) $x,\bar{z},\bar{y}$ (21) z,y,x (22) $z,\bar{y},\bar{x}$ (23) $\bar{z},y,\bar{x}$ (24) $\bar{z},\bar{y},x$	h,k,l permutable General: hkl : $h+k,h+l,k+l=2n$ $0kl$: $k,l=2n$ hhl : $h+l=2n$ $h00$: $h=2n$ Special: no extra conditions

48	h	$..m$	x,x,z $\bar{z},\bar{x},x$	$\bar{x},\bar{x},z$ $\bar{z},x,\bar{x}$	$\bar{x},x,\bar{z}$ x,z,x	$x,\bar{x},\bar{z}$ $\bar{x},z,\bar{x}$	z,x,x $x,\bar{z},\bar{x}$	$z,\bar{x},\bar{x}$ $\bar{x},\bar{z},x$
24	g	$2.mm$	$x,\tfrac{1}{4},\tfrac{1}{4}$	$\bar{x},\tfrac{3}{4},\tfrac{1}{4}$	$\tfrac{1}{4},x,\tfrac{1}{4}$	$\tfrac{1}{4},\bar{x},\tfrac{3}{4}$	$\tfrac{1}{4},\tfrac{1}{4},x$	$\tfrac{3}{4},\tfrac{1}{4},\bar{x}$
24	f	$2.mm$	$x,0,0$	$\bar{x},0,0$	$0,x,0$	$0,\bar{x},0$	$0,0,x$	$0,0,\bar{x}$
16	e	$.3m$	x,x,x	$\bar{x},\bar{x},x$	$\bar{x},x,\bar{x}$	$x,\bar{x},\bar{x}$		
4	d	$\bar{4}3m$	$\tfrac{3}{4},\tfrac{3}{4},\tfrac{3}{4}$					
4	c	$\bar{4}3m$	$\tfrac{1}{4},\tfrac{1}{4},\tfrac{1}{4}$					
4	b	$\bar{4}3m$	$\tfrac{1}{2},\tfrac{1}{2},\tfrac{1}{2}$					
4	a	$\bar{4}3m$	$0,0,0$					

Figure 1   (*continued*).

crystallographic notation, we will use Miller indices in parentheses $(h\,k\,l)$ and in braces $\{h\,k\,l\}$ to denote lattice planes or sets of lattice planes symmetrically equivalent under the action of a point group, respectively; vector components in brackets $[u\,v\,w]$ indicate lattice directions; lattice basis vectors are denoted by **a**, **b**, **c** in case of 3D periodic structures, and quasilattice vectors $\mathbf{a}_i$, $i = 1 \ldots n$ in case of quasiperiodic structures.

For the comparison of different crystal structure types, clear definitions of some terms are necessary (Lima-de-Faria et al., 1990):

● *Isopointal*: two structures are isopointal if they have the same space-group type, and the atomic (Wyckoff) positions, occupied either fully or partially at random, are the same in both structures; as there are no limitations on the values of the adjustable parameters of the Wyckoff positions or on the cell parameters, isopointal structures may have locally different geometric arrangements and atomic coordinations (AET) and may belong to different structure types.

An example for *isopointal* structure types are *tI*2-In and *tI*2-Pa (**Figure 2**).

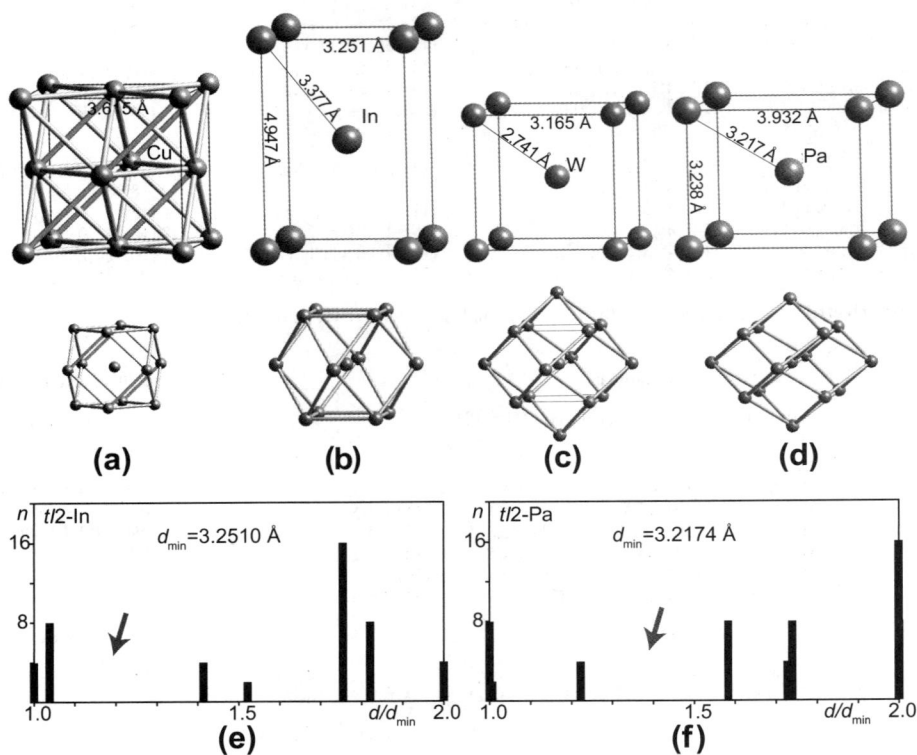

Figure 2 Comparison between the isopointal structures of (b) *tI*2-In and (d) *tI*2-Pa and their AET, (b) a cuboctahedron (coordination number CN12) and (d) a rhombdodecahedron (CN14). Due to the strongly differing axial ratios *c*/*a* of 1.522 for *tI*2-In, one for *cF*4-Cu and 0.824 for *tI*2-Pa, their AET differ significantly as well. As their AET show, the structure of *tI*2-In is closer to the *cF*4-Cu type (a), and that of *tI*2-Pa is closer to the *cI*2-W type (c). The distances histograms for (e) *tI*2-In and (f) *tI*2-Pa differ significantly from each other: the first big gap appears after 12 atoms in (e) and after 14 atoms in (f). Since the first 10 atoms in (f) have almost the same distances from the center, the first big gap opens after 14 atoms.

- *Isoconfigurational:* two structures are *configurationally isotypic* if they are isopointal and both the crystallographic point configurations (crystallographic orbits) and their geometrical interrelationships are similar; all geometrical properties, such as axial ratios, angles between crystallographic axes, values of corresponding adjustable positional parameters (x, y, z), and coordinations of corresponding atoms (AET) are similar. *Isoconfigurational structures belong to the same structure type.*

  For instance, the *isoconfigurational* metallic phase AlNi and the ionic compound CsCl are both representatives of the $cP2$-CsCl structure type although they strongly differ in their atomic interactions.

- *Isotypic:* two structures are *crystal chemically isotypic* if they are isoconfigurational and the corresponding atoms and bonds (interactions) have similar physical/chemical characteristics.

  For instance, $cP2$-CoAl and $cP2$-NiAl are *isotypic*.

- *Homeotypic:* two structures are *homeotypic* if one or more of the following conditions required for *isotypism* are relaxed:
  i. Identical or enatiomorphic space-group types, allowing for group/subgroup or group/supergroup relationships.
  ii. Limitations imposed on the similarity of geometric properties, i.e. axial ratios, interaxial angles, values of adjustable positional parameters, and the coordination of corresponding atoms.
  iii. Site occupancy limits, allowing given sites to be occupied by different atomic species.

  For instance, $cI2$-W and $cP2$-NiAl are *homeotypic* according to (i) and (iii).

- *Polytypic:* an element or compound is *polytypic* if it occurs in several structural modifications, each of which can be regarded as built up by stacking layers of (nearly) identical structure and composition, and if the modifications differ only in their stacking sequence. *Polytypism* is a special variant of *polymorphism.*

  For instance, the structure types $hP2$-Mg, $cF4$-Cu, $hP4$-La and $hR3$-Sm are all different stacking variants of closest packed layers (see **Figure 17** in Section 1.3.2.1); their structures are polytypic variants of dense sphere packings in general.

- *Polymorphic:* an element or compound is *polymorphic* if it occurs in several structural modifications as function of ambient conditions (temperature, pressure, external electric, magnetic, elastic fields, etc.). The different modifications are frequently called *polymorphs* and in case of the chemical elements, *allotropes.*

- *Aristotype:* high-symmetry basic structure that can be seen as idealized version of one or more different lower-symmetry *derivative structures,* the *hettotypes.*

  For instance, the $cP2$-CsCl structure can be seen as hettotype and that of $cI2$-W as its aristotype.

- *Prototype:* crystal structure of an element or compound used for the definition of a *structure type,* which represents the class of all materials with isoconfigurational structures.

1.2.2.1 *Sphere Packings and Space Filling, Polytypism*

Geometrically, crystal structures can be seen as densest possible packings of atoms and/or structural subunits under the constraints of chemical bonding. In case of covalent bonding, these constraints are rather hard and anisotropic; in case of typical metallic bonding, they can be quite soft and isotropic. Describing a structure in terms of structural subunits (AET, clusters, layers, frameworks) instead just by ball-and-stick models may give a better understanding of the spatial distribution of locally varying chemical bonding (covalent or polyionic bonding contributions) or odd stoichiometries. In most cases, however, it only serves as means for the visualization of complex crystal structures.

Due to the intrinsically isotropic nature of metallic bonding, the structure of typical metallic elements can often be described in terms of dense sphere packings. A *sphere packing* is an infinite set of noninterpenetrating hard spheres with the property that any pair of spheres is connected by a chain of spheres with mutual contact. A sphere packing is called *homogenous* for all spheres being symmetrically equivalent, otherwise it is called *heterogenous* (Koch and Fischer, 1992). In the latter case, the spheres of the different symmetrically non-equivalent subsets may have different radii and occupy positions of different crystallographic orbits. The number of types of heterogenous sphere packings is infinite, whereas it is finite for homogenous sphere packings. There exist, for instance, 199 different cubic and 394 different tetragonal homogenous sphere packings (Koch and Fischer, 1992).

The fractional packing densities, i.e. the fraction of space occupied by the spheres, are with $q = \pi/\sqrt{18} = 0.74048$, the highest for the well-known *hexagonal closest packing* (*hcp*) and *cubic closest packing* (*ccp*), respectively (**Figure 3**). In both cases, the *coordination numbers* are 12 and the distances to the nearest neighbors are the same. If we allow slightly different atomic radii, then the *topologically closest packings* (*tcp*), i.e. packings of face-sharing irregular tetrahedra, can have even higher packing densities (see Section 1.4.5). **Table 3** gives some examples for homogenous sphere packings with the highest and lowest densities and contact numbers. The number k of contacts per sphere amounts to $3 \leq k \leq 12$. It is remarkable that the monoclinic sphere packing with space group $C2/m$ and $k = 11$ shows only 3% lower packing density, then the highly symmetric cubic and hexagonal closest packings. **Table 4** shows space-filling values for the structures of a number of elements. Very low packing densities like that for $cF8$-C (diamond), for instance, indicate that a hard-sphere packing model is not an adequate

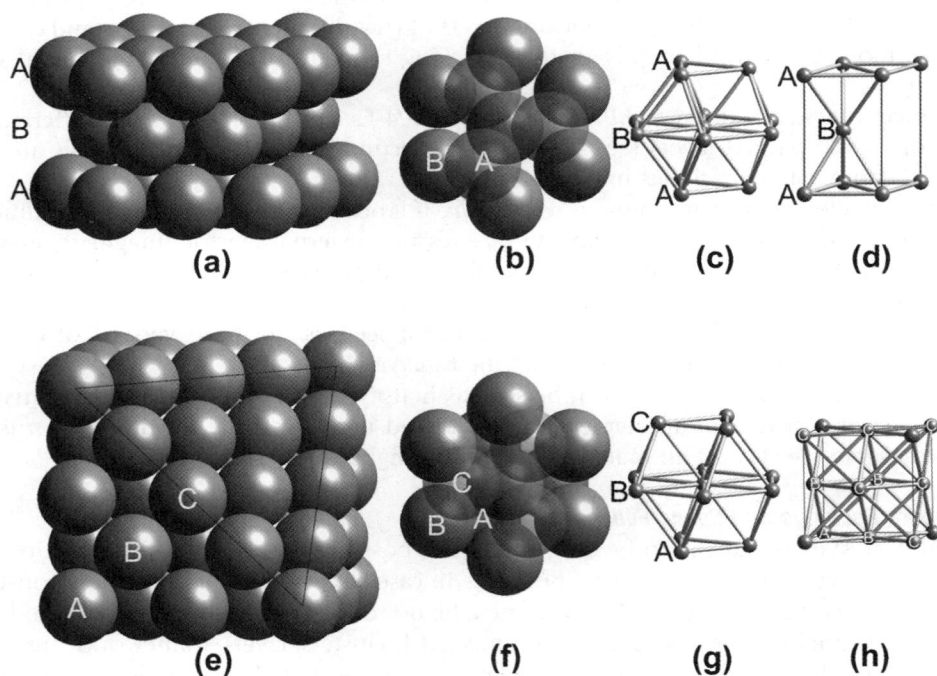

Figure 3 Characteristics of the hexagonal (a–d) and cubic (e–h) close sphere packings. The hexagonal and cubic unit cells are shown in (d) and (h), respectively. The layer sequences ..AB.. along the *c*-axis in the *hcp* case (a), and ..ABC.. along [111], the body diagonal of the cubic unit cell, in the *ccp* case (e) are marked. The AET correspond to a disheptahedron (b,c) and to a cuboctahderon (f,g), respectively.

Table 3   Characteristics of some homogenous sphere packings with high (low) contact numbers k and high (low) fractional packing densities q; d is the distance between the centers of spheres next to each other (Wilson and Prince, 1999)

k	Space group	Wyckoff position	Parameters	Distance d	Density q
12	$P6_3/mmc$	2c 1/3, 2/3, 1/4	$c/a = 2/3\sqrt{6} = 1.6330$	a	0.7405
12	$Fm\bar{3}m$	4a 0, 0, 0		$a/2\sqrt{2}$	0.7405
11	$C2/m$	4i x, 0, z	$x = 1/2\left(\sqrt{2}-1\right), z = 3\sqrt{2}-4$	b	0.7187
			$b/a = 1/3\sqrt{3}, c/a = 1/6\left(\sqrt{6}+2\sqrt{3}\right)$		
			$\cos\beta = 1/6\left(\sqrt{6}-2\sqrt{3}\right)$		
10	$I4/mmm$	2a 0, 0, 0	$c/a = \sqrt{6}/3 = 0.8165$	c	0.6981
3	$I4_132$	24h 1/8, y, 1/4-y	$y = 1/8\left(2\sqrt{3}-3\right)$	$a/4\left(2\sqrt{6}-3\sqrt{2}\right)$	0.0555

Table 4   Fractional packing densities q of elemental structures under the assumption of hard spherical atoms (Pearson, 1972)

Element	Pearson symbol, c/a	Space-filling value q	Element	Pearson symbol, c/a	Space-filling value q
Cu	cF4	0.740	Po	cP1	0.523
Mg	hP2, 1.63	0.740	Bi	hR2, 2.60	0.446
Zn	hP2, 1.86	0.650	Sb	hR2, 2.62	0.410
Pa	tI2	0.696	As	hR2, 2.80	0.385
In	tI2	0.686	Ga	oC8	0.391
W	cI2	0.680	Te	hP3	0.364
Hg	hR1	0.609	C (diamond)	cF8	0.340
Sn	tI4	0.535	P (black)	oC8	0.285
U	oC4	0.534			

description of such a structure. The hard constraints of directional covalent bonding dominate the structure formation overriding maximization of packing density.

The crystal structures of metallic elements adopt dense sphere packings as long as purely geometrical packing principles are dominating. Covalent bonding contributions and electronic band structure effects can give rise to the formation of more complex, lower symmetric structures.

Many crystal structures can be characterized topologically as stackings of atomic layers. The afore-mentioned *hcp* structure can be described as stacking of dense packed layers with period (stacking sequence) AB, the cubic closest packing (*ccp*) with period ABC (**Figure 3**). The atomic layers are denoted by A, B or C depending on their relative position to each other. Packing fractions as well as coordination numbers (CN = 12) are equal in both cases. The first-shell AET corresponds to a cuboctahedron in case of *ccp* and to a disheptahedron in case of *hcp*. The distribution of atomic distances does not become different until the third and higher coordination shells.

These two types of closest packed layer stackings are not the only possible ones. There exist infinitely many polytypic variants, with exactly the same coordination numbers and packing densities. Examples for such complex layer structures of metallic elements are *hP4*-La (ACAB) and *hR3*-Sm (ABABCBCAC) (see **Figure 17** in Section 1.3.2.1). There exist different notations for the description of layer stackings; frequently used is that of Jagodzinski (1954), which characterizes the layer sequence with regard to the either h(exagonal) or c(ubic) surrounding of each layer: in the stacking sequence …ABC…, B would be characterized by c; in the sequence …ABA…, B would be characterized by h. $hP4$-La (ACAB) and $hR3$-Sm (ABABCBCAC) would then be described by $chch$ and $chhchhchh$. Their hexagonality, i.e. the ratio of the number of h to the total number of h and c results in 0.5 and 0.67, respectively. In the less-detailed Ramsdell (1947) notation, these two sequences would be summarized to 4H and 9R, just listing the number of layers and the Bravais lattice type.

1.2.2.2   Coordination and Atomic Environment Types (AET)

A general technique to derive useful coordination polyhedra (AET) was suggested by Brunner and Schwarzenbach (1971): all interatomic distances around a particular atom are calculated up to a certain limit, and all atoms within a distance defined by the first maximum gap in a distances histogram belong to the coordination polyhedron. This is shown in the example of the isopointal element structures $tI2$-In and $tI2$-Pa (**Figure 2**). If there is no clear maximum gap observable, a second criterion may be the maximum-convex-volume rule: all coordinating atoms lying at the intersections of at least three faces should form a convex polyhedron (Daams and Villars, 1992).

The kind of coordination polyhedron formed by atoms B, with radius r_B, around an atom A, with radius r_A, strongly depends on the size ratio r_A/r_B of the constituents. The ideal size ratios for hard spheres touching each other are 0.155 in case of a triangular AET (coordination number CN = 3), 0.225 for a tetrahedral or square planar AET (CN = 4), $\sqrt{2} - 1 = 0.414$ for an octahedral AET (CN = 6), 0.528 for a trigonal prismatic AET (CN = 6), $\sqrt{3} - 1 = 0.732$ for a hexahedral AET (CN = 8), 0.902 for an icosahedral AET (CN = 12), one for a cuboctahedral or disheptahedral AET (CN = 12) and 1.803 for a dodecahedral AET (CN = 20).

1.2.2.3   Cluster-based Description of Structures

The symmetry of all crystal structures can be described by nD space groups, where $n = 3$ in case of periodic and $n > 3$ in case of aperiodic crystal structures. One has to keep in mind that a given space-group symmetry results from the optimal ordering of atoms under the constraints of the maximization of the packing density, of the attractive interactions, the configurational and vibrational entropy as well as of the minimization of repulsive interactions and not the other way round. In other words, the space-group symmetry of a crystal structure results from both the point-group symmetry of the AET and larger structural subunits (clusters) as well as the way these subunits pack.

The visualization of a crystal structure in terms of structural building blocks (structure motifs, fundamental building units, clusters, …) can be quite arbitrary. Although there are some conventions for the definition of AET, as mentioned above, there are no conventions for the derivation of larger subunits, if polyhedral frequently called *cluster*. Furthermore, even if one finds a good-looking cluster-based description, this does not mean that it has any meaning from a crystal-chemical point of view or that another cluster description would not do equally well.

Let us discuss these problems on the simple example of an fcc packing of hard spheres, for instance, as realized in the structure of $cF4$-Al. It is usually described as a packing of closest packed layers with sequence …ABC… (**Figure 3** in Section 1.2.2.1). Of course, an fcc packing of hard spheres is by no means a layer structure in the crystal-chemical meaning of the word. According to the cubic symmetry, the layer sequence …ABC… is realized in the four symmetrically equivalent [111] directions, i.e. along the threefold axes. So, if not a layer structure, what else would be a proper description of such a structure? The AET around each atom of fcc aluminum is a cuboctahedron (**Figure 4**). The next larger AET is an octahedron, etc. These "cluster shells" look quite convincing. However, the crucial point is that every single Al atom in the structure is surrounded by exactly the same AET and that the bond distances

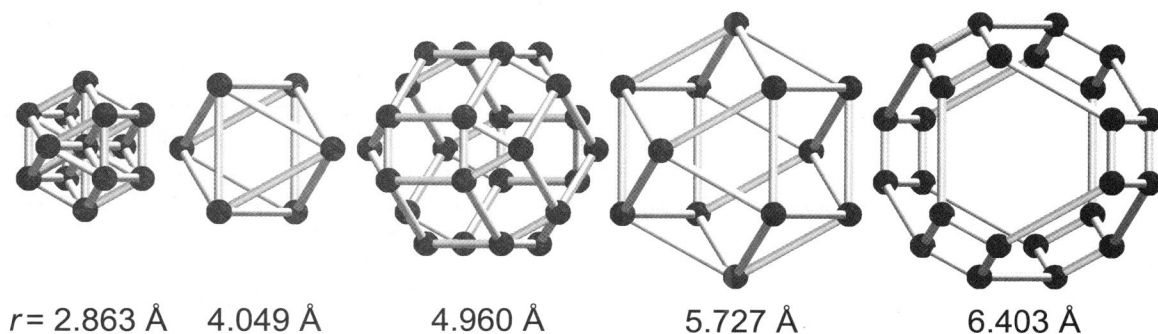

$r = 2.863$ Å   4.049 Å   4.960 Å   5.727 Å   6.403 Å

Figure 4 The first five "cluster shells" around any Al atom in *fcc* aluminum. The radii *r* of these coordination polyhedra (cluster shells) have a ratio of $1 : \sqrt{2} : \sqrt{3} : \sqrt{4} : \sqrt{5}$ (from Steurer, 2006).

between neighboring atoms are always exactly the same. Consequently, there is no crystal-chemical meaning at all in this kind of "cluster," it is just a geometrical construct.

Geometrical analysis of a complex crystal structure:

- Identify the content of subunit cells that are defined around high-symmetry points of the unit cell. Depending on site symmetry, useful subunit cells can be (semi)regular polyhedra that can be packed without gaps and overlaps (**Figure 10**). For instance, for cubic structures, the truncated octahedron (Kelvin polyhedron) has been suggested as subunit cell (Chieh, 1979, 1980, 1982). To give an example, in **Table 5**, the polyhedra (point configurations) are listed that are generated around the different Wyckoff positions of the space group $F\bar{4}3m$ (see **Figure 1** in Section 1.2.2).
- Identify recurrent structural subunits (AET, clusters) and the way they are linked. Choose those clusters that are closest to units, which are crystal chemically distinguishable from their environment.
- Analyze the packing of the subunit cells and clusters, respectively. A good starting point can be the well-known packing principles of uniform polyhedra or of tilings.
- The structure of clusters and/or their packing principles as well as structural relationships can sometimes be well described based on the higher dimensional approach for aperiodic structures.

Table 5   Polyhedra resulting from point configurations generated by the site symmetry groups at the different Wyckof positions of the space group $F\bar{4}3m$

Wyckoff position	Coordinates			Type of polyhedron	Center of polyhedron
	x	y	z		
4a	0	0	0	Large octahedron	4b
4b	1/2	1/2	1/2	Large octahedron	4a
4c	1/4	1/4	1/4	Large tetrahedron	4b
4d	3/4	3/4	3/4	Large tetrahedron	4b
16e	x	x	x	Tetrahedron	4c or 4d
24f	x	0	0	Octahedron	4a or 4b
24g	x	1/4	1/4	Octahedron	4c or 4d
48h	x	x	z	Octahedron	4b or 4c
96i	x	y	z	Truncated tetrahedron	4c or 4d
				Edge- and vertex-truncated tetrahedra	4a and 4b

1.2.2.4 Periodic and Quasiperiodic Tilings, Coverings

Many structures or substructures, periodic and quasiperiodic ones, can be described as decoration of 2D or 3D tilings by atoms or other structural subunits. A trivial case is taking the unit cell of a periodic structure as unit tile (prototile) and generating an infinite periodic tiling out of identical copies of this unit tile. It is less trivial if the prototiles are smaller than the unit cell because the resulting tilings can illustrate fundamental packing principles of more complex structures. Since quasiperiodic structures do not have lattice symmetry and a unit cell, the underlying quasiperiodic tilings are a proper way for the study of their symmetry and structural building principles. We will introduce a few basic tilings in the following. For a huge collection of tilings of all kinds, including periodic and quasiperiodic ones beside others, see Grünbaum and Shephard (1987).

Tilings that are frequently found in structures of intermetallics are the 11 Archimedean tilings, which were derived by Johannes Kepler in analogy to the Archimedean solids discussed below. The three *regular Archimedean tilings* consist of one type of regular polygons only and show just one type of vertex configuration. A regular polygon is convex with all its sides of equal length (equilateral) and all its angles equal (equiangular). The three regular Archimedean tilings are the triangle tiling 3^6—this is the tiling underlying a closest packed atomic layer, the square tiling 4^4 and the hexagon tiling 6^3. The *Cundy and Rollett symbol* of a vertex configuration n^m means that m n-gons meet at a vertex. The vertex configuration can also be written in the form of the Schläfli symbol $\{n,m\}$ or (n,m).

The eight *semiregular Archimedean tilings* are uniform. This means that they have only one type of vertex configuration, i.e. they are vertex transitive; they consist of two or more regular polygons as unit tiles (**Figure 5**). In case of layer structures, where one layer type corresponds to one of the Archimedean tilings, the layer next to it will preferentially be the respective dual tiling (*Catalan* or *Laves tiling*). The dual to a tiling can be obtained by putting vertices into the centers of the unit tiles and connecting them by lines. If the tiling is regular, then the dual tiling will be regular as well. The dual of the regular square tiling is a regular square tiling again, so this tiling is self-dual. The dual to the hexagon tiling is the triangle tiling (see **Figure 18** in Section 1.3.2.1). While the uniform semiregular tilings are described by their vertex configuration, their duals consist of just one type of polygon (are *isohedral*) but have more than one vertex configuration. Therefore, they are described by their face configuration, i.e. the sequential numbers of polygons meeting at each vertex of a face. For instance, the dual to the Archimedean snub square tiling, $3^2.4.3.4$, is the *Cairo pentagon tiling*, $V3^2.4.3.4$ (see **Figure 44** in Section 1.4.10.4). Its face configuration $V3^2.4.3.4$ means pentagonal unit tiles with corners, where 3, 3, 4, 3, 4 squashed pentagons meet.

In case of QCs, quasiperiodic tilings such as the 2D or 3D Penrose tiling (PT) are frequently underlying the observed structures. There are several versions of the 2D PT: a pentagon-based tiling (P1), a kite and dart version of it (P2), which may be further subdivided to give the Robinson triangle tiling, and a rhomb tiling (P3). Patches of these tilings are shown in **Figure 6**(a–d). All of them are mutually locally derivable and belong to the Penrose local isomorphism (PLI) class. According to its reciprocal space symmetry, the PT is a decagonal quasiperiodic tiling. The PLI-class tilings possess matching rules that force quasiperiodicity. If the matching rules are relaxed, other tilings become possible, which may be quasiperiodic, periodic, or all kinds of nonperiodic up to fully random.

The rhomb PT can be constructed from two unit tiles: a skinny (acute angle $\alpha = \pi/5$) and a fat rhomb (acute angle $\alpha = 2\pi/5$) with equal edge lengths (**Figure 7**). Their areas and frequencies in the PT are both in the ratio $1:\tau$. The construction has to obey matching rules, which can be derived from the scaling properties of the PT (see **Figure 6**(k)). The Ammann lines shown in **Figure 7** can be used as matching rules. In case of simpleton flips, i.e. jumps of one inner vertex in a squashed hexagon

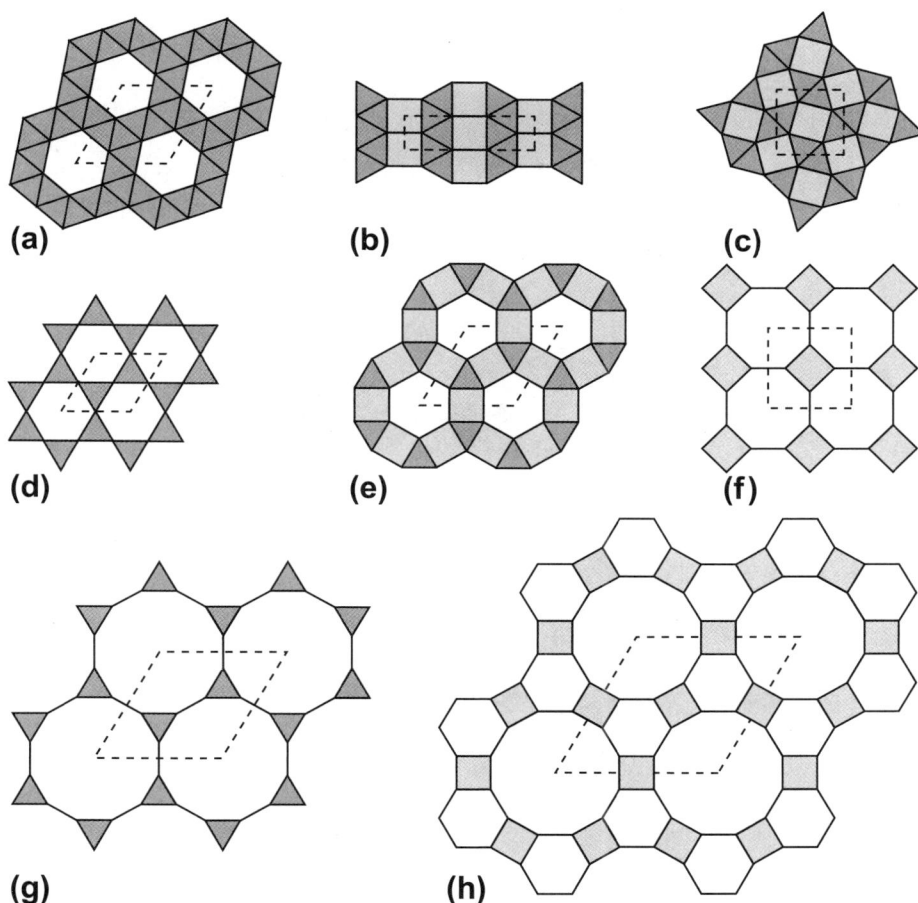

Figure 5 The eight semiregular Archimedean tilings, with their plane group symmetry and vertex configuration: (a) Snub hexagonal tiling, plane group $p6$, $3^4.6$; (b) elongated triangular tiling, $c2mm$, $3^3.4^2$; (c) snub square tiling, $p4gm$, $3^2.4.3.4$; (d) trihexagonal tiling (Kagomé net), $p6mm$, 3.6.3.6; (e) small rhombtrihexagonal tiling, $p6mm$, 3.4.6.4; (f) truncated square tiling, $p4mm$, 4.8^2; (g) truncated hexagonal tiling, $p6mm$, 3.12^2; (h) great rhombitrihexagonal tiling, $p6mm$, 4.6.12. The crystallographic unit cells of these tilings are outlined (dashed lines).

arrangement of three unit tiles, the Ammann lines are locally broken and displaced. Simpleton flips are seen as low-energy excitations in quasiperiodic structures.

A PT can be fully covered by partially overlapping copies of a covering cluster, which is called Gummelt decagon. The Gummelt decagon decorated with different variants of the PT is shown in **Figure 6**(a–d). The overlap rules are coded here in the "rocket decoration": for two overlapping decagons, the colors (light or gray shading) of the overlap areas must agree **Figure 6**(e–j). There are nine different allowed coordinations possible of a central Gummelt decagon by other decagons, so that all decagon edges are fully covered. The coordination numbers are 4, 5 or 6. The centers of the decagons form a pentagon PT when the overlap rules are obeyed. For more details on the PT, see Steurer and Deloudi (2009).

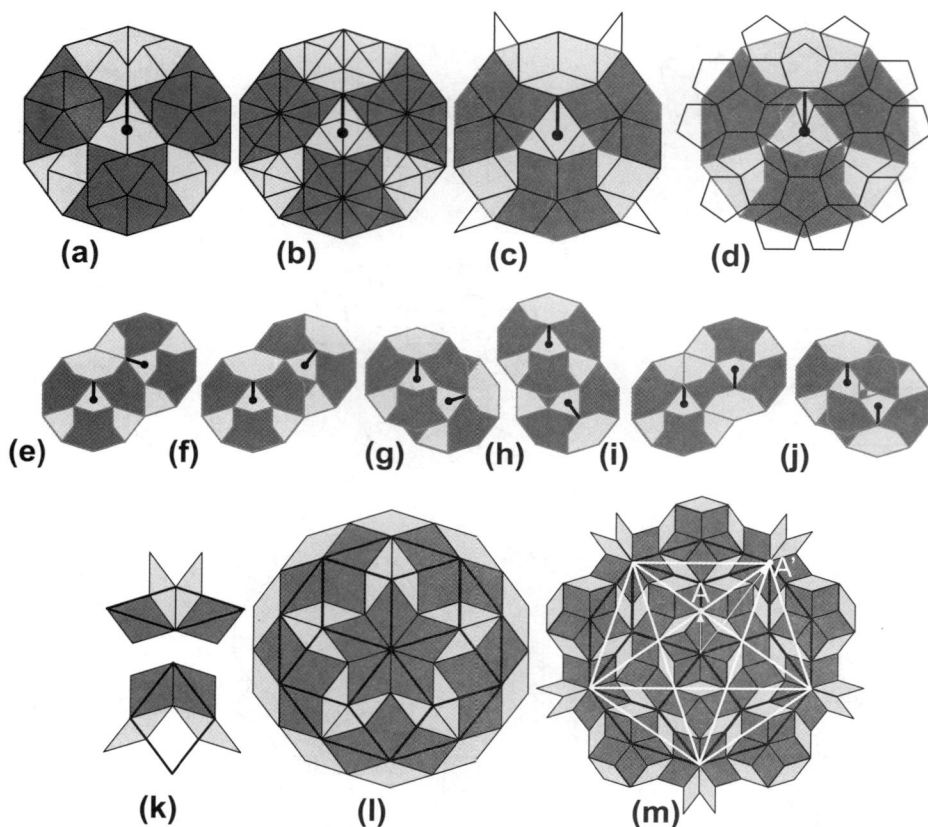

Figure 6 Gummelt-decagon covering patches of (a) kite-and-dart tiling (P2), (b) Robinson triangle tiling, (c) rhomb Penrose tiling (P3), and (d) Pentagon Penrose tiling (P1). The allowed overlaps of Gummelt decagons are shown in (e–j). For the shown rocket decoration, the colors of the overlap areas of two decagons must agree. The distances between decagon centers in (e,f,h,i) are τ times larger than those of (g,j). (k–l) scaling properties of the Penrose tiling. (k) The substitution (inflation) rule for the rhomb prototiles. In (l) a PT (*thin lines*) is superposed by another PT (*thick lines*) scaled by τ, in (m) scaling by τ^2 is shown. A subset of the vertices of the scaled tilings is the vertices of the original tiling. The rotoscaling operation, i.e. scaling by τ^2 and rotation around $\pi/10$, is also a symmetry operation of a pentagram (*white lines*), mapping each vertex of a pentagram onto another one, for instance, A to A' (m) (from Steurer and Deloudi, 2009).

PT show scaling symmetry by powers n of τ as illustrated in **Figure 6**(k–m). This means that the scaling operation maps each vertex of a PT onto another vertex of the PT, no new vertices are generated if n is a positive integer. In case of n being an even positive number, the scaling symmetry of the PT is also the scaling symmetry of the pentagramma. This is shown in **Figure 6**(m), where A is transformed into A' by a rotoscaling operation, i.e. scaling by τ^2 and rotation around $\pi/10$.

1.2.2.5 Convex Uniform Polyhedra and their Packing

A convex polyhedron is called *regular* if its faces are all equal and regular polygons are surrounding all vertices (corners) in the same way (with the same solid angles). In other words, *regular polyhedra* are *face-* and *vertex-transitive*. Without the latter condition, one obtains the *nonuniform face-transitive polyhedra*,

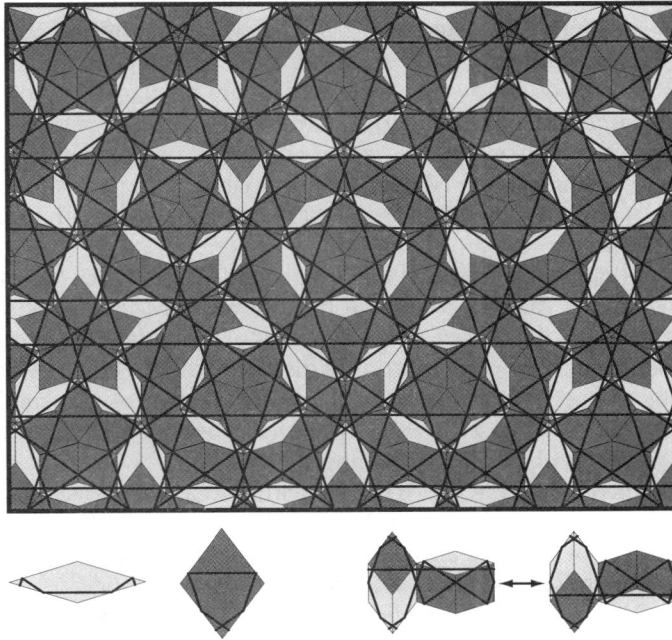

Figure 7 The Penrose rhomb tiling (P3) with Ammann lines drawn in. The decoration of the unit tiles by Ammann line segments and the action of simpleton flips are shown (from Steurer and Deloudi, 2009).

such as the rhombic dodecahedron, triacontahedron, or the pentagonal bipyramid, which are important AET or cluster shells in case of QC structures.

In 3D, there are five regular polyhedra, the *Platonic solids*: the tetrahedron, point group $\bar{4}3m$, 3^3; the octahedron, $m\bar{3}m$, 3^4; the hexahedron (cube), $m\bar{3}m$, 4^3; the icosahedron, $m\bar{3}\,\bar{5}$, 3^5; the pentagon-dodecahedron, $m\bar{3}\,\bar{5}$, 5^3. A uniform polyhedron can be characterized by its vertex configuration, which just gives the kind of polygons along a circuit around a vertex. The *Cundy and Rollet symbol* of a uniform polyhedron, p^q, denotes the type of face (p-gon), with q, the number of faces surrounding each vertex. Since all point groups of the Platonic solids contain a cubic point group as subgroup, there is always a defined orientation relationship to cubic symmetry (**Figure 8**).

Figure 8 The Platonic solids inscribed in cubic unit cells to show their orientation relationships to the two- and threefold axes of the cube: tetrahedron, point group $\bar{4}3m$, 3^3; octahedron, $m\bar{3}m$, 3^4; hexahedron (cube), $m\bar{3}m$, 4^3; icosahedron, $m\bar{3}\,\bar{5}$, 3^5; pentagon-dodecahedron, $m\bar{3}\,\bar{5}$, 5^3.

The dual q^p of any Platonic solid p^q is a Platonic solid again. The tetrahedron is its own dual, cube and octahedron are duals of each other, and so are the icosahedron and the pentagon-dodecahedron.

The other class of convex uniform polyhedra, i.e. with one type of vertex configuration only (*vertex-transitive*), is the *semiregular polyhedra*. Their faces all are regular polygons of at least two kinds. They include the 13 *Archimedean solids* (**Figure 9**) and an infinite number of prisms and antiprisms with n-fold rotational symmetry. The prisms, $4^2.n$, are constituted by two congruent n-gons plus n squares and show point symmetry N/mmm, with $N = n$. The antiprisms, $3^3.n$, consist of two congruent n-gons rotated by an angle of π/n against each other plus n equilateral triangles with point symmetry $\overline{(2N)}m2$ for $N = n$ and n even, and $\overline{N}2/m$ for $N = n$ and n odd.

The Archimedean solids can all be inscribed in a sphere and in one of the Platonic solids. Their duals are the *Catalan solids*, with faces that are congruent but not regular (*face-transitive*); instead of circumspheres like the Archimedean solids, they have inspheres. The midspheres, touching the edges, are common to both of them. The face configuration is used for the description of these face-transitive polyhedra. It is given by the sequential listing of the number of faces that meet around each vertex around a face. For instance, $V(3.4)^2$ describes the rhombic dodecahedron, where at the vertices around one rhombic face 3, 4, 3, 4 rhombs, respectively, meet.

The only uniform polyhedra that tile the 3D space without gaps and overlaps are the cube and the truncated octahedron (*Kelvin polyhedron*, Voronoi cell of the *bcc* lattice) (**Figure 10**(a)). Their packings have symmetry $Pm\overline{3}m$ in case of the cube and $Im\overline{3}m$ in case of the Kelvin polyhedron. In all other cases, at least two types of (semi)regular polyhedra are needed for space filling.

Truncated cubes can be packed sharing their octagonal faces, the remaining voids are filled by octahedra (**Figure 10**(b)). Octahedra are also needed to make the packing of square-sharing cuboctahedra space filling (**Figure 10**(c)). The gaps left in an edge-connected framework of octahedra can be filled by tetrahedra (**Figure 10**(d)). The same is true for a packing of hexagon sharing truncated tetrahedra (**Figure 10**(e)).

A *bcc* packing of truncated cuboctahedra, which touch each other with their hexagonal faces, need octagonal prisms for filling the gaps (**Figure 10**(f)). Three polyhedra are needed for the following six packings: square-sharing rhombicuboctahedra in a primitive cubic arrangement leave holes, which can be filled by cubes and cuboctahedra in the ratio 1:3:1 (**Figure 10**(g)). The gaps in an *fcc* packing of square-sharing rhombicuboctahedra can be filled by cubes and tetrahedra (**Figure 10** (h)).

Truncated cuboctahedra, in contact with their octagonal faces, form gaps to be filled with cubes and truncated octahedra (**Figure 10**(i)). Truncated octahedra are fully surrounded by cuboctahedra, sharing the square faces, and by truncated tetrahedra linked by the hexagonal faces (**Figure 10**(j)). Square-sharing truncated cuboctahedra form an *fcc* packing with voids, which can be filled with truncated cubes and truncated tetrahedra (**Figure 10**(k)). Finally, a packing that needs four types of uniform polyhedra to be space filling: truncated cubes linked via octagonal prisms form a primitive cubic tiling with rhombicuboctahedra in the center of the cubic unit cell and cubes filling the residual gaps (**Figure 10**(l)).

In case of quasiperiodic structures, polyhedra with icosahedral point-group symmetry play an important role. One of these, the triacontahedron, a zonohedron, and its packing will be discussed in the following. The rhombic triacontahedron (**Figure 9**(o)), a zonohedron with 30 faces, 60 edges and 32 vertices, belongs to the Catalan solids and is dual to the icosidodecahedron (**Figure 9**(h)). As dual to a vertex-transitive Archimedean solid, it is face-transitive and, additionally, also edge-transitive such as the rhombic dodecahedron (**Figure 9**(n)), another Catalan solid, and the Platonic solids. Subsets of its vertices form an icosahedron, a dodecahedron, a cube and a tetrahedron. The face configuration is given by the symbol $V3.5.3.5$; this means that each face is surrounded by vertices, where 3, 5, 3 and 5 faces meet.

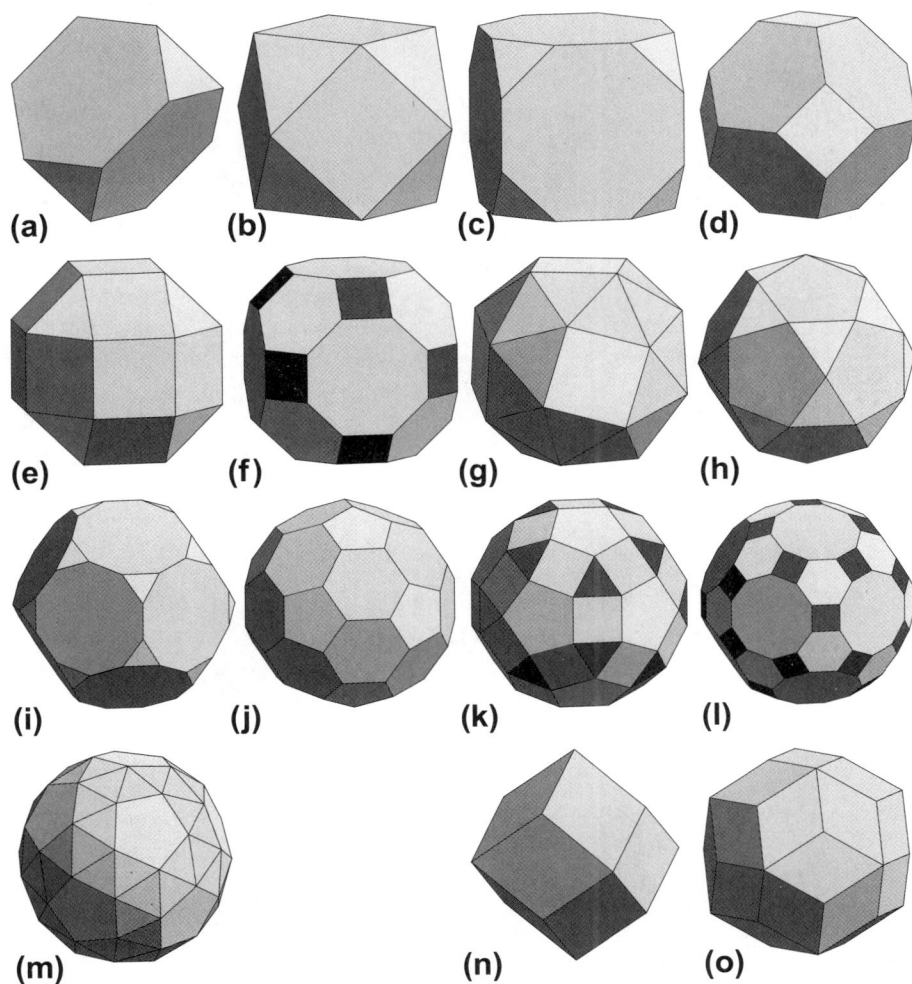

Figure 9 The 13 vertex-transitive Archimedean solids with their vertex configuration and point symmetry: (a) truncated (Friauf) tetrahedron, 3.6^2, point group $\bar{4}3m$; (b) cuboctahedron, $(3.4)^2$, $m\bar{3}m$; (c) truncated cube, 3.8^2, $m\bar{3}m$; (d) truncated octahedron, 4.6^2, $m\bar{3}m$; (e) small rhombicuboctahedron, 3.4^3, $m\bar{3}m$; (f) truncated cuboctahedron, $4.6.8$, $m\bar{3}m$; (g) snub cube, $3^4.4$, 432 (only one enantiomorph shown); (h) icosidodecahedron, $(3.5)^2$, $m\bar{3}\,\bar{5}$; (i) truncated dodecahedron, 3.10^2, $m\bar{3}\,\bar{5}$; (j) truncated icosahedron, 5.6^2, $m\bar{3}\,\bar{5}$; (k) small rhombicosidodecahedron, $3.4.5.4$, $m\bar{3}\,\bar{5}$; (l) truncated icosidodecahedron (great rhombicosidodecahedron), $4.6.10$, $m\bar{3}\,\bar{5}$; (m) snub dodecahedron, $3^4.5$, 235 (only one enantiomorph shown). The (n) rhombic dodecahedron, $V(3.4)^2$, $m\bar{3}m$, and the (o) rhombic triacontahedron, $V(3.5)^2$, $m\bar{3}\,\bar{5}$, are duals of the cuboctahedron (b) and the icosidodecahedron (h), respectively, and belong to the face-transitive Catalan solids.

In periodic and quasiperiodic structures, triacontahedra can partially overlap in well-defined ways (**Figure 11**). The overlap regions are zonohedra such as the rhombic icosahedron, the oblate rhombohedron and the rhombic dodecahedron. In case of rational approximants to quasiperiodic structures, the triacontahedral clusters pack in a way to form cubic unit cells as shown in **Figure 11**(d,e). In case of

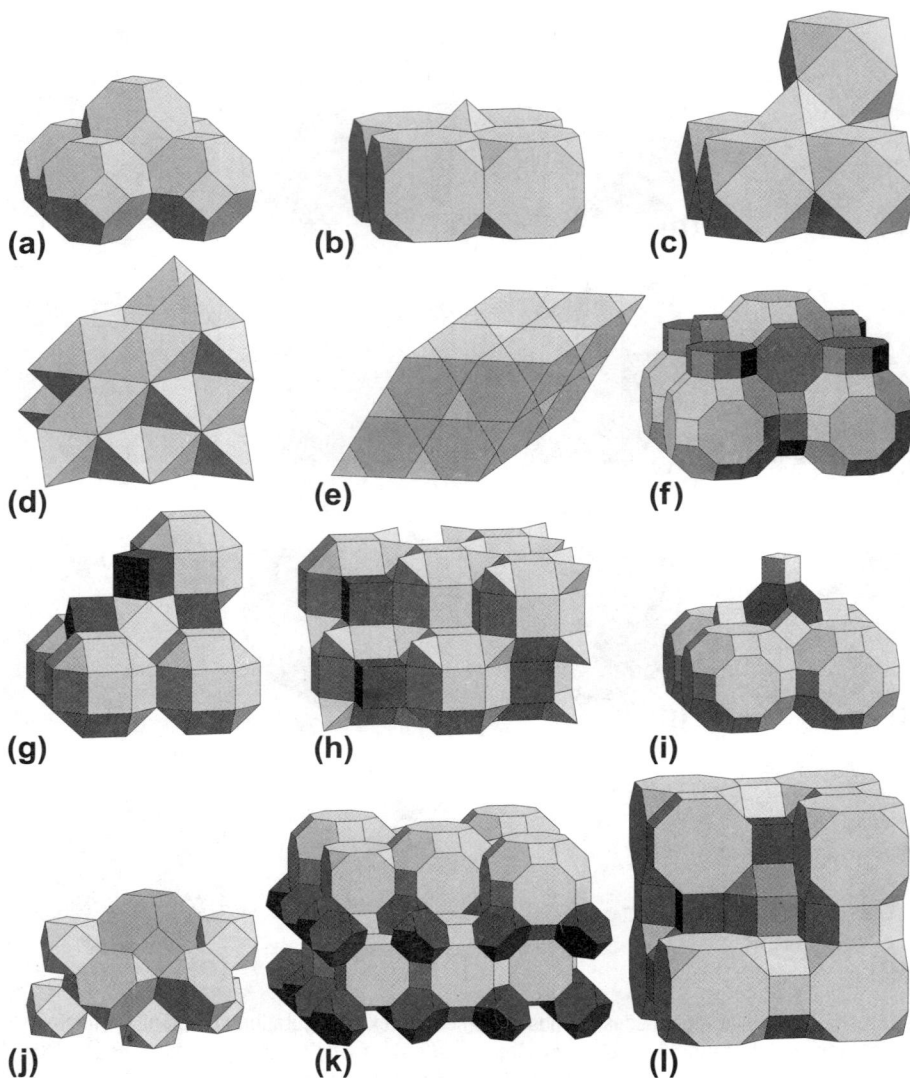

Figure 10 Packings of regular and semiregular polyhedra with resulting cubic space-group symmetry: (a) Truncated octahedra, $Im\overline{3}m$; (b) truncated cubes + octahedra, $Pm\overline{3}m$; (c) cuboctahedra + octahedra, $Pm\overline{3}m$; (d) octahedra + tetrahedra, $Fm\overline{3}m$; (e) truncated tetrahedra + tetrahedra, $Fm\overline{3}m$; (f) truncated cuboctahedra + octagonal prisms, $Im\overline{3}m$; (g) rhombicuboctahedra + cuboctahedra + cubes, $Pm\overline{3}m$; (h) rhombicuboctahedra + tetrahedra + cubes, $Fm\overline{3}m$; (i) truncated cuboctahedra + truncated octahedra + cubes, $Pm\overline{3}m$; (j) truncated octahedra + cuboctahedra + truncated tetrahedra, $Fm\overline{3}m$; (k) truncated cuboctahedra + truncated cubes + truncated tetrahedra, $Fm\overline{3}m$; (l) rhombicuboctahedra + truncated cubes + octagonal prisms + cubes, $Pm\overline{3}m$.

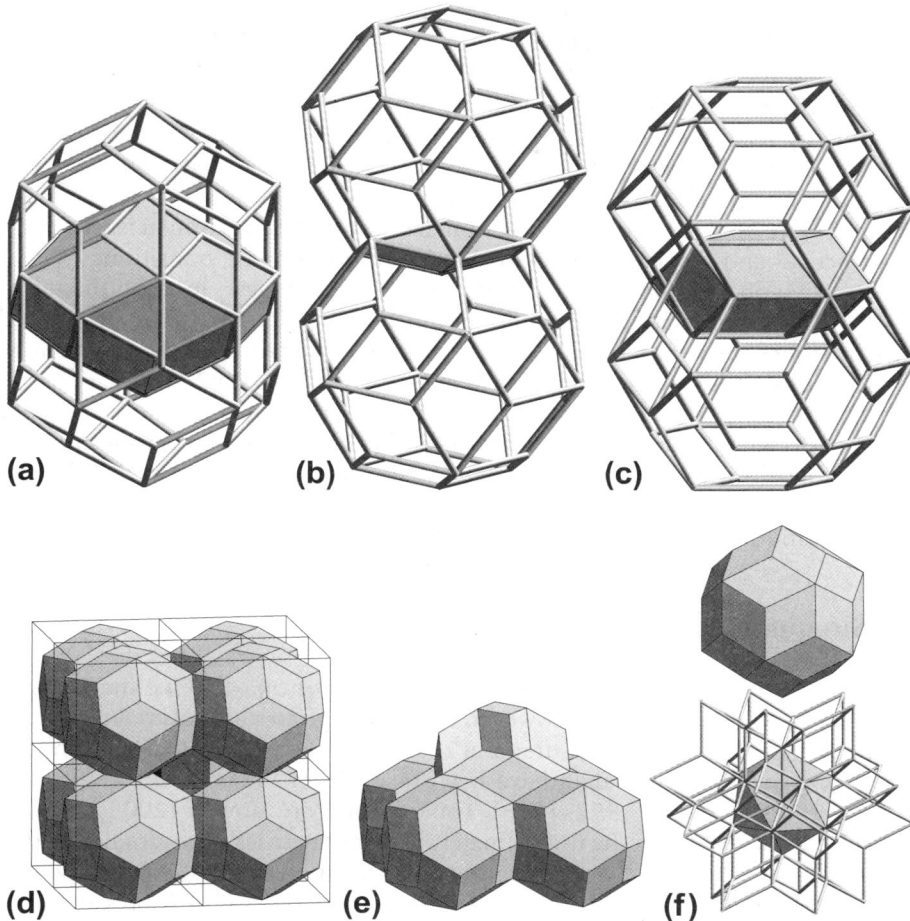

Figure 11 Triacontahedra overlapping along the (a) five-, (b) three- and (c) twofold directions. The shared volumes are zonohedra: (a) rhombic icosahedron, (b) oblate rhombohedron and (c) rhombic dodecahedron. The *bcc* packing of triacontahedra is shown in (d). The triaconahedra share faces along the twofold direction and one oblate rhombohedron in each of the eight threefold directions. Consequently, the empty space left in the center of eight triacontahedra has the shape of a dimpled triacontahedron (e). In (f), the packing of a triacontahedron into one of the 12 pentagonal dimples of a rhombic hexecontahedron is illustrated. A rhombic hexecontahedron can be formed by 20 prolate rhombohedra. The 12 inner vertices form an icosahedron (from Steurer and Deloudi, 2009).

quasiperiodic structures, at points of local high symmetry, arrangements like that shown in **Figure 11**(f) are frequently found.

1.2.3 Polymorphism

Most of the chemical elements and many intermetallics undergo structural (*polymorphic*) phase transformations from one modification to another as function of pressure and/or temperature. In case of the elements, the different modifications are also called *allotropes*. There are two basically different types of structural phase transformations: first-order transformations with a jumpwise change in the first derivatives of the Gibbs free energy, $G = H - TS$, i.e. in volume or entropy, and second-order transitions which show a jump in the second derivatives, i.e. heat capacity or compressibility. In both cases, the crystal structure changes at the transition point: for a first-order transition in a discontinuous way; generally no symmetry relationship needs to exist between the two modifications; for a second-order transition, a continuously changing order parameter can be defined and a group/subgroup relationship exists for the symmetry of the two modifications.

With regard to the structural changes resulting from a phase transformation, it is useful to distinguish between reconstructive- and displacive (dilatative, orientational)-phase transitions (see, e.g. Buerger, 1951). Reconstructive-phase transitions with essential changes in coordination numbers ($cP2$-Fe $\Leftrightarrow$ $cF4$-Fe, for instance, with coordination numbers CN = 8 and CN = 12, respectively), atomic positions and, sometimes, also in chemical bonding (nonmetallic gray $cF8$-Sn and metallic white $tI4$-Sn, for instance, with strongly differing minimum distances $d^{cF8\text{-}Sn} = 3.02$ Å and $d^{tI4\text{-}Sn} = 1.54$ Å). Reconstructive transformations are always of first order. Displacive-phase transitions are characterized by small atomic shifts not changing the first coordination shell (for instance, martensitic diffusionless lattice rearrangement). Order/disorder transitions are related to the ordered or disordered arrangement of structure elements (copper–gold system, for instance). Second-order (continuous) transitions can be characterized by continuously varying order parameters.

1.3 Crystal Structures of the Metallic Elements

In the following, the crystal structures of all modifications (allotropes) of the metallic elements (see **Table 1** in Section 1.2) will be discussed. However, only those phases are covered that are stable either at room temperature (RT) and given pressure or at room pressure (RP) and given temperature. If not indicated otherwise, crystal structure data have been taken from Villars and Calvert (1991), Young (1991), Massalski (1990), Tonkov (1992–1996), Tonkov and Ponyatovsky (2005) or *Pearson's Crystal Data* (Villars and Cenzual, 2011). In the unfortunately not-so-rare cases of contradicting data, the most recent and, hopefully, most reliable ones have been used, otherwise the Pearson symbol has been replaced by a question mark. Particularly, some of the structural data of high-pressure phases that are derived from very small and/or low-quality X-ray diffraction data sets may be revised once better data will be available.

1.3.1 Groups 1 and 2: Alkali and Alkaline Earth Metals

The alkali and alkaline earth metals (**Table 6**) can be seen as typical simple metals, conforming closest to the free electron-gas model of metals, at least at ambient temperature and pressure. The outer electrons occupy the ns orbitals, and ionization removes the electrons of a whole shell, in this way drastically reducing the atomic radius (e.g. $r_{Li} = 1.56$ Å, $r_{Li}^+ = 0.60$ Å). The absence of directional bonding leads to simple sphere packings. At ambient conditions, the alkali metals all crystallize in the *bcc* $cI2$-W structure type (see **Figure 27** (a) in Section 1.4.4). Generally, at higher temperatures, the *bcc* structure is assumed to be more stable than the *ccp* or *hcp* one because it allows larger vibrational entropy. At lower temperatures or higher pressures, the $cI2$-W-type structure transforms martensitically into the closest packed structure types $hR3$-Sm or $cF4$-Cu, respectively (see **Figure 17** in Section 1.3.2.1).

Under pressure and at RT, $cI2$-Li first transforms to a *ccp* phase due to s–p electron transfer; subsequently, above 39 GPa, it becomes liquid until it solidifies again at 67 GPa to semimetallic $oC40$-Li

Table 6 Structural information for the elements of groups 1 and 2, akali and alkaline earth metals, respectively. In the first line of each element box the element name, atomic number Z, and electronic ground-state configuration is given. Below, the different modifications with their prototype structure PT, if known, are listed together with the temperature T and pressure P limiting their stability range; if no temperature or pressure is given this means ambient conditions. Atomic volumes $V_{at} = V_{uc}/n_{at}$ are given where available, with V_{uc} and n_{at} the unit cell volume and the number of atoms per unit cell. Atomic volumes V_{at} are given at ambient conditions if no pressure P and temperature T is given in the form $T|P$. Only those HP and HT phases are listed that are stable at RT or RP, respectively. If no other reference is given, data are based on the compilations by Villars and Calvert (1991), Tonkov (1992–1996), Tonkov and Ponyatovsky (2005), Young (1991)

1	PT	T [K]	P [GPa]	V_{at} [Å³]	T|P	Ref.[a]
Li Lithium 3 [He]2s^1						
hR3	Sm	<74		20.99	20|RP	[1,2]
cI2	W			24.32		
cF4	Cu		>7.5	14.83	RT|8	[3]
liquid			>40		RT	[4]
oC40	Li		>67	6.7	240|75	[4-5]
oC24	Li		>86	6.5	RT|88	[4-6]
cP4	Li		>300	3.42	RT|300	[7]
Na Sodium 11 [Ne]3s^1						
hR3	Sm	<36		37.74	20|RP	
cI2	W		<65	39.50		
cF4	Cu		<105			
cI16	Li		<118	9.77	RT|115	[8]
oP8	MnP		<125	9.45	RT|119	[8]
tI19.3	inc[b]		<125	8.62	RT|147	[9]
hP4	La		>200	7.88	RT|200	[10]
K Potassium 19 [Ar]4s^1						
cI2	W		<11.4	75.72	301|RP	
cF4	Cu		<23	34.11	RT|12.4	[11]
tI19.2	inc[b]		>20	23.48	RT|22.2	[12]
oP8	MnP		>54	14.79	RT|58	[12]
tI4	Sn		>90			[12]
oC16			>96			[12]

2	PT	T [K]	P [GPa]	V_{at} [Å³]	T|P	Ref.[a]
Be Beryllium 4 [He]2s^2						
hP2	Mg	<1523		8.11		
cI2	W	>1523	<12	8.30	1528|RP	
oP4			>12	6.95	RT|28.3	
Mg Magnesium 12 [Ne]3s^2						
hP2	Mg		<50	23.24		
cI2	W		>50	12.88	RT|58	
Ca Calcium 20 [Ar]4s^2						
cF4	Cu	<721		43.63		
cI2	W	>721 or	>20	44.95	740|RP	
cP1	Po		>32	18.99	RT|42	[21,22]
tP8			>119			[23]
oC8			>143	10.50	RT|154	[23]
oP4			>158	9.94	RT|172	[24]

(Continued)

Table 6 Structural information for the elements of groups 1 and 2, akali and alkaline earth metals, respectively In the first line of each element box the element name, atomic number Z, and electronic ground-state configuration is given. Below, the different modifications with their prototype structure PT, if known, are listed together with the temperature T and pressure P limiting their stability range; if no temperature or pressure is given this means ambient conditions. Atomic volumes $V_{at} = V_{uc}/n_{at}$ are given where available, with V_{uc} and n_{at} the unit cell volume and the number of atoms per unit cell. Atomic volumes V_{at} are given at ambient conditions if no pressure P and temperature T is given in the form T|P. Only those HP and HT phases are listed that are stable at RT or RP, respectively. If no other reference is given, data are based on the compilations by Villars and Calvert (1991), Tonkov (1992–1996), Tonkov and Ponyatovsky (2005), Young (1991)—cont'd

1	PT	T [K]	P [GPa]	V_{at} [Å^3]	T|P	Ref.[a]
Rb Rubidium 37 [Kr]5s^1						
cI2	W	<7		87.10		
cF4	Cu	<13		37.63	RT|9.0	[13]
oC52	Rb	<16.6		31.42	RT|14.3	[14]
tI19.3	inc[b]	<20		28.85	RT|16.8	[15,16]
tI4	Sn	<48		25.96	RT|20.2	[15]
oC16	Cs	>48		17.97	RT|48.1	[17]
Cs Cesium 55 [Xe]6s^1						
cI2	W			110.45		
cF4	Cu		>2.3	53.57	RT|4.1	
oC84	Cs		>4.2	50.21	RT|4.3	[18]
tI4	Sn		>4.3	35.01	RT|8	
oC16	Cs		>10	26.09	RT|25.8	[19]
hP4	La		>72	11.48	RT|92	[20]
Fr Francium 87 [Rn]7s^1						

2	PT	T [K]	P [GPa]	V_{at} [Å^3]	T|P	Ref.[a]
Sr Strontium 38 [Kr]5s^2						
cF4	Cu	<504 or	<3.5	56.08		
hP2	Mg	<896		57.05	521|RP	
cI2	W	>896 or	>25	57.75	901|RP	
tI4	Sn		>24.4	22.42	RT|34.8	[25]
mC12			>37.7	20.53	RT|41.7	[26]
tI10.8	inc[b]		>46.3	17.74	RT|56	[27]
Ba Barium 56 [Xe]6s^2						
cI2	W		<5.5	62.99		
hP2	Mg		<12.6	38.44	RT|6.9	[28]
tI10.8	inc[b]		<45	31.17	RT|12.0	[29]
hP2	Mg		>45	20.34	RT|53	[29]
Ra Radium 88 [Rn]7s^2						
cI2	W			68.22		

[a][1] Ernst et al., [1986]; [2] Berliner & Werner [1986]; [3] Hanfland et al., [2000]; [4] Guillaume et al., [2011]; [5] Marqués et al., [2011]; [6] Rousseau et al., [2005]; [7] Ma et al., [2008]; [8] Gregoryanz et al., [2008]; [9] Lundegaard et al., [2009a]; [10] Ma et al., [2009]; [11] McMahon et al., [2006b]; [12] Lundegaard et al., [2009b]; [13] Takemura et al., [2000]; [14] Nelmes et al., [2002]; [15] Schwarz et al., [2001a]; [16] McMahon et al., [2001a]; [17] Schwarz et al., [1999b]; [18] McMahon et al., [2001b]; [19] Schwarz et al., [1998]; [20] Takemura et al., [2000]; [21] Mao et al., [2010]; [22] Yao et al., [2010]; [23] Fujihisa et al., [2008]; [24] Nakamoto et al., [2010]; [25] Allan et al., [1998]; [26] Bovornratanaraks et al., [2006]; [27] McMahon et al., [2000]; [28] Takemura [1994]; [29] Nelmes et al., [1999].

[b]Incommensurate host-guest structure.

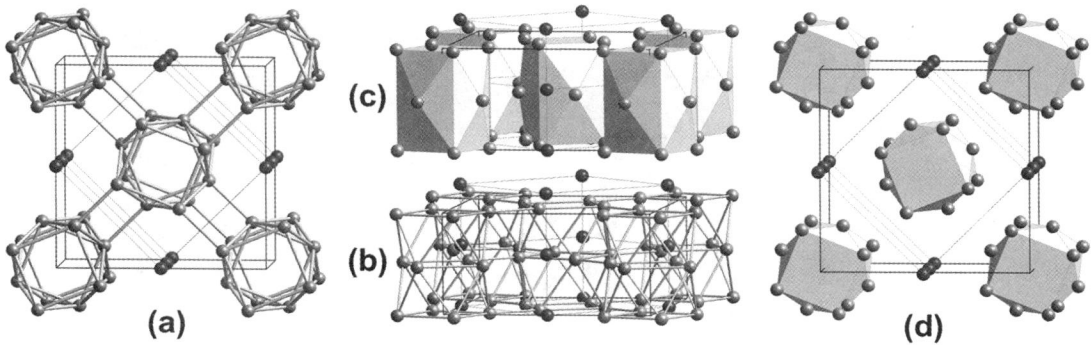

Figure 12 The structure of tI9.3-Rb in ball-and-stick as well as in polyhedral representation. The tetragonal body-centered host structure (bright atoms) contains a tetragonal body-centered guest structure (dark atoms). Both substructures are incommensurate to each other along the fourfold axis. The structure is closely related to the W substructure of tI32-W_5Si_3.

(Guillaume e al., 2011; Marqués et al., 2011). Calculations show that, with increasing pressure, electron density is pushed away from the atomic cores by Coulomb repulsion and accumulated at interstitial sites. At 86 GPa, again a poorly metallic modification forms, oC24-Li. Beyond 300 GPa, a cubic structure, cP4-Li, was predicted with sixfold coordinated Li. With $T_m \approx 190$ K, lithium has by far the lowest melting temperature at pressures around 50 GPa of any known material.

Sodium has a similar phase sequence as Li up to 125 GPa. Then, cI16-Na transforms, via semi-metallic oP8-Na and with a volume decrease of 0.5%, to tI19.3-Na, a tetragonal body-centered incommensurate host–guest structure (Lundegaard et al., 2009a). Remarkably, the guest sublattice is monoclinic. Around ≈ 200 GPa, sodium becomes an optically transparent wide-band-gap dielectric (Ma et al., 2009). This effect is explained by p–d hybridization of valence electrons and their repulsion by core electrons into interstitial sites of the compressed hP4-La type structure. The c/a ratio is with 1.391 at 320 GPa, for instance, extremely small compared to the ideal value of $4\sqrt{2/3} = 3.266$.

The phase sequence of potassium differs from that of sodium. In case of the heavier alkali metals, s–d electron transfer takes place under high pressure. At 23 GPa, cF4-K transforms into a tetragonal, incommensurately modulated host–guest structure, tI19.2-K, where the body-centered 16-atom host framework of square antiprisms contains C-centered linear chains of K atoms. Three further high-pressure structures have been observed, with one of them, oP8-K, also found in the phase sequence of sodium.

Rubidium shows a similar phase sequence as potassium. The transition from cF4-Rb to tI19-Rb (**Figure 12**) runs via a complex orthorhombic structure, oC52-Rb (**Figure 13**), however. The volume changes are 2.5% and 0.9%, respectively. The structure of oC52-Rb can be described as a six-layer stacking of 8- and 10-atom layers with a sequence 8-10-8-8-10-8 (Nelmes et al., 2002). At ≈ 17 GPa, oC52-Rb transforms into the body-centered tetragonal, incommensurate host–guest structure tI19.3-Rb (McMahon et al., 2001a). While sodium, potassium and rubidium all have the same type of host structures, they differ in their guest structures, which are monoclinic, C-centered and body-centered, respectively. Close to the pressure for the transition to tI4-Rb, the ratio of the periods of host and guest along the c-direction, c_H/c_G, reaches the commensurate value 5/3. The structural relationships and mechanisms of phase transitions between all six modifications of rubidium have been discussed by Katzke and Toledano (2005). Accordingly, oC52-Rb can be described as modulated structure with approximate modulation period of $13a$ along the initial [100] direction of cI2-Rb; tI4-Rb can be seen as lock-in phase of tI19.3-Rb.

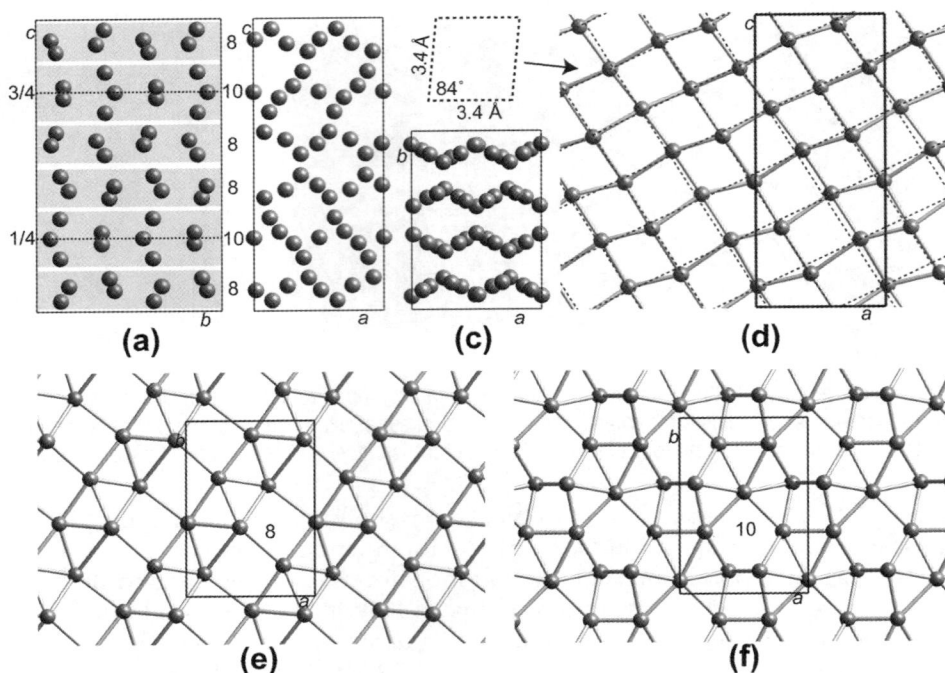

Figure 13   The structure of *oC*52-Rb in different views: (a–c) Projections along the *a*-, *b*- and *c*-directions, respectively. In (d), a puckered (010)-section with $0.3 \leq y \leq 0.5$ is depicted with an oblique dashed grid overlaid; its unit cell is shown at left of the subfigure. In (e–f), (001)-sections are shown; they correspond to the gray shaded regions marked in (a). These puckered sections contain eight or 10 atoms per unit cell as indicated.

Cesium shows a similar phase sequence as rubidium. The main difference is that the *ccp* modification of cesium transforms into an even more complex structure than *oc*52-Rb, namely *oC*84-Cs. This structure is also built from 8- and 10-atom layers, but now with the 10-layer sequence 8-8-10-8-8-8-8-10-8-8 (Nelmes et al., 2002) (**Figures 14–16**).

There is an extremely strong dependence of the atomic volume on pressure, which increases with increasing atomic number due to the shielding of the outer electrons by the increasing number of inner electron shells. In case of Cs, for instance, with increasing pressure, the valence electrons move from the 6s to the unfilled 5d band until the s–d electron transfer is complete at about 10 GPa. Due to the unique 5d (or 5d–5p hybridized) monovalent electronic states, directional bonding leads to the rather complex structures observed (Takemura et al., 1991). At higher pressure, 5p core repulsion dominates over 5d bonding resulting in the closest packed structure of *hP*4-Cs.

The alkaline earth metals behave to some extent similar as the alkali metals (**Table 6**). They crystallize under ambient conditions either in one of the two closest packed structures, *ccp* or *hcp*, or in the *bcc* structure type and also show several allotropic forms including incommensurate phases. The large deviation $c/a = 1.56$ from the ideal value of 1.633 for beryllium at ambient conditions indicates covalent bonding contributions. The structure of the high-pressure phase *oP*4-Be corresponds to slightly distorted *hcp*.

A long-lasting discussion has been led about the structure that calcium adopts above 32 GPa, which was experimentally observed as of the *cP*1-Po type (Gu et al., 2009 and references therein). By first

Crystal Structures of Metallic Elements and Compounds 29
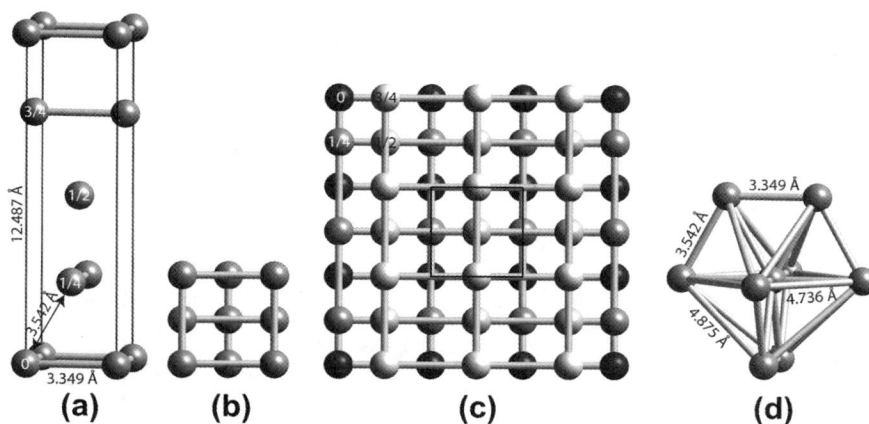

Figure 14 The structure of $tI4$-Cs in different views: Projections along (a) the a- and (b) c-direction. The packing of layers is illustrated in (c) with a projection of the layers at $z = 0$ (black atoms), $z = 1/4$ (dark gray atoms), $z = 1/2$ (gray atoms), $z = 3/4$ (light gray atoms). An AET is depicted in (d).

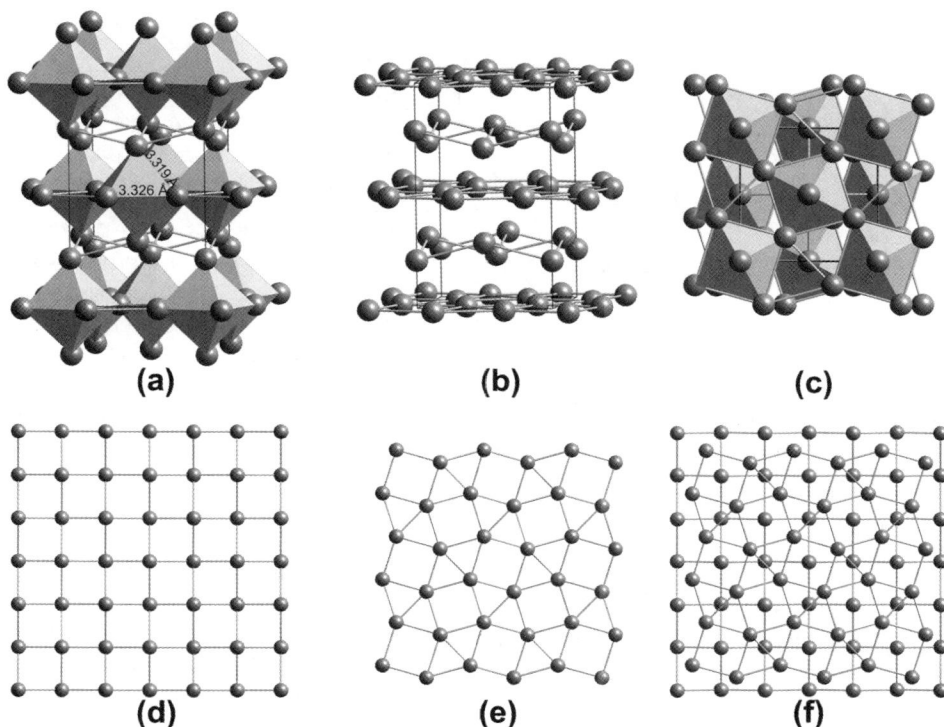

Figure 15 Different representations of the structure of $oC16$-Cs: (a,c) corner-connected octahedra in two projections; (b) packing of flat and puckered layers; layers in $x = 0$ (d) and $\approx 1/4$ (e), corresponding to Archimedean square tiling 4^4 and snub square tiling $3^2.4.3.4$, respectively, as well as their superposition (f).

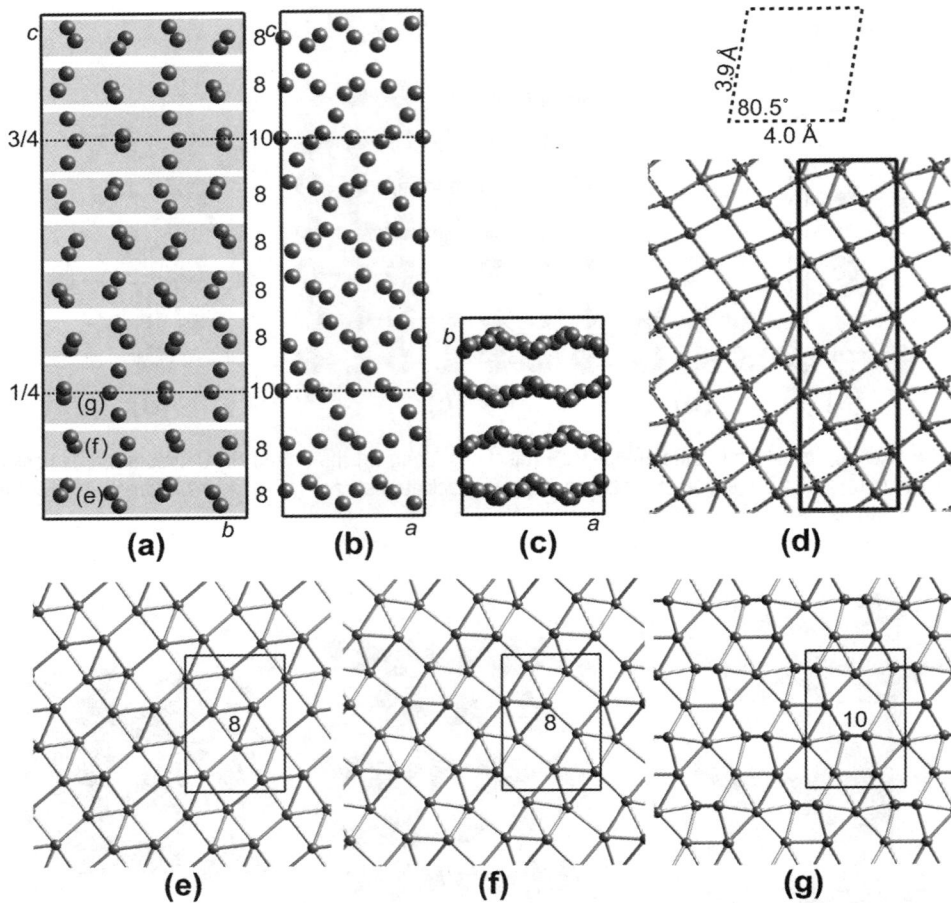

Figure 16 The structure of $oC84$-Cs in different views: (a–c) Projections along the a-, b- and c-directions, respectively. In (d), a puckered (010)-section with $0.07 \leq y \leq 0.17$ is depicted with an oblique dashed grid overlaid; its unit cell is shown at the top of the subfigure. In (e–g), (001)-sections are shown; they correspond to the gray shaded regions marked in (a). These puckered sections contain eight or 10 atoms per unit cell as indicated. The layers in (e) and (f) correspond to the Archimedean elongated triangular tiling $3^3.4^2$.

principles calculations, its dynamical stability at 300 K was confirmed (Yao et al., 2010), while a slightly rhombohedrally distorted version of it, $hR1$-Ca, was suggested by Mao et al. (2010).

At higher pressures, strontium runs through a phase sequence $tI4$-Sr $\rightarrow$ $mC12$-Sr $\rightarrow$ $tI10.8$-Sr. While $mC12$-Sr can be seen as monoclinic superstructure (helical distortion) of $tI4$-Sr, $tI10.8$-Sr corresponds to an incommensurate, tetragonal body-centered host–guest structure (with the guest being C-centered). Barium shows the same type of self-hosting structure, $tI10.8$-Ba, as intermediate phase between $hP2$-Ba below 12.6 GPa and above 45 GPa (Reed et al., 2000). Low-pressure $hP2$-Ba shows a strong decrease in the c/a ratio from 1.58 to 1.50 due to the s–d electron transfer, whereas that of high-pressure $hP2$-Ba is with a value of 1.575 (similar as $hP2$-Be at ambient conditions), pressure-independent in the studied range up to 90 GPa. This has been explained in the way that the s–d transfer is finished already at about 40 GPa (Takemura, 1994).

1.3.2 Groups 3–12: Transition Metals (TM)

The elements of groups 3–10 are typical metals that have in common that their d orbitals are partially occupied. These orbitals are only slightly screened by the outer s electrons leading to significantly different chemical properties of the transition elements going from left to right in the periodic table. The atomic volumes decrease rapidly with increasing number of electrons in bonding d orbitals due to cohesion, and increase with filling the antibonding d orbitals. The anomalous behavior of the 3d-transition metals Mn, Fe and Co may be explained by the existence of nonbonding d electrons (Pearson, 1972). Under pressure, sp bands rise in energy faster than d bands, resulting in an sp to d band electron transfer.

1.3.2.1 *Groups 3 and 4: Sc, Y, La, Ti, Zr, Hf*

Scandium, yttrium, lanthanum and actinium (**Table 7**) behave similar to some extent. They show similar phase sequences, where the high-pressure phases of the light elements occur as ambient pressure phases of the heavy homologs. However, there are some peculiarities for scandium. The first high-pressure modification has an incommensurate host/guest structure with the averaged Pearson symbol *tI*10.6; the *bcc* framework of the tetragonal host structure accommodates the *C*-face-centered guest structure (McMahon et al., 2006a). It can be described by the superspace group *I*4/*mcm*(00γ) and seems to be isostructural to the high-pressure modification of strontium found above 46.3 GPa (see **Table 6** in Section 1.3.1). Between 23 and 101 GPa, the wave vector component γ varies between 1.28 and 1.36, passing through the commensurate value of 4/3 at 72 GPa.

At highest pressures, above 240 GPa, *hP*6-Sc forms; the symmetry of its helical structure can be described by space group *P*6$_1$22 (Akahama et al., 2005). It was explained to result from a modulation of the stacking sequence of the (1,1,1) planes in an *fcc* arrangement due to 3d orbital interactions along the helical chains; the 3d orbital is considered to be hybridized with the 4s orbital. Due to the s–d electron transfer, the compressibility of Sc is rather high; at 297 GPa, the atomic volume amounts to only 32% of its value at ambient conditions.

Contrary to scandium, with increasing pressure, yttrium largely shows the typical rare-earth sequence of closest packed structures *hP*2-Mg → *hR*3-Sm → *hP*4-La → *cF*4-Cu. However, at highest pressure, due to strong electron–phonon coupling, yttrium adopts a hexagonal structure that can be seen as distorted *fcc*.

*hP*4-La is, with the sequence ..ACAB.., one of the simpler closest packed polytypic structures common for the lanthanoids (**Figure 17**). Another typical structure for lanthanoids is the *hR*3 phase of samarium with stacking sequence ..ABABCBCAC... The transition from *cF*4-La to *hR*8-La at 7 GPa is driven by pressure-induced softening of a transverse acoustic phonon mode distorting the *fcc* phase trigonally (Gao et al., 2007). At 60 GPa, the structure transforms back to the *cF*4-La (Porsch and Holzapfel, 1993).

Titanium, zirconium and hafnium (**Table 7**) crystallize in a slightly compressed *hcp* structure and transform to *bcc* at higher temperatures. At elevated pressures, the *hP*3-Ti phase is obtained by a martensitic transformation (**Figure 18**). Its packing density is with ∼0.57, slightly larger than that for simple-cubic *cP*1-Po (∼0.52) but substantially lower than for *cI*2-W (∼0.68) or *ccp* and *hcp* (∼0.74) structures. Calculations have shown that the *hP*3-Ti phase is stabilized by covalent bonding contributions from s–d electron transfer. At even higher pressures, zirconium and hafnium transform to the

Table 7   Structural information for the elements of groups 3 and 4. In the first line of each element box the element name, atomic number Z, and electronic ground-state configuration is given. Below, the different modifications with their prototype structure PT, if known, are listed together with the temperature T and pressure P limiting their stability range; if no temperature or pressure is given this means ambient conditions. Atomic volumes $V_{at} = V_{uc}/n_{at}$ are given where available, with V_{uc} and n_{at} the unit cell volume and the number of atoms per unit cell. Atomic volumes V_{at} are given at ambient conditions if no pressure P and temperature T is given in the form T|P. Only those HP and HT phases are listed that are stable at RT or RP, respectively. If no other reference is given, data are based on the compilations by Villars and Calvert (1991), Tonkov (1992–1996), Tonkov and Ponyatovsky (2005), Young (1991)

Group 3

PT		T [K]	P [GPa]	V_{at} [ų]	T|P	Ref.[a]
Sc Scandium 21 [Ar]3d¹4s²						
hP2	Mg			24.96		[1]
cl2	W	>1610		26.41	1623|RP	[1]
tl10.6	inc[b]		>20.5	18.65	RT|23	[2]
?			>104			[3]
?			>140			[3]
hP6	Sc		>240	7.96	RT|297	[3]
Y Yttrium 39 [Kr]4d¹5s²						
hP2	Mg			33.01		
cl2	W	>1755				
hR3	Sm		>12			
hP4	La		>25			
hP6 ?	Sc		>46			[4]
La Lanthanum 57 [Xe]5d¹6s²						
hP4	La			37.17		[5,6]
cF4	Cu	>533 or	>2.3	34.55	RT|2.3	[5]
cl2	W	>1153		38.65	1160|RP	
hR8	La		>7			
cF4	Cu		>60			
Ac Actinium 89 [Rn]6d¹7s²						
cF4	Cu			37.45		[7]

Group 4

PT		T [K]	P [GPa]	V_{at} [ų]	T|P	Ref.[a]
Ti Titanium 22 [Ar]3d²4s²						
hP2	Mg			17.65		
cl2	W	>1155		18.15	1193|RP	[8]
hP3	AlB₂		>2	17.23	RT|4	
oC4?	Ti		>116	10.30	RT|130	[8]
oP4?	Ti		>145	9.22	RT|178	[8]
Zr Zirconium 40 [Kr]4d²5s²						
hP2	Mg			23.28		
cl2	W	>1136				
hP3	AlB₂		>2.2			
cl2	W		>35			
Hf Hafnium 72 [Xe]4f¹⁴5d²6s²						
hP2	Mg			22.31		
cl2	W	>2013				
hP3	AlB₂		>38			
cl2	W		>71			
Rf Rutherfordium 104 [Rn]5f¹⁴6d²7s²						

[a][1] Kammler et al., [2008]; [2] McMahon et al. [2006a]; [3] Akahama et al., [2005]; [4] Grosshans et al., [1982]; [5] Porsch and Holzapfel, [1993]; [6] Seipel et al., [1997]; [7] Farr et al., [1961]; [8] Akahama et al., [2001].

[b]Incommensurate host-guest structure.

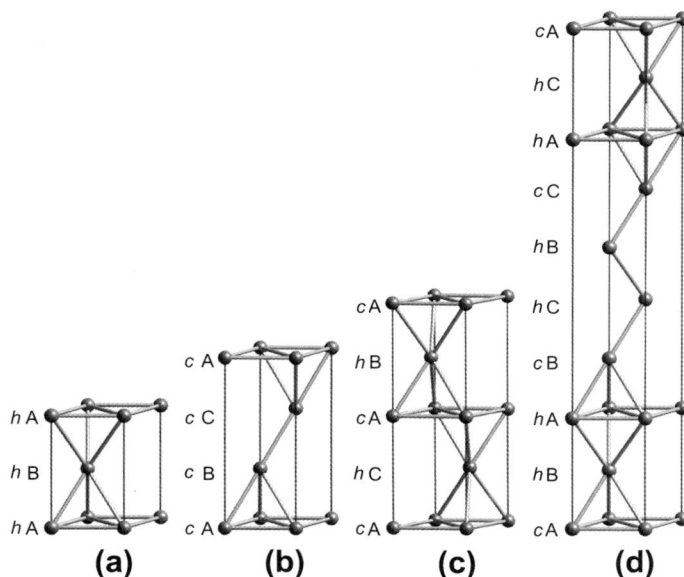

Figure 17 Schematic representation of the stacking sequences of the close packed structures (a) hP2-Mg, (b) cF4-Cu (in rhombohedral setting), (c) hP4-La and (d) hR3-Sm. The structure of hP4-La is frequently called *double hexagonal close packed* (*dhcp*).

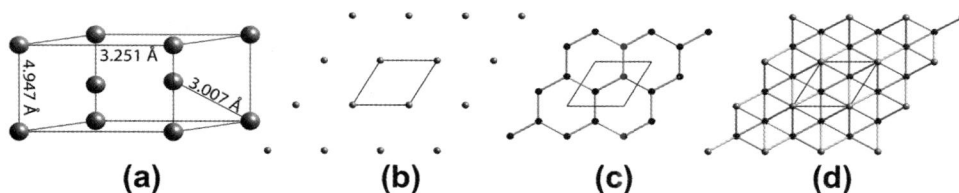

Figure 18 The hP3-Ti structure type: (a) one unit cell; layers in (b) $z = 0$ and (c) $z = 1/2$ and their projection on top of each other (d). The atomic arrangements in (b) and (c) correspond to the regular Archimedean triangle tiling 3^6 and its dual the hexagon tiling 6^3, respectively.

cI2-W type, while titanium first transforms to oC4-Ti and then to oP4-Ti (Akahama et al., 2001). While oC4-Ti can be seen as distorted *hcp* structure, oP4-Ti rather corresponds to a distorted *bcc* structure; however, these transitions are still controversially discussed (Verma et al., 2007), and a path-dependent direct transition hP3-Ti → cI2-W has been proposed (Ahuja et al., 2004).

1.3.2.2 Groups 5 and 6: V, Nb, Ta, Cr, Mo, W

At ambient pressure, vanadium, niobium, tantalum, molybdenum and tungsten have simple cI2-W-type structures at all temperatures (**Table 8**). For vanadium, a rhombohedral high-pressure phase of the hR1-Hg type has been found (Ding et al., 2007) in agreement with theoretical calculations, which explained the structural transition by a band Jahn–Teller mechanism (Verma and Modak, 2008). Chromium shows two antiferromagnetic phase transitions, which modify the structure only very slightly (Young, 1991).

Table 8   Structural information for the elements of groups 5 and 6. In the first line of each element box the element name, atomic number Z, and electronic ground-state configuration is given. Below, the different modifications with their prototype structure PT, if known, are listed together with the temperature T and pressure P limiting their stability range; if no temperature or pressure is given this means ambient conditions. Atomic volumes $V_{at} = V_{uc}/n_{at}$ are given where available, with V_{uc} and n_{at} the unit cell volume and the number of atoms per unit cell. Atomic volumes V_{at} are given at ambient conditions if no pressure P and temperature T is given in the form $T|P$. Only those HP and HT phases are listed that are stable at RT or RP, respectively. If no other reference is given, data are based on the compilations by Villars and Calvert (1991), Tonkov (1992–1996), Tonkov & Ponyatovsky (2005), Young (1991)

5	PT	T [K]	P [GPa]	V_{at} [Å³]	T|P	Ref.	6	PT	T [K]	P [GPa]	V_{at} [Å³]	T|P	Ref.
V Vanadium 23 [Ar]$3d^3 4s^2$							Cr Chromium 24 [Ar]$3d^4 4s^2$						
cI2	W			13.82			cI2	W			11.99		
hR1	Hg		>69	10.46	RT|90								
Nb Niobium 41 [Kr]$4d^4 5s^1$							Mo Molybdenum 42 [Kr]$4d^5 5s^1$						
cI2	W			17.98			cI2	W			15.58		
Ta Tantalum 73 [Xe]$4f^{14} 5d^3 6s^2$							W Tunsten 74 [Xe]$4f^{14} 5d^4 6s^2$						
cI2	W			18.02			cI2	W			15.85		

1.3.2.3 Groups 7 and 8: Mn, Tc, Re, Fe, Ru, Os

The high-temperature phases of manganese (**Table 9**), cF4-Mn and cI2-Mn, have typical metal structures, whereas cI58-Mn and cP20-Mn possess rather complicated structures (**Figure 19**). The paramagnetic cI58-Mn structure can be described as a $3 \times 3 \times 3$ superstructure of *bcc* unit cells, with 20 atoms slightly shifted and four atoms added, resulting in total 58 atoms. It transforms into an isostructural antiferromagnetic phase at the Neel temperature of 95 K. The structure of cP20-Mn is also governed by the valence electron concentration ("electron compound" or Hume-Rothery-type phase,

Table 9   Structural information for the elements of groups 7 and 8. In the first line of each element box the element name, atomic number Z, and electronic ground-state configuration is given. Below, the different modifications with their prototype structure PT, if known, are listed together with the temperature T and pressure P limiting their stability range; if no temperature or pressure is given this means ambient conditions. Atomic volumes $V_{at} = V_{uc}/n_{at}$ are given where available, with V_{uc} and n_{at} the unit cell volume and the number of atoms per unit cell. Atomic volumes V_{at} are given at ambient conditions if no pressure P and temperature T is given in the form $T|P$. Only those HP and HT phases are listed that are stable at RT or RP, respectively. If no other reference is given, data are based on the compilations by Villars and Calvert (1991), Tonkov (1992–1996), Tonkov & Ponyatovsky (2005), Young (1991)

7	PT	T [K]	P [GPa]	V_{at} [Å³]	T|P	Ref.	8	PT	T[K]	P [GPa]	V_{at} [Å³]	T|P	Ref.
Mn Manganese 25 [Ar]$3d^5 4s^2$							Fe Iron 26 [Ar]$3d^6 4s^2$						
cI58	Mn			12.21			cI2	W			11.78		
cP20	Mn	>980		13.62	1008|RP		cF4	Cu	>1185		12.13	1189|RP	
cF4	Cu	>1360					cI2	W	>1667		12.64	1712|RP	
cI2	W	>1410 or	>165	14.62	1422|RP		hP2	Mg		>13	10.66	RT|15	
Tc Technetium 43 [Kr]$4d^6 5s^1$							Ru Ruthenium 44 [Kr]$4d^7 5s^1$						
hP2	Mg			14.26			hP2	Mg			13.57		
Re Rhenium 75 [Xe]$4f^{14} 5d^5 6s^2$							Os Osmium 76 [Xe]$4f^{14} 5d^6 6s^2$						
hP2	Mg			14.71			hP2	Mg			13.99		

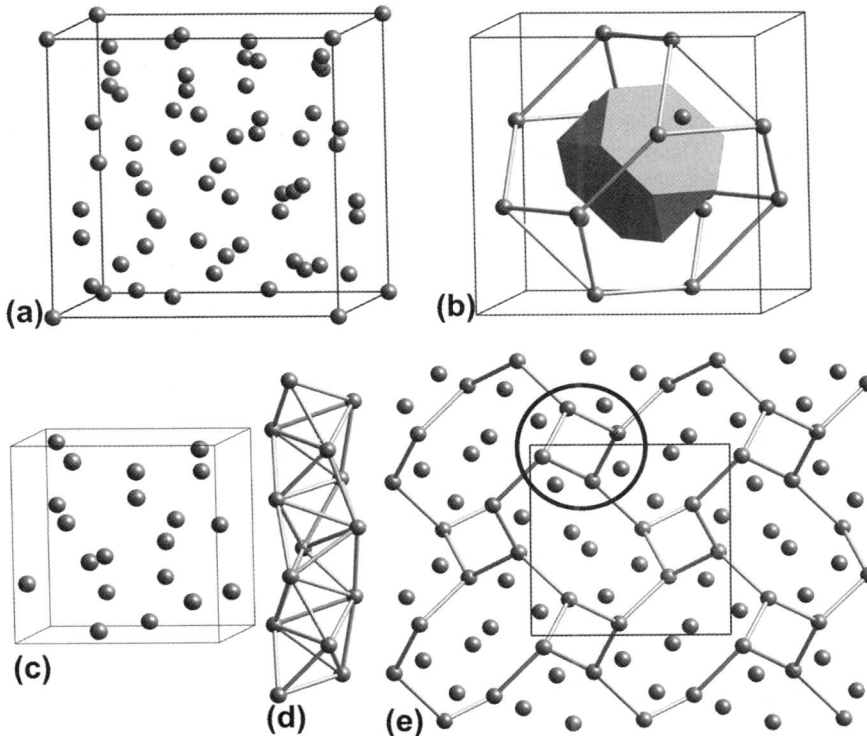

Figure 19 The complex structures of (a) *cI*58-Mn and (c) *cP*20-Mn. A fundamental 29-atom structural unit, which occupies the origin and body-center of the unit cell, it shown in (b): The Mn atom in the center is surrounded first by a small Friauf (CN16) polyhedron and then by a, by a factor $\approx \sqrt{2}$ larger truncated tetrahedron. The structure of (c) *cP*20-Mn can be seen as body-centered rod-packing of tetrahelices (d). The atoms generating such a tetrahelix are marked by a circle in the along a unit-cell edge projected structure (e).

see Section 1.4.4.2). It can be described by a *bcc* rod packing of Mn tetrahelices (Nyman et al., 1991), with the interstices filled by Mn-zigzag chains. The tetrahedra are slightly deformed to give a tetrahelix with a repeat period of eight tetrahedra. A tetrahelix consisting of undeformed tetrahedra is incommensurate and does not show rod periodicity. At 165 GPa, a phase transformation from *cI*58-Mn to *cI*2-Mn has been found, possibly due to the disappearance of magnetic moments (Fujisha and Takemura, 1995).

The technically most important element and main constituent of the Earth's core, iron (**Table 9**), shows four allotropic forms: ferromagnetic *cI*2-Fe, which transforms to a paramagnetic isostructural phase at a Curie temperature of 1043 K; at 1185 K *cF*4-Fe forms, while above 1667 K, iron adopts the *cI*2-Fe structure type again. However, taking into account the tetragonal distortion of the ferromagnetic structure by its exchange-coupled moments, the transition to the paramagnetic phase becomes a structural transition with a change in symmetry (see also Chapter 19, Volume II). There is only one high-pressure phase known at ambient temperature, nonmagnetic *hP*2-Fe, with a slightly compressed *hcp* structure. For technetium, rhenium, ruthenium and osmium, only simple *hcp* structures are known.

Table 10   Structural information for the elements of groups 9 and 10. In the first line of each element box the element name, atomic number Z, and electronic ground-state configuration is given. Below, the different modifications with their prototype structure PT, if known, are listed together with the temperature T and pressure P limiting their stability range; if no temperature or pressure is given this means ambient conditions. Atomic volumes $V_{at} = V_{uc}/n_{at}$ are given where available, with V_{uc} and n_{at} the unit cell volume and the number of atoms per unit cell. Atomic volumes V_{at} are given at ambient conditions if no pressure P and temperature T is given in the form $T|P$. Only those HP and HT phases are listed that are stable at RT or RP, respectively. If no other reference is given, data are based on the compilations by Villars and Calvert (1991), Tonkov (1992–1996), Tonkov & Ponyatovky (2005), Young (1991)

9	PT	T [K]	P [GPa]	V_{at} [Å³]	T|P	Ref.[a]	10	PT	T [K]	P [GPa]	V_{at} [Å³]	T|P	Ref.[a]
Co Cobalt 27 [Ar]3d⁷4s²							Ni Nickel 28 [Ar]3d⁸4s²						
hP2	Mg			11.08			cF4	Cu			10.94		
cF4	Cu	>673		11.36	793|RP								
			>105	7.40	RT|202	[1]							
Rh Rhodium 45 [Kr]4d⁸5s¹							Pd Palladium 46 [Kr]4d¹⁰						
cF4	Cu			13.75			cF4	Cu			14.72		
Ir Iridium 77 [Xe]4f¹⁴5d⁷6s²							Pt Platinum 78 [Xe]4f¹⁴5d⁹6s¹						
cF4	Cu			14.15			cF4	Cu			15.10		
hP14	Ir		>59	12.42	RT|65	[2]							

[a][1] Yoo et al., [2000]; [2] Cerenius and Dubrovinsky, [2000].

1.3.2.4   Groups 9 and 10: Co, Rh, Ir, Ni, Pd, Pt

Cobalt (**Table 10**) is dimorphous, *hcp* at ambient conditions and *ccp* at higher temperatures or pressures. By annealing it in a special way, stacking disorder can be generated: the *hcp* sequence ..ABAB.. is statistically replaced by a *ccp* sequence ..ABCABC.. yielding sequences like ..ABABABABCBCBCBC.. with a frequency of about one ..ABC.. among 10 ..AB... The high-pressure transition from magnetic *hP2*-Co to likely nonmagnetic *cF4*-Co occurs martensitically without any apparent volume change (Yoo et al., 2000).

Rhodium, iridium, nickel, palladium and platinum all crystallize in *ccp* structures at ambient conditions. For iridium, a high-pressure modification, *hP14*-Ir, is known (Cerenius and Dubrovinsky, 2000). It can be seen as 14-layer closest packed superstructure of *cF4*-Ir.

1.3.2.5   Groups 11 and 12: Cu, Ag, Au, Zn, Cd, Hg

The coinage metals—copper, silver and gold (**Table 11**)—are typical metals with *ccp* structure. Their single *ns* electron is less shielded by the filled d orbitals than the *ns* electron of the alkali metals by the filled noble gas shell. The d electrons also contribute to the metallic bond. These factors are responsible for the more noble character of these metals than of the alkali metals and that these elements are frequently grouped to the transition elements.

For zinc, cadmium and mercury (**Table 11**), covalent bonding contributions (hybridization of the filled d band to the conduction band) lead to deviations from the ideal ratio $c/a = 1.633$ for *hcp* to values of 1.856 (Zn) and 1.886 (Cd), respectively (Takemura, 1997). Consequently, the bonds within the *hcp* layers are shorter and stronger than between the layers. With increasing pressure, c/a approximates the ideal value with $c/a = 1.68$ at 30 GPa for Cd (Donohue, 1974) and $c/a = 1.76$ at 46.8 GPa for Hg (Schulte and Holzapfel, 1993). No phase transformations have been observed for pressures up to 126 GPa for Zn and up to 174 GPa for Cd.

Table 11   Structural information for the elements of groups 11 and 12. In the first line of each element box the element name, atomic number Z, and electronic ground-state configuration is given. Below, the different modifications with their prototype structure PT, if known, are listed together with the temperature T and pressure P limiting their stability range; if no temperature or pressure is given this means ambient conditions. Atomic volumes $V_{at} = V_{uc}/n_{at}$ are given where available, with V_{uc} and n_{at} the unit cell volume and the number of atoms per unit cell. Atomic volumes V_{at} are given at ambient conditions if no pressure P and temperature T is given in the form $T|P$. Only those HP and HT phases are listed that are stable at RT or RP, respectively. If no other reference is given, data are based on the compilations by Villars and Calvert (1991), Tonkov (1992–1996), Tonkov and Ponyatovsky (2005), Young (1991)

11	PT	T [K]	P [GPa]	V_{at} [Å³]	T|P	Ref.[a]	12	PT	T [K]	P [GPa]	V_{at} [Å³]	T|P	Ref.[a]
Cu Copper 29 [Ar]3d¹⁰4s¹							Zn Zinc 30 [Ar]3d¹⁰4s²						
cF4	Cu			11.81			hP2	Mg			15.20		
Ag Silver 47 [Kr]4d¹⁰5s¹							Cd Cadmium 48 [Kr]4d¹⁰5s²						
cF4	Cu			17.05			hP2	Mg			21.60		
Au Gold 79 [Xe]4f¹⁴5d¹⁰6s¹							Hg Mercury 80 [Xe]4f¹⁴5d¹⁰6s²						
cF4	Cu			16.96			hR1	Hg	<234.3		23.07	83|RP	
							tI2	Pa		>3.4	19.04	RT|15	
							mC6	Hg		>12	18.29	RT|20	[1]
							hP2	Mg		>37	16.91	RT|35.2	

[a][1] *Takemura et al. [2007].*

The rhombohedral structure of hR1-Hg can be derived from a *ccp* structure by compression along the threefold axis. In contrast to zinc and cadmium, the ratio $c/a = 1.457$ for a hypothetical distorted *hcp* structure is smaller than the ideal value. There also exist several high-pressure allotropes. The *hcp* modification, stable at highest pressures (at least up to 193 GPa), shows a ratio $c/a = 1.75$, comparable to those of Zn and Cd (Takemura et al., 2007). All mercury modifications can be seen as distorted *ccp* or *hcp* structures.

1.3.3   Groups 13–16: (semi)metallic Main Group Elements

Only aluminum, thallium and lead crystallize in closest packed structures characteristic for typical metals (**Table 12**). The s–d transfer effects, important for alkali- and alkaline-earth metals, do not appear for the heavier group 13 elements due to their filled d bands. Aluminum transforms from *ccp* to *hcp* (i.e. a change in the stacking sequence from ..ABC.. to ..AB..) at 217 GPa with a volume drop by 1.3%; no further changes have been observed up to 333 GPa (Akahama et al., 2006). The value of $c/a = 1.618(5)$ is only slightly smaller than the ideal value 1.633. The partial filling of the initially unoccupied d states may play a main role for this transition.

Gallium shows a more complex phase diagram. At ambient conditions, stable *oC*8-Ga forms a 6^3-network of distorted hexagons parallel to (100) at $x = 0$ and 1/2. The bonds between the layers are considerably weaker than within. At slightly elevated pressure, gallium adopts a complex orthorhombic modification, *oC*104 (Degtyareva et al., 2004a), which can be described as commensurate modulation of a simple *Fddd* eight-atom basis structure (Perez-Mato et al., 2006) (**Figure 20**). Three more modifications have been found to exist at higher pressures. *tI*2-Ga can be seen as a distorted *fcc* structure, which transforms to *cF*4-Ga at 120 GPa. In this transformation, the c/a value of the tetragonal phase changes continuously until it reaches $\sqrt{2}$ and an *fcc* structure is realized (Takemura et al., 1998).

Table 12   Structural information for the elements of groups 13 to 16. In the first line of each element box the element name, atomic number Z, and electronic ground-state configuration is given. Below, the different modifications with their prototype structure PT, if known, are listed together with the temperature T and pressure P limiting their stability range; if no temperature or pressure is given this means ambient conditions. Atomic volumes $V_{at} = V_{uc}/n_{at}$ are given where available, with V_{uc} and n_{at} the unit cell volume and the number of atoms per unit cell. Atomic volumes V_{at} are given at ambient conditions if no pressure P and temperature T is given in the form T/P. Only those HP and HT phases are listed that are stable at RT or RP, respectively. If no other reference is given, data are based on the compilations by Villars and Calvert (1991), Tonkov (1992–1996), Tonkov and Ponyatovsky (2005), Young (1991)

13/15	PT	T [K]	P [GPa]	V_{at} [Å³]	T|P	Ref.[a]
Al Aluminum 13 $[Ne]3s^2 3p^1$						
cF4	Cu			16.60		
hP2	Mg		>217	8.42		[1]
Ga Gallium 31 $[Ar]3d^{10} 4s^2 4p^1$						
oC8	Ga			19.58		
oC104	Ga		>2	17.62	RT|2.8	[2]
hR6	Ga		>10.5	15.76	RT|12.2	[2]
tI2	In		>14	15.27	RT|15.6	[2]
cF4	Cu		>120			[3]
In Indium 49 $[Kr]4d^{10} 5s^2 5p^1$						
tI2	In			26.16		
oF4	In		>45	15.00	RT|93	
Tl Thallium 81 $[Xe]4f^{14} 5d^{10} 6s^2 6p^1$						
hP2	Mg			28.59		
cI2	W	>503		29.00	523|RP	
cF4	Cu		>4	27.27	RP|6	
Sb Antimony 51 $[Kr]4d^{10} 5s^2 5p^2$						
hR2	As			30.21		
mI	inc[b]		>8.2	24.79	RT|6.9	[7]
tI	inc[b]		>8.6	23.95	RT|10.3	[7]
cI2	W		>28	20.45	RT|28.8	[7]
Bi Bismuth 83 $[Xe]4f^{14} 5d^{10} 6s^2 6p^3$						
hR2	As			35.39		
mC4	Bi		>2.55	31.37	RT|2.7	[7]
tI	inc[b]		>2.7	28.46	RT|6.8	[7]
cI2	W		>7.7	27.43	RT|8.5	[7]

14/16	PT	T [K]	P [GPa]	V_{at} [Å³]	T|P	Ref.[a]
Ge Germanium 32 $[Ar]3d^{10} 4s^2 4p^2$						
cF8	C			22.63		
tI4	Sn		>9	16.71	RT|12	[4]
oI4	Sn		>75	12.34	RT|81	
hP1	BiIn		>85	12.04	RT|83	
oC16	Si		>102	10.67	RT|135	[5]
hP2	Mg		>160	9.73	RT|180	[5]
Sn Tin 50 $[Kr]4d^{10} 5s^2 5p^2$						
cF8	C	<286.2		34.16	285|RP	
tI4	Sn	>286.2		27.05	298|RP	
tI2	Pa		>9.4	20.25	RT|24.5	[6]
cI2	W		>45	17.76	RT|53	
Pb Lead 82 $[Xe]4f^{14} 5d^{10} 6s^2 6p^2$						
cF4	Cu			30.32		
hP2	Mg		>13.9	28.02	RT|15.2	
cI2	W		>110	16.11	RT|127	
Po Polonium 84 $[Xe]4f^{14} 5d^{10} 6s^2 6p^4$						
cP1	Po			38.14	311|RP	
hR1	Hg	>348		36.61		

[a][1] Akahama et al., [2006]; [2] Degtyareva et al., [2004a]; [3] Takemura et al., [1998]; [4] Nelmes et al., [1996]; [5] Takemura et al., [1998]; [6] Desgreniers et al., [1989]; [7] Degtyareva

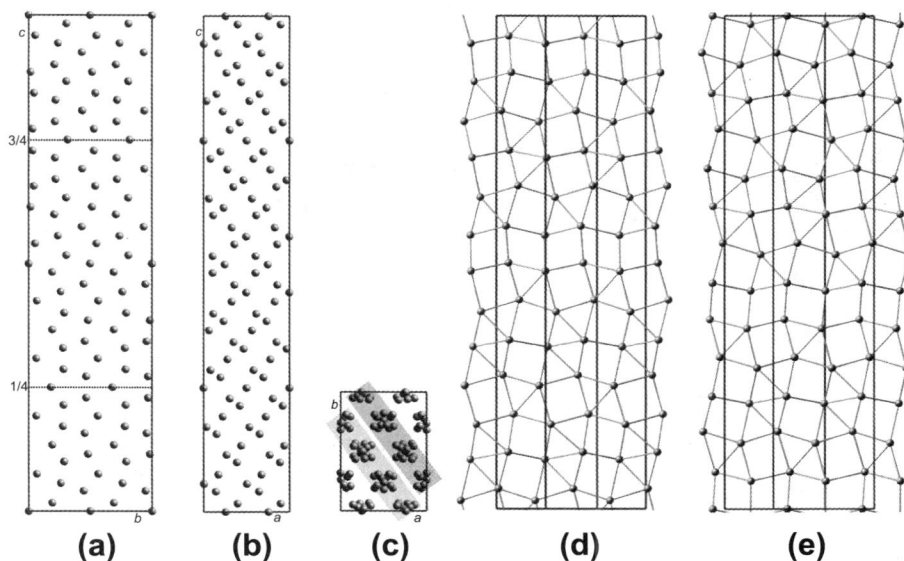

Figure 20 The structure of $oC104$-Ga in different views: (a–c) Projections along the a-, b- and c-directions, respectively. In (d,e), puckered (110)-bounded projections are shown; they correspond to the gray shaded regions marked in (c).

Indium, with the rather open ground-state structure $tI2$-In (see **Figure 2** in Section 1.2.2), transforms at 45 GPa by a simple orthorhombic distortion into the densely packed high-pressure phase $oF4$-In. The c/a ratio of $tI2$-In first increases with pressure from initially 1.522 to a maximum of 1.543 at 24 GPa until it decreases again. If a face-centered unit cell is chosen for $tI2$-In instead of the standard body-centered one, then the rather close relationship to fcc becomes obvious: the c/a ratio for $tF2$-In is 1.076 vs 1.0 for fcc.

For thallium, there is only one high-pressure phase transformation known, from hcp to ccp, with a volume decrease of only 0.75%.

Germanium (**Table 12**) crystallizes at ambient conditions in the diamond structure due to strong covalent bonding. At higher pressures, it transforms to the metallic white-tin ($tI4$-Sn) structure. This structure type can be regarded as being intermediate between the diamond structure of semiconducting $cF8$-Sn and $cF4$-Pb. For an ideal ratio of $c/a = 0.528$, one atom is sixfold coordinated. Increasing pressure, $oI4$-Ge is formed. This structure can be obtained by a slight distortion from $tI4$-Ge and transforms by further distortion into $hP1$-Ge. However, both phase transformations are of first order. $hP1$-Ge, with $c/a = 0.930$, has a quasi-eightfold coordination, the ideal ratio for CN = 8 would be $c/a = 1$. At 102 GPa, $hP1$-Ge transforms into $oC16$-Ge, where the coordination is increased to 10.5 on average (Takemura et al., 2001). At highest pressure, beyond 160 GPa, an hcp structure with 12-fold coordination is obtained. Thus, germanium runs with increasing pressure through phases with coordination numbers 4, 6, 8, 10.5 and 12.

The effective radii of tin in $tI4$-Sn and of lead in $cF4$-Pb are large compared with those of other typical metals with large atomic number due to incomplete ionization of the single ns electron. This means that in $cF8$-Sn, for instance, the electron configuration is $\ldots5s^15p^3$, allowing sp^3 hybridization and covalent tetrahedrally coordinated bonding, whereas in $tI4$-Sn, with $\ldots5s^25p^2$, only two p orbitals are available for covalent and one further p orbital for metallic bonding. $tI4$-Sn, at ambient conditions,

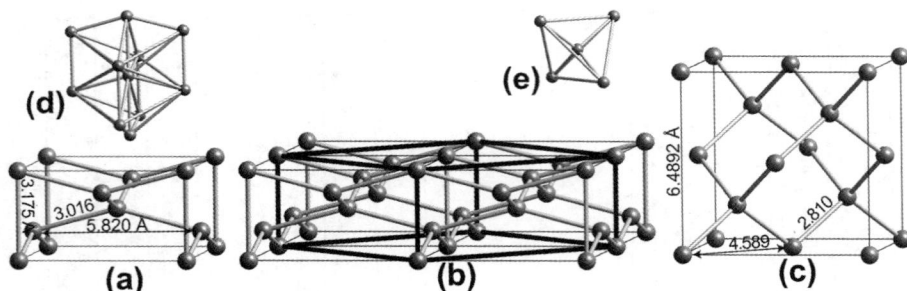

Figure 21   Relationships between the structures of the two tin allotrops: (a) nonmetallic gray tin, $cF8$-Sn, and (c) metallic white tin, $tI4$-Sn. In (b), the relationship between the two structures is shown. The thick black line marks the F-centered unit cell that becomes cubic if the c/a-ratio approaches $\sqrt{2}$. The AET of the structures in (a) and (c) are depicted in (d) and (e).

transforms at 9.4 GPa into $tI2$-Sn and finally, above 45 GPa, into $cI2$-Sn by continuously decreasing the tetragonal distortion of the structure. The transformation from low-temperature gray tin ($cF8$-Sn) to white tin ($tI4$-Sn) is accompanied by a decrease in atomic volume by 20% (**Figure 21**). Lead shows two high-pressure phases. It is amazing that the hcp to bcc transition takes place with just a volume reduction of 0.3% while that of the fcc to hcp transition amounts to 1.5% (Mao et al., 1990).

The structure of, at ambient conditions isotypic, antimony and bismuth (**Table 12**) consists of puckered layers of covalently bonded atoms stacked along the hexagonal axis. The structure, which is of the $hR2$-As type, can be regarded as a distorted primitive cubic structure ($cP1$-Po), in which the atomic distance d_1 within layers equals the distance d_2 between layers. The metallic character of these elements increases for d_2/d_1 approaching 1. At 8.2 GPa, $hR2$-Sb transforms, with a volume change of 5%, into a body-centered monoclinic, incommensurate host–guest structure, which can be seen as a slightly distorted version of the tetragonal host–guest structure stable at higher pressure (Degtyareva et al., 2004b). In particular, the guest atoms are arranged in zigzag chains. At 8.6 GPa, antimony transforms into a body-centered tetragonal, incommensurate host–guest structure, with the linear guest chains modulated only along the c-axis and the host framework perpendicular to it. The transformation is of first order and connected with a volume change of 0.5%, while the transition to $cI2$-Sb takes place with a volume change of 3.3%. At 2.55 GPa, $hR2$-Bi transforms into $mC4$-Bi, which can be considered as distorted simple cubic structure. Between 2.8 and 7.7 GPa, bismuth shows a similar body-centered tetragonal, incommensurate host–guest structure as antimony, with the linear guest chains modulated only along the c-axis and the host framework perpendicular to it (Degtyareva et al., 2004b). Both, antimony and bismuth transform to $cI2$-W type structures at highest pressures.

Polonium has an exceptionally simple crystal structure at ambient conditions; it is cubic primitive with one atom per unit cell, $cP1$-Po. This structure is stabilized by relativistic spin-orbit coupling, which leads to a hardening of most phonon frequencies. Above 348 K, it transforms to rhombohedral $hR1$-Po by a slight compression along [111]. The unusual lowering of point symmetry with increasing temperature has been explained by the growing influence of vibrational entropy (Verstrate, 2010).

1.3.4   Lanthanoids and Actinoids

Lanthanoids (**Table 13**) and actinoids (**Table 14**) are characterized by the fact that their valence electrons are increasingly filling the 4f and 5f orbitals, respectively. The term "lanthanoid contraction" refers to the decrease in atomic volumes with increasing number of electrons similar as it is the case for the

Table 13 Structural information for the Lanthanoides. In the first line of each element box the element name, atomic number Z, and electronic ground-state configuration is given. Below, the different modifications with their prototype structure PT, if known, are listed together with the temperature T and pressure P limiting their stability range; if no temperature or pressure is given this means ambient conditions. Atomic volumes $V_{at} = V_{uc}/n_{at}$ are given where available, with V_{uc} and n_{at} the unit cell volume and the number of atoms per unit cell. Atomic volumes V_{at} are given at ambient conditions if no pressure P and temperature T is given in the form $T|P$. Only those HP and HT phases are listed that are stable at RT or RP, respectively. If no other reference is given, data are based on the compilations by Villars and Calvert (1991), Tonkov (1992–1996), Tonkov and Ponyatovsky (2005), Young (1991)

PT		T [K]	P [GPa]	V_{at} [Å³]	T|P	Ref.[a]
Ce Cerium 58 $[Xe]4f^2 6s^2$						
cF4	Cu	<96		28.52	76|RP	
hP4	La	>96		34.78		
cF4	Cu	>220		34.37		
cI2	W	>999		34.97	1030|RP	
cF4	Cu		>0.76	28.00	RT|1.05	[1]
mC4	Ce		>5.1	23.59	RT|8.3	[2]
tI2	In		>12.2	12.76	RT|208	[3]
Pr Praseodymium 59 $[Xe]4f^3 6s^2$						
hP4	La			34.56		
cI2	W	>1069		35.22	1094|RP	
cF4	Cu		>4	29.05	RT|4	[4]
hR8	Pr		>7.4	25.09	RT|12.5	[4]
oI16			>13.7	21.99	RT|19	[4]
oC4	U		>20.5	19.06	RT|21.8	[4]
oP4			>147	11.82	RT|313	[5]
Nd Neodymium 60 $[Xe]4f^4 6s^2$						
hP4	La			34.15		
cI2	W	>1128		35.22	1156|RP	
cF4	Cu		>5	27.65	RT|5	[6]
hR8	Pr		>17			[4,6]
hP3	Nd		>35			[6]
mC4	U		>75	14.61	RT|89	[6,7]
oC4	U		>113	12.58	RT|155	[6]

PT		T [K]	P [GPa]	V_{at} [Å³]	T|P	Ref.[a]
Tb Terbium 65 $[Xe]4f^9 6s^2$						
oC4	Dy	<223		32.11	195|RP	
hP2	Mg	>223		31.97		
cI2	W	>1560				
hR3	Sm		>3	27.41	RT|6	[10]
hP4	La		>10			[10]
hR8	Pr		>30	18.61	RT|40.2	[10]
mC4	Ce		>51	12.48	RT|155	[10]
Dy Dysprosium 66 $[Xe]4f^{10} 6s^2$						
hP2	Mg			31.60		
cI2	W					
hR3	Sm		>7	27.34	RT|7	[11]
hP4	La		>17	21.67	RT|26.3	[11]
cF4	Cu		>39			[11]
oS8	Dy		>43	16.24	RT|87	[11]
Ho Holmium 67 $[Xe]4f^{11} 6s^2$						
hP2	Mg			31.09		
cI2	W	>1521				
hR3	Sm		>7	25.87	RT|8.5	
hP4	La		>19.5			
cF4	Cu		>54			
hR8	Pr		>58			

(Continued)

Table 13   Structural information for the Lanthanoides In the first line of each element box the element name, atomic number Z, and electronic ground-state configuration is given. Below, the different modifications with their prototype structure PT, if known, are listed together with the temperature T and pressure P limiting their stability range; if no temperature or pressure is given this means ambient conditions. Atomic volumes $V_{at} = V_{uc}/n_{at}$ are given where available, with V_{uc} and n_{at} the unit cell volume and the number of atoms per unit cell. Atomic volumes V_{at} are given at ambient conditions if no pressure P and temperature T is given in the form T|P. Only those HP and HT phases are listed that are stable at RT or RP, respectively. If no other reference is given, data are based on the compilations by Villars and Calvert (1991), Tonkov (1992–1996), Tonkov and Ponyatovsky (2005), Young (1991)—cont'd

PT		T [K]	P [GPa]	V_{at} [Å³]	T|P	Ref.[a]
Pm Promethium 61 [Xe]4f^{6}6s^2						
hP4	La			33.60		
cI2	W	>1163				
cF4	Cu		>10			
hR8?	Pr		>18			
Sm Samarium 62 [Xe]4f^{6}6s^2						
hR3	Sm			33.23		
hP2	Mg	>1007		33.79	980|RP	
cI2	W	>1195				
hP4	La		>6			
hR8?	Pr		>11			
hP3	Pr		>37.4	15.11	RT|77	
mC4	Nd		>105	13.74	RT|132	
Eu Europium 63 [Xe]4f^{7}6s^2						
cI2	W			48.07		
hP2	Mg		>12.5	27.39	RT|12.5	
hR8?	Pr		>18	23.58	RT|20	
Gd Gadolinium 64 [Xe]4f^{7}5d^{1}6s^2						
hP2	Mg			33.00		
hR3	Sm		>1.5	30.00	RT|3.5	[8]
hP4	La		>6.5	27.43	RT|10	[8]
cF4	Cu		>26			[9]
hR8	Pr		>33	20.64	RT|39	[9]
mC4	Pr		>60.5	17.28	RT|65	[9]

PT		T [K]	P [GPa]	V_{at} [Å³]	T|P	Ref.[a]
Er Erbium 68 [Xe]4f^{12}6s^2						
hP2	Mg			30.71		
hR3	Sm		>12.4			
hP4	La		>24			
cF4	Cu		>67.4			
Tm Thulium 69 [Xe]4f^{13}6s^2						
hP2	Mg			30.27		
cI2	W	>1800				
hR3	Sm		>9	25.01	RT|11.6	
hP4	La		>30			
Yb Ytterbium 70 [Xe]4f^{14}6s^2						
hP2	Mg			41.62	296|RP	[12]
cF4	Cu	>310 or		41.28	298|RP	[13]
cI2	W	>1065 or	> 3.5	43.76	1047|RP	
hP2	Mg		>26	18.98	RT|34	
cF4	Cu		>53	16.36	RT|53	
hP3	Nd		>98	10.85	RT|202	
Lu Lutetium 71 [Xe]4f^{14}5d^{1}6s^2						
hP2	Mg			29.90		
hR3	Sm		>25	21.13	RT|23	[14]
hP4	La		>45			[14]
hR8	Pr		>88	12.52	RT|142	[14]

[a] [1] Franceschi and Olcese [1969]; [2] McMahon and Nelmes, [1997]; [3] Vohra and Beaver [1999]; [4] Evans et al., [2009]; [5] Velisavljevic and Vohra [2004]; [6] Chesnut and Vohra [2000]; [7] Akella et al., [1999]; [8] Akella et al., [1988]; [9] Errandonea et al., [2007]; [10] Cunningham et al., [2007]; [11] Shen et al., [2007]; [12] Zhao et al., [1994]; [13] Chesnut and Vohra [1999]; [14] Chesnut and Vohra [1998].

Table 14 Structural information for the Actinoides. In the first line of each element box the element name, atomic number Z, and electronic ground-state configuration is given. Below, the different modifications with their prototype structure PT, if known, are listed together with the temperature T and pressure P limiting their stability range; if no temperature or pressure is given this means ambient conditions. Atomic volumes $V_{at} = V_{uc}/n_{at}$ are given where available, with V_{uc} and n_{at} the unit cell volume and the number of atoms per unit cell. Atomic volumes V_{at} are given at ambient conditions if no pressure P and temperature T is given in the form T|P. Only those HP and HT phases are listed that are stable at RT or RP, respectively. If no other reference is given, data are based on the compilations by Villars and Calvert (1991), Tonkov (1992–1996), Tonkov and Ponyatovsky (2005), Young (1991).

PT	T [K]	P [GPa]	V_{at} [Å³]	T|P	Ref.[a]
Th Thorium 90 [Rn]6d²7s²					
cF4	Cu		32.86		
cI2	W	>1673	34.01	1720|RP	
tI2	In	>90	17.56	RT|102	[1]
Pa Protactinium 91 [Rn]5f²6d¹7s²					
tI2	Pa		24.94		[2]
cI2	W	?			
oC4	U	>77	15.40	RT|130	[2]
U Uranium 92 [Rn]5f³6d¹7s²					
mP4	inc[b]	<43	20.58	4|RP	[3]
oC4	U		20.75		
tP30	CrFe	>941	21.81	955|RP	
cI2	W	>1048	22.06	1060|RP	
Np Neptunium 93 [Rn]5f⁴6d¹7s²					
oP8	Np	<554	19.22	586|RP	
tP4	Np	>554	20.31		
cI2	W	>850	21.81	873|RP	
Pu Plutonium 94 [Rn]5f⁶6d¹7s²					
mP16	Pu	<394	20.43		
mC34	Pu	>394	22.44	463|RP	
oF8	Pu	>478	23.14	508|RP	
cF4	Cu	>588	24.89	653|RP	
tI2	In	>741	24.78	750|RP	
cI2	W	>754	24.07	773|RP	
oP4	Pu	>37	14.56	RT|37	[4]

PT	T [K]	P [GPa]	V_{at} [Å³]	T|P	Ref.[a]
Bk Berkelium 97 [Rn]5f⁹7s²					
hP4	La		27.97		
cF4	Cu	>1203 or	>7		
oC4	U		>25	14.49	RT|45.9
Cf Californium 98 [Rn]5f¹⁰7s²					
hP4	La		27.27		
cF4	Cu				
oC4	U		>41	14.29	RT|46.6
Es Einsteinium 99 [Rn]5f¹¹7s²					
Fm Fermium 100 [Rn]5f¹²7s²					
Md Mendelevium 101 [Rn]5f¹³7s²					

(Continued)

Table 14 Structural information for the Actinoides In the first line of each element box the element name, atomic number Z, and electronic ground-state configuration is given. Below, the different modifications with their prototype structure PT, if known, are listed together with the temperature T and pressure P limiting their stability range; if no temperature or pressure is given this means ambient conditions. Atomic volumes $V_{at} = V_{uc}/n_{at}$ are given where available, with V_{uc} and n_{at} the unit cell volume and the number of atoms per unit cell. Atomic volumes V_{at} are given at ambient conditions if no pressure P and temperature T is given in the form T|P. Only those HP and HT phases are listed that are stable at RT or RP, respectively. If no other reference is given, data are based on the compilations by Villars and Calvert (1991), Tonkov (1992–1996), Tonkov and Ponyatovsky (2005), Young (1991)—cont'd

PT	T [K]	P [GPa]	V_{at} [Å³]	T|P	Ref.[a]	PT	T [K]	P [GPa]	V_{at} [Å³]	T|P	Ref.[a]
Am Americium 95 [Rn]5f^{7}7s^2						No Nobelium 102 [Rn]5f^{14}7s^2					
hP4	La	<1043		29.27							
cF4	Cu	>1043 or	>6.1	25.23	RT|7.8	[5]					
cI2	W	>1350									
oF8	Pu		>10.0	22.34	RT|10.9	[5]					
oP4	Am		>17	18.04	RT|17.6	[5]					
Cm Curium 96 [Rn]5f^{7}6d^{1}7s^2						Lr Lawrencium 103 [Rn]5f^{14}6d^{1}7s^2					
hP4	La	<1550		29.98							
cF4	Cu	>1550 or	>17			[6]					
mC4			>37			[6]					
oF8	Am		>56	16.23	RT|81	[6]					
oP4	Am		>95	13.65	RT|100	[6]					

[a][1] Vohra and Akella [1991]; [2] Haire et al., [2003]; [3] van Smaalen and George [1987]; [4] Sikka [2005]; [5] Heathman et al., [2000]; [6] Heathman et al., [2005].
[b]Incommensurate charge-density-wave based structure.

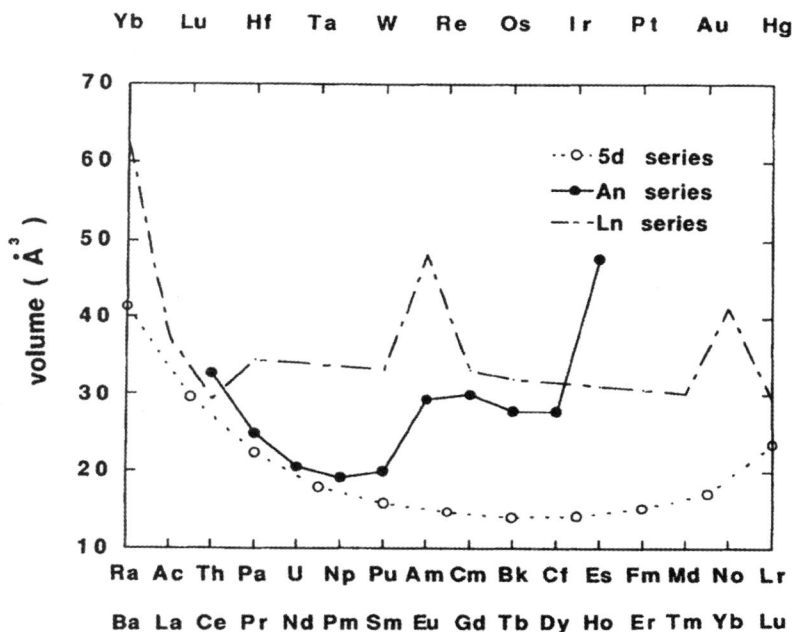

Figure 22   Atomic volumes of 5d transition elements (5d series) as well as of lanthanides (Ln series) and actinides (An series) (Wills and Eriksson, 1992).

transition metals (**Figure 22**). The chemical properties of the lanthanoids are rather uniform since their 4f orbitals are largely screened by their 5s and 5p electrons. All lanthanoids but Eu and Yb are trivalent. Eu and Yb are divalent at ambient conditions because of the favorable half- or fully filled f orbitals. The chemical behavior of the actinoids is somewhere in between that of the 3d transition metals and that of the lanthanoids, since the 5f orbitals are screened to a much lesser amount by the 6s and 6p electrons.

With the exception of Eu, all lanthanoids show at ambient conditions, closest packed structures: either a simple hcp structure with the standard stacking sequence ..AB.. or a $dhcp$ with a stacking sequence ..ACAB... Samarium has with ..ABABCBCAC.., an even more complicated stacking order with $n = 4.5$-fold superperiod. The ratio c/a is, in all cases, close to the ideal value of 1.633 or its multiple. With increasing pressure and decreasing atomic number, the sequence of closest packed phases $hP2$-Mg (..AB..) → $hR3$-Sm (..ABABCBCAC..) → $hP4$-La (..ACAB..) → $cF4$-Cu (..ABC..) → $hP6$-Pr appears (**Figure 17** in Section 1.3.2.1). The volume decrease can be explained by transfer of s and p electrons into the d band similar as for yttrium, which has no adjacent f states (Cunningham et al., 2007). At higher pressures, sometimes connected with a volume collapse up to 16%, a sequence of low-symmetry structures appears indicative of itinerant f electrons contributing to metallic bonding.

The phase sequence of lanthanum (see **Table 7** in Section 1.3.2.1) has already been discussed above. Cerium performs an isostructural transition at 0.76 GPa: the ccp structure is preserved but the volume collapses drastically by 16% due to a 4f–5d electron transfer. This collapsed ccp structure is also present below 96 K at ambient pressure. In case of the other lanthanoids, the isostructural transition is replaced by a ccp-to-distorted-ccp ($hR8$-Pr) transition. Further compression leads at 5.1 GPa to $oC4$-Ce and finally, beyond 12.2 GPa to $tI2$-Ce, which is stable at least up to 208 GPa (Vohra and Beaver, 1999).

Praseodymium is the first of the series of lanthanoids where the first-order transition $cF4$-Pr to the distorted ccp phase with the $hR8$-Pr type takes place. The subsequent transformations are connected with steep volume decrease ("volume collapse") due to delocalization of f electrons. Neodymium and samarium show a largely similar phase sequence at slightly higher pressures, however, no abrupt volume collapse has been observed due to f electron delocalization.

At ambient conditions, divalent europium shows quite a different behavior due to the stability of its half-filled 4f orbitals. Thus, it has more similarities to the alkaline earth metals, its phase diagram is rather comparable to that of barium, than to the other lanthanoids. A similar behavior is observed for ytterbium, which is divalent due to the stability of the completely filled 4f orbitals, and whose phase diagram resembles that of strontium. The anomalously large compression of both elements, Eu and Sm, is attributed to a mixing between di- and trivalent 4f configurations.

The c lattice parameter of Gadolinium exhibits an anomalous expansion when cooled below 298 K due to a change in the magnetic properties of the metal. Several other lanthanoids behave similarly. Terbium shows a similar behavior as cerium with some differences. The structure of the distorted ccp phase of Dysprosium was found to be orthorhombic, $oS8$-Dy. The c/a ratio for Ho increases with pressure from 1.570 for $hP2$-Ho to almost the ideal value of 1.633 for $hR8$-Ho at about 60 GPa.

At ambient conditions, divalent ytterbium shows two ccp phases in its phase sequence; the ccp phase at ambient pressure is a high-temperature phase, however. The high-pressure $hP3$-Yb phase, also found in trivalent Nd and Sm, can be considered as a distorted ccp structure. The rather high compression of 74% at 202 GPa is attributed to $4f^{14}$–$4f^{13}$ valence fluctuations and s–d electron rearrangements (Chesnut and Vohra, 1999).

According to their electronic properties, the actinoids (**Table 14**) can be divided into two subgroups: the elements from thorium to plutonium have itinerant 5f electrons, contributing to the metallic bonding, whereas in the elements from americium onward, the 5f electrons are localized similar to electrons in inner shells. This situation leads to superconductivity for thorium, protactinium and uranium, for instance, and to magnetic ordering for curium, berkelium and californium (Dabos-Seignon et al., 1993). The contribution of 5f electrons to bonding leads to low symmetry, small atomic volumes and high density in the case of the light actinoids while the heavier actinoids, as far as studied so far, crystallize in the hcp structure type at ambient conditions.

Thorium has, contrary to the other light actinoids, a high-symmetry ccp structure at ambient conditions behaving like a transition element. Its phase sequence resembles partly that of cerium indicating a similar electronic structure for these two elements at high pressure, what was confirmed by theoretical calculations (Hu et al., 2010). The population of the f band under pressure approaches that of the following element protactinium, which shows already at ambient conditions the $tI2$-Pa structure (Ce: >12.2 GPa; Th: >90 Gpa). Pa is the first actinoid element with f-bonding character already at ambient conditions. Compared to the $tI2$-In structure, which can be seen as uniaxially expanded bcc structure ($c/a = 1.66$ for Ce and 1.57 for Th), $tI2$-Pa is a compressed one ($c/a = 0.825$ for Pa). The AET is a cuboctahedron (CN = 12) in case of $tI2$-In and a rhombdodecahedron (CN = 14) in case of $tI2$-Pa (see **Figure 2** in Section 1.2.2).

For uranium ($p < 100$ GPa) and neptunium ($p < 50$ GPa), no phase transformations as function of pressure at ambient temperature have been found so far. Below 43 K, uranium shows an incommensurately modulated structure originating from a charge density wave state (van Smaalen and George, 1987). $oC4$-U, the modification stable at ambient conditions, can be seen as hcp structure with reduced displacement vector between neighboring layers and being contracted in the orthorhombic [100]- and [001]-directions by covalent bonding (**Figure 23**). In a packing model, the nonspherical shape due to covalent bonding leaves approximately 20% less interstitial space than in a hcp lattice explaining the missing primary solubilities of C, N and O in $oC4$-U (Blank, 1998).

The position of plutonium at the border of itinerant and localized 5f states (and itinerant 6d band) causes its unusually complex phase diagram as function of temperature, with structures typical for both

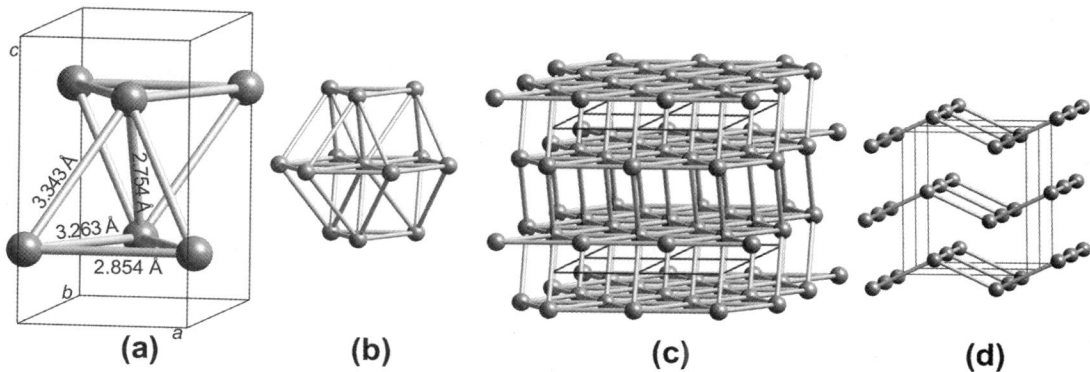

Figure 23 The structure of *oC*4-U in different views: (a) Unit cell and (b) one AET; projections along approximately [010] (c) and [100] (d) to visualize the stacking of distorted *hcp* layers; in (d) only contact distances <2.854 Å are shown.

cases. Thus monoclinic *mP*16-Pu can be considered as a distorted *hcp* structure with about 20% higher packing density than *cF*4-Pu due to covalent bonding contributions from 5f electrons (Ek et al., 1993). This ratio is quite similar to the aforementioned one of the isostructural transition in *ccp* Ce.

The phase diagram of americium is very similar to those of La, Pr and Nd. It is, due to the localization of 5f electrons, the first lanthanoid-like actinoid element. The f electron localization also leads to an approximately 50% larger atomic volume for Am compared to the preceding element Pu. Curium, with half-filled 5f band, shows a similar phase sequence as Am. The phase transitions are shifted to higher pressures, which are necessary to force the 5f electrons to take part in metallic bonding, in addition to the already itinerant 6d and 7s electrons.

1.4 Crystal Structures of Intermetallic Phases

While it is still possible to discuss in this chapter all crystal structures of the elements, this is no more feasible for the intermetallic phases. Due to space constraints, we have to focus on a properly selected small subset of the about 1900 structure types of the more than 40 000 intermetallic phases identified so far and listed in *Pearson's Crystal Data* (Villars and Cenzual, 2011). In our overview of characteristic structure types, their structural relationships and building principles, we will discuss some of those with the

- largest number of representatives,
- largest number of derivative structure types,
- highest importance for science and/or technical applications,
- best suitability for elucidating structural building principles in a didactic way.

After a general part on classification as well as some statistical information, we will discuss selected structure types, grouped to structure families wherever possible.

1.4.1 Statistics of Crystal Structure Types

The 50 most frequently occurring structure types are ranked in **Figure 24** according to the number of entries in *Pearson's Crystal Data* (Villars and Cenzual, 2011). Please keep in mind that the number of entries for a particular structure type may be higher than the number of chemically different

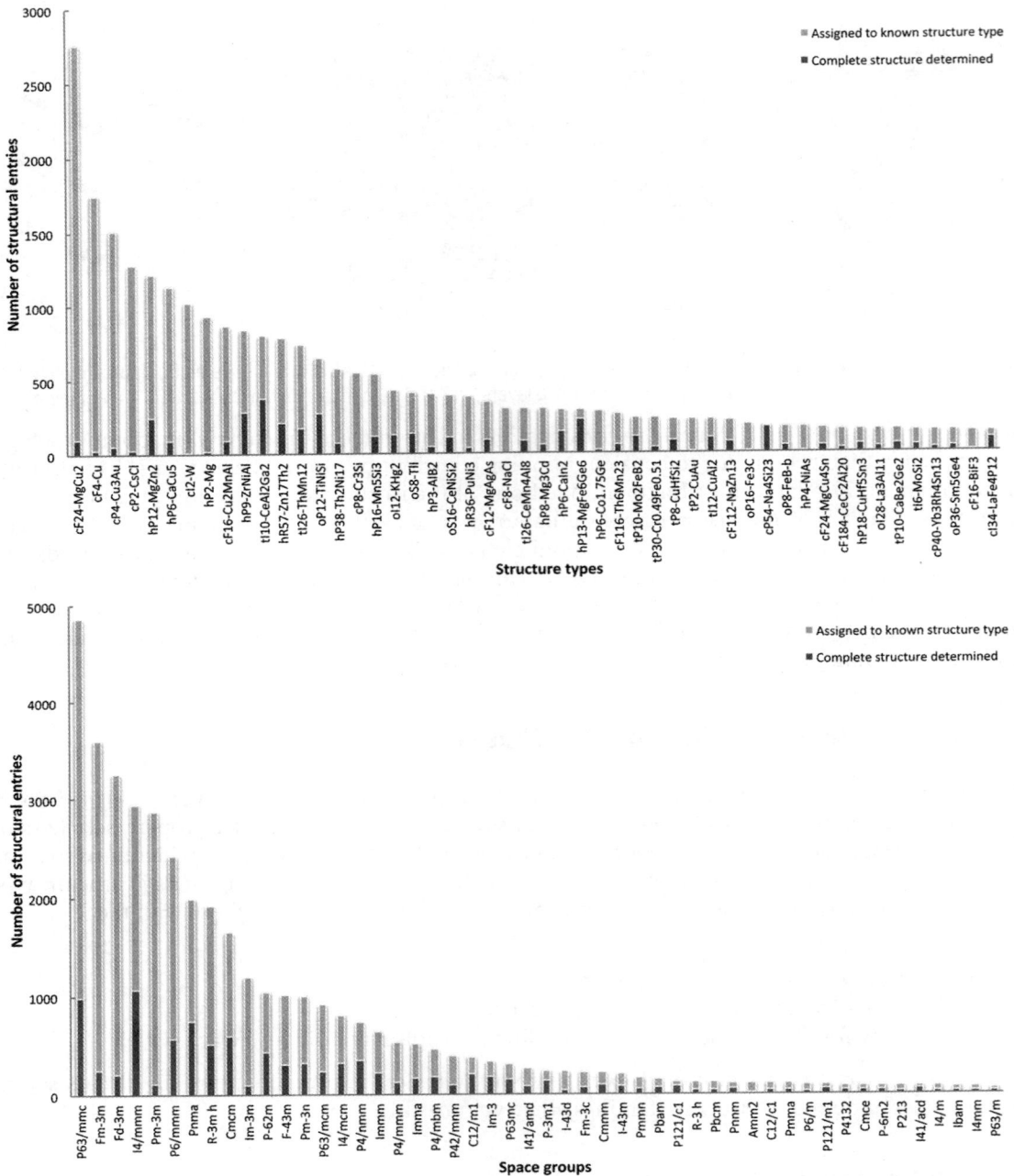

Figure 24   Frequency of the top 50 structure types (top) and of space-group symmetries (bottom) of intermetallics according to the number of entries listed in *Pearson's Crystal Data* (Villars and Cenzual, 2011) (light gray). Only a part of these entries is based on full structure analyses (marked dark gray in the histogram bars) (courtesy of J. Dshemuchadse).

representatives (compounds). The Laves phases belong to the structure family with most entries, approximately 2800 for $cF24$-Cu_2Mg and 1200 for $hP12$-$MgZn_2$. Second are closest packed structures such as $cF4$-Cu with 1700, its ordering variant $cP4$-$AuCu_3$ with 1500, and $hP2$-Mg with 900 entries. Also in the top group are $cI2$-W with 1000 and its superstructure $cP2$-CsCl with approximately 1300 entries in the data base. As well in the top 10 are the structure types of magnetic materials, $hP6$-$CaCu_5$ and $cF16$-Cu_2MnAl with 1100 and 900 entries.

Surprisingly, not only the highest symmetrical space groups are found in the group of the 10 most frequent space groups. The ranking is led by the two space groups of hexagonal and cubic closed sphere packings, $P6_3/mmc$ and $Fm\overline{3}m$ with close to 4900 and 3600 entries, respectively, followed by another cubic space group, $Fd\overline{3}m$ with 3300 entries; already at the fourth place comes the first lower symmetrical space group $I4/mmm$ with almost 3000 entries. The first monoclinic space group $C2/m$, ranked twenty-third, still accumulates close to 400 entries; the first triclinic space group appears as No. 52 with remarkable 61 entries. This distribution shows that although a large fraction of intermetallics adopts structures with symmetries of close sphere packings, the majority does not. This may be caused by anisotropic interactions such covalent bonding contributions but as well by odd stoichiometries preventing highly symmetric AET.

The frequency of AET of the chemical elements (Villars and Daams, 1993) and intermetallic phases has been extensively studied and published in a series of papers (Daams and Villars, 1992, 1993, 1994, 1997; Villars et al., 2004).

1.4.2   Classification of Structure Types and Families

The structures belonging to a particular structure type have to be isoconfigurational. This means that all geometrical properties, such as axial ratios, angles between crystallographic axes, values of corresponding adjustable positional parameters x, y, z, coordinations of corresponding atoms (AET) and entire configurations (clusters) of the two structures, are similar. Higher order (*derivative*) structure types can be obtained from lower order (*prototype* or *aristotype*) structures by systematic modifications:

- In case of *substitutional derivative structures*, a subset of atoms of one kind is replaced by atoms of another kind. *Interstitial (or "stuffed") derivatives* represent compounds in which unoccupied interstitial sites (voids) of the *parent basic structure* are (progressively) filled by atoms in the *derivative structure*. In general, the relationship between the unfilled parent (basic) structure and the derivatives based on filling one specific interstitial site approaches homeotypism.
  Example: Structure types derived by substitution from the *aristotype* $cI2$-W are $cP2$-CsCl, $cF16$-BiF_3, $cF16$-$MnCu_2Al$, $cF16$-NaTl and $cF16$-$TiCuHg_2$ (see **Figure 27** in Section 1.4.4).
- *Superstructures, commensurately* and *incommensurately modulated structures* are formed by periodic modification (substitutional and/or displacive modulation) of an underlying parent structure. A host structure can also be modified by a guest structure with a different period leading to a *composite (host/guest) structure*. If the period of the modulation wave or the guest structure and that of the parent or host structure are in an irrational ratio to each other, then an incommensurate structure results.
  Example: With substitutional and displacive modulation, we obtain $oI60$-EuAuSn as fivefold superstructure of the $oI12$-KHg_2 type (**Figure 37** in Section 1.4.7). The tetragonal body-centered host structure of $tI19.3$-Rb contains a tetragonal body-centered guest structure incommensurate to each other along the fourfold axis (**Figure 12** in Section 1.3.1).
- *Recombination structures* can be assembled when topologically simple *parent structures* are periodically divided into modules such as blocks, rods or slabs that, in turn, are recombined into derivative

structures by means of one or more structure building operations. *Modular structures* are a generalization of *recombination structures* when a derivative structure results from a combination of modules of different structure types.

Example: The μ-phase $hR13$-W_6Fe_7 can be seen as put together from a layer of the hexagonal bipyramids constituting the σ-phase $tP30$-CrFe and a layer of the truncated tetrahedra building the Laves phase $cF24$-Cu_2Mg (**Figure 32** in Section 1.4.5.2).

● *Hierarchical structures* are derivative structures formed by ordered substitution of a subset of atoms or of all atoms by larger structural units (cluster).

Example: Replacing in the $cF8$-ZnS structure, the Zn atoms at Wyckoff position $4a$ 0,0,0 by Mg and the S atoms at Wyckoff position $4c$ 1/4,1/4,1/4 by Cu tetrahedra, the structure type of the Laves phase $cF24$-Cu_2Mg (**Figure 31** in Section 1.4.5.1) is obtained.

1.4.3   Closest Packed Structures and Their Derivatives

By selective substitution of atoms forming closest packed structures and/or selective occupation of interstitial sites, a family of substitutionally ordered structures as well as of other derivative structures can be obtained. Some selected examples for derivative structures based on one unit cell of the fundamental $cF4$-Cu structure type are shown in **Figure 25**.

With substitutional ordering, we obtain from the basic *ccp* $cF4$-Cu ($Fm\bar{3}m$, Cu in $4a$ 0, 0, 0) structure type either $tP4$-AuCu ($P4/mmm$, Au in $1a$ 0, 0, 0 and $1c$ 1/2, 1/2, 0; Cu in $2e$ 0, 1/2, 1/2) or $cP4$-$AuCu_3$ ($Pm\bar{3}m$, Au in $1a$ 0, 0, 0; Cu in $3c$ 0, 1/2, 1/2) (**Figure 26(a–f)**).

Occupying half of the tetrahedral voids in $cF4$-Cu gives the $cF8$-C (diamond) structure type ($Fd\bar{3}m$, C in $8a$ 0, 0, 0). Allowing these atoms in 1/4, 1/4, 1/4 to be of a different kind, results in the $cF8$-ZnS (zincblende, sphalerite) structure type ($F\bar{4}3m$, Zn in $4a$ 0, 0, 0; S in $4c$ 1/4, 1/4, 1/4). Filling all tetrahedral voids with atoms of another kind leads to the $cF12$-CaF_2 (fluorite) structure type ($Fm\bar{3}m$, Ca in $4a$ 0, 0, 0; F in $8c$ 1/4, 1/4, 1/4).

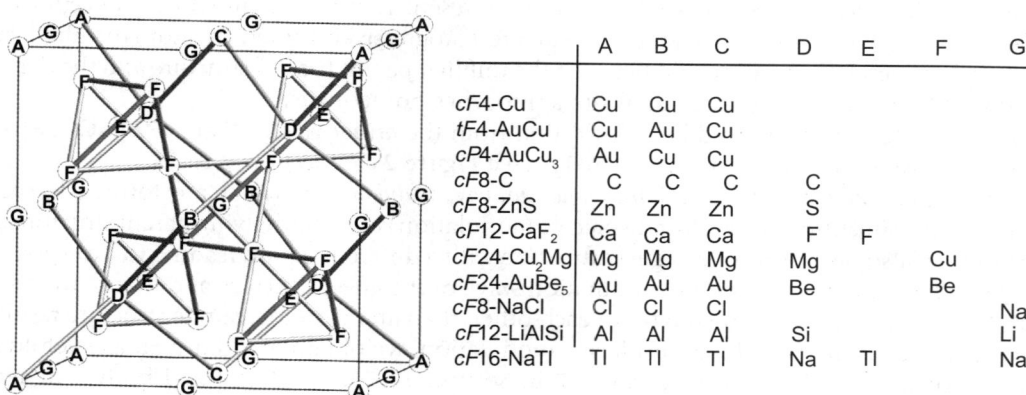

	A	B	C	D	E	F	G
$cF4$-Cu	Cu	Cu	Cu				
$tF4$-AuCu	Cu	Au	Cu				
$cP4$-$AuCu_3$	Au	Cu	Cu				
$cF8$-C	C	C	C	C			
$cF8$-ZnS	Zn	Zn	Zn	S			
$cF12$-CaF_2	Ca	Ca	Ca	F	F		
$cF24$-Cu_2Mg	Mg	Mg	Mg	Mg		Cu	
$cF24$-$AuBe_5$	Au	Au	Au	Be		Be	
$cF8$-NaCl	Cl	Cl	Cl				Na
$cF12$-LiAlSi	Al	Al	Al	Si			Li
$cF16$-NaTl	Tl	Tl	Tl	Na	Tl		Na

Figure 25   Site occupancies for some structures derived from the $cF4$-Cu type. Structures, where tetrahedral and octahedral sites A (0, 0, 0), B (1/2, 0, 1/2), C (1/2, 1/2, 0), D (1/4, 1/4, 1/4), E (1/4, 3/4, 1/4) and G (1/2, 1/2, 1/2) are occupied at the same time such as in $cF16$-NaTl can also be seen as derivatives of the $cP2$-CsCl type and will be discussed below.

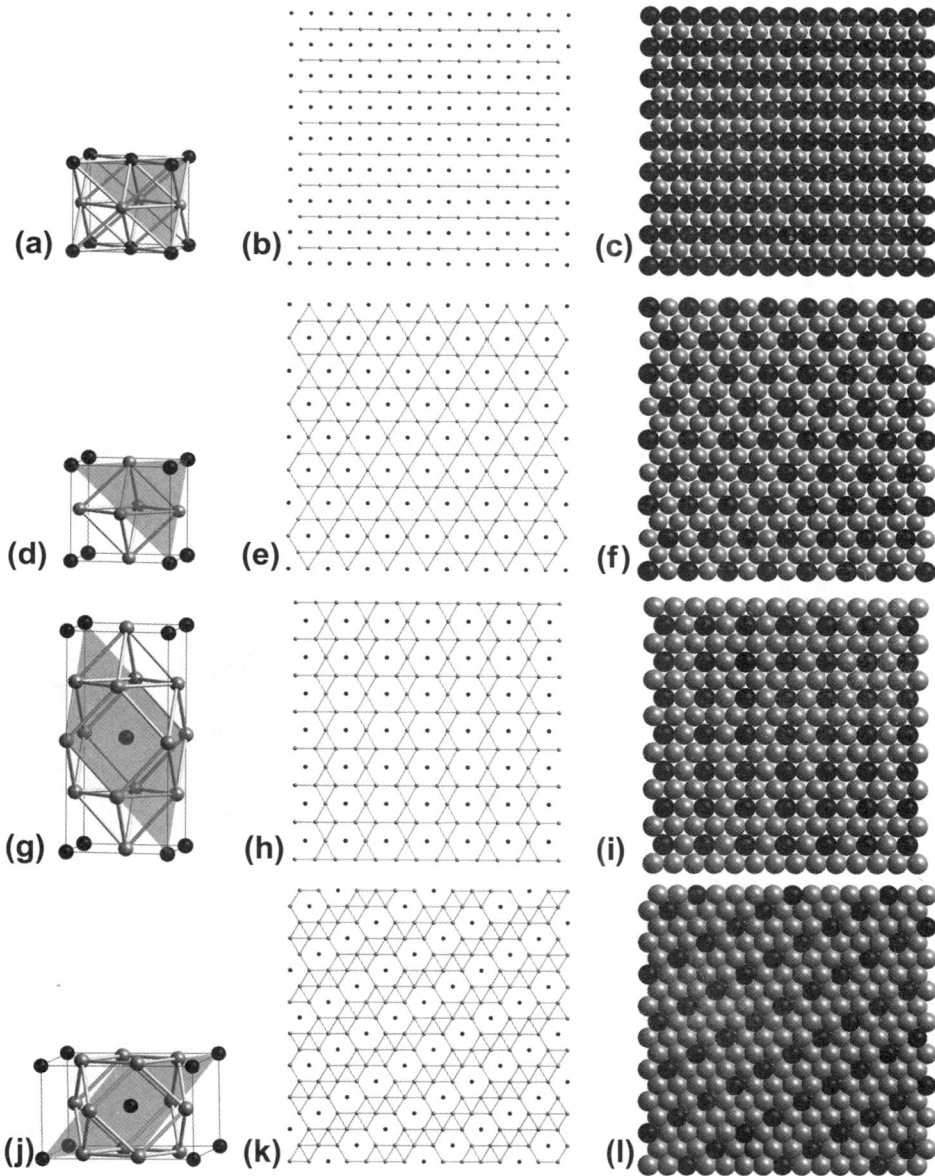

Figure 26 Seven structures are shown that can be considered to be *ccp* or *hcp* packings of hexagonal dense-packed (*hdp*) layers with a particular type of ordering. In each case, one unit cell is depicted, one *hdp* layer in the ball-and-stick mode as well with space filling atoms. The black spheres always correspond to the heavier atom. In the cases, where the orientation of the *hdp* layer is not perpendicular to a main lattice parameter, its orientation is indicated by a gray plane. (a–c) *t*P4-CuAu, (d–f) *c*P4-Cu₃Au, (g–i) *tI*8-TiAl₃, (j–l) *tI*10-MoNi₄, (m–o) *tI*16-ZrAl₃, (p–r) *o*P20-ZrAu₄ and (s–u) *h*P12-WAl₅.

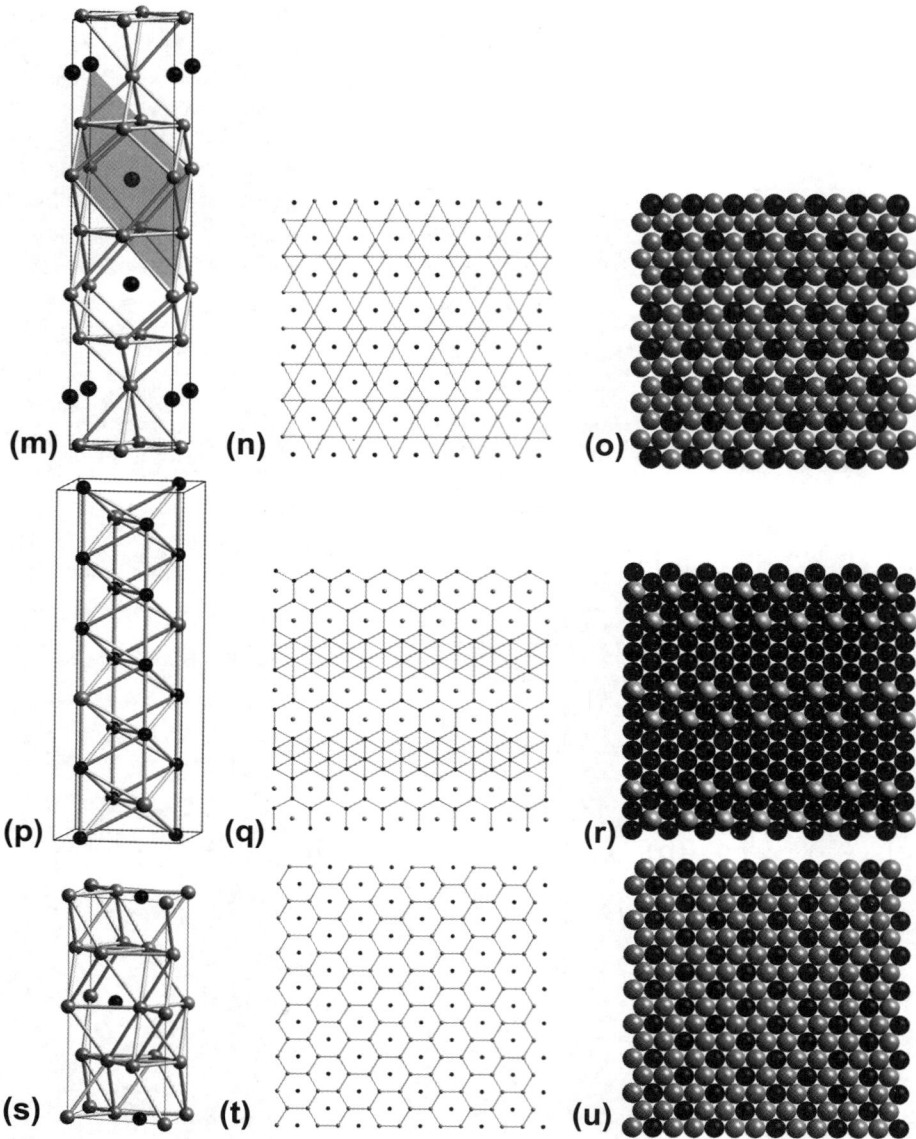

Figure 26 (*continued*).

If the empty tetrahedral voids in the diamond structure are occupied by tetrahedra of another atom species, then we obtain the cubic Laves phase $cF24$-Cu_2Mg $Fd\overline{3}m$, Mg in $8a$ 0, 0, 0; Cu in $16d$ 5/8, 5/8, 5/8. If the atoms in the tetrahedral voids are all of other kind than those at corners and face centers, the $cF24$-$AuBe_5$ structure type results ($F\overline{4}3m$, Au in $4a$ 0, 0, 0; Be in $4c$ 1/4, 1/4, 1/4, Be in $16e$ 5/8, 5/8, 5/8).

Occupying the octahedral voids yields the rocksalt (halite) structure $cF8$-NaCl ($Fm\overline{3}m$, Na in $4a$ 0, 0, 0; Cl in $4b$ 1/2, 1/2, 1/2). If, in addition, the tetrahedral voids are occupied as well, then we arrive at the

$cF16$-NaTl ($Fd\bar{3}m$, Tl in $8a$ 0, 0, 0; Na in $8b$ 1/2, 1/2, 1/2) structure type that is dealt with below, in Section 1.4.4, as $cI2$-W structure derivative.

Doubling the unit cell of $cF4$-Cu along one direction, lets us get the $tI8$-TiAl$_3$ structure type ($I4/mmm$, Ti in $2a$ 0, 0, 0; Al in $4d$ 0, 1/2, 1/4) (**Figure 26**(g–i)). Doing the same with the $cF8$-C (diamond) structure and allowing different occupancies yields the $tI16$-GaAs$_2$ type or in the ternary case, the $tI16$-FeCuS$_2$ (chalcopyrite) structure ($I\bar{4}2d$, Cu in $4a$ 0, 0, 0; Fe in $4b$ 0, 0, 1/2; S in $8d$ x, 1/4, 1/8 with $x = 1/4$).

Starting from the basic hcp $hP2$-Mg structure type ($P6_3/mmc$, Mg in $2c$ 1/3, 2/3, 1/4), we obtain $hP4$-C (lonsdaleite, hexagonal diamond) by occupying half of the tetrahedral voids with atoms of the same kind ($P6_3/mmc$, C in $2b$ 0, 0, 1/4 and in $2c$ 1/3, 2/3, 1/4); $hP4$-ZnS (wurtzite) results by occupying half of the tetrahedral voids with atoms of a different kind ($P6_3mc$, Zn in $2b$ 1/3, 2/3, 0; S in $2b$ 1/3, 2/3, 0.3748). Filling all octahedral interstices, one obtains the nickel-arsenide-type $hP4$-NiAs ($P6_3/mmc$, Ni in $2a$ 0, 0, 0; As in $2c$ 1/3, 2/3, 1/4).

For an ordered distribution of Ni and Sn in a $2 \times 2 \times 1$ superstructure of $hP2$-Mg, we obtain the $hP8$-Ni$_3$Sn structure type ($P6_3/mmc$, Sn in $2c$ 1/3, 2/3, 1/4; Ni in $6h$ 5/6, 2/3, 1/4). Other examples of ordering variants are shown in (**Figure 26**).

1.4.4   *cI2*-W-Based Structures and Their Derivatives

Sticking to cubic symmetry and allowing only substitutional ordering gives just one ordering variant of the bcc $cI2$-W ($Im\bar{3}m$, W in $2a$ 0, 0, 0) structure type without changing the unit cell size, $cP2$-CsCl ($Pm\bar{3}m$, Cl in $1a$ 1/2, 1/2, 1/2; Cs in $1b$ 0, 0, 0).

For the important class of derivative structures with F-centered $2 \times 2 \times 2$-fold supercells, we obtain five different structure types with four different stoichiometries AB, AB$_3$, ABC$_2$ and ABCD: $cF16$-NaTl ($Fd\bar{3}m$, Tl in $8a$ 0, 0, 0; Na in $8b$ 1/2, 1/2, 1/2), $cF16$-BiF$_3$ ($Fm\bar{3}m$, Bi in $4a$ 0, 0, 0; F in $4b$ 1/2, 1/2, 1/2; F in $8c$ 1/4, 1/4 1/4), $cF16$-MnCu$_2$Al (Heusler phase; $Fm\bar{3}m$, Al in $4a$ 0, 0, 0; Mn in $4b$ 1/2, 1/2, 1/2; Cu in $8c$ 1/4, 1/4, 1/4), $cF16$-TiCuHg$_2$ ($F\bar{4}3m$, Hg in $4a$ 0, 0, 0 and in $4d$ 3/4, 3/4, 3/4; Cu in $4b$ 1/2, 1/2, 1/2; Ti in $4c$ 1/4, 1/4, 1/4), $cF16$-LiPdMgSn ($F\bar{4}3m$, Sn in $4a$ 0, 0, 0; Mg in $4b$ 1/2, 1/2, 1/2; Pd in $4c$ 1/4, 1/4, 1/4; Li in $4d$ 3/4, 3/4, 3/4) (**Figure 27**(a–f)).

A representative of a $3 \times 3 \times 3$ superstructure, with two empty sites, is γ brass, $cI52$-Cu$_5$Zn$_8$ ($I\bar{4}3m$, Cu in $8c$ 0.828, 0.828, 0.828; Zn in $8c$ 0.11, 0.11, 0.11; Cu in 12 e .355, 0, 0; Zn in $24g$ 0.313, 0.313, 0.036) (**Figure 28** gamma brass near here).

Noncubic superstructures are known as well. Examples are $tI18$-V$_4$Zn$_5$ ($I4/mmm$, Zn in $2a$ 0, 0, 0 and $8h$ 0.328, 0.328, 0; V in $8i$ 0.348, 0, 0) and antiferromagnetic $tI6$-Cr$_2$Al ($T_N = 598$ K; $I4/mmm$, Al in in $2a$ 0, 0, 0 and Cr in $4e$ 0, 0, 0319; Atoji, 1965), with the $tI6$-MoSi$_2$ structure type (**Figure 27**(g–j)).

1.4.4.1   *Heusler Phases*

The more than 1500 Heusler phases known so far are ternary or quaternary intermetallic compounds that can be seen as ordered $2 \times 2 \times 2$ superstructures of the $cI2$-W structure type. One distinguishes between full- and half-Heusler phases with structure types $cF16$-MnCu$_2$Al ($Fm\bar{3}m$, Mn in 0,0,0; Cu in $8c$ 1/4,1/4,1/4; Al in 1/2,1/2,1/2) and $cF12$-LiAlSi ($F\bar{4}3m$, Al in $4a$ 0,0,0; Li in $4b$ 1/2,1/2,1/2; Si in $4c$ 1/4,1/4,1/4), respectively. $cF16$-MnCu$_2$Al can be considered a $cF16$-BiF$_3$ superstructure (**Figure 27**). Removing four of the eight Cu atoms in the eighth cube leads to the $cF12$-LiAlSi structure type. Full-Heusler phases have the composition A$_2$BC, with A and B transition elements, whereby B can be replaced, in some cases, by a rare-earth element or an alkali metal. For a recent review, see Trudel et al. (2010) or Graf et al. (2009), for instance.

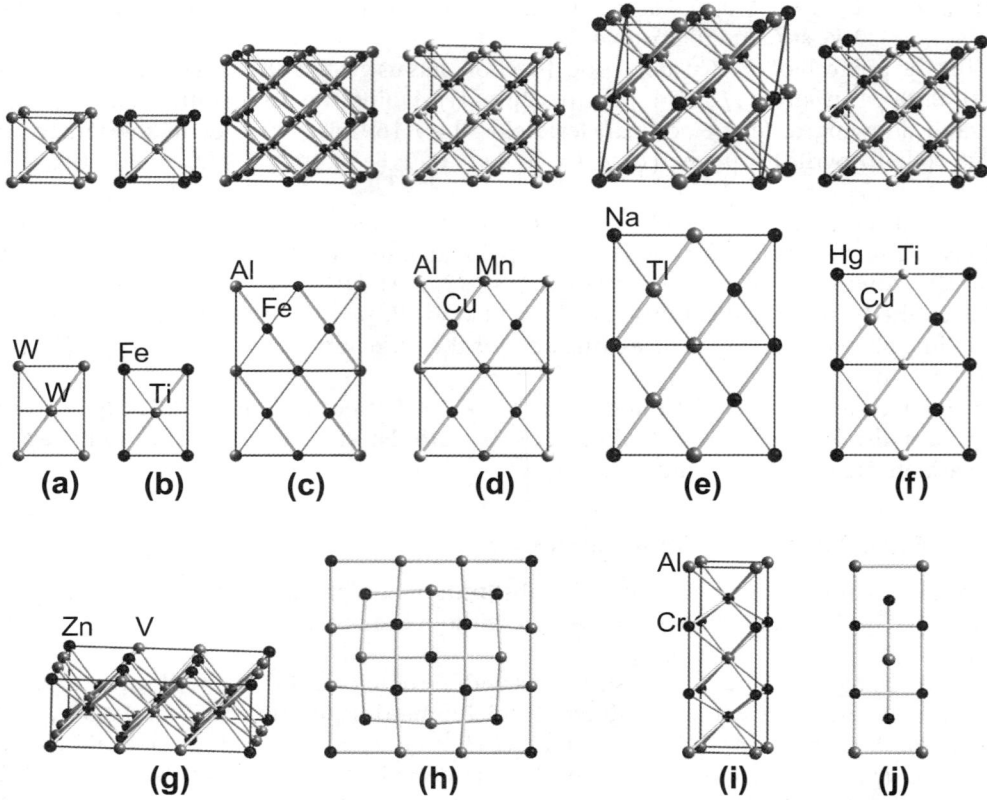

Figure 27 Examples of structures that can be described as superstructures of the $cI2$-W type. In each case one unit cell is shown as well as the (101) plane (as marked in (e), for instance). (a) $cI2$-W type; (b) TiFe, $cP2$-CsCl type; (c) Fe$_3$Al, $cF16$-BiF$_3$ type; (d) $cF16$-MnCu$_2$Al type; (e) $cF16$-NaTl type; (f) $cF16$-TiCuHg$_2$ type. The tetrahedral superstructures are shown with their unit cells as well as in projection along [010]: (g,h) $tI18$-V$_4$Zn$_5$ and (i,j) $tI6$-Cr$_2$Al with the $tI6$-MoSi$_2$ structure type.

Heusler phases can have interesting magnetic or thermoelectric properties, also superconducting representatives are known. Their structures are quite flexible concerning (partial) substitution of elements allowing fine-tuning of the valence electron concentration.

1.4.4.2   Hume-Rothery Phases on the Example of the Cu–Zn System

The brass phases in the system Cu–Zn are long-known typical examples for electronically stabilized crystal structures. With increasing Zn concentration, the electron concentration is increased, i.e. the average number of valence electrons per atom, e/a, which changes the diameter of the Fermi sphere. The crystal structure modifies and adjusts itself in order to bring the almost spherical Fermi surface, with radius k_F, in as close as possible contact with the respective Brillouin zone planes, with diffraction vector $\mathbf{H}$. It leads to a (pseudo)gap at the Fermi energy lowering the electronic energy and stabilizing the structure in this way. The underlying mechanism is based on interference phenomena leading to the formation of standing electron waves if the condition $2|k_F| = |\mathbf{H}|$ is fulfilled; this is the case for electron

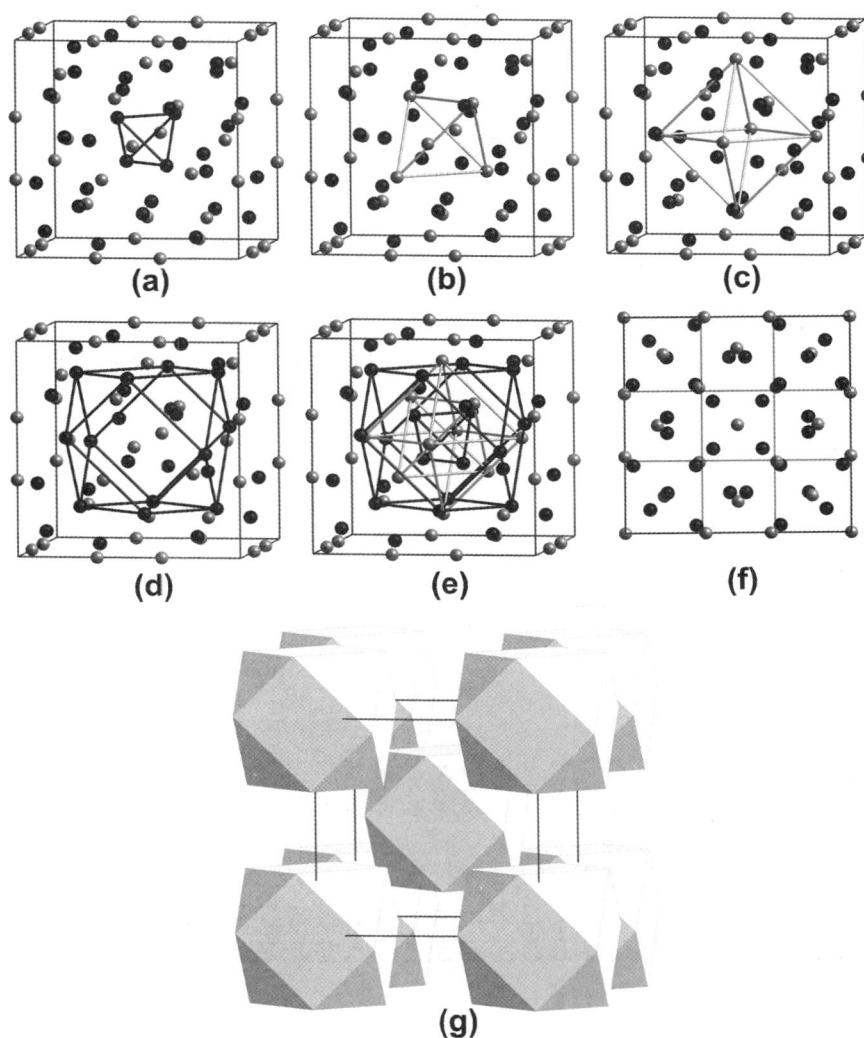

Figure 28 The structure of the I-cell γ brass, $cI52$-Cu$_5$Zn$_8$, $I\bar{4}3m$, in different representations (Cu…gray, Zn…black spheres). It can be described by a *bcc* arrangement of 26-atom clusters, each one consisting of (a) an inner Zn tetrahedron (IT) surrounded first by (b) an outer Cu tetrahedron (OT), then (c) by a Cu octahedron (OH) and, finally, (d,g) a Zn cuboctahedron (CO). The projected structure, with the *bcc* unit cell of the β phase drawn in, is shown in (f).

waves, with wave vector $|k_\mathrm{F}| = 1/\lambda$ and wave length $\lambda = 2d$, that are scattered on atomic planes (atoms on lattice planes, Bragg planes) with distance $d = 1/|\mathbf{H}|$. This mechanism is called Hume-Rothery mechanism, Fermi surface/Brillouin zone interaction or Fermi surface nesting. In **Table 15**, some examples for Hume-Rothery phases are listed.

The solid solution of Zn in *fcc* $cF4$-Cu, stable in the wide range $1 \leq e/a \leq 1.4$, is called α phase. Around $e/a \approx 3/2$, the β phase follows with a statistically disordered $cP2$-CsCl type structure. At lower

Table 15   Examples of Hume-Rothery phases (from Cahn and Haasen, 1996). The number of valence electrons is determined by the group in the periodic table of elements except for transition metals, where it was assumed zero here

Phases with cubic symmetry			Phases with hexagonal symmetry	
Disordered bcc structure β $1.36 \leq e/a \leq 1.59$ cl2-W	γ-brass structure γ $1.54 \leq e/a \leq 1.70$ cl52-Cu_5Zn_8	β-Mn structure μ $1.40 \leq e/a \leq 1.54$ cP20-Mn	ζ $c/a = 1.633$ $1.22 \leq e/a \leq 1.83$ hP2-Mg	ε $c/a = 1.57$ $1.65 \leq e/a \leq 1.89$ hP2-Mg
Cu–Be Ag–Zn Au–Al Cu–Zn Ag–Cd Cu–Al Ag–Al Cu–Ga Ag–In Cu–In Cu–Si Cu–Sn Mn–Zn	Cu–Zn Mn–Zn Cu–Cd Mn–In Cu–Hg Fe–Zn Cu–Al Co–Zn Cu–Ga Ni–Zn Cu–In Ni–Cd Cu–Si Ni–Ga Cu–Sn Ni–In Ag–Li Pd–Zn Ag–Zn Pt–Zn Ag–Cd Pt–Cd Ag–Hg Ag–In Au–Zn Au–Cd Au–Ga Au–In	Cu–Si Ag–Al Au–Al Co–Zn	Cu–Ga Cu–Si Cu–Ge Cu–As Cu–Sb Ag–Cd Ag–Hg Ag–Al Ag–Ga Ag–In Ag–Sn Ag–As Ag–Sb Au–Cd Au–Hg Au–In Au–Sn Mn–Zn	Cu–Zn Ag–Zn Ag–Cd Au–Zn Au–Cd Li–Zn Li–Cd

temperatures, Cu and Zn atoms order and the resulting phase is called β'. The γ phase (γ brass), with the $cI52$-Cu_5Zn_8 structure type, is stable for $e/a \approx 21/13$ (21 valence electrons and 13 atoms). Its structure can be seen as a $3 \times 3 \times 3$ superstructure of the β phase, with the corner atoms and the center atom removed (**Figure 28**). The hcp ε phase is found around $e/a \approx 7/4$. Depending on temperature, fcc $cF4$-Cu dissolves up to 38.3% Zn, while hcp $hP2$-Zn can only dissolve up to 2.8% Cu, forming a solid solution called η phase. By adding 2.8% Cu to Zn, the c/a ratio decreases from 1.856 to 1.805, still significantly deviating from the ideal ratio 1.633 due to covalent bonding contributions in the densely packed layers. The c/a ratio of the hcp ε phase, on the other hand, is with 1.568 smaller than the ideal value.

However, not all pseudogaps at the Fermi energy can be explained by the Hume-Rothery mechanism. They can also originate from structure-induced orbital hybridizations (covalent bonding contributions) resulting in bonding and antibonding energy levels around the Fermi energy. In this case, the electron concentration e/a is not the crucial parameter but the total number VEC of electrons per atom in the valence band (Mizutani, 2011). In case of transition metal atoms, the VEC includes the number of d electrons in the outermost electron shell, for instance.

The phases denoted as γ brass representatives may not only be of the $cI52$-Cu_5Zn_8 type ($I\bar{4}3m$, I-cell γ brass). If the two clusters A and B, occupying the corners and the center of the unit cell, are slightly different, then the P-cell γ brasses are obtained, $cP52$-Cu_9Al_4 type ($P\bar{4}3m$). This structure can be seen as cluster-decorated $cP2$-CsCl structure type. Furthermore, there exist F-cell γ brasses, $cF416$-$Cu_{41}Sn_{11}$ type ($F\bar{4}3m$), which can be described as $2 \times 2 \times 2$ superstructures of the P-cell type, now with four slightly different 26-atom clusters A, B, C and D (A in $4a$ 0, 0, 0, B in $4c$ 1/4, 1/4, 1/4, C in $4b$ 1/2, 1/2, 1/2 and D in $4d$ 3/4, 3/4, 3/4). By this kind of ordering, energetically unfavorable close Sn–Sn contacts can be avoided what would not be possible in P- or I cells at this composition (Booth et al., 1977). Finally, there are also pseudocubic R-cell γ brasses known that are of the $hR26$-Cr_8Al_5 type ($R3m$), with one 26-atom cluster per rhombohedral unit cell.

1.4.4.3 Complex Cluster-based Derivative Structures of the cl2-W Type

There is quite a large number of phases with structures that can be seen as $(p \times p \times p) = p^3$-fold superstructures of the $cF16$-NaTl type and therewith as $(2p)^3$-fold superstructures of the $cI2$-W type; for the structures known so far, p can adopt values of 3, 4, 7 and 11 (Dshemuchadse et al., 2011). Cluster-based structures with giant unit cells may have interesting physical properties because they contain different length scales: lattice period vs cluster diameter and therewith a spatially varying chemical composition. In the following, the structural building principles will be discussed on the example of $cF444$-$Ta_{36.4}Al_{63.6}$, with $p = 3$ (Weber et al., 2009; Conrad et al., 2009).

The structure of $cF444$-$Ta_{36.4}Al_{63.6}$ can be described as a ccp packing of endohedral fullerene-type $Ta_{57}Al_{102}$ clusters, which share their 12 pentagonal faces. The fundamental cluster, with diameter ≈ 13.5 Å, is centered by a Ta atom surrounded by an Al_{14} rhombic dodecahedron (first shell); Al atoms, stellating the faces of the Al_{14} rhombic dodecahedron, form a tetrahedrally distorted Al_{12} cuboctahedron, the corners of which center the pentagon faces of a Ta_{28} polyhedron (second shell); the third cluster shell conforms to an Al_{76} fullerene dual, the endohedral $Ta_{28}Al_{12}$ Frank–Kasper polyhedron, both with tetrahedral symmetry (**Figure 29**). Frank–Kasper (FK) polyhedra consist of triangular faces only (Section 1.4.5).

Generally, fullerenes are bounded by 12 p(entagon) faces, H h(exagon) faces and V vertices. The number of vertices and faces corresponds to $V = 20 + 2H$ and $F = 12 + H$; consequently, $H = 28$ for Al_{76}. The limiting polyhedron for $H = 0$ is the pentagonal dodecahedron. The dual to each fullerene F_V^F with V vertices and F faces is an FK polyhedron FK_F^V with F vertices and V faces (see also Alvarez, 2006).

It should be mentioned here that the structure of $cF444$-$Ta_{36.4}Al_{63.6}$ and related phases with $p = 3$ can also be described as F-cell γ brasses (see, for instance, Berger et al., 2007; and references therein).

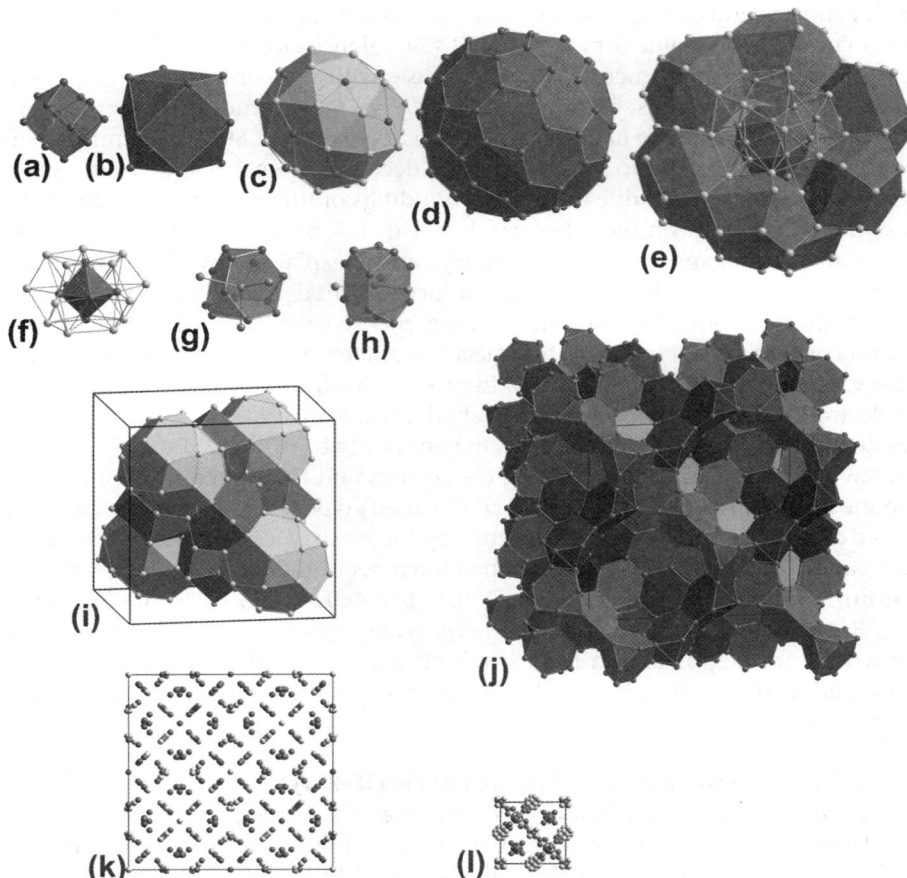

Figure 29   Structure of cF444-$Ta_{39}Al_{69}$, $F\bar{4}3m$. (a)–(d) Shells of the fundamental $Ta_{57}Al_{102}$ fullerene cluster: (a) Al_{14} rhombic dodecahedron around a central Ta atom; (b) distorted Al_{12} cuboctahedron merged with a Ta_{28} polyhedron (c); (d) Al_{76} fullerene shell consisting of 12 p(entagons), 28 h(exagons) and 76 v(ertices); (e) partially opened $Al_{102}Ta_{57}$ fullerene cluster with the mantle of bifrusta surrounding cluster (c) and centered by cluster (a); (f) $Ta_{15}Al_7$ pentagonal Ta_{15} bifrustum around an Al_7 pentagonal bipyramid; (g) Ta_5Al_{12} Friauf polyhedron (CN16 FK polyhedron): a truncated Al_{12} tetrahedron centered by a 4-connected Ta atom; (h) Ta_4Al_{12} CN15 FK polyhedron: a truncated Al_{12} trigonal prism centered by a 3-connected Ta atom. The packing of the $Ta_{28}Al_{12}$ cluster shells shown in (c) via pentagonal Ta bifrusta, drawn in (f), is illustrated in (i), that of the Al_{76} shells, depicted in (d), in (b). (b) All gaps in the cubic dense packing of Al_{76} fullerenes are filled with Friauf (see (g)) and CN15 FK polyhedra, shown in (h). Projections of the unit cells and the average structure are shown in (k) and (l), respectively.

1.4.5   Frank–Kasper (FK) Phases

The FK phases have *topologically close packed* (*tcp*) structures with exclusively tetrahedral interstitial sites, while *ccp* and *hcp* structures and their polytypic variants contain for each two tetrahedral voids one larger octahedral one. However, *tcp* structures are only possible as packings of hard spheres with at least two different diameters; they are based on the four FK polyhedra (Frank and Kasper, 1958, 1959) with

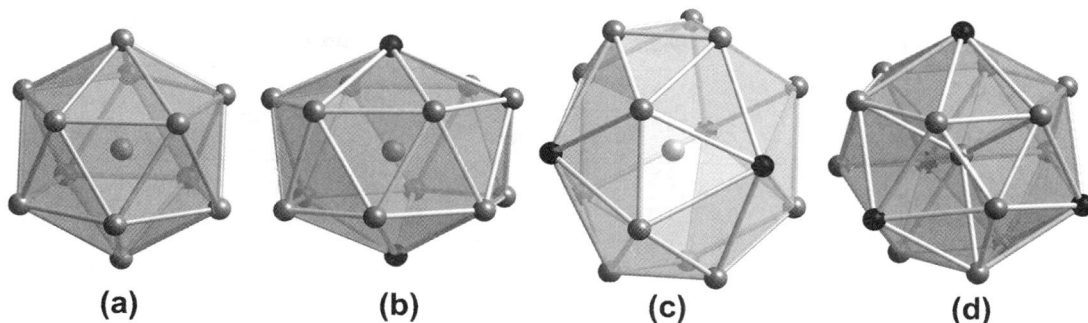

Figure 30   The four FK-polyhedra with their point symmetry: (a) CN12, icosahedron ($m\overline{3}\,\overline{5}$), (b) CN14, bicapped hexagonal antiprism ($\overline{12}2m$), CN15 ($\overline{6}m2$), CN16, Friauf polyhedron, hexagon-capped truncated tetrahedron ($\overline{4}3m$). They contain 20, 24, 26 and 28 irregular tetrahedra, respectively.

coordination numbers (CN) 12 (icosahedron; $m\overline{3}\,\overline{5}$), 14 (hexagonal antiprism with hexagon faces capped; $\overline{12}2m$), 15 ($\overline{6}m2$) and 16 (Friauf polyhedron; truncated tetrahedron with hexagon faces capped; $\overline{4}3m$). These simplicial polyhedra consist of irregular tetrahedra with one common vertex in the center and are bounded by equilateral triangles only (**Figure 30**).

The vertices of the FK polyhedra are either 5- or 6-coordinated. The number V_6 of 6-coordinated vertices of an FK-polyhedron, usually occupied by the larger atoms, is $V_6 = \text{CN} - V_5$, with $V_5 = 12$ in all cases. Their duals, resulting from replacing the faces by vertices and vice versa, are fullerenes with hexagon and pentagon faces, with the number of pentagon faces always 12. The *major ligand lines* connecting the 6-coordinated vertices with each other form the *major skeleton* of the structure.

tcp structures, exclusively built from FK-polyhedra, can be decomposed into flat atomic layers. The *primary layers* consist of triangles and pentagons (Laves phases, μ- and M-phase), or triangles and hexagons (σ-phase, $hP7$-Zr_4Al_3), or triangles, pentagons and hexagons (P-phase). They are accompanied by *secondary layers* of triangles and/or squares in a way that only tetrahedral voids are formed. This means that the vertices of the secondary layer cap the pentagons and hexagons only, and not the triangles. This distinguishes a secondary layer from one that is dual to the primary layer. The primary layers are usually located on mirror planes, the secondary layers in between. The pentagons/hexagons in two subsequent primary layers are arranged in a way two form antiprisms centered by an atom of the secondary layer.

Typical examples of FK phases and their composition regarding FK polyhedra are listed in **Table 16** and depicted in **Figures 31–33** (Sections 1.4.5.1 and 1.4.5.2).

1.4.5.1   Laves Phases and Polytypes

Friauf-Laves phases are binary or ternary intermetallic phases with more than 4000 entries in *Pearson's Crystal Data* (Villars and Cenzual, 2011). For a comprehensive review, see Stein et al. (2004, 2005). Structurally, Laves phases belong to the family of Frank–Kasper phases, crucial structure-determining parameters are size ratios, electronegativity difference and valence electron concentration. The structures of the different modifications of Laves phases, with general chemical composition AB_2, can be seen as different polytypic stacking variants of a fundamental layer consisting of four atomic planes: a Kagomé net (6.3.6.3) formed by the smaller B atoms is sandwiched between two triangle nets (3^6) of A atoms; on top of it, sits a triangle layer (3^6) of B atoms in one of

Table 16   Examples of Frank–Kasper phases with the frequencies of FK polyhedra building their structures (after Shoemaker et al, 1969)

Structure type	Frequency of			
	CN12	CN14	CN15	CN16
$cP8$-Cr_3Si	0.25	0.75	0	0
$hP7$-Zr_4Al_3	0.44	0.28	0.28	0
$cF24$-Cu_2Mg	0.67	0	0	0.33
$hP12$-$MgZn_2$	0.67	0	0	0.33
$hP24$-Ni_2Mg	0.67	0	0	0.33
$tP30$-$Cr_{46}Fe_{54}$	0.33	0.54	0.13	0
$hR13$-W_7Fe_6	0.55	0.15	0.15	0.15

two, by a rotation of π differing, variants. In the layer stacking sequence, one of these two variants is marked by a prime. From another point of view, one can illustrate the stacking sequence by describing the stacking of the layers packed from A-atom-centered truncated tetrahedra and tetrahedra formed by B atoms (**Figure 31**).

The prototype structures of binary and ternary Laves phases are the following:

- $hP12$-$MgZn_2$, $P6_3/mmc$, Mg in $4f$ 1/3, 2/3, 0.063; Zn in $2a$ 0, 0, 0 and $6h$ 0.830, 0.660, 1/4. The major skeleton is formed by the larger Mg atoms occupying the sites of the lonsdaleite (hexagonal diamond) structure. The Zn atoms constitute a 3D tetrahedral network around Mg atoms, consisting of stacked Kagomé nets connected via vertex-sharing tetrahedra. $hP12$-$MgZn_2$ can be seen as the basic two-layer variant AB (Ramsdell: 2H, Jagodzinski: hh).
- $cF24$-Cu_2Mg, $Fd\bar{3}m$, Mg in $8a$ 0, 0, 0; Cu in $16d$ 5/8, 5/8, 5/8. The major skeleton is formed by the larger Mg atoms occupying the sites of the diamond structure. The Cu atoms constitute a 3D tetrahedral network around Mg atoms, consisting of interpenetrating Kagomé nets connected by tetrahedra. $cF24$-Cu_2Mg can be seen as a three-layer polytypic stacking variant ABAC (3C, ccc).
- $hP24$-Ni_2Mg, $P6_3/mmc$, Mg in $4e$ 0, 0, 0.094 and $4f$ 1/3, 2/3, 0.094; Ni in $4f$ 1/3, 2/3, 0.844, $6g$ 1/2, 0, 0 and $6h$ 0.167, 0.334, 1/4. The major skeleton is formed by the larger Mg atoms occupying the sites of an intergrowth of diamond and lonsdaleite structures. The Ni atoms constitute a 3D tetrahedral network around Mg atoms, consisting of stacked Kagomé nets connected by tetrahedra. $hP24$-Ni_2Mg can be seen as a four-layer polytypic stacking variant ABAC (4H, $chch$).
- $hR42$-$Mg(Ag_{0.1}Zn_{0.9})_2$, $R\bar{3}m$. As example of a long-period Laves-phase polytype, we discuss the largest known so far, a 21-layer polytypic stacking variant ABC′B′ABC′BCA′C′BCA′CAB′A′CAB′ (21R, $(hchchch)_3$). There are also several other polytypes known (Komura and Kitano, 1977).

Whether a hexagonal or cubic Laves phase forms for a given chemical composition is influenced by the structure of the pure B element; if it is fcc, then in most cases, the cubic Laves phase forms. Phase transformations between different Laves phases have been observed as function of temperature, pressure and/or composition. Usually, the cubic Laves phase is the low-temperature (LT) modification. The phase fields of Laves phases range from very narrow to extended at either side of the stoichiometric composition. There are several binary and ternary systems known, where two or even all three types of Laves phases are observed, such as Co–Nb or Al–Cr–Ti. Although most Laves phases in ternary systems are just pseudobinaries, there exist several true ternaries such as $hP12$-Ta(Al,Ni)$_2$ or $hP12$-(Zr,Ti)(Ni,Ti)$_2$.

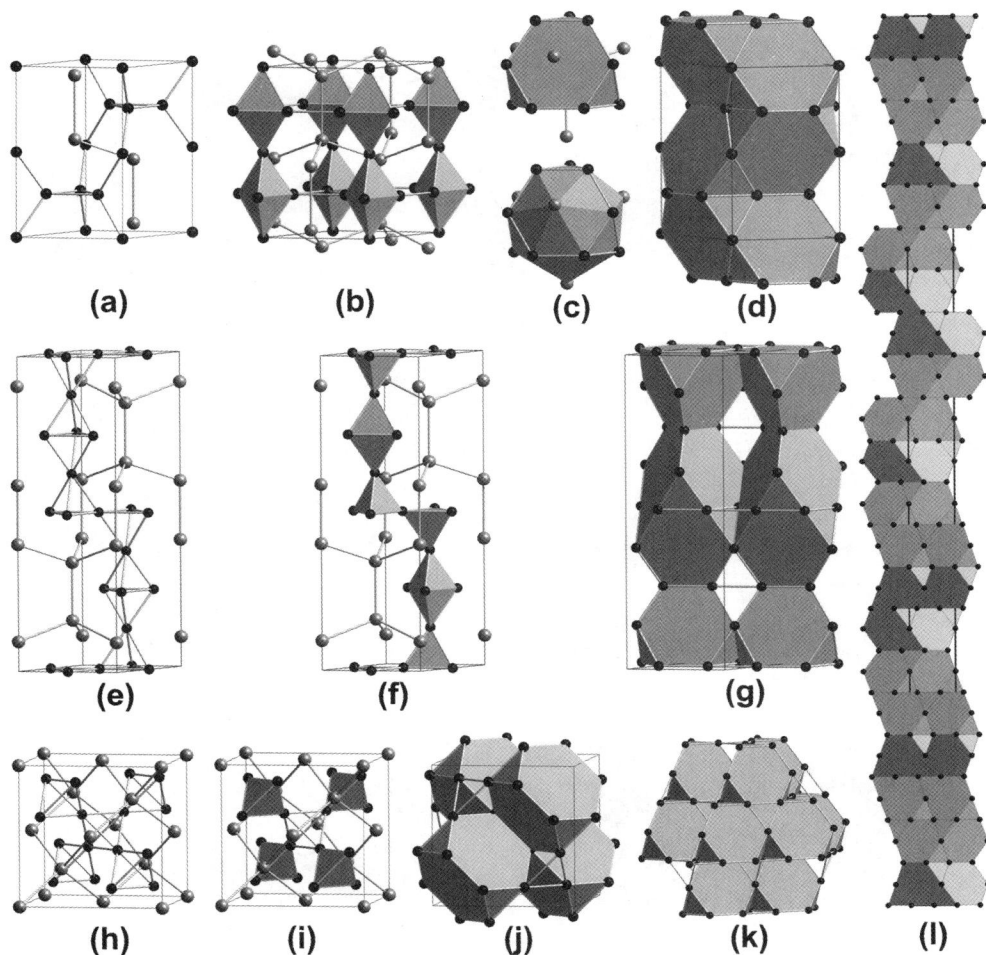

Figure 31 The structures of the three Laves phases MgTM$_2$ in different representations: (a–d) hP12-MgZn$_2$, (e–g) hP24-Ni$_2$Mg and (h–k) cF24-Cu$_2$Mg. The structures can be seen as different packings of TM tetrahedra (b) and Mg-centered truncated TM tetrahedra (c–d). The coordination polyhedron formed by the 12 TM and 4 Mg atoms (c) is called Friauf polyhedron or CN16 Frank–Kasper (FK) polyhedron, it is a hexagon face capped truncated tetrahedron. Not shown are the AET around B atoms, which are CN12 FK polyhedra, i.e. icosahedra formed by 6 A and 6 B atoms. (l) Polytypic variant of the Laves phase hR42-Mg(Ag$_{0.1}$Zn$_{0.9}$)$_2$. (Mg...gray, TM...black).

In the ideal case, only atoms of the same kind touch each other and the two subsets of homogenous sphere packings form an *interpenetrating heterogenous sphere packing*. In case of cF24-Cu$_2$Mg, for instance, the Mg atoms in $8a$ form a homogenous sphere packing with the shortest distance $d_1 = a\sqrt{3}/4$ and contact number $k = 4$; the Cu atoms in $16d$ constitute another sphere packing with shortest distance $d_2 = a\sqrt{2}/4$ and contact number $k = 6$. The shortest distance between Mg and Cu atoms amounts to $d_3 = a\sqrt{11}/8 > (d_1 + d_2)/2$ (Koch and Fischer, 1992). The ideal size ratio results to $r_A/r_B = \sqrt{3/2} = 1.225$. The experimentally observed radii ratios range between 1.05 and 1.70 (Stein et al., 2004).

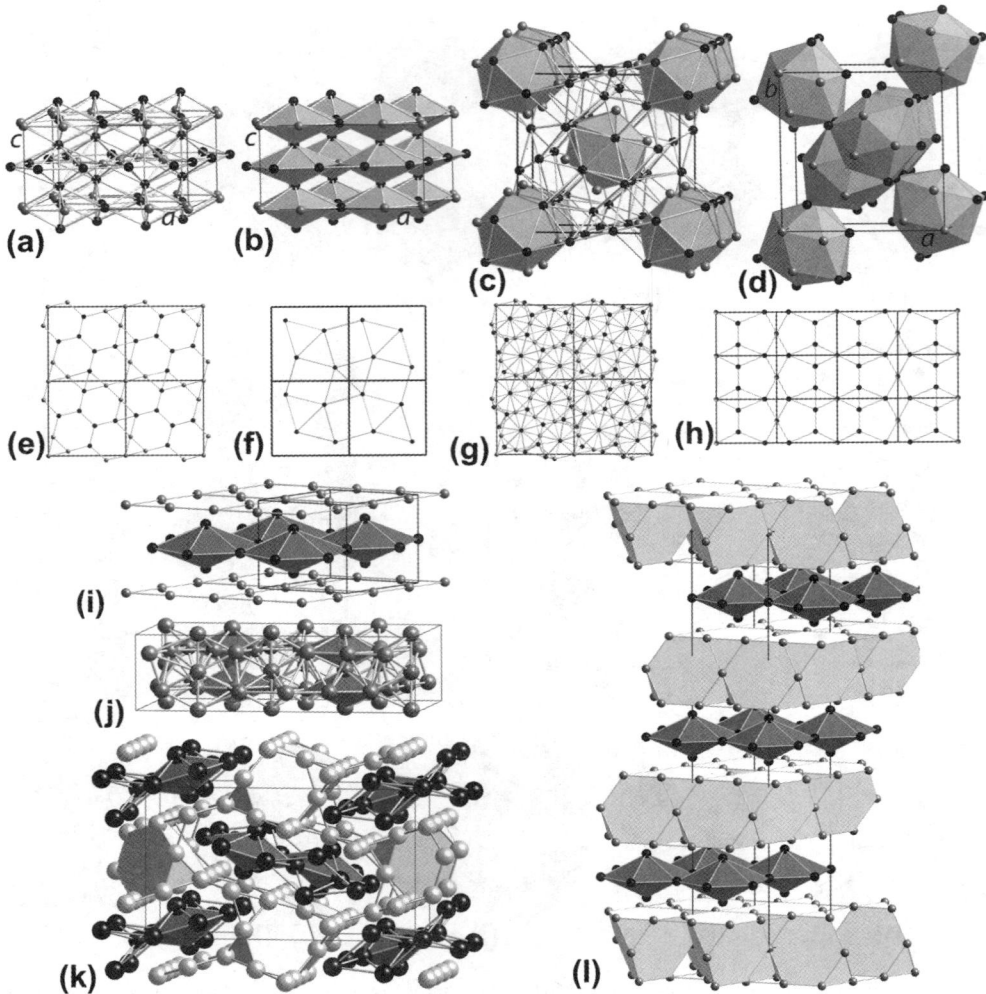

Figure 32 The structure of the σ-phase, *t*P30-CrFe in the system Al–Ta in different representations (Al...gray, Ta...black). It can be either described (b) by a packing of hexagonal bipyramids (*hbp*), edge-connected in the (110) planes and corner-sharing along the [001] direction, by a packing of partially overlapping CN12 icosahedra (only a subset shown) (c) or partially overlapping CN15 FK-polyhedra (d). The structure can also be decomposed into layers: a hexagon/triangle tiling in the primary layer with $z = 0$ (e), an Archimedean snub square tiling $3^2.4.3.4$ in the secondary layer $z = 1/4$ (f). In (g) the projection $0 \leq z \leq 1$ is shown and in (h) a (110) layer. Some phases, structurally related to the σ-phase are shown as well: (i) *h*P7-Zr_4Al_3, (j) P-phase *o*P56-$Mo_{21}Cr_9Ni_{20}$, (k) M-phase *o*P52-$Nb_{10}Ni_9Al_3$, (l) μ-phase *h*R13-W_6Fe_7 (heavier elements are shown in darker gray).

Depending on their chemical composition, Laves phases can show a variety of interesting physical properties:

- Superconducting materials: *c*F24-$(Zr, Hf)V_2$, *h*P12-$ZrRe_2$, *h*P24-Mo_2Hf, ...
- Magnetostrictive materials: $(Dy_{1-x}Tb_x)Fe_2$, ...

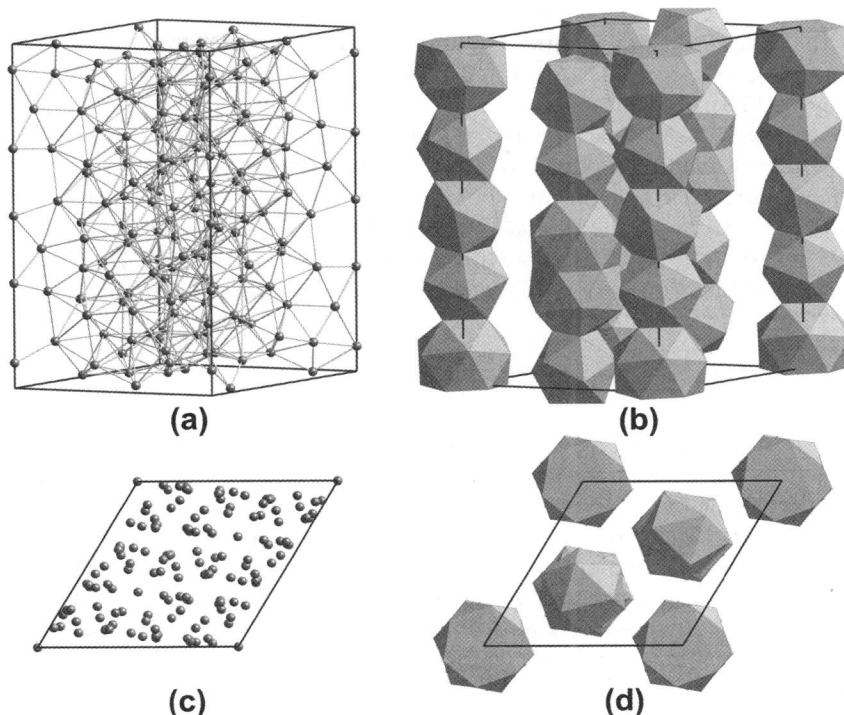

Figure 33 The R-phase, $hR53$-$Mo_3Cr_2Co_5$, in different representations: (a,c) perspective view, (b,d) projections along [001]. The structure can be described as columnar stackings of CN16–CN12–CN12–CN12–CN16 FK polyhedra, which are linked by tetrahedra.

- Magnetocaloric materials: $TbCo_2$, $Gd_5(Ge_xSi_{1-x})_4$, $Mn(Sb_xAs_{1-x})$, $La(Fe_{13-x}Si_x)$, ...
- Hydrogen storage materials: $Zr(V,Mn,Ni)_2$, ...
- HT structural materials: $Ta(Ni,Al)_2$, ...

1.4.5.2   σ, μ, M, P, and R Phase

The σ phase, $tP30$-$Cr_{46}Fe_{54}$ ($P4_2/mnm$), is assumed to be electronically stabilized with an electron/atom ratio in the range of $6.2 \leq e/a \leq 7$. Its structure contains three types of FK-polyhedra, CN12, CN14 and CN15 (**Figure 32**). In **Figure 32**(b), the CN14 FK-polyhedra are not shown as usual but as vertex-connected hexagonal bipyramids (*hbp*). This takes into account that the distances within each *hbp* are smaller than between *hbp*. In **Figure 32**(c), the arrangement of along [001] edge-connected CN12 FK-polyhedra (icosahedra) is illustrated on one subset. The other subset of icosahedra is strongly overlapping and fills the cell completely. The arrangement of CN15 FK-polyhedra is depicted in **Figure 32**(d).

The primary layers in $z = 0$, 1/2 are two-uniform hexagon/triangle tilings of the type $(6^2.3^2; 6.3.6.3)$ (**Figure 32**(e)), while the secondary layers in $z = 1/4$, 3/4 are two-uniform hexagon/triangle tilings of the type $3^2.4.3.4$ (**Figure 32**(f)). The pentagon/triangle tiling parallel to (110) is of the type $(5^2.3^2; 5.3^2.5.3; 5.3.5.3)$ (**Figure 32**(h)).

The μ phase, $hR13$-W_6Fe_7-type ($R\bar{3}m$, **Figure 32** sigma phase (l)), has been found mainly in systems containing Nb, Ta, Mo or W as one component and Fe, Co, Ni or Zn as the other (Joubert and Dupin, 2004). The structure can be seen as constituted from modules of $hP7$-Zr_4Al_3 ($P\bar{6}$, **Figure 32**(i)) and $cF24$-Cu_2Mg (**Figure 31**(k)).

While the σ phase can be seen as packing of CN14 FK polyhedra, the P-phase $oP56$-$Mo_{21}Cr_9Ni_{20}$ (*Pnma*; **Figure 32**(j)) alternately features modules of CN14 and of CN12 FK polyhedra. The M-phase $oP52$-$Nb_{10}Ni_9Al_3$ (*Pnma*; **Figure 32**(k)), on the other hand, contains modules of the Laves phase and the σ phase.

Another complex FK phases is the R phase, $hR53$-$Mo_3Cr_2Co_5$ ($R\bar{3}$; **Figure 33**), consisting of closest packed chains of face-sharing FK polyhedra with the sequence CN16–(CN12)$_3$–CN16. The R phase is one of the FK phases, which can only partially be described in terms of primary and secondary layers (Komura et al., 1960).

1.4.6 Zintl Phases

Zintl phases belong to the family of valence compounds. They are polyanionic compounds at the borderline between metallic and nonmetallic (ionic). Constituents of classical Zintl phases are electropositive elements of groups 1–2 on one side and electronegative elements of groups 13–16 on the other side. The compounds have to be electronically balanced such as covalently 2-center-2 electron bonded crystal structures. Consequently, they are stoichiometric line compounds, i.e. have a very narrow compositional stability range. Beside the aforementioned classical Zintl phases, there are also nonclassical Zintl phases known where transition metals (e.g. $cP2$-CsAu) or lanthanoids are involved instead of alkali or alkaline earth metals (Kauzlarich, 1996). Skutterudites, AB_3 (A … late TM, B … P, As, Sb), with the $cI32$-$CoAs_3$ structure type, can be seen as TM Zintl compounds (see Section 1.4.10.1).

The original concept has been introduced by Zintl (Zintl and Dullenkopf, 1932) on the example of $cF16$-NaTl. In this compound Na donates its electron to Tl, which then gets in its outermost electron shell isoelectronic to carbon. Therefore, the polyanionic substructure formed by the larger Tl^- ions should be that of diamond, while the small Na^+ ions are occupying the voids in the diamond (D-) net (a 3D four-connected net), making up a complementary D-net. The so-called *Zintl–Klemm concept* is a generalization of this example, saying that the anions in these compounds form polyanionic substructures that are the same as those of the neutral main group elements with the same number of valence electrons (pseudoatom concept). According to Miller (1996), Zintl phases are compounds whose bonding and nonbonding states are completely occupied and separated from the antibonding states by not more than 2 eV. This means that, generally, Zintl phases are semiconductors but they can also show metallic character if they have a few extra electrons or holes relative to those needed for the 2-center-2 electron bonds. An example for a metallic Zintl phase is $cP156$-$K_{29}NaHg_{48}$ (see **Figure 40** in Section 1.4.9), which is almost isostructural to the true Zintl phase $cP154$-$K_3Na_{26}In_{48}$, but is not electron balanced since In has three and Hg only two available valence electrons (Deiseroth and Biehl, 1999). The main structure motifs are two Na-centered icosahedral closo-cluster ($NaHg_{12}$) and six K-centered hexagonal antiprismatic arachno-clusters (KHg_{12}) per unit cell, embedded in a matrix of alkali metal atoms. The Hg atoms are close to each other and form a 3D network. Another example of a metallic Zintl phase is $hP8$-$BaSn_3$ and the series of polytypic superstructures, $(BaSn_3)_m[Ba(Sn_yBi_{1-y})_3]_n$, which are discussed in the next section.

Closed-shell clusters in Zintl phases are associated with delocalized bonding, i.e. the bonding electrons are shared by all atoms of the cluster. According to Wade's rule, $2n + 2$ bonding electrons are needed for so-called *closo* deltahedral clusters (all n vertices occupied), $2n + 4$ for *nido* (one vertex missing) and $2n + 6$ for *arachno* clusters (two vertices missing) (Wade, 1976). A deltahedron is a polyhedron whose faces are all equilateral triangles, similar to Greek capital deltas, Δ. The clusters can be isolated or connected. The delocalized intracluster (endo-)bonds are longer than the 2-center-2 electron (exo-)bonds between clusters.

1.4.7 Topological Layer Structures

Structures that can be topologically described as stackings of atomic layers are not necessarily layer structures in the crystal-chemical sense. This means that the chemical bonding between layers does not necessarily differ from that within the layers. However, the decomposition of a crystal structure into layers can simplify the description of the structure and reveal relationships to other structures.

It has been emphasized already by Samson (1964) that in each of the space groups $F23$, $P2_13$, $Fd3$, $Pa3$, $F432$, $F\bar{4}3m$, and $F\bar{4}3c$, every special position places at least one point on the (110) plane (see **Figure 29** in Section 1.4.4.3, for instance). One should keep in mind that atomic layers in cubic structures occur in at least three symmetrically equivalent orientations leading to a 3D periodic framework of interpenetrating atomic layers. This means that structures that crystallize with well-defined interpenetrating atomic layers will preferentially adopt one of these space group symmetries.

The $tP4$-AuCu ($P4/mmm$, Au in $1a$ 0, 0, 0 and $1c$ 1/2, 1/2, 0; Cu in $2e$ 0, 1/2, 1/2) structure type with representatives such as FePd, FePt, CoPt, etc. can be decomposed into (100) layers of 4^4 square tilings shifted against each other. $tI12$-CuAl$_2$ ($I4/mcm$, Cu in $4a$ 0, 0, 1/4; Al in $8h$ .1541, .6541, 0) can be described as stacking of $3^2.4.3.4$ triangle/square nets in $z = 0$, 1/2, with the squares capped by Cu in $z = 1/4$, 3/4 (**Figure 34**). This leads to edge-connected columns of face-sharing tetragonal Al antiprisms centered by Cu, which form chains along [001]. The space in-between has the shape of disorted Al tetrahedra.

Another stacking variant of $3^2.4.3.4$ nets is realized in the $tP10$-U$_3$Si$_2$ ($P4/mbm$, U in $2a$ 0, 0, 0 and $4h$ .181, .681, 1/2; Si in $4b$ .389, .889, 0) structure type (**Figure 34**). There, the nets are stacked on top of

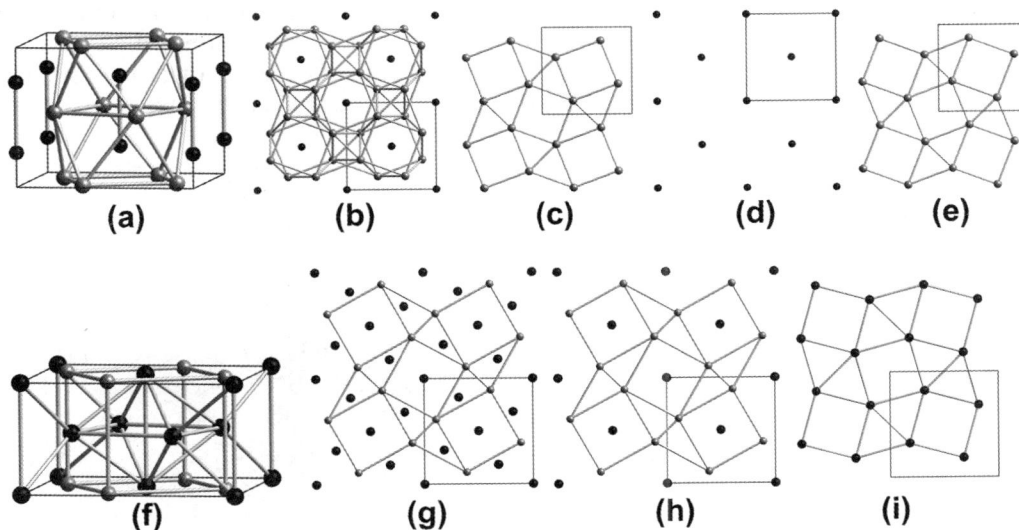

Figure 34 (a–e) The structure of $tI12$-CuAl$_2$: unit cell in perspective view (a) and projection along [001] (b); layers at $z = 0$ (c), $z = 1/4$ (d) and $z = 1/2$ (e) (Al…gray, Cu…black). (f–i) The structure of $tP10$-U$_3$Si$_2$: unit cell in perspective view (f) and projection along [001] (g); layers at $z = 0$ (h) and $z = 1/2$ (i) (Si…gray, U…black). The layers in (e) and (i) correspond to the Archimedean snub square tiling $3^2.4.3.4$.

each other separated by a square net of U atoms, which cap the faces of Si cubes and center the trigonal Si prisms. U can be seen to form a network of vertex-connected octahedra separated by Si-centered trigonal U prisms. The distance between Si atoms in neighboring trigonal U prisms is by far the shortest one. U–U and U–Si distances are shorter than all other Si–Si ones.

oC10-Fe_2AlB_2 ($Cmmm$, Al in 2a 0, 0, 0; B in 4i 0, 0.2071, 0; Fe in 4j 0, 0.3540, 1/2) (**Figure 35**) can be seen either as nonclose stacking of 4^4 nets or as prismatic stacking of 3^34^2 nets of Fe atoms. B atoms occupy the trigonal prismatic voids and Al centers the Fe cubes. The structure can also been seen as stacking of cubic and hexagonal slabs. The structure combines structure modules from the oC8-CrB ($Cmcm$, Cr in 4c 0, 0.146, 1/4; B in 4c 0, 0.440, 1/4) and the cP2-FeAl structure (Jeitschko, 1969). While the slabs of B-centered trigonal prisms, oC10-Fe_2AlB_2 have a sequence ABBA, those of oC8-CrB are ACCA. B means a shift of the slab by half a triangle edge, C by half the prism height.

In **Figure 26** (see Section 1.4.3), beside the aforementioned tP4-AuCu structure type, six more structures are shown that can be considered to be ccp or hcp packings of hdp layers with a particular type of ordering: cP4-$AuCu_3$ ($Pm\bar{3}m$, Au in 1a 0, 0, 0; Cu in 3c 0, 1/2, 1/2), tI8-$TiAl_3$ ($I4/mmm$, Ti in 2a 0, 0, 0; Al in 2b 0, 0, 1/2 and 4d 0, 1/2, 1/4), tI10-$MoNi_4$ ($I4/m$, Mo in 2a 0, 0, 0; Ni in 8h 0.2, 0.4, 0), tI16-$ZrAl_3$ ($I4/mmm$, Al in 4c 0, 1/2, 0, 4d 0, 1/2, 1/4 and in 4e 0, 0, 0.361; Zr in 4e 0, 0, 0.122), oP20-$ZrAu_4$ ($Pnma$, Au in 4c x, 1/4, z with x = 1/6 and z = 0, 1/5, 3/5, 4/5; Zr in 4c 1/6, 1/4, 2/5) and hP12-WAl_5 ($P6_3$, Al in 2a 0, 0, 0, 2b 1/3, 2/3, 0 and in 6c 1/3, 1/3, 1/4; W in 2b 1/3, 2/3, 1/2).

hP8-$BaSn_3$ (hP8-Ni_3Sn type) can be seen as a metallic Zintl phase. Partial substitution of Sn by Bi in the quasibinary homologous system $BaSn_{3-x}Bi_x$ ($0.4 \leq x \leq 1$) forms a series of polytypic superstructures, $(BaSn_3)_m[Ba(Sn_yBi_{1-y})_3]_n$, with periods in [001] direction of up to 39 atomic layers (**Figure 36**) (Ponou et al., 2008). These superstructure formation can be attributed to the partition of the structure in polar ($BaSn_3$) and nonpolar ($Ba(Sn_yBi_{1y})_3$) regions.

REME phases are a large class of intermetallics with RE:M:E = 1:1:1 stoichiometry, where RE is a rare-earth metal, actinoid or an element from groups 1–4; M is a late TM from groups 8–12, and E belongs to groups 13–15. While some of the REME phases are insulators or semiconductors, others have unusual magnetic and electronic properties. Some oP12-TiNiSi-type structures are known as heavy Fermion compounds. For comprehensive reviews, see Bojin and Hoffmann (2003a,b).

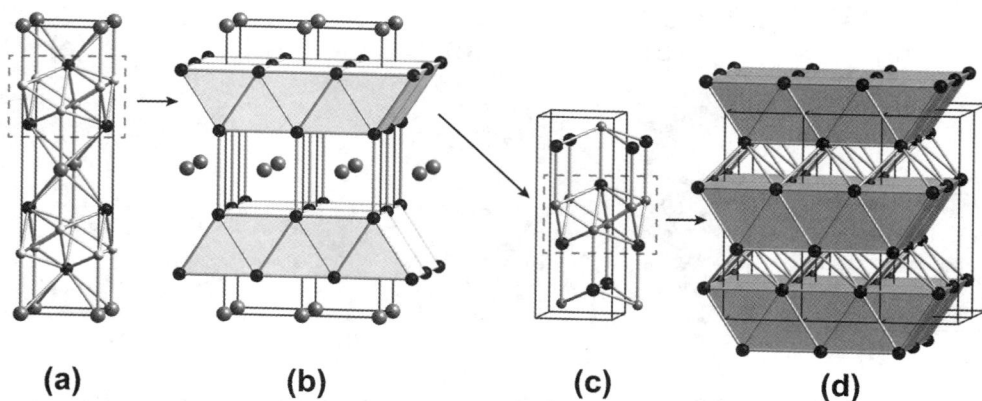

(a)   (b)   (c)   (d)

Figure 35   The structure of oC10-Fe_2AlB_2 (B…light gray, Al…medium gray, Fe…black) that contains modules of the oC8-CrB (marked by dashed rectangle; B…gray, Cr…black) and of the cP2-CsCl structure type. The B-centered triangular prisms are shaded.

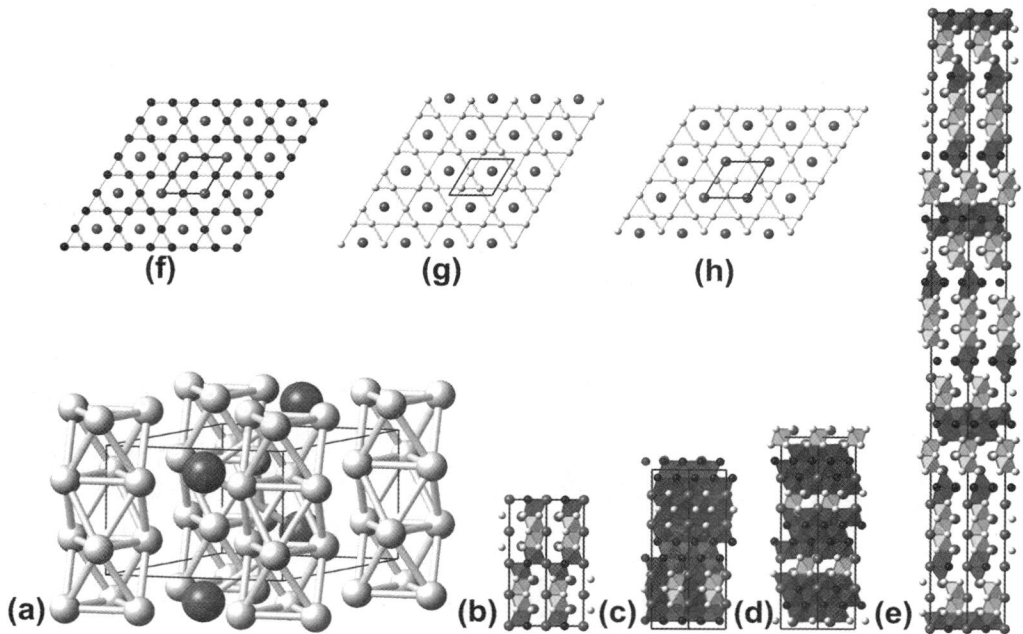

Figure 36 The structures of the compounds in the homologous series $(BaSn_3)_m[Ba(Sn_yBi_{1-y})_3]_n$, with (a) $y = 0$, $n = 0$, hh, $P6_3/mmc$, (b) $y = 0.43$, $m = 3$, $n = 1$, $(hhhc)_2$, $P6_3/mmc$, (c) $y = 0.39$, $m = 3$, $n = 2$, $(hhhcc)_2$, $P6_3/mmc$, (d) $x = 0.33$, $m = 2$, $n = 2$, $(hhcc)_3$, $R\bar{3}m$, (e) $y = 0.35$, $m = 10$, $n = 3$, $[(hhhc)_2(hhhhc)]_3$, $R\bar{3}m$ (Ponou et al., 2008). As example, are shown the *hdp* layers of the structure (c) at (f) $z = 0$, (g) 1/8, (h) 1/4. The stacking sequence of the *hdp* layers is indicated in this figure caption based on the Jagodzinski notation. In (b–e) are projections along [010] shown of $2 \times 2 \times 1$ unit cells in each case (Sn ... light gray, Ba ... dark gray, Bi black).

Although more than 2000 representatives of this class crystallize in about 40 structure types, they show common structural building principles. Most REME structures can be described as stackings of puckered ME layers consisting of eclipsed hexagonal structure elements with all possible lattice symmetries except triclinic. A typical representative of REME phases is *oI*60-EuAuSn. Its structure, with its polyanionic $AuSn^{2-}$ framework and Eu^{2+} cations in between, is shown in **Figure 37**. The structure can be seen as fivefold superstructure of the *oI*12-KHg$_2$ type, which itself is a fourfold superstructure of the *hP*3-AlB$_2$ type (**Figure 38**). For a review of equiatomic intermetallic Eu-compounds, see Pöttgen and Johrendt (2000).

1.4.8 Long-Period Structures

We have already seen several examples of long-period structures in the course of the discussion of the elements (Rb, Cs, Sr, Ba...), i.e. (incommensurately) modulated and/or composite structures. Further simple examples are long-range ordered antiphase domain structures in the system Cu–Au such as *oI*40-AuCu. Superstructures formed by different stacking sequences of *hdp* layers have been discussed above on the example of the polytypic superstructures, $(BaSn_3)_m[Ba(Sn_yBi_{1-y})_3]_n$.

Another example is the class of Nowotny chimney ladder structures that can be described as helical composite structures (Fredrickson et al., 2004ab; Sun et al., 2007). The compounds have the

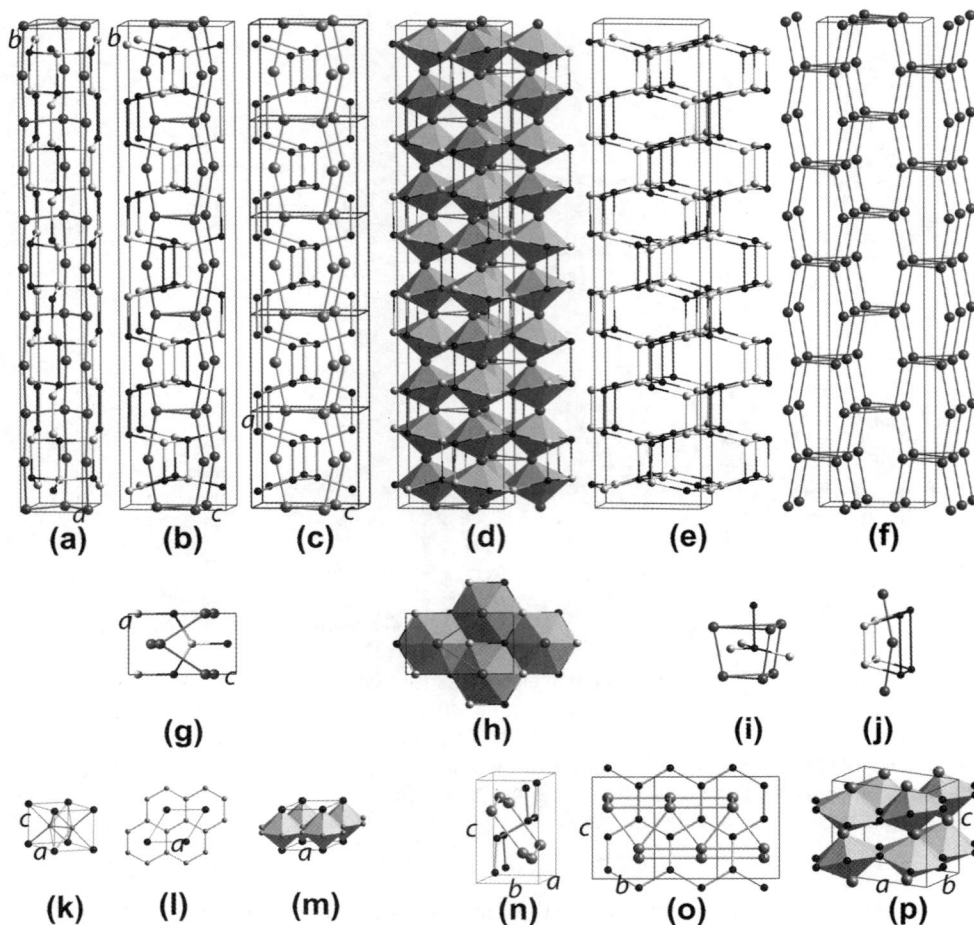

Figure 37 (a,b,d–j) The structure of *oI*60-EuAuSn in different views (Au black, Eu gray, Sn light gray). The polyanionic framework of Au and Sn atoms is shown in (e) and that of the Eu cations in (f). Projections of the unit cells (b,d) along [010] are shown in (g,h), respectively. The AET around Au and Eu are depicted in (i) and (j), respectively. The structure can be seen as fivefold superstructure of the *oI*12-KHg$_2$ type illustrated in (c, n-p) (K gray, Hg black), which itself is a fourfold superstructure of the *hP*3-AlB$_2$ type (k–m) (Al black, B gray).

general composition A$_t$B$_m$, with A, a TM of groups 4–9, and B, a main group element from groups 13–15. In the terminology of composite crystals, the host structure is formed by the TM atoms and the guest structure by the main group elements. With the underlying basis or average periodicity c_{av}, we obtain the actual period (lattice parameter) $c = (2t - m) c_{av}$; t and m are integers with a ratio $1.25 \leq m/t \leq 2$.

In **Figure 39**, the composite structures of *tP*32-Ir$_3$Ga$_5$, *tP*20-Ru$_2$Sn$_3$ and *oF*24-TiSi$_2$ are depicted, which show 3, 2 and 1 host and 5, 3, 2 guest periods, respectively, per unit cell. In **Figure 39**(b), one nicely sees the helical arrangement of the Ga atoms enclosed in the Ir diamond-like host framework; note the different periodicities of these substructures. Beside commensurate structures, with

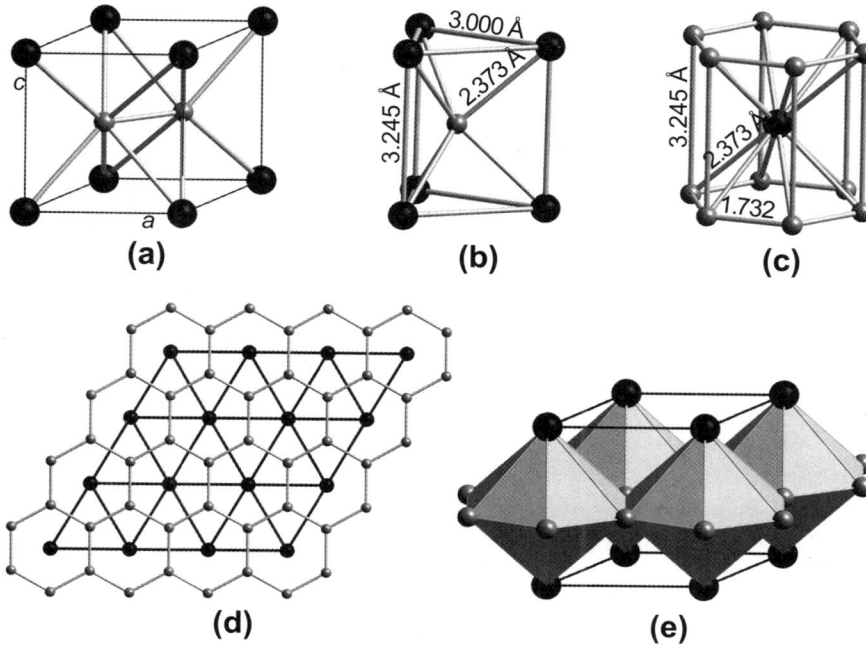

Figure 38 Example for different descriptions of $hP3$-AlB$_2$. Ball-and-stick model (a) with AET around (b) B and (c) Al. In (d), the mutually dual Al and B nets, regular Archimedean tilings 6^3 and 3^6, respectively, are shown in superposition. The polyhedron model (e) is based on the shortest interatomic distances.

periodicities of more than 300 Å, also a few incommensurate Nowotny chimney ladder structures have been described (Rohrer et al., 2000, 2001). (Mo/Rh)$_{11}$Ge$_{18}$, for instance, a mutually modulated composite structure, has satellite vectors $q_h = 0.364\ c^*$ for the host and $q_g = 0.389\ c^*$ for the guest substructure.

1.4.9 Hierarchical structures

Hierarchical structures can be formally derived from simpler structure types by replacing atoms by groups of atoms (clusters). This means that packing principles on smaller scale (atoms) are applied on a larger scale (clusters) analogously. Examples are (see also Bodak et al., 2006):

- $cP8$-Cr$_3$Si $\rightarrow$ $cP156$-K$_{29}$NaHg$_{48}$ (**Figure 40**(a,b));
- $cI2$-W $\rightarrow$ $cI44$-Ce$_6$Ni$_6$Si$_2$;
- $cP3$-CaTiO$_3$ $\rightarrow$ $cP39$-Mg$_2$Zn$_{11}$;
- $cF116$-Th$_6$Mn$_{23}$ $\rightarrow$ $cF1124$-Tb$_{117}$Fe$_{52}$Ge$_{112}$;
- quasiperiodic structures (**Figure 40**(c)).

Quasiperiodic (see Section 1.5) or fractal structures can also be seen as hierarchical structures. Due to their scaling symmetry, particular structural arrangements appear again and again on larger and larger scale. This is shown in **Figure 40**(c) on the example of a fractal pentagon tiling. It can be generated with a pentagon as initiator and the compound P5P as generator. Iterative application of the generator

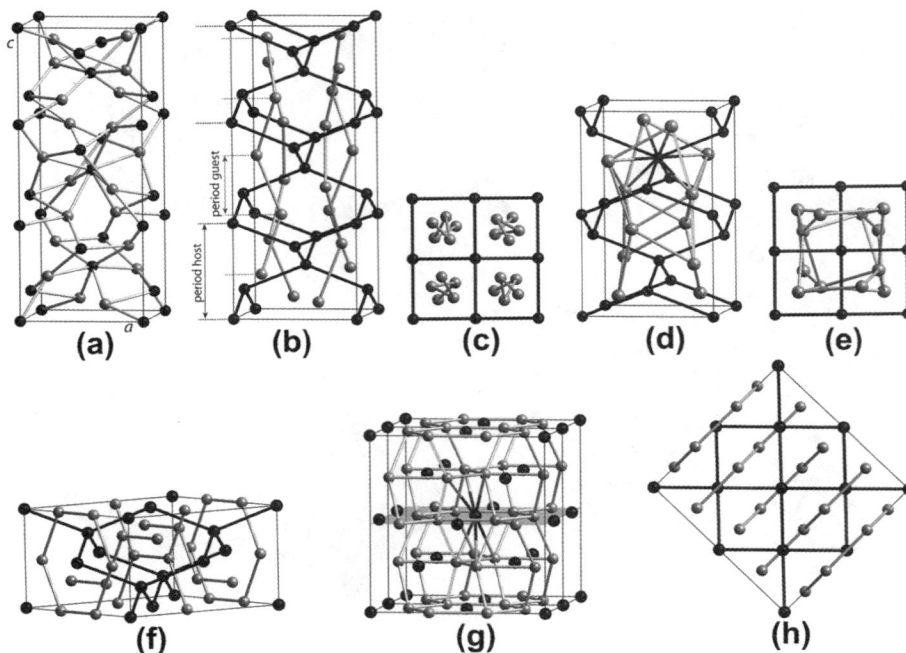

Figure 39   Examples for Nowotny chimney ladder structures: (a–c) $tP32$-Ir_3Ga_5, (d,e), $tP20$-Ru_2Sn_3 and (f–h) $oF24$-$TiSi_2$ with 3, 2 and 1 host and 5, 3, 2 guest periods per unit cell. Note that the bonds shown do not necessarily correspond to shortest atomic distances; they rather illustrate representative structural atomic arrangements. The lattice parameters a and b of $oF24$-$TiSi_2$ are each by a factor $\sqrt{2}$ larger than those of the other structures shown here. In (d) and (g), the very short Ru–Sn and Ti–Si bonds are marked by (black/gray) around one Ru and Ti atom, respectively.

creates larger and larger fractal tilings with particular pentagon arrangements appearing on every hierarchy level. If the gaps in the fractal tiling are properly filled, a quasiperiodic pentagon tiling is obtained. The centers of all pentagons then form a hexagon-boat-star (HBS) tiling.

1.4.10   Structures of Some Functional Intermetallics

Physical properties depend in very different ways on chemical composition, crystal structure and defects. For instance, hardness and elastic properties are mainly determined by bond strength and structural anisotropy while for transport properties, the kind of defects and their concentration can be crucial. Superconductivity strongly depends on the electronic band structure and valence electron concentration. In the following, the structures of a few examples of functional intermetallics are discussed.

1.4.10.1   Thermoelectric Materials

Thermoelectric materials can directly convert thermal into electrical energy and vice versa. The Seebeck effect describes the generation of an electrical voltage by a temperature gradient across a thermoelectric material. The inverse effect is called Peltier effect. The thermoelectric performance is described by the dimensionless figure of merit $ZT = (S^2\sigma/\kappa)T$, with S, the thermoelectric power ($S^2\sigma$, the power factor

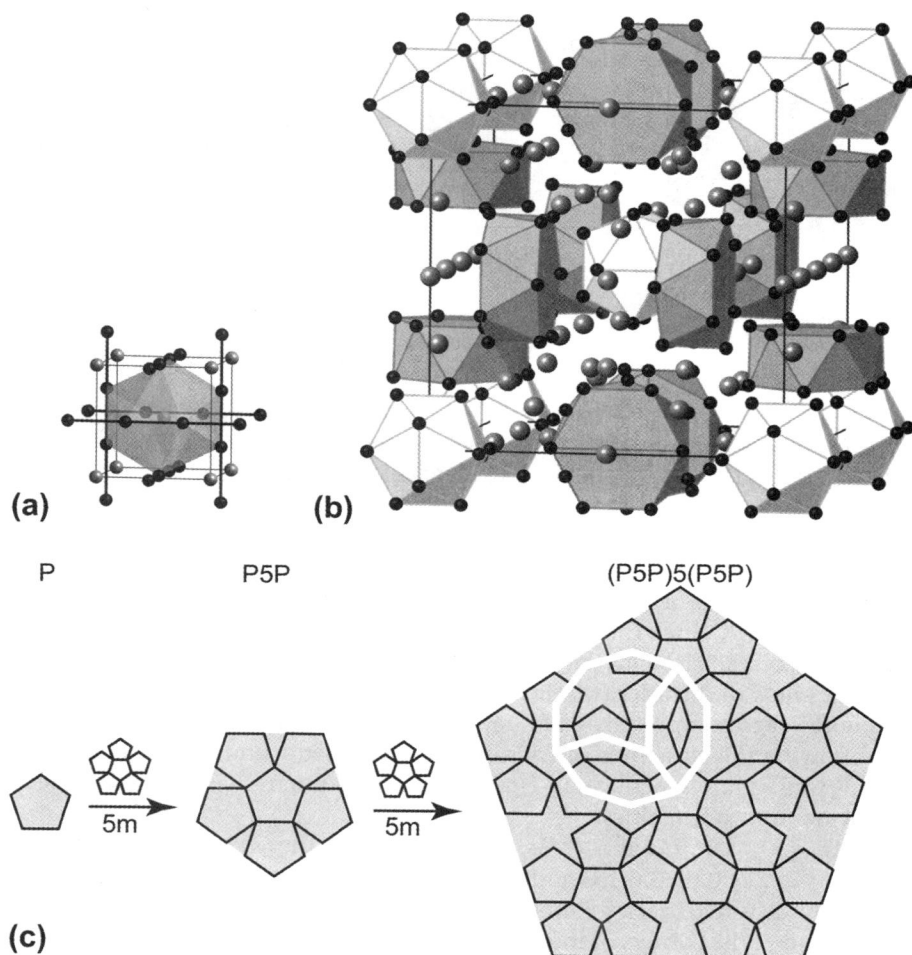

(a) P

(b) P5P

(P5P)5(P5P)

(c)

Figure 40 The structures of (a) $cP8$-Cr_3Si and (b) $cP156$-$K_{29}NaHg_{48}$ are in a hierarchical relationship to each other: the centers of the icosahedral $NaHg_{12}$ and the hexagonal antiprismatic KHg_{12} clusters correspond to the positions of Si and Cr atoms, respectively, in $cP8$-Cr_3Si. The interstitial sites between these clusters are occupied by K atoms. The gray levels of the atoms increase with the atomic number. In (c), the hierarchical relationships between different generations of scaled pentagon tilings are illustrated. The pentagon P gets on each of its edges twinned leading to the variant P5P; this is again fivefold twinned yielding (P5P)5(P5P). A typical cluster of an HBS tiling is outlined in white.

PF), σ, the electronic conductivity, $\kappa = \kappa_{el} + \kappa_{latt}$, the thermal conductivity with its electronic and lattice contribution, and T, the temperature. For a high-ZT (larger 1) material, the thermal conductivity should be as small as possible and the electronic conductivity as high as possible (PGEC, "phonon glass, electron crystal," concept); this is difficult to achieve because these parameters are partially directly related to each other. Consequently, the thermal conductivity can best be lowered by lowering the lattice contribution κ_{latt} by inhibiting the phonon propagation as far as possible. This can be achieved by introducing disorder, vacancies and/or nanoinclusions into structures and/or having "rattling

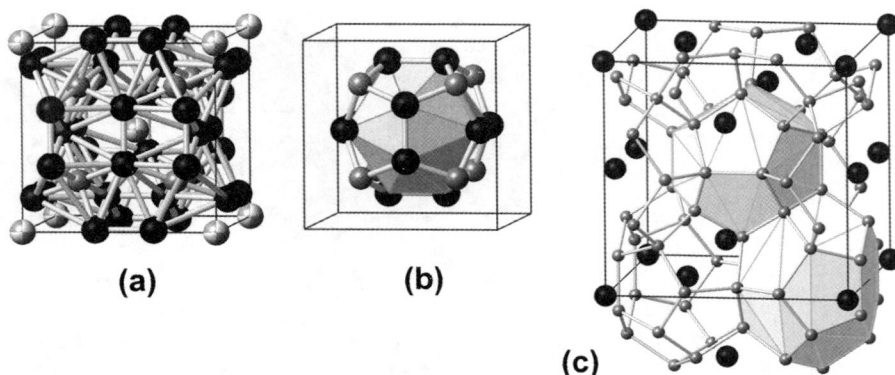

Figure 41   The structure of the (a,b) skutterudite $cI32$-CoAs$_3$ and the type I clathrate $cP54$-K$_4$Si$_3$. The icosahedral As$_{12}$-AET around the center is marked in (b). $cP54$-K$_4$Si$_{23}$ contains a Si$_{20}$-dodecahedron around the K atoms at the vertices and in the center, and around the other kind of K atoms a Si$_{24}$-tetrakaidecahedron, consisting of two hexagons and 12 pentagons.

atoms" in large voids such as it is the case for some skutterudites and clathrates. For reviews, see Sootsman et al. (2009) or Kleinke (2010), for instance. Materials with very complex crystal structures such as QCs, for instance, can also have large Seebeck coefficients due to their spiky electronic density of states and different kinds of disorder (Macia, 2001).

As an example for cage compounds with "rattling atoms" regarded as PGEC compounds, we will discuss in the following the structures of (narrow bandgap) semiconducting filled skutterudites and type I clathrates as well as intermetallic half-Heusler phases. Skutterudites are mainly antimonides and clathrates are silicides and germanides.

Skutterudites, AB$_3$ (A … late TM, B … P, As, Sb), crystallize in the $cI32$-CoAs$_3$-structure type ($Im\bar{3}$, Co in $8c$ 1/4,1/4,1/4; As in $24g$ 0,0.35,0.15) and can be seen as TM Zintl compounds (**Figure 41**). The Co atoms center slightly distorted, vertex-connected As$_6$-octahedra, which leave rather large icosahedral voids in the center and at the corners of the unit cell. In $cI32$-CoAs$_3$, the distance from the center to the As at the icosahedra vertices amounts to 3.118 Å. Together with the eight Co atoms in a distance of 3.546 Å from the center, the 12 As atoms form a slightly distorted dodecahedron. Filling this void with a heavy ion that is smaller than the void such as a trivalent RE ion or Ba^{2+}, for instance, inhibits long-wavelength phonon propagation by rattling. The general formula for filled skutterudites is M$_x$A$_4$B$_{12}$ ($x \leq 1$), such as Yb$_{0.19}$Co$_4$Sb$_{12}$ with a ZT $\approx$ 1 at 600 K (Nolas et al., 2000), for instance.

The type I clathrate structure, with general formula A$_2$B$_6$C$_{46}$ (A … Na, K, Ba; B,C … Al, Ga, In, Si, Ge, Sn), an ordered $cP54$-K$_4$Si$_{23}$ structure type ($Pm\bar{3}n$, K in $2a$ 0,0,0 and $6d$ 1/4,1/2,0; Si in $6c$ 1/4,0,1/2, $16i$ x,x,x with x = 0.185 and $24k$ 0,0.306,0.118), consists of a tetrahedral framework with cages that can incorporate large cations (**Figure 41**). The atoms in the cages donate their valence electrons to the framework. This leads to complete filling of the sp^3 orbitals of the tetrahedrally coordinated framework atoms and the clathrates, therefore, are expected to be semiconductors. These clathrates can be seen as semiconducting Zintl phases with A cations sitting in a polyanionic B$_6$C$_{46}$ framework. On the example of $cP54$-Ba$_8$In$_{16}$Ge$_{30}$, it was shown that the large Ba atoms in the In/Si cages are off-centered and show large atomic displacement parameters. Due to anharmonic potentials, the off-centering increases with temperature (Bentien et al., 2005).

Other interesting thermoelectric materials are half-Heusler phases (see Section 1.4.4.1), in particular, for HT applications at temperatures up to more than 1500 K. Examples with a ZT ≈ 0.8 at 1073 K are $cF12$-MNiSn (M = Ti, Zr, Hf) compounds doped by Sb at the Sn site (Culp et al., 2006). The empty sites in half of the eighth cubes of the unit cell give rise to narrow bands and resulting d orbital hybridization and semiconducting character.

1.4.10.2 Ferromagnetic Materials

The strongest permanent magnets contain RE elements because of their high magnetic moment and their stability against demagnetization (high coercitivity and remanence) (for a review, see Goll and Kronmüller, 2000). This is mainly achieved by their high magnetocrystalline anisotropy, which results from the coupling of their localized 4f electrons with the crystal electric field together with spin-orbit coupling. In case of $tP68$-$Nd_2Fe_{14}B$, the best permanent magnetic material so far (T_c = 585 K), Nd occupies two inequivalent sites 4f (with closer Fe distances) and 4g (with closer B distances), of which the latter has a strong preference for c-axis alignment and therewith defines the [001] easy-axis direction (Haskel et al., 2005).

The structure of $tP68$-$Nd_2Fe_{14}B$ ($P4_2/mnm$, Nd in 4f and 4g; Fe in 4c, 4e, 8j and 16k; B in 4f) (**Figure 42**) can be described as stacking of slabs of face- and vertex-sharing $Fe_{13}Nd_2$ CN14 FK polyhedra. The slabs are connected with each other via the Nd atoms, which themselves form flat layers together with the B atoms and Fe atoms outside the CN14 polyhedra. The B atoms, in the flat layers at $z = 0$ and $1/2$, center trigonal prisms formed by Fe atoms of the two adjacent layers. While the large Nd atoms need a much larger distance between adjacent Fe layers, Fe–B attraction leads to a much smaller distance. This contributes to the puckering of the Fe layers (Herbst et al., 1984). The rectangular faces of these edge-sharing prisms are capped by Nd atoms.

Another class of strong permanent magnets is constituted by $hP6$-$SmCo_5$ ($hP6$-$CaCu_5$ type; $P6/mmm$, Ca in 1a 0,0,0; Cu in 2c 1/3,2/3,0 and 3g 1/2,0,1/2) and $hR19$-Sm_2Co_{17} ($hR19$-Th_2Zn_{17} type; $R\overline{3}m$, Th in 6c 0,0,1/3; Zn in 6c 0,0,0.097, 9d 1/2,0,1/2, 18f 1/3,0,0 and 18h 1/2,1/2,1/6), which has the best performance at higher temperatures (T_c = 1189 K). Their structures can be described as stackings of alternating h(exagon) layers and K(agomé) layers (**Figure 43**). The h layers consist of only Ca-centered Cu-hexagons in case of $hP6$-$CaCu_5$ and of empty hexagons surrounded by Th-centered ones in case of $hR19$-Th_2Zn_{17}. The hexagons in the k nets are empty in both cases, however, a part of them is capped by Zn atoms forming hexagonal pyramids in $hR19$-Th_2Zn_{17}. These Zn atoms also cap the empty hexagons in the h layers, however, in a larger distance.

1.4.10.3 Magnetostrictive Materials

If a ferromagnetic material shows magnetostriction, then its shape/dimensions can be changed by the application of a magnetic field that changes the spontaneous magnetostrictive strain. There are many metals known that show a significant magnetostrictive effect. However, the commonly used material is Terfenol D, $cF24$-$Dy_{1-x}Tb_xFe_2$ ($x \sim 0.3$), with the structure of the $cF24$-Cu_2Mg-type Laves phase and a rather high Curie temperature of T_c = 653 K. While Terfenol D has a giant positive magnetostriction constant, $cF24$-$SmFe_2$, with the same structure type, has a giant negative one (Kawamura et al., 2006).

The magnetostrictive effect is also responsible for the Invar effect, originally discovered and used on $Fe_{65}Ni_{35}$ alloys (Guillaume, 1897). The negative magnetostrictive effect leads to an expansion of the crystal structure at ambient temperature. The thermal motion weakens the effect more and more with increasing temperature, thereby diminishing the magnetostrictive expansion what largely compensates the thermal expansion of the material. However, the main contributors to the invar effect are still controversially discussed (see Ruban et al., 2007 and references therein).

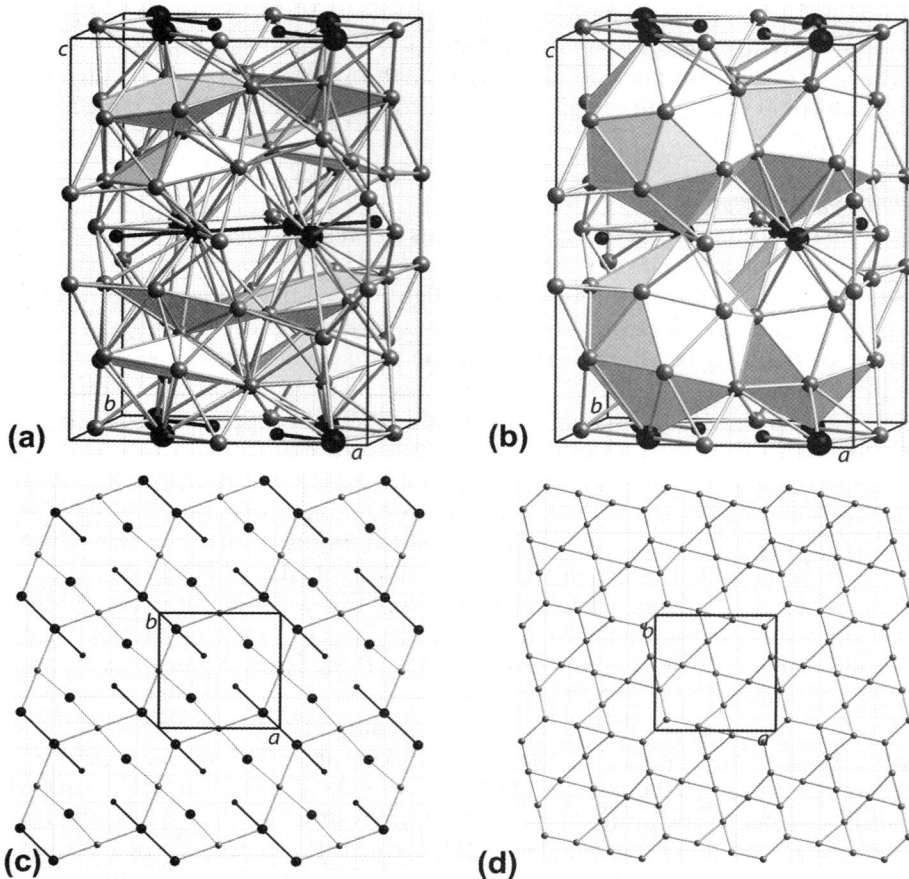

Figure 42 The structure of $tP68$-$Nd_2Fe_{14}B$ in different representations: (a) Fe-vertex-sharing hexagonal Fe-pyramids; (b) Fe-face- and Nd-vertex-sharing CN14 FK polyhedra; (c) flat and puckered atomic layers in $z = 0$ and (d) $0.1 \leq z \leq 0.26$ (Fe ... gray, Nd ... large, black, B ... small, black).

1.4.10.4 Magnetocaloric Materials

The magnetocaloric effect (MCE) can be used for refrigeration by adiabatic demagnetization. The magnetic part of the entropy S_M is lowered by adiabatic magnetization of a superparamagnetic material or a ferromagnetic material slightly above its Curie temperature (T_c). Thereby, the disorder of the magnetic spin system is greatly reduced. Since the total entropy remains constant in an adiabatic process, the vibrational entropy (lattice vibrations) and therewith the temperature has to increase by ΔT_{ad}. After the heat is removed from the system, the reverse process, which is similar to the expansion of a gas, adiabatic demagnetization restores the zero-field magnetic entropy again thereby decreasing lattice vibrations and therewith the temperature.

There are a couple of intermetallic phases known that show a giant magnetocaloric effect (GMCE), which would make them suitable as viable magnetic refrigerants. According to reviews by Gschneidner et al. (2005) and Brück (2005), intermetallics with large MCE include beside the mainly used Gd-based

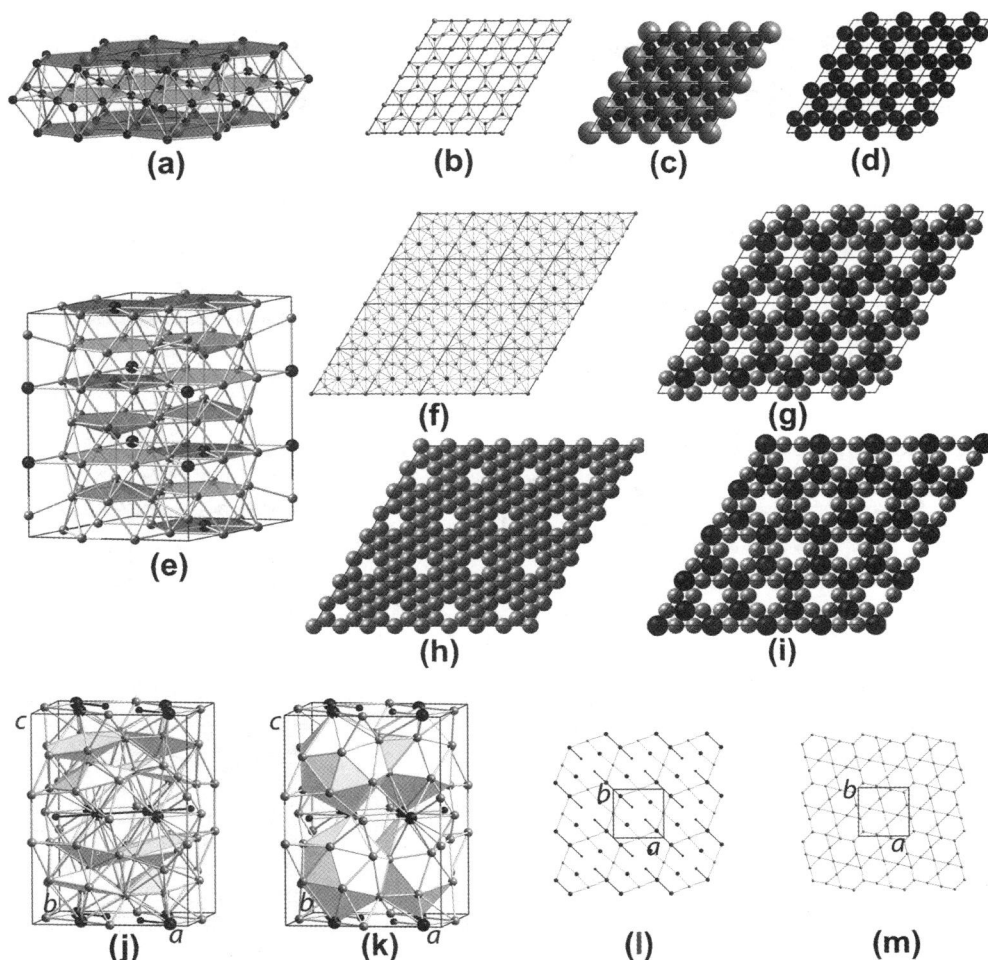

Figure 43 The structures of (a–d) the $hP6$-CaCu$_5$ type, (e–i) the $hR19$-Th$_2$Zn$_{17}$ type and (j–m) the $tP68$-Nd$_2$Fe$_{14}$B type in different representations (light atoms … gray, heavy atoms and B … black). In (b) and (f), the projections down [001] are depicted, in subfigures (c,d,g–i) cuts through the structures are shown with space-filling atomic spheres: (c) $z = 0$, (d) $z = 1/2$, (g) $z = 0$, (h) $0.05 \leq z \leq 0.25$, (i) $z = 1/3$. The Fe$_{13}$Nd$_2$ arrangements are illustrated in the form of (j) vertex connected hexagonal pyramids and as (k) face- and vertex-connected CN14 FK polyhedra. In (l), the flat atomic layer in $z = 0$ is depicted, and in (m), the puckered hexagon/triangle net of Fe atoms. The B atoms center trigonal Fe prisms.

solid solutions: the lanthanoid Laves phases (REM$_2$, with M … Al, Co, Ni; $cF24$-Cu$_2$Mg type), Gd$_5$(Si$_{1-x}$Ge$_x$)$_4$ ($oP36$-Sm$_5$Ge$_4$, $mP36$-Gd$_5$Ge$_2$Si$_2$ or $oP36$-Gd$_5$Si$_4$ type), La(Fe$_{13-x}$Si$_x$) ($cF112$-NaZn$_{13}$ type), Nd$_2$Fe$_{17}$ ($hR19$-Th$_2$Zn$_{17}$ type). Gd$_5$(Si$_{1-x}$Ge$_x$)$_4$ shows a first-order magnetic phase transition concentrating the large entropy change in a small temperature interval.

As example of a magnetocaloric material, the LT and HT structures of Gd$_5$(Si$_2$Ge$_2$) are illustrated in **Figure 44**. According to Choe et al. (2000), the structure can be described as stackings of slabs, which consist of Gd-centered Gd cubes. The slabs are shifted against each other in the (101) plane by half

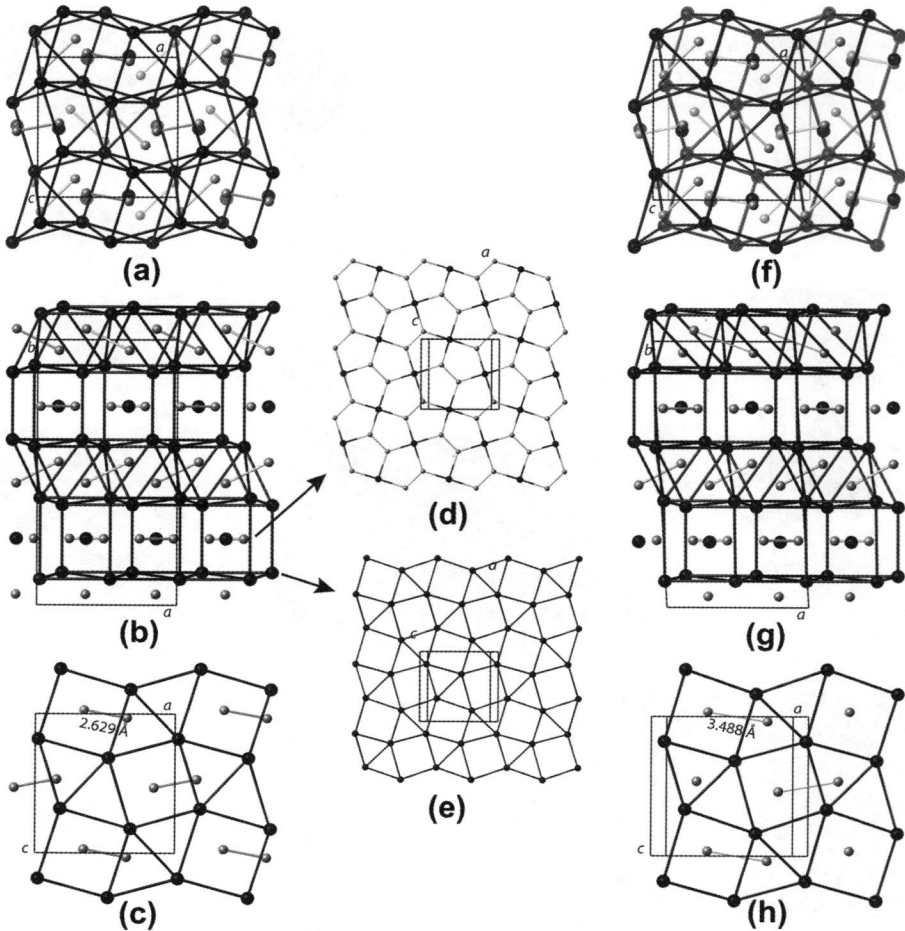

Figure 44 The structures of (a–e) $oP36$-Gd$_5$(Si$_2$Ge$_2$) at 243 K and (f–h) $mP36$-Gd$_5$(Si$_2$Ge$_2$) at 292 K: (a, f) projections onto (101), (b,g) along [001]; (c,h) bounded projections $-0.07 \leq y \leq 0.2$; (e) bounded projections of the slightly puckered Gd layer ($0.07 \leq y \leq 0.2$) and Si/Ge/Gd layer ($0.24 \leq y \leq 0.26$). The change in the dimer atom distance is indicated in (c,h) (Gd … black, Si/Ge … gray). The nets in (c,e,h) correspond to the Archimedean snub square tiling $3^2.4.3.4$, the net in (d) to its dual, the Catalan Cairo pentagonal tiling $V3^2.4.3.4$

a cube edge length, connected by dimers capping the cube faces of the two slabs. In the (101) plane, the cube faces are part of a $3^2.4.3.4$ snub square tiling. The Si/Ge atoms, centering the triangular Gd prisms, form dimers. These are part of a Cairo pentagon tiling (dual to the snub square tiling) together with the Gd in the cube centers. During the first-order phase transformation from the LT ferromagnetic to the HT paramagnetic phase at 276 K, the symmetry changes from $Pnma$ to $P112_1/a$. This structural transition can be controlled not only by temperature but also by the magnetic field, which can induce the transition even slightly above the transition temperature. The phase transition is also accompanied by a shear mechanism, which increases the atomic distance in half of the dimers by more than 30%, changing the electronic structure as well.

1.4.10.5 Magnetooptic Materials

The transmission of light through a magnetooptic material or the reflection on its surface can change its polarization (Faraday effect). It is not caused by the direct action of the magnetic field on the electromagnetic wave but by the changes in the crystal imposed by it.

Example: cF12-MnPtSb (half-Heusler type) (Antonov et al., 1997, **Figure 27** in Section 1.4.4.1).

The giant magnetooptic Kerr effect results from a unique combination of properties provided by the crystal structure and the chemical composition: the spin-orbit coupling strength, the magnitude of the $3d$-magnetic moment, the degree of hybridization in the bonding, the half-metallic character, or, equivalently, the Fermi-level filling of the band structure and the intraband plasma frequency.

1.4.10.6 Superconductors

Superconductivity of intermetallic phases strongly depends on the crystal structure as well as on the band structure and valence electron concentration. While the FK phase $cP8$-Cr_3Si is not superconducting, isoconfigurational Nb_3Sn and $TiNb_3$ are the most used materials for applications. The highest transition temperature in this structure class holds Nb_3Ge with $T_c = 23.2$ K. For the generation of very high magnetic fields, coils of Nb_3Sn ($T_c = 18.0$ K) are used.

The structure of $cP8$-Cr_3Si ($Pm\bar{3}n$, Si in $2a$ 0, 0, 0; Cr in $6c$ 1/4, 0, 1/2) can be described as packing of slightly distorted Si-centered, edge-sharing Cr-icosahedra (CN12), the edges of which form linear chains parallel to the axes of the unit cell (**Figure 45(a)**). These atomic chains are the crucial structural units for superconductivity in the respective superconducting compounds. Another description is based on CN14 FK-polyhedra (**Figure 45(b)**). In terms of the layer description, this structure can be seen as

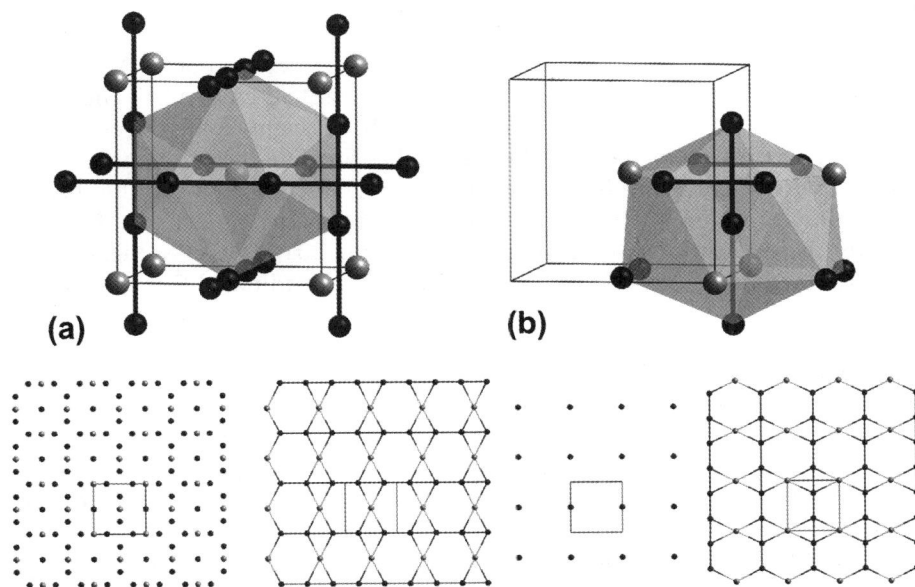

Figure 45 The structure of $cP8$-Cr_3Si in different representations: (a) emphasis on the Si-centered Cr-icosahedron (CN12) and the Cr-chains along {100}; (b) Cr-centered CN14 FK-polyhedron; (c) along [001] projected structure; sections at (d) $z = 0$, (e) $z = 1/4$ and (f) $z = 1/2$ (Si...gray, Cr...black).

stacking of four (001) layers per translation period: two-uniform hexagon-triangle tilings, $3^2.6^2$, at $z = 0$ and 1/2 (rotated by $\pi/2$), as well as regular square tilings, 4^4, at $z = 1/4$ and 3/4.

1.4.10.7 Highly Correlated Electron Systems

Intermetallics with highly correlated electron systems are of interest due to their magnetic and transport properties and potential applications in spintronics. Heavy fermion materials, with the strongest correlations, show at low-temperature enhanced interactions between f electrons and conduction electrons resulting in effective mass increase of the electrons by two order of magnitudes. Among other things, this leads to a large electronic specific heat such as indicated by the Sommerfeld coefficient $\gamma \approx 8000$ mJ mol^{-1} K^{-2}. A comprehensive review has been given by Thomas et al. (2006) listing a large number of heavy electron systems, grouped per structure type, with their physical properties. Most compounds have Ce, Yb or U as one of the constituents or, less frequently, also other rare-earth elements. Common structure types are tI10-ThCr$_2$Si$_2$ (CeCu$_2$Ge$_2$, YbCu$_2$Si$_2$, URu$_2$Si$_2$, ...), tP10-CaBe$_2$Ge$_2$ (CeCu$_2$Sb$_2$, CeIr$_2$Sn$_2$, UIr$_2$Si$_2$, ...), hP8-Ni$_3$Sn (CeAl$_3$, UPt$_3$, ...), hP9-ZrNiAl (YbNiAl, YbPtIn, ...) (**Figure 46**), oP12-TiNiSi (YbPtAl, YbNiSn, ...), cP4-AuCu$_3$ (CeIn$_3$, USn$_3$, ...), cF24-AuBe$_5$ (YbAgCu$_4$, UPdCu$_4$, ...), tP10-U$_3$Si$_2$ (U$_2$Ni$_2$In, U$_2$Pd$_2$Sn, ...) and cF112-NaZn$_{13}$ (UBe$_{13}$, NpBe$_{13}$, ...), cF184-CeCr$_2$Al$_{20}$ (YbFe$_2$Zn$_{20}$, ...), to name just a few.

The structure of tP10-U$_3$Si$_2$ ($P4/mbm$, U in $2a$ 0, 0, 0 and $4h$ 0.181, 0.681, 1/2) is shown in comparison with the related structure tI12-CuAl$_2$ ($I4/mcm$, Cu in $4a$ 0, 0, 1/4; Al in $8h$ 0.1541, 0.6541, 0; **Figure 34** in Section 1.4.7). Both structures are based on stackings of $3^2.4.3.4$ nets. oP12-TiNiSi ($Pnma$, Ti in $4c$ 0.021, 0.180, 1/4; Ni in $4c$ 0.142, 0.561, 1/4; Si $4c$ 0.765, 0.623, 1/4) and its aristotype oP12-CeCu$_2$ ($Imma$, Ce in $4e$ 0, 1/4, 0.538; Cu in $8h$ 0, 0.051, 0.165) can be described based on 3D four-connected polyanionic Ni–Si nets with Ti cations in the large channels (**Figure 47**(a–d)) (Landrum et al., 1998). The tI10-ThCr$_2$Si$_2$ type is an ordered tI10-BaAl$_4$ type ($I4/mmm$, Ba in $2a$ 0, 0, 0; Al in $4d$ 0, 1/2, 1/4 and $4e$ 0, 0, 0.38; **Figure 47**(e–h)).

The structure type cF112-NaZn$_{13}$ ($Fm\bar{3}c$, Na in $8a$ 1/4, 1/4, 1/4; Zn in $8b$ 0, 0, 0 and $96i$ 0, 0.181, 0.119; **Figure 48**) consists of face-sharing sodium-centered NaZn$_{24}$ snub cubes and Zn atoms in the holes left between. A subset of Zn atoms forms the corners of the eight eighth cubes; the other subset is arranged in inclined squares on the faces of these subcubes forming the snub cubes.

This structure type is of particular interest because it is frequently found in hard-sphere colloidal systems with size rations around 0.49–0.63. Calculations of the packing density of hard spheres with a size ratio of 0.58 yielded a packing density of 0.748 (Hudson, 2010). An even higher density of 0.771 could be obtained in a ternary system A$_{12}$BC, if icosahedrally coordinated C gets a smaller diameter.

1.5 Crystal Structures of Quasicrystals

QCs belong to the class of aperiodic crystals like incommensurately modulated structures or composite structures. They can be defined most generally in reciprocal (Fourier) space. Accordingly, the Fourier module of a quasiperiodic structure is of rank $n > d$, with d, the dimension of the quasiperiodic structure in physical space, and n, the number of basis vectors spanning the Fourier module.

A vector module of rank n is an infinite set of vectors resulting from all possible linear combinations of its n basis vectors. It is also called $\mathbb{Z}$-module to emphasize that the coefficients of all linear vector combinations are integers. An nD lattice is an example of a $\mathbb{Z}$-module of rank n. If the vector module is defined in reciprocal space, it is called Fourier module since direct (structure) and reciprocal (diffraction) spaces are related by the Fourier transform. The unit cell of a crystal structure is spanned by the

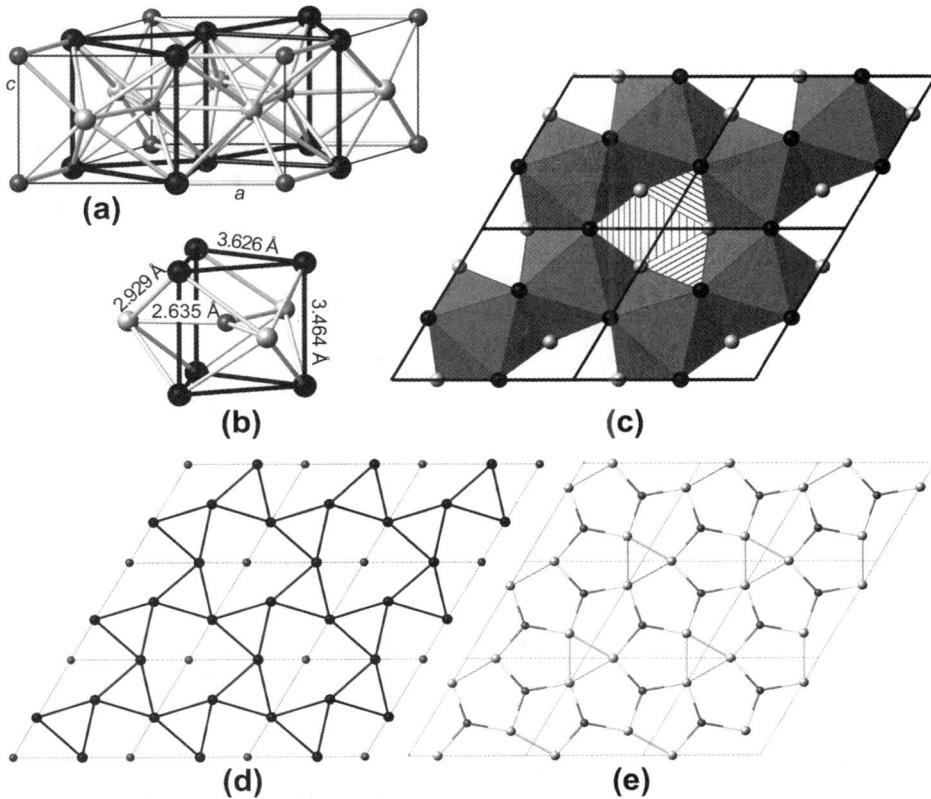

Figure 46   Example for different descriptions of $hP9$-ZrNiAl. Ball-and-stick model (a) of one unit cell and (b) the AET around Ni at Wyckoff position $2d$ 1/3,2/3,0, an Al-tricapped trigonal Zr prism. (Al ... light gray, Ni ... dark gray, Zr ... black). The polyhedron model (c) shows the face-sharing arrangment of these AET (dark gray) leaving open channels filled with columns of base-linked Zr-tricapped trigonal Al prisms (striped) centered around Ni at Wyckoff position $1a$ 0,0,0. The AET of Al corresponds to a distorted rhombic dodecahedron of composition $Zr_6Ni_4Al_4$ while Zr centers a pentagonal prism of composition Ni_4Al_6. The structure can be seen as ordered variant of the $hP9$-Fe_2P type (P $\rightarrow$ Ni, Fe at $3f$ x,0,0 $\rightarrow$ Zr, Fe at $3g$ x,0,1/2 $\rightarrow$ Al). The layers in (d) $z = 0$ corresponds to a shield/triangle tiling and that in (e) $z = 1/2$ to a pentagon/triangle tiling. The shields and triangles in the former are also constituents of a dodecagonal tiling, which has also square tiles in addition. In the latter, there are infinite wavy bands of pentagons, structure motifs resembling structural units of decagonal quasicrystals.

basis vectors of the direct lattice, the Bragg reflections of its diffraction pattern are located at the nodes of the reciprocal lattice.

A Fourier module of rank $n > d$ can be obtained as proper projection of an nD reciprocal lattice onto dD physical space. The nD embedding space $V = V^{\parallel} \oplus V^{\perp}$ of the nD reciprocal lattice consists of the dD physical (parallel, external) space $V^{\parallel}$ and the $(n - d)$D perpendicular (complementary, internal) space $V^{\perp}$ orthogonal to it.

Due to the properties of the Fourier transform, a projection in reciprocal space is related to a section in direct space and vice versa. Consequently, if the dD diffraction pattern of a QC can be described as projection of a Bragg-reflection-weighted nD reciprocal lattice, then the dD quasiperiodic structure can be obtained from a proper irrational physical-space section of an nD hypercrystal structure. Its nD unit

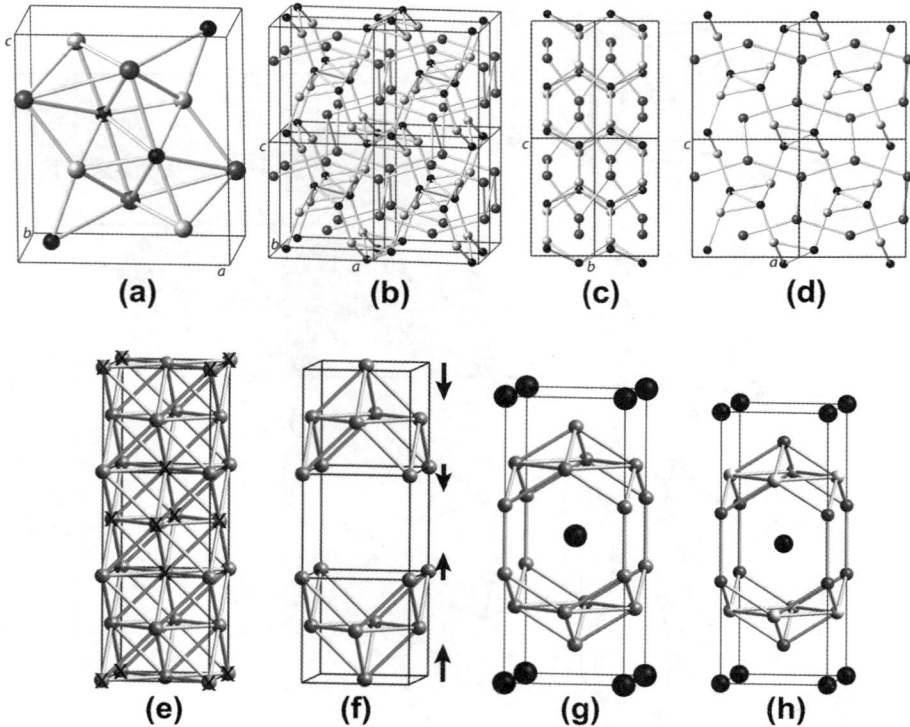

Figure 47 The structure of *oP*12-TiNiSi: (a) unit cell, (b) 2 × 2 × 2 unit cells and projections along [100] (c) and [010] (d). In (e–f), the relationship are shown between the structures of (e) *fcc* Al (*cF*4-Cu type), (g) *tP*10-BaAl$_4$ and (h) *tP*10-ThCr$_2$Si$_2$. Replacing the corner atoms of the tripled unit cell of *fcc* Al (e) by Ba atoms (black) and the central octahedron of crossed-out atoms by one further Ba atom and letting the structure relax along the arrows shown in (f), then we obtain the *tP*10-BaAl$_4$ structure type (g). Substituting Ba by Th, and Al by Cr (dark gray) and Si (light gray) in an ordered way, then we get the (h) *tP*10-ThCr$_2$Si$_2$ structure type.

cell contains hyperatoms whose perp(endicular)-space components are called *occupation domains* (atomic surfaces), while the par(allel)-space components refer to the atoms forming the QC structure. The big advantage of the higher dimensional approach is that quasiperiodic structures can be described as *n*D periodic hypercrystal structures allowing a closed, unit-cell-based description similar as for conventional 3D periodic crystal structures.

Alternatively, quasiperiodic structures can be described fully in par-space in terms of quasiperiodic tilings (quasilattices) decorated with atoms (see **Figure 7** in Section 1.2.2.4). Thereby, the unit tiles can be considered a kind of unit cells. Another, more crystal-chemically relevant, par-space description is based on one or more types of partially overlapping covering cluster(s) (see **Figure 6** in Section 1.2.2.4), whereby the centers of the clusters form again the vertices of a tiling.

The symmetry classification of QC is based on the corresponding Laue groups (point symmetry of the Bragg-intensity-weighted reciprocal lattice). Accordingly, the known QCs can be classified as octagonal ($8/m2/m2/m$ or $8/m$), decagonal ($10/m2/m2/m$ or $10/m$), dodecagonal ($12/m2/m2/m$ or $12/m$) and icosahedral ($\bar{5}\,\bar{3}2/m$). The full symmetry of QC can be described by *n*D space groups. Fortunately, only a very small subset out of them is needed due to the restriction that the projection of

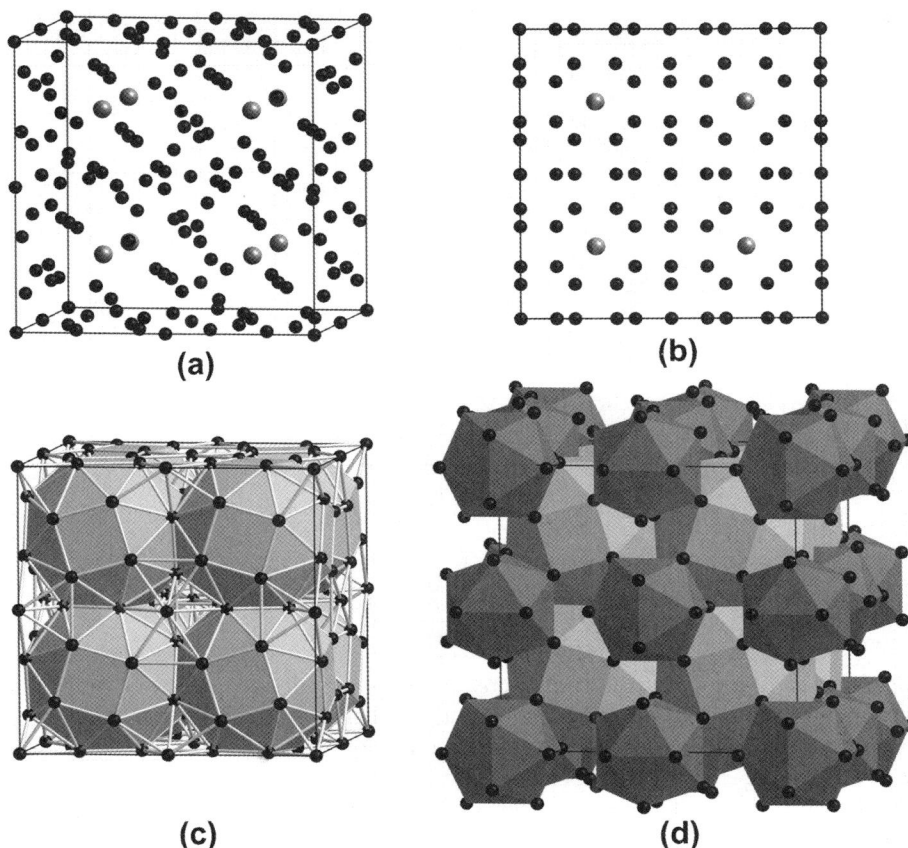

Figure 48 Different representations of the $cF112$-NaZn$_{13}$ structure type. One unit cell is shown in perspective view (a) and in projection (b) (Na ... gray, Zn ... black). The clusters around the Na atoms, both enantiomorphs of the snub cube $3^4 4$, and their packing is shown in (c). In (d), additionally, the icosahedral clusters are depicted around the Zn atoms in Wyckoff position 8 b 0 0 0. All clusters are slightly distorted.

the nD point group onto par-space has to be isomorphic to the point group of the actual quasiperiodic structure. This reduces drastically, for instance, the number of 6D point and space groups for icosahedral QCs from 7104 and 28 927 922 to 2 and 11, respectively (Souvignier, 2003). An alternative approach sticks to 3D par-space and describes all quasiperiodic space groups in reciprocal space (Rabson et al., 1991; Rokhsar et al., 1988).

1.5.1 The Fibonacci Sequence (FS) and the Higher Dimensional Approach

The simplest example of a quasiperiodic structure is based on the 1D Fibonacci sequence (FS). The FS can be generated by the substitution rule σ applied to the alphabet $\{S,L\}$: $S \Rightarrow L$, $L \Rightarrow LS$. This can be written employing the substitution matrix $\mathbf{S}$

Table 17 Generation of words $w_n = \sigma^n(L)$ of the Fibonacci sequence by repeated application of the substitution (inflation) rule $\sigma^n(L) = LS, \sigma^n(S) = L$; ν_n^L and ν_n^S denote the frequencies of the letters L and S in the words w_n; F_n are the Fibonacci numbers

n	$w_{n+2} = w_{n+1} w_n$	ν_n^L	ν_n^S
0	L	1	0
1	LS	1	1
2	LSL	2	1
3	LSLLS	3	2
4	LSLLSLSL	5	3
5	LSLLSLSLLSLLS	8	5
6	$\underbrace{\text{LSLLSLSLLSLLS}}_{w_5}\ \underbrace{\text{LSLLSLSL}}_{w_4}$	13	8
$\vdots$	$\vdots$	$\vdots$	$\vdots$
n		F_{n+1}	F_n

$$\sigma : \begin{pmatrix} L \\ S \end{pmatrix} \mapsto \underbrace{\begin{pmatrix} 1 & 1 \\ 1 & 0 \end{pmatrix}}_{S} \begin{pmatrix} L \\ S \end{pmatrix} = \begin{pmatrix} LS \\ L \end{pmatrix}.$$

Multiple application of the substitution rule creates longer and longer words w_n of the FS (**Table 17**). Since the number of letters is increased each time the substitution rule is applied, this operation is called inflation operation. The FS can also be generated by the concatenation rule $w_{n+2} = w_{n+1} w_n$. The frequencies $\nu_n^L = F_{n+1}$, $\nu_n^S = F_n$ of letters L, S in the words w_n are given by the Fibonacci numbers F_n, with $F_{n+2} = F_{n+1} + F_n$ and $F_0 = 0$, $F_1 = 1$. The Fibonacci numbers form a series with $\lim_{n \to \infty} F_{n+1}/F_n = \tau$. The definition of τ is given below.

The inflation operation does not change the already existing sequence (**Table 17**). With other words, the FS is invariant under the action of the substitution matrix $\mathbf{S}$, it shows scaling symmetry. The eigenvalues of the substitution matrix can be obtained from the eigenvalue equation

$$det|\mathbf{S} - \lambda\mathbf{I}| = 0$$

with the identity matrix $\mathbf{I}$. The evaluation of the determinant yields the characteristic polynomial

$$\lambda^2 - \lambda - 1 = 0$$

with the eigenvalues

$$\lambda_1 = \frac{1 + \sqrt{5}}{2} = 2\cos\pi/5 = \tau = 1.618...,$$

$$\lambda_2 = \frac{1 - \sqrt{5}}{2} = -2\cos 2\pi/5 = 1 - \tau = -\frac{1}{\tau} = -0.618...$$

and eigenvectors

$$\mathbf{v}_1 = \begin{pmatrix} \tau \\ 1 \end{pmatrix} \text{ and } \mathbf{v}_2 = \begin{pmatrix} -1/\tau \\ 1 \end{pmatrix}.$$

Then, we can explicitly write the eigenvalue equation $\mathbf{Sv}_i = \lambda \mathbf{v}_i$ for the first eigenvalue, for instance, in the form

$$\begin{pmatrix} 1 & 1 \\ 1 & 0 \end{pmatrix} \begin{pmatrix} \tau \\ 1 \end{pmatrix} = \begin{pmatrix} \tau + 1 \\ \tau \end{pmatrix} = \tau \begin{pmatrix} \tau \\ 1 \end{pmatrix}.$$

If the letters stand for a short distance S and a long distance $L = \tau \cdot S$, then the resulting quasiperiodic structure $s(\mathbf{r})$ shows scaling invariance under factors τ^n, $s(\mathbf{r}) = s(\tau^n \mathbf{r})$ (**Figure 49**). This means that the scaling operation maps each vector $\mathbf{r}$ to an already existing tiling vector $\tau \mathbf{r}$.

The FS can also be generated based on the higher dimensional approach. We need just one extra dimension along the perp-space $V^\perp$, because the 1D reciprocal space image (diffraction pattern in physical or par-space $V^\parallel$) of the FS corresponds to a $\mathbb{Z}$-module of rank 2. It can be seen as par-space projection of a 2D-weighted reciprocal lattice (**Figure 50**). Its Fourier transform is a 2D hypercrystal structure, from which the 1D FS results as cut. The occupation domains are straight-line segments of a length corresponding to the perp-space dimension of a single 2D unit cell.

If the 2D hypercrystal structure is sheared along the perp-space, then rational approximants are obtained when, in addition to the origin, one more reciprocal lattice point comes to lie on the par-space axis (**Figure 50(b)**). In case of an n/m approximant, this reciprocal lattice point has the lattice vector $m\mathbf{d}_1^{app} + n\mathbf{d}_2^{app}$.

What is the physical meaning of the nD approach? How to understand the concept of an occupation domain? The vertices of the FS are a subset of the set $\{mS + nL \,|\, m,n \in \mathbb{Z}\}$, i.e. a $\mathbb{Z}$-module of rank 2, where the selection criterion is defined by the occupation domain. The occupation domain projected onto par-space in a proper way defines the regions where vertices are allowed. This means that the FS has a *periodic average structure* (PAS) and the width of the projected occupation domains gives the maximum distance of a vertex from the closest lattice node of the PAS (**Figure 50(c)**). With other words, the PAS results if the FS is taken modulo the unit cell of the PAS. In case of the FS, the mapping of the vertices to the PAS is bijective.

The size and shape of an occupation domain can be derived by the strip-projection method (**Figure 51**). If we start from the condition that the vertices of the FS are a subset of the set $\{mS + nL \,|\, m,n \in \mathbb{Z}\}$, i.e. a $\mathbb{Z}$-module of rank 2, with a given ratio L/S of the lengths and frequencies of the line segments, then the most homogenous sequence can be obtained by proper projection of the content of

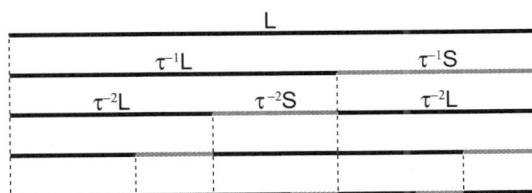

Figure 49 Graphical illustration of the substitution (inflation) rule applied to line segments L and S. Rescaling by a factor τ^{-1} at each step keeps the total length of the sequence constant. In this example, the number of tiles is inflated and their lengths are deflated.

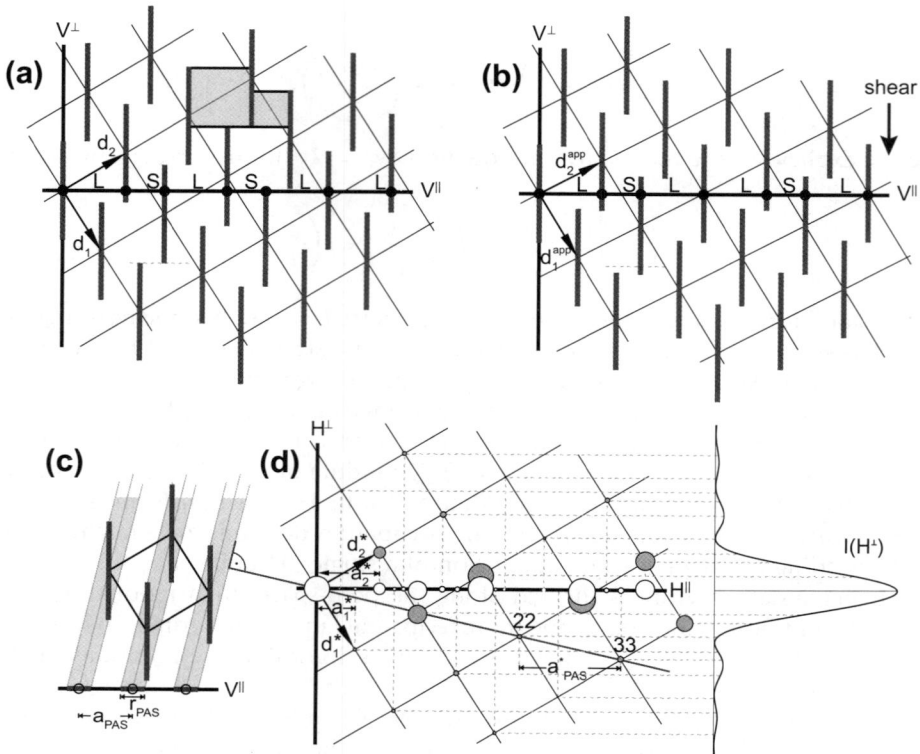

Figure 50   The FS in the nD description ($n = 2$). (a) The FS results as irrational section (slope τ^{-1}) of the 2D hypercrystal with basis $\mathbf{d}_1$, $\mathbf{d}_2$. The occupation domains are line segments parallel to $V^{\perp}$. Their length corresponds to the projection of the 2D unit cell onto $V^{\perp}$. (b) 2/1-approximant of the FS with period …LSL… obtained by shearing the hyperlattice parallel to $V^{\perp}$. (c) The PAS, with period a^{PAS}, is obtained by oblique projection of the hypercrystal structure along the gray-shaded strips onto $V^{\parallel}$. In reciprocal space (d), this oblique projection corresponds to a cut perpendicular to the projection direction. (d) The 1D reciprocal space with basis vectors $\mathbf{a}_2^* = \tau \mathbf{a}_1^*$ results from the projection of a 2D reciprocal lattice with basis vectors $\mathbf{d}_1^*$, $\mathbf{d}_2^*$. Bragg reflections located on the 2D reciprocal lattice are marked by filled circles, on $V^{\parallel}$ projected ones by open circles. The variation of the intensity of Bragg reflections as function of $\mathbf{H}^{\perp}$ is indicated at the righ-hand side of (d).

a strip cut out of a 2D lattice. The lattice and the strip, which is parallel to $V^{\parallel}$, have to include a proper angle in order to give the right ratio L/S. In case of the FS, the slope equals τ. The width of the strip, the window W (related to the occupation domain), is obtained by projecting the unit cell of the 2D lattice onto $V^{\perp}$. With other words, the higher dimensional description is just used in order to get the most homogenous packing of two unit tiles.

Another variant of generating the FS and its approximants is the *Klotz* (German word for block) construction method (Kramer and Schlottmann, 1989). Thereby, two squares with edge lengths L and S, respectively, are arranged to a fundamental domain (**Figure 52**(c)) already marked in (**Figure 50**(a)). The copies of this domain form a 2D uniform dihedral tiling under translation. The crystallographic unit cell (a parallelohedron in the general case) of this tiling can be obtained by connecting the centers of the large squares, for instance. The FS is generated by cutting this tiling along the par-space. The

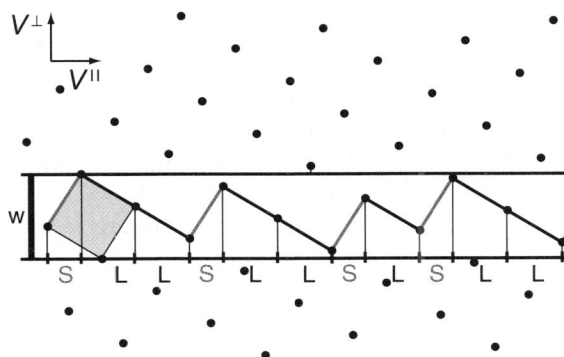

Figure 51 The strip-projection approach for the generation of the Fibonacci sequence. The unit cell of the 2D square lattice is shaded. The width of the strip, defining the window W, corresponds to the perp-space dimension of one unit cell. In our example, it includes the uppermost vertex but not the lowermost vertex of the unit cell.

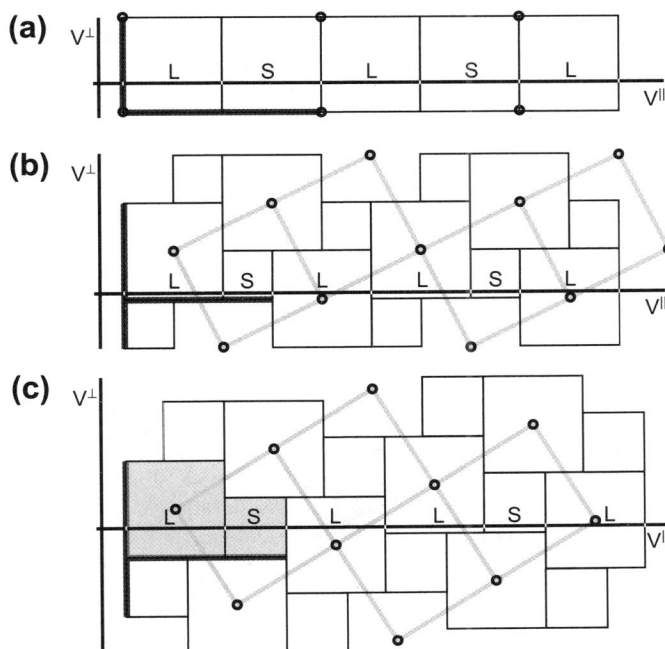

Figure 52 *Klotz* construction based on a fundamental domain (shaded) consisting of two squares. The ratio of their edge lengths is one in (a), two in (b) and τ in (c). Along the cutting line $V^{\parallel}$, this corresponds to 1D periodic approximant sequences (LS), (LSL), and the quasiperiodic FS, respectively. The thick lines mark the projections of the unit cell of the 2D lattice upon $V^{\parallel}$ and $V^{\perp}$, respectively. This gives the Delone (covering) cluster (LS) and the occupation domain, respectively. Vertices of 2D rectangular (b) or square lattices (a,c) are marked by open circles (from Steurer and Deloudi, 2008).

relative frequencies ν_L and ν_S of L and S tiles are proportional to the edge lengths of the corresponding squares.

The *Klotz* construction illustrates the property of the FS to be the most homogenous arrangement of unit tiles L and S with given ratio $\nu_L/\nu_S = L/S$, which is τ in case of the FS. This property follows from the fact that the 2D dihedral tiling is periodic and does not show any gaps. Since the area of the crystallographic square unit cell is equal to that of the dihedral domain, this is the densest possible arrangement of a homogenous structure with stoichiometry $L_\tau S$.

As shown in **Figure 52(c)**, the *covering cluster* of the FS corresponds to the word (LS). It covers the FS, with sometimes overlapping S tiles, in the following way:

$$(L(S)L)\ (LS)\ (L(S)L)\ (L(S)L)(LS)$$

The frequency of the word (L(S)L) is τ times larger than that of the word (LS). If, contrary to the word SS, the words LL, LS and LSL are energetically favorable configurations and $\nu_L/\nu_S = L/S = \tau$, then the FS is the structure with the lowest energy.

1.5.2   The Structure of Decagonal Phases

There are many stable and metastable decagonal QCs known, most of them Al-based (**Figure 53**). All are ternary phases and quite a few of them have a broad compositional stability range indicating entropic contributions to their stability by chemical disorder. Geometrically, their structures can be seen either as periodic stacking of quasiperiodic atomic layers or as covering of 2D quasiperiodic

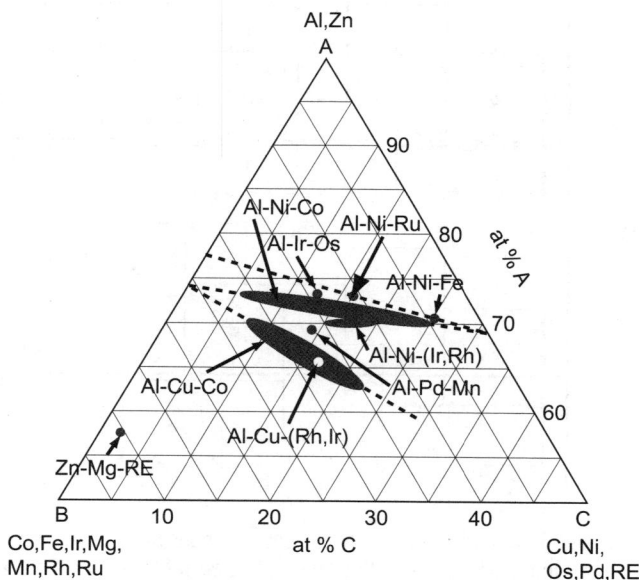

Figure 53 Stability regions of decagonal quasicrystals. RE denotes the rare-earth metals Y, Dy, Ho, Er, Tm, Lu. Note that only the A-rich part ($50 \leq A \leq 100$ at %) of the concentration diagram is shown (from Steurer and Deloudi, 2008).

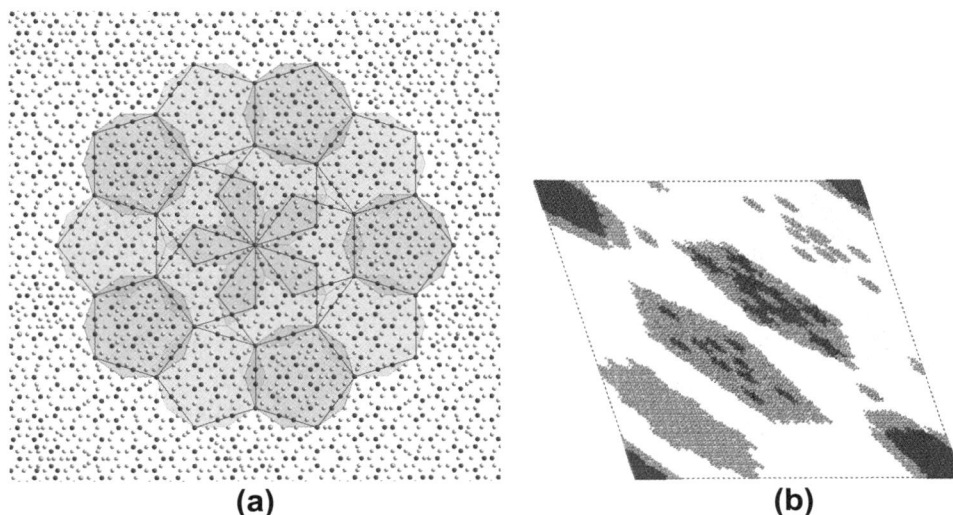

(a)　　　　　　　　　　　　　　　**(b)**

Figure 54　(a) 100×100 Å section of the structure of d-Co–Ni–Al based on a model of Deloudi and Steurer (2011) and (b) one unit cell of its monoclinic periodic average structure. Both figures show projections along the tenfold axis. The chemical decoration of the atomic surfaces is still visible after the oblique projection (Al gray, Co/Ni dark gray) (from Steurer and Deloudi, 2008).

tilings by partially overlapping columnar clusters (*cf.* Section 1.2.2.4). The columnar clusters, usually based on Gummelt decagons, can exhibit 2-, 4-, 6- or 8-layer periodicity. One has to keep in mind, however, that decagonal QCs are by no means layer structures and that the columnar clusters are mere structural building units. The tilings underlying most decagonal QC structures are closely related to the 2D PT in one of its variants: the romb PT, the pentagonal PT or the hexagon-boat-star (HBS) tiling.

The approximately 20 stable decagonal QCs known so far can be assigned to the following classes (Steurer, 2004):

1. Two-layer periodicity (sometimes with twofold superstructure along the periodic direction leading to a four-layer period).
   d-Co–Ni–Al type: Me-Cu–Al (Me = Co, Rh, Ir), Me–Ni–Al (Me = Co, Fe, Rh, Ru).
   d-Dy–Mg–Zn type: RE-Mg–Zn (RE = Y, Dy, Ho, Er, Tm, Lu).
2. Six-layer periodicity.
   d-Mn–Pd–Al type: Mn–Pd–Al, Mn–Fe–Al–Ge, Co–Cu–Ga, Cu–Fe–Ga–Si, V–Ni–Ga–Si.
3. Eight-layer periodicity.
   d-Os–Pd–Al type: Ru–Ni–Al, Me–Pd–Al (Me = Fe, Ru, Os), Os–Ir–Al.

An example of a decagonal QC structure is shown in **Figure 54**(a). The supercluster arrangement of the fundamental ≈ 20 Å diameter columnar clusters (shaded decagons) is indicated. If the infinite QC structure is taken modulo one unit cell of the PAS, the picture shown in **Figure 54**(b) is obtained. It results as well from the oblique projection of the 5D unit cell, it can be also seen as its shadow from hyperspace. In 5D space, the PAS distorted occupation domains have pentagonal symmetry ($5m$), the chemical decoration with Al and TM atoms is clearly visible. A detailed discussion of the structures of several Al-based decagonal QCs can be found in Deloudi and Steurer (2011).

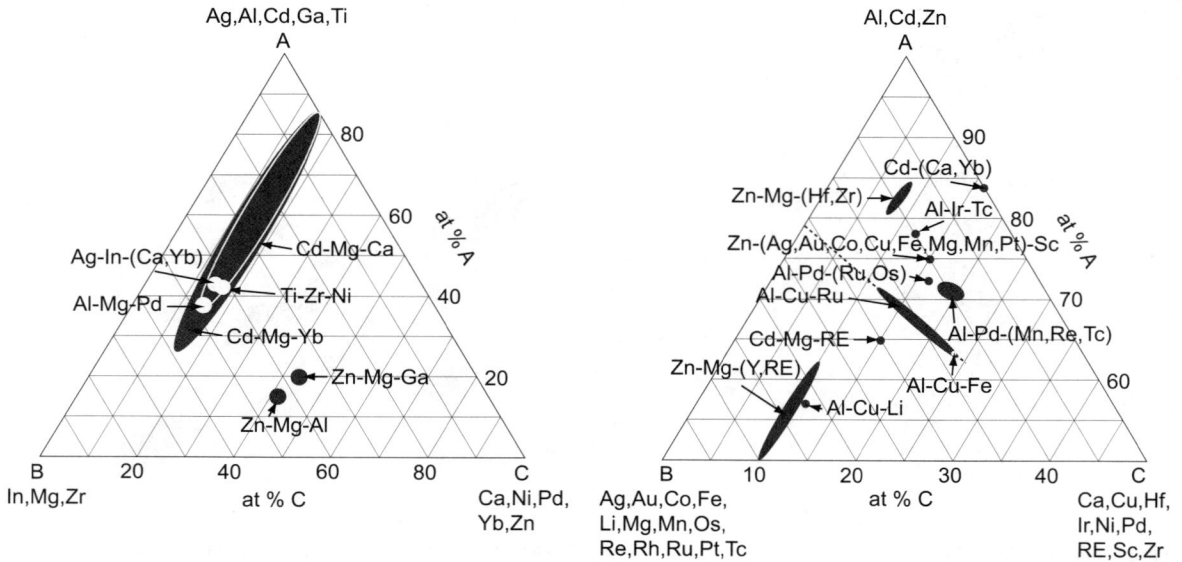

Figure 55 Stability regions of icosahedral quasicrystals. RE denotes the rare-earth metals Nd, Eu, Gd, Tb, Dy, Ho, Er, Tm, Yb, Lu in the case of i-RE–Mg–Cd; La, Ce, Pr, Nd, Gd, Tb, Dy, Ho, Er, Yb in the case of i-RE–Mg–Zn. Note that only the A-rich part ($50 \leq A \leq 100$ at.%) is shown in the concentration diagram at left (from Steurer and Deloudi, 2008).

1.5.3   The Structure of Icosahedral Phases

The majority of the about 60 stable QCs discovered so far shows icosahedral diffraction symmetry (**Figure 55**). Their structures can be classified according to the fundamental clusters they are built from: (A) Mackay cluster, (B) Bergmann cluster and (C) Tsai cluster. Another way to categorize them is according to their 6D lattice symmetry, P- or F-type and/or the quasilattice parameter a_r, i.e. the edge length of the corresponding Penrose rhombohedron.

The structures of all icosahedral QC can be described in a reasonably good approximation by cluster-decorated 3D PT with edge lengths $\tau^3 a_r$ of the unit rhombohedra. The quasilattice constant a_r and the lattice parameter a of the 6D hypercubic lattice are related by $a_r = a\sqrt{2}/2$ and $a_r = a\sqrt{2}/4$ in case of primitive and face-centered 6D unit cells, respectively.

1.5.3.1   *Mackay-Cluster-based Icosahedral Phases (Type A)*

The around 10 Mackay-cluster-based icosahedral phases known so far, all contain Al as main constituent, Cu or Pd as second component and a transition element of group 7 (Mn, Tc, Re) or group 8 (Fe, Ru, Os) as third constituent:

    i-Mn–Pd–Al type: $a_r = 4.541$-4.606 Å, $Fm\bar{3}\,\bar{5}$

    i-Me–Pd–Al (Me = Mn, Re, Ru, Os), *i-Me–Cu–Al* (Me = Fe, Ru, Os).

Sometimes, the Mackay-cluster-based icosahedral phases are called spd QCs due to the substantial hybridization between the p states of aluminum and the d states of the transition metal atoms (de Laissardiere et al., 2005). Their stability range is related to an electron concentration of $1.6 \geq e/a \geq 1.9$. There are about 10 1/1- and 2/1-approximants known so far.

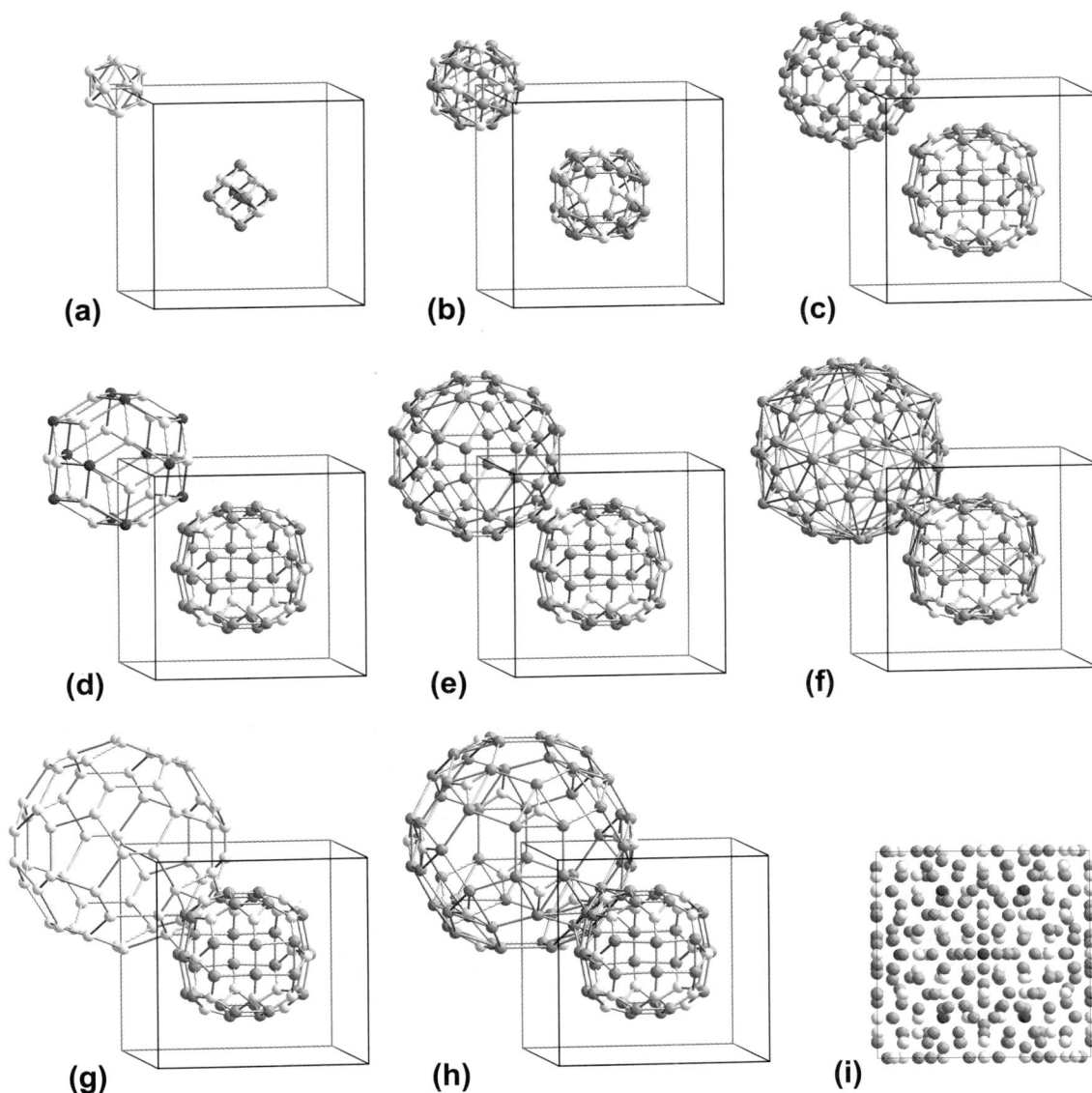

Figure 56 Shells of the clusters at the origin (a–h) and body center (a–c) of a Mackay-type cubic 2/1-approximant on the example of $Mn_6Pd_{23}Al_{70}Si$ ($a = 20.211$ Å, $Pm\bar{3}$; [Sugiyama et al., 1998]). (a) Pd_{12} icosahedron in the origin and Al_6Pd_8 rhombdodecahedron at the body-center; (b) $Al_{20}Pd_{12}$ origin-centered triacontahedron and $Al_{24}Pd_{12}$ cluster at the body center; (c) Al_{30} disordered icosidodecahedron with 60 partially occupied split positions and $Al_{36}Pd_{20}$ cluster at the body center; (d) $Mn_{12}Pd_{20}$ triacontahedron; (e) distorted Al_{60} rhombicosidodecahedron. (f) cluster shell of an Al_{60} truncated dodecahedron merged with an Al_{12} icosahedron linked via a joint tetrahedron to a Pd atom of the cluster in the body center; (g) truncated Pd_{60} icosahedron, capped by a Pd atom of the cluster in the body center; (h) cluster shell of an Al_{60} rhombicosidodecahedron merged with an Pd_{20} dodecahedron sharing a Pd atom with the cluster in the body center; (i) Projection of one unit cell along [100] (from Steurer and Deloudi, 2009).

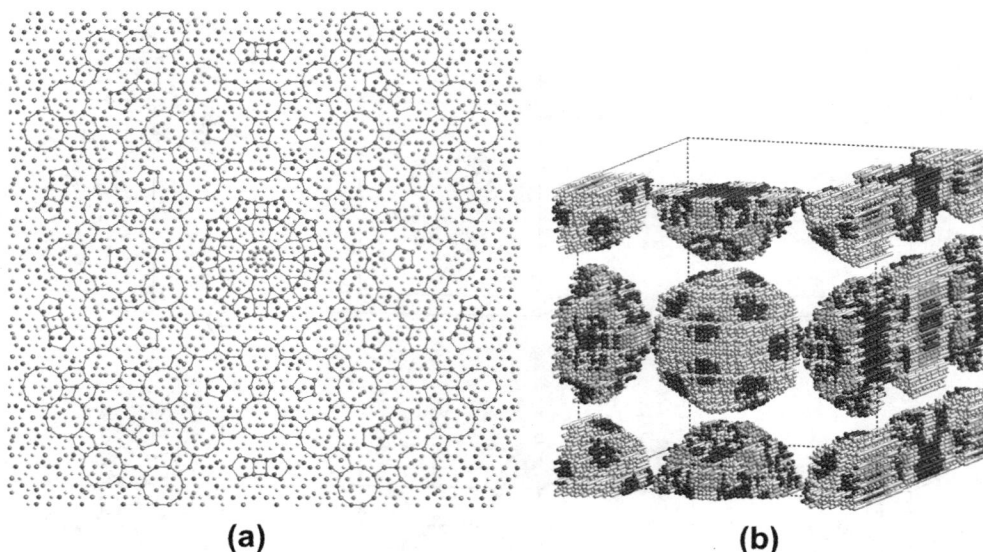

(a)   **(b)**

Figure 57   (a) Section of the structure of *i*-Fe–Cu–Al based on a model by Quiquandon and Gratias (2006) and (b) a part (for the sake of clarity) of the unit cell of its *fcc* periodic average structure, which is of the *cF*8-NaCl type. The chemical decoration of the atomic surfaces is still visible after the oblique projection (Al gray, Cu red, Fe green) (from Steurer and Deloudi, 2008).

The structure of the 2/1-approximant $Mn_6Pd_{23}Al_{70}Si$ (Sugiyama et al., 1998) is shown in **Figure 56**. This figure illustrates the shell structure of the pseudo-Mackay cluster, which is also the fundamental structural unit in the quasiperiodic paranet structures.

A section of the structure of i-Fe-Cu-Al, based on a model by Quiquandon and Gratias (2006), is shown in **Figure 57** together with its PAS. The centers of the projected occupation domains occupy the sites of the *cF*8-NaCl structure type.

1.5.3.2 *Bergmann-Cluster Based Icosahedral Phases (Type B)*

The more than 10 Bergmann-cluster-based icosahedral phases and their approximants known so far are with a few exceptions Mg-Zn based. They are frequently called Frank–Kasper QC because their approximants belong to the FK phases. Their stability range is related to an electron concentration of $2.1 \geq e/a \geq 2.4$.

  i-Ho–Mg–Zn type: $a_r = 5.01$-5.25 Å, $Pm\bar{3}\,\bar{5}$ or $Fm\bar{3}\,\bar{5}$

  i-RE-Mg–Zn–RE (RE = Y, Nd, Gd, Ho, Dy, La, Pr, Tb, Ce),

  i-Hf–Mg–Zn, *i*-Zr–Mg–Zn with symmetry $Fm\bar{3}\,\bar{5}$ and

  i-Li–Cu–Al, *i*-Mg–Zn–Ga, *i*-M-Mg–Al (M = Rh, Pd, Pt) with symmetry $Pm\bar{3}\,\bar{5}$ (i.e. these are disordered variants of the *F*-type QC).

  i-Zr–Ti–Ni.

The structure of the 2/1-approximant $Mg_{27}Zn_{47.3}Al_{10.7}$ (Lin and Corbett, 2006) is shown in **Figure 58**, illustrating the shell structure of the Bergmann-type cluster. The clusters share hexagon-faces of the fourth, fullerene-type, cluster shell. The fifth shells are triacontahedra sharing oblate rhombohedra with the overlapping neighboring clusters (*cf*. Section 1.2.2.5).

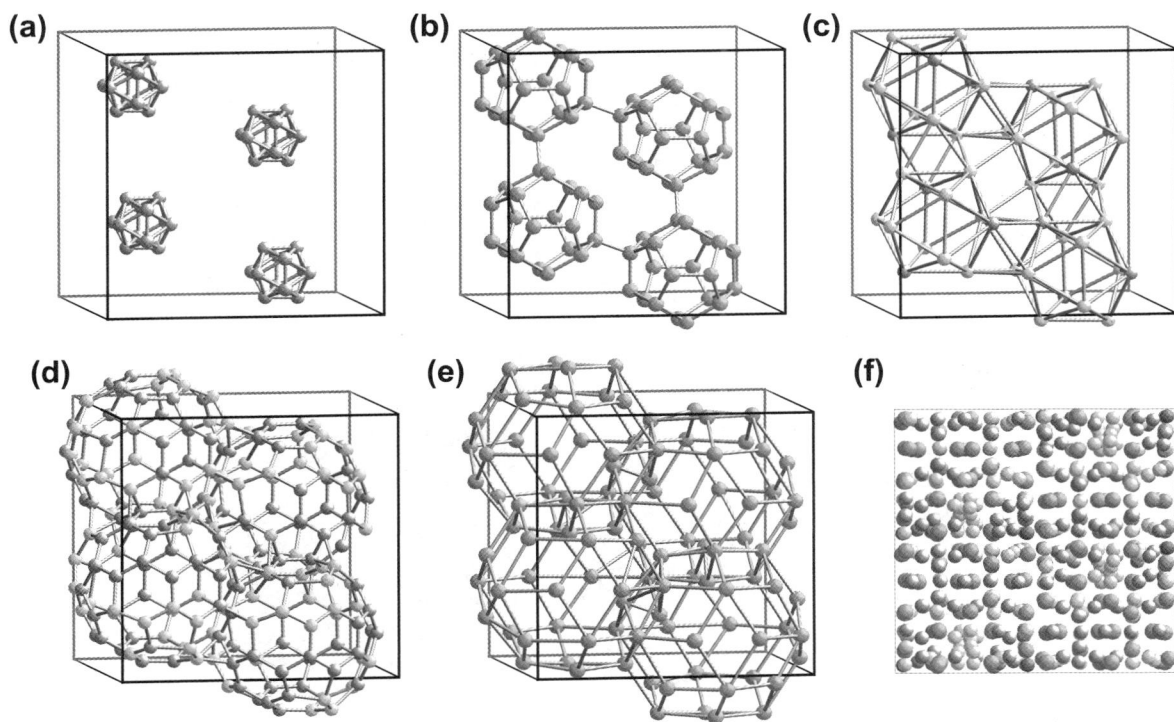

Figure 58 Shells of the characteristic cluster of a Bergmann-type cubic 2/1-approximant on the example of $Mg_{27}Zn_{47.3}Al_{10.7}$ ($a = 23.035$ Å, $Pa\bar{3}$, [Lin and Corbett, 2006]). (a) Zn_{12} icosahedron; (b) Mg_{20} dodecahedron linked to the neighboring cluster via two atoms with a distance of 3.176 Å; (c) $(Zn_{0.75}, Al_{0.25})_{12}$ icosahedron linked to the neighboring cluster via an octahedron; (d) distorted $(Zn_{0.77}, Al_{0.23})_{60}$ fullerene shell connected via a hexagon face to the neighboring cluster; (e) distorted Mg_{32} triacontahedron sharing oblate rhombohedra with the overlapping other clusters; (f) projection of one unit cell along [100] (from Steurer and Deloudi, 2009).

1.5.3.3 Tsai-Cluster-Based Icosahedral Phases (Type C)

The more than 30 ternary Tsai-cluster-based icosahedral phases and their more than 50 1/1- and 2/1-approximants known so far are mainly Cd or Sc based. There are also binary QCs known in the systems Cd–Ca and Cd–Yb, which are line compounds indicating that chemical disorder, and the entropy related to it, is not necessary for the stabilization of QCs.

i-Yb–Cd type: $a_r = 4.906 - 5.731$ Å, $Pm\bar{3}\,\bar{5}$ or $Fm\bar{3}\,\bar{5}$

i-Me–Cd (Me = Ca, Yb), *i*-Ca–Mg–Cd, RE–Mg–Cd (RE = Y, Nd, Eu, Gd, Tb, Dy, Ho, Er, Tm, Yb, Lu), *i*-Sc–Zn–Me (Me = Ag, Au, Co, Cu, Fe, Mg, Mn, Ni, Pd, Pt), *i*-Sc–Cu–Mg–Ga, *i*-Ti–Mg–Zn, *i*-Me–Ag-(Mg)–In (Me = Ca, Yb). Zn can be isoelectronically replaced by Cu–Ga in the case of *i*-Sc–Mg–Zn and Cd by Ag–In in the cases of *i*-Ca–Cd and *i*-Yb–Cd.

The structure of the 2/1-approximant $Ca_{13}Cd_{76}$ (Gomez and Lidin, 2001) is shown in **Figure 59**, illustrating the shell structure of the Tsai-type cluster. The outermost triacontahedral cluster shells share oblate rhombohedra with the overlapping neighboring clusters (*cf.* Section 1.2.2.5).

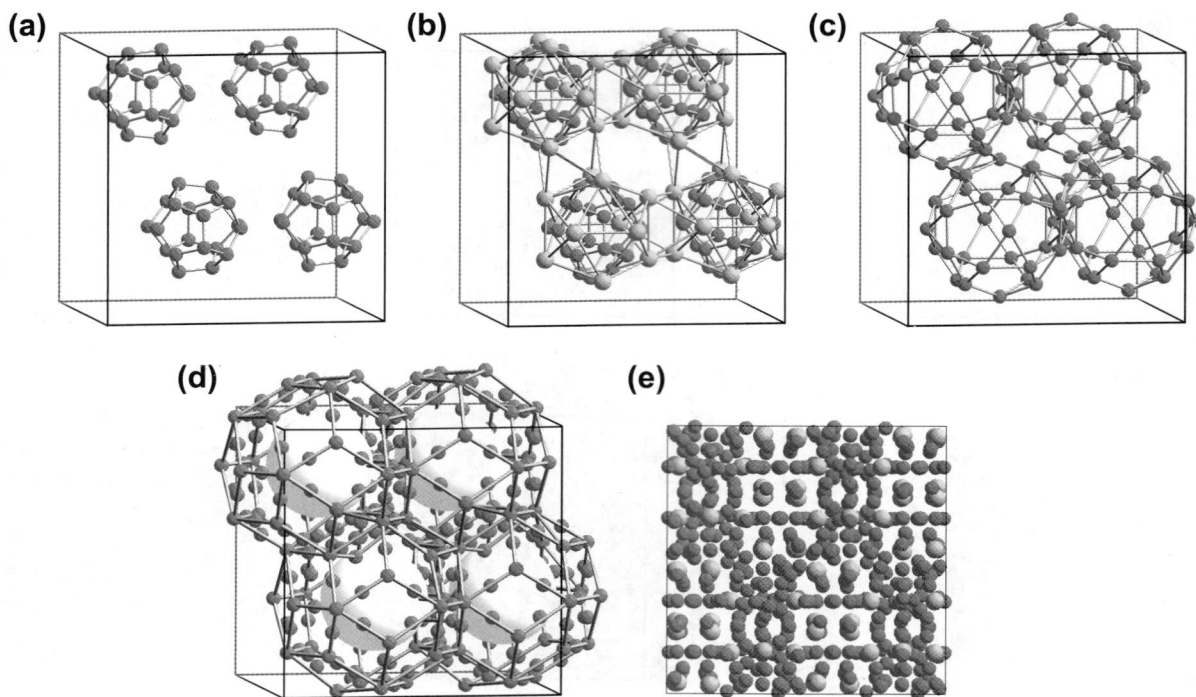

Figure 59 Shells of the Tsai-cluster in a cubic 2/1-approximant on the example of $Ca_{13}Cd_{76}$ ($a = 25.339$ Å, $Pa\bar{3}$; Gomez and Lidin, 2001). (a) Cd_{20} dodecahedron enclosing an orientationally disordered Cd_4 tetrahedron; (b) Ca_{12} icosahedron; (c) distorted Cd_{30} icosidodecahedron connected via a trigonal antiprism octahedron to the neighboring cluster; (d) Cd_{80} decorated triacontahedron sharing oblate rhombohedra with the overlapping other clusters; (e) Projection of one unit cell along [100] (from Steurer and Deloudi, 2009).

1.5.4   Remarks on Formation and Stability of Quasicrystals

While local atomic interactions and comparably simple packing principles are sufficient to grow periodic model structures, this does not seem to be the case for quasiperiodic ones. Quasiperiodicity of tilings can be forced by matching rules, which define the permitted and forbidden ways of assembling copies of a set of unit tiles. Consequently, a tiling obeying strict matching rules is a quasiperiodic tiling. However, matching rules are no growth rules. This means that even assembling the unit tiles according to the matching rules does not straightforwardly lead to a quasiperiodic tiling. Quite from the beginning, tile configurations may result that cannot be continued without leaving gaps or leading to overlaps or dead surfaces. Dead surfaces allow more than just one forced tile configuration, which can give rise to problems in a later assembling stage. Consequently, large defect-free quasiperiodic structures could only be grown by continuously disassembling wrong tile configurations and trying to reassemble the unit tiles in a better way. Basically, one would need global information in order to put tiles together properly from the very beginning.

An alternative approach for the description of quasiperiodic structures is based on coverings instead on tilings. Thereby, a patch of unit tiles of a particular quasiperiodic tiling acts as covering cluster, which

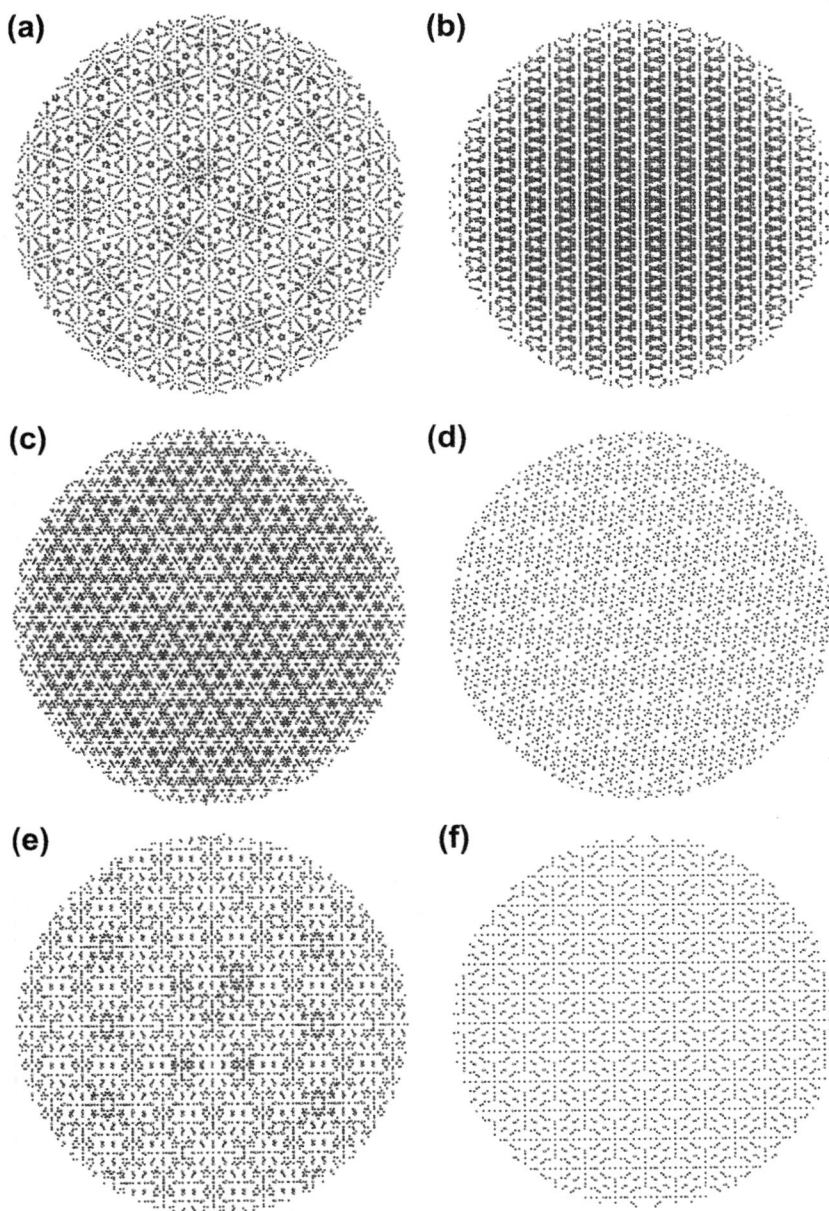

Figure 60 Projections of spherical sections (100 Å diameter) of the structure of *i*-Yb–Cd along (a) a fivefold axis, (c) a threefold axis and (e) a twofold axis. In (b), (d) and (f), the corresponding projections of 1/1-Cd–Yb are depicted, i.e., along the pseudo-fivefold as well as the three- and twofold directions, respectively. It is obvious that the atomic layers form a network compatible with fivefold symmetry only in (a) and not in (b). Almost all atoms are arranged in flat atomic layers that interpenetrate each other in a way, which is only possible in quasiperiodic structures (from Steurer and Deloudi, 2009).

can fully cover the tiling with well-defined overlaps. Consequently, there is a one-to-one relationship between coverings and tilings. Instead of by edge-by-edge (face-by-face) matching rules, 2D (3D) quasiperiodicity can be forced by overlap rules for shared areas (volumes). A well-known example used for the cluster-based description of DQC is the Gummelt decagon, which is based on the 2D PT (*cf.* Section 1.2.2.4). Furthermore, it has been shown that quasiperiodic ordering of low-energy Gummelt clusters can lead to a minimum in total energy (Steinhardt and Jeong, 1996). Coverings have also been discussed for icosahedral QC.

Growth rules based on the assemblage of fundamental building clusters are crystal-chemically more realistic than those based on fitting unit tiles together. The structural relevance of clusters as basic repeat units is corroborated by the fact that they are found in both QC and their periodic approximants. Furthermore, considering the high mobility and dynamics of atoms at QC formation temperatures, overlap rules are more robust than matching rules, because in a shared cluster volume, more atoms are directly involved than in a shared tile face. However, overlap rules alone are also no growth rules either, similar problems crop up as for matching rules.

Pseudogaps in the electronic density of states at the Fermi energy, caused by either Fermi surface nesting or spd hybridization of cluster atoms, have been found crucial for the stabilization of QC; however, electronic contributions of this kind cannot directly control the evolution of quasiperiodic order during QC growth. Anyway, QCs do exist, ergo growth rules must exist. However, these rules are not necessarily strict ones leading directly to perfectly quasiperiodic structures. They may be just constraining the way of quasiperiodic structure formation leading to only on-average quasiperiodic order. In this case, QC growth and structure optimization would be a two-step process. The first fast step would be dominated by steric and entropic factors, the second by energetic ones with major electronic contribution.

Consequently, one has to distinguish between the growth structure and the equilibrium structure, which can be obtained by annealing at suitably high temperature. While the former structure is just on-average quasiperiodic, the latter can be strictly quasiperiodic up to equilibrium phason fluctuations depending on temperature. The driving force for the formation of the equilibrium structure is mainly based on the electronic structure. Only a strictly quasiperiodic structure shows perfectly flat and well-ordered each other intersecting atomic layers causing perfectly sharp Bragg peaks (**Figure 60**). This framework may not be just a result of quasiperiodic order but assisting the growth of quasiperiodic structures. If the building clusters show icosahedral symmetry with the atoms located on a framework of interpenetrating atomic layers, then the quasiperiodic packing of these clusters must continue these atomic layers. This is a kind of extended or global growth rule coupled to icosahedral symmetry that only works if the structure grows in a quasiperiodic way.

References

Ahuja, A.K., Dubrovinsky, L., Dubrovinskaia, N., Osorio Guillen, J.M., Mattesini, M., Johansson, B., Le Bihan, T., 2004. Titanium metal at high pressure: Synchrotron experiments and ab initio calculations. Phys. Rev. B 69, 184102.

Akahama, Y., Fujihisa, H., Kawamura, H., 2005. New helical chain structure for scandium at 240 GPa. Phys. Rev. Lett. 94, 195503.

Akahama, Y., Kawamura, H., Le Bihan, T., 2001. New δ (Distorted-*bcc*) titanium to 220 GPa. Phys. Rev. Lett. 87, 275503.

Akahama, Y., Nishimura, M., Kinoshita, K., Kawamura, H., Ohisi, Y., 2006. Evidence of a *fcc-hcp* Transition in Aluminum at Multimegabar Pressure. Phys. Rev. Lett. 96, 045505.

Akella, J., Smith, G.S., Jephcoat, A.P., 1988. High-pressure phase-transformation studies in gadolinium to 106 Gpa. J. Phys. Chem. Sol. 49, 573–576.

Akella, J., Weir, S.T., Vohra, Y.K., Prokop, H., Catledge, S.A., Chesnut, G.N., 1999. High pressure phase transformations in neodymium studied in a diamond anvil cell using diamond-coated rhenium gaskets. J.Phys.: Cond. Matt. 11, 6515–6519.

Allan, D.R., Nelmes, R.J., McMahon, M.I., Belmonte, S.A., Bovornratanaraks, T., 1998. Structures and transitions in strontium. Rev. High Press. Sci. Technol. 7, 236–238.

Alvarez, S., 2006. Nesting of fullerenes and Frnk-Kasper polyhedra. Dalton Trans. 2045–2051.

Antonov, V.N., Oppeneer, P.M., Yaresko, A.N., Perlov, A.Ya., Kraft, T., 1997. Computationally based explanation of the peculiar magneto-optical properties of PtMnSb and related ternary compounds. Phys. Rev. B 56, 13012–13025.

Atoji, M., 1965. Antiferromagnetic structure of $AlCr_2$. J. Chem. Phys. 43, 222–225.

Bentien, A., Nishibori, E., Paschen, S., Iversen, B.B., 2005. Crystal structures, atomic vibration, and disorder of the type-I thermoelectric clathrates $Ba_8Ga_{16}Si_{30}$, $Ba_8Ga_{16}Ge_{30}$, $Ba_8In_{16}Ge_{30}$, and $Sr_8Ga_{16}Ge_{30}$. Phys. Rev. B 71, 144107.

Berger, R.F., Lee, S., Hoffmann, R., 2007. A quantum mechanically guided view of $Mg_{44}Rh_7$. Chem. Eur. J. 13, 7852–7863.

Berliner, R., Werner, S.A., 1986. The structure of the low-temperature phase of lithium metal. Physica B 136, 481–484.

Blank, H., 1998. Fractional packing densities and fast diffusion in uranium and other light actinoids. J. Alloys Comp. 268, 180–187.

Bodak, O., Demchenko, P., Seropegin, Y., Fedorchuk, A., 2006. Cubic structure types of rare-earth intermetallics and related compounds. Z. Kristallogr. 221, 482–492.

Bojin, M.D., Hoffmann, R., 2003a. The REME phases-I. An overview of their structural variety. Helv. Chim. Acta 86, 1653–1682.

Bojin, M.D., Hoffmann, R., 2003b. The REME phases-II. What's possible? Helv. Chim. Acta 86, 1683–1708.

Booth, M.H., Brandon, J.K., Brizard, R.Y., Chieh, C., Pearson, W.B., 1977. Gamma-brasses with F cells. Acta Crystallogr. B 33, 30–36.

Bovornratanaraks, T., Allan, D.R., Belmonte, S.A., McMahon, M.I., Nelmes, R.J., 2006. Complex monoclinic superstructure in Sr-IV. Phys. Rev. B 73, 144112.

Brück, E., 2005. Developments in magnetocaloric refrigeration. J. Phys. D: Appl. Phys. 38, R381–R391.

Brunner, G.O., Schwarzenbach, D., 1971. Limitation of coordination sphere and determination of coordination coefficient in crystal structures. Z. Kristallogr. 133, 127–133.

Buerger, M.J., 1951. Crystallographic aspects of phase transformations. In: Smoluchowski, R., Mayer, J.E., Weyl, W.A. (Eds.), Phase Transformations in Solids. John Wiley & Sons, New York, pp. 183–211.

Cerenius, Y., Dubrovinsky, L., 2000. Compressibility measurements on iridium. J. Alloys Comp. 306, 26–29.

Chan, R.W., Haasen, P., 1996. Physical Metallurgy. Vols. 1–3. North-Holland, Amsterdam.

Chesnut, G.N., Vohra, Y.K., 1998. Phase transformation in lutetium metal at 88 GPa. Phys. Rev. B 57, 10221–10223.

Chesnut, G.N., Vohra, Y.K., 2000. α-uranium phase in compressed neodymium metal. Phys. Rev. B 61, R3768–R3771.

Chesnut, G.N., Vohra, Y.K., 1999. Structural and electronic transitions in ytterbium metal to 202 Gpa. Phys. Rev. Lett. 82, 1712–1715.

Chieh, C., 1979. Archimedean truncated octahedron, and packing of geometric units in cubic crystal structures. Acta Crystallogr. A 35, 946–952.

Chieh, C., 1980. The Archimedean truncated octahedron. 2. Crystal structures with geometric units of symmetry 43m. Acta Crystallogr. A 36, 819–826.

Chieh, C., 1982. The Archimedean truncated octahedron. 3. Crystal structures with geometric units of symmetry m3m. Acta Crystallogr. A 38, 346–349.

Choe, W., Pecharsky, V.K., Pecharsky, A.O., Gschneidner Jr., K.A., Young Jr., V.G., Miller, G.J., 2000. Making and breaking covalent bonds across the magnetic transition in the giant magnetocaloric material $Gd_5(Si_2Ge_2)$. Phys. Rev. Lett. 84, 4617–4620.

Conrad, M., Harbrecht, B., Weber, T., Jung, D.Y., Steurer, W., 2009. Large, larger, largest–a family of cluster-based tantalum-copper-aluminides with giant unit cells. Part B: the cluster structure. Acta Crystallogr. B 65, 318–325.

Culp, S.R., Poon, S.J., Hickman, N., Tritt, T.M., Blumm, J., 2006. Effect of substitutions on the thermoelectric figure of merit of half-Heusler phases at 800 degrees C. Appl. Phys. Lett. 88, 042106.

Cunningham, N.C., Qiu, W., Hope, K.M., Liermann, H.P., Vohra, Y.K., 2007. Symmetry lowering under high pressure: structural evidence for f-shell delocalization in heavy rare earth metal terbium. Phys. Rev. B 76, 212101.

Daams, J.L.C., Villars, P., van Vucht, J.H.N., 1991. Atlas of Crystal Structure Types for Intermetallic Phases, vol. 4. American Society for Metals, USA.

Daams, J.L.C., Villars, P., 1992. Atomic-environment classification of the cubic "intermetallic" structure types. J. Alloys Comp. 182, 1–33.

Daams, J.L.C., Villars, P., 1993. Atomic-environment classification of the rhombohedral "intermetallic" structure types. J. Alloys Comp. 197, 243–269.

Daams, J.L.C., Villars, P., 1994. Atomic-environment classification of the hexagonal "intermetallic" structure types. J. Alloys Comp. 215, 1–34.

Daams, J.L.C., Villars, P., 1997. Atomic-environment classification of the tetragonal "intermetallic" structure types. J. Alloys Comp. 252, 110–142.

Dabos-Seignon, S., Dancausse, J.P., Gering, E., Heathman, S., Benedict, U., 1993. Pressure-induced phase-transition in alpha-Pu. J. Alloys Comp. 190, 237–242.

Desgreniers, S., Vohra, Y.K., Ruoff, A.L., 1989. Tin at high pressure: an energy-dispersive x-ray-diffraction study to 120 GPa. Phys. Rev. B 39, 10359–10361.

Degtyareva, O., McMahon, M.I., Allan, D.R., Nelmes, R.J., 2004a. Structural complexity in gallium under high pressure: relation to alkali elements. Phys. Rev. Lett. 93, 205502.

Degtyareva, O., McMahon, M.I., Nelmes, R.J., 2004b. High-pressure structural studies of group-15 elements. High Press. Res. 24, 319–356.

Deiseroth, H.-J., Biehl, E., 1999. $NaK_{29}Hg_{48}$: a contradiction to or an extension of theoretical concepts to rationalize the structures of complex intermetallics? J. Solid State Chem. 147, 177–184.

de Laissardiere, G.T., Nguyen-Manh, D., Mayou, D., 2005. Electronic structure of complex Hume-Rothery phases and quasicrystals in transition metal aluminides. Progr. Mater. Sci. 50, 679–788.

Deloudi, S., Steurer, W., 2011. Unifying cluster-based structure models of decagonal Al-Co-Ni, Al-Co-Cu and Al-Fe-Ni. Acta Crystallogr. B 67, 1–17.

Ding, Y., Ahuja, R., Shu, J., Chow, P., Luo, W., Mao, H.K., 2007. Structural phase transition of vanadium at 69 GPa. Phys. Rev. Lett. 98, 085502.

Donohue, J., 1974. The Structures of the Elements. John Wiley & Sons, New York.

Dshemuchadse, J., Jung, D.Y., Steurer, W., 2011. Structural building principles of complex fcc intermetallics with more than 400 atoms per unit cell. Acta Crystallogr. B 67, 269–272.

Ek, J. van, Sterne, P.A., Gonis, A., 1993. Phase stability of plutonium. Phys. Rev. B 48, 16280–16289.

Ernst, G., Artner, C., Blaschke, O., Krexner, G., 1986. Low-temperature martensitic phase transition of bcc lithium. Phys. Rev. B 33, 6465–6469.

Errandonea, D., Boehler, R., Schwager, B., Mezouar, M., 2007. Structural studies of gadolinium at high pressure and temperature. Phys. Rev. B 75, 014103.

Evans, S.R., Loa, I., Lundegaard, L.F., McMahon, M.I., 2009. Phase transitions in praseodymium up to 23 GPa: an x-ray powder diffraction study. Phys. Rev. B 80, 134105.

Farr, J.D., Giorgi, A.L., Bowman, M.G., Noney, R.K., 1961. The crystal structure of actinium metal and actinium hydride. J. Inorg. Nucl. Chem. 18, 42–47.

Franceschi, E., Olcese, G.L., 1969. A new allotropic form of cerium due to its transition under pressure to the tetravalent state. Phys. Rev. Lett. 22, 1299–1300.

Frank, F.C., Kasper, J.S., 1958. Complex alloy structures regarded as sphere packings. 1. Definitions and basic principles. Acta Crystallogr. 11, 184–190.

Frank, F.C., Kasper, J.S., 1959. Complex alloy structures regarded as sphere packing. 2. Analysis and classification of representative structures. Acta Crystallogr. 12, 483–499.

Fredrickson, D.C., Lee, S., Hoffmann, R., Lin, J., 2004a. The Nowotny chimney ladder phases: following the c_{pseudo} clue toward an explanation of the 14 electron rule. Inorg. Chem. 43, 6151–6158.

Fredrickson, D.C., Lee, S., Hoffmann, R., Lin, J., 2004b. The Nowotny chimney ladder phases: whence the 14 electron rule? Inorg. Chem. 43, 6159–6167.

Fujihisa, H., Nakamoto, Y., Shimizu, K., Yabuuchi, T., Gotoh, Y., 2008. Crystal structures of calcium IV and V under high pressure. Phys. Rev. Lett. 101, 095503.

Fujisha, H., Takemura, K., 1995. Stability and the equation of state of α-manganese under ultrahigh pressure. Phys. Rev. B 52, 13257–13260.

Gao, G.Y., Niu, Y.L., Cui, T., Zhang, L.J., Li, Y., Xie, Y., He, Z., Ma, Y.M., Zhou, G.T., 2007. Superconductivity and lattice instability in face-centered cubic lanthanum under high pressure. J. Phys. Condens. Matter 19, 425234.

Goll, D., Kronmüller, H., 2000. High-performance permanent magnets. Naturwissenschaften 87, 423–438.

Gomez, C.P., Lidin, S., 2001. Structure of $Ca_{13}Cd_{76}$: a novel approximant to the $MCd_{5.7}$ quasicrystals (M = Ca, Yb). Angew. Chem. Int. Ed. 40, 4037–4039.

Graf, T., Casper, F., Winterlik, J., Balke, B., Fecher, G.H., Felser, C., 2009. Crystal structure of new Heusler compounds. Z. Anorg. Allg. Chem. 635, 976–981.

Gregoryanz, E., Lundegaard, L.F., McMahon, M.I., Guillaume, C., Nelmes, R.J., Mezouar, M., 2008. Structural diversity of sodium. Science 320, 1054–1057.

Grosshans, W.A., Vohra, Y.K., Holzapfel, W.B., 1982. Evidence for a scft phonon mode and a new structure in rare-earth metals under pressure. Phys. Rev. Lett. 49, 1572–1575.

Grünbaum, B., Shephard, G.C., 1987. Tilings and Patterns. W. H. Freeman and company, New York.

Gschneidner Jr., K.A., Pecharsky, V.K., Tsokol, A.O., 2005. Recent developments in magnetocaloric materials. Rep. Prog. Phys. 68, 1479–1539.

Gu, Q.F., Krauss, G., Grin, Y., Steurer, W., 2009. Experimental confirmation of the stability and chemical bonding analysis of the high-pressure phases Ca-I, II, and III at pressures up to 52 GPa. Phys. Rev. B 79, 134121.

Guillaume, C.E., 1897. Recherches sur les aciers au nickel. Dilatations aux temperatures elevees; resistance electrique. CR Acad. Sci. 125, 235–238.

Guillaume, C.L., Gregoryanz, E., Degtyareva, O., McMahon, M.I., Hanfland, M., Evans, S., Guthrie, M., Sinogeikin, S.V., Mao, H.K., 2011. Cold melting and solid structures of dense lithium. Nat. Phys. 7, 211–214.

Hahn, T. (Ed.), 2002. International Tables for Crystallography, vol. A. Kluwer Academic Publishers, Dordrecht/Boston/London.

Haire, R.G., Heathman, S., Idiri, M., Le Bihan, T., Lindbaum, A., Rebizant, J., 2003. Pressure-induced changes in protactinium metal: importance to actinoid-metal bonding concepts. Phys. Rev. B 67, 134101.

Hanfland, M., Syassen, K., Christensen, N.E., Novikov, D.L., 2000. New high-pressure phases of lithium. Nature 408, 174–178.

Haskel, D., Lang, J.C., Islam, Z., Cady, A., Srajer, G., van Veenendaal, M., Canfield, P.C., 2005. Atomic origin of magnetocrystalline anisotropy in $Nd_2Fe_{14}B$. Phys. Rev. Lett. 95, 217207.

Heathman, S., Haire, R.G., Le Bihan, T., Lindbaum, A., Idiri, M., Normile, P., Li, S., Ahuja, R., Johansson, B., Lander, G.H., 2005. A high-pressure structure in curium linked to magnetism. Science 309, 110–113.

Heathman, S., Haire, R.G., Le Bihan, T., Lindbaum, A., Litfin, K., Méresse, Y., Libotte, H., 2000. Pressure induces major changes in the nature of Americium's $5f$ electrons. Phys. Rev. Lett. 85, 2961–2964.

Herbst, J.F., Croat, J.J., Pinkerton, F.E., Yelon, W.B., 1984. Relationships between crystal structure and magnetic properties in $Nd_2Fe_{14}B$. Phys. Rev. B 29, 4176–4178.

Hu, C.E., Zeng, Z.Y., Zhang, L., Chen, X.R., Cai, L.C., 2010. Phase transition and thermodynamics of thorium from first-principles calculations. Solid State Commun. 150, 393–398.

Hudson, T.S., 2010. Dense sphere packing in the $NaZn_{13}$ structure type. J. Phys. Chem. C 114, 14013–14017.

Jagodzinski, H., 1954. Polytypism in SiC crystals. Acta Crystallogr. 7, 300.

Jeitschko, W., 1969. The crystal structure of Fe_2AlB_2. Acta Crystallogr. B 25, 163–165.

Joubert, J.-M., Dupin, N., 2004. Mixed site occupancies in the μ phase. Intermetallics 12, 1373–1380.

Kammler, D.R., Rodriguez, M.A., Tissot, R.G., Brown, D.W., Clausen, B., Sisneros, T.A., 2008. In-situ time-of-flight neutron diffraction study of high-temperature α-to-β phase transition in elemental scandium. Met. Mater. Trans. A 39, 2815–2819.

Katzke, H., Toledano, P., 2005. Structural mechanisms and order-parameter symmetries for the high-pressure phase transitions in alkali metals. Phys. Rev. B 71, 184101.

Kauzlarich, S.M. (Ed.), 1996. Chemistry, Structure and Bonding of Zintl Phases and Ions. VCH Publishers, New York.

Kawamura, N., Taniguchi, T., Mizusaki, S., Nagata, Y., Ozawa, T.C., Samata, H., 2006. Functional intermetallic compounds in the samarium-iron system. Sci. Tech. Adv. Mater. 7, 46–51.

Kleinke, H., 2010. New bulk materials for thermoelectric power generation: clathrates and complex antimonides. Chem. Mater. 22, 604–611.

Koch, E., Fischer, W., 1992. Sphere packings and packings of ellipsoids. In: Wilson, A.J.C. (Ed.), International Tables for Crystallography. Kluwer Academic Publishers, Dordrecht.

Komura, Y., Kitano, Y., 1977. Long-period stacking variants and their electron-concentration dependence in the Mg-base Friauf-Laves phases. Acta Crystallogr. B 33, 2496–2501.

Komura, Y., Sly, W.G., Shoemaker, D.P., 1960. The crystal structure of the R phase. Mo-Co-Cr. Acta Crystallogr. 13, 575–585.

Kramer, P., Schlottmann, M., 1989. Dualization of Voronoi domains and *Klotz* construction-a general-method for the generation of proper space fillings. J. Phys. A: Math. Gen. 22, L1097–L1102.

Landrum, G.A., Hoffmann, R., Evers, J., Boysen, H., 1998. The TiNiSi family of compounds: structure and bonding. Inorg. Chem. 37, 5754–5763.

Leigh, G.J. (Ed.), 1990. Nomenclature of Inorganic Chemistry. Recommendations 1990. Blackwell Scientific Publications, Oxford.

Lima-de-Faria, J., Hellner, E., Liebau, F., Makovicky, E., Parthé, E., 1990. Nomenclature of inorganic structure types. Acta Crystallogr. A 46, 1–11.

Lin, Q.S., Corbett, J.D., 2006. New building blocks in the 2/1 crystalline approximant of a Bergman-type icosahedral quasicrystal. Proc. Natl. Acad. Sci. U.S.A. 103, 13589–13594.

Lundegaard, L.F., Gregoryanz, E., McMahon, M.I., Guillaume, C., Loa, I., Nelmes, R.J., 2009a. Single-crystal studies of incommensurate Na to 1.5 Mbar. Phys. Rev. B 79, 064105.

Lundegaard, L.F., Marqués, M., Stinton, G., Ackland, G.J., Nelmes, R.J., McMahon, M.I., 2009b. Observation of the $oP8$ crystal structure in potassium at high pressure. Phys. Rev. B 80, 020101.

Ma, Y.M., Eremets, M., Oganov, A.R., Xie, Y., Trojan, I., Medvedev, S., Lyakhov, A.O., Valle, M., Prakapenka, V., 2009. Transparent dense sodium. Nature 458, 182–185.

Ma, Y.M., Oganov, A.R., Xie, Y., 2008. High-pressure structures of lithium, potassium, and rubidium predicted by an ab initio evolutionary algorithm. Phys. Rev. B 78, 014102.

Macia, E., 2001. Theoretical prospective of quasicrystals as thermoelectric materials. Phys. Rev. B 64, 094206.

Marqués, M., McMahon, M.I., Gregoryanz, E., Hanfland, M., Guillaume, C.L., Pickard, C.J., Ackland, G.J., Nelmes, R.J., 2011. Crystal structures of dense liyhium: a metal-semiconductor-metal transition. Phys. Rev. Lett. 106, 095502.

Massalski, T.B., 1990. Binary Alloy Phase Diagrams, vols. 1–3. ASM International, USA.

Mao, H.K., Wu, Y., Shu, J.F., Hu, J.Z., Hemley, R.J., Cox, D.E., 1990. High-pressure phase-transition and equation of state of lead to 238 Gpa. Solid State Commun. 74, 1027–1029.

Mao, W.L., Wang, L., Ding, Y., Yang, W., Liu, W., Kim, D.Y., Luo, W., Ahuja, R., Meng, Y., Sinogeikin, S., Shu, J., Mao, H.K., 2010. Distortions and stabilization of simple-cubic calcium at high pressure and low temperature. Proc. Natl. Acad. Sci. 107, 9965–9968, corrections on p. 12734.

McMahon, R.J., Bovornratanaraks, T., Allan, D.R., Belmonte, S.A., Nelmes, R.J., 2000. Observation of the incommensurate barium-IV structure in strontium phase V. Phys. Rev. B 61, 3135–3138.

McMahon, Lundegaard, L.F., Hejny, C., Falconi, S., Nelmes, R.J., 2006a. Different incommensurate composite crystal structure for Sc-II. Phys. Rev. B 73, 134102.

McMahon, M.I., Nelmes, R.J., 1997. Different results for the equilibrium phases of cerium above 5 GPa. Phys. Rev. Lett. 78, 3884–3887.

McMahon, M.I., Nelmes, R.J., 2006. High-pressure structures and phase transformations in elemental metals. Chem. Soc. Rev. 35, 943–963.

McMahon, M.I., Rekhi, S., Nelmes, R.J., 2001a. Pressure dependent Incommensuration in Rb-IV. Phys. Rev. Lett. 87, 055501.

McMahon, M.I., Nelmes, R.J., Rekhi, S., 2001b. Complex crystal structure of cesium-iii. Phys. Rev. Lett. 87, 255502.

McMahon, M.I., Nelmes, R.J., Schwarz, U., Syassen, K., 2006b. Composite incommensurate K-III and a commensurate form: study of a high-pressure phase of potassium. Phys. Rev. B 74, 140102.

Miller, G.J., 1996. Structure and bonding at the Zintl border. In: Kauzlarich, S.M. (Ed.), Chemistry, Structure and Bonding of Zintl Phases and Ions, first ed. VCH Publishers, New York, pp. 1–59.

Mizutani, U., 2011. Hume-rothery Rules for Structurally Complex Alloy Phases. CRC Press, Taylor & Francis Group, Boca Raton, USA.

Nakamoto, Y., Sakata, M., Shimizu, K., 2010. Ca-VI: a high-pressure phase of calcium above 158 GPa. Phys. Rev. B 81, 140106.

Nelmes, R.J., Allan, D.R., McMahon, M.I., Belmonte, S.A., 1999. Self-hosting incommensurate structure of barium IV. Phys. Rev. Lett. 83, 4081–4084.

Nelmes, R.J., Liu, H., Belmonte, S.A., Loveday, J.S., Allan, D.R., McMahon, M.I., 1996. $Imma$ phase of germanium at ~80 GPa. Phys. Rev. B 53, R2907–R2909.

Nelmes, R.J., McMahon, M.I., Loveday, J.S., Rekhi, S., 2002. Structure of Rb-III: novel modulated stacking structures in alkali metals. Phys. Rev. Lett. 88, 155503.

Nolas, G.S., Kaeser, M., Littleton IV, R.T., Tritt, T.M., 2000. High figure of merit in partially filled ytterbium skutterudite materials. Appl. Phys. Lett. 77, 1855–1857.

Nyman, H., Carroll, C.E., Hyde, B.G., 1991. Rectilinear rods of face-sharing tetrahedra and the structure of β-Mn. Z. Kristallogr 196, 39–46.

Oganov, A.R., Lyakhov, A.O., Valle, M., 2011. How evolutionary crystal structure prediction works - and why. Acc. Chem. Res. 44, 227–237.

Oganov, A.R., Ma, Y.M., Xu, Y., Errea, I., Bergara, A., Lyakhov, A.O., 2010. Exotic behavior and crystal structures of calcium under pressure. Proc. Natl. Acad. Sci. 107, 7646–7651.

Pearson, W.B., 1972. The Crystal Chemistry and Physics of Metals and Alloys. Wiley-Interscience, New-York.

Perez-Mato, J.M., Elcoro, L., Aroyo, M.I., Katzke, H., Tolédano, P., Izaola, Z., 2006. Apparently complex high-pressure phase of gallium as a simple modulated structure. Phys. Rev. Lett. 97, 115501.

Pettifor, D.G., 1986. The structures of binary compounds. I. Phenomenological structure maps. J. Phys. C: Solid State Phys. 19, 285–313.

Pettifor, D.G., 1988. Structure maps for pseudobinary and ternary phases. Mater. Sci. Techn. 4, 675–691.

Ponou, S., Fässler, T.F., Kienle, L., 2008. Structural complexity in intermetallic alloys: Long-periodic order beyond 10 nm in the system $BaSn_3$/$BaBi_3$. Angew. Chem. Int. Ed. 47, 3999–4004.

Porsch, F., Holzapfel, W.B., 1993. Novel reentrant high pressure phase transtion in lanthanum. Phys. Rev. Lett. 70, 4087–4089.

Pöttgen, R., Johrendt, D., 2000. Equiatomic intermetallic europium compounds: syntheses, crystal chemistry, chemical bonding, and physical properties. Chem. Mater. 12, 875–897.

Quiquandon, M., Gratias, D., 2006. Unique six-dimensional structural model for Al-Pd-Mn and Al-Cu-Fe icosahedral phases. Phys. Rev. B 74, 214205.

Rabson, D.A., Mermin, N.D., Rokhsar, D.S., Wright, D.C., 1991. The space-groups of axial crystals and quasi-crystals. Rev. Mod. Phys. 63, 699–733.

Ramsdell, L.S., 1947. Studies on silicon carbide. Amer. Miner 32, 64–82.

Reed, S.K., Ackland, G.J., 2000. Theoretical and computational study of high-pressure structures in barium. Phys. Rev. Lett. 84, 5580–5583.

Rohrer, F.E., Lind, H., Eriksson, L., Larsson, A.-K., Lidin, S., 2000. On the question of commensurability - the Nowotny chimney-ladder structures revisited. Z. Kristallogr. 215, 650–660.

Rohrer, F.E., Lind, H., Eriksson, L., Larsson, A.-K., Lidin, S., 2001. Incommensurately modulated Nowotny Chimney-ladder phases $Cr_{1-x}Mo_xGe_{-1.75}$ with x=0.65 and 0.84. Z. Kristallogr 216, 190–198.

Rokhsar, D.S., Wright, D.C., Mermin, N.D., 1988. Scale Equivalence of Quasicrystallographic space-groups. Phys. Rev. B 37, 8145–8149.

Rousseau, R., Uehara, K., Klug, D.D., Tse, J.S., 2005. Phase stability and broken-symmetry transition of elemental lithium up to 140 GPa. Chem. Phys. Chem. 6, 1703–1706.

Ruban, A.V., Khmelevskyi, S., Mohn, P., Johansson, B., 2007. Magnetic state, magnetovolume effects, and atomic order in $Fe_{65}Ni_{35}$ Invar alloy: a first principles study. Phys. Rev. B 76, 014420.

Samson, S., 1964. A method for the determination of complex cubic metal structures and its application to the solution ofthe structure of $NaCd_2$. Acta Crystallogr. 17, 491–495.

Schulte, O., Holzapfel, W.B., 1993. Phase diagram for mercury up to 67 GPa and 500 K. Phys. Rev. B 48, 14009–14012.

Schwarz, U., Grzechnik, A., Syassen, K., Loa, I., Hanfland, M., 1999a. Rubidium-iv: a high pressure phase with complex crystal structure. Phys. Rev. Lett. 83, 4085–4088.

Schwarz, U., Syassen, K., Grzechnik, A., Hanfland, M., 1999b. The crystal structure of rubidium-VI near 50 GPa. Solid State Commun. 112, 319–322.

Schwarz, U., Takemura, K., Hanfland, M., Syassen, K., 1998. Crystal structure of cesium-V. Phys. Rev. Lett. 81, 2711–2714.

Seipel, M., Porsch, F., Holzapel, W.B., 1997. Characterization of the fcc-distorted fcc-structural transition in lanthanum in an extended pressure and temperature range. High Press. Res. 15, 321–330.

Shen, Y.R., Kumar, R.S., Cornelius, A.L., Nicol, M.F., 2007. High-pressure structural studies of dysprosium using angle-dispersive x-ray diffraction. Phys. Rev. B 75, 064109.

Shoemaker, C.B., Shoemaker, D.P., 1969. Structural properties of some σ-phase related phases. In: Giessen, B.C. (Ed.), Developments in the Structural Chemistry of Alloy Phases. Plenum Press, New York - London, pp. 107–139.

Sikka, S.K., 2005. A high pressure distorted α-uranium (Pnma) structure in plutonium. Solid State Commun. 133, 169–172. ann.

Sootsman, J.R., Chung, D.Y., Kanatzidis, M.G., 2009. New and old concepts in thermoelectric materials. Angew. Chem. Int. Ed. 48, 8616–8639.

Souvignier, B., 2003. Enantiomorphism of crystallographic groups in higher dimensions with results in dimensions up to 6. Acta Crystallogr. A 59, 210–220.

Stein, F., Palm, M., Sauthoff, G., 2004. Structure and stability of Laves phases. Part I. Critical assessment of factors controlling Laves phase stability. Intermetallics 12, 713–720.

Stein, F., Palm, M., Sauthoff, G., 2005. Structure and stability of Laves phases Part II - structure type variations in binary and ternary systems. Intermetallics 13, 1056–1074.

Steinhardt, P.J., Jeong, H.C., 1996. A simpler approach to Penrose tiling with implications for quasicrystal formation. Nature 382, 431–433.

Steurer, W., 2004. Twenty years of structure research on quasicrystals. Part 1. Pentagonal, octagonal, decagonal and dodecagonal quasicrystals. Z. Kristallogr. 219, 391–446.

Steurer, W., 2006. Stable clusters in quasicrystals: fact or fiction? Philos. Mag 86, 1105–1113.

Steurer, W., Deloudi, S., 2008. Fascinating quasicrystals. Acta Crystallogr. A 64, 1–11.

Steurer, W., Deloudi, S., 2009. Crystallography of Quasicrystals: Concepts, Methods and Structures. Springer, Heidelberg.

Sugiyama, K., Kaji, N., Hiraga, K., Ishimasa, T., 1998. Crystal structure of a cubic $Al_{70}Pd_{23}Mn_6Si$; a 2/1 rational approximant of an icosahedral phase. Z. Kristallogr. 213, 90–95.

Sun, J., Lee, S., Lin, J., 2007. Four-dimensional space groups for Pedestrians: composite structures. Chem. Asian J. 2, 1204–1229.

Takemura, K., 1994. High-pressure structural study of barium to 90 GPa. Phys. Rev. B 50, 16238–16246.

Takemura, K., 1997. Structural study of Zn and Cd to ultrahigh presures. Phys. Rev. B 56, 5170–5179.

Takemura, K., Christensen, N.E., Novikov, D.L., Syassen, K., Schwarz, U., Hanfland, M., 2000. Phase stability of highly compressed cesium. Phys. Rev. B 61, 14399–14403.

Takemura, K., Fujihisa, H., Nakamoto, Y., Nakano, S., Ohisi, Y., 2007. Crystal structure of the high-pressure γ phase of mercury: a novel monoclinic distortion of the close-packed structure. J. Phys. Soc. Jpn. 76, 023601.

Takemura, K., Kobayashi, K., Masao, A., 1998. High-pressure bct-fcc phase transition in Ga. Phys. Rev. B 58, 2482–2486.

Takemura, K., Schwarz, U., Syassen, K., Christensen, N.E., Hanfland, M., Novikov, D.L., Loa, I., 2001. High-pressure structures of Ge above 10 GPa. Phys. Sat. Sol.(b) 223, 385–390.

Takemura, K., Shimomura, O., Fujihisa, H., 1991. Cs(VI): a new high-pressure polymorph of cesium above 72 GPa. Phys. Rev. Lett. 66, 2014–2017.

Thomas, E.L., Millican, J.N., Okudzeto, E.K., Chan, J.Y., 2006. Crystal growth and the search for highly correlated intermetallics. Comm. Inorg. Chem. 27, 1–39.

Tonkov, E.Y., 1992–1996.. High Pressure Phase Transformations: A Handbook, Vols. 1-3. Gordon and Breach Science Publishers, Philadelphia.

Tonkov, E.Y., Ponyatovsky, E.G., 2005. Phase Transformations of Elements Under High Pressure. CRC Press, Boca Raton.

Trudel, S., Gaier, O., Hamrle, J., Hillebrands, B., 2010. Magnetic anisotropy, exchange and damping in cobalt-based full-Heusler compounds: an experimental review. J.Phys. D: Appl. Phys. 43, 193001.

van Smaalen, S., George, T.F., 1987. Determination of the incommensurately modulated structure of α-uranium below 37 K. Phys. Rev. B 35, 7939–7951.

Velisavljevic, N., Vohra, Y.K., 2004. Distortion of alpha-uranium structure in praseodymium metal to 311 Gpa. High Press. Res. 24, 295–302.

Verma, A.K., Modak, P., 2008. Structural phase transitions in vanadium under high pressure. Europhys. Lett. 81, 37003.

Verma, A.K., Modak, P., Rao, R.S., Godwal, B.K., Jeanloz, R., 2007. High-pressure phases of titanium: first-principles calculations. Phys. Rev. B 75, 014109.

Verstrate, M.J., 2010. Phases of Polonium via density functional theory. Phys. Rev. Lett. 104, 035501.

Villars, P., Brandenburg, K., Berndt, M., LeClair, S., Jackson, A., Pao, Y.-H., Igelnik, B., Oxley, M., Bakshi, B., Chen, P., Iwata, S., 2001. Binary, ternary and quaternary compound former/nonformer prediction via Mendeleev number. J. Alloys Comp. 317-318, 26–38.

Villars, P., Calvert, L.D., 1991. Pearson's Handbook of Crystallographic Data for Intermetallic Phases, vols. 1–4. ASM, USA.

Villars, P., Cenzual, K. (Eds.), 2011. Pearson's Crystal Data. ASM International, Materials Park, Ohio, USA.

Villars, P., Cenzual, K., Daams, J.L.C., Chen, Y., Iwata, S., 2004. Data-driven atomic environment prediction for binaries using the Mendeleev number. Part 1. Composition AB. J. Alloys Comp. 367, 167–175.

Villars, P., Daams, J.L.C., 1993. Atomic-environment classification of the chemical elements. J. Alloys Comp. 197, 177–196.

Vohra, Y.K., Akella, J., 1991. 5f bonding in thorium metal at extreme compressions: phase transitions to 300 GPa. Phys. Rev. Lett. 67, 3563–3566.

Vohra, Y.K., Beaver, S.L., 1999. Ultrapressure equation of state of cerium metal to 208 GPa. J. Appl. Phys. 85, 2451.

Wade, K., 1976. Structural and bonding patterns in cluster chemistry. Adv. Inorg. Chem. Radiochem. 18, 1–66.

Weber, T., Dshemuchadse, J., Kobas, M., Conrad, M., Harbrecht, B., Steurer, W., 2009. Large, larger, largest - a family of cluster-based tantalum-copper-aluminides with giant unit cells. Part A: structure solution and refinement. Acta Crystallogr. B 65, 308–317.

Wills, J.M., Eriksson, O., 1992. Crystal-structure stabilities and electronic structure for the light actinoids Th, Pa, and U. Phys. Rev. B 45, 13879–13890.

Wilson, A.J.C., Prince, E. (Eds.), 1999, International Tables for Crystallography, vol. C. Kluwer Academic Publishers, Dordrecht/Boston/London.

Yao, Y., Martonak, R., Patchkovskii, S., Klug, D.D., 2010. Stability of simple cubic calcium at high pressure: a first-principles study. Phys. Rev. B 82, 094107.

Yoo, C.S., Cynn, H., Söderlind, P., Iota, V., 2000. New β(fcc)-Cobalt to 210 GPa. Phys. Rev. Lett. 84, 4132–4135.

Young, D.A., 1991. Phase Diagrams of the Elements. University of California Press, Berkeley.

Zhao, Y.C., Porsch, F., Holzapfel, W.B., 1994. Irregularities of ytterbium under high pressure. Phys. Rev. B 49, 815–817.

Zintl, E., Dullenkopf, W., 1932. Über den Gitterbau von NaTl und seine Beziehung zu den Strukturen vom Typus des β-Messings. Z. Phys. Chem. B 16, 195–205.

Further Reading List

De Graef, M., McHenry, M.E., 2007. Structure of Materials: An Introduction to Crystallography, Diffraction, and Symmetry. Cambridge University Press, Cambridge.

Ferro, R. and A. Saccone, Intermetallic Chemistry (Pergamon Materials Series 13, Elsevier, Amsterdam).

O'Keeffe, M., Hyde, B.G., 1996. Crystal Structures. I. Patterns and Symmetry. Mineralogical Societey of America, Washington.

Pettifor, D.G., 1993. Electron theory of crystal structure. In: Cahn, R.W., Haasen, P., Kramer, E.J. (Eds.), Materials Science and Technology. A Comprehensive Treatment, vol. 1. VCH, Weinheim, pp. 61–122.

Vainshtein, B.K., 1994. Modern Crystallography I: Fundamentals of Crystals. Springer-Verlag, Berlin.

Vainshtein, B.K., Fridkin, V.M., Indenbom, V.L., 1982. Modern Crystallography II: Structure of Crystals. Springer-Verlag, Berlin.

Biography

Walter Steurer received his diploma and Ph.D. degrees in chemistry from the University in Vienna, Austria, in 1976 and 1979, respectively. Subsequently, he moved to the University of Munich, Germany, where he worked as a research associate and lecturer. In 1987, he concluded his habilitation thesis in the field of crystallography and mineralogy. For a short period, from 1992 until 1993, he was a professor of crystallography at the University of Hanover, Germany, and rejected calls to the Universities of Hamburg and Munich, Germany. Since 1993, he has been full Professor of Crystallography at the Laboratory of Crystallography of the ETH Zurich and University of Zurich, Switzerland. His main research topics comprise structural studies of complex intermetallic phases with focus on quasicrystals, their phase transformations, the modeling of order/disorder phenomena, and higher dimensional crystallography in general.

2   Electron Theory of Complex Metallic Alloys

U. Mizutani, Nagoya Industrial Science Research Institute, Nagoya, Japan
M. Inukai, Toyota Technological Institute, Nagoya, Japan
H. Sato, Aichi University of Education, Kariya-shi, Aichi, Japan
E.S. Zijlstra, Theoretical Physics, University of Kassel, Kassel, Germany

2.1   Introduction

2.1.1   What is the Electron Theory of Metals?

Each element exists either as a solid, a liquid, or a gas at ambient temperature and pressure. Alloys or compounds can be formed by assembling a mixture of different elements on a common lattice. Typically this is done by melting followed by solidification. Any material is, therefore, composed of a combination of the elements in the periodic table. A molar quantity of a solid contains as many as 10^{23} atoms. A solid is formed as a result of bonding among such a huge number of atoms. The entities responsible for the bonding are the electrons. The physical and chemical properties of a given solid are decided by how the constituent atoms are bonded through the interaction of their electrons among themselves and with the potentials of the ions. This interaction yields the electronic band structure characteristic of each solid: a semiconductor or an insulator is described by a filled band separated from other bands by an energy gap, and a metal by overlapping continuous bands. Its physical and chemical properties are determined by the resulting electronic structure. The electron theory of metals covers properties of electrons responsible for the bonding of solids as well as electron transport properties manifested in the presence of external fields or temperature gradients.

Studies of the electron theory of metals are also important from the point of view of application-oriented research and play a vital role in the development of new functional materials. The recent progress in the semiconducting devices like the integrated circuit or large-scale integrated circuit, as well as the development in magnetic and superconducting materials, certainly owes to the successful application of the electron theory of metals.

2.1.2   Historical Survey of the Electron Theory of Metals

In this section, the reader is expected to grasp only the main historical landmarks of the subject without going into details. The electron theory of metals has developed along with the development of quantum mechanics.

Heisenberg (1927) laid the mathematical foundation of quantum mechanics in 1925 and our most familiar Schrödinger equation was established in 1926 (Schrödinger, 1926). In 1926, Fermi (1926) and Dirac (1926) independently derived a new form of statistical mechanics based on the Pauli exclusion principle. In 1927, Pauli (1927) applied the newly derived Fermi–Dirac statistics to the calculation of the paramagnetism of a free-electron gas.

In 1928, Sommerfeld (1928) applied the quantum mechanical treatment to the electron gas in a metal. He retained the concept of a free-electron gas originally introduced by Drude (1900) and Lorentz (1905), but applied to it the quantum mechanics coupled with the Fermi–Dirac statistics. The specific heat, the thermionic emission, the electrical and thermal conductivities, the magnetoresistance, and the Hall effect were calculated quite satisfactorily by replacing the ionic potentials with a constant averaged potential equal to zero. The Sommerfeld free-electron model could successfully remove the difficulty associated with the electronic specific heat derived from the equipartition law based on classical statistics.

The Sommerfeld model was, however, unable to answer why the mean free path of electrons reaches 20 nm in a good conducting metal like silver at room temperature. Indeed, electrons in a metal are moving in the presence of strong Coulomb potentials due to ions. Therefore the success based on the concept of a free-electron behavior was received at that time with a great deal of surprise. The ionic potential is periodically arranged in a crystal. In 1928, Bloch (1928) could show that the wave function of a conduction electron in the periodic potential can be described in the form of the plane wave modulated by a periodic function with the period of the lattice, no matter how strong the ionic potential. The wave function is called the Bloch wave. The Bloch theorem provided the basis for the electrical resistivity; the entity that is responsible for the scattering of electrons is not a strong ionic potential itself but the deviation from its periodicity. Based on the Bloch theorem, Wilson (1931) was able to describe a band theory, which embraces metals, semiconductors, and insulators. The main frame of the electron theory of metals had been matured by about the middle of the 1930s. We can see it by reading the well-known textbooks by Mott and Jones (1936) and Wilson (1936).

Before ending this section, the most notable achievements since the 1940s in the field of electron theory of metals may be briefly mentioned. Bardeen and Brattain invented the point-contact transistor in 1948–1949 (Bardeen and Brattain, 1948). For this achievement the Nobel Prize was awarded to Bardeen, Brattain and Shockley in 1956. Superconductivity is a phenomenon in which the electrical resistivity suddenly drops to zero at its transition temperature T_c. The theory of superconductivity has been established in 1957 by Bardeen, Cooper and Schrieffer (BCS; Bardeen et al., 1957). The so-called BCS theory has been recognized as one of the greatest accomplishments in the electron theory of metals since the advent of the Sommerfeld free-electron theory. Naturally, the higher the superconducting transition temperature, the more likely are possible applications. A maximum superconducting transition temperature had been thought to be no greater than 30–40 K within the framework of the BCS theory. However, a new material, which undergoes the superconducting transition above 30 K, was discovered in 1986 (Bednorz and Müller, 1986) and has received intense attention from both fundamental and practical points of view. This was not an ordinary metallic alloy but a cuprate oxide with a complex crystal structure. More new superconductors in this family have been discovered successively and the superconducting transition temperature T_c has increased above 90 K in 1987, above 110 K in 1988 and almost 140 K in 1996. The electronic properties manifested by these superconducting oxides have become one of the most exciting and challenging topics in the field of the electron theory of metals.

The electron theory of metals has been originally constructed for crystals where the existence of a periodic potential was presupposed. Subsequently, an electron theory of a disordered system, where the periodicity of the ionic potentials is heavily distorted, was recognized to be also significantly important. Liquid metals are typical of such disordered systems. More recently, amorphous metals and semiconductors have received wide attention not only from the viewpoint of fundamental physics but also from many possible practical applications. In addition to these disordered materials, a nonperiodic yet highly ordered material known as a quasicrystal was discovered by Shechtman et al. (1984). The Nobel Prize in Chemistry was awarded to Shechtman in 2011. The icosahedral quasicrystal is now known to possess the two-, three-, and five-fold rotational symmetry, which is incompatible with the translational symmetry characteristic of an ordinary crystal. The electron theory should be extended to these nonperiodic materials as well and be cast into a more universal theory.

2.1.3 Outline of This Chapter

There are obviously numerous topics concerning the electron theory of metals, which include fields of electronics, magnetism, superconductivity, optics, and others not only in crystals but also in nonperiodic materials. It is certainly beyond the scope of the present chapter to cover such vast fields. Indeed, a number of excellent textbooks covering all these topics are available to the readers (Ashcroft and Mermin, 1976; Barrett and Massalski, 1966; Kittel, 1996; Mizutani, 2001; Smith and Jensen, 1989; Ziman, 1964). After the discovery of the Al–Mn quasicrystal in 1984 by Shechtman et al. (1984) and following extensive research, we have now become well aware that there exist a large number of intermetallic compounds, which crystallize into complex structure phases possessing >100 atoms per unit cell. One of the present authors (Uichiro Mizutani) has devoted himself for the last two decades or so to try to gain a deeper insight into the stabilization mechanism of such structurally complex metallic alloy phases including the family of quasicrystals, which are positioned at an extreme limit as having an infinitely large unit cell. The aim of the present chapter is to help graduate students in physics, chemistry, and materials science learn the fundamentals and usefulness in the electron theory of metals through reviewing the stabilization mechanism of complex metallic alloys (CMAs). Hopefully, the present work will be able to assist them to deepen their understanding on why nature is able to stabilize such a complex structure phase by repeating a large unit cell in a periodic manner.

It may be worthwhile mentioning what prior fundamental knowledge is required to read the present chapter. The reader is assumed to have taken an elementary course of quantum mechanics. We use in this chapter terminologies such as the wave function, the uncertainty principle, the Pauli exclusion principle, the perturbation theory, etc. without explanation. In addition, the reader is expected to have learned the elementary principles of classical mechanics and electromagnetic dynamics. It is also suggested that the reader would consult elementary and advanced excellent books on the electron theory of metals, whenever needed (Ashcroft and Mermin, 1976; Barrett and Massalski, 1966; Kittel, 1996; Mizutani, 2001; Smith and Jensen, 1989; Ziman, 1964).

The units employed in the present chapter are mostly those of the SI system, but the CGS units are often conventionally used, particularly in tables and figures. Practical units are also employed. For example, the resistivity is expressed in the units of ohms centimeter, which is the combined unit of CGS and SI.

2.2   Fundamentals in Alloy Phase Stability

2.2.1   Introduction

The topic on alloy phase stability had been considered as a long-lasting theme in the electron theory of metals in the twentieth century, which dates back to the 1920s, when Hume-Rothery (1926), Hume-Rothery et al. (1934) and Westgren and Phragmén (1928a, 1928b, 1929) pointed out that stable composition ranges for successively appearing fcc-α, bcc-β, complex cubic-γ, and hcp-ε and η phases obtained by adding polyvalent elements to noble metals are universally scaled in terms of electrons per atom ratio **e/a**. Mott and Jones (1936) interpreted this empirical rule as arising from the contact of the Fermi sphere with the Brillouin zone planes specific to the respective phases. Unfortunately, however, no significant progress had been made for many years to deepen the understanding of the Hume-Rothery electron concentration rule beyond the free electron model proposed by Mott and Jones.

Interest in the Hume-Rothery electron concentration rule had been revived in the late 1980s to early 1990s when many thermally stable quasicrystals were discovered by using the Hume-Rothery electron concentration rule as a guide (Tsai et al., 1988, 1990; Yokoyama et al., 1991). Since then, intense attention has been directed to the evidence for why a quasicrystal characterized by a highly ordered structure with the lack of a finite unit cell can be stabilized at a particular electron concentration **e/a**. Before describing details concerning this challenging theme in the electron theory of metals, we will discuss in Section 2.2 basic ideas and principles involved in alloy phase stabilization to help readers facilitate the understanding of the discussions developed in the following sections.

2.2.2   Cohesive Energy of a Solid

The cohesive energy of a solid refers to the energy required to separate constituent atoms apart from each other and to bring them to an assembly of neutral free atoms. Its value for elements in the periodic table is distributed over a few kJ/mol for inert gas elements to 837 kJ/mol for W having the highest melting point (Kittel, 1967).[1] There are typically three bonding types for a solid: ionic, covalent, and metallic bondings. In an ideally ionic crystal, its cohesive energy can be calculated by summing up the electrostatic energy for an assembly consisting of ions with unlike charges. This is known as the calculation of the Madelung constant (Kittel, 1967). Since ionic bonding originates from electrostatic interaction, the cohesive energy can arise even when overlap of wave functions between the neighboring atoms is absent and, hence, without the formation of the valence band. But, realistic ionic crystals are certainly not in such an ideal state but form the valence band through overlap of wave functions of the outermost electrons on neighboring atoms.

Degenerate orbital levels in the assembly of neutral atoms are lifted and split into the valence band through the orbital hybridization among neighboring atoms upon the formation of a solid. The cohesive energy in both metallic and covalent bonding types is gained by lowering the total energy relative to that of the assembly of free atoms (Mizutani, 2010).[2] In these cases, it is

---

[1] The cohesive energy, total energy, free energy of formation, and enthalpy of formation are expressed in units of kJ/mol or cal/mol or eV/atom. We use units of either eV/atom or kJ/mol (1 eV/atom = 96.44 kJ/mol) in the present chapter.

[2] The cohesive energy is defined relative to the ionization energy and hence represents a quantity positive in sign. The total energy given by Eqn 6 is defined relative to infinity and represents a quantity negative in sign. See more details in Mizutani, (2010).

meaningless to evaluate potential energies classically in the same manner as in ionic crystals discussed above. We need to calculate the total energy of a system in the presence of periodically (or nonperiodically) arranged ionic potentials of a solid. First-principles band calculations are indispensable to evaluate the cohesive energy in a solid, where metallic and covalent bondings are essential.

We will review briefly at this stage the physics of metallic bonding (Ashcroft and Mermin, 1976). We take as a starting point that the free electron gas with negative charges is uniformly distributed in an array of potentials due to periodically arranged positive ions and that the total charge of the electrons is just large enough to cancel that of the ions. The electrostatic energy per atom, when the array of potentials forms a bcc lattice, is reduced to

$$\varepsilon_0 = -\frac{2348}{(r_s/a_0)} \text{ kJ/mol,} \tag{1}$$

where r_s represents the averaged radius of a sphere that each free electron occupies in the space and a_0 is the Bohr radius equal to 0.0529 nm (Ashcroft and Mermin, 1976). In the case of Na, for example, the electrostatic energy amounts to $\varepsilon_0 = -600$ kJ/mol by inserting its appropriate ratio $r_s/a_0 = 3.91$ into Eqn 1.[3]

Let us suppose 1 mol of Na metal contains the Avogadro number of valence electrons. No more than two electrons can occupy the same quantum state due to the Pauli exclusion principle. As a result, the Fermi sphere is formed in the reciprocal space and the kinetic energy is inevitably increased. According to the free electron model, an average kinetic energy per electron is given by

$$\varepsilon_{kin} = \frac{3}{5}E_F = \frac{2903}{(r_s/a_0)^2} \text{ kJ/mol,} \tag{2}$$

where E_F is the Fermi energy (Ashcroft and Mermin, 1976).[4] It amounts to $\varepsilon_{kin} = 190$ kJ/mol for Na metal. The kinetic energy is a quantity with a positive sign. Hence, we see that, the smaller the r_s, the more unfavorable is the metallic bonding. The minimum appears at a particular distance r_s in the total energy given by the sum of Eqns 1 and 2. The system is stabilized, when the total energy is lower than that of the assembly of free atoms. This is the mechanism for metallic bonding. The difference between the two energies concerned is indeed what we call the cohesive energy. The value r_0 at the minimum corresponds to an equilibrium interatomic distance.[5] A variety of metallic phases are stabilized in nature by devising a mechanism unique to an individual system to lower the kinetic energy of electrons as much as possible (it is called the valence-band structure energy).

The cohesive energy in a solid with the covalent bonding is also gained in the same way as that with the metallic bonding and arises from orbital hybridizations of wave functions among constituent atoms. For instance, Si crystallizes in a diamond structure in its solid state, and is known to be typical of a covalently bonded semiconductor with an energy gap of about 1 eV, but it becomes metallic upon melting. Its heat of fusion is 50 kJ/mol and amounts to only 11% of the cohesive energy. This means

---

[3] The value of r_s turns out to be 0.207 nm for Na by inserting its lattice constant $a(= 0.422$ nm) into the relation $\frac{4\pi}{3}r_s^3 = \frac{a^3}{2}$.

[4] The Fermi energy refers to the energy of the highest occupied electronic states at absolute zero. The Fermi level or chemical potential at absolute zero coincides with the Fermi energy.

[5] Rigorously speaking, one needs to take into account the electron–electron interaction properly in the evaluation of the cohesive energy.

that the difference in the cohesive energy between metallic and covalent bondings is small, indicating that no essential difference exists between these two bonding mechanisms and that the energy involved in determining an ultimate structure is most likely at most several tens of kJ/mol or about 10% of the cohesive energy.

More details about the fundamental knowledge on bondings will be found in the literature (Ashcroft and Mermin, 1976; Kittel, 1967; Mizutani, 2001, 2010). Let us now discuss the stability of an alloy phase from the viewpoint of thermodynamic considerations in terms of the free energy of formation, enthalpy of formation, and entropy of formation.

2.2.3   Free Energy of Formation and Enthalpy of Formation

Suppose we prepare a solid solution (x_A, x_B) at absolute temperature T under a constant pressure P by mixing x_A of component A and x_B $(= 1 - x_A)$ of component B in an A–B alloy system. The free energy of formation or free energy of mixing ΔG_m is given by the relation

$$\Delta G_m = \Delta H_m - T\Delta S_m, \tag{3}$$

where ΔH_m is the enthalpy of formation or heat of mixing and ΔS_m is the entropy of formation or entropy of mixing (Swalin, 1972). Instead of the cohesive energy, we discuss now the energy and entropy relative to those of pure metals A or B. In other words, thermodynamic quantities obtained by a composition-weighted average of relevant values of constituent pure elements are taken as a reference state in Eqn 3. An equilibrium phase diagram can be understood by studying both temperature and composition dependences of these thermodynamic quantities. An alloy will be formed by mixing the atomic species A and B in proportion to respective concentrations x_A and x_B, provided that ΔG_m is negative under a given temperature and composition.

Figures 1 and 2 show phase diagrams in Au–Ni and Ni–Pt along with the composition dependences of the corresponding thermodynamic quantities at 1173 K and at 1625 K, respectively (Averbach et al., 1954; Okamoto, 2000). All thermodynamic quantities ΔG_m, ΔH_m and ΔS_m become zero at both

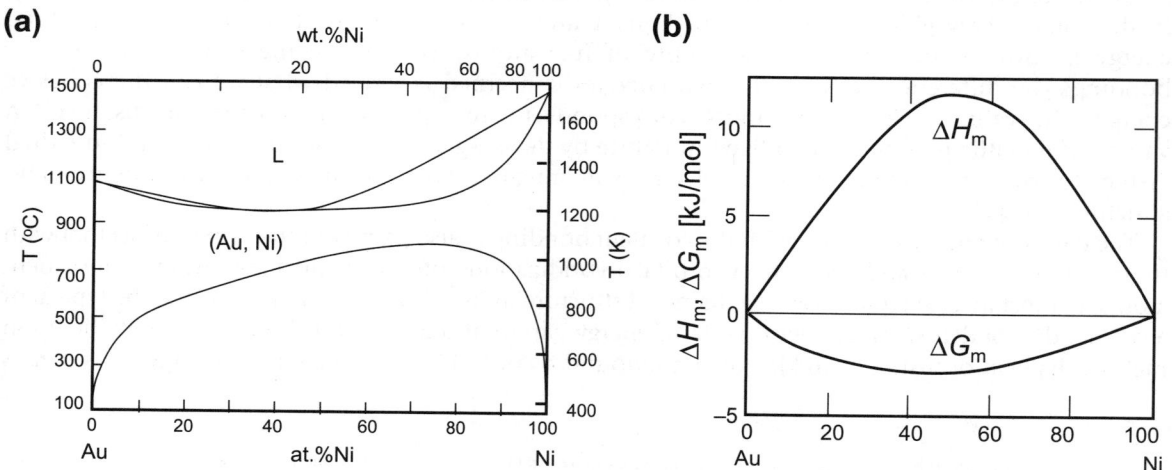

Figure 1   (a) Au–Ni phase diagram (Okamoto, 2000) and (b) heat of mixing ΔH_m and free energy of mixing ΔG_m at 1173 K (Averbach et al., 1954).

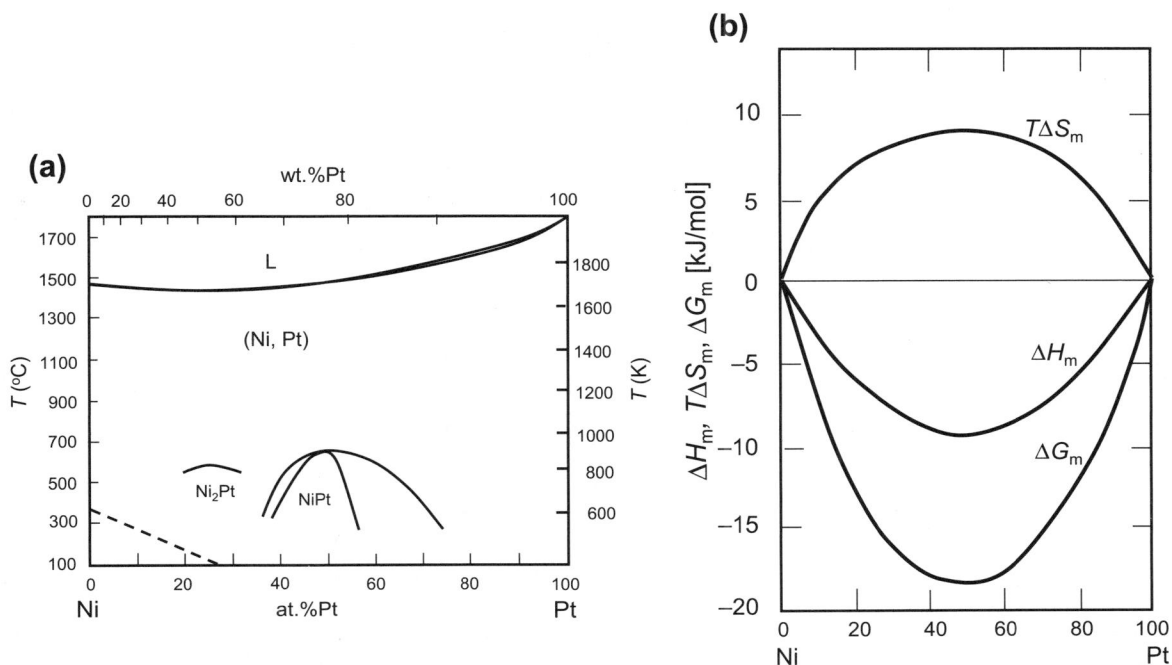

Figure 2 (a) Ni–Pt phase diagram (Okamoto, 2000) and (b) corresponding heat of mixing ΔH_m, entropy of mixing ΔS_m, and free energy of mixing ΔG_m at 1625 K (Swalin, 1972).

ends of the composition axis. The enthalpy of formation ΔH_m is positive in the Au–Ni system. At low temperatures, the entropy term in Eqn 3 is so small that ΔG_m is also positive. Hence, as can be seen in **Figure 1a**, the decomposition into two phases takes place at low temperatures. But a solid solution is formed when $\Delta G_m < 0$ at high temperatures as a result of an increase in the second term $T\Delta S_m$ in Eqn 3. The value of ΔH_m at $x = 0.5$ in **Figure 1b** takes its maximum of 11.7 kJ/mol, being merely 3% of the cohesive energy of pure elements like Au. In contrast, as shown in **Figure 2**, a complete solid solution is formed over a large temperature range in the Ni–Pt alloy system, since $\Delta H_m < 0$.

As far as the stability at absolute zero is concerned, we can discuss it in terms of ΔH_m, since no entropy term exists.[6] The value of ΔH_m has been evaluated by a large number of theoretical and experimental ways and distributed over a wide range covering from only ± a few up to ±100 kJ/mol, depending on the alloy system chosen. Miedema et al. (de Boer et al., 1988) examined binary phase diagrams consisting of two transition metal (TM) elements and pointed out that the number of intermetallic compounds existing in the resulting phase diagram increases from zero to five with increasing the heat of mixing of a 1:1 stoichiometric compound in a negative direction. As is clearly shown in **Figure 3**, we see that, the more negative the enthalpy of formation of the 1:1 compound, the more intermediate phases can exist (de Boer et al., 1988). **Figure 4** clearly demonstrates how the

---

[6] In principle, the formation of an alloy is possible at absolute zero when $\Delta H_m < 0$. According to the third law of thermodynamics, however, entropy, including the configuration entropy, should vanish at absolute zero. Hence, any alloy possessing chemical disorder or structural disorder is metastable at $T = 0$.

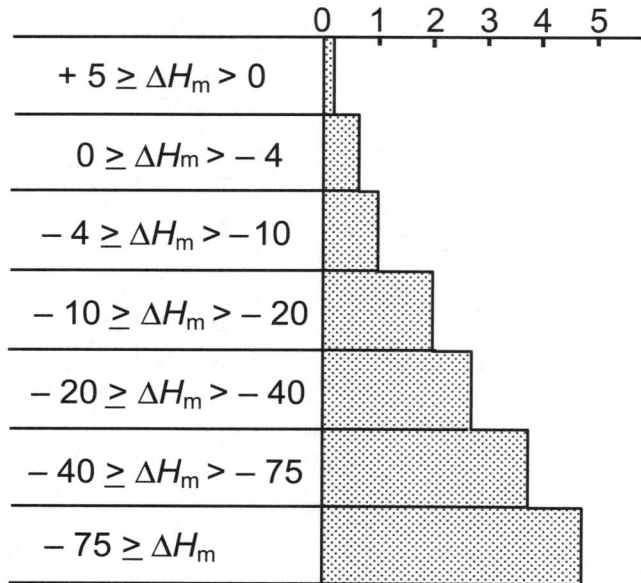

Figure 3   The number of intermetallic compounds existing in the phase diagram increases with increasing the heat of mixing ΔH_m to form an AB intermetallic compound in an A–B inter-TM alloy system (de Boer et al., 1988).

number of intermediate phases increases from one to three and to five, with increasing the heat of mixing in a negative direction as $\Delta H_m = 0$, -37 and -75 kJ/mol at $x_B = 0.5$. The most important message in this argument is that a difference in the heat of mixing amounts to only 5–20 kJ/mol, when several intermediate phases are competing with each other in a given alloy system, indicating that an energy difference involved in the competition between the two neighboring phases is fairly small.

2.2.4   Kinetic Energy of Electrons and the Role of the Pseudogap

The average kinetic energy per electron for pure Cu becomes $\varepsilon_{kin} = 405$ kJ/mol, if its Fermi energy of 7.0 eV is inserted into Eqn 2. Instead, a rough estimate of the potential energy using Eqn 1 amounts to -880 kJ/mol for pure Cu. Hence, we see that the kinetic energy of electrons should play a critical role in the cohesive energy. Since the kinetic energy, positive in sign, acts against gaining the cohesive energy, nature often devises a mechanism to lower the kinetic energy of a given system as much as possible.

We are ready to discuss the mechanism to lower the kinetic energy of electrons by forming a pseudogap across the Fermi level. We consider it most important to learn how much energy can be lowered through the formation of a pseudogap at the Fermi level. To grasp its essence, we assume the valence band to be approximated by a density of states (DOS) of a rectangle, as shown in **Figure 5**a (Mizutani, 2006a). The Fermi level E_F and the DOS at the Fermi level, $N(E_F)$, are set to be 7.0 eV and $N(E_F) = 0.21$ states/eV.atom appropriate to pure Cu in the free electron model, respectively. First-principles band calculations done for various approximants have revealed that the depth of a pseudogap at the Fermi level is generally about 20–60% that of the free electron DOS and that the width of the pseudogap is extended over the range $\Delta E = 500–1500$ meV. Let us assume that a pseudogap with the height H is formed at the Fermi level on the rectangular DOS with the height of H_0, as illustrated in **Figure 5**b. In order to create a pseudogap, we have to deplete electrons equal to $0.21 \times (1 - H/H_0) \cdot \Delta E$ states/atom

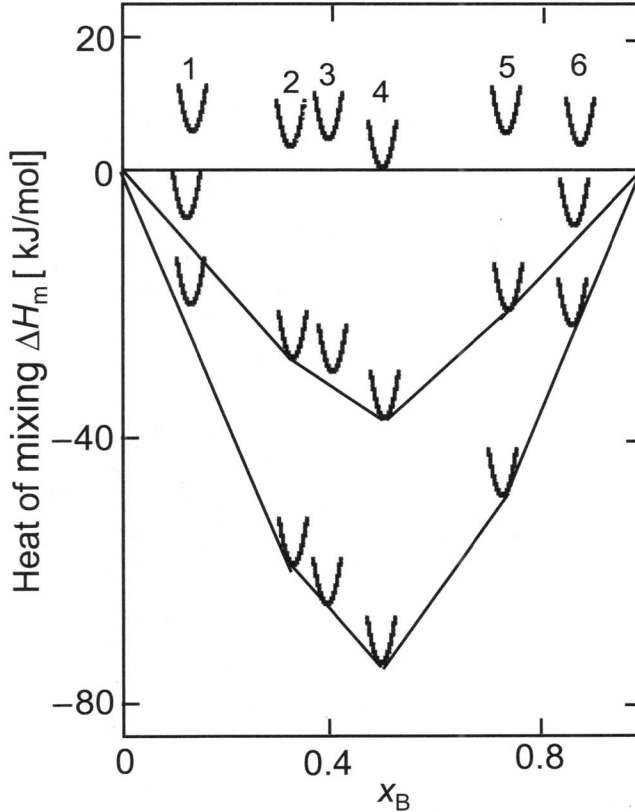

Figure 4 Composition dependence of heat of mixing ΔH_m in an A–B alloy system. The number of intermediate phases increases from unity up to five, as ΔH_m increases in a negative direction from zero, -37 to -75 kJ/mol (de Boer et al., 1988).

from the area (A). For the sake of simplicity, the depleted electrons are uniformly redistributed into the DOS below the pseudogap, as marked by the symbol (B). A resulting gain in the kinetic energy is calculated by using the relation shown in the footnote.[7]

As shown in **Figure 6**, the electronic energy is lowered by 30–60 kJ/mol, if a pseudogap 500–1000 meV wide is formed across the Fermi level. This is large enough to stabilize one phase relative to competing phase(s). However, if a pseudogap is formed near the bottom of the valence band, the reduction in the electronic energy would be limited to only a few kJ/mol. As is clear from the argument above, the electronic energy is most effectively lowered, when a pseudogap is formed at the Fermi level.

Before ending this section, several cases, in which a system is stabilized by forming either a true gap or a pseudogap across the Fermi level, are exemplified. Different mechanisms act to generate a pseudogap below.

[7] A reduction in the electronic energy ΔU is calculated by using the relation: $\Delta U = \int_0^{7-\Delta E} \{N_0(E) + \Delta N(E)\}EdE + \int_{7-\Delta E}^{7} N_{pg}(E)EdE - \int_0^{7} N_0(E)EdE$, where $N_0(E) = H_0 = 0.21$ states/eV.atom, $\Delta N(E)$ is an increment in the DOS due to transfer of depleted electrons (marked with B) and $N_{pg}(E) = 0.21(H/H_0)$ states/eV.atom. A simple manipulation leads to the relation ΔU [kJ/mol] $= 70.88\ \Delta E$ [eV] $\times (1 - (H/H_0))$.

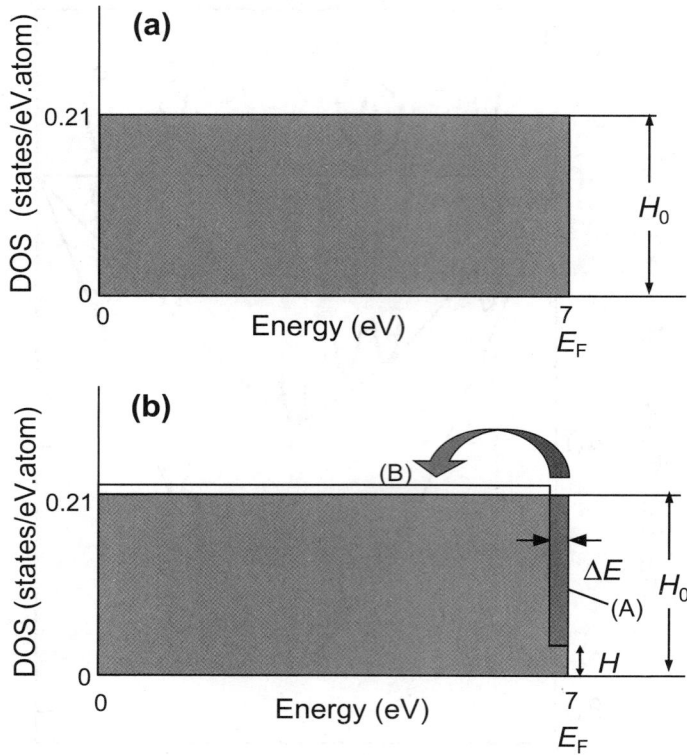

Figure 5 (a) DOS of Cu approximated by a rectangle with a height H_0. Free electron values are assigned to the Fermi level and the height of DOS at the Fermi level (Mizutani, 2006a). As a result, 1.4 electrons per atom are accommodated below the Fermi level. (b) A pseudogap with ΔE in width and H in height is created at the Fermi level. Depleted electrons in (A) are assumed to be uniformly redistributed over a whole valence band, as marked with (B).

1. The presence of four outermost electrons in the free atom of Si and Ge takes its full advantage of forming the covalent bonding upon solidification into a diamond structure. Namely, half-filled outermost electrons around a given atom tend to share the orbital with those of neighboring atoms. The resulting orbital hybridizations split the band into bonding and antibonding states and allow electrons to fill only the bonding states, resulting in an energy gap of about 1 eV across the Fermi level in Si. The formation of a true gap contributes to lowering the electronic energy and stabilizing the diamond structure. As mentioned above, a gain in the cohesive energy relative to that of liquid Si in metallic state is about 50 kJ/mol. Si can be regarded as being typical of an orbital hybridization-induced energy gap system. We will learn in the next section that a pseudogap can be also opened through the Fermi surface–Brillouin zone interactions particularly in structurally complex metallic alloy phases.

2. Some metals like Al, V, and Pb in the periodic table undergo the superconducting state at low temperatures by opening an energy gap of 0.5–1 meV at the Fermi level. This is brought about by the formation of the Cooper pair electrons. The superconducting energy gap is much smaller than the gap shown in **Figure** 5. According to the BCS theory, a gain in the cohesive energy for Al upon

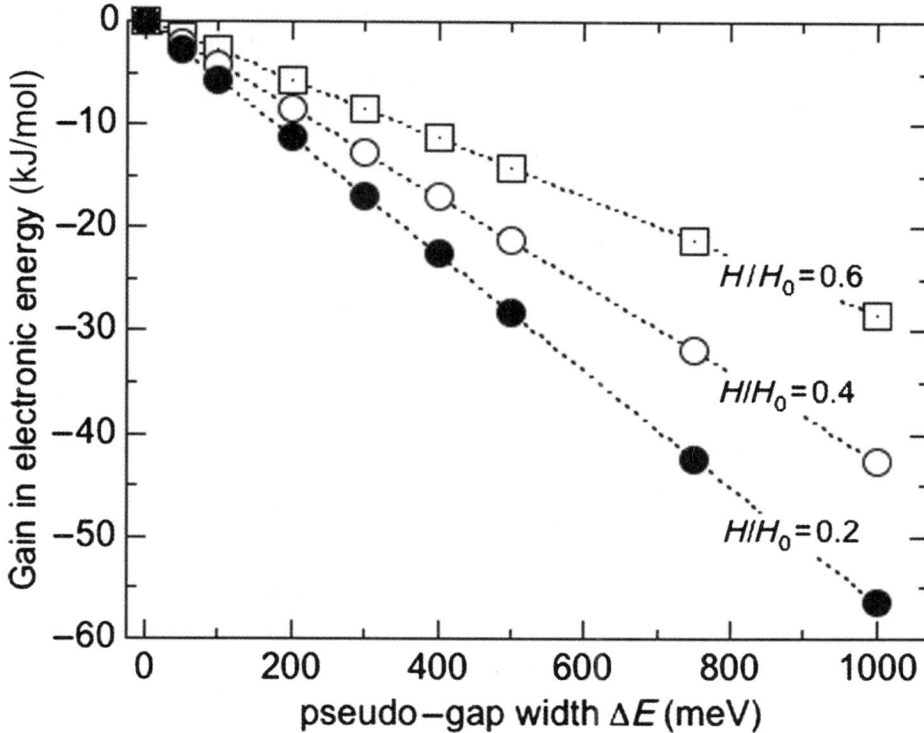

Figure 6 A gain in the electronic energy in the rectangle DOS model for Cu upon the formation of the pseudogap at the Fermi level under the conditions $H/H_0 = 0.2$, 0.4, and 0.6 (**Figure 5**b) as a function of the width of the pseudogap ΔE (Mizutani, 2006a).

superconducting transition amounts to $\dfrac{(k_B T_c)^2}{E_F} = 2.4 \times 10^{-9}$ eV/electron or 7.7×10^{-8} kJ/mol (Mizutani, 2001). A small energy involved upon transformation explains why the superconducting transition temperature is generally very low, for example, 1.19 K for pure Al. See more fundamental knowledge about superconductivity in Mizutani (2001).

3. A highly anisotropic organic molecular metal characterized as a pseudo-one-dimensional conductor is known to be stabilized through the so-called Peierls transition (Gruner, 1994). Charge-density waves or spin-density waves are excited by forming a periodic modulation through lattice deformation. As a result, a pseudogap is formed across the Fermi level and contributes to lowering the electronic energy. For example, $K_2Pt(CN)_4Br_{0.3} \cdot 3.2H_2O$ inorganic compound known as KCP opens a gap of 100 meV at the transition temperature of 189 K (Gruner, 1994).

2.3 Structure of Complex Metallic Alloys

2.3.1 What are Structurally Complex Metallic Alloys?

The structurally complex metallic alloys (CMAs) refer to a class of intermetallic compounds characterized by the possession of giant unit cells ranging from some tens up to >1000 atoms with well-defined atom clusters. Quasicrystals are included in this family as its extreme limit, since they are highly ordered but the most complex intermetallic compounds characterized by an infinitely large unit cell. Many CMAs possess a solid solution range so that the composition can be varied within a single-phase field while some are stable only as line compounds.

The aim of the present section is to help readers build up their own views of what quasicrystals and their approximants are and what other CMAs existing in equilibrium binary phase diagrams are.

2.3.2 Quasicrystals and Their Approximants

Many thermally stable quasicrystals were synthesized in the late 1980s through early 1990s following the discovery of the metastable Al–Mn quasicrystal by Shechtman et al. (1984). Since then, a large number of theoretical and experimental works have been accumulated to deepen our understanding of how atoms are built up to form a quasicrystal, why they can exist as a stable phase, and whether observed physical and chemical properties are different from those of ordinary crystals and are inherent to the quasiperiodicity of the lattice. The structure can be best described in terms of a six- or five-dimensional hypercubic lattice in the framework of the so-called cut-and-project method (Janot, 1994; Dubois, 2005). In the case of an icosahedral quasicrystal, an appropriate cut of a periodic density distribution $\rho_6(\mathbf{r})$ in the six-dimensional model space R_6 by the three-dimensional physical space $R_{3//}$ can generate a set of atom positions, provided that $\rho_6(\mathbf{r})$ is flat without any thickness in $R_{3//}$ and, hence, is completely embedded in the three-dimensional perpendicular space $R_{3\perp}$. These three-dimensional objects in the space $R_{3\perp}$ are called the atomic surfaces (ASs).

To visualize the situation, a square lattice is chosen as a hypercubic lattice in the two-dimensional space R_2 and the straight line segment with a length Δ is taken as the object AS, which is embedded in the space $R_{1\perp}$ and is perpendicular to the physical space $R_{1//}$, as shown in **Figure 7**. The position of $R_{1//}$ in the space R_2 is fixed by assigning its angle θ relative to the square lattice in R_2. The one-dimensional quasilattice or the Fibonacci chain[8] is created, if θ is set equal to $\tan^{-1}(1/\tau)$, where τ is the golden mean given by $\tau = \dfrac{\sqrt{5}+1}{2} \approx 1.6180339887\ldots$. Similarly, an icosahedral quasilattice is generated in $R_{3//}$, if $R_{3//}$ is fixed by the angle θ relative to R_6. Likewise, approximant lattices of different orders can be constructed by rationalizing the angle so as to conform with any Fibonacci ratio, $1/0$, $1/1$, $2/1$, $3/2$, $5/3$, $8/5\ldots\tau$.

A quasilattice is thus a structure obtained by setting an angle to $\tan^{-1}(1/\tau)$ whereas an approximant lattice is obtained by substituting the rationalized Fibonacci ratio for $1/\tau$. Both quasi and approximant lattices thus constructed are a mathematical product of the cut-and-project method from higher dimensions to three- or two-dimensional physical space. Real quasicrystals and their approximants can be obtained by decorating the resulting lattices with atoms. A cubic approximant can be obtained by assigning the same rationalized Fibonacci ratios for $\tan \theta_i$, where θ_i ($i = x$, y, and z) is an angle between the three principal x-, y-, and z-axes in the $R_{3//}$ space relative to those in the six-dimensional space R_6. All the approximants characterized by different Fibonacci ratios are, therefore, grouped into CMAs with a well-defined unit cell.

---

[8] A Fibonacci chain can be generated by an algorithm such that a segment of length L on a straight line is subdivided into a long segment L' and short segment S', inflate L' = L and S' = S, replace L by L + S and S by L, and so on. In its first generation, the set of LS is created from the 0th generation of the parent L. In the second, the third, fourth,... generations, sets of LSL, LSLLS, LSLLSLSL, ... appear. The number ratio of L over S is $1/0$, $1/1$, $2/1$, $3/2$, $5/3$, $8/5\ldots$ and eventually converges to an irrational number $\tau = 1.618 \ldots$, the golden mean τ. This is called the Fibonacci chain. The nth-order approximants can be constructed by terminating the operation at the nth generation.

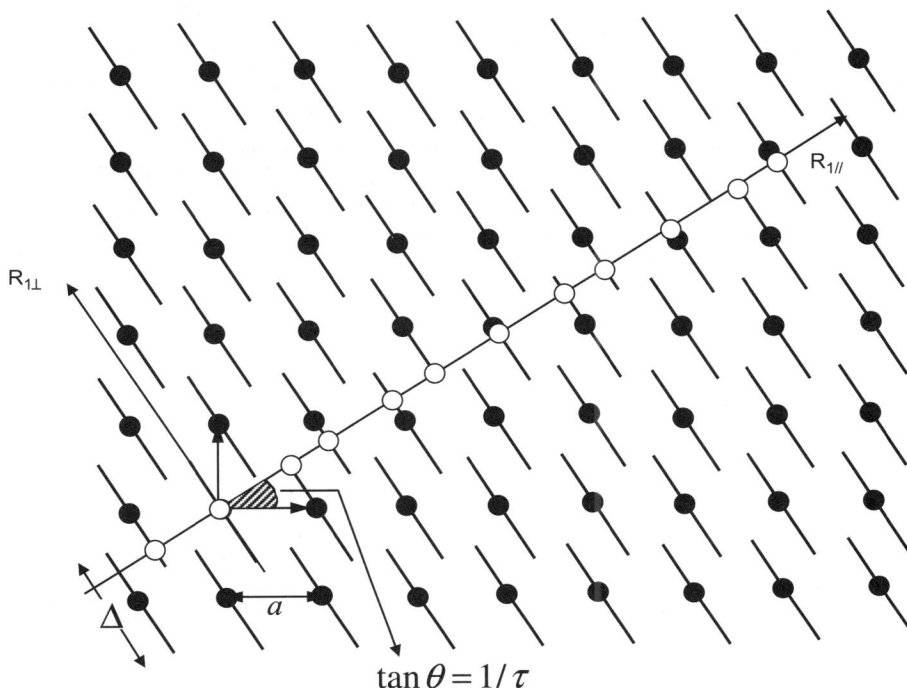

Figure 7 Construction of one-dimensional quasicrystal or Fibonacci lattice by the cut-and-project method from a two-dimensional square lattice (Mizutani, 2010). An angle θ relative to the two-dimensional square lattice in R_2 is chosen to be $\tan^{-1}(1/\tau)$, where $\tau = \dfrac{1 + \sqrt{5}}{2}$.

The structure of the 1/1–1/1–1/1 approximant is achieved by fixing $\tan\theta_x = \tan\theta_y = \tan\theta_z = 1/1$. A number of 1/1–1/1–1/1 approximants have been reported to exist as stable phases. For example, the $Al_xMg_{39.5}Zn_{60.5-x}$ ($20.5 \leq x \leq 50.5$) compounds known as the Bergman phase were identified as 1/1–1/1–1/1 approximant containing 160 atoms in its unit cell with space group $Im\bar{3}$ (Mizutani et al., 1997; Sun et al., 2000). Similarly, the $Al_{15}Mg_{43}Zn_{42}$ compound is identified as the 2/1–2/1–2/1 approximant with space group $Pa3$ and contains as many as 680 atoms in its cubic unit cell with the lattice constant of 2.303 nm (Takeuchi and Mizutani, 1995; Sugiyama et al., 2002; Lin and Corbett, 2006; Kreiner, 2009; see more details in Section 2.7.2.3). The family of 1/1–1/1–1/1 approximants can be handled in first-principles band calculations and has been chosen to comprehensively understand the stabilization mechanism of CMAs. However, the approximants having orders higher than 2/1–2/1–2/1 are at the moment too complex to perform first-principles band calculations. It must be kept in mind that even 1/1–1/1–1/1 approximants generally contain chemical disorder so that a model structure free from such disorder has to be constructed in order to make first-principles band calculations possible.

Icosahedral quasicrystals and their approximants can be classified according to atomic clusters building up its structure and the structure type (Mizutani, 2010). Three different atomic cluster types are known to exist in the literature: the first described is the rhombic triacontahedron (RT) containing 44 atoms in the cluster, the second is the Mackay icosahedron (MI) containing 54 atoms in the cluster, and the third is the Tsai-type cluster. The first one is denoted as the RT-type or the Frank Kasper-type cluster, since the Al–Mg–Zn 1/1–1/1–1/1 approximant mentioned above has been also referred to as the Frank-Kasper compound (Bergman et al., 1957). The RT-type cluster is found in systems like Al–Mg–X (X = Zn, Ag, Cu, and Pd) and Al–Li–Cu quasicrystals and their approximants (Mizutani et al., 2002).[9] The MI-type cluster involves a TM element as one of the major constituent elements. The MI-type cluster is found in systems composed of Al and TM elements like in Al–Mn, Al–Cu–Fe, Al–Cu–Ru, and Al–Pd–Re quasicrystals and their approximants. Tsai et al. (2000) discovered quasicrystals in the Cd–Yb alloy system and pointed out the possession of atomic clusters, which are different from the RT- and the MI-type clusters, as described below.

The successive shell structure of the RT-, MI-, and Tsai-type atomic clusters is illustrated in **Figures 8–10**, respectively (Mizutani, 2010). In both RT- and MI-type clusters, 12 atoms are located

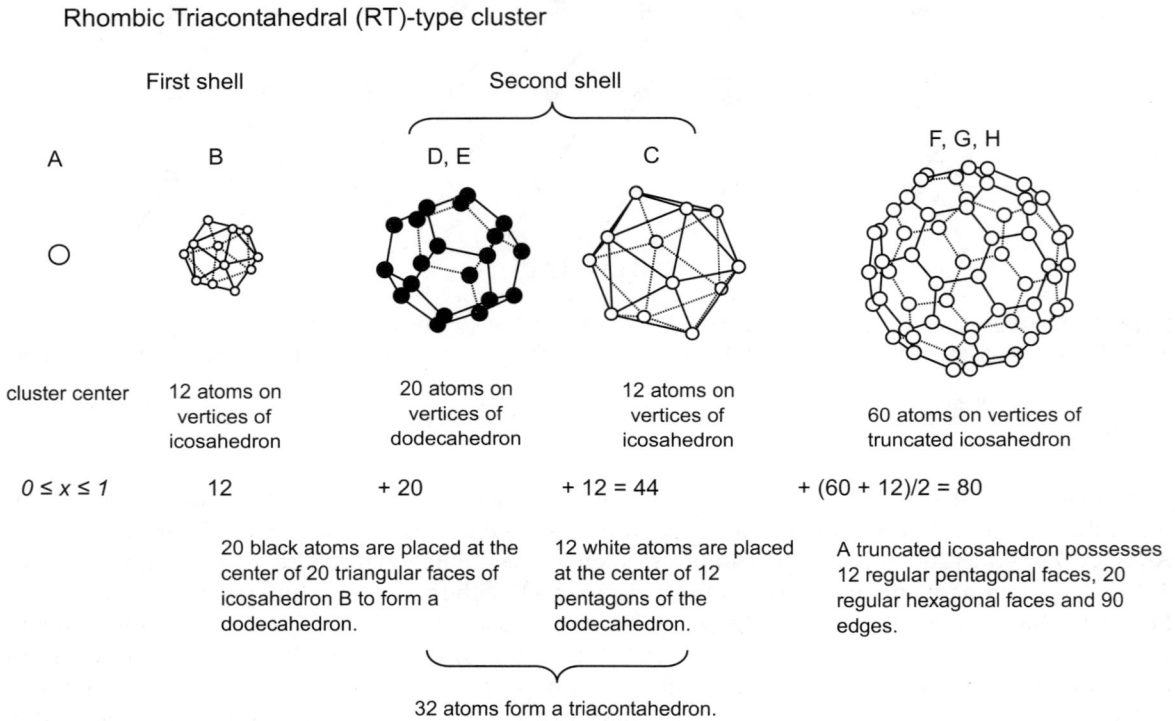

Rhombic Triacontahedral (RT)-type cluster

	First shell	Second shell		F, G, H
A	B	D, E	C	
cluster center	12 atoms on vertices of icosahedron	20 atoms on vertices of dodecahedron	12 atoms on vertices of icosahedron	60 atoms on vertices of truncated icosahedron
$0 \leq x \leq 1$	12	+ 20	+ 12 = 44	+ (60 + 12)/2 = 80
		20 black atoms are placed at the center of 20 triangular faces of icosahedron B to form a dodecahedron.	12 white atoms are placed at the center of 12 pentagons of the dodecahedron.	A truncated icosahedron possesses 12 regular pentagonal faces, 20 regular hexagonal faces and 90 edges.

32 atoms form a triacontahedron.

Figure 8   Successive shell structures of atoms in the atomic cluster of the RT-type quasicrystals and their approximants. Solid circles: largest atom like Mg and Li in Al–Mg–Zn and Al–Li–Cu, respectively (Mizutani, 2010). The assembly consisting of the first and second shells is called the RT cluster. Atoms on the sites F, G, H are shared by two neighboring Wigner–Seitz cells.

---

[9] Following the pioneering work by Bergman et al. (Bergman et al., 1957), the $Mg_{32}(Al, Zn)_{49}$ compound has been often referred to as possessing 162 atoms per unit cell, though the occupancy at the center of the cluster is less than unity, that is, 0.8 in their **Table 1**. Mizutani et al. (Mizutani et al., 1997) claimed from their powder diffraction Rietveld analysis for a series of $Al_xMg_{39.5}Zn_{60.5-x}$ 1/1–1/1–1/1 approximants (20.5 $\leq x \leq$ 50.5) that the fractional occupancies at the center of the cluster is at most 0.1 at 20.5 at.% Al and is decreased to zero when Al concentration exceeds 30 at.%. Hence, the total number of atoms in the unit cell is closer to 160.

around a given lattice site to form an icosahedron and constitute the first shell. The center of the icosahedron thus formed is fully or partially filled with an atom or more frequently completely vacant. In the Tsai-type atomic cluster, the first shell is a tetrahedron composed of three or four Cd or Zn atoms at its vertices around the cluster center, as shown in **Figure 10**. It is noteworthy that the presence of this unique tetrahedral unit internally breaks the icosahedral symmetry (Tsai et al., 2000).

There exist 20 triangular faces on an icosahedron. In the case of the RT-type cluster, a dodecahedron (see sites D and E in **Figure 8**) is formed by locating 20 larger atoms like Mg at the center of the triangular face of the icosahedron "B" of the first shell. The center of 12 pentagonal faces on the resulting dodecahedron is then filled with smaller atoms Al or X = Zn, Cu, Ag, and Pd in the case of the Al–Mg–X system. Totally 32 atoms constitute the RT as the second shell. The RT-type atomic cluster is, therefore, composed of 44 atoms including 12 atoms in the first shell and 32 atoms in the second shell. This is a characteristic feature of the atomic cluster found in both the RT-type quasicrystals and approximants.

The shell structure of the MI-type cluster is shown in **Figure 9**. Twelve atoms are located on vertices of an inner icosahedron (II) as the first shell. We put further 12 atoms at atomic sites immediately above the 12 atoms on the II. It forms a larger icosahedron called TM, since all the 12 vertices are filled with the TM element like Fe. Thirty midedge sites on the larger icosahedron TM are then filled with a mixture of atoms like Al or Cu atoms, which forms an icosidodecahedron. This constitutes the second shell together with atoms on the larger icosahedron TM. An assembly of these 54 atoms is called the MI-type cluster, as shown in **Figure 9**.

Mackay Icosahedral (MI)-type cluster

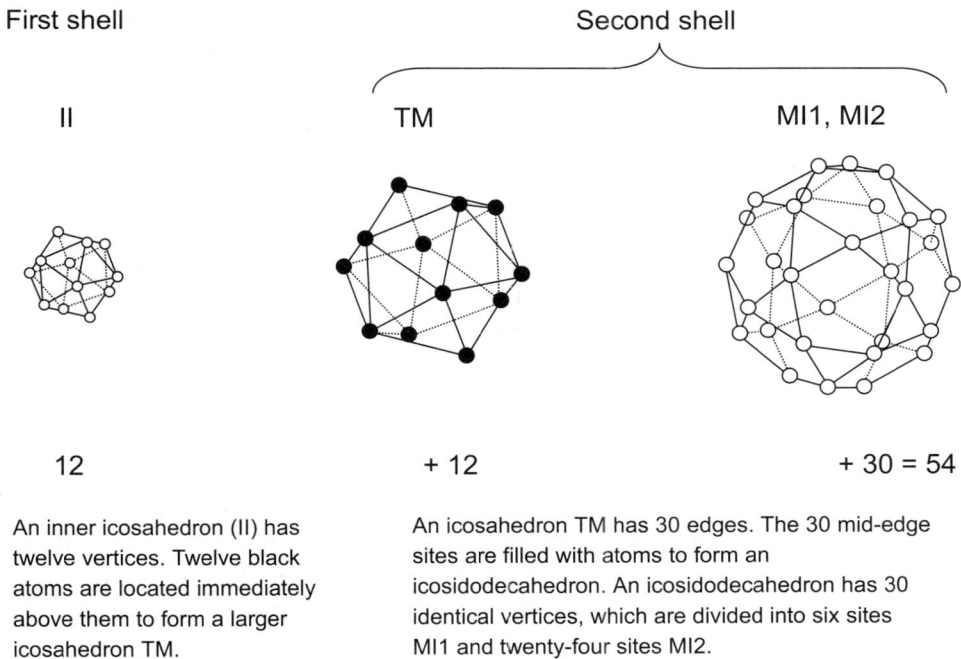

First shell	Second shell	
II	TM	MI1, MI2

12	+ 12	+ 30 = 54

An inner icosahedron (II) has twelve vertices. Twelve black atoms are located immediately above them to form a larger icosahedron TM.

An icosahedron TM has 30 edges. The 30 mid-edge sites are filled with atoms to form an icosidodecahedron. An icosidodecahedron has 30 identical vertices, which are divided into six sites MI1 and twenty-four sites MI2.

Figure 9 Successive shell structures of atoms in the atomic cluster of the MI-type quasicrystals and their approximants (Mizutani, 2010). Solid circles: the TM atoms like Fe.

Tsai-type atomic cluster

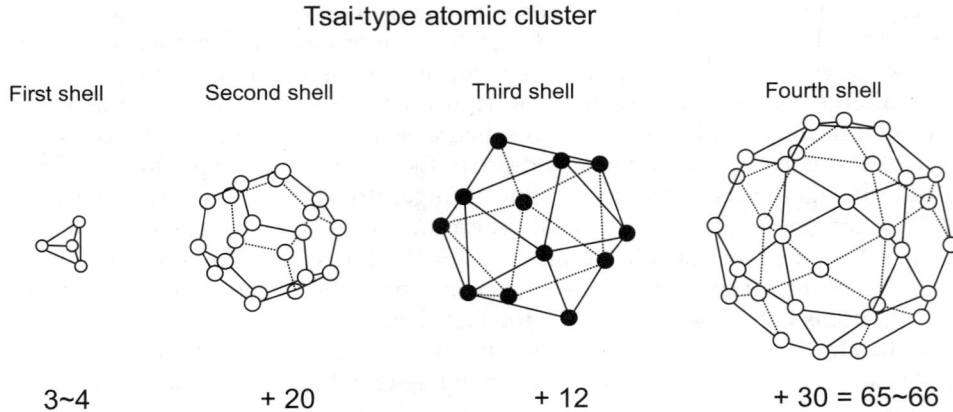

First shell	Second shell	Third shell	Fourth shell
3~4	+ 20	+ 12	+ 30 = 65~66

Figure 10   Successive shell structures of atoms in the atomic cluster of the Tsai-type quasicrystals and their approximants (Mizutani, 2010). Open circles: Cd or Zn, solid circles: Yb or Sc.

In the case of Cd_6Yb and Zn_6Sc approximants characterized by the Tsai-type cluster (Tsai et al., 2000), the second shell is represented by a dodecahedron with 20 Cd (Zn) atoms at its vertices, the third shell by an icosahedron with 12 Yb (Sc) atoms, and the fourth shell is a Cd (Zn) icosidodecahedron obtained by placing 30 Cd (Zn) atoms on the midedges of the Yb (Sc) icosahedron. In total, the icosahedral cluster consists of 65–66 atoms, as shown in **Figure 10**. In an icosahedral quasicrystal, the atomic cluster shown in **Figures 8–10** is distributed so as to satisfy an overall icosahedral symmetry on a quasiperiodic lattice. The structure of both RT- and MI-type clusters is characterized by the $m\overline{3}5$ symmetry but a slight distortion is always observed in 1/1–1/1–1/1 approximants. The RT-, MI-, and Tsai-type atomic clusters are located on the body center and corner of a cubic lattice to make a bcc packing in the 1/1–1/1–1/1 approximant, as described below.

The formation of a cubic unit cell may be explained by using the RT-type 1/1–1/1–1/1 approximant (Bergman et al., 1957). The triacontahedron in the RT-type cluster possesses 30 rhombic faces, each of which can be divided into two regular triangles, ending up with sixty triangles. Sixty atoms can be placed out from the centers of the 60 triangles on the faces of the triacontahedron to form a truncated icosahedron (see sites F, G and H in **Figure 8**). The truncated icosahedron has 60 vertices, 32 faces, and 90 edges. By adding further 12 atoms above the center of 12 regular pentagonal faces, one can form a truncated octahedron (OH) consisting of totally 72 atoms. Remember that a truncated OH is the Wigner–Seitz cell of a bcc structure (Mizutani, 2001). It can fill space without any overlap and/or void, when packed together in a body-centered cubic lattice, and assure the possession of a periodic lattice.

Figure 11 illustrates how atoms are grown into clusters and clusters into unit cells for both RT- and MI-type 1/1–1/1–1/1 approximants (Mizutani et al., 2002). For example, the $Al_{40.5}Mg_{39.5}Zn_{20}$ 1/1–1/1–1/1 approximant contains 160 atoms in the unit cell with the lattice constant of 1.4443 nm (Mizutani et al., 1997). Remember that 72 atoms on the truncated OH are shared with their neighboring ones so that 36 atoms together with 44 atoms in the RT cluster belong to each truncated OH. Therefore, we can say that each truncated OH contains 80 atoms and forms a bcc lattice in the RT-type 1/1–1/1–1/1 approximant.

The icosahedral quasicrystals and their approximants must be also distinguished according to the types of a quasilattice or approximant lattice involved. As shown below, we need to consider two

(a)

Icosahedron
of 12 atoms

Dodecahedron
of 20 Mg atoms

Rhombic Triacontahedron
or RT-cluster of 44 atoms

(b)   II   TM   MI1   MI2

Icosahedron
of 12 atoms

Larger Icosahedron
of 12 TM atoms

Mackay Icosahedron or
MI-cluster of 54 atoms

(c)

G1
G2

RT-type   MI-type

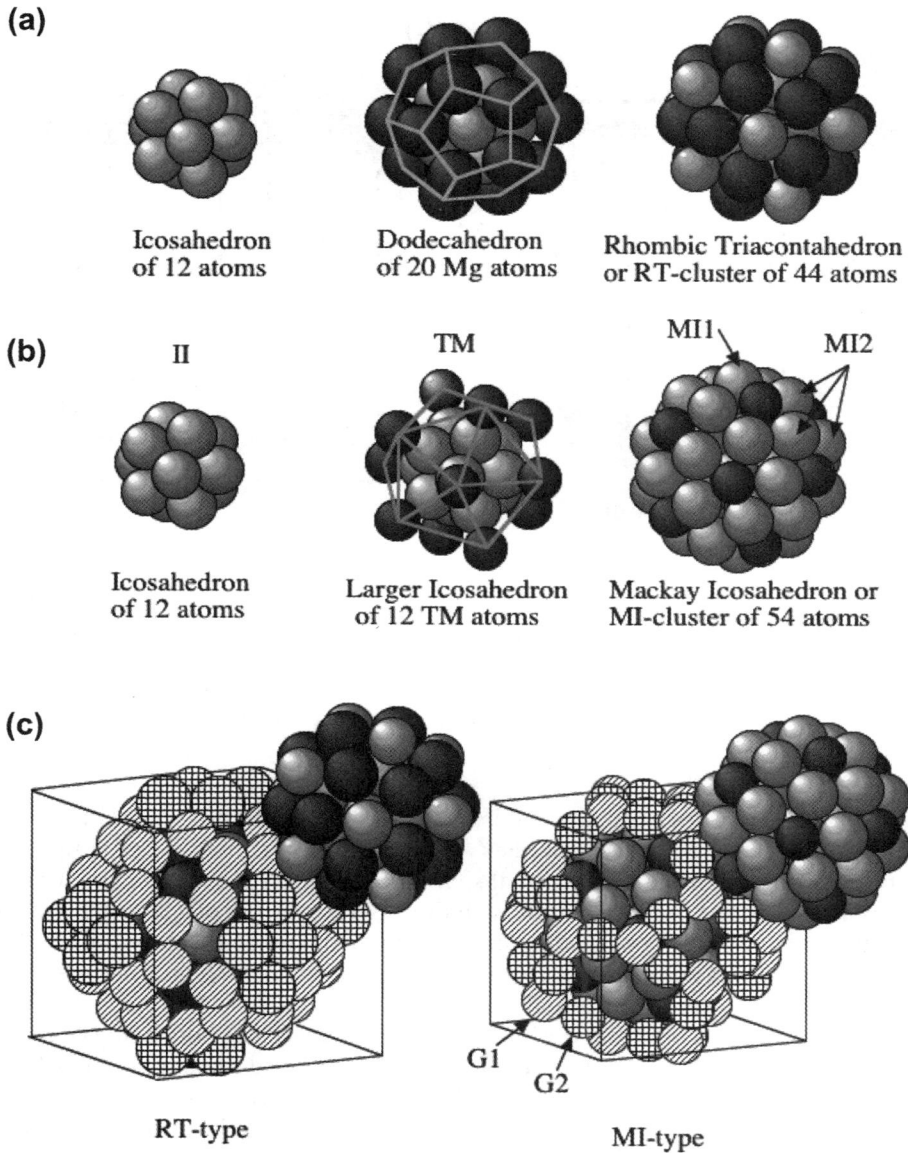

Figure 11   (a) RT- and (b) MI-type atomic clusters, and (c) unit cells of the RT- and MI-type 1/1–1/1–1/1 approximants (Mizutani, 2010). G1 and G2 refer to glue atoms.

different quasilattices of P- and F-types in real quasicrystals. In order to introduce two chemically different atomic clusters in an icosahedral quasicrystal, two families of lattice nodes, "+" or "−", are assigned for a simple cubic lattice in a six-dimensional space. Namely, the parity of either "+" or "−" is assigned, depending on whether the sum of the six corresponding coordinates is even or odd

(a)

(b)

F-type structure

P-type structure due to chemical disorder

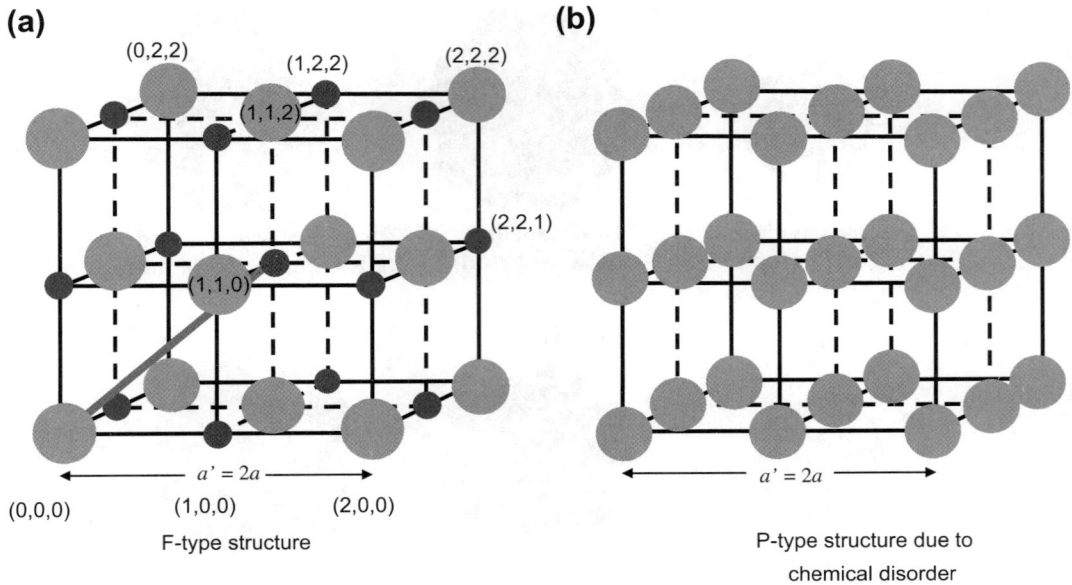

Figure 12   (a) F-type structure referring to the unit cell of NaCl (Mizutani, 2010). Small atoms are positioned at odd lattice nodes, where the sum of three coordinates is odd, while large atoms are at even nodes, where the sum of the coordinates is even. (b) P-type structure. Small and large atoms are not differentiated by X-ray diffraction measurements, when they are randomly distributed. (For color version of this figure, the reader is referred to the online version of this book.)

(Boudard et al., 1992). The superstructure can be generated by small differences in shapes, volumes, and chemical species in the AS located at sites "+" and "−". For the sake of simplicity, consider sodium chloride in ordinary three-dimensional space. As shown in **Figure 12**a (Mizutani, 2010), its Bravais lattice is a face-centered (F-type) cubic with the basis consisting of a large Cl atom and a small Na atom separated by one-half the body diagonal of a unit cube of lattice constant $a' = 2a$. As will be described below, this would help us understand the assignment of two families of lattice nodes in a simple cubic lattice in the six-dimensional space.

All indices in reflection lines observed for NaCl must be either even or odd, being characteristic of an fcc lattice with the lattice constant a'. However, the structure is reduced to a simple cubic lattice (P-type) of lattice constant a, provided that the two atoms happen to possess equal numbers of electrons like K^+ and Cl^- in KCl or two different atomic species cannot be differentiated because of their random distributions over the lattice. This is illustrated in **Figure 12**b. Now the crystal looks to X-rays as if it were a monatomic simple cubic lattice of lattice constant a. In other words, only even integers occur in the reflection indices when indexed with respect to a cubic lattice of lattice constant a'. They are called fundamental reflections. In the case of NaCl (F-type), all reflections of the fcc lattice of lattice constant a' must be present including weak lines with odd reflection indices, which are called superlattice reflections. In the case of the six-dimensional Bravais lattice discussed above, only fundamental reflections are observed in P-type quasicrystals, whereas superlattice reflections are additionally observed in F-type quasicrystals, which signifies the presence of two different atomic clusters arranged quasiperiodically.

Table 1 Classification of 1/1–1/1–1/1 approximants in terms of the type of lattice and the type of atomic cluster

Type of lattice	Type of atomic cluster	1/1–1/1–1/1 Approximant	Space group	Features	References
P	RT	$Al_xMg_{40}X_{60-x}$ (X = Zn, Cu, Ag, and Pd)	$Im\bar{3}$	Two identical clusters; chemical disorder in Al–X	(Mizutani et al., 1997; Takeuchi et al., 2000)
P	MI	$Al_{75}(Mn_{1-x}Fe_x)_{17}Si_8$ $(0.32 \leq x \leq 0.72)$	$Im\bar{3}$	Two identical clusters; chemical disorder in Mn–Fe and Al–Si	(Takeuchi et al., 2001)
P	MI	$Al_{68}Cu_7(Fe_{1-x}Ru_x)_{17}Si_8$ $(x = 0 \leq x \leq 1)$	$Im\bar{3}$	Two identical clusters; chemical disorder in Fe–Ru and Al–Si	(Takeuchi and Mizutani, 2002)
P	MI	$Al_{75}(Mn_{1-x}Fe_x)_{17}Si_8$ $(0 \leq x \leq 0.29)$	$Pm\bar{3}$	Two identical clusters; chemical disorder in Al–Si and Mn–Fe	(Cooper and Robinson, 1966; Takeuchi et al., 2001; Sugiyama et al., 1998)
F	MI	$Al_{54}Cu_{25.5}Fe_{12.5}Si_8$ $Al_{54}Cu_{25.5}Ru_{12.5}Si_8$	$Pm\bar{3}$	Two different clusters; chemical and structural disorder	(Takeuchi and Mizutani, 2002; Yamada et al., 1999)
P	Tsai	Cd_6M (M = Nd, Sm, Gd, Dy,Yb, Ca, Y)	$Im\bar{3}$	Two identical clusters; innermost disordered Cd tetrahedra	(Gomez and Lidin, 2003)
P	Tsai	Zn_6Sc	$Im\bar{3}$	Two identical clusters: innermost disordered Zn tetrahedra	(Lin and Corbett, 2004)

The argument above is extended to the generation of the two types of 1/1–1/1–1/1 approximants from the fcc lattice (F-type) in the six-dimensional space. The P-type 1/1–1/1–1/1 approximant is generated by assigning only the even lattice node with the lattice constant a' in the six-dimensional space. The F-type 1/1–1/1–1/1 approximant is likewise generated by assigning both even and odd lattice nodes with the lattice constant a'. The latter gives rise to two different atomic clusters, at the center and the corner of the cubic lattice, in the three-dimensional physical space.

The 1/1–1/1–1/1 approximants are thus classified by the type of lattice in six-dimensional space and the type of the atomic cluster in three-dimensional space. Typical examples are listed in **Table 1** (Cooper and Robinson, 1966; Gomez and Lidin, 2003; Lin and Corbett, 2004; Mizutani et al., 1997; Sugiyama et al., 1998; Sun et al., 2000; Takeuchi and Mizutani, 2002; Takeuchi et al., 2000, 2001; Yamada et al., 1999). They belong to the space group of either $Im\bar{3}$ or $Pm\bar{3}$. Space group of $Im\bar{3}$ is assigned, if all diffraction lines are indexed using the Miller indices, in which the sum $h + k + l$ is even. Space group $Pm\bar{3}$ is assigned, if extradiffraction lines appear, which are indexed with the sum of the Miller indices being odd.

2.3.3 Gamma Brasses

Gamma brasses containing 52 atoms per unit cell are typical of CMAs and are described in terms of a 26-atom cluster (Bradley and Jones, 1933). The advantage of working with gamma brasses as CMAs essentially stems from the following reasons: (1) the number of atoms in the unit cell is large enough to produce a sizable pseudogap at the Fermi level, which is important for understanding its stability, but is yet small enough to perform efficiently full-potential linearized augmented plane wave (FLAPW) band calculations, (2) a large number of combinations of elements in the periodic table give rise to stable

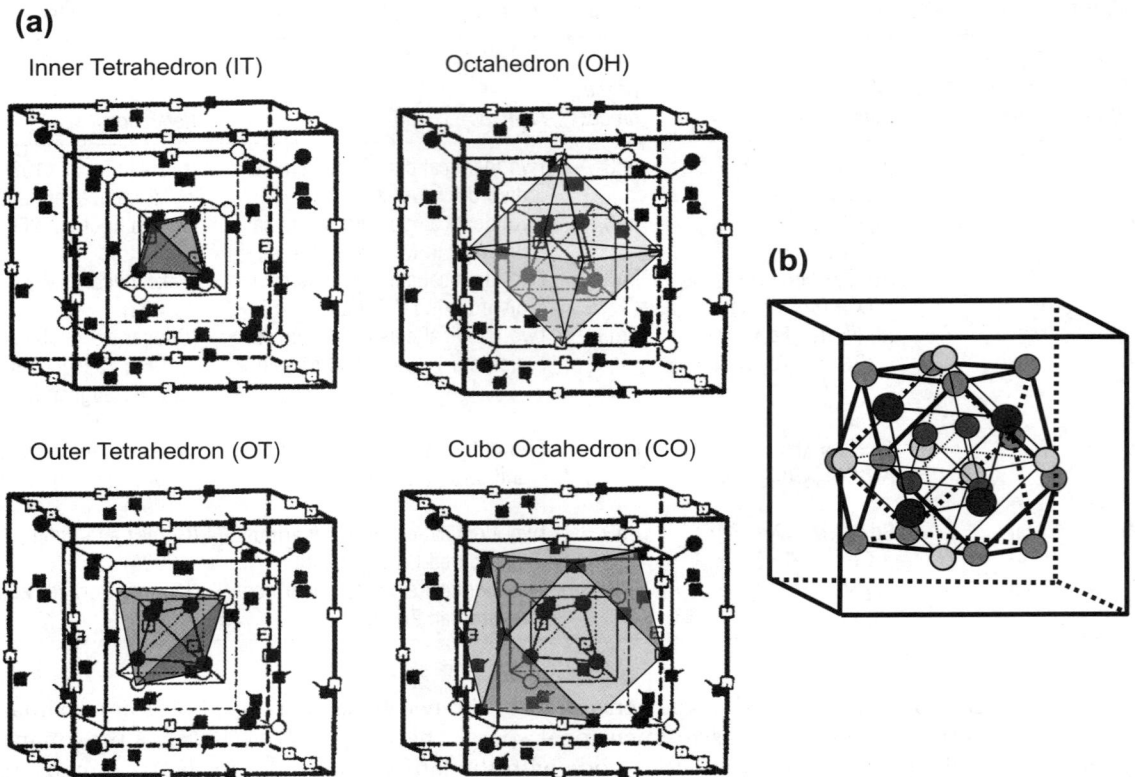

Figure 13 (a) Four shell structures consisting of an IT, OT, OH, and CO for the gamma brass. (b) 26-Atom cluster consisting of four atoms on IT (red), four atoms on OT (blue), six atoms on OH (yellow), and 12 atoms on CO (green) (Mizutani, 2006b). (For interpretation of the references to color in this legend, the reader is referred to the online version of this book.)

gamma brasses in more than 20 binary alloy systems, allowing us to carry out systematic studies, (3) many reliable structural data have already been accumulated and are available in the literature Villars (1997), and (4) the stabilization mechanism has been worked out for the first time by Mott and Jones in 1936 and has received intense attention since then by many researchers, as will be discussed in Section 2.5.

The 26-atom cluster is made up of four shells: four atoms are positioned on four vertices of the IT, four atoms on those of the outer tetrahedron (OT), six atoms on those of the OH, and 12 atoms on those of the cubooctahedron (CO), as illustrated in **Figure 13** (Mizutani, 2006b). Here the CO represents a polyhedron formed by meeting edges of the cube at the middle of octahedral faces.

Gamma brasses are divided into four families depending on the space group (Mizutani, 2010). The most abundant forms a bcc lattice with two identical 26-atom clusters with space group $I\overline{4}3m$. They are hereafter called I-cell gamma brasses. The second largest family forms a CsCl-type structure composed of two different 26-atom clusters. Since their space group is $P\overline{4}3m$, they are referred to as P-cell gamma brasses. Thus, both I- and P-cell gamma brasses normally contain 52 atoms in their cubic unit cells. As typical examples of the I- and P-cell gamma brasses, the 26-atom clusters in Cu_5Zn_8 and Al_4Cu_9 gamma brasses are illustrated in **Figure 14**, respectively (Mizutani, 2006b).

(a) Cu_5Zn_8 (I$\bar{4}$3m) **(b)** Al_4Cu_9 (P$\bar{4}$3m)

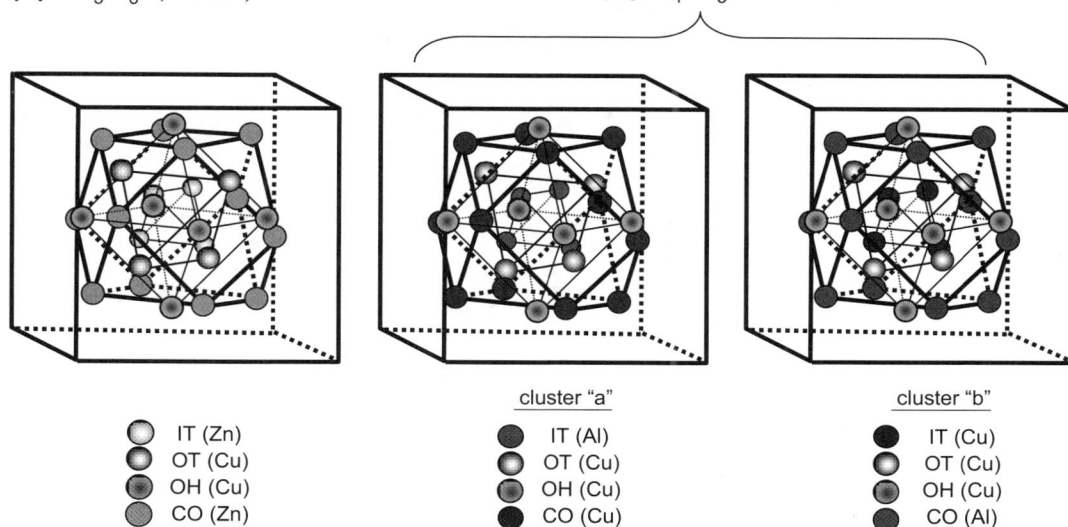

cluster "a" cluster "b"

○ IT (Zn)	● IT (Al)
◑ OT (Cu)	◐ OT (Cu)
◉ OH (Cu)	◉ OH (Cu)
◯ CO (Zn)	● CO (Cu)

● IT (Cu)
◐ OT (Cu)
◉ OH (Cu)
● CO (Al)

Figure 14 The 26-atom clusters in (a) Cu_5Zn_8 and (b) Al_4Cu_9 gamma brasses with space groups I$\bar{4}$3m and P$\bar{4}$3m, respectively (Mizutani, 2006b). (For color version of this figure, the reader is referred to the online version of this book.)

Though they are less frequent, we have two other families of gamma brasses. One includes systems that form superlattices with space group F$\bar{4}$3m, being referred to as F-cell gamma brasses. The fourth one is characterized by a rhombohedral unit cell with space group R3m, a subgroup of P$\bar{4}$3m, and is referred to as R-cell gamma brasses. Details are described elsewhere (Mizutani, 2010).

As listed in **Table 2**, we found I- and P-cell gamma brasses in 24 binary alloy systems (Mizutani, 2010). In most cases, the gamma-brass phase has a finite solid solution range (Okamoto, 2000). Since first-principles band calculations can be performed only for an ordered phase without involving any chemical and/or geometrical disorder, it is important to fill a specific atomic species into each shell in the 26-atom cluster without vacancies. Thus, the stoichiometric composition of an ideal gamma brass should fall in one of A_2B_{11}, A_4B_9, A_5B_8, A_7B_6, and $A_{10}B_3$ with space group I$\bar{4}$3m and a combination of these two with space group P$\bar{4}$3m. However, only the formula A_5B_8 has been experimentally found for I-cell gamma brasses, while the formula A_4B_9 for P-cell gamma brasses.[10] Such stoichiometric compounds are found within a solid solution range.

We have classified I- and P-cell gamma brasses into three groups I–III in terms of a combination of constituent elements in the periodic table, as listed in **Table 2** (Mizutani, 2010). Group I consists of a combination of monovalent noble metal and polyvalent element, whose valency is well-defined. The validity of the Hume-Rothery electron concentration rule of $e/a = 21/13$ has been claimed for group I. As a matter of fact, people in the 1920s have already been aware that the prototype Cu_5Zn_8 and Al_4Cu_9

---

[10] Gamma brasses with the composition TM_2Zn_{11} (TM = Fe, Co, Ni, Pd, etc.) are marginal but are still within a solid solution range. The gamma brasses with this composition generally involve chemical disorder as well as vacancies so that the number of atoms per unit cell is generally lower than 52.

Table 2   Classification of gamma brasses with space groups $I\bar{4}3m$ and $P\bar{4}3m$ in three different groups in binary alloy systems

	Gamma brass	e/a	Solid solution range (%)		Gamma brass	e/a	Solid solution range		Gamma brass	e/a	Solid solution range
I	Cu_5Zn_8 (I)	21/13	$57 < Zn < 70$	I	Cu_9In_4 (P)	21/13	$27.7 < In < 31.3$	II	Ni_2Be_{11} (I)	?	$11.5 < Ni < ?$
	Ag_5Cd_8 (I)	21/13	$57 < Cd < 62.5$		Ag_9In_4 (P)	21/13	$31.1 < In < 33.6$		Ni_2Cd_{11} (I)	?	$12 < Ni < 19.5$
	Ag_5Zn_8 (I)	21/13	$58 < Zn < 64.7$		Au_9In_4 (P)	21/13	$28.8 < In < 31.4$		Mn_2Zn_{11} (I)	?	$15.2 < Mn < 23$
	Cu_5Cd_8 (I)	21/13	$52.2 < Cd < 66$	II	Ni_2Zn_{11} (I)	?	$15 < Ni < 30$		Pt_2Zn_{11} (?)	?	$19 < Pt < 23$
	Au_5Cd_8 (I)	21/13	$61.2 < Cd < 67.6$		Pd_2Zn_{11} (I)	?	$14.5 < Pd < 24$		Al_8V_5 (I)	?	Al_8V_5
	Au_5Zn_8 (I)	21/13	$62.5 < Zn < 76$		Fe_2Zn_{11} (I)	?	$17 < Fe < 31$		Mn_3In (P)		Mn_3In
	Al_4Cu_9 (P)	21/13	$31.5 < Al < 37$		Co_2Zn_{11} (I)	?	$14.6 < Co < 31$	III	Ag_5Li_8 (I)	?	$63.5 < Li < 76$
	Cu_9Ga_4 (P)	21/13	$29.5 < Ga < 40$		Ir_2Zn_{11} (I)	?	$15.3 < Ir < 15.7$		$Li_{10}Pb_3$?	?	$22.5 < Pb < 23.5$

(I): space group $I\bar{4}3m$, (P): space group $P\bar{4}3m$, Solid solution range was taken from (Okamoto, 2000).

gamma brasses are stabilized at $e/a = 21/13$ in spite of different solute concentrations (Westgren and Phragmén, 1928a, 1928b, 1929). We will elucidate the Fermi surface–Brillouin zone (FsBz)-induced stabilization mechanism in Cu_5Zn_8 and Al_4Cu_9 by performing first-principles FLAPW band calculations in Section 2.5.3.

Group II includes totally 11 gamma brasses, which consist of 3d-transition metal elements like $TM = V$, Mn, Fe, Co, and Ni and either divalent elements like Be, Zn, and Cd or trivalent elements like Al and In. Since the e/a value for the TM element is not a priori known, it is not clear whether the gamma brass in group II is stabilized at $e/a = 21/13$ or not. This has been regarded for a long time as one of the most challenging themes in the electron theory of metals. We will show in Section 2.5.4 that the d state-mediated splitting or the d state-mediated-FsBz interaction plays a key role in the stabilization of group (II) gamma brasses and discuss a deviation from the empirical $e/a = 21/13$ law.

Group III includes gamma brasses consisting of a combination of two nontransition metal elements, for which a use of their nominal valencies does not yield the characteristic value of 21/13. The most notable example in this group is the $Ag_{100-x}Li_x$ ($x = 63.5$–76) gamma brass, a combination of two monovalent elements leading to $e/a = 1.0$, regardless of the Li concentration (Mizutani, 2010; Noritake et al., 2007). Hence, it has been wondered if Ag_5Li_8 gamma brass obeys the Hume-Rothery electron concentration rule or not. Its stability mechanism will be discussed in Section 2.5.5.

2.3.4 Other CMAs in Binary Alloy Systems

There are certainly a large number of CMAs other than families of approximants and gamma brasses even in binary alloy systems. At first, we point to the most notable three intermetallic compounds containing >1000 atoms per unit cell: β-Al_3Mg_2, Cu_3Cd_4, and $NaCd_2$, all their structures having been solved by Samson in 1960s (Samson, 1965, 1967, 1962). The β-Al_3Mg_2 compound contains 1168 atoms in the unit cell with clusters characterized by the Friauf polyhedra with space group $Fd\bar{3}m$ (Samson, 1965; Feuerbacher et al., 2007). According to the Al–Mg phase diagram (Okamoto, 2000), a solid solution range exists over 37.5 to 40.0 at.%Mg. In the case of Cu_3Cd_4, there are 1124 atoms in the unit cell with the lattice constant of 2.5871 nm (Samson, 1967). Its space group was deduced to be $F\bar{4}3m$. Dominant coordination shells in Cu_3Cd_4 are composed of Friauf polyhedra and icosahedra. The $NaCd_2$ with space group $Fd\bar{3}m$ contains 1120–1190 atoms in the unit cell with the lattice constant $a = 3.056$ nm (Samson, 1962). Unfortunately, however, we cannot execute first-principles band calculations for them by using experimentally reported atomic coordinates, because the unit cell is not only too large to perform efficient calculations but also all of them involve a large amount of vacancies and chemical disorder. Nevertheless, we will fairly reliably discuss FsBz interactions and the matching conditions for β-Al_3Mg_2 in Section 2.7.2.2.

We consider it practically more important to study the stabilization mechanism of CMAs having a lower number of atoms per unit cell than those of the Samson alloys discussed above. A close inspection into the literature (Villars, 1997) revealed that there exist a number of CMAs with a well-defined unit cell containing atoms over the range from more than 20 up to 500 atoms without involving any chemical disorder or vacancies in binary alloy systems. Their stabilization mechanism can be directly analyzed by first-principles band calculations.

Typical examples are picked up from Pearson's Handbook (Villars, 1997) and listed in **Table 3**. Among them, we direct our attention to the family containing more than two CMAs with a specific space group. First, we need to examine whether a deep pseudogap is commonly formed across the Fermi level in

Table 3   Typical CMAs other than gamma brasses and approximants in binary alloy systems

N	Space group	CMAs	N	Space group	CMAs
cF408	F$\bar{4}$3m	Ir$_7$Mg$_{44}$	cI52	I$\bar{4}$3m	Cu$_5$Zn$_8$
		Rh$_7$Mg$_{44}$			Ag$_5$Cd$_8$
		Ru$_7$Mg$_{44}$			Ag$_5$Li$_8$
		Os$_7$Sc$_{44}$			Ag$_5$Zn$_8$
oC276	Abm2	NiZn$_3$			Al$_8$V$_5$
cF176	Fd$\bar{3}$m	Al$_{10}$V			Al$_8$Cr$_5$
oC160	Amm2	Au$_3$Mg			Ni$_2$Zn$_{11}$
cI160	Im$\bar{3}$	Sc$_3$Zn$_{17}$	oF48	Fddd	CuMg$_2$
		Ru$_3$Be$_{17}$			CoIn$_2$
oI142	Immm	Ag$_{17}$Mg$_{54}$			CrSn$_2$
cP140	Pm$\bar{3}$	Fe$_6$Sc$_{29}$			Cr$_2$Sn$_3$
		Rh$_{13}$Sc$_{57}$			NbSn$_2$
cF116	Fm$\bar{3}$m	Li$_{23}$Sr$_6$			Sn$_3$V$_2$
		Mg$_{23}$Sr$_6$			Sn$_3$Ta$_2$
		Mn$_{23}$Th$_6$	cI40	Im$\bar{3}$m	Ge$_7$Ir$_3$
cF112	Fm$\bar{3}$c	NaZn$_{13}$			In$_7$Ni$_3$
		MgBe$_{13}$			In$_7$Pt$_3$
		SbBe$_{13}$			Sn$_7$Os$_3$
		RbZn$_{13}$	tI32	I4/mcm	Si$_3$W$_5$
		BaBe$_{13}$			Al$_3$Zr$_5$
		CaBe$_{13}$	oC28	Cmcm	Al$_6$Mn
		SrBe$_{13}$			Al$_6$Tc
		BaCu$_{13}$			Al$_6$Re
		LaCo$_{13}$			Al$_6$Fe
cF96	Fd$\bar{3}$m	NiTi$_2$			Al$_6$Ru
		CdNi	cI26	Im$\bar{3}$	Al$_{12}$Mo
		MgCo			Al$_{12}$W
		Nb$_{15}$Ni$_2$			Al$_{12}$Tc
		FeZr$_2$			Al$_{12}$Re
cP64	P$\bar{4}$3m	KGe	tI26	I4/mmm	Zn$_{12}$Sc
		RbGe			Be$_{12}$V
		CsGe			Be$_{12}$Cr
		KSi			Zn$_{12}$Y
		RbSi			Be$_{12}$Mo
		CsSi			Mn$_{12}$Y
					Zn$_{12}$Sm

them. If yes, we study as the next step whether it originates from FsBz interactions or orbital hybridizations or a combination of both. It is hoped that we are able to reveal that CMAs thus studied *do* obey the Hume-Rothery electron concentration rule by elucidating FsBz interactions involved. Indeed, we will deal with both Al$_6$TM (TM = Mn, Tc, Re, Fe, and Ru) ($N = 28$) with space group Cmcm and Al$_{12}$TM ($N = 26$) (TM = Mo, W, Tc, and Re) with space group Im$\bar{3}$ along this line in Section 2.7.2.1.

In the next section, we will describe the essence of first-principles band calculations with emphasis on why FLAPW band calculations are suitable to extract FsBz interactions. Basic ideas behind the

FLAPW–Fourier spectrum and Hume-Rothery plot analysis will then be explained in detail to facilitate readers to catch up with the electron theory of structurally CMA phases. Following the introduction of basic theories in Section 2.4, we will review the stabilization mechanism of CMAs by choosing diverse gamma brasses synthesized from different combinations of constituent elements in Section 2.5 and RT- and MI-type 1/1–1/1–1/1 approximants in Section 2.6 and other CMAs in Section 2.7. The theoretical interpretation of the Hume-Rothery electron concentration rule, together with **e/a** determination of transition metal elements, will be made to conclude the present chapter at the end of Section 2.7.

2.4   Electron Theory of Complex Metallic Alloys

2.4.1   First-Principles Band Calculations and the Hume-Rothery Electron Concentration Rule

As discussed in Section 2.2.1, the pioneering theory put forward by Mott and Jones in 1936 (Mott and Jones, 1936) certainly served as a milestone in establishing the basic idea for interpreting the Hume-Rothery electron concentration rule, which relates alloy phase stability to the Fermi sphere contact with the Brillouin zone planes. However, a scientific revolution was made in the late 1980s by Tsai et al. (1988, 1990), who could successfully synthesize a series of thermally stable quasicrystals by using the Hume-Rothery electron concentration rule as a guide. In the 1990s, a general consensus had been gradually built that the stability of structurally complex metallic alloys (CMAs) is most likely a consequence of lowering the electronic energy brought about by the development of a deep pseudogap at the Fermi level. As mentioned in Section 2.2.4, a pseudogap formation of depth and width comparable to those experimentally observed in quasicrystals and their approximants can gain an electronic energy of 10–40 kJ/mol, which is large enough to stabilize one such CMA relative to neighboring competing phases. First-principles band calculations must be employed as a tool to explore the pseudogap structure. However, its use is limited only to a crystal with a well-defined unit cell free from vacancies and chemical disorder. This means that we have to abandon quasicrystals and discuss their stabilization mechanism as an extreme limit from the data accumulated for CMAs with giant unit cells.

2.4.2   Origin of the Pseudogap: Orbital Hybridizations versus FsBz Interactions

The origin of the pseudogap at the Fermi level can be discussed from two different approaches: one from covalent bonding and the other from metallic bonding (Mizutani, 2010). Let us consider the former approach by considering an Al–Mn alloy. Here we assume a situation such that both Al and Mn atoms are placed a few tenths of a nanometer apart, corresponding to an average atomic distance in the alloy. If the Al-3p and Mn-3d energy levels are close to each other, the two atomic wave functions will overlap and mix with one another. This is called orbital hybridization and results in bonding and antibonding levels, as illustrated in **Figure 15**a (Mizutani, 2010). This also holds true upon forming a solid, i.e. an Al–Mn alloy phase. The bonding and antibonding levels will naturally be broadened into the respective bands, leaving a pseudogap in between bonding and antibonding subbands formed by Mn-3d states mixed with Al-3p states. **Figure 15**b shows the density of states (DOS) for the Al–Mn approximant containing 138 atoms in its cubic unit cell, as calculated by Fujiwara in 1989 (Fujiwara, 1989).[11] Gaussian curves are roughly drawn in **Figure 15**b as a guide to the eye to represent the

---

[11] A close inspection of **Figure 15**b reveals the existence of a sharp intervening peak right at the Fermi level inside a wide pseudogap caused by Al-sp/Mn-d orbital hybridizations (Fujiwara, 1989). This has been shown to originate from the use of the Elser–Henley model structure and the neglect of the combined correction term in the LMTO–ASA (linearized MTO-atomic sphere approximation) method (Zijlstra and Bose, 2003). A deep pseudogap without the intervening peak at the Fermi level was achieved by incorporating the combined correction term into the LMTO–ASA formalism for the structure experimentally deduced by Cooper and Robinson (Cooper and Robinson, 1966) (Section 2.6.5.1.1).

(a)

Anti-Bonding energy level (LUMO)

Al-3p ――――― <hexagon diagram> ――――― Mn-3d

Bonding energy level (HOMO)

(b)

3000 ┐   ┌ 800

DOS (states/Ry.cell)

Integrated DOS (states/cell)

Mn-3d/Al-3p
Bonding

Mn-3d/Al-3p
Anti-Bonding

0 └──┘
−1.0   Energy (Ry)   0 E_F 0.5

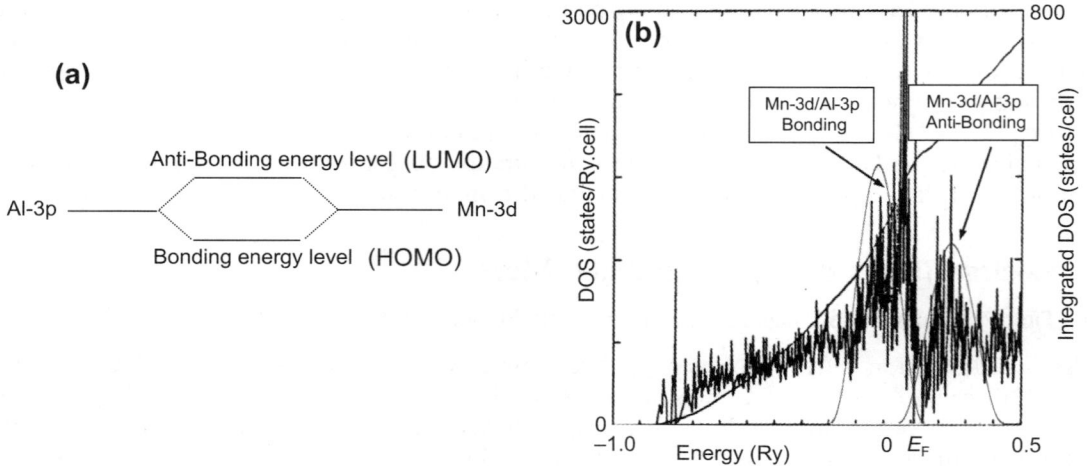

Figure 15 (a) Formation of bonding and antibonding levels due to orbital hybridization between neighboring Al-3p and Mn-3d atomic wave functions. (b) Formation of a pseudogap between bonding and antibonding subbands due to orbital hybridizations between Al and Mn atoms in an Al–Mn approximant (Mizutani, 2010). Gaussian curves are roughly fitted to the subband profiles as a guide to the eye. (For color version of this figure, the reader is referred to the online version of this book.)

resulting Al-3p/Mn-3d bonding and antibonding subbands. The Fermi level is found to fall inside the pseudogap created between the bonding and antibonding Mn-3d subbands.

The Al–Mn approximant mentioned above is favorably stabilized, since the bonding band is almost fully filled by electrons, whereas the antibonding band remains almost empty. This is the stabilization mechanism due to orbital hybridizations. Here it is kept in mind that it does not involve any parameter directly pertaining to the electron concentration **e/a**. It is also emphasized that an orbital hybridization-induced pseudogap has little to do with the size of the unit cell. In other words, the orbital-hybridization effect would not necessarily be enhanced, as the number of atoms in a unit cell is increased.

We have an alternative approach to generate a pseudogap at the Fermi level. This is due to the FsBz interaction and is approached from the metallic bonding picture. Let us consider itinerant electrons propagating throughout a periodically arranged ionic potential field. Stationary waves will be formed, when the wavelength of the electron wave matches the period of the ionic potential. The situation in real space is illustrated in **Figure 16**a (Mizutani, 2010). This is called the interference phenomenon and is equivalent to the fulfillment of the Bragg law. For the sake of simplicity, consider an electron propagating through a one-dimensional periodic potential field. Either a sine- or cosine-type stationary wave is formed, when the wave number of electron reaches $k = \pm\pi/a$, as shown in **Figure 16**b.[12] A closer look into **Figure 16**b indicates that the electronic energy of the cosine-type stationary wave must be lowered relative to the free electron value, since its charge density is the highest at the center of the

---

[12] The Bloch wave function in one-dimensional periodic array of ionic potentials is given by a linear combination of the unperturbed plane wave $A_0 e^{ikx}$ and the wave $A_1 e^{i[k-(2\pi/a)]x}$ perturbed by the set of lattice planes: $\psi(x) = \exp(ikx)[A_0 + A_1 \exp\{-i(2\pi/a)x\}]$. At $k = \pi/a$, the relation $A_0 = \pm A_1$ holds and the wave function is reduced to either $\sin(\pi x/a)$ or $\cos(\pi x/a)$. The wave of $k = \pi/a$ is reflected to the wave of $k' = -\pi/a$ by receiving a crystal momentum $G = -2\pi/a$ from the lattice planes and the reflected wave of $k = -\pi/a$ is again reflected to the wave of $k' = \pi/a$ by receiving a crystal momentum $G = 2\pi/a$ from the lattice planes. This process is infinitely repeated, resulting in a cosine- or sine-type stationary wave (Mizutani, 2010).

(a)

(b)

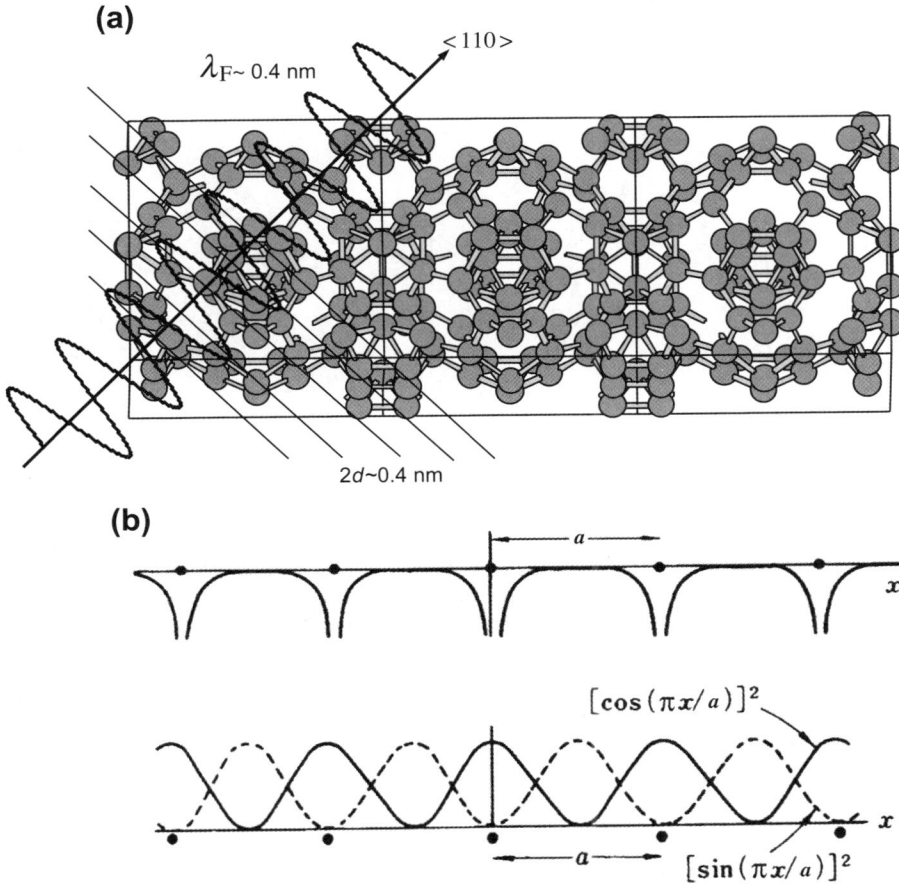

Figure 16 (a) Formation of stationary electron waves as a result of interference with periodically arranged ionic potentials, (b) formation of either cosine- or sine-type stationary waves in one-dimensional periodic ion potential field (Mizutani, 2010).

ion, where the potential is the lowest. The opposite is true for the sine-type stationary wave. This is the mechanism for the formation of an energy gap at the wave number $k = \pm\pi/a$ in the energy dispersion relation.

It is worthwhile to mention at this stage the interrelation between the set of lattice planes specified by the Miller indices $(h\,k\,l)$ in real space and the corresponding reciprocal lattice vector $\mathbf{G}$ in reciprocal space. For the sake of simplicity, let us consider a cubic system. A reciprocal lattice vector $\mathbf{G}$ having the components $(h\,k\,l)$ in units of $2\pi/a$ is perpendicular to the set of lattice planes with the same Miller indices. In the present chapter, we frequently use the square of the reciprocal lattice vector $|\mathbf{G}|^2$ in units of $(2\pi/a)^2$. Then, it is given by the squared sum of the Miller indices $|\mathbf{G}|^2 = h^2 + k^2 + l^2$ in the case of a cubic system.

Figure 17a illustrates the opening of an energy gap at the wave vector obtained by bisecting the reciprocal lattice vector $\mathbf{G}_{100}$, i.e. $k = \pm|\mathbf{G}_{100}|/2 = \pm\pi/a$ in a simple cubic lattice (Mizutani, 2010). This is certainly caused by the interference of electrons with the set of lattice planes $\{100\}$ with $|\mathbf{G}|^2 = 1$.

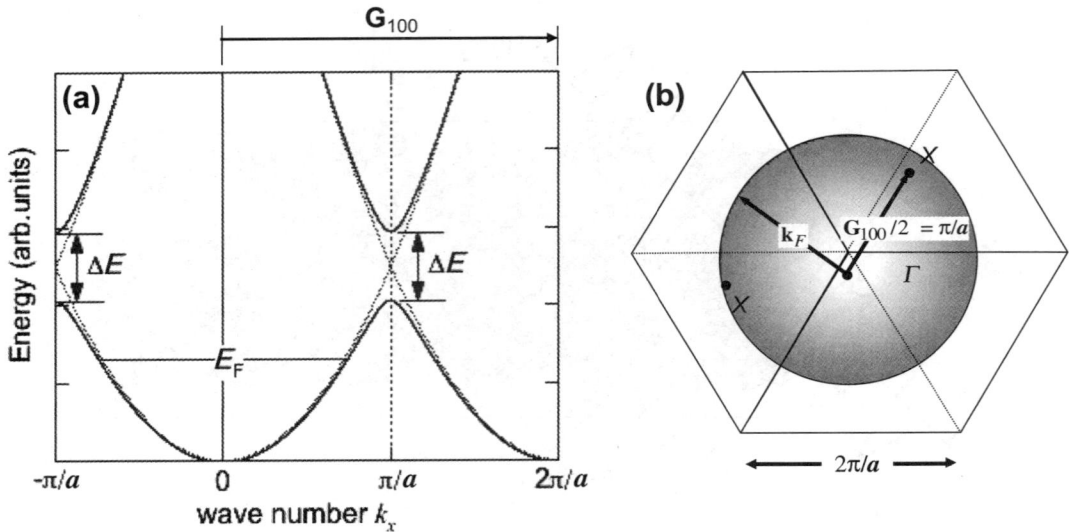

Figure 17   (a) Formation of an energy gap at the wave number $k = \pm\pi/a$ obtained by bisecting the reciprocal lattice vector $\mathbf{G}_{100}$ in a simple cubic lattice (Mizutani, 2010). The Fermi level is located well below the energy gap. (b) The Brillouin zone for a simple cubic lattice. The Fermi surface with the Fermi radius $\mathbf{k}_F$ corresponding to the electronic structure in (a) is schematically drawn.

As shown in **Figure 17**b, a cube with edge length $2\pi/a$ is constructed in the reciprocal space by perpendicularly bisecting six equivalent reciprocal lattice vectors. This is the Brillouin zone of a simple cubic lattice, across which an energy gap appears. In the case of a simple cubic lattice, the distance from the origin Γ to the center X of six square zone planes of its Brillouin zone is the same (**Figure 17**b).

The Brillouin zones in contact with the free electron Fermi sphere are illustrated in **Figure 18** for bcc, fcc, gamma brass, and 1/1–1/1–1/1 approximant. The Brillouin zone for the bcc lattice containing two atoms per unit cell is bounded by {110} rhombic dodecahedral planes with $|\mathbf{G}|^2 = 2$. The Brillouin zone for the fcc lattice containing four atoms per unit cell consists of eight {111} hexagonal zone planes with $|\mathbf{G}|^2 = 3$ plus six {200} square planes with $|\mathbf{G}|^2 = 4$, being totally bounded by 14 planes. The number of the Brillouin planes in contact with the free electron Fermi sphere increases with increasing the number of atoms per unit cell. In the case of the gamma brass with 52 atoms per unit cell, 12 {330} and 24 {411} zone planes are located at the same distance from the origin to form a 36-faced polyhedral Brillouin zone with $|\mathbf{G}|^2 = 18$, as shown in **Figure 18**c. The Fermi sphere contacting these zone planes can accommodate 1.538 electrons per atom. The number of the Brillouin zone planes further increases for the 1/1–1/1–1/1 approximant containing 160 atoms per unit cell. As shown in **Figure 18**d, the Brillouin zone is bounded by 48 {543}, 24 {710}, and 12 {550} zone planes, resulting in the 84-faced Brillouin zone with $|\mathbf{G}|^2 = 50$. The Fermi sphere touching all these zone planes accommodates 2.31 electrons per atom.

The argument above was made in the free electron model. In realistic systems, the same arguments hold true except that an energy gap across the Brillouin zone planes becomes finite so that the Fermi surface is no longer described by a sphere. The set of lattice planes interacting with electrons at the Fermi level plays a critical role in causing a pseudogap across the Fermi level. The corresponding reciprocal lattice vector $|\mathbf{G}|^2$ is hereafter called *critical*. Now one can realize that the *critical* $|\mathbf{G}|^2$ value increases with increasing the number of atoms per unit cell.

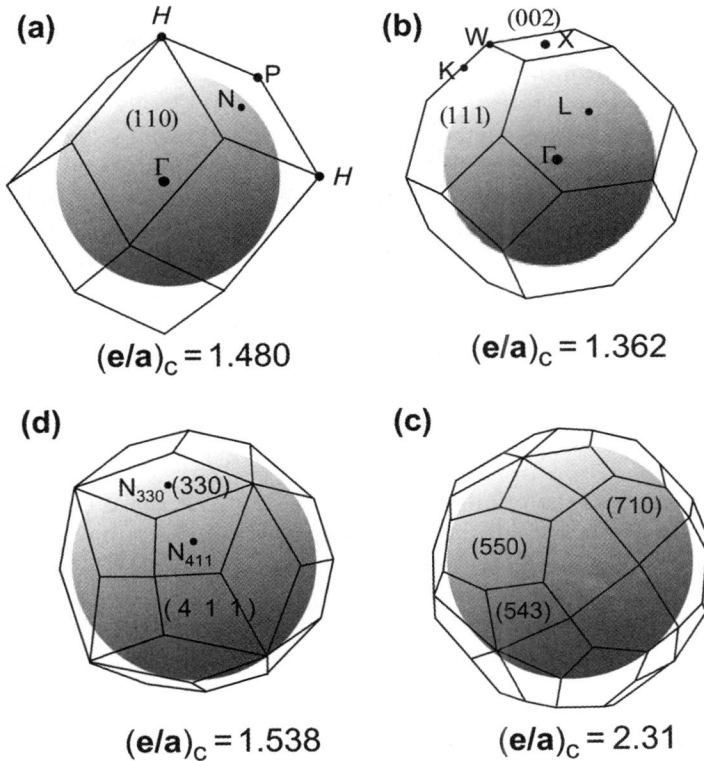

Figure 18 (a) Brillouin zone planes and inscribed free electron Fermi sphere in (a) bcc, (b) fcc, (c) gamma brass, and (d) 1/1–1/1–1/1 approximant phases. In the case of the fcc, the inscribed Fermi sphere first touches {111} zone planes with $|\mathbf{G}|^2 = 3$, since the distance $\Gamma L (= \sqrt{3}/2)$ is shorter than $\Gamma X (=$ unity) in units of $2\pi/a$. The $(e/a)_c$ denotes electrons per atom ratio obtained for inscribed Fermi sphere.

The contact of the Fermi surface with the equivalent Brillouin zone planes occurs simultaneously at each center. Hence, its effect on the DOS would be enhanced, as the number of equivalent zone planes is increased. An anomaly in the DOS caused by FsBz interactions is called the van-Hove singularity. We often encounter such a situation in CMAs that a deep valley is formed in the DOS, but the electronic states remain finite along this minimum. This is called a *pseudogap*. As emphasized in Section 2.2.4, its contribution to the stability of a solid would become most effective, when it is formed across the Fermi level. An FsBz-induced pseudogap plays a key role in stabilizing a CMA phase, in which the diffraction spectrum is characterized by a series of Bragg peaks. Included are quasicrystals, which also exhibit numerous sharp diffraction peaks and a more marked pseudogap the better ordered they are (Mizutani, 2010).

Let us suppose the interference phenomenon above to occur at the Fermi level and to be strong enough to cause a sizeable pseudogap there. Now the situation is envisaged such that the *effective* Fermi sphere with the diameter $2k_F$ is in contact with the Brillouin zone planes associated with the *critical* reciprocal lattice vector $|\mathbf{G}|$. This immediately leads to the condition

$$(2k_F)^2 = |\mathbf{G}|^2. \tag{4}$$

Equation 4 claiming the fulfillment of the interference condition has been often referred to as the Hume-Rothery matching condition (Mizutani, 2010). It is important to keep in mind that Eqn 4 is no longer based on the free electron model. A key point is how to define an effective Fermi sphere and deduce the value of $2k_F$ and how to extract the *critical* reciprocal lattice vector even for systems containing a large amount of transition metal elements. In the following sections, we will introduce an elegant technique on the basis of first-principles full-potential linearized augmented plane wave (FLAPW) band calculations to determine the *effective* Fermi diameter $2k_F$ and the *critical* reciprocal lattice vector $|G|$ involved in Eqn 4 in any intermetallic compounds, in which chemical disorder and defects are absent and a unit cell is well-defined. The matching condition given by Eqn 4 has been tested for many CMAs and will be discussed as the main subject in the following Sections.

A few more words may be added in relation to Eqn 4. The square of the Fermi diameter $(2k_F)^2$ of the *effective* Fermi sphere is expressed in the form

$$(2k_F)^2 = \left[\frac{3}{\pi}(e/a)N\right]^{2/3} = \left[\frac{3}{\pi}(e/uc)\right]^{2/3}, \tag{5}$$

in units of $(2\pi/a)^2$, where N is the number of atoms per unit cell and e/uc, a product of N and e/a, represents the number of electrons per unit cell. It is important to be reminded that $(2k_F)^2$ and *critical* $|G|^2$ both in units of $(2\pi/a)^2$ and e/uc as well increase with increasing N. This means that these three parameters $(2k_F)^2$, *critical* $|G|^2$ and e/uc can be defined only for crystals but not for a quasicrystal, since its unit cell is infinitely large. However, the parameter e/a can be still defined for quasicrystals, since it is a quantity independent of N. Note that the e/a value given as a composition average of those of constituent elements has been used for not only crystals but also quasicrystals, provided that the e/a values of the constituent elements are a priori known.

In the past two decades, expressions such as the "Hume-Rothery stabilization mechanism" and/or "Hume-Rothery-type stabilization" have been frequently employed, particularly upon discussing the stability of quasicrystals (Mizutani, 2010; Trambly de Laissardière et al., 2005). However, many people have used these phrases, whenever a pseudogap is experimentally found, or theoretically predicted, in a CMA without differentiating between the two origins: the FsBz and the orbital hybridizations. A very careful discussion is needed to extract the role of FsBz interactions on the formation of a pseudogap in systems, in which orbital hybridizations are predominant. Our goal in this chapter is to clarify this important issue.

2.4.3   First-Principles Electronic Structure Calculations

As emphasized in Section 2.1, we consider first-principles FLAPW electronic structure calculations to be an indispensable tool to discuss the phase stabilization mechanism of CMAs in terms of FsBz interactions. In this section, we devote ourselves to review what the FLAPW method is and why it is best suited to explore FsBz interactions and to reliably determine e/a. The expression "first-principles" in first-principles electronic structure calculations refers to the method of calculating the electronic structure by solving a one-electron Schrödinger equation without relying on any experimental results under the assumption that an electron experiences an effective potential within the framework of the density functional theory (DFT).

Itinerant electrons in the valence band move in a lattice while interacting with each other via the Coulomb force. Rigorously speaking, we must treat the electron–electron interaction in the context of the "many-body problem", which cannot be analytically solved. The motion of an electron in a metal has been treated in the so-called one-electron approximation, under which each electron independently propagates in an effective averaged potential. Both Hartree and Hartree–Fock approximations had been employed to construct an effective one-electron potential until 1964, when the DFT has been established (Hohenberg and Kohn, 1964; Kohn and Sham, 1965; Schlüter and Sham, 1982). This brought us a substantial progress in the reliability of one-electron electronic structure calculations. According to DFT, the total energy of an electron running in an effective potential field is given as a functional of the electron density. Indeed, Kohn and Sham (1965) provided the method of calculating the total energy of a system by treating the exchange-correlation energy of the electron in the local density approximation (LDA). Since details about the DFT–LDA theory will be found in the literature (Hohenberg and Kohn, 1964; Kohn and Sham, 1965; Schlüter and Sham, 1982), we simply note that the total energy of a system at absolute zero (Kohn and Sham, 1965) is expressed as

$$U_{\text{total}} = \sum_i \varepsilon_i - \frac{1}{2} \int\int \frac{n(\mathbf{r})n(\mathbf{r}')}{|\mathbf{r} - \mathbf{r}'|} \mathrm{d}\mathbf{r}\mathrm{d}\mathbf{r}' + \int n(\mathbf{r})\{\varepsilon_{\text{XC}}(n(\mathbf{r})) - \mu_{\text{XC}}(n(\mathbf{r}))\}\mathrm{d}\mathbf{r}, \tag{6}$$

where ε_i is the solution of an effective one-electron Schrödinger equation given by

$$\left\{ -\frac{h^2}{2m}\nabla^2 + v_{\text{eff}}(\mathbf{r}) \right\}\psi_i(\mathbf{r}) = \varepsilon_i\psi_i(\mathbf{r}). \tag{7}$$

The first term in Eqn 6 represents the one-electron band structure energy due to both valence and core electrons, whereas the second term is often referred to as the Hartree term representing the Coulomb potential energy due to the nucleus–electron interaction plus an average electron–electron interaction energy, and the third term represents the exchange-correlation energy derived in the LDA. Recent progress beyond the LDA has been made by adding gradient terms of the electron density to the exchange-correlation energy and its corresponding potential. This has led to the generalized gradient approximation developed by Perdew and Wang (1992) and Perdew et al. (1996).

There are two different approaches in first-principles electronic structure calculations: one is the pseudopotential method, which allows us to employ the set of plane waves as basis functions, and the other an all-electron method, which treats all electrons including core states explicitly. In the pseudopotential approach, the potential becomes so smooth that plane waves can be safely used as basis functions. But ignorance of the core states poses difficulties not only in discussing issues related to core states, such as core excitations and core level shifts, but also in accurately treating d-electrons having a higher tendency of localization in the valence band.

The all-electron method employs a one-electron effective potential determined by both nuclei and all electrons and, hence, s-, p-, d- and f-electrons are treated on the same ground. There are several all-electron first-principles band calculation methods, all of which basically divide up a crystal into regions inside atomic centered muffin-tin (MT) spheres, where Schrödinger's equation is solved numerically, and an interstitial (I) region. Both augmented plane wave (APW) and MT-orbital (MTO) methods are known to be typical of all-electron first-principles band calculation methods.

As emphasized in Section 2.4.2, the origin of a pseudogap across the Fermi level can be discussed in terms of orbital-hybridization effects and FsBz interactions. The respective formation mechanisms can be best studied by means of LMTO–ASA and FLAPW methods, respectively (Mizutani, 2010). Since the aim of the present chapter is to deepen our understanding of the Hume–Rothery stabilization mechanism in CMAs, we focus on only the FLAPW electronic structure calculations and point out how we can extract the *critical* $|\mathbf{G}|^2$ and square of the Fermi diameter $(2k_F)^2$, both of which are involved in Eqn 4.

2.4.4 APW Method

Prior to the discussion of the FLAPW method, we need to discuss the basic idea of the preceding relevant methods called the APW and LAPW methods (Mizutani, 2010). In this section, the essence of the APW method is described. A spherically symmetric potential is assumed inside the MT sphere with a radius a and, hence, wave functions can be rigorously solved as a product of radial wave functions and spherical harmonics. On the other hand, the potential outside the MT sphere is assumed to be constant so that plane wave can be taken as its solution:

$$\chi_{\mathbf{k}}(\mathbf{r}) = \sum_{\ell m}\{A^{\mathbf{k}}_{\ell m}u_{\ell}(E_{\ell}, \mathbf{r}) + B^{\mathbf{k}}_{\ell m}\dot{u}_{\ell}(E_{\ell}, \mathbf{r})\}Y_{\ell m}(\theta, \phi) \qquad (\mathbf{r} \leq a)$$
$$\chi_{\mathbf{k}}(\mathbf{r}) = \exp(i\,\mathbf{k}\cdot\mathbf{r}) \quad (\mathbf{r} > a) \tag{8}$$

For the sake of simplicity, a crystal consisting of a single atomic species is considered. In the region $r \leq a$, a trial function is expressed as

$$\chi^{\text{APW}}_{\mathbf{k}+\mathbf{G}}(E, \mathbf{r}) = \sum_{lm} A^{lm}_{\mathbf{k}+\mathbf{G}}R_{l}(E, \mathbf{r})Y_{lm}(\hat{\mathbf{r}}), \tag{9}$$

where $\mathbf{k}$ is the wave vector in the first Brillouin zone and $\mathbf{G}$ is an allowed reciprocal lattice vector. In the interstitial region $\mathbf{r} > a$, a trial function is expressed in the form of a plane wave with the wave vector $\mathbf{k} + \mathbf{G}$:

$$\chi^{\text{APW}}_{\mathbf{k}+\mathbf{G}}(\mathbf{r}) = \exp[i(\mathbf{k} + \mathbf{G})\cdot\mathbf{r}] \quad \mathbf{r} \in I. \tag{10}$$

Now Eqn 10 is expanded into the spherical harmonics:

$$\chi^{\text{APW}}_{\mathbf{k}+\mathbf{G}}(\mathbf{r}) = \exp[i(\mathbf{k} + \mathbf{G})\cdot\mathbf{r}] = 4\pi\sum_{\ell=0}^{\infty}\sum_{m=-\ell}^{\ell} i^{\ell}j_{\ell}(|\mathbf{k} + \mathbf{G}|r)Y^{*}_{\ell m}(\widehat{\mathbf{k} + \mathbf{G}})Y_{\ell m}(\hat{\mathbf{r}}), \tag{11}$$

where the symbol ⌢ over $\mathbf{k} + \mathbf{G}$ represents an angular variable (θ, φ) of the vector $\mathbf{k} + \mathbf{G}$. By imposing the continuity condition of Eqns 9 and 11 at the surface of the MT sphere, we can determine the coefficient $A^{\ell m}_{\mathbf{k}+\mathbf{G}}$ in Eqn 9 as follows:

$$A^{\ell m}_{\mathbf{k}+\mathbf{G}} = 4\pi i^{\ell}Y^{*}_{\ell m}(\widehat{\mathbf{k} + \mathbf{G}})\,\frac{j_{\ell}(|\mathbf{k} + \mathbf{G}|a)}{R_{\ell}(E, a)}. \tag{12}$$

In this way, we can construct the APW basis function, which is continuous across the MT sphere (Mizutani, 2010). Note that the derivative is not continuous at the surface of the MT sphere.

The function having the coefficient given by Eqn 12 is simply denoted as $\chi^{\text{APW}}_{\mathbf{k}+\mathbf{G}}(E, \mathbf{r})$. This is called the APW orbital. In order to guarantee $\chi^{\text{APW}}_{\mathbf{k}+\mathbf{G}}(E, \mathbf{r})$ to be the Bloch wave propagating throughout a crystal, we need to impose the Bloch condition:

$$\chi^{\text{APW}}_{\mathbf{k}+\mathbf{G}}(E, \mathbf{r} + \mathbf{R}) = e^{i(\mathbf{k}+\mathbf{G})\cdot\mathbf{R}}\chi^{\text{APW}}_{\mathbf{k}+\mathbf{G}}(E, \mathbf{r}), \tag{13}$$

where $\mathbf{R}$ is the lattice vector satisfying the relation $e^{i\mathbf{G}\cdot\mathbf{R}} = 1$. The wave function to describe the motion of an electron in a crystal is now expressed by summing APW orbitals as basis functions over the allowed reciprocal lattice vectors $\mathbf{G}$:

$$\psi_{\mathbf{k}+\mathbf{G}}(E, \mathbf{r}) = \sum_{\mathbf{G}} C(\mathbf{k} + \mathbf{G}) \chi_{\mathbf{k}+\mathbf{G}}^{\text{APW}}(E, \mathbf{r}). \tag{14}$$

The coefficients $C(\mathbf{k} + \mathbf{G})$ in Eqn 14 are determined as the solution of a set of equations by using the variational principle. Its secular determinantal equation is expressed as

$$\det\left\{\left(|\mathbf{k} + \mathbf{G}|^2 - E\right)\delta_{mn} + F_{mn}(E)\right\} = 0, \tag{15}$$

where coefficient $F_{mn}(E)$ is similar to the Fourier component of the ionic potential $V_{mn} \equiv V(\mathbf{G}_m - \mathbf{G}_n)$, that is, the form factor in nearly free electron (NFE) band calculations (Mizutani, 2010). However, in contrast to NFE band calculations, the energy to be solved is involved in $F_{mn}(E)$ through the terms $R_\ell(E,a)$ and $\dot{R}_\ell(E,a)$. Equation 15 must be linearized in order to perform a fast, but efficient computation.

2.4.5 LAPW Method

In this section, we study the essence of the LAPW method (Mizutani, 2010; Koelling and Arbman, 1975; Singh, 1994), which allows a fast computation by linearizing the APW method described in the preceding section. The determinant can be diagonalized, if energy-dependent logarithmic derivative of the radial wave function $\dfrac{\dot{R}_\ell(E, a)}{R_\ell(E, a)}$ in F_{mn} is made energy independent. For this purpose, we assume the trial radial wave function as follows (Mizutani, 2010):

$$\phi_\ell(E, r) = R_\ell(E_\nu, r) + \omega_\ell(E)\dot{R}_\ell(E_\nu, r), \tag{16}$$

where E_ν is called a linearization energy for each partial wave ℓ. In the region $\mathbf{r} \leq a$ in the Wigner–Seitz cell located at origin, the wave function is written as

$$\chi_{\mathbf{k}+\mathbf{G}}^{\text{LAPW}}(\mathbf{r}) = \sum_{\ell m}\left\{A_{\mathbf{k}+\mathbf{G}}^{\ell m}R_\ell(E_\nu, r) + B_{\mathbf{k}+\mathbf{G}}^{\ell m}\dot{R}_\ell(E_\nu, r)\right\}Y_{\ell m}(\hat{\mathbf{r}}) \tag{17}$$

instead of Eqn 9 (Koelling and Arbman, 1975). It must be noted that the ratio $\dfrac{B_{\mathbf{k}+\mathbf{G}}^{\ell m}}{A_{\mathbf{k}+\mathbf{G}}^{\ell m}}$ corresponds to $\omega(E)$ in Eqn 16 but is no longer energy dependent. The linearization energy E_ν in Eqn 16 is taken as the energy at the center of gravity in the band for the partial wave ℓ. In the same way as in the APW method, the wave function of an electron with the wave vector $\mathbf{k} + \mathbf{G}$ outside the MT sphere is given by Eqn 10 and is expanded into spherical harmonics in Eqn 11. However, there are two parameters $A_{\mathbf{k}+\mathbf{G}}^{\ell m}$ and $B_{\mathbf{k}+\mathbf{G}}^{\ell m}$ to be determined in Eqn 17. Hence, we can make not only the wave functions shown in Eqns 9 and 17 but also their derivatives to be continuous at the surface of the MT sphere.

Both $R_\ell(E_\nu, r)$ and $\dot{R}_\ell(E_\nu, r)$ are evaluated only for a linearization energy E_ν and $\chi_{\mathbf{k}+\mathbf{G}}^{\text{LAPW}}(\mathbf{r})$ provides sufficient basis for eigen functions in an energy range around this linearization energy. Thereby, the secular Eqn 15 becomes linear in energy, and all eigen energies can be found through one diagonalization of the secular matrix. However, this is made at the expense of an increased number of basis functions compared to the APW method. The convergence of this basis set is controlled by a cutoff parameter $R_{MT}K_{max}$ generally ranging over 6–9, where R_{MT} is the smallest MT sphere radius in the unit cell and K_{max} is the magnitude of the largest vector $\mathbf{K} = \mathbf{k} + \mathbf{G}$ in Eqn 14.

2.4.6　FLAPW Method

Both the APW and LAPW methods are constructed under the assumption of a spherically symmetrical potential and charge density. This rigid assumption may become less reliable for open structures such as semiconductors. Wimmer et al. (1981) were the first to remove the spherical shape approximation and could prove that the so-called FLAPW method they established is ideally suited to deal with the electronic structure of reduced symmetry systems like semiconductor surfaces.

In place of Eqn 8, the FLAPW method employs the following expressions for the total potential and, analogously, the total charge density:

$$V(\mathbf{r}) = \begin{cases} \sum\limits_{lm} V_{lm}(\mathbf{r})Y_{lm}(\hat{\mathbf{r}}) & (r \leq a) \\[2mm] \sum\limits_{\mathbf{K}} V_{\mathbf{K}}e^{i\mathbf{K}\cdot\mathbf{r}} & (r > a) \end{cases}, \tag{18}$$

where $\mathbf{K} = \mathbf{k} + \mathbf{G}$. Equation 18 implies that the potential and charge densities are no longer spherically symmetric both inside and outside the MT sphere but can reflect possible anisotropic distributions of charges around the centered atom. The MT approximation given by Eqn 8 in the APW and LAPW methods corresponds to retaining only the $l = 0$ component in $r \leq a$ and only the $\mathbf{K} = 0$ component in $r > a$ in Eqn 18.

The full-potential method is introduced to cope with a potential of arbitrary shape. The FLAPW method has become now most widely used as a tool capable of calculating the electronic structure with the highest accuracy among various all-electron first-principles band calculation methods.

2.4.7　"APW+lo" Method

The basis set of the FLAPW method is substantially larger than those for local function methods like the linear-MT-orbital (LMTO) method. This has been recognized for many years as a great disadvantage when applied to CMAs containing more than several tens of atoms per unit cell. Sjöstedt et al. (2000) presented a different way of linearizing the APW method that uses a rather small basis set for convergence and claimed that their newly developed "APW+lo" method yields numerically identical results to the conventional LAPW method.

There is not enough flexibility in variational freedom to find solutions when an energy-independent APW basis set is employed in FLAPW. This is the reason why the FLAPW method requires an increased number of basis functions. This difficulty was ingeniously circumvented by using a complementary basis set consisting of local orbitals for physically important l-quantum numbers, i.e. for $l \leq 3$ (Sjöstedt et al., 2000). The local orbitals impose no extra-condition on the APW basis set, and the number of plane waves in the interstitial is therefore unaffected.

Local orbitals are completely confined within the MT sphere and expressed in the following form:

$$\phi_{lm}^{lo}(\mathbf{r}, \mathbf{k}) = \begin{cases} R_{lm}^{lo}(r)Y_{lm}(\hat{\mathbf{r}}) & r \leq a \\ 0 & r > a \end{cases}, \tag{19}$$

where $R_{lm}^{lo}(r) = a_{lm}^{lo}u_l(r, E_\nu) + b_{lm}^{lo}\dot{u}_l(r, E_\nu)$, where E_ν is the linearization energy for all basis functions. The coefficient a_{lm}^{lo} is set to unity, while b_{lm}^{lo} is determined so as to make $\phi_{lm}^{lo} = 0$ at the MT boundary. The new basis functions are referred to as "APW+lo".

They tested the variational flexibility of a basis set by running both the "APW + lo" and LAPW methods self-consistently for pure Cu and Ce as a function of the cutoff parameter $R_{MT}K_{max}$. The complementary local orbitals for $l =$ s, p, and d, which renders nine extra basis functions to be added to the number of plane waves in the case of Cu. They revealed that "APW + lo" reaches the final total energy (within 1 mRy) for $R_{MT}K_{max} = 9$, corresponding to approximately $75 + 9$ basis functions, while LAPW needs $R_{MT}K_{max} = 10$ or about 100 basis functions. The two methods are confirmed to converge to numerically the same densities and total energies as well as individual eigen energies.

As shown by Madsen et al. (2001), this new scheme converges practically to identical results as the FLAPW method, but allows one to reduce $R_{MT}K_{max}$ by about one, leading to significantly smaller basis sets (up to 50%) and thus the corresponding computational time is drastically reduced (up to an order of magnitude). They concluded that the "APW+lo method" has opened a way to perform FLAPW electronic structure calculations even for CMAs that were previously inaccessible due to computational limitations (Madsen et al., 2001).

2.4.8 WIEN2k–FLAPW Program Package

The FLAPW band calculations can be performed by employing the commercially available WIEN2k–FLAPW program package (Blaha et al., 2012). The WIEN2k employs a mixed "FLAPW" and "APW + lo" basis within the framework of DFT. This code converges to results identical to the LAPW method with effective reduction in the cutoff parameter $R_{MT}K_{max}$ by one, leading to a significantly smaller basis set so that electronic structure calculations even for 1/1–1/1–1/1 approximants have been made possible. In the rest of Section 2.4, we will introduce basic ideas of the FLAPW–Fourier analysis and the Hume-Rothery plot methods by using the "Fourier coefficient" of the wave function outside the MT sphere at a given **k** in the Brillouin zone and given energy eigen value E, all of which can be generated by running WIEN2k as a function of allowed reciprocal lattice vectors **G**.

2.4.9 WIEN2k–Hume-Rothery Plot Method

The wave function shown in Eqn 14 outside the MT sphere is composed of plane waves summed over reciprocal lattice vectors allowed to a given structure. This is the unique feature of FLAPW method, from which FsBz interactions can be most effectively extracted. Running self-consistent cycles (SCFs) followed by the task "DOS" in WIEN2k can generate the so-called "case.output1" file, which lists the Fourier coefficients of wave functions outside the MT sphere at selected wave vector **k** in the first Brillouin zone and energy eigen values specified by the wave vector **k** + **G**, where **G** is the allowed reciprocal lattice vector. The default command "WFFIL" during the task "Initial calc." should be replaced by its optional command "WFPRI" to generate the "case.output1" file. The in-house Fortran-90 program described below was developed to perform the Hume-Rothery plot analysis by using the "case.output1" file.

The number of **k** points, N_k, in the first Brillouin zone is fixed to be 100–30,000, depending on the number of atoms per unit cell N involved. A value of N_k as high as 30,000 is needed for transition metal elements like bcc Mo, Ta, Re etc. with $N = 2$, while 100 is high enough for CMAs with $N > 300$.

In WIEN2k, the first Brillouin zone is automatically partitioned into N_k meshes and $\mathbf{k}_i$ is fixed at each corner, where i runs from one to N_k. The Fortran program (Inukai et al., 2011a) was constructed in such a way that the squared sum of the real and imaginary parts of the Fourier coefficient $C_{\mathbf{k}_i+\mathbf{G}}$ of the LAPW function is calculated at wave vector $\mathbf{k}_i$ and the energy eigenvalue E_j specified by $\mathbf{k}_i + \mathbf{G}$ in the dispersion relation. The Fortran program then seeks the set of LAPWs $\{2|\mathbf{k}_i + \mathbf{G}|\}^2_{E_j}$ having the largest Fourier coefficient $\sum_{\mathbf{G}} |C_{\mathbf{k}_i+\mathbf{G}}|^2$ for a given E_j and $\mathbf{k}_j$.[13] This is done for all $\mathbf{k}_i$ values over the range $1 \leq i \leq N_\mathbf{k}$ in the first Brillouin zone in an energy interval $E_j \leq E < E_j + \Delta E$, where E_j runs from the bottom of the valence band up to $+30$ eV above the Fermi level with an increment ΔE generally set to be 0.05 eV for all systems studied.

An average value of $\{2|\mathbf{k}_i + \mathbf{G}|\}^2_{E_j}$ over $i = 1$ to $N_\mathbf{k}$ is now calculated by using the relation

$$\langle \{2|\mathbf{k} + \mathbf{G}|\}^2 \rangle_E = \frac{\sum_{i=1}^{N_\mathbf{k}} \omega_i \{2|\mathbf{k}_i + \mathbf{G}|\}^2_E}{\sum_{i=1}^{N_\mathbf{k}} \omega_i}, \tag{20}$$

where ω_i represents degeneracies, possibly zero, of the selected electronic states $\{2|\mathbf{k}_i + \mathbf{G}|\}^2_E$ in a given energy interval and a subscript "j" in E_j is omitted. This gives rise to a single-valued energy dispersion relation, which describes the motion of itinerant electrons outside the MT sphere.

Since Eqn 20 involves an averaging procedure over selected $\mathbf{k}_i$ points in the first Brillouin zone, the quantity $\langle \{2|\mathbf{k} + \mathbf{G}|\}^2 \rangle_E$ thus obtained would naturally become less reliable, if the electronic structure becomes anisotropic. Inside the d-band, for example, the reciprocal lattice vector $\mathbf{G}$ in $\{2|\mathbf{k}_i + \mathbf{G}|\}^2_{E_j}$ selected as the electronic state having the maximum Fourier coefficient, would no longer be uniquely fixed but diversified from mesh to mesh.

The reliability of $\langle \{2|\mathbf{k} + \mathbf{G}|\}^2 \rangle_E$ may be quantitatively assessed by calculating the variance, which is defined as the mean of the square of the variable x_i or $\langle x_i^2 \rangle$ minus the square of its mean or $\langle x_i \rangle^2$:

$$\sigma^2 = \langle x_i^2 \rangle - \langle x_i \rangle^2 = \frac{\sum \omega_i x_i^2}{\sum \omega_i} - 2\langle x_i \rangle \frac{\sum \omega_i x_i}{\sum \omega_i} + \langle x_i \rangle^2$$

$$= \frac{\sum \omega_i (x_i - 2x_i \langle x_i \rangle + \langle x_i \rangle^2)}{\sum \omega_i} = \frac{\sum \omega_i (x_i - \langle x_i \rangle)^2}{\sum \omega_i}, \tag{21}$$

where ω_i represents degeneracies. In the present case, the variance is explicitly expressed as

$$\sigma^2(E) = \frac{\sum_{i=1}^{N_\mathbf{k}} \omega_i \left(\{2|\mathbf{k}_i + \mathbf{G}|\}^2_E - \langle \{2|\mathbf{k} + \mathbf{G}|\}^2 \rangle_E \right)^2}{\sum_{i=1}^{N_\mathbf{k}} \omega_i}. \tag{22}$$

Note that Eqn 22 is in units of $(2\pi/a)^4$. In order to make the variance in Eqn 22 to be independent of the unit cell size, we take its square root to reduce the units to $(2\pi/a)^2$ and divide it by $\langle \{2|\mathbf{k} + \mathbf{G}|\}^2 \rangle_{E_F}$ or $(2k_F)^2$. The resulting dimensionless standard deviation is expressed as

$$F(E) \equiv \frac{\sqrt{\sigma^2(E)}}{(2k_F)^2}, \tag{23}$$

which is distributed over the range from zero to about 0.5.

---

[13] The summation is taken under the condition $|\mathbf{k}_i+\mathbf{G}|$=constant and, hence, it counts the degeneracy of the state $\mathbf{k}_i+\mathbf{G}$, which is generally more than two at high symmetry points in the Brillouin zone.

The E versus $\langle\{2|\mathbf{k}+\mathbf{G}|\}^2\rangle_E$ data points would fall on a straight line passing through the origin and the standard deviation $F(E)$ defined in Eqn 23 would be extremely small, provided that the free electron model holds well. However, the value of $\langle\{2|\mathbf{k}+\mathbf{G}|\}^2\rangle_E$ would lose its physical meaning, if $F(E)$ becomes large. Hence, the data points having large $F(E)$ should be ignored. For this purpose, we draw a straight line by hand to fit only the data points with small $F(E)$ in the E versus $\langle\{2|\mathbf{k}+\mathbf{G}|\}^2\rangle_E$ plot. Once this is done, the square of the Fermi diameter $(2k_F)^2$ can be determined from the intercept at the Fermi level. The effective electrons per atom ratio $\mathbf{e}/\mathbf{a}$ can be calculated by inserting $(2k_F)^2$ thus obtained into the relation

$$(\mathbf{e}/\mathbf{a})_{\text{total}} = \frac{\pi\{(2k_F)^2\}^{3/2}}{3N}, \tag{24}$$

where N is the number of atoms per unit cell and $(2k_F)^2$ is in units of $(2\pi/a)^2$, where a is the lattice constant. The E versus $\langle\{2|\mathbf{k}+\mathbf{G}|\}^2\rangle_E$ plot is specifically called the Hume-Rothery plot, since it allows us to determine both $(2k_F)^2$ and $\mathbf{e}/\mathbf{a}$, the latter of which plays a key role in the Hume-Rothery electron concentration rule. For a more detailed account of this method, we refer the interested readers to journal articles (Asahi et al., 2005; Inukai et al., 2011a, 2011b; Mizutani et al., 2012, 2006, 2008, 2009, 2010, 2011) or the book (Mizutani, 2010).

2.4.10 WIEN2k–FLAPW-Fourier Method

The FLAPW-Fourier analysis is performed by summing the squared sum of real and imaginary parts of the Fourier coefficient outside the MT sphere over equivalent zone planes, which is hereafter denoted as $\sum|C_{\mathbf{k}+\mathbf{G}}|^2$, and plotting it as functions of both energy eigen value and square of the wave vector $\{2|\mathbf{k}+\mathbf{G}|\}^2$, where the wave vector $\mathbf{k}$ is generally taken at one of high symmetry points of the first Brillouin zone. In the case of the bcc lattice, the vector $\mathbf{k}$ is chosen as $\mathbf{k} = (1/2\ 1/2\ 0)$, $\mathbf{k} = (0\ 0\ 0)$, and $\mathbf{k} = (1\ 0\ 0)$ corresponding to symmetry points N, Γ, and H of the bcc Brillouin zone, respectively (Asahi et al., 2005; Inukai et al., 2011a, 2011b; Mizutani, 2010; Mizutani et al., 2012, 2006, 2008, 2009, 2010, 2011). In WIEN2k, the information above can be extracted from the "case.output1" file generated upon execution of the task "Band Structure" after the SCF cycles and the task "DOS" are completed. The energy dependence of $\sum|C_{\mathbf{k}+\mathbf{G}}|^2$ constructed for $\{2|\mathbf{k}+\mathbf{G}|\}^2$-specified plane waves at a particular symmetry point of the Brillouin zone is called the FLAPW–Fourier spectrum.

The electronic states $\{2|\mathbf{k}+\mathbf{G}|\}^2$ at such high symmetry points with a given $\mathbf{k}$ vector are denoted by $|\mathbf{G}|^2$. From the energy spectrum thus constructed, we can identify $|\mathbf{G}|^2$-specified plane waves dominating the electronic states at the Fermi level (Mizutani, 2010). The values of $|\mathbf{G}|^2$s thus extracted are identified as *critical* ones. This is nothing but the execution of the Fourier analysis of wave functions outside the MT sphere and can be used as a powerful tool to extract the FsBz interactions involved. This is called the FLAPW-Fourier method and its application to CMAs will be discussed in the following sections.

As a matter of fact, the Fourier method discussed above is obviously a standard approach in NFE band calculations. However, as emphasized in Mizutani (2010), the NFE model is not suitably applicable to a system involving d-states in its valence band. In contrast, the FLAPW–Fourier method chooses FLAPW orbitals as basis functions and allows us to perform accurate first-principles band calculations for any realistic crystal and to extract FsBz interactions in the same way as in NFE band calculations.

2.4.11 "Local" versus "NFE" Methods in $(2k_F)^2$ Determination

The determination of $(2k_F)^2$ by reading off the ordinate at the Fermi level in the Hume-Rothery plot is straightforward and reliable, provided that the standard deviation $F(E)$ is small in the neighborhood of

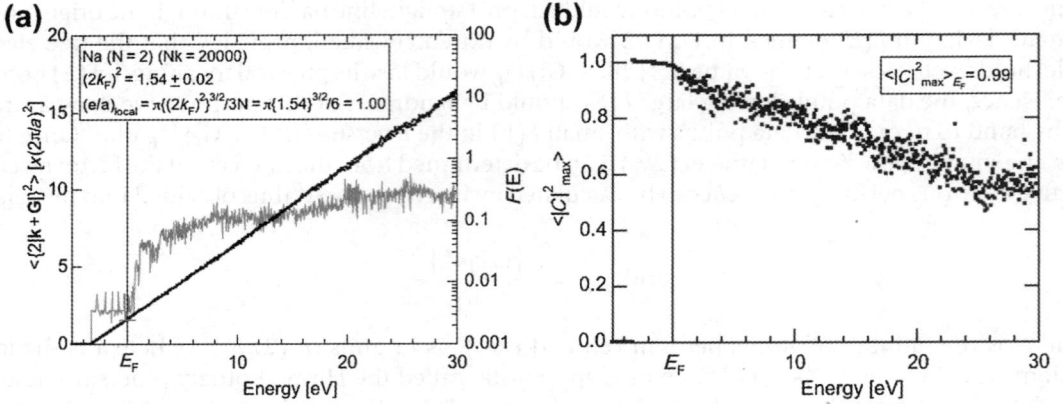

Figure 19 (a) Hume-Rothery plot and (b) energy dependence of $\langle|C|^2_{max}\rangle$ for Na. The "local" method is safely employed: a short horizontal red line is drawn at the Fermi level as a guide to the eye.

the Fermi level. As a typical example, the Hume-Rothery plot for Na is shown in **Figure 19**. The data points from the bottom of the valence band up to +30 eV almost perfectly fall on a straight line. The standard deviation is suppressed well below 0.1 below the Fermi level. Hence, the value of $(2k_F)^2$ can be safely determined by directly reading the ordinate at the Fermi level in **Figure 19a**. This may be hereafter called the "local" method, since "local" reading of data points across the Fermi level is validated. However, this is not always the case as we shall see.

At this stage, we introduce an additional parameter to judge the appropriateness of the "local" method. The maximum Fourier coefficient in wave function outside the MT sphere in energy range $E_j \leq E < E_j + \Delta E$ is picked up and averaged over all $\mathbf{k}_i$ values in the Brillouin zone. The resulting value at the Fermi level, which is denoted as $\langle \sum_{\mathbf{k+G}} |C_{\mathbf{k+G}}|^2_{max} \rangle_{E_F}$ or briefly $\langle |C|^2_{max} \rangle_{E_F}$, may be taken as an additional measure to judge the degree of itinerancy of electrons at the Fermi level. When the value is, say, below 0.1, we regard electrons at the Fermi level to be no longer itinerant. As shown in **Figure 19b**, the value for Na is almost unity, indicating that electrons at the Fermi level are almost perfectly itinerant.

Let us consider an opposite case showing a strongly localized character for electrons at the Fermi level. The Hume-Rothery plot and energy dependence of $\langle|C|^2_{max}\rangle$ for Fe are shown in **Figure 20a** and **b**, respectively. The data points in the Hume-Rothery plot heavily deviate from a linear behavior over a wide energy range across the Fermi level. The standard deviation well exceeds 0.1 over the energy range, where the Fe-3d band exists. The value of $\langle|C|^2_{max}\rangle_{E_F}$ is found to be only 0.036, indicating that electrons at the Fermi level are highly localized. Hence, we intentionally avoid data points over the energy range where the Fe-3d band exists and draw a red line by connecting data points below −6 eV and above about +20 eV, as shown in **Figure 20a**. The value of $(2k_F)^2$ is determined to be 1.6 ± 0.1 from the intersection of this line with the Fermi level. This is hereafter called the "NFE" method, since this hopefully allows us to cast the realistic electronic structure into the NFE framework without losing its underlying features. The accuracy in determining $(2k_F)^2$ for Fe is obviously poorer than that for Na obtained with the "local" method.

As is clear from the arguments above, we have to give up the "local" method and employ the "NFE" method, when the dimensionless standard deviation $F(E)$ across the Fermi level is >0.1 and the value of $\langle|C|^2_{max}\rangle_{E_F}$ is <0.1. The conditions above will be employed as a criterion in determining $(2k_F)^2$ from the Hume-Rothery plot in the rest of this chapter.

(a)

(b)

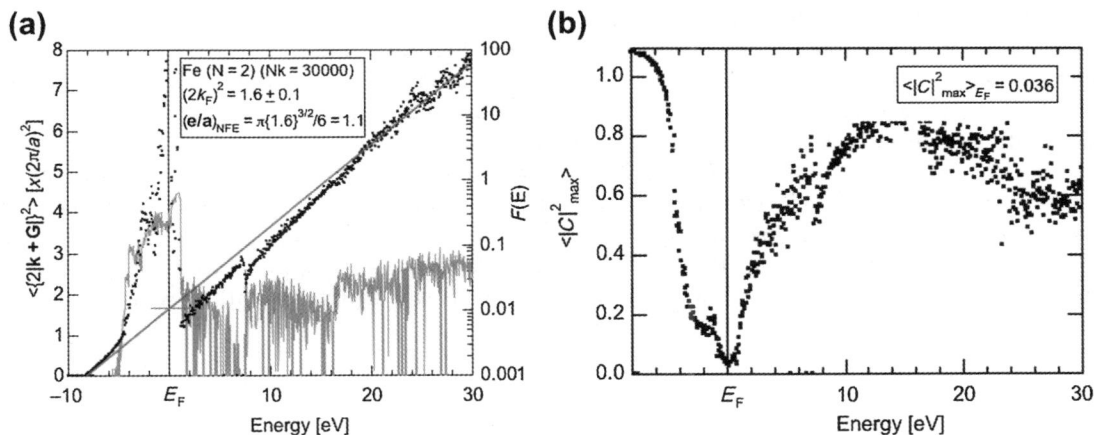

Figure 20 (a) Hume-Rothery plot and (b) energy dependence of $\langle |C|^2_{max} \rangle$ for nonmagnetic Fe. The NFE method has to be employed; a red NFE line is drawn by connecting data points, where the standard deviation $F(E)$ is suppressed below 0.1.

2.5 Stabilization Mechanism in a Series of Gamma-Brasses

2.5.1 Introduction

In the late 1920s Westgren and Phragmén (1928a, 1928b, 1929) pointed out for the first time that the gamma-brasses Cu_5Zn_8, Ag_5Cd_8, Al_4Cu_9, and $Cu_{31}Sn_8$ are all stabilized at a particular **e/a** value of 21/13 and obey the Hume-Rothery electron concentration rule in the same way as the β-brass.[14] Indeed, Mott and Jones (1936) were the first to discuss the **e/a** = 21/13 rule in terms of the contact of the free-electron Fermi sphere with the Brillouin zone of the gamma-brass, which is shown in **Figure 18**c. As emphasized in Section 2.4, however, no significant progress was made until recently to interpret the **e/a** = 21/13 rule for the Cu_5Zn_8 and Al_4Cu_9 gamma-brasses. As listed in **Table 2**, gamma-brasses can be classified into three groups. We will discuss in this section whether the Hume-Rothery stabilization mechanism holds universally or not in all gamma-brasses, regardless of the group classification, by making full use of the FLAPW–Fourier and Hume-Rothery plot methods.

In Section 2.5.2, we will guide the reader to get acquainted with the electronic structure of gamma-brasses by constructing a canonical gamma-brass electronic structure within the free-electron model. In Section 2.5.3, the stabilization mechanism in Group (I) gamma-brasses will be studied by choosing both Cu_5Zn_8 and Al_4Cu_9 gamma-brasses as representatives. In contrast to Group (I) gamma-brasses, those in Group (II) involve a large amount of the transition metal elements. Included in this group are Al_8V_5, Al_8Cr_5, Mn_3In, TM_2Zn_{11} (TM = Ni, Co, and Fe; Mizutani, 2010). Since the **e/a** value of the TM element is not a priori known, it has remained puzzling as to whether or not an ordered gamma-brass in Group (II) is also stabilized at **e/a** = 21/13 through the Hume-Rothery stabilization mechanism. Indeed, this has been regarded as one of the most challenging themes in

---

[14] The gamma-brass in the Cu–Sn system was later identified as forming a superlattice with space group F$\bar{4}$3m (Section 2.3.3).

the electron theory of metals in the twentieth century and efforts to elucidate its mechanism are a central issue of this chapter, with which we will deal in Section 2.5.4.

The Ag–Li gamma-brass is classified into Group (III). It has been identified to be isostructural to the prototype Cu_5Zn_8 (Arnberg and Westman, 1972; Noritake et al., 2007; Perlitz, 1933), though it consists of only monovalent elements Ag and Li without containing transition metal elements. Hume-Rothery (1962) himself mentioned in his book that "no combination of univalent elements can give the characteristic electron atom ratio of 21/13 although, if lithium were divalent, the above composition would be nearly that required for the 21/13 ratio." Hume-Rothery apparently tended to believe that the 21/13 rule can be universally applied to all gamma-brasses including the Ag–Li. Its stabilization mechanism as well as **e/a** determination will be discussed in Section 2.5.5.

2.5.2   Free-Electron Model for Gamma-Brasses

As noted in Section 2.5.1, we consider first the electronic structure of a gamma-brass constructed within the free-electron model. The energy dispersion relations along the direction ΓN for the bcc structure, to which gamma-brasses with space group $I\bar{4}3m$ belong, are shown in **Figure 21**.[15] Attention is now directed to the electronic states at the symmetry points N, at which many parabolic bands cross with each other. Crossing points can be indexed in terms of the square of the reciprocal lattice vector $|G|^2$. As discussed in Section 2.4.10, we are interested in the electronic state $2\{k+G\}$ of the plane wave in the FLAPW wave function outside the MT sphere. At symmetry points N or M, the wave vector **k** is chosen as ($^1/_2$ $^1/_2$ 0) so that $2\{k+G\}$ is expressed as ($1+2G_x$ $1+2G_y$ $2G_z$) or the set of odd, odd and even integers, since G_x, G_y, and G_z are integers. As a result, an insertion of reciprocal lattice vectors allowed to bcc lattice will give rise to $|G|^2$=2, 6, 10, 14, 18, 22, 26, 30, 34 ... at symmetry points N or M.

As already mentioned in relation to **Figure 18a**, the symmetry points N refer to the center of the {110} zone of the first Brillouin zone of a bcc lattice. Higher Brillouin zones can be equally constructed by bisecting perpendicularly reciprocal lattice vectors allowed to a given crystal structure (Mizutani, 2001). In the case of a bcc lattice, the center of higher zones, in which two integers in the set of Miller indices {hkl} are odd, always passes through the center of {110} zone planes upon reduction to the first zone. As shown in **Figure 21**, one can easily assign crossing points of free-electron parabolic bands at symmetry points N, which emerge one after another with increasing energy, as an increasing order of $|G|^2 = 2, 6, 10, 14, 18, 22, 26, 30, 34\cdots$ mentioned above. They obviously correspond to the center of {110}, {211}, {310}, {321}, {330}, {411} ... zone planes of the Brillouin zone. The Fermi level in **Figure 21** was drawn by filling electrons equal to e/a = 21/13 characteristic of the gamma-brass into the free-electron band. We see that electronic states near the Fermi level are dominated by plane waves specified by $|G|^2 = 18$ in the free-electron gamma-brass. This will serve as a key guide, when we discuss FsBz interactions in any realistic gamma-brasses in Group (I) to (III). It may be also noted that no electronic states exist over a wide energy range from −1 to about +2.5 eV at symmetry point Γ in **Figure 21**. This is a unique feature of the free-electron gamma-brass and has nothing to do with opening of a DOS pseudogap across the Fermi level.

---

[15] We focus on only gamma-brasses with space groups $I\bar{4}3m$ and $P\bar{4}3m$, the former being described as bcc and the latter as CsCl-type structure. The same dispersion relations as in **Figure 21** can be used for the CsCl-type gamma-brasses simply by replacing the symmetry point N by M (Mizutani, 2001).

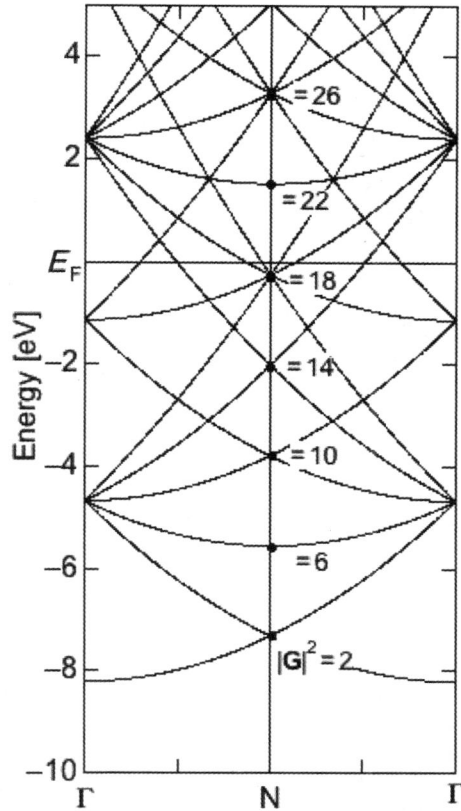

Figure 21   Energy dispersion relations along the direction ΓN for a canonical gamma-brass phase having **e/a** = 21/13 and N = 52 in the free-electron model (Mizutani, 2010).

2.5.3   Hume-Rothery Stabilization Mechanism in Cu_5Zn_8 and Al_4Cu_9 in Group (I)

Cu and Zn atoms possess 11 and 12 electrons outside the argon core, respectively. Hence, the Fermi level in both Cu_5Zn_8 and Al_4Cu_9 gamma-brasses is determined by filling $\dfrac{11 \times 5 + 12 \times 8}{13} = \dfrac{151}{13} =$ 11.615 and $\dfrac{11 \times 9 + 3 \times 4}{13} = \dfrac{111}{13} = 8.538$ electrons per atom into the respective valence bands. This is referred to as the valence electron concentration (**VEC**). Entirely different **VEC** values are assigned to Cu_5Zn_8 and Al_4Cu_9 gamma-brasses because Al donates 3 sp-electrons per atom while Zn donates 12 electrons per atom (2 sp- and 10 d-electrons) to the valence band. Instead, the number of itinerant electrons per atom, as already referred to as e/a, is concomitantly 21/13 $\left(= \dfrac{1 \times 5 + 2 \times 8}{13} = \dfrac{9 \times 1 + 3 \times 4}{13} \right)$, as Cu, Zn, and Al are monovalent, divalent, and trivalent elements, respectively. As is clear from the arguments above, the Hume-Rothery electron concentration rule is meaningful only when the parameter e/a is used as an electron concentration parameter. We will learn more about this in the rest of this section.

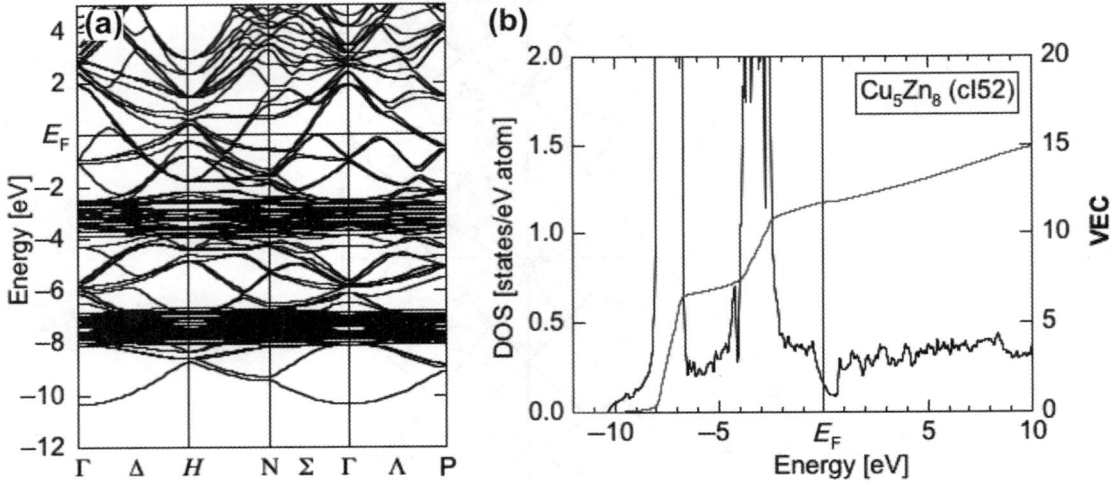

Figure 22   (a) Energy dispersion relations and (b) DOS calculated using WIEN2k for Cu_5Zn_8 gamma-brass.

To the best of our knowledge, an attempt to discuss the stability of the gamma-brass on the basis of first-principles band calculations was made for the first time by Paxton et al. (1997). They employed LMTO–ASA band calculations and claimed that not only {411} and {330} zones associated with $|G|^2 = 18$ but also {421}, {332}, and {422} zones contribute to the formation of the pseudogap at the Fermi level. Unfortunately, however, their conclusion is not consistent with that drawn from the FLAPW–Fourier analysis discussed below. We must say that information about FsBz interactions deduced from LMTO–ASA band calculations is fairly limited (Mizutani, 2010).

Asahi et al. (2005) were the first to discuss the Hume-Rothery stabilization mechanism in Cu_5Zn_8 and Al_4Cu_9 gamma-brasses by performing first-principles FLAPW band calculations with subsequent application of their newly developed FLAPW–Fourier method. The E–k relations and DOS derived from the FLAPW method for Cu_5Zn_8 are shown in **Figure 22 (a)** and **(b)**, respectively (Asahi et al., 2005; Mizutani, 2010). A large DOS around energies centered at -7.5 eV is due to the Zn-3d band, whereas that in the range from -2.5 to -4 eV is due to the Cu-3d band. A steep pseudogap is seen across the Fermi level. Its width is about 1.2 eV. The E–k relations and DOS for Al_4Cu_9 are similarly shown in **Figure 23 (a)** and **(b)**, respectively. The pseudogap is again observed across the Fermi level and its width is about 1.0 eV. The Cu-3d band is about 4 eV in width and is wider than that in Cu_5Zn_8, reflecting an increase in Cu concentration.

A closer look into E–k relations in **Figures 22a** and **23a** reveals the absence of electronic states below about -1 and above about $+1$ eV across the Fermi level at symmetry points N and M in Cu_5Zn_8 and Al_4Cu_9, respectively. As is clear from the arguments on **Figure 21**, symmetry points N and M correspond to the center of the {110}, {211}, {310}, {321}, {330}, {411} ... zone planes of the Brillouin zone of the bcc and CsCl structures, respectively. The DOS pseudogap of 1.2–1.0 eV in width in **Figures 22b** and **23b** almost coincides with the range of empty states at symmetry points N (or M) across the Fermi level in the corresponding dispersion relations shown in **Figure 22a**. Electronic states are also absent from -1 to $+2$ eV at the symmetry point Γ. However, as noted in Section 2.5.2, this is a characteristic feature of the free-electron model and makes no contribution to the formation of the DOS pseudogap. We are, thus, well convinced to

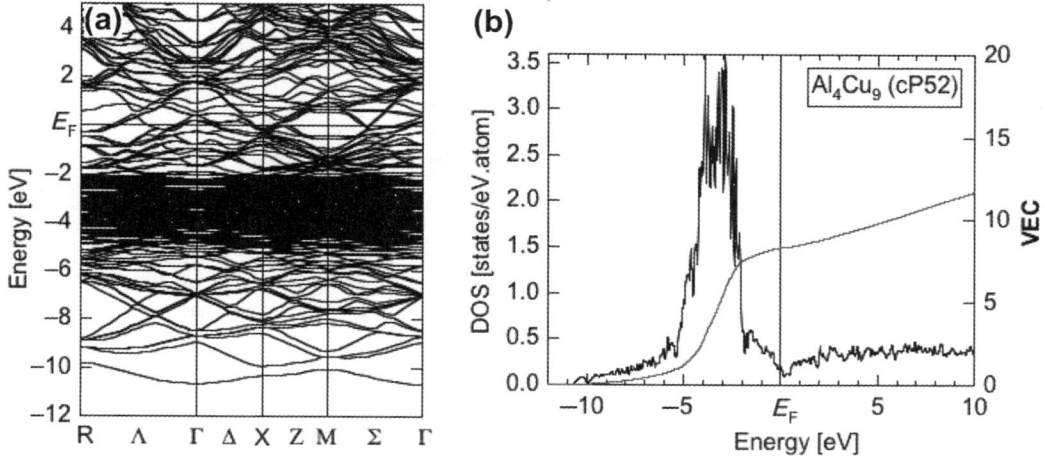

Figure 23 (a) Energy dispersion relations and (b) DOS calculated using WIEN2k for Al_4Cu_9 gamma-brass.

say that the DOS pseudogap in both Cu_5Zn_8 and Al_4Cu_9 gamma-brasses essentially originates from FsBz interactions at symmetry points N (or M). Now we are ready to index all electronic states appearing with increasing energies at symmetry points N and M in **Figures 22a** and **23a** by making full use of the FLAPW–Fourier method.

Figure 24 illustrates the matrix of the Fourier coefficient $\sum|C_{k+G}|^2$, or more precisely the square of the Fourier coefficients summed under the condition $|k_i+G|$=constant (see footnote 13). in the FLAPW wave function outside the MT sphere, as functions of energy eigenvalues E_1, E_2, E_3 .... in its column and $|G|^2$ in its row at symmetry points N for the gamma-brass. Suppose that E_2 in the matrix refers to the energy eigenvalue at the bottom of the pseudogap at symmetry points N. This corresponds to $E = -0.62$ eV below E_F in the case of Cu_5Zn_8, as can be seen from **Figure 22a**. One can construct either the $|G|^2$-dependence of the Fourier coefficient at selected energy-eigenvalue, say, at -0.62 eV for Cu_5Zn_8 or the energy dependence of the Fourier coefficient at selected $|G|^2$, as shown in **Figure 24**. In the rest of our discussions, we will employ the latter spectrum, that is, the energy dependence of the FLAPW–Fourier coefficients at selected $|G|^2$s.

Figure 25a and **b** display the FLAPW–Fourier spectra at symmetry points N (or M) over reciprocal lattice vectors $|G|^2$ from 2 to 26 for Cu_5Zn_8 and Al_4Cu_9, respectively. The DOS is superimposed. The $|G|^2 = 2$ states are concentrated near the bottom of the valence band and the center of gravity of each plane wave distribution increases with increasing $|G|^2$ or with increasing energy. This is quite consistent with the behavior expected from the free-electron model shown in **Figure 21**. Two important remarks should be addressed. First, electronic states are missing over about 1 eV across the Fermi level. It is clearly seen that electronic states corresponding to the bottom and top of the DOS pseudogap are exclusively dominated by $|G|^2 = 18$ in both Cu_5Zn_8 and Al_4Cu_9. This suggests the interference of electrons at the Fermi level with the set of {330} and {411} lattice planes to split the free-electron-like electronic states into bonding and antibonding states, leaving a pseudogap across the Fermi level. If this is indeed the case, the interference condition or the Hume-Rothery matching condition $(2k_F)^2 = |G|^2$ must hold. Its validity will be tested below.

Another important message is that plane waves are split into bonding and antibonding states at the bottom and top of Zn-3d and Cu-3d bands. In the case of Cu_5Zn_8, Zn-3d, and Cu-3d bands cause the

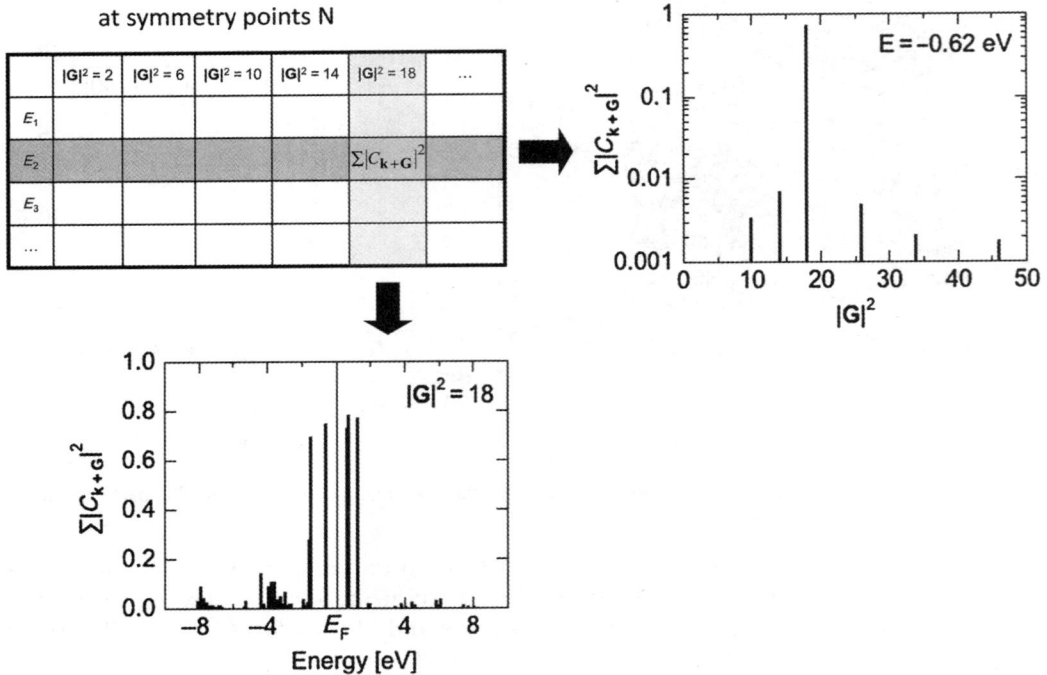

Figure 24 Illustration for construction of the FLAPW–Fourier energy spectrum and $|\mathbf{G}|^2$-dependence spectrum. The matrix lists the FLAPW–Fourier coefficient $\sum|C_{\mathbf{k+G}}|^2$ as functions of $|\mathbf{G}|^2$ in columns and energy eigenvalue in rows at symmetry points N for Cu_5Zn_8 gamma-brass.

splitting of $|\mathbf{G}|^2 = 6$ and 14 states, respectively. The splitting of $|\mathbf{G}|^2 = 14$ states is the most significant in Al_4Cu_9. The splitting is likely caused as a result of the orthogonality condition between sp- and d-like states. We hereafter call it *d states-mediated splitting*.

The Hume-Rothery plot for Cu_5Zn_8 is shown in **Figure 26a** along with the energy dependence of $\langle|C|^2_{max}\rangle$ in (b). The value of $\langle|C|^2_{max}\rangle_{E_F}$ is 0.57 and the standard deviation at the Fermi level $F(E_F)$ is <0.1. As noted in Section 2.4.11, we can safely determine the square of the Fermi diameter $(2k_F)^2$ by reading off the "local" value across the Fermi level, as shown in **Figure 26a**. The value is found to be 18.5 ± 0.1 in good agreement with $|\mathbf{G}|^2 = 18$ deduced from the FLAPW–Fourier analysis above. This confirms the validity of the Hume-Rothery matching condition. The **e/a** value encompassed by the Fermi sphere turned out to be 1.60.

Both Hume-Rothery plot and energy dependence of $\langle|C|^2_{max}\rangle$ for Al_4Cu_9 are shown in **Figure 27a and b**, respectively. Both $\langle|C|^2_{max}\rangle_{E_F} = 0.27$ and $F(E_F)$ <0.1 allow us to employ the "local" method to determine $(2k_F)^2$. The resulting value of 18.5 ± 0.1 is in good agreement with that for Cu_5Zn_8 within the accuracy of the present analysis. The **e/a** value turns out to be 1.60. All the results above verified that the DOS pseudogap across the Fermi level in both Cu_5Zn_8 and Al_4Cu_9 is induced by the FsBz interaction exclusively involving $|\mathbf{G}|^2 = 18$ and the Hume-Rothery electron concentration rule holds well with **e/a** = 21/13.

The same analysis has been confirmed to hold true in other gamma-brasses like Ag_5Zn_8 and Ag_9In_4 in Group (I) (Inukai et al., 2011b). Therefore, we conclude that the Hume-Rothery stabilization mechanism is fully validated for gamma-brasses in Group (I).

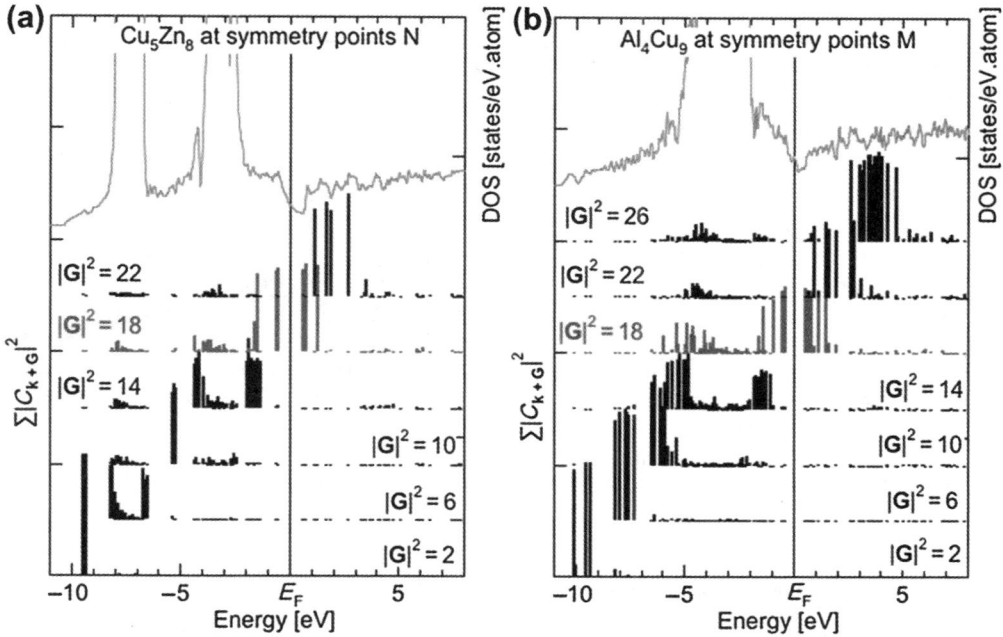

Figure 25 FLAPW–Fourier energy spectra (a) at symmetry points N for Cu_5Zn_8 gamma-brass and (b) at symmetry points M for Al_4Cu_9 gamma-brass. The respective total DOSs are superimposed. (For color version of this figure, the reader is referred to the online version of this book.)

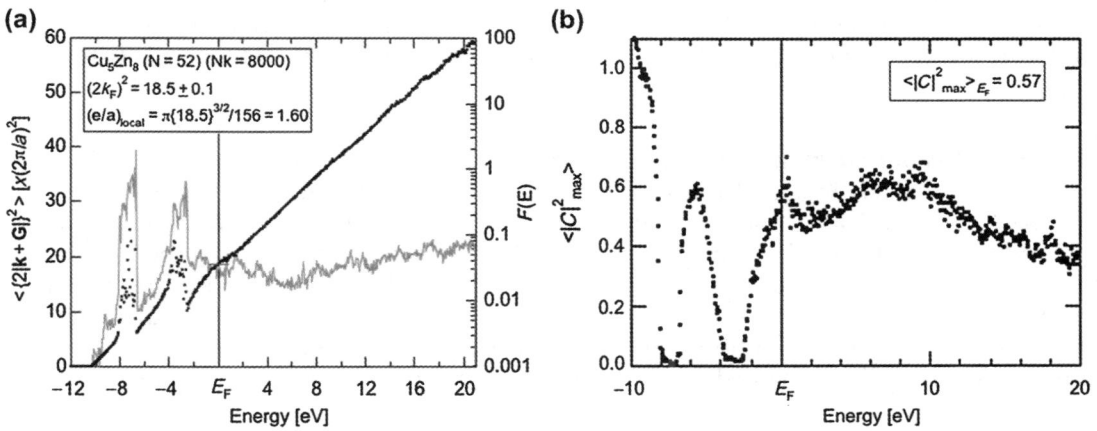

Figure 26 The Hume-Rothery plot and energy dependence of $\langle |C|^2_{max} \rangle$ for Cu_5Zn_8 gamma-brass.

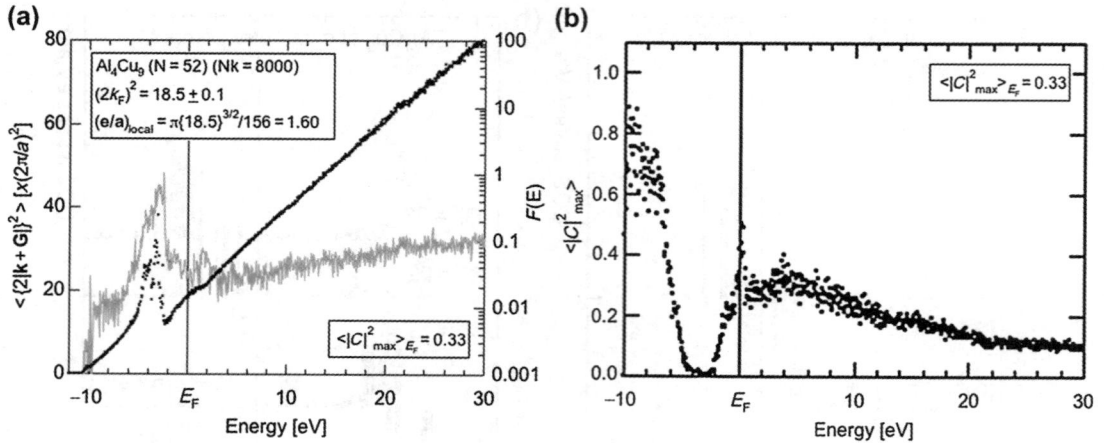

Figure 27   The Hume-Rothery plot and energy dependence of $\langle|C|^2_{max}\rangle$ for Al_4Cu_9 gamma-brass.

2.5.4   Stabilization Mechanism in Gamma-Brasses in Group (II)

Two gamma-brasses Al_8V_5 and Mn_3In are selected from Group (II) to analyze their stabilization mechanism in terms of the FLAPW–Fourier method.

2.5.4.1   *Construction of the Al_8V_5 Model Structure*

Brandon et al. (1977) revealed that Al_8V_5 gamma-brass is isostructural to Cu_5Zn_8 gamma-brass with space group $I\bar{4}3m$ and contains 52 atoms in the unit cell with the lattice constant $a = 0.9234$ nm. It is found that four sites on IT are shared by two Al and two V atoms, four sites on OT by four V atoms, six sites on octahedral (OH) by four V atoms, and two Al atoms and 12 sites on cuboctahedral (CO) by 12 Al atoms. For first-principles band calculations, four Al and six V atoms are exclusively filled into sites IT and OH, respectively. This is made possible without changing the overall composition of Al_8V_5. The experimentally determined fractional coordinates of all 52 atoms in the unit cell and the lattice constant (Brandon et al., 1977) are employed in FLAPW band calculations discussed below (Mizutani et al., 2006).

2.5.4.2   *Extraction of FsBz Interactions in Al_8V_5 Gamma-Brass*

The E–$\mathbf{k}$ dispersion relations and DOS for Al_8V_5 gamma-brass are shown in **Figure 28** (Mizutani, 2010; Mizutani et al., 2006). The V-3d band is extended over -3 to $+3$ eV across the Fermi level. It is apparently split into bonding and antibonding bands, leaving a deep DOS pseudogap at about $+0.5$ eV. We consider the V-3d/Al-3p orbital hybridizations to be mainly responsible for stabilizing Al_8V_5 gamma-brass, since the Fermi level falls at such a position that the bonding subband is almost fully filled while the antibonding subband is completely unoccupied. As a result, Al_8V_5 gamma-brass may well be regarded as being typical of an orbital hybridization-induced pseudogap system.

As emphasized in Sections 2.5.2–2.5.3, the reciprocal lattice vector $|\mathbf{G}|^2 = 18$ plays a specific role in forming an FsBz-induced pseudogap at the Fermi level and, thereby, contributing to the

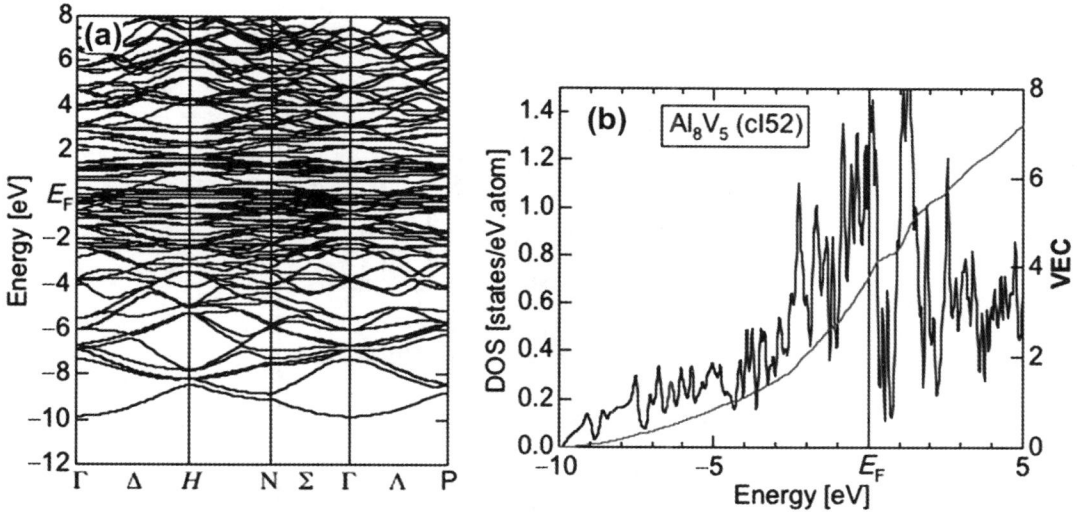

Figure 28   (a) Energy dispersion relations and (b) DOS calculated using the WIEN2k for Al_8V_5 gamma-brass (Mizutani et al., 2006).

stabilization of the gamma-brass phase in Group (I). It is of crucial importance to examine if the FsBz interaction involving $|G|^2 = 18$ remains important in Al_8V_5 gamma-brass, in which the V-3d band widely spreads across the Fermi level. The interference effect of electron waves with different sets of lattice planes in the presence of the V-3d band can be still studied by means of the FLAPW–Fourier method.

As shown in **Figure 29**, the energy spectrum of $\sum |C_{k+G}|^2$ at symmetry points N is constructed over $|G|^2$ values from 2 to 26 in the same manner as in **Figure 25**a and b for Cu_5Zn_8 and Al_4Cu_9. The total DOS is superimposed with gray in color. The V 3d-mediated splitting begins to set in when $|G|^2$ reaches 18. The intensity ratio of the resulting bonding states formed near the bottom of the V-3d band over the antibonding states formed near its top is extremely large at $|G|^2 = 18$ but gradually decreases with further increase in $|G|^2$. The intensity of bonding states becomes comparable to that of antibonding states at $|G|^2 = 26$, as can be seen in **Figure 29**. The intensity of antibonding states becomes stronger and stronger with further increase in $|G|^2$ beyond 26. Hence, the contribution of $|G|^2$ higher than 26 to the stability may well be ignored.

It is of crucial importance to realize that the $|G|^2 = 18$ states should have appeared a few electronvolts above the $|G|^2 = 14$ states or near the Fermi level, if the V-3d band were absent and the free-electron model were applied. In reality, we find that the $|G|^2 = 18$ states are pushed down to the bottom of the V-3d band as a result of the V 3d-mediated splitting. The same argument essentially holds true for $|G|^2 = 22$, which would have appeared immediately above the Fermi level, if the V-3d band were to be absent. From the arguments above, it is safely said that both $|G|^2 = 18$ and 22 states most significantly participate in the formation of bonding states near the bottom of the V-3d band and contribute to lowering the electronic energy. Among them, the $|G|^2 = 18$ waves characteristic of the gamma-brass structure make the largest contribution to lowering the electronic energy by forming bonding states below -2 eV near the bottom of the V-3d band and the $|G|^2 = 22$ waves are also important, since significant bonding states are formed below the Fermi level.

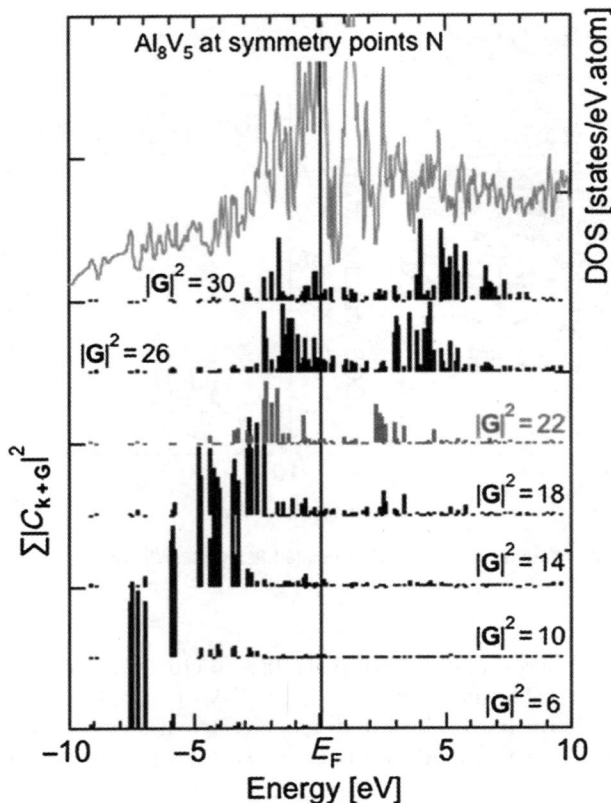

Figure 29   FLAPW–Fourier energy spectra at symmetry points N for Al_8V_5 gamma-brass (Mizutani et al., 2006). Its DOS is superimposed. (For color version of this figure, the reader is referred to the online version of this book.)

In contrast to gamma-brasses in Group (I), we revealed two *critical* reciprocal lattice vectors $|G|^2 = 18$ and 22, at least, at symmetry points N. The FLAPW-Fourier spectra at symmetry points Γ and H are shown in **Figure 30**a and b, respectively. We immediately see that the $|G|^2 = 24$ at symmetry point Γ and $|G|^2 = 20$ at symmetry points H also contribute to lowering the band structure energy through the V 3d states-mediated splitting. All the data above lead us to conclude that Al_8V_5 gamma-brass is stabilized through FsBz interactions involving $|G|^2 = 18$, 20, 22, and 24 as a result of the d states-mediated splitting. Thus, the stabilization mechanism for Al_8V_5 in Group (II) can be discussed in the framework of FsBz interactions. We revealed that it is different from that for gamma-brasses in Group (I) due certainly to the d states-mediated splitting in the presence of the V-3d band across the Fermi level. The *critical* reciprocal lattice vectors are no longer limited to solely $|G|^2 = 18$ but cover more than two in its neighborhood.

The Hume-Rothery plot for Al_8V_5 gamma-brass is shown in **Figure 31** along with the energy dependence of $\langle |C|^2_{max} \rangle$. The standard deviation $F(E)$ exceeds 0.1 over the energy range from -2 to $+2$ eV, where the V-3d band exists. Moreover, the value of $\langle |C|^2_{max} \rangle_{E_F}$ is merely 0.048 and is so small that electrons at the Fermi level should be regarded as being well localized. Hence, we inevitably choose the "NFE" approximation and draw a straight line as a guide, as shown in **Figure 31**a. The square of the

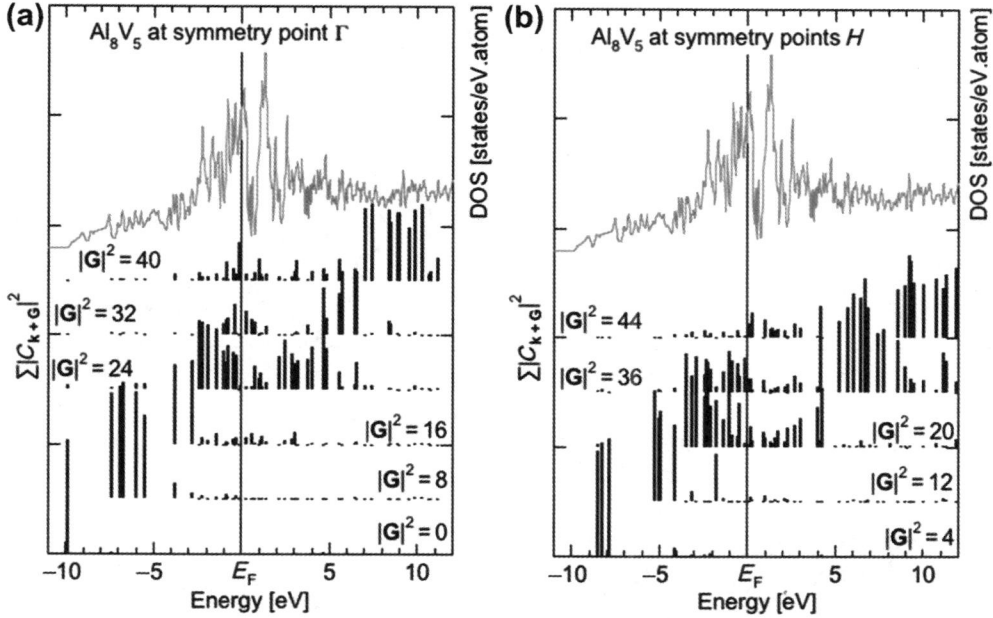

Figure 30 FLAPW–Fourier energy spectra at (a) symmetry point Γ and (b) symmetry points H for Al_8V_5 gamma-brass. Its DOS is superimposed.

Fermi diameter $(2k_F)^2$ is determined to be 22.8 ± 0.5 in units of $(2\pi/a)^2$.[16] The value is much higher than 18. The value of $(e/a)_{total}$ is deduced to be 2.19 and is no longer close to 21/13 (=1.615).

2.5.4.3 Construction of Mn₃In Model Structure

There exist significant mixtures of Mn and In atoms on both OH and CO sites in one of the 26-atom clusters "b" in Mn_3In with space group $P\bar{4}3m$ (Brandon et al., 1979). All six and 12 sites on b-OH and b-CO were fully filled with Mn and In atoms, respectively, to eliminate the chemical disorder (Inukai et al., 2011b). The resulting composition is shifted to $Mn_{40}In_{12}$ or 76.9 at.%Mn, being slightly higher in Mn concentration than that of Mn_3In or 75 at.%Mn.

2.5.4.4 Extraction of FsBz Interactions in Mn₃In Gamma-Brass

The E–$\mathbf{k}$ dispersion relations and DOS for the Mn_3In gamma-brass are shown in **Figure 32**a and b, respectively (Inukai et al., 2011b). The Mn-3d band is extended over -2 to $+1$ eV across the Fermi level. A deep pseudogap is located at about -1 eV below the Fermi level. As discussed for the Al_8V_5 gamma-brass in Section 2.5.4.2, Mn-3d/In-4sp orbital hybridizations are certainly responsible for splitting into bonding and antibonding Mn subbands and causing the pseudogap. The fact that the Fermi level is positioned in the middle of the antibonding Mn subband, no longer allows us to discuss its stability simply in terms of an orbital hybridization-induced pseudogap. Let us now try to extract FsBz interactions in the Mn_3In gamma-brass.

---

[16] In the original paper (Mizutani et al., 2006), the square of the Fermi diameter was determined to be 21.0 by taking an average of the two extrapolated straight lines: one from the data below -3 eV and the other from the data above $+3$ eV.

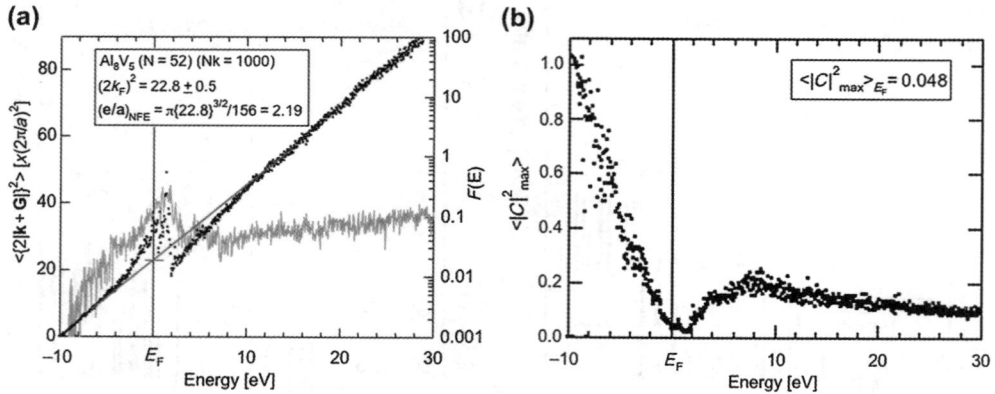

Figure 31 The Hume-Rothery plot and energy dependence of $\langle |C|^2_{max} \rangle$ for Al_8V_5 gamma-brass. (For color version of this figure, the reader is referred to the online version of this book.)

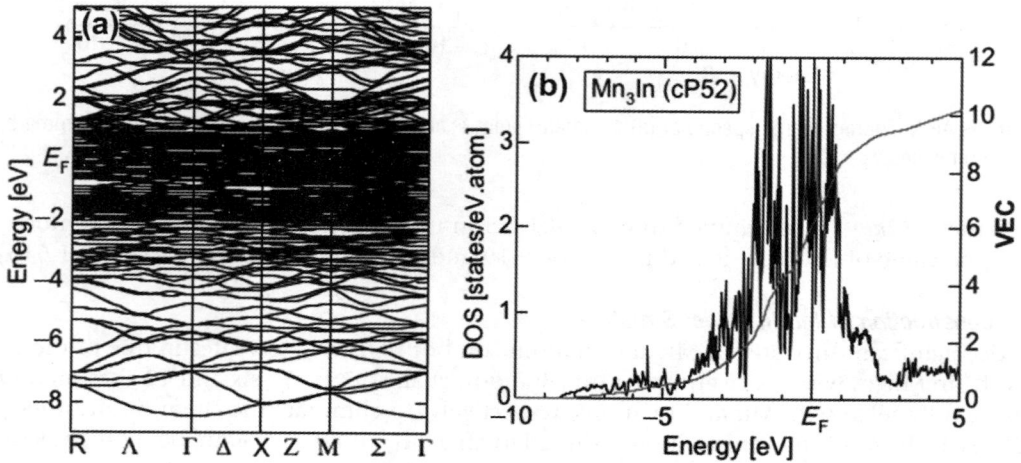

Figure 32 (a) Energy dispersion relations and (b) DOS calculated using the WIEN2k for Mn_3In gamma-brass (Inukai et al., 2011b).

Figure 33a and b show the FLAPW–Fourier spectra at symmetry points M and Γ for the Mn_3In gamma-brass, respectively. Owing to the Mn 3d-mediated splitting, the bonding states of the plane waves $|\mathbf{G}|^2 = 14$ and 18 are formed near the bottom of the Mn-3d band. They are obviously shifted to lower energies relative to the case, where Mn-3d band were absent like in the free-electron model shown in **Figure 21**. The shift is apparently the most significant in $|\mathbf{G}|^2 = 14$ and is also quite noticeable in $|\mathbf{G}|^2 = 18$ at symmetry points M and 16 and 20 at symmetry point Γ as well. They must be certainly responsible for contributing to lowering the electronic energy and, thereby, stabilizing the Mn_3In gamma-brass structure. The presence of more than two *critical* plane waves is again claimed to be a characteristic feature of the gamma-brasses in Group (II), where orbital hybridizations are substantial.

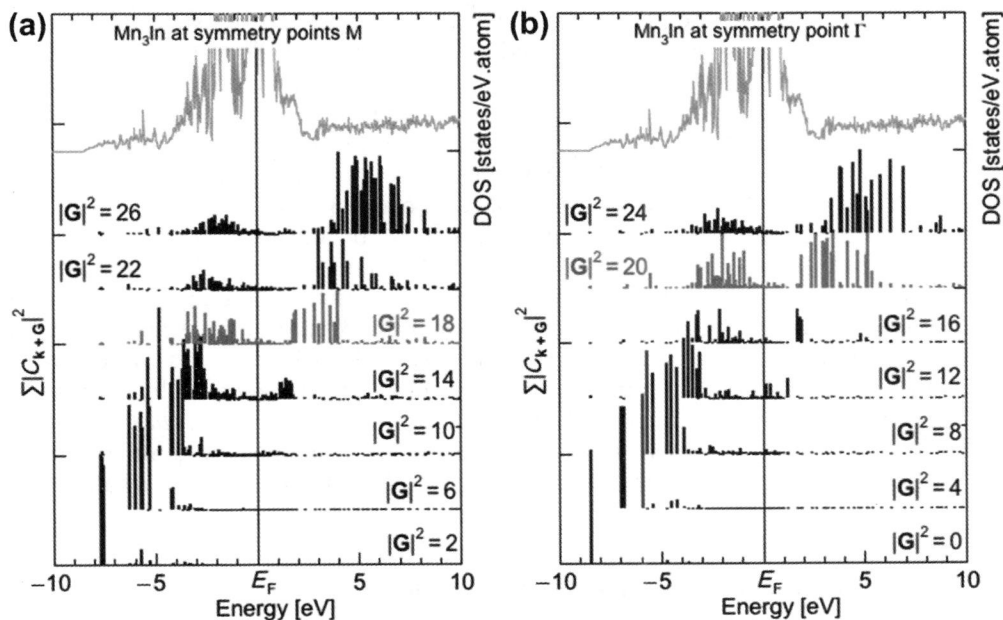

Figure 33 FLAPW–Fourier energy spectra at (a) symmetry points M and (b) symmetry point Γ for Mn₃In gamma-brass (Inukai et al., 2011b). Its DOS is superimposed. (For color version of this figure, the reader is referred to the online version of this book.)

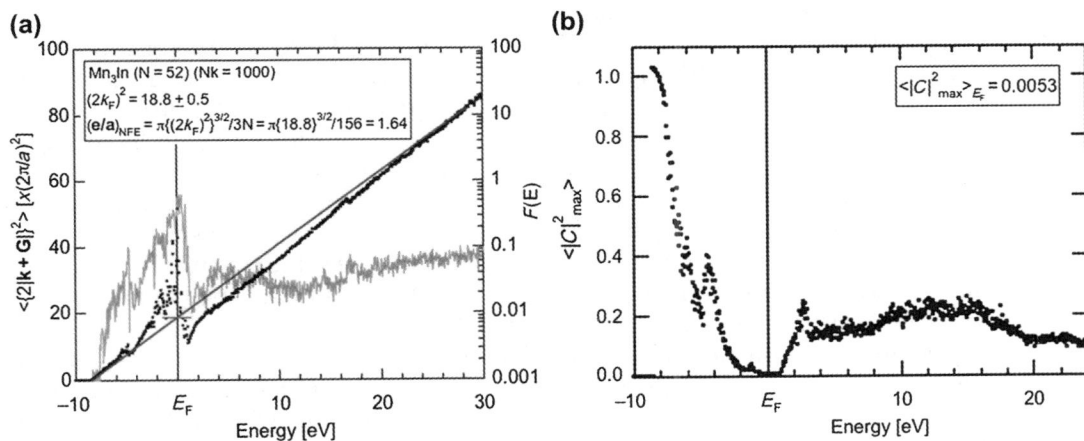

Figure 34 The Hume-Rothery plot and energy dependence of $\langle |C|^2_{max} \rangle$ for Mn₃In gamma-brass. (For color version of this figure, the reader is referred to the online version of this book.)

The Hume-Rothery plot and the energy dependence of $\langle |C|^2_{max} \rangle$ for the Mn_3In gamma-brass are plotted in **Figure 34a** and **b**, respectively. The standard deviation across the Fermi level is higher than 0.1 over energies, where the Mn-3d band exists, and $\langle |C|^2_{max} \rangle_{E_F}$ is only 0.0053. Therefore, the "NFE" approximation must be chosen. A straight line is drawn, as shown in **Figure 34a**, to pass through the data points, where the standard deviation is suppressed down to the level of 0.1. The value of $(2k_F)^2$ turns out to be 18.8 ± 0.5 and the resulting **e/a** value to be 1.64. The matching condition is again satisfied, since $(2k_F)^2$ is found in the range of *critical* $|G|^2$s over 16 to 20.

In summary, gamma-brasses in Group (II) are characterized by a deep DOS pseudogap induced by orbital hybridizations between transition metal elements like V and Mn and polyvalent elements like Al and In. The position of the pseudogap is, however, not necessarily located right at the Fermi level. Thus, the stability cannot be argued simply in terms of an orbital hybridization-induced pseudogap. Instead, the stability can be more consistently discussed on the basis of FsBz interactions. We could point out that d states-mediated splitting is responsible for lowering electronic energies by forming bonding states near the bottom of the TM-d band and that it involves more than two *critical* reciprocal lattice vectors centered at $|G|^2 = 18$, which serves as a crucial role in characterizing the gamma-brass structure. The values of $(2k_F)^2$ in Al_8V_5 and Mn_3In are deduced to be 22.8 and 18.8 and the resulting **e/a** values to be 2.19 and 1.64, respectively.

2.5.5 Stabilization Mechanism in Gamma-Brasses in Group (III)

In 1933, Perlitz (1933) identified the Ag_3Li_{10} compound to crystallize into the same structure as that of the prototype gamma-brass Cu_5Zn_8 and to possess its lattice constant of 0.994 nm. This is seemingly at variance with the Hume-Rothery electron concentration rule for the gamma-brass, as long as valencies of both Ag and Li are assumed to be unity. As mentioned earlier, Hume–Rothery (1962) believed that the Ag_3Li_{10} gamma-brass would be also stabilized near 21/13, if lithium were divalent. No theoretical verification for his postulate on the **e/a** value of the Ag-Li gamma-brass has been so far attempted. The "abnormal" Ag_5Li_8 gamma-brass is classified into Group (III) to analyze their stabilization mechanism in terms of the FLAPW–Fourier method (Mizutani et al., 2008).

2.5.5.1 *Construction of the Ag_5Li_8 Model Structure*

In 1972, Arnberg and Westman (1972) reported the structure of the $Ag_{30.2}Li_{69.8}$ gamma-brass by analyzing measured powder X-ray diffraction data. They identified its structure as a stacking of the 26-atom cluster to form the body-centered cubic lattice with space group $I\bar{4}3m$ (**Figure 14**). However, the reliability factor R_I was reduced to only about 10% because of the difficulty in its handling in air. More recently, Noritake et al. (2007) prepared the $Ag_{36}Li_{64}$ gamma-brass, the oxidation of which was minimized by conducting all operations in a glove box with flowing purified argon gas and reliably determined its crystal structure by analyzing the powder diffraction pattern taken with the use of synchrotron radiation beam. They found that Li atom enters exclusively into IT and CO sites, whereas the Ag atom into those on OT and OH sites in the 26-atom cluster but that small amounts of Li are mixed into OT and OH sites, resulting in chemical disorder.

An ordered structure was constructed simply by ignoring a small amount of Li atoms on OT and OH sites, where otherwise Ag atoms are occupied. This was made with a minimum sacrifice from the best-refined structure (Mizutani et al., 2008). The ordered Ag_5Li_8 gamma-brass with the lattice constant of 0.99066 nm was adopted. It may be noted that 38.5 at.%Ag concentration in Ag_5Li_8 is slightly off from the limiting concentration of 36.5 at.%Ag in the gamma-brass phase field in the equilibrium phase diagram (Okamoto, 2000).

2.5.5.2 Extraction of FsBz Interactions in Ag₅Li₈ Gamma-Brass

The energy dispersion relations and DOS for the Ag$_5$Li$_8$ gamma-brass are plotted in **Figure 35**a and b, respectively (Mizutani, 2010; Mizutani et al., 2008). The Ag-4d band is located near the bottom of the valence band. There is no measurable DOS pseudogap at the Fermi level. Instead, a deep pseudogap is formed at about +2 eV above the Fermi level. To identify its origin, we plotted in **Figure 36** the FLAPW–Fourier spectra at symmetry points N for the Ag$_5$Li$_8$ gamma-brass. The $|\mathbf{G}|^2 = 14$ states are centered at the Fermi level and exhibit no splitting to form any pseudogap. The distribution of the $|\mathbf{G}|^2 = 18$ states is centered in the vicinity of +2.0 eV in a sharp contrast to that formed across the Fermi level in the Cu$_5$Zn$_8$ gamma-brass in Group (I) (**Figure 25**). From this we can safely attribute the DOS pseudogap at about +2 eV to the FsBz-induced pseudogap associated with $|\mathbf{G}|^2 = 18$. The fact that the pseudogap appears at around +2 eV above the Fermi level, strongly suggests the electron concentration **e/a** to be much lower than 21/13, which serves as a key electron concentration parameter in the Hume-Rothery electron concentration rule for gamma-brasses in Group (I). It is therefore timely to determine the values of $(2k_F)^2$ and **e/a** by performing the Hume-Rothery plot analysis for the Ag$_5$Li$_8$ gamma-brass.

The Hume-Rothery plot and the energy dependence of $\langle |C|^2_{max} \rangle$ are shown in **Figure 37**a and b, respectively. Since the value of $\langle |C|^2_{max} \rangle_{E_F} = 0.42$ is large, we can apply the "local" method to determine $(2k_F)^2$ by reading off the ordinate at the Fermi level in **Figure 37**a. As indicated, the value turned out to be 13.5 ± 0.1 in a good agreement with the *critical* $|\mathbf{G}|^2$ equal to 14. Note that the matching condition $(2k_F)^2 = |\mathbf{G}|^2$ holds well by involving $|\mathbf{G}|^2 = 14$ but does not lead to the formation of a pseudogap at the Fermi level. The resulting **e/a** value turned out to be unity, being taken as the confirmation that both Ag and Li are indeed monovalent. As is clear from **Figure 37**, the ordinate reaches the value close to 18, if the Fermi level is raised to energies around +2.0 eV, where an FsBz-induced pseudogap exists. This means that the electron concentration of unity in Ag$_5$Li$_8$ is too small to meet the matching condition $(2k_F)^2 = |\mathbf{G}|^2$ involving 18, that is characteristic of gamma-brasses in Group (I).

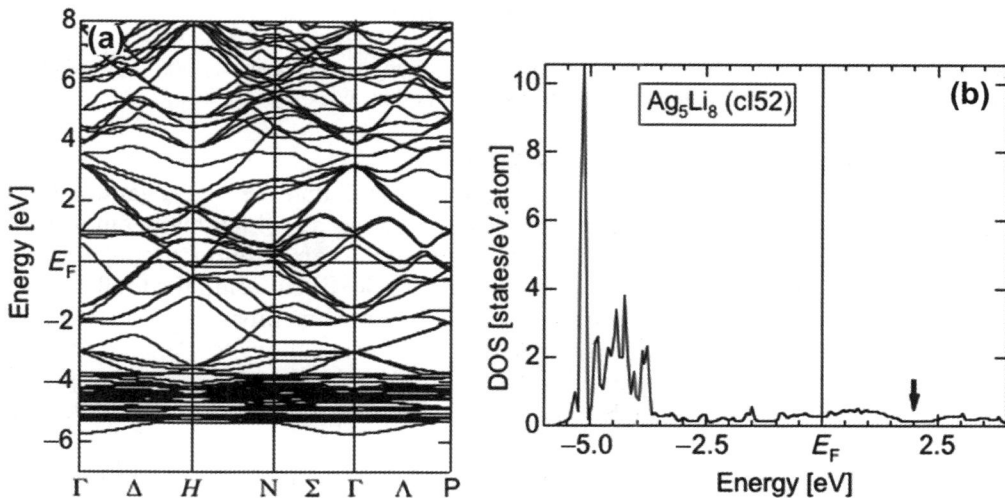

Figure 35 (a) Energy dispersion relations and (b) DOS calculated using the WIEN2k for Ag$_5$Li$_8$ gamma-brass (Mizutani, 2010). An arrow indicates the pseudogap.

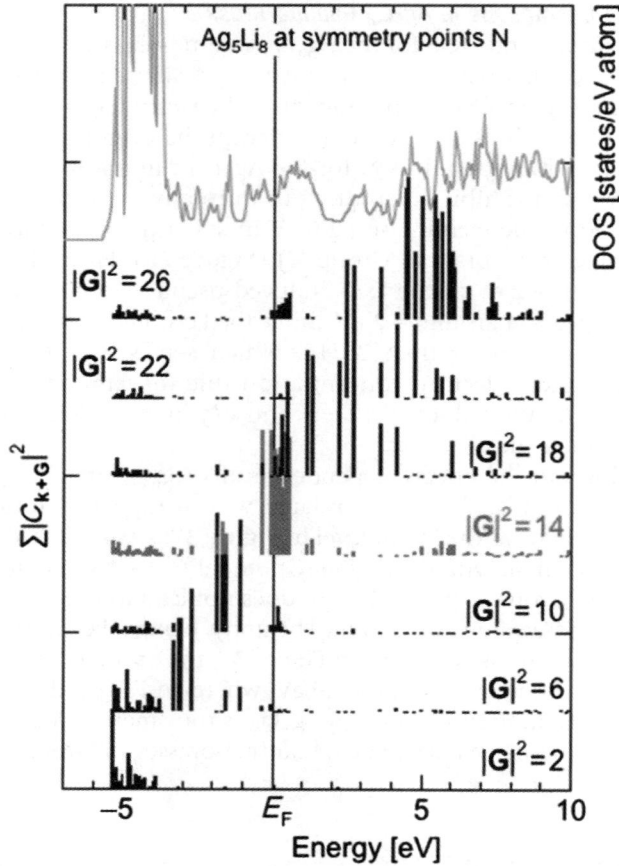

Figure 36 FLAPW-Fourier energy spectra at symmetry points N for Ag_5Li_8 gamma-brass. Its DOS is superimposed. (For color version of this figure, the reader is referred to the online version of this book.)

2.5.5.3 Stabilization Mechanism of Ag_5Li_8 Gamma-Brass

According to the Ag–Li phase diagram (Okamoto, 2000), we find that the Ag_5Li_8 gamma-brass phase containing 38.5 at.%Ag is competing with a bcc phase existing as a high-temperature phase. The stability of two competing phases has to be evaluated at the same composition, that is, 38.5 at.%Ag in the present case. However, it is difficult to calculate the electronic structure and the total energy of such a disordered bcc alloy with the same accuracy as those of Ag_5Li_8 gamma-brass. To circumvent this difficulty, Mizutani et al. (2008) performed the FLAPW band calculations for AgLi B2 compound, in addition to the gamma-brass, and constructed a DOS for the disordered 38.5 at.%Ag bcc phase simply by multiplying the DOS obtained for the B2-compound by a factor 4.85/6.0 so as to accommodate the same **VEC** as that in Ag_5Li_8.

In the remainder of this section, we discuss only the essence of a possible scenario for the stabilization mechanism of Ag_5Li_8 gamma-brass. **Figure 38**a shows the energy dependence of the **VEC** for the Ag_5Li_8 gamma-brass in comparison with a hypothetical bcc phase containing 38.5 at.%Ag. It is found that a sharp jump in **VEC** occurs at $E = -5.15$ eV for the Ag_5Li_8 gamma-brass, as marked with an arrow.

(a)

(b)

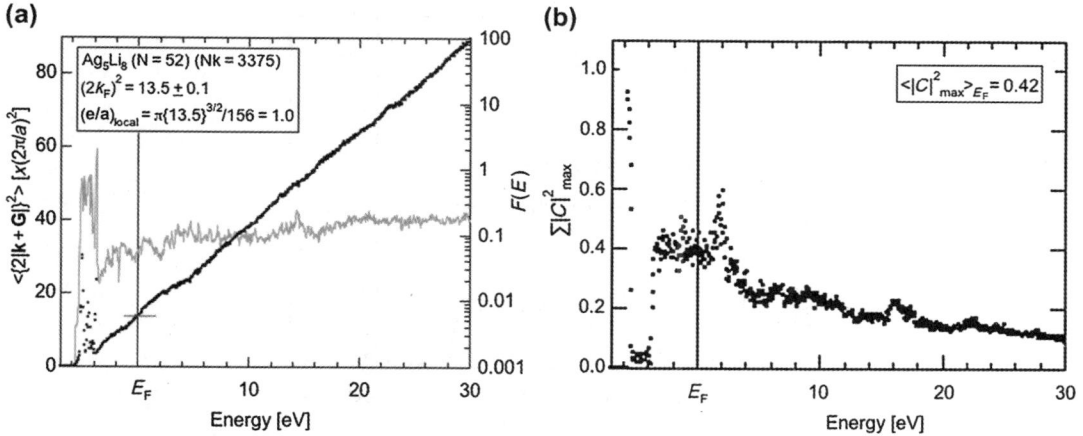

Figure 37 The Hume-Rothery plot and energy dependence of $\langle|C|^2_{max}\rangle$ for Ag_5Li_8 gamma-brass.

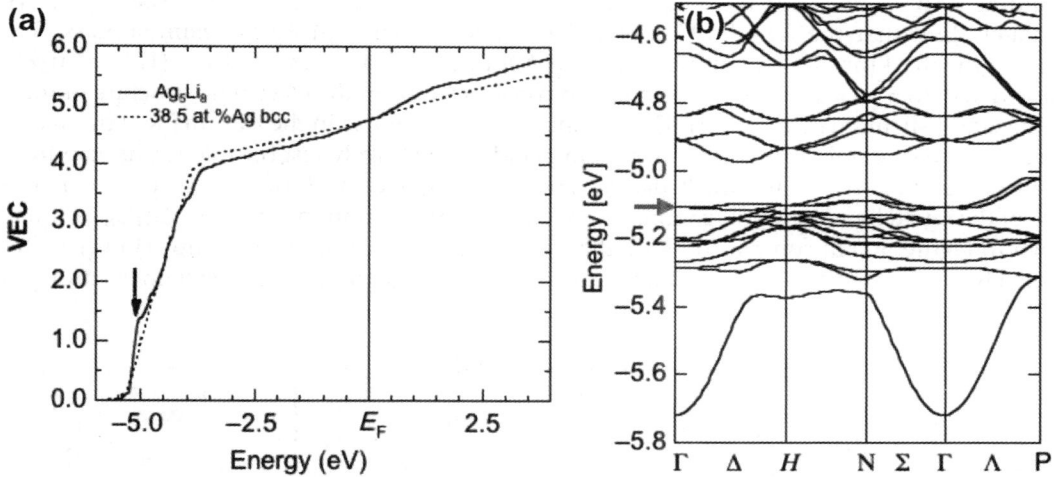

(a)

(b)

Figure 38 (a) **VEC** as a function of energy for Ag_5Li_8 gamma-brass (solid line) and 38.5 at.%Ag bcc model structure (dotted line) (Mizutani, 2010). An arrow indicates a steep jump in **VEC** in the gamma-brass. (b) Energy dispersion relations in the neighborhood of the Ag-4d states in Ag_5Li_8 gamma-brass. A red arrow indicates the presence of flat bands along ΓH and ΓN, which causes a jump in **VEC** in (a).

A jump in **VEC** occurs without changing energy and its magnitude reaches as large as 1.25. This large jump in **VEC** can be safely attributed to the existence of almost flat energy dispersions extending from points Γ to N at $E = -5.15$ eV, as is marked by red arrow in **Figure 38**b. Indeed, a δ-function-like sharp DOS is found near the bottom of the Ag-4d band in **Figure 35**b.

The **VEC** dependence of the DOS is plotted in **Figure 39** for the two phases. As is clear, Ag-4d bands for both phases can accommodate **VEC** roughly equal to 4.0 and are divided into bonding "B" and

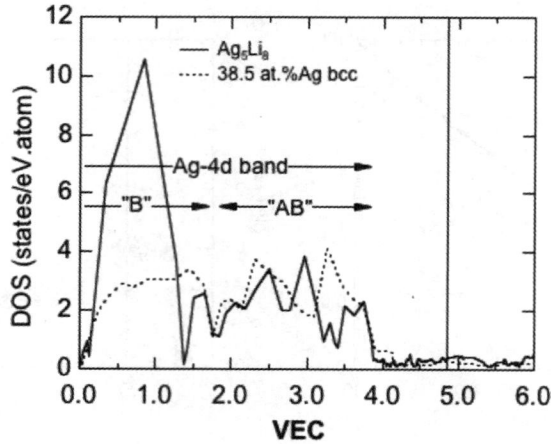

Figure 39   DOS as a function of **VEC** for Ag_5Li_8 gamma-brass (solid line) and 38.5 at.%Ag bcc model structure (dotted line). The vertical line represents the Fermi level, below which 4.85 electrons per atom are accommodated for both phases. Symbols "B" and "AB" represent the Ag-4d bonding and antibonding subbands, respectively (Mizutani, 2010).

antibonding "AB" states over the ranges $0 < VEC < 2.0$ and $2.0 < VEC < 4.0$, respectively. We can immediately find that the Ag-4d "B" subband is unusually large in Ag_5Li_8 gamma-brass. Such an abnormal growth of the "B" subband is found neither in gamma-brasses in Group (I) like Cu_5Zn_8 nor in a disordered 38.5 at.% bcc alloy, which is also constructed from the electronic structure of the CuZn B2 compound (Mizutani et al., 2009). This means that a sudden rise in the **VEC** marked by an arrow in **Figure 38**a is unique only to Ag_5Li_8 gamma-brass and is most likely responsible for its stabilization.

Before ending this section, we briefly discuss a possible origin for the formation of a flat band near the bottom of the Ag-4d band in Ag_5Li_8 gamma-brass. The X-ray diffraction spectrum for Ag_5Li_8 gamma-brass is shown in **Figure 40**a in comparison with that for Ag_5Zn_8 gamma-brass in Group (I) in (b). One can immediately realize that the (211) diffraction peak in Ag_5Li_8 gamma-brass is extremely strong and is

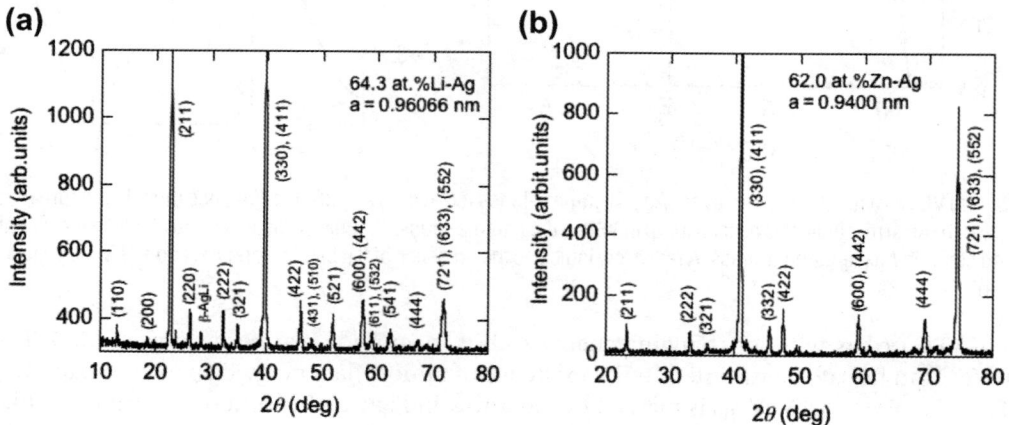

Figure 40   X-ray diffraction spectra taken with Cu-Kα radiation for (a) 64.3 at.%Li–Ag gamma-brass and (b) 62.0 at.%Zn–Ag gamma-brass (Mizutani, 2010).

comparable to the strongest peak $(330) + (411)$. This feature is unique only to Ag_5Li_8 and is not found in any other gamma-brasses. Noritake et al. (Noritake et al., 2007) attributed the occurrence of a huge (211) diffraction peak in Ag_5Li_8 to the predominant occupation of Ag atoms in the set of $\{211\}$ lattice planes.

Figure 41 reproduces the FLAPW–Fourier spectra for $|G|^2 = 6$ waves at symmetry points N along with the DOS for Ag_5Li_8 gamma-brass. The $|G|^2 = 6$ states are found to be split into bonding and antibonding states at the bottom, where the δ-function like DOS peak exists, and top of the Ag-4d band, respectively. This is typical of the Ag 4d states-mediated splitting. We show in **Figure 42** the s-, p-, d-states partial DOSs derived from WIEN2k for Ag_5Li_8 gamma-brass. Obviously, the flat band at $E=-5.15$ eV discussed above is mainly composed of Ag-4d states. However, as shown in the insert to **Figure 42**, both s- and p-states also participate in forming the flat band. Measuring from the bottom of the valence band, we can roughly estimate the kinetic energy of itinerant sp-electrons forming the flat band to be 2.5 eV. Its insertion into the free-electron equation $\lambda[nm] = 1.226/\sqrt{E[eV]}$ yields the wavelength $\lambda = 0.78$ nm. This is in good agreement with the lattice spacing $2d = 0.78$ nm in the set of $\{211\}$ lattice planes. In this way, the Bragg condition is satisfied and the stationary waves can be formed. We consider the Ag 4d states-mediated splitting to occur at such a low energy by involving the $|G|^2 = 6$ states and to result in a profound condensation of a large number of Ag-4d electrons at -5.15 eV in promoting the stability of this compound. We can definitely say that the stabilization of Ag_5Li_8 gamma-brass has little to do with the FsBz interaction involving the set of $\{330\}$ and $\{411\}$ lattice planes. The discussion on phase stability without involving a pseudogap at the Fermi level is quite difficult. But it may be possible that the stability

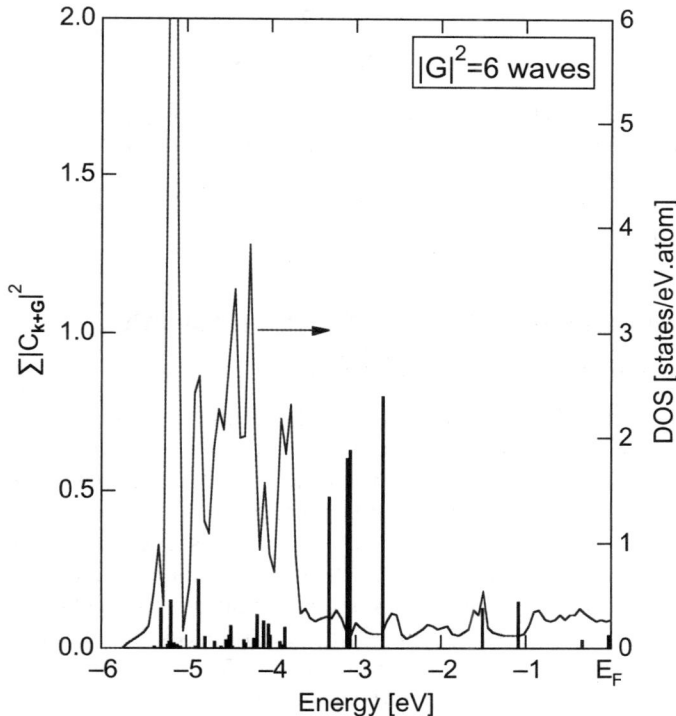

Figure 41 FLAPW–Fourier energy spectrum for $|G|^2 = 6$ waves for Ag_5Li_8 gamma-brass. The total DOS is superimposed. The Ag 4d-mediated splitting of $|G|^2 = 6$ waves is observed.

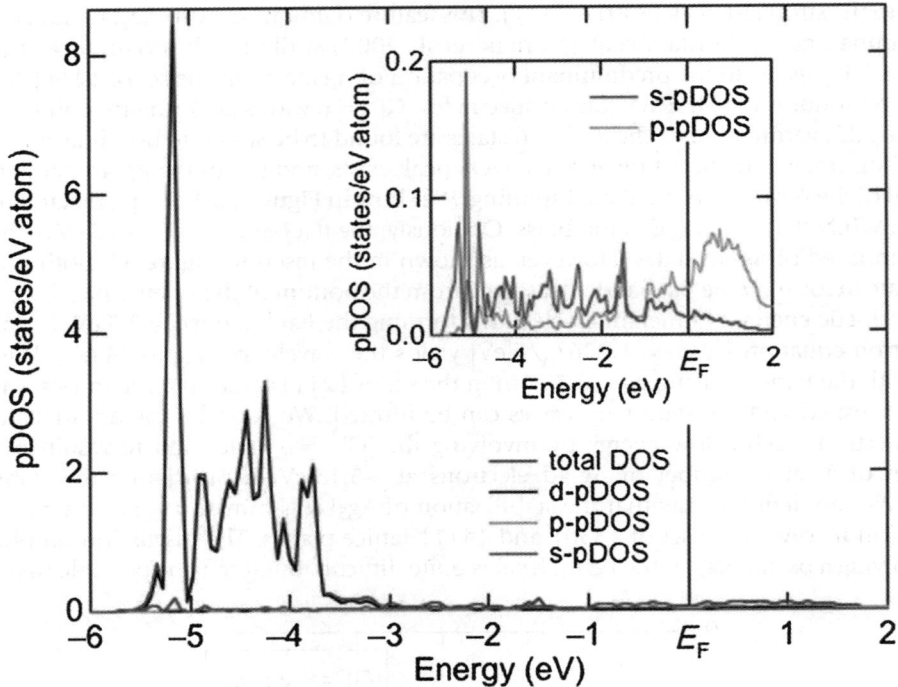

Figure 42   s-, p-, and d-partial DOSs and total DOS for Ag_5Li_8 gamma-brass. Insert shows s- and p-partial DOSs (Mizutani, 2010). (For color version of this figure, the reader is referred to the online version of this book.)

of this unique gamma-brass is essentially brought about by the set of Ag-rich {211} lattice planes, which is also characteristic of the gamma-brass structure.

2.6   Stabilization Mechanism in 1/1–1/1–1/1 Approximants

2.6.1   Electronic Structure of 1/1–1/1–1/1 Approximants

Following the discovery of the Al–Mn quasicrystal by Shechtman et al. (1984), quasicrystals have been established as a new family of compounds characterized by the so-called quasiperiodicity in combination with rotational symmetries forbidden in periodic crystals. Their stabilization mechanisms have become one of the most exciting recent topics in the electron theory of metals. However, first-principles band calculations based on the Bloch theorem in the reciprocal space are unfortunately not applicable because of the lack of translational symmetry, which is inherent in a perfectly periodic lattice of a crystal. Instead, one can perform first-principles band calculations for the family of *approximants*, since the lattice periodicity is assured, no matter how large the unit cell is. The electronic structure of the lowest order 1/1–1/1–1/1 approximant containing some 130–170 atoms in the unit cell has been extensively studied in the past. We consider this approach to be reasonable in order to deepen our understanding of the electronic structure of a quasicrystal, since we know from the cut-and-project method discussed in Section 2.3.2 that the local atomic structure of a quasicrystal and its approximant are essentially the same. In this section, we exclusively focus on the exploration of the stabilization mechanism in a series of 1/1–1/1–1/1 approximants as another class of CMAs in relation to the Hume-Rothery electron concentration rule.

Belin-Ferré et al. (Traverse et al., 1988) revealed in 1988 from the soft X-ray emission spectra a large depression in the DOS at the Fermi level in the Al–Mn quasicrystal relative to its amorphous and crystalline counterparts. Their pioneering work already hinted at the presence of a pseudogap in quasicrystals. At that time, however, people had intuitively expected that first-principles band calculations were beyond a level of computations, even for 1/1–1/1–1/1 approximants. Fujiwara (1989) could perform for the first time rudimentary LMTO–ASA band calculations for the Al–Mn approximant containing 138 atoms in the unit cell, though the neglect of the combined correction term led to the pseudogap structure as an artifact (see footnote 11 in Section 2.4.2). His work indicated a deep pseudogap at the Fermi level and suggested it to be most likely responsible for the stability of such CMAs containing icosahedral clusters similar to those in quasicrystals. Since then, first-principles LMTO–ASA electronic structure calculations have been extensively carried out to affirm if a pseudogap is a universal feature of all approximants. Readers may consult the recent developments on this topic in books and review articles (Laissardiere et al., 2005; Mizutani, 2010; Stadnik, 1999).

As briefly noted in Section 2.4.3 and explained in detail in Mizutani (2010) the LMTO–ASA method has been recognized as a fast but efficient scheme for first-principles electronic structure calculations. It owes its speed to a relatively small basis set, which consists essentially of atomic orbitals of constituent elements. This is the reason why the LMTO–ASA band calculation method is capable of determining the electronic structure even for CMAs with a giant unit cell, as in the case of 1/1–1/1–1/1 approximants. Moreover, it allows us to extract the effect of orbital hybridizations among neighboring atoms and their involvement in the formation of a pseudogap. On the other hand, the FLAPW method is more suited for the extraction of FsBz interactions, which serves as a key role in order to gain a deeper insight into the physics behind the Hume-Rothery electron concentration rule (Sections 2.4.3–2.4.11). Since the FLAPW wave function outside the MT sphere in a crystal is given as a sum of plane waves over >2000 reciprocal lattice vectors, solving a secular equation becomes a hard task for CMAs containing more than 100 atoms in the unit cell. This is probably the reason why, in the past, theoretical studies with respect to the Hume-Rothery electron concentration rule for CMAs have not made much progress beyond the free electron model. However, as emphasized in Sections 2.4.7–2.4.11, the incorporation of the "APW + lo" method developed after 2000 into the FLAPW formalism has made it possible to work on FsBz interactions even for CMAs containing more than 100 atoms in the unit cell.

Let us briefly review at this stage electronic structure calculations on 1/1–1/1–1/1 approximants available in literature. As mentioned in Section 2.3.2, icosahedral quasicrystals and their 1/1–1/1–1/1 approximants are classified with respect to atomic clusters building up its structure (Mizutani, 2010); the rhombic triacontahedron (RT)-type cluster containing 44 atoms, the Mackay Icosahedron (MI)-type cluster containing 54 atoms and the Tsai-type cluster containing 65–66 atoms. Representative works reported on first-principles electronic structure calculations for families of RT-, MI-, and Tsai-type 1/1–1/1–1/1 approximants are summarized in **Table 4** (Fujiwara, 1989; Fujiwara and Yokokawa, 1992; Hafner and Krajčí, 1993; Inukai et al., 2011a; Krajčí and Hafner, 1999; Krajčí et al., 1995; Mizutani et al., 2004, 2009, 2012; Nozawa and Ishii, 2008; Roche and Fujiwara, 1998; Sato et al., 2001; Sato et al., 2004; Takeuchi et al., 2003; Windisch et al., 1994; Zijlstra and Bose, 2003, 2004). As emphasized in Section 2.3.2, chemical disorder and defects were revealed at various sites in many of them. They must be removed to perform first-principles band calculations and, hence, the construction of a "model structure" is inevitable.

Efforts to probe more specifically the origin of the Hume-Rothery electron concentration rule were indeed fairly limited in the past. Among them, Trambly de Laissardière et al. (1995) discussed the Hume-Rothery electron concentration rule by constructing the Anderson Hamiltonian, which was

Table 4   First-principles band calculations for 1/1–1/1–1/1 approximants

Cluster type	System	Method	Reference
RT	$Al_{16}Mg_{39.5}Zn_{44.5}$	LMTO–ASA	(Hafner and Krajčí, 1993)
	$Al_{45}Mg_{40}Zn_{15}$	LMTO	(Roche and Fujiwara, 1998)
	$Al_{48}Mg_{64}Zn_{48}$	LMTO–ASA	(Sato et al., 2001)
	$Al_{60}Li_{32.5}Cu_{7.5}$	LMTO–ASA	(Fujiwara and Yokokawa, 1992)
	Al–Li–Cu	LMTO–ASA	(Windisch et al., 1994)
	$Al_{84}Li_{52}Cu_{24}$	LMTO–ASA	(Sato et al., 2004)
	$Al_{48}Mg_{64}Zn_{48}$ and $Al_{84}Li_{52}Cu_{24}$	FLAPW	(Inukai et al., 2011a)
MI	$Al_{82.6}Mn_{17.4}$	LMTO–ASA	(Fujiwara, 1989)
	Al–Cu–Fe–Si	LMTO–ASA	(Roche and Fujiwara, 1998)
MI	$Al_{68.8}Pd_{15.6}Mn_{15.6}$	LMTO–ASA	(Krajčí et al., 1995)
	$Al_{68.8}Pd_{15.6}Re_{15.6}$	LMTO–ASA	(Krajčí and Hafner, 1999)
	$Al_{73.6}Re_{17.4}Si_9$	LMTO–ASA	(Takeuchi et al., 2003)
	α-AlMn and AlPdMn	LMTO–ASA	(Zijlstra and Bose, 2003)
	$Al_{68}Cu_7TM_{17}Si_8$ (TM=Fe and Ru)	LMTO–ASA	(Mizutani et al., 2004)
	$Al_{68}Cu_7TM_{17}Si_8$ (TM=Fe and Ru)	FLAPW	(Mizutani et al., 2009)
	$Al_{114}Mn_{24}$, $Al_{114}Re_{24}$ $Al_{102}Re_{24}Si_{12}$ $Al_{108}Cu_6TM_{24}Si_6$ (TM=Fe and Ru)	FLAPW	(Mizutani et al., 2012)
Tsai	Cd_6Yb, Cd_6Ca Cd_6Ca	FLAPW	(Zijlstra and Bose, 2004) (Nozawa and Ishii, 2008)

composed of two terms describing the motion of nearly free sp-electrons on one hand and d-impurities on the other hand to treat electrons in Al–transition metal (TM) based alloys. Unfortunately, their approach is based on an empirical model Hamiltonian instead of first-principles band calculations. In this section, we aim at elucidating the stabilization mechanism of both RT- and MI-type 1/1–1/1–1/1 approximants on the basis of first-principles FLAPW band calculations with an emphasis on the clarification of the role of FsBz interactions in the formation of a pseudogap. They include Al–Mg–Zn and Al–Li–Cu approximants from the RT-type family and Al–Mn, Al–Re, Al–Re–Si, and Al–Cu–TM–Si (TM = Fe and Ru) 1/1–1/1–1/1 approximants from the MI-type family. As emphasized in Section 2.5 dealing with gamma-brasses, the free electron model and nearly free electron (NFE) band calculations will be again instructive enough for readers to grasp the physics on FsBz interactions in 1/1–1/1–1/1 approximants.

2.6.2 Free Electron Model for 1/1–1/1–1/1 Approximants

The free electron parabolic bands along the direction from Γ to the center of $\{710\}$ zone planes for a canonical 1/1–1/1–1/1 approximant are shown in **Figure 43** (Mizutani, 2010). The symmetry points N at the center of $\{710\}$ zone planes are located at $k = \dfrac{\sqrt{50}}{2} (=3.535)$ in units of $2\pi/a$ from the origin Γ in the extended Brillouin zone. In the same way as in **Figure 21** for gamma-brass, crossings of parabolic bands there can be indexed in terms of an ascending order of $|\mathbf{G}|^2$, as indicated in **Figure 43**. It is seen that degenerate states indexed as $|\mathbf{G}|^2 = 50$ fall closest to the Fermi level in the free electron model for a canonical 1/1–1/1–1/1 approximant, in which e/a = 2.3 is assumed. An increase in the number of atoms per unit cell from 52 in gamma-brasses to 160 in 1/1–1/1–1/1 approximants leads to an increase in *critical* $|\mathbf{G}|^2$ from 18 to 50 in the free electron model.

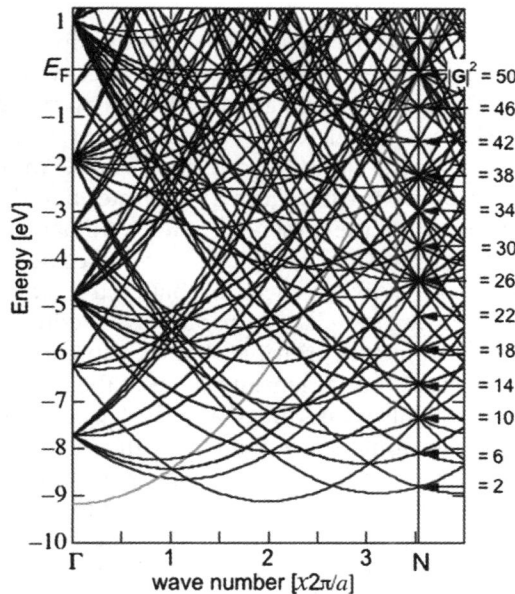

Figure 43 Energy dispersion relations along the direction ΓN for a canonical 1/1–1/1–1/1 approximant having **e/a** = 2.30 and $N = 160$ in the free electron model (Mizutani, 2010). A red curve indicates a parabolic band in the extended zone scheme.

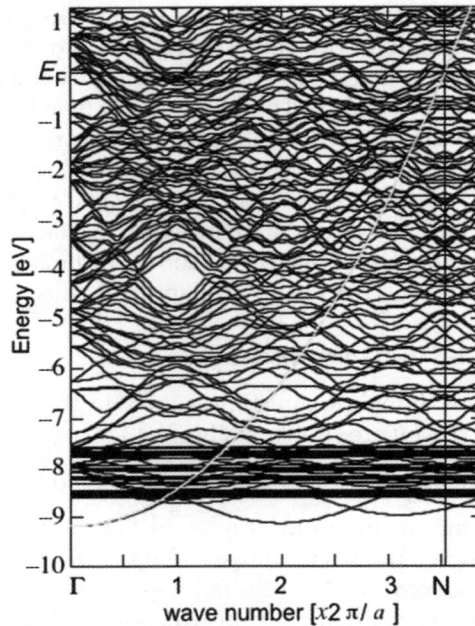

Figure 44   Energy dispersion relations along the direction <710> derived from LMTO–ASA method for the $Al_{48}Mg_{64}Zn_{48}$ 1/1–1/1–1/1 approximant (Sato et al., 2001). A red curve indicates a parabolic band in the extended zone scheme.

Now we are ready to compare free electron dispersion relations above with realistic ones calculated from first-principles band calculations. As will be described later, we will learn that the electronic structure of the RT-type Al–Mg–Zn approximant can be well approximated by the free electron model. **Figure 44** shows dispersion relations calculated by using the LMTO–ASA method for the RT-type $Al_{48}Mg_{64}Zn_{48}$ 1/1–1/1–1/1 approximant along the direction from Γ to the center of the {710} zone planes (Sato et al., 2001). Its comparison with **Figure 43** indicates that the overall feature of the band structures resembles each other except for the presence of the Zn-3d band in the latter. The lack of states at symmetry point Γ below the Fermi level and a large "hole" centered at -4 eV and wave number equal to unity in units of $2\pi/a$, for example, well reflect the band structures of the free electron model. Moreover, the red curve representing the free electron parabolic band in the extended zone scheme well traces dispersion relations derived from first-principles band calculations, as shown in **Figure 44**.

The analysis above strongly suggests that degenerate states at the center of {710} + {543} + {550} zones with $|G|^2=50$ in **Figure 43** are only weakly lifted as a result of the interaction with the set of these lattice planes and that resulting electronic states remain in the vicinity of the Fermi level. Another important remark should be addressed. We find from **Figure 43** that the frequency of crossings of parabolic bands in the free electron model exceeds 200 times along the direction ΓN over the energy range $-0.5 \leq E < 0.5$ eV across the Fermi level. All these degenerate states will be lifted and, as a result, energy dispersions in the neighborhood of crossings will be flattened, as shown in **Figure 44**. All of them would contribute to the formation of a DOS pseudogap of about 2 eV in width across the Fermi level. We will show below that *multizone effects* due to not only $|G|^2 = 50$ but also its neighboring ones are responsible for the formation of a sizable DOS pseudogap in CMAs having 160 atoms per unit cell.

2.6.3 NFE Approximation for 1/1–1/1–1/1 Approximants

As discussed in Section 2.4, first-principles band calculations essentially aim at solving a secular equation like Eqn 15. As mentioned in Section 2.4.10, one would naturally consider NFE band calculations to be the simplest and the most transparent in elucidating the origin of a DOS pseudogap in the Al–Mg–Zn approximant in terms of the form factor associated with each set of lattice planes. Let us assume the n-wave approximation in the framework of NFE band calculations (Mizutani, 2001). Since a secular determinantal equation is composed of an n-by-n determinant, one would end up solving an nth-order polynomial equation in the energy E:

$$a_n E^n + a_{n-1} E^{n-1} + \cdots + a_0 = 0, \tag{25}$$

where the coefficient a_n is composed of unperturbed free electron energies and nonzero form factors.[17] In the two-wave approximation as the simplest case, Eqn 25 is reduced to a quadratic equation and results in the well-known energy dispersion relation (Mizutani, 2001):

$$E(\mathbf{k}) = \left(\frac{1}{2}\right)\left\{-\left(\frac{a_1}{a_2}\right) \pm \sqrt{\left(\frac{a_1}{a_2}\right)^2 - 4\left(\frac{a_0}{a_2}\right)}\right\} = \left(\frac{1}{2}\right)\left\{(E_0 + E_n) \pm \sqrt{(E_0 - E_n)^2 + 4V_n^* V_n}\right\}, \tag{26}$$

where V_n is called the form factor or the Fourier component of the ionic potential associated with the reciprocal lattice vector $\mathbf{G} = (2\pi/a)\mathbf{n}$ and both $E_0 = \mathbf{k}^2$ and $E_n = (\mathbf{k} - \mathbf{G}_n)^2$ are unperturbed free electron energies centered at $\mathbf{G} = 0$ and $\mathbf{G} = \mathbf{G}_n$.

In the case of the two-wave approximation given by Eqn 26, we know that the perturbation due to the form factor V_n develops only in the vicinity of $E_0 = E_n$ and otherwise the free electron parabolic band is preserved. If the number of plane waves is increased above 1000, the NFE band calculations can reproduce well a realistic electronic structure, provided that the d-band is absent or is located far away from the Fermi level like in the Al–Mg–Zn approximant. Degenerate free electron parabolic bands indexed with $|\mathbf{G}|^2 = 50$ at symmetry points N near the Fermi level in **Figure 43**, for example, will be lifted as a result of perturbations due to many nonzero form factors involved in Eqn 25. We will learn below that the effect of each form factor on the dispersion relations in more than 1000-wave NFE approximation is still limited to only the vicinity of the crossing region of more than two parabolic bands, as long as the form factors are small enough to validate the NFE model.

Figure 45a and **b** compare wave number dependences of the form factor between Cu_5Zn_8 gamma-brass and the $Al_{48}Mg_{64}Zn_{48}$ 1/1–1/1–1/1 approximant. It is clear that only the $|\mathbf{G}|^2 = 18$ form factor is extremely large and located very near the Fermi level in Cu_5Zn_8. Its magnitude reaches 0.8 eV while the rest is lower than 0.1 eV in the range $0.9 \leq |\mathbf{G}|/2k_F \leq 1.0$. This is apparently responsible for the opening of a large gap reaching about 1 eV across the Fermi level at symmetry points N in its dispersion relations (**Figure 22a**) and allowed us to interpret the DOS pseudogap formed across the Fermi level (**Figure 22b**) solely in terms of the gap opening at symmetry points N. Instead, $|\mathbf{G}|^2 = 50$ associated with {543}, {710}, and {550} zone planes appears closest to the Fermi level in $Al_{48}Mg_{64}Zn_{48}$ 1/1–1/1–1/1 approximant but the magnitude of its form factor is only 0.2 eV. In addition, two form factors at $|\mathbf{G}|^2 = 46$ and 50, being comparable in magnitude, exist over the narrow range $0.9 \leq |\mathbf{G}|/2k_F \leq 1.0$.

---

[17] The form factor in the NFE model is defined as $V_G = \frac{1}{N}\sum_{i=1}^{N} V_{i(\alpha)}(\mathbf{r}_i)\exp(-i\mathbf{G}\cdot\mathbf{r}_i)$, where $V_{i(\alpha)}(\mathbf{r}_i)$ is the ionic potential due to an atomic species α at the position $\mathbf{r}_i$ in the unit cell and $\mathbf{G}$ is the reciprocal lattice vector in a given system.

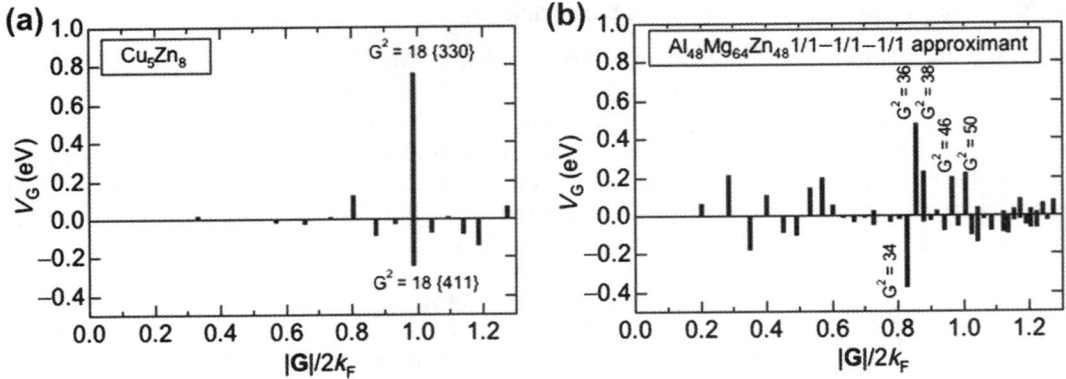

Figure 45 Form factors V_G in the NFE model as a function of the wave number normalized with respect to the Fermi diameter $2k_F$ for (a) Cu_5Zn_8 gamma-brass and (b) the $Al_{48}Mg_{64}Zn_{48}$ 1/1–1/1–1/1 approximant (Mizutani, 2010). Note that form factors take finite values only at wave numbers corresponding to the reciprocal vector $|G|$ allowed for each system. Sizable form factors appearing in the vicinity of the Fermi level are indexed in terms of $|G|^2$. The values of $(2k_F)^2$ for Cu_5Zn_8 gamma-brass and the $Al_{48}Mg_{64}Zn_{48}$ 1/1–1/1–1/1 approximant are taken to be 18.47 and 49.52, respectively, in units of $(2\pi/a)^2$, where a is the lattice constant.

To extract the role of form factors at $|G|^2 = 50$ and 46, which will be hereafter referred to as $V(50)$ and $V(46)$, in the formation of a pseudogap, we performed 1505-wave NFE electronic structure calculations for the $Al_{48}Mg_{64}Zn_{48}$ 1/1–1/1–1/1 approximant model structure (Section 2.6.4) in such a way that form factors other than the *critical* ones are set to zero. **Figure 46a and b** show energy

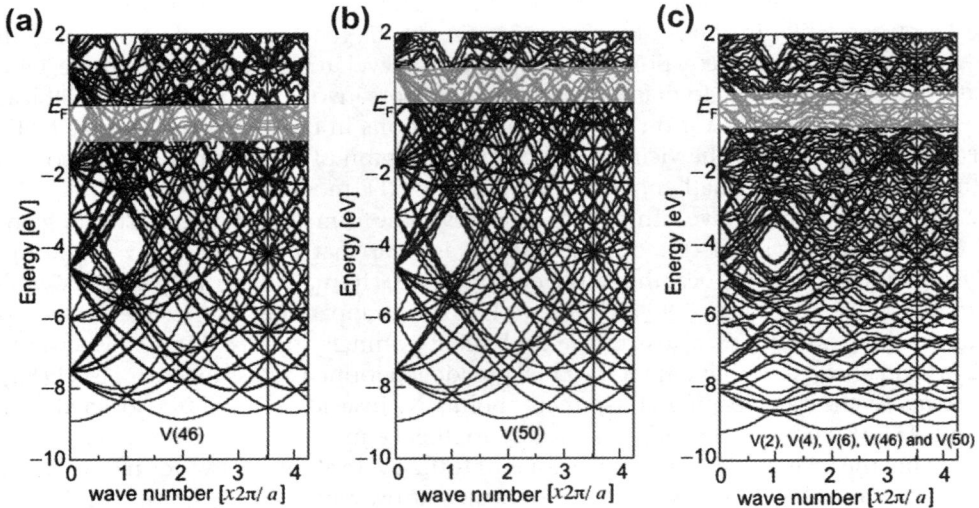

Figure 46 Energy dispersion relations along the direction <711> for the $Al_{48}Mg_{64}Zn_{48}$ 1/1–1/1–1/1 approximant obtained when (a) only the form factor at $|G|^2 = 46$ is nonzero and (b) only the form factor at $|G|^2 = 50$ is nonzero in 1505-wave NFE band calculations (Mizutani, 2010). In (c), form factors at $|G|^2 = 2, 4, 6, 46,$ and 50 are nonzero. An energy range, where sparse electronic states are produced by the respective form factors, is highlighted by a shaded rectangle.

dispersion relations derived when only either $V(46)$ or $V(50)$ is activated, respectively. It is clear that each form factor lifts degenerate states only in the energy region above the respective $|\mathbf{G}|/2k_F$ values shown in **Figure 45b** but preserves the free electron bands below them. As can be seen from dispersion relations inside shaded rectangles in **Figure 46a** and **b**, the form factor $V(50)$ causes electronic states to be less populated over energies centered immediately above the Fermi level, whereas $V(46)$ over those centered at -0.5 eV below the Fermi level. Sparse and flat dispersion relations become more evident in **Figure 46c**, where both form factors $V(50)$ and $V(46)$ are simultaneously activated, along with low energy-lying form factors $V(2)$, $V(4)$, and $V(6)$.

Figure 47a–**e** show the effect of different combinations of form factors on the DOS near the Fermi level in 1505-wave NFE band calculations. As shown in **Figure 47a**, we can confirm a pseudogap to be formed over the range -1 to $+1$ eV across the Fermi level in the DOS. The DOS in **Figure 47b**, where only the form factor $V(46)$ is taken into account, is characterized by a cusp at about -1 eV, followed by a sharply declining slope without clear formation of a pseudogap. It is similar to the van-Hove singularity derived in the model of Jones (Jones, 1937). A similar structure is formed but is shifted to the Fermi level in (c), where only the form factor $V(50)$ is introduced. A pseudogap structure, though its top edge is blurred, emerges across the Fermi level in (d), where both $V(50)$ and $V(46)$ are concomitantly activated. Finally, form factors in the range $|\mathbf{G}|^2 \leq 6$ or $|\mathbf{G}|/2k_F < 0.35$ are added to (d). As shown in (e), the DOS near the bottom of the valence band is heavily perturbed. More important is that the structure of the pseudogap becomes sharp and clear across the Fermi level.

From the analysis above we can say that two form factors $V(50)$ and $V(46)$, or the sets of the Brillouin zone planes associated with both $|\mathbf{G}|^2 = 50$ and 46, are essential in the formation of a DOS pseudogap across the Fermi level in the Al–Mg–Zn 1/1–1/1–1/1 approximant and its structure is matured through

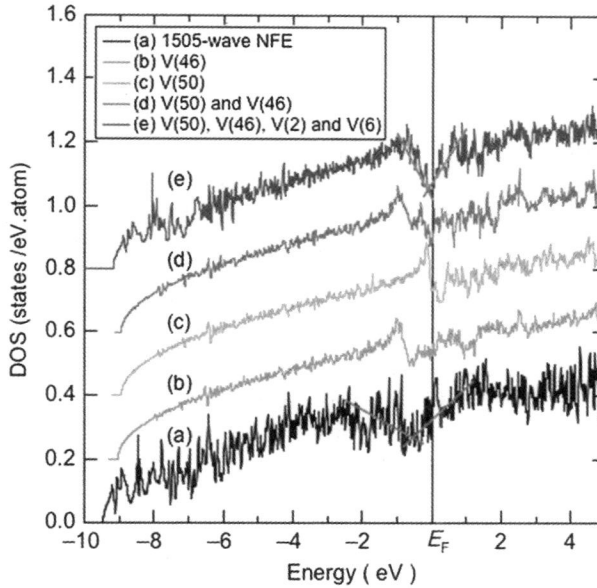

Figure 47 (a) DOS derived from 1505-wave NFE band calculations for the $Al_{48}Mg_{64}Zn_{48}$ 1/1–1/1–1/1 approximant. (b) DOS derived when only the form factor at $|\mathbf{G}|^2 = 46$ is nonzero. (c) DOS derived when only the form factor at $|\mathbf{G}|^2 = 50$ is nonzero. (d) DOS when both $|\mathbf{G}|^2 = 46$ and 50 are nonzero. (e) DOS derived when form factors at $|\mathbf{G}|^2 = 2, 4, 6, 46$, and 50 are nonzero (Mizutani, 2010). (For color version of this figure, the reader is referred to the online version of this book.)

assistance of low energy form factors in the range $|G|^2 \leq 6$. This can be viewed as the formation of a more or less spherical Brillouin zone net in the reciprocal space, thereby causing the Bragg reflections to occur in almost all directions. The analysis above points to the important conclusion that *multizone effects* or FsBz interactions involving *multizones* must play a key role in the formation of a pseudogap in the 1/1-1/1-1/1 approximant containing as many as 160 atoms per unit cell.

2.6.4   RT-Type 1/1-1/1-1/1 Approximants

2.6.4.1   *Construction of the $Al_{48}Mg_{64}Zn_{48}$ and $Al_{84}Li_{52}Cu_{24}$ Model Structures*

The atomic structure of the RT-type Al–Mg–Zn 1/1-1/1-1/1 approximants was experimentally determined by three independent groups (Bergman et al., 1957; Mizutani et al., 1997; Sun et al., 2000). The atomic structure involves a large amount of chemical disorder on sites B, C, and F in the RT-type cluster shown in **Figure 8**. Such chemical disorder has to be eliminated in order to carry out efficient band calculations. The site A is assumed to be vacant, since it is very weakly occupied by Al atoms (Mizutani et al., 1997; Sun et al., 2000). Experimentally, both Al and Zn atoms are randomly filled into sites B, C, and F with approximate ratios of 1:4, 3:2, and 3:2, respectively. In the model structure, only Zn atoms are filled into sites B and C, while Al atoms into sites F. Instead, Mg atoms are exclusively filled into sites D, E, G, and H without any chemical disorder, as indicated in **Figure 8**. The elimination of chemical disorder mentioned above leads to chemical formula $Al_{48}Mg_{64}Zn_{48}$ or $Al_{30}Mg_{40}Zn_{30}$ in atomic % with space group Im$\bar{3}$. The lattice constant reported by Mizutani et al. (1997) was 1.4355 nm for $Al_{48}Mg_{64}Zn_{48}$. This is the model structure employed in the FLAPW band calculations described below (Inukai et al., 2011a).

The atomic structure of the Al–Li–Cu 1/1-1/1-1/1 approximant was studied by the two different groups (Guryan et al., 1988; Audier et al., 1988). Both neutron and X-ray diffraction measurements were carried out on a powder sample with composition $Al_{94.6}Cu_{17}Li_{48.4}$ by Guryan et al. (1988) and on a single crystal with composition $Al_{88.62}Cu_{19.377}Li_{50.335}$ by Audier et al. (1988). Both sets of data are quite consistent with each other: the structure was successfully refined with space group Im$\bar{3}$ with the absence of atoms at the center of the cluster (sites A: 2a). Li atoms are filled into sites D, E, and H without measurable chemical disorder. Chemical disorder is the most substantial on sites C, where Al and Cu atoms are almost evenly distributed. Sites B and F are shared by Al and Cu atoms with a proportion of 89:11. In the model structure, Li atoms are exclusively filled into sites D, E, and H, Cu atoms into sites C and Al atoms into sites B, F, and G (Sato et al., 2004). This leads to the chemical formula $Al_{84}Li_{52}Cu_{24}$ per unit cell or $Al_{52.5}Li_{32.5}Cu_{15}$ in atomic % containing 160 atoms in the unit cell with space group Im$\bar{3}$. The lattice constant of 1.389 nm reported by Guryan et al. (1988) for $Al_{94.6}Cu_{17}Li_{48.4}$ was employed.

We have also studied the effect of the lattice relaxation on the electronic structure (Inukai et al., 2011a). But no discernible difference was obtained for both $Al_{48}Mg_{64}Zn_{48}$ and $Al_{84}Li_{52}Cu_{24}$ 1/1-1/1-1/1 approximants, indicating that our above-described model choices are at least consistent with the experimentally determined atomic positions. Hence, it suffices that we shall restrict ourselves to a discussion of the results obtained for the model structure described above using the experimentally obtained atomic coordinates.

2.6.4.2   *WIEN2k–FLAPW Band Calculations for $Al_{48}Mg_{64}Zn_{48}$ and $Al_{84}Li_{52}Cu_{24}$ 1/1-1/1-1/1 Approximants*

The WIEN2k–FLAPW band calculations with subsequent FLAPW–Fourier analysis were reported for two RT-type 1/1-1/1-1/1 approximants $Al_{48}Mg_{64}Zn_{48}$ and $Al_{84}Li_{52}Cu_{24}$ (Inukai et al., 2011a). The origin of the formation of the pseudogap in them was identified as resulting from FsBz interactions involved, as will be described below. Special care was exercised in order to carry out calculations in an efficient but reliable way for systems with giant unit cells, like the present approximants containing 160 atoms per unit cell. First, the radius of the MT sphere R_{MT} was chosen with care, in particular, in the case of Al–Li–Cu, where a proper description of the Cu-d states is the bottleneck for the number of plane waves needed. To reduce this problem, the value of R_{MT} for Cu was chosen about 14% larger than that of the elements without d states. In contrast, the choice of R_{MT} is less critical in the Al–Mg–Zn. Our final choice of R_{MT} ($R_{Al} = 2.50$, $R_{Mg} = 2.50$, and $R_{Zn} = 2.39$ for Al–Mg–Zn and $R_{Al} = 2.1824$, $R_{Li} = 2.1824$, and $R_{Cu} = 2.48$ for Al–Li–Cu in atomic units) was different from the automatic suggestion of the WIEN2k package.

There exist two more important parameters, which directly affect the computational time for executing self-consistent field (SCF) cycles for large systems. One is the cutoff parameter $R_{MT}K_{max}$ mentioned above, which determines the number of basis functions or size of the matrices. The smaller the cutoff parameter, the shorter is the SCF computational time. The other is the number of **k**-points in the first Brillouin zone, N_k. The total charge density and total spherical potential parameters in the MT sphere were first optimized by using $R_{MT}K_{max} = 6.0$ and a small $N_k = 64$, that is, $4 \times 4 \times 4$. Thirteen SCF cycles were needed to meet the total-energy convergence criterion of 0.0001 Ry. It took only 2 h and 43 min on an INTEL version (Linux OS: openSUSE 10.3, CPU: Intel Core 2 Quad 2.66 GHz, Memory: DDR2 4 GB (1333 MHz), Fortran compiler: Intel Fortran 10 + MKL 9 and compiler option: parallel). The resulting total charge density and total spherical potential in the MT sphere were saved and used in subsequent refinements of the two parameters N_k and $R_{MT}K_{max}$.

First, the value of N_k was increased from 64, 216, 1000, 1728 to 8000 under the condition $R_{MT}K_{max} = 6.0$ for the $Al_{48}Mg_{64}Zn_{48}$ approximant. A pseudogap across the Fermi level can hardly be resolved in the DOS with $N_k = 64$. However, it became slightly visible with $N_k = 216$. The reliable DOS consistent with those reported in the literature (Roche and Fujiwara, 1998; Sato et al., 2001) was achieved, when N_k was increased to 1000. Once the self-consistent total charge density and total spherical potentials are determined under $N_k = 64$, only three SCF cycles were needed for convergence with $N_k = 1000$. It took 5 h and 40 min for its completion. The dispersion relations and DOS obtained under the condition $R_{MT}K_{max} = 6.0$ and $N_k = 1000$ for the $Al_{48}Mg_{64}Zn_{48}$ approximants are shown in **Figure 48a** and **b**, respectively. The depth of the DOS pseudogap is shallow and is deduced to be $H/H_0 = 0.83$.

The dispersion relations and DOS are shown for the $Al_{84}Li_{52}Cu_{24}$ 1/1–1/1–1/1 approximant in **Figure 49a** and **b**, respectively. The calculations were made using the WIEN2k–FLAPW method under the condition $R_{MT}K_{max} = 6.0$ and $N_k = 1000$. It is clear that the depth of the pseudogap is much deeper than that in the Al–Mg–Zn approximant and that the depth ratio H/H_0 amounts to 0.28. This is consistent with the data in the literature (Fujiwara and Yokokawa, 1992; Windisch et al., 1994; Sato et al., 2004).

The effect of an increase in $R_{MT}K_{max}$ from 6.0 to 7.0 was studied for the $Al_{48}Mg_{64}Zn_{48}$ approximant. No measurable difference is observed over the whole valence band, except for a subtle difference in the DOS inside the Zn-3d band near the bottom of the valence band.

2.6.4.3 Extraction of FsBz Interactions in $Al_{48}Mg_{64}Zn_{48}$ 1/1–1/1–1/1 Approximant

The energy spectra of $\sum |C_{k+G}|^2$ for the $Al_{48}Mg_{64}Zn_{48}$ 1/1–1/1–1/1 approximant are shown in **Figure 50a–c** at symmetry points N over $38 \leq |G|^2 \leq 62$, at point Γ over $24 \leq |G|^2 \leq 72$ and at points H over $20 \leq |G|^2 \leq 84$, respectively, in the same manner as in **Figure 25a** and **b** for Cu_5Zn_8 and Al_4Cu_9. The total DOS is superimposed with gray in color. The range of $|G|^2$ of interest is more than two times higher than that in gamma-brasses because of an increase in the number of atoms per unit cell. Each plane wave specified by $|G|^2$ spreads over -2 and $+2$ eV around its center of gravity, which increases with increasing energy in accordance with the free electron behavior shown in **Figure 43**. Among them, the $|G|^2 = 50$ waves are densely and most evenly distributed across the Fermi level. Hence, $|G|^2 = 50$ can be designated to be most *critical* in the $Al_{48}Mg_{64}Zn_{48}$ 1/1–1/1–1/1 approximant. It must be

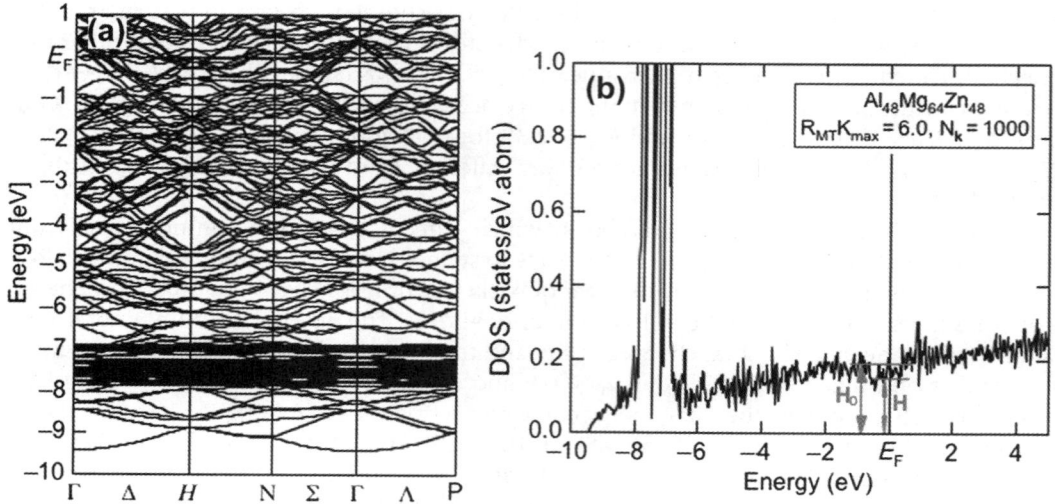

Figure 48 (a) Dispersion relations and (b) DOS calculated using WIEN2k for the $Al_{48}Mg_{64}Zn_{48}$ 1/1–1/1–1/1 approximant. The symbol H indicates an average height of the DOS at the Fermi level. H_0 is that in the absence of the pseudogap.

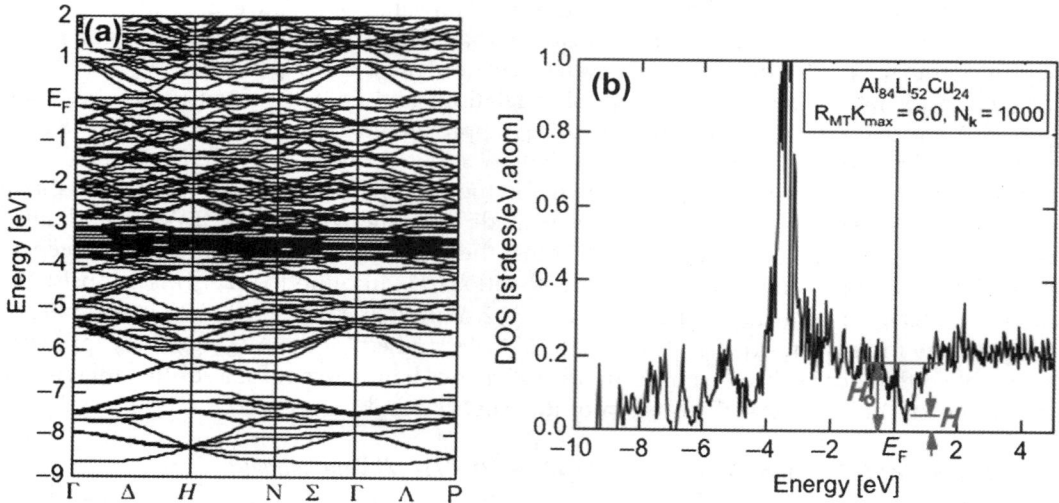

Figure 49 (a) Dispersion relations and (b) DOS calculated using WIEN2k for the $Al_{84}Li_{52}Cu_{24}$ 1/1–1/1–1/1 approximant. The symbol H indicates an average height of the DOS at the Fermi level. H_0 is that in the absence of the pseudogap.

reminded that, because of a relatively large full width of half maximum (FWHM), the electronic states at the Fermi level are contributed by not only the most *critical* $|\mathbf{G}|^2 = 50$ but also its neighboring ones extending over $|\mathbf{G}|^2$s from 44 to 54. This is taken as a theoretical proof for the presence of the *multizone effect* discussed in Section 2.6.3.

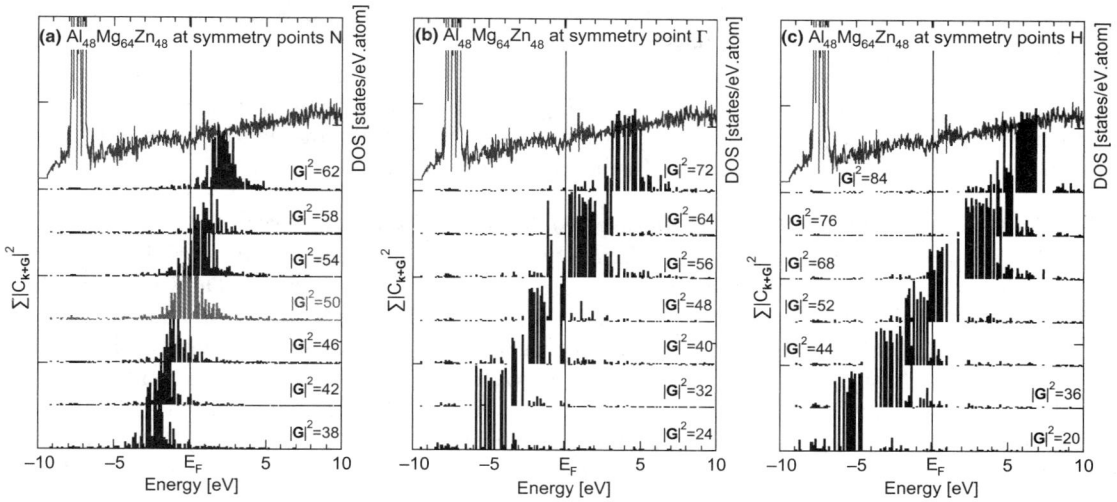

Figure 50 FLAPW–Fourier energy spectra at (a) symmetry points N, (b) symmetry point Γ and (c) symmetry points *H* for the $Al_{48}Mg_{64}Zn_{48}$ 1/1–1/1–1/1 approximant (Inukai et al., 2011a). Its DOS is superimposed. The most *critical* wave of $|G|^2 = 50$ is marked red in color. (For interpretation of the references to color in this legend, the reader is referred to the online version of this book.)

The Hume-Rothery plot and the energy dependence of $\langle|C|^2_{max}\rangle$ for the $Al_{48}Mg_{64}Zn_{48}$ 1/1–1/1–1/1 approximant are plotted in **Figure 51**a and b, respectively. The standard deviation is well below 0.1 over a wide energy range across the Fermi level, indicating that the free electron behavior holds well in this compound. The value of $\langle|C|^2_{max}\rangle_{E_F}$ at the Fermi level is 0.26. Therefore, the square of the Fermi diameter $(2k_F)^2$ can be reliably determined by locally reading off the ordinate at the Fermi level. It turns out to be 49.9±0.1 and the resulting **e/a** value to be 2.31. Therefore, we can confirm the validity of the

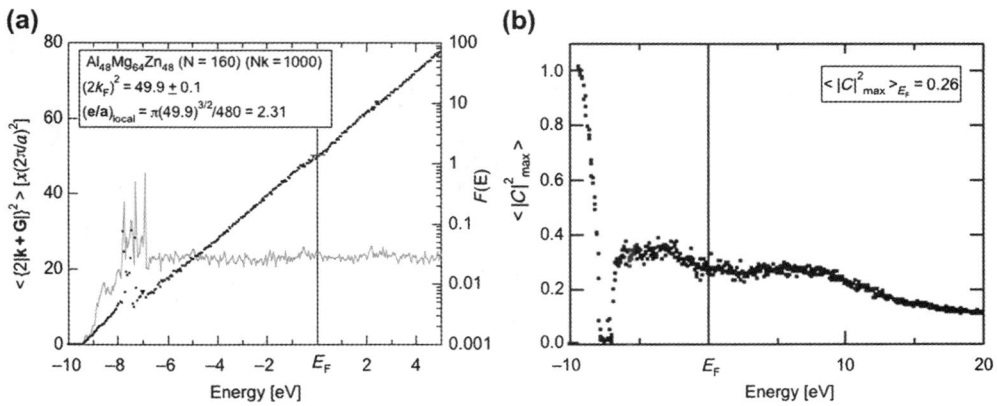

Figure 51 The Hume-Rothery plot and energy dependence of $\langle|C|^2_{max}\rangle$ for the $Al_{48}Mg_{64}Zn_{48}$ 1/1–1/1–1/1 approximant (Inukai et al., 2011a).

matching condition, since the most *critical* $|\mathbf{G}|^2$ was already deduced to be 50 from the FLAPW–Fourier spectrum shown in **Figure 50**a.

The FLAPW–Fourier analysis discussed above in combination with the NFE band calculations lead us to conclude that a shallow pseudogap shown in **Figure 48**b, being characterized by a relatively high pseudogap depth ratio H/H_0 equal to 0.83, originates solely from FsBz interactions involving the most *critical* $|\mathbf{G}|^2 = 50$ and its neighboring ones in the range from 44 to 54.

2.6.4.4 *Extraction of FsBz Interactions in Al$_{84}$Li$_{52}$Cu$_{24}$ 1/1–1/1–1/1 Approximant*

The FLAPW–Fourier spectra or energy dependences of $\sum|C_{\mathbf{k}+\mathbf{G}}|^2$ at symmetry points N, Γ and H for the Al$_{84}$Li$_{52}$Cu$_{24}$ 1/1–1/1–1/1 approximant are shown in **Figure 52**a–c along with its DOS, respectively. We can immediately see that the $|\mathbf{G}|^2 = 46$ wave interfering with the set of $\{631\}$ lattice planes at symmetry points N is the most evenly distributed across the Fermi level. Though the center of gravity for the $|\mathbf{G}|^2 = 50$ wave is located slightly above the Fermi level, its contribution is also almost equally important. Hence, both $|\mathbf{G}|^2 = 46$ and 50 may well be designated as being *critical* in this approximant. The FWHM is found to be 3.2 eV and is apparently larger than 2 eV in the Al$_{48}$Mg$_{64}$Zn$_{48}$ 1/1–1/1–1/1 approximant. As a result, electronic states at the Fermi level are decomposed into multiwaves ranging from $|\mathbf{G}|^2 = 40$ to 56.

The Hume-Rothery plot and energy dependence of $\langle|C|^2_{\max}\rangle$ for the Al$_{84}$Li$_{52}$Cu$_{24}$ 1/1–1/1–1/1 approximant are shown in **Figure 53**a and b, respectively. A "local" reading at the Fermi level is less accurate than that in the Al$_{48}$Mg$_{64}$Zn$_{48}$ 1/1–1/1–1/1 approximant because of the presence of anomalies associated with the formation of a large pseudogap. The value of $(2k_F)^2$ is deduced to be 47.1 ± 0.4. The effective $(e/a)_{\text{total}}$ value turns out to be 2.10, which is in a good agreement with an average e/a value of 2.05 obtained by assuming nominal valences of the constituent elements: $(e/a)_{\text{Al}} = 3.0$, $(e/a)_{\text{Li}} = 1.0$, and $(e/a)_{\text{Cu}} = 1.0$.

The square of the Fermi diameter $(2k_F)^2 = 47.1$ agrees well with the *critical* reciprocal lattice vector $|\mathbf{G}|^2 = 46$ deduced from the FLAPW–Fourier spectrum, demonstrating the fulfillment of the matching

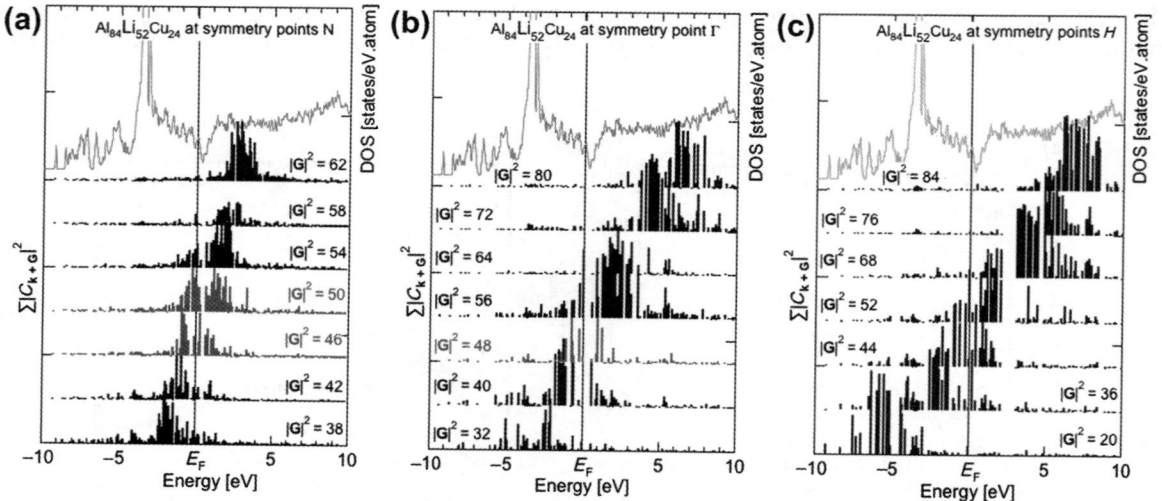

Figure 52 FLAPW–Fourier energy spectra at (a) symmetry points N, (b) symmetry point Γ, and (c) symmetry points H for the Al$_{84}$Li$_{52}$Cu$_{24}$ 1/1–1/1–1/1 approximant (Inukai et al., 2011a). Its DOS is superimposed. The most *critical* waves of $|\mathbf{G}|^2 = 46$, 48, and 50 are marked red in color. (For interpretation of the references to color in this legend, the reader is referred to the online version of this book.)

(a)   **(b)**

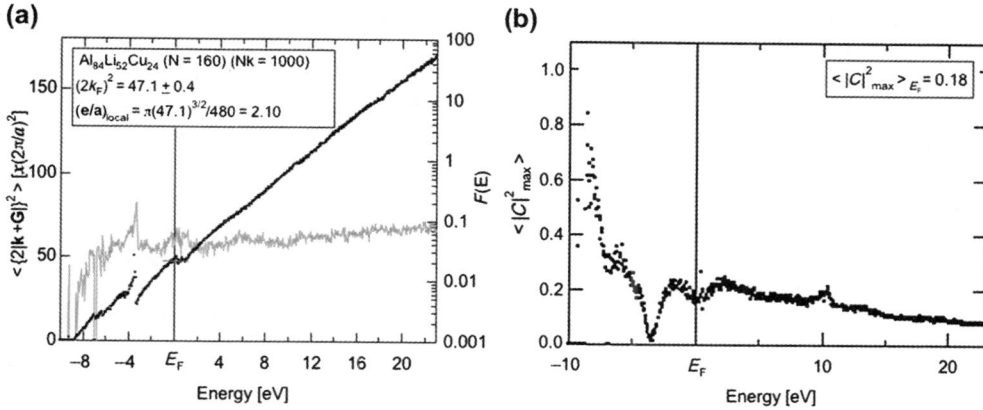

Figure 53 The Hume-Rothery plot and energy dependence of $\langle |C|^2_{max}\rangle$ for the $Al_{84}Li_{52}Cu_{24}$ 1/1–1/1–1/1 approximant (Inukai et al., 2011a).

condition $(2k_F)^2 = |\mathbf{G}|^2$ in the $Al_{84}Li_{52}Cu_{24}$ 1/1–1/1–1/1 approximant. We must stress that electronic states over $|\mathbf{G}|^2 = 40$–56 jointly participate in the formation of the pseudogap, confirming again the presence of the *multizone effect* unique to large systems like 1/1–1/1–1/1 approximants (Mizutani, 2010).

2.6.4.5   Why is the Pseudogap in $Al_{84}Li_{52}Cu_{24}$ Deeper than that in $Al_{48}Mg_{64}Zn_{48}$?

A difference in the pseudogap structure between these two RT-type approximants will be reviewed below from the point of view of both FsBz interactions and orbital hybridizations (Inukai et al., 2011a). The form factor for the $Al_{84}Li_{52}Cu_{24}$ model structure is shown in **Figure 54** as a function of the wave number normalized with respect to its Fermi diameter. It is clear that all form factors are comparable in magnitude to those in $Al_{48}Mg_{64}Zn_{48}$. The situation is, therefore, essentially the same as in the $Al_{48}Mg_{64}Zn_{48}$ 1/1–1/1–1/1 approximant discussed in Section 2.6.3.

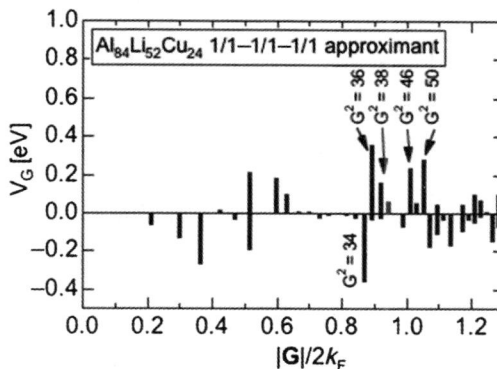

Figure 54 Form factor V_G in NFE band calculations as a function of the wave number normalized with respect to the Fermi diameter $2k_F$ for the $Al_{84}Li_{52}Cu_{24}$ 1/1–1/1–1/1 approximant (Sato et al., 2004). Sizable form factors appearing in the vicinity of the Fermi level are indexed in terms of $|\mathbf{G}|^2$. The value of $(2k_F)^2$ for $Al_{84}Li_{52}Cu_{24}$ is taken to be 45.20 in the units of $(2\pi/a)^2$, where a is the lattice constant.

The NFE model would be also reasonably applicable for $Al_{84}Li_{52}Cu_{24}$ approximant, since the Cu-3d band plays no essential role in the formation of a pseudogap (Sato et al., 2004). The 1061-plane wave approximation could reproduce well a pseudogap across the Fermi level and allowed us to identify zone planes involving $|\mathbf{G}|^2 = 46$ and 50 to be responsible for its formation. As is clear from the arguments above, the role of *multizone* effects in forming a pseudogap is essentially the same as that in the $Al_{48}Mg_{64}Zn_{48}$ 1/ 1-1/1-1/1 approximant discussed in Section 2.6.4.3. Indeed, the pseudogap depth ratio H/H_0 was deduced to be 0.66 from the 1061-plane wave NFE band calculations (Sato et al., 2004). This is much shallower than that of 0.28 calculated from the WIEN2k–FLAPW band calculations (**Figure** 49b).

Since the NFE band calculations take into account only FsBz interactions, we are naturally led to believe that not only FsBz interactions but also orbital hybridization effects play some key role in the $Al_{84}Li_{52}Cu_{24}$ approximant. As shown in **Figure 55**, Inukai et al. (2011a) revealed that strongly directional covalent-like bonding states are formed between Cu-4p and p-states of other constituent elements Al and Li over the energy range from -2 eV to the Fermi level, where the FsBz-induced pseudogap is present. Such sharp bonding states are formed at the expense of electronic states inside the pseudogap. They considered this to be effective enough to reduce the DOS background or to sharpen the declining slope in the pseudogap and to explain the reason why the pseudogap derived

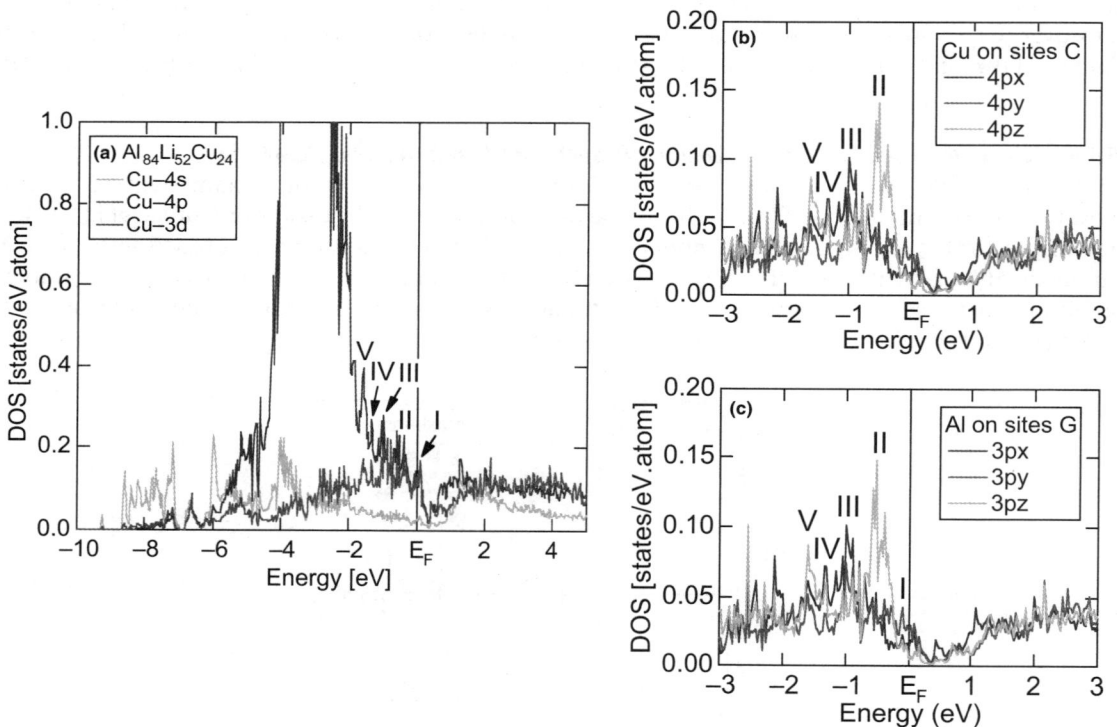

Figure 55   (a) 4s-, 4p-, and 3d-partial DOSs around Cu atoms on sites C for the $Al_{84}Li_{52}Cu_{24}$ 1/1−1/1−1/1 approximant (Inukai et al., 2011a). The partial DOS represents electronic states inside the MT sphere. Five successive peaks are present in both Cu-4p and Cu-3d partial DOSs and marked with numbers I–V. (b) p_x, p_y, and p_z-decomposed partial DOSs around Cu atoms on sites C and (c) that around Al atoms on sites G for the $Al_{84}Li_{52}Cu_{24}$ 1/1−1/1−1/1 approximant. Peaks are labeled with numbers I–VI. (For color version of this figure, the reader is referred to the online version of this book.)

from first-principles FLAPW band calculations is much deeper than that derived from the NFE model. Orbital hybridization effects are essentially absent in the $Al_{48}Mg_{64}Zn_{48}$ 1/1–1/1–1/1 approximant. This explains why the pseudogap in the latter is shallow and is caused by FsBz interactions alone.

In conclusion, the existence of more than two *critical* reciprocal lattice vectors around $|\mathbf{G}|^2 = 50$ is taken as a characteristic feature of the RT-type 1/1–1/1–1/1 approximants and is called the *multizone effect*. This is in sharp contrast to the situation in gamma-brasses, where dispersion relations are less crowded. The value of $(2k_F)^2$ for $Al_{48}Mg_{64}Zn_{48}$ and $Al_{84}Li_{52}Cu_{24}$ was determined from the Hume-Rothery plot to be 49.9 ± 0.1 and 47.1 ± 0.4, respectively, which led to the conclusion that the Hume-Rothery stabilization mechanism holds true with the fulfillment of the matching condition and that the *multizone effect* is taken as a unique feature in the RT-type 1/1–1/1–1/1 approximants.

2.6.5   MI-Type 1/1–1/1–1/1 Approximants

The MI-type quasicrystals and their approximants have received much more attention than the RT-type ones discussed above. To the best of our knowledge, however, no one has so far attempted to extract FsBz interactions in them because of the presence of TM elements as major constituent elements and a large number of atoms in the unit cell. The e/a determination of the TM element is certainly of crucial importance. It is further worthwhile studying whether the *multizone effect* discussed above for the RT-type approximants also applies to MI-type approximants. There is another phenomenon called *d state-mediated splitting*, which significantly perturbs the FLAPW–Fourier spectrum (Mizutani, 2010). It was revealed, for the first time, in gamma-brasses containing the TM element as a major constituent (Mizutani et al., 2006). One needs to study whether the *d state-mediated splitting* remains substantial and affects the argument on the Hume-Rothery stabilization mechanism in the MI-type approximants.

In the present chapter, we will review systematic studies on FLAPW–Fourier analyses for five MI-type 1/1–1/1–1/1 approximants (Mizutani et al., 2012): $Al_{114}Mn_{24}$, $Al_{114}Re_{24}$, $Al_{102}Re_{24}Si_{12}$, $Al_{108}Cu_6TM_{24}Si_6$ (TM = Fe and Ru), where orbital hybridizations are substantial between Al-sp and TM-d states. In order to clarify the role of FsBz interactions on the formation of a pseudogap across the Fermi level, we carried out the WIEN2k–FLAPW band calculations and, subsequently, FLAPW–Fourier analysis and the Hume-Rothery plot. Special care was again taken to optimize various parameters in the WIEN2k code in running the SCF cycles as efficiently as possible without any serious loss in accuracy and reliability of resulting data for such large systems.

2.6.5.1   *Atomic Structure of the MI-Type 1/1–1/1–1/1 Approximants and Construction of the Model Structures*

2.6.5.1.1   *$Al_{114}Mn_{24}$ 1/1–1/1–1/1 Approximant*

Cooper and Robinson (Cooper and Robinson, 1966) refined the X-ray diffraction data for an $Al_9Mn_2Si_{1.8}$ or $Al_{70.3}Mn_{15.6}Si_{14.0}$ single crystal and determined all 138-atom positions without distinguishing Si from Al. The space group was deduced to be Pm$\bar{3}$, since glue atoms filled outside the MI-cluster destroy the bcc symmetry and give rise to a simple cubic lattice. The composition for the refined structure was reduced to be $Al_{114}Mn_{24}$ by filling 24 Mn atoms on the vertices of a larger icosahedron and Al atoms onto remaining sites. The model structure was constructed by ignoring the presence of Si, as Cooper and Robinson did, and employed atomic coordinates and the lattice constant of $a = 1.268$ nm, which they refined.

2.6.5.1.2   *$Al_{102}Re_{24}Si_{12}$ 1/1–1/1–1/1 Approximant*

Takeuchi et al. (2003) determined 138 atom positions including positions of Si atoms in a series of Al–Re–Si 1/1–1/1–1/1 approximants by performing the Rietveld refinement analysis for powder diffraction data taken at SPring-8 synchrotron radiation facility, Japan, and revealed the complete ordering with space group $Pm\bar{3}$ at the composition $Al_{102}Re_{24}Si_{12}$. Their experimentally derived atomic structure with the lattice constant $a = 1.28603$ nm was directly employed for the WIEN2k–FLAPW electronic structure calculations.

2.6.5.1.3 $Al_{114}Re_{24}$ 1/1–1/1–1/1 Approximant

All Si atoms in the atomic structure of the $Al_{102}Re_{24}Si_{12}$ 1/1–1/1–1/1 approximant discussed above were intentionally replaced by Al atoms to make a direct comparison with the data for $Al_{114}Mn_{24}$. Note that both $Al_{114}Mn_{24}$ and $Al_{114}Re_{24}$ compounds are not existing in the equilibrium phase diagram but are intentionally constructed as a simplified model structure to grasp the essence of the MI-type 1/1–1/1–1/1 approximants and to examine whether the role of Mn and Re on the FsBz interactions is almost identical or not.

2.6.5.1.4 $Al_{108}Cu_6TM_{24}Si_6$ (TM=Fe or Ru) 1/1–1/1–1/1 Approximants

The atomic structure of the $Al_{108}Cu_6(Fe_{1-x}Ru_x)_{24}Si_6$ (x=0, 0.5 and 1) 1/1–1/1–1/1 approximants was experimentally determined by analyzing powder diffraction spectra taken by using synchrotron radiations at SPring-8, Japan, by means of the Rietveld method (Mizutani et al., 2001; Takeuchi and Mizutani, 2002). The space group is reduced to $Im\bar{3}$ as a result of random distributions of different chemical species in most equivalent sites in the unit cell. The exception is that the TM element Fe or Ru is exclusively filled into 12 sites on the larger icosahedron for both $Al_{108}Cu_6Fe_{24}Si_6$ and $Al_{108}Cu_6Ru_{24}Si_6$ (Section 2.3.2). The lattice constants were found to be 1.248 and 1.26832 nm, respectively. The experimentally derived total number of atoms per unit cell was deduced to be 139 in both cases (Mizutani, 2010).

To allow first-principles electronic structure calculations, we need to construct a model structure free from chemical disorder by a slight modification of the experimentally determined atomic structure discussed above (Mizutani, 2010; Mizutani et al., 2004, 2009). The presence of Cu atoms with occupancies less than about 10% on sites II, MI1, and MI2 is fully ignored. These sites together with glue sites G3 are filled with Al atoms. To compensate for the deficient Cu, the model assumes Cu atoms to fill into glue sites G1. Space group $Pm\bar{3}$ is intentionally employed to put six Cu atoms on glue sites G1 in one of the MI-type cluster and six Si atoms on G1 in the other one. We obtain the chemical formula $Al_{108}Cu_6TM_{24}Si_6$ (TM = Fe and Ru) for the model structure with the total number of atoms per unit cell equal to 144.

2.6.5.2 WIEN2k–FLAPW Band Calculations for MI-Type 1/1–1/1–1/1 Approximants

First-principles LMTO–ASA electronic structure calculations for MI-type 1/1–1/1–1/1 approximants have been made in the past: $Al_{114}Mn_{24}$ (Fujiwara, 1989; Zijlstra and Bose, 2003), $Al_{102}Re_{24}Si_{12}$ (Takeuchi et al., 2003), and $Al_{108}Cu_6TM_{24}Si_6$ (TM = Fe and Ru) (Mizutani et al., 2004). All of them were consistent with the finding of a deep DOS pseudogap across the Fermi level. The origin of the pseudogap was largely discussed from the viewpoint of orbital hybridization effects in the past. As noted above, attempts to elucidate FsBz interactions on the basis of first-principles band calculations have been so far very limited for MI-type 1/1–1/1–1/1 approximants. Present FLAPW band calculations were conducted for the five MI-type 1/1–1/1–1/1 approximants described above, using the WIEN2k program package (Mizutani et al., 2012). The exchange-correlation energy is calculated within the local density approximation (Perdew and Wang, 1992).

As discussed in Section 2.6.4.2, we need to carefully determine three important parameters in order to perform FLAPW band calculations with subsequent FLAPW–Fourier analysis in an efficient way for systems with giant unit cells: the radii of the MT spheres R_{MT} for constituent elements, the cutoff

Table 5   Key parameters in WIEN2k–FLAPW band calculations for MI-type 1/1–1/1–1/1 approximants

	R_{MT} (a.u.)	$R_{MT}K_{max}$	N_k
$Al_{114}Mn_{24}$	$r_{Al} = 2.06$ $r_{Mn} = 2.29$	5.52	1000
$Al_{102}Re_{24}Si_{12}$	$r_{Al} = 2.06$ $r_{Re} = 2.63$ $r_{Si} = 2.05$	5.41	1000
$Al_{114}Re_{24}$	$r_{Al} = 2.06$ $r_{Re} = 2.50$	5.44	1000
$Al_{108}Cu_6Fe_{24}Si_6$	$r_{Al} = 2.06$ $r_{Cu} = 2.12$ $r_{Fe} = 2.34$ $r_{Si} = 2.10$	5.60	1000
$Al_{108}Cu_6Ru_{24}Si_6$	$r_{Al} = 2.06$ $r_{Cu} = 2.16$ $r_{Ru} = 2.50$ $r_{Si} = 2.12$	5.50	1000

parameter $R_{MT}K_{max}$ and the number of **k**-points in the first Brillouin zone, N_k. They are listed in **Table 5** for the present five 1/1–1/1–1/1 approximants. The value of N_k was fixed to be 1000 for all five compounds containing 138–144 atoms per unit cell (Section 2.6.4.2). To facilitate the FLAPW–Fourier analysis, we lowered $R_{MT}K_{max}$ from the value 6.0 used for the RT-type 1/1–1/1–1/1 approximants (Inukai et al., 2011a). After careful repeated checks, we chose the value over the range 5.41–5.60, which appeared not to lead to a significant loss of accuracy for the electronic structure data used in the FLAPW–Fourier method. The need to reduce the size of the basis set used in the FLAPW–Fourier method and for the Hume-Rothery plot arose, because the number of independent atom coordinates is increased to 138 and 144 in the present MI-type approximants with space group Pm$\bar{3}$ compared to 80 in the RT-type approximants with space group Im$\bar{3}$. However, the higher value of $R_{MT}K_{max} = 6.0$ was invariably employed for structural relaxations.

The DOSs for the five MI-type 1/1–1/1–1/1 approximants are shown in **Figure 56**. The Fermi level is located in a deep pseudogap for all approximants studied. This is consistent with those reported in the

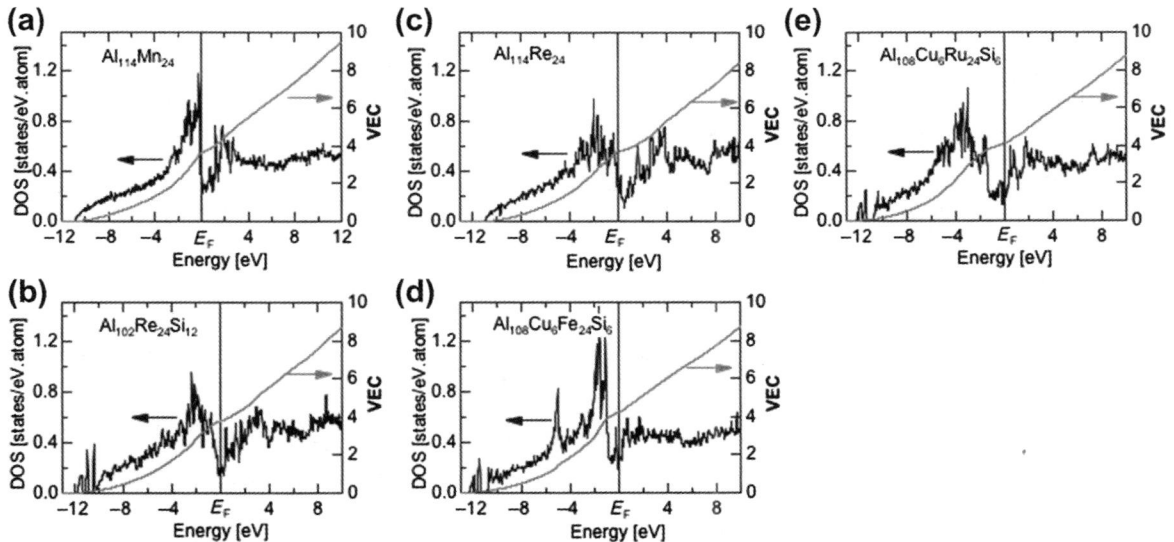

Figure 56   DOS calculated with WIEN2k using parameters listed in **Table 5** for the five MI-type 1/1–1/1–1/1 approximants (Mizutani et al., 2012). All of them are characterized by a deep pseudogap at the Fermi level. The valence electron concentration (VEC) represents the number of electrons per atom in the valence band.

literature (Fujiwara, 1989; Mizutani et al., 2004; Roche and Fujiwara, 1998; Takeuchi et al., 2003; Zijlstra and Bose, 2003). A comparison of DOSs between $Al_{102}Re_{24}Si_{12}$ and $Al_{114}Re_{24}$ tells us that Si serves as a means of shifting the Fermi level closer to the bottom of the pseudogap as a result of an increase in electron concentration.

Figure 57a–d compare Al- and TM-partial DOSs in both $Al_{114}Mn_{24}$ and $Al_{114}Re_{24}$ approximants. It is clear that the Mn and Re d-bands are split into bonding and antibonding subbands across the Fermi level. Similar splitting can be seen in Al-s and Al-p partial DOSs. The same argument has been already made for $Al_{114}Mn_{24}$ (Fujiwara, 1989), $Al_{102}Re_{24}Si_{12}$ (Takeuchi et al., 2003), $Al_{108}Cu_6Fe_{24}Si_6$, and $Al_{108}Cu_6Ru_{24}Si_6$ 1/1–1/1–1/1 approximants (Mizutani et al., 2004, 2009). Hence, we can safely take this as evidence that a pseudogap originates from orbital hybridization effects between Al-sp and TM-d states in all MI-type approximants containing Al and TM as major constituent elements. In the following sections, we will study the origin of a pseudogap in these five systems from the viewpoint of FsBz interactions by making full use of the FLAPW–Fourier analysis.

2.6.5.3 Extraction of FsBz Interactions in MI-Type 1/1–1/1–1/1 Approximants

The FLAPW–Fourier spectra were constructed at symmetry points M and X, and point Γ of a simple cubic lattice, since all the MI-type 1/1–1/1–1/1 approximants studied in the present work are

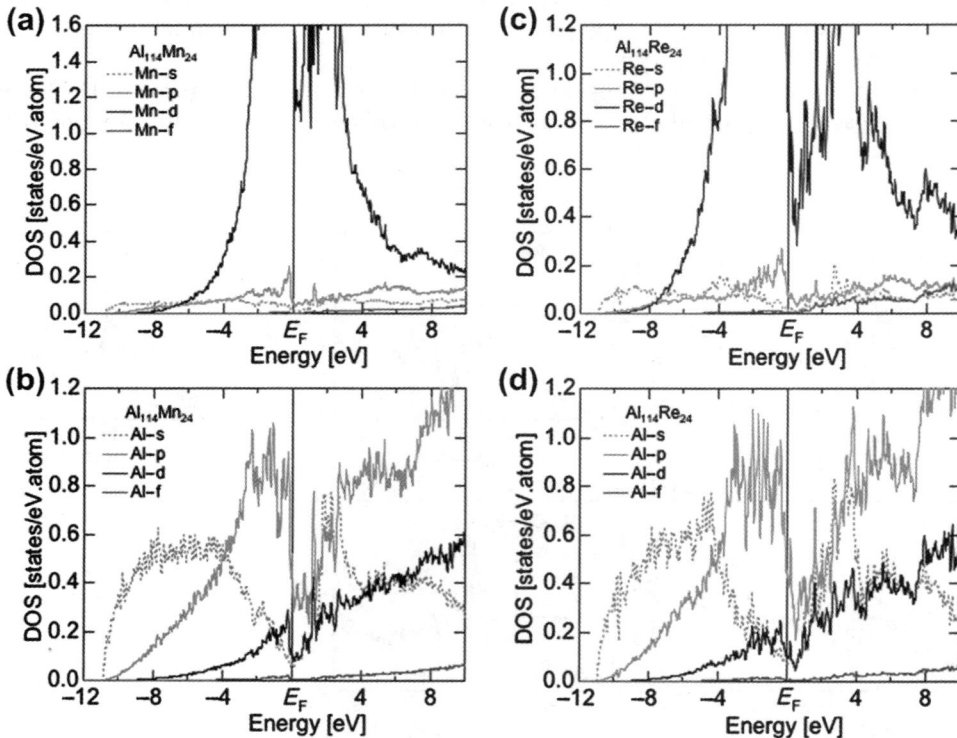

Figure 57 Al- and TM-partial DOSs for the $Al_{114}Mn_{24}$ and $Al_{114}Re_{24}$ 1/1–1/1–1/1 approximants (Mizutani et al., 2012). A deep pseudogap can be attributed to orbital hybridization effects between Al-sp and TM-d states. (For color version of this figure, the reader is referred to the online version of this book.)

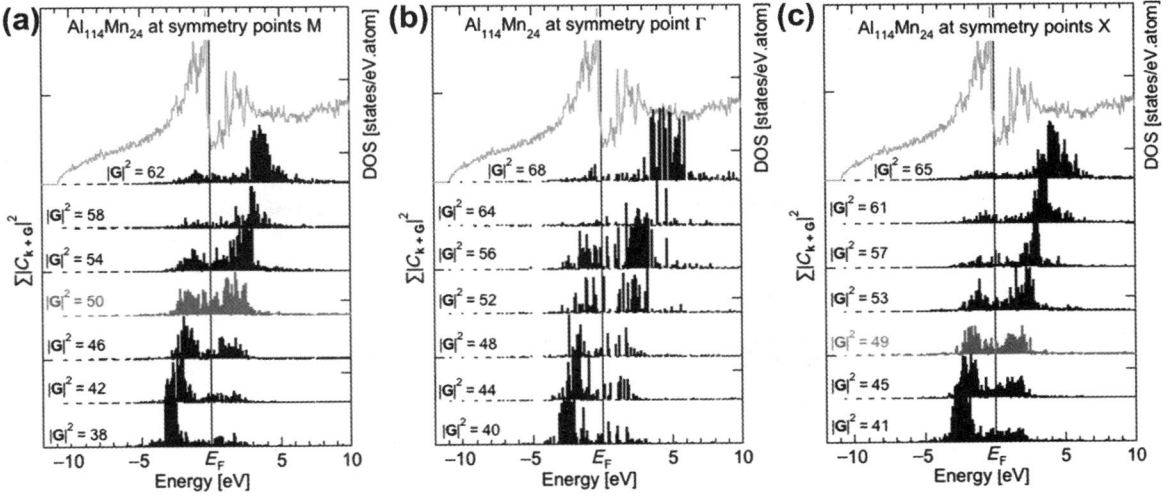

Figure 58 The FLAPW–Fourier spectra at (a) symmetry points M, (b) symmetry point Γ, and (c) symmetry points X for the $Al_{114}Mn_{24}$ 1/1–1/1–1/1 approximants (Mizutani et al., 2012). The DOS is superimposed. The most *critical* waves of $|G|^2 = 49$ and 50 are marked red in color.

characterized by space group $Pm\bar{3}$. Attention is now directed to only plane waves specified by $|G|^2$, whose Fourier coefficient $\sum_G |C_{k+G}|^2$ is sizable in the vicinity of the Fermi level. This is nothing but the way to extract the FsBz interactions involved.

The FLAPW–Fourier spectra, for example, energy dependences of $\sum_G |C_{k+G}|^2$ with $|G|^2$ in the range 38–68 for $Al_{114}Mn_{24}$ 1/1–1/1–1/1 approximant are shown at symmetry points M and X, and point Γ along with its DOS in **Figure 58a–c**, respectively. We can immediately see that all plane waves over $|G|^2 = 41$ to 56 possess sizable Fourier coefficients across the Fermi level. This is evidently taken as the confirmation of the *multizone effect* discussed for RT-type approximants in Section 2.6.4. Among them, the $|G|^2 = 50$ waves interfering with the set of {543} + {550} + {710} lattice planes are most evenly and densely distributed across the Fermi level. The FLAPW–Fourier spectra at symmetry points M for the $Al_{114}Re_{24}$ and $Al_{102}Re_{24}Si_{12}$ are shown in **Figure 59a** and **b**. They are quite similar to that for $Al_{114}Mn_{24}$ shown in **Figure 58**.

We show in **Figure 60a** and **b** the FLAPW–Fourier spectra at symmetry points M for the $Al_{108}Cu_6Fe_{24}Si_6$ and $Al_{108}Cu_6Ru_{24}Si_6$ 1/1–1/1–1/1 approximants, respectively. The sizable Fourier coefficients are observed at the Fermi level over the range $42 \leq |G|^2 \leq 54$. Among them, the $|G|^2 = 50$ waves are again most evenly and densely distributed around the Fermi level for both approximants. In summary, we could confirm not only the presence of the *multizone effect* involving $|G|^2$ from 41 to 56 at the Fermi level, but also identify $|G|^2 = 50$ to be the most *critical* for all MI-type approximants studied.

Figure 61a–c show the energy dependence of $\langle \{2|k+G|\}^2 \rangle_E$ and its standard deviation in units of $(2\pi/a)^2$ for $Al_{114}Mn_{24}$, $Al_{114}Re_{24}$, and $Al_{102}Re_{24}Si_{12}$ 1/1–1/1–1/1 approximants, respectively. The standard deviation exceeds 0.1 over the energy range, where the Mn and Re d-bands extend, respectively. Hence, we selected the NFE approximation and drew a straight line passing data points, as marked with red in color, while ignoring those with large standard deviations. We consider the straight line to represent well the dispersion relation for the motion of itinerant electrons outside the MT sphere.

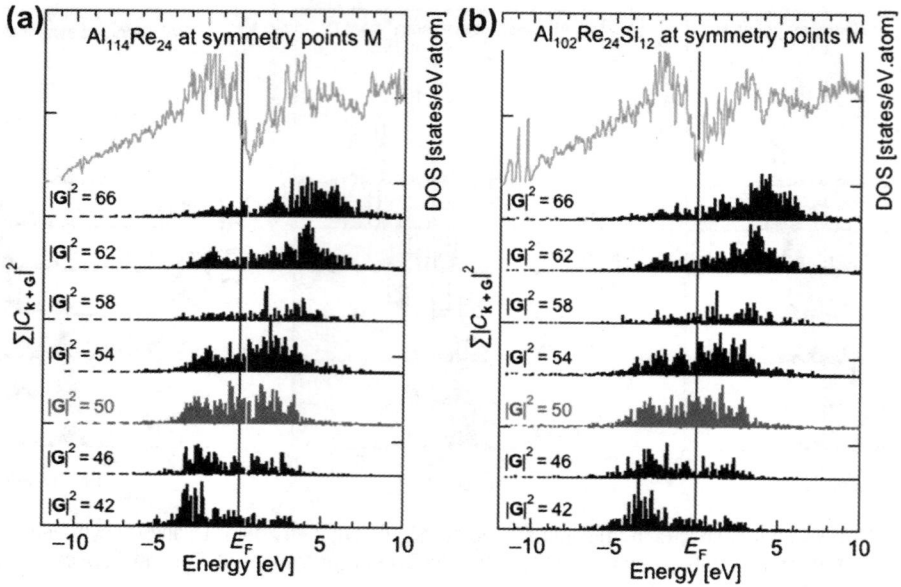

Figure 59   The FLAPW–Fourier spectra at symmetry points M for (a) the $Al_{114}Re_{24}$ and (b) the $Al_{102}Re_{24}Si_{12}$ 1/1–1/1–1/1 approximants. The DOS is superimposed. The most *critical* wave of $|\mathbf{G}|^2 = 50$ is marked red in color.

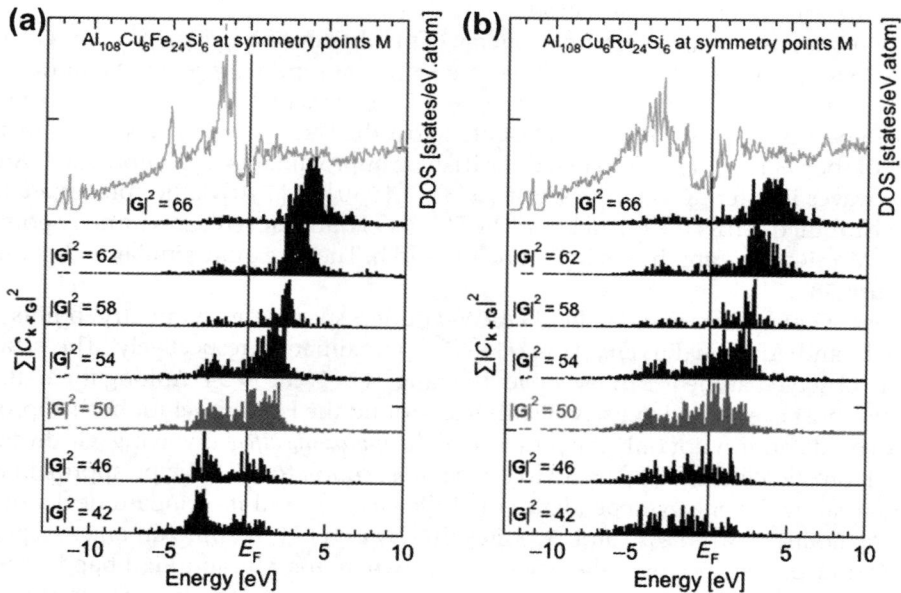

Figure 60   The FLAPW–Fourier spectra at symmetry points M for (a) the $Al_{108}Cu_6Fe_{24}Si_6$ and (b) the $Al_{108}Cu_6Ru_{24}Si_6$ 1/1–1/1–1/1 approximants. The DOS is superimposed. The most *critical* wave of $|\mathbf{G}|^2 = 50$ is marked red in color.

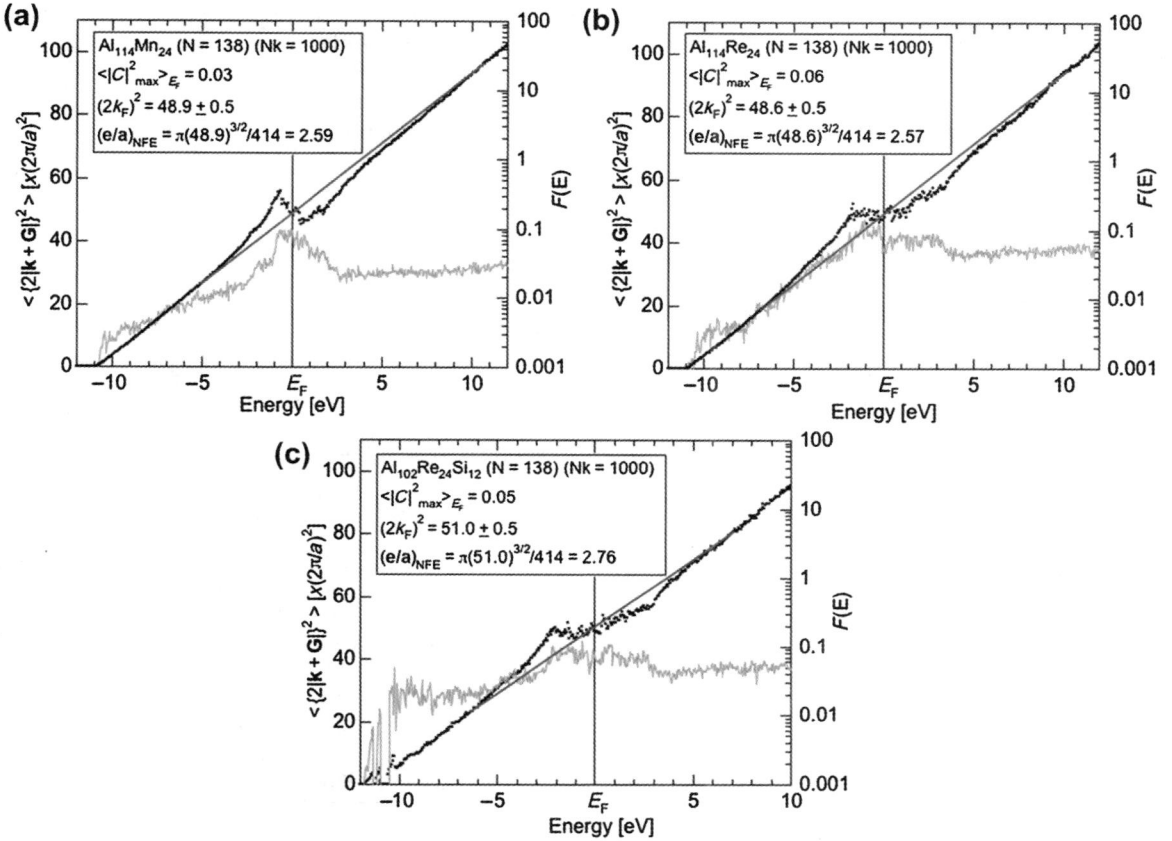

Figure 61   The Hume-Rothery plot for (a) the $Al_{114}Mn_{24}$, (b) the $Al_{114}Re_{24}$, and (c) the $Al_{102}Re_{24}Si_{12}$ 1/1–1/1–1/1 approximants (Mizutani et al., 2012).

In other words, we effectively cast the realistic electronic structure onto the free electron model without losing its basic features. The values of $(2k_F)^2$ for $Al_{114}Mn_{24}$, $Al_{114}Re_{24}$, and $Al_{102}Re_{24}Si_{12}$ 1/1–1/1–1/1 approximants can be determined to be 48.9 ± 0.5, 48.6 ± 0.5, and 51.0 ± 0.5 from the intercepts at the Fermi level, respectively. The replacement of some trivalent Al atoms by quadrivalent Si in $Al_{102}Re_{24}Si_{12}$ resulted in a slight increase in $(2k_F)^2$ to 51.0 ± 0.5. The effective $(e/a)_{total}$ value enclosed by the resulting Fermi sphere, can be easily calculated by inserting the value of $(2k_F)^2$ thus obtained and $N = 138$ into Eqn 24. The value turned out to be 2.59, 2.57, and 2.76 for $Al_{114}Mn_{24}$, $Al_{114}Re_{24}$, and $Al_{102}Re_{24}Si_{12}$ 1/1–1/1–1/1 approximants, respectively.

The Hume-Rothery plot was also made for the $Al_{108}Cu_6Fe_{24}Si_6$ and $Al_{108}Cu_6Ru_{24}Si_6$ 1/1–1/1–1/1 approximants. As shown in **Figure 62**, the role of Fe and Ru is essentially the same in these approximants. The resulting $(2k_F)^2$ and $(e/a)_{total}$ values, along with *critical* $|G|^2$ for all MI-type 1/1–1/1–1/1 approximants are summarized in **Table 6**. We can find that the matching condition $(2k_F)^2 = |G|^2$ holds well in the neighborhood of 50. It is important to be reminded that the matching condition involving $|G|^2 = 50$ in MI-type 1/1–1/1–1/1 approximants is the same as that in RT-type 1/1–1/1–1/1 approximants.

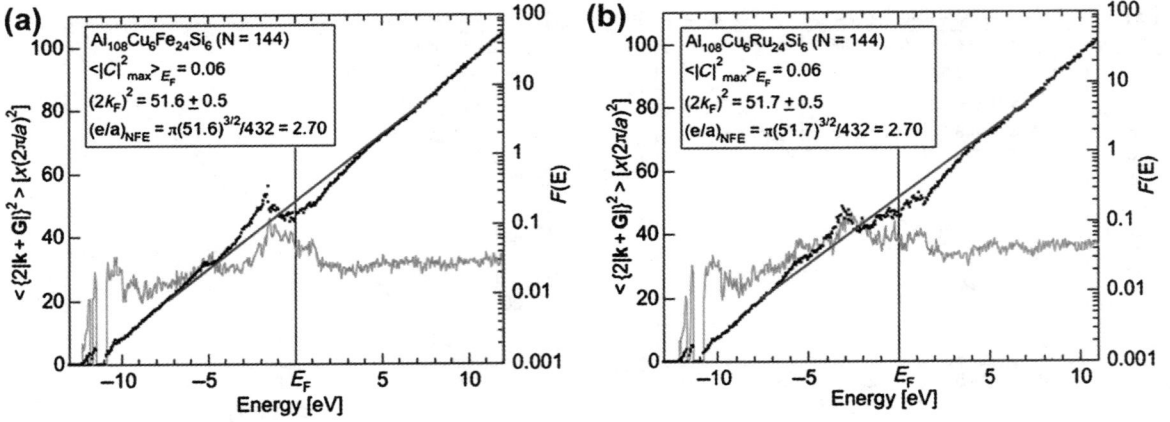

Figure 62   The Hume-Rothery plot for (a) the $Al_{108}Cu_6Fe_{24}Si_6$ and (b) the $Al_{108}Cu_6Ru_{24}Si_6$ 1/1–1/1–1/1 approximants.

Table 6   Square of the Fermi diameter, **e/a**, **e/uc**, and square of *critical* reciprocal lattice vectors $|G|^2$ for MI-type 1/1–1/1–1/1 approximants

	N *atoms per unit cell*	$(2k_F)^2 [x(2\pi/a)^2]$	*e/a*	*e/uc*	Critical $|G|^2 [x(2\pi/a)^2]$
$Al_{114}Mn_{24}$	138	48.9 ± 0.5	2.59	357	50
$Al_{102}Re_{24}Si_{12}$	138	51.0 ± 0.5	2.76	381	50
$Al_{114}Re_{24}$	138	48.6 ± 0.5	2.57	355	50
$Al_{108}Cu_6Fe_{24}Si_6$	144	51.6 ± 0.5	2.70	389	50
$Al_{108}Cu_6Ru_{24}Si_6$	144	51.7 ± 0.5	2.70	389	50
$Al_{48}Mg_{64}Zn_{48}$	160	49.9 ± 0.1	2.31	368	50

It is of great interest at this stage to consider if we can estimate the value of (e/a) of constituent TM elements Mn, Fe, Ru, and Re in the periodic table. We can evaluate from the Hume-Rothery plot not only $(e/a)_{element}$ of elements in the periodic table but also $(e/a)_{total}$ in intermetallic compounds existing in various binary alloy systems such as Al–TM alloy systems. As will be described in Section 2.7.4, we could determine the **e/a** values of relevant constituent elements as follows: $(e/a)_{Al} = 3.0$, $(e/a)_{Cu} = 1.0$, $(e/a)_{Si} = 4.0$, $(e/a)_{Mn} = 1.0$, $(e/a)_{Re} = 1.0$, $(e/a)_{Fe} = 1.1$, and $(e/a)_{Ru} = 1.0$. It is important to note that $(e/a)_{TM}$ is always positive in sign and distributed in the neighborhood of unity for all TM elements so far studied. Using these figures, we can easily calculate $(e/a)_{total}$ for $Al_{114}Mn_{24}$, $Al_{114}Re_{24}$, $Al_{102}Re_{24}Si_{12}$, $Al_{108}Cu_6Fe_{24}Si_6$, and $Al_{108}Cu_6Ru_{24}Si_6$ 1/1–1/1–1/1 approximants to be 2.65, 2.65, 2.74, 2.64, and 2.64, respectively. They are in good agreement with the data listed in **Table 6** within the accuracy of the present analysis. It should be emphasized that $(e/a)_{total}$ in all MI-type 1/1–1/1–1/1 approximants is distributed in the narrow range at around 2.6–2.7.

2.7   Hume-Rothery Electron Concentration Rule

2.7.1   The Physics Behind the e/uc versus $|G|^2$ Diagram

The Hume-Rothery electron concentration rule is the rule to link the family of a specific alloy phase with a particular electron per atom ratio **e/a**. **Figure 63** summarizes the energy dependence of the

Table 7   FWHM of $|G|^2 = 50$ plane wave for RT- and MI-type 1/1–1/1–1/1 approximants

	RT-type 1/1–1/1–1/1 approximants		MI-type 1/1–1/1–1/1 approximants				
System	$Al_{48}Mg_{64}Zn_{48}$	$Al_{84}Li_{52}Cu_{24}$	$Al_{114}Mn_{24}$	$Al_{114}Re_{24}$	$Al_{102}Re_{24}Si_{12}$	$Al_{108}Cu_6Fe_{24}Si_6$	$Al_{108}Cu_6Ru_{24}Si_6$
FWHM [eV]	2.0	3.2	5.0	6.8	7.1	5.3	6.6

$|G|^2 = 50$ waves for both rhombic triacontahedron (RT)- and Mackay icosahedron (MI)-type 1/1–1/1–1/1 approximants studied by means of the FLAPW–Fourier analysis (Mizutani et al., 2012). Though the full width at half maximum (FWHM) in the MI-type is definitely wider than that in the RT-type, $|G|^2 = 50$ can be designated as a *critical* reciprocal lattice vector and employed to describe an alloy phase characteristic common to both families of 1/1–1/1–1/1 approximants. Numerical values of FWHM for RT- and MI-type approximants are listed in **Table 7**. There exists a sizable discrepancy in e/a between the RT- and MI-type approximants: e/a $= 2.31$ and 2.10 for $Al_{48}Mg_{64}Zn_{48}$ and $Al_{84}Li_{52}Cu_{24}$,

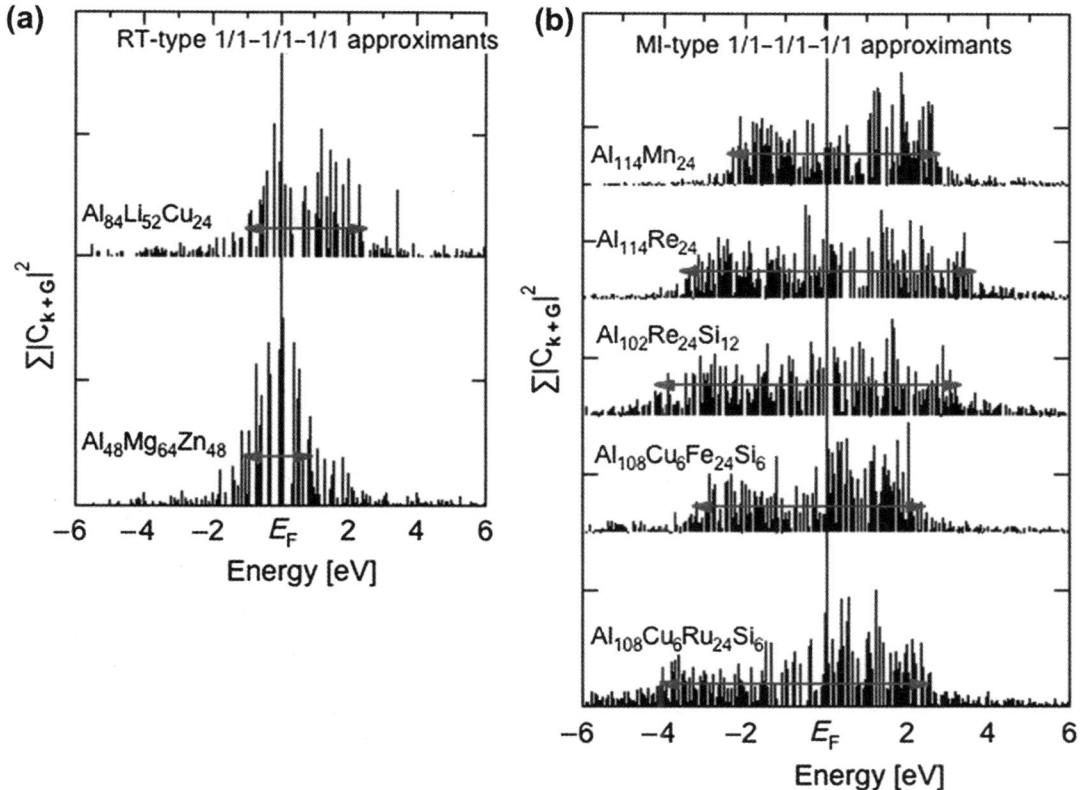

Figure 63   FLAPW–Fourier spectra for the $|G|^2 = 50$ waves for (a) RT-type $Al_{48}Mg_{64}Zn_{48}$ and $Al_{84}Li_{52}Cu_{24}$ 1/1–1/1–1/1 approximants and (b) five MI-type approximants (Mizutani et al., 2012). The FWHM is marked with an arrow in each energy spectrum.

respectively, in the RT-type 1/1–1/1–1/1 approximants, while $\mathbf{e/a} = 2.57–2.76$ for MI-type ones (**Table 6**). This means that $\mathbf{e/a}$, which represents the number of itinerant electrons per atom, may not be suitable as an electron concentration parameter to describe the Hume-Rothery electron concentration rule.

In the preceding sections, we have placed emphasis on the validity of the matching condition $(2k_F)^2 = |\mathbf{G}|^2$ for CMA phases including both gamma-brasses and 1/1–1/1–1/1 approximants. Here, it should be reminded that the matching condition itself represents the Hume-Rothery electron concentration rule, since information about the electron concentration enters through $(2k_F)^2$, while information about crystal structure through $|\mathbf{G}|^2$. It is also important to recall the statement in Section 2.4.2 such that both *critical* $|\mathbf{G}|^2$ and the square of the Fermi diameter $(2k_F)^2$ are cell size-dependent. Thus, one should use a cell size-dependent electron concentration parameter in place of $\mathbf{e/a}$ to link it with *critical* $|\mathbf{G}|^2$.

The number of atoms per unit cell, N, is plotted as a function of $\mathbf{e/a}$ in **Figure 64** for a variety of 1/1–1/1–1/1 approximants including the data for Tsai-type Cd_6Ca with $N = 168$ (Gomez & Lidin, 2003), to which $\mathbf{e/a} = 2.0$ may be safely assigned, since it consists of the two divalent elements Cd and Ca. (See comment 1 in Note Added in Proof). A decrease in N is observed with increasing $\mathbf{e/a}$ (Mizutani et al., 2012). From this we can expect the product of $\mathbf{e/a}$ and N, that is, $\mathbf{e/uc}$ to be kept almost constant for all

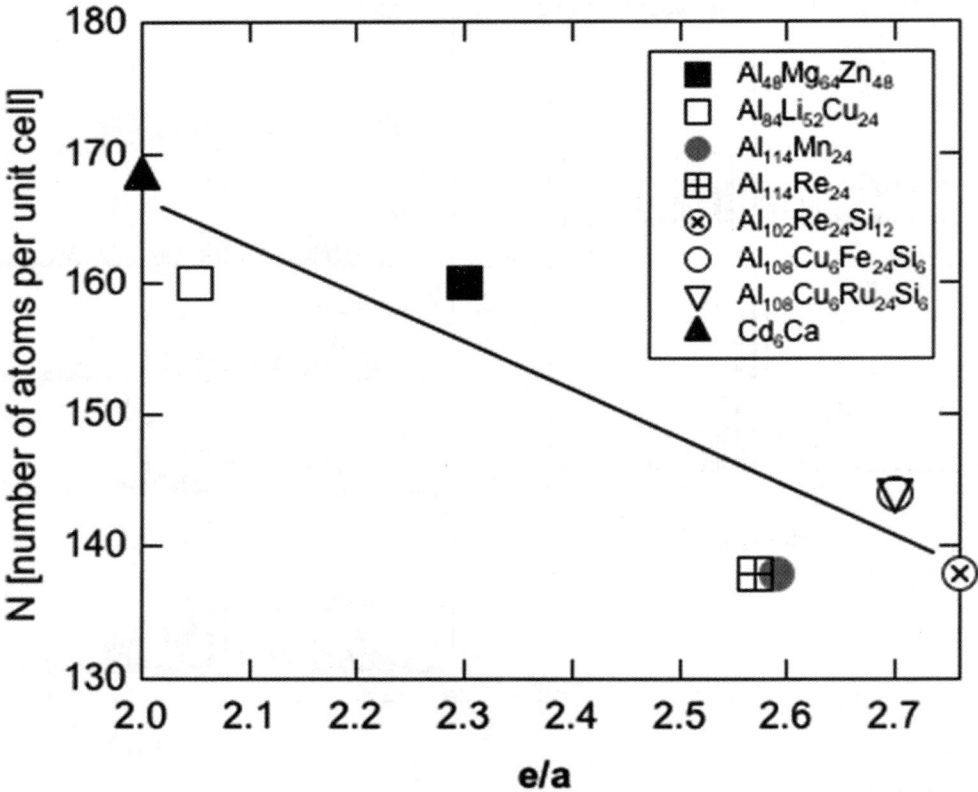

Figure 64 Number of atoms per unit cell as a function of **e/a** for RT-, MI-, and Tsai-type 1/1–1/1–1/1 approximants (Mizutani et al., 2012).

1/1–1/1–1/1 approximants, regardless of the atomic cluster types involved. We are, thus, encouraged to propose that the cell size-dependent **e/uc** may be better correlated with the *critical* reciprocal lattice vector $|\mathbf{G}|^2$ for CMAs characterized by a deep pseudogap at the Fermi level.

Once the FLAPW–Fourier analysis and the Hume-Rothery plot are performed, one can reliably determine both **e/uc** and *critical* $|\mathbf{G}|^2$ for CMAs, provided that needed structural information is available in the literature (Villars, 1997). As discussed in the previous sections, we have so far accumulated the relevant data for a series of gamma-brasses ($N=52$) and RT- and MI-type 1/1–1/1–1/1 approximants ($N=138$–144). In order to elucidate the physics of the Hume-Rothery electron concentration rule, we consider it to be of great importance to collect the set of **e/uc** versus *critical* $|\mathbf{G}|^2$ data in as many CMAs as possible. Let us now discuss the determination of **e/uc** and *critical* $|\mathbf{G}|^2$ data sets for some representative CMAs other than gamma-brasses and RT- and MI-type 1/1–1/1–1/1 approximants.

2.7.2 FLAPW–Fourier Analyses for CMAs Other than Gamma-Brasses and 1/1–1/1–1/1 Approximants

2.7.2.1 CMAs in Al–TM Alloy Systems

Three binary alloy systems Al–Mn, Al–Tc, and Al–Re, where the TM element is in Group 7 in the periodic table, are unique in the sense that there are more than two CMAs in a rather narrow Al-rich composition range; Al_6Mn ($N = 28$), $Al_{114}Mn_{24}$ ($N = 138$), and $Al_{10}Mn_3$ ($N = 26$) in the first, $Al_{12}Tc$ ($N = 26$) and Al_6Tc ($N = 28$) in the second and $Al_{12}Re$ ($N = 26$), Al_6Re ($N = 28$), and $Al_{114}Re_{24}$ ($N = 138$) in the third. Fortunately, we can perform FLAPW band calculations with subsequent FLAPW–Fourier and Hume-Rothery plot analyses for all of them, since a perfectly ordered structure free from any chemical disorder is available in the literature (Villars, 1997). Since the $Al_{114}TM_{24}$ (TM = Mn and Re) 1/1–1/1–1/1 approximants have already been discussed in Section 2.6, we deal with the remaining CMAs in the present section.

The Al_6Mn, Al_6Tc, and Al_6Re compounds, where Mn, Tc, and Re are in Group 7 in the periodic table, crystallize into an orthorhombic structure containing 28 atoms per unit cell (Villars, 1997). The FLAPW–Fourier spectra and Hume-Rothery plots for Al_6Mn and Al_6Re compounds are shown in **Figures 65 and 66**, respectively (Mizutani et al., 2012). A pseudogap is found across the Fermi level in both cases. The $|\mathbf{G}|^2 = 18$ is deduced to be commonly *critical* from the Fourier spectra, while $(2k_F)^2$ to be 17.7 ± 0.3 and 17.9 ± 0.3 from the Hume-Rothery plot. The data obtained for Al_6Tc were also found to be almost identical to those discussed above.

More surprisingly, the Al_6Fe and Al_6Ru compounds, where Fe and Ru are in Group 8 in the periodic table, are known to be isostructural to Al_6TM (TM = Mn, Tc, and Re) compounds discussed above (Villars, 1997).[18] The FLAPW–Fourier and Hume-Rothery plots were also made for them. Numerical data for all Al_6TM (TM = Mn, Tc, Re, Fe, and Ru) compounds are summarized in **Table 8**. Essentially, identical data were obtained for all of them. This indicates that the electronic structure due to electrons outside the MT sphere remains unchanged by interchange of relevant TM elements in Groups 7 and 8. The set of **e/uc** and $|\mathbf{G}|^2$ for all these Al_6TM compounds is commonly identified to be 79 ($= 2.83 \times 28$) and 18, respectively.

There is another family of isostructural CMAs in Al–TM alloy systems. The $Al_{12}TM$ (TM = Mo and W in Group 6) and $Al_{12}TM$ (TM = Tc and Re in Group 7) compounds are cubic with space group $Im\bar{3}$ (No. 204) and contain 26 atoms per unit cell (Villars, 1997). The FLAPW–Fourier spectra and

---

[18] Al_6Fe exists as a metastable phase in the Al–Fe alloy system (Kim and Cantor, 1994). All other compounds Al_6Mn, Al_6Tc, Al_6Re, Al_6Ru, $Al_{12}Tc$, $Al_{12}Re$, $Al_{12}Mo$, and $Al_{12}W$ exist as a stable phase.

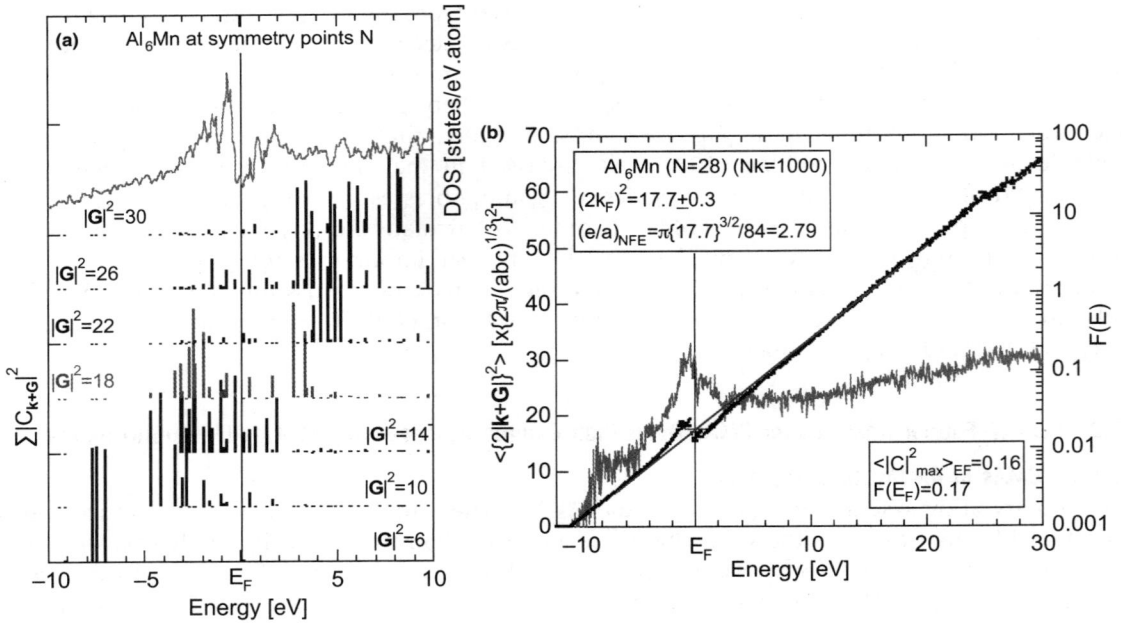

Figure 65 (a) FLAPW-Fourier energy spectra at symmetry points N and (b) Hume-Rothery plot for Al_6Mn compound. In (a), the total DOS is superimposed and the *critical* wave of $|G|^2 = 18$ is marked red in color. In orthorhombic crystal with lattice constants a, b and c, $|G|^2$ is expressed $\left[\left(\frac{bc}{a^2} \right)^{2/3} h^2 + \left(\frac{ca}{b^2} \right)^{2/3} k^2 + \left(\frac{ab}{c^2} \right)^{2/3} l^2 \right]$ in units of $\left(\frac{2\pi}{V_0^{1/3}} \right)^2$, where $V_0 = abc$ is the unit cell volume and $(h\ k\ l)$ is the set of Miller indices. Strictly speaking, the value of $|G|^2$ becomes non-integer, since $a \neq b \neq c$. For the sake of simplicity, however, $|G|^2 = h^2 + k^2 + l^2$ is employed in (a). The resulting error is found to be unimportant.

Hume-Rothery plots for $Al_{12}Tc$ and $Al_{12}Re$ are shown in **Figures 67 and 68**, respectively. A density of states (DOS) pseudogap at the Fermi level is clearly seen in both compounds. A perfect agreement in $(2k_F)^2$ and **e/a** is found between them. The relevant data for $Al_{12}Mo$ are shown in **Figure 69**. Its DOS is also characterized by a pseudogap at the Fermi level. As listed in **Table 8**, the resulting $(2k_F)^2$ and **e/a** values for all $Al_{12}TM$ compounds are in an excellent agreement with each other. The set of **e/uc** and *critical* $|G|^2$ is deduced to be 75 (= 2.87 × 26) and 18, respectively, for the $Al_{12}TM$ (TM = Mo, W, Tc, and Re) compounds.

As is clear from the arguments above, we revealed that the TM element can be selected from different columns or groups in the periodic table in both Al_6TM and $Al_{12}TM$ compounds. In connection with this statement, we remind the pioneering work by Tsai et al. (Tsai et al., 1988; Yokoyama et al., 1991), who discovered a series of Al–Cu–TM (TM = Fe, Ru, and Os) and Al–Pd–TM (TM=Mn and Re) quasicrystals. They intuitively assumed the Hume-Rothery electron concentration rule to hold in Al–TM alloys, as long as the TM element is selected from the same group in the periodic table. We have naturally supported their idea, since the electronic configuration outside the core of inert gas element is the same and thereby the chemical nature is similar to each other in a given group. Therefore, it is rather

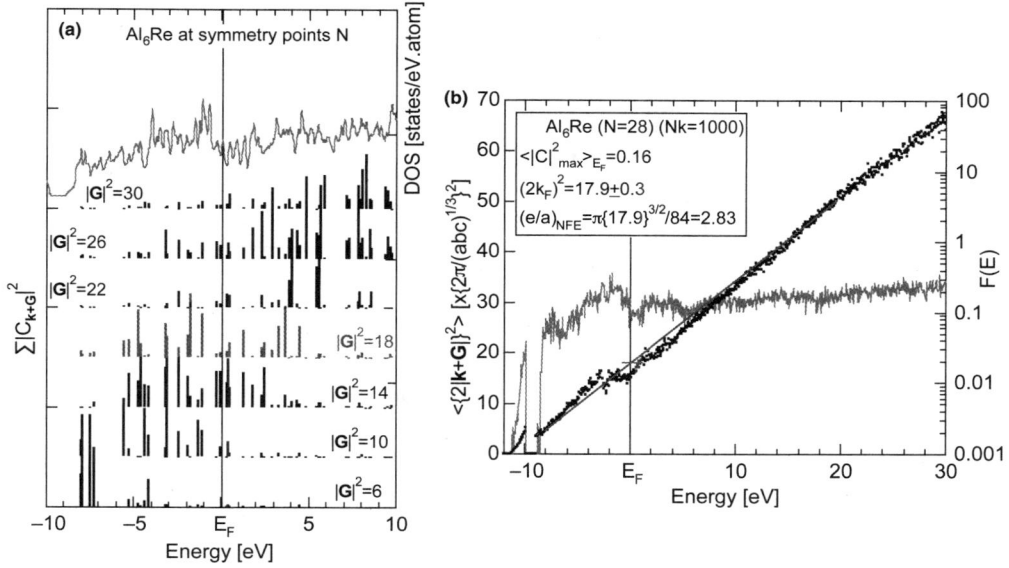

Figure 66 (a) FLAPW-Fourier energy spectra at symmetry points N and (b) Hume-Rothery plot for Al₆Re compound. In (a), the total DOS is superimposed, the *critical* wave of |**G²**| =18 is marked red in color and the same remark as in **Figure 65** is applied to the expression for |**G²**|.

surprising to find an identical electronic structure, as represented by $(2k_F)^2$ and **e/a**, in both $Al_{12}TM$ and Al_6TM compounds in spite of the involvement of TM elements over Groups 6–8. As will be discussed in Section 2.7.4, this will be taken as a strong indication that the effective **e/a** value for the TM elements involved must be very similar to each other.

Finally, the FLAPW–Fourier spectrum and Hume-Rothery plot are shown in **Figure 70** for the $Al_{10}Mn_3$ ($N=26$) compound, which is hexagonal with space group P63/mmc (No. 194) (Villars, 1997). A pseudogap happens to be heavily intervened by a sharp DOS peak at the Fermi level. The d state-mediated splitting is observed, particularly, for the $|G|^2 = 17$ waves. The values of $(2k_F)^2$ and **e/a** are deduced to be 16.0 ± 0.2 and 2.58, respectively. The set of **e/uc** and *critical* $|G|^2$ in $Al_{10}Mn_3$ was

Table 8 N, $(2k_F)^2$, **e/a**, **e/uc**, and *critical* $|G|^2$ for Al_6TM and $Al_{12}TM$ compounds

System	N	$(2k_F)^2$	e/a	e/uc	Critical $|G|^2$
Al₆Mn	28	17.7 ± 0.3	2.79	78	18
Al₆Tc	28	18.0 ± 0.3	2.85	80	18
Al₆Re	28	17.9 ± 0.3	2.83	79	18
Al₆Fe	28	17.9 ± 0.4	2.83	79	18
Al₆Ru	28	18.0 ± 0.3	2.85	80	18
Al₁₂Mo	26	17.2 ± 0.1	2.87	75	18
Al₁₂W	26	17.1 ± 0.1	2.85	74	18
Al₁₂Tc	26	17.4 ± 0.1	2.92	76	18
Al₁₂Re	26	17.4 ± 0.1	2.92	76	18

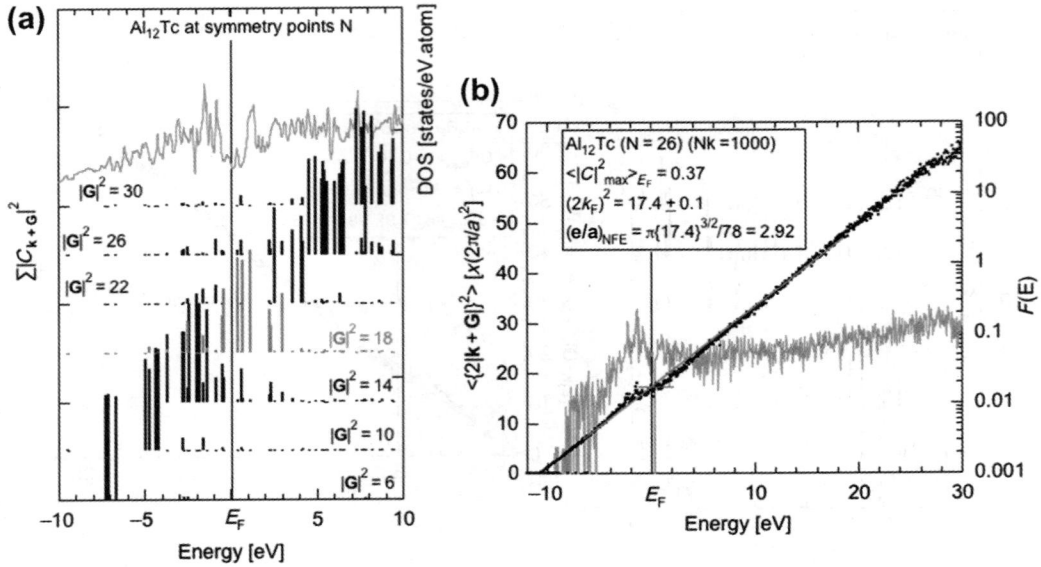

Figure 67   (a) FLAPW–Fourier energy spectra at symmetry points N and (b) Hume-Rothery plot for $Al_{12}Tc$ compound. Its DOS is superimposed. The *critical* wave of $|G|^2 = 18$ is marked red in color.

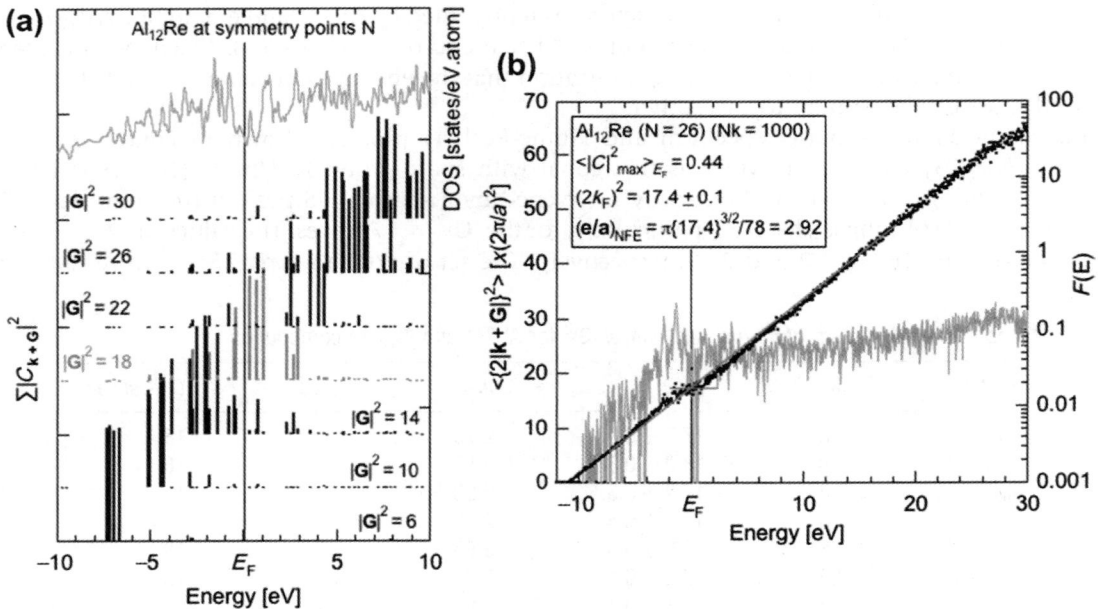

Figure 68   (a) FLAPW–Fourier energy spectra at symmetry points N and (b) Hume-Rothery plot for $Al_{12}Re$ compound. Its DOS is superimposed. The *critical* wave of $|G|^2 = 18$ is marked red in color.

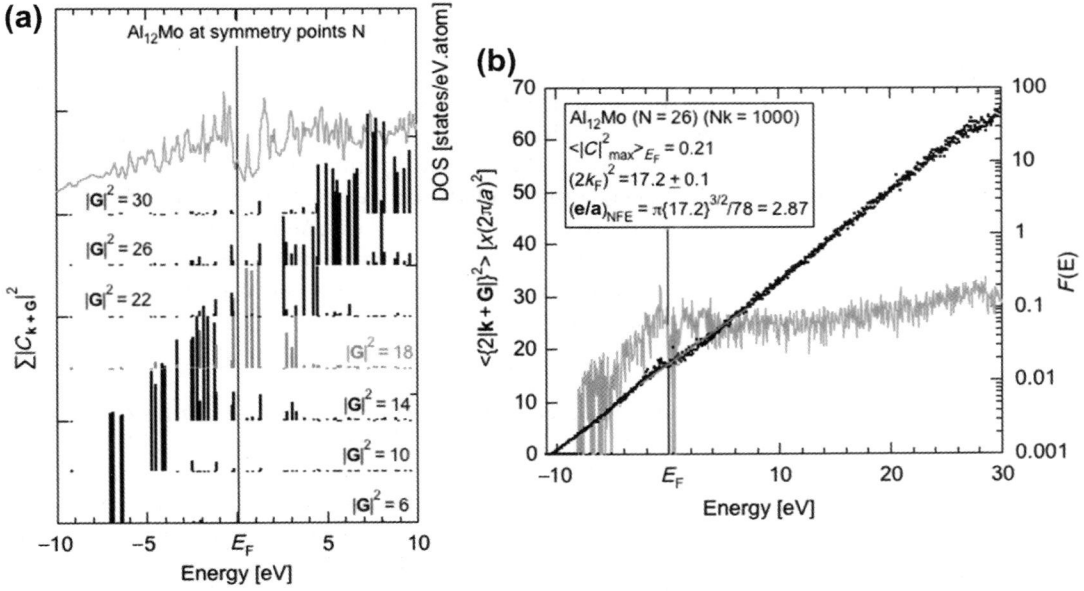

Figure 69   (a) FLAPW–Fourier energy spectra at symmetry points N and (b) Hume-Rothery plot for $Al_{12}Mo$ compound. Its DOS is superimposed. The *critical* wave of $|\mathbf{G}|^2 = 18$ is marked red in color.

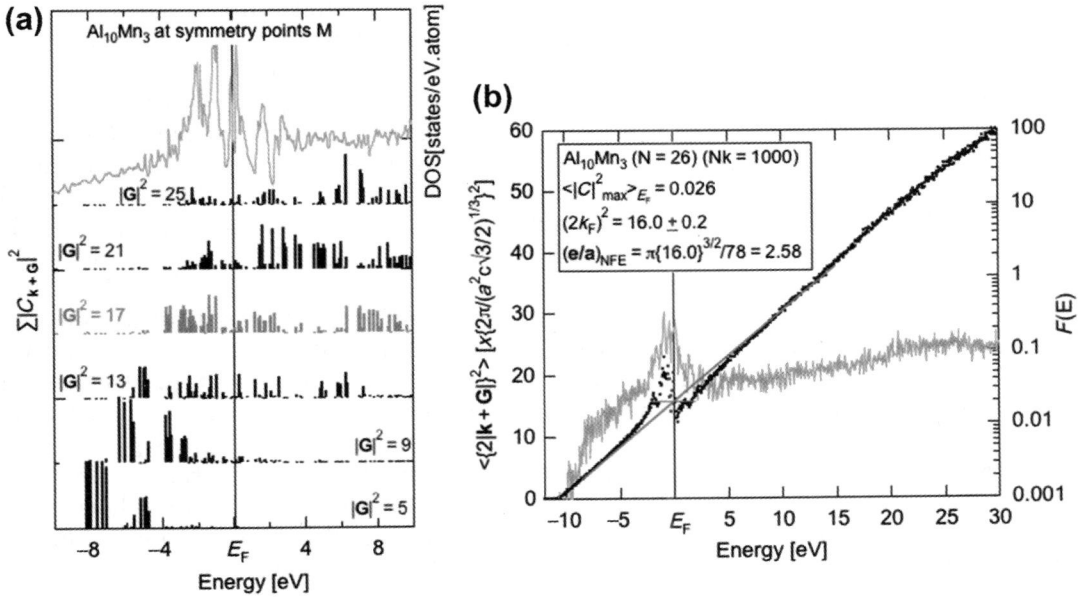

Figure 70   (a) FLAPW–Fourier energy spectra at symmetry points N and (b) Hume-Rothery plot for $Al_{10}Mn_3$ compound. Its DOS is superimposed. The *critical* wave of $|\mathbf{G}|^2 = 17$ is marked red in color.

determined to be 67 and 17, respectively. Thus, the $Al_{10}Mn_3$ compound can be distinguished in terms of **e/uc** and *critical* $|\mathbf{G}|^2$ from Al_6TM and $Al_{12}TM$ compounds discussed above.

Before ending this section, it is of crucial importance to note that the *critical* reciprocal vector $|\mathbf{G}|^2 = 18$ is commonly obtained not only for both Al_6TM (TM = Mn, Tc, Re, Fe, and Ru) and $Al_{12}TM$ (TM=Mo, W, Tc, Re, and Mo) compounds but also for gamma-brasses in spite of a large difference in **e/a** among Al_6TM (e/a=2.83), $Al_{12}TM$ (e/a=2.87), and gamma-brasses (e/a=1.6). However, a correlation emerges, if **e/uc** is used. Indeed, **e/uc** tends to converge into a common value around 80: 79 (=2.83 × 28), 75 (= 2.87 × 26), and 83 (= 1.6 × 52) for Al_6TM, $Al_{12}TM$, and gamma-brasses, respectively. As emphasized in Section 2.7.1, we consider the use of **e/uc** to be more powerful to discuss the Hume-Rothery electron concentration rule in a more systematic manner and to establish its firm basis in a more universal way.

2.7.2.2 β-Al₃Mg₂ and γ-Mg₁₇Al₁₂

Upon constructing the **e/uc** versus *critical* $|\mathbf{G}|^2$ diagram for CMAs, we consider studies of compounds with a further larger unit cell to be extremely important, since the relevant data would fill higher **e/uc** and $|\mathbf{G}|^2$ regions in the diagram. As described in Section 2.3.4, the β-Al_3Mg_2 compound has been known as one of the most notable CMAs possessing 1178 atoms in the unit cell (Feuerbacher et al., 2007; Samson, 1965). However, the experimental data on its electronic structure so far reported are rather limited. Dolinšek et al. (2007) measured the ^{27}Al Knight shift in β-Al_3Mg_2 and revealed the electron DOS at the Fermi level to be about 90% of that of fcc Al. Their work indicates the presence of a rather shallow pseudogap at the Fermi level. It is worthwhile exploring if it obeys the Hume-Rothery stabilization mechanism, since the alloy is free from TM elements and thereby orbital hybridization effects must be small. However, the number of atoms exceeding 1000 in the unit cell in addition to the presence of significant defects has so far prohibited us from the execution of first-principles band calculations. Thus, we are at the moment obliged to estimate its FsBz interactions, as described below, by carrying out WIEN2k–FLAPW band calculations for its substitute γ-$Mg_{17}Al_{12}$ existing next to β-Al_3Mg_2 in the Al–Mg binary phase diagram (Murray, 1988; Okamoto, 2000).

The γ-$Mg_{17}Al_{12}$ compound contains 58 atoms per unit cell with space group I$\bar{4}$3m and can be also regarded as being typical of CMAs. The FLAPW–Fourier spectrum at symmetry points N is shown in **Figure 71**a along with the DOS (Mizutani et al., 2010). A pseudogap is clearly formed at the Fermi level. The $|\mathbf{G}|^2 = 26$ wave is most evenly distributed across the Fermi level. Similarly, the $|\mathbf{G}|^2 = 24$ wave at symmetry point Γ is also found to be substantial at the Fermi level (Mizutani et al., 2010). Hence, both $|\mathbf{G}|^2 = 24$ and 26 were designated to be *critical* in γ-$Mg_{17}Al_{12}$. The value of $(2k_F)^2$ is deduced from the Hume-Rothery plot to be 25.9 ± 0.2 in units of $(2\pi/a)^2$, as shown in **Figure 71**b. The fulfillment of the matching condition in combination with the nearly free electron (NFE) band calculations led to conclude that the DOS pseudogap at the Fermi level originates from FsBz interactions involving $|\mathbf{G}|^2 = 24$ and 26 in γ-$Mg_{17}Al_{12}$. The value of **e/a** can be easily calculated to be 2.38 from $(2k_F)^2$ obtained above. This is in good agreement with its valence of 2.41 obtained by taking a composition-average of $(e/a)_{Al} = 3.0$ and $(e/a)_{Mg} = 2.0$. The value of **e/uc** is deduced to be 138 for γ-$Mg_{17}Al_{12}$.

The analysis on γ-$Mg_{17}Al_{12}$ above encouraged us to estimate the value of $(2k_F)^2$ for β-Al_3Mg_2 to be 204 by inserting the relation **e/uc** = (e/a) × N = 2.6 × 1178 = 3063 into Eqn 5, where e/a = 2.6 is obtained by using $(e/a)_{Al} = 3.0$ and $(e/a)_{Mg} = 2.0$ (Mizutani et al., 2010). **Figure 72** shows the powder X-ray diffraction spectrum taken with Cu-Kα radiation for β-Al_3Mg_2 (Mizutani et al., 2010). Representative diffraction lines were indexed by using the atomic structure data reported in Feuerbacher et al. (2007). The diffraction peak indexed as 12-fold {10 10 0}, 48-fold {10 86}, and 24-fold {14 20} corresponds to the set of lattice planes with $|\mathbf{G}|^2 = 200$, which best satisfies the

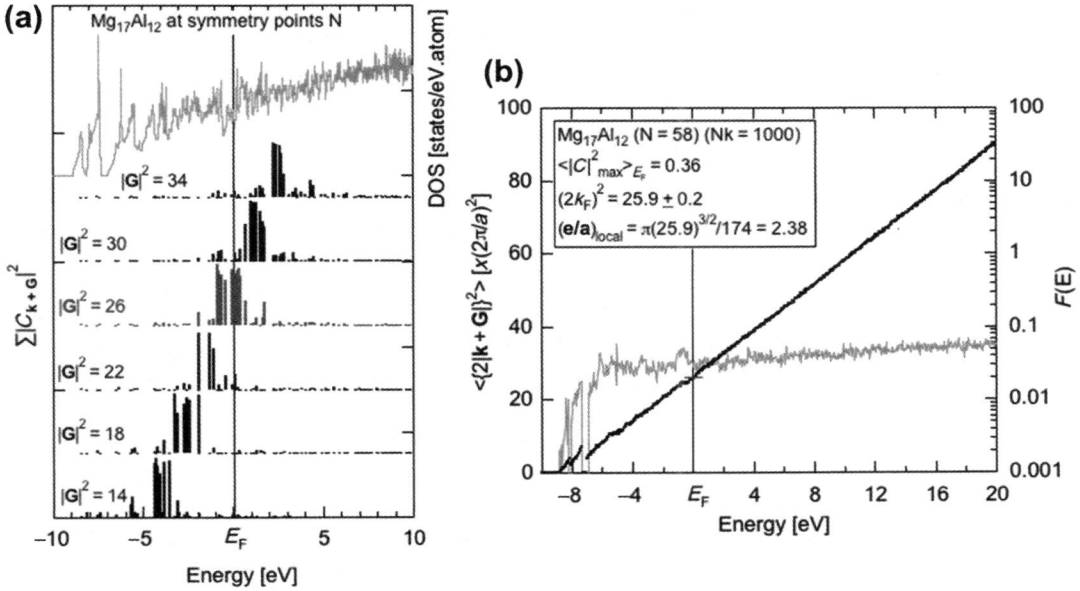

Figure 71　(a) FLAPW–Fourier energy spectra at symmetry points N and (b) Hume-Rothery plot for Mg₁₇Al₁₂ compound (Mizutani et al., 2010). Its DOS is superimposed. The *critical* wave of $|\mathbf{G}|^2 = 26$ is marked red in color.

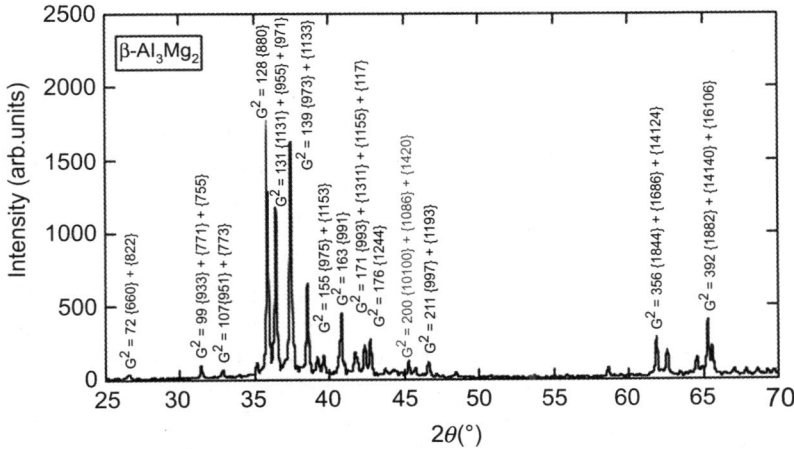

Figure 72　Powder X-ray diffraction spectrum taken with Cu-Kα radiation for β-Al₃Mg₂ (Mizutani et al., 2010).

matching condition with $(2k_F)^2 = 204$. Note that resulting 84 Brillouin zone planes are positioned at equidistance from the origin in the reciprocal space. Its neighboring zones would also contribute to the formation of the DOS pseudogap as a result of *multizone effects,* as discussed in Section 2.6 for 1/ 1-1/1-1/1 approximants. All of them would effectively encompass its Fermi surface in reciprocal space to produce a pseudogap across the Fermi level.[19] Though no first-principles electronic structure calculations are available at the moment, we have selected $|\mathbf{G}|^2 = 200$ as the most *critical* reciprocal lattice vector and $\mathbf{e/uc} = 3063$ as the electron concentration satisfying the matching condition for β-Al$_3$Mg$_2$ (Mizutani et al., 2010) and added it to the $\mathbf{e/uc}$ versus $|\mathbf{G}|^2$ diagram, which will be discussed in Section 2.7.3.

2.7.2.3 Al–Mg–Zn 2/1–2/1–2/1 Approximant

The 2/1–2/1–2/1 approximant in Al–Mg–Zn alloy system was first discovered by Takeuchi and Mizutani (1995) at the composition Al$_{15}$Mg$_{43}$Zn$_{42}$. The atomic structure was later studied by Sugiyama et al. (2002) at the composition Al$_{16}$Mg$_{42}$Zn$_{42}$ and by Lin and Corbett (September 12, 2006) at the composition Al$_{12.6}$Mg$_{31.8}$Zn$_{55.6}$. Lin and Corbett claimed the possession of 680 atoms per unit cell with space group Pa3. More recently, Kreiner (2009) reported that the 2/1–2/1–2/1 approximant exists at composition very close to Al$_{15}$Mg$_{43}$Zn$_{42}$ but not at Al$_{12.6}$Mg$_{31.8}$Zn$_{55.6}$ and that it contains 676 atoms per unit cell. A pseudogap across the Fermi level was confirmed to be present in the Al$_{15}$Mg$_{43}$Zn$_{42}$ 2/1–2/1–2/1 approximant through measurements of the electronic specific heat coefficient (Takeuchi and Mizutani, 1995). Though first-principles band calculations are not available because of substantial amounts of chemical disorder in addition to a large number of atoms exceeding 600 in the unit cell, we can safely assume its pseudogap to originate from FsBz interactions and to satisfy the matching condition.

The value of $(2k_F)^2$ for the Al$_{15}$Mg$_{43}$Zn$_{42}$ 2/1–2/1–2/1 approximant can be reliably estimated to be 125 by inserting the relation $\mathbf{e/uc} = N \times (\mathbf{e/a}) = 680 \times 2.15 = 1462$ into Eqn 5, where $\mathbf{e/a} = 2.15$ is obtained by using $(\mathbf{e/a})_{Al} = 3.0$ and $(\mathbf{e/a})_{Mg} = (\mathbf{e/a})_{Zn} = 2.0$. The set of 24-fold {11 20}, 24-fold {10 50}, 48-fold {10 43}, and 48-fold {865} lattice planes with $|\mathbf{G}|^2 = 125$ can be chosen as the one satisfying the matching condition and hence as being *critical* for the Al$_{15}$Mg$_{43}$Zn$_{42}$ 2/1–2/1–2/1 approximant (Mizutani, 2010).

2.7.3 New Hume-Rothery Electron Concentration Rule

Figure 73 shows $\mathbf{e/uc}$ as a function of *critical* reciprocal lattice vector $|\mathbf{G}|^2$ for CMAs so far studied (Mizutani, 2010). Relevant numerical data are listed in **Table 9**. All the $\mathbf{e/uc}$ versus *critical* $|\mathbf{G}|^2$ data are found to fall on a straight line with a slope of 3/2 on a log–log scale, being consistent with the relation:

$$\mathbf{e/uc} = \frac{\pi}{3}\left[(2k_F)^2\right]^{3/2} = \frac{\pi}{3}\left[|\mathbf{G}|^2\right]^{3/2}, \qquad (27)$$

where the last term is derived by inserting the matching condition $(2k_F)^2 = |\mathbf{G}|^2$ into the middle term in Eqn 27. As emphasized in Section 2.7.1, the cell size-dependence is clearly seen: the data points move up to higher $\mathbf{e/uc}$ and larger *critical* $|\mathbf{G}|^2$ values, as the number of atoms per unit cell N increases. We propose that **Figure 73** can be taken as a new Hume-Rothery electron concentration rule established on the basis of first-principles FLAPW band calculations coupled with the FLAPW–Fourier analysis and Hume-Rothery plot. It clearly links the electron concentration parameter $\mathbf{e/uc}$, instead of $\mathbf{e/a}$, and a crystal structure through *critical* $|\mathbf{G}|^2$. We confirmed, for the first time, that the Hume-Rothery electron concentration rule holds true even for strongly orbital-hybridizing systems as represented by the MI-type 1/1–1/1–1/1 approximants and other CMAs in the Al–TM alloy systems. This means that all the

---

[19] We consider the form factors in β-Al$_3$Mg$_2$ to be of the same magnitude as that in γ-Mg$_{17}$Al$_{12}$ (Mizutani et al., 2010). Judging from an increased multiplicity of zone planes, we tend to believe a pseudogap across the Fermi level in β-Al$_3$Mg$_2$ to be deeper than that in γ-Mg$_{17}$Al$_{12}$. The Knight shift data reported in Ref. (Dolinšek et al., 2007) might be influenced by a significant amount of defects in their synthesized sample, which would blur the DOS and contribute to shallowing the pseudogap.

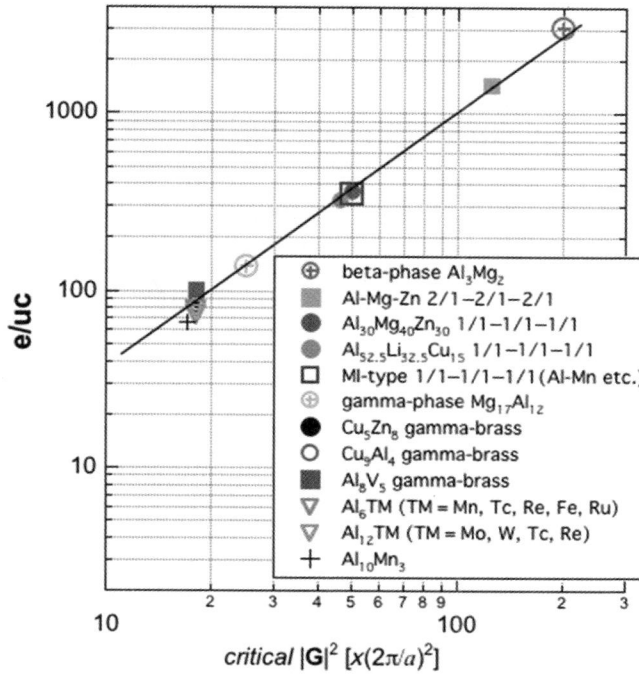

Figure 73 Hume-Rothery electron concentration rule concerning **e/uc** versus *critical* $|\mathbf{G}|^2$ on a log–log scale for a number of CMAs characterized by a pseudogap across the Fermi level. The critical $|\mathbf{G}|^2$ is expressed in units of $(2\pi/a)^2$ valid for cubic systems. The CMAs are cubic except for orthorhombic Al_6TM and hexagonal $Al_{10}Mn_3$. A straight line is referred to as "3/2-power law line" in the text. See the comment 2 in Note Added in Proof. (For color version of this figure, the reader is referred to the online version of this book.)

compounds falling onto the fitted line *do* obey the Hume-Rothery stabilization mechanism. Therefore, the new Hume-Rothery electron concentration rule can cover a wider range of alloys than the empirical Hume-Rothery rule, which was empirically established only for noble metal alloys (Mizutani, 2010).

It may be worthwhile noting, at this stage, that the extension of the present approach to the family of quasicrystals may not be straightforward. Obviously, the data points on the diagram would move up along the 3/2-power law line drawn in **Figure 73**, as the order of approximants increases. The data for

Table 9 *N*, **e/uc,** and *critical* $|\mathbf{G}|^2$ for representative CMAs

System	N	*e/uc*	Critical $|\mathbf{G}|^2$
β-Al_3Mg_2	1178	3063	200
$Al_{15}Mg_{43}Zn_{42}$ 2/1–2/1–2/1 approximant	680	1462	125
RT- and MI-type 1/1–1/1–1/1 approximants	138–160	355–389	50
γ-Phase $Mg_{17}Al_{12}$	58	138	24, 26
Group (I) gamma-brasses	52	83	18
Al_6Mn, Al_6Tc, Al_6Re, Al_6Fe, Al_6Ru	28	79	18
$Al_{12}Mo$, $Al_{12}W$, $Al_{12}Tc$, $Al_{12}Re$	26	75	18
$Al_{10}Mn_3$	26	67	17

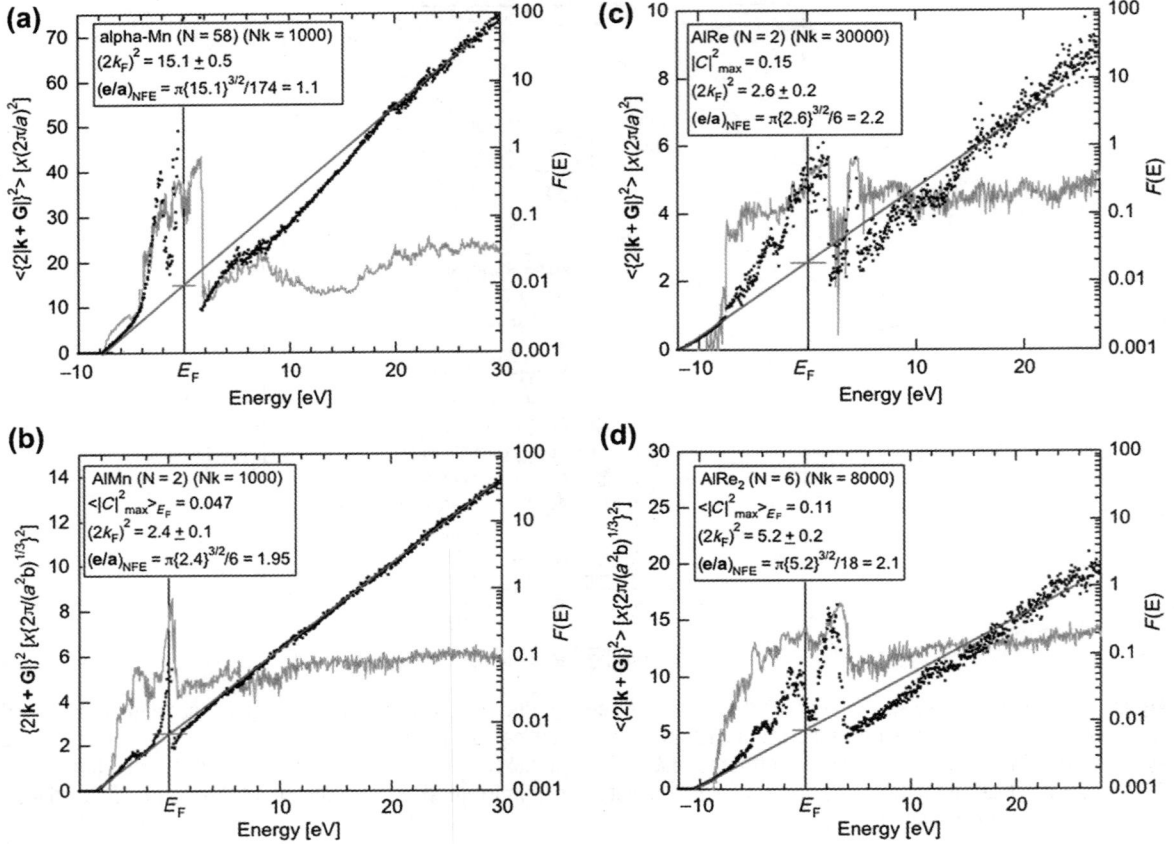

Figure 74   Hume-Rothery plot and its standard deviation $F(E)$ for (a) α-Mn ($N = 58$), (b) AlMn ($N = 2$), (c) AlRe ($N = 2$), and (d) AlRe$_2$ ($N = 6$) compounds.

quasicrystals, where the size of the unit cell is infinitely large, would be infinitely diverged. As noted in the Introduction, experimentalists have employed the **e/a**, instead of **e/uc**, and related it to the reciprocal lattice vector corresponding to one of the major diffraction lines to discuss the matching condition. This looks convenient, since **e/a** is a quantity independent of the size of the unit cell. However, the discussion along this line will breakdown, as soon as one attempts to calculate a physically more sound quantity like the Fermi diameter $2k_F$ by inserting **e/a** into Eqn 5. This is because an infinitely large unit cell prevents us to define the volume per atom V_a, which requires information about the dimension of a cell accommodating a given number of atoms.[20] We consider the basic difficulty in handling quasicrystals to stem from the breakdown of the Bloch theorem, on which the electron theory of metals including first-principles FLAPW band calculations totally rely. We are still too remote from establishing the electron theory of quasicrystals in order to describe motions of electrons in a quasiperiodic lattice.

---

[20] The volume per atom in a cubic system is obviously defined as $V_a = a^3/N$, where N is the number of atoms per cube with edge length a. Remember that the quasilattice constant defined from the cut-and-project method in Section 2.3 cannot be used to define V_a in a quasicrystal.

2.7.4 e/a Determination of TM Elements in the Periodic Table

It is also challenging, at the last stage of this chapter, to try to determine the effective **e/a** of the TM elements. Because of a limited space, we focus on the determination of effective **e/a** values of only Mn, Tc, and Re in Group 7 in the periodic table. For this purpose, we performed the FLAPW–Fourier and Hume-Rothery plot analyses for not only CMAs discussed above but also remaining compounds including pure elements: AlMn ($N = 2$) and α-Mn ($N = 58$) in Al–Mn alloy system, Al_3Tc_2 ($N = 5$), $AlTc_2$ ($N = 6$), and Tc ($N = 2$) in the Al–Tc alloy system and AlRe ($N = 2$), $AlRe_2$ ($N = 6$), and Re ($N = 2$) in the Al–Re alloy systems. Here we show only the Hume-Rothery plot data for α-Mn, AlMn, AlRe, and $AlRe_2$ compounds in **Figure 74** as representatives. The resulting **e/a** values are plotted as a function of the concentration of the TM element in **Figure 75**. The data points in the Al-rich composition range, where CMAs including quasicrystals and their approximants are formed, fall on a universal line connecting with $(e/a)_{Al} = 3.0$ and $(e/a)_{Mn} = (e/a)_{Tc} = (e/a)_{Re} = 1.0$.

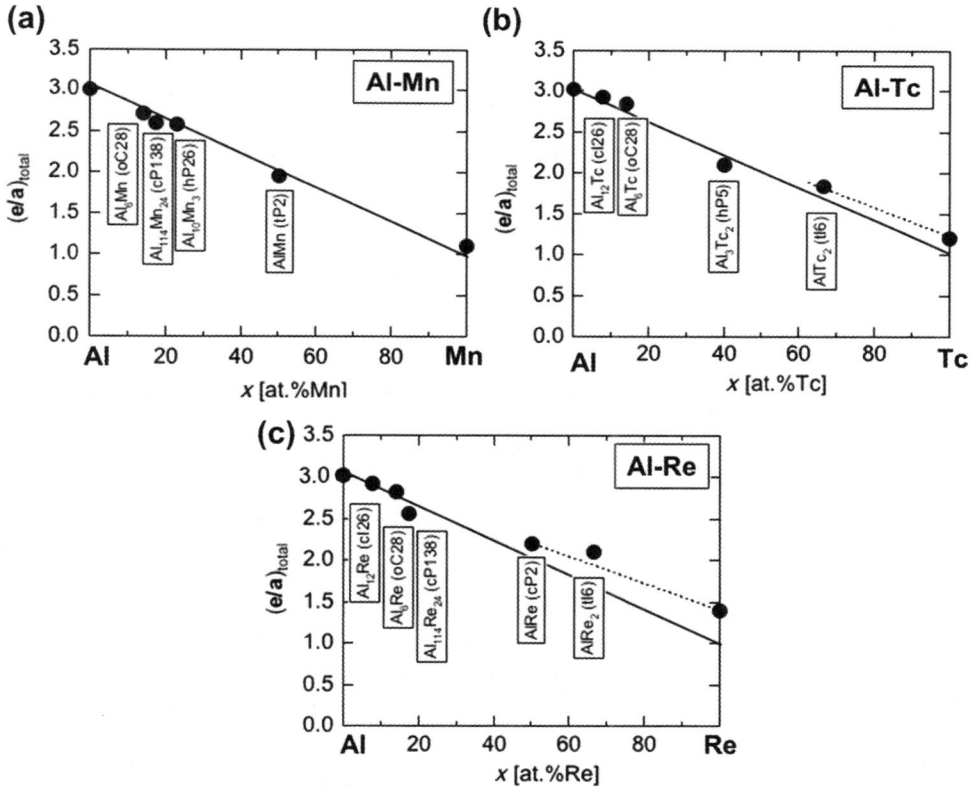

Figure 75 Composition dependence of $(e/a)_{total}$ for intermetallic compounds and pure elements in (a) Al–Mn, (b) Al–Tc, and (c) Al–Re alloy systems. A solid line was drawn by passing data points for Al-rich alloys, while a dotted line for Al–Tc and Al–Re alloy systems was drawn by passing data points only for Tc- or Re-rich alloys.

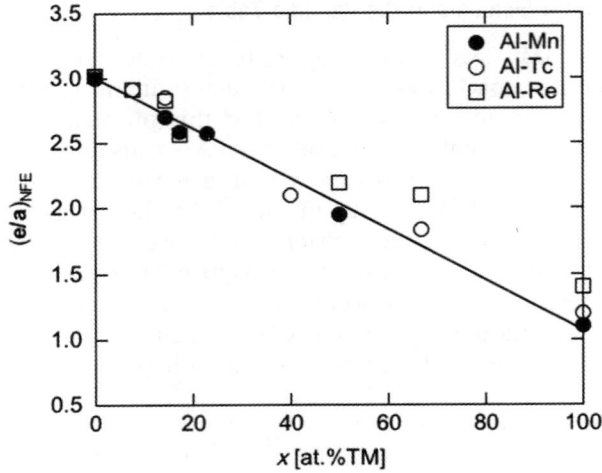

Figure 76 Composition dependence of $(e/a)_{total}$ for intermetallic compounds and pure elements in Al–Mn (●), Al–Tc (○), and Al–Re (□) alloy systems.

Table 10 e/a of 3d-, 4d-, and 5d-TM element embedded in the matrix of polyvalent elements like Al, Zn, and, etc

Ti	V	Cr	Mn	Fe	Co	Ni	Cu
0.9	0.9	0.9	1.0	1.1	1.1	0.0	1.0
Zr	Nb	Mo	Tc	Ru	Rh	Pd	Ag
1.4	1.2	1.3	1.0	1.0	1.2	0.0	1.0
Hf	Ta	W	Re	Os	Ir	Pt	Au
1.7	1.8	1.0	1.0	1.2	1.6	0.0	1.0

See the comment 3 in Note Added in Proof

To see its universality in the Al-rich composition range, all relevant data are plotted in a single diagram in **Figure 76**. A single straight line can be drawn through all the data points in the Al-rich composition range. The choice of $(e/a)_{TM} = 1.0$ for the three elements Mn, Tc, and Re is found to be quite reasonable. Similar analyses have been made for other 3d-, 4d-, and 5d-elements in the periodic table. They are summarized in **Table 10**. As already mentioned in Section 2.6.5.3, values of $(e/a)_{total}$ for $Al_{114}Mn_{24}$, $Al_{114}Re_{24}$, $Al_{102}Re_{24}Si_{12}$, $Al_{108}Cu_6Fe_{24}Si_6$, and $Al_{108}Cu_6Ru_{24}Si_6$ 1/1–1/1–1/1 approximants turn out to be 2.65, 2.65, 2.74, 2.64, and 2.64, respectively, and are distributed in the narrow range at around 2.7.

As repeatedly mentioned, Tsai et al. (Tsai et al., 1988; Tsai, 2005, 2008; Yokoyama et al., 1991) claimed that Al–Cu–TM (TM = Fe, Ru, and Os) and Al–Pd–TM (Mn and Re) quasicrystals they discovered are stabilized at $(e/a)_{total} = 1.8$, the value of which was derived by employing negative $(e/a)_{TM}$ values for TM elements Raynor proposed in 1949 (Raynor, 1949). The **e/a** values for typical

Table 11   e/a for typical icosahedral quasicrystals

Cluster-type	System	e/a
MI	$Al_{63}Cu_{25}Fe_{12}$	$3 \times 0.63 + 1 \times 0.25 + 1 \times 0.12 = 2.26$
	$Al_{63}Cu_{25}Ru_{12}$	$3 \times 0.63 + 1 \times 0.25 + 1 \times 0.12 = 2.26$
	$Al_{63}Cu_{25}Os_{12}$	$3 \times 0.63 + 1 \times 0.25 + 1 \times 0.12 = 2.26$
	$Al_{70}Pd_{20}Mn_{10}$	$3 \times 0.7 + 0 \times 0.2 + 0.1 \times 1 = 2.2$
	$Al_{70}Pd_{20}Tc_{10}$	$3 \times 0.7 + 0 \times 0.2 + 0.1 \times 1 = 2.2$
	$Al_{70}Pd_{20}Re_{10}$	$3 \times 0.7 + 0 \times 0.2 + 0.1 \times 1 = 2.2$
	$Al_{70}Pd_{20}V_5Co_5$	$3 \times 0.7 + 0 \times 0.2 + 0.9 \times 0.05 + 1.1 \times 0.05 = 2.2$
RT	$Al_{15}Mg_{44}Zn_{41}$	$3 \times 0.15 + 2 \times 0.44 + 2 \times 0.41 = 2.15$
	$Al_{55}Li_{35.8}Cu_{9.2}$	$3 \times 0.55 + 1 \times 0.358 + 1 \times 0.092 = 2.1$

Al-based MI-type icosahedral quasicrystals can be similarly calculated by a composition-average of e/a values listed in **Table 10** and are summarized in **Table 11** along with the data for RT-type quasicrystals. The resulting values of $(e/a)_{total}$ for MI-type quasicrystals are distributed in the neighborhood of 2.2, which is definitely lower than 2.7 for the corresponding $1/1-1/1-1/1$ approximants. More important is that the present studies based on first-principles FLAPW band calculations have decisively ruled out the empirical $(e/a)_{total} = 1.8$ rule and, instead, proposes that it should be replaced by the $(e/a)_{total} = 2.2$ rule for the MI-type icosahedral quasicrystals.

References

Arnberg, L., Westman, S., 1972. Acta Chemica Scandinavica 26, 1748.

Asahi, R., Sato, H., Takeuchi, T., Mizutani, U., 2005. Phys. Rev. B 72, 125102.

Ashcroft, N.W., Mermin, N.D., 1976. Solid State Physics. Saunders College, West Washington Square, Philadelphia, PA 19105. (Chapter 20), p. 410.

Audier, M., Pannetier, J., Leblanc, M., Janot, C., Lang, J.-M., Dubost, B., 1988. Physica B 153, 136.

Averbach, B.L., Flinn, A., Cohen, M., 1954. Acta Met 2, 92.

Bardeen, J., Brattain, W.H., 1948. Phys. Rev. 74, 230. Phys. Rev.. 75 (1949) 1208.

Bardeen, J., Cooper, L.N., Schrieffer, J.R., 1957. Phys.Rev 106, 162. Phys.Rev.108 (1957) 1175.

Barrett, C.S., Massalski, T.B. Structure of Metals, McGraw-Hill, New York. third edition 1966.

Bednorz, J.G., Müller, K.A., 1986. Z. Phys. B-Condensed Matter 64, 189.

Bergman, G., Waugh, J.L.T., Pauling, L., 1957. Acta Crystallogr 10, 254.

Blaha, P., Schwarz, K., Madsen, G., Kvasnicka, D. and Luitz, J.. http://www.wien2k.at/ (last accessed on June 17, 2013.).

Bloch, F., 1928. Z. Physik 52, 555.

de Boer, F.R., Boom, R., Mattens, W.C.M., Miedema, A.R., Niessen, A.K., 1988. Cohesion in Metals. North-Holland.

Boudard, M., de Boissieu, M., Janot, C., Herger, G., Beeli, C., Nissen, H.-U., Vincent, H., Ibberson, R., Audier, M., Dubois, J.M., 1992. J. Phys.Condens. Matter 4, 10149.

Bradley, A.J., Jones, P., 1933. J. Inst. Met 51, 131.

Brandon, J.K., Kim, H.S., Pearson, W.B., 1979. Acta Cryst B35, 1937.

Brandon, J.K., Pearson, W.B., Riley, P.W., Chieh, C., Stokhuyzen, R., 1977. Acta Cryst B33, 1088.

Cooper, M., Robinson, K., 1966. Acta Crystallog. 20, 614.

Dirac, P.A.M., 1926. Proc. Roy. Soc. (London) 112, 661.

Dolinšek, J., Apih, T., Jeglič, P., Smiljanić, I., Bilušić, A., Bihar, Ž., Smontara, A., Jagličić, Z., Heggen, M., Feuerbacher, M., 2007. Intermetallics 15, 1367.

Drude, P., 1900. Ann. Physik 1, 566.

Dubois, J.M., 2005. Useful Quasicrystals. World Scientific, Singapore.

Fermi, E., 1926. Z. Physik 36, 902.

Feuerbacher, M., Thomas, C., Makongo, J.P.A., Hoffmann, S., Carrillo-Cabrera, W., Cardoso, R., Grin, Y., Kreiner, G., Joubert, J.-M., Schenk, T., Gastaldi, J., Nguyen-Thi, H., Mangelinck-Noël, N., Billia, B., Donnadieu, P., Czyrska-Filemonowicz, A., Zielinska-Lipiec, A., Dubiel, B., Weber, T., Schaub, P., Krauss, G., Gramlich, V., Christensen, J., Lidin, S., Fredrickson, D., Mihalkovic, M., Sikora, W., Malinowski, J., Brühne, S., Proffen, T., Assmus, W., de Boissieu, M., Bley, F., Chemin, J.-L., Schreuer, J., Steurer, W., 2007. Z. Kristallogr 222, 259.

Fujiwara, T., 1989. Phys. Rev. B40, 942.

Fujiwara, T., Yokokawa, T., 1992. Phys.Rev. Letters 66, 333.

Gomez, C.P., Lidin, S., 2003. Phys. Rev. B 68, 024203.

Gruner, G., 1994. Density Waves in Solids. Addison-Wesley Longmans, Inc, U.S.A.

Guryan, C.A., Stephens, P.W., Goldman, A.I., Gayle, F.W., 1988. Phys. Rev. B 37, 8495.

Hafner, J., Krajčí, M., 1993. Phys. Rev. B 47, 11795.

Heisenberg, W., 1927. Z. Physik 43, 172.

Hohenberg, P., Kohn, W., 1964. Phys. Rev. 136, B864.

Hume-Rothery, W., 1926. J. Inst. Metals 35, 295.

Hume-Rothery, W., Mabbott, G.W., Channel-Evans, K.M., 1934. Phil. Trans. Roy. Soc. A 233, 1.

Hume-Rothery, W., 1962. Atomic Theory for Students of Metallurgy. Institute of Metals, Monograph and Report Series No.3. The Institute of Metals, London. p. 306.

Inukai, M., Zijlstra, E.S., Sato, H., Mizutani, U., 2011a. Phil. Mag. 91, 4247.

Inukai, M., Soda, K., Sato, H., Mizutani, U., 2011b. Phil.Mag 91, 2543.

Janot, C., 1994. Quasicrystals, second ed. Clarendon Press, Oxford.

Jones, H., 1937. Proc. Phys. Soc 49, 250.

Kim, D.H., Cantor, B., 1994. Phil. Mag. A 69, 45.

Kittel, C., 1967. Introduction to Solid State Physics, third ed. John Wiley & Sons, New York. (Chapter 3).

Kittel, C., 1996. Introduction to Solid State Physics, seventh ed. John Wiley & Sons, Inc, New York.

Koelling, D.D., Arbman, G.O., 1975. J. Phys.F Met. Phys. 5, 2041.

Kohn, W., Sham, L.J., 1965. Phys. Rev. 140, A1133.

Krajčí, M., Hafner, J., 1999. Phys. Rev. B 59, 8347.

Krajčí, M., Windisch, M., Hafner, J., Kresse, G., Mihalkovič, M., 1995. Phys. Rev. B 51, 17355.

Kreiner,G., Presented at Quasicrystal Meeting. Stuttgart, Germany, January 30, 2009.

Laissardiere, G.T., Nguyen-Manh, D., Mayou, D., 2005. Prog. Mat. Sci. 50, 679.

Lin, Q., Corbett, J.D., 2004. Inorg. Chem. 43, 1912.

Lin, Q., Corbett, J.D., 2006 September 12. Proc. Natl. Acad. Sci. U.S.A. 103 (37), 13589.

Lorentz, H.A., 1905. Arch. Neerllandaises des Sci. 10, 336. H.A.Lorentz, "The theory of electrons and its applications to the phenomena of light and radiant heat", (Dover Publications, New York, 1952).

Madsen, G.K.H., Blaha, P., Schwarz, K., Sjostedt, E., Nordstrom, L., 2001. Phys. Rev. B64, 195134.

Mizutani, U., 2006a. MATERIA 45 (8), 605–610 (in Japanese).

Mizutani, U., 2006b. Materia 45 (11), 803 (in Japanese).

Mizutani, U., Takeuchi, T., Sato, H., 2002. J. Phys.: Condens. Matter 14, R767.

Mizutani, U., Iwakami, W., Takeuchi, T., Sakata, M., Takata, M., 1997. Phil. Mag. Lett. 76, 349.

Mizutani, U., 2010. Hume-Rothery Rules for Structurally Complex Alloy Phases. CRC Press, Taylor&Francis Group, Boca Raton, FL.

Mizutani, U., 2001. Introduction to the Electron Theory of Metals. Cambridge University Press. Chap.5.

Mizutani, U., Inukai, M., Sato, H., 2011. Phil. Mag. 91, 2536.

Mizutani, U., Asahi, R., Sato, H., Takeuchi, T., 2006. Phys.Rev.B 74, 235119.

Mizutani, U., Inukai, M., Sato, H., Zijlstra, E.S., 2012. Phil. Mag. 92, 1691.

Mizutani, U., Asahi, R., Sato, H., Noritake, T., Takeuchi, T., 2008. J. Phys.: Condens. Matter 20, 275228.

Mizutani, U., Asahi, R., Takeuchi, T., Sato, H., Kontsevoi, O.Y., Freeman, A.J., 2009. Z. Kristallogr 224, 17.

Mizutani, U., Takeuchi, T., Sato, H., 2004. Prog. Mat. Sci. 49, 227.

Mizutani, U., Takeuchi, T., Banno, E., Fournee, V., Takata, M., Sato, H., 2001. Mater. Res. Soc. Symp. Proc, 643. [Materials Research Society, K.13.1.1.].

Mizutani, U., Kondo, Y., Nishino, Y., Inukai, M., Feuerbacher, M., Sato, H., 2010. J. Phys. Condens. Matter 22, 485501.

Mott, N.F., Jones, H., 1936. The Theory of the Properties of Metals and Alloys. Clarendon Press, Oxford. Dover 1958.

Murray, J.L., 1988. Al–Mg (aluminium–magnesium). In: Nayeb-Hashemi, A.A., Clark, J.B. (Eds.), Phase Diagrams of Binary Magnesium Alloys. ASM International, Metals Park, Ohio, pp. 17–34.

Noritake, T., Aoki, M., Towata, S., Takeuchi, T., Mizutani, U., 2007. Acta Cryst. B 63, 726.

Nozawa, K., Ishii, Y., 2008. J.Phys.: Condensed Matter 20, 315206.

Okamoto, H., 2000. Phase Diagrams for Binary Alloys. ASM International, Metals Park, Ohio. p.36.

Pauli, W., 1927. Z.Physik 41, 81.

Paxton, A.T., Methfessel, M., Pettifor, D.G., 1997. Proc.Roy.Soc. Lond. A 453, 1493.

Perdew, J.P., Wang, Y., 1992. Phys. Rev. B45, 13244.

Perdew, J.P., Burke, S., Ernzerhof, M., 1996. Phys. Rev. Let 77, 3865.

Perlitz, H., 1933. Z. Krist 86, 155.

Raynor, G.V., 1949. Prog. Met. Phys. 1, 1.

Roche, S., Fujiwara, T., 1998. Phys. Rev. B 58, 11338.

Samson, S., 1962. Nature 195, 259.

Samson, S., 1965. Acta Crystallogr. 19, 401.

Samson, S., 1967. Acta Crystallogr. 23, 586.

Sato, H., Takeuchi, T., Mizutani, U., 2001. Phys. Rev. B 64, 094207.

Sato, H., Takeuchi, T., Mizutani, U., 2004. Phys. Rev. B 70, 024210.

Schlüter, M., Sham, L.J., February, 1982. Phys. Today 35, 36.

Shechtman, D., Blech, I., Gratias, D., Cahn, J.W., 1984. Phys. Rev. Letters 53, 1951.

Schrödinger, E., 1926. Ann. Physik 79, 361, 489, 734.

Singh, D.J., 1994. Planewaves, Pseudopotentials and the LAPW Method. Kluwer Academic Publishers, Boston.

Sjöstedt, E., Nordström, L., Singh, D.J., 2000. Solid State Commun. 114, 15.

Smith, H., Jensen, H.H., 1989. Transport Phenomena. Clarendon Press, Oxford.

Sommerfeld, A., 1928. Z.Physik 47, 1. A.Sommerfeld and H.Bethe, "Elektronentheorie der Metalle" (Springer Verlag, 1967).

Stadnik, Z.M., 1999. Physical Properties of Quasicrystals. Springer Verlag.

Sugiyama, K., Kaji, N., Hiraga, K., 1998. Acta Cryst. C 54, 445.

Sugiyama., K., Sun, W., Hiraga, K., 2002. J. Alloys and Compounds 342, 139.

Sun, W., Lincoln, F.J., Sugiyama, K., Hiraga, K., 2000. Mat. Sci. Eng. 294-296, 327.

Swalin, R.W., 1972. Thermodynamics of Solids, second ed. John Wiley & Sons, New York.

Takeuchi, T., Mizuno, T., Banno, E., Mizutani, U., 2000. Mater. Sci. Eng. B 294-296, 522.

Takeuchi, T., Onogi, T., Banno, E., Mizutani, U., 2001. Mater. Trans. 42, 933.

Takeuchi, T., Mizutani, U., 2002. J. Alloys and Compounds 342, 416.

Takeuchi, T., Mizutani, U., 1995. Phys. Rev. B 52, 9300.

Takeuchi, T., Onogi, T., Otagiri, T., Mizutani, U., Sato, H., Kato, K., Kamiyama, T., 2003. Phys. Rev. B 68, 184203.

Trambly de Laissardière, G., Nguyen Manh, D., Magaud, L., Julien, J.P., Cyrot-Lackmann, F., Mayou, D., 1995. Phys. Rev. B 52, 7920.

Trambly de Laissardière, G., Nguyen-Manh, D., Mayou, D., 2005. Prog. Mat. Sci. 50, 679.

Traverse, A., Dumoulin, L., Belin, E., Sénémaud, C., 1988. In: Janot, Ch., Dubois, J.M. (Eds.), Quasicrystalline Materials. World Scientific, Singapore, pp. 399–408.

Tsai, A.P., 2005. The Science of Complex Alloy Phases. In: Massalski, T.B., Turchi, P.E.A. (Eds.), TMS (The Minerals, Metals & Materials Society, pp. 201–214.

Tsai, A.P., 2008. Sci. Technol. Adv. Mater 9, 013008.

Tsai, A.P., Guo, J.Q., Abe, E., Takakura, H., Sato, T.J., 2000. Nature 408, 537.

Tsai, A.P., Inoue, A., Masumoto, T., 1988. Jpn. J. Appl. Phys 27. L1587.

Tsai, A.P., Inoue, A., Yokoyama, Y., Masumoto, T., 1990. Mater.Trans.Jpn.Inst.Met 31, 98.

Villars, P., 1997. Pearson's Handbook. ASM International, Materials Park, OH.

Westgren, A., Phragmén, G., 1928a. Z.Anorg. Chemie 175, 80.

Westgren, A., Phragmén, G., 1928b. Metallwirtschaft 7, 700.

Westgren, A., Phragmén, G., 1929. Trans.Farad.Soc 25, 379.

Wilson, A.H., 1931. Proc. Roy. Soc. A133, 458.

Wilson, A.H., 1936. The Theory of Metals, first ed. Cambridge University Press. Second edition 1953.

Wimmer, E., Krakauer, H., Weinert, M., Freeman, A., 1981. Phys. Rev. B24, 864.

Windisch, M., Krajčí, M., Hafner, J., 1994. J. Phys.: Condens. Matter 6, 6977.

Yamada, H., Takeuchi T., Mizutani U., and Tanaka N., Mat. Res. Soc. Symp. Proc. vol. 553 (1999 Materials Research Society) Edited by J.M. Dubois, P.A. Thiel, A.P. Tsai and K. Urban, pp.117–122.

Yokoyama, Y., Tsai, A.P., Inoue, A., Masumoto, T., Chen, H.S., 1991. Mater. Trans. Jpn. Inst. Met. 32, 421.

Zijlstra, E.S., Bose, S.K., 2003. Phys. Rev. B 67, 224204.

Zijlstra, E.S., Bose, S.K., 2004. Phys. Rev. B 70, 184205.

Ziman, J.M., 1964. Principles of the Theory of Solids, first ed. Cambridge University Press. Second edition 1972.

Note Added in Proof

1. *Page 184:* The FLAPW-Fourier analysis and Hume-Rothery plot were carried out for Cd_6Ca approximant and several Cd-Ca compounds. The possession of **e/a**=2.0 was confirmed for both Ca and Cd. [U.Mizutani, M.Inukai, H.Sato, K.Nozawa and E.S.Zijlstra, "Aperiodic Crystals", pp.101-107, Chapter 14, edited by S.Schmid et al., Springer Dordrecht Heidelberg, 2013].

2. *Page 193*, caption to **Figure 73**: The validity of the new Hume-Rothery electron concentration rule has been confirmed in totally 39 CMAs characterized by a pseudogap across the Fermi level. [U.Mizutani, H.Sato, M.Inukai and E.S.Zijlstra, "e/a determination for 4d- and 5d-transition metal elements and their intermetallic compounds with Mg, Al, Zn, Cd and In", Phil.Mag. Phil.Mag. 93 (2013) 3353]

3. *Page 196:* Numerical data listed in **Table 10** are correct but are further refined in [U.Mizutani, H.Sato, M.Inukai and E.S. Zijlstra, "*e/a determination for 4d- and 5d-transition metal elements and their intermetallic compounds with Mg, Al, Zn, Cd and In*", Phil.Mag. 93 (2013) 3353] and [U.Mizutani, M.Inukai, H.Sato and E.S.Zijlstra, "*Hume-Rothery Stabilization Mechanism in Low-Temperature Phase Zn_6Sc Approximant and e/a Determination of Sc and Y in M-Sc and M-Y (M=Zn, Cd and Al) Alloy Systems*" in "*Aperiodic Crystals*", pp. 109–115, Chapter 15, edited by S. Schmid et al., Springer Dordrecht Heidelberg, 2013].

Biography

U.Mizutani studied the electronic structure of the Hume-Rothery alloy phases as a postdoctoral fellow under the guidance of Prof. T.B. Massalski, Carnegie-Mellon University, Pittsburgh, in late 1960s to 1970s. He wrote the review article on this subject in 1978 and on amorphous alloys in 1983 in Progress in Materials Science. His research field has then been shifted to studies of electronic structure and electron transport properties of quasicrystals and high-Tc superconducting oxides. He published the book entitled "Introduction to the Electron Theory of Metals" from Cambridge University Press in 2001, and the monograph entitled "Hume-Rothery Rules for Structurally Complex Alloy Phases" from CRC Press, Taylor & Francis Group in 2011.

He was a full professor of Nagoya University from 1989 until his retirement in 2005. He then served as a senior fellow at Toyota Physical & Chemical Research Institute over 2005 to 2009 and is now a senior fellow at Nagoya Industrial Science Research Institute.

Manabu Inukai is now a post doctoral fellow at Nagoya Institute of Technology. He has completed his bachelor course in engineering at Daido University and master course at Nagoya University. He has also completed his doctorate at Nagoya University. He has worked as Post doctoral fellow at Nagoya university (2009–2010), Japan Synchrotron Radiation Research Institute (2010–2011), Toyota Technological Institute (2011–2013).

Hirokazu Sato is now an Emeritus Professor of Aichi University of Education. He has been involved in both education and research in the field of solid state physics for almost 40 years and has many publications on the theory of the electron transport properties including the temperature dependence of electrical resistivity and Seebeck coefficient of amorphous metals. In the last 14 years, he has been contributing to elucidating the Hume-Rothery stabilization mechanism of intermetallic compounds in collaboration with Prof.U.Mizutani.

E.S. Zijlstra is a research associate at the University of Kassel. In 2001 he obtained his Ph.D. at the University of Nijmegen on the topic "Electronic properties of quasicrystals – A tight-binding study". During postdoctoral stays at the Max-Planck-Institute for Solid-State Research in Stuttgart (2000–2002) and Brock University in St. Catharines (2003–2005) he applied density functional theory to various electronic properties of complex metallic alloys. He also improved structural models of the most popular icosahedral quasicrystals. In Kassel he developed his own ab initio Code for Highly-exclted Valence Electron Systems (CHIVES), which he uses to simulate the structural response of solids and nanostructures to femtosecond-laser pulses.

3 Thermodynamics and Phase Diagrams

Arthur D. Pelton, Centre de Recherche en Calcul Thermochimique (CRCT), École Polytechnique de Montréal, Montréal, Canada

3.1 Introduction

An understanding of phase diagrams is fundamental and essential to the study of materials science, and an understanding of thermodynamics is fundamental to an understanding of phase diagrams. Knowledge of the equilibrium state of a system under a given set of conditions is the starting point in the description of many phenomena and processes.

A phase diagram is a graphical representation of the values of the thermodynamic variables when equilibrium is established among the phases of a system. Materials scientists are most familiar with phase diagrams that involve temperature, T, and composition as variables. Examples are T-composition phase diagrams for binary systems such as **Figure 1** for the Fe–Mo system, isothermal phase diagram sections of ternary systems such as **Figure 2** for the Zn–Mg–Al system, and isoplethal (constant composition) sections of ternary and higher order systems such as **Figures 3**a and **4**.

However, many useful phase diagrams can be drawn that involve variables other than T and composition. The diagram in **Figure 5** shows the phases present at equilibrium in the Fe–Ni–O_2 system at 1200 °C as the equilibrium oxygen partial pressure (i.e. chemical potential) is varied. The x-axis of this diagram is the overall molar metal ratio in the system. The phase diagram in **Figure 6** shows the equilibrium phases present when an equimolar Fe–Cr alloy is equilibrated at 925 °C with a gas phase of varying O_2 and S_2 partial pressures. For systems at high pressure, P–T phase diagrams such as the diagram for the one-component Al_2SiO_5 system in **Figure 7**a show the fields of stability of the various allotropes. When the pressure in **Figure 7**a is replaced by the volume of the system as the y-axis, the "corresponding" V–T phase diagram of **Figure 7**b results. The enthalpy of the system can also be a variable in a phase diagram. In the phase diagram in **Figure 3**b, the y-axis of **Figure 3**a has been replaced by the molar enthalpy difference $(h_T - h_{25})$ between the system at 25 °C and a temperature T. This is the heat that must be supplied or removed to heat or cool the system adiabatically between 25 °C and T.

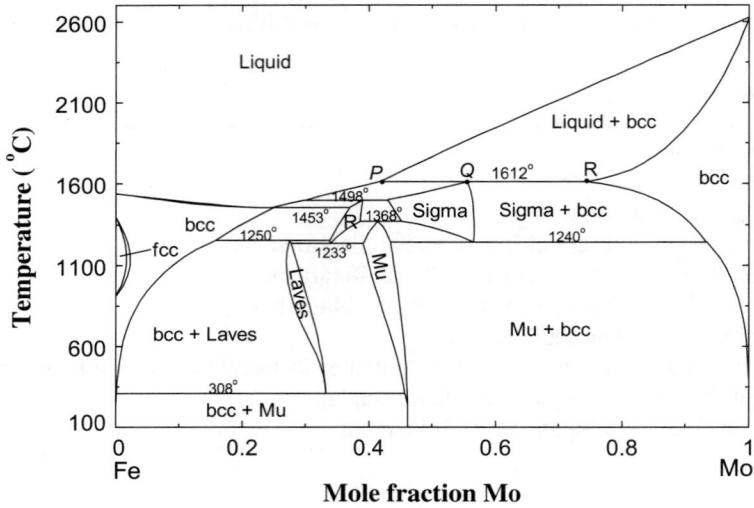

Figure 1   Temperature-composition phase diagram at $P = 1$ bar of the Fe–Mo system (see FactSage).

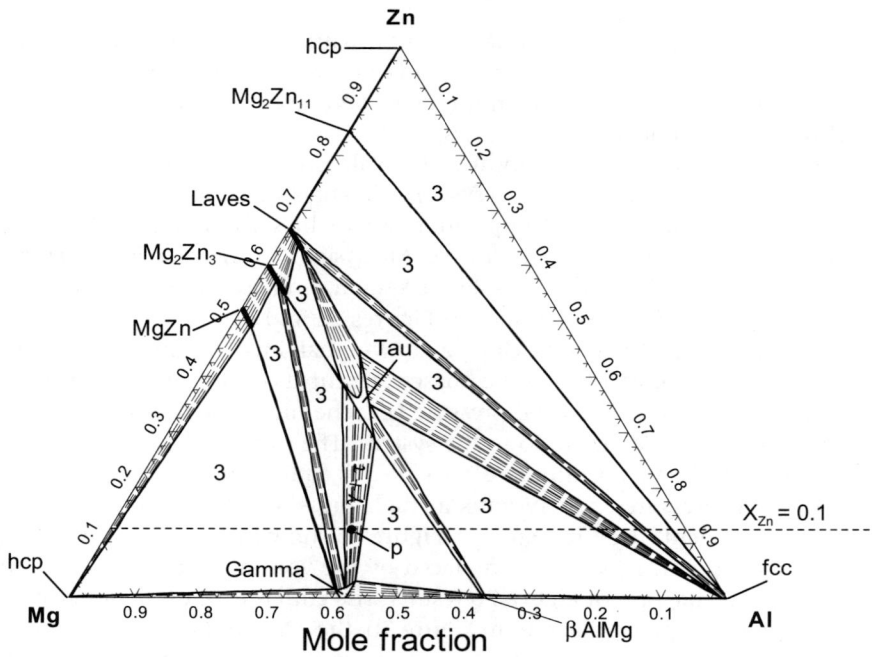

Figure 2   Isothermal section at 25 °C and $P = 1$ bar of the Zn–Mg–Al system (see FactSage).

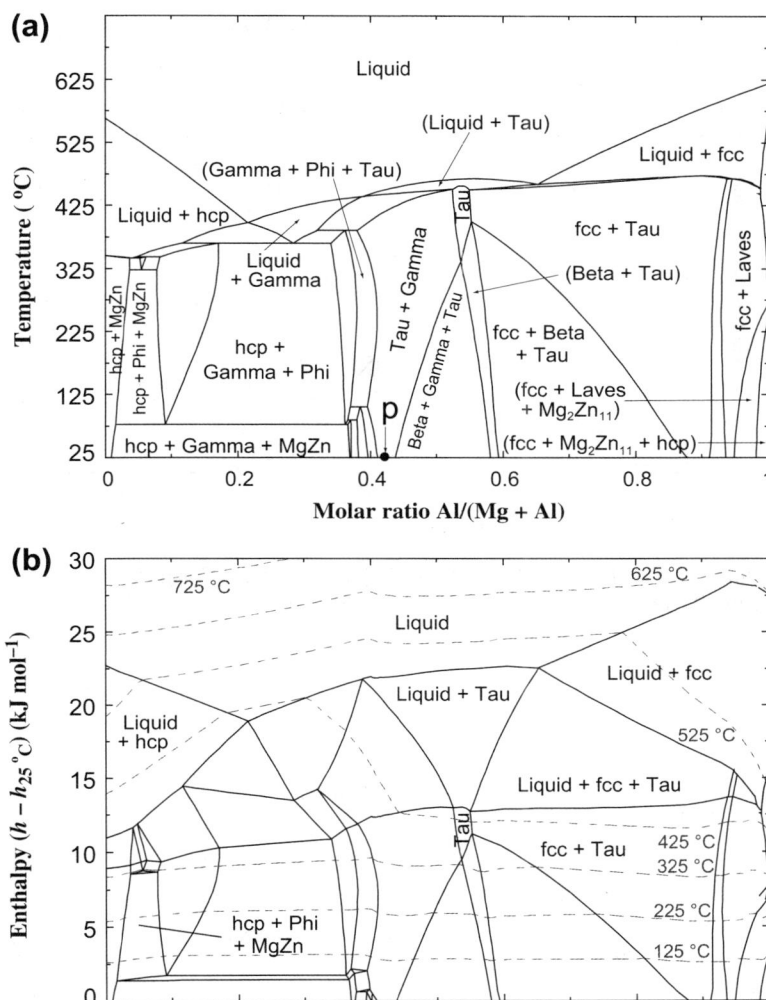

Figure 3 Isoplethal phase diagram section at $X_{Zn} = 0.1$ and $P = 1$ bar of the Zn–Mg–Al system. (a) Temperature *vs* composition; (b) Enthalpy *vs* composition (see FactSage). (For color version of this figure, the reader is referred to the online version of this book.)

The phase diagrams shown in **Figures 1–7** are only a small sampling of the many possible types of phase diagram sections. These diagrams and several other useful types of phase diagrams will be discussed in this chapter. Although these diagrams appear to have very different geometries, it will be shown that actually they all obey exactly the same geometrical rules.

The theme of this chapter is the relationship between phase diagrams and thermodynamics. To understand phase diagrams properly, it has always been necessary to understand their thermodynamic basis. However, in recent years, the relationship between thermodynamics and phase diagrams has taken on a new and important practical dimension. With the development of large evaluated databases

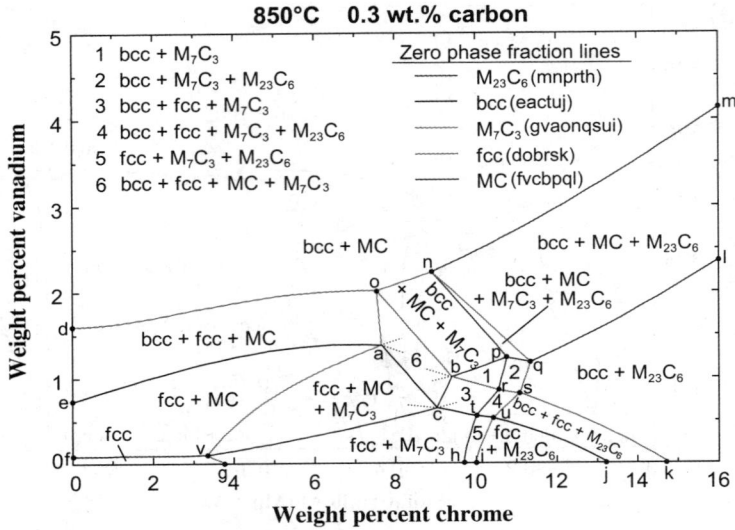

Figure 4 Phase diagram section of the Fe–Cr–V–C system at 850 °C, 0.3 wt.% C and $P = 1$ bar (see FactSage; SGTE). (For color version of this figure, the reader is referred to the online version of this book.)

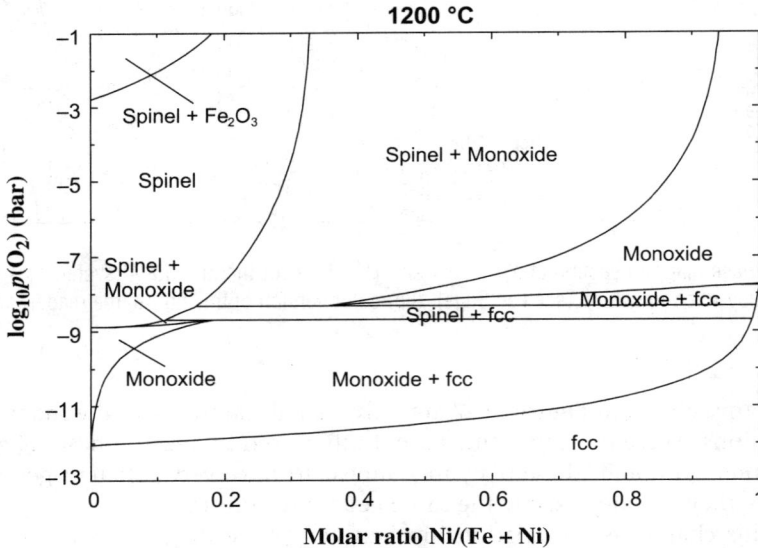

Figure 5 Phase diagram of the Fe–Ni–O$_2$ system at 1200 °C showing equilibrium oxygen pressure *vs* overall metal ratio (see FactSage).

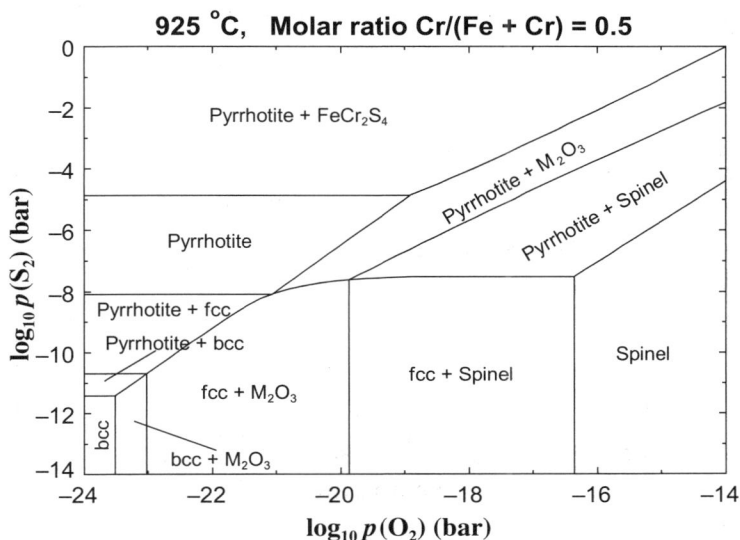

925 °C, Molar ratio Cr/(Fe + Cr) = 0.5

Figure 6 Phase diagram of the Fe–Cr–S_2–O_2 system at 925 °C showing equilibrium S_2 and O_2 partial pressures at constant molar ratio Cr/(Cr + Fe) = 0.5 (see FactSage).

of the thermodynamic properties of thousands of compounds and solutions, and of software to calculate the conditions for chemical equilibrium (by minimizing the Gibbs energy), it is possible to rapidly calculate and plot desired phase diagram sections of multicomponent systems. Most of the phase diagrams shown in this chapter were calculated thermodynamically with the FactSage software and databases (see FactSage in the list of websites at the end of this chapter).

Several large integrated thermodynamic database computing systems have been developed in recent years (see list of websites at the end of this chapter). The databases of these systems have been prepared by the following procedure. For every compound and solution phase of a system, an appropriate model is first developed giving the thermodynamic properties as functions of T, P and composition. Next, all available thermodynamic and phase equilibrium data from the literature for the entire system are simultaneously optimized to obtain one set of critically evaluated, self-consistent parameters of the models for all phases in two-component, three-component and, if data are available, higher order subsystems. Finally, the models are used to estimate the properties of multicomponent solutions from the databases of parameters of the lower order subsystems. The Gibbs energy minimization software then accesses the databases and, for given sets of conditions (T, P, composition...), calculates the compositions and amounts of all phases at equilibrium. By calculating the equilibrium state as T, composition, P, etc. are varied systematically, the software generates the phase diagram. As mentioned above, although the phase diagram sections shown in **Figures 1–7** are seemingly quite different topologically, they actually obey the same simple rules of construction and hence, can all be calculated by the same algorithm.

Section 3.2 provides a review of the fundamentals of thermodynamics as required for the interpretation and calculation of phase diagrams of all types. In Section 3.3, the Gibbs Phase Rule is developed in a general form suitable for the understanding of phase diagrams involving a wide range of variables including chemical potentials, enthalpy, volume, etc. Section 3.4 begins with a review of the

(a)

(b)

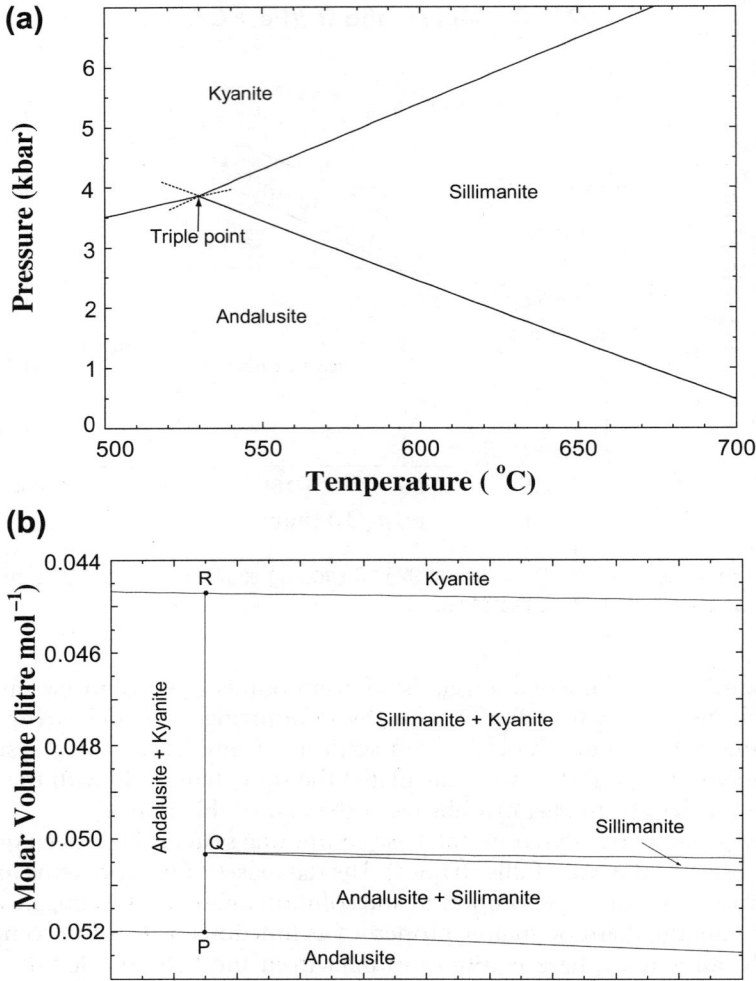

Figure 7   (a) $P-T$ and (b) $V-T$ phase diagrams of Al_2SiO_5 (see FactSage).

thermodynamics of solutions and continues with a discussion of the thermodynamic origin of binary T-composition phase diagrams, presented in the classical manner involving common tangents to curves of Gibbs energy. A thorough discussion of all features of T-composition phase diagrams of binary systems is presented in Section 3.5, with particular stress on the relationship between the phase diagram and the thermodynamic properties of the phases. A similar discussion of T-composition phase diagrams of ternary systems is given in Section 3.6. Isothermal and isoplethal sections as well as pol-ythermal projections are discussed.

Readers conversant in binary and ternary T-composition diagrams may wish to skip Sections 3.4–3.6 and pass directly to Section 3.7, where the geometry of general phase diagram sections is developed. In this section, the following are presented: the general rules of construction of phase diagram sections; the proper choice of axis variables and constants required to give a single-valued phase diagram section with each point of the diagram representing a unique equilibrium state; and a general algorithm for calculating all such phase diagram sections.

Section 3.8 outlines the techniques used to develop large thermodynamic databases through evaluation and modeling. Section 3.9 treats the calculation of equilibrium and nonequilibrium cooling paths of multicomponent systems. In Section 3.10, order/disorder transitions and their representation on phase diagrams are discussed.

3.2   Thermodynamics

This section is intended to provide a review of the fundamentals of thermodynamics as required for the interpretation and calculation of phase diagrams. The development of the thermodynamics of phase diagrams will be continued in succeeding sections.

3.2.1   The First and Second Laws of Thermodynamics

If the thermodynamic *system* under consideration is permitted to exchange both energy and mass with its *surroundings*, the system is said to be *open*. If energy but not mass may be exchanged, the system is said to be *closed*. The state of a system is defined by in*tensive properties*, such as temperature and pressure, which are independent of the mass of the system, and by *extensive properties*, such as volume and internal energy, which vary directly as the mass of the system.

3.2.1.1   *Nomenclature*

Extensive thermodynamic properties will be represented by upper-case symbols. For example, $G = $ Gibbs energy in joules (J). Molar properties will be represented by lower-case symbols. For example, $g = G/n = $ molar Gibbs energy in joules per mole (J mol^{-1}), where n is the total number of moles in the system.

3.2.1.2   *The First Law*

The *internal energy* of a system, U, is the total thermal and chemical bond energy stored in the system. It is an extensive state property.

Consider a closed system undergoing a change of state that involves an exchange of heat, dQ, and work, dW, with its surroundings. Since energy must be conserved:

$$dU_n = dQ + dW \tag{1}$$

This is the First Law. The convention is adopted whereby energy passing from the surroundings to the system is positive. The subscript on dU_n indicates that the system is closed (constant number of moles.)

It must be stressed that dQ and dW are not changes in state properties. For a system passing from a given initial state to a given final state, dU_n is independent of the process path since it is the change of a state property; however dQ and dW are, in general, path-dependent.

3.2.1.3 The Second Law

For a rigorous and complete development of the Second Law, the reader is referred to standard texts on thermodynamics. The *entropy* of a system, S, is an extensive state property which is given by Boltzmann's equation as:

$$S = k_B \ln t \qquad (2)$$

where k_B is the Boltzmann's constant and t is the multiplicity of the system. Somewhat loosely, t is the number of possible equivalent microstates in a macrostate, that is, the number of quantum states of the system that are accessible under the applicable conditions of energy, volume, etc. For example, for a system that can be described by a set of single-particle energy levels, t is the number of ways of distributing the particles over the energy levels, keeping the total internal energy constant. At low temperatures, most of the particles will be in or near the ground state. Hence, t and S will be small. As the temperature, and hence, U, increases, more energy levels become occupied. Consequently, t and S increase. For solutions, an additional contribution to t arises from the number of different possible ways of distributing the atoms or molecules over the lattice or quasilattice sites (Section 3.4.1.5). Again somewhat loosely, S can be said to be a measure of the disorder of a system.

During any spontaneous process, the total entropy of the universe will increase for the simple reason that disorder is more probable than order. That is, for any spontaneous process,

$$dS_{total} = (dS + dS_{surr}) \geq 0 \qquad (3)$$

where dS and dS_{surr} are the entropy changes of the system and surroundings, respectively. The existence of a state property S, which satisfies Eqn 3 is the essence of the Second Law.

Equation 3 is a necessary condition for a process to occur. However, even if Eqn 3 is satisfied, the process may not actually be observed if there are kinetic barriers to its occurrence. That is, the Second Law says nothing about the rate of a process, which can vary from extremely rapid to infinitely slow.

It should be noted that the entropy change of the system, dS, can be negative for a spontaneous process as long as the sum $(dS + dS_{surr})$ is positive. For example, during the solidification of a liquid, the entropy change of the system is negative in going from the liquid to the more ordered solid state. Nevertheless, a liquid below its melting point will freeze spontaneously because the entropy change of the surroundings is sufficiently positive due to the transfer of heat from the system to the surroundings. It should also be stressed that in passing from a given initial state to a given final state, the entropy change of the system dS is independent of the process path since it is the change of a state property. However, dS_{surr} is path-dependent.

3.2.1.4 The Fundamental Equation of Thermodynamics

Consider an open system at equilibrium with its surroundings and at internal equilibrium. That is, no spontaneous irreversible processes are taking place. Suppose that a change of state occurs in which S, V (volume) and n_i (number of moles of component i in the system) change by dS, dV and dn_i. Such a change of state occurring at equilibrium is called a *reversible* process, and the corresponding heat and work terms are dQ_{rev} and dW_{rev}. We may then write

$$dU = (\partial U / \partial S)_{v,n} \, dS + (\partial U / \partial V)_{s,n} dV + \sum \mu_i dn_i \qquad (4)$$

where

$$\mu_i = (\partial U / \partial n_i)_{S,V,n_{j \neq i}} \tag{5}$$

μ_i is the chemical potential of component i which will be discussed in Section 3.2.7. The absolute temperature is given as

$$T = (\partial U / \partial S)_{V,n} \tag{6}$$

We expect that temperature should be defined such that heat flows spontaneously from high to low T. To show that T as given by Eqn 6 is, in fact, such a thermal potential, consider two closed systems, isolated from their surroundings but in thermal contact with each other, exchanging only heat at constant volume. Let the temperatures of the systems be T_1 and T_2 and let $T_1 > T_2$. Suppose that heat flows from system 1 to system 2. Then $dU_2 = -dU_1 > 0$. Therefore, from Eqn 6,

$$dS = dS_1 + dS_2 = dU_1/T_1 + dU_2/T_2 > 0 \tag{7}$$

That is, the flow of heat from high to low temperature results in an increase in total entropy and hence, from the Second Law, is spontaneous.

The second term in Eqn 4 is clearly $(-P\,dV)$, the work of reversible expansion. From Eqn 6, the first term in Eqn 4 is equal to $T\,dS$, and this is then the reversible heat:

$$T\,dS = dQ_{rev} \tag{8}$$

That is, in the particular case of a process that occurs reversibly, (dQ_{rev}/T) is path-independent since it is equal to the change of a state property dS. Equation 8 is actually the definition of entropy in the classical development of the Second Law.

Equation 4 may now be written as

$$dU = T\,dS - P\,dV + \sum \mu_i\,dn_i \tag{9}$$

Equation 9, which results from combining the First and Second Laws, is called the *fundamental equation* of thermodynamics. We have assumed that the only work term is the reversible work of expansion (sometimes called "PV work"). In general, in this chapter, this will be the case. If other types of work occur, then non-PV terms, $dW_{rev(non-PV)}$, must be added to Eqn 9. For example, if the process is occurring in a galvanic cell, then $dW_{rev(non-PV)}$ is the reversible electrical work in the external circuit. Equation 9 can thus be written more generally as:

$$dU = T\,dS - P\,dV + \sum \mu_i\,dn_i + dW_{rev(non-PV)} \tag{10}$$

3.2.2 Enthalpy

Enthalpy, H, is an extensive state property defined as:

$$H = U + PV \tag{11}$$

Consider a closed system undergoing a change of state, which may involve irreversible processes (such as chemical reactions). Although the overall process may be irreversible, we shall assume that any work of expansion is approximately reversible (that is, the external and internal pressures are equal) and that there is no work other than work of expansion. Then, from Eqn 1:

$$dU_n = dQ - P\,dV \tag{12}$$

From Eqns 11 and 12, it follows that:

$$dH_n = dQ + V\,dP \tag{13}$$

Furthermore, if the pressure remains constant throughout the process, then

$$dH_p = dQ_p \tag{14}$$

Integrating both sides of Eqn 14 gives

$$\Delta H_p = Q_p \tag{15}$$

That is, the enthalpy change of a closed system in passing from an initial to a final state at constant pressure is equal to the heat exchanged with the surroundings. Hence, for a process occurring at constant pressure, the heat is path-independent since it is equal to the change of a state property. This is an important result. As an example, suppose that the initial state of a system consists of 1.0 mol of C and 1.0 mol of O_2 at 298.15 K at 1.0 bar pressure and the final state is 1 mol of CO_2 at the same temperature and pressure. The enthalpy change of this reaction is

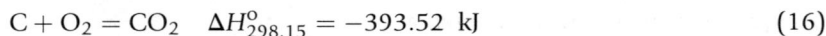

$$C + O_2 = CO_2 \quad \Delta H^o_{298.15} = -393.52 \text{ kJ} \tag{16}$$

(where the superscript on $\Delta H^o_{298.15}$ indicates the "standard state" reaction involving pure solid graphite, CO_2 and O_2 at 1.0 bar pressure). Hence, an exothermic heat of -393.52 kJ will be observed, independent of the reaction path, provided only that the pressure remains constant throughout the process. For instance, during the combustion reaction, the reactants and products may attain high and unknown temperatures. However, once the CO_2 product has cooled back to 298.15 K, the total heat that has been exchanged with the surroundings will be -393.52 kJ independent of the intermediate temperatures.

In **Figure 8** the standard molar enthalpy of Fe is shown as a function of temperature. The y-axis, $(h^o_T - h^o_{298.15})$, is the heat required to heat 1 mol of Fe from 298.15 K to a temperature T at constant pressure. The slope of the curve is the molar heat capacity at constant pressure:

$$c_p = (dh/dT)_p \tag{17}$$

From Eqn 17, we obtain the following expression for the heat required to heat 1 mol of a substance from a temperature T_1 to a temperature T_2 at constant P (assuming no phase changes in the interval):

$$(h_{T_2} - h_{T_1}) = \int_{T_1}^{T_2} c_p dT \tag{18}$$

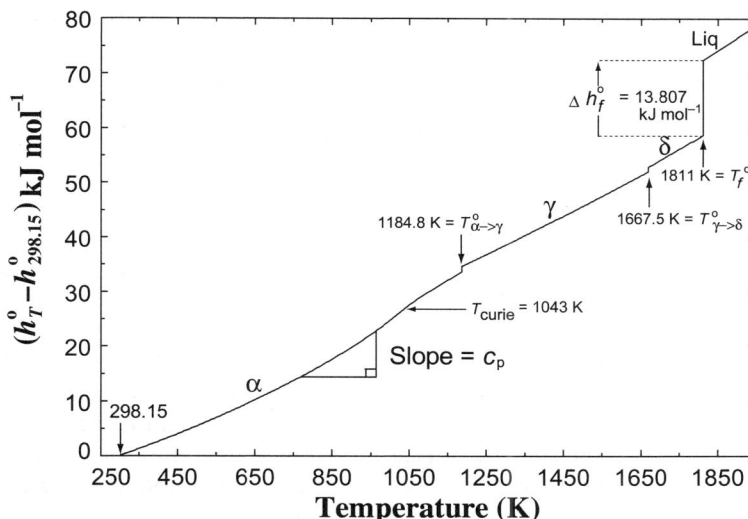

Figure 8 Standard molar enthalpy of Fe (see FactSage).

The enthalpy curve can be measured by the technique of drop calorimetry, or the heat capacity can be measured directly by adiabatic calorimetry.

The standard equilibrium temperature of fusion (melting) of Fe is 1811 K as shown in **Figure 8**. The fusion reaction is a *first-order phase change* since it occurs at constant temperature. The standard molar enthalpy of fusion Δh_f^o is 13.807 kJ mol^{-1} as shown in **Figure 8**. It can also be seen in **Figure 8** that Fe undergoes two other first-order phase changes, the first from α (bcc) to γ (fcc) Fe at $T_{\alpha \to \gamma}^o = 1184.8$ K and the second from γ to δ (bcc) at 1667.5 K. The enthalpy changes are, respectively, 1.013 and 0.826 kJ mol^{-1}. The Curie temperature, $T_{curie} = 1045$ K, which is also shown in **Figure 8**, will be discussed in Section 3.10.

3.2.3 Gibbs Energy

The Gibbs energy (also called the Gibbs free energy, or simply, the free energy), G, is defined as

$$G = H - TS \tag{19}$$

As given in Section 3.2.2, we consider a closed system and assume that the only work term is the work of expansion and that this is reversible. From Eqns 13 and 19,

$$dG_n = dQ + V\,dP - T\,dS - S\,dT \tag{20}$$

Consequently, for a process occurring at constant T and P in a closed system,

$$dG_{T,P,n} = dQ_P - T\,dS = dH_P - T\,dS \tag{21}$$

Consider the case where the "surroundings" are simply a heat reservoir at a temperature equal to the temperature T of the system. That is, no irreversible processes occur in the surroundings, which receive only a reversible transfer of heat $(-dQ)$ at constant temperature. Therefore, from Eqn 8,

$$dS_{surr} = -dQ/T \qquad (22)$$

Substituting into Eqn 21 and using Eqn 3 yields

$$dG_{T,P,n} = -T\,dS_{surr} - T\,dS = -T\,dS_{total} \leq 0 \qquad (23)$$

Equation 23 may be considered to be a special form of the Second Law for a process occurring in a closed system at constant T and P. From Eqn 23, such a process will be spontaneous if dG is negative.

For our purposes in this chapter, Eqn 23 is the most useful form of the Second Law.

An objection might be raised that Eqn 23 applies only when the surroundings are at the same temperature T as the system. However, if the surroundings are at a different temperature, we simply postulate a hypothetical heat reservoir, which is at the same temperature as the system, and which lies between the system and surroundings, and we consider the irreversible transfer of heat between the reservoir and the real surroundings as a second separate process.

Substituting Eqns 8 and 22 into Eqn 23 gives

$$dG_{T,P,n} = (dQ - dQ_{rev}) \leq 0 \qquad (24)$$

In Eqn 24, dQ is the heat of the actual process, while dQ_{rev} is the heat that would be observed were the process occurring reversibly. If the actual process is reversible, that is, if the system is at equilibrium, then $dQ = dQ_{rev}$ and $dG = 0$.

As an example, consider the fusion of Fe in **Figure 8**. At $T_f^o = 1811$ K, solid and liquid are in equilibrium. That is, at this temperature, melting is a reversible process. Therefore, at 1811 K:

$$dg_f^o = dh_f^o - T ds_f^o = 0 \qquad (25)$$

Therefore, the molar entropy of fusion at 1811 K is given by

$$\Delta s_f^o = \left(\Delta h_f^o / T\right) \quad \text{when } T = T_f^o \qquad (26)$$

Equation 26 applies only at the equilibrium melting point. At temperatures below and above this temperature, the liquid and solid are not in equilibrium. Above the melting point, $\Delta g_f^o < 0$ and melting is a spontaneous process (liquid more stable than solid). Below the melting point, $\Delta g_f^o > 0$ and the reverse process (solidification) is spontaneous. As the temperature deviates progressively further from the equilibrium melting point, the magnitude of Δg_f^o, that is, the magnitude of the driving force, increases. We shall return to this subject in Section 3.4.2.

3.2.4 Chemical Equilibrium

Consider first the question of whether silica fibers in an aluminum matrix at 500 °C will react to form mullite, $Al_6Si_2O_{13}$:

$$^{13}/_2 \ SiO_2 + 6Al = {}^{9}/_2 \ Si + Al_6Si_2O_{13} \qquad (27)$$

If the reaction proceeds with the formation of $d\xi$ moles of mullite, then from the stoichiometry of the reaction, $dn_{Si} = (9/2)\,d\xi$, $dn_{Al} = -6\,d\xi$, and $dn_{SiO_2} = -13/2\,d\xi$. Since the four substances are essentially immiscible at 500 °C, their Gibbs energies are equal to their standard molar Gibbs energies, g_i^o (also known as the standard chemical potentials μ_i^o as defined in Section 3.2.7), the standard state of a solid or liquid compound being the pure compound at $P = 1.0$ bar. The Gibbs energy of the system then varies as

$$dG/d\xi = g^o_{Al_6Si_2O_{13}} + 9/2\,g^o_{Si} - 13/2\,g^o_{SiO_2} - 6g^o_{Al} = \Delta G^o = -830 \text{ kJ} \tag{28}$$

where ΔG^o is called the standard Gibbs energy change of the reaction (Eqn 27) at 500 °C.

Since $\Delta G^o < 0$, the formation of mullite entails a decrease in G. Hence, the reaction will proceed spontaneously so as to minimize G. Equilibrium will never be attained and the reaction will proceed to completion.

3.2.4.1 Equilibria Involving a Gaseous Phase

An *ideal gas* mixture is one that obeys the ideal gas equation of state:

$$PV = nRT \tag{29}$$

where n is the total number of moles. Further, the partial pressure of each gaseous species in the mixture is given by

$$p_iV = n_iRT \tag{30}$$

where $n = \Sigma n_i$ and $P = \Sigma p_i$ (Dalton's Law for ideal gases). R is the ideal gas constant. The *standard state* of an ideal gaseous compound is the pure compound at a pressure of 1.0 bar. It can easily be shown that the partial molar Gibbs energy g_i (also called the chemical potential μ_i as defined in Section 3.2.7) of a species in an ideal gas mixture is given in terms of its standard molar Gibbs energy, g_i^o (also called the standard chemical potential μ_i^o) by

$$g_i = g_i^o + RT \ln p_i \tag{31}$$

The final term in Eqn 31 is entropic. As a gas expands at constant temperature, its entropy increases.

Consider a gaseous mixture of H_2, S_2 and H_2S with partial pressures p_{H_2}, p_{S_2} and p_{H_2S}. The gases can react according to

$$2H_2 + S_2 = 2H_2S \tag{32}$$

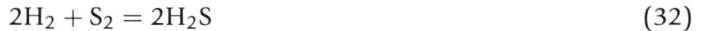

If the reaction (Eqn 32) proceeds to the right with the formation of $2d\xi$ moles of H_2S, then the Gibbs energy of the system varies as

$$
\begin{aligned}
dG/d\xi &= 2g_{H_2S} - 2g_{H_2} - g_{S_2} \\
&= \left(2g^o_{H_2S} - 2g^o_{H_2} - g^o_{S_2}\right) + RT\left(2\ln p_{H_2S} - 2\ln p_{H_2} - \ln p_{S_2}\right) \\
&= \Delta G^o + RT \ln\left(p^2_{H_2S}\ p^{-2}_{H_2}\ p^{-1}_{S_2}\right) \\
&= \Delta G
\end{aligned}
\tag{33}
$$

ΔG, which is the Gibbs energy change of the reaction (Eqn 32), is thus a function of the partial pressures. If $\Delta G < 0$, then the reaction will proceed to the right, so as to minimize G. In a closed system, as the reaction continues with the production of H_2S, p_{H_2S} will increase while p_{H_2} and p_{S_2} will decrease. As a result, ΔG will become progressively less negative. Eventually, an equilibrium state will be reached when $dG/d\xi = \Delta G = 0$.

For the equilibrium state, therefore,

$$\Delta G^\circ = -RT \ln K$$
$$= -RT \ln \left(p_{H_2S}^2 \; p_{H_2}^{-2} \; p_{S_2}^{-1} \right)_{\text{equilibrium}} \tag{34}$$

where K, the "equilibrium constant" of the reaction, is the one unique value of the ratio $(p_{H_2S}^2 \; p_{H_2}^{-2} \; p_{S_2}^{-1})$ for which the system will be in equilibrium at the temperature T.

If the initial partial pressures are such that $\Delta G > 0$, then the reaction (Eqn 32) will proceed to the left in order to minimize G until the equilibrium condition of Eqn 34 is attained.

As a further example, consider the possible precipitation of graphite from a gaseous mixture of CO and CO_2. The reaction is

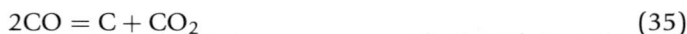

$$2CO = C + CO_2 \tag{35}$$

Proceeding as above, we can write:

$$dG/d\xi = g_C + g_{CO_2} - 2g_{CO}$$
$$= \left(g_C^\circ + g_{CO_2}^\circ - 2g_{CO}^\circ \right) + RT \ln \left(p_{CO_2} p_{CO}^{-2} \right)$$
$$= \Delta G^\circ + RT \ln \left(p_{CO_2} p_{CO}^{-2} \right) \tag{36}$$
$$= \Delta G = -RT \ln K + RT \ln \left(p_{CO_2} p_{CO}^{-2} \right)$$

If $(p_{CO_2} p_{CO}^{-2})$ is less than the equilibrium constant K, then precipitation of graphite will occur in order to decrease G.

Real situations are, of course, generally more complex. To treat the deposition of solid Si from a vapor of SiI_4, for example, we must consider the formation of gaseous I_2, I and SiI_2, so that three independent reaction equations must be written:

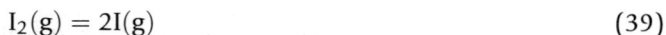

$$SiI_4(g) = Si(sol) + 2I_2(g) \tag{37}$$

$$SiI_4(g) = SiI_2(g) + I_2(g) \tag{38}$$

$$I_2(g) = 2I(g) \tag{39}$$

The equilibrium state, however, is still that which minimizes the total Gibbs energy of the system. This is equivalent to satisfying simultaneously the equilibrium constants of the reactions shown in Eqns 37–39.

Equilibria involving reactants and products that are solid or liquid solutions will be discussed in Section 3.4.1.4.

3.2.4.2 Predominance Diagrams

Predominance diagrams are a particularly simple type of phase diagram, which have many applications in the fields of hot corrosion, chemical vapor deposition, etc. Furthermore, their construction clearly illustrates the principles of Gibbs energy minimization.

A predominance diagram for the Cu–SO$_2$–O$_2$ system at 700 °C is shown in **Figure 9**. The axes are the logarithms of the partial pressures of SO$_2$ and O$_2$ in the gas phase. The diagram is divided into areas or domains of stability of the various solid compounds of Cu, S and O. For example, at point Z, where $p_{SO_2} = 10^{-2}$ and $p_{O_2} = 10^{-7}$ bar, the stable phase is Cu$_2$O. The conditions for coexistence of two and three solid phases are indicated, respectively, by the lines and triple points on the diagram. For example, along the line separating the Cu$_2$O and CuSO$_4$ domains, the equilibrium constant $K = p_{SO_2}^{-2} p_{O_2}^{-3/2}$ of the following reaction is satisfied:

$$Cu_2O + 2SO_2 + 3/2\, O_2 = 2CuSO_4 \tag{40}$$

Hence, along this line,

$$\log K = -\Delta G^\circ / RT = -2 \log p_{SO_2} - 3/2 \log p_{O_2} \tag{41}$$

This line is thus straight with slope $(-3/2)/2 = -3/4$.

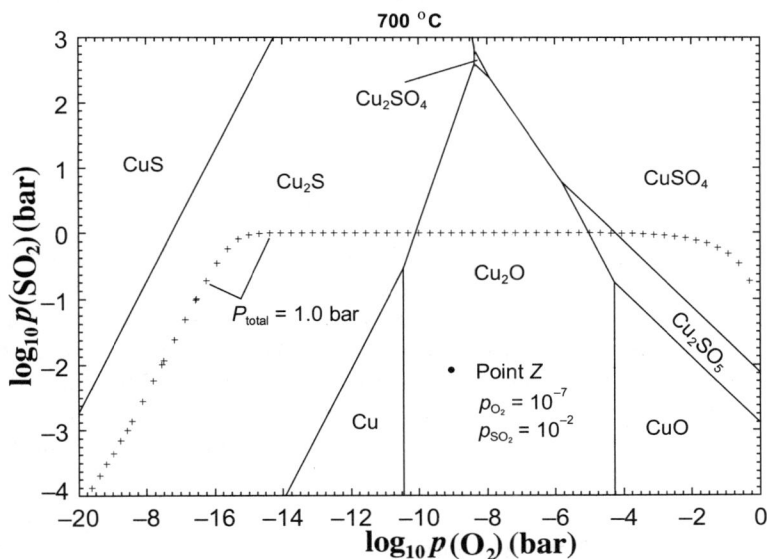

Figure 9   "Predominance" phase diagram of the Cu–SO$_2$–O$_2$ system at 700 °C showing the solid phases at equilibrium as a function of the equilibrium SO$_2$ and O$_2$ partial pressures (see FactSage).

We can construct **Figure 9** following the procedure of Bale et al. (1986). We formulate a reaction for the formation of each solid phase, always from 1 mol of Cu, and involving the gaseous species whose pressures are used as the axes (SO_2 and O_2 in this example):

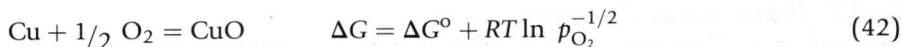

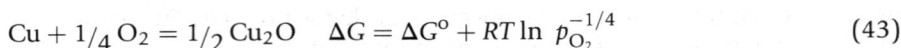

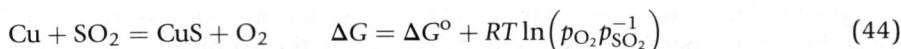

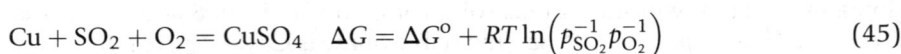

$$Cu + 1/_2\,O_2 = CuO \qquad \Delta G = \Delta G^\circ + RT \ln\, p_{O_2}^{-1/2} \tag{42}$$

$$Cu + 1/_4\,O_2 = 1/_2\,Cu_2O \quad \Delta G = \Delta G^\circ + RT \ln\, p_{O_2}^{-1/4} \tag{43}$$

$$Cu + SO_2 = CuS + O_2 \qquad \Delta G = \Delta G^\circ + RT \ln\left(p_{O_2}p_{SO_2}^{-1}\right) \tag{44}$$

$$Cu + SO_2 + O_2 = CuSO_4 \quad \Delta G = \Delta G^\circ + RT \ln\left(p_{SO_2}^{-1}p_{O_2}^{-1}\right) \tag{45}$$

and similarly for the formation of Cu_2S, Cu_2SO_4 and Cu_2SO_5.

The values of ΔG° are obtained from tables of thermodynamic properties. For any given values of p_{SO_2} and p_{O_2}, ΔG for each formation reaction can then be calculated. The stable compound is simply the one with the most negative ΔG. If all the ΔG values are positive, then pure Cu is the stable compound.

By reformulating Eqns 42–45 in terms of, for example, S_2 and O_2 rather than SO_2 and O_2, a predominance diagram with $\log\, p_{S_2}$ and $\log\, p_{O_2}$ as axes can be constructed. Logarithms of ratios or products of partial pressures can also be used as axes.

Along the curve shown by the crosses in **Figure 9**, the total pressure is 1.0 bar. That is, the gas consists not only of SO_2 and O_2, but of other species such as S_2 whose equilibrium partial pressures can be calculated. Above this line, the total pressure is >1.0 bar even though the sum of the partial pressures of SO_2 and O_2 may be <1.0 bar.

A similar phase diagram of $RT \ln\, p_{O_2}$ vs T for the Cu–O_2 system is shown in **Figure 10**. For the formation reaction,

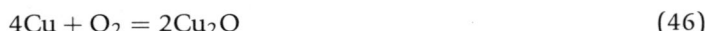

$$4Cu + O_2 = 2Cu_2O \tag{46}$$

we can write

$$\Delta G^\circ = -RT \ln K = RT \ln\left(p_{O_2}\right)_{equilibrium} = \Delta H^\circ - T\Delta S^\circ \tag{47}$$

The line between the Cu and Cu_2O domains in **Figure 10** is thus a plot of the standard Gibbs energy of formation of Cu_2O versus T. The temperatures indicated by the symbols M and $\boxed{M}$ are the melting points of Cu and Cu_2O, respectively. This line is thus simply a line taken from the well-known *Ellingham diagram or ΔG° vs T* diagram for the formation of oxides. However, by drawing vertical lines at the melting points of Cu and Cu_2O as shown in **Figure 10**, we convert the Ellingham diagram to a phase diagram. Stability domains for Cu(sol), Cu(liq), Cu_2O(sol), Cu_2O(liq) and CuO(sol) are shown as functions of T and of the imposed equilibrium p_{O_2}. The lines and triple points indicate conditions for two- and three-phase equilibria.

3.2.5   Measuring Gibbs Energy, Enthalpy and Entropy

3.2.5.1   *Measuring Gibbs Energy Change*

In general, the Gibbs energy change of a process can be determined by performing measurements under equilibrium conditions. For example, consider the reaction shown in Eqn 35. Experimentally,

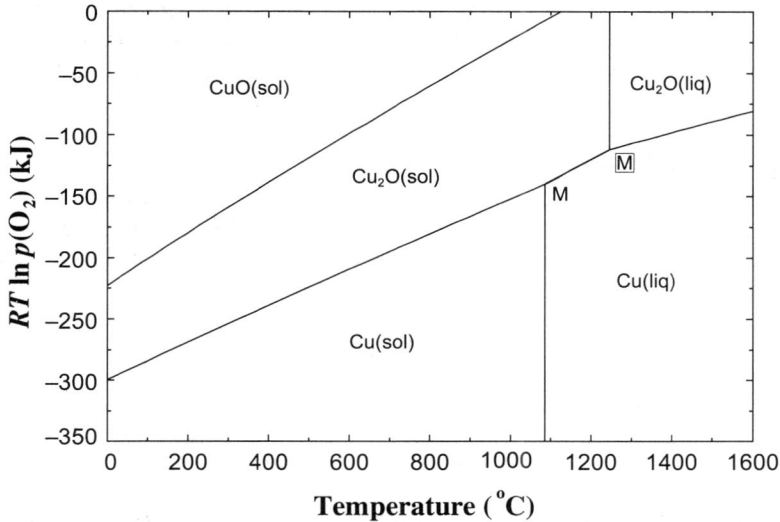

Figure 10 Predominance phase diagram (also known as a Gibbs energy—T diagram or an Ellingham diagram) for the Cu–O_2 system. Points M and M̄ are are the melting points of Cu and Cu_2O respectively (see FactSage).

equilibrium is first established between solid C and the gas phase and the equilibrium constant is then found by measuring the partial pressures of CO and CO_2. Equation 36 is then used to calculate $\Delta G°$.

Another common technique of establishing equilibrium is by means of an electrochemical cell. A galvanic cell is designed, so that a reaction (the cell reaction) can occur only with the passage of an electric current through an external circuit. An external voltage is applied which is equal in magnitude but opposite in sign to the cell potential, E (volts), such that no current flows. The system is then at equilibrium. With the passage of dz coulombs through the external circuit, Edz joules of work are performed. Since the process occurs at equilibrium, this is the reversible non-PV work, $dW_{rev(non\text{-}PV)}$. Substituting Eqn 11 in Eqn 19, differentiating, and substituting Eqn 10 gives:

$$dG = V\,dP - S\,dT + \sum \mu_l\,dn_i + dW_{rev(non-PV)} \tag{48}$$

For a closed system at constant T and P,

$$dG_{T,P,n} = dW_{rev(non-PV)} \tag{49}$$

If the cell reaction is formulated such that the advancement of the reaction by $d\xi$ moles is associated with the passage of dz coulombs through the external circuit, then from Eqn 49,

$$\Delta G\,d\xi = -E\,dz \tag{50}$$

Hence,

$$\Delta G = -FE \tag{51}$$

where ΔG is the Gibbs energy change of the cell reaction, $F = dz/d\xi = 96,500$ C mol^{-1} is the Faraday constant, which is the number of coulombs in a mole of electrons, and the negative sign follows the usual electrochemical sign convention. Hence, by measuring the open-circuit voltage, E, one can determine ΔG of the cell reaction.

3.2.5.2 Measuring Enthalpy Change

The enthalpy change, ΔH, of a process may be measured directly by calorimetry. Alternatively, if ΔG is measured by an equilibration technique as in Section 3.2.5.1, then ΔH can be determined from the measured temperature dependence of ΔG. From Eqn 48, for a process in a closed system with no non-PV work at constant pressure,

$$(dG/dT)_{P,n} = -S \qquad (52)$$

Substitution of Eqn 19 into Eqn 52 gives

$$(d(G/T)/d(1/T))_{P,n} = H \qquad (53)$$

Equations 52 and 53 apply to both the initial and final states of a process. Hence,

$$(d(\Delta G)/dT)_{P,n} = -\Delta S \qquad (54)$$

$$(d(\Delta G/T)d(1/T))_{P,n} = \Delta H \qquad (55)$$

Equations 52–55 are forms of the *Gibbs–Helmholtz equation.*

3.2.5.3 Measuring Entropy

As is the case with the enthalpy, ΔS of a process can be determined from the measured temperature dependence of ΔG by means of Eqn 54. This is known as the "second law method" of measuring entropy.

Another method of measuring entropy involves the *Third Law* of thermodynamics, which states that the entropy of a perfect crystal of a pure substance at internal equilibrium at a temperature of 0 K is zero. This follows from Eqn 2 and the concept of entropy as disorder. For a perfectly ordered system at absolute zero, $t = 1$ and $S = 0$.

Let us first derive an expression for the change in entropy as a substance is heated. Combining Eqns 52 and 53 gives:

$$(dH/dT)_{P,n} = T(dS/dT)_{P,n} \qquad (56)$$

Substituting Eqn 18 into Eqn 56 and integrating, then gives the following expression for the entropy change as 1 mol of a substance is heated at constant P from T_1 to T_2:

$$(s_{T_2} - s_{T_1}) = \int_{T_1}^{T_2} (c_p/T)\,dT \qquad (57)$$

The similarity to Eqn 18 is evident, and a plot of $(s_{T_2} - s_{T_1})$ versus T is similar in appearance to **Figure 8**.

Setting $T_1 = 0$ in Eqn 57 and using the Third Law gives

$$s_{T_2} = \int_0^{T_2} (c_p/T)\,dT \tag{58}$$

Hence, if c_p has been measured down to temperatures sufficiently close to 0 K, Eqn 58 can be used to calculate the absolute entropy. This is known as the "third law method" of measuring entropy.

3.2.5.4 Zero Entropy and Zero Enthalpy

The third law method permits the measurement of the absolute entropy of a substance, whereas the second law method permits the measurement only of entropy changes, ΔS.

The Third Law is not a convention. It provides a natural zero for entropy. On the other hand, no such natural zero exists for the enthalpy since there is no simple natural zero for the internal energy. However, by convention, the absolute enthalpy of a pure element in its stable state at $P = 1.0$ bar and $T = 298.15$ K is usually set to zero. Therefore, the absolute enthalpy of a compound is, by convention, equal to its standard enthalpy of formation from the pure elements in their stable standard states at 298.15 K.

3.2.6 Auxiliary Functions

It was shown in Section 3.2.3 that Eqn 23 is a special form of the Second Law for a process occurring at constant T and P. Other special forms of the Second Law can be derived in a similar manner.

It follows directly from Eqns 3, 12 and 22, for a process at constant V and S, that

$$dU_{V,S,n} = -T\,dS_{total} \leq 0 \tag{59}$$

Similarly, from Eqns 3, 13 and 22,

$$dH_{P,S,n} = -T\,dS_{total} \leq 0 \tag{60}$$

Equations 59 and 60 are not of direct practical use since there is no simple means of maintaining the entropy of a system constant during a change of state. A more useful function is the *Helmholtz energy*, A, defined as:

$$A = U - TS \tag{61}$$

From Eqns 12 and 40,

$$dA_n = dQ - P\,dV - T\,dS - S\,dT \tag{62}$$

Then, from Eqns 3 and 22, for processes at constant T and V,

$$dA_{T,V,n} = -T\,dS_{total} \leq 0 \tag{63}$$

Equations 23, 59, 60 and 63 are special forms of the Second Law for closed systems. In this regard, H, A and G are sometimes called *auxiliary functions*. If we consider open systems, then many other auxiliary functions can be defined as will be shown in Section 3.2.8.2.

In Section 3.2.3, the very important result was derived that the equilibrium state of a system corresponds to a minimum of the Gibbs energy G. This is used as the basis of nearly all algorithms for calculating chemical equilibria and phase diagrams. An objection might be raised that Eqn 23 only applies at constant T and P. What if equilibrium were approached at constant T and V for instance? Should we not then minimize A as in Eqn 63? The response is that the restrictions on Eqns 23 and 63 only apply to the paths followed during the approach to equilibrium. However, for a system at an equilibrium temperature, pressure and volume, it clearly does not matter along which hypothetical path the approach to equilibrium was calculated. The reason that most algorithms minimize G rather than other auxiliary functions is simply that it is generally easiest and most convenient to express thermodynamic properties in terms of the independent variables T, P and n_i.

3.2.7 The Chemical Potential

The chemical potential was defined in Eqn 5. However, this equation is not particularly useful directly since it is difficult to see how S can be held constant.

Substituting Eqn 11 into Eqn 19, differentiating and substituting Eqn 9 gives

$$dG = V \, dP - S \, dT + \sum \mu_i dn_i \qquad (64)$$

from which it can be seen that μ_i is also given by

$$\mu_i = (\partial G / \partial n_i)_{T,P,n_{j\neq i}} \qquad (65)$$

Differentiating Eqn 11 with substitution of Eqn 9 gives

$$dH = T \, dS + V \, dP + \sum \mu_i dn_i \qquad (66)$$

whence

$$\mu_i = (\partial H / \partial n_i)_{T,V,n_{j\neq i}} \qquad (67)$$

Similarly it can be shown that

$$\mu_i = (\partial A / \partial n_i)_{T,V,n_{j\neq i}} \qquad (68)$$

Among these equations for μ_i, the most useful is Eqn 65. The chemical potential of a component i of a system is given by the increase in Gibbs energy as the component is added to the system at constant T and P while keeping the numbers of moles of all other components of the system constant.

The chemical potentials of its components are intensive properties of a system; for a system at a given T, P and composition, the chemical potentials are independent of the mass of the system. That is, with reference to Eqn 65, adding dn_i moles of component i to a solution of a fixed composition will result in a change in Gibbs energy dG, which is independent of the total mass of the solution.

The reason that this property is called a chemical potential is illustrated by the following thought experiment. Imagine two systems, I and II, at the same temperature and pressure and separated by a membrane that permits the passage of only one component, say component 1. The chemical potentials of component 1 in systems I and II are $\mu_1^I = \partial G^I / \partial n_1^I$ and $\mu_1^{II} = \partial G^{II} / \partial n_1^{II}$. Component 1 is transferred across the membrane with $dn_1^I = -dn_1^{II}$. The change in the total Gibbs energy accompanying this transfer is then:

$$dG = d(G^I + dG^{II}) = -(\mu_1^I - \mu_1^{II})dn_1^{II} \tag{69}$$

If $\mu_1^I > \mu_1^{II}$, then $d(G^I + dG^{II})$ is negative when dn_1^{II} is positive. That is, the total Gibbs energy is decreased by a transfer of component 1 from system I to system II. Hence, component 1 is transferred spontaneously from a system of higher μ_1 to a system of lower μ_1. For this reason, μ_1 is called the chemical potential of component 1.

Similarly, by using the appropriate auxiliary functions, it can be shown that μ_i is the potential for the transfer of component i at constant T and V, at constant S and V, etc.

An important principle of phase equilibrium can now be stated. *When two or more phases are in equilibrium, the chemical potential of any component is the same in all phases.* This criterion for phase equilibrium is thus equivalent to the criterion of minimizing G. This will be discussed further in Section 3.4.2.

3.2.8 Some Other Useful Thermodynamic Equations

In this section, some other useful thermodynamic relations are derived. Consider an open system, which is initially empty. That is, initially S, V and n are all equal to zero. We now "fill up" the system by adding a previously equilibrated mixture. The addition is made at constant T, P and composition. Hence, it is also made at constant μ_i (of every component) since μ_i is an intensive property. The addition is continued until the entropy, volume and numbers of moles of the components in the system are S, V and n_i. Integrating Eqn 9 from the initial (empty) to the final state then gives:

$$U = TS - PV + \sum n_i \mu_i \tag{70}$$

It then follows that

$$H = TS + \sum n_i \mu_i \tag{71}$$

$$A = -PV + \sum n_i \mu_i \tag{72}$$

and most importantly

$$G = \sum n_i \mu_i \tag{73}$$

3.2.8.1 The Gibbs–Duhem Equation

By differentiation of Eqn 70,

$$dU = T\,dS + S\,dT - P\,dV - V\,dP + \sum n_i d\mu_i + \sum \mu_i dn_i \tag{74}$$

By comparison of Eqns 9 and 74, it follows that

$$S\,dT - V\,dP + \sum n_i d\mu_i = 0 \tag{75}$$

This is the general Gibbs–Duhem equation.

3.2.8.2   *General Auxiliary Functions*

It was stated in Section 3.2.6 that many auxiliary functions can be defined for open systems. Consider, for example, the auxiliary function:

$$G' = U - (-PV) - TS - \sum_1^k n_i\mu_i$$

$$= G - \sum_1^k n_i\mu_i \tag{76}$$

Differentiating Eqn 76 with substitution of Eqn 64 gives

$$dG' = V\,dP - S\,dT - \sum_1^k n_i d\mu_i + \sum_{k+1}^C \mu_i dn_i \tag{77}$$

where C is the number of components in the system. It then follows that

$$dG'_{T,P,\mu_i(1\leq i\leq k),\ n_i(k+1\leq i\leq C)} \leq 0 \tag{78}$$

is the criterion for a spontaneous process at constant T, P, μ_i (for $1 \leq i \leq k$) and n_i (for $k+1 \leq i \leq C$) and that equilibrium is achieved when $dG' = 0$. The chemical potential of a component may be held constant, for example, by equilibration with a gas phase as, for example, by fixing the partial pressure of O_2, SO_2, etc.

Finally, it also follows that the chemical potential may be defined very generally as

$$\mu_i = (\partial G'/\partial n_i)_{T,P,\mu_i(1\leq i\leq k),\ n_i(k+1\leq i\leq C)} \tag{79}$$

The discussion of the generalized auxiliary functions will be continued in Section 3.7.7.

3.3   The Gibbs Phase Rule

The geometry of all types of phase diagrams is governed by the *Gibbs Phase Rule*, which applies to any system at equilibrium:

$$F = C - P + 2 \tag{80}$$

The Phase Rule will be derived in this section in a general form along with several examples. Further examples will be presented in later sections.

In Eqn 80, C is the number of *components* in the system. This is the minimum number of substances required to describe the elemental composition of every phase present. As an example, for pure H_2O at 25 °C and 1.0 bar pressure, $C = 1$. For a gaseous mixture of H_2, O_2 and H_2O at high temperature, $C = 2$. (Although there are many species present in the gas including O_2, H_2, H_2O, O_3, etc., these species are all at equilibrium with each other. At a given T and P, it is sufficient to know the overall elemental composition in order to be able to calculate the concentration of every species. Hence, $C = 2$. C can never exceed the number of elements in the system.)

For the systems shown in **Figures 1–7**, respectively, $C = 2, 3, 3, 4, 3, 4$ and 1.

There is generally more than one way to specify the components of a system. For example, the Fe–Si–O system could also be described as the $FeO–Fe_2O_3–SiO_2$ system, or the $FeO–SiO_2–O_2$ system, etc. The formulation of a phase diagram can often be simplified by a judicious definition of the components as will be discussed in Section 3.7.5.

In Eqn 80, P is the number of *phases* present at equilibrium at a given point on the phase diagram. For example, in **Figure 1** at any point within the region labeled "Liquid + bcc", $P = 2$. At any point in the region labeled "Liquid," $P = 1$.

In Eqn 80, F is the *number of degrees of freedom* or the *variance* of the system at a given point on the phase diagram. It is the number of variables that must be specified in order to completely specify the state of the system. Let us number the components as $1, 2, 3, ..., C$ and designate the phases as $\alpha, \beta, \gamma, ...$ In most derivations of the Phase Rule, the only variables considered are T, P and the compositions of the phases. However, since we also want to treat more general phase diagrams in which chemical potentials, volumes, enthalpies, etc. can also be variables, we shall extend the list of variables as follows:

$$T, P$$

$$\left(X_1^\alpha, X_2^\alpha, ...X_{C-1}^\alpha\right), \left(X_1^\beta, X_2^\beta, ...X_{C-1}^\beta\right), ...$$

$$\mu_1, \mu_2, \mu_3, ...$$

$$V^\alpha, V^\beta, V^\gamma, ...$$

$$S^\alpha, S^\beta, S^\gamma, ...$$

where X_i^α is the mole fraction of component i in phase α, μ_i is the chemical potential of component i, and V^α and S^α are the volume and entropy of phase α. Alternatively, we may substitute mass fractions for mole fractions and enthalpies H^α, H^β, ... for S^α, S^β, ...

It must be stressed that the overall composition of the system is not a variable in the sense of the Phase Rule. The composition variables are the compositions of the individual phases.

The total number of variables is thus equal to $(2 + P(C - 1) + C + 2P)$.

(Only $(C - 1)$ independent composition variables X_i are required for each phase since the sum of the mole fractions is unity.)

At equilibrium, the chemical potential μ_i of each component i is the same in all phases (Section 3.2.7) as are T and P. μ_i is an intensive variable (Section 3.2.7), which is a function of T, P and composition. Hence,

$$\mu_i^\alpha\left(T, P, X_1^\alpha, X_2^\alpha, ...,\right) = \mu_i^\beta\left(T, P, X_1^\beta, X_2^\beta, ...,\right) = ... = \mu_i \tag{81}$$

This yields PC equations relating the variables. Furthermore, the volume of each phase is a function of T, P and composition, $V^\alpha = (T, P, X_1^\alpha, X_2^\alpha, ...,)$, as is the entropy of the phase (or, alternatively, the enthalpy). This yields an additional $2P$ equation relating the variables.

F is then equal to the number of variables minus the number of equations relating the variables at equilibrium:

$$F = (2 + P(C - 1) + C + 2P) - PC - 2P = C - P + 2 \tag{82}$$

Equation 80 is thereby derived.

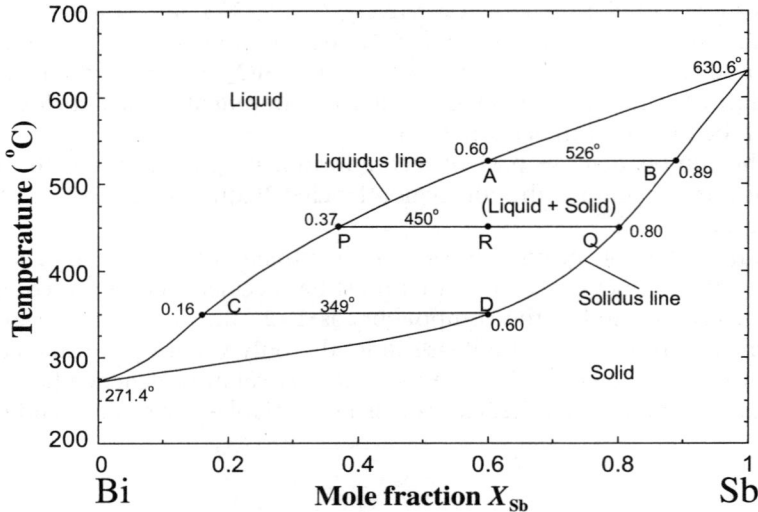

Figure 11   Temperature-composition phase diagram at $P = 1$ bar of the Bi–Sb system (see FactSage; SGTE).

3.3.1   The Phase Rule and Binary Temperature-Composition Phase Diagrams

As a first example of the application of the Phase Rule, consider the temperature-composition $(T–X)$ phase diagram of the Bi–Sb system at a constant applied pressure of 1.0 bar in **Figure 11**. The abscissa of **Figure 11** is the composition, expressed as mole fraction of Sb, X_{Sb}. Note that $X_{Bi}=1 - X_{Sb}$. Phase diagrams are also often drawn with the composition axis expressed as weight percentage.

At all compositions and temperatures in the area above the line labeled "liquidus," a single-phase liquid solution will be observed, while at all compositions and temperatures below the line labeled "solidus," there will be a single-phase solid solution. A sample at equilibrium at a temperature and overall composition between these two curves will consist of a mixture of solid and liquid phases, the compositions of which are given by the liquidus and solidus compositions at that temperature. For example, a sample of overall composition $X_{Sb} = 0.60$ at $T = 450$ °C (at point R in **Figure 11**) will consist, at equilibrium, of a mixture of a liquid of composition $X_{Sb} = 0.37$ (point P) and solid of composition $X_{Sb} = 0.80$ (point Q). The line PQ is called a *tie-line* or *conode*.

Binary temperature-composition phase diagrams are generally plotted at a fixed pressure, usually 1.0 bar. This eliminates one degree of freedom. In a binary system, $C = 2$. Hence, for binary isobaric $T–X$ diagrams the phase rule reduces to

$$F = 3 - P \tag{83}$$

Binary $T–X$ diagrams contain single-phase areas and two-phase areas. In the single-phase areas, $F = 3 - 1 = 2$. That is, temperature and composition can be specified independently. These regions are thus called *bivariant*. In two-phase regions, $F = 3 - 2 = 1$. If, say, T is specified, then the compositions of both phases are determined by the ends of the tie-lines. Two-phase regions are thus termed *univariant*. Note that although the overall composition can vary over a range within a two-phase region at constant T, the overall composition is not a parameter in the sense of the phase rule. Rather, it is the compositions of the individual phases at equilibrium that are the parameters to be considered in counting the number of degrees of freedom.

As the overall composition is varied at 450 °C between points P and Q, the compositions of the liquid and solid phases remain fixed at P and Q, and only the relative proportions of the two phases change. From a simple mass balance, we can derive the *lever rule* for binary systems: (moles of liquid)/(moles of solid) = QR/PR, where QR and PR are the lengths of the line segments. Hence, at 450 °C, a sample with overall composition $X_{Sb} = 0.60$ consists of liquid and solid phases in the molar ratio $(0.80 - 0.60)/(0.60 - 0.37) = 0.87$. Were the composition axis expressed as weight percent, then the lever rule would give the weight ratio of the two phases.

Suppose that a liquid Bi–Sb solution with composition $X_{Sb} = 0.60$ is cooled very slowly from an initial temperature above 526 °C. When the temperature has decreased to the liquidus temperature 526 °C (point A), the first solid appears, with a composition at point B ($X_{Sb} = 0.89$). As the temperature is decreased further, solid continues to precipitate with the compositions of the two phases at any temperature being given by the liquidus and solidus compositions at that temperature and with their relative proportions being given by the lever rule. Solidification is complete at 349 °C, the last liquid to solidify having composition $X_{Sb} = 0.16$ (point C).

The process just described is known as *equilibrium cooling*. At any temperature during equilibrium cooling, the solid phase has a uniform (homogeneous) composition. In the preceding example, the composition of the solid phase during cooling varies along the line BQD. Hence, in order for the solid grains to have a uniform composition at any temperature, diffusion of Sb from the center to the surface of the growing grains must occur. Since solid-state diffusion is a relatively slow process, equilibrium-cooling conditions are only approached if the temperature is decreased very slowly. If a sample of composition $X_{Sb} = 0.60$ is cooled rapidly from the liquid, concentration gradients will be observed in the solid grains, with the concentration of Sb decreasing toward the surface from a maximum of $X_{Sb} = 0.89$ (point B). Furthermore, in this case, solidification will not be complete at 349 °C since at 349 °C, the average concentration of Sb in the solid grains will be greater than $X_{Sb} = 0.60$. These considerations are discussed more fully in Section 3.9 and in Chapter 7 of this volume.

At $X_{Sb} = 0$ and $X_{Sb} = 1$ in **Figure 11**, the liquidus and solidus curves meet at the equilibrium melting points, or *temperatures of fusion* of Bi and Sb, which are $T^o_{f(Bi)} = 271.4$ °C and $T^o_{f(Sb)} = 630.6$ °C. At these two points, the number of components is reduced to 1. Hence, from the Phase Rule the two-phase region becomes *invariant* ($F = 0$) and solid and liquid coexist at the same temperature.

The phase diagram is influenced by the total pressure, P. Unless otherwise stated, T–X diagrams are usually presented for $P = $ constant $= 1.0$ bar. For equilibria involving only solid and liquid phases, the phase boundaries are typically shifted only by the order of a few hundredths of a degree per bar change in P. Hence, the effect of pressure upon the phase diagram is generally negligible unless the pressure is of the order of hundreds of bars. On the other hand, if gaseous phases are involved, the effect of pressure is very important. The effect of pressure will be discussed in Section 3.4.3.

3.3.1.1   *Three-phase Invariants in Binary Temperature-Composition Phase Diagrams*

When three phases are at equilibrium in a binary system at constant pressure, then from Eqn 83, $F = 0$. Hence, the compositions of all three phases as well as T are fixed and the system is said to be *invariant*. An example is the line PQR in **Figure 1** in the T–X diagram of the Fe–Mo system. At any overall composition of the system lying between points P and R at 1612 °C, three phases will be in equilibrium: liquid, sigma and bcc, with compositions at points P, Q and R, respectively. Several other invariants can also be seen in **Figure 1**. Binary T–X diagrams and invariant reactions will be discussed in detail in Section 3.5.

3.3.2   Other Examples of Applications of the Phase Rule

Despite its apparent simplicity, the Phase Rule is frequently misunderstood and incorrectly applied. In this section, a few additional illustrative examples of the application of the Phase Rule will be briefly presented. These examples will be elaborated upon in subsequent sections.

In the single-component Al_2SiO_5 system ($C = 1$) shown in **Figure 7**, the Phase Rule reduces to

$$F = 3 - P \tag{84}$$

Three allotropes of Al_2SiO_5 are shown in the figure. In the single-phase regions, $F = 2$; in these bivariant regions, both P and T (**Figure 7a**) or both V and T (**Figure 7b**) can be specified independently. When two phases are in equilibrium, $F = 1$. Two-phase equilibrium is, therefore, represented by the univariant lines in the P–T phase diagram in **Figure 7a**; when two phases are in equilibrium, T and P cannot be specified independently. When all three allotropes are in equilibrium, $F = 0$. Three-phase equilibrium can, thus, only occur at the invariant *triple point* (530 °C, 3859 bar) in **Figure 7a**.

The invariant triple point in **Figure 7a** becomes the invariant line PQR at $T = 530$ °C in the V–T phase diagram in **Figure 7b**. When the three phases are in equilibrium, the total molar volume of the system can lie anywhere on this line depending upon the relative amounts of the three phases present. However, the molar volumes of the individual allotropes are fixed at points P, Q and R. It is the molar volumes of the individual phases that are the system variables in the sense of the Phase Rule. Similarly, the two-phase univariant lines in **Figure 7a** become the two-phase univariant areas in **Figure 7b**. The lever rule can be applied in these regions to give the ratio of the volumes of the two phases at equilibrium.

As another example, consider the phase diagram section for the ternary ($C = 3$) Zn–Mg–Al system shown in **Figure 3a**. Since the total pressure is constant, one degree of freedom is eliminated. Consequently,

$$F = 4 - P \tag{85}$$

Although the diagram is plotted at constant 10 mol% Zn, this is a composition variable of the entire system, not of any individual phase. That is, it is not a variable in the sense of the Phase Rule, and setting it as a constant does not reduce the number of degrees of freedom. Hence, for example, the three-phase regions in **Figure 3a** are univariant ($F = 1$); at any given T, the compositions of all three phases are fixed, but the three-phase region extends over a range of temperature. Ternary temperature-composition phase diagrams will be discussed in detail in Section 3.6.

Finally, we shall examine some diagrams with chemical potentials as variables. As a first example, the phase diagram of the Fe–Ni–O_2 system in **Figure 5** is a plot of the equilibrium oxygen pressure (fixed, for example, by equilibrating the solid phases with a gas phase of fixed p_{O_2}) versus the molar ratio Ni/(Fe + Ni) at constant T and constant total hydrostatic pressure. For an ideal gas, it follows from Eqn 31 and the definition of chemical potential (Section 3.2.7) that

$$\mu_i = \mu_i^o + RT \ln \, p_i \tag{86}$$

where μ_i^o is the standard-state (at $p_i = 1.0$ bar) chemical potential of the pure gas at temperature T. Hence, if T is constant, μ_i varies directly as $\ln p_i$. The y-axis in **Figure 5** thus varies directly as μ_{O_2}. Fixing T and hydrostatic pressure eliminates two degrees of freedom. Hence, $F = 3 - P$. Therefore, **Figure 5** has the same topology as a binary T–X diagram like **Figure 1**, with single-phase bivariant areas, two-phase univariant areas with tie-lines, and three-phase invariant lines at fixed log p_{O_2} and fixed compositions of the three phases.

As a second example, in the isothermal predominance diagram of the Cu–SO_2–O_2 system in **Figure 9**, the x- and y-axis variables vary as μ_{O_2} and μ_{SO_2}. Since $C = 3$ and T is constant,

$$F = 4 - P \tag{87}$$

The diagram has a topology similar to the P–T phase diagram of a one-component system as in **Figure 7a**. Along the lines of the diagram, two solid phases are in equilibrium and at the triple points, three solid phases coexist. However, a gas phase is also present everywhere. Hence, at the triple points, there are really four phases present at equilibrium, while the lines represent three-phase equilibria. Hence, the diagram is consistent with Eqn 87.

It is instructive to look at **Figure 9** in another way. Suppose that we fix a constant hydrostatic pressure of 1.0 bar (for example, by placing the system in a cylinder fitted with a piston). This removes one further degree of freedom, so that now $F = 3 - P$. However, as long as the total equilibrium gas pressure is <1.0 bar, there will be no gas phase present, and so the diagram is consistent with the Phase Rule. On the other hand, above the 1.0 bar curve shown by the line of crosses in **Figure 9**, the total gas pressure exceeds 1.0 bar. Hence, the phase diagram calculated at a total hydrostatic pressure of 1.0 bar must terminate at this curve.

The Phase Rule may be applied to the four-component isothermal Fe–Cr–S_2–O_2 phase diagram in **Figure 10** in a similar way. Remember that fixing the molar ratio $Cr/(Fe + Cr)$ equal to 0.5 does not eliminate a degree of freedom since this composition variable applies to the entire system, not to any individual phase.

3.4   Thermodynamic Origin of Binary Phase Diagrams

3.4.1   Thermodynamics of Solutions

3.4.1.1   Gibbs Energy of Mixing

Liquid gold and copper are completely miscible at all compositions. The Gibbs energy of 1 mol of liquid solution, g^l at 1100 °C is drawn in **Figure 12** as a function of composition expressed as *mole fraction*, X_{Cu}, of copper. Note that $X_{Au} = 1 - X_{Cu}$. The curve of g^l varies between the standard molar Gibbs energies of pure liquid Au and Cu, g^o_{Au} and g^o_{Cu}. From Eqn 65, it follows that, for a pure component, g^o_i and μ^o_i are identical.

The function Δg^l_m shown in **Figure 12** is called the molar *Gibbs energy of mixing* of the liquid solution. It is defined as

$$\Delta g^l_m = g^l - \left(X_{Au}\mu^o_{Au} + X_{Cu}\mu^o_{Cu}\right) \tag{88}$$

It can be seen that Δg^l_m is the Gibbs energy change associated with the isothermal mixing of X_{Au} moles of pure liquid Au and X_{Cu} moles of pure liquid Cu to form 1 mol of solution according to the reaction:

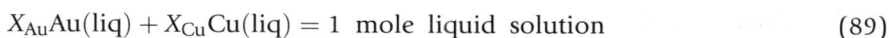

$$X_{Au}Au(liq) + X_{Cu}Cu(liq) = 1 \text{ mole liquid solution} \tag{89}$$

Note that for the solution to be stable, it is necessary that Δg^l_m be negative.

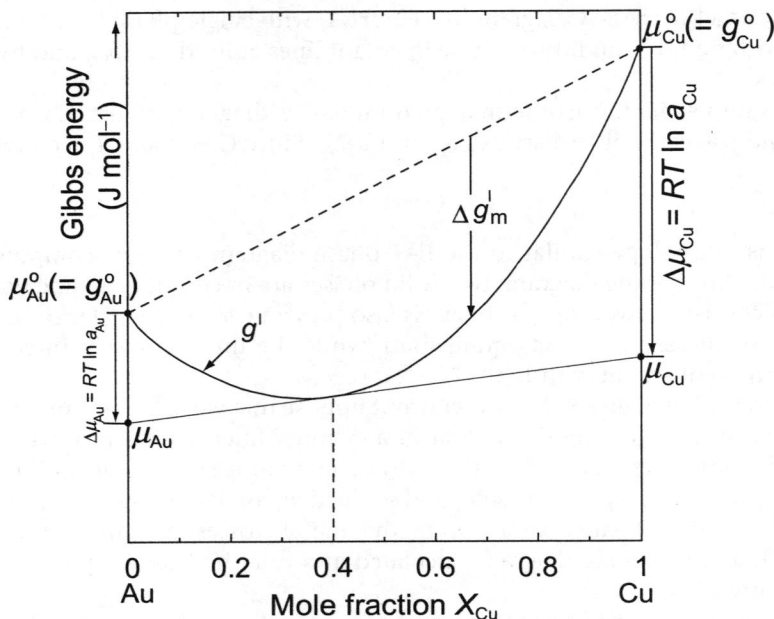

Figure 12 Molar Gibbs energy of liquid Au–Cu alloys at 1100 °C illustrating the tangent construction (see FactSage).

3.4.1.2 Tangent Construction

An important construction is illustrated in **Figure 12**. If a tangent is drawn to the curve of g^l at a certain composition ($X_{Cu} = 0.4$ in **Figure 12**), then the intercepts of this tangent on the axes at $X_{Au} = 1$ and $X_{Cu} = 1$ are equal to μ_{Au} and μ_{Cu}, respectively, at this composition.

To prove this, divide Eqns 64 and 73 by ($n_{Au} + n_{Cu}$) at constant T and P to obtain expressions for the molar Gibbs energy and its derivative:

$$g^l = X_{Au}\mu_{Au} + X_{Cu}\mu_{Cu} \tag{90}$$

and

$$dg^l = \mu_{Au}dX_{Au} + \mu_{Cu}dX_{Cu} \tag{91}$$

Since $dX_{Au} = -dX_{Cu}$, it can be seen that Eqns 90 and 91 are equivalent to the tangent construction shown in **Figure 12**.

These equations may also be rearranged to give the following useful expression for a binary system:

$$\mu_i = g + (1 - X_i)dg/dX_i \tag{92}$$

3.4.1.3 Relative Partial Properties

The difference between the chemical potential μ_i (also called the partial Gibbs energy g_i) of a component in solution and the chemical potential μ_i^o (or g_i^o) of the same component in a *standard state* is called the *relative chemical potential* (or *relative partial Gibbs energy*), $\Delta\mu_i$ (or Δg_i). It is usual to choose, as standard state, the pure component in the same phase at the same temperature. *The activity a_i of the component relative to the chosen standard state* is then defined in terms of $\Delta\mu_i$ by the following equation, as illustrated in **Figure 12**.

$$\Delta\mu_i = \mu_i - \mu_i^{\circ} = RT \ln a_i \tag{93}$$

From **Figure 12**, it can be seen that

$$\Delta g_m = X_{Au}\Delta\mu_{Au} + X_{Cu}\Delta\mu_{Cu} = RT(X_{Au}\ln a_{Au} + X_{Cu}\ln a_{Cu}) \tag{94}$$

The isothermal Gibbs energy of mixing can be divided into enthalpy and entropy terms, as can be the relative chemical potentials:

$$\Delta g_m = \Delta h_m - T\Delta s_m \tag{95}$$

$$\Delta\mu_i = \Delta h_i - T\Delta s_i \tag{96}$$

where Δh_m and Δs_m are the enthalpy and entropy changes associated with the isothermal mixing reaction shown in Eqn 89. It follows from Eqns 94–96 that

$$\Delta h_m = X_{Au}\Delta h_{Au} + X_{Cu}\Delta h_{Cu} \tag{97}$$

$$\Delta s_m = X_{Au}\Delta s_{Au} + X_{Cu}\Delta s_{Cu} \tag{98}$$

and so, tangent constructions similar to that of **Figure 12** can be used to relate the *relative partial enthalpies and entropies* Δh_i and Δs_i to the *integral molar enthalpy of mixing* Δh_m and *integral molar entropy of mixing* Δs_m, respectively.

3.4.1.4 Activity

The activity of a component in a solution was defined in Eqn 93. Since a_i varies monotonically with μ_i, it follows that *when two or more phases are in equilibrium, the activity of any component is the same in all phases*, provided that the activity in every phase is expressed with respect to the same standard state.

The use of activities in calculations of chemical equilibrium conditions is illustrated by the following example. A liquid solution of Au and Cu at $1100\,^{\circ}\mathrm{C}$ with $X_{Cu} = 0.6$ is exposed to an atmosphere, in which the oxygen partial pressure is $p_{O_2} = 10^{-3}$ bar. Will Cu_2O be formed? The reaction is

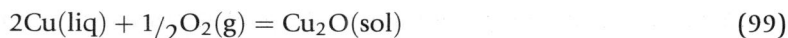

$$2Cu(\mathrm{liq}) + 1/2\,O_2(\mathrm{g}) = Cu_2O(\mathrm{sol}) \tag{99}$$

where the Cu(liq) is in solution. If the reaction proceeds with the formation of dn moles of Cu_2O, then $2dn$ moles of Cu are consumed, and the Gibbs energy of the Au–Cu solution changes by $-2(dG^l/dn_{Cu})$ dn. The total Gibbs energy then varies as

$$dG/dn = \mu^o_{Cu_2O} - 1/2\,\mu_{O_2} - 2(dG^l/dn_{Cu})$$

$$= \mu^o_{Cu_2O} - 1/2\,\mu_{O_2} - 2\mu_{Cu}$$

$$= (\mu^o_{Cu_2O} - 1/2\,\mu^o_{O_2} - 2\mu^o_{Cu}) - 1/2 RT \ln p_{O_2} - 2\,RT \ln a_{Cu}$$

$$= \Delta G^o + RT \ln\left(p^{-1/2}_{O_2}\,a^{-2}_{Cu}\right) = \Delta G \tag{100}$$

where Eqns 65,86 and 93 have been used.

For reaction shown in Eqn 99 at 1100 °C, $\Delta G^o = -68.73$ kJ (from FactSage—see list of websites at the end of this chapter). The activity of Cu in the liquid alloy at $X_{Cu} = 0.6$ is $a_{Cu} = 0.385$ (from FactSage). Substitution into Eqn 100 with $p_{O_2} = 10^{-3}$ bar gives

$$dG/dn = \Delta G = -7.50 \text{ kJ} \tag{101}$$

Hence, under these conditions, the reaction entails a decrease in the total Gibbs energy and so the copper will be oxidized.

3.4.1.5 Ideal Raoultian Solutions

An *ideal solution* or *Raoultian solution* is usually defined as one in which the activity of a component is equal to its mole fraction:

$$a^{ideal}_i = X_i \tag{102}$$

(With a judicious choice of standard state, this definition can also encompass ideal Henrian solutions, as discussed in Section 3.5.8.)

This Raoultian definition of ideality is generally only applicable to simple substitutional solutions. There are more useful definitions for other types of solutions such as interstitial solutions, ionic solutions, solutions of defects, polymer solutions, etc. That is, the most convenient definition of ideality depends upon the solution model. Solution models are discussed in Section 3.8.2. In the present section, Eqn 102 for an ideal substitutional solution will be developed with the Au–Cu solution as example.

In the ideal substitutional solution model, it is assumed that Au and Cu atoms are nearly alike, with nearly identical radii and electronic structures. This being the case, there will be no change in bonding energy or volume upon mixing, so that the enthalpy of mixing is zero:

$$\Delta h^{ideal}_m = 0 \tag{103}$$

Furthermore, and for the same reason, the Au and Cu atoms will be randomly distributed over the lattice sites. (In the case of a liquid solution, we can think of the lattice sites as the instantaneous atomic positions.) For a random distribution of N_{Au} gold atoms and N_{Cu} copper atoms over $(N_{Au} + N_{Cu})$ sites, Boltzmann's equation (Eqn 2) can be used to calculate the *configurational entropy* of the solution. This is the entropy associated with the spatial distribution of the particles:

$$S^{config} = k_B \ln(N_{Au} + N_{Cu})!/N_{Au}!\,N_{Cu}! \tag{104}$$

The configurational entropies of pure Au and Cu are zero. Hence, the configurational entropy of mixing, ΔS_m^{config}, will be equal to S^{config}. Furthermore, because of the assumed close similarity of Au and Cu, there will be no nonconfigurational contribution to the entropy of mixing. Hence, the entropy of mixing will be equal to S^{config}. Applying Stirling's approximation, which states that $\ln N! \approx [(N \ln N)-N]$ if N is large, yields

$$\Delta S_m^{ideal} = S^{config} = -k_B \left[N_{Au} \ln \frac{N_{Au}}{N_{Au} + N_{Cu}} + N_{Cu} \ln \frac{N_{Cu}}{N_{Au} + N_{Cu}} \right] \tag{105}$$

For 1 mol of solution, $(N_{Au} + N_{Cu}) = N^o$ where N^o = Avogadro's number. We also note that $(k_B N^o)$ is equal to the ideal gas constant R. Hence,

$$\Delta s_m^{ideal} = -R(X_{Au} \ln X_{Au} + X_{Cu} \ln X_{Cu}) \tag{106}$$

Therefore, since the ideal enthalpy of mixing is zero,

$$\Delta g_m^{ideal} = RT(X_{Au} \ln X_{Au} + X_{Cu} \ln X_{Cu}) \tag{107}$$

By comparing Eqns 94 and 107, we obtain

$$\Delta \mu_i^{ideal} = RT \ln a_i^{ideal} = RT \ln X_i \tag{108}$$

Hence, Eqn 102 has been demonstrated for an ideal substitutional solution.

3.4.1.6 Excess Properties
In reality, Au and Cu atoms are not identical, and so, Au–Cu solutions are not perfectly ideal. The difference between a solution property and its value in an ideal solution is called an *excess property*. The *excess Gibbs energy*, for example, is defined as

$$g^E = \Delta g_m - \Delta g_m^{ideal} \tag{109}$$

Since the ideal enthalpy of mixing is zero, the excess enthalpy is equal to the enthalpy of mixing:

$$h^E = \Delta h_m - \Delta h_m^{ideal} = \Delta h_m \tag{110}$$

Hence,

$$g^E = h^E - Ts^E = \Delta h_m - Ts^E \tag{111}$$

Excess partial properties are defined similarly:

$$\mu_i^E = \Delta \mu_i - \Delta \mu_i^{ideal} = RT \ln a_i - RT \ln X_i \tag{112}$$

$$s_i^E = \Delta s_i - \Delta s_i^{ideal} = \Delta s_i + R \ln X_i \tag{113}$$

(where μ_i^E is equivalent to g_i^E). Also,

$$\mu_i^E = h_i^E - Ts_i^E = \Delta h_i - Ts_i^E \tag{114}$$

Equations analogous to Eqns 94, 97 and 98 relate the integral and partial excess properties. For example, in Au–Cu solutions,

$$g^E = X_{Au} \; \mu_{Au}^E + X_{Cu} \; \mu_{Cu}^E \quad \text{and} \quad s^E = X_{Au} \, s_{Au}^E + X_{Cu} \, s_{Cu}^E \tag{115}$$

Tangent constructions similar to that of **Figure 12** can thus also be employed for excess properties, and an equation analogous to Eqn 92 can be written as

$$\mu_i^E = g^E + (1 - X_i) dg^E / dX_i \tag{116}$$

The Gibbs–Duhem Eqn 75 also applies to excess properties. At constant T and P,

$$X_{Au} \; d\mu_{Au}^E + X_{Cu} \; d\mu_{Cu}^E = 0 \tag{117}$$

In Au–Cu alloys, g^E is negative. That is, Δg_m is more negative than Δg_m^{ideal} and so, the solution is thermodynamically more stable than an ideal solution. We thus say that Au–Cu solutions exhibit *negative deviations from ideality*. If $g^E > 0$, then a solution is less stable than an ideal solution and is said to exhibit *positive deviations*.

3.4.1.7 Activity Coefficients

The *activity coefficient* of a component in a solution is defined as

$$\gamma_i = a_i / X_i \tag{118}$$

Hence, from Eqn 112,

$$\mu_i^E = RT \ln \gamma_i \tag{119}$$

In an ideal solution, $\gamma_i = 1$ and $\mu_i^E = 0$ for all components. If $\gamma_i < 1$, then $\mu_i^E < 0$ and from Eqn 112, $\Delta \mu_i < \Delta \mu_i^{ideal}$. That is, the component i is more stable in the solution than it would be in an ideal solution of the same composition. If $\gamma_i > 1$, then $\mu_i^E > 0$ and the driving force for the component to enter into solution is less than in the case of an ideal solution.

3.4.1.8 Multicomponent Solutions

The equations of this section were derived with a binary solution as an example. However, the equations apply equally to systems of any number of components. For instance, in a solution of components A–B–C–D..., Eqn 94 becomes

$$\Delta g_m = X_A \Delta \mu_A + X_B \Delta \mu_B + X_C \Delta \mu_C + \dots \tag{120}$$

3.4.2 Binary Temperature-Composition Phase Diagrams

In this section, we first consider the thermodynamic origin of simple "lens-shaped" phase diagrams in binary systems with complete liquid and solid miscibility. An example of such a diagram was given in **Figure 11**. Another example is the Ge–Si phase diagram in the lowest panel of **Figure 13**. In the upper

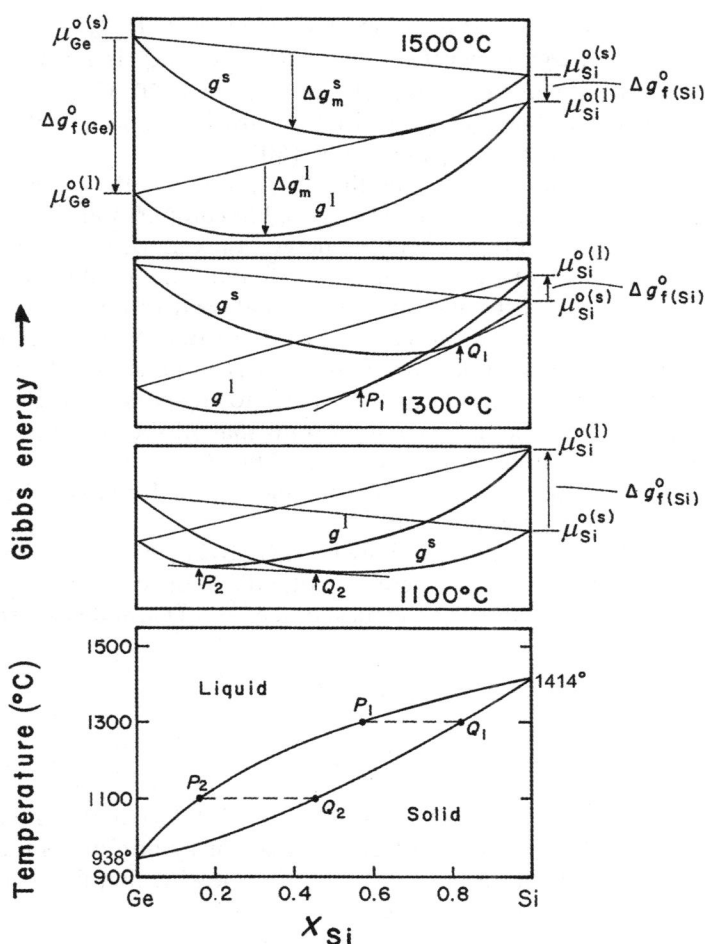

Figure 13 Ge–Si phase diagram at $P = 1$ bar (after Hansen, 1958) and Gibbs energy curves at three temperatures, illustrating the common tangent construction.

three panels of **Figure 13**, the molar Gibbs energies of the solid and liquid phases, g^s and g^l at three temperatures are shown to scale. As illustrated in the top panel, g^s varies with composition between the standard chemical potentials of pure solid Ge and of pure solid Si, $\mu_{Ge}^{o(s)}$ and $\mu_{Si}^{o(s)}$, while g^l varies between the standard chemical potentials of the pure liquid components, $\mu_{Ge}^{o(l)}$ and $\mu_{Si}^{o(l)}$. The difference between $\mu_{Ge}^{o(l)}$ and $\mu_{Ge}^{o(s)}$ is equal to the standard molar Gibbs energy of fusion (melting) of pure Ge, $\Delta g_{f\,(Ge)}^o = (\mu_{Ge}^{o(l)} - \mu_{Ge}^{o(s)})$. Similarly for Si, $\Delta g_{f\,(Si)}^o = (\mu_{Si}^{o(l)} - \mu_{Si}^{o(s)})$. The Gibbs energy of fusion of a pure component may be written as

$$\Delta g_f^o = \Delta h_f^o - T \Delta s_f^o \tag{121}$$

where Δh_f^o and Δs_f^o are the standard molar enthalpy and entropy of fusion.

Since, to a first approximation, Δh_f^o and Δs_f^o are independent of T, Δg_f^o is approximately a linear function of T. If $T > T_f^o$, then Δg_f^o is negative. If $T < T_f^o$, then Δg_f^o is positive. Hence, as seen in **Figure 13**, as T decreases, the g^s curve descends relative to g^l. At 1500 °C, $g^l < g^s$ at all compositions. Therefore, by the principle that a system always seeks the state of minimum Gibbs energy at constant T and P, the liquid phase is stable at all compositions at 1500 °C.

At 1300 °C, the curves of g^s and g^l cross. The line P_1Q_1, which is the *common tangent* to the two curves, divides the composition range into three sections. For compositions between pure Ge and P_1, a single-phase liquid is the state of minimum Gibbs energy. For compositions between Q_1 and pure Si, a single-phase solid solution is the stable state. Between P_1 and Q_1, a total Gibbs energy lying on the tangent line P_1Q_1 may be realized if the system adopts a state consisting of two phases with compositions at P_1 and Q_1 with relative proportions given by the lever rule (Section 3.3.1). Since the tangent line P_1Q_1 lies below both g^s and g^l, this two-phase state is more stable than either phase alone. Furthermore, no other line joining any point on the g^l curve to any point on the g^s curve lies below the line P_1Q_1. Hence, this line represents the true equilibrium state of the system, and the compositions P_1 and Q_1 are the liquidus and solidus compositions at 1300 °C.

As T is decreased to 1100 °C, the points of common tangency are displaced to higher concentrations of Ge. For $T < 938$ °C, $g^s < g^l$ at all compositions.

It was shown in **Figure 12** that if a tangent is drawn to a Gibbs energy curve, then the intercept of this tangent on the axis at $X_i = 1$ is equal to the chemical potential μ_i of component i. The *common tangent construction* of **Figure 13** thus ensures that the chemical potentials of Ge and Si are equal in the solid and liquid phases at equilibrium. That is,

$$\mu_{Ge}^l = \mu_{Ge}^s \tag{122}$$

$$\mu_{Si}^l = \mu_{Si}^s \tag{123}$$

This equality of chemical potentials was shown in Section 3.2.7 to be the criterion for phase equilibrium. That is, the common tangent construction simultaneously minimizes the total Gibbs energy and ensures the equality of the chemical potentials, thereby showing that these are equivalent criteria for equilibrium between phases as was discussed in Section 3.2.7.

If we rearrange Eqn 122, subtracting the Gibbs energy of fusion of pure Ge, $\Delta g_{f\,(Ge)}^o = \left(\mu_{Ge}^{o(l)} - \mu_{Ge}^{o(s)} \right)$, from each side, we obtain

$$\left(\mu_{Ge}^l - \mu_{Ge}^{o(l)} \right) - \left(\mu_{Ge}^s - \mu_{Ge}^{o(s)} \right) = -\left(\mu_{Ge}^{o(l)} - \mu_{Ge}^{o(s)} \right) \tag{124}$$

Using Eqn 93, we can write Eqn 124 as

$$\Delta \mu_{Ge}^l - \Delta \mu_{Ge}^s = -\Delta g_{f(Ge)}^o \tag{125}$$

or

$$RT \ln a_{Ge}^l - RT \ln a_{Ge}^s = -\Delta g_{f(Ge)}^o \tag{126}$$

where a_{Ge}^l is the activity of Ge (with respect to pure liquid Ge as standard state) in the liquid solution on the liquidus, and a_{Ge}^s is the activity of Ge (with respect to pure solid Ge as standard state) in the solid solution on the solidus. Starting with Eqn 123, we can derive a similar expression for the other component:

$$RT \ln a_{Si}^l - RT \ln a_{Si}^s = -\Delta g_{f(Si)}^o \tag{127}$$

Equations 126 and 127 are equivalent to the common tangent construction.

It should be noted that absolute values of chemical potentials cannot be defined. Hence, the relative positions of $\mu_{Ge}^{o(l)}$ and $\mu_{Si}^{o(l)}$ in **Figure 13** are arbitrary. However, this is immaterial for the preceding discussion since displacing both $\mu_{Si}^{o(l)}$ and $\mu_{Si}^{o(s)}$ by the same arbitrary amount relative to $\mu_{Ge}^{o(l)}$ and $\mu_{Ge}^{o(s)}$ will not alter the compositions of the points of common tangency.

The shape of the two-phase (solid + liquid) "lens" on the phase diagram is determined by the Gibbs energies of fusion, Δg_f^o, of the components and by the mixing terms, Δg^s and Δg^l. In order to observe how the shape is influenced by varying Δg_f^o, let us consider a hypothetical system A–B in which Δg^s and Δg^l are ideal Raoultian (Eqn 107). Let $T_{f(A)}^o = 800$ K and $T_{f(B)}^o = 1200$ K. Furthermore, assume that the entropies of fusion of A and B are equal and temperature-independent. The enthalpies of fusion are then given from Eqn 121 by the expression $\Delta h_f^o = T_f^o \Delta s_f^o$ since $\Delta g_f^o = 0$ when $T = T_f^o$ (Section 3.2.3). Calculated phase diagrams for $\Delta s_f^o = 3,\ 10$ and 30 J mol^{-1} K^{-1} are shown in **Figure 14**. A value of $\Delta s_f^o \approx 10$ J mol^{-1} K^{-1} is typical of most metals. When the components are ionic compounds such as ionic oxides, halides, etc., Δs_f^o can be significantly larger since there are several ions per formula unit. Hence, two-phase "lenses" in binary ionic salt or oxide phase diagrams tend to be wider than those encountered in alloy systems. If we are considering vapor–liquid equilibria rather than solid–liquid equilibria, then the shape is determined by the entropy of vaporization, Δs_v^o. Since Δs_v^o is usually an order of magnitude larger than Δs_f^o, two-phase (liquid + vapor) lenses tend to be very wide. For equilibria between two solid solutions of different crystal structure, the shape is determined by the entropy of solid–solid transformation, which is usually smaller than the entropy of fusion by approximately an order of magnitude. Therefore, two-phase (solid + solid) lenses tend to be narrow.

3.4.2.1 Effect of Grain Size and Strain Energy

The foregoing phase diagram calculations have assumed that the grain size of the solid phases is sufficiently large, so that surface (interfacial) energy contributions to the Gibbs energy can be neglected. It has also been assumed that strain energy is negligible. For very fine grain sizes in the submicron range, however, the interfacial energy can be appreciable and can have an influence on the phase diagram. This effect can be important with the development of materials with grain sizes in the nanometer range.

Interfacial and strain energies increase the Gibbs energy of a phase, and hence, decrease its thermodynamic stability. In particular, the liquidus temperature of a fine-grained solid will be lowered. If a suitable estimate can be made for the grain boundary or strain energies, then this can simply be added to the Gibbs energy expression of the phase for purposes of calculating phase diagrams.

3.4.3 Binary Pressure-Composition Phase Diagrams

Let us consider liquid–vapor equilibrium in a system with complete liquid miscibility, using, as example, the Zn–Mg system. Curves of g^v and g^l can be drawn at any given T and P, as in the upper panel

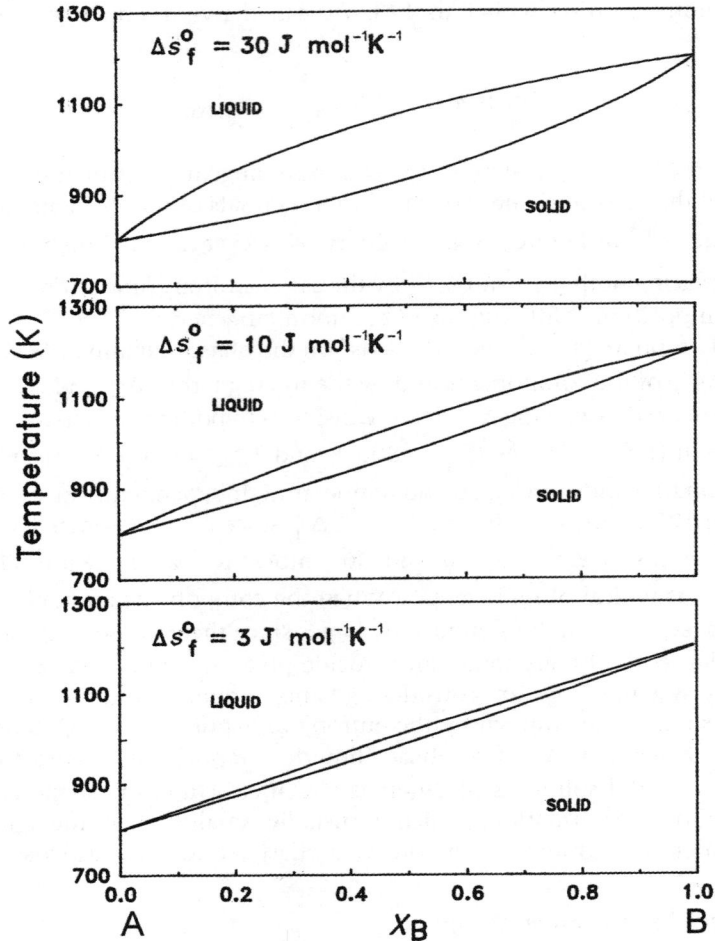

Figure 14 Phase diagram of a system A–B with ideal solid and liquid solutions. Melting points of A and B are 800 and 1200 K, respectively. Diagrams calculated for entropies of fusion $\Delta s^o_{f(A)} = \Delta s^o_{f(B)} = 3$, 10 and 30 J mol^{-1} K^{-1}.

of **Figure 15**, and the common tangent construction then gives the equilibrium vapor and liquid compositions. The phase diagram depends upon the Gibbs energies of vaporization of the pure components $\Delta g^o_{v\,(Zn)} = \left(\mu^v_{Zn} - \mu^{o(l)}_{Zn} \right)$ and $\Delta g^o_{v\,(Mg)} = \left(\mu^v_{Mg} - \mu^{o(l)}_{Mg} \right)$ as shown in **Figure 15**.

To generate the isothermal pressure-composition (P–X) phase diagram in the lower panel of **Figure 15**, we require the Gibbs energies of vaporization as functions of P. Assuming monatomic ideal vapors and assuming that pressure has negligible effect upon the Gibbs energy of the liquid, we can write, using Eqn 86:

$$\Delta g_{v(i)} = \Delta g^o_{v(i)} + RT \ln P \qquad (128)$$

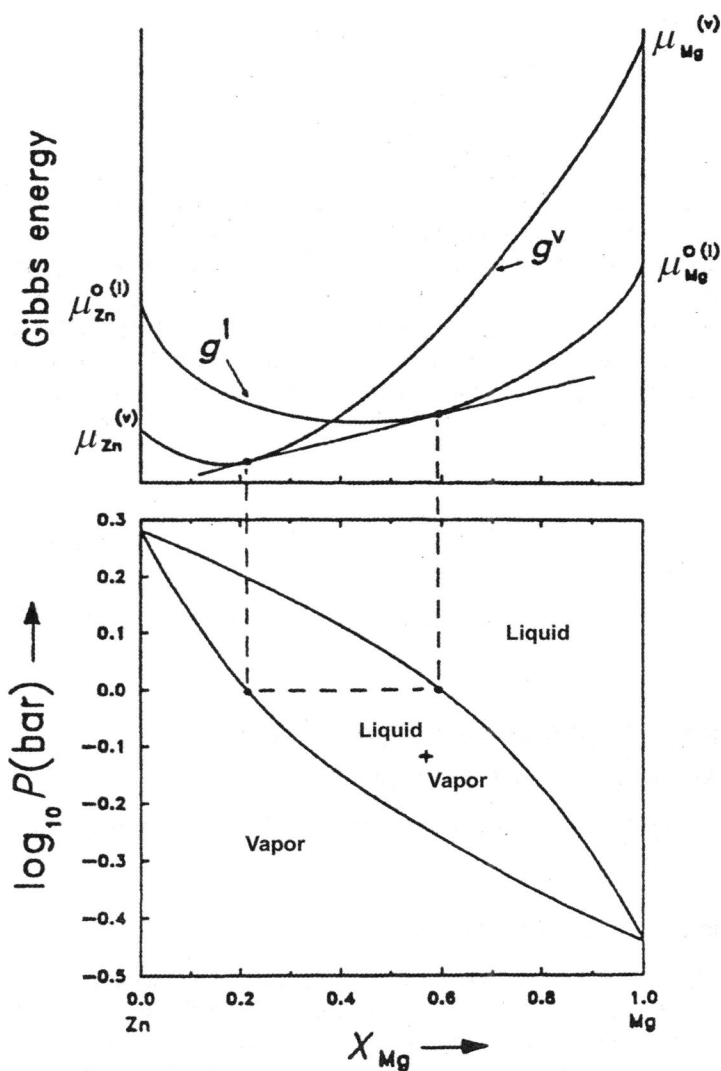

Figure 15 Pressure-composition phase diagram of the Zn–Mg system at 977 °C calculated for ideal vapor and liquid solutions. Upper panel illustrates common tangent construction at a constant pressure.

where $\Delta g^o_{v(i)}$ is the standard Gibbs energy of vaporization (when $P = 1.0$ bar), which is given by

$$\Delta g^o_{v(i)} = \Delta h^o_{v(i)} - T\Delta s^o_{v(i)} \tag{129}$$

For example, the standard enthalpy of vaporization of Zn is 115,310 J mol^{-1} at its normal boiling point of 1180 K (see FactSage). Assuming that $\Delta h^o_{v(Zn)}$ is independent of T, we calculate from Eqn 129 that $\Delta s^o_{v(Zn)} = 115310/1180 = 97.71$ J mol^{-1} K^{-1}. From Eqn 128, $\Delta g_{v(Zn)}$ at any T and P is thus given by

$$\Delta g_{v(Zn)} = (115310 - 97.71\ T) + RT \ln P \tag{130}$$

A similar expression can be derived for the other component, Mg.

At constant temperature then, the curve of g^v in **Figure 15** descends relative to g^l as the pressure is lowered, and the P–X phase diagram is generated by the common tangent construction. The diagram at 977 °C in **Figure 15** was calculated under the assumption of ideal liquid and vapor mixing ($g^{E(l)} = 0$, $g^{E(v)} = 0$).

P–X phase diagrams involving liquid–solid and solid–solid equilibria can be calculated in a similar manner through the following general equation, which follows from Eqn 64 and which gives the effect of pressure upon the Gibbs energy change for the transformation of 1 mol of pure component i from an α-phase to a β-phase at constant T:

$$\Delta g_{(i)\alpha \to \beta} = \Delta g^o_{(i)\alpha \to \beta} + \int_{P=1}^{P} \left(v^\beta_i - v^\alpha_i \right) dP \tag{131}$$

where $\Delta g^o_{(i)\alpha \to \beta}$ is the standard ($P = 1.0$ bar) Gibbs energy of transformation, and v^β_i and v^α_i are the molar volumes.

3.5 Binary Temperature-Composition Phase Diagrams

3.5.1 Systems with Complete Solid and Liquid Miscibility

Examples of phase diagrams of such systems were given in **Figures 11 and 13** and discussed in Sections 3.3.1 and 3.4.2.

3.5.2 Minima and Maxima in Two-Phase Regions

As discussed in Section 3.4.1.6, the Gibbs energy of mixing Δg_m may be expressed as the sum of an ideal term Δg_m^{ideal} and an excess term g^E. As shown in Section 3.4.2, if the solid and liquid phases are both ideal, or close to ideal, a "lens-shaped" two-phase region results. However, in most systems, even approximately ideal behavior is the exception rather than the rule.

Curves of g^s and g^l for a hypothetical system A–B are shown schematically in **Figure 16** at a constant temperature (below the melting points of pure A and B) such that the solid state is the stable state for both pure components. However, in this system, $g^{E(l)} < g^{E(s)}$ so that g^s presents a flatter curve than does g^l and there exists a central composition region in which $g^l < g^s$. Hence, there are two common tangent lines, P_1Q_1 and P_2Q_2. Such a situation gives rise to a phase diagram with a minimum in the two-phase region as observed in the quasibinary Na_2CO_3–K_2CO_3 system shown in **Figure 17**. At a composition and temperature corresponding to the minimum point, liquid and solid of the same composition exist in equilibrium.

A two-phase region with a minimum point as shown in **Figure 17** may be thought of as a two-phase "lens" which has been "pushed down" by virtue of the fact that the liquid is relatively more stable than the solid. Thermodynamically, this relative stability is expressed as $g^{E(l)} < g^{E(s)}$.

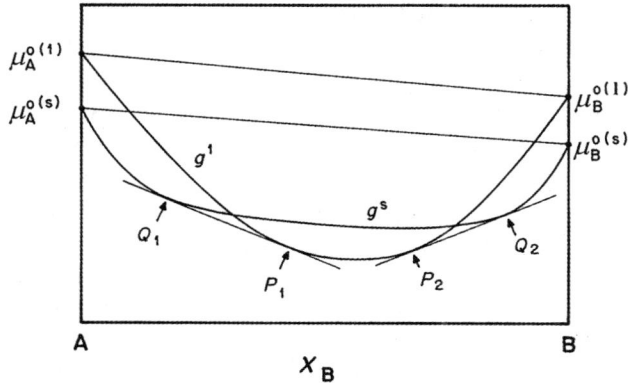

Figure 16 Isothermal Gibbs energy curves for solid and liquid phases in a system A–B, in which $g^{E(l)} < g^{E(s)}$. A phase diagram of the type of **Figure 17** results.

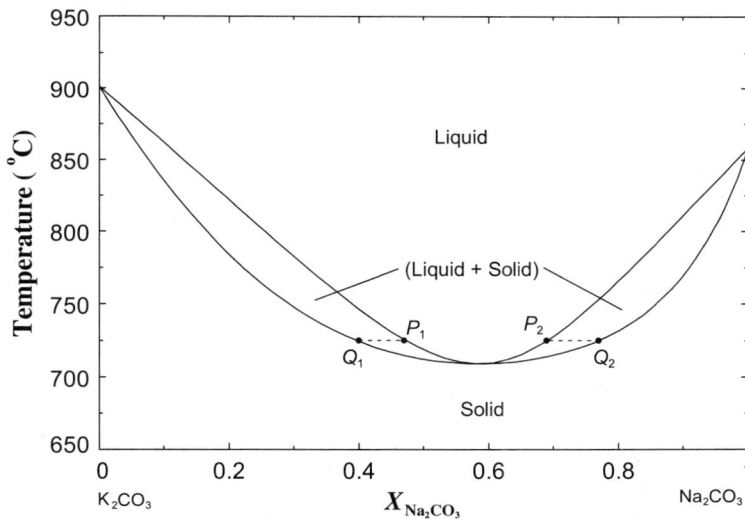

Figure 17 Temperature-composition phase diagram of the K_2CO_3–Na_2CO_3 system at $P = 1$ bar (see FactSage).

Conversely, if $g^{E(l)} > g^{E(s)}$ to a sufficient extent, then a two-phase region with a maximum will result. Such maxima in (liquid + solid) or (solid + solid) two-phase regions are nearly always associated with the existence of an intermediate phase, as will be discussed in Section 3.5.7.

3.5.3 Miscibility Gaps

If $g^E > 0$, then the solution is thermodynamically less stable than an ideal solution. This can result from a large difference in size of the component atoms, ions or molecules; or from differences in bond types, valencies, electronic structure; or from many other factors.

In the Au–Ni system, g^E is positive in the solid phase. In the top panel of **Figure 18**, $g^{E(s)}$ is plotted at 1200 K (Hultgren et al., 1973) and the ideal Gibbs energy of mixing, Δg_m^{ideal}, is also plotted at 1200 K. The sum of these two terms is the Gibbs energy of mixing of the solid solution, Δg_m^s, which is plotted at 1200 K as well as at other temperatures in the central panel of **Figure 18**. Now, from Eqn 107, Δg_m^{ideal} is always negative and varies directly with T, whereas g^E generally varies less rapidly with temperature. As a result, the sum $\Delta g_m^s = \Delta g_m^{ideal} + g^E$ becomes less negative as T decreases. However, the limiting slopes to the Δg_m^{ideal} curve at $X_{Au} = 1$ and $X_{Ni} = 1$ are both infinite, whereas the limiting slopes of g^E are always finite (Henry's Law). Hence, Δg_m^s will always be negative as $X_{Au} \to 1$ and $X_{Ni} \to 1$, no matter how low the temperature. As a result, below a certain temperature, the curve of Δg_m^s will exhibit two negative "humps." Common tangent lines P_1Q_1, P_2Q_2, P_3Q_3 to the two humps at different temperatures define the ends of tie-lines of a two-phase solid–solid *miscibility gap* in the Au–Ni phase diagram, which is shown in the lower panel in **Figure 18** (Hultgren et al., 1973). The peak of the gap occurs at the *critical* or *consolute* temperature and composition, T_c and X_c.

An example of a miscibility gap in a liquid phase will be given in **Figure 21**.

When $g^{E(s)}$ is positive for the solid phase in a system, it is usually also the case that $g^{E(l)} < g^{(s)}$ since the unfavorable factors (such as a difference in atomic dimensions), which cause $g^{E(s)}$ to be positive, will have less of an effect upon $g^{E(l)}$ in the liquid phase owing to the greater flexibility of the liquid structure to accommodate different atomic sizes, valencies, etc. Hence, a solid–solid miscibility gap is often associated with a minimum in the two-phase (solid + liquid) region, as is the case in the Au–Ni system.

Below the critical temperature, the curve of Δg_m^s exhibits two inflection points, indicated by the letter s in **Figure 18**. These are known as the *spinodal points*. On the phase diagram, their locus traces out the *spinodal curve*. The spinodal curve is not part of the equilibrium phase diagram, but it is important in the kinetics of phase separation, as discussed in Chapter 8 of this volume.

3.5.4 Simple Eutectic Systems

The more positive g^E is in a system, the higher is T_c and the wider is the miscibility gap at any temperature. Suppose that $g^{E(s)}$ is so positive that T_c is higher than the minimum in the (solid + liquid) region. The result will be a phase diagram such as that of the Ag–Cu system shown in **Figure 19**. The upper panel of **Figure 19** shows the Gibbs energy curves at 850 °C. The two common tangents define two two-phase regions. As the temperature is decreased below 850 °C, the g^s curve descends relative to g^l and the two points of tangency P_1 and P_2 approach each other until, at $T = 780$ °C, P_1 and P_2 become coincident at the composition E. That is, at $T = 780$ °C, there is just one common tangent line contacting the two portions of the g^s curve at compositions A and B and contacting the g^l curve at E. This temperature is known as the *eutectic temperature*, T_E, and the composition E is the *eutectic composition*. For temperatures below T_E, g^l lies completely above the common tangent to the two portions of the g^s curve and so, for $T < T_E$, a solid–solid miscibility gap is observed. The phase boundaries of this two-phase region are called the *solvus* lines. The word eutectic is from the Greek for "to melt well," since the eutectic composition E is a minimum melting temperature.

This description of the thermodynamic origin of simple eutectic phase diagrams is strictly correct only if the pure solid components A and B have the same crystal structure. Otherwise, a curve for g^s, which is continuous at all compositions, cannot be drawn. In this case, each terminal solid solution (α and β) must have its own Gibbs energy curve (g^α and g^β), and these curves do not join up with each other. However, a simple eutectic phase diagram as shown in **Figure 19** still results.

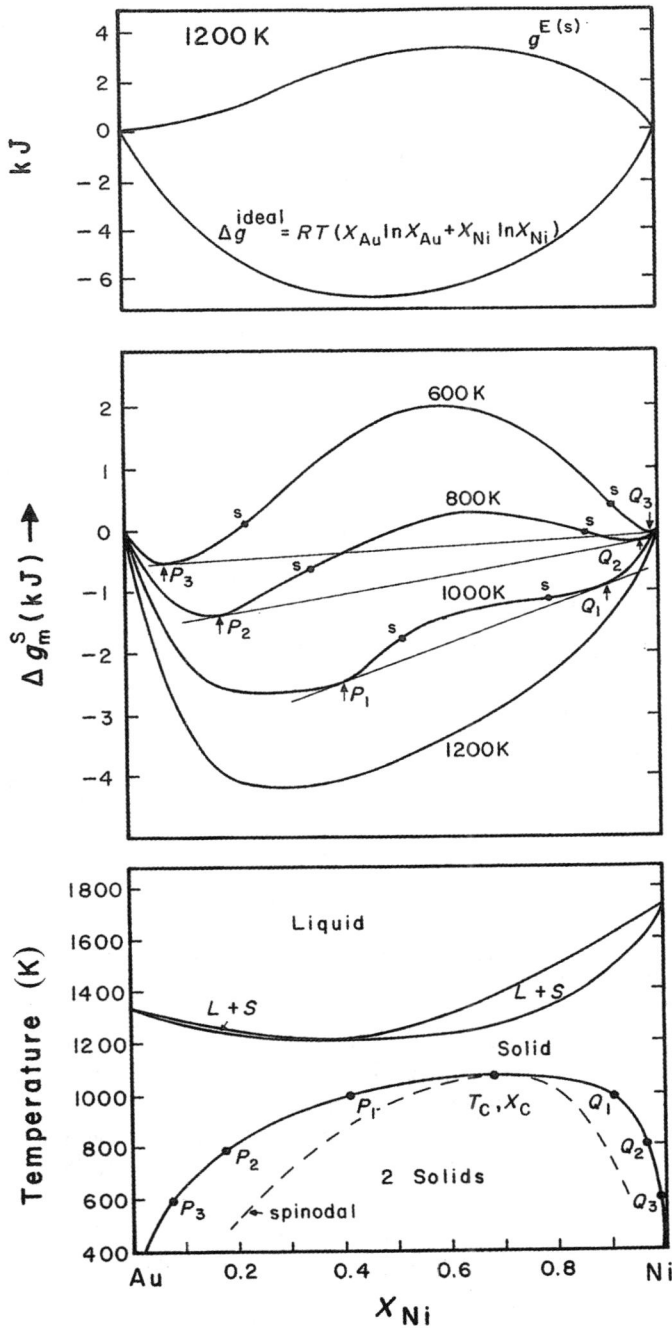

Figure 18 Phase diagram (after Hultgren et al., 1973) and Gibbs energy curves of solid solutions for the Au–Ni system at $P = 1$ bar. Letter s indicates spinodal points.

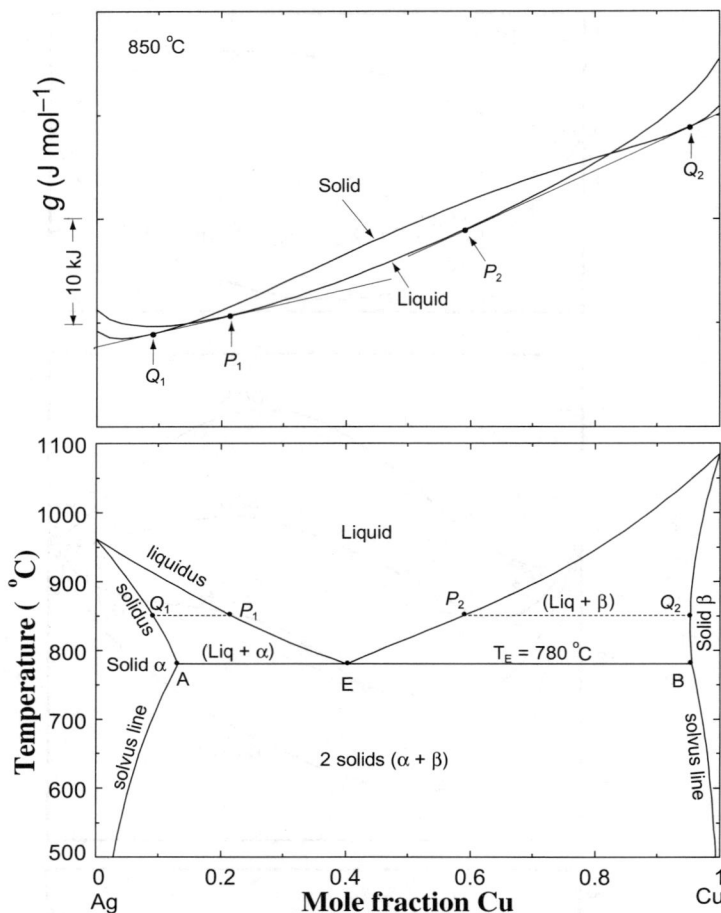

Figure 19   Temperature-composition phase diagram at $P = 1$ bar and Gibbs energy curves at 850 °C for the Ag–Cu system. Solid Ag and Cu have the same (fcc) crystal structure (see FactSage; SGTE).

Suppose a liquid Ag–Cu solution of composition $X_{Cu} = 0.22$ (composition P_1) is cooled from the liquid state very slowly under equilibrium conditions. At 850 °C, the first solid appears with composition Q_1. As T decreases further, solidification continues with the liquid composition following the liquidus curve from P_1 to E and the composition of the solid phase following the solidus curve from Q_1 to A. The relative proportions of the two phases at any T are given by the lever rule (Section 3.3.1). At a temperature just above T_E, two phases are observed: a solid of composition A and a liquid of composition E. At a temperature just below T_E, two solids with compositions A and B are observed. Therefore, at T_E, during cooling, the following *binary eutectic reaction* occurs:

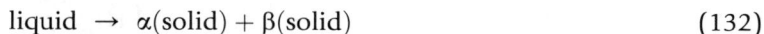

$$\text{liquid} \;\rightarrow\; \alpha(\text{solid}) + \beta(\text{solid}) \tag{132}$$

Under equilibrium conditions, the temperature will remain constant at $T = T_E$ until all the liquid has solidified, and during the reaction, the compositions of the three phases will remain fixed at A, B and E.

For this reason, the eutectic reaction is called an *invariant* reaction as was discussed in Section 3.3.1.1. More on eutectic solidification may be found in Chapter 7 of this volume.

3.5.5   Regular Solution Theory

Many years ago, Van Laar (1908) showed that the thermodynamic origin of a great many of the observed features of binary phase diagrams can be illustrated at least qualitatively by simple regular solution theory. A simple *regular solution* of components A and B is one for which,

$$g^{\mathrm{E}} = X_A X_B (\omega - \eta T) \tag{133}$$

where ω and η are parameters independent of temperature and composition. Substituting Eqn 133 into Eqn 116 yields, for the partial properties:

$$\mu_A^{\mathrm{E}} = X_B^2 (\omega - \eta T), \quad \mu_B^{\mathrm{E}} = X_A^2 (\omega - \eta T) \tag{134}$$

Several liquid and solid solutions conform approximately to regular solution behavior, particularly if g^{E} is small. Examples may be found for alloys, molecular solutions, and ionic solutions such as molten salts and oxides, among others. (The very low values of g^{E} observed for gaseous solutions generally conform very closely to Eqn 133.)

To understand why this should be so, we need only a very simple model. Suppose that the atoms or molecules of the components A and B mix substitutionally. If the atomic (or molecular) sizes and electronic structures of A and B are similar, then the distribution will be nearly random, and the configurational entropy will be nearly ideal as in Eqn 106. That is,

$$g^{\mathrm{E}} = \Delta h_{\mathrm{m}} - T s^{\mathrm{E(non\text{-}config)}} \tag{135}$$

Assuming that the bond energies ε_{AA}, ε_{BB} and ε_{AB} of first-nearest-neighbor pairs are independent of temperature and composition and that the average nearest-neighbor coordination number, Z, is also constant, we calculate the enthalpy of mixing resulting from the change in the total energy of first-nearest-neighbor pair bonds as follows.

In 1 mol of solution, there are $(N^{\circ}Z/2)$ nearest-neighbor pair bonds, where N° is Avogadro's number. Since the distribution is assumed to be random, the probability that a given bond is an A–A bond is equal to X_A^2. The probabilities of B–B and A–B bonds are, respectively, X_B^2 and $2X_A X_B$. The contribution to the molar enthalpy of mixing resulting from the change in total energy of first-nearest-neighbor bonds is then equal to the sum of the energies of the first-nearest-neighbor bonds in 1 mol of solution minus the energy of the A–A bonds in X_A moles of pure A and the energy of the B–B bonds in X_B moles of pure B:

$$\begin{aligned}
\Delta h^{\mathrm{FNN}} &= (N^{\circ}Z/2)\left(X_A^2 \varepsilon_{AA} + X_B^2 \varepsilon_{BB} + 2X_A X_B \varepsilon_{AB}\right) - (N^{\circ}Z/2)(X_A \varepsilon_{AA} + X_B \varepsilon_{BB}) \\
&= (N^{\circ}Z)[\varepsilon_{AA} - (\varepsilon_{AA} + \varepsilon_{BB})/2]X_A X_B \\
&= C_1 X_A X_B
\end{aligned} \tag{136}$$

where C_1 is a constant.

With exactly the same assumptions and arguments, we can derive equations for the enthalpy of mixing contributions resulting from the change in total energy of second-nearest-neighbor, Δh^{SNN}, third-nearest-neighbor, Δh^{TNN}, etc. pair bonds. These are clearly all of the same form, $C_i X_A X_B$. If we now assume that the actual observed enthalpy of mixing Δh_m, results mainly from these changes in the total energy of all pair bonds, then

$$\Delta h_m = \omega X_A X_B \tag{137}$$

where $\omega = C_1 + C_2 + C_3 + \ldots\ldots$

If we now define σ_{AA}, σ_{BB}, and σ_{AB} as the vibrational entropies of pair bonds and follow an identical argument to that just presented for the pair bond energies, we obtain

$$s^{\text{E(non-config)}} = (N^o Z/2)[\sigma_{AB} - (\sigma_{AA} + \sigma_{BB}/2)] = \eta X_A X_B \tag{138}$$

Equation 133 has thus been derived. If A–B bonds are stronger than A–A and B–B bonds, then $(\varepsilon_{AA} - \sigma_{AB}T) < [(\varepsilon_{AA} - \sigma_{AA}T)/2 + (\varepsilon_{BB} - \sigma_{BB}T)/2]$, so that $(\omega - \eta T) < 0$ and $g^E < 0$. That is, the solution is rendered more stable. If the A–B bonds are relatively weaker, then the solution is rendered less stable, $(\omega - \eta T) > 0$ and $g^E > 0$.

Simple nonpolar molecular solutions and simple ionic solutions such as molten salts often exhibit approximately regular behavior. The assumption of additivity of the energy of pair bonds is probably reasonably realistic for van der Waals or coulombic forces. For metallic alloys, the concept of a pair bond is, at best, vague, and metallic solutions tend to exhibit larger deviations from regular behavior.

In several solutions, it is found that $|\eta T| \langle (|\omega|$ in Eqn 133. That is, $g^E \approx \Delta h_m = \omega X_A X_B$, and so to a first approximation, g^E is independent of T. This is more often the case in nonmetallic solutions than in metallic solutions.

3.5.5.1 *Thermodynamic Origin of Simple Phase Diagrams Illustrated by Regular Solution Theory*

Figure 20 shows several phase diagrams calculated for a hypothetical system A–B containing a solid and a liquid phase with melting points of $T^o_{f(A)} = 800$ K and $T^o_{f(B)} = 1200$ K and with entropies of fusion of both A and B set to 10 J mol^{-1} K^{-1} which is a typical value for metals. The solid and liquid phases are both regular with temperature-independent excess Gibbs energies

$$g^{\text{E(s)}} = \omega^s X_A X_B \quad \text{and} \quad g^{\text{E(l)}} = \omega^l X_A X_B \tag{139}$$

The parameters ω^s and ω^l have been varied systematically to generate the various panels of **Figure 20**.

In panel (n), both phases are ideal. Panels (l)–(r) exhibit minima or maxima depending upon the sign and magnitude of $(g^{\text{E(l)}} - g^{\text{E(s)}})$, as has been discussed in Section 3.5.2. In panel (h), the liquid is ideal but positive deviations in the solid give rise to a solid–solid miscibility gap as discussed in Section 3.5.4. On passing from panel (h) to panel (c), an increase in $g^{\text{E(s)}}$ results in a widening of the miscibility gap, so that the solubilities of A in solid B and of B in solid A decrease. Panels (a)–(c) illustrate that negative deviations in the liquid cause a relative stabilization of the liquid with resultant lowering of the eutectic temperature. Panels (d) and (e) are discussed in Section 3.5.6.

3.5.6 Immiscibility—Monotectics

In **Figure 20e**, positive deviations in the liquid have given rise to a *liquid–liquid miscibility* gap. The quasibinary CaO–SiO$_2$ system, shown in **Figure 21**, exhibits such a feature. Suppose that a liquid of

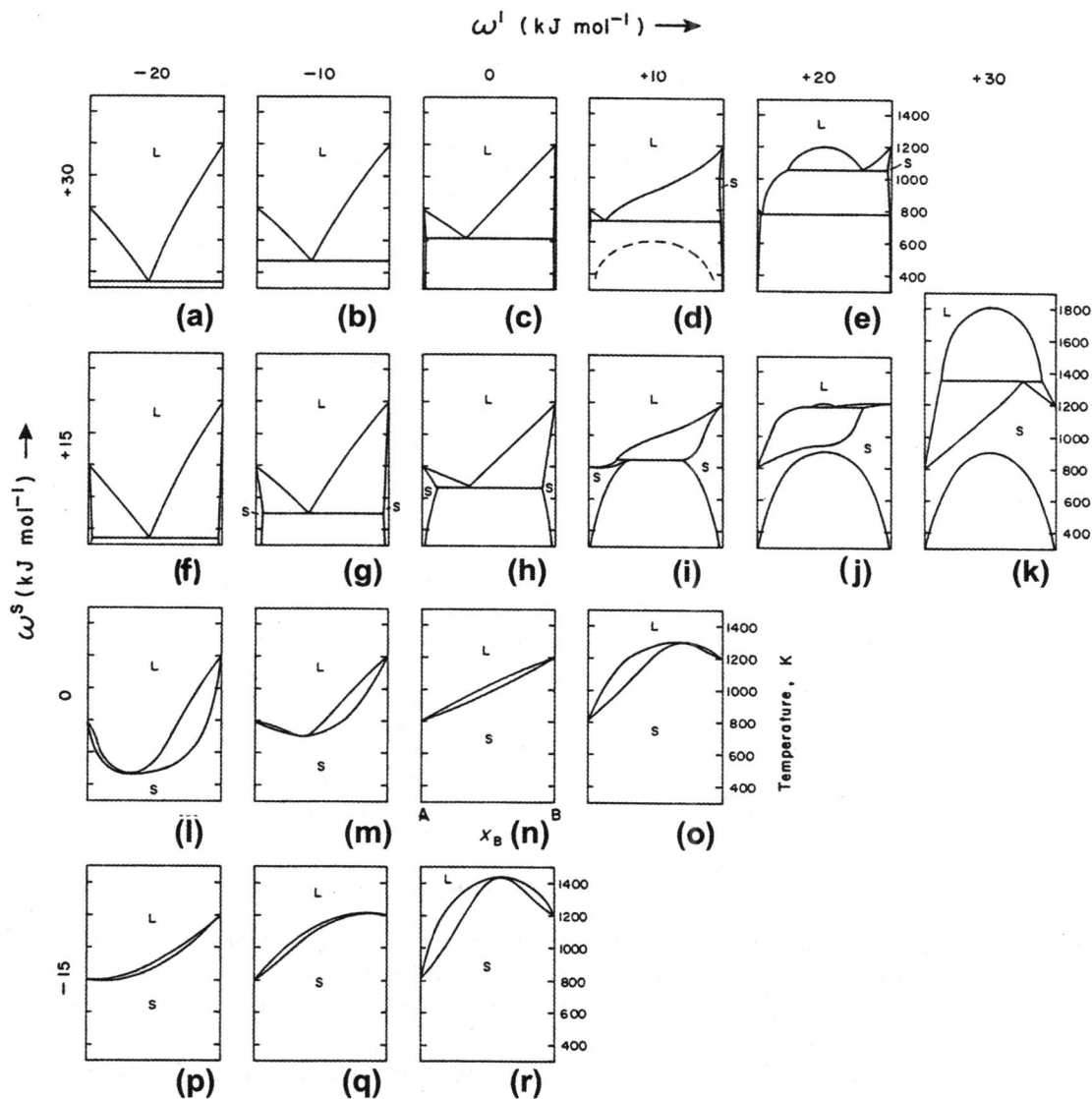

Figure 20 Topological changes in the phase diagram for a system A–B with regular solid and liquid phases brought about by systematic changes in the regular solution parameters ω^s and ω^l. Melting points of pure A and B are 800 K and 1200 K, respectively. Entropies of fusion of both A and B are 10.0 J mol^{-1} K^{-1} (Pelton and Thompson, 1975). Dashed curve in panel (d) is a metastable liquid miscibility gap.

composition $X_{SiO_2} = 0.8$ is cooled slowly from high temperatures. At $T = 1830\ °C$, the miscibility gap boundary is crossed and a second liquid layer appears with a composition of $X_{SiO_2} = 0.96$. As the temperature is lowered further, the composition of each liquid phase follows its respective phase boundary until, at $1689\ °C$, the SiO$_2$-rich liquid has a composition of $X_{SiO_2} = 0.988$ (point B), and in

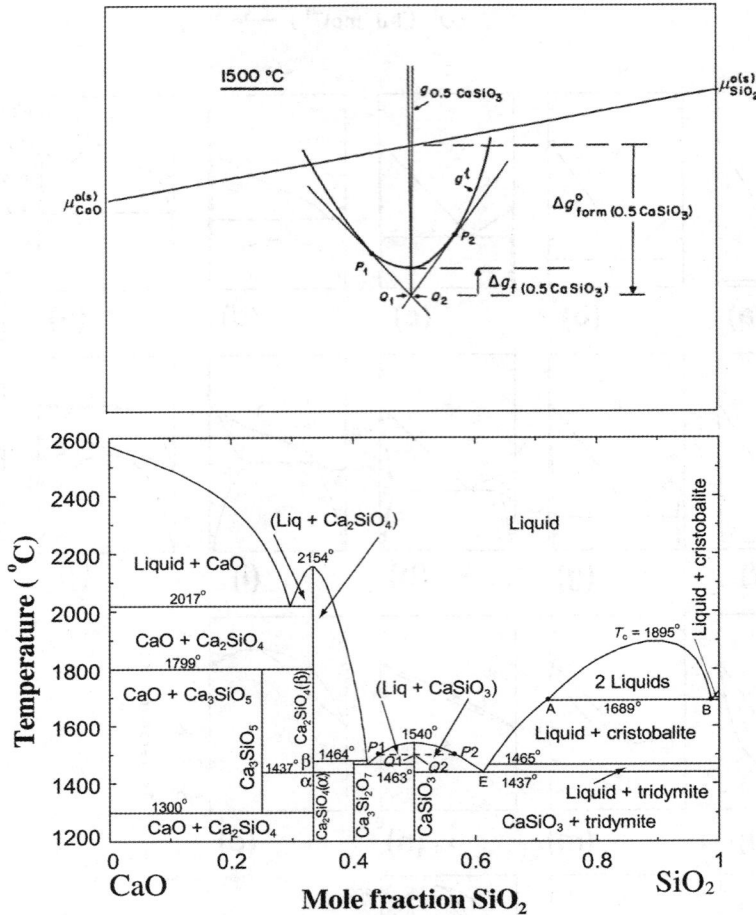

Figure 21   CaO–SiO$_2$ phase diagram at $P = 1$ bar and Gibbs energy curves at 1500 °C illustrating Gibbs energies of fusion and formation of the compound CaSiO$_3$ (see FactSage).(Reprinted with permission from Pelton, 2001a). The temperature difference between the peritectic at 1464 °C and the eutectic at 1463 °C has been exaggerated for the sake of clarity.

the CaO-rich liquid $X_{SiO_2} = 0.72$ (point A). At any temperature, the relative amounts of the two phases are given by the lever rule.

At 1689 °C, the following invariant *binary monotectic reaction* occurs upon cooling:

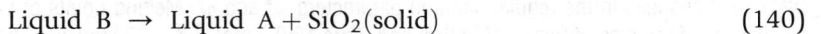

$$Liquid \; B \; \rightarrow \; Liquid \; A + SiO_2(solid) \tag{140}$$

The temperature remains constant at 1689 °C and the compositions of the phases remain constant until all of liquid B is consumed. Cooling then continues with precipitation of solid SiO$_2$ with the equilibrium liquid composition following the liquidus from point *A* to the eutectic *E*.

Returning to **Figure 20**, we see in panel (d) that the positive deviations in the liquid in this case are not large enough to produce immiscibility, but they do result in a flattening of the liquidus, which indicates a "tendency to immiscibility." If the nucleation of the solid phases can be suppressed by sufficiently rapid cooling, then a *metastable liquid–liquid miscibility gap* is observed as shown in **Figure 20**d. For example, in the Na_2O–SiO_2 system, the flattened (or "S-shaped") SiO_2-liquidus heralds the existence of a metastable miscibility gap of importance in glass technology.

3.5.7 Intermediate Phases

The phase diagram of the Ag–Mg system (Hultgren et al., 1973) is shown in **Figure 22**d. An intermetallic phase, β', is seen centered approximately about the composition $X_{Mg} = 0.5$. The Gibbs energy curve at 1050 K for such an intermetallic phase has the form shown schematically in **Figure 22**a. The curve $g^{\beta'}$ rises quite rapidly on either side of its minimum, which occurs near $X_{Mg} = 0.5$. As a result, the single-phase β' region appears on the phase diagram only over a limited composition range. This form of the Gibbs energy curve results from the fact that when $X_{Ag} \approx X_{Mg}$, a particularly stable crystal structure exists in which Ag and Mg atoms preferentially occupy different sites. The two common tangents P_1Q_1 and P_2Q_2 give rise to a maximum in the two-phase (β' + liquid) region of the phase diagram. (Although the maximum is observed very near $X_{Mg} = 0.5$, there is no thermodynamic reason for the maximum to occur exactly at this composition.)

Another intermetallic phase, the ε phase, is also observed in the Ag–Mg system (**Figure 22**d). The phase is associated with a *peritectic* invariant ABC at 744 K. The Gibbs energy curves are shown schematically at the peritectic temperature in **Figure 22**c. One common tangent line can be drawn to g^l, $g^{\beta'}$ and g^ε at 744 K.

Suppose that a liquid alloy of composition $X_{Mg} = 0.7$ is cooled very slowly from the liquid state. At a temperature just above 744 K, a liquid phase of composition C and a β' phase of composition A are observed at equilibrium. At a temperature just below 744 K, the two phases at equilibrium are β' of composition A and ε of composition B. The following invariant *binary peritectic reaction* thus occurs upon cooling:

$$\text{Liquid} + \beta' \text{ (solid)} \rightarrow \varepsilon \text{ (solid)} \tag{141}$$

This reaction occurs isothermally at 744 K with all three phases at fixed compositions (at points A, B and C). For an alloy with overall composition between points A and B, the reaction proceeds until all the liquid has been consumed. In the case of an alloy with overall composition between B and C, the β' phase will be the first to be completely consumed.

Peritectic reactions occur upon cooling generally with formation of the product solid (ε in this example) on the surface of the reactant solid (β'), thereby forming a coating which can reduce or prevent further contact between the reactant solid and liquid. Further reaction may thus be greatly retarded, so that equilibrium conditions can only be achieved by extremely slow cooling.

The Gibbs energy curve for the ε phase, g^ε, in **Figure 22**c rises more rapidly on either side of its minimum than does the Gibbs energy $g^{\beta'}$ of the β' phase. Consequently, the width of the single-phase region over which the ε phase exists (sometimes called its *range of stoichiometry* or *homogeneity range*) is narrower than that of the β' phase.

In the upper panel of **Figure 21** for the CaO–SiO_2 system, Gibbs energy curves at 1500 °C for the liquid and $CaSiO_3$ phases are shown schematically. $g_{0.5(CaSiO_3)}$ rises extremely rapidly on either side of its minimum. (We write $g_{0.5(CaSiO_3)}$ for 0.5 moles of the compound in order to normalize to a basis of 1 mol of components CaO and SiO_2.) As a result, the points of tangency Q_1 and Q_2 of the common tangents P_1Q_1 and P_2Q_2 nearly (although not exactly) coincide. Hence, the range of stoichiometry of the $CaSiO_3$ phase is very narrow (although never zero). The two-phase regions labeled

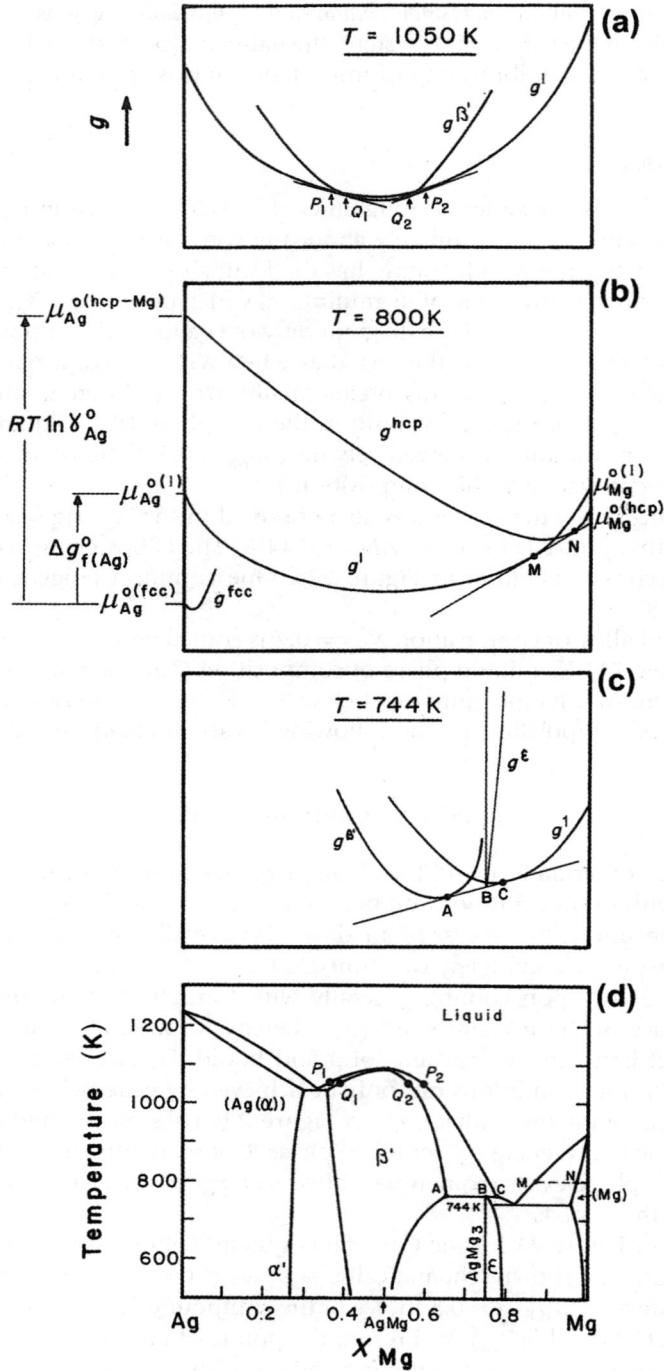

Figure 22   Ag–Mg phase diagram at $P = 1$ bar (after Hultgren et al., 1973) and Gibbs energy curves at three temperatures.

(liquid + CaSiO$_3$) in **Figure 21** are the two sides of a two-phase region that passes through a maximum at 1540 °C just as the (β' + liquid) region passes through a maximum in **Figure 22d**. Because the CaSiO$_3$ single-phase region is extremely narrow (although always of finite width), we refer to CaSiO$_3$ as a *stoichiometric compound*. Any deviation in composition from the stoichiometric 1:1 ratio of CaO to SiO$_2$ results in a very large increase in its Gibbs energy.

The ε phase in **Figure 22** is based on the stoichiometry AgMg$_3$. The Gibbs energy curve (**Figure 22c**) rises extremely rapidly on the Ag side of the minimum, but somewhat less steeply on the Mg side. As a result, Ag is virtually insoluble in AgMg$_3$, while Mg is sparingly soluble. Such a phase with a narrow range of homogeneity is often called a *nonstoichiometric compound*. At low temperatures, the β' phase exhibits a relatively narrow range of stoichiometry about the 1:1 AgMg composition and can properly be called a *compound*. However, at higher temperatures, it is debatable whether a phase with such a wide range of composition should be called a *compound*.

From **Figure 21**, it can be seen that if stoichiometric CaSiO$_3$ is heated, it will melt isothermally at 1540 °C to form a liquid of the same composition. Such a compound is called *congruently melting* or simply a *congruent compound*. The compound Ca$_2$SiO$_4$ in **Figure 21** is also congruently melting. The β' phase in **Figure 22** is also congruently melting at the composition of the liquidus/solidus maximum.

It should be noted with regard to congruent melting that the limiting slopes dT/dX of both branches of the liquidus at the congruent melting point (that is, at the maximum of the two-phase region) are zero since we are dealing with a maximum in a two-phase region (similar to the minimum in the two-phase region in **Figure 17**).

The AgMg$_3$ (ε) compound in **Figure 22** is said to *melt incongruently*. If solid AgMg$_3$ is heated, it will melt isothermally at 744 K by the reverse of the peritectic reaction as shown in Eqn 141, to form a liquid of composition C and another solid phase, β', of composition A.

Another example of an *incongruently melting compound* is Ca$_3$Si$_2$O$_7$ in **Figure 21**, which melts incongruently (or peritectically) to form liquid and Ca$_2$SiO$_4$ at the peritectic temperature of 1464 °C.

An incongruently melting compound is always associated with a peritectic. However, the converse is not necessarily true. A peritectic is not always associated with an intermediate phase. See, for example, **Figure 20i**.

For purposes of phase diagram calculations involving stoichiometric compounds such as CaSiO$_3$, we may, to a good approximation, consider the Gibbs energy curve, $g_{0.5(CaSiO_3)}$, in the upper panel of **Figure 21** to have zero width. All that is then required is the value of $g_{0.5(CaSiO_3)}$ at the minimum. This value is usually expressed in terms of the *Gibbs energy of fusion* of the compound, $\Delta g_{f(0.5CaSiO_3)}$ or of the *Gibbs energy of formation* $\Delta g_{form(0.5CaSiO_3)}$ of the compound from the pure solid components CaO and SiO$_2$ according to the reaction: 0.5 CaO (sol) + 0.5 SiO$_2$(sol) = 0.5 CaSiO$_3$(sol). Both these quantities are interpreted graphically in **Figure 21**.

3.5.8 Limited Mutual Solubility—Ideal Henrian Solutions

In Section 3.5.4, the region of two solids in the Ag–Cu phase diagram of **Figure 19** was described as a miscibility gap. That is, only one continuous g^s curve was assumed. If, somehow, the appearance of the liquid phase could be suppressed, the two solvus lines in **Figure 19**, when projected upward, would meet at a critical point, above which one continuous solid solution would exist at all compositions.

Such a description is justifiable only if the pure solid components have the same crystal structure, as is the case for Ag and Cu. However, consider the Ag–Mg system (**Figure 22**), in which the terminal (Ag) solid solution is face-centered-cubic (fcc) and the terminal (Mg) solid solution is hexagonal-close-packed (hcp). In this case, one continuous curve for g^s cannot be drawn. Each solid phase must have its own separate Gibbs energy curve, as shown schematically in **Figure** 22b at 800 K. In this figure, $\mu_{Mg}^{o(hcp)}$ and $\mu_{Ag}^{o(fcc)}$ are the standard molar Gibbs energies of pure hcp Mg and pure fcc Ag, while $\mu_{Ag}^{o(hcp\text{-}Mg)}$ is the standard molar Gibbs energy of pure (hypothetical) hcp Ag in the hcp-Mg phase.

Since the solubility of Ag in the hcp-Mg phase is limited, we can, to a good approximation, describe it as a *Henrian ideal solution*. That is, when a solution is sufficiently dilute in one component, we can approximate $\mu_{solute}^E = RT\ln \gamma_{solute}$ by its value in an infinitely dilute solution. That is, if X_{solute} is small, we set $\gamma_{solute} = \gamma_{solute}^o$, where γ_{solute}^o is the *Henrian activity coefficient* at $X_{solute} = 0$. Thus, for sufficiently dilute solutions, we assume that γ_{solute} is independent of composition. Physically, this means that in a very dilute solution, there is negligible interaction among solute particles because they are so far apart. Hence, each additional solute particle added to the solution makes the same contribution to the excess Gibbs energy of the solution and so $\mu_{solute}^E = dG^E/dn_{solute} = $ constant.

From the Gibbs–Duhem Eqn 117, if $d\mu_{solute}^E = 0$, then $d\mu_{solvent}^E = 0$. Hence, in a Henrian solution, $\gamma_{solvent}$ is also constant and equal to its value in an infinitely dilute solution. That is, $\gamma_{solvent} = 1$ and the solvent behaves ideally. In summary, then, for dilute solutions ($X_{solvent} \approx 1$), *Henry's Law* applies

$$\gamma_{solvent} \approx 1$$
$$\gamma_{solute} \approx \gamma_{solute}^o = \text{constant} \tag{142}$$

(Care must be exercised for solutions other than simple substitutional solutions. Henry's Law applies only if the ideal activity is defined correctly for the applicable solution model. Solution models are discussed in Section 3.8.2.)

Henrian activity coefficients can usually be expressed as functions of temperature:

$$RT\ln \gamma_i^o = a - bT \tag{143}$$

where a and b are constants. If data are limited, it can further be assumed that $b \approx 0$, so that $RT\ln \gamma_i^o \approx$ constant.

Treating the hcp-Mg phase in the Ag–Mg system (**Figure** 22b) as a Henrian solution, we write

$$g^{hcp} = \left(X_{Ag}\mu_{Ag}^{o(fcc)} + X_{Mg}\mu_{Mg}^{o(hcp)}\right) + RT\left(X_{Ag}\ln a_{Ag} + X_{Mg}\ln a_{Mg}\right)$$
$$= \left(X_{Ag}\mu_{Ag}^{o(fcc)} + X_{Mg}\mu_{Mg}^{o(hcp)}\right) + RT\left(X_{Ag}\ln\left(\gamma_{Ag}^o X_{Ag}\right) + X_{Mg}\ln X_{Mg}\right) \tag{144}$$

where a_{Ag} and γ_{Ag}^o are the activity and activity coefficient of silver with respect to pure fcc silver as standard state. Let us now combine terms as follows:

$$g^{hcp} = \left[X_{Ag}\left(\mu_{Ag}^{o(fcc)} + RT\ln \gamma_{Ag}^o\right) + X_{Mg}\mu_{Mg}^{o(hcp)}\right] + RT\left(X_{Ag}\ln X_{Ag} + X_{Mg}\ln X_{Mg}\right) \tag{145}$$

Since γ_{Ag}^o is independent of composition, let us define

$$\mu_{Ag}^{o(hcp\text{-}Mg)} = \mu_{Ag}^{o(fcc)} + RT\ln \gamma_{Ag}^o \tag{146}$$

From Eqns 145 and 146, it can be seen that, relative to $\mu_{Mg}^{o(hcp)}$ and to the hypothetical standard state, $\mu_{Ag}^{o(hcp\text{-}Mg)}$ defined in this way, the hcp solution is ideal. Equations 145 and 146 are illustrated in **Figure 22b**. It can be seen that as γ_{Ag}^{o} becomes larger, the point of tangency N moves to higher Mg concentrations. That is, as $\left(\mu_{Ag}^{o(hcp\text{-}Mg)} - \mu_{Ag}^{o(fcc)}\right)$ becomes more positive, the solubility of Ag in hcp-Mg decreases.

It must be stressed that $\mu_{Ag}^{o(hcp\text{-}Mg)}$, as defined by Eqn 146, is solvent-dependent. That is, $\mu_{Ag}^{o(hcp\text{-}Mg)}$ is not the same as, for example, $\mu_{Ag}^{o(hcp\text{-}Cd)}$ for Ag in dilute hcp-Cd solutions. However, for purposes of developing consistent thermodynamic databases for multicomponent solution phases, it is generally considered advantageous to select a set of values for the chemical potentials of pure elements and compounds in metastable or unstable crystal structures, such as $\mu_{Ag}^{o(hcp)}$, which are independent of the solvent. Such values may be obtained (1) from measurements at high pressures if the structure is stable at high pressure, (2) as averages of the values for several solvents, or (3) from first-principles quantum mechanical or molecular dynamics calculations. Tables (Dinsdale, 1991) of such values for many elements in metastable or unstable states have been prepared and agreed upon by the international community working in the field of phase diagram evaluation and thermodynamic database optimization (Section 3.8). The use of such solvent-independent values, of course, means that even in dilute solutions, nonzero excess Gibbs energy terms are usually required.

3.5.9   Geometry of Binary Temperature-Composition Phase Diagrams

In Section 3.3.1.1, it was shown that when three phases are at equilibrium in a binary system at constant pressure, the system is invariant ($F = 0$). There are two general types of three-phase *invariants* in binary phase diagrams. These are the *eutectic-type* and *peritectic-type* invariants as illustrated in **Figure 23**. Let the three phases concerned be called α, β and γ, with β as the central phase as shown in **Figure 23**. The phases α, β and γ can be solid, liquid or gaseous. At a eutectic-type invariant, the following invariant reaction occurs isothermally as the system is cooled:

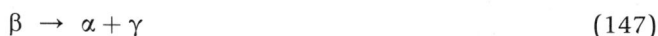

$$\beta \ \rightarrow \ \alpha + \gamma \tag{147}$$

while at a peritectic-type invariant, the invariant reaction upon cooling is

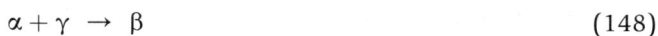

$$\alpha + \gamma \ \rightarrow \ \beta \tag{148}$$

Some examples of eutectic-type invariants are (1) *eutectics* (**Figure 19**), in which $\alpha = \text{solid}_1$, $\beta = \text{liquid}$, $\gamma = \text{solid}_2$; the eutectic reaction is $\text{liq} \rightarrow s_1 + s_2$; (2) *monotectics* (**Figure 21**), in which $\alpha = \text{liquid}_1$, $\beta = \text{liquid}_2$, $\gamma = \text{solid}$; the monotectic reaction is $\text{liq}_2 \rightarrow \text{liq}_1, +s$; (3) *eutectoids*, in which $\alpha = \text{solid}_1$, $\beta = \text{solid}_2$, $\gamma = \text{solid}_3$; the eutectoid reaction is $s_2 \rightarrow s_1 + s_3$; (4) *catatectics*, in which $\alpha = \text{liquid}$, $\beta = \text{solid}_1$, $\gamma = \text{solid}_2$; the catatectic reaction is $s_1 \rightarrow \text{liq} + s_2$.

Some examples of peritectic-type invariants are: (1) *peritectics* (**Figure 22**), in which $\alpha = \text{liquid}$, $\beta = \text{solid}_1$, $\gamma = \text{solid}_2$. The peritectic reaction is $\text{liq} + s_2 \rightarrow s_1$; (2) *syntectics* (**Figure 20k**), in which $\alpha = \text{liquid}_1$, $\beta = \text{solid}$, $\gamma = \text{liquid}_2$. The syntectic reaction is $\text{liq}_1 + \text{liq}_2 \rightarrow s$. (3) *peritectoids*, in which $\alpha = \text{solid}_1$, $\beta = \text{solid}_2$, $\gamma = \text{solid}_3$. The peritectoid reaction is $s_1 + s_3 \rightarrow s_2$.

An important rule of construction, which applies to invariants in binary phase diagrams is illustrated in **Figure 23**. This *extension rule* states that at an invariant, the extension of a boundary of a two-phase

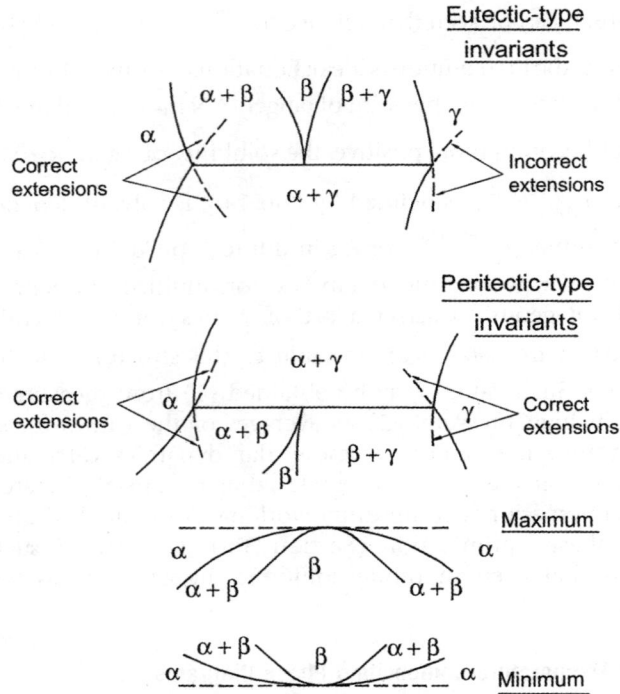

Figure 23 Geometrical units of binary phase diagrams, illustrating rules of construction.

region must pass into the adjacent two-phase region and not into a single-phase region. Examples of both correct and incorrect constructions are given in **Figure 23**. To understand why the "incorrect extension" shown is not correct, consider that the $(\alpha + \gamma)/\gamma$ phase boundary line indicates the composition of the γ phase in equilibrium with the α phase, as determined by the common tangent to the Gibbs energy curves. Since there is no reason for the Gibbs energy curves or their derivatives to change discontinuously at the invariant temperature, the extension of the $(\alpha + \gamma)/\gamma$ phase boundary also represents the stable phase boundary under equilibrium conditions. Hence, for this line to extend into a region labeled as single-phase γ is incorrect. This extension rule is a particular case of the general Schreinemakers' rule (Sections 3.7.4 and 3.7.7.1).

Two-phase regions in binary phase diagrams can terminate (1) on the pure component axes (at $X_A = 1$ or $X_B = 1$) at a transformation point of pure A or B; (2) at a critical point of a miscibility gap; (3) at an invariant. Two-phase regions can also exhibit maxima or minima. In this case, both phase boundaries must pass through their maximum or minimum at the same point as shown in **Figure 23**.

All the *geometrical units* of construction of binary phase diagrams have now been discussed. The phase diagram of a binary alloy system will usually exhibit several of these units. As an example, the Fe–Mo phase diagram was shown in **Figure 1**. The invariants in this system are peritectics at 1612, 1498 and 1453 °C; eutectoids at 1240, 1233 and 308 °C; and peritectoids at 1368 and 1250 °C. The two-phase (liquid + bcc) region passes through a minimum at $X_{Mo} = 0.2$.

Between 910 °C and 1390 °C is a two-phase $(\alpha + \gamma)$ γ-loop. Pure Fe adopts the fcc γ structure between these two temperatures but exists as the bcc α phase at higher and lower temperatures. Mo, on the other hand, is more soluble in the bcc than in the fcc structure. That is, $\mu_{Mo}^{o(bcc-Fe)} < \mu_{Mo}^{o(fcc-Fe)}$, as discussed in Section 3.5.8. Therefore, small additions of Mo stabilize the bcc structure.

In the CaO–SiO$_2$ phase diagram (**Figure 21**), we observe eutectics at 1437, 1463 and 2017 °C; a monotectic at 1689 °C; and a peritectic at 1464 °C. The compound Ca$_3$SiO$_5$ dissociates upon heating to CaO and Ca$_2$SiO$_4$ by a peritectoid reaction at 1799 °C and dissociates upon cooling to CaO and Ca$_2$SiO$_4$ by a eutectoid reaction at 1300 °C. Maxima are observed at 2154 and 1540 °C. At 1465 °C, there is an invariant associated with the tridymite $\rightarrow$ cristobalite transition of SiO$_2$. This is either a peritectic or a catatectic depending upon the relative solubility of CaO in tridymite and cristobalite. However, these solubilities are very small and unknown. At 1437 °C, Ca$_2$SiO$_4$ undergoes an allotropic transformation. This gives rise to two invariants, one involving Ca$_2$SiO$_4(\alpha)$, Ca$_2$SiO$_4(\beta)$ and Ca$_3$SiO$_5$, and the other Ca$_2$SiO$_4(\alpha)$, Ca$_2$SiO$_4(\beta)$ and Ca$_3$Si$_2$O$_7$. Since the compounds are all essentially stoichiometric, these two invariants are observed at almost exactly the same temperature of 1437 °C and it is not possible to distinguish whether they are of the eutectoid or peritectoid type.

3.6 Ternary Temperature-Composition Phase Diagrams

3.6.1 The Ternary Composition Triangle

In a ternary system with components A–B–C, the sum of the mole fractions is unity, $(X_A + X_B + X_C) = 1$. Hence, there are two independent composition variables. A representation of composition, symmetrical with respect to all three components, may be obtained with the equilateral "composition triangle" as shown in **Figure 24** for the Bi–Sn–Cd system. Compositions at the corners of the triangle correspond to the pure components. Compositions along the edges of the triangle correspond to the three binary subsystems Bi–Sn, Sn–Cd and Cd–Bi. Lines of constant mole fraction X_{Bi} are parallel to the Sn–Cd edge, while lines of constant X_{Sn} and X_{Cd} are parallel to the Cd–Bi and Bi–Sn edges, respectively. For example, at point a in **Figure 24**, $X_{Bi} = 0.05$, $X_{Sn} = 0.45$ and $X_{Cd} = 0.50$.

Similar equilateral composition triangles can be drawn with coordinates in terms of weight percentage of the three components.

3.6.2 Ternary Space Model

A ternary temperature-composition phase diagram at constant total pressure may be plotted as a three-dimensional "space model" within a right triangular prism with the equilateral composition triangle as base and temperature as vertical axis. Such a space model for a simple eutectic ternary system A–B–C is illustrated in **Figure 25**. On the three vertical faces of the prism are found the phase diagrams of the three binary subsystems, A–B, B–C and C–A which, in this example, are all simple eutectic binary systems. The binary eutectic points are e_1, e_2 and e_3. Within the prism, we see three *liquidus surfaces* descending from the melting points of pure A, B and C. Compositions on these surfaces correspond to compositions of liquid in equilibrium with A-, B- and C-rich solid phases.

In a ternary system at constant pressure, the Gibbs Phase Rule, Eqn 80, becomes

$$F = 4 - P \tag{149}$$

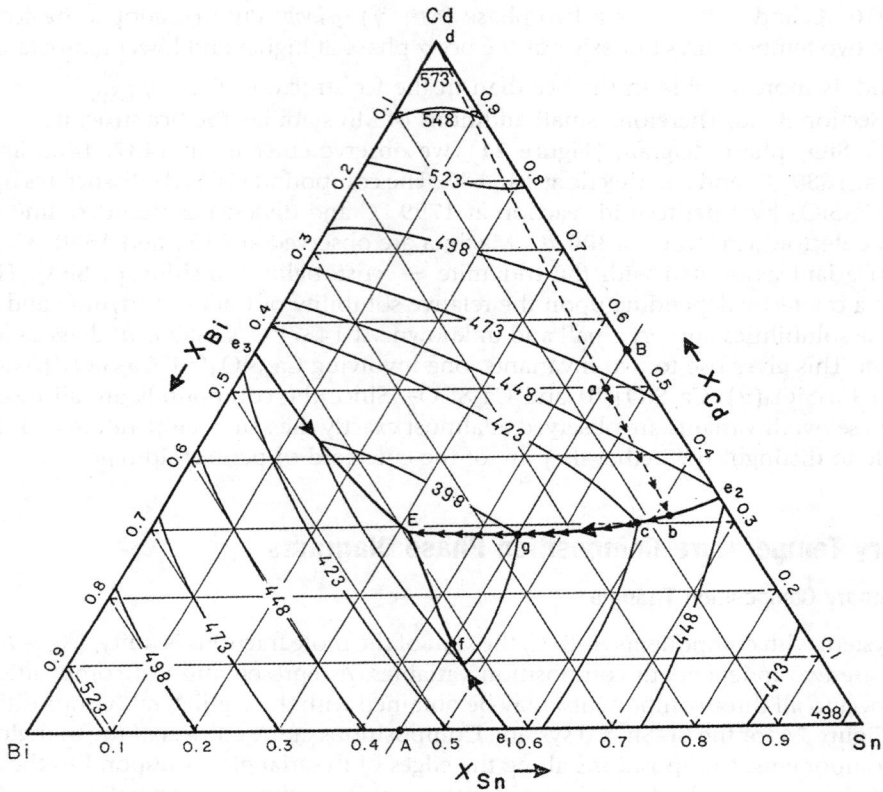

Figure 24 Projection of the liquidus surface of the Bi–Sn–Cd system onto the ternary composition triangle (after Bray et al., 1961–1962). Small arrows show the crystallization path of an alloy of overall composition at point *a* (*P* = 1 bar).

When the liquid and one solid phase are in equilibrium, $P = 2$. Hence, $F = 2$ and the system is bivariant. A ternary liquidus is thus a two-dimensional surface. We may choose two variables, say T, and one composition coordinate of the liquid, but then the other liquid composition coordinate and the composition of the solid are fixed.

The A- and B-liquidus surfaces in **Figure 25** intersect along the line e_1E. Liquids with compositions along this line are therefore in equilibrium with A-rich and B-rich solid phases simultaneously. That is, $P = 3$ and so, $F = 1$. Such "valleys" are thus called *univariant lines*. The three univariant lines meet at the *ternary eutectic point E* at which $P = 4$ and $F = 0$. This is an invariant point since the temperature and the compositions of all four phases in equilibrium are fixed.

3.6.3 Polythermal Projections of Liquidus Surfaces

A two-dimensional representation of the ternary liquidus surface may be obtained as an orthogonal projection upon the base composition triangle. Such a *polythermal projection* of the liquidus of the Bi–Sn–Cd system (Bray et al., 1961–1962) is shown in **Figure 24**. This is a simple eutectic ternary system with a space model like that shown in **Figure 25**. The constant temperature lines on **Figure 24**

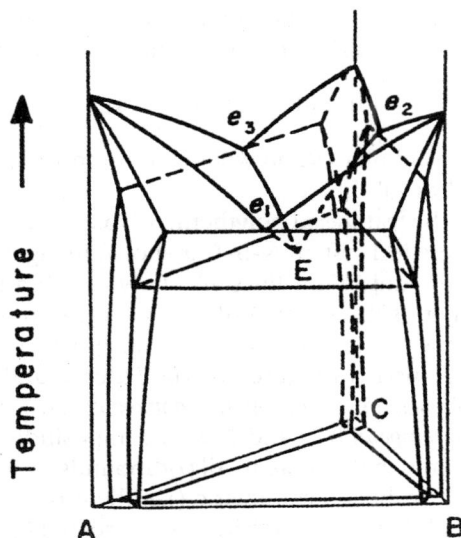

Figure 25 Perspective view of ternary space model of a simple eutectic ternary system. e_1, e_2 and e_3 are the binary eutectics and E is the ternary eutectic. The base of the prism is the equilateral composition triangle (pressure = constant).

are called *liquidus isotherms*. The univariant valleys are shown as heavier lines. By convention, the large arrows indicate the directions of decreasing temperature along these lines.

Let us consider the sequence of events occurring during the equilibrium cooling from the liquid of an alloy of overall composition a in **Figure 24**. Point a lies within the field of *primary crystallization* of Cd. That is, it lies within the composition region in **Figure 24**, in which Cd-rich solid will be the first solid to precipitate upon cooling. As the liquid alloy is cooled, the Cd-liquidus surface is reached at $T \approx 465$ K (slightly below the 473 K isotherm). A solid Cd-rich phase begins to precipitate at this temperature. Now, in this particular system, Bi and Sn are nearly insoluble in solid Cd, so that the solid phase is virtually pure Cd. (Note that this fact cannot be deduced from **Figure 24** alone.) Therefore, as solidification proceeds, the liquid becomes depleted in Cd, but the ratio X_{Sn}/X_{Bi} in the liquid remains constant. Hence, the composition path followed by the liquid (its *crystallization path*) is a straight line passing through point a and projecting to the Cd-corner of the triangle. This crystallization path is shown in **Figure 24** as the line ab.

In the general case in which a solid solution rather than a pure component or stoichiometric compound is precipitating, the crystallization path will not be a straight line. However, for equilibrium cooling, a straight line joining a point on the crystallization path at any T to the overall composition point a will extend through the composition, on the solidus surface, of the solid phase in equilibrium with the liquid at that temperature. That is, the compositions of the two equilibrium phases and the overall system composition always lie on the same tie-line as is required by mass balance considerations.

When the composition of the liquid has reached point b in **Figure 24** at $T \approx 435$ K, the relative proportions of the solid Cd and liquid phases at equilibrium are given by the *lever rule* applied to the tie-line dab: (moles of liquid)/(moles of Cd) = da/db, where da and ab are the lengths of the line segments. Upon further cooling, the liquid composition follows the univariant valley from b to E while Cd- and

Sn-rich solids coprecipitate as a binary eutectic mixture. When the liquidus composition attains the ternary eutectic composition E at $T \approx 380$ K, the invariant *ternary eutectic reaction* occurs:

$$\text{liquid} \rightarrow s_1 + s_2 + s_3 \tag{150}$$

where s_1, s_2 and s_3 are the three solid phases and where the compositions of all four phases (as well as T) remain fixed until all liquid has solidified.

In order to illustrate several of the features of polythermal projections of liquidus surfaces, a projection of the liquidus surface of a hypothetical system A–B–C is shown in **Figure 26**. For the sake of simplicity, isotherms are not shown; only the univariant lines are shown, with arrows to show the directions of decreasing temperature. The binary subsystems A–B and C–A are simple eutectic systems, while the binary subsystem B–C contains one congruently melting binary phase, ε, and one incongruently melting binary phase, δ, as shown in the insert in **Figure 26**. The letters e and p indicate binary eutectic and peritectic points. The δ and ε phases are called *binary compounds* since they have compositions within a binary subsystem. Two *ternary compounds*, η and ζ, with compositions within the ternary triangle, as indicated in **Figure 26**, are also found in this system. All compounds, as well as pure solid A, B and C (the α, β and γ phases), are assumed to be stoichiometric (i.e. there is no solid solubility). The fields of primary crystallization of all the solids are indicated in parentheses in **Figure 26**. The composition of the ε phase lies within its field since it is a congruently melting compound, while the composition of the

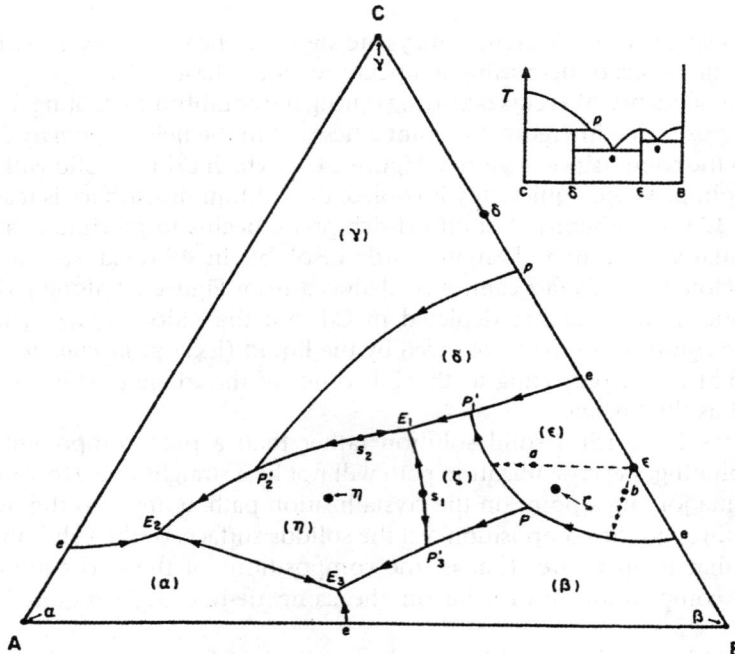

Figure 26 Projection of the liquidus surface of a system A–B–C. The binary subsystems A–B and C–A are simple eutectic systems. The binary phase diagram B–C is shown in the insert. All solid phases are assumed pure stoichiometric components or compounds. Small arrows show the initial parts of the crystallization paths of alloys of compositions at points *a* and *b*.

δ phase lies outside of its field since it is an incongruently melting compound. Similarly for the ternary compounds, η is a congruently melting compound while ζ is incongruently melting. For the congruent compound η, the highest temperature on the η liquidus occurs at the composition of η.

The univariant lines meet at a number of *ternary eutectics E* (three arrows converging on the point), a *ternary peritectic P* (one arrow entering, two arrows leaving the point), and several *ternary quasi-peritectics P'* (two arrows entering, one arrow leaving). Two *saddle points s* are also shown. These are points of maximum T along the univariant line but of minimum T on the liquidus surface along a section joining the compositions of the two solids. For example, s_1 is a maximum point along the univariant E_1P_3', but is a minimum point on the liquidus along the straight line $\zeta s_1 \eta$.

Let us consider the events occurring during equilibrium cooling from the liquid of an alloy of overall composition a in **Figure 26**. The primary crystallization product will be the ε phase. Since this is a pure stoichiometric solid, the crystallization path of the liquid will be along a straight line passing through a and extending to the composition of ε as shown in the figure.

Solidification of ε continues until the liquid attains a composition on the univariant valley. Thereafter, the liquid composition descends along the valley toward the point P_1' in coexistence with ε and ζ. At point P_1', the invariant *ternary quasi-peritectic reaction* occurs isothermally:

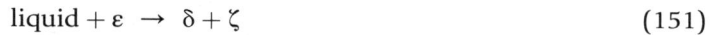

$$\text{liquid} + \varepsilon \;\rightarrow\; \delta + \zeta \tag{151}$$

Since there are two reactants, there are two possible outcomes: (1) the liquid is completely consumed before the ε phase and solidification will be complete at the point P_1'; (2) ε is completely consumed before the liquid and solidification will continue with decreasing T along the univariant line P_1E_1 with coprecipitation of δ and ζ until, at E_1, the liquid will solidify eutectically (liquid $\rightarrow \delta + \zeta + \eta$). To determine whether outcome (1) or (2) occurs, we use the mass balance criterion that, for three-phase equilibrium, the overall composition a must always lie within the *tie-triangle* formed by the compositions of the three phases. Now, the triangle joining the compositions of δ, ε and ζ, does not contain the point a, but the triangle joining the compositions of δ, ζ and liquid at P_1' does contain the point a. Hence, outcome (2) occurs.

An alloy of overall composition b in **Figure 26** solidifies with ε as primary crystallization product until the liquid composition contacts the univariant line. Thereafter, coprecipitation of ε and β occurs with the liquid composition following the univariant valley until the liquid reaches the peritectic composition P. The invariant *ternary peritectic reaction* then occurs isothermally:

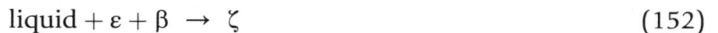

$$\text{liquid} + \varepsilon + \beta \;\rightarrow\; \zeta \tag{152}$$

Since there are three reactants, there are three possible outcomes: (1) the liquid is consumed before either ε or β and solidification terminates at P; (2) ε is consumed first and solidification then continues along the path PP_3'; or (3) β is consumed first and solidification continues along the path PP_1'. Which outcome actually occurs depends on whether the overall composition b lies within the tie-triangle (1) $\varepsilon\beta\zeta$, (2) $\beta\zeta P$, or (3) $\varepsilon\zeta P$. In the example shown, outcome (1) will occur.

A polythermal projection of the liquidus surface of the Zn–Mg–Al system is shown in **Figure 27**. There are 11 primary crystallization fields of nine binary solid solutions (with limited solubility of the third component) and of two ternary solid solutions. There are three ternary eutectic points, three saddle points, three ternary peritectic points and five ternary quasi-peritectic points.

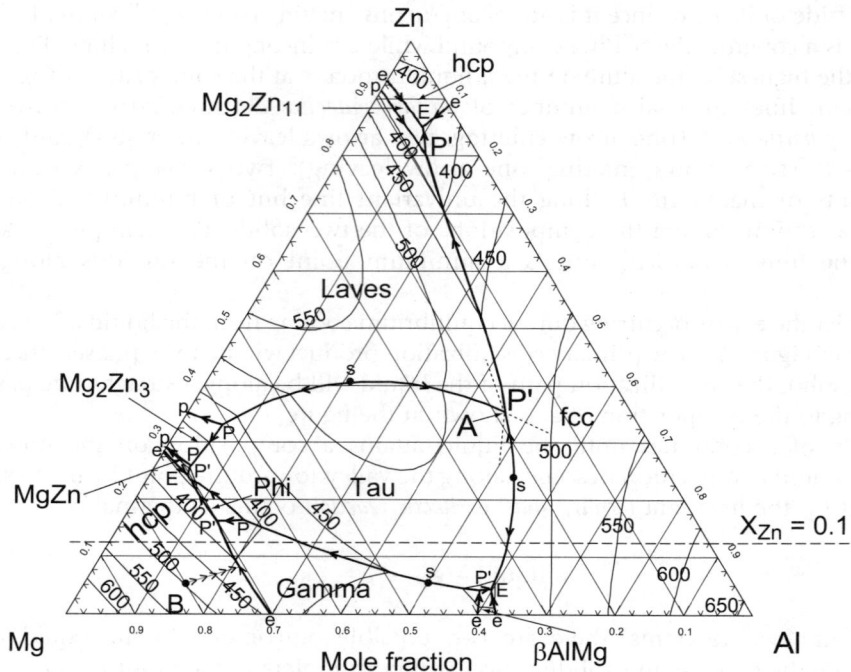

Figure 27   Polythermal projection of the liquidus surface of the Zn–Mg–Al system at $P = 1$ bar (see FactSage).

3.6.4   Ternary Isothermal Sections

Polythermal projections of the liquidus surface do not provide information on the compositions of the solid phases at equilibrium. However, this information can be presented at any one temperature on an *isothermal section* such as that shown for the Bi–Sn–Cd system at 423 K in **Figure 28**. This phase diagram is a constant temperature slice through the space model of **Figure 25**.

The liquidus lines bordering the one-phase liquid region of **Figure 28** are identical to the 423 K isotherms of the projection in **Figure 24**. Point c in **Figure 28** is point c on the univariant line in **Figure 24**. An alloy with overall composition in the one-phase liquid region of **Figure 28** at 423 K will consist of a single liquid phase. If the overall composition lies within one of the two-phase regions, then the compositions of the two phases are given by the ends of the *tie-line* that passes through the overall composition. For example, a sample with overall composition p in **Figure 28** will consist of a liquid of composition q on the liquidus and a solid Bi-rich alloy of composition r on the solidus. The relative proportions of the two phases are given by the lever rule:

(moles of liquid)/(moles of solid) $= pr/pq$,   where pr and pq are the lengths of the tie-line segments.

In the case of solid Cd, the solid phase is nearly pure Cd, so all tie-lines of the (Cd + liquid) region converge nearly to the corner of the triangle. In the case of the Bi- and Sn-rich solids, some solid solubility is observed. (The actual extent of this solubility is exaggerated in **Figure 28** for the sake of clarity of presentation.) Alloys with overall compositions rich enough in Bi or Sn to lie within the

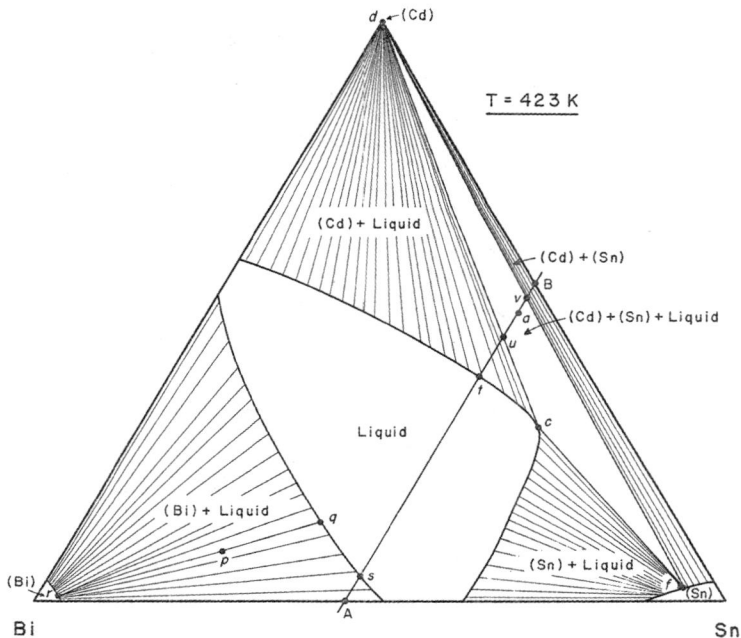

Figure 28   Isothermal section of the Bi–Sn–Cd system at 423 K at $P = 1$ bar (after Bray et al., 1961–1962). Extents of solid solubility in Bi and Sn have been exaggerated for clarity of presentation.

single-phase (Sn) or (Bi) regions of **Figure 28** will consist at 423 K of single-phase solid solutions. Alloys with overall compositions at 423 K in the two-phase (Cd + Sn) region will consist of two solid phases.

Alloys with overall compositions within the three-phase triangle *dcf* will, at 423 K, consist of three phases: Cd- and Sn-rich solids with compositions at *d* and *f* and liquid of composition *c*. To understand this better, consider an alloy of composition *a* in **Figure 28**, which is the same composition as the point *a* in **Figure 24**. In Section 3.6.3, we saw that when an alloy of this composition is cooled, the liquid follows the path *ab* in **Figure 24** with primary precipitation of Cd and then follows the univariant line with coprecipitation of Cd and Sn, so that at 423 K, the liquid is at the composition point *c*, and two solid phases are in equilibrium with the liquid.

3.6.4.1   Topology of Ternary Isothermal Sections

At constant temperature, the Gibbs energy of each phase in a ternary system is represented as a function of composition by a surface plotted in a right triangular prism with Gibbs energy as vertical axis and the composition triangle as base. Just as the compositions of phases at equilibrium in binary systems are determined by the points of contact of a common tangent line to their isothermal Gibbs energy curves (**Figure 13**), so the compositions of phases at equilibrium in a ternary system are given by the points of contact of a common tangent plane to their isothermal Gibbs energy surfaces (or, equivalently as discussed in Section 3.4.2, by minimizing the total Gibbs energy of the system). A common tangent plane can contact two Gibbs energy surfaces at an infinite number of pairs of points, thereby generating an infinite number of tie-lines within a two-phase region on an isothermal section. A common tangent

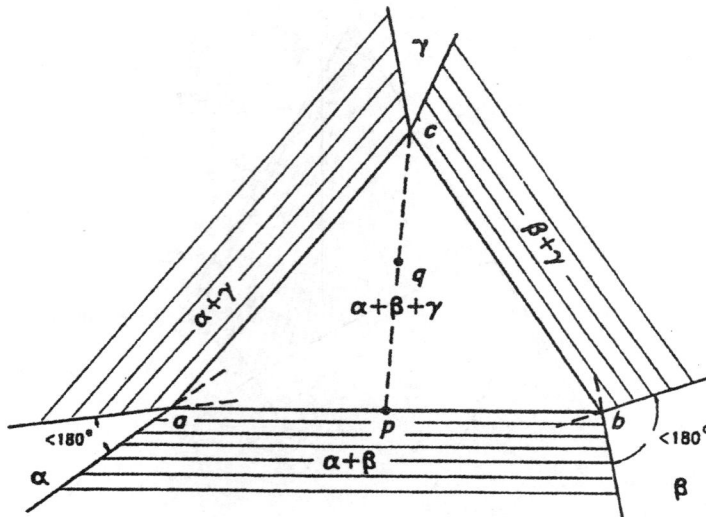

Figure 29 A tie-triangle in a ternary isothermal (and isobaric) section illustrating the lever rule and the extension rule (Schreinemakers' rule).

plane to three Gibbs energy surfaces contacts each surface at a unique point, thereby generating a three-phase tie-triangle.

Hence, the principal topological units of construction of an isothermal ternary phase diagram are three-phase ($\alpha + \beta + \gamma$) tie-triangles as in **Figure 29** with their accompanying two-phase and single-phase areas. Each corner of the tie-triangle contacts a single-phase region, and from each edge of the triangle, there extends a two-phase region. The edge of the triangle is a limiting tie-line of the two-phase region.

For overall compositions within the tie-triangle, the compositions of the three phases at equilibrium are fixed at the corners of the triangle. The relative proportions of the three phases are given by the *lever rule of tie-triangles*, which can be derived from mass balance considerations. At an overall composition q in **Figure 29**, for example, the relative proportion of the γ phase is given by projecting a straight line from the γ corner of the triangle (point c) through the overall composition q to the opposite side of the triangle, point p. Then, (moles of γ)/(total moles) $= qp/cp$ if compositions are expressed as mole fractions, or (weight of γ)/(total weight) $= qp/cp$ if compositions are in weight percent.

Isothermal ternary phase diagrams are generally composed of a number of these topological units. An example for the Zn–Mg–Al system at 25 °C is shown in **Figure 2**. At 25 °C, hcp-Mg, hcp-Zn and Mg_2Zn_{11} are nearly stoichiometric compounds, while Mg and Zn have very limited solubility in fcc-Al. The gamma and βAlMg phases are binary compounds, which exhibit very narrow ranges of stoichiometry in the Mg–Al binary system and in which Zn is very sparingly soluble. The Laves, Mg_2Zn_3 and MgZn phases are virtually stoichiometric binary compounds, in which Al is soluble to a few percent. Their ranges of stoichiometry are thus shown in **Figure 2** as very narrow lines. The Tau phase is a ternary phase with a small single-phase region of stoichiometry as shown. An examination of **Figure 2** will show that it is composed of the topological units of **Figure 29**.

An *extension rule*, a particular case of *Schreinemakers' Rule* (Sections 3.7.4 and 3.7.7.1), for ternary tie-triangles is illustrated in **Figure 29** at points a and b. At each corner, the extension of the boundaries of the single-phase regions, indicated by the broken lines, must either both project into the triangle as at

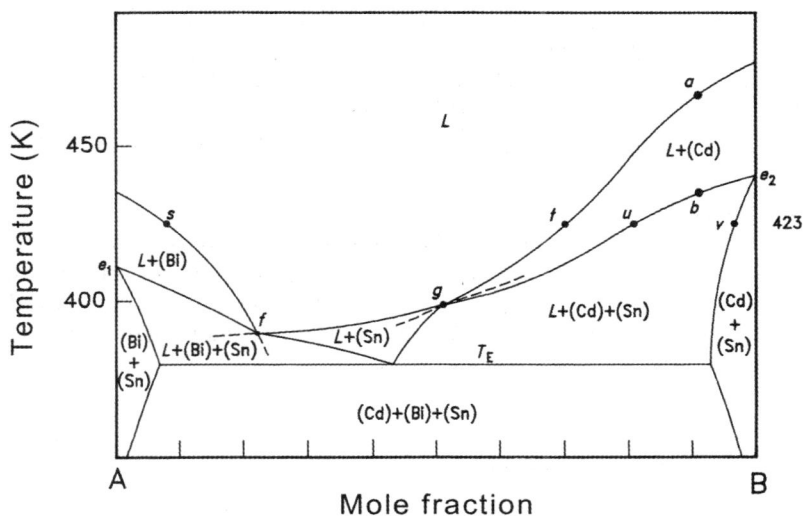

Figure 30 Isopleth (constant composition section) of the Bi–Sn–Cd system at $P = 1$ bar following the line AB of **Figure 24**. (Extents of solid solubility in the terminal solid solutions have been exaggerated for clarity of presentation.)

point a, or must both project outside the triangle as at point b. Furthermore, the angle between the boundaries must not be greater than $180°$.

Another important rule of construction, whose derivation is evident, is that within any two-phase region tie-lines must never cross one another.

3.6.5 Ternary Isopleths (Constant Composition Sections)

A vertical *isopleth*, or constant composition section, through the space model of the Bi–Sn–Cd system, is shown in **Figure 30**. The section follows the line AB in **Figure 24**.

The phase fields in **Figure 30** indicate which phases are present when an alloy with an overall composition on the line AB is equilibrated at any temperature. For example, consider the cooling, from the liquid state, of an alloy of composition a, which is on the line AB (**Figure 24**). At $T \approx 465$ K, precipitation of the solid (Cd) phase begins at point a in **Figure 30**. At $T \approx 435$ K (point b in **Figures 24 and 30**), the solid (Sn) phase begins to appear. Finally, at the eutectic temperature T_E, the ternary reaction occurs, leaving solid (Cd) + (Bi) + (Sn) at lower temperatures. The intersection of the isopleth with the univariant lines in **Figure 24** occurs at points f and g, which are also indicated in **Figure 30**. The intersection of this isopleth with the isothermal section at 423 K is shown in **Figure 28**. The points s, t, u and v of **Figure 28** are also shown in **Figure 30**.

It is important to note that on an isopleth, the tie-lines do not, in general, lie in the plane of the diagram. Therefore, the diagram provides information only on which phases are present, not on their compositions or relative proportions. The boundary lines on an isopleth do not, in general, indicate the phase compositions or relative proportions, only the temperature at which a phase appears or disappears for a given overall composition. The lever-rule cannot be applied on an isoplethal section.

An isoplethal section of the Zn–Mg–Al system at 10 mol% Zn was shown in **Figure 3a**. The liquidus surface of the diagram may be compared to the liquidus surface along the line where $X_{Zn} = 0.1$ in

Figure 27 and to the section at $X_{Zn} = 0.1$ in the isothermal section at 25 °C in **Figure 2**. Point p on the two-phase (Tau + Gamma) field in **Figure 3**a at 25 °C corresponds to point p in **Figure 2**, where the equilibrium compositions of the Gamma and Tau phases are given by the ends of the tie-line. These individual phase compositions cannot be read from **Figure 3**a.

The geometrical rules that apply to isoplethal sections will be discussed in Section 3.7.

3.6.5.1 Quasibinary Phase Diagrams

The binary CaO–SiO$_2$ phase diagram in **Figure 21** is actually an isopleth of the ternary Ca–Si–O system along the line $n_O = (n_{Ca} + 2n_{Si})$. However, all tie-lines in **Figure 21** lie within (or virtually within) the plane of the diagram because $n_O = (n_{Ca} + 2n_{Si})$ in every phase. Therefore, the diagram is called a *quasibinary* phase diagram. Similarly, the K$_2$CO$_3$–Na$_2$CO$_3$ phase diagram in **Figure 17** is a quasibinary isoplethal section of the quaternary K–Na–C–O system.

3.7 General Phase Diagram Sections

Phase diagrams involving temperature and composition as variables have been discussed in Sections 3.5 and 3.6 for binary and ternary systems. In the present section, we shall discuss the geometry of general phase diagram sections involving these and other variables such as total pressure, chemical potentials, volume, enthalpy, etc. for systems of any number of components. A few examples of such diagrams have already been shown in **Figures 4–7, 9 and 10**. It will now be shown that all these phase diagrams, although seemingly quite different geometrically, actually all obey one simple set of geometrical rules.

The rules which will be developed apply to two-dimensional phase diagram sections in which two thermodynamic variables are plotted as axes while other variables are held constant. We shall develop a set of sufficient conditions for the choice of axis variables and constants, which ensure that the phase diagram section is single-valued, with each point on the diagram representing a unique equilibrium state. Although these rules do not apply directly to phase diagram projections such as in **Figures 24, 26 and 27**, such diagrams can be considered to consist of portions of several phase diagram sections projected onto a common plane. Some geometrical rules specific to projections will be given in Section 3.7.13. Finally, since all single-valued phase diagram sections obey the same set of geometrical rules, one single algorithm can be used for their calculation as will be demonstrated in Section 3.7.4.1.

In Sections 3.7.1–3.7.6, we shall first present the geometrical rules governing general phase diagram sections without giving detailed proofs. Rigorous proofs will follow in Sections 3.7.7 and 3.7.9.

3.7.1 Corresponding Potentials and Extensive Variables

In a system with C components, we select the following $(C + 2)$ *thermodynamic potentials* (intensive variables): $T, P, \mu_1, \mu_2, \ldots\ldots, \mu_C$. For each potential ϕ_i, there is a *corresponding extensive variable* q_i related by the equation:

$$\phi_i = (\partial U/\partial q_i)_{q_j(j \neq i)} \tag{153}$$

The corresponding pairs of potentials and extensive variables are listed in **Table 1**. The corresponding pairs are found together in the terms of the fundamental Equation 9 and in the Gibbs–Duhem Equation 75.

Table 1   Corresponding pairs of potentials ϕ_i and extensive variables q_i

ϕ_i:	T	P	μ_1	$\mu_2\ldots$	μ_C
q_i:	S	$-V$	n_1	$n_2\ldots$	n_C

When a system is at equilibrium, the potential variables are the same for all phases (all phases are at the same T, P, μ_1, μ_2, ...) but the extensive variables can differ from one phase to the next (the phases can all have different volumes, entropies and compositions). In this regard, it is important to note the following. If the only variables which are held constant are potential variables, then the compositions of all phases will lie in the plane of the phase diagram section and the lever rule may be applied. Some examples are found in **Figures 1, 2 and 5**. However, if an extensive variable such as composition is held constant, then this is no longer necessarily true as was discussed, for example, in Section 3.6.5 regarding isoplethal sections such as **Figures 3, 4 and 30**.

3.7.2   The Law of Adjoining Phase Regions

The following *Law of Adjoining Phase Regions* (Palatnik and Landau, 1956) applies to all single-valued phase diagram sections: *"As a phase boundary line is crossed, one and only one phase either appears or disappears."*

    This law is clearly obeyed in any phase diagram section in which both axis variables are extensive variables such as **Figures 2, 3b, 4 and 28**. When one axis variable is a potential, however, it may at first glance appear that the law is not obeyed for invariant lines. For instance, as the eutectic line is crossed in the binary T composition phase diagram in **Figure 19**, one phase disappears and another appears. However, the eutectic line in this figure is not a simple phase boundary, but rather an infinitely narrow three-phase region. This is illustrated schematically in **Figure 31**, where the three-phase region has been "opened up" to show that the region is enclosed by upper and lower phase boundaries which are coincident. In passing from the ($\beta + L$) to the ($\alpha + \beta$) region, for example, we first pass through the ($\alpha + \beta + L$) region. Hence, the Law of Adjoining Phase Regions is obeyed. The three-phase region is infinitely narrow because all three phases at equilibrium must be at the same temperature. Another example is seen in the isoplethal section in **Figure 30**. At the ternary eutectic temperature, T_E, there is an

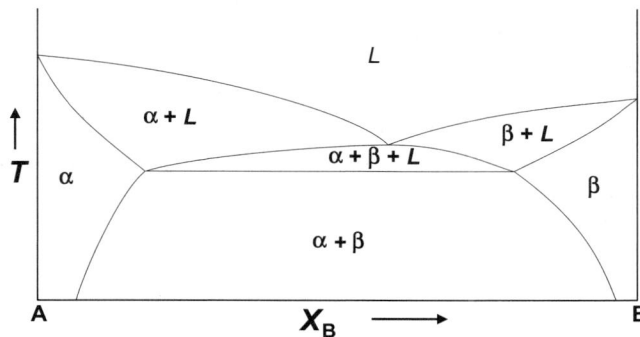

Figure 31   An isobaric temperature-composition phase diagram with the eutectic "opened up" to show that it is an infinitely narrow three-phase region. Reprinted with permission from Pelton (2001a).

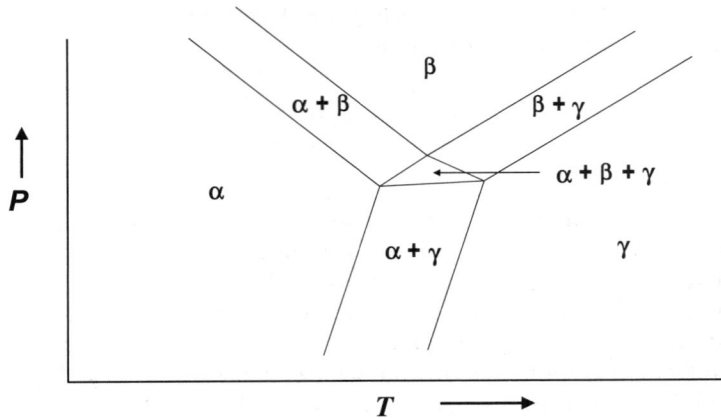

Figure 32 A *P–T* diagram in a one-component system "opened up" to show that the two-phase lines are infinitely narrow two-phase regions. Reprinted with permission from Pelton (2001a).

infinitely narrow horizontal four-phase invariant region (L + (Bi) + (Cd) + (Sn)). As yet another example, the *y*-axis of **Figure 5** is the chemical potential of oxygen, which is the same in all phases at equilibrium. This diagram exhibits two infinitely narrow horizontal three-phase invariant regions.

Similarly, if both axis variables are potentials, there may be several infinitely narrow univariant phase regions. For example, in the *P–T* diagram of **Figure 7**a, all lines are infinitely narrow two-phase regions bounded on either side by coincident phase boundaries as shown schematically in **Figure 32**, where all two-phase regions have been "opened up." The three-phase invariant triple point is then seen to be a degenerate three-phase triangle. Other examples of diagrams in which both axis variables are potentials are seen in **Figures 6, 9 and 10**. All lines in **Figures 9 and 10** are infinitely narrow two-phase univariant regions. In **Figure 6**, some lines are infinitely narrow three-phase univariant regions while others are simple phase boundaries.

3.7.3   Nodes in Phase Diagram Sections

All phase boundary lines in any single-valued phase diagram section meet at nodes where exactly four lines converge as illustrated in **Figure 33**. *N* phases ($\alpha_1, \alpha_2, ..., \alpha_N$), where $N \geq 1$, are common to all four regions. Two additional phases, β and γ, are disposed as shown in **Figure 33**, giving rise to two $(N + 1)$-phase regions and one $(N + 2)$-phase region. At every node, the generalized *Schreinemakers' Rule* requires that the extensions of the boundaries of the *N*-phase region must either both lie within the $(N + 1)$-phase regions as in **Figure 33**, or they must both lie within the $(N + 2)$-phase region.

Schreinemakers' Rule will be proven in Section 3.7.7.1.

An examination of **Figure 4**, for example, will reveal that all nodes involve exactly four boundary lines. Schreinemakers' Rule is illustrated on this figure by the dashed extensions of the phase boundaries at the nodes labeled *a*, *b* and *c* and in **Figure 30** by the dashed extension lines at nodes *f* and *g*.

For phase diagrams in which one or both axis variables are potentials, it might, at first glance, seem that fewer than four boundaries converge at some nodes. However, all nodes can still be seen to involve

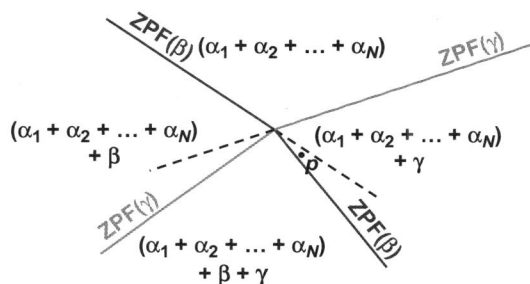

Figure 33 A node in a general phase diagram section showing the Zero Phase Fraction lines of the β and γ phases. Schreinemakers' Rule is obeyed. Reprinted with permission from Pelton (2001a). (For color version of this figure, the reader is referred to the online version of this book.)

exactly four phase boundary lines when all infinitely narrow phase regions are "opened up" as in **Figures 31 and 32**.

The extension rule illustrated in **Figure 29** for tie-triangles in ternary isothermal sections can be seemed to be a particular case of Schreinemakers' Rule, as can the extension rules shown in **Figure 23** for binary T-composition phase diagrams. Finally, an extension rule for triple points on P–T phase diagrams is illustrated in **Figure 7**a; the extension of each univariant line must pass into the opposite bivariant region. This rule, which can be seen to apply to **Figures 9 and 10** as well, is also a particular case of Schreinemakers' Rule.

An objection might be raised that a minimum or a maximum in a two-phase region in a binary temperature-composition phase diagram, as in **Figures 17, 21 and 22**, represents an exception to Schreinemakers' Rule. However, the extremum in such cases is not actually a node where four phase boundaries converge, but rather a point where two boundaries touch. Such extrema, in which two phase boundaries touch with zero slope, may occur for a C-phase region in a phase diagram of a C-component system when one axis is a potential. For example, in an isobaric temperature-composition phase diagram of a four-component system, we may observe a maximum or a minimum in a four-phase region separating two three-phase regions. A similar maximum or minimum in a $(C\text{-}n)$-phase region, where $n > 0$, may also occur along particular composition paths. This will be discussed in more detail in Section 3.9.4.

Other apparent exceptions to Schreinemakers' Rule (such as five boundaries meeting at a node) can occur in special limiting cases, as, for example, if an isoplethal section of a ternary system passes exactly through the ternary eutectic composition.

3.7.4   Zero Phase Fraction Lines

All phase boundaries on any single-valued phase diagram section are *Zero Phase Fraction (ZPF) Lines*, an extremely useful concept introduced by Gupta et al. (1986), which is a direct consequence of the Law of Adjoining Phase Regions. There is a ZPF line associated with each phase. On one side of its ZPF line, the phase occurs, while on the other side, it does not, as illustrated in **Figure 4**. There is a ZPF line for each of the five phases in **Figure 4**. The five ZPF lines account for all the phase boundaries on the two-dimensional section. Phase diagram sections plotted on triangular coordinates as in **Figures 2 and 28** similarly consist of one ZPF line for each phase.

With reference to **Figure 33**, it can be seen that at every node, two ZPF lines cross each other.

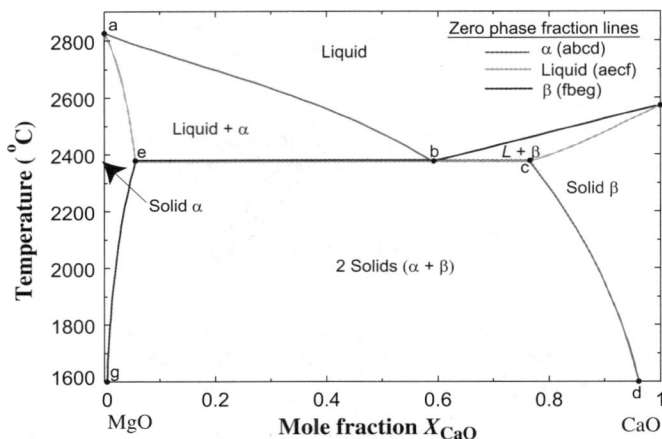

Figure 34 Binary *T*-composition phase diagram at $P = 1$ bar of the MgO–CaO system showing Zero Phase Fraction (ZPF) lines of the phases (see FactSage). (For color version of this figure, the reader is referred to the online version of this book.)

In the case of phase diagrams in which one or both axis variables are potentials, certain phase boundaries may be coincident as was discussed in Section 3.7.2. Hence, ZPF lines for two different phases may be coincident over part of their lengths as is illustrated in **Figure 34**. In **Figures 9 and 10**, every line in the diagram is actually two coincident ZPF lines.

3.7.4.1 General Algorithm to Calculate Any Phase Diagram Section

Since all single-valued phase diagram sections obey the same geometrical rules described in Sections 3.7.1–3.7.3, one general algorithm can be used to calculate any phase diagram section thermodynamically. As discussed in Section 3.2.3, the equilibrium state of a system (that is, the amount and composition of every phase) can be calculated for a given set of conditions (*T, P*, overall composition, chemical potentials, etc.) by software that minimizes the total Gibbs energy of the system, retrieving the thermodynamic data for each phase from databases. To calculate any phase diagram section, one first specifies the desired axis variables and their limits as well as the constants of the diagram. The software then scans the four edges of the diagram, using Gibbs energy minimization to find the ends of each ZPF line. Next, each ZPF line is calculated from one end to the other using Gibbs energy minimization to determine the series of points at which the phase is just on the verge of appearing. Should a ZPF line not intersect any edge of the diagram, it will be discovered by the program while it is tracing one of the other ZPF lines. When ZPF lines for all phases have been drawn, the diagram is complete. This is the algorithm used by the FactSage system software, which has been used to calculate most of the phase diagrams shown in this chapter (see FactSage).

3.7.5 Choice of Variables to Ensure that Phase Diagram Sections are Single-valued

In a system of *C* components, a two-dimensional phase diagram section is obtained by choosing two axis variables and holding $(C - 1)$ other variables constant. However, not every choice of variables will result in a single-valued phase diagram with a unique equilibrium state at every point. For example, on

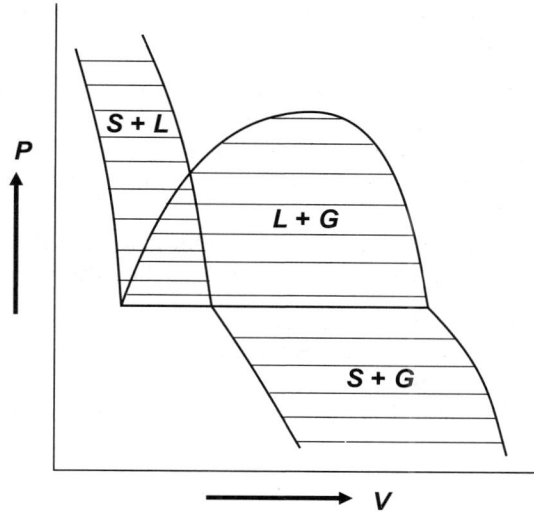

Figure 35 Schematic P–V diagram for H_2O. This is not a single-valued phase diagram section (reprinted with permission from Pelton (2001a)).

the P–V diagram for H_2O shown schematically in **Figure 35**, at any point in the area where the $(S + L)$ and $(L + G)$ regions overlap, there are two possible equilibrium states of the system. Similarly, the diagram of S_2 pressure versus the mole fraction of Ni at constant T and P in the Cu–Ni–S_2 system in **Figure 36** exhibits a region in which there are two possible equilibrium states.

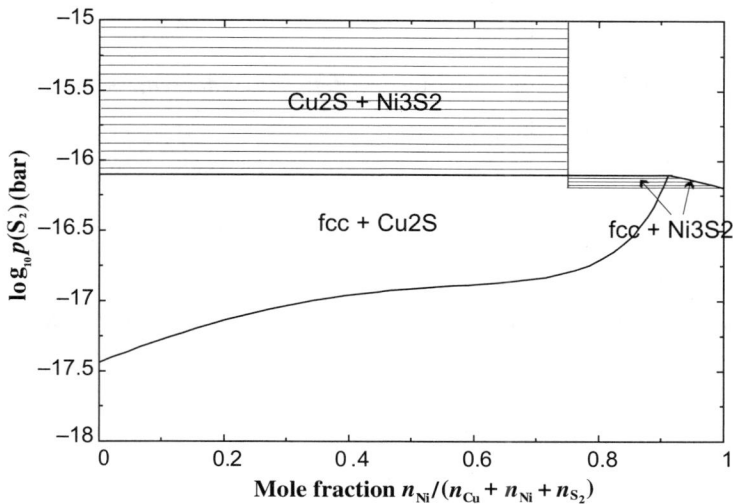

Figure 36 Equilibrium S_2 pressure versus mole fraction of Ni in the Cu–Ni–S_2 system at 427 °C. This is not a single-valued phase diagram (see FactSage).

In order to ensure that a phase diagram section is single-valued at every point, the following procedure for the choice of variables is sufficient:

(1) From each corresponding pair (ϕ_i, q_i) in **Table 1**, select either ϕ_i or q_i, but not both. At least one of the selected variables must be an extensive variable q_i.
(2) Define the size of the system as some function of the selected extensive variables (examples: $(n_1 + n_2) = $ constant; $n_2 = $ constant; $V = $ constant; etc.) and then normalize the chosen extensive variables.
(3) The chosen potentials and normalized extensive variables are then the $(C + 1)$ independent phase diagram variables. Two of these are selected as axes and the others are held constant.

Note that this procedure constitutes a set of sufficient (but not necessary) conditions for the diagram to be single-valued. That is, a diagram which violates these conditions may still be single-valued, but there is no assurance that this will be the case.

The procedure is illustrated by the following examples. In each example, the selected variables, one from each conjugate pair, are underlined.

System Fe–Mo (**Figure 1**):

$$\phi_i : \quad \underline{T} \quad \underline{P} \quad \mu_{Fe} \quad \mu_{Mo}$$
$$q_i : \quad S \quad -V \quad \underline{n_{Fe}} \quad \underline{n_{Mo}}$$
$$\text{Size} : (n_{Fe} + n_{Mo})$$
$$\text{Phase diagram variables} : T, P, n_{Mo}/(n_{Fe} + n_{Mo}) = X_{Mo}$$

System Fe–Cr–S_2–O_2 (**Figure 6**):

$$\phi_i : \quad \underline{T} \quad \underline{P} \quad \mu_{Fe} \quad \mu_{Cu} \quad \underline{\mu_{S_2}} \quad \underline{\mu_{O_2}}$$
$$q_i : \quad S \quad -V \quad \underline{n_{Fe}} \quad \underline{n_{Cu}} \quad n_{S_2} \quad n_{O_2}$$
$$\text{Size} : (n_{Fe} + n_{Cu})$$
$$\text{Phase diagram variables} : T, P, \mu_{O_2}, \mu_{S_2}, n_{Cr}/(n_{Fe} + n_{Cr})$$

System Zn–Mg–Al isopleth (**Figure 3a**):

$$\phi_i : \quad \underline{T} \quad \underline{P} \quad \mu_{Zn} \quad \mu_{Mg} \quad \mu_{Al}$$
$$q_i : \quad S \quad -V \quad \underline{n_{Zn}} \quad \underline{n_{Mg}} \quad \underline{n_{Al}}$$
$$\text{Size} : (n_{Mg} + n_{Al})$$
$$\text{Phase diagram variables} : T, P, n_{Al}/(n_{Mg} + n_{Al}), n_{Zn}/(n_{Mg} + n_{Al}) = X_{Zn}/(1 - X_{Zn})$$

System Al_2SiO_5 (**Figure 7b**):

$$\phi_i : \quad \underline{T} \quad \underline{P} \quad \mu_{Al_2SiO_5}$$
$$q_i : \quad S \quad \underline{-V} \quad n_{Al_2SiO_5}$$
$$\text{Size} : n_{Al_2SiO_5}$$
$$\text{Phase diagram variables} : T, V/n_{Al_2SiO_5}$$

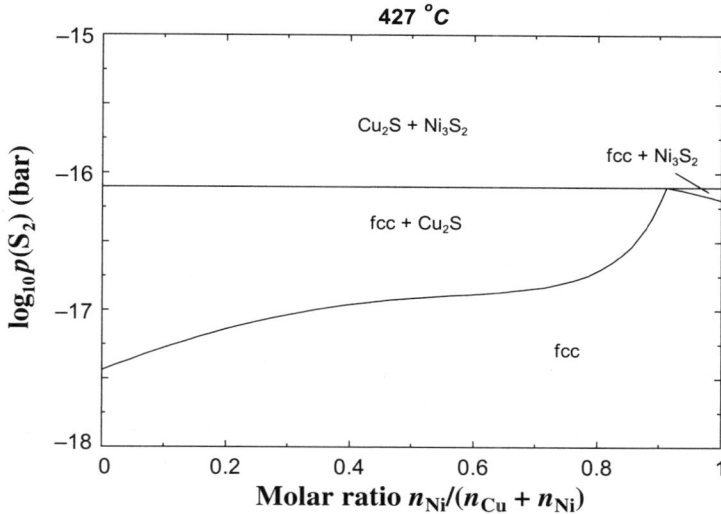

Figure 37 Equilibrium S_2 pressure versus molar ratio $n_{Ni}/(n_{Cu} + n_{Ni})$ in the Cu–Ni–S_2 system at 427 °C. This is a single-valued phase diagram (see FactSage).

System Cu–Ni–S_2 (**Figure 37**):

$$\phi_i : \quad \underline{T} \quad \underline{P} \quad \mu_{Cu} \quad \mu_{Ni} \quad \underline{\mu_{S_2}}$$

$$q_i : \quad S \quad -V \quad \underline{n_{Cu}} \quad \underline{n_{Ni}} \quad n_{S_2}$$

Size : $(n_{Cu} + n_{Ni})$

Phase diagram variables : $T, P, \mu_{S_2}, n_{Ni}/(n_{Cu} + n_{Ni})$

This choice of variables for the Cu–Ni–S_2 system gives the single-valued phase diagram in **Figure 37**. For the same system in **Figure 36**, an incorrect choice of variables was made, leading to a diagram which is not single-valued. The x-axis variable of **Figure 36** contains the term n_{S_2}. That is, both variables (μ_{S_2}, n_{S_2}) of the same corresponding pair were selected, thereby violating the first selection rule. Similarly, the selection of both variables $(P, -V)$ from the same corresponding pair results in the diagram of **Figure 35**, which is not single-valued. Similar examples of phase diagrams that are not single-valued have been shown by Hillert (1997).

Other examples of applying the procedure for choosing variables will be given in Sections 3.7.8–3.7.10.

In several of the phase diagrams in this chapter, log p_i or $RT \ln p_i$ is substituted for μ_i as an axis variable or constant. From Eqn 86, this substitution can clearly be made if T is constant. However, even when T is an axis variable of the phase diagram as in **Figure 10**, this substitution is still permissible since μ_i^o is a monotonic function of T. The substitution of log p_i or $RT \ln p_i$ for μ_i results in a progressive expansion and displacement of the axis with increasing T that preserves the overall geometry of the diagram.

3.7.6 Corresponding Phase Diagrams

If a potential axis, ϕ_i, of a phase diagram section is replaced by its normalized conjugate variable q_i, then the new diagram and the original diagram may be said to be *corresponding phase diagrams*. An example is seen in **Figure 7**, where the *P*-axis of **Figure 7a** is replaced by the *V*-axis of **Figure 7b**. Another example is shown in **Figure 38** for the Fe–O system. Placing corresponding diagrams beside each other as in **Figures 7 and 38** is useful because the information contained in the two diagrams is complementary. Note how the invariant triple points of one diagram of each pair correspond to the invariant lines of the other. Note also that as ϕ_i, increases, q_i increases. That is, in passing from left to right at constant *T* across the two diagrams in **Figure 38**, the same sequence of phase regions is encountered. This is a general result of the thermodynamic stability criterion which will be discussed in Section 3.7.7.

A set of corresponding phase diagrams for a multicomponent system with components 1,2,3,.........,*C* is shown in **Figure 39**. The potentials $\phi_4, \phi_5, \ldots, \phi_{C+2}$ are held constant. The size of the system (in the sense of Section 3.7.5) is q_3. The invariant triple points of **Figure 39b** correspond to the invariant lines of **Figure 39a** and d and to the invariant tie-triangles of **Figure 39c**. The *critical points* of **Figure 39b**, where the two-phase regions terminate, correspond to the critical points of the "miscibility gaps" in **Figure 39a**, c and d.

A set of corresponding phase diagrams for the Fe–Cr–O$_2$ system at $T = 1300\,°C$ is shown in **Figure 40**. **Figure 40b** is a plot of μ_{Cr} versus μ_{O_2}. **Figure 40a** is similar to **Figure 5**. **Figure 40c** is a plot of $n_O/(n_{Fe} + n_{Cr})$ versus $n_{Cr}/(n_{Fe} + n_{Cr})$, where the size of the system is defined by $(n_{Fe} + n_{Cr})$. The coordinates of **Figure 40c** are known as *Jänecke coordinates*.

Figure 40c may be "folded up" to give the more familiar isothermal ternary section of **Figure 40d** plotted on the equilateral composition triangle.

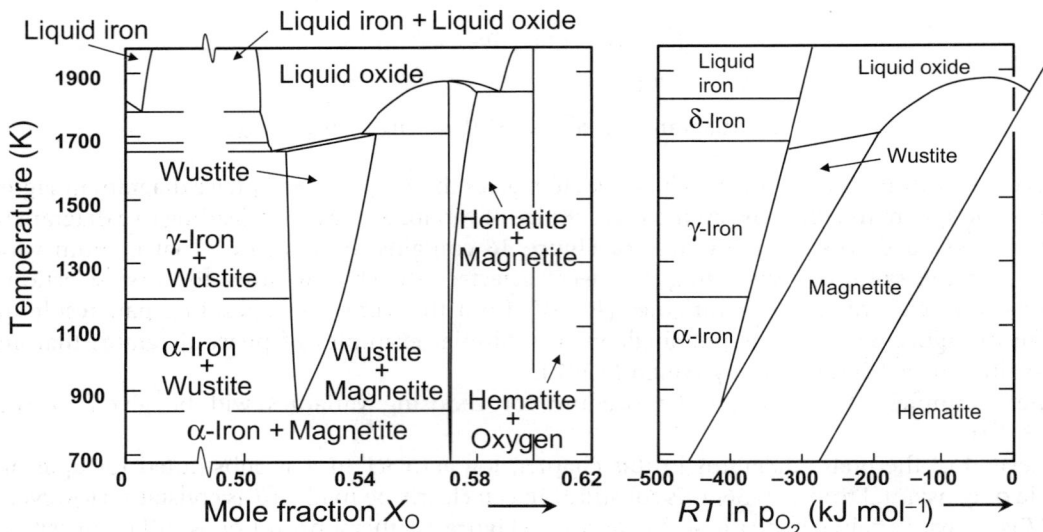

Figure 38 Corresponding T–X_O and $T - (\mu_{O_2} - \mu_{O_2}^0)$ phase diagrams of the Fe–O system (after Muan and Osborn, 1965).

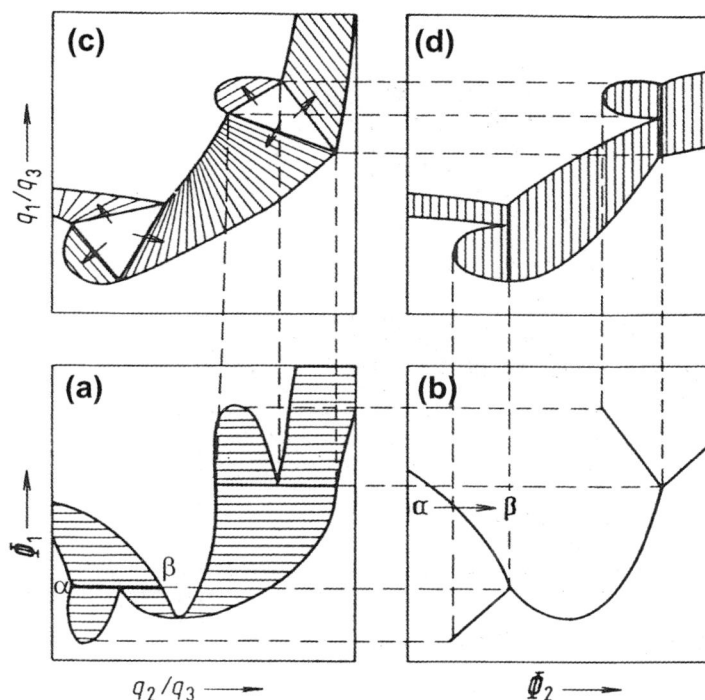

Figure 39 A corresponding set of phase diagrams for a C-component system when the potentials ϕ_4, ϕ_5, ..., ϕ_{C+2} are constant (Pelton and Schmalzried, 1973).

3.7.6.1 Enthalpy-Composition Phase Diagrams

The T-composition phase diagram of the Mg–Si system is shown in **Figure 41**a. A corresponding phase diagram could be constructed by replacing the T-axis by the molar entropy because (T,S) is a corresponding pair, although such a diagram would not be of much practical interest. However, just as $(dS/dT) > 0$ at constant P and composition, so also is $(dH/dT) > 0$. Hence, an H-composition phase diagram will have the same geometry as an S-composition diagram. (This will be demonstrated more rigorously in Section 3.7.12.)

Figure **41**b is the enthalpy-composition phase diagram corresponding to the T-composition diagram of **Figure 41**a. The y-axis is $(h - h_{25})$, where h_{25} is the molar enthalpy of the system in equilibrium at 25 °C. Referencing the enthalpy to h_{25} slightly distorts the geometry of the diagram because h_{25} is a function of composition. In particular, the tie-lines are no longer necessarily perfectly straight lines. However, this choice of reference renders the diagram more practically useful since the y-axis variable then directly gives the heat, which must be added, or removed, to heat, or cool, the system between 25 °C and any temperature T. The isothermal lines shown in **Figure 41**b are not part of the phase diagram but have been calculated and plotted to make the diagram more useful. It should be stressed that **Figure 41**b is a phase diagram section obeying all the geometrical rules of phase diagram sections discussed in Sections 3.7.1–3.7.4. It consists of ZPF lines and can be calculated using the same

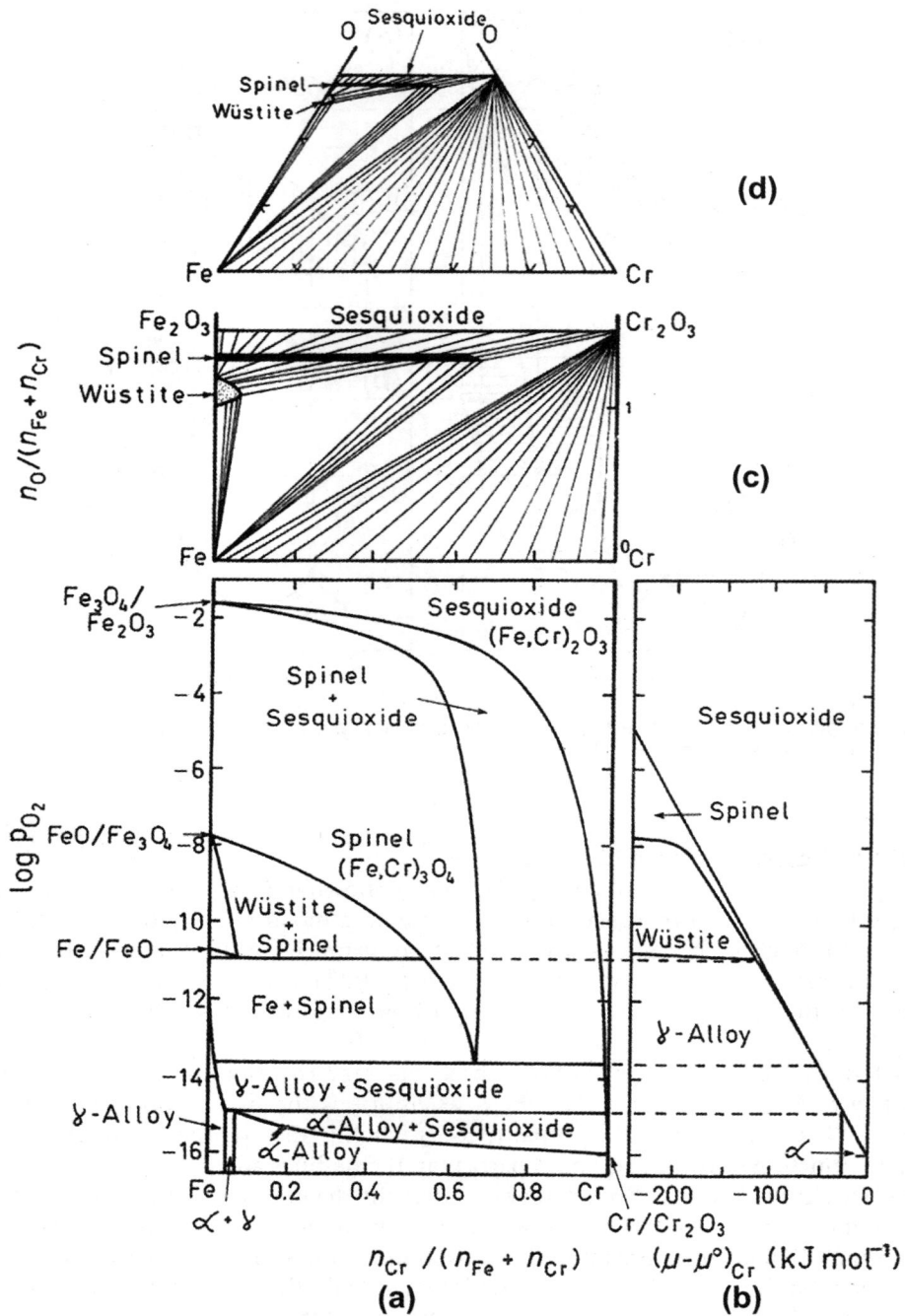

Figure 40   Corresponding phase diagrams of the Fe–Cr–O₂ system at 1300 °C (Pelton and Schmalzried, 1973).

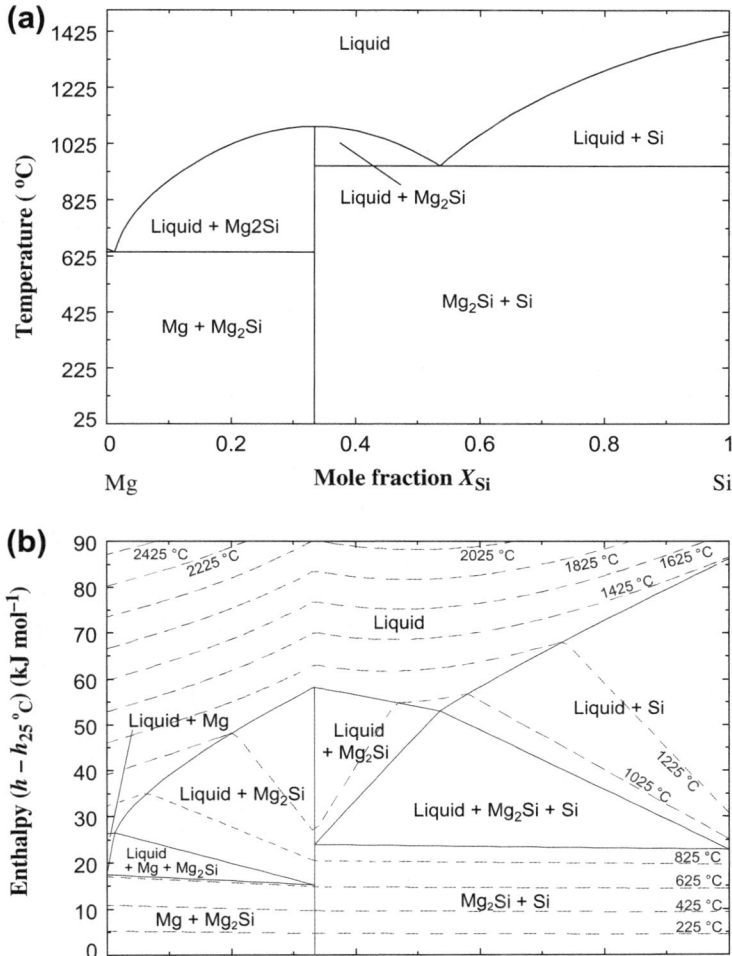

Figure 41 Corresponding T-composition and enthalpy-composition phase diagrams of the Mg–Si system at $P = 1$ bar. The enthalpy axis is referred to the enthalpy of the system at equilibrium at 25 °C (see FactSage). (For color version of this figure, the reader is referred to the online version of this book.)

algorithm as for any other phase diagram section as discussed in Section 3.7.4.1. Note that the three-phase tie-triangles are isothermal.

Another H-composition phase diagram is shown in **Figure** 3b for the Zn–Mg–Al system along the isoplethal section where $X_{Zn} = 0.1$.

3.7.7 The Thermodynamics of General Phase Diagram Sections

This section provides a more rigorous derivation of the geometry of general phase diagram sections that was discussed in Sections 3.7.2–3.7.6.

In Section 3.2.8.2, a general auxiliary function G' was introduced. Let us now define an even more general auxiliary function, B, as follows:

$$B = U - \sum_{i=1}^{m} \phi_i q_i \quad 0 \leq m \leq (C+2) \tag{154}$$

where the ϕ_i and q_i include all corresponding pairs in **Table 1** (if $m = 0$, then $B = U$). Following the same reasoning as in Section 3.2.8.2, we obtain for a system at equilibrium:

$$dB = -\sum_{1}^{m} q_i d\phi_i + \sum_{m+1}^{C+2} \phi_i dq_i \tag{155}$$

and the equilibrium state at constant ϕ_i ($0 \leq i \leq m$) and q_i ($m + 1 \leq i \leq C + 2$) is approached by minimizing B:

$$dB_{\phi_i(0 \leq i \leq m), q_i(m+1 \leq i \leq C+2)} \leq 0 \tag{156}$$

Figure 42 is a plot of the $\phi_1 - \phi_2 - \phi_3$ surfaces of two phases, α and β, in a C-component system when $\phi_4, \phi_5, \ldots, \phi_{C+2}$ are all held constant. For example, in a one-component system, **Figure 42** would be a plot of the $P-T-\mu$ surfaces. Let us now choose a pair of values ϕ_1 and ϕ_2 and derive the condition for equilibrium. Since all potentials except ϕ_3 are now fixed, the appropriate auxiliary function, from Eqns 154 and 70, is $B = q_3\phi_3$. (See also Eqns 73 and 76). Furthermore, let us fix the size of the system by setting q_3 constant. Then, from Eqn 156, the equilibrium state for any given pair of values of ϕ_1 and ϕ_2 is given by setting $dB = q_3 d\phi_3 = 0$, that is by minimizing ϕ_3. This is illustrated in **Figure 42**, where the base plane of the figure is the $\phi_1 - \phi_2$ phase diagram of the system at constant $\phi_4, \phi_5, \ldots, \phi_{C+2}$. For example, in a one-component system, if ϕ_1, ϕ_2, ϕ_3 are P, T, μ, respectively, then the $P-T$ phase diagram is given by minimizing μ, the molar Gibbs energy of the system. Clearly, therefore, a $\phi_1 - \phi_2$ diagram at constant $\phi_4, \phi_5, \ldots, \phi_{C+2}$ is single-valued.

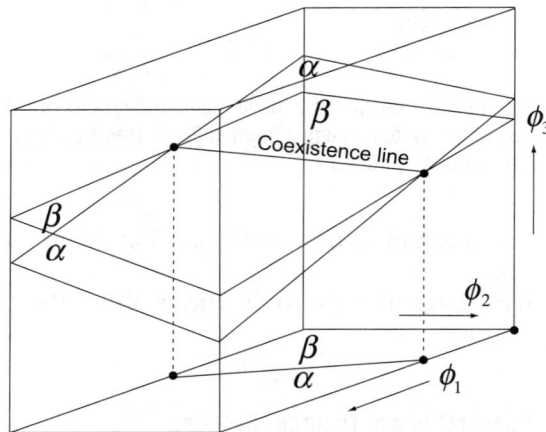

Figure 42 $\phi_1 - \phi_2 - \phi_3$ surfaces of α and β phases in a C-component system when $\phi_4, \phi_5, \ldots, \phi_{C+2}$ are constant (Reprinted with permission from Hillert, 2008).

As another example, consider the $T - \mu_{O_2}$ phase diagram of the Fe–O system at constant hydrostatic pressure in **Figure 38**. The potentials of the system are T, P, μ_{O_2} and μ_{Fe}. At constant P, and for any given values of T and μ_{O_2}, the equilibrium state of the system is given by minimizing μ_{Fe}. Similarly, the predominance diagram of the Cu–SO_2–O_2 system in **Figure 9** can be calculated by minimizing μ_{Cu}, which can be shown to be equivalent to the procedure described in Section 3.2.4.2.

A general ϕ_1–ϕ_2 phase diagram at constant ϕ_4, ϕ_5, ..., ϕ_{C+2} was shown in **Figure 39b**. The size of the system is defined by fixing q_3. We wish now to show that when **Figure 39b** is "opened up" by replacing ϕ_2 by q_2 to give the corresponding phase diagram in **Figure 39a**, the sequence of phase regions encountered in moving horizontally across **Figure 39a** is the same as in **Figure 39b**. For example, when the two-phase ($\alpha + \beta$) line in **Figure 39b** is "opened" up to form the two-phase ($\alpha + \beta$) region in **Figure 39a**, the α and β phases remain on the left and right sides, respectively, of the two-phase region.

We first derive the general *thermodynamic stability criterion* for a single-phase region. Imagine two bordering microscopic regions, I and II, within a homogeneous phase and suppose that a local fluctuation occurs in which an amount $dq_n > 0$ of the extensive property q_n is transferred from region I to region II at constant ϕ_i ($0 \leq i \leq m$) and constant q_i ($m + 1 \leq i \leq C + 2$), where $i \neq n$. Suppose that, as a result of the transfer, ϕ_n increases in region I and decreases in region II, that is, that $(d\phi_n/dq_n) < 0$. A potential gradient is thus set up with $\phi_n^I > \phi_n^{II}$. From Eqn 155, further transfer of q_n from region I to II then results in a decrease in $B : dB = (\phi^{II} - \phi^I)dqn < 0$. From Eqn 156, equilibrium is approached by minimizing B. Hence, the fluctuation grows, with continual transfer from region I to II. That is, the phase is unstable. Therefore, the criterion for a phase to be thermodynamically stable is:

$$d\phi_n/dq_n > 0 \tag{157}$$

for any choice of constant ϕ_i or q_i constraints ($i \neq n$). Since $\phi_n = (\partial B/\partial q_n)$, the stability criterion can also be written as:

$$\left(\partial^2 B/\partial q_n^2\right)\phi_{i(0 \leq i \leq m)}, q_{i(m+1 \leq i \leq C+2); i \neq n} > 0 \tag{158}$$

An important example is illustrated in **Figure 18** for the solid phase of the Au–Ni system. In this case, the appropriate auxiliary function is the Gibbs energy, G. In the central composition regions of the Gibbs energy curves in **Figure 18**, between the spinodal points indicated by the letter s, $(\partial^2 G/\partial n_{Ni}^2)_{T,P,n_{Fe}} < 0$. Hence, at compositions between the spinodal points, microscopic composition fluctuations will grow since this leads to a decrease in the total Gibbs energy of the system. Hence, the phase is unstable and phase separation can occur by spinodal decomposition as discussed in Chapter 7 of this volume. Other well-known examples of the stability criterion are $(dT/dS) > 0$ (entropy increases with temperature), and $(dP/d(-V)) > 0$ (pressure increases with decreasing volume).

Returning now to **Figure 39**, it can be seen from Eqn 157 that moving horizontally across a single-phase region in **Figure 39b** corresponds to moving in the same direction across the same region in **Figure 39a**. Furthermore, in the two-phase ($\alpha + \beta$) region, suppose that ϕ_2 is increased by an amount $d\phi_2 > 0$ keeping all other potentials except ϕ_3 constant, thereby displacing the system into the single-phase β region. From the Gibbs–Duhem Equation 75:

$$d\phi_3^\alpha = -(q_2/q_3)^\alpha d\phi_2 \quad \text{and} \quad d\phi_3^\beta = -(q_2/q_3)^\beta d\phi_2 \tag{159}$$

where $(q_2/q_3)^\alpha$ and $(q_2/q_3)^\beta$ are the values at the boundaries of the two-phase region.

Since, as has just been shown, the equilibrium state is given by minimizing ϕ_3, it follows that:

$$\mathrm{d}(\phi_3^\beta - \phi_3^\alpha)/\mathrm{d}\phi_2 = (q_2/q_3)^\alpha - (q_2/q_3)^\beta < 0 \tag{160}$$

Hence, the α-phase and β-phase region lie to the left and right, respectively, of the $(\alpha + \beta)$ region in **Figure 39a**, and **Figure 39a** is single-valued.

Since the stability criterion also applies when extensive variables are held constant, it can similarly be shown that when **Figure 39a** is "opened up" by replacing the ϕ_1 axis by q_1/q_3 to give **Figure 39c**, the sequence of the phase regions encountered as the diagram is traversed vertically is the same as in **Figure 39a**, and that **Figure 39c** is single-valued. Finally, it can be shown that the phase diagrams will remain single-valued if a constant ϕ_4 is replaced by q_4/q_3. **Figure 39a**, for example, is a section at constant ϕ_4 through a three-dimensional ϕ_1-ϕ_4-(q_2/q_3) diagram, which may be "opened up" by replacing ϕ_4 by (q_4/q_3) to give a ϕ_1-(q_4/q_3)-(q_2/q_3) diagram, which is necessarily single-valued because of the stability criterion. Hence, sections through this diagram at constant (q_4/q_3) will also be single-valued.

Consider now the node in a general phase diagram section shown in **Figure 33**. At any point on either phase boundary of the $(\alpha_1 + \alpha_2 + ... + \alpha_N)$ region, one additional phase, β or γ, becomes stable. Since there is no thermodynamic reason why more than one additional phase should become stable at exactly the same point, the Law of Adjoining Phase Regions is evident. These phase boundaries intersect at the node, below which there must clearly be a region in which β and γ are simultaneously stable. If the x-axis of the diagram is an extensive variable, then this region will have a finite width since there is no thermodynamic reason for the β-phase to become unstable at exactly the same point that the γ-phase becomes stable. It then follows that exactly four boundaries meet at a node. Of course, if the x-axis is a potential variable, then the two-phase boundaries of the $(\alpha_1 + \alpha_2 + ... + \alpha_N + \beta + \gamma)$ region will necessarily be coincident.

3.7.7.1   Schreinemakers' Rule
Schreinemakers' Rule (Schreinemakers, 1915; Pelton, 1995), which was discussed in Section 3.7.4, will now be proven for the general case. Consider the point p in **Figure 33**, which lies in the $(\alpha_1 + \alpha_2 + ... + \alpha_N + \gamma)$ region arbitrarily close to the node. If precipitation of γ were prohibited by kinetic constraints, point p would lie in the $(\alpha_1 + \alpha_2 + ... + \alpha_N + \beta)$ region since it lies below the dashed extension of the phase boundary line as shown in the figure. Let B be the appropriate auxiliary function. If precipitation of γ is prohibited, then $\mathrm{d}B/\mathrm{d}n^\beta < 0$, where $\mathrm{d}n^\beta$ is the amount of β-phase (of fixed composition) which precipitates. That is, precipitation of β decreases B. If γ is permitted to precipitate, then precipitation of β is prevented. Therefore,

$$\frac{\mathrm{d}}{\mathrm{d}n^\gamma}\left(\frac{\mathrm{d}B}{\mathrm{d}n^\beta}\right) > 0 \tag{161}$$

Since the order of differentiation does not matter:

$$\frac{\mathrm{d}}{\mathrm{d}n^\beta}\left(\frac{\mathrm{d}B}{\mathrm{d}n^\gamma}\right) > 0 \tag{162}$$

Hence, the extension of the other boundary of the $(\alpha_1 + \alpha_2 + ... + \alpha_N)$ region must pass into the $(\alpha_1 + \alpha_2 + ... + \alpha_N + \beta)$ region as shown in **Figure 33**. Similarly, it can be shown that if the extension of one boundary of the $(\alpha_1 + \alpha_2 + \ldots + \alpha_N)$ region passes into the $(\alpha_1 + \alpha_2 + \ldots + \alpha_N + \beta + \gamma)$ region, then so must be the extension of the other (Pelton, 1995).

3.7.8 Interpreting Oxide Phase Diagrams

Three phase diagrams for the Fe–Si–O$_2$ system are shown in **Figures 43–45**. Diagrams of these types are often a source of confusion.

Figure 43 is often called the "phase diagram of the Fe$_3$O$_4$–SiO$_2$ system." However, it is not a binary phase diagram, nor is it a quasibinary phase diagram. Rather, it is an isoplethal section across the

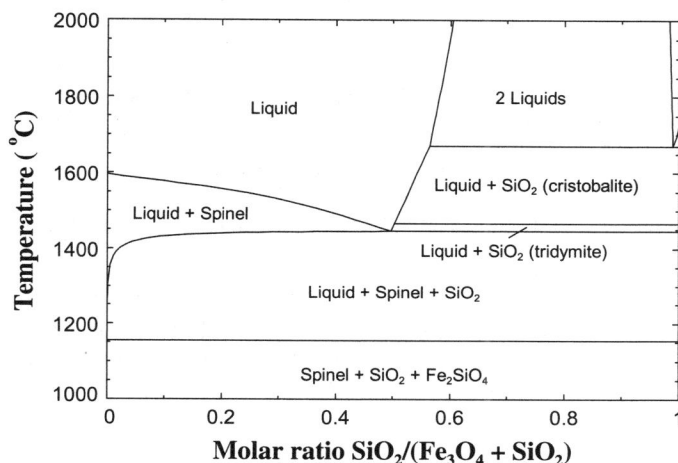

Figure 43   Isoplethal section across the Fe$_3$O$_4$–SiO$_2$ join of the Fe–Si–O$_2$ system. ($P = 1$ bar) (see FactSage).

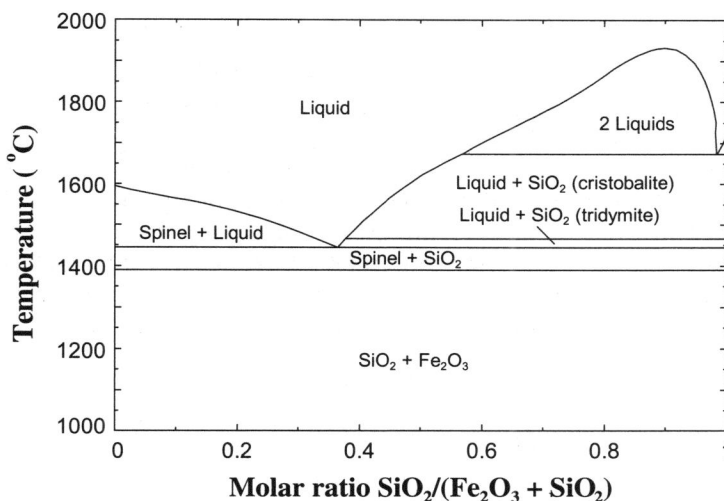

Figure 44   "Fe$_2$O$_3$–SiO$_2$ phase diagram in air ($p_{O_2} = 0.21$ bar)" (see FactSage).

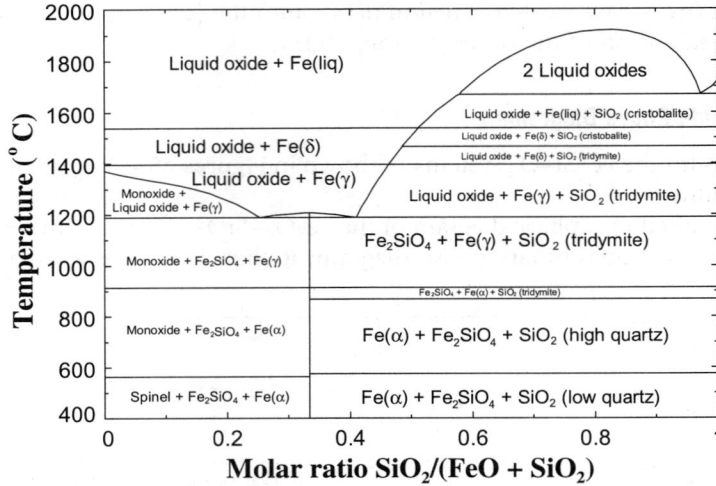

Figure 45 "FeO–SiO$_2$ phase diagram at Fe saturation" (see FactSage).

Fe$_3$O$_4$–SiO$_2$ joint of the Fe–Si–O$_2$ composition triangle. As with any isoplethal section, tie-lines in **Figure 43** do not necessarily lie in the plane of the diagram. That is, the x-axis gives only the overall composition of the system, not the compositions of individual phases. The lever rule does not apply. In the scheme of Section 3.7.5, **Figure 43** is described as follows:

$$\phi_i : \quad \underline{T} \quad \underline{P} \quad \mu_{Fe_3O_4} \quad \mu_{SiO_2} \quad \mu_{O_2}$$

$$q_i : \quad S \quad -V \quad \underline{n_{Fe_3O_4}} \quad \underline{n_{SiO_2}} \quad n_{O_2}$$

Size : $(n_{Fe_3O_4} + n_{SiO_2})$

Phase diagram variables : $T, P, n_{SiO_2}/(n_{Fe_3O_4} + n_{SiO_2}), n_{O_2}/(n_{Fe_3O_4} + n_{SiO_2})$

with $n_{O_2}/(n_{Fe_3O_4} + n_{SiO_2}) = 0$. The components have been defined judiciously as Fe$_3$O$_4$–SiO$_2$–O$_2$ rather than Fe–Si–O$_2$. While this is not necessary, it clearly simplifies the description.

Figure 44 is often called the "Fe$_2$O$_3$–SiO$_2$ phase diagram at constant p_{O_2}." This diagram is actually a ternary section along a constant p_{O_2} line as shown schematically in **Figure 46**. Note that the position of this line for any given p_{O_2} will vary with temperature. At lower temperatures, the line approaches the Fe$_2$O$_3$–SiO$_2$ join. Since p_{O_2} is the same in all phases at equilibrium, the compositions of all phases, as well as the overall composition, lie along the constant p_{O_2} line. Hence, **Figure 44** has the same topology as binary T-composition phase diagrams. All tie-lines lie in the plane of the diagram as illustrated in **Figure 46**, and the lever rule applies. In the scheme of Section 3.7.5, **Figure 44** is described as

$$\phi_i : \quad \underline{T} \quad P \quad \mu_{Fe_2O_3} \quad \mu_{SiO_2} \quad \underline{\mu_{O_2}}$$

$$q_i : \quad S \quad P \quad \underline{n_{Fe_2O_3}} \quad \underline{n_{SiO_2}} \quad n_{O_2}$$

Size : $(n_{Fe_2O_3} + n_{SiO_2})$

Phase diagram variables : $T, P, n_{SiO_2}/(n_{Fe_2O_3} + n_{SiO_2}), \mu_{O_2}$

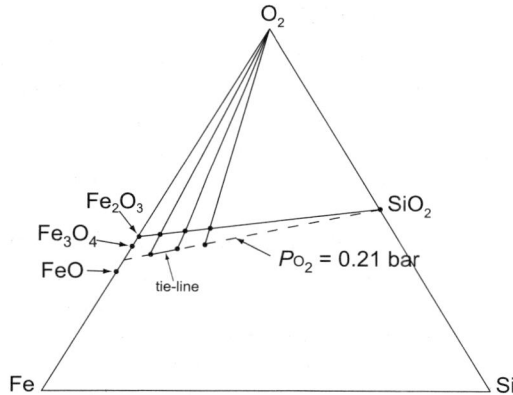

Figure 46   Constant p_{O_2} line in the Fe–Si–O$_2$ system illustrating the construction and interpretation of **Figure 44**.

The phase boundary lines in **Figure 44** give the ratio $n_{SiO_2}/(n_{Fe_2O_3} + n_{SiO_2})$ in the individual phases of the Fe$_2$O$_3$–SiO$_2$–O$_2$ system at equilibrium. However, they do not give the total composition of the phases, which may be deficient in oxygen. That is, some of the Fe may be present in a phase as FeO. The diagram may also be visualized as a projection of the equilibrium phase boundaries from the O$_2$-corner of the composition triangle onto the Fe$_2$O$_3$–SiO$_2$ join as illustrated in **Figure 46**. From an experimental standpoint, the x-axis represents the relative amounts of an initial mixture of Fe$_2$O$_3$ and SiO$_2$ that was charged into the apparatus at room temperature and that was subsequently equilibrated with air at higher temperatures.

Figure 45 is often called the "FeO–SiO$_2$ phase diagram at Fe saturation." Usually, this diagram is drawn without indicating the presence of Fe in each phase field although metallic Fe is always present as a separate phase. The horizontal lines in **Figure 45**, which indicate the allotropic transformation temperatures and melting point of Fe, are also generally not drawn. Diagrams such as **Figure 45** are best regarded and calculated as isoplethal sections with excess Fe. **Figure 45** was calculated by the following scheme:

$$\phi_i : \quad \underline{T} \quad \underline{P} \quad \mu_{FeO} \quad \mu_{SiO_2} \quad \mu_{Fe}$$

$$q_i : \quad S \quad -V \quad n_{FeO} \quad n_{SiO_2} \quad n_{Fe}$$

$$\text{Size} : (n_{FeO} + n_{SiO_2})$$

$$\text{Phase diagram variables} : T, \ P, n_{SiO_2}/(n_{FeO} + n_{SiO_2}), n_{Fe}/(n_{FeO} + n_{SiO_2})$$

with $n_{Fe}/(n_{FeO} + n_{SiO_2}) = 0.0001$. That is, the components of the system have been defined as FeO, SiO$_2$ and O$_2$ with a sufficient excess of Fe so as to always give a metallic phase at equilibrium. Note that the phase boundary compositions in **Figure 45** give only the ratio $n_{SiO_2}/(n_{FeO} + n_{SiO_2})$ in each phase but not the total compositions of the phases, which may contain an excess or a deficit of iron relative to the FeO–SiO$_2$ join. In this system, the solubility of Si in Fe is very small, and so has a negligible effect on the phase diagram. However, in other systems such as Fe–Zn–O$_2$, the equilibrium metal phase could actually contain more Zn than Fe, depending upon the conditions.

3.7.9   Choice of Components and Choice of Variables

As was discussed in Section 3.3, the components of a given system may be defined in different ways. A judicious selection can simplify the choice of variables for a phase diagram as has just been illustrated in Section 3.7.8. As another example, the selection of variables for the isothermal log p_{SO_2} − log p_{O_2} phase diagram of the Cu–SO$_2$–O$_2$ system in **Figure 9** according to the scheme of Section 3.7.5 is clearly simplified if the components are formally selected as Cu–SO$_2$–O$_2$ rather than, for example, as Cu–S–O.

In certain cases, different choices of variables can result in different phase diagrams. For the same system as in **Figure 9**, suppose that we wish to plot an isothermal predominance diagram with log p_{SO_2} as y-axis but with log a_{Cu} as the x-axis. The gas phase will consist principally of O$_2$, S$_2$ and SO$_2$ species. As the composition of the gas phase varies at constant total pressure from pure O$_2$ to pure S$_2$, the partial pressure of SO$_2$ passes through a maximum where the gas phase consists mainly of SO$_2$. Therefore, for the phase diagram to be single-valued, the log p_{SO_2} axis must encompass gaseous mixtures which consist either mainly of (O$_2$ + SO$_2$) or of (SO$_2$ + S$_2$), but not both. Let us define the components as Cu–SO$_2$–O$_2$ and choose the phase diagram variables as follows:

$$\phi_i : \quad \underline{T} \quad \underline{P} \quad \mu_{Cu} \quad \mu_{SO_2} \quad \mu_{O_2}$$

$$q_i : \quad S \quad -V \quad n_{Cu} \quad n_{SO_2} \quad \underline{n_{O_2}}$$

Size : n_{O_2}

Phase diagram variables : T, P, μ_{Cu}, μ_{SO_2}, n_{O_2}

As p_{SO_2} increases at constant T, P and n_{O_2}, p_{O_2} decreases while p_{S_2} always remains small. The phase diagram then essentially encompasses only the Cu–SO$_2$–O$_2$ subsystem of the Cu–S$_2$–O$_2$ system. A phase field for Cu$_2$O will appear on the diagram, for example, but there will be no field for Cu$_2$S. If, on the other hand, we choose the components as Cu–SO$_2$–S$_2$ with the size of the system defined as n_{S_2} = constant, then p_{S_2} varies along the p_{SO_2} axis while p_{O_2} always remains small. The phase diagram then encompasses only the Cu–SO$_2$–S$_2$ subsystem. A phase field for Cu$_2$S appears, but there will be no field for Cu$_2$O.

This example serves to illustrate again how the procedure for choosing variables given in Section 3.7.5 provides sufficient conditions for a phase diagram to be single-valued.

3.7.10   Phase Diagrams of Reciprocal Systems

A section at $T = 750\,°C$ and $P = 1.0$ bar of the *reciprocal salt system*, NaCl–CaCl$_2$–NaF–CaF$_2$ is shown in **Figure 47**. A reciprocal salt system is one consisting of two or more cations and two or more anions. Since excess Na, Ca, F and Cl are virtually insoluble in the ionic salts, this system is, to a very close approximation, a quasi-ternary isoplethal section of the Na–Ca–Cl–F system in which:

$$(n_F + n_{Cl}) = (2n_{Ca} + n_{Na}) \tag{163}$$

That is, the total cationic charge is equal to the total anionic charge. The compositions of all phases lie in the plane of the diagram and the lever rule applies.

The corners of the composition square in **Figure 47** represent the pure salts, while the edges of the square represent compositions in common-ion quasi-binary systems. The choice of phase diagram

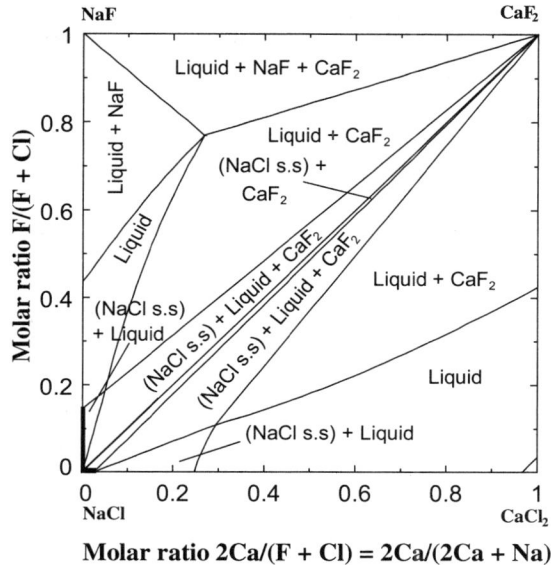

Figure 47 Phase diagram at $T = 750\ ^{\circ}C$ and $P = 1$ bar of the reciprocal system $NaCl–CaCl_2–NaF–CaF_2$ (see FactSage).

variables is most simply formulated, using the procedure of Section 3.7.5, as follows:

$$\phi_i : \quad \underline{T} \quad \underline{P} \quad \mu_{Na} \quad \mu_{Ca} \quad \mu_F \quad \mu_{Cl}$$

$$q_i : \quad S \quad -V \quad n_{Na} \quad n_{Ca} \quad n_F \quad n_{Cl}$$

Size : $(n_F + n_{Cl}) = (2n_{Ca} + n_{Na})$

Phase diagram variables : $T,\ P, n_F/(n_F + n_{Cl}), 2n_{Ca}/(2n_{Ca} + n_{Na}), (2n_{Ca} + n_{Na})/(n_F + n_{Cl}) = 1$

The axis variables of **Figure 47** are sometimes called "equivalent ionic fractions", in which the number of moles of each ion is weighted by its absolute charge. (Note that if Na, Ca, F or Cl were actually significantly soluble in the salts, the diagram could still be plotted as in **Figure 47**, but it would then no longer be a quasi-ternary section and tie-lines would not necessarily lie in the plane of the diagram, nor would the lever rule apply.)

3.7.11 Choice of Variables to Ensure Straight Tie-lines

Figure 47 could be drawn with the axis variables chosen instead as $n_F/(n_F + n_{Cl})$ and $n_{Ca}/(n_{Ca} + n_{Na})$, that is, as molar ionic fractions rather than equivalent ionic fractions.

In this case, the diagram would be a distorted version of **Figure 47**, but would still be a valid phase diagram obeying all the geometrical rules. However, tie-lines in two-phase regions would no longer be straight lines. That is, a straight line joining the compositions of two phases at equilibrium would not pass through the overall composition of the system. It is clearly desirable from a practical standpoint that tie-lines be straight.

It can be shown through simple mass balance considerations (Pelton and Thompson, 1975) that the tie-lines of a phase diagram, in which both axes are composition variables, will only be straight lines if the denominators of the two composition variable ratios are the same. This is the case in **Figure 47** where the denominators are the same because of the condition of Eqn 163. For another example, see **Figure 39c**. It can thus be appreciated that the procedure of Section 3.7.5 provides a sufficient condition for tie-lines to be straight by the stratagem of normalizing all composition variables with respect to the same system "size."

It can also be shown (Pelton and Thompson, 1975) that ternary phase diagram sections plotted on a composition triangle always have straight tie-lines.

3.7.12   Other Sets of Corresponding Variable Pairs

The set of corresponding variable pairs in **Table 1**, Section 3.7.1 is only one of the several possible sets. Substituting S from Eqn 71 into the Gibbs–Duhem Equation 75 and rearranging terms yields:

$$-Hd(1/T) - (V/T)dP + \sum n_i d(\mu_i/T) = 0 \qquad (164)$$

Hence, another set of corresponding variable pairs is that given in **Table 2**.

With this set of conjugate variable pairs replacing those in **Table 1**, the same procedure as described in Section 3.7.5 can be used to choose variables to construct single-valued phase diagrams. For example, the procedure for the H-composition phase diagram in **Figure 41b** is as follows:

$$\phi_i: \quad 1/T \quad \underline{P} \quad \mu_{Mg}/T \quad \mu_{Si}/T$$
$$q_i: \quad \underline{-H} \quad -V/T \quad n_{Mg} \quad n_{Si}$$

Size : $(n_{Mg} + n_{Si})$

Phase diagram variables : $H,\ P,\ n_{Mg}/(n_{Mg} + n_{Si})$

This demonstrates that **Figure 41b** is, in fact, a single-valued phase diagram. Although, in this particular case, an S-composition phase diagram would also be single-valued, and very similar in form to **Figure 41b**, H cannot simply be substituted for S in **Table 1** in all cases.

Many other sets of conjugate variable pairs can be obtained through appropriate thermodynamic manipulation. The set shown in **Table 2** and many other such sets have been discussed by Hillert (1997). For example, $(-A/T,T/P)$, $(-TU/P,1/T)$, $(\mu_i/T,n_i)$ is one such set. In conjunction with the selection rules of Section 3.7.5, all these sets of conjugate pairs can be used to construct single-valued phase diagram sections. However, these diagrams would probably be of very limited practical utility.

Table 2   Corresponding pairs of potentials ϕ_i and extensive variables q_i

$\phi_i:$	$1/T$	P	μ_i/T	μ_2/T	...	μ_C/T
$q_i:$	$-H$	$-V/T$	n_1	n_2	...	n_C

3.7.13 Extension Rules for Polythermal Projections

An extension rule for the univariant lines on polythermal projections is illustrated in **Figure 27** at point A. At any intersection point of three univariant lines, the extension of each univariant line passes between the other two. The proof, although straightforward, is rather lengthy and so will not be reproduced here.

3.8 Thermodynamic Databases for the Computer Calculation of Phase Diagrams

As discussed in Section 3.1, the past 25 years have witnessed a rapid development of large evaluated optimized thermodynamic databases and of software that accesses these databases to calculate phase diagrams by Gibbs energy minimization.

The databases for multicomponent solution phases are prepared by first developing an appropriate mathematical model, based upon the structure of the solution, which gives the thermodynamic properties as functions of temperature and composition for every phase of a system. Next, all available thermodynamic data (calorimetric data, activity measurements, etc.) and phase equilibrium data from the literature for the entire system are evaluated and simultaneously "optimized" to obtain one set of critically evaluated self-consistent parameters of the models for all phases in two-component, three-component and, if available, higher order systems. Finally, the models are used to estimate the properties of multicomponent solutions from the databases of parameters of lower order subsystems. Recently, experimental data are being supplemented by "virtual data" from first principles calculations; for a review, see Turchi et al. (2007).

The optimized model equations are consistent with thermodynamic principles and with theories of solutions. The phase diagrams can be calculated from these thermodynamic equations, and so one set of self-consistent equations describes simultaneously all the thermodynamic properties and the phase diagrams. This technique of analysis greatly reduces the amount of experimental data needed to fully characterize a system. All data can be tested for internal consistency. The data can be interpolated and extrapolated more accurately. All the thermodynamic properties and the phase diagram can be represented and stored by means of a relatively small set of coefficients.

This technique permits the estimation of phase diagrams of multicomponent systems based only upon data from lower order subsystems and allows any desired phase diagram section or projection to be calculated and displayed rapidly. In **Figure 48** is a calculated phase diagram section for a six-component system, which might be of interest in the design of new Mg alloys. This diagram was calculated on a laptop computer in approximately 2 min. Hence, many sections of a multicomponent system can be quickly generated for study.

The computer calculation of phase diagrams present many advantages. For example, metastable phase boundaries can be readily calculated by simply removing one or more stable phases from the calculation. An example is shown in **Figure 49** where the metastable liquid miscibility gap was calculated. Other metastable phase boundaries such as the extension of a liquidus below the eutectic temperature can be calculated similarly. Another advantage is the possibility of calculating phase compositions at points on an isoplethal section. For instance, compositions of individual phases at equilibrium cannot be read directly from **Figure 48**. However, with the calculated diagram on the computer screen, and with appropriate software (see FactSage), one can simply place the cursor at any desired point on the diagram and "click" in order to calculate the amounts and compositions of all phases in equilibrium at that point. Other software can follow the course of equilibrium cooling or of nonequilibrium "Scheil–Gulliver cooling" as will be discussed in Section 3.9. And, of course, the thermodynamic databases permit the

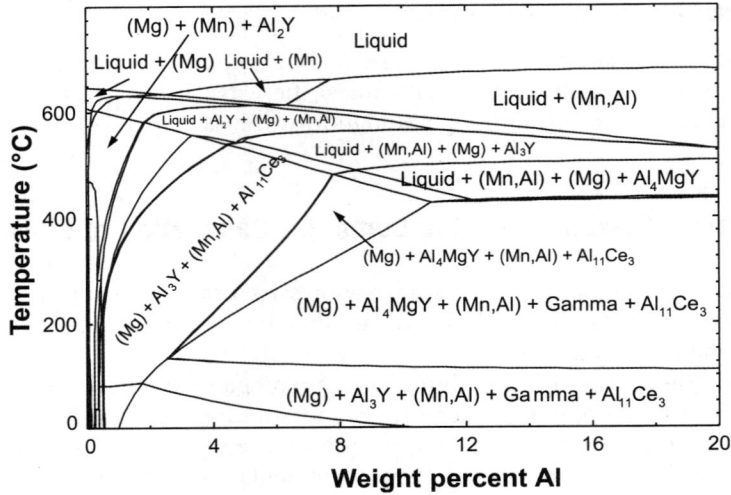

Figure 48   Isoplethal section of the Mg–Al–Ce–Mn–Y–Zn system at 0.05% Ce, 0.5% Mn, 0.1% Y and 1.0% Zn (weight percentage) at $P = 1$ bar (see FactSage).

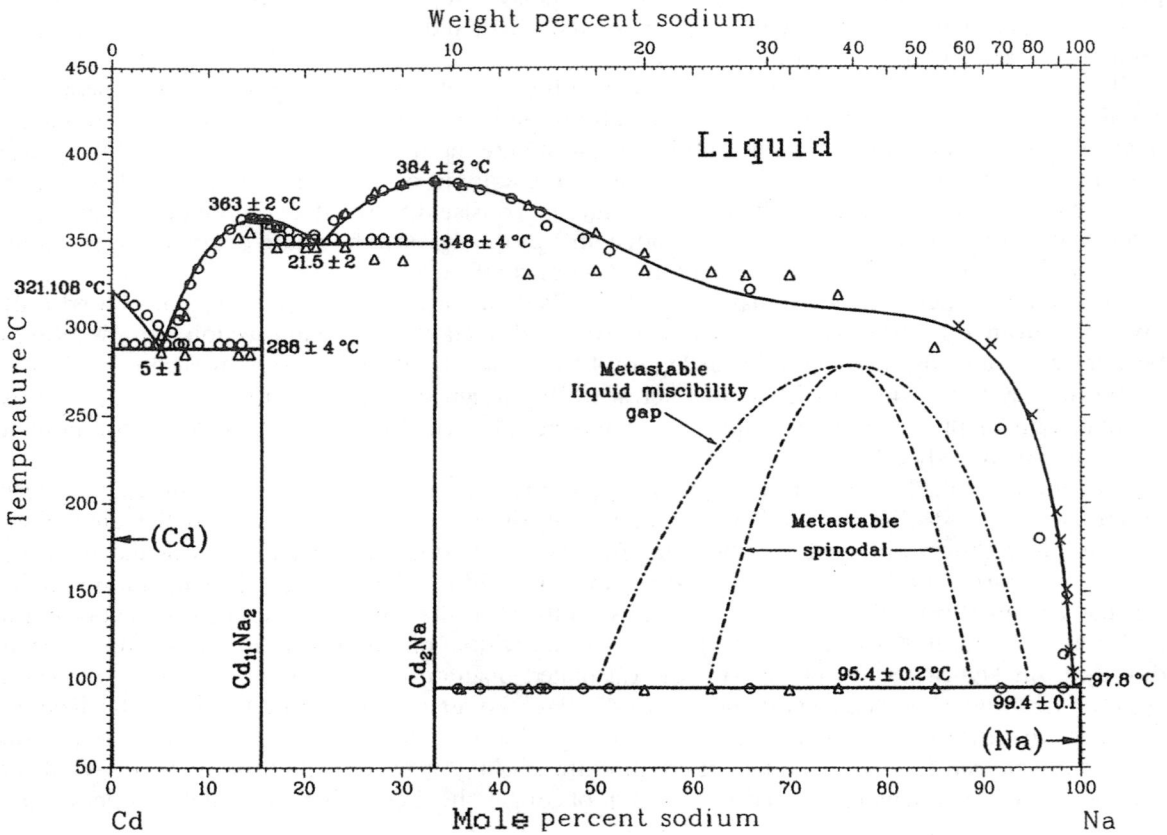

Figure 49   Cd–Na phase diagram at $P = 1$ bar calculated from optimized thermodynamic parameters showing experimental data points. O Kurnakow and Kusnetzow (1907), △ Mathewson (1906), X Weeks and Davies (1964).

calculation of chemical potentials, which are essential for calculating kinetic phenomena such as diffusion. Coupling thermodynamic databases and software with databases of diffusion coefficients and software for calculating diffusion (as with the Dictra program of Thermocalc (see Thermocalc)) provides a powerful tool for the study of the heat treatment of alloys for instance.

Among the largest integrated database computing systems with applications in metallurgy and material science are Thermocalc, FactSage, MTDATA and Pandat (see list of websites). These systems all combine large evaluated and optimized databases with advanced Gibbs energy minimization software. As well, the SGTE group (see SGTE) has developed a large database for metallic systems.

All these databases are extensive. For example, the SGTE alloy database contains optimized data for 78 components, involving hundreds of solution phases and stoichiometric compounds, 350 completely assessed and optimized binary systems and 120 optimized ternary systems. Equally extensive optimized databases have been developed for systems of oxides, salts, etc.

It is beyond the scope of this chapter to give a detailed account of the techniques of thermodynamic evaluation, optimization and modeling. The following sections are intended only to provide a general idea of what is involved in order to permit an appreciation of the advantages of the technique and of its limitations.

3.8.1 Thermodynamic/Phase Diagram Optimization of a Binary System

As a simple example of the optimization/evaluation of a binary system, we shall consider the Cd–Na system. The phase diagram, with points measured by several authors (Mathewson, 1906; Kurnakow and Kusnetzow, 1907; Weeks and Davies, 1964) is shown in **Figure 49**. From electromotive force measurements on alloy concentration cells, several authors have measured the activity coefficient of Na in liquid alloys. The data are shown in **Figure 50** at 400 °C. From the temperature dependence of $\mu_{Na}^E = RT\ln\gamma_{Na}$, the partial enthalpy of Na in the liquid was obtained via Eqn 114. The results are shown in **Figure 51**. Also, h^E of the liquid has been measured by Kleinstuber (1961) by direct calorimetry.

The system contains four solid phases, Cd, Na, $Cd_{11}Na_2$ and Cd_2Na, which are all assumed to be stoichiometric compounds. We seek expressions for the Gibbs energy of every solid phase and the liquid phase as functions of T and, in the case of the liquid, of composition. An expression for the Gibbs energy of a phase as a function of T and composition constitutes a complete thermodynamic description of the phase since all other thermodynamic properties can be calculated from G by taking the appropriate derivatives.

For pure solid elemental Cd, $h_{298}^o = 0.0$ by convention. From third law measurements (see Section 3.2.5.3), $s_{298}^o = 51.800$ J mol^{-1} K^{-1} (see SGTE). The heat capacity has been measured as a function of T and can be represented by the empirical equation:

$$c_P = 22.044 + 0.012548\ T + 0.139(10^5)T^{-2}\quad \text{J mol}^{-1}\text{K}^{-1} \tag{165}$$

Hence, from Eqns 18, 19 and 57, the following expression is obtained for the molar Gibbs energy of solid Cd:

$$\mu_{Cd}^{o(s)} = -7083 + 99.506\ T - 6.274(10^{-3})T^2 - 6966\ T^{-1} - 22.044\ T\ \ln\ T\quad \text{J mol}^{-1} \tag{166}$$

(where T is in Kelvin). A similar expression is obtained for $\mu_{Na}^{o(s)}$. The Gibbs energies of $Cd_{11}Na_2$ and Cd_2Na have not been measured and are obtained from the optimization procedure.

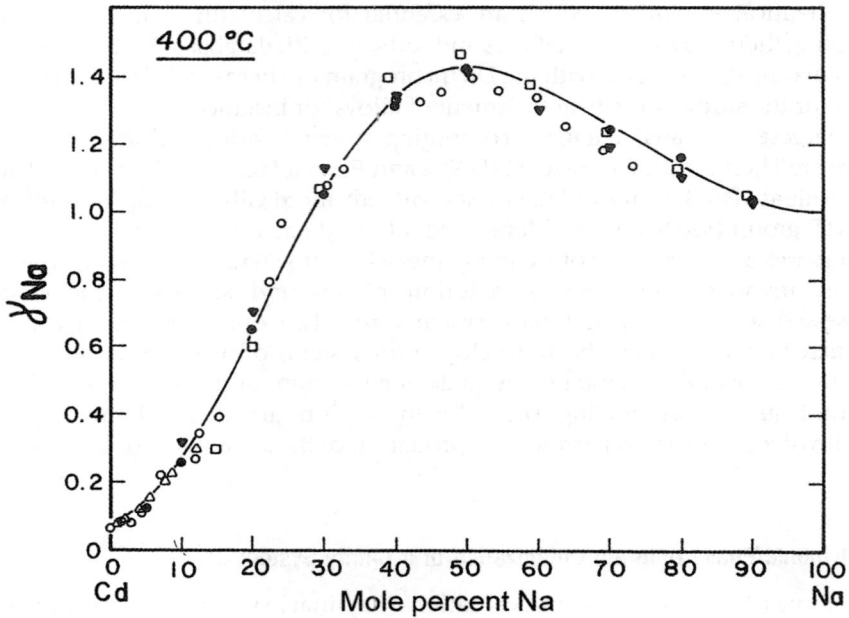

Figure 50   Sodium activity coefficient in liquid Cd–Na alloys at 400 °C. Line is calculated from optimized thermodynamic parameters. ☐ Hauffe (1940), ● Lantratov and Mikhailova (1971), △ Maiorova et al. (1976), ▼ Alabyshev and Morachevskii (1957), O Bartlett et al. (1970).

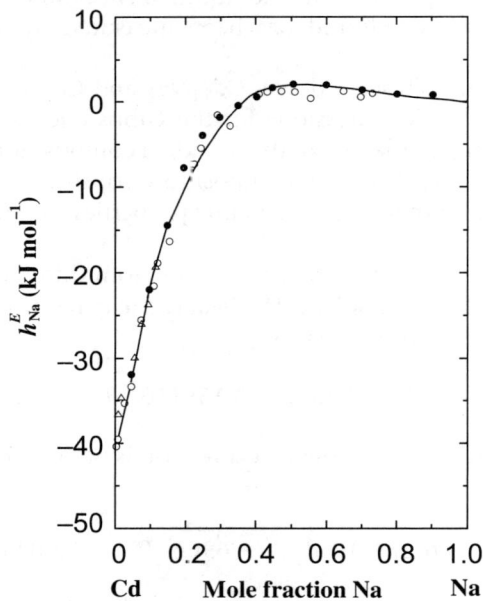

Figure 51   Partial excess enthalpy of sodium in liquid Cd–Na alloys. Line is calculated from optimized thermodynamic parameters. ● Lantratov and Mikhailova (1971), △ Maiorova et al. (1976), O Bartlett et al. (1970).

We now require an expression for the Gibbs energy of the liquid phase. From Eqns 88, 107 and 111,

$$g^l = X_{Cd}\mu_{Cd}^{o(l)} + X_{Na}\mu_{Na}^{o(l)} + RT(X_{Cd}\ln X_{Cd} + X_{Na}\ln X_{Na}) + \left(h^E - Ts^E\right) \tag{167}$$

Expressions similar to Eqn 166 can be obtained for $\mu_{Cd}^{o(l)}$ and $\mu_{Na}^{o(l)}$ from measurements of the melting points, enthalpies of fusion and heat capacities of the liquid phase of the pure elements. For many simple binary substitutional solutions, good representations of h^E and s^E are obtained by expanding the functions as polynomials in the mole fractions X_A and X_B of the components:

$$h^E = X_A X_B \left[h_0 + h_1(X_B - X_A) + h_2(X_B - X_A)^2 + h_3(X_B - X_A)^3 + ...\right] \tag{168}$$

$$s^E = X_A X_B \left[s_0 + s_1(X_B - X_A) + s_2(X_B - X_A)^2 + s_3(X_B - X_A)^3 + ...\right] \tag{169}$$

where the h_i and s_i are empirical parameters to be obtained during optimization. As many parameters are used as are required to represent the data in a given solution. For most solutions, it is a good approximation to assume that the coefficients h_i and s_i are independent of temperature.

If the series are truncated after the first term, then

$$g^E = h^E - Ts^E = X_A X_B(h_0 - Ts_0) \tag{170}$$

This is the equation for a regular solution discussed in Section 3.5.5. Hence, the polynomial representation can be considered to be an extension of regular solution theory. When the expansions are written in terms of the composition variable $(X_B - X_A)$, as in Eqns 168 and 169, they are said to be in *Redlich–Kister form*.

Differentiation of Eqns 168 and 169 and substitution into Eqns 116 and 114 yields the following expansions for the partial excess enthalpies and entropies:

$$h_A^E = X_B^2 \sum_{i=0} h_i \left[(X_B - X_A)^i - 2i X_A(X_B - X_A)^{i-1}\right] \tag{171}$$

$$h_B^E = X_A^2 \sum_{i=0} h_i \left[(X_B - X_A)^i - 2i X_B(X_B - X_A)^{i-1}\right] \tag{172}$$

$$s_A^E = X_B^2 \sum_{i=0} s_i \left[(X_B - X_A)^i - 2i X_A(X_B - X_A)^{i-1}\right] \tag{173}$$

$$s_B^E = X_A^2 \sum_{i=0} s_i \left[(X_B - X_A)^i - 2i X_B(X_B - X_A)^{i-1}\right] \tag{174}$$

Equations 168, 169 and 171–174 are linear in terms of the coefficients. Through the use of these equations, all integral and partial excess properties (g^E, h^E, s^E, μ_i^E, h_i^E, s_i^E) can be expressed by linear equations in terms of the one set of coefficients $\{h_i, s_i\}$. It is thus possible to include all available experimental data for a binary phase in one simultaneous linear optimization.

Several algorithms have been developed to perform such optimizations using least-squares, Bayesian, and other techniques. See, for example, Bale and Pelton (1983), Lukas et al. (1977), Dörner et al. (1980) and Köningsberger and Eriksson (1995).

All the data points shown in **Figures 49–51** along with the enthalpy of mixing data for the liquid (Kleinstuber, 1961) were optimized in one simultaneous operation (Pelton, 1988) to obtain the following expressions for h^E and s^E of the liquid:

$$h^{E(l)} = X_{Cd}X_{Na}\left[-12\,508 + 20\,316(X_{Na} - X_{Cd}) - 8\,714(X_{Na} - X_{Cd})^2\right]\mathrm{J\ mol}^{-1} \tag{175}$$

$$s^{E(l)} = X_{Cd}X_{Na}\left[-15.452 + 15.186(X_{Na} - X_{Cd}) - 10.062(X_{Na} - X_{Cd})^2 \\ - 1.122(X_{Na} - X_{Cd})^3\right]\mathrm{J\ mol}^{-1}\ \mathrm{K}^{-1} \tag{176}$$

During the same simultaneous optimization procedure, the optimized Gibbs energies of fusion of the two compounds were determined:

$$\Delta g^o_{f(1/13Cd_{11}Na_2)} = 6\,816 - 10.724\ T\quad \mathrm{J\ mol}^{-1} \tag{177}$$

$$\Delta g^o_{f(1/3Cd_2Na)} = 8\,368 - 12.737\ T\quad \mathrm{J\ mol}^{-1} \tag{178}$$

The above equations can be combined to give the Gibbs energies of formation of the two compounds from the solid elements. (See **Figure 21** for an illustration of the relation between the Gibbs energy of formation and the Gibbs energy of fusion of a compound.)

Equation 175 reproduces the calorimetric data within 200 J mol^{-1}. Equations 114, 119, 172 and 174 can be used to calculate h^E_{Na} and γ_{Na}. The calculated curves are compared to the measured points in **Figures 50 and 51**. The optimized enthalpies of fusion of the compounds of 6816 and 8368 J mol^{-1} agree within error limits with the values of 6987 and 7878 J mol^{-1} measured by Roos (1916).

The phase diagram in **Figure 49** was calculated from the optimized thermodynamic expressions. Complete details of the analysis of the Cd–Na system are given by Pelton (1988).

It can thus be seen that one simple set of equations can simultaneously and self-consistently describe all the thermodynamic properties and the phase diagram of a binary system.

The exact optimization procedure will vary from system to system depending upon the type and accuracy of the available data, the number of phases present, the extent of solid solubility, etc. A large number of optimizations have been published since 1977 in the *Calphad Journal* (Pergamon).

3.8.2 Solution Models

As discussed in Section 3.8.1, the polynomial model, Eqns 168 and 169, is an extension of regular solution theory (Section 3.5.5). Hence, the polynomial model can be expected to provide an acceptable description of the properties of a solution only if the physical assumptions of regular solution theory are approximately valid, that is, if the solution is approximately a random substitutional solution of atoms or molecules on a single sublattice with relatively weak interactions.

3.8.2.1 Short-Range-Ordering (sro) and the Quasichemical Model

Consider a binary solution of components A–B with a single sublattice (or quasilattice in the case of a liquid) and consider the formation of two nearest-neighbor A–B pairs from an A–A and a B–B pair by the following "quasichemical reaction":

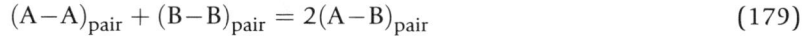

$$(A-A)_{pair} + (B-B)_{pair} = 2(A-B)_{pair} \tag{179}$$

Let the molar Gibbs energy change for this reaction be $(\omega - \eta T)$. This term has essentially the same interpretation in terms of bond energies as the regular solution parameter as in Eqns 136–138.

If $(\omega - \eta T)$ is negative, then reaction in Eqn 179 is displaced to the right, and at equilibrium, the fraction of (A–B) pairs will be greater than in an ideal randomly mixed solution. Conversely, if $(\omega - \eta T)$ is positive, reaction in Eqn 179 is shifted to the left and clustering of A and B particles occurs; if it is sufficiently positive, a miscibility gap will be observed. This sro can be modeled by the quasichemical model in the pair approximation (Guggenheim, 1935; Fowler and Guggenheim, 1939), which has been modified by Pelton and Blander (1984) and Blander and Pelton (1987). For the most recent and complete development of the modified quasichemical model (MQM) in the pair approximation, see Pelton et al. (2000) and Pelton and Chartrand (2001a).

In the MQM, expressions for the enthalpy and entropy of a solution are written in terms of the numbers of (A–A), (B–B) and (A–B) pairs. The expression for the entropy is obtained via the Helmholtz Equation 2 by randomly distributing the pairs over pairs of lattice sites rather than by distributing the atoms or molecules over lattice sites as in the regular solution theory. The parameter of the model, $(\omega - \eta T)$, can be expanded as an empirical polynomial in the component mole fractions similar to Eqns 168 and 169 or as a polynomial in the pair fractions. The numbers of each type of pair at equilibrium are then obtained by minimizing the Gibbs energy. This results in an equilibrium constant for the quasichemical reaction in Eqn 179. In the limit as $(\omega - \eta T)$ approaches zero, the MQM and the polynomial model become identical (Pelton et al, 2000). When $(\omega - \eta T)$ differs significantly from zero, the shapes of the curves of h^E and s^E versus composition predicted by the MQM differ significantly from the approximately parabolic shape (Eqn 133) predicted by the polynomial model, and generally correspond more closely to experimental data. The MQM has been used successfully (see FactSage) to model a large number of solutions, particularly liquid solutions, of metals, salts, sulfides and oxides.

Quasichemical theory has been further extended beyond the pair approximation to account for clusters of more than two particles. This model is known as the Cluster Variation Method (CVM). For a review of the CVM, see Inden (2001). A somewhat simplified version of the CVM, which is mathematically more tractable, is the Cluster Site Approximation (CSA) proposed by Oates and Wenzl (1996); see also Inden (2001).

3.8.2.2 Long-Range-Ordering (lro) and Sublattice Models

lro is generally treated by models which involve two or more sublattices. Ionic salts, for example, are modeled with one sublattice for cations and another for anions. As another example, in the spinal structure of certain ceramic oxides, the oxygen ions occupy one sublattice while some metal ions are distributed on a second sublattice with tetrahedral coordination and other metal ions are distributed on a third sublattice with octahedral coordination. Intermetallic solutions, such as Laves phases, are described by sublattice models. Interstitial solutions are modeled by introducing an interstitial sublattice on which some sites are occupied while others are vacant. Nonstoichiometric compounds A_xB_y with a range of homogeneity, as, for example, the compounds β' and ε in

Figure 22, are modeled as containing point defects such as substitutional defects (some A atoms on B-sublattice sites and B atoms on A-sublattice sites), interstitial atoms, or vacancies on one or more sublattices.

The simplest implementation of sublattice modeling is the Compound Energy Formalism (CEF) (Barry et al., 1992; Hillert, 2008), in which species are distributed randomly over each sublattice. The parameters of the model are, essentially, the site energies for the species on the different sublattices and the interaction energies between pairs of species on different sublattices and on the same sublattice. The equilibrium distribution of species among the sublattices is calculated as that which minimizes the Gibbs energy of the solution. The CEF has been applied successfully to model a very large number of solutions of metals, salts, oxides, sulfides, etc.

The CEF model accounts for lro, but not for sro. In some solutions, a satisfactory representation of the properties can only be obtained by modeling both lro and sro simultaneously. The CVM and CSA models permit this as does an extension of the MQM to systems with two sublattices (Pelton and Chartrand, 2001b). An approximate correction term for relatively small degrees of sro can also be added to the CEF equations (Hillert, 2008).

3.8.3   Estimating Thermodynamic Properties of Ternary and Multicomponent Solutions

Among 70 metallic elements, $70!/3!67! = 54\,740$ ternary systems and $916\,895$ quaternary systems are formed. In view of the amount of work involved in measuring even one isothermal section of a relatively simple ternary phase diagram, it is very important to have a means of estimating ternary and higher order phase diagrams.

For a solution phase in a ternary system with components A–B–C, one first performs evaluations/optimizations on the three binary subsystems A–B, B–C and C–A as discussed in Section 3.8.1 in order to obtain the binary model parameters. Next, the model is used, along with reasonable assumptions, to estimate the thermodynamic properties of the ternary solution.

Suppose that a phase, such as a liquid phase or an fcc phase, has been modeled in all three binary subsystems with the simple polynomial model of Eqns 168 and 169. The Gibbs energy of the ternary solution could then be estimated by the following equation first suggested by Kohler (1960):

$$g^{E} = (1 - X_{A})^{2}g^{E}_{B/C} + (1 - X_{B})^{2}g^{E}_{C/A} + (1 - X_{C})^{2}g^{E}_{A/B} \qquad (180)$$

In this equation, g^{E} is the excess molar Gibbs energy at a composition point in the ternary liquid phase and $g^{E}_{B/C}, g^{E}_{C/A}$ and $g^{E}_{A/B}$ are the excess Gibbs energies in the three binary systems at the same ratios $X_{B}/X_{C}, X_{C}/X_{A}$ and X_{A}/X_{B} as at the ternary point. If the ternary liquid phase as well as the three binary liquid phases are all regular solutions, then Eqn 180 is exact. In the general case, a physical interpretation of Eqn 180 is that the contribution to g^{E} from pair interactions between A and B species is constant at a constant ratio X_{A}/X_{B}, apart from the dilutive effect of the C species which is accounted for by the term $(1 - X_{C})^{2}$ taken from regular solution theory.

Ternary phase diagrams estimated in this way are often quite acceptable. The agreement between the experimental and calculated diagrams can be markedly improved by the inclusion of one or more "ternary terms" with adjustable parameters in the interpolation equations for g^{E}. For example, ternary terms $a_{ijk}X_{A}^{i}X_{B}^{j}X_{C}^{k}$ $(i \geq 1, j \geq 1, k \geq 1)$, which are zero in all three binary subsystems, could be added to Eqn 180 and the values of the parameters a_{ijk}, which gives the "best" fit to experimental ternary thermodynamic or phase equilibrium data could be found by optimization. This, of course, requires that ternary measurements be made, but only a few experimental points will generally suffice rather than the large number of measurements usually required for a fully experimental determination. In this way, the coupling of the thermodynamic approach with a few well-chosen experimental measurements greatly reduces the experimental effort involved in determining multicomponent phase diagrams.

Other equations, similar to the Kohler Equation 180 are often used to estimate the thermodynamic properties of ternary solutions from the model parameters of the binary subsystems when the polynomial model is used. For a discussion, and for extensions to estimating the thermodynamic properties of multicomponent solutions from binary and ternary model parameters, see Pelton (2001b). These equations, like the Kohler equation, are based on the polynomial model, which is an extension of regular solution theory, and they can only be expected to yield good estimates when the physical assumptions of the model are valid; that is, when the solution can reasonably be approximated as a random substitutional solution on a single sublattice with relatively weak interactions. For more complex solutions, involving sro or lro, a good estimate can only be expected when a proper physical model is employed. Therefore, if the solution in question is a liquid with appreciable sro, a model which takes sro into account must be used for the multicomponent solution, including all three binary subsystems, even though it might be possible to force fit a binary solution with a simpler model such as the polynomial model. If the solution exhibits lro, then an appropriate sublattice model must be used. As an example of the prediction of ternary phase diagrams from optimized model parameters for the binary subsystems, the liquid phase of the Zn–Mg–Al system was modeled using the MQM, and binary model parameters were obtained by evaluation/optimization of the three binary subsystems (see FactSage). The thermodynamic properties of the ternary liquid phase were then estimated with the MQM by assuming that the binary parameters remain constant in the ternary solution along lines of constant X_{Zn}/X_{Mg}, X_{Mg}/X_{Al} and X_{Al}/X_{Zn}. When the ternary liquidus projection was then calculated, based only on the binary parameters, a liquidus surface was calculated, which deviates from that shown in **Figure 27** by no more than 8° and 2 mol%. Since experimental data are available for this system, they were used to evaluate one small ternary model parameter. The resultant calculated diagram in **Figure 27** agrees with the available experimental data within the experimental error limits.

Such agreement is not exceptional. Phase diagrams calculated from the best thermodynamic databases generally agree with experimental measurements within or nearly within experimental error limits. In this regard, it may be noted that estimating the properties of a C-component system from the optimized parameters of its $(C - 1)$-component subsystems becomes increasingly more exact as C becomes larger. For example, if all the ternary subsystems of a four-component system have been experimentally investigated and optimized ternary model parameters have been obtained, then the estimation of the properties of the four-component solution are generally good.

For a general discussion of many solution models, see Hillert (2008). A great many models and their applications have been published since 1977 in the *Calphad Journal* (Pergamon).

3.9 Equilibrium and Nonequilibrium Solidification

3.9.1 Equilibrium Solidification

The course of equilibrium solidification can be followed through the use of isoplethal sections, such as that for the six-component Mg alloy shown in **Figure 48**. Polythermal liquidus projections are also useful in this respect. Consider an alloy with composition 80 mol% Mg, 15% Al and 5% Zn at point B in **Figure 27**. As this alloy is cooled at equilibrium, solid-hcp Mg begins to precipitate at the liquidus temperature, which is just below 500 °C. The liquid composition then follows the crystallization path shown in **Figure 27** until it attains a composition on the univariant line. The liquid composition then follows the univariant line toward the point P' with coprecipitation of hcp and the gamma phase. (For a discussion of equilibrium crystallization paths, see Section 3.6.3.) At point P', a quasiperitectic reaction occurs. All the foregoing information can be deduced from **Figure 27**. However, from this diagram alone, it is not possible to determine the amounts and compositions of the various phases at equilibrium because the extent of the solid solutions are not shown nor, for the same reason, is it possible to determine whether solidification is complete at the point P' or whether the liquid persists to lower temperatures.

However, with thermodynamic calculations, this information is readily obtained. Software is available, with tabular or graphical display, to calculate the amounts and compositions of every phase during equilibrium solidification as well as to display the sequence of reactions during solidification. In the current example, the reaction sequence, calculated using the FactSage software and databases, is

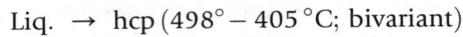

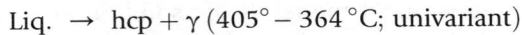

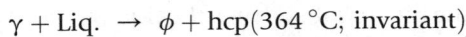

$$\text{Liq.} \rightarrow \text{hcp} \, (498° - 405 °\text{C; bivariant})$$

$$\text{Liq.} \rightarrow \text{hcp} + \gamma \, (405° - 364 °\text{C; univariant})$$

$$\gamma + \text{Liq.} \rightarrow \phi + \text{hcp} (364 °\text{C; invariant})$$

The solidification is calculated to terminate at 364 °C with complete consumption of the liquid by the quasiperitectic reaction. A graphical display of the calculated amounts of all phases during equilibrium cooling is shown in **Figure 52**.

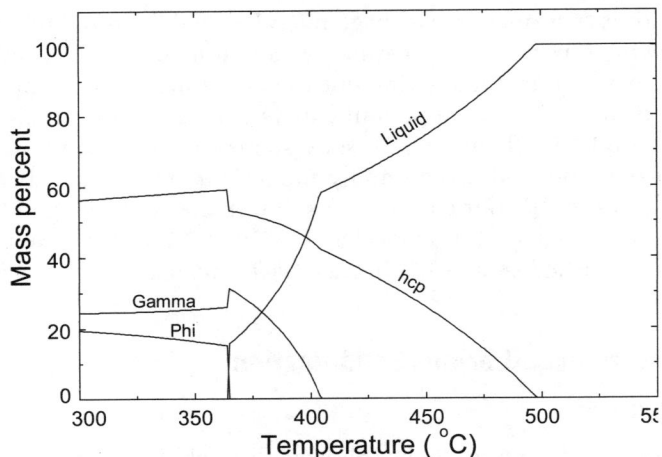

Figure 52   Calculated amounts of phases present during equilibrium cooling of an alloy of composition 85 mol% Mg, 15% Al, 5% Zn (point B in **Figure 27**) (see FactSage).

3.9.2   Nonequilibrium Scheil–Gulliver Solidification

Throughout equilibrium cooling, every solid-solution phase remains homogeneous with no concentration gradients. As was described in Section 3.3.1, during the equilibrium cooling of a Bi–Sb alloy (**Figure 11**) with $X_{Sb} = 0.6$, the first solid precipitate appears at 526 °C with a composition $X_{Sb} = 0.89$. As cooling continues, the composition of the solid phase follows the solidus to point D, at which its composition is $X_{Sb} = 0.60$, and solidification terminates at 349 °C. For all this to happen, diffusion must continually occur in the solid grains in order for their composition to remain uniform. Since solid-state diffusion is a relatively slow process, equilibrium cooling conditions are realized only for very slow cooling rates.

Consider now the limiting case of extremely rapid cooling such that no diffusion whatsoever takes place in the solid phase. However, the liquid is still assumed to be homogeneous, and equilibrium is assumed at the solid/liquid interface. Consequently, there will be a concentration gradient within the grains, with the concentration of Sb decreasing from a maximum of $X_{Sb} = 0.89$ at the points of initial nucleation. When the temperature has decreased to 349 °C, the concentration of Sb at the surface of the solid grains will be $X_{Sb} = 0.60$, but the average composition of the solid will lie between $X_{Sb} = 0.60$ and $X_{Sb} = 0.89$. Hence, a liquid phase is still present at 349 °C and will, in fact, persist down to the melting point of pure Bi, 271.4 °C.

This limiting case is called *Scheil–Gulliver cooling*. As another example, consider the cooling of the hypothetical ternary liquid with composition at point b in **Figure 26**. As discussed in Section 3.6.3, under equilibrium cooling conditions, solidification terminates at point P, where the liquid is completely consumed in the ternary peritectic reaction: Liquid + ε + β → ζ. During Scheil–Gulliver cooling, on the other hand, peritectic reactions (or any reactions involving solid reactants) cannot occur. Hence, the liquid continues to lower temperatures, its composition following the univariant lines past point P'_3 to the ternary eutectic composition E_3, where it is finally consumed in the ternary eutectic reaction. (The liquid does not follow the other path from P to P'_1 because tie-triangles εζX, where X is any point on the line PP'_1, do not contain the liquid composition at point P. That is, ε and ζ cannot coprecipitate from a liquid of composition P because of mass-balance considerations.)

Similarly, for the Zn–Mg–Al system, it was shown in Section 3.9.1 that a liquid of initial composition at point B in **Figure 27**, when cooled under equilibrium conditions, is completely consumed by the quasiperitectic reaction at point P'. Under Scheil–Gulliver cooling conditions, on the other hand, the liquid will continue to lower temperatures, following the univariant lines to the eutectic point E. Furthermore, the path followed by the composition of the liquid along the liquidus during precipitation of the primary hcp phase (shown by the small arrows in **Figure 27**) will be different in the cases of equilibrium and Scheil–Gulliver cooling, as will the compositions and relative amounts of the phases at any temperature. The course of Scheil–Gulliver cooling is easily followed through thermodynamic calculations using the following algorithm. The temperature is reduced from the initial liquid state in incremental steps, typically of the order of 1°–10°. At each step, the equilibrium state is calculated and then all solid phases are removed from the calculation, leaving only the liquid, which is then cooled by another incremental step and the process is repeated. At any temperature, the amount and average composition of every phase can thereby be calculated and presented in tabular or graphical form similar to **Figure 52**. The calculated (FactSage) reaction sequence in the present example is

$$\text{Liq.} \rightarrow \text{hcp} \, (498° - 405 \, °\text{C; bivariant})$$

$$\text{Liq.} \rightarrow \text{hcp} + \gamma \, (405° - 364 \, °\text{C; univariant})$$

$$\text{Liq.} \rightarrow \text{hcp} + \text{Phi} \, (364° - 343°C; \text{univariant})$$

$$\text{Liq.} \rightarrow \text{hcp} + \text{Tau} \, (343° - 341°C; \text{univariant})$$

$$\text{Liq.} \rightarrow \text{hcp} + \text{Tau} + \text{MgZn} \, (341°C; \text{invariant})$$

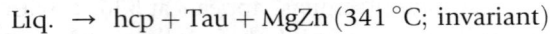

This reaction sequence may be compared to that given in Section 3.9.1 for equilibrium cooling of the same alloy.

3.9.3   General Nomenclature for Invariant and Other Reactions

As discussed in Section 3.5.9, in a binary isobaric temperature-composition phase diagram, there are two possible types of invariant reactions: "eutectic-type" ($\beta \rightarrow \alpha + \gamma$) and "peritectic type" ($\alpha + \gamma \rightarrow \beta$). In a ternary system, there are "eutectic-type" ($\alpha \rightarrow \beta + \gamma + \delta$), "peritectic-type" ($\alpha + \beta + \gamma \rightarrow \delta$), and "quasi-peritectic-type" ($\alpha + \beta \rightarrow \gamma + \delta$) invariants (Section 3.6.3). In a system of C components, the number of types of isobaric invariant reactions is equal to C. A reaction with one reactant, such as ($\alpha \rightarrow \beta + \gamma + \delta$) is clearly a "eutectic-type" invariant reaction but, in general, there is no standard terminology. These reactions may be conveniently described according to the numbers of reactants and products (in the direction which occurs upon cooling). Hence, the reaction ($\alpha + \beta \rightarrow \gamma + \delta + \varepsilon$) is a *2 → 3 reaction*; the reaction ($\alpha \rightarrow \beta + \gamma + \delta$) is a *1 → 3 reaction*; and so on.

A similar terminology can be used for univariant, bivariant, etc. isobaric reactions. For example, a *1 → 2* reaction ($\alpha \rightarrow \beta + \gamma$) is invariant in a binary system, but univariant in a ternary system. The ternary peritectic-type *3 → 1 reaction* ($\alpha + \beta + \gamma \rightarrow \delta$) is invariant in a ternary system, univariant in a quaternary system, bivariant in a quinary system, etc.

3.9.4   Quasi-Invariant Reactions

According to the Phase Rule, an invariant reaction occurs in a C-component system when $(C + 1)$ phases are at equilibrium at constant pressure. Apparent exceptions are observed where an isothermal reaction occurs even though fewer than $(C + 1)$ phases are present. However, these are just limiting cases.

An example was shown in **Figure 17**. A liquid with a composition exactly at the liquidus/solidus minimum in a binary system will solidify isothermally to give a single solid phase of the same composition. Similarly, at a maximum in a two-phase region in a binary system, the liquid solidifies isothermally to give a single solid phase, as, for example, at the congruent melting points of the β' phase in **Figure 22** and of the Ca_2SiO_4 and $CaSiO_3$ compounds in **Figure 21**. Such maxima are also associated with congruently melting compounds in ternary and higher order systems as, for example, the compound η in **Figure 26**.

In a ternary system, a minimum point may be observed on a univariant line as shown in **Figure 53**. In this hypothetical system, there are two solid phases, pure stoichiometric A and a solid solution of B and C, in which A is only slightly soluble. The A–B and C–A binary subsystems, thus, have simple eutectic phase diagrams, while the B–C system resembles **Figure 11**. A liquid with a composition at the minimum point m on the univariant line joining the two binary eutectic points e_1 and e_2 will solidify isothermally to give solid A (α) and a solid solution (β) of composition at point p. Point m is a local liquidus minimum, but the isothermal solidification at point m is not a ternary eutectic reaction since the latter involves three solid phases. Similar quasi-invariant reactions can also occur at liquidus minima in higher order systems.

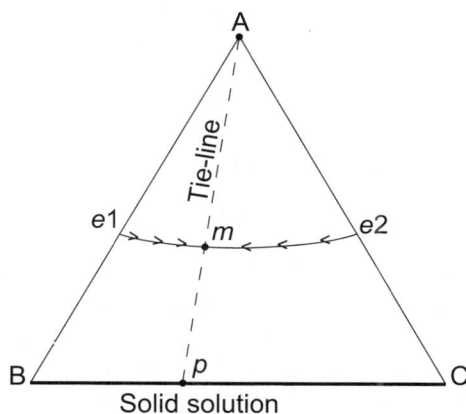

Figure 53 Liquidus projection of a system A–B–C with a minimum on the univariant line joining the binary eutectic points e_1 and e_2. Solid phases are stoichiometric A (α) and a binary solid solution B–C (β).

At point m in **Figure 53**, the eutectic quasi-invariant reaction Liq $\rightarrow \alpha + \beta$ occurs isothermally. A line joining two binary peritectic points p_1 and p_2 in a ternary system can also pass through a minimum at which an isothermal peritectic quasi-invariant reaction occurs: Liq $+ \alpha \rightarrow \beta$.

Another example of a quasi-invariant reaction is that which occurs at saddle points such as the point s in **Figures 26 and 27**. Saddle points occur at maxima on univariant lines. A ternary liquid with a composition exactly at a saddle point will solidify isothermally, at constant composition, to give the two solid phases with which it is in equilibrium. Higher order saddle points can be found in systems where $C \geq 4$.

3.10 Second-Order and Higher-Order Transitions

All the transitions discussed until now have been *first-order transitions*. When a first-order phase boundary is crossed, a new phase appears with extensive properties which are discontinuous with the other phases.

An example of a transition that is not first order is the ferromagnetic-to-paramagnetic transition of Fe, Co, Ni and other ferromagnetic materials that contain unpaired electron spins. Below its Curie temperature, $T_{curie} = 1043$ K, Fe is ferromagnetic since its spins tend to be aligned parallel to one another. This alignment occurs because it lowers the internal energy of the system. At 0 K at equilibrium, the spins are fully aligned. As the temperature increases, the spins become progressively less aligned since this disordering increases the entropy of the system, until at T_{curie}, the disordering is complete and the material becomes paramagnetic. There is never a two-phase region and no abrupt change occurs at T_{curie}, which is simply the temperature at which the disordering becomes complete.

The following model is vastly oversimplified and is presented only as the simplest possible model illustrating the essential features of a second-order phase transition.

Suppose that we may speak of localized spins and, furthermore, that a spin is oriented either "up" or "down." Let the fraction of "up" and "down" spins be ξ and $(1 - \xi)$. Suppose further that when two neighboring spins are aligned, a stabilizing internal energy contribution $\varepsilon < 0$ results. Assuming that the spins are randomly distributed, the probability of two "up" spins being aligned is ξ^2 and of two "down" spins being aligned is $(1 - \xi)^2$. The resultant contribution to the molar Gibbs energy is

$$g = RT(\xi \ln \xi + (1 - \xi) \ln (1 - \xi)) + \left[\xi^2 + (1 - \xi)^2\right]\varepsilon \tag{181}$$

The equilibrium value of ξ is that which minimizes g. Setting $dg/d\xi = 0$ gives

$$\xi/(1 - \xi) = \exp(2(1 - 2\xi)\varepsilon/RT) \tag{182}$$

At $T = 0$ K, $\xi = 0$ (or, equivalently, $\xi = 1$). That is, the spins are completely aligned. A plot of ξ versus T from Eqn 182 shows that ξ increases with T, attaining the value $\xi = 1/2$, when $T = -\varepsilon/R$. At all higher temperatures, $\xi = 1/2$ is the only solution of Eqn 182 above $T_{curie} = -\varepsilon/R$. As the spins become progressively more disordered with increasing temperature below T_{curie}, energy is absorbed, thereby increasing the heat capacity. Most of the disordering occurs over a fairly small range of temperature just below T_{curie}. This can be seen in **Figure 8** where the slope of the enthalpy curve increases just below T_{curie} and then quite abruptly decreases again above this temperature.

In this extremely simple model, there is a discontinuity in the slope $c_P = dh/dT$ at T_{curie}, that is, a discontinuity in the second derivative of the Gibbs energy. Such a transition is called a *second-order transition*.

For proper models of magnetism, see Chapter 19, Volume II. In reality, some ordering persists above T_{curie}, and the transformation is not exactly second order. Transitions involving discontinuities in the third, fourth, etc. derivatives of G are called third-order, fourth-order, etc. transitions.

The T-composition phase diagram of the Fe–Ni system is shown in **Figure 54**. The Curie temperature varies with composition as it traverses the bcc phase field as a line. Nickel also undergoes a magnetic

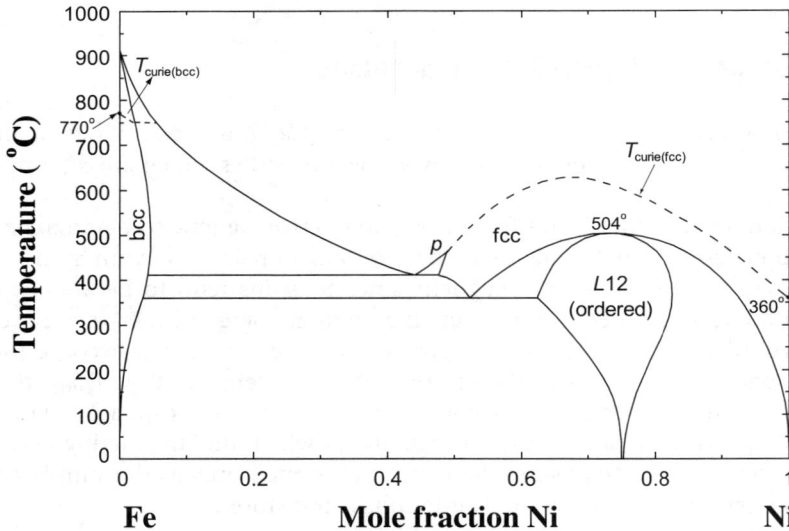

Figure 54 Temperature-composition phase diagram of the Fe–Ni system at $P = 1$ bar showing the magnetic transformations (see FactSage; SGTE; Massalski et al., 2001).

transition at its Curie temperature of 360 °C. The second-order transition line crosses the fcc phase field as shown.

At point *P* in **Figure 54**, the second-order transition line widens into a two-phase miscibility gap and the magnetic transition becomes first order. Point P is called a *tri-critical point*. (Other versions of the Fe–Ni phase diagram do not show a tri-critical point (Massalski et al., 2001).)

The magnetic transition is an example of an *order–disorder transition*. Another important type of order–disorder transition involves an ordering of the crystal structure. An example is seen in the Al–Fe system in **Figure 55**. At low temperatures, the body-centered phase exhibits lro. Two sublattices can be distinguished, one consisting of the corner sites of the unit cells and the other of the body-centered sites. The Al atoms preferentially occupy one sublattice while the Fe atoms preferentially occupy the other. This is the ordered B2 (CsCl) structure. As the temperature increases, the phase becomes progressively disordered, with more and more Al atoms moving to the Fe-sublattice and *vice versa* until, at the transition line, the disordering becomes complete with the Al and Fe atoms distributed equally over both sublattices. At this point, the two sublattices become indistinguishable, the lro disappears, and the structure becomes the A2 (bcc) structure. As with the magnetic transition, there is no two-phase region and no abrupt change at the transition line, which is simply the temperature at which the disordering becomes complete. As with the magnetic transition, most of the disordering takes place over a fairly small range of temperature just below the transition line. The heat capacity, thus, increases as the transition line is approached from below and then decreases sharply as the line is crossed. The transition is thus second order, or approximately second order.

The B2/A2 transition line may terminate at a tri-critical point like point P of **Figure 54**. Although it is not shown in **Figure 55**, there is most likely a tri-critical point near 660 °C.

Order–disorder transitions need not necessarily be of second- or higher order. In **Figure 54**, the L12 phase has an ordered face-centered structure with the Ni atoms preferentially occupying the

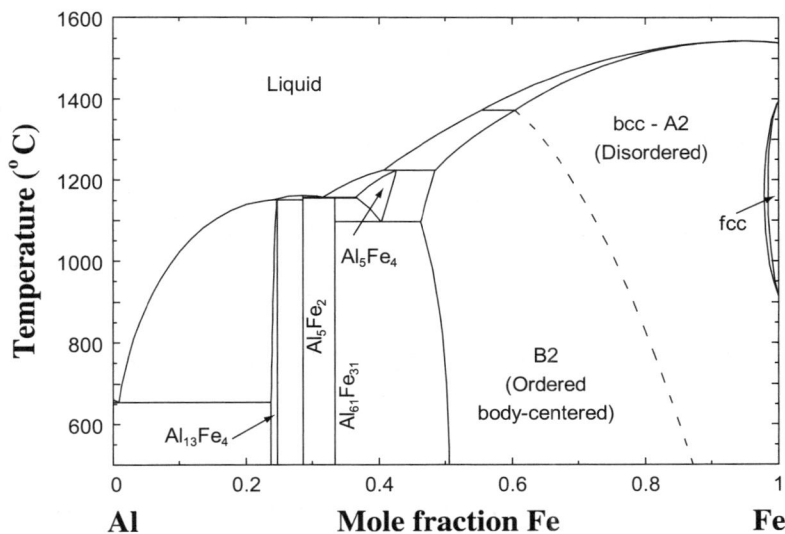

Figure 55 Temperature-composition phase diagram of the Al–Fe system at $P = 1$ bar showing the B2 → A2 order–disorder transition (see FactSage; SGTE).

face-centered sites of the unit cells and the Fe atoms in the corner sites. This structure transforms to the disordered fcc (A1) structure by a first-order transition with a discontinuous increase in enthalpy at the congruent point (504 °C). There is some controversy as to whether or not many reported second-order transition lines may actually be very narrow two-phase regions.

Although there is no lro in the high-temperature disordered structures, appreciable sro remains. For example, in the disordered A2 phase in the Al–Fe system, nearest neighbor (Fe–Al) pairs are still preferred as in the reaction in Eqn 179. Modeling of ordered phases and the order–disorder transition is a complex subject with a large literature. The CEF (Section 3.8.2.2) has been used with some success, but only through the introduction of a large number of adjustable parameters. A quantitative model most likely requires that the coupling of sro and lro be taken into account; the CVM and CSA models (Section 3.8.2.2) have been applied to this end. Ordered phases and order–disorder transitions are treated in Chapter 8 of this volume. See also Inden (2001).

3.11   Bibliography

3.11.1   Phase Diagram Compilations

From 1979 to the early 1990s, the American Society for Metals undertook a project to evaluate critically all binary and ternary alloy phase diagrams. All available literature on phase equilibria, crystal structures and, often, thermodynamic properties were critically evaluated in detail by international experts. Many evaluations have appeared in the *Journal of Phase Equilibria* (formerly *Bulletin of Alloy Phase Diagrams*), (ASM Intl., Materials Park, OH), which continues to publish phase diagram evaluations. Condensed critical evaluations of 4700 binary alloy phase diagrams have been published in three volumes (Massalski et al., 2001). The ternary phase diagrams of 7380 alloy systems have also been published in a 10-volume compilation (Villars et al., 1995). Both binary and ternary phase diagrams from these compilations are available on the website of the ASM alloy phase diagram center. Many of the evaluations have also been published by ASM as monographs on phase diagrams involving a particular metal as a component.

The SGTE (Scientific Group Thermodata Europe) group has produced several subvolumes within volume 19 of the Landolt–Börnstein series (SGTE, 1999–2009) of compilations of thermodynamic data for inorganic materials and of phase diagrams and integral and partial thermodynamic quantities of binary alloys.

Phase diagrams for over 9000 binary, ternary and multicomponent ceramic systems (including oxides, halides, carbonates, sulfates, etc.) have been compiled in the 14-volume series, *Phase Diagrams for Ceramists* (Levin et al, 1964–2005). Earlier volumes were noncritical compilations. However, more recent volumes included critical commentaries. See also the Am. Ceram. Soc./NIST website for the online version.

Hundreds of critically evaluated and optimized phase diagrams of metallic, oxide, salt and sulfide systems may be downloaded from the FactSage website.

3.11.2   Further Reading

A recent text by Hillert (2008) provides a thorough thermodynamic treatment of phase equilibria as well as solution modeling and thermodynamic/phase diagram optimization.

A classical discussion of phase diagrams in metallurgy was given by Rhines (1956). Prince (1966) presents a detailed treatment of the geometry of multicomponent phase diagrams. A series of five volumes edited by Alper (1970–1978) discusses many aspects of phase diagrams in materials science. Bergeron and Risbud (1984) give an introduction to phase diagrams, with particular attention to applications in ceramic systems.

In the *Calphad Journal*, (Pergamon) and in the *Journal of Phase Equilibria*, (ASM Intl., Materials Park, OH) are to be found many articles on the relationships between thermodynamics and phase diagrams.

The SGTE Casebook (Hack, 2008) illustrates, through many examples, how thermodynamic calculations can be used as a basic tool in the development and optimization of materials and processes of many different types.

Acknowledgments

It would have been impossible to write this chapter without the FactSage software. I am indebted to my colleagues Chris Bale, Gunnar Eriksson and Patrice Chatrand. Financial assistance over many years from the Natural Sciences and Engineering Research Council of Canada is gratefully acknowledged.

References

Alabyshev, A.F., Morachevskii, A.G., 1957. Z. Neorg. Khim. 2, 669.

Alper, A.M. (Ed.), 1970-1978. Phase Diagrams —J. Materials Science and Technology, vols. 1–5. Academic, New York.

Bale, C.W., Pelton, A.D., 1983. Metall. Trans. B 14, 77.

Bale, C.W., Pelton, A.D., Thompson, W.T., 1986. Canad. Metall. Quart. 25, 107.

Barry, T.I., Dinsdale, A.T., Gisby, J.A., Hallstedt, B., Hillert, M., Jansson, B., Jonsson, J., Sundman, B., Taylor, J.R., 1992. J. Phase Equilib. 13, 459.

Bartlett, H.E., Neethling, A.J., Crowther, P., 1970. J. Chem. Thermo. 2, 523.

Bergeron, C.J., Risbud, S.H., 1984. Introduction to Phase Equilibria in Ceramics. Columbus. Amer. Ceramic Soc, Ohio.

Blander, M., Pelton, A.D., 1987. Geochim. Cosmochim. Acta 51, 85.

Bray, H.F., Bell, F.D., Harris, S.J., 1961-1962. J. Inst. Metals 90, 24.

Dinsdale, A.T., 1991. Calphad J. 15, 317.

Dörner, P., Henig, E.-Th., Krieg, H., Lucas, H.L., Petzow, G., 1980. Calphad J. 4, 241.

Fowler, R.H., Guggenheim, E.A., 1939. Statistical Thermodynamics. Cambridge Univ. Press, Cambridge. p. 350.

Guggenheim, E.A., 1935. Proc. Roy. Soc. A148, 304.

Gupta, H., Morral, J.E., Nowotny, H., 1986. Scripta Metall. 20, 889.

Hack, K. (Ed.), 2008. SGTE Casebook-thermodynamics at Work, second ed. CRC Press, Boca Raton FL.

Hansen, M., 1958. Constitution of Binary Alloys, second ed. McGraw-Hill, New York.

Hauffe, K., 1940. Z. Elektrochem. 46, 348.

Hillert, M., 1997. J. Phase Equilib. 18, 249.

Hillert, M., 2008. Phase Equilibria, Phase Diagrams and Phase Transformations–their Thermodynamic Basis, second ed. Cambridge Univ. Press, Cambridge.

Hultgren, R., Desai, P.D., Hawkins, D.T., Gleiser, M., Kelly, K.K., Wagman, D.D., 1973. Selected Values of the Thermodynamic Properties of the Elements and Binary Alloys. Am. Soc. Metals, Metals Park, Ohio.

Inden, G., 2001. Atomic ordering. ch. 8. In: Kostorz, G. (Ed.), Phase Transformations in Materials. Wiley-VCH, Weinheim.

Kleinstuber, T., 1961. Ph.D. Thesis, Univ. Munich, Germany.

Kohler, F., 1960. Monatsh. Chem. 91, 738.

Köningsberger, E., Eriksson, G., 1995. A new optimization routine for ChemSage. CALPHAD J. 19, 207–214.

Kurnakow, N.S., Kusnetzow, A.N., 1907. Z. Anorg. Chem. 52, 173.

Lantratov, M.R., Mikhailova, A.G., 1971. Zh. Prikl. Khimii 44, 1778.

Levin, E.M., Robbins, C.R., McMurdie, H.F., 1964-2005. Phase Diagrams for Ceramists. Am. Ceramic Soc., Columbus, Ohio.

Lukas, H.L., Henig, E.-Th., Zimmermann, B., 1977. Calphad J. 1, 225.

Maiorova, E.A., Morachevskii, A.G., Kovalenko, S.G., 1976. Elektrokhimiya 12, 313.

Massalski, T.B., Okamoto, H., Subramanian, P.R., Kacprzak, L., 2001. Binary Alloy Phase Diagrams, second ed. Am. Soc. Metals, Metals Park, OH.

Mathewson, C.H., 1906. Z. Ang. Chem. 50, 180.

Muan, A., Osborn, F., 1965. Phase Equilibria Among Oxides in Steeimaking. Addison Wesley, Reading, MA.

Oates, W.A., Wenzl, H., 1996. Scripta Master. 35, 623.

Palatnik, L.S., Landau, A.I., 1956. Zh. Fiz. Khim. 30, 2399.

Pelton, A.D., 1988. Bull. Alloy Phase Diag. 9, 41.

Pelton, A.D., 1995. J. Phase Equilib. 16, 501.

Pelton, A.D., 2001a. Thermodynamics and phase diagrams of materials. ch. 1. In: Kostorz, G. (Ed.), Phase Transformations in Materials. Wiley-VCH, Weinheim.

Pelton, A.D., 2001b. A general geometric thermodynamic model for multicomponent solutions. Calphad J. 25, 319–328.

Pelton, A. D., Blander, M. 1984. Proc. AIME Symp. Molten Salts Slags. Warrendale, Pa.: The Metall. Soc.AIME, p. 281.

Pelton, A.D., Chartrand, P., 2001a. The modified quasichemical model II – multicomponent solutions. Met. Mat. Trans. 32A, 1355–1360.

Pelton, A.D., Chartrand, P., 2001b. The modified quasichemical model IV – two sublattice quadruplet approximation. Met Mat. Trans. 32A, 1409–1416.

Pelton, A.D., Degterov, S.A., Eriksson, G., Robelin, C., Dessureault, Y., 2000. The modified quasichemical model I binary solutions. Met. Mat. Trans. 31B, 651–660.

Pelton, A.D., Schmalzried, H., 1973. Metall. Trans. 4, 1395–1403.

Pelton, A.D., Thompson, W.T., 1975. Prog. Solid State Chem. 10 (3), 119.

Prince, A., 1966. Alloy Phase Equilibria. Elsevier, Amsterdam.

Rhines, F.N., 1956. Phase Diagrams in Metallurgy. McGraw-Hill, New York.

Roos, G.D., 1916. Z. Anorg. Chem. 94, 329.

Schreinemakers, F. A. H., 1915. Proc. K. Akad. Wetenschappen. Amsterdam (Section of Sciences) 18,116.

SGTE group 1999–2009, (Ed.), Lehrstuhl für Werkstoffchemie, RWTH, Aachen, Thermodynamic Properties of Inorganic Materials, Landolt-Börnstein Group IV, vol. 19. Springer, Berlin.

Turchi, P.E., Abrikosov, I.A., Burton, B., Fries, S.G., Grimvall, G., Kaufman, L., Korzhavyi, P., Manga, V.R., Ohno, M., Pisch, A., Scott, A., Zhang, W., 2007. Interface between quantum-mechanical-based approaches, experiments, and CALPHAD methodology. Calphad J. 31, 4–27.

Van Laar, J.J., 1908. Z. Phys. Chem. 63, 216, 64, 257.

Villars, R., Prince, A., Okamoto, H., 1995. Handbook of Ternary Alloy Phase Diagrams. Am. Soc. Metals, Metals Park, OH.

Weeks, J. R., Davies, H. A., 1964. AEC Report, Conf. 660712–1, BNL-10372.

List of Websites

American Ceramic Society/NIST, phase.ceramics.org

American Society for Metals, alloy phase diagram center, www1.asminternational.org/asmenterprise/apd/

FactSage, Montreal, www.factsage.com

NPL-MTDATA, www.npl.co.uk/advancedmaterials/measurement-techniques/modelling/mtdata

Pandat, www.computherm.com/pandat.html

SGTE, Scientific Group Thermodata Europe, www.sgte.org

Thermocalc, Stockholm, www.thermocalc.com

Biography

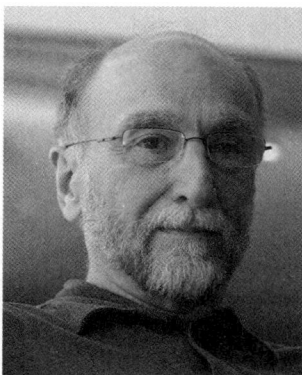

Arthur D. Pelton is Professor Emeritus in the Dep't.of Chemical Engineering and co-director of the Center for Computational Thermochemistry (CRCT) at the Ecole Polytechnique in Montreal, Quebec, Canada.

Dr. Pelton received his undergraduate and graduate degrees from the Dep't. of Metallurgy and Materials Science at the University of Toronto (PhD in 1970). Following post-doctoral studies at the Technical University in Clausthal, Germany and at MIT, he joined the faculty of the Ecole Poly-technique in the Dep't. of Materials Engineering in 1973. Dr. Pelton has co-authored over 250 technical papers and 14 book chapters. He is a Fellow of the Royal Society of Canada, the Canadian Academy of Engineering and of ASM. He is a recipient of the J. Willard Gibbs Phase Equilibria Award from ASM, the Hume-Rothery Prize for work in phase equilibria from the Institute of Metals, UK, and the Gibbs Triangle Award of Calphad. His primary field of interest is chemical thermodynamics in materials science. He is a co-founder of the FactSage thermodynamic database computing system.

4 Metallic Glasses

A.L. Greer, Department of Materials Science & Metallurgy, University of Cambridge, Cambridge, UK

4.1 Introduction

In the fourth edition of *Physical Metallurgy*, "metastable states of alloys" were covered in the chapter by Cahn and Greer (1996). That chapter included coverage of metallic glasses (MGs) as one example of the metastable states of interest at that time. Since that time, MGs have become one of the most studied classes of metallic material, at least when elevated-temperature applications are set aside. This chapter builds on Cahn and Greer (1996), but shifts the focus fully onto MGs. While MGs have some attractions as functional materials, the current interest is mainly in their potential as structural materials. This interest has been stimulated by the increasing range of alloy compositions that can be cast into fully glassy states without the need for high cooling rates. Samples of these so-called bulk metallic glasses (BMGs) have minimum cross-sections of 1 mm to 1 cm.

All alloys of technological importance are, in one way or another, *metastable*—that is, they depart from equilibrium at their composition by not having the lowest possible free energy. The development of new alloys with improved properties has often been a matter of increasing the departure from equilibrium, either in the final product or at an intermediate stage of its manufacture. In the 1950s with the application of techniques for achieving large departures from equilibrium, there emerged an entirely new class of metastable alloys, *metallic glasses*, with the defining characteristic that they lack the long-range atomic order associated with crystals. In this chapter, the nature of the glassy state is described, focusing on its manifestation in metallic systems (Section 4.2). Then the structures of MGs are examined (Section 4.3). The physical metallurgy of structural changes (Section 4.4) and of mechanical properties (Section 4.5) is covered, before ending with a brief overview of applications, actual and potential (Section 4.6).

4.2 Compositions, Thermodynamics and Kinetics

4.2.1 Fundamentals of the Glassy State

A *glass* has essentially the structure of a liquid, combined with the resistance to shear deformation of a solid. A glass is formed when a liquid that is cooled fails to crystallize and instead continuously and uniformly congeals into a solid. Noncrystalline, *amorphous*, alloys can also be obtained by a wide variety of other methods (**Table 1**); these alloys are not, in the strict sense, glasses. Thus glasses are members of the family of amorphous solids, but not all such solids are glasses.

Table 1   Production methods for glassy and amorphous alloys

Method	References
Rapid liquid cooling	
Melt-spinning, planar-flow casting	Liebermann (1980)
	Smith and Saletore (1986)
Atomization	Miller and Murphy (1979)
	Kim et al. (1989)
Wire formation in water	Hagiwara and Inoue (1993)
As a surface treatment:	
Scanned laser or electron beam	Schubert and Bergmann (1993)
	Breinan et al. (1980)
Pulsed laser beam	Lin and Spaepen (1986)
Supercooling of clean liquids	
Emulsion technique	Perepezko and Smith (1981)
Fluxing	Kui et al. (1984)
Solidification in free fall	Drehman and Turnbull (1981)
Physical vapor deposition	
Evaporation	Behrndt (1970)
Sputter-deposition	Leamy and Dirks (1977)
Chemical methods	
Electroless deposition	Watanabe and Tanabe (1976)
Electrodeposition	Brenner et al. (1950)
Precipitation	Wells et al. (1989)
Hydrogenation	Yeh et al. (1983)
Irradiation	
By light or heavy ions	Nastasi et al. (1986)
	Dunlop and Lesueur (1992)
By electrons	Luzzi and Meshii (1988)
By neutrons	Lesueur (1975)
Ion implantation	Hubler et al. (1985)
Ion mixing	Liu et al. (1983)
Mechanical methods	
Grinding	Schwarz et al. (1985)
Mechanical alloying	Koch et al. (1983)
Reactions	
Solid-state reaction of elements	Schwarz and Johnson (1983)
Decomposition of crystalline solid solution	Bormann (1994)

Adapted from Greer (1995) with permission.

It should be noted that amorphous and glassy states of a given composition can be quite distinct. For example, liquid silicon has a high density and is metallic. Crystalline silicon has tetrahedral covalent bonding. Amorphous silicon, which can be made by vapor deposition as well as by solidification from the liquid, also has local tetrahedral coordination with covalent bonding (Shao et al., 1998). In its bonding type, solid amorphous silicon is therefore quite distinct from the liquid and from the hypothetical glass that would be dense and metallic. A consequence is that the crystallization of amorphous silicon is a different transformation from crystallization of liquid silicon, with quite different kinetics (Stiffler et al., 1992).

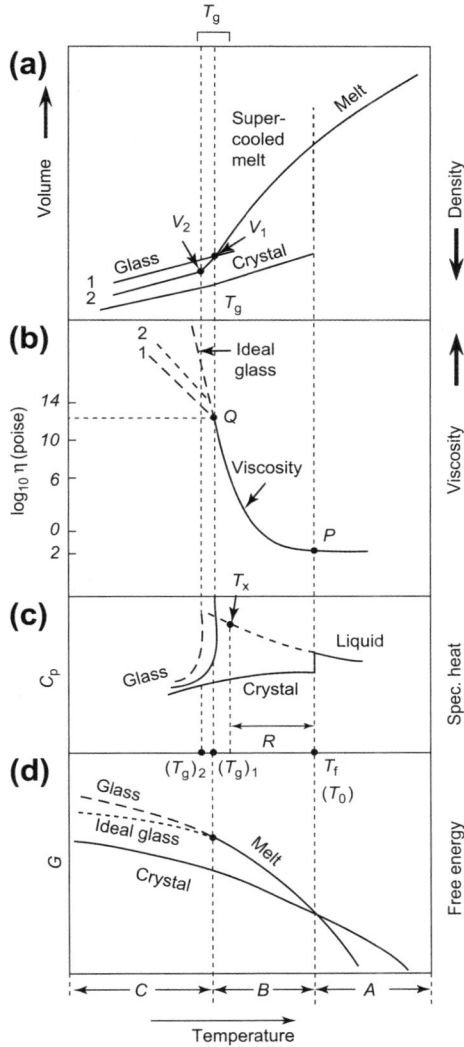

Figure 1   The glass transition: variation of several properties with temperature (4th edition, Figure 2).

Figure 1 shows schematically the changes in properties that occur when a liquid is cooled into the crystalline or glassy states. At the higher temperatures, range A, the melt is in equilibrium. It is difficult (though not impossible (Kelton and Greer, 2010a)) to superheat any crystal, especially metallic crystals above their thermodynamic melting point, so that metastable states in range A are very rare. In defining the boundary between ranges A and B, we use T_f (the equilibrium freezing temperature); for alloys with a solidus–liquidus gap, this is not the same as the melting temperature, but we ignore the complexities arising from an interposed solid + liquid range. In effect, **Figure 1** is for a system in which all the transformations are without solute partitioning.

In range B, the melt is thermodynamically supercooled and will crystallize rapidly if a critical nucleus is provided. At least at the higher temperatures in this range, the liquid has a low viscosity, and crystallization can be avoided only because the rate of cooling is such as to prevent nucleation of the crystalline phase. At the lowest temperatures in this range the liquid viscosity increases rapidly with falling temperature, and a glass can still be formed even in the presence of nuclei because the low atomic mobility stifles crystal growth. Crystallization studies (Section 4.4.3) provide evidence for copious nucleation occurring in the later stages of a quench without significant overall transformation of the liquid. In range B, the viscosity varies over some fifteen orders of magnitude, and the form of this variation has become progressively more measurable as compositions with greater glass-forming ability (GFA) (greater resistance to crystallization) have been developed. The form of viscosity change is discussed in Section 4.2.4. At all temperatures in range B the atomic mobility is adequate for the liquid to remain in configurational equilibrium (that is, in an internal equilibrium in which it has the lowest free energy of any state that it can reach by continuous change).

Crystallization, if it occurs, involves a discontinuous change in properties (as shown, for example, by volume in **Figure 1**). Crystallization can occur only in range B, and if it is avoided on cooling, range C is reached. Range C is defined by the glass-transition temperature T_g, below which the liquid configuration is congealed in a pattern that corresponds to equilibrium at T_g. Experimentally, T_g is close to the temperature at which the viscosity reaches 10^{12} Pa s. The nature of the transition at T_g can be seen by considering the temperature dependence of liquid/glass properties—for example volume and viscosity as shown in **Figure 1**. Above T_g the properties show a strong temperature dependence, reflecting the configurational changes in the liquid. Below T_g, however, atomic motions in the melt are so slow that during the cooling there is no time for configurational changes. Corresponding to the lack of configurational change in the glass, its properties have a rather weak temperature dependence, roughly matching the behavior of the crystal.

It is important to recognize that T_g is *not* a thermodynamically defined temperature; its location is determined wholly by kinetics. If the cooling rate is reduced (but not so much as to permit crystallization in range B), then the changing temperature-dependent liquid configurations can stay in equilibrium to a lower temperature, that is $(T_g)_1$ is lowered to, say, $(T_g)_2$. This slower cooling entails a smaller volume (higher density) and higher viscosity for the glass. Thus different glassy states can exist; Section 4.4.1 deals with the structural relaxation by which glassy states can change. On simple annealing, glasses densify and evolve toward an ideal glassy state represented by an extrapolation of the equilibrium properties, shown on **Figure 1** for the viscosity. How dense could an ideal glass be? This intriguing question was first addressed by Kauzmann (1948). He pointed out that extrapolation of liquid properties to lower temperatures would soon (that is, at a temperatures not far below the measured T_g) lead to what he termed an *apparent paradox* in which the characteristic difference between liquid and crystal properties would be inverted. For example, the liquid would become *denser* than the crystal and it would have *lower* entropy. While not thermodynamically forbidden, such an inversion seems unlikely, and Kauzmann proposed that some transformation, either an ideal (nonkinetic) glass transition or some type of spontaneous crystallization, must intervene to prevent the paradox being reached. He suggested that the isentropic point (at which the extrapolated liquid entropy matches that of the crystal) could be taken to represent the ideal glass-transition temperature and that the observed glass transition (found at finite rather than infinitely low cooling rates) occurs somewhat before the ideal transition. The fundamental nature of the transition into the glassy state remains an active research area. For example, there are explorations of how deposition from the vapor may allow access to stable amorphous states that are denser than could be obtained by cooling the liquid (Singh et al., 2013).

4.2.2 Production and Compositions

In connection with **Figure 1**, we have described the classical features of glass formation, well known for oxides and polymers. It is now established experimentally that metallic systems fit into the same picture. Around T_g, metallic glass-forming systems show all the characteristics (volumetric, rheological, thermal) of the conventional glass transition illustrated in the figure. MGs are classically the outcome of cooling an alloy liquid without the intervention of crystallization. But it is important to recognize that there has been much work on other ways of obtaining amorphous metallic states, often with the aim of expanding the range of glass-forming compositions. **Table 1** lists the range of techniques adopted, many of them involving the *amorphization*, in the solid state, of an initial crystalline phase. Measurements of properties and structures suggest that the amorphous metallic phases produced by other methods are similar to glasses produced by cooling the liquid and that on annealing the amorphous phases and the glasses would relax into essentially identical states.

In early work on MGs formed from the liquid, the focus was on the development of better techniques for rapid quenching. The most common methods, melt-spinning (Liebermann, 1980) and planar-flow casting (Smith and Saletore, 1986), remain of interest for the production of thin (10–100 μm thick) ribbons and foils, geometries that are useful for soft magnetic materials (Section 4.6). But current research focuses on expanding the range of compositions that do not require rapid quenching to form a glass, and which consequently permit glass formation in bulk (i.e. with minimum dimension exceeding 1 mm to 1 cm). There have been many attempts to understand and predict the GFA (quantified as the critical cooling rate for glass formation) of different compositions. The one that remains most generally applicable is that due to Turnbull (1969). Crystallization during cooling of the liquid is possible only in the temperature range from the equilibrium freezing temperature T_f down to the glass-transition temperature T_g. The closer these two temperatures are to each other, the greater the possibility that crystallization can be avoided on cooling, giving a glass.

Figure 2 shows, for a particular composition range in the binary system Pd–Si, how key quantities vary near a eutectic point. The liquidus (freezing) temperature T_f is strongly dependent on composition (**Figure 2(a)**), whereas T_g is almost constant (**Figure 2(b)**). Consequently, it is at the eutectic composition that the two temperatures are closest, conveniently characterized by the reduced glass-transition temperature, $T_{rg} = T_g/T_f$, which peaks at that point. The critical cooling rate for glass formation, calculated for example according to the methods of Davies (1976), is very sensitive to the value of T_{rg} (note that it is plotted on a logarithmic scale in the figure) and shows a sharp minimum near the eutectic point.

Thus the search for glass-forming systems is the search for deep eutectics, i.e. compositions at which the liquid is very stable compared to the crystalline phases in the given system. The chemical elements of interest are classified in **Table 2**, and their role in glass-forming systems is set out in **Table 3**. The classifications in these tables relate closely to those suggested by Cahn and Greer (1996), Inoue (2000), Takeuchi and Inoue (2005) and Cheng and Ma (2011). **Figure 3** shows equilibrium phase diagrams for binary systems representative of each category in **Table 3**. In each case the composition range(s) in which glass formation is possible by a technique such as melt-spinning can be seen to be correlated with eutectics.

In multicomponent systems, even deeper eutectics (i.e. with eutectic temperatures closer to T_g) can be achieved, and this is the basis for discovery of systems with higher T_{rg} and correspondingly lower critical cooling rates for glass formation and ultimately the capability to form BMGs. **Table 4** sets out a list of BMG-forming systems, highlighting the maximum diameter of rod that has been shown to be fully glassy when cast and the year of this demonstration. GFA is often discussed in terms of the "three empirical rules" proposed by Inoue (2000). These are that: (1) the alloy system should have three or more elements, (2) these elements should have a size (i.e. atomic diameter) mismatch exceeding 12% and (3) these

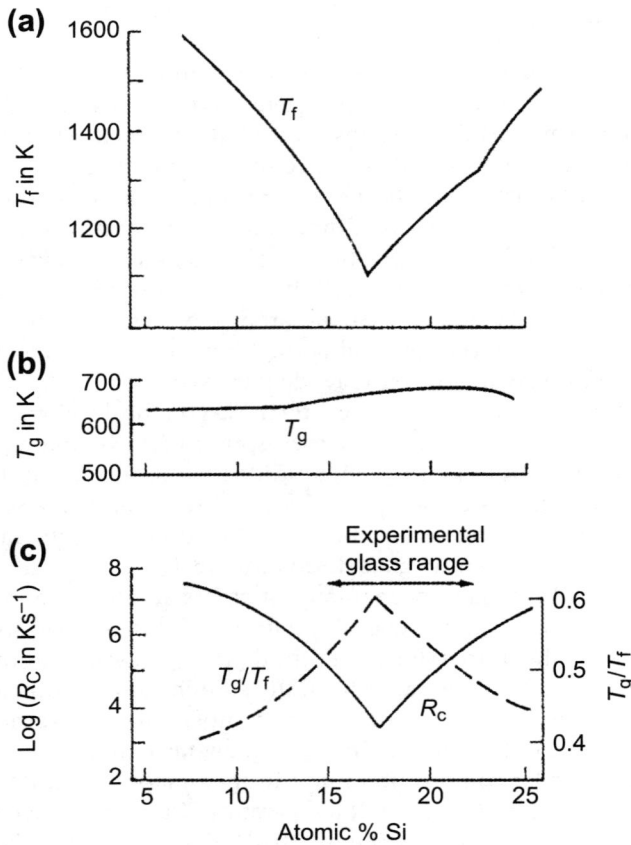

Figure 2 Equilibrium freezing (liquidus) temperature, T_f, glass-transition temperature, T_g, the ratio $T_{rg} = T_g/T_f$, and the calculated critical cooling rate for glass formation, R_c, for a range of compositions in the Pd–Si alloy system (4th edition, Figure 10).

Table 2 Classification of common constituent elements in MGs

Abbreviation	Description	Examples
AM	Alkali and alkaline-earth metals (groups IA and IIA)	Mg, Ca, Be*
SM	Semi- or simple metals (groups IIIA and IVA)	Al, Ga
ETM	Early transition metals (groups IVB to VIIB)	Ti, Zr, Hf, Nb, Ta, Cr, Mo, Mn
LTM	Late transition metals (groups IB, IIB and VIIIB)	Fe, Co, Ni, Cu, Pd, Pt, Ag, Au, Zn
RE	Rare-earth metals	Sc, Y, L, Ce, Nd, Gd
	Lanthanides + Sc, Y	
–	Actinide metals	U
NM	Metalloids and nonmetals	B, C, P, Si, Ge

*Although the element Be is an AM in this classification, its role in metallic glass formation is special, associated with its small size.
Adapted from Cheng and Ma (2011) with permission.

Table 3 Classification of typical MGs based on binary prototypes

Prototype	Base metal	Examples
LTM + NM	LTM	Ni–P, Pd–Si, Au–Si–Ge, Pd–Ni–Cu–P, Fe–Cr–Mo–P–C–B
ETM + NM	ETM	Ti–Si
ETM + LTM	ETM or LTM	Zr–Cu, Ni–Zr, Ni–Nb, Ti–Ni, Zr–Cu–Ni–Al, Zr–Ti–Cu–Ni–Be
RE + LTM	RE	Gd–Fe
SM + RE	SM or RE	Al–La, Ce–Al, Al–La–Ni–Co, La–(Al,Ga)–Cu–Ni
AM + LTM	AM	Mg–Cu, Ca–Mg–Zn, Ca–Mg–Cu
U + LTM	U	U–Co

Adapted from Cheng and Ma (2011) with permission.

elements should show negative heats of mixing with each other. These rules are all compatible with, and indeed are correlated with, eutectic compositions. It is increasingly recognized that GFA may be sharply dependent on composition, peaking dramatically at particular points in the multidimensional space describing alloys with many components. In this respect, ultrastable MGs may resemble stoichiometric intermetallic compounds. Although the glasses, unlike the compounds, lack crystallinity, they do have high degrees of order as seen thermodynamically (Section 4.2.3) and in their structure (Section 4.3).

4.2.3 Thermodynamics

By highlighting the importance of eutectics, the role of thermodynamics in determining GFA is already clear. But further aspects need to be considered. Garrone and Battezzati (1985) first drew attention to the fact that the heat of crystallization of a metallic glass is significantly less than the equilibrium heat of melting in the same system. That the enthalpy difference between the crystalline and the liquid/glassy states of a given system can be so temperature-dependent can be explained if the liquid, as it is cooled to approach T_g, has a large excess specific heat as shown in **Figure 1(c)**, i.e. a specific heat much larger than that of the crystal or glass at similar temperatures. With BMG-forming systems, it is possible to make direct measurements of the specific heat of the supercooled liquid, confirming that it is high (**Figure 4**) and, indeed, as T_g is approached is almost twice that of the crystal or glass. The large excess specific heat of the liquid means that, relative to the crystal, it is ordering significantly as it is cooled. The consequences can be seen in **Figure 5**. For pure metals, it is generally considered that the free-energy driving force ΔG for solidification rises linearly with supercooling below T_f. The curves in the figure show that for glass-forming systems, ΔG rises less than linearly with supercooling, a direct result of the excess specific heat of the liquid. The evident curvature reduces the driving force for crystallization and thereby facilitates glass formation.

A more striking point is that some of the curves for the glass-forming liquids in **Figure 5** have essentially zero slope at the lowest temperatures. As the slope of such plots is proportional to the entropy, this means that the liquid state about to form a glass has an entropy essentially equal to that of the crystal, the condition considered by Kauzmann (1948) and discussed in Section 4.2.1. How it can be that noncrystalline structures of glasses/liquids can show such high degrees of order is considered further in Section 4.3.

4.2.4 Kinetics

The viscosity of the supercooled liquid is the central kinetic parameter in considering glass formation. Different types of glass-forming system, network formers like SiO_2, ionic liquids, and those showing van der Waals bonding between molecules, do show different types of behavior, reflecting the different

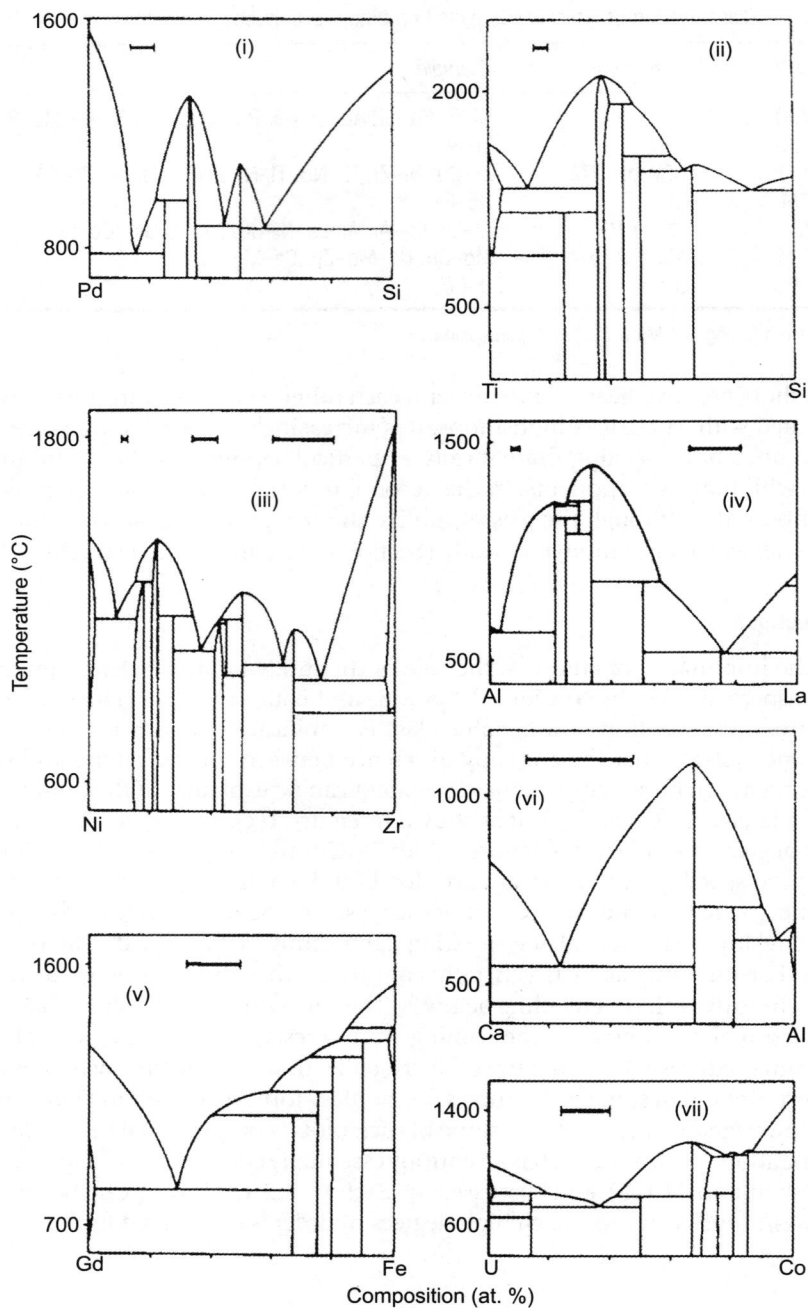

Figure 3   Representative equilibrium phase diagrams for each of the categories (i) to (vii) listed in **Table 3**. The glass-forming ranges are indicated by horizontal bars and show a good correlation with deep eutectics (4th edition, Figure 2).

Table 4   Representative bulk metallic glass compositions with the critical largest diameter of cylinders that can be cast fully glassy

Base metal	Composition (atomic %)*	Critical diam. (mm)	Production method	First report
Pd	$Pd_{40}Ni_{40}P_{20}$	10	Fluxing	Kui et al. (1984)
	$Pd_{40}Cu_{30}Ni_{10}P_{20}$	72	Water quenching	Inoue et al. (1997)
Zr	$Zr_{65}Al_{7.5}Ni_{10}Cu_{17.5}$	16	Water quenching	Inoue et al. (1993)
	$Zr_{41.2}Ti_{13.8}Cu_{12.5}Ni_{10}Be_{22.5}$	25	Copper mold	Peker and Johnson (1993)
Cu	$Cu_{46}Zr_{42}Al_7Y_5$	10	Copper mold	Xu et al. (2004)
	$Cu_{49}Hf_{42}Al_9$	10	Copper mold	Jia et al. (2006)
Rare earth	$Y_{36}Sc_{20}Al_{24}Co_{20}$	25	Water quenching	Guo et al. (2003)
	$La_{62}Al_{15.7}Cu_{11.15}Ni_{11.15}$	11	Copper mold	Tan et al. (2003)
Mg	$Mg_{54}Cu_{26.5}Ag_{8.5}Gd_{11}$	25	Copper mold	Ma et al. (2005)
	$Mg_{65}Cu_{7.5}Ni_{7.5}Zn_5Ag_5Y_5Gd_5$	14	Copper mold	Park and Kim (2005)
Fe	$Fe_{48}Cr_{15}Mo_{14}Er_2C_{15}B_6$	12	Copper mold	Ponnambalam et al. (2004)
	$(Fe_{44.3}Cr_5Co_5Mo_{12.8}Mn_{11.2}C_{15.8}B_{5.9})_{98.5}Y_{1.5}$	12	Copper mold	Lu et al. (2004)
	$Fe_{41}Co_7Cr_{15}Mo_{14}C_{15}B_6Y_2$	16	Copper mold	Shen et al. (2005)
Co	$Co_{48}Cr_{15}Mo_{14}C_{15}B_6Er_2$	10	Copper mold	Men et al. (2006)
Ti	$Ti_{40}Zr_{25}Cu_{12}Ni_3Be_{20}$	14	Copper mold	Guo et al. (2005)
Ca	$Ca_{65}Mg_{15}Zn_{20}$	15	Copper mold	Park and Kim (2004)
Pt	$Pt_{42.5}Cu_{27}Ni_{9.5}P_{21}$	20	Water quenching	Schroers and Johnson (2004a)

*The standard way of quoting MG compositions, and used throughout this chapter
Reprinted from Li et al. (2007) with permission.

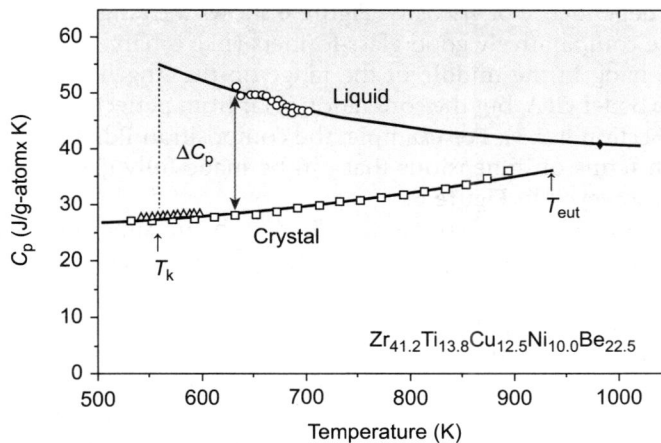

Figure 4   The specific heat of the equilibrium liquid (◆), supercooled liquid (○), glass (△) and crystal (□) states of the BMG-forming system Vitreloy 1. Redrawn from Busch et al. (1995). (For color version of this figure, the reader is referred to the online version of this book.)

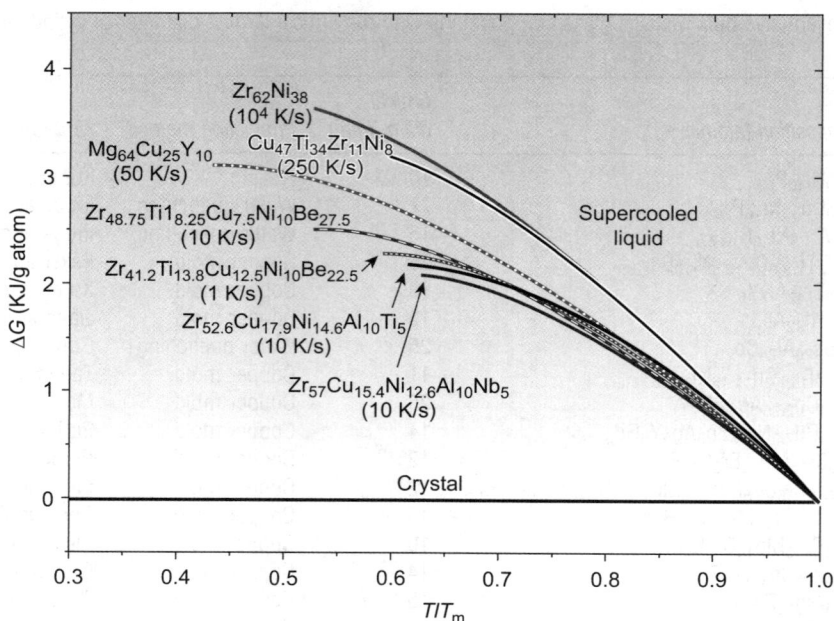

Figure 5 The difference in Gibbs free energy between the supercooled liquids of metallic glass-forming alloys and the crystalline state (usually a fine-scale mixture of several phases) of the same alloys. Redrawn from Busch et al. (2007). (For color version of this figure, the reader is referred to the online version of this book.)

sensitivities of the liquid structure to temperature changes. In the classification due to Angell (1995), *strong* network-forming liquids like SiO_2 have a near-Arrhenius dependence of viscosity. On the other hand, an organic liquid like o-terphenyl, composed of discrete molecules, is *fragile* with a strongly non-Arrhenius temperature dependence of viscosity. **Figure 6** shows the range of behavior that is observed. Metallic liquids that are comparatively good glass-formers (like binary Ni–Zr), or capable of forming BMGs, have viscosities lying in the middle of the range on the Angell plot. There is a tendency for stronger liquids to have better GFA, but the correlation is far from perfect, as the GFA is affected also by the thermodynamics (Section 4.2.3). For example, the composition $Pd_{40}Cu_{30}Ni_{10}P_{20}$ has the highest known GFA (at least in terms of dimensions that can be made fully glassy, **Table 4**), but is not the strongest of the liquids surveyed in **Figure 6**.

GFA is most directly linked to the crystal growth rate in the supercooled liquid, compared for different systems in **Figure 7**. Interestingly, fragile molecular systems such as o-terphenyl have growth rates (at a given reduced temperature) even lower than those of the strong oxide systems. Other fragile systems, such as the elements for which data are shown in the figure, especially the metal nickel, have growth rates so high that glass formation (suppression of crystallization) is very difficult if not impossible. The system $Zr_{50}Cu_{50}$ represents metallic glass-forming alloys, and shows that their behavior is intermediate (Wang et al., 2011). For systems in general, crystal growth rates are expected to be inversely proportional to the viscosity of the supercooled liquid. However, it has been suggested by Ediger et al. (2008) and confirmed by others (Nascimento and Zanotto, 2010) that this coupling of viscous flow and crystal-growth rates can break down as the temperature is lowered toward T_g, with the growth rate not decreasing as much as would be expected from the viscosity. The extent of this

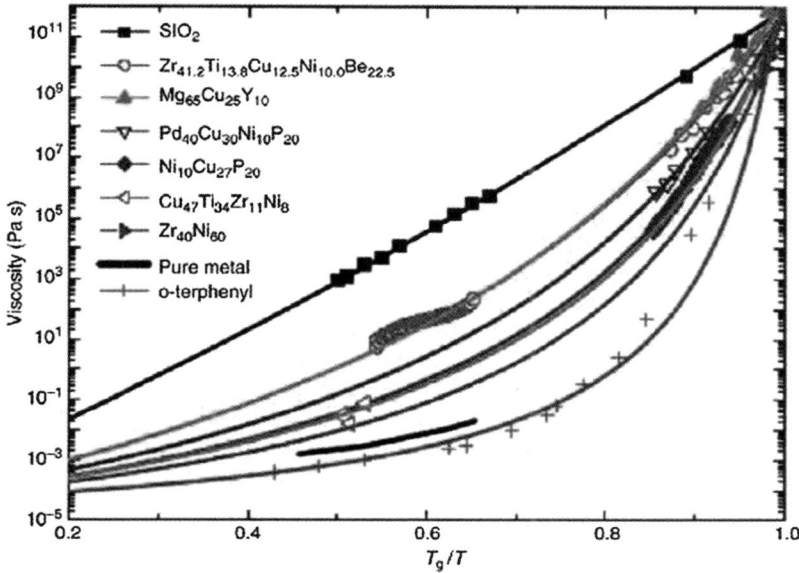

Figure 6 Viscosity of glass-forming liquids as a function of inverse temperature, normalized by the glass-transition temperature T_g. The metallic alloys are intermediate between strong network-forming liquids such as SiO_2, and fragile liquids such as o-terphenyl. Reprinted from Busch et al. (2007) with permission. (For color version of this figure, the reader is referred to the online version of this book.)

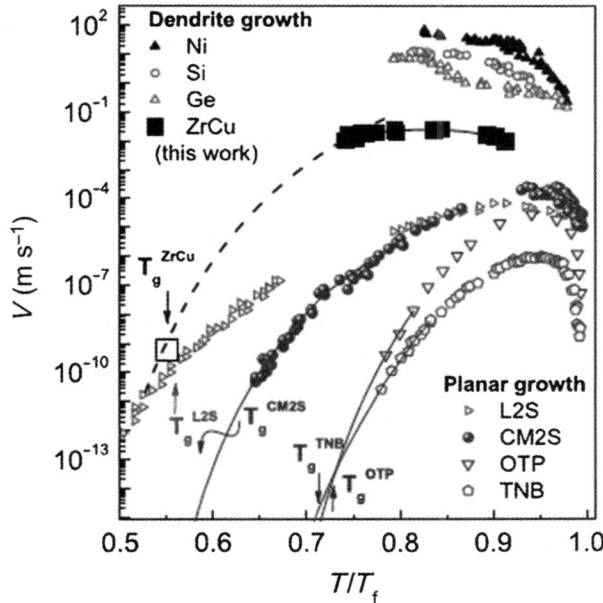

Figure 7 Crystal-growth rates as a function of reduced temperature (T/T_f) in liquids: elements (Ni, Si, and Ge), a glass-forming alloy ($Zr_{50}Cu_{50}$), molecular liquids (o-terphenyl (OTP) and tri-α-naphthylbenzene (TNB)), and oxides ($Li_2O \cdot 2SiO_2$ (L2S) and $CaO \cdot MgO \cdot 2SiO_2$ (CM2S)). Reprinted from Wang et al. (2011) with permission. (For color version of this figure, the reader is referred to the online version of this book.)

decoupling, negligible for strong liquids, is more evident for greater fragility. Crystal growth rates in some glass-forming chalcogenide liquids appear to be intermediate between those of the metal (Ni) and the metallic glass-former ($Zr_{50}Cu_{50}$) in **Figure 7**, and they show evidence for substantial decoupling (Orava et al., 2012; Sosso et al., 2012). Crystallization in MGs is also likely to show decoupling from viscous flow. Indeed, crystallization can be observed in MGs well below T_g, a phenomenon almost unknown in oxide glasses. Crystallization is considered further in Section 4.4.3.

4.3   Structure

4.3.1   Structure in the Absence of Crystallinity

In a single-phase solid alloy the distinctive feature of the glassy state is that it has no microstructure. There is no crystal lattice relative to which defects such as grain boundaries or dislocations can be defined. Though lacking periodicity, glasses in general, and MGs in particular, do have structure, and much effort has been expended to determine it. The existence of structure is consistent with the low entropy of MGs already noted in Section 4.2.4. Good progress has been made in elucidating the nature of metallic glasses' short-range order (SRO), i.e. the structure at the level of nearest-neighbor atomic coordination. The nature of their medium-range order (MRO), i.e. positional correlations over 1–2 nm, is more difficult to characterize and remains relatively poorly understood. While the structures of MGs can be distinctively different from one composition to another, there are many common features that will be the principal focus in this section. It is clear that the structure of an MG can strongly affect its properties. For example, annealing to obtain a slight increase in density, and barely detectable (by standard diffraction techniques) relaxation of its structure, is associated with a viscosity increase of some five orders of magnitude in a rapidly quenched metallic glass (Taub and Spaepen, 1979). While some links between structure and properties are now emerging, the understanding of such links in MGs is still very far behind that achieved for polycrystalline metals, for which the structural basis of properties is a materials-science paradigm. A review of all the work, over decades, on the structure of MGs lies outside the scope of this chapter. This section attempts to survey our current understanding and to highlight key contributions to it. A more comprehensive treatment can be found in the recent review by Cheng and Ma (2011).

The problem of determining the structure of a crystal consists of identifying the coordinates of all the atoms in the unit cell; though the task is very difficult for large and complex unit cells, in principle (if the "phase problem" can be solved) the structure can be determined precisely. Conversely, for a glass, the structure can be described only on a statistical basis: there are no unit cells and the environments of different atoms of a given species are necessarily diverse. In this point lies the difficulty of determining and describing the structure of any glass (or liquid). Almost all of the analyses of glass structure have been based on pair distribution functions (PDFs), because they can be determined from scattering experiments. PDFs give a measure of the probability of finding an atom center at a distance r from an average central atom; they are often given in a reduced form $G(r) = 4\pi r[\rho(r) - \rho_0]$, where $\rho(r)$ is the number of atoms per unit volume at a distance r and ρ_0 is the number of atoms per unit volume in the sample as a whole. The PDF gives statistical information about distances only, and this is insufficient to specify the structure. The angular distribution of interatomic vectors can readily be extracted from structural models, but cannot be determined from conventional scattering experiments.

4.3.2   Methods for Determination of Structure

The determination of glassy structures is based on the comparison of measured PDFs with those calculated from trial structural models. Especially with the acceleration in recent years in the possible rates of acquisition of X-ray and neutron scattering data, PDFs can be determined with good precision (Egami and Billinge, 2012), and can provide good discrimination between models. When an acceptable match is obtained between measured and calculated PDFs, there is still the problem that the solution is not unique; this is a more severe problem for glasses than for the analogous determination of crystal structures, because of the angular information missing from PDFs. The discussion which follows is based on PDFs determinable in diffraction experiments, but some other methods for structural analysis will be introduced; some of these methods are capable of yielding the angular information missing from PDFs.

The PDF of a glass is determined by Fourier inversion of scattering data, for example from normalized X-ray scattering curves of the kind shown in **Figure 8**. For an alloy glass one set of data is sufficient to determine a *radial distribution function* (RDF) in which the different types of atoms are not distinguished. Alternatively, a partial pair distribution function (PPDF) can be determined for each type of atom pair. For a binary alloy there are three types of pairs, and to determine the three PPDFs it is necessary to perform at least three scattering experiments in which the two types of atom have different relative scattering powers. An early example of this type of analysis is given in **Figure 9** showing the Ni–Ni, B–B and Ni–B PPDFs in a $Ni_{81}B_{19}$ glass. For a ternary alloy there would be six PPDFs, and an analysis of this or greater complexity has not yet been attempted. There are various ways of achieving the different relative scattering powers for the determination of PPDFs:

1. Different radiations: X-rays, neutrons and electrons are all possible.
2. Anomalous dispersion: For X-rays the effective atomic scattering factors can be altered by using a wavelength close to an absorption edge.

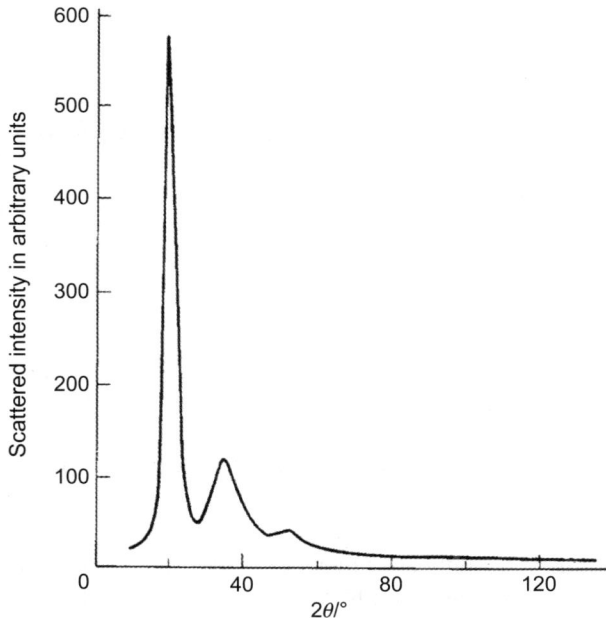

Figure 8   Scattered X-ray intensity as a function of scattering angle θ for amorphous $Ni_{81}B_{19}$. From Cargill, private communication (4th edition, Figure 25).

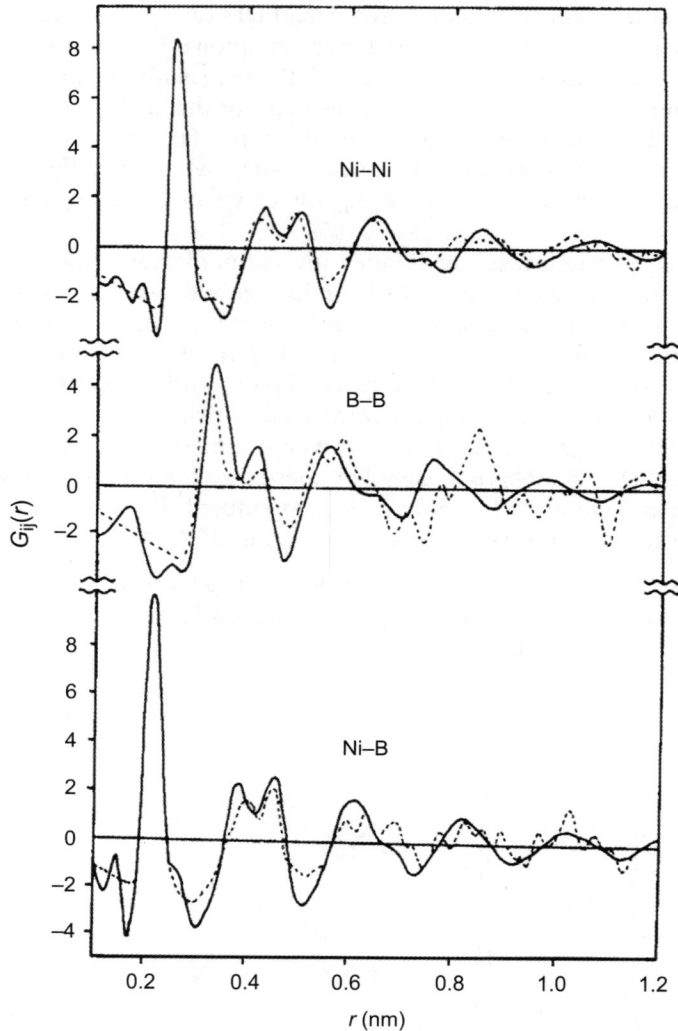

Figure 9 Partial pair distribution functions for glassy $Ni_{81}B_{19}$ as measured (full curves) by Lamparter et al. (1982) and as calculated (dashed curves) by Dubois et al. (1985) (4th edition, Figure 26).

3. Polarized neutrons: The scattering from ferromagnetic samples is different for different planes of polarization.
4. Isotopic substitution: Substitution of one isotope of an element for another leaves the structure of the glass unaffected, but may strongly alter neutron-scattering lengths. The PPDFs in **Figure 9** were determined in this way. Three $Ni_{81}B_{19}$ samples were prepared, one with natural nickel, one with ^{62}Ni and one with a mixture of ^{60}Ni and ^{62}Ni, to give an effective scattering length of zero. In this last case there was no scattering by nickel, and Fourier inversion of the neutron scattering data yielded the B–B PPDF directly.

5. Isomorphous substitution: It may be possible to substitute one element for another of similar type so that the structure is likely the same. X-ray scattering can then be used with the different atomic scattering factor. Recent work on systems such as Cu–Zr and Ni–Zr, in which the copper and nickel might be expected to play similar roles in the structure, however shows that the structures are rather different (Kaban et al., 2013). This suggests that isomorphous substitution should be a technique of last resort.

The experimental and computational procedures for the determination of RDFs and PPDFs have been extensively reviewed (Wagner, 1983; Egami, 1981; Suzuki, 1983). The basis for structure determination is the comparison between measured and calculated PDFs. Yet even without such comparison the measured PDFs reveal basic features of the glassy structure. The first peak in the RDF indicates the nearest neighbor separation and the area under the peak gives the average coordination number. The separation is as expected for a condensed phase. The coordination numbers are high, indicating the MGs are densely packed, as would be expected by extrapolation of the liquid properties. For a given atom, the number of atoms in its first-neighbor coordination shell must depend on the relative radii of the atom itself and those of the coordinating atoms. When PPDFs are examined, it is found that the distribution of the different elements in a glass can be far from random. This was first established by Sadoc and Dixmier (1976) who found that in a $Co_{81}P_{19}$ glass the relevant PPDF showed that P–P nearest-neighbor pairs were completely avoided. In glasses of this metal-metalloid type, this avoidance of metalloid neighbors seems to be a general feature (in **Figure 9**, for example, the first-neighbor peak for B–B is clearly missing). Of course, if the metalloid content were increased much above the approximately 20 at.% favored for glass formation, metalloid contact would become inevitable, but even then it is much less prevalent than in a random alloy (Cowlam et al., 1984).

It has been emphasized that pair distribution functions are limited in the information that they provide. For this reason there has been much interest in more local probes of atomic environments. X-ray absorption fine structure (XAFS) covers a variety of techniques in which the fine structure at or near an X-ray absorption edge is analyzed to obtain information specifically on the local environments of the particular species of absorbing atoms (Rehr and Albers, 2000). A structural model is still required, and the influences of atom positions and atom identities may be difficult to separate. Early studies on metal-metalloid glasses (Wong, 1981; Gurman, 1985) generally confirmed the conclusions from PPDF studies that the nearest neighbor environment in these glasses is well defined and can be similar to that in corresponding crystals. In more recent studies of the system Ni–Ag, in which there are no crystalline (or quasicrystalline) compounds, Luo et al. (2004) found strong icosahedral SRO in the amorphous phase.

Nuclear magnetic resonance (NMR) and Mössbauer spectroscopy, through the interaction of the nuclear quadrupole moment of the probe atom with the local field gradient, can provide information on the symmetry of the local environment (Slichter, 1990). As for XAFS, the environments probed are specific to one chemical element in the glass. The extracted information is unfortunately very model-dependent. NMR experiments on metal-metalloid glass again show local symmetry similar to that found in the corresponding crystalline phases. In Ni–B glasses, for example, Panissod et al. (1983) showed that the variation of the environments around [11]B nuclei is modeled well by a combination of the environments in crystalline Ni_3B and Ni_4B_3, each of which shows trigonal prismatic coordination of the boron. Interestingly, the environment of the boron in the glasses is not similar to that in crystalline Ni_2B, which does not have trigonal prisms. In cerium-based glasses containing no metalloids, Xi et al. (2007) interpreted their ^{27}Al NMR results in terms of a glassy structure dominated by aluminum-centered clusters with icosahedral symmetry. Mössbauer spectroscopy is a less discriminating test of structural models, but again there is evidence for crystal-like structural units (Whittle et al., 1985) and for local chemical order (Scholte et al., 1985).

Transmission electron microscopy (TEM) is very widely used to confirm the noncrystallinity of metallic-glass samples or to characterize the crystals or crystalline clusters that may be found within them. It has generally not been thought useful for quantitative study of glassy structure. Apart from the general difficulty of deriving a three-dimensional structure from a projected two-dimensional image, there are concerns about artifacts (such as surface contamination and crystallization) in thin-foil samples prepared for TEM examination. However, the recent work of Hirata et al. (2011) has shown that useful structural information can be obtained at the level of the nearest-neighbor shell around a given atom. Using an electron beam with a diameter (full width at half maximum) as small as 0.36 nm, the circular first diffraction halo expected in the diffraction pattern from a glass can break down into discrete spots (**Figure 10**). By matching the nanobeam electron diffraction (NBED) patterns thus obtained to simulated patterns, the structure of individual clusters can be inferred. By taking many such patterns the distribution of interatomic spacings was determined and found to be a reasonably

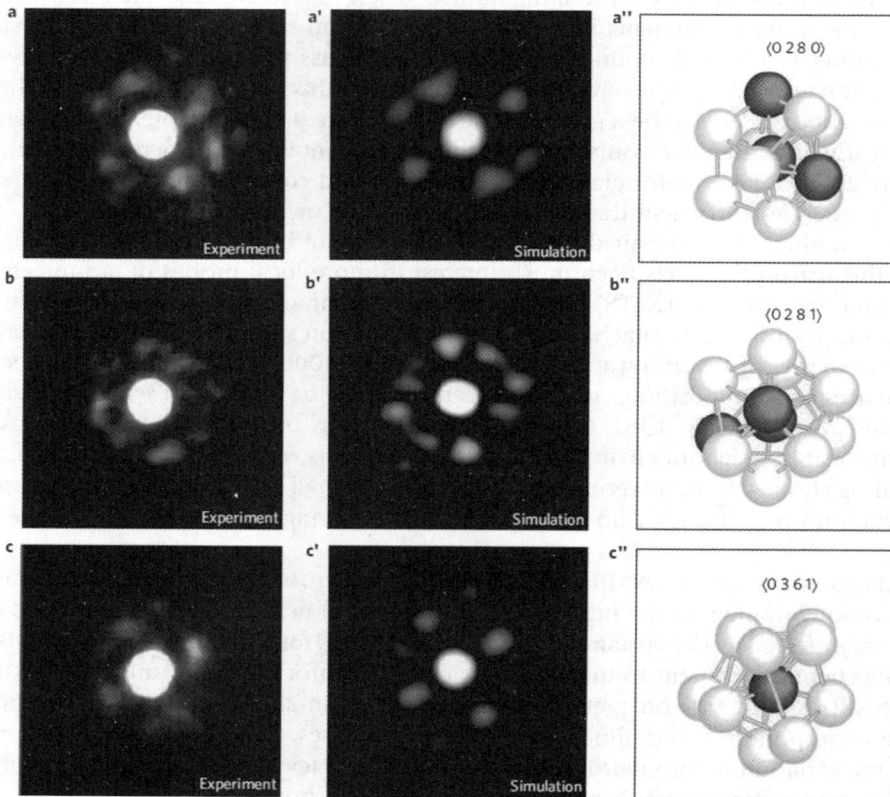

Figure 10   The structures of atom-centered clusters in $Zr_{66.7}Ni_{33.3}$ glass determined using simulation of NBED patterns obtained with a 0.36 nm beam: (a–c) measured patterns; (a′–c′) simulated NBED patterns calculated for the clusters shown in (a″–c″). Reprinted from Hirata et al. (2011) with permission. (For color version of this figure, the reader is referred to the online version of this book.)

good match to the distribution calculated from an ab initio molecular dynamics (MD) model of the glass, in this case $Zr_{66.7}Ni_{33.3}$.

Fluctuation electron microscopy (FEM) again uses a nanobeam of electrons, in this case scanned over the specimen (Treacy et al., 2005). In forming a dark-field image of the glassy phase, there is increased speckle contrast if there are relatively ordered regions. It has proved difficult to characterize the MRO in the glass itself, but the technique can detect crystalline order on a nanometer scale, and this has proved useful in studies of crystallization (Section 4.4.3).

Atom-probe tomography (APT) permits the imaging of individual atoms on the surface of a needle tip, and the identity of the atoms can be established by mass spectrometry when they are stripped from the tip by field evaporation (Miller, 2000). In recent years the data-acquisition rates have increased immensely, so that it is now possible to collect data on millions of atoms. The technique suffers, however, from relaxation of atomic configurations on the surface of the tip and cannot be relied on for detailed structural information. It is excellent, however, for measuring composition variation with essentially atomic resolution. Some results are examined later in Section 4.4.3.

Computer simulation plays important roles in developing and refining trial metallic-glass (and liquid) structures. The key methods are well developed and widely applied (Brázdová and Bowler, 2013). Molecular dynamics (MD) can be based on empirical interatomic potentials or on ab initio solution of the electronic structure. In MD the positions and velocities of the atoms are determined, and from an ensemble average of their instantaneous values the state of the material is determined. In the reverse Monte Carlo (RMC) method, a trial structure is iterated in a way simulating thermal annealing, with changes that improve the fit with measured data being automatically accepted while changes that degrade the fit are accepted with some probability (McGreevy, 2001). It should be noted that even if a good fit is achieved, it does not yield a uniquely defined structure, and the results will be most reliable when several types of experimental data are used as constraints.

These computational techniques can provide powerful support for experimental X-ray scattering studies. For example, the RMC method and MD simulations have been combined with scattering measurements on levitated droplets of equilibrium and supercooled liquids (Lee et al., 2008a). In this way, it has been shown that alloy liquids containing transition metals can develop strong icosahedral SRO consistent with good GFA.

4.3.3 Structural Models

The first structural model suggested for MGs was perhaps that based on "microcrystallites" (we would now term these nanocrystallites). The concept was that metallic solids appearing to be noncrystalline were not glasses, but simply polycrystalline solids in which the grain size is very small. Early work by Cargill (1970) showed clearly that the microcrystallite model does not provide quantitative fits to measured RDFs of glasses such as $Ni_{76}P_{24}$. There is no longer any doubt from very many studies of structure and properties that MGs are true glasses. Nevertheless, it has to be admitted that there are nanocrystalline solids that could easily be mistaken for glasses if scattering data of the kind in **Figure 8** were available only at low resolution. Calorimetric studies of such materials may be useful in showing that on annealing they transform to a polycrystalline aggregate of larger grains by a process of continuous grain growth rather than by nucleation and growth of crystals as in a glass (Chen and Spaepen, 1991).

It is intrinsic to a glass that its structure is inherited from the liquid. In this context the pioneering work is that of Bernal, who modeled the structure of pure metallic liquids as a packing of hard spheres all of the same diameter, realized physically by kneading a large population of steel balls in a bag

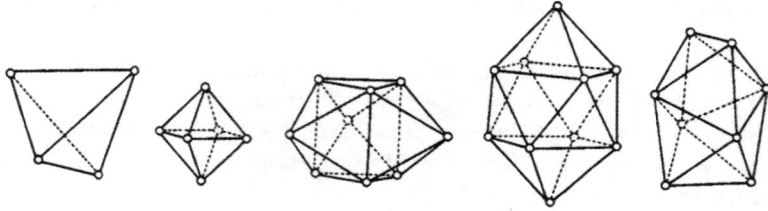

Figure 11 Polyhedra formed by dense random packing of hard spheres, according to Bernal (4th edition, Figure 27).

(Bernal, 1964). This *dense random packing* (DRP) is statistically reproducible, and the structure can be considered to be made up of only five simple polyhedra (canonical holes), shown in **Figure 11**, the smaller holes (tetrahedral and octahedral) being in the great majority. The topology of DRP is related to that of the continuous random network first proposed for tetrahedrally coordinated silicate glasses by Zachariasen (1932) (Connell, 1975; Chaudhari et al., 1976). While DRP is the starting point for considering the structure of MGs, it can certainly not be applied in its original form with infinitely hard spheres all of one size. Even the naming of this packing is problematic: if a packing is to be dense it cannot be random (Torquato et al., 2000). Nonetheless, the DRP model does capture the central feature of metallic-glass structure, namely that the packing is dense. Locally, the densest configuration of spheres of a given size is with their centers at the corners of a regular tetrahedron. Extending beyond the local scale, liquids and glasses approximate to polytetrahedral packing, but with considerable distortions as regular tetrahedra do not fill space.

DRP and its immediate variants fail to predict features such as the avoidance of metalloid–metalloid nearest neighbors noted in Section 4.3.2. This was first treated successfully in the *local coordination model* of Gaskell (1979). This suggests that the coordination of the metalloid atoms in metal-metalloid glasses is the same as in the corresponding crystalline compounds. The metalloid typically has nine neighboring metal atoms, the basic coordination polyhedron being a trigonal prism (giving six nearest neighbors, with three more slightly further away). The prisms can be linked in one of two ways (**Figure 12**), allowing them to be assembled into the nonperiodic structure of the glass. Computer models assembled in this way and then relaxed under appropriate atomic potentials gave good fits to measured distribution functions. The second nearest neighbor metal-metalloid distance in measured PPDFs can correspond

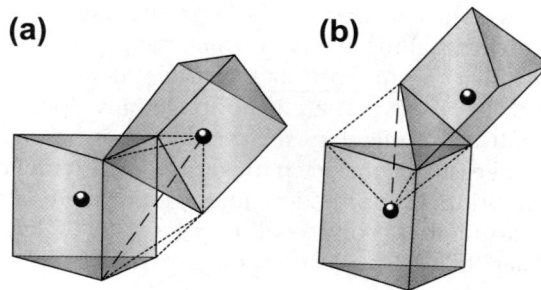

Figure 12 The packing of trigonal prisms as found in (a) Fe_3C and (b) Fe_3P. Each prism has a metalloid atom at its center. The long dashed line indicates the second-nearest metalloid–metal distance in each case. Redrawn from Gaskell (1983) (4th edition, Figure 28).

directly to one or the other of the distances expected from the types of linkage in **Figure 12**. Thus there is evidence for MRO in the glasses in which the packing of the prismatic units resembles that in particular crystal structures. More evidence for this is provided in the B–B PPDF in **Figure 9**, which shows structure out to large distances; this structure cannot be reproduced by a model based on randomly packed trigonal prisms. Gaskell suggested that the prisms themselves arise, not from any directed bonding, but from a compromise between the tendency to the closest possible packing and the need for space to insert metalloid atoms between the preponderant metal atoms (Gaskell, 1985).

The local-coordination model was taken further in the *chemical twinning model* of Dubois et al. (1985) who pointed out that similar prismatic coordination can be found in many metal–metal compounds. In this way, similar concepts can be applied to a very wide range of systems. Chemical twinning is the term used to describe the packing of coordination polyhedra in crystalline compounds. It is expected that GFA should be greatest when there are competing types of twinning in the alloy.

In many systems, it is useful to analyze structure in terms of atom-centered clusters (Shi et al., 2008). The ratio of the radii of the solute atom at the center and of the solvent atoms around it determines how many solvent atoms will fit in the coordination shell. Coordination numbers in typical MGs vary from 8 to 20. **Figure 13** shows the case of clusters centered on the solute atom boron in the simulated structure of $Ni_{81}B_{19}$ glass (**Figure 13**). In this case, as generally for metal-metalloid glasses as noted earlier, metalloid–metalloid nearest neighbors are not found, so the shells of the clusters are composed exclusively of nickel atoms. In other cases, for example, clusters in Cu–Zr glass, the shell may contain a mixture of atom types. Clearly, there are radius ratios and relative populations of atomic species of different size that will give particularly stable clusters and good packing. Such considerations underlie the *efficient cluster packing* model and related approaches that have been successful in rationalizing compositions favoring eutectics and good GFA (Miracle and Sanders, 2003; Miracle, 2004, 2006; Shi

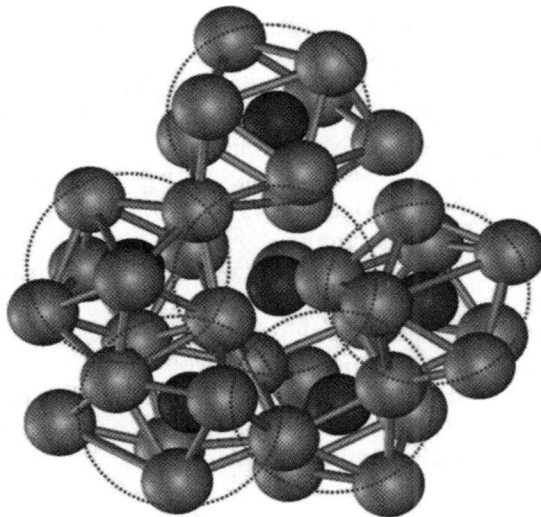

Figure 13   Structure of $Ni_{81}B_{19}$ metallic glass obtained from an ab initio molecular dynamics simulation, showing quasi-equivalent solute-centered clusters. The chemical ordering is such that solute boron atoms (purple) are fully coordinated by solvent nickel atoms (green) and do not make contact with each other. Reprinted from Shi et al. (2008) with permission.

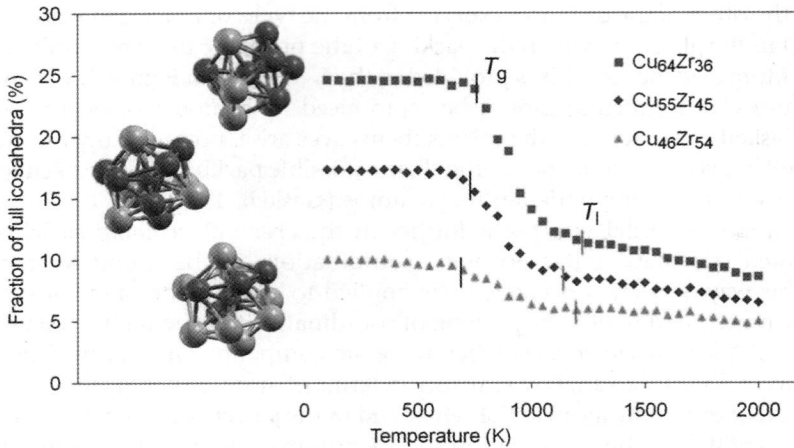

Figure 14   The fraction of Cu-centered clusters within which the coordination is icosahedral for three compositions of Cu–Zr glass simulated using molecular dynamics. The fraction is appreciable at the liquidus temperature T_l and rises on cooling toward the glass-transition temperature T_g. Representative icosahedra are shown. Reprinted from Cheng et al. (2008b) with permission. (For color version of this figure, the reader is referred to the online version of this book.)

et al., 2008). The clusters are not identical, but can be regarded as quasiequivalent (Sheng et al., 2006). A structure assembled from such clusters, predominantly vertex-sharing, has MRO to beyond 1 nm, and gives excellent fits to scattering data (Sheng et al., 2006). Importantly in this work, the structural modeling takes chemical effects into account and includes realistic many-body interactions; hard-sphere models would not be so effective.

Frank (1952) first proposed that the barrier to crystal nucleation in supercooled liquid metals arises mainly because atomic packing in the liquid is icosahedral and therefore incompatible with simple crystal structures. Studies of supercooled liquids are now facilitated by advances in containerless processing, enabling scattering experiments to be performed on levitated droplets. Such experiments reveal a variety of types of order developing as metallic liquids of different compositions are cooled, but *icosahedral* ordering is predominant (Lee et al., 2004, 2008a). Molecular-dynamics simulations are consistent with scattering data, and show (**Figure 14**) that icosahedral order, present even above the liquidus temperature, increases sharply on cooling toward the glass transition (Cheng et al., 2008a, 2008b). The ordering appears to be intimately associated with the increase in viscosity that leads to glass formation, as the simulations show that the least mobile atoms are those in icosahedral clusters. Furthermore, alloying (notably the addition of aluminum to Cu–Zr) that increases the fraction of such clusters correlates with improved GFA (Cheng et al., 2009c).

The cluster-based models are appropriate for systems in which the cluster-center species is a relatively dilute solute in the alloy (**Figure 13**). Other compositions have yet to be fully analyzed. It is certain, though, that all the structures will show significant ordering inherited from the liquid (**Figure 14**).

4.3.4   Challenges for Further Study

The nature of the glass transition, kinetic rather than thermodynamic, means that there is not a single glassy structure, but rather a range of structures mutually accessible through relaxation processes. The property changes associated with these are reviewed in Section 4.4.1. These changes can be substantial, yet the underlying structural changes can be very subtle and difficult to characterize. Structural relaxation of glasses in general, and MGs in particular, is often interpreted in terms of *free volume*. An as-quenched glass has distributed excess "free" volume that decreases on annealing. This description is useful in emphasizing the importance of volume in controlling atomic mobility in glasses (the case of viscous flow, for example, is considered in Section 4.5.2). At the atomic level, free volume is not uniformly distributed; where it is concentrated the atoms can be considered to be under hydrostatic tension. In the glass as a whole, there must, in compensation, be other atoms under hydrostatic compression. Egami (2006) has pioneered the description of glassy structures in terms of these atomic-level stresses. A more relaxed glass has fewer anomalously short or long interatomic separations and thus a narrower spread of atomic-level stresses (Dmowski et al., 2007). Consideration of topological fluctuations in the bonding network can lead to a general theory for glass formation, and Egami et al. (2007) have shown that based on such considerations the glass-transition temperature can be predicted from elastic constants.

Cheng and Ma (2008) have shown that MD modeling can be used to characterize changes in a structure on relaxation. In their model Cu–Zr–Al glass, a relaxed, denser state is one with more icosahedral order (Cheng and Ma, 2008). At present, there are rather few experimental studies of sufficient precision to provide a basis for testing such modeling.

After much study, the SRO in MGs is well characterized. It can be taken to be based on poly-tetrahedral packing and can often be usefully described in terms of quasiequivalent clusters. It is also notable that the SRO in the glasses is often very similar to that in the corresponding crystalline solids. In contrast, the poor understanding of MRO is particularly unfortunate, as the MRO may be a crucial area of difference between glassy and crystalline states, and important in determining glass properties. It is becoming increasingly apparent that MGs can be inhomogeneous on a nanometer scale, and that this inhomogeneity is intrinsic, inherited from the liquid. Ichitsubo et al. (2007) showed that the sound velocity in $Pd_{42.5}Ni_{7.5}Cu_{30}P_{20}$ glass was greater for nanometer wavelengths than for millimeter wavelengths. They attributed this to an elastic inhomogeneity with more and less stiff regions on a length scale of more than 5 nm. Wagner et al. (2011) used a variant of atomic-force microscopy (AFM) to map the contact resonance frequency at the surface of $Pd_{77.5}Cu_6Si_{16.5}$ glass and found inhomogeneity on the scale of about 20 nm. Strikingly, the statistical variation of resonance frequency from place to place on the glass was ten times greater than on comparable crystalline samples. The inhomogeneity can be akin to a nanoscale phase separation with local variations in composition and possibly regions of crystal-like order. It certainly includes regions of greater and lesser density. Such variations, still poorly characterized, are critical in affecting properties, and specific examples relating to crystallization (Section 4.4.3) and plastic deformation (Section 4.5.5) appear later.

4.4 Structural Evolution

4.4.1 Structural Relaxation

According to the analysis by Johari and Goldstein (1970), the atomic and molecular configurations in liquids and glasses change according to motions classified as primary (α) and secondary (β) relaxations. Primary relaxations describe the major rearrangements responsible for viscous flow. On cooling, the glass transition is reached when the decreasing mobility stifles these rearrangements. In the glass itself, then, there are no primary relaxations, but there are still possibilities for minor, localized and reversible rearrangements. These are the secondary relaxations that Johari and Goldstein suggested should be a "near-universal feature of the glassy state". Johari and Goldstein studied a variety of organic molecular

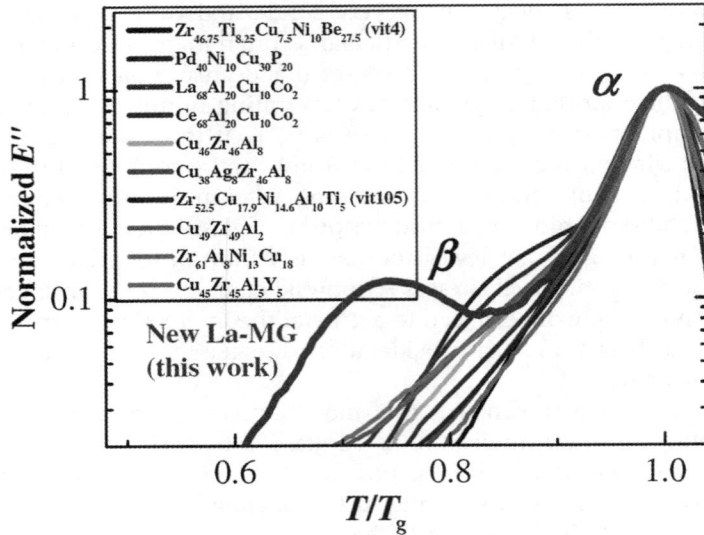

Figure 15 The loss modulus at 1 Hz (E'') as a function of temperature, showing varying degrees of β relaxation in La$_{68.5}$Ni$_{16}$Al$_{14}$Co$_{1.5}$ and in ten other MGs. The temperature scale is normalized with respect to T_g and E'' is normalized with respect to its maximum value at T_g. Reprinted from Yu et al. (2012) with permission. (For color version of this figure, the reader is referred to the online version of this book.)

liquids and fused salts, and made comparison with amorphous polymers. It is now becoming accepted that a description in terms of α and β relaxations can be applied also to MGs (Yu et al., 2013). The β relaxation in MGs has been studied mainly using dynamic mechanical analysis (DMA), measuring elastic constants and energy loss (mechanical damping, characterized by the loss modulus E'') in vibrating samples. **Figure 15** shows results for several BMGs and highlights a La-based glass that shows a strong β relaxation that, interestingly, appears to be correlated with greater plasticity. It is suggested that β relaxations underlie many of the mechanical responses (anelastic and plastic flow) of MGs as well as the changes of structure on annealing that are the focus of this section.

When a metallic glass is annealed close to its glass-transition temperature T_g, many physical and mechanical properties change as the glassy structure evolves toward the metastable equilibrium state of the supercooled liquid. The changes in structure and properties are generically called *relaxation*, and are not specific to MGs, There are, for example, significant relaxation-induced changes in the refractive index of oxide glasses for optical applications. At its most basic level (**Figure 1**) the evolution toward a more relaxed state involves an increase in density. One description of this is in terms of free volume (Section 4.3.4), the decrease of which implies a lowering of atomic mobility in the glass. As an example, **Figure 16** shows the results of sensitive interdiffusivity measurements in a deposited amorphous metal multilayer. On annealing, there is a decrease in volume (by less than 1%) and a substantial decrease in interdiffusivity (by more than three orders of magnitude). These results emphasize that while structural changes on relaxation may be slight, and indeed are difficult to characterize by the types of study outlined in Section 4.3.2, the associated property changes can be large. The reduction in atomic mobility on relaxation is particularly significant, and must of course impede the continuation of the relaxation itself.

Properties revealing relaxation effects include: density or length changes, elastic modulus, viscosity, diffusivity (**Figure 16**), Curie temperature (for ferromagnetic glasses), electrical resistivity, enthalpy, and

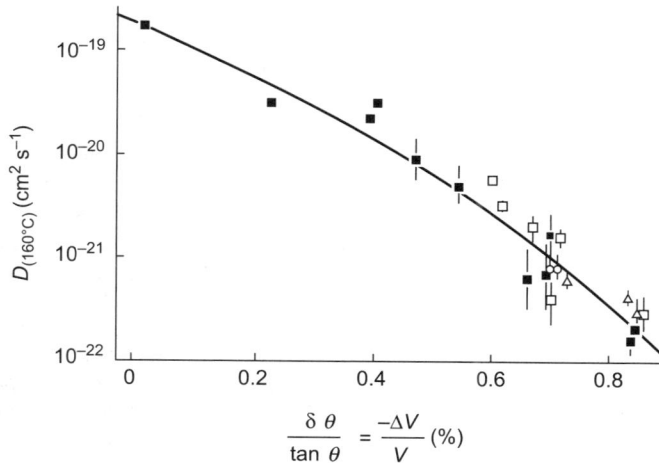

Figure 16 The interdiffusivity in an Fe–Ti amorphous multilayer at 160 °C, plotted against the relative decrease in atomic volume $\Delta V/V$. Adapted from Chason and Mizoguchi (1987) (4th edition, Figure 31(a)).

a number of others as shown in **Table 5**. **Figure 17** has examples of observed relaxation kinetics. On isothermal annealing, an as-quenched glass relaxes toward the metastable equilibrium state characteristic of the annealing temperature. While the glass remains far from equilibrium, property changes are often approximately linear with the logarithm of the anneal time. This is shown for the increase in Young's modulus in **Figure 17(a)**; such an evolution is necessarily monotonic. On annealing long enough, the metastable equilibrium states are reached, different for each annealing temperature. **Figure 17(b)** shows this behavior for the Curie temperature of a ferromagnetic glass. At higher annealing temperature, the change in property is initially faster, but the total extent of property change is less, as the higher-temperature equilibrium state is one of greater volume and in that sense is closer to the as-quenched state. These effects of temperature mean that the curves in **Figure 17(b)** intersect; this is the basis for experiments on the crossover effect (see below). If equilibrium states can be reached, then it is possible (if the glass is sufficiently stable, i.e. resistant to crystallization) to evolve from one equilibrium state to another. When evolving from the equilibrium attained on lower-temperature annealing toward the new equilibrium being established during a higher-temperature anneal, the property should change in the opposite direction to that seen when annealing an as-quenched glass. This is shown in the viscosity measurements in **Figure 17(c)**. That the same final property value can be reached from two directions helps to establish that this value is associated with a (metastable) equilibrium state. Reversible changes in structure near to T_g, evidenced in the property changes in **Figure 17(c)**, can be associated with α relaxations. At lower annealing temperatures, well below T_g, metastable equilibrium states are not attained in any reasonable time. Yet at such temperatures, there can be reversible property changes on successive anneals at two temperatures. Such effects were first observed for Curie temperature (Egami, 1978), and **Figure 17(d)** shows an example for Young's modulus. These reversible cycles cannot be from one metastable equilibrium state to another. There must be faster processes involving strictly local and reversible changes in structure; these can be associated with β relaxations.

That there is a spectrum of relaxation processes is also suggested by the discovery of the *crossover effect* in MGs, in respect of the Curie temperature (Greer and Leake, 1979) and elastic modulus (Scott and Kuršumović, 1982). The central phenomenon, also known as the *memory effect*, can be conveniently

Table 5   Structural relaxation in metallic glasses: properties affected and types of relaxation

Property affected by relaxation	Type of relaxation*					Selected references
	(i)	(ii)	(iii)	(iv)	(v)	
Density	I		Y			Gerling et al. (1988), Toloui et al. (1985)
Length changes	D		Y			Huizer and van den Beukel (1987)
Thermal expansion coefficient	D					Komatsu et al. (1983)
Enthalpy	D		Y		Y	Woldt (1988), Inoue et al. (1983)
Young's modulus	I		Y	Y	Y	Koebrugge and van den Beukel (1988), Chen et al. (1983)
Hardness/yield stress	I				Y	Bresson et al. (1982), Deng and Argon (1986)
Viscosity	I	Y		Y	Y	Volkert and Spaepen (1989)
Relaxation time for stress relief	I	Y		Y	Y	Williams and Egami (1976)
Diffusivity	D				Y	Akhtar and Misra (1986)
Curie temperature	I,D	Y	Y	Y	Y	Greer and Leake (1979), Egami (1978)
Electrical resistivity	I,D		Y			Kelton and Spaepen (1984), Balanzat et al. (1985)
Temperature coefficient of resistivity	D,I				Y	Kelton and Spaepen (1984), Balanzat et al. (1985)
Internal friction	D					Posgay et al. (1985)
Strain at fracture	D		Y		Y	Inoue et al. (1983), Mulder et al. (1983)
Saturation magnetization	D				Y	Nishi et al. (1986)
Magnetic after-effect	D		Y	Y	Y	Guo et al. (1986)
Induced magnetic anisotropy			Y	Y		Guo et al. (1986)
Structure factor (X-ray, neutron)	Y				Y	Caciuffo et al. (1989)
Extended X-ray absorption fine structure	Y					Baxter (1986)
Positron lifetime	D					de Vries et al. (1988)
Superconducting transition temperature	D		Y			Johnson (1981)
Mössbauer spectra	Y	Y			Y	Ström-Olsen et al. (1988)
Corrosion resistance	I					Masumoto et al. (1986)
Density of states at Fermi level	I,D				Y	Zougmoré et al. (1988)
Debye temperature	I				Y	Zougmoré et al. (1988)
Magnetostriction coefficient	I				Y	Cunat et al. (1988), Tarnóczi et al. (1988)

*The types of relaxation are: (i) monotonic relaxation; (ii) reversible relaxation near the glass transition temperature T_g; (iii) reversible well below T_g; (iv) memory effects; and (v) effects of production conditions. I and D indicate an increase or decrease, respectively, of the property value in monotonic relaxation; Y indicates occurrence of that type of relaxation.
Adapted from Greer (1993) with permission.

illustrated by an early study of changes in the refractive index of an oxide glass (Macedo and Napolitano, 1967). **Figure 18** and its caption explain how such an experiment is done. If there were only a single relaxation process, the plot would simply show a horizontal line when the temperature was raised from T_1 to T_2. Experiments like this show that there is not a one-to-one relationship between a property and the internal state of the glass, in the sense that the same value of the property can be associated with distinct structural states generated by different heat-treatment programs. There must be at least two distinct processes, but it is likely that there is a broad spectrum of processes with distinct activation energies (Leake et al., 1988). This outcome suggests the near-impossibility of associating

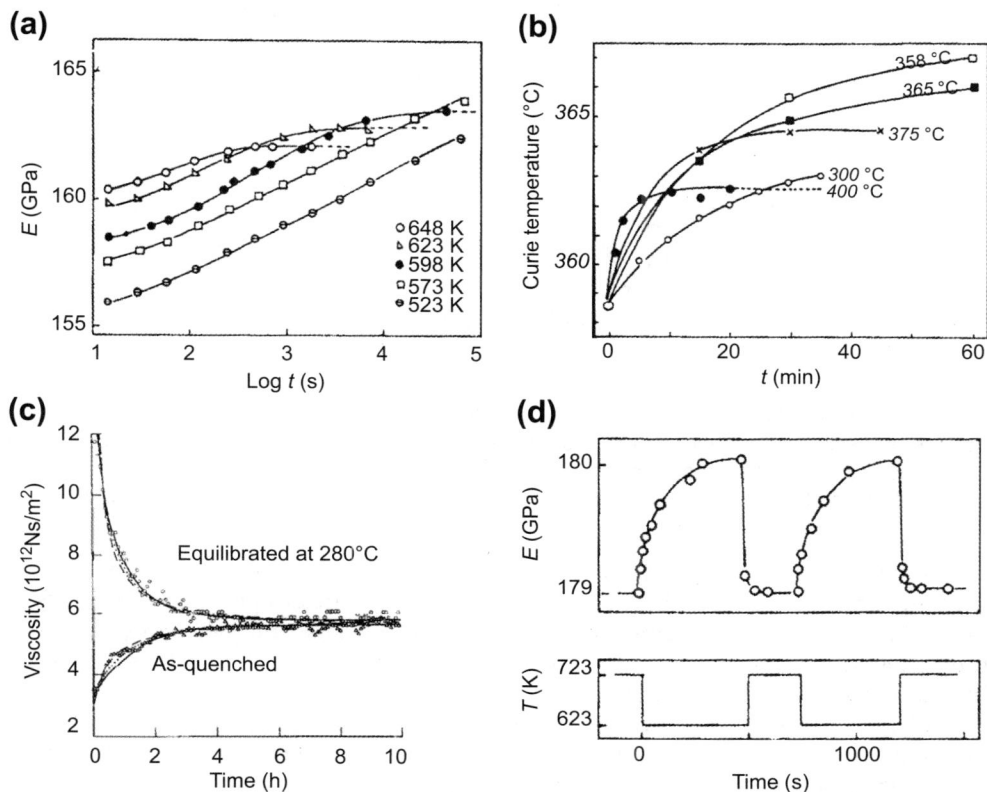

Figure 17 Different forms of property changes in MGs arising from structural relaxation: (a) irreversible change far from equilibrium in the early stages of annealing an as-quenched glass at selected temperatures—the Young's modulus of $Co_{58}Fe_5Ni_{10}B_{16}Si_{11}$; (b) the Curie temperature of $Fe_{80}B_{20}$ glass as a function of annealing time at selected temperatures; (c) the viscosity of glassy $Pd_{40}Ni_{40}P_{19}Si_1$ as a function of time at 300 °C for an as-quenched sample and for a sample previously equilibrated at 280 °C; (d) reversible changes in the Young's modulus of the same glass as in (a) when cycled between 623 and 723 K. Reprinted with permission from: (a,d) (4th edition, Figure 32(c) and Figure 32(b) respectively); (b) Greer and Leake (1979) (4th edition, Figure 32(d)); (c) Volkert and Spaepen (1989).

different aspects of relaxation with precisely defined structural changes. Modeling of relaxation kinetics has mostly been in terms of free volume, and rather quantitative approaches have been adopted (e.g. van den Beukel and Sietsma, 1990). If β relaxations underlie both flow processes and structural relaxation in MGs, then it may be of interest to consider the kinetics of stress relief. The results in **Figure 19** show the results of a wide range of measurements on a single Al-based glass at room temperature only. The form of the stress relief over time can be expressed as a relaxation time spectrum and, remarkably this shows an enormously wide range of relaxation time, from ~ 1 s to $\sim 3 \times 10^7$ s. It seems, then, that relaxation involves a very wide range of processes, and in this work, Ju et al. (2011) suggest that these can be associated with local shear transformation zones (STZs) of well defined size. STZs will be considered further in Section 4.5.2.

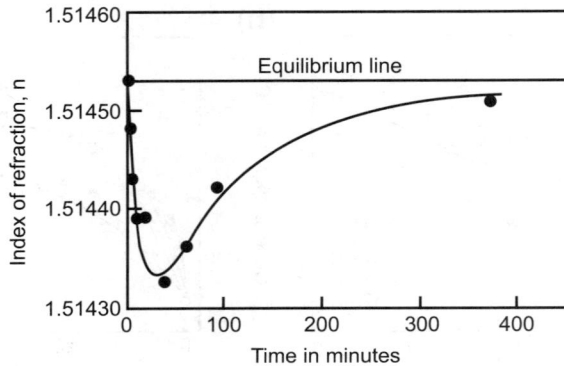

Figure 18 Refractive index vs time in a crossover experiment on a borosilicate glass held at temperature T_1 until the index evolved to a value corresponding to equilibrium at a second, higher, temperature T_2. The annealing temperature was then abruptly increased to T_2: instead of remaining constant, the index followed the curve shown, returning eventually to the T_2 equilibrium value. Reprinted from Macedo and Napolitano (1967) with permission (4th edition, Figure 33).

In this section we have so far considered only structural relaxation brought about by purely thermal treatments. Structural changes can also be induced by applying stresses and magnetic fields. Lee et al. (2008b) loaded $Ni_{62}Nb_{38}$ glass in compression at 95% of its macroscopic yield stress for 30 h at room temperature. They termed this treatment elastostatic compression as it should be safely in the elastic region. They found that samples treated in this way showed improved plasticity when subsequently loaded to beyond the yield stress in compression. This change was attributed to an increase in free volume, inferred from calorimetric measurements. Ke et al. (2011) performed a similar experiment on $Zr_{46.75}Ti_{8.25}Cu_{7.5}Ni_{10}Be_{27.5}$ BMG loaded at 80% of its macroscopic yield stress for up to 50 h at room temperature. In this case it was confirmed by direct measurement of density that the volume of metallic-glass samples subjected to such elastostatic compression does indeed increase. The small amount of creep at room temperature is associated with dilatation, the consequences of which will be explored further in Section 4.5.1. For the moment we note that the effect of this mechanical treatment is opposite to that of thermal annealing; if annealing is taken to induce *aging* of the sample, the elastostatic loading induces *rejuvenation*, giving a state that is less relaxed.

Ideally, glassy states are isotropic, but it has long been known that as-quenched glasses can show properties that are direction-dependent. Melt-spun ribbons in particular show strong anisotropy, as might be expected from the strong directional nature of the liquid shearing during ribbon production (Sinning et al., 1985; Tarumi et al., 2007). In contrast, BMGs tend to be isotropic. Here we focus on treatments to induce anisotropy in MGs. Mechanical treatments such as the elastostatic loading noted above are clearly directional and have the potential to induce anisotropy, as does the application of a magnetic field. The structural effects have been most studied for samples in which anisotropy has been induced by creep. Suzuki et al. (1987) were able to show directly by energy-dispersive X-ray diffraction from a $Fe_{40}Ni_{40}Mo_3Si_{12}B_5$ glass annealed under tensile stress that there was induced directional SRO that decayed gradually on subsequent stress-free annealing. The results of a similar experiment are shown in **Figure 20**. For an isotropic glass the diffraction pattern in transmission has circular halos, and when anisotropy is induced the halos are elliptical. Diffraction patterns were recorded for samples in two orientations, one rotated 90° relative to the other, the difference between the two patterns then highlighting any anisotropy. For a sample annealed without stress (**Figure 20(a)**) circular symmetry is

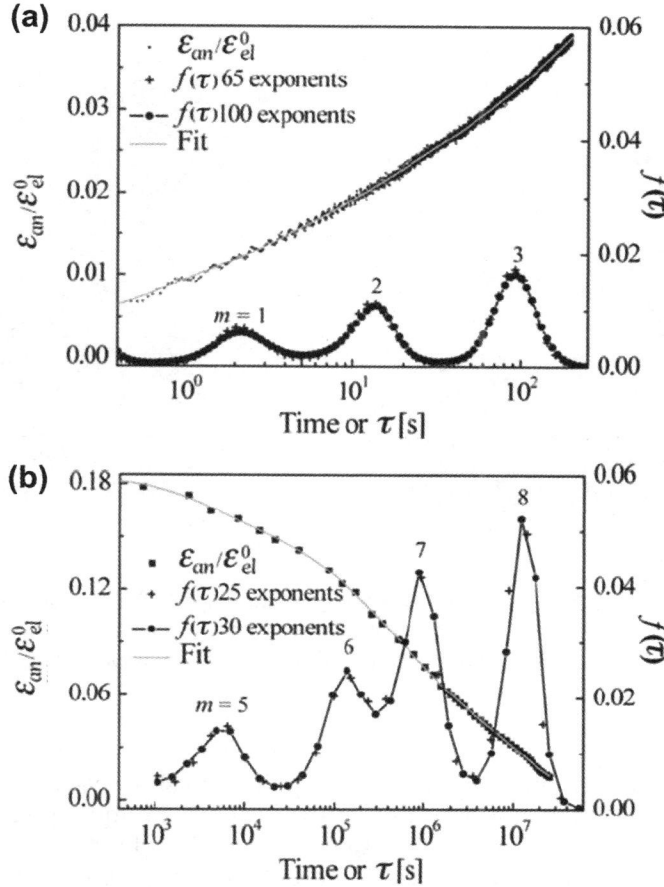

Figure 19 Sample relaxation curves and corresponding relaxation-time spectra for anelastic load relaxation of melt-spun $Al_{86.8}Ni_{3.7}Y_{9.5}$ ribbons in (a) nanoindentation cantilever and (b) mandrel measurements. In each case, two spectra are shown, obtained from fits with different numbers of fitting parameters. Reprinted from Ju et al. (2011) with permission.

maintained, but for the stress-annealed sample (now being measured stress-free at room temperature) the induced anisotropy is clear (**Figure 20(b)**). The induced ellipticity of the diffraction halos is large: the major and minor radii of the ellipse differ by up to 10%. Such a change is too large to be associated directly with a similar change in bond length. Rather the ellipticity of the halos is interpreted in terms of *bond-orientational anisotropy* (Tomida and Egami, 1993).

Resonant ultrasound spectroscopy has been used to show that MGs subjected to creep develop anisotropic elastic properties, with the normal stiffness coefficient parallel to the stress axis becoming different from that in the plane perpendicular to the axis. The anisotropy first has one sign, and then as creep continues it changes sign (Concustell et al., 2011). The first stage appears to be associated with reversible anelastic strain and the second with irreversible plastic strain (viscous flow). The same two stages of opposite sign also appear in magnetic anisotropy induced by annealing a metallic glass in tension (Nielsen and Nielsen, 1980).

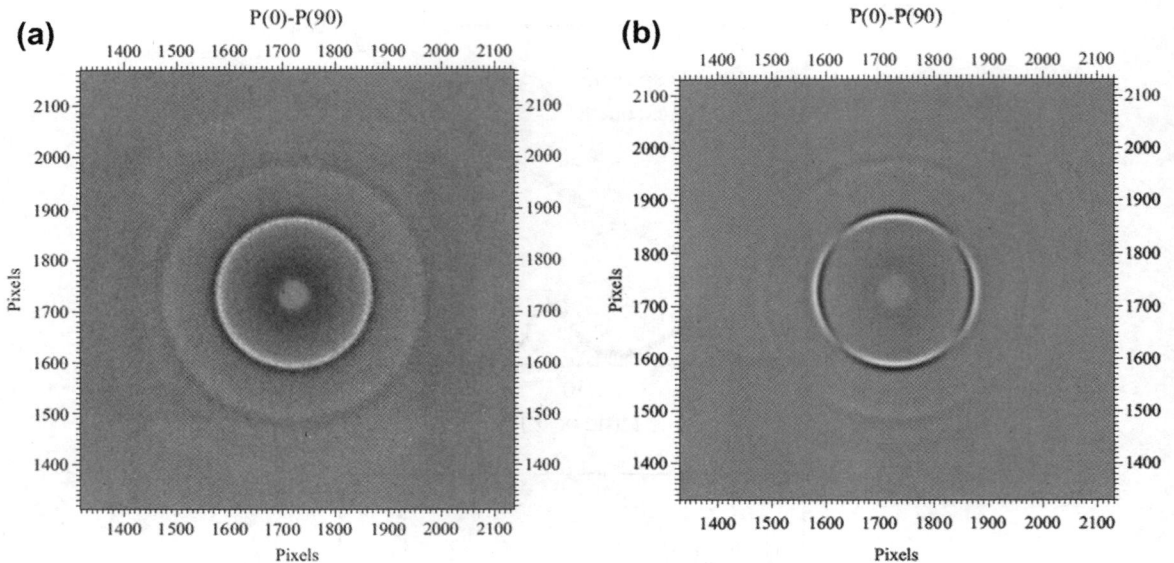

Figure 20 Difference between the 2D XRD patterns measured (at room temperature on stress-free samples) for two sample orientations P(0°)–P(90°) for an Fe-based metallic glass preannealed at 300 °C (a) without stress (b) under stress. The sample annealed under stress undergoes creep and shows anisotropy. Reprinted from Dmowski and Egami (2007) with permission.

4.4.2 Phase Separation

There are examples of metallic alloy liquids that in equilibrium show *phase separation*, i.e. decomposition into two liquids of differing composition. If the equilibrium liquid phase field is of sufficient extent, there is a fully developed *miscibility gap* (**Figure 21** shows this for a schematic binary system), in which two liquids coexist, bounded by the coexistence curve. Within the miscibility gap lies the spinodal curve that is the locus of points where the curvature of the Gibbs free energy as a function of composition $(\partial^2 G/\partial X^2)$ is zero. Inside the spinodal curve the liquid is unstable with respect to composition fluctuations however small, and undergoes *spinodal decomposition*. Between the coexistence and spinodal curves, the liquid is metastable to small fluctuations but can still decompose by nucleation and growth of a second liquid phase. The form of the miscibility gap illustrates that complete mixing, giving greater entropy, is favored at higher temperatures. Conversely demixing, if it can occur at all, is favored by lower temperatures. It is intrinsic to glass formation that the liquid is preserved down to the glass-transition temperature T_g, and the glass that forms is thermodynamically related to the liquid. Thus glass formation, by preserving the liquid and glassy states to low temperatures, is likely to reveal metastable miscibility gaps, i.e. gaps that do not appear in equilibrium diagrams because of the intervention of freezing to one or more crystalline phases.

Phase separation is well known in conventional oxide glasses (Zarzycki, 1991a), and indeed it is the basis of a number of industrial products. In particular, it is used to catalyze crystallization and to refine the crystalline microstructure in glass-ceramics (Zarzycki, 1991b; Kelton and Greer, 2010b). In MGs, however, the possibility of phase separation has been subject to some doubt. According to Inoue's empirical rules (Section 4.2.2), metallic glass formation is favored when the main component elements have negative heats of mixing. It is often considered that phase separation in metallic alloy

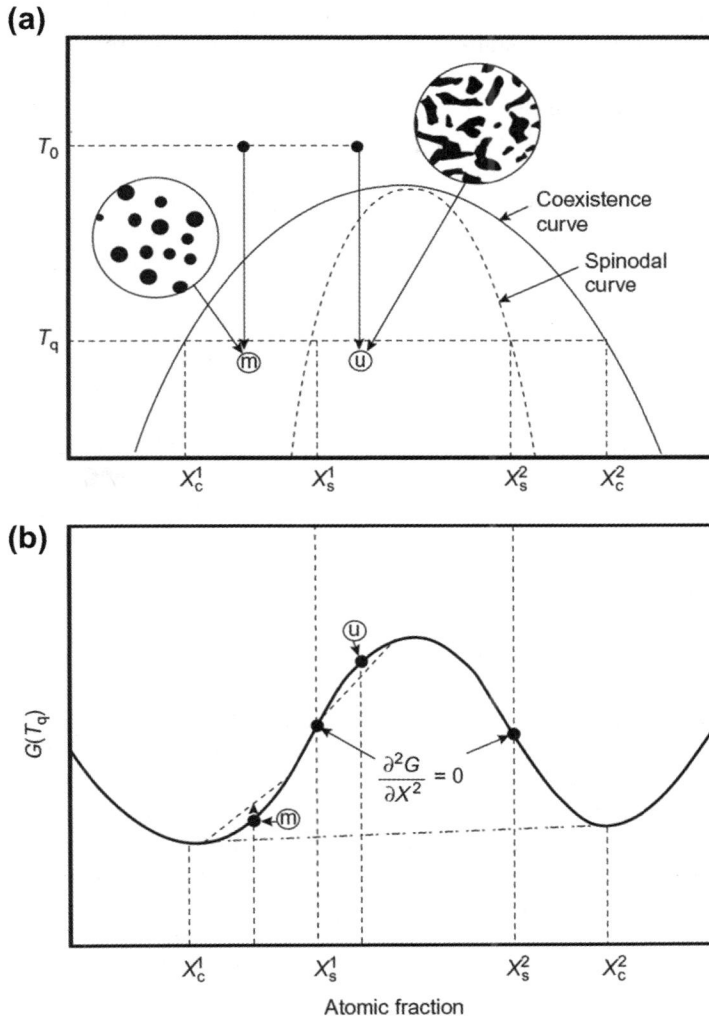

Figure 21 (a) A schematic phase diagram for a binary system showing phase separation with sketches of the initial morphologies obtained by cooling into the metastable (m) or unstable (u) regions. (b) The Gibbs free energy G of the system as a function of composition (atomic fraction X of one component) at the temperature T_q, showing how the coexistence and spinodal curves are determined. Reprinted from Kelton (1991) with permission.

liquids is associated with the demixing of two elements that have a positive heat of mixing. It then seems that any tendency to phase separation would be antithetical to glass formation. Elements with a positive heat of mixing, when included in a metallic glass-forming composition, are indeed very effective in inducing phase separation, and glass formation, even bulk glass formation, can still be achieved. It is also important to note, however, that a positive heat of mixing is not essential for phase separation. Even if all the elements in a given system have negative heats of mixing, there may be particular combinations of elements that are more favored (have heats of mixing that are more strongly negative) than others, and these can be the basis of the compositions that form on phase separation. Such compositions are sometimes considered to be defined by stable clusters in the liquid state. Perhaps because of doubts about the thermodynamic compatibility of phase separation and glass formation, early claims of phase separation in MGs were controversial, and there were certainly discrepancies between different studies (as reviewed by Lan et al., 2012). Now, however, as thermodynamic modeling of supercooled (i.e. metastable) liquid alloys has improved, CALPHAD techniques show clearly that there is a wide variety of glass-forming liquids and glasses in which phase separation can be expected. Furthermore, many experimental studies have not only revealed the phenomenon, but also demonstrated that it can be exploited for control of microstructure and potentially for optimization of properties.

Phase separation in MGs was first reported by Chen and Turnbull (1969), who noted that annealing-induced crystallization of Pd–Si based glasses with added gold was different below and above the glass-transition temperature T_g. The results for annealing just above T_g could best be explained by assuming that the homogeneous glass gave a homogeneous but unstable liquid that decomposed into two liquids that then showed crystallization to different phase mixtures. TEM provided further evidence for phase separation in suitably annealed samples. This work was taken further by Chou and Turnbull (1975) who found that at lower gold contents, for example in glassy $Pd_{80}Au_{3.5}Si_{16.5}$, phase separation occurred by nucleation and growth, while at higher contents, for example in $Pd_{74}Au_8Si_{18}$, it occurred by spinodal decomposition. In the latter case, they used small-angle X-ray scattering (SAXS) to study the kinetics and found that in the early stages of the separation, the logarithm of the SAXS intensity increased linearly with time, in accordance with Cahn's theory of spinodal decomposition. Chou and Turnbull (1975) noted that phase separation is of particular interest "because of its potential for generating structures in which two crystalline phases are finely interdispersed."

The phase separation in Pd–Au–Si and related glasses was seen on heating a glass formed on rapid quenching and presumed to be homogeneous. That phase separation could also occur on initial cooling of the liquid was shown clearly in the work of Tanner and Ray (1980) on Zr–Ti–Be glasses. Melt-spinning gives fully glassy ribbons, but it is evident that at intermediate compositions there are two calorimetric glass transitions rather than one. **Figure 22** shows differential scanning calorimeter (DSC) traces obtained on heating milligram metallic glass samples. In the DSC, the power necessary to heat the sample through any small temperature interval is automatically compared with the power necessary to heat a reference sample (undergoing no transformations) through the same temperature interval. As plotted in **Figure 22** (though the opposite convention is also common), an upward deflection corresponds to heat release from the sample. The glass transitions, occurring at around 400 °C or below, are indicated by downward steps in the baseline on heating. These steps correspond to the increase in specific heat when the glass is heated into the supercooled liquid state. On continued heating above the glass transition(s), there is crystallization as indicated by the exothermic peaks.

TEM (**Figure 23**) does indeed provide evidence for two phases in the as-quenched ribbons at intermediate compositions. The morphology in **Figure 23** is interconnected and sponge-like, a characteristic associated with spinodal decomposition, and different from the morphology of dispersed spheres associated with nucleation and growth (the contrasting morphologies associated with the early stages of decomposition in each case are shown schematically in **Figure 21(a)**). Others, though, have suggested that morphologies such as that in **Figure 23** could be an artifact of TEM specimen preparation (Nagahama et al., 2003).

It is evident (**Figure 21(a)**) that a miscibility gap widens as the temperature is lowered. Thus, as observed by Tanner and Ray (1980), the liquid phases formed on initial separation have compositions

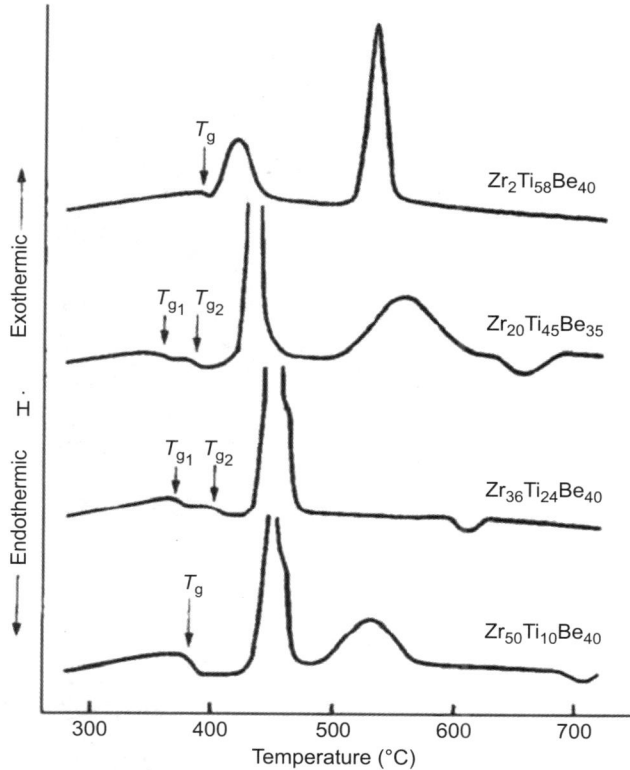

Figure 22 Differential scanning calorimetry (DSC) traces on heating (at 20 K min^{-1}) Zr–Ti–Be MGs produced by melt-spinning. One or two glass transitions, indicated by step-like increases in specific heat capacity, are followed by a sequence of crystallization exotherms. Reprinted from Tanner and Ray (1980) with permission.

that on continued cooling sit well inside the gap and are thus susceptible to further decomposition. In this way, there can be a sequence of phase separations as cooling proceeds, displaying a hierarchy of length scales, a phenomenon known in oxide glass-forming systems. Phase-separated microstructures formed at lower temperature are finer because of lower atomic mobility. Park et al. (2008) showed that phase separation can occur in the glassy state itself; they annealed a homogeneous melt-spun glass of $Zr_{28}Y_{28}Al_{22}Co_{22}$ (a composition showing phase separation in the liquid state on slower cooling) some 50 K below its T_g and found that phase separation developed on a nanometer scale. Similarly fine microstructures can also be found in as-quenched glasses. Kalay et al. (2012) used FEM and APT to characterize melt-spun $Al_{90}Tb_{10}$. This glass (**Figure 24**) shows clear inhomogeneity, with 2 nm clusters of nearly pure aluminum, that presumably act as nucleation centers for subsequent fine-scale devitri-fication (Section 4.4.3). The length scale of decomposition in **Figure 24** can be contrasted with scales of up to ~500 nm seen in other glass-forming systems showing separation in the liquid at higher temperatures. For a given system, the scale of the separated microstructure is finer for higher cooling rate, which can also alter the morphology from that due to nucleation and growth (on slow cooling, giving separation at small supercooling in the miscibility gap) to that due to spinodal decomposition (on fast cooling, separating at large supercooling) (Sohn et al., 2012).

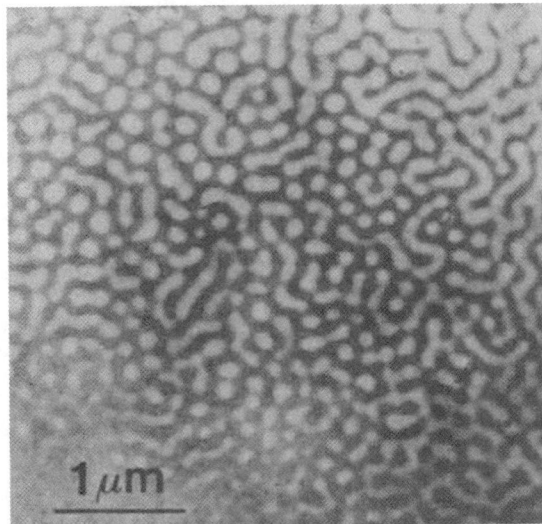

Figure 23   Bright-field transmission electron micrograph of as-quenched $Zr_{36}Ti_{24}Be_{40}$ metallic glass showing evidence for spinodal decomposition. Reprinted from Tanner and Ray (1980) with permission.

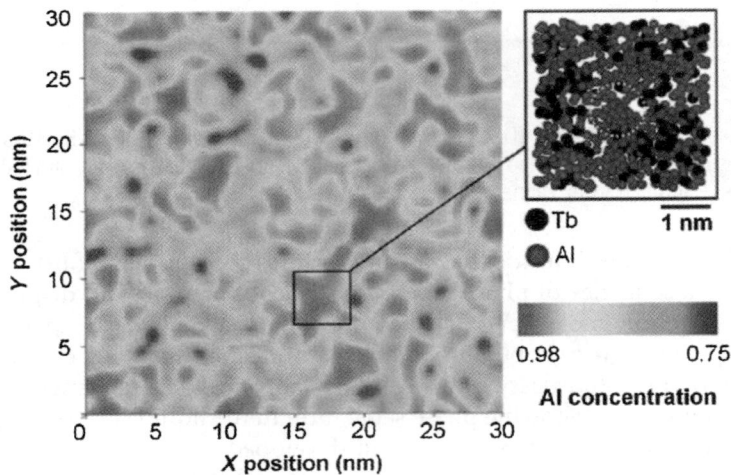

Figure 24   A map of aluminum concentration obtained from APT of a $30 \times 30 \times 4$ nm volume of melt-spun $Al_{90}Tb_{10}$ glass. The inset (volume $4 \times 4 \times 2$ nm) shows a cluster (circled) of pure Al. Reprinted from Kalay et al. (2012) with permission. (For color version of this figure, the reader is referred to the online version of this book.)

The liquid phases formed by separation on cooling may, depending on their GFA, form a glass or crystallize. In this way, a very wide variety of microstructures can form. **Figure 25**, from the work of Chang et al. (2010), shows some of the possibilities as revealed in the systems $Gd_{55}Al_{25}Co_{20}$–$Ti_{55}Al_{25}Co_{20}$ and $Gd_{55}Al_{25}Cu_{20}$–$Ti_{55}Al_{25}Cu_{20}$, in which phase separation is driven by

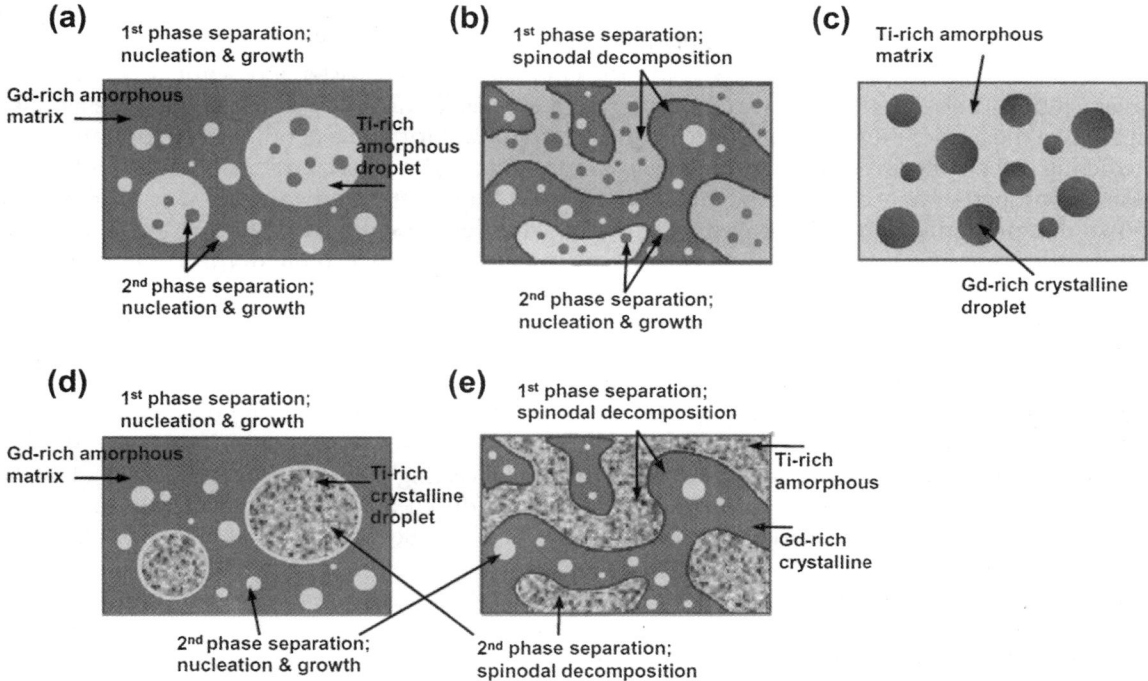

Figure 25 Schematic microstructures formed on melt-spinning of the systems $Gd_{55}Al_{25}Co_{20}$–$Ti_{55}Al_{25}Co_{20}$ and $Gd_{55}Al_{25}Cu_{20}$–$Ti_{55}Al_{25}Cu_{20}$. Sequential phase separation, driven by the positive heat of mixing of Gd and Ti, can occur by nucleation and growth or by spinodal decomposition, and the compositions thus formed may form a glass or crystallize. Reprinted from Chang et al. (2010) with permission. (For color version of this figure, the reader is referred to the online version of this book.)

the positive heat of mixing ($15 \ kJ \ mol^{-1}$) between gadolinium and titanium. **Figure 25(c)** shows (schematically) a microstructure of spherical crystals in a glassy matrix. In a different system (Co–Si–B–Cu), Nagase and Umakoshi (2010) found just such a microstructure with monocrystalline spheres. These crystals form in liquid droplets surrounded by a glassy matrix, and their shape is then not dictated by the crystal symmetry and anisotropic growth mechanisms as in the usual crystallization from a single uniform glassy phase (Section 4.4.3); this shows that it can be important to take phase separation into account in interpreting crystallized microstructures.

Can phase-separated microstructures be useful in property optimization?—this has been explored most for mechanical properties, as discussed in Section 4.5.5. Studies of the electrochemical behavior of two-phase glasses have a number of points of interest, including the exploration of possibilities for selective etching of one of the phases to obtain sponge-like structures of high specific surface area (Gebert et al., 2007). Phase separation also clearly influences magnetic properties (Han et al., 2012).

4.4.3   Crystallization

As noted in Section 4.2.4, MGs, like the supercooled liquids from which they form, are susceptible to crystallization. While there may be alloy compositions for which a metallic glass is more stable than any one crystalline phase, the glass seems always to have a higher free energy than a combination of crystalline phases. Crystallization of MGs has been widely studied, and a number of reviews are available, for example by Köster and Herold (1981), Köster (1984), Köster and Schünemann (1993), Greer (2001). There are several reasons for interest. A glass is formed only if crystallization is avoided on cooling the liquid; thus crystallization studies are important for understanding glass formation and predicting GFA. Once a glass is formed, it is important to characterize its stability in order that desirable properties may be preserved under service conditions. On the other hand, many studies have shown that properties can be improved by partial or complete crystallization. Indeed materials made by crystallization of a metallic glass precursor are now commercially exploited; this could be seen as analogous to conventional glass-ceramics in oxide systems. Crystallization of a glass (also known as *devitrification*) is very much slower than conventional solidification of a liquid, and it can readily be interrupted by lowering the temperature. This facilitates not only the development of optimized microstructures by controlled annealing, but also fundamental studies of the processes of nucleation and growth. Such studies have proved useful in understanding the glass-forming systems themselves, but it has also been possible to use the glasses as slow-motion analogs of metallic liquids and thereby to elucidate nucleation mechanisms in conventional metal casting (Schumacher et al., 1998). Finally, it is emerging that crystallization occurring during, and induced by, plastic deformation may make a significant contribution to enhancing the plasticity of MGs (Section 4.5.5, Hajlaoui et al., 2006; Pauly et al., 2010).

We compare first the basic kinetics of crystallization of the glassy state and solidification from the liquid state. **Figure 26** is a schematic illustration, for a glass-forming alloy, of the variation of the homogeneous nucleation frequency and the crystal growth rate over the temperature range from the liquidus down to below the glass-transition temperature T_g. As the supercooling is increased, the thermodynamic driving force for transformation to the crystalline state increases and the atomic mobility decreases. The forms of the temperature dependences of homogeneous nucleation and growth are readily derived from standard expressions (Greer, 1991); both show maxima at intermediate temperatures because of the lack of driving force at high temperature and the lack of mobility at low temperature. The maximum rate occurs at a lower temperature for nucleation than for growth, because of the need for the driving force to overcome the solid–liquid interfacial energy. As seen in **Figure 26**, conventional solidification occurs at rather small supercooling with moderate growth rates on heterogeneous nuclei. Crystallization of a glass occurs near T_g at very low growth rates on nuclei that could be heterogeneous (predominant at lower temperature) or homogeneous. In conventional solidification, the rate of latent heat production is matched by the rate of external heat extraction, and there are likely to be significant temperature gradients. As liquids are supercooled further, nucleation may occur at temperatures for which the atomic mobility is still large; crystal growth is then rapid and the release of latent heat is more rapid than the heat extraction, leading not only to temperature gradients but to a rapid overall temperature rise (*recalescence*). For crystallization of the glass, however, the crystal growth is so slow that the transformation can be isothermal, with negligible temperature gradients. In conventional casting, the solidification processes mostly occur close to the liquidus, at temperatures above the range in which recalescence is significant. The crystallization processes of concern in this chapter mostly occur near T_g at temperatures below the recalescence range. While solidification and crystallization can be viewed within one framework as indicated in **Figure 26**, significant differences can be expected. Taking dendritic growth as an example, the rate-limiting processes in solidification are solute and heat redistribution at the growing tip, whereas in crystallization the rate is limited by the mobility of the crystal–glass interface itself and by solute redistribution.

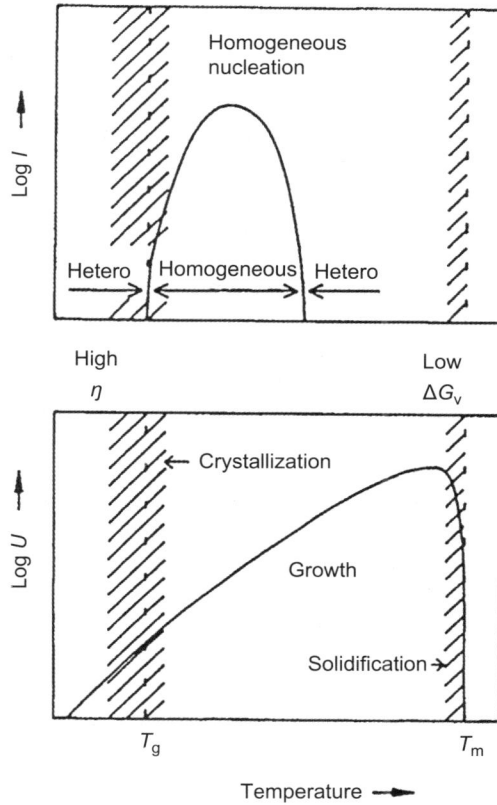

Figure 26 Schematic representation of the homogeneous nucleation frequency and the growth rate of crystals in a super-cooled melt or glass. The melting temperature, T_m, and glass-transition temperature T_g are shown, together with (shaded) the regions for conventional solidification and for crystallization of MGs. The temperature ranges in which homogeneous or heterogeneous nucleation dominate are marked, indicating that either can dominate for crystallization of MGs (4th edition, Figure 34).

Nevertheless, the common aspects of the processes in the two regimes have for example enabled heterogeneous nucleation in a metallic glass to be used to study grain refinement (*inoculation*) mechanisms in conventional metal casting. In such a study, Schumacher et al. (1998) embedded inoculant TiB_2 particles in an aluminum-based metallic glass. Crystallization of the alloy to α-Al occurs on the particle surfaces, and novel microscopical studies of the nucleation mechanism are possible because the crystallization proceeds so much more slowly than conventional solidification.

Typical DSC traces obtained on heating MGs have been shown in **Figure 22**. The trace for $Zr_{50}Ti_{10}Be_{40}$ shows a clear single glass transition, and then an interval of some 40 K before the onset of crystallization. Such a wide supercooled-liquid region indicates resistance to crystallization and implies good GFA of the kind associated with BMGs. In contrast, the trace for $Zr_2Ti_{58}Be_{40}$ shows little or no supercooled liquid region, and in many such cases it is not possible (in a simple DSC temperature scan) to detect the glass transition at all, as its calorimetric signature is overwhelmed by the onset of crystallization; such DSC traces are typical of glasses that require rapid quenching (e.g. by melt-spinning) for their formation. As shown in **Figure 22**, the crystallization following the glass transition often has distinct stages indicated by separate peaks in the DSC trace. The integrated area under each peak gives the associated enthalpy change. The different stages may correspond to successive crystallization of different parts of the sample (perhaps arising from composition changes induced by the crystallization processes themselves), or to transformations from metastable crystalline states toward the final equilibrium state. As the thermodynamic driving force for crystallization of the glass is high and the atomic mobility is low, it is likely that there will be metastable phases (or combinations of phases, for example a metastable eutectic) before equilibrium is reached.

In thermodynamic terms, three types of initial crystallization can be distinguished. In *polymorphic* crystallization the glass transforms to one crystalline phase of the same composition. The crystalline phase is not normally a solid solution of the base element, because glass formation would be difficult if such a crystallization were possible, and because any crystallization that did occur would likely involve solute partitioning. The phase formed in polymorphic crystallization is normally an intermetallic compound. In *primary* crystallization one crystalline phase is formed that has a composition different from that of the glass. Solute partitioning is involved, and the crystal growth is often dendritic. Complete transformation to the initial phase may not be possible, and the remaining glassy matrix of changed composition may undergo a distinctly different type of crystallization. The third type of crystallization is *eutectic*, in which two crystalline phases grow cooperatively. There is no overall composition difference between the glass and a eutectic colony growing within it. Within a colony straightforward lamellar structures are only rarely observed, and the microstructure is on a fine scale that does not vary strongly with crystallization conditions. Microstructures arising from the three types of crystallization are shown in **Figure 27**. Ternary eutectic crystallization is also known in which three crystalline phases grow cooperatively. The growth rate of eutectic colonies is particularly slow in such cases, which are usually associated with very good GFA (Garcia-Escorial and Greer, 1987). When phase separation (Section 4.4.2) occurs, the glassy phases of differing composition can crystallize distinctly. The driving forces for the various forms of decomposition are illustrated schematically in **Figure 28**.

Experimental studies of crystallization can be based on isothermal anneals or on continuous heating of the kind used for **Figure 22**. In either case, DSC (which in effect measures enthalpy changes in the sample) or measurement of other property changes (notably electrical resistivity) can be used to derive the volume fraction x transformed as a function of time t. Isothermal x vs t measurements are often analyzed according to Johnson–Mehl–Avrami–Kolmogorov kinetics, $x(t) = 1 - \exp[-(Kt)^n]$, where K is a temperature-dependent rate constant and n is an exponent that can be related to the modes of nucleation and growth. It must be pointed out, however, that such analyses are subject to errors of interpretation and must be supplemented by microscopical studies. In particular, if the transformation in a sample differs from place to place, quite erroneous conclusions may be reached (Holzer and Kelton, 1991). X-ray diffraction and related techniques are useful for identifying the crystalline products, but microstructural techniques are essential for understanding transformation kinetics. The scale of crystallites developing on annealing a metallic glass can vary enormously from a few nanometers (e.g. Kalay et al., 2012) up to as much as a millimeter in some ultraclean alloys with very high GFA (e.g. Drehman and Greer, 1984). Depending on the scale, TEM, scanning electron microscopy of polished and etched surfaces, or optical metallography may be appropriate. APT has proved useful in revealing composition profiles arising from solute partitioning in nanometer-scale primary crystallization (Sahu et al., 2010), while FEM is useful in characterizing MRO in the glass that may be a precursor for crystallization (Kalay et al., 2012). Intense radiation sources, such as X-rays from a synchrotron, can permit structural studies with very fine time resolution. Time-resolved studies

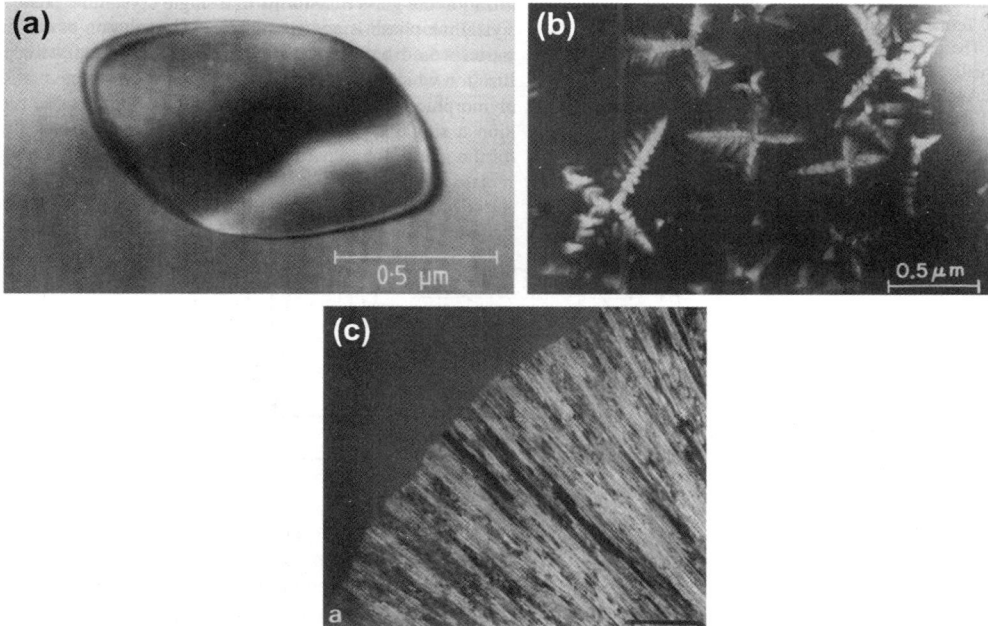

Figure 27 Transmission electron micrographs showing: (a) Polymorphic crystallization: $Fe_{76}B_{24}$ glass annealed at 400 °C for 10 min (reproduced with the kind permission of U. Köster). (b) Primary crystallization: $Fe_{81.5}B_{14.5}Si_4$ glass annealed at 445 °C for 20 min, showing α-iron dendrites. (c) Eutectic colony in $Fe_{40}Ni_{40}B_{20}$ glass, annealed at 390 °C for 1 h, showing γ-(Fe,Ni) + (Fe,Ni)$_3$B (marker = 10 μm) (4th edition, Figure 36).

of crystallization can reveal transient phases and other important aspects of the process (Brauer et al., 1992).

Figure 29 shows an example of a kinetic analysis, on melt-spun $Fe_{80}B_{20}$ glass, which shows eutectic crystallization to a mixture of α-Fe and the metastable boride, tetragonal Fe_3B. The traces in **Figure 29** are derived from DSC data and show the rate of change of transformed fraction during continuous heating. The data (main peaks) can be well matched by a numerical model that assumes eutectic growth from a fixed number of quenched-in nuclei. The population of these nuclei can be determined from the numerical fitting and was found to be of the order of 10^{18} m^{-3}, in agreement with direct TEM observations. The small high-temperature peaks in **Figure 29** arise from a thin layer on one side of the melt-spun ribbons, which contains no quenched-in nuclei.

Nucleation kinetics in MGs can be determined by direct counting of the numbers of crystals, by inference from overall transformation kinetics, or by measuring the crystal size distribution in a partially transformed sample. Examples of the latter are shown in **Figure 30**. The types of nucleation behavior that have been observed are illustrated schematically in **Figure 31**. Nucleation can occur during an anneal and can be homogeneous or heterogeneous, showing site saturation. In either case, transient effects may be evident. Alternatively, in some cases (e.g. as in the case in **Figure 29**) annealing merely leads to growth on preexisting nuclei.

While it appears that there can be homogeneous nucleation of crystals in MGs, there are few if any fully quantitative comparisons of measured kinetics with theoretical predictions. There are problems

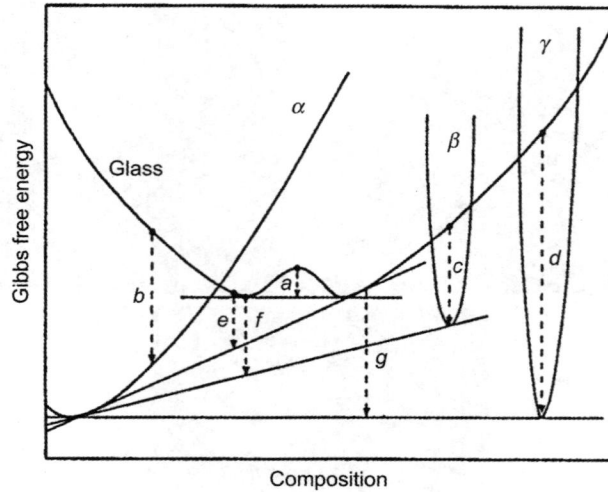

Figure 28 Schematic free energy curves for a binary alloy system, showing how a glass or amorphous phase can lower its free energy by undergoing: (a) separation into two amorphous phases (Section 4.4.2); (b) polymorphic crystallization into the supersaturated solid solution α; (c) polymorphic crystallization to the metastable compound β; (d) polymorphic crystallization to the equilibrium phase γ; (e) primary crystallization of α; (f) eutectic crystallization to α and metastable β; and (g) equilibrium eutectic crystallization to α and γ (4th edition, Figure 37).

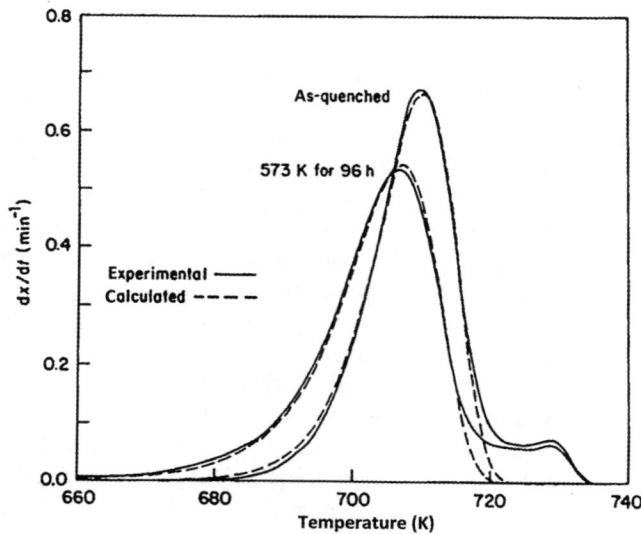

Figure 29 Crystallization kinetics, dx/dt, deduced from continuous DSC heating data, for $Fe_{80}B_{20}$ glass 32 μm thick (solid curves), compared with fitted curves derived from a model. The lower peak is for a preannealed sample. Reprinted from Greer (1982) with permission (4th edition, Figure 38).

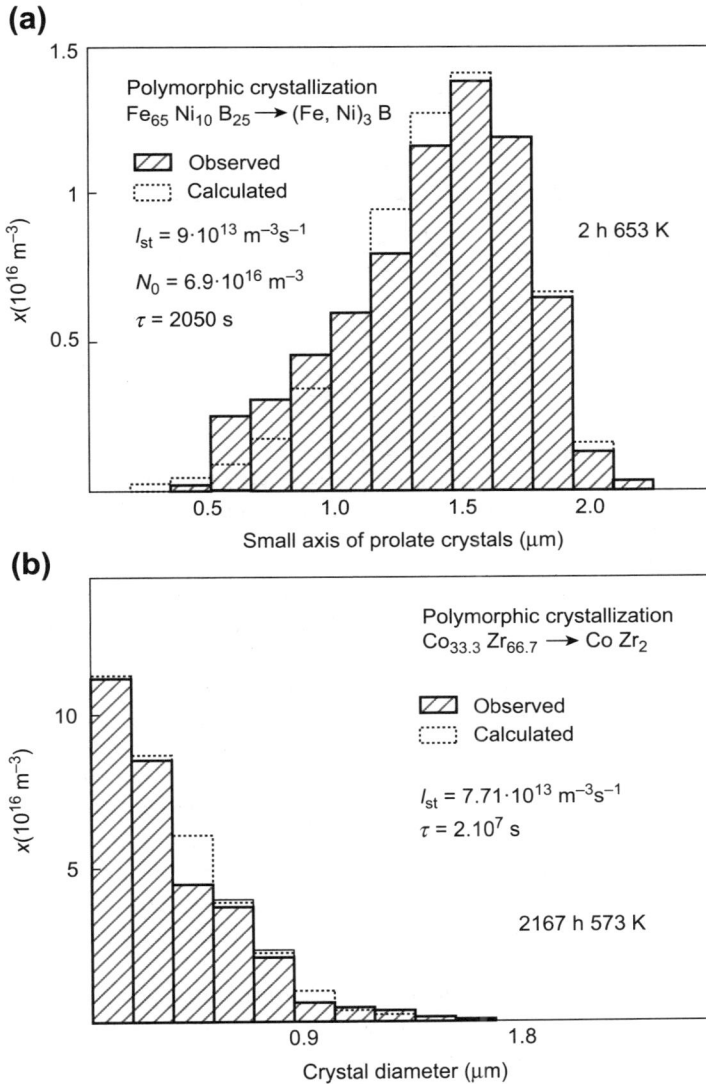

Figure 30 Typical histograms of crystal diameter distributions (observed compared with calculated) in partially transformed samples of polymorphically crystallizing MGs: (a) for $Fe_{65}Ni_{10}B_{25}$, assuming transient heterogeneous nucleation; (b) for $Co_{33}Zr_{67}$, assuming transient homogeneous nucleation. Redrawn from Köster and Blank-Bewersdorff (1987).

with the accuracy of both the measured kinetics and the input parameters for the theory. Detailed analysis of crystal nucleation in oxide glasses (Kelton and Greer, 1988) has shown that it is useful to have data on transient behavior. In such studies, the number of nuclei developing as a function of time is determined using two-stage annealing. The glasses are annealed at a temperature T_N, where the nucleation frequency is high but the growth rate is low, to produce a population of nuclei. An anneal at a higher temperature, T_G, where the nucleation rate is low but the growth rate is high, then allows these

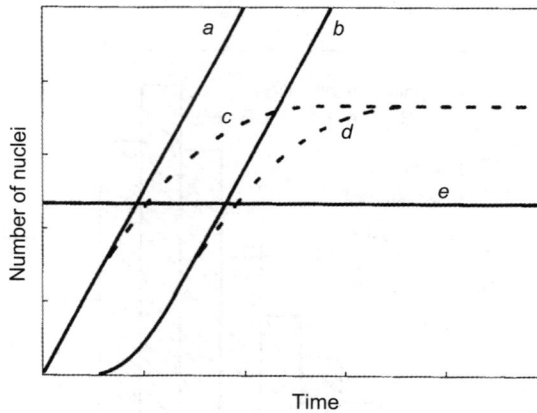

Figure 31   Schematic forms of the number of nuclei as a function of time for the following types of nucleation: (a) steady-state homogeneous, (b) transient homogeneous, (c) steady-state heterogeneous, (d) transient heterogeneous, (e) quenched-in active nuclei. Cases (c) and (d) show site saturation. Reprinted from Greer (1986) with permission (4th edition, Figure 40).

nuclei to grow to observable size. It has proved difficult to apply this technique to MGs, but **Figure 32(a)** shows the results of the first such study (Shen et al., 2009). Importantly, the product of the steady-state nucleation frequency and the induction time is independent of the atomic mobility, enabling the data for different nucleation temperatures T_N to be scaled to give a universal curve (**Figure 32(b)**), providing evidence both for the validity of classical nucleation theory in this case and the occurrence of transient homogeneous nucleation in this $Zr_{59}Ti_3Cu_{20}Ni_8Al_{10}$ glass. The form of the universal curve, in particular the product of the steady-state nucleation frequency and the induction time, enables the interfacial energy between the glass and the nucleating phase to be determined. In this case, the glass undergoes polymorphic crystallization to an icosahedral quasicrystalline phase, and the interfacial energy is remarkably low (0.010 ± 0.004 J m^{-2}), even when compared to values typical for liquid–quasicrystal interfaces. It is concluded that the liquid shows a strong increase in icosahedral SRO as it is cooled toward T_g. The low interfacial energy is then a consequence of the similar SRO in the glass and in the quasicrystalline phase.

The case just considered concerns homogeneous nucleation in the glass. There is also evidence that it can occur in the liquid during rapid quenching to the glassy state. As discussed in connection with **Figure 29**, the crystallization of $Fe_{80}B_{20}$ glass occurs by growth on a population of quenched-in nuclei. The large population of such nuclei and its strong dependence on quench rate suggest transient homogeneous nucleation during the rapid quenching of the alloy. Analysis suggests that in some cases inhibition of nucleation by transient effects is essential if glass formation is to be possible (Kelton and Greer, 1986). This is illustrated for the case of $Au_{81}Si_{19}$ in **Figure 33**. The dashed line in the lower figure indicates the low level of crystalline volume fraction (10^{-6}) that is conventionally taken as the criterion for successful formation of a (metallic) glass; it can be seen that this level is achieved with transient nucleation for a quench rate of 10^5 K s^{-1}, whereas if the nucleation rate were always in steady state the unrealistic rate of 10^8 K s^{-1} would be required. Thus crystallization studies reveal that for the formation of some MGs from the liquid, high quench rates are absolutely necessary.

In marked contrast, there are other glasses for which crystallization is clearly dominated by heterogeneous nucleation, the homogeneous nucleation frequency being very small. Examples are the

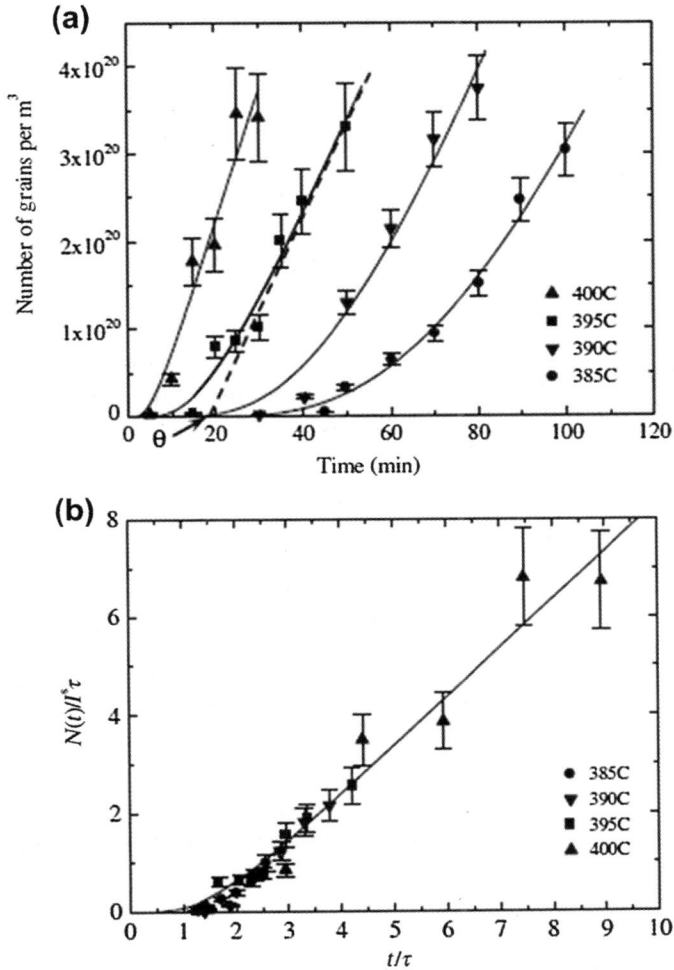

Figure 32 Two-stage annealing of the glass $Zr_{59}Ti_3Cu_{20}Ni_8Al_{10}$ gives nucleation and growth of quasicrystal grains of the same composition as the glass. (a) Number of grains per unit volume as a function of annealing time at the nucleation temperature T_N. The second anneal was at the growth temperature of 430 °C. The nucleation induction time θ is obtained by extrapolation, as shown by the dashed line for the 395 °C data. (b) Scaled data from (a) fit on a universal curve as predicted from classical theory of nucleation. Reprinted from Shen et al. (2009) with permission.

good glass-formers $Pd_{40}Ni_{40}P_{20}$ (Drehman and Greer, 1984) and $Pd_{77.5}Cu_6Si_{16.5}$ (Kiminami and Sahm, 1986). That the predominant nucleation is heterogeneous can be exploited, as processing under very clean conditions can permit bulk glass formation at very low cooling rates of less than 1 K s^{-1}. In such cases, crystallization of the glass proceeds typically from a low population of nucleation centers and very coarse microstructures result.

But many BMGs crystallize to very fine (even nanometer scale) microstructures, reflecting more complex behavior. For fundamental studies of crystallization kinetics, BMGs have the great

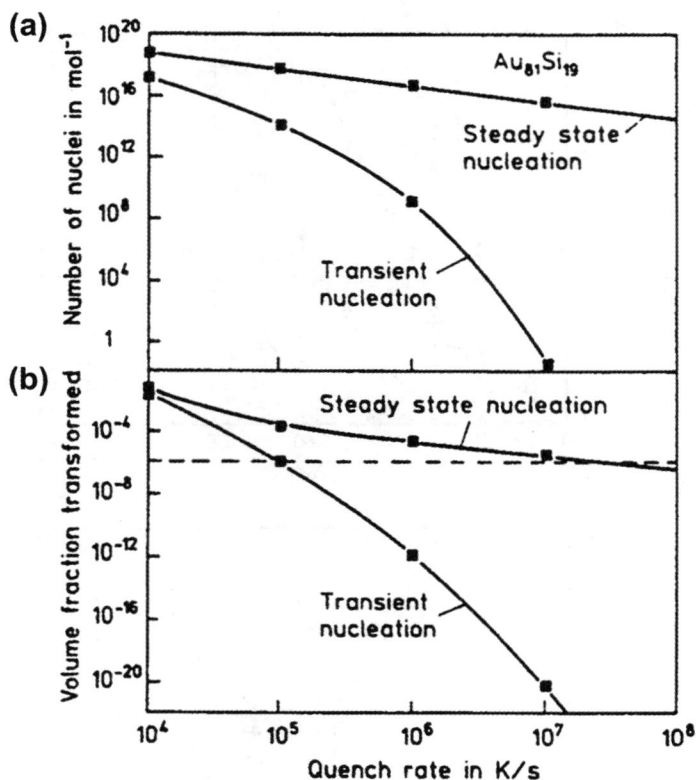

Figure 33   (a) The number of nuclei and (b) the transformed crystalline fraction calculated for quenching molten $Au_{81}Si_{19}$ at various rates, assuming either steady-state or transient homogeneous nucleation. Reprinted from Kelton and Greer (1986) with permission (4th edition, Figure 42).

advantage that their crystallization is sufficiently sluggish to permit the determination of the time–temperature–transformation (TTT) curve over the entire temperature range between T_g and the liquidus. **Figure 34** shows the TTT curve for $Zr_{41.2}Ti_{13.8}Cu_{12.5}Ni_{10.0}Be_{22.5}$, and such data have been acquired for several BMG-forming compositions. The characteristic C-curve is of the general shape for nucleation and growth kinetics of the classical forms shown in **Figure 26**. In some cases, for example for $Pd_{40}Cu_{30}Ni_{10}P_{20}$ (Löffler et al., 2000), the C-curve can be well fitted with classical nucleation and growth laws. But in most cases the simplicity of the C-curve is deceptive, masking great complexity; this is the case for the curve in **Figure 34**. The crystallization of $Zr_{41.2}Ti_{13.8}Cu_{12.5}Ni_{10.0}Be_{22.5}$ has been studied in detail, as reviewed by Schroers and Johnson (2000a, 2000b). The nature of the crystallization varies markedly with temperature. In the range 610–670 K, the primary crystallization yields a ccp phase; in the range 710–740 K, the primary phase is $MgZn_2$-type; and in the range 780 K to the liquidus, the primary phase is Zr_2Cu-type. In addition to the change of phase, there is a strong dependence of the scale of the microstructure on the devitrification temperature. Annealing at 610 K gives 10^{23} m^{-3} crystallites of diameter ~ 13 nm. In contrast, at 960 K, there are 10^8 m^{-3} crystallites of diameter ~ 1.5 mm. Thus the microstructural scale exhibited by this single alloy spans the entire range from a few

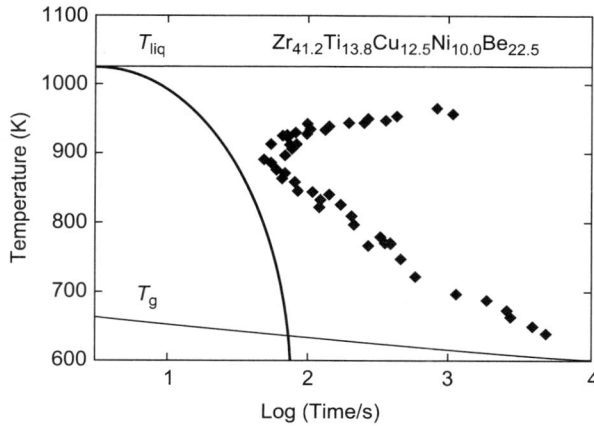

Figure 34   An isothermal (TTT) transformation diagram for the onset (10^{-4} fraction transformed) of primary crystallization of the bulk metallic glass-former $Zr_{41.2}Ti_{13.8}Cu_{12.5}Ni_{10.0}Be_{22.5}$ (Vitreloy 1). Crystallization occurs at temperatures between the liquidus at T_{liq} and the glass transition at T_g, and can be avoided by sufficiently rapid quenching of the liquid (dashed line). Adapted from Busch (2000) with permission.

nanometers to values characteristic of conventional alloy castings. The high-temperature crystallization kinetics can be fitted to classical nucleation and growth laws. But when extrapolated down to 610 K, the predicted population of nucleation centers is 10^2 m^{-3}, some $10^{21}\times$ less than that observed experimentally. The extrapolated behavior is consistent with the sparse nucleation that might be expected for BMG-forming systems and with the behavior observed in some systems, notably the $Pd_{40}Ni_{40}P_{20}$ and $Pd_{40}Cu_{30}Ni_{10}P_{20}$ alloys mentioned above. The discrepancy between this extrapolated behavior and the copious nucleation that is observed in low-temperature anneals can largely be attributed to prior phase separation on the glassy state. Evidence for phase separation in the particular case of $Zr_{41.2}Ti_{13.8-}Cu_{12.5}Ni_{10.0}Be_{22.5}$ has been reviewed by Löffler and Johnson (2001), and the phenomenon in MGs more generally has been discussed in Section 4.4.3. In any case, the complexity underlying the C-curve (**Figure 34**) explains why for systems of this type there may not be a straightforward correlation of GFA with the thermal stability of the glass (Waniuk et al., 2001).

Surfaces are often preferred sites for crystallization of a metallic glass. The main reason is likely to be modification of surface composition, for example by oxidation. There are clear possibilities to manipulate surface composition to enhance or impede nucleation, or to select crystallizing phases different from those that would be formed in the bulk. Comprehensive work in this area has been by Köster (1984). Surface crystallization can be used to improve the properties of a metallic glass (particularly magnetic properties) by inducing a compressive in-plane stress in a thin ribbon (Ok and Morrish, 1981). Wei and Cantor (1989) found that the surface crystallization of $Fe_{40}Ni_{40}B_{20}$ is enhanced if the alloy is first phase-separated by relaxation and then abraded, or else the surface is enriched in nickel by electroplating and the glass is then annealed. Sometimes, surface oxidation can have the opposite effect of inhibiting crystallization at the surface. Thus, with $Pd_{40}Ni_{40}P_{20}$, a thin NiO layer protects the glass from the local loss of phosphorus that is the prime cause of preferential surface nucleation (Garcia-Escorial and Greer, 1987). Removal of the original surface can drastically modify crystallization behavior. For example, $Ni_{66}B_{34}$ glass (which normally crystallizes from the surface with a strong [100] fiber texture) loses the quenched-in nuclei responsible for this if the surface is etched off (Köster and Schünemann, 1993).

Finally, in this section, we consider crystallization of MGs as a route to desirable microstructures. This concept, established for glass-ceramics, is of considerable interest for MGs. It may seem strange to adopt an extreme processing route (such as melt-spinning) or a complex composition (giving very high GFA) to obtain a metallic glass, in ribbon or bulk form, only then to crystallize it. The important point is that the crystalline product made from the glass can have a very fine and uniform microstructure, fine because the large effective supercooling favors nucleation, and uniform because of the uniformity of the glass itself and the lack of macrosegregation during solidification. Devitrification offers a particularly attractive way of making nanocrystalline material (with grain size less than 100 nm) inexpensively, controllably, and in comparative bulk without the need for consolidation of fine powders (an alternative route with contamination and safety issues).

Early work on the exploitation of devitrification was by Ray (1981) on alloys of Fe, Ni, Al, Cr, Mo, Co and W in multiple combinations, with 5–12 at.% of B or other metalloids as an aid to glass formation. Similar alloys, fully crystallized but derived from glassy precursors, were developed by Das et al. (1985) and others. Ni–Mo–B and Ni–Al–Ti–X–B and later other Ni–Mo–B alloys with added Cr were made by melt-quenching, comminution, and consolidation by extrusion or hot isostatic pressing. During the processing, ordered phases including Ni_4Mo, Ni_3Mo, Ni_2Mo and $Ni_3(Al,Ti)$ are precipitated from the crystallized matrix, together with stable boride precipitates. This family of alloys was developed commercially and related materials continue to be of interest for hard-facing coatings. Glasses such as Fe–Cr–Mo–C–B and Fe–Cr–Mo–C–V when devitrified give tool-steel materials with impressive high-temperature properties.

An important industrial application of a crystalline material of glass-forming composition is in the family of permanent magnet materials based on $Nd_2Fe_{14}B$. This phase combines high coercivity and high saturation magnetization, and is the basis for the strongest permanent magnets that are commercially available. The materials can be made by sintering an alloy powder, by melt-spinning the alloy to form a glass and then devitrifying, or by direct quenching to a nanocrystalline structure. The direct-quenching method was introduced by Croat et al. (1984), and has been fully commercialized. The annual production of 5500 tons is small, however, compared to that by the sintering route (nearly 50 000 tons per annum). **Figure 35** provides an overview of the production possibilities by direct quenching or by heating a glass. The hard magnetic phase is desired, the alternative soft magnetic phase must be completely avoided, and a fine grain size is desirable for high coercivity. In the materials of most interest, grain sizes range from 14 to 50 nm.

Interest in nanocrystalline magnetic materials also extends to the opposite extreme of very soft magnets. Compositions based on Fe–B–Si were well known in early studies of the soft magnetic properties of MGs. If the composition is modified by addition of copper (to enhance crystal nucleation) and niobium (to impede crystal growth), devitrification of such alloys leads to an isotropic micro-structure with a very fine (10 nm diameter) grains of bcc α-Fe(Si) embedded in the residual glass (Yoshizawa et al., 1988). At this scale the local magnetocrystalline anisotropies are averaged out, and soft magnetic properties result. These materials marketed under the trade name Finemet®, and other materials similarly derived from melt-spun ribbons of metallic glass, are attractive in comparison to fully glassy soft magnetic alloys because they can have very low magnetostriction and high saturation magnetization. The magnetostrictions of α-Fe crystallites and of the residual glassy phase in the microstructure are of opposite sign and substantially cancel (Herzer, 1991). APT studies (**Figure 36**) have been used to study the microstructural development in these materials. The as-quenched glass has the added copper distributed uniformly throughout, but it forms distinct clusters on annealing. These

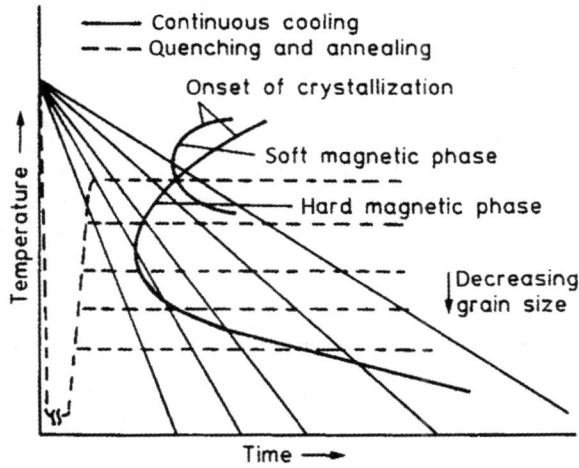

Figure 35 Schematic time–temperature–transformation diagram to rationalize the optimum conditions for producing Nd–Fe–B permanent magnets. Reprinted from Warlimont (1985) with permission (4th edition, Figure 43).

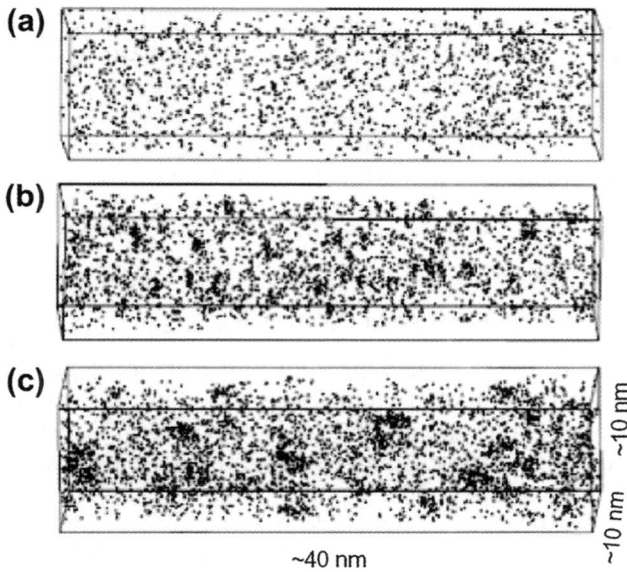

Figure 36 An APT map of copper in an $Fe_{73.5}Si_{13.5}B_9Nb_3Cu_1$ metallic glass: (a) as-quenched; after annealing at 400 °C for (b) 5 min and (c) 60 min. Reprinted with permission from (Hono et al., 1999).

then act as centers for heterogeneous nucleation, according to the sequence in **Figure 37**. When primary crystallization of α-Fe(Si) starts, the solutes niobium and boron are rejected into the residual glass and soon stop further growth when there is soft-impingement of the diffusion fields of the nearby crystallites (Hono et al., 1999).

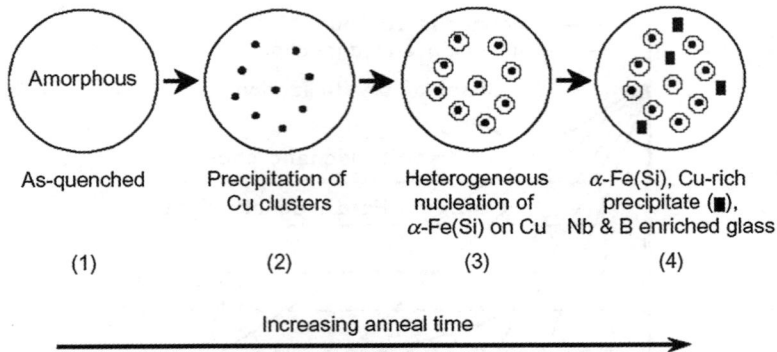

As-quenched   Precipitation of   Heterogeneous α-Fe(Si), Cu-rich
                        Cu clusters   nucleation of   precipitate (■),
                                           α-Fe(Si) on Cu   Nb & B enriched glass

   (1)   (2)   (3)   (4)

Increasing anneal time

Figure 37   Schematic diagram of the crystallization of Finemet® glass. Adapted from Hono et al. (1999) with permission.

Nanocrystalline materials made by devitrification of MGs may also be of interest for their mechanical properties. Aluminum-based alloys with rare-earth and late-transition-metal additions up to a total of 10–15 at.% can either by direct quenching or by devitrification yield microstructures on a nanometer scale with exceptional strength exceeding 1.5 GPa (Inoue et al., 1992). Consolidation of glassy or partially devitrified material may lead to a fully crystalline product, but one in which a fine-scale microstructure with desirable properties is retained. Superplasticity is of particular interest. For example, an alloy of the aluminum—rare earth—transition metal type has shown superplastic elongations of more than 500% at strain rates as high as 1 s^{-1} (Higashi et al., 1992). Even after consolidation and deformation, the mechanical properties (for example a tensile strength of 900 MPa) are superior to those of conventional high-strength aluminum alloys.

4.5   Mechanical Properties

4.5.1   Comparison with Conventional Engineering Materials

The availability of BMGs has greatly facilitated the study of mechanical properties and makes it possible to compare the glasses with conventional engineering materials. The comparisons presented here, mostly from Ashby and Greer (2006), use materials-selection maps in which ellipses enclose the range of property values associated with particular materials or groups of materials. An introduction to the use of such maps can be found in Ashby (2005).

Figure 38 is a map showing the elastic limit (yield stress) σ_y and Young's modulus E for a very wide range of metals, alloys and metal-matrix composites. The MGs, represented by the shaded ellipses and labeled with the compositions in atomic %, have high strengths that set them apart from the conventional materials. Their yield strengths scale with Young's modulus in a way that suggests a link with the ideal strength ($\sigma_y = E/20$). Indeed MGs approach this strength more closely than any other bulk metallic material. The σ_y values shown in **Figure 38** are for bulk samples; as noted in Section 4.5.3 higher values, even closer to the ideal strength, can be obtained in ultrasmall samples.

Maps such as that in **Figure 38** plotted with logarithmic scales make it easy to show contours giving the values of property combinations that are often useful *merit indices* for selection of materials for particular types of application (Ashby, 2005). **Figure 38** shows contours of the yield strain σ_y/E, and it

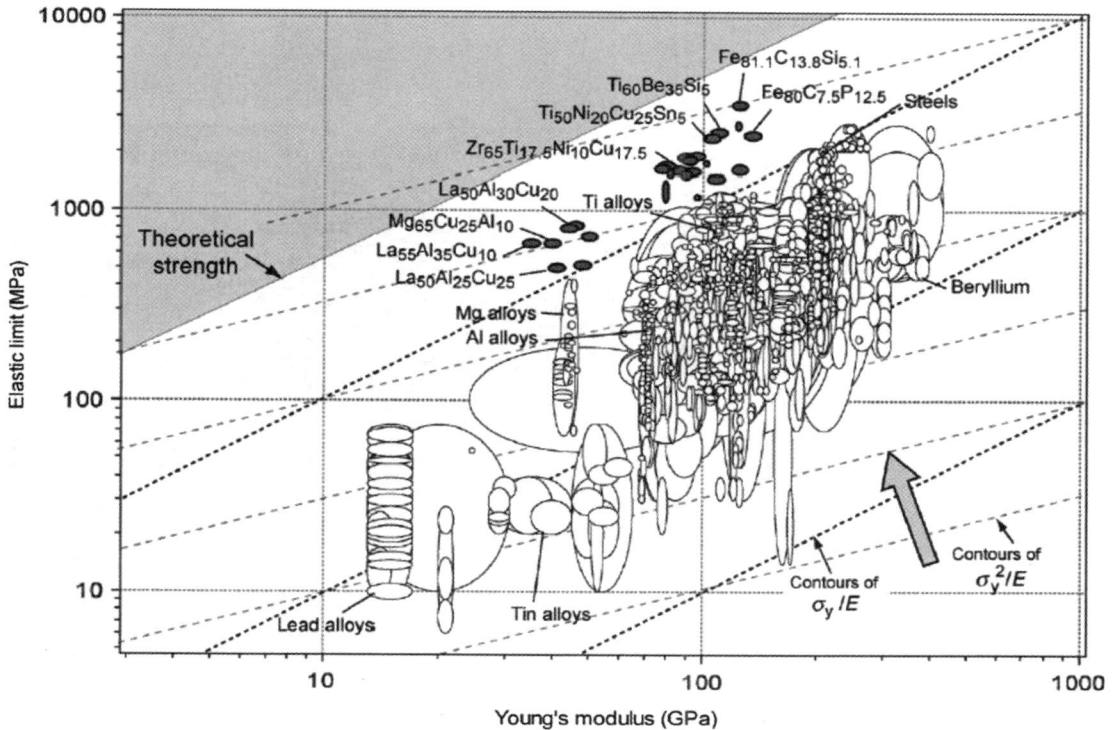

Figure 38 Materials-selection map showing yield stress σ_y and Young's modulus E for a selection of MGs (mostly BMGs) and for conventional metals, alloys and metal-matrix composites. Reprinted from Ashby and Greer (2006) with permission.

is clear that this quantity is much higher for the MGs than for any of the conventional metallic materials. The figure also shows contours of the quantity σ_y^2/E, which can be called *resilience* and is a measure of the capacity (per unit volume) of a material for reversible elastic storage of energy. The MGs also have exceptionally high values of this parameter, suggesting that they would be attractive for applications that exploit elasticity. It is then of interest to examine the mechanical damping, determining what fraction of the energy is lost in an elastic loading cycle. This is characterized by the *loss coefficient* η (Ashby, 2005). As shown in **Figure 39**, conventional metallic materials show a broad correlation between high resilience and low loss coefficient, indicating that plastic flow makes a major contribution to the energy loss. The MGs extend the range of this correlation and show exceptionally low η.

Conventional oxide glasses are characteristically brittle, and it is of fundamental interest and practical importance to explore whether MGs behave similarly. The resistance of a material to cracking can be characterized in various ways. In some cases the property of interest is the maximum load (stress) that a component (material) can support before fracture; this is related to the *fracture toughness* K_c. In other cases, the property of interest is the critical energy before fracture; this is related to the *toughness* $G_c(= K_c^2/E)$. **Figure 40** is a map of K_c and Young's modulus E, and shows contours of G_c. The values for MGs are compared with data for nearly 2000 conventional materials. The conventional materials fall into distinct classes. Whether characterized by K_c or G_c, metals are tough, while ceramics are brittle. The

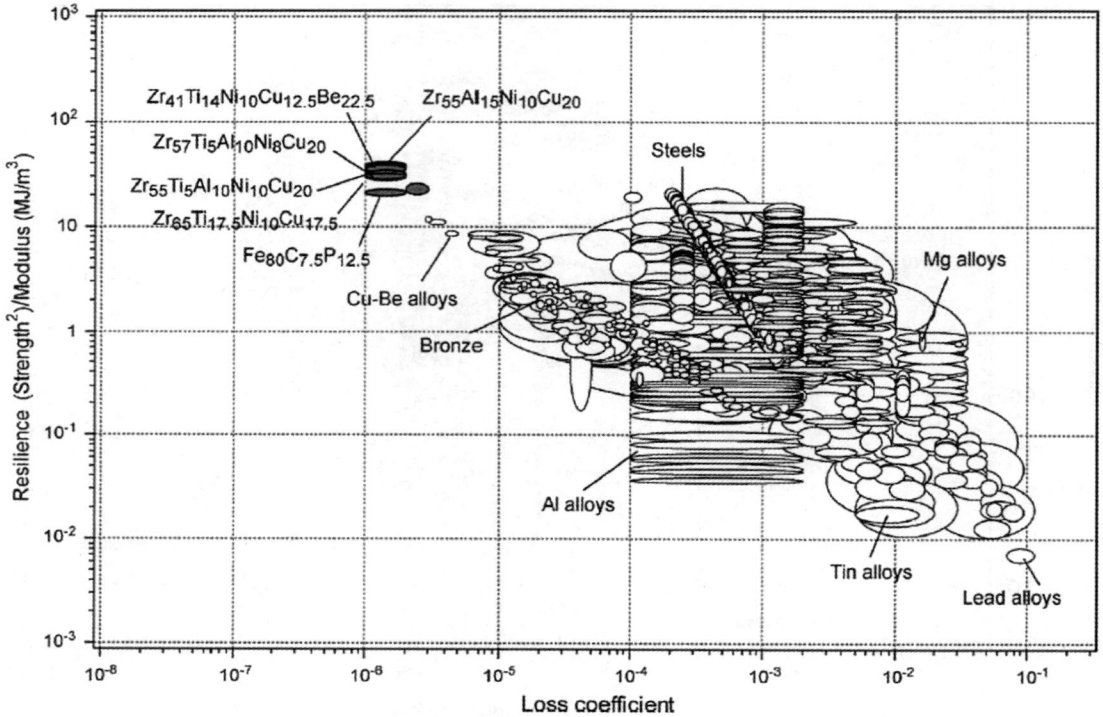

Figure 39 Resilience σ_y^2/E and loss coefficient η for MGs and conventional metallic materials (the same selection as in **Figure 38**). Reprinted from Ashby and Greer (2006) with permission.

toughness G_c cannot be less than the energy of the surfaces created when a material fails by cracking, defining an inaccessible (shaded) region on the map. The surface energies of materials scale with E (Ashby and Greer, 2006). Polymers have rather low K_c, but high values of G_c, giving them good impact resistance. In contrast, conventional oxide glasses have particularly low impact resistance. Quite remarkably, the metallic glasses do not form a clear class; their K_c and G_c values vary over some four orders of magnitude, from values approaching those of the toughest metals to values typical of ceramics. This behavior will be explored in Section 4.5.4.

At the tip of a sharp crack in a material there is a region, the *plastic* or *process zone*, in which plastic flow can occur. Plastic flow blunts the crack, and thus for a given material type (for example quenched and tempered steels) a lower yield stress σ_y, facilitating plastic flow, tends to increase toughness. The process-zone size, d, is given by

$$d = \frac{K_c^2}{\pi \sigma_y^2}.$$

(1)

If d is much smaller than the dimensions of a component, fast fracture by crack extension is the likely failure mode. On the other hand, if d is greater than the component dimensions, brittle failure is not expected. As already seen, metallic glasses have exceptionally high values of σ_y, and thus even

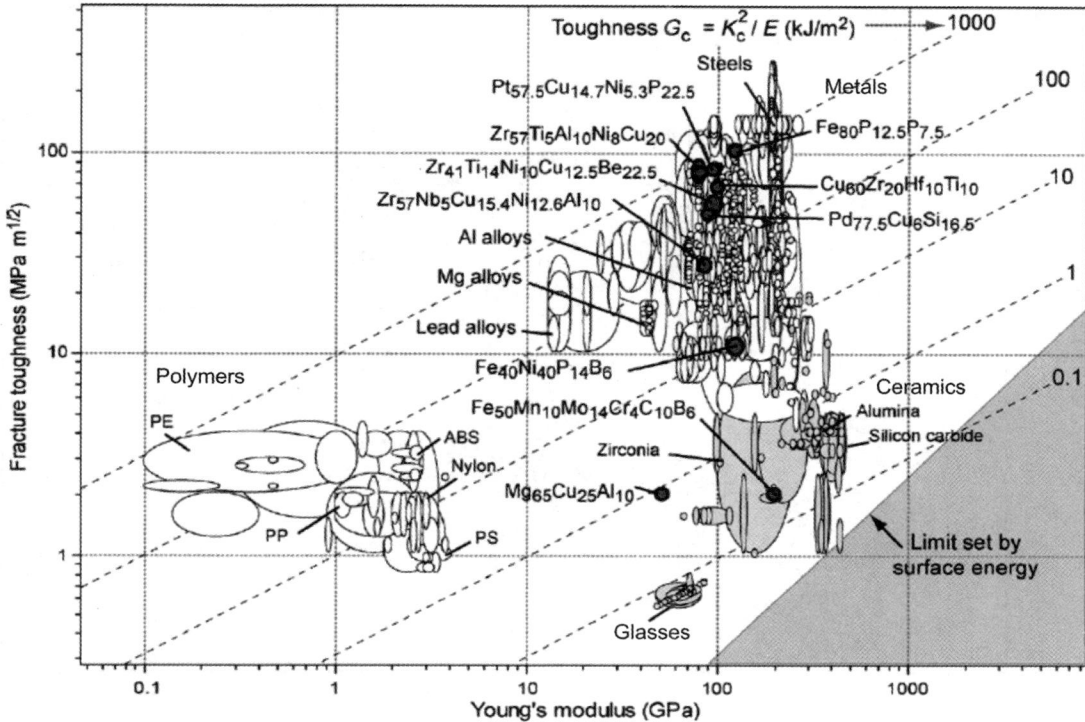

Figure 40 Fracture toughness K_c and Young's modulus E for MGs, compared with key conventional materials classes (metals, ceramics, polymers, oxide glasses). The dashed lines are contours of toughness (or fracture energy) G_c in kJ m^{-2}. Reprinted from Ashby and Greer (2006) with permission.

those with high fracture toughness K_c have rather small process-zone sizes. **Figure 41** is a map of K_c and σ_y showing contours of d. At best, the MGs have a d of a few millimeters, compared to values approaching 1 m in structural steels. The most brittle MGs have d values even smaller than those for oxide glasses.

A further area of concern for MGs is their fatigue behavior, found to be very variable as a result of intrinsic and extrinsic factors (Cameron and Dauskardt, 2006; Kruzic, 2011; Chuang et al., 2013). The fatigue life of a component under cyclic loading has two phases: crack initiation and crack propagation. For a typical high-strength conventional polycrystalline alloy, the time for initiation can be a significant, or even the dominant fraction of the total life. In contrast, initiation times in MGs appear to be very short indeed. The fatigue life of MGs is then essentially the time for crack propagation until catastrophic failure. The life under tensile loading is much shorter than under compression, and can be sensitive to specimen geometry. The threshold stress intensity for crack growth in MGs is ~ 1.3 MPa$\sqrt{m}$, at the low end of the range of values seen for polycrystalline alloys. The main problem, however, is the endurance limit for MGs: while this can be as high as 57% of the ultimate tensile strength it can also be as low as 8%, compared to $\sim 50\%$ for conventional high-strength alloys (Hess et al., 2006; Chuang et al., 2013).

From this brief survey, it is already clear that MGs have some properties that make them attractive for applications. We have highlighted the high elastic strain, high resilience and low loss coefficient that

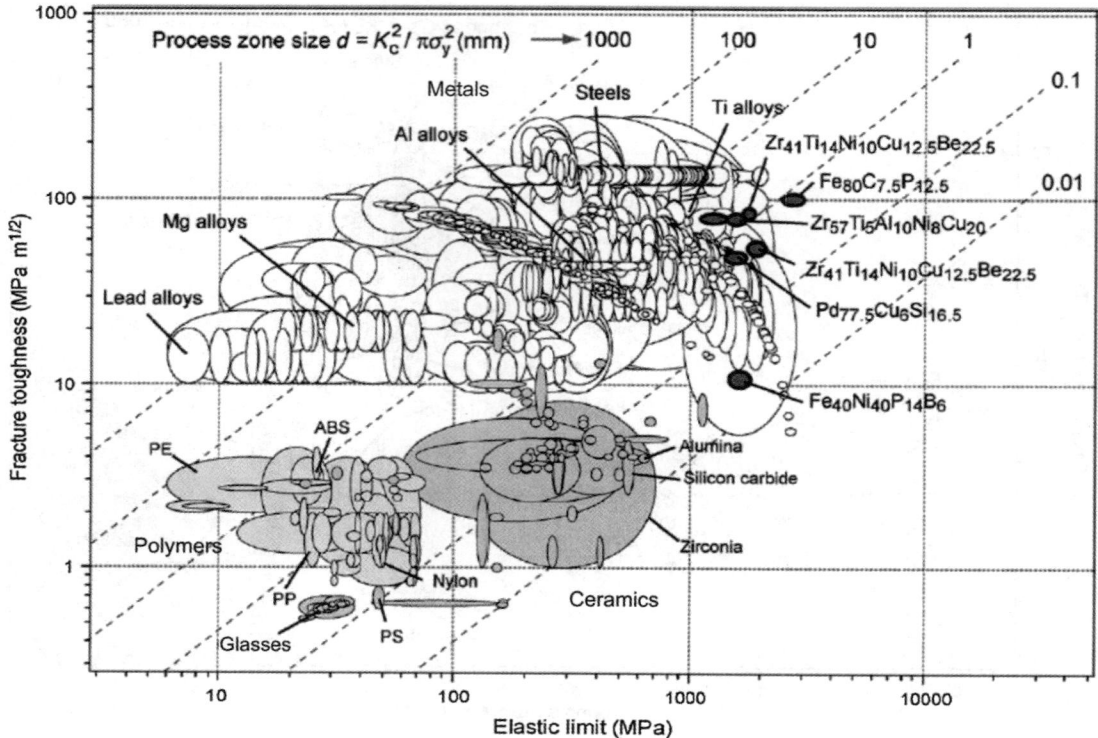

Figure 41 Fracture toughness and elastic limit (yield stress) for the same selection of materials as in **Figure 40**. The dashed lines are contours of process-zone size *d* in mm. Reprinted from Ashby and Greer (2006) with permission.

suggest application as springs and in some sporting goods. Applications of MGs are discussed in Section 4.6. On the other hand, MGs have some drawbacks. The mechanisms of plastic flow in MGs, underlying both good and bad aspects, are quite different from those of polycrystalline metals, and are explored next (Section 4.5.2).

4.5.2 Mechanisms of Deformation

In polycrystalline metals the characteristic elementary process governing plastic flow is dislocation glide. In MGs it was suggested by Argon (1979), on the basis of observations on bubble rafts, that under shear there can be local cooperative rearrangements of atoms. The spheroidal region in which this rearrangement occurs is a *shear transformation zone* (STZ). Unlike dislocations in crystals, microscopic imaging is not possible for STZs in glasses, and study of them has relied on modeling (Falk and Langer, 1998). Studies such as those of Zink et al. (2006) and Mayr (2006) on a simulated Cu–Ti glass suggest that STZs in MGs have a diameter of the order of 1 nm and that they contain from a few tens to a few hundreds of atoms. The activation volumes of STZs inferred from nanoindentation experiments tend to be rather larger, 2.5 nm^3 to 6.5 nm^3, and are bigger for glasses with larger Poisson's ratio (Pan et al., 2008). It has proved possible to detect STZs and their strain fields in colloidal glasses, which may be useful analogs for the MGs (Schall et al., 2007).

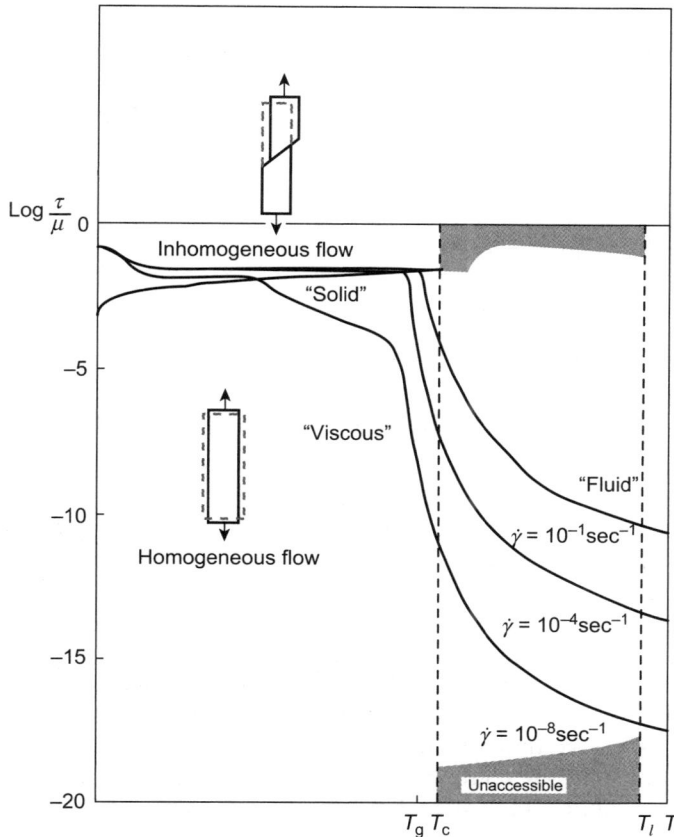

Figure 42 Spaepen's schematic deformation-mechanism map for a metallic glass/liquid, showing the conditions for homogeneous or inhomogeneous (shear-banding) flow. Reprinted from Spaepen (1977) with permission.

Figure 42 is the deformation-mechanism map for MGs proposed by Spaepen (1977). This shows that at low stresses the glasses show homogeneous viscous flow. Below the glass-transition temperature T_g this flow is very slow and can be regarded as creep. At T_g and above, i.e. in the supercooled liquid state, the flow can be rapid, and this is attractive for shaping MGs (Section 4.6). At high stresses, in a significant temperature range including room temperature, the plastic flow of the MGs is sharply localized into thin *shear bands*. The occurrence of shear bands is the most significant single hindrance to the wider structural application of MGs. Their nature is considered in Section 4.5.3.

As considered by Spaepen, the basis of the behavior in **Figure 42** is the shear-induced disordering of metallic-glass structure. This disordering is known to be accompanied by dilatation, and can most easily be described as the generation of free volume (Section 4.3.4). The availability of bulk samples of MGs has facilitated a wide range of quantitative studies of viscous flow, allowing the effects of shear-induced free volume to be characterized and analyzed. As an example we can take the work of Lu et al. (2003) on $Zr_{41.2}Ti_{13.8}Cu_{12.5}Ni_{10}Be_{22.5}$ (Vitreloy 1). **Figure 43** shows that when tested in uniaxial compression at

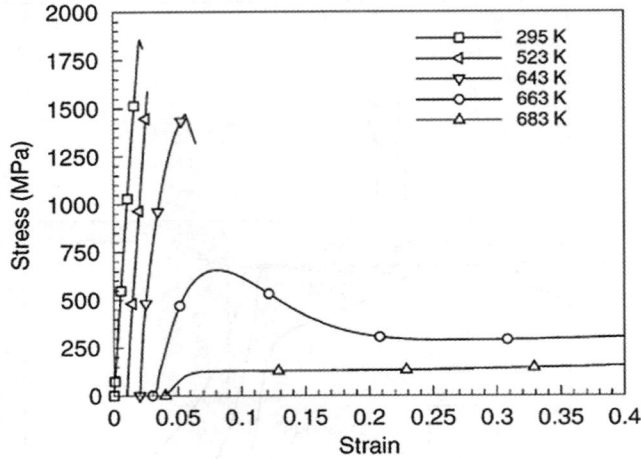

Figure 43 Stress–strain curves for Vitreloy 1 in uniaxial compression at a strain rate of $\dot{\varepsilon} = 0.1 \text{ s}^{-1}$ at the different temperatures indicated. For clarity the origin of successive curves has been shifted to the right. Reprinted from Lu et al. (2003) with permission.

room temperature, the glass yields at 1860 MPa. On yielding there is immediate catastrophic failure by shear along a single shear band. At higher temperatures, the yield stress is lower and catastrophic failure can be avoided. As flow continues, however, there is flow-thinning (most clearly seen in the curve for 663 K). This arises from the generation of free volume by the shear itself. Especially at elevated temperatures, any excess free volume of this kind will tend to decrease through structural relaxation toward equilibrium. Thus there are competing processes, and the overall behavior in the experiments falls into three regimes (**Figure 44**). At high temperature and low strain rate, the viscous flow is

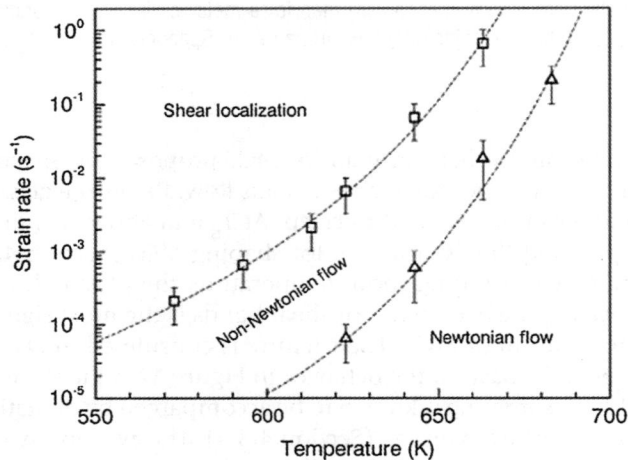

Figure 44 The boundaries between distinct modes of deformation for the BMG Vitreloy 1 as determined in uniaxial-compression tests. Reprinted from Lu et al. (2003) with permission.

homogeneous and Newtonian, i.e. the strain rate is linearly proportional to applied shear stress, implying a viscosity that is constant and specifically independent of strain rate. In this regime, structural relaxation dominates: the free-volume level and structure generally remain in equilibrium despite ongoing shear. At higher strain rate or lower temperature, the flow remains homogeneous, but is non-Newtonian. In this regime, the viscosity decreases with increasing strain rate: at any strain rate a nonequilibrium steady-state level of free volume is established and that level is higher at higher strain rate. At still higher strain rates or lower temperatures, the flow becomes inhomogeneous. The generation of free volume under shear permits the sheared volume to flow even faster and the result is shear localization (shear-banding).

The measured flow behavior of $Zr_{41.2}Ti_{13.8}Cu_{12.5}Ni_{10}Be_{22.5}$ glass summarized in **Figure 44** has been quantitatively modeled by Demetriou and Johnson (2004) in terms of free-volume production and annihilation.

4.5.3   Shear-Banding

The localization of plastic flow noted above is manifested as shear bands, giving unattractive surface markings on deformed samples. **Figure 45(a)** shows the surface steps on a bent wire, and also shows thin extrusions that presumably arise when the load is released and there is elastic spring-back of the bent sample. The close-up in **Figure 45(b)** shows the clean shear of the glass, the sample remaining coherent with no evidence of cracking. Shear-banding is important in MGs, and it has been very widely studied. The topic has been thoroughly reviewed by Greer et al. (2013); this section covers only the most prominent features.

It was noted in connection with **Figure 38** that the yield stress of MGs scales with elastic stiffness. This was examined more closely by Johnson and Samwer (2005) who found that, for a wide range of glasses tested in compression, the shear yield stress scales with the shear modulus (**Figure 46**), indicating a remarkably consistent value of shear yield strain of 2.67%. Thus, despite the unstable nature of flow in shear bands, the conditions for the onset of yielding are clear. As can be seen in **Figure 45(b)**,

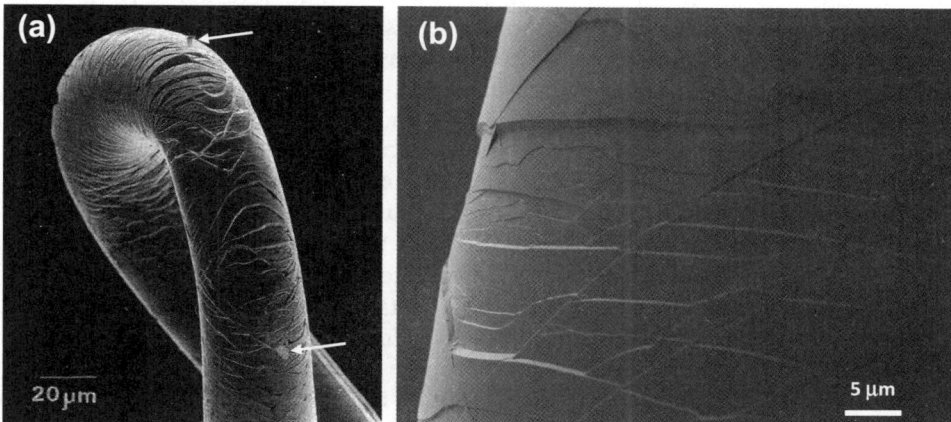

Figure 45   (a) Surface markings caused by shear bands in a bent wire (diam. 100 μm) of a $((Fe_{50}Co_{50})_{75}B_{20}Si_5)_{96}Nb_4$ (in at.%) metallic glass. The arrows indicate thin extruded plates of the glass. (b) A close-up of the shear offsets on the tensile side of the wire. Image credits: Dr Konstantinos Georgarakis, WPI Advanced Institute for Materials Research, Tohoku University).

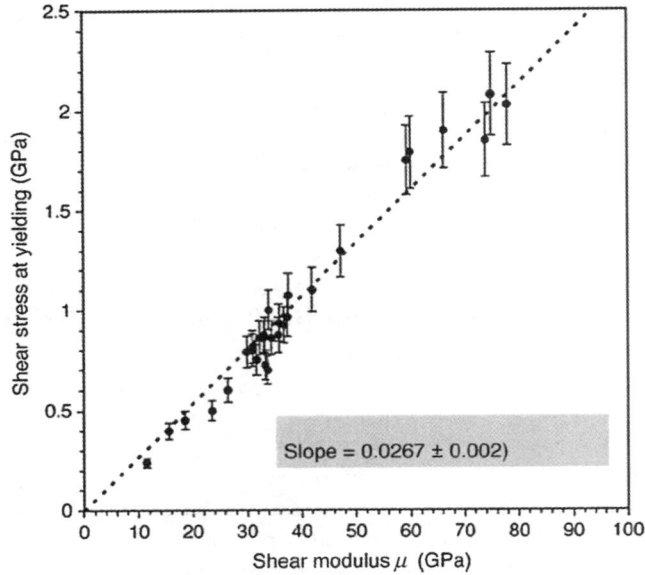

Figure 46 Measured shear stress at yielding, $\tau_y = \sigma_y/2$ as a function of shear modulus at room temperature for 30 bulk-metallic glasses. Reprinted from Johnson and Samwer (2005) with permission.

the shear offsets are sharp, and correspondingly, the bands must be thin. The thickness seen in TEM and in atomistic simulations is consistently in the range 10–20 nm (Zhang and Greer, 2006). The bands form close to the planes of greatest shear stress, i.e. at $\sim 45°$ to the principal stress axes. (Detailed discussion of this angle and of the deviations from 45° attributable, amongst other causes, to the normal-stress dependence of the shear flow stress, is given by Gao et al. (2011).) In tension, the shear softening usually leads to catastrophic failure on a single dominant shear band (**Figure 47**). Accordingly, even though there is extensive local plastic flow, the sample shows essentially zero ductility and can be regarded as macroscopically brittle. Such catastrophic failure is avoided in bending (**Figure 45(a)**) because the neutral (zero-stress) plane in the middle of the sample prevents a single shear band from spanning the sample cross-section. As shown in **Figure 47**, after some shear giving smooth shear offset surfaces, the sample under tension parts to form two fracture surfaces.

The characteristic vein pattern on these surfaces (**Figure 48**) suggests liquid-like flow of the sheared material. The onset of liquid-like flow is consistent with and explains the shear localization, but its origin has been controversial. On the one hand, the concentration of mechanical work done in thin shear bands is likely to lead to local heating, and indeed such heating is generally considered to be the origin of the adiabatic shear bands seen in some polycrystalline alloys (Wright, 2002). On the other hand, disordering in sheared material, usually characterized as an increase in free volume (Section 4.3.4, Section 4.5.2), would give a lowered viscosity without any necessary temperature rise. The debate on which is the dominant effect has been summarized by Lewandowski and Greer (2006), who also made measurements on notched beams of a Zr-based BMG loaded in bending to failure. The samples had been coated with a thin layer of tin, and it was found that this layer melted at shear bands close to the notch. From the width of the melted zone, Lewandowski and Greer were able to derive the likely evolution of temperature profiles around the shear bands. Although these measurements provided clear

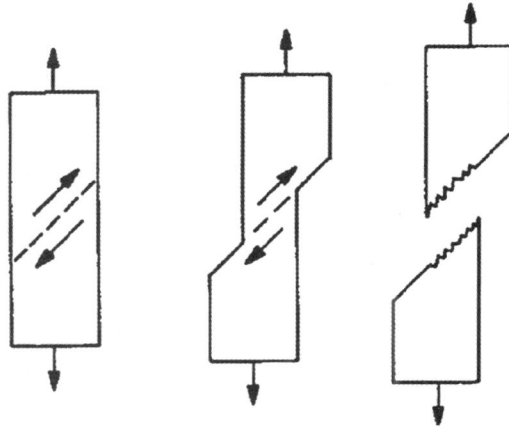

Figure 47   The flow and fracture process by which metallic-glass samples tested in tension have fracture surfaces with two distinct regions: smooth and veined. Reprinted from Leamy et al. (1972) with permission.

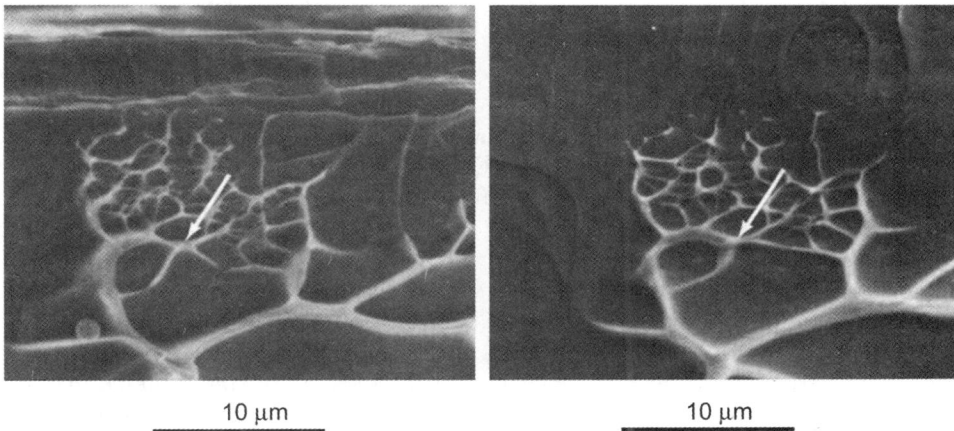

Figure 48   The vein pattern on the opposing segments of the fracture surface of a $Pd_{82}Si_{18}$ glassy sample. The left-hand image has been inverted to facilitate the comparison, with the arrows indicating corresponding points. Reprinted from Leamy et al. (1972) with permission.

evidence of heating, the width of the temperature profiles was much too great to be consistent with the 10–20 nm thickness of shear bands. It was concluded that the observed heating is a consequence of shear localization, not its cause. Subsequent continuum mechanics modeling by Jiang and Dai (2009) has confirmed that at the onset of shear-banding, local free volume increases occur before there is local heating.

If the free volume in a shear band is increased during shear, it is reasonable to expect that some of this might be quenched-in after the shear has ended. Evidence for this was provided by Pampillo (1972) who showed that shear bands are revealed when the polished surface of a deformed metallic glass is etched. This effect (**Figure 49**) suggests that the structural state of the material has been altered.

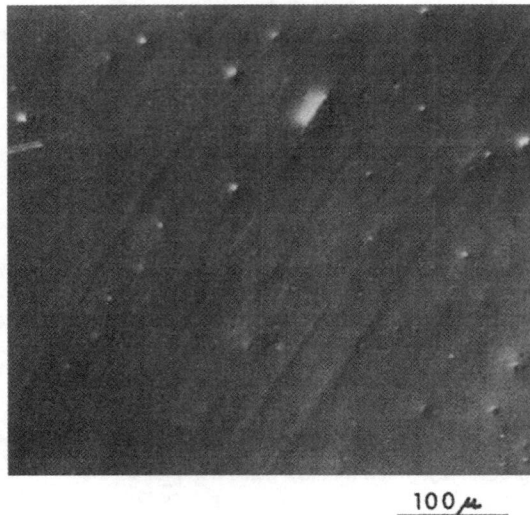

100 μ

Figure 49 Preferential etching of shear bands on a polished cross-section of deformed $Pd_{77.5}Cu_6Si_{16.5}$ glass. Reprinted from Pampillo and Chen (1974) with permission.

Subsequent annealing erased the preferential etching, as would be expected from structural relaxation toward equilibrium (Section 4.4.1). Pampillo and Chen (1974) showed that when a deformed sample was reloaded, shear resumed on existing shear bands (even when the sample was reshaped to avoid effects of stress concentration). This confirms that the material with higher free volume does indeed have a lower flow stress.

By using a very stiff loading machine, and small samples, catastrophic failure on a single shear band (**Figure 47**) can be avoided. For samples loaded in compression, there is a regime in which they show *serrated* flow and the duration of the load drops is clearly dependent on the sample temperature (Klaumünzer et al., 2010). This is clear evidence that shear bands can operate cold, i.e. at essentially the temperature of the body of the sample. In other cases, however, there is clear evidence for local heating and the temperature rises can be substantial. The fracture of MGs can, indeed, be accompanied by the ejection of liquid particles heated by several hundred degrees (Gilbert et al., 1999).

Cheng et al. (2009b) have analyzed the conditions under which shear bands may operate cold or hot. They modeled the conditions in a shear band during uniaxial compressive loading of cylindrical samples of aspect ratio (sample height to diameter) of 2:1. They considered quasistatic loading under displacement control. They show that when there is yielding, the extent of the shear offset before the load is relaxed (and shear stops) is, among other factors, proportional to the sample height. Their calculations (**Figure 50**) show the effect of sample size. In small samples (diameter $d = 1.0$, 1.5 and 2.0 mm, i.e. height = 2.0, 3.0 and 4.0 mm) the shear tends to a fixed offset (the vertical component of which is shown in **Figure 50(a)**), and correspondingly the temperature rise is small. This is the regime of serrated or *stick-slip* flow. For larger samples (diameter $d = 3.0$ and 4.0 mm, i.e. height = 6.0 and 8.0 mm), however, the temperature rise induces an extra softening effect such that the shear offset increases without limit (**Figure 50(a)**) and the shearing itself continues to accelerate (**Figure 50(b)**). As the greater offset produces still more heating, there is no steady state, but instead a *thermal runaway*

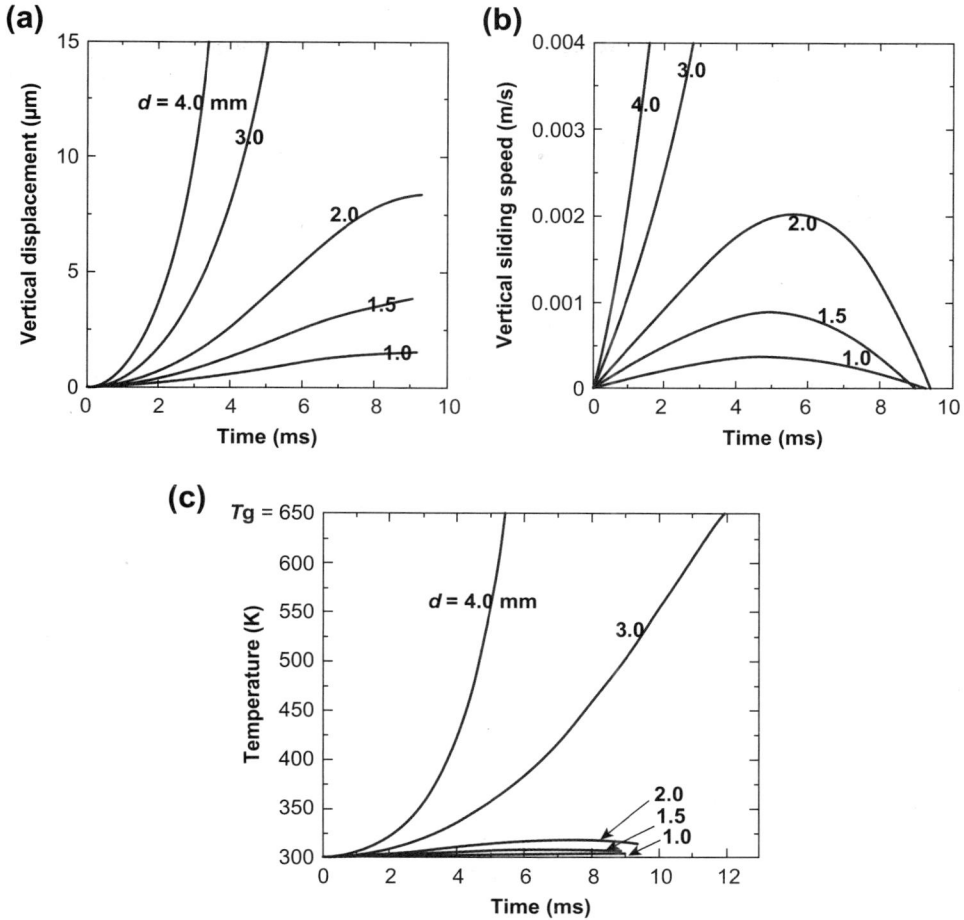

Figure 50 Continuum-mechanics modeling of a shear band in metallic-glass samples (of diameter in mm given by the labels) under uniaxial compression, as a function of shear duration: (a) axial displacement, (b) axial sliding speed, and (c) temperature in the band. Adapted from Cheng et al. (2009b) with permission. (For color version of this figure, the reader is referred to the online version of this book.)

(**Figure 50(c)**), associated with catastrophic failure. The conditions for the onset of such runaway phenomena may be a way to estimate the ultimate strength of materials in general (Braeck and Podladchikov, 2007).

We turn next to the vein pattern found on fracture surfaces. The standard model for its formation is *viscous fingering*, as analyzed by Saffman and Taylor (1958), of air into the liquid-like layer. As noted by Argon and Salama (1976), this mechanism requires the liquid-like layer to be 2–20 times thicker than the vein spacing. For the example shown in **Figure 48**, then, the liquid-like layer would have to be some 50 μm thick, much greater than the fundamental shear band thickness of 10–20 nm noted above. This can be explained by the sequence of events. Shear localization has its origin in free-volume generation and initially the shear bands are thin and operate cold. As the magnitude of shear offset increases, more mechanical work is done in the band, and there is significant heating. By the time the sample approaches failure (forming the veined fracture surfaces as shown in **Figure** 47) the heating may be considerable and the associated hot zone is the region within which the vein pattern forms.

We noted in Section 4.5.1 that MGs can have poor fatigue resistance compared to polycrystalline alloys. Shear banding appears to facilitate both the initiation and propagation of fatigue cracks. Shear bands can be nucleated easily at stress concentrations on a sample surface and they soon evolve into fatigue cracks. As the fatigue crack grows, further shear bands are generated and the crack propagates along these bands (Wang et al., 2010).

Finally in this section we consider the suppression of shear banding in small samples. There are now many experimental studies on metallic-glass specimens with diameters down to ~100 nm. These specimens have often been prepared by focused-ion-beam milling, and there have been concerns about artifacts caused by surface contamination and damage possibly associated with this technique. Nevertheless, a consistent trend has emerged that the flow stress rises as the sample diameter gets smaller, up to about two times the macroscopic flow stress at diameters of 100 nm. This "smaller-is-stronger" trend can be interpreted according to an energy-balance argument as first proposed by Volkert et al. (2008) and taken further by Wang et al. (2012). This is analogous to the Griffith analysis for crack propagation and treats a shear band as a shear crack that can propagate only if its length exceeds a critical value. In this way Wang et al. (2012) suggest that the flow stress should have a component proportional to $D^{-1/2}$, where D is the specimen diameter. Other approaches also suggest that shear-band nucleation must become more difficult, and ultimately impossible, as the sample size decreases. For example, Shimizu et al. (2006) use MD modeling to obtain a critical length scale of ~100 nm. It seems reasonable to suggest that the strength of a metallic glass should be size-dependent, with three regimes (**Figure 51**). In the macroscopic regime the strength is controlled by the conditions for shear-band propagation. At smaller sizes, shear-band nucleation becomes difficult and controls the strength. And at the smallest sizes, shear-banding must eventually be suppressed altogether. In that regime, the strength should approach the ideal value and plastic flow, if any, should be homogeneous.

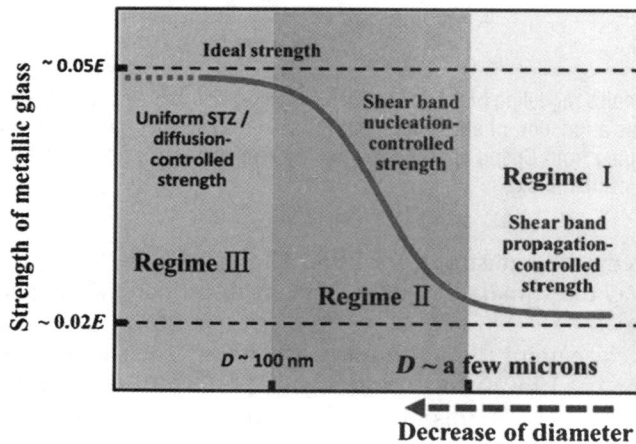

Figure 51   The "smaller-is-stronger" trend shown by MGs, interpreted in terms of three regimes with distinct controlling mechanisms. Reprinted from Wang et al. (2012) with permission. (For color version of this figure, the reader is referred to the online version of this book.)

Luo et al. (2010) have produced dramatic confirmation that homogeneous flow is possible in small samples. Using electropolishing to avoid the problems of focused-ion-beam milling, they prepared specimens with a cross-section less than 20 nm. In tension these showed homogeneous flow and elongations of up to 200%. This behavior was attributed to the suppression of shear banding and to the contribution of fast surface diffusion.

4.5.4 Toughness

The use of elastic constants to understand fracture was pioneered, for polycrystalline metals, by Pugh (1954). He noted that resistance to plastic shear is proportional to the shear modulus μ, and that resistance to dilatation and cracking is proportional to the bulk modulus B. A low value of μ/B should therefore favor plastic flow over cracking. Poisson's ratio ν shows a monotonic variation with μ/B, such that as μ/B is lowered ν increases toward its upper-bound value of 0.5 (characteristic of a liquid). Chen et al. (1975) noted "It is the high Poisson's ratio which is responsible for the ductile behavior of many MGs, " but this point was largely ignored until Schroers and Johnson (2004b) showed that a Pt-based metallic glass with an unusually high ν of 0.42 also showed exceptionally high plastic strain in compression ($\sim 20\%$) and exceptionally high fracture toughness (80 MPa m$^{1/2}$). Then Lewandowski et al. (2005) suggested a general correlation between ν and toughness (characterized in their work by the fracture energy G): the data points in **Figure 52** represent a variety of as-cast MGs based on Ce, Cu, Fe, Mg, Pd, Pt, or Zr, as well as a Zr-based glass (Vitreloy 1) in different annealing conditions. The MGs appear to show significant toughness only when ν exceeds a critical value of 0.31–0.32, whether that value is attained by choice of composition or annealing condition.

The elastic properties of MGs, including ν, can be estimated from the properties of the constituent metallic elements using simple rule-of-mixtures calculations (Zhang and Greer, 2007). While such estimates are crude, ignoring important factors such as the heats of mixing of particular pairs of elements, they are nevertheless sufficiently good to guide the development of MGs with higher ν. And, through the correlation in **Figure 52** MGs with higher plasticity and toughness are thus likely to be

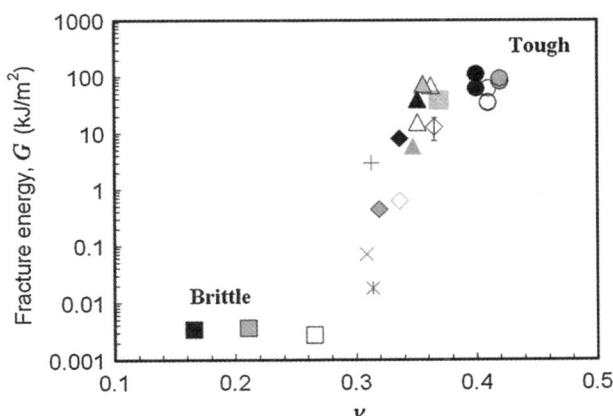

Figure 52 The correlation of fracture energy G with Poisson's ratio ν for MGs (as-cast and annealed) as well as for oxide glasses (■ fused silica, ▨ window glass, □ toughened, partially crystallized glass). The divide between the tough and brittle regimes is in the range $\nu_{crit} = 0.31$–0.32. Reprinted from Lewandowski et al. (2005) with permission. (For color version of this figure, the reader is referred to the online version of this book.)

obtained. The suggestion of a correlation between toughness and ν has been influential in the pursuit of MGs showing improved plasticity, and the nature of the correlation has been widely tested. The possible mechanism underlying the correlation has also been studied from the perspective of atomic-level structure, particularly the dependence on composition and processing history (Cheng et al., 2009a), and the principles thus inferred have been applied to guide the design of MGs with improved plasticity (Zhang et al., 2009) and toughness (He et al., 2011).

The work on plasticity or brittleness of MGs is now very extensive. Within simple series of glasses (varying composition or annealing treatment within a given system), it does seem that a higher plasticity or toughness is correlated with a higher ν. On the other hand, there is doubt about the existence and universality of a critical ν_{crit} above which MGs cease to be brittle. The universal correlation implied in **Figure 52** does not, for example, take into account how the testing temperature in each case compares with the glass-transition temperature of the glass. It has been suggested that ν_{crit} can be different for different alloy systems, even for those as closely related as Pd-based and Pt-based (Kumar et al., 2011). And studies of embrittlement on annealing also suggest that ν_{crit} is different for different systems (Kumar et al., 2013).

The toughness values in **Figure 52** are assumed to be intrinsic properties of the MGs. It has recently been noted, however, that they may be significantly influenced by extrinsic factors, such as the presence of brittle oxide inclusions (Madge et al., 2012). The presence or absence of such inclusions may well be correlated with the chemistry of the alloy system. When such extrinsic effects are excluded, it seems that there may not be a sharp transition such as that shown in **Figure 52**; rather there may just be a linear variation of $\log G$ with ν (as is seen in **Figure 52** when $G > 1$ kJ mol^{-1}). That does suggest that the pursuit of higher values of ν remains worthwhile to achieve greater toughness. Dramatic confirmation of this is provided by the development of a Pt-based BMG with $\nu = 0.43$ and $K_c \approx 125$ MPa m$^{1/2}$, beyond range of the MGs in **Figure 52**, (Demetriou et al., 2011a), and of a Pd-based BMG with $\nu = 0.42$ and $K_c \approx 200$ MPa m$^{1/2}$ (Demetriou et al., 2011b). In the latter case, the plastic zone diameter d (Section 4.5.1, Eqn (1)) is ~ 6 mm, much larger than for any other metallic glass, increasing the size of samples that could be made of this monolithic glass while still maintaining optimum properties (Ashby and Greer, 2006). This case is also particularly notable because the glass ($Pd_{79}Ag_{3.5}P_6Si_{9.5}Ge_2$) combines its exceptionally high toughness with a high σ_y of 1490 MPa, such that the product $K_c\sigma_y$ (a property termed *damage tolerance*) is the highest of any known engineering material (**Figure 53**). The key point, as illustrated for example by the ellipse for steels in the figure, is that a higher σ_y, other things being equal, usually implies a lower K_c (for the straightforward reason that crack-blunting through plastic flow at the tip is more difficult). It is remarkable, then, that this metallic glass (and later another Zr-based BMG (He et al., 2012)) can manage to combine a high σ_y with a high K_c, especially since a metallic glass can have no microstructural features (such as internal interfaces, grain boundaries) of the kind exploited in polycrystalline materials to improve toughness.

Can such an impressive property combination be linked to the nature of the shear-banding? The patterns of shear bands that form at a notch tip in the Pd-based glass with record-breaking damage tolerance are particularly finely spaced (Greer, 2011). A finer spacing, and consequently smaller shear offsets, obviously aid plasticity. The effects of sample size on shear-band spacing and shear offset, as reviewed by Miracle et al. (2011), may contribute to understanding the role of increasing ν. The analyses of Conner et al. (2003) and of Wei et al. (2013) both show that, for a given geometry of plate bending (i.e. given ratio of plate thickness H to radius of curvature R), the shear-band spacing λ and the shear offset Δu depend on ν, decreasing to zero as ν approaches its upper bound of 0.5 (**Figure 54**). These predictions have been quantitatively verified by experiment, albeit in a very limited number of tests (Wei et al., 2013). While these analyses are for the specific case of bent plates, it may more generally be

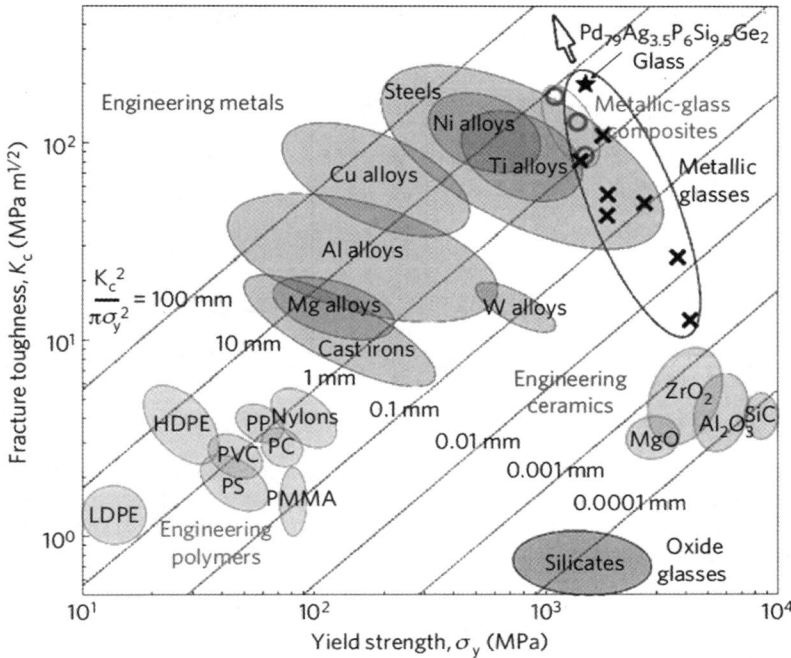

Figure 53 Materials-selection map comparing different materials classes in terms of the trade-off between fracture toughness (K_c) and yield stress (σ_y). Values are shown for monolithic BMGs ($\times$), for MG-based in situ dendritic composites of the kind described by Hays et al. (2000) ($\bigcirc$), and for the BMG $Pd_{79}Ag_{3.5}P_6Si_{9.5}Ge_2$ ($\star$) that shows the highest damage tolerance (the product $K_c\sigma_y$) of any known material. The contour lines indicate values of plastic zone diameter d (Eqn (1)). The arrow indicates the direction in which BMGs may potentially get access to still better combinations of K_c and σ_y. Reprinted from Demetriou et al. (2011b) with permission. (For color version of this figure, the reader is referred to the online version of this book.)

true that a higher ν favors smaller λ and Δu and therefore greater plasticity. The underlying analysis depends only on the geometry of flow as characterized by ν, while of course ν may be correlated with yet other factors such as the relative likelihood of shear-transformation and cavitation events as discussed by Demetriou et al. (2011b).

As briefly reviewed by Greaves et al. (2011), ν is also correlated with other factors: higher values of ν are associated with higher fragility of the liquid and lower GFA. The particularly damage-tolerant glass highlighted in **Figure 53** has a Pd-based composition that allows an unusually favorable combination of high ν and high GFA (Greer, 2011). In any case, it is clear that choice of composition can dramatically affect the pattern of shear-banding and the consequent mechanical properties of MGs. For greater plasticity and toughness, it is favorable to choose compositions showing higher ν, and those will typically have higher contents of elements that themselves show high ν (Zhang and Greer, 2007).

4.5.5 Improving Plasticity

Conventional polycrystalline alloys have many microstructural features that can be controlled in order to optimize their mechanical properties. In contrast, MGs have no microstructure, at least not at any

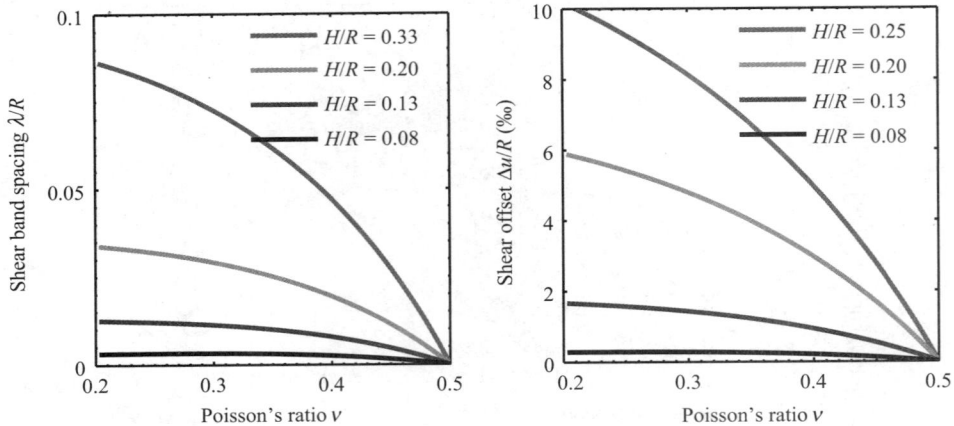

Figure 54   The predictions of Wei et al. (2013) for shear-band spacing λ and the shear offset Δu on the bands in metallic-glass plates of thickness H bent to a radius of curvature R. Both λ and Δu depend on Poisson's ratio ν, each tending to zero as $\nu \to 0.5$ (its upper bound, characteristic of a liquid). Reprinted from Wei et al. (2013) with permission. (For color version of this figure, the reader is referred to the online version of this book.)

length scale above that of MRO (Section 4.3.3). As noted above, particularly in connection with the damage-tolerant glass featured in **Figure 53**, the plasticity of MGs is improved if the pattern of shear bands is finer, giving more uniformly distributed flow and delaying the onset of catastrophically localized shear. To achieve this, the initiation of shear bands must be stimulated and their propagation must be hindered or blocked. The measures taken in essence involve introducing some inhomogeneity into the glassy structure.

Inhomogeneity can be induced by cold deformation, and this does improve subsequent plasticity. The methods used include: wire drawing (Inoue et al., 1982), cold-rolling (Yokoyama et al., 2001; Cao et al., 2010), uniaxial compression (He et al., 2008), channel-die compression (Scudino et al., 2010), shot-peening (Zhang et al., 2006) and imprinting (Scudino et al., 2011). The common feature is that the metallic glass develops harder (denser) and softer regions, detectable by TEM (Lee et al., 2009) and by direct hardness mapping (Song et al., 2011).

There has also been much interest in developing composite materials based on a metallic-glass matrix. These offer possibilities for microstructural design to permit control of relevant length scales. The most studied type of composite is that pioneered by Hays et al. (2000), in which dendrites of a ductile crystalline solid solution are embedded in a metallic-glass matrix. The dendrites form in situ as the primary solidification phase when the liquid is cooled, and the interdendritic liquid then forms a BMG. The volume fraction of the crystalline phase is substantial (typically 25%), and the properties of such composites are then often (according to a rule of mixtures) a volume-weighted average of the properties of the metallic glass (high yield stress, zero ductility) and the crystalline phase (lower yield stress, work-hardening and ductility). In this way it is hoped that an attractive compromise of properties can be achieved.

Hays et al. (2000) studied the metallic glass $Zr_{41.2}Ti_{13.8}Cu_{12.5}Ni_{10}Be_{22.5}$ (Vitreloy 1) with added zirconium, titanium and niobium to give a bcc (β) primary solidification phase with the approximate composition $Zr_3(Ti,Nb)$. The β-phase dendrites had a trunk length 50–150 μm, a trunk radius 1.5–2.0 μm and a secondary dendrite arm spacing 6–7 μm. The metallic glass itself has a yield stress of

1.9 GPa and a yield strain of 2%. Even under compression it shows little or no plasticity, failing catastrophically on or close to a single shear band. The composite, by comparison, has a lower yield stress of 1.3 GPa and yield strain of 1.2%. It shows significant plasticity and work-hardening and fails at an ultimate strength of 1.7 GPa and a total (elastic + plastic) strain exceeding 8%. The averaging of the properties of the two phases is evident. The composite is of particular interest because, in stark contrast to any monolithic metallic glass, it shows actual ductility: at failure the plastic strain in tension is 5% averaged over the gauge length of the specimen, and 15% in its necked region. Hays et al. (2000) found that the β-phase dendrite trunks in effect divide the glassy matrix into domains some 100 μm across. The dendrites promote the initiation of shear bands and restrict their propagation within these domains.

It is no doubt significant that the β-phase in these composites can show ductility and work-hardening. Further studies of similar composites has shown that twinning and martensitic transformation in the β-phase may also contribute to its plasticity (Szuecs et al., 2001).

It is of interest to explore what might be the optimum length scale of the microstructure in such composites. Eckert et al. (2002) showed that higher cooling rates give finer microstructures, and that when these are too fine (dendrite trunk length 2–6 μm, diameter 0.2–0.4 μm) the propagation of shear bands is not significantly hindered and there is little if any property benefit. Hofmann et al. (2008) suggested that a significant improvement in plasticity should be expected when the domain size between dendrite trunks is less than the plastic zone size d (Eqn (1)). They used isothermal holds in the two-phase (β + liquid) region during cooling to control the microstructural scale in a (Zr,Ti)-based BMG system, and were able to achieve a mode-I fracture toughness as high as ~ 170 MPa$\sqrt{\text{m}}$, as well as increased fracture energy and ductility. While the mode-I fracture energy (~ 341 kJ m^{-2}) is the highest known for any metallic material, it is interesting that even this optimized composite has a damage tolerance (the product of fracture toughness K_c and yield stress σ_y) less than that of the monolithic Pd-based BMG highlighted in **Figure 53**.

The domain size in the optimized dendritic composites is of the order of 100 μm. At shorter scales, when the metallic-glass domain size within a composite lies in Regime III (**Figure 51**), shear-banding is suppressed and the flow of the metallic glass is homogeneous. This length-scale can be accessed in laminates consisting of alternating layers of a metallic glass and a polycrystalline metal. With metallic-glass layers tens of micrometers thick, constraint can force the shear-band pattern to be fine and uniform. With layers a few nanometers thick, the metallic glass shows substantial homogeneous flow (e.g. thickness reductions of 75%) (Donohue et al., 2007).

It is expected that particles embedded within a metallic glass could stimulate the initiation and multiplication of shear bands and block their propagation. These particles can be added ex situ, or produced in situ by crystallization (Section 4.4.3). Micrometer-sized particles improve plasticity in compression, but not in tension (Choi-Yim et al., 1999). It appears that the particles are effective at initiating shear bands, but not at blocking catastrophic propagation. With nanometer-scale particles (from crystallization), there can be dramatic improvements in compressive plasticity (Inoue et al., 2005), and also marginal improvements in tensile plasticity (Fan et al., 1999). The latter effect is possibly because the particles modify the operation of shear bands by causing the flowing material to act as a semi-solid slurry (Hajlaoui et al., 2006). Such effects can be enhanced if the particles are formed, or are transformed, as a result of the deformation itself (Pauly et al., 2010), in some ways analogous to conventional transformation toughening of ceramics.

Phase separation (Section 4.4.2) can also be used to enhance plasticity. An example is the work of Park et al. (2012) on mixing the BMG compositions $Zr_{45}Cu_{50}Al_5$ and $Zr_{55}Co_{25}Al_{20}$ in varying proportions to form a Zr–Cu–Co–Al BMG that shows phase separation on a scale of a few nanometers.

A maximum compressive plasticity of 12% can be achieved (compared to 0.5 and 4% for the base glasses) when the two base alloys are mixed in 50:50 proportion. The improved plasticity can be attributed to the inhomogeneity in the glass forcing the branching and multiplication of shear bands. There are now many such demonstrations of improved plasticity arising from phase separation, and it is interesting that phase separation can be induced by rather small additions (as low as 1%), for example of iron in a Cu–Zr-BMGs (Pan et al., 2009) or vanadium in Fe-based BMGs (Kim et al., 2012).

It is clearly possible, through design of composites or mechanical treatments of monolithic MGs, to promote plasticity by attaining finer, more uniform shear-band patterns. While there are prospects for further improvements to the toughness and compressive plasticity of MGs, it has proved impossible, however, to raise their ductility (i.e. plasticity in tension) to rival that in polycrystalline metals.

4.6 Applications

Potential and actual applications of MGs have been reviewed by many authors (Johnson, 1999; Inoue, 2000; Wang et al., 2004; Salimon et al., 2004). **Table 6**, adapted from Ashby and Greer (2006), notes the distinctive characteristics of MGs compared to conventional metallic materials (i.e. their glassy nature and intrinsic lack of microstructural features such as grain and phase boundaries), and attempts to identify the attributes that make MGs attractive or unattractive for applications. Section 4.5 covered mechanical properties; other classes of property that have attracted interest are listed in the table. Of course, it is often a combination of properties that is important. For example, a magnetic read-head should have not only good soft-magnetic properties, but also good wear resistance; MGs having just this combination (unusual in conventional metallic magnetic materials) are attractive for this application (Kohmoto et al., 1989).

MGs were first widely used because of their excellent soft-magnetic properties, low coercivity and high permeability (Smith, 1993). Rapid solidification (with cooling rates 10^5–10^6 K s^{-1}) was necessary to achieve glass formation in the early iron-based compositions. Large-scale production was established using planar-flow casting to obtain sheets some tens of micrometers thick. This form of material is well suited for laminated transformer cores, for example. The high electrical resistivity of MGs (Nagel, 1977) is a further advantage in reducing eddy-current losses. The low losses in such cores are likely to generate further substantial use of MGs in electricity distribution networks. Other applications include magnetic shielding.

The ability to make MGs in larger cross-sections (**Table 4**) has allowed a broad range of structural applications to be considered. The high hardness of MGs combined with the lack of grain structure suggests a use in precision tooling, particularly knife edges, which can be sharpened to an exceptional edge, in principle down to the atomic limit. The hardness and ease of precision forming by viscous flow underlie the choice of a Zr-based MG for the gears in miniature electric motors (Niza et al., 2010). At present the cylindrical casing of the smallest of these motors has an outer diameter of 0.9 mm. Inside the casing, there is a sun-and-planet mechanism, with micro gears a few hundred micrometers in diameter. It is claimed that making the gears from an MG with a Vickers hardness of 520, rather than a carbon tool steel of hardness 500, extends their lifetime by a factor of 22. Precision shaping and wear resistance are presumably key factors favoring the MG. On a larger scale, related properties make MGs attractive for such items as watch cases and mobile-phone cases. With these, the ability to take a high polish, or to be finely patterned, and the scratch resistance and corrosion resistance are important. The use of noble metals and novelty may also be attractive. The practicalities of thermoplastic-forming

Table 6 Properties of MGs related to applications

Attributes	Attractive	Unattractive
General	• Absence of microstructural features such as grain and phase boundaries and of related composition variations (e.g. segregation). This allows components with features of near-atomic scale	• Present cost of components and processing • Optimization of composition for glass-forming ability prevents easy optimization for other properties, including low density
Mechanical	• High hardness, H, giving good wear and abrasion resistance • High yield strength, σ_y • Fracture toughness K_c and toughness G_c can be very high • High specific strength,* σ_y/ρ, $\sigma_y^{2/3}/\rho$ and $\sigma_y^{1/2}/\rho$ • High resilience per unit volume and mass,* σ_y^2/E and $\sigma_y^2/E\rho$ • Low mechanical damping	• Severe localization of plastic flow (shear-banding), giving zero ductility in tension • Fracture toughness K_c and toughness G_c can be very low • Can be embrittled by annealing • Small process-zone size ($d < 1$ mm) means that larger components may fail in a brittle manner
Thermal	• $T_g < T_c$ for some MGs, allowing processing as a supercooled liquid (T_g—glass-transition temp. T_c—temp. of crystallization onset)	• Instability above T_c limits high-temperature use
Electrical and magnetic	• High magnetic permeability • Resistivity is high and nearly independent of temperature	• Relatively high magnetostriction gives energy loss in oscillating field
Chemical	• Lack of grain structure and associated microstructural features (e.g. solute segregation) gives corrosion resistance • Selective etching of phase-separated glasses gives solids with high surface area of interest for catalysis	• Common (Zr, Ti)-based compositions are susceptible to take-up of oxygen, leading to crystallization and embrittlement
Environmental	• Some compositions biocompatible	• Not easily recycled once in a product (nonconventional compositions)
Processing	• Low solidification shrinkage and lack of grain structure give high precision and finish in castings • The high viscosity and low strain-rate sensitivity of the super-cooled liquid permit thermoplastic forming	• Current need for vacuum die-casting gives relatively slow production rate • High viscosity and susceptibility to crystallization give a narrow processing window with the need for high temperatures and high heating rates
Aesthetic	• Lack of grain structure allows a very high polish • High hardness and corrosion resistance give durability	
Potential markets	• Aesthetics, present novelty and rarity make MGs attractive for high-end "life-style" products • Properties and processing favor μm-to-mm scale structures	• Current high cost of material and processing limits applications to those with high added value

*Material (or merit) indices of this kind are discussed in Ashby (2005). The parameter ρ is the density.
Adapted with permission from Ashby and Greer (2006).

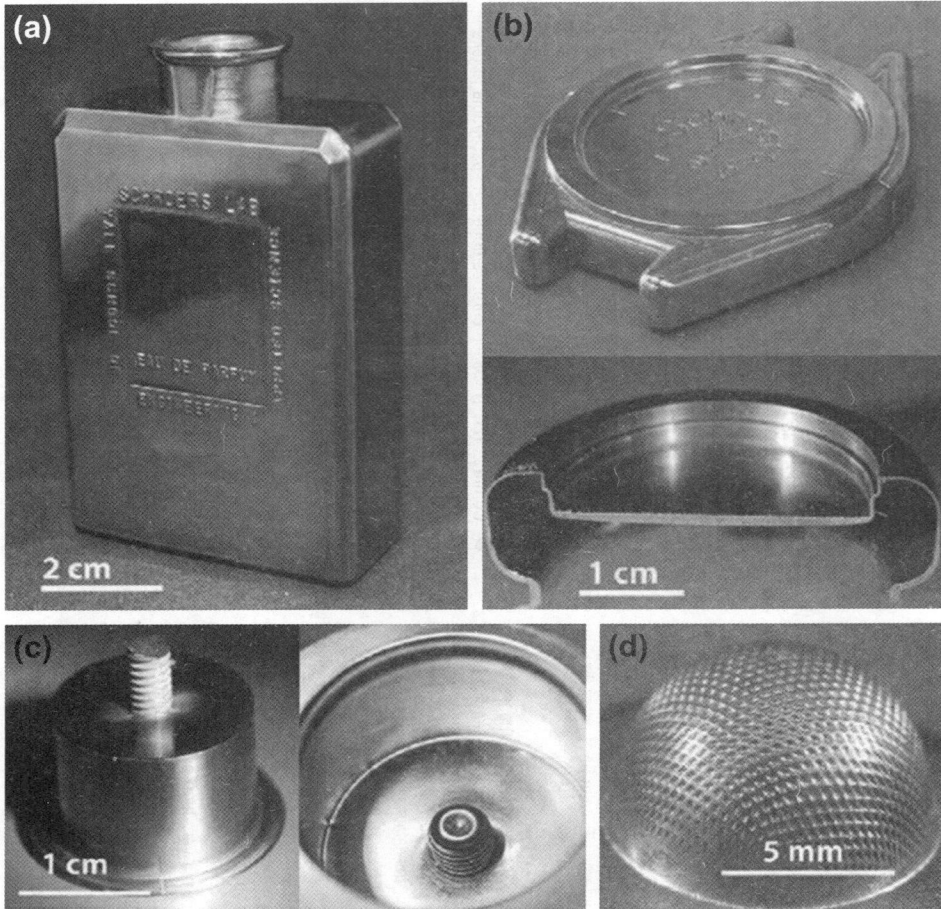

Figure 55 (a) and (b) Examples of objects made by blow-molding the metallic glass $Zr_{44}Ti_{11}Cu_{10}Ni_{10}Be_{25}$. The net-shape process exploits the flow of the material above the glass-transition temperature T_g and takes less than 1 min. (c) Joints created by blow-molding around a fastener site. (d) Surface patterning integrated into the blow-molding. Reprinted from Schroers et al. (2011) with permission. (For color version of this figure, the reader is referred to the online version of this book.)

processes such as blow-molding have been extensively explored (**Figure** 55) (Schroers et al., 2011), and it has been verified that atomically smooth surfaces (much flatter than can be achieved by polishing) can result from such thermoplastic processing (Kumar et al., 2010).

As noted in Section 4.4.3, MGs are susceptible to crystallization, and this imposes significant restrictions on the processing window for such techniques as blow-forming. Formability is improved at higher temperatures (Pitt et al., 2011). With suitable choice of composition, BMGs can be formed at low cooling rates. Similarly, the BMGs can be heated back to the liquid state without intermediate crystallization. Schroers et al. (2000) showed, however, that the critical heating rate to avoid crystallization is much higher (typically by two orders of magnitude) than the critical cooling rate, an asymmetry arising from nuclei quenched into the glass. This relative instability on heating is a problem for thermoplastic processing, and points to the need for high heating rates. Johnson et al. (2011) achieved rates around 10^6 K s^{-1} with excellent control by applying a millisecond electrical current pulse (from capacitor discharge) to a metallic-glass sample. This near-adiabatic process gives a few milliseconds in the liquid state, sufficient for injection molding. This technique ingeniously exploits the high electrical resistivity of MGs and their near-zero temperature coefficient of resistivity. The latter is crucial in ensuring that the pulse heating gives a remarkably uniform temperature distribution in the sample. **Figure 56** shows the processing window opened up by this promising technique.

A second category of structural application exploits the elastic properties of MGs. The exceptionally high elastic strain and elastic energy storage per unit volume and mass, and exceptionally low mechanical damping (all noted in Section 4.5.1), make MGs attractive for springs and for some sporting goods; golf-club heads and tennis-racket frames have been successfully marketed. The high elastic strain limit of MGs makes them attractive for use in strain-sensing devices such as pressure sensors, where their use gives greater sensitivity (a factor of four improvement) over the conventional devices using stainless steel (Nishiyama et al., 2007). MGs should also be attractive for a range of device applications requiring fast elastic response and transmission of elastic waves.

The ability to be patterned on a fine scale without interference from an internal microstructure has already been noted. Combined with high hardness, this should permit MG surfaces to be molded or etched on a near-atomic scale to serve as masters for reproduction of ultrahigh density digital data. The reasons for the attractiveness of MGs are shown in **Figure 57**; compared to other materials moldable in similar ways, such as the polymer PMMA, the MGs have higher strength at ambient temperature, have

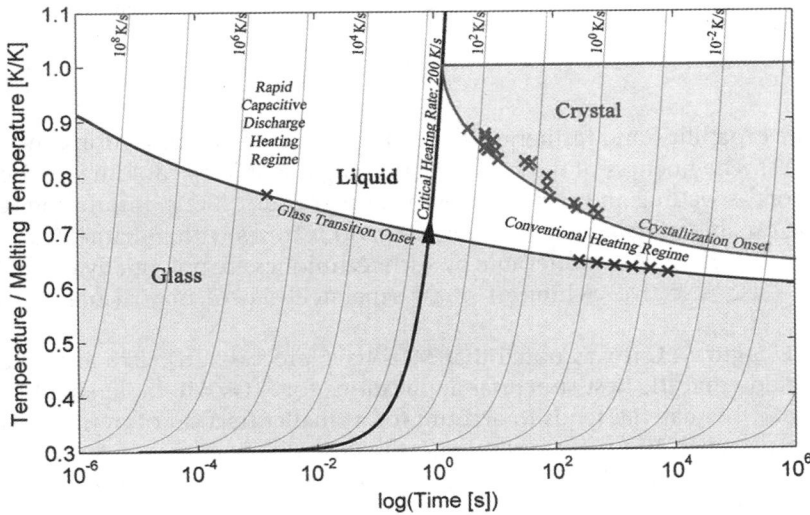

Figure 56 A map of the onset temperatures for the glass transition and for crystallization on continuous heating of Zr$_{41.2}$Ti$_{13.8}$Cu$_{12.5}$Ni$_{10}$Be$_{22.5}$ BMG (Vitreloy 1). A heating rate of $\sim$200 K s^{-1} is needed to bypass crystallization when heating from the glass into the liquid state. Heating at lower rates gives only a narrow temperature range (between T_g and the onset of crystallization) for thermoplastic forming. By heating at rates of the order of 10^6 K s^{-1}, achievable by applying an electrical pulse, there is easy access to the supercooled liquid state at any temperature above T_g. Reprinted from Johnson et al. (2011) with permission. (For color version of this figure, the reader is referred to the online version of this book.)

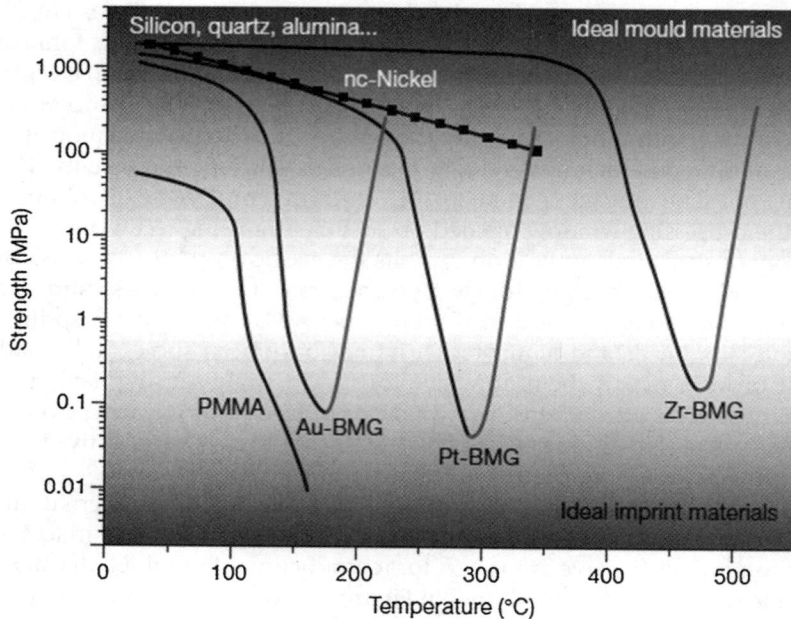

Figure 57 The temperature dependence of the strength of BMGs compared to the polymer PMMA. Below T_g the BMGs have high strength and can serve as mold materials. Above T_g they are soft and suitable for being imprinted. The red curves show the increase in BMG strength as they crystallize, a useful property after imprinting. Reprinted from Kumar et al. (2009) with permission.

higher temperature capability, and furthermore their hardening on crystallization can also be exploited (Kumar et al., 2009). The fineness of the surface patterning possible by molding has been studied and involves such factors as wetting and surface tension (Kumar et al., 2009). Imprinting (embossing) has been shown to be capable of generating high-aspect-ratio features with a diameter as small as 13 nm (**Figure 58**). The high surface areas achievable by such techniques are potentially useful for catalysis. As noted in Section 4.4.2, selective etching of phase-separated glasses may also be of interest in this context.

In Section 4.5.1, **Figure 41**, it was noted that MGs have process-zone sizes at best on a millimeter scale; we can conclude that the best structural applications for MGs will be in small components. For smaller components, the material tends to account for a smaller fraction of overall cost, and that may reduce or, with easier processing, even eliminate the cost disadvantage of MGs. MGs are, indeed, particularly attractive for use in MEMS (microelectromechanical systems) devices. Their good elasticity properties are highly relevant and their brittleness (low process-zone size) and negligible ductility are unimportant at small length scales. Importantly, MGs can be deposited as thin films and furthermore the GFA in deposition opens up a much wider range of compositions than is found in BMGs. An example of commercial application is the glassy Al_3Ti-based hinges supporting the micromirrors in data projectors using digital light processor (DLP) technology (Tregilgas, 2005). Devices have been manufactured with more than 10^6 mirrors on a single chip, and the sputter-deposited hinges tested to more than 10^{12} cycles without an onset of fatigue failure. A different type of potential MEMS application is shown in **Figure 59**.

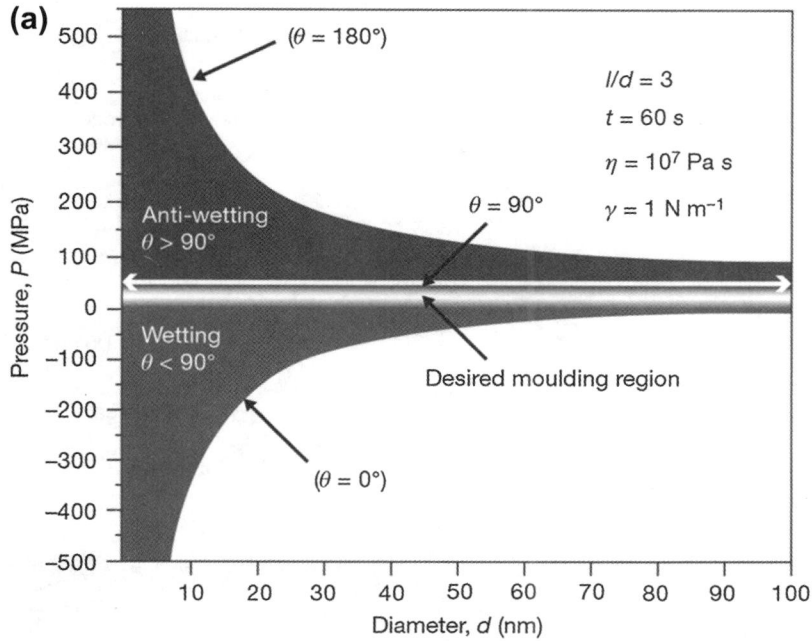

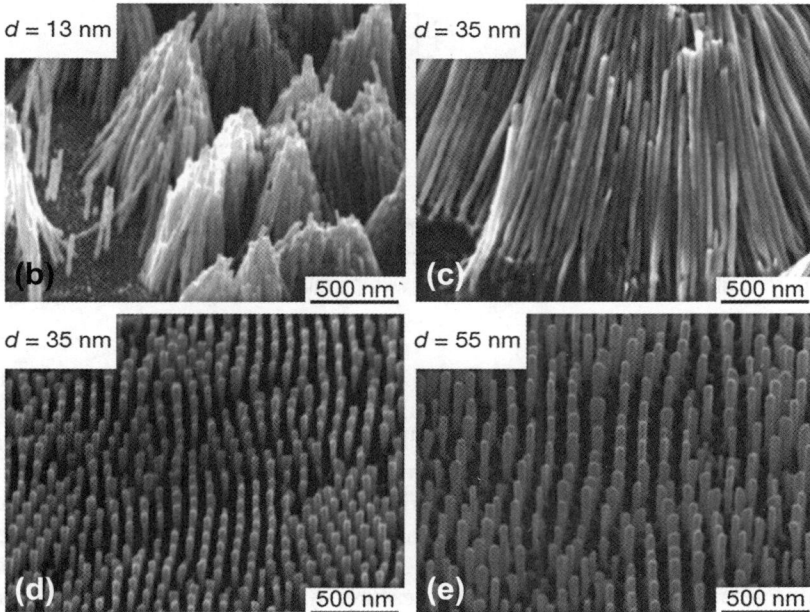

Figure 58   (a) The imprinting pressure necessary to fill a pore of diameter d is a function of d and of the wetting conditions. (b–e) SEM images of $Pt_{57.5}Cu_{14.7}Ni_{5.3}P_{22.5}$ BMG rods formed by pressing at temperatures above T_g into porous alumina. The rods of diameters 13 nm (b) and 35 nm (c) have an aspect ratio exceeding 50. Reprinted from Kumar et al. (2009) with permission. (For color version of this figure, the reader is referred to the online version of this book.)

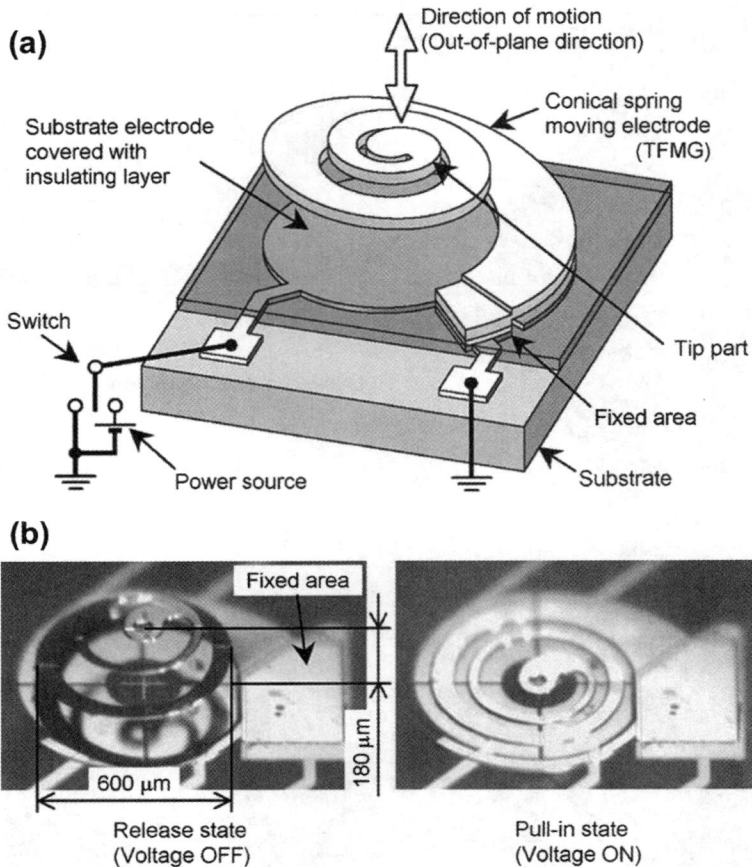

Figure 59 Conical spring microactuator made from a sputter-deposited Pd-based amorphous alloy. The thin-film spring is stretched out elastically and then transiently heated to fix it in the extended position, i.e. in the "release state" shown in the left-hand panel of (b). Reprinted from Fukushige et al. (2005) with permission.

A conical spring microactuator with a long stroke of 200 μm normal to the substrate has been fabricated from a 7.6 μm thick film of $Pd_{76}Cu_7Si_{17}$ glass (Fukushige et al., 2005).

The expected good mechanical properties of small cross-sections of metallic glass can be exploited in large components if the glass is a component of a composite material. The examples of laminates and of in situ dendritic composites were considered in Section 4.5.5. A further possibility is the application of MGs in foams (Brothers and Dunand, 2005). The narrow ligaments between the cells of the foam are expected to show good plasticity. Furthermore, the relatively high viscosity of glass-forming liquids decreases the draining rate of the liquid during processing, making it easier to achieve uniform cell size. There is no need for the addition of particulate dispersions, which are useful in slowing the drainage of low-viscosity conventional metallic liquids.

At present MGs are more expensive than conventional alloys, mostly because one or more of the elementary components is expensive. In some cases, (e.g. Pd), the element is intrinsically expensive; in other cases, the cost is increased because the elements (e.g. Zr) have to be of particularly high purity (in the case of zirconium is it difficult, but necessary, to have very low oxygen content). Even if the raw material costs are not so high, the lack of established mass production restricts availability and raises prices. Nonetheless, we can expect an ongoing reduction in the cost of BMG components as new compositions are developed that use inexpensive metals, and as improved processing permits the use of base metals of lower purity.

References

Akhtar, D., Misra, R.D.K., 1986. Effect of thermal relaxation on diffusion in a metallic glass. Scr. Metall. 20, 627–631.

Angell, C.A., 1995. Formation of glasses from liquids and biopolymers. Science 267, 1924–1935.

Argon, A.S., 1979. Plastic deformation in metallic glasses. Acta Metall. 27, 47–58.

Argon, A.S., Salama, M., 1976. The mechanism of fracture in glassy materials capable of some inelastic deformation. Mater. Sci. Eng. 23, 219–230.

Ashby, M.F., 2005. Materials Selection in Mechanical Design, third ed. Butterworth-Heinemann, Oxford.

Ashby, M.F., Greer, A.L., 2006. Metallic glasses as structural materials. Scr. Mater. 54, 321–326.

Balanzat, E., Stanley, J.T., Mairy, C., Hillairet, J., 1985. The nature of the atom species involved in the reversible relaxation of metallic glasses. Acta Metall. 33, 785–795.

Baxter, D.V., 1986. EXAFS studies of $La_{1-x}Ga_x$ metallic glasses. J. Non Cryst. Solids 79, 41–55.

Behrndt, K.H., 1970. Formation of amorphous films. J. Vac. Sci. Technol. 7, 385–398.

Bernal, J.D., 1964. The structure of liquids. Proc. Roy. Soc. Lond. A 280, 299–322.

Bormann, R., 1994. Thermodynamic and kinetic requirements for inverse melting. Mater. Sci. Eng. A 179–180, 31–35.

Braeck, S., Podladchikov, Y.Y., 2007. Spontaneous thermal runaway as an ultimate failure mechanism of materials. Phys. Rev. Lett. 98, 095504.

Brauer, S., Ström-Olsen, J.O., Sutton, M., Yang, Y.S., Zaluska, A., Stephenson, G.B., Köster, U., 1992. In situ x-ray studies of rapid crystallization of amorphous $NiZr_2$. Phys. Rev. B 45, 7704–7715.

Brázdová, V., Bowler, D.R., 2013. Atomistic Computer Simulations. Wiley-VCH, Weinheim.

Breinan, E.M., Snow, D.B., Brown, C.O., Kear, B.H., 1980. New developments in laser surface melting using continuous prealloyed powder feed. In: Mehrabian, R., Kear, B.H., Cohen, M. (Eds.), Rapid Solidification Processing, Principles and Technologies II. Claitor's, Baton Rouge, LA, pp. 440–452.

Brenner, A., Couch, D.E., Williams, E.K., 1950. Electrodeposition of alloys of phosphorus with nickel and cobalt. J. Res. Natl. Bur. Stand. 44, 109–122.

Bresson, L., Chevalier, J.P., Fayard, M., 1982. Bend testing metallic glasses. Effect of heat treatment on the mechanical properties of $Cu_{60}Zr_{40}$. Scr. Metall. 16, 499–502.

Brothers, A.H., Dunand, D.C., 2005. Ductile bulk metallic glass foams. Adv. Mater. 17, 484–486.

Busch, R., 2000. The thermophysical properties of bulk metallic glass-forming liquids. J. Metals 52 (7), 39–42.

Busch, R., Kim, Y.J., Johnson, W.L., 1995. Thermodynamics and kinetics of the undercooled liquid and the glass-transition of the $Zr_{41.2}Ti_{13.8}Cu_{12.5}Ni_{10.0}Be_{22.5}$ alloy. J. Appl. Phys. 77, 4039–4043.

Busch, R., Schroers, J., Wang, W.H., 2007. Thermodynamics and kinetics of bulk metallic glass. MRS Bull. 32, 620–623.

Caciuffo, R., Stefanon, M., Howells, W.S., Soper, A.K., Allia, P., Vinai, F., Melone, S., Rustichelli, F., 1989. Structural study of $Fe_{40}Ni_{40}B_{20}$ amorphous alloy. Physica B 156–157, 220–222.

Cahn, R.W., Greer, A.L., 1996. Metastable states of alloys. In: Cahn, R.W., Haasen, P. (Eds.), Physical Metallurgy, fourth ed. Elsevier, Amsterdam, pp. 1723–1830.

Cameron, K.K., Dauskardt, R.H., 2006. Fatigue damage in bulk metallic glass I: simulation. Scr. Mater. 54, 349–353.

Cao, Q.P., Liu, J.W., Yang, K.J., Xu, F., Yao, Z.Q., Minkow, A., Fecht, H.J., Ivanisenko, J., Chen, L.Y., Wang, X.D., Qu, S.X., Jiang, J.Z., 2010. Effect of pre-existing shear bands on the tensile mechanical properties of a bulk metallic glass. Acta Mater. 58, 1276–1292.

Cargill, G.S., 1970. Structural investigation of noncrystalline nickel-phosphorus alloys. J. Appl. Phys. 41, 12–29.

Chang, H.J., Yook, W., Park, E.S., Kyeong, J.S., Kim, D.H., 2010. Synthesis of metallic glass composites using phase separation phenomena. Acta Mater. 58, 2483–2491.

Chason, E.H., Mizoguchi, T., 1987. Structural relaxation and interdiffusion in amorphous Fe-Ti compositionally modulated films. Mater. Res. Soc. Symp. Proc. 80, 61–67.

Chaudhari, P., Graczyk, J.F., Henderson, D., Steinhardt, P., 1976. Transformations between random networks and dense random packed models for amorphous solids. Philos. Mag. 31, 727–732.

Chen, H.S., Krause, J.T., Coleman, E., 1975. Elastic constants, hardness and their implications to flow properties of metallic glasses. J. Non Cryst. Solids 18, 157–171.

Chen, H.S., Krause, J.T., Inoue, A., Masumoto, T., 1983. The effect of quench rate on the Young's modulus of Fe-, Co-, Ni- and Pd-based amorphous alloys. Scr. Metall. 17, 1413–1414.

Chen, H.S., Turnbull, D., 1969. Formation, stability and structure of palladium-silicon based alloy glasses. Acta Metall. 17, 1021–1031.

Chen, L.C., Spaepen, F., 1991. Analysis of calorimetric measurements of grain growth. J. Appl. Phys. 69, 679–688.

Cheng, Y.Q., Cao, A.J., Ma, E., 2009a. Correlation between the elastic modulus and the intrinsic plastic behavior of metallic glasses: the roles of atomic configuration and alloy composition. Acta Mater. 57, 3253–3267.

Cheng, Y.Q., Han, Z., Li, Y., Ma, E., 2009b. Cold versus hot shear banding in bulk metallic glass. Phys. Rev. B 80, 134115.

Cheng, Y.Q., Ma, E., Sheng, H.W., 2009c. Atomic level structure in multicomponent bulk metallic glass. Phys. Rev. Lett. 102, 245501.

Cheng, Y.Q., Ma, E., 2008. Indicators of internal structural states for metallic glasses: local order, free volume, and configurational potential energy. Appl. Phys. Lett. 93, 051910.

Cheng, Y.Q., Ma, E., 2011. Atomic-level structure and structure–property relationship in metallic glasses. Prog. Mater. Sci. 56, 379–473.

Cheng, Y.Q., Ma, E., Sheng, H.W., 2008a. Alloying strongly influences the structure, dynamics, and glass forming ability of metallic supercooled liquids. Appl. Phys. Lett. 93, 111913.

Cheng, Y.Q., Sheng, H.W., Ma, E., 2008b. Relationship between structure, dynamics, and mechanical properties in metallic glass-forming alloys. Phys. Rev. B 78, 014207.

Choi-Yim, H., Busch, R., Köster, U., Johnson, W.L., 1999. Synthesis and characterization of particulate reinforced $Zr_{57}Nb_5Al_{10}Cu_{15.4}Ni_{12.6}$ bulk metallic glass composites. Acta Mater. 47, 2455–2462.

Chou, C.-P.P., Turnbull, D., 1975. Transformation behavior of Pd–Au–Si metallic glasses. J. Non Cryst. Solids 17, 169–188.

Chuang, C.-P., Yuan, T., Dmowski, W., Wang, G.-Y., Freels, M., Liaw, P.K., Li, R., Zhang, T., 2013. Fatigue-induced damage in Zr-based bulk metallic glasses. Sci. Rep. 3, 2578.

Concustell, A., Godard-Desmarest, S., Carpenter, M.A., Nishiyama, N., Greer, A.L., 2011. Induced elastic anisotropy in a bulk metallic glass. Scr. Mater. 64, 1091–1094.

Connell, G.A.N., 1975. Dense random packings of hard and compressible spheres. Solid State Commun. 16, 109–112.

Conner, R.D., Johnson, W.L., Paton, N.E., Nix, W.D., 2003. Shear bands and cracking of metallic glass plates in bending. J. Appl. Phys. 94, 904–911.

Cowlam, N., Guoan, W., Gardner, P.P., Davies, H.A., 1984. $Ni_{64}B_{36}$—a transition metal-metalloid glass with first neighbour metalloid atoms. J. Non Cryst. Solids 61–62, 337–342.

Croat, J.J., Herbst, J.F., Lee, R.W., Pinkerton, F.E., 1984. Pr–Fe and Nd–Fe-based materials: a new class of high-performance permanent magnets. J. Appl. Phys. 55, 2078–2082.

Cunat, Ch, Hilzinger, H.R., Herzer, G., 1988. Relaxation parameters to simulate the changes in magnetostriction in amorphous magnetic alloys. Mater. Sci. Eng. 97, 497–500.

Das, S.K., Okazaki, K., Adam, C.M., 1985. Applications of rapid solidification processing to high temperature alloy design. In: Stiegler, J.O. (Ed.), High Temperature Alloys: Theory and Design. Met. Soc. AIME, Warrendale, pp. 451–471.

Davies, H.A., 1976. Formation of metallic glasses. Phys. Chem. Glasses 17, 159–173.

Demetriou, M.D., Floyd, M., Crewdson, C., Schramm, J.P., Garrett, G., Johnson, W.L., 2011a. Liquid-like platinum-rich glasses. Scr. Mater. 65, 799–802.

Demetriou, M.D., Johnson, W.L., 2004. Modeling the transient flow of undercooled glass-forming liquids. J. Appl. Phys. 95, 2857–2865.

Demetriou, M.D., Launey, M.E., Garrett, G., Schramm, J.P., Hofmann, D.C., Johnson, W.L., Ritchie, R.O., 2011b. A damage-tolerant glass. Nat. Mater. 10, 123–128.

Deng, D., Argon, A.S., 1986. Structural relaxation and embrittlement of $Cu_{59}Zr_{41}$ and $Fe_{80}B_{20}$ glasses. Acta Metall. 34, 2011–2023.

de Vries, J., Koebrugge, G.W., van den Beukel, A., 1988. Positron life time measurements during isothermal structural relaxation in amorphous $Fe_{40}Ni_{40}B_{20}$. Scr. Metall. 22, 637–641.

Dmowski, W., Egami, T., 2007. Observation of structural anisotropy in metallic glasses induced by mechanical deformation. J. Mater. Res. 22, 412–418.

Dmowski, W., Fana, C., Morrison, M.L., Liaw, P.K., Egami, T., 2007. Structural changes in bulk metallic glass after annealing below the glass-transition temperature. Mater. Sci. Eng. A 471, 125–129.

Donohue, A., Spaepen, F., Hoagland, R.G., Misra, A., 2007. Suppression of the shear band instability during plastic flow of nanometer-scale confined metallic glasses. Appl. Phys. Lett. 91, 241905.

Drehman, A.J., Greer, A.L., 1984. Kinetics of crystal nucleation and growth in $Pd_{40}Ni_{40}P_{20}$ glass. Acta Metall. 32, 323–332.

Drehman, A.J., Turnbull, D., 1981. Solidification behavior of undercooled $Pd_{83}Si_{17}$ and $Pd_{82}Si_{18}$ liquid droplets. Scr. Metall. 15, 543–548.

Dubois, J.M., Gaskell, P.H., Le Caër, G., 1985. A model for metallic glasses generated by chemical twinning. In: Steeb, S., Warlimont, H. (Eds.), Rapidly Quenched Metals. North-Holland, Amsterdam, pp. 567–572.

Dunlop, A., Lesueur, D., 1992. Irradiation of metallic targets with high-energy heavy-ions—high electronic excitation induced effects. Mater. Sci. Forum 97–99, 553–576.

Eckert, J., Kühn, U., Mattern, N., He, G., Gebert, A., 2002. Structural bulk metallic glasses with different length-scale of constituent phases. Intermetallics 10, 1183–1190.

Ediger, M.D., Harrowell, P., Yu, L., 2008. Crystal growth kinetics exhibit a fragility-dependent decoupling from viscosity. J. Chem. Phys. 128, 034709.

Egami, T., 1978. Structural relaxation in amorphous alloys—compositional short range ordering. Mater. Res. Bull. 13, 557–562.

Egami, T., 1981. Structural study by energy dispersive X-ray diffraction. In: Güntherodt, H.-J., Beck, H. (Eds.), Glassy Metals I. Springer, Berlin, pp. 25–44.

Egami, T., 2006. Formation and deformation of metallic glasses: atomistic theory. Intermetallics 14, 882–887.

Egami, T., Billinge, S.J.L., 2012. Underneath the Bragg Peaks: Structural Analysis of Complex Materials, second ed. Elsevier, Amsterdam.

Egami, T., Poon, S.J., Zhang, Z., Keppens, V., 2007. Glass transition in metallic glasses: a microscopic model of topological fluctuations in the bonding network. Phys. Rev. B 76, 024203.

Falk, M.L., Langer, J.S., 1998. Dynamics of viscoplastic deformation in amorphous solids. Phys. Rev. E 57, 7192–7205.

Fan, C., Takeuchi, A., Inoue, A., 1999. Preparation and mechanical properties of Zr-based bulk nanocrystalline alloys containing compound and amorphous phases. Mater. Trans. JIM 40, 42–51.

Frank, F.C., 1952. Supercooling of liquids. Proc. Roy. Soc. Lond. A 215, 43–46.

Fukushige, T., Hata, S., Shimokohbe, A., 2005. A MEMS conical spring actuator array. J. Microelectromech. Syst. 14, 243–253.

Gao, Y.F., Wang, L., Bei, H., Nieh, T.G., 2011. On the shear-band direction in metallic glasses. Acta Mater. 59, 4159–4167.

Garcia-Escorial, A., Greer, A.L., 1987. Surface crystallisation of melt-spun $Pd_{40}Ni_{40}P_{20}$ glass. J. Mater. Sci. 22, 4388–4394.

Garrone, E., Battezzati, L., 1985. The thermodynamic quantities frozen-in upon vitrification of metallic alloys. Philos. Mag. B 52, 1033–1045.

Gaskell, P.H., 1979. A new structural model for amorphous transition metal silicides, borides, phosphides and carbides. J. Non Cryst. Solids 32, 207–224.

Gaskell, P.H., 1983. Exploring the medium-range structure of binary amorphous metallic alloys and oxide glasses. In: Vitek, V. (Ed.), Amorphous Materials: Modeling of Structure and Properties. Met. Soc. AIME, Warrendale, pp. 47–64.

Gaskell, P.H., 1985. What do we need to know about the structure of amorphous metals? In: Wright, A.F., Dupuy, J. (Eds.), Glass, Current Issues. Martinus Nijhoff, Dordrecht, pp. 54–71.

Gebert, A., Mattern, N., Kühn, U., Eckert, J., Schultz, L., 2007. Electrode characteristics of two-phase glass-forming Ni–Nb–Y alloys. Intermetallics 15, 1183–1189.

Gerling, R., Schimansky, F.P., Wagner, R., 1988. Two-stage embrittlement of amorphous $Fe_{40}Ni_{40}P_{20}$ resulting from a loss of free volume and phase separation. Acta Metall. 36, 575–583.

Gilbert, C.J., Ager, J.W., Schroeder, V., Ritchie, R.O., Lloyd, J.P., Graham, J.R., 1999. Light emission during fracture of a Zr–Ti–Ni–Cu–Be bulk metallic glass. Appl. Phys. Lett. 74, 3809–3811.

Greaves, G.N., Greer, A.L., Lakes, R.S., Rouxel, T., 2011. Poisson's ratio and modern materials. Nat. Mater. 10, 823–837.

Greer, A.L., 1982. Crystallisation kinetics of $Fe_{80}B_{20}$ glass. Acta Metall. 30, 171–192.

Greer, A.L., 1986. Crystal nucleation and growth in metallic liquids and glasses. In: Haasen, P., Jaffee, R.I. (Eds.), Amorphous Metals and Semiconductors. Pergamon, Oxford, pp. 94–107.

Greer, A.L., 1991. Grain refinement in rapidly solidified alloys. Mater. Sci. Eng. A 133, 16–21.

Greer, A.L., 1993. Structural relaxation and atomic transport in amorphous alloys. In: Liebermann, H.H. (Ed.), Rapidly solidified alloys. Dekker, New York, pp. 269–301.

Greer, A.L., 1995. Metallic glasses. Science 267, 1947–1953.

Greer, A.L., 2001. From metallic glasses to nanocrystalline solids. In: Dinesen, A.R., Eldrup, M., Juul Jensen, D., Linderoth, S., Pedersen, T.B., Pryds, N.H., Schrøder Pedersen, A., Wert, J.A. (Eds.), Proceedings of the 22nd Risø International Symposium on Materials Science—Science of Metastable and Nanocrystalline Alloys: Structure, Properties and Modelling. Risø National Laboratory, Roskilde, pp. 461–481.

Greer, A.L., 2011. Damage tolerance at a price. Nat. Mater. 10, 88–89.

Greer, A.L., Cheng, Y.Q., Ma, E., 2013. Shear bands in metallic glasses. Mater. Sci. Eng. R 74, 71–132.

Greer, A.L., Leake, J.A., 1979. Structural relaxation and crossover effect in a metallic glass. J. Non Cryst. Solids 33, 291–297.

Guo, F., Poon, S.J., Shiflet, G.J., 2003. Metallic glass ingots based on yttrium. Appl. Phys. Lett. 83, 2575–2577.

Guo, F., Wang, H.-J., Poon, S.J., Shiflet, G.J., 2005. Ductile titanium-based glassy alloy ingots. Appl. Phys. Lett. 86, 091907.

Guo, H.Q., Kronmüller, H., Moser, N., Hofmann, A., 1986. Crossover phenomena of the induced anisotropy and of the magnetic after-effect in amorphous $Co_{58}Ni_{10}Fe_5Si_{11}B_{16}$. Scr. Metall. 20, 185–189.

Gurman, 1985. EXAFS and XANES studies of metallic glasses. In: Steeb, S., Warlimont, H. (Eds.), Rapidly Quenched Metals. North-Holland, Amsterdam, pp. 427–430.

Hagiwara, M., Inoue, A., 1993. Production techniques of alloy wires by rapid solidification. In: Liebermann, H.H. (Ed.), Rapidly Solidified Alloys. Dekker, New York, pp. 139–155.

Hajlaoui, K., Yavari, A.R., Doisneau, B., LeMoulec, A., Botta F., W.J., Vaughan, G., Greer, A.L., Inoue, A., Zhang, W., Kvick, Å., 2006. Shear delocalization and crack blunting of a metallic glasses containing nanoparticles: in situ deformation in TEM analysis. Scr. Mater. 54, 1829–1834.

Han, J.H., Mattern, N., Schwarz, B., Kim, D.H., Eckert, J., 2012. Phase separation and magnetic properties in Gd–(Hf, Ti, Y)–Co–Al metallic glasses. Scr. Mater. 67, 149–152.

Hays, C.C., Kim, C.P., Johnson, W.L., 2000. Microstructure controlled shear band pattern formation and enhanced plasticity of bulk metallic glasses containing *in situ* formed ductile phase dendrite dispersions. Phys. Rev. Lett. 84, 2901–2904.

He, L., Zhong, M.B., Han, Z.H., Zhao, Q., Jiang, F., Sun, J., 2008. Orientation effect of pre-introduced shear bands in a bulk-metallic glass on its "work-ductilising". Mater. Sci. Eng. A 496, 285–290.

He, Q., Cheng, Y.Q., Ma, E., Xu, J., 2011. Locating bulk metallic glasses with high fracture toughness: chemical effects and composition optimization. Acta Mater. 59, 202–215.

He, Q., Shang, J.Q., Ma, E., Xu, J., 2012. Crack-resistance curve of a Zr–Ti–Cu–Al bulk metallic glass with extraordinary fracture toughness. Acta Mater. 60, 4940–4949.

Hess, P.A., Menzel, B.C., Dauskardt, R.H., 2006. Fatigue damage in bulk metallic glass II: experiments. Scr. Mater. 54, 355–361.

Herzer, G., 1991. Magnetization process in nanocrystalline ferromagnets. Mater. Sci. Eng. A 133, 1–5.

Higashi, K., Mukai, T., Tanimura, S., Inoue, A., Masumoto, T., Kita, K., Ohtera, K., Nagahora, J., 1992. High strain rate superplasticity in an Al–Ni–misch metal alloy produced from its amorphous powders. Scr. Metall. Mater. 26, 191–196.

Hirata, A., Guan, P., Fujita, T., Hirotsu, Y., Inoue, A., Yavari, A.R., Sakurai, T., Chen, M., 2011. Direct observation of local atomic order in a metallic glass. Nat. Mater. 10, 28–33.

Hofmann, D.C., Suh, J.Y., Wiest, A., Duan, G., Lind, M.L., Demetriou, M.D., Johnson, W.L., 2008. Designing metallic glass matrix composites with high toughness and tensile ductility. Nature 451, 1085–1089.

Holzer, J.C., Kelton, K.F., 1991. Kinetics of the amorphous to icosahedral phase transformation in Al–Cu–V alloys. Acta Metall. Mater. 39, 1833–1843.

Hono, K., Ping, D.H., Ohnuma, M., Onodera, H., 1999. Cu clustering and Si partitioning in the early crystallization stage of an $Fe_{73.5}Si_{13.5}B_9Nb_3Cu_1$ amorphous alloy. Acta Mater. 47, 997–1006.

Hubler, G.K., Singer, I.L., Clayton, C.R., 1985. Mechanical and chemical properties of tantalum-implanted steels. Mater. Sci. Eng. 69, 203–210.

Huizer, E., van den Beukel, A., 1987. Reversible and irreversible length changes in amorphous $Fe_{40}Ni_{40}B_{20}$ during structural relaxation. Acta Metall. 35, 2843–2850.

Ichitsubo, T., Hosokawa, S., Matsuda, K., Matsubara, E., Nishiyama, N., Tsutsui, S., Baron, A.Q.R., 2007. Nanoscale elastic inhomogeneity of a Pd-based metallic glass: sound velocity from ultrasonic and inelastic X-ray scattering experiments. Phys. Rev. B 76, 140201(R).

Inoue, A., 2000. Stabilization of metallic supercooled liquid and bulk amorphous alloys. Acta Mater. 48, 279–306.

Inoue, A., Hagiwara, M., Masumoto, T., 1982. Production of Fe–P–C amorphous wires by in-rotating-water spinning method and mechanical properties of the wires. J. Mater. Sci. 17, 580–588.

Inoue, A., Hagiwara, M., Masumoto, T., Chen, H.S., 1983. The structural relaxation behavior of $Pd_{48}Ni_{32}P_{20}$, $Fe_{75}Si_{10}B_{15}$ and $Co_{72.5}Si_{12.5}B_{15}$ amorphous alloy wire and ribbon. Scr. Metall. 17, 1205–1208.

Inoue, A., Horio, Y., Kim, Y.H., Masumoto, T., 1992. Elevated temperature strength of an $Al_{88}Ni_9Ce_2Fe_1$ amorphous alloy containing nanoscale fcc-Al particles. Mater. Trans. JIM 33, 669–674.

Inoue, A., Nishiyama, N., Kimura, H., 1997. Preparation and thermal stability of bulk amorphous $Pd_{40}Cu_{30}Ni_{10}P_{20}$ alloy cylinder of 72 mm in diameter. Mater. Trans. JIM 38, 179–183.

Inoue, A., Zhang, T., Nishiyama, N., Ohba, K., Masumoto, T., 1993. Preparation of 16 mm diameter rod of amorphous $Zr_{65}Al_{7.5}Ni_{10}Cu_{17.5}$ alloy. Mater. Trans. JIM 34, 1234–1237.

Inoue, A., Zhang, W., Tsurui, T., Yavari, A.R., Greer, A.L., 2005. Unusual room-temperature compressive plasticity in nanocrystal-toughened bulk copper–zirconium glass. Philos. Mag. Lett. 85, 221–229.

Jia, P., Guo, H., Li, Y., Xu, J., Ma, E., 2006. A new Cu–Hf–Al ternary bulk metallic glass with high glass forming ability and ductility. Scr. Mater. 54, 2165–2168.

Jiang, M.Q., Dai, L.H., 2009. On the origin of shear banding instability in metallic glasses. J. Mech. Phys. Solids 57, 1267–1292.

Johari, G.P., Goldstein, M., 1970. Viscous liquids and the glass transition. II. Secondary relaxation in glasses of rigid molecules. J. Chem. Phys. 53, 2372–2388.

Johnson, W.L., 1981. Superconductivity in metallic glasses. In: Güntherodt, H.-J., Beck, H. (Eds.), Glassy Metals I. Springer, Berlin, pp. 191–223.

Johnson, W.L., 1999. Bulk glass-forming metallic alloys: science and technology. MRS Bull. 24 (10), 42–56.

Johnson, W.L., Kaltenboeck, G., Demetriou, M.D., Schramm, J.P., Liu, X., Samwer, K., Kim, C.P., Hofmann, D.C., 2011. Beating crystallization in glass-forming metals by millisecond heating and processing. Science 332, 828–833.

Johnson, W.L., Samwer, K., 2005. A universal criterion for plastic yielding of metallic glasses with a $(T/T_g)^{2/3}$ temperature dependence. Phys. Rev. Lett. 95, 195501.

Ju, J.D., Jang, D., Nwankpa, A., Atzmon, M., 2011. An atomically quantized hierarchy of shear transformation zones in a metallic glass. J. Appl. Phys. 109, 053522.

Kaban, I., Jóvári, P., Kokotin, V., Shuleshova, O., Beuneu, B., Saksl, K., Mattern, N., Eckert, J., Greer, A.L., 2013. Local atomic arrangements and their topology in Ni–Zr and Cu–Zr glassy and crystalline alloys. Acta Mater. 61, 2509–2520.

Kalay, Y.E., Kalay, I., Hwang, J., Voyles, P.M., Kramer, M.J., 2012. Local chemical and topological order in Al–Tb and its role in controlling nanocrystal formation. Acta Mater. 60, 994–1003.

Kauzmann, W., 1948. The nature of the glassy state and the behavior of liquids at low temperatures. Chem. Rev. 43, 219–256.

Ke, H.B., Wen, P., Peng, H.L., Wang, W.H., Greer, A.L., 2011. Homogeneous deformation of metallic glass at room temperature reveals large dilatation. Scr. Mater. 64, 966–969.

Kelton, K.F., 1991. Crystal nucleation in liquids and glasses. In: Ehrenreich, H., Turnbull, D. (Eds.), Solid State Physics. vol. 45, Academic Press, Boston, pp. 75–178.

Kelton, K.F., Greer, A.L., 1986. Transient nucleation effects in glass formation. J. Non Cryst. Solids 79, 295–309.

Kelton, K.F., Greer, A.L., 1988. Test of classical nucleation theory in a condensed system. Phys. Rev. B 38, 10089–10092.

Kelton, K.F., Greer, A.L., 2010b. Nucleation in Condensed Matter. Elsevier, Oxford, 512–517.

Kelton, K.F., Greer, A.L., 2010a. Nucleation in Condensed Matter. Elsevier, Oxford, 532–537.

Kelton, K.F., Spaepen, F., 1984. Kinetics of structural relaxation in several metallic glasses observed by changes in electrical resistivity. Phys. Rev. B 30, 5516–5524.

Kim, Y.W., Kim, H.M., Kelly, T.F., 1989. Amorphous solidification of pure metals in submicron spheres. Acta Metall. 37, 247–255.

Kim, H.-K., Lee, K.-B., Lee, J.C., 2012. Ductile Fe-based amorphous alloy. Mater. Sci. Eng. A 552, 399–403.

Kiminami, C.S., Sahm, P.R., 1986. Kinetics of crystal nucleation and growth in $Pd_{77.5}Cu_6Si_{16.5}$ glass. Acta Metall. 34, 2129–2137.

Klaumünzer, D., Maaß, R., Dalla Torre, F.H., Löffler, J.F., 2010. Temperature-dependent shear band dynamics in a Zr-based bulk metallic glass. Appl. Phys. Lett. 96, 061901.

Koch, C.C., Cavin, O.B., McKamey, C.G., Scarbrough, J.O., 1983. Preparation of "amorphous" $Ni_{60}Nb_{40}$ by mechanical alloying. Appl. Phys. Lett. 43, 1017–1019.

Koebrugge, G.W., van den Beukel, A., 1988. Free volume dependence of CSRO kinetics in amorphous $Fe_{40}Ni_{40}B_{20}$. Scr. Metall. 22, 589–593.

Kohmoto, O., Ohya, K., Ojima, T., 1989. Wear-resistant magnetic head using amorphous alloy material. IEEE Trans. Magn. 25, 4490.

Komatsu, T., Matusita, K., Yokota, R., 1983. Compositional dependence of thermal expansion coefficient of metallic glasses. J. Non Cryst. Solids 72, 279–286.

Köster, U., 1984. Micromechanisms of crystallization in metallic glasses. Z. Metallkd. 75, 691–697.

Köster, U., Blank-Bewersdorff, M., 1987. Transient nucleation in Co–Zr metallic glasses. Mater. Res. Soc. Symp. Proc. 57, 115–127.

Köster, U., Herold, U., 1981. Crystallization of metallic glasses. In: Güntherodt, H.-J., Beck, H. (Eds.), Glassy Metals I. Springer, Berlin, pp. 225–259.

Köster, U., Schünemann, U., 1993. Phase transformations in rapidly solidified alloys. In: Liebermann, H.H. (Ed.), Rapidly Solidified Alloys. Dekker, New York, pp. 303–337.

Kruzic, J.J., 2011. Understanding the problem of fatigue in bulk metallic glasses. Metall. Mater. Trans. A 42A, 1516–1523.

Kui, H.-W., Greer, A.L., Turnbull, D., 1984. Formation of bulk metallic glass by fluxing. Appl. Phys. Lett. 45, 615–616.

Kumar, G., Neibecker, P., Liu, Y.H., Schroers, J., 2013. Critical fictive temperature for plasticity in metallic glasses. Nat. Commun. 4, 1536.

Kumar, G., Prades-Rodel, S., Blatter, A., Schroers, J., 2011. Unusual brittle behavior of Pd-based bulk metallic glass. Scr. Mater. 65, 585–587.

Kumar, G., Staffier, P.A., Blawzdziewicz, J., Schwarz, U.D., Schroers, J., 2010. Atomically smooth surfaces through thermoplastic forming of metallic glass. Appl. Phys. Lett. 97, 101907.

Kumar, G., Tang, H.X., Schroers, J., 2009. Nanomoulding with amorphous metals. Nature 457, 868–872.

Lamparter, P., Sperl, W., Steeb, S., Blétry, J., 1982. Atomic structure of amorphous metallic $Ni_{81}B_{19}$. Z. Naturforsch. A 37, 1223–1234.

Lan, S., Yip, Y.L., Lau, M.T., Kui, H.W., 2012. Direct imaging of phase separation in $Pd_{41.25}Ni_{41.25}P_{17.5}$ bulk metallic glasses. J. Non Cryst. Solids 358, 1298–1302.

Leake, J.A., Woldt, E., Evetts, J.E., 1988. Gaussian activation energy spectra in reversible and irreversible structural relaxation. Mater. Sci. Eng. 97, 469–472.

Leamy, H.J., Chen, H.S., Wang, T.T., 1972. Plastic flow and fracture of metallic glass. Metall. Trans. 3, 699–709.

Leamy, H.J., Dirks, A.G., 1977. The microstructure of amorphous rare-earth/transition-metal thin films. J. Phys. D: Appl. Phys. 10, L95–L98.

Lee, G.W., Gangopadhyay, A.K., Hyers, R.W., Rathz, T.J., Rogers, J.R., Robinson, D.S., Goldman, A.I., Kelton, K.F., 2008a. Local structure of equilibrium and supercooled Ti–Zr–Ni liquids. Phys. Rev. B 77, 184102.

Lee, G.W., Gangopadhyay, A.K., Kelton, K.F., Hyers, R.W., Rathz, T.J., Rogers, J.R., Robinson, D.S., 2004. Difference in icosahedral short-range order in early and late transition metal liquids. Phys. Rev. Lett. 93, 037802.

Lee, M.H., Lee, J.K., Kim, K.T., Thomas, J., Das, J., Kühn, U., Eckert, J., 2009. Deformation-induced microstructural heterogeneity in monolithic $Zr_{44}Ti_{11}Cu_{9.8}Ni_{10.2}Be_{25}$ bulk metallic glass. Phys. Stat. Solidi RRL 3, 46–48.

Lee, S.-C., Lee, C.-M., Yang, J.-W., Lee, J.-C., 2008b. Microstructural evolution of an elastostatically compressed amorphous alloy and its influence on the mechanical properties. Scr. Mater. 58, 591–594.

Lesueur, D., 1975. Amorphisation par irradiation aux fragments de fission d'un alliage Pd–Si. Radiat. Effects 24, 101–110.

Lewandowski, J.J., Greer, A.L., 2006. Temperature rise at shear bands in metallic glasses. Nat. Mater. 5, 15–18.

Lewandowski, J.J., Wang, W.H., Greer, A.L., 2005. Intrinsic plasticity or brittleness of metallic glasses. Philos. Mag. Lett. 85, 77–87.

Li, Y., Poon, S.J., Shiflet, G.J., Xu, J., Kim, D.H., Löffler, J.F., 2007. Formation of bulk metallic glasses and their composites. MRS Bull. 32, 624–628.

Liebermann, H.H., 1980. The dependence of the geometry of glassy alloy ribbons on the chill block melt-spinning process parameters. Mater. Sci. Eng. 43, 203–210.

Lin, C.J., Spaepen, F., 1986. Nickel–niobium alloys obtained by picosecond pulsed laser quenching. Acta Metall. 34, 1367–1375.

Liu, B.-X., Johnson, W.L., Nicolet, M.-A., 1983. Structural difference rule for amorphous alloy formation by ion mixing. Appl. Phys. Lett. 42, 45–47.

Löffler, J.F., Johnson, W.L., 2001. Crystallization pathways of deeply undercooled Zr–Ti–Cu–Ni–Be melts. Scr. Mater. 44, 1251–1255.

Löffler, J.F., Schroers, J., Johnson, W.L., 2000. Time-temperature-transformation diagram and microstructures of bulk glass forming $Pd_{40}Cu_{30}Ni_{10}P_{20}$. Appl. Phys. Lett. 77, 681–683.

Lu, J., Ravichandran, G., Johnson, W.L., 2003. Deformation behavior of the $Zr_{41.2}Ti_{13.8}Cu_{12.5}Ni_{10}Be_{22.5}$ bulk metallic glass over a wide range of strain-rates and temperatures. Acta Mater. 51, 3429–3443.

Lu, Z.P., Liu, C.T., Thompson, J.R., Porter, W.D., 2004. Structural amorphous steels. Phys. Rev. Lett. 92, 245503.

Luo, J.H., Wu, F.F., Huang, J.Y., Wang, J.Q., Mao, S.X., 2010. Superelongation and atomic chain formation in nanosized metallic glass. Phys. Rev. Lett. 104, 215503.

Luo, W.K., Sheng, H.W., Alamgir, F.M., Bai, J.M., He, J.H., Ma, E., 2004. Icosahedral short-range order in amorphous alloys. Phys. Rev. Lett. 92, 145502.

Luzzi, D.E., Meshii, M., 1988. Chemical disordering in amorphization. J. Less Common Metals 140, 193–210.

McGreevy, R.L., 2001. Reverse Monte Carlo modelling. J. Phys. Condens. Matter 13, R877–R914.

Ma, H., Shi, L.L., Xu, J., Li, Y., Ma, E., 2005. Discovering inch-diameter metallic glasses in three-dimensional composition space. Appl. Phys. Lett. 87, 181915.

Macedo, P.B., Napolitano, A., 1967. Effects of a distribution of volume relaxation times in the annealing of BSC glass. J. Res. Natl. Bur. Stand. A 71A, 231–238.

Madge, S.V., Louzguine-Luzgin, D.V., Lewandowski, J.J., Greer, A.L., 2012. Toughness, extrinsic effects and Poisson's ratio of bulk metallic glasses. Acta Mater. 60, 4800–4809.

Masumoto, Y., Inoue, A., Kawasima, A., Hashimoto, K., Tsai, A.P., Masumoto, T., 1986. The effect of structural relaxation on the corrosion behaviour of amorphous $(Ni–Pd)_{82}Si_{18}$ alloys. J. Non Cryst. Solids 86, 121–136.

Mayr, S.G., 2006. Activation energy of shear transformation zones: a key for understanding rheology of glasses and liquids. Phys. Rev. Lett. 97, 195501.

Men, H., Pang, S.J., Zhang, T., 2006. Effect of Er doping on glass-forming ability of $Co_{50}Cr_{15}Mo_{14}C_{15}B_6$ alloy. J. Mater. Res. 21, 958–961.

Miller, M.K., 2000. Atom Probe Tomography: Analysis at the Atomic Level. Springer, New York.

Miller, S.A., Murphy, R.J., 1979. A gas-water atomization process for producing amorphous powders. Scr. Metall. 13, 673–676.

Miracle, D.B., 2004. A structural model for metallic glasses. Nat. Mater. 3, 697–702.

Miracle, D.B., 2006. The efficient cluster packing model—An atomic structural model for metallic glasses. Acta Mater. 54, 4317–4336.

Miracle, D.B., Concustell, A., Zhang, Y., Yavari, A.R., Greer, A.L., 2011. Shear bands in metallic glasses: size effects on thermal profiles. Acta Mater. 59, 2831–2840.

Miracle, D.B., Sanders, W.S., 2003. The influence of efficient atomic packing on the constitution of metallic glasses. Philos. Mag. 83, 2409–2428.

Mulder, A.L., van der Zwaag, S., van den Beukel, A., 1983. Embrittlement and disembrittlement in amorphous Metglas 2826 A. Scr. Metall. 17, 1399–1402.

Nagahama, D., Ohkubo, T., Hono, K., 2003. Crystallization of $Zr_{36}Ti_{24}Be_{40}$ metallic glass. Scr. Mater. 49, 729–734.

Nagase, T., Umakoshi, Y., 2010. Formation of dual-layer melt-spun ribbon through liquid phase separation. Intermetallics 18, 2136–2144.

Nagel, S.R., 1977. Temperature dependence of resistivity in metallic glasses. Phys. Rev. B 16, 1694–1698.

Nascimento, M.L.F., Zanotto, E.D., 2010. Does viscosity describe the kinetic barrier for crystal growth from the *liquidus* to the glass transition? J. Chem. Phys. 133, 174701.

Nastasi, M., Lilienfeld, D., Johnson, H.H., Mayer, J.W., 1986. Stability of metallic CsCl-structured alloys under ion irradiation. J. Appl. Phys. 59, 4011–4016.

Nielsen, O.V., Nielsen, H.J.V., 1980. Magnetic anisotropy in $Co_{73}Mo_2Si_{15}B_{10}$ and $(Co_{0.89}Fe_{0.11})_{72}Mo_3Si_{15}B_{10}$ metallic glasses induced by stress-annealing. J. Magn. Magn. Mater. 22, 21–24.

Nishi, Y., Kai, T., Tachi, M., Ishidaira, T., Yajima, E., 1986. Cooling condition dependence of saturated magnetic flux density in Fe-5at%Si-20at%B alloy glass. Scr. Metall. 20, 1099–1100.

Nishiyama, N., Amiya, K., Inoue, A., 2007. Recent progress of bulk metallic glasses for strain-sensing devices. Mater. Sci. Eng. A 449–451, 79–83.

Niza, M.E., Komori, M., Nomura, T., Yamaji, I., Nishiyama, N., Ishida, M., Shimizu, Y., 2010. Test rig for micro gear and experimental analysis on the meshing condition and failure characteristics of steel micro involute gear and metallic glass one. Mech. Mach. Theory 45, 1797–1812.

Ok, H.N., Morrish, A.H., 1981. Origin of the perpendicular anisotropy in amorphous $Fe_{82}B_{12}Si_6$ ribbons. Phys. Rev. B 23, 2257–2261.

Orava, J., Greer, A.L., Gholipour, B., Hewak, D.W., Smith, C.E., 2012. Characterization of supercooled liquid $Ge_2Sb_2Te_5$ and its crystallization by ultra-fast-heating calorimetry. Nat. Mater. 11, 279–283.

Pampillo, C.A., 1972. Localized shear deformation in a glassy metal. Scr. Metall. 6, 915–918.

Pampillo, C.A., Chen, H.S., 1974. Comprehensive plastic deformation of a bulk metallic glass. Mater. Sci. Eng. 13, 181–188.

Pan, D., Inoue, A., Sakurai, T., Chen, M.W., 2008. Experimental characterization of shear transformation zones for plastic flow of bulk metallic glasses. Proc. Natl. Acad. Sci. U.S.A. 105, 14769–14772.

Pan, J., Liu, L., Chan, K.C., 2009. Enhanced plasticity by phase separation in CuZrAl bulk metallic glass with micro-addition of Fe. Scr. Mater. 60, 822–825.

Panissod, P., Bakonyi, I., Hasegawa, R., 1983. Local boron environment in $Ni_{100-x}B_x$ metallic glasses—an NMR study. Phys. Rev. B 28, 2374–2381.

Park, B.J., Sohn, S.W., Kim, D.H., Jeong, H.T., Kim, W.T., 2008. Solid-state phase separation in Zr–Y–Al–Co metallic glass. J. Mater. Res. 23, 828–832.

Park, E.S., Kim, D.H., 2004. Formation of Ca–Mg–Zn bulk glassy alloy by casting into cone-shaped copper mold. J. Mater. Res. 19, 685–688.

Park, E.S., Kim, D.H., 2005. Formation of Mg–Cu–Ni–Ag–Zn–Y–Gd bulk glassy alloy by casting into cone-shaped copper mold in air atmosphere. J. Mater. Res. 20, 1465–1469.

Park, J.M., Han, J.H., Mattern, N., Kim, D.H., Eckert, J., 2012. Designing Zr–Cu–Co–Al bulk metallic glasses with phase separation mediated plasticity. Metall. Mater. Trans. A 43A, 2598–2603.

Pauly, S., Gorantla, S., Wang, G., Kühn, U., Eckert, J., 2010. Transformation-mediated ductility in CuZr-based bulk metallic glasses. Nat. Mater. 9, 473–477.

Peker, A., Johnson, W.L., 1993. A highly processable metallic glass: $Zr_{41.2}Ti_{13.8}Cu_{12.5}Ni_{10.0}Be_{22.5}$. Appl. Phys. Lett. 63, 2342–2344.

Perepezko, J.H., Smith, J.S., 1981. Glass formation and crystallization in highly undercooled Te–Cu alloys. J. Non Cryst. Solids 44, 65–83.

Pitt, E.B., Kumar, G., Schroers, J., 2011. Temperature dependence of the thermoplastic formability in bulk metallic glasses. J. Appl. Phys. 110, 043518.

Ponnambalam, V., Poon, S.J., Shiflet, G.J., 2004. Fe-based bulk metallic glasses with diameter thickness larger than one centimeter. J. Mater. Res. 19, 1320–1323.

Posgay, G., Kedres, F.J., Albert, B., Kiss, S., Haranguzó, I.Z., 1985. The effect of cooling conditions on the internal friction of a multicomponent metallic glass. In: Steeb, S., Warlimont, H. (Eds.), Rapidly Quenched Metals. North-Holland, Amsterdam, pp. 723–726.

Pugh, S.F., 1954. Relations between the elastic moduli and the plastic properties of polycrystalline pure metals. Philos. Mag. 45, 823–843.

Ray, R., 1981. High strength microcrystalline alloys prepared by devitrification of metallic glass. J. Mater. Sci. 16, 2924–2927.

Rehr, J.J., Albers, R.C., 2000. Theoretical approaches to X-ray absorption fine structure. Rev. Mod. Phys. 72, 621–654.

Sadoc, J.F., Dixmier, J., 1976. Structural investigation of amorphous CoP and NiP alloys by combined X-ray and neutron scattering. Mater. Sci. Eng. 23, 187–192.

Saffman, P.G., Taylor, G., 1958. The penetration of a fluid into a porous medium or Hele-Shaw cell containing a more viscous liquid. Proc. Roy. Soc. Lond. A 245, 312–329.

Sahu, K.K., Mauro, N.A., Longstreth-Spoor, L., Saha, D., Nussinov, Z., Miller, M.K., Kelton, K.F., 2010. Phase separation mediated devitrification of $Al_{88}Y_7Fe_5$ glasses. Acta Mater. 58, 4199–4206.

Salimon, A., Ashby, M.F., Bréchet, Y., Greer, A.L., 2004. Bulk metallic glasses: what are they good for? Mater. Sci. Eng. A 375–377, 385–388.

Schall, P., Weitz, D.A., Spaepen, F., 2007. Structural rearrangements that govern flow in colloidal glasses. Science 318, 1895–1899.

Scholte, P.M.L.O., Tegge, M., van der Woude, F., Buschow, K.H.J., Vincze, I., 1985. Mössbauer spectroscopy on amorphous Fe_xZr_{100-x} ($20 < x < 90$) alloys. In: Steeb, S., Warlimont, H. (Eds.), Rapidly Quenched Metals. North-Holland, Amsterdam, pp. 541–544.

Schroers, J., Hodges, T.M., Kumar, G., Raman, H., Barnes, A.J., Pham, Q., Waniuk, T.A., 2011. Thermoplastic blow molding of metals. Mater. Today 14, 14–19.

Schroers, J., Johnson, W.L., 2000a. Crystallization of $Zr_{41}Ti_{14}Cu_{12}Ni_{10}Be_{23}$. Mater. Trans. JIM 41, 1530–1537.

Schroers, J., Johnson, W.L., 2000b. History dependent crystallization of $Zr_{41}Ti_{14}Cu_{12}Ni_{10}Be_{23}$ melts. J. Appl. Phys. 88, 44–48.

Schroers, J., Johnson, W.L., 2004a. Highly processable bulk metallic glass-forming alloys in the Pt–Co–Ni–Cu–P system. Appl. Phys. Lett. 84, 3666–3668.

Schroers, J., Johnson, W.L., 2004b. Ductile bulk metallic glass. Phys. Rev. Lett. 93, 255506.

Schroers, J., Johnson, W.L., Busch, R., 2000. Crystallization kinetics of the bulk-glass-forming $Pd_{43}Ni_{10}Cu_{27}P_{20}$ melt. Appl. Phys. Lett. 77, 1158–1160.

Schubert, E., Bergmann, H.W., 1993. Rapidly solidified surface layers by laser melting. In: Liebermann, H.H. (Ed.), Rapidly Solidified Alloys. Dekker, New York, pp. 195–230.

Schumacher, P., Greer, A.L., Worth, J., Evans, P.V., Kearns, M.A., Fisher, P., Green, A.H., 1998. New studies of nucleation mechanisms in Al-alloys: implications for grain-refinement practice. Mater. Sci. Technol. 14, 394–404.

Schwarz, R.B., Johnson, W.L., 1983. Formation of an amorphous alloy by solid-state reaction of the pure polycrystalline metals. Phys. Rev. Lett. 51, 415–418.

Schwarz, R.B., Petrich, R.R., Saw, C.K., 1985. The synthesis of amorphous Ni–Ti alloy powders by mechanical alloying. J. Non Cryst. Solids 76, 281–302.

Scott, M.G., Kuršumović, A., 1982. Short-range ordering during structural relaxation of the metallic glass $Fe_{40}Ni_{40}B_{20}$. Acta Metall. 30, 853–860.

Scudino, S., Jerliu, B., Pauly, S., Surreddi, K.B., Kühn, U., Eckert, J., 2011. Ductile bulk metallic glasses produced through designed hetero-geneities. Scr. Mater. 65, 815–818.

Scudino, S., Surreddi, K.B., Khoshkhoo, M.S., Sakaliyska, M., Wang, G., Eckert, J., 2010. Improved room temperature plasticity of $Zr_{41.2}Ti_{13.8}Cu_{12.5}Ni_{10}Be_{22.5}$ bulk metallic glass by channel-die compression. Adv. Eng. Mater. 12, 1123–1126.

Shao, Y., Spaepen, F., Turnbull, D., 1998. An analysis of the formation of bulk amorphous silicon from the melt. Metall. Mater. Trans. A 29, 1825–1828.

Shen, J., Chen, Q.J., Sun, J.F., Fan, H.B., Wang, G., 2005. Exceptionally high glass-forming ability of an FeCoCrMoCBY alloy. Appl. Phys. Lett. 86, 151907.

Shen, Y.T., Kim, T.H., Gangopadhyay, A.K., Kelton, K.F., 2009. Icosahedral order, frustration, and the glass transition: evidence from time-dependent nucleation and supercooled liquid structure studies. Phys. Rev. Lett. 102, 057801.

Sheng, H.W., Luo, W.K., Alamgir, F.M., Bai, J.M., Ma, E., 2006. Atomic packing and short-to-medium range order in metallic glasses. Nature 439, 419–425.

Shi, L.L., Xu, J., Ma, E., 2008. Alloy compositions of metallic glasses and eutectics from an idealized structural model. Acta Mater. 56, 3613–3621.

Shimizu, F., Ogata, S., Li, J., 2006. Yield point of metallic glass. Acta Mater. 54, 4293–4298.

Singh, S., Ediger, M.D., de Pablo, J., 2013. Ultrastable glasses from *in silico* vapour deposition. Nat. Mater. 12, 139–144.

Sinning, H.R., Leonardsson, L., Cahn, R.W., 1985. Irreversible anisotropic length changes in $Fe_{40}Ni_{40}B_{20}$ and a search for reversible length changes in several metallic glasses. Int. J. Rapid Solidif. 1, 175–197.

Slichter, C.P., 1990. Principles of Magnetic Resonance, third ed. Springer, New York.

Smith, C.H., 1993. Applications of rapidly solidified soft magnetic alloys. In: Liebermann, H.H. (Ed.), Rapidly Solidified Alloys. Dekker, New York, pp. 617–663.

Smith, M.T., Saletore, M., 1986. Simple, low-cost planar flow casting machine for rapid solidification processing. Rev. Sci. Instrum. 57, 1647–1653.

Sohn, S.W., Yook, W., Kim, W.T., Kim, D.H., 2012. Phase separation in bulk-type Gd–Zr–Al–Ni metallic glass. Intermetallics 23, 57–62.

Song, K.K., Pauly, S., Zhang, Y., Scudino, S., Gargarella, P., Surreddi, K.B., Kühn, U., Eckert, J., 2011. Significant tensile ductility induced by cold rolling in $Cu_{47.5}Zr_{47.5}Al_5$ bulk metallic glass. Intermetallics 19, 1394–1398.

Sosso, G.C., Behler, J., Bernasconi, M., 2012. Breakdown of Stokes–Einstein relation in the supercooled liquid state of phase change materials. Phys. Stat. Solidi B 249, 1880–1885.

Spaepen, F., 1977. A microscopic mechanism for steady state inhomogeneous flow in metallic glasses. Acta Metall. 25, 407–415.

Stiffler, S.R., Evans, P.V., Greer, A.L., 1992. Interfacial transport kinetics during the solidification of silicon. Acta Metall. Mater. 40, 1617–1622.

Ström-Olsen, J.O., Brüning, R., Altounian, Z., Ryan, D.H., 1988. Structural relaxation in metallic glasses. J. Less Common Metals 145, 327–338.

Suzuki, K., 1983. Experimental determination of short-range structure of amorphous alloys by pulsed neutron scattering. In: Luborsky, F.E. (Ed.), Amorphous Metallic Alloys. Butterworths, London, pp. 74–99.

Suzuki, Y., Haimovich, J., Egami, T., 1987. Bond-orientational anisotropy in metallic glasses observed by X-ray diffraction. Phys. Rev. B 35, 2162–2168.

Szuecs, F., Kim, C.P., Johnson, W.L., 2001. Mechanical properties of $Zr_{56.2}Ti_{13.8}Nb_{5.0}Cu_{6.9}Ni_{5.6}Be_{12.5}$ ductile phase reinforced bulk metallic glass composite. Acta Mater. 49, 1507–1513.

Takeuchi, A., Inoue, A., 2005. Classification of bulk metallic glasses by atomic size difference, heat of mixing and period of constituent elements and its application to characterization of the main alloying element. Mater. Trans. 46, 2817–2829.

Tan, H., Zhang, Y., Ma, D., Feng, Y.P., Li, Y., 2003. Optimum glass formation at off-eutectic composition and its relation to skewed eutectic coupled zone in the La based La–Al–(Cu, Ni) pseudo ternary system. Acta Mater. 51, 4551–4561.

Tanner, L.E., Ray, R., 1980. Phase separation in Ti–Zr–Be metallic glasses. Scr. Metall. 14, 657–662.

Tarnóczi, T., Lovas, A., Kopasz, C., 1988. The influence of quenching rate on the relaxation processes in a nearly non-magnetostrictive metallic glass. Mater. Sci. Eng. 97, 509–513.

Tarumi, R., Shibata, A., Ogi, H., Hirao, M., Takashima, K., Higo, Y., 2007. Elastic anisotropy of an $Fe_{79}Si_{12}B_9$ amorphous alloy thin film studied by ultrasound spectroscopy. J. Appl. Phys. 101, 053519.

Taub, A.I., Spaepen, F., 1979. Isoconfigurational flow of amorphous $Pd_{82}Si_{18}$. Scr. Metall. 13, 195–198.

Toloui, B., Kuršumović, A., Cahn, R.W., 1985. Structural relaxation of neutron-irradiated $Fe_{40}Ni_{40}B_{20}$ metallic glass. Scr. Metall. 19, 947–952.

Tomida, T., Egami, T., 1993. Molecular dynamics study of structural anisotropy and anelasticity in metallic glasses. Phys. Rev. B 48, 3048–3057.

Torquato, S., Truskett, T.M., Debenedetti, P.G., 2000. Is random close packing of spheres well defined? Phys. Rev. Lett. 84, 2064–2067.

Treacy, M.M.J., Gibson, J.M., Fan, L., Paterson, D.J., McNulty, I., 2005. Fluctuation microscopy: a probe of medium range order. Rep. Prog. Phys. 68, 2899–2944.

Tregilgas, J.H., 2005. Amorphous hinge material. Adv. Mater. Process. 163 (1), 46–49.

Turnbull, D., 1969. Under what conditions can a glass be formed? Contemp. Phys. 10, 473–488.

van den Beukel, A., Sietsma, J., 1990. The glass transition as a free volume related kinetic phenomenon. Acta Metall. Mater. 38, 383–389.

Volkert, C.A., Donohue, A., Spaepen, F., 2008. Effect of sample size on deformation in amorphous metals. J. Appl. Phys. 103, 083539.

Volkert, C.A., Spaepen, F., 1989. Crossover relaxation of the viscosity of $Pd_{40}Ni_{40}P_{19}Si_1$ near the glass transition. Acta Metall. 37, 1355–1362.

Wagner, C.N.J., 1983. Experimental determination of atomic scale structure of amorphous alloys by scattering experiments. In: Luborsky, F.E. (Ed.), Amorphous Metallic Alloys. Butterworths, London, pp. 58–73.

Wagner, H., Bedorf, D., Küchemann, S., Schwabe, M., Zhang, B., Arnold, W., Samwer, K., 2011. Local elastic properties of a metallic glass. Nat. Mater. 10, 439–442.

Wang, C.-C., Ding, J., Cheng, Y.-Q., Wan, J.-C., Tian, L., Sun, J., Shan, Z.-W., Li, J., Ma, E., 2012. Sample size matters for $Al_{88}Fe_7Gd_5$ metallic glass: smaller is stronger. Acta Mater. 60, 5370–5379.

Wang, G., Liaw, P.K., Jin, X., Yokoyama, Y., Huang, E.-W., Jiang, F., Keer, L.M., Inoue, A., 2010. Fatigue initiation and propagation behavior in bulk-metallic glasses under a bending load. J. Appl. Phys. 108, 113512.

Wang, W.H., Dong, C., Shek, C.H., 2004. Bulk metallic glasses. Mater. Sci. Eng. R 44, 45–89.

Wang, Q., Wang, L.-M., Ma, M.Z., Binder, S., Volkmann, T., Herlach, D.M., Wang, J.S., Xue, Q.G., Tian, Y.J., Liu, R.P., 2011. Diffusion-controlled crystal growth in deeply undercooled $Zr_{50}Cu_{50}$ melt on approaching the glass transition. Phys. Rev. B 83, 014202.

Waniuk, T.A., Schroers, J., Johnson, W.L., 2001. Critical cooling rate and thermal stability of Zr–Ti–Cu–Ni–Be alloys. Appl. Phys. Lett. 78, 1213–1215.

Warlimont, H., 1985. New magnetic materials by rapid solidification. In: Steeb, S., Warlimont, H. (Eds.), Rapidly Quenched Metals. North-Holland, Amsterdam, pp. 1599–1609.

Watanabe, T., Tanabe, T., 1976. Formation and morphology of Ni–B amorphous alloy deposited by electroless plating. Mater. Sci. Eng. 23, 97–100.

Wei, G., Cantor, B., 1989. The effect of heat treatment and surface treatment on the crystallisation behaviour of amorphous $Fe_{40}Ni_{40}B_{20}$. Acta Metall. 37, 3409–3424.

Wei, Y.J., Lei, X.Q., Huo, L.S., Wang, W.H., Greer, A.L., 2013. Towards more uniform deformation in metallic glasses: the role of Poisson's ratio. Mater. Sci. Eng. A 560, 510–517.

Wells, S., Charles, S.W., Mørup, S., Linderoth, S., van Wonterghem, J., Larsen, J., Madsen, M.B., 1989. A study of Fe–B and Fe–Co–B alloy particles produced by reduction with borohydride. J. Phys. Condens. Matter 1, 8199–8208.

Whittle, G.L., Wells, P., Campbell, S.J., Calka, A., Stewart, A.M., 1985. Correlations between hyperfine parameters in Fe–B metallic glasses. In: Steeb, S., Warlimont, H. (Eds.), Rapidly Quenched Metals. North-Holland, Amsterdam, pp. 545–548.

Williams, R.S., Egami, T., 1976. Effects of deformation and annealing on magnetic amorphous alloys. IEEE Trans. Magn. 12, 927–929.

Woldt, E., 1988. The reversible enthalpy change of the metallic glass $Fe_{40}Ni_{40}B_{20}$—experiments and simulation in the activation energy spectrum model. J. Mater. Sci. 23, 4383–4391.

Wong, 1981. EXAFS studies of metallic glasses. In: Güntherodt, H.-J., Beck, H. (Eds.), Glassy Metals I. Springer, Berlin, pp. 45–77.

Wright, T.M., 2002. The Physics and Mathematics of Adiabatic Shear Bands. Cambridge University Press, Cambridge.

Xi, X.K., Li, L.L., Zhang, B., Wang, W.H., Wu, Y., 2007. Correlation of atomic cluster symmetry and glass-forming ability of metallic glass. Phys. Rev. Lett. 99, 095501.

Xu, D., Duan, G., Johnson, W.L., 2004. Unusual glass-forming ability of bulk amorphous alloys based on ordinary metal copper. Phys. Rev. Lett. 92, 245504.

Yeh, X.L., Samwer, K., Johnson, W.L., 1983. Formation of an amorphous metallic hydride by reaction of hydrogen with crystalline intermetallic compounds—a new method of synthesizing metallic glasses. Appl. Phys. Lett. 42, 242–243.

Yokoyama, Y., Yamano, K., Fukaura, K., Sunada, H., Inoue, A., 2001. Ductility improvement of $Zr_{55}Cu_{30}Al_{10}Ni_5$ bulk amorphous alloy. Scr. Mater. 44, 1529–1534.

Yoshizawa, Y., Oguma, S., Yamauchi, K., 1988. New Fe-based soft magnetic alloys composed of ultrafine grain structure. J. Appl. Phys. 64, 6044–6046.

Yu, H.B., Shen, X., Wang, Z., Gu, L., Wang, W.H., Bai, H.Y., 2012. Tensile plasticity in metallic glasses with pronounced β relaxations. Phys. Rev. Lett. 108, 015504.

Yu, H.-B., Wang, W.-H., Samwer, K., 2013. The β relaxation in metallic glasses: an overview. Materials Today 16, 183–191.

Zachariasen, W.H., 1932. The atomic arrangement in glass. J. Am. Chem. Soc. 54, 3841–3851.

Zarzycki, J., 1991a. Glasses and the Vitreous State. CUP, Cambridge, pp. 148–185.

Zarzycki, J., 1991b. Glasses and the Vitreous State. CUP, Cambridge, pp. 422–431.

Zhang, L., Cheng, Y.Q., Cao, A.J., Xu, J., Ma, E., 2009. Correlation between the elastic modulus and the intrinsic plastic behavior of metallic glasses: the roles of atomic configuration and alloy composition. Acta Mater. 57, 1154–1164.

Zhang, Y., Greer, A.L., 2006. Thickness of shear bands in metallic glasses. Appl. Phys. Lett. 89, 071907.

Zhang, Y., Greer, A.L., 2007. Correlations for predicting plasticity or brittleness of metallic glasses. J. Alloys Compd. 434, 2–5.

Zhang, Y., Wang, W.H., Greer, A.L., 2006. Making metallic glasses plastic by control of residual stress. Nat. Mater. 5, 857–860.

Zink, M., Samwer, K., Johnson, W.L., Mayr, S.G., 2006. Plastic deformation of metallic glasses: size of shear transformation zones from molecular dynamics simulations. Phys. Rev. B 73, 172203.

Zougmoré, F., Lasjaunias, J.C., Béthoux, O., 1988. Effect of structural relaxation on the thermodynamic properties of amorphous superconducting ZrCu alloys. J. Less Common Metals 145, 367–374.

Further Reading

Schuh, C.A., Hufnagel, T.C., Ramamurty, U., 2007. Mechanical behavior of amorphous alloys. Acta Mater. 55, 4067–4109.
Wang, W.H., 2009. Bulk metallic glasses with functional physical properties. Adv. Mater. 21, 4524–4544.
Wang, W.H., 2012. The elastic properties, elastic models and elastic perspectives of metallic glasses. Prog. Mater. Sci. 57, 487–656.
See also the collection of articles in *MRS Bull.* (2007) **32**(8) on "Bulk metallic glasses: at the cutting edge of metals research".

Biography

A. L. (Lindsay) Greer is Professor of Materials Science in the Department of Materials Science & Metallurgy at the University of Cambridge. He has also held faculty positions at Harvard University, Institut National Polytechnique de Grenoble, Washington University (St Louis) and the University of Vienna. He is an editor of *Philosophical Magazine* (founded in 1798, publishing papers on the structure and properties of condensed matter). His research interests are metallic glasses and crystal nucleation, grain refinement in casting, chalcogenide thin films for phase-change data storage, and electromigration. He has published more than 380 scientific papers, and is the author (with K. F. Kelton) of *Nucleation in Condensed Matter: Applications in Materials and Biology*. He is a Fellow of the Institute of Materials Minerals and Mining (UK).

5 Diffusion in Metals and Alloys

Zoltan Balogh and Guido Schmitz, Institute of Materials Physics, University of Münster, Münster, Germany

Table of Symbols

a lattice parameter

B magnetic field

c_i volume concentration of a given element

$c = c_A$

c_b tracer concentration in the grain bulk

c_v^{eq} equilibrium vacancy concentration per volume

c_{sc}; c_d; c_{gb} tracer concentration in the shortcut; in the dislocation; in the grain boundary

$\tilde{D}$ interdiffusion coefficient excluding thermodynamic factor

$\tilde{D}_D$ Darken interdiffusion coefficient

$\tilde{D}_{N.P.}$ Nernst–Planck interdiffusion coefficient

D_0 preexponential factor

D; D_i diffusion coefficient of a given species

D_i^* tracer diffusion coefficient of component i

D_{ij} diffusion coefficients according to the Onsager thermodynamics

D_b bulk-, lattice- or intragranular diffusion coefficient

D_V vacancy diffusion coefficient

D_{sc}; D_d; D_{gb} shortcut diffusion coefficient; dislocation diffusion coefficient; grain boundary diffusion coefficient

D_{ab}; D_{tj} aggregate boundary diffusion coefficient; triple junction diffusion coefficient

E Young modulus

$\mathrm{erf}(x)$; $\mathrm{erfc}(x)$ error function; complementary error function

f, f_i, $f_{ij}^{(i)}$ correlation factor; correlation factor for a given species; partial/collective correlation factor

G^B; G^F; G^M binding; formation; migration Gibbs energy

g; g'; g'' molar Gibbs energy; first; and second derivates of the molar Gibbs energy in respect to the mole fraction

H^B; H^F; H^M binding; formation; migration enthalpy;

ΔH activation enthalpy

$H(\nu)$ randomization frequency

$\hbar$ reduced Planck constant

J_z the magnetic quantum number

j_i diffusion flux density of a given species

$j = j_A$

j_v vacancy flux density

j^+; j^- diffusion flux density to the positive; negative direction

$\bar{j}$ interdiffusion flux density

k_B Boltzmann constant

k_c kinetic exponent

k_{IF} interface transport coefficient

l_c characteristic length for the linear–parabolic transition;

l_0 characteristic length for transition between Nernst–Planck regime—linear regime

l_1 characteristic length for transition between linear regime—Darken regime

M mobility coefficient, magnetization

M_{ij} mobility coefficients according to Onsager thermodynamics

m' diffusion asymmetry parameter

N number of atoms in the system

P shortcut double and triple product

p hydrostatic pressure

Q the momentum transfer in quasielastic neutron scattering experiment

ΔQ activation enthalpy

Q^{-1} the internal friction coefficient

R gas constant

RHS Right-hand side

S^B; S^F; S^M binding; formation; migration entropy

s segregation factor

T temperature

T_c critical temperature (of phase separation)

T_m melting temperature

ΔV activation volume

v_K Kirkendall velocity

x coordinate along the (main) diffusion direction

x_M the position of the Matano plane

z the number of nearest neighbors (coordination number)

z the coordinate along the grain boundary/ dislocation

Γ total jump rate

Γ_0 attempt frequency

δ interface width

γ volume fraction of the diffusion shortcut

δ the thickness of a diffusion shortcut (grain boundary, dislocation, etc.)

ε pair exchange parameter

ε_{ij} components of strain

ε_{ij} pair interaction energy between atoms of type i and j

η scaled "coordinate" of Boltzmann–Matano transformation

η viscosity

κ gradient energy term

Λ the density of diffusion shortcuts

Λ_0 wavelength of the neutron beam in a QNES experiment

λ the length of the jump vector (usually nearest neighbor distance)

μ; μ_i chemical potential; of a given element

v; v_i mole fraction; of a given element;

v_v^{eq} the equilibrium vacancy mole fraction in a crystal

Λ_{mod} the modulation length of a multilayer

σ Poisson ratio

σ_{ij} components of stress tensor

τ various characteristic times

Φ thermodynamic factor

Ω; Ω_i atomic volume; partial volume of component i

ω; ω_i jump rate between two selected sites; of a given species

ω_{exc} the excitation frequency in a mechanical spectroscopy measurement

ω_{Larm} the Larmor frequency

5.1   Introduction

Transport in materials via random individual atomic migration steps (jumps) is called 'diffusion'. General understanding and detailed theory of diffusional transport represent a classical field of material science. While in liquids and gases, diffusion rate ranges up to millimeters or even centimeters per second, transport in solid materials is rather slow. In densely packed metals near to the melting point, one can expect about a micrometer per second and this rate drops down to about a nanometer per second at half the melting temperature. At room temperature atomic migration is practically frozen, except for few exceptional cases of small interstitial impurities. Measurement of diffusion in solids needs therefore sensitive techniques and well-developed microscopy. Not surprising that a scientific proof of diffusion in solid metals came rather late in history of science (Roberts-Austen, 1896).

Although diffusion in solid state is slow, knowledge of diffusion processes is nevertheless fundamental to material scientists and engineers. Tailoring of microstructures by solid-state reactions is largely based on diffusion processes at elevated temperatures. Long-term stability and aging of materials, especially important in high-temperature applications, are controlled by diffusion. With downscaling of structural dimensions in current technology, control of diffusional transport on the nanometer scale gets decisive in electronic devices or MEMS (Micro Electro-Mechanical Systems), even in nanocrystalline construction materials for operation close to room temperature.

Understanding diffusion from a theoretical point of view is challenging as many different fields are involved. Complex boundary value problems of continuum diffusion equations found beautiful mathematic solutions in the past. A statistical description of independent atomic jumps requires profound thermodynamics making use of equilibrium and nonequilibrium concepts. Most ingenious physicists have contributed to this field. It was none other than Albert Einstein who provided the missing link between the continuum description of diffusion and the random walk of individual atoms (Einstein, 1905). Beginning with the second half of past century, the microscopic mechanisms of atomic jumps attracted increasing attention. Nowadays this field is particularly driven by the possibilities of computer simulation. Energetics and kinetics of atomic jumps are calculated by ab initio method, while the whole diffusion process is simulated by Monte Carlo (MC) or to greater detail by Molecular Dynamic (MD) simulation.

From the experimental point of view, one may state that the bulk of required diffusion data for metals are already known. But confronted to the actual task of discussing a given reaction or material, many scientists are suddenly surprised that just in their particular case required diffusion data are still missing. Continuous development of alloys and intermetallic phases motivates further new and more accurate measurement of respective diffusion properties. This is particularly true in the case of short circuit transport on the nanoscale. Coefficients of grain boundary diffusion in the dilute limit are partly known, but interdiffusion within grain boundaries in the technically interesting high concentration range, properties of interphase boundaries, triple junctions or higher order defects, and transport barriers at interfaces are still widely unexplored landscape.

Obviously in a general textbook like Physical Metallurgy some compromises have to be made as a result of the limited space dedicated to a given topic. While we tried our best to present as much useful information in the most didactic manner as possible, this chapter is not aimed to represent a complete coverage of all diffusion related phenomena. For those who would like to gain a deeper understanding or learn finer aspects of the atomic transport we advise trying some specialized textbooks or review articles. One of the most recent general textbook on the diffusion is Prof. Mehrer's excellent, *Diffusion in Solids, Fundamentals, Methods, Materials, Diffusion Controlled Processes* (Mehrer, 2006a). Prof. Gupta's

Diffusion Processes in Advanced Technological Materials gives insight on the application of diffusion processes (Gupta, 2004). We can suggest Prof. Gusak's *Diffusion-controlled Solid State Reactions in Alloys, Thin Films and Nanosystems* for everyone who would like to deepen his knowledge on reactive diffusion (Gusak, 2010). Those who are interested in the nanoscale aspects of diffusion could find the collection of reviews in Volume 19 of the Journal of Metastable and Nanocrystalline Materials quite useful (JMN, 2004).

The present authors are grateful to the members of the Institute of Materials Physics in Münster for providing useful information and helping with deep discussions. In particular, we would like to emphasize the valuable contribution in proofreading by Susann Nowak and Manuel Roussel, who provided the viewpoints of an "intelligent graduate student" and a "well-educated" post doc.

5.2   Atomistic Mechanism and Fundamental Relations of Diffusion

In this section, the basis for a fundamental understanding of diffusive transport should be laid. We will assume a microscopic point of view. Thus, from a principal knowledge of microscopic mechanisms of site exchange and the probability of this 'jump' event, we derive the phenomenological Fick laws. We demonstrate how phenomenological coefficients are related to the statistics of many independent jumps and will finally consider how this statistical process may be traced by methods of modern computer modeling.

5.2.1   Elementary Diffusion Mechanisms in Metals

Crystal structures of metals are distinguished by their dense packing. Hard to imagine, how atoms may exchange their lattice sites in a dense lattice. It seems to be natural and has been confirmed by experiment that diffusional transport in solid metals can only appear by point defects that act as vehicles of transport. The few possibilities of defect-mediated atomic migration that are realized in metals are illustrated in **Figure 1**. We have to mark an individual atom to follow the migration. In the figure this is simple: we use blue color. In real experiment, one can mark a tiny fraction of atoms by another isotope, possibly a radioactive one. Atoms marked in this way are called tracer atoms.

The migration of small solutes dispersed on interstitial sites of the host represents conceptually the simplest case of diffusion (**Figure 1(a)**). Provided low concentration, a solute on an interstitial site finds its neighbor interstices usually empty and so a spontaneous jump to the next free site is always possible. However, occupation of the interstitial site may lead to distortion of the surrounding host. Even more severe distortion is necessary to perform the jump to the next site through the small window offered by the neighboring host atoms. Since elastic energy required for distortion increases with the square of the solute size, only rather small atoms, such as hydrogen, carbon, or oxygen, are candidates to occupy interstitial sites in thermal equilibrium and to perform easy diffusion jumps. But if diffusion of interstitial species becomes possible, it is usually outstandingly fast.

By contrast, misplacing a host atom itself into an interstitial site (self-interstitial) requires in metals so large amount of energy, that a measurable fraction of self-interstitials is only expected after irradiation or severe plastic deformation. If self-interstitials are nevertheless present (e.g. as a consequence of irradiation with energetic particles), they allow an indirect transport mechanism, usually termed as the *interstitialcy* mechanism (**Figure 1(b)**). In a first step, a substitutional atom is kicked by a neighbored interstitial from its lattice position (1) into the next free interstice, while the former interstitial takes the

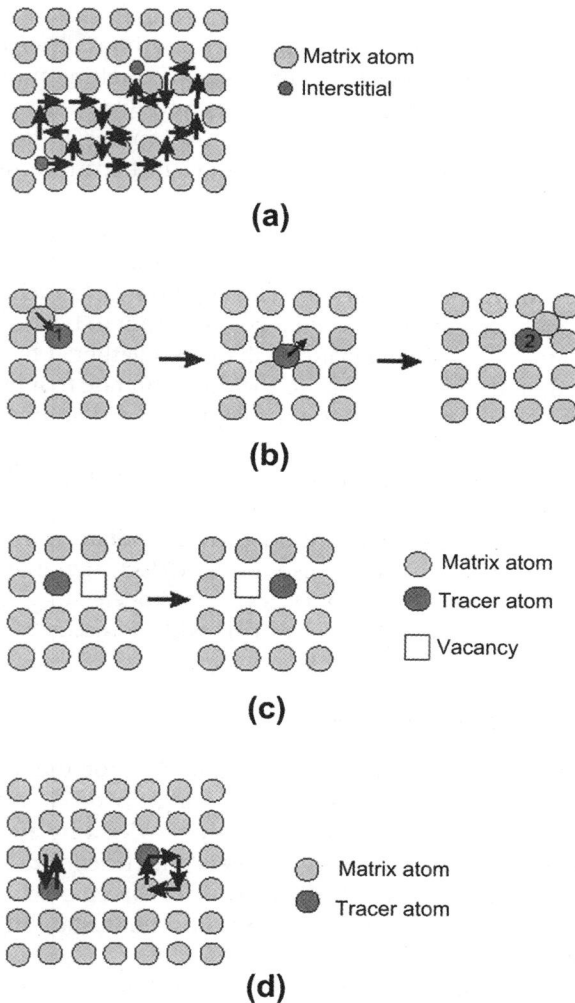

Figure 1 Site exchange in crystalline metals: (a) interstitial diffusion, (b) interstitialcy mechanism, (c) vacancy mechanism, (d) hypothetical direct site exchange (left) and ring exchange (right). (For color version of this figure, the reader is referred to the online version of this book.)

original lattice site. The newly formed interstitial repeats this step by substituting a third lattice atom, and so finds itself at the end at the new lattice position (2).

In contrast to self-interstitials, vacant lattice sites do belong to thermal equilibrium of a crystal in considerable concentrations. Near to the melting point, one may even expect a fraction of 10^{-4} to 10^{-3} of unoccupied lattice sites. Therefore, site exchange of atoms with neighbored vacancies (see **Figure 1(c)**) represents the most common mechanism of self or (substitutional) solute diffusion in metals. In comparison with small interstitials, this vacancy-mediated diffusion is rather slow, since atoms have to wait for a vacancy appearing in its neighborhood, before a jump becomes possible.

Intrinsic vacancies are formed by thermal activation. Therefore, temperature dependence of this kind of diffusion is quite pronounced. As a complication which is sometimes observed, pairs of vacancies may be formed in significant fraction if temperature approaches the melting point. In vicinity of a double vacancy, atomic jumps become more probable, since elastic distortion during the jump is easier accommodated by the larger empty volume nearby. If diffusion is significantly mediated by double vacancies, temperature dependency of the diffusion coefficient becomes more complex (see Section 5.5).

Alternative cooperative mechanisms of diffusion, such as the direct pair exchange or a closed ring replacement sequence (see **Figure 1(d)**) were also proposed initially to explain diffusion in crystalline metals. However, direct pair exchange in dense lattices requires too much elastic distortion and the probability for correlated sequences like a ring exchange turned out to be too small for being significant in crystalline metals. In contrast, convincing evidence has been gathered by computer simulation and also by measurements based on the isotope effect to the diffusion rate that the ring exchange or similar correlated replacement sequences play an important role in liquids and metallic glasses (Faupel et al., 2003).

In comparison with simple metals, the superlattice structure of intermetallic compounds gives rise to richer variety of possible mechanisms. If a substitutional atom jumps to a next neighbor site in a superlattice structure, it may find itself afterward on a wrong sublattice which is usually energetically unfavorable. Thus, for superlattice structures also point defect agglomerates (triple defects) and involved correlated jump sequences (6-jump cycles) have been suggested, which allow atomic migration without disturbing the ordered structure. This will be discussed in more detail in Section 5.5.5.

5.2.2 The Individual Jump: Transition State Theory

Understanding diffusion rates needs at first hand an idea to describe the probability of an individual jump. Although modern computer facilities allow following up the jump process in detailed MD simulations, real understanding in a classical sense is only provided by analytical descriptions which may be achieved by combining thermodynamic and general kinetic considerations. Quite a reasonable description delivers the so-called transition state theory which is based on classical work by Wert (1950) and Vineyard (1957). At least, this model provides a practical understanding of jump frequencies and their dependence on temperature.

The mechanisms sketched in **Figure 1** have in common that they require the jumping atom to pass a saddle on their pathway from initial to final equilibrium position. Obtaining the saddle position requires an elevated enthalpy in comparison with the equilibrium positions. The pathway itself may be a complex curvilinear as indicated in **Figure 2**, but can be certainly projected to a one-dimensional coordinate. Provided the jump rate is slow in comparison with phonon frequencies, one may assume that the crystal, apart from being constrained at the 1D coordinate along the transition path, establishes thermal equilibrium with respect to all other degrees of freedom of the phase space. Under this assumption, the jump rate ω between two selected lattice sites can be estimated by the thermodynamic probability of finding the jumping atom at the saddle times its Maxwell equilibrium velocity along the transition path (Schoeck, 1980). As most compact descriptions, the theory arrives at (Vineyard, 1957)

$$\omega = \Gamma_0 \exp\left(-\frac{H^{\mathrm{M}}}{k_{\mathrm{B}}T}\right) \tag{1a}$$

(a)

(b)

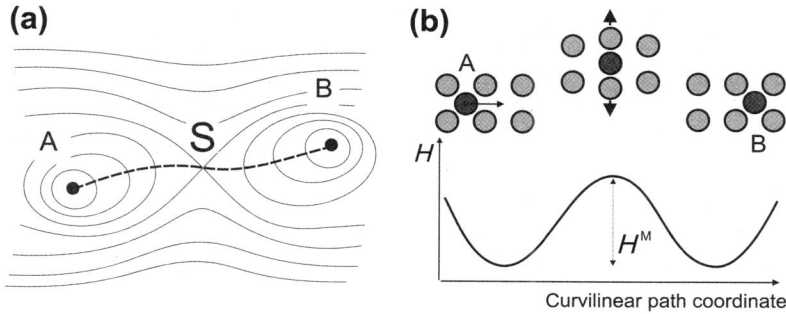

Figure 2 Energetic situation of the atomic jump: (a) 2D representation of the enthalpy landscape between two equilibrium positions (A and B) of the jumping atom. Both positions are linked by a curvilinear pathway. (b) Along this path, a clearly defined energy maximum appears at the saddle position. The respective migration enthalpy H^M may be imagined as elastic energy needed to push outward neighbored atoms that hinder the transition. (For color version of this figure, the reader is referred to the online version of this book.)

or

$$\omega = \tilde{\Gamma}_0 \exp\left(\frac{S^M}{k_B}\right) \cdot \exp\left(-\frac{H^M}{k_B T}\right) = \tilde{\Gamma}_0 \exp\left(-\frac{G^M}{k_B T}\right) \tag{1b}$$

for the transition rate of the atom. H^M denotes the migration enthalpy, which represents the enthalpy of the system in saddle position minus the enthalpy in the initial equilibrium position (A), while $G^M = H^M - TS^M$ contains already vibration parts of the entropy. The precise meaning of the two 'attempt' frequencies is

$$\Gamma_0 := \prod_{i=1}^{3N} \nu_i \bigg/ \prod_{i=1}^{3N-1} \nu_i^{(S)} \quad \text{and} \quad \tilde{\Gamma}_0 = \prod_{i=1}^{3N} \nu_i \bigg/ \prod_{i=1}^{3N-1} \nu_i^{(A)}; \tag{2}$$

in which ν_i, $\nu_i^{(S)}$, and $\nu_i^{(A)}$ denote the 3N Eigen frequencies of N atoms in the equilibrium configuration and the $3N-1$ frequencies of the system if constraint to the saddle (S) or the initial equilibrium (A) position on the pathway, respectively. In consequence of the second choice, the migration entropy S^M has to be defined as

$$S^M := k_B \ln\left[\prod_{i=1}^{3N-1} \nu_i^{(A)} \bigg/ \prod_{i=1}^{3N-1} \nu_i^{(S)}\right] \tag{3}$$

It is a matter of taste, which of the two formulations Eqn (1a) or (1b) should be better used. The advantage of the latter is that the attempt frequency $\tilde{\Gamma}_0$ contains only properties of the initial position, while migration entropy and enthalpy both comprise a comparison of the saddle to the initial equilibrium position.

Equations (1) and (2) may be criticized for their limited practical value, as exact phonon spectra, especially those of the saddle point configuration, are rarely known. However, computer techniques and numerical concepts have experienced tremendous progress so that nowadays the vibronic effects on the defect properties can be handled (Grabowski et al., 2009, 2011). Still, for easy estimation, Γ_0 or $\tilde{\Gamma}_0$ are identified by Debye's frequency (10^{12}–10^{13} Hz). In view of the dominating role of the energetic Boltzmann term in understanding the temperature dependence of jump rates, this practice is often well justified.

The transition state theory may also be blamed for using equilibrium concepts in describing a kinetic phenomenon. Indeed, it is questionable whether the short time, during which the jump appears, is sufficient to establish equilibrium at any constrained position along the transition path. Therefore, refined versions regarded as dynamical theories were suggested (Rice, 1958; Slater, 1959; Manley and Rice, 1960). However, the final formula for the jump rate turned out to be very similar to Eqn (1). Further, one may criticize that a system that has already overcome the saddle point must not necessarily end up in the final equilibrium position. Near to the saddle point, walk along the transition path may become itself diffusive in character, meaning it appears in small random forward and backward steps. In consequence, the absolute jump rate would be reduced similar as in classical theory of nucleation the formation rate of overcritical nuclei is reduced by the Zeldovich (Zeldovich, 1942) factor. The latter gets the more important the smaller the curvature of the Gibbs potential barriers at the saddle position, but can be taken into account quantitatively if the curvature of the enthalpy landscape along the pathway is known.

Current methods to determine the position and the barrier of the saddle points are briefly described in Section 5.2.9.2.

5.2.3   Continuum Diffusion Equations

Having clarified nature and rate of elementary jumps, we are now in the position to derive continuum equations which conveniently describe the atomic transport on a mesoscopic to macroscopic length scale. To avoid unnecessary complexity, we consider first diffusion of interstitial solutes in dilute limit for which no additional lattice defects are involved. A one-dimensional scheme as sketched in **Figure 3** is sufficient to obtain the desired fundamental relations. Imagine two adjacent planes of interstices spaced by the jump distance λ. Let them be occupied by different numbers $n(x)$ and $n(x + \lambda)$ of interstitial atoms per area being able to jump. The enthalpy landscape of an atom constrained to the diffusional pathway is sketched in the bottom row of the figure.

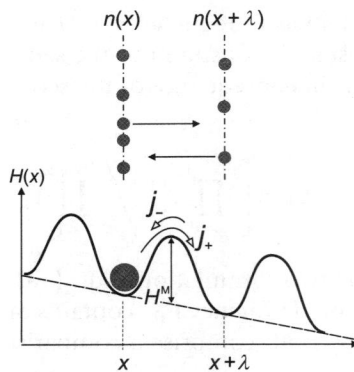

Figure 3   One-dimensional enthalpy profile assuming a constant driving force. Two adjacent layers of interstitial sites spaced by a distance λ are distinguished by local energy minima between which a barrier has to be overcome. Owing to a driving force, the barrier from left to the right gets slightly lower than that from right to left. As sketched at the top, the layers may be occupied by a different number $n(x)$ and $n(x + \lambda)$ of atoms per area.

In contrast to **Figure 2**, we allow now for an additional driving force (e.g. electrical field or chemical forces acting on the atoms in preferred direction). Therefore, the periodic potential of the energy landscape is superposed by a steady slope downward the x-direction. So, the jumps become slightly biased with respect to their direction. But, provided that additional driving forces are not too large, the individual migration steps are still controlled by thermal activation across the barrier. Therefore, the total current density of atoms along the x-axis results from the balance of atomic jumps from left to right minus those from right to left. Based on Eqn (1) this is expressed as

$$j = j^+ - j^-$$
$$= \Gamma_0 n(x)\left[\exp\left(-\frac{H^M + (H(x+\lambda) - H(x))/2}{k_BT}\right) - \frac{n(x+\lambda)}{n(x)}\exp\left(-\frac{H^M - (H(x+\lambda) - H(x))/2}{k_BT}\right)\right].$$

Transforming atomic numbers per area into concentrations $c(x)$ per volume yields

$$j = \Gamma_0 c(x)\lambda\,\exp\left[-\frac{H^M}{k_BT}\right]\cdot$$
$$\left(\exp\left[-\frac{(H(x+\lambda) - H(x))/2}{k_BT}\right] - \exp\left[-\frac{-(H(x+\lambda) - H(x))/2 - k_BT\ln[c(x+\lambda)/c(x)]}{k_BT}\right]\right),$$

where we have shifted for later convenience the second compositional pre-factor into the exponent. Driving forces and composition variation between neighbored planes are usually small with respect to k_BT, thus the exponentials can be expanded to linear order:

$$j \approx \Gamma_0 c\lambda\,\exp\left[-\frac{H^M}{k_BT}\right]\cdot\left(\frac{H(x) + k_BT\ln c(x) - H(x+\lambda) - k_BT\ln c(x+\lambda)}{k_BT}\right). \quad (4)$$

In the numerator of the last term at the right-hand side (RHS) of Eqn (4), the combination of the "enthalpy"[1] with the configurational entropy $k_BT\ln c$ are naturally identified by the chemical potential of the diffusor at the respective equilibrium sites. We arrive at the remarkably compact expression

$$j \approx -\underbrace{\frac{\Gamma_0 c\lambda^2}{k_BT}\exp\left[-\frac{H^M}{k_BT}\right]}_{=M}\cdot\frac{\partial\mu}{\partial x} = -M\cdot\frac{\partial\mu}{\partial x}, \quad (5)$$

which represents quite a natural linear relation: Atomic currents are generated by driving forces and controlled by the respective mobility M. In general, driving force to migration is represented by the negative gradient of the corresponding chemical potential. If this driving force stems exclusively from configurational entropy, we speak of diffusion in its original sense. In this case, the chemical potential fulfills

$$\frac{\partial}{\partial x}\mu = \frac{\partial}{\partial x}(\mu_0 + k_Bt\ln c) = \frac{k_BT}{c}\frac{\partial c}{\partial x}, \quad (6)$$

[1] In a strict sense, also the vibration parts of the entropy could be included by using G^M instead of H^M and the alternative definition of the attempt frequency (see Eqns (1) and (2)).

so that Eqn (5) may be formulated as the famous first Fick's law (Fick, 1855)

$$j = -D\frac{\partial c}{\partial x} \quad \text{with} \quad D := \Gamma_0\lambda^2 \exp\left[-H^M/(k_BT)\right] \tag{7}$$

A comparison of Eqn (5) with Eqn (7) discovers the fundamental relation between both mobility and diffusion coefficient:

$$D = \frac{k_BT}{c}M. \tag{8}$$

Continuity of matter requires for the concentration change (atoms per unit volume) with time:

$$\frac{\partial c}{\partial t} = -\frac{\partial}{\partial x}j = \frac{\partial}{\partial x}\left(D(c(x))\frac{\partial c}{\partial x}\right), \tag{9}$$

which is known as the second Fick's law in the case of a composition-dependent diffusion coefficient. If the coefficient is constant, we find the simpler linear form of Fick's second law:

$$\frac{\partial c}{\partial t} = D\frac{\partial^2 c}{\partial x^2}. \tag{10}$$

For the latter linear equation, a variety of mathematical solutions are available assuming different boundary and start conditions, while the nonlinear analog Eqn (9) knows only fewer and less general solutions and often requires numerical calculation.

Any driving force exceeding pure entropy, for example any excess term in the Gibbs energy of an alloy, leads to drift terms that are not included in the Fick laws. If entropic as well as drift forces act together, it is convenient to write Eqn (5) in a form that separates both influences

$$j = -D\frac{dc}{dx} + \frac{cD}{k_BT}F. \tag{11}$$

In this drift–diffusion equation, F stands for any driving force which may stem, for example from chemical, electric, magnetic, elastic, or temperature fields acting on migrating species. Again, the temporal evolution of local composition follows from continuity

$$\frac{\partial c}{\partial t} = \frac{\partial}{\partial x}\left(D\frac{\partial c}{\partial x}\right) - \frac{\partial}{\partial x}\left(\frac{cD}{k_BT}F\right). \tag{12}$$

In three-dimensional space the directions of the atomic current and the driving force are not necessarily parallel to each other anymore, as jump rates or distances may differ into the directions of different crystal axes. Obviously, the diffusion coefficient gets then a tensor character. Generalizing Eqn (7) to 3D space demonstrates that the diffusion coefficient has in general the form of a symmetrical tensor **D** of second rank,

$$\mathbf{j} = -\mathbf{D}\nabla\mathbf{x}. \tag{13}$$

As for any symmetrical tensor of second rank, three highly symmetric directions may be used as coordinate axes so that the tensor gets diagonal form. From this diagonal form, it is easily seen that cubic symmetry is already sufficient to degenerate the tensor **D** to the simple scalar diffusion coefficient D. In consequence, when measuring diffusion in a cubic lattice structure, one should always expect perfect isotropic behavior.

5.2.4 Generic Solutions of the Fick Laws

Since mathematical solutions of the stated differential equations depend on boundary and starting conditions, an almost unmanageable number of different functions are offered in respective textbooks to describe atomic currents and the evolution of composition profiles (Crank, 1975). The equations for atomic transport are mathematically equivalent to those that describe heat transfer. Thus, further useful solutions are available from textbooks addressing heat conduction (Carslaw and Jaeger, 1959). A few solutions of the second Fick's law in its linear form Eqn (10) are so important that they must be discussed here. On the one hand, they form the basis for the experimental determination of diffusion coefficients and on the other hand they are generic to derive further more specialized solutions.

Cases in which boundary conditions have to be fulfilled far at infinity must be distinguished from those with boundary conditions to be fulfilled in finite geometries. The former provide a reasonable approximation, if composition profiles are investigated on the length scale of micrometers in a macroscopic sample of millimeter dimensions. For this case, major solutions are based on the bell-shaped Gaussian function which broadens in time but always vanishes far from the diffusion zone:

$$c(x,t) = \frac{m}{2\sqrt{\pi Dt}} \exp\left(-\frac{x^2}{4Dt}\right) \tag{14}$$

Straightforward analysis demonstrates that this function indeed solves Eqn (10). In the presented form, Eqn (14) is normalized so that, when integrating along the x-axis, the total amount of diffusing species amounts to m. Furthermore, at the very beginning ($t \to 0$), this function is a representation of Dirac's δ function, meaning that all diffusing species are concentrated first in an infinitely thin slab and then diffuse outward to both sides (see **Figure 4(a)**). Remarkably, the width of the concentration profile increases proportional to the square root of Dt which is a characteristic feature of any diffusion process.

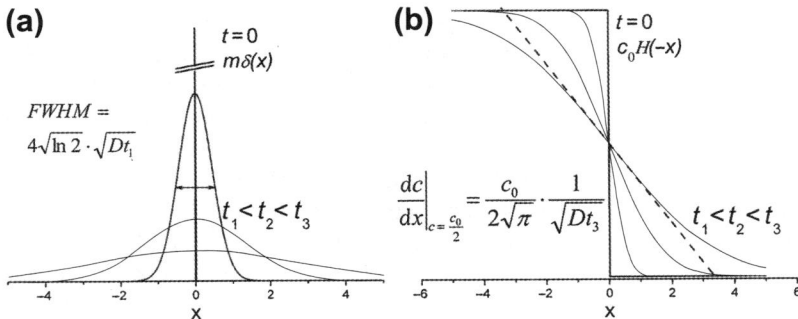

Figure 4 Evolution of the concentration profile starting (a) from an infinitesimally thin layer of the diffusor or (b) from a stepwise profile (A/B diffusion couple). The diffusion coefficient can be determined by measuring, for example the full width half maximum or the slope at the interface (dashed) at given time.

Since Fick's second law (Eqn (10)) is linear and Eqn (14) represents a δ function initially, the Gaussian solution can be used as a Green's function to construct solutions for any arbitrary initial composition profile.

The most important example is the *thick film solution* to be used for a diffusion couple that comprises two half spaces initially of different composition. Constructing the appropriate Heaviside-shaped starting profile from an infinite sum of δ functions

$$c(x) = c_0 \int_{-\infty}^{0} \delta(x - x') dx'$$

leads to a solution at later time of the form

$$c(x, t) = c_0 \int_{-\infty}^{0} \frac{\exp\left[-(x - x')^2/(4Dt) \right]}{2\sqrt{\pi Dt}} dx' \tag{15}$$

(see **Figure 4(b)**) which may be abbreviated by substituting $\eta := (x - x')/\sqrt{4Dt}$

$$= \frac{c_0}{\sqrt{\pi}} \int_{x/(2\sqrt{Dt})}^{\infty} \exp(-\eta^2) d\eta$$

Defining the so-called *Error Function*

$$\mathrm{erf}(x) := \frac{2}{\sqrt{\pi}} \int_{0}^{x} \exp(-\eta^2) d\eta, \tag{16}$$

Equation (15) is usually expressed as

$$c(x) = \frac{c_0}{2} \left[1 - \mathrm{erf}\left(\frac{x}{2\sqrt{Dt}} \right) \right], \tag{17}$$

which inherits from the Gaussian distribution the characteristic $\sqrt{Dt}$ dependence of the width of the mixed zone around the initial welding plane. Besides interdiffusion couples, the same solution is also useful in the description of diffusion into a "half space" sample (located on the positive x-axis) keeping the surface concentration (at $x = 0$) at the constant level $c_0/2$. The solution further plays a role in the quantitative description of short circuit transport along grain boundaries (see Section 5.6).

In contrast, solutions for finite geometries are frequently represented by a kind of Fourier series. For example, to describe out-diffusion (out gassing) from a thin slab bounded by surfaces at $x = 0$ and $x = l$ (thickness l) with an initial content c_0 while keeping the concentration c_s at the surfaces constant, the solution becomes

$$c(x, t) - c_s = (c_0 - c_s) \cdot \frac{4}{\pi} \sum_{n=0}^{\infty} \frac{(-1)^n}{2n+1} \exp\left(-\frac{(2n+1)^2 \pi^2}{l^2} Dt \right) \sin\left(\frac{(2n+1)\pi}{l} x \right). \tag{18}$$

For measurement of diffusion or permeation coefficients of a thin membrane, the solution for asymmetric concentrations on both sides of the slab becomes even more important (Ash et al., 1965):

$$c(x,t) - c_0 = c_1\left(1 - \frac{x}{l}\right) - c_1\frac{2}{\pi}\cdot\sum_{n=1}^{\infty}\frac{1}{n}\exp\left(-\frac{n^2\pi^2}{l^2}Dt\right)\cdot\sin\left(\frac{n\pi}{l}x\right) \tag{19}$$

which assumes that at zero time the concentration at one side ($x = 0$) suddenly jumps from c_0 to $c_0 + c_1$ while the concentration at the other side ($x = l$) is kept constant at c_0 (see **Figure 5**). After sufficient time, all exponential terms in the sum on the RHS vanish so that a linearly degrading composition profile remains in the steady state. Then, the steady permeation current through the membrane amounts obviously to

$$j_{\text{perm}} = -D\frac{\partial c}{\partial x} = c_1\cdot\frac{D}{l}. \tag{20}$$

Taking into account also the transition regime before reaching the steady state, the time integral of transported material Q reads

$$\frac{Q(t)}{l\cdot c_1} = \frac{Dt}{l^2} - \frac{1}{6} - \frac{2}{\pi^2}\sum_{1}^{\infty}\frac{(-1)^n}{n^2}\exp\left[-\frac{n^2\pi^2}{l^2}Dt\right] \xrightarrow[t\to\infty]{} \frac{D}{l^2}t - \frac{1}{6} \tag{21}$$

Again, all exponential terms vanish in the limit of long time. Thus, after a certain time lag, the permeated amount increases linearly with time as shown in **Figure 5**. The offset of the linear function from the origin represents the characteristic time lag which is determined by Eqn (21) to

$$t_{\text{lag}} = l^2/(6D). \tag{22}$$

Evaluation of permeation experiments by these equations is a common method to determine diffusion coefficients of fast diffusers (e.g. hydrogen in metals (Song and Pyun, 1990; Takano et al.,

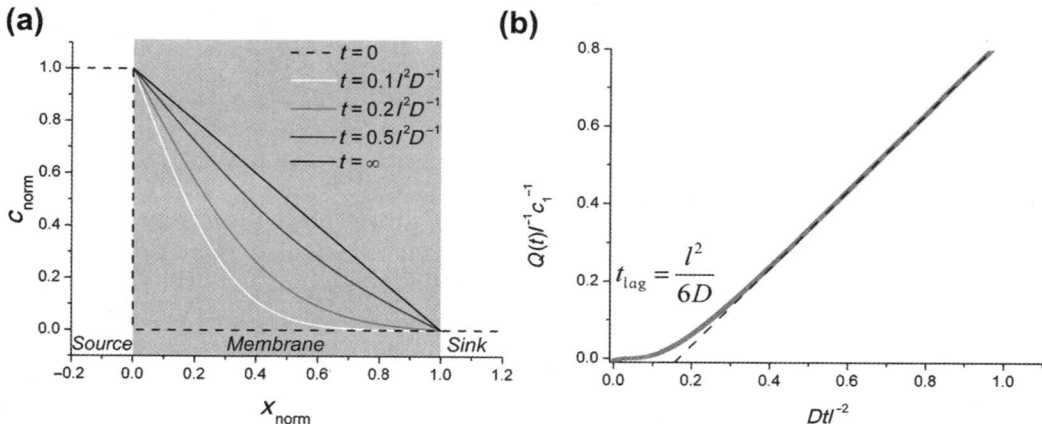

Figure 5 (a) Evolution of the concentration profile in a permeation experiment. (b) The total amount of transported material increases linearly with time but with a characteristic offset. The offset time lag contains the information on the diffusion coefficient, if the thickness of the membrane is known.

1995)). Similar procedures were also applied in the case of complex multilayer stacks (Schmitz et al, 1998). Remarkably by combining Eqns (20) and (22) both the coefficients of permeation ($\mu_P = c_1 \cdot D$) and of diffusion can be determined independently. Thus, besides the diffusion coefficient, also the concentration of mobile species within the membrane can be quantified.

5.2.5 The Statistical Interpretation of the Diffusion Coefficient

As it was already pointed out, diffusion is closely related to uncorrelated atomic jumps that obey general statistic principles. While already the early discovery of Brownian motion (Brown, 1828) has made obvious this statistical nature of the diffusion process and Fick's laws, Eqns (7) and (10) were formulated shortly later (Fick, 1855), it took surprisingly almost 50 further years before Einstein (1905) and Smoluchowski (1906) could provide independently the hidden link between the phenomenological diffusion coefficient and the mean square displacement of particles under chaotic statistical motion.

Consider an ensemble of particles being first concentrated in a sharply confined region at the origin and then starting wiggling around in uncoordinated manner. Naturally, after some time t the particles will be redistributed over a broader region and the width of this region increases with time. As particles move independently and nondeterministically, one can describe this process only by means of a time-dependent probability function $W(t,x)$. Let $W(t,x)dx$ represent the probability of finding a particle, initially located at the origin, in a small volume element dx located at position x at later time t. (As before, we restrict our considerations to the 1D case as generalization to three dimensions follows the same concepts.) By means of this probability function, we can obviously predict the temporal evolution of any particle distribution if it has been known at an earlier moment as

$$c(x, t+\tau) = \int_{x=-\infty}^{\infty} c(x-x', t)\, W(x', \tau)dx'. \tag{23}$$

Since this relation holds for arbitrary periods of evolution and concentration fields, we may consider the case of short evolution time τ and sufficiently smooth composition profiles. So we can expand with respect to time and diffusion length on the left-hand side and RHS of Eqn (14), respectively:

$$c(x,t) + \tau\frac{\partial c}{\partial t} + \ldots = \int_{x'=-\infty}^{\infty} \left[c(x,t) - x'\frac{\partial c}{\partial x} + \frac{1}{2}x'^2\frac{\partial^2 c}{\partial x^2} + \ldots \right] W(x', \tau)dx'. \tag{24}$$

Making use of the definition of averages or 'moments'

$$\langle x^n \rangle_\tau := \int_{x=-\infty}^{\infty} x^n \cdot W(x, \tau)dx,$$

suppressing higher order terms and comparing the remaining ones term by term, we approximate the latter equation by

$$\frac{\partial c}{\partial t} \approx \frac{\langle x^2 \rangle_\tau}{2\tau}\frac{\partial^2 c}{\partial x^2} - \frac{\langle x \rangle_\tau}{\tau}\frac{\partial c}{\partial x}, \tag{25}$$

which becomes exact in the limit of vanishing τ. Here and in the following, angle brackets denote a statistical averaging on all possible different jump sequences. Now we can compare with former Eqn (11) for the case of constant diffusion coefficient and driving force to find the fundamental equivalences (Einstein, 1905)

$$D = \lim_{\tau \to 0} \frac{\langle x^2 \rangle_\tau}{2\tau};$$

(26a)

$$\frac{DF}{k_B T} = \lim_{\tau \to 0} \frac{\langle x \rangle_\tau}{\tau}.$$

(26b)

Obviously, in the absence of a directed driving force the mean displacement of the particles Eqn (26b) is zero, while the mean *square* displacement is nevertheless finite and can represent the diffusion coefficient Eqn (26a). This relation represents the fundamental discovery of random walk theory. The analogous consideration is easily performed for the 3D case. Assuming isotropy (correct, e.g. for cubic symmetry) the mean square displacement reads

$$\langle \mathbf{R}^2 \rangle = \langle x^2 \rangle + \langle y^2 \rangle + \langle z^2 \rangle = 3\langle x^2 \rangle$$

so that the diffusion coefficient has to be identified by

$$D = \lim_{\tau \to 0} \frac{\langle \mathbf{R}^2 \rangle_\tau}{6\tau}$$

(27)

for isotropic diffusion in 3D space.

5.2.6 Random Walk on a Regular Lattice

The random walk of a diffusing particle is most easily discussed on a regular lattice in which each possible site is surrounded by z next neighbor sites to which the particle can jump as illustrated in **Figure 6**. Therefore, the possible jumps are characterized by microscopic vectors $\mathbf{r}_i$ which may point into z different directions as defined by the lattice coordination. Here, to keep our expressions simple, all of them shall have the same absolute length λ. The macroscopic displacement vector $\mathbf{R}$ joining the initial position of a particle and its final position after n individual jumps is therefore given as the vector sum

$$\mathbf{R} = \sum_{i=1}^{n} \mathbf{r}_i$$

and in consequence the mean square displacement must be calculated as

$$\langle \mathbf{R}^2 \rangle = \left\langle \sum_{i=1}^{n} \sum_{j=1}^{n} \mathbf{r}_i \mathbf{r}_j \right\rangle = \sum_{i=1}^{n} \langle \mathbf{r}_i^2 \rangle + 2 \sum_{i=2}^{n} \sum_{j<i} \langle \mathbf{r}_i \mathbf{r}_j \rangle.$$

(28)

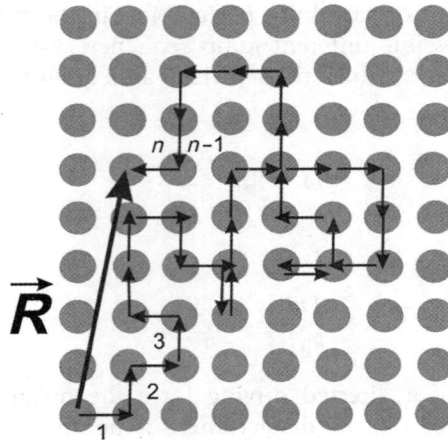

Figure 6 Random walk on a regular lattice. The elementary jumps (1...n) have the same length, but different (in this case four different) orientations. The net displacement after n jumps is marked by **R**.

As before, angle brackets symbolize the ensemble average on many different particle walks. At the RHS, we have conveniently split the double sum into its symmetric part with equal indices and its asymmetric part. Since in our consideration all microscopic jump vectors should have the same length λ, the first sum simply equals n times the square of this length. If different jumps happen to be indeed uncorrelated, as assumed for true random walk, the second part vanishes by averaging and so we find by virtue of Eqns (27) and (28) the convenient expression

$$D = \frac{n\lambda^2}{6n\,\bar{\tau}} = \frac{z\omega\lambda^2}{6} \tag{29}$$

that links the phenomenological diffusion coefficient to the microscopy of the lattice. Here $\bar{\tau}$ and z denote the average time between two successive jumps and the number of next neighbors (coordination number), respectively[2]. By taking into account individual jump distances into different lattice directions, generalization of the presented consideration to noncubic structures is easily possible.

Often, the assumption of uncorrelated migration steps is only a weak approximation, since memory effects between consecutive jumps appear in defect-mediated diffusion mechanisms. In such case, the asymmetric part of the sum in Eqn (28) does not vanish anymore. Usually, a correlation factor f is defined as

$$f := 1 + \lim_{n \to \infty} \frac{2\sum_{i=2}^{n}\sum_{j<i}\langle \mathbf{r}_i \mathbf{r}_j \rangle}{n\lambda^2}, \tag{30}$$

which quantifies the ratio between the real diffusivity and the value expected for truly random walk:

$$D = f \cdot \frac{z\omega\lambda^2}{6}. \tag{31}$$

---

[2] ω denotes the jump probability to a selected neighbor. So total jump probability is $\Gamma = \frac{\lambda}{\tau} = z.\omega$

In practice, an evaluation based on Eqn (31) can never include an infinite number of atomic jumps. When restricting the statistical set, sufficient convergence must be warranted so that at least reasonable averages are obtained. This needs particular care in the case of pronounced spatial or temporal heterogeneities. Under extreme assumptions (e.g. diffusion with orders of magnitude difference between jump lengths or residence times between successive jumps), the transport equation may even take a different form and the proportionality $\langle x^2 \rangle \sim t$ does not hold anymore (Philibert, 2009). Such heterogeneities may play an important role on the nanoscale which is discussed in Section 5.7.

5.2.7 Arrhenius Representation and Activation Parameters

When deriving a quantitative expression for diffusion by the vacancy mechanism, we have to take into account that a jump is only possible, if a vacancy appears in the neighborhood. Thus the generic jump rate ω in Eqn (31) must be multiplied by the probability of finding a vacancy on the next site in jump direction, which is approximately equal to the local vacancy fraction $v_V^{(eq)}$. Obeying furthermore Eqn (7), we find

$$D^* = \frac{\lambda^2}{6} \cdot \Gamma_0 \cdot z \cdot v_V^{eq} \cdot f \, \exp\left[-H^M/(k_B T) \right] \tag{32}$$

(Here and in the remaining part of this chapter, we designate compositions in terms of atomic fractions by the letter "v", while concentrations per volume are designated by the letter "c".) This kind of diffusion coefficient is called the tracer diffusion coefficient, which is often indicated by a star in the superscript. The equilibrium fraction of vacancies itself is known from basic thermodynamics so that a full description of the temperature dependence of the diffusion coefficient with regard to Eqn (1b) reads

$$D^* = \frac{\lambda^2}{6} \tilde{\Gamma}_0 zf \, \exp\left(-\frac{H^M + H^F}{k_B T} \right) \exp\left(\frac{S^F + S^M}{k_B} \right) = D_0 \, \exp\left(-\frac{\Delta H}{k_B T} \right) \tag{33}$$

The sum of the vacancy formation entropy S^F and the migration entropy S^M can be called as "diffusion entropy". The quantity ΔH is called the activation enthalpy of diffusion which may be split into the activation enthalpy of vacancy formation and their migration enthalpy.

Usually diffusion data are presented in the form of Arrhenius diagrams, meaning a plot of the logarithm of the diffusivity versus the inverse of temperature. For any simple diffusion mechanism as stated in Eqn (33), straight lines of negative slope are expected (see the many examples in Section 5.5).

The activation enthalpy can also be separated into two distinct parts by $\Delta H = \Delta E + p\Delta V$, where ΔE is the activation energy and ΔV is called the activation volume of diffusion. The activation volume can be measured by studying the pressure dependence of the diffusion coefficient. It can be also split into a formation and a migration part. From a physical point of view, the former describes the additional volume that is produced by formation of a vacancy, while the latter summarizes all the volume expansion which appears when hopping across the saddle position. Often the actual value of the activation volume allows decisive clues on the diffusion mechanism. For example, interstitial atoms require very little excess volume to be produced, while vacancies and divacancies are characterized by large excess volumes.

5.2.8 Correlation Factors

How can correlation as admitted by Eqn (30) appear in a statistical sequence of random individual migration steps? Indeed, considering the walk of a vacancy in a crystalline lattice, it is hard to imagine any kind of memory that 'reminds' the vacancy to the previously performed jumps before deciding for the next. But the situation becomes markedly different when focusing on the migration of the atoms. As will be shown below, jump statistics of the atoms needs detailed considerations already for a pure metal. Diffusion of a dilute impurity or even the case of concentrated alloys becomes even more complex and interesting.

5.2.8.1   Correlation in Self-Diffusion of Pure Metals

In a kind of thought experiment, we may mark a certain atom and follow its migration path. In experiment, such marking can be approximated by using a different isotope in very dilute concentration.[3] Obviously, the marked atom has first to wait for a vacancy to appear in the next neighbor shell before a site exchange becomes possible. Directly after the atom has performed a jump, the vacancy still remains in next-neighbor relationship. Thus after this initial jump, there is a significantly increased probability that in the next step atom and vacancy just back change their sites. Let us estimate more quantitatively. The probability that the back jump appears amounts to $1/z$, if z is the coordination number of the considered lattice. (Any of the z neighbors of the vacancy, among them the marked atom, may jump into the vacant site with equal probability.) If the back jump of the marked atom happens, two migration steps cancel and have in consequence no measureable effect on the diffusion process. Thus, in first approximation we expect for the correlation factor

$$f = 1 - 2 \cdot \frac{1}{z}. \tag{34}$$

Obviously, this formula cannot be exact, since more complex return paths of the vacancy that may comprise a sequence of several intermediate jumps are neglected. It allows nevertheless deducing general features of the correlation factor. Noteworthy, correlation factors of different lattice structures must reflect a tendency of decreasing with the coordination number of the host lattice. Lower coordination just increases the probability that the vacancy returns before the marked atom has 'left its memory' by exchanging with another independent vacancy. So, we expect for the example of an fcc lattice ($z = 12$) a correlation factor of 0.8333, while for a hypothetical 1D chain of atoms ($z = 2$) the correlation factor of Eqn (32) should even vanish. For this latter case long-ranged diffusive transport of atoms by the vacancy mechanism is impossible. On the other extreme, a dilute interstitial usually finds all its neighbored interstices empty and thus has always a free choice for the direction of the next jump. In consequence, no correlation appears for interstitial diffusion ($f = 1$) in any arbitrary lattice structure. Only when the interstitial sites become densely occupied, as it may happen for example for hydride phases in metals, the situation becomes quite similar to a vacancy mechanism and significant correlation effects are noted.

The analytical evaluation of Eqn (30) is complex as beside the direct follow-ups also all nonconsecutive pairs of jumps need to be considered. For a vacancy mechanism in cubic lattices it has been shown that Eqn (30) can be reduced to the correlation between consecutive pairs (see, e.g. (Mehrer, 2006a)):

$$f = \frac{1 + \langle \cos \vartheta \rangle}{1 - \langle \cos \vartheta \rangle}. \tag{35}$$

Here, ϑ denotes the angle between the directions of two consecutive jumps.

---

[3] Strictly speaking, such isotope markers represent only an approximation to ideal marking. In chemical sense isotopes certainly behave as the majority atoms, but different atomic mass may still have significant influence on the jumping rates (so-called isotope effect).

In literature, a variety of theoretical methods were suggested to calculate the correlation factor exactly. We cannot present or discuss these different concepts here, but only refer to some original articles (Barnes, 1952; Compaan and Haven, 1956; Le Claire and Lidiard, 1956; Mullen, 1961; Bakker, 1970; Benoist et al., 1977; Koiwa, 1977). The strict mathematical concepts as well as originality and accuracy in thinking demonstrated in these works must be acknowledged. However, with ongoing progress in computer simulation, the 'brute force' method by following the jump sequence in MC simulation and evaluating directly by means of Eqn (30) (Bruin and Murch, 1973; Murch and Thorn, 1977) becomes more and more promising and made alternative analytical methods almost obsolete (Belova and Murch, 2005). After defining appropriate jump probabilities, the MC method delivers direct results with high accuracy and reliability. Any possibly insufficient approximation in statistical averaging is avoided. So, it is not surprising that in prominent cases, MC simulation even discovered some weakness and inaccuracy in previous analytical concepts (Belova and Murch, 2001).

A summary of quantitative correlation factors for different lattice structures and diffusion mechanisms is given in **Table 1**. All values are calculated for atomic transport in pure materials (self-diffusion). The table provides at least an orientation on typical values. To our satisfaction, we find the rough estimate that the correlation factor should decrease with lattice coordination nicely confirmed. On the one hand, the quantitative data demonstrate that with the exception of the 'pathologic' case of the one-dimensional chain, the correlation factor will hardly slow down the diffusion coefficient by more than a factor 2. Thus, regarding the limited accuracy of diffusion measurements—often an order of magnitude accuracy means already a success—the influence of the correlation factor may simply be neglected in practical situations.

On the other hand, comparing the tabulated values for different jump mechanisms in the same lattice structure, it is seen that the correlation factor depends significantly on the defect that mediates migration. Thus, accurate measurements of the correlation factor may yield decisive evidence for identification of a particular migration mechanism; for example, divacancies that become potentially important near to the melting point should reflect in a significant decrease of the correlation factor by about 40% (from 0.7815 down to 0.4579). Typically, such investigations are corroborated by a study of the isotope effect to the diffusion rate. For example, Rothman et al. concluded from measurements of the isotope effect in self-diffusion of Ag that the curvature in the Arrhenius plot (see **Figure 42**) is most

Table 1 Correlation factors of self-diffusion in simple lattice structures. Values taken from (Belova and Murch, 2002; Montet, 1973; Murch, 1984; Compaan and Haven, 1956; Compann, 1958; Benoist et al., 1977; Mehrer, 2006a)

Lattice	Mechanism	Correlation factor f
1D chain	Vacancy	0.0
2D square	Vacancy	0.467
2D hexagonal	Vacancy	0.56006
3D simple cubic	Vacancy	0.6531
3D cubic bcc	Vacancy	0.7215
3D cubic fcc	Vacancy	0.7815
3D cubic diamond	Vacancy	0.5
3D cubic fcc	Divacancy	0.4579
3D cubic bcc	Divacancy	0.335–0.469
3D cubic fcc	Dumbbell interstitial ⟨001⟩	0.4395
3D cubic diamond	Collinear interstitialcy	0.727
3D any lattice	Interstitial	1.0

probably arising from an increased divacancy contribution (Rothman et al., 1970). Similarly, Mundy has shown by the pressure and temperature dependence of the isotope effect that the self-diffusion of Na requires a monovacancy and two different divacancy-type mechanisms (Mundy, 1971).

In contrast to pure metals, alloys require particular care with respect to correlation. If the binding between the vacancy and one component of the alloy gets particularly strong, quantitatively relevant correlation effects may appear that result in asymmetric diffusion rates between the alloy components. Also superlattice structures of intermetallic compounds can give rise to strong correlation effects: Imagine that a component A has a strong energetic preference for one sublattice, but all possible next-nearest neighbor jumps will transfer the A atom from this preferred sublattice to sites on another one. The energetic driving force for pushing back the atom to its initial site on the correct sublattice will become so strong that the direct back jump becomes the most probable alternative depressing the correlation factor close to zero.

5.2.8.2 Correlation in Impurity Diffusion

In very dilute alloys the impurity atoms (solutes) become insulated within the matrix of the solvent. In this case, diffusion of the impurities within the host can be studied without the disturbing influence of alloy thermodynamics (see also Section 5.3.1). Nevertheless, correlation effects are more complex than in the case of self-diffusion and can become more significant. Let us first consider the case of a densely packed fcc crystal.

Since we are treating a system in the dilute limit, we can neglect solute–solute interactions. But In comparison with self-diffusion, we have to take into account three important differences:

(1) According to the Lomer equation (Lomer, 1958), the probability of a vacancy occupying a nearest-neighbor site to a substitutional impurity is

$$p = z v_V^{eq} \exp\left(\frac{G^B}{k_B T}\right) = z \cdot \exp\left(\frac{S_{1V}^F - S^B}{k_B}\right) \exp\left(-\frac{H_{1V}^F - H^B}{k_B T}\right) \tag{36}$$

where G^B, H^B and S^B are the Gibbs enthalpy, the enthalpy and entropy of the solute–vacancy binding, v_V^{eq} is the equilibrium concentration, H_V^F and S_V^F are the enthalpy and entropy of formation of monovacancies in the pure matrix, respectively, and z is the coordination number. If G^B is positive, the solute–vacancy interaction is attractive, when negative, it is repulsive.

(2) Since the symmetry of the lattice is broken by random arrangement of the solutes, vacancy jumps are no longer equivalent. Instead of a unique vacancy jump frequency, one needs to distinguish five different frequencies (ω_0 to ω_4) for a complete characterization of the system when interaction is restricted to next-nearest neighbor distances (Lidiard, 1955, 1960). **Figure 7** illustrates the possible vacancy jumps for which different frequencies have to be distinguished.

(3) The impurity correlation factor is no longer a function of the lattice geometry alone. It also depends on the ratio between the different jump rates.

For the diffusion coefficient of the solute, we would like to use the same formalism as for the tracer diffusion but with a correlation factor f_2:

$$D_2^* = \frac{\lambda^2}{6} z f_2 \omega_2 v_V^{eq} \exp\left(\frac{G^B}{k_B T}\right) \tag{37}$$

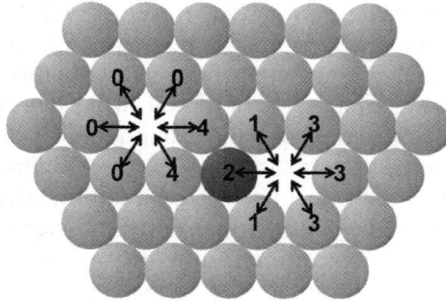

Figure 7 The five different jump possibilities for the vacancy in a next-neighbor dilute alloy model: (1) "rotation" around the solute atom, (2) exchange with the solute atom, (3) dissociation from the solute atom, (4) association with the solute atom, (0) exchange with a solvent atom without any interference from solute atoms. (For color version of this figure, the reader is referred to the online version of this book.)

in which z, λ, ω_2, v_V^{eq} denote the coordination number, jump distance, vacancy–solute exchange rate, and equilibrium fraction of vacant lattice sites, respectively. Manning calculated the correlation factor in the framework of the five frequency model (Manning, 1968)

$$f_2 = \frac{\omega_1 + \frac{7}{2}H_3\omega_3}{\omega_2 + \omega_1 + \frac{7}{2}H_3\omega_3}, \tag{38}$$

where H_3 represents the escape probability ($H_3 = \omega_3/\omega_0$). In thermal equilibrium, dissociation and association of the vacancy–solute complex are related to each other by the detailed balance

$$\frac{\omega_3}{\omega_4} = \exp\left(\frac{G^B}{k_B T}\right) \tag{39}$$

Comparing Eqn (37) with the analogous formula for self-diffusion (Eqn (33)) and using Eqn (39) one can identify three reasons why the impurity diffusion coefficient can differ from that of self-diffusion:

$$\frac{D_2^*}{D^*} = \frac{f_2}{f}\frac{\omega_2}{\omega_0}\exp\frac{G^B}{k_B T}. \tag{40}$$

The solute–vacancy binding may be non-zero, the rate of solute–vacancy exchange is different from that of solvent–vacancy exchange, or the correlation factors differ.

As a consequence of Eqn (38), self- and solute diffusion reveal different activation enthalpy. This difference obviously amounts to

$$\Delta Q = H_2^M - H^M - H^B + \Delta Q_f, \tag{41}$$

in which H_2^M, H^M, and H^B are the enthalpies of the vacancy–solute exchange, vacancy–solvent exchange, and vacancy–solute binding, respectively. ΔQ_f summarizes the activation enthalpy of the explicit temperature dependence of the correlation factor of solute diffusion. To estimate ΔQ_f, one can use an electrostatic model suggested by Lazarus (1954) and lCla (1962). In this model ΔQ_f is caused by the excess charge from the valence difference between the impurity and the matrix atoms (ΔZe). Since the vacancy itself is charged, electrostatic interaction appears between vacancy and the impurity. In metals free electrons screen these excess charges quite efficiently. Consequently, the H^B vacancy–solute binding is much smaller than the formation enthalpy H_V^F of the vacancies. This is the fundamental reason, why impurity diffusion coefficients fall into a narrow band around the self-diffusion coefficient in the case of normal impurity diffusion as presented in Section 5.5 (see **Figure 46**). A more comprehensive description of the effects related to the solute–vacancy binding will be given in the next section describing diffusion in concentrated alloys under thermodynamic driving forces (see Section 5.3).

The bcc crystal structure shows two prominent differences in comparison with the close-packed structures. (1) First neighbors of an atom are never first neighbors of each other. Accordingly, vacancy "rotation" around a given atom is not possible in a bcc lattice (see **Figure 8**). (2) Atoms in the second coordination shell are not much farther than atoms in the first shell; thus a next-nearest neighbor (1NN) approximation is likely an oversimplification. The specific bcc-related vacancy jumps are illustrated in **Figure 8**.

If the vacancy is in a nearest-neighbor position of the impurity atom, it can

- exchange with the impurity atom (ω_2),
- jump to a second next neighbor place of the impurity atom (ω_3),
- jump to a place in the third coordination sphere of the impurity atom ($\omega_{3'}$),
- jump to a place in the fifth coordination sphere of the impurity atom ($\omega_{3''}$).

If it is in the second nearest position of the impurity atom, it can

- jump to a nearest-neighbor position (ω_4),
- jump away from the proximity of the solute (ω_5).

In addition, frequencies ω_4, $\omega_{4'}$, $\omega_{4''}$ and ω_6 describe the rates of the reverse jumps of ω_3, $\omega_{3'}$, $\omega_{3''}$ and ω_5, respectively. Vacancy jumps far away from any solute atom may be characterized by the jump rate ω_0 for the pure metal.

Denoting the vacancy–solute binding Gibbs energies by G^{B_1} and G^{B_2} for the first and second nearest neighbors, respectively, we can formulate the equations of detailed balancing (similarly to Eqn (39)):

$$\frac{\omega_{3'}}{\omega_{4'}} = \frac{\omega_{3''}}{\omega_{4''}} = \exp\left(-\frac{G^{B_1}}{k_B T}\right) \tag{42a}$$

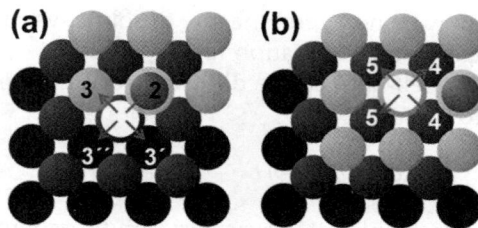

Figure 8 A bcc crystal with an impurity atom (dark orange, with lighter margins) and a vacancy (indicated by a white filled circle) in the first (a) and second (b) coordination shell. z-coordinates are represented by color coding (bottom: dark, middle: blue, topmost atoms: light). (For color version of this figure, the reader is referred to the online version of this book.)

$$\frac{\omega_5}{\omega_6} = \exp\left(-\frac{G^{B_2}}{k_B T}\right) \tag{42b}$$

Obviously realizing the escape by jumping first to the second neighbor shell (jump 3 in **Figure 8(a)**) and then escape (jump 5 in **Figure 8(b)**) or escape in one jump (jump 3' or jump 3" in **Figure 8(a)**) should have no effect on the detailed balancing, thus

$$\frac{\omega_5 \omega_3}{\omega_6 \omega_4} = \frac{\omega_{3'}}{\omega_{4'}} \tag{42c}$$

No analytical solution of the complete set of frequencies exists today. There are two alternatives to reduce the number of unknowns.

In the first model, $\omega_{4'} = \omega_{4''} = \omega_6 = \omega_0$, so jumps from higher coordination shells are unaffected by the presence of the impurity atom. From this model $\omega_3/\omega_{3'}$ and ω_3/ω_2 can be extracted. The second model only treats the nearest-neighbor interactions. Accordingly, $\omega_4 = \omega_{4'} = \omega_{4''}$ and $\omega_5 = \omega_6 = \omega_0$. The ω_3/ω_2 and ω_4/ω_0 ratios can be gained from this model. A theoretical treatment of the diffusion problem, within the framework of these models, has been carried out by Serruys and Brebec (1982).

5.2.9 Atomistic Simulation by MC Methods

With progress of available computing power, theoretical studies of diffusion by direct simulation of the random walk have become more and more important. Therefore, in the remaining part of this section we discuss major techniques of numerical simulation. The (kinetic) MC (KMC) method is a so-called state-to-state technique, that is only the equilibrium positions and the time spent in a given position are recorded. Second, the more realistic calculation of the microscopic atomic trajectories by MD will be reflected. The price for this more detailed information is however an increased complexity of the algorithms and hence higher demand for computing time. Finally, we shortly mention quantum mechanical based techniques (density functional theory) which can be used for a further improvement of the accuracy of the simulations, and will certainly become still more important in the next future.

In the MC method, physical problems are studied by the use of random numbers. The Teller–Metropolis method, probably the most famous MC algorithm appeared in 1953 (Metropolis et al., 1953). The usual goal of these simulations is to find the equilibrium state of a statistical ensemble and to investigate its properties.

Since random walk clearly obeys the laws of probability (see Section 5.2.6), MC methods can be used to model diffusion behavior too. The first attempts to describe a dynamic system by MC simulations were related to the field of radiation damage (e.g. (Beeler, 1966)). In these studies, the probability of creating a site exchange in a knock-on event was rather high owing to high particle energies. In normal diffusion, however, the ratio of successful jumps to attempted ones is almost negligible. For example, even for a very low jump barrier of 0.5 eV at room temperature only about 1 out of 10^7 vibration events would lead to a successful atomic exchange (Henkelman and Jónsson, 2001). Accordingly, the effectiveness of a naive MC calculation would be very low.

A solution of this problem was given by Bortz et al. by the "n-fold" mechanism or as currently termed by the KMC (Bortz et al., 1975). Every attempt leads to a successful event, which tremendously reduces the computing time for processes with low probability. The real dynamics of the system is computed by a second MC cycle which measures the time probably spent in the given configuration. This second element of the KMC method is called residence time algorithm. Let us focus to the vacancy mechanism although other diffusion mechanisms can be treated with the MC method too, see, for example (Gendt et al., 2001).

Figure 9 shows a two-dimensional example to illustrate the procedure. In a rectangular grid there is a single vacant location, with atoms 1–4 as nearest neighbors. The respective atom–vacancy exchange rates are ω_i ($i = 1...4$). In a first step one selects the exchange to be realized. The probability for each exchange is given by the simple formula:

$$P_i = \frac{\omega_i}{\sum_i \omega_i} \tag{43}$$

Accordingly, a first random number generation (R_1) chooses between the possible jumps, with respect to their probabilities (see **Figure 9**). Then in a second step, the residence time (τ) is calculated by a second random number (R_2) with respect to the total jump rate:

$$\tau = \frac{1}{\sum_i \omega_i} \ln(R_2) \tag{44}$$

In this way one can implement the vacancy jumps by very simple and fast code. Thus, the diffusion procedure can be investigated for a large cell size (above one million atoms) and long timescales (trillions of elementary jumps). The time range covered by KMC simulations is around six orders of magnitude larger than a MD simulation using similar computation time (Nato Science Series, 2007). Consequently processes like mixture/dissolution (Erdélyi et al., 2008), nucleation (Clouet et al., 2006; Erdélyi et al., 2010; Mao et al., 2012), ripening (Roussel and Bellon, 2006a), void formation (Gusak et al., 2005), surface diffusion (Pierre-Louis et al., 2009, Konwar et al., 2011), and so on, can be comfortably studied by MC simulation.

To get physically meaningful results, an accurate rate catalog (the set of ω_i) is required. This not only means that the jump rates ω_i are quantified correctly but also means that the set is complete, for example that in **Figure 9** second nearest-neighbor jumps (marked by *) do not take place. For example,

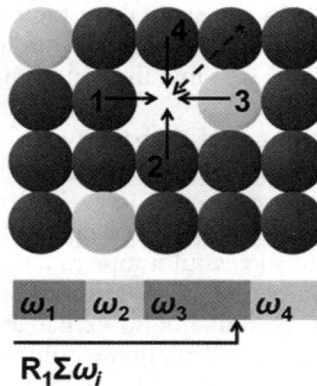

Figure 9 Selection of the jump in a KMC procedure. The ω_i jump rates for all possible escape paths (vacancy–atom exchanges 1–4) are summed up as a number line and a random number ($R_1 \in [0,1]$) is generated. Using this random number as a pointer, one of the jumps is selected (here jump direction #3). In this way, the selection is weighted by the probabilities of the potential jumps.

it is known that, while in most cases surface diffusion happens via adatom hopping (Kellogg et al., 1978), for W (Wrigley and Ehrlich, 1980) or Pt (Kellogg and Feibelman, 1990), the preferred mechanism is an adatom–substrate exchange. Thus, one can arrive at completely false conclusions if such peculiar paths are neglected. On the other hand, with a correct rate catalog the state to state accuracy of the KMC is well comparable with MD (Voter, 1986).

As a consequence, standard KMC is the most useful when the elementary jumps are well known and one is interested in the long(er) term evolution of the system. In terms of exactness and time window, KMC simulations are in between MD and phenomenological modeling. Since atom probe tomography (APT) delivers single atom analysis of a few million of atoms, this experimental method is especially well suited to be combined with MC simulations (Setna et al., 1994; Pareige et al., 1999; Mao et al., 2012).

5.2.9.1 *Obtaining the Jump Rate Catalog*
When the possible escape paths from a given state are known, one needs to calculate the respective escape rates. One of the most popular methods to determine the rate constants is the transition state theory (Vineyard, 1957) as was considered in Section 5.2.2.

If all the interatomic potentials are known, jump rates for each configuration can be computed. Usually further approximations are applied:

- The total energy of the system is calculated as the sum of pair interactions. Furthermore, only a fraction of those are affected by the jump, usually only the nearest neighbors. The vibration spectrum of the solid and the lattice positions remain unchanged during the jump event. The saddle point energy is often chosen to be independent of the configuration (Stolwijk, 1981; Martin, 1990), although in detailed cases, variation in the saddle point energy is necessary to correctly describe the diffusion kinetics (Pareige et al., 1999; Le Bouar and Soisson, 2002).
- The energy of each interaction is composed of pair and many-body interactions (Clouet et al., 2004; Mao et al., 2012). Higher order interactions are necessary if one would like to reproduce complex phase diagrams. These simulations still neglect the stress/strain-related effects.

In thermodynamic equilibrium, the total frequency of jumps from state i to state j must be equal to that of the reverse jumps (principle of detailed balance). Therefore a relation between the jump rates ($\omega_{i \rightarrow j}$) and the equilibrium population (χ_i) of states exists:

$$\omega_{i \rightarrow j}\chi_i = \omega_{j \rightarrow i}\chi_j \tag{45}$$

Since jump rates between two given states are invariable, their values found in equilibrium are applicable to the nonequilibrium state, too.

5.2.9.2 *Lattice-Free MC, Search for Saddle Points*
In lattice-free KMC methods, the possible transition pathways are not previously known. Assuming that the enthalpy landscape is established by ab initio calculations or empirical potentials, the task is to search for a minimum energy path, that is the saddle between the minima.

There are multiple algorithms suggested for this goal, a very successful implementation is called the nudged elastic band method (NEBM) (Mills and Jónsson, 1994; Mills et al., 1995). In this method the initial and final states (R_0 and R_N) are fixed and in between $N - 1$ intermediate states called replicas are added ($N \approx 5\text{--}20$) (see **Figure 10**). These replicas are connected by spring interactions to assure the continuity of the path. Accordingly, the resulting path can be imagined as an elastic band spread over

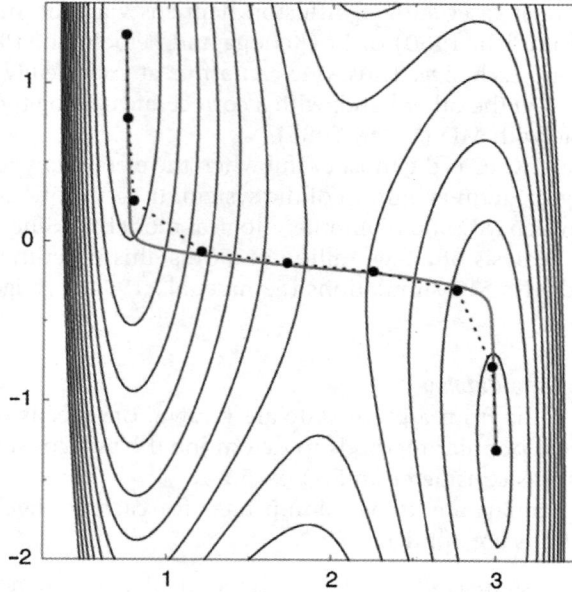

Figure 10 Finding the minimum energy path by the nudged band method. In this example 7 replicas are placed in between the fixed minima. By reducing the tangential components of the "true force" arriving from the potential landscape the position of the replicas converges to the minimum energy path (marked by a solid gray line). Reproduced with permission from Henkelman and Jónsson (2000).

the energy landscape. An optimization of the band, by minimizing the forces acting on the replicas will bring the band to the minimum energy path which is found by minimizing the action

$$
S(R_1, \ldots R_{N-1}) = \sum_{i=0}^{N} V(\vec{R_i}) + \sum_{i=1}^{N} \frac{Nk_i}{2} (\vec{R_i} - \vec{R_{i-1}})^2 \tag{46}
$$

where V is the potential from the enthalpy landscape, R_i are the position of the replicas and k_i are the spring constants. In the nudged band method, the minimization of the path is carried out in a way that components of the "true force" (i.e. of the gradient of V) which are parallel to path and components of the spring force which are perpendicular to the path are neglected. In this way, it can be assured that in the final configuration the band will fall to the minimum energy path (for which the forces from the potential perpendicular to the band are zero in each position). A refined version of the model has been presented by Henkelman and Jónsson (2000).

The NEBM requires the knowledge of the positions of the enthalpy minima. An alternative method, called the dimer method (Henkelman and Jónsson, 1999), can be used to search for the saddle point without this information. The method is based on the fact that the saddle point is characterized by a maximum along the lowest curvature and a minimum in the orthogonal direction (in all other orthogonal dimensions for an N-dimension space). In the dimer method two replicas with a very small separation are created and placed as a dimer first close to the assumed minimum. Then the dimer is moved slightly up-hill and rotated to find the lowest local curvature mode of the potential landscape. This process is repeated iteratively until the saddle point is found. This method has been further developed and found many applications (Henkelman and Jónsson, 2001; Xu and Henkelman, 2008). In so-called on the fly MC methods, the saddle energies are always recalculated, for example by the dimer method before a next jump is performed as shown in **Figure 11**.

Alternative 'on the fly' algorithms also exist. Some examples can be found, for example in Refs (Bocquet, 2002; El-Mellouhi et al., 2008; Konwar et al., 2011). With known saddle points, the KMC simulation can proceed either by solving Eqns (1) and (2) or by suitable model approximations (e.g. the jump rates only depend on the saddle height). These methods, though much slower than lattice KMC, still represent a significant advantage in computing time over MD calculations.

5.2.9.3 *Vacancy Concentration, Generation and Annihilation of Vacancies*

In a typical lattice KMC simulation only a single vacancy is used as diffusion vehicle. This is equivalent to a vacancy concentration of $\nu_V = 1/N$ where N is the system size. For a relatively large system, containing one million atoms, ν_V is 10^{-6}. This is still a rather high vacancy concentration, which can be only realistic at temperatures above half of the melting point (Siegel, 1978). Consequently, the time measured in the MC simulation is accelerated by a factor of $(\nu_V^{eq}(T)N)^{-1}$, where $\nu_V^{eq}(T)$ is the equilibrium vacancy fraction at temperature T.

This simple scaling holds only if the equilibrium vacancy concentration does not vary significantly in the process of interest. Solid-state reactions, phase separation or ordering transitions for example can create a product phase with different vacancy formation energies and different equilibrium vacancy

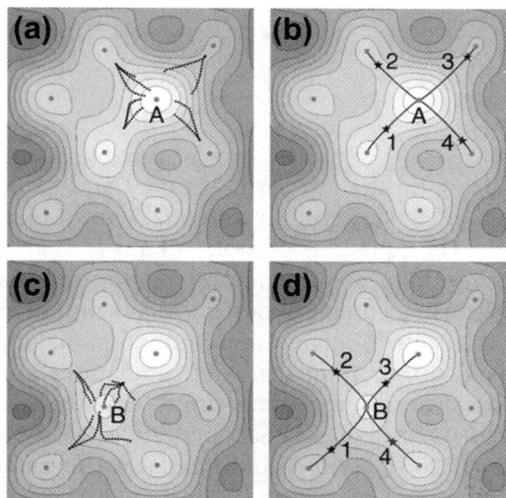

Figure 11 'On the fly' KMC simulation using the dimer search method. In the first step ten dimers are distributed randomly in the proximity of the minimum (A) to find the saddle points. (a) After finding the saddle point the system is allowed to slide down to the minimum along the gradient of the potential landscape. (b) Knowing the possible final configurations and the jump barriers, a KMC cycle selects the new position (B). The process is repeated again (c–d). Reproduced with permission from Henkelman and Jónsson (2001).

concentrations. Accordingly, as the system evolves, the proportionality factor between the simulated and the real time varies. This effect, if left untreated, can falsify the kinetics gained from the simulation. As the simplest way to compensate for this, the equilibrium vacancy concentration is calculated at regular intervals and the simulated time is calibrated with the determined vacancy supersaturation (Le Bouar and Soisson, 2002).

As an alternative possibility, vacancy sources/sinks are introduced which ensure that the real and the simulated vacancy concentrations are always proportional (Gendt et al., 2001; Hin et al., 2005). This can be achieved by introducing certain sites which have a chance to randomly generate a vacancy with the probability

$$P = \Gamma \exp\left(\frac{-H_V^F}{k_B T}\right) \tag{47}$$

in which Γ is the attempt frequency in the solvent and H_V^F is the vacancy formation enthalpy. The source also acts as a sink, that is if a vacancy returns to this position, it is annihilated so that a dynamical equilibrium for the vacancy concentration is established. This also means that the number of vacancies in the system is not fixed anymore and can fluctuate with time.

5.2.10 Atomic Simulation by MD

In MD the trajectories of interacting atomic/molecular species are calculated by a simultaneous (numerical) solution of the Newtonian equations of motion:

$$\frac{d^2 \vec{r_i}(t)}{dt^2} = m_i^{-1} \vec{F_i} \tag{48}$$

$$\vec{F_i} = \frac{\partial V\left(\vec{r_1}, \ldots, \vec{r_N}\right)}{\partial \vec{r_i}} \tag{49}$$

where F_i represents the force acting on the i-th atom at time t and r_i are the atomic coordinates. To calculate the force acting on the different atoms, the potential V must be a differentiable function with respect to all r_i atomic positions. The equations of motion Eqns (48) and (49) are integrated in small time steps Δt (typically from a few fs to ps). MD allows investigating the evolution of some tens of thousands atoms up to ms range using contemporary personal computers.

The main advantage of the MD is the full trajectory information making ad hoc assumptions on the diffusion jumps obsolete. It is also important that MD simulations are done in continuous space. Consequently, in principle, any structure, including simple liquids (Alder and Wainwright, 1959), amorphous materials (Teichler, 1997; Kluge and Schober, 2004), two-dimensional defects like grain boundaries (Suzuki and Mishin, 2005), solid–liquid interfaces (Rozas and Horbach, 2011) or one-dimensional defects such as dislocation (Fang and Wang, 2000), can also be treated without any problem.

One of the earliest applications of MD was made by Alder and Wainwright (1957) and Alder and Wainwright (1959) to describe the motion of rigid 2D disks in a liquid. Currently, MD is one of the most widely used technique to simulate various diffusion and chemical reaction-related processes.

5.2.10.1 Temperature and Kinetic Energy in MD

Since the equations of motion are not limited to equilibrium, MD is a very versatile tool to study atomic transport. To control the temperature of the system different options are available:

- The total energy of the system is conserved, that is the simulation is based on a microcanonical ensemble. If irreversible processes take place in the system, this condition will lead to a change of the kinetic energy/temperature.
- The velocities of the system are rescaled within some MD steps (Woodcock, 1971; Haile and Gupta, 1983) by a factor T_{act}/T_{set}. Where T_{set} is the desired system temperature and T_{act} is calculated from the actual velocities using the principle of equipartition

$$\sum_i^N \frac{1}{2} m_i v_i^2 = \frac{1}{2} N_{Df} k_B T_{act} \tag{50}$$

in which N_{Df} denotes the number of degrees of freedom of the system. As a main problem of this solution, the T_{set} temperature plays no role in the equation of motions. Thus fluctuations in temperature might occur. Too frequent rescaling might influence the trajectories themselves significantly.

- Defining a heat reservoir as a part of the system (Nose, 1984). By this, the degrees of freedom of the system are increased by 1 and energy is flowing back and forth between the heat reservoir and the system. This method is called extended system method. While it constructs a real canonical ensemble, energy oscillations can appear in the system because of the second-order coupling between the system and the reservoir. The period of these oscillations depends on the coupling factor. As an advantage, however, this treatment can be easily applied together with other extended system descriptions, such as the constant pressure condition (Andersen and Chem, 1980).
- In the weak coupling method (Berendsen et al., 1984) energy transfer between the system and a heat reservoir is realized by a stochastic and a friction term in the equation of motion (Langevin equation)

$$\frac{d^2 \vec{r_i}(t)}{dt^2} = m^{-1} \vec{F_i} - \gamma_i \frac{d\vec{r_i}}{dt} + R(t \cdot \gamma_i), \tag{51}$$

where the Gaussian stochastic term $R(t, \gamma_i)$ corresponds to frequent collisions with an ideal gas of light particles at temperature T_{set}. It has zero mean value. The damping constant γ_i controls the strength of the interaction with the heat reservoir. Through the Langevin equation both global thermalization and random local fluctuations appear in the system. Because of the first-order coupling between the reservoir and the system, temperature changes are nonoscillatory.

- The temperature of the system is rescaled by a Maxwellian rethermalization procedure (Andersen and Chem, 1980; Heyes, 1983). This method contains correction terms, which physically represent frequent collisions with light particles in thermal equilibrium. The resulting ensemble is isothermal and follows the Maxwell–Boltzmann distribution. Such thermalization can be achieved, for example by introducing an MC term as suggested by Heyes (1983).

"Heating up" the system is relatively easy in the MD method. Therefore it is well suited to investigate thermally activated processes. Because of constraints in computational time, however, the method is more appropriate to follow the evolution of the system at elevated temperatures. At low temperatures, at which the structure is reasonably stable, MD can be used to create a rate catalog for a KMC simulation (see, e.g. Ref. (Suzuki and Mishin, 2005)).

5.2.10.2 *Interatomic Potentials*

The most accurate way to reproduce the force field experienced by the moving atoms can be achieved by ab initio calculations based on quantum mechanical treatment of the electrons (Car and Parrinello, 1985). Not surprisingly, accuracy has its price in complexity and computational time. Therefore vast majority of contemporary MD is realized using semiempirical interatomic potentials.

Semiempirical interatomic potentials depend on the atomic positions (configuration) and have relatively simple functional form. The computational time scales linearly with the system size. Usually, the selected potentials contain some fitting parameters which are adjusted to reproduce the correct material behavior.

The most widely used interatomic potentials for metallic systems follow the embedded atom (Daw and Baskes, 1984) or the Finnis–Sinclair model (Finnis and Sinclair, 1984). In the embedded atom model, the potential is described by the equation

$$U = \frac{1}{2} \sum_{i,j(i \neq j)} \varphi_{S_i,S_j}\left(r_{ij}\right) + \sum_i F_{S_i}(\rho_i), \tag{52}$$

where $\varphi_{si,sj}(r_{ij})$ are pair interaction between the i-th and j-th atom in the system which depend on their distance r_{ij} and their chemical natures s_i and s_j. The embedding energy of atom i into a 'jellium' of electron density ρ_i at the site i is described by F_{si}. The electron density in turn can be calculated simply as the sum of the electron densities of all other atoms $j \neq i$ (Daw and Baskes, 1984).

$$\rho_i = \sum_{j \neq i} \rho_{S_j}\left(r_{ij}\right) \tag{53}$$

The second term on the RHS of Eqn (52) represents the many-body interactions which are responsible for a significant part of the binding energy in a metal. The embedding function itself is arbitrary in the model of Daw. In the Finnis–Sinclair method, it is postulated to be a negative function, and the absolute value is the square root of ρ_i (Finnis and Sinclair, 1984).

According to Eqns (52) and (53) a pure metal can be described by three functions ($\varphi(r)$, $F(\rho)$ and $\rho(r)$). Similarly, a binary A–B solution requires seven of them ($\varphi_{AA}(r)$, $\varphi_{AB}(r)$, $\varphi_{BB}(r)$, $F_A(\rho)$, $F_B(\rho)$, $\rho_A(r)$ and $\rho_B(r)$). However six of these functions are identical for those of the pure metals, so the only extra component is the $\varphi_{AB}(r)$ cross-interaction potential. For the sake of efficiency, the pair interaction and electron density functions are smoothly truncated after a cutoff distance. Usually, the cutoff distance is chosen to cover a few (3–5) coordination shells. Among other MD-related results, current developments in the potential generation procedures are summarized in a recent review by Mishin et al. (2010).

5.2.10.3 *MD Study of Diffusion in Amorphous Materials*

A decisive advantage of the MD method is its lattice-free nature. Therefore, diffusion in a noncrystalline medium, such as amorphous materials or structural defects, can be investigated. MD results on grain boundary diffusion will be discussed in Section 5.6.3.2. Here we consider exemplary findings on the diffusion in metallic glasses.

The first challenge in describing the diffusion in disordered media is the identification of the diffusion vehicles. Though, strictly speaking, even in crystalline solids there is no conservation law for defects. But assuming conservation of lattice sites limits possible defect types and accordingly the possible diffusion mechanisms. In a melt/amorphous material apart from a "full" vacancy or interstitial also partial defects as well as extended defects can exist. Consequently the distance between two equilibrium positions is less than the "lattice parameter". The atomic jump is often resulting in a local relaxation event. In simulating the atomic transport in Ni_xZr_{1-x} glasses, Teichler distinguished single atom hopping and cooperative motion (Teichler, 1997). According to his definition:

- atomic displacements that exceed half of atomic radii are considered as full hopping events
- if two hopping events take place no less than an atomic radius from each other, they are considered locally correlated; whether two jumps are locally correlated strongly depends on the observation time
- a series of locally correlated jumps are considered as a cooperative displacement

His results indicated that at Zr-poor alloys most of the Ni transport happens via cooperative jump events, while in Zr-rich alloys single atom hopping dominates the Ni transport.

In a so-called soft-sphere glass, low-frequency localized vibration modes (influencing clusters containing $N_s = 20$–100 atoms) exist. The activation energy of a diffusion hopping is the lowest along the direction of (a linear combination of) these vibrations. The hopping event will involve all N_s atoms with a maximal displacement equal to $dN_s^{-0.5}$ if d is the distance between the equilibrium position and the saddle point (Schober et al, 1997; Oligschleger and Schober, 1999). These collective jumps can also lead to a ring-exchange mechanism (Miyagawa et al., 1988).

For amorphous materials, MD-type simulations gave a good insight to the underlying diffusion mechanism and explained most of the features found in the experimental investigations (Faupel et al., 2003).

5.2.11 Density Functional Theory-Based Simulations

Solving quantum mechanical equations which deal with the movement of ions and electrons could explain a great deal of problems in material physics, chemistry or biology. There is no fundamental limitation to apply a complete quantum mechanical description to atomic migration. On the other hand, solving the related equations requires computational resources that scale by a steep N^3 with the system size N. Therefore quantum mechanics methods are mainly used to feed structural and energetic basic information into more resource efficient methods like the discussed MD or KMC simulation.

A particularly successful application of quantum mechanics to many particle systems is the density functional method. The obvious advantage of considering the electron density is that it represents a function of only three coordinates, much less than the $3N$ coordinates of the wave function of a system of N electrons.

The theory is based on the so-called Hohenberg–Kohn theorems (Hohenberg and Kohn, 1964). They state that the total energy can be uniquely expressed as a functional of the electron density. The minimum of the Kohn–Sham energy functional is equal to the ground-state energy of the system with a preset ion positions. For these discoveries Walter Kohn received a Nobel Prize in Chemistry in 1998.

The simplest realization of the density functional theory is called local density approximation. In this approximation, it is assumed that the exchange–correlation energy per electron at a given point is equal to that of a homogenous electron gas that has the same local density. Thus, in this method the effects of local electron density fluctuations are suppressed. Although this seems to be a rather crude simplification, the local density approximation is indeed proven to be highly successful in describing various material physics-related problems.

5.2.11.1 Calculating Point Defect Properties

Vacancy formation energies play a crucial role in diffusion as the activation energy for diffusion is the sum of the formation G_V^F and migration G^M Gibbs energies. The contributions of these two terms may vary from system to system. However as a rule of thumb, they can roughly be treated as equal to each other. Calculating defect properties represents a particular challenge, since the quantum mechanical equations have to be solved for a situation in which the translational symmetry is broken.

Owing to the highly efficient screening in metals, mean field approaches can rather accurately reproduce physical properties. Indeed, even the simple local or semi-local exchange–correlation functionals like local density approximation (Ceperley and Alder, 1980) or generalized gradient approximation (Perdew et al., 1992) are capable of providing a sufficiently accurate defect formation energy at $T = 0$ K (Gillan and Phys, 1989; Korhonen et al., 1995). This energy contains already the elastic contribution which is due to two major effects: a strong distortion near the defect itself and a long-range volumetric effect, which influences the lattice constant. First, the volume of the vacancy is treated as a variable. The volume of the defect must be chosen so that the free energy of the crystal is minimized (Grabowski et al., 2009).

In the recent decade efforts focused on the other terms appearing in the vacancy formation energy, namely the electronic, quasiharmonic and anharmonic contributions. The electronic contribution is nothing more than the contribution of the entropy of the electron gas appearing because of a $T \neq 0$ K treatment (Mermin, 1965). Quasiharmonic and anharmonic excitations are results of the noninteracting vibrations and interacting vibrations of the atomic core. A recent review on the $T \neq 0$ K properties of point defects is published by Grabowski et al. (2011).

Interaction between the impurities and the vacancies plays an important role, for example for the case of impurity diffusion in Al. Ab initio methods are more and more widely used to study the hopping mechanism and the vacancy–impurity interaction in this system (Sandberg and Holmestad, 2006; Simonovic and Sluiter, 2009).

5.2.11.2 Ab initio MD and KMC

In conventional MD, semiempirical interatomic potentials are used to establish the equation of motions for the atoms as described before. These potentials, while often successful for pure or binary systems, usually lack a strong theoretical foundation. Moreover for multicomponent systems their application becomes more and more questionable. An alternative approach could be the combination of density functional theory and MD. This was done successfully first by Car and Parrinello (1985).

In their approach they assumed that the Born–Oppenheimer approximation is valid (i.e. ionic and electronic wavefunctions are decoupled) and that the motion of ions can be described by classical mechanics. Since a detailed description of first principles MD is well beyond the scope of this chapter for more information consult some recent reviews such as Ref. (Mishin et al., 2010; Hutter, 2011).

Since KMC is a state-to-state technique, no equations of motion are solved, but hopping from one equilibrium site to another is modeled with predefined probabilities. The important task is to identify the possible escape routes and to determine their respective probability. Here, density functional theory can provide the required information to map the potential landscape (Fichthorn and Scheffler, 2000; Reuter and Scheffler, 2006). Especially challenging is the application of the density functional theory to lattice-free KMC. In such simulation, a preset rate catalog with a few atomic configurations is not applicable anymore and the escape paths and probabilities have to be calculated on the fly (Xu and Henkelman, 2008).

5.3 Interdiffusion in Concentrated Alloys and Thermodynamic Driving Forces

In the previous section, we discussed diffusive transport in the ideal situation of self- or impurity diffusion in the dilute limit as it is experimentally realized in a tracer study (see Section 5.4.1.1). In most practical situations of production and application of metallic structures however, atomic transport proceeds in the chemical regime of large concentration and pronounced concentration gradients. In this regime thermodynamic driving forces have to be taken into account. Furthermore, in concentrated alloys, mixing or demixing is usually a joint result of migration of all components. So, we have to discuss the simultaneous diffusion of the components, a process which is usually specified by the term *interdiffusion*.

5.3.1 Nonequilbrium Thermodynamics

We will start the discussion from the basic Eqn (5) found in the previous section. However, such equations have now to be formulated for each component involved. To keep our derivations simple but sufficiently general, we need to consider at least two alloy components A and B and, to address the most common case, the vacancies as moving species. Following the principles of nonequilibrium thermo- dynamics, we presume that any inhomogeneity in chemical potentials may drive mobile species. Thus, a minimum set of equations reads

$$j_A = -M_{AA}\frac{\partial \mu_A}{\partial x} - M_{AB}\frac{\partial \mu_B}{\partial x} - M_{AV}\frac{\partial \mu_V}{\partial x}$$

$$j_B = -M_{BA}\frac{\partial \mu_A}{\partial x} - M_{BB}\frac{\partial \mu_B}{\partial x} - M_{BV}\frac{\partial \mu_V}{\partial x} \qquad (54)$$

$$j_V = -M_{VA}\frac{\partial \mu_A}{\partial x} - M_{VB}\frac{\partial \mu_B}{\partial x} - M_{VV}\frac{\partial \mu_V}{\partial x}$$

(as before, we restrict considerations to 1D geometry). Aside, we note that these equations could be easily extended by further gradient terms that involve other driving forces, such as electrical fields or temper- ature gradients giving rise to electro- or thermomigration. The matrix of the mobility coefficients M_{kl} represents material properties which we would like to link below to diffusion coefficients measured in tracer experiments. The Onsager reciprocity relation states (Onsager, 1931) that this matrix must be symmetric ($M_{kl} = M_{lk}$). Often, it is argued in addition that the nondiagonal elements capturing the cross- interaction between different species can be neglected in first approximation. For vacancy diffusion, however, we do know that atomic transport happens by site exchange of atom–vacancy pairs, here A–V or B–V pairs. Thus, a driving force that pushes the vacancy to one side necessarily also drives the atom to the opposite side. On the other hand, A–V and B–V exchange may happen almost independently. Finally, the equations should merge into Eqn (5) if no gradient in the vacancy potential is present and so Eqn (8) is fulfilled. Putting all this together, Eqn (54) can be reasonably reduced to

$$j_A = -\frac{\nu_A D_A^*}{\Omega k_B T}\left(\frac{\partial \mu_A}{\partial x} - \frac{\partial \mu_V}{\partial x}\right)$$

$$j_B = -\frac{\nu_B D_B^*}{\Omega k_B T}\left(\frac{\partial \mu_B}{\partial x} - \frac{\partial \mu_V}{\partial x}\right) \qquad (55)$$

$$j_V = -(j_A + j_B)$$

To avoid unnecessary complexity, we have neglected here any effect of different or composition-dependent partial volumes, that is A and B atoms as well as the vacancy occupy the same constant atomic volume Ω. Here and in the following, we describe the composition of the alloy by atomic fractions ν_A and ν_B which have to be distinguished from concentrations c_A and c_B (measured in m^{-3} units). Furthermore, for the binary we tacitly define $\nu := \nu_A$ and so $\nu_B = 1 - \nu$.

The typical situation of interdiffusion in an A/B diffusion couple is illustrated by **Figure 12**. If at given position, the partial currents of the two components are not balanced, a compensating vacancy current is generated in regions free from vacancy sinks or sources. Where sinks and sources (here edge dislocation lines) are present, equilibration of vacancy density leads to nonconservative climb of dislocation and thus to variation of the number of available lattice sites. In consequence, volume shrinkage appears at the side of the faster, expansion at the side of the slower component. As net effect of both, the welding plan shifts by *convection* to the right with the velocity ν_K.

Being interested in diffusion on macroscopic scales, it is usually assumed that sufficient vacancy sources or sinks are available, so that the concentration of vacancies stays constant at its temperature-determined equilibrium. However, most prominent sink or source mechanisms are climb of edge dislocations or attachment and detachment at incoherent high-angle grain boundaries. In conventional metals, a typical density of dislocations (depending on the annealing and deformation history) amounts to about 10^6–10^9/cm^2 and so the average distance of the vacancies to the next sink may range to at least a micrometer. As a consequence, the assumption of a negligible vacancy gradient is not so clear anymore, if diffusion is studied on the nanometer scale. In the following, we will therefore distinguish two limiting cases: (1) so-called Darken interdiffusion: presence of vacancy equilibrium always and everywhere and (2) so-called Nernst–Planck interdiffusion: total absence of sinks or sources.

(1) *Darken interdiffusion*

By definition, density of vacancies is always in equilibrium. Thus, we can assume in Eqn (55)

$$\frac{\partial \mu_V}{\partial x} = 0. \tag{56}$$

If at a given reference plane (consider the initial welding plane of the diffusion couple, but the following is also valid for any other plane oriented perpendicular to the x-axis) the partial fluxes of the

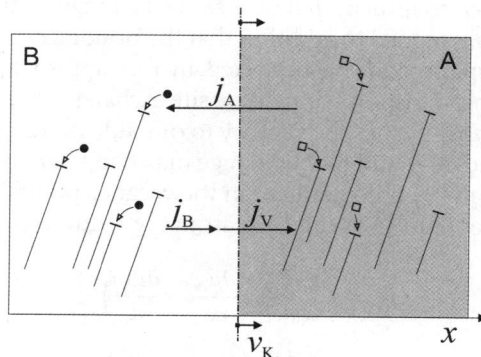

Figure 12 Interdiffusion in an A/B diffusion couple by the vacancy mechanism: A difference between fast A and slow B diffusion must be balanced by vacancies ($\square$) by climb of dislocations. Volume shrinkage (expansion) appears at the A-side (B-side) of the diffusion couple. In consequence, the original welding plane (dash-dotted) shifts to the A-side.

two components are not balanced (in the example of **Figure 12** flux of A is larger than flux of B), more atoms will cross the plane from right to left than from left to right. In consequence, the reference plane will shift to the right. Quantitatively, the velocity of this convection movement is given by

$$v_K = -\Omega(j_A + j_B) = \Omega\left(\frac{\nu_A D_A^*}{\Omega k_B T}\cdot\frac{\partial\mu_A}{\partial x} + \frac{\nu_B D_B^*}{\Omega k_B T}\cdot\frac{\partial\mu_B}{\partial x}\right) \tag{57}$$

Note: in a binary diffusion couple, j_A and j_B are directed opposite to each other. In the respective literature, Eqn (57) is known as the first Darken equation (Darken, 1948) and for reasons that will become transparent later (see Section 5.3.4) this velocity is usually termed as Kirkendall velocity. As an immediate consequence of the convective flow of matter, we have to distinguish from now the local lattice reference from the laboratory reference. In the former, we may apply Fick's equations for each component separately (atomic hops appear on the 'background' of the lattice). In the latter, convection shifts the lattice and therefore convection currents ($v_K\nu_A/\Omega$, respectively, $v_K\nu_B/\Omega$) must be added to the diffusion currents. In the following, we will mark variables in the laboratory system by a tilde. So we find for the current density in the laboratory reference

$$\begin{aligned}\tilde{j}_A &= j_A + \frac{\nu_A}{\Omega}v_K = -\frac{\nu_A D_A^*}{\Omega k_B T}\frac{\partial\mu_A}{\partial\tilde{x}} + \frac{\nu_A^2 D_A^*}{\Omega k_B T}\frac{\partial\mu_A}{\partial\tilde{x}} + \frac{\nu_A\nu_B D_B^*}{\Omega k_B T}\frac{\partial\mu_B}{\partial\tilde{x}}\\ &= -\frac{\nu_A\nu_B}{\Omega k_B T}\left(D_A^*\frac{\partial\mu_A}{\partial\tilde{x}} - D_B^*\frac{\partial\mu_B}{\partial\tilde{x}}\right) = -j_B - v_K\frac{\nu_B}{\Omega} = -\tilde{j}_B\end{aligned} \tag{58}$$

This formulation is absolutely symmetric with respect to the components. Obviously, there is no good reason anymore to distinguish in laboratory reference two partial currents. Both reveal the same absolute value. Therefore, we will call $\tilde{j} := \tilde{j}_A = -\tilde{j}_B$ simply the interdiffusion current that controls the evolution of the single measureable composition profile inside the sample. With regard to the Gibbs–Duhem relation for a binary system $\nu_A d\mu_A = -\nu_B d\mu_B$, we transform Eqn (58) into a more convenient form

$$\tilde{j} = -(\nu_B D_A^* + \nu_A D_B^*)\cdot\frac{\nu_A\nu_B}{\Omega k_B T}\left(\frac{\partial\mu_A}{\partial x} - \frac{\partial\mu_B}{\partial x}\right) \tag{59a}$$

$$\tilde{j} = -\underbrace{(\nu_B D_A^* + \nu_A D_B^*)}_{:=\tilde{D}\cdot}\cdot\underbrace{\frac{\nu_A\nu_B}{k_B T}\frac{\partial^2 g}{\partial\nu_A^2}}_{:=\Phi}\cdot\frac{\partial c_A}{\partial x}$$

$$\tilde{j} = -\tilde{D}\cdot\Phi\cdot\frac{\partial c_A}{\partial x} \tag{59b}$$

(g denotes the Gibbs potential per atom)

The latter Eqn (59b) gives rise to the definition of a new coefficient: the Darken interdiffusion coefficient

$$\tilde{D}_D = (\nu_B D_A^* + \nu_A D_B^*)\Phi \tag{60}$$

which includes the so-called thermodynamic factor

$$\Phi = \frac{\nu_A \nu_B}{k_B T} \frac{\partial^2 g}{\partial \nu^2}. \tag{61}$$

This definition of the interdiffusion coefficient is also called the second Darken equation. The coefficient allows expressing the joint mixing current in the form of a first Fick law as presented in Eqn (59b). It is a matter of taste whether the thermodynamic factor should be included into the interdiffusion coefficient or kept separately. Quite common in literature, partial intrinsic diffusion coefficients are defined by

$$D_A := \Phi \cdot D_A^*; \quad D_B := \Phi \cdot D_B^*. \tag{62}$$

With these, the interdiffusion coefficient is simply stated as

$$\tilde{D}_D = \nu_B D_A + \nu_A D_B \tag{63}$$

The Darken interdiffusion coefficient expresses the superposition of the diffusive transports of both atomic species. At intermediate compositions, its magnitude is essentially controlled by the faster of the two tracer diffusion coefficients. It should be furthermore noted that beside the explicit composition dependence expressed in Eqn (60), there is also an implicit one owing to the composition dependence of the tracer coefficients $D_A^*(\nu)$ and $D_B^*(\nu)$ and the thermodynamic factor $\Phi(\nu)$.

The thermodynamic factor captures the influence of excess (=energetic) driving forces. Pure diffusion can only be expected for an ideal solution for which this factor becomes indeed equal one. Let us illustrate the properties by using a simple symmetric regular solution model of a binary alloy:

$$g(\nu) = 2\nu(1-\nu)k_B T_c + k_B T[\nu \ln \nu + (1-\nu)\ln(1-\nu)]. \tag{64}$$

Here T_c denotes the critical temperature of the miscibility gap. The stronger a repulsive interaction between the alloy components, the higher is the critical temperature. On the other hand, a negative T_c in Eqn (64) may express attractive interaction (stronger mixing). Inserting Eqn (64) into the definition Eqn (61) we find

$$\Phi_{reg} = 1 - 4\nu_A \nu_B \frac{T_c}{T} \tag{65}$$

which is plotted for illustration in **Figure 13**. In view of Eqn (65) it is clear that the absolute value of the thermodynamic factor can hardly be larger than 10, except the case of very stable compounds for which the factor may range in the extreme to 100 or even 1000.

In decomposing systems, a finite composition range exists, for which the thermodynamic factor becomes even negative. As a consequence, diffusion currents become reversed. In this so-called spinodal regime, concentration fluctuations increase by 'up-hill' diffusion instead of being damped. Indeed, if calculating the evolution of local concentration on the basis of Eqn (59) with the same concept that led to Eqn (8), we find

$$\frac{\partial \nu}{\partial t} = \frac{\partial}{\partial x} \tilde{D}_D \frac{\partial \nu}{\partial x} \approx \tilde{D}_D \frac{\partial^2 \nu}{\partial x^2} \tag{66}$$

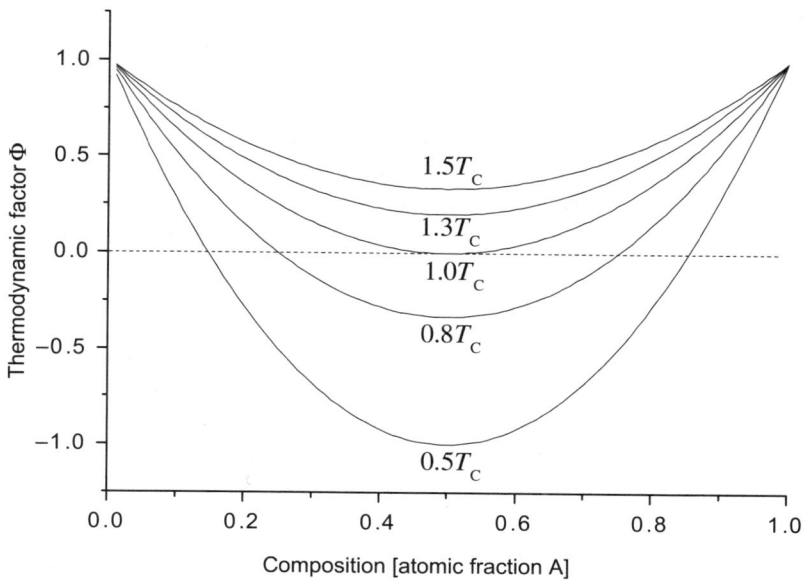

Figure 13   The thermodynamic factor as calculated by Eqn (65).

(The last approximation at the RHS is only valid if the composition dependence of the diffusion coefficient can be neglected. This can always be justified by sufficiently small amplitudes of composition fluctuation.) Inserting into Eqn (66) a periodic disturbance of wave vector k

$$\nu(x, t) := \nu_0 \exp(\beta t) \cdot \exp(ikx) \tag{67}$$

and solving the resulting equation for the growth constant β that quantifies the exponential increase rate of the disturbance, we find

$$\beta = -\tilde{D}_D k^2 \tag{68}$$

Visually β gets negative, fluctuations are obviously damped as expected in normal diffusion with a positive diffusion coefficient. But for a negative thermodynamic factor and thus negative $\tilde{D}_D$ and therefore positive β, the amplitude of the composition fluctuation will grow in time. The respective growth rate even increases with the square of the wave vector k, that is, for decreasing distances. This imposes serious difficulties with the stability of the system on small length scales, which can only be solved by introducing an additional gradient energy to the Gibbs potential of the inhomogeneous system, as was suggested by Hillert (1961) and Cahn and Hilliard (1958) (see also Section 5.7).

To understand diffusion, knowledge of atomic mobilities is essential. To determine the latter, diffusion has to be measured in situations which exclude the impact of discussed chemical driving forces. Usually this is achieved by applying a tiny amount of radiotracers which does not modify the chemistry of the sample. From Eqn (65) we can also learn that a measurement of dilute impurities by conventional analysis techniques, such as secondary ion mass spectrometry (SIMS), is also well suitable to determine tracer impurity diffusion coefficients. In the limit $v_A \to 1$; $v_B \to 0$ the thermodynamic factor becomes equal one and thus, energetic forces no longer influence the diffusive transport.

(2) Nernst–Planck case

In studying diffusion on short-length scales, we cannot be sure anymore that sufficient vacancy sinks or sources are available. Let us consider the other extreme: absolute absence of sinks and sources. Since in most situations the partial atomic currents are not equal, one has to take into account an additional flux of vacancies. Local density fluctuations of the vacancies are built up and so the gradient of vacancy density becomes an efficient additional driving force. With time, this driving force first increases. However, generally this force is opposed to the atomic driving force acting on the faster component. Therefore, soon after starting the diffusion process the opposing 'vacancy pressure' becomes that large that it depresses the velocity of the fast component down to that of the slow one. Them both partial currents balance exactly. The establishment of this balanced situation takes normally very short time in comparison with the timescale of atomic diffusion since vacancy redistribution appears a factor $1/v_V^{eq}$ faster than that of atoms. So after a short transition time, we may set

$$j_A = -j_B \tag{69}$$

Reversely we can use the requirement of Eqn (69) to determine the gradient of vacancy potential that is required to balance atomic currents:

$$\frac{\partial \mu_V}{\partial x} = \left(v_A D_A^* \frac{\partial \mu_A}{\partial x} + v_B D_B^* \frac{\partial \mu_B}{\partial x} \right) \Big/ \left(v_A D_A^* + v_B D_B^* \right) \tag{70}$$

Since atomic currents have become equal, no convection appears. So, one can directly use Eqn (55) to calculate the current in the laboratory reference:

$$\tilde{j} = j_A = -\frac{D_A^* D_B^*}{v_A D_A^* + v_B D_B^*} \cdot \frac{v_A v_B}{\Omega k_B T} \left(\frac{\partial \mu_A}{\partial x} - \frac{\partial \mu_B}{\partial x} \right) = -\underbrace{\frac{D_A^* D_B^*}{v_A D_A^* + v_B D_B^*}}_{=:D_{N.P.}} \cdot \underbrace{\frac{v_A v_B}{k_B T} \frac{\partial^2 g}{\partial v_A}}_{\Phi} \cdot \frac{\partial c_A}{\partial x}$$

$$\tilde{j} = -\tilde{D}_{N.P.} \cdot \frac{\partial c_A}{\partial x} \tag{71}$$

We obtain a Fick law that is formally identical to former Eqn (59b) of the Darken case. But now the interdiffusion coefficient reads

$$D_{N.P.} = \Phi \frac{D_A^* D_B^*}{v_A D_A^* + v_B D_B^*} = \frac{D_A D_B}{v_A D_A + v_B D_B} \tag{72}$$

This so-called *Nernst–Planck interdiffusion* coefficient is essentially controlled by the slower component. The faster component has to wait for the slower one since vacancy transport does not allow compensation of asymmetric atomic currents anymore. In consequence, a diffusive process in Nernst–Planck regime appears to be slower than in the Darken regime. So, the density of vacancy sinks and sources becomes a decisive factor in understanding reaction rates, as it controls whether fast Darken or slow Nernst–Planck transport appears on the studied length scale. In a transition regime of finite sink efficiency, one can even expect a superposition of both regimes.

The Nernst–Planck diffusion coefficient was first introduced in studies of atomic transport in ionic crystals (Mehrer, 2006b). In this case equal ionic fluxes j_A^+ and j_B^- are enforced by the requirement of charge neutrality. Whether the same Nernst–Planck regime could be observed in metals too, though for other reasons, has been discussed quite controversially until recently. In current experiments with reactions on the nanoscale, now almost convincing evidence has been collected that indeed this regime becomes important in understanding atomic transport (for further details see Section 5.7).

5.3.2 A Summary: The 'Zoo' of Diffusion Coefficients

In previous sections not a single but a variety of different diffusion coefficients were defined, which bears the risk of getting lost. Let us present a comprehensive synopsis in form of a table that hopefully can help in keeping the clear view and summarizes under which conditions this various types of diffusion coefficients can be experimentally observed (see **Table 2**).

5.3.3 Concentration Dependence of Diffusion Coefficients

Since a composition dependence of the interdiffusion coefficient is a prominent feature which has a pronounced impact on the outcome of experiments, we will present some general considerations on measurement and understanding of this dependency.

Table 2   The systematic between diffusion coefficients and respective measurement conditions

Random walk: $D_{\text{rand}} = z\omega\frac{\lambda^2}{6}$	Theoretical coefficient for truly random walk in ideal Brownian motion. Only entropic driving forces.	Eqn (29)
Tracer self-diffusion: $D_A^*(\nu) = z\omega_A(\nu)\cdot\frac{\lambda^2}{6}f_A(\nu)\cdot\nu_V^{\text{eq}}$ (vacancy mechanism)	Usually assumed to be the diffusion in pure metals, but can be determined in alloys as well. Characteristic is the absence of energetic driving forces which is achieved in experiment by using radiomarkers. In general, the tracer coefficient depends on the composition ν of the host since jump rates and correlation factor depend on local composition.	Eqn (32) with (7)
Tracer impurity diffusion: $D_B^*(\nu) = \frac{\lambda^2}{6}f_2(\nu)\omega_2(\nu)c_V^{\text{eq}}e^{G^B/k_BT}$	Usually assumed to be the diffusion of a diluted impurity in a practically pure host, but can be determined in alloys as well, if radiomarkers are used.	Eqn (37)
Intrinsic diffusion coefficients: $D_A(\nu) = \Phi(\nu)\cdot D_A^*(\nu),\ D_B(\nu) = \Phi(\nu)\cdot D_B^*(\nu)$	Component diffusion coefficients including thermodynamic driving forces. Measurement by interdiffusion experiments if movement of the reference plane is recorded in addition (see Section 5.3.4).	Eqn (62)
Interdiffusion coefficients: $\tilde{D}_D := \nu_B D_A + \nu_A D_B,$ $\tilde{D}_{\text{N.P.}} := \frac{D_A D_B}{\nu_A D_A + \nu_B D_B}$	Two interdiffusion coefficients describe the joint effect of the partial currents. Evaluation of composition profiles in A/B diffusion couples delivers this composition-dependent diffusion coefficient. The composition dependence is three fold: (1) jump rates depend on composition of matrix, (2) thermodynamic factor is composition dependent, (3) superposition of partial currents is weighted by local composition.	Eqns (63) and (72)

5.3.3.1 The Boltzmann–Matano Analysis

If the diffusion coefficient does not depend on concentration, interdiffusion in a binary diffusion couple should produce simple error-function-shaped profiles which could be evaluated by direct fitting (Section 5.2.4). However, the theoretical considerations presented before make clear that such ideal behavior is a rare case, while significant composition dependence must be expected as the rule. To evaluate a composition profile in terms of a concentration-dependent diffusion coefficient, the so-called Boltzmann–Matano analysis is the established procedure.

Given a concentration dependence, Fick's second law in the form of the nonlinear partial differential Eqn (9) is the right starting point. This differential equation has no general analytical solution. However, under the assumption of a step-like starting profile and boundary conditions to be fulfilled at infinity (far from the diffusion zone), it is possible to transform this equation into an ordinary differential equation (Matano, 1933). By substituting

$$\eta := \frac{x - x_{\mathrm{M}}}{\sqrt{t}} \tag{73}$$

Equation (9) transforms into

$$-\frac{\eta}{2} \cdot \frac{\partial c}{\partial \eta} = \frac{\partial}{\partial \eta}\left(\tilde{D} \cdot \frac{\partial c}{\partial \eta}\right) \tag{74}$$

which may be integrated and then evaluated at fixed time to find the relation

$$-\frac{1}{2t} \cdot \int_{\nu_{\mathrm{L}}}^{\nu*} (x - x_{\mathrm{M}})\mathrm{d}\nu = \tilde{D}(\nu)\frac{\mathrm{d}\nu}{\mathrm{d}x}\bigg|_{x*} \tag{75}$$

in which ν_{L} denotes the terminating composition at the left side (see **Figure 14**). The parameter x_{M} defines the position of the so-called Matano plane. By the definition of Eqn (73) it is clear that this plane is initially located at the welding plane of the diffusion couple. However, the latter may shift during the diffusion process so that the position of the Matano plane cannot be localized by optical inspection of the diffusion zone at later stages. To derive a condition that fixes its position, we note that at infinity ($x \to \infty$) the concentration gradient on the RHS of Eqn (75) vanishes. So performing the integration over the whole composition range requires

$$\int_{\nu_{\mathrm{L}}}^{\nu_{\mathrm{R}}} (x - x_{\mathrm{M}})\mathrm{d}\nu = 0 \tag{76}$$

which defines the position x_{M} of the Matano plane.

Figure 14 illustrates how the different terms needed to evaluate Eqn (75) are determined from a measured composition profile. After positioning the Matano plane so that the hatched areas to the left and right become equal, one can evaluate the integral I^* and the local slope S^* of the concentration profile at arbitrary composition ν^*. $\tilde{D}(\nu^*)$ is then simply $-I^*/(2t \cdot S^*)$ (mind I^* is negative).

For reactive systems that form intermetallic phases in the diffusion zone, the composition dependence of the diffusion coefficient becomes quite complex due to the major influence of the thermodynamic factor. Distinguished reaction products are reflected in the composition profile by a sequence of composition

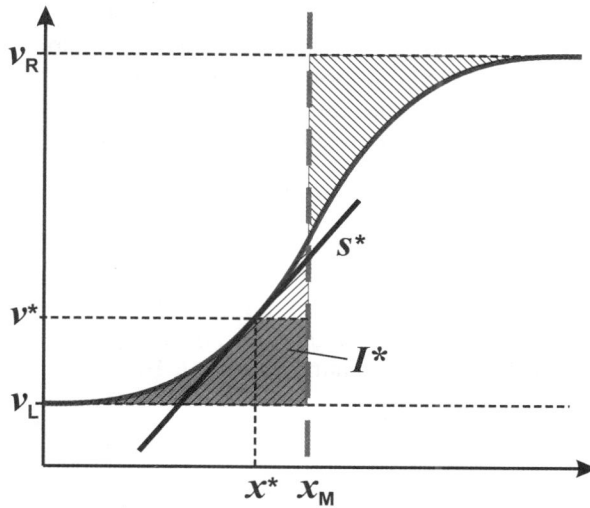

Figure 14 Evaluation of the interdiffusion coefficient according to Boltzmann–Matano. The measured composition profile is shown in red. The terminating concentrations (v_L and v_R) are indicated by horizontal dashed lines, while the position of the Matano plane (x_M) is marked by the vertical dashed line. This line has been positioned so that the hatched areas to both sides become equal. The composition at the given position x^* is v^*; the black line marks the gradient of composition at this position, while the green shaded area represents the integral $I^* = \int_{v_L}^{v^*} (x - x_M)\mathrm{d}v$. (For color version of this figure, the reader is referred to the online version of this book.)

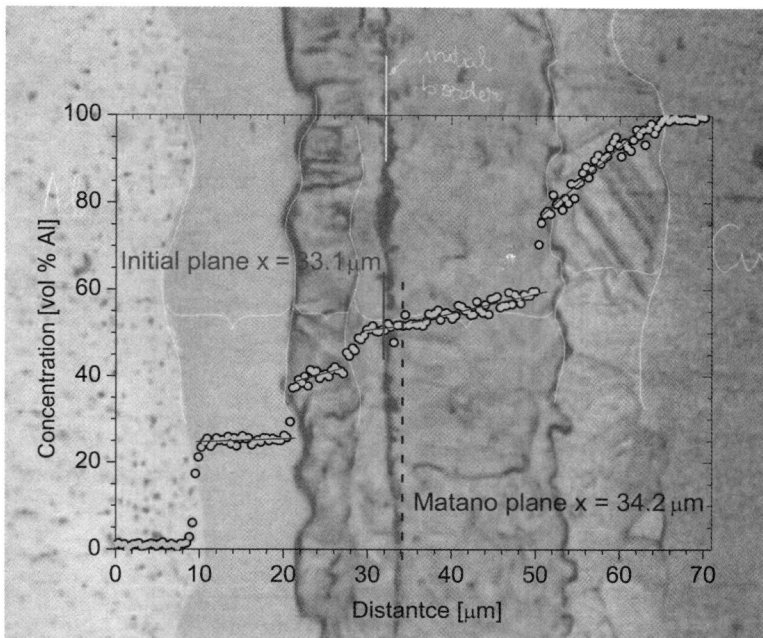

Figure 15 Cross-section through an Al/Cu interdiffusion couple after diffusion annealing (scanning electron microscopy). The corresponding composition profile as determined by energy-dispersive spectrometry (EDX) is overlaid. Different intermetallic phases reflect by regions of reduced concentration slope. The Matano plane has shifted by about 1 μm from the intitial welding plane. (For color version of this figure, the reader is referred to the online version of this book.)

plateaus and discontinuous concentration steps at the phase boundaries, as exemplified in **Figure 15**. Since the presented mathematical derivation does not rely on any particular shape of the developing composition profile, except the step-like starting condition, the Boltzmann–Matano method can naturally be applied to also determine interdiffusion coefficients of the various intermetallic reaction products.

So, the Matano method is only limited by two important restrictions:

(1) The Boltzmann–Matano equation (Eqn (74)) describes an infinite system. It is only applicable if the boundaries of the specimen are far from the diffusion zone.
(2) Interdiffusion must not induce any local volume change. This may appear if the interdiffusion species have a different partial volume. In such case a refined method derived by Sauer and Freise should be applied (Sauer and Freise, 1962). In addition the partial volumes, may even become composition dependent, which is reflected in deviation from Vegard's law $V_m(v) = vV_A + (1 - v)V_B$. Then $V_m(v)$ and the partial volumes are related by **Figure 16**.

$$V_A(v) = V_m(v) + \frac{\partial V_m}{\partial v}(1 - v); \quad V_B(v) = V_m(v) - \frac{\partial V_m}{\partial v}v \tag{77}$$

Diffusion couples with negative deviation from Vegard's law shrink, while couples with positive deviation expand during diffusion. Sauer and Freise (1962) deduced the required extension to the Matano analysis by the expressions

$$\tilde{D}(v) = \frac{V_A}{V_m}vD_B + \frac{V_B}{V_m}(1 - v)D_A \tag{78}$$

$$\tilde{D}(v^*) = \frac{V_m}{2t}\left(\frac{dY}{dx}\bigg|_*\right)^{-1} \cdot \left[(1 - Y^*)\int_{-\infty}^{x*}\frac{Y}{V_m}dx + Y^*\int_{x*}^{\infty}\frac{1 - Y}{V_m}dx\right] \tag{79}$$

in which $Y := (v(x) - v_R)/(v_L - v_R)$ and v_L and v_R are the mole fraction at the left and right sides far from the diffusion zone, respectively.

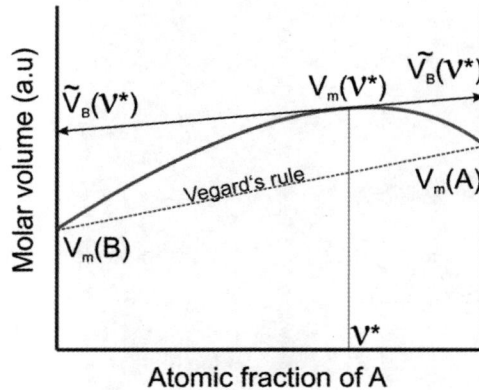

Figure 16 Illustration of partial molar volume change during interdiffusion. The composition-dependent partial molar volumes are determined from the tangent sections with the ordinate axes.

5.3.3.2 Concentration Dependence of Jump Rates

Since the activation energy of self-diffusion scales in metals roughly with the melting point (see the discussion of homologous temperatures in Section 5.5.1), jumping rates in different pure metals vary enormously when compared at equal absolute temperatures. Therefore, when studying a miscible diffusion couple with terminating components of different melting points, it cannot be a surprise that the diffusion coefficients of the alloy may vary across the composition range by 8–10 orders of magnitudes (Erdélyi and Beke, 2011). Interestingly, the consequences of this dominating effect have not been studied until recently (Erdélyi et al., 1999), while attempts to describe the comparably small influence by the partial correlation factors have attracted interest for decades (see Section 5.3.3.3).

With regard to the dominant dependence on melting point and therefore on binding energy, a rough guideline for the composition dependence of the diffusion coefficients can be derived (Kosevich et al., 1994). Let us consider the tracer diffusivity of component A in an AB alloy. Motivated by experimental observation, we presume that for any diffusivity the approximate relation

$$D = D' \cdot \exp\left[-\frac{E_{\text{coh}}(\nu)}{k_{\text{B}}T} \right] \qquad (80)$$

holds with suitable chosen prefactor D'. We approximate cohesive energy by a regular solution

$$E_{\text{coh}}(\nu) = -\left(\nu \frac{\varepsilon_{\text{AA}}}{2} z + (1-\nu) \frac{\varepsilon_{\text{BB}}}{2} z + \nu(1-\nu)\varepsilon z \right). \qquad (81)$$

E_{coh} is defined here as a positive quantity; ε_{AA}, ε_{BB}, and ε_{AB} denote the (negative) binding energies between respective atomic pairs and $\varepsilon = \varepsilon_{\text{AB}} - (\varepsilon_{\text{AA}} + \varepsilon_{\text{BB}})/2$. With these definitions, we calculate the tracer diffusivity of A atoms in pure A or B metal as

$$D_{\text{A}}^{(\text{A})^*} = D_{\text{A}}^0 \exp\left[\frac{\varepsilon_{\text{AA}} z}{2k_{\text{B}}T} \right]; \quad D_{\text{A}}^{(\text{B})^*} = D_{\text{A}}^0 \exp\left[\frac{\varepsilon_{\text{BB}} z}{2k_{\text{B}}T} \right] \qquad (82)$$

and the diffusivity at intermediate compositions as

$$D_{\text{A}}^*(\nu) = D_{\text{A}}^0 \exp\left[\frac{\varepsilon_{\text{AA}} z}{2k_{\text{B}}T} \nu \right] \exp\left[\frac{\varepsilon_{\text{BB}} z}{2k_{\text{B}}T}(1-\nu) \right] \exp\left[\frac{\varepsilon z}{k_{\text{B}}T} \nu(1-\nu) \right] = \left(D_{\text{A}}^{(\text{A})^*} \right)^{\nu} \left(D_{\text{B}}^{(\text{B})^*} \right)^{1-\nu} \exp\left[\frac{z\varepsilon}{k_{\text{B}}T} \nu(1-\nu) \right] \qquad (83)$$

This simple model may be criticized for conceptional weakness. But the major interest here is to clarify that exponential interpolation between terminating diffusion coefficients is probably the best idea, if detailed data for intermediate compositions are not available. In logarithmic scale, diffusion coefficients are consequently interpolated by a straight line as illustrated in the sketch **Figure 17**, eventually modified by additional interaction terms which stabilize or destabilize the alloy.

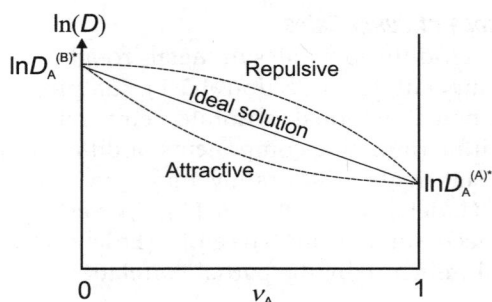

Figure 17   General composition dependence of alloy diffusivities.

With availability of sufficient computing power, MC simulations became a suggested possibility and the need for models that could at least qualitatively describe thermodynamic properties and correlation of diffusion in alloys did arise. Usually these models use the pair interaction energies in the stable configuration (ε_{AA}, ε_{BB}, and ε_{AB}) and similar at the saddle point (ε_{AA}^*, ε_{BB}^*, ε_{AB}^*) to describe the atomic jumps. One of the first of such studies was performed by Radelaar (1970) with the aim of characterizing the role of short-range ordering in diffusion.

Since then, other MC studies were more interested in general trends of statistics, and its dependence on general thermodynamic parameters. An important step was the integration of the pair exchange parameter $\varepsilon = \varepsilon_{AB} - (\varepsilon_{AA} + \varepsilon_{BB})/2$ to include the ordering tendencies of the alloys (Stolwijk, 1981; Murch, 1982). In this way, a general description of diffusion behavior using a well-measurable thermodynamic parameter became available. Regarding a vacancy mechanism, the total binding energies of an A or B atom with z_A A neighbors are

$$E_A = z_A \varepsilon_{AA} + (z - z_A - 1)\varepsilon_{AB} \tag{84a}$$

$$E_B = z_A \varepsilon_{AB} + (z - z_A - 1)\varepsilon_{BB}. \tag{84b}$$

Saddle point energies and attempt frequencies were chosen to be equal for both atomic species. Using this model, for example correlation factors (Stolwijk, 1981; Murch, 1982) and tracer diffusion coefficients (Murch, 1982) were calculated in dependence on the pair exchange parameter and the asymmetry parameter $(\varepsilon_{AA} - \varepsilon_{BB})/\varepsilon$.

In their aim to understand the consequences of strong diffusion "asymmetry", that is significantly different mobilities at both terminating sides of the diffusion couple, Erdélyi et al. (2010) also pointed out the importance of the kinetic parameter $m = (\varepsilon_{AA} - \varepsilon_{BB})/2$ which they used in a normalized form

$$m' = \frac{2zm}{k_B T \log_{10} e} \tag{85}$$

to describe the difference of the diffusivities in orders of magnitude units. (The factor $\log_{10} e$ in the denominator is only required to express the difference in the more convenient order of magnitude scale.) In this way, the parameter m is connected to a well-known experimental quantity. From a mathematical point of view, their MC model is equivalent to the previously mentioned work of Stolwijk (1981) and Murch (1982).

Inserting pair exchange and asymmetry parameters (ε and m), Eqn (84a,b) transform to

$$E_A = -z_A(\varepsilon - m) + (z - 1)\varepsilon_{AB} \tag{86a}$$

$$E_B = -(z - 1 - z_A)(\varepsilon + m) + (z - 1)\varepsilon_{AB} \tag{86b}$$

and the jump rates read (compare Eqn (2.1)):

$$\omega_A(z_A) = \Gamma_0 \exp\left(-\frac{\tilde{E}_0 + z_A(\varepsilon - m)}{k_B T}\right)$$
$$\omega_B(z_A) = \Gamma_0 \exp\left(-\frac{\tilde{E}_0 + (z - 1 - z_A)(\varepsilon + m)}{k_B T}\right). \tag{87}$$

The constant $\tilde{E}_0$ is connected to the saddle point energy E^S by $\tilde{E}_0 = -E^S + (Z - 1)\varepsilon_{AB}$.

The main advantages of this MC ansatz are the following: (1) Atomic jump rates $\omega_A(c)$ and $\omega_B(c)$ can be different for any given concentration. (2) Both jump rates are concentration dependent, even if the alloy is completely miscible. (3) Clustering (ordering or phase separating) tendencies are included. (4) Both the m and ε parameters can be quantified from simply measurable quantities, for example diffusion asymmetry and critical temperature of miscibility gap.

On the other hand, important limitations of the model are the following: (1) Vacancy migration and formation are not decoupled. The composition of the higher jump rate also reveals the higher vacancy concentration (see **Figure 18**). In view of the controlling impact of the melting point this is not necessarily unphysical. (2) The preexponential factor $\Gamma_0 \times \exp(-\tilde{E}_0/k_B T)$ is a free parameter and has no relation to either atomistic or phenomenological quantities.

Models using similar assumptions and formulations are widespread in current MC-type simulations of diffusion (e.g. (Soisson and Martin, 2000; Roussel and Bellon, 2006b; Potvoce and Tréglia, 2012)). More complex parameterizations, using, for example variable saddle points or higher order solute–solute interactions are also available in the literature (Mao et al., 2012) and may be used to match the model to realistic properties.

5.3.3.3 Correlation Factors for Concentrated Alloys

As shown by Eqns (31) and (33), a correlation factor of the tracer self-diffusion coefficient can be easily defined as deviation from ideal random walk of the atoms. A similar strategy must be followed in the case of concentrated alloys. However, here the correlations are more complex. Composition-dependent tracer correlation factors for both component ($f_A(\nu)$, $f_B(\nu)$) and remarkably also for the vacancy $f_v(\nu)$ have to be distinguished. Following the treatment of Murch and Qin (1994) we briefly summarize the way to obtain these different factors.

Atomic transport in concentrated alloys can be described by coupled equations such as shown by Eqn (54). For a correct decription one needs a correlation factor for all the D_{ij} phenomenological coefficients. Statistical treatment, similar to the Einstein formula on these coefficients, was done by Allnatt and Phys (1982). If we assume equidistant jumps of distance λ and follow the reasoning of Section 5.2.6, the moments in the denominators are simply $n_i\lambda^2$ ($i = A,B$) with n_i as the total number

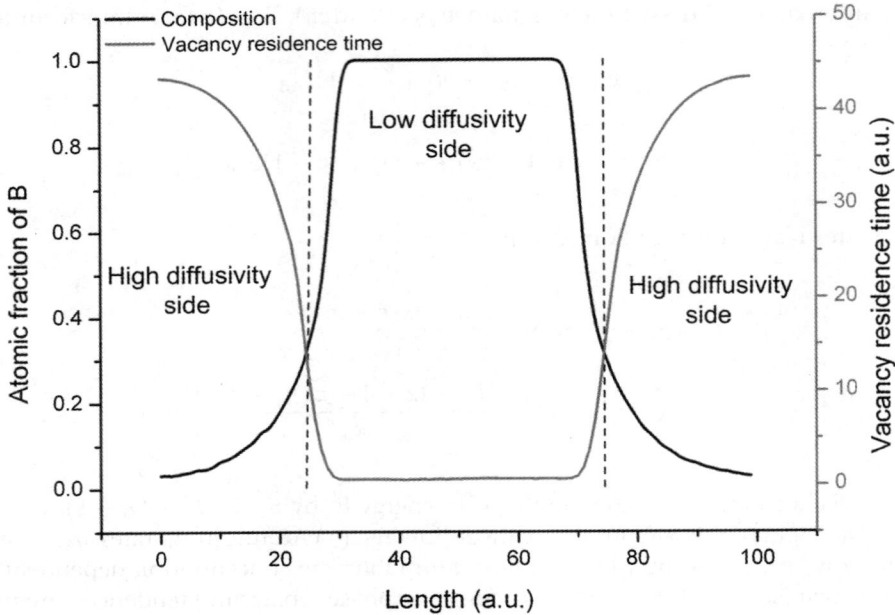

Figure 18 Cumulated vacancy residence time (corresponding to local vacancy concentration) and composition profile obtained by a KMC simulation of interdiffusion in an A/B diffusion couple with $m' = 2$. The asymmetry of the composition profile induced by diffusion asymmetry is clearly visible (intermixing appears mostly on the high diffusivity side). The vacancy itself spends most of the time on the high diffusivity side since higher diffusivity corresponds to lower vacancy formation energy. Dashed lines mark the original interfaces.

of jumps atoms of species i made within time t and R_i is the total jump vector of species i. Consequently

$$f_{AA}^{(A)} = \frac{\langle \vec{R}_A^2 \rangle}{n_A \lambda^2}; \quad f_{BB}^{(B)} = \frac{\langle \vec{R}_B^2 \rangle}{n_B \lambda^2}; \quad f_{AB}^{(A)} = \frac{\langle \vec{R}_A \vec{R}_B \rangle}{n_A \lambda^2}; \quad f_{AB}^{(B)} = \frac{\langle \vec{R}_A \vec{R}_B \rangle}{n_B \lambda^2} \tag{88}$$

these equations are exactly analogous to Eqn (30) above and allow the determination of the partial correlation factors via a KMC simulation. From these partial correlation factors one can derive further correlation factors.

A detailed work shows that Eqn (62) for the intrinsic diffusion and Eqn (60) for the interdiffusion are somewhat inaccurate and a more precise form would be

$$D_i(v) = \Phi(v) D_i^*(v) r_i(v) = \Phi(v) D_{i,\,\mathrm{random}} f_i(v) r_i(v) \tag{89}$$

$$D_{DM} = \left[(1 - v) D_A^* + v D_B^* \right] \Phi(v) S \Phi(v) S \tag{90}$$

The new correction factors ($r_A(v)$, $r_B(v)$ and S) are known as vacancy wind, or Manning factors. Eqn (90) is also called the Darken–Manning equation. The correlation factors for the intrinsic diffusion can be expressed as a function of the collective ($f_{AA}^{(A)}$, $f_{AB}^{(A)}$, $f_{BB}^{(B)}$ and $f_{AB}^{(B)}$) (Murch and Qin, 1994):

$$f_A r_A = f_{AA}^{(A)} - \frac{v f_{AB}^{(A)}}{1 - v} \quad \text{and} \quad f_B r_B = f_{BB}^{(B)} - \frac{(1 - v) f_{AB}^{(B)}}{v} \tag{91}$$

The actual effect of the above vacancy wind factors, on the other hand, is usually well below the measurement uncertainty. Neglecting them will usually not cause a great problem in the evaluation of experimental data.

5.3.3.4 Analytical Expressions for the Correlation Factors

With current computational facilities the most straightforward way to obtain the correlation factors are KMC simulations. Nevertheless, because of their historical importance and possible educational value, it is still useful to briefly expound an analytical model of the mentioned correlation factors. Here we present the basic results of the so-called Manning model (Manning, 1967, Manning, 1971). It was shown later by Belova and Murch (2000a) that more refined models such as that from Holdsworth and Elliot (1986) or Moleko et al. (1989) show better agreement with the results of MC simulations which are believed to be statistically correct. Nonetheless, the Manning model qualitatively still contains the main effects of the more refined models while being simpler to be understood.

Manning considers a (binary) random alloy in a kind of lattice gas model. The considered alloy is ideal; accordingly no driving force other than the entropy is present for the atomic transport. Moreover each lattice site is equivalent and characterized by mean occupation of the species (v A atoms and $(1 - v)$ B atoms). The vacancy-A atom and the vacancy-B atom exchanges have a frequency ω_A and ω_B, respectively. The net jump rate Γ_v of the vacancy is therefore

$$\Gamma_v = [\omega_A v + \omega_B (1 - v)] z = \omega z \tag{92}$$

in which z is the coordination number.

As summarized in **Table 2**, the tracer coefficients for the species $i = A, B$ have the form

$$D_i^*(v) = \frac{1}{6} \lambda^2 v_v^{eq} z \omega_i f_i(v) \tag{93}$$

A formally similar equation can be derived for the diffusion of the vacancies

$$D_V^{eff}(v) = \frac{1}{6} z \lambda^2 \omega f_v(v) = \frac{1}{6} z \lambda^2 [\omega_A v + \omega_B (1 - v)] f_v(v) \tag{94}$$

In a pure metal the vacancy correlation factor f_v is equal to unity, however in an alloy, where $\omega_A \neq \omega_B$ it can deviate from one.

Diffusion of the vacancy happens via exchanges with A or B atoms, thus it can be expressed as a sum of two partial/collective diffusion coefficient.

$$D_V^{eff}(v) = v D_V^{eff,A} + (1 - v) D_V^{eff,B} = -\frac{1}{6} z \lambda^2 \left[v \omega_A f_{AV}^{(A)}(v) + (1 - v) \omega_B f_{VB}^{(B)}(v) \right] \tag{95}$$

where the negative sign arrives from the fact that the vacancy and atomic currents flow in the opposite direction (see, e.g. Ref. Mehrer, 2006c). From Eqns (94) and (95) the vacancy correlation factor can be expressed with the partial correlation factors $f_{vi}^{(i)}(v)$

$$f_v(v) = -\frac{v\omega_A f_{VA}^{(A)}(v) + (1-v)\omega_B f_{VB}^{(B)}(v)}{v\omega_A + (1-v)\omega_B} \tag{96}$$

The tracer correlation factors f of pure metals depend only on the lattice structure (see **Table 1**). For a pure metal with the same structure as the considered random alloy characterized by the exchange rate ω, Manning defines an escape or randomization frequency H_0 in the form (Manning, 1971)

$$H_0 := \frac{2f\omega}{1-f}. \tag{97}$$

This is interpreted as the rate at which the system loses its memory after a given jump (the vacancy reaches the so-called randomization plane), that is jumps separated by a time H_0^{-1} are always uncorrelated. Since in the random alloy the vacancy diffusion rate is reduced by the factor $f_v(v)$, the escape frequency of the alloy has to be reduced accordingly

$$H(v) = H_0 f_v(v) \tag{98}$$

So, by interpreting Eqn (97) in reversed direction using the modified randomization frequency, he finds the individual correlation factor $f_i(v)$ for species i

$$f_i(v) = \frac{H_0 f_v(v)}{2\omega_i + H_0 f_v(v)} \tag{99}$$

Now only the link between the tracer $f_i(v)$ and the partial correlation factors $f_{vi}^{(i)}(v)$ is missing, which in the Manning formalism (Manning, 1971) is simply

$$f_{vi}^{(i)}(v) = -f^{-1} \cdot f_i(v) \tag{100}$$

Substituting into Eqn (99), the f_v vacancy correlation factors from Eqn (96), the $f_{vi}^{(i)}(v)$ partial correlation factors from Eqn (100) and H_0 from Eqn (97) (and restricting ourselves to a given constant v) one gets

$$f_i = \frac{-2\omega_i f_i + (M_0 + 2)(\omega_A f_A v + \omega_B f_B (1-v))}{(M_0 + 2)(\omega_A f_A v + \omega_B f_B (1-v))} \tag{101}$$

where the constant $M_0 = 2f(1-f)^{-1}$.

In this equation all terms $\omega_i f_i(v)$ can be replaced by the tracer coefficient [Eqn (93)] which leads to the most well-known form of the correlation factors in the Manning formalism.

$$f_A = f\left(1 - \frac{2(1-v)(D_A^* - D_B^*)}{M_0(vD_A^* + (1-v)D_B^*)}\right) \tag{102a}$$

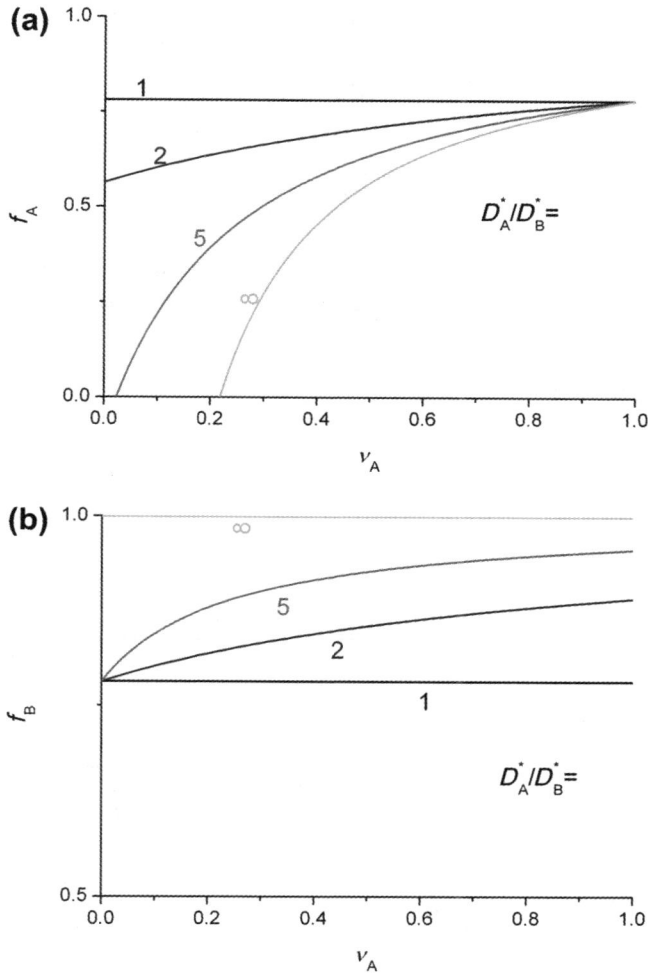

Figure 19 Correlation factors according to the Manning model for the faster (a) and the slower (b) components. The slower B atoms are slightly accelerated in A-rich alloys, while A atoms are decelerated in B-rich ones. Presumably unphysical, the correlation factors for the A atoms reach zero for some combinations of D_A^*/D_B^* and v.

$$f_B = f\left(1 + \frac{2v(D_A^* - D_B^*)}{M_0(vD_A^* + (1-v)D_B^*)}\right) \qquad (102b)$$

The remaining free parameters are the ratio of the tracer diffusivities and the composition. **Figure 19** shows the dependence of f_A and f_B on them. Note that, since D_i^* already contains the correlation factors a constant D_A^*/D_B^* does not mean a constant ω_A/ω_B.

A similar derivation yields the vacancy wind factors, which in the Manning model are

$$r_A = \frac{1}{f} - \frac{D_B^*(1-f)}{f\left(vD_A^* + (1-v)D_B^*\right)} \tag{103a}$$

$$r_B = \frac{1}{f} - \frac{D_A^*(1-f)}{f\left(vD_A^* + (1-v)D_B^*\right)} \tag{103b}$$

Since $D_A^* > D_B^*$ this means that in the case of *intrinsic diffusion* the vacancy wind effect helps the faster component while hinders the slower one as it can be seen in **Figure 20**.

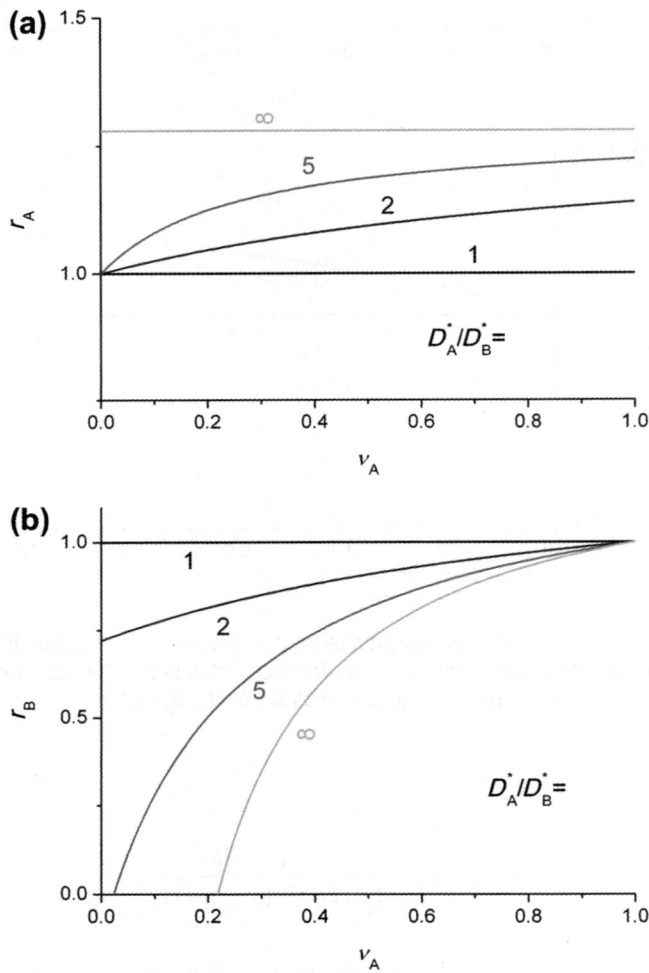

Figure 20 Vacancy wind factors (correlation factors for the intrinsic diffusion) according to the Manning model for the faster component (a) and the slower component (b). Unlike to the correlation factors the vacancy wind helps the faster component.

As can be seen from these graphs, the Manning model, although providing qualitative information on the behavior of correlation factors, suffers from a principal weakness, as it excludes a certain composition range if too large a tracer diffusivity ratio. This weakness is shared with more refined models too (Holdsworth and Elliot, 1986; Moleko et al., 1989), although the latter can reproduce better the correlation factors obtained by MC simulation for a given set of D_A^*/D_B^* and v.

5.3.4 The Kirkendall Effect

The physical relevance of the vacancy mechanism, on which the derivation of the Darken equations in Section 5.3.1 is based, has been controversially discussed historically for rather long time. Only the marker experiments by Smigelskas and Kirkendall (1947) presented first clear evidence that the partial intrinsic fluxes of the alloy components are indeed decoupled from each other, which could not be understood by a direct site or ring exchange. In the famous experiments, inert Mo wires were put at the interface between the Cu and the α-brass half spaces of a diffusion couple. By interdiffusion, the Mo markers shifted away from the original interface. This is naturally explained by unbalanced intrinsic diffusion currents that generate volume shrinkage at one side and expansion at the other side of the interface (see illustration of **Figure 12**).

The theoretical basis to understand this effect quantitatively was already laid in Section 5.3.1. The migration rate of the individual atoms with respect to the local lattice frame is characterized by the intrinsic diffusion coefficients D_A and D_B. These may differ and be concentration-dependent as well. If so, unbalanced intrinsic diffusion currents will lead to a convection shift of the lattice reference. The local velocity of this movement is quantified by Eqn (57). Inert markers that cannot diffuse are practically fixed to the surrounding lattice. They must therefore shift together with the local lattice reference.

Soon after the first experiments, the Kirkendall effect was observed in many other examples (da Silva and Mehl, 1951; Barnes, 1952; Heumann and Walther, 1957). Whenever markers were put at the interface of the diffusion couples and seen later still concentrated in a thin layer, this layer had moved from the original interface. As an experimental fact, these layers stay in an environment of constant local composition and move parabolically in time, that is $\Delta x = Kt^{0.5}$, where K is a temperature-dependent constant.

Powder particles of rather stable oxides with sufficient mass contrast (e.g. ThO_2) represent more reliable fiducial markers which allow accurate experiments. Embedding these particles at various depths in a multifoil diffusion sample (Heumann and Walther, 1957), the velocity of lattice convection, also called Kirkendall drift (Cornet, 1974; van Loo et al., 1979) can be measured at arbitrary depth positions within the diffusion zone. According to Eqn (57), any position of the reference lattice can move with its own velocity. Remarkably, only markers that are put at the central interface (at the Matano plane) between the homogeneous half spaces of the diffusion couple can shift coherently, staying concentrated in a thin layer that is called a Kirkendall plane (Cornet, 1974).

5.3.4.1 *Stability of the Kirkendall Plane*

The uniqueness and the stability of the Kirkendall plane were not questioned for decades. The few indications that the marker wires/particles did not remain in a single narrow zone were mostly rejected as experimental errors. Theoretical investigations by van Loo et al. (1990) concluded that vacancies can be created or annihilated in the bulk or at the interfaces depending upon the intrinsic diffusion coefficients in the product and parent phases and the compositions at the interfaces. If the interface becomes an important vacancy source/sink, then the vacancy flux can differ to both sides of the sample. This complication may result in multiple Kirkendall layers in the specimen. However, a clear picture was derived only recently after bifurcation of the Kirkendall planes was proven and analyzed by systematic experimental studies of well-selected diffusion couples (van Dal et al., 2000; van Dal et al., 2001a; Paul et al., 2004). These experiments widened our understanding of formation and stability of the Kirkendall marker planes. For a demonstration of the bifurcation of the Kirkendall plane, see **Figure 21**.

When putting marker particles at the contact interface (Matano plane), they are first microscopically distributed over the full composition range owing to unavoidable fluctuations in the sharp composition transition (for illustration see **Figure 22**). Why are these particles later found nicely concentrated in thin layers which move coherently with the diffusion profile at a position of constant composition?

After preparation, each marker starts moving with its own Kirkendall velocity. On the other hand, according to the Boltzmann–Matano analysis, the concentration profile spreads to both sides of the Matano plane with a parabolic kinetics. Thus, a position of constant concentration moves with the

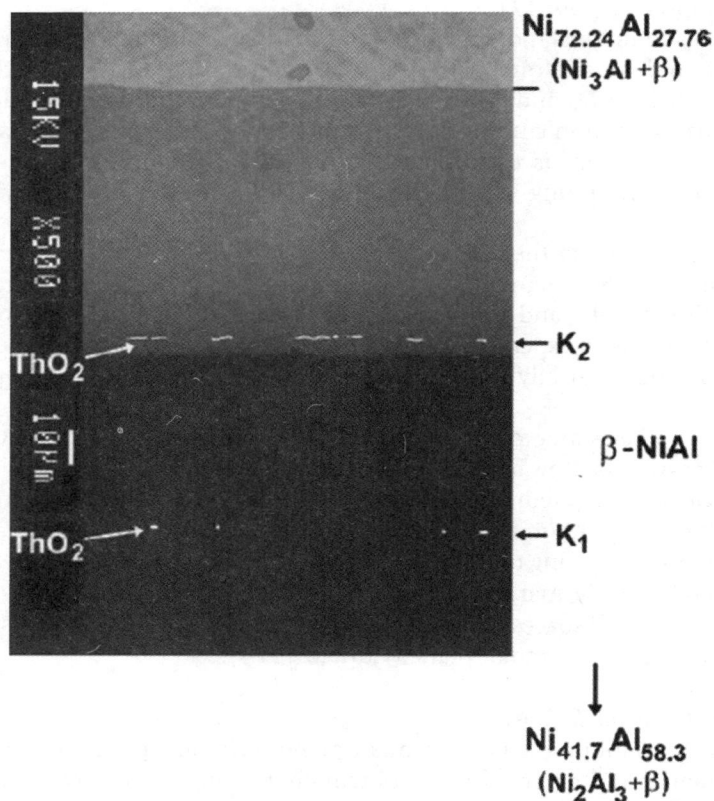

Figure 21 Reaction layer in an Ni/Al diffusion couple. Inert ThO_2 markers are observed at two well-defined planes: K_1 and K_2. Reproduced with permission from Paul et al. (2004).

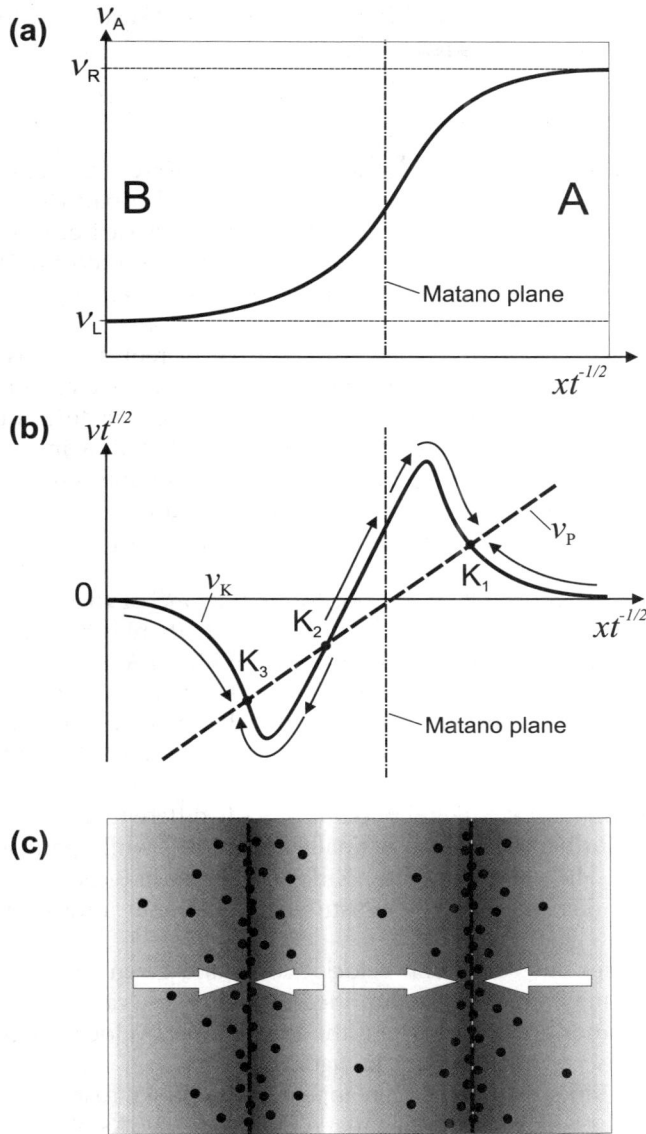

Figure 22 Schematic composition profile (a), velocity diagram (b) and inert marker distribution in the diffusion zone of an A/B couple. By scaling the x and v axes by $\sqrt{t}$ and $\sqrt{t}^{-1}$, respectively, the two upper diagrams become time independent. The Matano plane marks the position of the initial interface. At the A-rich side, the Kirkendall velocity (solid line) is assumed to be positive, while negative at the B-rich side. The spreading velocity of the composition profile is plotted as dashed line. Positions of the profile farther from the Matano plane depart with higher velocity. Possible positions of Kirkendall plane are marked by the cross-points K_1, K_2, K_3. Arrows in (b) and (c) indicate the attraction fields of stable Kirkendall planes. Effective migration of the markers moves them toward these attraction planes.

local velocity v_P of the profile that is proportional to its present distance x from the Matano plane and inversely proportional to the square root of time:

$$v_P(x) = \kappa_P \frac{x}{\sqrt{t}}. \qquad (104)$$

Imagine the marker particle is located at a position, where Kirkendall velocity and the velocity of the composition profile are identical. Then, shortly later the marker particle will obviously find itself still in the same surrounding. Kirkendall and profile velocity decrease both inversely proportional to the square root of time. So, this equality of velocities will be preserved. Obviously, we have to postulate just this equality of Kirkendall and profile velocity as the generic property of the Kirkendall marker plane.

The situation of a typical diffusion couple at given time may be represented as shown in **Figure 22**. In the A-rich part, the partial diffusivity of A atoms may be higher and vice versa in the B-rich side. Thus, the Kirkendall velocity will change sign somewhere in the middle of the diffusion couple. Furthermore, all concentration gradients vanish far from the diffusion zone. Following from Eqn (57), v_K will therefore go to zero for both ends of the couple. In agreement to Eqn (104) the profile velocity at the different points within the diffusion zone is reflected by a linear function. Obviously, this line (dashed) cuts the curve of the Kirkendall velocity (solid) at three points K_1, K_2, K_3, allowing in principle three different Kirkendall planes of concentrated markers.

What happens to marker particles that are not located at one of these ideal positions? Consider a particle slightly left from position K_1. The Kirkendall velocity there is higher than the velocity of the diffusion profile. Thus, the particle will move to positions of higher A concentration until it catches the Kirkendall plane K_1. The opposite situation holds for particles slightly right of K_1. They remain behind until meeting the Kirkendall plane. In this way, the Kirkendall plane obviously acts as attractor which collects sooner or later all marker particles. Analogous considerations hold for the Kirkendall plane at K_3.

The situation of the hypothetical Kirkendall plane at K_2 is different, as in this case a particle that is slightly ahead will move further away until it finally merges K_1, while particles slightly behind will completely move toward K_3. In consequence, the Kirkendall plane at K_2 is unstable against fluctuations and cannot survive. The markers there will be redistributed and finally attracted to the stable Kirkendall planes at K_1 and K_3.

In dependence on the specific Kirkendall velocity curve and given initial position of the Matano plane, different situations may appear as exemplified in **Figure 23** (van Dal et al., 2001a). As before, locations of Kirkendall planes are found where the local marker velocity is equal to the Kirkendall velocity, that is at the intersections shown in the figures.

One, two, or even three Kirkendall planes can appear in the shown cases. Not all of these Kirkendall planes are "stable" in the sense that they attract marker particles. The K_2 plane in **Figure 23(b and c)** is characterized by a positive gradient in the Kirkendall velocity and is therefore unstable. The planes at K_1 and K_3 intersect the velocity curve where the gradient is negative. Therefore, they attract the inert markers and accordingly they form a concentrated layer of markers after some time.

The ideal diffusion couple to observe this effect forms a reaction product with a rather broad existence range in composition so that marker velocity plots like those in **Figure 23** can be realized (Paul et al., 2004). If more than one intermetallic reaction product exist, multiple Kirkendall planes can be stable; for example, trifurcation of the marker particles was observed in Ti/Al (Paul et al., 2005a).

A recent work by Ghosh and Paul (2007) suggested that whenever a stable Kirkendall plane is found in a reaction layer, the grain morphology is different on the two sides of the Kirkendall plane.

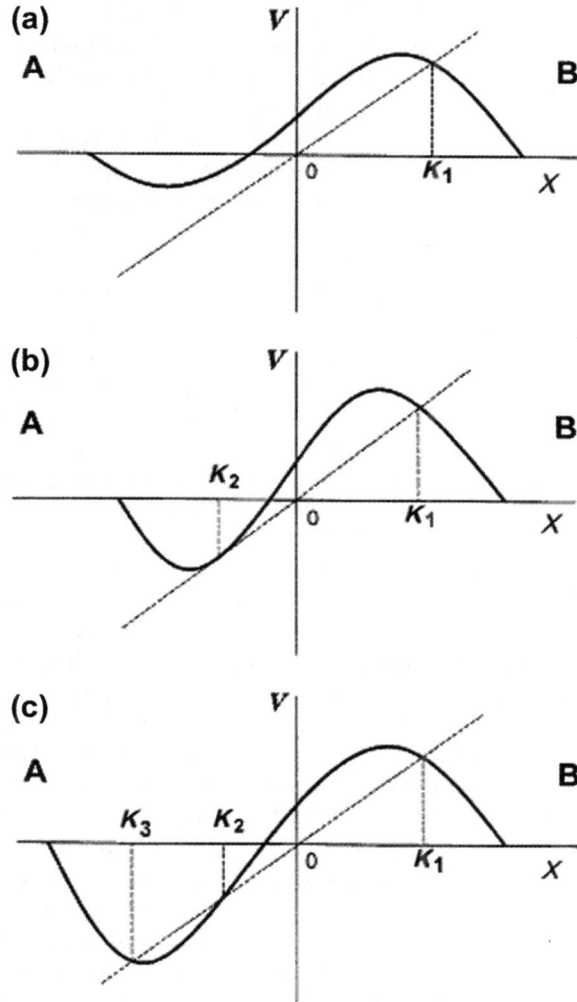

Figure 23   Local Kirkendall velocity (solid) and spreading velocity of the composition profile (dashed). Depending on the specific situation, one (a), two (b) or three (c) Kirkendall planes can appear. Stable Kirkendall planes are only formed when the gradient of the Kirkendall velocity is negative (here at K_1 and K_3). Reproduced with permission from van Dal et al. (2000).

The abrupt transition between different grain structures may be even used to localize Kirkendall plane(s) without using inert markers. The reason for such duplex morphology is proposed to lie in the fact that normally the layer of a product phase grows at both its interfaces. For a β phase between an α and a γ phase, for example, both the α/β and the β/γ interfaces act as a source for new β volume; however the different elementary reaction processes at the two interfaces lead to different microstructures. According to the model of Ghosh and Paul, these two growth zones meet each other exactly at the Kirkendall plane (Ghosh and Paul, 2007).

5.3.4.2 Kirkendall Voiding

The previous considerations based on the Darken equations assumed that the vacancy sources and sinks are sufficient to maintain equilibrium vacancy concentration, and the accommodation of new lattice planes does not generate any stress or at least this stress relaxes in a very short time. This idealized situation is rarely achieved. Porosity formation associated with the Kirkendall effect was documented as early as 1953 by Seitz (1953). The total volume of voids amounts up to half of the material lost on the side of the faster diffusing species. This process of porosity formation is called Kirkendall or Frenkel voiding. Later, Balluffi had shown that as little as only 1% relative excess in vacancy concentration can initiate void nucleation (Balluffi, 1954).

Vacancy supersaturation and elastic stress are obviously interconnected. For example, if vacancies are annihilated at edge dislocations, these dislocations will climb with a direction component perpendicular to the diffusion direction. At a relative excess vacancy concentration of about 10%, the stress caused by annihilating the vacancies is comparable with the yield strength of the material (Brinkmann, 1955). External stress, on the other hand, influences the vacancy annihilation, concentration and pore formation, and can even suppress pore formation (Barnes and Mazey, 1958). Kirkendall voiding is usually considered as disadvantageous for most applications, as it negatively influences mechanical stability of devices or disturb electrical or magnetic properties.

Nowadays however, porous nanoparticles which may be used in catalysis, ion-exchange or medical application attracted interest. The creation of hollow nanoparticles by Kirkendall voiding opened a new production route (Yin et al., 2004). Kirkendall voids appeared after the reaction of metallic nanoparticles with O or S (see **Figure 24**) (Wang et al., 2005; Nakamura et al., 2007; Fan et al., 2007; Railsback et al., 2010). The first demonstration of the analogous effect was found at much larger core–shell BeNi and BeCo particles, where Kirkendall voiding (Aldinger, 1974) and subsequent Ostwald ripening of the pores produced partially hollow particles. Most recently, this effect was also observed in the completely miscible metallic Au/Ag core–shell system (Glodán et al., 2010), which confirmed that nanovoids can be formed as a result of a pure Kirkendall effect. In the past few years, intensive theoretical and simulation effort was invested to clarify the details of this nanovoid formation and some progress was indeed achieved (e.g. Ref. (Gusak and Tu, 2009)), although the complete understanding is missing, for example nucleation of the voids under the eventual effect of elastic stress has been neglected so far. There is common agreement that, owing to energetic contribution of the surfaces, hollow nanoparticles are metastable structures. Therefore in the final state, a dense particle has to be reformed again. Indeed, the respective driving forces that lead finally to the shrinkage of the voids are already acting at the void formation stage too. Typically a fast void formation is observed, followed by a slower densification (Nakamura et al., 2007). According to A. Gusak's explanation the shrinkage is controlled by the diffusion of the slower species (Gusak et al., 2005). Summarizing, the main processes that influence the behavior of the core–shell system are

(1) Kirkendall voiding: By uncompensated intrinsic transport, a vancancy flux is directed toward the center. As a result of the finite almost defect free geometry, it is impossible to equilibrate the vacancy supersaturation.
(2) Gibbs–Thomson effect: The equilibrium vacancy concentration in a curved system must be higher at the inner surface than at the outer surface. This leads to an outward vacancy flux, consequently to the shrinkage of the void. This effect is continuously acting on the particle, and if it is strong enough it may suppress the formation of the void at all.
(3) Inverse Kirkendall effect: dealloying and segregation of the faster diffusing component happen at the inner surface as a result of outward flux of vacancies. This effect slows down the shrinkage (Gusak et al., 2005).

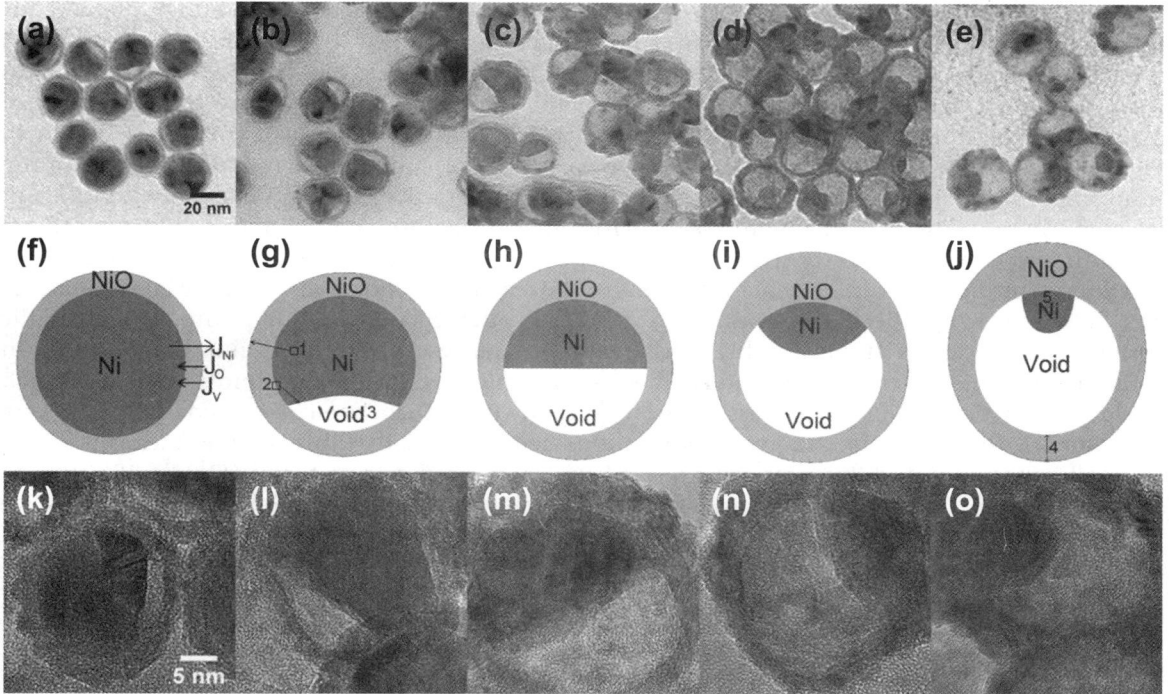

Figure 24 High-resolution transmission electron microscopy (HRTEM) investigations and the skech of a hollow nanoparticle formation during the oxidation of Ni. Reproduced with permission from Railsback et al. (2010).

(4) As a result of the restricted size, curved geometry and possible large specific volume changes, stress should play a significant role. A recent model developed by Erdélyi and Schmitz (2012) to describe the stress effects during solid-state reactions in curved geometry may be applied to characterize these phenomena.

5.3.5 Reactive Diffusion

Interdiffusion in metallic diffusion couples is closely related to solid-state reactions in layer-like geometries. Indeed the reaction and growth of an intermetallic product can be understood as interdiffusion with a quite particular composition dependence of the diffusion coefficient. In this case, the thermodynamic factor contains all the required information to describe phase stablilities and thus the driving forces to atomic transport and reaction. So no wonder that formation of a new phase of intermediate composition is often called *reactive diffusion*. Studying reactive diffusion became popular and attracted intensive research since the 1970s when miniaturization in microelectronics required a detailed understanding on formation and growth rates of silicides. But even today, diffusive reactions left many unsolved problems when it comes to the nanometric dimension and new geometries such as wires and spheres. Tiny part of this current research is discussed in Sections 5.3.4 and 5.7 of this review. For a comprehensive overview on the physics of solid-state reactions by diffusion, the reader is referred to a recent textbook (Gusak, 2010).

When the phase diagram of a binary system shows at studied temperature intermetallic compounds, one can expect that all these compound phases appear in the diffusion zone, started one after the other ordered according to their respective composition (see for illustration the Al/Cu example shown in **Figure 15.**). Let us discuss as the most simple generic case the formation of a single product phase. A schematic composition profile across such reaction zone is shown in **Figure 25(a)**. To bring A and B atoms together for further reaction, either A atoms have to diffuse from right to left or B atoms from left to the right side. In practice both will happen with different weights. Thus growth requires interdiffusion across the already formed layer. If this diffusion process determines the growth rate, we speak about diffusion control to be distinguished from interface or reaction-controlled growth. The concentrations of the reaction product (ν_L, ν_R) at the layer interfaces are determined by the phase-boundaries stated in the phase diagram. Often intermetallic compounds reveal an unmeasurably small existence range $\Delta \nu$ (so-called line compound). In consequence, the concentration gradient across this intermetallic phase almost vanishes which misleads sometimes to the quick assumption that growth of such line compounds might be very sluggish. The following quantitative considerations demonstrates that this is by no means the case, as small concentration differences can still lead to pronounced driving forces (Gusak, 2010).

In a very stable intermetallic one can neglect to first approximation configuration entropy. Thus, we use a parabolic representation to model the Gibbs energy of the intermetallic phase:

$$g(\nu) = g_I + g'(\nu - \nu_I) + \frac{g''}{2}(\nu - \nu_I)^2. \tag{105}$$

A prime or double prime denote in the usual way first or second derivation with respect to the atomic fraction ν. Higher order terms can certainly be neglected within a sufficiently small composition window $\Delta \nu$. Since the composition range is narrow, composition in the intermetallic compound is

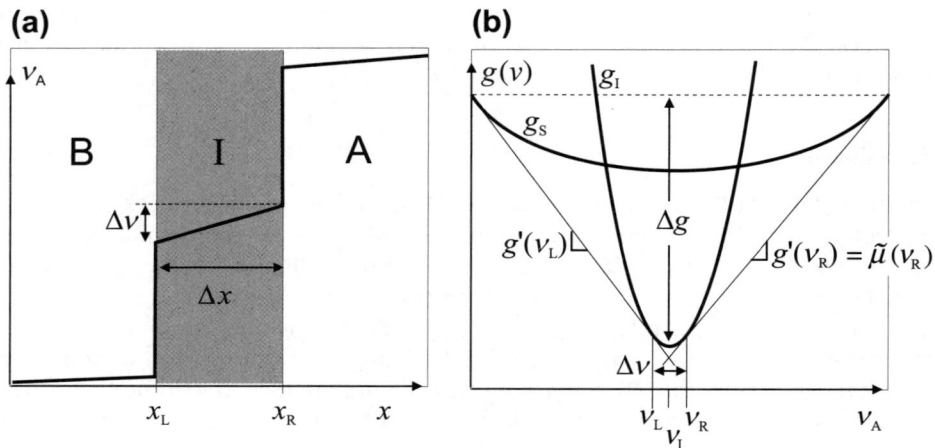

Figure 25 (a) Schematic composition profile across an interdiffusion couple with formation of a single product phase. (b) Gibbs energy diagram of the solution phase (g_s) and the intermetallic compound (g_I). Equilibrium compositions at the phase boundaries are determined by 'double tangents' (thin solid).

practically constant. So, divergence of the flux becomes negligible and flux itself must be almost constant across the intermetallic phase:

$$dc/dt = -\partial j/\partial x \approx 0$$

$$\text{const} \approx \left|\Omega\cdot\tilde{j}\right| = \tilde{D}(\nu)\frac{\partial \nu}{\partial x} \tag{106}$$

One can freely draw a constant into an integral, so that with regard to Eqn (106) the expression

$$\left|\Omega\cdot\tilde{j}\right| = \frac{\int\limits_{x_L}^{x_R} \tilde{D}(\nu)\frac{\partial \nu}{\partial x}\,dx}{\int\limits_{x_L}^{x_R} dx} = \frac{\int\limits_{\nu_L}^{\nu_R} \tilde{D}(\nu)\,d\nu}{\Delta x} \tag{107}$$

is obviously correct. For further interpretation, we insert the interdiffusion coefficient of the Darken case Eqns (60) and (61)

$$\tilde{D}(\nu) = \frac{\nu_I(1-\nu_I)}{k_B T}g''\left(\nu_I D_B^*(\nu) + (1-\nu_I)D_A^*(\nu)\right).$$

Since the composition range is narrow, any composition variation may be neglected, setting everywhere $\nu = \nu_I$. We have to be more careful with the composition dependence of the tracer diffusion coefficients. Due to the complex superlattice structure of intermetallic compounds this can be rather pronounced. Therefore, we interpret Eqn (107) as

$$\left|\Omega\cdot\tilde{j}\right| = \frac{1}{\Delta x}\cdot\frac{\nu_I(1-\nu_I)}{k_B T}g''\cdot\underbrace{\int\limits_{\nu_L}^{\nu_R}\left(\nu_I D_B^*(\nu) + (1-\nu_I)D_A^*(\nu)\right)d\nu}_{:=\overline{D}_I\cdot\Delta\nu} = \frac{1}{\Delta x}\cdot\frac{\nu_I(1-\nu_I)}{k_B T}\overline{D}_I\cdot g''\Delta\nu$$

$$= \frac{1}{\Delta x}\cdot\frac{\nu_I(1-\nu_I)}{k_B T}\overline{D}_I\cdot\left(\tilde{\mu}(\nu_R) - \tilde{\mu}(\nu_L)\right)$$

in which $\overline{D}_I$ represents the interdiffusion coefficient averaged on the composition range of the intermetallic compound. (The integral

$$\overline{D}_I\Delta\nu := \int\limits_{\nu_L}^{\nu_R} \tilde{D}(\nu)\,d\nu \tag{108}$$

is also called the Wagner interdiffusivity (Wagner, 1969; Gusak, 2010).) With regard to the Gibbs potentials sketched in **Figure 25(b)**, the double tangents at both sides of the intermetallic phase, which are equivalent to the exchange potentials ($\tilde{\mu}(\nu_R) = \mu_A(\nu_R) - \mu_B(\nu_R)$ and $\tilde{\mu}(\nu_L) = \mu_A(\nu_L) - \mu_B(\nu_L)$), can be geometrically approximated. This leads to the remarkably compact result

$$\left|\Omega\cdot\tilde{j}\right| = \frac{1}{\Delta x}\cdot\frac{\nu_I(1-\nu_I)}{k_B T}\overline{D}_I\cdot\left(\frac{\Delta g}{1-\nu_I} + \frac{\Delta g}{\nu_I}\right) = \frac{\Delta g}{k_B T}\cdot\frac{\overline{D}_I}{\Delta x} \tag{109}$$

The interdiffusion current that controls growth of the product layer is obviously just proportional to the average interdiffusion coefficient, proportional to the chemical driving force of phase formation and inversely proportional to the thickness that has been already reached. Noteworthy, the composition range Δv drops out of the formula which avoids any difficulties in quantification of a too narrow concentration range. This fact reflects once again that instead of concentrations the physically correct driving forces for atomic transport are the gradients in the respective chemical potentials.

For further growth of the product layer, the diffusion current has to deliver the missing A atoms at the left and the missing B atoms at the right side. So using the current of Eqn (109), the growth rate of the intermetallic product is calculated as

$$\frac{d\Delta x}{dt} = \frac{dx_R}{dt} - \frac{dx_L}{dt} = \Omega \tilde{j} \cdot \left(\frac{1}{1 - \nu_I} + \frac{1}{\nu_I} \right)$$

$$\frac{d\Delta x}{dt} = \frac{1}{\nu_I (1 - \nu_I)} \cdot \frac{\Delta g}{k_B T} \cdot \frac{\overline{D}_I}{\Delta x} \tag{110}$$

The latter expression has to be integrated to determine the phase width. We obtain

$$\Delta x^2(t) = \underbrace{\frac{2}{\nu_I (1 - \nu_I)} \cdot \frac{\Delta g \overline{D}_I}{k_B T}}_{=:k^2} \cdot t$$

$$\Delta x(t) = k \cdot t^{1/2} \tag{111}$$

Shown here for the explicit example, the parabolic growth proportional to square root of time is a general characteristic for any diffusional growth of layer-like products in solid-state reactions. In any case, the respective growth constant k combines thermodynamic driving forces, diffusion coefficients and the solubility limits given in the phase diagram.

Finally the case of multiple product layers has to be considered, since it represents the typical situation in reactive binary systems. The problem was solved first by Kirkaldy (1958) and by Kidson (1961) assuming that compositions at the interfaces match those of local equilibrium. Furthermore, each phase is characterized by a composition-independent interdiffusion coefficient (e.g. defined by averaging over the phase range as done before). Conservation of matter must be obeyed at the interfaces. Hence, the movement of any interface is determined by the balance of related currents in accordance to

$$\frac{dx_k}{dt} = \frac{j^{(k)} - j^{(k-1)}}{c_i^{(k)} - c_f^{(k-1)}} \tag{112}$$

(For a definition of variables see **Figure 26.**) Growth of a layer is controlled by shift of two interfaces. In consequence, the growth of a phase depends not only on the transport through itself but also on that through the neighbor, layers which indicates a competitive nature of phase growth. Although stable in equilibrium, a phase may shrink inside the diffusion zone because of the overwhelming growth of its neighbors.

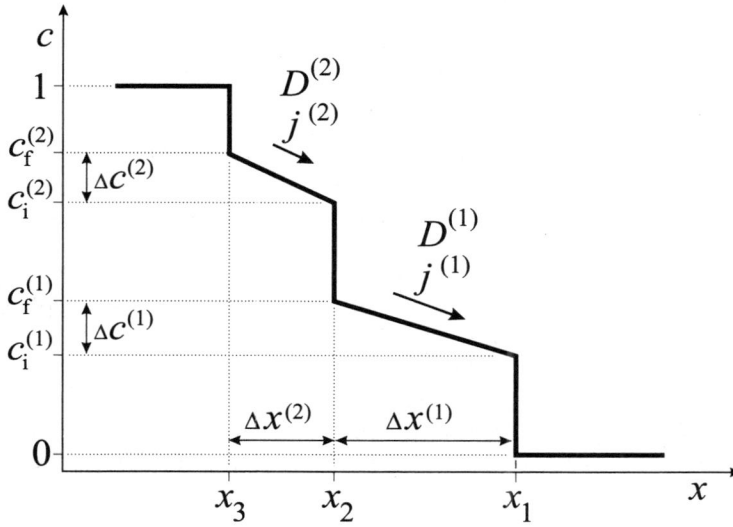

Figure 26 Schematic composition profile through a reaction zone with two intermetallic layers (1) and (2) forming between the components A and B. Definition of the variables as required for the quantitative considerations.

Let us consider the situation sketched in **Figure 26** to discuss this competition. Two intermetallic phases have formed. The solubility in the terminating phases is assumed to be negligible to avoid lengthy formulas. In this case, the growth of the compound layers is given by

$$\frac{d\Delta x^{(1)}}{dt} = \frac{dx_1}{dt} - \frac{dx_2}{dt} = C_{11} \cdot j^{(1)} + C_{12} \cdot j^{(2)}$$

$$\frac{d\Delta x^{(2)}}{dt} = \frac{dx_2}{dt} - \frac{dx_3}{dt} = C_{21} \cdot j^{(1)} + C_{22} \cdot j^{(2)} \tag{113}$$

where

$$C_{11} = \frac{1}{c_i^{(2)} - c_f^{(1)}} + \frac{1}{c_i^{(1)}}$$

$$C_{22} = \frac{1}{c_i^{(2)} - c_f^{(1)}} + \frac{1}{1 - c_f^{(2)}} \cdot \tag{114}$$

$$C_{12} = C_{21} = -\frac{1}{c_i^{(2)} - c_f^{(1)}}$$

A similar set of linear equations with a tridiagonal matrix (C_{ij}) is formulated in the general case of an arbitrary number of compounds (Kidson, 1961). The growth competition between both compounds is conveniently expressed by the current ratio $r := j^{(1)}/j^{(2)}$. It is seen from Eqn (113) that growth of compound (1) requires

$$r > r_1 := -C_{12}/C_{11} \tag{115}$$

Likewise

$$r > r_2 := -C_{22}/C_{12} \tag{116}$$

is needed for the growth of compound (2).

The most important point to notice here is that diffusion-controlled kinetics will never allow a phase to shrink away. As r_1 is always smaller than r_2, we face a situation as sketched in **Figure 27**. Given initially a current ratio $r < r_1$ (situation I. in the figure), at first compound (1) will shrink. But by this, also $j^{(1)}$ increases according to Ficks law $j^{(1)} = -D^{(1)}\Delta c^{(1)}/\Delta x^{(1)}$, respectively, Eqn (109). Thus, before the compound can disappear, the current ratio will reach r_1, at which point the compound (1) starts growing again. Finally, a steady state of constant current ratio between both phases is reached. The analogous is expected for the opposite initial situation (II.).

In consequence, all compounds will grow in steady state simultaneously proportional to a parabolic time law. Apart from geometrical factors of the order of unity, which take into account the peculiarities of the underlying phase diagram, the relative thicknesses of the compounds are fixed by their characteristic products $D^{(k)}\Delta c^{(k)}$ (Kidson, 1961) if only concentration gradients are considered as driving force, or by $D^{(k)} \cdot \Delta g^{(k)}/k_B T$, if the more general formalism of Eqn (109) is used for the different phases.

This prediction of concomitant diffusional growth is certainly one of the best confirmed laws of material physics in late reaction stages as observed in macroscopic interdiffusion couples. It is of similar generality as the Ostwald ripening, for the late stages of decomposition processes. However remarkably, it stands in stark contrast to many experimental observations in thin film reactions. Quite common here, only one or at most a few of the in principle possible phases appear in the diffusion zone (Bene, 1982; Pretorius et al., 1991). Till today a proper and complete understanding of this phase selection problem in reactive diffusion has not achieved yet, although progress in theoretical understanding of the phase nucleation has been achieved (Desré and Yavari, 1990; Gusak, 1990; Pasichnyy et al., 2005; Gusak et al., 2011). For further contributions and discussions to this topic see Section 5.7.

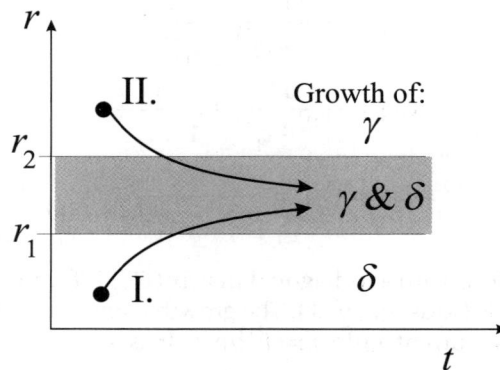

Figure 27 Development of the ratio r of transport currents. A unique steady state is established irrespective of different starting conditions.

5.4 Experimental Techniques to Study Atomic Transport

Experimental methods for diffusion research can be grouped into two broad categories. The first category covers techniques which evaluate composition profiles on various length scales on the basis of the Fick laws. Probably the most well known of these techniques is the radiotracer measurement. However, other depth profiling methods Secondary ion mass spectrometry (SIMS), Auger electron spectroscopy (AES), Rutherford backscattering (RBS), etc.) and (electron) microscopy, atom probe tomography (APT) and some diffraction-based techniques belong to this category as well. The second category contains so-called microscopic techniques. These are based on the investigation of phenomena that are directly related to the presence of point defects or the diffusional jumps. Mechanical spectroscopy and specialized nuclear techniques belong to this category.

In this section, we cover some of the currently available experimental options to gain diffusion data, with a special emphasis given to the developments of the past 20 years. In the first subsection we will summarize the depth profiling methods. The second part offers an outlook to X-ray or neutron diffraction-based techniques, the third part will contain the microscopy-based applications while the fourth and last will summarize the indirect techniques that determine individual jumps and mobilities.

5.4.1 Depth Profiling Methods

5.4.1.1 Radiotracer Techniques

In a radiotracer experiment, the diffusion of a small amount of radioactive material is characterized. The usage of radioactive isotopes was pioneered by the Nobel Laureate György Hevesy (Groh and von Hevesy, 1920; Groh and von Hevesy, 1921). Since the concentration of tracer atoms is very small, ideally infinitesimal, diffusion of a tracer atom is not influenced by other tracer atoms. The diffusion coefficient one can measure in a radiotracer experiment is called *tracer diffusion coefficient* (see Section 5.3.2 about the meaning of the different diffusion coefficients). One can of course measure impurity diffusion accurately with the radiotracer method. However, the main advantage of this technique is the ability to measure self-diffusion in pure metals, alloys or even in ordered phases.

In the first step the surface of the sample is polished. To gain a concentration profile the source of the radiotracer atoms A^* is placed on top of the specimen by dripping off a tracer-containing solution, by evaporation or by electrodeposition. Homogeneity of the deposit is not necessary as long as remaining irregularities of the tracer layer are negligible in comparison with the diffusion length within the specimen. The sample is then (isothermally) heat treated under given annealing conditions (e.g. temperature and time) to allow for the diffusion of the tracer atoms. For most experiments the sample is protected by either inert gas or a vacuum environment.

Finally, a composition depth profile is created by sectioning (see the methods below) and measuring the activity of each slice. The main steps of the radiotracer experiment are summarized in **Figure 28**. To evaluate the diffusion coefficient, one has to fit the measured composition profile with an appropriate theoretical model. Since nuclear counting methods are highly sensitive and reliable, major amount of high-quality diffusion data is measured by the radiotracer method (see, e.g. the collection of data in (Landolt–Börnstein, 1990a–c)). The available range and the accuracy of the diffusion coefficient depend almost exclusively on the sectioning methods applied. Since these cover a large variety, from mechanically cutting disks of few mm thickness to sputter-etching layers of a few nm, radiotracer-based investigations can be used from near melting point T_m with macroscopic diffusion distances of centimeters, down to even below $T_m/2$ with diffusion lengths in atomic dimensions (see **Figures 42** and **44**). Remarkably, this corresponds to more than 12 orders of magnitude in the diffusion coefficients.

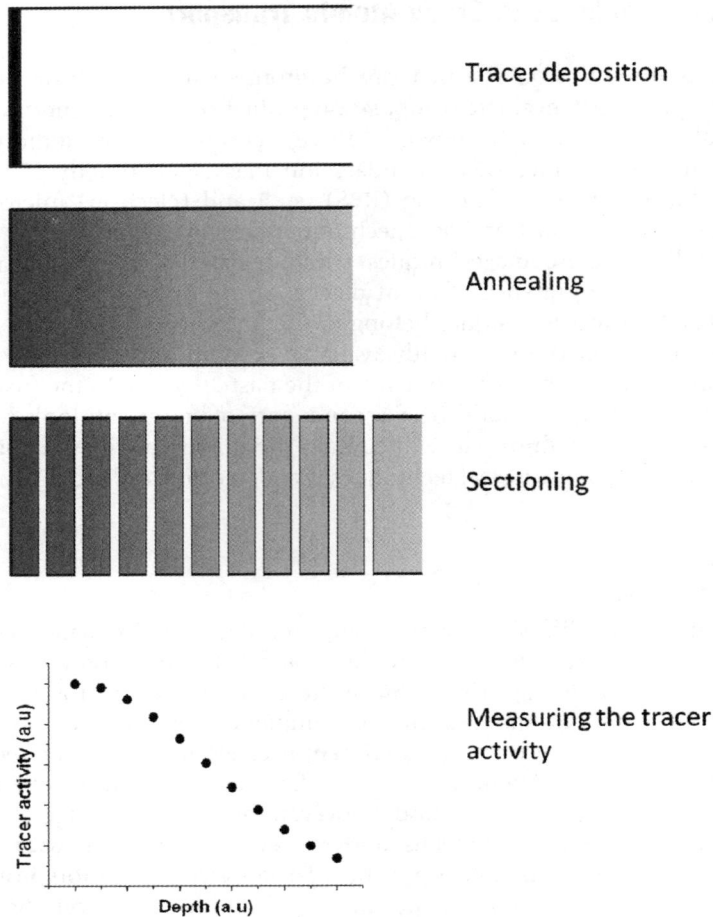

Figure 28   The main steps of a radiotracer diffusion experiment. (For color version of this figure, the reader is referred to the online version of this book.)

In the case of homogeneous lattice diffusion, the tracer isoconcentration surfaces are parallel to the original surface of the sample. Therefore, the main requirement of the sectioning technique is the parallelism of the sections. Methods can be grouped into mechanical and nonmechanical sectioning. The main methods of mechanical sectioning are

- For diffusivities in the range of 10^{-15} m^2 s^{-1} and above (even slower if very long annealing times are possible) sectioning with a precision lathe can produce adequate resolution.
- Application of microtomes allows the measurement of diffusivities down to the 10^{-17} m^2 s^{-1} range. Both the lathe and the microtome sectioning require ductile materials.
- For brittle materials, the appropriate technique is mechanical grinding since by this conservation of the material can easily be ensured. Since the depth resolution of the grinding is somewhat better than of the previous methods, the achievable diffusivity range is extended down to 10^{-18} m^2 s^{-1}. The depth scale of grinding is usually calibrated by precisely weighting the sample before and after grinding.

After sectioning, either the activity of the removed material or the residual activity (Gruzin, 1952) of the specimen is measured. The latter technique assumes that the original alignment of the sample can be restored after measuring the activity.

Given a diffusion coefficient of 10^{-18} m^2 s^{-1}, the diffusion length $(Dt)^{0.5}$ amounts after 1 year annealing to only 5.6 µm. Consequently, nanoscale resolution in sectioning is required to quantify even lower diffusivities.

● Electrochemical etching to study tracer diffusion was pioneered by Pawel et al. (Pawel and Lundy, 1964; Pawel, 1964). In this method a surface oxide was grown by electrochemical techniques on the specimen. The oxide layer was homogeneous on W, Ta and Nb surfaces and its thickness could be controlled by the applied voltage. Using an adhesive tape a complete removal of the tracer-containing oxide was possible. The resolution of this method is better than 10 nm and diffusion coefficients down to 10^{-23} m^2 s^{-1} could be reliably measured.

● Ion beam sputtering (IBS) can be used in principle for any material that has a sufficiently low vapor pressure. The technique was first applied to measure tracer diffusion coefficients by Gupta and Tsui (1970). In this method, the sample is bombarded with low-energy ions (<1000 eV) and the sputtered material is deposited onto an adequate substrate, for example a tape moving with a constant velocity. That way, the sectioning depth is converted to a lateral scale. The resolution of this method is limited by surface roughening. However, a few nanometer resolution is routinely achieved (e.g. Maier et al., 1976; Mundy et al., 1981) so that a diffusivity down to the 10^{-24} m^2 s^{-1} range can be measured.

In principle, the mentioned techniques can be used to also study shortcut diffusion along dislocations and grain boundaries (see Section 5.6) if the macroscopic homogeneity of the specimen can be assured. Radiotracer measurements can even be carried out in specially produced bicrystals that contain only a single well-defined grain boundary (e.g. Ref. (Divinski et al., 2012)). The limiting factor in these studies is the sensitivity for small amounts of tracer atoms, since the active volume around dislocations and grain boundaries is very small. This becomes especially critically if desired depth resolution requires very fine sectioning.

Appropriate tracer isotopes exist for most elements, with a few exceptions which are unfortunately technologically important, such as Al or Si (^{26}Al has a low specific activity, while ^{31}Si has a very short half-life of only 2.5 h). To get at least approximate information, a chemically similar substitute, such as In or Ga for Al (Herzig et al., 1999a; Minamino et al., 2002) or Ge for Si, may be studied (Salamon and Mehrer, 1999, 2003). Some investigations, for example Ref. (Gude and Mehrer, 1997), compared results using ^{31}Si and ^{71}Ge and found a very good agreement.

As a further alternative one can combine an interdiffusion study with the measurement of tracer diffusion of one constituent. Based on the Darken equation (Eqn (60)), one can then calculate the tracer diffusion coefficient of the other constituent, if the thermodynamic factor Φ is known from thermodynamic measurements or theoretical calculations.[4]

By measuring the Kirkendall shift in interdiffusion, one can get a further relation between the intrinsic diffusion coefficients:

$$v_{\mathrm{K}} = \Omega(D_{\mathrm{B}} - D_{\mathrm{A}}) \cdot \frac{dc_{\mathrm{B}}}{dx} \qquad (117)$$

---

[4] In principle, the Darken–Manning Equation (Eqn (90)) should be used. But in view of the typical accuracy of diffusivity measurements, the minor modification by the vacancy wind factor can usually be neglected.

In this way, both tracer coefficients can be calculated from a single interdiffusion experiment (e.g. (van Dal et al., 2001b)). The accuracy of such investigations, however, is certainly less than that of direct radiotracer measurements.

5.4.1.2   Sectioning of Liquid Metals

Diffusion in liquids is orders of magnitude faster than in solids and the Brownian motion of the atoms can cover macroscopic distances. Thus, a useful composition profile can be established by rather crude sectioning techniques (either after solidification or by purpose built instruments like shear cells). One of the pioneers of the diffusion, Sir Roberts-Austen, has measured the diffusion coefficient of gold and platinum in liquid lead in 1896 with an acceptable accuracy even by modern standards (Roberts-Austen, 1896).

On the other hand, Brownian motion of single atoms is hardly the dominant transport mechanism in a normal liquid. The main challenge for a diffusion experiment in liquids is to suppress convective flows.

- A possible solution to suppress the thermal convective flow is the usage of capillary-shaped specimen holders of 1–2 mm diameter; this technique is called long capillary method (Ozelton and Swalin, 1968).
- Specific volume change during melting and solidification can also initiate an undesired convection. To avoid these problems a full liquid stage treatment, including the sectioning, can be carried out. The required experimental setup is called a shear cell (Nachtrieb and Petit, 1956; Masaki et al., 2005). It consists of a series of disks that can rotate coaxially. Small holes are bored into the disk, which can be joined into a long capillary by rotating the disks appropriately. After the diffusion experiment, the disks can be rotated to another position separating the liquid column into different compartments (see **Figure 29**).
- Different densities of the two liquids are also a source of convective flow. For precision measurement, microgravity conditions are needed. The first of these measurements was carried out in 1973 in the Skylab space station by Ukanwa (1974). Measuring diffusion in liquids is still one of the important branches of space-based scientific programs (e.g. Ref. (Masaki et al., 2005; Griesche et al, 2007)).

5.4.1.3   Surface Analytical Techniques (AES and X-ray Photoelectron Spectroscopy)

Core electrons remain localized in (metallic) atoms and are therefore characterized by well-defined, element-specific energy levels instead of broad bands. It is possible to excite core electrons to a free state (**Figure 30(a)**). The kinetic energy of the resulting free electron will be the difference of the exciting energy and the bonding energy of the electron ($E_{kin} = E_{exc} - E_{bond}$). If the exciting energy is known, one can determine the bonding energy by measuring the kinetic energy. Characteristic X-rays (usually Al K_α or Mg K_α sources) have both the required energy and the narrow energy width to produce and evaluate these photoelectrons with sufficient accuracy. X-ray photoelectron spectroscopy (XPS) was introduced as a surface analytical instrument by Siegbahn and Edvarson (1956).

Also, core electrons may be ejected by an energetic electron beam. After the ejection, the atom possesses a free state in one of the inner core states. This will be shortly filled by an electron from an outer orbital. The excess energy of this follow-up process can be released in two ways, either as characteristic X-rays (which is utilized by electron microprobe analysis) or via the Auger process

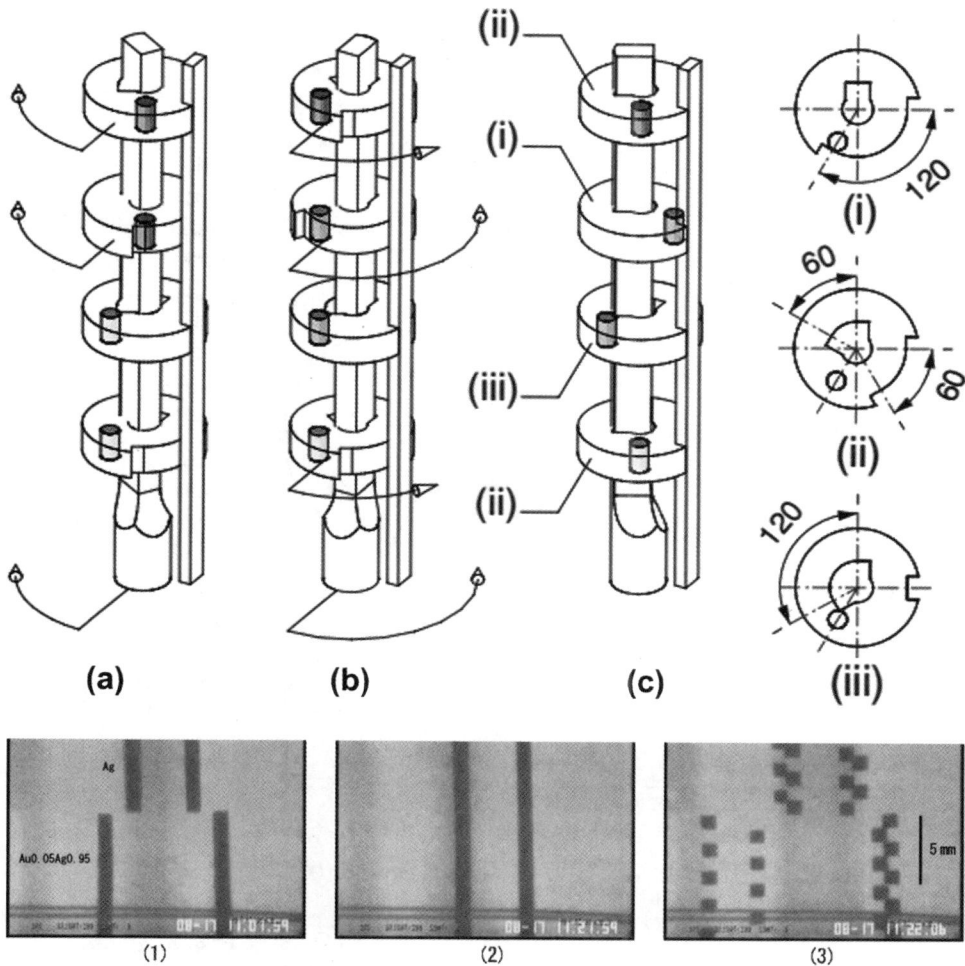

Figure 29 Top: The rotation mechanism of a shear cell. Bottom: X-ray image of the liquid column in a shear cell experiment. The liquid is seen as darker region. Before joining the column (1) the two phases are well separated. During the experiment (2) the cylinders are rotated to join the columns, thus allowing the diffusion to proceed. At the end of the diffusion the cylinders are rotated once more allowing a sectioning in the liquid phase (3). Reproduced with permission from Masaki et al. (2005). (For color version of this figure, the reader is referred to the online version of this book.)

(Meitner, 1922; Auger, 1923). In the latter, the released energy is transferred to another electron in a way similar to an internal photoexcitation. As can be seen in **Figure 30(b)**, this process involves three electronic orbitals of (usually) well-defined energies. Accordingly, the kinetic energy of the released electron is again characteristic to the element. AES gained popularity as a practical surface analytical tool after the pioneering work of Harris (1968).

Both the photo- and Auger electrons are distinguished by very low inelastic mean free paths. These certainly depend on the host material and the electron energy, but generally they fall in the range of

(a)

Photoeffect

$E_{kin} = h\nu - E_{bin}{}^{1s}$

Electron bombardment

Atom in ground state

(b)

Auger transition(KLL)

2p
2s ———— L

1s ———— K

X-ray emission (K_α)

2p
2s ———— L

1s ———— K

2p
2s ———— L

1s —o———— K

Excited state

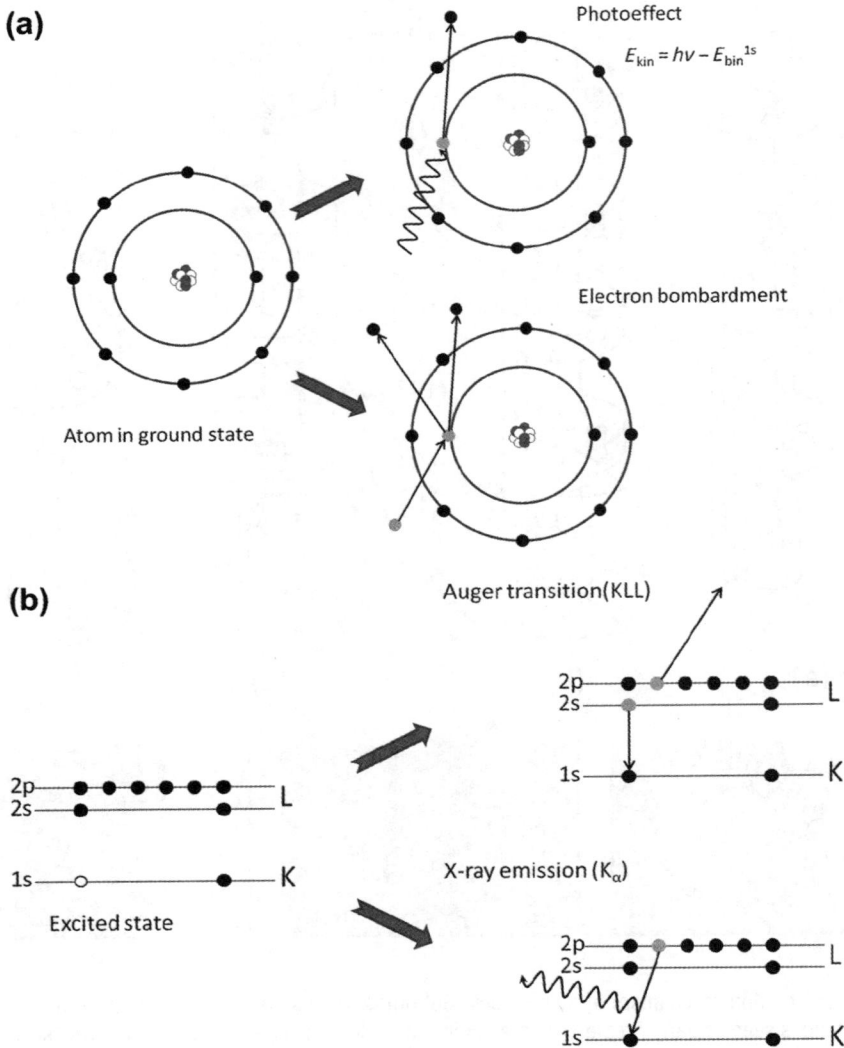

Figure 30 (a) Ejecting an electron by photoeffect or electron bombardment. (b) Scheme of the relaxation by emission of a characteristic X-ray photon or by emission of an Auger electron. The photo- and Auger electrons as well as the characteristic X-ray can be used for local compositional analysis. (For color version of this figure, the reader is referred to the online version of this book.)

a few nanometers at most (e.g. Ref. (Tanuma et al., 1988)). Consequently, Auger- and photoelectrons deliver information on the composition of the (near) surface layer. By knowing the attenuation length of the electrons and the probability of certain Auger transitions or photo excitation processes quantitative depth information can be derived.

The described techniques are element selective, but not isotope sensitive; therefore, they are unsuitable to investigate self-diffusion. Since photoelectron and Auger peaks are measured on a rather high background of inelastically scattered electrons, performing a reliable measurement with concentrations below 1 at% is difficult. As an advantage, however, any element with $Z > 2$ can be detected if it is present in sufficient concentration. Therefore, these techniques are best suited to study diffusion in concentrated alloys.

The following routes can be taken to measure a diffusion coefficient:

- Depth profiling: As the attenuation length of the electrons is small, they are coming only from the near surface layer. Electron spectroscopy requires ultrahigh vacuum conditions. Therefore, usually a depth profile is achieved with the help of IBS for which contemporary AES or XPS devices are often equipped with an ion gun. A given amount of material is removed, and then the fresh surface is analyzed. This elementary cycle can be repeated many times, although surface roughening, ballistic mixing and preferential sputtering may deteriorate the resolution for long sputtering times.
- For deeper penetration profiles up to few hundred μm, a mechanical technique called ball cratering can be used (Thompson et al., 1979). In this technique, a steel ball covered with a diamond paste is rotated on the surface of the sample, which produces a small hemispherical crater on the specimen surface. The geometry of the crater is set by the curvature of the ball. By means of scanning Auger microscopy, one can construct a depth profile simply by performing a line scan across the dimple.
- The permeability of defects (dislocations or grain boundaries) can be measured by the surface accumulation or Hwang–Balluffi method (Hwang and Balluffi, 1979). In this method, the source of the diffusant is covered by a thin film specimen and the surface concentration of the film is monitored by following the characteristic Auger or photoelectron signals (see **Figure 31**). For grain boundary diffusion in the so-called C-kinetics regime (see Section 5.6.1.1) and constant source of concentration c_0, the evolution of the surface concentration c_s is given as

$$c_s(t) = c_0 \frac{k'}{k''} \left(1 - \exp\left(\frac{\delta D_{GB} \Lambda}{\delta_s h k'} t^* \right) \right) \tag{118}$$

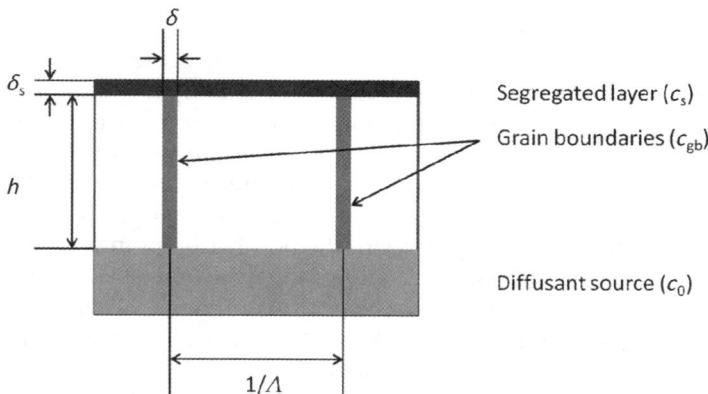

Figure 31 Specimen geometry in a surface accumulation experiment. (For color version of this figure, the reader is referred to the online version of this book.)

where δ is the effective diffusion width of the grain boundary, δ_s is the thickness of the segregated layer at the surface, Λ is the density of the grain boundaries, h is the thickness of the film, $k' = c_s/c_{gb}$ is the relative segregation factor between the surface and the grain boundary, while $k'' = c_{gb}/c_0$ is the relative segregation factor between grain boundary and source. The time t^* is corrected for the effect of the transient before reaching steady-state conditions in the grain boundary (similarly to the case of permeation experiments, Eqn (21)). This transient time can also be used for a crude estimation of the diffusion coefficient. The Hwang–Balluffi method can be used to study grain boundary transport in the case of very thin films (few tens of nanometers), for which reasonable sectioning cannot be carried out. The diffusant must show a definite surface segregation tendency. Since the surface accumulation experiment registers the kinetics, the method requires an in situ heat treatment.

5.4.1.4   *SIMS and Secondary Neutral Mass Spectrometry*

In sputter sectioning, removed ions themselves can be analyzed by mass spectrometry. All kind of mass spectrometers (magnetic sector, quadrupole and time of flight) can be used. Time-of-flight mass spectrometry has the advantage that the complete mass spectrum is measured simultaneously. It needs, however, a well-defined trigger signal; so a pulsed ion source is required. Since mass spectrometry is able to distinguish between different isotopes, it is possible to also study self-diffusion. One can use radioactive or stable tracers in the case of isotope heterostructures (e.g. Ref Kube et al., 2010). As for all methods based on mass spectrometry any element can be detected (unlike, e.g. with Auger electron spectrometry), although the separation of species with similar mass/charge (e.g. $^{28}Si^+$, $^{12}C^{16}O^+$, $^{14}N_2^+$, $^{56}Fe^{2+}$) ratio is not always straightforward and requires expensive instruments with high mass resolution.

To generate the ions for the mass spectrometer, two approaches are possible with their respective advantages and disadvantages. In the first, ion sputtering is applied to the specimen surface by which only a fraction of the desorbed species (not more than a few percent) is ionized. The mass spectrometer analyzes only this ionized fraction. This technique is called Secondary ion mass spectrometry (SIMS). For the second approach, plasma etching is used and desorbed neutral atoms/molecules are postionized by laser radiation or electron cyclotron wave resonance plasma. This technique is called secondary neutral mass spectrometry (SNMS).

As with any IBS technique, the depth resolution of the SIMS is excellent and in the range of 1 nm; though surface roughening and ion-induced mixing causes a degradation of the resolution with increasing sputter depth. A lateral resolution below 100 nm is achievable with focused ion beams (Benninghoven, 1994). The sputtering yield, and more importantly the ratio of ions to neutral atoms, depends however strongly on the nature of the primary ions, the bombarding energy and the specimen composition. It can cover a range of 10^{-5} to 10^{-1}. The influence of the specimen composition on ionization is called matrix effect (Deline et al., 1978). Therefore, quantitative analysis of specimens with noticeable concentration variations or interfaces is challenging (Ferrari et al., 2003; Mansilla and Wirtz, 2012). On the other hand, an appropriate choice of the bombarding ions and energy can significantly increase the ion yield for a given element. This allows the detection of some impurities down to the 10^{12}–10^{13} atoms $\times$ cm^{-3} range by modern commercial instruments. Accordingly the method is well suited to study diffusion of very dilute impurities, for example Ref. (Bracht et al., 2007).

In SNMS, desorption and ionization are separated. Since 90–99.999% of the sputtered species are ionized, matrix effects have much less influence. Therefore, reconstructing a composition depth profile from the measured intensities versus time is easier than in the case of the SIMS (Oeschner et al, 2009). Because of the post ionization process, the signal to noise ratio is however worse than for

SIMS. Consequently SNMS is better suited to study diffusion in concentrated alloys or investigating solid–state reactions (Lakatos et al., 2010).

5.4.1.5 Ion Beam-Based Methods

The first ion scattering experiment was performed by Rutherford in his famous experiment that has proven the existence of the atomic nucleus (Rutherford, 1911). In a modern surface analysis-oriented approach, a monoenergetic beam of ions, usually α-particles of few MeV energy, is used for depth profiling. The energetic beam particles are scattered by the atomic nuclei of the specimen. The kinematic factor K of the backscattering process is a monotonically decreasing function of the target nucleus mass, that is energetic particles loose more energy, if they are reflected by a lighter nucleus. Before and after the scattering event, the beam particles suffer an energy loss, which is proportional to the depth. Therefore, the final energy of the backscattered particle depends on the type of atom it had knocked-on and the depth of the collision event. The yield at a given energy is therefore related to the composition profile.

As can be seen from **Figure 32**, this technique is most useful, when the diffusants or atoms on the top (in this case Pt) are heavier than the matrix atoms or atoms on the bottom (in this case Ge), since this avoids overlap of the contributions of the components. The penetration depth of an α particle with a few MeV energy lies in the μm range. Therefore, RBS can only measure relatively small diffusion

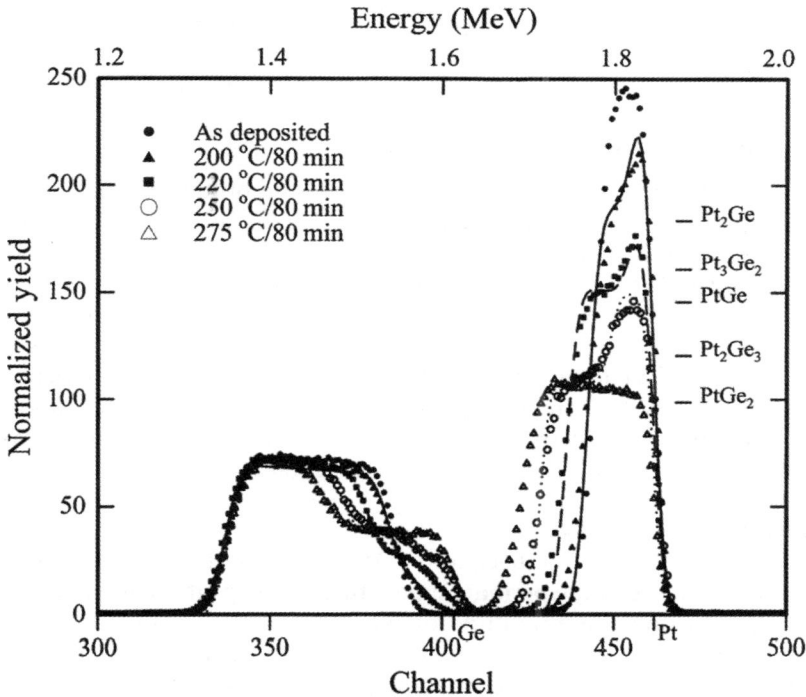

Figure 32 RBS spectrum to trace the formation of different platinum germanides. For a given target species, depth decreases with energy, i.e. atoms at the surface produce the highest recoil energies which are marked for Ge and Pt on the x-axis. Ge or Pt atoms located deeper in the sample are seen left from these two limits. Reproduced with permission from Nemutudi et al. (2000).

lengths and, accordingly, small diffusion coefficients. Depth resolution is in the range of a few nanometers, depending on the material.

RBS carried out with an ion energy of ~ 100 keV is called medium energy ion scattering (MEIS). This method has an excellent mass and depth resolution (<0.5 nm), however, the penetration depth is very small, so only the region within a few nanometers from the surface is accessible. MEIS is furthermore isotope sensitive, thus it is well suited to study, for example, the transport of oxygen through a thin surface oxide layer (Gusev et al., 1995).

If the cross-section of the target nuclei shows a resonance at a specific energy, this process can dominate the yield of the backscattered particles. So, by changing the primary beam energy one can carry out a depth profiling, since the penetration energy loss determines the depth, where the resonant scattering has happened. The main advantage of this method is the higher yield for light elements, which are difficult to analyze by standard RBS. The important light elements possess a strong (α,α) resonance useful for depth profiling, C at 4.262 MeV (Feng et al., 1994), O at 3.05 MeV (Leavitt et al., 1990) and N at 3.576 MeV (Herring et al., 1958). A recent collection of references about the low-energy cross-sections of various light elements can be found in Ref. (Abriola et al., 2011).

Resonant nuclear reactions can be used in similar manner. Indeed, resonant nuclear reaction analysis has a further advantage over resonant backscattering: the detected particle after the reaction differs from the primary particles (e.g. deuteron $\rightarrow$ proton). Accordingly, the background is much reduced. A list of useful resonant nuclear reactions for elements between H and F are listed in a review article by Demortier (2003). The resulting proton energies produced by various light elements after a bombardment by 1800 keV deuterons at 150 °C can be read from Figure 1 of Ref. (Vickridge, 1988).

5.4.2   Diffraction-Based Techniques

As scattering cross-sections for X-rays or neutron beams depend on the chemical nature of the atoms, measured diffraction patterns will be characteristic not only for the lattice structure but also for the given distribution of different atomic species. Phase growth/shrinkage can be followed by the change of the intensity of the characteristic X-ray diffraction (XRD) reflexes, for example (Singh et al., 2009). With high-intensity synchrotron or free electron laser sources, this can be done even in situ (Cserháti et al., 2008).

Furthermore, the penetration depth of the X-rays/neutrons, for small angles, strongly depends on the incident angle. Therefore, diffraction with this radiation is inherently capable of depth profiling. Since reconstructing the depth profile from the measured intensity is practically impossible, the usual approach is to simulate the intensity curve for a given structure and modify structural parameters, until a sufficient agreement with the experimental data is found (Stearns, 1988).

5.4.2.1   X-Ray and Neutron Reflectivity

The origin of this method can be traced back to the production of the first multilayers (DuMond and Youtz, 1940). As a consequence of the periodic superstructure, Bragg reflections can be observed in the small-angle XRD. By diffusion the amplitude of the composition modulation decreases which in consequence reduce the intensity of these reflections (see **Figure 33**). If the diffusion coefficient is concentration independent, and the concentration profile is harmonic, then the intensity of the reflection will decrease according to

$$\frac{d}{dt}\left[\ln\left(\frac{I}{I_0}\right)\right] = -\frac{8\pi}{\Lambda_{mod}{}^2}D \tag{119}$$

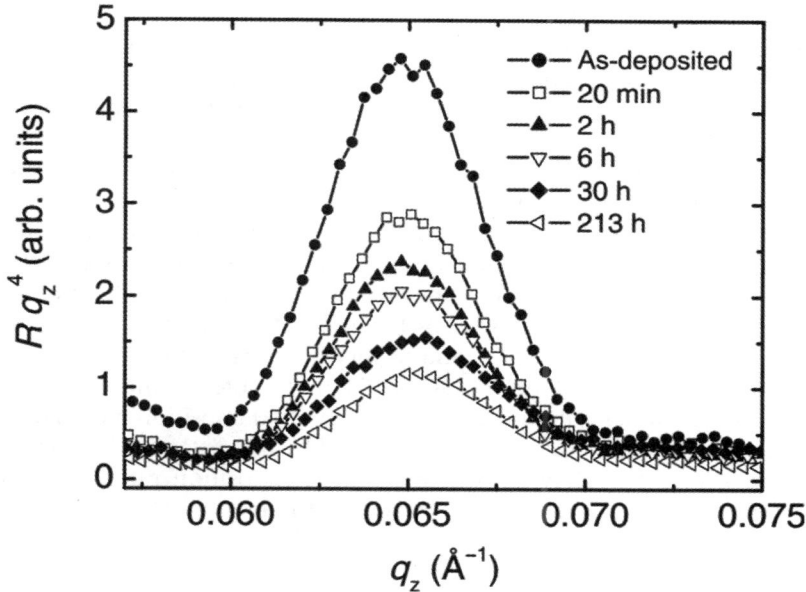

Figure 33 Decrement of the first-order Bragg peak during annealing of an $^{nat}Fe_{7nm}$ / $^{57}Fe_{3nm}$ isotope multilayer at 633 K. The image was taken from Chakravarty et al. (2009).

where Λ_{mod} is the modulation length of the multilayer and D is the volume interdiffusion coefficient (DuMond and Youtz, 1940) (compare Sections 5.3.1 and 5.7). This technique is especially suited to study low-diffusion coefficients, for example for a multilayer of 10 nm modulation length, a diffusion coefficient of 10^{-21} m^2 s^{-1} would mean that the intensity would decrease to half of the original value in a day. As any periodic structure can be modeled by a Fourier series, the method can be used for arbitrary composition profiles. Indeed, if the diffusion coefficient is concentration independent, then the higher order terms would decrease more quickly, thus after a given time the composition profile will become harmonic (sinusoidal) anyway. This effect, however, can be hampered by non-Fickian diffusion phenomena in nanoscale (see Section 5.7).

The homogeneity of the diffusion coefficient can be guaranteed in an isotope multilayer structure, however, X-ray reflectometry normally cannot distinguish between different isotopes. One possibility is the application of the Mössbauer resonance technique, by that it is possible to characterize the diffusion in isotope multilayers by X-ray reflectivity for some elements (Gupta et al., 2005; Gröstlinger et al., 2012).

As neutrons are scattered by the atomic nuclei with isotope contrast neutron reflectivity offers a far more general solution to this problem. Neutron reflectivity can be carried out using two different modes. In the first, the reflectivity of a monochromatic neutron beam is measured as a function of the incidence angle similar to an XRD experiment. In the second, the incidence angle is fixed and the reflectivity is measured as a function of the neutron wavelength (Mâaza et al., 1992). The latter is called time-of-flight method, since the wavelength of the neutrons is inversely proportional to their velocities. Neutron reflectivity allows the determination of diffusion coefficients in the 10^{-25}–10^{-26} m^2 s^{-1} range (Schmidt et al, 2008); thus, it is one of the most sensitive techniques to measure self-diffusion. With the growing number of high brightness spallation sources worldwide, this kind of measurement method is expected to gain more importance in the future.

5.4.2.2   X-Ray Standing Wave Method

When an X-ray beam is reflected from a reasonably flat interface between two domains of different refraction index, it will produce a standing wave pattern. The nodal planes of the standing wave are parallel to the diffracting planes and they have a similar periodicity (Batterman, 1969; Andersen et al., 1976). The position of the nodal planes can be controlled by the incident angle. If the X-ray energy is furthermore tuned close to resonant fluorescence energy of a selected species, then composition profiling with subnanometer resolution is possible.

The fluorescence intensity will be the product of the intensity of local wave field and the local absorber density. Since the field penetration and the standing wave pattern also depend on the refraction properties, composition profiles are determined by comparing the measured fluorescence intensity plots with simulated ones (see **Figure 34**) (Erdélyi et al., 2009).

To increase the fluorescence intensity/contrast, multilayer specimens (Gupta et al., 2010) or a "waveguide" geometry can be used (Erdélyi et al., 2009). In the latter technique, the specimen is placed between two high-refraction index reflector layers (e.g. 6th period transition metals from Hf to Au). The periodicity of the standing waves in this case is determined by the distance of the waveguide layers.

X-ray standing wave methods allow a nondestructive composition profiling with subnanometer resolution and accordingly the ability to investigate atomic transport properties down to the 10^{-24} m^2 s^{-1} diffusivity range.

5.4.3   Microscopy-Based Techniques

Fabricating a cross-section parallel to the diffusion direction and analyzing the resulting surface or thin film via some chemical/isotope-sensitive microscopy is probably the most straightforward way to establish a composition profile. A decisive advantage of the microscopy is the ability to correlate structural and chemical information. The two- or three-dimensional (in the case of APT) field of view represents another important advantage important in the study of heterogeneous short circuit transport (compare Section 5.6). Depending on the respective technique, the resolution of these methods varies from μm down to the nm range, although usually the practical detection limit and the signal to noise ratio are not particularly impressive.

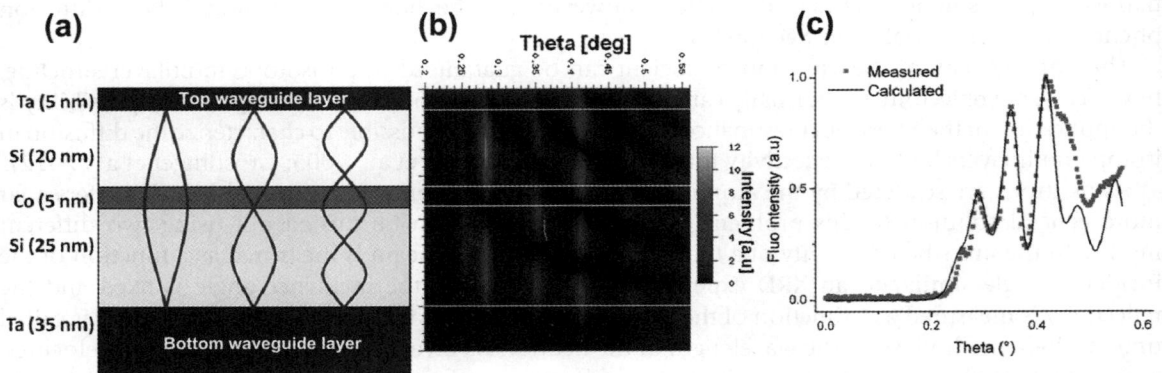

Figure 34   Exemplary structure (a), the calculated wave intensity for irradiation by 7.9 keV X-rays with varying incident angle (b), the resulting fluorescence intensity (c). Data provided by Erdélyi et al. (2009). (For color version of this figure, the reader is referred to the online version of this book.)

5.4.3.1   Microanalysis Electron Probe Microanalysis (EPMA)/Electron Microprobe Analysis (EMPA) Scanning Electron Microscope

High kinetic energy electrons are capable of ejecting a strongly bound core electron, with the consequence of producing characteristic X-ray photons or Auger electrons similar as previously discussed (see **Figure 30**). Electron microprobe analysis uses an electron beam of 10–50 keV energy to excite characteristic X-ray photons. By appropriate electrostatic or magnetic lenses, the electron probe can be focused down to a radius in the nanometer range. Since, however, electron trajectories are quite long within the bulk material, the interaction volume, which emits X-rays, is significantly larger than the probe. This leads to a spatial resolution near to a micrometer. By lowering the incident electron energy an improvement in the resolution can be achieved at the cost of lower signal/noise ratio (Gauvin et al., 2012). A composition profile can be derived by scanning the electron beam over the specimen.

The first electron probe microanalyzer was produced by Castaing (1951). The physical principles and technical requirements for an electron beam microanalyzer and a scanning electron microscope (SEM) are similar. Therefore, contemporary SEMs are often equipped with X-ray detectors, and act as electron microprobes too.

Two methods are being used to analyze the characteristic X-rays: wavelength-dispersive X-ray spectrometry and EDX. In the former, diffraction by a single crystal is utilized to disperse the collimated X-ray beam into different angles according to their wavelength. Scintillation or proportional counters are used to collect the X-ray photons at the different angles. In a modern wavelength-dispersive spectrometer, even a series of single crystals is used to cover a wavelength interval as broad as possible. In the latter method, the energy of the X-rays is analyzed by a solid-state detector (e.g. Li-doped Si, or more recently high-purity Si drift detectors). The main advantage of the first method is the better wavelength resolution of the single crystal analyzers, which allows a better separation of elements with small Z difference (e.g. S and P) and the better detection limit (down to 0.1 at% as compared with the ~ 1 at% for energy-dispersive spectrometers). Wavelength-dispersive spectrometers can also usually analyze a broader range of elements (with the exception of H, He and Li). The main advantage of the energy-dispersive spectrometers is their simpler realization and operation. Furthermore, the spectrometer detects the whole energy spectrum simultaneously, allowing a significantly faster analysis. Normally the method can detect elements from Na using a conventional Be window. Previous windowless designs made the detection of lighter element from B available. Current state of the art, large area Si drift detectors allows the quantitative analysis of Li and Be too.

SEMs also offer a third possibility to create a composition profile, by the so-called Z-contrast. This technique uses the elastically backscattered electrons. The backscattering yield is higher for elements with higher atomic number Z. Using calibration points of given composition, the determination of the local averaged atomic number with a $\Delta Z \sim 0.1$ precision can be carried out on well-prepared specimen. For a binary (or quasibinary) system this allows the determination of the local composition. The Z-contrast is larger for elements with lower atomic numbers (Ball and McCartney, 1981).

Electron probe microanalysis is capable of constructing a chemical map with submicrometer resolution. It can be mainly used to study interdiffusion of species. Combined with electron microscopy, to correlate structural properties (e.g. the position of the inert markers or crystal structure) highly accurate characterization of the atomic transport properties is possible (see, e.g. Ref. (Paul et al., 2005b)).

5.4.3.2   Analytical Scanning Transmission Electron Microscope

The realization of a practical scanning transmission electron microscope became possible in the late 1960s (Crewe et al., 1969, 1970). In this technique, a sharply focused electron beam is scanned over a thin sample. By interaction with the specimen, a part of the electrons are scattered, and so local information can be obtained by counting the nonscattered electrons, or the electrons scattered to a given angle. Since the scatter cross-section is a monotonous function of the atomic number, the images contain information on the local atomic number (at least for samples of homogenous thickness). Therefore, Z-contrast imaging using a high-angle annular dark field detection mode is possible in this method too (Pennycook and Jensen, 1991).

This arrangement allows the same chemical mapping opportunities as in an electron microanalyzer. Because of the thin specimen, there are virtually no multiple electron scattering events. Therefore, the X-ray photons are coming only from the illuminated spot, allowing a much better spatial resolution than in an electron microprobe/SEM.

Inelastic electron scattering events contain additional information about the target (the electrons in the specimen). Characteristic energy losses can be followed by an electron spectrometer. The method is called electron energy loss spectrometry (EELS); an example is shown in **Figure 35**. The measured spectrum contains the zero-loss peak (electrons that suffered no inelastic scattering), the plasmon peaks (collective excitations of the specimen electron system), and the characteristic atomic absorption edges. Evaluation of the absorption edges allows a quantitative chemical mapping. The electron energy loss spectrum not only contains information about the chemical nature of the atoms but also can be used to identify the different chemical surroundings of the measured atom (e.g. oxidation states) (Laffont et al., 2006). As can be seen in **Figure 35**, the intensity falls rapidly with the amount of energy loss; therefore, this method provides better results in the analysis of lighter elements. Accordingly, EELS is a good complement to energy-dispersive X-ray spectrometry, since the latter is more accurate for heavier elements.

5.4.3.3   Atom Probe Tomography

APT evolved from field ion microscopy, which was the first microscopy technique capable of resolving atomic lattice planes (Müller, 1951). If an external field in the range of $\sim 50\,\text{Vnm}^{-1}$ is applied to a specimen, onset of field evaporation of ionized atoms is observed. Such high fields can be achieved by a convenient voltage of a few kV, if the apex radius of used tip-shaped samples is below 100 nm. The

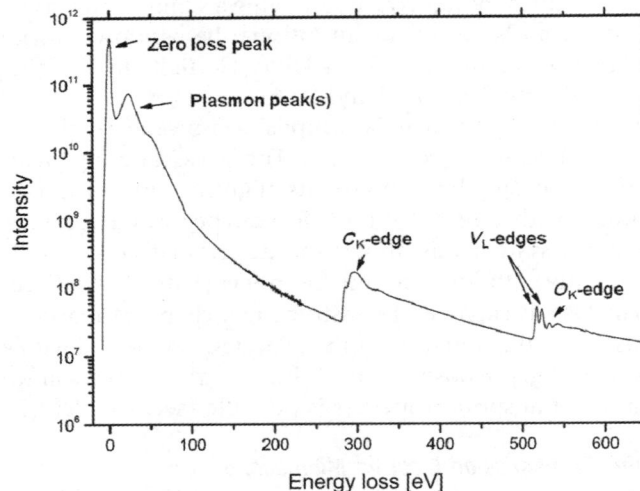

Figure 35   EELS spectrum on V_2O_5 obtained in a modern energy-filtered TEM with the most important features (zero-loss peak, plasmon peaks and the characteristic edges). Note the logarithmic scale. (Tobias Gallasch, Univ. of Muenster, 2011 (Gallash et al., 2011)).

desorbed ions follow a trajectory determined by the shape of the external field, and the impact position is detected by a 2D detector system of single ion sensitivity. A very recent textbook about the technique was written by Gault et al. (2012a).

If the voltage is kept below the so-called evaporation threshold, and additional short electric pulses or thermal spikes via laser irradiation are applied, it is possible to trigger the field evaporation. By recording the flight time between the trigger and the impact on the detector, a time-of-flight mass spectrometry becomes feasible. Furthermore by calculating the ion trajectories, a full tomographic reconstruction of the original sample is possible applying appropriate reconstruction algorithms. Modern wide-angle atom probe instruments are capable of analyzing specimens of a few micrometers in length and up to 100 nm radius.

The decisive advantages of the APT are

● Depending on the investigated material, it can have lattice resolution in the depth scale and at least subnanometer lateral resolution. Consequently, three-dimensional chemical maps can be produced with similar resolution in all directions. This allows identifying the composition field of linear defects such as dislocations (Blavette et al., 1999) or triple junctions (Stender et al., 2011). The single atom sensitivity is outstanding. In suitable cases limited structural information can also be gained (Gault et al., 2012b), so correlating structures with composition fields is possible.

● APT is "projection free", and evaluation is done after reconstructing a virtual representation of the specimen (see **Figure 36**). Therefore, "sectioning" to produce a local composition profile can be done at any region and any alignment, so local analysis of even curved grain boundaries is well possible (Chellali et al., 2012). Furthermore, many different tools, such as the isoconcentration surfaces shown in **Figure 36**c (which highlights the volumes above a given concentration threshold), could be used to better designate nanometric regions of interests.

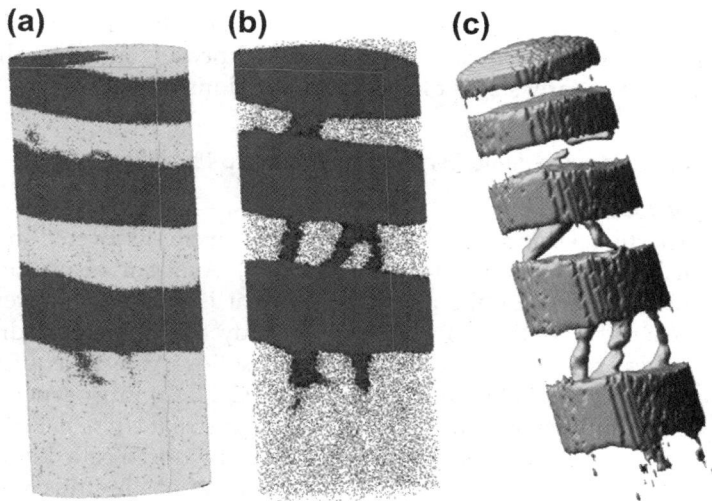

Figure 36   (a) Part of an atom probe analysis of an Fe/Cr multilayer. Fe atoms are represented by red dots while Cr is green. (b) Plotting only the Fe atoms, pipe-like structures appear in the Cr layers that indicate triple junction transport. (c) Isoconcentration surfaces are one of the visualization tools helping to designate regions of interest (Stender et al., 2011).

- In field evaporation, only a fraction of the surface atoms can be desorbed at a given time, since the evaporation probability strongly depends on the effective local field. The evaporation sequence is therefore to its major part deterministic. Accordingly, no surface roughening happens during depth profiling. Furthermore, the evaporation process is gentle, that is it induces no rearrangement of atoms. Thus, 'ballistic' mixing can also be avoided. Owing to these beneficial properties, the depth resolution of the atom probe does not deteriorate with depth, unlike the ion bombardment-based methods (Moutanabbir et al., 2011).
- Since the species are identified by mass spectrometry, atom probe is isotopically sensitive, so self-diffusion can be studied (Moutanabbir et al., 2011; Shimizu et al., 2011).
- Diffusion of low mass elements down to hydrogen can be studied as well. Only the residual atmosphere of the typical turbomolecular pumps and cryopumps used by the atom probe causes possible restrictions which may be circumvented by using isotopes like deuterium to achieve an adequate signal to noise ratio (Gemma et al., 2009).

On the other hand main challenges and disadvantages are

- The method can only investigate sharp, needle-shaped specimens. While by using focused ion beam cutting methods it is nowadays possible to fabricate atom probe tips from almost any solid, this still limits the investigated volume to less than a cubic micrometer.
- The chemical detection limit is rather high, around 100 ppm. In unfortunate cases it is even higher.
- The reconstruction method itself is uncalibrated; without the help of good distance markers (e.g. lattice periodicity) or an accurate atomic density map, the lateral and depth scale can have relatively large distortions.
- Based on projective geometry, the specimen itself represents the "microscope". Therefore, irregular specimen shape would cause distortions in the reconstructed sample. Different correlative microscopic techniques are actively researched to minimize this problem (e.g. (Dimitrieva et al., 2011; Haley et al., 2011)).

In summary, if the reconstruction and evaluation are done in a careful manner, APT offers a powerful measurement technique in studying nanoscale diffusion. It is especially suited for the study of diffusion along internal interfaces or the early stages of solid-state reactions.

5.4.4 Indirect Methods to Determine Jump Geometries, Rates and Mobilities

5.4.4.1 Mechanical Spectroscopy

Thermally activated anelastic relaxation phenomena in solids were discovered in the first half of the twentieth century (Gorski, 1935; Snoek, 1941; Zener, 1943). Anelastic relaxation happens after the instantaneous elastic response, but unlike plastic behavior it is completely reversible. Measured as internal friction, the delayed relaxation can be used to study the elementary diffusion steps, since constant or oscillating stress can induce atomic motion in solids. As a prerequisite, studied solutes, solute–defect complexes, and especially interstitials (e.g. H, C, N, O) must generate local unisotropic strain in the lattice, which can interact with the macroscopic external stress field.

A quantitative description of the anelastic behavior is given by the so-called linear response model. In this, the Hooke law is modified to include the time derivatives of the stress σ and the strain ε

$$\sigma + \tau_\varepsilon \dot{\sigma} = M_R(\varepsilon + \tau_\sigma \dot{\varepsilon}) \tag{120}$$

where M_R is the relaxed elastic modulus and τ_σ and τ_ε are the stress and strain relaxation time, respectively. If the time derivates are vanishing, Eqn (120) represents exactly the Hooke law. Under application of uniaxial stress, the appropriate M_R is the Young modulus, while for shear it is the shear modulus. If the rate of change of the stress is high, then

$$\dot{\sigma} \approx M_R \frac{\tau_\sigma}{\tau_\varepsilon} \dot{\varepsilon} = M_U \dot{\varepsilon} \tag{121}$$

The new modulus M_U is called the unrelaxed elastic modulus.

If periodic stress is applied to the material, a phase shift between stress and strain is observed, which results in a hysteresis loop at intermediate frequencies in the stress strain diagram. This behavior is called internal friction and can be characterized by the Q^{-1} coefficient which measures the ratio of the anelastically dissipated energy and the peak-stored elastic energy. The internal friction is maximal (**Figure 37** shows the dependence of the Q^{-1} from ω_{exc}) for the exciting frequency

$$\omega_{exc} = \frac{1}{\sqrt{\tau_\varepsilon \tau_\sigma}} = \tau^{-1} \tag{122}$$

where τ is called mean relaxation time. τ is usually thermally activated according to an Arrhenius temperature dependence. In a typical measurement, the excitation frequency is kept at a given value and the energy loss is measured as a function of the temperature. Accordingly, the loss peak shifts to higher temperatures as the excitation frequency increases.

The activation enthalpy can be calculated as

$$\Delta H = -k_B \frac{d(\ln \omega_{exc})}{d(T^{-1})} \tag{123}$$

Point defects can act as elastic dipoles in a solid. Reorienting them parallel to the stress results in anelastic relaxation. If this relaxation is achieved by a diffusion jump, then a simple relation between diffusion coefficient and relaxation time exists. The Snoek effect (Snoek, 1941) is the prime example of this behavior, which is related to the stress-induced migration of interstitial C, N and O in bcc materials. The octahedral sites occupied by these impurities reveal a tetragonal symmetry (distance to the nearest matrix atoms is smaller in the $\langle 100 \rangle$ direction than in the $\langle 110 \rangle$ direction). Consequently, the strain fields of the interstitials are also characterized by tetragonal symmetry. For an unstrained material all octahedral sites are energetically equivalent, however, if uniaxial stress is applied to the specimen (other than along the $\langle 111 \rangle$ direction), then a degeneration of the octahedral sites is induced. This causes a change of occupation of the possible octahedral sites, if the interstitials have enough mobility. So, the relaxation time of the internal friction measurement contains information on the mean residence time of the solutes at a given interstitial site, and evidently on the diffusion coefficient:

$$D = \frac{1}{36} \frac{a^2}{\tau_R} \tag{124}$$

where τ_R is the relaxation time and a is the lattice parameter. (For the derivation of this diffusion equation consult, e.g. Ref. (Mehrer, 2007)). The Snoek effect is an extremely sensitive technique, which allows studying interstitial diffusion in bcc and hcp materials down to the 10^{-24} m^2 s^{-1} diffusivity range. As a main drawback of the technique interstitials in fcc materials cannot be studied, since the octahedral sites have a cubic symmetry there (except those of a dumbbell structure).

The Zener effect (Zener, 1943) is similar to the Snoek effect; however, the elastic dipoles are formed by interstitial pairs, substitutional solute pairs, solute–vacancy complexes, and so on. This means that

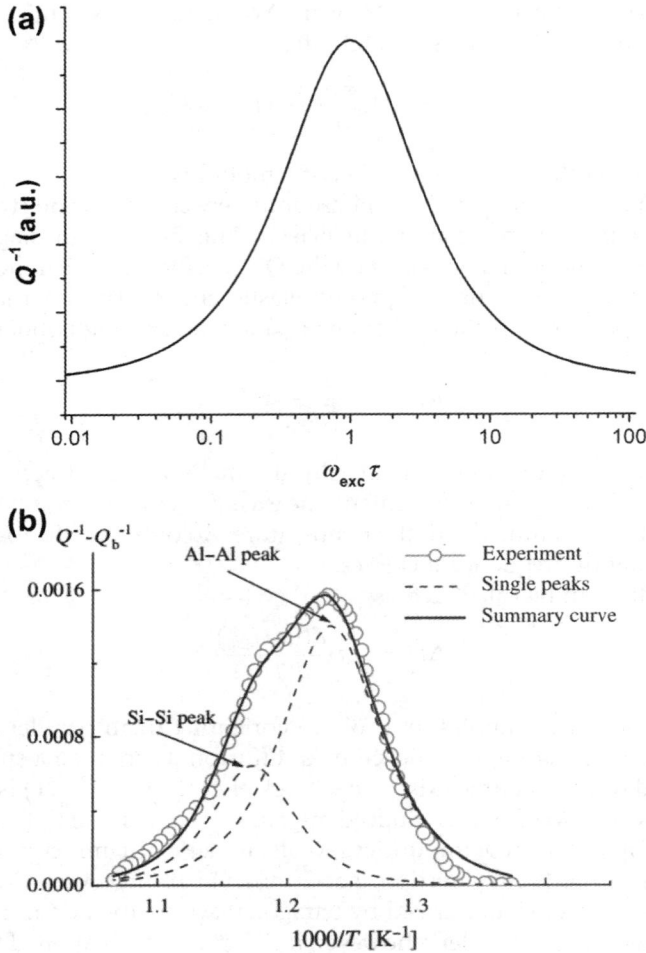

Figure 37 (a) Theoretical shape of the internal friction as a function of the exciting frequency, the Q^{-1} shows a maximum at $\omega_{exc} = \tau^{-1}$. (b) Experimental internal friction curve measured in a ternary $Fe_{0.75}Al_{0.20}Si_{0.5}$ alloy. Figure taken from Pavlova et al. (2006). The internal friction peak contains the contribution both from the Al and Si.

the reorientation of the elastic dipole is usually realized by series of elementary jumps; therefore, correlating the diffusion coefficient with the relaxation time is not as straightforward as before.

By bending a specimen of thickness d, a macroscopic strain gradient is established, which produced a gradient in the chemical potential for the solutes. This could cause a macroscopic transport of the solutes as the net volume transport reduces the strain. This effect was first observed by Gorski (1935), and unlike the Snoek or the Zener relaxation, it is a result of a real long-range diffusion transport. The τ_G Gorski relaxation time can be expressed as

$$\tau_G = \frac{d^2}{\pi^2 \Phi D_B} \tag{125}$$

where D_B is the diffusion coefficient of species B and Φ is the thermodynamic factor, which is present because Gorski relaxation produces an "up-hill" diffusion leading to a composition gradient. The diffusion must be rather fast to measure the Gorski relaxation, which only allows investigating H diffusion. In that field, Gorski relaxation studies are quite successful, information on the grain boundary diffusion can be even gained by this method (Sinning, 2000).

5.4.4.2 Nuclear Magnetic Resonance and Nuclear Magnetic Relaxation

In a static magnetic field of field strength B_0, nuclei with nonvanishing magnetic moment are precessing by the Larmor frequency

$$\omega_{Larm} = \frac{qgB_0}{2m} = \gamma B_0 \tag{126}$$

where q, m and g are the charge, mass and g-factor of the nucleus, respectively, and γ is called gyromagnetic ratio. The latter parameters are known for the nuclei of interest. Due to the Zeeman effect, a degeneration into $2J + 1$ energy levels arises, $U_M = J_z \hbar \omega_{Larm}$, in which J_z is an integer quantum number between $-J$ and J. At thermal equilibrium the magnetic moments are distributed according to the Boltzmann statistics. As the degeneration of quantum states is small for magnetic fields below 1 T, the population of the states shows little variation. For a system containing N nuclei per volume unit, a magnetic field of B_0 leads to an equilibrium magnetization of

$$M^{eq} = N\gamma^2 \frac{\hbar^2}{3k_B T}(J+1)JB_0. \tag{127}$$

If a radio frequency magnetic field with a frequency close to the Larmor frequency and field strength B_1 is applied transversally to B_0, transitions between the degenerated Zeeman levels are excited (**Figure 40(a)**). Experimentally, this is realized by wrapping a coil around the specimen (see **Figure 38**). If the frequency of the B_1 field fulfills the resonance condition (Eqn. (126)), the magnetic moment will be tilted in the y–z plane with an angular velocity of γB_1. By applying a suitable pulse length t_p of B_1, the magnetic moments can be inverted $(t_p = \pi/(\gamma B_1))$ or tilted into the x–y plane $(t_p = \pi/(2\gamma B_1))$. If the magnetic moment M gets a component perpendicular to the

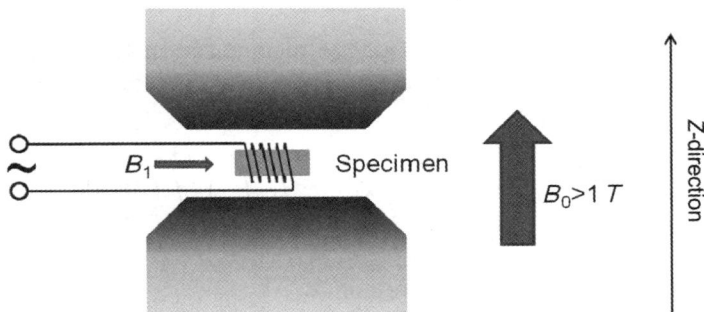

Figure 38 Geometry of a nuclear magnetic resonance experiment. (For color version of this figure, the reader is referred to the online version of this book.)

external field B_0, a precession in the x–y plane will occur. This precession will induce an a.c. voltage in the coil.

The evolution of the magnetic moment of the sample within the combined field is described by the Bloch equations modified by Torrey (1956):

$$\frac{\partial M_z}{\partial t} = \gamma\left(M \times B\right)_z - \frac{M_z - M_z^{eq}}{T_1} + \nabla\left[D\nabla\left(M - M_{eq}\right)_z\right] \tag{128a}$$

$$\frac{\partial M_x}{\partial t} = \gamma\left(M \times B\right)_x - \frac{M_x}{T_2} + \nabla\left[D\nabla\left(M - M_{eq}\right)_x\right] \tag{128b}$$

$$\frac{\partial M_y}{\partial t} = \gamma\left(M \times B\right)_y - \frac{M_y}{T_2} + \nabla\left[D\nabla\left(M - M_{eq}\right)_y\right] \tag{128c}$$

B in these equations is the vectorial sum of B_1 and B_0. The first term on the RHSs of these equations represents the precession of the spins around the magnetic field. The second terms contain the phenomenological constants T_1 and T_2. If no transverse field is present, then T_1 determines the time necessary for the system to return to the equilibrium magnetization from a given M_z value. This represents a coupling between the spin system and any other degree of freedom in the material, which is called "lattice" in this technique; the T_1 coefficient is therefore called 'spin-lattice' relaxation time. T_2 represents the relaxation of the transverse magnetic moments, that is the relaxation of the ensemble of the nuclear magnetic moments toward isotropic distribution laterally. This term is important, since T_2 is smaller than T_1, and accordingly, the nuclear magnetic moments can reach a quasi-equilibrium state in the lateral plane before the system of all moments equilibrates with the "lattice". The coefficient T_2 is called spin–spin relaxation and is strongly related to the measured width of the nuclear magnetic resonance (NMR) signal.

The third term on the RHSs of Eqn (128), introduced by Torrey (1956), contains information about the diffusion/transport of the magnetic moments in a gradient field. Magnetic moment diffusion can happen by the diffusion of the atomic nucleus carrying magnetic moment or by spin exchange. Spin exchange, however, is almost insignificant in comparison with atomic diffusion. Assuming that D represents practically the atomic diffusion coefficient is therefore well justified. Corrections for the case of nonnegligible spin exchange can be found in the original paper of Torrey (1956). Indeed, solving Eqn (128) for a measurement in a gradient magnetic field allows evaluation of diffusion coefficients without any further assumption.

Having prepared initially the z component of the magnetization to a value M_0 (the transversal component is zero), it will be later after time t_{echo}:

$$M(t_{echo}) = M_0 \exp\left(-\frac{t_{echo}}{T_2}\right)\exp\left[-\gamma^2 D \int\limits_0^{t_{echo}}\left(\int\limits_0^{t'} G(t'')dt''\right)^2 dt'\right] \tag{129}$$

if a field gradient in z-direction $G := \partial B/\partial z$ is present.

The most simple realization of a so-called spin echo experiment (Hahn, 1950) is the $\pi/2 - \tau - \pi - \tau$ experiment, where $\tau = \frac{1}{2}t_{echo}$. By a first $\pi/2$ pulse, the magnetic moment is tilted to the x–y plane. Let us first neglect spin–spin relaxation and atomic diffusion. The spins start precessing, however, due to the field gradient with different angular velocities. So, a loss of coherency appears which results in a decrement of the total transversal magnetic moment. After a time $\tau = {}^1/_2\, t_{echo}$ however, an inversion π pulse is applied. The spin system is mirrored in the x–y plane; consequently the moments with lower precession frequency will find themselves "ahead" of the faster ones. Therefore, the whole spin system will become coherent again after exactly t_{echo} (twice the time τ) and the transversal magnetic moment reaches the initial value M_0. The echo signal is measured by the sensor coil. In reality the transverse magnetic moment decreases with time due to the spin–spin relaxation (as stated by the first exponential factor on the RHS in Eqn (129)) and due to diffusion of the magnetic moments in the gradient field (as stated by the second exponential factor). By varying the gradient G and t_{echo}, the diffusion coefficient of the atoms can be evaluated.

The decay due to the atomic diffusion, however, has to be quicker than the decay due to the spin–spin relaxation. Thus, this so-called spin echo technique is applicable to study very fast atomic transport processes such as H-diffusion. Experiments are usually performed with a pulsed field gradient to achieve the highest possible field gradient (Stejskal and Tanner, 1965).

Alternatively nuclear magnetic relaxation techniques evaluate the diffusion coefficient by analyzing the relaxation times T_1 and T_2. The time T_1 can be measured when no transversal magnetic field is present. Then Eqn (128a) simplifies to

$$\frac{\partial M_z}{\partial t} = -\frac{M_z - M_z^{eq}}{T_1} \tag{130}$$

Figure 39 illustrates the most simple measurement sequence to determine T_1. First, an inversion pulse is applied, which is followed by a relaxation period during a time interval τ. Then, using a $\pi/2$

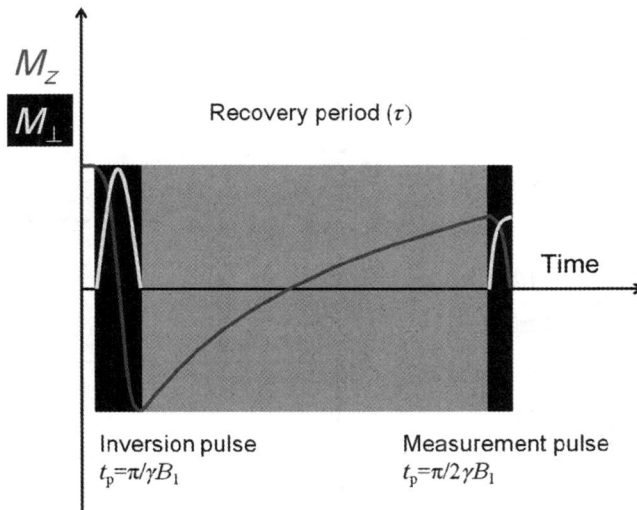

Figure 39 Schematic illustration of an inversion–recovery measurement for measuring T_1. First the magnetization M_z is switched to the inverted state by the so-called inversion pulse. M_z starts relaxing with characteristic time T_1. Then after a relaxation time τ, it is tilted into the transversal direction by the measurement pulse and the absolute value of the momentary M is determined.

pulse, the magnetic moment is tilted into the x–y plane so that its value can be measured. Repeating this sequence for different τ, the relaxation curve $M_z(\tau)$ is determined and T_1 can be evaluated. The main component which determines the spin–lattice relaxation is the dipole–dipole interaction of the magnetic moments, and for nuclei with $J > 1/2$, the quadrupole interactions (where J is the angular momentum of the nucleus). Both of these are influenced by the atomic jumps, for example Ref (Torchia and Szabo, 1982).

For nuclei with $J = 1/2$, only dipole spin–spin interactions take place. The nuclei precess in a magnetic field, which is a combination of the external B_{ext} and the local field fluctuation B_{local} produced by the dipoles of the neighboring nuclei (since a dipole field decays with r^{-3}, only the very near nuclei have to be taken into account). Since these local fields vary randomly, a dispersion appears in the Larmor frequency, which leads to a broadening of the resonance line. This broadening is proportional to the local field fluctuations

$$\Delta\omega_{Larm} = \frac{1}{T_2} \sim \gamma \Delta B_{local}. \tag{131}$$

The effective field sensed by any nucleus is the time average of the local fields experienced within the T_2 relaxation time. If the local fields are randomly distributed, the average fluctuation is large for a nonmoving nucleus and smaller for a diffusing one which samples and so averages many different neighborhoods. In consequence, the resonance line becomes sharper as result of the diffusion (**Figure 40**). This effect is called motional narrowing. In consequence, the T_2 spin–spin relaxation time contains the desired information about the diffusion length.

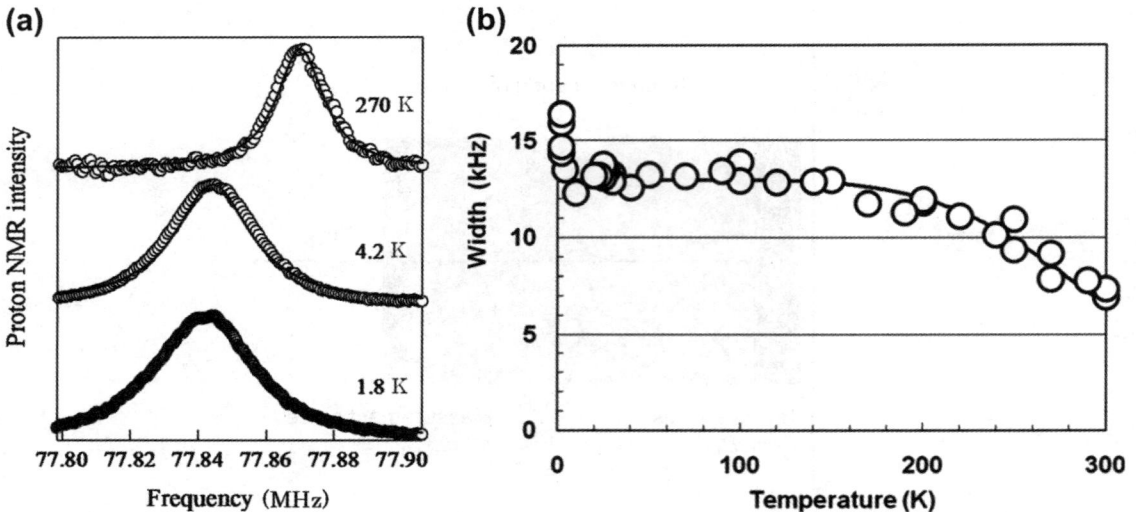

Figure 40   (a) Proton NMR spectra in an $Ni_{32.4}Nb_{21.6}Zr_{36}H_{10}$ metallic glass at 1.8 K, 4.2 K and at 270 K. (b) At temperatures above 150 K the onset of motional narrowing of the resonance can be observed. Reproduced with permission from Niki et al. (2012).

5.4.4.3 Mössbauer Spectroscopy and Quasielastic Neutron Scattering

The Mössbauer effect means the recoilless emission and absorption of γ-quantums by atomic nuclei. The momentum and the energy of the recoil are absorbed by the whole crystal, which results in a γ-photon with extremely narrow line width. It was detected by Mössbauer (1958). Mössbauer spectroscopy (MBS) uses two specimens; one of them serves as the γ-source emitting quanta of energy E_γ, while the other acts as absorber. The source is moved with a velocity v with respect to the absorber, which produces a Doppler shift $\Delta E = E_\gamma v/c$. The line width of the absorption spectra can then be analyzed in dependence on the relative velocity.

Diffusion jumps cause a broadening of the Mössbauer linewidth (Singwi and Sjölander, 1960). Without diffusion only the natural line width Γ_0, which is determined by the lifetime of the excited atomic state τ_N, can be observed ($\Gamma_0 \approx \hbar/\tau_N$). If the mean residence time τ_R is comparable with the τ_N lifetime, then the wave packet is cut into a series of shorter wave packets. If $\tau_R \ll \tau_N$, then the linewidth is increased by $\Delta\Gamma = \Gamma - \Gamma_0 \approx \hbar/\tau_R$.

Neglecting the correlation effects, the diffusion coefficient on a Bravais lattice with jump distance λ can be expressed as

$$D \approx \frac{\lambda^2}{12} \frac{\Delta\Gamma}{\hbar} \tag{132}$$

Only four isotopes exist which allow diffusion investigations via MBS: ^{57}Fe, ^{119}Sn, ^{151}Eu and ^{161}Dy. Not surprisingly, most investigations were carried out by using ^{57}Fe. The diffusion process itself must be fast enough, which means D is above 10^{-15} m^2 s^{-1}.

In an elastic neutron scattering event, similarly to the Mössbauer effect, the momentum is absorbed by the lattice itself (Vogl, 1996). In a neutron diffraction spectrum, the elastic scattering event appears as a narrow Bragg line. Similarly to the Mössbauer effect, atomic migration produces a broadening effect (**Figure 41**). This phenomenon is called quasielastic neutron scattering (QENS). Using this method, diffusion jumps comparable with the neutron wavelength can be studied. Between line width and the diffusion coefficient the relation

$$\Delta E = 2\hbar D Q^2 \tag{133}$$

holds, in which $Q = 4\pi \sin(\Theta/2)/\Lambda_0$ is the momentum transfer, Θ the scattering angle, Λ_0 the wavelength of the incident neutron beam and ΔE represents the line width.

The dispersion relation of the neutrons offers a unique capability to investigate the structural and dynamical properties of solids, since the frequencies of the neutron wave are comparable with lattice vibration frequencies, while their wavelength is comparable with the lattice parameter. Furthermore, neutrons are uncharged and they interact with the nuclei only; consequently, their penetration depth can be very high.

Measurement of diffusion by QENS requires rather high diffusion coefficients. It is mainly used to study H-diffusion (Hempelmann, 2000) but also applicable to study diffusion in liquids (Brillo et al., 2008).

5.4.4.4 Positron Annihilation

Positrons are the antiparticles of the electrons. They are created, for example by a β$^+$ nuclear decay event, The emitted positrons are very quickly thermalized in a lattice. A collision of a thermal positron and electron will produce two γ-photons, each with a characteristic energy of 511 keV.

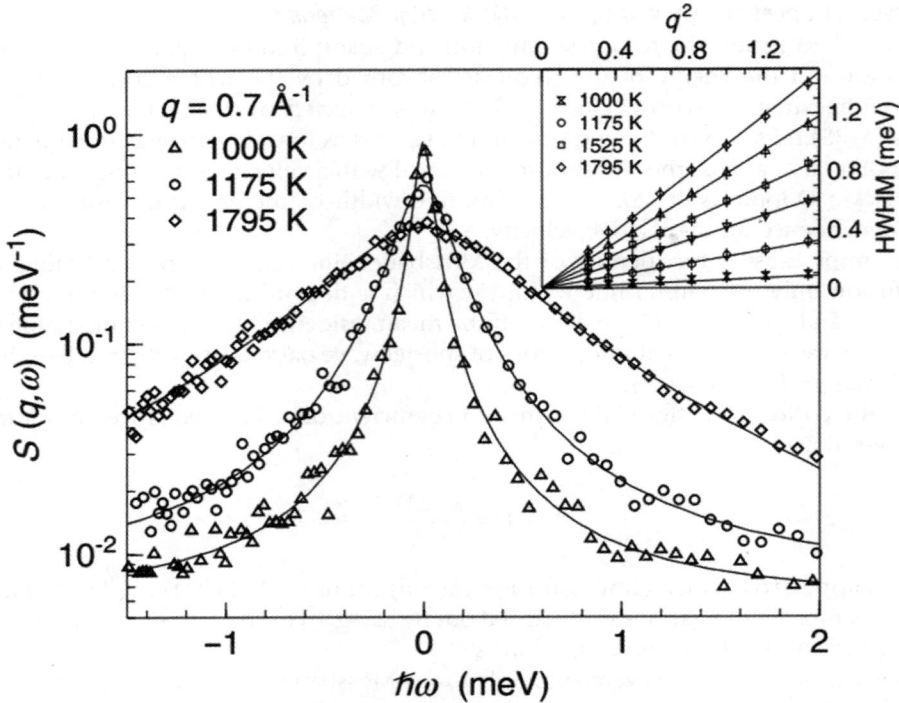

Figure 41 Broadening of the QENS peak as result of the increasing diffusion in an $Al_{80}Cu_{20}$ melt. The inset shows the dependence of the peak width on the square of the momentum transfer q^2. Figure taken from Brillo et al. (2008).

Positron annihilation (PA) spectroscopy itself is not a tool to measure the diffusion coefficient. However, by characterizing vacancy properties it provides very valuable input parameters for under-standing of the diffusion mechanism. This is especially important in intermetallics, where the point defect concentration in the different sublattices can show dramatic differences and possibly strong composition dependence. The two main methods of PA spectroscopy are measurements of the positron lifetime and the analysis of the Doppler broadening. For an introductory review of the techniques the reader may consult, for example Ref. (West, 1973).

The lifetime of the positrons depends on the probability of electron–positron collision, and there-fore on the local electron density. Vacancies are negatively charged lattice defects that attract and trap positrons. Furthermore, the electron density around them is smaller; meaning the lifetime τ_T of trapped positrons is larger than the lifetime of free positrons inside the lattice τ_F. Thus, the mean positron lifetime τ can be expressed as

$$\tau = \tau_F \frac{1 + \tau_T c_V \sigma}{1 + \tau_F c_V \sigma}, \tag{134}$$

In which c_V is the monovacancy concentration and σ the cross-section of trapping. Since $\tau_T > \tau_F$, this function increases monotonously with the monovacancy concentration. If c_V is sufficiently large, then all positrons are annihilated at a vacancy. Consequently, the method is not applicable to very high vacancy concentration above the 10^{-4} range. Vacancy migration can be studied in systems with nonequilibrium vacancy concentrations. By measuring the time required to reach equilibration one can derive the vacancy diffusivity (Würschum et al, 1995).

To measure the lifetime of the positron, a suitable radioisotope as positron source (e.g. ^{22}Na) can be used, where the β^+ decay is accompanied by the emission of a characteristic γ-photon. In this way both creation and annihilation of the positrons can be observed by detecting the corresponding γ radiation. As an alternative, also a pulsed positron beam may be used to implant positrons into the solid, for example (Hugenschmidt et al, 2005). Usually only the mean lifetime can be determined, however for high-quality spectra, containing a large number of individual events (e.g. 2×10^7 in Ref. Becvář et al., 1995), a deconvolution and the evaluation of the different lifetimes can be possible.

Doppler broadening spectroscopy is based on the fact that the kinetic energy of a thermalized positron is negligible; therefore any shift observed in the energy of the annihilation γ-photons is attributed to the electron momentums. If the annihilation event happens as a result of a delocalized conduction band electron, the Doppler shift will be less than 2 keV, while high momentum core electrons produce a much higher Doppler shift. Open volume defects, such as vacancies, drastically reduce the probability of core electron–positron annihilation events; therefore, a sharpening of the annihilation line can be observed. Using a coincidence detection system consisting of two diagonally placed high-purity Ge detectors, the γ-background can be very efficiently reduced, which allows the detection of the highly Doppler-shifted part of the annihilation peak. So, the properties of embedded layers, impurity clusters (Pikart et al., 2011), or the chemical neighborhood of defects can be studied (Takagiwa et al., 2006).

5.5 Elementary Diffusion Properties of Metals and Intermetallic Compounds

In this section an overview on the diffusive behavior of metallic materials is presented. Observing the general trends among different elements and lattice structures, it may become possible to quantitatively estimate diffusivities even in those cases in which no appropriate experimental data are available. Self-diffusion is certainly the most fundamental process which takes place in any solid metal and it allows conclusions on the typical jumping rate of the lattice atoms. Thus, it is not surprising that this property has been the most widely studied. In the following section we first summarize general properties of self-diffusion for major metallic lattice structures and present simple empirical rules to estimate quantitative parameters of self-diffusion. Second, we briefly describe the most important characteristics of impurity diffusion in close-packed metals. We address the remarkably fast transport in the case of open metals and of interstitial solutes. Finally, the complex situation of intermetallic compounds is reflected and exemplified by experimental results.

5.5.1 Self-Diffusion in Pure Metals

In a self-diffusion measurement usually radioactive or stable isotopes are used as tracers. Although not the most current source, the shear amount of data summarized in Landolt–Börnstein (1990a) and NIMS gives a good impression about the work carried out in the past. The temperature range from the melting point (T_m) down to half the melting point is covered for a wide range of metallic systems. This means a diffusivity range from 10^{-11} m^2 s^{-1} to 10^{-24} m^2 s^{-1}. Contemporary nuclear methods, such as neutron reflectometry (Schmidt et al, 2008), may allow the extension of this scale to 10^{-26} m^2 s^{-1}. Owing to the vast experimental work we nowadays have a good understanding of the general properties of self-diffusion in pure metals.

- Diffusion vehicles for self-diffusion in metals are vacancy-type defects, either mono- or divacancies (Adda and Philibert, 1966; Seeger et al., 1970; Heumann, 1992). This has been supported by a wide variety of experimental evidence. (1) The pioneering work of Kirkendall had shown that different atom species in substitutional alloys have different diffusivities (Smigelskas and Kirkendall, 1947). (2) Vacant lattice sites are the dominant (point) defects in pure metals in thermal equilibrium. This can be confirmed by various measurement techniques from dilatometry to PA spectroscopy (Bergersen and Scott, 1969). (3) Measured activation volumes in high-pressure experiments are in agreement with the values predicted by vacancy diffusion models (Mehrer, 1996). (4) The sum of the vacancy formation and migration enthalpies is in the range of the activation enthalpy of diffusion (Peterson, 1978; Mehrer, 1978).
- Diffusivities near the melting point are similar in metals of similar crystal structures, indeed diffusivities plotted in a homologous temperature scale (T_m/T) lie in a more or less narrow band for both bcc and fcc metals.
- The diffusion usually, although not exclusively, obeys the Arrhenius rule $D(T) = D_0\exp(-Q/k_BT)$, see Section 5.2, Eqn (32). The exact cause of possible deviations is still debated.

5.5.1.2 Self-Diffusion in fcc and hcp Metals

Self-diffusion coefficients of some fcc metals in a homologous temperature scale are shown in **Figure 42**. As a general characteristic, it can be stated that diffusion coefficients almost exclusively follow a strong Arrhenius rule. A slight upward curvature can be observed for some fcc metals (as shown in the inset of **Figure 42**). The most widely accepted explanation of the latter effect is an increased divacancy contribution (Backus et al., 1978; Anderson and Simak, 2004) although alternative theories to describe the observed deviation also exist (Varotsos and Alexopoulos, 1986; Sandberg and Grimwall, 2001). Thus a generally accepted theory is still lacking. Furthermore, Carling et al. claimed that vacancy–vacancy interactions are repulsive in Al, thus divacancies are fundamentally unstable (Carling et al., 2000). Uesugi et al. however found that while first nearest-neighbor vacancy pairs are indeed unstable in Al, second nearest-neighbor divacancies are energetically favored (Uesugi et al., 2003).

With the exception of Pb, the diffusivities at the melting point lie in the 10^{-13} m^2 s^{-1} and 10^{-12} m^2 s^{-1} range. And again with the exception of Pb, the difference between self-diffusivities does not exceed three orders of magnitudes at 0.5 T_m. Consequently the activation enthalpy ΔQ for self-diffusion shows a strong correlation with the melting point:

$$\Delta Q = 17RT_m \tag{135}$$

where R is the gas constant and T_m is the melting point of the considered fcc metal. This correlation is called the van Liempt rule (van Liempt, 1935).

The preexponential factor D_0 lies in the range from several 10^{-6} m^2 s^{-1} to few 10^{-4} m^2 s^{-1}. This implies that the diffusion entropy (the sum of vacancy formation entropy and migration entropy (Eqn (3))) falls in between 1 and 5 k_B. Within a given group of the periodic table, the diffusion coefficients (in the homologous scale) become larger for the heavier elements.

Hcp metals (such as Zn, Be, Cd, and Mg) are characterized by anisotropic transport properties. Diffusion along the hexagonal axis (c-axis) is slightly faster than perpendicular to it. Other than that, with the exception of Be, they are very similar to the fcc metals. Diffusivities at the melting point are near 10^{-12} m^2 s^{-1}, and the van Liempt rule estimates the activation enthalpy well. Activation enthalpies predicted by the van Liempt rule are compared with the measured values in **Figure 43**.

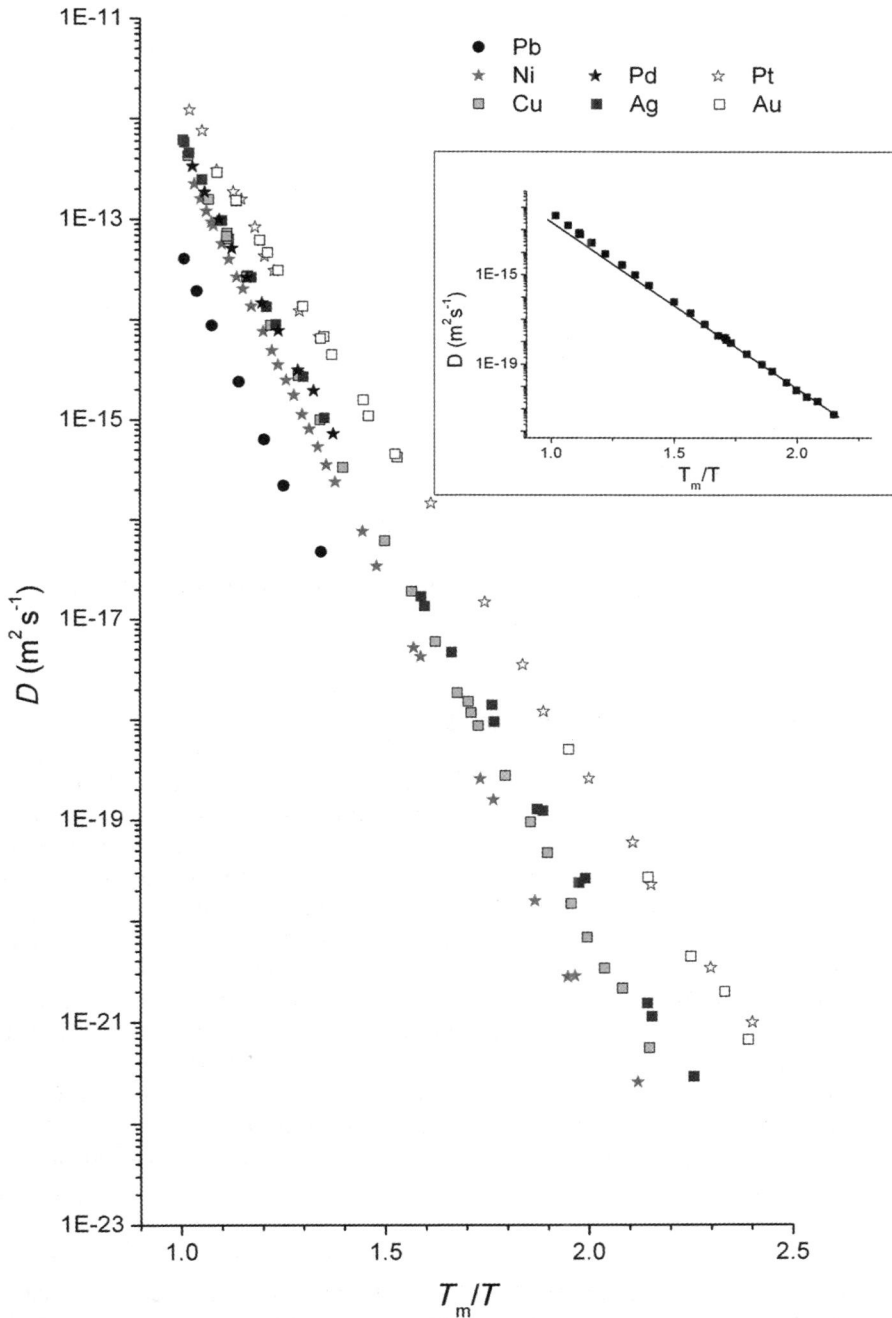

Figure 42 Self-diffusion coefficient in homologous temperature scale for fcc metals (source of the data Pb: (Nachtrieb and Handler, 1955), Ni: (Bakker, 1968; Maier et al., 1976), Pd: (Peterson, 1964), Pt: (Cattaneo et al., 1962; Rein et al., 1978), Cu: (Rothman and Peterson, 1969; Maier et al., 1973), Ag: (Rothman et al., 1970; Lam et al., 1973), Au: (Okkerse, 1956; Rupp et al., 1969)). The inset exemplifies the Cu self-diffusion for which a slight, but nevertheless clearly visible deviation from pure Arrhenius behavior appears.

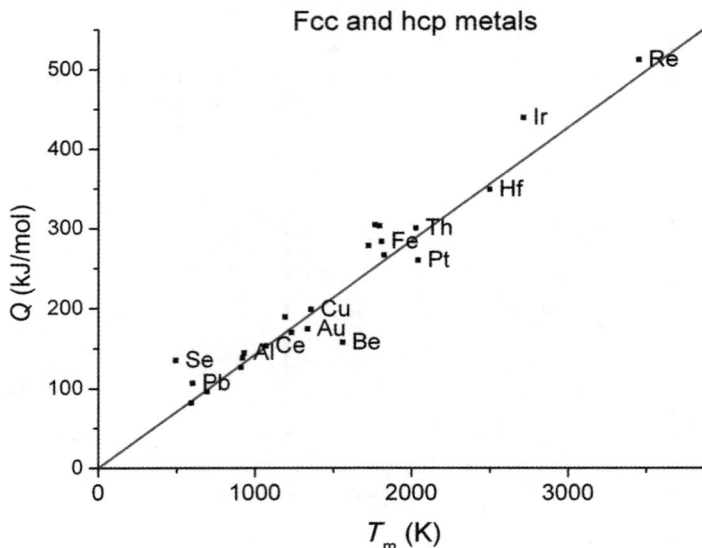

Figure 43 Activation enthalpy of self-diffusion as a function of the melting temperature in fcc and hcp metals. The line represents the van Liempt rule, a simple phenomenological relation.

5.5.1.3 Self-Diffusion in bcc and Tetragonal Lattice Structures

Self-diffusion coefficients in homologous temperature scale of some bcc metals are plotted in **Figure 44**. The scatter in the activation energies is larger than for the fcc metals, and some metals possess a very prominent upward curvature in the Arrhenius diagram. To a certain extent, the van Liempt rule holds for bcc metals too (see **Figure 45**). In spite of the larger scatter at lower temperatures, the diffusivities near the melting point are very similar and lie in the range of $10^{-12}\,\mathrm{m^2\,s^{-1}}$ and $10^{-11}\,\mathrm{m^2\,s^{-1}}$. So, diffusion near melting point is somewhat faster in bcc than in fcc metals. Diffusivities are similar for elements from the same group in the periodic system. Diffusion is the fastest in the alkali metals and the slowest in group VI metals (Cr, Mo, W). It is usually true, similarly to fcc metals, that the heavier elements are characterized by faster diffusion coefficients within one group of the periodic table. A notable exception from this rule is potassium. Group IV transition metals (Ti, Zr, Hf) undergo a phase transition and as a result their diffusion behavior is more complex.

It is generally true for bcc metals that the migration enthalpy H^{M} of vacancies is much smaller than their formation enthalpy H^{F}. $H^{\mathrm{M}}/H^{\mathrm{F}} \approx 0.2$ as compared with 0.8–0.9 in fcc metals. In bcc, material relaxation around vacancies plays a more important role than in close-packed structures; up to 13% relaxation toward the vacancies can happen (Willaime, 2001). This explains the large structural

Figure 44 Self-diffusion coefficients in homologous temperature scale for bcc metals. The scatter in diffusivities is much larger than in the case of close-packed structures. Note the metals from the same group of the periodic system (Na, Li, K or Ta, Nb, V or Cr, Mo, W) have similar diffusivities. The plotted diffusion coefficient for Li is the ^{7}Li tracer in an ^{6}Li matrix. Source data (Li: (Lodding et al., 1970), Na: (Mundy, 1971), K: (Mundy et al., 1971), V: (Pelleg, 1974), Nb: (Einziger et al., 1978), Ta: (Weiler et al., 1982), Cr: (Mundy et al., 1976; Mundy et al., 1981), Mo: (Maier et al., 1979), W: (Mundy et al., 1978)).

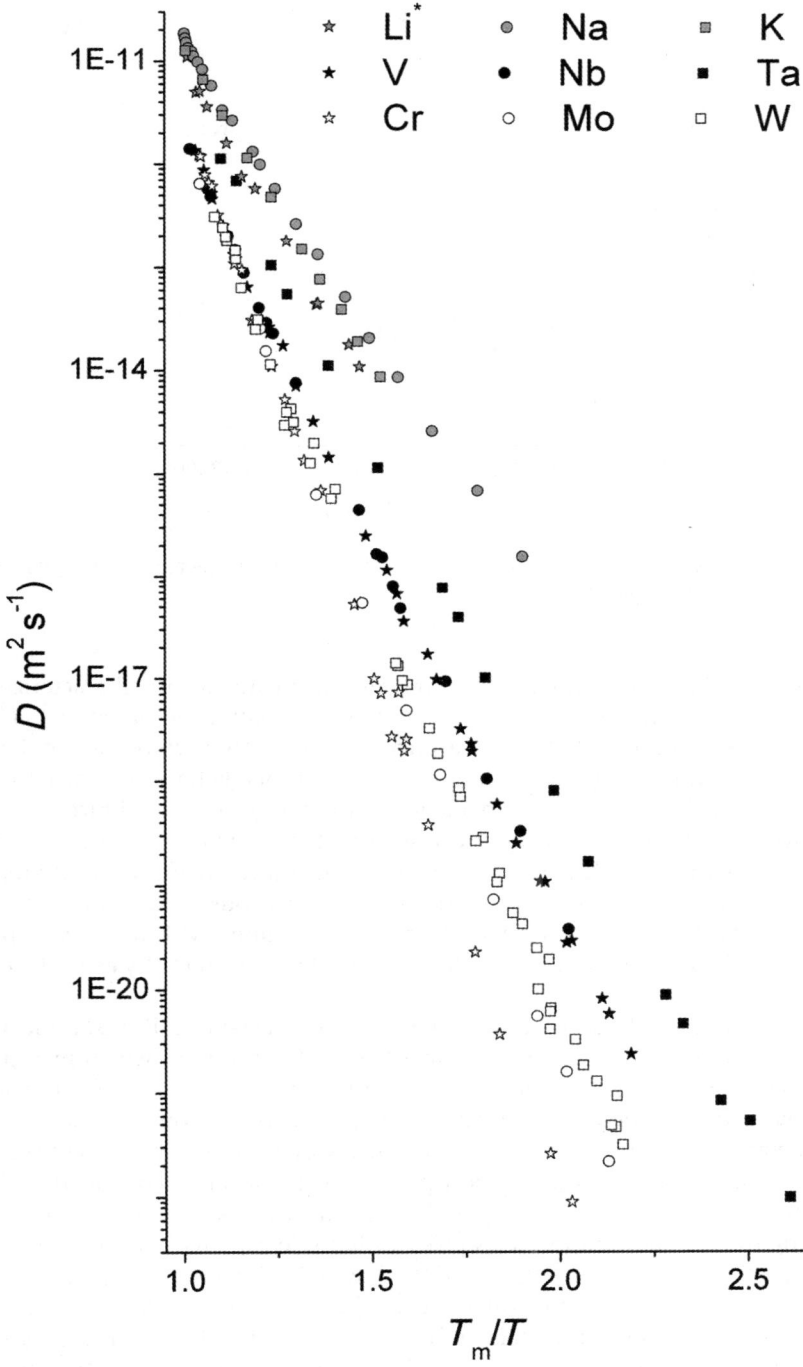

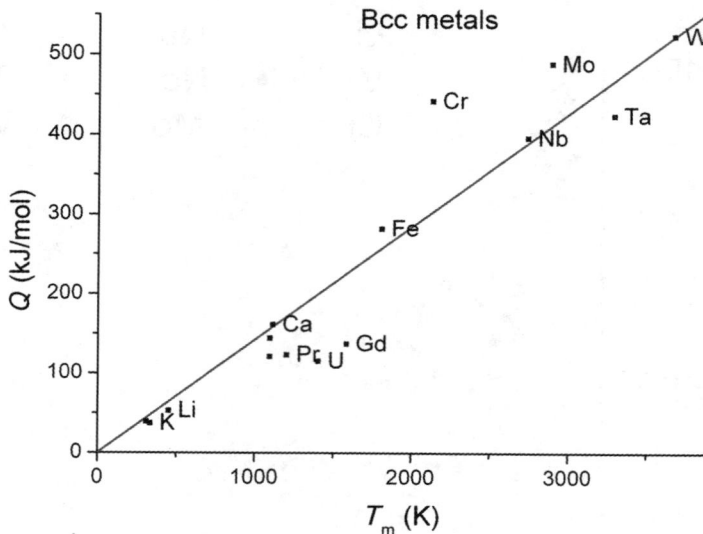

Figure 45 The activation enthalpy of self-diffusion as a function of the melting temperature in bcc metals. The line represents the van Liempt rule, a simple phenomenological relation.

relaxation and lower migration barriers in bcc metals. Alkali metals are characterized by an exceptionally small migration enthalpy as compared with the formation enthalpy $H^M/H^F \approx 0.1$ (Schulz, 1991). The calculated activation enthalpy for vacancy formation can even be greater than the measured activation enthalpy for diffusion. On the contrary, Group VI metals have somewhat higher migration enthalpies $H^M/H^F \approx 0.5$. This explains the behavior of the diffusivities in **Figure 44**.

In the past, multiple concepts have been proposed to address the problem of the stronger curvature in the Arrhenius plot. These included strong contribution of short circuits, impurity-related effects and divacancies. However, all of them have to be discarded for various failures in describing the experimental observations. So far, a monovacancy description combined with the special properties of the electronic structure and lattice dynamics provided the best description (Petry et al., 1991a; Herzig et al., 1999b).

The bcc structure is intrinsically soft against certain shear deformations. For example, Group IV metals have unusually low frequencies for the longitudinal L $2/3\langle 111 \rangle$ phonons which may promote vacancy jumps in this direction. Köhler and Herzig (1988) claimed that this effect results in vanishing vacancy migration enthalpy in the said metals. The vacancy formation enthalpy, on the other hand, is scaled with the melting temperature, with Group IV transition metals being a possible exception (Schulz, 1991). Furthermore, Ho et al. found that the phonon properties are strongly correlated with the electronic structure (Ho et al., 1983, 1984) as the filling of the d-electron shells increases the frequency of the L $2/3\langle 111 \rangle$ via a strong directional bonding. This gives a plausible explanation for the relatively high activation energy of Cr (Trampenau et al., 1993) and the low activation energy of β-Ti (Petry et al., 1991b).

Owing to their low symmetry, tetragonal metals (like In, Sn) are characterized by anisotropic diffusivities. However unlike the hcp metals, diffusion is slower parallel to the tetragonal axis (c-axis). Activation energy is above what is predicted by the van Liempt rule and diffusion is slower in the homologous scale as compared with cubic or hcp metals.

5.5.1.4 *Estimation of Self-Diffusion Coefficients*

Although the tracer self-diffusion coefficient is already measured for most metallic systems (see Ref. (Landolt–Börnstein, 1990a; NIMS)), methods to estimate the diffusion coefficient or related parameters are sometimes helpful. These range from empirical estimations to full theoretical characterization based on ab initio models. In this section, we only give a short summary of the former.

The most well-known empirical formulas are

(1) The Varotsos formula (Varotsos, 1978; Varotsos and Alexopoulos, 1986) is based on the idea that the activation enthalpy has a form $\Delta Q = CB\Omega$, where Ω is the atomic volume, B is the bulk modulus and C a lattice-dependent constant (for the explanation of this constant see (Varotsos, 1978)). For cubic material Eqn (33) then will have a form like

$$D \sim \exp\left(-\frac{CB\Omega}{k_B T} \right) \tag{136}$$

The agreement between the diffusivity values predicted by the formula and the experimental data is rather good.

(2) The van Liempt rule (Eqn (135)) was already mentioned. A similar relation exists between the latent heat of melting and the activation enthalpy $\Delta Q = 15.2 L_M$ (Nachtrieb rule; (Nachtrieb et al., 1952)).

(3) There is a correlation between the activation volume (ΔV) and the activation enthalpy: $\Delta V = 4\Delta Q \chi$, where χ is the compressibility (Keyes rule; (Keyes, 1958)).

(4) The Zener formula (Zener, 1951) allows the calculation of the migration entropy. In this model, the migration Gibbs energy is equal to the elastic energy required to strain the lattice for the jump.

$$\Delta S = \frac{\Lambda \beta H^M}{T_m}$$

$$\beta = \frac{d(E/E_0)}{d(T/T_m)} \tag{137}$$

where ΔS is the migration entropy (see Eqn (3)), H^M is the migration enthalpy, E is the Young modulus at considered temperature, E_0 is the Young modulus at 0 K, T_m is the melting point and Λ is a crystal structure-related constant (1 for bcc and 0.55 for fcc). This formula accurately describes the measured diffusion entropy.

5.5.2 Diffusion in Dilute Alloys

This section deals with diffusion in binary alloys where one constituent has a significantly smaller atomic fraction than the other. The majority component, which contributes more than 99% of the alloy, is called, in the usual sense, matrix or solvent. The minority component is called solute. Low concentration of solute atoms means that most solutes become insulated in the matrix. Accordingly, solute–solute interactions are negligible. Since radiotracer and mass spectrometry-based techniques have an excellent sensitivity, measuring concentrations below 1 ppm is a possibility. Diffusion of the solute measured under this situation is often called impurity diffusion.

We first deal with substitutional alloys. Here, the vacancy mechanism dominates atomic diffusion as in pure metals. Vacancy diffusion in close-packed metals is grouped into two broad categories. As normal diffusion, one may designate any behavior similar to that observed for noble metals. Everything else, from the particularly slow diffusion in the trivalent Al, to the very fast impurity diffusion in the "open metals" (like Pb) is grouped as exceptional diffusion. In the last part, we discuss the diffusion of interstitial solutes (H, C, N and O).

5.5.2.1 Vacancy-Mediated Diffusion in Substitutional Solutions

Figure 46 shows the diffusion of various substitutional impurities in Ag (together with the self-diffusion coefficient of Ag). This Arrhenius diagram reveals the typical characteristics of so-called normal impurity diffusion.

(1) The diffusivities of the solutes lie in a relatively narrow band around the self-diffusion coefficients. Above two-thirds of T_m (melting temperature) the relation $0.01 < D^{solute}/D^{self} < 100$ holds.
(2) For most preexponential factors $0.1 < D_0^{solute}/D_0^{self} < 10$ applies.
(3) The activation enthalpies are close to each other $0.75 < \Delta Q^{solute}/\Delta Q^{self} < 1.25$.

Substitutional impurities in fcc (Au, Cu, Ni, ..., etc.) and hcp (Zn) metals behave similarly. See Landolt–Börnstein (1990b) and Neumann and Tuijn (2002) for a collection of relevant diffusion data.

5.5.2.2 Diffusion in Al

Although Al is an fcc metal, like the ones associated with the "normal" impurity diffusion, the impurity diffusion in Al (see **Figure** 47) is markedly different from that, for example, in Ag (**Figure** 46). Most transitional metal impurities have extremely high activation enthalpies as well as very low prefactors D_0, so none of the common characteristics of impurity diffusion is fulfilled for this metal.

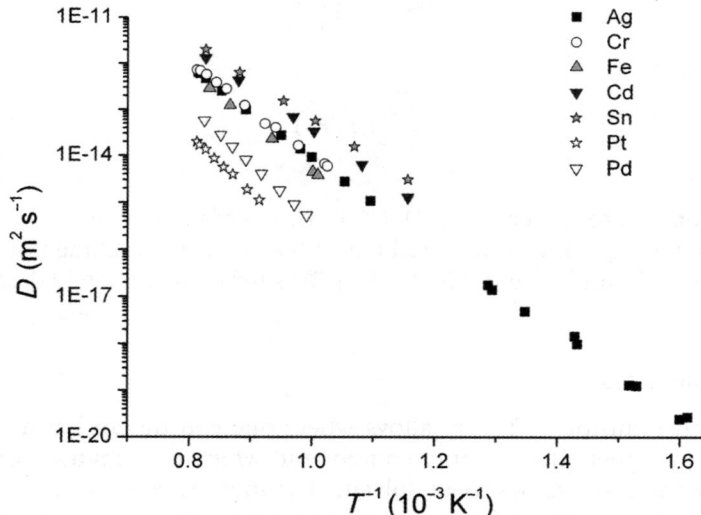

Figure 46 Diffusion coefficients of some impurities in Ag as compared with self-diffusion. Data source: Ag: (Rothman et al., 1970; Lam et al., 1973), Cr: (Neumann et al., 1981), Fe: (Mullen, 1961), Cd and Sn: (Tomizuka and Slifkin, 1954), Pt: (Neumann et al., 1982), Pd: (Peterson, 1963).

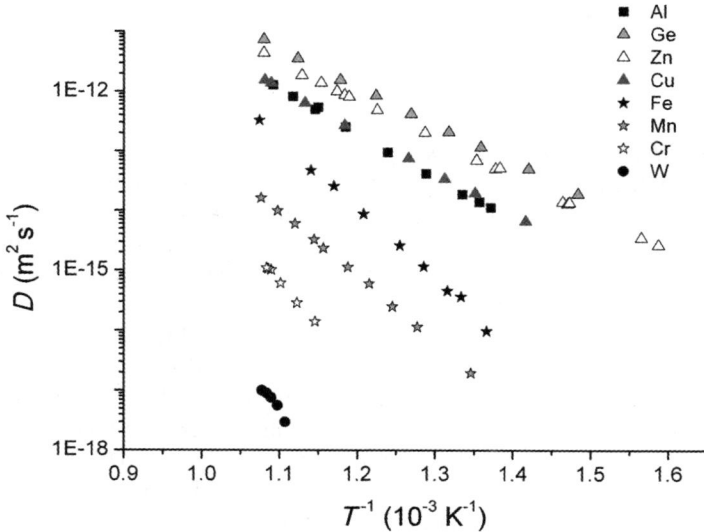

Figure 47 Impurity and self-diffusion in Al. Nontransition metal solutes (e.g. Ge, Zn, Cu) show normal impurity diffusion, however, some transition metals (such as Fe, Mn) migrate extremely slowly in Al. Source of data: Al: (Lundy and Murdock, 1962), Ge, Zn and Cu: (Peterson and Rothman, 1970), Fe, Mn and Cr: (Rummel et al., 1995a), W: (Chi and Bergner, 1982).

Nontransition elements have diffusivities similar to self-diffusion and their activation volume is close to that of the self-diffusion (Beyeler and Adda, 1968). For transition metal impurities, however, a very high activation volume (up to 2.7 Ω_{Al}) was reported (Rummel et al., 1995b). These findings are nevertheless in agreement with the empirical Keyes formula (Keyes, 1958), according to which high activation volume is correlated to high activation enthalpy.

Ab initio calculations have demonstrated that 3d and 4d impurities have a strong repulsive solute–vacancy interaction, while 4sp and 5sp impurities possess a weakly attractive solute–vacancy interaction (Hoshino et al., 1996). Recently, an extensive theoretical characterization of impurity diffusion in Al was carried out by Simonovic and Sluiter (2009) using electronic density functional theory and transition state theory in harmonic approximation. They were able to reproduce the experimental values with a reasonable accuracy (although the calculated values for transition metals are still somewhat below the experimental ones). Their main findings can be summarized in the following:

(1) For nontransition impurities there is a solute–vacancy binding, the strength of this binding correlates to the relative molar volume of the impurity in comparison with the Al host. The activation enthalpy of diffusion is close to that of the self-diffusion, usually it is slightly lower.
(2) The rate-determining step for nontransition metal impurities is the Al–vacancy exchange, the prefactor is approximately equal to that of the self-diffusion.
(3) Early transition metals strongly repel vacancies regardless of their molar volume. The effect is the strongest for Group V metals and very strong for Group IV and VI metals too. Activation energy for diffusion is therefore significantly higher than that of the self-diffusion.
(4) The rate-determining step for diffusion of transition metals is the solute–vacancy exchange. So the prefactors for impurity diffusion show a larger scatter.

(5) Lanthanides with their nondirectional bonds to Al have an attractive vacancy–solute interaction. Since they have large molar volumes, this attraction is relatively strong, which causes a low diffusion activation energy.

5.5.2.3 Fast Impurity Diffusion in Open Metals

One of the earliest systematic investigations of atomic transport in solid matter was that of the diffusion of Au impurities in Pb by Roberts-Austen (1896). At that time the peculiarity of Au diffusion in lead was still not realized. After investigating the self-diffusion and other impurity elements in Pb with the radiotracer technique (Groh and von Hevesy, 1920, 1921), the exceptional characteristic of Au impurity diffusion in Pb was appreciated.

Figure 48 summarizes the impurity diffusion in Pb. Normal diffusion was encountered for some elements, but noble metals for example reveal extremely high diffusion coefficients, more than three orders of magnitude above self-diffusivity. Similar fast noble metal diffusion can be observed in the matrices of Sn, In and Tl. These metals, which are characterized by large interatomic distances despite small atomic radii, are often called "open" metals (Hood, 1988). Small solute atoms in these materials proceed by enhanced diffusion mechanism, which may combine substitutional and interstitial character. Hg and Cd diffuse via an interstitial vacancy pair (atomic exciton) and Au and Ag as free interstitials (Warburton and Turnbull, 1975a). On the other hand, impurity–impurity interaction leads to the formation of low mobile solute pairs. For example solute diffusion becomes slower in the case of higher concentration of Au (Cohen et al., 1977) or Ni (Anenzoubadrour et al., 1988).

A theoretical description of the atomic transport has been given by Warburton and Turnbull (1975b), called dissociation theory. Solutes are distributed among substitutional and interstitial places.

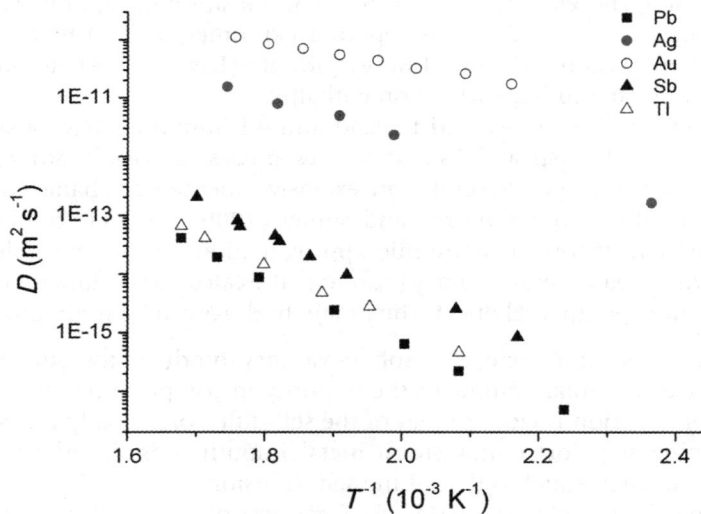

Figure 48 Impurity and self-diffusion in the "open metal" Pb. Impurity diffusion is orders of magnitude faster than self-diffusion for elements with smaller atomic radii (Ag and Au). Larger atoms (Sb and Tl) show no exceptional impurity diffusion. Source of data: Pb: (Nachtrieb and Handler, 1955), Ag: (Hu and Huntington, 1982), Au: (Ascoli et al., 1956), Sb: (Nishikawa and Tsumuraya, 1971), Tl: (Resing and Nachtrieb, 1961).

The substitutional species are considered immobile in this picture (their mobility is close to that of the slow self-diffusion). To become mobile, they have to be dissociated according to

$$S_i + V \leftrightarrow S_s \tag{138}$$

where S_i is a solute atom on interstitial places, V is a vacancy and S_s is a solute on a substitutional place. The solute diffusion then contains three steps: first it dissociates to a vacancy and an interstitial, these two mobile species take some diffusion jumps and finally the vacancy and the interstitial recombine in a new position. Solutes dissolving via this mechanism are called hybrid solutes. In local equilibrium, the law of mass action is fulfilled:

$$\frac{c_i c_V}{c_s} = \frac{c_i^{eq} c_V^{eq}}{c_s^{eq}} = K(T) \tag{139}$$

where c_i, c_s and c_V are the concentrations of interstitial solutes, substitutional solutes and vacancies, respectively. $K(T)$ is the dissociation constant, a parameter which depends only on the temperature. The superscript "eq" indicates the respective concentrations measured in thermal equilibrium. In a normal metal the density of vacancy sources and sinks is high enough that thermal equilibrium is reached almost any time. The concentration of vacancies is therefore set by the thermodynamics of the host matrix, while concentration of the interstitial and substitutional solutes depends on the total impurity concentration and $K(T)$.

The measured effective diffusivity is then the sum of the vacancy and interstitial diffusivities (Frank et al., 1984):

$$D_{eff} = \frac{c_i^{eq}}{c_i^{eq} + c_s^{eq}} D_i + \frac{c_V^{eq}}{c_i^{eq} + c_s^{eq}} D_V \tag{140}$$

where D_i is the diffusion coefficient for interstitial transport and D_V is the mobility of the impurity–vacancy complex. If $c_i^{eq} D_i >> c_V^{eq} D_V$, the substitutional transport is negligible. If a solute is dominantly interstitially solved, then it can be seen trivially that $D_{eff} \approx D_i$. For solutes which are solved mostly substitutionally Eqn (140) will take the form

$$D_{eff} \approx D_i \frac{c_i^{eq}}{c_s^{eq}} \tag{141}$$

which contains the ratio of interstitially and substitutionally solved impurities. This is the reason for the relatively wide range of diffusivities for fast impurities.

Si and Ge also belong to the Group IV such as Pb and Sn. So they can also be treated as an open metal for the diffusion of various late transition elements.

5.5.3 Interstitial Diffusion of H, C, N, and O

Solutes with considerably smaller size than matrix atoms migrate via interstitial sites. These interstices can be of tetrahedral or octahedral symmetry depending on the geometry of the host lattice. Since migrating atoms already occupy the interstitial sites, no further defect formation is necessary for diffusion. For the same reason, a term analogous to solute–vacancy binding is also lacking. Accordingly, only the migration enthalpy plays a role in the activation energy. Diffusion of the heavier C, N, and O can be treated by the classical diffusion theories. However, for H isotopes, quantum effects influence the diffusion properties.

5.5.3.1 "Heavy" Interstitial Diffusers (C, N, and O)

Because of the small size of these atoms as compared with the matrix atoms, the migration barrier between two interstitial sites is rather low. Consequently, diffusion of these impurities is very fast in comparison with diffusion of substitutional solutes (Landolt–Börnstein, 1990c). Indeed, it can be as high as in liquids. **Figure 49** shows the diffusion coefficient of C, N and O in V in comparison with self-diffusion. The activation enthalpies are 116 kJ/mol, 148 kJ/mol, 123 kJ/mol, and 309 kJ/mol, respectively.

Because of such high diffusivity and also because O and N are gases, experimental methods to determine interstitial diffusion differ from the conventional tools:

- The steady-state method is applicable for fast diffusers. Steady-state composition profiles can be measured, while the flux can be easily controlled by gas flow or electrochemical methods.
- C has a suitable radiotracer in form of the long lifetime ^{14}C isotope. For N and O stable tracers (^{18}O and ^{15}N) exist accordingly methods like SIMS (Han and Helms, 1986) can be applied.
- Nuclear reaction analysis is a natural complement for ion implantation in the same specimen.
- Outgassing of O and N or decarburization in the case of C. In these processes the sample is usually embedded in a primary solid solution.
- Most metals have a strong tendency to form oxides, nitrides or carbides. In a diffusion experiment this can lead to the formation of a reaction layer. By detecting the concentration profile, interdiffusion coefficients can be evaluated.
- The jump rate of the interstitials is so high that mechanical spectroscopy measurements (Snoek relaxation, see Section 5.4.4.1) can provide diffusion coefficient values (Weller, 2006) down to ambient temperature.

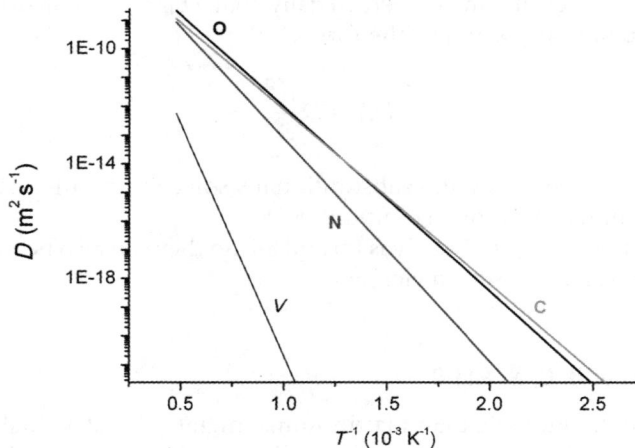

Figure 49   The diffusivity of "heavy" interstitials in comparison with self-diffusion in V. Source: Schmidt and Warner (1972) and Pelleg (1974).

Fast interstitial diffusants can be measured over a very wide temperature (and diffusivity) range combining macroscopic and microscopic measurements. In this way, they provide an excellent testing ground to check the linearity of the Arrhenius relationship. **Figure 50** shows the diffusion coefficient of C and N in Fe measured by different techniques. As can be seen the diffusivity covers almost 16 orders of magnitude (da Silva and McLellan, 1976).

Diffusion of interstitial solutes migrating by the interstitial mechanism in a cubic lattice can be described by the following equation:

$$D = ga^2 \exp\left(\frac{S^M}{k_B}\right)\exp\left(-\frac{H^M}{k_B T}\right) \tag{142}$$

where a is the lattice parameter, g is a geometric factor which depends on the lattice and the interstitial site (tetra- or octahedral). The activation enthalpy for the diffusion is simply the migration enthalpy of the interstitial solute while the migration entropy corresponds to the analog of Eqn (3).

As can be seen in **Figure 50**, N diffusion shows very little deviation from linearity, while there is a slight upward curvature in the C diffusion in α-Fe. Several attempts were made to clarify the behavior of C in Fe. Belova and Murch (2005) claimed that rotation of C interstitials around each other has a higher jump rate than migration of a singular impurity. The tendency to form C–C pairs is responsible for the enhanced diffusion of C at higher concentrations and/or temperatures.

5.5.3.2 Interstitial Diffusion of H

The smallest activation enthalpies and largest isotope effects are found for H transport in metals. Furthermore, because of the low mass of H, quantum effects can appear in the H diffusion. We restrict

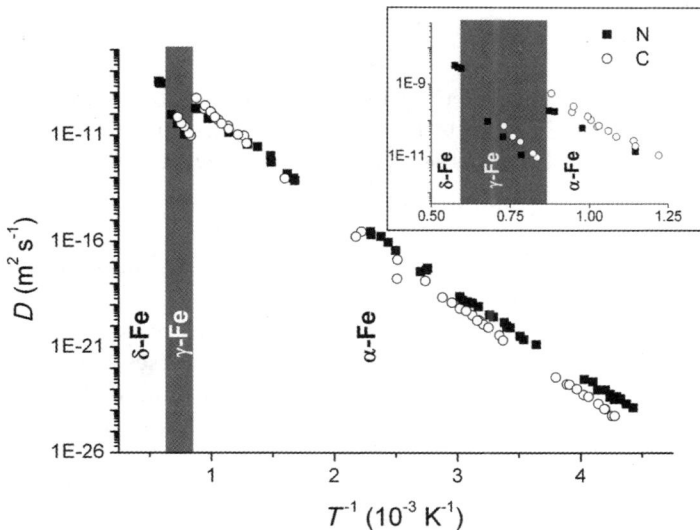

Figure 50 N and C diffusion in α-Fe (Source of data: (Grieverson and Turkdogan, 1964a, 1964b; Smith, 1964; Lord and Beshers, 1966). The diffusion is slower in the γ-phase, but in the case of N recovers to extrapolated value in the δ-phase. The inset shows the region near the phase transitions.

ourselves to give a brief summary of these effects. Comprehensive reviews about H diffusion can be found in Alefeld and Vökl (1978, 1979) and Wipf (1997).

The ability of H to penetrate into metals was first observed by Graham (1866). He also concluded that metals can form nonstoichiometric alloys with hydrogen, in contrast to nonmetals which produce stoichiometric hydrides. In 1914 Sieverts found that the H concentration in metals is proportional to the square root of the H_2 pressure (Sieverts, 1914). This can be considered as direct evidence of the dissociation of the H_2 molecule. Since these early pioneering investigations vast scientific work was devoted to study H diffusion in metals. A very recent motivation for such studies is the fact that metals can store hydrogen more efficiently than compressed gases or liquids (Sakintuna et al., 2007). Also the fast diffusion of H allows reasonably short charge/discharge times. Data about H diffusion in metals can be found in Landolt–Börnstein (1990d).

H occupies octahedral sites if the size of the host atom is below 0.134 nm and tetrahedral sites if it is larger than 0.137 nm (Somenov and Shilstein, 1979). For V, both tetrahedral and octahedral sites are allowed. The small mass of the H isotopes allows the observation of quantum effects in diffusion. By having an even lower mass than the proton, the anti-muon (μ^+) can provide particular clues about quantum diffusion (Stoneham, 1979). The possible diffusion regimes for H are

- Coherent tunneling: the interstitial propagates through the material like a free electron. Since this regime requires a very regular structure, it is only achievable (if possible at all) with high-purity materials at low temperature.
- Incoherent or phonon-assisted tunneling: the ground states of occupied and unoccupied interstitial sites are different. Phonons are used to equalize the energies of neighboring sites. This regime is observed in many H diffusion experiments (Stoneham, 1978).
- Classical regime: as the name suggests, this is the well-known hopping mechanism of diffusion mostly discussed throughout this chapter. It has been observed for many hydrogen-metal systems too.
- High-temperature regime: the residence time at the equilibrium sites becomes comparable with the transition time between sites.

H diffusion can be investigated by a variety of methods. Some of them are especially useful for this particular diffusor. Other standard methods, however, are less effective in studying H diffusion. In the following, we give a short list of the possible techniques and their eventual drawbacks.

- Methods optimized for fast diffusing interstitials such as adsorption–desorption methods can be used to investigate H diffusion, too.
- Mechanical spectroscopy is applicable (e.g. Sinning, 2000); however, there is no unambiguous experimental evidence for an H-related Snoek effect.
- While tritium is a suitable radiotracer material (half-life: 12.32 years), isotope effects on the diffusivity are very strong owing to the large change in relative mass (see **Figure 51**).
- Permeation methods can be used to measure diffusion of H through a thin foil. In the steady-state method, the partial pressure of H_2 is kept constant on both sides of the metal (Schmitz et al, 1998; Yamakawa et al., 2001, 2005).
- The favorable gyromagnetic ratio of the proton and the absence of the quadrupole coupling make NMR (Cotts, 1972) especially suitable to study H diffusion.
- Since the neutron scattering cross-section of the proton is more than an order of magnitude larger than for any other nuclei, QENS allows the investigation of H transport (Hempelmann, 2000).

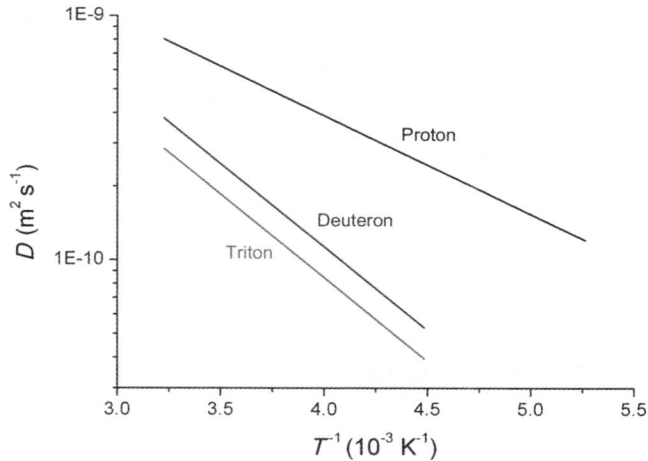

Figure 51   Diffusivity of the H isotopes in Nb according to the data of Matusiewicz (1977). Unlike for higher mass diffusors, the activation enthalpies of the different isotopes show significant variation too.

The basic idea of QENS is that interaction with moving nuclei results in a broadening of the neutron energy spectrum (see Section 5.4.4.3).

Some metals (e.g. Ti and Nb) reveal a negative enthalpy of solution and very high diffusion coefficients for H. For example at room temperature, the diffusion coefficient of H in Nb is 10^{-9} m^2 s^{-1}. So H is more than 10 orders of magnitude faster than the heavy interstitials.

Figure 51 shows the diffusion of proton, deuteron, and triton in Nb (Matusiewicz and Birnbaum, 1977). Since the diffusion is fast and the temperature dependence is small, the Gorski technique (see Section 5.4.4.1) can be used for ^{1}H and D in the whole temperature range. Protons diffuse faster than deuterium, while tritium is the slowest of the hydrogen isotopes. However, in stark contrast to the classical isotope effect, which predicts a difference only in the D_0 prefactor (Wert, 1950), the activation enthalpies are also different for the H isotopes.

In fcc metals, the prefactor follows the classical rule. However, there is an inverse relation for the activation enthalpy $\Delta H_{\mathrm{proton}} > \Delta H_{\mathrm{deuteron}}$. Accordingly, at low temperatures ^{1}H diffuses slower than D in Pd (Bomhold and Wicke, 1967). This effect potentially allows the separation of H isotopes.

5.5.4   Diffusion in Concentrated Binary Alloys

The practical usage of alloys exceeds that of pure metals by far. It is not surprising that self- and impurity diffusion measurements in different alloys are very numerous (see Refs. (Landolt–Börnstein, 1990e) and (Landolt–Börnstein, 1990f)). Systematic measurements to explore the concentration dependence, on the other hand, are rare. These include Ag–Au (Mallard et al., 1963), Au–Ni (Reynolds et al., 1957), Co–Ni (Million and Kučera, 1969; Million and Kučera, 1971), Cu–Ni (Monma et al., 1964), Fe–Ni (Million et al., 1980), Fe–Pd (Fillon and Calais, 1977), Ge–Si (Silvestri et al., 2006; Kube et al., 2010),

Nb–Ti (Gibbs et al., 1963) and Pb–Tl (Resing and Nachtrieb, 1961). Two important trends are indicated by these measurements:

- For a given composition v, the diffusion coefficients of species A ($D_A(v)$) and B ($D_B(v)$) usually do not show a difference larger than a factor 10. When they do so, it indicates that the diffusion mechanisms are probably different for the two tracers.
- The activation enthalpies (ΔQ) and the pre-factors (D_0) are scaled by the melting temperature of the alloy as in the case of self-diffusivities. This means that for fully miscible alloys the concentration dependence of the activation enthalpy is linear. Accordingly, the diffusion coefficients are characterized by exponential concentration dependence (see Section 5.3.3.2).

5.5.5 Diffusion Properties of Long-Range Ordered Intermetallic Compounds

Intermetallics or ordered alloys are compound phases of two (or more) metals or semimetals, which possess a crystal structure that is different from their constituents. Their physical and chemical properties cover a wide range of possible characteristics. Owing to advantageous properties (such as high hardness, wear corrosion and heat resistance) their technical application has a long history. The role of aluminides and silicides as lightweight, corrosion and heat-resistant materials has grown significantly in the recent decades. In addition, silicides play an ever increasing role in modern micro- and nanoelectronics.

Ab initio calculations based on density functional theory explain why certain combinations of metals would prefer a structure with long-range order (preferential occupation sites on sublattices). These ordered phases exist in a wide variety. They can have different structures, thermal stability and existence range. Some intermetallics are real "line phases", while others can exist as off-stoichiometric compound too. Moreover there are intermetallics which are only observed in off-stoichiometric form. An eventual off-stoichiometry can be compensated in two main ways (without lacking generality we assume an A-rich side):

- Antisite atoms: A atoms occupy β sites (which normally belong to B atoms), they are denoted by A_β
- Structural vacancies: Since there are not enough B atoms to occupy each designated sublattice site some of them remain unoccupied. This effect could produce extreme high vacancy densities up to a few percent. Vacancies in the β sites are denoted by V_β.

Apart from these structurally necessary defects, thermally excited defects appear too. This also implies that the order parameter at $T > 0$ deviates from unity. Local defects from the perfect chemical order are necessarily present. It is important to note that vacancies at different sublattices usually see a different atomic environment, and therefore different site energy. Accordingly the concentrations of thermal vacancies are usually not equal in the two sublattices. This difference can sometimes, like in the case of $MoSi_2$, be rather pronounced (Zhang et al., 2004).

Intermetallics are characterized by short- and long-range order. Both affect the diffusion. An important consequence is that the diffusion mechanism must conserve the order state, at least in a statistical sense. This can lead to a significant difference in the diffusivities of the different species. Furthermore, significantly different vacancy density on the sublattices and high energies of antisite formation lead to pronounced driving forces for a direct back jump after an atom has left the correct sublattice by a diffusion jump. Therefore, correlation factors of diffusion may become very low. With increasing temperature disorder increases and so does the correlation factor. Consequently, a nonnegligible part of the temperature dependence of the diffusivity is related to the correlation factor (Belova and Murch, 2000b).

Diffusion mechanisms which preserve the local order include the six-jump mechanism (Elcock and McCombie, 1958), the triple-defect mechanism (Stolwijk et al., 1980) and the anti-structure bridge (Kao and Chang, 1993). Assuming particular mechanisms allows the calculation of activation energies and entropies using ab initio or MD methods. If one constituent forms a sublattice that allows (like in the case of Ni_3Al) nearest-neighbor jumps, the diffusion of the majority atoms A and the minority atoms B can be decoupled. If the B antisite requires not to high-energy, impurity diffusion of B on the A sublattice can even dominate the diffusion of B (Numakura et al., 1998).

In the following section we briefly overview the most important intermetallic structures. Then, we continue with the proposed diffusion mechanisms in the ordered alloys.

5.5.5.1 Long-Range Ordered Structures of Intermetallics
Figures 52–58 show some of the most important intermetallic superstructures.

The B2 or CsCl lattice is among the simplest intermetallic structures. It is composed of two interpenetrating simple cubic lattices, each α site is surrounded by 8 β sites and vice versa, thus no nearest-neighbor jumps within a coherent sublattice are possible. As can be seen in **Figure 52**, this structure can be derived from a bcc lattice. The composition is approximately AB, examples of B2-type intermetallics are FeAl, NiAl, CoGa, CuZn.

The $D0_3$ or Fe_3Si lattice is also derived from the bcc lattice, however unlike for B2, here the second sublattice contains alternating darker α and lighter β sites. Accordingly α sites have four or eight α sites among their nearest neighbors, so the transport of the A atoms does not necessarily need complex mechanisms. The approximate composition is A_3B, examples of $D0_3$-type intermetallics are Fe_3Si, Fe_3Al, Cu_3Sn.

The $L1_2$ or Cu_3Au structure can be derived from the fcc lattice. A atoms, occupying the darker sites of **Figure 53** have 4 B and 8 A nearest neighbors, while B atoms are surrounded by 12 A atoms. Like in the $D0_3$, the A atoms can migrate using α sites only. Composition is approximately A_3B, and typical examples are Ni-based compounds like Ni_3Ge, Ni_3Al or Ni_3Ga.

The $L1_0$ or CuAu structure can be derived from the body-centered tetragonal structure with a c/a ratio close to that of the fcc structure (see **Figure 53**). A and B atoms occupy alternating planes. As a result of this layer-like geometry, $L1_0$ intermetallics are characterized by two different diffusion coefficients, $D^{\parallel}$ parallel and $D^{\perp}$ perpendicular to the tetragonal axis. The approximate composition is AB, while β-TiAl and magnetic materials CoPt and FePt are typical representatives of this structure.

The $D0_{19}$ is an ordered hexagonal structure which can be derived from the hcp lattice (see **Figure 54**). The composition is approximately A_3B, an example of this structure is Ti_3Al. Compounds of Ni and several semimetals (Ge, As, Sb) also crystallize in this structure.

Figure 52 The B2 and the $D0_3$ structure.

Figure 53   The L1$_2$ and the L1$_0$ structure.

The B8 or NiAs structure has an approximate composition from AB to A$_2$B. Various Ni$_x$Sn$_y$ alloys appear in certain variants of this structure. The B (Sn; lighter) atoms form a compressed hcp lattice with a c/a ratio of about 1.3. The A (Ni; darker) atoms can be found at the octahedral (a) or double tetrahedral interstitial sites (b) as presented in **Figure 55**.

In the B3$_2$ or NaTl structure the composition is approximately AB. The B3$_2$ structure can be visualized as two interpenetrating diamond sublattices (see **Figure 56**). β-LiAl and β-LiIn are among the materials of this structure.

The B20 or FeSi structure is cubic. Each A atom is surrounded by seven B "nearest-neighbor atoms" although the atomic distances slightly vary, and a second nearest-neighbor shell with six A atoms. In FeSi, for example these atomic distances are the following: one B atom at 0.511a, three B atoms at 0.521a, and three B atoms at 0.56a in the first shell, six A atoms at 0.613a for Fe and 0.620a for Si central atoms, respectively; *a* is the length of the unit cell. Transition metal silicides like FeSi, CrSi and CoSi crystallize in this structure.

The A15 or Cr$_3$Si structure is characterized by an approximately A$_3$B composition. B (darker) atoms form a bcc lattice while the A (lighter) atoms form chains of split interstitials along $\langle 100 \rangle$, $\langle 010 \rangle$ and

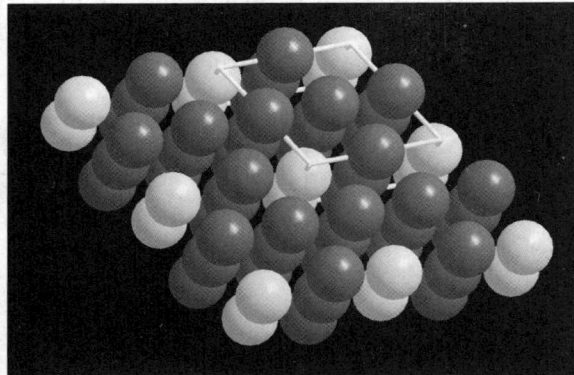

Figure 54   The D0$_{19}$ structure. Note, while the dark gray A atoms can diffuse on their own sublattice both in plane and normal, the light gray B atoms have an easy diffusion direction along the c-axis, but have to use antisites for in-plane diffusion.

(a)   **(b)**

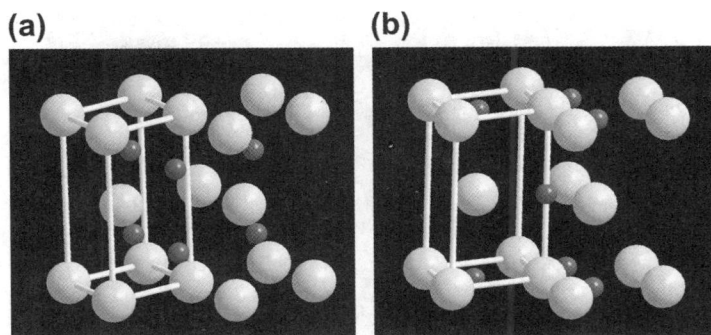

Figure 55   The octahedral (a) and the double tetrahedral (b) sites in B8 intermetallics.

$\langle 001 \rangle$ planes (see **Figure 57**). Along these chains fast diffusion of the majority component is possible. Type II superconductors like Ni_3Ti, Ni_3Sn or V_3Ga crystallize in this structure.

The $C11_b$ or $MoSi_2$ structure has an approximate composition of A_2B. The unit cell of this structure can be seen in **Figure 57**. The $C11_b$ structure can be derived from the body-centered tetragonal structure, B (darker) atoms have 10 A (lighter) nearest neighbors while A atoms have 5 A and 5 B neighbors. $MoSi_2$ and WSi_2 are typical examples of this crystalline structure.

The cubic C15 or Mg_2Cu structure belongs to a family of structures called Laves phases (Sauthoff, 1995). The structure is illustrated in **Figure 58**. The approximate composition is A_2B. B (darker) atoms are forming a diamond structure while A (lighter) atoms are forming a network of tetrahedral joining at their vertices. To form a binary C15 structure, a ratio $r_B/r_A > 1.25$ is required in the relative atomic radii.

5.5.5.2 Sublattice Diffusion Mechanism

If one of the components belong to a sublattice which allows internal nearest-neighbor jumps (e.g. A component in $L1_2$), migration in this sublattice does not affect local order. If the concentration of B_α antisite atoms is sufficiently high, coherent diffusion on the majority sublattice can be the dominant mechanism for minority atoms too (Numakura et al., 1998). If transport of the two components happens on the different sublattices, the diffusion coefficients are usually de-coupled. As the system gets more and more ordered, the diffusivity of the minority atoms becomes significantly reduced, while the migration of the majority component is much less affected. This effect is called "the ordered Cu_3Au

Figure 56   The $B3_2$ and the B20 structure.

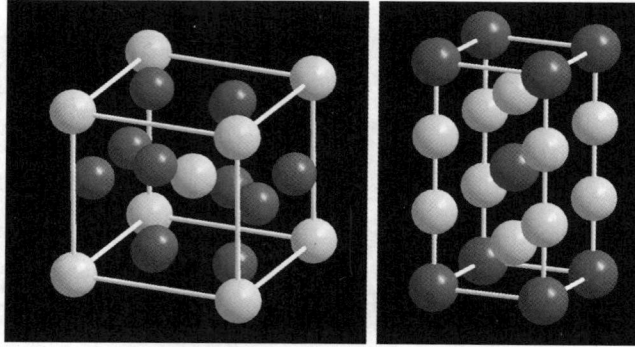

Figure 57   The A15 and the C11$_b$ structure.

effect" by d'Heurle and Gas (1986) and Schoijet and Girifalco (1968), and the shorter term "Cu$_3$Au-rule" is also in use.

In the model proposed by Numakura et al. (1998) to describe the diffusion in an L1$_2$ structure, the authors assumed that vacancy concentration is similar to that of pure metals and diffusion happens by the monovacancy mechanism and nearest-neighbor atom–vacancy exchange. Consequently, the diffusion of the majority element can be treated as self-diffusion in the α sublattice. Because of similar geometry, this sublattice is often called the NbO lattice (Koiwa and Ishioka, 1983). The diffusion coefficient of the majority element is therefore

$$D_A^* = \frac{2}{3}\lambda^2 v_{v,\alpha} \omega f \tag{143}$$

where λ is the lattice parameter, ω is the jump rate, f is the correlation factor (which is 0.689 according to Koiwa and Ishioka (1983)) and $v_{v,\alpha}$ is the vacancy fraction in the α-sublattice. Comparing Eqn (143)

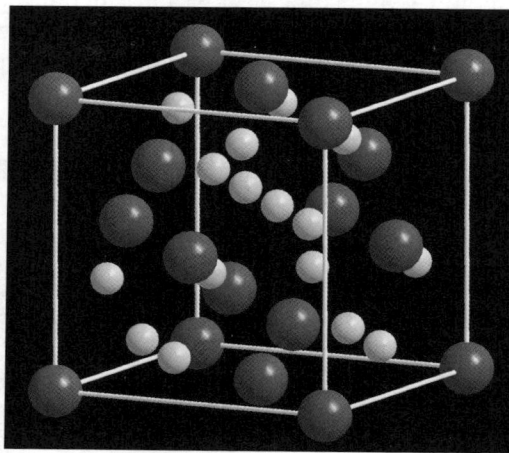

Figure 58   The C15 structure.

with Eqn (32) no difference in the diffusion on an NbO lattice and the fcc lattice exists apart from the 2/3 factor and a different correlation factor.

Similarly, if a B atom is already located on the α-sublattice $B_\alpha - V_\alpha$ exchanges does not promote further disorder. Consequently impurity diffusion of the B atoms in the α-sublattice can be the dominant mechanism, if the order in the alloy is not especially high. In this case, a treatment similar to the five-frequency model of Lidiard (1955) (see Section 5.2.8.2) can be used to calculate the impurity diffusion coefficient (Numakura et al., 1993):

$$D_B^* = \frac{2}{3}\lambda^2 v \; _{V,\alpha} \frac{w_4}{w_3} w_2 f p_\alpha \tag{144}$$

In which the definition of the vacancy jump rates w_i follow the systematic used in Section 5.2.8.2. f is the correlation factor while p_α measures the probability of finding a B atom in an antisite position.

5.5.5.3 Six-Jump Cycle

The six-jump cycle (6JC) or Elcock loop mechanism was first proposed to interpret diffusion in the B2 structure (Elcock and McCombie, 1958) but was adapted to other superstructures later. In six well-correlated jumps, the vacancy first destroys and then reconstructs local order as illustrated in **Figure 59**. The vacancy uses both the α- and β-sublattices during the migration process. Therefore, the tracer diffusivites of the two constituents in a stoichiometric alloy, taking correlation effects correctly into account, must not differ more than a factor of 2.034 (Arita et al., 1989). The diffusion in, for example, β-AgMg (Domian and Aaronson, 1965), β-AuCd (Gupta et al., 1967), β'-AuZn (Gupta and Lieberman, 1971) and β-CuZn (Kuper et al., 1956) was found to obey this rule indicating the validity of such mechanism. However, if deviations from the perfect stoichiometry occur, a large number of structural defects appear. Interaction with these defects can seriously influence the D_A/D_B ratio, thus ratios larger than 2 does not necessarily disprove the six-jump cycle mechanism (Belova and Murch, 2002).

Since the six-jump mechanism requires a series of coordinated jumps for the vacancy to proceed, the correlation factors are, not surprisingly, rather small. One can distinguish between two different correlation types in the six-jump mechanism. The first type of correlation factor treats the whole six successful vacancy–atom exchanges as one effective jump. For example, this correlation factor is in NiAl $f = 0.782$ and 0.860 for Ni and Al, respectively (Divinski and Herzig, 2000). The second type of correction measures the chance of the first vacancy–atom exchange to initiate the six-jump mechanism; for NiAl there is a considerable difference in this kind of correlation factor: 0.445 for Ni and 0.022 for Al (Divinski and Herzig, 2000).

Computer simulations have shown that the six-jump mechanism can be the key mechanism at low temperatures and highly stoichiometric alloys. However, other loops gain importance with increasing disorder (Athènes et al., 1997). Indeed the six-jump mechanisms in their pure form such as shown in **Figure 59** are restricted to a rather narrow temperature range (Belova and Murch, 2000b, 2000c).

5.5.5.4 Triple-Defect Mechanism

The triple-defect mechanism was first proposed to describe the diffusion in the CoGa system (Stolwijk et al., 1980). Triple defects are a combination of a divacancy and an antisite atom as can be seen in **Figure 60** (Wasilewski, 1968). From pure stoichiometric point of view, the triple defect is equivalent to a $V_\alpha + V_\beta$ vacancy pair (as can be seen in the second stage of **Figure 60**) and can be thermodynamically preferred if vacancy formation energy differs strongly on the two sublattices.

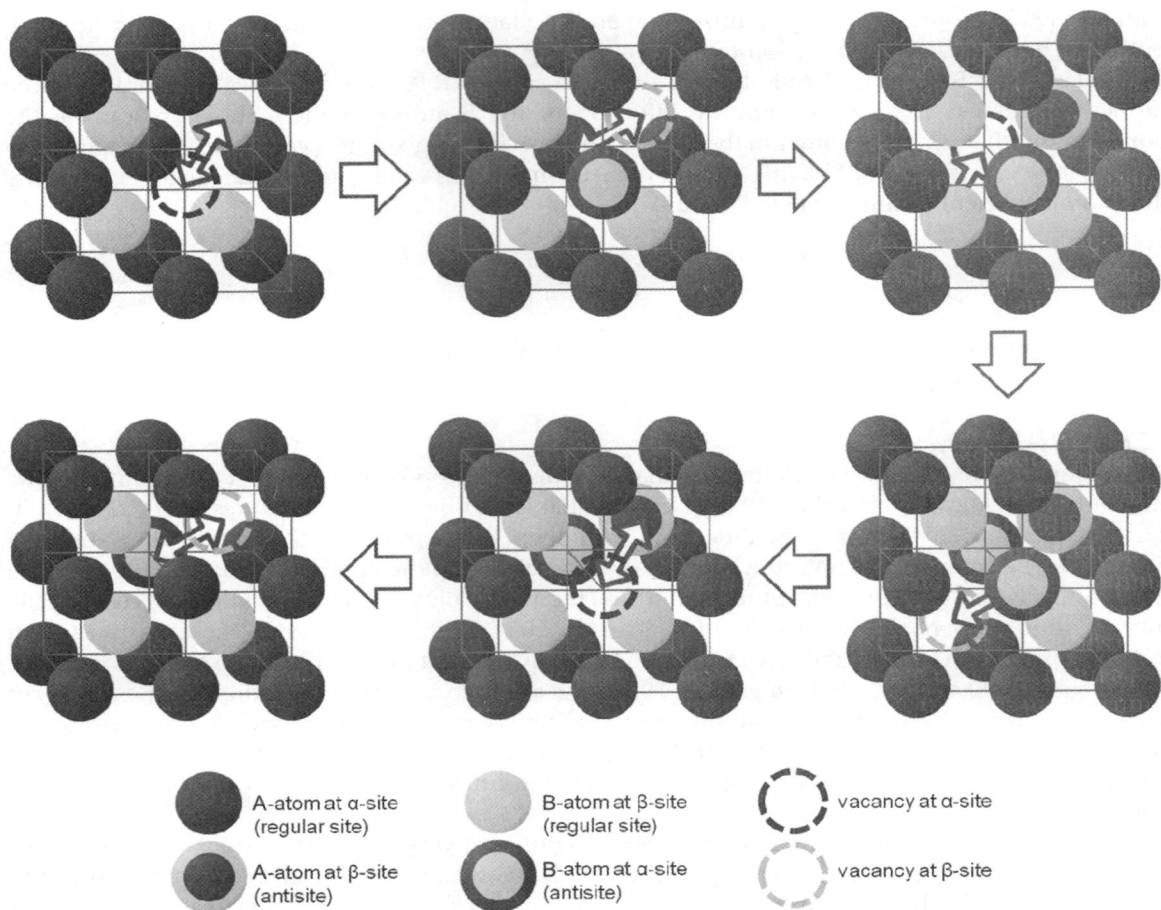

Figure 59 The so-called ⟨100⟩ straight cycle in B2 intermetallics, one of the possible realization of the six-jump mechanism in this structure. (For color version of this figure, the reader is referred to the online version of this book.)

Figure 60 shows the migration of the triple defect which can happen by two nearest-neighbor jumps of the B atom (e.g. NiAl (Frank et al., 2001)) or by two nearest-neighbor jumps of the A atom accompanied by a second nearest-neighbor jump of the B atom (e.g. CoGa (Stolwijk et al., 1980)). Like the six-jump mechanism, the triple-defect mechanism also includes highly correlated jumps involving both species. Consequently the diffusivities are coupled and according to Bakker et al. D_A/D_B cannot exceed 13.3:1 (Bakker et al., 1981). A recent MC study carried out by Belova et al. arrived to a somewhat stricter criterion for the ratio of diffusivities. According to them, it cannot exceed 11.2:1 (Belova et al., 2009).

The intermediate configuration during the diffusion jump is called nearest-neighbor vacancy pair (the second stage on **Figure 60**). In principle this defect can also act as a diffusion vehicle in an ordered alloy. The influence of such vacancy pairs is nonetheless minimal in intermetallics, although in ionic crystals they can play an important role (Schottky pairs). Triple-defect and vacancy pair mechanisms are closely related (Belova and Murch, 2004a).

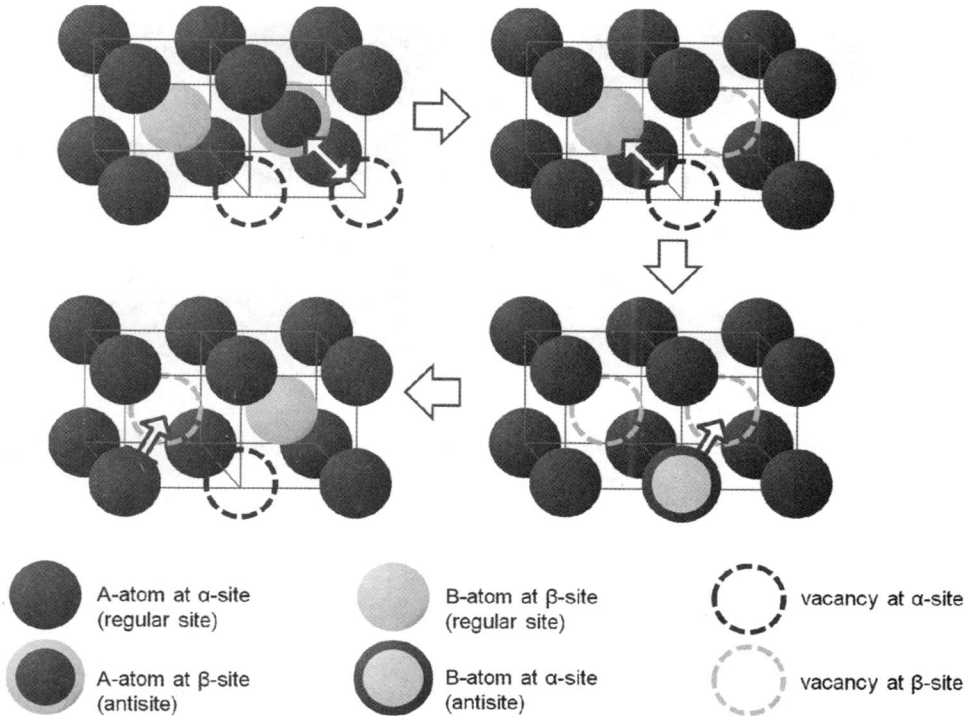

Figure 60 The migration of the triple defect via two nearest-neighbor jumps of both the A and B atoms in a B2 superlattice structure. (For color version of this figure, the reader is referred to the online version of this book.)

5.5.5.5 Antistructure Bridge Mechanism

This mechanism was originally proposed by Kao and Chang to describe diffusion in the B2 superlattice (Kao and Chang, 1993) and it was extended to the L1$_2$ system shortly after by (Divinski and Larikov (1997). As it can be seen in **Figure 61** the antistructure bridge mechanism in B2 requires a vacancy in the second nearest-neighbor position to an antisite atom. Then in the first jump, the vacancy creates a new antisite atom, while it annihilates the old one with the second jump. As a net effect, the vacancy and the antisite atom exchanged their position.

Because of the above requirement, long-range diffusion can only proceed by the antistructure bridge mechanism, if the number of antisite atoms is sufficiently high. Then, the vacancy can always make hops between different antisite atoms. According to the MC simulations carried out by Divinski and Larikov, this percolation threshold is around 5% B$_\alpha$ concentration in B2, while 11% in L1$_2$ structure (Divinski and Larikov, 1997). These seemingly high concentrations of antistructure atoms are far from being unlikely. While this mechanism, in its pure form, can only work if the percolation criterion is fulfilled, it can nevertheless also contribute to the atomic transport working in tandem with other diffusion mechanisms (usually the sublattice mechanism) if the criterion is not hold. This contribution cannot be neglected for the diffusion in, for example, NiAl (Frank et al., 2001) or TiAl (Mishin and Herzig, 2000).

As the antistructure bridge mechanism aids only one of the contributing species, diffusion coefficients of the constituents are completely decoupled.

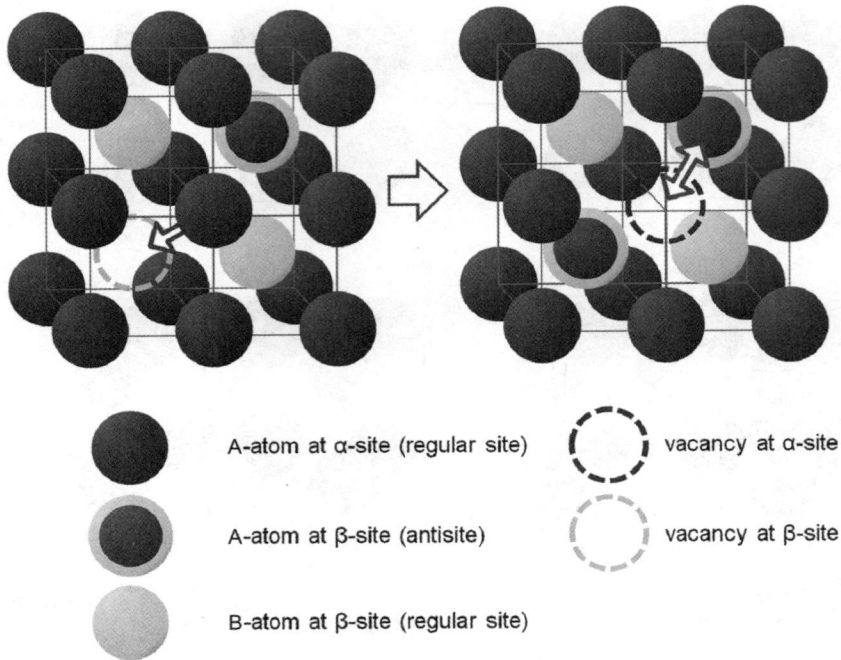

A-atom at α-site (regular site)   vacancy at α-site

A-atom at β-site (antisite)   vacancy at β-site

B-atom at β-site (regular site)

Figure 61   The antistructure bridge mechanism in B2 intermetallics. (For color version of this figure, the reader is referred to the online version of this book.)

5.5.5.6 *Order–Disorder and Order–Order Transformations*

Order–disorder and order–order transitions change the characteristic distribution of atoms. Thereby the importance of various diffusion mechanism and jump sequences can be altered at the phase transition. In general, the density of antisite atoms increases as the critical temperature of the order–disorder transformation is approached. As a consequence, the contribution of the antistructure bridge mechanism may also increase.

One of the pioneering work in studying the effect of order–disorder transition on the diffusion was carried out by Kuper et al. in the beta brass (Kuper et al., 1956). As it is shown in **Figure 62**, a break point in the Arrhenius curve can be observed for both constituents at the A2 → B2 transition temperature. In the case of CuZn, activation energy for diffusion of both components is higher in the ordered phase. In the disordered bcc FeCo alloy the self-diffusion of Fe is somewhat faster than Co. In the ordered B2 state however, it is the activation energy of Co which is the smaller. As a result, not much below the critical temperature of order–disorder the migration rate of Co surpasses that of Fe (Nitta et al., 2004).

In multicomponent alloys quite many possibilities for order–order-type transitions exist. For example in the magnetic shape memory alloy Ni_2MnGa, the high-temperature phase is characterized by B2′-type order (Ni atoms occupy α sites and β sites are occupied by randomly distributed Mn and Ga atoms) while the lower temperature phase is characterized by $L1_2$ or Heusler-type structure (all three atomic species have their own sublattices). The diffusion rates of all three components were measured by radiotracer method (Erdélyi et al., 2007). Ga was found to be the slowest diffusing species in both phases and its diffusion is characterized by single activation energy. Ni and Mn, on the contrary, are

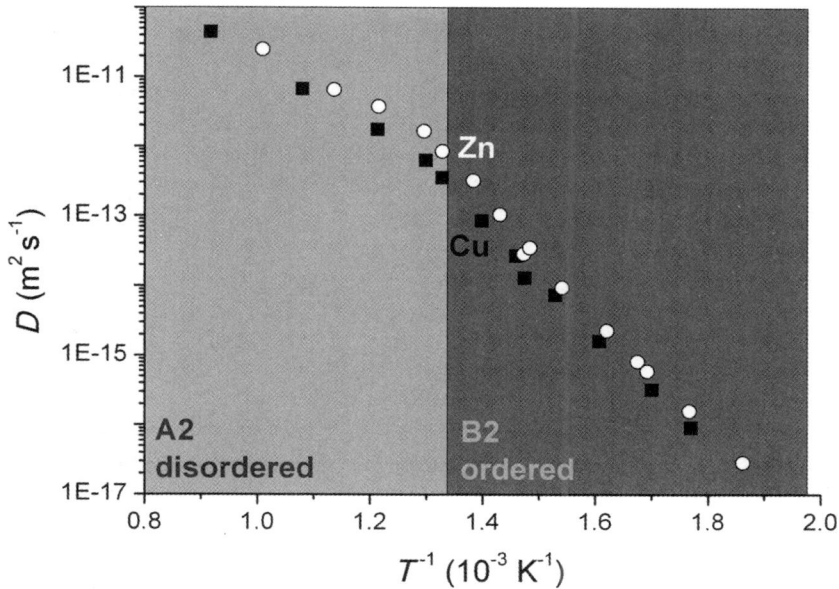

Figure 62 Tracer diffusion of both constituent in beta brass (Data source: Ref. (Kuper et al., 1956)). The A2 → B2 transition at 468 °C causes an abrupt change in the diffusion properties.

characterized by diffusivities which are higher than that of the Ga and close to each other (see **Figure 63**). The activation energy of the three constituents is similar in the B2' phase; however in the L1$_2$ phase the activation energy of the transition metals is significantly smaller than that of the Ga.

The authors explained this behavior by introducing Ga-containing and Ga-free cycles within the framework of the six-jump mechanism. Since the jump rates of the transition metals are much higher, the second sequence is characterized by a lower barrier. Therefore in the more ordered phase, the diffusion of Ni and Mn is dominated by the Ga-free jumps. In the less ordered B2' phase, the above two jump sequences are indistinguishable (since the β sites contain randomly distributed Mn and Ga atoms). Therefore the diffusion of the three elements becomes strongly coupled.

5.5.6 Experimental Results of Diffusion Studies in Intermetallics

The diffusion mechanisms in intermetallics are much more complex than in disordered alloys, as defect formation and migration mechanisms can show a remarkable difference from structure to structure or in between the different sublattices. Not surprisingly, this gives rise to a wide range of diffusion behavior. Large anisotropy or orders of magnitude diffusivity difference between the two constituents and/or impurities are fairly common in ordered alloys. In the following we give some examples on the diffusion phenomena in different intermetallic structures. This selection concentrates mostly on the technologically important aluminides and silicides.

5.5.6.1 B2 Intermetallics: NiAl and FeAl

NiAl is an ordered alloy which is characterized by triple-defect-type disorder (two Ni vacancies and an Ni antisite atom), and the defect migration takes place by two nearest-neighbor jumps in both

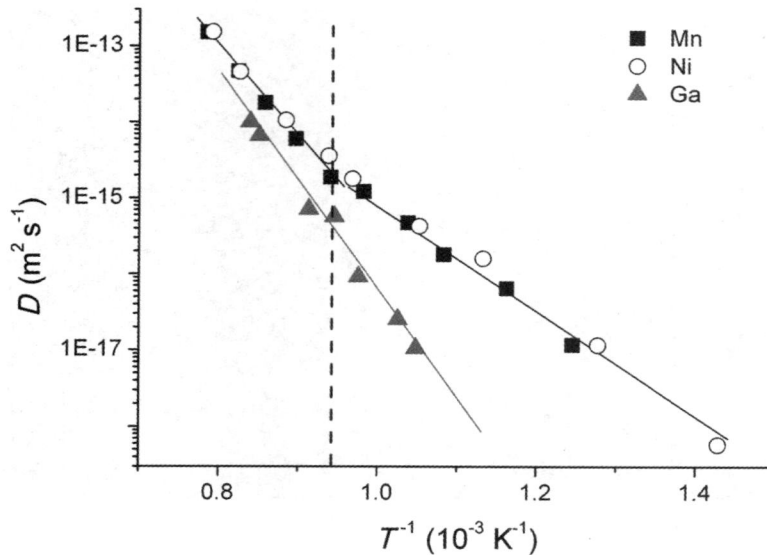

Figure 63 Tracer diffusion coefficients in the Ni_2MnGa magnetic shape memory alloy (Source of data: Ref. (Erdélyi et al., 2007)). The dashed line marks the $B2' \rightarrow L1_2$ transition. In the more ordered low-temperature phase the diffusion of the transition metals is decoupled from the Ga diffusion.

sublattices. Off-stoichiometry to the Al-rich side is compensated by an increasing number of structural vacancies (triple defects) and antisite atoms to the Ni-rich side.

Interdiffusion and Ni tracer diffusion were investigated in this system. An up-to-date and comprehensive collection of references can be found, for example in a recent simulation-related article by Zhang et al. (2011). Ni tracer diffusivities show an upward curvature from the Arrhenius relation at all compositions above 1500 K (Frank et al., 2001). But the Ni diffusion coefficient does not show a remarkable increase when increasing Al content above stoichiometry. This indicates that next-nearest neighbor jumps on the Ni sublattice have a negligible contribution. On the other hand, increasing Ni concentration causes a decrease in the Ni activation enthalpy which indicates a stronger influence from the antistructure bridge mechanism.

Al diffusion can be calculated from the interdiffusion coefficients using the Darken approach (Kim and Chang, 2000; Paul et al., 2005b). Partial molar volumes are concentration dependent in the NiAl system; therefore the Boltzmann–Matano analysis would produce artifacts, to obtain correct values, the Sauer–Freise treatment (see Eqn (79)) has to be used. Interdiffusion coefficients show a significant drop around the stoichiometric concentration. The minimum was found in a slightly Ni-enriched alloy (e.g. (Paul et al., 2005b)), with the exception of the results by Kim and Chang (2000), who determined a sharp minimum exactly at the stoichiometric composition. The role of the vacancy wind effect in NiAl is debated. Divinski and Herzig reproduced the experimental data of Kim and Chang by MC simulations (Divinski and Herzig, 2002) and claimed that the sharp drop around the stoichiometric composition is a consequence of the very small Manning factor, which is characteristic for the triple-defect mechanism. Paul et al., on the other hand, claimed that S is close to unity (Paul et al., 2005b) and neglecting the Manning factor would not cause a difference larger than the typical experimental error.

As a substitute for Al, the tracer diffusion of In was investigated in NiAl (Lutze-Birk and Jacobi, 1975; Minamino et al., 2002). Compared with Ni tracer diffusion, In diffuses with approximately the same diffusion coefficient as Ni on the Ni-rich side, while In diffusivity is about an order of magnitude faster on the Al-rich side. These results are consistent with the diffusivity ratio of the triple-defect mechanism.

FeAl has a wide existence range in composition, although shifted to the Fe-rich side (from 22 to 55 at % Al). Far from the stoichiometric composition the transition to the DO_3 phase is possible. Structural defects are antisite atoms both in the Fe- and Al-rich alloys, and thermal defects consist of antisite atoms and vacancies in the Fe sublattice (Fähnle et al., 1999). Up to now, no unequivocal diffusion mechanism is known for the B2 FeAl.

Fe tracer diffusion in the B2 phase is characterized by a linear Arrhenius dependence at compositions corresponding to $Fe_{52}Al_{48}$, $Fe_{67}Al_{33}$ and $Fe_{74.5}Al_{25.5}$ (Eggersmann and Mehrer, 2000) and $Fe_{75}Al_{25}$ (Tokei et al., 1997). Activation enthalpy increases as the composition approaches stoichiometry. Furthermore, the more ordered DO_3 phase shows a higher, the disordered A2 phase a lower activation enthalpy than the B2 ordered phase.

Interdiffusion measurement carried out by Salamon and Mehrer (2005a) indicates that intrinsic diffusivities of the Al are somewhat higher than for Fe; however both diffusion transports are very probably coupled.

5.5.6.2   L1$_2$ Intermetallics: Ni$_3$Al

The Ni$_3$Al is one of the most thoroughly studied intermetallic systems from a diffusion point of view (see Ref. (Zhang et al., 2011) for a recent collection of references). The Ni atoms migrate using the sublattice mechanism (Shi et al., 1995; Frank et al., 1995). The diffusion of Ni is concentration independent at higher temperatures while it might show a weak minimum around the stoichiometric concentration at lower temperatures (Hoshino et al., 1988). In the same experiment a highly curved Arrhenius plot was found. Hoshino et al. explained these effects by the presence of structural vacancies, which have their minimal concentration around the stoichiometric composition. Radiotracer and SIMS investigations by Frank et al. revealed no curvature in the Arrhenius plots (Frank et al., 1995). They made an increased contribution of grain boundary transport responsible for the results of Hoshino et al. since grain boundary diffusion of Ni also shows notable composition dependence and so it could reproduce both of these effects (Frank and Herzig, 1997). Computer simulation indicated (Athènes and Bellon, 1999; Duan, 2007) however that antisite atoms can interact with vacancies and increase the diffusivity of both components.

Interdiffusion coefficients in Ni$_3$Al are characterized by a linear Arrhenius dependence and do not show a significant composition dependence similarly to the Ni self-diffusion. To derive the Al tracer diffusivity one has to calculate the thermodynamic and vacancy wind factors, which are considerable sources of errors. There is a general agreement that Al diffusion has insignificant composition dependence and that the diffusivity and activation enthalpy of Al diffusion are not much different from Ni. Shi et al. observed a crossover in the Al and Ni diffusivities at 1223 K. Above this temperature Al is the faster diffuser, below Ni (Shi et al., 1995). Cserháti et al. found that Al diffusion coefficient is always below that of Ni, even though Al self-diffusion has a slightly lower activation energy (243 kJ/mol as compared with 303 kJ/mol for Ni). The slower Al diffusivity is a consequence of the orders of magnitude smaller prefactor (Cserháti et al., 2003). Ikeda et al. on the other hand concluded that Al diffusion is characterized by significantly higher activation energy (ranging to about 400 kJ/mol) (Ikeda et al., 1998).

Tracer diffusion measurements were carried out using Ga and Ge radiotracer to simulate the Al atoms. Both of the Al substitutes have an activation energy of about 60 kJ/mol above that of Ni self-diffusion (363 kJ/mol and 368 kJ/mol, respectively) (Divinski et al., 1998). Higher activation energy for the minority component in an $L1_2$ phase is consistent with the "Cu_3Au rule" (Schoijet and Girifalco, 1968). Al migrates as an impurity in the Ni sublattice at lower temperatures; at higher temperatures the increased disorder also allows the antisite bridge mechanism to operate. Owing to that Al diffusivity could approach that of Ni.

5.5.6.3 DO₃ Intermetallics: Fe₃Al, Fe₃Si and Cu₃Sn

Fe_3Al has a DO_3 structure at low temperatures; the existence range of this phase is relatively wide. Major structural defects are antisite atoms in this system. The main thermally activated defects in this system are monovacancies which can be formed in both sublattices, unlike the B2 FeAl phase (Mayer et al., 1997). The relatively small formation energy of antisite atoms and the existence of vacancies in both sublattices mean that Fe atoms can also diffuse with the aid of the Al vacancies. As can be seen in **Figure 64**, the activation energy for Fe diffusion is higher in the DO_3 phase than in the B2 phase of similar composition (278 kJ/mol to 232 kJ/mol) (Eggersmann and Mehrer, 2000). It should be noted however that the B2 phase for Fe_3Al is highly disordered, which alone reduces the activation enthalpy. Compared with the stoichiometric B2 phase, the difference is less significant (262 kJ/mol). The tracer diffusion coefficients of Al substitutes Zn and In do show a similar kink at the B2 → DO_3 transition like the Fe. The change in the activation energy is less pronounced, which cause the diffusion of the Al substitutes to decouple from Fe (Eggersmann and Mehrer, 2000).

The DO_3 ordered phase in the Fe_3Si is stable from close to ambient temperature till up to the melting point in the appropriate composition range. Dominant defects in off-stoichiometric alloys are the antisite atoms both in the Fe- and Si-rich sides (Dennler and Hafner, 2006). In the Fe sublattice vacancy formation enthalpy is relatively low at stoichiometric composition (≈ 0.75 eV) and somewhat higher

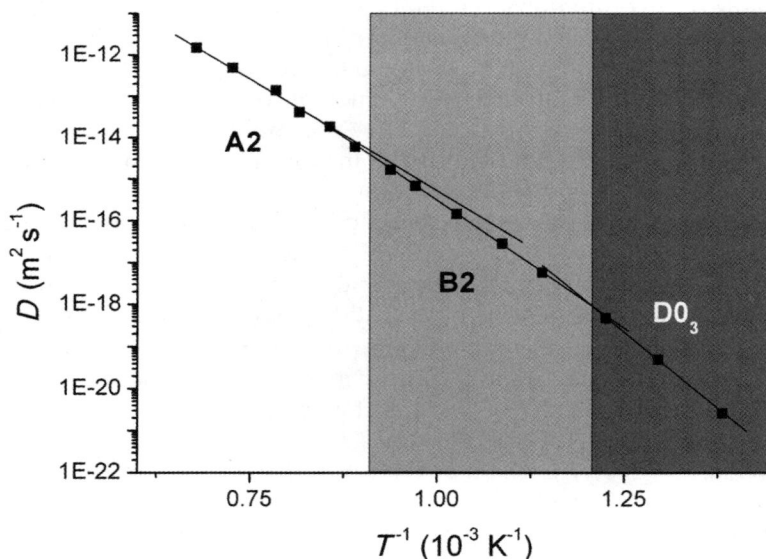

Figure 64 Diffusion of Fe in Fe_3Al (source of the data: (Eggersmann and Mehrer, 2000)), the activation energy increases with increasing order.

(≈ 1.1 eV) in $Fe_{79}Si_{21}$ (Kümmerle et al., 1995). The formation enthalpy of Si vacancies is significantly higher.

Fe atoms diffuse by the sublattice mechanism in Fe_3Si (Gude and Mehrer, 1997). The temperature dependence can be characterized by a linear Arrhenius behavior for the stoichiometric compound. The activation energy is very small (158 kJ/mol), well below that of the pure Fe or of the disordered Fe_xSi_{1-x} alloys. For Fe-rich alloys, the Arrhenius plot shows a downward curvature and the activation enthalpy increases (Gude and Mehrer, 1997). This effect is most probably a consequence of a ferromagnetic–paramagnetic transition in the material. Since the stoichiometric alloy has the lowest Curie point and saturation magnetization, this effect does not change the diffusion behavior in a significant way. MBS has shown that Fe diffusion is fastest in the stoichiometric alloy and slower in both the Fe- and Si-rich sides (Thiess et al., 2001).

The diffusion of the Si was modeled by ^{71}Ge tracer. As a check, the diffusion of ^{71}Ge and short lifetime ^{31}Si show a good agreement. The activation enthalpy of the Ge diffusion is about twice as large as for Fe diffusion, while the composition dependence of the diffusivities is negligible. The minority component is up to five orders of magnitude slower than the majority component in the Fe_3Si system (Gude and Mehrer, 1997).

Cu_3Sn is an important ordered alloy as it formed by the reaction of major solder alloys with the Cu base material. Therefore its properties can determine the strength of the solder joint. Diffusion of Cu in this system appears most probably by the sublattice mechanism, while Sn diffuses as an impurity on the Cu sublattice. A recent measurement carried out by Paul et al., using the Kirkendall marker technique, has found that both Cu and Sn migrate with an extremely low activation energy (78.8 kJ/mol and 79.7 kJ/mol) and the diffusion of Cu is about 30 times faster than that of Sn (Paul et al., 2011). The interdiffusion coefficient determined in this work is in a good agreement with other measurements reported in the literature (Onishi and Fujibuchi, 1975; Viancio et al., 1994; Dreyer et al., 1995).

To the contrary, Thiess et al. (2003) found by means of QENS that Sn diffusion is faster than Cu in the ordered Cu_3Sn and is most probably aided by the anti-structure bridge mechanism. Ab initio calculations performed by Chen et al. have shown that Cu–Sn bonds are much weaker than Cu–Cu bonds; therefore the migration of Sn is accelerated (Chen et al., 2008). A very recent MD simulation by Gao and Qu reached a different conclusion, and they proposed that the activation energy of Cu diffusion is about 10% less than that of Sn. Consequently Cu is the dominant diffusing species (Gao and Qu, 2012).

5.5.6.4 *$L1_0$ Intermetallics: TiAl and FePt*

$L1_0$ intermetallic γ-TiAl has attracted much interest because of its lightweight and good mechanical properties at high temperatures. As a consequence of the $L1_0$ structure, the in-plane (perpendicular to the tetragonal axis) and out-of-plane (parallel to the tetragonal axis) diffusion mechanisms can be significantly different. Therefore, such a system must be characterized by two diffusion coefficients and measurements in single crystalline γ-TiAl specimens are imperative. Antisite atoms represent the major structural defects, while PA spectroscopy was not able to detect structural vacancies in γ-TiAl (Shirai and Yamaguchi, 1992). Thermal vacancies are preferably formed on the Ti sublattice (Raju et al., 1996).

Ti self-diffusion was measured in both directions by Ikeda et al. using the radiotracer technique (Ikeda et al., 2001a). In-plane diffusion is significantly faster, while the activation energy is smaller (311 kJ/mol and 371 kJ/mol, respectively). The difference is caused by the larger formation energy of the Al sublattice vacancies as well as the smaller correlation factor for the out-of-plane jumps. The role of the correlation factor can be clearly observed when the diffusion of an impurity element with no site

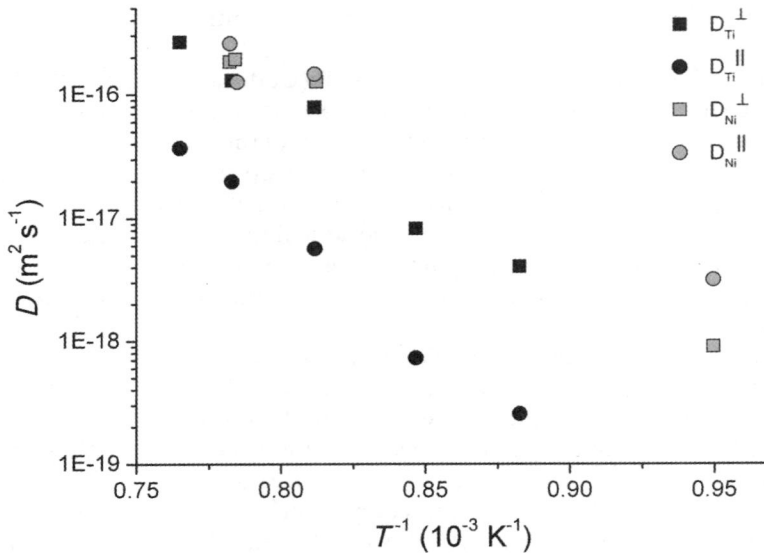

Figure 65 Diffusion of Ti and Ni impurities in TiAl. The anisotropy is opposite for the two diffusants revealing a strong correlation effect. Data taken from Ikeda et al. (2001b).

preference (such as Ni) is studied (Ikeda et al., 2001b). These results are shown in **Figure 65**. By contrast, a repetition of this measurement by Nosé et al. revealed a reversed anisotropy for Ni (Nosé et al., 2006), that is in-plane diffusion of the impurity atoms is slower than out of plane.

Diffusion of the Al substitute In was also analyzed by SIMS in the same work (Nosé et al., 2006). The anisotropy of the In diffusion is similar in character to the Ti, although not so pronounced. Activation energy for In (Al) diffusion is 285 kJ/mol and 248 kJ/mol for parallel and perpendicular to the tetragonal axis (in plane), respectively. This is significantly less than the above-mentioned values for Ti diffusion. Al is supposed to diffuse as an impurity in the Ti planes.

FePt-ordered alloys are interesting as magnetic storage materials because of their high magnetocrystalline anisotropy. The range of existence for the $L1_0$ ordered phase is rather broad from 33 to 55 at % Pt at 873 K.

Nosé et al. carried out a complete radiotracer investigation using ^{59}Fe and ^{103}Pd (as a substitute for Pt) tracer (Nosé et al., 2005). They found that the diffusion of Fe and Pt agrees within one order of magnitude; however the Fe diffusion is anisotropic with the faster diffusion in in-plane direction.

In-plane Fe self-diffusion was investigated at low temperature (773–873 K) by Rennhofer et al. using nuclear resonant (or Mössbauer) X-ray scattering (Rennhofer et al., 2006). Their activation enthalpy (160 kJ/mol) and preexponential factor were surprisingly low (3.45×10^{-13} m^2 s^{-1}) which, according to the explanation of the authors, is a result of a highly correlated diffusion mechanism such as the six-jump cycle. In a recent experiment, the same group investigated diffusion perpendicular to the $\langle 100 \rangle$ planes (Gröstlinger et al., 2012). The activation energy was found to be somewhat higher (183 kJ/mol) and overall diffusion was a factor of 20 slower as compared with the in-plane diffusion. The authors explained the anisotropy by strong ordering tendency of FePt, which seriously limits the number of possible diffusion paths when both sublattices have to be used.

5.5.6.5 DO$_{19}$ Intermetallics: Ti$_3$Al

In technologically important TiAl alloys not only γ-TiAl but also α$_2$-Ti$_3$Al appear and the two phases form a lamellar microstructure. α$_2$-Ti$_3$Al has a relatively broad existence range from 24 to 37 at% Al, depending on the temperature. PA spectrometry revealed that no structural vacancies are observable in α$_2$-Ti$_3$Al and off-stoichiometry is compensated by antisite atoms (Würschum et al, 1996). The same study has also shown that thermally activated vacancies appear preferably on the Ti sublattice. Self-diffusion of Ti proceeds via vacancy mechanism in the ordered alloy. Al atoms are most probably migrating as impurities in the Ti sublattice.

Diffusion of the [44]Ti tracer and interdiffusion was investigated by Rüsing and Herzig (1996). Ti tracer diffusion was found to be slower in the intermetallic phase than in pure α-Ti. The difference is caused by the smaller prefactor as the activation enthalpies are almost equal (288 kJ/mol and 303 kJ/mol, respectively). The activation energy shows no clear composition dependence on the Ti-rich side of the alloy, and the differences are comparable with the measurement error.

The interdiffusion is faster than Ti tracer diffusion, which is hardly surprising in a system with $\Phi > 1$. The interdiffusion coefficient shows slight composition dependence on the Ti-rich side, it becomes slower when getting farther from the stoichiometric composition. Al tracer coefficients have been determined from the analysis of the interdiffusion using the Darken–Manning equation (Eqn (90)). They were found to be smaller than those of Ti, while the activation enthalpy is significantly higher (394 kJ/mol) (Rüsing and Herzig, 1996).

[69]Ga tracer as an Al substitute was used by Herzig et al. to gain more direct information about the Al diffusion (Herzig et al., 1999a). The activation enthalpy of the Ga diffusion was slightly above the Ti self-diffusion (315 kJ/mol), however well below the estimated value from the Darken–Manning treatment. Slower diffusion of the minority element is in agreement with the Cu$_3$Au rule (d'Heurle and Gas, 1986).

5.5.6.6 C11$_b$ Intermetallics: MoSi$_2$

Molybdenum disilicide is an ordered alloy with a rather narrow existence range. Because of its excellent electrical properties as well as oxidation and temperature resistance, it is widely used in high-temperature heating elements. Structural vacancies in well-annealed Mo$_2$Si do not exist according to PA studies (Zhang et al., 2004). Thermal vacancies are formed on the Si sublattice; no vacancies on the Mo sublattice are detected within the experimental accuracy of the method.

Si diffusion was studied in-plane (or perpendicularly to the tetragonal axis) and parallel to the tetragonal axis by both the short lifetime [31]Si and the Si substitute [71]Ge tracer (Salamon and Mehrer, 2003; Salamon et al., 2004). Diffusion coefficients agree to an Arrhenius law in both cases. Since in the in-plane direction, diffusion may appear insulated in the Si sublattice, the activation enthalpy is smaller (186 kJ/mol versus 225 kJ/mol).

The diffusion of Mo has been studied by Salamon and Mehrer by radiotracer technique using the [99]Mo isotope (Salamon and Mehrer, 2005b). The diffusion of Mo was found to be extremely slow in comparison with Si. At 1450 K this difference is more than six orders of magnitude (see **Figure 66**). Also the anisotropy is enhanced as compared with Si. The slower diffusion is found parallel to the tetragonal axis. The activation enthalpies for both components show a remarkable difference, too (586 kJ/mol parallel and 468 kJ/mol perpendicular to the tetragonal axis). This extreme difference indicates that Mo diffusion proceeds via defects in the Mo sublattice.

5.5.6.7 B20 Intermetallics: FeSi and CoSi

Iron silicides are investigated because of their magnetic properties and applications as metal–semiconductor connections. Because of the B20 structure, nearest-neighbor jumps within one sublattice are impossible.

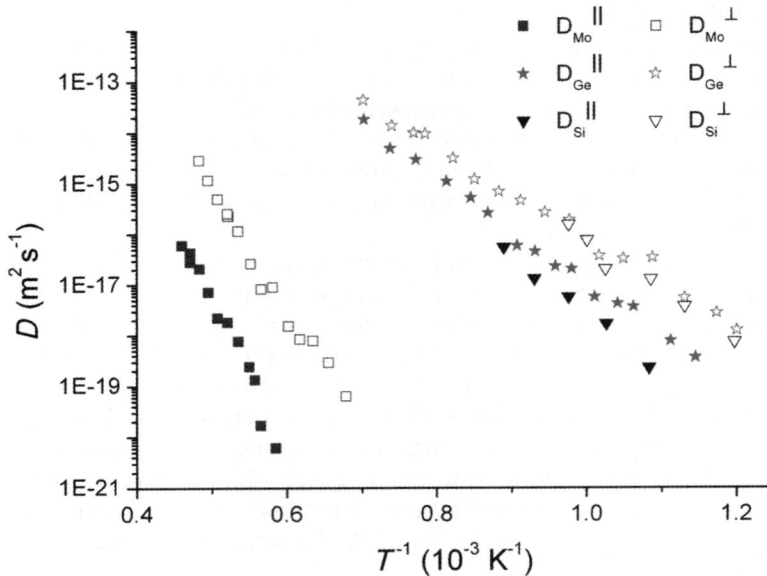

Figure 66   Diffusion of [99]Mo, [71]Ge and [31]Si parallel and perpendicular (in-plane) to the tetragonal axis in the Mo_2Si system according to Refs. (Salamon and Mehrer, 2003; Salamon et al., 2004). Because of the geometry (see the C11b structure in **Figure 58**) and vastly different vacancy formation enthalpies, a very significant difference in the diffusivities is present for the two constituents. The anisotropy is also noteworthy.

Atoms in the second nearest-neighbor shell, however, are not significantly farther, so second nearest-neighbor vacancy jumps cannot be excluded.

Tracer diffusion measurements using ^{59}Fe, [71]Ge (as a Si substitute) (Salamon and Mehrer, 1999) and the short lifetime [31]Si istope (Riihimäki et al., 2008) revealed an uncoupled diffusion behavior. The diffusion of both species is very slow indicating a strong covalent bonding of the constituents. Diffusion of Fe is one to four orders of magnitude slower than Ge diffusion in the temperature range of 1000–1500 K. The activation enthalpy of Fe self-diffusion is almost twice as large as for Ge diffusion (431 kJ/mol and 234 kJ/mol, respectively). This strongly suggests that both species diffuse independently on their respective sublattices.

Si diffusion is somewhat slower than Ge diffusion and characterized by a slightly higher activation enthalpy. Riihimäki et al. explained this result by an attractive interaction between the larger Ge tracer atoms and the Si sublattice vacancies (Riihimäki et al., 2008).

Cobalt–silicide phases are intensively researched because of their application as interconnect in very-large-scale integration technology. Van Dal et al. investigated the growth of the B20-structured CoSi at the interface of a $Co_2Si/CoSi_2$ reaction couple using the Kirkendall marker technique (van Dal et al., 2001b). By this, both the interdiffusion coefficient as well as the two tracer coefficients can be determined (For details see Sections 5.3.1 and 5.3.4). The reaction zone is characterized by a duplex morphology (see **Figure 67**); the Co-rich side of CoSi is characterized by very large grains, while the Si-rich side has significantly refined grain structure. A large number of pores also appeared near to the $CoSi/CoSi_2$ interface. From the position of the Kirkendall plane, the ratio of the diffusivities was calculated and Si diffusion was found to be at least 20 times faster than Co diffusion in the CoSi.

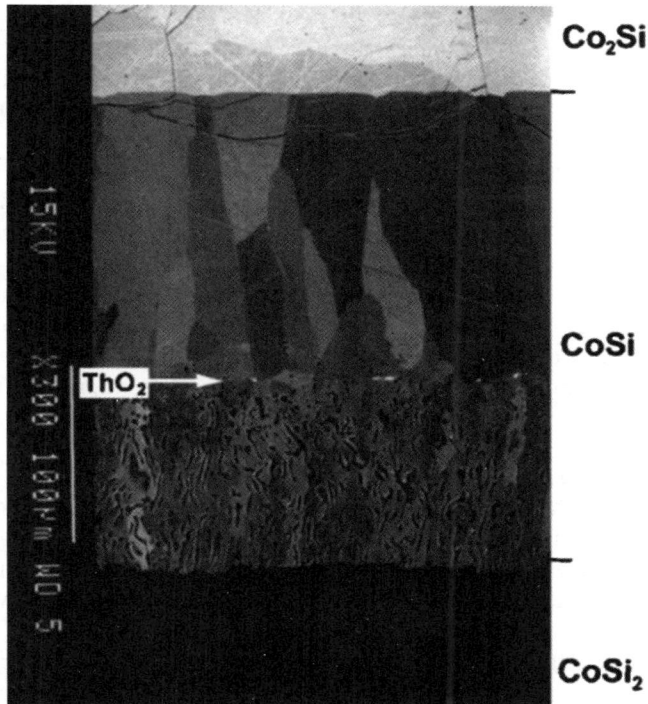

Figure 67 Interdiffusion between Co$_2$Si and CoSi$_2$: The growing reaction product CoSi spreads to both sides of the original interface that is marked by inert ThO$_2$. Microstructures on both sides of the Kirkendall plane differ significantly. Picture is taken from van Dal et al. (2001b).

The growth of the phase was found to be parabolic and the activation enthalpy of the Si diffusion was 192 kJ/mol.

5.5.6.8 Laves Phases: the C15 Co$_2$Nb

Co$_2$Nb has interesting magnetic properties and represents a good balance between high strength, oxidation resistance, melting point and low density which is a promising combination of advantages for aerospace applications. The Co$_2$Nb phase mainly exists on the Co-rich side of the stoichiometry. The off-stoichiometry is mostly compensated by Co antisite atoms according to gravimetry measurements (Saito and Beck, 1960).

Denkinger and Mehrer investigated the tracer diffusion of both components by ^{57}Co and ^{95}Nb tracers (Denkinger and Mehrer, 2000). Their main findings can be summarized as the following: Co diffusion is faster than Nb diffusion at a given temperature and composition, although the activation enthalpies are quite similar (295 kJ/mol and 292 kJ/mol, respectively, in Nb$_{31}$Co$_{69}$). With increasing Co content, the activation enthalpy of Co diffusion decreases, while the diffusivity increases. A faster Co diffusion is in agreement with both the Cu$_3$Au rule and with the much smaller size of the Co atoms. Diffusion of both components is likely to be driven by a vacancy mechanism.

5.5.6.9   B8 Intermetallics: Ni–Sn

B8 intermetallics have a very broad existence range as extra Ni atoms can occupy part of the double tetrahedral sites in the structure. Moreover thermal excitation can also cause Ni atoms to change their normal octahedral interstitial sites to the double tetrahedral ones. Ni atoms can occupy antisite positions too, although it is less important in Ni–Sn (Schmidt et al, 1992a). Because of symmetry, the diffusivity differs along and perpendicular to the hexagonal axis.

Diffusion of ^{63}Ni and ^{113}Sn tracer atoms was investigated by Schmidt et al. (1992a, 1992b). Ni migration is realized by jumps from octahedral Ni chains to double tetrahedral interstitial sites (see **Figure 55**). The diffusion coefficient of Ni is characterized by a linear Arrhenius dependence. Diffusion perpendicular to the hexagonal axis is slightly faster than parallel to it. The activation energy is practically similar for both directions and composition independent (≈ 210 kJ/mol). The diffusion of Sn also proceeds by using the double tetrahedral sites. Sn diffusion is also characterized by a linear Arrhenius dependence, and the activation enthalpy is again independent of the orientation (290 kJ/mol). Diffusion parallel to the hexagonal axis is faster than perpendicular to it.

5.5.6.10   B3$_2$ Intermetallics: β-LiAl

β-LiAl is a mixed ionic and electronic conductor, which can be used, among other applications, to produce tritium for fusion reactions. Off-stoichiometry is either compensated by Li structural vacancies or Li–antisite atoms; moreover there is an attractive interaction between Li–antisite atoms and vacancies on the Li sublattice (Tarczon et al., 1988). Diffusion of Li proceeds via vacancy mechanism on the Li sublattice.

Tarczon et al. studied the diffusion of Li by pulsed field gradient NMR measurements (Tarczon et al., 1988). Li diffusion was found to follow a linear Arrhenius behavior. Both the pre-factor (D_0) and the activation enthalpy (ΔH) decreased with increasing Li content. Li diffusion coefficient is as high as 10^{-11} m^2 s^{-1} at ambient temperature. Assuming two different jump times allowed a better description of the experimental data.

The diffusion of the Li was also investigated by 0.84 s (!) half life ^{8}Li isotope at Tokai Radioactive Ion Accelerator Complex (Jeong et al., 2005). In this experiment, the ^{8}Li isotopes produced in nuclear reaction are implanted to the preheated sample with 4 MeV final energy, which corresponds to an average implanting depth of 10 μm. ^{8}Li ion decays into two α particle, which have an average escape distance of 8 μm. Therefore one can observe the diffusion of Li from measuring the time evolution of the rate of α-particles. The implantation was carried out in 1.5 s, while the subsequent diffusion experiment lasted 4.5 s. The evaluation of the spectra can be done by comparing measured and simulated time evolutions. These results are comparable with the measurements of Tarczon et al., thought a 30% systematic difference was observed.

5.5.6.11   A15 Intermetallics: V$_3$Ga and V$_3$Si

V$_3$Ga is often used in the high field inner coils of superconducting magnets. The T_c is 14.2 K, while the upper critical field (H_{c2}) is over 19 T.

The diffusion of ^{48}V and ^{67}Ga tracer was analyzed by van Winkel et al. (1984). Bulk diffusion of Ga was found to be minimal, and the analysis of the penetration profiles has shown grain boundary diffusion only. Vanadium bulk diffusion is characterized by a linear Arrhenius plot, with an activation energy of 415 kJ/mol. This value is rather high as compared with pure metals.

V$_3$Si is a well-known superconductor with a T_c of 17 K. The existence range of this compound is very narrow. The defect concentration has not been investigated yet. A similar A15 Nb$_3$Sn is characterized by antisite-type structural defects. Structural vacancy concentration is negligible according to ab initio calculations (Besson et al., 2007). Nearest-neighbor jumps on their own sublattice are possible for V atoms, but not for Si atoms.

Diffusion in this system was analyzed using the Kirkendall marker technique by Kumar et al. (2009). They encountered an almost negligible Si diffusion, while V diffusion is characterized by a linear Arrhenius dependence. The activation enthalpy for V tracer diffusion was calculated to be 272 kJ/mol. The immeasurable Si diffusion rate can be explained by the lack of sufficient Si antisite atoms (which could migrate as an impurity in the V sublattice).

5.6   Short Circuit Transport

Unlike gases, liquids, or amorphous materials, crystalline solids contain structural defects which may represent alternative diffusion paths. Usually, dislocations and grain boundaries offer an environment of less order, more free volume and possibly weaker chemical bonds. They are therefore often distinguished by enhanced diffusivity (at least for pure metals, where the different local chemistry plays no role). As *"diffusion shortcuts"*, they lead to a fast *short circuit* transport of atoms.

In the following section, we briefly introduce the phenomenological treatment of fast diffusivity paths, and offer an outlook on their fundamental and practical relevance. Next, we will describe dislocations and the related diffusion mechanisms, summarize the general characteristics of grain boundary (GB) diffusion, and finally mention some modern results on shortcut diffusion (e.g. nonequilibrium grain boundaries, triple junctions).

5.6.1   Diffusion along Fast Paths

Figure 68 shows the so-called diffusion spectrum of Ni. From this example it is immediately clear that self-diffusion in dislocations and grain boundaries is many orders of magnitude faster than in a regular lattice. Furthermore the diffusion activation enthalpy is also lower in the diffusion shortcuts, thus the difference in diffusivity increases with decreasing temperature. It is interesting to note, that while diffusion data for well-defined volume close to equilibrium state show nearly no scatter, measurements of shortcut diffusion are characterized by more pronounced fluctuations. This partly stems from experimental/methodical uncertainties; however the nonuniformity of the defect structures also contributes to this effect. Indeed notions like "grain boundary diffusion coefficient" are simplifications with more or less justification depending on the exact reproducible preparation of experimental structures.

To illustrate the importance of these shortcuts in atomic transport, let us assume a crude slab model, consisting of grains with 500 μm and grain boundaries of 0.5 nm width. No material exchange between the grain interior and grain boundaries should be allowed. Then, using the data of **Figure 68**, it can be estimated that the effective material transport within the grain boundaries will exceed the lattice transport if the temperature is below ~1100 K. If the temperature is even below ~850 K (approximately half of the melting temperature T_m) gross grain boundary diffusion would exceed the volume-related transport by a factor of 100. Indeed this means that, even in coarse-grained materials, lattice diffusion is negligible as compared with shortcut diffusion for a relatively wide and technologically important temperature domain. For finer grained, especially for nanocrystalline materials, this temperature limit is even higher.

5.6.1.1   Kinetic Regimes for Self-Diffusion
In a more realistic approach, the diffusing species can walk through the whole sample volume, spend some time in defects then enter the grain bulk and possibly visit another defect. The general diffusion

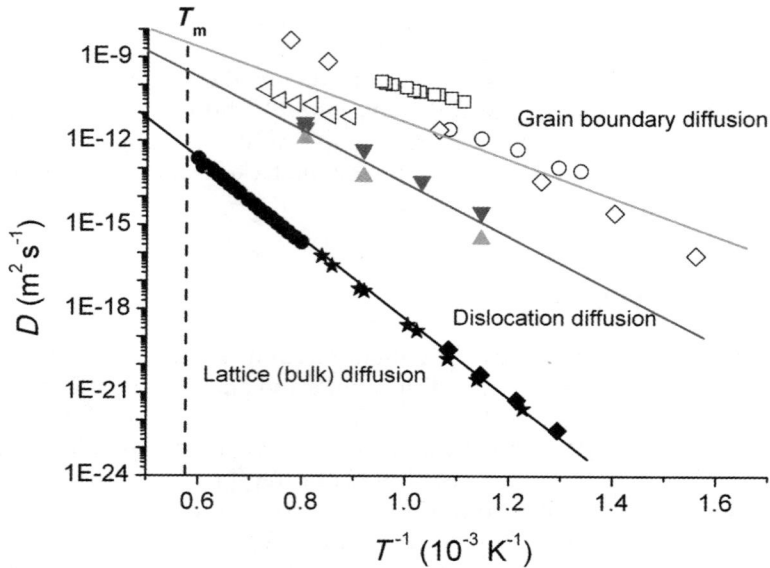

Figure 68   Diffusion spectrum of self-diffusion of Ni with exemplary lattice/bulk (◆ (Wazzan, 1965), ● (Bakker, 1968) and ★ (Maier et al., 1976)), dislocation (▼ edge and ▲ screw dislocation, both (Canon and Stark, 1969)), and grain boundary (◁ (Lange et al., 1964), ○ (Wazzan, 1965), □ (de Reca and Pampillo, 1975) and ◇ (Neuhaus and Herzig, 1988)) data. It is immediately seen that lattice diffusion is orders of magnitude slower than the diffusion along "shortcuts", especially at lower temperatures. The dashed line indicates the melting temperature T_m of Ni. Solid lines are just to guide the eyes.

behavior in such a system can be interpreted by kinetic regimes. The most simple, yet still the most widely used classification was suggested by Harrison (1961). This work described diffusion with the aid of a single static grain boundary or a dislocation network. Transport along the defects is characterized by a single diffusion coefficient D_{sc} which is order of magnitude larger than the bulk diffusion coefficient D_b (to avoid any misunderstanding lattice diffusion will also be indexed by a subscript in this section). As a further simplification, let us discuss first tracer self-diffusion which excludes any impact of chemical segregation.

According to Harrison three fundamental regimes can be distinguished (see **Figure 69**). In the so-called *A-kinetic regime*, the bulk diffusivity is large so that the diffusant atoms visit multiple shortcut channels and the residence time in the different transport paths is proportional to the volume fraction

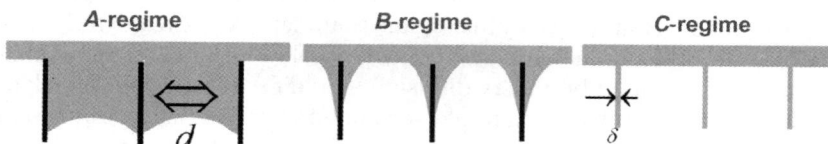

Figure 69   Kinetic regimes according to the Harrison classifications (Harrison, 1961). In the *A*-kinetics the lattice diffusion length $(D_b t)^{0.5}$ is much larger than the average distance between the fast diffusion paths (grain diameter d). The *C*-kinetics is characterized by exclusive transport along the diffusion shortcuts, that is the lattice diffusion length is less than the width of the shortcut (δ). In between one can find the *B* regime. (For color version of this figure, the reader is referred to the online version of this book.)

of the given structural element. The macroscopic sample as a whole obeys the Fick law and is characterized by a single effective diffusion coefficient D_{eff}:

$$D_{eff} = \gamma D_{sc} + (1 - \gamma)D_b \tag{145}$$

where γ is the volume fraction of the diffusion shortcuts. This kinetic regime is established if

$$\Lambda \sqrt{D_b t} \gg 1 \tag{146}$$

where t is the annealing time and Λ is the typical reciprocal distance between diffusion shortcuts (grain boundaries, dislocations, ..., etc.). This regime can be established if the diffusion coefficient (in other words the annealing temperature) is high, the annealing time is long, or if the characteristic distance between the defects is small. The latter is easily achieved in nanocrystalline materials, with grain sizes below 100 nm. Therefore this regime is very important in nanoscaled materials.

Equation (145) is derived for a slab model as it was shown in **Figure 69** accordingly the shortcuts 'communicate' using the perfect crystalline lattice as medium. A realistic material however is characterized by a statistic network of grain boundaries/dislocations. Belova and Murch analyzed the effect of the different geometries by MC simulation (Belova and Murch, 2004b) and they concluded that Eqn (145) overestimates the diffusion coefficient by 10–50%. The most correct description was achieved by using the so-called Maxwell–Garnett equation (Maxwell–Garnett, 1904).

$$D_{eff} = \frac{D_{sc}[(3 - 2\gamma)D_b + 2\gamma D_{sc}]}{\gamma D_b + (3 - \gamma)D_{sc}} \tag{147}$$

The *C-kinetics regime* represents the opposite of the above-mentioned A-regime, as the bulk diffusion is practically frozen and so transport is restricted to the shortcuts. However, from experimental point of view, both regimes share an important similarity. The measured composition profile is characterized by a single diffusion coefficient, in this case the shortcut diffusion coefficient D_{sc}. The condition to establish the C-regime is

$$\sqrt{D_b t} < \frac{\delta}{20} \tag{148}$$

where δ is the characteristic width of the given diffusion shortcut. Since δ is comparable with or less than a nanometer, even for $D_{sc} = 10^{12} \times D_b$ the shortcut diffusion penetration profile $((D_{sc}t)^{0.5} = 10^6(D_bt)^{0.5})$ would not exceed a few μm. Accordingly, the sectioning in experiments must be comparably fine. Furthermore since the diffusion is restricted to the shortcuts, tracer atoms enter only a small fraction of the specimen. This leads to very low tracer quantities per section and so C-kinetics tracer diffusion measurements require high sensitivity and care. Not surprisingly, accurate self-diffusion measurements in C-kinetics are extremely rare, for example (Sommer and Herzig, 1992; Divinski et al., 2010). On the other hand, since these measurements directly provide the D_{sc} shortcut diffusion coefficient their importance cannot be overestimated.

The third type of diffusion regime, called *B-kinetics regime*, represents a complex situation. Neither of the limiting conditions before is fulfilled. Consequently leakage of material from the diffusion shortcut to the bulk is present; however, transport at each shortcut happens still independently. The tracer penetration profile depends on the actual geometry; therefore this regime will be discussed in detail in the respective sections of dislocation (Section 5.6.2) and grain boundary diffusion

(Section 5.6.3). In general, experiments in the B regime do not allow a direct determination of the D_{sc} diffusion coefficient; instead a double product P of the transport cross-section or size of the shortcut δ and D_{sc} ($P = \delta D_{sc}$) can be evaluated. By a combined B- and C-kinetics measurement the width of the shortcut can be estimated at least at the temperature range where both regimes can be measured (Gas et al., 1992; Divinski et al., 2010).

The transitions between the above kinetic regimes are not abrupt, but continuous. Therefore transition regimes (AB, BC) can be also defined for a more accurate description. Other possible classification diffusion kinetics, including nonstationary grain boundaries can be found, for example in the review article of Bernardini et al. (2004).

5.6.1.2 Segregation Effects
Grain boundaries and dislocations are usually characterized by weaker bonds and more free volume. Since in most cases A–B (matrix atom–impurity atom) pair bonds are either stronger or weaker than the A–A (matrix atom–matrix atom) bonds, removing impurity atoms from the bulk and placing them at the defects will either result in a Gibbs energy loss or gain. In both cases, however, the local composition at the defects differs from the bulk. To understand shortcut diffusion in alloys this segregation effect has to be taken into account. For low concentration impurities, Henry's law is fulfilled for the segregation:

$$c_{sc} = sc_b \tag{149}$$

where c_{sc} and c_b are the concentration of impurities in the defect and the bulk, respectively, and s is the segregation factor.

For C-type diffusion, since the bulk diffusion is frozen, the defect-bulk system is chemically not equilibrated. Since there is so no communication between the bulk and the grain boundaries, segregation has no effect, the measurements directly deliver the shortcut diffusion coefficient D_{sc} irrespectively of the diffusant type. For the B-type diffusion however, instead of a double product, the P parameter becomes a triple product containing the segregation factor in addition:

$$P = \delta s D. \tag{150}$$

Indeed it can be said, that the double product is simply the $s = 1$ case of the triple product. Measuring both in the B- and C-kinetics, if δ is known can provide the segregation factor and the segregation enthalpy, for example Ref. (Surholt et al., 1994; Divinski et al., 2007).

For A-type diffusion, the residence time of the diffusant in different structural elements is no more proportional to the simple volume fraction. Therefore, a correction has to be introduced into Eqn (145) which reads

$$D_{eff} = \tau D_{sc} + (1 - \tau)D_b = \gamma \frac{c_{sc}}{\gamma c_{sc} + (1 - \gamma)c_b} D_{sc} + \left(1 - \gamma \frac{c_{sc}}{\gamma c_{sc} + (1 - \gamma)c_b}\right)D_b \tag{151a}$$

where τ is the fraction of the time the diffusant spends in the diffusion shortcuts. In coarse-grained materials it is still true for even quite high segregation factors that $(1 - \gamma) >> s\gamma$. Under these conditions the above equation simplifies to

$$D_{eff} = s\gamma D_{sc} + (1 - s\gamma)D_b \tag{151b}$$

This formula was first suggested by Mortlock (1960).

As for the case of self-diffusion, the transition between the regimes cannot be abrupt. This leads to the appearance of transition regimes, which contain the segregation effects too (Belova et al., 2012). Even more, as the tracer is usually supplied from the free surface, tracer atoms can enter with a different probability to the grain interior and the grain boundary (Belova et al., 2012). This could lead to initial tracer distribution, which does not necessarily respect the segregation equilibrium between the grain boundary and the bulk. As a result transients in the apparent segregation factor might arise in a B-kinetics measurement.

5.6.2 Dislocation Pipe Diffusion

Dislocations are crystal imperfections, which in the simplest case can be imagined as the border of an extra semiplane introduced to the crystal. A detailed description of the dislocation and their characteristics can be found in chapter 16, Volume II. In this section, we only consider properties that are related to their diffusion shortcut behavior.

Dislocations are linear (one-dimensional) defects. Smoluchowski proposed a model, which assumes dislocations as pipes of $\delta \approx 0.5$ nm radius that are characterized by a dislocation diffusion coefficient D_d, larger than the D_b bulk diffusion coefficient (Smoluchowski, 1952). This model is inspired by the Fisher model of grain boundary diffusion (Fisher, 1951) (see Section 5.6.3.3). The density of dislocations in well-annealed metals is around 10^{10} m^{-2}, which corresponds to a volume fraction of 10^{-8}. For highly deformed materials, however, the dislocation density can be many orders of magnitude higher. Combined with a strong segregation to dislocations, this means that dislocation pipe diffusion can be the dominant mode of material transport in extraordinary cases. For example dislocation pipe diffusion is assumed to explain the accelerated segregation kinetics found by Chang et al. in investigating Bi segregation to Cu grain boundaries (Chang et al., 1999).

Since the exact nature (especially the geometry) of the diffusion shortcuts plays no role in the A- or C-type kinetic regime, the conditions (Eqns (146) and (148)) and the formulas (Eqn (151b)) hold for the case of dislocation pipe diffusion. In the following we will describe the diffusion in an isolated dislocation in type B-regime.

Using cylindrical coordinates r (distance from the dislocation core) and z (the depth along the dislocation line) is the straightforward choice. The basic equations to be solved are the following: the Fick equations within ($r \leq \delta$) and outside of the dislocation ($r > \delta$)

$$\frac{\partial c_b}{\partial t} = D_b \left[\frac{1}{r} \frac{\partial}{\partial r} \left(r \frac{\partial c_b}{\partial r} \right) + \frac{\partial^2 c_b}{\partial z^2} \right] \quad \text{for} \quad r > \delta \tag{152a}$$

$$\frac{\partial c_d}{\partial t} = D_d \left[\frac{1}{r} \frac{\partial}{\partial r} \left(r \frac{\partial c_d}{\partial r} \right) + \frac{\partial^2 c_d}{\partial z^2} \right] \quad \text{for} \quad r \leq \delta \tag{152b}$$

where D_b and D_d are the diffusion coefficients in the bulk and dislocation, and $c_b(r,z,t)$ and $c_d(r,z,t)$ are the respective concentrations. Flux continuity and segregation equilibrium must be fulfilled at the dislocation–bulk interface:

$$\left(D_b \frac{\partial c_b}{\partial r} \right)_{r=\delta^+} = \left(D_d \frac{\partial c_d}{\partial r} \right)_{r=\delta^-} \tag{153a}$$

$$sc_b\left(r = \delta^+\right) = c_d\left(r = \delta^-\right). \tag{153b}$$

Finally the composition gradient must vanish far from the dislocation. Usually two main initial conditions are analyzed in the literature: constant source and instantaneous source. The first means that a constant c_0 concentration is maintained on the surface, while in the second case an initial amount of M atoms per unit area is deposited on the surface (see Section 5.2.4 on the solution of the diffusion equation). Detailed theoretical description of the diffusion along dislocations can be found in a series of articles by Le Claire and Rabinovitsch (1981), Le Claire and Rabinovitsch (1982), Le Claire and Rabinovitsch (1983), and Le Claire and Rabinovitsch (1984). In the following we restrict ourselves to present the mean concentrations (e.g. the total activity measured in a section cut perpendicular to the dislocation pipes) and only for the simplest $s = 1$ case. Let us introduce the normalized variables

$$\eta := \frac{z}{\sqrt{D_b t}}; \quad \alpha := \frac{\delta}{\sqrt{D_b t}}; \quad \Delta := \frac{D_d}{D_b}; \quad \beta := \frac{(\Delta - 1)\delta}{\sqrt{D_b t}} \approx \frac{D_d \delta}{D_b \sqrt{D_b t}}. \tag{154}$$

From these variables η scales the effect of direct surface diffusion, while α scales the outdiffusion from the dislocation. β is an enhancement factor, although its meaning for the dislocation pipe model is less transparent than in the case of grain boundary diffusion (Section 5.6.3). For constant source, the mean composition follows a complementary error function part that captures bulk transport from the source and an additional term called "dislocation tail".

$$\overline{(c_{\text{const}})(\eta)} = c_0 \left(\text{erfc}\, \frac{\eta}{2} + \pi \delta^2 \rho_d Q'(\eta) \right) \tag{155}$$

where ρ_d is the dislocation density. For the case of an instantaneous source, the solution is equal to the so-called thin film solution (Eqn (14)) plus a "dislocation tail".

$$\overline{(c_{\text{inst}})(\eta)} = \frac{M}{\sqrt{\pi D_b t}} \left(\exp\left(-\frac{\eta^2}{4}\right) + \pi \delta^2 \rho_d Q''(\eta) \right) \tag{156}$$

where Q' and Q'' are quite complex integrals containing Bessel functions (see Refs. (Le Claire and Rabinovitsch, 1981, 1982) for more information) which among other parameters depend on the normalized variable η. To evaluate the dislocation diffusion coefficient from a measured profile, however, no detailed knowledge about these parameters is required. The most important information is that Q' and Q'' are roughly exponential functions of η, if η becomes larger than 4–5. The slope of both Q' and Q'' is equal to

$$\frac{\partial Q'}{\partial \eta} = \frac{\partial Q''}{\partial \eta} = -\frac{A(\alpha)}{\sqrt{\alpha \beta}} \tag{157}$$

where $A(\alpha)$ is close to unity, varies slowly with α and has a very weak dependence on $\alpha\beta$. This function is plotted in **Figure 70**.

Combining either Eqn (155) or Eqn (156) with Eqn (157) one can get the following equation for the slope of the concentration profile:

$$\frac{\partial \ln(\bar{c})}{\partial z} = -\frac{-A(\alpha)}{\delta \sqrt{\frac{D_d}{D_b} - 1}} \approx \frac{-A(\alpha)}{\delta} \sqrt{\frac{D_b}{D_d}} \tag{158}$$

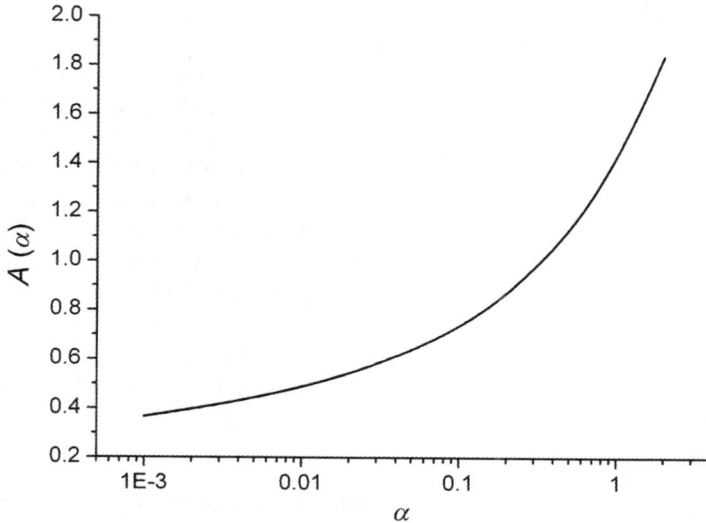

Figure 70   Plot of the function A(α), note the logarithmic scale for α

Thus, the logarithm mean concentration of a section depends linearly on the depth (unlike the famous $\sim z^{6/5}$ dependence in the case of grain boundary diffusion discussed in the next section). Furthermore, time dependence is introduced through A(α), which is a slowly varying function of α (see **Figure 70**) and so the slopes are practically independent of the time for dominant dislocation diffusion. This allows a clear separation of the dislocation tail from the grain boundary tail in experiment, since the latter has a $\sim t^{-0.25}$ time dependence. Since the volume fraction of dislocations is quite small in well-annealed metals, dislocation diffusion measurements are rare. A couple of experimental data is collected in (Landolt–Börnstein, 1990g).

5.6.3   Grain Boundary Transport

Since quite a large fraction of the technical materials are polycrystalline, diffusion along grain boundaries plays a significant role for many applications. Grain boundary diffusion is many orders of magnitude faster than lattice diffusion especially at lower temperatures (see **Figure 68**). Accordingly, it is very often the dominant atomic transport mechanism. Intensive, systematic investigation of grain boundary diffusion started in the 1950s (Barnes, 1950; Fisher, 1951; Hoffmann and Turnbull, 1951; Le Claire, 1951). The first accurate grain boundary diffusion measurement was carried out by Hoffmann and Turnbull (1951). They investigated the self-diffusion of Ag by radiotracer technique in single crystalline and polycrystalline specimens. When they plotted the activity as a function of the depth, they observed a deep penetration tail of the tracer for polycrystalline samples which was not present in the single crystals. This effect was attributed to shortcut diffusion along the grain boundaries. Based on the wide variety of possible grain boundary structures, it is indeed quite remarkable that properties of grain boundary transport can be often characterized by a single parameter, the grain boundary diffusion coefficient D_{gb}.

After these early investigations quite an amount of experimental work was invested to study this phenomenon in more detail (see, e.g. Refs. (Landolt–Börnstein, 1990h or Kaur et al., 1989)). Grain boundary transport is especially important in the fine-grained (submicro- or nanocrystalline) materials for which the grain boundary fraction can reach a few percents. For these kind of materials, substructures of the grain boundary network like triple junctions (Palumbo et al., 1990) or certain "fast grain boundaries" (Divinski et al., 2004) can gain importance. Highly disordered so-called nonequilibrium grain boundaries are especially prominent in materials produced by severe plastic deformation (Valiev, 2004; Sauvage et al., 2012). Because of its wide variety, this field is very actively researched nowadays, as crucial experimental data and/or appropriate theoretical models are often still lacking.

There are two general possibilities of experimental investigation of grain boundary diffusion: either by taking a polycrystalline specimen to characterize some mean properties of the ensemble (e.g. Refs. (Lange et al., 1964; Neuhaus and Herzig, 1988)) or by using carefully designed bicrystal specimens with a single grain boundary of known properties (e.g. Ref. (Divinski et al., 2012)). Measurements using APT represent a middle ground between these two extremes, since in this method single grain boundaries can be examined and that provides not only the mean grain boundary diffusion coefficient but the typical scatter among grain boundaries too (e.g. Ref. (Ene et al., 2005; Chellali et al., 2012)).

5.6.3.1   *Characterization and Structure of Grain Boundaries; Atomic Transport Mechanisms*

Grains in a nontextured material are oriented randomly; therefore a wide variety of orientation relation between adjacent grains must be expected. The properties of the grain boundary, including the atomic transport characteristics too, depend on the misorientation of the grains and also on the orientation of the grain boundary plane relative to the grains. In total five macroscopic grain boundary parameters are required: the three rotation angles to transform the grains into each other and the normal to the grain boundary plane.

If the misorientation angle between the two grains is less than $\sim 15°$, the grain boundary can be described as a series of discrete dislocations. These kinds of grain boundaries are *called low-angle grain boundaries*. The dislocation pipe model of Smoluchowski (1952) is designated to treat grain boundary diffusion in this particular situation. In the following discussion, it is implicitly assumed that considered grain boundaries do not belong to this class.

Among the high-angle grain boundaries one finds a distinguished class, called special grain boundaries. These boundaries represent specific misorientations for which the coincidence site lattice (CSL) contains a notable fraction of the total atoms. The CSL can be constructed by extending the crystalline structures of the different grains into each other and marking those atoms which belong to both lattices (see **Figure 72**).

These special grain boundaries are called ΣN boundaries, where N^{-1} is the fraction of atoms on the CSL, for example, $\Sigma 5$ means that every fifth atom belongs to the CSL. Special grain boundaries are characterized by higher order and a periodic structure (see **Figure 71(a)**). Probably the best known of them is the $\Sigma 3$ or coherent twin boundary. Not surprisingly these more ordered structures are usually characterized by lower grain boundary energy and lower diffusivity than ordinary high-angle grain boundaries.

Grain boundaries for which no such periodic arrangement can be defined are called general high-angle grain boundaries. In this case atomic arrangement appears largely disordered. Contemporary MD simulations have shown (see (Mishin et al., 2010) for a recent review) that the disordered zone around the grain boundaries grows as temperature increases (see **Figure 73**). Some simulations also predicted (e.g. (Ciccotti et al., 1983; Frolov and Mishin, 2009)) that a grain boundary premelting can happen close to, but well below the bulk melting point. Naturally, these effects may seriously affect diffusion and segregation properties of grain boundaries.

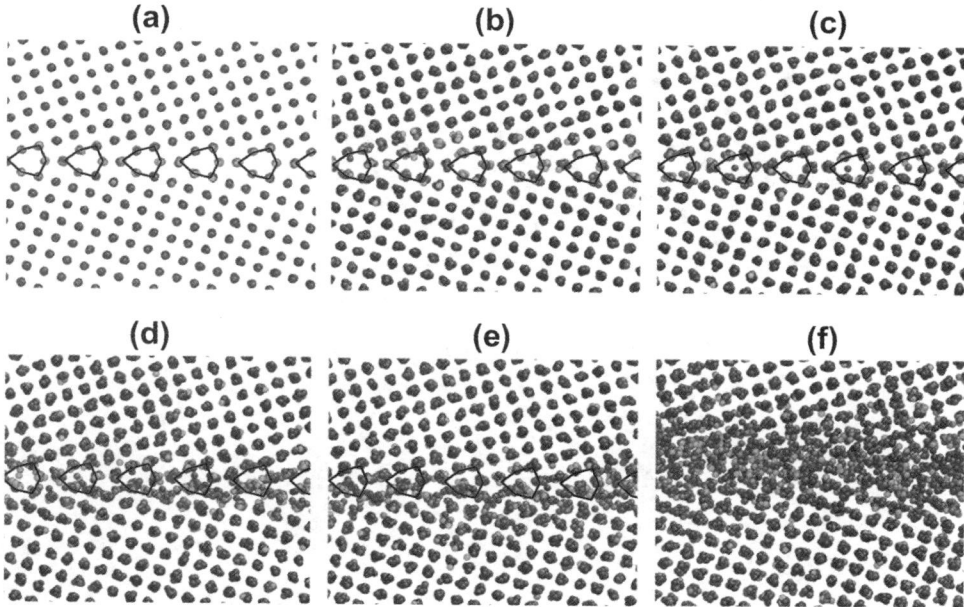

Figure 71 MD simulation of a Σ5 (310) grain boundary in Cu after 10 ns equilibration at 200 K (a), 700 K (b), 800 K (c), 900 K (d), 1000 K (e) and 1300 K (f). At low temperature open channels are visible in the structure causing a significant anisotropy in the diffusion coefficients. As the temperature increases, deviations from this ordered configuration appear and near the melting point, the GB becomes liquid like. Reproduced with permission from Divinski et al. (2012). (For color version of this figure, the reader is referred to the online version of this book.)

Atom probe measurements on the Ni diffusion in Cu thin films by Chellali et al. (2012) indicated that the width of the Ni "segregation zones" around the Cu grain boundaries shows a considerable variation with the temperature, even though the conditions of the C-kinetics were always fulfilled. Furthermore, to establish a consistent description of the material transport between grain boundaries and triple junction, the authors had to assume also a temperature-dependent variation of the transport width δ.

The point defect formation and migration energies of GBs show remarkable differences from that of the bulk. First of all, because of the larger free volume, the formation energy is generally smaller for both vacancy and interstitial type of defects. However, as grain boundaries do not have lattice, symmetry these formation energies show a strong variation with the local environment (Suzuki and Mishin, 2005). Moreover, the formation energy of the vacancy and interstitial-like defects are close to each other, so both of them can contribute to the atomic transport. As a further complication, both the vacancy and the interstitial-like defects can split into partials (Suzuki and Mishin, 2005). Not surprisingly, there is a rich variation of diffusion mechanisms for grain boundary diffusion and the measured total atomic transport is the net effect of possible mechanisms. At lower temperatures and defect concentrations, single defect mechanism seems still to be likely, while at higher temperatures jumps involving multiple defects have significant contribution.

Under these conditions, it is not a small wonder that grain boundary diffusion coefficients, at least for a fixed grain boundary orientation follow the Arrhenius dependence usually quite well (see the

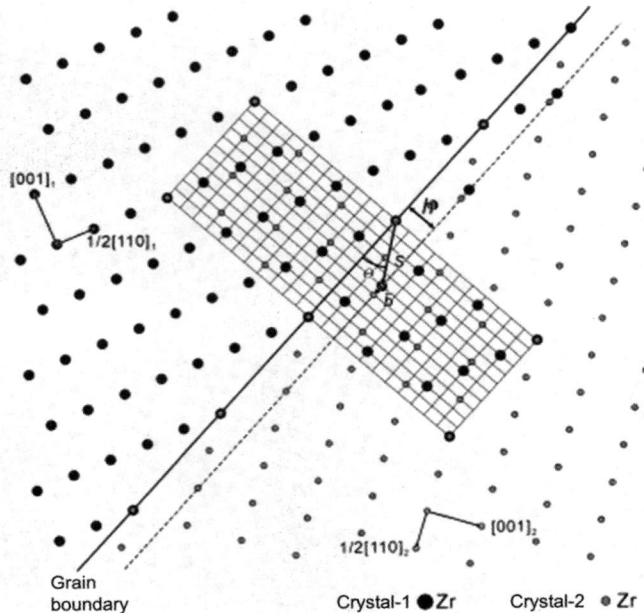

Figure 72 Constructing the CSL in the case of Zr Σ11 (113) grain boundary. The atomic planes are expanded to the neighboring grains. If a given site belongs to both lattices, it is called coincidence site (large atoms marked in gray at the corners of the rectangle and along the grain boundary plane). The unit cell of the CSL is naturally larger than the unit cell of the crystal itself (in this case 11 times). Redrawn after Yoshida et al. (2004).

experimental results in Landolt–Börnstein, 1990h). This indicates, that for a given grain boundary, there is a dominant mechanism of diffusion or that multiple diffusion mechanisms are characterized by activation energies lying close enough. Nevertheless, the diffusion coefficient and even the mechanism can show significant variation from grain boundary to grain boundary.

5.6.3.2 MD Simulations on Grain Boundaries

The lattice-free nature of MD combined with its still large enough simulation volume make this method ideal to study the equilibrium structure, thermodynamic properties and diffusion mechanisms of grain boundaries.

As a further advantage, asymmetrical and/or nonspecific grain boundaries can be simulated by MD too. According to experimental studies, these types of grain boundaries have a significantly higher share than the symmetric ones in realistic polycrystalline materials (Rohrer et al., 2004). As an illustration of the capabilities of the method, a relatively recent MD study by Brown and Mishin (2007) revealed that the asymmetric tilt Σ11 grain boundary in Cu stabilizes itself by creating nanoscale facets composed of high- and low-angle grain boundary types. Surprisingly, a significant portion of these facets cannot be characterized by any CSL group, which indicates the limitations of the CSL description. Another interesting phenomenon, namely the formation of a thin bcc layer at the Σ3 (211) grain boundaries of Cu was predicted too by MD simulations. This effect is confirmed by a high-resolution TEM study (Schmidt et al, 1995).

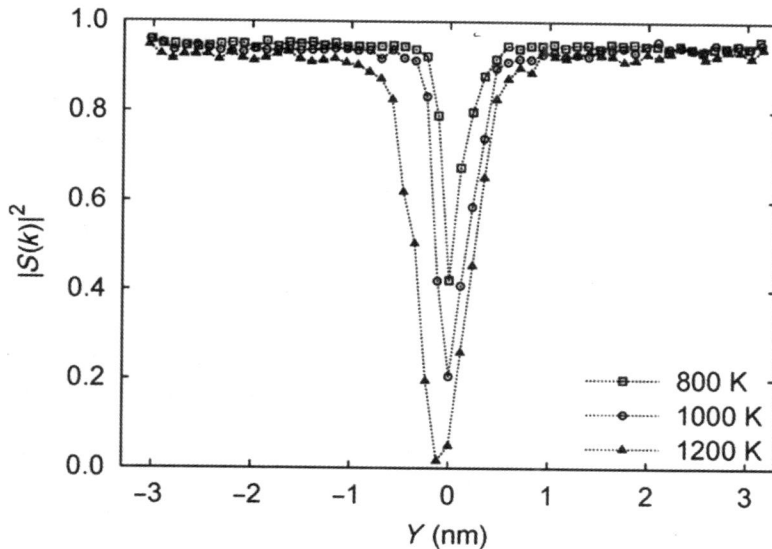

Figure 73 Variation of the static structure factor $|S(k)|^2$ near a $\Sigma 5$ (310) grain boundary of Cu at different temperatures. $|S(k)|^2 = 1$ means perfect crystalline order, while $|S(k)|^2 = 0$ represents the random arrangement of a liquid (see, e.g. Ref. Lutsko et al., 1988 for the definition). The width of the disordered zone increases with increasing temperature. At 1200 K (much below the 1358 K melting point of Cu) already very little order is preserved in the center of the grain boundary. Reproduced with permission from Suzuki and Mishin (2005).

Under well-chosen conditions, also structural transitions within the grain boundary (including the so-called grain boundary premelting) or different diffusion mechanisms can be studied in detail.

A combined radiotracer experiment and MD study on Ag diffusion in asymmetric $\Sigma 5$ (310) grain boundaries of Cu by Divinski et al. found that the diffusion coefficient is strongly anisotrop below 826 K, while this anisotropy disappears above that temperature (Divinski et al., 2012). The authors explained the measured anisotropy at lower temperature as a result of open channels along the [001] tilt axis (see **Figure 71**) (Sørensen et al., 2000), while at higher temperature significant deviation from ordered structure appears. Near the melting point significant amorphization happens in the grain boundary zone, which rapidly spreads out as the temperature increases, although the width of the amorphous grain boundary zone remains finite below T_m. A recent overview on the atomistic modeling of the grain boundary premelting was given by Williams and Mishin (2009).

To derive quantitative predictions on diffusion coefficients, a given defect (vacancy or interstitial) is introduced at certain positions in the grain boundary core. The formation energies of these defects are computed using molecular statics in harmonic approximation for the atomic vibrations. The probabilities of finding a defect in those positions are calculated from these formation free energies (Sørensen et al., 2000; Suzuki and Mishin, 2005). To gain information about the migration properties, the single defects are allowed to walk in an MD simulation. Analysis of the atomic trajectories has shown that defect migration in the grain boundaries is often a complex event that involves the displacement of multiple atoms. According to these simulations the dominant defect type and diffusion mechanism varies from grain boundary to grain boundary. Vacancies and interstitials in the more open structure of the grain boundaries are often found to be delocalized (Suzuki and Mishin, 2005).

5.6.3.3 B-*Kinetics Diffusion in Grain Boundaries*

The *B*-kinetics regime is specified by two conditions. (1) The bulk diffusion is fast enough to produce a considerable leakage of diffusors from the grain boundary to the bulk. (2) On the other hand it is low enough to prevent the overlapping of the composition fields of different grain boundaries. Thus $s\delta << D_b t << d$ must be fulfilled, where δ is the width of the grain boundary, s is the grain boundary segregation factor, D_b is the bulk diffusivity, t is the time and d is the grain diameter. This means that the atomic transport problem is reduced to the solution of the diffusion along and around a single grain boundary embedded in a semi-infinite bulk (when the uniformity of grain boundaries can be assumed). The first attempt to solve this problem was the Fisher model (Fisher, 1951).

Figure 74 shows a sketch of the geometry in the Fisher model: a grain boundary with a width δ and a diffusion coefficient D_{gb} is embedded in a semi-infinite bulk, which is characterized by a diffusion coefficient D_b. It is assumed that $D_{gb} >> D_b$. The depth z is measured from the free surface, while x measures the distance from the core of the grain boundary. Along the third (y) direction translational symmetry holds, thus diffusion is essentially two dimensional. The mathematical problem is very similar to the dislocation-pipe model of Smoluchowski (see Section 5.6.2). Indeed historically it was the success of the Fisher model which inspired a respective model for dislocation diffusion.

$$\frac{\partial c_b}{\partial t} = D_b \left[\frac{\partial^2 c_b}{\partial x^2} + \frac{\partial^2 c_b}{\partial z^2} \right] \quad \text{for} \quad |x| > \frac{\delta}{2} \tag{159a}$$

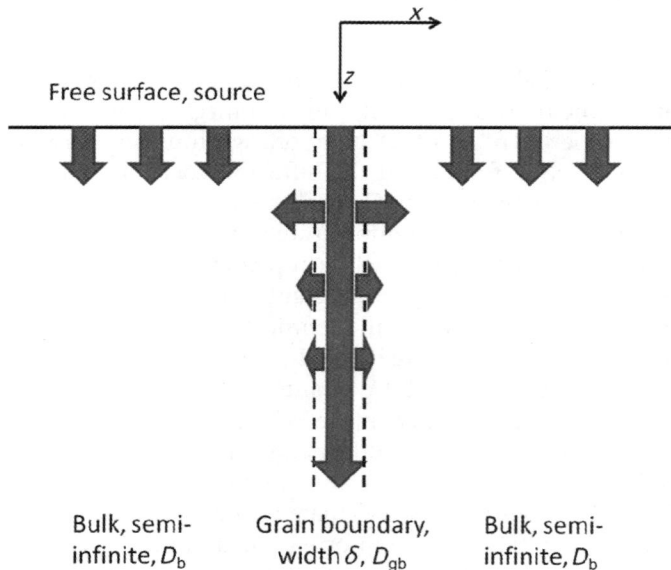

Figure 74   Sketch illustrating the basic idea of the Fisher model. (The length of the arrows approximately represents the amount of diffusant transported.) Far from the surface, the dominant diffusion process is the grain boundary transport and the leakage of diffusants from the grain boundary into the grain interior. (For color version of this figure, the reader is referred to the online version of this book.)

$$\frac{\partial c_{gb}}{\partial t} = D_{gb}\left[\frac{\partial^2 c_{gb}}{\partial x^2} + \frac{\partial^2 c_{gb}}{\partial z^2}\right] \quad \text{for} \quad |x| \leq \frac{\delta}{2} \tag{159b}$$

Similarly to the case of dislocations, flux continuity and continuity of concentration (with possible segregation effects taken into account) must be fulfilled between grain boundary and bulk. Since the extent of the grain boundary in the x direction is about or below a nanometer and $D_{gb} \gg D_b$, it is reasonable to assume a full homogenization across the grain boundary (c_{gb} does not depend on x). Then, the only relevant process in the x-direction is the leakage of material from the grain boundary into the two adjoining grains. Therefore Eqn (159b) can be simplified into

$$\frac{\partial c_{gb}}{\partial t} = D_{gb}\frac{\partial^2 c_{gb}}{\partial z^2} + \frac{2D_b}{\delta}\left(\frac{\partial c_b}{\partial x}\right)_{x \to \frac{\delta}{2}+}. \tag{159c}$$

The second term on the RHS describes the escape of diffusant into the grains (gradient of composition is negative). To solve the equations, the following normalized parameters are convenient:

$$\eta := \frac{z}{\sqrt{D_b t}}; \quad \xi := \frac{x - \delta/2}{\sqrt{D_b t}}; \quad \Delta := \frac{D_{gb}}{D_b}; \quad \beta := \frac{(\Delta - 1)\delta}{2\sqrt{D_b t}} \approx \frac{D_{gb}\delta}{2D_b\sqrt{D_b t}} \tag{160}$$

η has its usual meaning and it is exactly the same as in the case of dislocation pipe diffusion, ξ measures the outdiffusion from the grain boundary, while the β (or LeClaire parameter) measures the enhancement of transport due to the grain boundary. It can be imagined as the ratio of the transport capacities of the grain boundary (D_{gb} diffusion coefficient times δ width) and the growing near grain boundary zones at both sides (D_b diffusion coefficient times $(D_b t)^{0.5}$ width per side after time t). A large β means a pronounced deep grain boundary tail as compared with the lattice diffusion dominated surface zones. **Figure 75** shows the effect of β on the shape of the isoconcentration contours. Clearly for smaller β values, the grain boundary fringes are strongly reduced; as a rule of thumb, a $\beta > 10$ is required to evaluate the double $P = D_{gb}\delta$ or triple product $P = D_{gb}\delta s$ reliably.

The two limiting cases for the diffusion problem are as discussed before the constant source and the thin film source. An approximate solution to the constant source case was already given by Fisher (1951), who stated that far from the surface the composition decreases exponentially with the depth z and follows a complementary error function dependence on the distance from the grain boundary x (compare general solutions of Fick laws in Section 5.2.4):

$$c(z,x) = c_0 \exp(-\lambda_1 z)\,\text{erfc}\left(\frac{x}{2\sqrt{D_b t}}\right) \quad \lambda_1 = \left(\frac{2}{D_{gb}\delta}\right)^{0.5}\left(\frac{D_b}{\pi t}\right)^{0.25} \tag{161}$$

The rather complex exact solution for the case of a constant source was derived by Whipple (1954), while Suzouka provided an exact solution for the thin film or instantaneous source case (Suzouka, 1961; Suzouka, 1964). Here, we restrict ourselves to some qualitative remarks as the full interpretation of these results is beyond the scope of this summary. The most important difference between the two cases is that the fast grain boundary transport quickly depletes the limited source in the latter case (assuming absence of surface diffusion). This results in a lower concentration of diffusant within the grain boundary than in the bulk for the near surface region. There is an inward diffusion current from

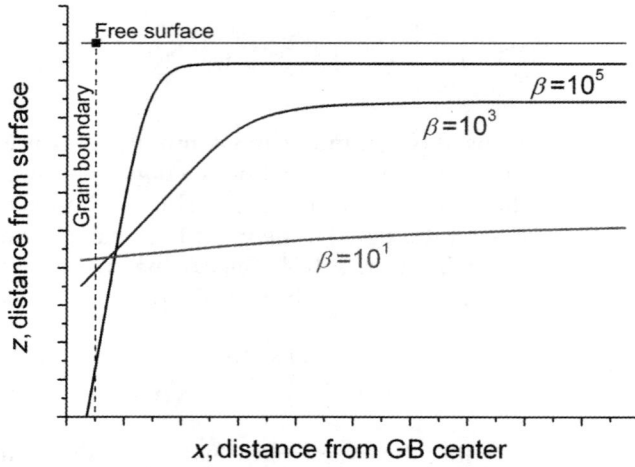

Figure 75 Tracer isoconcentration curves near a grain boundary calculated for $\beta = 10$, 10^3 and 10^5 (D_{gb} was kept constant while D_b was increased) and a constant source at the surface. With decreasing β, the relative contribution of the lattice diffusion increases resulting in a less pronounced grain boundary cusp (Finite differentia simulation).

bulk to grain boundaries in this zone. Far from the surface, the grain boundary behaves as "normal", and there is a leakage of material to the grains.

For most diffusion experiments one-dimensional sectioning is used. Accordingly, it is not the local composition, but the mean composition of a thin layer that is measured. This can be achieved in the model by integrating the composition in a x-y plane parallel to the surface. In B-kinetics the amount of tracer in the grain boundary itself usually can be neglected. The result is, similarly to the dislocation case (Eqns (155) and (156)), the amount of bulk diffusion from the surface plus a grain boundary tail.

The main difference between the dislocation and grain boundary diffusion is the exact functional form of this composition tail. While for dislocations, the logarithm of the mean concentration depends linearly on the depth z, for grain boundaries $\ln(c(z)) \sim z^{6/5}$ holds. The triple product can be calculated as

$$sD_{gb}\delta = 1.308\sqrt{\frac{D_b}{t}} \cdot \left(\frac{\partial \ln(\bar{c}(z))}{\partial z^{6/5}}\right)^{-5/3} \tag{162}$$

for constant source, while

$$sD_{gb}\delta = 1.322\sqrt{\frac{D_b}{t}} \cdot \left(\frac{\partial \ln(\bar{c}(z))}{\partial z^{6/5}}\right)^{-5/3} \tag{163}$$

for the thin film source condition.

These solutions can be applied if the conditions

$$\beta > 10 \quad \text{and} \quad \frac{s\delta}{2\sqrt{D_b t}} < 0.1 \tag{164}$$

are fulfilled. A very detailed analysis and discussion of the grain boundary diffusion can be found in the textbook of Kaur et al. (1995).

Comparing the second inequality of Eqn (164) with Eqn (148) it is clear that there is a finite transition in between a pure C and pure B regimes. Evaluating the penetration profile at this transition by either method will result in the underestimation of the diffusion coefficient.

To comprehend this statement, one has to consider the evaluation preconditions. In a C-regime evaluation, it is expected that the tracer remains in the grain boundary and the flux is determined by D_{gb}. If there is unexpected outdiffusion from the grain boundary, then the tracer atoms spend some time in the lattice zones, characterized by the smaller D_b diffusion coefficient, that is the effective flux will be reduced in comparison with the expectation of the C kinetics. On the other hand, for a B-regime evaluation, the amount of tracer in the grain boundary itself is assumed to be negligible. If grain boundary transport is too pronounced in comparison with the bulk, most of the tracer atoms penetrate deeply in the material and will not be found at the depth postulated by Eqn (162) or similar equations of Whipple (1954), Suzuoka (1961), and Suzuoka (1964). Again the real flux will be underestimated. Thus negative deviations in Arrhenius plots of grain boundary diffusivity will always point at the appearance of the transition regime (for an example see (Divinski et al., 2007)).

Nevertheless a careful measurement combined with accurate evaluation in this transition regime could provide both P, δ and D_{gb} (Szabo et al., 1990).

5.6.4 Diffusion in Systems with Multilevel Hierarchy

As it was mentioned in the previous sections, the diffusivity strongly depends on the properties of the actual grain boundary. Accordingly the scatter between grain boundary diffusion measurements is much larger than in the case of lattice/bulk diffusion (see **Figure 68**). Nevertheless for a coarse-grained material, taking a single grain boundary diffusion coefficient seems to deliver an adequate description of the material transport. In fine-grained nanocrystalline or highly deformed materials, the volume fraction of different types of shortcuts other than relaxed high-grain boundaries can reach a rather high level. Part of them offers an even faster diffusion path, thus they can significantly influence, if not dominate, the atomic transport.

In this section we will briefly describe three recent studies of such cases: (1) diffusion in a material with double-scale microstructure, (2) diffusion in triple junctions and (3) diffusion in nonequilibrium grain boundaries (see **Figure 76** for sketches). We will present the evaluation of the diffusion data as well as some recent experimental result in this new and expanding field of diffusion studies.

5.6.4.1 *Diffusion in Agglomerates of Grains*

An established route to produce nanocrystalline materials is mechanical alloying and a subsequent sinter treatment. The resulting material often has two characteristic length scales: first the size of the single crystallites and second the size of compacted aggregates/agglomerates (see **Figure 76(a)**). The difference between these two can reach many orders of magnitude (Rabkin et al., 2011). Grains within one aggregate are usually characterized by moderate misorientation and the grain boundaries in between are relaxed, representing equilibrium grain boundaries as can be found in the common material. Diffusion along these grain boundaries is characterized by a diffusion coefficient, which corresponds to the D_{gb} of the coarse-grained materials. The diffusion coefficient D_{ab}, describing the transport along "aggregate boundaries", is much larger however. Consequently, the agglomerate boundaries behave as shortcuts as compared with the standard grain boundaries. To describe such

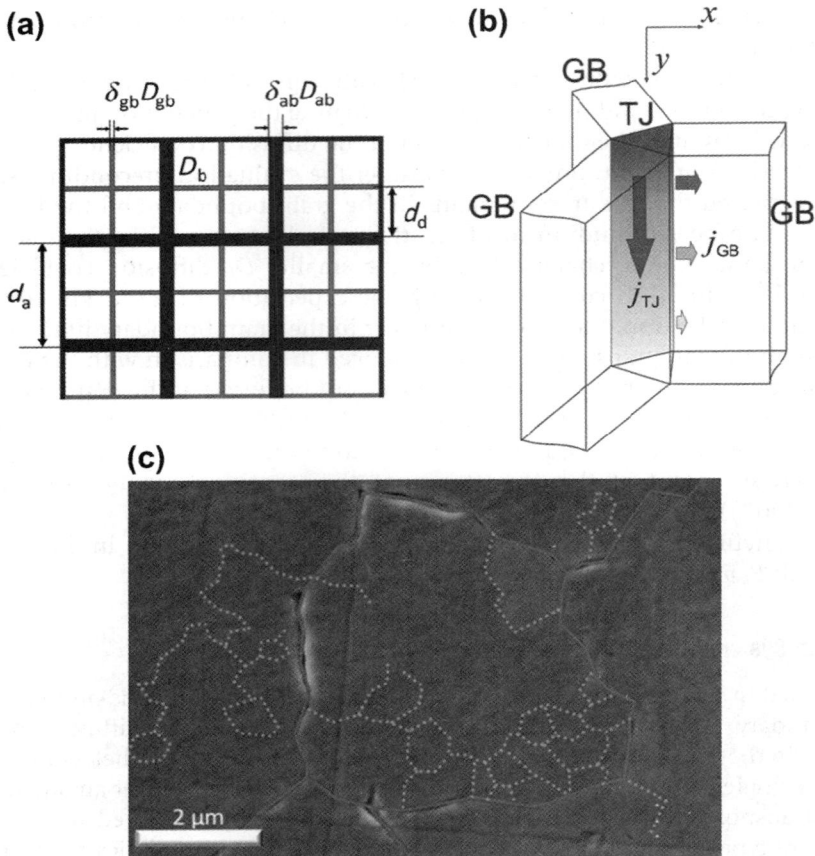

Figure 76 Examples of multilevel hierarchies: (a) aggregates of nanograins, (b) triple junctions, (c) nonequilibrium grain boundaries in a severely deformed material (Ribbe, 2012). (For color version of this figure, the reader is referred to the online version of this book.)

system, obviously three diffusion coefficients are needed ($D_{ab} >> D_{gb} >> D_b$) and in consequence the systematic of the kinetic regimes becomes more complicated.

Analyzing the diffusion of [110 m]Ag in sintered γ-Fe-40 wt% Ni powder, Divinski et al. found exactly the described situation (Divinski et al., 2004). With a straightforward extension of the Harrison classification (Harrison, 1961), they defined two-level kinetic regimes. In their nomenclature, a first letter designates the relation between the bulk/grain boundary system, while the second designates the grain boundary/agglomerate boundary system. The meaning of these partial regimes corresponds to the systematic of the Harrison scheme. For example *C–B* regime means that the bulk transport is frozen (*C* regime for the bulk/grain boundary system); while grain boundary migration is possible. However the diffusion length along the grain boundaries must be less than the half of the typical dimension of the agglomerates (*B* regime for the grain boundary/agglomerate boundary system). Depending on the ratio of the diffusivities and the size of grains and agglomerates, every combination of the regimes is possible. Divinski et al. (2004) contains the derivation of some important formulas to evaluate the relevant diffusion coefficients from experimental concentration profiles.

5.6.4.2 Diffusion in Triple Junctions

Triple junctions are topologically necessary, line-shaped defects which are formed at the merging of three grain boundaries (see **Figure 76(b)**). Their distinguished properties are already mentioned by McLean (1957) who assumed a positive line tension as compared with normal grain boundaries. This was indeed proven experimentally by a meticulous atomic force microscopy analysis of Cu tricrystals by Gottstein et al. (2010) and Zhao et al. (2010). Segregation of Cr into Fe triple junctions have shown that they represent preferred locations of segregation. Triple junction segregation enthalpy was found to be different from but in between that of the surface and grain boundary (Stender et al., 2011).

The above results indicate that triple junctions are characterized by weaker bonding between the atoms and/or a higher amount of free volume than grain boundaries. This hints that triple junctions should be diffusion paths with higher diffusivity and lower activation enthalpy. Using a simple model of tetradecahedra-shaped grains, Palumbo et al. estimated the volume fraction of the triple junctions (Palumbo et al., 1990) and they concluded that for a fine-grained nanocrystalline material (~ 10 nm grain size), the triple junction contribution can dominate the materials properties. This is especially true for the case of diffusion, where a dramatically higher triple junction diffusion coefficient D_{tj} could lead to significant triple junction contribution even for still relatively small volume fractions (Chen and Schuh, 2007a, 2007b).

From a theoretical point of view, the diffusion in a system composed of triple junctions, grain boundaries and bulk is similar to the previous situation of agglomerate boundaries. The kinetic regimes can also be described using the same nomenclature. If volume diffusion is frozen, then the situation is simplified to a two-level problem containing a fast diffusion path (triple junction) and slow diffusion path (grain boundary) (Chellali et al., 2011). This is quite similar to the classical Fisher problem of grain boundary diffusion, only the geometry is slightly different. In the grain boundary–bulk case, the diffusant can leave the grain boundary (fast diffusion path) into two neighboring grains (slow diffusion path), while for the triple junction–grain boundary case the triple junction (fast diffusion path) is connected to three grain boundaries (slow diffusion path). Therefore the composition profiles in the C–B regime can be approximated by the solution

$$c(z,x) = c_0 \exp(-\lambda_1 z)\,\mathrm{erfc}\left(\frac{x}{2\sqrt{D_{gb}t}}\right) \quad \lambda_1 = \left(\frac{3\delta}{D_{tj}q}\right)^{0.5}\left(\frac{D_{gb}}{\pi t}\right)^{0.25} \tag{165}$$

where z and x measures the distance from the free surface and triple junction core, respectively, δ is the effective transport width of the grain boundary, while q is the effective transport cross-section of the triple junction. This equation is practically a carbon copy of Eqn (161), only replacing the geometry-related constants. Earlier solution for the more complex B–B case can be found in Ref. (Klinger et al., 1997).

Systematic experimental studies of triple junction diffusion are rare. To directly investigate the triple junction transport, either special specimen geometry (tricrystals) (Bokstein et al., 2001) or local methods, such as APT (Chellali et al., 2011, 2012) (see **Figure 77**) are needed. The few analysis carried out up to now show the same general tendency, that the triple junction diffusion coefficient is few hundred times larger than that of the grain boundaries at $\sim T_m/2$. Chellali et al. also measured the activation enthalpy for the diffusion of Ni in the triple junctions of Cu and they found as demonstrated in **Figure 77** that it is around two-thirds of the grain boundary activation enthalpy (Chellali et al., 2012). Nevertheless, in view of the mentioned and a few other experiments, enhanced diffusion along triple junctions can nowadays judged as being confirmed as a realistic possibility (e.g. Refs. (Mikhailovskii et al., 1991; Schmitz et al, 2006; Portavoce et al., 2010)).

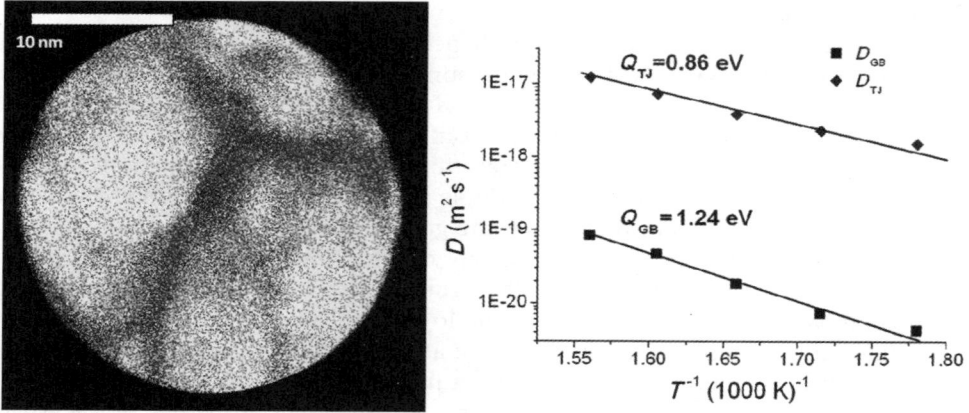

Figure 77 Left: Atom probe volume reconstruction showing the Ni atoms (blue) concentrated along grain boundaries of Cu (yellow). Triple junctions are clearly identified. Evaluating a large number of similar atomic maps provides diffusion coefficients for individual grain boundaries and triple junctions. Right: diffusion coefficients in an Arrhenius plot. Triple junction diffusion is characterized by higher diffusion coefficients and lower activation enthalpy (Chellali et al., 2012).

5.6.4.3 Diffusion in Nonequilibrium Grain Boundaries

Ultrafine-grained materials have shown promising properties, such as having simultaneously high yield strength and ductility (Valiev et al., 2000; Valiev, 2004). Severe plastic deformation is an established way to produce materials with grain sizes in the 0.1–1 μm range. In severely deformed material, grain boundaries with a broad spectrum of properties may appear, although controlling their characteristics is nevertheless possible to a certain degree (grain boundary engineering (Valiev et al., 2000)). From a material transport point of view, the so-called nonequilibrium grain boundaries, that is, grain boundaries with high level of stress, higher density of grain boundary and lattice dislocations than required geometrically and higher free volume (Kolobov et al., 2001; Wilde et al., 2010; Sauvage et al., 2012), are especially important, since they are characterized by an enhanced diffusivity. Recent high-resolution electron microscopy investigations revealed that the structural width of these nonequilibrium grain boundaries can be also higher (~2 nm) which may also be a source of increased atomic transport (Sauvage et al., 2012). On the other hand, nonequilibrium grain boundaries are expected to quickly relax to ordinary high-angle grain boundaries according to Nazarov (2000) with the relaxation time

$$\tau = \frac{k_B T d^3}{A \delta G \Omega D_{gb}},$$ (166)

in which d, δ, Ω, G, and D_{gb} denote grain size, grain boundary width, atomic volume, shear modulus, and the grain boundary diffusion coefficient, respectively. A is a numerical constant which is 143 for grain boundaries with glissile and 163 for sessile dislocation. According to the Nazarov model $\tau \approx 45$ min for Cu at 398 K, thus nonequilibrium grain boundaries relax into normal high-angle grain boundaries quickly. This model, however, assumes that the absorption of the lattice dislocations into the high-angle grain boundaries does not feel any barrier. Therefore realistic relaxation times might be much longer (Divinski and Wilde, 2008). Indeed some experiments indicate an increment in the diffusivity in fine-grained materials well after the supposed relaxation time (Kolobov et al., 2001; Fujita et al., 2002). Nonetheless, great care is needed in the interpretation of experiments concerning severely plastic deformed materials, as high distortion could produce percolating porosity (Ribbe et al., 2009).

If a specimen contains both nonequilibrium and relaxed grain boundaries, it produces a situation similar to the aggregates (Section 5.6.4.1). A three-stage hierarchy of fast nonequilibrium grain boundaries, "normal" grain boundaries and bulk volume is present (Divinski and Wilde, 2008). In investigating various ultrafine-grained systems, Divinski and Wilde (2008) found that typical tracer penetration plots contain three domains (**Figure 78(a)**). The tracer concentration falls quickly near to the surface. This is followed by Gaussian decay of concentration characteristic for a grain boundary C-kinetics experiment. Surprisingly, however, this region is followed by a third domain of Gaussian decay, much less steep than in the second region. Evaluating the second domain, they found that this material transport process is characterized by a diffusion coefficient which is close to the D_{gb} values of coarse-grained materials. The third domain is characterized by a diffusion coefficient more than two orders of magnitude above that of relaxed grain boundaries.

Recently Divinski et al. measured the diffusion of ^{63}Ni tracer in severely deformed Ni and found that some relaxation of nonequilibrium grain boundaries did take place; however this process is quite slow (Divinski et al., 2011). The transition to the relaxed state starts above 400 K (**Figure 78(c)**) and significant amount of ultrafast diffusion paths is present in the material after being annealed at 500 K for 72 h (roughly 100 times the τ of Eqn (167)). This is a clear indication that nonequilibrium grain boundaries can have a serious and longer lasting influence on the atomic transport in severely deformed materials. More information about this topic can be found, for example, in a recent review by Sauvage et al. (2012).

5.7 Diffusion in Nanometric Dimensions

A general overview on diffusion in metals cannot be complete without discussing the particular aspects that appear in diffusion on very short-length scales. On the one hand, recent progress in experimental techniques allows focusing even on a subnanometer distances in diffusion research and on the other hand, emergence of new technologies and even already application of nanostructured materials require exactly this knowledge. One can certainly state that 'nano-transport' has motivated a major fraction of current theoretical and experimental research in the supposedly old field of diffusion.

Here we can address only a few selected topics. First, the general limitations of the Fick equations when approaching atomic dimensions are pointed out. Second, the current understanding of interfacial transport barriers and vacancy distribution will be presented. Finally, we will consider the influence of elastic stress and its relaxation, which become particularly obvious in problems of nanometric dimensions.

5.7.1 The Limitations of the Fick Equation on Short Length Scales

In Sections 5.2 and 5.3, the fundamental continuum equations of diffusion were obtained under assumptions which are not valid anymore when it comes to short dimensions or large composition gradients. Let us study a few examples of such possible limitations:

(1) Impact of gradient energies

When considering decomposition by diffusion in the spinodal regime (see Section 5.3.1), we already pointed out that composition fluctuations become emphasized by up-hill diffusion, since the negative

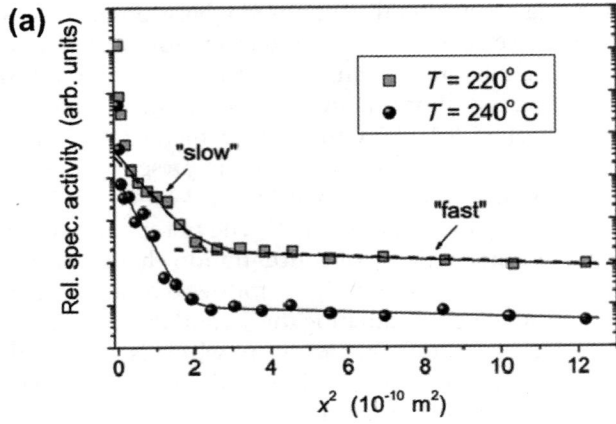

(a)

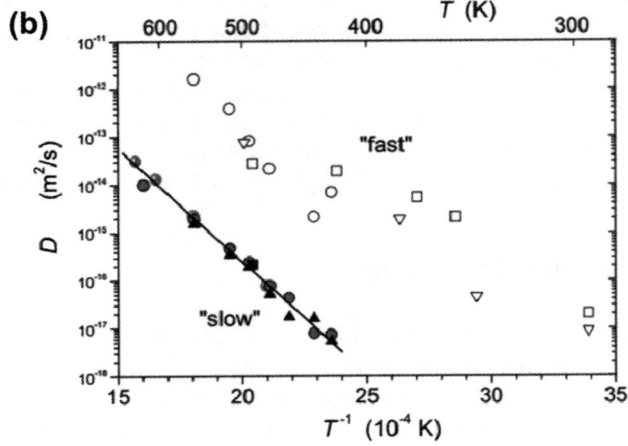

(b)

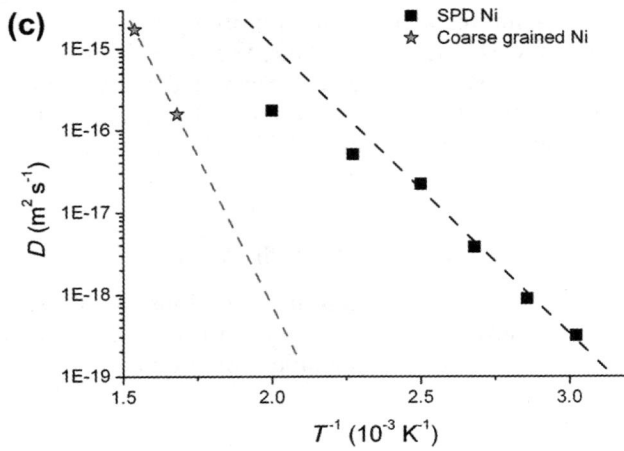

(c)

thermodynamic factor inverses the direction of transport. Based on Eqn (68), the growth of a compositional perturbation becomes even faster with decreasing wave length. We would end up in an unrealistic infinitely fine-scaled microstructure of A- and B-rich phases.

Cahn and Hilliard (1958), Cahn (1965) and independently Hillert (1961) were the first to notice that this shortcoming in theoretical description is a consequence of neglecting interfacial energies. To avoid this, they suggested treating the Gibbs energy beside composition also as a function of composition gradients and higher derivations $g(v,v',v'',\dots)$ and so to expand this functional around the homogeneous alloy. In leading order, the thermodynamic potential is extended by an extra energy that is proportional to the square of the local concentration gradient. In other words, this so-called gradient energy normally 'punishes' too steep concentration gradients. Since chemical potentials represent derivatives of the Gibbs energy, a corresponding gradient term must also appear when calculating chemical potentials. For the particular case of the interdiffusion exchange potential this reads

$$\tilde{\mu} = \mu_A - \mu_B = g'(v) - 2\kappa \frac{\partial^2 v}{\partial x^2} \tag{167}$$

The second term on the RHS represents the additional thermodynamic driving force stemming from gradient energy. The parameter κ represents the so-called gradient energy coefficient, while $g(v)$ is the Gibbs energy of the homogeneous phase, as considered in bulk alloy thermodynamics. (As before, prime or double prime indicate first or second derivative with respect to v.)

Inserting this exchange potential into Eqn (59a) a modified second Fick law is derived:

$$\frac{\partial v}{\partial t} = \tilde{D}_D \cdot \left[\frac{\partial^2 v}{\partial x^2} - \frac{2\kappa}{g''} \frac{\partial^4 v}{\partial x^4} \right]. \tag{168}$$

Here, we need to recall that g'' and in consequence also $\tilde{D}_D$ may change sign. A negative g'' and possibly negative $\tilde{D}_D$ indicate the case of up-hill diffusion (see Section 5.3.1). Let us determine the influence of the additional gradient term by studying the growth rate of a sinusoidal perturbation. Inserting Eqn (67) this in Eqn (168) we now obtain for the rate coefficient of exponential growth

$$\beta = -\tilde{D}_D k^2 \left(1 + \frac{2\kappa}{g''} k^2 \right), \tag{169}$$

which is the original solution (Eqn (68)) of the macroscopic case modified by a term that becomes important in proportion to the square of the wave vector k. Regarding the sign of this modification, the present authors tend to assume that in continuum descriptions the gradient coefficient has to be

Figure 78 (a) Typical penetration profile of an ultrafine-grained material that reveals three segments of different diffusion rate. (b) Diffusion coefficient characterizing the two faster segments, the "slow" diffusion coefficient is equal to the D_{gb} coefficient of relaxed grain boundaries. (Figures taken from (Divinski and Wilde, 2008), reproduced by permission.) (c) Relaxation of nonequilibrium grain boundaries begins at moderate temperatures in Ni, however they still have a significant contribution even for 500 K and long annealing times. Data for severe plastic-deformed Ni taken from Divinski et al. (2011) while data for coarse-grained Ni from Divinski et al. (2010).

positive. Otherwise serious issues with the stability of homogeneous alloys would arise. Still one has to distinguish the cases of a positive or negative interdiffusion coefficient. In the former, the gradient energy term accelerates mixing; while in the latter, it slows down the diffusional built-up of composition oscillations (hinders demixing).

(2) The impact of discrete lattices

Taking an increasingly microscopic point of view, one finally has to regard the discrete nature of the lattice. In discrete formulation, the diffusional flux of species i between the atomic planes n and $n + 1$ by the vacancy mechanism may be written as

$$j_i^{n \to n+1} = \frac{z_v \left(v_i^n v_V^{n+1} \Gamma^{n \to n+1} - v_i^{n+1} v_V^n \Gamma^{n+1 \to n} \right) N}{A}. \tag{170}$$

Parameter z_v denotes the number of nearest neighbors in an adjacent plane (vertical coordination number), v_i^n and v_V^n are the fractions of species i and vacancies in the nth plane. N is the number of lattice sites per plane, A the area of the considered cross-section, and $\Gamma^{n \to n+1}$ the jump frequency from layer n to layer $n + 1$. This kind of microscopic model considering average concentrations on atomic planes is also called kinetic mean field or Martin model (Martin, 1990).

Comparing Eqn (170) to the continuum equations of Section 5.2 (Eqns (31) and (32) but here only diffusion in one dimension, so replacing $1/6$ by $1/2$), the diffusion coefficient D_i of species i is related to the jump frequencies by

$$D_i(x) = \frac{1}{2} z_v \lambda^2 \left(\Gamma^{n \to n+1} + \Gamma^{n+1 \to n} \right) v_V(x) \tag{171}$$

with λ being the component of jump distance in x-direction. As can be seen, continuum description and the discrete equations of Martin contain the corresponding quantities. However, this does not necessarily mean that they are quantitatively identical. Indeed, a systematic deviation appears, if composition variations become too steep.

To demonstrate this, we may assume an alternative, purely mathematical point of view. In simply discretizing the second derivative in the second Fick law (Eqn (10)) we obtain

$$\frac{\partial v^{(n)}}{\partial t} = \frac{\tilde{D}_D}{\lambda^2} \left(v^{(n+1)} - 2v^{(n)} + v^{(n-1)} \right), \tag{172}$$

a formulation that captures discrete concentrations $v^{(n)}$ on adjacent lattice planes right away.

As done before, we illustrate the properties of this diffusion equation by inserting a periodic perturbation, in this case a discrete version of Eqn (67), to get

$$\beta v_0 e^{\beta t} \cdot e^{ik\lambda n} = \frac{\tilde{D}_D}{\lambda} v_0 e^{\beta t} \left(e^{ik\lambda(n+1)} - 2e^{ik\lambda n} + e^{ik\lambda(n+1)} \right)$$

$$\beta = -\frac{2\tilde{D}_D}{\lambda^2} (1 - \cos k\lambda) \tag{173}$$

The latter expression may be expanded to leading order for ease of comparison with Eqn (169)

$$\beta \approx - \tilde{D}_\mathrm{D} k^2 \left(1 - \frac{2\lambda^2}{4!} k^2 + \dots \right). \tag{174}$$

Similar to Eqn (169) for the gradient energy, we find again a deviation from the macroscopic expression that is proportional to the square of k. In contrast to the previous however, the influence of the discrete lattice is expected to decelerate interdiffusion in any case.

(3) Impact of nonlinear terms in the Fick equation:

In the general derivation of the conventional Fick laws in Eqn (4), a linearization of exponential functions was performed, which cannot be valid anymore in the limit of strong composition gradients. In this limit, instead of Eqn (59b), the exact expression of the interdiffusion flux reads

$$\tilde{j} = -2 \frac{\tilde{D}\nu(1-\nu)}{\Omega\lambda} \cdot \sinh\left[\frac{\lambda}{2k_\mathrm{B}T} \frac{\partial\tilde{\mu}}{\partial x} \right]. \tag{175}$$

By expanding the sinh function to first order beyond the Fick approximation, we find as interdiffusion current

$$\tilde{j} = -\frac{\tilde{D}}{\Omega} \left(\frac{\partial\nu}{\partial x} + \frac{1}{24} \left(\frac{\lambda}{k_\mathrm{B}T} g'' \right)^2 \left(\frac{\partial\nu}{\partial x} \right)^3 \right) \tag{176}$$

and by using the equation of continuity

$$\frac{\partial\nu}{\partial t} = \tilde{D}_\mathrm{D} \frac{\partial^2\nu}{\partial x^2} \left(1 + \frac{1}{8} \left(\frac{\lambda}{k_\mathrm{B}T} g'' \right)^2 \cdot \left(\frac{\partial\nu}{\partial x} \right)^2 \right). \tag{177}$$

In contrast to both small distance corrections (cases i and ii) introduced before, this correction is nonlinear in the composition variables. Therefore, the analysis in terms of a constant damping rate of Fourier components is not possible anymore. To compare nevertheless the different short-distance corrections, exemplary numerical solutions are shown in **Figure 79**. The temporal evolution of a sinusoidal perturbation in an ideal alloy has been calculated by Eqns (68), (169) and (174), or Eqn (177). The wavelength amounted to 2π lattice constants (say $k = \lambda^{-1}$). Clearly, the corrections to the Fick law depend significantly on the model. Most important, they are even of different sign, possibly leading to acceleration or deceleration when it comes to small distances. Furthermore, the contribution of a correction depends sensitively on material parameters and naturally also on the period length of the modulation. In general one has to expect a superposition of the different effects. In view of this situation, it is hard to identify the origin of experimental deviations from Fick laws in short-range diffusion. Premature interpretation of older experiments (Cook and Hilliard, 1969) demonstrating a negative gradient coefficient may even have to be reconsidered in this regard.

With state of the art experimental techniques, diffusion can be better and better observed on length scales comparable with the atomic distances. But owing to various experimental difficulties, the

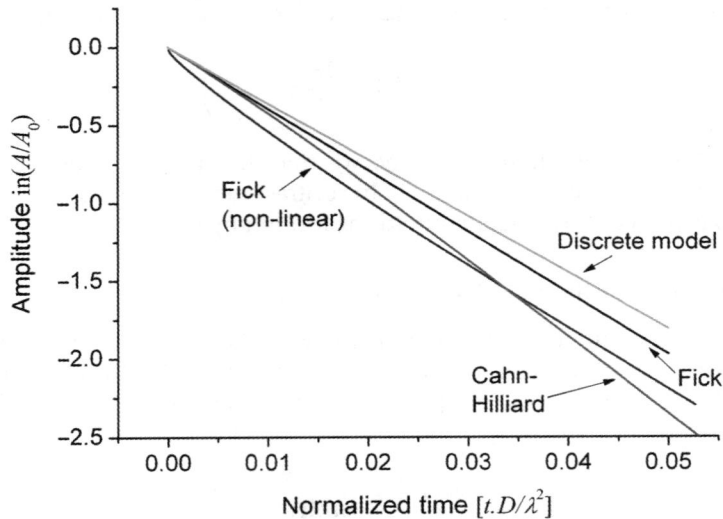

Figure 79 Damping of a small-period sinusoidal concentration modulation as calculated by the conventional Fick law (Eqn (68)), the nonlinearized Fick (Eqn (177)), the Cahn–Hilliard thermodynamics of inhomogeneous systems (Eqn (169)), or the discrete lattice model (Eqn (174)) (ideal alloy, $\lambda = 0.2$ nm, $T = 600$ K, $\kappa = 1.6 \times 10^{-22}$ J nm^2, wavelength of concentration modulation: $2\pi/k = 1.3$ nm).

investigation of the validity limit of the Fick equations and a correct interpretation of the earliest stage of interdiffusion are an exciting and certainly not finished area of experimental research.

The influence of gradient energies was checked shortly after proposition by using XRD analysis of man-made superlattice structures of miscible Au–Ag metallic layers (Cook and Hilliard, 1969). Studying the fading of superlattice reflections with annealing time, the effective diffusion coefficient was determined for different multilayer periods. Significant deviations from macroscopic diffusion were proven for a superlattice period below about 1 nm, not more than only 4–5 jump distances. A slightly larger critical length of about 2 nm was found by Cammarata and Greer for multilayers of metallic glasses (Cammarata and Greer, 1984). Both results were interpreted as a quantitative agreement with the Cahn–Hilliard theory, although a reevaluation of the results might be recommended in view of the possible impact of described alternative short-range corrections.

Interestingly in current work performing analogous experiments with neutron irradiation and multilayers of pure isotope contrast (chemically homogeneous) a particular fast interdiffusion rate is observed initially that slows down significantly during longer annealing (Schmidt et al, 2008). This behavior could possibly be explained by the nonlinearized Fick equation. But initial supersaturation of vacancies stemming from layer deposition, and their annihilation may represent a natural alternative to understand fast diffusion rates at the very beginning of annealing.

The difference between a discrete and a continuum representation of the Cahn–Hilliard concept was early investigated theoretically by Cook et al. (1969). Assuming a concentration-independent diffusion coefficient, the authors demonstrated that the discrete version Eqn (174) and the continuum description Eqn (68) begin to differ, when the wavelength of the perturbation falls short of about 6 lattice distances.

For combinations of metals with repulsive interaction (e.g. Fe/Cu, Ag/Cu), the impact of the gradient energy was tested recently based on analysis by APT (Stender et al., 2008). By using this accurate 3D-resolved chemical analysis (see also Section 5.4.3.3), possible disturbing influence of microstructural artifacts such as shortcut diffusion and segregation along grain boundaries or local interface roughness can be safely ruled out. So the chemical structure of the interfaces can be measured with an accuracy of 2–3 Å. In agreement with the Cahn–Hilliard theory, it was demonstrated that even for nonmixing metals, short-ranged interdiffusion takes place at initially sharp interfaces, until an equilibrium concentration gradient is reached. The typical mixing depth amounts to up to 1–1.5 nm. The expected temperature dependence of the equilibrium concentration gradient was nicely confirmed in experiment (see **Figure 80**). This allowed quantification of the gradient energy coefficients. Noteworthy, the quantitative value of the gradient coefficient was up to a factor of 10 higher than a theoretical prediction by Cahn and Hilliard (1958) which was based on purely energetic arguments. In consequence, effects of the gradient energy may become visible already at a coarser length scale than originally expected.

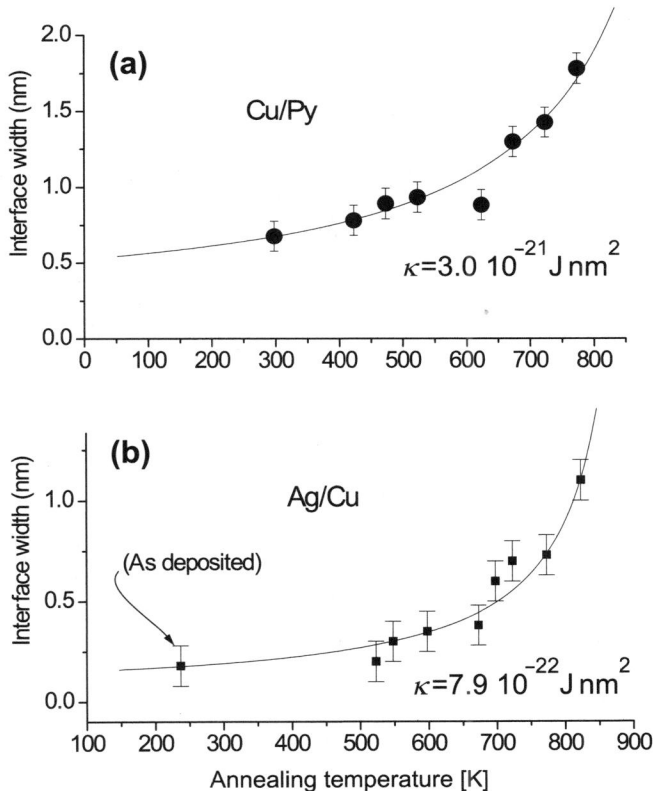

Figure 80 Equilibrium chemical thickness of the layer interface in immiscible Cu/Ni$_{80}$Fe$_{20}$ (a) and Ag/Cu (b) multilayers versus temperature of annealing. Measurements by APT (mind the extremely short-length scale of analysis). Solid lines represent the prediction of the Cahn–Hilliard theory. Stated gradient energy coefficients are derived by fitting to the experimental data. Reproduced from data published in Stender et al. (2008).

Regarding this rare experimental evidence, the practical importance of discussed short-distance corrections to diffusion could be doubted as long as they appeared only on the length of 2–4 lattice constant. However, in a series of several contemporary studies, Erdélyi et al. (1999) investigated the particular influence of a strong concentration dependence of the diffusion coefficients. They assume exponential dependence, which is an acceptable approximation (at least for ideal alloy systems, see Section 5.3.3.2) and thus model the tracer diffusion coefficients of the alloy components by interpretation of Eqn (83) as

$$D_i(v) = D_{0,i} \cdot \exp[m_i(1 - v)] = D_{0,i} \cdot 10^{m_i'(1-v)} \qquad (178)$$

(Parameter m' describes the change of diffusivity in order of magnitude units. Here diffusivity is chosen to be higher in the B-rich alloy.) By using MC simulation and the kinetic mean field model (Eqn (170)) they demonstrated that strong composition dependence of the diffusion coefficients leads to phenomena which are at first sight quite peculiar (see **Figure 81**). Naturally, a symmetric error-function-shaped composition profile (**Figure 81(a)**) can only be expected if the diffusion coefficient is reasonably constant. By contrast, if $m' = 4$, mixing appears qualitatively different: the A layer of low atomic mobility remains pure and shrinks, while the B layer of high mobility reveals practically a homogeneous distribution of in-diffused A atoms (see **Figure 81(b)**). Remarkably, the interface between the two layers remains rather sharp, even though the two metals are fully miscible. Such asymmetrical intermixing was experimentally observed, for example in the Si/Ge system (Csik et al., 2001; Tripathi, 2009). Since the sharp interface is a consequence of the kinetic conditions, the chemical transition at the interface may represent a kind of an optimum steady-state shape. It is a reasonable expectation that even resharpening could appear by diffusion, if the interface is initially broader than this steady-state value. (Note that this is a kinetically driven transient effect, unlike sharpening due to chemical separation in decomposing alloys.)

Let us assume an interface with a linear concentration transition (**Figure 82**, gray data points). Since the gradient of concentration dc/dx or atomic fraction dv/dx is constant in the interface, the atomic flux

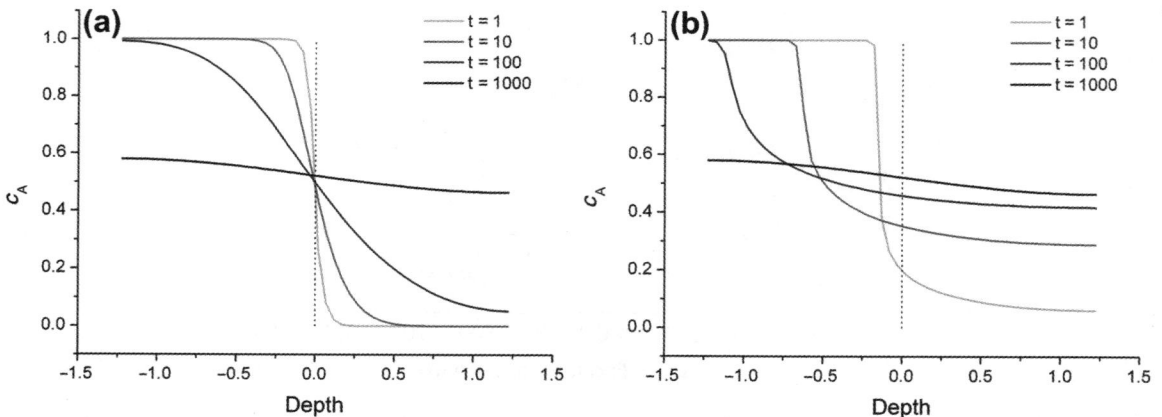

Figure 81 Evolution of the composition profile in the case of $m' = 0$ (a) and $m' = 4$ (b). The dotted line marks the position of the original interface. Intermixing takes an alternative route, if atomic mobilities show significant concentration dependence. (Calculations by kinetic mean field simulation, grid spacing 0.05 nm. Data kindly provided by Erdélyi and Beke (2011).

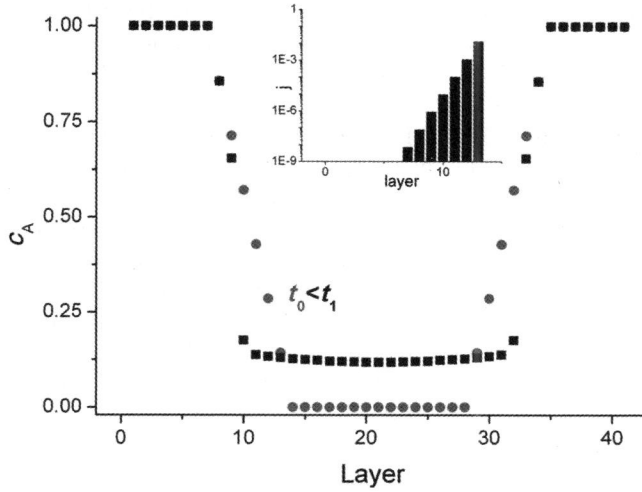

Figure 82   Interface sharpening as a result of a composition-dependent diffusion coefficient ($m' = 7.3$). The inset shows the diffusion current at the time t_0 in a logarithmic scale. Data kindly provided by Erdélyi et al. (2004b).

varies only with the local diffusion coefficient. Considering a finite volume element $(x, x + \Delta x)$, the inward and outward fluxes of A atoms are

$$j_{A,in}(x) = -\frac{\Delta c}{\Delta x} D_0 \exp(m(1 - v(x)))$$

$$j_{A,out}(x + \Delta x) = -\frac{\Delta c}{\Delta x} D_0 \exp(m(1 - v(x + \Delta x))). \tag{179}$$

Balancing both and making Δx infinitesimally small yields

$$\frac{dc_A}{dt} = \frac{\Delta c}{\Delta x} D_0 \exp(m(1 - v(x))) m \frac{dv}{dx} = -KD(v) \tag{180}$$

with a suitable constant K. Since the latter is positive, the interface region is characterized by a net loss of A atoms. However, since $D(v)$ increases with the local B content, B-rich regions lose A atoms faster than B-poor regions and so shortly later, the composition profile has transformed into the shape that is represented by the black data points in **Figure 82**. In agreement to this theoretical prediction, sharpening of the interface was experimentally observed in Mo/W by in situ synchrotron XRD experiments (Erdélyi et al., 2004b) and in Ni/Cu by APT (Balogh et al., 2011).

The sharpening itself is a general effect of the concentration dependence of the diffusion coefficient and is therefore not restricted to nanometric dimensions of the diffusion zone. It can be already predicted in continuum descriptions. A detailed analysis of the interface sharpening in a continuum framework is, for example, given in Ref. (Wan et al., 2012).

However going beyond the expectation of continuum modeling, kinetic mean field (Erdélyi et al., 2002; Erdélyi et al., 2004a) and MC simulation (Erdélyi and Beke, 2003; Roussel and Bellon, 2006b) have discovered a peculiar behavior of the kinetically stabilized sharp interface which can obviously only be expected for diffusion on nanometric distances: the interface does not shift parabolically anymore (compare for illustration, e.g. the positions of the steep transitions in **Figure 81(a) and (b)** after the same time). This clearly contradicts to what has been said before in the context of the Boltzmann transformation (Section 5.3.3.1) in continuum description.

In classical random walk theory of Einstein (1905), a normal distribution of jump lengths and residence times between atomic jumps is assumed. Under these conditions, the $x \sim t^{0.5}$ rule can be derived for the mean displacement of the particles. As immediate consequence any plane of constant composition must shift by parabolic time dependence (see Boltzmann transformation in Section 5.3.3.1 and the discussion of the Kirkendall planes in Section 5.3.4.1). This "normal diffusion", however, covers only part of the possible random walk mechanisms. Diffusion processes with higher (superdiffusion) or lower exponents (subdiffusion) are also known in different fields (Klafter and Sokolov, 2005). General transport can be discussed with the help of a fractional diffusion coefficient D_α and by a general Fick law

$$\frac{\partial c}{\partial t} = D_\alpha \frac{\partial^\alpha c}{\partial x^\alpha} \tag{181}$$

in which α is an arbitrary number which represents the reciprocal time exponent $k_c = 1/\alpha$ in the kinetic relation $\sim t^{k_c}$. (The physical unit of the diffusion coefficient must be properly chosen to fulfill the equation.) In analogy to the Fick equations, now planes of constant composition in the diffusion zone shift proportional to $\sim t^{1/\alpha}$ with respect to the initial Matano plane. To the bottom end, such deviations from Fickian kinetics result from the fact that the residence times of the diffusing species strongly vary with position.

As stated previously, Cook et al. demonstrated in their experiments with superlattice structures that deviations from Fickian behavior can be expected, if the wavelength of concentration variations falls short of about 6 lattice constants. Now the remarkable point is that in the case of strong concentration dependence of diffusion, part of the complete composition profile reveals a pretty steep slope even though the total depth of the intermixed zone remains still appreciable. Thus, the minimum depth of the diffusion zone, at which first deviations from the continuum approach may appear, can be much larger, even a few 100 lattice constants if the diffusion coefficient strongly depends on the concentration ($m' > 4$) (Erdélyi et al., 1999).

Kinetic mean field simulations using a regular solution model were carried out by Erdélyi et al. to explore the behavior of the kinetic exponent k_c (Erdélyi et al., 2004a). It was found (see **Figure 83**) that the value of k_c increases and approaches even a value of 1 at large diffusion asymmetry (parameter m'), while larger pair exchange parameters ε pushes the kinetics toward $k_c = 0.25$. The former resembles the limiting case of a linear barrier-limited transport which will be the topic of the following subsection, while the latter may reflect the dominance of the fourth-order term in the diffusion equation (Eqn (168)) of the Cahn–Hilliard case or in the growth exponent (Eqn (174)) of the discrete formulation (Sokolov et al., 2002). Such non-Fickian dissolution kinetics was indeed already observed in the phase separating Au/Ni (Katona et al., 2005), and in the completely miscible amorphous and crystalline Si/Ge (Balogh et al., 2008, 2010) system. In the latter study, experimental and simulated kinetic exponents were found in reasonable agreement (see **Figure 83**).

Although the curves of **Figure 83** were extracted from atomic simulation, we try to provide a qualitative picture. Imagine the atomic transport being characterized by three different currents: J^A in the low-mobility A-rich part, J^I in the interface region and J^B in the high-mobility B-rich part

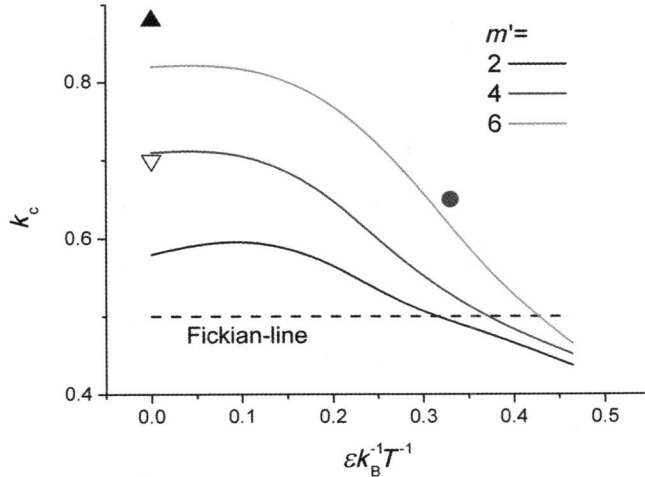

Figure 83 Simulated values of the kinetic exponent k_c for the dissolution of a thin film (~ 2 nm) as a function of $\varepsilon/k_B T$ and m' parameters as stated. (Data provided by Erdélyi et al. (2004a).) The data points represent experimental data: ▲ c-Si/c-Ge at 750 K ($m' \approx 9$, $\varepsilon/k_B T \approx 0$) (Balogh et al., 2010), ▽ a-Si/a-Ge at 603 K ($m' \approx 4$, $\varepsilon/k_B T \approx 0$) (Balogh et al., 2008), ● Ni/Au at 680 K ($m' \approx 6$, $\varepsilon/k_B T \approx 0.33$) (Katona et al., 2005).

(see **Figure 84(a)**). Since the mobility in the A-rich domain is orders of magnitude slower than in the other regions of the sample, $J^A \approx 0$ can be assumed. Since the diffusion coefficient is strongly concentration dependent, in the early stages $J^I \ll J^B$ although the interface region has the steepest slope. At this stage, the process is controlled by slow almost constant J^I with the consequence of linear kinetics. With proceeding diffusion, J^I decreases only slowly, since the sharp interface shape is conserved, while the B-rich part is filled with A-atoms, which decreases the diffusivity there significantly. After a critical time, which depends on the m' parameter, J^B will become the rate-limiting step, and thus Fickian kinetics is restored (**Figure 84(b)**). In possible experimental confirmation, studying the dissolution of amorphous Si into amorphous Ge, a transition from $k_c = 0.7$ to $k_c = 0.5$ (**Figure 84(c)**) was observed (Balogh et al., 2008), while in the case of crystalline Si/Ge (Balogh et al., 2010), in the same time window of measurement still no transition to lower k_c was found. This is nicely in line with the higher diffusion asymmetry of the crystalline system (Prokes, 1986; Kube et al., 2010) which let expect the transition even later.

5.7.2 Interfacial Transport Barriers

Already in Section 5.3.5, it was stated that in reactive thin film interdiffusion couples most expected phases predicted by the bulk phase diagram are missing first and will appear only later when the depth of the diffusion zone reached a few micrometers. There must be obviously a fundamental shortcoming with volume diffusion concepts when it comes to thin reaction couples. As seen from Eqn (110), diffusion let expect infinitely fast growth rates at the very beginning of phase formation. In view of the possible diffusion mechanisms which are all bound to finite jump rates, such prediction is clearly unphysical. It is a natural suggestion that additional transport barriers at the interfaces somehow limit the transport currents. Similar to diffusion, the driving force for transport across the interface stems from a difference $\Delta\tilde{\mu}_{IF}$ in chemical potentials at both sides of the interface. Thus, one may postulate for

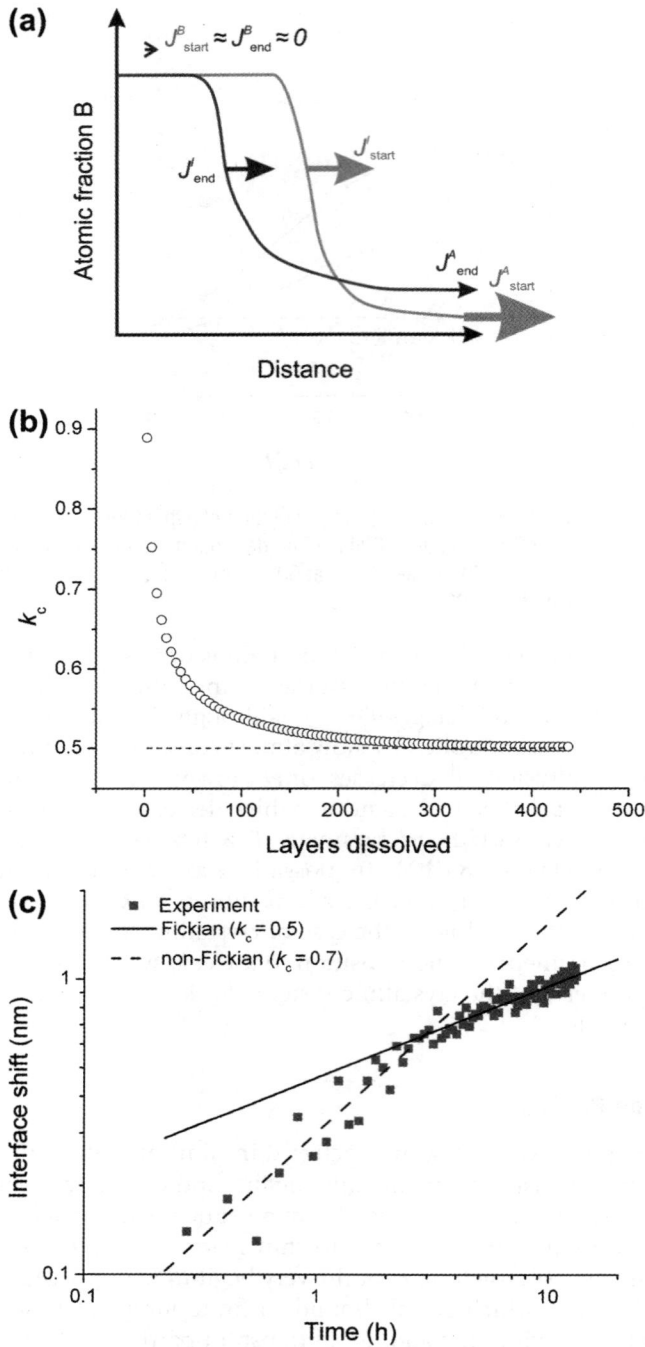

Figure 84 a) Sketch of the three distinguished kinetic domains of the diffusion zone. (b) Transition from nanoscale superdiffusion toward the microscale conventional Fickian diffusion for a system with $m' = 7$ and $\varepsilon/k_B T = 0.09$ (data from simulation kindly provided by Erdélyi et al. (2004a)). (c) Interface shift measured during the dissolution of a-Si into a-Ge representing an experimental demonstration of the transition from a non-Fickian toward a Fickian regime (according to the data of (Balogh et al., 2008)).

the migration velocity of the interface

$$v_{IF} = k_{IF}\frac{\Delta\tilde{\mu}_{IF}}{\Omega} \tag{182}$$

in which k_{IF} takes a similar role in interface transport as the diffusion coefficient in volume transport. As before, a tilde marks the interdiffusion exchange potential $\tilde{\mu} = \mu_A - \mu_B$ taking into account the combined driving force on both species. Following a concept which has been suggested by Deal and Grove (1965), we consider the growth of a single product (Section 5.3.5), but now under the additional influence of the interfacial transport barriers. For simple reasoning, we assume the phase symmetrically in the middle of the composition range. Both interfaces should have identical kinetic properties.

Part of the driving force is consumed to push atoms across the interfaces. So that we have to modify Eqn (110) in describing the growth of the product layer

$$\frac{d\Delta x}{dt} = \frac{1}{4}\frac{\overline{D}_I}{\Delta x}\frac{1}{k_B T}(\tilde{\mu}_R - \tilde{\mu}_L - 2\Delta\mu_{IF}), \tag{183}$$

Here, we have assumed $\nu_I = 0.5$. Inserting $\Delta\tilde{\mu}_{IF}$ from Eqn (182) into (183) and solving for the growth rate, we get

$$\frac{d\Delta x}{dt} = \frac{1}{4}\frac{\overline{D}_I\Delta g}{k_B T}\left(\Delta x + \frac{1}{4}\frac{\overline{D}_I\Omega}{k_B T \cdot k_{IF}}\right)^{-1} \tag{184}$$

The last term in the bracket on the RHS represents a characteristic length

$$l_c := \frac{1}{4}\frac{\overline{D}_I\Omega}{k_B T \cdot k_{IF}}. \tag{185}$$

With this, the interpretation of Eqn (184) is obvious. For sufficiently small thickness of the product phase ($\Delta x \ll l_c$), the growth is controlled by reaction at the interfaces and so it appears with a constant rate of

$$\frac{d\Delta x}{dt} = \frac{\Delta g}{\Omega}k_{IF}, \tag{186}$$

and consequently the integrated thickness is expected to increase linearly with time. With further increasing thickness however, bulk diffusion transport across the product layer slows down and so when the total thickness of the product layer exceeds l_c, the process comes under diffusion control. A transition into parabolic growth happens.

In general, one may expect a linear growth regime at the beginning of any reactive diffusion process. There was also hope that growth limitation by interfacial barriers could explain the suppression of certain phases in the early stages of reactive diffusion (Gösele and Tu, 1982; Gösele and Tu, 1989). However quite disappointing, although the linear–parabolic transition is widespread in textbooks on diffusion, it has been rarely observed experimentally in solid-state reaction (Millares et al., 1990). Clear examples are regularly found for oxidation reactions (Poate et al., 1978) and some reliable experiments with solid-state reactions were carried out in Si/Me systems (d'Heurle and Gas, 1986; Cheng and Chen, 1991; Nemouchi et al., 2005). Interestingly, the linear regime is quite often only found for phases that appear as second or third in the reaction zone, for example (Cserháti et al., 2008). For these, the driving force is already significantly reduced which gives room that interface barriers become relatively more important. In purely metallic diffusion couples, no undisputed clear example of interface control is known to the authors. Grain boundaries, rough interfaces, and undesired oxide impurities at the interface are factors which may explain observed deviations from parabolic growth as well. In addition, the influence of a strong concentration dependence of diffusivity opens an alternative to explain linear growth as has been explained previously.

According to arguments by Gusak (2010), an observable control of kinetics by interfacial barriers in metals should be doubted in general. With respect to atomic mobility, heterophase interfaces in metals are considered to be rather similar to large-angle grain boundaries. The total duration of transferring an atom from one phase to the other comprises therefore the time of transfer across the interface, the time for detachment and the time for attachment. A realistic estimation for the time of transfer is $\tau_1 \sim \delta^2/D_{GB}$, where δ describes the thickness of the open interface (a few Angstroem at most). Attachment may need some search for the right lattice site by random walk along the interface. Thus, $\tau_2 \sim d^2/D_{GB}$ in which d may represent a distance of a few tens of lattice constants. For conservative estimation, the time of detachment could be at most the same as for attachment. So in total, for the transfer time $\tau_T = (\delta^2 + 2d^2)/D_{GB} \approx d^2/D_{GB}$. On the other hand, the typical time for transporting an atom through the intermetallic product layer amounts to

$$\tau_D = \frac{\Delta x}{v} = \frac{\Delta x\, v_I}{j}\frac{1}{\Omega} = \frac{\Delta x^2 v_I}{\overline{D}\Delta v} \geq \frac{\Delta x^2}{\overline{D}} \tag{187}$$

Reaction control at the interfaces would require that τ_T is larger than τ_D. So, it can only be expected, if the layer thickness is smaller than

$$\Delta x \leq \sqrt{2d^2\overline{D}/D_{GB}} = l_c. \tag{188}$$

The important point here is that diffusivity in grain boundaries is typically four to eight orders of magnitude larger than volume diffusivity. In consequence, the critical thickness l_c below which growth is controlled by the interfaces could hardly be larger than a single lattice constant. In consequence parabolic growth is practically always observed under realistic experimental conditions.

On the other hand Gusak et al. (Gusak, 2010) demonstrated that linear growth kinetics may be explained by a completely new aspect which is also bound to the particular conditions of a nanometric diffusion length. Since the basic idea of this reasoning found recently first experimental evidence, it should be repeated here.

In Section 5.3.1, the important difference between the Darken and Nernst–Planck regime of interdiffusion was already worked out. In Darken regime, different diffusivities of the atomic components must be compensated by a directed flux of vacancies. Given a pretty short diffusion length across the product layer, only a few nanometers, vacancies will not find any possibility of equilibration inside the product layer. Next sinks and sources are only found at the incoherent interphase boundaries, but their density and thus their efficiency may be limited. A quantitative description of this situation is derived in the following manner (see also **Figure 85**):

Let A atoms be faster than B. With

$$\frac{D_i^*}{k_B T}\frac{\partial \mu_i}{\partial x} \approx \frac{D_i}{v_i}\frac{\partial v_i}{\partial x}$$

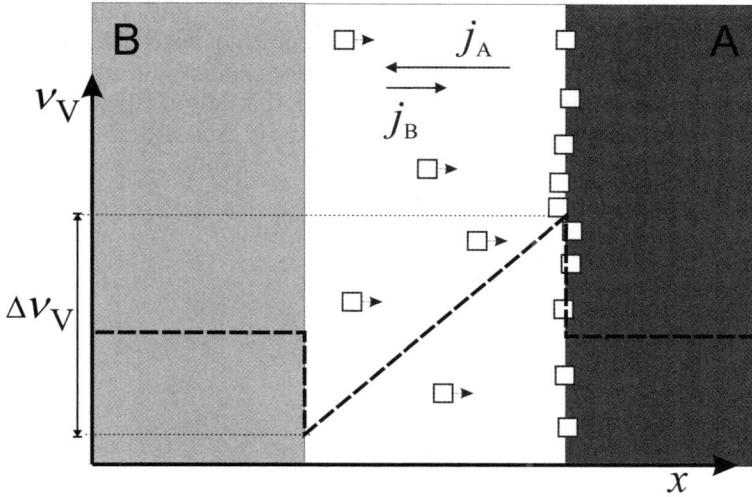

Figure 85 Growth of intermetallic product AB (white) by reactive diffusion of two metals A (dark) and B (light) on the basis of the vacancy mechanism (vacancies are symbolized by open squares). Since A diffuses faster than B, vacancies are pushed to the right. A gradient in vacancy density (dashed line) is generated across the formed intermetallic, if sinks and sources are only present at the interfaces and have insufficient efficiency for full equilibration.

and the convenient definition

$$D_V := \left(\nu D_A^* + (1 - \nu) D_B^* \right) / \nu_V, \tag{189}$$

we derive from basic Eqn (55) the simplified versions

$$\Omega j_A = -D_A \frac{\Delta \nu}{\Delta x} + \frac{\nu}{\nu_V} D_A^* \frac{\Delta \nu_V}{\Delta x} \tag{190}$$

$$\Omega j_V = (D_A - D_B) \frac{\Delta \nu}{\Delta x} - D_V \frac{\Delta \nu_V}{\Delta x}. \tag{191}$$

The current expressed by Eqn (191) withdraws vacancies from the interface at the B side (left in **Figure 85**) and pushes them continuously toward the A side (right). Thus, vacancy depletion at the former and vacancy enrichment at the latter interface would appear, if no sinks and sources counteracted. If sinks and sources are active however, the temporal evolution of the local vacancy densities at the interfaces $\nu_{V,L}$ and $\nu_{V,R}$ is described as

$$\begin{aligned} \frac{\partial \nu_{V,L}}{\partial t} &= -\frac{\Omega j_V}{\delta} - \frac{\nu_{V,L} - \nu_V^{eq.}}{\tau} \approx 0 \\ \frac{\partial \nu_{V,R}}{\partial t} &= +\frac{\Omega j_V}{\delta} - \frac{\nu_{V,R} - \nu_V^{eq.}}{\tau} \approx 0 \end{aligned} \tag{192}$$

Here, the parameter δ denotes the effective thickness of the interface. Parameter τ represents the characteristic time of vacancy annihilation or generation and so measures the efficiency of the sinks/sources. At vanishing τ, efficiency of the sink/sources becomes infinitely high and no deviation of vacancy density from equilibrium would happen. Then reaction appeared in the Darken regime. But in the opposite case of τ going to infinity, Nernst–Planck conditions were established.

However here, it is the intermediate case of a finite τ that should be considered. Under these circumstances, we expect that a steady-state vacancy density is established, so that the RHSs of Eqn (192)) approximately vanish. As a consequence, we calculate the finite difference between the vacancy densities at the opposite sides of the intermetallic layer:

$$\Delta\nu_V = \nu_{V,R} - \nu_{V,L} = \frac{2\tau}{\delta}\Omega j_V, \tag{193}$$

which is further quantified by combining Eqns (192) and (194) to get

$$\Delta\nu_V = \frac{2\tau}{\delta}\frac{(D_B - D_A)\Delta\nu}{\Delta x + 2D_V\tau/\delta} = \frac{l_1}{D_V}\frac{(D_A - D_B)\Delta\nu}{\Delta x + l_1}. \tag{194}$$

To obtain the second equality on the RHS, we have defined the characteristic length $l_1 := 2D_V\tau/\delta$. Inserting the vacancy difference of Eqn (194) into Eqn (190), one obtains after lengthy, but standard algebra

$$\Omega\tilde{j}_A = -\tilde{D}_D\frac{\Delta\nu}{\Delta x}\cdot\frac{1 + (\tilde{D}_{NP}/\tilde{D}_D)(l_1/\Delta x)}{1 + l_1/\Delta x} = -\tilde{D}_D\frac{\Delta\nu}{\Delta x}\cdot\frac{1 + l_0/\Delta x}{1 + l_1/\Delta x}. \tag{195}$$

Here another characteristic length $l_0 := (\tilde{D}_{N.P.}/\tilde{D}_D)\cdot l_1$ has been defined. If partial diffusivities of the components are different, l_0 is necessarily smaller than l_1. In estimating l_1, we note that $2D_V\tau$ just corresponds to the square of the free diffusion length of a vacancy before becoming annihilated. This roughly corresponds to the width of the intermetallic product layer. By contrast δ is only a few Angstroem. Thus, l_1 will probably range to about 100 nm, while a reasonable estimate of l_0 could be $l_0 \approx l_1/100$.

The remarkable interpretation of Eqn (195) is the following: The two characteristic lengths allow distinguishing three stages of reaction. In the first stage, when the thickness is still thinner than l_0, the number of vacancies arriving at the interfaces is so large that action of the sinks can be neglected. In consequence, the conditions of Nernst–Planck transport are fulfilled and so, the interdiffusion current is

$$\Omega\tilde{j}_A = -\tilde{D}_{N.P.}\frac{\Delta\nu}{\Delta x}. \tag{196}$$

Since it is inversely proportional to the layer width, growth appears parabolic and is controlled by the slow diffusivity of the Nernst–Planck regime. However under realistic assumptions, since l_0 equals at most a few nanometers, this stage can hardly be discovered in experiment.

The second stage is defined by $l_0 << \Delta x < l_1$. Here eqn (195) yields

$$\Omega\tilde{j}_A = -\tilde{D}_D\frac{\Delta\nu}{l_1} \tag{197}$$

Surprisingly, the interdiffusion current remains constant during this regime, controlled by the fast Darken coefficient. So, linear growth kinetics can appear in a significant thickness range, up to about

100 nm. This linear regime is not due to a reaction barriers at the interfaces but is caused by limited annihilation of vacancies at the interfaces.

Finally in the third stage, thickness way exceeds both characteristic lengths ($l_1 << \Delta x$) and so

$$\Omega \tilde{j}_A = -\tilde{D}_D \frac{\Delta v}{\Delta x}. \tag{198}$$

Growth is inversely proportional to the thickness. The kinetics will become parabolic again, but now it is controlled by the fast Darken interdiffusion coefficient.

The described theory was selected here for presentation, since it demonstrates impressively that transport in reactive diffusion on short dimensions may become quite complex, as many different mobile species are involved and local equilibrium cannot be established anymore. Such complexity on the nanometer scale has not been studied experimentally with sufficient accuracy to demonstrate the practical relevance of models like the presented one. Sure, to experimentalists, who aim at the measurement of diffusion coefficients, one can only give the advice to avoid such complex situations. But on the other hand, exactly these complex situations of reactive diffusion are of particular importance for any technology based on nanoscaled devices. Here, migration of only a handful of atoms can be already sufficient to degenerate the functionality. So beside scientific curiosity, technological development provides good reasons enough to focus on reactive diffusion in the absolute nanoscale.

In particular, the presented theoretical study points out that significant deviation from vacancy equilibrium must be expected, since in nanoscale transport, vacancy currents may become too high to be compensated by dilute sources and sinks. Interestingly, recent experimental observations by Ene et al. (2009) could be indeed interpreted as evidence that vacancy distribution and in particular the Nernst–Planck interdiffusion regime plays an important role on nanometric length scales. In spherical or hemispherical core–shell nanostructures, the reaction rate depends decisively on the stacking order of the reaction partners. This was, for example, demonstrated for the reaction of Cu/Al/Cu triple layers which were deposited on the apex of nanometric needles. The growth rate of the intermetallic product Al_2Cu was measured by APT. At both reactive interfaces clear parabolic growth were found, but interestingly with significantly different rates that were interpreted as interdiffusion by fast Darken versus slow Nernst–Planck interdiffusivity (see **Figure 86**).

However, the interpretation of these experiments has been furthermore related to another important effect, the generation of diffusion-induced elastic stress. This phenomenon becomes particularly important on nanometric dimensions, since plastic relaxation by dislocation nucleation and glide is hindered. Remarkably, the efficiency of vacancy sinks and sources can be controlled by internal stress and thus the appearance of the Nernst–Planck transport regime may be bound to generation of sufficient stress. Therefore, we cannot finish this section on distinguished nanoscale effects without presenting at least a few general remarks on diffusion-induced stress.

5.7.3 Diffusion and Elastic Stress

The interdependence of diffusion and stress has so many different facets that it is difficult not to lose the track of all important aspects. An overview chart as in **Figure 87** might be helpful for a discussion. It is a modified version of a scheme suggested by Beke et al. (2004).

In interdiffusion in the high-concentration chemical regime, imbalanced diffusion currents, different partial volumes of the components, and reaction-induced volume excess can lead to local variation of

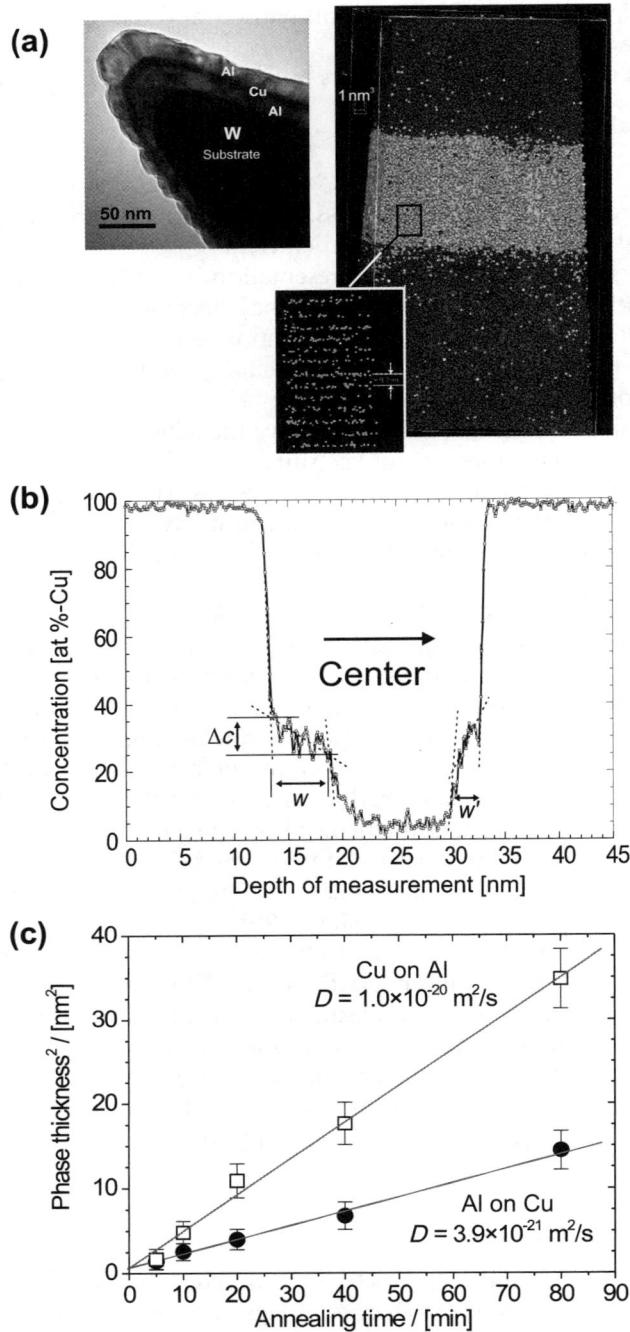

Figure 86 Reactive diffusion of a Cu/Al/Cu triple layer deposited upon a hemispherical tip apex of about 25 nm curvature radius. (a) Sample geometry and exemplary tomographic reconstruction by APT. (b) Composition profile after 40 min annealing at 110°C demonstrates asymmetric thickness of the reaction product. (c) Square of measured thickness of the intermetallic products versus annealing time. Reproduced from data published in Schmitz et al. (2009).

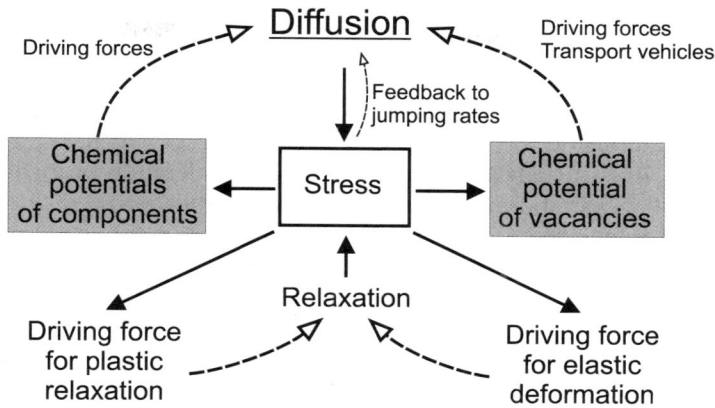

Figure 87 Scheme illustrating the interdependence of diffusion-related stress phenomena at the example of vacancy diffusion in crystalline lattices. The situation is characterized by complex feedback loops that modify atomic transport. Adapted after Beke et al. (2004).

the sample volume. Stress-free expansion in one and contraction in another part of the sample necessarily leads to elastic coherency stress. This stress modifies in turn the chemical potentials of the atomic species and also that of the vacancies. (Let us consider as the most common case, a vacancy mechanism in crystalline host.) Since the resulting stress tensor is usually inhomogeneous, new driving forces to atomic transport appear which in general are opposed to the initial chemical forces and so will retard the overall mixing process. As a rule of trend, build-up stress decelerates the faster diffusing species and may accelerate the slower one. As a consequence, a transition from the fast Darken-type interdiffusion to the slower Nernst–Planck regime may happen (Beke et al., 2004; Erdélyi and Schmitz, 2012) (see below).

Furthermore, vacancies are nonconserved species. So, hydrostatic stress modifies their local density which also causes a variation of the atomic possibilities to jump. Finally, in nanostructures of closed geometries, such as cylinders or spheres, induced stress can become much higher than the macroscopic yield strength. It may range up to several GPa, since nucleation of dislocations and thus plastic relaxation is prevented. Such high stress may even couple directly to the activation volume of diffusion and so can reduce the probability of jumps. In **Figure 87**, these direct feedback loops to the diffusion rate are indicated by dashed arrows in the upper part of the figure. To make the situation even more complex, there are further feedback mechanisms to the generated elastic stress as the latter can partly relax by plastic or elastic deformation (Opposits et al., 1998) (leading to the Gorski effect). These feedback mechanisms are indicated in the bottom part of the figure.

In correlation to grain or phase boundaries further surprising phenomena are observed in thin film diffusion couples which are probably also correlated to diffusion-induced stress. By diffusion of too big or too small atoms along grain boundaries, these boundaries themselves start migrating, leaving behind an intermixed bulk zone. This so-called diffusion-induced grain boundary migration (DIGM) (Yoon, 1989; Penrose, 2004) is closely related to a second process which is regularly observed in size-mismatched thin films interdiffusion couples: By so-called diffusion-induced recrystallization (DIR), new grains are nucleated even in epitaxial single crystalline films (Hartung and Schmitz, 2001). The resulting dynamical process of interdiffusion and grain growth induces exciting stepwise concentration profiles, which fundamentally differ from Fick interdiffusion profiles and are still only partly under-stood (Schmitz et al, 2010; Eich et al., 2012).

In the following we will try to present a more quantitative picture so that the coupling between the different phenomena becomes more concrete. Historically, one of the earliest trials to describe the interrelation of stress and diffusion was made by Larché and Cahn (1982, 1985). However their model had still some weakness. In particular it did predict even an acceleration of mixing by induced stress which contradicts general physical arguments. The first successful model including all of the basic phenomena was given by Stephenson (1988). A recent summary of the actual developments can be found in Svoboda and Fischer (2011). Here we will present the equations in analogy to the Stephenson concept. The reader can find a quite general derivation of the complete set of equations in Erdélyi and Schmitz (2012).

A general difficulty arises from the fact that elastic fields are long ranged and so, boundary conditions get a decisive influence. The following considerations are based on the very common geometry of a thin film interdiffusion couple that is clamped to a rigid substrate surface, as sketched in **Figure 88**. We define axes and directions as stated in this figure.

Although in principle interaction with nondiagonal terms of the stress tensor could appear via the diaelastic effect (interaction due to composition dependence of elastic constants), we presume that only the direct interaction (parelastic effect) of the hydrostatic stress part with the atomic volumes provides the dominant driving force. Dealing with strain effects, we have to skip from now the assumption of a constant atomic volume, but need to distinguish partial volumes (Ω_A, Ω_B, and Ω_V). These can furthermore be dependent on composition to also include a possible volume excess of reactions.

Since the thin film diffusion couple is fixed to the substrate (no lateral displacement possible), but is free to expand in perpendicular direction, stress will build-up exclusively in the in-plane directions. Strain is caused by different reasons which need to be sharply distinguished: (1) elastic deformation

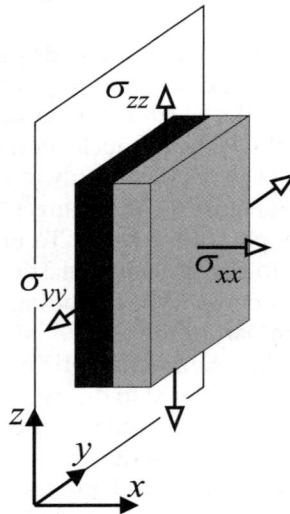

Figure 88   A thin film diffusion couple fixed to a substrate. Stress may develop in the lateral y- and z-directions while perpendicular stress in x-direction vanishes due to the free surface boundary condition.

(connected to stress by Hook's law), (2) stress-free expansion (due to exchange of atoms of different size or excess volume of reactions), and (3) plastic deformation. The following relations hold for the diagonal components of the stress and strain tensors:

$$\sigma_{xx} = 0; \quad \sigma_{yy} = \sigma_{zz} = -\frac{E}{1-\sigma}\left(\varepsilon^{(SF)} + \varepsilon_{yy}^{(P)}\right) = -\frac{3p}{2},$$
(199)

in which E, σ, $\varepsilon^{(SF)}$, $\varepsilon_{yy}^{(P)}$, and p denote Young modulus, Poisson's ratio, the isotropic diagonal components of the stress-free expansion, the in-plane component of plastic deformation strain and the hydrostatic pressure $p = -(\sigma_{yy} + \sigma_{zz})/3$, respectively. Derivation with respect to time (in the lattice reference frame) gives the temporal variation of the in-plane stress:

$$\frac{d\sigma_{yy}}{dt} = -\frac{E}{1-\sigma}\left(\frac{d\varepsilon^{(SF)}}{dt} + \frac{d\varepsilon_{yy}^{(P)}}{dt}\right).$$
(200)

The stress-free expansion summarizes all diffusion effects which change the relative volume:

$$\frac{d\varepsilon^{(SF)}}{dt} = -\frac{1}{3}\left\{\frac{\partial}{\partial x}[(\Omega_A - \Omega_V)j_A] + \frac{\partial}{\partial x}[(\Omega_B - \Omega_V)j_B] - S\right\}.$$
(201)

(Mind that Ω_i can be composition dependent.) In a single jump, vacancy and atom exchange their sites. So, comparably small volume change appears by the first two terms on the RHS. The last term however captures the effect of generating ($S > 0$) or annihilating vacancies ($S < 0$). These processes change the local number of lattice sites and have therefore possibly a strong contribution to stress-free expansion. Equilibration of the vacancy density appears at sinks/sources by a reaction of first order

$$S(x) = k_S\left[c_{V,0}^{(eq.)}\exp\left(-\frac{\Omega_V p(x)}{k_B T}\right) - c_V(x)\right],$$
(202)

where we have regarded the fact that the equilibrium density of vacancies is modified by local hydrostatic pressure. The reaction constant k_S controls the efficiency (i.e. the density) of sinks or sources.

Plastic deformation by dislocation glide is only driven by shear components of the stress, which yields the tensor equation

$$\frac{d}{dt}\hat{\varepsilon}^{(P)} = \frac{1}{2\eta}\left(\hat{\sigma} - \frac{1}{3}\text{tr}\hat{\sigma}\right).$$
(203)

In this equation, the material parameter η represents a suitable Newton viscosity. For the given geometry Eqn (203) can be reduced to

$$\frac{d\varepsilon_{yy}^{(P)}}{dt} = \frac{d\varepsilon_{zz}^{(P)}}{dt} = -\frac{\sigma_{yy}}{6\eta}.$$
(204)

By continuous integration of Eqns (200), (201) and (204), the development of local hydrostatic pressure during the interdiffusion/reaction process is calculated. This pressure interacts with the respective partial volumes of the different species. Thus, any gradient of the pressure presents an additional driving force to migration and so the first Fick law is modified by drift terms to (compare with Eqn (11))

$$j_i = -D_i \frac{\partial c_i}{\partial x} - \frac{D_i^* c_i}{k_B T} \Omega_i \frac{dp}{dx} \tag{205}$$

to calculate the intrinsic currents of the components $i(=A,B)$.

As in Section 5.3.1, the final step in calculating the interdiffusion current is the transformation into the laboratory reference system. For this, we need the local convection velocity (Kirkendall velocity). We regard that the divergence of this velocity field is coupled to the evolution of local volume and so

$$\frac{\partial v_K}{\partial x} = \frac{1}{V}\frac{dV}{dt} = 3\frac{d\varepsilon^{(SF)}}{dt} - \frac{3(1-2\sigma)}{E}\frac{dp}{dt}. \tag{206}$$

Here, the last term on the RHS includes the elastic volume change by compression. Knowing thereby the convection velocity, the concentration change by interdiffusion in the laboratory frame is finally calculated by adding intrinsic diffusion and convective currents:

$$\frac{d\widetilde{c_A}}{dt} = -\frac{\partial}{\partial x}(j_A + c_A \cdot v_K) - c_A S \tag{207}$$

Unexpectedly, at the RHS a further new term arises. This term is necessary, since formation of a vacancy (i.e. introducing a new empty lattice site) dilutes the volume concentration of the remaining atoms.

The set of equations given by Eqns (199)–(207) represent basically the 'Stephenson model'. It yields a full description of the main diffusion-related phenomena summarized in **Figure 87**. Thus, beside mere Fick interdiffusion of the atoms, it also takes into account stress generation, plastic relaxation, vacancy formation and migration. Still it is a quite rough model, as it assumes isotropic elasticity and a very simple viscosity model of plastic deformation. Also any microstructural effect, such as inhomogeneous distribution of vacancy sinks and sources or migration along grain boundaries are not considered. Last but not least it holds only for the assumed simple thin film layer geometry and boundary conditions. Nevertheless, the model has, for example been successfully applied to diffusion-induced sample bending (Daruka et al., 1996) and to interdiffusion in NiZr multilayers (Greer, 1996).

Recently, a kind of Stephenson model was also modified and extended to be applied to spherical geometries, so that the case of core–shell nanoparticles and curved substrates in general, such as has been shown in **Figure 86**, could be discussed. Due to the particular geometrical constraint, and formation of distinct product phases, the model becomes complex but stays fully analytically (Erdélyi and Schmitz, 2012). Nicely, the calculations predicted the observed asymmetry in growth rate at the two interfaces, as it is demonstrated in **Figure 89**.

The situation of a spherical core shell nanoparticle (a stack of $Al_{15\ nm}/Cu_{15\ nm}/Al_{15\ nm}$ was coated on an inert elastically hard core of 30 nm radius) was adapted by suitable model assumptions. The figure shows the situation after some annealing that gave rise to the formation of intermetallic product layers (marked by gray color in the geometry sketch) of a few nanommeters in thickness. Reaction to the product Al_2Cu is linked to a volume excess of about 6%. As a consequence, beside the immediate compressive stress in the reaction products, simply because they are too large for the available space, also compressive hydrostatic stress is build-up in the center of the sample, while tensile stress appears

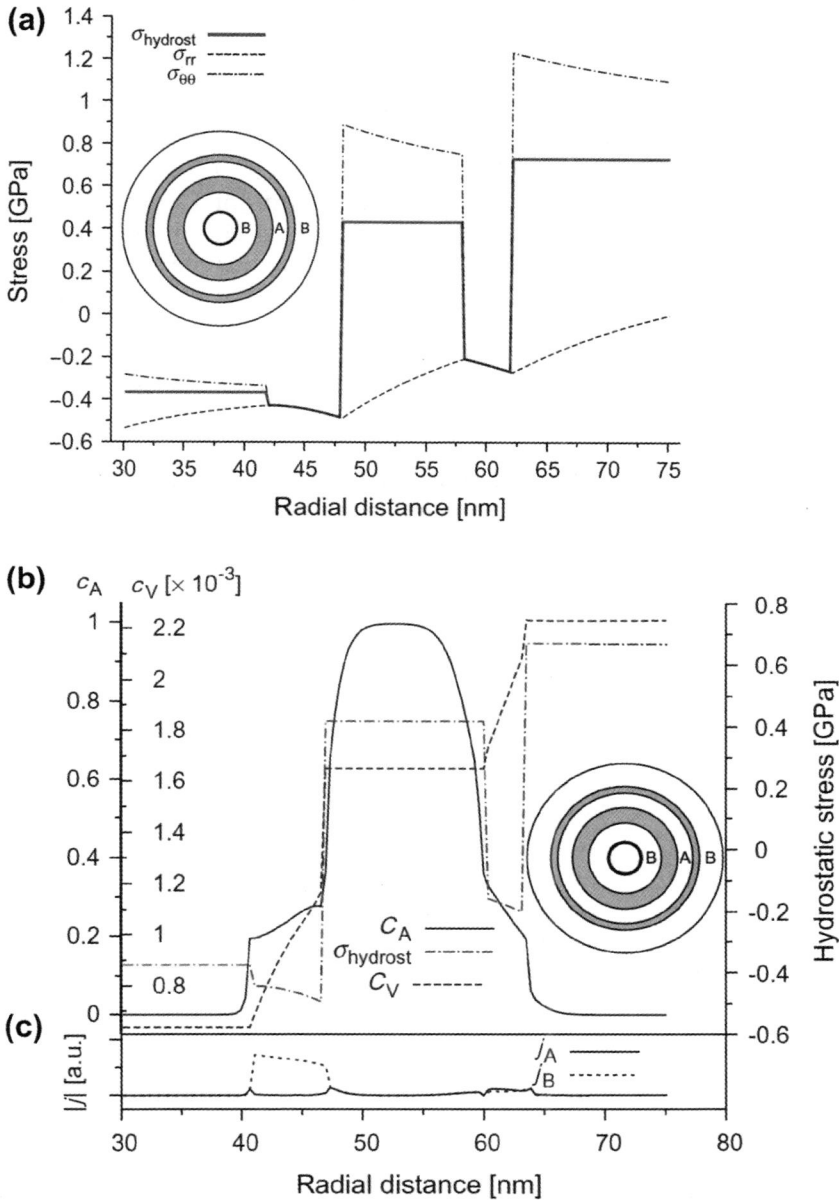

Figure 89 Solid-state reaction in a core–shell nanostructure: Using the Stephenson model for spherical geometry (Erdélyi and Schmitz, 2012), the reaction of a triple layer Al/Cu/Al deposited on a spherical substrat of 30 nm curvature radius has been calculated. Plots represent a sample stage after given annealing time. (a) Radial profiles of hydrostatic (solid), radial (dashed) and tangential (dash-dotted) stress component. (b) Composition profiles of Cu (solid) and vacancies (dashed) in comparison with hydrostatic stress (dash-dotted). For the calculation, stress-free equilibrium concentration of vacancies was exaggerated to 10^{-3}. (c) Partial intrinsic currents of Cu (solid) and Al (dashed). Reproduced with permission from Erdélyi and Schmitz (2012). (For color version of this figure, the reader is referred to the online version of this book.)

in the outer shell (**Figure 89(a)**, solid line). In consequence, across both product layers a considerable difference in stress appears which gives rise to enormous gradients in the vacancy concentration (see **Figure 89(b)**, dashed line).

In the case that the faster component Al(=B) diffuses outward (appears here at the inner Al/Cu interface) vacancies have to be transported inward. This vacancy transport is additionally driven by the vacancy gradient which has established across the product layer. By contrast, the faster component needs to diffuse inward for the opposite stacking of Cu and Al (here at the outer interface). Vacancies would have to migrate outward against the driving force of the vacancy gradient. This depresses the fast transport completely, as demonstrated in the bottom row of the figures (**Figure 89(c)**) which presents the quantitative calculation of the partial currents. Clearly the partial current of Al(=B) is much faster than that of Cu(=A) across the inner intermetallic product layer. The conditions of Darken interdiffusion are established. By contrast, at the outer product layer, both partial currents become almost equal and small. The conditions of the slow Nernst–Planck regime are fulfilled there.

As a consequence the respective parabolic growth rates of the reaction layers become indeed significantly different, in perfect agreement to the experimental observation. This close correlation of theoretical analysis and experiments with nanometric hemispherical substrates yields strong evidence that indeed dramatic deviations from vacancy equilibrium appear on the nanoscale and that this may lead indeed to establishment of Nernst–Planck interdiffusion conditions.

References

Abriola, D., Barradas, N.P., Bognanović-Radović, I., Chiari, M., Gurbich, A.F., Jeynes, C., Kokkoris, M., Mayer, M., Ramos, A.R., Shi, L., Vickridge, I.C., 2011. Nucl. Instrum. Methods Phys. Res. Sect. B 269, 2972.
Adda, Y., Philibert, J., 1966. La diffusion dans les Solides. Presse Universitaires de France.
Alder, B.J., Wainwright, T.E., 1957. J. Chem. Phys. 27, 1208.
Alder, B.J., Wainwright, T.E., 1959. J. Chem. Phys. 31, 459.
Aldinger, F., 1974. Acta Metall. 22, 923.
Alefeld, G., Vökl, J., 1978. Hydrogen in Metals I—Application-oriented Properties. In: Topics in Applied Physics, vol. 28. Springer Verlag.
Alefeld, G., Vökl, J., 1979. Hydrogen in Metals II—Basic Properties. In: Topics in Applied Physics, vol. 28. Springer Verlag.
Allnatt, A.R., 1982. J. Phys. C: Solid State Phys. 15, 5605.
Anenzoubadrour, H., Moya, G., Bernardini, J., 1988. Acta Metall. 36, 767.
Andersen, S.K., Golovchenko, J.A., Mair, G., 1976. Phys. Rev. Lett. 37, 1141.
Andersen, H.C., 1980. J. Chem. Phys. 72, 2384.
Anderson, D.A., Simak, S., 2004. Phys. Rev. B 70, 115108.
Arita, M., Koiwa, M., Ishioka, S., 1989. Acta Metall. 37, 1363.
Ascoli, A., Germagnoli, E., Mongini, L., 1956. Nouvo Cimento 4, 123.
Ash, R., Barrer, R.M., Palmer, D.G., 1965. Br. J. Appl. Phys. 16, 873.
Athènes, M., Bellon, P., Martin, G., 1997. Philos. Mag. A 76, 565.
Athènes, M., Bellon, P., 1999. Philos. Mag. A 79, 2243.
Auger, P., 1923. C.R.A.S 177, 169.
Backus, J.G.E.M., Bakker, H., Mehrer, H., 1978. Phys. Stat. Sol. (B) 64, 151.
Bakker, H., 1968. Phys. Stat. Sol. 28, 569.
Bakker, H., 1970. Phys. Stat. Sol. 38, 167.
Bakker, H., Stolwijk, N.A., Hoetjeseijkel, M.A., 1981. Philos. Mag. 43, 251.
Balluffi, R.W., 1954. Acta Metall. 2, 194.
Ball, M.D., McCartney, D.G., 1981. J. Microsc. 124, 57.
Balogh, Z., Erdélyi, Z., Beke, D.L., Langer, G.A., Csik, A., Boyen, H.G., Wiedwald, U., Ziemann, P., Portavoce, A., Girardeaux, C., 2008. Appl. Phys. Lett. 92, 143104.
Balogh, Z., Erdélyi, Z., Beke, D.L., Wiedwald, U., Pfeiffer, H., Tschetschetkin, A., Ziemann, P., 2010. Thin Solid Films 519, 952.

Balogh, Z., Chellali, M.R., Greiwe, G.H., Schmitz, G., Erdélyi, Z., 2011. Appl. Phys. Lett. 99, 181902.
Barnes, R.S., 1950. Nature 166, 1032.
Barnes, R.S., 1952. Proc. Phys. Soc. B 65, 512.
Barnes, R.S., Mazey, D.J., 1958. Acta Metall. 6, 1.
Batterman, B.W., 1969. Phys. Rev. Lett. 22, 703.
Becvář, F., Nowotný, I., Procházka, I., Rafaja, D., Kern, J., 1995. Appl. Phys. A 61, 335.
Beeler, J.R., 1966. Phys. Rev. 150, 470.
Beke, D.L., Szabó, I.A., Erdélyi, Z., Opposits, G., 2004. Mat. Sci. Eng. A 387, 4.
Belova, I.V., Murch, G.E., 2000a. Philos. Mag. A 80, 1469.
Belova, I.V., Murch, G.E., 2000b. J. Phys. Chem. Solids 61, 1755.
Belova, I.V., Murch, G.E., 2000c. Philos. Mag. A 80, 2073.
Belova, I.V., Murch, G.E., 2001. Diff. Def. For. 194–199, 533.
Belova, I.V., Murch, G.E., 2002. Philos. Mag. A 82, 269.
Belova, I.V., Murch, G.E., 2004a. J. Metastable Nanocryst. Mater. 19, 25.
Belova, I.V., Murch, G.E., 2004b. Philos. Mag. 84, 3637.
Belova, I.V., Murch, G.E., 2005. Philos. Mag. 85, 4515.
Belova, I.V., Shaw, D., Murch, G.E., 2009. J. Appl. Phys. 106, 113707.
Belova, I.V., Fiedler, T., Kulkarni, N., Murch, G.E., 2012. Philos. Mag. 92, 1748.
Benoist, P., Bocquet, J.-L., Lafore, P., 1977. Acta Metall. 25, 265.
Bene, R.W., 1982. Appl. Phys. Lett. 41, 529.
Benninghoven, A., 1994. Surf. Sci. 299–300, 246.
Bergersen, B., Scott, M.J., 1969. Solid State Commun. 7, 1703.
Berendsen, H.J.C., Postma, J.P.M., van Gunsteren, W.F., DiNola, A., Haak, J.R., 1984. J. Phys. Chem. 81, 3684.
Bernardini, J., Girardeaux, C., Erdelyi, Z., Lexcellent, C., 2004. J. Metastable Nanocryst. Mater. 19, 35.
Besson, R., Guyot, S., Legris, A., 2007. Phys. Rev. B 75, 054105.
Beyeler, M., Adda, Y., 1968. J. Phys. 29, 345.
Blavette, D., Cadel, E., Fraczkiewicz, A., Menand, A., 1999. Science 286, 2317.
Bocquet, J.L., 2002. Diff. Def. For. 203–205, 81.
Bokstein, B., Ivanov, V., Oreshina, O., Peteline, A., Peteline, S., 2001. Mat. Sci. Eng. A 302, 151.
Bomhold, G., Wicke, E., 1967. Z. Phys. Chem. 56, 133.
Bortz, A.B., Kalos, M.H., Lebowitz, J.L., 1975. J. Comput. Phys. 17, 10.
Bracht, H., Silvestri, H.H., Sharp, I.D., Haller, E.E., 2007. Phys. Rev. B. 75, 035211.
Brinkmann, J.A., 1955. Acta Metall. 3, 141.
Brillo, J., Chathoth, S.M., Koza, M.M., Meyer, A., 2008. Appl. Phys. Lett. 93, 121905.
Brown, R., 1828. Philos. Mag. 4, 161.
Brown, J.A., Mishin, Y., 2007. Phys. Rev. B 76, 134118.
Bruin, H.J., Murch, G.E., 1973. Philos. Mag. 27, 1475.
Cahn, J.W., Hilliard, J.E., 1958. J. Chem. Phys. 28, 258.
Cahn, J.W., 1965. J. Chem. Phys. 42, 93.
Cammarata, R.C., Greer, A.L., 1984. J. Non-cryst. Solids 61–62, 889.
Canon, R.F., Stark, J.P., 1969. J. Appl. Phys. 40, 4366.
Carslaw, H.S., Jaeger, J.C., 1959. Conduction of Heat in Solids. Oxford University Press.
Car, M., Parrinello, R., 1985. Phys. Rev. Lett. 55, 2471.
Carling, K., Wahnstrom, G., Mattsson, T.R., Mattsson, A.E., Sandberg, N., Grimwall, G., 2000. Phys. Rev. Lett. 85, 3862.
Castaing, R., 1951. Ph.D. thesis, Univ. Paris.
Cattaneo, F., Germagnoli, E., Grasso, F., 1962. Philos. Mag. 7, 1373.
Ceperley, D.M., Alder, B.J., 1980. Phys. Rev. Lett. 45, 566.
Chang, L.S., Rabkin, E., Straumal, B.B., Baretzky, B., Gust, W., 1999. Mater. Sci. Forum 294–296, 585.
Chakravarty, S., Schmidt, H., Tietze, U., Lott, D., Lalla, N.P., Gupta, A., 2009. Phys. Rev. B 80, 014111.
Cheng, J.Y., Chen, L.J., 1991. J. Appl. Phys. 69, 2161.
Chen, Y., Schuh, C.A., 2007a. Scr. Mater. 57, 256.
Chen, Y., Schuh, C.A., 2007b. J. Appl. Phys. 101, 063524.
Chen, J., Lai, Y.S., Ren, C.Y., Huang, D.J., 2008. Appl. Phys. Lett. 92, 081901.
Chellali, M.R., Balogh, Z., Zheng, L., Schmitz, G., 2011. Scr. Mater. 65, 343.
Chellali, M.R., Balogh, Z., Bouchikhaoui, H., Schlesiger, R., Stender, P., Zheng, L., Schmitz, G., 2012. Nano Lett. 12, 3448.

Chi, N.V., Bergner, D., 1982. DIMETA 82. In: Proc. Int. Conf. on Diffusion in Metals and Alloys, p. 334.

Ciccotti, G., Guillôpé, M., Pontikis, V., 1983. Phys. Rev. B 27, 5576.

Clouet, E., Nastar, M., Sigli, C., 2004. Phys. Rev. B 69, 064109.

Clouet, E., Hin, C., Gendt, D., Nastar, M., Soisson, F., 2006. Adv. Eng. Mater. 8, 1210.

Cohen, B.M., Turnbull, D., Warburton, W.K., 1977. Phys. Rev. B 16, 2491.

Compaan, K., Haven, Y., 1956. Trans. Faraday Soc. 52, 786.

Cook, H.E., Hilliard, J.E., 1969. J. Appl. Phys. 40, 2191.

Cook, H.E., de Fontaine, D., Hilliard, J.E., 1969. Acta Metall. 17, 765.

Cornet, J.F., 1974. J. Phys. Chem. Solids 35, 1247.

Compaan, K., Haven, Y., 1958. Trans. Faraday Soc. 54, 1498.

Cotts, R.M., 1972. Ber. Bunsenges. Phys. Chem. 76, 760.

Crank, J., 1975. The Mathematics of Diffusion, second ed. Oxford University Press, Oxford.

Crewe, A.V., Isaacson, M., Johnson, D., 1969. Rev. Sci. Instrum. 40, 241.

Crewe, A.V., Wall, J., Langmore, J., 1970. Science 168, 1338.

Cserháti, C., Paul, A., Kodentsov, A.A., van Dahl, M.J.H., van Loo, F.J.J., 2003. Intermetallics 11, 297.

Cserháti, C., Balogh, Z., Csik, A., Langer, G.A., Erdélyi, Z., Glodán, G., Katona, G.L., Beke, D.L., Zizak, I., Darowski, N., Dudzik, E., Feyerherm, R., 2008. J. Appl. Phys. 104, 024311.

Csik, A., Langer, G.A., Beke, D.L., Erdélyi, Z., Menyhárd, M., Sulyok, A., 2001. J. Appl. Phys. 89, 804.

Darken, L.S., 1948. Trans. AIME 175, 184.

Daruka, I., Szabo, I.A., Beke, D.L., Cserhati, C.S., Kodensov, A., van Loo, F.J.J., 1996. Acta Mater. 44, 4981.

Daw, M.S., Baskes, M.I., 1984. Phys. Rev. B 29, 6443.

Deal, B.E., Grove, S., 1965. J. Appl. Phys. 36, 3770.

Deline, V.R., Katz, W., Evans Jr., C.A., 1978. Appl. Phys. Lett. 33, 832.

Demortier, G., 2003. J. Electron. Spectrosc. Relat. Phenom. 129, 243.

Denkinger, M., Mehrer, H., 2000. Philos. Mag. A 80, 1245.

Dennler, S., Hafner, J., 2006. Phys. Rev. B 73, 174303.

de Reca, N.W., Pampillo, C.A., 1975. Scr. Metall. 9, 1355.

Desré, P., Yavari, A.R., 1990. Phys. Rev. Lett. 64, 1533.

d'Heurle, F.M., Gas, P., 1986. J. Mater. Res. 1, 205.

Dimitrieva, O., Choi, P., Gerstl, S.S.A., Ponge, D., Raabe, D., 2011. Ultramicroscopy 111, 623.

Divinski, S.V., Larikov, L.N., 1997. J. Phys. Condens. Matter 9, 7873.

Divinski, S.V., St. Frank, Södervall, U., Herzig, C., 1998. Acta Mater. 46, 4369.

Divinski, S.V., Herzig, C., 2000. Intermetallics 8, 1357.

Divinski, S.V., Herzig, C., 2002. Diff. Def. For. 203–205, 177.

Divinski, S.V., Hisker, F., Kang, Y.S., Lee, J.S., Herzig, C., 2004. Acta Mater. 52, 631.

Divinski, S.V., Ribbe, J., Schmitz, G., Herzig, C., 2007. Acta Mater. 55, 3337.

Divinski, S., Wilde, G., 2008. Mater. Sci. Forum 584–586, 1012.

Divinski, S.V., Reglitz, G., Wilde, G., 2010. Acta Mater. 58, 386.

Divinski, S.V., Reglitz, G., Rösner, H., Estrin, Y., Wilde, G., 2011. Acta Mater. 59, 1974.

Divinski, S.V., Edelhoff, H., Prokofjev, S., 2012. Phys. Rev. B. 85, 144104.

DuMond, J., Youtz, J.P., 1940. J. Appl. Phys. 11, 357.

Domian, H.A., Aaronson, H.I., 1965. Diffusion in body centered metals. Am. Soc. Metals, 209.

Dreyer, K.F., Neils, W.K., Chromik, R.R., Grossman, D., Cotts, E.J., 1995. Appl. Phys. Lett. 67, 2795.

da Silva, L.C.C., Mehl, R.F., 1951. Trans. AIME 191, 155.

da Silva, J.R.G., McLellan, R.B., 1976. Mat. Sci. Eng. 26, 83.

Duan, J., 2007. J. Phys. Condens. Matter 9, 086217.

Eggersmann, M., Mehrer, H., 2000. Philos. Mag. A 80, 1219.

Eich, S.M., Kasprzak, M., Schmitz, G., Gusak, A., 2012. Acta Mater. 60, 3469.

Einstein, A., 1905. Ann. Phys. 322, 549.

Einziger, R.E., Mundy, J.N., Hoff, H.A., 1978. Phys. Rev. B 17, 440.

Elcock, E.W., McCombie, C.W., 1958. Phys. Rev. B 109, 605.

El-Mellouhi, F., Mousseau, N., Lewis, L.J., 2008. Phys. Rev. B 78, 153202.

Ene, C.B., Schmitz, G., Kirchheim, R., Hütten, A., 2005. Acta Mater. 53, 3383.

Ene, C., Nowak, Oberdorfer, C., Schmitz, G., 2009. Appl. Surf. Sci. 109, 660.

Erdélyi, Z., Beke, D., Nemes, P., Langer, G., 1999. Philos. Mag. 79, 1757.

Erdélyi, Z., Szabó, I.A., Beke, D.L., 2002. Phys. Rev. Lett. 89, 165901.

Erdélyi, Z., Beke, D.L., 2003. Phys. Rev. B 68, 092102.

Erdélyi, Z., Katona, G.L., Beke, D.L., 2004a. Phys. Rev. B 69, 113407.

Erdélyi, Z., Sladacek, M., Stadler, L.M., Zizak, I., Langer, G.L., Kis-Varga, M., Beke, D.L., Sepiol, B., 2004b. Science 306, 1913.

Erdélyi, G., Mehrer, H., Imre, A.W., Lograsso, T.A., Schlagel, D.L., 2007. Intermetallics 15, 1078.

Erdélyi, Z., Beke, D.L., Taranovskyy, A., 2008. Appl. Phys. Lett. 92, 133110.

Erdélyi, Z., Cserháti, C., Csik, A., Daróczi, L., Langer, G.A., Balogh, Z., Varga, M., Beke, D.L., Zizak, I., Erko, A., 2009. X-ray Spectrom. 38, 338.

Erdélyi, Z., Balogh, Z., Beke, D.L., 2010. Acta Mater. 58, 5639.

Erdélyi, Z., Beke, D.L., 2011. J. Mater. Sci. 46, 6465.

Erdélyi, Z., Schmitz, G., 2012. Acta Mater. 60, 1807.

Fähnle, M., Mayer, J., Meyer, B., 1999. Intermetallics 7, 315.

Fang, Q.F., Wang, R., 2000. Phys. Rev. B 62, 9317.

Fan, H.J., Gösele, U., Zacharias, M., 2007. Small 3, 1660.

Faupel, F., Frank, W., Macht, M.P., Naundorf, V., Rätzke, K., Schober, H.R., Sharma, S.K., Teichler, H., 2003. Rev. Mod. Phys. 73, 237.

Feng, Y., Zhou, Z., Zhou, Y., Zhao, G., 1994. Nucl. Instrum. Methods Phys. Res. B86, 225.

Ferrari, S., Perego, M., Fanciulli, M., 2003. Appl. Surf. Sci. 203–204, 52.

Fichthorn, K.A., Scheffler, M., 2000. Phys. Rev. Lett. 84, 5371.

Fick, A., 1855. Ann. Phys. 94, 59.

Fillon, J., Calais, D., 1977. J. Phys. Chem. Solids 38, 81.

Finnis, M.W., Sinclair, J.E., 1984. Philos. Mag. A 50, 45.

Fisher, J.C., 1951. J. Appl. Phys. 22, 74.

Frank, W., Gösele, U., Mehrer, H., Seeger, A., 1984. Diffusion in Silicon and Germa-nium. In: Murch, G.E., Nowick, A.S. (Eds.), Diffusion in Crystalline Solids. Academic Press, p. 63.

Frank, St., Södervall, U., Herzig, C., 1995. Phys. Stat. Sol. B 191, 45.

Frank, St., Herzig, C., 1997. Mat. Sci. Eng. A 239-240, 882.

Frank, St., Divinski, S.V., Södervall, U., Herzig, C., 2001. Acta Mater. 49, 1399.

Frolov, T., Mishin, Y., 2009. Phys. Rev. B 79, 174110.

Fujita, T., Horita, Z., Langdon, T.G., 2002. Philos. Mag. A 82, 2249.

Gallash, T., Stockhoff, T., Baither, D., Schmitz, G., 2011. J. Power Sources. 196, 428.

Gao, F., Qu, J., 2012. Mater. Lett. 73, 92.

Gas, P., Beke, D.L., Bernardini, J., 1992. Philos. Mag. Lett. 65, 133.

Gault, B., Moody, M.P., Cairney, J.M., Ringer, S.P., 2012a. Atom Probe Microscopy, Springer Series in Material Science, vol. 160. Springer Verlag.

Gault, B., Moody, M.P., Cairney, J.M., Ringer, S.P., 2012b. Mater. Today 15, 378.

Gauvin, R., Brodusch, N., Michaud, P., 2012. Def. Diff. Forum 323–325, 61.

Gemma, R., Al-Kassab, T., Kirchheim, R., Pundt, A., 2009. Ultramicroscopy 109, 631.

Gendt, D., Maugis, P., Martin, G., Nastar, M., Soisson, F., 2001. Diff. Def. For. 194–199, 1779.

Ghosh, C., Paul, A., 2007. Acta Mater. 55, 1927.

Gibbs, G.B., Graham, D., Tomlin, D.H., 1963. Philos. Mag. 8, 1269.

Gillan, M.J., 1989. J. Phys. Condens. Matter 1, 689.

Glodán, G., Cserháti, C., Beszeda, I., Beke, D.L., 2010. Appl. Phys. Lett. 97, 113109.

Gorski, W.S., 1935. Z. Phys. Sowjetunion 8, 457.

Gösele, U., Tu, K.N., 1982. J. Appl. Phys. 53, 3252.

Gösele, U., Tu, K.N., 1989. J. Appl. Phys. 66, 2619.

Gottstein, G., Shvindlerman, L.S., Zhao, B., 2010. Scr. Mater. 62, 914.

Graham, T., 1866. Philos. Trans. R. Soc. 156, 399.

Grabowski, B., Ismer, L., Hickel, T., Neugebauer, J., 2009. Phys. Rev. B 79, 134106.

Grabowski, B., Hickel, T., Neugebauer, J., 2011. Phys. Status Solidi B 248, 1295.

Greer, A.L., 1996. Def. Diff. Forum 129–130, 163.

Grieverson, P., Turkdogan, E.T., 1964a. Trans. AIME 230, 407.

Grieverson, P., Turkdogan, E.T., 1964b. Trans. AIME 230, 1605.

Griesche, A., Macht, M.P., Suzuki, S., Kraatz, K.H., Frohberg, G., 2007. Scripta Mater. 57, 477.

Groh, J., von Hevesy, G., 1920. Ann. Physik 63, 85.

Groh, J., von Hevesy, G., 1921. Ann. Physik 65, 216.

Gröstlinger, F., Rennhofer, M., Leitner, M., Partyka-Jankowska, E., Sepiol, B., Laenens, B., Plackaert, N., Vantomme, A., 2012. Phys. Rev. B 85, 134302.

Gruzin, P.L., 1952. Dokl. Akad. Nauk. SSSR 86, 289.

Gude, A., Mehrer, H., 1997. Philos. Mag. A 76, 1.
Gupta, D., Lazarus, D., Lieberman, D.S., 1967. Phys. Rev. 153, 863.
Gupta, D., Tsui, R.T.C., 1970. Appl. Phys. Lett. 17, 294.
Gupta, D., Lieberman, D.S., 1971. Phys. Rev. B 4, 1070.
Gupta, D., 2004. Diffusion Processes in Advanced Technological Materials. William Andrew Publ., Norwich (USA).
Gupta, A., Gupta, M., Chakravarty, S., Rüffner, R., Wille, H.C., Leupold, O., 2005. Phys. Rev. B 72, 012104.
Gupta, A., Kumar, D., Phatak, V., 2010. Phys. Rev. B 81, 155402.
Gusak, A.M., 1990. Ukr. Phys. J. 35, 725.
Gusev, E.P., Lu, H.C., Gustafsson, T., Garfunkel, E., 1995. Phys. Rev. B 52, 1759.
Gusak, A.M., Zaporozhets, T.V., Tu, K.N., Gösele, U., 2005. Philos. Mag. 85, 4445.
Gusak, A.M., Tu, K.N., 2009. Acta Mater. 57, 3367.
Gusak, A.M., 2010. Diffusion-controlled Solid State Reactions. Wiley-VCH Verlag, Weinheim (Germany).
Gusak, A.M., Hodaj, F., Schmitz, G., 2011. Philos. Mag. Lett. 91, 610–620.
Hahn, E.L., 1950. Phys. Rev. 80, 580.
Haile, J.M., Gupta, S., 1983. J. Chem. Phys. 79, 3067.
Haley, D., Petersen, T., Ringer, S.P., Smith, G.D.W., 2011. J. Microsc. 244, 140.
Han, C.J., Helms, C.R., 1986. J. Appl. Phys. 59, 1767.
Harrison, L.G., 1961. Trans. Faraday Soc. 57, 1191.
Harris, L.A., 1968. J. Appl. Phys. 39, 1419.
Hartung, F., Schmitz, G., 2001. Phys. Rev. B 64, 245418.
Hempelmann, R., 2000. Quasielastic Neutron Scattering and Solid State Diffusion. Oxford Science Publication.
Henkelman, G., Jónsson, H., 1999. J. Phys. Chem. 111, 7010.
Henkelman, G., Jónsson, H., 2000. J. Phys. Chem. 113, 9978.
Henkelman, G., Jónsson, H., 2001. J. Phys. Chem. 115, 9657.
Herring, D.F., Chiba, R., Gasten, B.R., Richards, H.T., 1958. Phys. Rev. 112, 1210.
Herzig, C., Friesel, M., Derdau, D., Divinski, S.V., 1999a. Intermetallics 7, 1141.
Herzig, C., Köhler, U., Divinski, S.V., 1999b. J. Appl. Phys. 85, 8119.
Heumann, T., Walther, G., 1957. Z. Metallkd. 48, 151.
Heumann, T., 1992. Diffusion in Metallen. Springer Verlag, Berlin.
Heyes, D.M., 1983. Chem. Phys. 82, 285.
Hillert, M., 1961. Acta Metall. 9, 525.
Hin, C., Soisson, F., Maugis, P., 2005. Diff. Def. For. 237–240, 721.
Ho, K.M., Fu, D.L., Harmon, B.N., 1983. Phys. Rev. B 28, 6687.
Ho, K.M., Fu, D.L., Harmon, B.N., 1984. Phys. Rev. B 29, 1575.
Hoffmann, R.E., Turnbull, D., 1951. J. Appl. Phys. 22, 634.
Hohenberg, P.C., Kohn, W., 1964. Phys. Rev. 136, B864.
Holdsworth, P.C.W., Elliot, R.J., 1986. Philos. Mag. A 54, 601.
Hood, G.M., 1988. J. Nucl. Mater. 159, 149.
Hoshino, K., Rothman, S.J., Averbeck, R.S., 1988. Acta Metall. 36, 1271.
Hoshino, T., Zeller, R., Dederichs, P.H., 1996. Phys. Rev. B. 53, 8971.
Hu, C.-K., Huntington, H.A., 1982. Phys. Rev. B 26, 2782.
Hugenschmidt, C., Schreckenbach, K., Stadlbauer, M., Straßer, B., 2005. Nucl. Instrum. Methods Phys. Res. Sect. A 554, 384.
Hutter, J., 2011. WIREs Comput. Mol. Sci. 2, 604.
Hwang, J.C.M., Balluffi, R.W., 1979. J. Appl. Phys. 50, 1339.
Ikeda, T., Almazouzi, A., Numakura, H., Koiwa, M., Sprengel, W., Nakajima, A., 1998. Acta Mater. 46, 5369.
Ikeda, T., Kadowaki, H., Nakajima, H., Inui, H., Yamaguchi, M., Koiwa, M., 2001a. Mat. Sci. Eng. A 312, 155.
Ikeda, T., Kadowaki, H., Nakajima, H., 2001b. Acta Mater. 49, 3475.
Jeong, S.C., Katayama, I., Kawakami, H., Watanabe, Y., Ishiyama, H., Miyatake, H., Sataka, M., Okayasu, S., Sugai, H., Ichikawa, S., Nishio, K., Nakanoya, T., Ishikawa, N., Chimi, Y., Hashimoto, T., Yahagi, M., Takada, K., Kim, B.C., Watanabe, M., Iwase, A., Hashimoto, T., Ishikawa, T., 2005. Nucl. Instrum. Methods Phys. Res. Sect. B 230, 596.
J. Metastable Nanocryst. Mater., 2004. vol. 19: Nanodiffusion, Transtech Publ., (Switzerland).
Kao, C.R., Chang, Y.A., 1993. Intermetallics 1, 237.
Katona, G.L., Erdélyi, Z., Beke, D.L., Dietrich, C., Weigl, F., Boyen, H.G., Koslowski, B., Ziemann, P., 2005. Phys. Rev. B 71, 115432.
Kaur, I., Gust, W., Kozma, L., 1989. Handbook of Grain and Interphase Boundary Diffusion. Ziegler Press, Stuttgart (Germany).
Kaur, I., Mishin, Y., Gust, W., 1995. Fundamentals of Grain and Interphase Boundary Diffusion. John Wiley and Sons Ltd.

Kellogg, G.L., Tsong, T.T., Cowan, P., 1978. Surf. Sci. 70, 485.

Kellogg, G.L., Feibelman, P.J., 1990. Phys. Rev. Lett. 64, 3143.

Keyes, R.W., 1958. J. Chem. Phys. 29, 467.

Kidson, G.V., 1961. J. Nucl. Mat 3, 21.

Kim, S., Chang, Y.A., 2000. Metall. Mater. Trans. A 31, 1519.

Kirkaldy, J.S., 1958. Can. J. Phys. 36, 917.

Klafter, J., Sokolov, I.M., 2005. Phys. World 18, 29.

Klinger, L.M., Levin, L.A., Petelin, A.L., 1997. Diff. Def. For. 143–147, 1523.

Kluge, M., Schober, H.R., 2004. Phys. Rev. B 70, 224209.

Köhler, U., Herzig, C., 1988. Philos. Mag. A 58, 769.

Koiwa, M., 1977. Philos. Mag. 36, 893.

Koiwa, M., Ishioka, S., 1983. Philos. Mag. 48, 1.

Kolobov, Yu.R., Grabovetskaya, G.P., Ivanov, M.B., Zhilyaev, A.P., Valiev, R.Z., 2001. Scr. Mater. 44, 873.

Konwar, D., Bhute, V.J., Chatterjee, A., 2011. J. Phys. Chem. 135, 174103.

Korhonen, T., Puska, M.J., Nieminen, R.M., 1995. Phys. Rev. B 51, 9526.

Kosevich, V.M., Gladkikh, A.N., Karpovskyi, M.V., Klimenko, V.N., 1994. Interface Sci. 2, 247.

Kube, R., Bracht, H., Hansen, J.L., Larsen, A.N., Haller, E.E., Paul, S., Lerch, W., 2010. J. Appl. Phys. 7, 073520.

Kümmerle, E.A., Badura, K., Sepiol, B., Mehrer, H., Schaefer, H.E., 1995. Phys. Rev. B 52, R6947.

Kumar, A.K., Laurila, T., Vourinen, V., Paul, A., 2009. Scr. Mater. 6, 377.

Kuper, A.B., Lazarus, D., Manning, J.R., Tomizuka, C.T., 1956. Phys. Rev. 104, 1536.

Laffont, L., Wu, M.Y., Chevallier, F., Poizot, P., Morcrette, M., Tarascon, J.M., 2006. Micron 37, 459.

Lakatos, A., Langer, G.A., Csik, A., Cserhati, C., Kis-Varga, M., Daroczi, L., Katona, G.L., Erdélyi, Z., Erdelyi, G., Vad, K., Beke, D.L., 2010. Appl. Phys. Lett. 97, 233103.

Lam, N.Q., Rothman, S.J., Mehrer, H., Nowicki, L.J., 1973. Phys. Stat. Sol. (B) 57, 225.

Lange, W.A., Hässner, A., Mischer, G., 1964. Phys. Stat. Sol. 5, 63.

Landolt–Börnstein, 1990a. New Series III/26. Springer Verlag, Berlin (chapter 2), pp. 34–84.

Landolt–Börnstein, 1990b. New Series III/26. Springer Verlag, Berlin (chapter 3), pp. 88–203.

Landolt–Börnstein, 1990c. New Series III/26. Springer Verlag, Berlin (chapter 8), pp. 471–503.

Landolt–Börnstein, 1990d. New Series III/26. Springer Verlag, Berlin (chapter 9), pp. 504–573.

Landolt–Börnstein, 1990e. New Series III/26. Springer Verlag, Berlin (chapter 4), pp. 213–276.

Landolt–Börnstein, 1990f. New Series III/26. Springer Verlag, Berlin (chapter 5), pp. 279–366.

Landolt–Börnstein, 1990g. New Series III/26. Springer Verlag, Berlin (chapter 11), pp. 626–629.

Landolt–Börnstein, 1990h. New Series III/26. Springer Verlag, Berlin (chapter 12), pp. 630–716.

Larché, F.C., Cahn, J.W., 1982. Acta Metall. 30, 1835.

Larché, F.C., Cahn, J.W., 1985. Acta Metall. 33, 331.

Lazarus, D., 1954. Phys. Rev. 93, 973.

Leavitt, J.A., McIntyre Jr., L.C., Ashbaugh, M.D., Oder, J.G., Lin, Z., Dezfouly-Arjomandy, B., 1990. Nucl. Instrum. Methods Phys. Res. B44, 260.

Le Bouar, Y., Soisson, F., 2002. Phys. Rev. B 65, 094103.

Le Claire, A.D., 1951. Philos. Mag. 42, 468.

Le Claire, A.D., 1962. Philos. Mag. 7, 141.

Le Claire, A.D., Lidiard, A.B., 1956. Philos. Mag. 1, 518.

Le Claire, A.D., Rabinovitsch, A., 1981. J. Phys. C: Solid State Phys. 14, 3863.

Le Claire, A.D., Rabinovitsch, A., 1982. J. Phys. C: Solid State Phys. 15, 3455.

Le Claire, A.D., Rabinovitsch, A., 1983. J. Phys. C: Solid State Phys. 16, 2087.

Le Claire, A.D., Rabinovitsch, A., 1984. J. Phys. C: Solid State Phys. 17, 991.

Lidiard, A.B., 1955. Philos. Mag. 46, 1218.

Lidiard, A.B., 1960. Philos. Mag. 5, 1171.

Lodding, A., Mundy, J.N., Ott, A., 1970. Phys. Stat. Sol. 38, 559.

Lomer, W.M., 1958. Vacancies and Other Point Defects in Metals and Alloys. Institute of Metals.

Lord, A.E., Beshers, D.N., 1966. Acta Metall. 14, 1659.

Lundy, T.S., Murdock, J.F., 1962. J. Appl. Phys. 33, 1671.

Lutze-Birk, A., Jacobi, H., 1975. Scr. Metall. 9, 761.

Lutsko, J.F., Wolf, D., Yip, S., Philpott, S.R., Nguyen, T., 1988. Phys. Rev. B 38, 11572.

Mâaza, M., Sella, C., Kâabouchi, M., 1992. Appl. Surf. Sci. 60-61, 573.

Maier, K., Bassani, C., Schüle, W., 1973. Phys. Lett. 44A, 539.

Maier, K., Mehrer, H., Lessmann, E., Schüle, W., 1976. Phys. Stat. Sol. B 78, 689.

Maier, K., Mehrer, H., Rein, G., 1979. Z. Metallkd. 70, 271.

Mallard, W.C., Gardner, A.B., Bass, R.F., Slifkin, L.M., 1963. Phys. Rev. 129, 617.

Manley, O.P., Rice, S.A., 1960. Phys. Rev. 117, 632.

Manning, J.R., 1967. Acta Metall. 15, 817.

Manning, J.R., 1968. Diffusion Kinetics for Atoms in Crystal. van Nostrand Comp., Princeton.

Manning, J.R., 1971. Phys. Rev. B 4, 1111.

Mansilla, C., Wirtz, T., 2012. Appl. Surf. Sci. 258, 4813.

Mao, Z., Booth-Morrison, C., Sudbrack, C.K., Martin, G., Seidman, D.N., 2012. Acta Mater. 60, 1871.

Martin, G., 1990. Phys. Rev. B 41, 2279.

Masaki, T., Fukazawa, T., Matsumoto, S., Itami, T., Yoda, S., 2005. Meas. Sci. Technol. 16, 327.

Matano, C., 1933. Jpn. J. Phys. 8, 109.

Matusiewicz, J., Birnbaum, H.K., 1977. J. Phys. F.: Met. Phys. 7, 2285.

Maxwell-Garnett, J.C., 1904. Philos. Trans. Royal Soc. London 203, 385.

Mayer, J., Meyer, B., Oehrens, J.S., Bester, G., Börnsen, N., Fähnle, M., 1997. Intermetallics 5, 597.

McLean, D., 1957. Grain Boundaries in Metals. Clarendon Press, Oxford, pp. 49–50.

Mehrer, H., 1978. J. Nucl. Mat. 69–70, 38.

Mehrer, H., 1996. Def. Diff. Forum 129, 57.

Mehrer, H., 2006a. Diffusion in Solids, Fundamentals, Methods, Materials, Diffusion-controlled Processes. In: Springer Series in Solid State Sciences, vol. 155. Springer-Verlag, Berlin.

Mehrer, H., 2006b. Diffusion in Solids, Fundamentals, Methods, Materials, Diffusion-controlled Processes. In: Springer Series in Solid State Sciences, vol. 155. Springer-Verlag, Berlin (chapter 11).

Mehrer, H., 2006c. Diffusion in Solids, Fundamentals, Methods, Materials, Diffusion-controlled Processes. In: Springer Series in Solid State Sciences, vol. 155. Springer-Verlag, Berlin (chapter 12).

Mehrer, H., 2007. Diffusion in Solids, Fundamentals, Methods, Materials. In: Diffusion-controlled Processes, Springer Series in Solid State Sciences, vol. 155. Springer-Verlag, Berlin (chapter 14).

Meitner, L., 1922. Z. Phys. 9, 131.

Mermin, N.D., 1965. Phys. Rev. 137, A1441.

Metropolis, N., Rosenbluth, A.W., Rosenbluth, M.N., Teller, A.H., Teller, E., 1953. J. Phys. Chem. 21, 1087.

Mikhailovskii, I.M., Rabukin, V.B., Velikodnaya, O.A., 1991. Phys. Status Solidi A 125, K65.

Million, B., Kučera, J., 1969. Acta Metall. 17, 339.

Million, B., Kučera, J., 1971. Czech. J. Phys. B. 21, 161.

Million, B., Ruzickova, J., Velisek, J., Vrestal, J., 1980. Mat. Sci. Eng. 50, 43.

Millares, M., Pieraggi, B., Lelièvre, E., 1990. Scr. Metall. Mater. 27, 1777.

Mills, G., Jónsson, H., 1994. Phys. Rev. Lett. 72, 1124.

Mills, G., Jónsson, H., Schenter, G.K., 1995. Surf. Sci. 324, 305.

Minamino, Y., Koizumi, Y., Inui, Y., 2002. Diff. Def. For. 194–199, 517.

Mishin, Y., Herzig, C., 2000. Acta Mater. 48, 589.

Mishin, Y., Asta, M., Li, J., 2010. Acta Mater. 58, 1117.

Miyagawa, H., Hiwatari, Y., Bernu, B., Hansen, J.P., 1988. J. Chem. Phys. 88, 3879.

Moleko, L.K., Allnatt, A.R., Allnatt, E.L., 1989. Philos. Mag. 59, 149.

Monma, K., Suto, H., Oikawa, H., 1964. J. Jpn. Inst. Met. 128, 192.

Mortlock, A.J., 1960. Acta Metall. 8, 132.

Mössbauer, R.L., 1958. Z. Phys. 151, 124.

Moutanabbir, O., Isheim, D., Seidman, D.N., Kawamura, Y., Itoh, K.M., 2011. Appl. Phys. Lett. 98, 013111.

Montet, G., 1973. Phys. Rev. B 7, 650.

Müller, E., 1951. Z. Phys. 131, 136.

Mullen, J.G., 1961. Phys. Rev. 121, 1649.

Mundy, J.N., 1971. Phys. Rev. B. 3, 2431.

Mundy, J.N., Miller, T.E., Porte, R.J., 1971. Phys. Rev. B 3, 2445.

Mundy, J.N., Tse, C.W., McFall, W.D., 1976. Phys. Rev. B 13, 2349.

Mundy, J.N., Rothman, S.J., Lam, N.Q., Hoff, H.A., Nowicki, L.J., 1978. Phys. Rev. B 18, 6566.

Mundy, J.N., Hoff, H.A., Pelleg, J., Rothman, S.J., Nowicki, L.J., Schmidt, F.A., 1981. Phys. Rev. B 24, 658.

Murch, G.E., 1984. J. Phys. Chem. Sol. 45, 451.

Murch, G.E., Thorn, R.J., 1977. J. Phys. Chem. Sol. 38, 789.

Murch, G.E., 1982. Philos. Mag. A 45, 941.

Murch, G.E., Qin, Z., 1994. Def. Diff. For. 109–110, 1.

Nachtrieb, N.H., Weil, J.A., Catalano, E., Lawson, A.W., 1952. J. Chem. Phys. 20, 1189.

Nachtrieb, N.H., Handler, G.S., 1955. J. Chem. Phys. 23, 1569.

Nachtrieb, N.H., Petit, J., 1956. J. Chem. Phys. 24, 746.

Nakamura, R., Tokozakura, D., Nakajima, H., Lee, J.G., Mori, H., 2007. J. Appl. Phys. 101, 074303.

Nato Sci. Ser., 2007. vol. 235, Radiation effects in solids, (chapter 1)., pp. 1–24.

Nazarov, A.A., 2000. Interface Sci. 8, 315.

Nemutudi, R.S., Comrie, C.M., Chrums, C.L., 2000. Thin Solid Films 358, 270.

Nemouchi, F., Mangelinck, D., Bergmann, C., Gas, P., 2005. Appl. Phys. Lett. 86, 041903.

Neumann, G., Pfundstein, M., Reimers, P., 1981. Phys. Stat. Sol. 64, 225.

Neumann, G., Pfundstein, M., Reimers, P., 1982. Philos. Mag. A 45, 499.

Neuhaus, P., Herzig, C., 1988. Z. Metallkd. 79, 595.

Neumann, G., Tuijn, C., 2002. Impurity Diffusion in Metals. Scitec Publication Ltd, Uetickon-Zürich, Switzerland.

Niki, H., Okuda, H., Oshiro, M., Yogi, M., Seki, I., Fukuhara, M., 2012. J. Appl. Phys. 111, 124308.

NIMS, Internet site: www.diffusion.nims.go.jp.

Nishikawa, S., Tsumuraya, T., 1971. Philos. Mag. 26, 941.

Nitta, H., Iijima, Y., Tanaka, K., Yamazaki, Y., Lee, C.G., Matsuzaki, T., Watanabe, T., 2004. Mat. Sci. Eng. A 382, 243.

Nose, S., 1984. Mol. Phys. 52, 255.

Nosé, Y., Ikeda, T., Nakajima, H., Numakura, H., 2005. Diff. Def. For. 237–240, 450.

Nosé, Y., Terashita, N., Ikeda, T., Nakajima, H., 2006. Acta Mater. 54, 2511.

Numakura, H., Yamada, T., Koiwa, M., Szabó, I.A., Hono, K., Sakurai, T., 1993. Diff. Def. For. 95-98, 869.

Numakura, H., Ikeda, T., Koiwa, M., Almazouzi, A., 1998. Philos. Mag. A 77, 887.

Oeschner, H., Getto, R., Kopnarski, M., 2009. J. Appl. Phys. 105, 063523.

Okkerse, B., 1956. Phys. Rev. 103, 1246.

Oligschleger, C., Schober, H.R., 1999. Phys. Rev. B 59, 811.

Onishi, M., Fujibuchi, H., 1975. Trans. Jpn. Inst. Met. 16, 539.

Onsager, L., 1931. Phys. Rev. 37, 405.

Opposits, G., Szabó, S., Beke, D.L., Guba, Z., Szabó, I.A., 1998. Scr. Mater. 39, 977.

Ozelton, M.W., Swalin, R.A., 1968. Philos. Mag. 18, 441.

Palumbo, G., Thorpe, S.J., Aust, K.T., 1990. Scr. Mater. 24, 1347.

Pareige, C., Soisson, F., Martin, G., Blavette, D., 1999. Acta Mater. 47, 1789.

Pasichnyy, M.O., Schmitz, G., Gusak, A.M., Vovk, V., 2005. Phys. Rev. B 72, 014118.

Paul, A., Kodentsov, A.A., van Loo, F.J.J., 2004. Acta Mater. 52, 4041.

Paul, A., Kodentsov, A.A., van Loo, F.J.J., 2005a. Def. Diff. For. 237–240, 813.

Paul, A., Kodentsov, A.A., van Loo, F.J.J., 2005b. J. Alloys. Compd. 403, 147.

Paul, A., Ghosh, C., Boettinger, W.J., 2011. Metall. Mater. Trans. A 42, 952.

Pavlova, T.S., Golovin, I.S., Sinning, H.R., Golovin, S.A., Siemers, C., 2006. Intermetallics 14, 1238.

Pawel, R.E., Lundy, T.S., 1964. J. Appl. Phys. 35, 435.

Pawel, R.E., 1964. Rev. Sci. Instrum. 35, 1066.

Pelleg, J., 1974. Philos. Mag. 29, 383.

Pennycook, S.J., Jensen, D.E., 1991. Ultramicroscopy 37, 14.

Penrose, O., 2004. Acta Mater. 52, 3901.

Perdew, J.P., Chevary, J.A., Vosko, S.H., Jackson, K.A., Pederson, M.R., Singh, D.J., Fiolhais, C., 1992. Phys. Rev. B 46, 6671.

Peterson, N.L., 1963. Phys. Rev. 132, 2471.

Peterson, N.L., 1964. Phys. Rev. 136, A568.

Peterson, N.L., Rothman, S.J., 1970. Phys. Rev. B. 1, 3264.

Peterson, N.L., 1978. J. Nucl. Mat. 69–70, 3.

Petry, W., Heiming, A., Herzig, C., Trampenau, J., 1991a. Def. Diff. For. 75, 211.

Petry, W., Heiming, A., Trampenau, J., Alba, M., Herzig, C., Schober, H.R., Vogl, G., 1991b. Phys. Rev. B 43, 10993.

Philibert, J., 2009. Int. J. Mat. Res. 100, 6.

Pierre-Louis, O., Chame, A., Saito, Y., 2009. Phys. Rev. Lett. 103, 195501.

Pikart, P., Hugenschmidt, C., Horisberger, M., Matsukawa, Y., Hatekayama, M., Toyama, T., Nogai, Y., 2011. Phys. Rev. B 84, 014106.

Poate, J.M., Tu, K.N., Mayer, J.W., 1978. Thin Films—Interdiffusion and Reactions. New York.

Portavoce, A., Chow, L., Bernardini, J., 2010. Appl. Phys. Lett. 96, 214102.

Potvoce, A., Tréglia, G., 2012. Phys. Rev. B 85, 224101.

Pretorius, R., Vredenberg, A.M., Saris, F.W., de Reus, R., 1991. J. Appl. Phys. 70, 3636.

Prokes, S.M., 1986. PhD thesis, Harvard University.

Rabkin, E., Gottstein, G., Shvindlerman, L.S., 2011. Scr. Mater. 65, 1101.

Radelaar, S., 1970. J. Phys. Chem. Solids 31, 219.

Railsback, J.G., Johnston-Peck, A.C., Wang, J., Tracey, J.B., 2010. ACS Nano 4, 1913.

Raju, S., Mohandas, E., Raghunathan, V.S., 1996. Scr. Mater. 34, 585.

Rein, G., Mehrer, H., Maier, K., 1978. Phys. Stat. Sol. (A) 45, 253.

Rennhofer, M., Sepiol, B., Sladecek, M., Kmiec, D., Stankov, S., Vogl, G., Koslowski, M., Kozubski, R., Vantomme, A., Meersschaut, J., Rüffer, B., Gupta, A., 2006. Phys. Rev. B 74, 104301.

Resing, H.A., Nachtrieb, N.H., 1961. J. Phys. Chem. Solids 21, 40.

Reuter, K., Scheffler, M., 2006. Phys. Rev. B 73, 045433.

Reynolds, J.E., Averbach, B.L., Cohen, M., 1957. Acta Metall. 5, 29.

Ribbe, J., Baither, D., Schmitz, G., Divinski, S.V., 2009. Phys. Rev. Lett. 102, 165501.

Ribbe, J., 2012. Dissertation, Atomarer Transport in ultrafeinkörnigen Legierungen Univ. of Münster, Germany.

Rice, S.A., 1958. Phys. Rev. 112, 804.

Riihimäki, I., Virtanen, A., Pusa, P., Salamon, M., Mehrer, H., Räisänen, J., 2008. EPL 82, 66005.

Roberts-Austen, W.C., 1896. Philos. Trans. R. Soc. 187, 383.

Rohrer, G.S., Saylor, D.M., el Dasher, B., Adams, B.L., Rollet, A.D., Wynblatt, P., 2004. Z. Metallkd. 95, 197.

Rothman, S.J., Peterson, N.L., 1969. Phys. Stat. Sol. 35, 105.

Rothman, S.J., Peterson, N.L., Robinson, J.T., 1970. Phys. Stat. Sol. 39, 635.

Roussel, J.M., Bellon, P., 2006a. Phys. Rev. B 63, 184114.

Roussel, J.M., Bellon, P., 2006b. Phys. Rev. B 73, 085403.

Rozas, R.E., Horbach, J., 2011. EPL 93, 26006.

Rummel, G., Zumkley, Th., Eggersman, M., Freitag, K., Mehrer, H., 1995a. Z. Metallkd. 86, 122.

Rummel, G., Zumkley, Th., Eggersman, M., Freitag, K., Mehrer, H., 1995b. Z. Metallkd. 86, 131.

Rupp, W., Ermert, U., Sizmann, R., 1969. Phys. Stat. Sol. 33, 509.

Rüsing, J., Herzig, C., 1996. Intermetallics 4, 647.

Rutherford, E., 1911. Philos. Mag. 21, 669.

Saito, S., Beck, P.A., 1960. Trans. Metall. Soc. AIME 218, 670.

Sakintuna, B., Lamari-Darkrim, F., Hirscher, M., 2007. Int. J. Hydrogen Energ. 32, 1121.

Salamon, M., Mehrer, H., 1999. Philos. Mag. A 79, 2137.

Salamon, M., Mehrer, H., 2003. Diff. Def. For. 216–217, 161.

Salamon, M., Strohm, A., Voss, T., Laitinen, P., Rihiihimäki, I., Divinski, S., Frank, W., Räisänen, J., Mehrer, H., 2004. Philos. Mag. A 84, 737.

Salamon, M., Mehrer, H., 2005a. Z. Metallkd 96, 4.

Salamon, M., Mehrer, H., 2005b. Z. Metallkd 96, 833.

Sandberg, N., Grimwall, G., 2001. Phys. Rev. B 63, 184109.

Sandberg, N., Holmestad, R., 2006. Phys. Rev. B 73, 014108.

Sauer, F., Freise, V., 1962. Z. Elektrochem. 66, 353.

Sauthoff, G., 1995. Intermetallics. VCH, Weinheim.

Sauvage, X., Wilde, G., Divinski, S.V., Horita, Z., Valiev, R.Z., 2012. Mat. Sci. Eng. A 540, 1.

Schmidt, F.A., Warner, J.C., 1972. J. Less-common Met. 26, 325.

Schmidt, H., Frohberg, G., Wever, H., 1992a. Acta Metall. Mater. 40, 3105.

Schmidt, H., Frohberg, G., Miekeley, W., Wever, H., 1992b. Phys. Stat. Sol. B 171, 29.

Schmidt, C., Ernstm, F., Finnis, N.W., Vitek, V., 1995. Phys. Rev. Lett. 75, 2160.

Schmitz, G., Kesten, P., Kirchheim, R., 1998. Phys. Rev. B 58, 7333.

Schmitz, G., Ene, C.B., Lang, C., Vovk, V., 2006. Adv. Sci. Technol. 46, 126.

Schmidt, H., Gupta, M., Gutberlet, T., Stahn, J., Burns, M., 2008. Acta Mater. 56, 464.

Schmitz, G., Ene, C.-B., Nowak, C., 2009. Acta Materialia 57, 2673.

Schmitz, G., Baither, D., Kasprzak, M., Kim, T.H., Kruse, B., 2010. Scripta Mater. 63, 484–487.

Schoijet, M., Girifalco, L.A., 1968. J. Phys. Chem. Solids 29, 911.

Schoeck, G., 1980. Thermodynamics and thermal activation of dislocations. In: Nabarro, F.R.N. (Ed.), Dislocations in Solids, vol. 3. North-Holland, Amsterdam.

Schober, H.R., Gaukel, C., Oligschleger, C., 1997. Def. Diff. Forum 143–147, 723.

Schulz, H., 1991. Mat. Sci. Eng. A 141, 149.

Seeger, A., Schumacher, D., Schilling, W., Diehl, J., 1970. Vacancies and Interstitials in Metals. Noorth Holland, Amsterdam.
Seitz, W., 1953. Acta Metall. 1, 355.
Serruys, Y., Brebec, G., 1982. Philos. Mag. A 46, 661.
Setna, R.P., Cerezo, A., Hyde, J.M., Smith, G.D.W., 1994. Appl. Surf. Sci. 76–77, 203.
Shirai, Y., Yamaguchi, M., 1992. Mat. Sci. Eng. A. 152, 173.
Shi, Y., Frohberg, G., Wever, H., 1995. Phys. Stat. Sol. A 152, 361.
Shimizu, Y., Kawamura, Y., Uematsu, M., Tomita, M., Kinno, T., Okada, N., Kato, M., Uchida, H., Takahashi, M., Ito, H., Ishikawa, H., Ohji, Y., Takamizawa, H., Nagai, Y., Itoh, K.M., 2011. J. Appl. Phys. 109, 036102.
Sieverts, A., 1914. Z. Phys. Chem. 88, 451.
Siegbahn, K., Edvarson, K., 1956. Nucl. Phys. 1, 137.
Siegel, R.W., 1978. J. Nucl. Mater. 69–70, 117.
Silvestri, H.H., Bract, H., Hansen, J.L., Larsen, A.N., Haller, E.E., 2006. Semicond. Sci. Technol. 21, 758.
Simonovic, D., Sluiter, M.H.F., 2009. Phys. Rev. B 79, 054304.
Singwi, K.S., Sjölander, S., 1960. Phys. Rev. 120, 1093.
Sinning, H.R., 2000. Phys. Rev. Lett. 85, 3201.
Singh, S., Basu, S., Bhatt, P., Poswal, A.K., 2009. Phys. Rev. B 79, 195435.
Slater, N.B., 1959. Theory of Unimolecular Reactions. Cornell Univ. Press, Ithaca N.Y.
Smigelskas, A.D., Kirkendall, E.O., 1947. Trans. AIME 171, 130.
Smith, R.P., 1964. Trans. AIME 230, 476.
Smoluchowski, M., 1906. Ann. Phys. 326, 756.
Smoluchowski, R., 1952. Phys. Rev. 87, 482.
Snoek, J.L., 1941. Physica 8, 711.
Soisson, F., Martin, G., 2000. Phys. Rev. B 62, 203.
Sokolov, I., Klafter, J., Blumen, A., 2002. Phys. Today 55, 48.
Somenov, V.A., Shilstein, S.S., 1979. Prog. Mater. Sci. 24, 267.
Sommer, J., Herzig, C., 1992. J. Appl. Phys. 72, 2758.
Song, R.H., Pyun, S., 1990. J. Electrochem. Soc. 137, 1051.
Sørensen, M.R., Mishin, Y., Voter, A.F., 2000. Phys. Rev. B 62, 3658.
Stejskal, E.O., Tanner, J.E., 1965. J. Chem. Phys. 42, 288.
Stearns, M.B., 1988. Phys. Rev. B 38, 8109.
Stephenson, G.B., 1988. Acta Metall. 36, 2663.
Stender, P., Ene, C.B., Galinski, H., Schmitz, G., 2008. Intern. J. Mat. Res. 99, 480.
Stender, P., Balogh, Z., Schmitz, G., 2011. Phys. Rev. B 83, 121407(R).
Stoneham, A.M., 1978. J. Nucl. Mater. 69, 109.
Stoneham, A.M., 1979. Hyperfine Interact. 6, 211.
Stolwijk, N.A., van Gend, M., Bakker, H., 1980. Philos. Mag. A 42, 783.
Stolwijk, N.A., 1981. Phys. Stat. Sol. B 105, 223.
Surholt, T., Mishin, Y.M., Herzig, C., 1994. Phys. Rev. B 50, 3577.
Suzuoka, T., 1961. Trans. Jap. Inst. Metal. 2, 25.
Suzuoka, T., 1964. J. Phys. Soc. Jpn. 19, 839.
Suzuki, A., Mishin, Y., 2005. J. Mater. Sci. 40, 3155.
Svoboda, J., Fischer, F.D., 2011. Acta Mater. 59, 1212.
Szabo, I.A., Beke, D.L., Kedves, F.J., 1990. Philos. Mag. A 62, 227.
Takano, N., Murakami, Y., Terasaki, F., 1995. Scr. Metall. Mater. 32, 401.
Takagiwa, Y., Kanazawa, I., Sato, K., Murakami, H., Kobayashi, Y., Tamura, R., Takeuchi, S., 2006. Phys. Rev. B 73, 092202.
Tanuma, S., Powell, C.J., Penn, D.R., 1988. Surf. Interf. Anal. 11, 577.
Tarczon, J.C., Halperin, W.P., Chen, S.C., Brittain, J.O., 1988. Mat. Sci. Eng. A 101, 99.
Teichler, H., 1997. Def. Diff. For. 143–147, 717.
Thiess, H., Sepiol, B., Rüffer, R., Vogl, G., 2001. Diff. Def. For. 194–199, 363.
Thiess, H., Baron, A., Ishikawa, D., Ishikawa, T., Miwa, D., Sepiol, B., Tsutsui, S., 2003. Phys. Rev. B 67, 184302.
Thompson, V., Hintermann, H.E., Chollet, L., 1979. Surf. Technol. 8, 421.
Tokei, Zs., Bernardini, J., Gas, P., Beke, D.L., 1997. Acta Mater. 45, 541.
Tomizuka, C.T., Slifkin, L., 1954. Phys. Rev. 96, 610.
Torrey, H.C., 1956. Phys. Rev. 104, 563.
Torchia, D.A., Szabo, A., 1982. J. Magn. Reson. 49, 107.

Trampenau, J., Petry, W., Herzig, C., 1993. Phys. Rev. B 47, 3132.
Tripathi, S., Sharma, A., Shripathi, T., 2009. Appl. Surf. Sci. 256, 489.
Uesugi, T., Kohyama, M., Higashi, K., 2003. Phys. Rev. B 68, 184103.
Ukanwa, A.O., April 30–May 1, 1974. M558 radioactive tracer diffusion. In: Proceedings of the Third Space Processing Symposium-Skylab Results, vol. I. NASA Marshall Space Flight Center, Alabama, pp. 425–456.
Valiev, R.Z., Islamgaliev, R.K., Alexandrov, I.V., 2000. Prog. Sci. Mater. 45, 103.
Valiev, R., 2004. Nat. Mater. 3, 511.
Varotsos, P.A., 1978. J. Phys. F 8, 1373.
Varotsos, P.A., Alexopoulos, K.D., 1986. Thermodynamics of Point Defects and Their Relation with Bulk Properties. North Holland.
van Dal, M.J.H., Pleumeekers, M.C.L.P., Kodentsov, A.A., van Loo, F.J.J., 2000. Acta Mater. 48, 385.
van Dal, M.J.H., Gusak, A.M., Cserháti, C., Kodentsov, A.A., van Loo, F.J.J., 2001a. Phys. Rev. Lett. 86, 3352.
van Dal, M.J.H., Huibers, D.G.G.M., Kodentsov, A.A., van Loo, F.J.J., 2001b. Intermetallics 9, 409.
Viancio, P.T., Erickson, K.L., Hopkins, P.T., 1994. J. Electron. Mater. 22, 721.
Vickridge, I.C., 1988. Nucl. Instrum. Methods Phys. Res. Sect. B 34, 470.
van Liempt, J.A.M., 1935. Z. Physik 96, 534.
Vineyard, G.H., 1957. J. Phys. Chem. Solids 3, 121.
van Loo, F.J., Bastin, G.F., Rieck, G.D., 1979. Sci. Sintering 11, 9.
van Loo, F.J.J., Pieraggi, B., Rapp, R.A., 1990. Acta Metall. Mater. 38, 1769.
Vogl, G., 1996. Physica B 226, 135.
Voter, A.F., 1986. Phys. Rev. B 34, 6819.
van Winkel, A., Lemmens, M.P.H., Weeber, A.W., Bakker, H., 1984. J. Less-common Met. 99, 257.
Wagner, C., 1969. Acta Metall. 17, 99.
Wang, Y., Cai, L., Xia, Y., 2005. Adv. Mater. 17, 473.
Wan, H., Shen, Y., Yin, X., Chen, Y., Sun, J., 2012. Acta Mater. 60, 2539.
Warburton, W.K., Turnbull, D., 1975a. Thin Solid Films 25, 71.
Warburton, W.K., Turnbull, D., 1975b. In: Nowick, A.S., Burton, J.J. (Eds.), Diffusion in Solids, Recent Development. Academic press, p. 171.
Wasilewski, R.J., 1968. J. Phys. Chem. Solids 29, 39.
Wazzan, A.R., 1965. J. Appl. Phys. 36, 3596.
Weiler, D., Maier, K., Mehrer, H., 1982. Dimeta 82. In: Proc. Int. Conf. on Diffusion in Metals and Alloys, p. 342.
Weller, M., 2006. Mat. Sci. Eng. A. 442, 21.
Wert, C.A., 1950. Phys. Rev. 79, 601.
West, R.N., 1973. Adv. Phys. 22, 263.
Whipple, R.T.P., 1954. Philos. Mag. 45, 1225.
Willaime, F., 2001. Adv. Eng. Mater. A 3, 283.
Williams, P.L., Mishin, Y., 2009. Acta Mater. 57, 3786.
Wilde, G., Ribbe, J., Reglitz, G., Wegner, M., Rösner, H., Estrin, Y., Zehetbauer, M., Setman, D., Divinski, S., 2010. Adv. Eng. Mater. 12, 758.
Wipf, H., 1997. Hydrogen in Metals III—Properties and Applications. In: Topics in Applied Physics, vol. 79. Springer Verlag.
Woodcock, L.V., 1971. Chem. Phys. Lett. 10, 257.
Wrigley, J.D., Ehrlich, G., 1980. Phys. Rev. Lett. 44, 661.
Würschum, R., Grupp, C., Schaefer, H.E., 1995. Phys. Rev. Lett. 75, 97.
Würschum, R., Kümmerle, E.A., Badura-Gergen, K., Seeger, A., Herzig, C., Schaefer, H.E., 1996. J. Appl. Phys. 80, 724.
Xu, L., Henkelman, G., 2008. J. Phys. Chem. 129, 114104.
Yamakawa, K., Ege, M., Luderscher, B., Hirscher, M., Kronmüller, H., 2001. J. Alloys Compd. 321, 17.
Yamakawa, K., Ege, M., Hirscher, M., Luderscher, B., Kronmüller, H., 2005. J. Alloys Compd. 393, 5.
Yin, Y., Rioux, R.M., Erdonmez, C.K., Hughes, S., Somorjai, G.A., Alivisatos, A.P., 2004. Science 304, 711.
Yoon, D.N., 1989. Annu. Rev. Mater. Sci. 19, 43.
Yoshida, H., Yokoyama, K., Shibata, N., Ikuhara, Y., Sakuma, T., 2004. Acta Mater. 52, 2349.
Zeldovich, J., 1942. J. Exp. Theor. Phys. 12, 525.
Zener, C., 1943. Trans. AIME 152, 122.
Zener, C., 1951. J. Appl. Phys. 22, 372.
Zhang, X.Y., Sprengel, W., Staab, T.E.M., Inui, H., Schaefer, H.E., 2004. Phys. Rev. Lett. 92, 155502.
Zhao, B., Verhasselt, J.C., Shvindlerman, L.S., Gottstein, G., 2010. Acta Mater. 58, 5646.
Zhang, L.J., Du, Y., Chen, Q., Steinbach, I., Huang, B.Y., 2011. Int. J. Mat. Res. 101, 1461.

Biography

Dr. Zoltán Balogh studied Physics at the University of Debrecen, Hungary. In this period he took part EU supported research cooperation with the University of Ulm, Germany and the Paul Cézanne University in Marseille, France. He completed his Ph.D. thesis entitled "Nanoscale effects caused by diffusion asymmetry" in 2009 under the guidance of Prof. Dr. Dezső Beke and Dr. Zoltán Erdélyi. Shortly after he has joined the research Group of Prof. Guido Schmitz at University of Münster as a post doc fellow. Recently this group has relocated to the University of Stuttgart. In his career path Zoltán Balogh specialized in investigating diffusion and solid state reaction at very short length and time scale using various high resolution techniques.

Prof. Dr. Guido Schmitz studied Physics and Theology at the University of Freiburg, Germany. Continuing his studies at the University of Göttingen in 1988, Peter Haasen became his mentor, encouraging him to focus his further career on the science of metals and materials. In 2002, after completing research assignments at UC Los Angeles, USA and earning his habilitation in Göttingen, he was called upon as Professor of Materials Physics in Münster, Germany. Prof. Schmitz received the Werner Köster award by the Germany Society for Material Science (DGM) and holds a honorary doctorate of the National University of Cherkasy, Ukraine. Most recently, he followed a call to the University of Stuttgart to head the Chair of Material Physics. In his career path of engaged research and teaching, Guido Schmitz has developed a particular expertise in atomic transport and nanoanalysis of solid state reactions via atom probe tomography.

6   Defects in Metals

NUMAKURA Hiroshi, Department of Materials Science, Osaka Prefecture University, Naka-ku, Sakai, Japan

6.1   Introduction

Real crystals contain a variety of imperfections, or defects, by nature. Some of them are of atomic dimensions, and are called atomic or point defects, while others are extended in one, two, or three dimensions, producing linear, planar, or volume defects. Many of the physical properties of crystalline solids are strongly influenced by those imperfections, thus being described as "microstructure-sensitive". The well-known examples are mechanical properties of metals: strength, ductility, and toughness. Also, the kinetic processes, i.e. diffusion and phase transformations, which control the microstructure formation, are governed by the density and mobility of the defects present in the material. Point defects mediate movements of atoms, dislocations provide ductility, grain boundaries often make metals brittle but also strengthen them by impeding dislocation motion, for example. Quantitative knowledge on the properties and behavior of those defects is therefore important.

In this chapter, we first give an overview of defects observed in common metals and alloys, from vacancies, interstitial atoms, and impurity (or solute) atoms, to dislocations, stacking faults, and various interfaces and boundaries. Detailed discussion on dislocations and stacking faults are given in Chapter 16, Volume II in the context of plastic deformation, and the complex extended defects, which are elements of "microstructure", are accounted for in other chapters dedicated to the topic. We therefore focus on point defects in the following part of this chapter. After explaining standard experimental techniques, we discuss the structure and thermodynamic properties of elementary point defects (vacancies and self-interstitial atoms), and the statistical thermodynamics of the formation of point defects and their interaction in metals and alloys.

6.2   Overview

6.2.1   Point Defects

Figure 1 is a schematic illustration of common point defects in crystals. Atomic vacancies (a) and self-interstitial atoms (b, c) are elementary intrinsic defects. A pair of a vacancy and a self-interstitial atom is called a Frenkel defect. Foreign atoms (e, f) are extrinsic defects, and are referred to as impurity or solute atoms, depending on the context. A foreign atom may be present in a space between the host atoms, or in a place where a host atom is normally present. The former is an interstitial solute atom (e), and the latter is a substitutional solute atom (f). Complex defects may be formed by association of these defects, such as pairs of vacancies (divacancies, d), pairs of a vacancy and a solute atom (g, h), triplets of

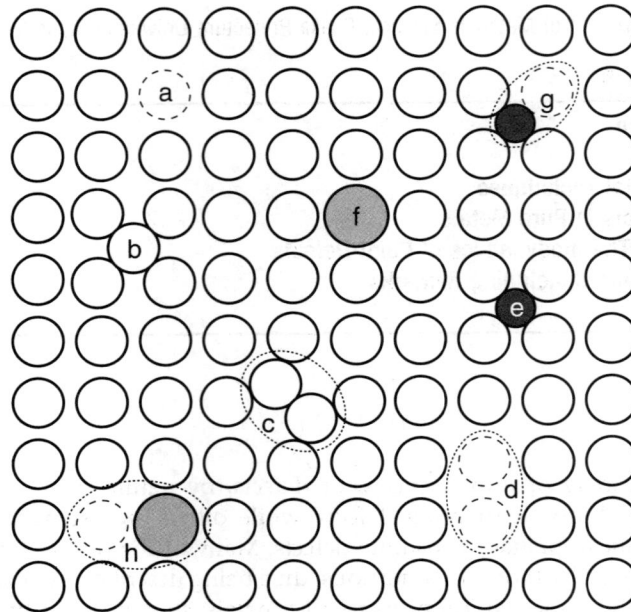

Figure 1   Point defects in an elemental crystal. a: an atomic vacancy, b: a self-interstitial atom at an interstitial position, c: a self-interstitial atom in the split (or "dumb-bell") configuration, d: a pair of vacancies (divacancy), e: an interstitial foreign atom, f: a substitutional solute atom, g: a pair of a vacancy and an interstitial solute atom, h: a pair of a vacancy and a substitutional solute atom.

vacancies or larger clusters, and so on, because the association most often lowers the free energy of the defect-containing crystal.

6.2.1.1   *Vacancies and Self-interstitial Atoms*

6.2.1.1.1   Thermal Formation

Intrinsic atomic defects are spontaneously produced at high temperatures and exist in thermal equilibrium owing to the configurational entropy associated with their presence. For example, when n vacancies are introduced into a crystal made up of N atoms, the number of possible ways to arrange them over $N + n$ sites is

$$W = \frac{(N+n)!}{N!\,n!}.$$ (1)

This gives the configurational entropy

$$S_c = k \ln W \approx k[(N+n) \ln (N+n) - N \ln N - n \ln n],$$ (2)

where k is the Boltzmann constant. Writing the increase in the free energy of the crystal by introduction of one vacancy as Δg_f and assuming that $n \ll N$ (so that interactions between vacancies can be ignored), the total change in the free energy is given by

$$\Delta G = n\Delta g_f - TS_c,$$ (3)

where T is temperature. The competition of the two terms results in the equilibrium concentration

$$c = \frac{n}{N+n} = \exp\left(-\frac{\Delta g_f}{kT}\right), \tag{4}$$

which is obtained from the condition $\partial \Delta G/\partial n = 0$.

The free energy of formation Δg_f is defined as the change in the free energy of the solid by introduction of a single defect. It can be written as

$$\Delta g_f = \Delta h_f - T\Delta s_f \tag{5}$$

with the enthalpy and entropy of formation, Δh_f and Δs_f, and the former is further divided to the change in the internal energy and the work associated with the formation,

$$\Delta h_f = \Delta e_f + P\Delta v_f. \tag{6}$$

Here, Δe_f is the formation energy, Δv_f is the formation volume, and P is the external pressure. These parameters of formation are characteristic of the defect species and determine their thermodynamic behavior.

The formation energy Δe_f of a vacancy in metals is of the order of 1 eV,[1] varying from 0.7 eV in aluminum to 3.6 eV in tungsten. Since it is due largely to the loss of chemical bonding, the formation energy is high in strongly bonded solids. It is in fact known to be about 1/3 of the cohesive energy of the solid. The formation volume Δv_f of a vacancy is one atomic volume if no relaxation occurs, but in reality it is somewhat smaller, by 10 to 40%. Under normal conditions the $P\Delta v_f$ term is several orders of magnitude smaller than the formation energy. Therefore, Δh_f and Δe_f are practically equal to each other. The vacancy formation enthalpy in pure metals is known to be correlated with the melting temperature of the material, T_m. The relation for face-centered cubic (fcc) metals is displayed in **Figure 2** (a), which is expressed as $\Delta h_f \approx 10\,kT_m$. In the same figure is shown the relation between the cohesive energy and the melting temperature. With another empirical knowledge on the formation entropy, Δs_f is commonly 1 to 2 times k (see below), the vacancy concentration at the melting point can be estimated, from Eqn (4), as $\exp(-10 + (1 \sim 2)) \approx 10^{-4}$. For the vacancy formation energy of fcc metals, there is another correlation: $(\Delta h_f/m)^{1/2}\,h/\Omega^{1/3}$ is proportional to kT_D, where m is the atomic mass, h is the Planck constant, Ω is the atomic volume, and T_D is the Debye temperature (Mukherjee, 1965). This relation is shown in **Figure 2(b)**.

The formation entropy Δs_f is the change in the entropy of the solid by the introduction of a defect, other than the configurational entropy. It is known to be due mostly to atomic vibrations.[2] In the harmonic approximation, the vibrational entropy of the solid consisting of N atoms is, under the condition $h\nu_i \ll kT$ (high-temperature approximation), written as

$$S_{vib} = k \sum_{i=1}^{3N-3} \left(1 - \ln\frac{h\nu_i}{kT}\right), \tag{7}$$

---

[1] 1 eV is equal to 1.60×10^{-19} J and corresponds to 96.5 kJ mol^{-1}. As the energies associated with defects in solids are commonly of this order of magnitude, we use this unit throughout this chapter.
[2] See, e.g., Chapter 4 of Flynn (1972) or Chapter 15 of Girifalco (2000).

(a)

(b)

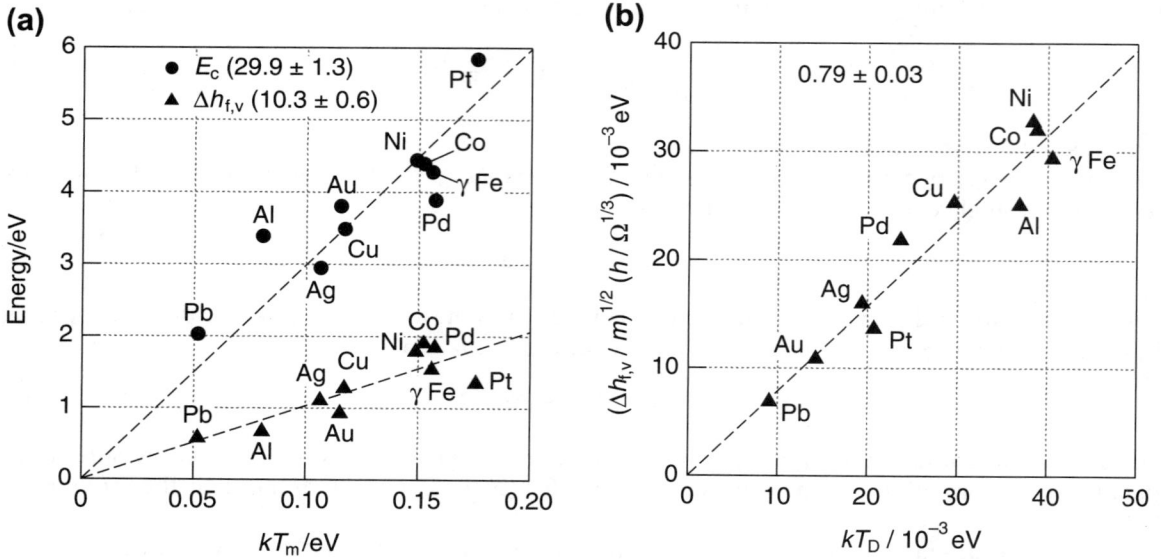

Figure 2 (a) Relations of cohesive energy E_c and vacancy formation enthalpy $\Delta h_{f,v}$ with melting temperature T_m in fcc metals. (b) Relation between the vacancy formation enthalpy and the Debye temperature T_D in fcc metals. The numbers are proportionality constants to kT_m or kT_D.

where ν_i are the frequencies of normal modes of vibration. When the frequencies of the perfect crystal are changed to ν_i' by the introduction of a defect, the formation entropy is given by

$$\Delta s_f = k \sum_{i=1}^{3N-3} \ln \frac{\nu_i}{\nu_i'}. \tag{8}$$

We consider the simplest case where only the force constants associated with the nearest neighbor atoms of the defect are affected. With the Einstein model, where all the normal modes are assumed to have the same vibration frequency, only z of the total $3N - 3$ frequencies are changed from ν to ν', where z is the coordination number. The formation entropy is then simplified to

$$\Delta s_f = zk\ln \frac{\nu}{\nu'}. \tag{9}$$

Around a vacancy, the force constants are smaller than in the regular structure. They lower some of the vibration frequencies and therefore lead to a positive Δs_f. The formation entropy of a vacancy in metals is known to be $(1 \sim 2) \times k$ in both theory and experiment (see Section 6.4).

Self-interstitial atoms in metals were assumed in the early years to be an atom located at an interstitial site (Huntington and Seitz, 1942), but later it turned out that the "split", or "dumb-bell" configuration, originally envisaged as the saddle-point configuration, was the preferred form (Huntington, 1953). Defect "c" in **Figure 1** depicts it: the extra atom displaces one of the adjacent atoms and the two atoms are positioned symmetrically about the regular position. The atomic structures conceived for face-centered cubic (fcc) and body-centered cubic (bcc) metals are illustrated in **Figure 3**.

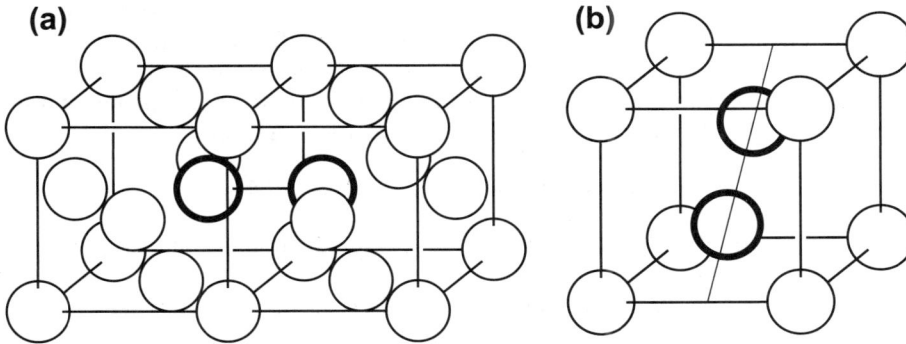

Figure 3 Atomic structure of a self-interstitial atom in the split (dumb-bell) configuration. (a) The $\langle 100 \rangle$ dumb-bell in fcc metals. (b) The $\langle 110 \rangle$ dumb-bell in bcc metals.

The equilibrium concentration is given by a formula similar to Eqn (4), but with a numerical factor, z', stemming from the multiplicity of possible configurations, i.e. equivalent interstitial positions and/or orientations per atomic site,[3]

$$c \approx z' \exp\left(-\frac{\Delta g_f}{kT} \right). \tag{10}$$

In either configuration the formation enthalpy Δh_f of an self-interstitial atom is considerably higher than a vacancy, by a factor of two or three. The formation entropy is not known experimentally but is suggested by theory to be larger, at least a few times, than that of a vacancy, because of the characteristic vibration modes (Ram, 1991a, see also Section 6.4.1.1). The concentration of self-interstitial atoms can therefore be roughly estimated at the square or the cube of that of vacancies, i.e. negligibly low. However, the large formation entropy indicates relative importance at high temperatures, and together with the fact that they cause elastic softening, particularly in the shear modulus (Schilling, 1978), a theory of melting by an increased population of self-interstitial atoms at high temperature has been proposed (Granato, 1994a), against the traditional idea that melting occurs when the vacancy concentration exceeds a critical value (Frenkel, 1937). Nevertheless, the absolute number of the self-interstitial atoms present in thermal equilibrium is so small that their influences on the material properties can usually be ignored.

6.2.1.1.2 Migration (Diffusion)
Point defects move around in the crystal by jumping between neighboring sites. The jump, which is the elementary process of atomic diffusion, is usually thermally activated, and the rate is expressed as

$$w = \nu \exp\left(-\frac{\Delta g_m}{kT} \right), \tag{11}$$

whose theoretical foundation was established by Vineyard (1957). The prefactor ν, called the attempt frequency, is essentially the vibration frequency of atoms, and is in the range from 10^{13} to 10^{14} s^{-1} in

---

[3] For the dumb-bell interstitial atoms in **Figure 3**(a) and (b), z' is 3 and 6, respectively.

common metallic solids. The activation energy for the jump, or the free energy of migration, Δg_m, is the difference in the free energy between the initial state in which the diffusing atom is at the stable position and the activated intermediate state in which the atom is in a "saddle point" configuration in the jump path. Similar to the free energy of formation, it is written by the enthalpy and entropy of migration, Δh_m and Δs_m, and the former by the energy and volume of migration, Δe_m and Δv_m:

$$\Delta g_m = \Delta h_m - T\Delta s_m = \Delta e_m + P\Delta v_m - T\Delta s_m. \qquad (12)$$

Again, Δh_m is practically equal to Δe_m because the $P\Delta v_m$ is much smaller than Δe_m under the ambient pressure.

The migration parameters of vacancies and self-interstitial atoms in metals have been determined from the kinetics of recovery of the point defects produced by quenching or low-temperature irradiation (Section 6.3.2). For a vacancy in metals, the migration enthalpy $\Delta h_{m,v}$ is similar in magnitude to, but somewhat smaller than, the formation enthalpy $\Delta h_{f,v}$. **Figure 4** shows the migration enthalpy in fcc metals and the activation energy of self-diffusion, Q. As atomic diffusion in metallic crystals is mediated by vacancies, the sum of the formation enthalpy and migration enthalpy of a vacancy is expected to be equal to the activation energy of diffusion. This is in fact the case: for fcc metals, $\Delta h_f \approx 10\ kT_m$ (**Figure 2(a)**), $\Delta h_m \approx 7\ kT_m$, and $Q \approx 18\ kT_m$. The last is the relation found by Brown and Ashby (1980).

On the other hand, the migration energy of a self-interstitial atom is commonly of the order of 0.1 eV or even less, being much lower than that of a vacancy. The elementary process of the migration of a $\langle 100 \rangle$ dumb-bell interstitial atom in fcc metals is illustrated in **Figure 5**. With such a low migration energy and the high formation energy, self-interstitial atoms in metals are, therefore, essentially unstable defects. When self-interstitial atoms are introduced accidentally or artificially in the solid, they rapidly disappear by recombining with vacancies or by migrating to surfaces and internal defects. Note, however, that in solids of more open structure, self-interstitial atoms may not be as unfavorable as in

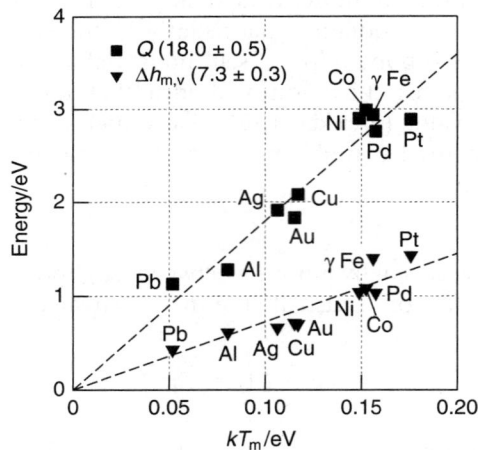

Figure 4 The vacancy migration enthalpy $\Delta h_{m,v}$ and the activation energy of self-diffusion Q in fcc metals. The values of $\Delta h_{m,v}$ for γ iron and cobalt are the differences between Q and $\Delta h_{f,v}$. The numbers to the legends are proportionality constants to kT_m (T_m: melting temperature).

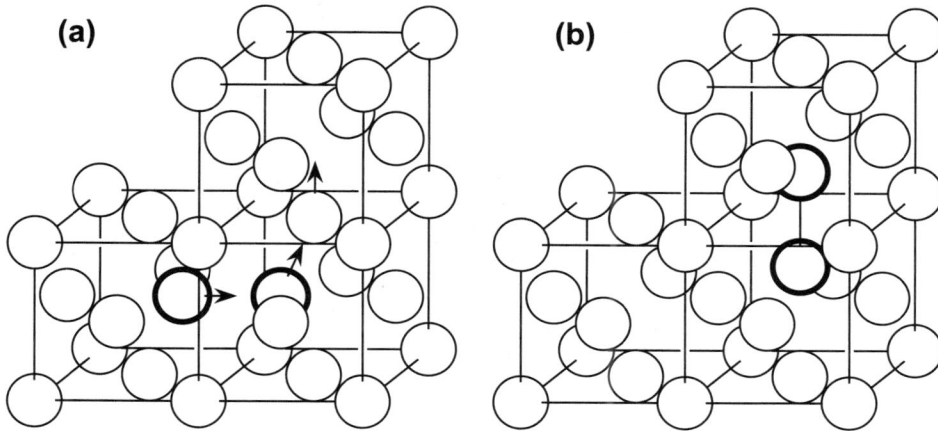

Figure 5   Motion of a ⟨100⟩ dumb-bell interstitial atom in an fcc crystal.

the densely-packed metallic crystals; the formation energy may be of similar magnitude to, or even lower than, that of a vacancy. In crystalline silicon, of the diamond cubic structure, for example, self-interstitial atoms are known to be present in an appreciable number at high temperatures and contribute to self-diffusion to a comparable extent with vacancies (Mehrer, 2007, Chapter 23).

6.2.1.1.3 Non-equilibrium Production

Vacancies and self-interstitial atoms of concentrations far above the equilibrium levels can be produced by several methods: rapid quenching from high temperatures, mechanical working, and irradiation with energetic particles. In the first case, the defects produced at a high temperature may be retained at the low temperature, where their migration is very slow. The defects are then "frozen-in" and remain supersaturated. This method is effective for vacancies, and was employed for basic studies of their properties. However, application to self-interstitial atoms in metals is virtually impossible, since their equilibrium concentration is too low and, moreover, their mobility is too high to retain them. In the second case, both vacancies and self-interstitial atoms are created by interaction between dislocations (see Chapter 16, Volume II). For quantitative studies, however, this method is not convenient because controlling the amount of the defects is difficult, and other defects, which are inevitably produced together, complicate the experiment.

The third case, high energy irradiation, is an effective means to introduce Frenkel defects. When a solid material is irradiated with electrons, ions, or neutrons, collisions of incident particle with atoms in the solid occur. The target atom may be displaced out of the regular site, leaving a vacancy behind, if the energy transferred from the incident particle is sufficiently high. The critical value for the "knock-on" event to take place is a few ten electron-volts in most metals. **Figure 6** shows the experimental data of the displacement threshold energy, E_d, of common metals (Vajda, 1977; Jung, 1991). The maximum energy transferred to the target atom of mass M on the elastic collision of an incident particle of mass m with kinetic energy E is

$$E_{p,max} = 4 \frac{mM}{(m+M)^2} E \tag{13}$$

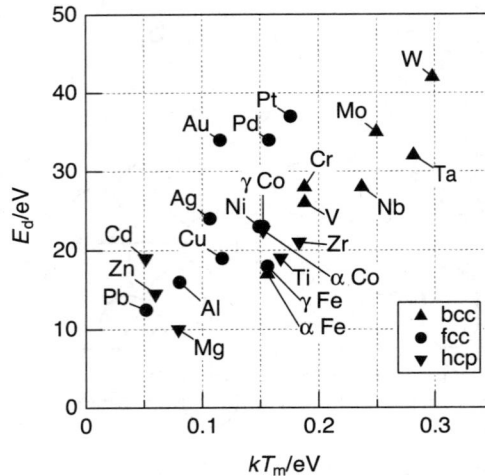

Figure 6 Displacement threshold energy of pure metals plotted against kT_m, where T_m is the melting temperature. Data are from Vajda (1977) and Jung (1991).

(see, e.g., Vajda 1977). If the incident particle is much lighter than the target atom, i.e. $m \ll M$, which is the case for electron or neutron irradiation, this formula is simplified to $E_{p,max} \approx 4(m/M)E$. **Figure 7** shows the maximum energy transfer from electrons[4] and neutrons calculated from Eqn (13). The horizontal line indicates the typical value of the displacement threshold energy, 25 eV. Neutrons of 1 keV cause displacements of atoms in most metals, while for electrons, energies of several hundred keV are necessary for a knock-on to take place. With any particle, much higher energy is required to displace atoms in heavier target materials, i.e. of large M, because of the scattering law. In addition, the displacement threshold energy is generally high in such heavy, strongly bonded metals as Mo and W, which have high melting temperatures, as shown in **Figure 6**.

When the kinetic energy of the displaced atom, called a primary knock-on atom (PKA), is high, it may act as an incident particle and may cause further knock-ons. A chain of multiple atomic displacements triggered by a high energy PKA is called a displacement cascade. **Figure 8** illustrates displacement cascades, which were found in computer simulation (Yoshida 1961). How many defects are produced by a PKA depends on its kinetic energy. According to the simple model for displacement of atoms by Kinchin and Pease (1955), above E_d the average number of displacements is proportional to the energy of PKA. Taking nickel as an example, whose displacement threshold is 23 eV, with 1 MeV electrons the maximum energy transfer is about 70 eV but the average kinetic energy of a PKA is of similar magnitude to E_d (in this particular case it is about 4 eV). Therefore, the defects introduced by electrons of this energy are mostly closely spaced self-interstitial atoms and vacancies. This is the technique that has widely been used in experimental studies of Frenkel defects in metals. On the other hand, fast neutrons of the same energy may transfer up to 66 keV to a target atom in the same material. The average PKA energy is about half of

---

[4] For high-energy electrons, the incident energy is to be multiplied by a factor due to the relativistic effect, $[1 + E/(2mc^2)]$, where c is the speed of light.

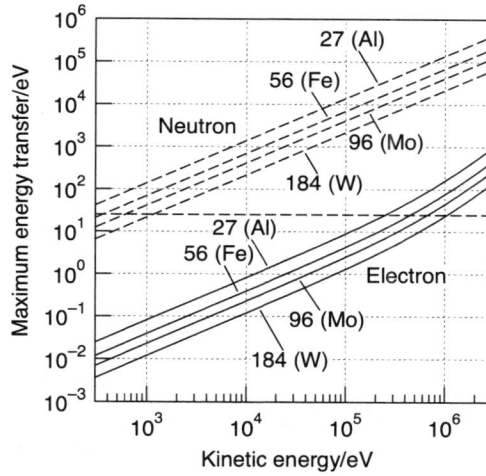

Figure 7 The maximum energy transfer from an incident electron (solid curves) or neutron (dashed curves) to a target atom of mass number *M* (indicated by the tag) by an elastic collision.

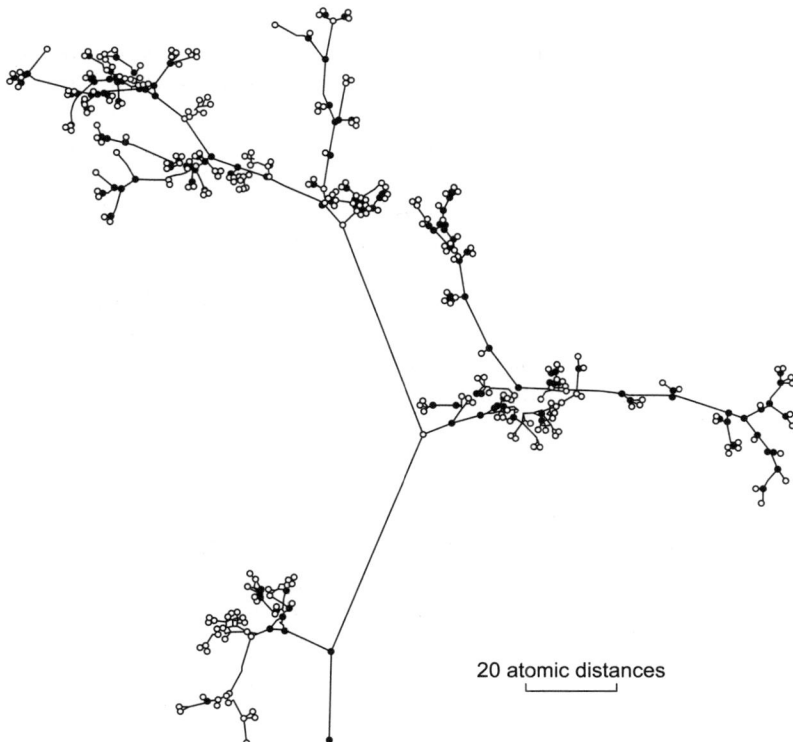

Figure 8 Displacement cascades in germanium caused by a PKA of 10 keV observed in computer simulation (Yoshida 1961). Open and filled circles denote self-interstitial atoms and vacancies, respectively.

the maximum energy transfer,[5] and the number of displacements expected for this case from the Kinchin–Pease model is over 700.

Irradiation with heavy particles, in particular of very high energy, produces significant disorder. When their concentrations are not too high, those nonequilibrium defects decrease their numbers to the standard levels at the temperature, by migrating to various sinks (dislocations, internal boundaries, external surfaces, etc), or by recombining with their counterparts. Defects of the same species often agglomerate, however, to form pairs, triplets, or larger clusters of various shapes and sizes, and those clusters become new sinks for the remaining defects. The dynamics of the production of defects, and the mechanism and kinetics of their redistribution and annihilation form the basis for understanding radiation damage of metals.

6.2.1.2 Foreign Atoms

Foreign species of atoms are sometimes added as solute atoms expecting a beneficial role, but in other cases they enter the solid as impurities. The configurational entropy associated with the introduction of impurity atoms reduces the free energy as in the case of intrinsic defects by the same $-TS_c$ contribution, where S_c may alternatively be called the entropy of mixing. While formation of an intrinsic point defect is always an endothermic process, dissolution of the foreign atoms is either endothermic or exothermic, depending on the combination of the guest and host atom species.[6] The former is characterized by a positive heat of solution, Δh_s, and its equilibrium concentration (the solubility) increases with temperature, and the reverse is true for the case in which Δh_s is negative. The solubility is not determined solely by temperature but depends also on the thermodynamic activity of the impurity species (the vapor pressure, in practice) in the environment. This is the situation of, for example, hydrogen or carbon in iron or nickel. On the other hand, the latter is the case where foreign atoms lower the enthalpy when they dissolve into the material, usually forming stable chemical bonds with the host atoms. Examples of this case are hydrogen or oxygen in titanium or vanadium. Here, introduction of the foreign atoms is favored by both the enthalpy and entropy effects, so that they are absorbed spontaneously from the environment. This is the reason why metals that are chemically active to light elements are difficult to purify; elimination of such elements as hydrogen, carbon, nitrogen, and oxygen, which are abundantly present in the atmosphere, is a process against the free energy.

The atoms of those light elements are small and usually occupy interstitial positions when dissolved in metals, forming an interstitial solution. The octahedral and tetrahedral interstitial sites in the body-centered cubic, face-centered cubic and hexagonal close-packed structures are illustrated in **Figure 9**. Heavier foreign atoms generally form substitutional solutions. The case of boron is marginal; it is not yet well known if B atoms occupy interstitial sites or the same sites as the host atoms in some metals in which the element plays a useful role, i.e. iron, nickel, and their alloys (Fors and Wahnström, 2008; McLellan, 1995).

6.2.1.3 Interaction between Point Defects

When two vacancies come next to each other, as "d" in **Figure 1**, the internal energy is lowered, since the number of missing bonds is decreased from $2z$ (for two isolated vacancies) to $2z - 1$, where z is the coordination number. This energy gain is the primary origin of stabilization of the pair, i.e. a divacancy.

---

[5] The elastic scattering of a neutron by an atom occurs nearly isotropically in the center-of-mass system of the two-particle collision mechanics, and therefore the energy spectrum of a PKA is almost uniform from E_d to $E_{p,max}$, giving the average PKA energy $\bar{E}_p \approx (1/2)E_{p,max}$. In contrast, forward scattering is predominant in the elastic scattering of a charged particle, which makes a PKA energy spectrum much enhanced at the low energy end, leading to a low average energy. See Billington and Crawford (1961), Corbett (1966), or Thompson (1969) for details.
[6] Practical data are found in the compilation by Fromm and Gebhardt (1976).

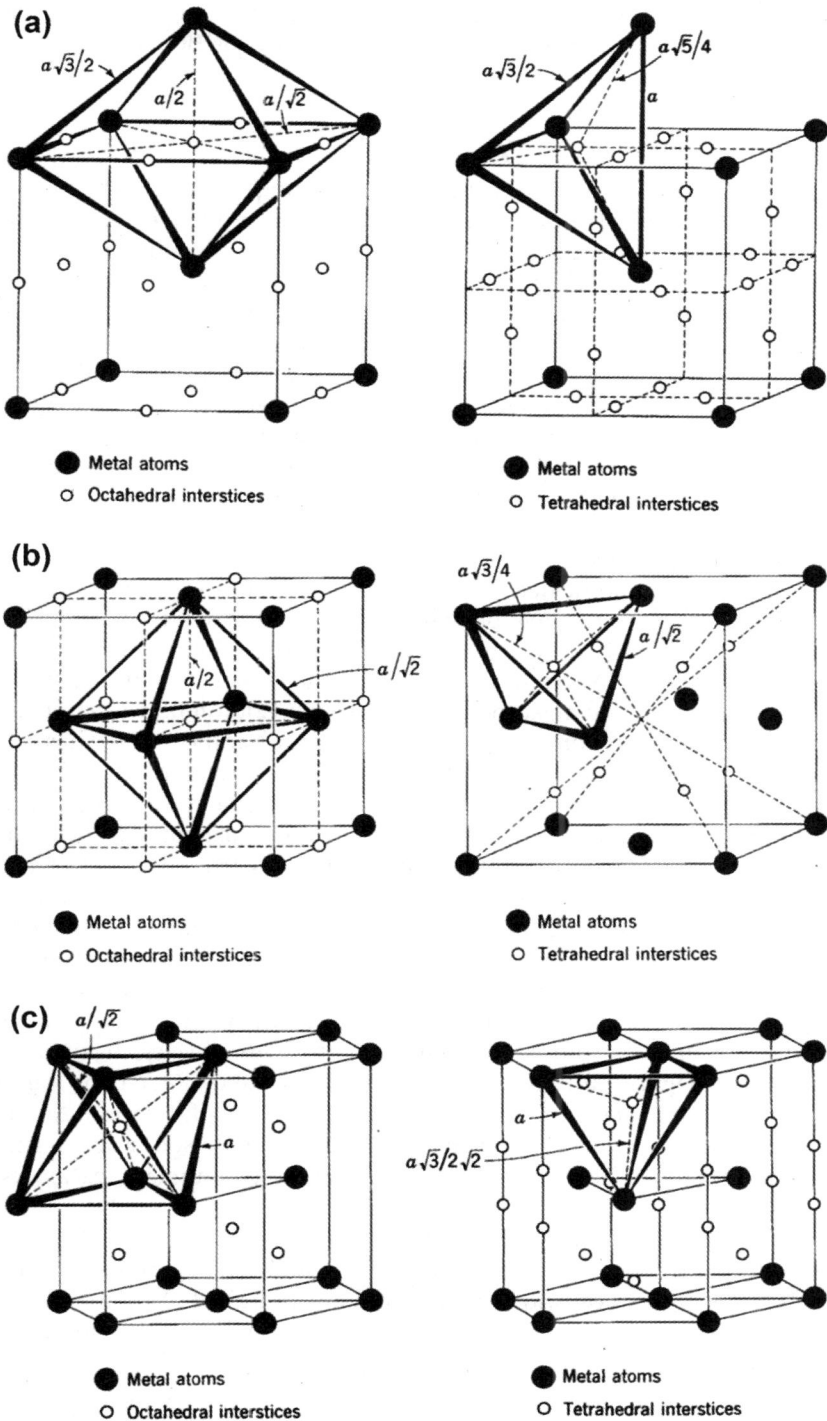

Figure 9 Octahedral (left) and tetrahedral (right) interstitial sites in (a) bcc, (b) fcc, and (c) hcp structures. After Barrett and Massalski (1980).

Because of the stabilization energy, or "binding energy", the concentration of divacancies, c_{2v}, is higher than that expected for random association of single vacancies by the Boltzmann factor, i.e.

$$c_{2v} = \frac{z}{2} c_v^2 \exp\left(\frac{B}{kT}\right), \qquad (14)$$

where c_v is the concentration of single vacancies. The binding energy B is a small fraction of the formation energy, and is typically 0.2 eV. Formation of divacancies thus increases the total number of vacant sites, and the fraction of paired vacancies increases with temperature. If c_v is 10^{-4} in Eqn (14) and $B/(kT) = 2.5$ (for such a combination as $B = 0.2$ eV and $T = 930$ K, for example), c_{2v} is of the order of 10^{-6}. Since divacancies in metals are usually more mobile than single vacancies, their formation more or less enhances atomic diffusion, in particular at high temperatures. On the other hand, when vacancies formed at a high temperature are brought to a low temperature by rapid quenching and are left to migrate under the condition that their number is maintained, the fraction of the pairs increases at the expense of the singles as their redistribution proceeds, since the factor $\exp[B/(kT)]$ favors pairs at low temperatures. Supersaturated vacancies, as well as self-interstitial atoms, tend to agglomerate to form clusters by the same principle: association of the point defects almost always reduces the internal energy, so that clustering and growth occur spontaneously. This can be the primary mode of elimination of surplus defects in crystals in which not many sinks are available.

Foreign atoms and intrinsic point defects often form pairs or, more generally, "complexes". For example, an interstitial C atom is known to be bound to a vacancy in bcc iron, with a binding energy of about 0.5 eV (Becquart et al., 2007). Interaction of substitutional solute atoms with vacancies has also been studied (Hautojärvi, 1987; Wolverton, 2007). In ionic crystals, mutual attraction or repulsion between an aliovalent impurity and a vacancy occurs through Coulomb interaction: a divalent cation impurity in an alkali halide crystal, for example, attracts a cation vacancy, which has an effective charge $-e$, while it repels an anion vacancy, with an effective charge $+e$. Even though charge differences are screened in metallic solids and cannot give rise to such long-range interaction as in ionic crystals, solute atoms in metals may interact with vacancies by an analogous effect at short distances. In addition to the electrostatic interaction, mechanical distortion accompanying each point defect also gives rise to attraction or repulsion between them. Since a vacancy produces a local reduction in volume, interstitial and oversized substitutional solute atoms, which cause dilatation, tend to associate themselves with a vacancy. When a substitutional solute atom interact attractively with a vacancy, it finds more often a vacancy in the immediate vicinity than does a solvent atom. The solute–vacancy interaction is therefore an important factor in their mobility. There is a microscopic model for the relationship between the interaction and the diffusion coefficient of solute species, using five distinct jump frequencies illustrated in **Figure 10** (Lidiard, 1955). Solute–vacancy interaction can affect the equilibrium concentration of vacancies, similarly to the case of divacancy formation. Attractive interaction effectively decreases the vacancy formation energy at the sits around a solute atom, and thus enhances the equilibrium concentration. This is the origin of the super-abundant vacancies in hydrogen-containing metals (Section 6.4.2.3).

The example of a C-solute bound to a vacancy mentioned above is one of the "trapping" of a mobile species to a less mobile one, resulting in reduction of the mobility. Trapping of light foreign atoms dissolved in metals is one of the most widely known and much studied phenomena among a variety of solute–defect and solute–solute interaction.[7] The trapping center can be either an intrinsic point defect

---

[7] For reviews, see, e.g., Wert (1978) and Fukai (2005) for H, Numakura and Koiwa (1996, 1997) and Numakura (2012) for C, N and O.

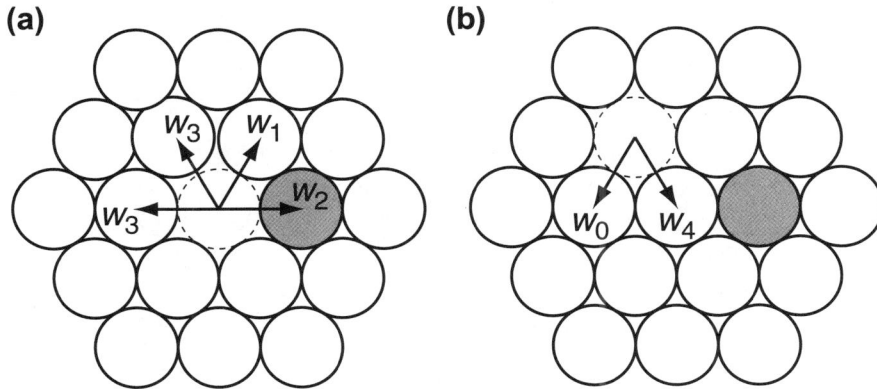

Figure 10 Vacancy jumps in the five-frequency model of solute–vacancy interaction in an fcc crystal. The shaded circle and the dashed circle indicate the solute atom and the vacancy, respectively. w_0 is the frequency of all vacancy–solvent exchange jumps that occur at places away from the solute atom.

or a solute atom, but is not restricted to point defects. C and N atoms are attracted to dislocation cores to form "Cottrell atmospheres" (Cottrell and Bilby, 1949), and the same solutes, as well as H atoms, segregate to grain- and interphase boundaries (Lejček, 2010). When solute atoms are trapped, their diffusion mobility is reduced, and at the same time their solubility may be enhanced because of the attraction. Trapping influences mechanical properties of metals through these effects on the solubility and diffusivity of the solute species, and is therefore an important problem.

6.2.1.4 Point Defects in Alloys

Defect phenomena pertaining to dilute solid-solution alloys can be discussed in terms of the solute–defect and solute–solute interactions outlined above. Thermal formation of intrinsic defects in concentrated solid-solution alloys may be understood as being apparently the same as the case of pure metals. The equilibrium concentration of vacancies, for example, can be expressed in the same form as Eqn (4). However, the formation energy $\Delta g_{f,v}$ is a function of the composition, because the energy required to create a vacancy depends on the environment of the atomic site, i.e. the numbers A and B atoms at the adjacent sites. According to a simple pair interaction model with the mean-field approximation (the Bragg–Williams model), the effective formation energy of a vacancy is given by a fractional average of those in pure A and pure B solids with an additional term stemming from the interaction between A and B components (Section 6.5.2).

In ordered alloys, where two (or more) different species of atoms are arranged on their own sublattice sites, the situation is rather different. The point defects specific to the ordered structure are illustrated in **Figure 11**: vacancies in each sublattice, atoms occupying the sites for the other species, called antisite atoms or antistructure defects, and self-interstitial atoms of each species. A set of vacancies corresponding to the chemical formula, which increases the volume of the crystal by one unit cell, is called a Schottky defect. It consists of a pair of A and B vacancies in a compound AB, or three A vacancies and one B vacancy in a compound A_3B. Antisite atoms are common defects in ordered alloys that exist over a certain range of composition around the stoichiometry. All of these defects are produced at high temperatures by the effect of the configurational entropy. Usually the antisite atoms are formed in much larger numbers than the others. In ionic crystals, the numbers of

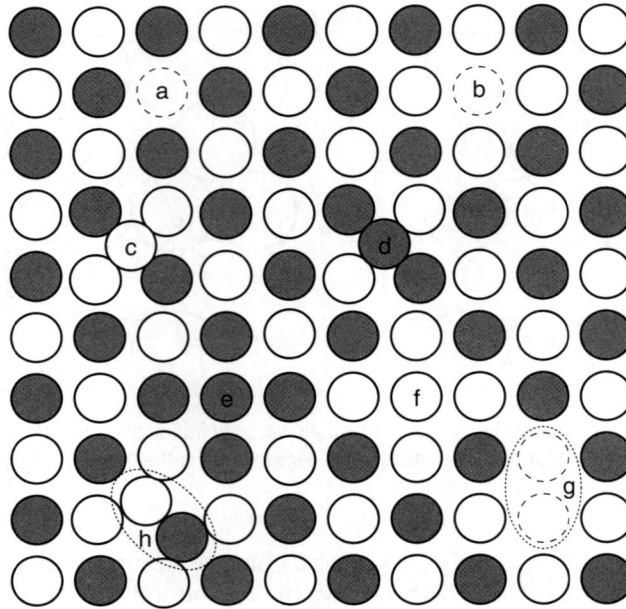

Figure 11 Point defects in an ordered AB alloy, where A atoms (open circles) and B atoms (shaded circles) occupy their own sublattice sites α and β, respectively. a: a vacancy at an α sublattice site, V_α, b: a vacancy at a β sublattice site, V_β, c: a self-interstitial A atom, A_I, d: a self-interstitial B atom, B_I, e: an antistructure B atom, B_α, f: an antistructure A atom, A_β, g: a pair of V_α and V_β (Schottky defect), h: a self-interstitial A atom of a "mixed dumb-bell" configuration.

the defects must be in proper proportions to maintain the charge neutrality, but in metallic alloys the restriction appears not too strong. In any case, as the concentrations of these defect species are important factors in the properties of ordered alloys and intermetallic compounds, quantitative understanding is required.

The concentrations of the antisite atoms in an ordered alloy are related to the degree of order, which is a function of temperature and composition. **Figure 12** shows the experimental observation on β-brass, viz., CuZn alloy of the B2 (CsCl) structure (Norvell and Als-Nielsen, 1970). If no vacancies are considered, the long-range order (LRO) parameter η is defined from the populations of the antisite atoms as follows:

$$\eta = 1 - \left(\frac{N_{B\alpha}}{N_\alpha} + \frac{N_{A\beta}}{N_\beta}\right), \qquad (15)$$

where $N_{B\alpha}$ and $N_{A\beta}$ are the numbers of the antisite atoms B in the α sublattice and A in the β sublattice, and N_α and N_β are the numbers of those sublattice sites, respectively. Each term in the parentheses is the site fraction of the antisite atoms, and is related in turn to the LRO parameter, for the simplest case of an AB compound of the stoichiometric composition, as

$$\frac{N_{B\alpha}}{N_\alpha} = \frac{N_{A\beta}}{N_\beta} = \frac{1}{2}(1 - \eta). \qquad (16)$$

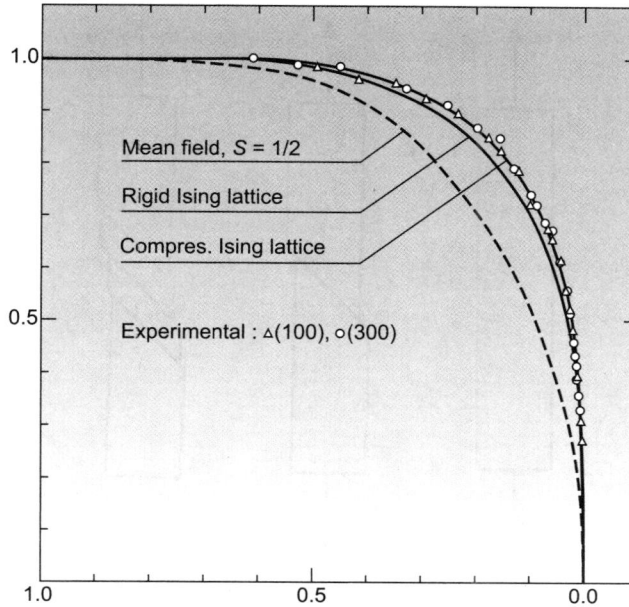

Figure 12 The degree of long-range-order in β brass as a function of temperature determined by neutron diffraction (Norvell and Als-Nielsen 1970). The horizontal axis shows the reduced temperature $(T_c - T)/T_c$, where T_c is the critical temperature of order–disorder transition. The lines are predictions from various theories.

The relations become slightly more complicated if the composition deviates from the stoichiometry, and much more when vacancies are taken into consideration. Computation of the vacancy concentrations also becomes complicated even if they are assumed to be low, since the concentrations of vacancies on the different sublattices are linked to each other and also to the concentrations of the antisite atoms. The statistical thermodynamics of the Schottky defects and antisite atoms are discussed in detail in Section 6.5.3.

6.2.2 Dislocations

Metals deform irreversibly by a mechanical force exceeding the elastic limit, or the yield strength. If a rod of a single crystal of a metal is subjected to a tensile test, it deforms by slip as illustrated in **Figure 13**, which is recognized by steps on the surface. By the Laue method of X-ray diffraction, one would find that the slip took place along a particular crystallographic plane in a particular direction. However, the slip along each plane does not occur at once throughout from one side of the surface to the other, but proceeds step by step in the body, by the motion of line defects called dislocations. **Figure 14** schematically illustrates the deformation process, which is driven by the shear component of the tensile load exerted on the slip plane in **Figure 13**. Theoretical estimates of the stress required for shifting two half crystals along the slip plane far exceed the experimentally observed critical shear stress. Dislocations were later found to move at much lower stress; it turned out that slip deformation

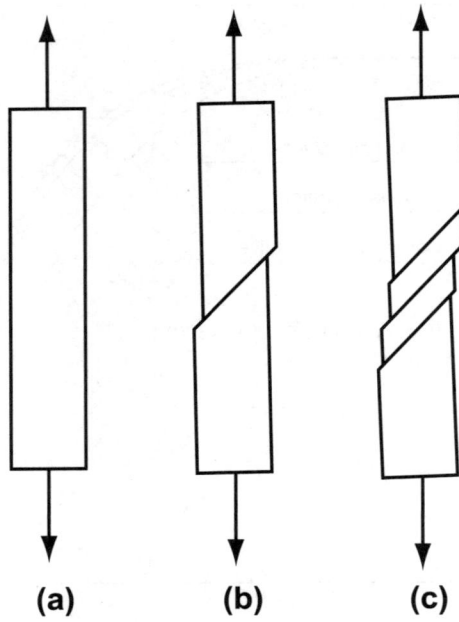

Figure 13   Slip deformation of a single crystal in a tensile test.

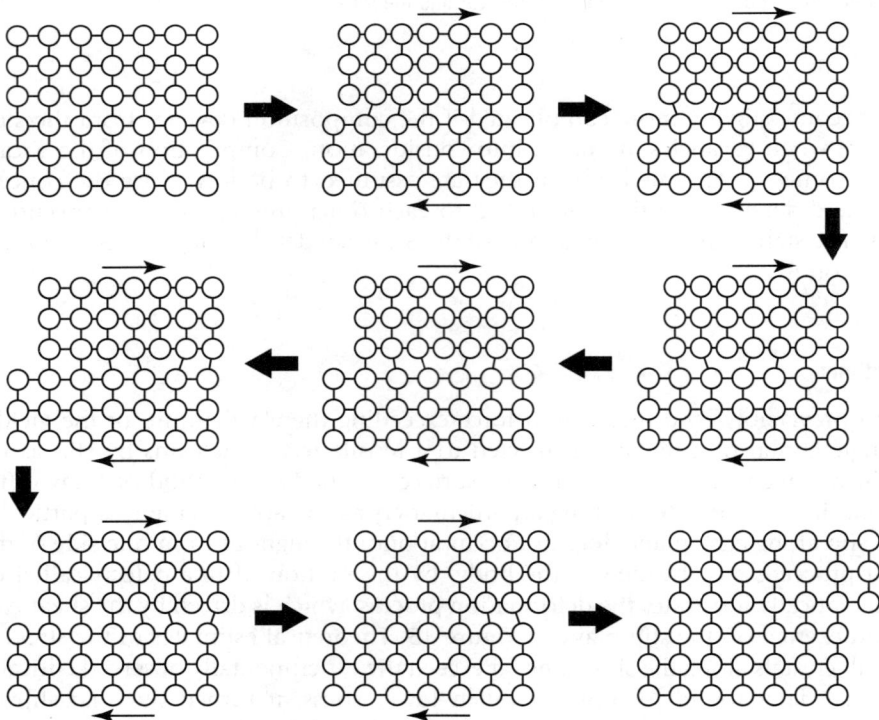

Figure 14   Slip deformation by a motion of an edge dislocation.

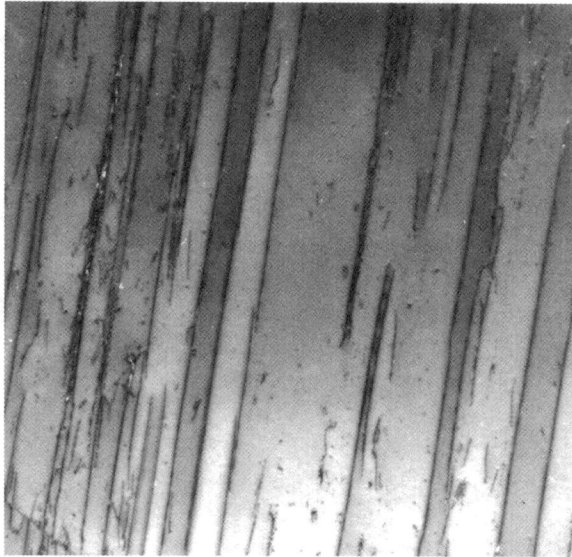

Figure 15 Transmission electron micrograph of screw dislocations in titanium deformed at room temperature (Numakura H., unpublished). Frame width is about 3.5 μm.

occurred by the motion of a large number of dislocations.[8] **Figure 15** shows dislocations in deformed titanium observed by transmission electron microscopy.

Dislocations are geometrical defects that can be envisaged in an elastic medium. An edge dislocation and a screw dislocation are illustrated in **Figure 16(a) and (b)**. First a planar cut is made to the middle of the body, and then the material above and below are shifted from each other along the cut-plane. The boundary line of the cut-plane, which is an elastic singularity, is the dislocation, and the displacement vector is the Burgers vector. In a crystalline solid, if the magnitude of the shift is equal to the translation vector of the crystal, the shifted part recovers the periodic structure. The atomic arrangement is disturbed at and along the dislocation line, and there is a characteristic strain field around it. If the shift is given normal to the dislocation line, an edge dislocation is created, which has an extra half atomic plane in the compressed side. On the other hand, a shift parallel to the dislocation line creates a screw dislocation, around which the periodic structure of the crystal is changed to a spiral. The dislocation line and the displacement vector may not be normal nor parallel, but can make any angle, defining the "character" of the dislocation. **Figure 16(c)** shows a curved dislocation. The character is edge at one end and screw at the other, but is "mixed" everywhere else.

The line of an edge dislocation and the Burgers vector define a plane on which the dislocation moves. The glide motion of dislocations tends to occur on a particular crystallographic plane in a particular crystallographic direction. The preferable combination of plane and the direction of the dislocation motion is the slip system observed in experiment. The slip direction is most often the nearest neighbor direction, and the slip plane is the most widely separated atomic planes. Commonly observed Burgers vectors and the glide planes of dislocations are listed in **Table 1**.

---

[8] For details, see Chapter 16 of volume II, or Chapters 1 and 8 of Hirth and Lothe (1982).

(a)

(b)

(c)

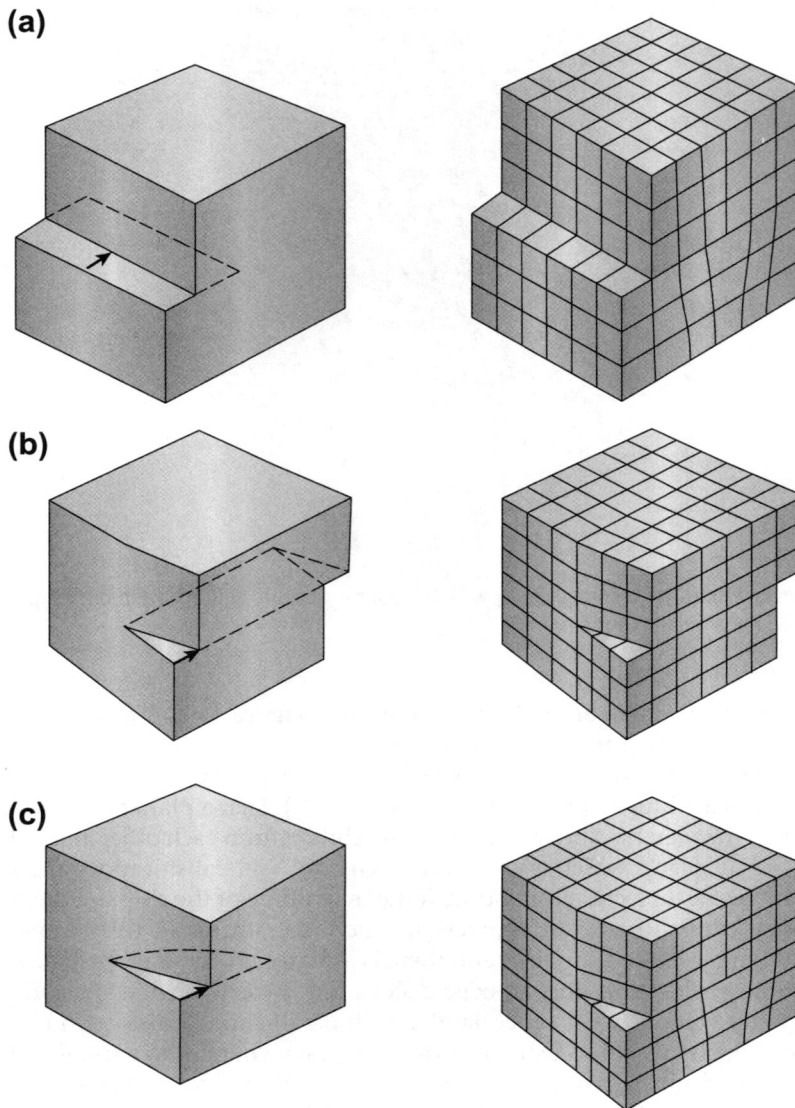

Figure 16 Schematic illustrations of an edge dislocation (a), a screw dislocation (b), and a mixed dislocation (c). The arrow indicates the Burgers vector.

In ordered alloys, the translation vector retaining the periodic structure is generally longer than in pure metals, and so is the Burgers vector of a dislocation. **Figure 17** shows the translation vectors in an fcc crystal and in an ordered alloy of the Cu_3Au structure. To retain the correct atomic arrangement, the Burgers vector in the latter must be twice as long as in a disordered alloy. Since the atomic and elastic distortions associated with a dislocation are much more extensive in ordered

Table 1 The Burgers vectors and glide planes of dislocations commonly observed in metals. The magnitude of the Burgers vectors is given in units of the relevant lattice parameter.

Structure	Burgers vector	Glide plane
fcc	$\frac{1}{2}\langle 1\bar{1}0\rangle$	$\{111\}$
bcc	$\frac{1}{2}\langle 11\bar{1}\rangle$	$\{1\bar{1}0\}, \{112\}$
hcp	$\frac{1}{3}\langle 11\bar{2}0\rangle$	$\{0002\}, \{1\bar{1}00\}$

alloys or intermetallic compounds, the glide motion of a dislocation is harder and more complicated than in pure metals.

The formation energy of a dislocation is much higher than point defects because of the long-range elastic strain field around it. The energy of a dislocation may be divided into the core energy and the elastic strain energy. The latter can be evaluated by the strain energy stored in a hollow cylinder surrounding a dislocation line. It is given by linear elasticity theory as

$$E = \frac{Gb^2}{4\pi}\ln\frac{r}{r_c},\tag{17}$$

where G is the shear modulus, b is the magnitude of the Burgers vector, and r_c and r are the inner radius and the outer radius of the cylinder, respectively (Hirth and Lothe, 1982, Chapter 3). The core of a dislocation is the part in which the distortion is so large that linear elasticity is not applicable. Here we assume that $r_c \approx b$. According to careful estimations, the core energy is roughly given by $Gb^2/(4\pi)$. The outer radius r can be taken as half the average distance between dislocations, as dislocations tend to arrange themselves in a configuration in which their strain fields cancel out each other. The dislocation density in well-annealed metals is usually 10^{11} to 10^{12} per square meter, so that the average distance is about a few micrometers, or $10^4 b$, which gives $\ln(r/r_c) \approx 9$. Then the total energy is about $10Gb^2/(4\pi)$

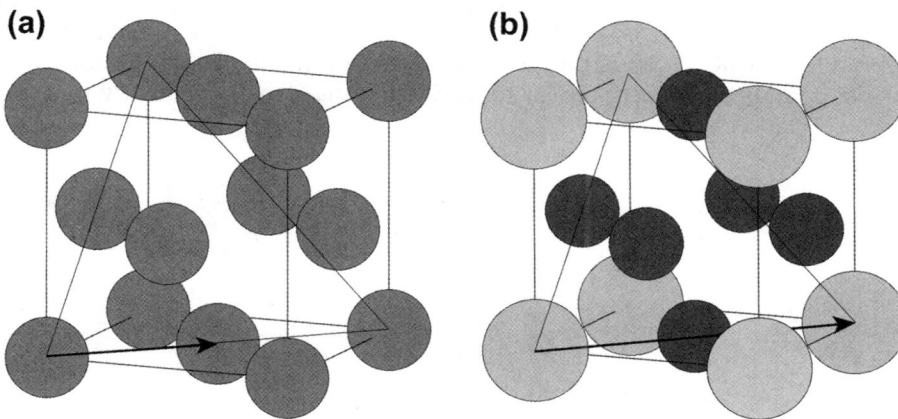

Figure 17 The Burgers vector of a perfect dislocation in an fcc crystal (a) and in an ordered alloy of the Cu$_3$Au (L1$_2$) structure (b).

per unit length, or $10Gb^3/(4\pi)$ per length of atomic distance. Taking G and b for copper, 50 GPa and 0.26 nm, we obtain 3×10^{-9} J m^{-1} as the self energy, or 5 eV per b.

With this estimate, the energy required to generate a dislocation penetrating a crystallite of 10 μm in diameter amounts to be of the order of 10^{-14} J, or 10^5 eV. Remembering that the thermal energy is merely 1/40 eV at room temperature and still much smaller than 1 eV even at 3000 K, the probability of finding a dislocation generated by thermal agitation in crystal grains in conventional polycrystalline metals is practically zero, even though there are as many possible places for a dislocation line as 10^{10}. Dislocations are nonequilibrium defects introduced into the crystal by accident during solidification or phase transformation, or by mechanical forces. Mechanisms of dislocation generation are discussed in Chapter 16, Volume II.

6.2.2.1 *Disclinations*

There is another class of one-dimensional defects, called disclinations. While dislocations have a characteristic shear displacement (translation), disclinations are defects associated with local rotation. They are known to occur in liquid crystals, but their energy is expected to be very high in metallic crystals and are rarely observed. An exceptional example was reported, however, in heavily deformed iron by high-resolution transmission electron microscopy (Murayama et al., 2002). Interested readers are referred to comprehensive reviews (Romanov and Vladimirov, 1992; Romanov, 2003; Kleman and Friedel, 2008).

6.2.3 Stacking Faults

Many metals crystallize into the close-packed structures, fcc and hexagonal close packed (hcp), in which the atomic planes of triangular lattice are stacked on the positions illustrated in **Figure 18** in a sequence …ABCABC… (fcc) or …ABAB… (hcp). Staking faults are planar faults that are created by either removing an atomic layer or inserting an extra layer, which are called an intrinsic fault, I, or extrinsic fault, E, respectively. The intrinsic fault in an fcc crystal has a stacking sequence …ABC|BC…, where the bar indicates the missing layer. The sequence around the extrinsic fault is …ABCBABC…, where B at the center is the extra plane. In both the cases, the close-packing is preserved, so that no nearest-neighbor bonds are disturbed. There are, however, irregularities in the second-neighbor relations across the fault, two for both I and E. Stacking faults can also be created by shear displacements. In the structure …ABCABC…, if the right-half part is shifted as a rigid body in such a way that A changes to B, then the sequence becomes …ABCBCA…. It is the same fault as the intrinsic fault I introduced above. Twin boundaries in fcc crystals are defects similar to stacking faults. A twin boundary separates two twin

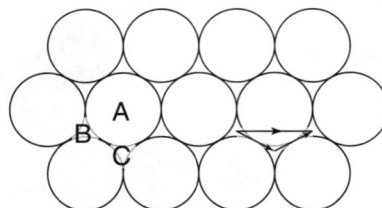

Figure 18 Schematic illustration of a close-packed atomic plane. Atoms in the next layer may be placed either at position B or C. The horizontal arrow on the right indicates the translation vector between adjacent B positions, and the short vectors indicate the translation in two steps through position C.

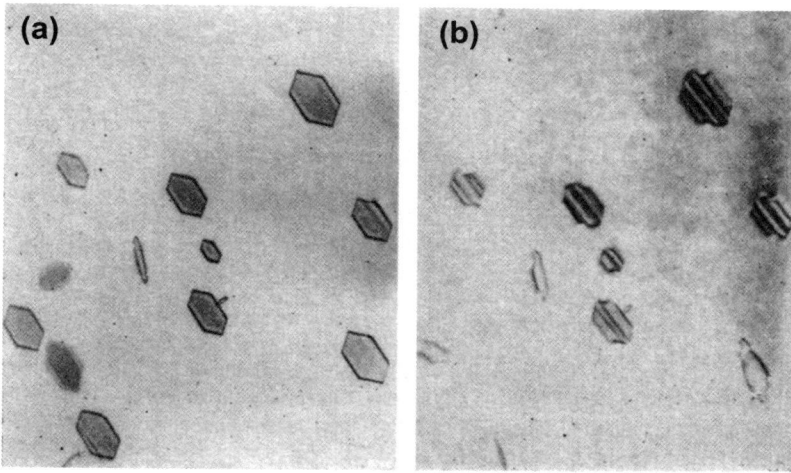

Figure 19 (a) Dislocation loops with a stacking fault in pure aluminum quenched from 923 K. (b) The same area as (a), showing unfaulting of some of the loops by thermal stresses due to the electron beam. After Cotterill and Segall (1963).

crystals, each of which corresponds either to a mirror reflection or 180° rotation of the other. The stacking sequence across a twin boundary in a fcc crystal is ...ABCACBA..., in which the A layer at the center is shared by the two crystals. Here, improper arrangement occurs between the second neighbor C–C layers only, so that its energy must be about half of those of I and E faults.

In the hcp structure, removing a single layer, for example B, results in a sequence ...ABAB|BAB..., which is unstable and must transform to ...ABAB|CBC... by itself. This is an intrinsic fault called I_1, as there is only one set of improperly placed second neighbor layers. An extrinsic fault is created by inserting a layer C between B and A, giving a stacking sequence ...ABABCABAB.... In this sequence there are three pairs of wrong second neighbor layers. The translational shift along the close-packed plane that changes the sequence from ...ABABABAB... to ...ABABCACA... produces another stacking fault in the hcp structure. This fault includes two wrong second-neighbor relations and is called an intrinsic fault I_2. The energies of I_1, I_2, and E are roughly in proportions of 1:2:3.

Stacking faults are observed in quenched and irradiated metals. Supersaturated vacancies and self-interstitial atoms often agglomerate on a close-packed plane, forming a disc with a dislocation loop at its perimeter. **Figure 19(a)** shows transmission electron micrographs of intrinsic stacking faults formed by condensation of vacancies in aluminum quenched from a high temperature (Cotterill and Segall, 1963). The fringes indicate the presence of a stacking fault on the loop plane, which is {111}, with the dislocation line segments parallel to ⟨110⟩ type directions. The dislocation loop bounding the fault is called a faulted loop or a Frank loop.[9] These vacancy discs often collapse with the facing atomic planes being sheared to attain a regular ...ABCABC... sequence, only leaving a loop of a perfect dislocation, which is called a perfect loop or a prismatic loop.[10] Some of the loops in **Figure 19(a)** are found to have been transformed to perfect loops, no longer with the fringe contrast, in **Figure 19(b)**.

---

[9] The dislocation is a Frank sessile dislocation, whose Burgers vector, $(1/3)\langle 111\rangle$, is normal to the fault plane.
[10] Having changed the Burgers vector to $(1/2)\langle 1\bar{1}0\rangle$, the dislocation segment can slip out onto the prismatic plane, to which the Burgers vector is parallel.

(a) a b a b a b a b   **(b)** a b a b a b a b

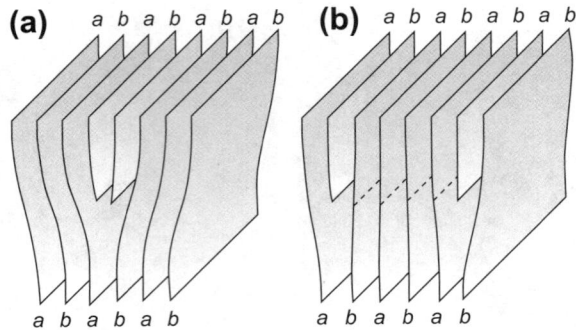

a b a b a b   a b a b a b

Figure 20 Schematic illustration of the structure of a perfect dislocation (a) and an extended dislocation (b). Between the partial dislocations each of the planes above the extension plane is connected to a plane below of the other type. This disorder forms a stacking fault in elemental crystals such as pure metals, or a antiphase boundary for a super-dislocation in an ordered alloy (**Figure 17**(b)). After Haasen (1996).

The intrinsic faults created by the mutual displacement are closely related to dislocations in fcc and hcp metals. The Burgers vector of an ordinary dislocation in these metals is the vector connecting the nearest neighbor atoms, from B to the neighboring B in **Figure 18**, $(1/2)\langle 1\bar{1}0\rangle$ in the case of the fcc structure. The displacement often occurs in two steps, by splitting the dislocation in two "partial" dislocations, with displacement vectors $(1/6)\langle 2\bar{1}\bar{1}\rangle$ and $(1/6)\langle 1\bar{2}1\rangle$, which correspond, respectively, to the displacement that first brings atoms on the next layer, at position B, to position C, and that brings to the next B, as indicated by vectors. In between the partial dislocations, which are called Shockley partial dislocations, a planar fault is created by the incomplete shift. In the fcc crystal, it is the intrinsic fault I, and in the hcp crystal it is I_2. The structures of a perfect dislocation and an extended dislocation formed by the splitting are illustrated in **Figure 20**. In the well-annealed fcc metals and alloys of low stacking-fault energies, such extended dislocations are commonly observed, in addition to grown-in stacking faults and twins. **Figures 21** shows examples in Cu–Ge solid-solution alloys.

The stacking fault energy, SFE in short, is an important parameter determining the dissociation of a dislocation, which has a significant influence on the plastic deformation behavior. Their values have

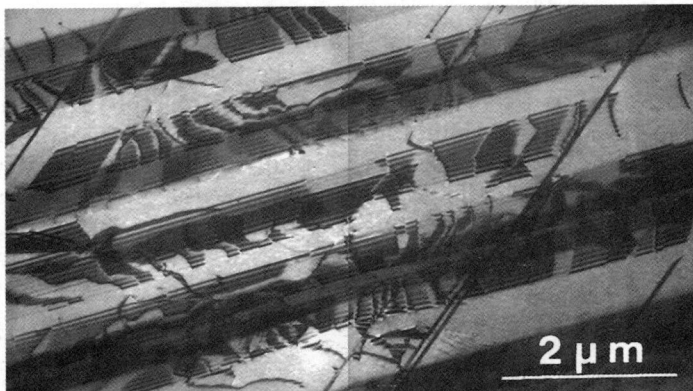

2 µm

Figure 21 Transmission electron micrograph of extended dislocations in Cu–8 at.% Ge solid-solution alloy deformed by 10%.

been determined mostly by transmission electron microscopy from the width of extended dislocations and other configurations of dislocations. While it is over 100 mJ m^{-2} for the intrinsic fault in common fcc and hcp metals, it is markedly lower in noble metals, 45 in copper, 16 in silver, and 32 in gold, in mJ m^{-2} (Hirth and Lothe, 1982, Appendix 2). These low values in these metals decrease further by alloying (Gallagher, 1970), including the Cu–Ge alloys shown in **Figure 21**. The SFE of austenitic stainless steels is known to be similarly low (e.g., Schramm and Reed, 1975).

Stacking faults on non-close-packed planes and those in non-close-packed structures, e.g., bcc, are also of much interest, as dissociation of a dislocation into partial dislocations with a stacking fault, if possible, facilitates its glide motion. In the bcc structure, no stacking faults are expected from a hard sphere model. Dislocations in bcc metals are barely extended because of the absence of low-energy stacking faults, and this is believed to be the major reason why dislocations, in particular screw dislocations, in bcc metals are much less mobile than those in fcc metals (Vítek, 1974). This aspect is particularly important, as mentioned earlier, in ordered alloys and intermetallic compounds, where the structure of the stacking faults are complicated because of the ordered atomic arrangement. Other types of planar faults arising from the ordered structure (e.g., antiphase boundaries—see below) may also be involved in extension of a dislocation. The geometry and energies of planar faults in this class of materials have therefore been studied extensively by experiment, theory and computer simulation (Paidar and Vitek, 2002).

6.2.4   Domain Boundaries

6.2.4.1   Antiphase Boundaries

One of the other types of planar faults that occurs in ordered alloys are interfaces where the structure is regular with respect to atomic positions but is irregular with respect to chemical species. **Figure 22** illustrates this defect, called an antiphase boundary, or APB. It can happen in ordered compounds as grown-in stacking faults or twins in elemental crystals: two ordered regions, or "domains" begin to grow in a single crystal of a disordered alloy and eventually they meet where the atomic positions find themselves out-of-phase. On the other hand, if a super-dislocation in an ordered compound is split into partial dislocations and one of the latter passes through, an APB is left behind. Dislocations in some class of intermetallic compounds are decomposed into partial dislocations bounding an APB and stacking faults, and their mobility is strongly dependent on the energies of the planar faults. Evaluation of the energies of the variety of planar defects is therefore crucial in understanding the deformation behavior. Transmission electron microscopy is a most useful tool (Sun, 1995; Veyssyère and Douin, 1995) but theoretical calculations can also give good estimates (see, e.g., Mishin, 2004).

6.2.4.2   Magnetic Domain Boundaries

While APBs are interfaces between chemically ordered domains, there are similar interfaces between magnetically ordered domains in magnetic materials. In the ferromagnetic state of bcc iron, below 770 °C, the magnetic moment of each atom is all aligned in one of the three ⟨100⟩ directions. In a single crystal, therefore, six physically distinguishable domains can exist, whose magnetization direction is in parallel or antiparallel to [100], [010] or [001]. Their interfaces are magnetic domain boundaries, in which the local magnetic moment is smoothly rotated from one orientation to the other, as illustrated in **Figure 23** (Kittel, 1996). These domain boundaries increase the internal energy, but are naturally formed to make internal loops of spontaneous magnetization, by which the magnetic field produced outside is minimized.

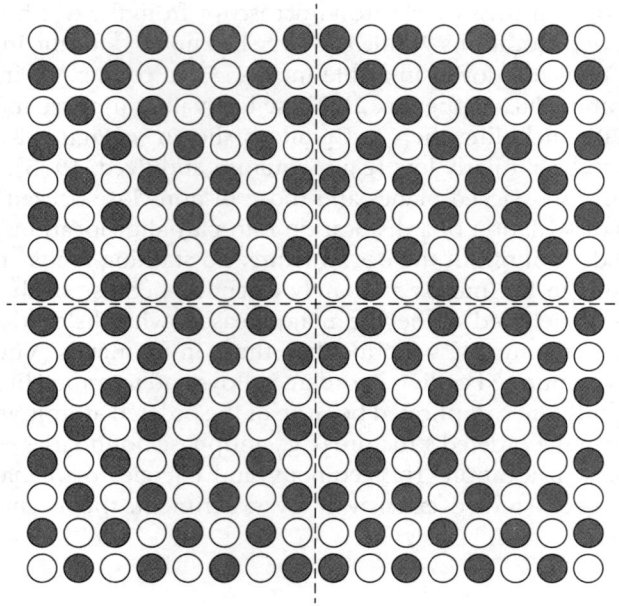

Figure 22   Antiphase boundaries (dashed lines) separating four ordered domains in a binary ordered alloy.

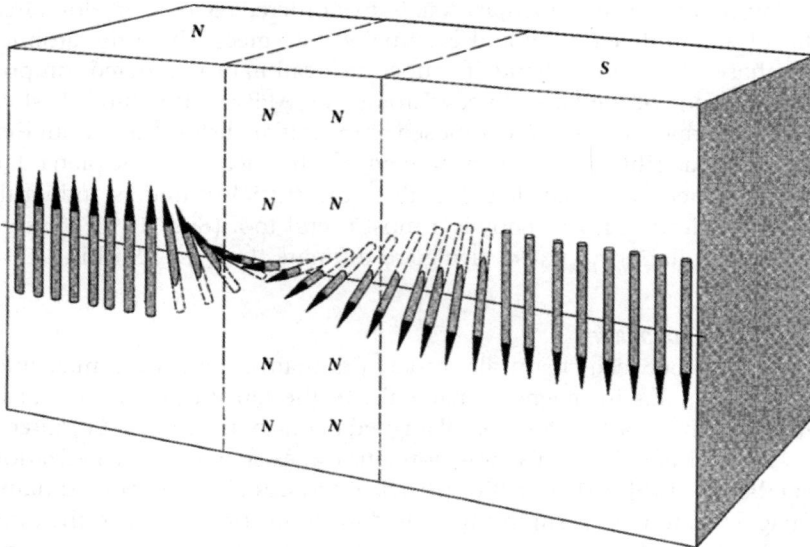

Figure 23   A magnetic domain boundary, or a Bloch wall, in which the direction of magnetization is rotated by 180°. After Kittel (1996).

A technical magnetization process occurs by movements of magnetic domain boundaries, which are driven by an external magnetic field (Chikazumi, 1997, Chapter 18). If the external field is applied in parallel to one of the possible magnetization directions, the domains which are favorably oriented grow at the expense of unfavorable ones. The magnetization curve goes up until it reaches the saturation level, when the domain boundaries are swept out from the sample piece. On decreasing the external field, domain boundaries are reintroduced from the surface and the original domain configuration, or a configuration similar to it, is recovered at zero field. In soft magnetic materials, smooth movements of domain boundaries are desired for better performance. Since other crystal defects (point defects, dislocations, impurity atoms) interfere the motion of domain boundaries, it is essential to remove defects as much as possible. Single crystals are favorable in this respect, as passage of domain boundaries through differently oriented crystal grains must be difficult. It is for this reason that very coarse-grained steels are used for the core of voltage transformers.

6.2.4.3 Variants

Both chemical and magnetic ordering are accompanied by distortion of the crystal. For example, when an alloy of Fe and Pt of equiatomic composition is cooled down from a high-temperature solid-solution, a chemical ordering begins to occur in the fcc structure at 1280 °C to form the layer-by-layer arrangement shown in **Figure 24**, the L1$_0$ structure. As this structure is of tetragonal symmetry, the lattice parameter in the direction normal to the layers may be different from the other two. In fact, it becomes appreciably shorter as the chemical order develops, and the ratio of the lattice parameters c/a is about 0.95 at room temperature (Kudielka and Runow, 1976). The three differently oriented ordered domains are distinguishable simply from the dimensions, with no need of looking into the atomic arrangement. These domains are referred to as "variants", not "grains", since the domains can vanish to recover the fcc single crystal if heated above the order–disorder transition temperature.

The distortion accompanying magnetic order, called "magnetostriction", is in general much smaller than that associated with chemical order. In terms of linear strain, it is $+2 \times 10^{-5}$ for iron in $\langle 001 \rangle$, and -2×10^{-5} for nickel in $\langle 111 \rangle$. Some alloys of iron, for example Fe–Ga and Fe–Co, exhibit magnetostriction several or ten times larger than iron. Those materials are potentially useful in such applications as magnetic actuators (Clark et al., 2000). Magnetostriction is desired to be small for soft magnetic applications, because it causes energy loss under alternating magnetic field, called hysteresis loss. Amorphous ferromagnets are excellent soft magnetic materials because they have no anisotropy in

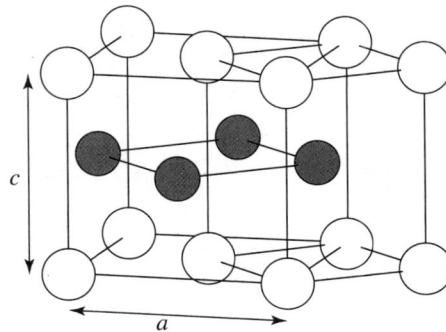

Figure 24 The L1$_0$ (CuAu) structure.

magnetization, small magnetostriction, and as mentioned earlier no grain boundaries nor other crystalline defects.

6.2.5   Grain and Twin Boundaries

Grain boundaries are interfaces where crystal grains of different orientations join in a polycrystalline solid. The atomic arrangement is irregular and the atomic density is generally low in grain boundaries. Similarly to dislocations they are defects accidentally produced during phase transformation or mechanical working. Grain boundaries in real materials are often curved, but to reduce the internal energy they have a tendency to become flat, and further rearrange themselves to change to a boundary that better matches the two grains, of many atomic bonds and less extraneous space. One class of such low-energy boundaries is twin boundaries mentioned previously.

There are two types of simple boundaries, tilt and twist boundaries, which are illustrated in **Figure 25**. The energy of a grain boundary increases with increasing the misorientation angle, but at large angles it begins to decrease as they become close to other special low-energy boundaries, displaying a diagram such as **Figure 26** (Hasson et al., 1972). This is the basic information for quantitatively characterizing grain boundaries. The cusps in the energy diagram correspond to boundaries of good matching, including twin boundaries that occur at some angles specific to the crystal geometry. Twins occur not only on crystal growth but also by plastic deformation. **Figure 27** compares shear deformation achieved by slip (a) and formation of a twin crystal (b). Twinning deformation, or deformation twinning, is important at low temperatures, particularly in low-symmetry crystals in which not many slip systems are available (Mahajan and Williams, 1973; Narita and Takamura, 1992). Also, it is relevant to martensitic transformation discussed in Chapter 8 of this volume.

Grain boundaries accommodate impurity atoms (grain boundary segregation), often causing the boundaries brittle, provide pathways for fast diffusion, or allow sliding of grains along the boundary giving rise to superplastic deformation. They also serve as sinks and sources for point defects and dislocations. Corrosion proceeds at grain boundaries, and this is why amorphous alloys are more corrosion-resistant. In so many ways, grain boundaries play important roles in material properties that they are discussed in many parts in later chapters. Nanocrystals fabricated by vapor quenching and

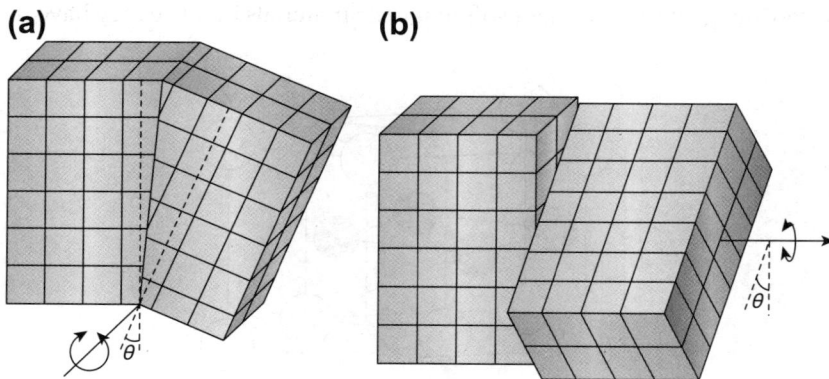

Figure 25   Geometry of grain boundaries. (a) Tilt boundary. (b) Twist boundary. θ is the misorientation angle. After Mittemeijer (2010).

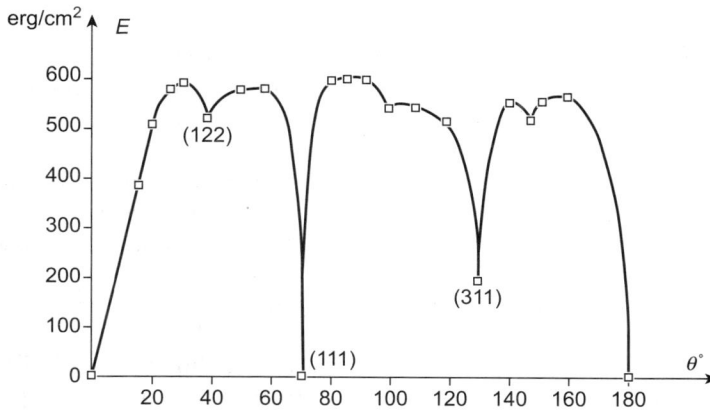

Figure 26 Energy of symmetrical ⟨110⟩ tilt boundaries in aluminum calculated using a Morse potential. After Hasson et al. (1972).

other nonequilibrium methods display interesting properties such as high mechanical damping, fast diffusion, stabilization of nonequilibrium structures and phases, realization of high-temperature phases at low temperatures, and so on, where many of the findings are attributable to the high density of grain boundaries. Those nanocrystalline materials used to be obtained only in the form of powders or thin films. However, with the advent of severe cold-work techniques, metals of very fine microstructure, consisting of submicrometer grains, are now being produced in bulk form, thus being called bulk nano(structured)-metals. They often exhibit unusual mechanical properties such as surprisingly high yield strength and not-too-poor ductility. The mechanism of grain refining—production of grain boundaries—and the roles of the copious grain boundaries are one of the current topics in physical metallurgy (Zehetbauer and Zhu, 2009; Huang and Langdon, 2013).

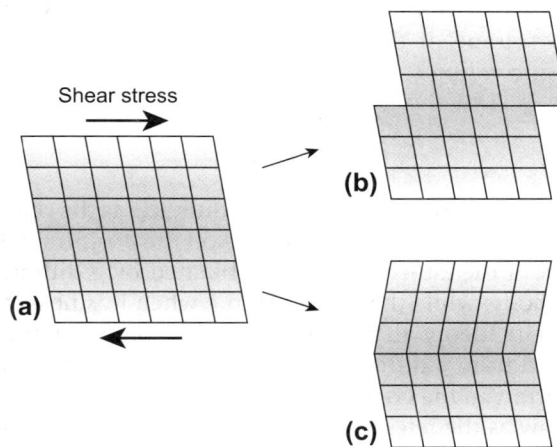

Figure 27 Deformation of a crystal subjected to shear stress (a) by slip (b) and by twinning (c).

6.2.6   Three-dimensional Defects

Supersaturated point defects tend to agglomerate and form large clusters by the same reason as diva-cancy formation. Examples of planar clusters are discs of vacancies and self-interstitial atoms discussed earlier, which evolve to stacking faults and/or dislocation loops. In quenched fcc metals, defects of a shape of a tetrahedron with stacking faults on the faces are observed by electron microscopy (Silcox and Hirsch, 1959; Loretto et al., 1965). They have the same geometry as the Thompson tetrahedron[11] and are believed to be formed by agglomeration of vacancies. A simplest form of such defects may be generated from three nearest-neighbor vacancies arranged as a triangle in a close-packed plane, on which an atom in the next plane falls to form a tetrahedron of vacancies with an atom at the center. A larger tetrahedron may be created by collapse of vacancy discs on the close-packed planes intersecting each other. No such defects formed by self-interstitial atoms are known.

A larger number of vacancies may segregate to form a three-dimensional hollow space, or a void. Voids in metal crystals assume polyhedral shapes, with the internal surfaces parallel to low-index crystallographic planes whose surface energy is low. For example, those in quenched fcc metals are in the form of an octahedron bounded by {111} planes (Brimhall and Mastel, 1969). Voids are commonly observed in metals and alloys heavily irradiated at moderate temperatures (0.3 to 0.4 times the melting temperature), and are the origin of "void swelling" of the material. Helium and other rare gas elements are produced in nuclear fusion and fission reactors, and they form gas bubbles in struc-tural components because their solid-solubility is very low. Their formation mechanism is closely related to void formation, and the effects of bubbles on mechanical properties are also a critical issue in degradation of structural materials in nuclear technology (Wolfer, 1984; Odette and Lucas, 2001).

6.3   Experimental Techniques

6.3.1   Differential Dilatometry

As vacancies are formed with increasing temperature, the volume of a piece of metal increases more than that caused by thermal expansion since new atomic sites are added to the crystal. The absolute concentration of vacancies can be determined by precisely measuring the difference between macro-scopic and microscopic changes in volume with temperature. Introduction of vacancies and/or self-interstitial atoms affect the volume of the crystal, $V = N\Omega$, in two ways: changing the number of atomic sites, N, and the atomic volume, Ω, i.e.

$$\frac{\Delta V}{V} = \frac{\Delta N}{N} + \frac{\Delta \Omega}{\Omega}. \tag{18}$$

A point defect slightly displaces the atoms around it, and thus acts as the center of elastic distortion. The volume change of the solid due to the distortion depends on the position of the defect in the body: it is greater than that expected from the lattice parameter obtained by a diffraction experiment when the defect is at the center of the body, while the reverse is true when it is near the surface. Eshelby (1954) showed, by continuum elasticity theory, that the macroscopic volume change, $\Delta V/V$, is identical to the microscopic $\Delta \Omega/\Omega$ determined from a diffraction experiment when a large number of defects are dis-tributed randomly and uniformly in the body. By taking a sufficiently low temperature where the defect concentration is negligibly low as the reference state, the change in the number of atomic sites can be

---

[11] With the four faces parallel to {111} and the six edges parallel to ⟨110⟩.

obtained by measuring the relative changes of the macroscopic and microscopic volumes from the relation

$$\frac{N(T) - N_0}{N_0} = \frac{V(T) - V_0}{V_0} - \frac{\Omega(T) - \Omega_0}{\Omega_0}, \tag{19}$$

where the symbols with subscript 0 indicate those at the reference temperature. For the case of vacancies the left-hand side is their concentration c_v, and for self-interstitial atoms it is $-c_i$. When both the two defect species are present, it is $c_v - c_i$. Since the concentration of self-interstitial atoms is usually very low, the absolute concentration of vacancies can be determined by this experiment.

Figure 28(a) shows the experimental result for pure aluminum (Simmons and Balluffi, 1960a). For crystals of cubic symmetry, the changes in the macroscopic and microscopic volumes can be obtained from relative changes in the linear dimension of the sample, L, and the lattice parameter, a, respectively. The absolute concentration of vacancies is then obtained as

$$c_v \approx 3\left(\frac{\Delta L}{L_0} - \frac{\Delta a}{a_0}\right). \tag{20}$$

The deviation of $\Delta L/L_0$ from $\Delta a/a_0$ at the temperatures close to the melting temperature (660 °C) demonstrates that vacancies are produced in appreciable numbers. **Figure 28(b)** is a plot of the vacancy concentration calculated by Eqn (20) against reciprocal temperature. The formation enthalpy and the formation entropy (see the next section) are obtained from the slope and the intercept, respectively. This method was successfully applied to aluminum, silver, gold, and copper by Simmons and Balluffi (1960a, 1960b, 1962, 1963).

The mobility of vacancies can be determined by tracing transient changes of the volume after a rapid change in temperature. As illustrated in **Figure 29**, if a sample is first equilibrated at temperature T_1 and is then heated up abruptly to T_2 and kept constant, the volume of the sample first increases without

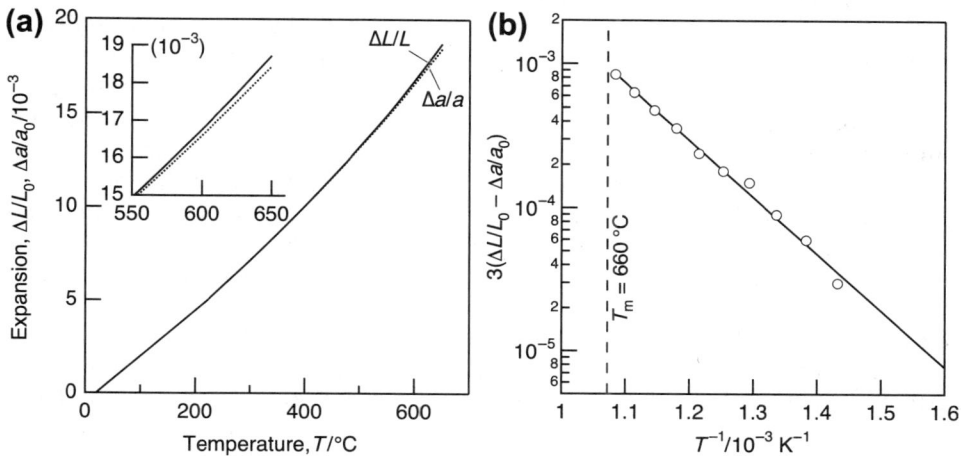

Figure 28 (a) Changes in macroscopic length, $\Delta L/L_0$, and lattice parameter, $\Delta a/a_0$, of pure aluminum. (b) Arrhenius plot of the vacancy concentration derived from (a) according to Eqn (20). After Simmons and Balluffi (1960a).

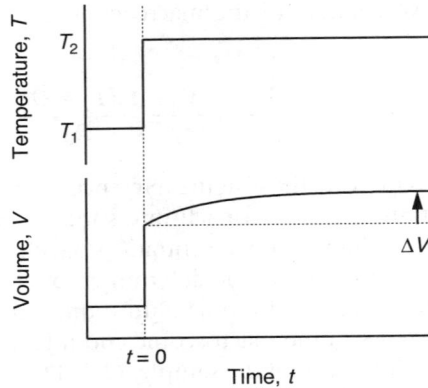

Figure 29 Gradual increase in volume (lower) arising from production of vacancies on an abrupt increase in temperature from T_1 to T_2 (upper).

significant delay by thermal expansion of the lattice, and then it gradually changes with time. This behavior comes from evolution of the vacancy concentration to approach a new equilibrium. Since the volume of the crystal containing n vacancies at temperature T can be written as

$$V(T) = N\,\Omega^\circ(T) + n\,\Delta v_{f,v}(T), \tag{21}$$

where $\Omega^\circ(T)$ is the atomic volume of a defect-free crystal, the variation of the volume with the holding time t is expressed as

$$\Delta V(T_1, T_2; t) = [n(T_2, t) - n(T_1, t = 0)]\Delta v_{f,v}(T_2). \tag{22}$$

While being kept at temperature T_2, for vacancies to attain the new equilibrium concentration with uniform spatial distribution, they must be generated from internal sources (or annihilate at sinks if temperature is lowered), which require their migration. The kinetics of the volume change is thus controlled by their diffusion, and its rate constant is characterized by the migration enthalpy $\Delta h_{m,v}$. This method was first applied to an intermetallic compound CoGa by van Ommen and de Miranda (1981), who measured the length change by mechanical probing. Schaefer and coworkers improved the method by utilizing a laser interferometer and evaluated the migration enthalpy of vacancies in FeAl and NiAl (Schaefer et al., 1999; Sprengel et al., 2002). Change in volume due to changes in the defect concentration over a limited temperature range is generally small and is difficult to detect. These experiments were successful owing to the unusually high vacancy concentrations in the materials studied.

In the above techniques, knowledge on the formation volume, $\Delta v_{f,v}$, was not required. The formation volume is, however, one of the important defect parameters, and can be determined by measuring the lattice parameter of samples with and without the defects at the same temperature. The volume of the crystal consisting N atoms and n vacancies are written as Eqn (21). Considering that the number of sites is changed from N to $N + n$, it can alternatively be expressed using the relaxation volume, $\Delta v_{rel,v}$, as

$$V = (N + n)\Omega^\circ + n\,\Delta v_{rel,v}. \tag{23}$$

This equation may be taken as the definition of $\Delta v_{\mathrm{rel,v}}$. The relation between the two parameters is readily found from the two equations:

$$\Delta v_{\mathrm{f,v}} = \Omega^\circ + \Delta v_{\mathrm{rel,v}}. \tag{24}$$

Since the left-hand side of Eqn (23) can be written as $(N+n)\Omega$, where Ω is the atomic volume of the defect-containing crystal, we have

$$\frac{\Omega - \Omega^\circ}{\Omega^\circ} = c_{\mathrm{v}}\frac{-\Omega^\circ + \Delta v_{\mathrm{f,v}}}{\Omega^\circ} = c_{\mathrm{v}}\frac{\Delta v_{\mathrm{rel,v}}}{\Omega^\circ}. \tag{25}$$

Therefore, the magnitude of the relaxation volume can be determined by measuring the relative change in the atomic volume, $\Delta\Omega/\Omega^\circ$, and the defect concentration c_{v}. The method works also for self-interstitial atoms with the same formula as Eqn (25), with c_{v} replaced by c_{i}. Since the number of atomic sites is decreased when self-interstitial atoms are created, the relation between the relaxation volume and the formation volume is different from that for vacancies, i.e.

$$\Delta v_{\mathrm{f,i}} = -\Omega^\circ + \Delta v_{\mathrm{rel,i}}. \tag{26}$$

The relaxation volume of a vacancy is negative because of the inward relaxation of neighboring atoms. In fcc metals it is known to be $-(0.2 \sim 0.4) \times \Omega$. It follows that the formation volume is somewhat smaller than one atomic volume. On the other hand, the relaxation volume of a self-interstitial atom is positive and is much larger in magnitude; values over one atomic volume have been reported for many metals, leading to a positive formation volume.

6.3.2 Electrical Resistivity

The presence of point defects can be detected sensitively by electrical resistivity at low temperatures. The concentration can be determined simply by comparing the resistivity values with and without the defects, since point defects (including solute atoms) contribute to the resistivity in proportion to their concentration (Rossiter, 1987). Measurements are to be made at low temperatures to suppress the disturbances from thermal vibration, for the changes in resistivity due to the defects are usually small. In fact, the ratio of the residual resistivity at liquid helium temperature to the room temperature value, called residual resistivity ratio (RRR), is used as a most reliable indication of the purity of ultra-pure metals. Relative changes in the defect concentration, for example during irradiation and subsequent annealing, can be monitored precisely, whereas the absolute concentration can be determined only if the magnitude of the specific contribution is known. The method can be applied equally to point defects in pure metals (Lucasson and Walker, 1962), disordered and ordered alloys (Sharma et al., 1978; Vaessen et al., 1984; Karsten et al., 1991; Gilbert et al., 1973; Rivière et al., 1980; Alamo et al., 1986; Dimitrov et al., 1992a, 1992b; Sattonnay et al., 1997). As the contributions from various defects are integrated into a single value, however, discrimination of multiple defect species, if present, is impossible. Electrical resistivity due to dislocations, stacking faults, and grain boundaries have also been studied (Brown, 1982), but they do not usually allow quantitative analysis because interpretation is not straightforward.

6.3.2.1 Formation and Migration Enthalpies

The formation enthalpy of vacancies in various metals has been determined by measuring the electrical resistivity of samples quenched from high temperatures (Kauffman and Koehler, 1955; Bradshaw and

Pearson, 1956; Bauerle and Koehler, 1957). A sample equilibrated at temperature T_q is rapidly cooled to a temperature that is low enough to prevent any motion of vacancies, and the resistivity ρ is measured at a cryogenic temperature T_r. The procedure is repeated for several quenching temperatures, and the values of the excess resistivity

$$\Delta\rho(T_q; T_r) = \rho(T_q; T_r) - \rho_0(T_r) \tag{27}$$

due to retained vacancies are plotted against the reciprocal of T_q. Here, $\rho_0(T_r)$ is the resistivity of a reference sample that is free from excess vacancies. The plot would fall on a straight line, similarly to **Figure 28(b)**, whose slope gives the enthalpy of formation, $\Delta h_{f,v}$. Many of the experimental data of $\Delta h_{f,v}$ in metals were obtained by this method (Siegel, 1978; Ehrhart and Schultz, 1991). The formation volume can be evaluated by a similar set of experiments through the effect of external pressure on the concentration (Huebener and Holmes, 1963; Grimes, 1965; Emrick and McArdle, 1969). It is not impossible to detect the extra resistivity by vacancies in thermal equilibrium directly at high temperatures: attempts for aluminum and tungsten have been reported (Simmons and Balluffi, 1960c; Kraftmakher, 1996). However, the increase due to the defects is relatively small and cannot readily be distinguished from other effects, in particular the large contribution of lattice anharmonicity.

Resistivity measurements have been applied most successfully to determination of the migration enthalpy of point defects in annealing experiments (Kauffman and Koehler, 1955; Bauerle et al., 1956; Bradshaw and Pearson, 1957). After freezing-in the defects formed at a high temperature by quenching, or producing them by irradiation at a low temperature, the resistivity of the sample is measured during the course of isothermal or isochronal annealing at an intermediate temperature where the defects move at a measurable rate. The supersaturated defects tend to annihilate to sinks or precipitate to form clusters, and the decay of the resistivity due to the reduction in the number of the defects is monitored through resistivity. Since the kinetics of the decay is controlled by the migration of the defects, its activation enthalpy can be determined from the variation of the rate constant with temperature. For example, if the defects annihilate at unsaturable sinks and its rate is controlled by diffusion of individual defects, the kinetics of annihilation is described by a simple rate equation

$$\frac{dc}{dt} = -Kc, \tag{28}$$

where c denotes the excess concentration of the defects. The rate constant K is proportional to the diffusion coefficient of the defect and the density of the sinks. The decay of the excess concentration thus follows a simple exponential time law,

$$c(t) = c(0) \exp(-Kt). \tag{29}$$

It can be traced by measuring the resistivity and, as K is proportional to the diffusion coefficient and therefore to the jump frequency, one can determine the migration enthalpy from the temperature dependence of the decay rate. By studying changes of the migration enthalpy with external pressure, the migration volume, i.e. the activation volume of migration, can also be evaluated (Emrick, 1961).

6.3.2.2 *Analysis of Annealing Curves*

The kinetics of the decay may not always be as simple as the example above. If the concentration of defects decreases by formation of pairs, its rate would be proportional to the square of the concentration. The kinetic equation may be, depending on the mechanism, of the form

$$\frac{dc}{dt} = -Kc^n, \tag{30}$$

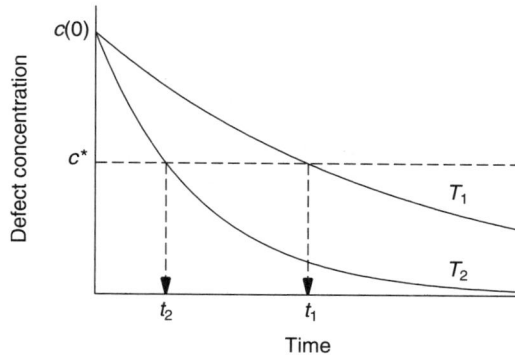

Figure 30 Schematic diagram of the cross-cut method for determining the activation energy from isothermal annealing curves at different temperatures.

like a chemical reaction of order n. More generally, it can be written as

$$\frac{dc}{dt} = -K f(c), \tag{31}$$

with an arbitrary function of concentration $f(c)$. It follows that

$$-\int_{c(0)}^{c(t)} \frac{1}{f(c)} dc = Kt. \tag{32}$$

The constant K is governed by the mobility of the defect, and if it obeys the Arrhenius law, determination of its activation energy is useful for identifying the rate-controlling process.

Several methods are commonly employed for this purpose, for example, analyzing two (or more) isothermal annealing curves, an isothermal annealing with a sudden change of temperature, and annealing during a continuous heating at a constant rate[12] (Damask and Dienes, 1963, Chapter 3). In the first, which is called the cross-cut method, two identically prepared samples, which must contain the same number of defects, are annealed isothermally at different temperatures T_1 and T_2, and the decay of the defect concentration is measured as a function of time, as illustrated in **Figure 30**. The time at which the decay curve cuts an arbitrary chosen level of concentration, c^*, must be different for the two decay curves: t_1 at temperature T_1, and t_2 at temperature T_2. They are related from Eqn (32) as

$$-\int_{c(0)}^{c^*} \frac{1}{f(c)} dc = K_0 \exp\left(-\frac{Q}{kT_1}\right) t_1 = K_0 \exp\left(-\frac{Q}{kT_2}\right) t_2. \tag{33}$$

---

[12] These methods of analysis can also be used with other experimental techniques, provided that relative changes in the defect concentration are traced.

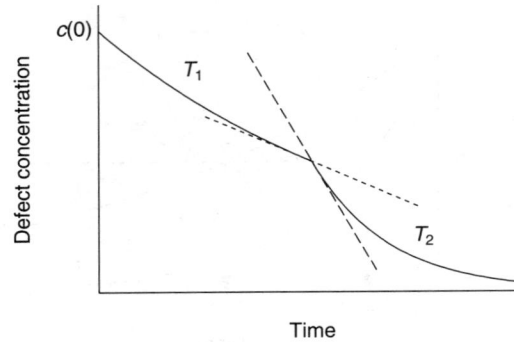

Figure 31 Schematic diagram of the slope-change method for determining the activation energy from an annealing curve with an abrupt change in temperature from T_1 to T_2.

The activation energy Q can therefore be determined as

$$Q = k \frac{\ln(t_1/t_2)}{1/T_1 - 1/T_2}. \tag{34}$$

In the second, called the slope-change method, the annealing temperature is raised from T_1 to T_2 in the course of annealing of a single sample. The slope of the c versus t curve changes, as illustrated in **Figure 31**. The slopes of the two segments of the annealing curve are both given by Eqn (31), with common c and thus $f(c)$ at the instant of the temperature change. Therefore, if the annealing is controlled by a single thermally activated process, the ratio of the slopes immediately before and after the temperature change is given by

$$\frac{(dc/dt)_1}{(dc/dt)_2} = \frac{K(T_1)}{K(T_2)} = \exp\left[-\frac{Q}{k}\left(\frac{1}{T_1} - \frac{1}{T_2}\right)\right], \tag{35}$$

from which the activation energy Q is readily obtained. While determining the slope from experimental curves may not be easy, this method has an advantage that no knowledge on $f(n)$ is required.

6.3.3 Positron Annihilation

Positrons injected into a solid are rapidly thermalized, migrate through the solid, and after some time they annihilate with electrons, producing characteristic two γ rays of 511 keV. The most common positron source ^{22}Na emits a γ ray of 1.28 MeV when it undergoes the β^+ decay. The lifetime can readily be measured as the time interval between the emission of the 1.28 MeV ray, which signals the injection of a positron, and the two 511 keV rays. The lifetime of positrons in solids, which is of the order of 100 ps, depends on the electron density of their environment. Positrons injected into the solid tend to be trapped, or localized, at vacancies or their clusters, i.e. vacancy-type defects, where the electron density is low. Dislocations and grain boundaries are also known to act as trap sites. The lifetime of positrons trapped at these defects is extended to a value characteristic of the defect species, and the values for common defects are known by experiment or by theoretical calculations. The type and density of trapping centers can therefore be determined from a spectrum of the lifetime. This is the technique called positron annihilation spectroscopy, PAS, or positron lifetime spectroscopy, PLS.

The lifetime changes according to the numbers of positrons annihilating at various sites. If we consider only one defect species, e.g., monovacancies in an elemental crystal, a simple two-state trapping model gives the trapping rate κ_1 as

$$\kappa_1 = \sigma_1 c_1 = I_1 \left(\frac{1}{\tau_0} - \frac{1}{\tau_1} \right), \tag{36}$$

where σ_1 is the specific trapping rate and c_1 is the defect concentration (see, e.g., Siegel, 1978). On the right-hand side, I_1 is the relative intensity of the extended-lifetime component in the spectrum, and τ_0 and τ_1 are lifetimes of free and trapped positrons, respectively, all of which can be evaluated from an observed lifetime spectrum. The specific trapping rate σ_1 is generally unknown, but if it is assumed to be independent of temperature one can determine the formation enthalpy of the defect from an Arrhenius plot of κ_1. For materials for which data of absolute defect concentrations are available from some other experiments (most probably by differential dilatometry), one may compare them with the measured values of κ_1 to evaluate σ_1, which may then be adopted for other materials of similar electronic structures to estimate the concentration of the particular defect species from PAS experiments.

The two γ rays emitted at each of the annihilation event provide additional methods of analysis: angular correlation of annihilation radiation (ACAR), and Doppler broadening. Since the momentum of the positron–electron pair is conserved before and after the annihilation, the two γ rays are emitted slightly deviating from the exact antiparallel relation. By the same effect, the energy of the emitted γ rays is shifted by the Doppler effect. The magnitudes of the angular deviation and the Doppler shift both reflect the energy of the electron with which the positron annihilated. Positrons trapped at a vacancy-type defect suffer less from the Doppler effect because they are less likely to encounter high-energy, inner-shell electrons of an atom. This leads to characteristic sharpening of the overall photon-energy spectrum, from which the defect concentration may be evaluated.

Measurements of the lifetime and Doppler broadening are now standard methods for studying vacancy-type defects owing to their high sensitivity: vacancy concentrations down to 1 ppm levels can be accurately determined. These techniques have been applied successfully to a variety of materials: pure metals (Seeger, 1973; Siegel, 1978; West, 1979; Siegel, 1982; Schaefer, 1982; Hautojärvi, 1987; Schaefer, 1987), intermetallic compounds (Schaefer et al., 1999; Sprengel et al., 2002), and semiconductors and ceramics (Rempel et al., 2002). Since PAS experiments under extreme conditions are not too difficult, they have served to determine the formation volume of vacancies, in which the effect of pressure on the vacancy concentration is to be measured (Dickman et al., 1977; Wolff et al., 1997; Müller et al., 2001). Migration of vacancies can also be studied by PAS, through the kinetics of equilibration on abrupt changes in temperature similar to **Figure 29** (Schaefer and Schmid, 1989; Würschum et al., 1995).

6.3.4   Calorimetry

The enthalpy of the solid increases when defects are introduced by an amount proportional to the number of defects and their formation enthalpy. This enthalpy stored in the solid, commonly referred to as "stored energy", is released on annihilation of the defects. Therefore, if the energy release is measured during the course of an annealing experiment, the decrease in the defect density can be determined. Conversely, if the changes in the population of the defects are simultaneously monitored by, for example, electrical resistivity measurements, the enthalpy per defect can be evaluated (DeSorbo, 1960).

This technique is used for estimating the formation enthalpy of Frenkel defects (Losehand et al., 1969). A sample is first irradiated by electrons at a low temperature to introduce vacancies and self-interstitial atoms, and then subjected to annealing. Recovery of the intrinsic defects in pure metals occurs in characteristic stages: migration and annihilation of self-interstitial atoms take place first and those of vacancies follow at much higher temperatures. If the contribution per Frenkel pair to the electrical resistivity is known, the enthalpy of per pair can be evaluated. As the formation enthalpy of a vacancy is known in many cases, the formation enthalpy of a self-interstitial atom may be obtained as the difference between the formation enthalpy of a Frenkel pair and that of a vacancy. This is an important means of experimentally evaluating the formation enthalpy of a self-interstitial atom.

6.3.5   Mechanical and Magnetic Relaxation

6.3.5.1   Anelastic Relaxation
While vacancies and substitutional solute atoms are expected to induce isotropic distortion, defects of lower point symmetry than the host crystal, such as pairs of point defects, some interstitial solute atoms (e.g., those in bcc metals) and also self-interstitial atoms produce anisotropic distortion. Low-symmetry defects can be in several crystallographically equivalent but distinguishable orientations. For example, a dumb-bell self-interstitial atom in fcc metals, shown in **Figure 3(b)**, is of tetragonal symmetry and thus can be in either of the three orientations, [100], [010], and [001]. Normally they are distributed equally over all possible orientations, but if external stress is applied in such a way as to interact differently with the defects in different orientations, some orientations may be stabilized over the others. This results in an increase in the population of the favored orientation(s), producing a macroscopic, anisotropic deformation. This is the phenomenon known as anelastic relaxation, which manifests itself as either a delayed deformation under a constant stress (elastic after-effect), stress relaxation under a constant strain, strain lagging from the stress under forced vibration, decay of the amplitude of free vibration, or attenuation of sound waves traveling the solid.

All of these are dissipation of the mechanical energy due to stress-induced redistribution or re-orientation of defects, and provide a useful means of studying the defect concentration and mobility, through the magnitude of the energy dissipation and the rate of the redistribution, respectively. Measurements of the energy dissipation, or internal friction, as a function of vibration frequency or temperature allow identification of various defects, as well as determination of their density and mobility (Nowick, 1978). Below we give a brief summary of the principle concerning point defects. For a thorough account of the principle and practice of mechanical relaxation, readers are referred to Nowick and Berry (1972) and Schaller et al. (2001).

6.3.5.2   Relaxation Strength—Thermodynamics
The characteristic distortion associated with a defect is described by "λ tensor", which is defined as the strain ε produced by the unit concentration of the defect in orientation p,

$$\lambda_{ij}^{(p)} \equiv \frac{\mathrm{d}\varepsilon_{ij}}{\mathrm{d}c_p}. \tag{37}$$

While the volume change due to the defect is given by the trace of the λ tensor, the magnitude of the relaxation is determined by the dispersion with respect to the stress axis. When the partial concentrations c_p deviate from the average value, c/n (n is the number of possible orientations) by external stress σ, one observes an anleastic strain[13]

$$\varepsilon^{an} = \sum_{p=1}^{n} \left(c_p - \frac{c}{n} \right) \lambda^{(p)}. \tag{38}$$

It can readily be shown that the relaxation of the elastic compliance, $\delta J = \varepsilon^{an}/\sigma$, under small stress σ is given by

$$\delta J = \frac{c\,\Omega}{kT} \left[\frac{1}{n} \sum_p \left(\lambda^{(p)} \right)^2 - \left(\frac{1}{n} \sum_p \lambda^{(p)} \right)^2 \right], \tag{39}$$

where Ω is the atomic volume (Nowick and Berry, 1972, Chapter 8).

   Figure 32 illustrates the characteristic strain produced by a tetragonal defect in a cubic crystal, such as an interstitial solute atom at the octahedral site or the $\langle 100 \rangle$ dumb-bell self-interstitial atom. In this case, there are three equivalent but orientationally distinguishable configurations: the tetragonal axis is parallel to [100], [010] and [001], which may be labeled $p = 1$, 2, and 3, respectively. Under a uniaxial tensile stress σ_{33}, for example, the relevant component λ_{33} is λ_1 for $p = 1$ and 2 (knowing that $\lambda_1 = \lambda_2$), and is λ_3 for $p = 3$. Eqn (39) then gives the relaxation magnitude of the compliance, which is the reciprocal of Young's modulus in the $\langle 001 \rangle$ direction, as follows:

$$\delta E_{\langle 001 \rangle}^{-1} = \frac{2}{9} \frac{c\Omega}{kT} |\lambda_1 - \lambda_3|^2. \tag{40}$$

If either the concentration c or the so-called "shape factor" $|\lambda_1 - \lambda_3|^2$ is known from another experiment, the other can be determined by measuring the relaxation magnitude. Carbon and nitrogen atoms dissolved in α iron give rise to a relaxation by stress-induced redistribution, called the Snoek relaxation

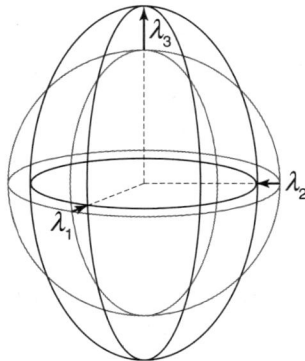

Figure 32   Schematic illustration of the elastic distortion due to a defect of tetragonal symmetry in a cubic crystal. The strain produced by the defect is represented by an ellipsoid whose principal axes measure $1 + \lambda_i$, where λ_i are principal values of the λ tensor.

---

[13] For simplicity, we consider a uniaxial stress and the corresponding strain component and omit the subscripts.

(Nowick and Berry, 1972, Chapter 9). With the concentration c of the solute determined by chemical analysis, the shape factor evaluated from the relaxation magnitude is known to agree well with the tetragonal distortion of Fe–C and Fe–N martensite. It indicates that in both dilute solid solution and the supersaturated martensite the C and N atoms occupy the same positions, the octahedral interstitial sites. The Snoek relaxation is observed for the light solute species (C, N, O) in bcc metals.

A useful feature of anelastic relaxation is that its intensity depends on the orientation of the external stress. For example, it is apparent from **Figure 3(a)** that a tetragonal defect in a cubic crystal would respond to a tensile stress if the stress is applied in either of $\langle 100 \rangle$ directions, but no reorientation, and thus no relaxation, would occur if the stress is applied parallel to $\langle 111 \rangle$ directions, for in the latter case the stress interacts with the defects in the three orientations identically. In general, how a point defect responds to external stress is determined by its symmetry, and is an important property governing the relaxation behavior. Defects of tetragonal symmetry in cubic crystals respond only to $\{110\}\langle 1\bar{1}0 \rangle$ shear stress, while those of trigonal symmetry (having a principal axis in $\langle 111 \rangle$) only to $\{001\}\langle \bar{1}00 \rangle$ shear stress, producing relaxation in the compliance S' $(=2(S_{11} - S_{12}))$ and S $(=S_{44})$, respectively. Such rules, called "selection rules", have been derived from group theory and summarized by Nowick and Heller (1965). For cubic crystals, the relaxation of the compliance J under tensile stress of an arbitrary orientation is given by

$$\delta J = \left(\frac{1}{3} - \Gamma \right) \delta S' + \Gamma \, \delta S. \tag{41}$$

Here, Γ is defined as $\Gamma \equiv \gamma_2^2 \gamma_3^2 + \gamma_3^2 \gamma_1^2 + \gamma_1^2 \gamma_2^2$ with γ_i being the directional cosines between the stress axis and the crystal axes i, and ranges from 0 for the stress axis in $\langle 100 \rangle$ to 1/3 for $\langle 111 \rangle$. The dependence of the relaxation intensity on the stress direction is thus the opposite between tetragonal defects and trigonal defects, as illustrated in **Figure 33**. Examination of the relaxation strength in single crystals thus allows identification of the defect symmetry, which can be an important clue to determining the atomic configuration of the defect.

6.3.5.3 *Relaxation Rate—Kinetics*

Measurements of the relaxation rate provide information on the mobility of the defect. As illustrated in **Figure 34**, if a stress is applied abruptly to a defect-containing solid and is kept constant, an extra deformation, called anelastic strain, appears gradually with time above the instantaneous elastic strain. The time-dependent deformation in this quasi-static experiment is described by a simple exponential time law,

$$\varepsilon(t) = \sigma_0 \left\{ J + \delta J \left[1 - \exp\left(-\frac{t}{\tau} \right) \right] \right\}, \tag{42}$$

with the relaxation time τ. If the stress is removed, the strain drops but not immediately to zero, and the remnant strain disappears exponentially with the same relaxation time. This behavior is called elastic after-effect. On the other hand, if a sinusoidal stress is applied the strain lags behind the stress. The tangent of the phase lag, ϕ, is a measure of the loss of mechanical energy, which is expressed by the Debye equation

$$\tan \phi = \frac{\delta J}{J} \frac{\omega \tau}{1 + (\omega \tau)^2}, \tag{43}$$

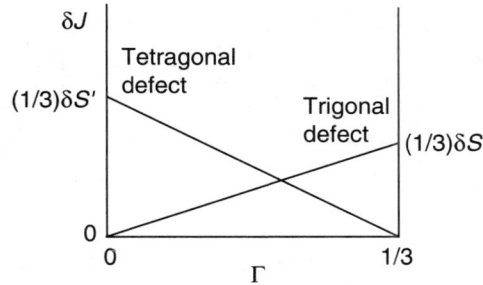

Figure 33 Effects of the direction of the external tensile (or compressive) stress on the magnitude of anelastic relaxation in compliance J due to tetragonal and trigonal defects in cubic crystals. $\Gamma = 0$ and 1/3 correspond respectively to a stress parallel to $\langle 100 \rangle$ and $\langle 111 \rangle$, respectively.

where ω is the angular frequency of the applied stress (Nowick and Berry, 1972, Chapter 3). As this function exhibits a maximum at $\omega\tau = 1$, one can readily determine the relaxation time by measuring the phase lag as a function of vibration frequency. The relaxation rate, τ^{-1}, is in simple cases directly proportional to the frequency of the reorientation jump of the defect.[14] Since the atomic jumps controlling the relaxation is often the elementary process of diffusion, the diffusion coefficient is obtained from the relaxation rate.

Long-range diffusion of defects under stress gradient gives rise to a mechanical relaxation effect called the Gorsky effect, or Gorsky relaxation (Nowick and Berry, 1972, Chapter 11), which is apparently similar to effects due to short-range reorientation or redistribution. If defects that produce a volume dilatation are present in, for example, a plate-shaped specimen and the plate is bent as shown in **Figure 35**, the defects migrate from the compressed side to the expanded side. This

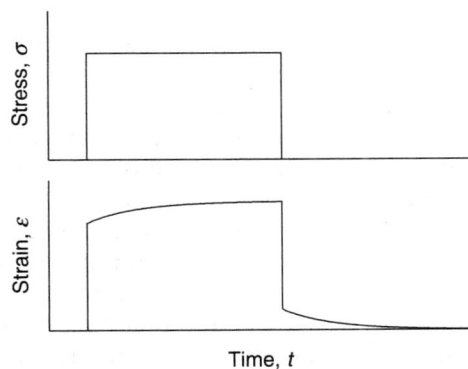

Figure 34 Elastic after-effect, in which strain appears or disappears gradually with time on abrupt changes in external stress.

[14] In cases where more than one type of atomic jumps are involved, the relaxation rate is a function of their frequencies.

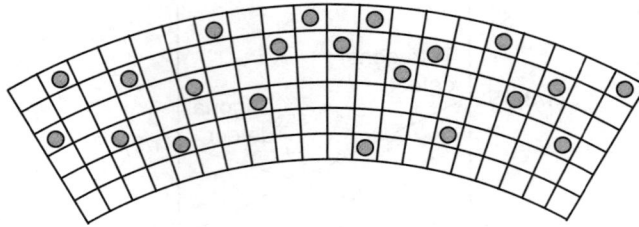

Figure 35 Schematic illustration of the Gorsky effect. When a plate containing dilatational defects is bent by external stress the defects migrate from the compression side to the tension side.

redistribution gives rise to a retarded but reversible elastic deformation; the stress–strain response exhibits either elastic after-effect or stress relaxation. For a specimen of a rectangular cross section, the relaxation time is given by

$$\tau = \frac{d^2}{\pi^2 D},$$ (44)

where d is the thickness of the specimen. This method has been used efficiently for measuring hydrogen diffusion in metals. In either case of short-range reorientation and long-range migration, measurements of the rate of mechanical relaxation allow determination the diffusion coefficient out of the ranges usually accessible by conventional techniques such as interdiffusion and radiotracer experiments.

6.3.5.4 Magnetic Relaxation
Magnetic relaxation techniques, whose principles are similar to the mechanical relaxation methods described above, are available to ferromagnetic materials (Kronmüller, 1970; Rathenau and de Vries, 1969). One of them is magnetic anisotropy relaxation, which is based on the directional preference of low-symmetry defects with respect to the magnetization direction, e.g., parallel to or perpendicular to it. A single crystal specimen is held in an external magnetic field and kept at saturation magnetization. If the direction of the field is abruptly changed to another direction, for example from [100] to [010] with α iron, the low-symmetry defects change their orientation at a rate limited by their jump frequency. The gradual change in their orientation distribution is reflected to the macroscopic magnetization, which can be monitored by a torque magnetometer. One would thus observe transient behavior similar to the elastic after-effect experiment, **Figure 34**.

Another technique, called magnetic disaccommodation, is based on interaction of defects and domain walls, and is to be done under demagnetized conditions. First the sample is demagnetized by an alternating magnetic field with its amplitude gradually being increased to a level at which the domain structure and the distribution of defects are randomized. After demagnetization, the AC susceptibility is measured at a small field amplitude continuously with time. If the defects are of anisotropic nature and have a preferred orientation with respect to the direction of the local magnetic moment, the defects in the domain walls change their orientation within the wall to be in the preferred configuration. The domain walls become less mobile because the directions of local magnetization have been stabilized in turn by the reoriented defects. This results in a decrease of susceptibility with time, usually in the common form $1 - \exp(-t/\tau)$ with a rate constant τ. This effect was known in the 1930s as a drop of initial permeability (disaccommodation) of nominally pure carbonyl iron (Richter, 1937a, 1937b), which later turned out to be caused by redistribution of C and N impurities (Snoek, 1939a, 1939b; Brissonneau, 1958). The kinetics of the reorientation of defects can be studied by this technique together with variations with temperature, but it is not easy in general to determine other information, such as their concentration, in a quantitative manner.

6.3.6 X-ray and Neutron Scattering

Point defects in crystals give rise to characteristic effects on scattering intensities of X-rays and neutrons, in addition to the shift of Bragg peak positions due to changes in interatomic distances. While a regular periodic array of atoms produces sharp diffraction peaks at particular directions, no intensity is observed everywhere else because of the destructive interference of scattered waves. Atomic displacements due to defects disturb the latter, and cause diffuse scattering in various ways, depending on the character of the distortion. The displacements at the immediate vicinity of the defect lead to diffuse intensities in regions far from Bragg peaks, whereas the long-range strain field of the defect, in contrast, produces strong diffuse scattering near the Bragg peaks (Huang scattering), with their intensities proportional to the defect concentration and the square of the scattering amplitude characteristic of the defect species (Dederichs, 1973; Ehrhart et al., 1974). Measurements of diffuse scattering intensities can, therefore, give detailed information on the defects; atomic configuration, and symmetry and strength of the strain field can be studied, which can be used in turn for identification of defect species. Evolution of metastable defects can also be monitored, since clusters of point defects also produce significant diffuse scattering. On the other hand, local strains by dislocations cause broadening of the diffraction peak, and by using this effect the density of dislocations can be evaluated by precisely analyzing the profiles of diffraction peaks (Ungár and Tichy, 1999).

The principle of a scattering experiment and the diffuse scattering from a defect-containing crystal are schematically illustrated in **Figure 36** (Robrock, 1990). The wave vectors of the incident and the

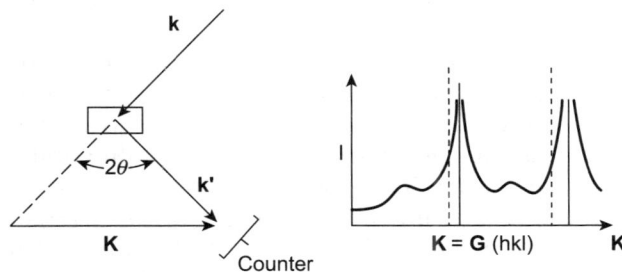

Figure 36 The geometry of the scattering experiment (upper) and a schematic diffraction pattern (lower). **k** and **k′** are wave vectors of the incident and diffracted waves, **K** = **k′** − **k** is the scattering vector, and θ is the scattering angle. The vertical dashed and solid lines in the diffraction pattern indicate the position of the Bragg peaks from a defect-free crystal and a defect-containing crystal, respectively. After Robrock (1990).

scattered photons or neutrons are $\mathbf{k}$ and $\mathbf{k}'$, respectively, and the scattering vector is defined as $\mathbf{K} = \mathbf{k}' - \mathbf{k}$. Bragg peaks occur when $\mathbf{K}$ equals to any of the reciprocal lattice vectors $\mathbf{G}$, and diffuse scattering appears in between. If the concentration of the defects is low and their spatial distribution is random, as explained above the amplitude of diffuse scattering at regions where $\mathbf{K}$ is far from $\mathbf{G}$ is dominated by displacements of atoms in the vicinity of the defect, while that of Huang scattering (at $\mathbf{K}$ regions close to $\mathbf{G}$) is determined by the long-range displacement field. Low-symmetry defects induce anisotropic displacements in the crystal, and contribute to anisotropic diffuse scattering through orientation-dependent Fourier components of the displacement field. By measuring how diffuse scattering intensities depend on $\mathbf{K}$ (i.e. orientation and distance in the reciprocal space), therefore, the symmetry of the defect may be identified. To determine the nature of the defect, the measured intensity profiles are compared with those calculated theoretically by assuming a particular atomic configuration, and the one that agrees best with the experimental observation is chosen (Ehrhart, 1978). In practice, however, extremely precise measurements are required to evaluate differences between diffuse scattering intensities with and without defects, and the defect concentration must be at least 100 ppm. The intensity of Huang scattering, on the other hand, increases as $|\mathbf{K} - \mathbf{G}|^{-2}$ on approaching a Bragg peak (i.e. $|\mathbf{K} - \mathbf{G}| \to 0$), which is advantageous for experimental observation. These techniques played important roles in determining the symmetry and structure of self-interstitial atoms in metals in the 1970s (Ehrhart, 1978).

6.3.7 Transmission Electron Microscopy

Since the early years of commercial electron microscopes, study of defects in metals has been one of the most useful applications of transmission electron microscopy. The first observations of dislocations in metals were reported in mid 1950s (Hirsch et al., 1956; Bollmann, 1956), and those of stacking faults and various secondary defects in quenched metals immediately followed. Direct observation of single point defects is still difficult (it is possible only with field-ion microscopes), but very small clusters of point defects (containing about ten of them) can be identified owing to well-developed imaging techniques with the aid of the refined theory of diffraction contrast (Hirsch et al., 1965; Amelinckx et al., 1970). With continued developments of instruments and new methods, dislocation cores and grain boundaries are imaged routinely in atomic resolution, serving for better understanding of their properties. Moreover, by combining with chemical analysis methods (energy dispersive X-ray spectroscopy, EDS, and/or electron energy loss spectroscopy, EELS), it has recently become possible to quantitatively analyze segregation of a trace amount of impurity atoms to grain boundaries (e.g., Shigesato et al., 2012).

As the acceleration voltage of conventional transmission electron microscopes is a few hundred kV or above, electron microscope observations may have unwanted damage effects. For studies of point defects in solids, however, it is not always a drawback, as it provides a means of in situ observation of electron irradiation. Electrons injected to the thin foil specimen stochastically make elastic collisions with the atoms in the solid and may cause knock-on events mentioned earlier. Making use of this effect, dynamic behavior of point defects—growth, shrinkage, and/or coalescence of dislocation loops, voids, and other forms of clusters—can be studied by in situ observation under defect production. Also, evolution of defect clusters during annealing can be monitored by using specimen holders with which the temperature can be changed. Recent observations of rapid one-dimensional motion of nanometer-sized dislocation loops are examples attracting renewed interests to dynamics of defect clusters (Arakawa et al., 2007; Matsukawa and Zinkle, 2007).

6.4 Point Defects in Pure Metals

6.4.1 Vacancies and Self-interstitial Atoms

In **Table 2** are summarized experimental data for vacancies and self-interstitial atoms in pure metals of fcc, bcc, and hcp structures. The formation enthalpy and entropy of a vacancy have been obtained mostly from equilibrium measurements at high temperatures by differential dilatometry or positron annihilation. The migration enthalpy of a vacancy $\Delta h_{m,v}$ can be derived from self-diffusion activation energy Q and the formation enthalpy $\Delta h_{f,v}$ as far as the diffusion behavior is normal, i.e. the self-diffusion occurs by a simple monovacancy mechanism and the diffusion coefficient obeys the Arrhenius law. Many of the other data have been obtained from experiments on non-equilibrium defects. As mentioned earlier, excess point defects can be introduced by quenching, plastic deformation and particle irradiation. The standard method for studying the properties of self-interstitial atoms is electron irradiation, in which pairs of a vacancy and a self-interstitial atom (Frenkel pairs) can be introduced in a controlled manner, by adjusting the energy and the fluence of the incident electrons. The mobilities of the elementary defects can be studied by measuring the electrical resistivity in the course of annealing, where characteristic recovery stages appear in the resistivity versus annealing time or temperature curve (Lucasson and Walker, 1962; Schilling and Sonnenberg, 1973; Balluffi, 1978; Young, 1978). **Figure 37** shows, as a typical example, the case of electron-irradiated pure copper (Ehrhart and Schultz, 1991). Each of the stages corresponds to a particular process such as annihilation of close vacancy–interstitial pairs, clustering of vacancies and of interstitial atoms, annihilation to sinks, etc. With proper interpretation and analyses, information on the mobility of the relevant defects can be obtained. A comprehensive account of annealing theories is given in Chapter 2 of Damask and Dienes (1963).

The formation enthalpy of an self-interstitial atom is evaluated as the difference between those of a Frenkel pair and of a vacancy, where the former is commonly obtained from the energy release on annealing using calorimetry. The formation volume Δv_f of a vacancy and that of a self-interstitial atom have been determined directly by diffuse X-ray scattering, or indirectly from the pressure dependence of the equilibrium concentration. Virtually no experimental data are found of the volume and entropy of migration of a vacancy, nor of the entropy of formation, the volume and entropy of migration of a self-interstitial atom. Theoretical studies are expected to supply those pieces of information: calculation of equilibrium concentrations of point defects at non-zero temperatures are now being attempted (Grabowski et al., 2011). Also, recovery of radiation induced point defects, various defect reactions and annihilation processes by annealing can be studied by computation combining ab initio calculations and Monte Carlo simulation (Fu et al., 2005).

6.4.1.1 In fcc Metals

In early computer simulation studies the stable atomic structure of a vacancy was found to be a simple vacant atomic site, with the neighboring atoms being displaced, or "relaxed", inwards to some extent (Johnson and Brown, 1962; Johnson, 1966). This is supported by the agreement between theory and experiment of positron annihilation on the formation enthalpy (Schaefer, 1987), not only for fcc metals but also for metals of other structures. The formation volume of a vacancy is close to one atomic volume, i.e. the relaxation is small in magnitude. The same conclusion was derived by later simulation studies using more sophisticated interatomic potentials (Finnis and Sinclair, 1984; Daw and Baskes, 1984; Daw et al., 1993). In the end of 1980s it became possible to calculate the formation energy of a vacancy fairly accurately from electron theory (Gillan, 1989; De Vita and Gillan, 1991; Dederichs et al., 1991; Mehl and Klein, 1991). Owing to the rapid development of "ab initio" calculations based on the density-functional theory since then, formation and migration energies of point defects are now

Table 2 Experimental data for the properties of vacancies and self-interstitial atoms and related parameters in pure metals. E_c: cohesive energy, T_m: melting temperature, T_D: Debye temperature, Q: activation energy of self-diffusion, Δh_f: formation enthalpy, Δv_f: formation volume, Δs_f: formation entropy, Δh_m: migration enthalpy. Ω is the atomic volume, and k is the Boltzmann constant. The properties of point defects are mostly from the compilations by Schaefer (1987), Ehrhart and Schultz (1991), Schultz (1991), Schober et al. (1992), and Wollenberger (1996), and those from other sources are as indicated. The data of cohesive energy and Debye temperature are due to Kittel (1996), and those of self-diffusion to Mehrer et al. (1990)

Class	Material	E_c/eV	T_m/K	T_D/K	Q/eV	Temperature of recovery stages/K			Vacancy				Self-interstitial atom		
						Stage I	Stage III	Stage V	Δh_f/eV	$\Delta v_f/\Omega$	$\Delta s_f/k$	Δh_m/eV	Δh_f/eV	$\Delta v_f/\Omega$	Δh_m/eV
fcc metals	Al	3.39	934	428	1.28	37	220	420	0.67	0.95	0.7	0.61	3.0, 3.6	0.9	0.12
	Pb	2.03	601	105	1.13	4	160	280–300	0.58		0.7, 2.6	0.43			0.01
	γ Fe	4.28	1811[a]		2.94				1.4–1.7						
	α Co	4.39	1768		2.99				1.91[f]						
	Ni	4.44	1728	450	2.90	56	340	760	1.79	0.8		1.04		0.8	0.15
	Pd	3.89	1828	274	2.76	35	300–350		1.85			1.03			
	Pt	5.84	2041	240	2.89	22	500	<870	1.35	0.38–0.76	0.4, 1.3	1.43	1.7[l]	0.6–0.8	0.06–0.07
	Cu	3.49	1358	343	2.08	38	250	550–600	1.28	0.75	1.5–2.8	0.70	3.4–4.7[l]		0.12
	Ag	2.95	1235	225	1.76	28	230	540–600	1.11		1.5	0.66			0.09
	Au	3.81	1337	165	1.83	<0.01	290	650	0.93	0.85	0.72	0.80			
bcc metals	V	5.31	2183	380	3.2	≤3.8	170		2.1, 2.2			0.5			
	Nb	7.57	2750	275	4.1	<6.3	200–270		2.7, 3.1			0.55			
	Ta	8.10	3269	240	4.4	4–6	260–300		3.1			0.7			
	Cr	4.10	2180	630	4.6	36	350		2.1			0.95			
	Mo	6.82	2896	450	5.0	≈40	450–640		3.1	0.9	1.6	1.35		0.1	0.08
	W	8.90	3459	400	6.5		620–900		3.6		3.2	1.7		0.1	0.05, 0.09
	α Fe (para)	4.28	1811[a]		2.5				1.4–1.85			0.65			
	α Fe (ferro)	4.28	1811[a]	470	3.0	≤110	220–278		1.6–2.0	0.95		0.55	4.7, 5.0	0.1	0.25–0.30

hcp metals													
Mg	1.51	922	400	1.43	<13	130–200	≤400	0.6–0.8		0 ± 0.3	0.45–0.6		(0.03)
Zn	1.35	693	327	0.98	13	120–160	160–280	0.54	0.4	0.5, 1.0	0.42	2.5	0.015
Cd	1.16	594	209	0.83	≤3.6	120	<300	0.46	0.46	0.3, 0.5	0.4		
α Ti	4.85	1941[b]	420	3.14[d]	120–130	250–300		1.55[g] 1.27[h]					
α Zr	6.25	2128[b]	291	3.17[e]	≈102	250–300		1.6[i]	0.95		0.54–0.7	−0.4	0.26, 0.30
α Hf	6.44	2506[b]	252	3.84, 3.61				1.8–1.9[j] 2.45[k]					
Re	8.03	3459	430	5.30	90	630	≈1100				2.2	2.5	(≈0.16)
ε Co	4.39	1768[c]	445		45–60	250–350	>500		0.95		0.72		0.10–0.15

[a]T_m of bcc δ phase.
[b]T_m of bcc β phase.
[c]T_m of fcc α phase.
[d]Köppers et al. (1997).
[e]Hood et al. (1997).
[f]Schulte and Campbell (1979).
[g]Shestopal (1966).
[h]Hashimoto et al. (1984).
[i]Hood (1986).
[j]Hood (1988).
[k]Hood and Schultz (1995).
[l]Difference between the formation enthalpy of a Frenkel pair and that of a vacancy.

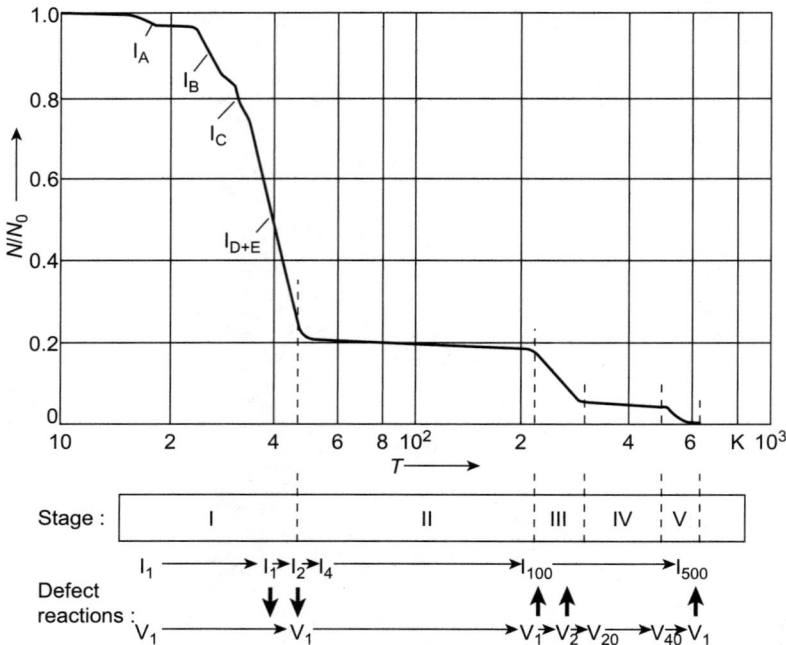

Figure 37 Recovery stages observed in annealing of electron-irradiated pure copper. I_1 and V_1 denote single interstitial atoms and vacancies, and I_2 and V_2 di-interstitials and di-vacancies, respectively. After Ehrhart and Schultz (1991).

computed readily and reliably (e.g., Korzhavyi et al., 1999; Hoshino et al., 1999, 2001; Nazarov et al., 2012).

The stable structure of a self-interstitial atom was found to be the $\langle 100 \rangle$ dumb-bell configuration by theoretical calculations (Huntington, 1953; Seeger et al., 1962; Johnson and Brown, 1962; Johnson, 1966; Schober, 1977; Jesson et al., 1997; Mishin et al., 2001a). The formation energy is roughly a few times that of a vacancy, with minor differences from other configurations. The presence of $\langle 100 \rangle$ dumb-bell interstitials has been evidenced by diffuse X-ray scattering, and diaelastic[15] and anelastic effects (for references, see Ehrhart (1978) and Schilling (1978)). Self-interstitial atoms are known to produce large volume dilatation: the formation volume is known to be as large as one atomic volume. This means that the relaxation volume around a single self-interstitial atom is about twice the atomic volume.[16] On the other hand, the anisotropy in the strain field, which is of tetragonal symmetry, is known to be rather weak (Ehrhart, 1978; Schilling, 1978; Robrock, 1990).

The strong distortion associated with self-interstitial atoms gives rise to characteristic effects on the dynamical properties of the crystal (Scholz and Lehmann, 1972; Dederichs et al., 1978). **Figure 38** illustrates the localized and resonance vibration modes associated with a $\langle 100 \rangle$ dumb-bell interstitial atom. They give a local vibrational spectrum that is very different from the ideal spectrum of a perfect crystal, as shown in **Figure 39**. The two resonance modes, A_{2u} and E_g, were detected in a single-crystal of

---

[15] Diaelasticity is a mechanical relaxation caused by stress-induced generation of elastic dipoles, while anelasticity is due to stress-induced reorientation or redistribution (Granato, 1994b).
[16] See Eqn (26).

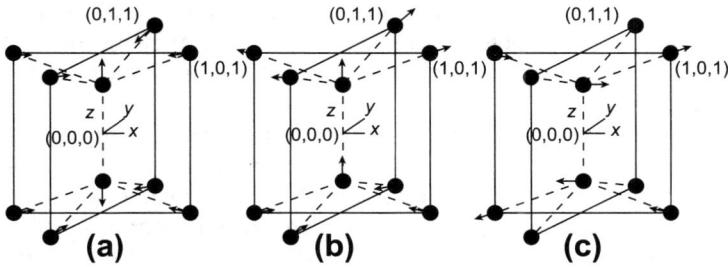

Figure 38 Localized and resonance modes of a $\langle 100 \rangle$ dumb-bell interstitial atom in an fcc crystal. (a) localized mode A_{1g}, (b) translational resonance mode A_{2u}, (c) vibrational resonance mode E_g. After Robrock (1990).

aluminum by inelastic diffuse scattering of neutrons (Urban et al., 1987). The low-frequency resonance modes significantly increase the formation entropy (Ram, 1984). Granato (1992, 1993) pointed out that this entropy could raise the concentration self-interstitial atoms at high temperatures to a level that might significantly affect various physical and thermodynamic properties.

As discussed in Section 6.2.1.1, the activation energy of self-diffusion and the vacancy formation enthalpy in pure metals are roughly proportional to the melting temperature. The relations for fcc metals were shown in **Figures 2(a)** and **4**. They are expressed as

$$Q = (18.0 \pm 0.5)kT_m, \tag{45}$$

$$\Delta h_{f,v} = (10.3 \pm 0.6)kT_m, \tag{46}$$

where platinum and gold markedly deviate in the latter. The relation for the migration enthalpy of a vacancy, which has been independently determined by recovery experiments (see below), is

$$\Delta h_{m,v} = (7.3 \pm 0.3)kT_m. \tag{47}$$

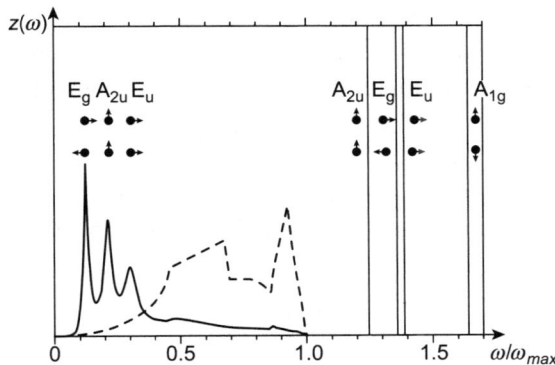

Figure 39 Local vibrational spectrum of a $\langle 100 \rangle$ dumb-bell interstitial atom in an fcc crystal calculated using a Morse potential (Dederichs et al. 1978). The dashed line shows the spectrum of a defect-free crystal. After Robrock (1990).

As mentioned earlier, the sum of the formation enthalpy and the migration enthalpy is close to the activation energy of self-diffusion, which is consistent with the vacancy mechanism[17] (Balluffi, 1978). The volumes of formation and migration also sum up to the activation volume of self-diffusion (Emrick, 1961; Dickerson et al., 1965).

The most reliable data on the migration enthalpy of a vacancy and of a self-interstitial atom have been obtained from annealing kinetics after quenching or electron irradiation. An exemplary annealing curve of electron-irradiated pure copper was shown in **Figure 37**: the electrical resistivity, which represents the number of defects, decreases with the progress of annealing. The recovery curve of copper, considered as a prototype, exhibits three characteristic stages. Stage I, ending at about 50 K, stage III between 200 K and 300 K, and stage V around 600 K. There used to be a long-standing controversy in interpretation of defect reactions occurring at each of the recovery stages (Balluffi, 1978; Young, 1978). Many experimental observations seem to support the "one-interstitial" recovery model, over the "two-interstitial" model, attributing annihilation of self-interstitial atoms by correlated or free migration to stage I, and that of freely migrating vacancies to sinks to stage III. Clusters formed by agglomeration of defects are more stable than isolated point defects, and survive through higher temperatures. At stage II interstitial clusters grow and eventually form small interstitial loops, which continue to grow and remain up to stage V. Similar growth of vacancy clusters occur at stage IV. At stage V, the vacancy and/or interstitial clusters dissociate and the vacancies annihilate at interstitial loops, whereby all the defects are removed.

The temperature where stage I begins in fcc metals is empirically related to the Debye temperature T_D as

$$T_I = 0.14\, T_D \tag{48}$$

(Lucasson et al., 1987). The correlation is rationalized by the role of phonons in the motion of a self-interstitial atom through its resonant vibration modes. The only exception is gold, whose stage I temperature is too low to be detected in experiment. The very high mobility of self-interstitial atoms indicated by the extremely low stage I temperature is attributed to the anomalous phonon dispersion (Lucasson et al., 1987). The temperatures of the three prominent stages in fcc metals are related also with the melting temperature T_m, as follows:

$$
\begin{array}{lll}
\text{Stage I} & : & 0.020 \pm 0.015 \\
\text{Stage II} & : & 0.20 \pm 0.02 \\
\text{Stage III} & : & 0.45 \pm 0.03
\end{array}
$$

where the values are in T/T_m.

The migration enthalpy of a vacancy in fcc metals is well described by the theory developed by Flynn (1968, 1972), which relates the migration enthalpy to elastic constants through lattice vibrations as

$$\Delta h_{m,v} = \delta^2 \langle C \rangle \Omega, \tag{49}$$

where δ^2 is a numerical parameter (≈ 0.1 for fcc crystals), $\langle C \rangle$ is the average elastic constant, which is given for cubic crystals from the anisotropic elastic stiffness constants C_{ij} by

$$\frac{1}{\langle C \rangle} = \frac{2}{15}\left(\frac{3}{C_{11}} + \frac{1}{C'} + \frac{1}{C_{44}}\right), \tag{50}$$

---

[17] It is not the case for some bcc transition metals in which the Arrhenius plot of the self-diffusion coefficient is markedly curved upwards (the next subsection).

where $C' = (C_{11} - C_{12})/2$ is the shear stiffness constant for $\{110\}\langle 1\bar{1}0\rangle$ shear deformation. A generalization of Flynn's theory, which does not include the parameter δ, was developed later by Schober et al. (1992) on the basis on phonon dispersion curves using lattice Green's function. The values of the migration enthalpy predicted by the theory satisfactorily agree with experimental values in most fcc metals.

6.4.1.2 In bcc Metals

The structure of a vacancy is a simple vacant atomic site also in bcc metals, with small shifts in the positions of neighboring atoms, retaining the cubic symmetry. Johnson (1964) devised a short-ranged central-force interatomic potential for bcc iron, the Johnson potential, and calculated energies and volumes of formation and migration energies of vacancies and self-interstitial atoms. The stable form of a self-interstitial atom was suggested to be the $\langle 110 \rangle$ dumb-bell configuration, **Figure 3(b)**. Later simulation studies found the same for vanadium, α iron, molybdenum, and tungsten (Guinan et al., 1977; Miller, 1981; Taji et al., 1989). For the latter three metals, the structure was confirmed by mechanical relaxation and diffuse X-ray scattering experiments (DiCarlo et al., 1969; Hivert et al., 1970; Okuda and Mizubayashi, 1973, 1976; Townsend et al., 1976; Mizubayashi and Okuda, 1977, 1981; Ehrhart, 1978; Grasse et al., 1984; Wallner et al., 1987). Unlike the case of fcc metals, the distortion produced by the $\langle 110 \rangle$ dumb-bell interstitial, which is of orthorhombic symmetry, is strongly anisotropic (Schilling, 1978). This is advantageous to the experimental techniques cited above. In Johnson's calculation the $\langle 111 \rangle$ dumb-bell configuration was less stable than the $\langle 110 \rangle$ counterpart, but the energy difference was small (~ 0.3 eV). Recent ab initio calculations suggest that these two configurations are almost equally stable, when the $\langle 111 \rangle$ dumb-bell is extended in the direction of its axis to form a crowdion[18] (Derlet et al., 2007, and references therein). The dumb-bell interstitial atoms have similar effects on the lattice dynamical properties to the case of those in fcc metals. The local vibrational spectrum of a $\langle 110 \rangle$ dumb-bell interstitial consists of low-frequency resonance modes and high-frequency localized modes (Ram, 1991b), which explains the enhanced thermal displacements observed in experiment.

The relations of the activation energy of self-diffusion and the vacancy formation enthalpy with the melting temperature in bcc transition metals are shown in **Figure 40**, together with the cohesive energy and the migration enthalpy. The former two are written as

$$Q = (19.0 \pm 1.1)kT_{\mathrm{m}}, \tag{51}$$

$$\Delta h_{\mathrm{f,v}} = (11.7 \pm 0.2)kT_{\mathrm{m}}. \tag{52}$$

Here, while $\Delta h_{\mathrm{f,v}}$ is well correlated to the melting temperature as in fcc metals, scatter is apparent for Q. It is more evident for the migration enthalpy $\Delta h_{\mathrm{m,v}}$ shown in the same figure. As can be seen also in **Table 2**, the vacancy migration enthalpy of bcc transition metals does not systematically scale to the melting temperature (Ehrhart and Schultz, 1991; Schultz, 1991) but depends on the group in the periodic table. The ratio $\Delta h_{\mathrm{m,v}}/(kT_{\mathrm{m}})$ is above 5 for group 6 metals (Cr, Mo, W), while it is roughly 2.5 for group 5 metals (V, Nb, Ta). The migration enthalpy is even smaller in the high-temperature bcc phases of group 4 metals, Ti and Zr, to which the anomalously high self-diffusivity with strong upward curvature is attributed (Herzig and Köhler, 1987). Similar trends are also found in the mobility of

---

[18] A crowdion (named after "crowd of ions") is a long chain of atoms whose number (typically 10 to 20) is larger by one than the number of regular sites.

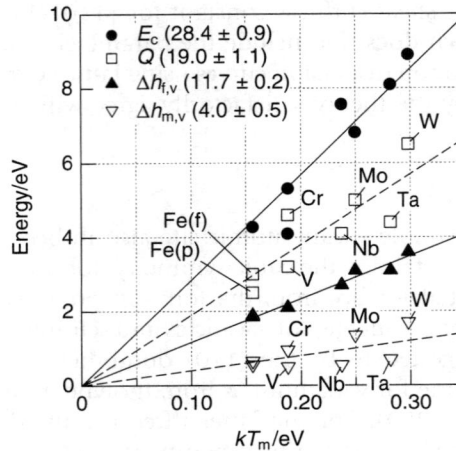

Figure 40 Relations of cohesive energy, activation energy of self-diffusion, formation enthalpy and migration enthalpy of a vacancy in bcc metals with melting temperature T_m. The numbers are proportionality constants to kT_m. "Fe(p)" and "Fe(f)" indicate the paramagnetic and ferromagnetic states of bcc iron, respectively.

self-interstitial atoms: the temperature of stage I recovery is markedly lower in group 5 metals than in group 6 and, accordingly, so is the migration enthalpy of a self-interstitial atom. These group-specific properties are understood as originating from the lattice dynamical properties: the characteristic low-frequency phonon mode, of longitudinal acoustic $(2/3)\langle 111 \rangle$, in these bcc metals (Herzig, 1993). Flynn's theory (1968, 1972) accounts for the trend in the migration enthalpy, and even quantitative agreement is obtained with $\delta^2 = 0.041$. The general theory by Schober et al. (1992) satisfactorily explains the experimental observations.

Resistivity recovery experiments on electron-irradiated materials revealed distinctive features of bcc metals in the mobilities of self-interstitial atoms and vacancies. Interpretation of the complex recovery behavior was difficult in the early years of investigation, but later it turned out to be due in part to interstitial impurities (H, C, N, O). With efforts to prepare high purity materials by efficiently reducing those impurity contents, stages I and III are now understood analogously to the case of fcc metals. As mentioned above, the migration of self-interstitial atoms begins at temperatures below 10 K in group 5 metals, while it starts only above 30 K in group 6 metals. The free migration stage of self-interstitial atoms in group 6 metals is suggested to occur by planar migration, which may be the case for other bcc metals as well (Schultz, 1991). In α iron, however, three-dimensional migration, which is the mechanism in fcc metals, may be dominant (Lucasson et al., 1987).

6.4.1.3 In hcp Metals

Studies of defects in hcp metals have been behind those in cubic metals, even though there are such important materials as titanium, zirconium, and hafnium in this class, particularly in nuclear technology. Reviews of studies up to the late 1980s are available for atomic structure and mobility of point defects (Bacon, 1988; Frank, 1988), diffusion in α zirconium (Hood, 1988), and microstructure evolution in zirconium during irradiation (Griffiths, 1988). There is also a more recent review focussing on collective behavior of point defects in hcp metals under irradiation (Woo, 2000), which is a key for understanding "irradiation growth", the dimensional changes due to anisotropic diffusion and agglomeration of point defects (Fidleris, 1988; Holt, 2008). At the time when the two landmark compilations of the data of diffusion and point defects were made (Mehrer et al., 1990; Ehrhart and Schultz, 1991), some of the essential parameters were still lacking in hcp metals. During the last two decades, reliable values of the activation energy of self-diffusion and the vacancy formation energy in group 4 metals have become available, as seen in **Table 2**. Moreover, theoretical calculations of atomic configurations, and formation and migration energies are actively being made also for hcp metals (Raji et al., 2009; Samolyuk et al., 2013).

The correlations of the cohesive energy, the activation energy of self-diffusion and the vacancy parameters with the melting temperature are shown in **Figure** 41. Except for the cohesive energy, the divalent metals (Mg, Zn, Cd) and transition metals (Ti, Zr, Hf, Re, Co) exhibit similar systematic correlation with melting temperatures. They are written as

$$Q = (17.7 \pm 0.2)kT_{\mathrm{m}}, \tag{53}$$

$$\Delta h_{\mathrm{f,v}} = (10.3 \pm 0.3)kT_{\mathrm{m}}, \tag{54}$$

$$\Delta h_{\mathrm{m,v}} = (7.4 \pm 0.4)kT_{\mathrm{m}}. \tag{55}$$

The relations are quantitatively similar to those for fcc metals.

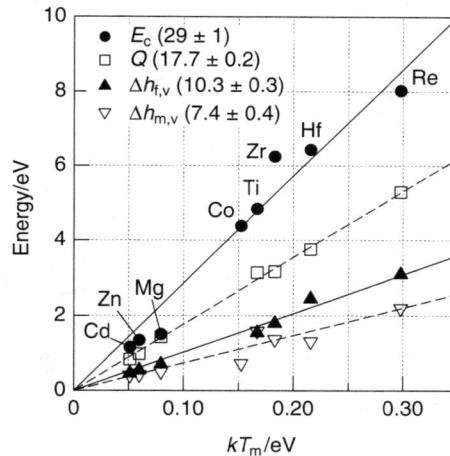

Figure 41 The cohesive energy E_{c}, activation energy of self-diffusion Q, vacancy formation enthalpy $\Delta h_{\mathrm{f,v}}$, and vacancy migration enthalpy $\Delta h_{\mathrm{m,v}}$ in hcp metals. The values of $\Delta h_{\mathrm{m,v}}$ for titanium, zirconium and hafnium are $Q - \Delta h_{\mathrm{f,v}}$, while $\Delta h_{\mathrm{f,v}}$ for rhenium is $Q - \Delta h_{\mathrm{m,v}}$. The numbers to the legends are proportionality constants to kT_{m}. For Ti, Zr, Hf, and Co, T_{m} of the high-temperature phases is used.

The atomic structure of point defects in hcp metals was first studied by Johnson and Beeler (1981). According to their simulation using a short-ranged pair potential, the energy and volume of vacancy formation were about 1.5 eV and 0.5 Ω, respectively, and a divacancy is stable with a binding energy of 0.6 eV for the in-plane configuration and 0.4 eV for the out-of-plane configuration. In **Figure 42** are illustrated configurations of a self-interstitial atom in the hcp structure. Relative stabilities of various configurations of a self-interstitial atom, as well as multiple vacancies and interstitials, have been studied by computer simulation, and more recently by ab initio calculations. The stable configuration of a self-interstitial atom is often B_O or C in the former, depending on the interatomic potentials used. For example, configuration C is most stable with a model in which $c/a = 1.63$, but configuration B_O is favored for $c/a = 1.86$ (Oh and Johnson, 1989).[19] In ab initio calculations, on the other hand, the O configuration is of the lowest formation energy (Raji et al., 2009; Samolyuk et al., 2013). In both the cases the energies of the various configurations are not significantly different. The properties found in these studies are generally similar to those in fcc metals: the migration energy of a vacancy is appreciably smaller than the formation energy, the association energy of a divacancy is of about -0.1 eV, the formation energy of a self-interstitial atom is roughly twice as large as that of a vacancy, and the migration energy is much smaller than that of a vacancy.

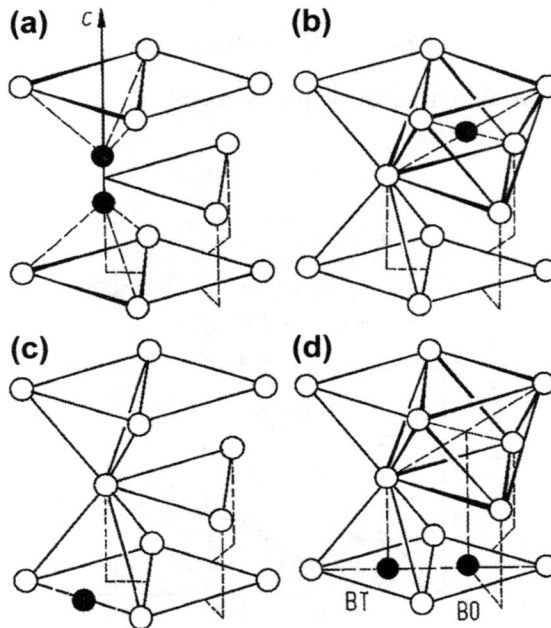

Figure 42 Configurations of a self-interstitial atom in hcp crystals. (a) $\langle 0001 \rangle$-split (dumb-bell) configuration, C, (b) octahedral, O, (c) crowdion in the basal plane, B_C, (d) interstitials in the basal plane at a projection of the tetrahedral site, B_T, and octahedral site, B_O. After Ehrhart and Schultz (1991).

---

[19] The latter is close to Zn and Cd, while the former to the other hcp metals.

6.4.2 Interaction between Point Defects

6.4.2.1 Defect Complexes in General

Complexes of point defects may form by association of individual defects, and the associated defects are often energetically more favored. The balance between the associated and separated defects may be described by a chemical reaction between the defect α and defect β:

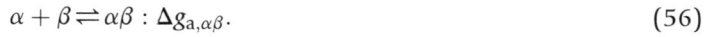

$$\alpha + \beta \rightleftharpoons \alpha\beta : \Delta g_{a,\alpha\beta}. \tag{56}$$

The free energy of association $\Delta g_{a,\alpha\beta}$ is defined as the difference in the formation energy between the isolated and associated states:

$$\Delta g_{a,\alpha\beta} = \Delta g_{f,\alpha\beta} - \left(\Delta g_{f,\alpha} + \Delta g_{f,\beta} \right). \tag{57}$$

Association is favored when $\Delta g_{a,\alpha\beta}$ is negative, and vice versa. The equilibrium concentration of the complexes is expressed in the form

$$c_{\alpha\beta} \approx A\, c_\alpha c_\beta \exp\left(-\frac{\Delta g_{a,\alpha\beta}}{kT} \right), \tag{58}$$

where A is a number of the order of 10 depending on the geometrical configuration of the associated defect. The free energy of binding, $\Delta g_{b,\alpha\beta} \equiv -\Delta g_{a,\alpha\beta}$, is often used to characterize the preference for the complex.

The volume change is given by the trace of the λ tensor and can be evaluated, for cubic crystals, by the relative change in the lattice parameter caused by the addition of the solute,

$$\frac{\Delta V}{V} = \mathrm{tr}\left(\lambda_{ij} \right) = \frac{3}{a}\frac{da}{dc}. \tag{59}$$

According to anisotropic elasticity theory, two centers of isotropic distortion interact with each other through the elastic anisotropy of the host crystal (Eshelby, 1956; Leibfried and Breuer, 1978). The interaction energy in cubic crystals is expressed as

$$E_{\mathrm{int}} = \frac{15}{4\pi}\frac{C' - C_{44}}{[3(1-\nu)/(1+\nu)]^2}\frac{\Delta V_1 \Delta V_2}{r^3}\Lambda, \tag{60}$$

where ν is Poisson's ratio, ΔV_1 and ΔV_2 are the changes in volume associated with defect 1 and defect 2, respectively, and r is the distance between them. The last factor Λ is defined by the orientation of the pair as

$$\Lambda \equiv \alpha_1^4 + \alpha_2^4 + \alpha_3^4 - \frac{3}{5}, \tag{61}$$

with α_i being the directional cosines between the pair axis and the three principal crystal axes. According to this formula, in crystals whose elastic anisotropy $A = C_{44}/C'$ is greater than 1, for example, two point defects producing volume changes of the same sense attract each other if aligned in $\langle 100 \rangle$ but repel in $\langle 111 \rangle$. A defect pair thus formed may produce, in turn, uniaxial distortion along the axis of the pair, and cause an anelastic relaxation effect, which is called Zener relaxation for the case of substitutional atom pairs (Nowick and Berry, 1972, Chapter 10). A similar relaxation effect is anticipated for divacancies and interstitial solute atom pairs. The latter is in fact observed for carbon in fcc metals (Numakura et al., 2000).

6.4.2.2 Divacancies

At elevated temperatures where the equilibrium concentration of vacancies is fairly high, vacancies at adjacent sites, i.e. divacancies, trivacancies, or higher order clusters, may occur. By considering the reduction in the number of missing bonds and also in the strain energy, association of two vacancies is expected to reduce the enthalpy of the crystal, as discussed in Section 6.2.1.3. The free energy of association is defined from the free energies of formation as

$$\Delta g_{a,vv} = \Delta g_{f,vv} - 2\Delta g_{f,v}. \tag{62}$$

The concentration of divacancies, defined as the ratio of the number of divacancies to the number of lattice sites, is expressed as

$$c_{vv} \approx \frac{z}{2} \exp\left(-\frac{2\Delta g_{f,v}}{kT}\right) \exp\left(-\frac{\Delta g_{a,vv}}{kT}\right) = \frac{z}{2} \exp\left(-\frac{\Delta g_{f,vv}}{kT}\right). \tag{63}$$

The total concentration of vacant lattice sites is given by

$$c_{v,total} = c_v + 2c_{vv} \approx \exp\left(-\frac{\Delta g_{f,v}}{kT}\right)\left[1 + z\exp\left(-\frac{\Delta g_{f,v} + \Delta g_{a,vv}}{kT}\right)\right]. \tag{64}$$

It increases if there is strong binding between two vacancies, i.e. the association energy $\Delta g_{a,vv}$ is negative and its magnitude is not negligible in comparison to $\Delta g_{f,v}$. For fcc metals, the enthalpy of association is estimated to be -0.1 to -0.3 eV (Balluffi, 1978).[20]

Divacancies in fcc metals are suggested to be more mobile than single vacancies, by computer simulation (Johnson, 1965) and also by experiment (Franklin and Birnbaum, 1971). In bcc metals, the stable configuration of a divacancy is two vacant sites at the second-neighbor positions, and its migration energy is, on the other hand, similar in magnitude to that of the single vacancy (Johnson, 1964). In both fcc and bcc metals, the self-diffusion coefficient often exhibits positive deviation from the Arrhenius law at high temperatures. It was assumed to be due to an increasing contribution of divacancies to diffusion with increasing temperature (e.g., Seeger and Mehrer, 1970). In 1975 Gilder and Lazarus (1975) proposed that the upward curvature could be explained without considering divacancies, but from temperature-dependent formation and migration enthalpies of monovacancies, which arise from anharmonic atomic vibrations. This idea was later substantiated experimentally for the bcc transition metals, as mentioned in Section 6.4.1.2. In this context, the role of divacancies in the self diffusion in fcc metals has been questioned and critically re-analyzed (Mundy, 1987, 1992). More recently, ab initio calculations suggest, for aluminum, that divacancies are even energetically unfavorable, and attribute the non-Arrhenius temperature

---

[20] Theoretical values (Foiles et al., 1986; Klemradt et al., 1991) are more or less similar to the experimental estimates

dependence of the vacancy concentration to lattice anharmonicity (Carling and Wahnström, 2000; Carling et al., 2003).

6.4.2.3 Vacancy–Solute Complexes

In dilute solid-solution alloys, complexes of vacancies and solute atoms may be formed, owing to electronic and/or elastic interaction between the two defect species. Formation of pairs of a vacancy and a solute atom in metals has been studied theoretically and experimentally (Le Claire, 1978; March, 1978; Bérces and Kovács, 1983; Hautojärvi, 1987), and the diffusion behavior of the solute atoms, as well as the solvent atoms, has been analyzed (Le Claire, 1978, 1993). As the concentration of solute atoms can be, and usually is, much higher than that of vacancies and may have significant effects on the properties of the material, we discuss the formation of vacancy–solute atom pairs in slightly more detail (Howard and Lidiard, 1964).

Let us consider a dilute binary A–B alloy consisting of N_A and N_B of A and B atoms, respectively. The number of lattice sites next to a B atom is zN_B and the number of other sites is $N_A - zN_B$. The probabilities of finding a vacancy at these sites are given respectively by $\exp(-\Delta g_{f,v}/kT) \times \exp(-\Delta g_{a,vs}/kT)$ and $\exp(-\Delta g_{f,v}/kT)$. The numbers of isolated vacancies and those associated with a solute atom are thus written as

$$n_v = (N_A - zN_B) \exp\left(-\frac{\Delta g_{f,v}}{kT}\right),$$
$$(65)$$

$$n_{vs} = zN_B \exp\left(-\frac{\Delta g_{f,v} + \Delta g_{a,vs}}{kT}\right).$$
$$(66)$$

Their concentrations are obtained by dividing by the total number of lattice sites, $N_A + N_B + n_v + n_{vs}$. Assuming that the concentrations of the solute atoms and vacancies are low, we obtain the expression for the total concentration of vacancies, called the Lomer formula:

$$c_{v,total} \approx \exp\left(-\frac{\Delta g_{f,v}}{kT}\right)\left[1 - zc_B + zc_B \exp\left(-\frac{\Delta g_{a,vs}}{kT}\right)\right],$$
$$(67)$$

where $c_B \equiv N_B/(N_A + N_B)$ is the mole fraction of the solute.

Figure 43 shows Arrhenius plots of calculated concentrations of isolated vacancies, c_v, vacancy-solute atom pairs, c_{vs}, and their sum, $c_{v,total}$, for the case of $z = 12$, $c_B = 1 \times 10^{-2}$, and $\Delta g_{a,vs} = -0.15\,\Delta g_{f,v}$. Since the solute concentration is low in this example, c_v is virtually equal to the vacancy concentration in the absence of the solute atoms. The attractive interaction between a vacancy and the solute atom gives rise to vacancies bound to the solute atoms at low temperatures. With decreasing temperature the number of vacancy–solute atom pairs progressively increases and becomes larger than the number of isolated vacancies, leading to higher total concentration of vacancies than without the solute atoms. If we take aluminum for practice, for which $\Delta h_{f,v} \approx 0.7$ eV and $\Delta s_{f,v}/k \approx 0.7$, the melting temperature (934 K) and room temperature correspond respectively to 8 and 26 in the horizontal axis, $\Delta g_{f,v}/kT$. The association energy of -0.15 in terms of $\Delta g_{f,v}$ is equivalent to about -0.1 eV, which applies to such solute atoms as In and Sn (Wolverton, 2007). In this case, below $\Delta g_{f,v}/kT = 15$ (or 520 K), where c_{vs} and c_v cross, the majority of vacancies are bound to the solute atoms.

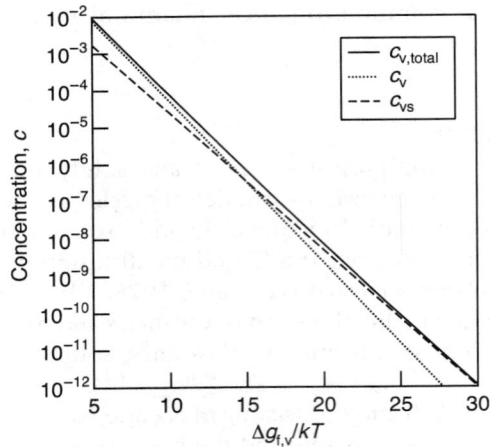

Figure 43 Arrhenius plots of the concentrations of isolated vacancies, c_v (dotted line), of vacancy–solute atom pairs, c_{vs} (dashed line), and the total vacancy concentration, $c_{v,total}$ (solid line), in a dilute solid-solution alloy with coordination number 12.

For some electropositive solutes in copper and silver, experimental values of $\Delta g_{a,vs}$ ranging from -0.1 to -0.2 eV are reported (Köstler et al., 1987). In Section 6.2.1.3 was mentioned that there is strong attraction between interstitial C atom and a vacancy in bcc iron, for which $\Delta g_{a,vs}$ is estimated at -0.5 eV or even larger in magnitude. For the case of hydrogen as the solute species, an unusually large number of vacancies, amounting to 10 at.%, has been known to be produced (Fukai and Ōkuma, 1993; Fukai, 2003). These vacancies, called "super-abundant vacancies", are an example of the enhancement of the vacancy concentration by the attractive solute–vacancy interaction described by the Lomer formula. The high concentration of vacancies has immediate impact on substitutional diffusion (Hayashi et al., 1998; Iida et al., 2005) and is therefore of practical importance. There is ample evidence that interstitial hydrogen atoms attract vacancies to form clusters, with a binding energy of 0.3 to 0.4 eV, and considerably reduce the effective formation energy of vacancies (Harada et al., 2005; Hiroi et al., 2005; Fukai et al., 2006). The entire picture is supported by ab initio calculations (Tateyama and Ohno, 2003; Nazarov et al., 2010).

6.4.2.4 Self-interstitial–Solute Complexes

Interaction between self-interstitial atoms and solute atoms is also known. In irradiated metals with solute (or impurity) atoms, self-interstitial atoms are often trapped to a solute atom, by which recovery is retarded (Maury et al., 1985, and references therein). The complex sometimes takes a "mixed" dumb-bell configuration, which consists of a host atom and a solute atom both displaced from regular positions. The properties of mixed dumb-bell interstitials are discussed by Dederichs et al. (1978). Mechanical relaxation is a convenient tool for studying the behavior of such anisotropic point defects. The monograph on this technique by Robrock (1990) describes the principle and practical applications to the study of normal and mixed dumb-bell interstitials in pure metals and dilute alloys.

6.5   Statistical Thermodynamics of Point Defects

As briefly outlined in Section 6.2.1.1, point defects may be present in thermal equilibrium owing to the configurational entropy, and their concentration is expressed by a formula of the form of Eqn (4). Here we derive the formula using a formalism based on virtual chemical potentials of defects (Allnatt and Lidiard, 1993), which can be adopted consistently for point defects in ordered alloys (Section 6.5.3).

6.5.1   Vacancies and Self-interstitial Atoms in Pure Metals

6.5.1.1   Vacancies

First we consider vacancies in an elemental crystal. The free energy of a crystal containing N_a atoms and N_v vacancies at temperature T and pressure P is written as

$$G(N_a, N_v) = G^\circ(N) + N_v g_v^\infty - TS_c, \tag{68}$$

where $G^\circ(N)$ is the free energy of a perfect crystal consisting of N sites, and N is taken to be equal to $N_a + N_v$. The quantity g_v^∞ is the work required to remove one atom from inside of the perfect crystal, take it away to an infinite distance, and leave it there at rest. It includes the loss of internal energy associated with chemical bonding, changes in the vibration entropy, effects of atomic displacements and elastic distortion on the internal energy, entropy and volume of the crystal. We assume that the concentration of defects is low, and interaction between defects is ignored. Accordingly the second term on the right-hand side is made to be proportional to N_v. The configurational entropy is given by

$$S_c = k \ln \frac{N!}{N_a! N_v!}. \tag{69}$$

Using Stirling's approximation, the free energy is written as

$$G(N_a, N_v) = G^\circ(N) + N_v g_v^\infty - kT \ln(N \ln N - N_a \ln N_a - N_v \ln N_v). \tag{70}$$

Now we introduce chemical potentials of atoms and vacancies:

$$\mu_a = \left(\frac{\partial G}{\partial N_a}\right)_{T,P,N_v}, \tag{71}$$

$$\mu_v = \left(\frac{\partial G}{\partial N_v}\right)_{T,P,N_a}. \tag{72}$$

From Eqn (70), the following expressions are derived:

$$\mu_a = \mu^\circ + kT \ln \frac{N_a}{N} = \mu^\circ + kT \ln(1 - c_v) \tag{73}$$

and

$$\mu_v = \mu^\circ + g_v^\infty + kT \ln \frac{N_v}{N} = \Delta g_{f,v} + kT \ln c_v, \tag{74}$$

in which

$$\mu^\circ \equiv \left(\frac{\partial G^\circ}{\partial N}\right)_{T,P} \tag{75}$$

is the free energy per atomic site in the perfect crystal,

$$c_v \equiv \frac{N_v}{N} = \frac{N_v}{N_a + N_v} \tag{76}$$

is the site fraction of vacancies, and

$$\Delta g_{f,v} \equiv \mu^\circ + g_v^\infty \tag{77}$$

is the free energy of vacancy formation. The equilibrium concentration is obtained from the condition $\mu_v = 0$,

$$c_v^{eq} = \exp\left(-\frac{\Delta g_{f,v}}{kT}\right). \tag{78}$$

The free energy of the vacancy-containing crystal can be expressed using the chemical potentials, i.e.

$$G(N_a, N_v) = N_a\mu_a + N_v\mu_v. \tag{79}$$

Its change associated with introduction of vacancies is found, by substituting the expressions for the chemical potentials, to be

$$\Delta G \equiv G(N_a, N_v) - N_a\mu^\circ = N_a kT \ln(1 - c_v) + N_v\left(\Delta g_{f,v} + kT \ln c_v\right)$$
$$= N_v\Delta g_{f,v} + (N_a + N_v)kT[(1 - c_v)\ln(1 - c_v) + c_v\ln c_v]. \tag{80}$$

The first term is the linearly increasing energy penalty, and the second term is the entropy of mixing, which lowers the free energy very steeply on introducing vacancies. If we simplify $(1 - c_v)\ln(1 - c_v) \approx -c_v$ by neglecting second-order terms, the above equation can be rewritten, using the relation $\Delta g_{f,v} = -kT \ln c_v^{eq}$, as

$$\Delta G = kT\left[N_a\ln(1 - c_v) + N_v(-\ln c_v^{eq} + \ln c_v)\right] \approx kTN_v\left(\ln\frac{N_v}{N_v^{eq}} - 1\right), \tag{81}$$

with N_v^{eq} the equilibrium number of vacancies at temperature T. The variation of ΔG with N_v is drawn using this expression in **Figure** 44. The diagram shows that the value of ΔG is $-kT$ times the number of vacancies at the equilibrium concentration, which is evident also in Eqn (81). It indicates that the average Gibbs energy is $-kT$ per point defect in equilibrium (Johnson, 1994).

6.5.1.2 Self-interstitial Atoms
The formalism for self-interstitial atoms is essentially the same as for vacancies described above. There is a difference in W, which is in this case the number of ways to arrange N_i self-interstitial atoms in N' configurations available in the crystal:

$$W = \frac{N'!}{(N' - N_i)!N_i!}. \tag{82}$$

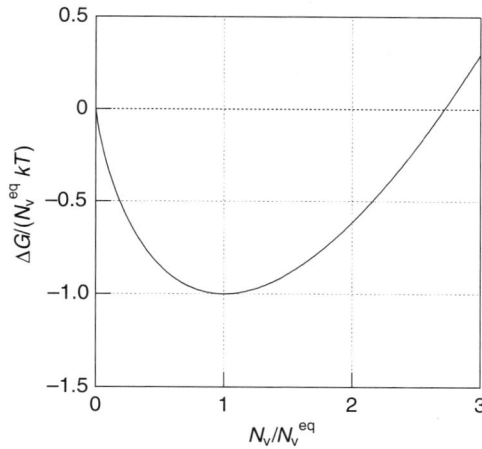

Figure 44 Change in the Gibbs energy on introduction of vacancies.

Letting the number of distinct positions and/or distinguishable orientations for an interstitial atom be z' per atomic site, the number of available configurations, N', in a crystal consisting of N_a atoms can be written as $z'N_a$. Following the procedure for the vacancy concentration, it is straightforward to arrive at the expression for the equilibrium site fraction of self-interstitial atoms,

$$\frac{N_i}{z'N_a - N_i} = \exp\left(-\frac{\Delta g_{f,i}}{kT}\right), \tag{83}$$

with the formation energy

$$\Delta g_{f,i} \equiv g_i^\infty. \tag{84}$$

The parameter g_i^∞ is the work required to bring an atom from outside and insert it to one of the interstitial sites in a perfect crystal. By defining the concentration of self-interstitial atom as N_i/N_a, its equilibrium value is written, for low concentrations, as

$$c_i^{eq} \approx z' \exp\left(-\frac{\Delta g_{f,i}}{kT}\right). \tag{85}$$

This appears essentially the same as the expression for vacancies, Eqn (78), but the difference in the definition of the Gibbs energy of formation is to be noted.

6.5.2 Vacancies in Solid-Solution Alloys

The equilibrium concentration of vacancies in a concentrated binary alloy, which was briefly discussed in Section 6.2.1.4, can be formulated by statistical thermodynamics of a ternary alloy, with atoms A and B and vacancies V. Below is given a simple model based on nearest-neighbor pair interaction, which is essentially equivalent to the well-known regular solution model. The first task is to express the free energy of a ternary alloy consisting of A and B atoms and vacancies whose numbers are N_A, N_B, and N_V, respectively. The enthalpy of the alloy can be written as the sum of the pair interaction energies e_{pq},

$$H = \sum_{p,q} N_{pq} e_{pq}, \tag{86}$$

where p and q are either A, B, or V, and N_{pq} is the number of p–q pairs. It can readily be obtained by counting the numbers of the nearest neighbor p–q bonds about each of the A and B atoms and vacancies, assuming that the probability of finding a particular species at an atomic site is equal to its mole fraction. The result is

$$H = \frac{Nz}{2} \left[c_A^2 \, e_{AA} + c_B^2 \, e_{BB} + c_V^2 \, e_{VV} + 2(c_A \, c_B \, e_{AB} + c_B \, c_V \, e_{BV} + c_V \, c_A \, e_{VA}) \right], \tag{87}$$

with z the coordination number. Here we have introduced the total number of sites

$$N \equiv N_A + N_B + N_V \tag{88}$$

and concentrations

$$c_p \equiv \frac{N_p}{N}, \tag{89}$$

where p is either A, B, or V. On the same assumption of random distribution of atoms and vacancies, the configurational entropy is given by

$$S_c = k \ln \frac{N!}{N_A! \, N_B! \, N_V!} \approx N \ln N - N_A \ln N_A - N_B \ln N_B - N_V \ln N_V$$
$$= -N(c_A \ln c_A + c_B \ln c_B + c_V \ln c_V). \tag{90}$$

The chemical potentials of the constituent species are described by differentiating the free energy $G = H - TS_c$ with respect to N_A, N_B, or N_V. The results are

$$\mu_A = \frac{z}{2} \left[e_{AA} - (1 - c_A)(c_B \, v_{AB} + c_V \, v_{VA}) + c_B \, c_V \, v_{BV} \right] + kT \ln c_A, \tag{91}$$

$$\mu_B = \frac{z}{2} \left[e_{BB} - (1 - c_B)(c_V \, v_{BV} + c_A \, v_{AB}) + c_V \, c_A \, v_{VA} \right] + kT \ln c_B, \tag{92}$$

$$\mu_V = \frac{z}{2} \left[e_{VV} - (1 - c_V)(c_A \, v_{VA} + c_B \, v_{BV}) + c_A \, c_B \, v_{AB} \right] + kT \ln c_V, \tag{93}$$

where v_{pq} are the effective pair interaction (EPI) energies defined as

$$v_{pq} \equiv e_{pp} + e_{qq} - 2e_{pq}. \tag{94}$$

The equilibrium concentration of vacancies is obtained by setting $\mu_V = 0$. It gives formally the same expression as Eqn (4), but with an effective free energy of formation in place of Δg_f,

$$\Delta g_f^{\text{eff}} \equiv (1 - c_V)\left(c_A\, \Delta g_f^A + c_B\, \Delta g_f^B\right) + \frac{z}{2} c_A\, c_B\, v_{AB}. \quad (95)$$

In deriving this expression e_{VV} is set to zero, which may be a reasonable choice. The quantities

$$\Delta g_f^A \equiv -\frac{z}{2} v_{VA} \quad (96)$$

$$\Delta g_f^B \equiv -\frac{z}{2} v_{BV} \quad (97)$$

are the vacancy formation energies at $c_A \to 1$ and $c_B \to 1$, respectively. If the A–B interaction is attractive (favoring ordering), v_{AB} is positive and the formation energy in the alloy rises from the fractional average of Δg_f^A and Δg_f^B, while it lowers if the A–B interaction is repulsive (favoring phase separation). **Figure 45** shows a practical example of the variation of the effective formation energy of a vacancy with composition according to Eqn (95) but neglecting c_V on the right-hand side. The values of the parameters are those evaluated for the Ni–Al system (Numakura et al., 1998): $\Delta g_f^A = 1.76$ eV (for nickel), $\Delta g_f^B = 0.73$ eV (for aluminum), and $v_{AB} = 0.232$ eV (for Ni$_3$Al), with $z = 12$. The formation energy is increased by alloying because of the introduction of stable A–B bonds, exhibiting a maximum at $c_B \approx 0.13$.

6.5.3 Vacancies and Antisite Atoms in Ordered Alloys

There is growing interest in ordered alloys and intermetallic compounds as novel structural and functional materials. In particular, transition-metal aluminides, gallides and silicides are attractive

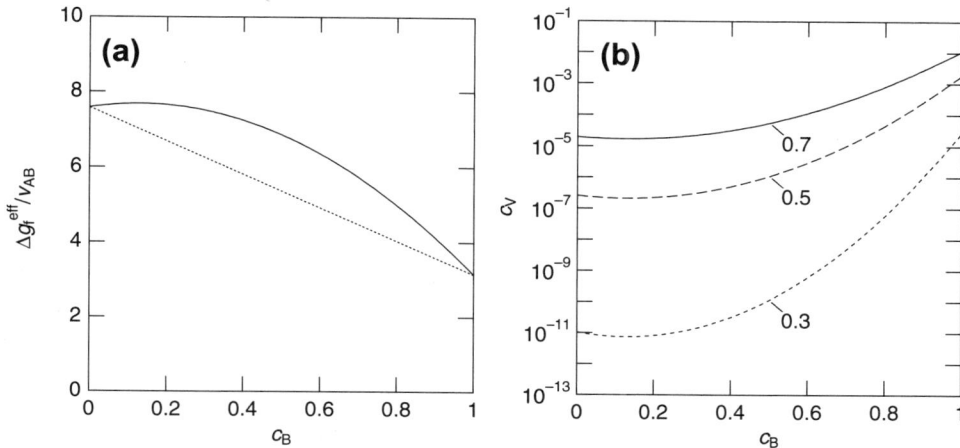

Figure 45 (a) Effective vacancy formation energy Δg_f^{eff} (solid curve) in disordered A–B alloy calculated by Eqn (95) with $z = 12$ and $v_{AB} = 0.232$ eV. The dotted line is the arithmetic average of Δg_f^A ($= 7.59\, v_{AB}$) and Δg_f^B ($= 3.15\, v_{AB}$). (b) Vacancy concentration c_V calculated by Eqn (4) using the Δg_f^{eff} for Δg_f. Tags to the three curves indicate kT/v_{AB}.

as high-strength lightweight alloys for high-temperature use. The interesting but somewhat exotic properties of these materials often originate from point defects in the ordered crystal structure (de Novion, 1995). Ordered alloys and intermetallic compounds exhibit a certain range of stability in composition around the stoichiometry, in which the deviation must be realized by intrinsic point defects, i.e. antisite atoms and vacancies. These defects influence, often strongly, the properties of the material. For instance, in the aluminides of iron and nickel of the B2 structure, FeAl and NiAl, the mechanical strength and diffusion rates vary sensitively with deviation of the composition from the stoichiometry (Pike et al., 1997; Kim and Chang, 2000). Detailed knowledge of point defects, viz., the types, concentrations, and mobilities, is important for understanding these properties.

To provide a theoretical basis, we discuss in this section the statistical thermodynamics of point defects in binary ordered alloys (Cheng et al., 1967; Schapink, 1969; Püschl et al., 2007). The theory will show what governs the balance of vacancies and antisite atoms in thermal equilibrium at a given composition. As a practical example, defect concentrations in an L1$_2$ ordered alloy calculated with the Bragg–Williams approximation are presented.

6.5.3.1 *Definitions*

We consider a binary ordered alloy A$_m$B$_n$ containing vacancies and antisite atoms, but no interstitial atoms. The sublattices for the species A and B are referred to as α and β, respectively. The constituents are vacancies of α and β sublattice sites V$_\alpha$ and V$_\beta$, antisite atoms B$_\alpha$ and A$_\beta$, and regular atoms A$_\alpha$ and B$_\beta$. For convenience, they are indicated by the following labels:

$$\begin{array}{cccc} 1 & V_\alpha & 2 & V_\beta \\ 3 & B_\alpha & 4 & A_\beta \\ 5 & A_\alpha & 6 & B_\beta \end{array}$$

Let a unit cell be made up of m α sites and n β sites, and the system consist of N unit cells. The numbers of A and B atoms in the system, N_A and N_B, define the composition of the alloy:

$$x_A = \frac{N_A}{N_A + N_B} = \frac{c_A}{c_A + c_B}, \tag{98}$$

$$x_B = \frac{N_B}{N_A + N_B} = \frac{c_B}{c_A + c_B}, \tag{99}$$

where $c_A \equiv N_A/N$ and $c_B \equiv N_B/N$ are referred to as site fractions. The problem is to determine the equilibrium numbers of the four defect species at a given composition and temperature. The number of components in the system is two (atoms A and B), but is formally three with vacancies as the tertiary component; it is a statistical thermodynamics of a ternary system, where we are to find equilibrium distributions of the three components in a crystal comprising two sublattices.

Let the numbers of the six species be N_i ($i = 1, 2, ..., 6$). Indeed not all of them are independent. First, the numbers of atoms are conserved:

$$N_4 + N_5 = N_A, \tag{100}$$

$$N_3 + N_6 = N_B. \tag{101}$$

On the other hand, the numbers of vacancies are variable. So is the size of the system, N, which increases or decreases with the numbers of vacancies. The ordered structure is nevertheless to be maintained. The ratio of the numbers of the sublattice sites must therefore satisfy

$$N_1 + N_3 + N_5 = mN, \tag{102}$$

$$N_2 + N_4 + N_6 = nN. \tag{103}$$

These four constraints reduce the number of independent variables from seven (N_i and N) to three. Alternatively, if the last two conditions, (102) and (103), are combined to a single relation

$$\frac{N_1 + N_3 + N_5}{m} = \frac{N_2 + N_4 + N_6}{n}, \tag{104}$$

there are six formal variables (N_i) and three constraints, again leaving only three variables independent.

6.5.3.2 Defect Equilibria

To derive equations governing the equilibria of the four defect species, we employ the same argument as Section 6.5.1 for vacancies in elemental crystals, using virtual chemical potentials (Allnatt and Lidiard, 1993). First we write the free energy of a solid consisting of N unit cells and containing vacancies and antisite atoms under pressure P and at temperature T:

$$G(N_i) = G^\circ(N) + \sum_{i=1}^{4} N_i g_i^\infty - TS_c. \tag{105}$$

The first term on the right-hand side is the free energy of a defect-free crystal, the second term is the changes due to the introduction of defects, and the last term is the contribution of the configurational entropy. Each of the parameters g_i^∞ is defined, as before, as the work required to introduce a defect i into a defect-free crystal of the same number of unit cells. On the assumption that the defects are distributed randomly in each of the sublattices, the configurational entropy S_c is given by

$$S_c = k \ln \frac{(mN)!}{N_1! \, N_3! \, N_5!} \frac{(nN)!}{N_2! \, N_4! \, N_6!}. \tag{106}$$

The chemical potentials of the constituent species are defined as

$$\mu_i \equiv \left(\frac{\partial G}{\partial N_i} \right)_{T,P,N_{j\neq i}}. \tag{107}$$

They are found by differentiating Eqn (105) and using the Stirling formula to be

$$\mu_i = \frac{\mu^\circ}{m+n} + g_i^\infty - kT \left(\frac{m \ln m + n \ln n}{m+n} - \ln c_i \right) \tag{108}$$

for $i = 1$ to 4, and

$$\mu_i = \frac{\mu^\circ}{m+n} - kT \left(\frac{m \ln m + n \ln n}{m+n} - \ln c_i \right) \tag{109}$$

for $i = 5$ and 6, where

$$\mu^\circ \equiv \left(\frac{\partial G^\circ}{\partial N} \right)_{T,P} \tag{110}$$

is the free energy per unit cell of the defect-free crystal, and c are defined as

$$c_i \equiv \frac{N_i}{N}. \quad (i = 1, 2, ..., 6) \tag{111}$$

In deriving these expressions, a formula

$$\left(\frac{\partial N}{\partial N_i} \right)_{T,P,N_{j \neq i}} = \frac{1}{m+n} \tag{112}$$

is used, which is obtained from the relation

$$\sum_{i=1}^{6} N_i = (m+n)N. \tag{113}$$

Using the expressions for the chemical potentials, the free energy can be rewritten as

$$G(N_i) = \sum_{i=1}^{6} N_i \mu_i. \tag{114}$$

Now we incorporate the three constraints, Eqns (104), (100), and (101), into the free energy using Lagrange multipliers $\lambda_1, \lambda_2, \lambda_3$:

$$G(N_i) = G^\circ(N) + \sum_{i=1}^{4} N_i g_i^\infty - kT \left[(mN)\ln(mN) + (nN)\ln(nN) - \sum_{i=1}^{6} N_i \ln N_i \right]$$
$$+ \lambda_1 \left(\frac{N_1 + N_3 + N_5}{m} - \frac{N_2 + N_4 + N_6}{n} \right) + \lambda_2 (N_4 + N_5 - N_A) + \lambda_3 (N_3 + N_6 - N_B). \tag{115}$$

Applying the condition for equilibrium

$$\mu_i = 0 \quad (i = 1, 2, ..., 6) \tag{116}$$

gives a set of six simultaneous equations for the six variables N_i. By eliminating λ_1, λ_2 and λ_3 from those equations, one obtains

$$0 = m \, \mu_1 + n \, \mu_2, \tag{117}$$

$$\mu_1 + \mu_6 = \mu_2 + \mu_3, \tag{118}$$

$$\mu_2 + \mu_5 = \mu_1 + \mu_4, \tag{119}$$

which correspond respectively to the reactions

$$(\text{null}) \;\leftrightarrow\; m\,V_\alpha + n\,V_\beta, \tag{120}$$

$$V_\alpha + B_\beta \;\leftrightarrow\; B_\alpha + V_\beta, \tag{121}$$

$$A_\alpha + V_\beta \;\leftrightarrow\; V_\alpha + A_\beta. \tag{122}$$

Substitution of Eqns (108) and (109) for μ_i and some rearrangements give the following equations:

$$\left(\frac{N_1}{mN}\right)^m \left(\frac{N_2}{nN}\right)^n = \exp\left(-\frac{\Delta g_S}{kT}\right), \tag{123}$$

$$\frac{N_2 N_3}{N_1 N_6} = \exp\left(-\frac{\Delta g_{XB}}{kT}\right), \tag{124}$$

$$\frac{N_1 N_4}{N_2 N_5} = \exp\left(-\frac{\Delta g_{XA}}{kT}\right), \tag{125}$$

with

$$\Delta g_S \equiv \mu^\circ + m\,g_1^\infty + n\,g_2^\infty, \tag{126}$$

$$\Delta g_{XB} \equiv g_2^\infty + g_3^\infty - g_1^\infty, \tag{127}$$

$$\Delta g_{XA} \equiv g_1^\infty + g_4^\infty - g_2^\infty. \tag{128}$$

Equations (123)–(125), which are called respectively the Schottky product equation, B-antisite disorder equation and A-antisite disorder equation, determine the equilibrium defect concentrations. Combining Eqns (124) and (125) yields

$$\frac{N_3 N_4}{N_5 N_6} = \exp\left(-\frac{\Delta g_{XB} + \Delta g_{XA}}{kT}\right), \tag{129}$$

which determines the equilibrium concentration of a pair of antisite atoms. This equation is apparently identical to the formula of Bragg and Williams (1934, 1935) governing the order–disorder transition without considering vacancies. The theory of Hagen and Finnis (1998) is virtually identical to what is described here, with only minor differences in the definitions of the parameters and variables.

6.5.3.3 *Practice*

The equilibrium defect concentrations at a given composition and temperature can be obtained by solving the three Equations, (123) to (125), for N_i. The controlling parameters, Δg_S, Δg_{XB}, and Δg_{XA}, may be evaluated from atomistic calculations of g_i^∞, if the defect concentrations are low enough. In general, however, the concentrations of the antisite atoms may be as high as 10^{-1}, so that one cannot assume a dilute-limit condition, as in the model of Wagner and Schottky (1931), in which no other defects are assumed to be present around a defect. Since the energy to create a particular defect at a particular atomic site depends on the atomic arrangement around the site, the formation energy of any defect is a function of the concentrations of all the defect species. A simple method to deal with this problem is the pair interaction model for computing the enthalpy of an alloy. Below we discuss the case of L1$_2$ ordered alloys as an example. Hagen and Finnis (1998) present a different method for evaluating the three energy parameters, which may be convenient for importing numerical values of defect energies obtained by ab initio calculations.

For simplicity we consider pair interactions between nearest neighbor atoms only, and adopt the point approximation for the configurational entropy (i.e. the Bragg–Williams model). We denote the interaction energy between a neighboring pair of species p and q by e_{pq}, where p and q are either atom A, atom B or a vacancy V. In the A$_3$B alloy of the L1$_2$ structure, atoms A and B are placed at the face-center sites and cube-corner sites, respectively, in the unit cell of the fcc structure, **Figure 17(b)**. The enthalpy of the alloy is obtained by summing the pair interaction energies e_{pq}, and the configurational entropy is given by Eqn (106) with $m = 3$ and $n = 1$. The following expression for the free energy is obtained:

$$
\begin{aligned}
G(N_i) = -\frac{2}{N}&\left[\left(\frac{2}{3}N_3 N_5 + N_3 N_4 + N_5 N_6\right)v_{AB} + \left(\frac{2}{3}N_1 N_5 + N_1 N_4 + N_2 N_5\right)v_{AV}\right.\\
&\left.+ \left(\frac{2}{3}N_1 N_3 + N_2 N_3 + N_1 N_6\right)v_{BV}\right] + 6[N_A\,e_{AA} + N_B\,e_{BB} + (N_1 + N_2)e_{VV}]\\
&- kT\left[(3N)\ln(3N) + N \ln N - \sum_{i=1}^{6} N_i \ln N_i\right].
\end{aligned}
\tag{130}
$$

Here we have introduced the effective pair interaction (EPI) energies defined in the previous subsection.[21] To derive defect-equilibria equations, here we choose conveniently N, N_3, and N_4 as independent variables without recourse to Lagrange multipliers, and apply the conditions for equilibrium,

$$
\left(\frac{\partial G}{\partial N}\right)_{T,P,N_3,N_4} = 0,
\tag{131}
$$

$$
\left(\frac{\partial G}{\partial N_3}\right)_{T,P,N,N_4} = 0,
\tag{132}
$$

$$
\left(\frac{\partial G}{\partial N_4}\right)_{T,P,N,N_3} = 0.
\tag{133}
$$

---

[21] Note that $e_{pq} = e_{qp}$, and also $v_{pq} = v_{qp}$.

These equations give respectively the Schottky product equation, B-antisite disorder equation, and A-antisite disorder equation. The energies controlling the defect equilibria turn out to be

$$\Delta g_S = 2\left[\left(\frac{2}{3}c_3c_5 + c_3c_4 + c_5c_6\right)v_{AB} + \left(\frac{2}{3}c_1c_5 + c_1c_4 + c_2c_5 - 3c_A\right)v_{AV}\right.$$
$$\left. + \left(\frac{2}{3}c_1c_3 + c_2c_3 + c_1c_6 - 3c_B\right)v_{BV}\right] + 24e_{vv}, \tag{134}$$

$$\Delta g_{XB} = 2\left[\left(\frac{c_5}{3} - c_4\right)(v_{AB} - v_{AV}) + \left(c_6 - \frac{c_3}{3} + \frac{c_1}{3} - c_2\right)v_{BV}\right], \tag{135}$$

$$\Delta g_{XA} = 2\left[\left(c_6 - \frac{c_3}{3}\right)(v_{AB} - v_{BV}) + \left(\frac{c_5}{3} - c_4 - \frac{c_1}{3} + c_2\right)v_{AV}\right]. \tag{136}$$

The vacancy–vacancy interaction energy e_{vv} remaining in Eqn (134) is taken to be zero by definition. The defect concentrations at given temperature and composition are therefore determined completely by the three EPI energies, v_{AB}, v_{AV}, and v_{BV}. For a given set of their values, the concentrations c_i that satisfy the equations for defect equilibria can be found by iteration or trial and error. **Figures 46 and 47**

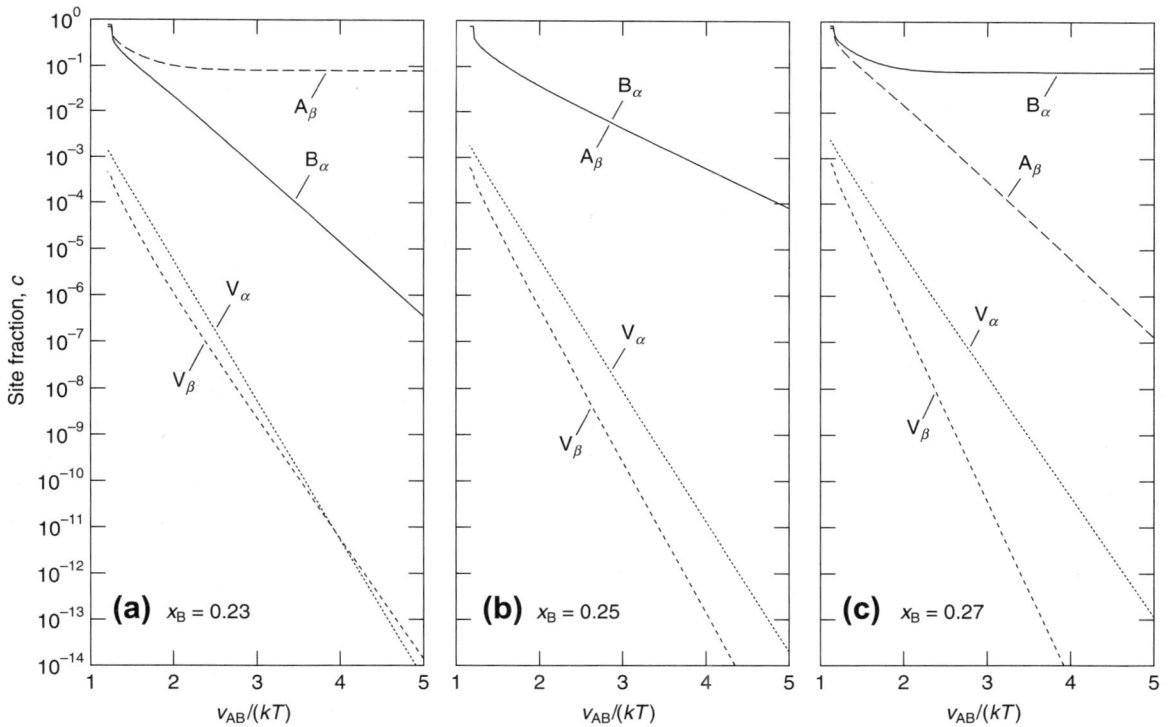

Figure 46 Site fractions of vacancies and antisite atoms in an A_3B alloy of the $L1_2$ structure as functions of reciprocal temperature. x_B is the atom fraction of component B. See text for details.

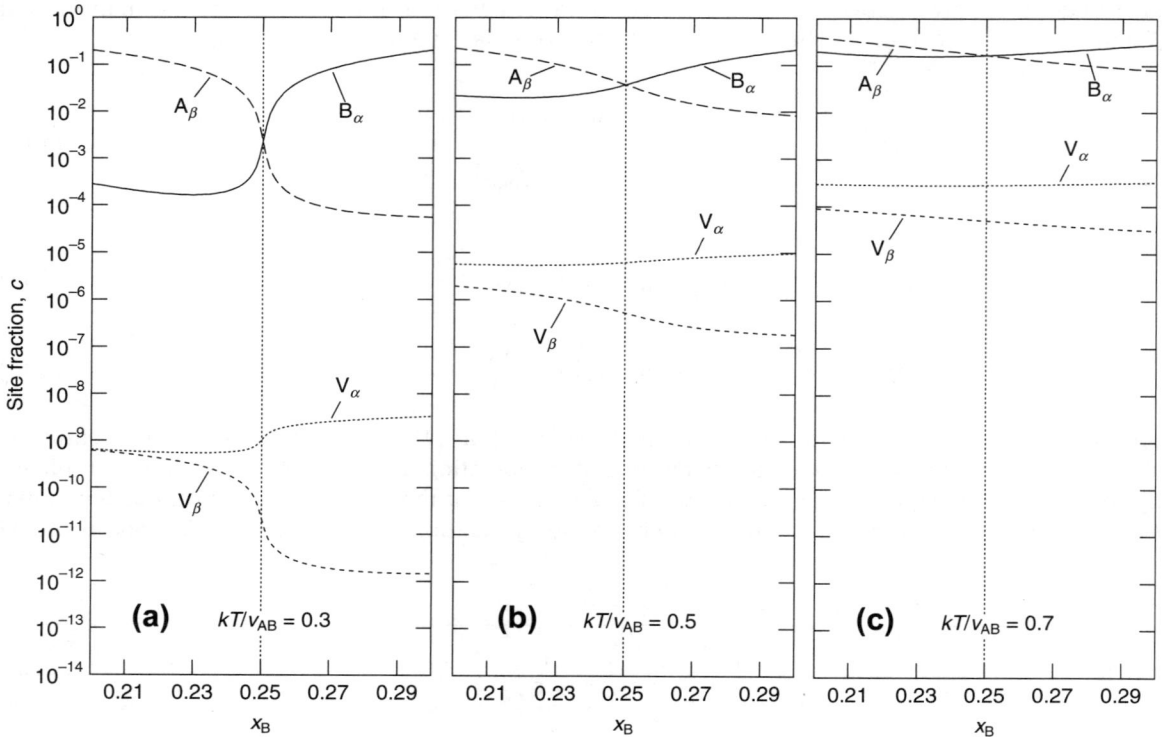

Figure 47 Same as **Figure 46**, but as functions of composition.

show the equilibrium concentrations of vacancies and antisite atoms calculated for $v_{AV}/v_{AB} = -1$ and $v_{BV}/v_{AB} = -0.5$. It is a model of an A_3B alloy where the vacancy formation enthalpy of the pure solid of A is about twice as large as that of B. Similar calculations on model A–B alloys of the B2 and $L1_0$ structures were presented in a previous work (Püschl et al., 2007).

Comparison of those defect concentrations calculated for different structures reveals that the defect equilibria depend as much on the structure, i.e. the stoichiometry m and n, and the coordinations numbers z_i (the number of j sublattice sites around an i sublattice site), as on the EPI energies v_{pq}; there are patterns characteristic of a particular ordered structure. Unusually high concentrations of vacancies observed for aluminides and gallides of iron-group transition metals with the B2 structure are a real example (de Novion, 1995). It occurs, for instance, in NiAl, considerably on the Ni sublattice at Al-rich compositions. In Ni_3Al ($L1_2$), in contrast, although the same trend is known (Numakura et al., 1998) and is also found in **Figures 46** and **47**, the vacancy concentration stays at a normal level. The pair interaction model is useful for understanding the origin of these properties. Asymmetrical EPI energies for A–V and B–V give rise to unbalanced vacancy concentrations on the two sublattices, and if either of them is small in magnitude, the concentration of vacancies can be comparable to those of the antisite atoms. The extent of the enhancement of vacancy formation depends on geometrical arrangements of sublattice sites, i.e. the coordination numbers; it is conspicuous in the B2 structure but is not in $L1_0$, nor in $L1_2$. This topic was discussed in more detail in the previous work (Püschl et al., 2007).

6.6 Summary and Concluding Remarks

Defects in metals are one of the most extensively studied topics of physical metallurgy in the last 50 years. Since 1950s theories of elasticity, statistical thermodynamics and then electron theory for defects were developed, and with the advent of new experimental techniques and their rapid developments the fundamentals of the nature of point defects were established, with particular emphasis on fcc metals, which laid the basis for defect-controlled phenomena such as diffusion and phase transformation. With strong needs of basic research for nuclear technology, studies on refractory bcc metals were actively performed in the 1970s and continued through 1980s, with much efforts on preparing high purity materials and careful experiments such as high temperatures and ultrahigh vacuum conditions. The extensive experimental data on point defects in metals acquired by the end of 1980s were compiled in one volume of Landolt–Börnstein series, "Atomic defects in metals", published in 1991 (Ullmaier, 1991).

Since then, it appears that efforts to fulfill the needs for the missing pieces of information are not actively growing, but are rather diminishing these days. Instead, theoretical and computational studies have been rapidly developing, owing to the significant advances in computational resources and techniques. Reproducing the energy of a vacancy in simple metals was a difficult problem until the late 1980s, but now the energies of formation and migration of point defects, as well as their interaction, are computed routinely from first principles using state-of-the-art packages. In addition to electronic total-energy calculations, theories and practical computations of lattice-dynamical properties are relevant to the study of defects (Pierron-Bohnes and Mehaddene, 2007). Flynn's theory connecting diffusional jumps of atoms and lattice vibrations was developed and generalized to be capable of predicting the migration energy of vacancies in metals (Schober et al., 1992) and further for ordered alloys (Kentzinger and Schober, 2000). If phonon dispersion relations, which are inputs to the theory, are not available, they can be calculated from first principles. The formation entropy of a point defect can also be evaluated from the vibrational spectrum of the defect (Mishin et al., 2001b). Today, first-principle calculations of the behavior of point defects at finite temperatures are being developed (see, e.g. Hickel et al., 2012). On this trend the experimental data in **Table 2** may serve as the standard reference for theoretical calculations to check their reliability, and the missing information will, sooner or later, be supplied from theory.

Still much work is needed on interaction between defects, including solute atoms, dislocations, and interfaces. Defect–defect, solute–defect, and solute–solute interactions have strong impact on the properties of structural materials. The problems are more complex than those for single defect species; experiments are not straightforward, and neither are theoretical treatments. We hope to see progress in this field by concerted experimental and theoretical investigations in the years to come.

References

Alamo, A., de Novion, C.H., Desarmot, G., 1986. Production and annealing of point defects in electron-irradiated copper-gold alloys. Radiat. Eff. 88, 69–91.

Allnatt, A.R., Lidiard, A.B., 1993. Atomic Transport in Solids. Cambridge University Press, Cambridge.

Amelinckx, S., Gevers, R., Remaut, G., Van Landuyt, J. (Eds.), 1970. Modern Diffraction and Imaging Techniques in Material Science. North-Holland, Amsterdam.

Arakawa, K., Ono, K., Isshiki, M., Mimura, K., Uchikoshi, M., Mori, H., 2007. Observation of the one-dimensional diffusion of nanometer-sized dislocation loops. Science 318, 956–959.

Bacon, D.J., 1988. A review of computer models of point defects in hcp metals. J. Nucl. Mater. 159, 176–189.

Balluffi, R.W., 1978. Vacancy defect mobilities and binding energies obtained from annealing studies. J. Nucl. Mater. 69 & 70, 240–263.

Barrett, C., Massalski, T.B., 1980. Structure of Metals—Crystallographic Methods, Principles, and Data, third revised ed. Pergamon Press, Oxford.

Bauerle, J.E., Koehler, J.S., 1957. Quenched-in lattice defects in gold. Phys. Rev. 107, 1493–1498.

Bauerle, J.E., Klabunde, C.E., Koehler, J.S., 1956. Resistivity increase in water-quenched gold. Phys. Rev. 102, 1182.

Becquart, C.S., Raulot, J.M., Bencteux, G., Domain, C., Perez, M., Garruchet, S., Nguyen, H., 2007. Atomistic modeling of an Fe system with a small concentration of C. Comput. Mater. Sci. 40, 119–129.

Bérces, G., Kovács, I., 1983. Vacancies and vacancy complexes in binary alloys. Philos. Mag. A 48, 883–901.

Billington, D.S., Crawford Jr, J.H., 1961. Radiation Damage in Solids. Princeton University Press, Princeton.

Bollmann, W., 1956. Interference effects in the electron microscopy of thin crystal foils. Phys. Rev. 103, 1588–1589.

Bradshaw, F.J., Pearson, S., 1956. Quenching vacancies in platinum. Philos. Mag. 1, 812–820.

Bradshaw, F.J., Pearson, S., 1957. Quenching vacancies in gold. Philos. Mag. 2, 379–383.

Bragg, W.L., Williams, E.J., 1934. The effect of thermal agitation on atomic arrangement in alloys. Proc. Roy. Soc. Lond. A 145, 699–730.

Bragg, W.L., Williams, E.J., 1935. The effect of thermal agitation on atomic arrangement in alloys. II. Proc. Roy. Soc. Lond. A 151, 540–566.

Brimhall, J.L., Mastel, B., 1969. Stability of voids in neutron irradiated nickel. J. Nucl. Mater. 33, 186–194.

Brissonneau, P., 1958. Contribution à l'étude quantitative du traînage magnétique de diffusion du carbone dans le fer α. J. Phys. Chem. Solids 7, 22–51.

Brown, A.M., Ashby, M.F., 1980. Correlations for diffusion constants. Acta Metall. 28, 1085–1101.

Brown, R.A., 1982. The interaction of conduction electrons with dislocations and grain boundaries in metals. Can. J. Phys. 60, 766–778.

Carling, K., Wahnström, G., 2000. Vacancies in metals: from first-principles calculations to experimental data. Phys. Rev. Lett. 85, 3862–3865.

Carling, K.M., Wahnström, G., Mattsson, T.R., Sandberg, N., Grimvall, G., 2003. Vacancy concentration in al from combined first-principles and model potential calculations. Phys. Rev. B 67, 054101.

Cheng, C.Y., Wynblatt, P.P., Dorn, J.E., 1967. Vacancy models for concentrated binary alloys—II Long-range ordered alloys. Acta Metall. 15, 1045–1056.

Chikazumi, S., 1997. Physics of Ferromagnetism. Oxford University Press, New York.

Clark, A.E., Restorff, J.B., Wun-Fogle, M., Lograsso, T.A., Schlagel, D., 2000. Magnetostrictive properties of body-centered cubic Fe–Ga and Fe–Ga–Al alloys. IEEE Trans. Magn. 36, 3238–3240.

Corbett, J.W., 1966. Electron radiation damage in semiconductors and metals. In: Seitz, F., Turnbull, D. (Eds.), Solid State Physics. Academic Press, New York (Suppl. 7).

Cotterill, R.M., Segall, R.L., 1963. The effect of quenching history, quenching temperature and trace impurities on vacancy clusters in aluminium and gold. Philos. Mag. 8, 1105–1125.

Cottrell, A.H., Bilby, B.A., 1949. Dislocation theory of yielding and strain ageing of iron. Proc. Phys. Soc. A 62, 49–62.

Damask, A.C., Dienes, G.J., 1963. Point Defects in Metals. Gordon and Breach, New York.

Daw, M.S., Baskes, M.I., 1984. Embedded-atom method: derivation and application to impurities, surfaces, and other defects in metals. Phys. Rev. B 29, 6443–6453.

Daw, M.S., Foiles, S.M., Baskes, M.I., 1993. The embedded-atom method: a review of theory and applications. Mater. Sci. Rep. 9, 251–310.

de Novion, C., 1995. Point defects. In: Westbrook, J.H., Fleischer, R.L. (Eds.), Intermetallic Compounds: Principles and Practice, vol. 1: Principles. John Wiley & Sons, Chichester, pp. 559–583 (Chapter 23).

De Vita, A., Gillan, M.J., 1991. The ab initio calculation of defect energetics in aluminium. J. Phys. Condens. Matter 3, 6225–6237.

Dederichs, P.H., Lehmann, C., Schober, H.R., Scholz, A., Zeller, A., 1978. Lattice theory of point defects. J. Nucl. Mater. 69 & 70, 176–199.

Dederichs, P.H., Hoshino, T., Drittler, B., Abraham, K., Zeller, R., 1991. Total-energy calculations for point defects in metals. Physica B 172, 203–204.

Dederichs, P.H., 1973. The theory of diffuse x ray scattering and its application to the study of point defects and their clusters. J. Phys. F Met. Phys. 3, 471–496.

Derlet, P.M., Nguyen-Manh, D., Dudarev, S.L., 2007. Multiscale modeling of crowdion and vacancy defects in body-centered-cubic transition metals. Phys. Rev. B 76, 054107.

DeSorbo, W., 1960. Calorimetric studies on annealing quenched-in defects in gold. Phys. Rev. 117, 444–450.

DiCarlo, J.A., Snead Jr., C.L., Goland, A.N., 1969. Stage-I interstitials in electron-irradiated tungsten. Phys. Rev. 178, 1059–1072.

Dickerson, R.H., Lowell, R.C., Tomizuka, C.T., 1965. Effect of hydrostatic pressure on the self-diffusion rate in single crystals of gold. Phys. Rev. 137, A613–A619.

Dickman, J.E., Jeffery, R.N., Gustafson, D.R., 1977. Vacancy formation volume in indium from positron-annihilation measurements. Phys. Rev. B 16, 3334–3337.

Dimitrov, C., Sitaud, B., Zhang, X., Dimitrov, O., Dedek, U., Dworschak, F., 1992a. Radiation-induced defects in solid solutions and intermetallic compounds based on the Ni-Al system: I. Low-temperature electron-irradiation damage. J. Phys. Condens. Matter 4, 10199–10210.

Dimitrov, C., Sitaud, B., Zhang, X., Dimitrov, O., Dedek, U., Dworschak, F., 1992b. Radiation-induced defects in solid solutions and intermetallic compounds based on the Ni–Al system: II. Recovery of radiation damage. J. Phys. Condens. Matter 4, 10211–10226.

Ehrhart, P., Schultz, H., 1991. Properties and interactions of atomic defects in metals and alloys. In: Ullmaier, H. (Ed.), Atomic Defects in Metals, Landolt-Börnstein New Series, Group III: Crystal and Solid State Physics, vol. 25. Springer-Verlag, Berlin Heidelberg, pp. 88–379 (Chapter 2).

Ehrhart, P., Haubold, H.-G., Schilling, W., 1974. Investigation of point defects and their agglomerates in irradiated metals by diffuse X-ray scattering. Adv. Solid State Phys. 14, 87–110.

Ehrhart, P., 1978. The configuration of atomic defects as determined from scattering studies. J. Nucl. Mater. 69 & 70, 200–214.

Emrick, R.M., McArdle, P.B., 1969. Effect of pressure on quenched-in electrical resistance in gold and aluminum. Phys. Rev. 188, 1156–1162.

Emrick, R.M., 1961. Pressure effect on vacancy migration rate in gold. Phys. Rev. 122, 1720–1733.

Eshelby, J.D., 1954. Distortion of a crystal by point imperfections. J. Appl. Phys. 25, 255–261.

Eshelby, J.D., 1956. The continuum theory of lattice defects. Solid State Phys. 3, 79–144.

Fidleris, V., 1988. The irradiation creep and growth phenomena. J. Nucl. Mater. 159, 22–42.

Finnis, M.W., Sinclair, J.E., 1984. A simple empirical N-body potential for transition metals. Philos. Mag. A 50, 45–55.

Flynn, C.P., 1968. Atomic migration in monatomic crystals. Phys. Rev. 171, 682–698.

Flynn, C.P., 1972. Point Defects and Diffusion. Oxford University Press, Oxford.

Foiles, S.M., Baskes, M.I., Daw, M.S., 1986. Embedded-atom-method functions for the fcc metals Cu, Ag, Au, Ni, Pd, Pt, and their alloys. Phys. Rev. B 33, 7983–7991.

Fors, D.H.R., Wahnström, G., 2008. Nature of boron solution and diffusion in α-iron. Phys. Rev. B 77, 132102.

Franklin, D.G., Birnbaum, H.K., 1971. An anelastic study of quenched gold. Acta Metall. 19, 965–971.

Frank, W., 1988. Intrinsic point defects in hexagonal close-packed metals. J. Nucl. Mater. 159, 122–148.

Frenkel, J., 1937. On the liquid state and the theory of fusion. Trans. Faraday Soc. 33, 58–65.

Fromm, E., Gebhardt, E. (Eds.), 1976. Gase und Kohlenstoff in Metallen. Springer-Verlag, Berlin Heidelberg.

Fu, C.-C., Dalla Torre, J., Willaime, F., Bocquet, J.-L., Barbu, A., 2005. Multiscale modelling of defect kinetics in irradiated iron. Nat. Mater. 4, 68–74.

Fukai, Y., Ōkuma, N., 1993. Evidence of copious vacancy formation in Ni and Pd under a high hydrogen pressure. Jpn. J. Appl. Phys. 32, L1256–L1259.

Fukai, Y., Yokota, S., Yanagawa, J., 2006. The phase diagram and superabundant vacancy formation in Co–H alloys. J. Alloys Compd. 407, 16–24.

Fukai, Y., 2003. Formation of superabundant vacancies in M–H alloys and some of its consequences: a review. J. Alloys Compd. 356–357, 263–269.

Fukai, Y., 2005. The metal-hydrogen system—Basic bulk properties, second ed. Springer-Verlag, Berlin Heidelberg.

Gallagher, P.C.J., 1970. The influence of alloying, temperature, and related effects on the stacking fault energy. Metall. Trans. 1, 2429–2461.

Gilbert, J., Herman, H., Damask, A.C., 1973. Electron irradiation of Cu$_3$Au. Radiat. Eff. 20, 37–42.

Gilder, H.M., Lazarus, D., 1975. Role of vacancy anharmonicity on non-Arrhenius diffusion behavior. Phys. Rev. B 11, 4916–4926.

Gillan, M.J., 1989. Calculation of the vacancy formation energy in aluminium. J. Phys. Condens. Matter 1, 689–711.

Girifalco, L.A., 2000. Statistical Mechanics of Solids. Oxford University Press, New York.

Grabowski, B., Hickel, T., Neugebauer, J., 2011. Formation energies of point defects at finite temperatures. Phys. Status Solidi B 248, 1295–1308.

Granato, A.V., 1992. Interstitialcy model for condensed matter states of face-centered-cubic metals. Phys. Rev. Lett. 68, 974–977.

Granato, A.V., 1993. Thermodynamic properties of liquid and amorphous monatomic metals from an interstitialcy model. J. Non Cryst. Solids 156–158, 402–406.

Granato, A.V., 1994a. Self-interstitials as basic structural units of liquids and glasses. J. Phys. Chem. Solids 55, 931–939.

Granato, A.V., 1994b. The temperature-independent component of the diaelastic effect. J. Alloys Compd. 211/212, 503–508.

Grasse, D., von Guérard, B., Peisl, J., 1984. Fast neutron-irradiation of molybdenum studied by diffuse X-ray scattering. J. Nucl. Mater. 120, 304–306.

Griffiths, M., 1988. A review of microstructure evolution in zirconium alloys during irradiation. J. Nucl. Mater. 159, 190–218.

Grimes, H.H., 1965. Vacancy formation in gold under high pressure. J. Phys. Chem. Solids 26, 509–516.

Guinan, M.W., Stuart, R.N., Borg, R.J., 1977. Fully dynamic computer simulation of self-interstitial diffusion in tungsten. Phys. Rev. B 15, 699–710.

Haasen, P., 1996. Physical Metallurgy, third ed. Cambridge University Press, Cambridge.

Hagen, M., Finnis, M.W., 1998. Point defects and chemical potentials in ordered alloys. Philos. Mag. A 77, 447–464.

Harada, S., Yokota, S., Ishii, Y., Shizuku, Y., Kanazawa, M., Fukai, Y., 2005. A relation between the vacancy concentration and hydrogen concentration in the Ni–H, Co–H and Pd–H systems. J. Alloys Compd. 404–406, 247–251.

Hashimoto, E., Smirnov, E.A., Kino, T., 1984. Temperature dependence of the Doppler-broadened lineshape of positron annihilation in α-Ti. J. Phys. F Met. Phys. 14, L215–L217.

Hasson, G., Boos, J.-Y., Herbeuval, I., Biscondi, M., Goux, C., 1972. Theoretical and experimental determinations of grain boundary structures and energies: correlation with various experimental results. Surf. Sci. 31, 115–137.

Hautojärvi, P., 1987. Vacancies and vacancy-impurity interactions in metals studied by positrons. Mater. Sci. Forum 15–18, 81–98.

Hayashi, E., Kurokawa, Y., Fukai, Y., 1998. Hydrogen-induced enhancement of interdiffusion in Cu–Ni diffusion couples. Phys. Rev. Lett. 80, 5588–5590.

Herzig, C., Köhler, U., 1987. Anomalous self-diffusion in bcc IVB metals and alloys. Mater. Sci. Forum 15–18, 301–322.

Herzig, C., 1993. Soft phonons and diffusion systematics in bcc transition metals and alloys. Defect Diffusion Forum 95–98, 203–220.

Hickel, T., Grabowski, B., Körmann, F., Neugebauer, J., 2012. Advancing density functional theory to finite temperatures: methods and applications in steel design. J. Phys. Condens. Matter 24, 053202.

Hiroi, T., Fukai, Y., Mori, K., 2005. The phase diagram and superabundant vacancy formation in Fe–H alloys revisited. J. Alloys Compd. 404–406, 252–255.

Hirsch, P.B., Horne, R.W., Whelan, M.J., 1956. Direct observations of the arrangement and motion of dislocations in aluminium. Philos. Mag. 1, 677–684.

Hirsch, P.B., Howie, A., Nicholson, R.B., Pashley, D.W., Whelan, M.J., 1965. Electron Microscopy of Thin Crystals. Butterworths, London.

Hirth, J.P., Lothe, J., 1982. Theory of Dislocations, second ed. John Wiley & Sons, New York.

Hivert, V., Pichon, R., Bilger, H., Bichon, R., Verdone, J., Dautreppe, D., Moser, P., 1970. Internal friction in low temperature irradiated bcc metals. J. Phys. Chem. Solids 31, 1843–1855.

Holt, R.A., 2008. In-reactor deformation of cold-worked Zr–2.5Nb pressure tubes. J. Nucl. Mater. 372, 182–214.

Hood, G.M., Schultz, R.J., 1995. The temperature dependence of positron annihilation in α-Hf. Mater. Sci. Forum 175–178, 375–378.

Hood, G.M., Zou, H., Schultz, R.J., Matsuura, N., Roy, J.A., Jackman, J.A., 1997. Self- and Hf diffusion in α-Zr in dilute, Fe-free Zr(Ti) and Zr(Nb) alloys. Defect Diffusion Forum 143–147, 49–54.

Hood, G.M., 1986. Diffusion and vacancy properties of α-Zr. J. Nucl. Mater. 139, 179–184.

Hood, G.M., 1988. Point defect diffusion in α-Zr. J. Nucl. Mater. 159, 149–175.

Hoshino, T., Papanikolaou, N., Zeller, R., Dederichs, P.H., Asato, M., Asada, T., Stefanou, N., 1999. Comput. Mater. Sci. 14, 56–61.

Hoshino, T., Mizuno, T., Asato, M., Fukushima, H., 2001. Full-potential KKR calculations for point defect energies in metals, based on the generalized-gradient approximation: I. Vacancy formation energies in fcc and bcc metals. Mater. Trans. 42, 2206–2215.

Howard, R.E., Lidiard, A.B., 1964. Matter transport in solids. Rep. Prog. Phys. 27, 161–240.

Huang, Y., Langdon, T.G., 2013. Advances in ultrafine-grained materials. Mater. Today 16, 85–93.

Huebener, R.P., Holmes, C.G., 1963. Pressure effect on vacancy formation in gold. Phys. Rev. 129, 1162–1168.

Huntington, H.B., Seitz, F., 1942. Mechanism for self-diffusion in metallic copper. Phys. Rev. 61, 315–325.

Huntington, H.B., 1953. Mobility of interstitial atoms in a face-centered metal. Phys. Rev. 91, 1092–1098.

Iida, T., Yamazaki, Y., Kobayashi, T., Iijima, Y., Fukai, Y., 2005. Enhanced diffusion of Nb in Ni–H alloys by hydrogen-induced vacancies. Acta Mater. 53, 3083–3089.

Jesson, B.J., Foley, M., Madden, P.A., 1997. Thermal properties of the self-interstitial in aluminum: an ab initio molecular-dynamics study. Phys. Rev. B 55, 4941–4946.

Johnson, R.A., Beeler Jr, J.R., 1981. Point defects in titanium. In: Lee, J.K. (Ed.), Interatomic Potentials and Crystalline Defects. The Metallurgical Society of AIME, Warrendale, PA, pp. 165–177.

Johnson, R.A., Brown, E., 1962. Point defects in copper. Phys. Rev. 127, 446–454.

Johnson, R.A., 1964. Interstitials and vacancies in α iron. Phys. Rev. 134, A1329–A1336.

Johnson, R.A., 1965. Point defect calculations for copper. J. Phys. Chem. Solids 26, 75–80.

Johnson, R.A., 1966. Point-defect calculations for an fcc lattice. Phys. Rev. 145, 423–433.

Johnson, R.A., 1994. Average Gibbs energy per lattice defect. Phys. Rev. B 50, 13799–13800.

Jung, P., 1991. Production of atomic defects in metals. In: Ullmaier, H. (Ed.), Atomic Defects in Metals, Landolt-Börnstein New Series, Group III: Crystal and Solid State Physics, vol. 25. Springer-Verlag, Berlin Heidelberg, pp. 1–87.

Karsten, K., Dedek, U., Dworschak, F., Schilling, W., 1991. Resistivity recovery in dilute AuFe alloys after low temperature electron irradiation. Radiat. Eff. Defects Solids 116, 315–328.

Kauffman, J.W., Koehler, J.S., 1955. Quenching-in of lattice vacancies in pure gold. Phys. Rev. 97, 555.

Kentzinger, E., Schober, H.R., 2000. Migration energies in $L1_2$ intermetallic compounds. J. Phys. Condens. Matter 12, 8145–8158.

Kim, S., Chang, Y.A., 2000. An interdiffusion study of a NiAl alloy using single-phase diffusion couples. Metall. Mater. Trans. A 31A, 1519–1524.

Kinchin, G.H., Pease, R.S., 1955. The displacement of atoms in solids by radiation. Rep. Prog. Phys. 18, 1–51.

Kittel, C., 1996. Solid State Physics, seventh ed. John Wiley & Sons, New York.

Kleman, M., Friedel, J., 2008. Disclinations, dislocations, and continuous defects: a reappraisal. Rev. Mod. Phys. 80, 61–115.

Klemradt, U., Drittler, B., Hoshino, T., Zeller, R., Dederichs, P.H., 1991. Vacancy–solute interactions in Cu, Ni, Ag, and Pd. Phys. Rev. B 43, 9487–9497.

Köppers, M., Derdau, D., Friesel, M., Herzig, C., 1997. Self-diffusion and group III (Al, Ga, In) solute diffusion in hcp titanium. Defect Diffusion Forum 143–147, 43–48.

Korzhavyi, P.A., Abrikosov, I.A., Johansson, B., 1999. First-principles calculations of the vacancy formation energy in transition and noble metals. Phys. Rev. B 59, 11693–11703.

Köstler, C., Faupel, F., Sander, L., Hehenkamp, T., 1987. Equilibrium vacancies in Cu and Ag alloys with electropositive solutes. Philos. Mag. A 56, 831–840.

Kraftmakher, Y., 1996. On equilibrium point defects in metals. Philos. Mag. A 74, 811–822.

Kronmüller, H., 1970. Studies of point defects in metals by means of mechanical and magnetic relaxation. In: Seeger, A., Schumacher, D., Schilling, W., Diehl, J. (Eds.), Vacancies and Interstitials in Metals. North-Holland, Amsterdam, pp. 667–728.

Kudielka, H., Runow, P., 1976. Einflüsse der Temperatur auf Symmetrie, Ordnungs- und Schwingungszustand des FePt-Gitters. Z. Metallkd. 67, 699–703.

Le Claire, A.D., 1978. Solute diffusion in dilute alloys. J. Nucl. Mater. 69 & 70, 70–96.

Le Claire, A.D., 1993. Some aspects of diffusion in dilute alloys. Defect Diffusion Forum 95–98, 19–40.

Leibfried, G., Breuer, N., 1978. Point Defects in Metals I—Introduction to the Theory. Springer-Verlag, Berlin Heidelberg.

Lejček, P., 2010. Grain Boundary Segregation in Metals. Springer-Verlag, Berlin Heidelberg.

Lidiard, A.B., 1955. Impurity diffusion in crystals (mainly ionic crystals with the sodium chloride structure). Philos. Mag. 46, 1218–1237.

Loretto, M.H., Clarebrough, L.M., Segall, R.L., 1965. Philos. Mag. 11, 459–465.

Losehand, R., Rau, F., Wenzl, H., 1969. Stored energy and electrical resistivity of Frenkel defects in copper. Radiat. Eff. 2, 69–74.

Lucasson, P.G., Walker, R.M., 1962. Production and recovery of electron-induced radiation damage in a number of metals. Phys. Rev. 127, 485–500.

Lucasson, P., Maury, F., Lucasson, A., 1987. On the migration of interstitial defects in irradiated cubic metals. Mater. Sci. Forum 15–18, 231–236.

Mahajan, S., Williams, D.F., 1973. Deformation twinning in metals and alloys. Int. Metall. Rev. 18, 43–61.

March, N.H., 1978. Point defect–solute interactions in metals. J. Nucl. Mater. 69 & 70, 490–520.

Matsukawa, Y., Zinkle, S.J., 2007. One-dimensional fast migration of vacancy clusters in metals. Science 318, 959–962.

Maury, F., Lucasson, A., Lucasson, P., Moser, P., Loreaux, Y., 1985. Interstitial migration in dilute FeSi and FeAu alloys. J. Phys. F Met. Phys. 15, 1465–1484.

McLellan, R.B., 1995. The diffusion of boron in nickel. Scr. Metall. Mater. 33, 1265–1267.

Mehl, M.J., Klein, B.M., 1991. All-electron first-principles supercell total-energy calculation of the vacancy formation energy in aluminium. Physica B 172, 211–215.

Mehrer, H., Stolica, N., Stolwijk, N.A., 1990. In: Mehrer, H. (Ed.), Diffusion in Solid Metals and Alloys, Landolt-Börnstein New Series, Group III: Crystal and Solid State Physics, vol. 26. Springer-Verlag, Berlin Heidelberg, pp. 32–84 (Chapter 2).

Mehrer, H., 2007. Diffusion in Solids—Fundamentals, Methods, Materials, Diffusion-controlled Processes. Springer-Verlag, Berlin Heidelberg.

Miller, K.M., 1981. Point defect–dislocation interactions in molybdenum. J. Phys. F Met. Phys. 11, 1175–1189.

Mishin, Y., Mehl, M.J., Papaconstantopoulos, D.A., Voter, A.F., Kress, J.D., 2001a. Structural stability and lattice defects in copper: ab initio, tight-binding, and embedded-atom calculations. Phys. Rev. B 63, 224106.

Mishin, Y., Sørensen, M.R., Voter, A.F., 2001b. Calculation of point-defect entropy in metals. Philos. Mag. A 81, 2591–2612.

Mishin, Y., 2004. Atomistic modeling of the γ and γ'-phases of the Ni–Al system. Acta Mater. 52, 1451–1467.

Mittemeijer, E.J., 2010. Fundamentals of Materials Science—The Microstructure–property Relationship Using Metals as Model Systems. Springer-Verlag, Berlin Heidelberg.

Mizubayashi, H., Okuda, S., 1977. Elastic after-effect studies of lattice defects in Mo after fast neutron irradiation at 5 K. Radiat. Eff. 33, 221–235.

Mizubayashi, H., Okuda, S., 1981. Elastic after-effect studies of self-interstitials in tungsten after fast neutron irradiation at 5 K. Radiat. Eff. 54, 201–216.

Mukherjee, K., 1965. Monovacancy formation energy and Debye temperature of close-packed metals. Philos. Mag. 12, 915–918.

Müller, M.A., Sprengel, W., Major, J., Schaefer, H.-E., 2001. Activation volume and chemical environment of atomic vacancies in intermetallic compounds. Mater. Sci. Forum 363–365, 85–87.

Mundy, J.N., 1987. Diffusion mechanism in f.c.c. metals. Phys. Status Solidi B 144, 233–241.

Mundy, J.N., 1992. Self diffusion in pure metals. Defect Diffusion Forum 83, 1–18.

Murayama, M., Howe, J.M., Hidaka, H., Takaki, S., 2002. Atomic-level observation of disclination dipoles in mechanically milled nanocrystalline Fe. Science 295, 2433–2435.

Narita, N., Takamura, J., 1992. Deformation twinning in f.c.c. and b.c.c. metals. In: Nabarro, F.R.N. (Ed.), Dislocations in Solids, vol. 9. North-Holland, Amsterdam, pp. 135–189 (Chapter 9).

Nazarov, R., Hickel, T., Neugebauer, J., 2010. First-principles study of the thermodynamics of hydrogen-vacancy interaction in fcc iron. Phys. Rev. B 82, 224104.

Nazarov, R., Hickel, T., Neugebauer, J., 2012. Vacancy formation energies in fcc metals: influence of exchange-correlation functionals and correlation schemes. Phys. Rev. B 85, 144118.

Norvell, J.C., Als-Nielsen, J., 1970. Long-range order in β brass. Phys. Rev. B 2, 277–282.

Nowick, A.S., Berry, B.S., 1972. Anelastic Relaxation in Crystalline Solids. Academic Press, New York.

Nowick, A.S., Heller, W.R., 1965. Dielectric and anelastic relaxation of crystals containing point defects. Adv. Phys. 14, 101–166.

Nowick, A.S., 1978. Anelastic studies of intrinsic atomic defects. J. Nucl. Mater. 69 & 70, 215–227.

Numakura, H., Koiwa, M., 1996. The Snoek relaxation in dilute ternary bcc alloys: a review. J. Phys. IV Colloq. C8, Suppl. J. Phys. III 6, 97–106.

Numakura, H., Koiwa, M., 1997. Snoek relaxation in ternary body-centered cubic alloys. In: Wolfenden, A., Kinra, V.K. (Eds.), M3D III: Mechanics and Mechanisms of Material Damping, ASTM Standard Technical Publication 1304. American Society for Testing and Materials, West Conshohocken, PA, pp. 383–393.

Numakura, H., Ikeda, T., Koiwa, M., Almazouzi, A., 1998. Self-diffusion mechanism in Ni-based L1$_2$ type intermetallic compounds. Philos. Mag. A 77, 887–909.

Numakura, H., Kashiwazaki, K., Yokoyama, H., Koiwa, M., 2000. Anelastic relaxation due to interstitial solute atoms in face-centred cubic metals. J. Alloys Compd. 310, 344–350.

Numakura, H., 2012. Interaction between interstitial and substitutional solute atoms in α iron. In: Furuhara, T., Numakura, H., Ushioda, K. (Eds.), Proceedings of the 3rd International Symposium on Steel Science (ISSS 2012). The Iron and Steel Institute of Japan, Tokyo, pp. 19–28.

Odette, G.R., Lucas, G.E., 2001. Embrittlement of nuclear reactor pressure vessels. JOM 53 (7), 18–22.

Oh, D.J., Johnson, R.A., 1989. Relationship between c/a ratio and point defect properties in hcp metals. J. Nucl. Mater. 169, 5–8.

Okuda, S., Mizubayashi, H., 1973. Study on the relaxation peaks due to the stage I defects in neutron irradiated molybdenum. Cryst. Lattice Defects 4, 75–82.

Okuda, S., Mizubayashi, H., 1976. Anelasticity study of self-interstitials in tungsten. Phys. Rev. B 13, 4207–4216.

Paidar, V., Vitek, V., 2002. Stacking-fault-type interfaces and their role in deformation. In: Westbrook, J.H., Fleischer, R.L. (Eds.), Intermetallic Compounds: Principles and Practice, Vol. 3: Progress. John Wiley & Sons, Chichester, pp. 437–467 (Chapter 22).

Pierron-Bohnes, V., Mehaddene, T., 2007. Lattice statics and lattice dynamics. In: Pfeiler, W. (Ed.), Alloy Physics: A Comprehensive Reference. Wiley-VCH, Weinheim, pp. 119–171 (Chapter 4).

Pike, L.M., Chang, Y.A., Liu, C.T., 1997. Point defect concentrations and hardening in binary B2 intermetallics. Acta Mater. 45, 3709–3719.

Püschl, W., Numakura, H., Pfeiler, W., 2007. Point defects, atom jumps and diffusion. In: Pfeiler, W. (Ed.), Alloy Physics: A Comprehensive Reference. Wiley-VCH, Weinheim, pp. 173–280 (Chapter 5).

Raji, A.T., Scandolo, S., Mazzarello, R., Nsengiyumva, S., Härting, M., Britton, D.T., 2009. Ab initio pseudopotential study of vacancies and self-interstitials in hcp titanium. Philos. Mag. 89, 1629–1645.

Ram, P.N., 1984. Green's-function calculation of entropy of formation of self-interstitials in Cu. Phys. Rev. B 30, 6146–6153.

Ram, P.N., 1991a. Vibrational properties of self-interstitials in metals. Radiat. Eff. Defects Solids 118, 1–93.

Ram, P.N., 1991b. Dynamics of self-interstitial atoms in bcc metals. Phys. Rev. B 43, 6977–6985.

Rathenau, G.W., de Vries, G., 1969. Diffusion. In: Berkowitz, A.E., Kneller, E. (Eds.), Magnetism and Metallurgy. Academic Press, New York, pp. 749–814 (Chapter XVI).

Rempel, A.A., Sprengel, W., Blaurock, K., Reichle, K.J., Major, J., Schaefer, H.-E., 2002. Identification of lattice vacancies on the two sublattices of SiC. Phys. Rev. Lett. 89, 185501.

Richter, G., 1937a. Über die magnetische Nachwirkung am Carbonyleisen. Ann. Phys. 29, 605–635.

Richter, G., 1937b. Über die mechanische und die magnetische Nachwirkung des Carbonyleisens. Ann. Phys. 32, 683–700.

Rivière, J.P., Junqua, N., Beretz, D., 1980. Recovery of a Fe–40 at.% Al ordered alloy after 20 K neutron or electron irradiation. Radiat. Eff. Lett. 57, 75–79.

Robrock, K.-H., 1990. Mechanical Relaxation of Interstitials in Irradiated Metals. Springer-Verlag, Berlin Heidelberg.

Romanov, A.E., Vladimirov, V.I., 1992. Disclinations in crystalline solids. In: Nabarro, F.R.N. (Ed.), Dislocations in Solids, vol. 9. North-Holland, Amsterdam, pp. 191–402.

Romanov, A.E., 2003. Mechanics and physics of disclinations in solids. Eur. J. Mech. A/Solids 22, 727–741.

Rossiter, P.L., 1987. The Electrical Resistivity of Metals and Alloys. Cambridge University Press, Cambridge.

Samolyuk, G.D., Golubov, S.I., Osetsky, Y.N., Stoller, R.E., 2013. Self-interstitial configurations in hcp Zr: a first principles analysis. Philos. Mag. Lett. 93, 93–100.

Sattonnay, G., Ma, F., Dimitrov, C., Dimitrov, O., 1997. Radiation-induced defects in electron-irradiated γ-TiAl compounds: the effect of composition. J. Phys. Condens. Matter 9, 5527–5542.

Schaefer, H.-E., Schmid, G., 1989. Vacancy migration at high temperatures in Au studied by positron annihilation measurements in fast pulse-heating experiments. J. Phys. Condens. Matter 1 (Suppl. A), SA49–SA54.

Schaefer, H.-E., Frenner, K., Würschum, R., 1999. Time-differential length change measurements for thermal defect investigations: intermetallic B2-FeAl and B2-NiAl compounds, a case study. Phys. Rev. Lett. 82, 948–951.

Schaefer, H.E., 1982. Thermal equilibrium studies of vacancies in metals by positron annihilation. In: Coleman, P.G., Sharma, S.C., Diana, L.M. (Eds.), Positron Annihilation. North-Holland, Amsterdam, pp. 369–380.

Schaefer, H.-E., 1987. Investigation of thermal equilibrium vacancies in metals by positron annihilation. Phys. Status Solidi A 102, 47–65.

Schaller, R., Fantozzi, G., Gremaud, G. (Eds.), 2001. Mechanical spectroscopy Q^{-1} 2001. Trans Tech Publications, Zürich.

Schapink, F.W., 1969. Statistical Thermodynamics of Vacancies in Binary Alloys. (PhD thesis), Delft: Delft University of Technology.

Schilling, W., Sonnenberg, K., 1973. Recovery of irradiated and quenched metals. J. Phys. F Met. Phys. 3, 322–350.

Schilling, W., 1978. Self-interstitial atoms in metals. J. Nucl. Mater. 69 & 70, 465–489.

Schober, H.R., Petry, W., Trampenau, J., 1992. Migration enthalpies in FCC and BCC metals. J. Phys. Condens. Matter 4, 9321–9338.

Schober, H.R., 1977. Single and multiple interstitials in FCC metals. J. Phys. F Met. Phys. 7, 1127–1138.

Scholz, A., Lehmann, C., 1972. Stability problems, low-energy-recoil events, and vibrational behavior of point defects in metals. Phys. Rev. B 6, 831–816.

Schramm, R.E., Reed, R.P., 1975. Stacking fault energies of seven commercial austenitic stainless steels. Metall. Trans. A 6A, 1345–1351.

Schulte, C.W., Campbell, J.L., 1979. Positron trapping in cobalt and trapping threshold temperature correlations. Appl. Phys. 19, 269–273.

Schultz, H., 1991. Defect parameters of b.c.c. metals: group-specific trends. Mater. Sci. Eng. A141, 149–167.

Seeger, A., Mehrer, H., 1970. Analysis of self-diffusion and equilibrium measurements. In: Seeger, A., Schumacher, D., Schilling, W., Diehl, J. (Eds.), Vacancies and Interstitials in Metals. North-Holland, Amsterdam, pp. 1–58.

Seeger, A., Mann, E., von Jan, R., 1962. Zwischengitteratome in kubisch-flächenzentrierten kristallen, insbesondere in kupfer. J. Phys. Chem. Solids 23, 639–656.

Seeger, A., 1973. Investigation of point defects in equilibrium concentrations with particular reference to positron annihilation techniques. J. Phys. F Met. Phys. 3, 248–294.

Sharma, B.D., Sonnenberg, K., Antesberger, G., Kesternich, W., 1978. Electrical resistivity of electron-irradiated concentrated Fe–Cr–Ni alloys during isochronal annealing. Philos. Mag. A 37, 777–788.

Shestopal, V.O., 1966. Specific heat and vacancy formation in titanium at high temperatures. Soviet Phys. Solid State 7, 2798–2799.

Shigesato, G., Fujishiro, T., Hara, T., 2012. Boron segregation to austenite grain boundary in low alloy steel measured by aberration corrected STEM–EELS. Mater. Sci. Eng. A 556, 358–365.

Siegel, R.W., 1978. Vacancy concentrations in metals. J. Nucl. Mater. 69 & 70, 117–146.

Siegel, R.W., 1982. Positron annihilation spectroscopy of defects in metals—an assessment. In: Coleman, P.G., Sharma, S.C., Diana, L.M. (Eds.), Positron Annihilation. North-Holland, Amsterdam, pp. 351–368.

Silcox, J., Hirsch, P.B., 1959. Direct observations of defects in quenched gold. Philos. Mag. 4, 72–89.

Simmons, R.O., Balluffi, R.W., 1960a. Measurements of equilibrium vacancy concentrations in aluminum. Phys. Rev. 117, 52–61.

Simmons, R.O., Balluffi, R.W., 1960b. Measurement of the equilibrium concentration of lattice vacancies in silver near the melting point. Phys. Rev. 119, 600–615.

Simmons, R.O., Balluffi, R.W., 1960c. Measurements of the high-temperature electrical resistance of aluminum: resistivity of lattice vacancies. Phys. Rev. 117, 62–68.

Simmons, R.O., Balluffi, R.W., 1962. Measurement of equilibrium concentrations of lattice vacancies in gold. Phys. Rev. 125, 862–872.

Simmons, R.O., Balluffi, R.W., 1963. Measurement of equilibrium concentrations of vacancies in copper. Phys. Rev. 129, 1533–1544.

Snoek, J.L., 1939a. Magnetic after effect and chemical constitution. Physica 6, 161–170.

Snoek, J.L., 1939b. Mechanical after effect and chemical constitution. Physica 6, 591–592.

Sprengel, W., Müller, M.A., Schaefer, H.-E., 2002. Thermal defects and diffusion. In: Westbrook, J.H., Fleischer, R.L. (Eds.), Intermetallic Compounds: Principles and Practice, Vol. 3: Progress. John Wiley & Sons, Chichester, pp. 275–293 (Chapter 15).

Sun, Y.-Q., 1995. Structure of antiphase boundaries and domains. In: Westbrook, J.H., Fleischer, R.L. (Eds.), Intermetallic Compounds: Principles and Practice, Vol. 1: Principles. John Wiley & Sons, Chichester, pp. 495–517 (Chapter 21).

Taji, Y., Iwata, T., Yokota, T., 1989. Molecular-dynamical calculations of irradiation-produced point defects in bcc metals. Phys. Rev. B 39, 6381–6387.

Tateyama, Y., Ohno, T., 2003. Stability and clusterization of hydrogen-vacancy complexes in α-Fe: an ab initio study. Phys. Rev. B 67, 174105.

Thompson, M.W., 1969. Defects and Radiation Damage in Metals. Cambridge University Press, Cambridge.

Townsend, J.R., Schildcrout, M., Reft, C., 1976. Mechanical studies of irradiation-induced defects in Cu and W. Phys. Rev. B 14, 500–516.

Ullmaier, H., 1991. Atomic defects in metals. In: Landolt-Börnstein New Series, Group III: Crystal and Solid State Physics, vol. 25. Springer-Verlag, Berlin Heidelberg.

Ungár, T., Tichy, G., 1999. The effect of dislocation contrast on X-ray line profiles in untextured polycrystals. Phys. Status Solidi A 171, 425–434.

Urban, R., Ehrhart, P., Schilling, W., Schober, H.R., Lauter, H., 1987. Phys. Status Solidi B 144, 287–304.

Vaessen, P., Lengeler, B., Schilling, W., 1984. Recovery of the electrical resistivity in electron-irradiated, concentrated silver-zinc alloys. Radiat. Eff. 81, 277–292.

Vajda, P., 1977. Anisotropy of electron radiation damage in metal crystals. Rev. Mod. Phys. 49, 481–521.

van Ommen, A.H., de Miranda, J., 1981. A dilatometric study of vacancy mobility in the intermetallic compound CoGa. Philos. Mag. A 43, 387–407.

Veyssyère, P., Douin, J., 1995. Dislocations. In: Westbrook, J.H., Fleischer, R.L. (Eds.), Intermetallic Compounds: Principles and Practice, Vol. 1: Principles. John Wiley & Sons, Chichester, pp. 519–558 (Chapter 22).

Vineyard, G.H., 1957. Frequency factors and isotope effects in solid state rate processes. J. Phys. Chem. Solids 3, 121–127.

Vítek, V., 1974. Theory of the core structures of dislocations in body-centred-cubic metals. Cryst. Lattice Defects 5, 1–34.

Wagner, C., Schottky, W., 1931. Theorie der geordneten Mischphasen. Z. Phys. Chem. B 11, 163–210.

Wallner, G., Franz, H., Rauch, R., Schmalzbauer, A., Peisl, J., 1987. Diffuse X-ray scattering from defects after low temperature neutron irradiation. Mater. Sci. Forum 15–18, 907–912.

Wert, C.A., 1978. Trapping of hydrogen in metals. In: Alefeld, G., Völkl, J. (Eds.), Hydrogen in Metals II—Application-oriented Properties. Springer-Verlag, Berlin Heidelberg, pp. 305–330 (Chapter 8).

West, R.N., 1979. Positron studies of lattice defects in metals. In: Gautojärvi, P. (Ed.), Positrons in Solids, pp. 89–144.

Wolfer, W.G., 1984. Advances in void swelling and helium bubble physics. J. Nucl. Mater. 367–378.

Wolff, J., Broska, A., Franz, M., Köhler, B., Hehenkamp, Th., 1997. Defect analysis in intermetallic alloys with positron annihilation. Mater. Sci. Forum 255–257, 593–595.

Wollenberger, H.J., 1996. Point defects. In: Cahn, R.W., Haasen, P. (Eds.), Physical Metallurgy, fourth, revised and enhanced ed. Elsevier Science BV, Amsterdam, pp. 1621–1721 (Chapter 18).

Wolverton, C., 2007. Solute–vacancy binding in aluminum. Acta Mater. 55, 5867–5872.

Woo, C.H., 2000. Defect accumulation behaviour in hcp metals and alloys. J. Nucl. Mater. 276, 90–103.

Würschum, R., Grupp, C., Schaefer, H.-E., 1995. Simultaneous study of vacancy formation and migration at high temperatures in B2-type Fe aluminides. Phys. Rev. Lett. 75, 97–100.

Yoshida, M., 1961. Distribution of interstitials and vacancies produced by an incident fast neutron. J. Phys. Soc. Jpn. 16, 44–50.

Young Jr, F.W., 1978. Interstitial mobility and interactions. J. Nucl. Mater. 69 & 70, 310–330.

Zehetbauer, M.J., Zhu, Y.T., 2009. Bulk Nanostructured Materials. Wiley-VCH, Weinheim.

Biography

NUMAKURA Hiroshi studied nuclear engineering at Tohoku University in Sendai and obtained Ph.D. in 1986. He joined Kyoto University and worked there first as assistant professor and then as associate professor at Department of Materials Science and Engineering for 21 years. In 1996-97 he spent 10 months at Ecole Nationale Supérieur de Mécanique et Aerotechnique in Futuroscope, France, as visiting researcher. He moved in 2007 to Osaka Prefecture University, where he currently holds the position of professor at Department of Materials Science. His research aims at understanding the behaviour of defects in solid materials from an atomic level, integrating theory and experiment. His recent interests are on solute-solute interaction in metals and point defects and diffusion in intermetallics and related compounds.

7 Solidification

William J. Boettinger, Materials Science and Engineering Division, Materials Measurement Laboratory, NIST, Gaithersburg, MD, USA

Dilip K. Banerjee, Materials Science and Engineering Division, Materials Measurement Laboratory, NIST, Gaithersburg, MD, USA

7.1 Introduction

A general view of the formation of the solid from its melt is given. The macroscopic process is driven by the extraction of heat from the melt, but fluid flow is also relevant. Thus the first section deals with transport phenomena. Transport, including species diffusion, also plays a role in microscopic processes. Microstructural development is then explored through an examination of thermodynamics with a description of phase diagrams and their associated thermodynamic databases and computational methods. Following sections treat nucleation, interface attachment kinetics, solute distribution, cellular and dendritic growth, and polyphase solidification. Subsequently, the effect of fluid flow on the grain structure of castings, macrosegregation and porosity are treated. Finally we provide a discussion of developing processes employing various aspects of solidification.

It is to be noted that this document builds on the previous versions of this chapter authored by William Tiller and Herald Biloni in earlier editions of *Physical Metallurgy*. To them we are indebted. The reader is referred to several new monographs on the subject of solidification for different perspectives on this topic: Dantzig and Rappaz (2009), Stefanescu (2010), Kelton and Greer (2010) and Glicksman (2011). Articles by Boettinger et al. (2000) and Asta et al. (2009) summarize progress and future challenges in the area of solidification.

7.2 Transport Phenomena during Solidification

Solidification is driven primarily by heat transfer from the liquid through the solid being formed and through the mold into the surroundings to remove liquid metal superheat and heat of fusion. Occasionally, part of the heat of fusion flows into the liquid if it is supercooled. Unlike solid-state metallurgical phase transformations that generally involve much smaller heats of transformation, solidification involves important considerations of temperature gradients and heat flow. Because the liquid and solid have different concentrations, gradients of concentration and the resultant diffusion of alloying additions also must be considered in both the liquid and the solid. Understanding is further complicated by the resultant density variations that occur in the liquid that lead to buoyancy-driven convection. Motion of the mold and contraction in the solidifying metal also contribute to the liquid movement. Therefore, solidification includes energy, mass, species, and momentum transport. The formal treatment of this problem involves additional complexity as a consequence of the continuous generation of latent heat at the moving liquid–solid (L–S) interface, and the variation of the thermophysical properties of the metal–mold system with temperature. Transport processes are important not only on the scale of the casting, but also on the scale of the microscopic shape of the moving liquid–solid interface. This interface can be smooth, cellular, or dendritic features. These features produce the microstructure of the solidified alloy. The coupling of the macroscopic and microscopic transport processes is an essential feature of solidification.

In this section we describe (1) the general transport equations and extract simpler equations used to obtain analytical expressions for heat transfer and solute diffusion in the solid/liquid metal system; (2) the complex situation of the heat transfer coefficient between metal and mold wall; (3) examples of heat flow analysis in one-dimensional (1-D) as well as the use of controlled (directional) solidification for research purposes and (4) general aspects of software packages used to numerically treat the complex casting shapes involved in industrial foundry applications.

7.2.1   General Transport Equations

7.2.1.1   Energy Transport

The transport equation for thermal energy H^V(enthalpy/volume) is given by

$$\frac{\partial}{\partial t}\left(H^V\right) + \nabla \cdot \left(\vec{v}\, H^V\right) = \nabla \cdot (k\nabla T), \tag{1}$$

where $\vec{v}$ is the medium velocity, t is the time, k is the thermal conductivity and T is the temperature. There are two approaches to solving this equation: one treats the liquid, solid and mold separately with matching boundary conditions on the interfaces. The other treats the liquid–solid system as a single domain.

Separate treatment of the phases is only practical when the shape of the interface between the solid and the liquid is relatively smooth, as in some cases for the solidification of pure substances or dilute alloys. To obtain the simplest equation for heat flow one assumes constant density in each phase and the absence of fluid flow. Using $H^V = H_0^V + C^V T$, where H_0^V is a reference enthalpy and C^V is the heat capacity per unit volume, the pure conduction heat diffusion equation is obtained separately for the solid and liquid, namely,

$$\frac{\partial T}{\partial t} = \alpha \nabla^2 T, \tag{2}$$

if C^V is assumed constant and where $\alpha = k/C^V$ is the thermal diffusivity. The matching boundary conditions at the interface between the solid and the liquid region are given by

$$T_L^* = T_S^* = T^*,$$

(3)

$$k_S G_S|^* - k_L G_L|^* = VL_V.$$

(4)

Here $T_L^* =$ liquid temperature at the interface, $T_S^* =$ solid temperature at the interface, and $T^* =$ interface temperature that in general depends on liquid composition, interface curvature and the normal component of the interface velocity, V (see Section 7.5); $k_S =$ solid thermal conductivity, $k_L =$ liquid thermal conductivity, $G_L|^*$ and $G_S|^*$ are the normal components of the temperature gradients in the liquid and solid and $L_v =$ volumetric latent heat. This form of the heat flow equation and matching conditions will be used to develop the various microstructure models in Sections 7.4–7.8.

The single-domain treatment is used for most numerical software packages for the modeling of casting of alloys especially those involving dendritic solidification. It solves the transport in a single computational domain that includes fully liquid, mushy zone and fully solid regions. The mushy dendritic region is characterized by a volume fraction of liquid g_L, and a volume fraction of solid g_S that sum to unity. In general these fractions are functions of position and time and represent a coarse-grained view of the mush. Early work on single-domain methods was presented by Bennon and Incropera (1987a,b), Rappaz and Stefanescu (1988) and Rappaz (1989). The enthalpy for a solid–liquid system can be described as

$$H^V = H_L^V(1 - g_S) + H_S^V g_S.$$

(5)

The enthalpies of the liquid phase and solid phase are functions of temperature and composition that also depend on position and time just like the fraction solid. If we assume that the liquid and solid enthalpies are only functions of temperature and that in addition the solid fraction is only a function of temperature, then one obtains

$$H^V = H_L^V - g_S[H_L^V - H_S^V] = (H_0 + C_L^V T) - g_S L_V.$$

(6)

Neglecting the advection term, eqn (1) becomes

$$\frac{\partial}{\partial t}(C_L^V T) - L_v \frac{dg_S}{dT}\frac{\partial T}{\partial t} = \nabla \cdot (k \nabla T).$$

(7)

The numerical implementation of single-domain methods is subdivided into two main approaches: the effective (modified) specific heat method and the enthalpy method. Rappaz (1989) discusses the advantages and disadvantages of each. In the modified specific heat method, eqn (7) is written as

$$\frac{\partial}{\partial t}(C_e^V T) = \nabla \cdot (k \nabla T),$$

(8)

where C_e^V is the modified specific heat given by

$$C_e^V = C_L^V - L_V \frac{dg_s}{dT}. \tag{9}$$

Although, the modified specific heat is easy to implement in a numerical code, it offers some numerical challenges. This is because of the fact that there is typically a strong variation of the modified specific heat at the liquidus or eutectic temperature because of the large value of the $\frac{dg_s}{dT}$ term. Nevertheless most commercial casting software typically employs the modified specific heat method.

In standard macroscopic modeling of solidification the relationship $g_S(T)$ or $C_e^V(T)$ can be deduced either from differential thermal analysis (DTA) measurements or from solute microsegregation models (such as that of Scheil (1942)) or its modifications for dendrite tip kinetics or solid diffusion to be discussed in Section 7.7. These models can be combined with a thermodynamic alloy database (described in Section 7.3) to provide a detailed description of the enthalpy of each phase and consequently the heat capacity and heat of fusion. The experimental interpretation of DTA-type measurements is complicated by the thermal lags of standard instruments. These lags are because of the spatial separation of the temperature sensors and the melt. An analysis of these lags and the use of enthalpy–temperature information obtained from a CALPHAD-type database to simulate the DTA response of complex alloys is given by Boettinger and Kattner (2002).

In the enthalpy method the change in enthalpy is the independent variable, which is solved at each time step. In numerical approaches, eqn (1) is solved (ignoring advection for the time being for the sake of explanation only) by tracking both enthalpy and temperature at each computational node. In an explicit formulation, the enthalpy at each node of the finite element (FE) model at time t_{n+1} is obtained by knowing the nodal enthalpies at time t_n and evaluating the right side of eqn (1) at time t_n. Then, nodal temperatures at time t_{n+1} can be updated from an enthalpy versus temperature relationship such as the one given in eqn (6). However, the accuracy of the explicit scheme can be questionable because of its strong dependence on the time step increment. Implementation of the enthalpy method is also difficult in the implicit form. This is because it is difficult to ascertain if the proper relation between enthalpy and temperature is maintained during the solution. To address some of the challenges with the implementation of the enthalpy method, Dantzig and Rappaz (2009) introduced the enthalpy-specific heat method, which involves evaluation of the right side of eqn (1) at time t_{n+1} and manipulation of the left-hand side of eqn (1) using the effective specific heat. A predictor–corrector approach is used to obtain the nodal temperatures at time t_{n+1}. Nodal temperatures are the independent variables in this formulation. A useful variation of the enthalpy method is called temperature recovery method (Tszeng et al., 1989). A recent article by Felice et al. (2009) suggests that the enthalpy method converges faster than the direct temperature solution approach.

More complex methods, where g_S is not simply a function of T, must be considered if the details of nucleation and growth are to be coupled to the macroscopic heat flow (Rappaz, 1989). For example, for columnar dendritic growth, g_S depends on the local isotherm velocity and temperature gradient, as well as the temperature. For solidification involving equiaxed microstructures, the growth of solidifying grains depends on the local undercooling. In addition, the solidification path is dependent on the number of nucleated grains in the undercooled melt. The microstructure at a given location is obtained by coupling the macroscopic heat diffusion equation with a microscopic model that allows for nucleation and growth of grains based on local cooling conditions.

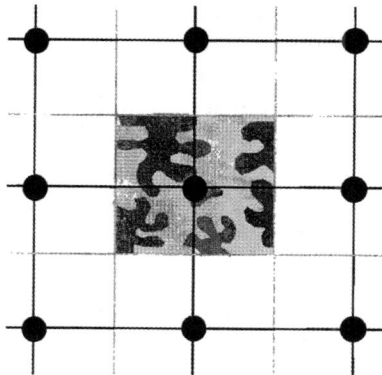

Figure 1 2-D illustration of the unit volume of material associated with an FEM (finite element model) node point (center black dot). The FEM mesh is shown as thin black lines, and the boundaries of the volumes associated with each node are represented by gray lines.

There are various approaches available for modeling the evolution of microstructure. In the simplest form, deterministic models are used to compute nucleation and growth of grains that are assumed to be motionless. The output of these microscopic models is the change in volume fraction of solid, dg_s, at each time step, dt. The microscopic models are coupled with the macroscopic heat diffusion eqn (7) through dg_s. A representative volume is assumed at each macroscopic node (in a typical finite element formulation) in which the microscopic model equations are solved (see **Figure 1**). As shown in the figure, each computational node in the macroscopic computation domain is associated with a representative volume. This volume associated with a macroscopic node is essentially the locus of points that lie closer to the node of interest than to a neighboring node. A simple model often assumes that this representative volume is closed to mass and momentum transport. This means that transport of grains (nucleated elsewhere) through advection into this representative volume is ignored. Details of coupling issues and their effects on time step and computational efficiency are discussed by Rappaz (1989) and Stefanescu (2009).

7.2.1.2 Species (solute) Transport
Species (solute) transport during solidification is important in determining the segregation in the final solidified part. The species transport equation is obtained for concentration, C mass/volume, using a solute diffusivity, D, in a similar manner to energy transport as

$$\frac{\partial C}{\partial t} + \nabla \cdot (C \, \vec{v}) = \nabla \cdot (D \nabla C). \tag{10}$$

Again, this equation can be solved by treating the liquid and solid as different regions with a moving interface or as a single domain using average quantities. For the two domain method we can again obtain a simple diffusion equation only by neglecting fluid flow

$$\frac{\partial C}{\partial t} = D \nabla^2 C \tag{11}$$

that is solved separately in the liquid and solid with matching conditions on the liquid–solid interface. The values of the concentrations on the liquid and solid sides of the interface, C_L^* and C_S^*, are generally different. The possible choices for the local values for these concentrations will be described in Section 7.3 and are related to the phase diagram at the temperature of the interface or to more complex velocity-dependent description described in Section 7.5. The flux boundary conditions at the L–S interface is

$$D_S G_S^c\big|^* - D_L G_L^c\big|^* = V\big(C_S^* - C_L^*\big), \tag{12}$$

where D_L and D_S are the liquid and solid solute diffusivities, $G_L^C\big|^*$ and $G_S^C\big|^*$ are the normal components of the concentration gradients at the interface in the liquid and solid, and V is the normal component of the interface velocity.

For the single-domain method, the concepts of volume averaging proposed by Ni and Beckermann (1991) can be employed. Because it is too time consuming to solve exact conservation equations on a microscopic scale over an entire casting, macroscopic models are developed over a finite-sized representative volume element (RVE) by using volume averaging (see **Figure 2**). This representative volume element will contain both solid and liquid phases in the mushy zone. This volume is selected to be larger than individual phase particles in the mixture but small enough to retain spatial variations of properties in the mixture. Typically the size of this volume is of the order of several dendrite arm spacings but less than the characteristic length of the mushy zone in the heat flow direction.

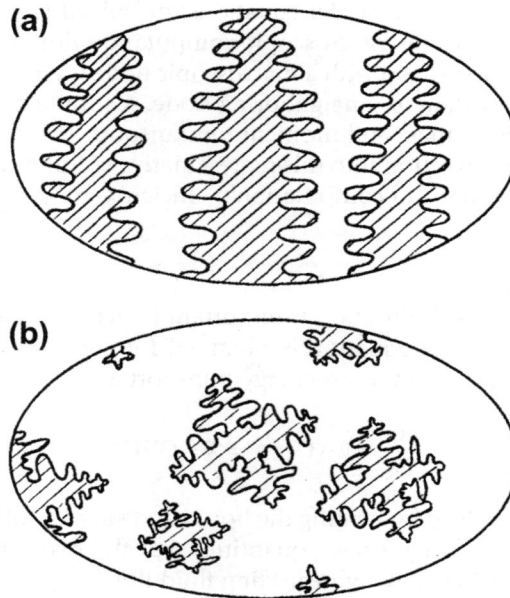

Figure 2 Schematic illustration of the averaging volume (e.g. RVE) containing (a) columnar dendritic crystals and (b) equiaxed dendritic crystals.

Average quantities of species concentration are computed over this RVE. For a two-phase L–S mixture involving two components, the average concentration in the RVE is given by

$$\overline{C} = \int_{V_{RVE}} [g_S C_S + g_L C_L] dV \approx g_S \langle C_S \rangle + g_L \langle C_L \rangle, \tag{13}$$

where $\langle C_L \rangle$ are $\langle C_S \rangle$ are the average concentrations of the liquid and solid in the RVE. These averages as well as g_L and g_S vary as functions of position and time in the different RVEs that represent the system. Defining $\vec{v}_S$ and $\vec{v}_L$ as velocities of solid and liquid phase and using eqn (10), the solute conservation equation can be written as

$$\frac{\partial (g_S \langle C_S \rangle + g_L \langle C_L \rangle)}{\partial t} + \nabla \cdot (g_S \langle C_S \rangle \vec{v}_S + g_L \langle C_L \rangle \vec{v}_L) = \nabla \cdot (g_L D_L \nabla C_L + g_S D_S \nabla C_S). \tag{14}$$

To recover models like the Scheil approach to be described in Sections 7.5 and 7.7, the liquid is assumed to be uniform in concentration on the scale of the RVE and $\langle C_L \rangle$ is replaced by C_L^*.

Equation (14) takes the general form of a conservation equation: accumulation of solute is described by the first term; advection of solute because of solid and liquid motion is given by the second term. The right-hand term is the diffusive term. Equation (14) can be written in advection–diffusion form for mixture composition by substituting ϕ for the average mixture composition, $\overline{C}$, where ϕ is the transported quantity as

$$\frac{\partial \phi}{\partial t} + \nabla \cdot (\phi \vec{v}) = \nabla \cdot (D \nabla \phi) + S_\phi, \tag{15}$$

where S_ϕ is the source term. Equations written in advection–diffusion form are suitable for easy implementation in time-implicit solution in numerical algorithms employed in commercial software.

7.2.1.3 Mass and Momentum Transport

Convection within the melt is important and influences solidification at both the macroscopic and the microscopic levels. At the macroscopic level it can change the shape of the isotherms, reduce the thermal gradients within the liquid region and alter the mode of species transport from diffusive to convective. Convection can cause long-range transport of individual species that leads to macrosegregation. The orientation of columnar dendritic structure, the occurrence of the columnar-to-equiaxed transition (CET) and nucleation initiated by dendrite arm detachment are all influenced by convection (see Section 7.9). During the casting process fluid flow occurs both during mold filling and due to natural convection. Once mold filling is complete, natural convection may occur because of thermal and solutal gradients. In addition, surface tension gradients at the free surface can produce shear stresses which affect convection. In the mushy region, volume change in solidification can alter the fluid flow.

The influence of convection on the thermal field can be considered in several ways. In the simplest approach, the advection term in eqn (1) is replaced by an increased thermal conductivity of the metal. In the most comprehensive approach, the fluid flow is solved by fully coupling the solution of the energy eqn (1) with the solution of equations for momentum conservation and mass conservation.

As done for energy and species transport, we will first describe the mass and momentum conservation equation for the two-domain approach and then we will write such equations for the one-domain approach. In the two-domain approach, the fluid flow equations are solved only in the liquid region and the velocity is usually assigned a zero value at the liquid–solid interface and everywhere in the solid region. The solution of the fluid dynamics equations provides time-dependent values of liquid velocity and pressure over the computation domain. The Navier–Stokes equations describe the conservation of linear momentum, which, along with mass conservation (continuity equation), are needed to describe the flow and pressure fields. General forms of these equations for a Newtonian fluid are given by

$$\rho \frac{\partial \vec{v}}{\partial t} + \rho \vec{v} \cdot \nabla \vec{v} = \nabla \cdot (\eta \nabla \vec{v}) - \nabla P + \rho \vec{b}, \tag{16}$$

$$\frac{\partial \rho}{\partial t} + \nabla \cdot \rho \vec{v} = 0. \tag{17}$$

The first two terms on the left side of eqn (16) describe the local and convective acceleration of the liquid and the first term on the right is the viscous shear force, the second term is the pressure gradient, and the third term on the right contains body forces. Note that ρ and η are density and viscosity, respectively.

In the two-domain approach, matching conditions for the continuity and Navier–Stokes equations need to be derived at the solid–liquid interface moving at a velocity of $\vec{v}^*$. Here we establish the matching (jump) condition only for the continuity equation as it is relevant to solidification shrinkage. Let us assume a control volume containing solid–liquid region having an interface area of S. Also, let us assume that the interface has a finite thickness. The velocities of the solid and liquid at the interface are denoted as $\vec{v}_S^*$ and $\vec{v}_L^*$. The integral form of eqn (17) can be written as

$$\int_V \frac{\partial \rho}{\partial t} + \int_V \nabla \cdot \rho \vec{v} = 0. \tag{18}$$

Using Gauss's theorem, the above equation can be written as

$$\int_V \frac{\partial \rho}{\partial t} + \int_S \rho \vec{v} \cdot \vec{n} \, dS = 0. \tag{19}$$

Note $\vec{n}$ is a unit normal vector.

In the limit when the thickness of the interface tends to zero (assuming a sharp L–S interface), the volume integral in the above equation becomes zero and the following expression can be written (note that the difference in velocity between the liquid- or solid-phase velocity and the interface velocity $\vec{V}$ needs to be considered as the L–S interface itself is moving),

$$\rho_L (\vec{v}_L^* - \vec{V}) \cdot \vec{n} - \rho_S (\vec{v}_S^* - \vec{V}) \cdot \vec{n} = 0. \tag{20}$$

If the velocity of the solid is taken as zero (e.g. solid does not move once it forms), then the fluid flow required to feed the change in density at the interface is obtained by rearranging the above equation and is given by

$$\overrightarrow{v}_L^* \cdot \overrightarrow{n} = -\beta \overrightarrow{V} \cdot \overrightarrow{n}, \tag{21}$$

where β is the solidification shrinkage factor defined as $(\rho_S - \rho_L)/\rho_L$. Because the solid is usually more dense than the liquid, the liquid flow velocity is negative for a solidifying liquid ($\overrightarrow{V} \cdot \overrightarrow{n} > 0$); that is, the liquid flows toward the freezing L–S interface to feed the shrinkage.

In the single-domain approach, average equations for the liquid phase in the mushy region eqns (16) and (17) can be written following the volume-averaging technique proposed by Ni and Beckermann (1991) as

$$g_L \rho_L \frac{\partial \overrightarrow{v}_L}{\partial t} + g_L \rho_L \nabla \cdot \overrightarrow{v}_L \overrightarrow{v}_L = \nabla \cdot (g_L \eta_L \nabla \overrightarrow{v}_L) - g_L \nabla P + g_L \rho_L \overrightarrow{b} - g_L^2 \eta_L \frac{\overrightarrow{v}_L}{K}, \tag{22}$$

$$\frac{\partial \overline{\rho}}{\partial t} + \nabla \cdot (g_L \rho_L \overrightarrow{v}_L) = 0, \tag{23}$$

where the solid has been assumed to be rigid and stationary, $\overline{\rho}$ is the volume-averaged density in the RVE and K is the permeability in the mushy zone. Note also that the right-hand side of eqn (22) contains extra terms that result from the volume-averaging process (included in the body force term) and forces because of flow through porous media governed by Darcy's law (last term on right-hand side) as described below by eqn (25). This term drives the velocities to zero when the material is fully solid. These equations can only be solved using a numerical approach for an arbitrary casting geometry. Accurate modeling of flow and pressure fields requires consideration of density variations in the solidification process, turbulence modeling (especially during the filling process), and consideration of momentum transfer between liquid and solid phases through suitable source terms.

Modeling the free surface during mold filling is a very difficult problem. One method that is widely used is the one introduced by Hirt and Nichols (1981) and is called the volume of fluid method (VOF). The VOF method can be described by an equation similar to the continuity eqn (17),

$$\frac{\partial(\rho_L F)}{\partial t} + \nabla \cdot (\rho_L \overrightarrow{v} F) = 0. \tag{24}$$

Here, volume function, F, is assigned a value of 1 in the fully liquid region, while it has a value of 0 in gas (or air) region. Equation (24) is solved along with all the other basic conservation equations to track the free surface at each time.

Fluid flow models need to be capable of describing flow through the mushy zone. Fluid flow through a solidifying region will be influenced by the morphology of the dendrite network in the mushy zone. Most numerical models treat flow in the mushy zone as a flow through a porous media obeying Darcy's law (Poirier, 1987) as

$$\overrightarrow{v}_L = -\frac{K}{\eta g_L}(\nabla P - \rho \overrightarrow{g}), \tag{25}$$

where $\vec{v}_L$ is liquid velocity, K is the permeability in the mushy zone, η is the viscosity, g_L is the liquid volume fraction. The permeability of the dendrite network is empirically determined with either the Hagen–Poiseuille or the Blake–Kozeny model (Poirier, 1987). For example, the permeability in the Blake–Kozeny model is given by

$$K = \frac{Cd^2 g_L^3}{(1 - g_L^2)}, \tag{26}$$

where C is a constant that depends on the geometry of the flow channel and d is a characteristic dimension of the solid microstructure usually related to the dendrite arm spacing. We will revisit the issue of permeability in dendritic structures in Section 7.9.3.

7.2.2 Heat Transfer at the Metal–Mold Interface

The heat flow from the solidifying metal is often limited by the metal–mold thermal resistance. This resistance is quantified by the value of the heat transfer coefficient, h_i, defined by

$$q = h_i(T_{iS} - T_{iM}), \tag{27}$$

corresponding to a Newtonian heat transfer model where q is the heat flux across the interface, T_{iS} is the metal temperature and T_{iM} the mold temperature, both at the metal–mold interface. When the melt first enters into contact with the mold wall, the mold surface is usually at a lower temperature than the liquid. The thermal contact is not perfect and the h_i value depends on the complex nature of the contact between metal and mold as shown in **Figure 3a**. Also important are the thermal resistance imposed by any coating present on the mold surface (Biloni, 1977) and the "air gap" that often develops between the mold surface and the solidifying metal because of metal shrinkage (Das and Paul, 1993). Consequently, the physical nature of the thermal contact can change with time and from point to point and may also depend on the wetting capacity of the melt, existence of oxides, grease, and so on.

Ho and Pehlke (1984, 1985) and Campbell (1991a,b) give a clear picture of the heat flow at the metal–mold interface including the origin, development, and nature of the so-called air gap. The following facts must be considered:

(1) When the metal enters the mold, good contact exists between the molten metal and the mold surface as Prates and Biloni (1972) and Morales et al. (1979) proved through the analysis of the casting surface structure. Contact occurs at the peaks of the mold surface roughness where large supercooling creates predendritic nuclei (Biloni and Chalmers, 1965). The application of pressure enhances the contact and the h_i value can be increased dramatically (Campbell, 1991a).
(2) After the creation of a solidified layer with sufficient strength, both the casting and the mold deform because of thermal contraction and the contact is reduced to isolated points at greater separations than that determined by the surface roughness. The interface gap starts to open and the conduction across the interface decreases drastically. Consequently, the h_i value falls by more than one order of magnitude.

When radiation is neglected, the important mechanism becomes the conduction of heat through the gas phase in the gap. In this case, $h_i = k_i/e$, where e is the equivalent thickness averaged over the metal–mold interface and k_i an effective thermal conductivity of the gas. Then, h_i corresponds to an average value represented by the equivalent model of **Figure 3b** but local values can be considerably different. The important aspects for modeling h_i are the identity of the gas in the gap and the gap

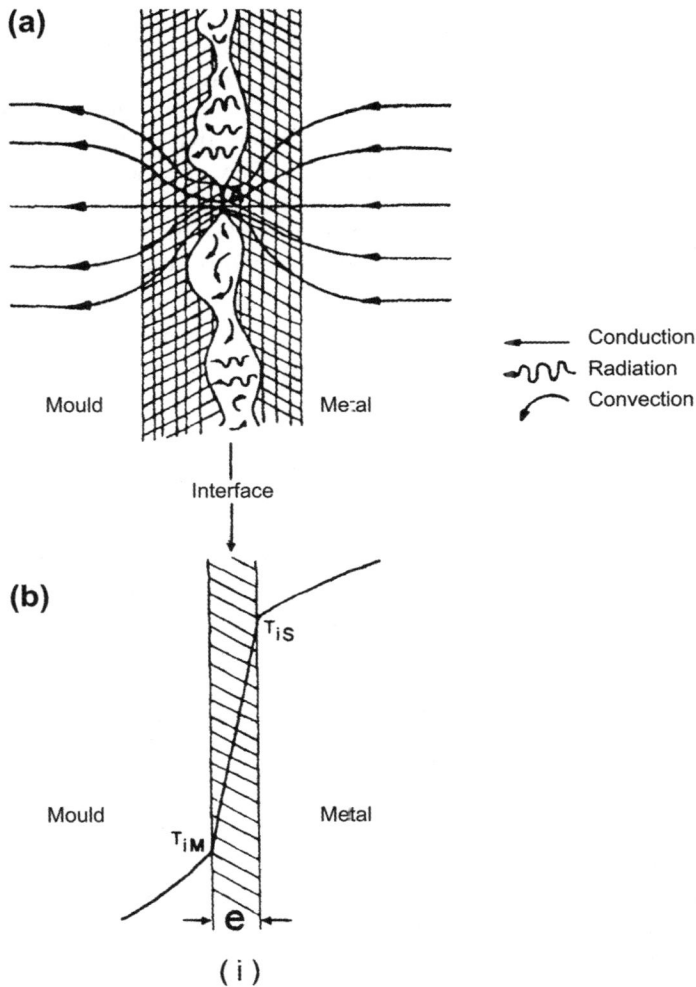

Figure 3 Nature of the thermal contact between the metal and the mold and equivalent Newtonian model of the metal–mold interface. (a) The complex nature of the contact and different types of heat transfer occurring at the interface are shown schematically. At point A good local contact assures a higher heat conduction. As a consequence, h_i will locally be higher than at the rest of the metal–mold interface. (b) The equivalent Newtonian model is based on an effective value of the gap, e, to yield an average value of h_i.

thickness. When radiation effects dominate, an effective heat transfer coefficient pertaining to the radiative heat transport can be written as

$$h_{\mathrm{rad}} = \frac{\sigma \left(T_{iS}^2 + T_{iM}^2 \right)\left(T_{iS} + T_{iM} \right)}{\varepsilon_{iS}^{-1} + \varepsilon_{iM}^{-1} - 1}, \tag{28}$$

where ε is the emissivity and σ is the Stefan–Boltzmann constant.

Ho and Pehlke (1984) and Campbell (1991a,b) discuss the nature of the gas in the gap. For iron and steel castings in sand molds, the gas present is likely to be hydrogen. This is significant, because of the high thermal conductivity of hydrogen, which is of the order of seven times greater than that for air. For metallic molds the lower H_2 gas content in the gap will result in a lower thermal conductivity, but the h_i value may still be twice that of air (Campbell, 1991b). These conclusions must be considered seriously when h_i values are estimated for heat transfer calculations.

The width of the gap is treated as a thermal expansion effect of the casting and the mold (Campbell, 1991a). If the mold expansion is considered homogeneous, transient heat flow consideration yields that

$$\frac{d}{L_0} \sim \alpha_C(T_m - T) + \alpha_M(T_{Mi} - T_0), \tag{29}$$

where d = gap size, L_0 = casting diameter, T_m = freezing point, T_{Mi} = temperature of the mold interface, T_0 = original temperature of the mold, α_C = casting thermal diffusivity, α_M = mold thermal diffusivity.

However, in general, there will be powerful geometrical effects and the gap thickness can change differently in various parts of the mold. Therefore, the situation in shaped casting is complicated as Campbell (1991b) affirms and has yet to be tackled successfully by theoretical models. Indeed Ho and Pehlke (1984) and (1985) demonstrate the difference in h_i value obtained for chilling surfaces located at the top or bottom of a cylindrical Al casting. Additional details on implementing the heat transfer across the casting/mold interface are described later in this section.

Mehrabian (1982) reported the measurement of h_i in splat cooling, in pressurized aluminum against a steel mold and in liquid die casting against a steel mold. He estimated that an upper limit exists for practically achievable heat transfer between liquid and substrates of about $h_i = 10^5$ J m^{-2} K^{-1} s^{-1} to $h_i = 10^6$ J m^{-2} K^{-1} s^{-1}. **Table 1** gives the order of magnitude of h_i for different conditions of a metal in contact with a mold.

In practice, those performing engineering modeling of castings employ careful measurements and analysis to obtain accurate heat transfer coefficients. Using thermocouple measurements and numerical solution of the inverse heat conduction problem (IHCP), Ho and Pehlke (1985) have obtained h_i values that show the onset of gap formation, and its time evolution.

Table 1   Order of magnitude of heat transfer coefficient, h_i, for different processes (Mehrabian, 1982)

Process	h_i (W/m^2 K)
Massive mold, polished	4×10^3
Massive mold, coated	7.5×10^2
Cooled mold, polished	5×10^3
Cooled mold, coated	10^3
Pressure cast	3×10^3–3×10^4
Die cast	5×10^4
Drop smash	10^4–10^5
Splat cooling	10^5–10^6

Hao et al. (1987) performed experiments with ductile iron and discussed the effect on h_i of the expansion of graphite precipitated during the solidification period. Sharma and Krishman (1991) discussed the effect of the microgeometry of molds considering several combinations of V-shaped grooves upon the mold or chill surface. Das and Paul (1993) determined h_i in castings and quenching using a solution technique for inverse problems based on the boundary element method (BEM).

Recently, Ferreira et al. (2005) studied the effect of melt temperature profile on the transient metal/mold heat transfer coefficient during solidification using numerical modeling of the IHCP. They performed solidification experiments with Sn-5 wt% Pb, Al-10 wt% Cu, and Al-4.5 wt% Cu alloys and used experimentally obtained temperatures in the numerical IHCP model to determine transient metal/mold heat transfer coefficients, h_i. It was shown that h_i profiles can be affected significantly by the initial melt temperature distribution. The numerical model used to simulate segregation profiles was based on an approach proposed by Voller (1998) but with improvement such as the ability to include temperature- and concentration-dependent thermophysical properties, variable metal/mold interface heat transfer coefficient, and to account for a space-dependent initial melt temperature profile. Ferreira et al. (2005) used an iterative approach and a systematic sensitivity study to arrive at an optimum value of the heat transfer coefficient. The sensitivity of this interfacial heat transfer coefficient to initial melt superheat variations was shown to depend on the alloy thermal properties, the wetting of the mold surface by the melt, and on how large the metal/mold heat transfer coefficient is relative to the thermal conductivity of the materials (e.g. the Biot number).

Spinelli et al. (2005) studied the heat transfer coefficients during upward and downward transient directional solidification (DS) of Al–Si alloys (5 wt% Si, 7 wt% Si, and 9 wt% Si). They also used a combination of experimentally recorded data and results from a numerical inverse method to obtain the heat transfer coefficient. They showed that for both upward and downward growth directions, the transient metal/coolant heat transfer coefficient, h_i, can be expressed as a power function of time and h_i increases with decrease in silicon content of the alloy. Additionally, they found that h_i values obtained for the upward solidification direction were significantly higher than those for the downward growth direction because of melt convection.

In commercial FE software, generally a pair of coincident nodes is created at the metal/mold interface at each location. Normally, these nodes have the same physical coordinates. These nodes are used to form the so-called interfacial heat transfer "element" or "gap element". Often a feature called "multi-point constraint" is used to form elements when the nodes at the boundaries of metal and mold materials do not align. In this method, the overall effect of the gap is included through its influence on the interfacial heat transfer coefficient.

In a numerical heat transport model that is also coupled with a thermal stress calculation model, the gap at the metal/mold interface is obtained at each time from the deformation of each material. Some advanced numerical codes have implemented this approach where the interfacial heat transfer will vary as the gap dimensions change at different locations. Lewis and Ransing (2000) used a thermo-elasto-viscoplastic material model to simulate the air gap at metal/mold interface. Shrinkage of metal at the interface is computed by the mechanical model and this information is used to compute the air gap. A contact algorithm was developed for modeling the interaction at the interface. The interaction between casting and mold occurs in the following ways: (a) perfect contact: no relative motion occurs (this condition is used until the start of the solidification), (b) slippage and contact: no movement occurs in normal direction, but both frictional and frictionless slipping can occur in the tangential direction, and (c) air gap: relative motion in both tangential and normal directions. They proposed the following equation for interfacial heat transfer coefficient,

$$h_i = \frac{k_a}{d + k_a/h_{i,\text{init}}},\qquad(30)$$

where d is the air gap width, k_a is the air thermal conductivity, $h_{i,\text{init}}$ is the initial value of h_i. Lewis and Ransing (1998) also proposed an empirical equation which is a function of casting interface temperature and has three arbitrary constants. Suitable values can be assigned to these constants to match an experimental variation of h_i with respect to casting interface temperature. The following equations describe their approach:

$$h_i = e^{a_1} e^{-a_2/x^2} \frac{1}{x^{a_3}},\qquad(31)$$

$$x = \sqrt{2a_2/a_3} + \max(0, T_L - T_{\text{int}}),\qquad(32)$$

where T_L is the liquidus temperature and T_{int} is the temperature of the casting at the interface with the mold and a_1, a_2 and a_3 are the adjusted constants to fit the data. They proposed maintaining calibrated values of these constants in a database for pairs of casting/mold material combinations.

7.2.3   Experimental Methods Involving Controlled Solidification

The difficulties associated with predicting external heat transfer in casting have led to the development of two research techniques of controlled solidification: unidirectional solidification and solidification with prescribed bulk supercooling. Much of the understanding of solidification laws comes from unidirectional solidification experiments based on the simple principle that the extraction of latent heat must be achieved without allowing the melt to supercool sufficiently to permit the nucleation of crystals ahead of the solidification front. In practice this requires a heat sink that removes heat from the solid and a heat source that supplies heat to the melt. Extensive use of such techniques pioneered by Chalmers (1964, 1971) produced the basis of the modern understanding of solidification microstructure. The basic heat flow objectives are to obtain a unidirectional thermal gradient across the interface and to move it so that the interface translates at a controlled rate. For a planar L–S interface the gradients are related by eqn (4). Techniques have also been developed based on this approach to obtain single crystals. Flemings (1974) gives details of the different techniques used.

Although a number of variations in the construction of DS equipment have been described in the literature, in many cases, the process used is the Bridgman technique where a cylindrical crucible is moved through a fixed temperature gradient with a constant translation velocity. It allows for the growth velocity of the interface and the temperature gradient to be separately controlled. If the pulling speed is not too high and the sample not too wide, the heat flow and the solidification can be assumed to be unidirectional. It is often assumed that the interface will remain stationary with respect to the furnace during most of the growth period and that the growth velocity of the interface is equal to that of the crucible or the moving furnace. Clyne (1980a,b) combined experimental solidification of commercial purity aluminum with a finite difference mathematical model to determine the relationship between the interface and the pulling speeds and the conditions under which the departure from ideal behavior would be significant.

Another directional controlled process corresponds to the extraction of heat via a bottom chill. Growth occurs in a direction parallel and opposite to the heat flux direction. In this situation a better control of microstructure and properties is obtained in comparison with conventional casting. However the microstructure is not as uniform as in the Bridgman method, because V and G_L decrease with the distance from the chill. It is possible to improve the microstructure uniformly by programed furnace temperature and withdrawal rates.

More recently, Xu and Naterer (2004) developed a new technique for controlling the liquid–solid interface motion for pure materials by using a combination of experimental/numerical technique. An inverse method was used to determine the interfacial velocity by matching the time-dependent position of the calculated and experimental interface position. They observed that an accelerating interface required higher interfacial temperature gradients over time, while these gradients became nearly constant when the interface moved with a constant velocity. They contended that with control strategies such as closed-loop feedback control, the proposed method could provide a better controlled solidification process.

The above examples of controlled solidification do not involve bulk liquid supercooling. However, when a crystal is nucleated at a specified temperature and grows freely into a super-cooled liquid, the bulk supercooling $\Delta T = T_m - T_{\text{bulk liquid}}$ plays a major role in determining the structure observed during the solidification process. This type of study has been performed with both low and high supercooling and important structural information has been obtained in organic material analogs (Glicksman et al., 1976) and metallic alloys (Flemings and Shiohara, 1984; Willnecker et al., 1989, 1990). An external needle has been used to stimulate nucleation at desired levels of bulk liquid supercooling. Because solidification is typically dendritic in these cases, a calculation of the heat flow is intimately entwined with the model for dendritic growth as described in Section 7.7.

7.2.4   Heat Flow in Simple Solidification Geometries

Two examples of the analysis of 1-D heat flow during solidification for pure metals undergoing planar (nondendritic) solidification are given here. The first is applicable in slow solidification into a superheated melt where the L–S interface temperature can be taken as the pure metal melting point. The second treats the case where bulk supercooling is present and interface kinetics is included because of the rapid solidification. These examples are followed by a discussion of powder solidification.

7.2.4.1   *Freezing at Mold Wall*

When a molten metal is suddenly placed in contact with a cold planar mold surface, solidification will ensue. The calculation of the rate of solidification is the subject of this section. The basic problem was addressed by Clyne and García (1980), who expanded on prior work of Pires et al. (1974), as follows: (i) conductive heat flow is unidirectional with semi-infinite metal and mold regions; (ii) the Newtonian interface resistance is represented by a constant heat transfer coefficient, h_i; (iii) the metal solidifies with a planar L–S interface that remains at the equilibrium melting point; (iv) the liquid metal has zero superheat; (v) convection currents and radiation losses are assumed to be negligible; and (vi) the thermal properties of the metal and mold do not change with temperature.

Only an approximate analytical solution to this problem is possible. It was called the "virtual adjunct method" (VAM) by Clyne and García (1980). Their basic assumption is that Newtonian resistance at the metal/mold interface is considered equivalent to the previously solidified layer of metal. In the so-called virtual system, thermal conduction as described by Fourier's law is used (the thermal contact between metal and mold is assumed perfect and is represented by an infinite heat transfer coefficient) (see **Figure 4**). Heat flow is considered in two regimes (metal and mold), separated by a hypothetical plane of constant temperature at the interface. Metal and mold side contributions to the thermal resistance at the interface are modeled using "preexisting adjuncts of solid material". Niyama and Anzai (1992) illustrate the approximate nature of the solution. They argue that this idea works nicely as long as the temperature distribution in the solidified layer is assumed to be linear.

Following Clyne and Garcia, the parameters of the real and virtual system (primed) are related by

$$x' = S_0 + x, \tag{33a}$$

$$S' = S_0 + S, \tag{33b}$$

$$t' = t_0 + t, \tag{33c}$$

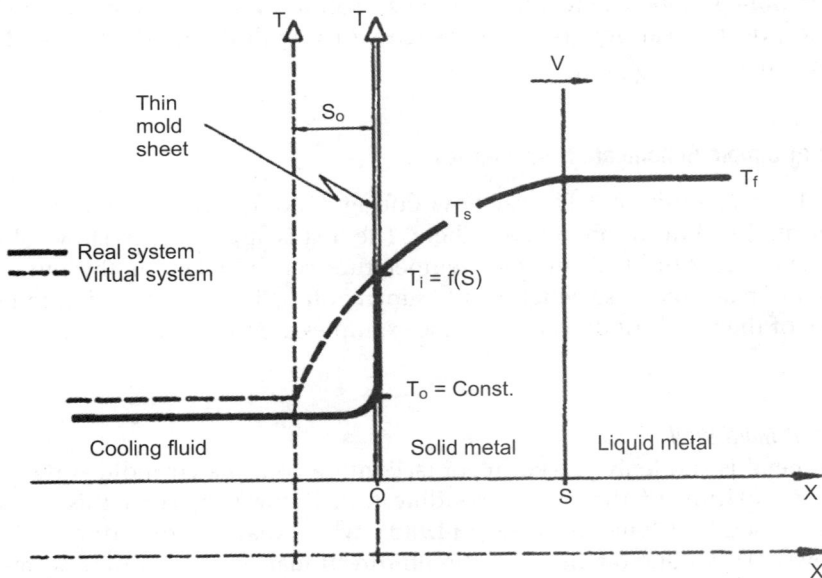

Figure 4 Temperature versus distance plot during metal solidification in a cooled mold for the real system (solid line) and for the virtual system (dashed line) used in the approximate heat flow calculation (Clyne and García, 1980).

where t_0 is the time needed to form solidified thickness S_0 in the virtual system. The boundary conditions are

$$T_S = \begin{cases} T_0 & \text{at} & x' = 0 \\ T_f & \text{at} & x' = S' \end{cases}. \tag{34}$$

Then the solution to eqn (2) is given by

$$T_S = A + B \ \text{erf}\left(\frac{x'}{2\sqrt{\alpha_S t'}}\right), \tag{35}$$

where A and B are constants obtained using the boundary conditions. If φ is defined as

$$\varphi = \frac{S'}{2\sqrt{\alpha_S t'}}, \tag{36}$$

then

$$t' = \frac{S'^2}{4\alpha_S \varphi^2}, \tag{37}$$

and similarly the solidified thickness S_0 at time t_0 is

$$t_0 = \frac{S_0^2}{4\alpha_S \varphi^2}. \tag{38}$$

We can write equation (37) as

$$t + t_0 = \frac{(S_0 + S)^2}{4\alpha_S \varphi^2}. \tag{39}$$

Using the expression for t_0 (eqn (38)), after manipulation, the solution for the solidification time, t_S, as the function of the distance solidified, X, has the form,

$$t_S(X) = AX^2 + BX, \tag{40}$$

where

$$A = \frac{1}{4\varphi^2 \alpha_S}, \tag{41}$$

$$B = \frac{L^V}{h_i(T_m - T_0)} \tag{42}$$

with φ evaluated numerically from the equation

$$\varphi \exp\left(\varphi^2\right)[M + \text{erf}(\varphi)] = \frac{\text{Ste}}{\sqrt{\pi}}. \tag{43}$$

Here, α_S is the thermal diffusivity of the solid metal, $M = \sqrt{k_S \rho_S C_p^S / k_M \rho_M C_p^M}$ = mold constant, subscripts S and M refer to solidified metal and mold, respectively, C_p^i are the specific heats, L_{mass} is the heat of fusion all on a per unit mass basis and Ste is the dimensionless Stefan number, Ste = $C_p^S(T_m - T_0)/L_{mass}$. The temperature in the solid is given by

$$T_S(x,t) = T_{ms} + \frac{T_m - T_{ms}}{\text{erf}(\varphi)} \, \text{erf}\left(\frac{X}{2\sqrt{\alpha_S t}}\right), \tag{44}$$

where T_{ms} represents mold temperature at mold–solid interface.

Equation (40) can be used to obtain the well-known equation (Chvorinov's rule) for the solidification time for a casting solidifying in an insulated mold such as a sand mold. If one assumes that there is no temperature discontinuity across the metal/sand mold interface (e.g. perfect thermal contact), the interfacial heat transfer coefficient tends to infinity. From eqn (42), B becomes zero and eqn (40) reduces to the well-known Chvorinov's rule

$$t_f = C\left(\frac{V}{A}\right)^2, \tag{45}$$

where t_f is the total solidification time for a casting of volume V and surface area A. C is a constant for a given metal–mold material and mold temperature. The second term in eqn (40) will dominate when the heat transfer coefficient at the metal/mold interface is finite and thermal conductivities of metal and mold are very large. In this case, the term A given by eqn (41) will be very small and hence solidification time will be governed by the metal/mold interfacial heat transfer coefficient.

Pires et al. (1974) and Garcia and Prates (1978) successfully checked eqn (40) for the particular case of efficient refrigeration of the chill mold in unidirectional solidification of Sn, Pb, Zn and Al. In this case, $M \simeq 0$, which gives a more simplified form of the general solution. Garcia et al. (1979) extended the experiments to molds where $M \neq 0$. These authors calculated the h_i value from the experimental curves, $X = f(t_S)$, through eqn (42).

Garcia and Clyne (1983) consider that in the case of Al–Cu alloys, having an appreciable temperature range of solidification, and concluded that this method can be used without major modification. An analytical treatment that takes account of the mushy zone has been developed by Lipton et al. (1982). In addition, the VAM has been extended to freezing processes involved in certain types of splat cooling (Clyne and Garcia, 1981). When the effects of superheat are considered, numerical methods are necessary, as shown for example by Hills et al. (1975).

7.2.4.2   Rapid Freezing in Contact with a Cold Substrate with Initial Melt Supercooling

In contrast to the previous example, splat cooling is a method in which liquid is suddenly placed in contact with a cold substrate. The liquid may appreciably supercool below the melting point before any solid forms. The interplay of liquid volume to be frozen, melt supercooling and external heat transfer to the cold substrate controls the solidification speed. A major complication occurs for heat flow analysis when high solidification velocities are involved even for pure materials. The temperature of the L–S interface T^* cannot be treated as a constant, equal to the melting point, T_m, but rather it is a function of the interface velocity. As a result, the heat flow analysis depends on the details of this function. Clyne (1984) treats the case of a pure metal freezing with a smooth (nondendritic) L–S interface governed by a kinetic law for the interface velocity given by $v = \mu(T_M - T^*)$ where μ is the linear interface attachment coefficient (see Section 7.5).

Calculations show the importance of the initial supercooling ΔT on the development of high solidification velocities. Large values of ΔT can develop if nucleation on the substrate is difficult. **Figure 5** (from Clyne, 1984) shows temperature–time plots at two positions inside a 50-μm thick layer of an Al melt cooling in contact with a substrate with $h_i = 10^6$ W m^{-2} K^{-1}, for $\mu = 4$ cm s^{-1} K^{-1} and nucleation at a dimensionless supercooling, a figure that might be expected for homogeneous

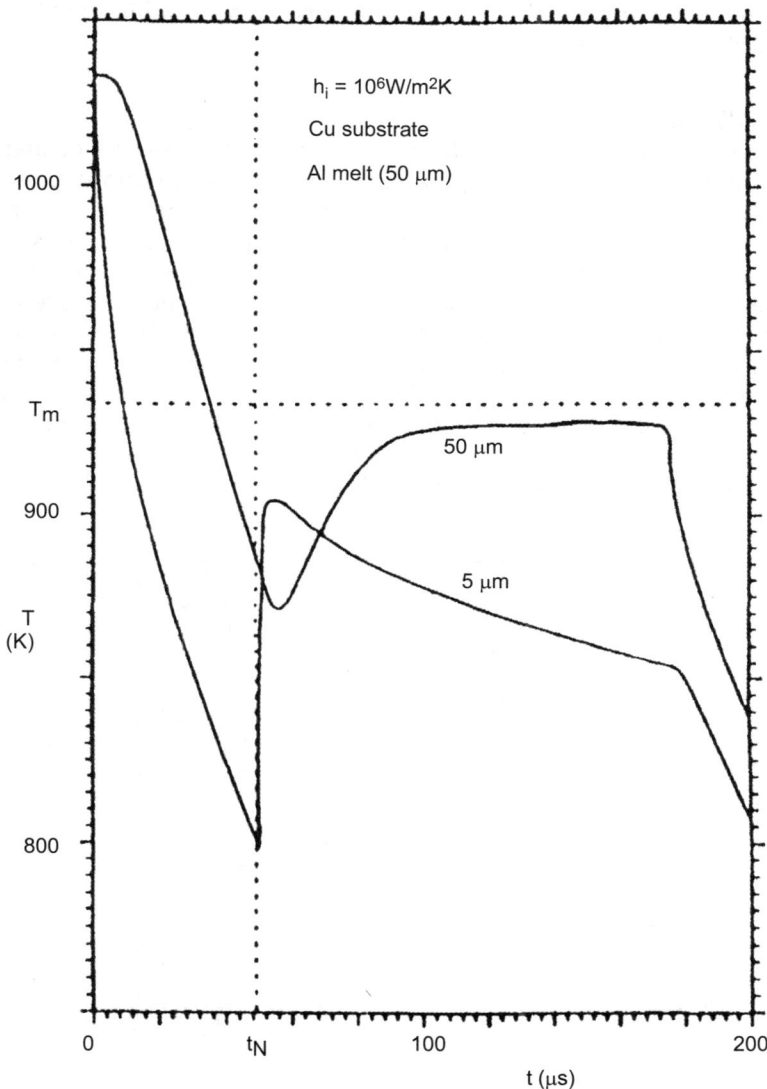

Figure 5   Calculated temperature–time histories for two positions within a liquid layer 5 and 50 μm from a chilling substrate. Nucleation occurs at the substrate surface at time, t_N, at an undercooling of ~ 0.4 L/C_p^L. The recalescence after the passing of the liquid–solid interface is evident at both positions (Clyne, 1984).

nucleation (see Section 7.4). At a position 5 μm from the substrate, for example, the temperature in the liquid drops until nucleation occurs. The temperature at this position then rises rapidly because of latent heat evolution as the L–S interface proceeds from the substrate toward the 5 μm position. After the interface passes the 5 μm position, the temperature falls slowly. A plot for the temperature at the 50 μm position shows a smaller initial drop in temperature followed by a gradual rise as the interface approaches. The supercooling of the interface when it passes the 5 and 50 μm positions is approximately 25 K and 5 K, respectively, which correspond to interface velocities of 100 and 20 cm s^{-1}, respectively. Thus, the initial supercooling produces a much higher solidification rate close to the chill than would be possible without initial supercooling.

7.2.4.3 Powder Solidification

Levi and Mehrabian (1982) examined the heat flow during rapid cooling of metal droplets. They developed a numerical solution based on the enthalpy method for simulating the solidification process from a single nucleation event occurring at the powder surface and a solid growing with a smooth (nondendritic) L–S interface fanning out from this point. Their results are compared with the trends predicted from a Newtonian cooling model. They also discuss the implications of single versus multiple nucleation events. **Figure 6** adapted by Boettinger and Perepezko (1985) from Levi and Mehrabian (1982) shows the interface temperature and interface velocity as solidification proceeds from one side of the droplet to the other, increasing the fraction solid. The curves show the case of an initial supercooling of $\Delta\theta = 0.5 (\sim 182$ K for Al) for various values of h_i. The velocity starts at a high value (>3 m/s) and slows as the interface moves across the droplet. This decrease in velocity is because of the evolution of the latent heat at the L–S interface and the resultant reduction in the interface supercooling. The effect of changing the heat transfer coefficient by two orders of magnitude

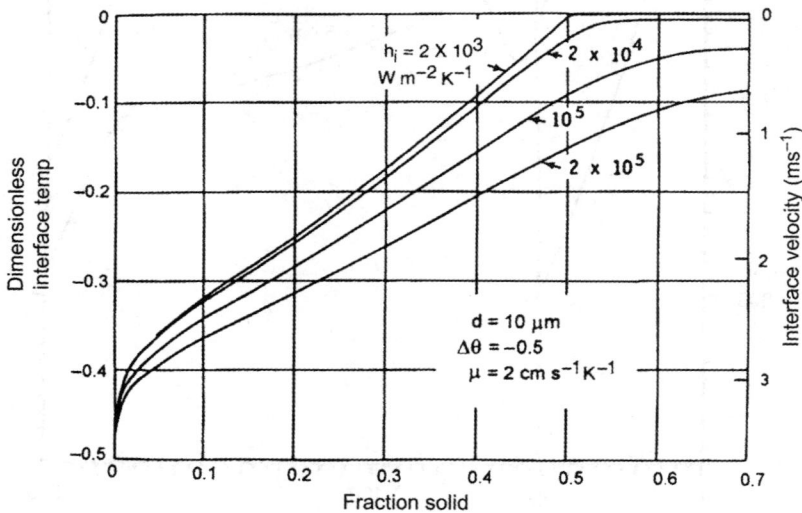

Figure 6 Calculated interface temperature (Levi and Mehrabian, 1982) for the solidification of a powder particle initially undercooled by 1/2 L/C_p^l. The temperature rises and the velocity falls as growth proceeds from the point of nucleation on the powder surface across the powder particle. The effect of various values of the heat transfer coefficient h_i is also shown. The velocity scale on the right was added by Boettinger and Perepezko (1985).

is primarily to alter the velocity after the fraction solid exceeds the dimensionless initial supercooling (0.5 in this case). Growth at small fraction solid is controlled primarily by heat flow inside the powder particle, while growth at large fraction solid is controlled by external heat flow. If no initial supercooling were present, the growth velocities across the entire particle are typical of those seen at high fraction solid in **Figure 6**; that is, less than 10 cm s^{-1}.

The concepts of understanding powder solidification is also utilized in spray forming where near-net shape preforms are manufactured with refined microstructure (Leatham et al., 1991). In an atomization spray process, the molten metal stream is disintegrated into a dispersion of micron-sized droplets. The droplets are rapidly cooled at a maximum rate of 10^5 K s^{-1} to 10^7 K s^{-1}. The spray can be described by a two-phase flow. There are interchanges of momentum and energy between the gas and the metal droplets. The atomization of the melt stream into droplets of diameter 10–500 μm with associated rapid cooling to the solid and semisolid state provides a large number of growth centers for the residual liquid lying on the substrate. This leads to a fine equiaxed microstructure, typically in the range 10–100 μm. Such fine microstructure offer advantages in material strength without requiring any further heat treatment. See Section 7.10 for further details on spray forming.

7.2.5 Heat Flow in Complex Geometries

In more complicated geometries there are several aspects that lead to modeling difficulties, including the nonlinearity of the heat transfer problem because of the latent heat of fusion, the geometric complexity of the shaped castings (see **Figure 7**), the disparities in thermal properties between the metal

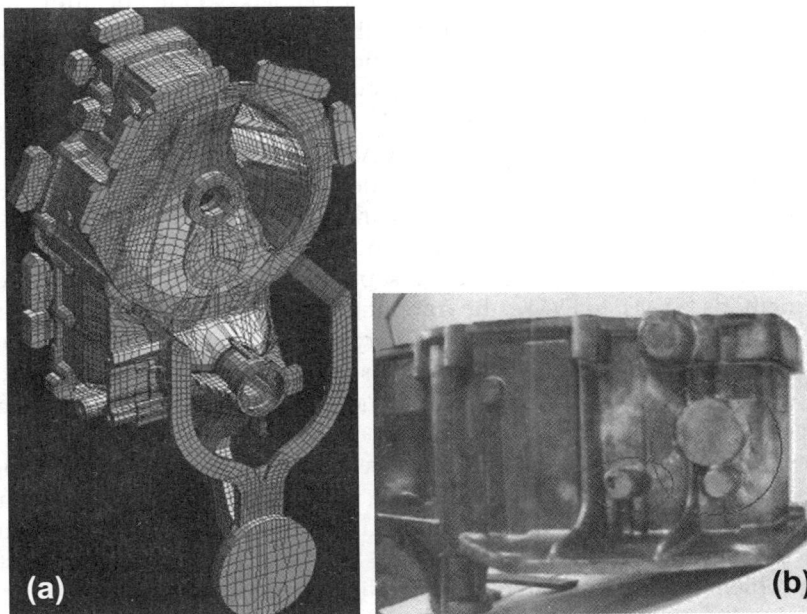

Figure 7 (a) A shaded finite element model of an improved design of a casting for an aluminum transfer case housing; (b) early trial casting with serious mis-runs and other defects. Courtesy: Mahaney and Kim, EKK Inc.

and mold, temperature-dependent thermophysical properties, and heat transfer coefficient. In addition, a combination of fluid flow, heat transfer, and mass transport modeling is required to adequately solve a casting problem involving complex geometries. In this framework, the techniques of numerical modeling are absolutely necessary. For example, an automotive die cast aluminum transfer case housing in **Figure 7** shows the nature of the complexity of the casting geometry and the corresponding finite element method (FEM) model which includes the casting, runner and gating system with biscuit, overflows, main inserts, operator and helper slides, main bores, cores, and water lines.

Since the pioneering work of Henzel and Keverian (1965) for a heavy steel casting using a finite difference technique, an explosive growth of computer modeling techniques has arisen, especially in the past 25 years (Berry and Pehlke, 1988). As a result macroscopic modeling of solidification processes is well developed and different processes can be modeled.

7.2.5.1   *General Numerical Modeling Approaches*

Various numerical methods have been used to treat solidification. Rappaz (1989) discusses five main computational techniques: (i) the finite difference method (FDM) with or without the alternative implicit direction (ADI) time-stepping scheme, (ii) the FEM, (iii) the BEM, (iv) the direct finite difference method, and (v) the control volume element method. He discussed the basic advantages and/or inconveniences of these methods using schematic two-dimensional (2-D) enmeshments that are associated with the five main computational techniques. Ohnaka (1991) analyzed solidification with a thermal conduction model and critically reviewed these computational techniques. Berry and Pehlke (1988) give a comprehensive view of the steps to be taken when solidification modeling is used, stressing the fact that the thermophysical properties and details about mold material are often poorly known. Indeed, conditions such as moisture, property dependence on temperature, would require an almost limitless database. Software packages permit, through suitable interactions, the generation of maps displaying the variation of specific criteria functions that affect casting soundness, such as local temperature gradient, freezing time, front speed or cooling rate.

Many casting modeling software are commercially available. Commercial software[1] dedicated to casting modeling include CAPCAST, JSCast, MAGMASOFT, MAVIS-FLOW, ProCAST, SIMTEC, and SOLIDCast. Furrer and Semiatin (2010) provide a detailed summary of Computational Materials Modeling and Simulation software. The choice of computational grid can determine preference for a particular software package. All computational methods for solidification are based on either structured/regular grid or unstructured/irregular grid. Most popular structured grid-based methods are FDM or finite volume methods (FVM). In FVM's, the computational domain is subdivided into a grid of cells just like FDMs. But in FVMs, the cells can be arbitrary quadrilaterals in 2-D or hexahedra in three-dimensional (3-D) (or can represent any shape enclosed by nodes). In FDM, the grids are obtained using curvilinear coordinate systems. A drawback of FVM is that it is difficult to formulate higher order FVMs (O'Rourke et al., 1997).

The FEM is based on unstructured grids. In structured grid-based methods, algebraic equations are solved using the regular order of the grid resulting in minimum memory requirements. Typical 1-D problems result in tridiagonal matrices when implicit algorithms are used. For multidimensional problems the fractional step method (Anderson et al., 1984) or the ADI method mentioned above is used to split the problem into different steps for each dimension of the problem. In the structured grid approach, complicated geometries typically involved in thin-walled casting are difficult to discretize

---

[1] Product names are given as examples and do not constitute an endorsement by the National Institute of Standards and Technology (NIST).

with a regular grid. Also, large variations in section thicknesses can result in a huge number of cells if a uniform cell size were used for discretization.

In the unstructured grid-based method, it is usually more judicious to store the assembly of matrix elements. Iterative methods such as the conjugate gradient method are used for symmetric matrices. For nonsymmetric matrices, a generalization of the conjugate gradient method can be used. The FEM allows modeling a complex shaped casting quite accurately. Automatic meshing is often employed but elements of poor quality can arise for complex geometries that require manual intervention. Memory requirements for FEM can be large and the solution can take large computational time.

A mesh type between a structured and an unstructured mesh is called block-structured mesh. Here the computational domain is developed using "blocks" of structured grid (Cross et al., 2006).

7.2.5.2 Specific Procedures for Numerical Modeling of Castings

7.2.5.2.1 Geometry Building and Meshing

The geometry includes cast part, mold, gates, cores, and runners. In most casting software, an electronic data interface is provided to retrieve the geometry. Typically, the solid geometry is built using a computer-aided design (CAD) or computer-aided manufacturing (CAM) software. Meshing is important as the accuracy of computationally predicted results depends on the quality of mesh. In general, the higher is the mesh density, the higher is the computational accuracy and CPU time (as storage and memory requirements increase).

7.2.5.2.2 Description of Material Properties

The material properties are assigned to a region comprising cells or elements based on the material type associated with that region. Some of the typical material properties include thermal conductivity, enthalpy, emissivity, viscosity. Some of these properties could be functions of temperature or composition. Most commercial casting software has a built-in materials database composed of properties of common casting alloys and mold materials.

7.2.5.2.3 Assigning Initial and Boundary Conditions

For a typical heat transfer analysis, initial metal and mold temperatures need to be provided. These can be provided in terms of constant values or a distribution of values for the material domain. The initial temperature distribution can be obtained from the output of a mold filling software package for a sequential analysis (first fluid, then thermal). For mold filling, the initial flow velocity, such as velocity at the inlet, is provided.

Different boundary conditions are applied at different locations. Typically, Dirichlet conditions or convective heat flow conditions are applied to the exterior sand mold surface. Radiative heat flow conditions are prescribed on the exterior ceramic shell surfaces such as those used in investment casting applications of high-temperature alloys. At the metal–mold surface, Newtonian cooling is often prescribed that requires knowledge of metal–mold heat transfer coefficient.

7.2.5.2.4 Selecting Calculation Parameters

Some of the calculation parameters include simulation time, time step values, convergence criteria, relaxation factors, output result frequency. Appropriate choices of these parameters are needed to obtain accurate results in a reasonable computational time. In most software, a minimum and a maximum time step values are specified. For an explicit time integration process, the maximum time step value needs to satisfy the value from the stability criterion for numerical convergence. Since an implicit formulation is usually stable, a large value of the maximum time step can be selected. Convergence criteria are needed for an iteration sequence associated with solution of a nonlinear system of equations encountered in an implicit time integration scheme. An optimum value of the convergence criteria is needed to obtain an accurate solution in a reasonable time. In implicit solution scheme, a user may have to use an appropriate value of under-relaxation factors, which vary depending on variables (e.g., temperature, velocity components, pressure etc.).

7.2.5.2.5 Solution
In this step, the user chooses the type of analysis, for example, heat transfer, heat transfer/fluid flow, heat transfer/stress analysis. A heat conduction model may be sufficient if mold filling is quick and the effect of filling on metal and mold temperatures is not significant. Thermal stress calculation is desired in applications (e.g. single crystal or DS airfoils) where stress-related defect prediction is important.

7.2.5.2.6 Postprocessing
Most software has postprocessing capability that allows direct visualization of computational results. These may include time-dependent temperature distribution, flow velocity, stress

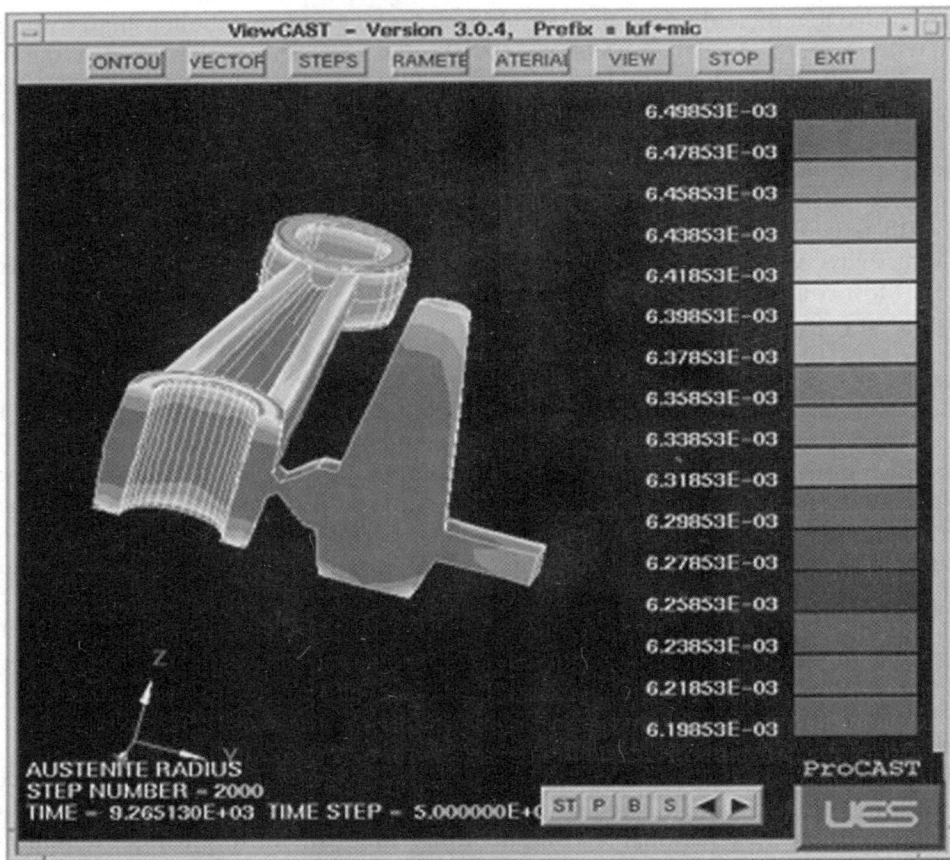

Figure 8   A map of final eutectic grain size (austenite) of a ductile iron casting computed using macro-microscopic modeling approach in ProCAST 3.0. UES Software, Inc., Dayton, OH—Current owner: ESI North America, Columbia, MD, USA.

distribution. Additionally, derived quantities can be displayed that can provide insight into the solidification-related defects. Some of these quantities are isochrons (an isochron is a contour plot that illustrates regions of constant time for castings to cool to a given temperature or a fraction of solid), temperature gradient, solidification velocity, temperature gradient/solidification velocity ratios, local cooling rate, criteria functions (e.g. Niyama criterion for porosity; Section 7.9.4.2.1) and void fraction. An example showing the austenite grain size prediction in a ductile iron casting is given in **Figure 8**. The calculation involved coupling macroscopic heat flow equations with a model for computing microstructure in the casting by applying proper nucleation and growth laws as discussed in Section 7.2.1.1.

7.3 Thermodynamics of Solidification

The application of thermodynamics to solidification is described. Topics include the various levels of equilibrium that can be employed, the construction of equilibrium interface conditions for curved interfaces, T_0 curves, and the use of thermodynamic databases for solidification. Finally because of its derivation from thermodynamics and "gradient flow" kinetic arguments, the basics of the phase field approach to modeling solidification are outlined.

7.3.1 Hierarchy of Equilibrium

The process of solidification cannot occur at equilibrium. However it is clear that different degrees of departure from full equilibrium occur and constitute a hierarchy which is followed with increasing solidification rate. This hierarchy is outlined in **Table 2** (Boettinger and Perepezko, 1985).

The conditions required for global equilibrium (i) are usually obtained only after long-term annealing. Chemical potentials and temperature are uniform throughout the system. Under such conditions no changes occur with time. Global equilibrium is invoked for descriptions of solidification where the temperature is considered to be spatially uniform but slowly decreasing with time. The compositions of the entire liquid and entire solid phases follow the liquidus and solidus of the phase

Table 2 Hierarchy of equilibrium

(1) Full Diffusional (Global) Equilibrium
  (a) No chemical potential gradients (compositions of phases are uniform)
  (b) No temperature gradients
  (c) Applied to entire system at each instant during temperature changes
  (d) Lever rule applicable
(2) Local Interfacial Equilibrium
  (a) Chemical potential for each component continuous across the interface
  (b) Phase diagram gives compositions and temperatures only at solid–liquid interface
  (c) Correction made for interface curvature (Gibbs–Thomson Effect)
(3) Metastable Local Interfacial Equilibrium
  (a) Important when stable phase cannot nucleate or grow fast enough
  (b) Metastable phase diagram (a true thermodynamic phase diagram missing the stable phase or phases) gives the interface conditions
(4) Interfacial Nonequilibrium
  (a) Phase diagram fails to give temperature and compositions at interface
  (b) Chemical potentials are not equal at interface
  (c) Free energy functions of phases still lead to criteria for the "impossible" (Baker and Cahn, 1971)

diagram. The fractions of the system that are liquid and solid are given by the *lever rule* at each temperature during cooling. The lever rule is a consequence of a simple solute balance. This situation is only realized during solidification for certain combinations of slow cooling, small sample size or very fast diffusing interstitial solutes in the solid phases.

During most solidification processes, gradients of temperature and composition must exist within the phases. However one can often accurately describe the overall kinetics using diffusion equations to describe the changes in temperature and composition within each phase and using the equilibrium phase diagram to give the allowable combinations of temperature and compositions of the liquid phase and solid phase just adjacent to the solidification interface. This is called local interface equilibrium [(ii) in **Table 2**].

For a binary alloy, if the liquidus and solidus of the phase diagram can be approximated as straight lines, the local equilibrium condition can be written as

$$T^* = T_m + m_L^E C_L^* \tag{46}$$

and

$$C_S^* = k^E C_L^*, \tag{47}$$

where T^* and T_m are the interface temperature and pure solvent melting temperature, m_L^E is the liquidus slope, C_L^* and C_S^* are the compositions at the interface of the liquid and solid and k^E is the equilibrium partition coefficient (ratio of the slope of the liquidus to the slope of the solidus). Although dimensionless, k^E depends on the units of the composition employed.

The phase diagram is measured for coarse microstructures where the curvatures of interfaces between phases are negligible and/or zero. When phase diagram data are used as local conditions on curved interfaces between phases, the data must be modified for the normally small but important Gibbs–Thomson effect. For isotropic interface energy the temperature of the interface between liquid and solid is given by

$$T^* = T_m + m_L^E C_L^* - T_m \Gamma K_m, \tag{48}$$

where $\Gamma = \frac{\gamma_{LS}}{L^V}$, the ratio of the liquid–solid surface energy to the latent heat per unit volume, K_m is the sum of the principle curvatures of the interface (defined as positive when the center of curvature is in the solid). For a metallic alloy, the melting point is depressed ~ 0.1 K for a solid particle with radius of curvature of 1 µm. In the usual case where the liquid–solid interface energy is anisotropic, the γ_{LS} in the parameter Γ is replaced by the interface stiffness, $\gamma_{LS} + \partial^2 \gamma_{LS}/\partial \theta_1^2 + \partial^2 \gamma_{LS}/\partial \theta_2^2$. This quantity includes the second derivative of the interface energy with respect to the change of the normal to the interface in the two principle (θ_1 and θ_2) directions. In extreme cases of anisotropy, it is this factor that permits the existence of a sharp corner on a faceted crystal in contact with a liquid at equilibrium.

One often employs a fluid–fluid interface analogy to replace the fluid–solid interface to obtain the essence of the Gibbs–Thomson effect for isotropic interface energy. The effect is then considered to be due to the pressure difference that would exist across a boundary with an interface tension. For mechanical equilibrium, the pressure on the concave inside of the interface must be higher than that pressure on the convex outside by an amount $\Delta p = \gamma_{LS} K_m$. Even though the pressures are unequal across the interface, the temperature and chemical potential for each component must be equal across the interface for equilibrium. For a pure material the chemical potential is just the molar free energy and we thus have values on the sides of the interface

$$\begin{cases} T^L = T^S = T^*, \\ p^L = p^S - \gamma_{LS} K_m, \\ G_m^L(T^L, p^L) = G_m^S(T^S, p^S). \end{cases} \tag{49}$$

To use molar free energies for the liquid and solid obtained at a common pressure, say p^L (used for measuring the phase diagram), the free energy of the solid can be expanded about the pressure of the liquid and then rewritten as

$$G_m^S(T^S, p^S) = G_m^S(T^L, p^L) + \frac{\partial G_m^S(T, p)}{\partial p}\bigg|_{p=p_L} (p^S - p^L) \tag{50}$$

$$= G_m^S(T^L, p^L) + V_m^S \gamma_{LS} K_m,$$

where V_m^S is the molar volume of the solid. Thus if one uses a molar free energy for the liquid and solid obtained at a common pressure p^L, the term $V_m^S \gamma_{LS} K_m$ must be added to the molar free energy of the solid phase before finding the temperature where the free energies of the liquid and solid are equal. Then

$$G_m^L(T^L, p^L) = G_m^S(T^L, p^L) + V_m^S \gamma_{LS} K_m \tag{51}$$

leads to the expression for the melting point modification, eqn (48).

Local equilibrium is never strictly valid, but it is based on the notion that interfaces will equilibrate much more quickly than will bulk phases. The conditions described in (ii) of **Table 2** are widely used to model the majority of solidification processes that occur in castings. For example, under the assumptions of fast diffusion in the liquid phase, negligible diffusion in the solid phase, and local equilibrium at the interface neglecting the curvature correction, the Scheil equation (see Sections 7.6.2.2 and 7.7.3.2) gives a reasonable first approximation to the dendritic coring or microsegregation in conventional castings. Clearly phase diagrams constitute an essential part of the database for the modeling and analysis of solidification problems. Although they are typically measured under the assumptions of global equilibrium, phase diagrams may be applied locally to describe solidification interfaces with the Gibbs–Thomson effect added when interface curvature is large.

Metastable equilibrium, (iii) in **Table 2**, can also be used locally at interfaces. One can understand the microstructural change of cast iron from the stable gray form (austenite and graphite) to the metastable white form (austenite and cementite) with increasing solidification rate (and interface supercooling) using information from the stable and the metastable phase diagrams combined with a kinetic analysis using local equilibrium concepts (Jones and Kurz, 1980). The eutectic temperature and composition for white cast iron are well-defined thermodynamic quantities just as they are for gray cast iron. As with stable equilibrium, metastable equilibrium is represented by a common tangent construction to the molar free energy versus composition curves for the liquid, austenite, and cementite phases and thus minimizes the free energy as long as graphite is absent. When solidification is complete, a two-phase mixture of austenite and cementite can exist in

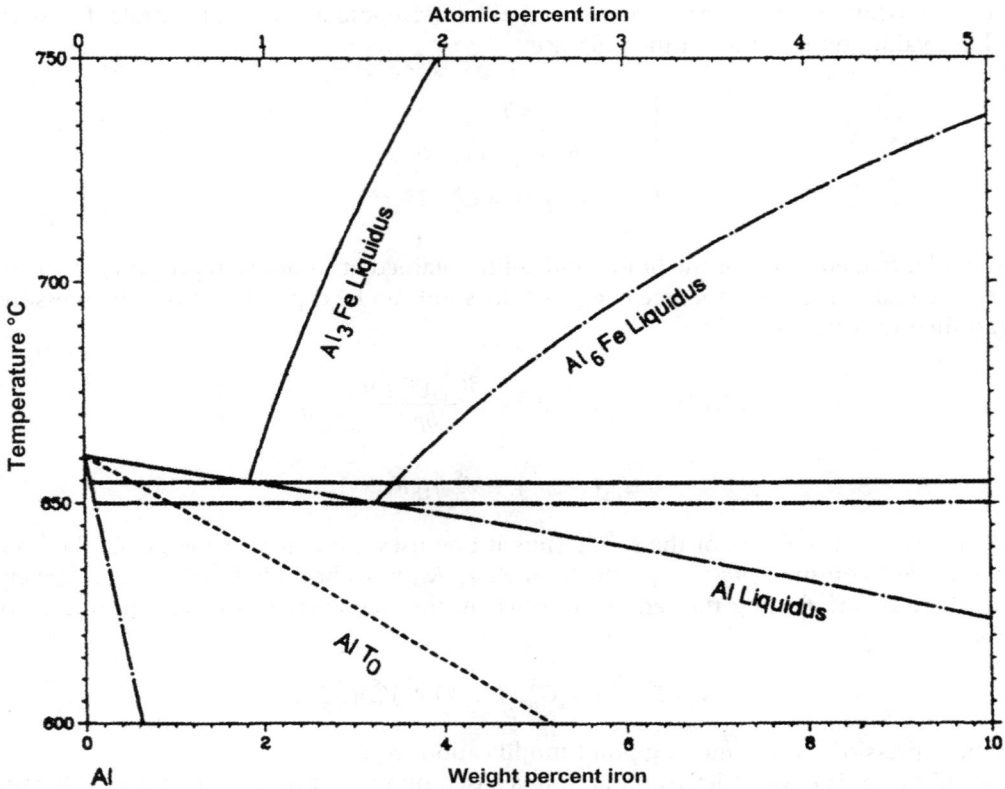

Figure 9 Calculated Al-rich portion of the stable Al–Fe phase diagram is shown by solid lines. The T_0 for the solidification of Al solid solution is shown dashed. The solvus curves are omitted for clarity (Murray, 1983a).

a global metastable equilibrium. The concept of local metastable equilibrium is especially important during rapid solidification (Perepezko and Boettinger, 1983; Perepezko, 1988) because some equilibrium phases, especially those with complex crystal structures, have sluggish nucleation and/ or interface attachment kinetics and may be absent in rapidly solidified microstructures.

An example of a metastable phase diagram superimposed on a stable phase diagram is given in **Figure 9** for the Al–Fe system. If Al_3Fe is absent, phase boundaries involving a metastable phase, Al_6Fe, which is isomorphous with Al_6Mn, are obtained. In particular a metastable eutectic, $L \rightarrow Al + Al_6Fe$, occurs. Transitions of microstructures involving Al_3Fe to those involving Al_6Fe have been observed with increasing solidification speed by Adam and Hogan (1972) and by Hughes and Jones (1976). The competitive growth kinetics of the two can also be analyzed using the stable and metastable phase diagrams.

For local equilibrium, whether stable or metastable, the chemical potentials of the components for the liquid and solid are equal across the interface. In **Table 2**, however, another situation is described in (iv), and relates to the situation where the temperature of the interface remains continuous but the chemical potentials cannot be approximated as being equal across an interface growing at a high solidification rate and large interface supercooling. These rapid growth rates can trap the solute into the freezing solid at levels exceeding the equilibrium value for the corresponding liquid composition present at the interface. Thus the chemical potential of the solute increases upon being incorporated in the freezing solid in a process called *solute trapping*. This increase in chemical potential of the solute across the interface must be balanced by a decrease in chemical potential of the solvent in order for crystallization to occur; that is, to yield a net decrease in free energy (Baker and Cahn, 1971). The free energy change during solidification, ΔG (product minus reactant; i.e. solid minus liquid) is given by

$$\Delta G = \left(\mu_S^A - \mu_L^A\right)\left(1 - C_R^*\right) - \left(\mu_S^B - \mu_L^B\right)C_R^*, \tag{52}$$

where μ_S^A and μ_S^B are the chemical potentials for species A and B in the solid, and μ_L^A and μ_L^B are the chemical potentials in the liquid. These potentials are functions of the temperature and solid or liquid composition at the interface during solidification (C_S^* or C_L^*). Baker and Cahn call C_R^* the composition of the *reactant*. It is the composition of the increment of alloy whose change from liquid to solid is being considered. As described in the section on interface kinetics, C_R^* is thought to be between the values of C_L^* and C_S^* depending on the kinetic model. We note that for solidification of pure A eqn (52) reduces to just the difference in molar free energy between pure solid and liquid which is often approximated as

$$\Delta G = L_m \frac{(T - T_m)}{T_m}, \tag{53}$$

where L_m is the molar heat of fusion.

Despite the loss of interface equilibrium during rapid solidification, the free energy functions of the solid and liquid phases and their associated chemical potentials can be used to define the possible range of compositions that can exist at the interface at various temperatures. This restriction is obtained by the requirement that ΔG be negative. For $C_R^* = C_S^*$, **Figure 10** shows the region of allowable values of solid compositions C_S^* at the interface for a fixed liquid composition, C_L^*, at the interface as a function of interface temperature (Boettinger, 1982). Such allowable regions can be calculated from a thermodynamic model of the system of interest.

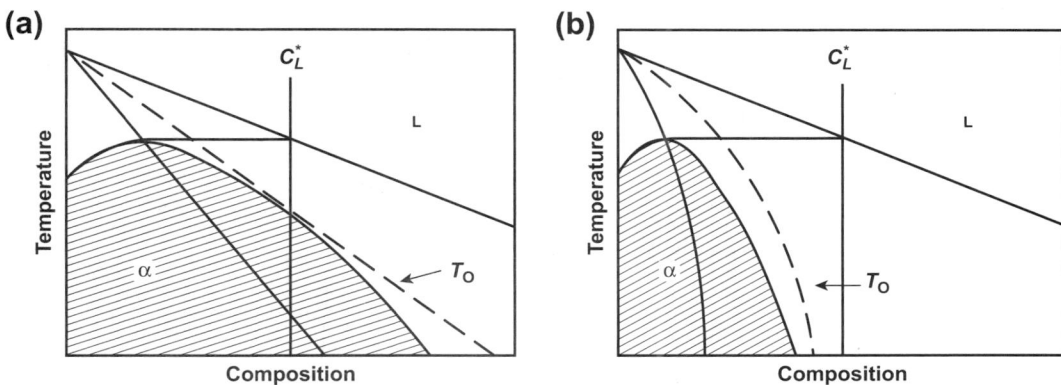

Figure 10 The shaded regions indicate thermodynamically allowed solid compositions C_S^* that may be formed from liquid of composition C_L^* at various temperatures. In (a) partitionless solidification is possible for a liquid composition of C_L^* whereas in (b) the T_0 temperature plunges and partitionless solidification is impossible for liquid of composition C_L^* (Boettinger, 1982).

7.3.2 T_0 Curves

As shown in **Figure 10**, a thermodynamic temperature can be described for any selected pair of liquid and solid compositions that is the highest temperature at which the crystallization can occur. A limiting case can be considered where the composition of the solid formed at the interface, C_S^*, equals the composition of the liquid at the interface, C_L^*. This is called *partitionless solidification,* and is favored at very high solidification rate. The T_0 temperature is defined as the highest interface temperature where this can occur (Aptekar and Kamenetskaya, 1962; Biloni and Chalmers, 1965). This temperature traces a curve as a function of interface composition that lies between the liquidus and solidus curves of the phase diagram. The curve is found thermodynamically by finding the temperature where the molar free energies of the liquid and solid phases are equal for the interface composition of interest. It is the temperature where $\Delta G = 0$ for $C_R^* = C_S^* = C_L^*$ in eqn (52).

T_0 curves exist for the liquid with stable or metastable phases, and lie between the liquidus and solidus for those phases. In fact for dilute alloys the slope of the T_0 curve is $m_L^E[(\ln k^E)/(k^E - 1)]$. **Figure 11** shows schematically, possible T_0 curves for three eutectic phase diagrams (Boettinger, 1982). An important use of these curves is to determine whether a bound exists for the extension of solubility by rapid melt quenching. If the T_0 curves plunge to very low temperatures as in **Figure 11a**, single-phase α or β crystals with composition beyond their respective T_0 curves cannot be formed from the melt. In fact, for phases with a retrograde solidus, the T_0 curve plunges to absolute zero at a composition no greater than the liquidus composition at the retrograde temperature, thus placing a bound on solubility extension (Cahn et al., 1980). A retrograde solidus is a solidus whose slope becomes infinite at some temperature. Experiments on laser-melted doped Si alloys seem to confirm this bound (White et al., 1983). Eutectic systems with plunging T_0 curves

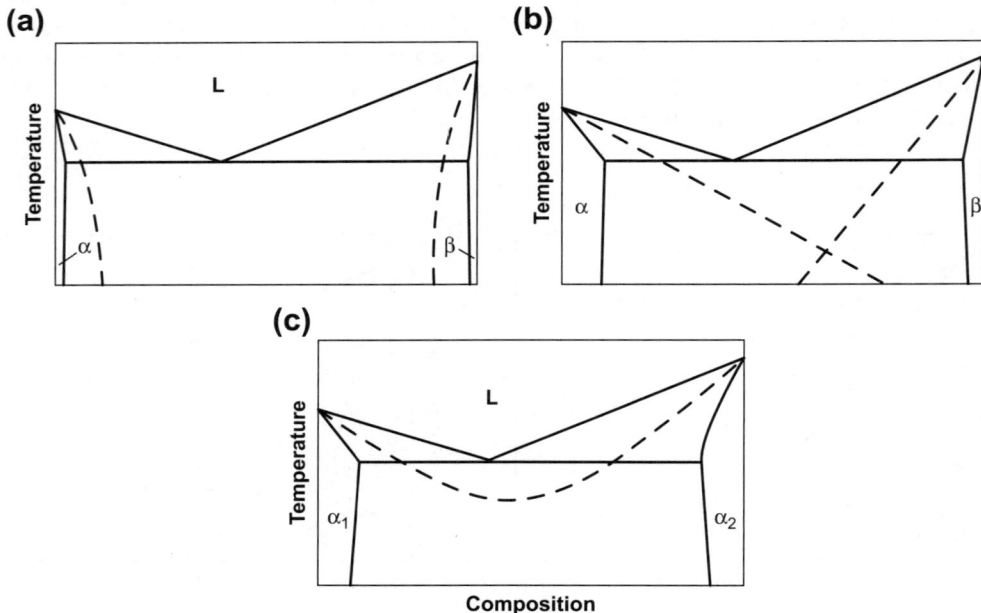

Figure 11   Schematic representation of (a) plunging, (b) crossing and (c) continuous. T_0 curves for liquid to crystal transformations in three types of eutectic systems (Boettinger, 1982).

are good candidates for easy metallic glass formation. An alloy in the center of such a phase diagram can only *crystallize* into a mixture of solid phases with compositions different from each other and from the liquid regardless of the departure from equilibrium. The diffusional kinetics of creating these solid phases from the liquid phase may depress the solidification temperature to near the glass transition temperature, T_g, where an increased liquid viscosity effectively halts crystallization and an amorphous solid is formed.

In contrast, alloys with T_0 curves that are only slightly depressed below the stable liquidus curves, as in **Figure 11**b and c, make good candidates for extension of the solubility limit of the solid (crystalline) phase and unlikely ones for glass formation. In **Figure 11**b the crystal structures of α and β are different and the T_0 curve cross, whereas in **Figure 11**c the crystal structures are the same and the T_0 curve is continuous across the diagram. At temperatures and liquid compositions below the T_0 curves, partitionless solidification is thermodynamically possible. Ni–Cr and Ag–Cu are examples of the behavior in **Figure 11**b and c, respectively. Solubility limit extension is useful metallurgically because of the possibility of the development of larger fractions of strengthening precipitates (dispersiods) during subsequent thermal processing (although limited to only once such thermal cycle).

7.3.3 Thermodynamic Databases

In addition to experimental work to determine phase diagrams, researchers have sought to determine the underlying free energy functions of phases by two methods. The first method uses first principle calculation of interaction energies between atoms in the various phases and statistical mechanics (Asta et al., 2001; van de Walle and Ceder, 2002). Progress with this method has been impressive over the past 15 years and is providing much needed formation energies for the second method. However improvements of the accuracy of the temperature scale are required for engineering metallurgical processing. The second method, which is referred to as the CALPHAD method (*cal*culation of *phase diagrams*), employs simple, mathematically analytical, free energy functions of the phases that approximate physical models for the atomic interactions in each phase (e.g. see Andersson et al., 2002). The regular solution model, its generalization, and its extension to multiple sublattices are examples of this approach. Sets of free energy functions for phases are published in the literature for many binary and ternary systems providing a compact data set for complex alloy constitution. Numerical parameters for these analytical functions are obtained by fitting the available thermodynamic data (calorimetry, vapor pressure, emf, etc., and first principle-derived formation energies) and the experimental phase diagram. **Figure 12** shows the calculated phase diagram by Ansara et al. (1998) for the Co–Mo system compared with the experimental data. For higher order systems, considerable success has been achieved using thermodynamic treatment of the ternary subsystems and employing thermodynamic extrapolation methods to treat the quaternary and all higher order systems.

The thermodynamic calculation approach to phase diagrams permits estimates to be made for the solidification of complex practical materials. As an example, the constituent ternary subsystems of the Ni–Cr–Co–Ti–Ta–Al–Mo–W system have been modeled and a thermodynamic interpolation method has been applied to produce predictions for an eight-component alloy CMSX-2, a commercial single crystal aero-turbine alloy (Boettinger et al., 1998). For solidification of multicomponent alloys, partition coefficients, k_i^E and liquidus slopes m_{Li}^E are defined for each alloying addition to the major component. For a liquid concentration of the alloy, C_{Li}^* in **Table 3** the liquidus (interface) temperature is computed as 1606 K and the solid concentrations at that temperature, C_{Si}^*, the partition ratios, k_i^E and

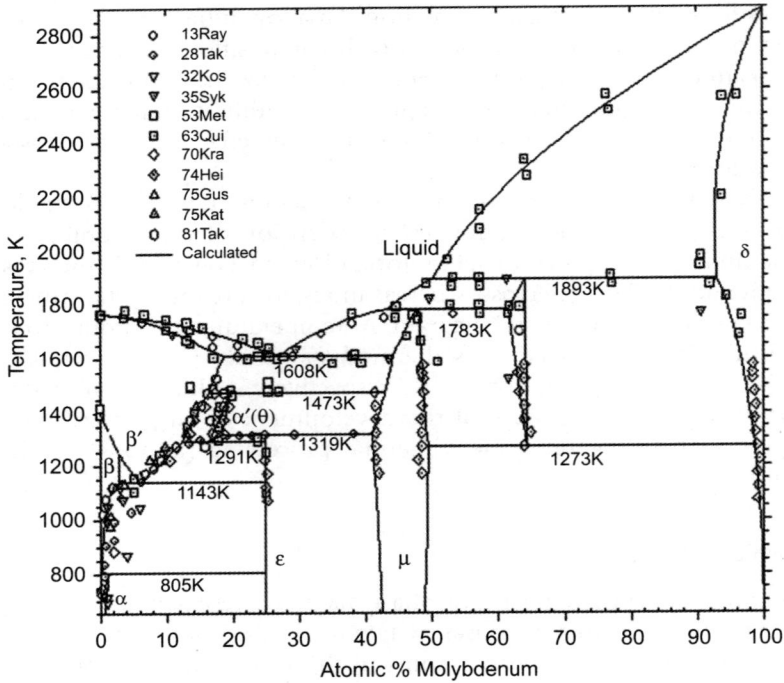

Figure 12 Calculated Co–Mo phase diagram (lines) and experimental points (Ansara et al., 1998).

the liquidus surface slopes, m_{Li}^E are given for each solute. Note that, unlike binary systems, the sign of $m_{Li}^E(1 - k_i^E)$ need not be negative in multicomponent alloys.

As solidification proceeds, the same determination can be performed as solute rejected by the growing solid enriches the remaining liquid. Thus output from thermodynamic software can be used to deliver phase diagram information to the various solidification models to be described in the remainder of this chapter. The ability to output the concentration dependence of parameters is a major asset of this method. For heat flow modeling, the enthalpy of the phases as a function of phase composition and temperature can also be obtained from thermodynamic databases.

Table 3   Example of output from the thermodynamic subroutine. The units of concentration are wt% and liquidus slope are K/wt%. Ni is treated as the solvent. The liquidus temperature is computed as 1606 K. From Boettinger et al. (1998)

CMSX-2	Ni	Cr	Co	Ti	Ta	Al	Mo	W
C_{Li}^*	65.8	8.0	5.0	1.0	6.0	5.6	0.6	8.0
C_{Si}^*	68.2	8.45	3.49	0.492	3.27	5.11	0.403	10.6
m_{Li}^E	–	−1.80	−5.73	−16.3	−5.00	−15.44	−5.72	−0.641
k_i^E	–	1.06	0.699	0.492	0.545	0.913	0.672	1.32

7.3.4 The Phase Field Method

In the 1990s a method was refined to treat solidification and other phase transformations that avoids mathematical tracking of the interface between the phases. This method is particularly useful for modeling of solidification with complex interface shapes and multiphase solidification processes. The basis of the phase field model for solidification of pure materials is found in efforts by Langer (1986) and Calginalp (1986), with related ideas in the context of antiphase boundary migration by Allen and Cahn (1979). Such models treat the presence of an interface by the introduction of a gradient correction to the thermodynamic potentials. It employs a phase field variable, for example, ϕ, which is a function of position and time, to describe whether the material is liquid or solid. The variable has some features in common with the fraction of liquid used for mushy zone descriptions and the enthalpy method described in Section 7.2 but is a much more detailed variable that represents the interface transition. The behavior of this variable is governed by an equation that is coupled to equations for heat and solute transport. Interfaces between liquid and solid are described by smooth but highly localized changes of this variable between fixed values that represent solid and liquid (here 0 and 1, respectively). The approach is described here and reference will be made to dynamic calculations throughout the chapter. References for alloy solidification are Wheeler et al. (1992), Calginalp and Xie (1993), Warren and Boettinger (1995), Steinbach et al. (1996), Beckermann et al. (1999) and Kim et al. (1999).

7.3.4.1 The Phase Field Variable

A single scalar order parameter can be used to model solidification of a single-phase material. However, to employ such a simple description of the liquid–solid transition necessarily requires a number of approximations. **Figure 13** adapted from Mikeev and Chernov (1991) shows one possible physical interpretation of a single scalar phase field variable. The interfacial region and its motion during solidification are depicted by a damped wave that represents the probability of finding an atom at a particular location. On the left, the atoms tend to be located at discrete atomic lattice points corresponding to the crystal. As the liquid is approached, the probability has the same average value but becomes less localized to lattice points as indicated by the reduced amplitude of the wave. Finally the probability achieves a constant value in the liquid indicating the absence of localization of atoms to

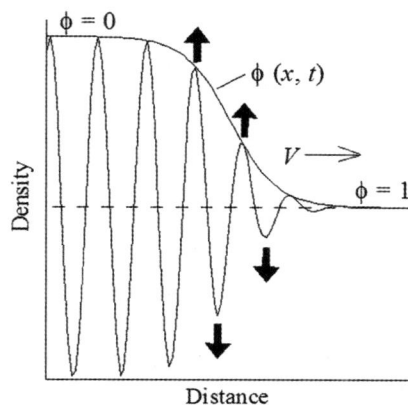

Figure 13 Schematic representation of a possible physical interpretation of the phase field variable. Adapted from Mikeev and Chernov (1991).

specific sites, that is, a liquid. The *amplitude* of the wave might be related to the phase field ϕ. Multiple scalar phase field variables can be used to treat situations where more than two solid phases appear as in eutectic and peritectic reactions. Polycrystalline material can also be treated using multiple phase field variables (Chen, 1995) or using an angular variable in addition to the phase field variable (Warren et al., 2003).

7.3.4.2 *Thermodynamic Derivation of the Phase Field Model*

There are two approaches to phase field modeling: those that use a thermodynamic treatment with gradient energies and gradient flow dynamics and those that are mostly concerned with reproducing the traditional sharp interface approach. Here we only examine a thermodynamic treatment with a single scalar order parameter. This approach is found in many papers on the phase field method and determines the evolution equations from a few basic concepts. The following is extracted from the review of Boettinger et al. (2002).

By demanding that the entropy always increases locally for a system where the internal energy and concentration are conserved, restrictions on the relationships between the fluxes of internal energy and concentration can be obtained. The simplest postulates yield generalizations of Fourier's and Fick's laws of diffusion. Since the phase field variable is not conserved, a separate relationship governing ϕ is required to guarantee that the entropy increases.

To treat cases containing interfaces, an entropy functional S is defined over the system volume V as

$$S = \int_V \left[s(e, c, \phi) - \frac{\varepsilon_e'^2}{2}|\nabla e|^2 - \frac{\varepsilon_c'^2}{2}|\nabla c|^2 - \frac{\varepsilon_\phi'^2}{2}|\nabla \phi|^2 \right] dV, \tag{54}$$

where s, e, c and ϕ are the entropy density, internal energy density, concentration and phase field, respectively, with ε_e', ε_c' and ε_ϕ' being the associated gradient energy coefficients. The entropy density s must contain a *double well* in the variable ϕ that distinguishes the liquid and solid. The quantity S also includes the entropy associated with gradients (interfaces). From such an approach follows naturally equations for energy (heat) diffusion, solute diffusion and phase field evolution. A formulation using this entropy functional especially with all three gradient energy contributions is quite general (Bi and Sekerka, 1998).

We will present a simpler isothermal formulation to which we append a heat flow equation. This gives essentially equivalent results to the entropy formulation if $\varepsilon_e' = 0$. The enthalpy per unit volume is expressed as

$$H = H_0 + C^V T + L^V \phi, \tag{55}$$

yielding an equation for thermal diffusion with a source term given by

$$C^V \frac{\partial T}{\partial t} + L^V \frac{\partial \phi}{\partial t} = \nabla \cdot (k \nabla T), \tag{56}$$

where H_0 is a constant, T is the temperature, C^V is the heat capacity per unit volume that in general depends on temperature, L_V is the latent heat per unit volume, and k is the thermal conductivity. From the second term in eqn (56), it can be seen that the latent heat evolution occurs where ϕ is changing with time, that is, near a moving interface. This has close similarity with eqn (7); however the meaning of g_s and ϕ is quite different, the former representing a dendritic mushy zone and the latter representing a transition between liquid and solid on the nanometer scale.

An isothermal treatment forms the free energy functional F, which must decrease during any process, as

$$F = \int_V \left[f(\phi, c, T) + \frac{\varepsilon_C^2}{2} |\nabla c|^2 + \frac{\varepsilon_\phi^2}{2} |\nabla \phi|^2 \right] dV, \tag{57}$$

where $f(\phi, c, T)$ is the free energy density, and where the gradient energy coefficients have different units than the (primed) gradient energy coefficients used in eqn (54).

For equilibrium, the variational derivatives of F must satisfy the equations,

$$\frac{\delta F}{\delta \phi} = \frac{\partial f}{\partial \phi} - \varepsilon_\phi^2 \nabla^2 \phi = 0, \tag{58}$$

$$\frac{\delta F}{\delta c} = \frac{\partial f}{\partial c} - \varepsilon_C^2 \nabla^2 c = \text{constant}, \tag{59}$$

where the gradient energy coefficients are assumed constant. The constant in eqn (59) occurs because of the constraint that the total amount of solute in the volume V is constant; that is, concentration is a conserved quantity.

For time-dependent situations, the simplest equations that guarantee a decrease in total free energy with time (an increase in entropy in the general formulation) are given by

$$\frac{\partial \phi}{\partial t} = -M_\phi \left[\frac{\partial f}{\partial \phi} - \varepsilon_\phi^2 \nabla^2 \phi \right], \tag{60}$$

$$\frac{\partial c}{\partial t} = \nabla \cdot \left[M_C c (1 - c) \nabla \left(\frac{\partial f}{\partial c} - \varepsilon_C{}^2 \nabla^2 c \right) \right]. \tag{61}$$

The parameters M_ϕ and M_C are positive mobilities related to the interface kinetic coefficient and solute diffusion coefficient as described below. Equations (60) and (61) have different forms because composition is a conserved quantity and the phase field is not. Equation (60) is called the Allen–Cahn equation when applied to antiphase boundary motion. Equation (61) is the Cahn–Hilliard equation. In this formulation the pair are coupled through the energy function $f(\phi, c, T)$. In the second equation, we set $\varepsilon_C = 0$ for the remainder of this paper. With $\varepsilon_C \neq 0$ and a double well (miscibility gap) in $f(c)$, eqn (61) describes spinodal decomposition.

7.3.4.3 The Alloy Free Energy Function

The free energy density uses two functions: a double-well function and an interpolating function. Here we chose the two functions, $g(\phi) = \phi^2(1 - \phi)^2$ and $p(\phi) = \phi^3(6\phi^2 - 15\phi + 10)$, respectively. Note that by design, $p'(\phi) = 30g(\phi)$, ensuring that $\partial f / \partial \phi = 0$ when $\phi = 0$ and 1, for all temperatures. In this document $\phi = 0$ represents solid and $\phi = 1$ represents liquid, although the opposite convention is often used. Plots of these functions are given in **Figure 14**.

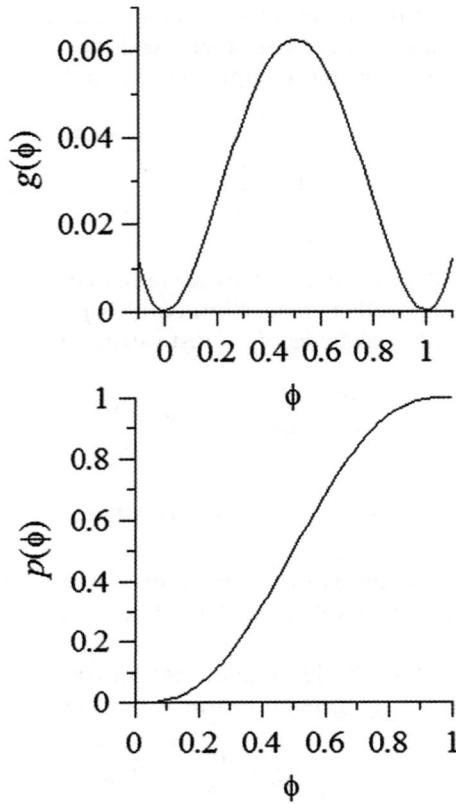

Figure 14 The functions $g(\phi)$ and $p(\phi)$.

The alloy free energy density $f(\phi, c, T)$ can be constructed in several ways. One method can be broken into three steps:

(1) Start with the ordinary free energy of the pure components as liquid and solid phases, $f_A^L(T), f_A^S(T), f_B^L(T)$ and $f_B^S(T)$. They are functions of temperature only.

(2) Form a function, $f_A(\phi, T)$ that represents both liquid and solid for pure A as

$$
\begin{aligned}
f_A(\phi, T) &= (1 - p(\phi))f_A^S(T) + p(\phi)f_A^L(T) + W_A g(\phi) \\
&= f_A^S + p(\phi)\left[f_A^L(T) - f_A^S(T)\right] + W_A g(\phi),
\end{aligned}
\tag{62}
$$

This function combines the free energies of the liquid and solid with the interpolating function $p(\phi)$ and adds an energy hump, W_A, between them. A similar expression can be obtained for component B.

(3) Form the function, $f(\phi, c, T)$ that represents, for example, a regular solution of A and B,

$$
\begin{aligned}
f(\phi, c, T) &= (1 - c)f_A(\phi, T) + cf_B(\phi, T) + R'T\{(1 - c)\ln(1 - c) + c \ln c\} \\
&\quad + c(1 - c)\{\Omega_S(1 - p(\phi)) + \Omega_L p(\phi)\},
\end{aligned}
\tag{63}
$$

where Ω_L and Ω_S are the regular solution parameters of the liquid and solid that again are combined with the interpolating function $p(\phi)$. The gas constant R' and the regular solution parameters are described on a unit volume basis.

A second method to form $f(\phi, c, T)$ is convenient if the free energies of the solid phase and liquid phase, $f^L(c,T)$ and $f^S(c,T)$, are already available; for example, from a CALPHAD thermodynamic modeling of the phase diagram. The free energy density is

$$f(\phi, c, T) = \{(1 - c)W_A + cW_B\}g(\phi)$$
$$+ (1 - p(\phi))f^S(c, T) + p(\phi)f^L(c, T). \tag{64}$$

Such a construction will lead to the same form as eqn (63) if regular solution models are used for the liquid and solid phases.

A simplification is often made for the pure elements, namely,

$$f_S^A(T) = 0,$$

$$f_L^A(T) - f_S^A(T) = \frac{L_A^V(T_m^A - T)}{T_m^A}. \tag{65}$$

This takes the solid as the standard state and expands the difference between liquid and solid free energies around the melting point. The pure component melting points are T_m^A and T_m^B, and the latent heats are L_A^V and L_B^V Then the pure element free energy can be given by

$$f_A(\phi, T) = W_A g(\phi) + L_A^V \frac{T_m^A - T}{T_m^A} p(\phi). \tag{66}$$

The effect of the second term is to raise or lower the minimum in the free energy at $\phi = 1$ (liquid) depending on the whether the temperature is above or below the melting point with the free energy at $\phi = 0$ (solid) being fixed at zero.

The choice for the form of the functions $g(\phi)$ and $p(\phi)$ is quite arbitrary and others are used in various publications. In the limit in which the interface thickness is small, the choice does not matter.

7.3.4.4   Pure Material
Consider the case where $c = 0$; that is, pure component A. Combining eqns (60) and (66),

$$\frac{\partial \phi}{\partial t} = M_\phi^A \varepsilon_\phi^2 \left[\nabla^2 \phi - \frac{2W_A}{\varepsilon_\phi^2} \phi(1 - \phi)(1 - 2\phi) \right]$$

$$- \frac{30 M_\phi^A L_A^V}{T_m^A} (T_m^A - T)\phi^2(1 - \phi)^2. \tag{67}$$

Equilibrium solutions, $(\partial \phi/\partial t = 0)$, are obtained if ϕ is constant with values equal to zero or one. These situations correspond to a single-phase solid or a single-phase liquid, respectively. The equilibrium equation, eqn (58) permits these states to exist at any temperature including those for a metastable supercooled liquid or a superheated solid.

An equilibrium 1-D solution $(\partial \phi / \partial t = 0)$ to eqn (67) also exists only at $T = T_{\mathrm{m}}^{\mathrm{A}}$ for a planar transition zone between liquid ($\phi = 1$) and solid ($\phi = 0$) where ϕ varies only in the x-direction normal to the interface. The solution is

$$\phi(x) = \frac{1}{2}\left[1 + \tanh\left(\frac{x}{2\delta_{\mathrm{A}}}\right)\right], \tag{68}$$

where δ_{A} is measure of the liquid–solid interface thickness of pure A given by

$$\delta_{\mathrm{A}} = \frac{\varepsilon_\phi}{\sqrt{2 W_{\mathrm{A}}}}. \tag{69}$$

The value of the interface thickness is a balance between two opposing effects. The interface tends to be sharp to minimize the volume of material where ϕ is between 0 and 1 and $f(\phi, T)$ is large (as described by W_{A}). The interface tends to be diffused to reduce the energy associated with the gradient of ϕ (as described by ε_ϕ).

Figure 15 shows the variation of ϕ and the integrand of eqn (57), $f(\phi, T_M) + \frac{1}{2}\varepsilon_\phi^2|\nabla\phi|^2$, with distance across an interfacial region for the equilibrium solution, eqn (68). The area of the hatched

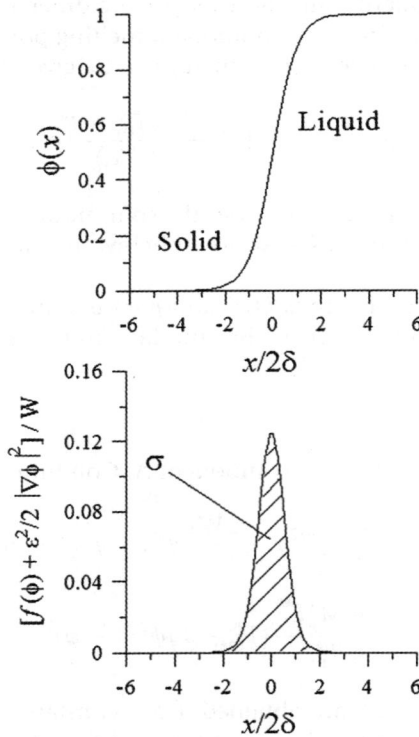

Figure 15 Variation of the phase field parameter, ϕ, and the energy/unit volume, $f(\phi) + \frac{1}{2}\varepsilon_\phi^2(\nabla\phi)^2$, with distance, x, across a stationary flat liquid–solid interface at the melting point in a pure material.

region corresponds to the surface free energy and is given by direct integration of the solution, eqn (38), as

$$\gamma_{LS}^A = \frac{\varepsilon_\phi \sqrt{W_A}}{3\sqrt{2}} . \tag{70}$$

Note that it is possible to take the limit, $\delta \to 0$, while keeping γ_{LS}^A fixed. Equations (69) and (70) can be used to calculate the values of the parameters ε_ϕ and W_A to match with selected values for γ_{LS}^A and δ_A, namely,

$$\varepsilon_\phi = \sqrt{6\gamma_{LS}^A \delta_A},$$

$$W_A = 3\frac{\gamma_{LS}^A}{\delta_A} . \tag{71}$$

In the planar solution above, the Laplacian, $\nabla^2\phi$ in eqn (60), is simply the second derivative of ϕ with respect to x. To examine a curved equilibrium interface (eqn (67) with $\partial\phi/\partial t = 0$), the simplest mathematics is to transform to spherical coordinates with no angular dependence. Then the Laplacian includes an extra term, in addition to the second derivative with respect to r, namely, $(2/r)(\partial\phi/\partial r)$. With this extra term, no solution exists when $T = T_m^A$. However for an interface with a specified radius, R, a solution does exist at a temperature below T_m^A, namely, $T = T_m^A - \Gamma/R$, where $\Gamma = \gamma_{LS}^A T_M^A / L_A^V$. This selection of one interface radius for each (interface) temperature is consistent with the Gibbs–Thomson effect for a sharp interface model as described in Section 7.3.1.

To examine a moving flat steady-state interface, the simplest mathematics is to transform eqn (60) in 1-D to a coordinate frame moving at constant velocity V. Then $\partial\phi/\partial t$ changes to $-V(\partial\phi/\partial x)$. With this change, no solution exists if $T = T_M^A$. However, a solution does exist if the temperature is given by

$$T = T_m^A - \frac{V}{\mu_A}, \tag{72}$$

where

$$\mu_A = \frac{6M_\phi^A \varepsilon_\phi L_A^V}{T_m^A \sqrt{2W_A}}, \tag{73}$$

if δ_A is small

This selection of one interface velocity for each (interface) temperature is consistent with the classical approach to linear interface attachment kinetics (see Section 7.5; viz., eqn (97)). Equation (73) can be used to determine a value of M_ϕ^A from a knowledge (or estimate) of μ_A, namely,

$$M_\phi^A = \frac{\mu_A T_m^A}{6\delta L_A^V} . \tag{74}$$

In numerical solutions, the minimum mesh spacing must scale with the chosen value of δ_A so that the interface structure can be resolved mathematically. Physically for metallic systems, the interface thickness is of the order of a few atomic dimensions. Such fine mesh spacing is only practical in 1-D at the current time. Thus one would prefer to use a larger value of δ_A and retain the accuracy of eqn (72). By treatment of the temperature variation that must exist across a diffuse interface, Karma and Rappel (1996) have calculated the relation between μ_A and M_ϕ^A that is correct to second order in δ_A (if the thermal conductivity k of the liquid and solid are equal); viz.,

$$\frac{1}{M_\phi^A} = \frac{6\delta_A L_A}{T_M^A}\left[\frac{1}{\mu_A} + A\left(\frac{\delta L_A}{k}\right)\right], \tag{75}$$

where $A \approx 5/6$. If $\delta_A \to 0$, eqn (75) recovers eqn (74).

As pointed out by Karma and Rappel, if one wants to treat the case of local interface equilibrium where $\mu_A = \infty$, then for a given value of δ_A, eqn (75) gives the appropriate value of M_ϕ^A. This is useful because many researchers are interested in comparing the predictions of the phase field method with mathematical solutions of the sharp interface formulation that assume local equilibrium (i.e. ignoring interface kinetics). Such modifications have been also performed for alloys by Karma (2001). On the other hand, some researchers are interested in high solidification speed where interface kinetics plays a very important role.

7.3.4.5 Alloys
We recall the general form of the free energy, $f(\phi, c, T)$, eqn (63), for the case of an alloy. The general shape is shown in **Figure 16a** for $\Omega \leq 0$ and at a fixed temperature between the pure component melting points. For fixed temperature, simultaneous solution of eqns (58) and (59) in one dimension shows that ϕ will vary from 1 to 0 across a stationary interface between liquid and solid as C varies between two specific compositions C_L and C_S as shown by the dotted line on the (ϕ, c) base of the plot. The values far from the interface are given by a plane tangent at two points to the free energy surface. The tangent plane has the properties that

$$\frac{\partial f}{\partial \phi} = 0 \quad \text{and} \quad \frac{\partial f}{\partial c} = \text{constant} \tag{76}$$

as required by eqns (58) and (59) whenever ϕ and c have no spatial variation; that is, far from the interface. This construction is the same as that obtained by classical thermodynamic methods applied to equilibrium between bulk phases; namely, the common tangent construction to the liquid and solid free energies at the selected temperature. The common tangent line to $f(0, c, T)$ and $f(1, c, T)$ at fixed T is the projection of the tangent plane to the surface $f(\phi, c, T)$ as shown in **Figure 16b**.

The phase field mobility and the diffusion mobility can be taken as

$$M_\phi = (1 - c)M_\phi^A + cM_\phi^B,$$

$$M_C = \frac{(1 - p(\phi))D_S + p(\phi)D_L}{R'T}, \tag{77}$$

where D_L and D_S are the diffusion coefficients in the bulk liquid and solid phases. Also we note that the interpolating function $p(\phi)$ need not be the same as the interpolating function introduced in eqn (62).

(a)

(b)

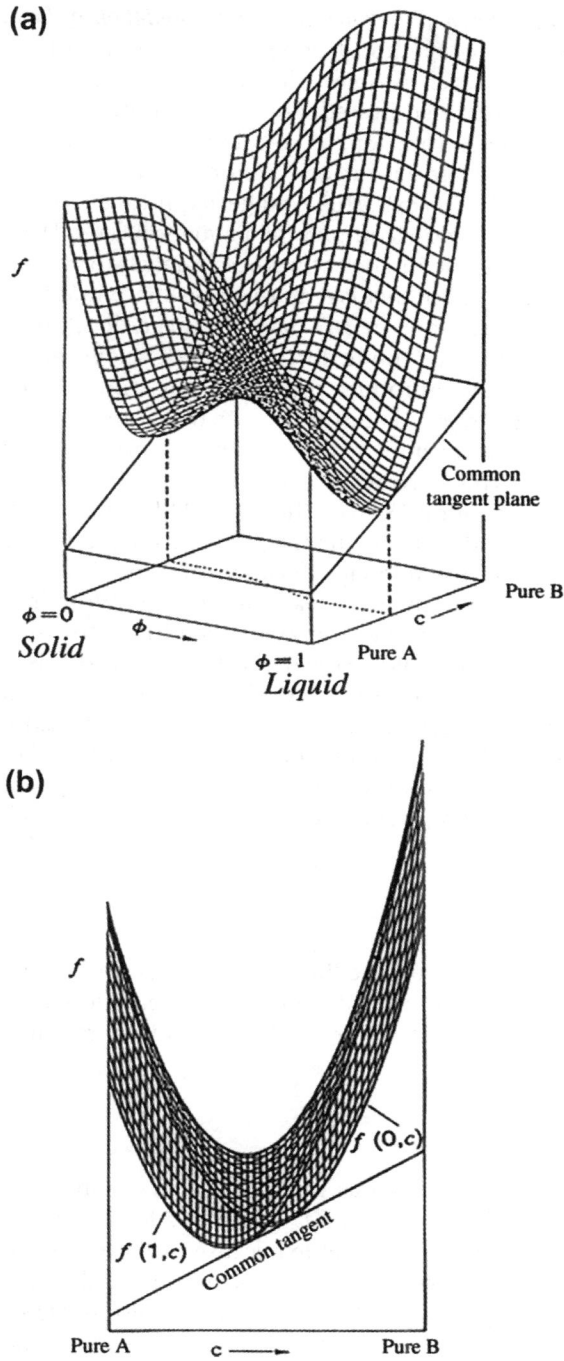

Figure 16   (a) The alloy free energy function $f(\phi,c,T)$ for a temperature, T, between the pure component melting points $(T_M^B < T < T_M^A)$, the locus of ϕ and c through the thin interfacial region is shown dotted. (b) Projection of (a) along the ϕ axis showing the common tangent line. The free energy surface is only shown for $0 < \phi < 1$.

7.3.4.6 *Anisotropy, Noise and Numerical Considerations for Simulation of Dendritic Growth*

A major triumph of the phase field method has been its application to dendritic growth as will be described in Section 7.7.3.1. However to obtain "realistic" simulations of dendrites from the above equations, three other issues must be addressed: anisotropy, noise and computation times. The most widely used method to include anisotropy for 2-D calculations is to assume that ε_ϕ in eqn (57) depends on an angle θ (Kobayashi, 1994). Here θ is the orientation of the interface with respect to the frame of reference given by $\tan\theta = (\partial\phi/\partial y / \partial\phi/\partial x)$. This change requires a recomputation of the variational derivative in eqn (58) resulting in a more complex field equation for ϕ. Using this method of including anisotropy, formal asymptotics, as the interface thickness is taken to zero, yields the same form of the anisotropic Gibbs–Thompson equation that is employed for sharp interface theories (McFadden et al., 1993). Methods for including anisotropy in 3-D have also been developed.

Calculations performed using a coarse finite difference mesh will usually exhibit side branching as an artifact because discretization errors introduce numerical noise into the calculation (a coarse square mesh can also induce a synthetic fourfold anisotropy). As the computational mesh is refined however, side branches disappear. Therefore, to study realistic structures, noise is introduced at controlled levels, which induces side branching. This is typically implemented using random fluctuations of a source term added to the phase field equation (Kobayashi, 1994; Elder et al., 1994).

Computational issues associated with solving the phase field equations are quite challenging, being primarily focused on the value of the interface thickness and the finite difference mesh spacing. The central difficulty is having sufficient numerical resolution of the diffuse interface and at the same time having sufficient nodes/elements to model the entire dendritic structure. This often limits simple solution methods to high supercooling where the dendrite is small. This is an active area of research. In particular, techniques using adaptive FE calculations have been useful to enable more detailed simulations (Provatas et al., 1998). Hybrid methods are also being developed that solve the thermal diffusion equation using random walkers rather than a numerical solution to the differential equation (Karma et al., 2000). Examples of microstructural predictions using phase field methods will be interspersed throughout the following sections on dendrites, eutectics, and peritectics.

7.4 Nucleation

The classical picture of nucleation from pure liquids and alloys and the important effects of catalytic particles are described. The important topic of dendrite fragmentation as a source of growth centers is deferred to Section 7.9.2. Kelton and Greer (2010) provide a recent comprehensive treatise on nucleation.

7.4.1 Nucleation in Pure Liquids

Nucleation during solidification is defined as the formation of a small crystal from the melt that is capable of continued growth. Below T_m the solid phase has a lower free energy than the liquid phase, but a small solid particle is not necessarily stable because of the free energy associated with the L–S interface. The change in free energy corresponding to the liquid–solid transition must therefore include not only the change in free energy between the two phases but also the free energy of the L–S interface.

To arrive at a description of nucleation it is also important to realize that, at any supercooling, there exists within the melt a statistical distribution of atom clusters or embryos of different sizes having the character of the solid phase. The probability of finding an *embryo* of a given size increases as the temperature decreases. *Nucleation* occurs when the supercooling is such that there are sufficient embryos with a radius larger than a *critical radius* (Hollomon and Turnbull, 1953).

7.4.1.1 *Calculation of the Critical Radius and Energy Barrier*

The change in the free energy per unit volume, ΔG (solid minus liquid), to form a solid embryo of spherical shape of radius, r, from liquid of a pure material involves a sum of the volume free energy change and the surface free energy associated with the L–S interface given by

$$\Delta G = \Delta G_v + \Delta G_i = -\frac{4}{3}\pi r^3 \frac{L^V \Delta T}{T_m} + 4\pi \gamma_{SL} r^2, \tag{78}$$

where ΔG_v is the change in free energy on a volume basis from equation (53) on solidification associated with the volume and ΔG_i is the free energy associated with the interface, γ_{SL} is the L–S interfacial free energy, L^V is the latent heat per unit volume and ΔT is the supercooling with respect to the bulk melting temperature. The critical radius, r^*, occurs when ΔG has a maximum given by the condition, $d(\Delta G)/dr = 0$, as

$$r^* = 2\gamma_{SL}/T_m L^V \Delta T. \tag{79}$$

The critical radius for a given temperature is identical to what would be predicted using the Gibbs–Thomson Effect described in Section 7.3 for the manner in which the liquid–solid interface curvature changes the equilibrium liquid–solid interface temperature.

Figure 17, taken from Kurz and Fisher (1989), gives a comprehensive picture of the variation of the free energy of an embryo as a function of its radius and ΔT: (a) at temperatures T greater than T_m both ΔG_v and ΔG_i increase with r. Therefore the sum ΔG increases monotonically with r. (b) At the melting point, $\Delta G_v = 0$ but ΔG_i still increases monotonically. (c) Below the equilibrium temperature the sign of ΔG_v is negative because the liquid is less stable than the solid while the behavior of ΔG_i is the same as in (a) and (b). At large values of r, the cubic dependence of ΔG_v dominates over the quadratic dependence of ΔG_i and ΔG passes through a maximum at the critical radius, r^*. When a thermal fluctuation causes an embryo to become larger than r^*, growth will occur as a result of the decrease in the total free energy. Nucleation in a homogeneous melt is called *homogeneous nucleation* and from eqn (78) the *critical energy of activation* for an embryo of radius r^* is given by

$$\Delta G^* = \frac{16\pi}{3} \frac{\gamma_{SL}^3 T_m^2}{(L^V \Delta T)^2}. \tag{80}$$

The unlikelihood that statistical fluctuations in the melt can create crystals with a large radius is the reason why nucleation is so difficult at small values of the supercooling. However small contamination particles in the melt, oxides on the melt surface or contact with the walls of a mold may catalyze nucleation at a much smaller supercooling and with fewer atoms required to form the critical nucleus. This is known as *heterogeneous nucleation*.

In **Figure 18**, homogeneous nucleation and heterogeneous nucleation are compared for a flat catalytic surface and isotropic interface energies. For this simple case, the embryo is a spherical cap that makes an angle θ with the substrate given by

$$\gamma_{cL} - \gamma_{cS} = \gamma_{SL} \cos \theta, \tag{81}$$

(a)

(b)

(c)

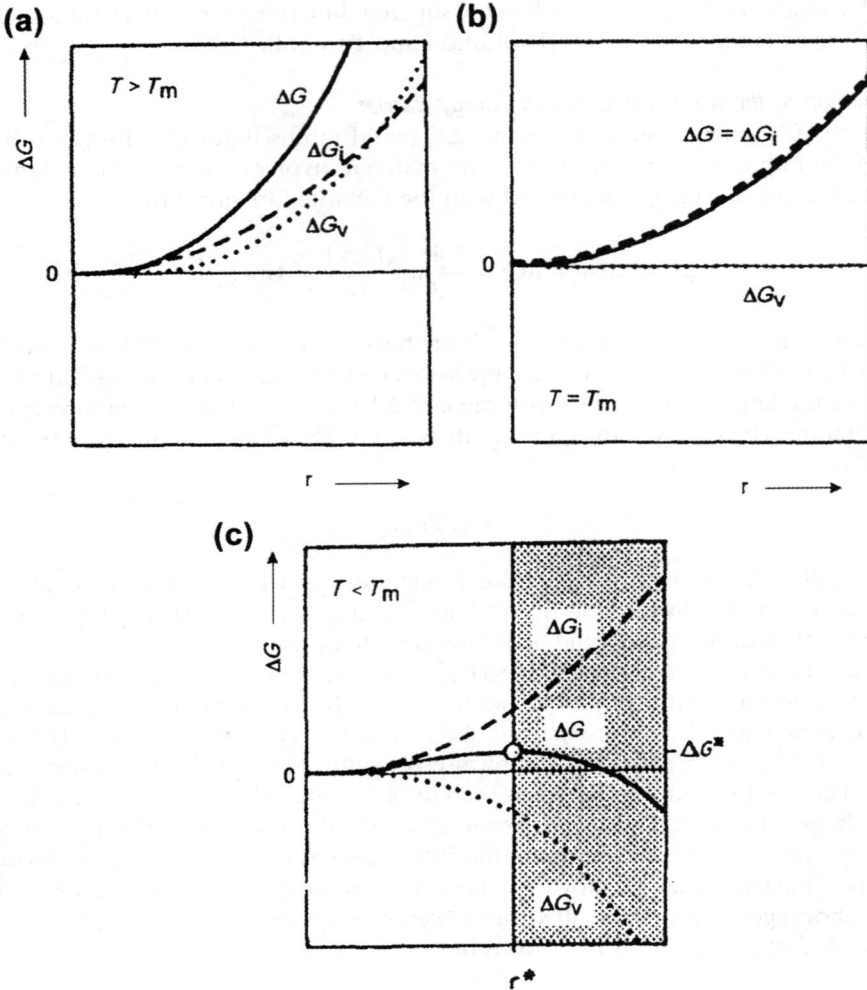

Figure 17 Volume, surface and total values of the free energy of a crystal cluster as a function of radius, r, at three temperatures: (a) $T > T_m$, (b) $T = T_m$, and (c) $T < T_m$ (Kurz and Fisher, 1989).

where γ_{cL} is the catalyst–liquid interfacial free energy and γ_{cS} the catalyst–solid interfacial free energy. At a supercooling, ΔT, the critical radius of the spherical cap is again given by eqn (79), but the number of atoms in the critical nucleus is smaller than that for homogeneous nucleation as a consequence of the catalytic substrate. Indeed the thermodynamic barrier to nucleation ΔG^* is reduced by a factor $f(\theta)$ to

$$\Delta G^* = \frac{16\pi}{3} \frac{\gamma_{SL}^3 T_m^2}{(L^V \Delta T)^2} f(\theta), \tag{82}$$

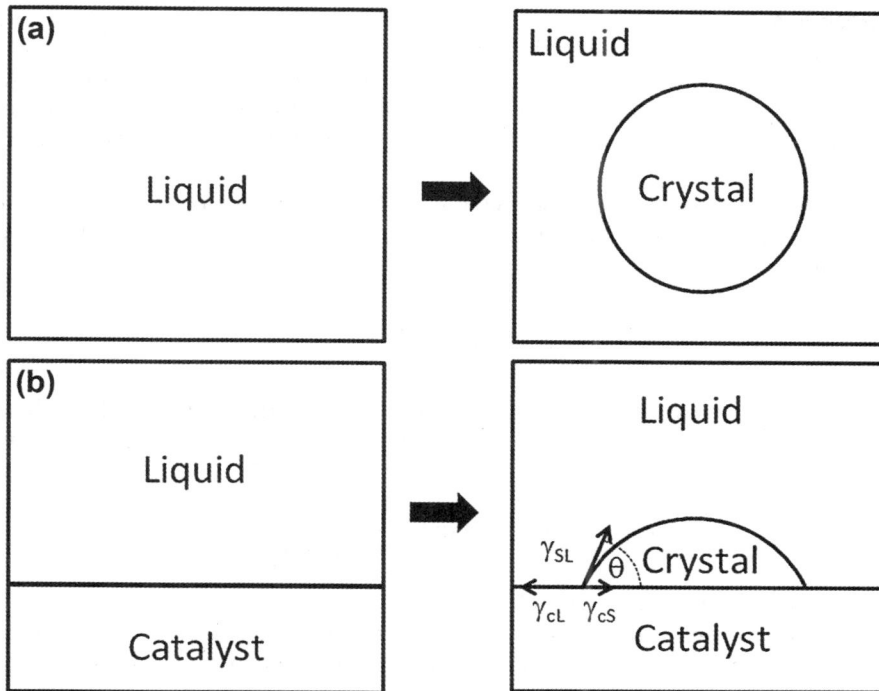

Figure 18 Schematic comparison of (a) homogeneous and (b) heterogeneous nucleation of a crystal in a supercooled liquid. The interface energies are assumed to be isotropic and in (b) the catalytic surface is assumed to be flat.

where

$$f(\theta) = \frac{(2 + \cos \theta)(1 - \cos \theta)^2}{4}. \tag{83}$$

If nucleation occurs in a scratch or a cavity of the catalytic substrate, the number of atoms in a critical nucleus and the value of ΔG^* can be reduced even more (Turnbull, 1950). For a planar catalytic surface, the reduction in the free energy barrier compared with that for homogeneous nucleation depends on the contact angle. Any value of θ between 0 and 180° corresponds to a stable angle. When $\theta = 180°$, the solid does not interact with the substrate, $f(\theta) = 1$ and the homogeneous nucleation result is obtained. When $\theta = 0°$, the solid "wets" the substrate, $f(\theta) = 0$, and $\Delta G^* = 0$. As a result, solidification can begin immediately when the liquid cools to the freezing point. From the classical heterogeneous nucleation point of view, a good nucleant corresponds to a small contact angle between the nucleating particle and the growing solid. According to eqn (81), this implies that γ_{cS} must be much lower than γ_{cL}. However, in general, the values of γ_{cS} and γ_{cL} are not known and, therefore it is rather difficult to predict the potential catalytic effectiveness of a nucleant. Factors that determine θ are largely unknown but likely involve lattice disregistry between the substrate and the stable phase, the topography of the catalytic substrate surface, the chemical nature of the catalytic surface and absorbed films on the catalytic substrate surface.

7.4.1.2 Nucleation Rate

The rate of homogeneous nucleation, I, is the number of embryos formed with a size that just exceeds the critical value per unit time per unit volume of liquid. Similarly, the heterogeneous nucleation rate is considered per unit area of active catalytic site. To determine the rate of nucleation, it is necessary to find expressions for the number of embryos that have critical size and the rate at which atoms or molecules attach to the critical nucleus.

By considering the entropy of mixing between a small number, N_n, of crystalline clusters, each of which contains n atoms, and N_L atoms of the liquid, an expression for the equilibrium number of clusters with n atoms can be obtained as,

$$\frac{N_n}{N_L} = \exp\left(\frac{-\Delta G_n}{k_B T}\right), \tag{84}$$

where k_B is Boltzmann's constant, ΔG_n is the value of ΔG, obtained from eqn (78) for a cluster of radius, r, containing n atoms. Using eqn (82), the number, N_n^*, of clusters of critical radius, r^*, is given by

$$\frac{N_n^*}{N_L} = \exp\left(\frac{-\Delta G_n^*}{k_B T}\right) = \exp\left(\frac{16\pi}{3k_B T}\frac{\gamma_{SL}^3 T_m^2}{(L^V \Delta T)^2}f(\theta)\right), \tag{85}$$

where ΔG_n^* corresponds to the critical size cluster.

If one can assume that an equilibrium number of critical nuclei can be maintained in the melt during the nucleation process, the homogeneous nucleation rate (#nuclei/time/volume of liquid) is then given by

$$I = K_1 \nu \frac{N_n^*}{N_L}, \tag{86}$$

where K_1 is the number of atoms per unit volume of liquid ($\sim 10^{29} m^{-3}$) and ν is the frequency at which atoms attach to the critical embryos. This is called the steady-state nucleation rate. The value of ν for metallic (monoatamic) melts scales with the atomic vibration frequency ($\sim 10^{13} s^{-1}$). For melts with (molecular) structural units, ν scales with D_L/a_0^2, where D_L is the diffusion coefficient in the liquid and a_0 is the distance between liquid structural units. For metals, this attachment rate is fairly independent of temperature and so

$$I = K_2 \exp\left(-\frac{K_3}{T\Delta T^2}\right), \tag{87}$$

where K_2 typically has a value of $10^{42}/m^3$ s and where K_3 contains the temperature-independent quantities in the exponential term in eqn (85). For nonmetallic melts, where the diffusion coefficient in the liquid can depend strongly on temperature,

$$I = K_4 \exp\left(-\frac{\Delta G_d}{k_B T}\right)\exp\left(-\frac{K_3}{T\Delta T^2}\right), \tag{88}$$

where K_4 includes the preexponential factor for diffusion, and ΔG_d is the activation energy for diffusion. For heterogeneous nucleation, similar expressions can be developed but is described per unit area of catalytic surface.

An evaluation of eqn (87) for homogeneous nucleation ($f(\theta) = 1$) shows that as the supercooling is increased, I increases very rapidly at a critical supercooling in the range of $\Delta T = 0.2T_m$ to $\Delta T = 0.4T_m$. Changes in the preexponential term in eqn (87) by orders of magnitude do not appreciably affect the calculated supercooling for sensible nucleation rates. This rapid rise in nucleation rate with decreasing temperature effectively defines a *nucleation temperature*. Thus for a metal with a melting point of 1000 K, homogeneous nucleation would occur for supercoolings of 200 K–400 K.

During rapid cooling of the melt, especially to large supercoolings in glass-forming alloys, atomic transport may be too slow to maintain an equilibrium number of clusters. This requires the examination of transient nucleation theory and effectively introduces a delay time into nucleation kinetics that can be important during glass formation or during devitrification as described by Kelton and Greer (2010).

Another assumption of the classical nucleation theory is that the free energy per unit volume and free energy per unit surface area are independent of the size of the embryo. Since the interface between solid and liquid is usually considered to be diffuse on the level of a few atomic dimensions, embryos that are a few atomic dimensions in radius cannot be described classically as above. This leads to a radius (or temperature) dependence of the surface energy as shown by Larson and Garside (1986) and Spaepen (1994). Indeed Tolman (1949) proposed that for small particle of radius r that the interface energy itself would have a first-order correction given by

$$\gamma_{LS} = \gamma_{LS}^0 \left(1 - 2\frac{\delta}{r}\right), \qquad (89)$$

where γ_{LS}^0 is the standard value of the interfacial energy and δ is a parameter related to the L–S interface thickness.

For heterogeneous nucleation, Perepezko (1988) has pointed out that if θ approaches zero, the thickness of the spherical cap can approach atomic dimensions, even when the cap radius is much larger. This fact would also necessitate a nonclassical approach to heterogeneous nucleation because the liquid–solid interface structure may be influenced by the liquid–substrate interface structure. Diffuse interface theoretical methods have been used to explore situations where the substrate induces ordering in the liquid (Granasy et al., 2007; Warren et al., 2009).

7.4.2 Effect of Melt Subdivision

When a volume of liquid metal is converted into an array of liquid droplets, large supercoolings before solidification are often obtained in many of the droplets. This fact leads to a method for the study of nucleation and to understand the large supercooling and refined microstructures obtained in metal powders created by industrial gas atomization and plasma spray processes. For the study of nucleation, this method was pioneered by Turnbull and Cech (1950) and continued most notably by Perepezko and Anderson (1980). An example of the effect of supercooling on microstructure development for gas-atomized Al-8 wt% Fe alloys is given by Boettinger et al. (1986).

If the heterogeneous nucleating sites contained within a given liquid volume are distributed randomly, the arrangement of nucleants among the droplets may be described by a Poisson distribution. For this case the nucleant-free droplet fraction, X, is represented by $X = \exp(-mV_d)$ where m is the number of nucleants per volume in the melt and V_d is the droplet volume. Based on experience with droplet emulsion samples (Perepezko and Anderson, 1980), supercooling effects become measurable for size refinement below about 100 μm diameter and can become appreciable for powder sizes less than about 10 μm. This suggests that typical values for nucleant densities within the volume of a melt must be in the range from about 10^6 to 10^9 cm^{-3}. The relatively sharp selection of a given X value (e.g. $X = 0.9$) with the droplet volume indicates the important role of size refinement in achieving large supercooling. Similar relationships can be developed for surface nucleant distributions.

7.4.3   Experiments on Nucleation in Pure Metals—the Liquid–Solid Interface Energy

Perepezko and Anderson (1980) have also summarized the principal methods for nucleation experiments conducted at slow cooling rates as shown in **Figure 19**. The most common method corresponds to the dispersion of a pure metal into droplets within a suitable medium. For metals that melt below 500 °C, organic fluids with added surfactants are used to form the droplet dispersion. In addition to the isolation of nucleants discussed above, the surfactant may play a role in rendering some nucleates inactive. For systems with high melting points, molten salts and glasses have been employed. In both cases, independence and separation of droplets are maintained by a thin inert coating which must be noncatalytic to nucleation. Such dispersions of droplets can be thermally cycled in a DTA or DSC (differential scanning calorimeter) to determine the supercooling of the majority of droplets before nucleation. Perepezko (1984) and Kelton (1991) have summarized the maximum supercoolings obtained by various workers. In particular Kelton (1991) has used the maximum obtained supercooling to extract values for γ_{LS} for 30 pure metals using the classical nucleation theory. This is possible because the nucleation temperature depends on γ_{LS}^3, and the results are relatively insensitive to the poorly known prefactor of the exponential in eqn (87). The results are well fit by the expression

$$\gamma_{SL} = \alpha L_m V_m^{-2/3} N_a^{-1/3}, \tag{90}$$

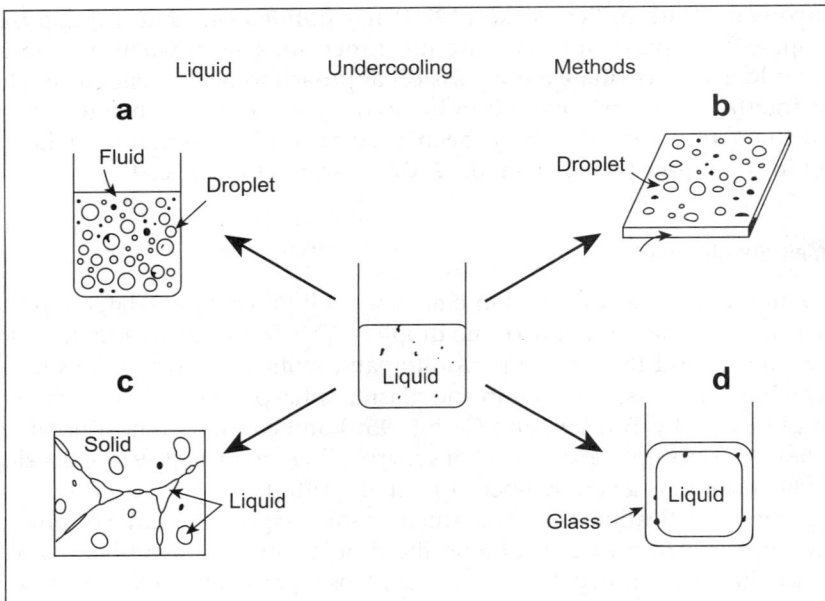

Figure 19   Sample configuration for different supercooling methods (a) droplet-emulsion, (b) substrate isolation, (c) liquid inclusion and (d) bulk encapsulation by glass (Perepezko and Anderson, 1980).

where N_a is Avogadro's number, V_m is the molar volume and with the constant $\alpha = 0.44$. This procedure may provide only a lower bound on the value of the interface energy, unless it is proven that the nucleation is indeed homogeneous. It may be that heterogeneities still limit the observed maximum supercoolings in most metals. Other methods of measuring the interface energy are noted that use grain boundary groove experiments (Gunduz and Hunt, 1985; Hardy et al., 1991; Marasli and Hunt, 1996).

Closely aligned with the correlation above is the theoretical model of Spaepen (1975) and Spaepen and Meyer (1976) that estimate the surface energy based on the configurational entropy of a structural transition that enforce polyhedral atomic packing in the interfacial region. Their result is equivalent to the expression

$$\gamma_{SL} = \alpha(T/T_m)L_m V_m^{-2/3} N_a^{-1/3}, \tag{91}$$

where α is a numerical factor related to the atomic packing (0.866 for face-centered cubic (FCC) and hexagonal close pack (HCP) and 0.71 for body-centered cubic (BCC) metals).

7.4.4 Atomistic Calculation of Interface Energy

Because experimental measurements of γ_{LS} are difficult and because of its importance not only to nucleation prediction but also to the Gibbs–Thomson effect, molecular dynamics (MD) simulations are being used to obtain information about γ_{LS} and its anisotropy. The increasing use of MD is largely because of the availability of more realistic embedded atom potentials to treat metals (Daws and Baskes, 1984). **Figure 20**, from Hoyt et al. (2006), shows a summary of results for several metals using different potentials. The scaling of the interface energy with the heat of fusion is again demonstrated with values of α of 0.55 and 0.3 in eqn (90) for FCC and BCC metals, respectively.

The anisotropy of the interface energy can also be determined with MD using either cleavage methods or the capillary fluctuation method. The latter monitors small equilibrium undulations in the

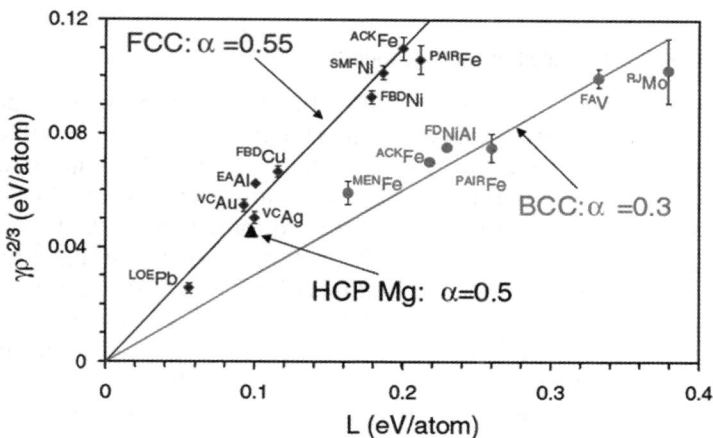

Figure 20 Liquid–solid interface energy for pure BCC and FCC metals calculated by the capillary fluctuation method using embedded atom method (EAM) potentials; the plot shows the correlation of the normalized interface energy with heat of fusion for the two structures (Hoyt et al., 2006).

interface shape from which the interface stiffness used in the anisotropic form of the Gibbs–Thomson coefficient (Section 7.3.1) can be directly obtained (Hoyt et al., 2003).

7.4.5 Alloy Nucleation

For a pure material, ΔG_v depends only on temperature. However for a binary alloy ΔG_v depends on the temperature and also on the composition of the liquid and of the solid nuclei. Thus for a given liquid composition, critical values of nucleus composition as well as size are required to determine ΔG^*. If the surface energy and $f(\theta)$ are independent of temperature, the smallest value of r^* (hence easiest nucleation) is obtained if the composition of the critical cluster maximizes ΔG given in eqn (52) with $C_R^* = C_S^*$. It is apparent from eqn (52) that ΔG (and hence ΔG_v) would be maximized for a composition of the solid with $\mu_S^A - \mu_L^A = \mu_S^B - \mu_L^B$; that is, by a parallel tangent construction as shown in **Figure 21**. This maximum driving force condition has been proposed (Hillert, 1953; Thompson and Spaepen, 1983) to find the favored nucleus composition for a given temperature and liquid composition. To use this condition, one must have a thermodynamic model for the alloy of interest; that is, the free energy functions for the liquid and solid phases must be known. For simple analysis, regular solution models are often employed for the liquid and solid phases. More precise models that fit the measured phase diagram and other thermodynamic data are often available in the literature using the CALPHAD method. By including a simple model of the composition dependence of the surface energy, Ishihara et al. (1985) have shown that the critical nucleus composition can approach the bulk liquid composition at large supercoolings.

Experience with alloy supercooling indicates that the composition dependence of the nucleation temperature, T_N, often reflects the composition dependence of liquidus temperature T_L. For example in the Pb–Sb system, the supercooling results shown in **Figure 22** show that T_N follows a similar trend to T_L even for different T_N levels resulting from catalytic sites of different potency, that is, different surface coatings (Richmond et al., 1983). The maximum ΔG_v condition to determine nucleus composition has been used to successfully predict the composition dependence of measured values of T_N in various alloy systems (Thompson and Spaepen, 1983).

7.4.6 Formation of Metastable Phases by Supercooling

One of the most dramatic effects of large supercoolings before solidification is the possibility of forming metastable phases. An elegant yet simple example occurs for pure Bi (Yoon et al., 1986). A dispersion of Bi droplets was cooled from above the Bi melting temperature of 271 °C to approximately 50 °C where nucleation took place. Upon reheating the dispersion, melting occurred at 174 °C, well below the Bi melting point indicating that a metastable phase had been formed at 50 °C. If one examines the pressure–temperature diagram for Bi and extrapolates the melting curve for the high pressure Bi(II) phase to atmospheric pressure, one obtains a metastable melting point for Bi(II) very close to 174 °C. In addition, performing the supercooling experiment with an increase in ambient pressure modified the melting point of the metastable phase in a manner consistent with it being Bi(II). Thus the formation of metastable Bi(II) occurred rather than the stable Bi(II) phase at large supercooling.

The bulk free energy change for solidification, ΔG_v, is always largest for the stable phase. However in the context of the heterogeneous nucleation theory described above, a metastable phase may make a smaller contact angle with a particular catalytic site than does the stable phase. Thus the barrier for nucleation of a metastable phase may be smaller than the barrier for the stable phase. Of course one must always supercool below the melting point of the metastable phase in order for ΔG_v for the

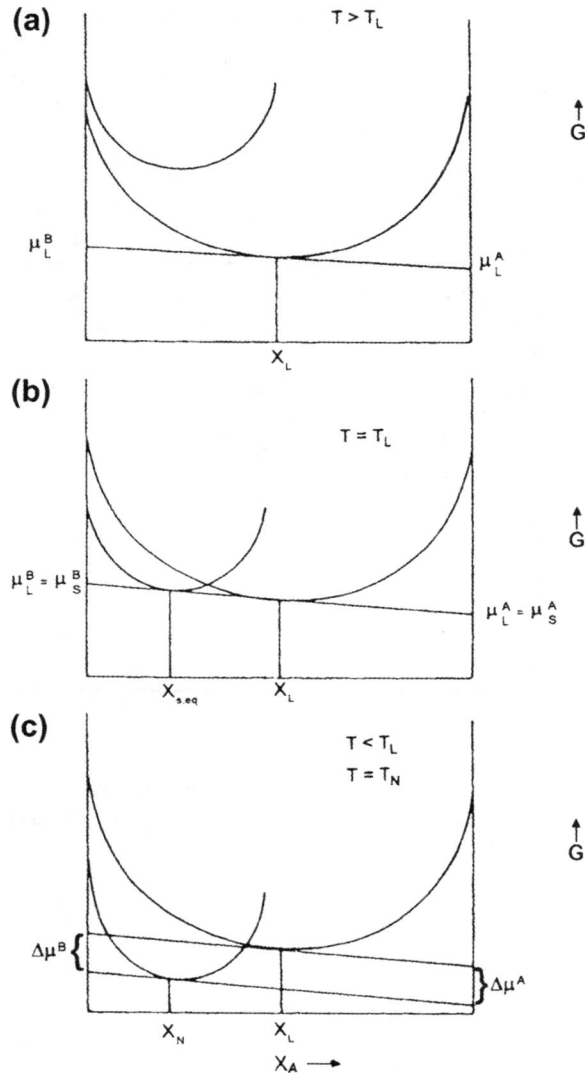

Figure 21 Schematic free energy versus composition curves for liquid and solid at three temperatures: (a) above the liquidus for composition X_L, (b) at the liquidus, and (c) below the liquidus at an arbitrary nucleation temperature. The composition of a nucleus, X_N, that maximizes the free energy change at the temperature given in (c) is given by the parallel tangent construction (Thompson and Spaepen, 1983).

metastable phase to be negative. Similarly metastable phases have been formed in alloy systems. In fact in the Pb–Sn system, by avoiding the nucleation of the stable Sn phase, the metastable Pb liquidus and solidus curves have been measured more than 80 K below the Pb–Sn eutectic temperature as shown in **Figure 23** (Fecht and Perepezko, 1989). When nucleation did occur in this supercooled state, a metastable phase was formed.

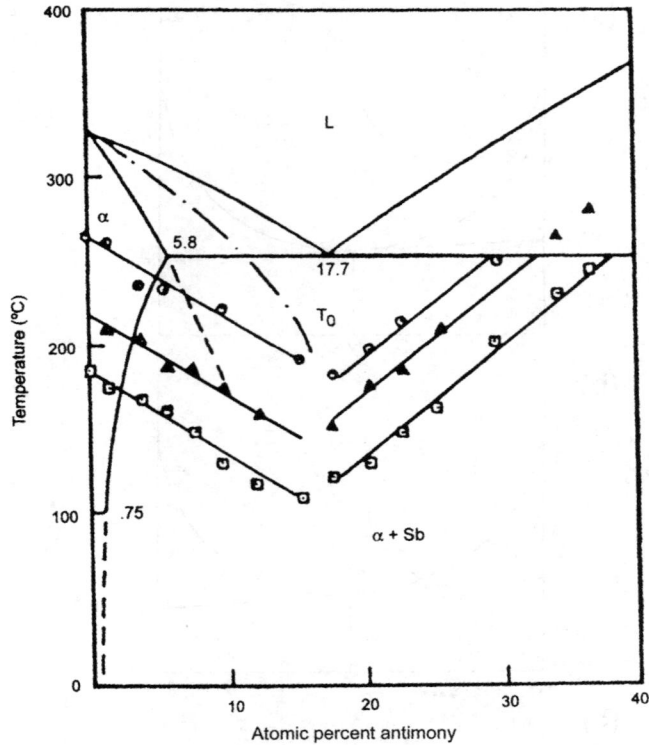

Figure 22   Summary of nucleation temperatures for Pb–Sb alloys which generally follow the liquidus slope (Richmond et al., 1983). Supercooling trends at different levels are produced by different droplet surface coating treatments.

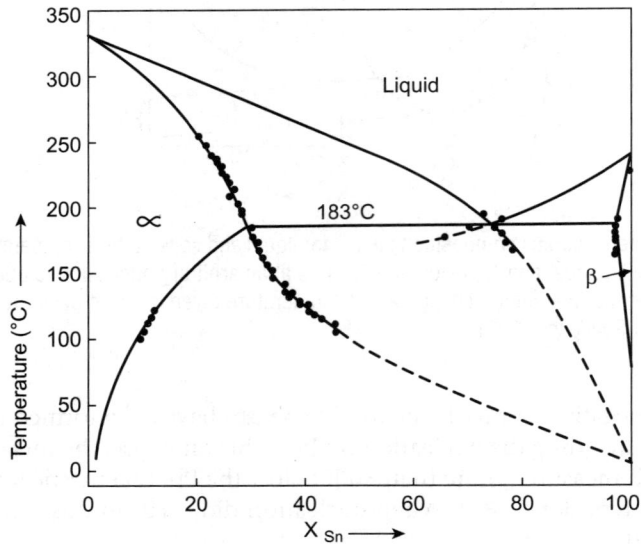

Figure 23   The stable Pb–Sn phase diagram (solid line), including measured and calculated metastable extensions (dashed line) of liquidus and solidus curves (Fecht and Perepezko, 1989).

Another example of metastable phase formation occurs in the Fe-Ni system. Kim et al. (1988a,b) have used Spaepen's model of the interface energy with $\alpha = 0.866$ for FCC and 0.71 for BCC metals to calculate homogeneous nucleation rates for the FCC and BCC phases for all compositions in Fe–Ni alloys. Due primarily to the higher α value for FCC (hence more difficult nucleation using eqn (91)), nucleation of BCC phase is favored in these alloys even for compositions where FCC has a higher liquidus. Indeed the presence of BCC was confirmed in their own experiments with 3–30 µm size powders as well as in the earlier work of Cech (1956).

7.4.7 Innoculants

One of the most important examples of structure modification in industry is the grain refinement of Al alloys using inoculants that increase heterogeneous nucleation. A fine grain size in shaped castings ensures uniformity and isotropy of mechanical properties by promoting a fine distribution of second phases and microporosity if present. The grain-refining inoculants used in the aluminum industry employ so-called master alloys containing Al with Ti, B and C. This is a complex subject beyond the scope of this chapter and excellent reviews are available (Perepezko, 2008; Kelton and Greer, 2010).

One important idea on this subject is found in the work of Greer et al. (2000) defining the free growth model of heterogeneous nucleation. In Section 7.4 it has been assumed that once the critical size is obtained the growth will proceed practically uninhibited. One can model the heat flow and growth of a spherical particle in a pure material starting from a radius slightly larger than the critical radius quite rigorously. Since the temperature of the bulk liquid is at the nucleation temperature and the interface radius is slightly larger than the critical value, the interface temperature through the Gibbs–Thomson effect is slightly above the nucleation temperature. Thus latent heat can flow into the supercooled liquid. As growth proceeds, the radius gets larger, the interface temperature rises (through the Gibbs–Thomson effect) and the heat flow in the bulk liquid becomes more rapid and growth continues. However for the growth of a spherical cap on a catalytic substrate, growth cannot proceed uninhibited.

As catalytic sites have limited lateral extent, other considerations are necessary. Greer et al. (2000) considered a disk-like catalytic particle whose catalytic face is of diameter d (**Figure 24**). The spherical cap can only grow until its base diameter reaches the limit of the face diameter of the catalyst. The triple junction between the liquid, solid and catalyst becomes spatially pinned and the contact angle calculated from eqn (81) becomes irrelevant. For the solid metal to continue to grow, the radius of the liquid–solid interface must actually decrease. The minimum radius is equal to d. The Gibbs–Thomson effect would then require the bulk undercooling value for further growth (free growth, denoted by subscript fg) to be at least

$$\Delta T_{fg} = \frac{4T_m \gamma_{LS}}{L^V d} \tag{92}$$

to effect dissipation of the heat of fusion. If this level of bulk supercooling is available, continued growth of the spherical cap solid would then cause the liquid–solid interface radius to increase and it would be *free* to grow controlled by the temperature of the melt undergoing recalescence and by the limits of solute diffusion for alloy melts. With this model the statistical variation of nucleation temperatures predicted by the classical approach would disappear. A fixed start of sensible solidification (nucleation) would be obtained if all of the substrates had the same size. However when this model is considered within the context of the distribution of inoculant sizes that are present in real grain refiners, a stochastic distribution of initiation temperatures is obtained.

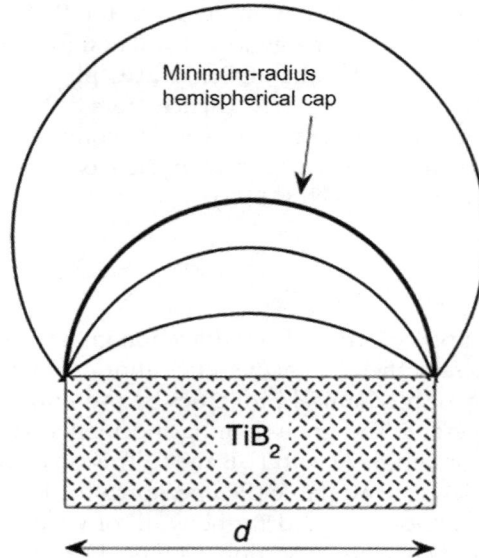

Figure 24 Free growth model of Greer et al. (2000), which shows that heterogeneous nucleation on a catalytic surface of limited lateral extent cannot produce growth until the triple junction between liquid, solid and heterogeneity reaches the edge of the catalytic surface and the size. When the solid reaches the critical hemispherical condition it is free to effectively produce solidification.

7.5   Interface Kinetics

As mentioned in Section 7.3, local equilibrium is often a good approximation for interface conditions during solidification of metals and alloys under casting conditions. Here we quantify the degree of nonequilibrium (interface supercooling) required to move an interface between a crystal and a melt at a given velocity. First we describe pure materials and then describe alloy effects focusing on the nonequilibrium incorporation of solute into a crystal growing at high solidification velocity.

7.5.1   Pure Materials

The nature of the L–S interface and the rate at which atoms join a crystal from the melt have a decisive influence on the kinetics and morphology of crystal growth. For solidification of a pure material, the parameter that governs the atomic or molecular attachment kinetics is the interface supercooling, ΔT_k, which is the difference between the thermodynamic melting point and the interface temperature. The dependence of ΔT_k on growth velocity is the subject of this section. Interface supercooling is a distinct topic from the topic of bulk supercooling and/or bulk supersaturation. Bulk supercooling is the difference between the thermodynamic melting point and the temperature in the liquid far away from the solid–liquid interface. The response to bulk supercooling is controlled by the transport of heat and solute by conduction/diffusion and/or convection in the liquid to be described in Section 7.7 for dendritic growth.

7.5.1.1   Interface Structure

There are two approaches involved in the description of the transition from a liquid to a crystal across an interface. In the first, atoms are considered to belong to either the crystal or the liquid and the interface is considered to be a *sharp interface*. The geometry of the surface that separates the two types of atoms may be smooth or meandering on an atomic scale. The former is called a facetted interface while the later is called a rough interface. In either case the atomic position of atoms in the crystal at the interface is considered to be in perfect crystallographic positions. The second approach includes the additional possibility of a gradual transition in atomic position from the randomness associated with a liquid to the perfect registry of the crystal. The later is called a *diffuse interface*.

Jackson (1958) has considered a sharp interface model and estimated the conditions when a facetted or a rough interface will occur between liquid and solid. Using a near-neighbor bond model and assuming that a random arrangement of atoms are added to an atomically planar crystal surface, he obtained an expression for the change in free energy as a function of the fraction, x, of the possible sites occupied by "added solid atoms" as

$$\frac{\Delta G}{RT_m} = \alpha^* x(1 - x) + x \ln x + (1 - x)\ln(1 - x), \tag{93}$$

where

$$\alpha^* = \left\{ \frac{L_m}{RT_m} \right\} \xi, \tag{94}$$

and where R is the gas constant, L_m the molar latent heat and ξ is a factor depending on the crystal-lography of the interface. This factor is always less than unity, is usually greater than 0.5 and is largest for close-packed planes. This theory has been successfully used to classify and categorize atomic scale interface morphologies (Jackson, 1971). When $\alpha^* < 2$, the minimum value of ΔG occurs at $x = 1/2$; that is, when half the sites are full. This represents a rough interface. In these circumstances, from a more macroscopic point of view the L–S interface is, in general, nonfaceted and may exhibit smooth changes in orientation on a scale larger than atomic dimensions. When $\alpha^* > 2$, minima in $\Delta G/RT_m$ occur at two values of x, one near zero and one near unity indicating an interfacial layer with mostly empty site or mostly full sites. This represents a smooth (facetted) interface. Planes that are not close packed have smaller values of ξ and, thus for some materials, can exhibit roughness while close-packed planes may be facetted.

Another approach to interface roughness comes from the consideration of how thermal vibrations affect the surface energy of an atomic step on an otherwise facetted interface (Chernov, 1984). It is found that the step energy vanishes when L_m/RT_m falls below a critical value of order unity. When the step energy goes to zero there is no barrier to surface roughening. This analysis gives the same qualitative result as Jackson's approach. Various statistical multilevel models of interface structure and Monte Carlo simulations (Temkin, 1964, 1969; Leamy and Jackson, 1971; Jackson, 1974) also indicate the importance of the ratio L_m/RT_m. A common feature of all these models is that the roughness of the interface increases with decreasing L_m/RT_m. **Figure 25** shows simulations of an interface at several values of L_m/RT_m (Leamy and Gilmer, 1974).

As mentioned in the Section 7.4 on nucleation, MD simulations have been used to model interface structure. Simulations show that the transition between liquid and solid for a material with a Lennard–Jones interatomic potential takes place over several atomic layers (Broughton et al., 1981).

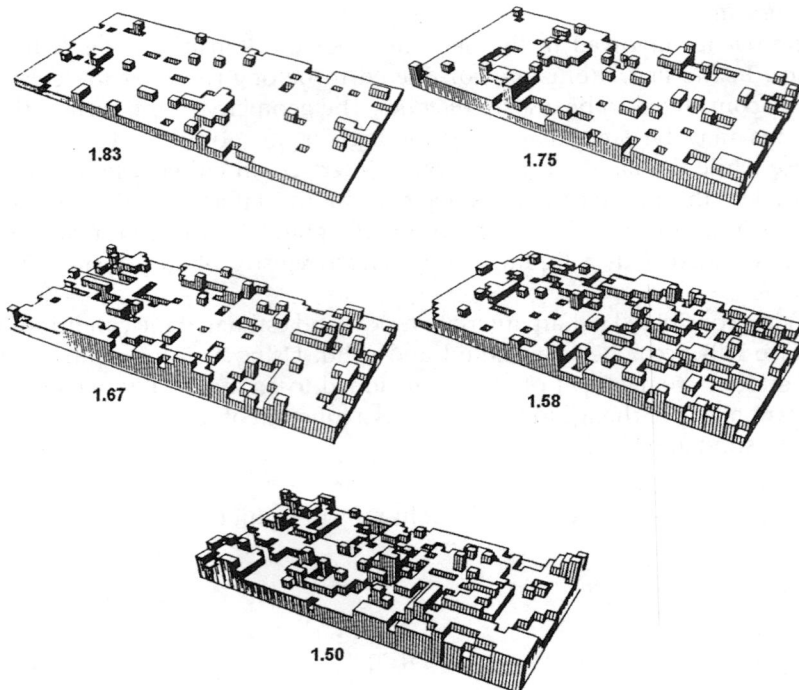

Figure 25 Surface configuration showing increasing roughness with decreasing values of L_m/RT_m (Leamy and Gilmer, 1974).

This potential approximates nondirectional metallic-like bonding. **Figure 26** shows the calculated structure of successive (111) layers between the liquid and the crystal.

In another technique, density functional theory, superposition of ordering waves is employed to represent the local atomic density (Oxtoby and Haymet, 1982). This method also shows the interface to be several atom layers thick as sketched in **Figure 13**. The expansion relating the free energy to the local density uses order parameters that describe the amplitude of the ordering waves through the interfacial region. This is a generalization of the gradient energy approach of Cahn (1960) except that liquid structure factor data are used to determine the interface thickness and gradient energy coefficient.

Whether a sharp or diffuse interface model is used, a facetted interface must grow by layer or lateral growth. In contrast L–S interfaces that are rough can grow by continuous growth because there are so many sites for easy attachment. As we see below continuous growth is faster than faceted growth at the same level of interface supercooling.

7.5.1.2 Continuous Growth

The growth of a rough interface is called continuous or normal growth because the interface can propagate normal to itself in a continuous manner because of the large number of sites for easy atom attachment. To obtain the velocity–supercooling function for continuous growth of single-component

(111) face

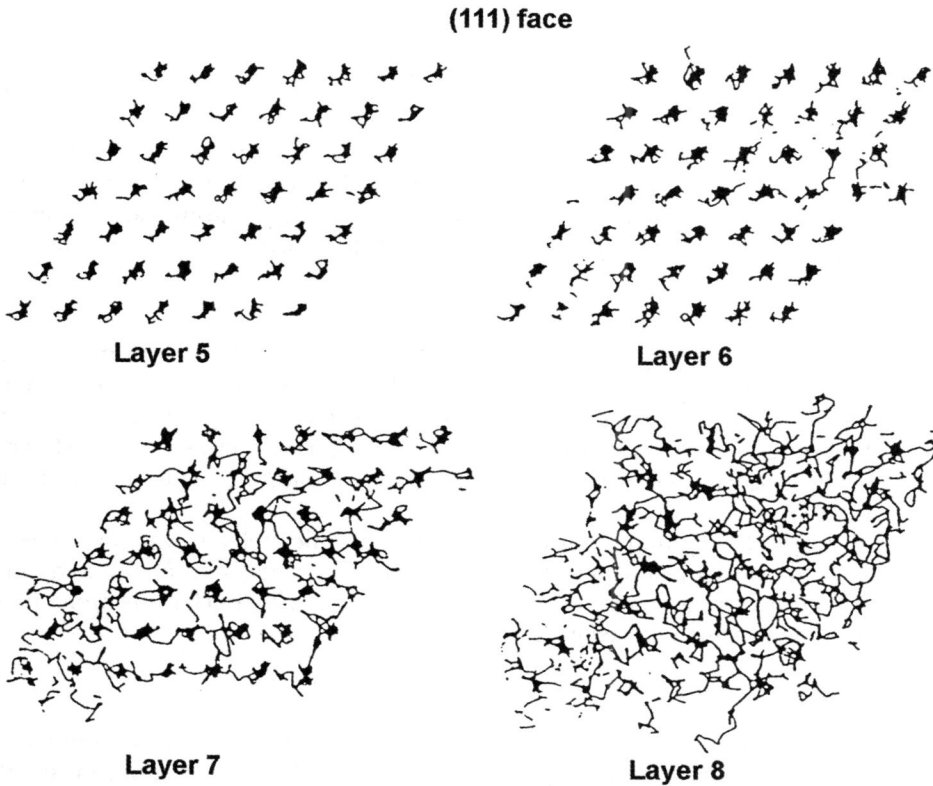

Layer 5 · Layer 6

Layer 7 Layer 8

Figure 26 Trajectories of the molecules in layers parallel to a (111) interface going from solid (layer 5) into the liquid (layer 8) that were obtained by MD simulations by Broughton et al. (1981).

melts, the growth velocity is typically expressed as a product of a factor involving the thermodynamic driving force for solidification and a kinetic prefactor involving the interface mobility:

$$V(T^*) = V_C(T^*)[1 - \exp(\Delta G/RT^*)], \tag{95}$$

where T^* is the interface temperature and ΔG is the Gibbs free energy change per mole of material solidified (defined to be negative for solidification). The bracketed term in eqn (95) represents a difference between the "forward flux" (liquid → solid) and the "backward flux". The kinetic prefactor, $V_c(T)$, is the rate of the forward flux alone, and corresponds to the hypothetical maximum growth velocity at infinite driving force. Near equilibrium ($\Delta G \ll RT$), the exponential can be expanded, and using eqn (53) for ΔG results in a linear relation between velocity and interface supercooling:

$$V(T_i) = V_C \frac{L_m \Delta T_k}{RT_m^2}. \tag{96}$$

This is often written as

$$V = \mu \Delta T_k, \tag{97}$$

where μ is called the linear interface kinetic coefficient.

In conventional modeling of interface kinetics (Wilson, 1900; Frenkel, 1932; Turnbull, 1962; Jackson, 1975) it is assumed that the rate of the forward reaction, that is, the rate at which atoms can jump across the interface to join the solid, is similar to the rate at which atoms can diffuse in the melt. Consequently, the kinetic prefactor is assumed to scale with the diffusivity in the liquid

$$V_C = f_1 D_L / a_0, \tag{98}$$

where a_0 is an interatomic spacing and f_1 is a geometrical factor of order unity. Because of the temperature dependence of D_L, V will first increase linearly, then go through a maximum and finally decrease as the supercooling increases. This relation has extensive experimental support for the crystallization of oxide glasses and other covalent materials (Jackson et al., 1967). However a prefactor that scales with viscosity has never been verified for monatomic melts such as liquid metals (Broughton et al., 1982).

Turnbull and Bagley (1975) pointed out that for simple atomic melts in which the intermolecular potential is largely directionally independent, crystallization events may be limited only by the impingement rate of atoms with the crystal surface and therefore can be much more rapid than diffusive events. According to their *collision-limited growth model*,

$$V_C = f_3 V_S, \tag{99}$$

where V_S is the velocity of sound in the liquid metal (~ 1 km s^{-1}) and f_3 is another numerical factor of order unity. The important consequence is that V_S is about three orders of magnitude greater than D_L/a_0 for typical metallic melts, resulting in a correspondingly more mobile crystal/melt interface. Hence for a given velocity, the kinetic undercooling is much smaller. In addition no maximum is expected in the velocity–supercooling curve. The collision-limited growth model has been confirmed as an upper bound on V_c by the analysis of the velocities of rapidly growing dendrites growing into pure Ni melts (Coriell and Turnbull, 1982), by MD calculations on Lennard–Jones systems (Broughton et al., 1982) and by pulsed laser melting experiments on Cu and Au (McDonald et al., 1989).

In the MD simulation of growth, Broughton et al. (1982) approximate their results using eqn (96) but with

$$V_C = f_4 \sqrt{k_B T_m / m_w}, \tag{100}$$

where k_B is the Boltzmann constant, m_w is the atomic weight and f_4 is of order unity. This speed is the average thermal velocity and corresponds to the velocity at which atoms can strike the lattice sites because of thermal vibrations. For Ni at its melting point, this velocity is 8.6×10^4 cm s^{-1}, which is less than the speed of sound for liquid Ni estimated by Coriell and Turnbull (1982). For Ni at small supercoolings, this value of V_C yields a kinetic coefficient, μ, of 200 cm s^{-1} K^{-1} corresponding to negligible interface supercoolings under ordinary solidification conditions. In addition, Rodway and Hunt (1991) using the Seebeck effect to measure the velocity–interface supercooling relation for Pb have obtained a value of 28 cm s^{-1} K^{-1} for μ that agrees well with that predicted by Broughton et al. (1982) using the average thermal speed for this lower melting point material. Mikeev and Chernov

(1991) and Chernov (2004) employed density functional theory and data for the structure factor of a Lennard–Jones liquid determined by scattering experiments to estimate a value of $f_4 = 0.72$ for $\{100\}$ interfaces.

Since the work of Broughton et al., MD simulations have been using more realistic embedded atom potentials to compute interface kinetic coefficients. As an example, μ values for Mo and V are found to lie in the range 9–16 cm s^{-1} K^{-1} (Hoyt et al., 2006). The values are greater, by a factor of roughly 3/2, than those predicted by the density functional theory (DFT) based model of Mikeev and Chernov (1991), but the magnitude of the model–simulation discrepancy is similar to that observed in previous MD studies for FCC using embedded atom potentials. Monk et al. (2010) show the difficulty of extracting interface attachment kinetic coefficients from MD simulations when the kinetics is fast compared with the heat flow. They find 48 cm s^{-1} K^{-1} for Ni.

Mendelev et al. (2010) examine values of μ for different orientations using a variety of embedded atom potentials for the FCC metals. For Al, Cu and Ni values averaged over the various potentials are 100 cm s^{-1} K^{-1}, 70 cm s^{-1} K^{-1} and 70 cm s^{-1} K^{-1} for $\{100\}$, respectively. All potentials give $\mu^{100} > \mu^{110} > \mu^{111}$ for the high-symmetry $\{100\}$, $\{110\}$ and $\{111\}$ planes, respectively. However, this anisotropy is sufficiently small that no missing growth orientations are usually seen; that is, no facets are exposed because of growth kinetics.

7.5.1.3 *Facetted Crystal Growth*

Most metals of primary interest solidify by a continuous growth mechanism with very small interface supercooling and no facets. Examples include Al, Cu, Fe, Ni, and Pb and Sn. Metals that grow with a facetted interface include Bi, Ga and Sb. Intermetallic compounds and Si typically grow with facetted interfaces. Intermetallic compounds often freeze in combination with nonfacetted phases to form eutectics in many practical alloys (see Section 7.8.1.3). Phases that grow with facetted interfaces have highly anisotropic growth kinetics by definition. A brief description of their growth mechanisms is given.

If the interface is atomically smooth and free of any defects, the growth rate is limited by the nucleation of surface clusters. These clusters must form on the interface to create the necessary surface steps for lateral growth. The lateral spreading rate is assumed to occur quite rapidly at a speed determined by the continuous growth law described above. The classical theory of 2-D nucleation was developed by Volmer and Marder (1931). The growth law (for the formation of cylindrical surface clusters) has the form

$$V \sim \exp \left\{ \frac{-\pi \gamma_e^2 h T_m V_m}{k_B T L_m \Delta T_k} \right\}, \tag{101}$$

where γ_e is the ledge energy per unit area and h is the step height. According to eqn (101) the growth rate is effectively zero at small supercooling and increases sharply at some critical supercooling.

If one or more screw dislocations emerge at the L–S interface it is not necessary to nucleate new layers to provide the sites for lateral attachment. The step generated by each dislocation moves one plane each time it sweeps around the dislocation (Frank, 1949). It was shown by Hillig and Turnbull (1956) that the distance between neighboring turns of the spiral is inversely proportional to ΔT_k, and therefore the total length of step is directly proportional to ΔT_k. For small supercooling, the rate of growth, therefore, will be

$$V \sim (\Delta T_k)^2, \tag{102}$$

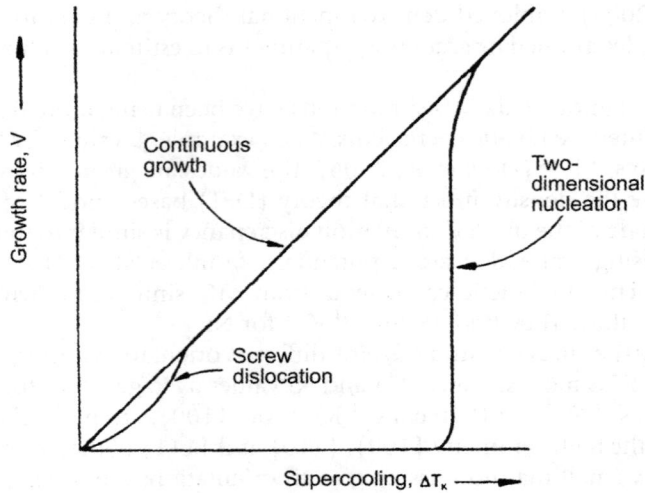

Figure 27   Growth rate versus interface supercooling according to the three classical laws of interface kinetics.

because the rate of growth per unit length of step should also be proportional to ΔT_k (Chalmers, 1964). **Figure 27** schematically shows the growth laws for continuous, 2-D nucleation, and screw dislocation-assisted growth laws. It is seen that interface undercooling required for step growth is considerable higher than for continuous growth. Another source of steps at the L–S interface is the reentrant angle resulting from the emergence of twin planes at the L–S interface. This mechanism is found to be important for Si and Ge (Hamilton and Seidensticker, 1960). See Flemings (1974). Twins are frequently found at high growth rates in many Al alloys (Salgado-Ordorica and Rappaz, 2008).

The Jackson α^* classification scheme (see eqn (94)) while extremely useful cannot predict transitions found in growth kinetics. The models of Cahn (1960), Cahn et al. (1964) and Chernov (1984) predict for a given material a kinetic roughening transition from lateral growth at low supercooling to continuous growth at high supercooling. An experiment showing the transition has been performed by Peteves and Abbaschian (1986) on Ga. The transition to continuous growth occurs at a supercooling of about 4 K and a growth rate of about 0.5 cm s^{-1} for the (111) face.

7.5.2   Binary Alloys

Baker and Cahn (1971) described the general formalism for the interface conditions for solidification of a *binary* alloy in terms of two response functions. One choice for these functions describes the interface temperature, T^*, and the composition of the solid at the interface, C_S^*. Neglecting orientation and curvature effects, these response functions can be written as follows:

$$T^* = T(V, C_L^*) \tag{103}$$

and

$$C_S^* = C_L^* k(V, C_L^*), \tag{104}$$

where V is the local interface velocity. The functions $T(V, C_L^*)$ and $k(V, C_L^*)$ must be determined by a detailed kinetic model for the interface. They are applied at the interface as boundary conditions for the solution of the appropriate diffusion equations for heat and solute. At zero velocity they are very simply related to the phase diagram: $T(0, C_L^*)$ is the equation for the phase diagram liquidus and $k(0, C_L^*)$ is the equation for the equilibrium partition coefficient k^E, which can depend on composition. The dependence of k on curvature is thought to be negligible (Flemings, 1974). The possible forms for the functions T and k for solidification are constrained by the condition that $\Delta G < 0$ as described by eqn (52) and shown in **Figure 10**. In the limit of very high rate, the interface partition coefficient may approach unity only if the interface temperature decreases below the T_0 temperature as described also in Section 7.3. The kinetic partition coefficient can also depend on the crystallographic orientation of the growing interface. If a crystal grows with an L–S interface having regions that are curved and facetted, the incorporation of solute into the crystal behind the facet can be quite different from the rest of the crystal. Brice (1973) calls this the *facet effect*.

For solidification, the process that accomplishes the formation of the crystal structure from the liquid and the process that establishes the compositions of that crystal at the interface are considered as distinct. Indeed for metals, the former is only limited by the rate at which atoms hit the interface (collision limited growth) whereas the latter process requires diffusive interchanges between liquid and solid to reach equilibrium solute partitioning. It is this separation of time scales that permits solute trapping for crystal growth at high velocity; there is insufficient time for the diffusive rearrangements before the solute is buried under additional crystalline material. For this reason the $k(V)$ expression will involve a diffusion coefficient, whereas the interface temperature equation can be obtained in an identical manner to a pure material, eqn (95), employing eqn (100). It is however necessary to use the value for ΔG obtained for alloys, eqn (52) rather than the simple expression eqn (53) that is proportion to ΔT_k.

7.5.2.1 Interface Temperature

For dilute solutions, an analytical expression for the interface temperature when solute trapping is present can be obtained. Boettinger and Coriell (1986) evaluated ΔG from eqn (52) for $C_R^* = C_S^*$ for dilute alloys. A more general expression that used a selectable value of C_R between C_S^* and C_L^* depending on a parameter β

$$C_R^* = \beta C_L^* + (1 - \beta)C_S^* \tag{105}$$

was expressed by Aziz and Boettinger (1994). The parameter β spans the limiting cases with and without solute drag (Hillert, 1999). Solute drag represents a diminution of the driving energy for insertion into eqn (95). Using the β parameter, eqn (52) expressed for dilute alloys is

$$\frac{\Delta G}{RT^*} = \frac{1 - k^E}{m_L^E}\left(T_m + m_L^E C_L^* - T^*\right) + C_L^*\left[k^E - k + (1 + (1 - \beta)(k - 1))\ln\frac{k}{k^E}\right]. \tag{106}$$

One can then write the response function for the interface temperature of a flat interface as follows:

$$T^* = T_m + m_L(V)C_L^* - \frac{RT_m^2}{L_m}\frac{V}{V_C}, \tag{107}$$

where $m_L(V)$ is given by

$$m_L(V) = m_L^E \left\{ 1 + \frac{k^E - k + [k + \beta(1-k)]\ln k/k^E}{1 - k^E} \right\}. \tag{108}$$

This expression is valid regardless of the particular model chosen for the function $k(V)$. As $k(V)$ goes from k^E to unity, $m_L(V)$ changes from m_L^E to $m_L^E[(\ln k^E)/(k^E - 1)]$, which is the slope of the T_0 curve. The last term in eqn (107) can be identified as the interface kinetic supercooling necessary to drive the formation of the lattice. For a curved interface, eqn (107) must include an additional term for the Gibbs–Thomson effect.

7.5.2.2   Partition Coefficient—Solute Trapping

Several models for the dependence of the partition coefficient on velocity, eqn (104), have been formulated for continuous growth. The model of Baker (1970) is quite general permitting the possibility of non-monotonic dependence of k on V. Other theories predict that the partition coefficient changes monotonically from its equilibrium value to unity as the growth velocity increases. In these models the interface partition coefficient is significantly changed from the equilibrium value, k^E, only when a dimensionless growth velocity, V/V_D approaches one. Here V_D is the diffusive speed for atom exchange between the crystal and the liquid. V_D is the ratio of a diffusion coefficient in the interfacial region, D_i, for that exchange to the interatomic distance, a_0 or the interface width. The value of this diffusion coefficient is likely bounded by those of the liquid and the solid. Using the liquid value for the interface diffusion coefficient typical of metals (2.5×10^{-5} cm^2/s) and a length scale of 0.5 nm, V_D should be less than 5 m/s. For ordinary solidification (castings) $V \ll 1$ m/s and the local equilibrium assumption is valid.

The functional form of the models of Aziz (1982) and of Jackson et al. (1980) for nonfacetted growth is given by

$$k = \frac{k^E + V/V_D}{1 + V/V_D}. \tag{109}$$

At a velocity of V_D, the partition coefficient is the arithmetic mean of the equilibrium partition coefficient and unity (see **Figure 35a**). Experimental evidence (Smith and Aziz, 1994) using eqn (109) suggests that V_D is in the range between 6 and 38 m/s. Because eqn (109) has no dependence on composition, it cannot treat the situation shown in **Figure 11b**, in which partitionless solidification is impossible for some compositions and a nondilute numerical model is required. Generalization of eqn (109) to treat these situations is found in Aziz and Kaplan (1988).

The local nonequilibrium model of Galenko and Sobolev (1997) makes use of a modified Fick's law, accounting for the finite relaxation time of diffusion profiles in the bulk liquid and adds another critical speed, $V_B > V_D$, giving

$$k(V) = k^E \frac{\left[1 - (V/V_B)^2 \right] + (V/V_D)}{1 - (V/V_B)^2 + (V/V_D)}, \tag{110}$$

that approaches unity more quickly than the Aziz model. Note that eqn (109) is recovered if $V_B \to \infty$. Another model for solute trapping has been proposed by Jackson et al. (2004), which for rough interfaces is:

$$k(V) = k_E^{1/(1+AV)}, \tag{111}$$

where, A is a constant that depends on the square root of the diffusion coefficient.

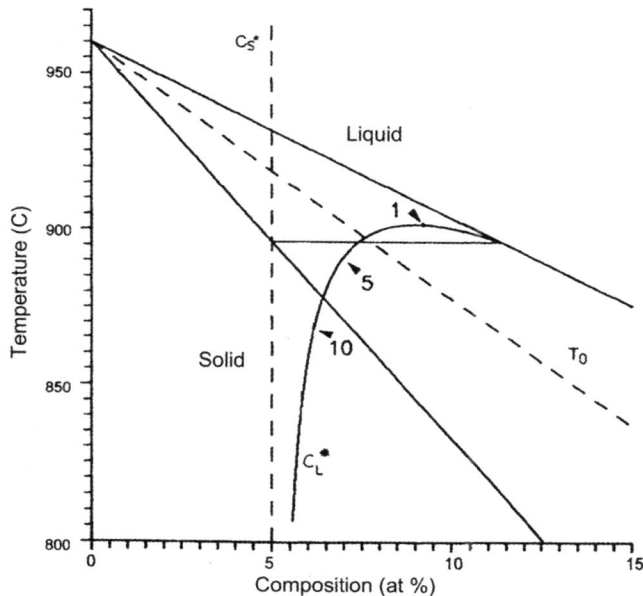

Figure 28 Plot of interface temperature versus liquid composition C_L^* at the interface for a fixed solid composition, ($C_S^* = 5$ at%), forming at the indicated velocities (in m/s) superimposed on the phase diagram.

Figure 28 shows a plot of the response functions obtained using eqns (107)–(109) superimposed on a phase diagram including the liquidus, solidus, and T_0 curves. The plot gives the liquid composition at the interface and the interface temperature as a function of interface velocity for a fixed solid composition as would be appropriate for steady-state solidification. This would be the case for directional steady-state solidification at different but constant velocities. The figure is based on a phase diagram for Ag–Cu with $T_m = 960\,°\text{C}$, $k^E = 0.44$ and $m_L^E = -5.6$ K (at%)$^{-1}$, $\beta = 0$, $C_S^* = 5$ at% Cu, $V_D = 5$ m s^{-1}, and $V_C = 2 \times 10^3$ m s^{-1}. At zero velocity, the composition of the liquid lies on the liquidus curve, a situation that corresponds to local equilibrium. At intermediate velocities (about 0.1 m s^{-1}) the composition of the liquid moves toward the composition of the solid but with an increased interface temperature. At high velocities the liquid composition is near the solid composition and the interface temperature is near, but slightly below, the T_0 curve. At still higher velocities, where the partition coefficient is essentially unity, the temperature drops with increasing velocity. The increase in temperature with increasing velocity seems strange but is a result of fixing the solid composition.

This analysis provides a pair of thermodynamically consistent response functions for dilute alloys. For concentrated alloys no analytical expressions can be written because k^E depends on composition. However, given a thermodynamic description of the liquid and solid phases and values for the two kinetic parameters V_D and V_C, the response functions can be calculated numerically (Aziz and Kaplan, 1988). **Figure 29** taken from Aziz and Kaplan (1988) shows the result for all compositions of an ideal binary alloy when the solute drag is not considered. These curves give the combinations of C_L^*, C_S^*, and T^* possible for the indicated velocities. Extensive experimental research has been focused on testing these models (Kittl et al., 2000). They conclude that inclusion of the solute drag effect is not necessary.

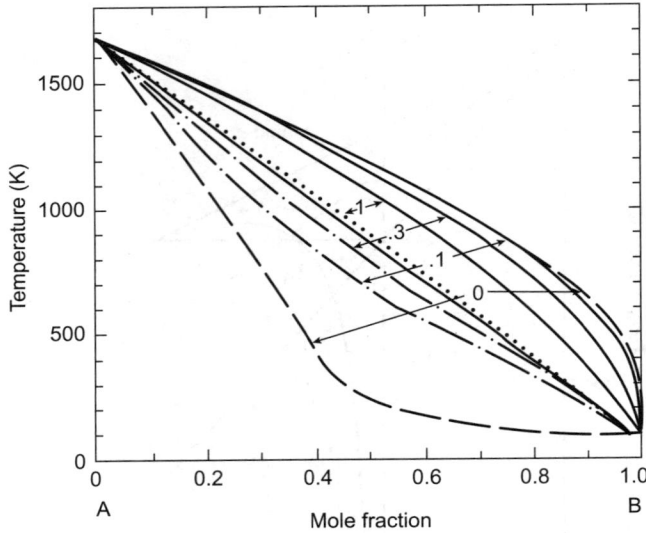

Figure 29 Kinetic interface condition diagram for ideal liquid and solid with pure component melting points as shown, pure component entropies of fusion equal to R, and $V_C/V_D = 100$. Dashed lines: equilibrium liquidus and solidus labeled zero, dash-dot lines: kinetic liquidus and solidus at different interface velocity given as values of V/V_D, dotted line: T_0 curve (Aziz and Kaplan, 1988).

Ahmad et al. (1998) have compared the results of a phase field model of rapid solidification with eqns (107)–(109). For a diffusion coefficient interpolated through the interface according to $D_I = D_S\phi + D_L(1 - \phi)$ where ϕ is the phase field, a good fit to the Aziz model and a value of $\beta = 24/35$ for the drag coefficient were obtained. In the limit of high velocity, they obtained the result that

$$V_D = \frac{3}{8}\left[\frac{D_I}{l_A}\right]\frac{\ln\left(1/k^E\right)}{1 - k^E},\tag{112}$$

where D_I is the solute diffusivity in the diffuse interface and l_A is the interface thickness. Ahmad et al. (1998) shows that a positive correlation of experimental values of V_D obtained by Aziz (1996) for systems with different values of k^E following this expression has been observed experimentally. However it should be noted that other results were obtained with the phase field model if a different interpolation scheme was employed; namely, $D_I = D_S^\phi D_L^{1-\phi}$. Multistep models that approach a phase field model have also been considered by Smith and Aziz (1994).

Yang et al. (2011) have recently calculated the kinetic properties of atomically rough liquid–solid interfaces in two model alloy systems with FCC-based crystal structures using MD with embedded atom potentials. Results for solute trapping at high V are most consistent with the two-parameter solute trapping theory of Galenko and Sobolev (1997) and the temperature equation with an intermediate value of β.

7.5.2.3 Long-Range Order–Disorder Trapping

Solute trapping ideas were extended to chemically ordered intermetallic phases by Boettinger and Aziz (1989). Rapid solidification experiments indicate that some compounds, which are normally ordered at the solidus, can be forced to solidify into the chemically disordered form of the crystal structure, for example, B2 → BCC or L1$_2$ → FCC. Examples are found in Inoue et al. (1984), Huang et al. (1986),

Boettinger et al. (1988a), Huang and Hall (1989). Assadi et al. (1998) has applied the model of Boettinger and Aziz (1989) to the rapid solidification kinetics of the L1$_2$ and B2 phases in the Ni-rich part of the Ni–Al alloy system. Excellent agreement with the modeling is found with their experimental studies.

The theory treats the trapping of disorder by a consideration of solute trapping on each sublattice of the chemically ordered phase. At high rates, there is insufficient time to proportion the solute onto each sublattice and a chemically disordered crystal with the same lattice can be formed. Often however, this chemically disordered phase reverts to the equilibrium ordered phase during solid-state cooling with a resultant microstructure consisting of a high density of antiphase domains. An approximate expression giving the solidification velocity at which the long-range order parameter, η, at the L–S interface goes to zero is given by

$$V_{\eta \to 0} = V_D \left[\frac{T_C}{T_m} - 1 \right],$$ (113)

where T_m is the melting point of the ordered phase and T_C is the temperature where the solid phase would disorder during heating if melting could be prevented; that is, the metastable critical temperature for the order–disorder reaction. Clearly the closer T_C is to the melting point, the lower is the velocity to obtain disorder trapping. Strongly ordered compounds cannot usually be disordered by rapid solidification techniques because $T_C \gg T_m$.

7.6 Solidification of Alloys with Planar and Nearly Planar L–S Interfaces

The analysis of the shape of the L–S interface on a scale larger than atomic dimensions begins with a consideration of plane front growth. Plane front growth is assumed in simple analytical models that treat cases of segregation with different sets of assumptions about solute diffusion. Consideration of plane front growth is also used as a starting point to understand the more complex interface shapes and nonuniform solid composition that accompany cellular and dendrite growth most common in castings. In actual practice, plane front growth is necessary for crystal growth of materials for electronic applications to achieve best properties and for zone refining to achieve high purity. To experimentally achieve planar growth, it is necessary to obtain an L–S interface that is both macroscopically and microscopically planar. Macroscopic planarity is achieved by heat and fluid flow management to produce a 1-D temperature profile near the location of the liquid–solid interface. This is called directional solidification (DS) and requires good furnace design and avoiding convection in the melt. Microscopic planarity is achieved by avoiding interface instabilities because of constitutional supercooling (CS).

7.6.1 General Formulation of 1-D Growth

In general, the transport of heat and solute during DS in the absence of convection with a planar L–S interface growing in the z-direction is described by the 1-D thermal and solute diffusion equations. The simplest forms are obtained from eqns (2) and (11) for constant and equal solid and liquid densities,

$$\frac{\partial T}{\partial t} = \alpha \frac{\partial^2 T}{\partial z^2}$$ (114)

$$\frac{\partial C}{\partial t} = D \frac{\partial^2 C}{\partial z^2},$$ (115)

where α and D take on different values in the liquid and solid. If we place the origin of the spatial coordinate system at the moving interface and assume it is moving at constant speed, the diffusion equations become

$$\frac{\partial T}{\partial t} - V\frac{\partial T}{\partial z} = \alpha\frac{\partial^2 T}{\partial z^2}, \tag{116}$$

$$\frac{\partial C}{\partial t} - V\frac{\partial C}{\partial z} = D\frac{\partial^2 C}{\partial z^2}. \tag{117}$$

These equations must be solved for the temperature and composition profiles $T(z, t)$ and $C(z, t)$ in the liquid and solid subject to local equilibrium conditions at the interface given by

$$T_S^* = T_L^* = T_m + m_L^E C_L^* \tag{118}$$

and

$$C_S^* = k^E C_L^*, \tag{119}$$

where the subscripts L and S of the variables T and C denote values in the liquid and solid sides, respectively, and the asterisk denotes vales at the interface. A balance of heat and solute fluxes at the interface also requires

$$k_S\frac{\partial T_S}{\partial z}\bigg|^* - k_L\frac{\partial T_L}{\partial z}\bigg|^* = V\,L_V \tag{120}$$

and

$$D_S\frac{\partial C_S}{\partial z}\bigg|^* - D_L\frac{\partial C_L}{\partial z}\bigg|^* = V\big(C_L^* - C_S^*\big). \tag{121}$$

For rapid solidification, nonequilibrium interface conditions at the interface can replace eqns (118) and (119).

If one uses a characteristic length, L_0 and a characteristic time t_0 and defines $z' = z/L_0$ and $t' = t/t_0$ eqns (116) and (117) become

$$\left(\frac{L_0^2}{\alpha t_0}\right)\frac{\partial T}{\partial t'} - \left(\frac{VL_0}{\alpha}\right)\frac{\partial T}{\partial z'} = \frac{\partial^2 T}{\partial z'^2}, \tag{122}$$

$$\left(\frac{L_0^2}{D t_0}\right)\frac{\partial C}{\partial t'} - \left(\frac{VL_0}{D}\right)\frac{\partial C}{\partial z'} = \frac{\partial^2 C}{\partial z'^2}. \tag{123}$$

Four dimensionless numbers $\dfrac{VL_0}{\alpha}, \dfrac{VL_0}{D}, \dfrac{L_0^2}{\alpha t_0}, \dfrac{L_0^2}{D t_0}$ are apparent. To within a numerical factor, these are common dimensionless numbers defined in many heat and solute diffusion problems. The first two are called thermal and solute *Peclet* numbers respectively. The reciprocals of the last two are called *Fourier* numbers. In the liquid and solid phases, the corresponding values of the thermal diffusion and solute diffusion coefficients are employed for analysis of processes in the liquid or solid, respectively. Fourier numbers are most important for unsteady processes and Peclet numbers are important for steady-state (constant V) processes. These parameters arise in many models of solidification. In this section we consider events on the scale of the liquid–solid interface and treat mostly steady processes ($\partial/\partial t = 0$). Thus the *Peclet* numbers take on prime importance. However we will employ Fourier numbers in the models of microsegregation described in Section 7.7.3.

Typical values (in $m^2\,s^{-1}$) for the thermal and solute diffusivity for alloys are $\alpha_S = \alpha_L = 10^{-3}$, $D_L = 10^{-9}$ and $D_S = 10^{-14}$. For a conventional solidification process, $V = 10^{-4}\,m\,s^{-1}$ and a length scale typical of the L–S mushy zone is $L_0 < 10^{-2}$ m. Hence, the thermal Peclet numbers (for liquid and solid) are much smaller than unity and therefore, the thermal diffusion equation in the liquid and solid can be approximated for steady-state processes as

$$\frac{\partial^2 T}{\partial z^2} = 0. \tag{124}$$

This implies that the temperature profile for 1-D heat flow has a linear profile in each phase with gradients that satisfies eqns (118) and (120) at the L–S interface. The solute Peclet numbers are not small and eqn (123) must be used to determine how the concentration varies in the z-direction. Thus in the analysis of solidification microstructure, one often views the heat flow and the solidification velocity rather trivially, as, for example, might be obtained in a DS furnace and under the control (or imagination) of the investigator. In complex castings, often the local liquidus isotherm velocity computed by a macroscopic heat flow code is used as an approximation to the solidification velocity for the purposes of microstructure analysis. In either case we often consider the microstructure as having been formed under an "imposed" solidification velocity as presented in subsequent sections on dendrites and eutectics.

7.6.2 Solute Redistribution during 1-D Solidification

Four cases of the diffusional transport of solute and the resultant solute distribution in the frozen solid can be distinguished for solidification of a 1-D sample (**Figure 30**): *equilibrium* freezing (global equilibrium assumed), and *Scheil* freezing, *steady-state* freezing and freezing with partially stirred liquid where local interface equilibrium is assumed. Equilibrium and Scheil freezing will constitute limiting cases of dendritic microsegregation to be described in Section 7.7.3. Steady-state freezing provides the

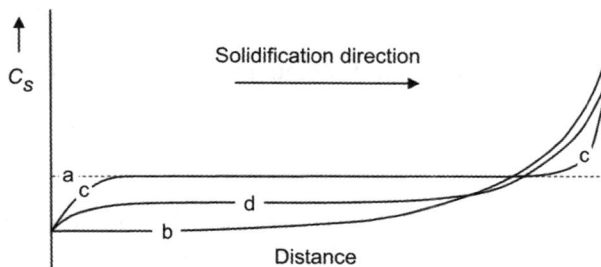

Figure 30 Solute distribution remaining in the solid *after* 1-D solidification: (a) equilibrium freezing; (b) complete liquid mixing with no solid diffusion; (c) diffusion in liquid only; (d) partial mixing in liquid, including convection.

starting point for the description of interface shape stability theory that explains why cells and dendrites are so common in cast structure. We note that the term Scheil freezing is sometimes referred to as Scheil–Gulliver freezing to more properly reflect historical contributions. If the sample has macroscopic dimensions, these limiting cases can describe simple examples of macrosegregation.

7.6.2.1   Equilibrium Freezing

Here the L–S interface advances so slowly that solute diffusion maintains spatial uniformity but slow temporal changes of composition in both the liquid and the solid (global equilibrium, see Section 7.3). The phase compositions follow the liquidus and solidus of the phase diagram during cooling. If the phase diagram predicts a single-phase solid structure for the bulk alloy composition after cooling to low temperatures, the difference in solid and liquid composition occurring during the solidification process will disappear after solidification is complete (**Figure 30a**). In more quantitative terms, for this case to be applicable, the solute *Peclet numbers* for each phase must be much be small and tend toward zero. Only under unusual circumstances, extremely small V and small sample length L_0, might this be expected to apply during solidification. One such case is Fe–C growing at very low velocity because of the high solid-state diffusivity of interstitial C in Fe. In substitutional alloys, equilibrium freezing does not occur in any practical casting or crystal growth situations. Only for these cases where the solid and liquid compositions are spatially uniform at each instant is the lever rule expected to yield correct results during the freezing process.

7.6.2.2   Complete Liquid Mixing, with no Solid Diffusion

The conditions considered in this limiting case are (i) complete mixing of the liquid at each instant; (ii) no solid-state diffusion; (iii) local equilibrium at the L–S interface. The assumption of complete liquid mixing and no solid diffusion is often made because diffusion in the liquid is typically orders of magnitude faster than in the solid (Scheil freezing). One may consider that the solute Peclet number for the liquid to be extremely small but the solute Peclet number for the solid to be near infinity. In this case there is no need to solve the diffusion equation, eqn (123). Only a statement of solute conservation is necessary. When a fraction of the sample, f_S, is solidified, the concentration of the remaining liquid is found to obey the following differential equation

$$C_L\left(1 - k^E\right)\frac{df_S}{dt} - \left(1 - f_S\right)\frac{dC_L}{dt} = 0$$

or   (125)

$$\frac{dC_L}{C_L} = \frac{1 - k^E}{1 - f_S}\,df_S,$$

where the asterisk is dropped because the liquid is spatially uniform in composition. The later equation is directly integrated for constant k^E to give

$$C_L = C_0(1 - f_S)^{k^E - 1},$$ (126)

where C_0 is the nominal alloy composition. The temperature of the interface as the interface progresses can be simply obtained by substituting eqn (126) into eqn (118).

$$T^* = T_m + m_L^E C_0(1 - f_S)^{k^E - 1}.$$ (127)

After freezing is complete, this process leaves a solute distribution in the solid (**Figure 30b**) given by

$$C_S = k^E C_0 (1 - f_S)^{k^E - 1}, \tag{128}$$

where now f_S is taken as the fractional distance down the 1-D sample. Because we have generally neglected density differences in this simple discussion, we have been careless about the definition of f_S. We note that if the concentration variable C is described in mass fraction, then f_S is actually mass fraction. If C is described in atomic fraction then f_S is atomic fraction.

The Scheil approach will be applied as an approximation to the description of microsegregation produced by dendritic solidification in Section 7.7.3. Indeed the rod geometry with a planar interface is not a necessary assumption if the Gibbs–Thomson effect can be neglected, and the equations are valid for any interface shape but f_S will not be a linear distance fraction.

Gulliver (1922), Hayes and Chipman (1938), Scheil (1942) and Pfann (1952) all developed this equation, also called the "nonequilibrium lever rule". Equation (126) can only be considered a limiting case since most real systems have at least some solid-state diffusion and some level of incomplete liquid mixing. For alloys containing a eutectic (see Section 7.8) the liquid composition may reach the eutectic composition before freezing is complete. At this point, the remaining liquid will freeze as a eutectic mixture. For cases when k^E depends on composition, eqn (125) can be numerically integrated.

7.6.2.3 Steady-State Diffusion-controlled Freezing

Another practically important limiting case of 1-D solidification occurs when all the assumptions described in Section 7.6.2.2 apply, except item (i). In this case mixing in the liquid is not complete and is governed by eqn (123). **Figure 30c** shows the resulting solute distribution along the rod after solidification. If the full time-dependent diffusion equation eqn (115) is solved, three distinct regions occur: an initial transient, a steady-state region and a terminal transient. The first is required to establish the steady-state boundary layer of solute ahead of the interface and the third arises from the interaction of the boundary layer with the end of the specimen. The diffusion boundary layer in the liquid ahead of the L–S interface is a region of the system that transports the solute missing from the initial transient in the solid and maintains a constant solid composition in the central region of the rod. The moving boundary layer changes the liquid interface composition from C_0 to C_0/k^E, and disappears at the end of solidification by "depositing" its solute in the final transient. **Figure 31** (Kurz and Fisher, 1989) shows the distribution of solute in the liquid and solid along the rod during unidirectional solidification.

When the steady-state condition has been reached in the central region of the bar (**Figure 30c**), the solution to eqn (117) for solute distribution in the liquid in front of the interface is given by

$$C_L = C_0 \left[1 + \frac{1 - k^E}{k^E} \exp\left(-\frac{V}{D_L} z\right) \right], \tag{129}$$

where D_L is the solute diffusion coefficient in the liquid and z the distance from the interface. Note that the liquid concentration at the interface ($z = 0$) is C_0/k^E producing a solid composition C_0. In eqn (129), the thickness of the solute rich layer is given by the characteristic distance, D_L/V. It is quite important to know the extent of each of the three regions shown in **Figure 30c** The reader is referred to Verhoeven et al. (1988, 1989) for a summary of analysis of the initial and final transients.

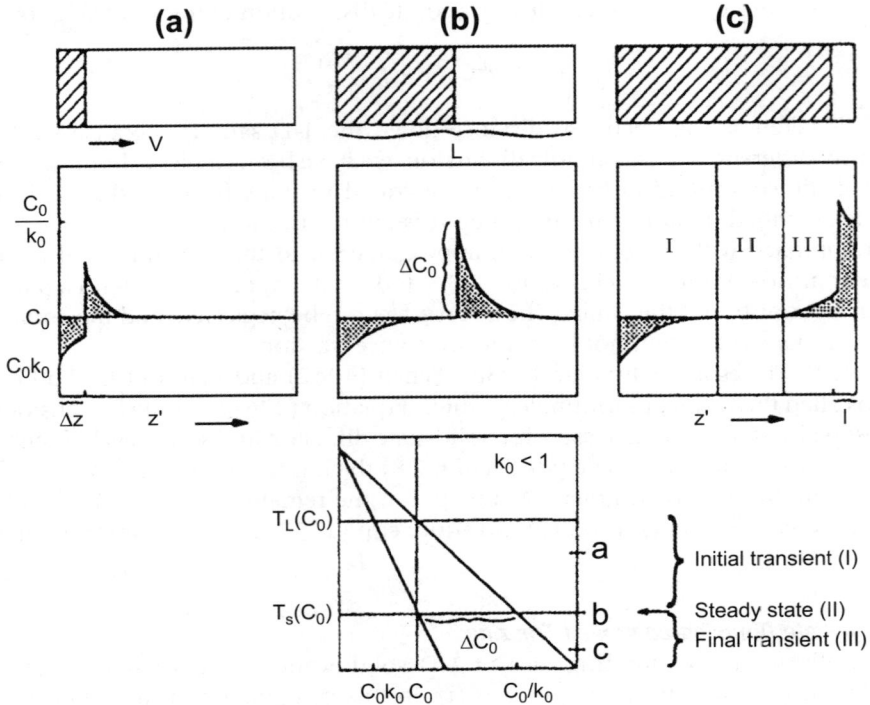

Figure 31 Development of initial and final transient during steady-state diffusion controlled planar solidification for an alloy of composition C_0 (Kurz and Fisher, 1989).

7.6.2.4 Convection Effects–Freezing with Partial Mixing in the Liquid (Boundary Layer Approach)

Free convection, because of solute or thermal gradients in the liquid, or forced convection, because of crystal rotation or electromagnetic forces, strongly influence segregation. This subject has been reviewed by various authors: Hurle (1972), Carruthers (1976), Glicksman et al. (1986), Brown (1988), Favier (1990). Fluid flow has important technological consequences for the processing of electronic materials where the solute distribution during nominally planar growth conditions determines the quality of the devices. Complex fluid flow within the mushy zone during dendritic solidification is also important. This topic is deferred until Section 7.9.

For planar growth, the interval between the extreme cases of complete mixing and diffusion controlled freezing for planar growth was bridged by the pioneering work of Burton et al. (1953) (BPS). In this simple approach, a diffusion layer of thickness δ_F is assumed near the interface outside of which the liquid composition is maintained uniform by convection. An expression for solute distribution (**Figure 30d**) after freezing is complete in the sample is

$$C_S = k_{eff} C_0 (1 - f_S)^{k_{eff} - 1}.$$

(130)

with an effective distribution coefficient given by

$$k_{eff} = \frac{k^E}{k^E + (1 - k^E) \exp\left[\left(-\frac{V \delta_F}{D_L}\right) \frac{\rho_S}{\rho_L}\right]}.$$

(131)

Fluid flow affects the solute distribution through the parameter $V\delta_F/D_L$ in eqn (131). For vigorous convection in the liquid, $\delta_F \to 0$, $k_{eff} \to k^E$, and eqn (131) is the same as the Scheil result. For negligible convection, $\delta_F \to \infty$, $k_{eff} \to 1$, and eqn (130) gives a constant solute profile. The model has been quite successful in describing 1-D segregation in the presence of laminar and turbulent convection during plane front growth. The parameter k_{eff} should not be confused with the function $k(V)$ that describes solute trapping at high solidification velocity.

In the metallurgical context, macrosegregation produced during directional dendritic solidification can sometime be interpreted following the BPS approach. One example is found in Dupouy et al. (1989), who measured the macrosegregation produced during dendritic solidification for upward and downward (with respect to gravity) solidification. Boettinger et al. (1981) observed similar macrosegregation down the length of a directionally solidified sample when convection disrupts the D_L/V solute boundary layer required to support off-eutectic planar growth (see Section 7.8).

The BPS approach, while particularly easy to use, neglects many factors which have been more recently considered. A major effort has been made to include time dependence in the BPS model when Czochralski and Bridgman crystal growth is considered (Wilson, 1978, 1980; Favier, 1981a,b; Favier and Wilson, 1982). Camel and Favier, (1984a,b) and Favier and Camel (1986) used an order of magnitude analysis and scaling to examine different flow regimes in terms of dimensionless numbers in Bridgman crystal growth. Priede and Gerbeth (2005) show a breakdown of the BPS approach for flow patterns that radially converge near the liquid–solid interface.

7.6.2.5 Zone Melting

If instead of melting the entire sample, a molten zone with planar interface is passed down sample, purification of metals can be achieved. The most important variables in the zone-melting process are (i) zone length; (ii) charge length; (iii) initial distribution of solute in the charge; (iv) vapor pressure and (v) zone travel rate (constant or variable). Manipulation of these variables can produce a large variety of impurity distributions in the solid charge. A multipass zone-refining device provides a more efficient system than the single-pass system originally developed. The reader is referred to the important contributions of Pfann (1966) concerning this technique as well as the *zone-leveling* and the *temperature gradient zone melting* techniques reviewed by Biloni (1983).

Rodway and Hunt (1989) established a criterion for optimizing the zone length during multipass zone refining. The technique has been applied numerically, to model the redistribution in a rod, for various values of k^E. The important conclusions are the following: (i) a variable zone size (VZS) along the bar during the process causes a considerable increase in the rate at which the ultimate distribution is approached, compared with a fixed zone size process. This leads to a significant improvement in the usable fraction of the rod. (ii) The optimum zone length at any stage in the process is independent of the k_{eff} value. Consequently for a material containing many impurities with different k^E values ($k^E < 1$ or $k^E > 1$), the VZS is optimum for all of them.

7.6.3 Lateral Segregation

In the previous discussion we have assumed the interface to be planar. In the event that the thermal distribution of the crystal growth apparatus is not perfect, macroscopic curvature of the interface can develop. If convective mixing can be neglected, Coriell and Sekerka (1979), and Coriell et al. (1981) have modeled the lateral segregation that will be present for a given shape. Their numerical and analytical results treat the segregation in terms of the distance that the interface deviates from planarity (δ_p), the sample width, the characteristic diffusion distance (D_L/V) and the partition coefficient, k^E. The radial segregation is greatest when k^E is small and when $\delta_p/(D_L/V) \gg 1$. Detailed calculations of lateral segregation because of convection driven by longitudinal and radial gradients, typical of Bridgman growth upward and downward (with respect to gravity), are described by Chang and Brown (1983). Experiments performed by Schaefer and Coriell (1984) in the transparent succinonitrile–acetone system show the effect of radial gradients at an L–S interface.

7.6.4 Morphological Stability of a Planar Interface

In Section 7.6.2 it was assumed that the L–S interface was microscopically planar during solidification. Thus the composition profile induced in the solid varied only in the direction of growth. However even if the heat flow is controlled to be unidirectional and the isotherms are planar, a planar interface may be unstable to small changes in shape. Lateral composition variations can then be induced in the solid on a scale much smaller than the sample width. The morphological stability theory defines the conditions under which this instability can occur. Instability of the interface ultimately leads to the development of cellular and dendritic structures. A review of the stability theory by Coriell and McFadden (1993) is available.

7.6.4.1 *Theory*

The temperature and solute profiles for $k^E < 1$, as a function of the distance, z, from a planar L–S interface for growth at constant velocity V are shown in **Figure 32**. A linear stability analysis was first applied to a growing sphere and later to the plane by Mullins and Sekerka (1963,1964) and Sekerka (1967) under the conditions of local interface equilibrium and isotropic surface energy. One of the first extensions of this approach came when Cahn (1967) treated anisotropic interface kinetics and interface energy. Many other assumptions of the original theory have been relaxed over the years as summarized by Sekerka (1986).

The analysis begins by perturbing the shape of a planar L–S interface initially located at $z = 0$ in the moving frame of reference to

$$z = \delta \exp(\sigma t) \sin(2\pi x/\lambda), \tag{132}$$

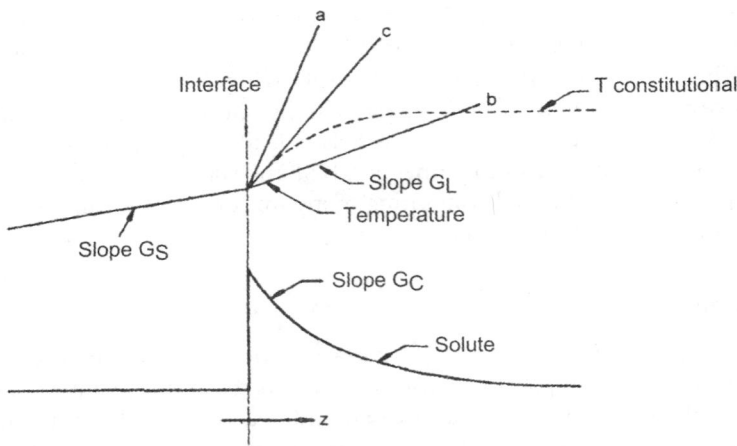

Figure 32 Temperature and solute profile for $k_E < 1$ as a function of distance z measured from the L–S interface when constrained growth at a velocity V occurs for a binary alloy. The concept of CS is shown ahead of the interface.

where δ is the perturbation amplitude, λ is the wavelength, and σ is the growth (or decay) rate of the perturbation. The value for σ is determined by solving the steady-state heat flow and diffusion equations with appropriate boundary conditions for small values of δ (linear theory). The planar interface shape is stable if the real part of σ is negative for all values of λ. The conditions giving the stability–instability demarcation ($\sigma = 0$) reduce to an equation with three terms corresponding to the three factors contributing to the overall stability of the interface, the thermal field, the solute field and the capillarity forces. The stability equation is

$$G - m_L^E G_c \xi_c + \frac{4\pi^2 T_m \Gamma}{\lambda^2} = 0. \tag{133}$$

where Γ is the Gibbs–Thomson constant defined in Section 7.3.1. The parameter G is the conductivity-weighted temperature gradient given by

$$G = \frac{k_S G_S + k_L G_L}{2\bar{k}}, \tag{134}$$

where $\bar{k} = (k_S + k_L)/2$ is the mean of the liquid and solid thermal conductivities. The parameter G_c is the composition gradient in the liquid, which, for a planar interface moving at constant velocity, is obtained from eqn (129) and is given by

$$G_c = \frac{V C_0 (k^E - 1)}{k^E D_L}. \tag{135}$$

The parameter ξ_c can usually be set equal to unity if $V\lambda/2D_L \ll 1$. However ξ_c may deviate significantly from unity under rapid solidification conditions. ξ_c can be computed from the following equation

$$\xi_c = 1 + \frac{2k^E}{1 - 2k^E - \left[1 + \left(\frac{4\pi D_L}{V\lambda}\right)^2\right]^{1/2}}. \tag{136}$$

Technically eqn (115) is correct only when $V\lambda/2\alpha_L \ll 1$, where α_L is the liquid thermal diffusivity. This condition is almost never violated even during rapid solidification. See Kurz and Fisher (1989) for a complete description of this detail.

If the left-hand side of eqn (133) is positive, the interface is stable. The first term is stabilizing for positive temperature gradients; if the temperature gradient is negative (growth into a supercooled melt), this term is destabilizing. If a pure material is considered, this is the only possible destabilizing term. Thus a planar interface in a pure material is only unstable for growth into a supercooled melt. The second term represents the effect of solute diffusion in the liquid and, being negative, is always destabilizing. The third term, involving capillarity, has a stabilizing influence for all wavelengths, though its effect is largest at short wavelengths. This is the sort of stabilizing effect to be expected from surface energy which tends to promote an interface shape with the least area, namely a plane.

Figure 33 shows a plot summarizing the stability of a planar interface for dilute Al–Cu alloys. **Figure 33a** shows the value of σ versus λ for selected values of G_L, V, and C_0 (200 K cm^{-1}, 0.1 cm s^{-1} and 0.1 wt% Cu). Under these conditions a range of wavelengths have a positive value for σ, and are therefore unstable. The smallest unstable wavelength is usually referred to as the marginal wavelength and the

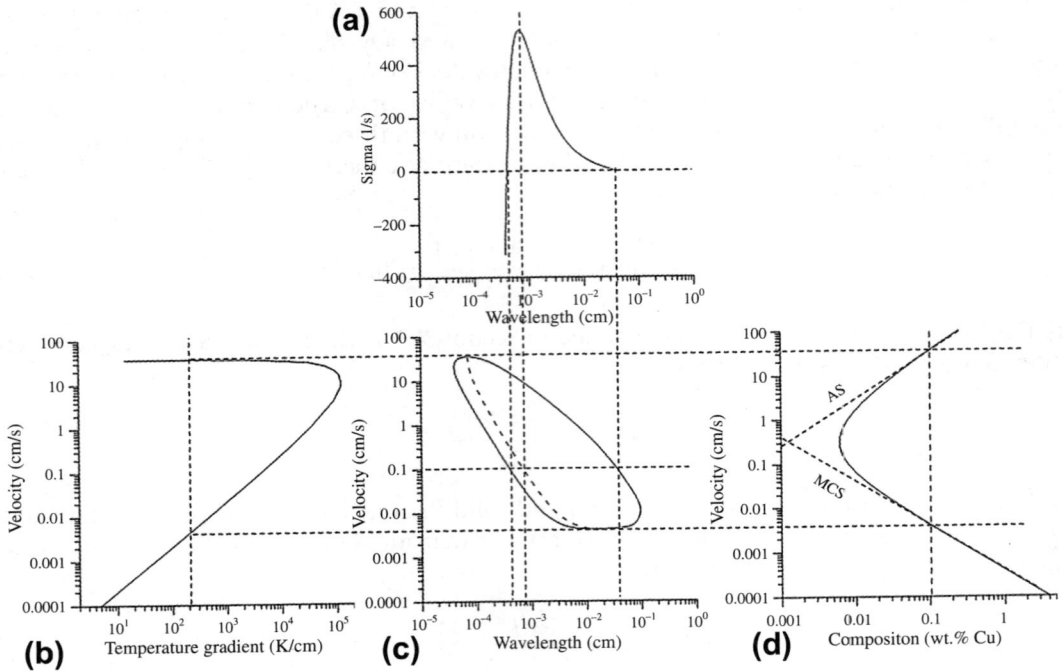

Figure 33 Summary of the results of interface stability theory for Al-rich Al–Cu alloys (a) perturbation growth rate vs. perturbation wavelength, (b) velocity vs. liquid temperature gradient, (c) velocity vs. wavelength and (d) velocity versus composition (Coriell and Boettinger, 1994).

wavelength with the largest value of σ is called the fastest growing wavelength. For some velocities, $\sigma < 0$ for all λ, and the interface is said to be stable. **Figure** 33c shows a closed curve of λ and V values for fixed G_L and C_0 (200 K/cm, 0.1 wt% Cu) for which $\sigma = 0$. At values of V and λ inside the closed curve the interface is unstable. Note that instability only exists over a range between two critical velocities. Outside this range, for all λ and the interface is always stable. Also shown dashed is the wavelength that corresponds to the fastest growing wavelength. **Figure** 33d shows the two critical velocities as a function of C_0 for fixed G_L (200 K/cm). The curves for the two velocities merge into a smooth curve for low C_0. Instability occurs to the right of the curve. **Figure** 33b shows the critical velocities as a function of G_L for fixed C_0 (0.1 wt% Cu). Note the insensitivity of the upper velocity stability limit to the value of G_L.

The upper and lower critical velocities may be approximated by two different limiting cases:

(i) If the interface grows at slow velocities, unstable wavelengths are large. Thus the capillarity forces are small and can be neglected; the stability criterion becomes: $G \geq m_L G_c$. If, in addition, eqn (134) is coupled with eqn (120), and $VL^V \ll 2k_L G_L$, the stability criterion becomes

$$\left(k_L/\bar{k}\right) G_L \geq m_L^E G_c, \tag{137}$$

which is called the *modified CS criterion*. For $k_S = k_L$ this reduces to the *CS criterion* (to be examined in more detail in Section 7.6.4.2).

(ii) If the interface grows at high velocity, unstable wavelengths become small and capillarity dominates. In this case, stabilization because of the temperature gradient is negligible and one obtains the *absolute stability condition*:

$$m_L^E G_c \leq k^E T_m \Gamma (V/D_L)^2 \tag{138}$$

or

$$V \geq \frac{m_L^E C_0 (k^E - 1) D_L}{k^{E^2} T_m \Gamma}. \tag{139}$$

The modified CS criterion and the absolute stability criterion serve as asymptotes to the exact result at low and high velocity, respectively, shown in **Figure 33d**.

7.6.4.2 Relationship to CS

The CS criterion obtained by Tiller et al. (1953) before the morphological stability theory was developed serves as a model to understand the major cause of instability. The approach determines if any part of the liquid ahead of a moving interface is supercooled with respect to its local liquidus temperature. The solute distribution in front of the interface given by eqn (129) and the corresponding liquidus temperature for the composition at each point in front of the interface given by

$$T_L(z) = T_m + m_L^E C_0 \left[1 + \frac{(1 - k^E)}{k^E} \exp\left(-\frac{V}{D_L} z \right) \right]. \tag{140}$$

The actual temperature in the liquid ahead of the unperturbed interface because of the temperature gradient, G_L, is

$$T(z) = T_m + \frac{m_L^E C_0}{k^E} + G_L z. \tag{141}$$

Figure 32 considers three possible values of the actual temperature. For case (b), the actual temperature is less than the local liquidus temperature $T_L(z)$ (labeled $T_{constitutional}$) for a range of values of z and the liquid is said to be constitutionally supercooled; as a consequence the L–S interface is unstable. Case (a), where the actual temperature exceeds $T_L(z)$, corresponds to a stable L–S interface. Case (c) is the critical condition. It can easily be demonstrated that the interface will be stable for

$$\frac{G_L}{V} \geq \frac{m_L C_0}{D_L} \frac{k_E - 1}{k_E}. \tag{142}$$

Even though the stability criterion derived from this simple method is very similar to that derived from the more complex treatment, it does not yield any information about the size scale of the instability.

7.6.4.3 Experiments

Low V—From eqn (142), the decrease of the parameter G_L/VC_0 controls the evolution of the L–S interface from plane to a corrugated form called cells and eventually to dendrites in some cases.

Experimental techniques used for metals can be based on the fact that the interface instabilities produce a redistribution of solute that can reveal the origin and development of the instabilities in metallographic sections. For example, Biloni et al. (1965a) used anodic oxidation to produce interference colors to detect the presence or absence of cells in extremely dilute alloys such as 99.993 wt% and 99.9993 wt% purity Al. Quenching can also be used to delineate the shape of the liquid–solid interface in alloys and transparent organic metal analogs have also been successfully employed for *in situ* observation of instability processes.

Many authors have determined experimentally that the CS criterion corresponds reasonably well to the transition from plane to unstable interfaces (Chalmers, 1964; Flemings, 1974). Biloni et al. (1966), through critical experiments with Sn–Pb ($k^E < 1$) and Sn–Sb ($k^E > 1$) alloys, were among the first to establish that depressions at the L–S interface rather than projections are the first sign of interface instability. Current knowledge of the origin of the instability and its evolution can be summarized as follows:

(i) The first sign of instability is segregation associated with depressions at the interface: grain boundaries, striation boundaries and isolated depressions or *nodes*. These nodes occur in an ordered arrangement in tetragonal Sn base alloys (Biloni et al., 1966) and FCC Pb–Sb alloys (Morris and Winegard, 1969). However, in Zn–Cd hexagonal close-packed alloys, the first array of nodes is disordered (Audero and Biloni, 1973). Alloy crystallography as well as crystal orientation have a large influence on the morphology of the interface formed after the breakdown of the planar interface.

(ii) The grooves associated with grain boundaries and striation boundaries act as built-in distortions of the plane front, and interface breakdowns begin here, spreading outward to other portions of the crystal (Schaefer and Clicksman, 1970). The same effect occurs adjacent to the container surface (Sato and Ohira, 1977).

(iii) The evolution from nodes or depressions at the interface into a regular or hexagonal substructure is obtained as the CS increases. This occurs by the formation of interface depressions that connect the nodes to initially form elongated cells and finally a hexagonal arrangement. This process depends on the alloy crystallography (Morris and Winegard, 1969; Biloni et al., 1965b; Biloni et al., 1967; Audero and Biloni, 1973). **Figure 34** corresponds to the evolution from a planar interface to a cellular interface shown by the shape of the decanted interface and by metallographic sectioning slightly behind the interface.

High V—Rapid solidification can produce microstructures free of cellular and dendritic segregation. Absolute stability predicts that by increasing the interface velocity, a transition to plane front growth is possible and thus microsegregation-free solidification. Narayan (1982) summarized the morphological stability results obtained with pulsed laser melting and resolidification in Si alloys. The transition velocities observed were in the m/s range and good agreement with the predictions of the absolute stability velocity if the velocity-dependent value of the partition coefficient was used (see Section 7.6.4.4). Boettinger et al. (1984a) showed that in Ag-1 and -5 at% Cu alloys, a transition from cellular structures to plane front growth could be obtained by increasing the growth velocity beyond 0.3 and 0.6 m/s, respectively. These velocities are a factor of two larger than the calculated absolute stability values. However in view of the uncertain materials parameters, the agreement is reasonable. Hoaglund et al. (1991) performed a detailed study using pulsed laser melting and resolidification of Si–Sn alloys. **Figure 35** shows the values of $k(V)$ determined and the fit to the morphological stability theory using these measured values of $k(V)$. Ludwig and Kurz

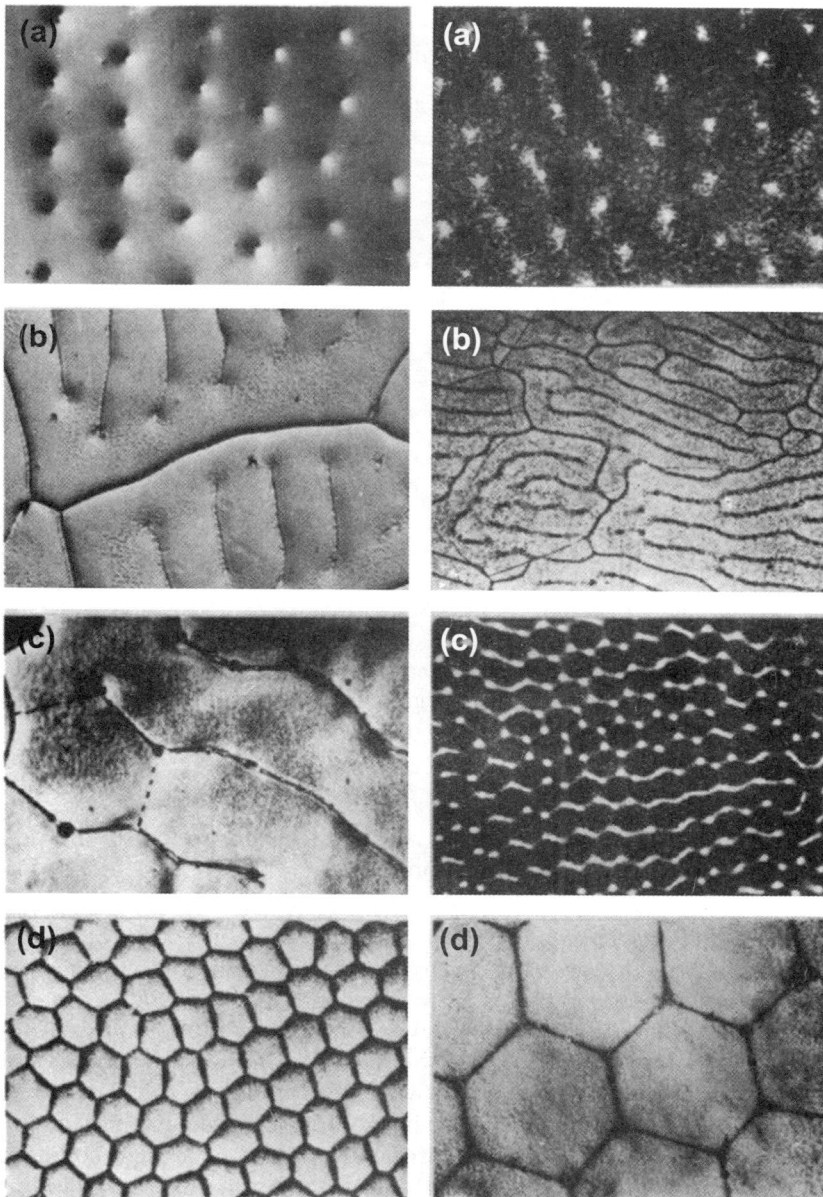

Figure 34 Evolution of the segregation substructure as a function of CS shown by decanting the interface (left) and slightly behind it by metallographic sectioning (right). The growth direction is normal to the page. The amount of CS increases from (a) to (d). Magnifications: (a) left: ×100, right: ×100; (b) left: ×100, right: ×50; (c) left: ×150, right: ×50; (d) left: ×50, right: ×150 (Biloni, 1970).

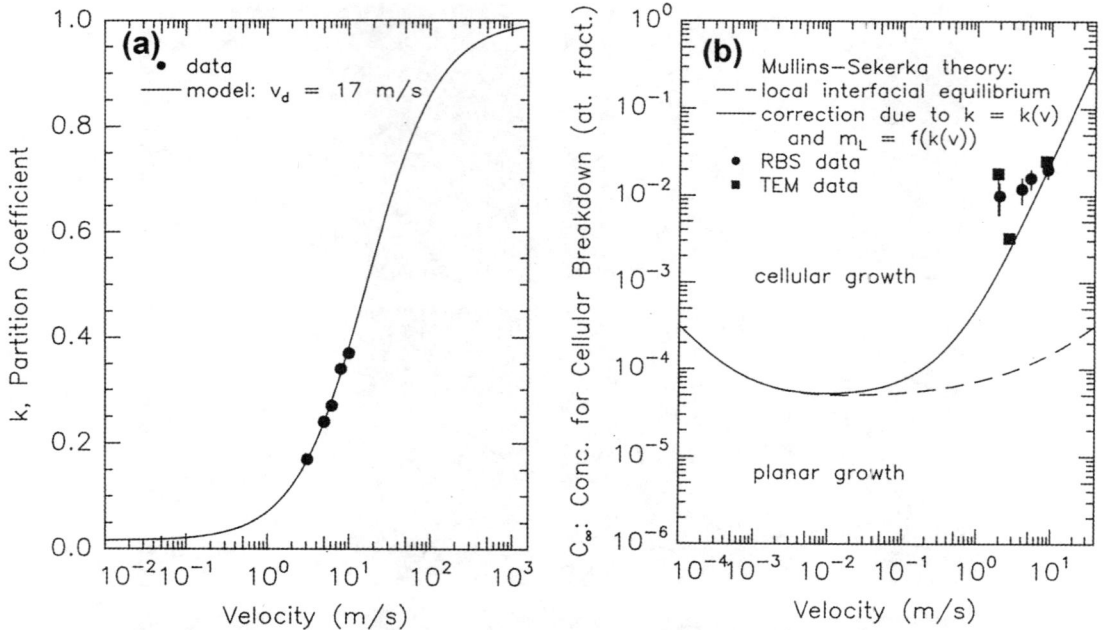

Figure 35   (a) Velocity dependence of measured partition coefficient of Sn in (100) Si. $k_E = 0.016$. Data are fit with $V_D = 17$ m/s. (b) Critical concentration above which cellular breakdown occurs: dashed curve stability theory using k_E; solid line, stability theory using velocity dependent $k(V)$ and $m_L(V)$ (Hoaglund et al., 1991).

(1996) directly observed the stabilization of a planar interface as the solidification velocity was increased in a transparent succinonitrile–argon alloy.

7.6.4.4 Further Theoretical Developments

The morphological stability theory was extended to include nonequilibrium solute trapping by Coriell and Sekerka (1983). Complex temporally oscillatory instabilities are found. A first-order approximation of the results suggests that the absolute stability condition is modified by substituting the $k(V)$ expression from eqn (109) into eqn (139) and solving implicitly for the critical velocity. This effect reduces the absolute stability velocity compared with what would be calculated on the basis of an equilibrium partition coefficient. The oscillatory instabilities exposed in this analysis and in more detailed studies are related to the banded microstructures observed at high solidification velocities (Boettinger et al., 1984a) (see Section 7.8.1.7). Other modifications to the linear stability theory include extension to multicomponent alloys (Coriell et al., 1987), anisotropic thermal conductivities (Coriell et al., 1990), and modification of the latent heat because of heat of mixing effects (Nandapurkar and Poirier, 1988).

Methods for dealing with finite amplitude perturbations and the transition to cellular structures near the lower critical velocity corresponding to the CS condition have been summarized by Coriell et al. (1985). Two types of behavior are found depending primarily on the value of the partition coefficient (Wollkind and Segal, 1970; Caroli et al., 1982). If k^E is near unity, the transition (bifurcation) is supercritical; that is, a small increase in the velocity past the critical value leads to interface shapes that are only slightly deformed. The wavelength of the instability is close to the fastest growing wavelength of the linear theory. For small values of k^E, the bifurcation is subcritical; that is, a small increase above the critical value of the linear theory leads to interfaces deformed by a large amount. The wavelength is two to four times larger than the fastest growing wavelength predicted by linear theory (Somboonsuk and Trivedi, 1985). The available data on the planar to cellular transition have been reviewed by Cheveigne et al. (1988). Agreement with the above bifurcation concepts was demonstrated. However it has been argued by Lee and Brown (1993) that the range of velocity where the subcriticallity exists may be too small for experimental observation. They attribute the observation of spacings much smaller than the fastest growing wavelength to the fact that, at velocities only a few percent above the critical velocity, wavelengths up to factors of four smaller than the critical are also unstable. The instability of these smaller wavelengths is evident in the flatness of the bottom of the closed curve in **Figure 33**c. Indeed the behavior of the spacing and amplitude of cells that form very close to the critical velocity is very complex (Eshelman et al., 1988). However, few practical situations involve velocities that are so carefully controlled.

An important theoretical contribution by Warren and Langer (1990, 1993) analyzed the development of the interface shape instability during the initial transient of solidification (see initial portion of **Figure 31**c) rather than its development from the steady-state situation where the solute profile is described by eqn (129). The analysis predicts a spacing for the growing perturbations that is larger than the fastest growing wavelength of the steady-state analysis in agreement with the observations of Somboonsuk and Trivedi (1985). Further experimental confirmation of the approach was obtained by Losert et al. (1998a,b). A more detailed discussion of cell and dendrite shapes is deferred to Section 7.7.

7.6.5 Coupled Interface and Fluid Flow Instabilities

Fluid flow is often present during solidification and upsets the diffusion of solute and heat assumed above. Thus fluid flow can alter the morphological stability of planar interfaces. Flow in the dendritic mushy zone will be deferred to Section 7.9.

Hurle (1969) and Coriell et al. (1976) first treated the impact of fluid flow on interface stability using the simple boundary layer approach described in Section 7.6.2.4. The thickness of the diffusion boundary layer, δ_F, in front of the planar interface was reduced because of flow of an unspecified origin. A reduction in the boundary layer thickness increases the value G_c and alters the stability criterion simply through eqn (137) if δ_F remains larger than the interface perturbation wavelength. Favier and Rouzaud (1983) improved this approach with a deformable boundary layer.

Complete coupling of the fluid flow and diffusion phenomena near an L–S interface lead to many complex results that influence morphological and fluid stability. These have been summarized by Coriell and McFadden (1990). When DS is performed with the liquid above the solid, rejection of a less dense solute can cause a liquid density that increases with height near the interface if the temperature gradient is not sufficiently large. This may cause fluid flow to occur. However even if the overall density decreases with height, convection can still occur when a light solute is rejected. This is termed *double-diffusive convection* because of the fact that heat and solute diffuse with vastly different rates during the flow. A physical argument for the instability can be obtained by considering the forces that act on a small packet of liquid that is given a displacement upward away from the interface into a region of hotter fluid that contains less solute. Because heat transfer is more rapid than solute transfer, the packet will become hotter but remain approximately of the same composition. The displaced packet then finds itself surrounded by liquid of the same temperature but with less solute. The packet is thus less dense than its surroundings and continues to rise. Generally the wavelength of an unstable interface that forms because of the fluid instability is much larger than those of the ordinary morphological instability. However under some conditions, the wavelengths are comparable and complex time-dependent oscillations can occur. Fully developed flow caused by density gradients during dendritic growth is described in Section 7.9.

Another case of instability can occur even when the rejected solute is denser than the solvent. **Figure 36** shows experiments by Burden et al. (1973) in which a macroscopic deformation of the L–S interface was observed during growth at very slow rates (1 μm s^{-1}). Although these observations were made during cellular and dendritic growth, an analysis by Coriell and McFadden (1989) for a noncellular interface seems to apply in this case. Instabilities occurred for wavelengths of the order of millimeters in this case and were determined to be due primarily to the difference in thermal conductivity of the liquid and solid phases. A physical explanation involves the creation of a slight radial temperature gradient when a long wavelength perturbation is present. This induces a flow that ultimately leads to the denser solute accumulating in the depressions in the interface. In the experiments of Burden et al. (1973), these depressions occurred at the walls of the container and the wavelength was approximately twice the container diameter.

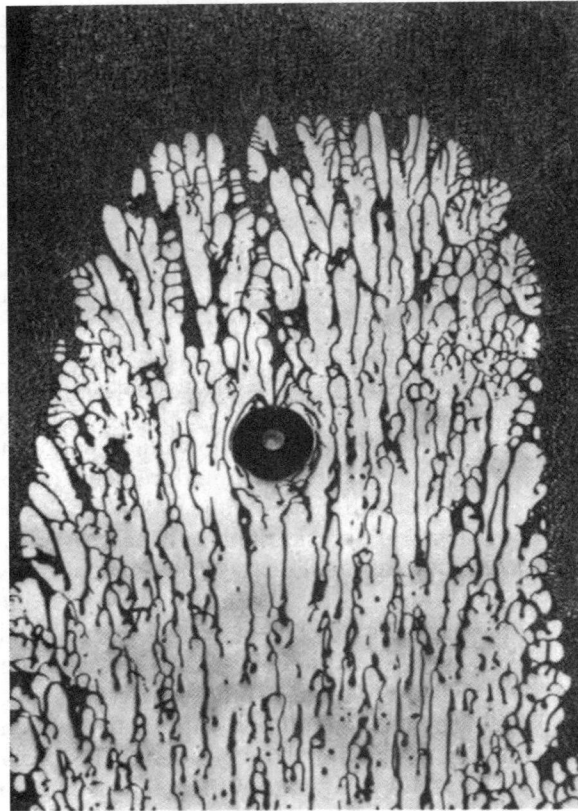

Figure 36 Macroscopic interface shape across the full diameter of 4 mm for an Al-10 wt% Cu alloy grown at $V = 2.8 \times 10^{-4}$ cm/s and $G_L = 60$ K/cm because of solute convection (Burden et al., 1973).

Murray et al. (1991) studied the effect of time variations of the gravitational force during solutal convection. In the context of experiments in microgravity, this research permits an assessment of the effects of so-called *g-jitter* that occurs during space flight. In space, surface tension-driven flows can also become important.

7.7 Cellular and Dendritic Solidification

The instability of the planar shape of the liquid–solid interface described in Section 7.6.4 leads to solidification by a cellular mechanism and, at conditions further from stability, by a dendritic mechanism. After the passage of the solidification front, a variation of composition remains in the solid on a length scale characteristic of the cellular or dendritic growth that is called microsegregation. This microsegregation pattern typically remains frozen in the solid because of the small ratio of the solute diffusion coefficients in the solid and liquid ($\sim 10^{-4}$ for substitutional solid solutions). In many cases the composition variation is so severe that a second solid phase solidifies in the intercellular or interdendritic regions even though none would be predicted based on a consideration of global equilibrium. The focus of this section is the prediction of the spacings associated with cellular and dendritic growth and the degree of microsegregation produced by that growth. These spacings are important in the selection of heat treatment times and temperatures for the homogenization of ingots as well as the properties of as-cast materials. The prediction of the microsegregation pattern is a fundamental goal of solidification modeling. Control of practical casting defects such as macrosegregation, porosity and hot tearing must start from an understanding of cellular and dendritic growth.

We first consider the growth of an isolated *alloy* dendrite growing into a melt that is bulk supercooled below the liquidus temperature. The relations between dendrite tip velocity, tip radius, tip composition and bulk supercooling will be established. Application of these relations is also made to constrained dendritic growth where no bulk supercooling exists but the velocity is specified. The growth of arrays of cells and dendrites is then considered to develop an understanding of the factors that control the primary and secondary spacings of fully developed cells and dendrites. We then describe microsegregation, the application of the Scheil equation and corrections made to include incomplete liquid mixing, dendrite tip kinetics, and solid diffusion. Finally we described solidification path analysis and resultant microsegregation for multicomponent alloys.

7.7.1 Alloy Dendritic Growth

7.7.1.1 *Theory of the Tip Region*

Only the tip region of a growing dendrite is modeled with the following approach. Two cases are distinguished: free growth and constrained growth. The model for free dendritic growth treats the situation where an isolated nucleus initiates growth into a uniformly supercooled melt. This situation has application in the growth of an equiaxed grains in castings. The melt temperature far from the dendrite can be taken as the nucleation temperature, T_N, the latent heat flows into the supercooled liquid and the temperature gradient at the tip is negative. The goal is to predict the dendrite tip velocity, V, tip radius, r, tip solid composition, C_S^*, and tip temperature, T^*, as functions of bulk liquid composition C_0 and the bulk supercooling, ΔT, below the liquidus temperature, given by

$$\Delta T = T_m + m_L^E C_0 - T_N. \tag{143}$$

The heat flow away from the tip into the liquid must be computed as outlined below.

The model for constrained growth most accurately treats the situation that occurs during DS, where the heat flow and solidification velocity are controlled by a moving furnace. Constrained growth approximates the situation of dendritic growth of columnar grains in a casting where melt superheat and heat of fusion flow to a cold mold wall or already frozen alloy behind the growing tips. The solidification velocity is approximately "imposed" by the isotherm velocity computed from a macroscopic heat flow considerations and the temperature gradient is positive. The goal of models is to predict the dendrite tip radius, tip solid composition and tip temperature for a given liquid alloy composition as functions of velocity and temperature gradient.

For both free and constrained growth, the diffusion of solute into the liquid ahead of the growing dendrite is treated by assuming that the dendrite tip region has the shape of a paraboloid of revolution. By solving the diffusion equation in the liquid and neglecting solid diffusion, Ivantsov (1947) has shown that the liquid composition at the dendrite tip C_L^* is given by

$$C_L^* = \frac{C_0}{1 - [1 - k(V)]Iv(P_c)}, \tag{144}$$

where $k(V)$ is the interface partition coefficient described in Section 7.5.2.2 and $P_c = Vr/2D_L$ is the solute Peclet number. The Ivantsov function, $Iv(P)$, is equal to $Pexp(P)E_1(P)$ where $E_1(P)$ is the first exponential integral of P, a function that is easily available from mathematics tables or software subroutine libraries. (The case of a 2-D (plate) dendrite has also been treated and involves the complementary error function.) The composition gradient in the liquid at the tip, G_c^*, is given from the conservation of solute flux condition at the tip as

$$G_C^* = -\frac{V}{D_L}[1 - k(V)]C_L^*. \tag{145}$$

The dendrite tip temperature, T^*, must be given by the local interface condition equation evaluated at the velocity and liquid composition present at the tip, namely,

$$T^* = T_m + m_L(V)C_L^* - \frac{2T_m\Gamma}{r} - \frac{RT_m^2}{L_m}\frac{V}{V_C}. \tag{146}$$

This expression is the same as eqn (107) except for the inclusion of the Gibbs–Thompson supercooling $2T_m\Gamma/r$ for a tip radius of r. It also includes interfacial nonequilibrium solute trapping effects that are important at high velocity as described in Section 7.5. Letting $m_L(V) = m_L^E$, $k(V) = k^E$ and $V_C = \infty$ recovers the usual local equilibrium condition valid for slow solidification.

The treatment of the temperature field is handled differently for free and constrained growth. For free dendritic growth, heat flow in the solid can be neglected, and an Ivantsov solution for the temperature field can be obtained in the liquid. The temperature T^* at the dendrite tip given by eqn (146) is thus higher than the bulk liquid temperature far from the tip and must be consistent with

$$T^* = T_N + (L_V/C^V)Iv(P_t), \tag{147}$$

where $P_t = Vr/2\alpha_L$ is the thermal Peclet number. The temperature gradient in the liqlpha;uid at the tip, G_L^*, is given from the conservation of heat flux at the tip as

$$G_L^* = -(2/r)(L_V/C^V)P_t = -\left(\frac{L_V}{C^V}\right)\left(\frac{V}{\alpha_L}\right). \tag{148}$$

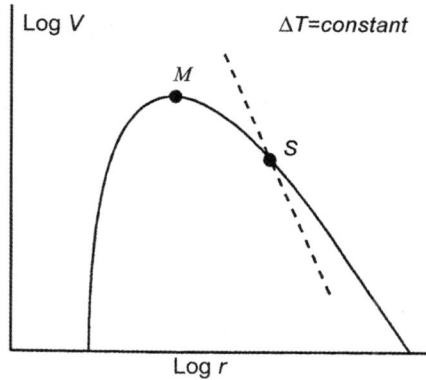

Figure 37 Schematic relationship between growth velocity and tip radius for dendritic growth for a fixed bulk supercooling. Point *M* corresponds to the maximum growth rate hypothesis and point *S* is the operating point of the dendrite corresponding to the marginal stability hypothesis, eqn (150), shown dashed.

For constrained growth, the temperature field and temperature gradient are assumed to be imposed by the macroscopic heat flow (e.g. imposed by a DS furnace).

For free growth, a combination of eqns (143), (144), (146), and (147) yields

$$\Delta T = \frac{L_V}{C^V} Iv(P_t) + m_L^E C_0 \left[1 - \frac{m_L(V)/m_L^E}{1 - [1 - k(V)]Iv(P_c)}\right] + \frac{2 T_m \Gamma}{r} + \frac{RT_m^2}{L_m} \frac{V}{V_C}. \tag{149}$$

This equation connects the values of dendrite growth rate and tip radius that are possible for each value of bulk supercooling ΔT. A plot of the relationship is shown in **Figure 37**. The curve shows that the growth velocity would be low for small and large values of radius and exhibits a maximum velocity at intermediate values of the radius. At small values of r, the creation of a dendrite tip with high curvature (high surface area) retards growth. At large values of radius, the buildup of heat and solute at the blunt tip is so severe that growth is retarded. Thus intermediate values of radius permit higher growth rates. Experiment has shown that a unique value of velocity and tip radius occur for each value of supercooling. Thus an additional condition is necessary to uniquely specify V and r for each value of bulk supercooling.

For constrained growth, prescribed values of G_L and V also fail to specify the tip temperature and radius of the dendrite tip as seen in eqn (146).

7.7.1.2 The Radius Versus Velocity Selection Criterion

The additional condition required to select the operating point of the dendrite has been the subject of much research. The above consideration only specifies the combined effect of the product of velocity and tip radius. For many years a maximum growth rate hypothesis was employed corresponding to point *M* in **Figure 37**. Careful experiments for a pure material by Glicksman et al. (1976) showed that this hypothesis gave dendrite tip radius values that were too small. Ideas concerning the stability of the tip were first considered by Oldfield (1973). The marginal stability condition of Langer and Mueller-Krumbhaar (1978) is now commonly used and has been the subject of some experimental validation. It equates the operating tip radius with the minimum unstable wavelength prediction from linear stability theory for a planar interface (Section 7.6.4) growing at the same velocity with temperature and composition gradients that are present at the dendrite tip. (These values of the gradients will not correspond to those present for planar growth of an alloy of composition C_0 growing at the same velocity.) The detailed predictions of the marginal stability hypothesis will be explored in Section 7.7.1.3.

Despite the success of the marginal stability argument, the approach has been met with concern among theoreticians particularly because the effect of anisotropic interface energy is not included. Clearly anisotropy is important in dendritic growth. Dendrites in cubic materials grow in [100] directions. Blodgett et al. (1974) measured the anisotropy of the liquid–solid surface energy of suc-cinonitrile. They determined that the surface energy had maxima in the [100] directions. To determine the Gibbs–Thomson coefficient, $T_m\Gamma$, for an anisotropic material with interface energy γ_{SL} that depends on two orientation angles θ_1 and θ_2 in the principle curvature directions, the surface energy γ_{SL} is replaced by the interface stiffness, $\gamma_{SL} + \dfrac{\partial^2 \gamma_{SL}}{\partial \theta_1^2} + \dfrac{\partial^2 \gamma_{SL}}{\partial \theta_2^2}$. This quantity is smallest when γ_{SL} is largest, that is, in the [100] for cubic systems.

Further analysis by many researchers has been summarized by Billia and Trivedi (1993). Lacombe et al. (1995) have performed careful measurements of the fluted or "Phillips screwdriver" shaped dendrite tip region for pure succinonitrile. McFadden et al. (2000) have matched this shape by employing an assumption that the cross-section of the dendrite stalk can be approximated by the equilibrium shape for the cubic fourfold equilibrium crystal shape. They find systematic deviations from the $T^* = T_N + (L_V/C^V)I_v(P_t)$, solution; namely, that the tip temperature is lower for a fixed value of P_t.

An approach to the radius selection problem that fully includes the anisotropy of the interface energy is called *microscopic solvability*. Microscopic solvability indicates that by including the effect of aniso-tropic surface energy, only one shape preserving solution is found to the diffusion equation, thus specifying the velocity versus tip radius condition uniquely. Microscopic solvability predicts that the tip radius will follow essentially the same relationship given by marginal stability analysis except that the value of σ^* (see Section 7.7.1.3) depends on the degree of anisotropy in the surface energy.

Karma and Rappel (1997) have found solvability predictions to generally agree with the more accurate phase field results and to experiments. The basics of the phase field method were described in Section 7.3.4. The method succeeds in obtaining realistic dendritic growth shapes for pure materials (Calginalp, 1986; Kobayashi, 1991, 1992, 1993) and for alloys Warren and Boettinger (1995) with no extra condition required to specify the operating state of the dendrite tip. The velocity and tip radius are naturally selected by the solution to the differential equations. A full summary of phase field modeling of dendritic growth is omitted here for brevity. Some results for alloys will be given in Section 7.7.3.1. A review of phase field research is given by Boettinger et al. (2002).

7.7.1.3 *Details of the Marginal Stability Criterion*
Following only the marginal stability approach, an equation for the tip radius, r, valid for small and large Peclet numbers (Boettinger and Coriell, 1986; Lipton et al., 1987; Boettinger et al., 1988b) is given by

$$r^2 = \frac{T_m\Gamma/\sigma^*}{m_L(V)G_C^*\xi_c - G\xi_t}, \tag{150}$$

where

$$\xi_t = 1 - \frac{1}{\left(1 + (\sigma^* P_t^2)^{-1}\right)^{1/2}}, \tag{151}$$

and

$$\xi_c = 1 + \frac{2k(V)}{1 - 2k(V) - \left(1 + \left(\sigma^* P_c^2\right)^{-1}\right)^{\frac{1}{2}}}. \tag{152}$$

Here the parameter σ^* is taken as $1/4\pi^2$. The parameters ξ_t and ξ_c can be set equal to unity if P_t and $P_c \ll 1$, but may deviate from unity as the velocity approaches either the CS or absolute stability condition. The parameter G is the conductivity-weighted temperature gradient given by

$$G = \frac{k_S G_S + k_L G_L}{k_S + k_L}. \tag{153}$$

If the conductivities (liquid and solid) are equal, G is simply the mean of the liquid and solid temperature gradients. For free dendritic growth, $G_S = 0$, and

$$G = G_L^*/2. \tag{154}$$

For growth into the supercooled melt G is determined from eqn (148). For constrained growth, G is determined by the imposed temperature gradient of the furnace. Either way the operating condition given by the marginal stability hypothesis gives a significantly larger tip radius in comparison to that determined by the maximum growth rate hypothesis as seen in **Figure 37**.

Using the marginal stability condition, eqn (150), a complete specification of the dendrite tip can be obtained for free growth or for constrained growth. In the former case, eqns (144), (145) and (148) are used in eqn (150) and solved simultaneously with eqn (149) to numerically determine the tip radius and velocity for a given value of ΔT. We refer to the results of this theory as *IV/MS* after the Ivantsov solution using the marginal stability criterion for tip radius selection. A simplified version of this theory for free growth that assumes local interface equilibrium is given by Lipton et al. (1984).

An example of the results of the full theory for free growth into a supercooled melt including the high Peclet number corrections (eqns (151) and (152)) and the nonequilibrium interface conditions for temperature and compositions is shown in **Figure 38** for an Ag-15 wt% Cu alloy (Boettinger et al., 1988b). **Figure 38a** shows that the dendrite growth velocity increases with supercooling. Note the more rapid increase in velocity near $\Delta T = 100$ K. **Figure 38b** shows how the nonequilibrium interface partition coefficient at the tip increases very sharply toward unity because of solute trapping at this same level of supercooling. This indicates the transition to partitionless solidification for the tip. In fact, for supercoolings greater than 200 K the dendritic growth rate is the same as that for a pure metal with a melting point equal to the T_0 temperature for an Ag-15 wt% Cu alloy. **Figure 38c** shows how the composition of the solid at the dendrite tip depends on supercooling. The composition rises from 5.5 wt% Cu $(=k^E C_0)$ for low supercooling to 15 wt% Cu for $\Delta T = 100$ K. This plot shows how microsegregation in dendritic structures can be reduced by increased supercooling. Finally, **Figure 38d** gives the dendrite tip radius. The general decline in radius with increased supercooling is sharply arrested at $\Delta T = 100$ K as the dendrite tip changes from solute-controlled length scales to the larger thermally controlled length scales corresponding to $k(V) \rightarrow 1$.

For constrained growth, prescribed values of G and V are used in eqns (144), (145) and (150) to implicitly give the radius. The radius and velocity are then used to determine C_L^* using eqn (144) and T^* using eqn (146). An example of the results of this theory for constrained growth during DS of Ag-Cu alloys assuming local equilibrium is shown in **Figure 39** from Kurz et al. (1986). The calculated tip radius and tip temperature as a function of imposed growth rate for a temperature gradient of 10^5 K/cm

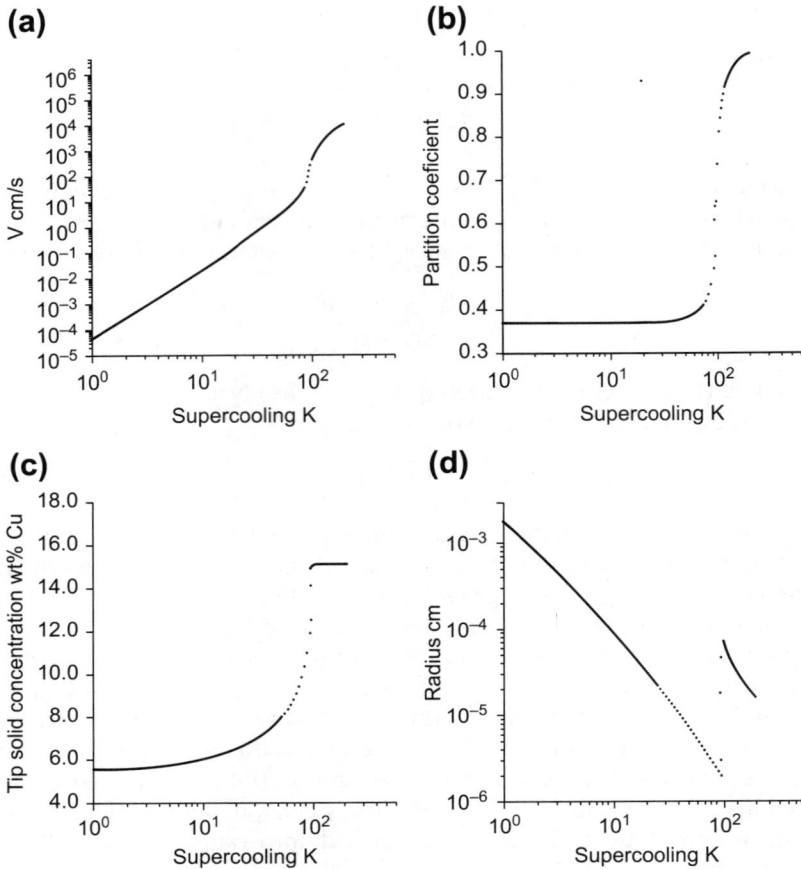

Figure 38   Plots of calculations for free dendritic growth of Ag-15 wt% Cu for various supercoolings below the liquidus, ΔT: (a) growth velocity, V; (b) nonequilibrium partition coefficient, $k(V)$; (c) solid composition at the dendrite tip, C_S^*; and (d) tip radius, r (Boettinger et al., 1988b).

are given. The general trend for the radius follows $Vr^2 = $ constant for each alloy. However near the CS and absolute stability velocities the tip radius approaches infinity (planar interface) and deviates from this simple law. The tip temperature is closest to the liquidus at intermediate velocities and approaches the solidus temperatures at low and high velocity. A very important result from this theory is that the solid composition at the center of a dendrite, $k^E C_L^*$, is only slightly greater than $k^E C_0$ at intermediate velocity but increases toward C_0 as the velocity decreases toward the CS limit or increases toward the absolute stability limit. This result impacts the discussion of microsegregation below.

7.7.1.4   Approximate Theory for Low Supercooling
For low supercooling, the model presented above can be simplified. For low values of P_c, local equilibrium can be assumed and $Iv(P_c)$ can be approximated by

$$Iv(P_c) \simeq P_c, \tag{155}$$

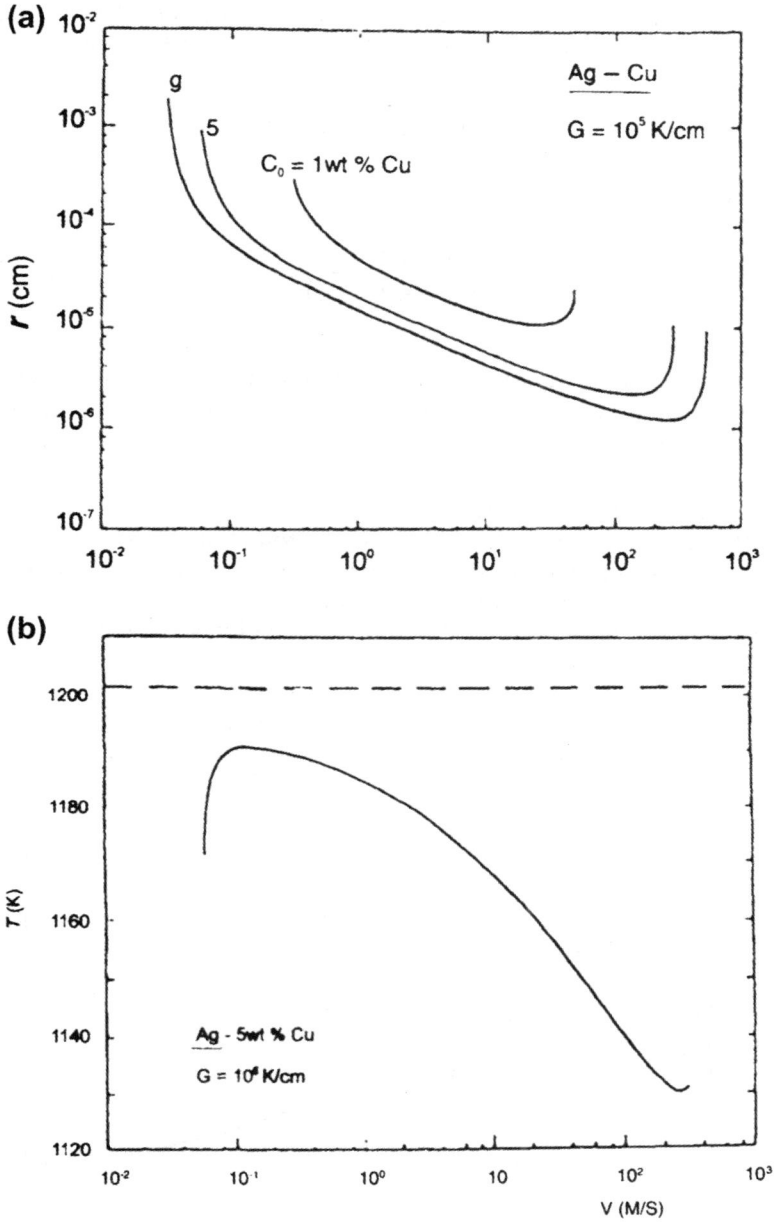

Figure 39 Calculated dendrite tip radius (a) and tip temperature (b) for Ag–Cu alloys of the noted compositions as a function of velocity for constrained growth at a temperature gradient of 10^5 K/cm. In (b) the liquidus temperature is shown dashed (Kurz et al., 1986).

an expression called the hemispherical approximation (Kurz and Fisher, 1981). This approximation leads to a particularly simple expression for the dendrite velocity as a function of supercooling

$$V = \frac{D_L}{\pi^2 m_L^E C_0 (k^E - 1) T_m \Gamma} \Delta T^2, \tag{156}$$

with a tip radius given by

$$r = \left(\frac{D_L T_m \Gamma / \sigma^*}{m_L^E C_0 V (k^E - 1)} \right)^{1/2}. \tag{157}$$

These relations can be applied to free growth or to constrained growth. However for the latter, the expressions are only valid when the temperature gradient is small. In this case the supercooling, ΔT, in eqn (156) is the difference between the liquidus temperature for the bulk alloy composition and the dendrite tip temperature.

7.7.1.5 *Experiments on Alloy Dendritic Growth*

Chopra et al. (1988) have compared the results of the Lipton et al. (1984) theory with experiments using succinonitrile–acetone transparent alloys. Their results are shown in **Figure 40**. One of the most interesting results of the dendrite growth theory and experiments is that the addition of a small amount of solute actually increases the growth rate of the dendrites above that for the pure material

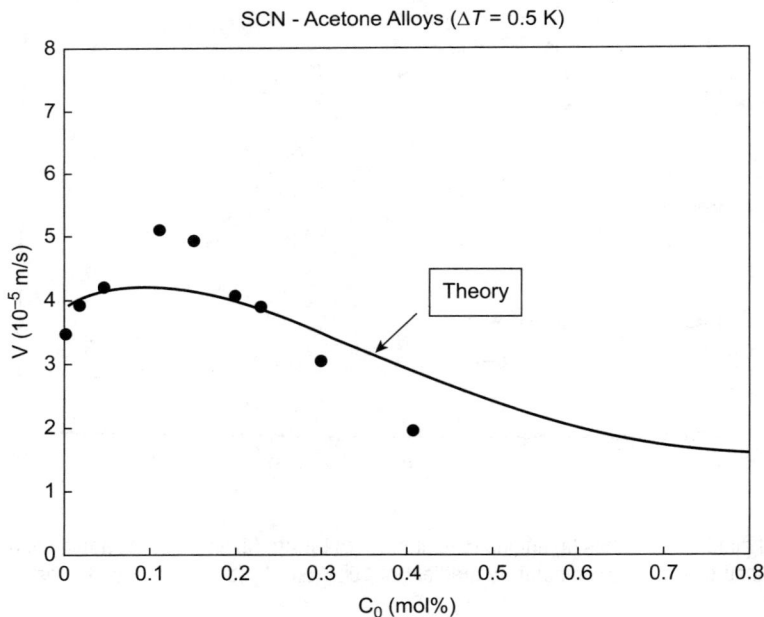

Figure 40 Comparison of theoretical and measured dendrite growth velocities as a function of alloy composition for a fixed bulk supercooling below the liquidus of 0.5 K for succinonitrile–acetone alloys (Chopra et al., 1988).

for fixed bath supercooling below the liquidus. This effect is because of the destabilizing influence (through G_c^* in eqn (150)) of the solute on the dendrite tip. The solute pushes the marginal wavelength and consequently the tip radius to a smaller value and permits the dendrite to grow at a higher speed compared with the pure material. With further increased levels of solute, the transport difficulties associated with the solute begin to dominate and the dendritic growth rate slows for a fixed supercooling below the liquidus.

For high supercoolings (100–300 K), several groups have obtained alloy dendritic growth rate data as a function of bulk supercooling for alloys using carefully designed experiments in levitated samples (Wu et al., 1987; Eckler et al., 1992). Good agreement between theory and experiment has been obtained. **Figure 41** shows a summary of experimental data and theoretical predictions of the model that incorporates nonequilibrium interface kinetics for pure Ni and for Ni–B alloys by Eckler et al. (1992). The abrupt increase in velocity at high supercoolings is associated with the transition from solute controlled dendritic growth to thermally controlled growth brought about by the solute trapping (velocity-dependent partition coefficient).

The predictions of the theory for constrained growth presented above have been compared by Kurz et al. (1988) to measurements of cell compositions for Ag-15 wt% Cu alloy samples prepared by moving electron beam melting performed by Boettinger et al. (1987). **Figure 42** shows bars corresponding to the observed range of composition across cells of the Ag-rich phase as a function of growth velocity as measured by energy dispersive x-ray analysis in scanning transmission electron microscope (TEM). The solid curve shows the results of the theory. The curve marked IV/MS is the same theory without the inclusion of the velocity dependence of the partition coefficient. The curve BH is the older dendritic growth theory of Burden and Hunt (1974a,b) which uses the maximum growth rate hypothesis rather than the marginal stability analysis. It can be seen that the formulation is closer to the observed experimental data.

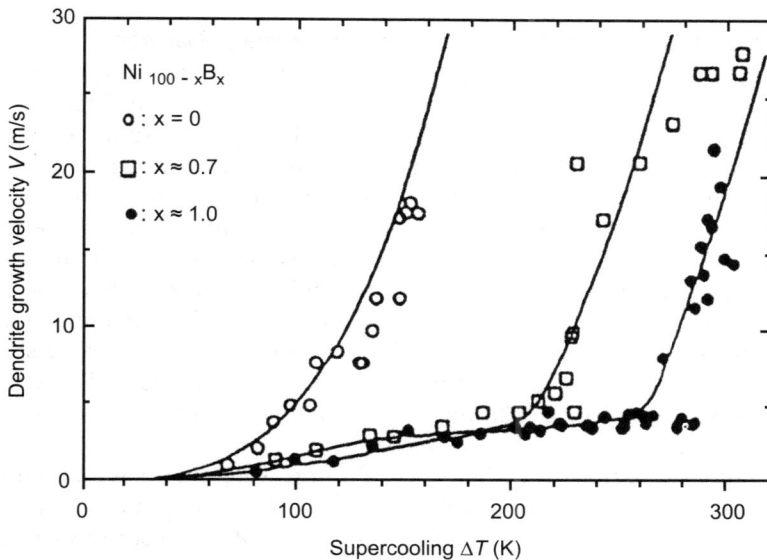

Figure 41 Comparison of theoretical and measured dendrite growth velocities as a function of bulk supercooling below the liquidus for pure Ni and two Ni–B alloys (Eckler et al., 1992).

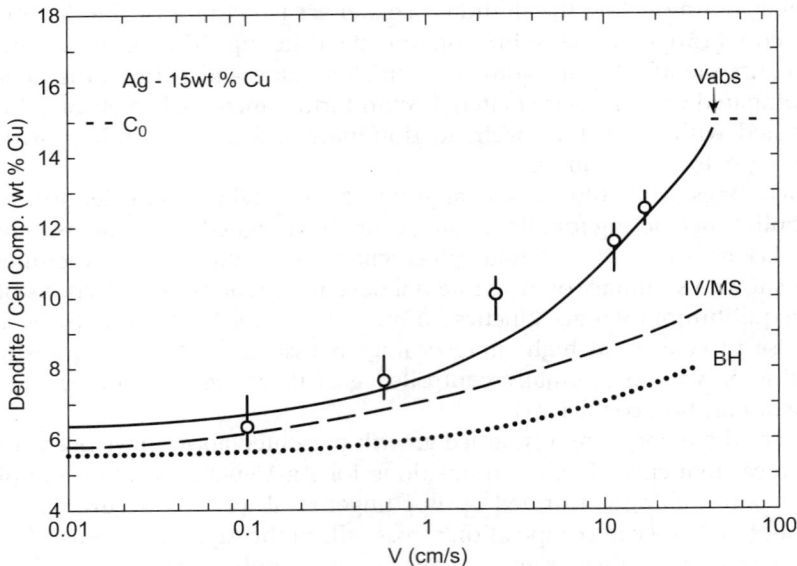

Figure 42 Measured average composition (circles) of cells of the Ag-rich phase in an Ag-15 wt% Cu alloy as a function of solidification velocity (Boettinger et al., 1987). The three curves are (i) the predictions of dendritic growth theory with velocity-dependent partition coefficient (solid curve), (ii) without the velocity-dependent partition coefficient, curve labeled IV/MS, (iii) the older Burden–Hunt model labeled BH (Kurz et al., 1988).

7.7.2 Cell and Dendrite Spacings

7.7.2.1 Numerical Calculations of Arrayed Cell and Dendrite Primary Spacings

During constrained growth, cells and dendrites form arrays with a characteristic spacing transverse to the growth direction called the primary spacing. To model the primary spacing, interactions with neighboring cells/dendrites must be considered. Hunt and Lu (1996) summarize their efforts using finite difference calculations with a sharp interface model to study cell/dendrite shapes and primary spacings during constrained growth. The diffusion equation for solute in the liquid is solved. Computations are performed for single cells and dendrites with radial symmetry in a cylindrical computational domain of various diameters. This geometry is used to approximate a hexagonal array of cells or dendrites with a primary spacing equal to the cylinder diameter. Two classes of solutions were obtained, sometimes for the same composition, growth velocity and temperature gradient: those with rounded nearly hemispherical interface shapes and those with parabolic interface shapes. The authors associate these two shapes with cells and dendrites, respectively. Side branching was not simulated in the dendrite calculations. Multicell calculations were also performed to examine the stability of arrays of cells and dendrites with various primary spacings.

For a very wide primary spacing, only dendritic solutions were obtained and a unique value of tip radius was determined for a given set of growth conditions. This computational regime is thought to approximate isolated dendrites. They confirmed the important role of anisotropy on the selection of the radius of the tip as developed separately by the microscopic solvability approach.

For smaller primary spacings, both cell and dendrite solutions could be obtained for the same growth conditions. For each, solutions could be obtained only over a range of primary spacings. The

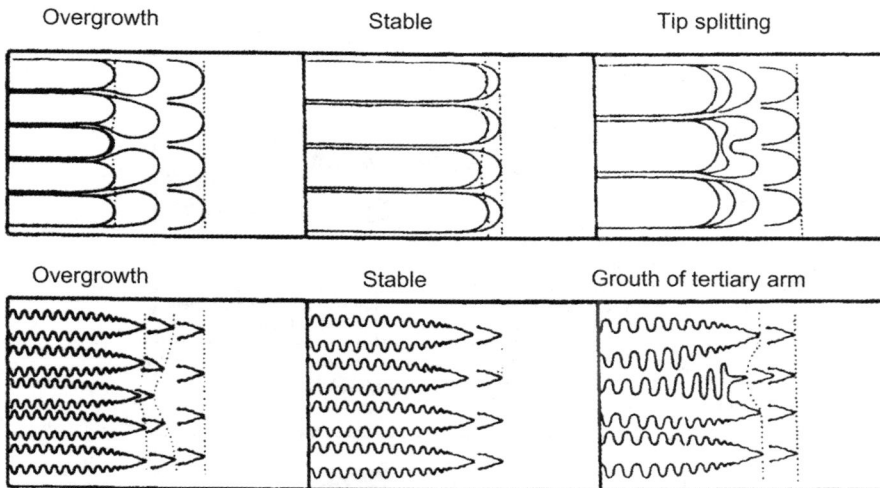

Figure 43 Schematic illustration of primary spacing adjustment mechanism for cells and dendrites (Hunt and Lu, 1996).

physical processes that define the adjustment mechanisms to the spacings are shown in **Figure 43**. The minimum stable cell spacing is determined by overgrowth of one cell by the adjacent cells. The maximum cell spacing occurs when the cell tips become very flat and on the verge of splitting. For dendrites, the minimum spacing is also determined by overgrowth. The maximum spacing is assumed to be twice the minimum spacing based on an argument regarding the transition of tertiary arms into new primary arms. An example of the range of primary spacing that could be calculated is shown in **Figure 44** along with experimental data. Also the existence of a range of spacings for cells for given growth conditions agrees with observations of Eshelman and Trivedi (1988).

Employing phase field calculations of 3-D steady-state growth shapes during DS, Gurevich et al. (2010) also confirmed the co-existence of cell and dendrite growth forms (fingers and needles in their terminology) for small and large spacings, respectively, as shown in **Figure 45**. They also found that the two types of shapes merged smoothly as the growth velocity increased. The existence of the two computed growth shapes for identical velocity and temperature gradient is confirmed.

The computations described above do not describe the conditions at the root of cells. The walls of adjacent cells never join and a thin layer of liquid persists far behind the cell tips. Ungar and Brown (1985) have performed FE calculations of deep cells using a composite coordinate technique that allows cell walls to fold over and join at the cell roots. In fact small drop-like structures are found to form at the cell roots that may be indicative of the pinch-off and periodic shedding of liquid droplets. Such shedding of liquid droplets may be the cause of spherical second phase dispersoid particles that occur in some rapidly solidified alloys (Boettinger et al., 1988c).

7.7.2.2 Cell to Dendrite Transition

As the solidification velocity is increased for a given alloy composition and temperature gradient, one observes a transition in structure: planar → cellular → dendritic → cellular → planar. The lower and upper transitions between planar and cellular structures are given by the modified CS and the absolute stability criteria, respectively (Section 7.6.4). The transitions between cellular structures and

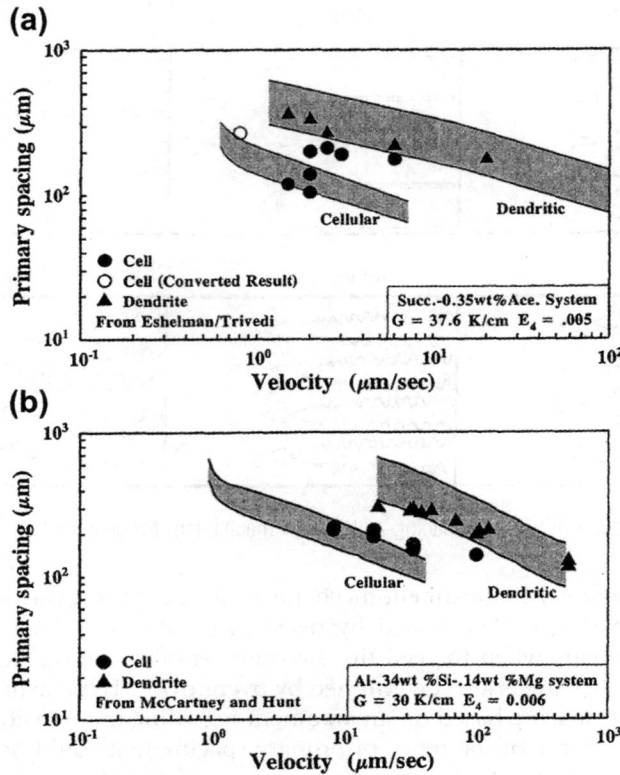

Figure 44 Comparison of the predicted ranges of cell/dendrite primary spacings with experimental results: (a) succinonitrile–0.35 wt% acetone; (b) Al-0.34 wt% Si-0.14 wt% Mg (Hunt and Lu, 1996).

dendritic structures at intermediate velocities are not well understood. Experimental data for the *low velocity transition* were obtained by Tiller and Rutter (1956) and Chalmers (1964) and were fit by the expression

$$G_L V^{1/2} = \text{constant} \cdot C_0 / k^E, \tag{158}$$

with a constant of proportionality that depends on crystallographic growth direction. Later Kurz and Fisher (1981) proposed the condition

$$G_L / V = \frac{m_L^E C_0 (k^E - 1)}{D_L}, \tag{159}$$

which is k^E times the critical value of G_L/V for the plane to cell transition given by eqn (142)).

Despite this early work, experiments by Somboonsuk et al. (1984) and Trivedi et al. (2003) clearly show that cells and dendrites can coexist at the same velocity and that the conditions required for the transition are more difficult to calculate (see Billia and Trivedi 1993). Local variations in primary spacing are seen to occur and when sufficient space exists adjacent to a cell, it will develop side

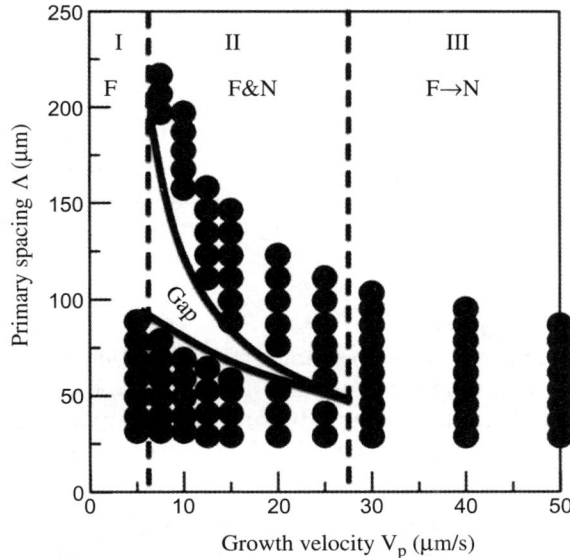

Figure 45   Range of spacings for the existence of finger and needle axisymmetric steady-state growth solutions as a function of growth velocity for $G = 38$ K/cm, $C_0 = 0.7$ wt%, and anisotropy $\varepsilon = 0.01$. Three different growth regimes can be distinguished (I) with only fingers for low velocity, (II) distinct branches of finger and needle solutions for intermediate velocities, and (III) merging of these branches into a single branch with a continuous shape evolution from finger to needle with increasing spacing (Gurevich et al., 2010).

branches. These observations confirm the findings of calculated cell and dendrite structures at the same velocity described in Section 7.7.2.1. Hunt postulated that the cell to dendrite transition might occur when the tip temperature for the cell becomes lower than that for the dendrite. This postulate also predicts an upper transition velocity from dendrites back to cells with increasing solidification velocity in the rapid solidification regime. This upper transition is observed experimentally but has received no theoretical treatment.

7.7.2.3   Analytical Expressions for Primary Spacing
Analytical theories of primary spacing often yield a power law expression over a range of solidification conditions that are not close to the CS or absolute stability velocities given by

$$\lambda_1 = A_1 G_L^{-m} V^{-n}. \tag{160}$$

The result of the model by Hunt (1979) coincides closely with the expression derived by Kurz and Fisher (1981) using quite different assumptions about the dendrite geometry. The values $m = 0.5$ and $n = 0.25$ were obtained by both with the constant, A_1, being proportional to the fourth root of the alloy composition for dilute alloys. Thus for a given V and G_L, the primary spacing is larger for concentrated alloys than for dilute alloys. Traditionally many authors have considered that experimental values of λ_1 are better correlated to the local cooling rate, given by the product $G_L V$, with $m = n \approx 0.5$ (Flemings, 1974; Okamoto et al., 1975; Okamoto and Kishitake, 1975; Young and Kirkwood, 1975). On the other hand, in steels a broad discrepancy exists for the values of m and n, although many of the experimental studies have not been performed under controlled solidification conditions. Jacobi and Schwerdtefeger (1976), controlling G_L and V separately, obtained values of $m = 0.25$ and $n = 0.72$. The complexity of modeling primary spacing is indicated by the fact that values of m and n fit to the numerical computations of Hunt and Lu (1996) depend on the anisotropy of the surface energy.

Equation (160) is often used in conjunction with microstructural measurements to estimate the freezing conditions for various rapid solidification processing techniques. In view of the complexity of modeling λ_1 and the disparity in experimental values even at slow cooling rates, such an approach should be used with great caution.

7.7.2.4 Secondary Dendrite Arm Spacing
During dendritic growth of cubic materials, the paraboloid-shape dendrite tip flutes laterally in the four {100} longitudinal planes containing the [100] growth direction. Each flute then develops perturbations down the length of the dendrite shaft that become secondary arms as shown in **Figure 46**. The initial spacing of the secondary arms near the tip has been observed in transparent pure materials and

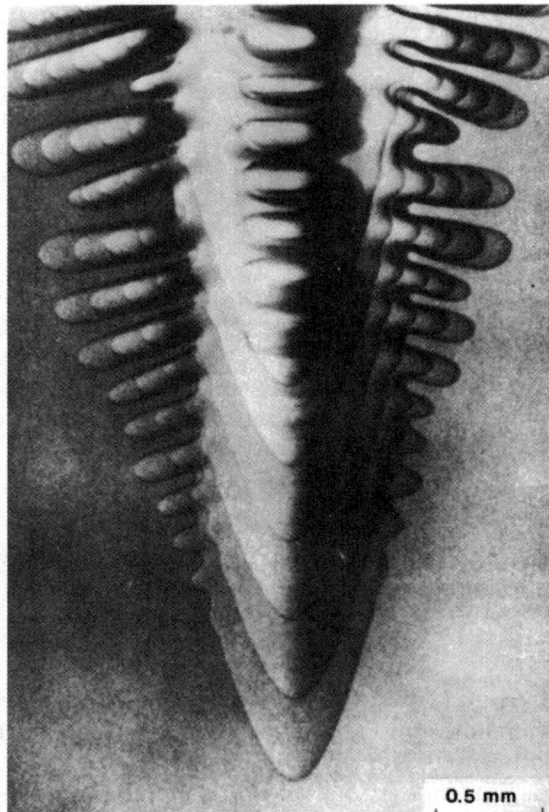

0.5 mm

Figure 46 Superposition of time-lapse photographs of the growth of a succinonitrile dendrite. Side branch evolution on the {100} branching sheets is evident (Huang and Glicksman, 1981).

alloys (Huang and Glicksman, 1981; Trivedi and Somboonsuk, 1984). For both constrained and free growth, the spacing near the tip is approximately 2.5 times the tip radius over a broad range of growth conditions. Theoretical work by Langer and Mueller-Krumbhaar (1981) has also confirmed this relationship, which is related to the question of the tip stability discussed previously.

Theoretical questions remain over the physical mechanism of the secondary arms. Two views have emerged, that of Glicksman et al. (2007), Glicksman (2012) and that of Echebarria et al. (2010). The former, considers only growth into an undercooled melt. They argue that there exists a natural limit process cycle whereby the tip velocity changes slightly with a periodicity that spews out the secondary arms. The latter authors find that this mechanism as well as amplification of naturally occurring thermal noise can cause secondary arms in specific cases.

For cast alloys, secondary arm spacing is practically important because it sets the length scale associated with the final microsegregation pattern that relates directly to mechanical properties. The final spacing is usually much coarser than the one formed near the tip and results from coarsening of secondary dendrite arms during residence in the mushy zone. The observed mechanism of coarsening is the melting or dissolution of smaller arms at the expense of larger arms. Through the Gibbs–Thompson effect, local differences in curvature give rise to slight temperature and/or composition variations along the liquid–solid interface. Diffusion of heat and/or solute in response to these differences cause the dissolution of small arms and the growth of others effectively increasing the average spacing of the secondary arms.

For all but the most dilute alloys, the rate of coarsening is determined by the diffusion of solute in the liquid (Voorhees, 1990) rather than by heat. Models were obtained for isothermal coarsening of the secondary arms held in the mushy zone by Kattamis et al. (1967) and Feurer and Wunderlin (1977). The latter obtained an expression for the secondary arm spacing, λ_2, given by

$$\lambda_2 = 5.5 \left(M^* t_f\right)^{1/3}, \tag{161}$$

where

$$M* = \frac{T_m \Gamma D_L \ln(C_b/C_0)}{m_L^E (k^E - 1)(C_b - C_0)}, \tag{162}$$

where C_b is the final composition of the liquid at the base of the dendrite. When solidification is completed by a eutectic reaction, C_b is equal to the eutectic composition. The numerical factor in eqn (161) depends on the details of the geometry of the coarsening process and the dendritic structure and should be viewed as approximate. For DS the local solidification time can be approximated by

$$t_f = \frac{\Delta T_f}{G_L V}, \tag{163}$$

where ΔT_f is the temperature difference between the dendrite tip and the base of the dendrite. Thus for a given growth velocity and temperature gradient λ_2 decreases with increasing composition. Equation (162) has been extended to multicomponent alloys by Rappaz and Boettinger (1999).

Kirkwood (1985) obtained the above result only when the cooling rate is small compared with $m_L C_0/t_f$. At larger cooling rates, the strict dependence $\lambda_2 \sim t_f^{1/3}$, was not obtained and he ascribed the scatter in λ_2 versus t_f data to this effect. In real solidification processes, coarsening takes place simultaneously with the subsequent increase in the fraction solid during cooling. Mortensen (1991) presents a model that includes this effect.

Recent experimental work has focused on a more complete measurement of the geometry of the dendritic structure (Fife and Voorhees, 2009; Genau et al., 2009; Aagesen et al., 2010). Mean and Gauss curvature measurements of the 3-D structures from point to point along the entire liquid–solid interface as well as surface area per unit volume replace the simple linear secondary arm spacing. These curvatures, which can be positive and/or negative, change with time during coarsening. These data have been compared with theory and models and a clearer picture of mushy zone coarsening is emerging. Advances in direct observation of coarsening in metallic alloys using synchrotron radiation are also shedding new information on secondary arm coarsening (Rivulcaba et al., 2009).

7.7.3   Microsegregation

Microsegregation is the pattern of composition variation that remains in a solidified alloy. It includes the composition variation across cells or dendrites as well as the formation of other phases in the intercellular or interdendritic regions. One of the major goals of microsegregation analysis is the prediction of the volume fraction of eutectic or other secondary phases that may form between the cells and/or dendrites of the primary phase.

7.7.3.1   *Phase Field Approach to Modeling Microsegregation during Dendritic Growth*

The microsegregation that occurs during dendritic solidification is extremely complex with details that can best be predicted using the phase field method. **Figure 47** from Warren and Boettinger (1995) shows snap shots of 2-D calculations of the dendrite tip region (a) and (b) and the interdendritic region (c) and (d) for two different values of the solid diffusion coefficient. The different colors indicate the composition. The rejection of solute into the liquid around the tip region is clearly seen. Less obvious are the small changes in the shades of blue of the liquid in the interdendritic liquid that governs secondary arm coarsening deep in the mushy zone. The complex microsegregation patterns in the solid are apparent with larger gradients evident for the case with the small solid diffusion coefficient.

Microsegregation can be quantified with modern energy dispersive X-ray measurement of composition. Pixel by pixel quantitative analysis is now possible. A histogram of the number of pixels with composition within predefined bins can be obtained. In a binary alloy where monotonic solid composition versus fraction solid behavior is expected, the histogram can be integrated (accumulated) to provide a composition versus (area) fraction measurement. An example of this procedure was performed by Boettinger and Warren (1996) on microsegregation patterns predicted by the phase field method for a rapidly growing dendrite growing into an initially undercooled melt and also subject to external heat extraction. **Figure 48** summarizes the result and makes comparison to the simple Scheil prediction. One notes in (c) that the composition at low fraction solid is well above the value predicted by the Scheil equation. This effect is exaggerated here because of the very rapid dendrite tip growth in the initially undercooled melt. One also notes the flatness of the composition versus fraction solid curve during the initial portion of fast growth. The modest effect of the variation of external heat extraction during recalescence and a variation of the ratio of liquid to solid diffusivity is also shown.

With this as a backdrop, we now proceed to describe much simpler methods to model microsegregation, The simplest approaches to describe microsegregation considers a volume in the solidifying material small enough to be isothermal at any instant but large enough to capture the cell or dendrite spacing. Such a volume is shown in **Figure 49**. We discuss the Scheil approach and then proceed to modify the approach.

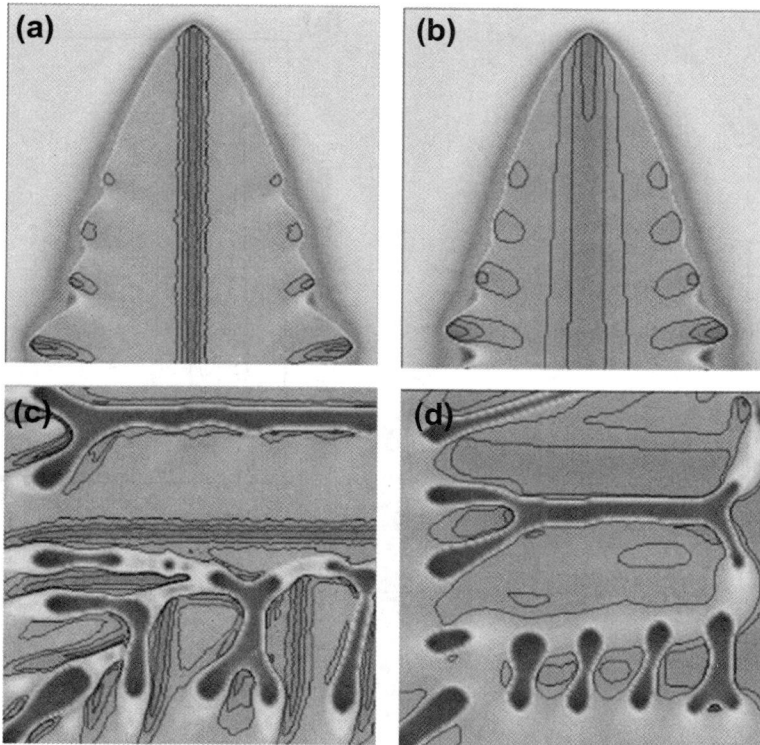

Figure 47 Phase field simulations of dendritic growth showing the variation of composition in the liquid and in the solid. Plots showing the effects of variations in the solid diffusivity: (a) and (c) show the solute profile for $D_S/D_L = 10^{-4}$, and (b) and (d) show the solute profile for $D_S/D_L = 10^{-1}$. (a) and (b) show the tip region and (c) and (d) show the interdendritic region. The bulk alloy concentration is 0.408 and three contours for 0.397, 0.399 and 0.4005 are shown in the solid as thin black lines (Warren and Boettinger, 1995).

7.7.3.2 Scheil Approximation

The simplest approximation for the prediction of microsegregation during dendritic solidification uses the Scheil equation described in Section 7.6.2.2 but applied to the small volume considered in **Figure 49**. With this approach no assumptions are actually necessary concerning the geometry of the freezing solid until one desires to convert from fraction solid to distance in a cast microstructure. For example, the dependence of volume fraction solid on distance, r_1 from the center of a cell or dendrite with radial symmetry with a primary spacing, λ_1, can be obtained with the expression,

$$f_S = 1 - \left(\frac{r_1}{\lambda_1}\right)^2. \tag{164}$$

Other geometries require other relationships between fraction solid and distance.

A comparison of the Scheil solidification of alloys with three different binary phase diagrams is shown in **Figure 50**. The volume under consideration (shown as a rectangular box) can be thought of as moving downward through the structure as the solidification occurs. Solidification will always start at

(a)

(b)

(c)

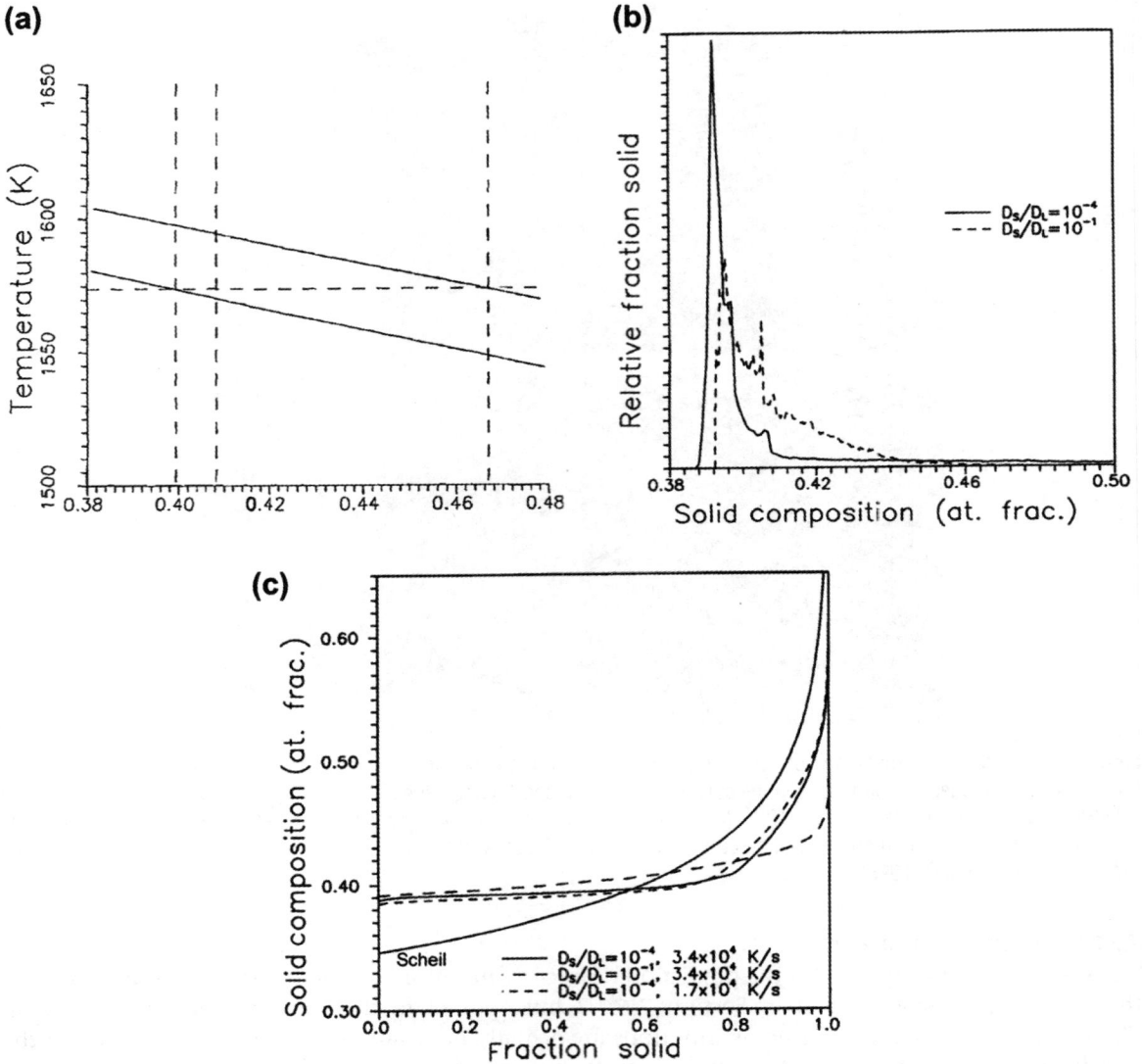

Figure 48 Details of the microsegregation computed for dendritic growth Boettinger and Warren (1996). (a) Portion of phase diagram showing the bulk alloy composition of 0.408 and the liquid and solid compositions at the nucleation temperature; (b) composition histogram; and (c) concentration versus solid fraction in the solidified material compared with a Scheil prediction.

the liquidus temperature under the assumptions of the Scheil equation. The first phase to form is the α phase in all three cases and is depicted as a large cell in the microstructure sketch even though it might be dendritic in reality. As solidification continues, the box moves downward and the fraction of solid in the box increases. The entire liquid portion of the box becomes enriched uniformly in component B following the liquidus curve. However the entire solid composition does not follow the solidus curves. Only the increment of solid forming at each instant follows the solidus curve leaving behind

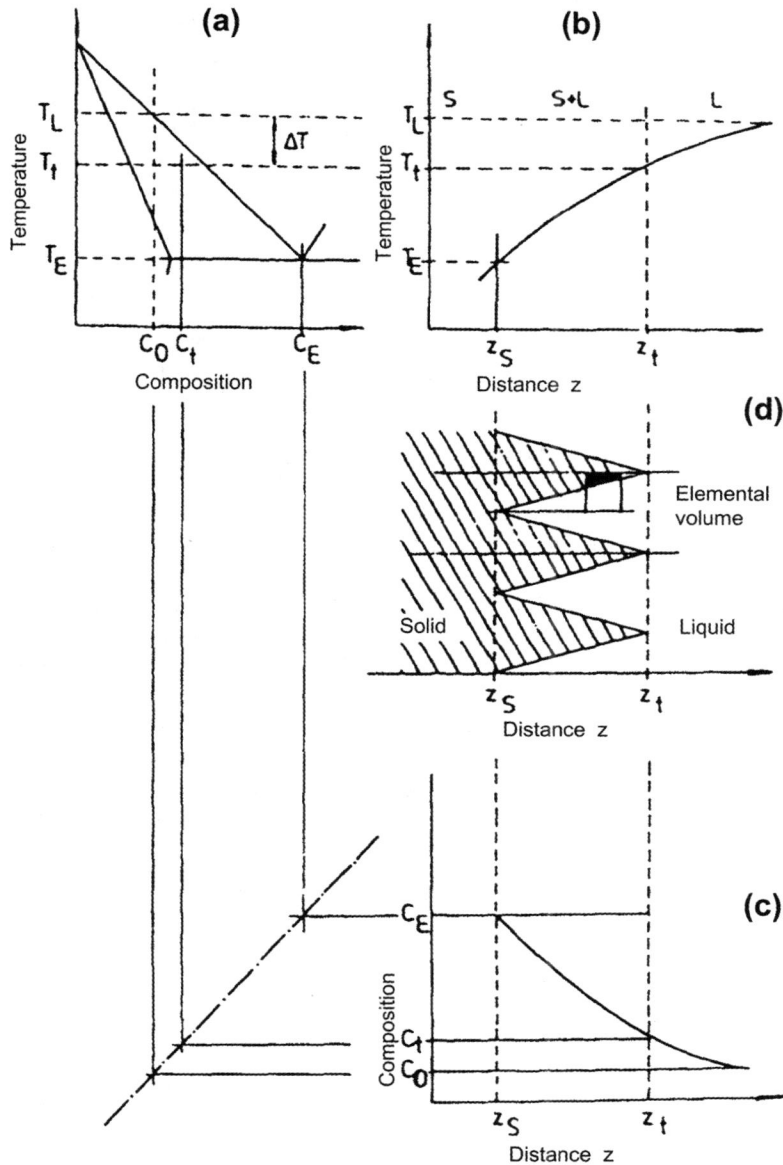

Figure 49 Conceptual model of cellular/dendritic freezing showing: (a) portion of the phase diagram; (b) schematic temperature distribution in the melt; (c) distribution of solute within the liquid in the mushy zone; (d) representation of plate-like, unbranched dendrites showing the position of a characteristic elemental volume (Brody and Flemings, 1966).

a microsegregated solid. The Scheil equation, or the other analytical approaches described below, is used to describe only the dendritic stage of solidification. As the fraction of solid increases (the box moves down further), the liquid composition and the temperature may approach (a) a eutectic point or (b, c) a peritectic point where the liquidus curve for a second phase is intersected. In the former case,

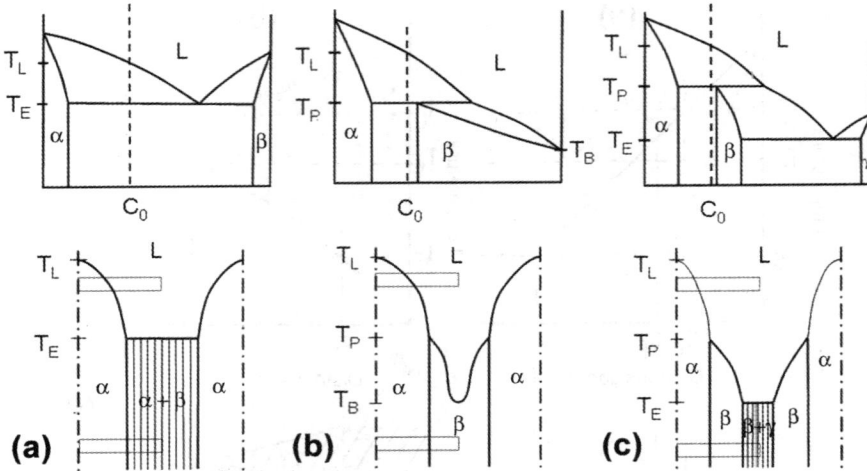

Figure 50 Schematic of the interface shape and microstructure produced during dendritic solidification of three binary alloys with different phase diagrams, (a) eutectic, (b) peritectic and (c) combined peritectic and eutectic.

one ceases to employ the Scheil description and the remaining liquid usually freezes along a planar interface (i.e. at one temperature) as two phase eutectic mixture of α and β phases. If a peritectic feature is present in the phase diagram, solidification of the peritectic β phase will begin as the primary α or first phase ceases. This leads to a coating of the second phase over the dendrite. Under the assumption of the Scheil equation, no solid diffusion occurs and the boundary between the α and β in the microstructure sketch remains vertical. That is, no solid-state reaction occurs. In (b), solidification is completed with the coating of the β phase while in (c) the final liquid freezes as a two phase eutectic mixture of β and γ phases. More details about eutectic and peritectic solidification will be presented in Section 7.8.

For the case of **Figure 50a**, the fraction of eutectic that forms can be evaluated by determining the amount of liquid that remains when the liquid composition reaches the eutectic composition. This fraction of liquid would then be converted to a fraction of eutectic, f_E given by

$$f_E = \left(\frac{C_E}{C_0} \right)^{\frac{1}{k^E - 1}}. \tag{165}$$

That the Scheil equation gives a good first approximation to dendritic microsegregation in the columnar zone of ingots has been proven, among other authors, by Kattamis and Flemings (1965) for low alloy steels and Weinberg and Teghtsoonian (1972) for Cu-base alloys.

7.7.3.3 Modification of the Scheil Approach

The predictions of the Scheil equation establish an upper bound on the severity of microsegregation, for example, as determined by the predicted fraction of eutectic. Battle (1992) and Kraft and Chang (1997) present a comparison of various microsegregation models. Effects that alter the predicted Scheil microsegregation are (a) the variation in cell or dendrite tip composition with growth conditions; (b) the degree of lateral liquid mixing between the cells or dendrites; and (c) diffusion in the solid. All of these corrections are made in a single comprehensive analysis by Tong and Beckermann (1998) that uses solute *Fourier* numbers in the liquid and solid. They point out that with the exception of a solute species like C in Fe, one rarely needs to correct for solid and liquid diffusion under the same growth conditions for the same alloy. Correction for limited liquid diffusion is needed at high growth rates whereas corrections for solid diffusion are important for slower solidification.

7.7.3.3.1 Liquid Diffusion

The simplest method to break the assumption of complete liquid mixing was proposed by Flood and Hunt (1987b) and used by Gandin et al. (1999) is called the *truncated Scheil* approach. This was modified by Boettinger et al. (1998) to ensure solute conservation as follows. The composition at the dendrite tip, C_L^{tip} based on a local estimate of the bulk supercooling or the isotherm velocity and temperature gradient is obtained using eqn (144). The lever rule (conservation of solute) is then employed using this liquid tip composition, the corresponding solid composition at the tip of $k^E C_L^{tip}$, and the bulk alloy composition C_0 to compute a fraction of solid at the tip f_S^{tip} from a solute balance,

$$C_L^{tip}\left(1 - f_S^{tip}\right) + k^E C_L^{tip} f_S^{tip} = C_0 \tag{166}$$

giving

$$f_S^{tip} = \frac{1}{1 - k^E}\left[1 - \frac{C_0}{C_L^{tip}}\right]. \tag{167}$$

This liquid tip composition and value of the fraction solid at the tip are then used as initial values for the integration of the Scheil equation in differential form (eqn (125)) to predict the remainder of the microsegregation profile. In a sense, this treats the dendrite tip as if it is squared off. The result is

$$C_L^* = \begin{cases} C_L^{tip} & 0 \leq f_S < f_S^{tip} \\ C_L^{tip}\left(\dfrac{1 - f_S}{1 - f_S^{tip}}\right)^{k^E - 1} & f_S \geq f_S^{tip} \end{cases}. \tag{168}$$

The corresponding fraction of eutectic for a phase diagram as in **Figure 50a** is given by

$$f_E = \left(1 - f_S^{tip}\right)\left(\frac{C_E}{C_L^{tip}}\right)^{\frac{1}{k^E - 1}}. \tag{169}$$

A method that patches the two parts of the truncated Scheil prediction into a smoothly varying result was obtained by Giovanola and Kurz (1990). They employed a method that uses the IV/MS prediction for the tip composition without assuming complete mixing between the dendrites near the tip region. They employ a polynomial to connect the tip composition to a composition well behind the tips where the complete mixing assumption is valid. The solid composition versus fraction solid relation is determined by numerically solving algebraic equations. The semi-analytical method of Tong and Beckermann (1998) is a later approach to this problem.

Regardless of the exact model used, the most important metallurgical feature of microsegregation models that employs dendrite tip kinetics is the alteration of the composition of solid that forms at $f_S = 0$.

The Scheil equation implicitly assumes that no solute is built up in front of the dendrite tip. Thus the first solid to form has composition, $k^E C_0$. Increasing levels of solute in the liquid at the tip correspondingly increases the amount of solute in the solid along the central spine of the dendrite trunk (and secondary arms) leaving less solute to accumulate in the interdendritic liquid regions at high fraction solid. Thus in a simple eutectic type system, for example, the amount of eutectic that forms in interdendritic regions will be reduced compared with the Scheil prediction. Alternately, for a peritectic phase diagram, a smaller fraction of peritectic phase would coat the primary dendritic structure.

In the rapid solidification regime, other effects such as the variation of the partition coefficient with tip velocity can be included. Microsegregation profile must be measured in the TEM because of the small primary spacings. Bendersky and Boettinger (1985) and Boettinger et al. (1987) measured the composition profiles across very fine cells of Ag-15 wt% Cu alloy. The cores of the cells were found to have relatively flat solute profiles in general agreement with the simple truncated Scheil prediction. Their results at a growth velocity of 0.12 m s^{-1} has been used for comparison to various models as shown in **Figure 51** taken from Tong and Beckermann (1998).

The computation method of Lu et al. (1994) or phase field calculations yield the most detailed information about the details of microsegregation over a broad range of growth conditions.

7.7.3.3.2 Solid Diffusion

Brody and Flemings (1966) were the first to account for diffusion in the solid as a correction to the Scheil approach. They retained the assumption of complete liquid mixing, but permitted diffusion in the solid. The solute balance eqn (125) is modified to

$$C_L(1 - k^E)\frac{df_S}{dt} - (1 - f_S)\frac{dC_L}{dt} = \frac{1}{L_0}D_S\frac{\partial C_S}{\partial x}\bigg|^*, \tag{170}$$

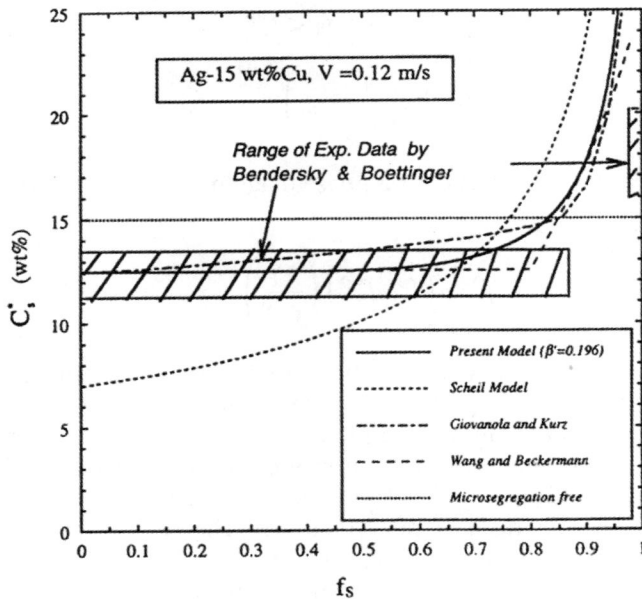

Figure 51 Comparison by Tong and Beckermann (1998) of various microsegregation models with the data experimental data of Bendersky and Boettinger (1985).

where the term on the right-hand side involves the concentration gradient in the solid at the liquid–solid interface and L_0 is the length of the volume element under consideration (related to the dendrite arm spacing). They then approximated the solute concentration gradient in the solid at the L–S interface and derived the following expressions:

$$C_L^* = C_0\left[1 - \frac{f_S}{(1 + \alpha_\theta k^E)}\right]^{k^E-1} \tag{171}$$

and

$$C_L^* = C_0\left[1 - f_S(1 - 2\alpha_\theta k^E)\right]^{(k^E-1)/(1-2\alpha_\theta k^E)}. \tag{172}$$

Equation (171) considers the case where the growth rate is constant and eqn (172) considers the case where the growth is parabolic, that is, $V \sim t^{-1/2}$. The solid solute Fourier number $\alpha_\theta = D_S t_f/\lambda_1^2$ is a measure of the extent of diffusion of solute in the solid where t_f = local solidification time. Brody and Flemings (1966) also obtained the more exact numerical solutions of the solid diffusion equation.

Equations (130) and (131) are good approximations only when $\alpha_\theta \ll 1$, that is, with limited solid diffusion during the solidification process. Kurz and Clyne (1981) and Ohnaka (1986) modified this approach to obtain the two proper limiting cases corresponding to the Scheil equation (no solid diffusion, $\alpha_\theta = 0$) and to the lever rule (complete solid diffusion, $\alpha_\theta = \infty$) by modifying the parameter α_θ. The equation of Ohnaka for parabolic growth is:

$$C_L^* = C_0[1 - \gamma f_S]^{(k^E-1)/\gamma}, \tag{173}$$

with

$$\gamma = 1 - \frac{2\alpha_\theta}{1 + 2\alpha_\theta}. \tag{174}$$

The fraction of eutectic in this case is given by

$$f_E = 1 - \frac{1}{\gamma}\left[1 - \left(\frac{C_E}{C_0}\right)^{\frac{\gamma}{k^E-1}}\right], \tag{175}$$

which recovers the Scheil expression for f_E if $\gamma = 1(\alpha_\theta = 0)$.

Another approximation for back diffusion has been given by Wang and Beckermann (1993) who find a differential equation for solute balance

$$\frac{dC_L}{df_S} + \left[\frac{k^E(1 + 6\alpha_\theta) - 1}{1 - f_S} + \frac{6\alpha_\theta}{f_S}\right]C_L = \frac{6\alpha_\theta C_0}{f_S(1 - f_S)}, \tag{176}$$

where their value of $\alpha_\theta = 4D_S t_f/\lambda_1^2$. Equation (176) has an integral solution

$$C_L = C_0\frac{6\alpha_\theta(1 - f_S)^{[k^E(1+6\alpha_\theta)-1]}}{f_S^{6\alpha_\theta}}\int_0^{f_S}\left[f^{6\alpha_\theta-1}(1 - f)^{-(1-6\alpha_\theta)k^E}\right]df, \tag{177}$$

that must be evaluated numerically to give the liquid composition as a function of fraction solid. Beckermann and Wang's approach also has the proper limits of global equilibrium and Scheil freezing for $\alpha_\theta \to \infty$ and $\alpha_\theta \to 0$, respectively. Corrections because of solid diffusion are very important for extremely slow solidification or when solid diffusion is very rapid as in the case of interstitial solutes, as for carbon in steels (Schneider and Beckermann, 1995).

While the above methods do permit one to calculate the fraction of solid versus temperature curve that is required for heat flow modeling for use in eqn (9) or (10), they only give the value of C_S at the L–S interface and do not give the solute profile that remains in the solid after solidification. Full numerical solutions can overcome these limitations.

7.7.4   Solidification of Ternary Alloys

Practical alloys typically contain many components and a discussion of solidification in ternary alloys is appropriate and the study of ternary phase diagrams is essential. Much of the complexity of systems containing more than three components can be understood conceptually with the study of ternary systems. The essential difference from binary systems is the need to specify tie-line information. Tie-lines are isothermal lines that connect the concentrations of the two phases in two phase regions of the 3-D ternary phase diagram, for example for the liquid and solid. The concept of tie lines need not be geometrical however. They represent a functional relationship between concentration values that describe the liquid and those that specify the solid; thus they can be generalized to any number of components. 2-D temperature versus composition sections of ternary systems only indicate the identity of the phase present and a temperature and average composition, but not their individual phase compositions because tie-lines rarely lie in the plane of the section shown.

Many texts are available for the study of ternary phase diagrams. Examples include Rhines (1956), Massing and Rodgers (1960), Prince (1966), and West (1982). One difficulty with these standard texts is that discussions of solidification always assume global equilibrium during cooling. The greatest value comes from learning the structure of the phase diagrams but then using them to describe conditions of local interface equilibrium for understanding and modeling solidification.

Depending on the alloy system, primary solidification can be followed by secondary products. Three phase equilibrium between a liquid and two solid phases occurs routinely in ternary systems and is represented graphically by isothermal tie-triangles, for example, binary eutectic reactions ($L \to \alpha + \beta$) and binary peritectic reactions ($L + \alpha \to \beta$) in a ternary system. Such three phase equilibria can occur over a range of temperature in contrast to a binary system and are called monovariant in a ternary system at constant pressure. Lower temperature phase diagram features may occur at a fixed temperature and involve four phases. In a ternary system these are invariant reactions. There are three types, $L \to \alpha + \beta + \gamma$ (ternary eutectic) and $L + \alpha \to \beta + \gamma$ (quasi-peritectic) and the rare $L + \alpha + \beta \to \gamma$ (ternary peritectic)[2]. However, in analogy with a peritectic invariant in a binary alloy, the last two indicate merely a liquid composition and temperature where a "switching" of the products of solidification occur in most solidification situations. A discussion of the microstructure produced by some of the phase diagram features is described forthwith.

For solidification of a single solid phase from a ternary liquid, the tie lines can be described by a pair of partition coefficients, $k_i^E = C_{Si}^*/C_{Li}^*$ for the relationship between the solid and liquid concentration for each alloying addition. In general these are not constants and depend on temperature and/or

---

[2] There is no universally accepted terminology for these equilibria.

concentration themselves. Use of a CALPHAD-type approach as described in Section 7.3.3 is best in these situations to quantify how the composition of liquid and solid phases are related.

For the dendritic growth of the primary phase, the Scheil approach can be utilized, and if desired, a dendrite tip model constructed. For a treatment of the kinetics of the dendrite tip in a multicomponent alloy, Ivantsov solutions are obtained for each solute to determine the composition of each solute at the dendrite tip. The marginal stability criterion for the multicomponent alloy is applied to determine the tip radius as shown by Bobadilla et al. (1989) and Rappaz et al. (1990). Rappaz and Boettinger (1999) have extended both dendrite tip equations as well as the secondary arm spacing prediction to multicomponent alloys and to situations with unequal liquid diffusion coefficients for the components.

Use of the Scheil approach to determine the solidification path is quite useful and will be explored in some detail. The solidification path for an n-component alloy is the composition of liquid (and thus solid) as a function of fraction solid obtained from the set of $n-1$ equations

$$\frac{dC^*_{Li}}{df_S} = \frac{\left(1 - k^E_i\right)C^*_{Li}}{1 - f_S},\qquad(178)$$

that are uncoupled if the $k^E_i (= C^*_{Si}/C^*_{Li})$ do not depend on the composition of the other species. If the m^E_{Li} and k^E_i are constants, then for an alloy of initial composition C_{0i},

$$C^*_{Li} = C_{0i}(1 - f_S)^{k^E_i - 1}\quad i = 1, 2, \ldots, n-1$$
$$T^* = T_m + \sum_{i=1}^{n-1} m^E_{Li}C^*_{Li},\qquad(179)$$

during primary solidification. For a ternary alloy a plot of C^*_{L1} versus C^*_{L2} on the liquidus surface of the phase diagram shows the solidification path. One then determines the fraction solid at which the liquid path crosses a monovariant line. If the line is of eutectic character, simultaneous solidification of two solid phases occurs and the liquid composition follows this path of the line during eutectic solidification as the temperature decreases. Analytical description of monovariant eutectic solidification path is given by Mehrabian and Flemings (1970). If the solidification path for primary solidification crosses a monovariant peritectic line, solidification switches to single-phase growth of the new phase. The Scheil equation is then applied using values of m^E_{Li} and k^E_i for the new solid phase but with the appropriate starting value for f_S.

A comparison of the Scheil solidification of alloys with three different ternary phase diagrams is shown schematically in **Figure 52**. As in **Figure 50**, the volume under consideration can be thought of as moving down through the structure as the solidification occurs. The α, β, and γ phases are solid solutions based on pure A, B, and C respectively. The δ phase is an intermetallic based in the A–B system. In **Figure 52(b)**, the intermetallic forms from the melt congruently, whereas in **Figure 52(c)** it forms by a peritectic reaction.

Solidification starts at the liquidus temperature. The first phase to form is the α phase in all cases and is depicted as a large cell in the microstructure sketch even though it might be dendritic in reality. As the solidification continues, the box moves downward until the entire liquid portion of the box becomes enriched in component B and C following the liquidus surface. If the solid solution α were pure A ($k^E_B = k^E_C = 0$), the liquid composition would follow a linear path directly away from the A corner of the phase diagram. If $k^E_B \neq k^E_C \neq 0$, the prediction of the Scheil equation would induce curvature in the path. Only the increment of α phase solid forming at each instant follows the solidus surface (not shown)

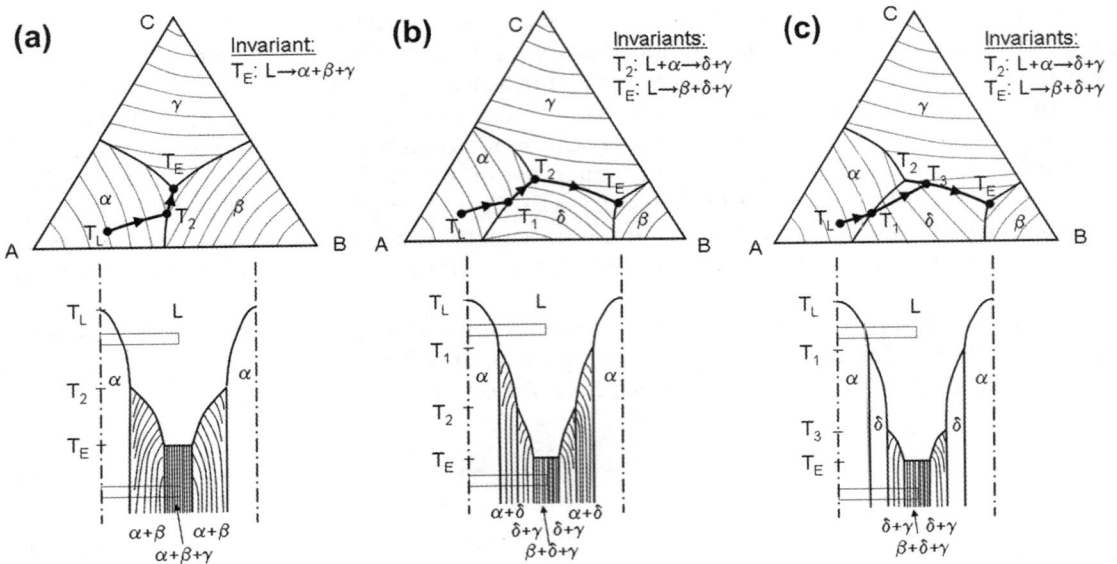

Figure 52 Schematic of the interface shape and microstructure produced during dendritic solidification of three ternary alloys with different phase diagrams.

leaving behind a microsegregated solid. As the fraction of solid increases (the box moves down further), the liquid composition and the temperature approach the monovariant lines on the liquidus surface of the phase diagrams.

For cases (a) and (b) the monovariant lines are of eutectic character and constitute valleys on their respective liquidus surfaces. However unlike a eutectic in a binary system, this eutectic forms over a range of temperature and the remaining liquid in the box follows the valley. The solid forming during this process is a two-phase mixture composed of $\alpha + \beta$ (**Figure 52(a)**) or $\alpha + \delta$ (**Figure 52(b)**) as shown schematically. The phase compositions as well as the phase fraction of the eutectic left in the solid microstructure will in general vary as the remaining liquid descends the liquidus valley. The case where the liquid path crosses a peritectic monovariant line is shown **Figure 52(c)**. Here, solidification switches from α to δ phase. The δ phase forms a coating on the α phase.

In **Figure 52(a)** the liquid at lower temperature finally reaches the ternary eutectic composition and forms a three phase microstructure at a single temperature with a planar L–S interface. In **Figure 52(b)**, the liquid encounters a special point where two eutectic valleys meet to from a third that continues down in temperature. This point represents the liquid composition of the invariant reaction, $L + \alpha \rightarrow \delta + \gamma$. When the liquid reaches this point on the phase diagram, the solidification switches from leaving a two phase coating of $\alpha + \delta$ on the original dendrite to leaving an addition coating of a different two phase eutectic of $\delta + \gamma$. Solidification of this alloy is finally completed at a ternary eutectic point where β, δ, and γ phases form at a single temperature (planar liquid–solid interface). For the case of **Figure 52(c)**, the path traverses the δ liquidus surface, between temperatures T_1 and T_3 until it intersects the monovariant eutectic valley $L \rightarrow \delta + \gamma$. The path then follows this valley forming a two-phase mixture of $\delta + \gamma$ until the freezing is completed at the ternary eutectic temperature.

This is an idealized description even beyond that imposed by the Scheil approach. The drawings assume that the various eutectic solidification processes all form fine multiphase structures accomplished by coupled growth to be described in Section 7.8.1.2. In particular cases, the phases may form divorced eutectics (Section 7.8.1.6) especially if the space remaining between the dendrites is small compared with the spacing of the eutectic. Then the phases may form in the interdendritic regions in isolation.

7.7.5 Example of Ternary Solidification Path Analysis Derived from a Thermodynamic Data Base

Calculation of solidification paths using full integration of phase diagram tie-lines with the Scheil approach (including solid diffusion) have been conducted by Chen and Chang (1992).

Another example of such an analysis describes the cast microstructure of a Ni-Cr-Zr based alloy being studied for anode materials in improved Ni-metal-hydride batteries (Boettinger et al., 2010). The competition between the cubic ($C15$) and hexagonal ($C14$) forms for the Laves phase in the solidified material is of interest. Computation of Scheil and equilibrium solidification paths was performed with Thermo-Calc software (Andersson et al., 2002) with the thermodynamic database of Ansara et al. (1998). The composition of the liquid phase during Scheil solidification of a Ni-20.6 at% Cr-33.5 at% Zr alloy is shown superimposed on the liquidus surface in **Figure 53a**. An isothermal section at 1473 K is shown in **Figure 53b** to indicate the various solid phases. The details of the predicted temperature versus fraction of solid curve as well as the Cr versus Zr paths of the liquid and solid phases are shown in **Figure 54a** and **b**, respectively. A common lettering system is used to relate points on the two graphs. Occurring between temperatures a and b is the dendritic solidification of $C14$. At point b the liquid composition has reached the monovariant peritectic line, $L + C14 \rightarrow C15$ (at the border between the $C14$ and $C15$ regions on the liquidus surface in **Figure 54a**). In the Scheil limit the solidification process switches from the freezing of $C14$ to the freezing of $C15$. Between temperatures b and c occurs the continued freezing of the C15 phase. Between temperatures c and d, monovariant eutectic solidification $L \rightarrow C15 + Ni_{10}Zr_7$ is predicted to begin (following the border between the $C15$ and $Ni_{10}Zr_7$ phases on

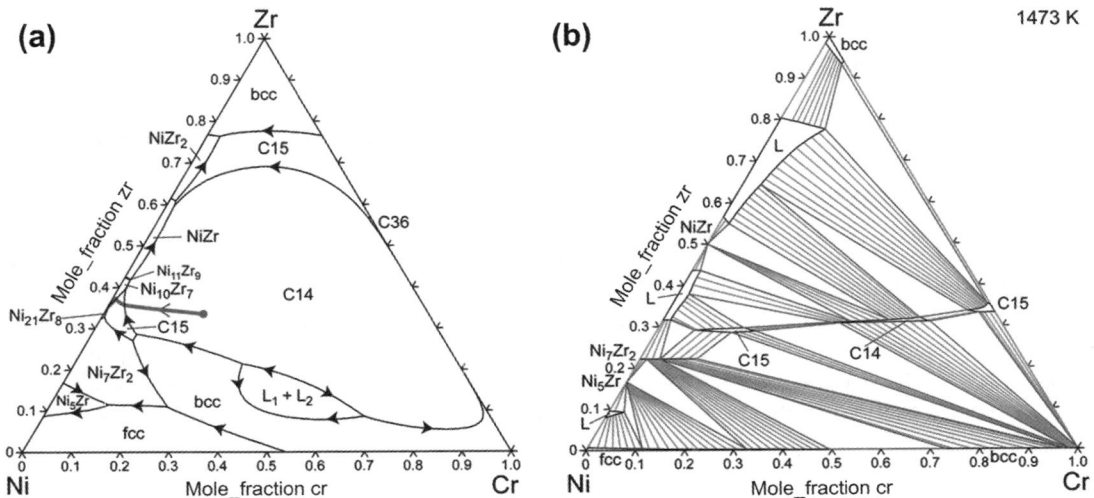

Figure 53 Ni–Cr–Zr; (a) liquidus surface; (b) isothermal section at 1473 K. The red circle shows the location of the initial composition and the red arrow the path of the liquid phase composition during Scheil solidification (Boettinger et al., 2010).

(a)

(b)

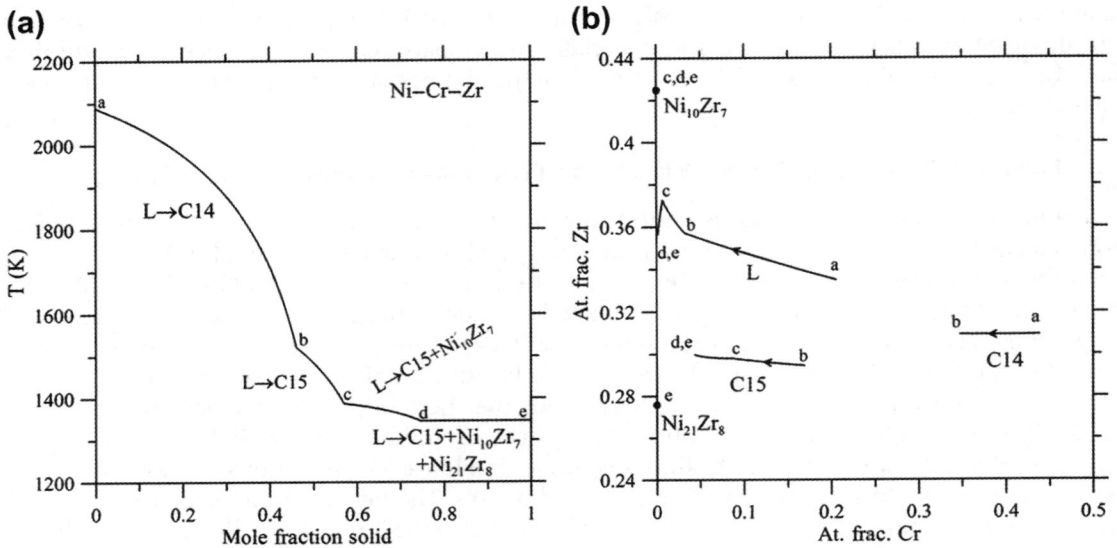

Figure 54 Ni-20.6 at% Cr-33.5 at% Zr alloy; (a) fraction solid versus temperature plot and (b) Cr and Zr composition of liquid and solid phases during solidification. The relationship during solidification among temperature, fraction solid and phase compositions are indicated with letters (a) through (e).

the liquidus surface in **Figure 54a)** and at temperature e, final freezing is accomplished by ternary eutectic solidification $L \rightarrow C15 + Ni_{10}Zr_7 + Ni_{21}Zr_8$. Using this ternary alone, the sequence of phases forming from the melt observed experimentally in a much more complex multicomponent alloy are generally reproduced.

7.8 Polyphase Solidification

Eutectic, peritectic, and monotectic solidification involve the freezing of an alloy at or near a special liquid composition. This special composition is defined thermodynamically as the liquid composition that can be in equilibrium with two other phases at the same temperature. This equilibrium can only occur at a single temperature in a binary system. For the eutectic and peritectic cases, the liquid is in equilibrium with two solid phases and for the monotectic case, the liquid is in equilibrium with a solid phase and another liquid phase. The monotectic case involves alloys that exhibit a miscibility gap in the liquid phase. In systems with more than two components, polyphase solidification can involve two phases in equilibrium with a liquid over a range of temperatures and can involve more than two phases. The following deals primarily with binary alloys except for Section 7.8.1.4, where the influence of a ternary addition on eutectic solidification is described. Aspects of the solidification path involving multicomponent eutectic and peritectic solidification were described in Sections 7.7.4 and 7.7.5.

7.8.1 Eutectic Solidification

In a binary alloy, eutectic solidification converts a liquid into two solid phases in close spatial proximity. Alloys near eutectic compositions are very important in the casting industry because of several characteristics: (i) low liquidus temperatures compared with the pure component melting points that simplify melting and casting operations; (ii) zero or small freezing ranges that effectively eliminate the dendritic mushy zone thereby reducing macrosegregation and shrinkage porosity while promoting excellent mold filling; and (iii) possibilities of forming "in situ" composites. Among the most common eutectic or near-eutectic alloys of industrial importance are cast irons, Al–Si alloys, wear-resistant alloys and solders. In many other practical alloys that freeze dendritically, secondary phases are formed near the end of freezing by eutectic solidification.

Eutectic solidification involves the following stages: eutectic liquid is supercooled and one of the solid phases nucleates, causing solute enrichment in the surrounding liquid and sympathetic nucleation of the second solid phase. Repeated nucleation and/or overgrowth of one solid phase by the other produces a growth center that defines an individual eutectic grain. For many eutectics, solidification proceeds by simultaneous growth of two interspersed solid phases at a common liquid–solid interface. The solute rejected into the liquid by each phase is taken up by the adjacent phase particles. As solidification proceeds, spatial and crystallographic rotation of the solid phases together with competitive overgrowth of adjacent eutectic grains leads to a stable solidification front, with the surviving eutectic grains having, as much as possible, maximized the solidification rate, minimized the α–β interfacial energy and oriented the α–β interfaces in the heat flow direction. The ability of one solid phase in the eutectic to stimulate nucleation of the other varies widely for different α–β phases and, in general, experimental knowledge about the nucleation process in eutectic solidification has been sparse (Mondolfo, 1965). This complex nucleation process has been the subject of simulation by phase field methods (e.g. Elder et al., 1994; Gránásy and Pusztai, 2002). The following focuses primarily on the growth aspects of eutectics. Two excellent summaries of the early literature on eutectic solidification can be found in reviews by Chadwick (1963) and Hogan et al. (1971).

7.8.1.1 Eutectic Classification

When a eutectic liquid solidifies, the resulting material generally consists of a dispersed two-phase microstructure that is approximately 10 times finer than cells or dendrites formed under the same conditions. The exact arrangement of the two phases in the eutectic microstructure can vary widely, depending on the solidification conditions and the particular eutectic alloy being solidified. Hunt and Jackson (1966) provided a simple classification scheme according to the interface kinetics of the component phases that remains useful. A correlation was found between eutectic morphology and the entropies of fusion of the two solid phases, concepts discussed in Section 7.5.1. The classification refers to nonfaceted–nonfaceted eutectics (nf–nf); nonfaceted–faceted eutectics (nf–f); and faceted–faceted eutectics (f–f). Very little is known about the solidification of f–f eutectics but extensive research has been performed on nf–nf and nf–f eutectics. Kerr and Winegard (1967) argued that the faceted or nonfaceted nature of the liquid–solid interface of the phases is better described by the entropies of solution of the individual phases because the (eutectic) liquid composition is generally quite different from the solid-phase compositions. Croker et al. (1973) examined a large number of eutectic microstructures and in addition found that the volume fraction of the solid phases in the eutectic structure and the growth velocity must also be considered to obtain a good classification scheme.

7.8.1.2 Nonfaceted–Nonfaceted Eutectics

Eutectic mixtures of two nonfaceted phases tend to form regular microstructures consisting of either alternate lamellae of α and β or rods of α embedded in a β matrix that grow in a continuous edgewise manner into the melt. Experimentally, it is found that eutectics in which one phase has a very low volume fraction tend to grow in a rod-like manner. When the volume fraction of one of the phases is less than about 0.3, a rod-like structure has lower total $\alpha-\beta$ surface energy than a lamellar structure. This assumes that the surface energies are isotropic. However a 25% anisotropy of $\alpha-\beta$ interface energy can stabilize lamellar structures at practically any value of volume fraction. In marginal cases, lamellar-to-rod transitions can occur for a given alloy by changing the growth conditions. 3-D microstructural analysis of eutectic structures shows that both phases are continuous in the growth direction over large regions that are called eutectic grains. A eutectic grain can be described approximately as two inter-penetrating single crystals of the two component phases having almost constant crystallographic orientation. Hogan et al. (1971) give a thorough discussion of eutectic grains and a summary of the orientation relationships between phases and the habit plane of the $\alpha-\beta$ interfaces. The orientation relationship and the orientation of the interfaces usually allow for a high degree of fit between the two crystal structures that minimizes the $\alpha-\beta$ surface energy.

In lamellar or rod eutectic solidification, the two phases, α and β, solidify side by side with an approximately planar and isothermal L–S interface, supercooled ΔT below the equilibrium eutectic temperature. During solidification of an A–B alloy, the A-rich α phase rejects B atoms into the liquid and the B-rich β phase rejects A atoms. The interaction of the diffusion fields in the liquid in front of the two phases gives rise to the term *coupled growth*, which is commonly used to describe eutectic solidi-fication. At a given solidification rate V, the spacing of the lamellae or rods, λ_E and the interface supercooling below the eutectic temperature, ΔT, are controlled by a balance between: (i) the necessity for lateral diffusion of excess A and B in the liquid just ahead of the L–S interface, which favors a small interlamellar or interrod spacing, and (ii) the necessity to create $\alpha-\beta$ interfacial area, which tends to favor large λ_E (fewer interfaces).

Over the years the theory of lamellar or rod growth for the slow solidification velocities when $\lambda_E V/2D_L \ll 1$ has been developed by Zener (1946), Tiller (1958), Hillert (1957), Jackson and Hunt (1966), Magnin and Trivedi (1991). The basis of the analysis is illustrated in **Figure 55**. The interface temperature at each interface point is composed of three contributions controlled by the velocity on the interface at each point, the composition of the liquid at the interface at each point and the curvature of the interface at each point. The supercooling below the eutectic temperature, T_E, for each point on the L–S interface (α–L and β–L) can be described by

$$\Delta T = \Delta T_k + \Delta T_D + \Delta T_c. \tag{180}$$

The first term is the interface attachment kinetic supercooling (described in Section 7.5) and is usually neglected compared with the other terms. Thus it is assumed that each point on the α–L and β–L interfaces is at local equilibrium, which is described by the metastable extension of the two liquidus and solidus lines below the T_E. ΔT_D is the supercooling below the eutectic temperature because of the local variation in composition from the eutectic composition. This local variation is approximated by solving the steady-state diffusion equation in the liquid for a planar $\alpha-\beta$ interface growing at velocity V and spacing λ_E. ΔT_c is the supercooling because of curvature (Gibbs–Thomson effect).

It is then assumed that the total supercooling, ΔT, given by eqn (180) is constant across the interface; that is, the interface is isothermal. Thus across the interface, any variation in ΔT_D must be balanced by a variation in ΔT_c, giving a constant value for $\Delta T = \Delta T_D + \Delta T_c$ at each point of the interface. Because of the solution of the diffusion equation, ΔT_D has a minimum value near the $\alpha-\beta$–L groove, and thus ΔT_c has a maximum value there. The radius of curvature of each solid phase interface is, therefore, smallest near the triple junctions and leads to an interface shape similar to that shown in **Figure 55c**. Jackson

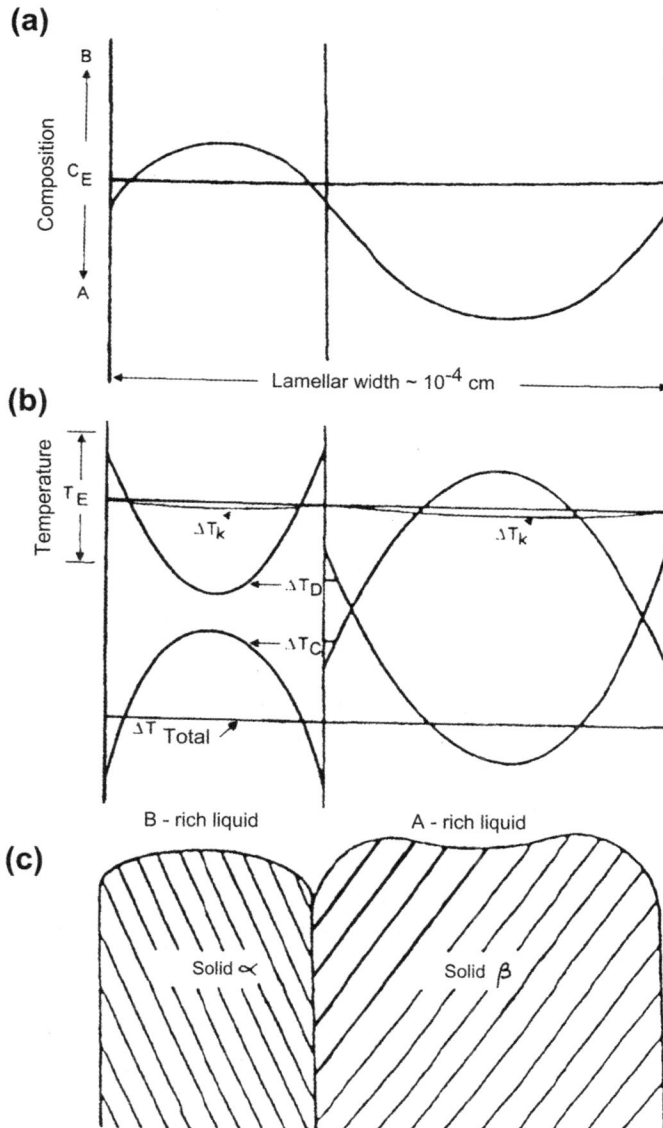

Figure 55 (a) Liquid composition (% B) across an α–β interface. (b) Contributions to the total supercooling (ΔT) existing at the L–S interface, ΔT_D, ΔT_c, and ΔT_K are the solute, curvature and kinetic supercoolings, respectively. (c) Shape of the lamellar L–S liquid interface (Hunt and Jackson, 1966).

and Hunt (1966) have shown that the predicted interface shape agrees very well with the interface shape observed in the transparent hexachloroethane–carbon bromide model system.

By averaging the composition deviation from the eutectic composition in the liquid in front of each phase and averaging the interface curvature of each phase (which depends on the width of the phase and the angle the interface makes with the α–β interface), the total supercooling is found to be (using Jackson and Hunt's approach)

$$\Delta T = K_5 \lambda V + K_6/\lambda, \tag{181}$$

where K_5 and K_6 are constants. For lamellar growth they are given by

$$K_5 = \frac{\overline{m}_L^E C_0^* P_E}{D_L f_\alpha f_\beta}, \tag{182}$$

and

$$K_6 = 2\overline{m}_L^E \left(\frac{\Gamma_\alpha \sin \theta_\alpha}{m_{L\alpha}^E f_\alpha} + \frac{\Gamma_\beta \sin \theta_\beta}{m_{L\beta}^E f_\beta} \right), \tag{183}$$

with

$$\overline{m}_L^E = \frac{m_{L\alpha}^E m_{L\beta}^E}{m_{L\alpha}^E + m_{L\beta}^E}. \tag{184}$$

The parameters $m_{L\alpha}^E$ and $m_{L\beta}^E$ are the liquidus slopes for the alpha and beta phases (both defined positive and for their respective solutes), C_0^* is the difference in composition between the solid phases, D_L is the liquid diffusion coefficient, f_α and f_β are the volume fraction of the solid phases in the eutectic, P_E is a series function of the phase fractions and can be approximated (Trivedi and Kurz, 1988) by

$$P_E = 0.3383 \left(f_\alpha f_\beta \right)^{1.661}. \tag{185}$$

Γ_α and Γ_β are the Gibbs–Thomson coefficients (surface energy/entropy of fusion per unit volume), and θ_α and θ_β are the angles that the alpha and beta interfaces make with a plane perpendicular to the α–β interface. Because a balance of tensions must exist at the α–β–L triple point, the solid–solid energy exerts its influence through its effect on these angles. A low value of $\gamma_{\alpha\beta}$ leads to small values for θ_α and θ_β and hence a small value of ΔT_c. Very similar expressions are obtained for rod eutectic growth by Jackson and Hunt (1966).

The individual terms of eqn (181) are shown schematically for different growth rates in **Figure 56**. It can be seen that ΔT is biggest for large lamellar spacings because diffusion is difficult, and also for small spacings where curvature effects are dominant. Clearly, the values of λ_E and ΔT are not fixed uniquely by V, yet in experiments it is well established that the value of λ_E generally decreases with increasing V. Hence an additional condition is required to specify the operation point on each λ_E versus ΔT curve.

The simplest additional condition is obtained by assuming that growth occurs at the minimum ΔT for a given V or, equivalently, a maximum V for a given ΔT. This condition is called the *extremum condition*. Using this condition,

$$\lambda_E^2 V = K_6/K_5, \tag{186}$$

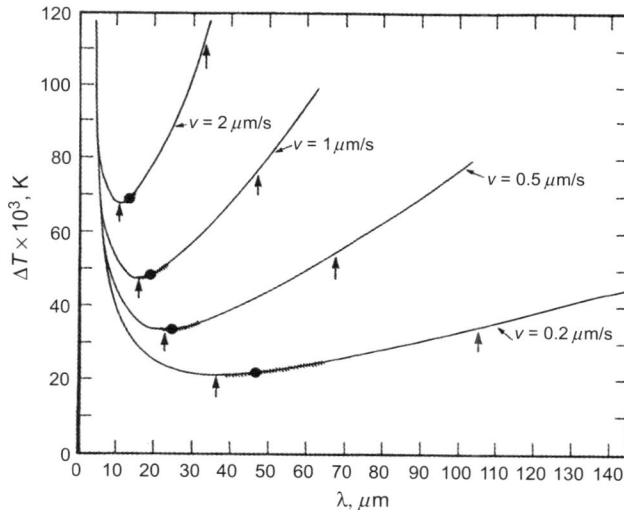

Figure 56   Interface supercooling, ΔT, as a function of lamellar spacing, λ_E, for different growth rates, V using eqn (181) for CBr_2–C_2Cl_6 organic eutectic. The vertical arrows show the theoretical minimum (extremum) and maximum spacings for stable lamellar growth. Also shown are the experimentally observed range of spacings (hatch marks on curve) and the mean spacing (filled circles) (Seetharaman and Trivedi, 1988).

and

$$\Delta T^2 / V = 4 K_5 K_6. \tag{187}$$

Many investigators (e.g. Jordan and Hunt, 1971, 1972; Tassa and Hunt, 1976) have found the average spacing and interface supercooling to be close to the values given by these equations using the extremum condition. Small spacing adjustments occur during continued growth by the motion of faults (lamellar edges) perpendicular to the growth direction.

The discovery that the dendrite tip radii could not be described by a maximum growth rate hypothesis (see Section 7.7) led to further consideration of whether the extremum condition for eutectic growth was valid. Experiments on eutectics (Jordan and Hunt, 1972; Seetharaman and Trivedi, 1988) have shown experimentally that a small range of spacings are observed for a given growth rate. They find that the minimum spacing observed is close to that given by the extremum condition but that the average spacing is somewhat larger than that given by eqn (186). Much research has been focused on defining the allowable range.

The basic concepts were first proposed by Jackson and Hunt (1966), who indicated that only spacings within a certain range were stable to fluctuations in the shape of the L–S interface as shown in **Figure 57**. They quoted unpublished work by J. W. Cahn that argued that spacings smaller than that given by the extremum condition are inherently unstable. This instability is because of the pinching of an individual lamella or rod that locally increases the spacing (**Figure 57a**). Detailed theoretical analysis of this aspect has been performed by Langer (1980), Dayte and Langer (1981) and Dayte et al. (1982).

That the allowable range of spacing has a maximum value at any velocity was first proposed by Jackson and Hunt (1966). This maximum spacing also follows a $\lambda_E^2 V = $ constant law. If the spacing exceeds the extremum value by a critical factor, the larger volume fraction phase develops a pocket that drops

(a)   **(b)**

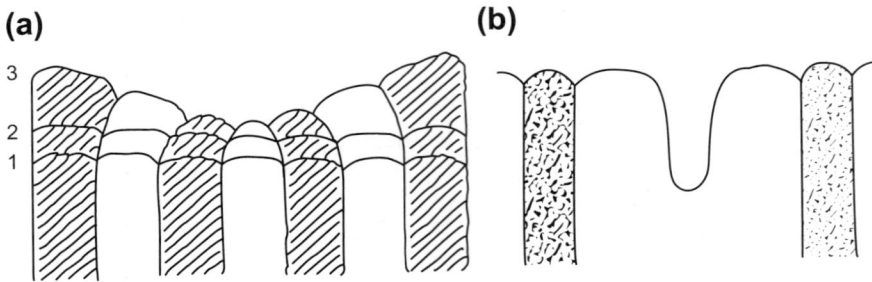

Figure 57 (a) Schematic illustration of the instability of lamellae with λ_E less than the extremum value. The lamella in the center will be pinched off with time. (b) The shape instability of the interface on one phase that occurs when the spacing becomes too large. A new lamella may be created in the depressed pocket (Jackson and Hunt, 1966). Figure taken from Trivedi and Kurz (1988).

progressively back from the interface until growth of the other phase ultimately occurs in it (**Figure 57b**). They took this condition to occur when the slope of the L–S interface in the pocket became infinite. For example, this condition yields a maximum values for the $\lambda_E^2 V$ constant of 10 and two times the extremum value for volume fractions of 0.5 and 0.1, respectively. As a consequence of the formation of the new lamella, the local spacing is abruptly reduced by a factor of two. However, the careful experiments of Seetharaman and Trivedi (1988) show that the maximum observed spacing is much smaller than this estimate (see **Figure 56**), giving an average spacing that is only ~ 20% larger than the extremum value (or minimum stable spacing). Thus the maximum value of spacing occurs before the pocket depression attains infinite slope. While further research is required on this topic, the extremum value is often taken as a good approximation for nf–nf growth. Other stability issues of eutectic growth involving the solidification of alloys with compositions different from the eutectic composition are described below.

Convection in the liquid near the interface of a growing eutectic has been found to increase the value of λ_E (Junze et al., 1984; Baskaran and Wilcox, 1984). Flow parallel to the interface distorts the liquid concentration profile in front of the lamellae slightly and alters the diffusion controlled growth. The effect is greatest when the dimensionless parameter, $G_u \lambda_E^2 / D_L$, is large, where G_u is the gradient normal to the interface of the fluid flow velocity parallel to the interface. Vigorous stirring is required to alter the spacing significantly.

The phase field method has been used to simulate eutectic growth. Karma (1994) showed the correspondence to the sharp interface approach described above. Wheeler et al. (1996) developed various models appropriate for eutectics with different solid thermodynamic descriptions. Simulations have been performed by Nestler and Wheeler (2000), Lewis et al. (2003), Kim et al. (2004), and Plapp and Karma (2002). 3-D simulations by Plapp (2007) of eutectic growth enabled the exploration of the transition between rod and lamellar structures, their disordering and the different undercooling versus spacing relations for different rod packings. Parisi and Plapp (2010) showed that other steady-state structures were possible in addition to rods and lamellae. These simulations shed light on realistic imperfect eutectic structures. Phase field work on ternary eutectics is summarized by Hecht et al. (2004).

7.8.1.3 Nonfaceted–Faceted Eutectics

The modeling of nf–f eutectics is quite important given the fact that eutectics of technological importance such as Al–Si and Fe–C belong to this class. Fisher and Kurz (1979) and Kurz and Fisher

(1979) summarize the main features of (nf–f) eutectic growth. When (nf–f) eutectics are compared with (nf–nf) eutectics, several characteristics can be noted:

(i) The degree of structural regularity is much lower and a wide distribution of local spacings is observed.

(ii) For a given growth rate and fraction of phases, the average spacing and the average interface supercooling below the eutectic composition for growth of an nf–f eutectic are much larger than for an nf–nf eutectic.

(iii) For a given growth rate, the supercooling and the spacing for nf–f eutectic decrease as the temperature gradient is increased. No such effect is seen for nf–nf eutectics.

To understand the increased average supercooling below the eutectic temperature, early investigations considered interface attachment supercooling (first term in eqn (180)) for the faceted phase. However Steen and Hellawell (1975) and Toloui and Hellawell (1976) showed that the kinetic supercooling of Si in Al–Si eutectic is too small to explain the increased eutectic front supercooling. Indeed, the Si phase in Al–Si and the graphite phase in Fe–C contain defect planes parallel to the plate growth direction that enable easy growth (twins in Si and rotation boundaries in graphite). Toloui and Hellawell (1976) suggested that the large supercoolings were because of the difficulties of adjusting the spacing to minimize the diffusion distance. These difficulties are related to the anisotropy of growth of the faceted phase.

Measurement of spacing and supercooling on the model system camphor–naphthalene by Fisher and Kurz (1979) permitted important results to be obtained. The system exhibits two distinct eutectic growth forms: one regular and the other irregular. By assuming that the measured spacing and supercooling for the regular growth were given using the nf–nf theory with the extremum condition, the various materials parameters for this system were obtained. When ΔT versus λ_E was plotted for the relevant growth rate, the spacing and supercooling values for irregular growth fell on the same derived theoretical curve for regular eutectic growth but with spacings much larger than the extremum value. Thus the coarseness of the structure is the cause of the large supercooling of growing nf–f eutectics. Indeed theoretical analysis of the λ_E versus ΔT curves by Magnin and Kurz (1987) that relax the assumption of an isothermal two-phase interface made in the nf–nf theory show a deviation from the nf–nf theory only at very slow growth rates where the constants in eqn (181) become functions of G_L. Thus for growth at more normal speeds, the theory turns to an analysis of why the spacing is so big for nf–f eutectics.

The general argument employed to understand why the average spacing of f–nf eutectics is large focuses on determining the stable range for eutectic spacings at a given velocity. Important contributions have been made by Fisher and Kurz (1980), Magnin and Kurz (1987) and Magnin et al. (1991). For irregular eutectics the growth directions of different lamellae are not parallel. Thus as growth proceeds, the local spacing decreases between converging lamellae and increases between diverging lamellae (**Figure 58**). For converging lamellae, when their separation decreases below the extremum value, one of the lamellae is pinched off, just as for nf–nf growth. For diverging lamellae, when the local spacing increases beyond a critical value, Fisher and Kurz (1980) have suggested that the faceted phases branches into two diverging lamella. The formation of the new lamella decreases the local spacing. The anisotropic growth kinetics of the faceted phase leads to what is termed *branching-limited growth*. Several criterion have been proposed to determine the maximum value of spacing where the branching takes place: Fisher and Kurz (1980) and Magnin and Kurz (1987). Magnin and Kurz (1987) suggest that this branching instability occurs when the faceted phase interface develops a depression of some characteristic depth; for example, when it drops below a line joining the two triple points for the lamella. The

(a)

(b)

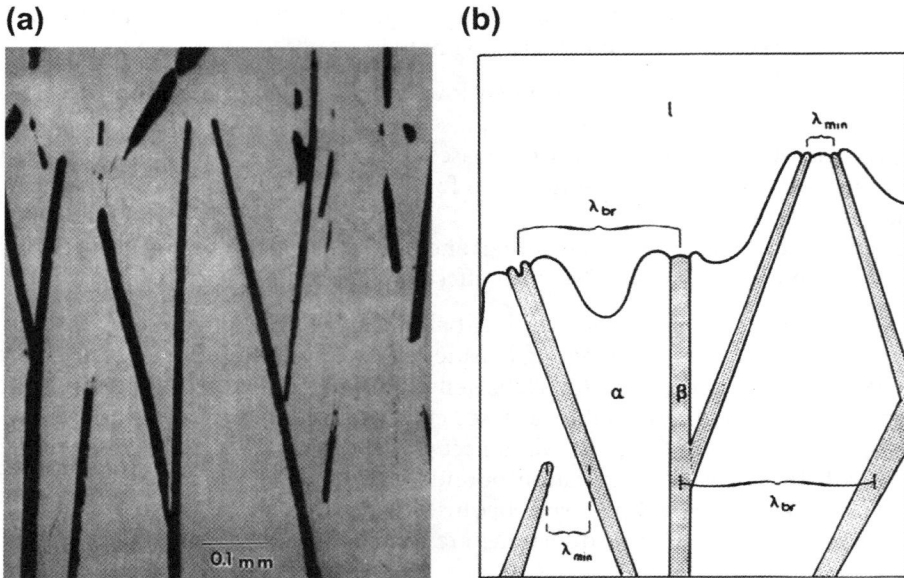

Figure 58 Proposed growth behavior of irregular eutectics, showing branching at λ_{br} and termination at λ_{ex}. (a) Fe-graphite eutectic growth at $V = 1.7 \times 10^{-2}$ μm/s and (b) schematic representation of solid–liquid interface during growth (Magnin and Kurz, 1987).

average spacing lies between the minimum spacing and the spacing that cause the branching instability. Magnin et al. (1991) argue that the mechanism that establishes the minimum and maximum spacings remains undetermined and that the inherently unsteady solidification of nf–f eutectics plays a fundamental role. Many issues remain to be studied in this area especially those regarding orientation relationships and their relationship to the branching mechanism.

7.8.1.4 Eutectic cells and Eutectic Dendrites

In addition to consideration of the stability of the eutectic spacing, two other instabilities can influence the microstructure of alloys at or near the eutectic composition. These involve a ternary component addition to a binary eutectic and, as described in Section 7.8.1.5, the deviation of the average composition of the eutectic microstructure in a binary system from the thermodynamic eutectic composition.

A ternary component added to a binary eutectic can lead to a cellular structure (Chadwick, 1963). The mechanism is similar to the cellular breakdown in single-phase solidification treated in Section 7.6.4; for a critical value of G_L/V the average planar L–S interface of the eutectic structure can become unstable and the solidification front becomes corrugated. The cells (often called eutectic colonies) are quite large containing many (10–100) eutectic spacings with the lamellae curving to remain approximately normal to the liquid–solid interface. Thus cells are most noticeable for nf–nf eutectics. Near the edge of eutectic colonies there is often a transition to a rod structure. Bertorello and Biloni (1969) propose that the inception of the instability occurs at depressions in the interface because of eutectic grain boundaries or at fault terminations at the L–S interface. Phase field simulations of colony formation have been performed by Plapp and Karma (2002) as shown in **Figure 59**.

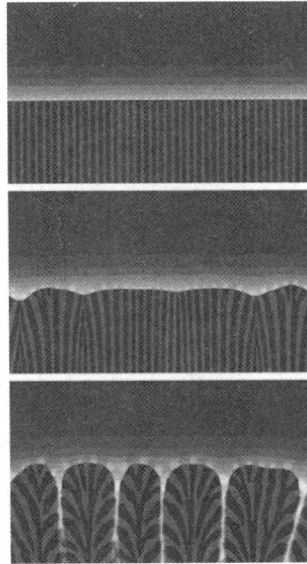

Figure 59 Phase field calculation results showing the time evolution of colony formation for an alloy with a small amount of third element (impurity) added to a binary eutectic. The smooth contours ahead of the front represent isoconcentration lines of the ternary impurity; the small "halos" just in front of the growing lamellae are a visualization of the interlamellar eutectic diffusion field (Plapp and Karma, 2002).

If an excessive amount of a ternary element is added, the eutectic colony can actually evolve into a two-phase dendrite with secondary arms. In this case, ternary eutectic is usually found between the two-phase dendrite Sharp and Flemings (1974). **Figure 60** shows a micrograph of a quenched liquid–solid interface of such a growth form in Pb–Sn–Cd system (Boettinger, 1973). Spiral two-phase dendrites have

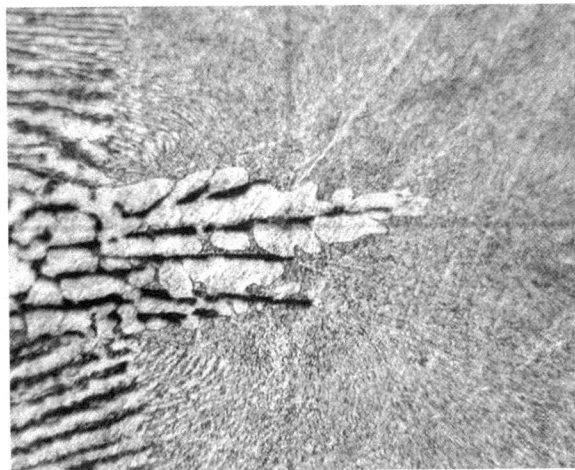

Figure 60 Two phase dendrite formed when a large quantity of a third element is added to a binary eutectic. The alloy lies on the Sn-Cd two phase valley in the Sn–Pb–Cd ternary system (Boettinger, 1973).

recently been observed by Akamatsu et al. (2010). A parabolic shape composed of the two solid phases form a double helix. This growth mechanism permits the eutectic spacing to maintain a near constant value.

7.8.1.5 Competitive Growth-Coupled Zone

As a binary alloy with a composition different from the eutectic composition cools from the liquidus to the eutectic temperature, dendritic growth of the primary phase followed by eutectic growth of the remaining interdendritic liquid is expected. However, there is the *range* of alloy composition, temperature gradient and growth rate (or interface supercooling) where it is possible to freeze these liquids as eutectic microstructures without dendrites. Thus it is important to distinguish between the concepts of a thermodynamic eutectic point and a eutectic microstructure. This range of conditions is called the coupled zone.

Pioneering investigations in this field were those of Tammann and Botschwar (1926) and Kofler (1950) in organic systems, which established that at low or zero temperature gradients, the range of alloy compositions for coupled growth widened with increasing growth velocity. Later, Mollard and Flemings (1967) showed that a widening of the composition range for coupled growth was not restricted to large growth rates but could also be widened at small growth rates as long as the G_L/V ratio was large enough. Solidification with a high value of G_L/V suppresses the dendritic growth of the primary phase. Other milestones in the development of the knowledge of the coupled zone were the investigations by Hunt and Jackson (1967), Jackson (1968), Burden and Hunt (1974c), Tassa and Hunt (1976) and Kurz and Fisher (1979).

For a binary system, the coupled zone concept is most easily understood using an approach that employs three steps (Kurz and Fisher, 1979): for each overall liquid composition, (i) consider all the growth forms possible, that is, α dendrites, β dendrites, and eutectic of α and β. (ii) Analyze the growth kinetics of each of these forms and determine the interface (or tip) temperatures of the growth forms as a function of V, and possibly of G_L. (iii) Apply the competitive growth criterion, for example, that the morphology having the highest interface temperature for a given growth rate, or the highest growth rate for a given temperature will dominate. The range of temperatures and compositions within which eutectic growth is fastest defines the coupled zone and is sometimes plotted on the phase diagram. The composition range of the coupled zone can also be plotted versus velocity because each value of interface temperature corresponds to a known value of the growth velocity for the dominant growth structure. For growth conditions where the eutectic is not dominant, the microstructure consists of a mixture of dendrites and eutectic.

Figure 61 shows an example of a competitive growth analysis for a system involving an nf–f eutectic (the β phase is faceted). The kinetic curves are shown for α dendrites, β dendrites and (planar) eutectic. The curves for the dendrites depend on the value of the temperature gradient and this dependence leads to the decreased interface (tip) temperatures at low velocity and the widening of the coupled zone at high G_L/V ratio. The skewed nature of the coupled zone about the eutectic composition for this nf–f system is because of two factors (Kurz and Fisher, 1979): the nf–f eutectic and the faceted dendritic phase require higher supercooling for a given growth rate than an nf–nf eutectic and a nonfaceted dendrite, respectively. The former is because of the branching difficulties already discussed and the latter is because of the fact that faceted phase dendrites usually grow as a plate or 2-D dendrite rather than a paraboloid or 3-D dendrite. Diffusion of solute away from the tip region of a plate dendrite is more difficult and leads to increased supercooling. Thus for the alloy shown, one expects with increasing growth rate (or supercooling): eutectic, β dendrites (with eutectic), eutectic, and α dendrites (with eutectic). Thus α dendrites form from a composition on the "wrong side" of the eutectic. For

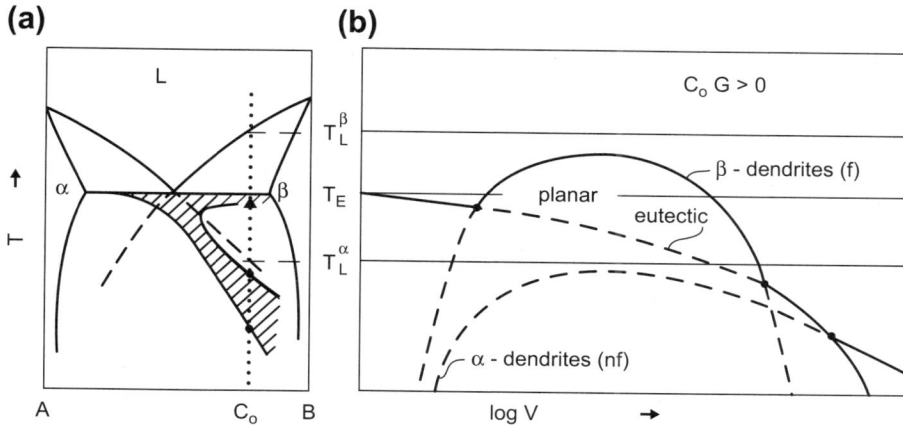

Figure 61 The origin of the coupled zone (hatched) is understood by considering the variation in eutectic interface temperature and dendrite tip temperatures for an off-eutectic alloy. The dominant microstructure for any composition at a given velocity (or supercooling) is that which grows with the highest temperature (or fastest growth rate) (Trivedi and Kurz, 1988).

a system with an nf–nf eutectic the coupled zone is symmetric about the eutectic composition and the formation of dendrites of a particular phase is not observed on the "wrong side" of the eutectic. The behavior of an nf–f system leads to much confusion if a simplistic, purely thermodynamic view of solidification is employed. It can also lead to difficulty in determining the thermodynamic eutectic compositions by purely metallographic methods. It is further noted that near the growth rate where a microstructural transition from dendritic to eutectic structure takes place, the interdendritic eutectic will not have an average composition equal to the thermodynamic eutectic composition (Sharp and Flemings, 1973).

The methodology of competitive growth outlined above provides an adequate framework to understand the major features of the transition from eutectic to dendritic growth. Stability analyses using perturbation methods were performed by Jordan and Hunt (1971) and Hurle and Jakeman (1968). More subtle variations in eutectic microstructure occur under conditions close to the stability/instability transitions that required more analysis. Jackson and Hunt (1966) observed a tilting of lamellae when the growth rate was suddenly increased. Zimmermann et al. (1990) have observed oscillations where the width of the Al lamellae varies at the expense of the Al_2Cu lamellae in the growth direction in Al-rich $Al–Al_2Cu$ off-eutectic alloys. Gill and Kurz (1994) observed another type of instability where the width of both lamellae varies in the growth direction. Han and Trivedi (2000) documented the formation of alternating bands of the primary phase and eutectic near the dendrite to eutectic transition. Karma (1987) succeeded in simulating these instabilities using Monte Carlo methods. He related the appearance of the instability with increasing velocity to critical values of the concentration gradient in the liquid ahead of the interface.

7.8.1.6 Divorced Eutectics

When the liquid remaining between a primary dendritic phase reaches the eutectic composition, eutectic solidification usually occurs. Typically one observes the same eutectic microstructures already described between the dendrite arms especially if the fraction of liquid remaining between the dendrites when the liquid reaches the eutectic composition is large compared with the eutectic spacing. If however the fraction of liquid remaining is so small that the width is comparable to the eutectic spacing, the characteristic two-phase structure may not be observed. It is easier for the second solid phase to form as individual particles or a layer between the dendrites with the additional primary phase merely thickening the dendrite. This occurs more often for a faceted second phase because coupled nf–f eutectics grow with larger spacings and hence require more space to develop their characteristic morphology. Thus the final solidified microstructure consists of dendrites or cells with interdendritic single phase. This microstructure is sometimes misinterpreted as resulting from a peritectic reaction in complex alloys where the phase diagram is unknown. An example of a divorced eutectic involving a facetted phase is shown in **Figure 62**.

7.8.1.7 *Rapid Solidification of Eutectic Alloys*

Rapid solidification produces a very rich variety of microstructures for alloys near eutectic composi-tions. As limiting cases at high rates of solidification rate, glass formation or extended crystalline solubility is expected depending on the thermodynamic structure of the T_0 curves, as described in Section 7.3.2, **Figure 11**. How microstructures and phase distributions evolve from the classical microstructures described above as the solidification velocity is increased is now described.

7.8.1.7.1 *Modification of Eutectic Theory*

In Section 7.7, the general theory presented for dendritic growth included the modifications necessary to treat high growth rates; namely, modifications of the tip stability condition for high Peclet numbers (of order unity) and the inclusion of nonequilibrium interface conditions (solute trapping). For eutectic growth the Peclet number, $\lambda_E V/2D_L$ becomes large at relatively low velocity (~ 10 cm/s) where the effects of solute trapping are not too important. Thus Trivedi et al. (1987) recomputed the solute field in the liquid in front of a growing eutectic when the Peclet number is large while retaining the local interface equilibrium assumption. The theoretical results are similar to those for nf–nf growth at slow

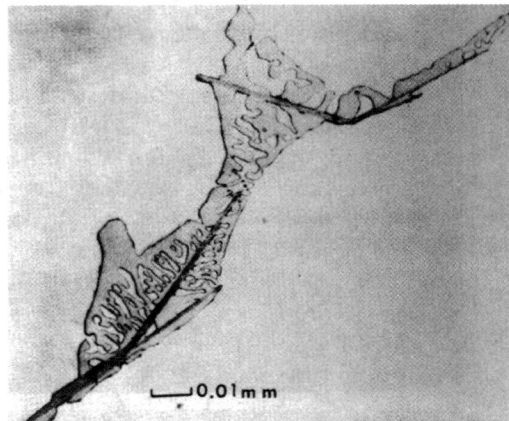

Figure 62 Micrograph of interdendritic region of an as-cast 2219 Aluminum alloy. Parts of the needles of the Al_7Cu_2Fe phase are imbedded in Al dendritic matrix showing that it formed along with the Al as a divorced (uncoupled) L $\rightarrow$ Al + Al_7Cu_2Fe eutectic. Later freezing shows a fine two-phase mixture of Al + Al_2Cu with additional needle-shaped Al_7Cu_2Fe formation consistent with the fact that the phase diagram calls for a ternary eutectic of L $\rightarrow$ Al + Al_7Cu_2Fe + Al_2Cu (Boettinger, 1981).

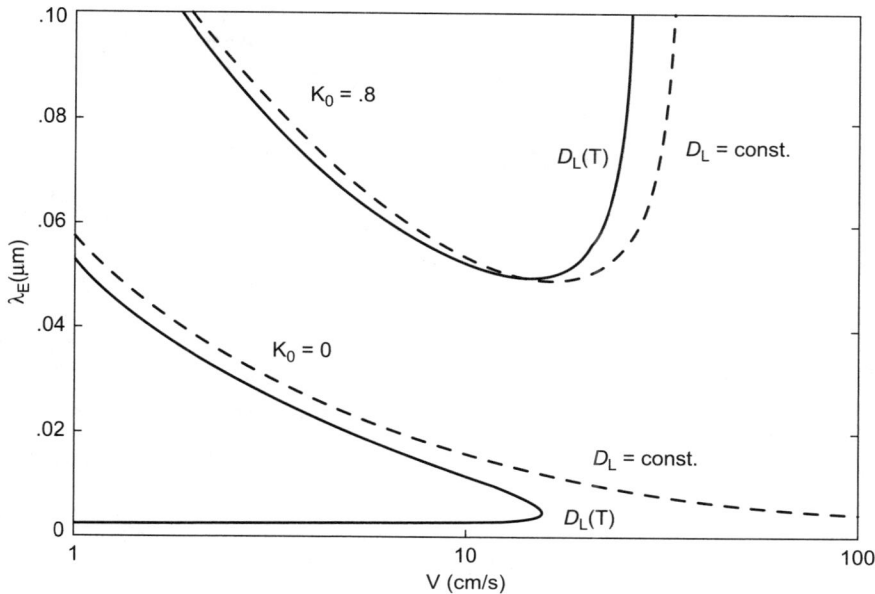

Figure 63 Plots showing variation of λ_E with V when D_L is assumed constant and when D_L depends on temperature. At high solidification velocities the $\lambda_E^2 V$ "constant" depends on the Peclet number and the relationship between λ_E and V is altered (Trivedi et al., 1987).

velocity except that the function P_E in eqn (185) depends not only on the volume fractions of the solid phases but also on the Peclet number, the shape of the metastable extensions of the liquidus and solidus curves below the eutectic temperature, and the partition coefficients. Also at high velocity the supercooling can become sufficiently large that the temperature dependence of the liquid diffusion coefficient must be considered (Boettinger et al., 1980).

These considerations alter the $\lambda_E^2 V$ "constant" at high speed and the spacing versus velocity relation (**Figure 63**) in a way that depends strongly on the equilibrium partition coefficients of the two phases, taken to be equal in the Trivedi et al. (1987) analysis. Two cases are distinguished depending on whether (a) the partition coefficients are close to unity or (b) they are close to zero. In case (a), the eutectic interface temperature is found to approach the metastable solidus temperature of one of the constituent solid phases as the velocity is increased. During this approach, the eutectic spacing actually *increases* with increasing velocity. Indeed eutectic solidification is replaced by single-phase microsegregation-free (planar) growth at high velocities. In case (b) the interface temperature cannot reach a metastable solidus curve of either phase. The supercooling becomes so large that the temperature dependence of the diffusion coefficient has a major influence and the spacing decreases with velocity faster than predicted by a constant $\lambda_E^2 V$ value. In both cases there exists a maximum velocity for coupled eutectic growth. In case (a) the eutectic is replaced by single-phase growth of one of the phases whereas in case (b) glass formation is likely if the interface temperature reaches the glass transition temperature where the melt viscosity increases substantially (diffusion coefficient plummets). In fact, the cases where the partition coefficients are close to zero are those that would exhibit plunging T_0 curves and lead to glass formation as described in

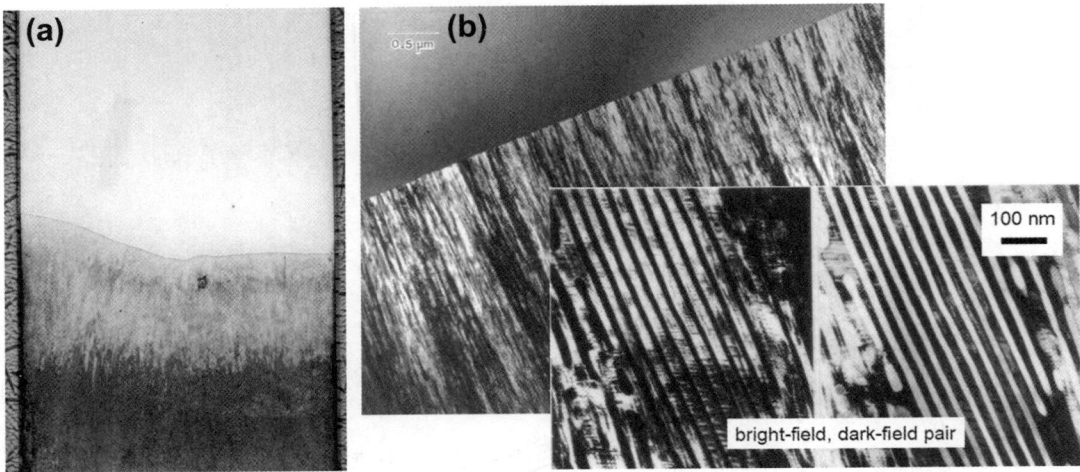

Figure 64   (a) Transition from fine eutectic growth to metallic glass when attempting DS at 2.5 mm/s. (a) Optical micrograph showing full diameter of directional solidified sample. (b) TEM micrographs at low and high magnification proving that the growth just before the transition to glass is a fine two-phase eutectic (Boettinger et al., 1980; Boettinger, 1982).

Section 7.3. This would have occurred naturally in the Trivedi et al. (1987)'s approach if nonequilibrium interface conditions had been included.

The asymmetry of the coupled zone for f–nf eutectics has an impact on these considerations. Glass formation often occurs in systems involving phase diagrams with f–nf eutectics; that is, between a solid solution and an intermetallic compound. Thus the composition with the smallest maximum growth rate for the eutectic (and hence easiest glass formation) may be shifted toward the direction of the faceted intermetallic phase. Case (a) above may also lead to the formation of a metastable crystalline phase if the eutectic interface temperature drops below the (metastable) solidus for such a phase. The concept of growth limitations of eutectic growth was explored in detail by Boettinger (1982).

7.8.1.7.2   Experiments—Glass Formation and Microsegregation Free Crystalline Solidification
An example of glass formation when a eutectic structure cannot grow fast enough is found in DS experiments on Pd–Cu–Si alloys in thin-walled narrow tubes with liquid metal cooling. It was shown that metallic glass could be formed by exceeding a critical growth velocity (about 2 mm s^{-1}) for eutectic solidification as shown in **Figure 64** (Boettinger et al., 1980; Boettinger, 1982). In this case ample nuclei were present but the eutectic alloy could not grow fast enough and glass was formed.

It is clear from the above that solidification velocity plays a dominant role in controlling microstructure. To control solidification velocity at high rates, surface melting and resolidification employing a moving heat source have been used to create small trails of material that are solidified at speeds close to the scan speed (Boettinger et al., 1984a,b). This technique is useful for speeds up to α/d where α is the thermal diffusivity and d is the diameter of the focused electron or laser beam. For higher speeds, surface melting and resolidification employing a pulsed laser or electron beam must be used to melt a thin surface layer. Maps giving the predominant microstructure as a function of speed and alloy composition are then produced. **Figure 65** shows such a map for Ag–Cu alloys. Similar maps have been constructed for Al–Al$_2$Cu (Gill and Kurz, 1994).

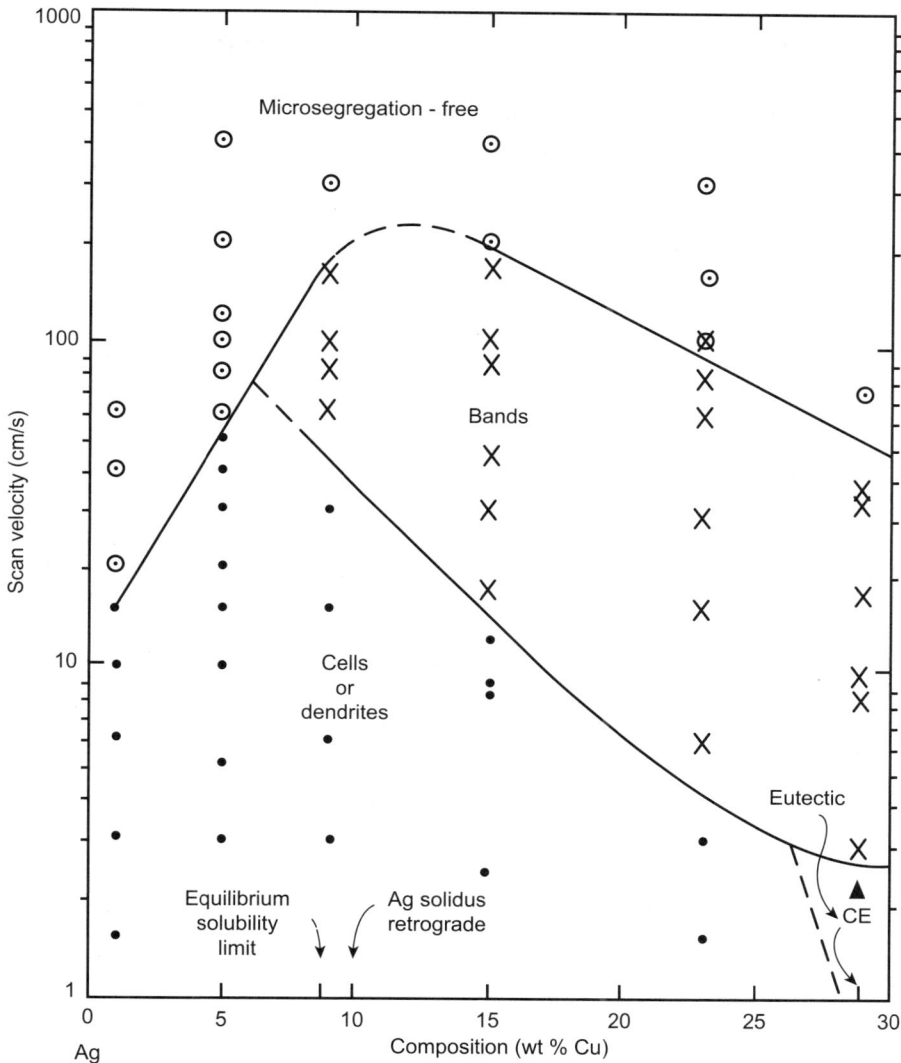

Figure 65 Experimental results for the variation in microstructure observed for Ag–Cu alloys depending on solidification velocity (Boettinger et al., 1984a).

In **Figure 65** four microstructural domains are obtained: cells/dendrites and eutectic microstructure at slow speed, bands at intermediate speed and microsegregation-free single-phase FCC at high speed. The boundary on the left is because of absolute stability and was described in Section 7.6.4. Eutectic growth ceases at approximately 2.5 cm/s generally following the description above except that an intermediate structure of bands is observed before single-phase growth dominates at high velocity.

This banded microstructure consist of thin (1 μm) regions *parallel* to the growth front that alternate between cellular solidification and cell-free solidification. The general character of this structure is because of details of solute trapping (Merchant and Davis, 1990; Braun and Davis, 1991; Gremaud et al., 1991; Carrad et al., 1992). Ordinarily, interface kinetics requires that the temperature of a planar single-phase growth front decreases with increasing velocity. However over the range of velocity where the partition coefficient is approaching unity the interface temperature actually increases with increasing velocity. This reversed behavior is the basic cause of an instability that leads to the banded microstructure. At high speeds in the near eutectic Ag–Cu alloys, the microsegregation-free structures are caused by the fact the partition coefficient has gone to unity.

7.8.1.7.3 *Metastable Crystalline-Phase Formation*

An analysis of growth competition has also been highly successful at explaining observed transitions from microstructures involving stable phases to those involving metastable phases. In the Fe–C system the transition from gray cast iron (Fe-graphite) to white cast iron (Fe-Fe_3C) has been extensively studied (Jones and Kurz, 1980). This transition occurs at relatively slow speeds not normally considered to be rapid. However the same principles can be employed at higher rates for other alloy systems using appropriately modified kinetic laws. The competitive growth analysis must include the dendritic and eutectic growth involving the stable and the possible metastable phases. **Figure 66** shows the coupled zones for Al–Al_3Fe (the stable eutectic) and Al–Al_6Fe (the metastable eutectic) summarizing experimental and theoretical work of several groups (Adam and Hogan, 1972; Hughes and Jones, 1976; Gremaud et al., 1987). The metastable Al_6Fe phase forms at an increased solidification velocity when the interface temperature drops below about 920 K. One of the most striking results of this diagram is the fact that alloys with compositions far on the Fe-rich side of either eutectic can form a microstructure consisting of primary Al at large supercooling and increased velocity; that is, the Al phase is the first to freeze. Indeed a determination of the powder size dependence of microstructural transitions in Fe-rich

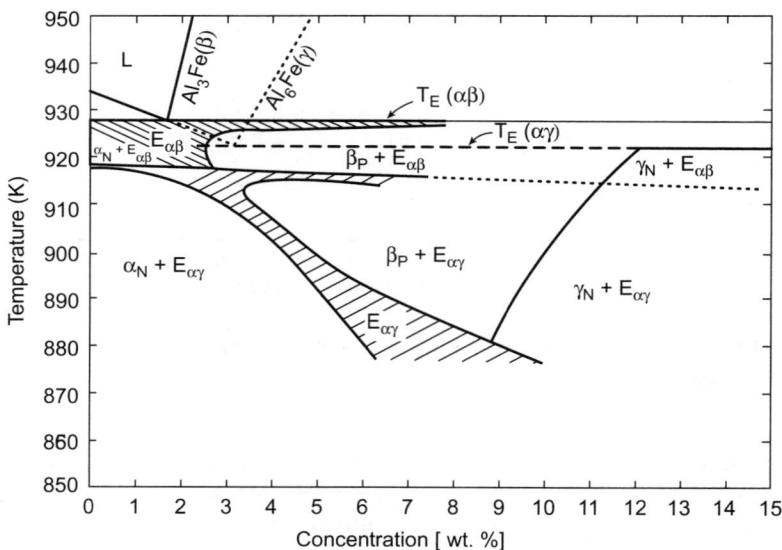

Figure 66 Theoretical coupled zones for the stable Al–Al_3Fe and metastable Al–Al_6Fe eutectics (hatched). The microstructure present at any value of interface supercooling and average alloy composition has been determined by a competitive growth analysis. Subscripts P and N refer to plate (2-D) and needle (3-D) dendrites, respectively (Trivedi and Kurz, 1988).

Al–Fe eutectic alloys from primary Al_3Fe, to eutectic $Al + Al_6Fe$, to primary Al as the powder size decreases is consistent with increasing velocities (supercoolings) calculated for the different size powders (Boettinger et al., 1986). Similar microstructural transitions have been observed for other Al-based transition metal alloys. The ability to form a matrix phase of an nf (usually ductile) phase for alloys with a large excess of alloying additions has been a major motivation for alloy development through rapid solidification processing. These alloys from a class of dispersoid strengthened materials.

An example of how one obtains phase diagram information for metastable phases is available for the Al–Fe system (Perepezko and Boettinger, 1983). An examination of the Al–Fe–Mn ternary system shows that the $Al_6(Fe,Mn)$ phase has a (stable) liquidus surface and an $Al-Al_6(Fe,Mn)$ monovariant eutectic valley. Extrapolation of this surface and the valley in the 3-D phase diagram to the Al–Fe binary reveals an estimate for the Al_6Fe liquidus and the $Al-Al_6Fe$ eutectic composition and temperature. A CALPHAD analysis of the ternary system would permit a more precise estimate of the metastable Al–Fe system.

7.8.1.8 *Eutectic Casting Alloys*

In this section we will discuss two most popular foundry alloys, Al–Si and Cast Iron. These eutectics are of the nonfaceted–faceted type. In these alloys, the low volume fraction phase in the eutectic (Si or graphite) has high entropy of melting and the microstructure is often termed irregular. The low volume fraction phase usually grows as fibers when its volume fraction is low (typically less than 0.25). However, lamellar structure can be formed even if the volume fraction of the second phase is very low if the specific interface energy between the phases is highly anisotropic (e.g. Fe–C alloys). It may be mentioned here that the graphite phase in ductile iron does not grow as a fiber. Ductile iron eutectic is an exception. In the cast iron literature, spheroidal graphite (SG) cast iron (i.e. ductile iron) eutectic is often called a "divorced eutectic," a usage different from that described in Section 7.8.1.6. There is no coupled growth of the austenite and graphite phase in this eutectic.

The mechanical properties of these metal–nonmetal (nonfaceted–faceted) eutectics are dominated by the morphologies in which the nonmetals solidify. As nonfaceted–faceted eutectics, the asymmetry of the couple zone must always be considered in the interpretation of microstructure regarding the presence or absence of primary phases as described in Section 7.8.1.5. In both systems the structure can be modified either by rapid cooling or by controlled addition of specific elements. The use of elemental additions has an advantage because their effect is essentially independent of the casting section thickness. Thus we shall only discuss modification by additives. The modification of the structure of these alloys and the resultant effect on the mechanical properties are clear examples of the manipulation of the structure based on the application of fundamental principles.

7.8.1.8.1 *Aluminum–Silicon Alloys*

Many studies have been focused on the mechanism of modification in Al–Si alloy and several reviews are available (Chadwick, 1963; Smith, 1968; Hellawell, 1970; Granger and Elliot, 1988). Most important is the change of the morphology of the silicon phase in the eutectic mixture. Day and Hellawell (1968) studied the morphology of the eutectic in directionally solidified alloys. By varying the temperature gradient in the liquid at the solid–liquid interface (G_L) and solidification velocity (V), they obtained five different types of microstructure. Massive silicon particles were observed at low V and high G_L values (when the solidification interface was almost planar). At lower G_L values and low V values, a variety of microstructures were observed. At higher growth rates (exceeding approximately 6 $\mu m\ s^{-1}$), a flake-like, interconnected silicon phase was observed. Most chemically unmodified Al–Si alloys exhibit this type of irregular eutectic.

In unmodified alloys, silicon particles appear more or less spherical in shape at low magnification although fibrous morphology can be observed at higher magnification. During the growth of unmodified Al–Si eutectic, the Si flakes contain widely spaced {111} twins that provide for easy growth in the [111] direction and difficult growth normal to [111]. This is called twin-plane reentrant edge (TPRE) growth. In unmodified alloys, the Si phase in the eutectic appears as coarse flakes that grow more or less independently of the Al phase.

With small additions of alkaline or alkaline earth metals (especially Na and Sr), the Si phase takes on a somewhat finer branched fibrous form that grows at a common liquid–solid interface with the Al phase to form a composite-like structure with improved properties. The easy branching of the modified Si leads to a more regular and finer structure as described in Section 7.8.1.3. Modifiers also change the morphology of primary Si in hypereutectic alloys. Although nucleation studies have been performed (Crossley and Mondolfo, 1966; Ross and Mondolfo, 1980), the modifying effects of Na and Sr are now thought to be growth related (Hanna et al., 1984). Nucleation remains important however, through the addition of Al–Ti–B master alloys to control the Al grain size in hypoeutectic alloys and through the addition of P (usually Cu–P) to promote heterogeneous nucleation (on AlP) and refinement of primary Si in hypereutectic alloys.

In modified alloys, the Si fibers contain a much higher density of twins that exhibit an internal zigzag pattern (Lu and Hellawell, 1987, 1988). Since both Na and Sr are concentrated in the Si phase, these authors proposed a mechanism whereby the modifying elements are adsorbed on the growth ledges spewing out from the reentrant corner. These adsorbed atoms cause the formation of new twins because of stacking errors on the growing interface caused by size mismatch of the Si and the modifier. A hard sphere model for atomic packing was used to define a critical ratio of 1.65 of modifier to Si atomic size that promotes twinning. The growth of the Si is then thought to occur by repeated twin formation in a more isotropic manner than by TPRE governed growth. This more isotropic growth permits the Si fibers to branch and adjust eutectic spacing. The formation of the internal zigzag twin structure is also consistent with the observation of microfaceting on the Al–Si interface (Lu and Hellawell, 1987).

Major and Rutter (1989) proposed that a certain concentration of Sr is required at the interface to achieve modification (Clapham and Smith, 1988). Below a critical concentration, growth of the Si is by the TPRE mechanism typical of unmodified eutectic. If growth occurs for a sufficient distance to accumulate Sr concentration at the solid–liquid interface above a critical level, the reentrant edges are poisoned. Then, new twins form as described above. A continuous cycle of twin formation, TPRE growth, poisoning, new twin formation and so on can occur. Qiyang et al. (1991) confirmed the adsorption of Na on {111} Si in agreement with the poisoning of the twin reentrant edges. The phenomenon of overmodification can be explained as complete suppression of the TPRE mechanism resulting from elevated quantities of the modifying addition. In this way, formation of Al bands in overmodified structures (Fredriksson et al., 1973) may be explained.

The efficacy of the modification can be evaluated by cooling curve analysis of the modified alloy. Often a higher eutectic undercooling is observed for the solidification of a modified Al–Si eutectic than that for an unmodified eutectic. However, this is not the sole indicator of the efficiency of the modification as unmodified alloy can exhibit similar or even larger eutectic undercooling. The modified eutectic growth has a quadratic dependence on the undercooling, which includes the kinetic undercooling at the eutectic front. The growth constant has been evaluated by experiment (Degand et al., 1996). This growth constant was found to decrease with the increase in the degree of modification. The traditional Jackson–Hunt law of regular eutectic growth can be applied for the case of modified Al–Si eutectic utilizing this growth constant for computing the liquid–solid transformation kinetics.

7.8.1.8.2 Cast Iron

It is known that cast iron, belonging to the family of high-carbon Fe alloys, can solidify according to either the stable iron–graphite system (gray iron) or the metastable Fe–Fe_3C system (white iron). As a consequence, the eutectic may be austenite–graphite or austenite–cementite (ledeburite). Furthermore, the complex chemical composition of the material has important and powerful effects on the structure of cast iron. Commercial alloys usually contain Si, minor additions of S, Mn and P and trace elements such as Al, Sn, Sb and Bi as well as the gaseous elements H, N and O. Both forms of cast iron (white and gray) have technological importance. Several comprehensive reviews and books have been published both from fundamental and from technological points of view (Morrog, 1968a,b; Minkoff, 1983; Elliot, 1988; Stefanescu, 1988; Craig et al., 1988; Hughes, 1988; Stefanescu et al., 1988). Gray iron is the most interesting because of the different morphologies that the graphite can achieve and the resulting differences in mechanical and physical properties. Although semantic problems have confused scientists and foundrymen in the past, a general understanding of the mechanisms of nucleation, growth and modification of the graphite phase has occurred in the last couple of decades. In this section the present status of knowledge in the area will be discussed briefly.

It is known that the growth of the stable Fe-graphite eutectic is favored over the metastable Fe–Fe_3C eutectic at low solidification velocities or by the addition of elements such as Si and Al. These elements increase the temperature difference between the stable Fe-graphite and metastable Fe–Fe_3C eutectic temperatures. In addition a wide variety of compounds have been claimed to serve as nuclei for graphite, including oxides, silicates, sulfides, nitrides, carbides and intermetallic compounds. Most of the nucleation mechanisms are connected with impurities existing in the melt or with inoculants that promote heterogeneous nucleation of the graphite. Although other inoculants are used, Fe–Si inoculants are the most powerful and popular (Elliot, 1988; Stefanescu, 1988; Skaland et al., 1993). The liquid–solid transformation in cast iron can be understood by considering how either the primary austenite or the primary graphite/cementite phases nucleate and grow before the onset of the eutectic solidification.

Tian and Stefanescu (1993) performed quenching experiments to study the nucleation of austenite grains in hypoeutectic cast irons. With an increase in cooling rate, the number of austenite dendrites that nucleated from the melt increased. The number of austenite grains nucleated had a quadratic dependence on the melt undercooling. Because of the relatively symmetric coupled zone in the white cast iron eutectic compared with the gray iron, the primary austenite growth does not have much influence on the eutectic ledeburite solidification (Mampey, 2001). Since there is asymmetry in the stable gray iron-coupled eutectic zone, effects of cooling will be more pronounced on the amount of austenite that forms in comparison with the amount predicted by equilibrium solidification (solidification under global equilibrium conditions is a good approximation for the solidification of Fe–C because of the fast interstitial diffusion rate of C in Fe).

A comprehensive review of the mechanism of formation of graphite from liquid is provided by Banerjee and Stefanescu (1991) and Stefanescu (2009). They postulated that SG is the natural form of graphite that forms from liquid. The presence of impurities such as S and O alters this natural shape to a flake graphite form possibly because of a surface adsorption mechanism. Growth of intermediate graphite shapes such as compacted graphite may possibly result from a mixed growth mechanism (Zhu et al., 1985).

The morphology and characteristic of the eutectic, whether stable or metastable, with or without modification, are very important in determining the physical and thermal properties. Thus, it is worthwhile to consider the most important eutectic structures observed. The microstructures of these major forms are shown, for example, by Stefanescu (1988).

White irons: the metastable unalloyed $Fe-Fe_3C$ ledeburite eutectic is classified as quasi-regular. Hillert and Subbarao (1968) described the mode of growth of the eutectic as well as the orientations arising between Fe_3C and γ-austenite. Powell (1980) has shown that the eutectic structure can be modified by quenching. By adding Cr or Mg, a plate-like Fe_3C structure associated with equiaxed grains can be achieved (Stefanescu, 1988).

Gray irons: for high-purity Fe–C–Si alloys, the structure of the Fe–G eutectic is *spheroidal* (Sadocha and Gruzlesky, 1975). However as stated above the presence of impurities in the melt causes the graphite to take a flake morphology and gray flake iron is considered to be the *characteristic* form from a practical point of view. Modification of this structure gives different graphite morphologies: *nodular, compact* or *vermicular*, and *coral*. We shall be concerned only with the growth of eutectic structures without a primary phase and we will refer mainly to the three structures widely used in industry: flake, compact or vermicular, and nodular or spheroidal cast iron.

Gray flake irons: the growth of the flake structure is well understood. Once graphite has nucleated, the eutectic cell or colony grows in an approximately radial manner and each flake is in contact with austenite up to the growing edge. The crystals of graphite grow in the close-packed strong bonding "*a*" crystallographic direction using steps created by rotation boundaries. These boundaries are defects in the crystals in the form of rotations of the lattice around the <0001> axis. According to Minkoff (1990), the screw dislocations on $\{10\bar{1}0\}$ planes, which have been proposed as an alternative growth mechanism, are inactive (**Figure 67**a). Elliot (1988), Stefanescu (1988) and Skaland et al. (1993) discuss the effect of S and O as promoters of the flake graphite morphology on the basis of their adsorption on the high-energy ($10\bar{1}0$) plane. Thus, growth becomes predominant in the "*a*" direction. The result is a *plate-like* or *flake* graphite.

Nodular or SG irons: until recently it has been widely accepted that the growth of this eutectic begins with nucleation and growth of graphite in the liquid, followed by early encapsulation of these graphite spheroids in austenite shells. The result is eutectic cells presenting a single nodule (Wetterfall et al., 1972). Thus, it is common practice in the foundry industry to associate the number of nodules to the number of eutectic cells. However recent research by Sikora et al. (1990) and Banerjee and Stefanescu (1991) indicated the existence of simultaneous nucleation of both the dendritic austenite and the SG. Banerjee and Stefanescu (1991) proposed the following theory for the nodular iron eutectic: (i) graphite nodules and austenite nucleate independently in the liquid at the beginning of solidification. Austenite forms because of the asymmetry of the coupled zone; (ii) very limited growth of nodules occurs in liquid; (iii) the interaction between both phases during solidification is aided possibly by convection in the liquid; (iv) graphite nodule encapsulation in austenite occurs immediately after these interactions. This mechanism essentially gives rise to the formation of eutectic cells presenting several nodules in one austenite grain. This is a fact to be taken into account when micromodelling of the structure is attempted; (v) further growth of graphite nodules occur by diffusion of carbon through the austenite shell only after the nodules are attached in austenite; and (vi) austenite dendrites grow partly because of carbon diffusion and also because of melt undercooling and supersaturation. **Figures 68** and **69** show a schematic of this mechanism for both directional and multidirectional solidification. This proposed mechanism was proven by Sikora et al. (1990). Their study of microsegregation pattern around eutectic aggregate demonstrated that solidification in SG iron begins with independent formation of graphite nodules and austenite. Subsequently, graphite nodules are incorporated in growing austenite and grow further by diffusion of carbon through austenite shell.

Regarding the spheroidal growth of the graphite, several theories exist in the literature and they have been reviewed by Minkoff (1983), Elliot (1988) and Stefanescu (1988). Minkoff (1990) considers that the relationship among supercooling, melt chemistry and crystalline defects determines the spheroidal

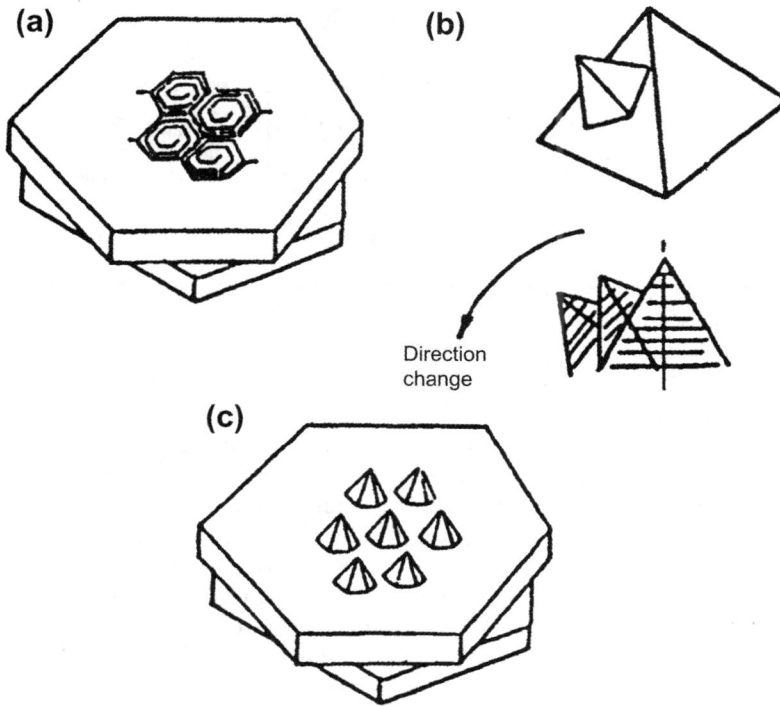

Figure 67 (a) Mechanism of growth of flake graphite from a rotation boundary which provides steps for the nucleation of (1010) faces. (b) Mechanism of growth of SG by repeated instability of pyramidal surfaces forming a radial array. (c) Mechanism of growth of compacted graphite by development of pyramidal forms on the crystal surface at steps because of screw dislocations (Minkoff, 1990).

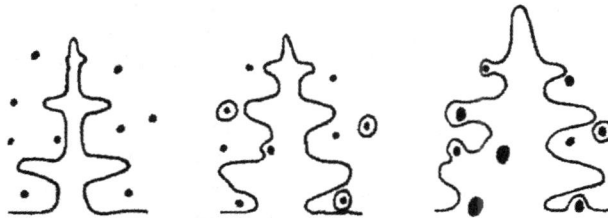

Figure 68 Schematic illustration of the progression of growth of the austenite–SG eutectic in DS (Banerjee and Stefanescu, 1991).

growth of the graphite. In this case the screw dislocation mechanism is considered dominant in causing repeated instability of the pyramidal surfaces, so that a radial array of pyramids is formed (**Figure 67b**).

Compact or vermicular graphite irons: this intermediate graphite morphology has been studied extensively because of its technological importance (Riposan et al., 1985). The graphite is interconnected within the eutectic cell but its growth differs from flake graphite. As in the case of spherulitic growth, several theories exist (Stefanescu, 1988; Elliot, 1988). The influence of the melt chemistry is very important. The occurrence of compact graphite form requires a balance between flake-promoting

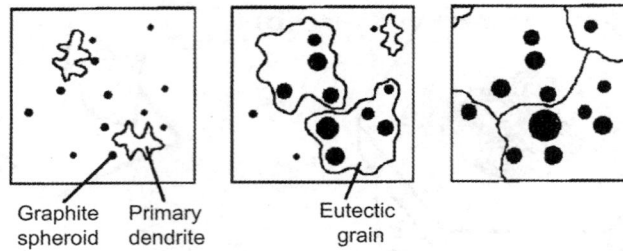

Graphite   Primary   Eutectic
spheroid   dendrite   grain

Figure 69   Schematic illustration of the sequence of formation of eutectic grains in sphedoidal graphite iron for the case of multiDS (Banerjee and Stefanescu, 1991).

elements, such as S and O, spheroidizing elements such as Mg, Ce and La, and antispheroidizing elements such as Ti and Al (Subramanian et al., 1985).

Minkoff (1990), in his general approach to the interdependence of supercooling, chemical composition and crystalline defects, considered that compact or vermicular graphite forms are intermediate between flake and spherulitic formation, and that the rounded morphology of the structure is a result of the thickening of graphite crystals at small values of supercooling by growth from the steps of screw dislocations, which have Burgers vector in the <0001> direction (**Figure 67c**). We consider that the mechanism of graphite growth as well as modification by proper chemical agents is an open research field. Unidirectional solidification of flake and compact graphite irons with and without "in situ" modification in front of the L–S interface open questions about the forms of crystalline growth of the graphite phase, as well as the influence of the impurities and chemical agents on the surface tension of the Fe–G eutectic phases (Roviglione and Hermida, 1994).

The complete structural transition in hypoeutectic Cerium-treated nodular cast iron has been extensively studied by Banerjee and Stefanescu (1991). They varied G/V ratio and level of Cerium impurity in DS experiments. They showed that the metastable to stable transition depends mainly on the G_L/V ratio. However, the transition from one graphite shape to another (e.g. flake or lamellar to compacted to nodular) depended mostly on the Cerium level in the melt (see **Figure 70**). Values of the shape factor (nodularity) indicated that only quasi-spheroidal nodules were possible in these cerium-treated cast irons. **Figure 71** shows the typical microstructure obtained for each type of graphite morphology in these experiments.

Based on the experimental work by Rickert and Engler (1985) and the work by Banerjee and Stefanescu (1991), a sequence of changes in the eutectic morphology in directionally solidified cast iron is proposed in **Figure 72**. As G_L/V decreases or C_0 increases (reactive impurity level, e.g. Ce, Mg), the solid/liquid interface changes from planar to cellular to equiaxed while the graphite shape remained largely flakes. Cooperative growth of austenite and graphite occurs. Further decrease of G_L/V and increase in C_0 results in a formation of an irregular interface, with austenite dendrites protruding in the liquid. Graphite becomes compacted and then spheroidal. Eutectic growth is divorced.

Gray to white eutectic transition: the unexpected occurrence of white iron (chill) in gray iron castings is a major cause of scrap and economic loss. The tendency for the formation of white iron depends on metallurgical quality of the melt and also the cooling conditions prevailing in the castings before the onset of eutectic solidification. The difference between the stable and metastable eutectic temperatures is about 6 °C in cast iron. This difference can be increased or decreased by addition of alloying elements. Graphitizers such as Si and Ni increase the stable eutectic temperature and reduce the metastable temperature, while carbide promoters such as Cr and V have opposite effect on the eutectic

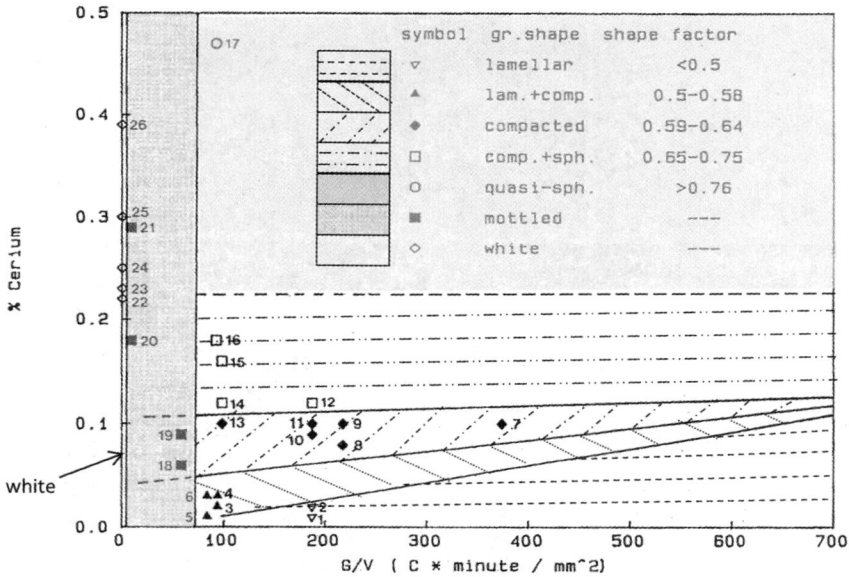

Figure 70 Influence of *G/V* ratios and wt% Ce level on structural transition in cast iron (Banerjee and Stefanescu, 1991).

temperatures. The physical model proposed for the evaluation of stable/metastable eutectic transition is based on a concept of a critical cooling rate. At cooling rates smaller than the critical cooling rate, stable eutectic is formed while at cooling rates higher than the critical cooling rate white iron eutectic is formed. Upadhya et al. (1990) applied this concept to a test casting with varying section thicknesses. They experimentally determined the value of critical cooling rate for a hypoeutectic gray iron casting. Subsequently they used this value of the critical cooling rate in a solidification modeling algorithm to predict where gray and white regions will form. Their numerical predictions compared reasonably well with experimental data (see **Figure 73a,b**). Magnin and Kurz (1985) proposed a similar concept based on the existence of critical solidification velocity. However, nucleation potential of the melt and influence of melt chemistry on the transformation temperatures also need to be considered. Fras and Lopez (1993) proposed a concept of *chilling equivalent* in an empirical expression that included the amount of melt superheat, nucleation and growth coefficients, and alloy specific heat. It was shown that chilling tendency decreases with an increase in pouring temperature, eutectic grain density and eutectic growth rates. It may be emphasized here that this type of analysis of chilling behavior can be applied in a foundry only where melting and inoculation practices are accurately monitored and controlled. Under these conditions, this type of approach for prediction of structural transition can be used as a process control tool.

7.8.2 Monotectic Solidification

In some metallic systems, the liquid separates into two distinct liquid phases of different composition during cooling before any solidifcation. On the phase diagram, the range of temperature and average composition where this separation occurs, as well as the compositions of the two liquid phases are given by a dome-shaped curve that defines the miscibility gap. The maximum temperature of the

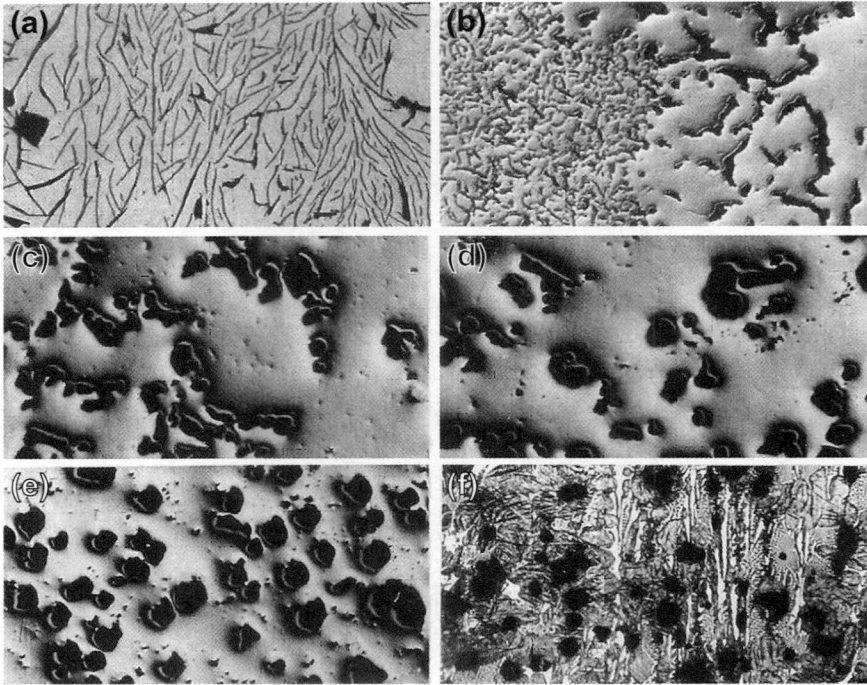

Figure 71 Typical microstructures for various structural regions in Ce-treated hypoeutectic irons, after DS: (a) lamellar graphite, (b) lamellar + compacted, (c) compacted, (d) compacted + spheroidal, (e) quasi-spereoidal, and (f) mottled. 50× (Bandyopadhyay (Banerjee) et al., 1990).

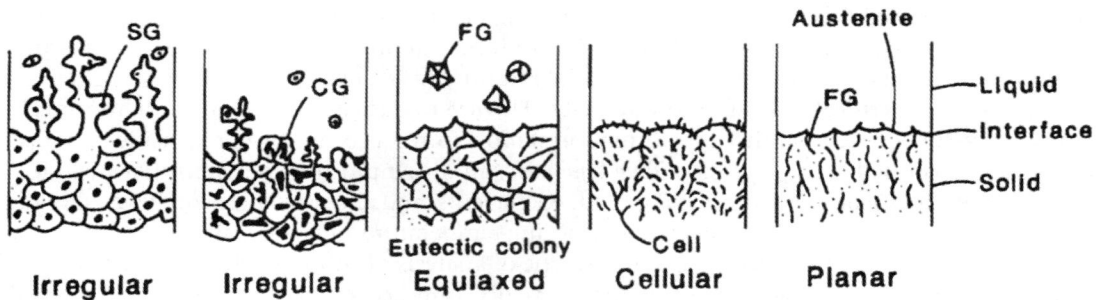

Figure 72 Possible morphologies of the solid–liquid interface and of the graphite in eutectic cast iron (Banerjee and Stefanescu, 1991).

miscibility gap is called the critical temperature. An example of a miscibility gap is shown in **Figure 74** for the Al–In system.

Even for alloys outside the miscibility gap, such as for Al-rich alloys in **Figure 74**, a consideration of the miscibility gap is important in developing an understanding of solidification microstructure. Under ordinary conditions, solidification of these alloys begins with the formation of dendrites of the Al solid

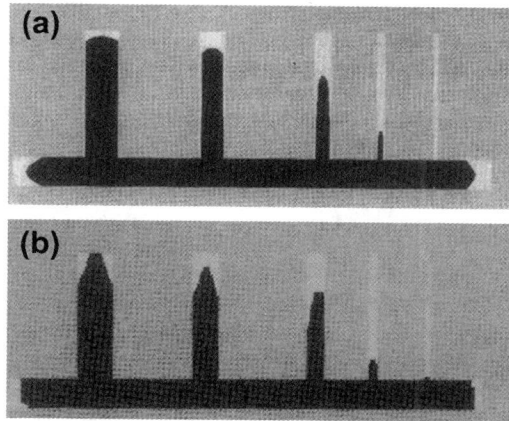

Figure 73   Maps of gray and white cast iron structural regions in a casting: (a) experimental, (b) calculated (Upadhya et al., 1990).

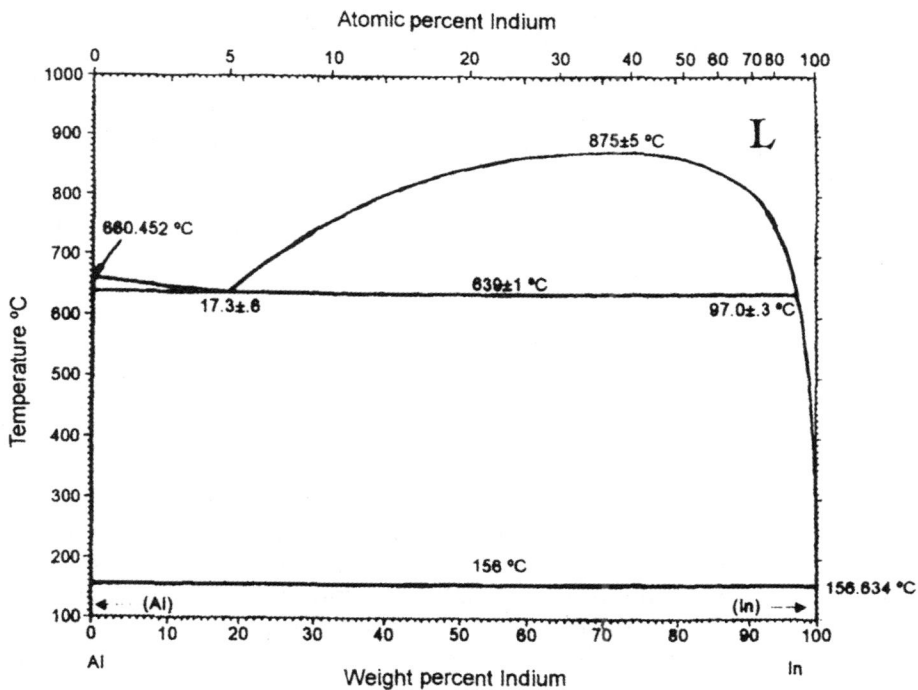

Figure 74   Al-In monotectic-type diagram (Murray, 1983b).

phase and enrichment of the liquid remaining between the dendrites with In until the composition reaches the edge of the miscibility gap. This composition (17.3 wt% In, **Figure 74**) defines the monotectic composition and temperature where the "reaction" liquid $L_1 \rightarrow$ solid S_1 + liquid L_2. Formally this reaction is the same as a eutectic reaction except that on cooling, one of the product phases is a liquid, the liquid defined by the other In-rich side of the miscibility gap at 97% In. At much lower temperatures this liquid may solidify in a terminal eutectic reaction $L_2 \rightarrow S_1 + S_2$.

Some sulfide and silicate inclusions in commercial Fe-based alloys are thought to form by monotectic solidification (Flemings, 1974). Free-machining Cu alloys containing Pb also involve this reaction. Considerable research has been focused on DS for fundamental reasons but also because of the potential for producing aligned growth of composites, or (with selective removal of one phase) thin fibers or microfilters (Grugel and Hellawell, 1981). For this latter purpose it is important to describe the possibilities of coupled growth of the S_1 and L_2 phases from the L_1 phase of monotectic composition.

7.8.2.1 Directional Solidification of Monotectic Alloys

As in the case of eutectic solidification, a wide variety of microstructures can be produced by DS of monotectic alloys. Lamellar microstructures are not observed in monotectic systems because the volume fraction of the L_2 phase is usually small. Three types of structures are observed. The first and most interesting and useful, typified by Al–In (Grugel and Hellawell, 1981), is a regular fibrous or composite structure that consists of closely packed liquid cylinders of a uniform diameter embedded in a matrix of the solid phase. These liquid cylinders solidify at much lower temperatures to solid rods of In by a divorced eutectic reaction. If the growth rate is increased, the distance between cylinders decreases and the structure gives way to a second type of microstructure that consists of discrete droplets of L_2 embedded in a solid matrix. The third type, typified by Cu–Pb (Livingston and Cline, 1969), is more irregular consisting of interconnected globules that take on some degree of alignment as the growth rate is increased. Although the L_2 cylinders formed at a monotectic reaction are susceptible to ripening and spheroidization during subsequent cooling, the droplet and irregular structures are not thought to form by coarsening (Grugel and Hellawell, 1981).

One of the most important considerations for understanding the different microstructures comes from a consideration of whether a stable triple junction can exist between L_1, L_2, and S_1 (Chadwick, 1965). This condition can only occur if

$$\gamma_{S_1 L_1} + \gamma_{L_1 L_2} > \gamma_{S_1 L_2}. \qquad (188)$$

If the inequality is satisfied, regular fibrous structures can be obtained. The eutectic theory for rod growth can then be applied although some modifications are required to treat the increased diffusion in the L_2 phase (Grugel and Hellawell, 1981). When this inequality is not satisfied, L_2 does not "wet" S_1, and L_1 will tend to coat the interface between L_2 and S_1; that is, L_1 preferentially wets S_1 to the exclusion of L_2. Cahn (1979) calls this the *perfect wetting case* and during monotectic growth, the L_2 phase will form droplets in the L_1 phase just ahead of the growing S_1 interface. As growth proceeds, the droplets are pushed by the interface and the size of the droplets increases until they reach a critical size where they are engulfed into the growing S_1 solid. The critical size for engulfment is determined by microscopic fluid flow around the droplet. As the solidification velocity is increased, irregular semicontinuous liquid rods can be partially engulfed in the solid as shown schematically in **Figure 75**. This irregular engulfment is believed to be the origin of the irregular globular microstructure typified by Cu–Pb.

Cahn (1979) showed that in monotectic systems, the perfect wetting case should be expected if the temperature difference between the critical temperature of the miscibility gap and the monotectic temperature is small. Thus, irregular composite structures are formed in these systems. When the temperature difference is large, perfect wetting does not occur, a stable triple junction can exist, and regular composite growth is expected. This idea was confirmed by the addition of a ternary element to

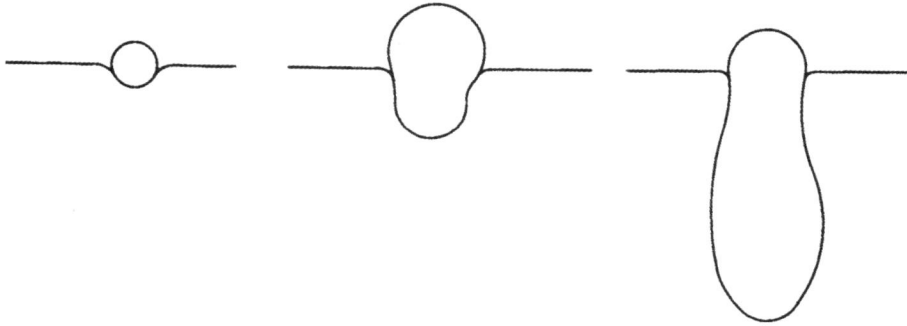

Figure 75 Schematic sequence to show liquid particle pushing, growth, and engulfment during irregular monotectic growth (Grugel et al., 1984).

a binary monotectic alloy by Grugel and Hellawell (1981). This addition altered the height of the miscibility gap and hence the wetting behavior. Grugel et al. (1984) found experimentally that the border between systems with regular and irregular composite structures occurred when the ratio of the monotectic temperature to the critical temperature on an absolute temperature scale is approximately 0.9.

For regular fibrous growth, the spacing varies inversely with the square root of the velocity. Grugel et al. (1984) showed that the "(spacing)^2 velocity" constant was about an order of magnitude larger for irregular growth than for regular growth. Derby and Favier (1983) have presented a different model for the occurrence of regular and irregular structures similar to those used for irregular eutectics. Kamio et al. (1991) have shown that the value of the temperature gradient has a large effect on the transitions between aligned growth and the formation of droplets.

Grugel and Hellawell (1981) also examined whether composites could be grown for compositions different than the exact monotectic composition. They found that the dendritic growth of the S_1 phase could be suppressed by sufficiently large values of G_L/V to permit planar composite growth just as for off-eutectic alloys. Attempts to grow composites with compositions on the other side of the monotectic (within the miscibility gap) failed because of convective instabilities. Reduced gravity experiments have been employed by Andrews et al. (1992) to avoid these difficulties.

Nestler et al. (2000a) performed phase field calculations of monotectic growth including fluid flow in the two liquid phases. They were able to show the conditions under which a stable two phase composite could be grown as well as the conditions that lead to irregular microstructures and liquid droplet pushing.

7.8.2.2 Rapid Solidification of Monotectic Alloys

Some alloys whose phase diagrams do not contain a miscibility gap or a monotectic reaction form microstructures consisting of droplets embedded in a matrix of a primary phase after rapid solidification. If the liquidus curve has a portion where the slope is close to zero, a metastable miscibility gap lies just beneath the liquidus curve. In fact the temperature difference between the liquidus and the metastable critical point is proportional to the liquidus slope (Perepezko and Boettinger, 1983). Thus with the supercooling inherent in many rapid solidification processes, alloy microstructure can be influenced by the presence of the metastable miscibility gap and its associated metastable monotectic reaction. The microstructure of rapidly solidified Al–Be alloys, which consists of fine Be particles in an

Al matrix, have been interpreted in this manner (Elmer et al., 1994). In fact even some slowly cooled alloys can exhibit microstructures characteristic of monotectic solidification even though there is no apparent miscibility gap. Verhoeven and Gibson (1978) showed that oxygen impurities raise the metastable miscibility gap in the Cu–Nb system so that it becomes stable and produces droplet microstructures.

7.8.3   Peritectic Solidification

The phase diagram for the Pb–Bi system is shown in **Figure 76a**. If a liquid with 33 wt% Bi is cooled, and global equilibrium could be maintained (see Section 7.3), the alloy would be composed of $L + \alpha$ at a temperature just above the peritectic temperature of 184 °C, denoted T_p, and would be composed of single-phase β just below T_p. This gives rise to the notion of a peritectic "reaction" that occurs on cooling that is written as $L + \alpha \rightarrow \beta$. However the solid-state diffusion required to accomplish this "reaction" during the solidification process greatly reduces the amount of the β phase formed with the exception of interstitial alloys.

7.8.3.1   *Peritectic Solidification during Dendritic Growth*

Under conditions where the α phase grows dendritically, the β phase will usually begin to form along the surface of the α phase as shown in the two drawings shown in **Figure 50b** and c. Although the β phase can be formed by three mechanisms, the most important during continuous cooling is the formation of β directly from the melt. The simplest way to treat this situation is to employ the Scheil approach in a small volume of the interdendritic liquid with the usual assumptions: local equilibrium at the solid–liquid interface, uniform liquid composition at each instant (temperature) and no solid diffusion. At small fraction solid, solidification of α phase occurs in the normal way with build up of solute in the liquid between the dendrites following the Scheil equation. When the liquid composition reaches 36 wt% Bi (see **Figure 76**), denoted C_p, solidification switches from the α phase to the β phase.

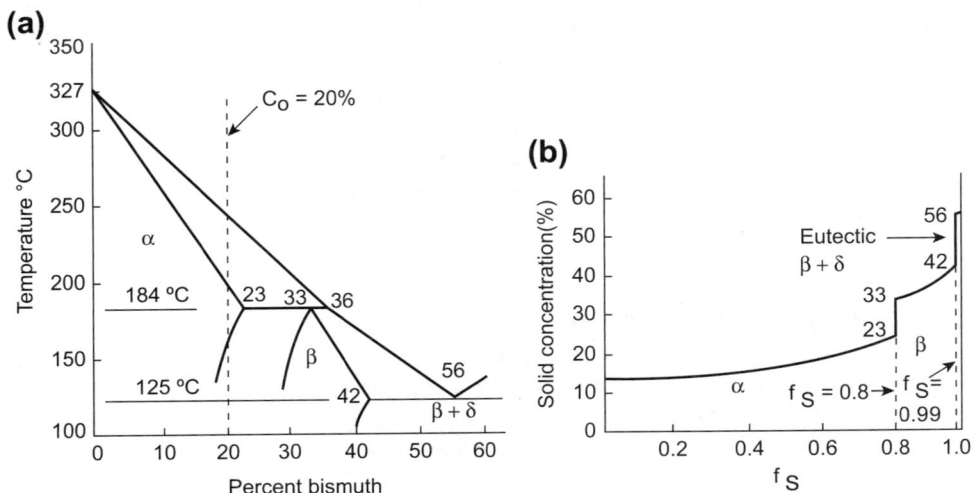

Figure 76   (a) Pb–Bi peritectic phase diagram. (b) For the solidification of a 20% Bi alloy, concentration (wt% Bi) in the solid according to a Scheil model of solidification (Flemings, 1974).

A different value of the partition coefficient given by the β liquidus and solidus must then be employed in the Scheil equation to follow the continued enrichment of the liquid composition in the component Bi. Often one must employ a concentration dependent partition coefficient for the β phase in peritectic systems, a situation that may require numerical solution of the differential form of the Scheil equation. The solid composition and the fraction of α and β phases formed by this mechanism using the Scheil model are shown in **Figure 76b**. For Pb–Bi alloys the final solidification product is eutectic. One notes that using the Scheil approach, *any* alloy composition to the left of 36 wt% (in **Figure 76a**) will contain α phase in the solidified microstructure. Many systems involve a cascade of peritectic reactions, with solidification switching from phase to phase forming separate layers around the initial α dendrite.

The second and third mechanisms for the formation of β are less important and more difficult to model. The geometry and connectivity of the L, α and β phases determine their relative importance. Both decrease the fraction of α phase from that predicted by a Scheil analysis. They have been referred to as the *peritectic reaction* and the *peritectic transformation* by Kerr et al. (1974). The peritectic reaction requires that all three phases be in contact with each other. This occurs in the vicinity of the liquid–α–β triple junction and involves partial dissolution of the α phase and solidification of the β by diffusion of solute through the liquid from the L–β boundary to the L–α boundary (**Figure 77a**). Hillert (1979) gives an approximate analysis of this process.

The third way that β phase can form, the peritectic transformation, involves solid-state diffusion and the motion of the α–β interface during cooling as shown in **Figure 77b**. This mechanism is

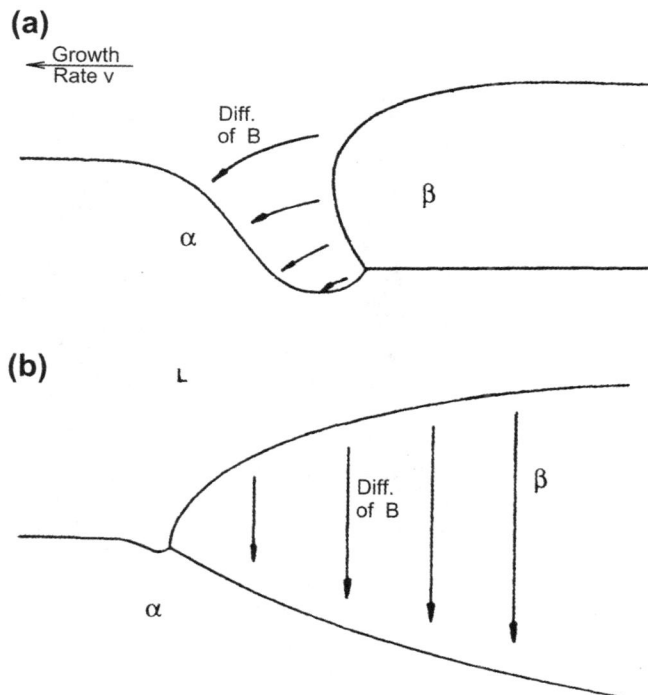

Figure 77 Peritectic reaction and peritectic transformation. (a) In the peritectic reaction a second solid phase, β, grows along the surface of the primary phase α by diffusion through the liquid. (b) In the peritectic transformation, diffusion of B atoms through the already formed solid phase β occurs (Hillert, 1979).

very important when the solid diffusion coefficient is large; for example, for interstitial solutes such as carbon in Fe. Indeed the peritectic reaction in low-carbon steels ($L + \delta$-Fe $\rightarrow \gamma$-Fe) seems to go to completion; that is, no δ-Fe is observed in these alloys. The diffusion problem that governs the motion of the α–β interface involves long-range (scale of dendrite spacing) transport of solute from the liquid across the β phase to the α–β interface. At this interface, this flux of solute causes the α phase to dissolve at the expense of the growing β phase. The analysis must use compositions for the interfaces that are given by the $L + \beta$ and $\alpha + \beta$ two phase fields on the phase diagram. Hillert (1979) has given an approximation for the growth of the β phase by solid-state diffusion. Because of the long-range diffusion, the thickness of the β phase increases with the square root of time if the interface compositions and diffusion coefficients can be assumed to be independent of temperature.

Fredriksson and Nylen (1982) have measured the fraction of L, α, and β phases as a function of distance behind the peritectic isotherm by quenching various alloys during DS. The relative contributions of the three mechanisms are analyzed and compared with the measurements. In one alloy (Al–Mn) the β phase did not grow along the L–α interface but grew independently from the melt.

A phase field simulation of solidification by Hecht et al. (2004) of an Fe–C–Mn alloy (**Figure 78**) shows the formation of the peritectic sheath of austenite forming over the primary ferrite dendrite. The solid-state peritectic transformation is shown far behind the dendrite tip region where the ferrite phase is being eaten by the austenite phase.

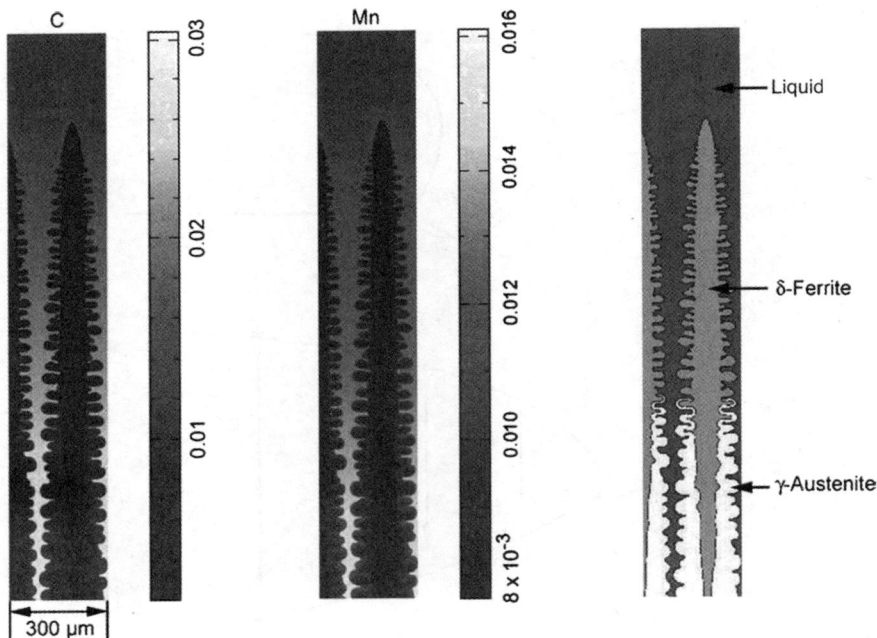

Figure 78 Phase field simulation of solidification of Fe-1 at% C-1 at% Mn at a cooling rate of 1 K s^{-1} in a temperature gradient of 2×10^4 km^{-1}. Shown are the composition maps of carbon and manganese in atom fractions as well as the phase map (Hecht et al., 2004).

7.8.3.2 Aligned Peritectic Growth

A two phase solid–solid region exists at low temperatures in binary peritectic phase diagrams. Of interest are attempts to achieve coupled growth of the α and β phases to produce a fine two-phase composite structure. It was thought that if the G_L/V ratio were large enough to suppress dendritic and cellular solidification of the α phase and force a planar solidification front, then coupled growth of the two solid phases might be possible. This would also require that the composition of the liquid near the interface be maintained near C_p, a liquid composition from which both α and β could form. When the G_L/V ratio was high enough to suppress dendritic and cellular growth of the primary phase Boettinger (1974) and Brody and David (1979) both obtained banded structures with alternating layers of the α and β phases rather than coupled growth. The systems were found to be extremely sensitive to minor growth rate fluctuations that lead to the solidification of the alternating bands of α and β. These banded structures have been studied theoretically by Trivedi (1995) for possible use to determine heterogeneous nucleation temperature. However a subsequent attempt by Vandyoussefi et al. (2000) was successful in obtaining couple growth in the Fe–Ni system. Work by Lo et al. (2003), Dobler et al. (2004), and Dobler and Kurz (2004) have elucidated this interesting solidification phenomenon.

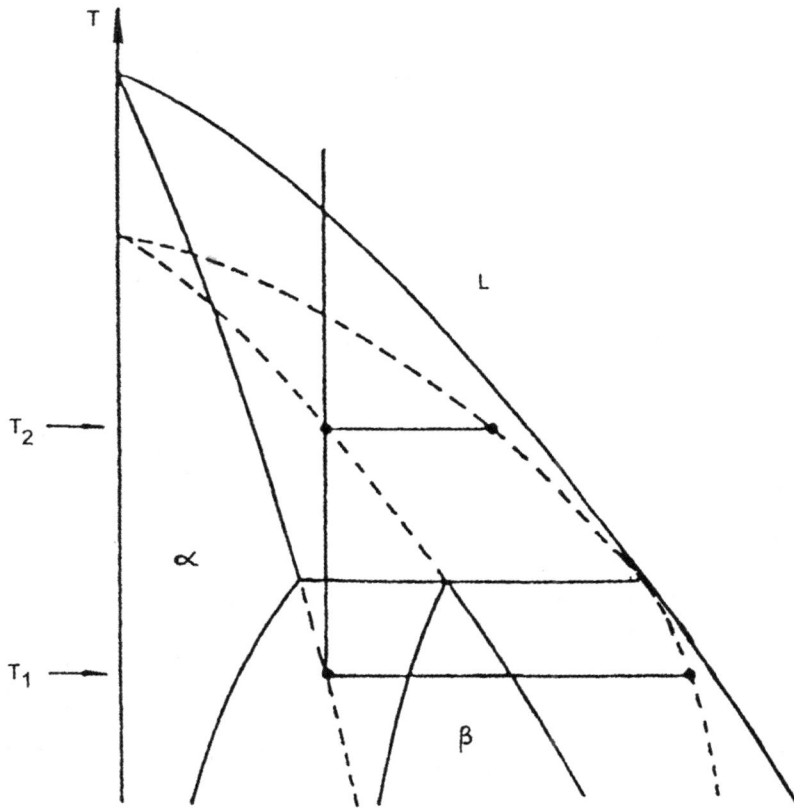

Figure 79 Peritectic phase diagram with metastable extensions of α and β liquidus and solidus curves below and above the peritectic temperature, respectively. For an alloy of the composition of the vertical line, planar steady-state growth of either α or β is possible at the temperatures T_1 and T_2, respectively (Hillert, 1979).

7.8.3.3   Rapid Solidification of Peritectic Systems

As shown in **Figure 79**, the metastable extension of the liquidus curve of the α phase below the peritectic temperature must lie below the stable β liquidus. However when compared with a eutectic system, the metastable α liquidus is relatively close to the stable β liquidus. Thus there exists considerable opportunity for the formation of the α phase directly from the melt at modest levels of supercooling at compositions where it is not expected. Whether or not this happens depends on the competition of nucleation and growth kinetics for the α and β phases. An example of this kind of behavior is found in the classic experiments of Cech (1956) on solidification of small droplets (3–30 μm) of Fe-30 wt% Ni alloys in a drop tube. This system contains a peritectic reaction, L + BCC → FCC. Experiments showed that BCC was formed from the melt (and not by solid-state transformation) at compositions with high Ni content where only FCC should have formed. In contradiction to the experimental facts, analysis of the dendritic growth kinetics of the competing FCC and BCC structures showed that FCC would be the favored product phase at all supercoolings for Fe-30 wt% Ni (Boettinger, 1988). This result is due primarily to the fact that the partition coefficient for BCC is much larger than for FCC. On the other hand, analysis of the nucleation behavior indicates that BCC is favored over FCC if the nucleation is homogeneous or if the contact angle on heterogeneities is greater than about 45° (Kelly and Vandersande, 1987). Thus only nucleation can explain the observed BCC structure.

In alloys containing a cascade of peritectic reactions, the close proximity of stable and metastable liquidus curves can also explain why one or more phases may be skipped in the layer structure that coats the initial dendritic phase in rapidly solidified alloys.

7.9   Cast Structure and Fluid Flow

The flow of molten metal or alloy is an important consideration during casting. Although the requirements for mold filling especially in narrow cross sections have been considered for centuries, the importance of flow near the freezing interface has been recognized much more recently. In this section we consider (i) the general origins of fluid flow in castings; (ii) the development of casting macrostructure (grain structure); (iii) the macrosegregation observed in ingots and castings because of flow in the mushy zone; and (iv) the formation of porosity and inclusions. Finally, the foundry concept of *fluidity* associated with the mold filling capability of alloys will be briefly described.

7.9.1   General Aspects of Fluid Flow in Castings

There are many aspects of material transport that can occur during solidification: (i) residual flow because of mold filling, (ii) thermal and solute-driven buoyancy convection, (iii) convection because of expansion or contraction of alloys upon solidification, (iv) convection driven by the variation of free surface energy with temperature and concentration, (v) effect of external forces (rotation, magnetic fields, etc.), (vi) floating or settling of free crystals because of density differences, (vii) dendritic breakage and transport, and (viii) pushing of equiaxed crystals and/or inclusions by the columnar solidification front.

Experimental observation of liquid metal motion is difficult. Cole (1971) and Weinberg (1975) extensively studied the flow in the fully liquid part of castings using radioactive tracers. They overcame the handicap of other experiments performed with transparent model liquids having much lower thermal conductivity than metals. Convection has the largest effect on thermal transport and macrostructure when the L–S interface (position of dendrite tips) is parallel to the gravity vector. Thus horizontal solidification is dramatically affected by convection because of the horizontal temperature gradients.

A situation where fluid flow makes a large change in the heat flow is found in welding (Kou and Sun, 1985). Here capillary flow on the free surface of a weld puddle can reverse the circulation direction depending on surface active additions and significantly alter the penetration depth. Limmaneevichitr and Kou (2000) have employed transparent analog materials to visualize these flows.

Flows in metals and the heat transfer because of the flow can be reduced by the application of a magnetic field resulting in an induced eddy current that exerts a body force on the fluid. Rotation of ingots gives an effect similar to the application of a magnetic field; here the body force is a Coriolis force that deflects particles of fluid in a direction normal to the axis of rotation, and normal to the direction of fluid motion. On the other hand, an increase in heat transfer can be accomplished by vigorous fluid motion near the L–S interface. Rotation or oscillations of the crucible, a rotating magnetic field, or electromagnetic field interactions can be used for this purpose (Cole, 1971).

Liquid alloys have viscosities similar to water and consequently, the flow is rapid and liquid alloys are usually treated as Newtonian fluids. Indeed turbulent flow occurs quite frequently during rapid mold filling increasing the possibilities for entrapment of oxide and for mold erosion. Thus smooth transitions between different mold segments are needed to reduce turbulence. Filters are typically used to choke liquid flow and reduce turbulence in the runners. Campbell (2003) suggests a maximum flow velocity of 0.5 m/s to avoid major turbulence.

Current mathematical models of flow used for castings usually assume laminar flow. When turbulence modeling is required (for high-pressure die casting [HPDC]), the flow is treated in terms of time-averaged quantities, since a direct numerical approach is considered impractical. The instantaneous quantities are described as the sum of a mean time-averaged value and a fluctuating quantity. As a consequence, the resulting fluid flow equations lead to unknown correlations between fluctuating part of the velocity and temperature. Turbulence models describe these correlations in terms of mean-flow quantities (Schlichting, 1979). One common model that is widely used for turbulence modeling is called k-ε model. In this model, turbulence kinetic energy, k and viscous dissipation rate of turbulent kinetic energy, ε define the turbulence field. The turbulence viscosity can be directly related to k and ε. Equations for turbulent quantities are solved along with basic conservation equations of mass, momentum, and energy to fully describe the flow field.

7.9.2 Effect of Flow on the Grain Size of Castings

The classical grain structure of ingots will be described, followed by descriptions of the source and survival of centers for growth of grains.

7.9.2.1 Description of Classical Ingot Structure

The classical representation of ingot macrostructure shows three distinct zones: the chill zone, which is a peripheral region near the mold surface, composed of small equiaxed grains, the columnar zone composed of grains elongated in the heat flow direction and a central equiaxed zone. Inside each grain a substructure of cells, dendrites, and/or eutectic exists. Fluid flow during solidification affects the origin and development of the three zones significantly. Extensive research has been performed because of the important influence of macro- and microstructure of ingots and castings upon subsequent as-cast mechanical properties or forming properties.

7.9.2.1.1 Chill Zone

The formation of the chill zone structure involves complex interactions of liquid metal flow, metal–mold heat transfer, heterogeneous nucleation and dendritic growth, a subject reviewed by Biloni

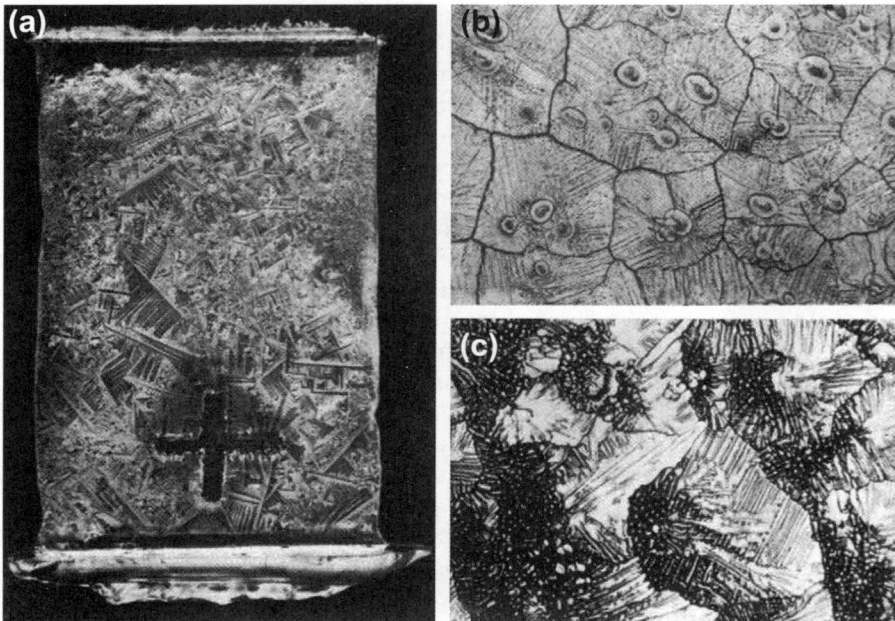

Figure 80   (a) Al-1 wt% Cu cast in a cold copper mold coated with lampblack, except in the area of the cross. (b) Substructure corresponding to the uncoated region of (a). Each grain has a predendritic region as origin. (c) Substructure corresponding to the coated region; notice the "cells," probably produced by a multiplication mechanism, as origin of the dendrites (Biloni, 1980).

(1980). When the liquid melt is rapidly cooled in the vicinity of a cold mold, heterogeneous nucleation on the mold wall or on other heterogeneous sites occurs producing a large number of initiation sites. The growth from these sites is limited by impingement with neighboring crystals formed at almost similar times. Therefore, sizes of these grains are usually very small and uniform with random orientations. Bower and Flemings (1967) experimentally simulated the thermal conditions existing in the chill zone using thin samples filled quickly by a vacuum technique. They found a dendritic substructure in the chill grains and established that a *grain multiplication* (*fragmentation*) mechanism induced by melt turbulence during pouring was quite important for formation of this substructure.

The effect of the mold coating modifications shows the importance of local values of the heat transfer in determining the chill zone structure. Biloni and Morando (1968) coated a copper mold with lampblack except for a small region. (cross in **Figure 80**). The chill grains in the region without the lampblack were smaller and contained a different predendritic substructure with solute rich cores than the region coated with lampblack. These solute-rich cores are an indication of solute trapping at the dendrite tips because of higher local solidification rate than was present in chill grains formed in the coated regions. Morales et al. (1979) showed how a tailored coating microprofile can influence the columnar grains originating from the chill zone (**Figure 81**). Mizukami et al. (1991) showed that the predendritic nuclei density per unit area for an Al-5 wt% Cu alloy poured in a bare copper mold decreased with an increase in the waviness of the surface (rugosity).

In an attempt to mimic the conditions occurring when the liquid metal first encounters a chill surface, Mizukami et al. (1993) studied the initial solidification of 18–8 stainless steel droplets ejected

Figure 81 Longitudinal section of an ingot poured from the bottom, after macroetch. The difference in grain size is due to mold walls with different microgeometries. The small columnar grains start at the asperities of an alumina mold coating presenting a controlled microgeometry; the very large grain started from a wall coated with a very smooth film of lampblack (Morales et al., 1979).

into cold Cu substrates using pressurized inert gas. They investigated both the microstructure and undercoolings in the vicinity of the substrate surface by varying parameters such as substrate roughness, inert gas pressure, and substrate and liquid steel droplet temperatures. They used a novel technique to measure the steel droplet temperatures at the contact surface of the substrate. A 0.5 mm diameter hole was made in the substrate. Through this hole, a silicon photodiode was used to measure the surface temperature with reasonable accuracy at about 1 μs time intervals. The cooling rate was computed before the onset of solidification by analyzing the recorded cooling curve. The cooling rate of steel droplets was influenced by the substrate material, both substrate and liquid droplet temperatures, and gas pressure. As expected, measured undercoolings increased as the cooling rate was enhanced. Also, as the cooling rate was increased, surface grain density was found to increase. Electron microprobe analysis was used to characterize the microstructure. Above a critical supercooling of about 50 K (equivalent in their experiments to a cooling rate of 5000 K/s), metastable cellular γ formed at the chill face that changed to the stable δ phase farther from the chill face. Below a supercooling of about 20 K–30 K, stable δ phase with dendritic microstructure formed at the chill face. Between these two supercooling ranges, a mixture of cellular and dendritic microstructure was observed. Thus, for this stainless steel, the conditions in the chill zone not only influence the grain density but also the phase selection between austenite or ferrite growth.

The control of chill surface quality during continuous casting of steel is a major focus for the determination of proper process parameters. Bobadilla et al. (1993) noted that the surface of vertically cast billets develop parallel depressions perpendicular to the casting direction. These depressions are called "oscillation marks" and were found to occur at regular spacings. These marks were associated with surface defects. Detailed analysis revealed that these oscillation marks were suitable for entrapment of inclusions during early solidification of steel. It was assumed that these oscillation marks resulted from the freezing of the curved part of liquid meniscus in contact with the cold mold. In summary, mold surface geometry, casting process parameters, surface defects, nucleation and growth of grains control the microstructure of the chill zone.

The fineness of the chill zone microstructure formed with high initial supercooling has been examined to obtain enhanced mechanical properties. Research by Yavari et al. (2008) showed that it is possible to obtain chill zone copper alloys of about 2–4 mm thickness with fracture strengths of about 1.9 GPa. Such high strengths result from the formation of nanocrystalline surface layer next to the mold contact surface. They showed that the melt can be significantly supercooled when the alloy composition is chosen such that the contact surface of the copper mold used does not serve as preferred nucleation sites for solidification. By reducing the casting section thickness, a copper alloy casting composed of chill zone alone is possible with strengths similar to those of stainless steels. These principles are increasingly being used to develop "chill-zone" ingots of various other alloy systems with superior mechanical strengths.

7.9.2.1.2   Columnar Zone

Further from the mold surface, the grain structure evolves into columnar grains growing roughly parallel to the heat flow direction. Each grain is typically composed of many primary dendrite stalks. Walton and Chalmers (1959) proposed the competitive mechanism through which favorably oriented grains eliminate other grains such that a crystallographic texture arises. In FCC and BCC alloys, a preferred <100> orientation is characteristic of the structure.

In conventional ingots, columnar growth may not be perpendicular to the mold wall if convection sweeps past the L–S interface because of horizontal temperature gradients. If convection is diminished through magnetic fields or mold rotation, perpendicular columnar growth can be restored (Cole, 1971). For the twin-roller strip casting of steels, Takatani et al. (2000) characterized the grain orientation using electron backscattered diffraction, and showed that the <100> texture that develops near the central region of the strip was inclined by about 15° from the casting direction (toward the upstream direction of the flow).

In general, the growth of the columnar front in a casting can be modeled using a macroscopic heat code that solves eqn (7). As described in Section 7.2, this requires a fraction solid versus temperature (and other variables) relationship obtained from one of the various dendritic microsegregation models presented in Section 7.7.3. The various computational methods are summarized by Rappaz and Stefanescu (1988) and Rappaz (1989). In principle, a fully iterative numerical scheme is required; that is, one estimates a fraction solid versus (and hence enthalpy) versus temperature relation, computes the thermal field and then resets the fraction solid versus temperature relation based on local values of isotherm velocity, temperature gradient and/or local freezing time. Iteration is required until convergence is obtained. This is almost never done because of computational constraints. The Scheil approach is normally used for the heat flow calculation and other details like the dendrite arm spacing or dendrite tip temperatures are recovered in a postprocessing step.

To model the developing grain shape in the columnar zone, Rappaz and Gandin (1993) developed an approach that includes an orientation variable for each grain and uses a cellular automaton (CA) technique. **Figure 82** shows a schematic example of this type of modeling for the competitive columnar growth of three grains. (Some equiaxed grains that nucleate and grow in front of the columnar zone are also depicted which will be discussed in a later section.) The columnar grains on the left and on the right contain dendrites whose <100> crystallographic orientations are nearly perpendicular to the liquidus isotherm. These dendrites grow with the same velocity V_L, as that of the isotherms. The grain in the middle of the figure, having a deviation of its <100> crystallographic direction from the heat flow direction, has dendrite tips growing with a velocity $V_\theta = V_L/\cos\theta$ that is larger than V_L. According to the growth kinetic model of the dendrite tip for constrained growth (Section 7.7), the faster growing (misoriented) dendrites are characterized by a larger tip supercooling. Thus the tips lag behind those of the better oriented grains and thus the central misoriented grain is pinched off by the left and right grains. Also shown in **Figure 82** is how the left hand grain boundary is parallel to the nominal heat flow direction while the right-hand boundary is tilted because of the details of tertiary arm formation. This effect was also recovered in the modeling. In complementary work, Gandin et al. (1995) measured

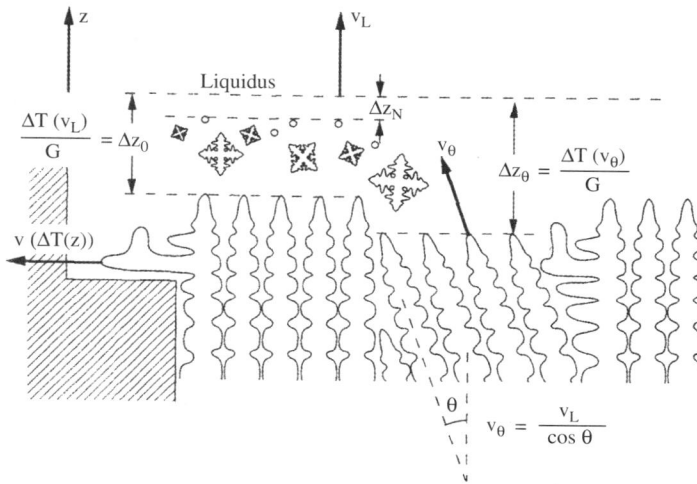

Figure 82 Competing processes during directional dendritic growth: development of preferred orientation in the columnar region, formation of equiaxed grains ahead of the columnar front. Three grains are shown (Rappaz and Gandin, 1993).

columnar grain diameters and the crystallographic texture in the columnar zone of a commercial superalloy casting. The orientation distribution of the columnar grains narrows and the grain size increases as the distance from the mold surface increases. The probabilistic model produces reasonable agreement with experiment. The method is also useful for modeling the grain competition in the grain selector during DS for the production of modern turbine blades (Rappaz and Gandin, 1994).

7.9.2.1.3 Central Equiaxed Zone

In many ingots a central zone of equiaxed grains is formed. Typically this grain structure is more desirable than the columnar structure described above because of improved mechanical properties. Equiaxed grains grow ahead of the columnar dendrites as shown in **Figure 82**. The columnar to equiaxed transition (CET) occurs when growth centers formed by the mechanisms described below can survive and grow to occupy sufficient space to impede the advance of the columnar front. Some ingots have almost no equiaxed zone while others have almost no columnar zone. The major challenges to predict the CET and the size of the equiaxed zone in castings involve an accurate description of the source of the nuclei (growth sites) and an accurate description of the competing growth rates of the columnar and equiaxed crystals under the prevailing conditions.

7.9.2.2 Source of Growth Sites for Equiaxed Solidification

The three sources of growth sites for equiaxed grains in the bulk liquid zone of a casting are heterogeneous nucleation on heterogeneities, nucleation on intentionally added grain refiners, and dendritic fragments produced and transported from the columnar zone.

7.9.2.2.1 Heterogeneous Sites and Grain Refiners

In alloys without grain refiners, Thevoz et al. (1989) and Stefanescu et al. (1990) found it necessary to postulate the existence of unknown catalytic sites to properly predict grain size. To predict accurately the grain size of castings, the former employed a distribution of nucleation temperature for the sites while the latter employed a cooling rate dependence for the number of active sites. The identification of the nature of the heterogeneous sites was not addressed.

The mechanisms by which grain refiners operate to produce viable growth centers has been described by Perepezko (2008) and Kelton and Greer (2010). Aspects of lattice matching, adsorption and the control of inoculant fading are subjects beyond the scope of this chapter.

7.9.2.2.2 Dendrite Fragments

There is mounting evidence that dendrite fragments are the growth initiation centers for equiaxed grains in non-grain refined castings as well as for spurious grain formation in directionally solidified single crystal aero-turbine blades. Over the years, some researchers had argued for a mechanical breakage model of fragmentation because of the existence of vigorous fluid flow and have thus limited the idea of fragmentation to highly convecting situations. But Pilling and Hellawell (1996) analyzed the pressure exerted on a fragile dendrite network by buoyancy-driven fluid flow and estimated the stress in the solid dendrite. They concluded that dendrite breakage (fracture) was unlikely to occur especially considering the ductile nature of metals near their melting point. They argued for a dendrite remelting mechanism because of deceleration of the growth front caused by convection. Indeed such an effect had already been discovered by Jackson et al. (1966). They showed that a modest deceleration of the growth rate of a dendritic array could cause dendrite arm separation during DS of transparent organic metal/alloy analogs.

Conclusive evidence for fragmentation was obtained by Hellawell et al. (1993) who performed studies of solute convection caused by liquid density gradients in the mushy zone. Using a transparent analog system undergoing vertical solidification they showed the formation of rising convective plumes of solute rich and less dense liquid within the fully molten zone. These plumes caused the formation of hollow channels in the dendritic array (**Figure 83**). A gentle radial flow from the surrounding mushy zone collected into the hollow channels to feed to the strong upward plume that propagates into the fully molten zone. Often dendrite fragments were directly observed to be ejected from the mushy zone into the plumes that rose into the fully liquid region. Other fragments were not ejected but merely settled deeper into the mushy zone within the channel.

Further convincing evidence for a fragmentation mechanism was obtained in metals by Mathiesen et al. (1999, 2006) using synchrotron generated X-rays to produce time resolved microradiographs of dendritic fragmentation. **Figure 84** shows a sequence of snap shots of an Al–Cu alloy during DS.

In castings, where growth happens inwards from the mold walls, both the temperature gradients and growth rates decrease in a continuous manner. Indeed Liu et al. (2002) further studied the mechanism of the detachment of dendrite side arms induced by deceleration in the NH_4Cl–H_2O and $[CH_2 CN]_2$–H_2O systems. They noted that detachment of dendrite arms rarely occurs during steady-state growth and often is promoted by deceleration. They noted that the fraction of detached arms increased as a function of time and rate of deceleration. Since there was no evidence of thermal recalescence as a deceleration mechanism in their experimental setup, they concluded that solutal effects arising from the slow adjustment of the spacings of the primary dendrites was the main cause for detachment. They contended that increased interdendritic solute concentration cause necking where secondary arms attach to the primary stems. The rates of detachment at these locations were found to be related to solute diffusion.

Esaka et al. (2003) studied the formation of equiaxed grains because of forced convection by conducting in situ observation using a transparent organic substance. They noticed emission of spherical particles from the columnar zone while studying the transport behavior using a fiber scope installed in the experimental setup. These particles were identified as secondary dendrite arms that were detached

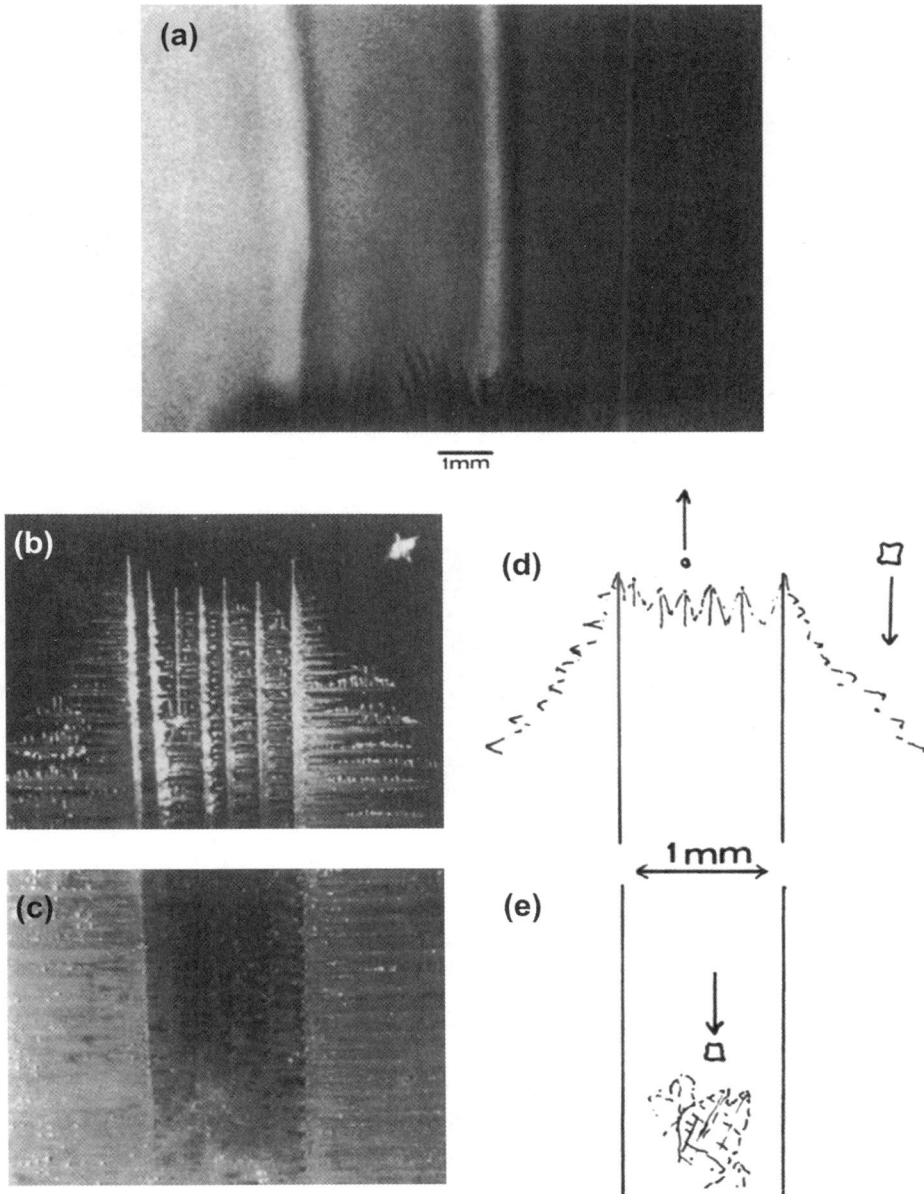

Figure 83 (a) Low magnification view of solute-rich plumes rising from a mushy zone of NH$_4$Cl dendrites growing from an NH$_4$Cl–H$_2$O solution being cooled from the bottom. (b) High magnification view of upper region of channel formed in the mushy zone with a single fragment that has been ejected, grown and is falling back to the advancing dendritic front causing the formation of a new grain. (c) High magnification view deep in the channel where dendritic fragments have fallen back into the channel (forming the characteristic freckle appearance seen in micrographs after solidification of superalloys). (d) and (e) Shown sketches of (b) and (c). (Adapted from Hellawell et al., 1993).

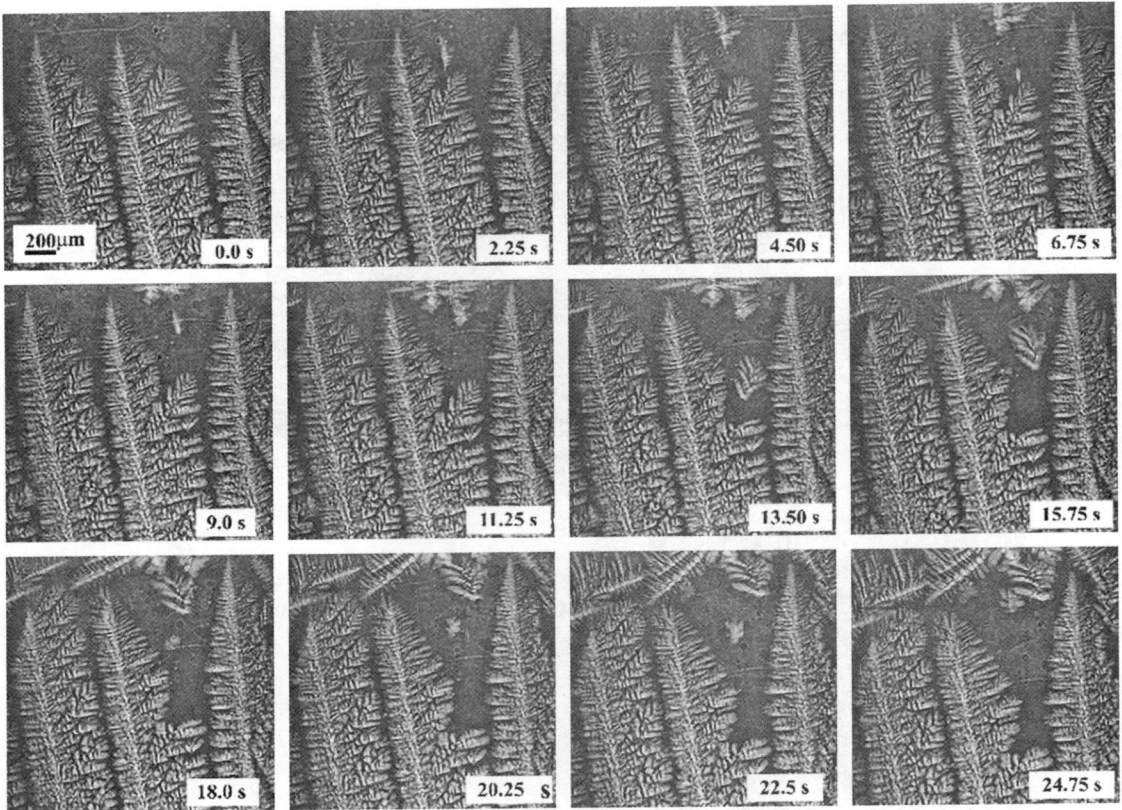

Figure 84 Dendrite fragmentation and columnar front blockage in metallic Al–Cu alloy observed by synchrotron radiation (Mathiesen et al., 2006).

by the fluctuating thermal and/or solutal field or by the process mechanical stirring used. The multiplication rate of equiaxed grains was found to increase with an increase in flow velocity.

Rivullcaba et al. (2007) using the synchrotron observation of fragmentation in Al–Cu observed a localized deceleration of single dendrites and the subsequent solute enrichment locally near the site of fragmentation. Such small fluctuations are likely because of convection and are much more subtle than those that occur in the solute rich plumes shown in **Figure 83**.

An extreme case of deceleration is found during recalescence from deeply bulk undercooled melts. Recalescence is reheating because of rapid heat of fusion release after fast solidification that cannot be suppressed by the rate of external heat extraction. Small grain sizes were observed experimentally only at intermediate initial (bath) undercooling (between 80 K and 175 K for Ni–Cu). Schwartz et al. (1994) proposed a model for dendritic breakup to provide the growth centers for the observed fine grains at the intermediate initial undercooling. Their idea is based on a simple model that treats the dendrite stalk as a cylinder undergoing a breakup into spheroidal particles at a temperature near the liquidus. The breakup is driven by the possibility to reduce interface area and the rate is governed by diffusion kinetics. They proposed that if the estimated time for breakup is shorter than the quasi-isothermal time period after the reheating and during subsequent freezing, fragmentation is predicted to occur. The quasi-isothermal time period was measured by pryometry on levitation melted and freezing samples. Because these measured times scale inversely with the cooling rate of the melt before nucleation (a measure of the external cooling conditions), a predictive model was obtained that agreed with the grain size transitions observed.

The accumulating evidence presented above suggests that a dendrite detachment mechanism is the most plausible explanation for the formation of equiaxed zone in castings with no grain refiners. Quantitative prediction of fragment yield (number per volume per time) as a function of local flow conditions in the mushy zone must await further research.

7.9.2.3 Viability of Growth Sites for Equiaxed Solidification

Regardless of the source of sites, the formation of either spurious grains in single crystal blade production or the formation of the equiaxed zone in a casting requires that the sites must survive and be able to grow. We first describe the simplest description of survival as presented by a discussion of the CET and then follow with a discussion more specific to dendrite fragments. The former example treats the growth of a seed with no fluid flow and in the simple situation of a linear temperature gradient in the liquid ahead of a columnar front. In the latter example, the motion and growth/dissolution of seeds in the complex thermal and solute environment of a flowing melt is considered.

7.9.2.3.1 Columnar to Equiaxed Transition

Early research on this topic was performed by Maxwell and Hellawell (1975). The model of Hunt (1984) and Flood and Hunt (1987a) use more refined dendritic growth and nucleation models. The structure will be fully equiaxed or fully columnar depending on whether the liquid temperature gradient is smaller or larger respectively than a critical value given by

$$G_{\mathrm{L}} = 0.617\, N_0^{1/2} \left[1 - \left(\frac{\Delta T_{\mathrm{n}}}{\Delta T_{\mathrm{c}}} \right)^3 \right] \Delta T_{\mathrm{c}}, \tag{189}$$

where N_0 = density of nucleating sites, ΔT_{n} = supercooling required for heterogeneous nucleation and ΔT_{c} = supercooling of the dendrite tips below the liquidus temperature of the columnar front. The experimental results for Al-3 wt% Cu by Ziv and Weinberg (1989) are in close agreement with eqn (189). Factors which promote a CET by this mechanism are: low temperature gradient, which increases the size of the constitutionally supercooled region in front of the dendritic tips, large solute content (increases the value of ΔT_{c} for fixed growth speed), a small value for ΔT_{n} (potent nucleation sites), and a large number of nuclei. Hunt's model ignores many complexities of the dendritic growth of equiaxed grains and nucleation was assumed to take place at a single temperature rather than over a range of temperature. It therefore cannot predict the effect of solidification conditions on equiaxed grain size (Kerr and Villafuerte, 1992). More detailed models of equiaxed growth employing empirical nucleation laws have been combined with numerical solutions of the heat flow to predict the grain size of fully equiaxed structures (Thevoz et al., 1989; Rappaz, 1989; Stefanescu et al., 1990). These more detailed analyses could be used to predict the CET. Flood and Hunt (1988) have critically reviewed the models and experiments of several researchers.

Rappaz and Gandin (1993), employing the cellular automata (CA) method, simulated the CET. **Figure 85**a corresponds to the simulation of the final grain structure of an Al-5 wt% Si casting when cooled at 2.3 K s^{-1}. **Figure 85**b corresponds to an Al-7 wt% Si casting cooled at 2.3 K s^{-1} and **Figure 85**c to an Al-7 wt% Si casting cooled at 7.0 K s^{-1}. Comparisons among the three figures show the

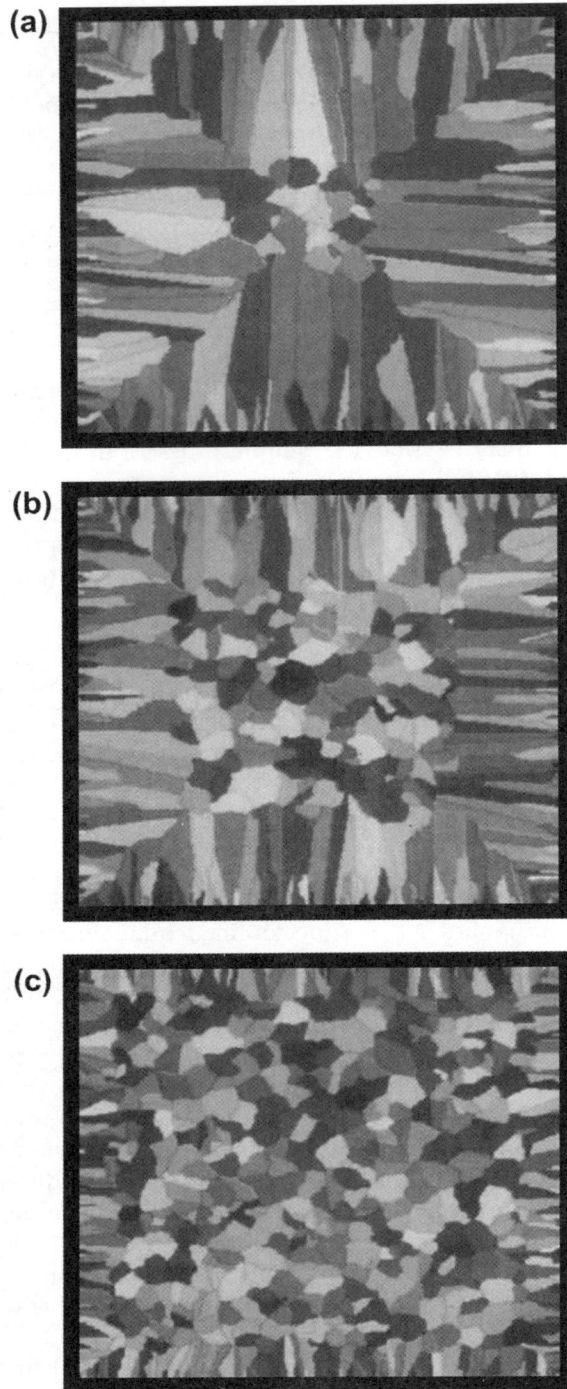

Figure 85 Simulation of columnar and equiaxed structures by Rappaz and Gandin (1993). (a) Al-5 wt% Si cooled at 2.3 K s^{-1}; (b) Al-7 wt% Si cooled at 2.3 K s^{-1}, and (c) Al-7 wt% Si cooled at 7.0 K s^{-1}.

effect of the alloy composition and cooling rate upon the grain structure. Rappaz and Gandin (1994) present a very comprehensive review of the modeling of grain structure formation in solidification processes.

The above analysis assumed that a physical blocking of the columnar front by equiaxed grains causes the CET. However around every growing equiaxed dendritic grain is a boundary layer of solute. Martorano et al. (2003) propose that solutal interactions between the equiaxed grains and the advancing columnar front are important. The resulting differences in the CET prediction are demonstrated and a revised isotherm velocity versus temperature gradient map for the CET is presented.

Kurz et al. (2001) proposed an approach based on development of processing maps by combining a model for CET with local solidification conditions computed numerically. They relaxed some of the simplifications in the original Hunt approach. Their dendritic growth model included nonequilibrium effects, which showed marked changes from Hunt's model in terms of values of processing conditions (G_L and V) where the CET will occur. They demonstrated the utility of development of processing maps in prediction of CET for welding and laser treatment of castings. Such maps could help in avoiding or promoting CET based on specific requirements. Such a microstructure selection map is shown in **Figure 86** for laser repair of single crystal turbine blades of CMSX-4 alloy. The first transition in the figure at low velocities is for planar to cellular/dendritic breakdown, while the next transition at higher growth velocity is for CET.

7.9.2.3.2 Details of Fragment Survival

The question of whether a fragment or for that matter any potential growth site survives might begin by applying the analysis of the growth of a spherical solid particle or a free dendrite into an undercooled melt (the latter presented in Section 7.7). The difficulty however is that the freely growing dendrite model assumes a surrounding that is quiescent and has uniform temperature and composition far from the growing crystal. In reality, a potential growth site finds itself in a moving fluid with nonuniform temperature and composition. To answer the survival question, many approximations and complicated analyses are necessary.

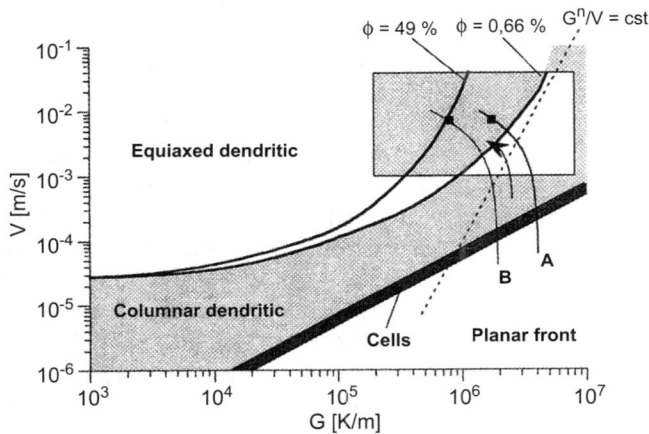

Figure 86 Microstructure selection map for CMSX-4. The rectangular insert represents range of conditions that are typical for laser process. The evolution of G and V is calculated for initial substrate temperatures of 20 °C (curve A) and 500 °C (curve B). The squares represent average values. φ is the volume fraction of equiaxed grains. cst is a function of alloy constants and nucleation density (Kurz et al., 2001).

Hellawell et al. (1997) experimentally studied the dendrite fragmentation mechanism and the influence of fluid flow in the transport of grain fragments. The observed dendrite fragments had sizes between 10 and 50 μm. At temperatures equal to or above the liquidus temperature, dissolution of particles occurs rapidly because of solute diffusion in liquid. They estimated that the survival time of these fragments to be less than 1 s once the liquidus temperature is reached. They proposed a possible approach to quantify the particle density by subtracting the rate of dissolution of particles or "death rate" from the supply of particles (fragmentation) or "birth rate".

Computation of fragment survival is a difficult problem. Because the density of the solid is normally higher than the liquid, they tend to sink once the upward convecting currents decrease. How well they are carried by the local flow depends on their size through their drag coefficients. Melting and/or freezing occur depending on the local temperature and liquid concentration. Gu et al. (1997) performed heroic calculations of how fragments might survive under some conditions during vertical solidification as they flow up into the bulk liquid above the columnar zone and as they fall back down to the columnar growth front and form a growth center for an equiaxed grain (or to stray grains in the case of single crystal growth of superalloy blades).

7.9.3   Macrosegregation

Macrosegregation is defined as variations in composition that exist over large dimensions, typically from millimeters to the size of an entire ingot or casting. Macrosegregation can be a significant problem for large castings. Macrosegregation typically results from relative movement of fluid and solid during the solidification process. We have already considered forms of macrosegregation in the initial and final transients during planar growth of a rod sample (Section 7.6.2) and during zone refining. However to define or measure macrosegregation for dendritically solidified samples, it is necessary to determine an average composition over a volume element that contains several dendrite arms. As we will see, changes in the dendritic microsegregation profile in such a volume element because of flow of solute-rich liquid in or out of the volume element during solidification will change the average composition of the volume element away from the nominal composition of the alloy.

Figure 87 shows a drawing of a large steel ingot showing some of the major types of macrosegregation commonly found. Positive and negative macrosegregation refer to solute content greater or less than the average. A cone of negative segregation forms at the bottom region of the ingot, where equiaxed grains formed early in the solidification process are thought to settle. Heavier inclusions may also settle in this region (Flemings, 1974). Positive segregation at the centerline and the hot top segregation result from buoyancy-driven and solidification shrinkage induced flow toward the end of the solidification process. Channel segregates forming a V pattern in the central region often contain fine equiaxed grains with increased levels of microporosity. The A-segregates are formed in the upper and outer regions of the ingot. These segregates are typically solute rich.

In describing models that predict macrosegregation we proceed from the simple to the complex. First is an approach appropriate for a columnar structure where the solid in the mushy zone remains attached. Situations where the solid can move as free-floating equiaxed grains are described second. The former also includes a discussion when the mushy zone is deformed as in continuous casting.

7.9.3.1   *Macrosegregation Models for Columnar Structures*
7.9.3.1.1   Modification of the Scheil Equation to Include Flow

The first mathematical treatment of macrosegregation was provided by Flemings and Nereo (1967) and Mehrabian et al. (1970) and has been summarized by Flemings (1974, 1976). Using a volume element

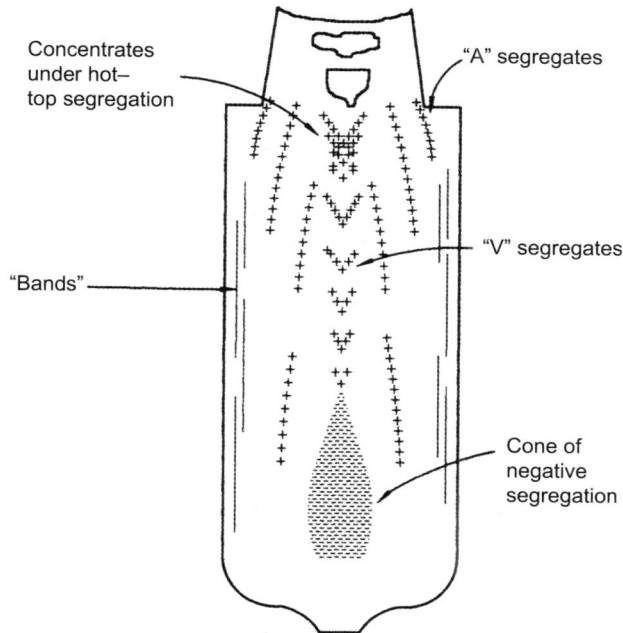

Figure 87 Different types of macrosegregation in an industrial steel ingot (Flemings, 1974).

similar to that chosen in **Figure 49** that was used to describe microsegregation in Section 7.7, a mass balance is performed under the additional possibility that flow of liquid in or out of the volume elements can occur and that the liquid and solid can have different densities. Thus the necessity for flow to feed solidification shrinkage is treated. The result is a modified form of the local solute redistribution, LSRE (Scheil), equation

$$\frac{dg_S}{dC_L^*} = \frac{(1-\beta)}{(1-k)}\frac{g_L}{C_L^*}\left(1 - \frac{\vec{v}\cdot\vec{n}}{|V_T|}\right), \tag{190}$$

where $\beta = (\rho_S - \rho_L)/\rho_S$ = solidification shrinkage; $\vec{v}\cdot\vec{n}$ = velocity vector component of interdendritic liquid in the direction normal to the isotherm; V_T = isotherm velocity; g_S and g_L are volume fractions of solid and liquid respectively.

This expression assumes: (i) local equilibrium without curvature correction, (ii) uniform liquid composition in the small volume of interest, (iii) no solid diffusion, (iv) constant solid and liquid density, (v) no solid motion, and (vi) absence of voids. In this approach, the appropriate values for $\vec{v}$ and V_T at each location must be determined from a separate calculation involving thermal analysis and flow in the mushy zone. However given these values for each small volume element, C_L^* and hence C_S^* as a function of g_S can be determined along with the fraction of eutectic if appropriate for the phase diagram of the alloy. The average value of C_S^* from $g_S = 0$ to 1 gives the average composition at each location in the casting. The average composition will not in general be equal to the nominal alloy composition. Indeed one hallmark of macrosegregation is a varying amount of eutectic from one region of the casting to another.

If one defines a flow factor, ξ, as

$$\xi = (1 - \beta)\left(1 - \frac{\vec{v} \cdot \vec{n}}{|V_T|}\right),$$ (191)

and assumes it to be a constant, eqn (190) can be integrated to yield

$$\frac{C_L^*}{C_0} = (1 - g_S)^{\frac{k-1}{\xi}}.$$ (192)

If $\xi = 1$, eqn (192) reverts to the standard Scheil equation and the average composition is equal to the nominal composition resulting in no macrosegregation. Note that $\xi = 1$ when the liquid flow occurring is exactly that required for the solidification shrinkage; that is,

$$\vec{v} \cdot \vec{n} = -\frac{\beta}{(1 - \beta)}|V_T|.$$ (193)

We now discuss specific cases depending on the value of ξ where $\beta > 0$ and $k < 1$ are assumed:

1. $\xi > 1$; i.e., $\left(\frac{\vec{v} \cdot \vec{n}}{|V_T|} < -\frac{\beta}{(1 - \beta)}\right)$: at the same solid fraction C_L^* is *smaller* than the value computed with

   the Scheil equation resulting in *negative* macrosegregation. This happens when the absolute value of the flow velocity toward regions of higher solid fraction is greater than is required to feed the shrinkage.

2. $0 < \xi < 1$; i.e., $\left(1 > \frac{\vec{v} \cdot \vec{n}}{|V_T|} > -\frac{\beta}{(1 - \beta)}\right)$: at the same solid fraction C_L^* becomes *larger* than the value

   computed with the Scheil equation resulting in *positive* macrosegregation. This happens when the absolute value of the flow velocity toward regions of higher solid fraction is less than is required to feed the shrinkage). A particularly simple case occurs at the chill face of a casting. Here $\vec{v} \cdot \vec{n}$ must be zero because there can be no flow into the chill face. This clearly produces positive macrosegregation. This is commonly observed in ingots and is termed inverse segregation because it is reversed from what one would expect based on the initial transient of plane front growth (**Figure 30**).

3. $\xi < 0$; i.e., $(\vec{v} \cdot \vec{n} > |V_T|)$: this condition can occur if liquid flows toward the region of lower solid fraction with a velocity higher than the isotherm velocity. Using eqn (192) results in solid fraction decreasing with decreasing temperature. This results in localized remelting. This is the case for channel macrosegregation discussed below.

The LSRE approach had shed significant light on macrosegregation. However a separate analysis of the flow velocity must be performed. In simple cases the flow can be computed by treating the mushy zone as a porous media and *D'Arcy's Law* is used. The pressure gradient and the body force because of gravity control the fluid velocity according to

$$\vec{v} = -\frac{K}{\eta g_L}(\nabla P - \rho_L \vec{g}),$$ (194)

where K = specific permeability; η = viscosity of the interdendritic liquid; ∇P = pressure gradient; $\vec{g}$ = acceleration vector because of gravity. Determining an accurate expression for the permeability of a mushy zone is a difficult problem since the value of K depends on interdendritic channel size and geometry. The Blake–Kozeny form for the permeability is widely used and is given by

$$K = \frac{C\lambda^2 g_L^3}{(1 - g_L^2)},\tag{195}$$

where C is a constant with a value of about 0.2 and λ a characteristic dimension. It is either the primary or secondary dendrite arm spacing. A better measure is the L–S interface area per unt volume.

Erdmann et al. (2010) provide a comprehensive review of permeability of the mushy zone for both columnar and equiaxed microstructures. For equiaxed dendrites, permeability is treated as a scalar quantity. For a columnar microstructure it is a tensor. The tensor is treated as transversely isotropic with two unique values: K_{zz} parallel to the primary dendrite trunks, and $K_{xx} = K_{yy}$ transverse to the dendrite trunks. Variation of the permeability as a function of liquid fraction is also summarized by these authors for both columnar and equiaxed microstructures using both experimental and numerical results. Numerical results were necessary for large liquid fractions as it is difficult to obtain such values experimentally.

Another topic that has received attention is the change in macrosegregation because of deformation of the mushy zone. Such deformation can suck or squeeze liquid in or out of the dendritic mush altering the macrosegregation profile. Using a simple theoretical approach, Lesoult and Sella (1988) employed a modified form of the LSRE in which they considered the influence of interdendritic strains and deformation of the solid network on the volume of the mushy zone element. This model allowed a simple prediction of the centerline segregation. Other more complicated approaches to treat deformation of the mushy zone are described in the next section.

7.9.3.1.2 *Solution of the Transport Equations for Macrosegregation during Columnar Solidification*
One of the major shortcomings of using the LSRE approach discussed above to predict macrosegregation is that these models neglect flow in the fully liquid region ahead of the mushy region. In addition, some types of macrosegregation shown in **Figure 87** cannot be predicted using the LSRE method; namely, V-segregates, negative cone segregation and channel segregates.

Ridder et al. (1981) were first to model the influence of flow in bulk liquid on the flow in the mushy zone. They investigated a case appropriate to electro-slag remelting and employed a three domain model with sharp boundaries between the fully liquid zone, the mush and the fully solid zone. They determined for the case examined that the fluid flow in the liquid metal pool, because of natural convection, had little effect on interdendritic fluid flow and the resulting macrosegregation.

With the wider use of one domain models over the past decade because of improving computational capabilities, a comprehensive treatment of solidification phenomena by solving coupled energy, mass, momentum, and species conservation equations have allowed a more complete method to model complex macrosegregation patterns. The reader is referred to the review provided by Beckermann (2002). Some of the progress is described below.

Gu and Beckermann (1999) conducted a 2-D simulation of a steel ingot casting using single-domain volume-averaged model. Their model showed that strong buoyancy-driven flow conditions exist throughout the solidification process. They computed macrosegregation profiles of the solute elements carbon and sulfur at the ingot centerline along the vertical direction. Results agreed well with the experimentally measured values (**Figure 88**). The model did not predict A-segregates (freckles) or V-segregates: the former because an insufficiently refined mesh was employed to resolve the localized channel flow and the later because the model did not include movement of equiaxed crystals according to the authors (see below).

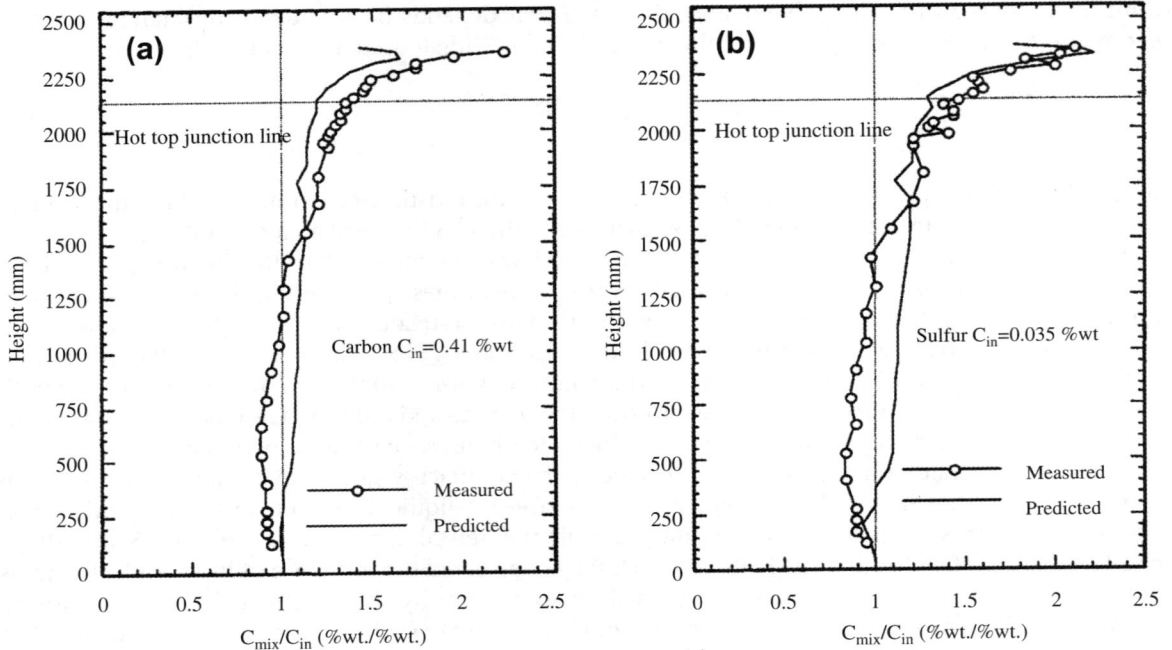

Figure 88   Comparison of measured and predicted macrosegregation variations along vertical centerline of a steel ingot (C_{mix} and C_{in} are local solute mixture concentration and initial solute concentration, respectively) (Gu and Beckermann, 1999).

Channel segregation or freckles are a particularly troublesome defect in the upward, DS of single crystal superalloy castings and electroslag remelted ingots. After solidification these defects consist of abrupt and large variations in composition consisting of chains of solute-rich grains. We have described the formation of solute rich plumes that can lead to dendrite fragmentation in Section 7.9.2.2. They result from movement of interdendritic liquid because of buoyancy from thermal and solutal density gradients that opens localized vertical channels in the liquid–solid region. Experimental work by Hellawell (1990) proves that the initiation of the channels is at the dendritic growth front itself. As growth proceeds, the localized depression in the dendritic front deepens and collects even more solute rich liquid to feed the upward moving plume. Thus the defect forms when the upward velocity of the liquid in the mushy zone exceeds the upward isotherm velocity producing positive macrosegregation. During upward DS of Ni-base superalloys, the solutal buoyancy forces result from the fact that light elements such as Al and Ti are rejected into the liquid in front of the L–S interface ($k_i < 1$) while, heavy elements such as W and Re ($k_i > 1$) are depleted in the liquid in front of the L–S interface. During subsequent solidification, the channels often become filled with dendrite fragments that appear in the final casting as freckle-like chains of dendritic equiaxed crystals under microscopic examination (see **Figure 89**).

To establish the conditions that lead to the localized channel flow and subsequently to freckles, Schneider et al. (1997) developed a 2-D micro/macrosegregation model, in which the solidification path was computed at each time step by conducting a phase equilibrium calculation with a multi-component thermodynamic database. Several other researchers have modeled freckle formation in directional casting of superalloys. Felicelli et al. (1998) reported 3-D calculations of macrosegregation using a similar approach as described above. The localized channel flow was computed. They found that freckles often appear on vertical sidewalls of castings. However, such calculations in 3-D are

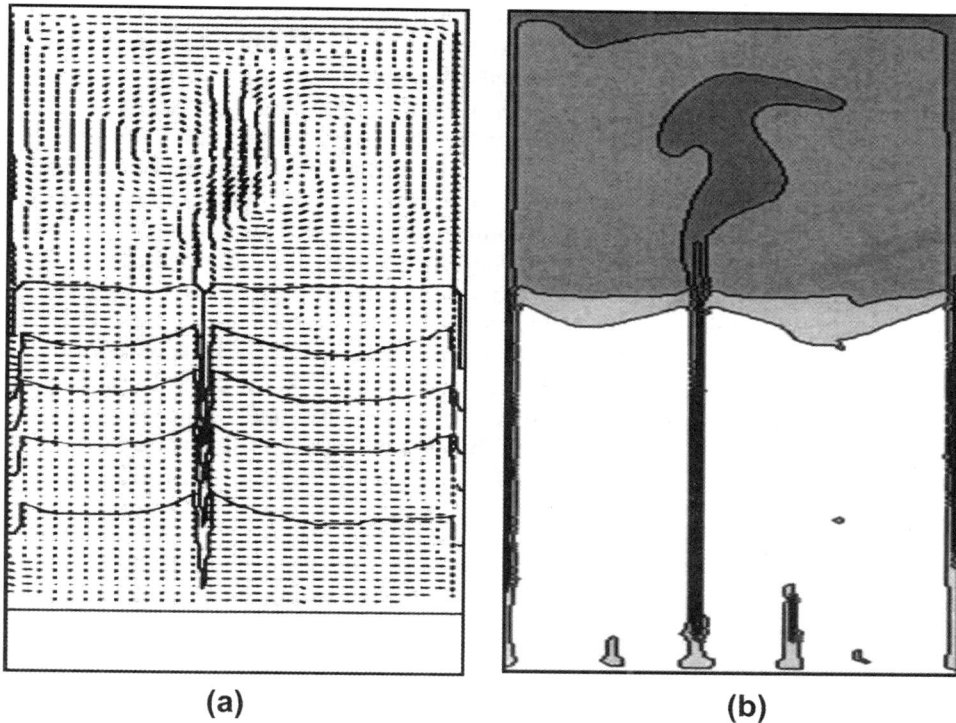

(a)   **(b)**

Figure 89 (a) Predicted velocity vectors (largest represents 6.3 mm/s) and contours of solid fraction (in increments of 20%); (b) macrosegregation pattern (Ti concentration normalized by initial concentration shown in equal increments between 0.87 and 1.34) showing freckle formation during upward DS of a single crystal Ni-base superalloy in 5 × 15 cm rectangular computational domain (Schneider et al., 1997).

extremely computationally intensive for practical casting configurations. Thus a simple freckle predictor concept has been developed for vertical upward solidification. The dimensionless Rayleigh number compares the buoyancy forces to the viscous force and is used in many natural convection circumstances. Worster (1992) and Beckermann et al. (2000) have introduced a "mushy zone" Rayleigh number

$$Ra_h = \frac{g\Delta\rho(h)\overline{K}(h)h}{\mu\alpha}, \tag{196}$$

where g is the acceleration of gravity, $\Delta\rho(h)$ is the density difference between the bulk liquid density and the liquid density at depth h in the mushy zone behind the dendrite tips, μ is the liquid viscosity, $\overline{K}(h)$ is the permeability averaged between the tips and the position h, and α is the thermal diffusivity. This Rayleigh number is a function of position in the mush, yet averages over the mush above the position of interest. Use of this Rayleigh number predictor requires finding the maximum local Rayleigh number in the mushy region and determining whether it exceeds a critical value and therefore predicts freckle formation. The authors determined the value of this critical Rayleigh number to be approximately 0.25 for directionally solidified Ni-base superalloy based on analysis of experimental data. Application of this criterion depends a great deal on an accurate knowledge of local thermal conditions, liquid densities, and solidification path. Another point to note is that Rayleigh number criterion alone is not a sufficient condition to predict whether freckles will form in a given casting. Despite these shortcomings this method provides guidelines for composition selection and processing conditions to avoid freckle defects in single crystal superalloy turbine airfoil casting.

In a typical continuous casting of steel slabs, bulging often occurs in the region between two consecutive rolls. This bulging causes relative liquid–solid velocities. Compression of the solidifying network releases solute enriched liquid that moves toward the central region causing positive macrosegregation. Kajitani et al. (2001) implemented an innovative approach in their model of macrosegregation because of mush deformation that combined uncoupled computation of the temperature field and the slab bulging profile. With this information as input, they obtained the solid velocity and then, using mass balance and D'Arcy's Law, computed the fluid flow and pressure fields. The predicted carbon macrosegregation profile is shown in **Figure 90**. It can be seen that bulging produced increased positive macrosegregation. Additionally, the carbon macrosegregation profile across the thickness of the slab after six rolls showed substantial positive macrosegregation in the central region, while negative macrosegregation was obtained in the peripheral regions. Computed results showed the opposite behavior for the case of shrinkage alone. Although no comparison was provided with experimental data, the modeling efforts established the role of bulging and shrinkage in producing complex macrosegregation pattern in continuous casting.

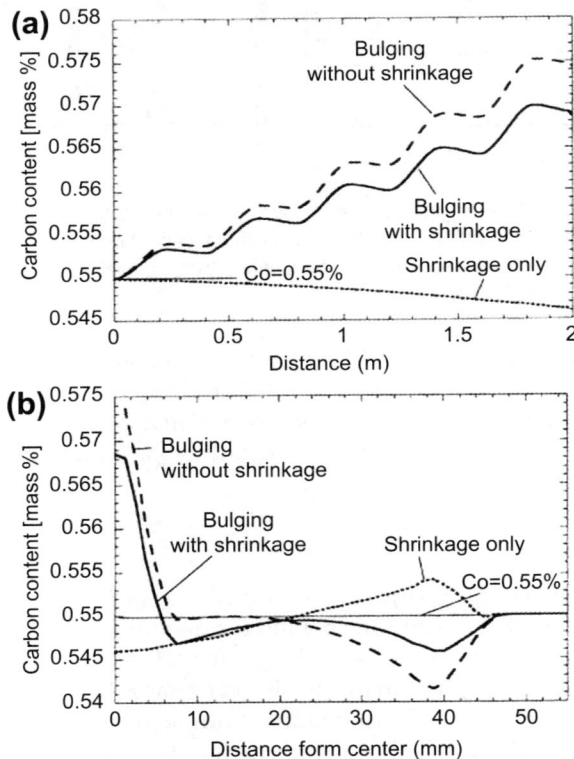

Figure 90   Predicted macrosegregation patterns during continuous casting of steel; C_0 is initial carbon content. (a) Along centerline (distance between rolls is 400 mm); (b) across slab thickness after six rolls ($f_s = 0.37$ at centerline, maximum bulging of 0.1 mm) (Kajitani et al., 2001).

A full coupling of stress and strain behavior in the solid with thermal, fluid, species transport of the mushy zone is extremely complex and computationally time consuming. This topic is beyond the scope of the present document. The topic of hot tearing however is amenable to some level of approximation as described for example by Mathier et al. (2009).

7.9.3.2 Macrosegregation Models for Equiaxed Structures

Until now the research discussed has not included the advection and solidification of freely floating dendritic grains. Yet such aspects of solidification have a significant role on certain types of macrosegregation. In particular negative cone macrosegregation is generally considered to be because of the settling of grains in the bottom portion of large ingots.

Macrosegregation in castings depends on the microstructure. Finn et al. (1992) studied macrosegregation in Al-4.5 wt% Cu castings that were direct chill cast with melts with and without grain refinement. They obtained positive centerline macrosegregation for the grain refined alloy ingot but negative centerline macrosegregation for the unrefined alloy. Another indication of the importance of freely floating equiaxed grains was observed by Choudhary and Gosh (1994), who reported the presence of strong macrosegregation pattern near the CET region.

Wang and Beckermann (1993) applied their multiphase model to simulate macrosegregation where both melt convection and equiaxed grain movement were considered. They considered the influence of different nucleation rates on the macrosegregation profile. Their model predictions for Al-4.5 wt% Cu alloy system (2-D simulation of a 5 cm × 10 cm rectangular cavity cooled from the left sidewall, which is aligned parallel to the gravity vector) predicted the following: positive macrosegregation along the bottom wall, A-segregates in the lower central region, no channel segregates, and less overall macrosegregation for higher nucleation rates.

Lee et al. (2000) solved a single-domain model with coupled energy, mass, momentum, and species conservation equations. Motion of equiaxed grains was allowed to occur until a critical solid fraction was reached. Their prediction of macrosegregation pattern matched experimentally measured patterns for continuous casting of steel. However, these studies did not include cases where microstructure showed a CET.

Guillemot et al. (2006) developed a model for predicting solid and liquid flow induced macrosegregation, which was coupled with a CA-FE model for computing microstructure during solidification. They included the effects of: (a) undercooling ahead of a columnar dendrite front, (b) nucleation of equiaxed grains in the undercooled melt, and (c) macroscopic transport and sedimentation of equiaxed grains. The CA model accounts for the transport of grains because of liquid flow and sedimentation. They applied this approach to model the solidification of a Pb-48 wt% Sn alloy in a rectangular cavity cooled from one of the vertical boundaries. They obtained good correlation between computed macrosegregation profiles and experimental measurements available in the literature. **Figure 91** (S) shows the final grain density for the case of transport of grains because of liquid movement and sedimentation, which was much higher at the bottom of the casting than was obtained for the case where no grain movement was considered. Their model also predicted the formation of channels (top left region of **Figure 91(d)**), which resulted from instability of the dendritic growth front. Equiaxed grains are shown to accumulate at the root of this channel as in typical freckle defects.

Wu and Ludwig (2009) developed predictive capability to model equiaxed dendrite solidification with melt convection and grain transport and sedimentation. Their modeling approach is based on a modified form of the volume-averaged single-domain approach proposed by Wang and Beckermann (1996). On a microscopic level, their model includes nucleation and growth of either globular–dendritic or dendrite grains. Inside each grain, the nonuniform solute distribution is modeled. The numerical

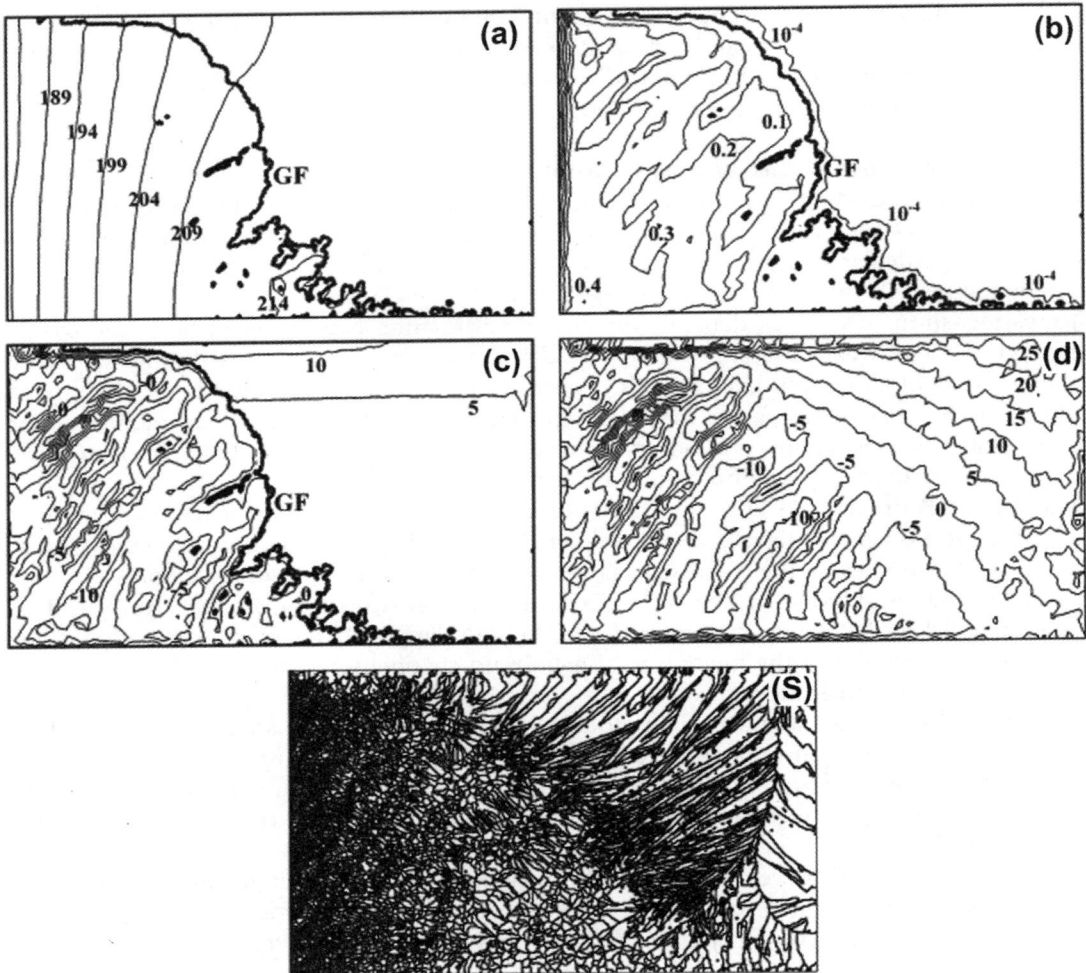

Figure 91 Predicted results for a CA-FE simulation using realistic parameters for the growth kinetics of the structure and nucleation of equiaxed grains which are free to move in the liquid. (a) Temperature in °C; (b) solid fraction; (c) relative average Sn composition after 100 s of solidification. The thick line labeled "GF" corresponds to the position of the growth front determined from the CA model; (d) relative composition map when the ingot is fully solidified; (S) final simulated grain structure (Guillemot et al., 2006).

model predictions for the fraction of phases were validated with experimental data reported by Nielsen et al. (2001) for Al–Cu alloys. **Figure 92** shows the final macrosegregation and microstructure patterns in a square casting of Al-4.7 wt% Cu alloy. The model predictions show positive segregation in the upper region, while negative segregation was predicted in the middle and lower regions of the casting. Their analysis showed that sedimentation of grains and melt convection induced by the sedimentation process play an important role in the final macrosegregation pattern during equiaxed solidification. They identified two main mechanisms for macrosegregation: (a) negative macrosegregation was caused by replacement of solute-rich melt by solute-depleted grains and (b) positive macrosegregation is caused by

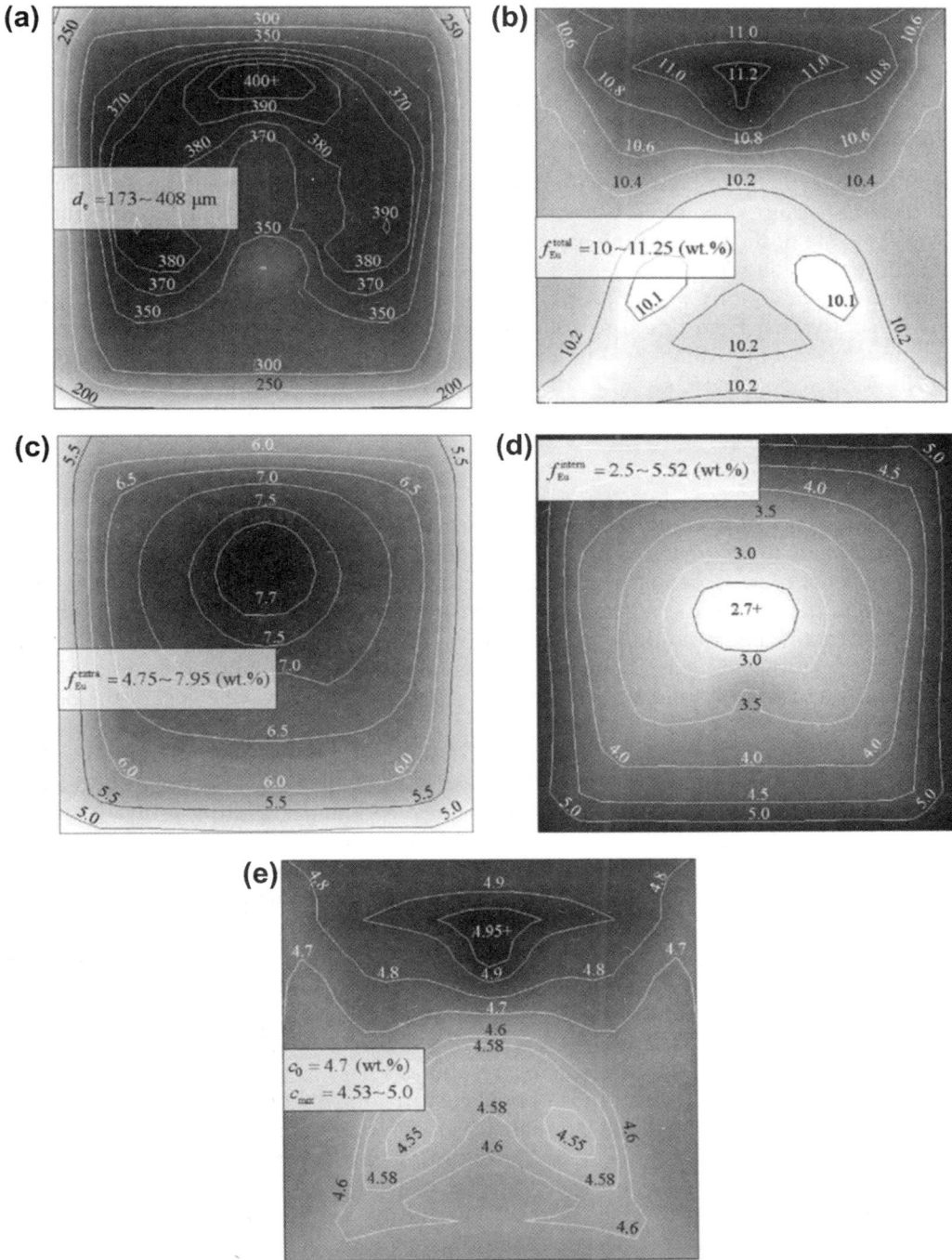

Figure 92 Final microstructure and macrosegregation pattern in a square casting of Al-4.7 wt% Cu alloy. All quantities are shown using gray scale, with dark color for highest value and light color for the lowest value of solute (Wu and Ludwig, 2009).

the replacement of the solute depleted grains by solute-rich melt. Additionally, their analyses indicated more macrosegregation for pure "globular" equiaxed solidification than that for dendritic equiaxed solidification for identical casting conditions.

7.9.4   Porosity

Porosity in castings can occur because of a number of factors. Failure to feed the mass deficit resulting from contraction because of solidification can result in porosity. Porosity related to difficulties of mass feeding on the scale of the casting dimensions is called macroporosity. On the other hand, microporosity is exhibited on the scale of the dendritic structure and is because of insufficient feeding of liquid into the interdendritic regions. Microporosity can also result from the decrease in solubility of gaseous components (if present) from liquid to solid phases. The morphology of microporosity because of shrinkage and gas is often different, but the formation of gas porosity is also dependent on the shrinkage flow. Microporosity has a strong deleterious effect on the mechanical properties of cast alloys, especially on fatigue and ductility and its reduction is of primary concern in the production of premium grade castings.

7.9.4.1   *Macroporosity*

Much of foundry practice is involved with the placement of chills and risers that maintain proper temperature gradients to retain an open path of liquid metal from the riser to the solidification front. Indeed the major use of macroscopic heat flow modeling of castings is to identify potential locations in the casting where the solidifying regions are cut off from the risers. Macroporosity results from the thermal contraction of the liquid during cooling as well as the density change on solidification. Macroporosity typically forms in the region of the casting that solidifies last if liquid feeding is insufficient. The external heat transfer conditions at the casting surface on the macroscale govern the directionality of solidification internally.

Macroporosity can be avoided by proper casting design using the right combinations of risers, ingates, vents, and chill placements. Pelini (1953) and Campbell (1991a) provide detailed descriptions of this topic. The primary role of even the simplest commercial casting software is to help the casting engineer properly rig the casting. For example most casting software has the capability to predict macroporosity once an accurate assessment of external heat transfer conditions is established. For example plots of the time evolution of a particular isotherm or fraction of solid can identify the locations of closed loops where mass feeding is difficult. The void fraction can be computed at these locations using the density versus temperature curve of the alloy in the mushy region and knowing the fraction solid at which feeding is effectively shut off. This fraction solid is called "critical" fraction solid and it depends on the alloy and in particular, the metallurgical quality of the melt. A typical number that is often used for ductile iron castings is 0.8. However, a value as low as 0.41 has been used for Al alloys. Often, a value of 0.7 provides reasonable results.

Figure 93(a) shows plots of macroporosity predicted by ProCAST™[3] software for the HPDC of an aluminum alloy. The predicted locations compare reasonably with those obtained in the actual casting. **Figure 93(b)** shows a similar comparison for a steel casting in which measured porosity in the casting was determined by X-ray analysis.

In conclusion, the modeling of macroporosity can greatly reduce the trial and error involved in developing proper casting rigging. However, the casting engineer must determine proper heat transfer

---

[3] Trade names are used for completeness only and do not constitute an endorsement of the NIST.

(a)

(b)

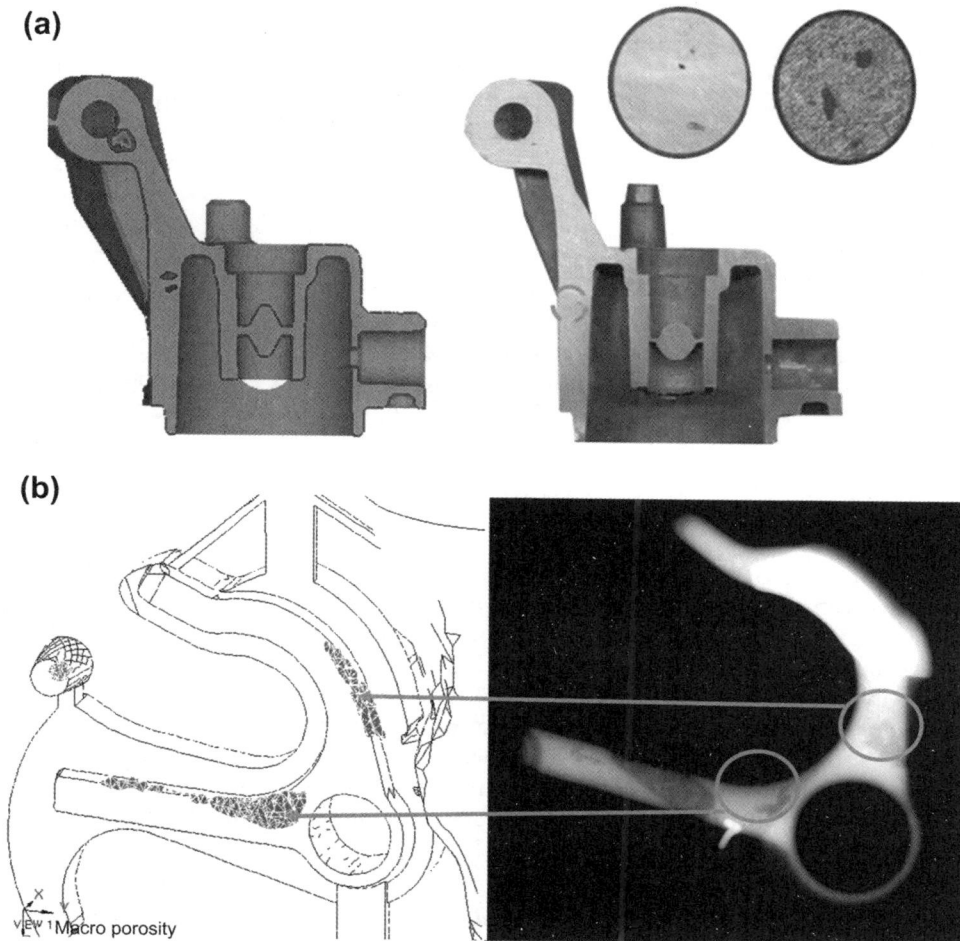

Figure 93 Comparison between predicted and experimentally obtained shrinkage porosities for (a) HPDC of aluminum alloys; (b) casting for a steel locking jaw. Courtesy: ESI Group, Columbia, MD.

coefficients on the external surfaces of the casting and obtain accurate materials data for this type of modeling to be of use.

7.9.4.2 Microporosity

Even if a path of liquid metal remains substantially open to the riser, porosity on the scale of the dendritic structure can still form. When liquid metal flows through the mushy zone to feed solidification shrinkage, the liquid metal pressure in the mushy zone drops below the external atmospheric pressure. Microporosity forms when the local pressure in the mushy zone drops below a critical value. Thus detailed prediction of microporosity requires a rather complete description of fluid flow in the mushy zone as does the prediction of macrosegregation developed in Section 7.9.3. Clearly the larger the freezing range of an alloy and the smaller the temperature gradient, the more tortuous the liquid channels are in the mushy zone. This leads to greatly increased difficulty of feeding the shrinkage and a greater reduction of the liquid metal pressure deep in the mushy zone (far behind the dendrite tips). Indeed one of the major reasons that alloys near eutectic compositions are so commonly used as casting alloys is their small freezing ranges and thus the avoidance of microporosity.

The critical reduction of the liquid pressure required for the formation of micropores depends on many factors. Indeed the initial formation of a pore is a heterogeneous nucleation problem (Campbell, 1991a) that requires the same consideration as described in Section 7.4, but with pressure substituted for temperature. It involves some of the same considerations as the prediction of cavitation during complex fluid flows. In principle one needs to know the contact angle of the pore on the solid–liquid interface. It is common to postulate that the critical pore radius (in the sense of nucleation) above which the pore can grow is related to the scale of the dendrite structure. Kubo and Pehlke (1985) let the critical radius be equal to the primary dendrite spacing, λ_1, whereas Poirier et al. (1987) relate the critical radius to the space remaining between dendrites. They obtain an expression for the liquid metal pressure where a void can form as

$$P \leq P_G - \frac{4\gamma_{LG}}{g_L\lambda_1},\tag{197}$$

where P_G is the pressure of gas in the pore if any, and γ_{LG} is the surface energy of the liquid–gas (vacuum) interface. If no dissolved gas is present and if the surface energy were zero, porosity would form at locations in the casting where the liquid metal pressure drops to zero (technically the partial pressure of the metallic vapor in equilibrium with liquid and solid). Dissolved gas increases the likelihood of porosity formation whereas the inclusion of the surface energy effect makes it more difficult to nucleate a pore, necessitating a negative liquid pressure to form a void. It is evident that microporosity because of solidification shrinkage and dissolved gas if present are coupled.

The value used for P_G in eqn (197) also depends on the fraction of liquid, g_L. Dissolved gas in a liquid alloy causes porosity because the solubility of the gas species in liquid metal usually exceeds the solubility in the solid. Indeed one defines a partition coefficient for the gas, k_{gas}^E as the ratio of the equilibrium solubilities of the gaseous species in the solid and the liquid. This definition is identical to that used for nongaseous solutes in an alloy. Considering again a small region of the dendritic array as depicted in **Figure 49**, as an alloy solidifies, the dissolved gas is rejected into the remaining liquid where its level increases. The exact amount of increase depends on the details of diffusion of the gaseous species in the liquid and solid phase in a manner identical to that for a metallic solute. However, because the diffusion rate of the gaseous species in solid metal is usually quite high, the gas content of the solid phase is usually assumed to be uniform within the volume element being considered. Thus the lever rule can be applied to compute the concentration of gaseous species in the liquid as a function of fraction solid. This increase in the concentration of gaseous species in the liquid leads to an expression for the equilibrium pressure that would exist in a pore if it formed given by

$$P_G = \frac{P_G'}{\left(g_L\left(1 - k_{gas}^E\right) + k_{gas}^E\right)^2},\tag{198}$$

where P_G' is the partial pressure of the gas in equilibrium at the free surface of the melt (given by the initial concentration of gas in the melt (Brody, 1974). Here Sievert's Law has been used in which the partial pressure of a diatomic gas in equilibrium with a saturated condensed phase is proportional to the square of the partial pressure. Thus it can be seen how the initial content of dissolved gas in the melt (through its effect on P'_G), its lack of solubility in the solid (through k^E_{gas}) and the pressure drop in the liquid metal required to feed solidification shrinkage all contribute to the formation of dendritic microporosity deep in the dendritic array. This combined effect is particularly important for aluminum castings where the solubility of hydrogen in the solid is only one tenth of that in the liquid. The assumption that the concentration of the gaseous species in the solid is uniform within the volume element in the dendritic array is relaxed in work described below.

Stefanescu (2005) and Lee and Wang (2010) provide summaries of the various approaches that have been used to predict the propensity for porosity formation in castings. While success has occurred in predicting the location and even the volume fraction of porosity, predicting the size distribution and shape of the pores is a much more difficult problem, yet it is vital to the prediction of critical flaws for mechanical property degradation prediction. As with our discussion of macrosegregation, the modeling of microporosity is easiest for columnar structures and more difficult for equiaxed structures.

7.9.4.2.1 Criteria Functions

To provide a simple noncomputationally intensive tool for incorporation into casting modeling software, so-called criteria functions have been developed. Using the computed local values of liquid isotherm velocity, temperature gradient at the liquidus temperature, local freezing time, and so on, a function is evaluated at each position within a casting to predict areas where microporosity is most likely. Such criteria functions are most useful for columnar structures.

The most popular criteria function that is widely used in casting software as a predictive tool for microporosity because of shrinkage was developed by Niyama et al. (1982). The critical value for the criteria function where microporosity develops was determined experimentally for low-carbon steel, but the general approach makes sense for columnar structures in general. The criteria function is closely related to the calculation of the pressure drop in the mushy region as follows.

One assumes dendrite trunks growing in the positive x-direction, and a liquid flow in the mushy zone in the same direction and governed by Darcy's law ignoring the gravity term (eqn (25)). The liquid flow velocity in the mushy zone is given by

$$v = -\frac{K(g_L)}{\eta g_L}\frac{dP}{dx}.$$ (199)

Recall from eqn (21) that the fluid flow required to feed the change in density at the liquid–solid interface is given by

$$v = -\beta V_T,$$ (200)

where the interface velocity has been assumed to be the isotherm velocity V_T. Combining the above two equations one obtains

$$\frac{dP}{dx} = \frac{\eta g_L \beta V_T}{K(g_L)}.$$ (201)

This equation shows the value of the pressure gradient that will exist for adequate flow to feed the shrinkage. Note that the pressure gradient is positive; that is, the pressure drops as one proceeds more deeply into the mushy zone away from the dendrite tips (-x direction). The above expression can be integrated between the liquidus temperature and the final eutectic temperature, taking G_L as the temperature gradient ($G_L = dT/dx$) to get the pressure difference between the dendrite tips and the location of complete solidification

$$\Delta P = \frac{\eta\beta V_T}{G} \int_{T_L}^{T_E} \frac{g_L}{K(g_L)} dT. \tag{202}$$

The volume fraction of liquid, g_L can be written as a function of temperature, for example using the Scheil assumption. Thus, the integrand is a function of temperature and the integral is a constant for a given alloy and dendrite spacing. Letting the integral be C and noting that the cooling rate and velocity of the liquidus isotherm are related by $\dot{T} = G V_T$, the pressure drop can be expressed as

$$\Delta P = C\eta\beta \frac{\dot{T}}{G^2}. \tag{203}$$

It can be seen that the pressure drop is higher with lower temperature gradient and higher cooling rate. Therefore, the propensity for porosity formation increases with lower temperature gradient and higher cooling rate. While simple, this approach contains the essential factors that govern microporosity formation. After a choice of the constant C either by selection of a g_L and K models or from experiment, maps of the criteria function at each position over the entire casting can be exhibited by commercial software packages to assist the casting engineer in refining the rigging system.

Various other criteria functions have been developed with varying degrees of success. Lee et al. (1990) proposed a criteria function that was applicable for wide freezing range alloys. This function included temperature gradient, local solidification time, and solidification velocity separately. Suri et al. (1994) proposed a criteria function called the feeding resistance number. This is based on the concept that interdendritic feeding is dependent on the specific surface area of the solid structure and thermal parameters. A high value of this number at a location indicates that feeding is difficult and hence microporosity is expected to occur. Beech et al. (1998) proposed feeding criteria function based on a drag force coefficient, which is a function of the local solid fraction.

7.9.4.2.2 Fluid Flow Calculations of Microporosity in Columnar Structures

To actually calculate the size and fraction of porosity after solidification is complete, a more complex analysis is required that solve the energy, mass, and momentum conservation equations. Focusing on hydrogen in Al–Cu alloys, three papers performed this type of analysis: Kubo and Pehlke (1985), Poirier et al. (1987) and Zhu and Ohnaka (1991). They assumed respectively that the maximum pore size was equal to the secondary dendrite arm spacing, $1/2$ of residual liquid space between primary arms and $1/2$ of residual liquid space between secondary arms. Permeability was calculated using the Blake–Kozeny model in all. Kubo and Pehlke further simplified the permeability calculation by using the Blake–Kozeny model only between 0.01 and 0.7 liquid fraction. Poirier et al. included the anisotropic character of the permeability in their model. In all of these papers, the effect of the initial H content, solidification velocity, temperature gradient (and cooling rate), and ambient pressure were simulated. The methods all employ the assumption that the gaseous species diffuse rapidly and only treat the pore size in terms of an equivalent spherical pore. In essence, none of these models can predict the time-dependent pore growth and pore size distribution in castings.

The details regarding size and shape constraints on gas filled pores in the interdendritic regions require consideration. In a columnar–dendritic mushy zone of an alloy, the interdendritic spaces between primary dendritic arms are greater than those between secondary arms. Hence, voids tend to form in the primary interdendritic space rather than in the spaces between secondary arms.

Lee and Hunt (1995) were first to include an analysis of diffusion of the gaseous species in the solid and thus relaxed the lever assumption for the gaseous species. Thus it incorporates ideas similar to those employed in the "back diffusion calculations" in Section 7.7. In the 2-D calculations a pore size distribution was computed using experimental input of nucleation data from X-ray microradiography measurements.

7.9.4.2.3 *Calculations of Microporosity in Equiaxed Structures (no grain movement)*

In equiaxed alloys, Poirier et al. (1987) postulated simple models for estimating the largest diameters of pores that can form and discussed simple mechanisms of pore formation in intergranular and inter-dendritic spaces. If the pores form intergranularly, the main parameter which controls the pore size and the excess pressure is the grain size. A finer grain size will require a greater local liquid pressure drop and larger dissolved gas content to generate pores. Because there is no well-defined primary spacing for equiaxed microstructure, the secondary dendrite arm spacing (SDAS) is the key parameter for inter-dendritic pore formation.

Sung et al. (2001) modeled microporosity in IN718 equiaxed investment castings. Their multi-component model calculated convection and macrosegregation and also included the transport of gas solutes. This model predicted the regions of possible intergranular porosity by comparing computed local liquid pressure value with that obtained using the Sievert's law. They also studied the influence of several variables on the propensity of microporosity formation. These variables are: the mass transfer of hydrogen and nitrogen from the casting to the casting/mold gap, the final grain size, a grain-shape parameter and casting section thickness. The mass transfer coefficient across the casting/mold gap was found to be the most important factor. This indicated that the microporosity in equiaxed investment castings depends strongly the on the development of the gap at the casting/ceramic shell mold interface, gas pressures in the gap region, and the transport of gases through the shell mold during solidification and subsequent cooling.

Sabau and Viswanathan (2002) provided a comprehensive approach for prediction of microporosity in equiaxed aluminum alloy castings. Their computational methodology included solidification, interdendritic fluid flow, hydrogen precipitation, and porosity formation. Such an approach can be easily implemented in a comprehensive casting simulation software. However, a rigorous model was not provided for computation of pore radius. Instead, pore radius was taken to be proportional to dendrite spacing, which changes with local solidification time.

Following up on earlier work on porosity formation in columnar grain structures, the diffusion of the gaseous species was combined with a CA method by Atwood and Lee (2003). This approach deals directly with the tortuous shape of hydrogen porosity in Al–Si alloys. The model predicts the percentage of porosity, the average pore size, pore size distribution, and maximum chord length of porosity. Their results showed good qualitative agreement with experimental data obtained for an Al-7 wt% Si alloy. However, this model does not allow for modeling porosity in the case of any grain movement during the solidification process.

7.9.4.2.4 *Calculations of Microporosity in Equiaxed Structures (with grain movement)*

In 2004 Lee et al. proposed a multiscale approach for prediction of microporosity in Al–Si–Cu alloy castings that includes the motion of freely floating grains. Despite being computationally prohibitive

(a)

(b)

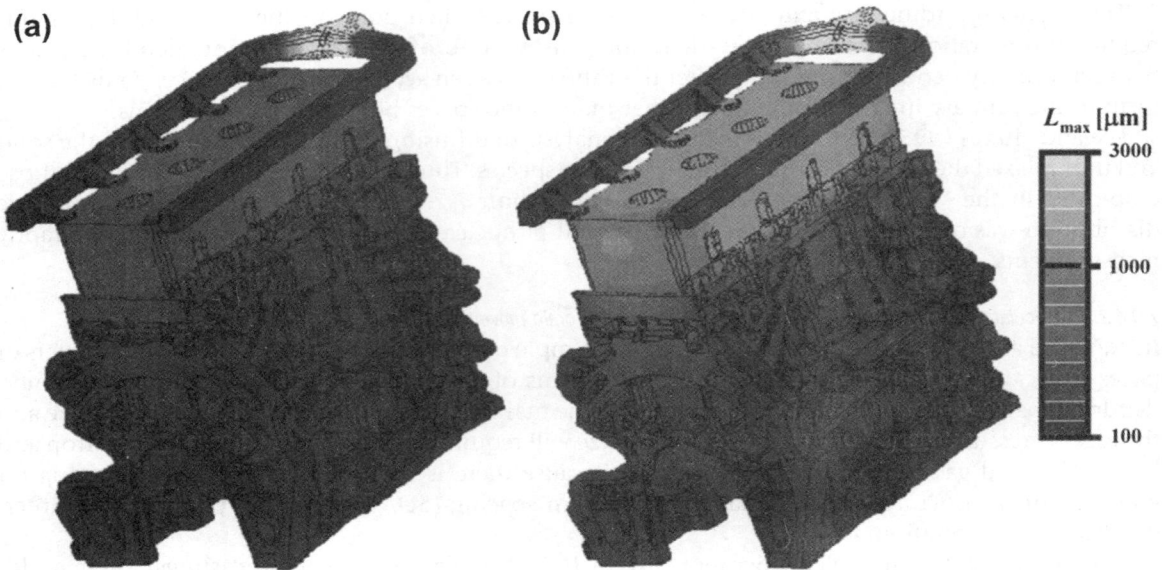

Figure 94 Application of multiscale model to predict maximum pore length (L_{max}) in an engine block casting made of W319 alloy. (a) $C_H = 0.1$ ml/100 g; and (b) $C_H = 0.2$ ml/100 g (Lee et al., 2004).

for an industrial scale casting, this approach is the most comprehensive for modeling microporosity in castings. This approach computes porosity by addressing both solidification shrinkage and gas porosity.

The thermal and flow field are obtained by solving the energy, momentum, and continuity equations on the macroscopic scale. The solid phase nucleation and growth were computed on a meso-scale using a stochastic nucleation model in conjunction with a CA-FD (cellular automata – finite difference) model that is linked with the macroscopic model through the temperature field. The growth of pores is computed by solving the diffusion of hydrogen on microscopic scale while nucleation is considered using a stochastic approach. Nucleation of pores is dependent on the hydrogen supersaturation in the liquid phase. The pore growth model is coupled with the macroscopic model through the pressure field. This model included growth restriction of pores that impinge on the solid phase resulting in the possibility of nonspherical pore growth. They predicted maximum pore size and distribution in an engine block made of commercial W319 alloy and obtained a good match with experiment (see **Figure 94**).

This model is complex and the microscopic model used for pore nucleation and growth needs to be validated by using experimental tools. Fortunately, recent advances in the use of in situ visualization tools such as X-ray radiography and X-ray tomography have allowed comprehensive studies of pore formation and growth. A resolution below 1 µm allows use of X-ray tomography to study pore formation and growth as shown in **Figure 95** (Mathiesen and Arnberg, 2007).

7.9.5 Inclusions

At present it is very clear that inclusions exert an important influence on fracture behavior of commercial materials. As a result, this portion of the field of solidification is receiving much greater attention. Inclusions are typically metal oxides or sulfides. Those that are solid in the melt above the

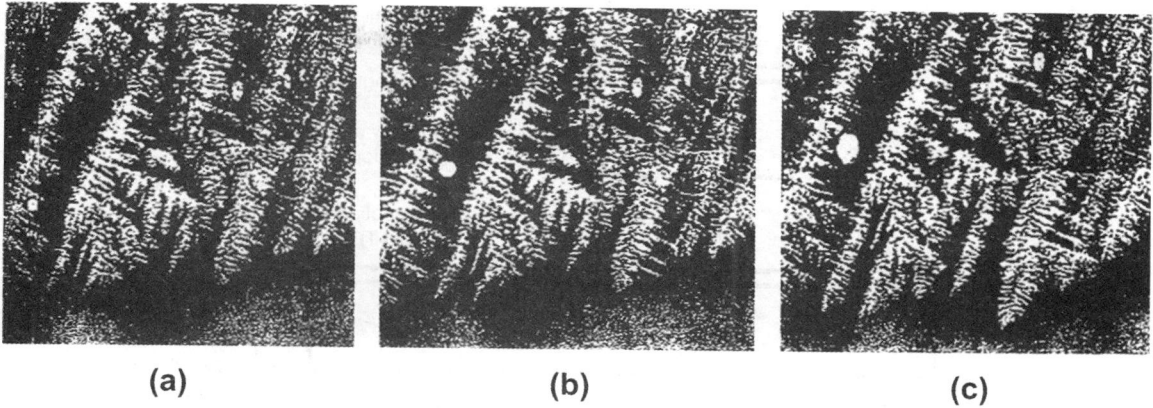

(a)   **(b)**   **(c)**

Figure 95 Time sequence, (a), (b) (c), of X-ray microradiographs showing the formation of gas porosity in a directionally solidified Al-30 wt% Cu alloy, and its subsequent entrapment by the eutectic front (Mathiesen and Arnberg, 2007).

liquidus temperature of the first metallic phase to form are called primary and those that form later and often interdendritically are called secondary.

Control of inclusions is necessary to prevent degradation of mechanical properties. Some of the controls that are practiced in the metal casting industry are chemistry control, mold/metal interface and refractory control, separation techniques, and inclusion shape control. Trojan and Fruehan (2008) provide a review of these control mechanisms. They highlight that inclusion shape control is a key control mechanism needed to prevent degradation of the casting in strain rate sensitive applications. This shape control can be achieved by controlling the surface energy between inclusions and the liquid metal through proper modification of the chemistry of the alloy. Campbell (1991a) and Trajan (1988) treat extensively both primary and secondary inclusions and their effect on mechanical properties in ferrous and nonferrous alloys.

With the increasing availability of multicomponent thermodynamic databases for various alloy systems, thermodynamic modeling of inclusion formation is becoming more common. Thermodynamic calculations on the multicomponent alloy systems can be conducted to study the formation of different types of inclusions during melt and ladle processing (e.g. inoculation, deoxidation practice etc.). When a full thermodynamic database is available that includes the metallic and nonmetallic phases the approach can treat the formation of the inclusion as it would a metallic phase. Appropriate solidification models (e.g. Lever, Scheil, or back diffusion) can be used to compute the solidification path, solute distribution, and amount of each phase (including inclusions) that forms from the liquid at each temperature as the melt cools.

7.9.5.1   Primary Inclusions

Primary inclusions correspond to: (i) exogenous inclusions (slag, dross, entrapped mold material, refractories); (ii) fluxes and salts suspended in the melt as a result of a prior melt treatment process; and (iii) oxides of the melt which are suspended on top of the melt and are entrapped within by turbulence. In the steel industry a significant reduction of inclusions is obtained by their floating upward and adhering to or dissolving in the slag at the melt surface. In the aluminum industry filtering has become a common practice and the development of better filters is an important area of research (Ross and

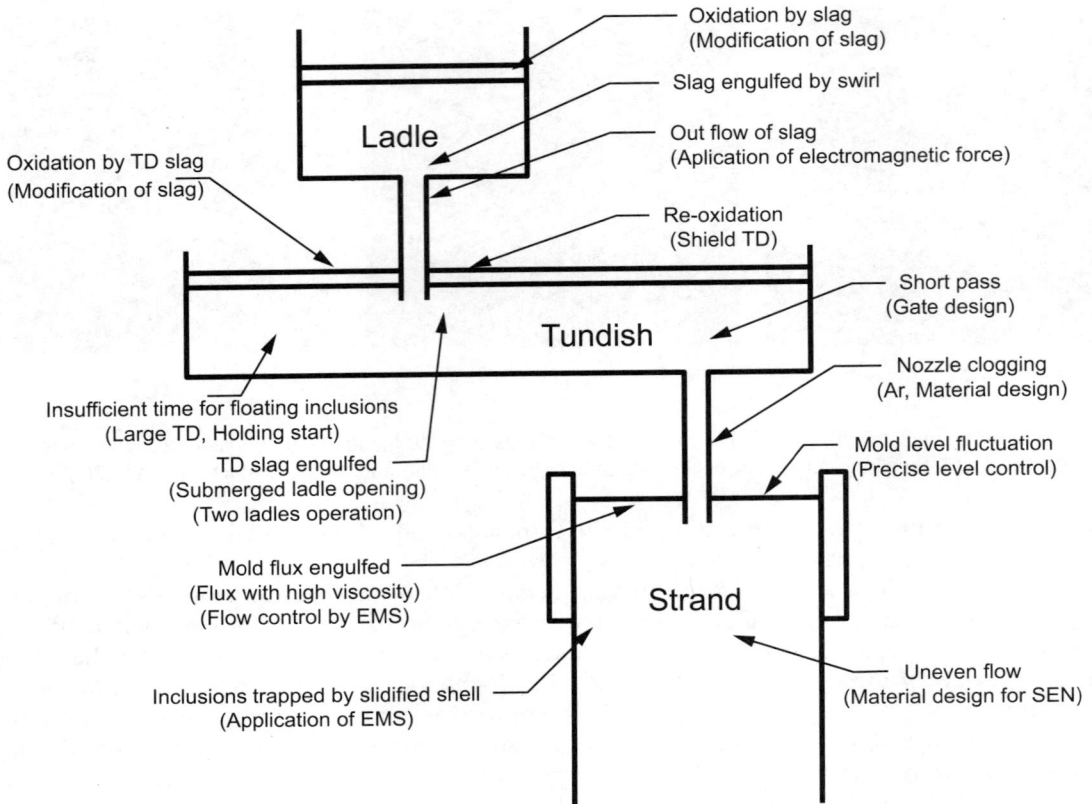

Figure 96 Schematic drawing of measures to reduce inclusion defects (Esaka, 1999).

Mondolfo, 1980; Apelian, 1982). Examples of approaches to prevent inclusions during the continuous casting of steel are shown in **Figure 96** (Esaka, 1999).

An example of the use of a thermodynamic data base to guide ladle processing is found in the work of Vayrynen et al. (2009), who studied the control and removal of inclusions in continuous casting of steel. In one case study, they investigated the possibility of inclusion formation during ladle treatment and casting of high-carbon and silicon spring steels. Their goal was to have effective silicon deoxidation (without aluminum deoxidation) by bringing the melt into contact with a well controlled slag. Their thermodynamic calculations showed that it was feasible to obtain steel with desired chemistry, for example, low O_2, without alumina and Ca–Al–silicate inclusions. Their thermodynamic calculations were in good agreement with experimental data. Zhang et al. (2004) used the thermodynamic approach to control inclusions during the steelmaking process.

Machinability of low-carbon free-cutting steel at moderate and high cutting speeds can be improved by engineering certain indigenous glassy inclusions. These researchers used a thermodynamic model for deoxidation control of steel during melt/slag and melt/oxide inclusion equilibration. Their study found that these inclusions improved machinability by providing the needed lubrication at the tool–chip interface during high-speed machining operations. The validity of the thermodynamic model is shown

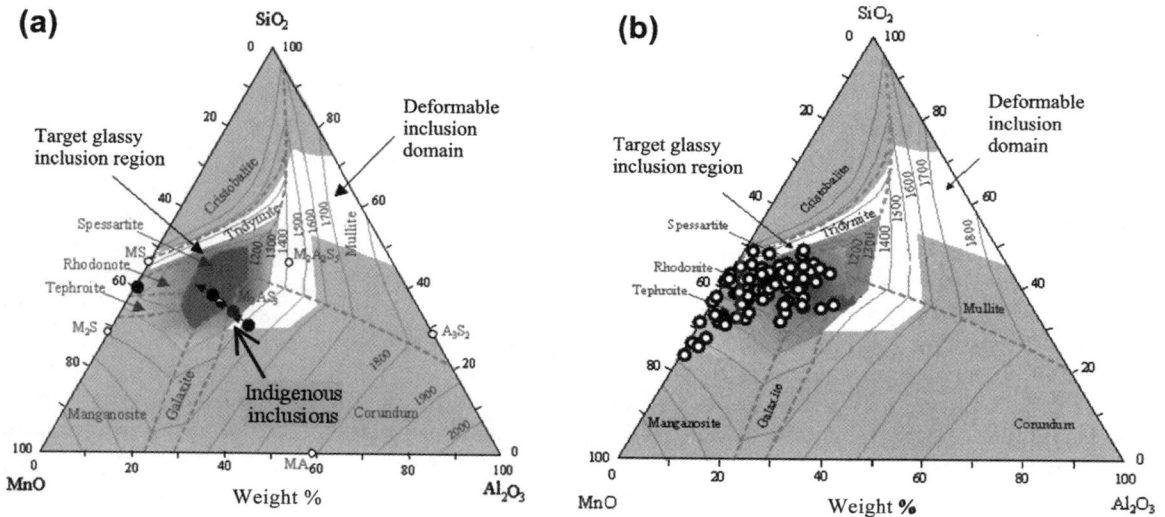

Figure 97   (a) Final step in the design of deoxidation in the continuous caster: prediction of indigenous inclusions (based on melt chemistry and thermodynamic equilibrium) after solidification in the caster; and (b) energy dispersive x-ray chemical analysis of indigenous oxide inclusions of MnO–Al$_2$O$_3$–SiO$_2$ in inclusion engineered 11SMn37 steel as-rolled (Zhang et al., 2004).

in **Figure** 97. The thermodynamic model computed the inclusion composition for particular melt chemistries. It can be seen that the composition of the measured inclusions in the ternary diagram showed a good match with the predicted composition of the inclusions. Note that the results are shown for the final step in the three step inclusion engineering process. Inclusions are shown in **Figure 97(b)** in the dark regions of the target inclusion areas. Thus thermodynamic modeling was successfully used to design a controlled deoxidation process and melt/slag and melt–oxide inclusion equilibration during the steelmaking process to obtain the desired glassy oxide inclusions.

7.9.5.2 Secondary Inclusions

Secondary inclusions are those which form concurrently with the solidification of the major metallic phase. Although in industrial practice commercial alloys involve multicomponent systems, a first approach to the understanding of the formation of secondary inclusions has been achieved through the considerations of ternary diagrams involving the most important impurity elements under consideration (Flemings, 1974). The solidification reactions occurring during the process, together with the values of the various partition coefficients of the impurity elements in the metallic phase play an important role in the type, size and distribution of inclusions in the final structure. Important ternary systems to be considered are Fe–O–Si, Fe–O–S, and Fe–Mn–S from which the formation of silicates, oxides and sulfides results. Fredriksson and Hillert (1972), through carefully controlled solidification, were able to determine the formation of four types of MnS inclusions formed by different reactions.

Nakama et al. (2009) studied the formation of sulfides as a normal part of the solidification path of the Fe–Cr–Ni–Mn–S alloy system. Predictions of the path using a multicomponent phase diagram provided by a thermodynamic calculation of the type described in Section 7.3.3 was compared with the observed formation of the sulfide inclusions. Depending on the phase sequence encountered on the path of the different alloys studied, the sulfides were located in different positions within the dendritic microstructure.

7.9.5.3 Inclusion Pushing/Entrapment

An important effect to consider when a moving solidification front intercepts an insoluble particle is whether the inclusion is pushed or engulfed. If the solidification front breaks down into cells, dendrites or equiaxed grains, two or more solidification fronts can converge on the particle. In this case, if the particle is not engulfed by one of the fronts, it will be pushed in between two or more solidification fronts and will be entrapped in the solid at the end of local solidification. Stefanescu and Dhindaw (1988) reviewed the variables of the process as well the available theoretical and experimental work for both directional and multidirectional solidification. Subsequently, Shangguan et al. (1992) presented an analytical model for the interaction between an insoluble particle and an advancing L–S interface. There exists a critical velocity for the pushing–engulfment transition of particles by the interface. The critical velocity is a function of a number of materials parameters and processing variables, including the melt viscosity, the wetting behavior of the particle and the matrix, the density difference as well as the thermal conductivity difference between the particle and the matrix, and the particle size. Qualitatively the theoretical predictions compare favorably with experimental observations.

Shibata et al. (1998) studied the interaction of the solid–liquid interface with inclusions during the casting of steel. They studied Al_2O_3 clusters floating in a low-carbon Al-killed steel melt ahead of the liquid–solid interface. **Figure 98** shows that fine inclusions were pushed while large inclusions were engulfed. They noticed that pushing was enhanced when the solidification velocity was reduced. They obtained experimentally a relationship where the maximum pushing velocity was inversely proportional to the inclusion radius (**Figure 99**). They noticed that inclusions in the melt ahead of the liquid–solid interface are almost always engulfed at solidification velocities typically encountered

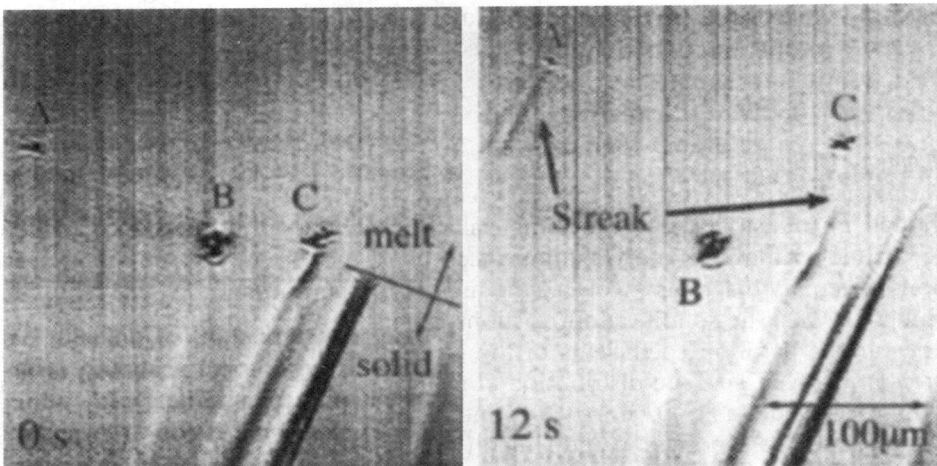

Figure 98   Pushing of small aggregates of Al_2O_3 inclusions by the advancing solid–liquid interface in a low-carbon Al-killed steel. Note that inclusions A and C are pushed leaving grooves behind them, while inclusion B is engulfed (Shibata et al., 1998).

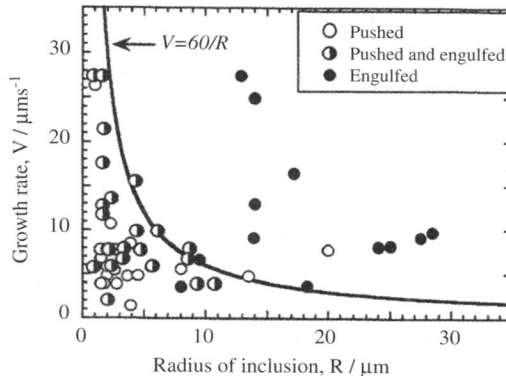

Figure 99 Critical condition for the engulfment of Al_2O_3 inclusions during solidification of low-carbon Al-killed steel at 1800 K (Shibata et al., 1998).

during continuous casting of steels. To avoid this engulfment, the inclusions need to be prevented from coming into close proximity of the L–S interface. This can be achieved by sweeping or washing the melt/shell boundary by clean melt flow controlled by an electromagnetic field.

7.9.6 Fluidity

Over the years the foundryman has found it useful to employ a quantity called *fluidity*. The concept arises from practical concerns regarding the degree to which small section sizes can be filled with metal during castings with various alloys. This property is measured through one of several types of fluidity tests. Hot metal is caused to flow into a long channel of small cross section and the *maximum length* that the metal flows before it is stopped by solidification is a measure of fluidity. Fluidity depends on many factors. They fall in the following three categories:

Alloy variables: alloy composition, viscosity of melt, pouring temperatures, and thermophysical properties as functions of temperatures.
Mold variables: mold temperature, thermophysical properties, metal/mold heat transfer coefficient and surface tension.
Casting variables: metallostatic head, channel dimensions, types and amount of inclusions.

In addition, fluidity also depends on whether the alloy freezes with plane front or with columnar or equiaxed dendrites. There are several techniques available for measuring fluidity (Sabatino, 2005). However, one major shortcoming is that data from various fluidity tests cannot be directly compared. Most popular tests are vacuum fluidity test, fluidity spiral in sand mold, and multichannel fluidity test in metal mold.

Flemings (1974) reviewed the field and his contributions to the study of fluidity. Later, Campbell (1991a,b) gives a general view of this property and stresses the importance of the factors that influence and limit the validity of the fluidity test. It is worthwhile to follow the approach of Campbell (1991a) who considers three cases: (i) *maximum fluidityl length*, L_f, determined by an experiment where the cross-sectional area of the channel is large enough that the effect of surface tension is negligible; (ii) L_f, when surface tension is important, and (iii) *continuous fluidity length*, L_c.

Measured fluidity can be compared with predictions of commercial casting software that solves fluid dynamics, heat transfer, and solidification kinetics in castings. Using accurate data for thermophysical properties and boundary conditions, the fluidity length can be computed assuming that flow is significantly reduced or stopped when dendrite coherency is reached.

When the channel section becomes thinner than a critical value (~ 0.30 cm for most alloys), the resistance to liquid flow increases because of surface tension (Flemings, 1974). This is particularly critical in technologies such as the casting of aerofoils, propellers, and turbine blades (Campbell, 1991a). Campbell and Olliff (1971) distinguish two aspects of filling thin sections: flowability, essentially following the rules discussed above and fillability limited by surface tension.

7.10   Developing and Emerging Processes

Solidification remains an important topic in new and emerging manufacturing processes and technologies such as spray forming, clean metal nucleated casting (CMNC), laser-based additive manufacturing, metal foam processing, semisolid forming, and liquid melt processing. These are described along with a few new concepts that have been introduced in high-pressure die casting.

7.10.1   Spray Forming

Spray forming (also known as spray casting) is a process for a near-net shape materials processing approach. It utilizes melting, atomization of droplets, and subsequent deposition and consolidation of droplets to produce near fully dense preforms. Brennan (1958) proposed modern spraying methods to produce semifinished products more than 50 years ago. However, this process received major attention when the Osprey process was proposed in the 1970s (Leatham et al., 1991). Considerable research and development over the past couple of decades have led to significant enhancement of this technology.

In the recent past, a large number of studies have been conducted to understand the spray deposition process. Many researchers chose to use modeling to understand the spray forming process. These models can be applied in several stages, for example, metal and gas delivery, atomization, spray, deposition, and consolidation. A brief summary of the modeling approaches is provided by Liu (2002).

In the spray forming process, molten metal is disintegrated into a stream of droplets by subjecting the melt to atomization. Typically, this disintegration is achieved by gas atomization, centrifugal atomization, or ultrasonic atomization (Liu, 2002). Gas atomization is by far is the most viable option.

Figure 100 shows a schematic of a spray forming process employing gas atomization. The melt is typically kept at a very low superheat. The melt chamber is purged with an inert gas and a slight over pressure of gas is maintained to prevent oxidation of the melt. The melt is discharged from the bottom to an impinging atomization spray (inert gas jets) that disintegrates the melt to droplets in the range of 10–500 µm in diameters. The droplets are rapidly cooled in flight and can reach a velocity of 50–100 m/s. The droplets reach the substrate below at varying level of fraction solid and cool further on hitting the substrate. Most of the latent heat of fusion (60–80%) is removed in several milliseconds following gas impingement while the rest is removed over a time period as long as 300 s. As was described in Section 7.4, isolation of potent nucleation sites to a small fraction of the droplet distribution promotes supercooling. Varying shapes of preforms can be employed. The absence of macrosegregation and the presence of fine grains result in enhanced mechanical properties of the formed part compared with those parts developed using normal ingot processing approach.

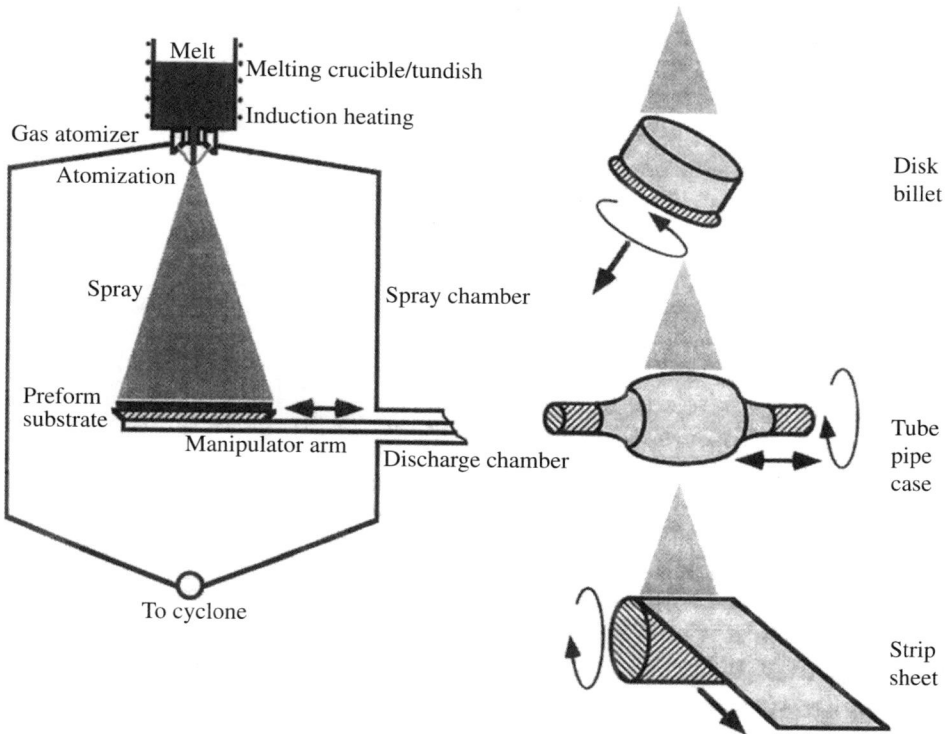

Figure 100 A schematic of a spray forming process. Courtesy: W.T. Carter, GE CRD, Schenectady, NY, 2004.

Spray forming offers several advantages: (a) low processing temperatures coupled with high heat extraction reduces segregation, and coarsening; (b) inert atmosphere maintained during processing reduces the potential for defect formation in the microstructure; (c) a relatively homogeneous distribution of nonmetallic phases (e.g. oxides, nitrides, silicides) may result in dispersion strengthened materials; (d) potential for producing alloys with compositions beyond those that can be produced using conventional processing methods; (e) low tooling costs coupled with reduced energy requirement and high throughput could make this an attractive rapid prototyping tool; (f) microstructural homogeneity possible in this process allows it to be used, for example, to make sputtering targets (e.g. aluminum–neodymium alloys); and (g) fine grain sizes, reduced size of secondary phases, and metastable phases.

Spray forming has been used quite extensively for the processing of structural materials (Lewis and Lawley, 1991; Leatham and Lawley, 1993), highly reactive, for example, Mg alloys (Savage and Froes, 1984), Al alloys (Baram, 1991), Fe-base alloys (Mathur et al., 1989) high-temperature materials (Bricknell, 1986), intermetallic compounds (Liu et al., 1994), and reinforced metal matrix composites (Singer, 1991). Specialty steels, superalloys, aluminum alloys, and copper alloys have been processed commercially using spray forming. Spray-processed Al alloys include Al–Si alloys for automotive applications, Al–Li alloys for aerospace and marine applications. Copper alloys such as Cu–Cr, Cu–Fe, and Cu–Zr have been used to produce strips with enhanced properties. High-Cr cast iron rolls have been produced by spray forming (Leatham, 1996). Ni-base superalloy billets have also been commercially produced by spray forming (Leatham, 1996).

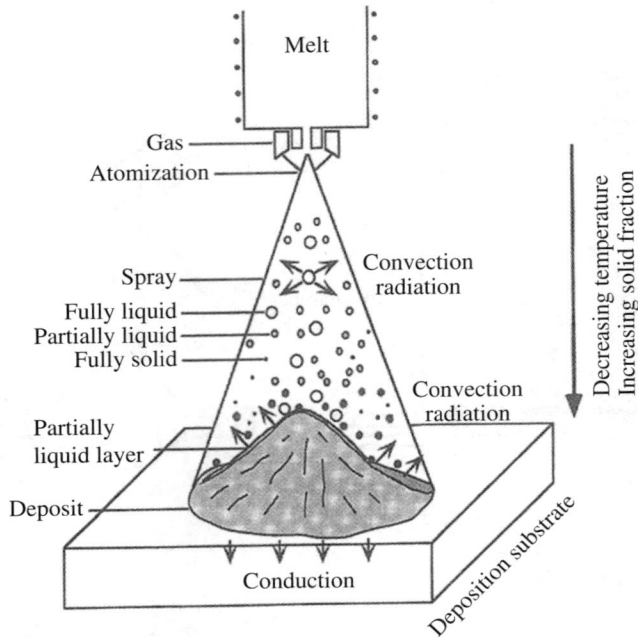

Figure 101 Heat transfer and solidification in spray and deposit (Liu, 2002).

Some of the common disadvantages of spray forming are as-cast porosity that requires further remedial steps such as rolling, extrusion, or hot isostatic pressing; metal losses because of overspray (e.g. atomized droplets that do not arrive at the preform).

7.10.1.1 *Atomization*
During spray forming, a high kinetic energy gas is used to impact the liquid metal to produce fine droplets. Several modes of atomization have been outlined (Liu, 2000): liquid jet breakup, liquid sheet breakup, liquid film breakup, liquid ligament breakup, and so on. Simultaneous occurrence of two or more of these mechanisms is likely. The two primary modes of breakup during conventional spray forming are liquid jet and liquid film breakup. Details of the atomization mechanisms are described (Liu, 2000). A wide range of droplet diameters are possible in gas atomization, which can be described as a stochastic process. Often the droplet diameter distribution has been approximated with a log-normal distribution.

7.10.1.2 *Process Parameters*
Geometric parameters that are of interest are dimensions of the preform; dimensions of nozzle; substrate configuration, angle of deposition. Types of atomizer gas chosen are also important. The most common atomization gas is argon. However, nitrogen, helium, and air have also been used for certain alloy systems and products. Other variables that are important are superheat of the melt, melt flow rate, gas pressure and flow rate, distance between the atomizer and the substrate, relative motion of the spray and substrate. A list of these variables is provided (Liu, 2002; page 674–675).

7.10.1.3 Solidification

Figure 101 shows a schematic of cooling and solidification during spray forming. During spray forming, the solidification of the droplets occurs in two stages: (a) during flight in the spray and (b) in the deposit on the substrate. The cooling in flight is mainly because of forced convection and is estimated to be in the range of 10^5 to 10^7 K s^{-1}. Following nucleation, latent heat is released rapidly because of high growth rates resulting in recalescence. Often temperatures nearly equal to the equilibrium melting temperatures are attained following recalescence within a few microseconds in flight. Following recalescence, the cooling rate is reduced (10^3 to 10^6 K s^{-1}). Typically small droplets are fully solidified upon arrival at the substrate, while large droplets could be fully liquid. Intermediate droplets arrive at the substrate at varying levels of fraction solid. Heat extraction is mainly because of conduction at the substrate bottom, while heat is lost because of convection at the substrate top. Cooling rates typically range from 1 to 10 K s^{-1} on the substrate.

7.10.1.4 Microstructure

The final microstructure depends on the process parameters used in the spray forming. Microstructure varies both in morphology and in scale among droplets depending on supercooling and thermal histories in the droplets during the entire cooling process. Microsegregation-free microstructure is observed in fine droplets that were fully solidified before impact, while larger droplets show dendritic microstructure with varying secondary dendritic arm spacings. The mushy droplets undergo abrupt change in cooling rate from a high value in the spray to a much lower value at the preform. The abrupt change in cooling rate and consequent high shear forces during impact and spreading of droplets on hitting the preform destabilize the dendritic network in the droplets resulting in fragmentation. These large numbers of broken dendrite fragments serve as efficient nuclei resulting in fine equiaxed microstructure. See Section 7.9 for details on fragmentations and development of equiaxed microstructure. Some homogenization of the microsegregation may take place in the substrate following solidification. Thus a microstructure may contain regions with a very fine scale microsegregation and/or grain size because of the undercooling, but may also contain regions solidified on the substrate where solid-state diffusion and coarsening mechanisms are more dominant.

7.10.1.5 Porosity

Spray forming is a complex process. Impact, spreading, and consolidation of droplets are very important in determining the properties of the deposits. Two-phase fluid flow, heat transfer, and solidification occur during droplet impingement and consolidation process. On a macroscopic level, the droplets may break up or bounce after impinging the substrate. The interaction among droplets and the droplet/substrate interaction can dictate the properties of the sprayed part.

A common problem in this process is the propensity for porosity formation. Several process parameters such as impact velocity, angle of impingement, droplet size, droplet temperature, the heat transfer rate, and substrate temperature can have influence on the porosity formation. Following Delplanque et al. (1997), major categories of porosity are chemical porosity, physical porosity, and dynamic porosity. Chemical porosity results because of possible dissolved gas in the alloy or from dissolution because of a particular foaming agent used. Physical porosity results from shrinkage because of solidification. The dynamic porosity results from the complex droplet/droplet and/or droplet/substrate, droplet/gas interactions. The porosity can form both at the surface and at the internal region of the deposition. The dynamic porosity can consist of both macroporosity and microporosity.

Macroporosity typically occurs at locations around solidified particles at interparticle boundaries. Nonplanarity of the surface in the solidified layer could impede free flow of mush or liquid from subsequent deposition. This could happen if these droplets cannot physically reach the void region. These porosities can have dimensions similar to the droplets.

Microporosity occurs because of separation/breakup of liquid in flight, which could result in voids being trapped during subsequent impingement of liquid at the substrate. Such porosity is smaller than droplet size. These types of microporosity are called transgranular microporosity. Intragranular microporosity (smaller in size than the droplets) results from entrapped gas bubbles trapped within droplets during atomization. Shorter flight time as well as high droplet velocities may allow these voids to be retained within droplets during completion of solidification following impingement at the substrate. Such microporosity can also develop from the presence of inclusions or dispersoids in the droplets, which could serve as potential sites for voids to form during subsequent solidification.

7.10.2   Clean Metal Nucleated Casting

The electrical power production industry has been working diligently to improve thermodynamic efficiency and fuel economy of land-based gas turbines made of superalloys. One key requirement to achieve these goals is to have higher operating temperatures. Additionally, continued higher electrical output demand necessitated the design and development of large turbines. To meet these performance demands at elevated temperatures, it is necessary to have a homogeneous, defect-free microstructure. As the part size is increased, it has become a real challenge to produce such large parts using conventional casting technology that meets these challenging performance demands at higher operating temperatures. Current casting technology for superalloys already has been pushed to the limits and could not be enhanced further to make the next generation of large ingots needed for new high-efficiency, higher output turbines. The clean metal nucleated casting (CMNC) approach has recently emerged as a commercially viable approach that can address some of these challenges faced by conventional casting technology. CMNC is a variation of the original nucleated casting proposed by Sankaranarayanan et al. (1995) and Tyler and Watson (1995) in which a stream of metal is atomized by a gas and the droplets are collected in a mold. This is a variation of the spray forming process discussed above. In this particular process, the thermal conditions are adjusted so that the deposit metal is approximately 25–35% solid; therefore the metal is inviscid and requires a collection mold. If the conditions are adjusted so that the deposit is approximately 25% liquid, the metal is quite viscous and therefore, a collection mold is not necessarily required, which is the spray forming process as discussed above.

General Electric Company has successfully used this technique to develop a turbine disk for the land-based gas turbine with a Ni-base 718 alloy. The turbine disk is the rotating hub on which turbine blades are mounted. Because the metal is cast as a semisolid as opposed to a superheated liquid in a conventional casting process, large ingots can be produced without major segregation defects. Therefore, CMNCs of large turbine wheels facilitate the production of freckle-free and inclusion-free parts that allow higher operating temperatures than those obtained with conventional cast part. The technique allows production of large, fine-grained, homogeneous superalloy ingots at a rate 6–10 times faster than the current state of the art, and cuts the number of melting steps needed to make a superalloy by a third. The fine-grained structure obtained directly from CMNC makes it possible to sharply reduce or eliminate the thermo-mechanical processing steps needed to convert the ingot into forgeable billets. CMNC changes the process for the casting of superalloys from one of liquid processing to one of casting in the semisolid state, thereby enabling large gains in productivity, small grain size and compositional uniformity.

Figure 102 is a schematic of the CMNC process (Carter and Jones, 2004). In this process, a consumable electrode is fed into the furnace from above. The bottom surface of the electrode is kept immersed in a hot liquid slag. The electrode melts at the bottom and the liquid metal is bottom fed into

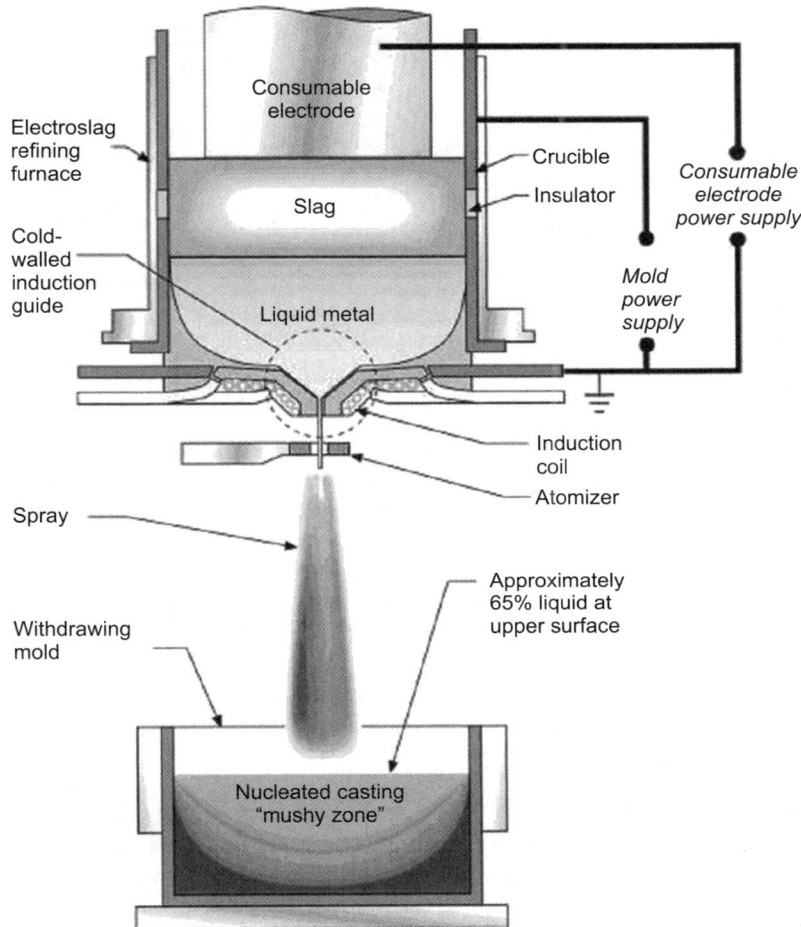

Figure 102   Schematic of CMNC system (Carter and Jones, 2004).

the gas atomizer via a cold-walled induction guide (CIG) nozzle. The CIG nozzle is a segmented, water-cooled copper funnel with induction heating designed to ensure that the liquid metal stream does not freeze. The use of a copper nozzle ensures that no ceramic inclusions are reintroduced into the liquid, as would normally occur with the ceramic nozzles typically used in traditional spray forming processes. The pouring system relies on active pressure control above the melt to keep the pouring rate constant and assure a constant solidification rate. The melt is gas atomized in a similar manner to the one described for the spray forming process. Process parameters such as the melt superheat, the gas-to-metal ratio, and the spray distance are adjusted to achieve a metal pool where the top surface of the deposited layer contains approximately 25–35% solid fraction. Therefore, a semisolid metal is cast into the collection mold. Many small particles will potentially remelt after deposition. However, a sufficient number of solid particles should be present to serve as growth sites for further solidification. The existence of a sufficient number of sites will ensure that a fine-grained structure is obtained. The

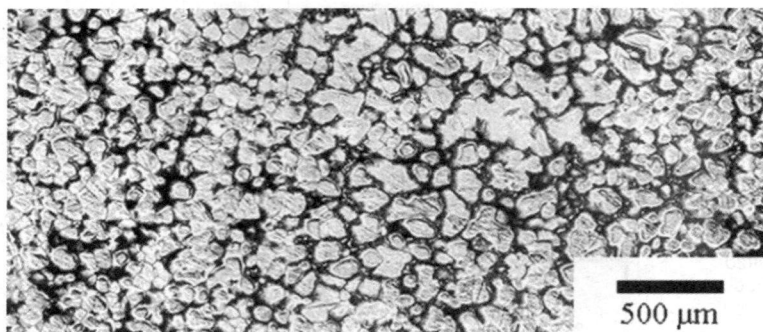

Figure 103   Microstructure of alloy 718 as cast in a nucleated casting trial (Carter and Jones, 2004).

collection mold plays a critical role in this regard. At steady state, the sprayed ingot is withdrawn through the stationary sidewalls so as to maintain a constant metal spray distance at all times.

One of the goals of the CMNC process is to replace traditional electroslag refining and vacuum arc remelting with the single CMNC processing step that results in large energy savings. The most significant advantage of this process is a macrosegregation-free ingot with a grain size of about 0.075 mm (see **Figure 103**). This grain structure is substantially finer than that of the standard triple melt ingot. It is possible that a small CMNC ingot may be forged into a disk immediately after casting without any traditional intermediate steps involving billet conversion. However, larger diameter ingots will still require metalworking to develop an acceptable billet structure.

7.10.3   Laser-Based Additive Manufacturing

Today, additive manufacturing (AM) process is used as an advanced technology tool to manufacture new products or repair existing products. AM is essentially a rapid prototyping tool that uses an approach where metal (melted by a laser beam) is deposited layer by layer according to the desired geometry. Each layer essentially is a section of the desired geometry of the final product. Today there are several approaches for AM such as selective laser sintering/melting (SLS/M), laser engineered net shaping (LENS) and its variant, direct metal deposition (DMD), laminated object manufacturing (LOM), electron beam melting and ultrasonic consolidation (Sanjay and Sisa, 2011).

The wide uses of laser-based processes (SLS/M, LENS/DMD, and LOM) shows a broader adoption of this technique as an AM approach. Although most of these processes use powder as the source of metal, the SLS/M approach uses a powder bed while in LENS, the blown powder is melted with a laser beam. LOM uses metallic foil as the building unit. This section will mainly focus on LENS and DMD techniques for AM.

A schematic of the LENS process is shown in **Figure 104**. In the LENS process, powders are fully melted by laser beam. Typically, an industrial-grade laser beam is focused on a work piece to produce a moving melt pool. A small amount of powder is injected into the melt pool where it becomes molten and solidified at a high cooling rate to produce a layer of the metal. The motion of the laser beam is controlled according to the CAD geometry of the part and subsequent layers are deposited on previously solidified layers. This type of processing typically results in fine equiaxed microstructure depending on the composition of the powders and processing conditions used. Such microstructure

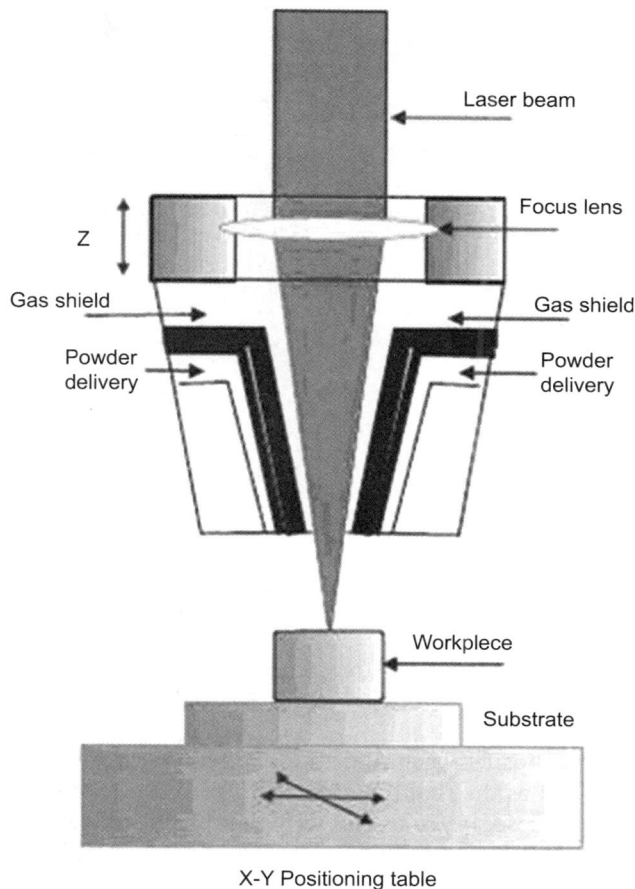

Figure 104 Schematic of the LENS process (Kumar, 2008).

may result in superior mechanical properties than those obtained in corresponding wrought parts. This process results in high dimensional accuracy and surface roughness of less than 10 microns (Sanjay and Sisa, 2011).

Commercially, mainly iron-, titanium- and nickel-based alloys have been deposited using LENS process. This process has been used to make highly complex parts which could not be produced by conventional means. Additionally, repair/refurbishment of parts has been conveniently done by this process. LENS has found wide use in medical, tooling, aerospace, and automotive industries (Kumar, 2008). **Figure 105** shows a titanium hip stem produced with LENS. The LENS process finds wide use in producing functionally graded materials (Sanjay and Sisa, 2011), where the composition of the deposit can vary as a function of location.

LENS machines are equipped with feedback control which allows a precise control on dimensional tolerance and process parameters thereby providing the ability to produce parts with more predictable mechanical properties. Monitoring of the temperature of the melt pool, cooling rate of the deposited

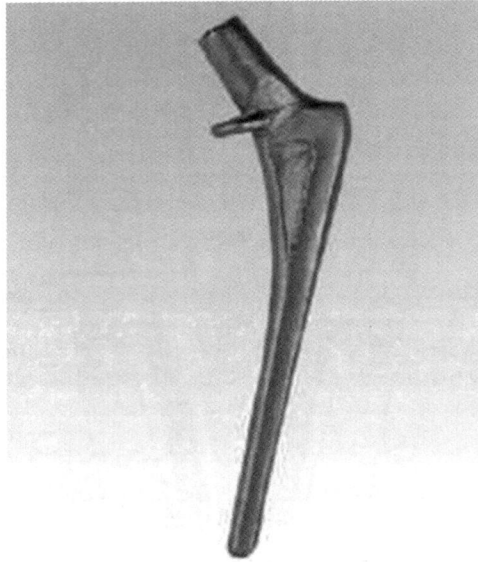

Figure 105   Titanium hip stem produced with LENS process. Courtesy: Optomec on http://www.optomec.com/.

layer, and composition monitoring via real-time spectroscopic analysis of the plasma plume are facilitating more accurate control of the process. Such feedback control is enabling wider commercialization of this technology.

In the LENS process, each deposited layer undergoes a thermal cycling because of the deposition of the subsequent layers on it. Just as the mold in a die casting process, the peak temperature reached in a layer gradually reduces as the laser beam moves to higher positions. Laser power, scanning velocity, powder mass flow rate, and time elapsed between deposition of consecutive layers have large effect on the thermal behavior and resulting microstructure and mechanical properties of the deposit. The microstructure in the deposited part depends on the rapid solidification when a layer is formed and the subsequent heating encountered because of deposition of additional layers. Cellular and dendritic microstructures were obtained in the deposited layers (Collins et al., 2003; Liu and Dupont, 2003). Syed et al. (2005) and Dinda et al. (2009) reported that dendritic structure was present in the top layer and dendritic/cell microstructure was present in the previously solidified bottom layers. Bi et al. (2006) reported columnar grain structure in the top layer of a stainless steel 316 alloy part. These studies clearly show the evidence of dendritic microstructure in the deposited part.

The mechanical properties of the LENS parts depend on the microstructure. The grain size has a marked influence on the recrystallization process. Katayama and Matsunawa (1984) reported the following relationship for secondary dendrite arm spacing (SDAS) for laser welding of 310 stainless steels:

$$\lambda_2 = 25 \, \dot{T}^{-0.28} \tag{204}$$

where λ_2 is in μm and cooling rate $\dot{T}$ is in $K\,s^{-1}$. Soodi et al. (2010) reported surface hardness measurements for six alloys that were formed using DMD. **Figure 106** compares the measured values against values for annealed samples. The DMD process results in much higher values than those obtained in the annealed samples.

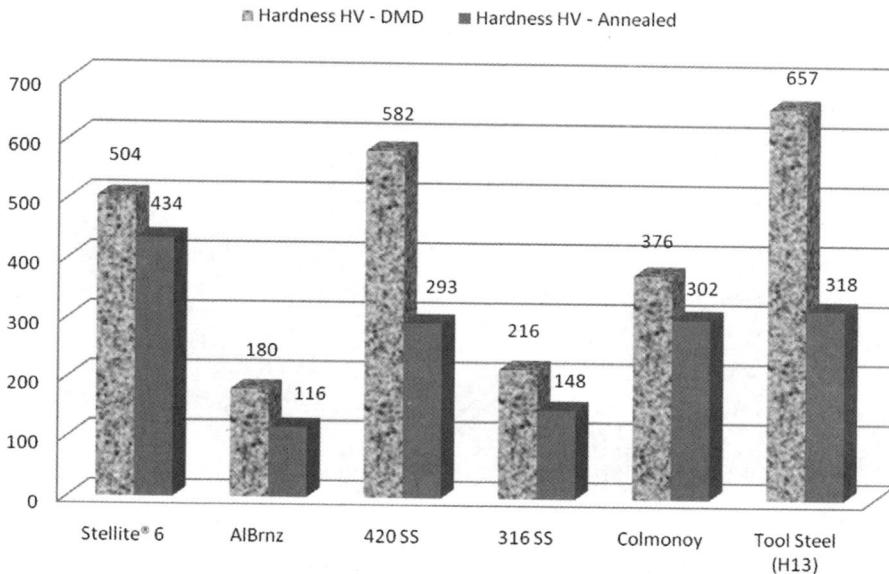

Figure 106 Hardness comparison between DMD versus annealed samples for six alloys (Soodi et al., 2010).

Figure 107 shows a 3-D isometric view of the microstructure at three locations through the thickness of the deposited layers for an IN 625 deposits using DMD (Dinda et al., 2009). Columnar dendrites are present through the entire deposited layer, although there was a change in dendrite orientation from horizontal at the top to near vertical at the bottom. The dendrite orientation follows the heat flux direction as dictated by the laser scan direction.

Microhardness measurement of samples deposited under wide variation of laser processing parameters showed that hardness values did not vary much (254 ± 6VHN) (Dinda et al., 2009). This is because most of the strengthening elements (Nb and Mo) remained in the γ-Ni matrix because of the very high cooling rate (10^3 to 10^5 K s^{-1}). Deposited samples were subsequently annealed at 700, 800, 900, 1000, 1100, and 1200 °C for 1 h. Sample annealed at 700 °C showed some age hardening effect because of precipitation of metastable, coherent γ'' [Ni$_3$Nb] in the γ-matrix. This γ'' phase transforms to stable, coherent δ-phase above 700 °C, which provides less strengthening. Further decrease in hardness results from the coarsening effect and gradual dissolution of the δ-phase above 1000 °C. **Figure 108** shows the plot of hardness as a function of annealing temperatures.

There have been a number of modeling studies in recent times to model the solidification in the laser deposition process. One such study was by Yin and Felicelli (2010) in which they used a 2-D FEM model in conjunction with CA technique to simulate growth of dendrites in the molten pool in the LENS process. They studied the effect of processing conditions (e.g. laser moving speed, layer thickness, and substrate size) on the solidification microstructure. The study concluded that dendritic growth occurred at high cooling rates associated with LENS, while the dendrite orientation and SDAS varied as a function of the location in the melt pool and cooling rates. The study also concluded that the numerical models should consider dendrite coarsening and solid-state transformation as a result of deposition of additional layers on top of a solidified layer.

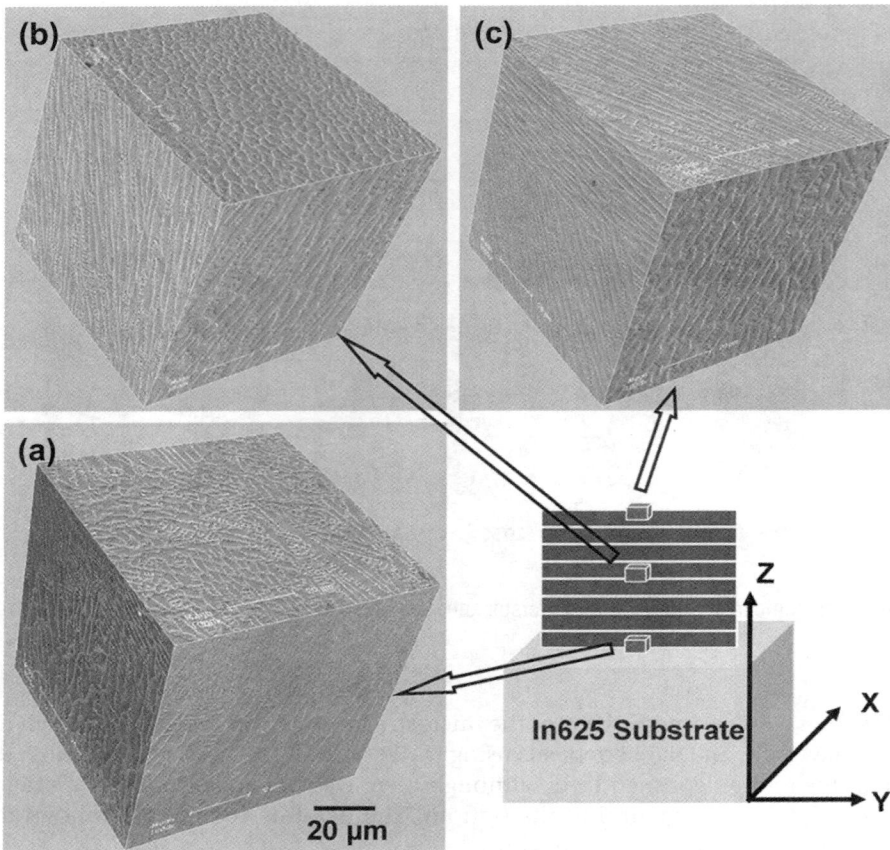

Figure 107 3-D view of microstructures at three locations of the deposit obtained with DMD of an IN 625 alloy on substrate: (a) first layer, (b) middle of the sample (between first and top layer), and (c) top layer (Dinda et al., 2009).

7.10.4 Metal Foam

Metal foam is a porous structure consisting of a solid metal containing a large fraction of gas-filled pores. Typically, 40–95% of the total volume consists of voids or pores. These pores can be isolated (closed foam), or they can form an interconnected network (open-cell foam), or a combination of both. Metallic foams retain some of the properties of the base metal used to form the foam structure. Recently, a novel structure, which is based on a lotus-type growth (Hyun et al., 2001; Hyun and Nakajima, 2003), has been developed. Such structure is characterized by having long cylindrical pores aligned in one direction.

In recent times, metal foams are finding more use as a new material for automobiles. Metallic foams can increase sound dampening while reducing the overall weight of the automobiles. They can also help absorb impact energy from vehicular crashes. Titanium alloy foams offer superior mechanical strength coupled with high corrosion resistance, low density and offer enhanced biocompatibility. These properties make

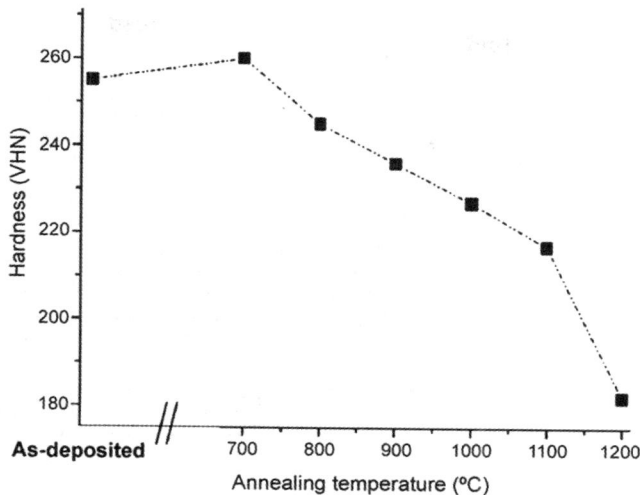

Figure 108 Variation of hardness with aging temperature of laser deposited IN 625 Alloy (Dinda et al., 2009).

them suitable for use as bone replacement materials in medical implants (Shapovalov, 1994; Dunand, 2004). Al alloy-based foams are finding increased use because of cheaper cost of production compared with conventional technology and improved properties. Metal foam technology has also been applied in the treatment of the automotive exhaust gas (Alantum). Compared with the traditional catalytic converter, the metal foam substrate can offer better heat transfer and exhibits excellent mass transport properties offering possibilities for using less platinum catalyst (SAE International).

Since the inception of the metal foam, several processes have been developed to introduce uniformly distributed large size pores in the metallic materials. These processes involve (a) liquid metal route, (b) powdered metal technique, and (c) methodologies where direct sintering of foamed elements is used. Metallic foams have tremendous potential of being a cost-effective engineered material with improved properties. One concern during processing has been the stability of the cell structure in terms of having proper mechanical strength to retain the network of porous structure.

7.10.4.1 Processing

A schematic showing an overview of the processing of metal foam is shown in **Figure 109** (Srivastava and Sahoo, 2007) and shows two main approaches: liquid metal- and powder metal-based routes. The intent of the foam processing is to introduce large and uniformly sized pores. Dissolved gases precipitate from the liquid melt during solidification and help develop a foamed structure if they are not allowed to escape. Commonly used external gases used to foam liquid melt are air, argon, nitrogen, and so on. Sometimes used is a blowing agent, which is a chemical that decomposes during processing to produce pores.

7.10.4.2 Processing of Liquid Metals

Liquid metal foaming using an external gas was patented by Alcan International (Jin et al., 1990). Rotating impellers are used to stir gas injected into the metal pool and the bubbles develop as the metal solidifies to form the foam. These impellers create fine, uniformly distributed bubbles. Sufficiently fine

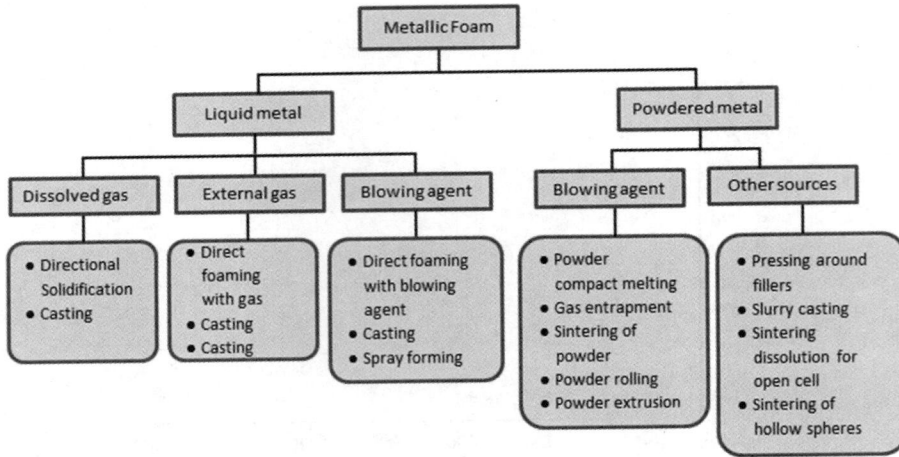

Figure 109   A listing of various approaches for processing of metal foams (Srivastava and Sahoo, 2007).

bubbles are needed to create foam of satisfactory quality. The floating foam can be continuously pulled off of the surface of the melt using various techniques (see **Figure 110** which shows the use of a conveyor belt as one technique to obtain sheets). In another example for the production of foamed composite, the melt after proper treatment can be cast in a permanent mold (Daoud et al., 2007). For the liquid metal processing using an external gas source, a two-step process is used. In the first step, the viscosity of the melt is brought to an optimum level by introducing ceramic particles of proper size (often 5–10 μm). The volume fraction of particles can range up to 20%. Most common particles are SiC, Al_2O_3, and MgO. Selection of these particles depends on the quality of the wetting of the particles by the liquid metal. External gas is then injected in the second step to obtain uniform dispersion of gas bubbles. As an alternative to the use of external gas, certain blowing agents are also used. These are

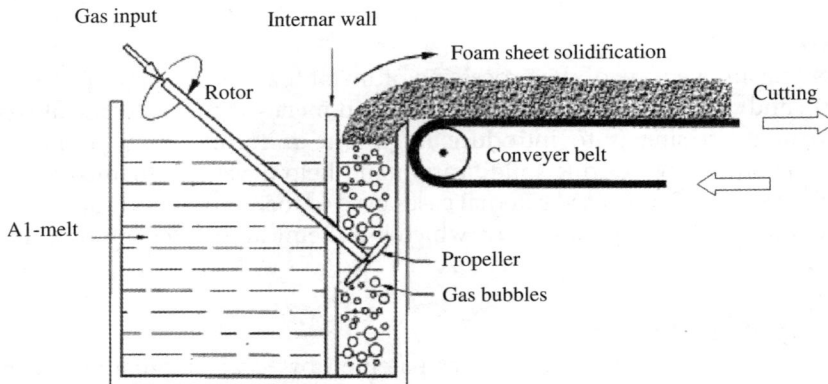

Figure 110   Apparatus for melt gas injection and foam sheet solidification (Asholt, 1999).

typically compounds that decompose on heating thereby releasing needed gases into the melt. Shinko Wire[4] (Japan) has commercialized this process in which the melt is continuously stirred following the introduction of particulate additives into the melt to control the viscosity. During this step, the temperature of the melt is also controlled. After the desired foaming is achieved, a foamed block can be produced using a standard casting process.

Spray forming has been discussed in detail previously and is also another approach to develop metal foam product. Decomposition of the blowing agent during deposition can lead to large amount of voids in the sprayed preform. Liquid metal foaming with dissolved gas sources takes advantage of typical gases present in alloys (e.g. hydrogen in Al, nitrogen in Fe). Control of temperature and applied pressure are important factors in ensuring a particular level of dissolved gases in the melt. As temperatures are lowered, the liquid undergoes a eutectic-type reaction and gives out the dissolved gas as a second phase, which is entrapped during solidification to generate a porous foam structure (Hyun et al., 2001, Simone and Gibson, 1997). Typically, DS at controlled cooling and at proper gas partial pressure will lead to gases that are entrapped in the solidified phase. Ikeda et al. (2005) used this approach to develop lotus-type porous stainless steel structure. It was shown that porosity increases with the increase in the partial pressure of the gas, while an increase in solidification rate reduced the porosity size. Hyun et al. (2001) reported that about 50% increase in tensile strength was achieved along the direction of pore orientation in comparison with that obtained in the transverse direction. Porosity levels between 5% and 75% can be achieved in this process. Foaming can also be achieved by casting liquid metal in the interstitial spaces around properly stacked holding materials (Banhart, 2001; Rabiei and O'Neill, 2005). These holding materials can be organic or hollow spheres, which can be removed later by leaching or dissolution. Such a technique has been used to produce foamed products of Al, Mg, Pb, and Zn alloys. DS of monotectic or hypermonotectic alloys can also be done to develop an aligned structure and a selective removal of the second phase will lead to a foamed product. Yasuda et al. (2004) removed the In phase in solidified hypermonotectic Al–In alloy to form aligned foam product.

7.10.4.3 Processing of Powdered Metals

The use of solid powder metal to make foam structures has been studied extensively by researchers. A detailed review of the processing can be found by Srivastava and Sahoo (2007). For the case of powder compact melting using a blowing agent as a gas source, alloy powders are mixed with the foaming agent, then compacted using processes such as hot pressing and extrusion, and then heated to a temperature below the melting point of the matrix metal to initiate the foaming process (Zeppelin et al., 2003). TiH_2 and ZrH_2 were used as foaming agent for Al and Zr foam parts, respectively. The quality and the stability of the cellular foamed structure produced depend on the processing conditions used. Other foaming routes have also been adopted. In one such process, sintering of metal powders can be used to develop a foam structure with porosity as high as 50%. Sintering of hollow spheres has been used to form both closed and open foam products. In this approach, pore size distributions depend on the size and arrangement of the spheres and hence, the properties of the foamed part can be controlled. Zhao and Sun (2001) proposed a technique where they produced an Al foam structure using spheres of NaCl. Sintering was done at a temperature of 680 °C (much below the melting point of NaCl). Sintered parts were subsequently immersed in warm running water to dissolve the NaCl. Metal foams can be produced by making powder metal slurries and mixing chemicals that produce the gases

---

[4] The inclusion of a specific trade name is for information only and does not constitute and endorsement by NIST.

as reaction products at an elevated temperature which lead to the foamed structure. Foamed products are then sintered to develop the necessary strength. This process has been successfully used to produce foams of stainless steels and Inconel 625 alloy (Bleck, 2004).

Recently, Fife and Voorhees (2009) used directional freeze-casting to create an aligned pore structure. In this process, metal powders are typically mixed with water to make slurry. The slurry is directionally solidified to create an aligned pore structure between the ice dendrites. The metal powders are rejected into the interdendritic space. The solidified structure is then dried in vacuum, which sublimes the ice leaving the large aligned and elongated porous structure. Sintering of the structure is used to create dense walls that separate the voids in the product. The process can be used to make a porous structure using variable powder size. A binder is often used in the slurry to prevent powder collapse after ice is removed during the sublimation process. The structure is controlled by maintaining a rather slow growth of solidification (about 3 μm s^{-1}).

7.10.4.4 Foam Structure Stabilization

One particular concern during processing is the collapse of foam, often because of instability of the cell wall under pressure differences. Growth of foam beyond a certain diameter will lead to thinning of cell walls leading ultimately to collapse. Surface tension is a key parameter in this regard. A good wetting of the particles with matrix can reduce the pressure in the cell wall. Gergely and Clyne (2004) studied the physics of the foaming process and modeled aqueous foam formation. Strengthening of cell walls will help in combating this collapse. Cell walls are often strengthened with the addition of particles in the melt thereby enhancing the viscosity of the melt (Srivastava and Sahoo, 2007). The ceramic particles move to cell boundaries thereby strengthening the cell walls (see **Figure 111**).

Pretreatment of blowing agents has been effective in better foaming. For example, TiH$_2$ particles have been heated at elevated temperatures for long duration leading to the formation a layer of oxide on the particle surfaces. This apparently aids in a delayed decomposition for better foaming control. Matijasevic and Banhart (2006) developed an enhanced, uniform cell wall structure using this technique. In situ studies have been conducted to understand the mechanism of foaming process. Processing conditions during foaming also play a role in strengthening of the cell walls. For example, the heating

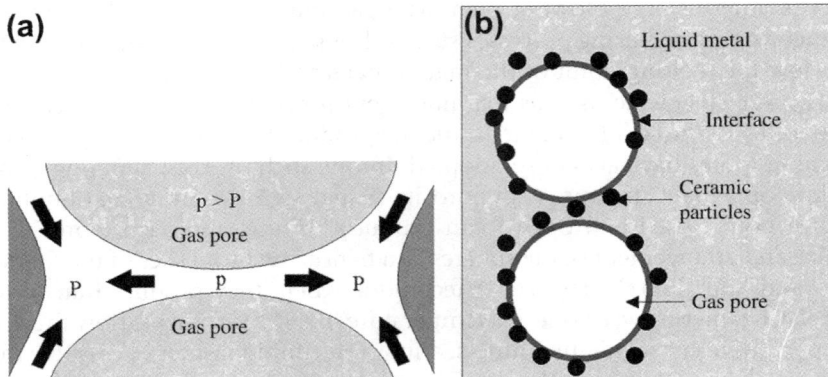

Figure 111 (a) Cell wall structure and related phenomena during foaming process (Wuebben et al., 2000); (b) the effect of ceramic particles for stabilization during the foaming process (Wuebben et al., 2003).

rate during foaming and compaction conditions can have a strong influence on the foam stabilization (Srivastava and Sahoo, 2007).

7.10.4.5 *Microstructure and Morphological Analysis*

The microstructure of the metal in the foam wall will vary based on composition and processing conditions. Daoud et al. (2007) developed foam composites using ZC63 Mg alloy using liquid metal foaming technique. They studied the microstructure of both unreinforced and reinforced (with fly ash) foamed composites. Optical microscopy of the cast base alloy revealed α-Mg dendrite, $Mg(Zn,Cu)_2$ eutectic phase, and Cu_5Zn_8 and CuMnZn intermetallics in the interdendritic regions. While the unreinforced composite showed a typical dendritic structure, the reinforced composite showed a cellular dendritic structure, with eutectic involving intermetallic compounds at the cell boundaries. The average dendrite arm/cell spacing in the reinforced composite was 10 μm compared with 100 μm in unreinforced alloy.

Morphological analysis is important to understand the structure of the metal foam. Recently, X-ray tomography has been successfully used to quantitatively characterize the structure of titanium pores created using directional freeze-casting and subsequent sintering of titanium powders (Fife et al., 2009). Such studies can help determine pore connectivity, volume fraction of pores, and interface shape distribution, and how pores evolve during the processing. Interface shape distributions and interface normal distributions can help assess variations in pore shape and orientation among foams produced using varying powder sizes and processing conditions. Fife et al. (2009) concluded that foams produced using smaller Ti powder size showed more complete sintering and desirable macroporosity with reduced level of undesirable microporosity.

7.10.5 Semisolid Metal Forming

Semisolid metal forming (SSM) was first introduced in the 1970s when Flemings and his coworkers studied flow behavior of semisolid metallic alloys. Soon thereafter, companies such as Alumax and ITT-TEVES commercialized this process to produce parts for automotive applications. Today, SSM is being used as a low-cost method to produce parts in large quantities. Essentially, SSM is a variation of die casting process, in which near-net shape parts can be produced. After melt processing, shaped casting is produced using conventional die casting process. In this process, die life is extended as the temperatures are lower (temperature is typically in the mushy or solid + liquid region) than that in conventional pressure die casting. Also, SSM forming has the potential to significantly reduce production costs because of the near-net shape character of the process, the ability to achieve high production rates, and reduced machining and finishing costs. There are three major variations of this process: thixocasting, rheocasting, and thixomolding. A detailed description of each of the methods is provided by Pan et al. (2008). In the thixocasting process (a two-step casting process), first a feedstock is prepared which is subsequently reheated in the mushy region and then cast to produce the SSM part. The feedstock or the bar making process is typically done using any of the four processes: mechanical stirring, strain-induced melt activation, grain-refined direct chill casting, and magnetohydrodynamic (MHD) casting. Among these processes, the most popular method is MHD, which results in a melt of exceptional quality. However, thixocasting is not often cost-effective and therefore, this is often used for niche markets for specialty applications.

Rheocasting involves preparation of a thixotropic slurry which is cooled and solidified to produce SSM parts. In rheocasting, nucleation and growth of grains in the mushy region are controlled to maintain a globular dendrite structure. Thermal management is very important to achieve the optimum melt condition before casting often in a vertical casting machine. Rheocasting is often a much cheaper option than thixocasting.

Thixomolding is a commercial process for semisolid processing of magnesium alloys. It is a process similar to injection molding for polymeric materials. This process involves feeding of Mg pellets into a heated injection system with the aid of reciprocating screw. Semisolid slurry is produced by using the shear action of this screw. An argon gas environment is maintained to reduce any chance of oxide formation. This slurry is then injected into a mold to produce the SSM part.

SSM slurry consists of a mixture of liquid metal and globular solid particles. One of the unique behaviors that are attributed to the SSM materials is that on reheating to the semisolid state, they can be handled much like solid (SSM material has a yield stress just like solid), yet they can flow like a viscous fluid if the applied stress exceeds yield stress. This is why SSM materials can be used to form a variety of complex shapes. Flemings (1991a) and Diewwanit and Flemings (1996) pointed out that a variety of alloys can be processed as SSM materials, for example, Al–Cu, Al–Si, Cu-base alloys, Mg- and Ti-base alloys, cast irons, steels, low-alloy and stainless steels, Ni- and Co-base superalloys. Flemings (1991a) pointed out that the flow resistance of these materials depends on their processing history through the grain size and degree of particle agglomeration in the mixture. The unusual flow behavior is attributed to the disagglomeration of particle clusters that occurs during shearring. Today SSM materials are finding increased use in automotive, aerospace, and defense applications. Many alloys of Cu, Fe, Zn, and Ti are being used in commercial applications (Brown and Flemings, 1993; Tims, 2002). Cost and performance of SSM products drive their increased use across all these industries.

Several mechanisms have been proposed to describe how the globular structure forms in SSM processing. It is generally agreed that forced convection associated with turbulent flow contributes to dendrite fragmentation as discussed in Section 7.9. These fragments serve as potential nucleation sites. If these nuclei do not grow beyond a certain point, it is possible to obtain large number of spherical particles in this process.

Defects associated with die casting of SSM materials are decreased because the flow is typically laminar (less entrapped gas, surface oxidation, etc.) and less solidification shrinkage (and fewer hot tears) because of use of mushy state of the alloys. Other significant advantages are less macro-segregation, less variation of grain size because of section size changes, and so on. Some of the defects associated with SSM are blisters, flow-related defects (e.g. laps, nonfill, cold shuts), shrinkage porosity, and so on. Blisters can result after heat treatment because of entrapped hydrogen gas generated from die lubricants. Blisters can be controlled by having suitable die lubricants. Most of the flow-related defects can be significantly reduced by increasing the metal temperature before die casting and having an improved gating design that ensures a laminar flow. Solidification shrinkage can be reduced by maintaining a directionality of solidification and maintaining sufficient pressure from the ram to generate enough force to drive the semisolid material. The solidification process needs to be carefully monitored during processing as a large amount of the material can be solidified during the filling process depending on the casting geometry. Nonuniform temperatures in the die from repeated use can affect the cooling and hence, the solidification pattern.

7.10.5.1 Material Models
Since its introduction four decades ago, the understanding of the fundamental behavior of SSM materials is still evolving. An appropriate constitutive model is needed to describe the viscosity of the SSM material for use in mathematical models for processing of these materials into shaped parts for various applications. The behavior of SSM materials depends a great deal on the fraction solid, shear

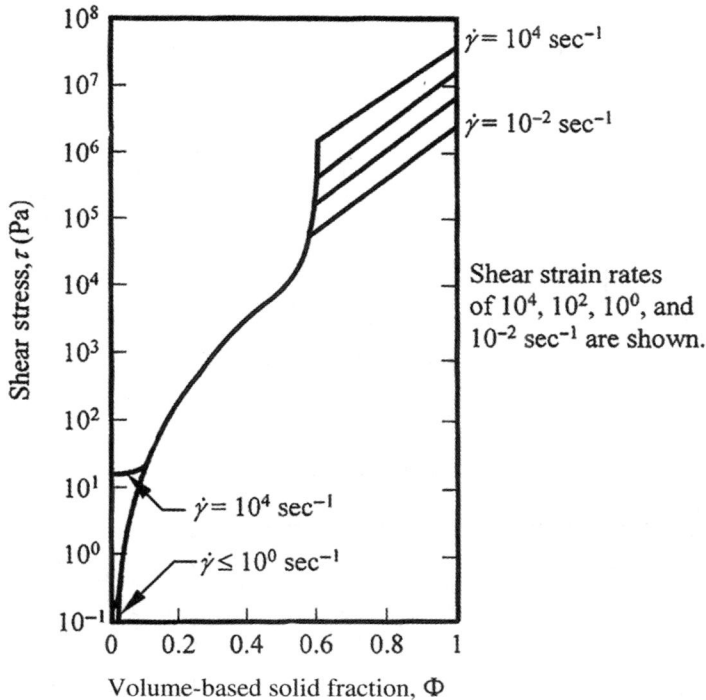

Figure 112   Shear stress for Al–Si-based SSM alloys at various solid fractions and shear strain rates (Sigworth, 1996).

strain rate, and history of processing. **Figure 112** shows the variation of shear stress as a function of solid fraction for varying strain rates for Al–Si SSM alloys (Sigworth, 1996). It can be seen that shear stress has a strong dependence on solid fraction and shear strain rates. Therefore, a relatively small variation in fraction solid can have a large impact on material response. The rheological behavior of SSM materials needs to be more fully understood in order for SSM processing to be commercially successful.

Empirical relationships have often been used to describe slurry viscosity as a function of solid fraction. Sigworth (1996) provided a summary of such models, yet he urges caution because these models do not include the dependence of viscosity on shear strain rate. Additionally, most of these relationships fail to model Bingham fluid behavior usually observed in the SSM slurries. The shear stress has a nonlinear relationship with strain rate for SSM materials. **Figure 113** shows the apparent viscosity as a function of fraction solid for two steel alloys (Flemings, 1991a). It is clear that the cooling rate has a strong influence on viscosity: a higher viscosity results from a high cooling rate, possibly because of formation of smaller grains in the solidified microstructure. Most popular viscosity models for SSM slurries use a power law behavior and are developed from characterization of experimental data as given by the equation

$$\mu_a = K\dot{\gamma}^{n-1},\tag{205}$$

(a)

440C stainless steel

(b)

AISI 4340 low alloy steel

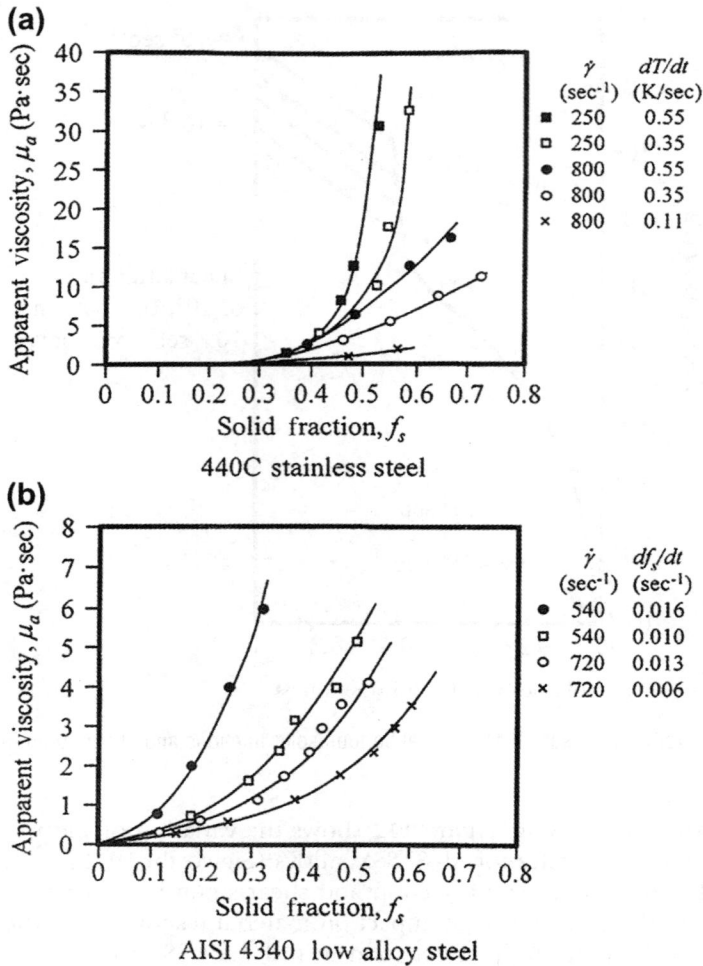

Figure 113 Fluid mechanical behavior of SSM steel alloy slurries over prolonged shearing at constant shear strain rates (Flemings, 1991a).

where n is the power law exponent and K is a power law index. For Newtonian flow, $n = 1$, and the viscosity does not depend on shear rate. The data in **Table 4** have been reported by Flemings (1991a) for Al alloy for several processing conditions.

It may be noted here that both K and n are strong functions of the alloy composition and fraction solid and have a strong dependence on grain size and prior history of processing. An extensive review of the various material models available for SSM slurries is provided by Tims (2002).

7.10.6 Liquid Melt Processing—New Techniques for Mg Alloys

Traditionally, grain refiners are added to liquid melts before pouring to produce fine-grained equiaxed microstructures in conventional casting processes. Although Zr has been an efficient grain refiner for

Table 4   Parameters determined by Flemings (1991a) for eqn (205)

Condition	N	K(Pa s^n)
Steady state	0.1	30
Continuous cooling at constant shear strain rate	−0.5	2300
Solidified and partially remelted at low shear strain rates	0.32	39 800

Mg-alloys with no Al, potent grain refiners (in the form of master alloys) for Al-containing Mg-base alloys are yet to be found. However a few innovative techniques have been developed for liquid melt processing of AM series (Mg–Al–Mn) and AZ series (Mg–Al–Zn) alloys that have potential for increased application in the automotive industry. These techniques are based on intense melt shearing that uniformly disperses MgO particles (present in the oxide films and skins) in the melt and these fine particles act as potent nucleation sites for solidification. The process remains effective as long as the dispersion of MgO oxide particles is maintained.

In 2009, Fan et al. proposed a melt shearing approach in which a twin-screw machine was used. They used this approach for treating both liquid and semisolid alloys of light metals. They obtained superior microstructure and mechanical properties using this technology. Recently, Fan et al. (2011) proposed a new technology based on the same principle of melt shearing but with simplified, high shear equipment.

The high shear equipment for melt shearing is basically a rotor–stator unit as shown in **Figure 114** (Fan et al., 2011). While in operation at a speed as high as 10^4 r.p.m., the rotor shears the liquid metal in the region between the rotor and stator and in the openings at the stator. This operation (a) creates a well-mixed flow in the melt and (b) shears the melt near the tip region of the shear device as shown in **Figure 114**. Degassing can be done during the melt shearing process. During the degassing process, argon is introduced into the melt underneath the shear unit shown in **Figure 114**. The shearing process creates well-dispersed and uniformly distributed solid particles and gas bubbles and uniform temperature and composition field in the liquid. Note that the solid particles are MgO for the case of AZ and AM alloys.

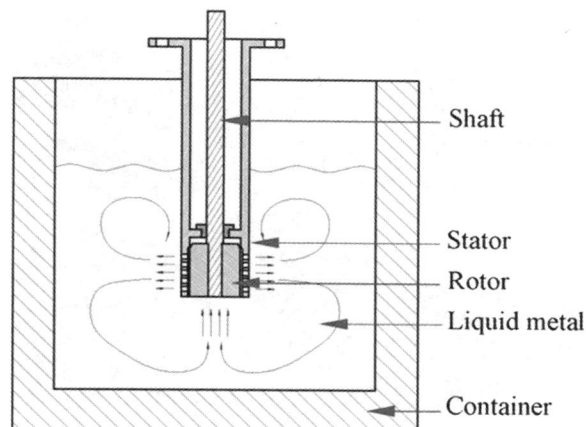

Figure 114   Schematic of the high shear device for liquid metal treatment (Fan et al., 2011).

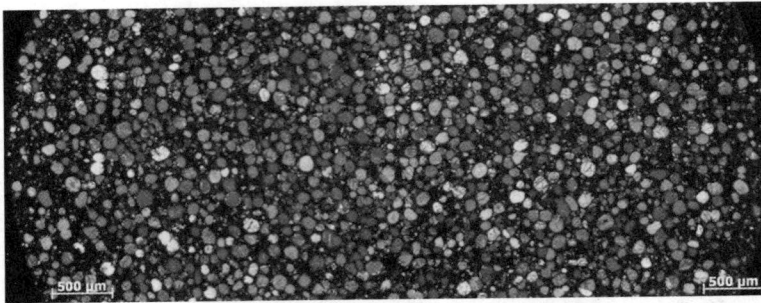

Figure 115   Microstructure of semisolid AZ91D magnesium alloy prepared by intensive melt shearing (Fan et al., 2011).

Figure 115 shows microstructure of a semisolid AZ91D magnesium alloy prepared by intensive melt shearing with the new high shear device. The high shear treatment results in uniformly distributed α-Mg grains whose sizes have a narrow distribution. Such melt treatment before DC casting showed much smaller, uniformly distributed, globulitic grains (**Figure 116**). This is in contrast to the microstructure consisting of large dendritic grains that are obtained in a conventionally cast ingot. Therefore, this process can help produce significant grain refinement.

In a similar process, Fan et al. (2010) developed another melt shear technology called melt conditioning by advanced shear technology. In this process, liquid metal is fed into a twin-screw device with an accurate temperature control. A pair of corotating and intermeshing screws subjects the liquid/semisolid material to melt shearing. This process can be used before die casting or strip casting both in batch and continuous manner. **Figure 117** shows an example of how mechanical properties obtained from this process (melt conditioning before die casting) compare with those in the conventional die casting process. Both ultimate tensile strength (UTS) and ductility showed marked improvement and results show very little dependence on melt superheat in the MC-HPDC (melt sheared-high-pressure die casting)

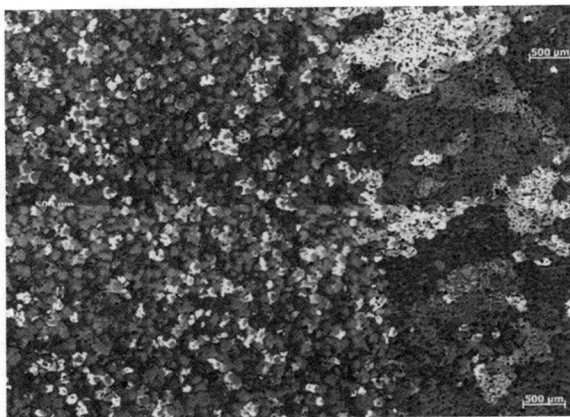

Figure 116   The transition zone of DC cast AZ31 magnesium alloy from nonshearing to shearing (right to left) (Fan et al., 2011).

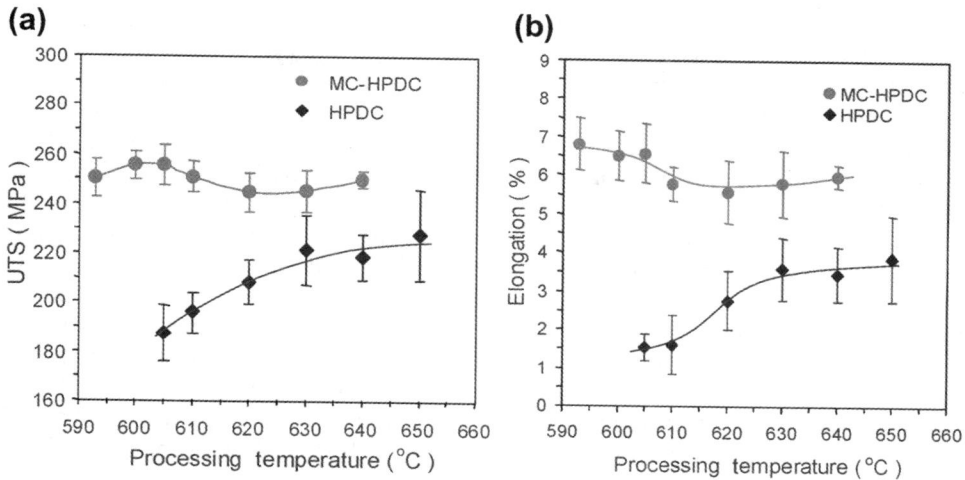

Figure 117 Mechanical properties of AZ91D alloy produced by the HPDC and the MC-HPDC processes as a function of the processing temperature (Fan et al., 2010).

process. In contrast, the conventional die cast samples showed decreasing values of UTS and ductility with reduced superheat. This suggests that melt superheat control is much more critical in the conventional HPDC process.

The effect of melt shearing combined with a continuous strip production using a twin roll caster (TRC) has also been investigated. The conventional TRC production of an AZ91D alloy showed large columnar grain structure. The microstructure showed typical centerline segregation in the mid-depth region of the strip and other defects such as hot cracks and surface bleeds (Fan et al., 2010). With melt conditioning by shearing, a fully equiaxed microstructure was obtained through the depth of the strip. Additionally, no centerline segregation was noticed (**Figure 118**) and other defects were greatly

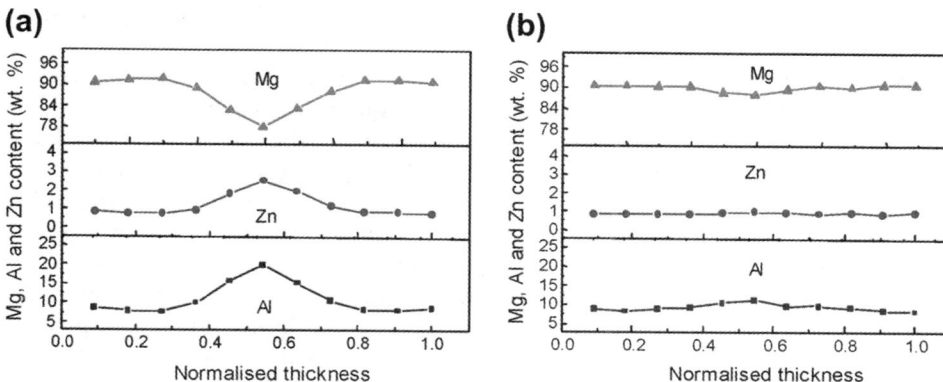

Figure 118 Through-thickness variation of chemical composition in AZ91D alloy strip (4 mm thickness) produced by (a) the conventional TRC and (b) the MC-TRC processes (Fan et al., 2011).

reduced. As mentioned earlier, melt shearing disperses the MgO present in oxide films into large number of fine MgO particles, which act as potent nucleation sites for the resultant equiaxed solidification microstructure.

More details were uncovered by Fan et al. (2009) regarding sheared AZ91D alloy. Using a pressurized filtration technique the potential nucleating agents were concentrated for further electron microscopic study. The pressurized filtration experiment showed that oxide films consisted of densely populated nano-sized MgO particles in a liquid matrix. These MgO particles were effectively dispersed into individual particles of 100–200 nm because of the action of the melt shearing process. Both MgO/α-Mg and MgO/Al$_8$Mn$_5$ interfaces were found to be semi-coherent with very small crystallographic misfit, while the Al$_8$Mn$_5$/α-Mg interface with the imbedded intermetallics was incoherent because of poor crystallographic matching. A coherent or semi-coherent interface between nucleating particles and the nucleated phase is a required condition for the nucleating particles to act as heterogeneous nuclei because such interfaces provide much lower nucleation barrier. Therefore, for melt sheared AZ91D alloys, MgO particles can act as potential nucleating sites for both the primary α-Mg and Al$_8$Mn$_5$ intermetallics, while it is very unlikely that Al$_8$Mn$_5$ particles could act as potent nucleating sites for the α-Mg phase.

7.10.7 Die Casting—Recent Advances

Castability of an alloy for die casting applications focuses on two major issues: hot tearing and die soldering. Hot tearing usually results from stresses that develop during feeding and is not commonly found to occur for HPDC of alloys such as Al alloys. However, controlled solidification study with the aid of coupled numerical codes is assisting engineers in understanding how stresses develop during the feeding process. Vacuum-assisted die filling and proper thermal management during the die cycles have been used to enhance castability and improve mechanical properties and surface quality of the cast parts. Such approaches have been proven to work for Mg alloys. Die soldering or "die sticking" occurs when molten alloy being cast "welds" to the die surface. This results in damage to the die after multiple cycles and leads to poor surface quality of the casting including loss of adequate dimensional control in the cast part. This occurs in Al-die casting in steel dies because of intermetallic compound that forms at the interface because of affinity between Al and Fe. Die soldering can be controlled by adjusting melt chemistry, controlling process conditions, and proper die surface condition. Addition of Fe and Sr has been shown to reduce the die soldering in Al alloys. High temperatures and increased melt velocity aggravate die soldering. SSM processing discussed above reduces die soldering because of lower temperatures used during casting.

Die coatings have been used as a barrier between steel dies and Al alloys. The coatings need to be able to withstand harsh conditions imposed by the liquid melt at increased velocity. Common coatings that have been used include (TiAl)N, CrN, and so on (Lin et al., 2006). Dies are also subjected to treatments such as nitriding or nitro-carburizing to combat die soldering by preventing surface erosion. The die coating should have the following properties: (a) resistance to erosion, wear, and oxidation; (b) nonwetting with the liquid melt; (c) good adherence to die substrate during shot cycling; (d) ability to accommodate residual stresses because of thermal effects; and (e) ability to delay thermal fatigue induced cracking. Lin et al. (2006) proposed that an optimized, multilayer, and graded coating system needed to be developed in which the overall coating system exhibits synergistic properties. Their proposed optimized system includes in sequence: (a) surface modification of the die substrate using, for example, ion nitriding or ferritic nitrocarburizing that increases hardness and provides increased mechanical support for subsequent coating layers; (b) adhesion layer comprising a thin layer (100–200 nm) of Cr or Ti on the modified substrate; (c) intermediate layer or graded layer where

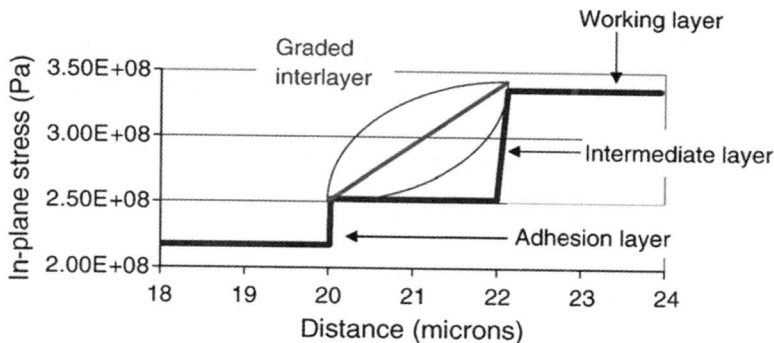

Figure 119   Computed variation of in-plane stress in the optimized coating system as applied on the H13 steel substrate (Lin et al., 2006).

composition, microstructure, and properties change continuously to better accommodate and withstand thermal residual stresses produced during process cycling; (d) nonwetting working layer that needs to be wear and oxidation resistant.

Finite Element (FE) modeling of the intermediate layer of the coating system (with Al_2O_3 as a working layer) showed that a graded TiAlN intermediate layer demonstrated an acceptable in-plane stress (**Figure 119**). The in-plane stress in the graded intermediate layer was between that at the adhesion layer of Ti and the outer working layer of Al_2O_3. The working layer shows a much higher residual stress than that at the substrate. The graded working layer appears to accommodate this apparent mismatch in residual stresses between the adhesion layer and the working layer. Therefore, this intermediate layer will essentially accommodate the residual thermal stress induced by the forming cycle. Such optimized die coating methodology has been successfully developed for aluminum pressure die castings. A similar approach can be developed for other metal/die combinations.

References

Aagesen, L.K., Fife, J.L., Lauridsen, E.M., Voorhees, P.W., 2010. The evolution of interfacial morphology during coarsening: a comparison between 4D experiments and phase-field simulations. Scr. Mater. 64, 394–397.

Adam, C.M., Hogan, L.M., 1972. J. Aust. Inst. Met. 17, 81.

Ahmad, A., Wheeler, A.A., Boettinger, W.J., McFaddden, G.B., 1998. Solute trapping and solute drag in a phase field model of rapid solidification. Phys. Rev. E 58, 3436–3450.

Akamatsu, S., Perrut, M., Bottin-Rousseau, S., Faivre, G., 2010. Spiral two-phase dendrites. Phys. Rev. Lett. 104, 056101.

Allen, S.M., Cahn, J.W., 1979. Acta Metall. 27, 1085.

Andersson, J.O., Helander, T., Höglund, L., Shi, P., Sundman, B., 2002. Thermo-Calc and DICTRA, computational tools for materials science. CALPHAD 26, 273–312.

Anderson, D.A., Tannehill, J.C., Pletcher, R.H., 1984. Computational Fluid Mechanics and Heat Transfer. McGraw Hill/Hemisphere. New York.

Andrews, J.B., Schmale, A.L., Sandlin, A.C., 1992. J. Cryst. Growth 119, 152.

Ansara, I., Dupin, N., Joubert, J.M., Latroche, M., Percheron-Guégan, A., 1998. J. Phase Equilib. 19, 6–10. errata 19, 98.

Apelian, D., 1982. In: Pampillo, C.A., Biloni, H., Mondolfo, L., Sacchi, F. (Eds.), Aluminum Transformation Technology and Applications 1981. ASM, Metals Park, OH, p. 423.

Aptekar, J.L., Kamenetskaya, D.S., 1962. Fiz. Met. Metalloved. 14, 123.

Asholt, P., 1999. In: Banhart, J., Ashby, M.F., Fleck, N. (Eds.), Proc. Int. Conf. Bremen, Germany, June 14–16. MIT Press-Verlag, p. 133.

Assadi, H., Barth, M., Greer, A.L., Herlach, D.M., 1998. Kinetics of solidification of intermetallic compounds in the Ni-Al system. Acta Mater. 46, 491–500.

Asta, M., Ozolins, V., Woodward, C., 2001. A first-principles approach to modeling alloy phase equilibria. J. Met. 16–21.

Asta, M., Beckermann, C., Karma, A., Kurz, W., Napolitano, R., Plapp, M., Purdy, G., Rappaz, M., Trived, R., 2009. Solidification microstructures and solid-state parallels: recent developments, future directions. Acta Mater. 57, 941–971.

Atwood, R.C., Lee, P.D., 2003. Simulation of the three dimensional morphology of solidification porosity in an aluminium-silicon alloy. Acta Mater. 51, 5447–5466.

Audero, M.A., Biloni, H., 1973. J. Cryst. Growth 18, 257.

Aziz, M.J., 1982. Model for solute redistribution during rapid solidification. J. Appl. Phys. 53, 1158.

Aziz, M.J., 1996. Interface attachment kinetics in alloy solidification. Metall. Mater. Trans. A 27, 671.

Aziz, M.J., Boettinger, W.J., 1994. On the transition from short-range diffusion-limited to collision-limited growth in alloy solidification. Acta Metall. Mater. 42, 527.

Aziz, M.J., Kaplan, T., 1988. Continuous growth model for alloy solidification. Acta Metall. Mater. 36, 2335.

Baker, J.C., 1970, Interfacial Partitioning During Solidification, Ph.D. Thesis, Massachusetts Institute of Technology, Chapter V, (see also Cahn et al. (1980)).

Baker, J.C., Cahn, J.W., 1971. In: Solidification. ASM, Metals Park, OH, p. 23.

Bandyopadhyay (Banerjee), D.K., Stefanescu, D.M., Minkoff, I., Biswal, S.K., 1990. Structural transitions in directionally solidified spheroidal graphite cast iron. Proc. Fourth Int. Symp. Phys. Metall. Cast Iron, Mater. Res. Soc. 27–34.

Banerjee, D.K., Stefanescu, D.M., 1991. Structural transitions and solidification kinetics of SG cast iron during directional solidification experiments. AFS Trans. 104, 747–759.

Banhart, J., 2001. Manufacture, characterisation and application of cellular metals and metal foams. Prog. Mater. Sci. 46, 559.

Baram, J., 1991. Metall. Trans. 22A (10), 2515–2522.

Baskaran, V., Wilcox, W.R., 1984. Influence of convection on lamellar spacing of eutectics. J. Cryst. Growth 67, 343.

Battle, T.P., 1992. Int. Mater. Rev. 37, 249.

Beckermann, C., 2002. Modelling of macrosegregation: applications and future needs. Int. Mater. Rev. 47, 243–261.

Beckermann, C., Diepers, H.-J., Steinbach, I., Karma, A., Tong, X., 1999. Modeling melt convection in phase-field simulations of solidification. J. Comput. Phys. 154, 468–496.

Beckermann, C., Gu, J.P., Boettinger, W.J., 2000. Development of a freckle predictor via Rayleigh number method for single-crystal nickel-base superalloy castings. Metall. Mater. Trans. A 31, 2545–2557.

Beech, J., Barkhudarov, M., Chang, K., Chin, S.B., 1998. In: Thomas, B.G., Beckermann, C. (Eds.), Modeling of Casting, Welding, and Advanced Solidification Processes VIII. The Minerals, Metals, and Materials Soc. Warrendale, PA, p. 1071.

Bendersky, L.A., Boettinger, W.J., 1985. Cellular microsegregation in rapidly solidified Ag-15 wt% Cu alloys. In: Steeb, S., Warlimont, H. (Eds.), Rapidly Quenched Metals. Elsevier Science Publishers, B.V. pp. 887–890.

Bennon, W.D., Incropera, F.P., 1987a. A continuum model for momentum, heat and species transport in binary solid-liquid phase change systems—I. Model formulation. Int. J. Heat Mass Transfer 30, 2161–2170.

Bennon, W.D., Incropera, F.P., 1987b. A continuum model for momentum, heat and species transport in binary solid-liquid phase change systems—II. Application to solidification in a rectangular cavity. Int. J. Heat Mass Transfer 30, 2171–2187.

Berry, J.T., Pehlke, R.D., 1988. Modeling solidification heat transfer. In: Metals Handbook Casting, ninth ed. 15. ASM, Metals Park, OH, p. 850.

Bertorello, H.R., Biloni, H., 1969. On the origin of cellular substructure in A1-CU eutectic. Trans. Metall. Soc. AIME 245, 1375.

Bi, Z., Sekerka, R.F., 1998. Phase-field model for solidification of a binary alloy. Physica A 261, 95.

Bi, G., Gasser, A., Wissenbach, K., Drenker, A., Poprawe, R., 2006. Characterization of the process control for the direct laser metallic powder deposition. Surf. Coat. Technol. 201, 2676.

Billia, B., Trivedi, R., 1993. (North-Holland, Amsterdam). Pattern formation in crystal growth. In: Hurle, D.T.J. (Ed.), Handbook of Crystal Growth 1B: Fundamentals, Transport and Stability, p. 899.

Biloni, H., 1970. Metall. ABM (Ass. Brasilera de Metais) 26, 803.

Biloni, H., 1977. Cienc. Interam. 18 (3–4), 3.

Biloni, H., 1980. In: Pampillo, C.A., Biloni, H., Embury, D.E. (Eds.), Aluminum Transformation Technology and Applications 1979. ASM, Metals Park, OH, p. 1.

Biloni, H., 1983. Solidification (North Holland, Amsterdam). In: Cahn, R.W., Haasen, P. (Eds.), Physical Metallurgy, third ed. p. 478.

Biloni, H., Chalmers, B., 1965. Trans. Metall. Soc. AIME 233, 373.

Biloni, H., Morando, R., 1968. Trans. Metall. Soc. AIME 242, 1121.

Biloni, H., Bolling, G.F., Domian, H.A., 1965a. Trans. Metall. Soc. AIME 233, 1926.

Biloni, H., Bolling, G.F., Cole, G.S., 1965b. Trans. Metall. Soc. AIME 233, 251.

Biloni, H., Bolling, G.F., Cole, G.S., 1966. Trans. Metall. Soc. AIME 236, 930.

Biloni, H., Di Bella, R., Bolling, G.F., 1967. Trans. Metall. Soc. AIME 239, 2012.

Bleck, W., 2004. Met. Powder Rep. March, 18.

Blodgett, J.A., Schaefer, R.J., Glicksman, M.E., 1974. A holographic system for crystal growth studies: design and applications. Metallography 7, 453.

Bobadilla, M., Lacaze, J., Lesoult, G.J., 1989. Influence des conditions de solidification sur le deroulement de la solidification des aciers inoxydables austenitiques. J. Cryst. Growth 89, 531.

Bobadilla, M., Jolivet, J.M., Lamant, J.Y., Larrecq, M., 1993. Continuous casting of steel: a close connection between solidification studies and industrial process development. Mater. Sci. Eng. A 173, 275.

Boettinger, W.J., 1973. NBS Gaithersburg, MD, unpublished research.

Boettinger, W.J., 1974. The structure of directional solidified two-phase Sn-Cd peritectic alloys. Metall. Trans. 5, 2023.

Boettinger, W.J., 1981, unpublished research.

Boettinger, W.J., 1982. Growth kinetic limitations during rapid solidification. In: Kear, B.H., Giessen, B.C. (Eds.), Rapidly Solidified Crystal Line and Amorphous Alloys. Elsevier, North Holland, NY, p. 15.

Boettinger, W.J., 1982. The effect of alloy constitution and crystallization kinetics on the formation of metallic glass. In: Masumoto, T., Suzuki, K. (Eds.), Rapidly Quenched Metals IV. The Japan Inst. Metals, p. 99.

Boettinger, W.J., 1988, NIST, Gaithersburg, MD, unpublished research.

Boettinger, W.J., Aziz, M.J., 1989. Growth kinetic limitations during rapid solidification. Acta Metall. 37, 3379.

Boettinger, W.J., Coriell, S.R., 1986. Science and technology of the supercooled melt. In: Sahm, P.R., Jones, H., Adam, C.M. (Eds.), 1986. Nato ASI Series, E-No. 114. Martinus Nijhoff, Dodrecht, p. 81.

Boettinger, W.J., Kattner, U.R., 2002. On DTA curves for the melting and freezing of alloys. Metall. Mater. Trans. A 33, 1779–1794.

Boettinger, W.J., Perepezko, J.H., 1985. In: Das, S.K., Kear, B.H., Adam, C.M. (Eds.), Rapidly Solidified Crystalline Alloys. TMS-AIME, p. 21.

Boettinger, W.J., Warren, J.A., 1996. The phase-field method: simulation of alloy dendritic solidification during recalescence. Metall. Mater. Trans. A 27, 657–669.

Boettinger, W.J., Biancaniello, F.S., Kalonji, G.M., Cahn, J.W., 1980. Eutectic solidification and the formation of metallic glasses. In: Mehrabian, R., Kear, B.H., Cohen, M. (Eds.), 1980. Rapid Solidification Processing: Principles and Technologies, vol. 2. Claitor's, Baton Rouge, p. 50.

Boettinger, W.J., Biancaniello, F.S., Coriell, S.R., 1981. Solutal convection induced macrosegregation and the dendrite to composite transition in off-eutectic alloys. Metall. Trans. A 12, 321–327.

Boettinger, W.J., Coriell, S.R., Sekerka, R.F., 1984a. Mechanisms of microsegregation-free solidification. Mater. Sci. Eng. 65, 27–36.

Boettinger, W.J., Shechtman, D., Schaefer, R.J., Biancaniello, F.S., 1984b. Metall. Trans. 15A, 55.

Boettinger, W.J., Bendersky, L.A., Early, J.G., 1986. An analysis of the microstructure of rapidly solidified Al-8 wt. % Fe powder. Metall. Trans. 17A, 781.

Boettinger, W.J., Bendersky, L.A., Coriell, S.R., Schaefer, R.J., Biancaniello, F.S., 1987. Rapidly solidified amorphous and crystalline alloys. J. Cryst. Growth 80, 17–25.

Boettinger, W.J., Bendersky, L.A., Biancaniello, F.S., Cahn, J.W., 1988a. Rapid solidification of ordering of B2 and L2 (sub 1) phases in the NIAl-NITI system. Mater. Sci. Eng. 98, 273.

Boettinger, W.J., Coriell, S.R., Trivedi, R., 1988b. Rapid solidification processing: principles and technologies. In: Mehrabian, R., Parrish, P.A. (Eds.). Claitor's Publishing, Baton Rouge, p. 13.

Boettinger, W.J., Bendersky, L.A., Schaefer, R.J., Biancaniello, F.S., 1988c. On the formation of dispersoids during rapid solidification of an Al–Fe–Ni alloy. Metall. Trans. A 19, 1101–1107.

Boettinger, W.J., Kattner, U.R., Banerjee, D.K., 1998. Analysis of solidification path and microsegregation in multicomponent alloys. In: Thomas, B., Beckermann, C. (Eds.), 1998. Modeling of Casting, Welding and Advanced Solidification Processes, VIII. TMS, Warrendale, PA, pp. 159–170.

Boettinger, W.J., Coriell, S.R., Greer, A.L., Karma, A., Kurz, W., Rappaz, M., Trivedi, R., 2000. Solidification microstructures: recent developments, future directions. Acta Mater. 48, 43–70.

Boettinger, W.J., Warren, J.A., Beckermann, C., Karma, A., 2002. Phase-field simulation of solidification. Annu. Rev. Mater. Res. 32, 163–194.

Boettinger, W.J., Newbury, D.E., Wang, K., Bendersky, L.A., Chiu, C., Kattner, U.R., Young, K., Chao, B., 2010. Examination of multiphase (Zr, Ti)(V, Cr, Mn, Ni)2 Ni–MH electrode alloys. Part 1: dendritic solidification structure. Metall. Mater. Trans. A 41, 2033–2047.

Bower, T.F., Flemings, M.C., 1967. Trans. Metall. Soc. AIME 239, 1629.

Braun, R.J., Davis, S.H., 1991. Rapid directional solidification: Bifurcation theory. J. Cryst. Growth 112, 670.

Brennan, J.B., 1958. USA Patent No. 2639490 and 2864137.

Brice, J.C., 1973. The growth of crystals from liquids. In: The Growth of Crystals from Liquids, p. 120. (North Holland, Amsterdam).

Bricknell, R.H., 1986. The structure and properties of a Nickel-based superalloy produced by Osprey atomization-deposition. Metall. Trans. 17A, 583–591.

Brody, H.D., 1974. In: Burke, J.J., Flemings, M.C., Quorum, A.E. (Eds.), Solidification Technology. Brook Hill Pub., Chenut Hills, MA, p. 53.

Brody, H.D., David, S.A., 1979. Controlled solidification of peritectic alloys. In: Solidification and Casting of Metals. The Metals Society, London, p. 144.

Brody, H.D., Flemings, M.C., 1966. Trans. Metall. Soc. AIME 236, 615.

Broughton, J.Q., Bonissent, A., Abraham, F.F., 1981. The fcc (111) and (100) crystal–melt interfaces: A comparison by molecular dynamics simulation. J. Chem. Phys. 74, 4029.

Broughton, J.Q., Gilmer, G.H., Jackson, K.A., 1982. Crystallization rates of a Lennard-Jones liquid. Phys. Rev. Lett. 49, 1496.

Brown, R.A., 1988. Theory of transport processes in single crystal growth from the melt. AIChE J. 34, pp. 881–911.

Brown, S.B., Flemings, M.C., 1993. Adv. Mater. Processes 36–40.

Burden, M.H., Hunt, J.D., 1974a. J. Cryst. Growth 22, 99.

Burden, M.H., Hunt, J.D., 1974b. J. Cryst. Growth 22, 109.

Burden, M.H., Hunt, J.D., 1974c. J. Cryst. Growth 22, 328.

Burden, M.H., Hebditch, D.J., Hunt, J.D., 1973. J. Cryst. Growth 20, 121.

Burton, J.A., Prim, R.C., Schlichter, W.P., 1953. J. Chem. Phys. 21, 1987.

Cahn, J.W., 1960. Acta Metall. 8, 554.

Cahn, J.W., 1967. In: Peiser, H.S. (Ed.), Crystal Growth. Pergamon Press, Oxford, p. 681.

Cahn, J.W., 1979. Metall. Trans. 10A, 119.

Cahn, J.W., Hillig, W.B., Sears, G.W., 1964. Acta Metall. 12, 1421.

Cahn, J.W., Coriell, S.R., Boettinger, W.J., 1980. Rapid solidification. In: White, C.W., Peercy, P.S. (Eds.), Laser and Electron Beam Processing of Materials. Academic Press, NY, p. 89.

Calginalp, G., 1986. Arch. Ration. Mech. Anal. 92, 205.

Calginalp, G., Xie, W., 1993. Phase-field and sharp-interface alloy models. Phys. Rev. E 48, 1897–1909.

Camel, D., Favier, J.J., 1984a. Thermal convection and longitudinal macrosegregation in horizontal Bridgman crystal growth: I. Order of magnitude analysis. J. Cryst. Growth 67, 42.

Camel, D., Favier, J.J., 1984b. Thermal convection and longitudinal macrosegregation in horizontal bridgman crystal growth: II. Practical laws. J. Cryst. Growth 67, 57.

Campbell, J., 1991a. Castings. Heinemann, London.

Campbell, J., 1991b. Mater. Sci. Technol. 7, 885.

Campbell, J., 2003. Castings, second ed. Butterworth-Heinemann, London, 32.

Campbell, J., Olliff, I.D., 1971. AFS Cast Met. Res. J. ASM Metals Park, OH, May, 55.

Caroli, B., Caroli, C., Roulet, B., 1982. On the emergence of one-dimensional front instabilities in directional solidification and fusion of binary mixtures. J. Phys. 43, 1767.

Carrad, M., Gremaud, M., Zimmermann, M., Kurz, W., 1992. Acta Metall. Mater. 40, 983.

Carruthers, J.R., 1976. Thermal convection instabilities relevant to crystal growth from liquids. In: Preparation and Properties of Solid State Materials, vol. 2. (Marcel Recker, NY).

Carter, W.T., Jones, R.M.F., 2004. In: Proceedings of TMS Annual Meeting, Charlotte, NC, March 14–18.

Cech, R.E., 1956. Trans. Metall. Soc. AIME 206, 585.

Chadwick, G.A., 1963. In: Chalmers, B. (Ed.), 1963. Progress in Materials Science, vol. 12. Pergamon Press, Oxford, p. 97.

Chadwick, G.A., 1963. Prog. Mater. Sci. 12, 2.

Chadwick, G.A., 1965. Br. J. Appl. Phys. 16, 1095.

Chalmers, B., 1964. Principles of Solidification. Wiley, NY.

Chalmers, B., 1971. In: Solidification. American Society for Metals, ASM, Metals Park, OH, p. 295.

Chang, Ch.J., Brown, R.A., 1983. J. Cryst. Growth 63, 343.

Chen, L.Q., 1995. Scr. Metall. Mater. 32, 115.

Chen, S.W., Chang, Y.A., 1992. Microsegregation in solidification for ternary alloys. Metall. Mater. Trans. A 23, 1038–1043.

Chernov, A.A., 1984. Modern Crystallography III: Crystal Growth. Springer-Verlag, Berlin.

Chernov, A.A., 2004. Notes on interface growth kinetics 50 years after Burton, Cabrera and Frank. J. Cryst. Growth 264, 499–518.

Cheveigne, De. S., Guthmann, C., Kuroski, P., Vicente, E., Biloni, H., 1988. J. Cryst. Growth 92, 616.

Chopra, M.A., Glicksman, M.E., Singh, N.B., 1988. Dendritic Solidification in Binary Alloys. Metall. Trans. 19A, 3087.

Choudhary, S.K., Gosh, A., 1994. Morphology and Macrosegregation in Continuously Cast Steel Billets. ISIJ Int. 34, 338–345.

Clapham, L., Smith, R.W., 1988. Segregation behaviour of strontiumin modified and unmodified Al–Si alloys. J. Cryst. Growth 92, 263.

Clyne, T.W., 1980a. J. Cryst. Growth 50, 684.

Clyne, T.W., 1980b. J. Cryst. Growth 50, 691.

Clyne, T.W., 1984. Numerical treatment of rapid solidification. Metall. Trans. 15B, 369.

Clyne, T.W., García, A., 1980. Assessment of a new model for heat flow during unidirectional solidification of metals. Int. J. Heat Mass Transfer 23, 773.

Clyne, T.W., Garcia, A., 1981. The application of a new solidification heat flow model to splat cooling. J. Mater. Sci. 16, 1643.

Cole, G.S., 1971. In: Solidification. ASM, Metals Park, OH, p. 201.

Collins, P.C., Banerjee, R., Banerjee, S., Fraser, H., 2003. Mater. Sci. Eng. A 352, 118.

Coriell, S.R., Boettinger, W.J., 1994, NIST, unpublished results.

Coriell, S.R., McFadden, G.B., 1989. J. Cryst. Growth 94, 513.

Coriell, S.R., McFadden, G.B., 1990. In: Koster, J.N., Sani, R.L. (Eds.), 1990. Low Gravity Fluid Dynamics and Transport Phenomena, vol. 130. Progress in Astronautics and Aeronautics AIAA, Washington DC, p. 369.

Coriell, S.R., McFadden, G.B., 1993. In: Hurle, D.T.J. (Ed.), 1993. Handbook of Crystal Growth, vol. 1. Elsevier Science Publishers, p. 785.

Coriell, S.R., Sekerka, R.F., 1979. Lateral Solute Segregation During Unidirectional Solidification of a Binary Alloy with a Curved Solid-Liquid Interface. J. Cryst. Growth 46, 479.

Coriell, S.R., Sekerka, R.F., 1983. Oscillatory Morphological Instabilities Due to NonEquilibrium Segregation. J. Cryst. Growth 61, 499.

Coriell, S.R., Turnbull, D., 1982. Acta Metall. 30, 2135.

Coriell, S.R., Hurle, D.T.J., Sekerka, R.F., 1976. Interface stability during crystal growth: The effect of stirring. J. Cryst. Growth 32, 1.

Coriell, S.R., Boisvert, R.F., Rehm, R.G., Sekerka, R.F., 1981. J. Cryst. Growth 54, 167.

Coriell, S.R., McFadden, G.B., Sekerka, R.F., 1985. Annu. Rev. Mater. Sci. 15, 119.

Coriell, S.R., McFadden, G.B., Voorhees, P.W., Sekerka, R.F., 1987. J. Cryst. Growth 82, 295.

Coriell, S.R., McFadden, G.B., Sekerka, R.F., 1990. J. Cryst. Growth 100, 459.

Craig, D.B., Nornung, M.J., Cluhan, T.K., 1988. Metals Handbook Casting, ninth ed. 15. ASM, Metals Park, OH, 629.

Croker, M.N., Fidler, R.S., Smith, R.W., 1973. The characterization of eutectic structure. Proc. R. Soc. London, Ser. A 133, 15.

Cross, M., Pericleous, K., Croft, T.N., McBride, D., Lawrence, J.A., Williams, A.J., 2006. Computational Modelling of Mold Filling, and Relative Free Surface Flows in Shape Casting: An Overview of Challenges. Metall. Mater. Trans. B 37, 879–885.

Crossley, P.A., Mondolfo, L.F., 1966. Mod. Cast. 49, 89.

Dantzig, J.A., Rappaz, M., 2009. Solidification. EPFL Press (CRC Press Taylor & Francis Group).

Daoud, A., AbouEl-Khair, M.T., Abdel-Aziz, M., Rohatgi, P., 2007. Compos. Sci. Technol. 67, 1842–1853.

Das, S., Paul, A.J., 1993. Metall. Trans. 24B, 1073.

Daws, M.S., Baskes, M.I., 1984. Embedded-atom method – derivation and application to impurities, surfaces, and other defects in metals. Phys. Rev. Lett. B29, 6443–6453.

Day, M.G., Hellawell, A., 1968. Proc. R. Soc. London, Ser. A 305, 473.

Dayte, V., Langer, J.S., 1981. Phys. Rev., B 24, 4155.

Dayte, V., Mathur, R., Langer, J.S., 1982. J. Stat. Phys. 29, 1.

Degand, C., Stefanescu, D.M., Laslaz, G., 1996. In: Ohnaka, I., Stefanescu, D.M. (Eds.), Solidification Science and Processing. TMS, Warrendale, PA, p. 55.

Delplanque, J.P., Lavernia, E.J., Rangel, R.H., 1997. Numerical Investigation of Multi-phase Flow Induced Porosity Formation in Spray Deposited Materials. TMS. Annual Meeting, Orlando, FL.

Derby, B., Favier, J.J., 1983. Acta Metall. 31, 1123.

http://vcc-sae.org/abstracts/1703-development-metal-foam-based-aftertreatment-diesel-passenger-car.

Diewwanit, I., Flemings, M.C., 1996. Semi-solid forming of hypereutectic Al-Si alloys. In: Hale, W. (Ed.), Light Metals. TMS, Warrendale, PA, pp. 787–793.

Dinda, G.P., Dasgupta, A.K., Mazumder, J., 2009. Laser aided direct metal deposition of Inconel 625 superalloy: microstructural evolution and thermal stability. Mater. Sci. Eng. A 509, 98–104.

Dobler, S., Kurz, W., 2004. Phase and microstructure selection in peritectic alloys under high G–V ratio. Z. Metallkd. 7, 592–595.

Dobler, S., Lo, T.S., Plapp, M., et al., 2004. Peritectic coupled growth. Acta Mater. 52, 2795–2808.

Dunand, D.C., 2004. Processing of Titanium Foams. Adv. Eng. Mater. 6, 369.

Dupouy, M.D., Camel, D., Favier, J.J., 1989. Natural convection in directional dendritic solidification of metallic alloys—I. Macroscopic effects. Acta Mater. 37, 1143–1157.

Echebarria, B., Karma, A., Gurevich, S., 2010. Onset of sidebranching in directional solidification. Phys. Rev. E 81, 021608.

Eckler, K., Cochrane, R.F., Herlach, D.M., Feuerbacher, B., Jurisch, M., 1992. Phys. Rev. B 45, 5019.

Elder, K.R., Drolet, F., Kosterlitz, J.M., Grant, M., 1994. Stochastic eutectic growth. Phys. Rev. Lett. 72, 677.

Elliot, R., 1988. In: Cast Iron Technology. Butterworths.

Elmer, J.W., Aziz, M.J., Tanner, L.E., Smith, P.M., Wall, M.A., 1994. Formation of bands of ultrafine beryllium particles during rapid solidification of Al-Be alloys: Modeling and direct observations. Acta Metall. Mater. 42, 1065.

Erdmann, R.G., Poirier, D.R., Hendrick, A.G., 2010. Permeability in the mushy zone. Mater. Sci. Forum 649, 399–408.

Esaka, H., 1999. In: Cantor, B., 'Reilly, K.O. (Eds.), Solidification and Casting. Institute of Physics Publishing, Bristol UK, p. 56.

Esaka, H., Wakabayashi, T., Shinozuka, K., Tamura, M., 2003. Origin of Equiaxed Grains and their Motion in the Liquid. Phase. ISIJ Int. 43 (9), 1415–1420.

Eshelman, M.A., Trivedi, R., 1988. Scr. Metall. 22, 893.

Eshelman, M.A., Seetharaman, V., Trivedi, R., 1988. Acta Metall. 36, 1165.

Fan, Z., Wang, Y., Xia, M., Arumuganathar, S., 2009. Enhanced heterogeneous nucleation in AZ91D alloy by intensive melt shearing. Acta Mater. 4891–4901.

Fan, Z., Xia, M., Bian, Z., Bayandorian, I., Cao, L., Li, H., Scamans, G.M., 2010. Refinement of Solidification Microstructure by the MCAST Process. Mater. Sci. Forum 649, 315–323.

Fan, Z., Zuo, Y.B., Jiang, B., 2011. Mater. Sci. Forum 690, 141–144.

Favier, J.J., 1981a. Acta Metall. 29, 197.

Favier, J.J., 1981b. Acta Metall. 29, 205.

Favier, J.J., 1990. Recent advances in Bridgman growth modelling and fluid flow. J. Cryst. Growth 99, 18.

Favier, J.J., Camel, D., 1986. J. Cryst. Growth 79, 50.

Favier, J.J., Rouzaud, A., 1983. J. Cryst. Growth 64, 387.

Favier, J.J., Wilson, L.O., 1982. J. Cryst. Growth 58, 103.

Fecht, H.J., Perepezko, J.H., 1989. Metall. Trans. 20A, 785.

Felice, V., de Jardy, A., Combeau, H., 2009. In: Lee, P.D., Mitchell, A., Williamson, R. (Eds.), Int. Symp. on Liquid Metal Processing and Casting. TMS publishing, Warrendale, PA, p. 97.

Felicelli, S.D., Poirier, D.R., Heinrich, J.C., 1998. Modeling freckle formation in three dimensions during solidification of multicomponent alloys. Metall. Mater. Trans. B 29, 847–855.

Ferreira, I.L., Spinelli, J.E., Pires, J.C., Garcia, A., 2005. The effect of melt temperature profile on the transient metal/mold heat transfer coefficient during solidification. Mater. Sci. Eng. A 408, 317.

Feurer, U., Wunderlin, R., 1977. DGM Fachber, 38. See also Kurz and Fisher (1989).

Fife, J.L., Voorhees, P.W., 2009. The morphological evolution of equiaxed dendritic microstructures during coarsening. Acta Mater. 57, 2418–2428.

Fife, J.L., Li, J.C., Dunand, D.C., Voorhees, P.W., 2009. Morphological analysis of pores in directionally freeze-cast titanium foams. J. Mater. Res. 24 (1), 117.

Finn, T.L., Chu, M.G., Bennon, W.D., 1992. In: Beckermann, C., et al. (Eds.), Micro/macroscale Phenomena in Solidification. ASME, New York, pp. 17–26.

Fisher, D.J., Kurz, W., 1979. In: Solidification and Casting of Metals. The Metals Soc. London, p. 57.

Fisher, D.J., Kurz, W., 1980. Acta Metall. 28, 777.

Flemings, M.C., 1974. Solidification Processing. McGraw Hill, New York.

Flemings, M.C., 1976. Scand. J. Metall. 5, 1.

Flemings, M.C., 1991. Metall. Trans. 22A, 957–981.

Flemings, M.C., Nereo, G.E., 1967. Trans. Metall. Soc. AIME 239, 1449.

Flemings, M.C., Shiohara, Y., 1984. Mater. Sci. Eng. 65, 157.

Flood, S.C., Hunt, J.D., 1987a. J. Cryst. Growth 82, 552–560.

Flood, S.C., Hunt, J.D., 1987b. Appl. Sci. Res. 44, 27.

Flood, S.C., Hunt, J.D., 1988. Metals Handbook. In: Casting, ninth ed. 15. ASM, Metals Park, OH, p. 130.

Frank, F.C., 1949. Discussions of the Faraday Society, 5, 48.

Fras, E., Lopez, H.F., 1993. Trans. AFS 101, 355.

Fredriksson., H., Hillert, M., 1972. Scand. J. Metall. 2, 125.

Fredriksson, H., Nylen, T., 1982. Met. Sci. 16, 283.

Fredriksson, H., Hillert, M., Lange, N., 1973. J. Inst. Met. 101, 285.

Frenkel, J., 1932. Phys. Z. Sowjetunion 1, 498.

Furrer, D.U., Semiatin, S.L., 2010. Metals process simulation. In: Asm Handbook, vol. 22B. ASM, Materials Park, OH, pp. 649–656.

Galenko, P., Sobolev, S., 1997. Phys. Rev. E 55, 343.

Gandin, C.A., Rappaz, M., West, D., Adams, B.L., 1995. Grain texture evolution during the columnar growth of dendritic alloys. Metall. Mater. Trans. A 26, 1543–1551.

Gandin, C.A., Desbioles, J.L., Rappaz, M., Thevoz, P., 1999. A three dimensional cellular automaton-finite element model for the prediction of solidification grain structure. Metall. Mater. Trans. A 30, 3153–3165.

Garcia, A., Clyne, T.W., 1983. In: Charles, J.A. (Ed.), Solidification Technology in the Foundry and Casthouse. The Metals Society, London, p. 33.

Garcia, A., Prates, M., 1978. Metall. Trans. 9B, 449.

Garcia, A., Clyne, T.W., Prates, M., 1979. Metall. Trans. 10B, 85.

http://www.alantum.com/en/gastreatment.html.

Genau, A.L., Voorhees, P.W., Thornton, K., 2009. The morphology of topologically complex interfaces. Scr. Mater. 60, 301–304.

Gergely, V., Clyne, T.W., 2004. Drainage in standing liquid metal foams: modelling and experimental observations. Acta Mater. 52, 3047.

Gill, S.C., Kurz, W., 1994. Rapidly solidified Al-Cu Alloys—Experimental determination of the microstructure selection map. Acta Metall. Mater. 41, 3563.

Giovanola, B., Kurz, W., 1990. Modeling of microsegregation under rapid solidification conditions. Metall. Trans. 21A, 260.

Glicksman, M.E., 2011. Principles of Solidification. Springer, New York.

Glicksman, M.E., 2012. Mechanism of Dendritic Branching. Metall. Mater. Trans. 43A, 391–404.

Glicksman, M.E., Schaefer, R.J., Ayers, J.D., 1976. Metall. Trans. 7A, 1747.

Glicksman, M.E., Coriell, S.R., McFadden, G.B., 1986. Interaction of flows with the crystal-melt interface. Annu. Rev. Fluid Mech. 18, 307.

Glicksman, M.E., Lowengrub, J.S., Li, S., Li, X., 2007. A deterministic mechanism for dendritic solidification kinetics. J. Met. 27.

Gránásy, L., Pusztai, T., 2002. Diffuse interface analysis of crystal nucleation in hard-sphere liquid. J. Chem. Phys. 117, 10121.

Granasy, L., Pusztai, T., Saylor, D., Warren, J.A., 2007. Phase field theory of heterogeneous crystal nucleation. Phys. Rev. Lett. 98, 035703.

Granger, D.A., Elliot, R., 1988. Metals Handbook. In: Casting, ninth ed. 15. ASM, Metals Park, OH, p. 15.

Greer, A.L., Bunn, A.M., Tronche, A., Evans, P.V., Bristow, D.J., 2000. Modelling of inoculation of metallic melts: application to grain refinement of aluminium by Al–Ti–B. Acta Mater. 48, 2823–2835.

Gremaud, M., Kurz, W., Trivedi R., 1987. Unpublished work. See Trivedi and Kurz (1988).

Gremaud, M., Carrad, M., Kurz, W., 1991. Acta Metall. Mater. 39, 1431.

Grugel, R.N., Hellawell, A., 1981. Metall. Trans. 12A, 669.

Grugel, R.N., Lograsso, T.A., Hellawell, A., 1984. Metall. Trans. 15A, 1003.

Gu, J.P., Beckermann, C., 1999. Simulation of convection and macrosegregation in a large steel ingot. Metall. Trans. 30A, 1357–1366.

Gu, J.P., Beckermann, C., Giamei, A.F., 1997. Motion and remelting of dendrite fragments during directional solidification of a nickel-base superalloy. Metall. Mater. Trans. A 28, 1533–1542.

Guillemot, G., Gandin, C.A., Combeau, H., 2006. ISIJ Int. 46 (6), 880–895.

Gulliver, G.H., 1922 (London). In: Griffin, Charles (Ed.), Metallic Alloys, p. 397.

Gunduz, M., Hunt, J.D., 1985. Acta Metall. 33, 1651.

Gurevich, S., Karma, A., Plapp, M., Trivedi, R., 2010. Phase-field study of three-dimensional steady-state growth shapes in directional solidification. Phys. Rev. E 81, 011603.

Hamilton, D.R., Seidensticker, R.G., 1960. J. Appl. Phys. 31, 1165–1168.

Han, S.H., Trivedi, R., 2000. Banded microstructure formation in off-eutectic alloys. Metall. Mater. Trans. A 31, 1819–1832.

Hanna, M.D., Lu, Shu-Zu, Hellawell, A., 1984. Metall. Trans. 15A, 459.

Hao, S.W., Zhang, Z.Q., Chen, J.Y., Liu, P.C., 1987. AFS Trans. 95, 601.

Hardy, S.C., McFadden, G.B., Coriell, S.R., Voorhees, P.W., Sekerka, R.F., 1991. Measurement and analysis of grain boundary grooving by volume diffusion. J. Cryst. Growth 114, 467.

Hayes, A., Chipman, J., 1938. Mechanism of solidification and segregation in a low carbon. Trans. Metall. Soc. AIME 135, 85.

Hecht, U., Gránásy, L., Pusztai, T., Böttger, B., Apel, M., Witusiewicz, V., Ratke, L., De Wilde, J., Froyen, L., Camel, D., Drevet, B., Faivre, G., Fries, S.G., Legendre, B., Rex, S., 2004. Multiphase solidification in multicomponent alloys. Mater. Sci. Eng. R (46–49).

Hellawell, A., 1970. The growth and structure of eutectics with silicon and germanium. Progress in Material Science 15, 3.

Hellawell, A., 1990. In: Weinberg, F., Lait, V.E., Samarasekera, I.V. (Eds.), International Symposium on Solidification Processing. Pergamon Press, p. 395.

Hellawell, A., Sarazin, J.R., Steube, R.S., 1993. Channel convection in partially solidified alloys. Philos. Trans. R. Soc. A 345 (1677), 507–544.

Hellawell, A., Liu, S., Lu, S.Z., 1997. Dendrite Fragmentation and the Effects of Fluid Flow in Castings. JOM, 18–20.

Henzel Jr, J.G., Keverian, J., 1965. The Theory and Application of Digital. Computer in Predicting Solidification Patterns. J. Met. 17, 561.

Hillert, M., 1953. Nuclear composition—A factor of interest in nucleation. Acta Metall. 1, 764.

Hillert, M., 1957. Jernkontorets Ann. 141, 757.

Hillert, M., 1979. In: Solidification and Casting of Metals. The Metals Society, London, p. 81.

Hillert, M., 1999. Acta Mater. 47, 4481.

Hillert, M., Subbarao, V.V., 1968. In: Solidification of Metals. Iron and Steel Inst. Publication no 110, London, p. 204.

Hillig, W.D., Turnbull, D., 1956. J. Chem. Phys. 24, 219.

Hills, A.W.D., Malhotra, S.L., Moore, M.R., 1975. The solidification of pure metals (and eutectics) under uni-directional heat flow conditions: II. Solidification in the presence of superheat. Metall. Trans. 6B, 131.

Hirt, W., Nichols, B.D., 1981. Volume of fluid (VOF) method for the dynamics of free boundaries. J. Comput. Phys. 39, 201–225.

Ho, K., Pehlke, R.D., 1984. Transient methods for determination of metal–mold interfacial heat transfer. AFS Trans. 92, 587.

Ho, K., Pehlke, R.D., 1985. Metal-mould interfacial heat transfer. Metall. Trans. 16B, 585.

Hoaglund, D.E., Aziz, M.E., Stiffler, S.R., Thomson, M.O., Sad, J.Y., Peercy, P.S., 1991. Effect of Nonequilibrium Interface Kinetics on Cellular Breakdown of Planar Interface During Rapid Solidification of Si-Sn. J. Cryst. Growth 109, 107.

Hogan, L.M., Kraft, R.W., Lemkey, F.D., 1971. In: Herman, H. (Ed.), 1971. Advances in Materials Research, vol. 5. Wiley, New York, p. 83.

Hollomon, J.H., Turnbull, D., 1953 (Interscience, New York). Prog. Met. Phys. vol. 4, 333.

Hoyt, J.J., Asta, M., Karma, A., 2003. Atomistic and continuum modeling of dendritic solidification. Mater. Sci. Eng. R 41, 121–163.

Hoyt, J.J., Asta, M., Sun, D.Y., 2006. Philos. Mag. 86, 3651–3664.

Huang, S.C., Glicksman, M.E., 1981. Acta Metall. 29, 717.

Huang, S.C., Hall, E.L., 1989. Mater. Res. Soc. Symp. Proc. 133, 373.

Huang, S.C., Hall, E.L., Chang, K.M., Laforce, R.P., 1986. Metall. Trans. 17A, 1685.

Hughes, I.C.H., 1988. Metals Handbook. In: Casting, ninth ed. 15. ASM, Metals Park, OH, p. 647.

Hughes, I.R., Jones, H., 1976. J. Mater. Sci. 11, 1781.
Hunt, J.D., 1979. In: Solidification and Casting of Metals. Metals Society, London, p. 3.
Hunt, J.D., 1984. Mater. Sci. Eng. 65, 75.
Hunt, J.D., Jackson, K.A., 1966. Trans. Metall. Soc. AIME 236, 843.
Hunt, J.D., Jackson, K.A., 1967. Trans. Metall. Soc. AIME 239, 864.
Hunt, J.D., Lu, S.-Z., 1996. Numerical modeling of cellular/dendritic array growth: spacing and structure predictions. Metall. Mater. Trans. A 27, 611–623.
Hurle, D.T.J., 1969. J. Cryst. Growth 5, 162.
Hurle, D.T.J., 1972. J. Cryst. Growth 13/14, 39.
Hurle, D.T.J., Jakeman, E., 1968. J. Cryst. Growth 3–4, 574.
Hyun, S.K., Nakajima, H., 2003. Mater. Lett. 57, 3149.
Hyun, S.K., Murakami, K., Nakajima, H., 2001. Mater. Sci. Eng. A 299, 241.
Ikeda, T., Aoki, T., Nakajima, H., 2005. Metall. Mater. Trans. A 36, 77.
Inoue, A., Masumoto, T., Tomioka, H., Yano, N., 1984. Int. J. Rapid Solidif. 1, 115.
Ishihara, K.N., Maeda, M., Shingu, P.H., 1985. Acta Metall. 33, 2113.
Ivantsov, G.P., 1947. Dokl. Akad. Nauk SSSR 58, 567.
Jackson, K.A., 1958. In: Liquid Metals and Solidification. ASM, Metals Park, OH, p. 174.
Jackson, K.A., 1968. Trans. Metall. Soc. AIME 242, 1275.
Jackson, K.A., 1971. In: Solidification. ASM, Metals Park, OH, p. 121.
Jackson, K.A., 1974. J. Cryst. Growth 24/25, 130.
Jackson, K.A., 1975. In: Hannay, N.B. (Ed.), 1975. Treatise on Solid State Chemistry, vol. 5. Plenum, NY, p. 1.
Jackson, K.A., Hunt, J.D., 1966. Trans. Metall. Soc. AIME 236, 1129.
Jackson, K.A., Hunt, J.D., Uhlmann, D.R., Seward III, T.P., 1966. Trans. Metall. Soc. AIME 236, 149.
Jackson, K.A., Uhlmann, D.R., Hunt, J.D., 1967. J. Cryst. Growth 1, 1.
Jackson, K.A., Gilmer, G.H., Leamy, H.J., 1980. In: White, C.W., Peercy, P.S. (Eds.), Laser and Electron Beam Processing of Materials. Academic Press, NY, p. 104.
Jackson, K.A., Beatty, K.M., Gudgel, K.A., 2004. J. Cryst. Growth 271, 481.
Jacobi, H., Schwerdtefeger, K., 1976. Metall. Trans. 7A, 811.
Jin, I., Kenny, L.D., Sang, H., 1990, US patent 4,973,358.
Jones, H., Kurz, W., 1980. Metall. Trans. 11A, 1265.
Jordan, R.M., Hunt, J.D., 1971. J. Cryst. Growth 11, 141.
Jordan, R.M., Hunt, J.D., 1972. Metall. Trans. 3, 1386.
Junze, J., Kobayashi, K.F., Shingu, P.H., 1984. Metall. Trans. 15A, 307.
Kajitani, T., Drezet, J.-M., Rappaz, M., 2001. Metall. Mater. Trans. A 32, 1479–1491.
Kamio, A., Kumai, S., Tezuka, H., 1991. Mater. Sci. Eng. A 146, 105.
Karma, A., 1987. Phys. Rev. Lett. 59, 71.
Karma, A., 1994. Phase-field model of eutectic growth. Phys. Rev. E 49, 2245–2250.
Karma, A., 2001. Phase-field formulation for quantitative modeling of alloy solidification. Phys. Rev. Lett. 87, 115701.
Karma, A., Rappel, W.-J., 1996. Phys. Rev. E 54, R3017.
Karma, A., Rappel, W.-J., 1997. J. Cryst. Growth 174, 54.
Karma, A., Lee, Y.H., Plapp, M., 2000. Phys. Rev. E 61, 3996.
Katayama, S., Matsunawa, A., 1984. In: Proceedings of the Materials Processing Symposium, vol. 44. Laser Institute of America, ICALEO, p. 60.
Kattamis, T.Z., Flemings, M.C., 1965. Trans. Metall. Soc. AIME 233, 992.
Kattamis, T.Z., Couglin, J.C., Flemings, M.C., 1967. Trans. Metall. Soc. AIME 239, 1504.
Kelly, T.F., Vandersande, J.B., 1987. Int. J. Rapid Solidif. 3, 51.
Kelton, K.F., 1991. Crystal Nucleation Liquids Glasses. Solid State Phys. Adv. Res. Appl. 45, 75–177.
Kelton, K.F., Greer, A.L., 2010. Nucleation in Condensed Matter. Elsevier, Amsterdam.
Kerr, H.W., Villafuerte, J.C., 1992. In: Cieslak, H.J., Perepezko, J.H., Kang, S., Glicksman, M.E. (Eds.), The Metal Science of Joining. TMS Pub. p. 11.
Kerr, H.W., Winegard, W.C., 1967. International Conference on Crystal Growth, suppl. to the Physics and Chemistry of Solids, 179.
Kerr, H.W., Cisse, J., Bolling G.F., 1974. 22, 677.
Kim, Y.W., Lin, H.M., Kelly, T.E., 1988a. Acta Metall. 36, 2525.
Kim, Y.-W., Lin, H.-M., Kelly, T.E., 1988b. Acta Metall. 36, 2537.
Kim, S.G., Kim, W.T., Suzuki, T., 1999. Phase-field model for binary alloys. Phys. Rev. E 60, 7186–7197.
Kim, S.G., Kim, W.T., Suzuki, T., et al., 2004. Phase-field modeling of eutectic solidification. J. Cryst. Growth 261, 135–158.

Kirkwood, D.H., 1985. Mater. Sci. Eng. 73, L1.

Kittl, J.A., Sanders, P.G., Aziz, M.J., Brunco, D.P., Thompson, M.O., 2000. Acta Mater. 48, 4797–4811.

Kobayashi, R., 1991. J. Jpn. Assoc. Cryst. Growth 18 (2), 209 (in Japanese).

Kobayashi, R., 1992. In: Kai, S. (Ed.), Pattern Formation in Complex Dissipative Systems. World Science, Singapore, p. 121.

Kobayashi, R., 1993. Modeling and numerical simulations of dendritic crystal-growth. Physica D 63, 410–423.

Kobayashi, R., 1994. Exp. Math. 3, 59.

Kofler, A., 1950. Z. Metallkd. 41, 221.

Kou, S., Sun, D.K., 1985. Fluid-flow and weld penetration in stationary arc welds. Metall. Trans. 16A, 203–213.

Kraft, T., Chang, Y.A., 1997. Predicting microsegregation. J. Met. 49, 20–28.

Kubo, K., Pehlke, R.D., 1985. Metall. Trans. 16B, 959.

Kumar, S., Leuven, K.U., 2008. Ph.D. thesis, Belgium.

Kurz, W., Clyne, T.W., 1981. Metall. Trans. 12A, 965.

Kurz, W., Fisher, D.J., 1979. Int. Mater. Rev. 5–6, 177.

Kurz, W., Fisher, D.J., 1981. Acta Metall. 29, 11.

Kurz, W., Fisher, D.J., 1989. Fundamentals of Solidification, third ed. Trans. Tech. Publication, Switzerland.

Kurz, W., Giovanola, B., Trivedi, R., 1986. Acta Metall. 34, 823.

Kurz, W., Giovanola, B., Trivedi, R., 1988. J. Cryst. Growth 91, 123.

Kurz, W., Bezencon, C., Gaumann, M., 2001. Sci. Technol. Adv. Mater. 2, 185–191.

Lacombe, J.C., Koss, M.B., Fradkov, V.E., Glicksman, M.E., 1995. Phys. Rev. E 52, 2778.

Langer, J.S., 1980. Phys. Rev. Lett. 44, 1023.

Langer, J.S., 1986. In: Grinstein, G., Mazenko, G. (Eds.), Directions in Condensed Matter Physics. World Scientific, Philedelphia, Pa, p. 164.

Langer, J.S., Mueller-Krumbhaar, H., 1978. Acta Metall. 26, 1681.

Langer, J.S., Mueller-Krumbhaar, H., 1981. Acta Metall. 29, 145.

Larson, M.A., Garside, J., 1986. J. Cryst. Growth 76, 88.

Leamy, H.J., Gilmer, G.H., 1974. J. Cryst. Growth 24/25, 499.

Leamy, H.J., Jackson, K.A., 1971. J. Appl. Phys. 42, 2121.

Leatham, A.G., 1996. Adv. Mater. Processes 150 (2), 31–34.

Leatham, A.G., Lawley, A., 1993. Int. J. Powder Metall. 29 (4), 321–329.

Leatham, A.G., Brooks, R.G., Coombs, J.S., Ogilvy, A.G.W., 1991. In: Proceedings of 1st International Conference on Spray Forming. Osprey Metals Ltd, Neath, U.K.

Lee, J.T.C., Brown, R.A., 1993. Phys. Rev. B 47.

Lee, P.D., Hunt, J.D., September, 1995. In: Proceedings of Modeling of Casting, Welding, and Advanced Solidification Processes VII. London, UK.

Lee, P.D., Wang, J., 2010. In: Furrer, D.U., Semiatin, S.L. (Eds.), 2010. Asm Handbook, vol. 22B, p. 254.

Lee, Y.W., Chang, E., Chieu, C.F., 1990. Modeling of feeding behavior of solidifying Al-7Si-0.3 Mg alloy plate casting. Metall. Trans. B 21, 715.

Lee, S.Y., Chung, S.I., Hong, C.P., 2000. In: Sahm, P.R., et al. (Eds.), Modeling of Casting, Welding and Advanced Solidification Processes IX. Shaker Verlag, Aachen, pp. 648–655.

Lee, P.D., Chirazi, A., Atwood, R.C., Wang, W., 2004. Multiscale modeling of solidification microstructures, including microsegregation and microporosity, in an Al-Si-Cu alloy. Mater. Sci. Eng. A 365, 57–65.

Lesoult, G., Sella, S., 1988. Solid State Phenom. 3–4, 167–178.

Levi, C.G., Mehrabian, R., 1982. Metall. Trans. 13A, 221.

Lewis, R.E., Lawley, A., 1991. Spray forming of metallic materials: an overview. In: Froes, F.H. (Ed.), Powder Metallurgy in Aerospace and Defense Technologies. MPIF, Princeton, NJ, pp. 173–184.

Lewis, R.W., Ransing, R.S., 1998. A correlation to describe interfacial heat transfer during solidification simulation and its use in the optimal design of castings. Metall. Mater. Trans. B 29 (2), 437–448.

Lewis, R.W., Ransing, R.S., 2000. The optimal design of interfacial heat transfer coefficients via a thermal stress model. Finite Elem. Anal. Des. 24, 193–209.

Lewis, D.J., Boettinger, W.J., Warren, J.A., 2003. In: Stefanescu, D.M., Warren, J.A., Jolly, M.R., et al. (Eds.), Three Dimensional Phase Field Modeling of Binary Eutectics, in Modeling of Casting, Welding and Advanced Solidification Processes-X, pp. 5–12.

Limmaneevichitr, C., Kou, S., 2000. Visualization of Marangoni convection in simulated weld pools containing a surface-active agent. Weld. J. 79, 324S–330S.

Lin, J., Carrera, S., Kunrath, A.O., Zhong, D., Myers, S., Mishra, B., Ried, P., Moore, J.J., 2006. Design methodology for optimized die coatings: the case for aluminum pressure die casting. Surf. Coat. Technol. 201, 2930–2941.

Lipton, J., A.Garcia, Heinemann, W., 1982. Arch. Einsenhüttenw. 53, 489.

Lipton, J., Glicksman, M.E., Kurz, W., 1984. Mater. Sci. Eng. 65, 57.

Lipton, J., Kurz, W., Trivedi, R., 1987. Acta Metall. 35, 957.

Liu, H., 2000. Science and Engineering of Droplets. William Andrew Publishing, Norwich, NY.

Liu, H., 2002. In: Yu, Kuang-O. (Ed.), Modeling for Casting and Solidification Processing. Marcel Dekker, New York, p. 655.

Liu, W.P., Dupont, J.N., 2003. In-situ reactive processing of nickel aluminides by laser-engineered net shaping. Metall. Mater. Trans. A 34, 2633–2641.

Liu, H., Rangel, R.H., Lavernia, E.J., 1994. Modeling of reactive atomization and deposition processing of Ni$_3$Al. Acta Metall. Mater. 42 (10), 3277–3289.

Liu, S., Lu, S., Hellawell, A., 2002. J. Cryst. Growth 234, 740–750.

Livingston, J., Cline, H., 1969. Trans. Metall. Soc. AIME 245, 351.

Lo, T.S., Dobler, S., Plapp, M., et al., 2003. Two-phase microstructure selection in peritectic solidification: from island banding to coupled growth. Acta Mater. 51, 599–611.

Losert, W., Shi, B.Q., Cummins, H.Z., 1998a. Evolution of dendritic patterns during alloy solidification. Proc. Natl. Acad. Sci. USA. 95, 431–438.

Losert, W., Shi, B.Q., Cummins, H.Z., 1998b. Proc. Natl. Acad. Sci. USA. 95, 439–442.

Lu, S.Z., Hellawell, A., 1987. Metall. Trans. 18A, 1721.

Lu, S.Z., Hellawell, A., 1988. In: Proc. Solidification Processing 1987.

Lu, S.Z., Hunt, J.D., Gilgien, P., Kurz, W., 1994. Acta Metall. Mater. 42, 1653.

Ludwig, A., Kurz, W., 1996. Direct observation of solidification microstructures around absolute stability. Acta Mater. 44, 3643–3654.

Magnin, P., Kurz, W., 1985. In: Fredriksson, H., Hiller, M. (Eds.), The Physical Metallurgy of Cast Iron, vol. 34. Proceedings of the MRS, North Holland, p. 263.

Magnin, P., Kurz, W., 1987. An analytical model of irregular eutectic growth and its application to Fe-C. Acta Metall. 35, 1119.

Magnin, P., Trivedi, R., 1991. Acta Metall. Mater. 39, 453.

Magnin, P., Mason, J.T., Trivedi, R., 1991. Acta Metall. Mater. 39, 469.

Major, J.F., Rutter, J.W., 1989. Mater. Sci. Technol. 5, 645.

Mampey, F., 2001. In: Proceedings of Cast Iron Division, AFS 105th Casting Congress. AFS, Des Plaines, IL, p. 51.

Marasli, N., Hunt, J.D., 1996. Solid–liquid surface energies in the Al–CuAl$_2$, Al–NiAl$_3$, and Al–Ti systems. Acta Mater. 44, 1085–1096.

Martorano, M.A., Beckermann, C., Gandin, C.A., 2003. A solutal interaction mechanism for the columnar-to-equiaxed transition in alloy solidification. Metall. Mater. Trans. A 34, 1657–1674.

Massing, G., Rodgers, B.A., 1960. Ternary Dystems. Dover, New York.

Mathier, V., Vernede, S., Jarry, P., Rappaz, M., 2009. Two-phase modeling of hot tearing in aluminum alloys: applications of a semicoupled method. Metall. Mater. Trans. A 40, 943–957.

Mathiesen, R., Arnberg, L., 2007. J. Met. 59, 20–26.

Mathiesen, R.H., Arnberg, L., Mo, F., Weitkamp, T., Snigirev, A., 1999. Time resolved X-ray imaging of dendritic growth in binary alloys. Phys. Rev. Lett. 83, 1562–1565.

Mathiesen, R.H., Arnberg, L., Bleuet, P., Somogyi, A., 2006. Crystal fragmentation and columnar-to-equiaxed transitions in Al–Cu studied by synchrotron X-ray video microscopy. Metall. Mater. Trans. A 37, 2515–2524.

Mathur, P., Apelian, D., Lawley, A., 1989. Analysis of the spray deposition process. Acta Metall. Mater. 37 (2), 429–443.

Matijasevic, B., Banhart, J., 2006. Improvement of aluminium foam technology by tailoring of blowing agent. Scr. Mater. 54, 503.

Maxwell, I., Hellawell, A., 1975. An analysis of the peritectic reaction with particular reference to Al-Ti alloys. Acta Metall. 23, 901.

Mcdonald, C.A., Malvezzi, A.M., Spaepen, F., 1989. J. Appl. Phys. 65, 129.

McFadden, G.B., Wheeler, A.A., Braun, R.J., Coriell, S.R., Sekerka, R.F., 1993. Phase-field models for anisotropic interfaces. Phys. Rev. E 48, 2016.

McFadden, G.B., Coriell, S.R., Sekerka, R.F., 2000. Effect of surface free energy anisotropy on dendrite tip shape. Acta Mater. 48, 3177–3181.

Mehrabian, R., 1982. Int. Mater. Rev. 27, 185.

Mehrabian, R., Flemings, M.C., 1970. Macrosegregation in ternary alloys. Metall. Trans. A 1, 455–464.

Mehrabian, R., Keane, N., Flemings, M.C., 1970. Metall. Trans. 1, 1209.

Mendelev, M.I., Rahman, M.J., Hoyt, J.J., Asta, M., 2010. Modell. Simul. Mater. Sci. Eng. 18, 074002.

Merchant, G.J., Davis, S.H., 1990. Acta Metall. Mater. 38, 2638.

Mikeev, L.V., Chernov, A.A., 1991. J. Cryst. Growth 112, 591.

Minkoff, I., 1983. The Physical Metallurgy of Cast Iron. Wiley.

Minkoff, I., 1990. In: Weinberg, F., Lait, J.E., Samarasekera, L.V. (Eds.), Int. Symposium on Solidification Processing. Pergamon Press, p. 255.

Mizukami, H., Suzuki, T., Umeda, T., 1991. Tetsu to Hagane 77 (10), 1672.

Mizukami, H., Suzuki, T., Umeda, T., Kurz, W., 1993. Mater. Sci. Eng. A 173, 363–366.

Mollard, F., Flemings, M.C., 1967. Growth of composites from the melt (Part 2). Trans. Metall. Soc. AIME 239, 1534.

Mondolfo, L., 1965. J. Aust. Inst. Met. 10, 169.

Monk, J., Yang, Y., Mendelev, M.I., Asta, M., Hoyt, J.J., Sun, D.Y., 2010. Modell. Simul. Mater. Sci. Eng. 18, 015004.

Morales, A., Glicksman, M.E., Biloni, H., 1979. In: Solidification and Casting of Metals. The Metals Society, London, p. 484.

Morris, L.R., Winegard, W.C., 1969. J. Cryst. Growth 5, 361.

Morrog, H., 1968a. J. Iron Steel Inst. 206, 1.

Morrog, H., 1968b. In: The Solidification of Metals. Iron and Steel Institute, Publication no 110, London, p. 238.

Mortensen, A., 1991. On the rate of dendrite arm coarsening. Metall. Trans. 22A, 569–574.

Mullins, W.W., Sekerka, R.F., 1963. J. Appl. Phys. 34, 323.

Mullins, W.W., Sekerka, R.F., 1964. J. Appl. Phys. 35, 444.

Murray, J.L., 1983a. Mater. Res. Soc. Symp. Proc. 19, 249.

Murray, J.L., 1983b. Bull. Alloy Phase Diagrams 4, 271.

Murray, B.T., Coriell, S.R., McFadden, G.B., 1991. J. Cryst. Growth 110, 713.

Nakama, K., Haruna, Y., Nakano, J., Sridhar, S., 2009. The effect of alloy solidification path on sulfide formation in Fe–Cr–Ni alloys. ISIJ Int. 49, 355–364.

Nandapurkar, P., Poirier, D.A., 1988. J. Cryst. Growth 92, 88.

Narayan, J.J., 1982. J. Cryst. Growth 59, 583.

Nestler, B., Wheeler, A.A., 2000. A multi-phase-field model of eutectic and peritectic alloys: numerical simulation of growth structures. Physica D 138, 114–133.

Nestler, B., Wheeler, A.A., Ratke, L., Stoecker, C., 2000. Phase-field model solidification a monotectic alloy convection. Physica D 141, 133–154.

Ni, J., Beckermann, C., 1991. A volume-averaged 2-phase model for transport phenomena during solidification. Metall. Trans. 22B, 349–361.

Nielsen, O., Appolaire, B., Combeau, H., Mo, A., 2001. Metall. Mater. Trans. A 32, 2049.

Niyama, E., Anzai, K., 1992. An analysis of unidireactional solidification of pure metals cooled through an interface resistence. Metall. Trans. 23B, 881–882.

Niyama, E., Uchida, T., Morikawa, M., Saito, S., 1982. AFS Cast Met. Res. J. 7, 52.

Ohnaka, I., 1986. Trans. ISIJ 26, 1045.

Ohnaka, I., 1991. Freezing and melting heat transfer in engineering (Chapter 21). In: Ching, K.C., Seki, H. (Eds.), Solidification Analysis of Casting. Hemispher Pub.

Okamoto, T.K., Kishitake, K., 1975. J. Cryst. Growth 29, 137.

Okamoto, T.K., Kishitake, K., Bessho, I., 1975. J. Cryst. Growth 29, 131.

Oldfield, W., 1973. Mater. Sci. Eng. 11, 211.

Optomec on http://www.optomec.com/.

Oxtoby, D.W., Haymet, A.D.J., 1982. J. Chem. Phys. 76, 6262.

O'Rourke, P.J., Haworth, D.C., Ranganathan, R., 1997. Asm Handbook. ASM, Materials Park, OH, 20.

Pan, Q.Y., Apelian, D., Jorstad, J.L.J., 2008. Asm Handbook, 15. ASM International, Materials Park, OH, 761.

Parisi, A., Plapp, M., 2010. Defects and multistability in eutectic solidification patterns, EPL, 90 26010.

Pelini, W.S., 1953. Trans. AFS 61, 61.

Perepezko, J.H., 1984. Nucleation Undercooled Liquids. Mater. Sci. Eng. 65, 125–135.

Perepezko, J.H., 1988. Metals Handbook (ASM, Metals Park, OH). In: Casting, 15. ASM, Metals Park, OH, p. 101.

Perepezko, J.H., 2008. Nucleation kinetics and grain refinement. In: Asm Handbook. Casting, vol. 15. ASM International, pp. 276–287.

Perepezko, J.H., Anderson, I.E., 1980. In: Machlin, E.S., Rowland, T.S. (Eds.), Synthesis and Properties of Metastable Phases. TMS-AIME, Warrendale, PA, p. 31.

Perepezko, J.H., Boettinger, W.J., 1983. Mater. Res. Soc. Symp. Proc. 19, 223.

Peteves, S.D., Abbaschian, G.J., 1986. J. Cryst. Growth 79, 775.

Pfann, W.G., 1952. Trans. Metall. Soc. AIME 194, 747.

Pfann, W.G., 1966. Zone Melting, second ed. Wiley, New York.

Pilling, J., Hellawell, A., 1996. Mechanical deformation of dendrites by fluid flow. Metall. Trans. A 27, 229–232.

Pires, S., Prates, M., Biloni, H., 1974. Z. Metallkd. 65, 143.

Plapp, M., 2007. Three-dimensional phase-field simulations of directional solidification. J. Cryst. Growth 303, 49–57.

Plapp, M., Karma, A., 2002. Eutectic colony formation: a phase-field study. Phys. Rev. E 66, 061608.

Poirier, D.R., 1987. Permeability for flow of interdendritic liquid in columnar-dendritic alloys. Metall. Trans. 18B, 245–255.

Poirier, D.R., Yeum, K., Mapples, A.L., 1987. Metall. Trans. 18A, 1979.

Powell, G.L.F., 1980. Met. Forum 3, 37.

Prates, M., Biloni, H., 1972. Metall. Trans. 3, 1501.

Priede, J., Gerbeth, G., 2005. Breakdown of Burton Prim Schlichter approach and lateral solute segregation in radially converging flows. J. Cryst. Growth 285, 261–269.

Prince, A., 1966. Alloy, Phase Equilibria. Elsevier, Amsterdam.

Provatas, N., Goldenfeld, N., Dantzig, J., 1998. Phys. Rev. Lett. 80, 3308.

Qiyang, L., Qing Chun, L., Qifu, L., 1991. Acta Metall. Mater. 39, 2497.

Rabiei, A., O'Neill, A.T., 2005. Mater. Sci. Eng. A 404, 159.

Rappaz, M., 1989. Int. Mater. Rev. 34, 93.

Rappaz, M., Boettinger, W.J., 1999. On dendritic solidification of multicomponent alloys with unequal liquid diffusion coefficients. Acta Mater. 47, 3205.

Rappaz, M., Gandin, Ch.A., 1993. Acta Metall. Mater. 41, 345.

Rappaz, M., Gandin, C.H.A., 1994. Mater. Res. Bull. no 1, 20.

Rappaz, M., Stefanescu, D.M., 1988. Metals Handbook. In: Casting, ninth ed. 15. ASM, Metals Park, OH, 883.

Rappaz, M., David., S.A., Vitek, J.M., Boatner, L.A., 1990. Metall. Trans. 21A, 1767.

Rhines, F., 1956. Phase Diagrams in Metallurgy: Their Development and Application. McGraw-Hill.

Richmond, J.J., Perepezko, J.H., Lebeau, S.E., Cooper, K.P., 1983. In: Mehrabian, R. (Ed.), Rapid Solidification Processing: Principles and Technologies III. NBS, Washington, DC, p. 90.

Rickert, A., Engler, S., 1985. In: Fredriksson, H., Hiller, M. (Eds.), The Physical Metallurgy of Cast Iron, Proceedings of the MRS, North Holland, 34, p. 165.

Ridder, S.D., Kou, S., Mehrabian, R., 1981. Effect of fluid-flow on macrosegregation in axi-symmetric ingots. Metall. Mater. Trans. B 12, 435–447.

Riposan, T., Chisamera, M., Sofroni, L., Brabie, V., 1985 (North Holland). In: Fredricksson, H., Hillert, M. (Eds.), Physical Metallurgy of Cast Iron, p. 131.

Rivulcaba, D., Mathiesen, R.H., Eskin, D.G., Arnberg, L., Katgerman, L., 2009. In-situ analysis of coarsening during directional solidification experiments in high-solute aluminum alloys. Metall. Mater. Trans. B 40, 312.

Rivullcaba, D., Mathiesen, R.H., Eskin, D.G., Arnberg, L., Katgerman, L., 2007. In-situ observations of dendritic fragmentation due to local solute-enrichment during directional solidification of an aluminum alloy. Acta Mater. 55, 4287–4292.

Rodway, G.H., Hunt, J.D., 1991. Thermoelectric investigation of solidification of lead I. Pure lead. J. Cryst. Growth 112, 554.

Rodway, G.H., Junt, J.D., 1989. J. Cryst. Growth 97, 680.

Ross, A.B. De, Mondolfo, L.F., 1980. In: Pampillo, C.A., Biloni, H., Embury, D.E. (Eds.), Aluminum Transformation Technology and Applications. ASM, Metals Park, OH, p. 81.

Roviglione, A., Hermida, H.D., 1994. Mater. Charact. 32, 127.

Sabatino, M.D., 2005. "Fluidity of Aluminum Foundry Aloys", NTNU doctoral thesis, p. 185.

Sabau, A.S., Viswanathan, S., 2002. Metall. Mater. Trans. B 33 (2), 243–255.

Sadocha, J.P., Gruzlesky, 1975. In: Metallurgy of Cast Iron. Georgi Pub.Co, St.Saphorin, Switzerland, p. 443.

Salgado-Ordorica, M.A., Rappaz, M., 2008. Acta Mater. 56, 5708–5718.

Sanjay, K., Sisa, P., 2011. Adv. Mater. Res. 227, 92–95.

Sankaranarayanan, A., Tyler, D.E., Watson, W.G., Cheskis, H.P., List, G.A., 1995. "Vertical Casting Process", US. Patent 5,381,847.

Sato, T., Ohira, G., 1977. J. Cryst. Growth 44, 78.

Savage, S.J., Froes, F.H., 1984. J. Met. 36 (4), 20–33.

Schaefer, R.J., Clicksman, M.G., 1970. Metall. Trans. 1, 1973.

Schaefer, R.J., Coriell, S.R., 1984. Convect1on-Induced Distortion of a Solid-Liquid Interface. Metall. Trans. 15A, 2109.

Scheil, E., 1942. Z. Metallkd. 34, 70.

Schlichting, H., 1979. Boundary Layer Theory, seventh ed. McGraw Hill, New York.

Schneider, M.C., Beckermann, C., 1995. Simulation of micro-/macrosegregation during the solidification of a low-alloy steel. ISIJ Int. 35, 665–672.

Schneider, M.C., Gu, J.P., Beckermann, C., Boettinger, W.J., Kattner, U.R., 1997. Modeling of micro-/macrosegregation and freckle formation in single-crystal nickel-base superalloy directional solidification. Metall. Mater. Trans. A 28, 1517–1531.

Schwartz, M., Karma, A., Eckler, K., Herlach, D.M., 1994. Physical mechanism of grain refinement in solidification of undercooled melts. Phys. Rev. Lett. 73, 1380–1384.

Seetharaman, V., Trivedi, R., 1988. Eutectic growth: Selection of interlamellar spacings. Metall. Trans. 19A, 2955.

Sekerka, R.F., 1967. In: Peiser, H.S. (Ed.), Crystal Growth. Pergamon press, Oxford, p. 691.

Sekerka, R.F., 1986. Am. Assoc. Cryst. Growth Newslett. 16, 2.

Shangguan, D., Ahuja, S., Stefanescu, D.M., 1992. Metall. Trans. 23A, 669.

Shapovalov, V., 1994. MRS. Bull. 19, 24.

Sharma, D.G.R., Krishman, 1991. AFS Trans. 99, 429.

Sharp, R.M., Flemings, M.C., 1973. Metall. Trans. 4, 997.

Sharp, R.M., Flemings, M.C., 1974. Metall. Trans. 4, 823.

Shibata, H., Yin, H., Yoshinaga, S., Emi, T., Suzuki, M., 1998. In-situ observation of engulfment and pushing of non-metallic inclusions in steel melt by advancing melt/solid interface. ISIJ Int. 38, 149–156.

Sigworth, G.K., 1996. Can. Metall. Q 35, 101–122.

Sikora, J.A., Rivera, G.L., Biloni, H., 1990. In: Lait, J.E., Samarasekera, L.V. (Eds.), Proceedings of the F. Weinberg Symposium on Solidification Processing. Pergamon press, p. 255.

Simone, A.E., Gibson, L.J., 1997. J. Mater. Sci. 32, 451.

Singer, A.R.E., 1991. Mater. Sci. Eng. A 135, 13–17.

Skaland, T., Grong, O., Grong, T., 1993. Metall. Trans. A 24, 2321.

Smith, R.W., 1968. In: The Solidification of Metals. Iron and Steel Inst. Public. 110, London, p. 224.

Smith, P.M., Aziz, M.J., 1994. Acta Metall. Mater. 42, 3515.

Somboonsuk, K., Trivedi, R., 1985. Acta Metall. 33, 1061–1068.

Somboonsuk, K., Mason, J.T., Trivedi, R., 1984. Metall. Trans. 15A, 967.

Soodi, M., Brandt, M., Masood, S., 2010. Adv. Mater. Res. 129–131, 648–651.

Spaepen, F., 1975. Acta Metall. 23, 729.

Spaepen, F., 1994. Homogeneous nucleation and the temperature dependence of the crystal-melt interfacial tension. In: Ehrenreich, H., Turnbull, D. (Eds.), 1994. Solid State Physics, 47, pp. 1–32.

Spaepen, F., Meyer, R.B., 1976. Scr. Metall. 10, 257.

Spinelli, J.E., Ferreira, I.L, Garcia, A., 2005. In: 6th World World Congresses of Structural and Multidisciplinary Optimization, (Rio de Janeiro, Brazil).

Srivastava, V.C., Sahoo, K.L., 2007. Mater. Sci. Poland 25 (3), 733–753.

Steen, H.A.H., Hellawell, A., 1975. Acta Metall. 23, 529.

Stefanescu, D.M., 1988. Metals Handbook. In: Casting, ninth ed. 15. ASM, Metals Park, OH, 168.

Stefanescu, D.M., 2005. Computer simulation of shrinkage related defects in castings - a review. Int. J. Cast Met. Res. 18, 129–143.

Stefanescu, D.M., 2009. Science and Engineering of Casting Solidification, second ed. Springer, NY.

Stefanescu, D.M., 2010. Science and Engineering of Casting Solidification, second ed. Springer, New York.

Stefanescu, D.M., Dhindaw, B.K., 1988. Metals Handbook. In: Casting, ninth ed. 15. ASM, Metals Park, OH, p. 142.

Stefanescu, D.M., Hummer, R., Nechtelberger, E., 1988. Metals Handbook. In: Casting, ninth ed. 15. ASM, Metals Park, OH, 667.

Stefanescu, D.M., Upadhya, G., Bandyopadhyay (Banerjee), D., 1990. Metall. Trans. 21A, 993.

Steinbach, I., Pezzolla, F., Nestler, B., Sesselberg, M., Prieler, R., Schmitz, G.J., Rezende, J.L.L., 1996. A phase field concept for multiphase systems. Physica D 94, 135–147.

Subramanian, S.V., Kay, D.A.R., Purdy, G.R., 1985 (North Holland). In: Fredriksson, H., Hillert, M. (Eds.), Physical Metallurgy of Cast Irons, p. 47.

Sung, P.K., Poirier, D.R., Felicelli, S.D., Poirier, E.J., Ahmed, A., 2001. Simulations of microporosity in IN718 equiaxed investment castings. J. Crystal Growth 226, 363–377.

Suri, V.K., Paul, A.J., El-Kaddah, N., Berry, J.T., 1994. Determination of correlation factors for prediction of shrinkage in castings. Part I: prediction of microporosity in castings: a generalized criterion. AFS Trans. 138, 861–867.

Syed, W., Pinkerton, A.J., Lin, L., 2005. Appl. Surf. Sci. 247, 268.

Takatani, H., Gandin, C., Rappaz, M., 2000. Acta Mater. 48, 675–688.

Tammann, G., Botschwar, A.A., 1926. Z. Anorg. Chem. 157, 27.

Tassa, M., Hunt, J.D., 1976. J. Cryst. Growth 34, 38.

Temkin, D.E., 1964. Crystallization Processes. (Tranl. by Consultants Bureau, New York, 1966) p. 15.

Temkin, D.E., 1969. Sov. Phys. Crystallogr. 14, 344.

Thevoz, Ph., Desbiolles, J.L., Rappaz, M., 1989. Metall. Trans. 20A, 311. See also M. Rappaz, 1989, Int. Mater. Rev. 34, 93.

Thompson, C.V., Spaepen, F., 1983. Acta Metall. 31, 2021.

Tian, H., Stefanescu, D.M., 1993. In: Piwonka, T.S., Voller, V., Katgerman, L. (Eds.), Modeling of Casting, Welding, and Advanced Solidification Processes-vi. TMS, Warrendale, PA.

Tiller, W.A., 1958. In: Liquid Metals and Solidification. ASM, Metals Park, OH, p. 276.

Tiller, W.A., Rutter, J.W., 1956. Can. J. Phys. 34, 96.

Tiller, W.A., Jackson, K.A., Rutter, J.W., Chalmers, B., 1953. Acta Metall. 1, 453.

Tims, M.L., 2002. In: Yu, Kuang-O. (Ed.), Modeling for Casting and Solidification Processing. Marcel Dekker, New York, p. 417.

Tolman, R.C., 1949. J. Chem. Phys. 17, 333.

Toloui, B., Hellawell, A., 1976. Acta Metall. 24, 565.

Tong, X., Beckermann, C., 1998. A diffusion boundary layer model of microsegregation. J. Cryst. Growth 187, 289–302.

Trajan, P.K., 1988. Metals Handbook. In: Casting, ninth ed. 15. ASM, Metals Park, OH, p. 88.

Trivedi, R., 1995. Theory of layered-structure formation in peritectic systems. Metall. Mater. Trans. A 26, 1583–1590.

Trivedi, R., Kurz, W., 1988. In: Stefanescu, D.M., Abbaschian, G.J., Bayuzick, R.J. (Eds.), Solidification Processing of Eutectic Alloys. The Metallurgical Society, Warrendale, PA, p. 3.

Trivedi, R., Somboonsuk, K., 1984. Mater. Sci. Eng. 65, 65.

Trivedi, R., Magnin, P., Kurz, W., 1987. Acta Metall. 35, 971.

Trivedi, R., Shen, Y., Liu, S., 2003. Cellular-to-dendritic transition during the directional solidification of binary alloys. Metall. Mater. Trans. A 34, 395–401.

Trojan, P.K., Fruehan, R., 2008. In: Asm Handbook, vol. 15, pp. 74–83.

Tszeng, T.C., Im, Y.T., Kobayashi, S., 1989. Thermal-analysis of solidification by the temperature recovery method. Int. J. Mach. Tools Manuf. 29, 107–120.

Turnbull, D., 1950. Kinetics of heterogeneous nucleation. J. Chem. Phys. 18, 198–203.

Turnbull, D., 1962. J. Chem. Phys. 66, 609.

Turnbull, D., Bagley, B.G., 1975. In: Hannay, N.B. (Ed.), 1975. Treatise on Solid State Chemistry, vol. 5. Plenum, NY, p. 513.

Turnbull, D., Cech, R.E., 1950. J. Appl. Phys. 21, 804.

Tyler, D.E., Watson, W.G., 1995. In: Proceedings of the 1995 International Conference & Exhibition on powder Metallurgy and Particulate Materials, pp. 7-99–7-107.

Ungar, L.H., Brown, R.A., 1985. Phys. Rev. B 31, 5931.

Upadhya, G., Banerjee, D.K., Stefanescu, D.M., Hill, J.L., 1990. Heat treatment-solidification kinetics modeling of structural transitions: chill formation in gray iron. AFS Trans. 156, 699–706.

van de Walle, A., Ceder, G., 2002. Automating first-principles phase diagram calculations. J. Phase Equilib. 23, 348–359.

Vandyoussefi, M., Kerr, H.W., Kurz, W., 2000. Two-phase growth in peritectic Fe–Ni alloys. Acta Mater. 48, 2297–2306.

Vayrynen, P., Wang, S., Louhenkilpi, S., Holappa, L., October 25–29, 2009. Mater. Sci. Technol. (Pittsburgh, PA).

Verhoeven, J.D., Gibson, E.D., 1978. J. Mater. Sci. 13, 1576.

Verhoeven, J.D., Gill, W.N., Puszynski, J.A., Ginde, R.M., 1988. J. Cryst. Growth 89, 205.

Verhoeven, J.D., Gill, W.N., Puszynski, J.A., Ginde, R.M., 1989. J. Cryst. Growth 97, 254.

Voller, V.R., 1998. Can. Metall. Q 37, 169.

Volmer, M.I., Marder, M., 1931. Z. Phys. Chem. A 154, 97.

Voorhees, P.W., 1990. Metall. Trans. 21A, 27.

Walton, D., Chalmers, B., 1959. Trans. Metall. Soc. AIME 188, 136.

Wang, C.Y., Beckermann, C., 1993. Mater. Sci. Eng. A 171, 199.

Wang, C.Y., Beckermann, C., 1996. Metall. Mater. Trans. A 27, 2765–2783.

Warren, J.A., Boettinger, W.J., 1995. Prediction of dendritic growth and microsegregation patterns in a binary alloy using the phase-field method. Acta Metall. Mater. 43, 689–703.

Warren, J.A., Langer, J.S., 1990. Phys. Rev. A 42, 3518–3525.

Warren, J.A., Langer, J.S., 1993. Phys. Rev. E 47, 2702–2712.

Warren, J.A., Kobayashi, R., Lobovsky, A.E., Carter, W.C., 2003. Extending phase field models of solidification to polycrystalline materials. Acta Mater. 51, 6035–6058.

Warren, J.A., Pusztai, T., Kornyei, L., Granasy, L., 2009. Phys. Rev. B 79, 014204.

Weinberg, F., 1975. Metall. Trans. 6A, 1971.

Weinberg, F., Teghtsoonian, E., 1972. Metall. Trans. 3, 93.

West, D.F.R., 1982. Ternary Equilibrium Diagrams, second ed. Chapman and Hall, London.

Wetterfall, S.E., Fredricksson, H., Hillert, M., 1972. J. Iron Steel Inst. 210, 323.

Wheeler, A.A., Boettinger, W.J., McFadden, G.B., 1992. A phase field model for isothermal phase transitions in binary alloys. Phys. Rev. A 45, 7424–7239.

Wheeler, A.A., McFadden, G.B., Boettinger, W.J., 1996. Phase-field model for solidification of a eutectic alloy. Proc. R. Soc. London, Ser. A 452, 495–525.

White, C.W., Zehner, D.M., Campisano, S.U., Cullis, A.G., 1983. In: Poate, J.M., Foti, G., Jacobson, D.C. (Eds.), Surface Modification and Alloying by Lasers, Ion, and Electron Beams. Plenum Press, NY, p. 81.

Willnecker, R., Herlach, D.M., Feuerbacher, B., 1989. In: Proc. 7th. Europ. Symp. on "Materials and Fluid Sciences under Microgravity, SP-295. ESA, Oxford, p. 193.

Willnecker, R., Herlach, D.M., Feuerbacher, R., 1990. Appl. Phys. Lett. 56, 324.

Wilson, H.A., 1900. Philos. Mag. 50, 238.

Wilson, L.O., 1978. J. Cryst. Growth 44, 371.

Wilson, L.O., 1980. J. Cryst. Growth 48, 363.

Wollkind, D., Segal, L., 1970. Philos. Trans. R. Soc. London 268, 351.

Worster, M.G., 1992. Instabilities of the liquid and mushy regions during solidification of alloys. J. Fluid Mech. 23, 649–669.

Wu, M., Ludwig, A., 2009. Acta Mater. 57, 5632–5644.

Wu, Y., Piconne, T.J., Shiohara, Y., Flemings, M.C., 1987. Metall. Trans. 18A, 915.

Wuebben, T., Odenbach, S., Banhart, J., 2000. In: Zitha, P., Banhart, J., Verbist, G. (Eds.), Proc. Eurofoam 2000. MIT-Verlag, Bremen, pp. 98–103.

Wuebben, T., Stanzick, H., Banhart, J., Odenbach, S., 2003. J. Phys. Condens. Matter 15, S427.

Xu, R., Naterer, G.F., 2004. Int. J. Heat Mass Transfer 47 (22), 4821.

Yang, Y., Humadi, H, Buta, D, Laird, B.B, Sun, D, Hoyt, J.J, Asta, M, 2011. Atomistic simulations of non-equilibrium crystal-growth kinetics from alloy melts, Phys. Rev. Lett. 107, 025505.

Yasuda, H., et al., 2004. Mater. Lett. 58, 911.

Yavari, A.R., Ota, K., Georgarakis, K., LeMoulec, A., Charlot, F., Vaughn, G., Greer, A.L., Inoue, A., 2008. Chilled zone copper with the strength of stainless steel and tailorable color. Acta Mater. 56, 1830–1839.

Yin, H., Felicelli, S.D., 2010. Acta Mater. 58, 1455–1465.

Yoon, W., Paik, J.S., Lacourt, D., Perepezko, J.H., 1986. J. Appl. Phys. 60, 3489.

Young, K.P., Kirkwood, D.H., 1975. Metall. Trans. 6A, 197.

Zener, C., 1946. Trans. Metall. Soc. AIME 167, 550.

Zeppelin, F., Hirsher, M., Stanzick, H., Banhart, J., 2003. Compos. Sci. Technol. 63, 2293.

Zhang, X., Roelofs, H., Lemgen, S., Urlau, U., Subramanian, S.V., 2004. Steel Res. Int. 75 (5), 314–321.

Zhao, Y.Y., Sun, D.X., 2001. Scr. Mater. 44, 105.

Zhu, J.D., Ohnaka, I., 1991. In: Rappaz, M., Dabu, M.R., Mahin, K.W. (Eds.), Modelling of Casting, Welding and Advanced Solidification Processes, p. 435.

Zhu, P., Sha, R., Li, Y., 1985. The physical metallurgy of cast iron. In: Fredriksson, H., Hiller, M. (Eds.), Proceedings of the MRS, North Holland, 34, p. 3.

Zimmermann, M., Karma, A., Carrard, M., 1990. Phys. Rev. B 42, 833.

Ziv, I., Weinberg, F., 1989. Metall. Trans. 20B, 731.

Biography

William J. Boettinger is a NIST Fellow (Emeritus) in the Materials Science and Engineering Division of the Material Measurement Laboratory at NIST in Gaithersburg, MD. Before joining the permanent staff at NBS/NIST, he was a NRC/NAE postdoctoral research associate at NBS from 1972–1974. From 1977 until 1995 he held a concurrent position as a professorial lecturer at the George Washington University where he taught a graduate course in phase transformations.

His expertise includes the thermodynamics and kinetics of metallurgical systems especially the relationship of alloy microstructure to processing conditions. He has over one hundred seventy publications on topics including dendritic, eutectic and peritectic solidification, rapid solidification, phase diagrams, diffusion, intermetallic compounds, measurement of crystal perfection using x-rays, soldering and Sn whisker growth.

Dr. Boettinger was awarded the Department of Commerce Bronze Medal in 1980, Silver Medals in 1983 and 1994, and Gold Medals in 1999 and 2003. He received the Materials Science Division Award of the American Society for Materials (ASM) in 1989 and was made a fellow of that society in 1994. He co-chaired the 1994 Physical Metallurgy Gordon Research Conference with J. H. Perepezko. He has received several honors from The Minerals, Metals and Materials Society (TMS): the 1999 Champion Mathewson Best Paper Award, the 2001 TMS Bruce Chalmers Award for research in solidification and a Fellow in 2006. He was named a NIST Fellow in 2001, elected a member of the National Academy of Engineering in 2006 and received the federal government's Presidential Rank Award in 2007.

Dr. Dilip K. Banerjee is a research engineer in the Mechanical Performance Group of the Materials Measurement Laboratory (MML) at the National Institute of Standards and Technology (NIST). He received his Ph.D. in Materials Engineering from The University of Alabama. He received his MBA from the Robert H. Smith School of Business at the University of Maryland, College Park.

Dr. Banerjee worked as a senior scientist in UES, Inc. where he was a developer of the commercial finite element casting simulation software, ProCAST. He later joined the General Electric Company as a leader of the software quality group in its Information Systems Division.

Dr. Banerjee joined NIST as a program manager in the Advanced Technology Program (ATP). He served as a co-chair of the ATP's proposal selection board in 2004 and later served as a senior program manager in the Standards Services Division at NIST. Subsequently, he worked in the Engineering Laboratory at NIST where his research work involved studying the effects of fire in structures. Dr. Banerjee has an extensive background in numerical modeling. He worked in the technical team to investigate the failure of the World Trade Center Building 7.

His current research effort is focused on developing accurate computational models in support of efforts of NIST Center of Automotive Lightweighting (NCAL) to develop the measurement methodology and analysis necessary for the U.S. auto industry to transition to advanced lightweighting materials. Dr. Banerjee received the "Outstanding Achievement Award" from the General Electric Company. He is also the recipient of the "The John Deere Outstanding Graduate Student Award" from The University of Alabama. He was recognized for outstanding services by the NIST/ATP Director.

Dr. Banerjee was a U.S. Department of Commerce Science and Technology Fellow and is an American Society for Quality Certified Software Quality Engineer. He serves on the ASCE (American Society of Civil Engineers) Fire Protection Technical Committee. He is also a General Electric Company certified "Six Sigma Green Belt."

8 Diffusional Phase Transformations in the Solid State

W.A. Soffa, Department of Materials Science and Engineering, University of Virginia, USA
David E. Laughlin, Department of Materials Science and Engineering, Carnegie Mellon University, Pittsburgh, PA, USA

8.1 Introduction

8.1.1 General Concepts: Classification of Phase Transformations

This chapter is about diffusional phase transformations. It is appropriate to first define the terms **phase** and **phase transformation**. A phase is a physically distinct homogeneous portion of a thermodynamic system delineated in space by a bounding surface, called an interphase interface, and distinguished by its state of aggregation (solid, liquid or gas), crystal structure, composition and/or degree of order. Each phase in a material system generally exhibits a characteristic set of physical, mechanical and chemical properties and is, in principle, mechanically separable from the whole.

A phase transformation in a material system occurs when one or more of the phases in a system changes their state of aggregation, crystal structure, degree of order or composition resulting from a reconfiguration of the constituent particles (atoms, molecules, ions, electrons, etc.) comprising the phase. This reconfiguration is a change in the thermodynamic state leading to a more stable condition described by appropriate thermodynamic potentials such as a decrease in the Gibbs free energy (G) at constant temperature (T) and pressure (P). Whether describing the freezing of a metal or the onset of ferromagnetism in iron (Fe), a change in phase is indicated when small changes in relevant thermodynamic variables produce marked changes and sometimes dramatic qualitative changes in the nature of the system. These changes can occur abruptly (discontinuously) or gradually (continuously) at critical values of certain thermodynamic variables. The decrease in free energy accompanying the reconfiguration is often referred to as the thermodynamic "driving force" for the phase change.

In his well-known collected works Gibbs formulated the fundamental conditions for equilibrium of phases in a thermodynamic system based on a simple principle of mathematics. If a system has a total of x variables which are related by r equations, the number of independent variables which can be altered without changing the state of equilibrium is $x - r$. We call the parameter $F = x - r$ the *variance* or *degrees of freedom of the system*. If the state of an individual phase is completely determined by its temperature, pressure and composition then the total number of variables associated with a system composed of P phases is $P(C+1)$ wherein $(C-1)$ composition variables must be specified to describe the composition of a phase. If the temperature and pressure are uniform throughout the system and the chemical potentials of the individual chemical species are equal within the coexisting phases, the number of equations relating the variables is $(P-1)(C+2)$; thus, the variance of the system is $F = P(C+1) - (P-1)(C+2) = C+2-P$. This is the famous *Gibbs phase rule* which is usually written as $P + F = C + 2$. Care must be exercised in properly specifying the number of components C; it is not always simply the number of elements present in the system. The variance or degrees of freedom can be understood as the number of thermodynamic variables that can be changed arbitrarily (within limits) by an experimenter without causing the disappearance of phases or the appearance of new phases while probing the system. It should be mentioned that if the chemical species are reactive and there are R independent reaction equilibria among the N species, the variance is then given by $F = C - R + 2 - P$. Finally, it should be clear that the phase rule $P + F = C + 2$ will be altered if other thermodynamic variables are essential in the description of a phase such as electric and magnetic fields as well as stress.

8.1.2   Diffusional and Nondiffusional Transformations

Phase transformations in the solid state may be divided into two broad categories: (1) those requiring the movement of atoms over distances of the order of atomic spacings or greater effected by essentially stochastic thermally activated jumps down a chemical potential gradient often accompanied by significant redistribution of solute and composition changes and (2) those whose atoms at the boundary of the new growing phase are displaced in a synchronous and coordinated fashion, over fractions of atomic spacings. The former are called diffusional transformations and are the subject of this chapter. The latter are often called displacive transformations (or sometimes martensitic). Following Frank, Professor Christian has referred to the former type of transformations as civilian (they occur with little or no atomic coordination) and the latter as military transformation, since the atomic motions occur cooperatively and in step with each other.

8.1.3   Replacive and Reconstructive Transformations

Diffusional transformations can be classified as either replacive or reconstructive (Buerger, 1948, 1951). In the former, the atomic jumps merely rearrange the positions of the species on the underlying lattice whereas in the latter the new phase that arises generally has a very different crystal structure. Replacive diffusional transformations are of two types: those in which the solute atoms cluster together to form a solute-enriched phase and those in which the solute and solvent atoms form an atomically ordered arrangement which is crystallographically related to the parent phase. (Of course, ordering of the phase could occur after the clustering step or vice versa. See Section 8.6.3 subsequently.) Replacive transformations may occur with complete coherency between the new phase and its matrix phase. On the other hand, reconstructive diffusional transformations involve an atom by atom disassembling of the parent phase and an atom by atom assembling of a new phase. The new phase generally forms with an interface that is incoherent with that of the matrix. An incoherent interface is one in which there is no systematic matching of atomic planes at the interface boundary.

   Diffusional transformations can differ from each other with respect to their thermodynamic character, resultant microstructures and kinetics. In the following section we give an overview of such classifications.

8.1.4 Discontinuous and Continuous Transformations: Gibbs (1875, 1876, 1877, 1878)

In well-known (but seldom read) papers Gibbs (1875, ff) distinguished between two fundamentally distinct types of phase transformations, namely those that are initiated by fluctuations that are large in degree but small in spatial extent and those that are initiated by fluctuations that are small in degree and relatively large in spatial extent.

(A phase) "...may be capable of continued existence in virtue of properties which prevent the commencement of *discontinuous* changes. But a phase which is unstable in regard to *continuous* changes is evidently incapable of permanent existence on a large scale except in consequence of passive resistances to change..." (Gibbs, 1875, ff, our emphasis).

Here, Gibbs contrasts changes in what we call today an order parameter, such as the difference in composition, atomic order or magnetization within a phase, with changes that result in the formation of an *entirely* new phase. A single phase may not be able to continuously change via infinitesimal fluctuations of order parameter, but could change if a large enough fluctuation in the order parameter were to appear. This kind of transformation is called by Gibbs a *discontinuous change* (transformation). This kind of transformation contrasts with *continuous* changes or *transformations*. Continuous transformations require only infinitesimal changes in order parameter to proceed and initially occur within the existing phase. *Discontinuous transformations* are those which require large fluctuations in order parameter and hence they form a new phase clearly distinguishable from the initially existing one.

Thus, Gibbs introduces two types of phase transformations: those which are initiated by fluctuations that are small in spatial extent but large in degree (*discontinuous*) and those which are initiated by fluctuations that are small in degree but large in spatial extent (*continuous*).

From the time of this article by Gibbs, until 1956, nearly all research on the initiation of phase transformations was done in terms of *discontinuous transformations*, the so-called nucleation and growth transformations. In 1956, Mats Hillert, working on his graduate degree at MIT, ushered in the spinodal era by discussing decomposition by a *continuous transformation* which was later termed by Cahn as "spinodal decomposition" (Cahn, 1961).

The distinction between continuous and discontinuous transformations includes aspects of thermodynamics, kinetics and microstructure. **Thermodynamics** is included in the use of such terms as stable and unstable for the phases under consideration. Also, the changes in the thermodynamic functions may be continuous or discontinuous with the advancement of the transformation. The **kinetics** of the two types of transformation are different in the sense that in a continuous transformation there is no sharp change in the kinetics as a function of time, whereas for discontinuous transformations (those that are diffusional) the reaction often starts off slowly and speeds up with time. See Section 8.3 subsequently. The **microstructural differences** are included in the idea of the extent of the spatial range of the fluctuations. For continuous changes the initiation of the change occurs within a single phase, that is, there is no sharp interface between two incipient phases. Also it occurs in all regions of the sample, not just localized regions as in nucleation events.

8.1.5 Homogeneous and Heterogeneous Transformations: Christian (2002)

Another classification of phase transformations has been given by Professor J. W. Christian in his classic treatise (Christian, 2002). *Heterogeneous* transformations involve the spatial partitioning of the system into regions that *have transformed* and regions which *have not transformed*. Such regions are separated by an interphase interface. *Homogeneous* transformations occur uniformly throughout the entire system and the changes occur continuously in time as well as uniformly in space. Therefore, all *homogeneous* transformations are *continuous according to this viewpoint.*

This classification is similar to that of Gibbs discussed earlier. All nucleation and growth processes are heterogeneous transformations, just as they are discontinuous transformations. It should be noted that homogeneous nucleation is a heterogeneous transformation. Homogeneous transformations are continuous.

8.1.6 First- and Higher Order Phase Transitions: Ehrenfest (1933)

In the early 1930s, Ehrenfest (1933) suggested a very useful scheme for distinguishing different types of phase transformations/transitions based on the behavior of certain thermodynamic variables in the vicinity of the phase change. The ORDER of a phase transition in this thermodynamic classification is defined according to the lowest differential/derivative of the relevant thermodynamic potential such as the Gibbs free energy which is discontinuous. The discontinuous first derivatives of the Gibbs free energy which define a FIRST-ORDER transition are as follows:

$$\left(\frac{\partial G}{\partial T}\right)_P = -S \quad \left(\frac{\partial G}{\partial P}\right)_T = V \quad \left(\frac{\partial\left(\frac{G}{T}\right)}{\partial\left(\frac{1}{T}\right)}\right)_P = H$$

for which it emerges that the entropy S is discontinuous at the transition temperature T_t and the phase change occurs isothermally with a discontinuity in H and a so-called latent heat of transformation ΔH_t. It follows from the basic definition of a heat capacity that the heat capacity or specific heat C_P will exhibit an infinite discontinuity at the transition temperature. The following second derivatives of the free energy are discontinuous at the onset of a SECOND-ORDER transition:

$$\left(\frac{\partial^2 G}{\partial T^2}\right)_P = -\frac{C_P}{T} \quad \left(\frac{\partial^2 G}{\partial P^2}\right)_T = -\beta V \quad \left(\frac{\partial H}{\partial T}\right)_P = C_P$$

where $-\frac{1}{V}\left(\frac{\partial^2 G}{\partial P^2}\right)_T = \beta$ is the compressibility. Such transitions have no latent heat nor coexistence of phases at the transition temperature. The specific heat/heat capacity at the transition temperature shows a finite discontinuity according to the Ehrenfest thermodynamic classification. SECOND-ORDER transitions are rare in nature: however, the transition from the normal state to the superconducting state in zero magnetic field is the textbook example of such a transition. The majority of phase changes in materials are FIRST ORDER but there is a plethora of transitions which are clearly *not* FIRST ORDER or SECOND ORDER; that is, they are third order or higher. However, in practice, it is extremely difficult to analyze behavior explicitly associated with these higher order derivatives; thus, we find it useful to talk about FIRST-ORDER, SECOND-ORDER and HIGHER ORDER transitions (sometimes called λ transitions). All higher order transitions occur without a latent heat similar to second-order behavior but display markedly different behaviors with respect to their signature discontinuities in C_P. The nature of these singularities is yet to be fully resolved theoretically.

8.1.7 Landau Classification

In Section 2.5 a detailed discussion of the Landau approach and classification of phase transitions will be presented. In the Landau method the free energy is written as a Taylor series expansion with respect to one or more order parameters which characterize the system and from this analysis two distinct types of transformation behavior become evident: those in which the curvature of the free energy of the high-temperature phase becomes zero at the transition temperature T_C and negative immediately below and those for which the curvature is positive at the transition temperature but may become negative at temperatures well below the equilibrium transition temperature. The former are identical to the second-order (continuous) transitions of Ehrenfest whereas the latter are the same as first-order (discontinuous) transitions discussed earlier. Far below the equilibrium transition T_C the first-order transition may exhibit thermodynamic (and kinetic) behavior associated with a continuous transformation under these nonequilibrium conditions since at a temperature $T_i^- < T_C$ the curvature of the free energy functional of the disordered phase becomes negative rendering the disordered state unstable. The temperature T_i^- is an instability temperature and it should be noted that for second-order transitions $T_C = T_i^-$.

8.1.8 Summary

This chapter deals with solid-state transformations in which atomic diffusion plays a major role. In this brief introductory summary we have distinguished between transformations that are continuous and homogeneous and those which are discontinuous and heterogeneous in nature. More details of these transformation modes will be discussed in the following sections. All second/higher order transitions are expected to proceed continuously and homogeneously. All first-order transitions might be expected to occur discontinuously and heterogeneously at and in the vicinity of the equilibrium transition temperature. However, well below the equilibrium transition temperature the intrinsically first-order transition can occur continuously and homogeneously under these nonequilibrium conditions.

8.2 Energetics

In the thermodynamics of phase stability the Gibbs free energy is of major importance and utility. This thermodynamic state function generally defined by

$$G = E + \mathrm{PV} - \mathrm{TS} = H - \mathrm{TS} \tag{1}$$

provides a tool for evaluating the relative stability of competing phases and quantitatively evaluating the thermodynamic "driving forces" for phase transformations as a function of relevant thermodynamic parameters such as temperature (T), pressure (P), composition (X) and, in some instances, can be generalized to include applied electric and/or magnetic fields. The internal energy (E) and entropy (S) of a phase can often provide a direct connection with factors relating to atomic structure and interatomic interactions. Initially, we will restrict ourselves to $G = G(T,P,X)$. If $G(T,P,X)$ is known then virtually all the thermodynamic properties of a material system can be deduced, in principle, from familiar derivatives of the Gibbs free energy and related quantities. For this reason the free energy is often called the characteristic function of the system. (The Helmholtz free energy $F = E - \mathrm{TS}$ differs from the Gibbs free energy by the familiar PV term which is only significant at relatively high pressures in the case of condensed phases). At constant T and P systems tend to evolve toward a state of lower free energy and stable equilibrium is characterized by a minimum in the free energy functional. Of course, local minima in the free energy of the system can occur, producing metastable states.

8.2.1 Allotropic/Polymorphic Transformations

One can write the Gibbs free energy of a solid crystalline phase α as

$$G^\alpha(T,P) = G^\alpha(T_0,P_0) + \int_{P_0}^{P} V^\alpha dP + \int_{T_0}^{T} S^\alpha dT \tag{2}$$

using the thermodynamic identities, $\left(\dfrac{\partial G^\alpha}{\partial P}\right)_T = V^\alpha$ and $\left(\dfrac{\partial G^\alpha}{\partial T}\right)_P = -S^\alpha$ where V^α and S^α are the molar volume and entropy of the α phase, respectively; both molar quantities generally are a function of T and P. $G^\alpha (T_0, P_0)$ is the free energy of the α phase in some arbitrary reference state defined by T_0 and P_0. This free energy functional can be used to calculate a free energy surface in the thermodynamic space $G(T.P)$ for the α phase and similarly for the vapor and liquid phases as well as an alternative solid phase β. These free energy surfaces intersect and that portion of each surface which represents the lowest free energy corresponds to the stable phase in a particular P–T range. The intersection of these surfaces produces two or more phases in equilibrium shown in the familiar P–T diagram such as the coexistence of a liquid and solid phase over a range of temperature and pressure along the melting curve. See **Figure 1**.

Along this curve/line in the P–T the free energy of the solid phase $G^S = G^L$, that is, the molar free energies of the solid and liquid phases are equal. Thus, $dG^S = dG^L$ for any movement along the melting curve giving $dG^S = V^S dP - S^S dT = V^L dP - S^L dT = dG^L$ resulting in the well-known Clapeyron equation

$$\frac{dP}{dT} = \frac{S^L - S^S}{V^L - V^S} \tag{3}$$

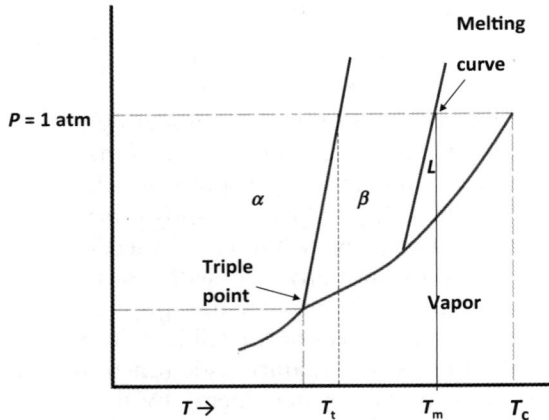

Figure 1   Schematic Pressure–Temperature phase diagram of a material displaying two allotropic solid phases. T_t is the transition temperature from the low-temperature α phase to the higher temperature β phase and T_m is the melting point of β at $P = 1$ atm.

where $(S^L - S^S)$ is the entropy of melting (fusion) and for any point on the curve can be written as

$$\Delta S^{S \to L} = \frac{\Delta H^{S \to L}}{T_m} \tag{4}$$

where $\Delta H^{S \to L}$ is the latent heat of melting (fusion) at the melting point T_m; The quantity $(V^L - V^S)$ is the volume change $\Delta V^{S \to L}$ accompanying the phase change. Clearly, the slope of the melting curve is determined by the relative molar volumes or densities of the liquid and solid phases along the coexistence curve. A similar analysis can be applied to the line representing two-phase equilibrium between allotropic or polymorphic modifications in the $P-T$ diagram as discussed subsequently as well to the sublimation and vaporization curves giving rise to the Clausius–Clapeyron equation.

If we restrict our discussion to isobaric conditions, say $P = 1$ atm, we can write the Gibbs free energy of a phase as a function of temperature as follows:

$$G(T) = H(T) - TS(T) = H_0 + \int_0^T C_P dT - T \int_0^T \frac{C_P}{T} dT \tag{5}$$

or

$$G(T) = H_0 - \int_0^T \left[\int_0^T \frac{C_P}{T} dT \right] dT$$

using the thermodynamic relationship $(\partial G/\partial T) = -S$ with H_0 essentially an integration constant. Of course, H_0 is the enthalpy of the phase at $T = 0$ K which for a solid phase is related to the cohesive energy recalling that $H \approx E$ for condensed phases. It is evident that various contributions to the heat capacity or specific heat play a central role in determining the variation of the free energy of a phase with temperature and thus the relative stability of the phases of the system.

Many materials including a number of important pure metals can exist in different stable crystalline forms at different temperatures and pressures. This phenomenon is called *polymorphism* or *allotropy* and the various crystalline forms are called *polymorphs* or *allotropes*. When the substances are pure elements (Ti, Zr, Sn, Fe, C) the terms allotropy and allotropes are preferred whereas when referring to compounds (SiO_2, ZrO_2, TiO_2, BN) the terms polymorphism and polymorphs are more appropriate. *Allotropic* and *polymorphic transformations* are important in a wide variety of engineering materials but, of course, from a technological point of view the occurrence of allotropy in iron (Fe) is most prominent giving rise to the eutectoid reaction in Fe–C–X alloys which is the basis for the heat treatment of engineering steels. Our discussion of allotropy/polymorphism will deal primarily with pure metals but we shall arrive at a number of important general conclusions.

Let us start our thermodynamic analysis of allotropy with a metal like zirconium (Zr) which is rather straightforward compared with iron (Fe) which will be addressed subsequently because magnetic effects are essentially absent in Zr. At one atmosphere pressure ($P = 1$ atm), pure Zr exhibits a stable hcp (α) structure from 0–1143 K and transforms to a bcc structure (β) which is stable from 1143 K to its melting point at 2125 K. The heat of transformation $\Delta H_t^{\alpha \to \beta}$ is 1040 cal mol^{-1} and the heat of fusion ΔH_f is approximately 4900 cal mol^{-1}. These latent heats are characteristic of the first-order nature of the $\alpha \to \beta$ and $\beta \to L$ transformations. In this one-component system the α, β and L phases are essentially

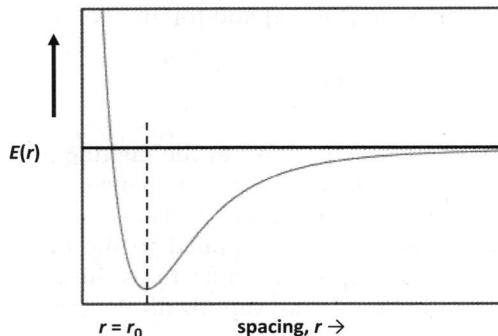

Figure 2 Schematic of the energy vs. atomic separation for an assembly of atoms showing a minimum in the energy at r_0, the equilibrium spacing. This curve is the result of long-range attractive interactions and short-range repulsive interactions.

the thermodynamically "competing" phases; we shall focus our attention on the relative stability of the solid α (hcp) and β (bcc) phases. Using the foregoing results for the free energies of the solid phases we write

$$G^{\alpha}(T) = H_0^{\alpha} + \int_0^T C_P^{\alpha}\, \mathrm{d}T - T \int_0^T \frac{C_P^{\alpha}}{T}\, \mathrm{d}T \tag{6}$$

and

$$G^{\beta}(T) = H_0^{\beta} + \int_0^T C_P^{\beta}\, \mathrm{d}T - T \int_0^T \frac{C_P^{\beta}}{T}\, \mathrm{d}T \tag{7}$$

for the α and β allotropes, respectively. The lead terms are the enthalpies $H_0^{\alpha} \approx E_0^{\alpha}$ and $H_0^{\beta} \approx E_0^{\beta}$ at 0 K of these phases and are related to the cohesive energy of the solids. Imagine that the atoms of the metal are placed on the lattice sites of a bcc lattice with a lattice spacing that is so large that the atoms do not interact. If the lattice spacing is then progressively reduced to the order of the size of the atoms themselves, the energy of the assembly will vary essentially like that shown in **Figure 2** and a minimum will appear in the energy versus interatomic spacing which will determine the equilibrium spacing of the crystal in the bcc arrangement; similarly, for the hcp structure. In general, the nature of the energy versus spacing curve is determined by long-range attractive and short-range repulsive interactions and can often be described by an expression of the form

$$E(r) = -\frac{A}{r^n} + \frac{B}{r^m} \tag{8}$$

where r is a interatomic spacing and A, B, m and n ($m > n$) are constants; $E(r)$ is the change in energy relative to the noninteracting assembly. The minimum in the curve $E(r = r_0) = E_C$ is approximately the cohesive energy of the solid related to the energy required to vaporize the solid into its constituent atoms. The energy of the solid at $T = 0$ K given earlier also includes a zero point vibrational energy, thus

$$E_0 = E_C + E_0^{VIB}$$

where E_0 is a composite term of the cohesive energy term (negative) and the zero point vibrational energy (positive). The cohesive energy of zirconium is approximately 125–150 kcal mol^{-1} and the zero point vibrational energy is estimated to be less than one percent the cohesive or binding energy.

When considering the thermodynamics of allotropy in a metal like Zr, it is clear from the foregoing formulation of the free energies of the α and β phases that the heat capacity of a particular phase is critical in determining the variation of its free energy with temperature. The heat capacity of a solid phase, in general, can be considered to be composed of three parts: C_P^L the so-called lattice heat capacity associated with the thermal vibrations of the constituent atoms or ions about their equilibrium positions, C_P^e the electronic heat capacity arising from thermal excitation of free or conduction electrons in the vicinity of the Fermi level of a metallic solid and C_P^m the magnetic contribution associated with disordering or decoupling of electron spins in ferromagnetic or antiferromagnetic solids. Thus, the total heat capacity of the α and β phases can be written as

$$C_P = C_P^L + C_P^e + C_P^m$$

with the different components varying in a manner specific to the phase in question (α or β). In a paramagnetic metal such as zirconium (Zr), the magnetic contribution to the heat capacity or free energy is considered to be negligible ($C_P^m = 0$).

For our purposes, the lattice or vibrational component of the heat capacity is adequately described by the Debye theory of the specific heat or heat capacity. Theoretical approaches to the vibrational heat capacity generally calculate C_V rather than C_P but they are readily related through the thermodynamic relationship

$$C_P - C_V = \frac{\alpha^2 VT}{\beta}$$

where α is the coefficient of thermal expansion, β the compressibility and V the molar volume, and to a good approximation can be written as

$$C_P = C_V \left(1 + 10^{-4}\ T\right) \text{cal deg}^{-1}\ \text{mol}^{-1}.$$

According to the Debye theory the expression for the lattice heat capacity C_V^L as a function of temperature is given as follows

$$C_V^L = 9R \left(\frac{T}{\theta_D}\right)^3 \int_0^{x_m} \frac{x^4 e^x}{(e^x - 1)^2}\ dx \tag{9}$$

where $x = \dfrac{h\nu}{k_B T}$ and $x_m = \dfrac{h\nu_{max}}{k_B T}$ where ν is a vibrational frequency and ν_{max} is the "cut-off" frequency or maximum vibrational frequency in the vibrational or phonon spectrum of the normal modes characterizing the solid; the parameter $\theta_D = \dfrac{h\nu_{max}}{k_B}$ is called the Debye characteristic temperature and h, k_B

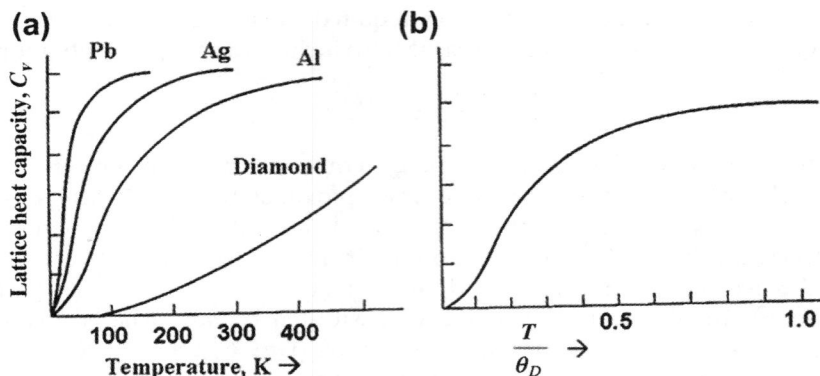

Figure 3 (a) Lattice heat capacity, C_V as a function of the temperature for different materials with different Debye temperatures. For Pb, $\theta_D = 88$ K; Ag, $\theta_D = 215$ K, Al, $\theta_D = 385$ K and Diamond, $\theta_D = \sim 2000$ K. (b). Master lattice heat capacity curve. After Smallman (1963).

and R have their usual meanings. This integral cannot be evaluated analytically; however, extensive tables have been computed numerically for C_V^L as a function of $\frac{T}{\theta_D}$. Also, values of the Debye energy, entropy and free energy have been compiled as a function of $\frac{T}{\theta_D}$. **Figure 3**a and b shows the Debye heat capacity as a function of the temperature for different materials with different Debye temperatures and versus reduced temperature, $\frac{T}{\theta_D}$. When plotted as a function of $\frac{T}{\theta_D}$ the data essentially lie on a master curve in this approximation. At low temperatures $T < 0.02\ \theta_D$, the lattice heat capacity is to a very good approximation given by

$$C_V^L = 234R \left(\frac{T}{\theta_D} \right)^3 \tag{10}$$

in agreement with experiment. This T^3 law is a major triumph of the Debye theory. At high temperatures C_V^L approaches the classical Dulong–Petit limit of $3R$. Note that at the Debye temperature, C_V^L essentially reaches the classical limit (about 0.96 the value of $3R$). Also, it is readily shown that the zero point vibrational energy in the Debye approximation is given by $\frac{9R\theta_D}{8}$.

The Debye frequency $\nu_{max} = \nu_D$ ($\sim 10^{12}$ to $10^{14}\ s^{-1}$) and the Debye temperature $\theta_D = \frac{h\nu_D}{k_B}$ are related to the strength of the bonding and the elastic properties of the solid. If one thinks roughly in terms of the atoms of atomic mass m being coupled by springs of stiffness λ, the characteristic vibrational frequency might be expected to vary as $\left(\frac{\lambda}{m} \right)^{1/2}$. Carbon, which is elastically stiff and composed of relatively light atoms, has a Debye temperature of about 1860 K whereas lead is soft and heavy with a Debye temperature of 102 K. We will see that the Debye temperatures of different allotropic forms can

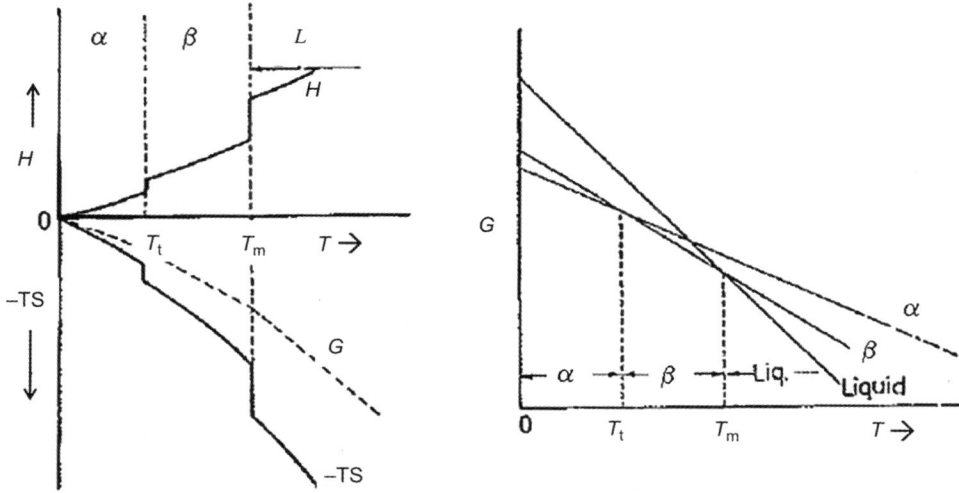

Figure 4   Schematic diagram showing the variation of G, H and −TS as a function of temperature for the α, β and liquid phases in a system exhibiting an allotropic phase change at $T = $ Tt. After Smallman (1963).

play a central role in the relative stability of phases. The Debye temperatures for the allotropes of Zr are approximately θ_D (hcp) = 260 K and θ_D (bcc) = 212 K.

The electronic contribution to the heat capacity for a metallic solid deriving from thermal excitation of the conduction electrons is predicted from the quantum theory of solids to be dependent on the density of states near the Fermi level and given by $C_V^e = \Gamma T$ where Γ is a constant and for most metals is typically about 10^{-3} to 10^{-4} cal deg^{-2} mol^{-1}. Transition metals generally show somewhat higher values because of their relatively high density of states. For example, Γ for Cu and Ag is about 1.5×10^{-4} whereas for γ-Fe and α-Fe, Γ is 8×10^{-4} and 12×10^{-4}, respectively. For zirconium (Zr; 4 d^2 5 s^2), Γ_α(hcp) is 7.1×10^{-4} and Γ_β(bcc) is 4.4×10^{-4} cal deg^{-2} mol^{-1}. It should be clear that the electronic contribution is expected to be important only at very low and very high temperatures.

With the thermodynamic machinery and data described briefly here it is possible to plot the energies (E, H), entropies (S) and free energies (G) of the α and β phases as a function of temperature as shown schematically in **Figure 4**. The thermodynamic properties of the liquid phase can also be computed (the heat capacity of the liquid phase is approximately $C_P = 8.0$ cal mol^{-1} deg^{-1}). The reader is reminded that the slopes of the free energy curves are directly related to the entropies of the competing phases, S^α, S^β and S^L through the thermodynamic identity $\left(\dfrac{\partial G}{\partial T}\right)_P = -S$. It is left to the reader to show how one can estimate the cohesive energy of the β phase in the thermodynamic analysis described earlier.

This thermodynamic analysis of phase stability and allotropy in zirconium (Zr) leads to some very important general conclusions. At low temperatures the more close-packed (hcp) α phase with stronger bonding (greater cohesive energy) tends to be favored because of the relative importance of the energy and entropic terms in the expression for the free energy $G = E + PV - TS$. The more open (bcc) β phase with higher internal energy is the phase of higher vibrational entropy as the temperature is increased. This is made clear by examining **Figure 5a** where the relative heat capacities of the bcc (β) and hcp (α) phases are shown schematically. The heat capacity of the more weakly bonded β phase with its lower

(a)

(b)

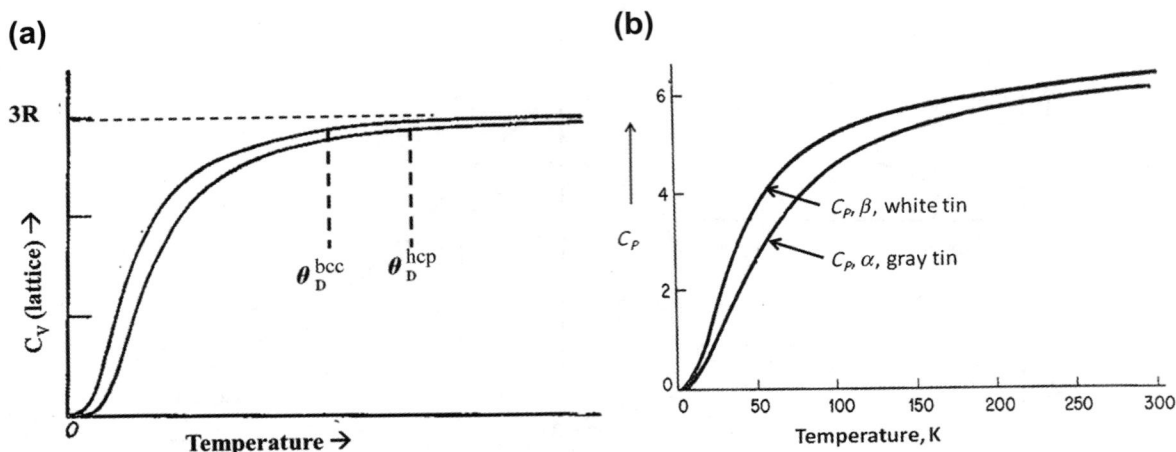

Figure 5 (a) Schematic showing relative lattice heat capacities and Debye temperatures for α (hcp) and β (bcc) allotropes in metals such as titanium (Ti) and Zirconium (Zr). (b) Schematic of the relative heat capacities of gray tin ($\theta_D = 260$ K) and white tin ($\theta_D = 212$ K)

characteristic temperature θ_D (and lower ν_D) lies above that of the more strongly bonded α phase (higher θ_D and ν_D) as they approach the Dulong–Petit limit. The leads to a higher entropy of the β phase eventually stabilizing it as its steeper G versus T curve falls below that of the α phase at high temperatures. Indeed, in many cases, the structure (in metals and nonmetals) of closest packing is favored at low T and a more open structure, one with weaker bonding and greater vibrational entropy, becomes stable at high T. See Th, Ti, Sr, Hf, Tl, Li, Na as well as NaCl-type to CsCl-type transformations. Importantly, these general conclusions cannot be applied in such a straightforward manner when directional bonding and magnetic effects become major considerations.

An interesting case in this regard is allotropy in tin (Sn) where at one atmosphere pressure ($P = 1$ atm) below about 286 K this element is stable in a diamond cubic structure called gray Sn (α) and above this transition temperature the stable form is ordinary Sn (β) possessing the usual metallic properties and exhibiting a body-centered tetragonal structure. The gray Sn is quite brittle and is a semiconductor. The densities of these two allotropes are quite different with $\rho(\alpha)$ and $\rho(\beta)$ being about 5.75 g cm^{-3} and 7.28 g cm^{-3}, respectively; therefore, there is a large volume expansion (approximately 27%) accompanying the $\beta \rightarrow \alpha$ transformation. This large volume expansion renders the transformation very sluggish. Importantly, the lower density diamond cubic form has a Debye temperature $\theta_D(\alpha) = 230$ K compared with $\theta_D(\beta) = 200$ K for the higher density metallic white Sn. See **Figure 5b**. The gray Sn with a coordination $z = 4$ is more strongly bonded than the white Sn $z = 6$; the nearest-neighbor distance in the directionally bonded diamond cubic structure is 0.280 nm compared with 0.302 nm in the metallic bct phase with space group $I\frac{4_1}{a}md$. The metallic phase is the phase which is stabilized by the vibrational entropy above 286 K. The difference in bonding is of central importance here.

The allotropic behavior of iron (Fe) is of paramount importance not only because of its technological significance—it might be said to essentially form the basis of an entire industrial and technological era—but also because it represents a most important example of where magnetic contributions to the free energy have a profound effect on the stability of the competing phases. At one atmosphere pressure, Fe is stable as a body-centered cubic (bcc) structure from 0 K to about 1183 K where it

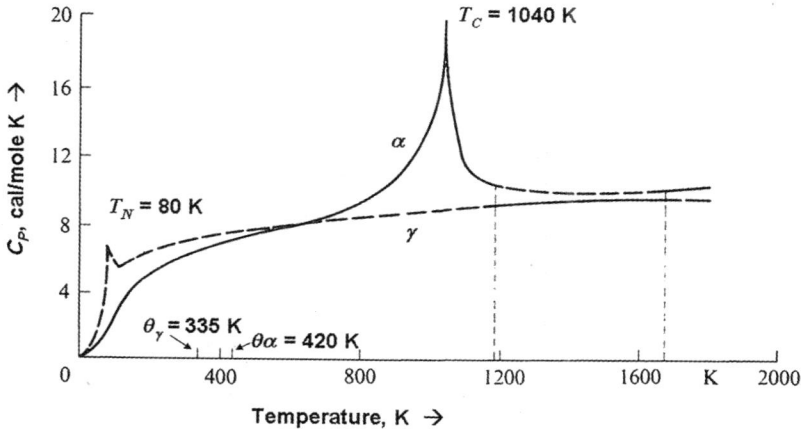

Figure 6 Schematic of the heat capacities of the competing phases in Fe at 1 atm pressure. After Haasen (2001). Note the peak of the γ phase heat capacity at low temperature represents its Neel temperature.

transforms to a face-centered cubic (γ) allotrope which is stable to approximately 1673 K wherein it reverts back to a stable bcc (δ) structure before melting at about 1808 K.

Zener (1955) was one of the first to attempt rigorously to understand the allotropy of Fe in terms of the physical properties and behavior of the solid phases emphasizing the importance of magnetism (C_P^m). Weiss and Tauer (1956) refined these ideas emphasizing the division of the heat capacity explicitly into lattice (C_P^L), electronic (C_P^e) and magnetic (C_P^m) components as discussed earlier. **Figure 6** shows a representation of the heat capacities of the competing phases in Fe at one atmosphere pressure. These authors then took the existing data and calculated the thermodynamic properties (H, S and G) of the α(bcc) and γ(fcc) phases. (Of course, the high temperature δ(bcc) phase is the reemergence thermodynamically of the α phase.) This was followed by an elaborated analysis by Kaufman et al. (1963) emphasizing a more rigorous description of the magnetic properties, particularly of the suggested low-temperature magnetic properties of the γ(fcc). These two works represent the seminal approaches to the subject. At low temperatures where entropy effects are expected to be less important, the ferromagnetic α(bcc) is the phase of lowest internal energy, the magnetic ordering lowers the energy by $\approx RT_C$ (≈ 2000 cal mol^{-1}), where T_C is the Curie temperature (1141 K) of the ferromagnetic phase, but as the temperature increases a magnetic transition (~ 50–80 K) occurs in the competing γ(fcc) phase associated with the antiferromagnetic $\rightarrow$ paramagnetic state at an Néel temperature T_N or a Schottky-like two-level (high-spin/low-spin) specific heat anomaly. This magnetic effect in the γ(fcc) imparts significant entropy to the phase and the $-TS$ term in the free energy stabilizes the γ(fcc) phase with respect to the magnetically ordered ferromagnetic bcc phase above 1183 K (this effect dominates differences in Debye temperature or electronic contributions to the heat capacity). However, between about 775 and 1141 K disordering of the spins in the ferromagnetic state leads to a ferromagnetic $\rightarrow$ paramagnetic transition in the α-Fe and this increased magnetic entropy stabilizes the α phase with respect to the γ phase at elevated temperatures to the melting point. The thermodynamic results are summarized schematically in the Gibbs free energy versus temperature curves depicted in **Figure 7**. Importantly, it is the magnetic disordering or Schottky-like anomaly in the γ phase at low temperatures which is postulated to give rise to a stable fcc allotrope over a restricted temperature range and if it were not for this transition the γ phase would not appear. The range of stability of γ-Fe is determined by

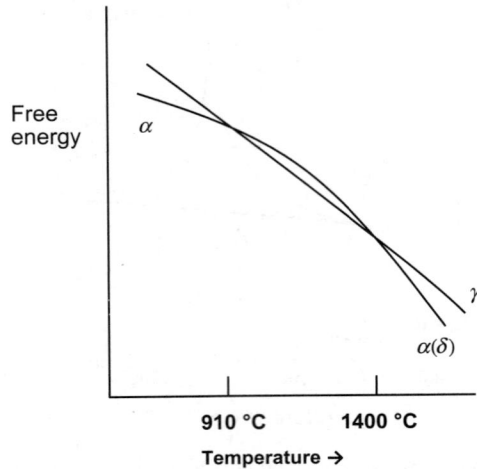

Figure 7 Schematic of the free energies of α and γ iron vs. temperature showing the return of α at higher temperature.

a delicate balance of relatively large contributions to the thermodynamics of the phases and the maximum value of $(G^\alpha - G^\gamma)$ is only about 15 cal mol^{-1}; thus, it is not surprising that alloying elements can selectively stabilize the α and γ phases as is well known in ferrous physical metallurgy. More recent investigators (Massalski and Laughlin, 2009) have focused on elucidating and amplifying these conclusions with the latter discussing the old "β-iron" controversy and the definition of a phase and phase transition. Finally, the $T - P$ for Fe is shown in **Figure 8** showing that allotropic phase

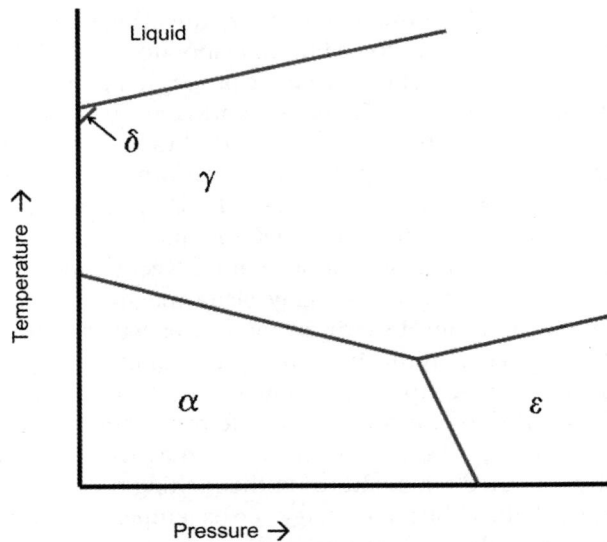

Figure 8 Schematic plot of the temperature vs. pressure phase diagram of iron. Note that at low pressure only the α, γ, δ and liquid phases are stable, whereas at high pressures ε (hcp) becomes a stable phase.

changes can be induced in the system by pressure as exemplified by the appearance of ε-Fe at high pressures.

8.2.2 Thermodynamics of Alloys: Energetics and Stability of Solutions

The Gibbs free energy of an alloy phase at constant temperature (T) and pressure (P) can be written as

$$G' = \sum_{i}^{n} \mu_i n_i \tag{11}$$

where G' is the extensive free energy and n_i is the number of moles of species i contained in the phase and μ_i is the chemical potential of species i in solution defined by

$$\mu_i = \left(\frac{\partial G'}{\partial n_i}\right)_{T,P,n_j} \tag{12}$$

wherein μ_i is also called the partial molar free energy $\overline{G}_i$. This fundamental definition applies to substitutional and interstitial solute elements in crystalline phases. From basic solution thermodynamics the chemical potential of a species in solution also can be written as

$$\mu_i(T,P) = \mu_i^0(T,P) + RT \ln a_i \tag{13}$$

where $\mu_i^0(T,P)$ is the chemical potential in a specified standard state (e.g. the free energy per mole of pure i at the given T and P) and a_i is the activity of the species i in solution. The chemical potential or partial molar free energy defined above can be understood as the free energy per mole of species i contributed to the total free energy of a solution of given composition or as the change in free energy when one mole of i is added to a large quantity of solution at constant composition. Partial molar quantities related to the enthalpy ($\overline{H}_i$) and entropy ($\overline{S}_i$) can be defined in a similar fashion and we can write $\overline{G}_i = \overline{H}_i - T\overline{S}_i$.

In multicomponent and polyphase systems we will find it convenient to write the free energy of a solution as a molar quantity, that is, G^α is taken to be the free energy per mole of the α phase and given by

$$G^\alpha(T,P) = G(T,P,X_A^\alpha,X_B^\alpha,\dots X_M^\alpha) \tag{14}$$

with X_M^α referring to the mole fractions of the components A, B, ...M, wherein $\sum_{i}^{M} X_M^\alpha = 1$, that is, only $M-1$ mole fractions specifying the composition of a phase are independent. Thus, at constant T and P,

$$G^\alpha = X_A^\alpha \mu_A^\alpha + X_B^\alpha \mu_B^\alpha + \dots \tag{15}$$

A binary or two-component system A–B can be described thermodynamically by $G^\alpha(T,P,X)$ where X is the mole fraction of B in the α phase, and at constant T and P we can write

$$G^\alpha = (1-X)\mu_A^\alpha + X\mu_B^\alpha \tag{16}$$

In a multicomponent polyphase system composed of α, β, γ, ... phases, the criterion for chemical equilibrium is that $\mu_i^\alpha = \mu_i^\beta = \mu_i^\gamma = \dots$ for all components $i = 1,2, \dots, C$ across the system. (Note that the number of components may be less than the number of species in systems exhibiting ionic or covalent bonding.) For two phases α and β in thermodynamic equilibrium in a binary system $(C=2)$, $\mu_A^\alpha = \mu_A^\beta$ and $\mu_B^\alpha = \mu_B^\beta$. A material system is composed of C components when $(C-1)$ composition variables are required to describe a phase.

The activity of a chemical species introduced in Eqn (13) is more rigorously defined as the ratio of fugacities $a_i = \frac{f_i}{f_i^0}$ where f_i is the fugacity of the species in solution and f_i^0 is the fugacity in a specified standard state such as the pure component at a given temperature (T) and total pressure (P). In this case, the activity of the pure substance is taken as unity. The fugacity f_i of component i is, to a good approximation, equal to the partial vapor pressure p_i of i over the solution. Thus, the activity of component i can be written as $a_i = \frac{p_i}{p_i^0}$ where p_i^0 is the vapor pressure of pure i. Other standard states can be invoked for convenience such as the infinitely dilute solution of species i. An ideal solution is defined by setting $a_i = X_i$ (Raoult's law) where X_i is the mole fraction or atomic fraction of i in solution, as discussed earlier. Thus, for an ideal solution

$$\mu_i - \mu_i^0 = RT \ln X_i \tag{17}$$

at constant total pressure P. Generally, solutions are not ideal and the deviation from ideal behavior is taken into account by writing $a_i = \gamma_i X_i$ where γ_i is called the activity coefficient. Dilute solutions often obey Henry's law and $\gamma_i = \gamma_0 = $ constant over a restricted composition range. The deviation from ideality expressed by γ_i can be greater or less than unity depending on the atomic interactions of the various species in solution, as will be discussed subsequently.

Let us consider forming a binary liquid or solid solution by dissolving B in A and consider the energetics of this process or reaction, namely,

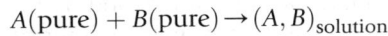

$$A(\text{pure}) + B(\text{pure}) \rightarrow (A, B)_{\text{solution}}$$

The free energy change accompanying the formation of the solution is called the free energy of mixing, $\Delta G_M = \Delta H_M - T\Delta S_M$, where ΔH_M is the heat (enthalpy) of mixing and ΔS_M is the entropy of mixing. In this brief review of the thermodynamics of alloys we will make extensive use of an approach often called **graphical thermodynamics** (Gibbs, 1873), particularly the use of free energy–composition diagrams in numerous contexts. **Figure 9** shows schematically the free energy per mole of solution, G, as a function of composition, X, for a binary system exhibiting complete solubility in the liquid and/or solid state. The prominent dashed line shows the free energy of a mechanical mixture of the components A and B as a function of composition and the free energy of mixing, ΔG_M, is indicated graphically with respect to the mixture of the components. The free energy per mole of solution can be written as

$$G = (1-X)G_A + XG_B + \Delta G_M \tag{18}$$

where $G_A = \mu_A^0$ and $G_B = \mu_B^0$ are the free energies per mole of the pure components A and B, respectively, and ΔG_M is the change in the free energy on mixing.

For an ideal solution,

$$\Delta G_M^{id} = RT\{(1-X)\ln(1-X) + X \ln X\} \tag{19}$$

whereas, in general, ΔG_M can be expressed as

$$\Delta G_M = RT\{(1-X)\ln a_A + X \ln a_B\} = RT\{(1-X)(\mu_A - \mu_A^0) + X(\mu_B - \mu_B^0)\} \tag{20}$$

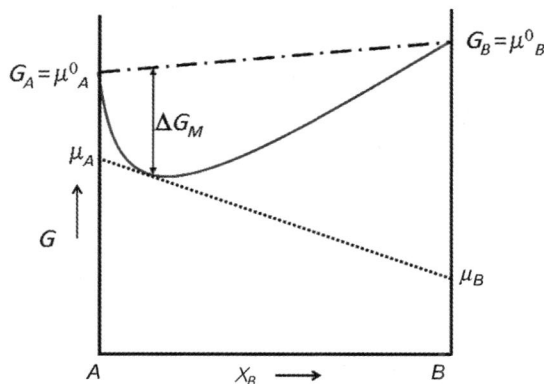

Figure 9 Free energy vs. composition curve of a binary system showing the chemical potentials of the A and B components in the alloy phase as well as those in their pure state.

or

$$\Delta G_M = RT\{(1 - X)\ln \gamma_A + X \ln \gamma_B\} + RT\{(1 - X)\ln(1 - X) + X \ln X\} \tag{21}$$

It is readily shown that if the solute of a dilute solution obeys Henry's law ($a_B = \gamma^0 X_B$), then the solvent obeys Raoult's law ($a_A = X_A$) using the Gibbs–Duhem equation $(1 - X)d\mu_A + Xd\mu_B = 0$ which stems from the property that the Gibbs free energy is homogeneous function of first order. Writing $G = (1 - X)\mu_A + X\mu_B$ for a binary solution and differentiating with respect to X and employing the Gibbs–Duhem relation leads to two very important relations widely used in the graphical thermodynamics of solutions:

$$\mu_A = G - X\frac{dG}{dX} \tag{22}$$

and

$$\mu_B = G + (1 - X)\frac{dG}{dX} \tag{23}$$

as shown in **Figure 9**. We shall see that this provides a basis for the common tangent construction which illustrates graphically chemical equilibrium in phase mixtures. If the binary A–B phase represented by the G vs. X in **Figure 9** is a solid phase then clearly the stable components A and B have the same crystal structure.

If we employ a ΔG_M vs. X curve to denote the energetics of the solid solutions, the intercepts represent relative partial molar quantities $\mu_A - \mu_A^0$ and $\mu_B - \mu_B^0$. See **Figure 10**. Since

$$\Delta G_M = \Delta H_M - T\Delta S_M \tag{24}$$

one can write the heat of mixing as

$$\Delta H_M = (1 - X)(\overline{H}_A - H_A^0) + X(\overline{H}_B - H_B^0) \tag{25}$$

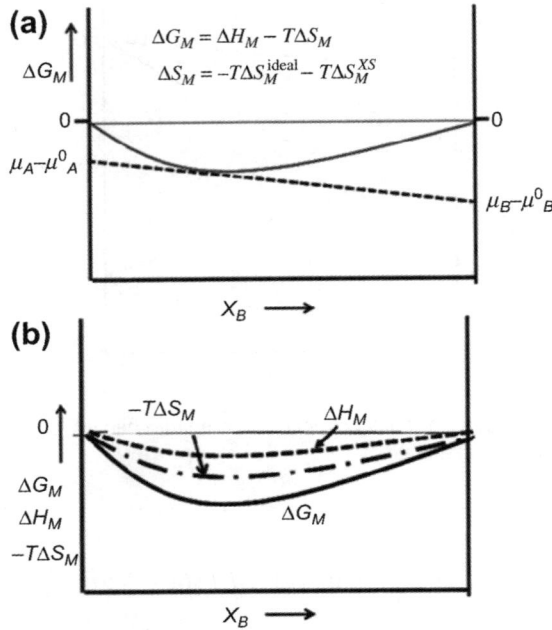

Figure 10   (a) ΔG of mixing plot for a binary solution. (b) Plot showing ΔG, ΔH and ΔS of mixing of the solution phase.

where $\overline{H}_A$ and $\overline{H}_B$ are the partial molar enthalpies of A and B in solution; H_A^0 and H_B^0 are the enthalpies of pure A and B per mole, respectively. Here we extend our definition of an ideal solution wherein

$$\Delta H_M = 0 \quad \text{and} \quad \Delta S_M = \Delta S_M^{id} = -R[(1-X)\ln(1-X) + X\ln X] \tag{26}$$

The later of which is called the ideal entropy (configurational) of mixing, calculated using elementary statistical thermodynamics applied to the random placement of the A and B atoms on the lattice sites. The ideal entropy of mixing has a maximum value of 1.38 cal mol^{-1} at $X = 0.5$ and is symmetric about this value. A reference state or solution model for the energetics of the A–B solution is the *regular solution* wherein ΔH_m is taken to be non-zero and $\Delta S_M = \Delta S_M^{id}$ (which has a built-in contradiction as discussed subsequently).

If the energetics of the solid solution are modeled in terms of short-range pairwise interactions or interatomic bonds between the A and B atoms on the lattice sites (assuming in the zeroth approximation interactions between first nearest-neighbors only (see **Figure 11**) described by bond energies E_{AA}, E_{BB} and E_{AB} (taken to be negative), then assuming random occupation of the N lattice sites ($N = N_A + N_B$, N_A and N_B being the number of A and B atoms, respectively) one finds that the heat of mixing ΔH_m can be written in a straightforward manner as

$$\Delta H_M = (n_{AA}E_{AA} + n_{BB}E_{BB} + n_{AB}E_{AB}) - N\frac{z}{2}[(1-X)E_{AA} + XE_{BB}] \tag{27}$$

where n_{AA}, n_{BB} and n_{AB} are the number of A–A, B–B and A–B bonds in a random solid solution (statistical distribution) and z is the first nearest-neighbor coordination of a lattice site. Clearly, the first

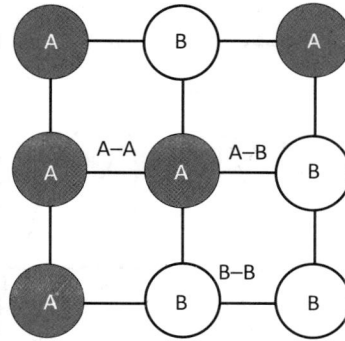

Figure 11 Schematic showing A–A, B–B and A–B first nearest-neighbor bonds.

term represents the enthalpy/energy of the solution and the second term is the enthalpy/energy of a mechanical mixture of the pure components. Importantly, it can be shown that ΔH_M can also be expressed as

$$\Delta H_M = NzX(1 - X)\left[E_{AB} - \frac{(E_{AA} + E_{BB})}{2}\right] = NzX(1 - X)V \tag{28}$$

indicating that if $\Delta H_M > 0$ ($V > 0$) A–A and B–B bonds are favored (there is a tendency for the atoms to surround themselves with like nearest-neighbors or clustering; if $\Delta H_M < 0$ ($V < 0$), A–B first nearest-neighbor bonds are favored indicating a tendency for ordering.

At high temperatures the configurational entropy will tend to produce a random distribution of the species in solution; however, as the temperature is lowered the mutual interactions of the A and B atoms will give rise to clustering ($\Delta H_M > 0$) or ordering ($\Delta H_M < 0$) effects referred to earlier. In the case of $\Delta H_M > 0$ the atomic interactions can lead to the appearance of a miscibility gap and phase separation below a temperature critical T_C as shown in **Figure 12a**. In the simple regular solution model developed here, $T_C = 2\dfrac{\Delta H_M}{R}$ (at $X = 0.5$). The two-phase field derives from the appearance of a double-well free energy $G(X) = H(X) - TS(X)$ versus composition (X) curve depicted in **Figure 12b** showing a common tangent construction establishing chemical equilibrium between two phases α_1 and α_2 having different compositions but the same crystal structures. We call attention to the two inflection points $\left(\dfrac{\partial G^2}{\partial X^2} = 0\right)$ in the $G(X)$ vs. X curve since it can be shown that supersaturated solutions lying between these points where $\dfrac{\partial G^2}{\partial X^2} < 0$ are thermodynamically unstable with respect to diffusional processes leading to phase separation. Supersaturated solutions on the $G(X)$ vs. X curve where $\dfrac{\partial G^2}{\partial X^2} > 0$ between the inflection points (the so-called chemical spinodes) and the equilibrium compositions are metastable with respect to phase separation. Small composition fluctuations in a metastable solution tend to decay because they entail a local increase in free energy whereas fluctuations in unstable solutions will tend to amplify spontaneously (see the chord construction in **Figure 12c**) and lead to phase separation. Furthermore, the diffusion coefficient in

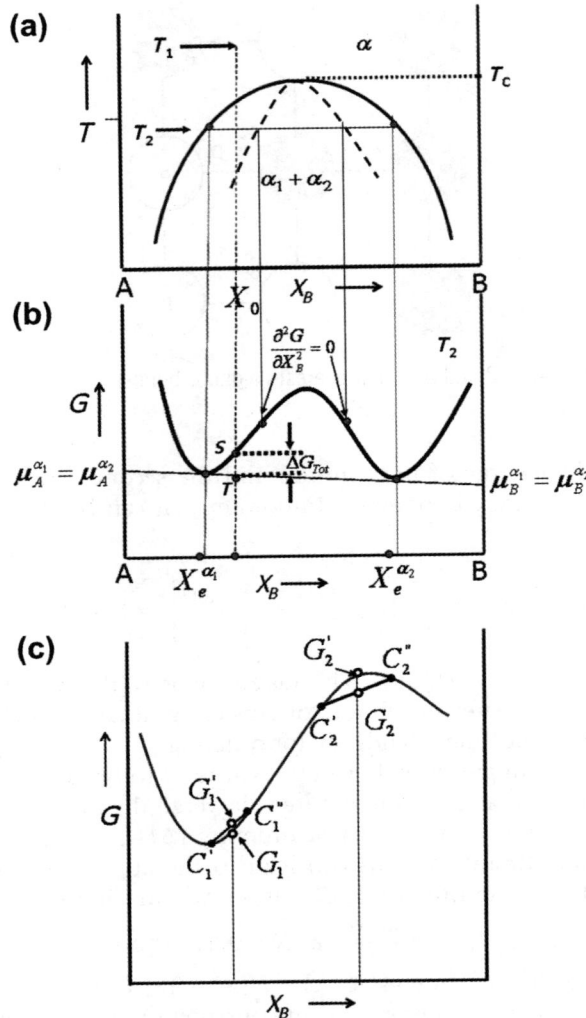

Figure 12 (a) Symmetric miscibility gap of an A–B alloy. (b) Free energy vs. composition curve of the alloy shown in (a) at temperature T_2. (c) Free energy vs. temperature curve for an alloy with an asymmetric miscibility gap, displaying by the cord construction the stability at 1 and instability at 2.

a binary system is proportional to $\dfrac{\partial G^2}{\partial X^2}$ which indicates that diffusive flow will occur up the concentration gradient (uphill diffusion) leading to spontaneous unmixing in thermodynamically unstable supersaturated solid solutions. This diffusional instability is often referred to as spinodal decomposition of the supersaturated state.

For the simple case of a miscibility gap let us explore further the properties of the $G(X)$ vs. X curve and introduce the concept of the overall "thermodynamic driving force" for a precipitation reaction within a metastable supersaturated solid solution, that is, the free energy change ΔG_{Total} for the phase reaction:

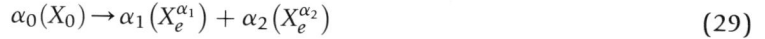

$$\alpha_0(X_0) \rightarrow \alpha_1\left(X_e^{\alpha_1}\right) + \alpha_2\left(X_e^{\alpha_2}\right) \tag{29}$$

wherein the supersaturated α_0 phase of composition X_0 quenched from temperature T_1 to T_2 decomposes or phase separates within the miscibility gap into an equilibrium phase mixture of phases α_1 and α_2 of compositions $X_e^{\alpha_1}$ and $X_e^{\alpha_2}$, respectively, the relative proportions given by the lever rule. Within the graphical thermodynamics of **Figure 12b** the decrease in the free energy accompanying phase separation can be shown to be the segment –ST (free energy change per mole of solution α_0) where S is a point on the tangent to the G vs. X curve at X_0 and T is the point on the common tangent at the overall composition X_0. However, during the initial stages of formation of the two-phase mixture in a supersaturated matrix of composition X_0 (nucleation), the formation of a small region of composition X^N (nucleus) is accompanied by a release of free energy per mole of the nucleus indicated by $-PQ$ in **Figure 13**. The matrix is effectively infinite in extent with respect to a small fluctuation and acts as a chemical potential reservoir, that is, the small fluctuation forming a potential nucleus occurs without disturbing the matrix composition or chemical potential. The free energy change (per mole of the nucleus or embryo) $-PQ$ is given by

$$\Delta G = X^N\left[\mu_B^N - \mu_B^M\right] + \left(1 - X^N\right)\left[\mu_A^N - \mu_A^M\right] \tag{30}$$

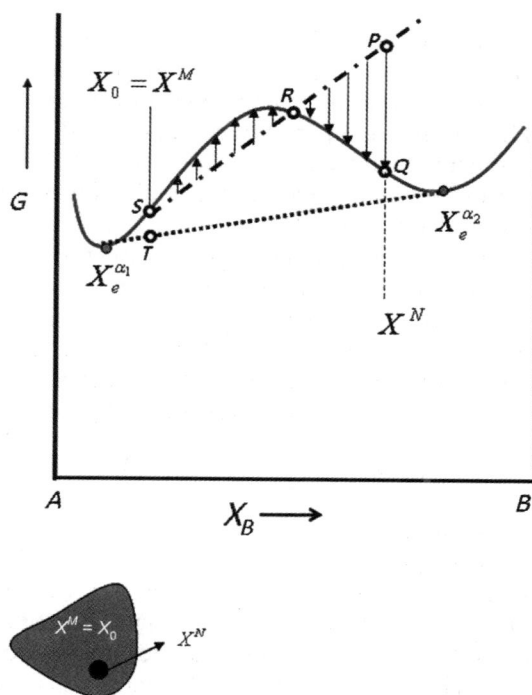

Figure 13 Free energy vs. composition diagram of an alloy with a miscibility gap, showing the driving force (PQ) for the nucleation of B-enriched phase from an alloy of composition X_0. X^M is the composition of the initial nuclei.

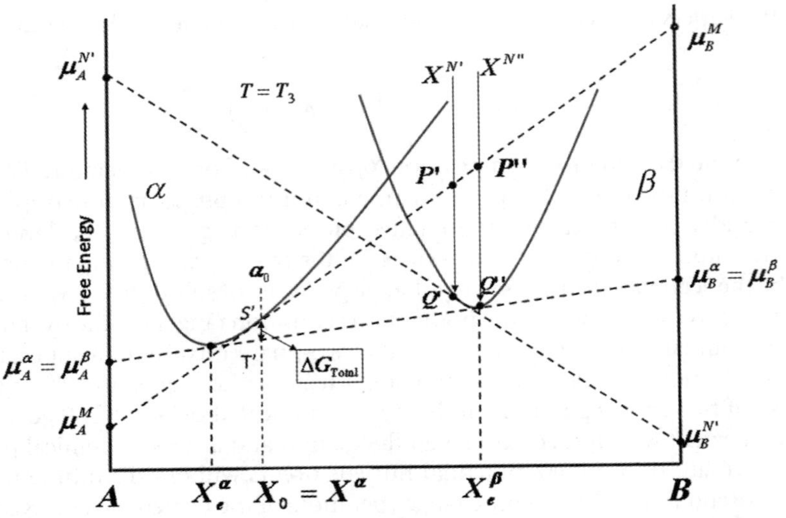

Figure 14   Free energy vs. composition diagram of an alloy with α and β phases of different crystal structure, showing the driving force for the nucleation of β phase from an alloy of composition X_0. See text.

where μ_B^N and μ_A^N refer to the chemical potentials of B and A in the nucleus/embryo of composition X^N and μ_B^M and μ_A^M refer to the chemical potentials of B and A in the matrix of composition X_0. The free energy release is the "effective" thermodynamic driving force for the formation of a nucleus/embryo of second phase of composition X^N within the supersaturated matrix. It should be noted that a fluctuation beyond R where the tangent to the free energy vs. composition (G vs. X) curve at the matrix composition intersects the G vs. X curve is needed to release free energy; smaller fluctuations than R actually increase the free energy locally and tend to decay.

Two-phase equilibrium between phases of different crystal structures is depicted in **Figure 14** using the graphical thermodynamic representation and common tangent construction. The common tangent construction establishes the equilibrium compositions of the conjugate phases X_e^α and X_e^β and the equality of the chemical potentials $\mu_A^\alpha = \mu_A^\beta$ and $\mu_B^\alpha = \mu_B^\beta$. If a solid solution α of composition $X^\alpha = X_0$ is rapidly cooled/quenched from to T_3 to produce a supersaturated state α_0, there is an overall thermodynamic driving force ΔG_{Total} (per mole of solution) for the precipitation reaction:

$$\alpha_0(X_0) \rightarrow \alpha_e(X_e^\alpha) + \beta_e(X_e^\beta) \tag{31}$$

which is the free energy release accompanying the formation of the equilibrium two-phase mixture as discussed earlier and represented graphically by $-S'T'$ in the associated G vs. X diagram. Again the segment $-P'Q'$ is the effective driving force or free energy released per mole of a small β nucleus or embryo of composition X^N which might form during the initial stages of precipitation (as above) and $P''Q''$ is the free energy released if the nucleus/embryo has the composition X_e^β.

Assuming Henry's law is obeyed by the solute (X) and that Raoult's law is obeyed by the solvent, the expression for the overall free energy change (ΔG_{Total}; see **Figure 14**) for the precipitation reaction

$$\alpha_0 \rightarrow \alpha_e + \beta_e$$

can readily be shown to be

$$\Delta G_{\text{Total}} = RT \left[X^\alpha \ln \frac{X_e^\alpha}{X^\alpha} + (1 - X^\alpha) \ln \frac{(1 - X_e^\alpha)}{(1 - X^\alpha)} \right] \tag{32}$$

However, the free energy released per mole of a β nucleus of the equilibrium composition X_e^β $(-P''Q'')$ is shown to be

$$\Delta G_N = RT \left[X_e^\beta \ln \frac{X_e^\alpha}{X^\alpha} + \left(1 - X_e^\beta \right) \ln \frac{(1 - X_e^\alpha)}{(1 - X^\alpha)} \right] \tag{33}$$

If the β nucleus is dilute in A, $(1 - X_e^\beta) \approx 0$ and thus the effective driving force governing the formation of a β nucleus can be approximated by

$$\Delta G_N = -RT \ln \frac{X^\alpha}{X_e^\alpha} = -RT \ln(SS) \tag{34}$$

where SS is the supersaturation of the α_0 matrix.

It should be pointed out that in an in-depth treatment of the nucleation problem in supersaturated solid solutions ΔG_N is related to the important parameter ΔG_V which is the free energy released per unit volume of the nucleus or embryo and given by $\Delta G_V = \Delta G_N / V_N$ where V_N (V^β) is the molar volume of the nucleus.

8.2.2.1 Coherent Phase Equilibria

The effect of elastic stress on phase equilibria in coherent multiphase systems represents an important but complex problem in the thermodynamics of solids. Cahn (1962b) in his seminal work on spinodal decomposition clearly showed the major influence that stress and elastic misfit can have on thermodynamic stability and microstructural evolution in cubic crystals leading to the concept of a coherent spinodal and a fundamental understanding of the occurrence of crystallographically aligned modulated structures in spinodally decomposing systems. Importantly, he recognized that the thermodynamics of stressed solids had not been addressed rigorously and essentially represented an inadequately solved problem in phase equilibria. Larche and Cahn (1973) and Robin (1974) in the early 1970s revealed that to describe the thermodynamics of coherent phase equilibria some fundamentally new questions had to be addressed and some basic concepts modified to properly describe the nature of the conjugate phases comprising metastable phase mixtures with elastic strain energy as a major contribution to the free energy of the system. Williams (1980, 1984) addressed the problem in a rather straightforward manner combining elasticity and solution thermodynamics applying his analysis to basic free energy–composition schemes showing clearly that the elastic energy of coherent phase mixtures can lead to novel and sometimes subtle changes in our thermodynamic description of heterogeneous phase equilibria. For example, two-phase fields are found to contract markedly and sometimes disappear completely as well as showing discontinuities in volume fraction as phase boundaries are crossed. Furthermore, the study of coherent phase equilibria has shown that thermodynamic equilibrium within a phase does not require the uniformity of the individual chemical potentials of the components but a constant so-called diffusion potential related to the difference in chemical potentials $\mu_A - \mu_B$. One also finds the equilibrium compositions of the metastable phases can be greater or less than the compositions of the unconstrained incoherent phases. This pioneering work mentioned earlier was followed by rigorous generalized treatments of the thermodynamics of elastically stressed crystals exemplified by the work of Voorhees and Johnson (2004), Pfeifer and Voorhees (1991), Johnson and Voorhees (1987), Johnson and Mueller (1991) and Ardell and Maheshwari (1995). A perusal of this literature shows that many of the results are strongly dependent upon the model used to represent the two-phase mixture, boundary conditions and assumptions regarding the variation of lattice parameters with composition in the conjugate phases. Khachaturyan in his classic treatise (1978) and subsequent work (1983) has developed a formalism emphasizing the central role of elastic energy in phase transformations occurring in the solid state.

For pedagogical reasons let us look at some relatively simple analyses which illustrate many of the salient and novel effects alluded to above. Essentially following Williams (1980, 1984), consider two cubic phases α and β with lattice parameters a_α and a_β within an incoherent two-phase state which are isotropically strained to have identical lattice parameters in a coherent metastable phase mixture; the lattice parameters are assumed composition independent. In this first approximation the compositions of the metastable conjugate phases remain the same as the unstressed equilibrium compositions established by the common tangent construction. The total free energy of the coherent phase mixture is written as a sum of the chemical free energies of the α and β phases in the unstressed state plus an elastic coherency strain energy which is a quadratic function of both the misfit strain and the volume fraction of the β phase. These free energies are shown graphically in **Figure 15a**. The nonlinear elastic term extends from point A to B passing through the intersection points X' and X''. Clearly a two-phase mixture is only stable between X' and X'' but the lever rule is applied to compositions A and B in this case. The elastic energy has contracted the extent of the coherent two-phase field compared with the incoherent equilibrium state and results in discontinuities in the metastable volume fractions of α and β phases as the compositions X' and X'' are crossed. If the elastic energy curve rises above the point W the metastable two-phase region disappears. This point W is called the Williams point. A more general case is shown in **Figure 15b** which allows for a composition dependence of the lattice parameters and a variation in the compositions of the metastable coherent conjugate phases. The metastable coherent two-phase field is contracted to the region between X' and X'' and the metastable phases are X'_α and X'_β, respectively; the lever rule is applied using these compositions. (Note: the morphology assumed has been a random array of spheres.)

The case of a miscibility gap can be addressed applying the same methodology. Allowing the lattice parameters of the conjugate phases to vary linearly with composition and assuming constant elastic properties yields a modified suppressed miscibility gap similar to result in the earlier work of Cahn (1962b). See **Figure 15c**. Here the metastable equilibrium compositions of the phases in the coherent state are delineated by the tangency of the elastic energy curve to the free energy–composition curve.

8.2.2.2   *Magnetism:Phase Equilibria and Phase Diagrams*

In our discussion of the allotropy of iron (Fe) we called attention to the critical role of magnetism (ferromagnetism and antiferromagnetism) in the relative stability of phases and cited early seminal papers on the subject. Here we wish to briefly examine a general approach to describing the contribution of magnetic effects to the heat capacity or specific heat specifically related to the occurrence of ferromagnetism in metals and alloys due primarily to Hillert and Jarl (1978) and extended by Inden (1976, 1981). Although the exact nature of the singularity of the heat capacity C_P^{mag} in the vicinity of the higher order ferromagnetic $\rightarrow$ paramagnetic transition is yet to be resolved theoretically, Inden introduced an accurate semiempirical operational description of the magnetic contribution to the heat

(a)

(b)

(c)

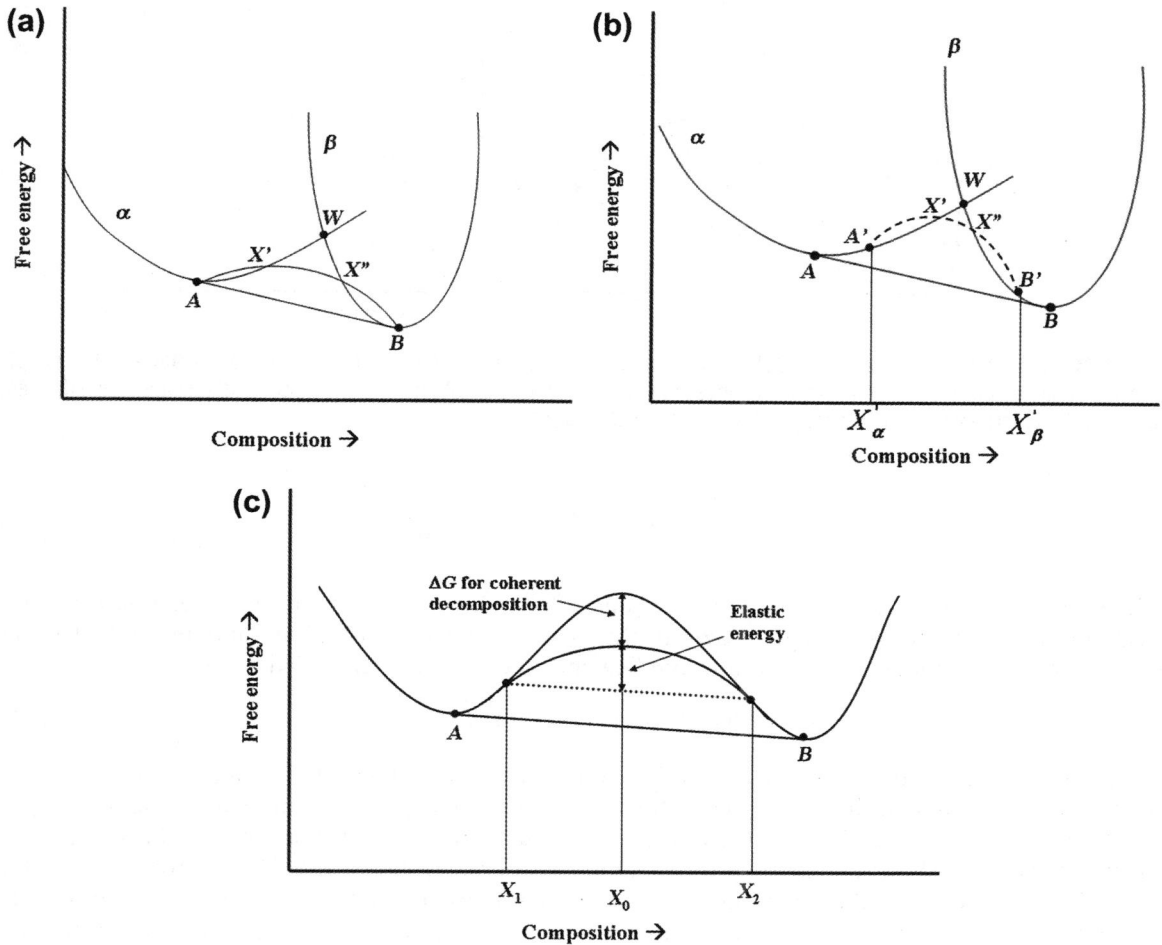

Figure 15 (a) Free energy–composition scheme showing superposition of elastic energy extending from A to B wherein the compositions of the conjugate phases remain the same as in the unstressed state. The coherent two-phase field has contracted to X'–X" but the lever rule is applied to the compositions A and B of the conjugate phases. The point W where the coherent two-phase field would disappear is called the Williams point. (b) More general case of coherent two-phase equilibrium where the compositions of the conjugate phases are changed to A' and B' and the two-phase field has contracted to X'–X" with the lever rule applied to compositions A' and B'. (c) Coherent two-phase equilibrium with attendant elastic energy applied to a miscibility gap.

capacity or specific heat taking into account the thermal disruption of both the LRO and SRO of the spins associated with the phase transition which is written as

$$C_P^{\text{mag}}(\text{LRO}) = K^{\text{LRO}} R \ln\left(\frac{1 + \tau^3}{1 - \tau^3}\right) \quad \text{for} \quad \tau = \frac{T}{T_C} \leq 1. \tag{35}$$

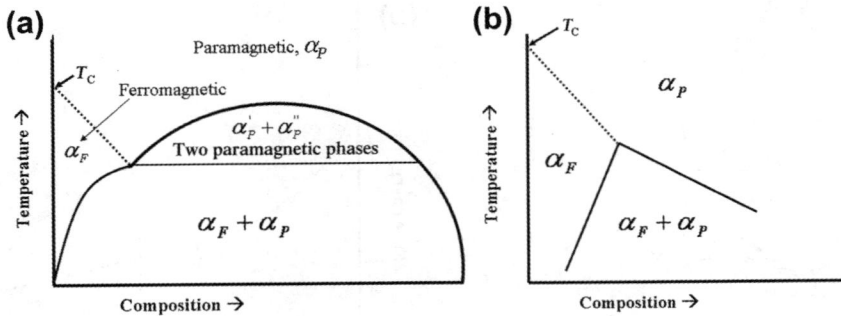

Figure 16   (a) Schematic phase diagram showing the intersection of a line of critical points (Curie temperature) with a simple miscibility gap delineating different magnetic phases α_P, α'_P, α''_P and α_F. (b) A line of critical points in a binary phase diagram showing the emergence of a tricritical point and two-phase region.

and

$$C_P^{mag}(\text{SRO}) = K^{\text{SRO}} R \ln\left(\frac{1 + \tau^{-5}}{1 - \tau^{-5}}\right) \quad \text{for} \quad \tau \geq 1 \tag{36}$$

where T_C is the Curie temperature with K^{LRO} and K^{SRO} being two constants characterizing the ferromagnetic and paramagnetic states, respectively. With several simplifications and the aid of series expansions the magnetic contribution to the free energy can be approximated by

$$G^{mag} = RTf(\tau, p)\ln(\beta + 1) \tag{37}$$

where $f(\tau, p)$ is a complex function of τ and includes a parameter p which is a constant defined as the fraction of the total spin disordering enthalpy absorbed above the Curie temperature (destruction of short-range order/SRO) and is given as $p = .28$ for fcc metals and 0.40 for bcc metals; β is the average magnetic moment (Bohr magnetons) of the atoms comprising the system. To effectively apply this formalism to alloys the salient parameters must be carefully evaluated as a function of composition. This remains an important and challenging problem in alloy physics and computational thermodynamics (Lukas et al., 2007).

Meijering (1963a) analyzed the interaction of a line of higher order transition such as a Curie temperature locus within a binary phase diagram with a miscibility gap as shown in **Figure 16**a. The magnetic transition can significantly distort the shape of the miscibility gap compared with the usual topology but also can markedly distort the associated spinodal curve. Later Nishizawa et al. (1979) discussed similar effects specifically in ferromagnetic α–Fe systems including the technologically important Fe–Cr system. Later Inden (1981 and 1982) analyzed the general behavior of second-/higher order lines in binary phase diagrams including atomic ordering. An important feature of the analysis is that a higher order line can terminate at a sharp critical point (tricritical point) within the interior of the binary diagram giving rise to a two-phase region, for example equilibrium between ferromagnetic and paramagnetic phases of different compositions in a two-phase region. See **Figure 16**b.

8.2.3   Metastability

A phenomenon of major importance in the thermodynamics and kinetics of phase transformations, particularly in the area of precipitation from supersaturated solid solution, is the formation of

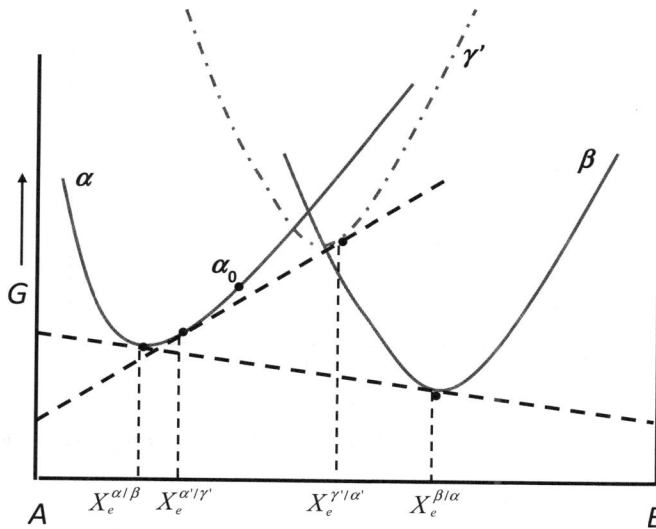

Figure 17 Free energy vs. composition curve of an alloy that has α and β as equilibrium phases at the temperature shown, but also has a metastable phase γ'. The compositions of the stable and metastable equilibria are shown.

metastable phases in the phase reaction in preference to the nucleation and growth of the thermodynamically stable precipitate. Indeed, in the heat treatment of age hardening alloys we find that the decomposition of the supersaturated state often proceeds through a series of metastable states before the emergence of the stable equilibrium precipitate and the optimum physical and mechanical properties invariably develop in association with the formation of fine-scale metastable precipitates within the parent matrix. We will see subsequently that this occurs as a result of favorable nucleation kinetics despite the fact that the thermodynamic driving forces discussed earlier are greatest for the formation of the equilibrium second phase. In **Figure 17** we show a rather straightforward graphical representation of a competing γ' phase that can form in a supersaturated α_0 establishing a metastable equilibrium phase mixture composed of α' and γ'; stable equilibrium is the $\alpha + \beta$ two-phase mixture. The metastable equilibrium gives rise to a metastable solvus shown in **Figure 18** deriving from a metastable common tangent construction with respect to the γ' phase. We see that the metastable γ' phase has a greater solubility in the α phase than the equilibrium precipitate. Furthermore, it is clear from the graphical thermodynamics that the overall driving force ΔG_{Total} is less for the phase reaction $\alpha_0 \rightarrow \alpha' + \gamma'$ than for the formation of the stable equilibrium phase mixture $\alpha + \beta$. Also, it can be shown by graphical thermodynamic construction that the "effective driving force" for nucleation of the β is greater than that for the nucleation of the metastable phase, γ'.

8.2.3.1 Eutectoid Decomposition

An important solid-state transformation in alloys is the case of eutectoid decomposition which has played a central role in governing the heat treatment of engineering steels. Also the proeutectoid reaction (a precipitation reaction) has been engineered to tailor the microstructure of hypoeutectiod compositions to achieve extraordinary combinations of properties (strength, ductility, etc.) for a wide range of technological applications. The eutectoid reaction per se is associated with a phase diagram configuration as shown in **Figure 19** and can be written generally as $\gamma \rightarrow \alpha + \beta$, that is,

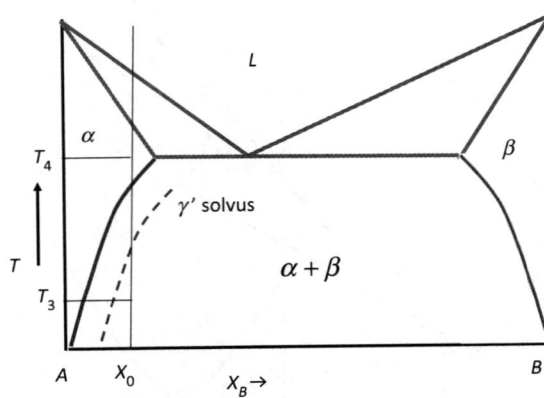

Figure 18 A simple eutectic phase diagram of a binary A–B alloy. The dotted line represents the metastable solvus line for the metastable γ' solid phase.

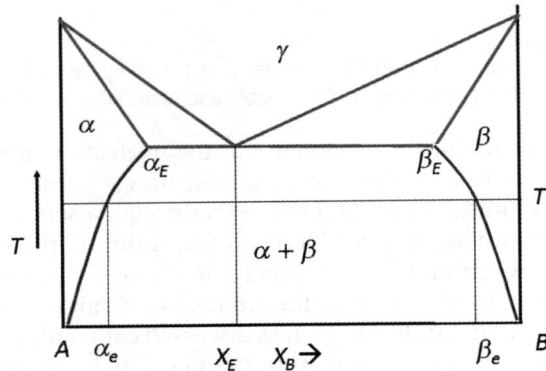

Figure 19 A simple eutectoid phase diagram for the alloy A–B. On cooling, the γ phase transforms to α and β at the invariant temperature.

a high-temperature γ phase decomposes into two phases α and β having different crystal structures. A schematic free energy–composition diagram for the eutectoid decomposition of an undercooled γ phase of eutectoid composition X_E at some temperature below the equilibrium eutectoid reaction isotherm is depicted in **Figure 20** where $\Delta G = \Delta G_{Total}$ for the eutectoid reaction is given by SE. The eutectoid reaction involves a codeposition of two phases often exhibiting a characteristic morphology. The eutectoid microconstituent in carbon and low-alloy steels generally exhibits a classic lamellar morphology similar to that arising in eutectic freezing and this mode of transformation is often called cellular phase separation as will be discussed later.

An interesting feature of the free energy scheme depicted in **Figure 20** is that a γ composition of S' can lower its free energy by transforming to a metastable α' phase without a change in composition. This alternative reaction path can involve either of two transformation mechanisms, namely, the well-known diffusionless martensitic transformation or diffusional massive transformation. The free energy change $S'E'$ is the thermodynamic driving force for these compositionally invariant transformation

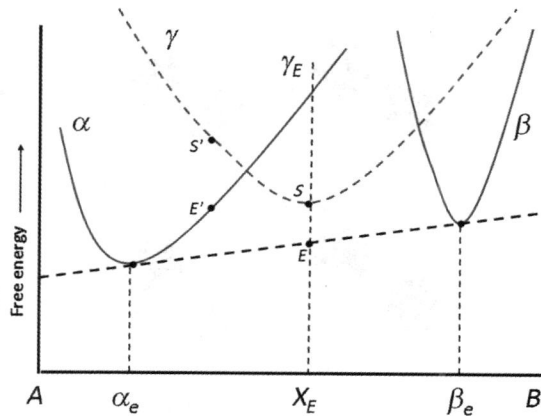

Figure 20   Free energy vs. composition curves for the three phases α, β and γ at the temperature T'.

modes. The *martensitic transformation* involves a characteristic distortion or strain which converts the parent phase (γ) to the product phase (α') often at very high velocities. The atoms within the unit cell of the parent phase generally move small distances relative to the unit cell dimensions essentially shearing from one crystal structure to another in the absence of thermally activated atomic jumps of individual atoms but rather involving a cooperative or synchronous shearing of groups of atoms. The *massive transformation* on the other hand occurs by a diffusional nucleation and growth mechanism involving short-range thermally activated and uncorrelated atomic jumps across a migrating transformation front or interphase interface (Massalski, 1958, 1970).

8.2.4 Atomic Ordering in Alloys: Superlattices

Solid solution formation is a basic phenomenon in modern physical metallurgy and materials science. The introduction of different atomic species onto the sites of a crystalline solid to form an alloy gives rise to a structural modification which can markedly influence the properties of the metallic material. (Note: the solute may enter the solid solution interstitially or substitutionally; in this discourse we will focus our attention on atomic arrangements in subtitutional solid solutions.)

The mixing of atomic species on the sites of a solid solution may not be random even under equilibrium conditions, that is, the probability of a pair of sites being occupied by specific atoms is not simply equal to the probability obtained by multiplying their respective atomic fractions. These nonrandom distributions of atomic species derive from different interatomic interactions between, say, A–A, B–B and A–B pairs of atoms in a binary solution. Indeed, randomness is the exception rather than the rule in real alloy systems. When the A and B atoms in a binary alloy have a preference for like atoms, that is, A–A and B–B pairings, the behavior is termed **clustering**. If the energetics of the solution favor unlike A–B pairings, the deviation from randomness is called **ordering**. If the preference for A–B pairs persists only over a few to several interatomic distances, the solid solution is said to exhibit **short-range order (SRO)**; whereas, if these correlations persist over large distances compared with the unit cell dimensions, the ordering is denoted **long-range order (LRO)** and the crystal structure and atomic arrangements can be described in terms of interpenetrating sublattices occupied preferentially by A and B atoms creating an ordered solid solution or superlattice. See **Figure 21**. In this figure the gray atoms represent average occupancy by white and black atoms: that is, the disordered binary phase.

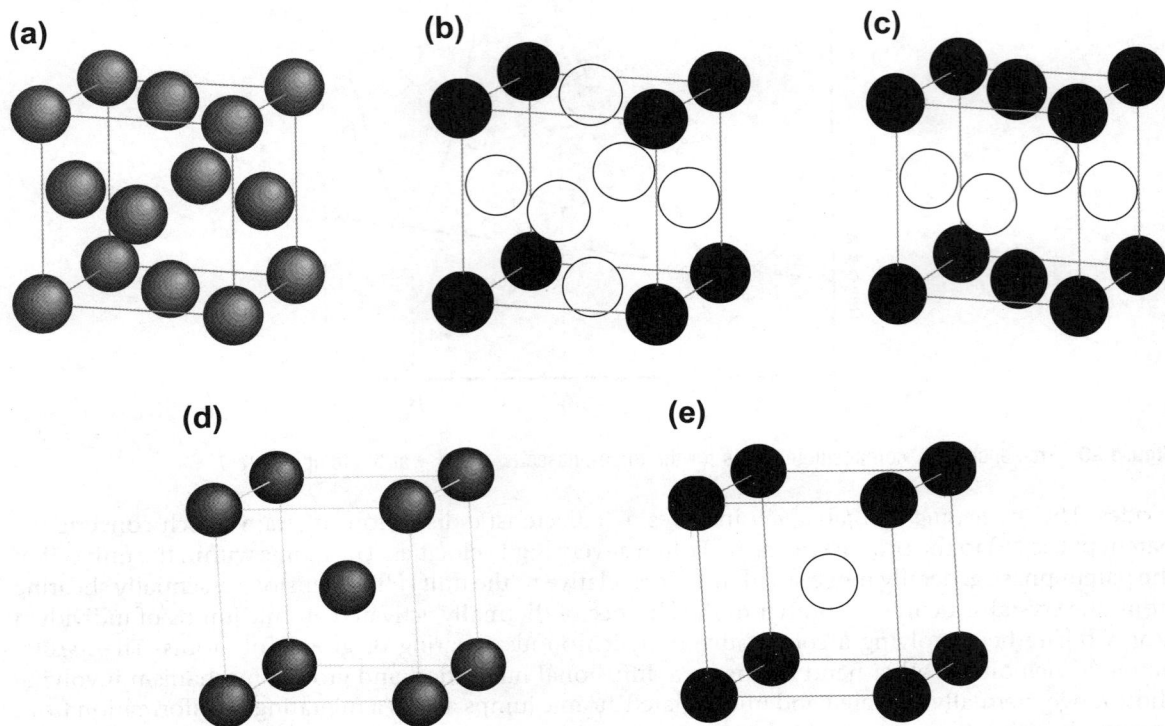

Figure 21   Disordered atomic structures and ordered atomic structures. (a) A1 (fcc, Cu), (b) L1$_2$ (Cu$_3$Au), (c) L1$_0$ (CuAu), (d) A2 (bcc, W), (e) B2 (CsCl).

These superlattices tend to form in the vicinity of stoichiometric ratios, for example, A_3B, AB, and are often referred to as intermetallic compounds, particularly when the ordered phase melts congruently or when the intermediate solid solution exists over a narrow composition range. We are going to be most interested in ordered (LRO) solid solutions which become disordered at elevated temperatures, that is, the LRO breaks down above a critical temperature but some SRO may persist. In the ordered state, the degree of LRO is not necessarily perfect, but is disrupted by thermal energy (temperature) and deviations from stoichiometry. Subsequently we shall quantify the description of the degree of order in a solid solution, but before doing this let us look briefly at the history of this phenomenon.

Tammann (1919) suggested that LRO could develop in a substitutional metallic solid solution. Earlier work by Kurnakow et al. (1916) indicated intriguing behavior in Cu–Au alloys of properties such as electrical resistivity in the vicinity of Cu$_3$Au and CuAu compositions, depending on the thermal treatment of these alloys. Bain (1923) and Johannson and Linde (1925) first reported "superlattice diffraction lines" in X-ray powder diffraction patterns. Indeed, because new periodicities appear in the structure of the ordered solution compared with the "disordered" parent phase, new "reflections" appear in the diffraction pattern. For example, in the A$_3$B superlattice viewed as a *crystallographic derivative* of a parent disordered FCC solid solution wherein the atomic sites are statistically occupied by "average" atoms, the distance between identical planes is doubled along the $\langle 100 \rangle$ directions as a result of the atomic or chemical ordering. New superlattice lines appear as a result of the new and larger d-spacings. The disordered FCC solid solution is properly viewed as a crystal based on an FCC Bravais lattice with the basis being a "statistical" or "average atom" occupying the points of the Bravais lattice. The A_3B ($L1_2$) superlattice is a crystal structure based on the simple cubic Bravais lattice with an A_3B arrangement decorating the lattice points. See **Figures 21** a, b. Importantly, this change in Bravais lattice constitutes a thermodynamic phase change or phase transformation.

These exciting new ideas and results percolated in the 1920s and stimulated a great deal of theoretical attention. In the 1930s a truly classic paper by Bragg and Williams (1934) appeared based on a rather simple statistical thermodynamic approach defining the underlying energetics of the solid solution in terms of an ordering energy related to the work done, W, in interchanging A and B atoms in an ordered structure from "right" sites to "wrong" sites among the sublattice sites characterizing the superlattice. They introduced an LRO parameter based on the occupancy of the appropriate sublattices and formulated the underlying energetics of the solution in terms of an ordering energy due to unspecified long-range forces acting on the individual atoms, the strength of which is proportional to the degree of atomic order; the (configurational) entropy of the ordered solution was formulated assuming random mixing of the different atomic species on the aforementioned sublattices of the emerging ordered structure. Importantly, the original Bragg–Williams theory makes no explicit use of short-range forces or pairwise interaction energies as incorporated later in the so-called quasichemical approaches. The assumption of random mixing on the sublattices fails to account for any local correlations or SRO. The degree of ordering is defined solely in terms an occupancy of the sublattices or the LRO parameter defined explicitly below. However, as we shall see the Bragg–Williams approach captured salient features of the ordering transformation in AB (CuZn) and A_3B (Cu$_3$Au) compositions in cubic solid solutions and is referred to as the zeroth approximation in the scheme of quasichemical descriptions, which generally write the solution energetics in terms of AA, BB and AB bond energies (pairwise interaction energies) (Bethe, 1935; Guggenheim, 1952). Indeed, the original Bragg–Williams formulation of the problem is found to be basically equivalent to a quasichemical approach incorporating AA, BB and AB bond energies in its underlying physics when the assumption of random mixing on the sublattices (regular solution) is invoked and is essentially homologous to the Weiss molecular field theory of ferromagnetism (Weiss, 1907). In the modern parlance of cluster variation methods (CVM), Bragg–Williams theory is a point cluster approximation (de Fontaine, 1973). The quasichemical approach of Bethe, emphasizing pairs of atom is a mean field theory in which pairs are immersed in a mean field in contrast to the Bragg–Williams formulation wherein single atomic species are effectively immersed in the background of the mean field derived from an "average" environment.

The Bragg–Williams theory introduces an LRO parameter based on occupancy of the appropriate sublattices, A atoms on α sublattice sites, B atoms on β sublattice sites ("right" sites) and B atoms on α sites, A atoms on β sites ("wrong" sites), and so on. We first will examine the equiatomic AB alloy undergoing the A2(bcc) → B2(sc) ordering transition. See **Figure 21**d and e. The degree of LRO is formulated quantitatively most generally as

$$\eta = \frac{(r_\alpha - X_A)}{Y_\beta} = \frac{(r_\beta - X_B)}{Y_\alpha} \tag{38}$$

where r_α is the fraction of α-sites occupied by A atoms ("right" atoms) and r_β is the fraction of β-sites occupied by B atoms ("right" atoms); Y_α is the fraction of α-sites and Y_β is the fraction of β-sites in the ordered superstructure, respectively. X_A and X_B are the atomic fractions of A and B in the alloy. This expression for η is applicable to stoichiometric and nonstoichiometric compositions and has a maximum value of unity in the perfectly ordered stoichiometric alloy. Clearly for non-stoichiometric compositions the maximum possible value is less than one. Using this description the order parameter varies from $\eta = 0$ in the disordered state (random solid solution) to $\eta = 1$ in the perfectly ordered state and most importantly the intensities of the "superlattice reflections" in diffraction patterns are generally found to vary as η^2 for both stoichiometric and nonstoichiometric compositions.

As mentioned earlier, the Bragg–Williams ordering energy or interchange energy is assumed to be related to the work done, W, in interchanging A and B atoms in the ordered structure from "right" sites to "wrong" sites among the sublattice sites characterizing the superlattice or superstructure. In this theory the interchange energy is assumed to be linearly related to the degree of order in the emerging superlattice as $W = \eta W_0$ where W_0 is the interchange energy when the alloy exhibits a state of perfect order ($\eta = 1$). An expression for the equilibrium degree of order η as a function of temperature was developed in terms of a simple kinetic equation describing the atomic transfer between sublattice sites (atomic transfer from "right" sites to "wrong" sites and the reverse); the forward and back reaction rates were set equal at equilibrium. The ratio of the rate constants (equilibrium constant) was set equal to a Boltzmann-like term $\exp\left(-\dfrac{W}{k_BT}\right)$, where $W = \eta W_0$ and k_B is the familiar Boltzmann constant in accord with basic chemical thermodynamics. When the solution energetics are written in terms of AA, BB and AB nearest-neighbor bond energies (E_{AA}, E_{BB} and E_{AB}) the interchange energy term W_0 is readily shown to be equal to $W_0 = -z(2E_{AB} - E_{AA} - E_{BB})$ where z is the nearest-neighbor coordination. When $2E_{AB} < E_{AA} + E_{BB}$ unlike pairs are favored, that is, the system tends to show ordering. This Bragg–Williams formulation gives the variation of η with temperature as

$$\ln\left(\frac{1-\eta}{1+\eta}\right) = -\frac{W_0\eta}{2k_BT} \tag{39}$$

or,

$$\eta = \tanh\left(\frac{W_0\eta}{4k_BT}\right) \tag{40}$$

which can be solved numerically. See **Figure 22**.

The results for the $A2 \rightarrow B2$ disorder–order show a continuous change in the degree of order from $\eta = 1$ at low temperatures to $\eta = 0$ at the critical temperature T_C, which can be shown to be given by $T_C = \dfrac{W_0}{4k_B}$. There is no latent heat associated with the phase transition and thermodynamic analysis reveals a finite discontinuity in the specific heat or heat capacity (C_v or C_P) at the critical temperature T_C as depicted in **Figure 23**.

This behavior shows all the earmarks of an Ehrenfest transition of Second Order (Ehrenfest, 1933). The Bragg–Williams theory does describe, at least qualitatively, the general behavior of systems such as CuZn (β-brass); however, the experimental results in terms of the behavior of the specific heat in the vicinity of the critical temperature are more complex. The neglect of local correlations (SRO) is clearly a major shortcoming of this zeroth approximation wherein SRO is found to persist even above T_C. See **Figure 24**. Indeed, a complete solution to this singularity has not yet been achieved. See Pippard (1966).

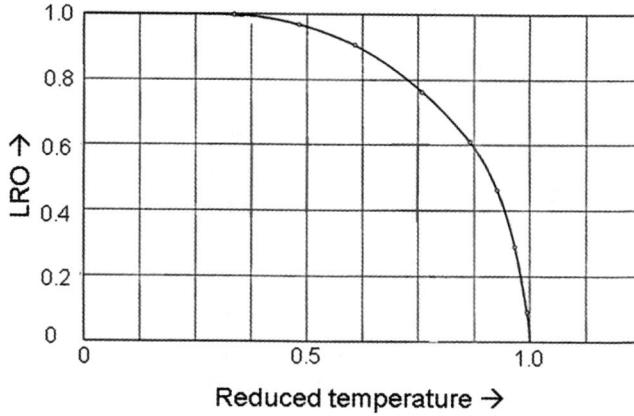

Figure 22 The order parameter vs. temperature plot for a Bragg–Williams second-order transition. Note the infinite slope as T approaches T_C. After Nix and Shockley (1938).

If the solution thermodynamics are formulated in terms of pairwise interaction energies for the A–B binary alloy assuming random mixing on the sublattices, the free energy of mixing of the ordering alloy can be written as

$$F_M = E_M - TS_M = NzV\left\{c(1-c) + \frac{\eta^2}{4}\right\} - k_B T \ln \omega \tag{41}$$

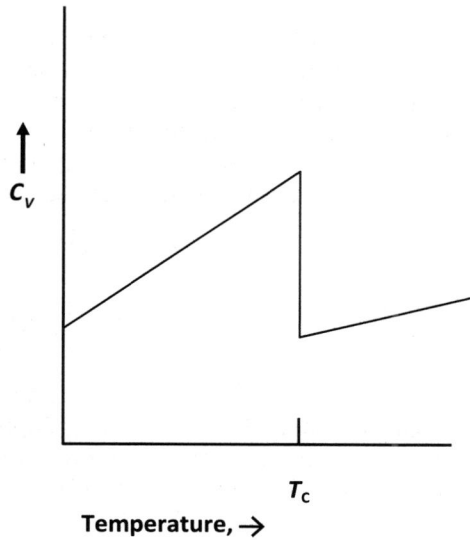

Figure 23 Plot of heat capacity vs. temperature for a Bragg–Williams second-order transition. Note the finite discontinuity at the critical temperature.

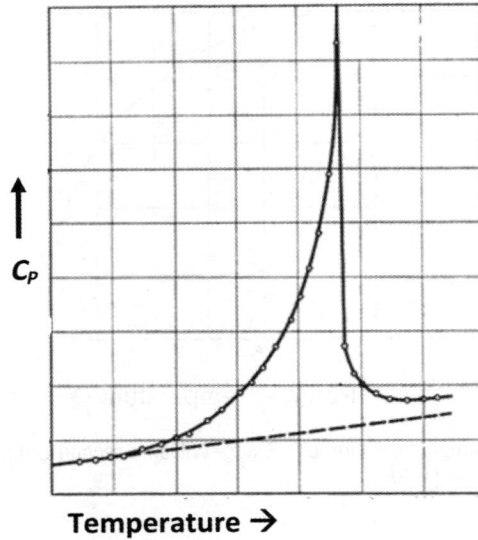

Figure 24　Variation of C_P vs. temperature plot for CuZn. Notice that the shape looks like the Greek letter *lamda*. This curve is different than that predicted by the Bragg–Williams treatment. After Nix and Shockley (1938).

where N is the total number of atoms in the alloy, $z = z_1 = 8$ is the coordination number of the first nearest-neighbor shell for the B2 structure, c is atomic fraction of B atoms,

$$V = \frac{1}{2}\left(2E_{AB}^1 - E_{AA}^1 - E_{BB}^1\right),$$

where the superscripts refer specifically to first or nearest-neighbor (*n–n*) interactions (later we will generalize the Bragg–Williams model to include 2nd *n–n* interactions), and $k_B \ln \omega$ is the configurational entropy term (the vibrational entropy contribution is neglected). The ω term is the number of possible arrangements or microstates corresponding to thermodynamically equivalent macrostates or distributions of A and B atoms on the sites of the sublattices (thermodynamic probability). The configurational entropy derived assuming random mixing on the sublattices can be written as

$$S_M = k_B \ln \omega = k_B \ln \frac{\left(\frac{N}{2}\right)!\left(\frac{N}{2}\right)!}{(N_A^\alpha)!(N_B^\alpha)!\left(N_A^\beta\right)!\left(N_B^\beta\right)!} \tag{42}$$

where N_A^α and N_B^α are the number of A and B atoms on the α sites and N_A^β and N_B^β are the number of A and B atoms on the β sites, respectively. Using the usual Stirling's approximation, and counting the number of AA, BB and AB bonds in terms of the concentration c and the sublattice occupancies, the free energy of mixing can be written as a function of c and η as follows:

$$F_M(c, \eta) = Nz_1 V\left\{c(1-c) + \frac{\eta^2}{4}\right\} + \frac{Nk_B T}{2}\left\{\left(1 - c + \frac{\eta}{2}\right)\ln\left(1 - c + \frac{\eta}{2}\right) + \left(c - \frac{\eta}{2}\right)\ln\left(c - \frac{\eta}{2}\right)\right.$$

$$\left. + \left(1 - c - \frac{\eta}{2}\right)\ln\left(1 - c - \frac{\eta}{2}\right) + \left(c + \frac{\eta}{2}\right)\ln\left(c + \frac{\eta}{2}\right)\right\} \tag{43}$$

the first term being the heat of mixing $H_m = Nz_1 V c (1 - c)$ associated with forming a disordered solid solution of A and B atoms statistically occupying the sites of a bcc solid solution and $V < 0$ for the ordering system. Taking $\dfrac{\partial F_M}{\partial \eta} = 0$ and $\dfrac{\partial^2 F_M}{\partial^2 \eta} = 0$ yields $\eta = \tan h\left(-\dfrac{4V\eta}{k_B T}\right)$ and $T_C = -\dfrac{4V}{k_B}$, consistent with the classic Bragg–Williams results outlined earlier.

The Bragg–Williams approach or zeroth approximation of the quasichemical models can readily be extended to the case of A1(fcc) $\rightarrow$ A$_3$B (L1$_2$, sc) ordering where

$$F_M = Nz_1 V \left\{ c(1 - c) + \frac{\eta^2}{16} \right\} - k_B T \ln \omega \dots \tag{44}$$

with $S_M = k_B \ln \dfrac{\left(\dfrac{3N}{4}\right)! \left(\dfrac{N}{4}\right)!}{(N_A^\alpha)!(N_B^\alpha)!}$ with the α and β sublattices for the A$_3$B superstructure shown in **Figure 21b**. The expanded entropy term becomes

$$S_M = -\frac{Nk_B}{4} \left\{ 3\left(c - \frac{\eta}{4}\right)\ln\left(c - \frac{\eta}{4}\right) + 3\left(1 - c + \frac{\eta}{4}\right)\ln\left(1 - c + \frac{\eta}{4}\right) + \left(c + \frac{3\eta}{4}\right)\ln\left(c + \frac{3\eta}{4}\right) \right.$$

$$\left. + \left(1 - c - \frac{3\eta}{4}\right)\ln\left(1 - c - \frac{3\eta}{4}\right) \right\} \tag{45}$$

Taking $\dfrac{\partial F_M}{\partial \eta} = 0$ yields the equilibrium value of the order parameter η for the A$_3$B composition as a function of temperature given by the transcendental equation as follows:

$$\ln\left(\frac{(1 + 3\eta)(3 + \eta)}{3(1 - \eta)^2}\right) = -\frac{2z_1 V \eta}{3k_B T} \tag{46}$$

For $F_M = F_M(\eta, T)$ there exists a critical temperature T_C such that $F_M(\eta^*, T_C) = F_M(0, T_C)$ with $\dfrac{\partial F_M(\eta^*, T_C)}{\partial \eta} = \dfrac{\partial F_M(0, T_C)}{\partial \eta} = 0$. At the critical temperature T_C, one finds by successive approximation, $\eta^* = 0.463$ and the critical temperature approximately given by $T_C = -\dfrac{0.137 z_1 V}{k_B}$. The variation of η with temperature is depicted in **Figure 25** and clearly displays a markedly different behavior than the A2 $\rightarrow$ B2 transition. At the critical temperature T_C the order parameter η undergoes a discontinuous change from $\eta = \eta^*$ to $\eta = 0$ and represents an equilibrium between an ordered phase ($\eta = \eta^*$) and a disordered phase ($\eta = 0$) with an associated latent heat ΔH_t given approximately by 0.78 T_C cal^{-1} mol^{-1} (Nix and Shockley, 1938). These thermodynamic features indicate a first-order transition according to the Ehrenfest classification.

The Bragg–Williams approximation does predict a first-order transition for the A1 $\rightarrow$ A$_3$B (L1$_2$) ordering in agreement with experiment, but the quantitative limitations are strikingly evident when one compares the value of $\eta^* = 0.463$ with the generally observed values of 0.7–0.8. Higher level theoretical approximations, for example cluster methods, predict values nearer 0.9. Furthermore, the more rigorous theoretical treatments predict lower critical temperatures by as much 50% (Christian, 2002).

As demonstrated earlier, the Bragg–Williams model captures essential features of AB and A$_3$B ordering (the B2 and L1$_2$ superstructures being crystallographic derivatives of bcc and fcc parent phases, respectively) showing the occurrence of two-phase equilibrium between a disordered phase

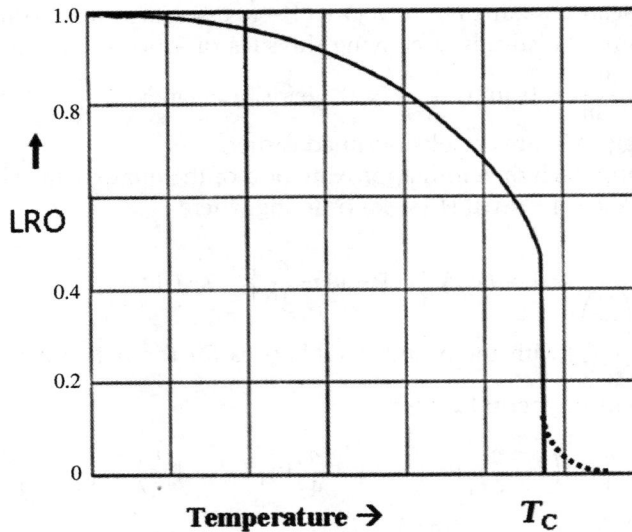

Figure 25 The Bragg–Williams LRO parameter vs. temperature plot for the fcc to L1$_2$ transition. The discontinuity at the critical temperature signals that this is a first-order (Ehrenfest) transition. After Nix and Shockley (1938).

($\eta = 0$) and an imperfectly ordered phase ($0 < \eta < 0$) at $T = T_C$ for the L1$_2$ ordering but no such two-phase equilibrium for the B2 ordering occurs which shows a continuous transition from $\eta = 1$ to $\eta = 0$ at T_C. The phase transitions are identified as first-order and second order according to the well-known Ehrenfest criteria. The free energies of the phases formulated earlier included a composition dependence although in the thermodynamic analysis we emphasized the stoichiometric compositions. Computational thermodynamic analysis using the general expressions for the free energies F_m (c, η, T) allows one to generate a temperature (T) versus composition (C) phase diagram and the ordering transitions are mapped into a conventional binary phase diagram in **Figure 26**. In the late 1930s Nix and Shockley (1938) compiled an extensive review of the subject including various early theoretical approaches.

8.2.4.1 First and Second Nearest–Neighbor Interactions

The energetics of binary substitutional metallic solid solutions based on pairwise interaction energies or A–A, B–B and A–B bonds within a zeroth approximation quasichemical approach can be readily extended to include both first nearest-neighbor (1st nn) and second nearest-neighbor (2nd nn) interactions by distinguishing two so-called interchange energies:

$$V = \frac{1}{2}\left(2E_{AB}^1 - E_{AA}^1 - E_{BB}^1\right)$$

$$U = \frac{1}{2}\left(2E_{AB}^2 - E_{AA}^2 - E_{BB}^2\right) \tag{47}$$

where E_{AA}^i, E_{BB}^i and E_{AB}^i refer to the ith nearest-neighbor interaction energies—all taken to be negative reflecting the "strength" of the A–A, B–B and A–B bonds, respectively. In the usual approaches

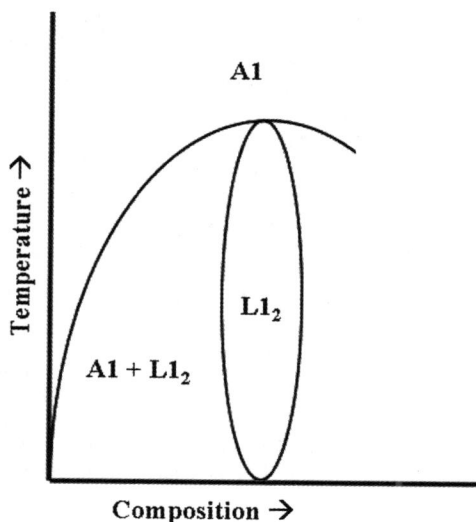

Figure 26 The Bragg–Williams phase diagram for A1(fcc) and L1$_2$ phases. Note the two-phase region (A1 + L1$_2$) between the ordered L1$_2$ phase and the disordered fcc phase, since this transition is first order.

incorporating only 1st nn interactions, $V < 0$ favors unlike A–B bonds and the system is classified as having a tendency for *ordering*. For $V > 0$, like A–A and B–B pairs are favored within the solid solution and the system has a tendency for *clustering* and *phase separation*. The heat or enthalpy of mixing can be written as $\Delta H_m = NzC(1-c)V$ for a bcc or fcc assembly of N atoms where $(1-c)$ and c are the atomic fractions of the species A and B, respectively; z is the 1st nn coordination of the structure. Clearly, $\Delta H_m < 0$ indicates a tendency for ordering and $\Delta H_m > 0$ is indicative of a clustering system. Importantly, in the 1st nn approximation ordering and clustering effects are essentially mutually exclusive—the system either tends to order (SRO or LRO) or shows clustering and a tendency to phase separate. However, extending this pairwise interaction model to include 2nd nn interactions, the model becomes richer in terms of possible behaviors since the 1st nn and 2nd nn (and higher) interactions may be of opposite sign. Ordering and phase separation are no longer mutually exclusive behaviors and the association of a miscibility gap within an ordering system emerges and this interplay of ordering and clustering tendencies can markedly influence the thermodynamic stability of a solution with respect to ordering and phase separation. Spinodal decomposition can actually be involved in the formation of ordered precipitates during the decomposition of supersaturated solid solutions (see e.g. Soffa and Laughlin, 1982, 1988, 1989).

Following the straightforward approach of Ino (1978) in the case of A2 → B2 ordering, the free energy of mixing is written as a function of the composition c and order parameter η as follows:

$$F_M(c, \eta) = Nc(1-c)[z_1 V + z_2 U] + \frac{N\eta^2}{4}[z_1 V - z_2 U] + \text{entropy terms} \tag{48}$$

wherein we have included the 2nd nn intereaction energies with $z_1 = 8$ and $z_2 = 6$ referring to coordination of the 1st nn and 2nd nn shells for the A2 → B2 ordering; η is the usual Bragg–Williams (LRO)

(a)

(b)

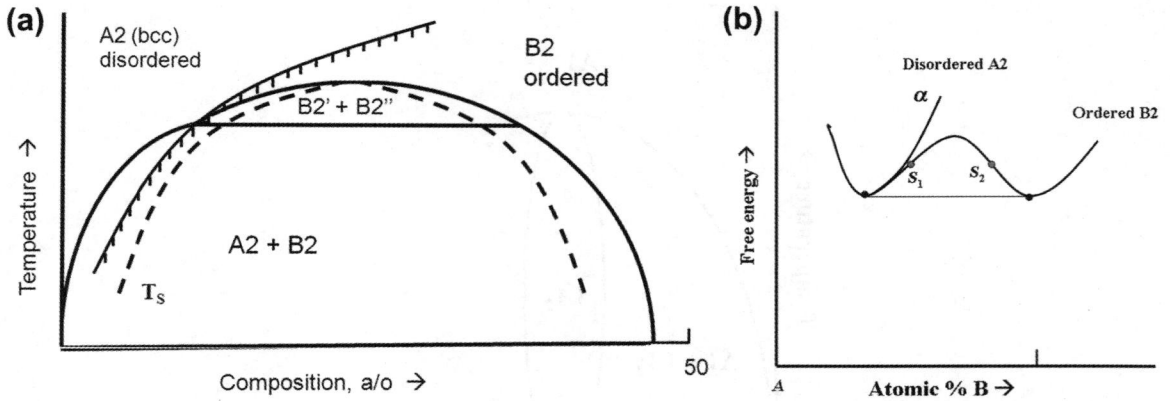

Figure 27   (a) Interplay of ordering and phase separation in an A2/B2 ordering system showing phase separation of the partially ordered B2 phase and associated spinodal locus. (b) Free energy–composition curves showing ordering and phase separation at low temperatures.

order parameter which in the case of the A2 → B2 transition in a stoichiometric A–B alloy can be written simply as

$$\eta = \frac{R - W}{R + W} \tag{49}$$

with R being the total number of "right"atoms and W the total number of "wrong" atoms occupying the conventional α and β sublattices of the B2 superlattice denoted previously. The entropic terms in this generalized Bragg–Williams model are the same as those in Eqn (40) assuming random mixing on the sublattices.

This free energy functional can lead to a phase diagram configuration shown schematically in **Figure 27** for the case $U \approx \frac{|V|}{3}$ with $V < 0$ and $U > 0$. In this situation a miscibility gap appears in the system below the order → disorder line of critical points along with an associated chemical spinodal locus $\left(\frac{\partial^2 F_M}{\partial c^2} = 0 \right)$. An alloy of composition $c \approx 0.25$ cooled from the α (A1; disordered bcc) single-phase field will order upon crossing the A2 → B2 critical temperature giving rise to an imperfectly ordered B2 superlattice and as the temperature is further decreased the alloy will phase separate into two imperfectly ordered B2 phases—B2′ + B2″—of different compositions. At still lower temperatures one phase will become disordered and the other will increase in composition and degree of order resulting in an equilibrium two-phase mixture of solute-depleted α (A1; disordered bcc) + solute-enriched B2 (ordered; sc). An associated free energy–composition diagram is depicted in **Figure 27b** showing the solution energetics and instabilities deriving from the expanded model which includes higher order interactions.

The generalized Bragg–Williams model including 1st and 2nd nn interactions can be extended to A1(fcc) → L1$_2$(A$_3$B; sc) ordering systems discussed earlier. (The reader is reminded that A2 → B2 ordering is generally a second-order/higher order transition whereas A1 → L1$_2$ is first order according to the Ehrenfest classification.) The free energy of mixing of the fcc-based solid solution relative to the pure components A and B can be written as

$$F_M(c, \eta) = Nc(1 - c)[12V + 6U] + \frac{N\eta^2}{16}[12V - 18U] + \text{entropic terms} \tag{50}$$

and again the entropic terms assuming random mixing on the conventional sublattices α_1, α_2, α_3 (the three designated α sublattices are crystallographically equivalent) and β are identical to those formulated previously. For $\dfrac{U}{|V|} \approx 0.4$ with $V < 0$ and $U > 0$ one finds a phase diagram configuration and stability loci depicted in **Figure 28a**. This description of the phase boundaries and stability limits includes a locus T_i^- which represents instability with respect to ordering and a chemical spinodal (T_S) locus instability with respect to phase separation contingent on prior ordering (condition spinodal instability). **Figure 28a** and b shows a summary of these results in graphical form including a free energy–composition diagram. This will be discussed, in detail, in a subsequent section on precipitation of ordered phases within supersaturated solid solutions. What is clear is that in both cases (the A2 → B2 and the A1 → L1$_2$ ordering systems) the inclusion of 1st and 2nd nn interactions leads to a substantially more

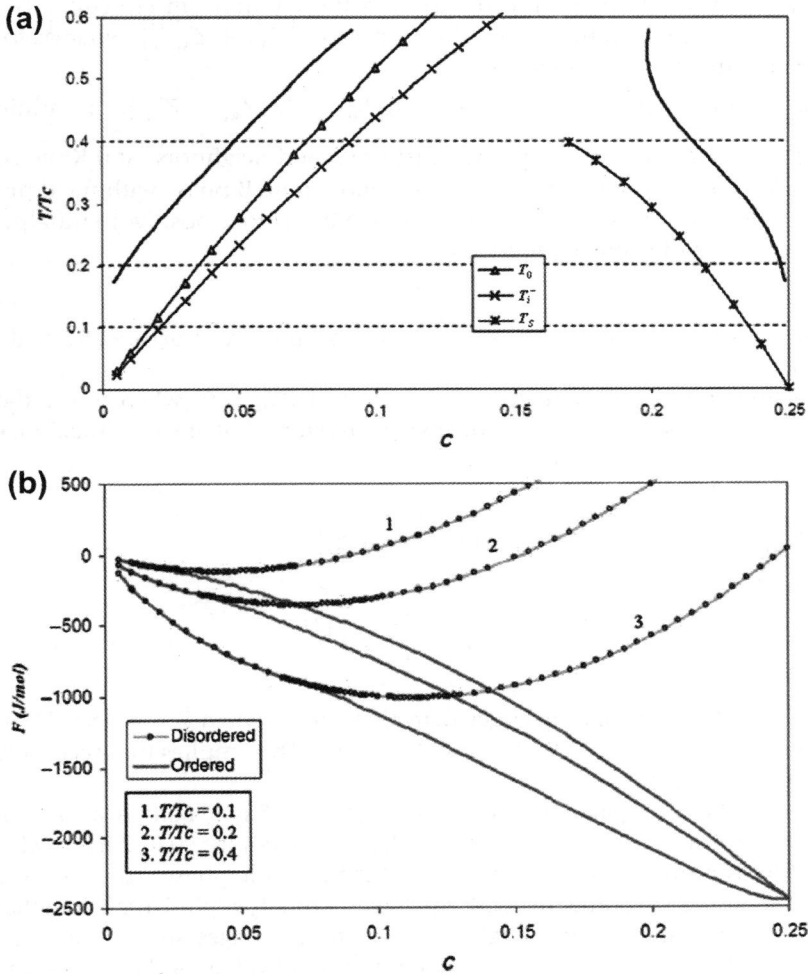

Figure 28 (a) The phase boundaries and instability loci and (b) the free energy vs. composition curves at three temperatures for $\frac{U}{|V|} \approx 0.4$ for the L1$_2$ ordering. After Soffa et al. (2010).

complex solution behavior which can allow for varied diffusional paths involving ordering and clustering in the transformation of the system as it approaches equilibrium (Soffa et al., 2010).

8.2.4.2   The Ground State of Binary Alloys with V < 0

Following Richards and Cahn (1971) we can write the enthalpy of mixing of an A–B alloy up to second neighbors as

$$\Delta H_M = NV\left[Z_{AB}^{(1)} + Z_{AB}^{(2)}\frac{U}{V}\right] \tag{51}$$

where U and V are defined above, Z_{AB}^i is the number of ith neighbor A-B bonds per atom and N is the total number of atoms in the alloy. The values of Z_{AB}^i can be easily calculated by considering the number of A-B bonds in the unit cell and dividing by the number of atoms in the unit cell. At 0 K (the ground state) the structure which minimizes ΔH_M is the equilibrium structure. Thus if $V < 0$, the equilibrium state is the one which maximizes the expression $\left[Z_{AB}^{(1)} + Z_{AB}^{(2)}\frac{U}{V}\right]$, whereas for $V > 0$ the state which minimizes it is the equilibrium state.

If both V and U are positive, the expression $\Delta H_M = NV\left[Z_{AB}^{(1)} + Z_{AB}^{(2)}\frac{U}{V}\right]$ is minimized for both $Z_{AB}^{(1)}$ and $Z_{AB}^{(2)}$ equal to zero, that is no opposite first or second neighbors. At 0 K the equilibrium state of A–B alloys would be composed of a pure A phase and a pure B phase with fractions determined by the lever rule. Any other configuration would have a larger (more positive) enthalpy of mixing and therefore not be in a state of stable equilibrium.

8.2.4.2.1   BCC Ground States

Let us look at the two ordered phases of 50% B based on the bcc structure (B2 and B32) shown in **Figure 29**a and b.

It can be seen that for the B2 structure, $Z_{AB}^{(1)} = 4$ and $Z_{AB}^{(2)} = 0$, whereas for the B32 structure $Z_{AB}^{(1)} = 2$ and $Z_{AB}^{(2)} = 3$. Thus, if we compare the two expressions that are to be maximized to determine the stable state:

$$\text{for B2}: \left[4 + 0\frac{U}{V}\right]$$

$$\text{for B32}: \left[2 + 3\frac{U}{V}\right]$$

it is seen that the expression for B32 is larger than that for B2 when $\frac{U}{V} > \frac{2}{3}$. That is, B32 is the stable ordered phase when $\frac{U}{V} > \frac{2}{3}$: otherwise B2 is the stable phase. This implies that second neighbors are of importance in the B32 structure.

One other case could be compared with these two ordered phases, namely a completely disordered 50% B alloy. For this case the average $Z_{AB}^{(1)} = 2$ and $Z_{AB}^{(2)} = 1.5$. For $\frac{U}{V} > 0$ B32 has the lower enthalpy of mixing and therefore is more stable. But for $\frac{U}{V} < \frac{2}{3}$ B2 is the stable phase. Thus, if $V < 0$, the homogeneously disordered phase is never the one with the lowest enthalpy of mixing. Note that for the case of negative $\frac{U}{V}$ the use of first and second neighbor interaction energies shows that a configurationally disordered phase is not stable at 0 K in agreement with the third law of thermodynamics.

At 25% B for $\frac{U}{V} > 0$ the phase with the D0$_3$ structure is the stable one since all of the B atoms have opposite first and second neighbors, which is favored when both U and $V < 0$. Between the composition 0% B and 25% B, a two-phase mixture of pure A and fully ordered stoichiometric

(a) **(b)** **(c)**

(d) **(e)**

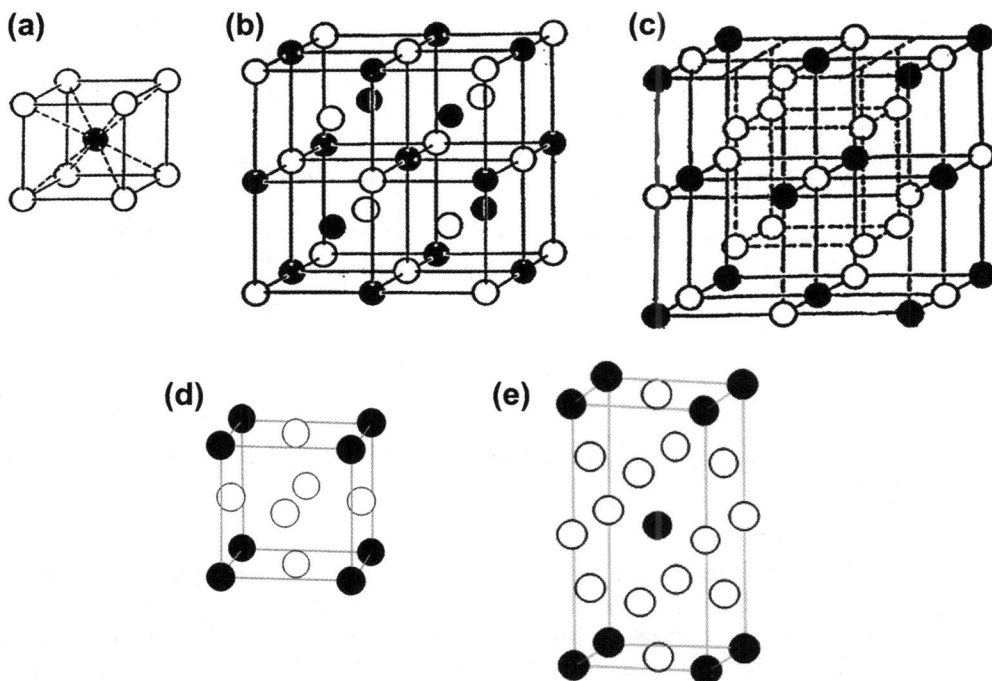

Figure 29 (a) The B2 (CsCl) structure. (b) The B32 (NaTl) structure. (c) The DO_3 (BiF$_3$) structure. (d) The $L1_2$ (fcc derivative) structure. (e) The DO_{22} (fcc derivative) structure.

DO_3 has the lowest enthalpy of mixing and is therefore the stable configuration. Between the composition 0% B and 25% B, a homogeneously ordered DO_3 phase (order parameter less than unity) may have the same configurational enthalpy of mixing (it is degenerate with the two-phase mixture) but since it has configurational entropy, it is not a stable phase at 0 K according to the third law of thermodynamics.

A diagram showing the stable phases for bcc derivative ordered structures is shown in **Figure 30**a.

8.2.4.2.2 FCC Ground States

Figure 29d and e displays two possible ordered phases that have derivative structures of the FCC (A1) structure, namely the $L1_2$ (Cu$_3$Au) and DO_{22} (Ti$_3$Al) structures. It can be seen that for the $L1_2$ structure, $Z_{AB}^{(1)} = 3$ and $Z_{AB}^{(2)} = 0$, whereas for the structure DO_{22}, $Z_{AB}^{(1)} = 3$ and $Z_{AB}^{(2)} = 0.5$. Thus, if we compare the two expressions that are to be maximized to determine the stable state:

$$\text{for } L1_2 : \left[3 + 0\frac{U}{V}\right]$$

$$\text{for } DO_{22} : \left[3 + 0.5\frac{U}{V}\right]$$

it is seen that the expression for DO_{22} is larger than that for $L1_2$ when $\frac{U}{V} > 0$. That is, DO_{22} is the more stable ordered phase whenever there is a tendency for next near neighbors to prefer to be opposite.

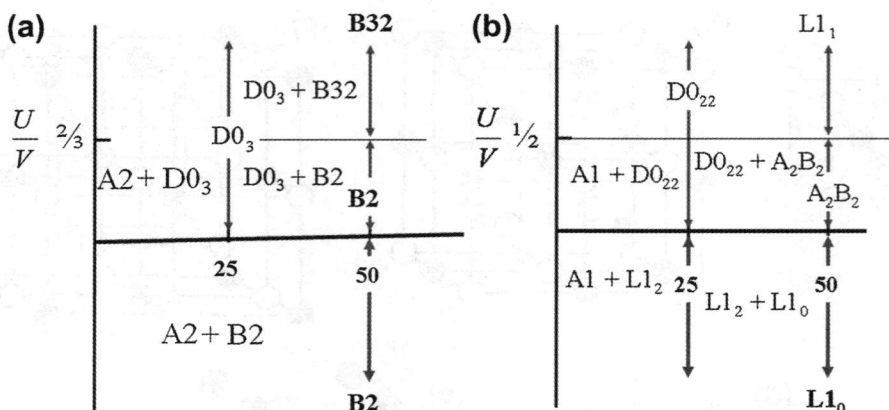

Figure 30 (a) A partial ground state diagram for bcc derivative ordered phases. (b) A partial ground state diagram for fcc derivative ordered phases.

Other cases of FCC derivative structures could be explored. For example the AB alloy (50% each) has three possible ordered phases: $L1_0$, $L1_1$ and the so-called A_2B_2 structure (space group $I\frac{4_1}{a}md$). By similar reasoning as shown above it can be shown that for $\frac{U}{V} < 0$ $L1_0$ is the ground state phase: for $0 < \frac{U}{V} < \frac{1}{2}$ the A_2B_2 structure is the ground state and for $\frac{U}{V} > \frac{1}{2}$ $L1_1$ is the ground state structure. A partial ground state phase diagram for FCC derivative structures is shown in **Figure 30**b.

More complete ground state diagrams for BCC and FCC can be found in Richards and Cahn (1971). An HCP ground state diagram is found in Singh and Lele (1991).

8.2.4.3   k-Space Solution Energetics

Let us now briefly introduce an extension of the familiar Bragg–Williams quasichemical treatment employed earlier using the language of Khachaturyan's static concentration wave (SCW) formalism (Khachaturyan, 1978, 1983) which has been employed extensively by Khachaturyan and others.

In this "k-space energetics" the site occupancies and interchange energies are formulated in k-space (reciprocal space) through the use of Fourier analysis and the discrete Fourier transform. In this approach the site occupation probability $n(\mathbf{r})$ is given by

$$n(\mathbf{r}) = c + \sum Q(\mathbf{k})\exp(ik\cdot r) \tag{52}$$

where $Q(\mathbf{k})$ is the amplitude of the Fourier component with wave vector $\mathbf{k}$ and c is the average composition; $\mathbf{r} = x\mathbf{a_1} + y\mathbf{a_2} + z\mathbf{a_3}$ is a vector in real space locating the atomic positions (x,y,z) within the conventional unit cell. Here the atomic arrangement is viewed as a superposition of concentration waves with wave vectors $\mathbf{k}$ and amplitude $Q(\mathbf{k})$ localized in the first Brillouin zone. Furthermore, the interchange energies are represented by a Fourier transform $V(k)$ as follows:

$$V(k) = \sum V(r)\exp(ik\cdot r) \tag{53}$$

where the summation is over the 1st and 2nd nn shells. For the case of fcc-based $L1_2$ ordering the superstructure can be generated by the emergence of a set of concentration waves $\{k_0\}$ comprising a so-called "star" of the structure (dictated by symmetry) and written as

$$k_0 = 2\pi\left(a_1^* + a_2^* + a_3^*\right)$$

and the occupation probability written as

$$n(r) = c + \sum (\gamma\eta)\exp(ik_0 \cdot r)$$

where γ is a constant depending on the crystal structure and equal to $1/4$ for the $L1_2$ superstructure and η is the order parameter. Thus, for the evolution of the $L1_2$ superstructure:

$$n(r) = c + \frac{\eta}{4}\sum \exp\left(2\pi\frac{x}{a}\right) + \exp\left(2\pi\frac{y}{a}\right) + \exp\left(2\pi\frac{z}{a}\right) \tag{54}$$

resulting from the growth of three symmetrically related concentration waves along [100], [010] and [001], respectively. The Fourier transforms of the interchange parameter for 1st and 2nd nn interactions yield

$$V(k_0) = -4V + 6U \tag{55}$$

$$V(0) = 12V + 6U \tag{56}$$

where V and U are the 1st and 2nd nn interchange parameters defined above; the term $V(0)$ for $k=0$ (Brillouin zone center) is associated with the disordered reference solid solution. In this formalism the free energy functional for the fcc solid solution is written as

$$F_m(c, \eta) = Nc(1 - c)V(0) - \frac{3}{16}N\eta^2 V(k_0) + \text{entropic terms} \tag{57}$$

wherein the entropic terms are identical to those derived above assuming random mixing on the sublattices of the superstructure.

8.2.4.4 k-Space Solution Energetics/L1₀

The k-space formulation of fcc-based solution energetics and stability analysis has been extended to the A1 → $L1_0$ ordering transition by Cheong and Laughlin (1994) and Soffa et al. (2011) employing the generalized Bragg–Williams approach (1st nn and 2nd nn interactions) including an elastic relaxation term in the free energy. The free energy of mixing of an A–B binary solution is written as

$$F_m(c, \eta) = Nc(1 - c)[12V + 6U] + \frac{N\eta^2}{4}[4V - 6U] + \text{entropic terms} \tag{58}$$

where the entropic terms are given by

$$S_m = -\frac{Nk_B}{2}\left\{\left[(1 - c) + \frac{\eta}{2}\right]\ln\left[(1 - c) + \frac{\eta}{2}\right] + \left[c - \frac{\eta}{2}\right]\ln\left[c - \frac{\eta}{2}\right] + \left[c + \frac{\eta}{2}\right]\ln\left[c + \frac{\eta}{2}\right]\right.$$
$$\left. + \left[(1 - c) - \frac{\eta}{2}\right]\ln\left[(1 - c) - \frac{\eta}{2}\right]\right\} \tag{59}$$

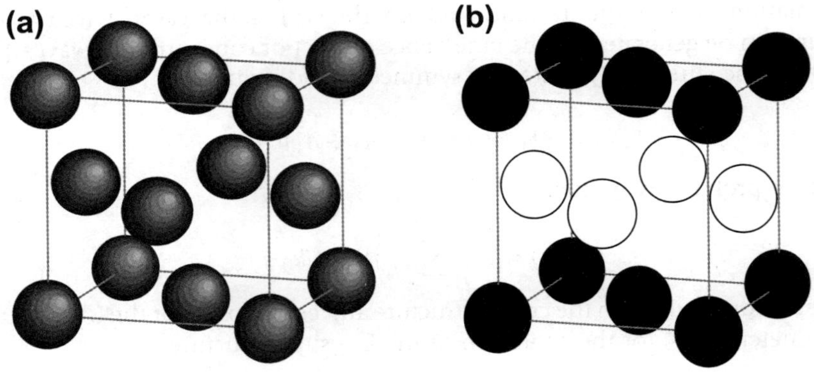

Figure 31   (a) FCC unit cell and (b) the L1$_0$ unit cell based on the FCC unit cell.

assuming random mixing on the L1$_0$ sublattice sites within a conventional fcc unit cell. See **Figure 31**. These can be then be transposed to the k-space representation as follows:

$$F_m(c, \eta) = Nc(1 - c)[V(0)] - \frac{N\eta^2}{4} V(k_0) + \text{entropic terms} \qquad (60)$$

where $k_0 = 2\pi a_3^*$ and $V(0) = 12V + 6U$ and $V(k_0) = -4V + 6U$. This model based on a rigid cubic lattice will lead to a second-order/higher order A1 $\rightarrow$ L1$_0$ transition similar to the result arrived at in the classic work of Nix and Shockley (1938). Guggenheim in 1952 using a quasichemical approach and a tetrahedron approximation was able to capture the first-order character of the A1 $\rightarrow$ L1$_2$ transition within this precursor of the modern CVM. Interestingly, Larikov et al. (1975) asserted that the phase change is first order if changes in lattice dimensions are incorporated into the free energy functional within a modified Bragg–Williams model. If an elastic energy term deriving from the elastic relaxation stemming from the cubic to tetragonal transformation strain is grafted on to the free energy expression in Eqn (60) in the form $E_{\text{ELASTIC}} = -Ne^4\eta^4$ where e is an elastic strain coupling term which is a function of the transformation strain and elastic constants, the ordering transition becomes first order. (See Cheong and Laughlin, 1994).

8.2.5 Landau Theory of Phase Transformations

In the late 1930s, Landau (1937) proposed that all second order transitions (including the ferromagnetic to paramagnetic transition in iron) and many first order phase transitions such as atomic ordering in alloys can be characterized by one or more so-called generalized order parameters (η). The order parameter describes salient properties of an assembly of constituent particles, that is electrons, atoms, ions, spins, and changes systematically as a critical temperature T_c (at constant pressure), for example, is approached. The order parameter describes the evolution of the system in terms of measurable physical parameters and has an equilibrium value for a given set of relevant thermodynamic variables ($T, P, E, H, \ldots$). The behavior of the order parameter in the vicinity of the critical point serves as a useful basis for classifying the nature of the phase transition/transformation. The order parameter η can be formulated to describe the

(1) magnetization of a ferromagnet
(2) polarization of a dielectric

(3) occupancy of sublattices in an alloy superstructure
(4) fraction of superconducting electrons in a metal and
(5) atomic displacements associated with structural phase changes

When normalized, $\eta = 0$ at high temperatures in the disordered state and becomes finite in the ordered state at low temperatures with $\eta \to 1$ as $T \to 0$.

The essential hypothesis of the Landau theory is that the free energy difference between the ordered ($\eta > 0$) and disordered states ($\eta = 0$), $G(\eta)$ can be expanded in a power series in the order parameter in the neighborhood of the critical point, that is, the free energy is assumed to be an analytic function of η in the vicinity of $\eta = 0$. Landau recognized that this may not be rigorously true but suggested that this would not affect the general character of the transformation arising in the model. Ginzburg, Levanyuk and Sobyanin (1987) quantified the region in which the expansion is not valid and showed that when long-range fields (electric, magnetic, strain) and interactions are involved, the region in which the Landau theory fails is small, of the order of a degree or two. Furthermore, it should be mentioned that Landau's phenomenological theory is essentially a "mean field" theory and basically a generalization of the Weiss molecular field theory approximation to ferromagnetism (Weiss, 1907).

Landau wrote the free energy difference

$$G(\eta) = G(\eta \neq 0) - G(\eta = 0)$$

between states of finite order parameter (ordered states) and states $\eta = 0$ (disordered states) as

$$G(\eta) = A\eta^2 + B\eta^3 + C\eta^4 + D\eta^5 + E\eta^6 + \ldots \tag{61}$$

where the coefficients $A, B, C, \ldots$ are generally functions of temperature (T) and pressure (P), that is, $A(T,P)$, $B(T,P)$, $C(T,P)$. At constant pressure, A can be taken as a linear function of temperature given by

$$A = a(T - T_0) \tag{62}$$

with B, C, essentially constants, in the first approximation. Importantly, the coefficient $A = a(T - T_0)$ is such that the parameter a is a positive constant for the case where the high-temperature phase is the high-symmetry phase. In general, symmetry considerations play an important role in analyses based on the Landau theory. The coefficient A clearly represents the curvature of the $G(\eta)$ versus η plot at $\eta = 0$ and changes sign at $T = T_0$. Furthermore, T_0 will be identified as an instability temperature wherein the high-temperature phase at $\eta = 0$ becomes thermodynamically unstable with respect to a low-temperature ordered phase.

When truncating the expansion beyond fourth order, namely,

$$G(\eta) = A\eta^2 + B\eta^3 + C\eta^4 \tag{63}$$

there are two broad classes of interest, namely, where $B = 0$ and $B \neq 0$. (Note: for $B = 0$ there are also two cases of interest, $C > 0$ and $C < 0$; however, for $C < 0$ we must include a sixth order term to keep the free energy positive at large values of the order parameter.)

For the case $B = 0$, $G(\eta)$ is an even function of η, that is

$$G(\eta) = G(-\eta) \tag{64}$$

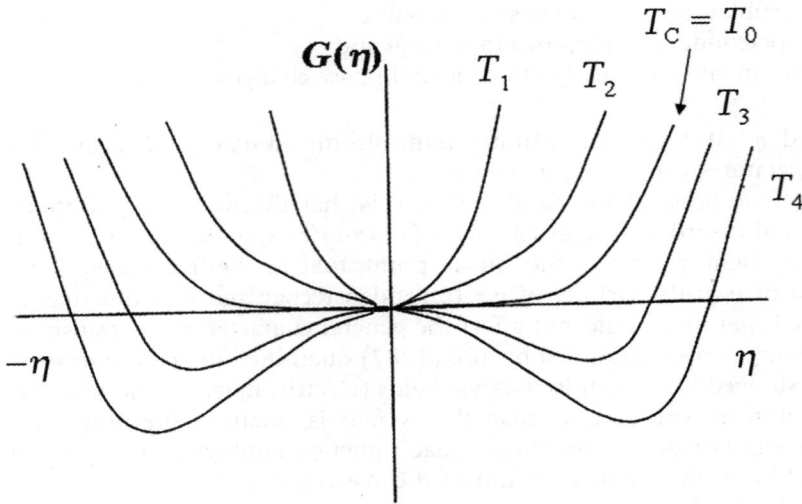

Figure 32 Gibbs free energy for the case of the Landau expansion with A and $C \neq 0$. At T_C the disordered phase becomes unstable. $T_1 > T_2 > T_C = T_0 > T_3 > T_4$. This represents a higher order transition.

and $A = a(T - T_0)$ changes sign at $T = T_0$. The behavior of $G(\eta)$ as a function of temperature and the variation of the order parameter with temperature are shown in **Figure 32**.

This is the signature of a SECOND-ORDER transition in the Landau theory. The equilibrium order parameter varies continuously from zero to a finite value on cooling through the transition or critical temperature $T = T_C = T_0$. (Note that the physical states of the system for $+\eta$ and $-\eta$ are essentially identical.) At equilibrium, for $T < T_C$, taking $\frac{\partial G}{\partial \eta} \equiv G_\eta = 0$ yields the following equation:

$$\eta = \left(\left(\frac{a}{2C} \right) (T_C - T) \right)^{\frac{1}{2}} \tag{65}$$

and inserting $\eta = 1$ at $T = 0$ K gives $\left(\frac{a}{2C} \right) = \frac{1}{T_C}$, leading to the well-known result

$$\eta = \left(\frac{(T_C - T)}{T_C} \right)^{\frac{1}{2}} = \left(1 - \frac{T}{T_C} \right)^{\frac{1}{2}} \tag{66}$$

which is identical to the result deriving from mean field approaches with a so-called critical exponent of $1/2$. This SECOND-ORDER or continuous transition is virtually identical to the behavior predicted by Ehrenfest (1933). In this case, $T = T_C = T_0 = T_i^-$, the critical temperature T_C can be identified with the parameter T_0 which is an instability temperature T_i^- for the disordered phase on cooling. Below $T_C = T_i^-$ small fluctuations about the disordered state ($\eta = 0$) experience no intrinsic thermodynamic restoring force and the free energy is monotonically decreasing to equilibrium ordered states $+\eta^*$ and $-\eta^*$ with the magnitude of η^* increasing continuously as T is decreased below T_C. At T_C, $G_\eta = 0$ and $G_{\eta\eta} = 0$ at $\eta = 0$, that is, just above or at $T = T_C$ finite fluctuations away from the disordered state ($\eta = 0$) lead to small changes in the free energy of the system and tend to appear profusely throughout the system because of the flattening of the $G(\eta)$ versus η curve near $\eta = 0$, virtually as premonitory signaling of the phase transition. Just below the critical temperature T_C, minima appear in the free energy at small values of η and the free energy decreases monotonically to these equilibrium values.

This model describes the salient features of a second-order phase transition and the behavior of relevant thermodynamic properties of the system in a straightforward manner.

The difference in Gibbs free energy between the ordered state and disordered state, G_O is found by substituting the equilibrium value of η in 2.65 into 2.63 (with $B = 0$). One obtains

$$G_O = -\frac{a(T_C - T)^2}{2T_C} \tag{67}$$

The entropy S of the system varies continuously through the transition and the entropy change can be found from $S_O = -\dfrac{\partial G_O}{\partial T}$ as

$$S_O = -a\frac{(T_C - T)}{T_C} \tag{68}$$

The heat capacity relative to the disordered state at constant pressure is thus

$$\Delta C_P^O = T\left(\frac{\partial S_O}{\partial T}\right) = \frac{aT}{T_C}. \tag{69}$$

which produces a finite discontinuity in the heat capacity at $T = T_C$ equal to the coefficient a. See **Figure 22**.

Finally, the enthalpy difference H_O is equal to $= G_O + TS_O$:

$$H_O = \frac{a}{2}\frac{\left(T^2 - T_C^2\right)}{T_C} \tag{70}$$

This equation shows an absence of a latent heat of transformation at T_C. However, the slope of the variation of the order parameter η with temperature as $\eta \to 1$ at $T = 0$ K is negative in this mean field model, which is at variance with the predicted slope of zero according to the requirements of the Third Law of Thermodynamics.

The SECOND-ORDER transition behavior predicted earlier occurs very rarely in real materials. However, there is one important case that shows the earmarks of a textbook second-order transition, namely, the normal $\to$ superconducting transition in zero field (Stanley, 1971). Most phase transitions/transformations occurring in nature are FIRST ORDER according to Ehrenfest's thermodynamic classification and will be discussed subsequently, in detail, but there are important transitions in materials which are not FIRST ORDER or SECOND ORDER. For example, the magnetic transition in iron (Fe) at its Curie temperature (1041 K)—paramagnetic→ferromagnetic transition—is at least third order (Pippard, 1966). The specific heat of many ferromagnets in the vicinity of the Curie temperature (T_C) exhibits singular behavior quite different than predicted for a second-order transition as discussed earlier. Similarly, a number of ordering transitions and superlattice formation in alloys which are clearly not first-order transitions show similar behaviors, for example the A2 $\to$ B2 ordering in Cu–Zn alloys (β-brass). See **Figure 24**. Indeed, the classic Weiss molecular approach to the magnetic transition in iron (Fe) and the Bragg–Williams approach to superlattice formation are mean field theories and predict second-order behavior. The complexities of the behavior near the critical temperature for these transitions which are neither first or second order are lumped into a category called HIGHER ORDER or λ-TRANSITIONS (the latter description based on the shape of the specific heat near T_C, see **Figure 24**) circumventing the need to differentiate discontinuities in third, fourth, and so on, derivatives of the free energy functional. The complex behavior mentioned earlier derives from short-range ordering effects, correlations and fluctuations near the critical point and remains a challenge in modern statistical physics.

If in the Landau expansion the series is again truncated beyond fourth order, but we take $B < 0$ and $C > 0$ with $A = a(T - T_0)$ as above, we reveal another type of behavior of major importance—a FIRST-ORDER transition. In this case, A remains finite at $T = T_C$ and two minima appear at $\eta = 0$ and $\eta = \eta^*$, respectively. This situation represents two-phase equilibrium at T_C between a disordered phase ($\eta = 0$) and an ordered phase ($\eta = \eta^*$). This behavior is indicative of a FIRST-ORDER transition characterized by a discontinuous change in the entropy (S) and enthalpy (H) at the transition temperature T_C associated with a latent heat ΔH_t and $\Delta S_t = \Delta H_t/T_C$. For $T < T_C$, a minimum in the free energy develops with increasing order parameter as the temperature is decreased. A single-phase ordered state becomes the thermodynamically preferred state of the system with the order parameter continuously increasing toward unity after the discontinuous jump at $T = T_C$. See **Figure 33**.

Note also that the free energy functional $G(\eta)$ is no longer an even function of η, that is, $G(\eta) \neq G(-\eta)$ for this case, that is, it is an odd function of η.

This thermodynamic analysis has implications regarding the possible mechanisms of the ordering transformation. Here we see in this FIRST-ORDER case that at the higher temperature regime $T \leq T_C$ there is local maximum between the equilibrium ordered state $\eta = \eta^*$ and the disordered state $\eta = 0$ including $T = T_C$. This represents an intrinsic thermodynamic restoring force within the disordered state with respect to small fluctuations away from the disordered state $\eta = 0$; that is, relatively small fluctuations tend to be damped or decay, thus to trigger the transformation, a relatively large fluctuation away from the initial state is required which releases free energy and allows the system to seek the lower free energy equilibrium ordered state. This is not the nucleation barrier Δg^*, per se to form the ordered

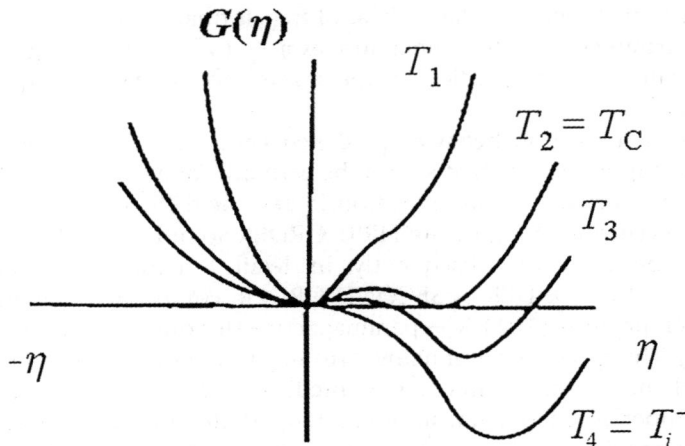

Figure 33   Gibbs free energy for the case of the Landau expansion with A, B and $C \neq 0$. At $T_2 = T_C$ the disordered phase has the same free energy as an ordered phase. At T_3 an ordered phase is stable. At T_0, if the disordered phase is retained it becomes unstable with respect to ordering. This represents a first-order transition. Here, $T_1 > T_2 = T_C > T_3 > T_i^-$.

state but a thermodynamic condition which imposes a requirement that the ordered state can only be achieved by large fluctuations away from $\eta = 0$—a nucleation event. At a temperature $T = T_0 = T_i^- < T_C$, the local maximum disappears and the free energy $G(\eta)$ versus η curve is monotonically decreasing away from $\eta = 0$ rendering the disordered state thermodynamically unstable with respect to the equilibrium ordered state $\eta = \eta^*$. It can be shown that

$$T_C - T_i^- = \frac{B^2}{4aC} \tag{71}$$

for this case.

This thermodynamic instability appearing when the system is displaced far from equilibrium suggests that the kinetic behavior will mimic the second-order/higher order transition whereby it is possible that ordering will occur continuously or homogeneously throughout the undercooled system without the initial partitioning of the system into transformed and untransformed regions, as is the case for nucleation and growth processes. In the special case of atomic ordering and superlattice formation in alloys this involves uniformly occurring preferential atomic jumping into sites which locally increase the order parameter.

Another case of interest involves the symmetrical FIRST-ORDER behavior wherein $A = a(T - T_0)$, $B = 0$, $C < 0$, $D = 0$ and $E > 0$; that is, the $G(\eta)$ functional is symmetric about $\eta = 0$ with two phases in thermodynamic equilibrium at $T = T_C$ and $T_0 = T_i < T_C$. See **Figure 34**.

As above, only the quadric term is given a temperature dependence and $C < 0$ and $E > 0$ are taken as constants. The sixth order term insures the proper thermodynamic behavior at low temperatures yielding real solutions for the order parameter. This specific behavior would apply to the A1(fcc) $\rightarrow$ L1$_0$ superlattice formation in equiatomic AB alloys (Cheong and Laughlin, 1994).

The Landau approach is increasingly entering the metallurgical phase transformation literature and providing valuable insight in various contexts. For example, it was argued for a number of years that the well-known BCC $\rightarrow$ ω-phase transformation in Zr-base and Ti-base alloys might be second/higher order. However, since the atomic displacement (which is the order parameter in this case) character- izing the transformation produces different structures for $\eta+$ versus $\eta-$ (atomic displacements in opposite directions), the free energy functional $G(\eta)$ is an odd function and therefore the transition is first order (de Fontaine, 1973). In the materials field, in general, the Landau theory has been a partic- ularly valuable tool in the area of ferroelectric behavior and in analyzing certain displacive trans- formations (International Tables for Crystallography, 2006). Finally, as mentioned earlier, symmetry considerations and restrictions emerge quite naturally in the Landau theory and one important result is that if the transformation is second order or continuous, the symmetry group of the low-temperature phase must be a subgroup (lower symmetry) of the group of the high-temperature phase.

8.2.6 Surface Energy and Capillarity Effects

A wide variety of effects related to surface or interfacial (free) energy and interface curvature play a central role in phase transformations and microstructural development. These effects include most prominently the nucleation and growth of phases, coarsening and the shape and distribution of phases in polyphase materials. The term capillarity refers to a composite of phenomena related to the ener- getics and kinetic behavior of surfaces and interfaces and their role in fundamental processes governing the evolution of material structure during processing and heat treatment. (The term surface is some- times used to refer strictly to the boundary between a condensed phase and a vapor phase, whereas the

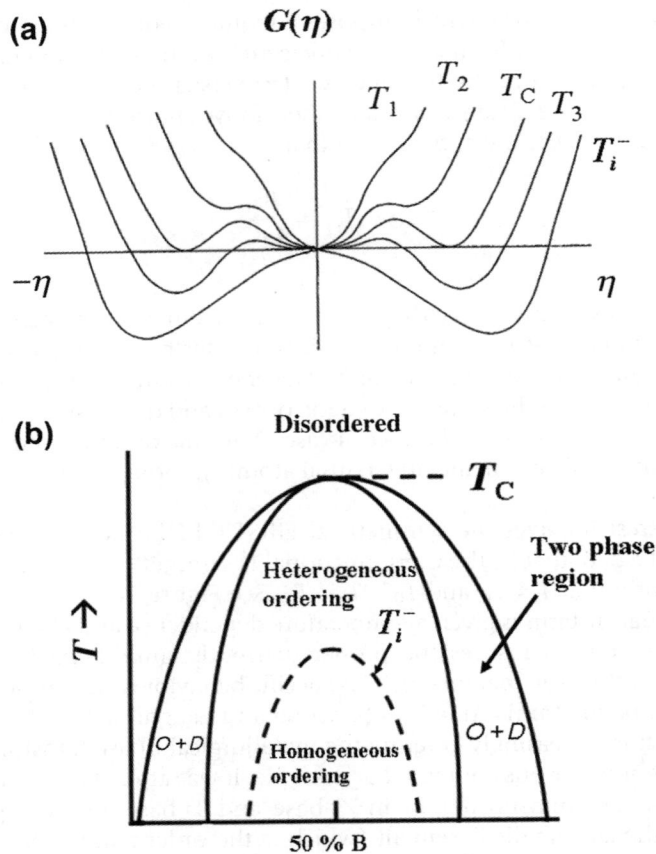

Figure 34 (a) Gibbs free energy for the case of the Landau expansion with A, C and $E \neq 0$, but $C < 0$. At T_C the disordered phase is in equilibrium with an ordered phase ($|\eta| \leq 1$). The disordered phase becomes unstable with respect to ordering at T less than or equal to T_i^-. This is the case of a symmetrical first-order transition. In the figure $T_1 > T_2 > T_C > T_3 > T_i^-$. (b) Schematic of the phase diagram that corresponds to the free energy curves of the 50%B plot in 34a. Note the region of homogeneous/spinodal ordering below T_i^-.

term interface is considered more general, referring to the region of contact between any two phases as well as grain boundaries within a single-phase polycrystalline solid. However, these terms are more often than not used synonymously in the literature and textbooks.)

Surface energy and interfacial effects are of paramount importance when the surface area-to-volume ratio of a phase is relatively large and/or when the interface has a large curvature. Generally, the effects of surface energy and curvature derive from the fact that the atomic environments of the atoms at an interface are different than those residing in the bulk phase. As a result, the energy of the surface atoms can be quite different than those located in the bulk and, indeed, the energy to add an atom to a surface or boundary location is generally higher than adding an atom to the bulk; this is essentially the origin of surface or interfacial energy as an *excess thermodynamic quantity* and is often described in terms of unsaturated/dangling bonds or perturbed electronic charge density.

When discussing the energetics and properties of interfaces three terms arise which must be carefully distinguished and understood. The *surface (free) energy* (σ) is the quantity associated with the reversible work required to create a unit area of new surface under prescribed thermodynamic constraints. If the area of an existing surface is increased by an increment dA, σdA is the surface work expended as the surface is extended without changing the nature or intrinsic structure of the surface. The *surface tension* (γ) and *surface stress* (f_{ij}) relate to forces acting within an interface resisting stretching the surface layer and which determine the forces against which work must be expended in extending the surface; the surface tension and surface stress essentially act as restoring forces resisting stretching the surface or interface. Importantly, the surface stress (f_{ij}) generally is a second rank tensor quantity describing the plane stress state of the surface or interface and the surface tension (γ) can be viewed as the average of the surface stresses in two mutually perpendicular directions within the surface. The surface stress at any point in a surface is the force acting across a line which passes through this point in the limit as the line length goes to zero. The surface (free) energy (σ) is a scalar quantity which has the same fundamental units as the surface stress and surface tension ($J\ m^{-2}/N\ m^{-1}/ergs\ cm^{-2}/dyne\ cm^{-1}$) and for liquids (which are isotropic) $\sigma = \gamma$, that is, the surface free energy and the surface tension of liquids are identical in the absence of adsorption or desorption effects. An important feature of the liquid state is that atoms have sufficient mobility to rearrange themselves during the extension of the surface maintaining a constant surface structure which is not necessarily the case for a crystalline solid. Furthermore, for crystalline solids, which are generally anisotropic, σ varies with orientation and the equilibrium shape of a crystal is found to be nonspherical, as discussed more fully subsequently. In the following brief review of the thermodynamics of interfaces we will emphasize the role of the specific interfacial free energy σ associated with a surface, internal interface or interphase interface and will neglect any adsorption/desorption effects.

Let us now briefly elucidate the relationship between surface (free) energy σ and the surface stress f (f_{ij}) in a more fundamental thermodynamic approach. The presence of a surface or interface of area A in a system contributes an excess Gibbs free energy $G^s = \sigma A$. If we now extend the surface at constant T and P by an amount dA, there can be two contributions to the surface work required as follows:

$$dG^S = d(\sigma A) = \sigma dA + Ad\sigma \tag{72}$$

where the term $\sigma\,dA$ is simply the increase in free energy due to the new interfacial area created maintaining a constant interface structure or constant σ in the process, whereas the second term, $Ad\sigma$, allows that an existing surface may be strained or distorted thereby modifying σ. We can use this result to define the surface stress f as

$$dG^S = fdA = \sigma dA + Ad\sigma = \sigma dA + A\frac{d\sigma}{dA}dA$$

$$f = \sigma + A\frac{d\sigma}{dA} \tag{73}$$

or, more rigorously,

$$f_{ij} = \sigma\delta_{ij} + \frac{\partial\sigma}{\partial\varepsilon_{ij}} \tag{74}$$

wherein $dA = A\delta_{ij}d\varepsilon_{ij}$ and $d\varepsilon_{ij}$ is a surface elastic strain tensor (assuming the modification of σ derives from elastic distortion of the surface structure). For a crystal surface with threefold or higher rotational symmetry the normal stresses within the surface are equal and no shear stresses act within the surface.

This follow from Hermann's theorem, which states that "…if an r rank tensor has an N fold symmetry axis and $r < N$, then this tensor also has a symmetry axis of infinite order along N." (Hermann, 1934; Sirotin and Shaskolskaya, 1982). Thus the property is isotropic in the plane perpendicular to N.

Therefore, when the surface stress is isotropic it can be taken as a scalar quantity wherein

$$f - \sigma = \frac{d\sigma}{de} \tag{75}$$

and $e = \dfrac{dA}{A}$ is an elastic strain within the surface. Only when $\dfrac{d\sigma}{de} = 0$ does the surface (free) energy σ equal the surface stress f (Howe, 1997; Trivedi, 1999).

The most straightforward thermodynamic definition of σ can be written as

$$\sigma = \left(\frac{dW}{dA}\right)_{T,P,n_i} = \left(\frac{dW}{dA}\right)_{T,V,n_i} \tag{76}$$

where dW is the reversible surface work required to increase the surface area(A) of a surface or interface by an amount dA at constant T,P or T,V in a thermodynamic system at constant composition (n_i). From the fundamental relationships for reversible changes in the Gibbs (G) and Helmholtz (F) free energies:

$$dG = -SdT + VdP + \sigma dA + \sum_i \mu_i dn_i \tag{77}$$

and

$$dF = -SdT - PdV + \sigma dA + \sum_i \mu_i dn_i \tag{78}$$

we can also write

$$\sigma = \left(\frac{\partial G}{\partial A}\right)_{T,P,n_i} = \left(\frac{\partial F}{\partial A}\right)_{T,V,n_i} \tag{79}$$

where σ is viewed as a specific interfacial free energy associated with an excess free energy stemming from the presence of surfaces or interfaces. Sometimes when dealing with processes unfolding at constant T,V and μ_i (such as nucleation) the surface (free) energy can be defined as

$$\sigma = \left(\frac{dW}{dA}\right)_{T,V,\mu_i} \tag{80}$$

which will be discussed subsequently. Finally, in a crystalline solid, σ can be highly anisotropic varying markedly with the crystallographic direction of the normal to the crystal plane in question. This anisotropy is readily illustrated using the well-known polar plot shown in **Figure 35**. The polar plot can be used to establish the equilibrium shape of a crystal through the Gibbs–Wulff construction as discussed later in this section.

Let us now consider a closed thermodynamic system (constant composition) at equilibrium composed of two phases α and β at constant T, V and uniform chemical potentials μ_i; the β phase is in the form of a small sphere of radius r embedded in the α phase and the $\alpha–\beta$ interface is characterized by a specific surface or interfacial free energy σ (assumed to be isotropic). See **Figure 36**. If a small virtual

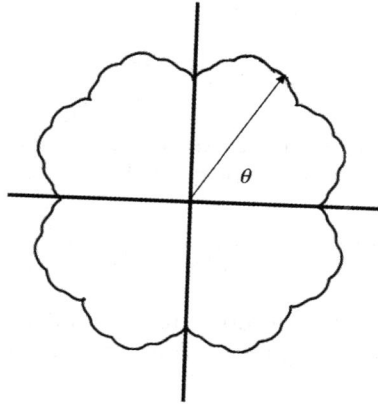

Figure 35 Schematic polar plot of a crystal with a fourfold axis of symmetry perpendicular to the plane of the diagram. After Christian (2002).

reversible displacement of the interface occurs expanding normal to itself into the α phase, the change in Helmholtz free energy is given by:

$$dF = -SdT - P^\alpha dV^\alpha - P^\beta dV^\beta + \sigma dA + \sum_i \mu_i dn_i \tag{81}$$

where σdA is the reversible surface work expended in the operation. Under isothermal $(dT = 0)$ and isochoric $(dV = dV^\alpha + dV^\beta = 0)$ conditions, at constant composition $(dn_i^\alpha = -dn_i^\beta)$ and uniform chemical potential of all species $\mu_i^\alpha = \mu_i^\beta$ at equilibrium $dF = 0$ and it follows that

$$P^\beta - P^\alpha = \sigma \frac{dA}{dV^\beta} \tag{82}$$

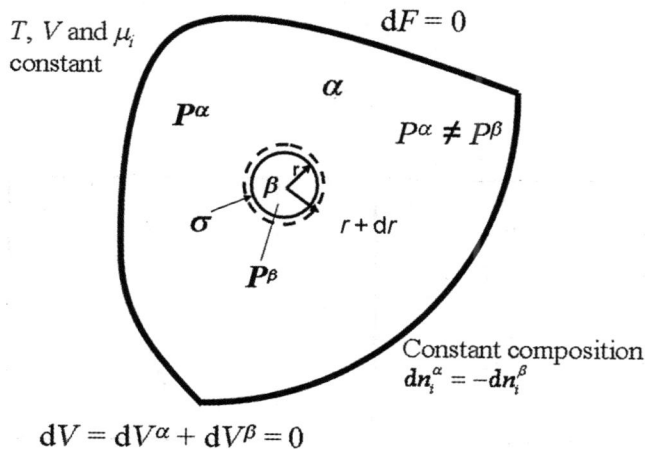

Figure 36 Small spherical β phase particle of radius r in equilibrium with the matrix α. A thermodynamic analysis of the α/β equilibrium shows that $P^\beta - P^\alpha = \dfrac{2\sigma}{r}$ where σ is the interfacial free energy.

Since for a spherical surface $\dfrac{dA}{dV^\beta} = \dfrac{2}{r}$ we find that

$$P^\beta - P^\alpha = \frac{2\sigma}{r} \tag{83}$$

that is, under equilibrium conditions there is a pressure difference across the α–β interface with $P^\beta > P^\alpha$. This is the famous Laplace equation which can be generalized for an arbitrary interface wherein locally

$$P^\beta - P^\alpha = \sigma \frac{dA}{dV^\beta} = \sigma \left(\frac{1}{r_1} + \frac{1}{r_2} \right) \tag{84}$$

where r_1 and r_2 are the local principal radii of curvature at a point defined within two orthogonal planes intersecting the surface at that point. See **Figure 37**. This is a fundamental result of the thermodynamics of capillarity theory.

The Laplace equation can be used to derive the important Gibbs–Thomson relation using the thermodynamic relation $d\mu_T = VdP$. Consider the α phase in the preceding discussion to be the vapor phase in equilibrium with a condensed phase (solid or liquid) β for a single-component system.

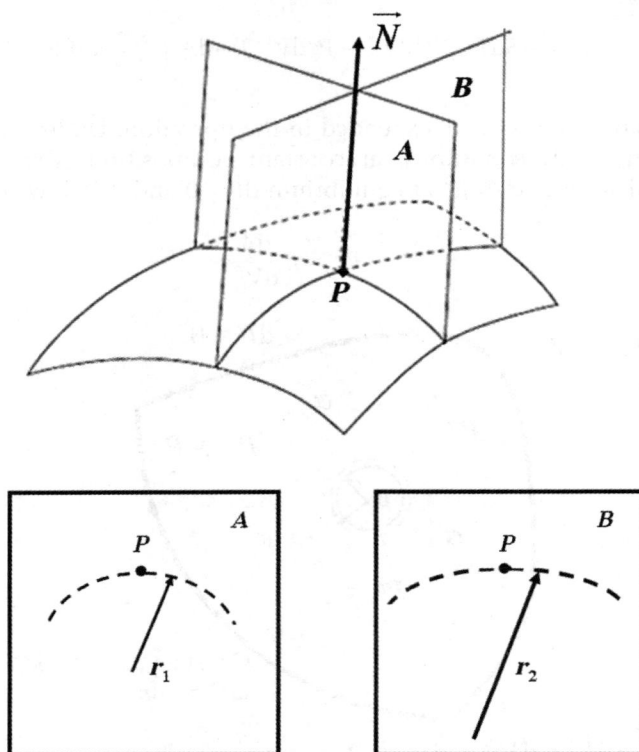

Figure 37   Principal radii of curvature of a surface element defined in two orthogonal planes A and B passing through the point P with local normal $\vec{N}$.

The change in chemical potential of the β phase as a result of the pressure change at constant temperature within the sphere compared with P_0, the equilibrium vapor pressure in equilibrium with β across a flat or plane interface is

$$\mu^\beta\left(T, P^\beta\right) - \mu^\beta(T, P_0) = \int_{P_0}^{P^\beta} V^\beta dP \tag{85}$$

where V^β is the molar volume of the β phase and $P^\beta = P^\beta(r)$ according to the Laplace equation. To maintain thermodynamic equilibrium the vapor pressure will change from P_0 to $P^\alpha(r) = P(r)$. Using the approximation $\int_{P_0}^{P^\beta} V^\beta dP \approx V^\beta \Delta P$ with $\Delta P = P^\beta - P_0 \sim P^\beta - P$ since $(P^\beta - P) >> (P - P_0)$, we can now derive the vapor pressure $P^\alpha(r) = P(r) = P$ in equilibrium with the small particle or droplet as a function of the size (r) of the β phase. Since $\mu^\beta\left(T, P^\beta\right) - \mu^\beta\left(T, P_0\right) = \mu^\alpha(T, P) - \mu^\alpha(T, P_0)$ it follows that

$$\mu^\alpha(T, P) - \mu^\alpha(T, P_0) = RT \ln \frac{P(r)}{P_0} = \frac{2\sigma V^\beta}{r} \tag{86}$$

after inserting $(P^\beta - P_0) \sim (P^\beta - P) = \frac{2\sigma}{r}$ from the Laplace equation. This is a version of the Gibbs–Thomson equation which derives from the influence of the capillarity pressure $(P^\beta - P^\alpha)$ across the curved interface.

Importantly, if we are dealing with solutions of condensed phases (solid or liquid), the change in chemical potential of a species i within the β phase resulting from interface curvature can be viewed to first order as a change in free energy $\Delta P \overline{V_i}$ where $\overline{V_i}$ is the partial molar volume of the component i in solution and ΔP is the capillarity pressure as above. Thus, the change in chemical potential of i in solution in the condensed β phase beneath the curved interface can be written as

$$\mu_i^\beta(r) = \mu_i^\beta(\infty) + \frac{2\sigma \overline{V_i}}{r} \tag{87}$$

where $\mu_i^\beta(\infty)$ is the chemical potential i beneath a planar interface $(r = \infty)$ and $\mu_i^\beta(r)$ is the chemical potential of i within the small spherical β phase of radius r. This is just another version of the Gibbs–Thomson equation. When the two phases α and β are in equilibrium separated by a planar interface $\mu_i^\beta(r) = \mu_i^{\alpha/\beta(\infty)}$ whereas with β in the form of a small spherical particle $\mu_i^\beta(r) = \mu_i^{\alpha/\beta(r)}$ where $\mu_i^{\alpha/\beta(\infty)}$ and $\mu_i^{\alpha/\beta(r)}$ are the chemical potentials of i in the α phase in equilibrium with the β phase across a planar interface and a curved interface, respectively. (Note that when $r >> 2\sigma \overline{V_i}$ capillarity effects are unimportant and bulk thermodynamics prevails.)

Let us look at the influence of capillarity effects on the liquid $\rightarrow$ solid transformation in a single-component system, for example the melting of a pure metal at ambient pressure. The thermodynamics of the transition can be represented graphically in a G vs. T (molar free energy versus temperature) plot shown in **Figure 38**. The melting temperature T_m of the bulk solid phase is the intersection of the free energy curves for the bulk solid and liquid phases where $G^L = G^S$ (molar free energies of the bulk liquid and solid phases, respectively). Now consider an equilibrium between

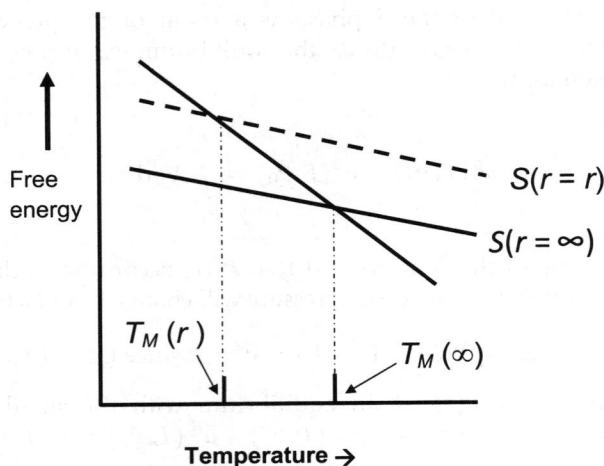

Figure 38 Schematic of free energy vs. temperature curve for the liquid phase and a large solid particle and a small solid particle (dotted line, showing that T_M is lowered).

a small spherical crystallite of radius r with the bulk liquid phase. The molar free energy (chemical potential) of the solid is now altered by capillarity and is a function of size given by $G^S(r) = G^S(\infty) + \frac{2\sigma_{S-L}V^S}{r}$ where the notation (∞) refers to the bulk solid. Clearly this Gibbs–Thomson effect modifies the equilibrium between liquid and solid raising the curve for the solid phase and a melting temperature $T'_m(r)$ defined by $G^S(r) = G^L = G^L(\infty)$ is established effectively lowering the melting point of the small crystallite compared with that of the bulk solid. This lowering of the melting point of the small crystallite contains the seeds of a Gibbsian capillarity-based nucleation theory. Indeed, the small crystallite is in unstable equilibrium with the bulk liquid having an equal probability of growing or melting at the temperature $T_M(r)$ and size r. This concept will be explored more fully in Section 8.4 of this chapter on nucleation.

Let us consider a two-phase mixture in the simple binary phase diagram shown in **Figure 39** and the corresponding free energy vs. composition (X) diagram at a temperature T_A. The common tangent construction depicted in **Figure 39b** establishes the equality of the chemical potentials of the components A and B in the conjugate α and β phases in thermodynamic equilibrium. We consider that the dispersed phase (β) is essentially a nearly stoichiometric compound phase existing over a very restricted composition range (as indicated in the phase diagram) about the composition A_aB_b; the α matrix phase is a solid solution with appreciable solid solubility of B in A. Again the notation (∞) will be used to denote bulk phases or equilibrium in the absence of capillarity effects. We assume that the overall composition of the alloy X_0 is in the two phase field and that the β phase essentially has a composition of X_0^β which remains unchanged when capillarity pressure effects are present at small particle sizes. The dispersed β phase in the α matrix is assumed to be in the form of small spherical particles on the submicron/nanoscale and are of uniform size with radius r; the α–β interface is characterized by an interfacial (free energy) $\sigma_{\alpha\beta}$ (isotropic). In **Figure 40** the common tangent construction shows the change in the composition of the α matrix in equilibrium with the β phase from

(a)

(b)

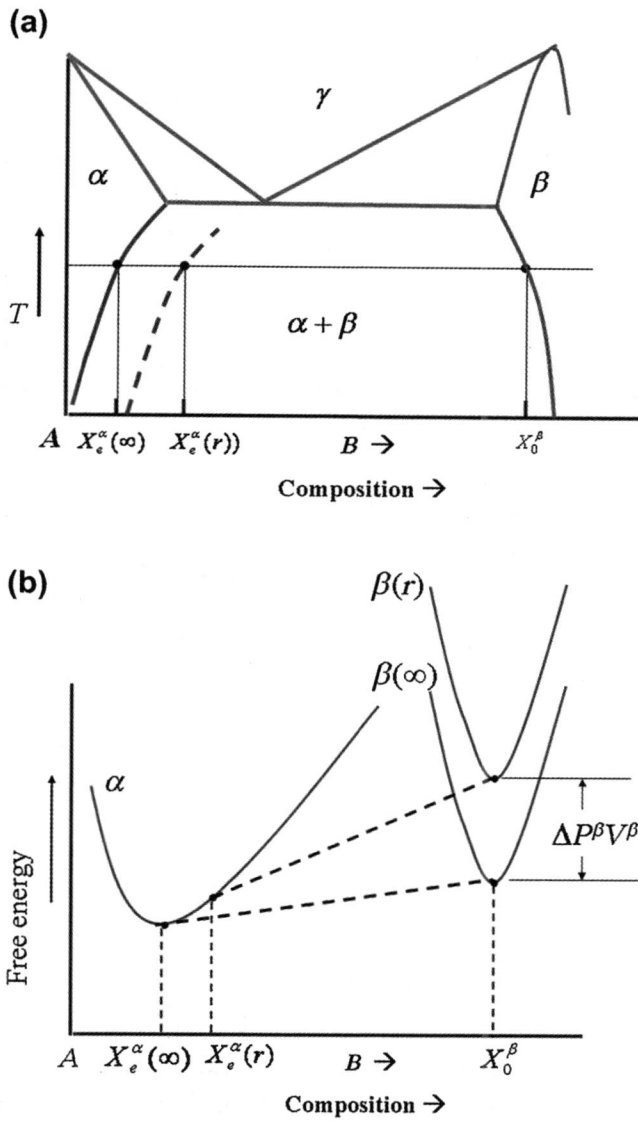

Figure 39 (a) Binary *A–B* phase diagram showing equilibrium between a precipitate phase β (A_aB_b) in a form of small particles and a matrix phase α. This schematic shows the change in solvus for small particles vs. equilibrium between bulk phases α and β. (b) Free energy composition curves for the α/β equilibrium in Figure 39a showing the effect of capillarity on the solid solubility.

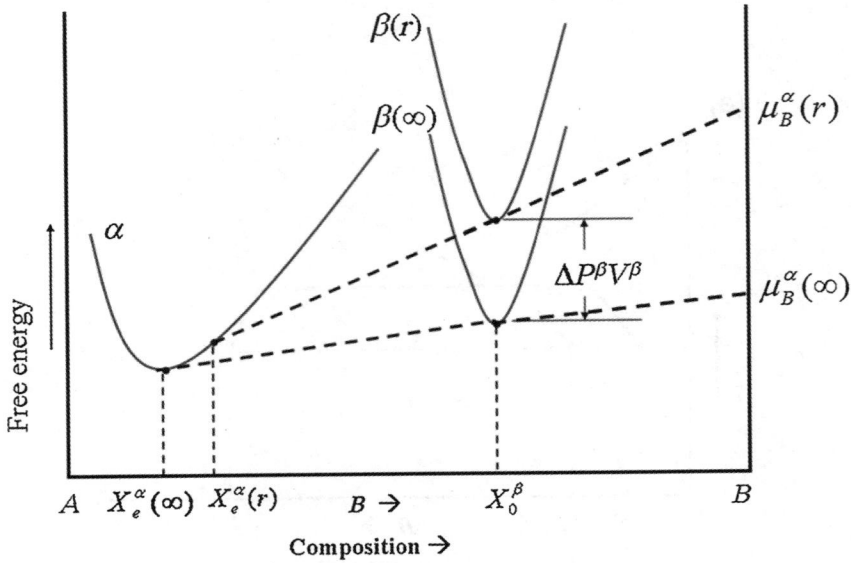

Figure 40 Influence of capillarity on the equilibrium between two phases α (matrix) and β (precipitate) for the general case wherein the capillarity effects change the equilibrium composition of the matrix and precipitate phases.

$X_e^\alpha(\infty)$ to $X_e^\alpha(r)$ as a result of the change in pressure within the β phase resulting from the small particle size. From construction, this change in solid solubility with particle size can be approximated as follows:

$$\frac{\Delta P^\beta V^\beta}{\left[X_0^\beta - X_e^\alpha(\infty)\right]} = \frac{RT \ln\left[\dfrac{X_e^\alpha(r)}{X_e^\alpha(\infty)}\right]}{\left[1 - X_e^\alpha(\infty)\right]} \tag{88}$$

and since $\Delta P^\beta V^\beta = \dfrac{2\sigma_{\alpha\beta}}{r}$

$$RT \ln\left[\frac{X_e^\alpha(r)}{X_e^\alpha(\infty)}\right] = \left(\frac{2\sigma_{\alpha\beta}}{r}\right) \frac{\left[1 - X_e^\alpha(\infty)\right]}{\left[X_0^\beta - X_e^\alpha(\infty)\right]} \tag{89}$$

where V^β is the molar volume of the β phase. This Gibbs–Thomson equation is an excellent approximation assuming $[(X_e^\alpha(r) - X_e^\alpha(\infty))]$ is small and that the relevant activity coefficients do not change significantly over this composition range. This Gibbs–Thomson effect essentially defines a new size-dependent solvus as depicted in **Figure 39a**. The two-phase equilibrium can be generalized to the case where the β phase is a solution phase and this is depicted in **Figure 40**; now the composition of the dispersed phase in equilibrium with the α matrix can change when the capillarity pressure shifts the free energy–composition curve and a new common tangent is established. This can be worked out analytically as well when the change in chemical potential of the β phase associated with this change in composition is properly taken into account in addition to the change in pressure within the small particles.

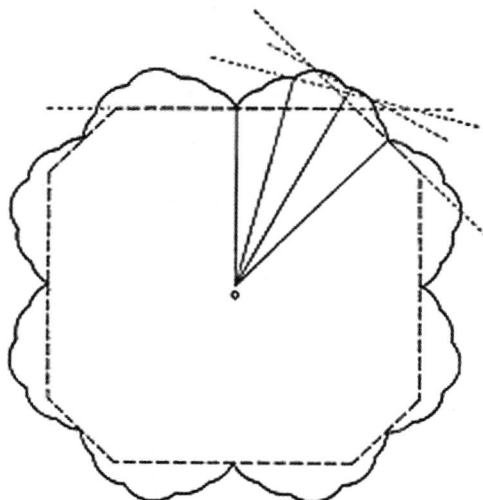

Figure 41 Wulff construction using the surface free energy polar plot. The inner envelope of the perpendiculars to the vectors drawn to each point on the plot gives rise to the equilibrium shape of the crystal. After Christian (2002).

In our preceding discussion we have briefly mentioned that the surface (free) energy σ of a crystalline solid can be highly anisotropic, that is, the energy of the $\{111\}$ faces can have a very different energy than the $\{100\}$ or $\{110\}$ faces in a cubic crystal. This is illustrated in the $\sigma(\mathbf{n})$ versus $\mathbf{n}$ plot where $\mathbf{n}$ is the normal to a crystallographic plane in the material and the distance from the origin in this polar plot (the magnitude of $\mathbf{n}$) is proportional to the value surface energy characterizing the particular plane; this so-called Gibbs–Wulff construction has the property that a radius vector $\mathbf{n}$ locates a point on the locus generated and the tangent plane normal to the radius vector at that point is an image of the crystal plane—the Wulff plane. Thus, one generates an envelope of Wulff planes as a function of $\mathbf{n}$ and the *inner envelope* can be shown to form a figure which minimizes the surface energy $\int \sigma dA$ and defines the equilibrium shape of the crystal. See **Figure 41**. Depending on the nature of the Wulff plot the crystal can be bounded by continuously curved regions or extended flat regions. In general, cusps in the $\sigma(\mathbf{n})$—plot result in finite flat regions.

The equilibrium shape can be used to formulate a generalized Gibbs–Thomson equation based on the construction shown in **Figure 42**. It is found that the parameter $\sigma(\mathbf{n})/\lambda(\mathbf{n})$ is an invariant of the equilibrium shape where $\lambda(\mathbf{n})$ is the normal distance defined in the two-dimensional representation. This condition insures that the capillarity pressure difference across all surface elements of the equilibrium shape is the same and thus the chemical potential is invariant along the surfaces and faces defining the equilibrium shape of the crystal (Johnson, 1965).

The anisotropy of the surface energy of a crystal can be understood structurally in terms of "broken bonds" associated with so-called terrace, ledge and kink densities for a general orientation as depicted in **Figure 43**. Certain atomic configurations will minimize the surface free energy for an arbitrary orientation and distorting these surface structures during stretching contributes to the relationship between surface stress and surface energy discussed earlier. Furthermore, the orientation dependence of $\sigma = \sigma(\mathbf{\theta})$ where $\mathbf{\theta}$ is an orientation parameter can lead to a significant torque acting on a surface tending to rotate the surface plane to a new orientation of lower energy. As a consequence of the orientation dependence of σ nominally flat surfaces can break up into two or more prominent areas of low-index planes called

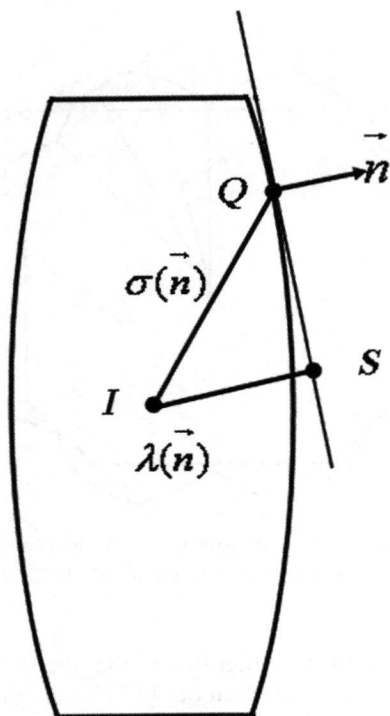

Figure 42 Schematic of generalized Gibbs–Thomson relation defining the vectors **n** and λ (**n**) with respect to the plane S. The parameter $\sigma(\mathbf{n})/\lambda(n)$ is a constant along the locus defining the equilibrium shape. After Johnson (1965).

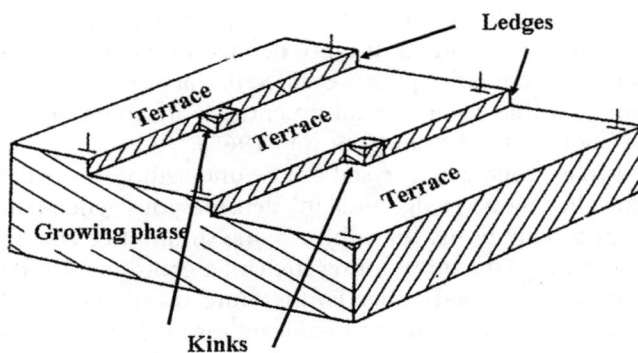

Figure 43 Terrace–Ledge–Kink (TLK) representation of a typical crystal interface.

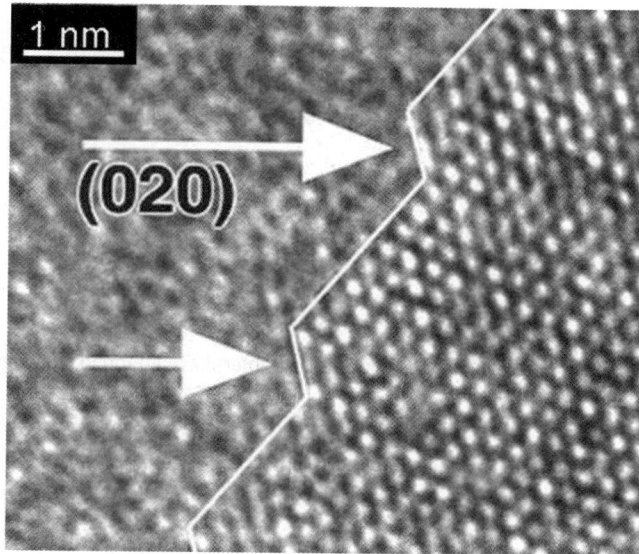

Figure 44 HREM image of facets between a product and parent phase. After Yanar et al. (2002).

Table 1 Typical surface energies in materials (metallic)

Liquid–solid	100–2000 erg cm^{-2} = mJ m^{-2}
Vapor–solid	500–3000 erg cm^{-2} = mJ m^{-2}
Solid–solid	
Coherent	10–200 erg cm^{-2} = mJ m^{-2}
Semicoherent	200–400 erg cm^{-2} = mJ m^{-2}
Incoherent	400–1000 erg cm^{-2} = mJ m^{-2}
Grain boundary	400–600 erg cm^{-2} = mJ m^{-2}

facets which can markedly lower the overall surface energy even though the surface area is increased. This occurs readily at higher temperatures where atomic transport is rapid. See **Figure 44**. This behavior is described most clearly using the so-called inverse σ-plot ($1/\gamma$ plot; Meijering, 1963b).

Table 1 shows a compilation of typical surface energies for various types of interfaces relevant to our discourse on phase transformations in subsequent sections. Note that solid–liquid interfacial energies (σ_{S-L}) are typically in the range 100–2000 erg cm^{-2} whereas solid–solid interphase interfacial or surface energies ($\sigma_{\alpha\beta}$) can range from 10 to 1000 erg cm^{-2} depending on the degree of coherency or crystallographic matching.

In describing the thermodynamics of interphase interfaces Gibbs (1875, ff) pointed out that the density, energy, entropy, and so on, of the individual phases in contact are not likely to be uniform up to the plane of contact resulting in a sharp discontinuity of thermodynamic properties. Cahn and Hilliard (1958, 1959) in their generalized treatment of the thermodynamics of inhomogeneous systems formulated a thermodynamic treatment of a diffuse interface characterized by a transition layer as shown in **Figure 45** in contrast to the so-called sharp interface. The composition and free energy in the interfacial region are considered to be continuous functions through transition region. The free

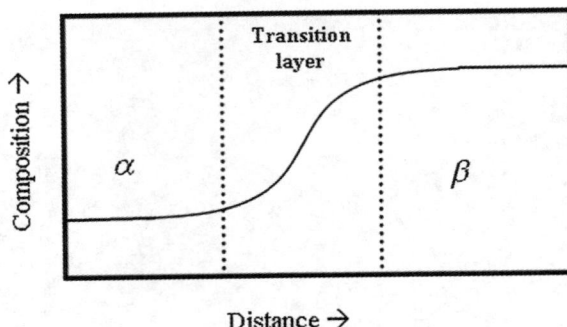

Figure 45 Schematic of composition vs. distance plot through an interphase interface showing the transition layer.

energy of the two-phase system incorporating the energetics of the transition layer can be written as

$$G = An_V \int_{-\infty}^{+\infty} \left\{ g(C) + \kappa \left(\frac{dC}{dx}\right)^2 \right\} dx \tag{90}$$

where n_V is the number of atoms/molecules per unit volume and $g(C)$ is the free energy per atom/molecule of a homogeneous solution of composition C; the second term $\kappa \left(\frac{dC}{dx}\right)^2$ associated with the local gradient in composition is called the gradient energy with the gradient energy coefficient κ assumed to be a constant. (This term derives from a truncated Taylor expansion of the free energy functional.) The surface free energy per unit area is calculated by subtracting from this expression the free energy of the inhomogeneous system the free energy that it would have if the phases actually were homogeneous up to a fiducial sharp interface. This leads to an expression for the surface free energy σ (per unit area) given by

$$\sigma = n_V \int_{-\infty}^{+\infty} \left\{ \Delta g(C) + \kappa \left(\frac{dC}{dx}\right)^2 \right\} dx \tag{91}$$

where $\Delta g(C) = g(C) - [(1 - C)\mu_A^e + C\mu_B^e]$ with μ_A^e and μ_B^e being the equilibrium chemical potentials of species A and B in the two phases. Clearly the $\Delta g(C)$ term is an excess free energy stemming from the species being at nonequilibrium compositions with respect to the bulk phases in crossing the interfacial region. This interfacial free energy functional is a minimum when $\Delta g(C) = \kappa \left(\frac{dC}{dx}\right)^2$ which produces a sigmoidal composition profile through the interface. The interfacial free energy with a change in the variable of integration can alternatively be written as

$$\sigma = 2n_V \int_{C_\alpha}^{C_\beta} [\kappa \Delta g(C)]^{\frac{1}{2}} dC \tag{92}$$

For further details of this continuum description of a coherent interface the reader is directed to the seminal papers of Cahn and Hilliard cited earlier and to the text of Howe (1997). Importantly, this approach to interphase interfaces is central to our modern understanding of spinodal decomposition which will be treated, in detail, in subsequent sections. It also has proven very useful in the modeling of microstructural evolution using the phase field approach.

8.3   Rate Processes in Solids

8.3.1   Basic Concepts

During the latter part of the nineteenth century a number of scientists became interested in the fundamental phenomena governing the rate of chemical reactions. This was a time when thermodynamics and the kinetic theory of gases were reaching maturity along with a developing statistical mechanics and the emergence of physical chemistry as a scientific and academic discipline. The Dutch chemist van't Hoff in 1884 presented a treatise on reaction rates (Etudes de dynamique chimique). He asserted that a wide variety of chemical reactions showed a temperature dependence described by an exponential function with the rate $\propto \exp\left(-\dfrac{T_R}{T}\right)$ where T_R is a constant characteristic of the reaction over a broad temperature range and T is the absolute temperature. In the 1880s Svante Arrhenius on an academic travel grant from the Swedish Academy of Sciences for a time associated with van't Hoff in Amsterdam and in 1889 offered an interpretation of the exponential dependence of reaction rates on temperature by introducing the concept of activation energy as fundamental to understanding the rate at which a reaction proceeds. The basic idea is that the rate of a reaction is not governed by the thermodynamic difference between reactants and products but by an intermediate state of higher (free) energy than both reactants and products which constitutes a barrier—activation barrier—that must be overcome when the reactant and product species encounter each other through gaseous collisions or within solutions. It should be pointed out that during his scientific sojourn Arrhenius studied with Ludwig Boltzmann in Austria and indeed the "Boltzmann distribution" and Boltzmann factor indicated that in collisions the number of atomic or molecular events which are likely to achieve some threshold energy to allow reaction should be proportional to $\exp\left(-\dfrac{E_A}{RT}\right)$ where E_A is related to the T_R parameter in van't Hoff's exponential expression.

During the twentieth century an Arrhenius equation of the type rate $= A \exp\left(-\dfrac{Q}{RT}\right)$ has been found to apply to a plethora of rate processes from chemical reactions to diffusion in solids. The rate of chemical reactions is often formulated as rate $= k_R C_A^\alpha C_B^\beta$ where k_R is called a specific rate constant and C_A^α and C_B^β are concentrations of the reactants and α and β are characteristic exponents; the rate constant generally obeys an Arrhenius relation. In chemical reactions the activated state or "activated complex" is seen as a critical configuration or distortion of the participating species on a potential surface in a hyperspace and the reaction proceeds along a "reaction coordinate" from reactants to products over a saddle point on the potential surface. In modern theoretical statistical mechanical treatments of the fundamental process the "activated complex" is considered to be a valid thermodynamic state that is in quasi-equilibrium with the reactant state and the reaction rate is controlled by the concentration of activated complexes and the rate at which this extraordinary state at the top of the activation barrier can decompose to the product state. This theoretical superstructure and its various formulations is often referred to as Absolute Reaction Rate Theory as is comprehensively reviewed in the classic treatise by Glasstone et al. (1941). A rigorous extension of this formalism particularly applicable to atomic migration in solids is that due to Vineyard (1957).

The activated state for an elementary diffusional jump from one site to another is viewed as an intermediate state at the top of the (free energy) barrier separating lattice sites and the thermally activated jumps can be seen in terms of Absolute Reaction Rate Theory by considering an assembly of harmonic oscillators where at any time in the system an equilibrium number of these oscillators is at the top of the barrier and this species crosses through the activated state with a certain frequency $\frac{1}{\tau}$ where τ is the mean time it takes for the migrating species to pass through a region of spatial extent δ where the potential energy $V(x)$ is assumed to be constant and equal to E_A. In general, the potential energy of the oscillator in the well is described in the harmonic approximation as $V(x) = E_0 + \frac{1}{2} Kx^2$. Staying with the basic tenets of Absolute Reaction Rate Theory, the probability that a given oscillator will at any instant of time be at the top of the activation barrier can be written as a ratio of partition functions (the ratio of the number of oscillators in the "activated state" compared with the number of harmonic oscillators vibrating about their equilibrium positions within the potential well or "reactants"). The ratio of partition functions essentially relates to an equilibrium constant for the equilibrium between the atoms in normal vibrational states and the extraordinary species in the activated state. The ratio of partition functions can be written as

$$\frac{Z_A^*}{Z_A} = \frac{\int_{X^*-\delta}^{X^*+\delta} \exp\left(-\frac{V(x)}{k_B T}\right) dx}{\int_{-\infty}^{+\infty} \exp\left(-\frac{V(x)}{k_B T}\right) dx} = \frac{\delta \exp\left(-\frac{E_M}{k_B T}\right)}{\sqrt{\frac{2\pi k_B T}{K}} \exp\left(-\frac{E_0}{k_B T}\right)} = \delta \sqrt{\frac{K}{2\pi k_B T}} \exp\left(-\frac{E_M - E_0}{k_B T}\right) \qquad (93)$$

where X^* is the coordinate locating the activated state along the reaction coordinate. This ratio of partition functions also can be interpreted as the ratio of the time spent at the top of the barrier τ^* compared with the time in a normal state τ' which is approximately the total time under consideration ($\tau' \gg \tau^*$). From classical statistical mechanics the time for the activated species to pass through the activated state is given by $\frac{\delta}{v^*} = \frac{\delta}{\sqrt{\frac{k_B T}{2\pi m^*}}}$ (v^* is the average velocity along the jump direction and m^* is the mass of the migrating species). Therefore, we can write that the average number of crossings per unit time or the average jump frequency of the migrating species as follows:

$$\omega = \frac{\tau^*}{\left(\frac{\delta}{v^*}\right)} \text{ times } \frac{1}{\tau'} = \frac{\sqrt{\frac{k_B T}{2\pi m^*}}}{\delta} \text{ times } \frac{\tau^*}{\tau'} = \frac{\sqrt{\frac{k_B T}{2\pi m^*}}}{\delta} \delta \sqrt{\frac{K}{2\pi k_B T}} \exp\left(-\frac{E_M - E_0}{k_B T}\right)$$

$$= \frac{1}{2\pi} \sqrt{\frac{K}{m^*}} \exp\left(-\frac{E_M - E_0}{k_B T}\right)$$

$$\omega = \nu \exp\left(-\frac{\Delta E}{k_B T}\right) \qquad (94)$$

where ν is the vibrational frequency (Einstein frequency $\approx 10^{12} - 10^{14} \text{ s}^{-1}$) and $\Delta E = E_M - E_0$ is the height of the activation barrier separating equilibrium sites. This is the rate at which diffusing atoms in a crystal jump into adjacent vacant sites based on this elementary version of the statistical mechanics of thermally activated atomic jumping based on a treatment by Girifalco (1971). See **Figure 46**. (If the

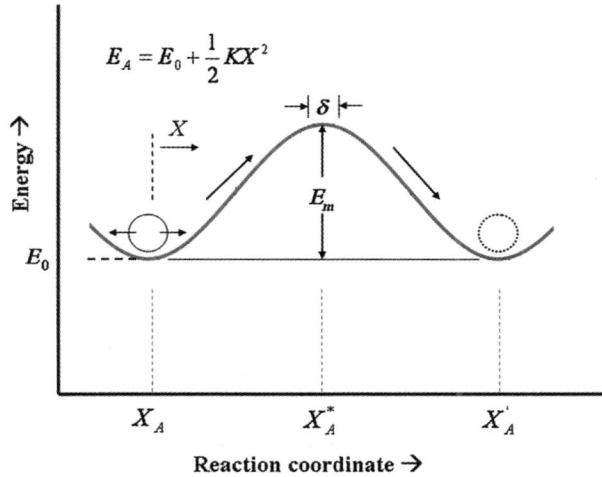

Figure 46 Schematic diagram showing the energetics of an elementary thermally activated atomic jump from X_A to X'_A. The migrating atom must overcome the activation barrier, E_m through thermal fluctuation. In this simple analysis the migrating atom is viewed as a simple harmonic oscillator within the equilibrium energy well having energy equal to E_A. The activated state is assumed to exist over a region δ at the top of the barrier.

atom is migrating through a periodically varying *free* energy field then the jump rate can be written as $\omega = \nu \exp\left(-\dfrac{\Delta g_m}{k_B T}\right)$ where Δg_m is a free energy barrier for migration: see **Figures 47** and **48**).

Let us briefly return to the essential features of the classic Eyring formulation of reaction rate theory which was more specifically directed at chemical reaction rates such as $A \leftrightarrows A^* \rightarrow B$. The rate of reaction was described explicitly in terms of the concentration of "activated complexes" and the frequency with which this species A^* decomposes along the reaction coordinate to produce the product state B. The reactant state is assumed to be in equilibrium with the activated state as mentioned earlier. The concentration of activated species is proportional to $\exp\left(-\dfrac{\Delta G^*}{k_B T}\right)$ where ΔG^* is considered a standard free energy of formation of the activated complex associated with an equilibrium constant

$$\frac{C_A^*}{C_A} = K^* = \exp\left(-\frac{\Delta G^*}{k_B T}\right) \tag{95}$$

where C_A^* and C_A are the equilibrium concentrations of A^* and A per unit volume. Writing the equilibrium constant as a ratio of molecular partition functions per unit volume including translational, vibrational, rotational, and so on, degrees of freedom and again assuming that the activated state exists over a distance δ along the reaction coordinate the specific rate constant for a simple first-order reaction $\left(-\dfrac{dC_A}{dt} = k_R C_A\right)$ is $K^* C_A$ times $\dfrac{\sqrt{\dfrac{k_B T}{2\pi m^*}}}{\delta}$; this last term represents the frequency with which A^* passes through the activated state and decomposes to produce the products similar to the treatment earlier. If

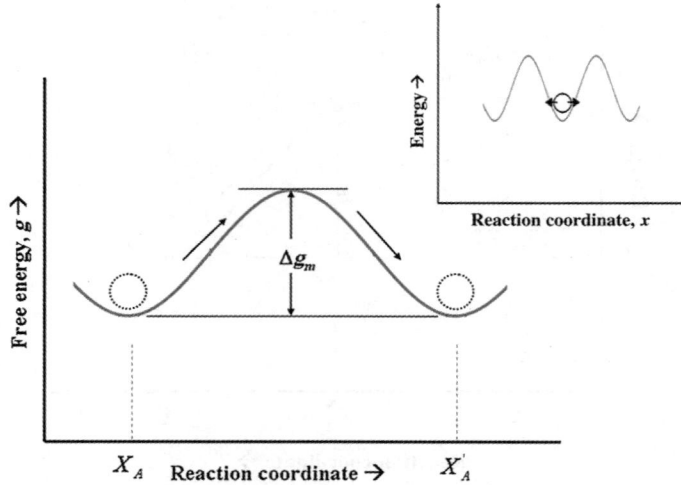

Figure 47 Elementary thermally activated jump viewed as a migration over free energy barriers Δg_m.

one factors out the translational degree of freedom along the reaction coordinate from the total partition function of the activated complex writing $Z_{A^*} = \sqrt{\dfrac{2\pi m^* k_B T}{h^2}}$ times Z'_{A^*}, then the expression for the rate constant becomes $k_R = \dfrac{k_B T}{h} \exp\left(-\dfrac{(\Delta G^*)'}{k_B T}\right)$ where $\Delta G^{*'}$ is the standard free energy of formation of an activated species which has no translational degree of freedom along the reaction

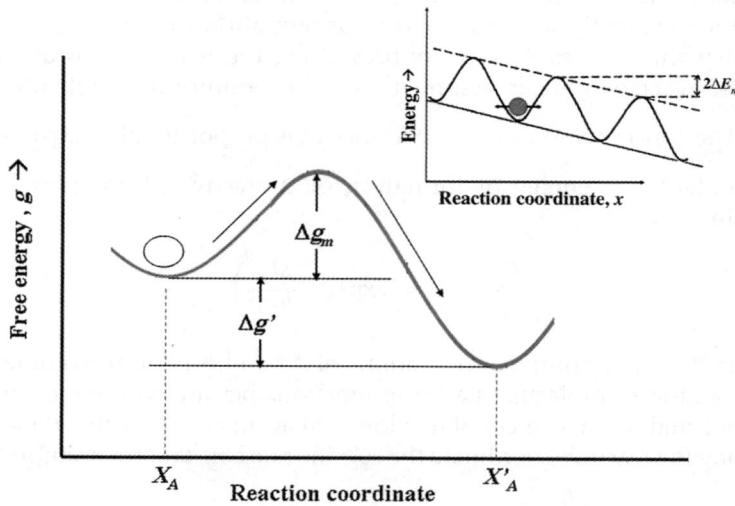

Figure 48 Thermally activated atomic migration under the influence of a driving force $\Delta g'$ which biases the jumps since the activation barrier is not the same for jumps from left to right compared with right to left.

coordinate corresponding to decomposition of the state; $\dfrac{k_B T}{h}$ is essentially a universal frequency $\approx 10^{12}$ to 10^{14} s^{-1} governing the rate process; the rate constant for the reaction can also be written as

$$k_R = \frac{k_B T}{h} \exp\left(\frac{(\Delta S^*)'}{k_B}\right) \exp\left(-\frac{(\Delta H^*)'}{k_B T}\right)$$

separating out the entropic and enthalpic components of the activation free energy. Clearly the rate process is expected to show an Arrhenius behavior as a function of temperature (the pre-exponential term $\frac{k_B T}{h}$ shows a weak temperature dependence compared with

$$\exp\left(-\frac{(\Delta H^*)'}{k_B T}\right)$$

agreeing with the virtually universal experimental result

$$k_R = A \exp\left(-\frac{Q}{k_B T}\right)$$

Returning to thermally activated atomic jumping during diffusion in solids, let us look at the so-called Wert–Zener modification. In applying reaction rate theory to their studies of solid-state diffusion Wert and Zener (1950) were theoretically uncomfortable with the idea of not conserving the number of degrees of freedom in passing from the reactant state to an activated state and thus suggested factoring out a vibrational degree of freedom from the partition function of an atom in a normal site, namely $\dfrac{k_B T}{h\nu}$ leading to a modified expression

$$k_R = \nu \exp\left(-\frac{(\Delta G^*)''}{k_B T}\right)$$

where $\Delta G^{*''}$ can be interpreted as the work done in taking the migrating atom from its normal state in an atomic site to the activated state constraining it to vibrate normal to the reaction path.

The Arrhenius concept of an activated state of higher (free) energy separating an initial state from a final state is universally applied to a wide range of thermally activated processes from chemical reactions, nucleation and growth, diffusion, coarsening, plastic deformation, and so on, and the Arrhenius plot—applied to rate constants, diffusivities, growth rates, and so on—generally involves plotting the natural log of a rate parameter versus $\dfrac{1}{T}$ with the slope equal to $-\dfrac{Q_{exp}}{R}$ and generally one attempts to relate the magnitude of this experimental activation energy Q_{exp} to elementary processes controlling the rate phenomenon in question.

8.3.2 Diffusion Kinetics

Diffusion is the flow of matter from one region to another in a material system generally driven by a concentration gradient. However, a more rigorous description of diffusional processes identifies the "driving force" as a gradient in the chemical potential of the migrating species. This is obvious from the fact that two phases of vastly differing compositions can coexist in thermodynamic equilibrium but the chemical potentials of the components are uniform throughout the phase mixture and no diffusive flow occurs within the compositionally inhomogeneous system. Also, during an allotropic trans-formation a more stable phase grows into the less stable phase often mediated by short-range diffu-sional jumps across an advancing interface at constant composition and the net atomic flux to the more stable phase derives from the local gradient in chemical potential.

Adolf Fick in the middle of the nineteenth century formulated a set of differential equations—Fick's First and Second laws—describing chemical diffusion analogous to Fourier's earlier treatment of heat flow. Fick's laws of chemical diffusion relate the flow of matter explicitly to the concentration gradient and are written as

$$J_x = -D\frac{\partial c}{\partial x} \qquad \text{(1 – Dimensional)} \tag{96}$$

and

$$\vec{J} = -D\nabla c \qquad \text{(3 – Dimensional)} \tag{97}$$

and

$$\frac{\partial c}{\partial t} = \frac{\partial\left(D\frac{\partial c}{\partial x}\right)}{\partial x} = D\frac{\partial^2 c}{\partial x^2} \qquad (D \text{ constant and } c = c(x,t))$$

and

$$\frac{\partial c}{\partial t} = -\text{div}\,\vec{J} = D\nabla^2 c \qquad (D \text{ constant and } c = c(x,y,z,t)) \tag{98}$$

where D is a phenomenological parameter (like the thermal conductivity in Fourier's equations) called the *diffusion coefficient* or *diffusivity*; J_x and $\mathbf{J}$ (J_x, J_y, J_z) are the fluxes (atoms cm^{-2} s^{-1} or mol cm^{-2} s^{-1}) or mass flow rates across a unit cross-sectional area in a local concentration gradient. Importantly, employing the local concentration gradient as the driving force for mass flow within a phase is acceptable operationally if the absence of a concentration gradient leads to a uniformity of the chemical potentials and chemical equilibrium; therefore, in many diffusion problems Fick's laws are invoked. Fick's First law relates the concentration gradient to the local concentration gradient and the solution of Fick's Second law (the diffusion equation) gives the time evolution of the concentration $c(x,y,z,t)$ for various initial and boundary conditions.

When describing diffusion fluxes in a material system it is important to define the frame of reference or coordinate system (which might be moving relative to a fixed reference frame or laboratory frame) in which the diffusion fluxes are prescribed. In the conventional diffusion couple setup between two pure metals A and B which interdiffuse across an initial weld interface between the two metal blocks the concentration profile evolves typically as shown in **Figure 49**. If the diffusion fluxes of A and B are described in terms of a coordinate system fixed relative to the ends of the couple—a laboratory frame—the intermixing can be analyzed in a mathematically convenient manner employing a single diffusion coefficient D_{AB}—the interdiffusion coefficient—wherein the fluxes are written as

$$J_A = -D_{AB}\frac{\partial c_A}{\partial x} \text{ and } J_B = -D_{AB}\frac{\partial c_B}{\partial x} \tag{99}$$

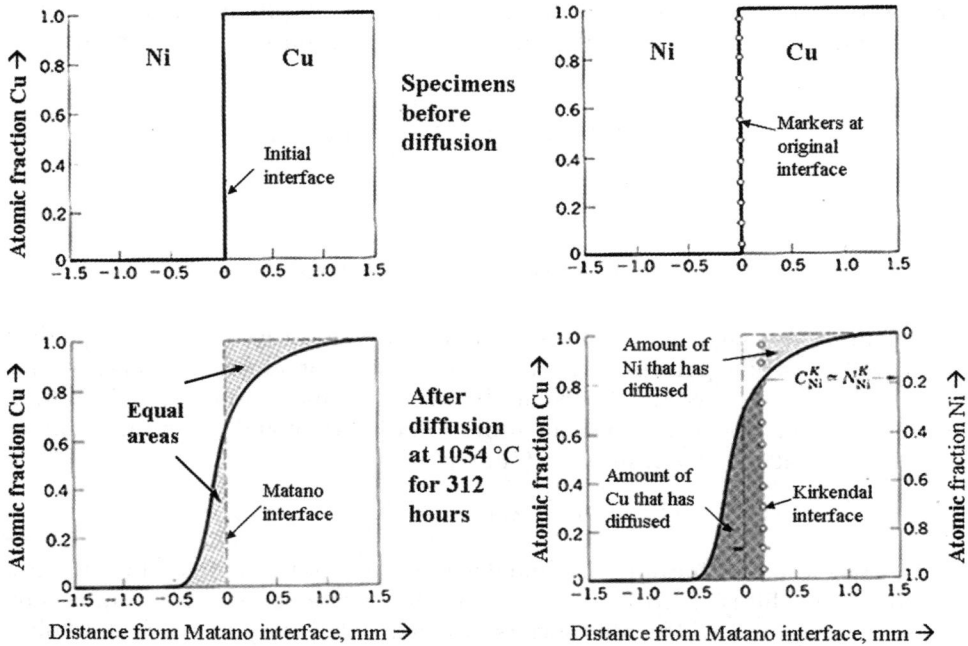

Figure 49 Schematic of interdiffusion between two metals A and B showing the Matano and Kirkendall interfaces wherein the intrinsic diffusivities are unequal ($D_B > D_A$). The Matano interface corresponds to the initial weld interface and the Kirkendall interface is the plane of the markers which initially were located at the weld interface but moved as a result of the unequal diffusive flows of the A and B atoms. After Guy and Hren (1974).

and in this frame $J_A + J_B = 0$ or $J_A = -J_B$. In this description the initial weld interface is the so-called *Matano interface* across which equal numbers of A atoms have crossed from left to right as B atoms from right to left. However, the classic Kirkendall–Smigelskas effect dealing with interdiffusion in substitutional metallic solid solutions indicated that the atomic migration occurs via a vacancy mechanism and that the A and B atoms do not diffuse down their local concentration gradients at the same rate, that is, the interdiffusion process actually involves two diffusion coefficients D_A and D_B called the intrinsic diffusivities or diffusion coefficients of A and B, respectively. The intrinsic diffusivities relate the diffusional fluxes of A and B to their local concentration gradients in a frame of reference fixed on a lattice plane in the diffusion-affected zone, that is,

$$J'_A = -D_A\frac{\partial c_A}{\partial x} \text{ and } J'_B = -D_B\frac{\partial c_B}{\partial x} \tag{100}$$

where $J'_A \neq J'_B$.

The unequal rates of migration of A and B lead to a net flux of vacancies across the initial weld interface and causes the destruction of lattice planes on one side and the creation of lattice planes on the other side as evidenced by the marker movement in the Kirkendall–Smigelskas experiment (1947). There is effectively a bulk flow or convective motion in the diffusion-affected zone and lattice planes are essentially moving with this motion relative to the ends of the diffusion couple. (The bulk flow is basically a plastic regression in the solid involving dislocation climb.) In the elegant Darken analysis of the interdiffusion process it is clear that in the laboratory frame the fluxes J_A and J_B are composed of two components:

$$J_A = J'_A + c_A V_m = -D_A \frac{\partial c_A}{\partial x} + c_A V_m \tag{101}$$

and

$$J_B = J'_B + c_B V_m = -D_B \frac{\partial c_B}{\partial x} + c_B V_m \tag{102}$$

where the first term on the right is a purely diffusive flow down the local concentration gradient and the second term is a bulk flow giving rise to an apparent flux in the laboratory frame where $J_A + J_B = 0$; V_m is the local velocity of the alloy medium relative to the laboratory frame and revealed by the velocity of markers in the diffusion-affected zone. The interdiffusion coefficient and the intrinsic diffusivities can be shown to be related at any composition as follows:

$$D_{AB} = X_B D_A + X_A D_B \tag{103}$$

where X_A and X_B are the atomic fractions of A and B with D_A and D_B the intrinsic diffusivities as defined earlier. Note that these diffusion coefficients are generally a function of composition although the assumption $D_{AB} =$ constant is widely used successfully in a wide variety of practical problems.

The use of the interdiffusion coefficient incorporates the bulk flow into the diffusion fluxes and allows us to describe the intermixing and time evolution of the concentration versus distance profile without concerning ourselves with sorting out these effects (bulk flow versus diffusive flow across the reference plane and $D_A \neq D_B$). If one wants to calculate the time to homogenize an alloy casting or dissolve a second phase during solution treatment of an age hardening alloy one can just use the interdiffusion coefficient at the temperature of interest and solve Fick's Second law.

The simplest expression relating the local chemical potential gradient to the rate of flow/flux of a diffusing species i for one-dimensional flow along the x-axis can be written as

$$J_i = -L_i \frac{\partial \mu_i}{\partial x} \tag{104}$$

where L_i is a phenomenological coefficient and $\frac{\partial \mu_i}{\partial x}$ is the chemical potential gradient of i along the x-axis. We can also write the diffusion flux in terms of a local diffusion velocity of the migrating species V_i along the x-axis and a mobility M_i as

$$J_i = c_i V_i = -M_i c_i \frac{\partial \mu_i}{\partial x} \tag{105}$$

where M_i is the velocity per unit "driving force" (the gradient in chemical potential) and equal to $\frac{L_i}{C_i}$. This flux is a purely diffusive flow of the migrating species down its local chemical potential gradient. We now relate the diffusion flux to the concentration gradient according to Fick's First law

$$J'_i = -D_i \frac{\partial c_i}{\partial x}$$

where D_i is clearly an intrinsic diffusion coefficient or intrinsic diffusivity. Writing

$$J_i = -M_i c_i \frac{\partial \mu_i}{\partial c_i} \frac{\partial c_i}{\partial x} \tag{106}$$

and recalling that $\mu_i = \text{constant} + RT \ln a_i$ and $a_i = \gamma_i c_i$ we arrive at the relation

$$D_i = M_i RT \left(1 + \frac{\partial \ln \gamma_i}{\partial \ln c_i}\right) = M_i RT \left(1 + \frac{\partial \ln \gamma_i}{\partial \ln X_i}\right) \tag{107}$$

wherein the term $\left(1 + \dfrac{\partial \ln \gamma_i}{\partial \ln X_i}\right)$ is called the "thermodynamic factor" which relates the solution thermodynamics to aspects of the atomic migration.

We shall elucidate the role of the thermodynamic factor by reviewing the relationship between atomic jumps from one lattice site to another and between lattice planes in a concentration gradient. If c_1 and c_2 are the concentrations (atoms cm^{-3}) of the diffusing species on adjacent planes 1 and 2 ($c_1 > c_2$) normal to the concentration gradient and of spacing α, the number of atoms of the diffusing species per unit area of these planes is $n_1 = \alpha c_1$ and $n_2 = \alpha c_2$, respectively. If these atoms are jumping randomly from site to site with a frequency Γ s^{-1} wherein one-sixth of their jumps are down or up the concentration gradient, on an atomic level there will be a net flux (atoms $\text{cm}^{-2}\,\text{s}^{-1}$) down the concentration gradient given by $\frac{1}{6}\alpha\Gamma(c_1 - c_2)$. The local concentration gradient is essentially $-\dfrac{(c_1 - c_2)}{\alpha}$ and thus the flux can be related to the concentration gradient as follows:

$$J = \frac{1}{6}\Gamma\alpha^2(c_1 - c_2) = -\frac{1}{6}\Gamma\alpha^2 \frac{dc}{dx} \tag{108}$$

which is Fick's First law if the intrinsic diffusivity is taken to be $D = \frac{1}{6}\Gamma\alpha^2$. This is an important result relating a macroscopic phenomenological parameter—D in Fick's First law—to the underlying atomistics of the diffusion process, namely an atomic jump frequency Γ and an elementary jump distance α. If the jumps are restricted to nearest-neighbor sites, $\alpha = \dfrac{\sqrt{2}}{2}a_0$ and $\alpha = \dfrac{\sqrt{3}}{2}a_0$ for FCC and BCC lattices, respectively, where a_0 is the lattice parameter of the conventional cubic unit cell. If the atomic jumps are mediated by exchange with an adjacent vacancy–vacancy mechanism—as is the case of diffusion in substitutional metallic solid solutions (and self-diffusion in pure metals)—the above expression must be modified to include what is called a "correlation factor" f that accounts for the fact that when the vacancy mechanism is operative the atomic jumps are not completely random since after executing an elementary atomic jump the diffusing species statistically is most likely to jump back into the vacant site from whence it came. Thus, we write

$$D = \frac{f}{6}\Gamma\alpha^2$$

where $f = 0.78$ and 0.72 for FCC and BCC lattices, respectively.

Using radioactive tracers one can monitor the atomic migration of the tracer into a pure metal or into a homogeneous alloy in the absence of a concentration gradient, that is, one can analyze the tracer diffusion process and measure self-diffusion coefficients or tracer diffusivities (D_M^*, D_A^*, D_B^*) that describe the nearly random (but including correlation effects) jumping of the tracer species in a pure metal or alloy. Let us return to the migration of A and B in a binary alloy in a concentration gradient and recall that the simple analysis presented earlier for diffusion in the concentration gradient was based on random jumping and flow down the gradient is essentially a statistical flow from plane 1 to plane 2 because there are more atoms on plane 1 than on plane 2 (there is no bias to their individual jumps up or down the gradient). Therefore, the variation of the free energy of the diffusing atom from site to site looks like that shown in **Figure 47**. The activation barrier for jumps from left to right is the same as that for jumps from right to left leading to random jumping. However, if the diffusing atom "sees" a free energy profile like that in **Figure 48** because of the effects of different atomic environments, for example, the jumps will be biased producing what effectively amounts to a drift velocity along the concentration gradient. The diffusive flow in the concentration gradient can essentially be considered to be composed of two components, namely, a statistical flow and flow deriving from a "driving force" which produces a drift velocity.

Let us return to the incorporation of the thermodynamic factor into the expression for the intrinsic diffusivities that is

$$D_i = M_i RT \left(1 + \frac{\partial \ln \gamma_i}{\partial \ln X_i} \right).$$

First we note that for dilute solutions ($\gamma_i = \gamma_i^\circ = $ constant) and ideal solutions ($\gamma_i = 1$) this becomes $D_i = M_i RT$ which is the famous Nernst–Einstein equation.

Assuming that $M_i = M_i^*$, we write $D_i^* = M_i^* RT = M_i RT$ where D_i^* is a tracer diffusivity. We then write

$$D_i = D_i^* \left(1 + \frac{\partial \ln \gamma_i}{\partial \ln X_i} \right)$$

relating the intrinsic diffusivity to the corresponding tracer diffusivity at a given concentration. Thus, the diffusion flux in a concentration gradient in a binary alloy can be separated into a statistical flow and a drift term (nonrandom jumping) in the concentration gradient as follows:

$$J_i = -D_i^* \frac{\partial c_i}{\partial x} + c_i V_m^D \tag{109}$$

with the drift velocity $V_m^D = -M_i \frac{\partial \mu_i'}{\partial x}$ and $\mu_i' = RT \ln \gamma_i$ which is the nonideal contribution to the chemical potential of the diffusing species i.

At this point in our review we remind the reader that like the thermal conductivity the diffusivities and mobilities are tensor properties of a crystalline solid (Second Rank polar matter tensors). Thus, for diffusion the fluxes along the x-, y- and z-axes of a species should be written in terms of the components of the second rank polar tensor D_{ij} as follows:

$$J_1 = -D_{11} \frac{\partial c}{\partial x_1} - D_{12} \frac{\partial c}{\partial x_2} - D_{13} \frac{\partial c}{\partial x_3}$$

$$J_2 = -D_{21} \frac{\partial c}{\partial x_1} - D_{22} \frac{\partial c}{\partial x_2} - D_{23} \frac{\partial c}{\partial x_3} \tag{110}$$

$$J_3 = -D_{31} \frac{\partial c}{\partial x_1} - D_{32} \frac{\partial c}{\partial x_2} - D_{33} \frac{\partial c}{\partial x_3}$$

where the notation $x \to 1$, $y \to 2$ and $z \to 3$ is used to relabel the coordinate axes; J_1, J_2 and J_3 are diffusion fluxes along the $1(x)$, $2(y)$ and $3(z)$ axes, respectively. In this generalized description of diffusion in an anisotropic crystal we see that composition gradients along orthogonal axes can, in principle, induce flows along the 1, 2 and 3 coordinate axes. Thus, the net flux vector $\mathbf{J}(J_1, J_2, J_3)$ is not necessarily parallel to the vector ∇c, i.e. $\left(\frac{\partial c}{\partial x_1}, \frac{\partial c}{\partial x_2}, \frac{\partial c}{\partial x_3} \right)$. However, if the second rank tensor is referred to the so-called principal axes the diffusivity tensor D_{ij} takes the following diagonal form

$$D_{ij} = \begin{pmatrix} D_{11} & 0 & 0 \\ 0 & D_{22} & 0 \\ 0 & 0 & D_{33} \end{pmatrix} \qquad (111)$$

wherein gradients along the principal axes always lead to flows parallel to these axes; however, if the gradient ∇c is along a direction which is not a principal axis the net flux vector $\mathbf{J}(J_1, J_2, J_3)$ is not parallel to ∇c in the general anisotropic case. The nature of the diffusivity tensor is strongly dependent on crystal symmetry and embodied in Neumann's Principle: *The symmetry elements of any physical property of a crystal must include the symmetry elements of the point group of the crystal.* The symmetry of cubic crystals requires that $D_{11} = D_{22} = D_{33} = D$ rendering cubic crystals isotropic with respect to diffusion: in fact, cubic crystals are isotropic with respect to all physical properties which are second rank tensorial properties of the material including the thermal conductivity. For tetragonal and hexagonal crystals $D_{11} = D_{22} \neq D_{33}$ and for the orthorhombic case $D_{11} \neq D_{22} \neq D_{33}$. The principal axes for cubic, tetragonal and ortho-rhombic crystals are along the conventional orthogonal crystal axes of these systems. For hexagonal crystals the c-axis is a principal axis and orthogonal directions within the basal plane are principal axes.

Let us now revisit the relationship of thermally activated atomic jumps of individual atoms and macroscopic diffusion behavior. As pointed out earlier virtually all diffusion coefficients empirically exhibit a temperature dependence of the Arrhenius form

$$D = D_0 \exp\left(-\frac{Q}{RT}\right)$$

where Q is called the experimental activation energy for diffusion which is typically ≈ 45–65 kcal mol^{-1} for diffusion in substitutional metallic solid solutions and self-diffusion in pure metals; the pre-exponential D_0 is generally ≈ 0.1–1.0 cm^2 s^{-1}. The activation energy for interstitial diffusion, for example C in α-Fe and γ-Fe is ≈ 20–30 kcal mol^{-1}. We derived a relationship between the diffusivity (D as in Fick's laws) of a migrating species and atomistic parameters through the basic equation

$$D = \frac{f}{6} \Gamma \alpha^2 \qquad (112)$$

where Γ and α are the average jump frequency and the elementary jump distance, respectively. The correlation factor, f, is unity for diffusion of interstitial solutes jumping among the atomic sites of a dilute interstitial solid solution and less than one for diffusion of a substitutional solute in solid solution migrating via a vacancy mechanism as discussed earlier. The average jump frequency can be addressed using an elementary interpretation of thermally activated jumping as follows: the diffusing atom is vibrating about its equilibrium position within a potential well with some frequency $\nu \approx 10^{13}$ s^{-1} and the probability that on any oscillation against the barrier separating atomic sites the atom will be energetic enough to move over the barrier and jump to an adjacent site (if the site is available to accommodate the migrating species) is from elementary statistical thermodynamics $\exp\left(-\dfrac{\Delta g_m}{k_B T}\right)$ where Δg_m is the height of the (free) energy barrier or the (free) energy of activation for the elementary diffusional jump. (The subscript m explicitly identifies this as an activation free energy for *atomic migration*.) See **Figure 47**. For diffusion occurring via a vacancy mechanism the probability that the site is vacant is $\exp\left(-\dfrac{\Delta g_v}{k_B T}\right)$ where Δg_v is the (free) energy of formation of a vacant site ≈ 1 eV in a metallic system. If there are z equivalent sites surrounding the migrating atom the average jump frequency will be $vz \exp\left(-\dfrac{\Delta g_m + \Delta g_v}{k_B T}\right)$ with $z = 12$ (fcc) and $z = 8$ (bcc) for nearest-neighbor jumps. For interstitial diffusion, for example C and N in Fe, the probability that an adjacent interstitial site is vacant is essentially one so that the jump frequency is given by $vz \exp\left(-\dfrac{\Delta g_m}{k_B T}\right)$. Thus, the diffusivity of a species controlled by thermally activated jumps mediated by a vacancy mechanism can be written as

$$D = \frac{f\alpha^2 vz}{6}\exp\left(-\frac{\Delta g_m + \Delta g_v}{k_B T}\right) \tag{113}$$

and for interstitial diffusion

$$D = \frac{\alpha^2 vz}{6}\exp\left(-\frac{\Delta g_m}{k_B T}\right) \tag{114}$$

and are often written as

$$D = \frac{f\alpha^2 vz}{6}\exp\left(-\frac{\Delta g_m + \Delta g_v}{k_B T}\right) = \frac{f\alpha^2 vz}{6}\exp\left(\frac{\Delta S_m + \Delta S_v}{R}\right)\exp\left(-\frac{\Delta H_m + \Delta H_v}{RT}\right)$$

and

$$D = \frac{\alpha^2 vz}{6}\exp\left(-\frac{\Delta g_m}{k_B T}\right) = \frac{\alpha^2 vz}{6}\exp\left(\frac{\Delta S_m}{R}\right)\exp\left(-\frac{\Delta H_m}{RT}\right) \tag{115}$$

respectively. Clearly, these expressions for the diffusivity are of the Arrhenius form $D = D_0 \exp\left(-\dfrac{Q}{RT}\right)$ and from a plot of $\ln D$ vs. $\frac{1}{T}$ the slope $-\dfrac{Q}{R}$ is $\left(-\dfrac{\Delta H_m + \Delta H_v}{R}\right)$ or $\left(-\dfrac{\Delta H_m}{RT}\right)$, respectively.

The intercept of the plot $\ln D_0$ is related to the preexponential parameters. The experimental activation energy Q is a composite term when a vacancy mechanism is operative composed of the enthalpic part of the free energy of activation for migration ΔH_m and the enthalpic contribution to the free energy of formation of vacant sites. In the case of interstitial diffusion Q is related directly to the enthalpic contribution to the free energy of activation for an elementary thermally activated jump from one interstitial site to another adjacent site.

Various aspects of atomic migration and thermally activated atomic jumping are often related to the basic tenets of what is called random walk theory (Shewmon, 1963). The root mean squared distance a migrating atom will travel from its initial position in time t after $n = \Gamma t$ jumps is shown to be $R = n^{1/2}\alpha$ and thus can be related to the diffusivity as $R = (6Dt)^{1/2}$ from our discussion earlier. If the diffusion is isotropic then $R^2 = X^2 + Y^2 + Z^2$ and for an arbitrary set of axes is such that $R^2 = 3X^2$ since all directions X, Y and Z are equivalent and thus the root mean squared travel distance along a particular direction is $X = \sqrt{2Dt}$. This relation is useful for "back of the envelope" estimates of diffusion distances in a particular context, for example approximately how thick is a carburized layer after a case-hardening treatment of steel ball bearings for some time t.

The discussion of mass transport to this point has focused on what we call bulk or lattice diffusion involving atomic jumps among the normal lattice sites or interstitial sites of the crystalline solid.

However, we know that most metallic (and ceramic) solids are polycrystalline and contain lattice defects such as dislocations. Atomic migration along grain boundaries and dislocations can occur much more rapidly than within the bulk crystal lattice with an activation energy of the order of one-half that for lattice diffusion. However, although diffusion is more rapid along these short-circuit paths the fraction of atoms associated with grain boundaries and dislocations is small of order 10^{-6} to 10^{-8} depending on the grain size and dislocation density. The contribution of short-circuit paths to overall mass transport depends on the ratio $\dfrac{D_{gb}f}{D_L}$ or $\dfrac{D_P f}{D_L}$ where D_L is the ordinary lattice or bulk diffusivity and D_{gb} and D_P are the diffusivities characterizing the grain boundary and dislocation "pipe" high-diffusivity paths; f is the fraction of atomic sites associated with the short-circuit paths. When these ratios approach 1 the short-circuit paths start to make a significant contribution to the overall diffusive flow and the apparent or effective diffusion coefficient is enhanced compared with that deriving from bulk diffusion alone. This occurs for grain boundary diffusion below about two-thirds the melting point $\frac{2}{3}T_M$ and for dislocations below about one-half the melting point $\frac{1}{2}T_M$, that is, short paths become more important at lower temperatures.

Numerous practical problems involving diffusion require a solution to the diffusion equation (Fick's Second law) for various initial and boundary conditions and a number of solutions relevant to materials problems are discussed in texts such as Shewmon (1963), Glicksman (2000), Wilkinson (2000) and Kirkaldy and Young (1987). It should be pointed out that useful solutions to the diffusion equation can often be extracted from books on heat flow such as Carslaw and Jaeger (1946) since the heat flow equation and its solutions can be converted to solutions to diffusion problems through a change in variables.

8.3.3.1 *Johnson–Mehl–Avrami–Kolmogorov (JMAK) Kinetics*

In the previous sections the kinetic mechanisms involved in the change of phase have been reviewed. In this section a general phenomenological approach to the kinetics of phase change is discussed. The following discussion is based on a series of papers published between 1937 and 1942 (Kolmogorov, 1937; Avrami, 1938, 1940, 1941; Johnson and Mehl, 1939). For a recent commentary see Barmak (2010) and Hillert (2011).

It is important to be able to understand and delineate how fast a transformation occurs, and to be able to display this on a simple plot of fraction transformed, X, versus time, t. A typical plot of the volume fraction transformed, X versus time, t for a first-order phase transformation is shown in **Figure 50**.

It can be seen that such transformations start off slowly, increase in rate and then slow down as the transformation nears completion. From the figure it can be seen that the equation for this curve should display the conditions

$$X = 0 \ \text{ at } \ t = 0$$

$$X = 1 \ \text{ at } \ t = \infty$$

$$\frac{dX}{dt} = 0 \ \text{ at } \ t = 0$$

$$\frac{dX}{dt} = 0 \ \text{ at } \ t = \infty$$

An equation which satisfies these conditions is

$$x = 1 - \exp(-(kt)^n) \tag{116}$$

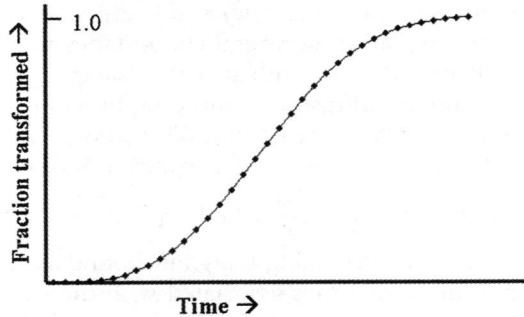

Figure 50 Typical volume fraction transformed, X, vs. time for a first-order phase transformation at a given temperature.

(Note: some authors place the temperature-dependent constant k outside of the parenthesis in the above equation. We prefer to keep it within so that it always has the units of inverse time, regardless of the value of the exponent, n.)

If sufficient data are available, the two constants k and n can be determined from a plot of $\ln\ln\frac{1}{(1-X)}$ vs. $\ln t$. The slope of the plot will be equal to n and the intercept will be $n\ln k$. These values can then be used to determine the fraction transformed at any time at the temperature in question. The effect of temperature comes into play via the temperature-dependent $k(T)$.

At a given temperature, the maximum rate of transformation occurs at the point of inflection of the curve, that is, when the second derivative of X with respect to time is equal to zero. The value of X_{max} in terms of the value of n is determined as follows:

$$X = 1 - \exp(-(kt)^n)$$

$$\frac{dX}{dt} = nk(kt)^{n-1}(1 - X)$$

$$\frac{d^2X}{dt^2} = (n-1)nk^2(kt)^{n-2}(1-X) + nk(kt)^{n-1}\left(-\frac{dX}{dt}\right)$$

$$set = 0 \ \text{and} \ \text{solve}:$$

$$(n-1)nk^2(kt)^{n-2}(1-X) = nk(kt)^{n-1}(1-X)nk(kt)^{n-1}$$

$$(kt)^n = \frac{n-1}{n} \ \text{or} \ X_{max} = 1 - \exp\left(-\frac{n-1}{n}\right)$$

Thus, if the inflection point can be determined, the value of n can be found. This is a second way to obtain the value of n for a given transformation. This can only be utilized if there is sufficient data to obtain the point of inflection of the X vs. t plot.

8.3.3.2 Initial Slopes of X vs. Time Curves
From the derivative of Eqn (116) we obtain

$$\frac{dX}{dt} = nk(kt)^{n-1}(1 - X) \tag{117}$$

The following can be seen to obtain

(1) $n > 1 \left(\dfrac{dX}{dt}\right)_{t=0} = 0$

(2) $n = 1 \left(\dfrac{dX}{dt}\right)_{t=0} = \text{constant} = nk^n = k$

(3) $0 < n < 1 \left(\dfrac{dX}{dt}\right)_{t=0} = \infty$

(4) $n < 0$ (not physical)

It can be seen that the expression 116 also allows for initially rapid transformations, if n is less than or equal to 1.

8.3.3.3 Models for Predicting Values of n and k

From the above equations we have

$$X = 1 - \exp[-(kt)^n]$$
$$\frac{dX}{dt} = \exp[-(kt)^n]nk(kt)^{n-1} \tag{118}$$
$$\frac{dX}{dt} = (1-X)nk^n t^{n-1}$$

For short times, by expanding the exponential in the first expression and dropping higher order terms we obtain

$$X = (kt)^n$$
$$\frac{dX}{dt} = nk^n t^{n-1} \tag{119}$$

The difference between equations 118 and 119 is the factor $(1-X)$, which is the fraction of the sample that is untransformed. This term has been called the "impingement factor" and it takes into account the fact that the volume fraction available to transform decreases with time.

We have

$$\left(\frac{dX}{dt}\right)_{\text{corrected for impingement}} = (1-X)\left(\frac{dX}{dt}\right)_{\text{not corrected for impingement}} \tag{120}$$

The fraction not corrected for impingement is called the extended volume fraction, X_x.

Note, by rearranging Eqn (120) and integrating, the following is obtained:

$$\frac{dX}{(1-X)} = dX_X$$

$$-\ln(1-X) = X_x$$

$$(1-X) = \exp(-X_X)$$

$$X = 1 - \exp(-X_X) \tag{121}$$

Thus if the transformation rate can be modeled in terms of the extended volume fraction transformed (that is, not taking into account impingement) we can obtain the actual transformation fraction transformed by Eqn (121).

8.3.3.4 Modeling the Nucleation Process

The rate of nucleation, I, is the number of transformation centers, N, coming into existence per unit time per unit volume of untransformed phase. Thus

$$I = \frac{1}{(1-X)} \frac{dN}{dt}$$

Avrami (1939) has written

$$\frac{dN}{N} = -v_1 dt$$

$$N = N_0 \exp(-v_1 t)$$

where N is the number of possible nuclei sites remaining and N_0 is total number of possible nuclei sites. (Note that the rate of change of the remaining sites is equal to but opposite in sign to the rate of change of sites where nucleation has occurred.)

The number of possible sites for nucleation decreases with time. Thus, using the above assumption of Avrami, we obtain for the rate of nucleation

$$\frac{dN}{dt} = N_0 v_1 \exp(-v_1 t)$$

$$I = \left(\frac{1}{1-X}\right) N_0 v_1 \exp(-v_1 t)$$

Two extreme cases may be considered:

(1) v_1 is very *large*:
$\frac{dN}{dt} = 0$: all nucleation has occurred at $t = 0$.
(2) v_1 is very *small*:
$\frac{dN}{dt} \approx N_0 v_1 =$ constant nucleation rate.

8.3.3.5 Modeling the Growth Process

The growth rate, G, of a linear dimension r of the new phase is defined as

$$G = \frac{dr}{dt}$$

There are two cases of interest for solid-state diffusional transformations, those whose rate of growth is constant

(1) $G =$ constant,

thus $r \propto t$ (linear)

(2) $G \propto t^{-1/2}$,

hence $r \propto t^{1/2}$ (parabolic).

8.3.3.6 Modeling the Volume Fraction Transformed
We will look at four special cases for the prediction of the values of k and n.

Case 1a	$I = 0$	$G =$ constant
Case 1b	$I = 0$	$G =$ parabolic
Case 2a	$I =$ constant	$G =$ constant
Case 2b	$I =$ constant	$G =$ parabolic

Case 1a
(1) All nucleation at $t = 0$
(2) Particles grow as spheres
(3) $G = \frac{dr}{dt} =$ constant.
Thus: $X_X = N \frac{4}{3} \pi G^3 t^3$ and $X = 1 - \exp\left(-\frac{4\pi}{3} N G^3 t^3\right)$

For this case we obtain $n = 3$ and $k = \left(\frac{4\pi N}{3}\right)^{\frac{1}{3}} G$

Case 1b
(1) All nucleation at $t = 0$
(2) Particles grow as spheres
(3) $G = k_r t^{-1/2}$

Thus $dr = k_r t^{-1/2} dt$
$r - r_0 = 2 k_r t^{1/2}$

$r \approx 2 k_r t^{1/2}$.

Now $X_X = \frac{32}{3} \pi k_r^3 \; N t^{3/2}$

$X = 1 - \exp\left(-\frac{32\pi}{3} k_r^3 N t^{3/2}\right)$

For this case we obtain $n = 3/2$ and $k = \left(\frac{32\pi}{3} N\right)^{\frac{2}{3}} k_r^2$

Case 2a (The original Johnson–Mehl Equation)
Assumptions:
(1) I and G not functions of x, t
(2) Random nucleation in untransformed regions
(3) Particles grow as sphere

At time t, the volume of a spherical particle that formed at $\tau(0 < \tau < t)$ is

$$\text{Vol} = \frac{4}{3}\pi G^3 (t - \tau)^3$$

$$\text{Thus } X_X = \int_{\tau=0}^{\tau=t} \frac{4}{3}\pi G^3 (t - \tau)^3 I d\tau = \frac{\pi}{3} G^3 I t^4$$

$$\text{and } X = 1 - \exp\left(-\frac{\pi}{3} I G^3 t^4\right)$$

For this case $n = 4$ and $k = \left(\frac{\pi I G^3}{3}\right)^{\frac{1}{4}}$

Case 2b

Same assumptions as 2a, but the growth is parabolic in time.

$$G = k_r t^{-1/2}$$

$$\text{Vol} = \frac{4}{3}\pi k_r^3 (t - \tau)^{3/2}$$

$$X_X = \int_{t=0}^{\tau=t} \frac{4}{3}\pi k_r^3 (t - t)^{3/2} I d t = \frac{4}{3}\frac{2}{5}(\pi k_r^3 I) t^{5/2}$$

$$X = 1 - \exp\left(-\frac{8\pi}{15} k_r^3 I t^{5/2}\right)$$

For this case $n = 5/2$ and $k = \left(\frac{8\pi k_r^3 I}{15}\right)^{\frac{2}{5}}$

From these cases it can be seen that the time exponent n for the growth of the new phase shaped as spheres is made up of two terms, p and q.

$p = 0$	If all nucleation occurs at $t = 0$
$p = 1$	If nucleation occurs with time
$q = 3$	If growth is linear
$q = 3/2$	If growth is parabolic

Thus, n can be $\frac{3}{2}, 3, \frac{5}{2}$ or 4 for the nucleation and growth of spheres.

Other values of n can be obtained. For example, if the nucleation rate is a function of time, n can be greater than 4.

If the new phase is not spherical in shape, other values of n can be found. For example, suppose the particle grows linearly as a plate in two dimensions but the third-dimension parabolic. Then, q is determined to be 5/2. The value of n could be 5/2 or 7/2, or greater, depending on the nucleation conditions.

It can be seen that the same value of n could be determined for cases with very different nucleation, growth or shapes of the new phase. It is always wise to supplement these kinetic studies with actual metallographic observations of the shape of the new phase.

Other complications can also arise. For example the diffusion fields of the growing particles could overlap, causing a slowing down of the kinetics. This is called soft impingement and the values of n and k will no longer be constant for a given temperature.

8.4   Classical Nucleation

Nucleation is basically a fluctuation phenomenon within an undercooled or supersaturated phase that produces regions of a new phase (or combination of phases) that can grow spontaneously dissipating the excess free energy and changing the phase constitution of the material system. The rate of this basic process during phase transformations occurring in materials systems often determines the microstructural scale of the transformation products and thus can ultimately influence the structure–properties relationships in engineering materials. The emergence of the new phase generally occurs at distinct sites within the parent phase and the number of these transformation centers or nuclei appearing in a unit volume per unit time essentially defines a nucleation rate. If the nuclei appear at random sites within the parent phase, the nucleation process is termed *homogeneous*. The formation of the new phase may be catalyzed by the presence of singularities in the system such as container walls or lattice defects and occurs preferentially at these special sites; this nonrandom appearance of the new phase or phases is called *heterogeneous* nucleation. The rate of heterogeneous nucleation is clearly limited by the density of these special sites.

Let us look briefly at the elementary thermodynamics of fluctuation behavior in an isolated system. Writing the combination of the First and Second Laws of Thermodynamics as

$$\Delta E = T\Delta S + W \tag{122}$$

where T is the temperature, W is the work done on or within the system and ΔS is the entropy change associated with some process occurring within the system; ΔE is the associated change in internal energy of the system. If an appreciable fluctuation occurs locally within the isolated system, $\Delta E = 0$, one finds

$$\Delta S = -\frac{W}{T} \tag{123}$$

Recalling that ΔS can be written according to elementary statistical thermodynamics as

$$\Delta S = k_B \ln\left(\frac{P_2}{P_1}\right) \tag{124}$$

where P_2 is the thermodynamic probability of the fluctuated state and P_1 is the thermodynamic probability of the initial metastable state. We see that the less probable fluctuated state entails $\Delta S < 0$ and a local heterophase fluctuation occurs with a probability related to $\exp\left(-\frac{W}{k_B T}\right)$, where W is essentially the work that must be expended, in principle, to create the fluctuation or perturbation. Such thermal fluctuations are associated with the initiation of phase transformations in a metastable parent phase and in the case of a nucleation event the thermodynamic driving force for the phase change can contribute to the work of formation of the fluctuation or nucleus of the new phase. We therefore expect that the rate of nucleation will be proportional to a term $\exp\left(-\frac{W}{k_B T}\right)$, where W is the work required to assemble a critical nucleus as discussed subsequently.

The foundations of what metallurgists and materials scientists, physicists, and chemists call classical nucleation theory (CNT) were essentially laid by Gibbs (1875, ff) in the late nineteenth century. He clearly showed that the initiation of a phase transformation in an undercooled or supersaturated metastable phase produced by crossing a phase boundary in the relevant phase diagram, generally encounters a "nucleation barrier" inhibiting formation of the thermodynamically preferred phase or phases. This kinetic barrier or inhibition arises because the embryos of the new thermodynamically more stable phase(s) involve the formation of small regions separated by an interphase interface requiring an expenditure of (free) energy or work in their creation. This surface (free) energy or work presents a primary barrier to the formation of the new phase; however, in the case of solid–solid phase transformations, the appearance of embryos/nuclei of the new phase may be accompanied by local misfit stresses/strains within the parent and emerging phase as a result of a difference in molar volume between phases or due to crystallographic mismatch. The associated elastic energy contributes to the work involved in the formation of the heterophase fluctuations which trigger the transformation of the system. In the language of the chemist the nucleation barrier is an activation barrier that controls the rate of the reaction or rate of formation of the "seeds" of the more stable phase.

Based on the seminal ideas of Gibbs, Volmer and Weber (1926), Farkas (1927) and Becker and Doering (1935) in the 1920s and 1930s formulated the approach of CNT primarily addressing the formation of nuclei in supersaturated vapor phases. This VWBD theory embodied the capillarity thermodynamics of Gibbs and was later modified and refined by Becker (1938), Turnbull and Fisher (1949), Russell (1970) and Aaronson et al. (2010) and applied to nucleation in condensed phases. In spite of the uncomfortable assumptions of the classical theory regarding extrapolation of bulk thermodynamic properties to the small embryos/nuclei and assumptions regarding the nature of the interphase interfaces between the parent and emerging phase, CNT has provided a useful semi-quantitative basis for systematizing the transformation behavior in a wide variety of material systems, and, importantly, for understanding the development of microstructure during the thermomechanical processing of engineering materials.

8.4.1   Basic Tenets of CNT

8.4.1.1   *Fundamentals: Homogeneous Nucleation*
In the first approximation to calculating the rate of nucleation we follow the quasi-equilibrium approach of Volmer and Weber (1926) which writes the nucleation rate as the product of the equilibrium concentration of critical nuclei times the rate at which single atoms/molecules (monomers) can join the nuclei rendering them supercritical and capable of spontaneous growth as transformation centers of the new phase. Any back flux of supercritical nuclei is neglected. We address the equilibrium distribution of embryos of various sizes up to the critical size using the usual Gibbs free energy approximation to describe the energetics of their formation which is not rigorous when the compressibility of the phases cannot be ignored. For condensed phases this is a useful approximation and the error entailed is quite small.

Whether in a vapor phase or condensed phase it is assumed that the embryos/nuclei of the incipient phase form through a series of bimolecular reactions of the form

$$\alpha_n + \alpha_1 \rightarrow \alpha_{n+1}$$

$$\ldots$$

$$\ldots$$

$$\ldots$$

$$\alpha_{n-1}^* + \alpha_1 \rightarrow \alpha_n^*.$$
$$\alpha_n^* + \alpha_1 \rightarrow \alpha_{n+1}^*$$

where the α_n is a cluster or embryo containing n atoms/molecules (monomers) and α_n^* is a cluster of critical size; the last step is considered to be irreversible and governs the effective nucleation rate. The other bimolecular reactions are in metastable equilibrium and these equilibria determine the distribution of embryos of a given size at any time in the system. The irreversible crossing of clusters of critical size to the supercritical state is assumed to not perturb the metastable equilibrium distribution. This quasi-equilibrium approach to the nucleation problem is fundamentally similar to the so-called Absolute Reaction Rate Theory (Glasstone et al., 1941) applied to the description of the rate of chemical reactions with the critical nucleus (cluster of critical size α_n^*) playing the role of the so-called "activated complex" or "transition state".

The overall reaction for cluster formation can be considered to be

$$n\alpha_1 = \alpha_n$$

and is assumed to be amenable to a treatment based on standard equilibrium chemical thermodynamics in terms of an equilibrium constant and the van't Hoff isotherm yielding an expression for the equilibrium concentration of embryos of various sizes n given by

$$\frac{N_n}{N_0} = \exp\left(-\frac{\Delta G_n}{k_B T}\right) \tag{125}$$

where N_n is the equilibrium number of embryos containing n monomers, N_0 is the number of unassociated atoms or molecules (essentially equal to the number of atoms or molecules in the system) and ΔG_n is a standard free energy of formation of embryos of size n. The number of clusters of critical size or critical nuclei at any time is then

$$\frac{N_{n^*}}{N_0} = \exp\left(-\frac{\Delta G_n^*}{k_B T}\right) = \exp\left(-\frac{\Delta G^*}{k_B T}\right) \tag{126}$$

(It should be noted that the critical nuclei are in unstable thermodynamic equilibrium with the supersaturated or undercooled parent phase.) See **Figure 51a**. The undercooled parent phase is characterized by an equilibrium distribution of embryos and nuclei at any time and the nucleation rate is simply governed by the concentration of critical nuclei C^* (number per unit volume) and the rate (ω^*) at which single atoms/molecules (monomers) can join these clusters of critical size taking them over the nucleation barrier and into the realm of spontaneous growth to form regions of the new more stable phase, that is, $I_v = C^*\omega^*$, where I_v is a widely used nomenclature designating the rate of homogeneous nucleation as the number of nuclei crossing the barrier per unit volume per unit time. In the case of condensation from a vapor phase ω^* can be estimated from the kinetic theory of gases. In condensed systems (liquid $\rightarrow$ solid and solid $\rightarrow$ solid transformations), the ω^* term involves a thermally activated diffusional jump from the parent phase to the critical nucleus and will be discussed in detail subsequently. See **Figure 51**.

Let us now look at the free energy of formation of embryos/nuclei using the conventional approach where the Gibbs free energy is used to estimate the work of formation of the critical nucleus. It is assumed that the transformation ($\alpha \rightarrow \beta$) is occurring at constant composition in a condensed phase and that bulk thermodynamic properties of the parent and product phases can be extrapolated to the small length scales involved in the nucleation process including the relevant surface energies and, furthermore, the interface between the parent and incipient phases is assumed to be sharp. The free energy of formation of a cluster containing n atoms/molecules, ΔG_n, written as a sum of a volume term and surface term, neglecting any strain energy accompanying the transformation as follows:

$$\Delta G_n = nv(\Delta G_v) + n^{\frac{2}{3}}\Sigma a_i \sigma_i \tag{127}$$

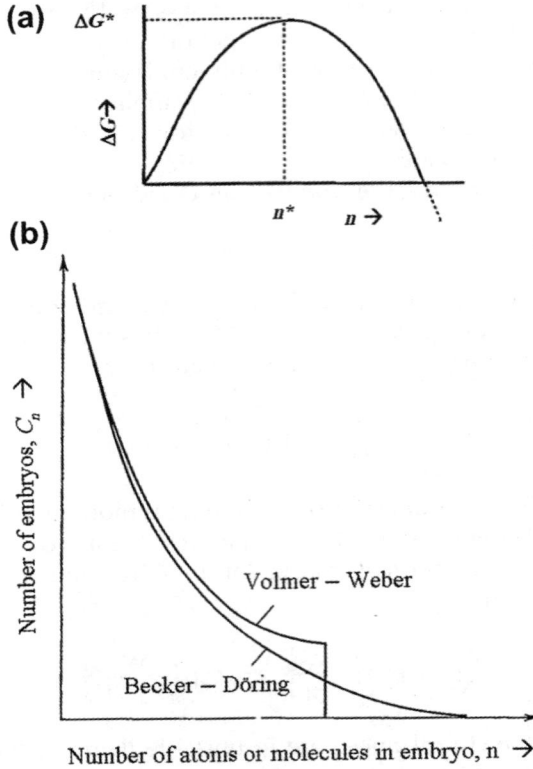

Figure 51   (a) Free energy of formation of clusters/embryos as a function of n, the number of atoms/molecules in the cluster. The cluster n^* is the cluster of critical size and the free energy of formation is ΔG^*. (b) The number of embryos as a function of cluster size n, according to Volmer–Weber and Becker–Doring approaches. After Reed-Hill (1972).

where v is the atomic/molecular volume, ΔG_V is the free energy released per unit volume of the phase change $\alpha \rightarrow \beta$, effectively the thermodynamic "driving force" for the transformation, and $n^{\frac{2}{3}} \sum a_i \sigma_i$ is the surface energy associated with the cluster/matrix interface expressed as a summation over facets i, where a_i is a geometric parameter, with units of area. Henceforth, this is written as simply as $\Sigma a\sigma$ following the notation of Turnbull (1956). Assuming that the embryos and nuclei develop with an optimized shape, one finds that ΔG_n as a function of size n, goes through a maximum at $n = n^*$. See **Figure 51a**. These critical values are given by

$$n^* = -\left(\frac{2(\Sigma a\sigma)}{3v\Delta G_V}\right)^3 \tag{128}$$

and

$$\Delta G_{n^*} = \Delta G^* = \frac{4(\Sigma a\sigma)^3}{27(v\Delta G_V)^2} \tag{129}$$

The activation barrier ΔG^* is the minimum work required through thermal fluctuations to create a heterophase fluctuation that can spontaneously decrease its free energy by growth to macroscopic dimensions. The cluster containing n^* monomers is the critical nucleus and is in unstable equilibrium with the undercooled parent phase; the critical nucleus has equal probability of growing or shrinking to subcritical size whereas embryos with $n < n^*$ are more likely to shrink in size. For a spherical nucleus r^*, the following well-known textbook equations result

$$\Delta G^* = \frac{16\pi\sigma^3}{3\Delta G_V^2} = \frac{4\pi(r^*)^2\sigma}{3} \tag{130}$$

and

$$r^* = -\frac{2\sigma}{\Delta G_V} \tag{131}$$

where σ is the isotropic surface energy of the α–β interface. We see that in this simple formulation of the nucleation barrier that the work to create the critical nucleus through thermal fluctuations is one-third the actual work expended in forming the α–β interface, the other two-thirds is supplied by the thermodynamic driving force.

If elastic strains accompany the appearance of the new phase (solid–solid transformation) the associated strain energy must be considered in the energetics of cluster/embryo formation. Assuming that the strain energy scales with the volume of the cluster and is independent of shape, in the first approximation, it is straight forward to write a modified expression for the standard free energy of formation of a cluster containing n atoms/molecules as

$$\Delta G_n = nv(\Delta G_V + \Delta G_S) + n^{\frac{2}{3}}\Sigma a\sigma \tag{132}$$

where ΔG_S is the attendant strain energy per unit volume of the cluster. Again the free energy of formation passes through a maximum at ΔG^* and n^* given by

$$\Delta G^* = \frac{4(\Sigma a\sigma)^3}{27v^2(\Delta G_V + \Delta G_S)^2} \tag{133}$$

and

$$n^* = -\left(\frac{2(\Sigma a\sigma)}{3v(\Delta G_V + \Delta G_S)}\right)^3 \tag{134}$$

Since ΔG_V is negative and ΔG_S is positive, the strain energy reduces the effective driving force and can suppress transformation. Generally, strain energy will increase the undercooling or supersaturation required for nucleation of the new phase. The strain energy only affects the equilibrium shape of the critical nucleus when it becomes a significant fraction of the driving force, that is, a critical level of strain energy is required to markedly influence the nucleus shape.

If the surface and strain energies are functions of the shape and strain, the ΔG^* appears at a saddle point in a hyperspace $\Delta G_n = f(n, s, \varepsilon)$ where s is a shape parameter or set of shape parameters and ε is a set of strains in the matrix and nucleus and n is the number of monomers in the nucleus as before. The size and shape of the nucleus optimize the surface and strain energies at the saddle point n^*, s^*, ε^*. In crystalline solids the energetics of nucleation are complicated further by the fact that various orientation relationships between the parent and product phases can be established and various types of interfaces (coherent, semicoherent, and incoherent) and elastic misfits can develop as discussed previously (see **Figure 43**). In **Figure 52** a coherent quasi-spherical precipitate is depicted showing attendant coherency strains.

Numerous investigators have approached the estimation of the strain energy attendant to nucleation in solid–solid transformations. We first mention the classic work of Nabarro (1940) which is strictly applicable to an incoherent nucleus where the strain energy introduced into the parent phase by an incompressible nucleus in the shape of a prolate or oblate spheroid is calculated. The results have been presented in numerous textbook treatments and treatises over the years (e.g. Christian, 2002) showing a maximum for a sphere and minimum for a disc and an intermediate level for a rod-like geometry. However, it is generally agreed today that incoherent nucleation is likely to be strain free if the diffusion of vacancies is possible and therefore these well-known results are not actually applicable to diffusional nucleation in solids (Russell, 1970).

Eshelby (1957) considered the formation of coherent nuclei of one cubic phase within another where the strain associated with the transformation is a uniform expansion or contraction, the nucleus and matrix are assumed to be elastically isotropic and have the same elastic constants. The strain energy per unit volume of the nucleus, E_S, is found to be given by

$$E_S = \frac{E(\varepsilon_{11}^T)^2}{(1-v)} \tag{135}$$

where $3\varepsilon_{11}^T = \Delta V/V$, the fractional volume misfit; E and v are the Young's modulus and Poisson's ratio related to the shear modulus $G = \dfrac{E}{2(1+v)}$ in the isotropic approximation. The transformation strain ε_{11}^T is defined as $\varepsilon_{11}^T = \delta = \dfrac{a_m^0 - a_p^0}{a_m^0}$, the so-called disregistry or stress-free linear transformation strain, with a_m^0 and a_p^0 being the lattice parameters of the (unconstrained) matrix (m) and precipitate (p)

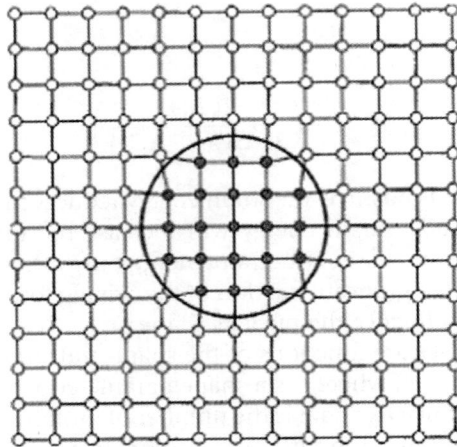

Figure 52   Coherent nucleus/embryo with attendant coherency strain.

phases, the term precipitate being used to generically refer to the nucleus/embryo of the new phase. This strain energy per unit volume of the nucleus/embryo is independent of shape. For volume misfits of a few to several percent the strain energy can reach values which are a significant fraction of the thermodynamic driving force (10–100 erg/cm^3).

Eshelby also considered the case of a rigid incompressible matrix and a nucleus with finite elastic constants (assuming elastic isotropy) and showed that the strain energy per unit volume of the nucleus is about three times that for the case of equal stiffnesses and given by

$$E_S = \frac{3E_P\delta^2}{2(1 - 2\nu_P)} \tag{136}$$

and for an incompressible nucleus the result is

$$E_S = \frac{3E_M\delta^2}{(1 + \nu_M)} \tag{137}$$

which is approximately 1.5 times that for equal stiffnesses of the matrix and nucleus. If the transformation has associated shear strain, the Eshelby analysis assuming elastic isotropy and equal elastic stiffness of the matrix and nucleus predicts a level of elastic shear strain energy per unit volume of the precipitate which is a function of particle shape:

$$E_S = \frac{E(\epsilon_{13}^T)^2}{(1 + \nu)} \frac{\pi(2 - \nu)}{4(1 - \nu)} \frac{c}{r} \tag{138}$$

where ϵ_{13}^T is a shear component of transformation strain tensor of an oblate spheroid lying in the $x_1 - x_2$ plane and c/r the shape parameter of the oblate spheroid of major axis r and minor axis c (c being along x_3 normal to the plane $x_1 - x_2$). The strain energy now becomes a function of the particle shape described by c/r even in the isotropic approximation and equal elastic stiffness. This result generally favors a plate-like morphology but at the expense of surface energy indicating a nucleus shape which optimizes the surface energy and strain energy expenditures in the free energy of formation. Finally, Aaronson et al. (2010) report on the volume strain energy for coherent nuclei/precipitates with differing but finite elastic constants and considered the role of elastic anisotropy. For "soft" particles discs are favored; for "hard" particles spheres are the energetically favorable morphology. When the problem of anisotropy is addressed, clearly the orientation relationship becomes a new variable. It is found that generally if any anisotropy exists the strain energy is shape dependent.

Before formulating the supposedly more rigorous steady-state theory (Becker-Doering, 1935) description of the nucleation problem let us derive an expression for the rate of homogeneous nucleation based on the quasi-equilibrium approach developed earlier. Recall that the nucleation rate is considered to be given by the product of the concentration of critical nuclei times the rate at which this activated state can cross over the barrier, that is, the rate at which atoms/molecules (monomers) can attach themselves to the nuclei taking them into the regime of spontaneous growth accompanied by a progressive decrease in the free energy of the cluster or heterophase fluctuation. In condensed phases this latter step of atomic attachment will generally involve a thermally activated diffusional jump across the interface from parent phase to the nucleus and the rate, $\omega^{*'}$ expected to be controlled the number of

monomers in the vicinity of the matrix–nucleus interface times an Arrhenius-type term $\exp\left(-\dfrac{q_m}{k_B T}\right)$

where q_m is an activation energy for the diffusional jump or atomic migration. Thus, the nucleation rate can be expressed as

$$I_V = \omega^* C^* = K_{\text{homo}} \exp\left(-\frac{(\Delta G^* + q_m)}{k_B T} \right) \tag{139}$$

where the ω^* term is the rate of atomic jumping across the interface and C^* is the concentration of critical nuclei in unstable thermodynamic equilibrium with the parent phase. The result is typically

$$I_V = (10^{30} \text{ to } 10^{40}) \exp\left(-\frac{(\Delta G^* + q_m)}{k_B T} \right) \tag{140}$$

and a sensible nucleation rate is expected for $\Delta G^* = (60–70)\, k_B T$ in most material systems. An important feature of this result is that the onset of nucleation is so sharp that the nucleation rate can be expected to vary over as much as five or six orders of magnitude in a temperature range of only a few degrees, that is, there is virtually a critical undercooling or supersaturation required to initiate nucleation (see **Figure 53**). This precipitous rise or burst slows down and passes through a maximum and eventually decays exponentially to zero at 0 K. This drop-off stems from the limited atomic mobility at low temperatures (the $\exp\left(-\frac{q_m}{k_B T} \right)$ term).

The steady-state theory of nucleation developed by Becker and Doering (1935) and others (e.g. Zeldovich, 1943) attempts to formulate the nucleation problem more fundamentally as a kinetic process defining the nucleation rate as a steady-state current through the embryo size distribution or along the n axis of the ΔG_n versus n-plot employed earlier. In this treatment where the nucleation rate is essentially a steady-state current or diffusion flux in "n-space" account is taken that embryo/clusters which actually exceed the critical nucleus size described earlier may decompose back to the subcritical regime and furthermore takes into account that the effective concentration of clusters in the vicinity of the top of the barrier is perturbed from the metastable equilibrium value. See **Figure 51**. The current through the size distribution is written as a difference equation:

$$j = \beta(n)C(n) - \alpha(n+1)C(n+1) \tag{141}$$

where $\beta(n)$ is the rate at which monomers are absorbed by embryos of size n and $\alpha(n+1)$ is the rate of decomposition of embryos of size $n+1$; $C(n)$ nd $C(n+1)$ are the steady-state concentrations of

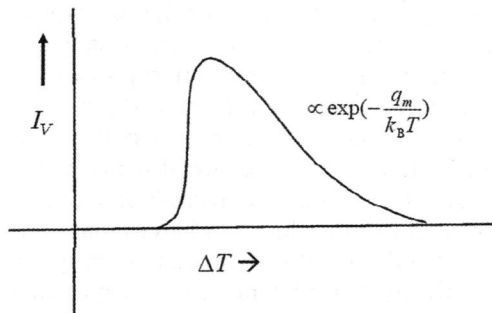

Figure 53 Predicted variation of the rate of homogeneous nucleation as a function of undercooling.

embryos of sizes n and $n + 1$, respectively (C_n^0 and C_{n+1}^0 are then taken as the equilibrium concentrations). See **Figure 51**b. Solving for the steady-state flux yields the following result:

$$J = Z\beta^* C_{n^*}^0 = Z\omega^* C^* = Z\omega^* N \exp\left(-\frac{\Delta G^*}{k_B T}\right) \tag{142}$$

wherein Z is called the Zeldovich nonequilibrium factor, $\beta^* = \omega^*$ is the rate at which single atoms/molecules (monomers) can join the critical nucleus, $C_{n^*}^0 = C^*$ is the equilibrium concentration of clusters of size $n = n^*$ (critical nuclei), N and ΔG^* have their usual meaning previously defined earlier. The Zeldovich factor is shown to be related to the curvature of the ΔG_n versus n curve at $n = n^*$. It should be pointed out that $Z \approx 10^{-1}$ to 10^{-2} for most systems of interest, thus, the nonequilibrium factor reduces the predicted nucleation rate by a factor of about 100 and given that the preexponential for condensed systems is $\approx 10^{30}$ to 10^{40} this correction is not likely to be resolved experimentally. However, the theory indicates that a time-dependent nucleation rate should be observed given by

$$J(t) = Z\omega^* N \exp\left(-\frac{\Delta G^*}{k_B T}\right) \exp\left(-\frac{\tau}{t}\right) \tag{143}$$

where τ is essentially an incubation time characterizing the approach to steady-state nucleation and approximately given by $\tau \approx \dfrac{1}{\omega^* Z}$. See **Figure 54**.

8.4.1.2 Heterogeneous Nucleation

It must be emphasized again that in the vast majority of phase transformations occurring in metallurgical (materials) systems involving nucleation and growth of new phases, homogenous nucleation is extremely rare and heterogeneous or catalyzed nucleation generally dominates the phase change at significantly reduced undercoolings or supersaturations compared with those required to induce homogeneous nucleation. It is extremely difficult to isolate systems whether undercooled melts, supersaturated vapors or solid solutions from the influence of catalyzing singularities such as container/mold walls, impurity particles or lattice defects. Some ingenious experimental approaches such as the classic small droplets experiments used in the study of solidification (Vonnegut, 1948; Turnbull and Fischer, 1949) have been designed to isolate the influence of nucleation catalysts. Large undercoolings compared with those observed for the freezing of bulk melts were observed but the validity of the results and interpretation of

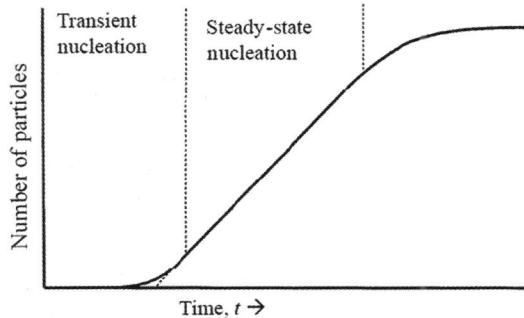

Figure 54 Number of particles nucleated as a function of time, showing an interval of steady-state nucleation.

these famous experiments has been called into question as to whether the realm of homogeneous nucleation was actually observed (Kelton and Greer, 2010). However, the Gibbsian capillarity-based CNT outlined earlier for homogeneous nucleation provides a framework for constructing an analysis of the energetics and kinetics of heterogeneous nucleation. (It should be mentioned that Gibbs also addressed the energetics of heterogeneous nucleation compared with homogeneous nucleation.)

Heterogeneous nucleation preferentially at "special sites" occurs because the interaction of the embryos/nucleus with the singularity, for example mold wall during freezing, reduces the free energy barrier ΔG^*, allowing the formation of the new phase at substantially reduced undercoolings. Nucleation of the solid phase during the freezing of undercooled melts is known to take place on impurity particles and container walls and preferential nucleation in solid–solid transformations is well established to occur at grain boundaries and dislocations (Christian, 2002). Generally, catalyzing the nucleation process on a planar singularity or surface essentially depends on a net reduction of surface energy required to form a nucleus (in the absence of significant strain energy effects).

Following a simple surface energy/capillarity approach, the free energy of formation of an embryo/nucleus of a β phase on a surface/substrate(s) within an undercooled α phase can be written as

$$\Delta G_s = A_{\beta s} W_s + A_{\alpha\beta} \sigma_{\alpha\beta} + V^\beta \Delta G_V \tag{144}$$

where $A_{\beta s}$ is the area of the β–s interface created, $A_{\alpha\beta}$ is the area of the α–β interface created and V^β is the volume of the spherical cap (see **Figure 55**); $\sigma_{\alpha\beta}$ is the isotropic surface energy of the α–β interface and ΔG_v has its usual meaning, that is, the free energy released per unit volume of the β phase formed. The parameter $W_s = \sigma_{\beta s} - \sigma_{\alpha s}$ is the free energy change accompanying the replacement of a unit area of the α–s interface with a unit area of β–s interface with surface energies $\sigma_{\alpha s}$ and $\sigma_{\beta s}$, respectively. There are three relevant interfacial free energies or surface energies $\sigma_{\alpha\beta}$, $\sigma_{\alpha s}$ and $\sigma_{\beta s}$ and can be related by $\sigma_{\alpha\beta}\cos\theta + \sigma_{\beta S} = \sigma_{\alpha S}$ assuming an incompressible substrate and mechanical equilibrium at the junction of the σ, β and s phases, where θ is called the contact angle from surface chemistry describing the "wetting" of the substrate by the β phase. Also, the contact angle can be written in terms of W_s as $\cos\theta = -\dfrac{W_s}{\sigma_{\alpha\beta}}$. For a spherical cap of radius r, $A_{\beta s} = \pi r^2 \sin^2\theta$, $A_{\alpha\beta} = 2\pi r^2(1-\cos\theta)$ and $V^\beta = \dfrac{\pi r^3(2 - 3\cos\theta - \cos^3\theta)}{3}$. Inserting these relations into Eqn (144) and setting $\dfrac{\partial \Delta G_s}{\partial r} = 0$, the nucleation barrier is found to be

$$\Delta G_s^* = \frac{16\pi\sigma_{\alpha\beta}^3 f(\theta)}{3\Delta G_V^2} \tag{145}$$

where

$$f(\theta) = \frac{(2 - 3\cos\theta + \cos^3\theta)}{4} \tag{146}$$

whereby $f(\theta)$ reduces the nucleation barrier for $\theta < \pi$; that is, the "wetting" of the substrate surface by the β phase catalyzes the nucleation process. Also, it is found that

$$r^* = -\frac{2\sigma_{\alpha\beta}}{\Delta G_V} \tag{147}$$

and therefore the critical radius (of curvature) is identical to that for homogeneous nucleation but the volume is decreased. Indeed, an interesting result of this analysis is that ΔG_s^* is proportional to the

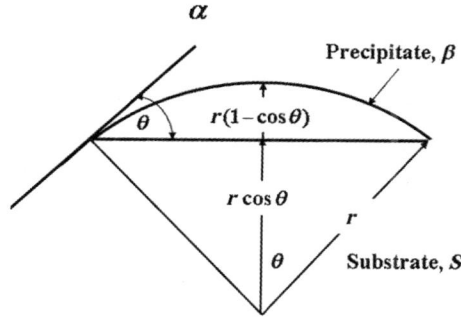

Figure 55 A schematic of a spherical cap embryo showing the relevant geometric parameters.

volume of the critical nucleus V^*, and it can be shown that for $\theta = \pi/2$ the nucleation barrier ΔG_s^* is reduced to exactly one-half that for homogeneous nucleation. Also, note that $\Delta G_s^* \to 0$ as $\theta \to 0$, that is, the nucleation barrier vanishes in the case of "perfect wetting".

The treatment earlier can readily be used to address the heterogeneous nucleation of new phases on grain boundaries in solid–solid transformations (Christian, 2002). It is well established experimentally that grain boundaries (surfaces, edges and corners) act as favorable sites for nucleation in metallic and ceramic materials in conjunction with polymorphic transformations, massive transformations, precipitation and eutectoid decomposition. A straightforward extension of the foregoing analysis is the case of an incoherent allotriomorph forming on a grain boundary surface in the shape of a doubly spherical lens (see **Figure 56**b). The energetics of formation of an embryo/nucleus on the boundary surface can be formulated as

$$\Delta G_s(B) = V^\beta \Delta G_\gamma + A_{\alpha\beta}\sigma_{\alpha\beta} - A_{\alpha\alpha}\sigma_{\alpha\alpha} \tag{148}$$

where V^β is the volume of the β embryo/nucleus, $A_{\alpha\beta}$ is the area of the α–β interface created and $A_{\alpha\alpha}$ is the area of the grain boundary removed and ΔGv is the free energy released per unit volume accompanying the $\alpha \to \beta$ transformation as above. The geometric terms for this shape are given by

$$V^\beta = \frac{2\pi r^3 \left(2 - 3\cos\theta - \cos^3\theta\right)}{3} \tag{149}$$

$$A_{\alpha\alpha} = \pi r^2 \sin^2\theta \tag{150}$$

$$A_{\alpha\beta} = 4\pi r^2 (1 - \cos\theta) \tag{151}$$

where r is the radius of curvature of the spherical cap and θ is the contact angle characterizing the interaction of the embryo/nucleus with the planar grain boundary surface. Clearly, in this case, $2\cos\theta = \dfrac{\sigma_{\alpha\alpha}}{\sigma_{\alpha\beta}}$, where $\sigma_{\alpha\alpha}$ is the grain boundary energy and $\sigma_{\alpha\beta}$ is the interfacial free energy associated with the new α–β interface. Setting $\dfrac{\partial \Delta G_S}{\partial r} = 0$, as earlier, one finds that

$$\frac{\Delta G_S^*(B)}{\Delta G_H^*} = \frac{2 - 3\cos\theta + \cos^3\theta}{2} \tag{152}$$

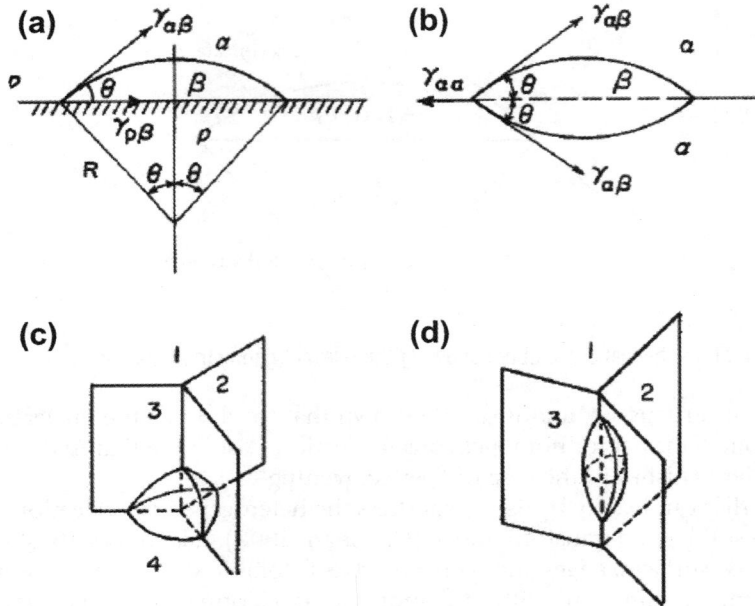

Figure 56 Nucleus/embryo geometries for nucleation at grain boundary faces, corners and edges. After Jena and Chaturvedi (1992).

where $\Delta G_S^*(B)$ is the reversible work to create the grain boundary nucleus on the boundary surface and $\Delta G_H^* = \dfrac{16\pi\sigma_{\alpha\beta}^3}{3\Delta G_V^2}$ is the nucleation barrier for homogeneous nucleation of β within the undercooled α phase. The catalytic potency of the grain boundary surface increases as the ratio $\sigma_{\alpha\alpha}/\sigma_{\alpha\beta}$ increases and the nucleation barrier vanishes as $\sigma_{\alpha\alpha} \to 2\,\sigma_{\alpha\beta}$ or as $\theta \to 0$.

Given the well-known topology of polycrystalline aggregates we know that special grain boundary sites are located at the intersection of grain boundaries such as three-grain junctions (edges) and four-grain junctions (corners) and a similar analysis can be carried out for these more complex embryo/nucleus shapes (see **Figure 56**c and d). The results indicate a greater reduction in ΔG_S^* for the grain boundary edge and corner sites, that is, $\Delta G_S^*(C) < \Delta G_S^*(E) < \Delta G_S^*(B) < \Delta G_H^*$ as shown in **Figure 57**; $\Delta G_S^*(C)$, $\Delta G_S^*(E)$ and $\Delta G_S^*(B)$ refer to the nucleation barriers at the corner (C), edge (E) and planar (B) boundary sites, respectively. It should be mentioned that $\Delta G_S^* \to 0$ at different values of $\dfrac{\sigma_{\alpha\alpha}}{\sigma_{\alpha\beta}}$ for the corner and edge sites, namely, $\Delta G_S^*(C) \to 0$ as $\sigma_{\alpha\alpha}/\sigma_{\alpha\beta} \to \dfrac{2\sqrt{2}}{\sqrt{3}}$ and $\Delta G_S^*(E) \to 0$ as $\dfrac{\sigma_{\alpha\alpha}}{\sigma_{\alpha\beta}} \to \sqrt{3}$. Importantly, heterogeneous nucleation at grain boundary sites can markedly reduce the nucleation barrier for the formation of the new more stable phase allowing transformation to be initiated at substantially reduced undercoolings or supersaturations; however, the contribution of these sites to an overall nucleation rate is always limited by the density of these sites in the parent phase, as mentioned earlier. The density of the various sites (surfaces/faces (B),

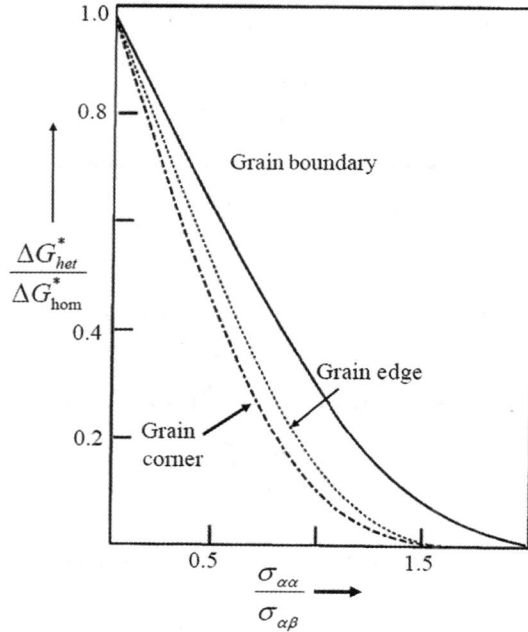

Figure 57 Nucleation barriers for heterogeneous nucleation on grain boundary sites compared with homogeneous nucleation. After Cahn (1956).

edges (E), corners (C)) is expected to scale as $\frac{\delta}{d}$, $\left(\frac{\delta}{d}\right)^2$ and $\left(\frac{\delta}{d}\right)^3$, respectively, where δ is the grain boundary thickness and d is the average grain diameter $\left(\frac{\delta}{d} \approx 10^{-5} - 10^{-6}\right)$. Cahn (1956) has analyzed this aspect of the problem, in detail, and shows that corner and edge sites become important with respect to the overall nucleation rate primarily at very small undercoolings and when $\frac{\sigma_{\alpha\alpha}}{\sigma_{\alpha\beta}}$ is relatively small. See **Figure 57**.

Other factors can influence the energetics of nucleation at grain boundaries such as the possibility of grain boundary nuclei establishing a low-energy coherent or semicoherent interface and an orientation relationship with one of the grains and an incoherent interface with respect to an adjoining grain with subsequent grow governed primarily by migration of the more mobile incoherent interphase interfaces. Also, facets can create low-energy regions along an interphase interface at special orientations (see **Figure 58**a and b).

Many investigators dating back to the 1940s suggested that dislocations might be expected to catalyze nucleation of phases in solid-state transformations. Cahn (1957) presented the first quantitative treatment of heterogeneous nucleation at dislocations addressing the energetic of nucleation of incoherent nuclei. Gomez-Ramirez and Pound (1973) also subsequently analyzed the case of incoherent nuclei, whereas Larche' (1979), Lyobov and Solov'yev (1965) and Dollins (1970) considered the case of coherent nuclei. In the first approximation, Cahn assumed that an incoherent nucleus/embryo of cylindrical shape of radius r, forming along a straight dislocation, effectively removes the strain energy

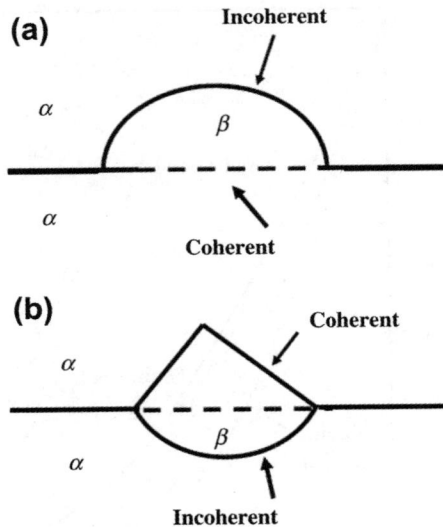

Figure 58 Grain boundary nuclei showing an interplay between coherent (semicoherent) and incoherent segments of the interphase interface.

of the dislocation over a unit length. The free energy of formation of the embryo per unit length is written as

$$\Delta G(r) = \pi r^2 \Delta G_V - A \ln r + 2\pi r \sigma + \text{constant} \tag{153}$$

where A is taken as $\dfrac{\mu b^2}{4\pi(1-\nu)}$ for an edge and $\dfrac{\mu b^2}{4\pi}$ for a screw dislocation; μ is the shear modulus and b is the Burgers vector with all other parameters having their usual meanings.

Taking $\dfrac{\partial \Delta G(r)}{\partial r} = 0$, it is found that, in general, $\Delta G(r)$ passes through a minimum at small r and then a maximum at intermediate values of r as a function of r. See **Figure 59**. A parameter $\alpha_D = -\dfrac{2A\Delta G_V}{\pi \sigma^2}$ emerges as an important variable governing the behavior. For $\alpha_D > 1$ the driving force/strain energy term dominates the surface energy term and there is no barrier to nucleation on the dislocation and spontaneous growth of the new phase occurs along and out from the line defect. If $\alpha_D < 1$, a subcritical metastable cylinder along the line defect is predicted to form along the dislocation (Cottrell atmosphere) at small r and a nucleation barrier appears at larger $r = r^*$ as shown in **Figure 59**. Cahn then addressed the optimum shape of a nucleus and calculated the free energy barrier to form a nucleus on the dislocation compared with that for homogeneous nucleation. See **Figure 59b and 60**. Given that the total number of dislocation sites essentially scales as $\approx N^{1/3}\rho$ where N is the total number of atomic sites in the system and ρ is the dislocation density, heterogeneous nucleation at dislocations is expected to become important when $\alpha_D \approx 0.5$ for $\rho \approx 10^8 – 10^{10}$ cm^{-2}. Clearly, α_D reflects the catalytic potency of dislocations to induce heterogeneous nucleation in this treatment. Gomez-Rameriz and Pound (1973) attempted to investigate the dislocation strain field effects in more detail assuming that the nucleation barrier is relatively insensitive to the nucleus shape. In both cases, dislocations are predicted to lower the barrier for incoherent nucleation of the new phase. The works of

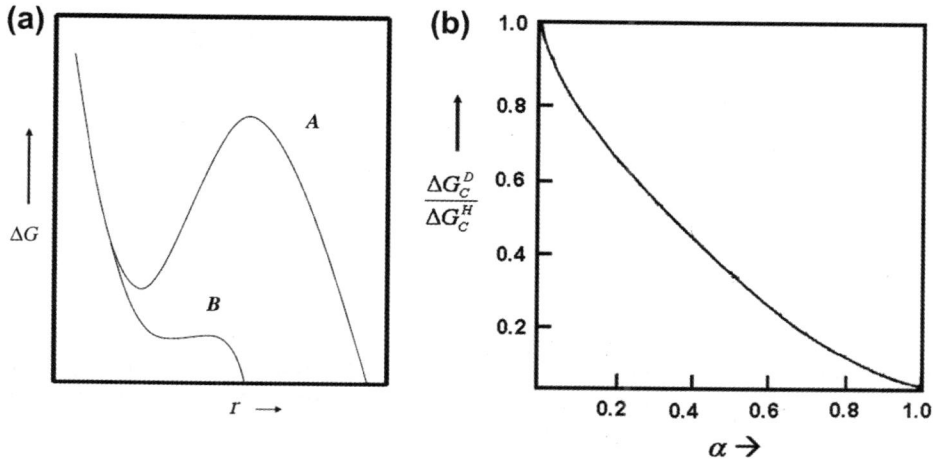

Figure 59   (a) Free energy of formation of a nucleus on a dislocation for $\alpha_D < 1$ and $\alpha_D > 1$ (after Cahn, 1957). (b) The nucleation barrier to form a nucleus on a dislocation as a function of α_D compared with the barrier for homogeneous nucleation of a spherical precipitate. After Cahn (1957).

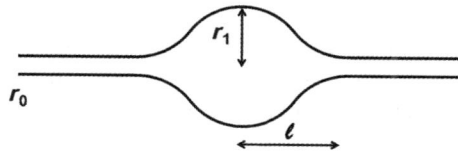

Figure 60   A schematic representation of a nucleus forming along a dislocation line. After Cahn (1957).

Lyubov and Solovyov (1965), Rollins (1970) and Larche' (1979) on the nucleation of coherent precipitates suggest that the elastic interaction of the nuclei and dislocation strain fields significantly relieves the usual strain energy term in the nucleation energetics. The nuclei form preferentially in the compressive or expansive regions of the dislocation strain field, that is, in regions that relieve the largest proportion of the dilatational transformation strains (edge). However, elastic analysis indicates that nucleation can be suppressed near dislocations if the nucleus has a small misfit strain and if the new phase has a much higher shear modulus than that of the matrix. The analysis of Larche' of the coherent nucleation problem suggests that for an edge dislocation the interaction of a misfitting spherical nucleus/embryo with the elastic field of the dislocation can be folded into a correction to the effective interfacial free energy term $\sigma_{\alpha\beta}$ wherein the usual expression for $\sigma_{\alpha\beta}$ is replaced by $\sigma_{\alpha\beta} - \dfrac{\mu b^2 (1 + \nu) \varepsilon^T}{9\pi(1 - \nu)}$ where ε^T is the stress-free transformation strain. This correction can be very significant ($\approx 30\text{–}40\%$) in systems with a large $e^T \approx 0.05$. These analyses for incoherent and coherent nuclei indicate a significant catalytic potency of dislocations on the nucleation process of similar magnitude relative to that of homogeneous nucleation, in general.

From this discussion of heterogeneous nucleation in solid–solid transformations it must be emphasized again that the impact of catalyzed nucleation at singularites in the system with respect to the

Table 2 Nucleation site densities (after Nicholson, 1970)

	Grain boundary		Grain edge		Dislocation density		Vacancy N_V
Defect density	Grain size, 5×10^{-4} m	Grain size, 5×10^{-6} m	Grain size, 5×10^{-4} m	Grain size, 5×10^{-6} m	$10^7/\text{cm}^2$	$10^{10}/\text{cm}^2$	10^{-6}
$\dfrac{N_S^{het}}{N_S^{hom}}$	10^{-6}	10^{-4}	10^{-13}	10^{-9}	10^{-8}	10^{-5}	10^{-6}

overall nucleation rate and transformation behavior is always limited by the density of these special sites. **Table 2** after Nicholson (1970) tabulates the typical density of sites relative to that for homogeneous nucleation. Under circumstances where there is an actual competition between heterogeneous and homogeneous nucleation, the rate of heterogeneous compared with that for homogeneous nucleation will scale to a good approximation as ratio of available sites times $\exp\left(\dfrac{\Delta G^*_{HET} - \Delta G^*_{HOMO}}{k_B T}\right)$ where $\Delta G^*_{HET} < \Delta G^*_{HOMO}$ and the ratio of sites $N_{HET}/N_{HOMO} \ll 1$. This analysis clearly indicates that heterogeneous nucleation will be most significant at small undercoolings or supersaturations. These conclusions are in excellent agreement with a plethora of experimental results.

Stowell (2002) has subjected the classical (capillarity) theory of nucleation to a critical analysis incorporating some ideas of Kashchiev (2000) in the evaluation and reevaluation of the data of LeGoues and Aaronson (1984) and Kirkwood (1970) from studies of well-known precipitation systems. He found reasonable agreement with the form of the equations of the capillarity model when noise associated with heterogeneous nucleation during some heat treatments was sorted out. A number of investigators (Kampmann and Kahlweit, 1970; Wagner and Kampmann, 1991) have called attention to competitive coarsening during the early stages of nucleation and growth and this will be discussed in the section on precipitation. Kelton and Greer (2010) in their recent book focused on nucleation in condensed systems present an extraordinary overview of nucleation applied to a broad range of topics from materials to biology; however, the treatment of nucleation in Christian's treatise on transformations [2002] remains the most rigorous treatment of the subject.

We close this section on CNT by pointing out that a treatment of nucleation behavior can be developed wherein the ad hoc assumptions of a sharp interface between the parent and emerging phases as well as a uniform composition within the effective nuclei are relaxed. This nonclassical formulation sometimes referred to as Cahn–Hilliard (1958 and 1959) nucleation will be addressed in the precipitation section as well.

8.5 Diffusional Growth of Phases

8.5.1 Parent and Product Phases Have the Same Composition

After *nucleation* of a more stable phase within a metastable phase the nuclei moving over the nucleation barrier generally enter the stage wherein their *growth* leads to a continuous decrease in the free energy of the system. The first case we address is where the growing phase has the same composition as the parent phase. This analysis is relevant to allotropic/polymorphic and massive transformation in the solid state. The migration of the interphase interface is assumed to be essentially

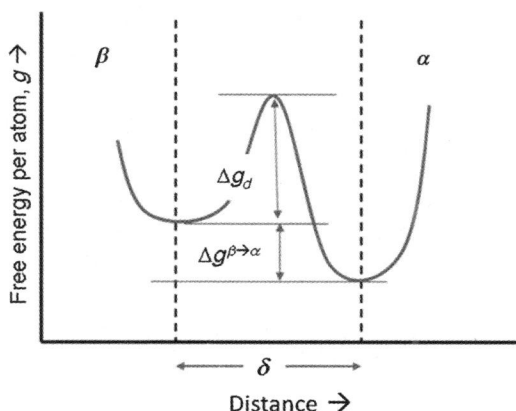

Figure 61   Variation of the free energy per atom, g, in the vicinity of the α/β interface during the transformation $\beta \rightarrow \alpha$; $\Delta g^{\beta \rightarrow \alpha}$ is the thermodynamic driving force and Δg_d is the free energy of activation for an elementary diffusional jump of an atom from β to α at the interface. The effective thickness of the interface is δ.

isotropic and controlled by thermally activated atomic jumps across the interface involving continuous, nearly random jumping of individual atoms or atomic attachment mediated by a ledge mechanism. We begin by describing essentially random atomic detachment and attachment at the interface between two phases α and β where α is the more stable growing phase. The interphase boundary is assumed to be disordered or incoherent and strain energy effects are assumed negligible. In the vicinity of the α–β interface the atoms see a free energy profile like that shown in **Figure 61**. The atoms are jumping from β to α and from α to β but the energetics of these elementary atomic processes give rise to a net transfer of atoms from β to α leading to growth of the α phase and the consumption of the β phase.

Let us assume that the effective number of atoms per unit area N_S at the boundary is the same at both sides of the interface, that is, $N_S^\alpha = N_S^\beta = N_S$ and the vibrational frequency (attempt frequency) of these atoms is taken to be $\nu \approx 10^{12} \text{ s}^{-1}$. From elementary rate theory we expect the rate of jumping of atoms from the β phase to the α phase to be described essentially by

$$\frac{dN^{\beta \rightarrow \alpha}}{dt} = N_S \nu \exp\left(-\frac{\Delta g_d}{k_B T} \right) \tag{154}$$

wherein the atoms at the interface "attack" the activation barrier $\Delta g_d \nu$ times per second and the probability that an atom will have enough (free) energy to surmount the barrier on any attempt is $\exp\left(-\dfrac{\Delta g_d}{k_B T} \right)$. Similarly, we write for the jumping of atoms from α to β

$$\frac{dN^{\beta \rightarrow \alpha}}{dt} = N_S \nu \exp\left(-\frac{\Delta g_d + \Delta g^{\beta \rightarrow \alpha}}{k_B T} \right) \tag{155}$$

where Δg_d is the free energy barrier controlling thermally activated jumps from β to α : the effective barrier for reverse jumping $\alpha \rightarrow \beta$ is larger by $\Delta g^{\beta \rightarrow \alpha}$, the thermodynamic "driving force" (per atom) for the $\alpha \to \beta$ transformation. The net flux of atoms (atoms cm^{-2} s^{-1}) across the interface is given by the difference in these jump rates and can be written as

$$\frac{dN}{dt} = \frac{dN^{\alpha \to \beta}}{dt} - \frac{dN^{\beta \to \alpha}}{dt} = N_S \nu \exp\left(-\frac{\Delta g_d}{k_B T}\right)\left(1 - \exp\left(-\frac{\Delta g^{\beta \to \alpha}}{k_B T}\right)\right) \qquad (156)$$

This net flux can be converted to a linear growth rate or interface velocity since a net transfer of N_S atoms across a unit area of interface advances the interface by a distance δ, the thickness of the interfacial region; thus, we can write the velocity of the interface (cm s^{-1}) as

$$V = \delta \nu \exp\left(-\frac{\Delta g_d}{k_B T}\right)\left(1 - \exp\left(-\frac{\Delta g^{\beta \to \alpha}}{k_B T}\right)\right) \qquad (157)$$

The velocity of the interface or growth rate of the more stable low temperature phase α into the undercooled high-temperature β phase generally will vary directly as the difference in free energy as a function of undercooling below the equilibrium transformation temperature $T_t = T_{\alpha\beta}$. See **Figure 62**. At temperatures where $k_B T \gg \Delta g^{\beta \to \alpha}$ taking $\left(1 - \exp\left(-\frac{\Delta g^{\beta \to \alpha}}{k_B T}\right)\right)$ after expanding in a Taylor series gives a velocity $V \approx \delta \nu \frac{\Delta g^{\beta \to \alpha}}{k_B T} \exp\left(-\frac{\Delta g_d}{k_B T}\right)$ essentially proportional to the thermodynamic driving force. Recall that $\frac{\Delta g^{\beta \to \alpha}}{k_B T}$ is proportional to ΔT at small undercoolings. At low temperatures $k_B T \ll \Delta g^{\beta \to \alpha}$, the $\exp\left(-\frac{\Delta g^{\beta \to \alpha}}{k_B T}\right)$ term is small and the velocity $V \approx \delta \nu \exp\left(-\frac{\Delta g_d}{k_B T}\right)$, that is, the rate of growth is limited by the atomic mobility at low temperatures decreasing exponentially as $T \to 0$ K. See **Figure 63**. This description appears to apply very well to the temperature dependence of the growth rate of the transformation α-Sn (white) $\to \beta$-Sn (gray) below $T_{\alpha\beta} \approx 286$ K ($13\,^\circ$C). See **Figure 64a and b**. Note that for the reverse transformation β-Sn (gray) $\to \alpha$-Sn (white) the growth rate is monotonically increasing with temperature since the atomic mobility is increasing continuously as the transformation temperature is raised.

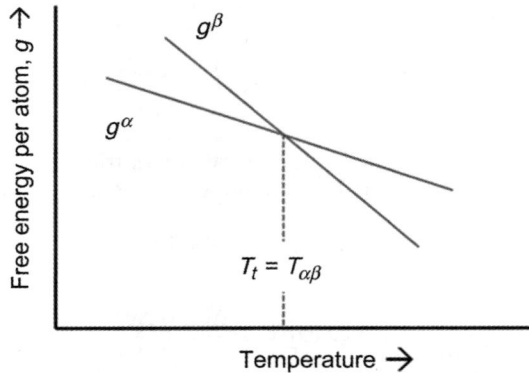

Figure 62   The free energy per atom of the α and β phases as a function of temperature; $T_t = T_{\alpha\beta}$ is the equilibrium transition temperature for the allotropic/polymorphic phase change.

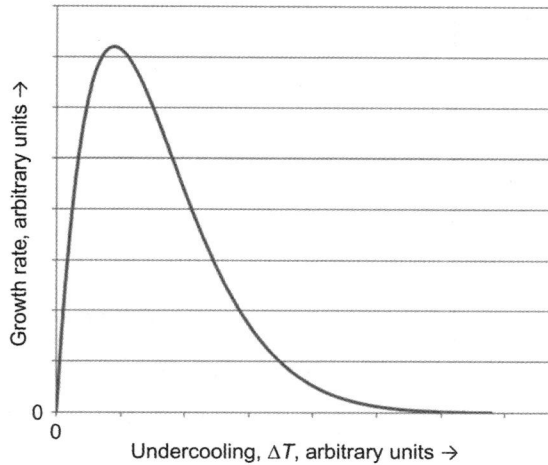

Figure 63 The growth velocity of the α/β interface as a function of ΔT ($\Delta T = T_{\alpha\beta} - T$), predicted by Eqn (156). Note that the velocity varies approximately linearly with ΔT for small ΔT and decreases exponentially at large ΔT and low temperatures.

This description of the growth rate for a compositionally invariant transformation is often referred to in the literature as the Burke–Turnbull analysis (see Burke and Turnbull, 1952) which can be extended to include the case wherein the atomic attachment and detachment processes involve ledges moving within the interphase interface. See **Figure 65**. A modified Burke–Turnbull equation is often invoked where the velocity of the interface is written as follows:

$$V = \frac{h}{\lambda} \delta v \exp\left(-\frac{\Delta g_d}{k_B T}\right)\left(1 - \frac{\Delta g^{\beta \to \alpha}}{k_B T}\right) \tag{158}$$

with h being the average ledge height and λ the average ledge spacing. This modified version has been applied to the massive transformation occurring in several systems as discussed subsequently.

8.5.2 Parent and Product Phases Having Different Compositions

Let us consider diffusional growth of a spherical precipitate phase β within a supersaturated solid solution α of average composition C_0. See **Figure 66**. We assume that during the early and intermediate stages (times) of growth the matrix composition far from the growing particle remains C_0 and a solute-depleted region exists primarily in the near vicinity of the particle. See **Figure 66**. In our initial approximation we assume that growth is diffusion-controlled and that local equilibrium exists at the α–β interface, that is, the composition of α is $C_\alpha^\beta = C_\alpha^\beta(\infty) = C_\alpha^E$ which is in thermodynamic equilibrium with the β phase of composition $C_\beta^\alpha = C_\beta^\alpha(\infty) = C_\beta^E$ given by the phase diagram. If one employs a so-called quasi-steady-state solution to the diffusion field (this assumes that the concentration profile in the matrix changes very slowly with time), the solution to Fick's second law is approximately

$$C(\rho) = \frac{A}{\rho} + B \tag{159}$$

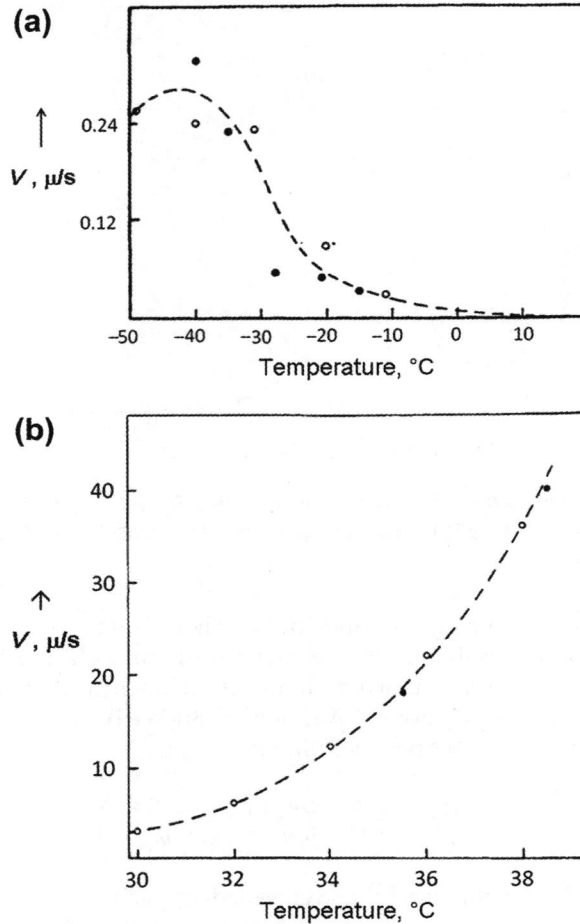

Figure 64   Experimental results for the α to β and the β to α transformations in Sn, showing the variation of the growth rate V as a function of undercooling (a) and superheating (b). After Burgers and Groen (1957).

where ρ is the distance from the center of the particle; the instantaneous radius of the particle is taken to be $\rho = R$. Since at $\rho \gg R$ the composition is taken to be C_0 and at the interface $\rho = R$ the composition is fixed at C_α^E the constants A and B can readily be determined and the solution for the concentration profile in the matrix is given by

$$C(\rho) = C_0 - \frac{R}{\rho}\left(C_0 - C_\alpha^E\right) \tag{160}$$

and the concentration gradient at the advancing interface is given by

$$\left(\frac{\partial C(\rho)}{\partial \rho}\right)_{\rho=R} = \frac{\left(C_0 - C_\alpha^E\right)}{R} \tag{161}$$

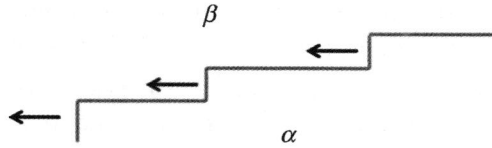

Figure 65   Schematic of the operation of a ledge growth mechanism wherein atomic attachment occurs at the ledge risers.

Applying a mass balance to the advancing α–β spherical interface requires that

$$\left(C_\beta^E - C_\alpha^E\right)\frac{dV}{dt} = 4\pi R^2 \left(\frac{D\left(C_0 - C_\alpha^E\right)}{R}\right) \tag{162}$$

where $\dfrac{dV}{dt} = 4\pi R^2 \dfrac{dR}{dt}$ and D is an appropriate diffusion coefficient (note: the concentrations C are given as atoms cm^{-3} or mol cm^{-3}). Here the growth rate is completely determined by the atomic flux arriving at the interphase interface via diffusive flow down the concentration gradient in the matrix. The growth rate is given by

$$\frac{dR}{dt} = \frac{D}{R}\frac{\left(C_0 - C_\alpha^E\right)}{\left(C_\beta^E - C_\alpha^E\right)} \tag{163}$$

and integration from $t = 0$ to $t = t$ yields

$$R^2(t) - R^2(0) = 2Dt\frac{\left(C_0 - C_\alpha^E\right)}{\left(C_\beta^E - C_\alpha^E\right)} \tag{164}$$

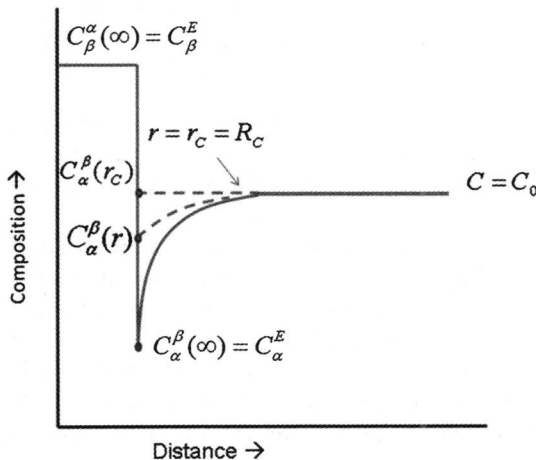

Figure 66   Composition profiles in the vicinity of a growing spherical β precipitate within a matrix of initial composition C_0. The solid line represents the diffusion field in the absence of interface curvature or capillarity effects. The dashed curves show the effects of capillarity on the concentration profiles.

for the time evolution of the particle radius during diffusion-controlled growth with $R(0)$ being the particle radius at the onset of the growth process. Assuming $R(t) >> R(0)$, the radius $R(t)$ increases as $t^{\frac{1}{2}}$ and the volume of the particle V_P varies as $t^{\frac{3}{2}}$. Ham (1959) showed that provided the shape of a nonspherical particle (defined by the eccentricity for prolate and oblate spheroids) remains constant during diffusional growth, the volume will increase as $t^{\frac{3}{2}}$ since each dimension will grow as $t^{\frac{1}{2}}$. It should be pointed out that if the growing particle is small enough where capillarity effects come into play, the concentration profile is changed since the matrix concentration in local equilibrium with the small particle of radius r will be enhanced approximately to

$$C_\alpha(r) = C_\alpha^E\left(1 + \frac{2\sigma V_m^\beta}{rRT}\right) \tag{165}$$

where σ is the interfacial free energy or surface energy of the α–β interface and V_m^β is the molar volume of the β phase. We have used the simplifying approximation that the atomic fraction $X_\beta^E \approx 1$ and that $\exp\left(\frac{2\sigma V_m^\beta}{rRT}\right)$ can be approximated $\left(1 + \frac{2\sigma V_m^\beta}{rRT}\right)$. The instantaneous growth rate $\frac{dR}{dt}$ is modified to

$$\frac{dR}{dt} = \frac{D}{R}\frac{(C_0 - C_\alpha^E)}{(C_\beta^E - C_\alpha^E)}\left(1 - \frac{R_C}{R}\right) \tag{166}$$

where at $R = R_C$, the equilibrium concentration C_α^β at the interface reaches C_0 and $\frac{dR}{dt}\to 0$ according to Eqn (166). Note in this limit the particle is in thermodynamic equilibrium with the supersaturated matrix. See **Figure 66**.

Note: The reader is reminded that the capillarity correction for the α–β equilibrium is more rigorously given by

$$X_\alpha^\beta(r) = X_\alpha^\beta(\infty)\exp\left(\frac{2\sigma V_m^\beta(1 - X_\alpha^\beta(\infty))}{rRT\left(X_\beta^\alpha - X_\alpha^\beta(\infty)\right)}\right) \tag{167}$$

where $X_\alpha^\beta(r)$ is the matrix composition (atomic fraction) or composition of the α phase in equilibrium with a spherical β precipitate of composition $X_\beta^\alpha = X_\beta^\alpha(\infty) = X_\beta^E$ (assumed to be constant) and $X_\alpha^\beta(\infty)$ is the composition of the α phase in equilibrium with β across a flat interface $r = \infty$. Clearly as $X_\beta^E \to 1$ the usual approximations obtain.

Let us consider the case where a steady-state obtains at the interface during diffusional growth as a concentration C_I is established at the interface such that an interfacial reaction governed by a rate equation $K_I(C_\beta - C_I)$ where K_I is a rate constant describing the rate at which atoms can join the growing phase at the interface is equal to the rate at which atoms arrive at the matrix–precipitate interface by diffusion down a modified concentration gradient in the matrix. If the interfacial reaction is very sluggish ($K_I << D$) the interface concentration C_I approaches C_0 and when C_I reaches C_0 the reaction is said to be *interface controlled* and the depleted region in the vicinity of the growing particle virtually vanishes. At intermediate steady-state values $C_\alpha^E < C_I < C_0$ growth is said to governed by *mixed control*. Note that in the limiting case $C_I \to C_0$ the rate of growth becomes a constant value and $R \propto t$. See **Figure 67**.

Another case of interest is where a precipitate phase nucleates copiously along the grain boundaries of a parent phase forming chunks or semicontinuous slabs which grow out into the

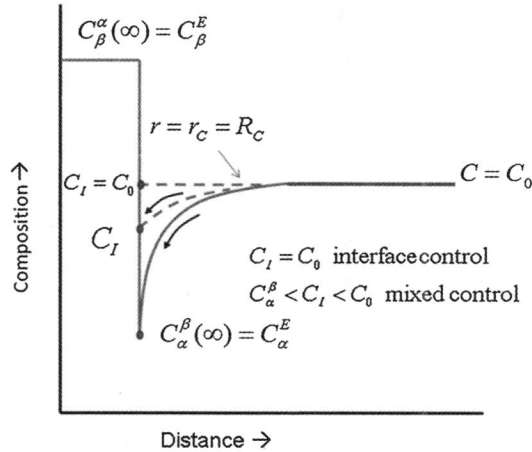

Figure 67 Composition profiles in the vicinity of a growing spherical β precipitate when an interfacial reaction leads to mixed and interface controlled growth.

matrix behind a virtually planar reaction front or interphase interface. These grain boundary chunks are often referred to as grain boundary allotriomorphs. See **Figure 68**. A classic example of this situation is the formation of proeutectoid ferrite in steels (Fe–C–X alloys). Upon cooling a plain carbon steel containing about the 0.4 wt.% C from the austenite phase field, the ferrite (bcc) pro-eutectoid phase (~ 0.02 wt.% C) nucleates primarily at the austenite grain boundaries exhibiting two morphologies, namely "chunky" allotriomorphs and sharp, blade-like growths emanating from the grain boundary regions into the parent austenite referred to as Widmanstatten "sideplates". See **Figure 68**. Importantly, the grain boundary ferrite nuclei generally establish an orientation relationship, for example the Kurdjumov–Sachs relationship wherein the closest packed planes and directions of the two phases are parallel with one of the adjoining grains and subsequent growth occurs primarily into the other grain where no bicrystallographic relationship exists. The migrating ferrite–austenite interface is essentially a disordered incoherent phase boundary. We will focus our attention first on the ferrite slabs (α) which grow into the parent austenite (γ) via diffusion of carbon in front of an advancing planar interphase interface into the austenite. See **Figure 69**. In the first approximation we again assume that the carbon concentration at large distances (much larger than the ferrite allotriomorph thickness) remains $C_0 \approx 0.4$ wt. % in our analysis and that local equilibrium prevails at the ferrite–austenite interface, that is, C^α and C^γ are the equilibrium compositions of ferrite and austenite given by the Fe–C phase diagram at the transformation temperature. The carbon concentration in the austenite in the vicinity of the growing ferrite normal to the reaction front can be assumed to be adequately described by an error function solution to Fick's second law based on the boundary conditions in spite of the moving interface. The carbon distribution in the austenite can be written as

$$C = C_0 - (C_0 + C^\gamma)\left[\left(1 - erf\left(\frac{z - r}{2\sqrt{Dt}}\right)\right)\right] \qquad (168)$$

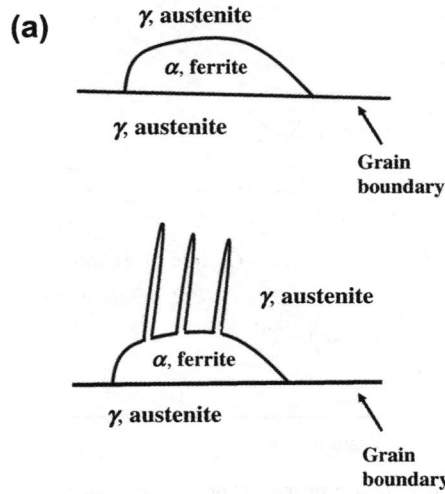

Figure 68   (a) Schematics of ferrite morphologies showing the allotriomorphs and sideplates typical in Fe–C–X alloys. (b) micrographs displaying morphologies of the allotriomorphs and sideplates typical in Fe–C–X alloys. After Shewmon (1963).

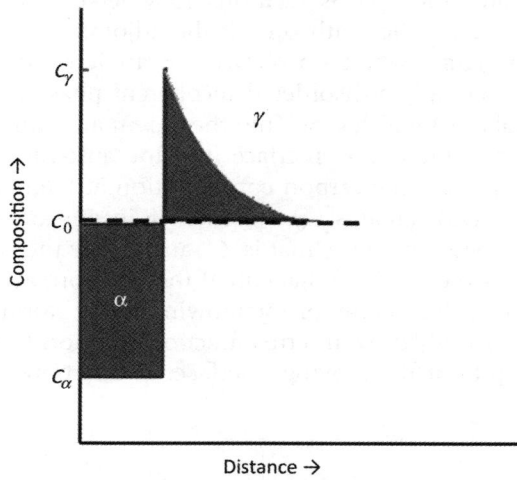

Figure 69   Carbon concentration profile in the parent austenite in front of a growing ferrite slab.

where the variable z is taken as the distance from the grain boundary and $z = r$ is the location of the advancing ferrite–austenite interphase interface. At the interface $\left(\frac{\partial C}{\partial z}\right)_{z=r} = \frac{C_0 - C^\gamma}{\sqrt{\pi D t}}$, and we can formulate a planar growth rate $\frac{dr}{dt}$ as above, equating the flux of carbon rejected into the austenite to the flux of carbon down the concentration gradient into the austenite as follows:

$$\frac{dr}{dt} = \left(\frac{D}{\pi t}\right)^{\frac{1}{2}} \frac{C^\gamma - C_0}{C^\gamma - C^\alpha} \tag{169}$$

and integrating show that the thickness of the ferrite slab follows a rate law $r \propto t^{\frac{1}{2}}$.

Let us now briefly address the appearance of the striking ferrite morphology called Widmanstatten sideplates wherein blades or spikes of the α phase shoot out into the parent austenite. The first approximation to understanding this morphological transformation is to use the simple analysis applied to the formation of dendrites during the freezing of an alloy. The advance of the solid–liquid interface generally involves the rejection or enhancement of solute as the solid forms from the liquid and a concentration gradient develops in the liquid phase in front of the interface. If a thermal fluctuation produces a small bump on an otherwise planar solid–liquid interface the isoconcentration lines in the liquid will be compressed leading to an enhanced diffusion of solute to or away from the interface allowing a spike of solid to grow out rapidly ahead of the generally plane front. This "point effect of diffusion" leads to a breakdown of the planar interface and amounts to an interfacial instability leading to dendritic growth. See **Figure 70**. The growth of such spikes in the solid state can be analyzed using a modified version of the approaches employed earlier for the case of ferrite sideplates assuming that the sideplates emerge from interfacial instability at the ferrite–austenite interface resulting from the point effect of diffusion of carbon within the parent austenite. The emergence of a blade-like protuberance at the interface will change the local equilibrium concentration of carbon and the effective concentration gradient at the blade edge or tip because of capillarity. The gradient at the edge with an effective radius of curvature ρ can be approximated by

$$\left(\frac{\partial C}{\partial r}\right)_{edge} = \frac{C^\gamma - C_0}{k\rho}\left(1 - \frac{\rho_C}{\rho}\right) \tag{170}$$

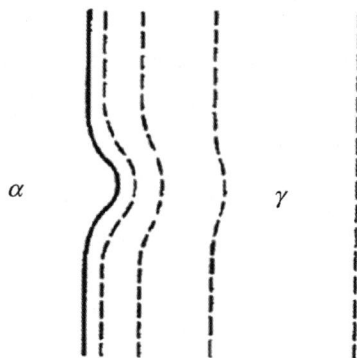

Figure 70 Compression of the isoconcentration lines in the austenite resulting from a local protuberance developing at the interface through fluctuations. After Shewmon (1963).

and the edge velocity written as

$$V_{edge} = D \frac{C^\gamma - C_0}{k\rho(C^\gamma - C^\alpha)} \left(1 - \frac{\rho_C}{\rho}\right) \tag{171}$$

where, as above, ρ_C defines a critical curvature that renders the ferrite in local equilibrium with the supersaturated austenite of composition C_0 and $V \to 0$; k is a constant of order unity. The growth velocity of the edge or tip controlled by the diffusion of carbon in the austenite under the influence of the point effect of diffusion is a maximum for $\rho = 2\rho_C$. This expression for the growth rate of the blade or sideplate is the well-known Zener–Hillert equation (Zener 1946 and Hillert, 1957). A more detailed and rigorous treatment of the problem has been formulated by Trivedi (1970, 1975).

In the case of Widmanstatten sideplates considerations of crystallography and interface structure as well as mechanism must be folded into a rigorous treatment. A number of investigators using various approximations have addressed these issues.

The analyses of precipitate growth discussed earlier did not address the case where the diffusion fields of the growing particles impinge or where the growth of many particles simultaneously leads to a decrease in the average composition of the matrix well away from the depleted shells surrounding the growing particles during the latter stages. The treatments earlier were developed pedagogically to show how a fundamental approach to precipitate growth requires an interplay of thermodynamics and diffusion kinetics with capillarity or surface curvature often entering the problem as a major factor when describing microstructural evolution on a fine scale. One approach to the impingement problem modifies the growth rate of a spherical particle as follows:

$$\frac{dR}{dt} = \frac{D}{R} \left(\frac{C_0 - C_\alpha^E}{C_\beta^E - C_\alpha^E}\right)(1 - Y) \tag{172}$$

where Y is the fraction of the second phase precipitated at time t and given by

$$Y = \frac{C_m(t) - C_\alpha^E}{C_0 - C_\alpha^E} \tag{173}$$

with $C_m(t)$ being the average concentration throughout the matrix at time t. If $(1 - Y)$ is known as a function of time t the rate equation can be integrated as above (Wert and Zener, 1950).

8.5.3 Cellular Phase Separation: Eutectoid Decomposition and Discontinuous Precipitation

Cellular phase separation refers to phase reactions wherein a single phase decomposes into two phases via a mechanism involving the concomitant formation of the new phases behind a reaction front. The formation of the new phases generally produces a characteristic morphology, for example, lamellar phase mixtures, common in eutectoid (and eutectic) decomposition as well as in precipitation systems. The phase reactions can be described as

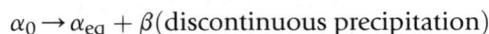

$$\gamma \to \alpha + Fe_3C \,(\text{eutectoid decomposition})$$

$$\alpha_0 \to \alpha_{eq} + \beta \,(\text{discontinuous precipitation})$$

noting in the case of discontinuous precipitation the product phase α_{eq} has the same crystal structure as the parent phase, α_0. The two-phase duplex microconstituent resulting from the cellular reaction can exhibit various morphologies, for example, lamellar, rod like, depending on the specificities of the surface energies and nucleation and growth behavior of the two phases comprising the two-phase mixtures. These resultant structures are examples of self-organization of the phases forming the microconstituents.

8.5.3.1 Eutectoid Decomposition

Let us now consider a classic metallurgical problem in microstructural evolution involving the diffusional growth of technological significance relating to the efforts in the mid-twentieth century to establish a sound scientific understanding of practical issues in the heat treatment of steels such as hardenability (Zackay and Aaronson, 1962). The eutectoid reaction in Fe–C–X alloys involves the decomposition of austenite into two phases α (ferrite) and Fe_3C (cementite) below the eutectoid reaction isotherm ($\sim$723 °C). See the Fe–C binary diagram in **Figure 71**. Between approximately 723 °C and 550 °C the eutectoid decomposition $\gamma \rightarrow \alpha + Fe_3C$ occurs in a manner which produces a self-organized duplex structure composed of alternating lamellae of the α and Fe_3C phases called pearlite after early descriptions by optical metallographers. See **Figure 72**. The blades of the phases apparently conjugate along low-energy $\alpha - Fe_3C$ interphase interfaces and the characteristic spacing of

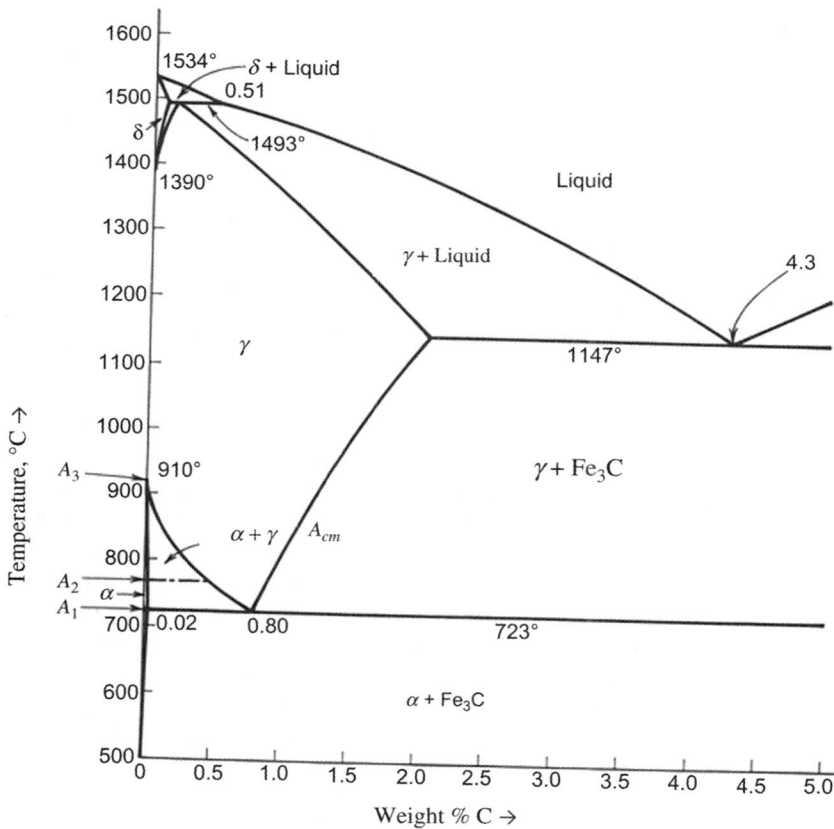

Figure 71 Binary metastable Fe–Fe$_3$C phase diagram. After Shewmon (1963).

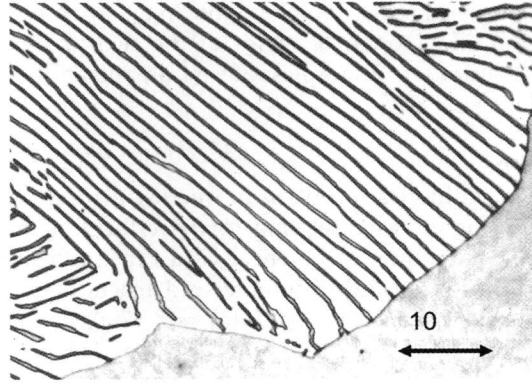

Figure 72   Microstructure of pearlite which has grown into an austenite (γ) grain. After Shewmon (1963).

the lamellar microconstituent is found to vary systematically with the transformation temperature. At temperatures between 550 and 650 °C the interlamellar spacing is much smaller than the spacing observed when the transformation occurs just below the eutectoid temperature. We also find that the growth rate of the lamellar colonies increases rapidly with the degree of undercooling below the equilibrium eutectoid temperature.

Let us try to develop a systematics to understand the salient features of this well-known solid-state transformation based on a fundamental thermodynamic framework and basic diffusion kinetics. Our approach will not be the most rigorous but will again emphasize pedagogy and principles governing the phenomena and which are generally applicable to understanding the microstructural evolution in conjunction with phase transformations.

The lamellar pearlite colonies emerge at the grain boundaries of the parent austenite phase and grow behind a reaction front as shown in **Figure 72**. As the two-phase aggregates grow a partitioning of carbon occurs between a solute-depleted ferrite (α) phase and a solute-enriched cementite (Fe_3C) phase via diffusional processes. The diffusional redistribution of carbon can occur within the austenite just ahead of the reaction front, within the austenite–pearlite boundary or even locally within the ferrite.

We will first approach the problem assuming that volume diffusion within the parent austenite in the near vicinity of the reaction front on a length scale of the order of S controls the edgewise growth.

Before developing the diffusional kinetics governing the cooperative growth of the $\alpha + Fe_3C$ duplex microconstituent let us briefly address the thermodynamics of the austenite $\rightarrow$ pearlite reaction and the observation that the lamellar aggregates exhibit a constant spacing S during isothermal growth wherein S becomes smaller as the undercooling ΔT below the eutectoid temperature increases, that is, the pearlite becomes finer at lower transformation temperatures. We begin our analysis by noting that the lamellar aggregates produced by the eutectoid reaction $\gamma \rightarrow \alpha + Fe_3C$ store free energy within the transformation product stemming from the creation of α-Fe_3C interfaces in the two-phase duplex microconstituent. The net free energy change accompanying the formation of pearlite must include this surface energy expenditure. Thus, the net free energy released per unit volume of austenite transformed, ΔG_V, is written as

$$\Delta G_V(S) = \Delta G_V(\infty) + \frac{2\sigma_{\alpha-Fe_3C}}{S} \tag{174}$$

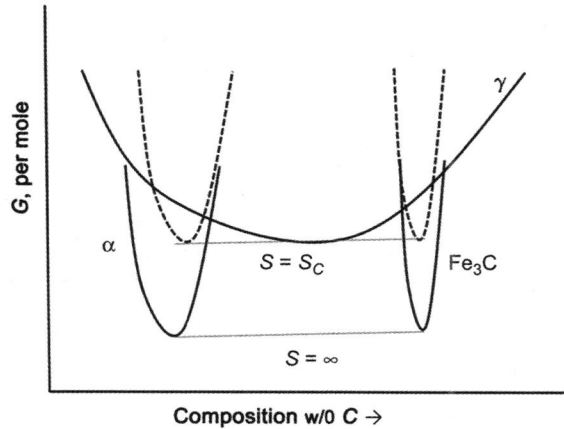

Figure 73 Schematic free energy–composition diagrams for the α, γ and Fe_3C phases showing the influence of capillarity on the phase equilibrium; at a critical spacing of the pearlite $S = S_C$ the two-phase microconstituent pearlite is in thermodynamic equilibrium with the parent austenite phase. $S = \infty$ refers to a spacing of the pearlite where capillarity or interface curvature effects are negligible.

where $\Delta G_V(\infty)$ is the bulk free energy change per unit volume or for a spacing so large that the interfacial free energy associated with the $\alpha - Fe_3C$ interfaces is negligible and the second term $\dfrac{2\sigma_{\alpha-Fe_3C}}{S}$ is the interfacial free energy stored per unit volume when forming the lamellae of finite spacing S. Since the thermodynamic driving force $\Delta G_V(\infty)$ is < 0 below the eutectoid temperature and $\dfrac{2\sigma_{\alpha-Fe_3C}}{S} > 0$, there is a critical spacing $S = S_C$ such that $\Delta G_V(S_C) = 0$, that is, the austenite (γ) is in equilibrium with the pearlite ($\alpha + Fe_3C$) phase mixture. See **Figure 73**.

Note: The shape of the ferrite and cementite tips at the advancing interface are dictated by the surface (free) energies and these curvatures give rise to significant capillarity effects raising the free energy curves of these phases and changing the local equilibrium concentrations along the interface depending on the spacing S. The critical spacing is given by

$$S_C = -\frac{2\sigma_{\alpha-Fe_3C}}{\Delta G_V(\infty)} \qquad (175)$$

This result will be used to formulate an approximate capillarity correction of the form $1 - \dfrac{S_C}{S}$ as shown in the following discussion.

Let us now derive a first approximation to an expression for the growth rate or velocity of the pearlite cells into the austenite assuming that the cooperative growth is controlled by (volume) diffusion in the near vicinity of the reaction front as mentioned earlier. As a ferrite blade moves into the austenite with a velocity V, a lateral flux of carbon must accompany this advance given by $V(C_\gamma^\alpha - C_\alpha^\gamma)_I$ where C_γ^α is the concentration of carbon in the austenite (γ) in local equilibrium with the ferrite (α) and C_α^γ is the concentration of carbon at the ferrite tip in local equilibrium with γ. (The subscript I refers to local equilibrium at the interface.) The carbon in the austenite diffuses down

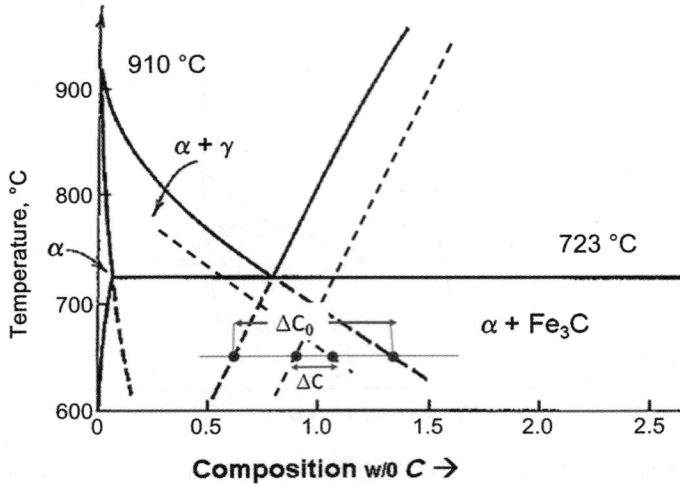

Figure 74 Portion of the Fe–C phase diagram showing shift of phase boundaries resulting from capillarity or interface curvature effects which modify the effective concentration gradient relevant to the diffusional austenite → pearlite reaction. ΔC_0 refers to local equilibria in the absence of capillarity effects whereas ΔC is the "corrected" concentration gradient.

a gradient in the austenite which goes as $\frac{\Delta C}{S}$ where $\Delta C = (C_\gamma^\alpha - C_\gamma^{Fe_3C}) = (C_\gamma^\alpha(\infty) - C_\gamma^{Fe_3C}(\infty))(1 - \frac{S_C}{S})$ often written as

$$\Delta C = \Delta C_0 \left(1 - \frac{S_C}{S}\right) \text{where } \Delta C_0 = \left(C_\gamma^\alpha(\infty) - C_\gamma^{Fe_3C}(\infty)\right)$$

so ΔC_0 refers to bulk phases and the $\Delta C_0 \left(1 - \frac{S_C}{S}\right)$ is the capillarity correction for interface curvature of the ferrite and cementite lamellae. See **Figures 73** and **74**. Equating these fluxes one arrives at an expression for the growth velocity as follows:

$$V = \frac{D_C^\gamma \left(C_\gamma^\alpha - C_\gamma^{Fe_3C}\right)}{\left(C_\gamma^\alpha - C_\alpha^\gamma\right)_I \eta S} = \frac{D_C^\gamma \Delta C_0 \left(1 - \frac{S_C}{S}\right)}{\left(C_\gamma^\alpha - C_\alpha^\gamma\right)_I \eta S} \tag{176}$$

where D_C^γ is the diffusivity of carbon in γ and η is a geometric parameter of order unity. However, we do not have a solution in closed form because as shown earlier the system can utilize any spacing $S < S_C$ at a given undercooling ΔT below the eutectoid reaction isotherm. What spacing does the system select? Some variational principle must be invoked. Let us suppose we assume that S is the spacing that maximizes the growth rate $V(S)$. Setting $\frac{dV(S)}{dt} = 0$, we find that $V = V_{MAX}$ when $S = 2S_C$; the growth rate then becomes

$$V = V_{max} = \frac{D_C^\gamma \Delta C_0}{\left(C_\gamma^\alpha - C_\alpha^\gamma\right)_I 4\eta S_C} \tag{177}$$

and since $\Delta C_0 \sim \Delta T$ and $S_C \sim \Delta T^{-1}$ the growth rate might be expected to vary as $V \sim D_C^{\gamma} (\Delta T)^2$. This predicted dependence of V on ΔT or temperature appears to describe the growth rate of pearlite in a variety of steels. Also, the observed pearlite spacing apparently obeys the relation $S_C \sim \Delta T^{-1}$ as predicted earlier. Eutectoids with a pearlitic-like or lamellar morphology occur in alloy systems with substitutional solute elements but the growth velocities generally vary as $D(\Delta T)^3$ because the solute redistribution does not occur by volume diffusion in the parent phase but is controlled by boundary diffusion within the interface between the growing cells and the parent phase. This results from the fact that in the case of boundary diffusion control the effective cross-section area through which the diffusion flux passes is reduced by a factor of the order of $\dfrac{\delta}{S}$ giving a growth velocity $V \sim \dfrac{D_B \Delta C_0 \delta}{S_c^2}$ where D_B is a boundary diffusivity and δ is the thickness of the boundary region.

8.5.3.2 Discontinuous Precipitation

Discontinuous precipitation is essentially a mode of precipitation similar to the austenite–pearlite reaction described earlier in that it is a form of cellular phase separation involving heterogeneous nucleation of a phase along the grain boundaries of an alloy and the formation of colonies of a two-phase mixture exhibiting a characteristic morphology which grow through a cooperative codeposition of phases behind a migrating high-angle grain boundary. See **Figure 75**. However, the phase reactions involved are somewhat different most generally we could have

Type 1:	$\alpha_0 \rightarrow \alpha_{eq} + \beta$
Type 2:	$\alpha' + \beta' \rightarrow \alpha_{eq} + \beta$
Type 3:	$\alpha' + \delta' \rightarrow \alpha_{eq} + \beta$

where in the first case α_0 is an initial supersaturated solid solution and α_{eq} is a solute-depleted α phase (same crystal structure) which is in equilibrium with the β precipitate. In the second case, $\alpha' + \beta'$ is a metastable (often coherent) two-phase mixture which initially has formed throughout the matrix by a uniform continuous precipitation from supersaturated solid solution and then subsequently consumed by the more stable two-phase mixture $\alpha_{eq} + \beta$ through the growth of the cellular colonies. Here the coherent metastable β' phase has the same structure as the equilibrium β phase and thus this reaction can be considered essentially a coarsening reaction driven by a reduction of surface free energy and strain energy. The third case involves continuous precipitation of a coherent metastable δ' phase which has a different crystal structure than the equilibrium β phase.

Figure 76 shows proposed mechanisms for the genesis of the self-assembled colonies each of which have been observed to operate in various systems. Cellular or discontinuous precipitation is ubiquitous in precipitation systems often leading to a degradation of engineering properties.

Once assembled the cells grow by boundary diffusion control and the growth rate generally varies as $\dfrac{D_B \delta}{S^2}$ where D_B is a grain boundary diffusivity, S is the spacing of the lamellar morphology and δ is an effective boundary thickness. The thermodynamic driving force for the cellular reaction is the chemical free energy often augmented by coherency strain energy ($\alpha' + \beta' \rightarrow \alpha_{eq} + \beta$) but always reduced by the interfacial free energy stored in the interphase interfaces incorporated in the cellular morphology. In thermomechanically processed alloys the stored energy of cold work in the matrix or phase mixture being consumed by the colonies can provide additional driving force for the migrating grain boundaries similar to recrystallization of a cold-worked alloy. It is interesting to note that cellular or discontinuous

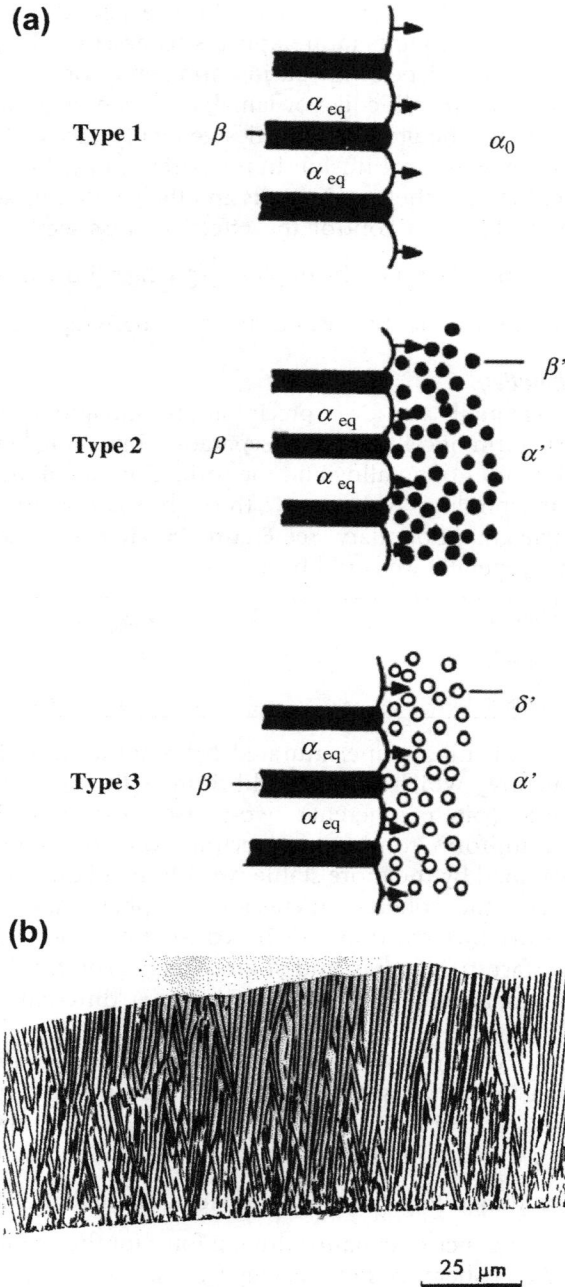

Figure 75 (a) Schematic showing various grain boundary discontinuous reactions. After (Williams and Butler, 1981).

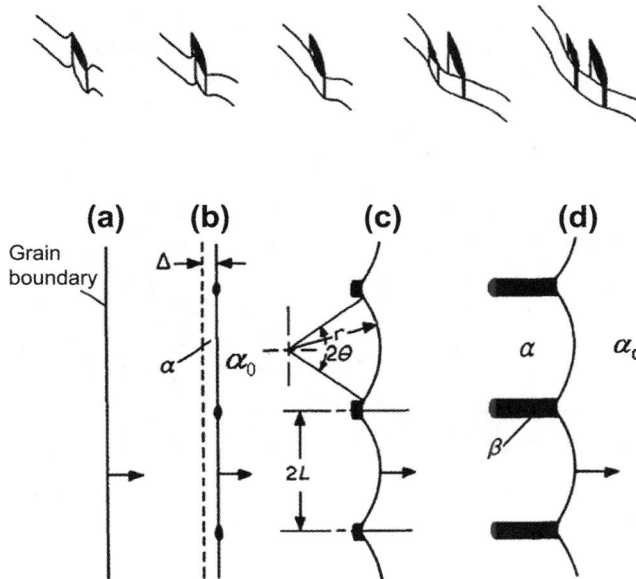

(a) **(b)** **(c)** **(d)**

Figure 76 Schematics showing possible mechanisms involved in the genesis of cellular/discontinuous precipitation. After Williams and Butler (1981), Fournelle and Clarke (1972) and Turnbull and Tu (1970).

precipitation in age hardening alloys was often referred to as a "recrystallization reaction" because the microstructural changes involved the sweeping of the structure by migrating high-angle grain boundaries as occurs in recrystallization. It is important to mention that the lamellar products of discontinuous or cellular precipitation show a characteristic spacing at a given degree of supersaturation or driving force determined by similar thermodynamic and kinetic factors discussed in eutectoid decomposition and the edgewise growth of pearlite (Gust, 1979; Williams and Butler, 1981; Manna et al., 2001).

8.5.4 Growth of Widmanstatten Morphologies

A classic precipitate growth morphology is that of Widmanstatten platelets which were first discovered by Count Alyois von Widmanstatten in his early nineteenth century studies of meteorites (Barrett and Massalski, 1966). See **Figure 77**. The Widmanstatten patterns in meteorites were coarse enough to be viewed by the naked eye and derived from precipitation of α (bcc) platelets within a γ (fcc) matrix in essentially Fe–Ni alloys under conditions of extraordinarily slow cooling rates ($\sim 1\,°C$ per 10^6 years) during their journey through space. The α platelets form parallel to the {111} planes of the γ matrix matching the {111} *habit planes* of the matrix and the {110} planes of the α precipitates across the broad faces of the Widmanstatten plates. Similar microstructures have been revealed on the microscopic scale in numerous alloy systems wherein matching of planes and directions of the matrix and precipitate phases across the habit plane occurs. In the case of the Cu–Si alloy shown in **Figure 78** the precipitation occurs along the {111} planes of an fcc matrix which are parallel to the (0001) of the basal planes of the hexagonal platelet precipitates with the close-packed directions in the conjugate phases parallel across the interface plane as well. The phases remain fully coherent across the broad faces until the very late stages of aging when the misfit strains eventually lead to a loss of coherency.

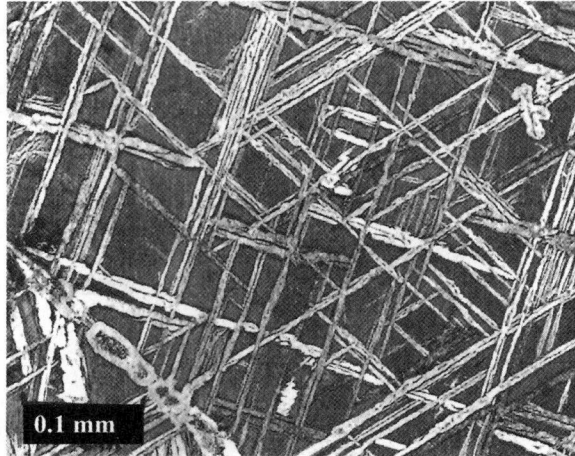

Figure 77 Widmanstatten precipitates in an Fe–Ni meteorite showing platelets of α(bcc) forming along the {111} planes of the γ(fcc) phase. After Finniston (1971).

This striking plate-like morphology appears to be primarily the result of a growth anisotropy directly related the structure of the interphase interfaces bounding the precipitate phase. However, there still persists some controversy as to the degree to which thermodynamic factors are folded into the formation and propagation of these low-energy interfaces. The broad faces of the Widmanstatten plates are virtually always coherent or semicoherent whereas the sides are generally disordered or incoherent. This difference in interfacial structure markedly influences the effective mobility of the interfaces during diffusional growth. The broad faces are structurally inhibited and almost certainly grow normal to themselves (plate thickening) via a ledge mechanism (see **Figure 41**) where atoms are added to the disordered risers of the ledges leading to lateral motion of the ledges within the interphase boundary. The sides are relatively

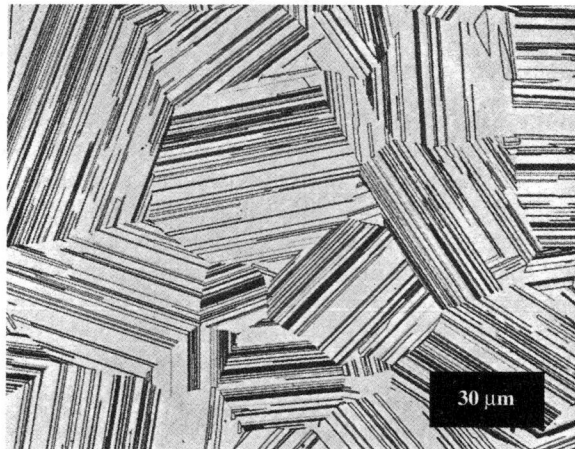

Figure 78 Widmanstatten platelets of a hexagonal phase precipitated along the {111} matrix (fcc) planes in a Cu–Si alloy. After Barrett and Massalski (1966).

much more mobile compared with the coherent/semicoherent faces of the platelets and in principle can grow (lengthen) at a rate controlled by volume diffusion whereas the thickening will depend on the supply and migration of the ledges within the interface (Aaronson et al., 2010).

8.5.5 Coarsening (Ostwald Ripening) of a Two-Phase Mixture in a Binary Alloy

8.5.5.1 *Particle Coarsening in a Matrix*

We now address a very important aspect of microstructural behavior from both a scientific and technological point of view. Consider the formation of a fine dispersion of second-phase particles (β) distributed uniformly within a matrix (α) resulting from nucleation and growth within an initially supersaturated solid solution during heat treatment. In the latter stages of this precipitation process the supersaturation of the matrix is substantially reduced because the growth of the precipitates drains solute from the surrounding phase wherein the matrix composition approaches $C_\alpha^\beta(\infty)$, the composition of the parent matrix in thermodynamic equilibrium with the β precipitates. The precipitates generally have nucleated at different times and grown to produce a range of particle sizes about some mean radius $\bar{r}$. Because of this size distribution even after the supersaturation has virtually vanished and no new particles appear there are local chemical potential and concentration gradients in the system associated with the particles because of the variation of solid solubility with particle size (Gibbs–Thomson or capillarity effects). See **Figure 79**. The concentration of solute in the matrix in local equilibrium in the near vicinity of a particle differs from particle to particle and importantly differs from the concentration in equilibrium with the mean particle size. The larger particles in the distribution tend to grow and the smaller tend to shrink with time as solute flows by diffusion in the matrix leading to an increase in the average particle size. The overall thermodynamic driving force for this microstructural change is the decrease in the total interfacial (free) energy per unit volume associated with the α–β interphase interfaces mediated by capillarity effects. This process is called (particle) *coarsening* or sometimes referred to as *Ostwald ripening*.

The classic approach to describing this diffusion-controlled coarsening of a two-phase mixture was carried out by Greenwood (1956) and later extended to include particle size distribution effects by Lifshitz and Slyozov (1961) and Wagner (1961). In the literature the theory is often called the L–S–W theory of particle coarsening, but should explicitly include Greenwood's seminal contribution and thus appropriately be referred to as the G–L–S–W theory (Balluffi et al., 2005). Greenwood's versions (1956, 1978) are somewhat simpler treatments of the kinetic analysis and will be substantially followed here for heuristic and pedagogical reasons. The approach is essentially what is referred to as a "mean field" solution to the diffusion problem wherein the growing or shrinking particles do not "see" the other particles except through a change in the average solute concentration remote from any

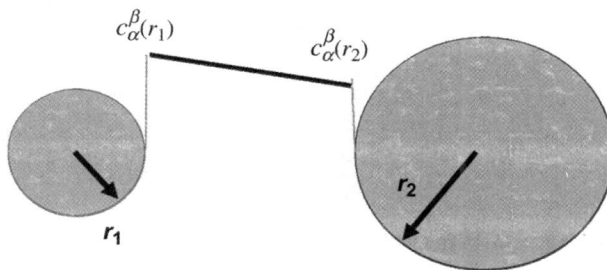

Figure 79 Schematic drawing showing concentration gradient between two particles of different sizes, $r_2 > r_1$, leading to a flow of solute from the vicinity of r_1 to r_2.

individual particle, that is, a particle in question is immersed in a mean or average environment which is perturbed only locally in the near vicinity of a particle (Gibbs–Thomson effect). This approximation essentially monitors the behavior of a particle effectively at zero volume fraction.

We start our discussion by assuming the volume fraction of the second-phase particles β is constant, that is

$$\sum_{i=1}^{i=N_p} V_i = \text{constant} \tag{178}$$

and

$$\sum_{i=1}^{i=N_p} \frac{dV_i}{dt} = \sum_{i=1}^{i=N_p} 4\pi r_i^2 \frac{dr_i}{dt} = 0 \tag{179}$$

where N_P is the number of particles at time t and V_i and r_i are the volume and radius of the ith particle, respectively.

The concentration of solute in the α matrix in local equilibrium with the ith particle of radius r_i is given by the familiar Gibbs–Thomson expression:

$$C_\alpha^\beta(r_i) = C_\alpha^\beta(\infty)\left(1 + \frac{2\sigma_{\alpha\beta}V^\beta}{r_i RT}\right) \tag{180}$$

assuming dilute or ideal solution behavior for the solute in α. We also are assuming that in this first approximation that the precipitate phase β is essentially pure B in an A–B binary solid solution; this simplification will be modified later in the formulation. If the concentration remote from the particle is taken to be $C^\wedge$ we can use the quasi-steady-state solution to describe the instantaneous rate of growth (or shrinkage) of the ith particle as follows:

$$\frac{dr_i}{dt} = \frac{D}{r_i}\left(\frac{[C^\wedge - C_\alpha^\beta(r_i)]}{[C_\beta^\alpha(\infty) - C_\alpha^\beta(r_i)]}\right) \tag{181}$$

where $C_\beta^\alpha(\infty) - C_\alpha^\beta(r_i)$ can be written as $C_\beta^\alpha(\infty) - C_\alpha^\beta(\infty)$ by replacing $C_\alpha^\beta(r_i)$ by $C_\alpha^\beta(\infty)$ in the difference compared with $C_\beta^\alpha(\infty)$. A bit of algebra shows that $C^\wedge = \overline{C}$, the mean concentration in the matrix or the concentration in equilibrium with the average particle size r. Noting also that in this approximation $[C_\beta^\alpha(\infty) - C_\alpha^\beta(\infty)] \approx \frac{1}{V^\beta}$ since $[C_\beta^\alpha(\infty) >> C_\alpha^\beta(\infty)]$ when the β phase approaches pure B. Thus, we can write

$$\frac{dr_i}{dt} = \frac{2D}{r_i}\left[C_\alpha^\beta(\infty)\frac{\sigma_{\alpha\beta}(V^\beta)^2}{RT}\right]\left[\frac{1}{\overline{r}} - \frac{1}{r_i}\right] \tag{182}$$

for the growth rate of a particle relative to the average particle size and find that $\frac{dr_i}{dt} > 0$ for $r_i > \overline{r}$ and $\frac{dr_i}{dt} < 0$ for $r_i < \overline{r}$, that is, particles with radii greater than the average particle radius increase in size while particles with radii smaller than the average tend to shrink. The "big" particles are growing at the expense of the "smaller" ones even leading to the disappearance of some of the smaller members of the ensemble. It is useful to focus on the fastest growing particles for some radius $r_i = r_m$. Setting $\frac{d}{dr_i}\left(\frac{dr_i}{dt}\right) = 0$ one finds that $r_m = 2\bar{r}$ leading to the expression:

$$\frac{dr_m}{dt} = \frac{2D}{r_m^2}\frac{C_\alpha^\beta(\infty)\sigma_{\alpha\beta}(V^\beta)^2}{RT} \tag{183}$$

for the growth rate of the fastest growing particle size. Integrating and inserting the relationship $r_m = 2\bar{r}$ we arrive at a kinetic law for the time evolution of the average particle size during diffusion-controlled coarsening given by

$$\bar{r}^3(t) - \bar{r}^3(0) = \frac{3}{4}DC_\alpha^\beta(\infty)\frac{\sigma_{\alpha\beta}(V^\beta)^2}{RT}t = K_1 t \tag{184}$$

where $\bar{r}(0)$ is the average particle size at the onset of the coarsening process. If $\bar{r}(t) \gg \bar{r}(0)$ this is often written as $\bar{r}(t) \approx K't^{\frac{1}{3}}$. The exact solution emanating from the Lifshitz and Slyozov (1961) theory when more rigorous attention is given to the nature of the behavior of the size distribution gives a modified K_1 as $K_1 = \frac{8}{9}D\left(\frac{C_\alpha^\beta(\infty)\sigma_{\alpha\beta}(V^\beta)^2}{RT}\right)$.

Coarsening is fundamentally a process rooted in capillarity effects expressed in the Gibbs–Thomson equation. However, most precipitates are not terminal solid solutions and the variation of solid solubility with particle size should be corrected for the precipitate composition, as mentioned earlier in the discussion of particle growth, that is, for intermediate phase precipitates, for example A_3B, the capillarity term that should be used in the treatment is that of Eqn (167) which includes a composition term which is essentially 4 for Al_3Li and 3 for Al_2Cu precipitate phases. Another consideration is that of nonspherical shapes and anisotropy of the interfacial or surface free energy σ. As discussed earlier a generalized Gibbs–Thomson equation can be formulated for the nonspherical equilibrium shapes wherein a parameter $\frac{\sigma_i}{\lambda_i}$ is invariant around the particle; σ_i is the surface energy of the ith surface element on the particle interface and λ_i is the distance of this element from the center of the precipitate projected along the normal to the surface element. The result is that the chemical potential is constant around the surface of the particle and thus the solubility is constant about the equilibrium shape.

Let us briefly return to the diffusion-controlled coarsening rate equations 181 and 182 and insert the general capillarity term mentioned earlier (Eqn (167)). Furthermore, we will forego the approximation $[C_\beta^\alpha(\infty) - C_\beta^\beta(\infty)] \approx 1/V^\beta$ for the more general case where the dispersed phase is not a terminal solid solution or pure B but an intermediate composition A_aB_b. Eqn (183) then becomes

$$\frac{dr_m}{dt} = \left(\frac{2DX_\alpha^\beta(\infty)}{r_m^2}\right)\left(\frac{\sigma_{\alpha\beta}V^\beta}{RT}\right)\left(\frac{1 - X_\alpha^\beta(\infty)}{\left(X_\beta^\alpha(\infty) - X_\alpha^\beta(\infty)\right)^2}\right) \tag{185}$$

wherein we have assumed $[X_\beta^\alpha(\infty) - X_\alpha^\beta(r_i)] \approx [X_\beta^\alpha(\infty) - X_\alpha^\beta(\infty)]$ as discussed earlier. Also, again we find $r_m = 2\bar{r}$ and, thus, integrating as before gives

$$\bar{r}^3(t) - (\bar{r}(0))^3 = K_2 t$$

where $K_2 = \left(\frac{3DX_\alpha^\beta(\infty)}{4}\right)\left(\frac{\sigma_{\alpha\beta}V^\beta}{RT}\right)\left(\frac{1 - X_\alpha^\beta(\infty)}{(X_\beta^\alpha(\infty) - X_\alpha^\beta(\infty))^2}\right)$

Interjecting the LSW refinements, K_2 becomes

$$K_2 = \left(\frac{8DX_\alpha^\beta(\infty)}{9}\right)\left(\frac{\sigma_{\alpha\beta}V^\beta}{RT}\right)\left(\frac{1 - X_\alpha^\beta(\infty)}{\left(X_\beta^\alpha(\infty) - X_\alpha^\beta(\infty)\right)^2}\right)$$

This more rigorous capillarity correction was used by Novotny and Ardell (2001) in their analysis of coarsening in binary Al–Sc alloys.

As stated earlier, the classic GLSW theory is essentially a mean field theory virtually for zero volume fraction of second-phase particles. However, as pointed out back in the 1960s and 1970s by Asimow (1963) and Ardell (1972) the volume fraction of second phase is expected to become important at volume fractions where the distances between precipitates is of the order of the mean particle size altering the effective diffusion fields controlling the coarsening process. As a first approximation Ardell suggested that the basic result $r^3(t) - r^3(0) = Kt$ derived earlier in the GLSW approximation be multiplied by a factor $K'(f_V)$ which is a function of the volume fraction f_V which ranges from 1 ($f_V = 0$) to about 10 ($f_V = 0.25$) increasing the coarsening rate significantly. He also pointed out that the steady-state particle size distribution for bulk diffusion-controlled coarsening was expected to be markedly different than that predicted by the GLSW theory. However, the experimental results seem quite complicated and system specific wherein composition changes and changes in coherency strains may play a role. It is important to point out that in spite of the simplifications of the original treatment a kinetic law of the form $r^3(t) - r^3(0) = Kt$ seems to hold for the coarsening or ripening of two-phase microstructures across a wide variety of systems (Ardell, 1987).

In this discussion we did not delve into grain boundary diffusion-controlled coarsening of precipitates or coarsening of precipitates associated with dislocation arrays involving pipe diffusion as a rate-controlling process (Martin et al., 1997).

8.5.5.2 *Discontinuous Coarsening*

In the coarsening reactions discussed earlier, particles of a single phase distributed in a solid solution matrix increase their average size with time. Another type of coarsening occurs in eutectoid and cellular type microstructures as shown in **Figure 80**. Here both phases increase their size (and therefore spacing) discontinuously. Here the word discontinuous is used because the coarsening is not uniform in space but occurs at an advancing reaction front or interface driven by a reduction in surface energy of the lamellar two-phase mixture. This parallels the use of the term in Section 8.1.4 for discontinuous transformations. In the case shown in **Figure 80** the two phases are incoherent with each other (they need not be). Another point of interest is that the diffusion process for this coarsening is along the boundaries separating the coarsened region from the uncoarsened region, as opposed to the matrix diffusion in the particle coarsening reactions.

8.5.5.3 *The Effect of Strain on Coarsening*

So far we have considered only the role of surface energy on the coarsening of precipitates. Another important thermodynamic energy to consider is the elastic energy due to misfit between the crystal lattices of the particles and the matrix. Such energy depends on the shape, habit, configuration and volume of the precipitates (Khachaturyan et al., 1988).

The elastic energy of an isolated particle in a matrix plays a role in determining the shape of the particle. Indeed shape changes occur as the particle size increases to take into account the elastic

Figure 80 Co–Si alloy displaying discontinuous coarsening of a cellular microstucture. After Livingston and Cahn (1975).

anisotropies of the matrix and precipitate. For example a precipitate with isotropic surface energy still may have a cuboidal shape to lower its elastic energy.

Another consideration is the role of the interaction between the stress fields of the precipitates, commonly called an interaction energy. Ardell and Nicholson (1966) have shown that such interaction energy can cause precipitates to align along elastically soft directions of the matrix. Khachaturyan and Airapetya (1974) also showed that a large enough precipitate can decompose into smaller ones and align themselves along elastically soft directions of the matrix. This was later observed experimentally by Miyazaki et al. (1982) and Doi et al. (1984) for cuboidal precipitates of γ' in an Ni alloy. See **Figure 81**.

In such cases, the increase in surface area (and hence surface energy) is off set by the decrease in total elastic energy because of the negative elastic interaction term. Such a process produces smaller precipitates and has been termed "inverse coarsening".

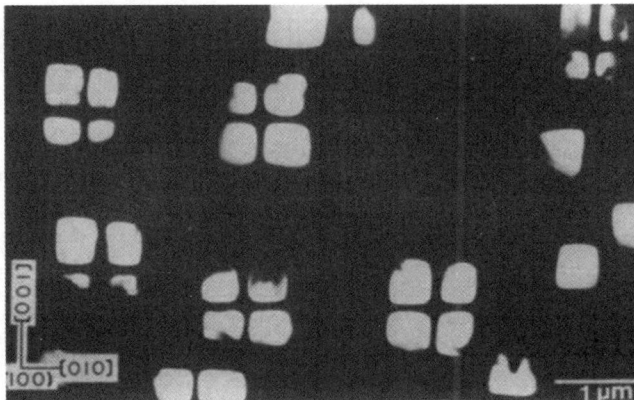

Figure 81 Elastic effects during coarsening leading to particle splitting. After Doi (1994)

8.6 Precipitation from Solid Solution

Precipitation from solid solution produces a phase mixture from the decomposition of an initial supersaturated phase resulting in a matrix phase whose crystal structure is similar to that of the parent phase, but of different composition and usually different lattice parameter(s), as well as an essentially dispersed phase (precipitate) that may differ in crystal structure, composition and/or degree of order. The physical, mechanical and chemical properties of the resultant two-phase alloy can vary markedly with the size, shape (morphology) and distribution of the precipitate phase within the matrix and this solid-state reaction provides the basis for one of the most powerful and versatile means available to the physical metallurgist and materials engineer for tailoring the properties of high-strength alloys through heat treatment and thermomechanical processing called *age hardening* or *precipitation hardening*. Precipitation effects can influence magnetic and superconducting properties as well as mechanical strength. Sometimes precipitation can cause unwanted effects in alloy applications and must be accounted for and controlled.

Precipitation from supersaturated solid solution as a phase transformation is ubiquitous in metallic and ceramic systems and in **Figure 82** we show various phase diagram configurations which can give rise to precipitation reactions in the solid state. Precipitation phenomena embody a range of fundamental processes central to understanding the role of transformation behavior in controlling microstructural development during processing and heat treatment, namely *nucleation*, *growth*, and *coarsening*

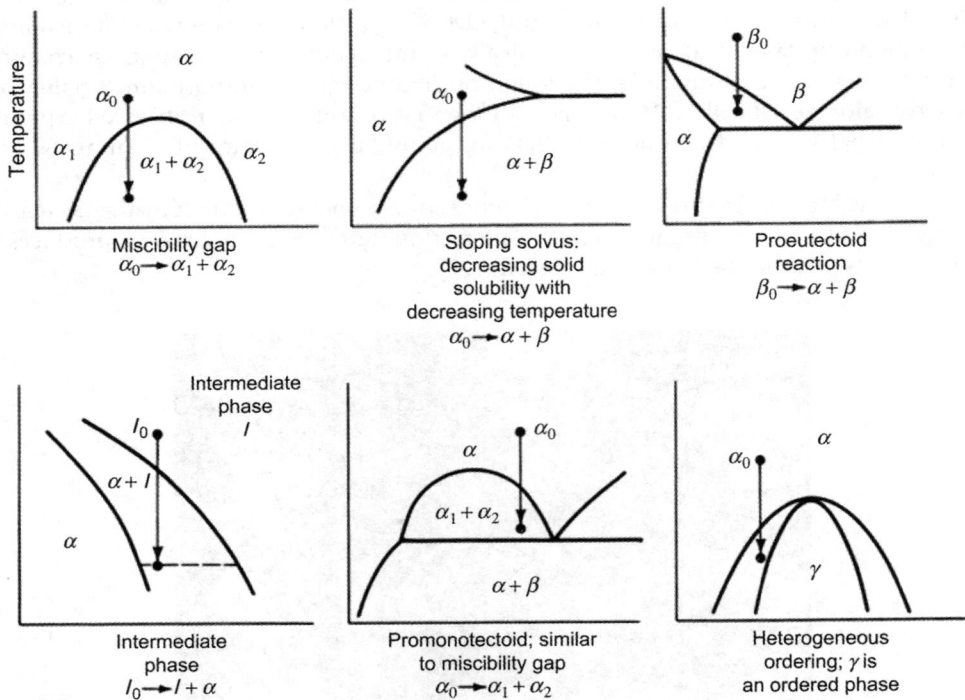

Figure 82 Equilibrium phase diagram configurations illustrating various conditions for precipitation of a second phase in a binary alloy. After Soffa (1985).

and competition among them. Furthermore, this solid-state transformation involves some of the most challenging fundamental issues at the forefront of thermodynamics and statistical mechanics such as nonequilibrium, irreversibility, the theory of rate processes, fluctuation behavior and self-organization in material systems.

8.6.1   Historical Background

During the first decade of the twentieth century A. Wilm in Germany wondered whether the emerging aluminum alloys could be strengthened employing methods used in other alloys, such as the quench hardening approach that was so successful with steels, or what is now called substitutional solution hardening, employed in Cu alloys. Both of these methods of strengthening had been used dating back to antiquity. The as-quenched aluminum alloys, however, were found to be soft much to Wilm's surprise. Furthermore, often the fastest cooling rates appeared to be softer yet. Fortunately following a set of Saturday morning "failures", he reexamined the disappointing results on Monday of the next week and found that the hardness and tensile properties had increased markedly while the specimens sat for 2 days at room temperature. They had "aged" over the weekend and these somewhat misguided experiments led to the discovery of age hardening. (Fortuitously, trace elements in the alloy micro-chemistry, e.g. Mg rendered these alloys "naturally aging", that is, unknown to Wilm at the time significant precipitation occurred at room temperature within the supersaturated alloys produced by quenching.) Wilm then more systematically explored the aging response in alloys containing ~ 4 wt.% Cu plus about 0.5 wt. % Mg along with traces of Mn, Fe and Si and patented his findings in 1906. His first archival publication was not contributed until 1911 (Wilm, 1911). This genre of alloy was christened DURALUMIN and was the primary structural material for the famous Zeppelins that were flying about between 1910 and 1920. The strengthening from the aging behavior increased the yield strength of the alloys by a factor of 2 or more compared with the as-quenched condition.

However, full exploitation of this phenomenon and extension to other alloys lay dormant for a decade until 1919 when Merica et al. (1921) in the United States identified the fundamental basis for the aging response; they recognized that this was the result of decreasing solid solubility with temperature resulting in precipitation from supersaturated solid solution of a second phase during the aging process at ambient (naturally aging) or moderately elevated temperature in a two-phase field. Quenching from a single-phase field at high temperature produces the supersaturated state, for example excess Cu dissolved in Al and the aging allows the nonequilibrium state to relax toward equilibrium (stable or metastable) dictated by the relevant solvus line in the phase diagram. Archer and Jeffries (1925) put forth a slip interference theory or "keying effect" as the mechanism for the enhanced mechanical strength wherein they hypothesized that the increased resistance to plastic flow derived from submicroscopic particles of a second phase which inhibit the elementary slip process of glide along crystallographic planes.

The association of this strengthening mechanism with decreasing solid solubility with decreasing temperature and the formation of fine-scale "pre-precipitates" or precipitate phases during aging by Merica et al., (1921) was the major breakthrough since following this revelation numerous age hardening alloys were developed not restricted to DURALUMIN-type aluminum alloys including copper-base and nickel-base alloys. Here we have an excellent example of a modest fundamental understanding of a phenomenon providing the foundation for the development of a new alloy technology beyond the specific context of the seminal discovery. Today *age hardening* or *precipitation hardening* is indeed one of the most effective approaches to designing high-strength alloys for a plethora of applications in modern technology.

The details of the nanoscale phenomena often involved in the precipitation process were not elucidated until the advent of sophisticated X-ray diffraction methods in the 1930s (Guinier 1938b; Preston 1938c) and ultimately through direct observation with the emergence of transmission electron microscopy in the 1950s and 1960s (Nicholson and Nutting, 1958; Nicholson et al., 1959). It was found that the precipitation process in the original DURALUMIN-type alloys generally exhibited a complex multistage decomposition of the supersaturated solid solution before the formation of the equilibrium phase indicated by the phase diagram, for example $CuAl_2$ (θ) in the binary Al–Cu system involving the appearance of nanoscale copper-rich "zones". Mehl and Jetter (1940) called these early-stage zones or precipitates Guinier–Preston (G.P.) zones after the early X-ray investigators in France and Great Britain (Guinier 1938a, 1938b; Preston 1938a, 1938b, 1938c.). Most importantly, it is now recognized that effective age hardening in most alloy systems generally derives from the formation of fine-scale metastable phases (sometimes referred to as "transition precipitates") before the appearance of the stable equilibrium precipitate phase; the formation of the equilibrium phase often leads to a degradation of properties developed with the metastable precipitates. A quantitative approach to correlating the precipitation microstructures with mechanical properties had to wait for the resolution of the nature of these ultrafine scale dispersions of second-phase particles and the development of a mature dislocation theory of plastic flow. Most notable is the theory of the yield strength due to Orowan (1948). Approaches to a quantitative description of the flow stress of particle hardening up to 1960 are summarized in the review by Kelly and Nicholson (1963) including the early work of Mott and Nabarro (1940). Books by Martin (1998) and Nembach (1996) provide excellent overviews of progress in the field. We shall not treat this topic in any depth in this chapter.

8.6.2 Nucleation and Spinodal Decomposition

In this section we focus on the initial stages of formation of a second phase through homogeneous decomposition within a supersaturated solid solution. Fundamental to understanding the breakdown of a supersaturated solid solution produced by quenching from a single-phase region into a two-phase field is the concept of the isothermal free energy–composition diagram and its relation to the phase diagram as discussed earlier. We will first consider precipitation within a miscibility gap such as depicted in **Figure 12**a. The associated free energy–composition curve is generally depicted showing a characteristic double well with a region of negative curvature in between the local minima delineated by the so-called chemical spinodal points $\frac{\partial^2 G}{\partial X^2} = 0$. The common tangent construction establishes the compositions of the equilibrium phases $X_e^{\alpha_1}$ and $X_e^{\alpha_2}$, respectively. Between the two inflection points (spinodal points) the free energy curve has a negative curvature and supersaturated states lying in this region are intrinsically unstable with respect to diffusional processes (recall the chord construction and that the effective diffusion coefficient $D \propto \frac{\partial^2 G}{\partial X^2}$, as discussed in an earlier section) and will tend to spontaneously phase separate. Supersaturated states lying on the curve where the curvature (point S in **Figure 12**) is positive are metastable and require a relatively large fluctuation in composition locally to initiate the precipitation reaction (nucleation). Gibbs recognized the possibility of these different kinetic paths leading to the breakdown of the supersaturated state and the formation of the two-phase mixture indicated by the phase diagram through the phase reaction $\alpha_0 \rightarrow \alpha_1 + \alpha_2$, where α_0 is the initial supersaturated solution and α_1 and α_2 are the equilibrium conjugate phases in the two-phase field, respectively. According to Gibbs, decomposition of the supersaturated state can be triggered by large localized fluctuations in composition which can grow spontaneously by diffusion down the concentration gradient which develops in the vicinity of these "critical nuclei" or by the continuous growth of

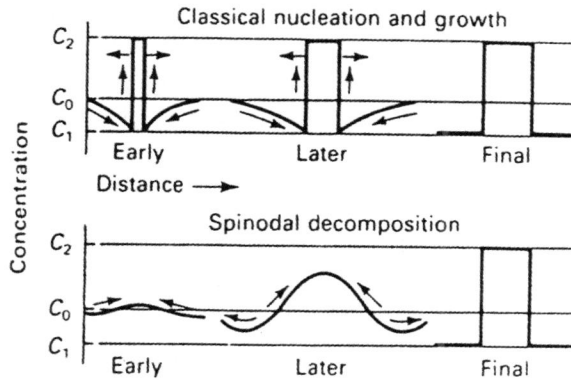

Figure 83 Classical nucleation and growth contrasted with spinodal decomposition as alternative modes of diffusional transformation leading to the formation of a second phase. After Soffa and Laughlin (1985).

initially low-amplitude composition fluctuations which are more spatially extended and exhibit rather diffuse incipient interphase interfaces during the early stages of phase separation gradually evolving into a distinct two-phase mixture. See **Figure 83**. Clearly the latter case involves "up-hill diffusion" or an effective negative diffusion coefficient. The first alternative refers to essentially **classical** nucleation of the new phase and the latter describes that which in the modern lexicon is called spinodal decomposition. The overall rate of these processes depends on the rate of atomic migration and the diffusion distances involved (undercooling). It is important to point out that the "spinodal line" depicted in **Figure 12**a is not a phase boundary but a demarcation indicating a difference in thermodynamic stability of supersaturated states and essentially a limit of metastability (**Figure 84**).

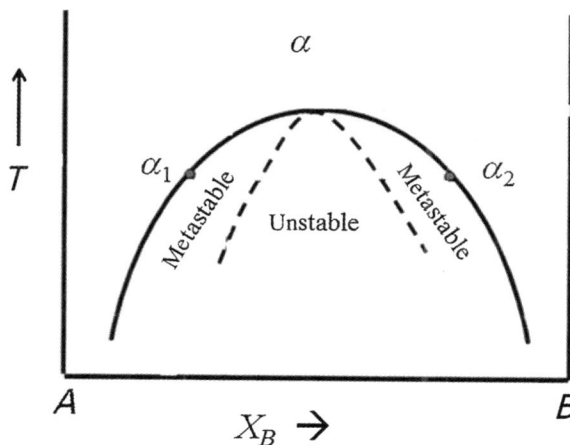

Figure 84 Schematic delineating metastable and unstable supersaturated states within a simple miscibility gap.

It is generally well established that at low to moderate undercoolings or supersaturations CNT as formulated earlier is an effective operational description of the system behavior and can be applied semiquantitatively depending on available data. The rate of homogeneous nucleation is expected to vary as $\frac{4\pi(r^*)^2}{a^4} D N_0 \exp\left(-\frac{\Delta G^*}{k_B T}\right)$ for a spherical nucleus, where a is an atomic dimension $\sim 0.2\text{--}0.4$ nm and the other terms as defined previously. The nucleation barrier ΔG^* is composed of the surface energy and strain energy terms and varies roughly as $\Delta G^* = \frac{A\sigma^3}{(\Delta T)^2 T}$ where ΔT is the undercooling below the relevant solvus and is a measure of the degree of supersaturation (driving force); A is essentially a constant and σ and T have their usual meanings. If one defines the start of the nucleation and growth precipitation reaction by a parameter τ that is the time to observe 1% transformation (generally controlled by the nucleation rate), then one can generate a locus in a classic TTT (time–temperature–transformation) diagram for 1% transformation yielding a well-known C-curve in the TTT transformation map. See **Figure 85**. The parameter $\frac{1}{\tau}$ essentially is a measure of the rate of reaction (nucleation rate) and scales as $\exp\left(-\frac{Q_D}{k_B T}\right) \exp\left(-\frac{A\sigma^3}{(\Delta T)^2 T}\right)$ Q_D is the activation energy for the thermally activated diffusional jumps from the parent phase to the critical nucleus. This approximate description shows the central role of the nature of the interphase interface and the associated interfacial free energy σ of a precipitate phase in determining the rate of formation of the phase during isothermal aging. Clearly coherent phases with low σ

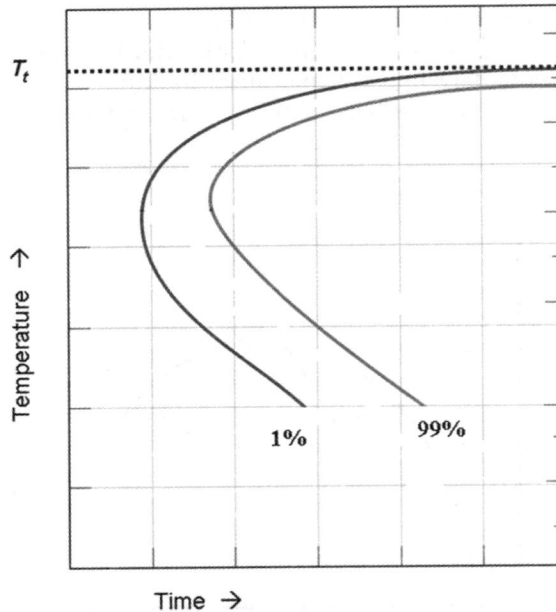

Figure 85 Typical Time—Temperature—Transformation (TTT) diagram exhibited by diffusional phase transformations. The start of transformation is labeled 1% and the finish 99%.

($\sim$50–200 erg cm^{-2}) are favored over incoherent phases with significantly higher σ ($\sim$400–1000 erg cm^{-2}) if the attendant strain energy for coherent nucleation is not prohibitive. This will be a major consideration when considering more complex phase diagrams and phases (stable or metastable) which do not have the same crystal structure as we find with the simple miscibility gap. Importantly, the equation for the initial nucleation rate shows the fundamental basis for the ubiquitous C-curve kinetics; at small undercoolings the nucleation barrier is high because the "driving force" is low and at large undercoolings the kinetics are sluggish because of restricted atomic mobility (high value of the activation energy for diffusional jumps in substitutional solid solutions).

Let us now consider a supersaturated state which lies within the "spinodal region" $\left(\frac{\partial^2 G}{\partial X^2} < 0\right)$. As mentioned earlier, the unstable solution tends to spontaneously unmix or phase separate without the requirement of a distinct nucleation step but through the amplification of initially extended low-amplitude fluctuations and this spinodal reaction involving diffusional clustering allows for a continuous evolution of the equilibrium $\alpha_1 + \alpha_2$ phase mixture. See **Figure 83**. The essential features of the spinodal process can be understood by considering this diffusional unmixing as the inverse of the homogenization of a nonuniform solid solution exhibiting a sinusoidal (or cosinusoidal) variation in composition with distance. In metastable solutions these small deviations from the average concentration, C_0, will generally decay according to $\Delta C = C(y,t) - C_0 = \Delta C_0 \exp\left(-\frac{t}{\tau}\right)$, where $\Delta C_0(y) = A_0 \sin\frac{\pi y}{\lambda}$ describes the initial composition fluctuation as a function of distance y and wavelength λ. The relaxation time of decay $\tau \approx \dfrac{\lambda^2}{D}$ where D is an appropriate diffusion coefficient and in a binary system is related to the curvature of the free energy–composition diagram $\dfrac{\partial^2 G}{\partial X^2}$ as mentioned in preceding discussions. In a metastable solution D is positive and the fluctuation decays exponentially toward the state of uniform composition (homogenization). However, if the solution is unstable, D is negative, and "uphill" diffusion occurs, that is diffusive flow of solute up the concentration gradient, and the amplitude of the concentration fluctuation grows with time, that is, $\Delta C = \Delta C_0 \exp\left(-\frac{t}{\tau}\right)$ becomes $\Delta C = \Delta C_0 \exp(R(\beta)t)$ where $R(\beta)$ is an amplification factor and a function of the wave number $\beta = \frac{2\pi}{\lambda}$. (This treatment of the behavior of a simple sinusoidal concentration wave in a heterogeneous solution is more general than it appears since through Fourier analysis an arbitrary composition heterogeneity can be considered as a superposition of Fourier components of various amplitudes and wavelengths and independently subject to a stability analysis.) This simple treatment captures important aspects of the spinodal process; however, the behavior of concentration fluctuations in an unstable solution requires that rigorous attention be paid to the energetics of the inhomogeities or concentration waves. We will find that long wavelength components will tend to grow sluggishly but short wavelength components are suppressed by a so-called gradient energy or incipient interfacial/surface free energy associated with the diffuse interfaces which evolve during the continuous unmixing process. We will address this important issue in the following overview of the modern theory of spinodal decomposition.

Becker (1937) and Dehlinger (1937, 1939) apprehended in the 1930s that the spinodal points on the free energy–composition curve of a binary system delineated a regime of supersaturated states wherein the diffusion coefficient in Fick's laws was negative and anticipated some unique behavior. Later Borelius (1937) attempted to develop a theory of localized fluctuation behavior contrasting behavior inside and outside the spinodal region, but never captured the essence of the problem. The core of the modern theory emerged when Hillert (a former student of Borelius) using a one-dimensional discrete lattice model based on regular solution energetics analyzed the behavior of composition fluctuations using a modified diffusion equation (Hillert, 1956, 1961). In his analysis a term related to composition differences between adjacent planes arises in describing the chemical potential gradient governing the diffusion flux across the discrete lattice. This term in the thermodynamics of an inhomogeneous binary solution is equivalent to the gradient energy referred to above. Hillert's formulation produced a nonlinear diffusion equation which he solved numerically showing selective growth of certain wavelengths of the composition fluctuations or modulations. The result clearly indicated a spontaneous periodic clustering or phase separation without the occurrence of a classical nucleation event. Initially low-amplitude extended composition fluctuations on a certain length scale were found to be amplified through "uphill" diffusion leading to continuous phase separation now universally known as spinodal decomposition. Here in Hillert's doctoral thesis at MIT submitted in 1956 a numerical solution to the diffusion equation reveals the essential behavior of unstable solutions exhibiting Gibbs' alternative to classical nucleation (fluctuations of the second kind) as a mode of breakdown of the supersaturated solution and continuous phase separation resulting in quasi-periodic concentration waves evolving toward the formation of the equilibrium phase mixture. This model also provided a fundamental basis for the observations of Bradley (1940) and analyses of Daniel and Lipson (1943) in the 1940s of periodic precipitation in the "sideband alloys" such as Cu–Ni–Fe alloys. The X-ray results were in accord with periodic modulation of the composition along the $\langle 100 \rangle$ directions of the decomposing cubic parent phase with a wavelength of 100 Å, the same order predicted by Hillert. In addition, Hillert showed that outside the spinodal region a nucleation barrier appears, in addition to discussing G. P. zone formation, diffuse interfaces as well as continuous ordering. This revolutionary thesis work was belatedly published in 1961 (Hillert et al., 1961) but its impact preceded the formal publication in an archival journal (see Cahn, 2006 for this evaluation of the impact of the paper which he says greatly influenced his work). Cahn and Hilliard at the General Electric research Laboratory were impressed and intrigued by Hillert's work and began to work on this topic and addressed some of the fundamental issues embodied in it. This team of outstanding thermodynamicists with excellent mathematical facility first focused on the proper thermodynamic description of the diffuse interfaces as well as the general thermodynamic description of a compositionally inhomogeneous system. The work was followed by a treatment of the nucleation problem outside the spinodal but relaxing the ad hoc assumptions of CNT regarding the nature of the interphase interface separating the nucleus and parent phase and the composition profile of the nucleus (Cahn and Hillard, 1958, 1959). They showed that indeed the work to form the critical nucleus vanishes at the spinodal as well (Cahn and Hillard, 1958, 1959). This Cahn–Hilliard or nonclassical nucleation will be discussed in more detail subsequently after discussing Cahn's theory of the spinodal process.

The Cahn theory of spinodal decomposition[1] is based on a stability analysis of concentration waves (Fourier components) in an inhomogeneous system and a solution of a generalized or modified diffusion equation in a three-dimensional continuum wherein the gradient energy is incorporated explicitly using a Ginzburg-Landau (1950) approximation. Using the Cahn–Hilliard (Cahn and Hillard, 1958, 1959) result to describe the thermodynamics of a nonuniform solution (a solution exhibiting concentration fluctuations about the mean concentration) one writes the total free energy of the system as

$$G = \int [g(c) + \kappa(\nabla c)^2] dV \tag{186}$$

where the first term under the integral is the local free energy density $g(c)$ (the local free energy per unit volume of a homogeneous region of composition c) and the second term $\kappa(\nabla c)^2$ is the excess free

---

[1] van der Waals (1908) first used the term spinodal (Cahn 1968) and Cahn coined the term spinodal decomposition (Cahn, 1961).

energy per unit volume due to the local concentration gradient ∇c or *gradient energy* (κ is called the gradient energy coefficient). Consider a one-dimensional composition wave along the y-axis of a rectangular block of uniform cross-sectional area A' and length L ($V = A'L$) of wavelength λ described by

$$c(y) - c_0 = A\cos(\beta y) \tag{187}$$

where c_0 is the mean concentration, A is the amplitude of the modulation and $\beta = \dfrac{2\pi}{\lambda}$ is the wave number as above. Expanding the free energy density about the average composition gives

$$g(c) = g(c_0) + (c - c_0)\left(\frac{\partial g}{\partial c}\right)_{c=c_0} + \frac{1}{2}(c - c_0)^2\left(\frac{\partial^2 g}{\partial c^2}\right)_{c=c_0} + \text{higher order terms} \tag{188}$$

yielding to second order the following expression for the free energy difference between the inhomogeneous solution and an initial homogeneous solution

$$\frac{\Delta G}{V} = \frac{A^2}{4}\left(\frac{\partial^2 g}{\partial c^2} + 2\kappa\beta^2\right) \tag{189}$$

which indicates that if the second derivative is positive the fluctuated or inhomogeneous state is a higher free energy state and will tend to decay back toward the state of uniform composition (κ is assumed to be positive giving rise to an incipient surface energy). However, if the second derivative $\dfrac{\partial^2 g}{\partial c^2}$ is negative (inside the spinodal region) periodic fluctuations with wavelengths greater than λ_c or $\dfrac{2\pi}{\beta_C}$ given by

$$\lambda_C = \left[-\frac{8\pi^2\kappa}{\dfrac{\partial^2 g}{\partial c^2}}\right]^{\frac{1}{2}} \tag{190}$$

lead to a decrease in free energy and tend to grow spontaneously; the supersaturated state is unstable with respect to such composition fluctuations. Clearly the effect of the gradient energy is to prohibit decomposition on too fine a scale in spite of the kinetic advantage the short wavelength fluctuations have in terms of shorter diffusion distances.

The stability analysis discussed earlier must be amended if the molar volume of the solution varies with composition and coherency strains develop between adjacent regions of different composition in the inhomogeneous system. For an isotropic system this misfit strain energy (per unit volume) associated with the concentration wave can be approximated by the expression

$$E_S = \frac{A^2\eta^2 E}{2(1 - \nu)} \tag{191}$$

where η is the linear expansion per unit composition change, E is Young's modulus and ν is Poisson's ratio. For a crystalline solid (cubic) η can be taken to be $\dfrac{d\ln(a)}{dc}$ where a is the lattice parameter. The excess free energy of the inhomogeneous solution with a quasi-sinusoidal composition modulation is then written as

$$\frac{\Delta G}{V} = \frac{A^2}{4}\left(\frac{\partial^2 g}{\partial c^2} + 2\kappa\beta^2 + \frac{2\eta^2 E}{1 - \nu}\right) \tag{192}$$

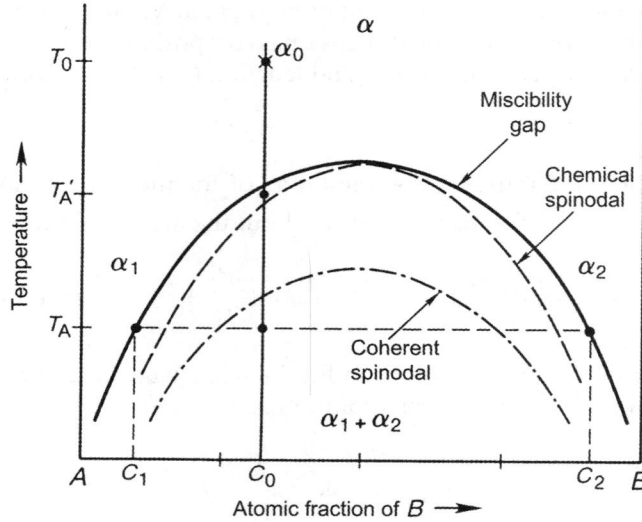

Figure 86 Schematic showing miscibility gap in the solid state and associated spinodal lines (chemical and coherent). After Soffa and Laughlin (1985).

which essentially defines a new stability criterion or a new spinodal region delineated by a locus called the *coherent spinodal* lying beneath the conventional *chemical spinodal*. See **Figure 86**. The strain energy has the effect of stabilizing the solution with respect to extended fluctuations of all wavelengths. The limit of stability is now given by

$$\left(\frac{\partial^2 g}{\partial c^2} + \frac{2\eta^2 E}{1 - \nu} \right) = 0 \tag{193}$$

resulting from the influence of coherency strain energy on the energetics of the inhomogeneous solution which can totally suppress the spinodal process in some systems.

Crystalline solids can be highly anisotropic in terms of their elastic properties including cubic crystals. This introduces a new and important consideration with respect to the behavior of the concentration waves which tend to evolve during decomposition of the unstable supersaturated state. As a result of elastic anisotropy the elastic energy attendant to the formation of concentration waves differs depending on the crystallographic direction along which the composition modulation occurs and this will be shown to impact the nature of the resultant microstructures stemming from the spinodal process.

A cubic crystal generally has three independent elastic constants C_{11}, C_{12} and C_{44} whereas in the isotropic case $2\,C_{44} - C_{11} + C_{12} = 0$, thus in an isotropic material there are only two independent elastic constants. In the case of a cubic crystal the strain energy term takes the form $2\eta^2 Y$ where

$$Y[100] = \frac{(C_{11} + 2C_{12})(C_{11} - C_{12})}{C_{11}} \quad \text{and} \quad Y[111] = \frac{6C_{44}(C_{11} + 2C_{12})}{4C_{44} + C_{11} + 2C_{12}}$$

for concentration waves along $\langle 100 \rangle$ and $\langle 111 \rangle$ directions, respectively. The elasticity parameter Y is a minimum for modulations along $\langle 100 \rangle$ when $2C_{44} - C_{11} + C_{12} > 0$ and a minimum for modulations along $\langle 111 \rangle$ when $2C_{44} - C_{11} + C_{12} < 0$. Thus, concentration waves will tend to develop preferentially along these directions cannibalizing the other Fourier components in the spectrum characterizing the initial composition inhomogeneity depending on the elastic anisotropy of the system (Cahn, 1961, 1962a, 1968 and Hillard, 1970).

Spinodal decomposition involves the selective amplification of concentration waves during the early stages of decomposition. We now focus our attention on the essential kinetic features of this process as a diffusional growth process occurring within the unstable system which represents the inverse of the homogenization problem involving "uphill diffusion" (flow against the concentration gradient; however, the diffusive flow is always down the chemical potential gradient). For diffusion in the binary system in one dimension along a y-axis the fluxes (moles cm^{-2} s^{-1}) of the components A and B can be related to the local gradients of chemical potential of these species as follows: the driving

$$J_A' = -C_A M_A \left(\frac{\partial \mu_A}{\partial y} \right) = -(1 - X)\rho_m M_A \left(\frac{\partial \mu_A}{\partial y} \right) \tag{194a}$$

$$J_B' = -C_B M_B \left(\frac{\partial \mu_B}{\partial y} \right) = -X\rho_m M_B \left(\frac{\partial \mu_B}{\partial y} \right) \tag{194b}$$

where C_A and C_B are the local concentrations (mol cm^{-3}), $(1 - X)$ and X are the corresponding atomic fractions, ρ_m the molar volume of the solution assumed to be constant; M_A and M_B are the mobilities of the diffusing species (diffusion velocities per unit driving force) with $\frac{\partial \mu_A}{\partial y}$ and $\frac{\partial \mu_B}{\partial y}$ the gradients in chemical potential (the driving forces for diffusion). These fluxes J_A' and J_B' are purely diffusive flows down the local chemical potential gradients of A and B, respectively, measured with respect to the lattice frame (a coordinate system fixed on a lattice plane) that moves with the bulk or convective flow as observed through a local marker movement with respect to a laboratory or Matano frame (Kirkendall effect). With respect to the Matano frame the fluxes are written as

$$J_A = J_A' + C_A v_m \text{ and } J_B = J_B' + C_B v_m \tag{195}$$

where v_m is a local marker velocity and noting that $J_A + J_B = 0$ in this frame. Eliminating v_m between the flux equations one can write the flux J_B as follows:

$$J_B = -\rho_m M \frac{\partial (\mu_B - \mu_A)}{\partial y} \tag{196}$$

where $(\mu_B - \mu_A)$ is sometimes referred to as the *diffusion potential* for the interdiffusion and $M = X(1 - X)[(1 - X) M_B + X M_A]$ is essentially an interdiffusion mobility. We recall from solution thermodynamics that $(\mu_B - \mu_A) = \frac{\partial G}{\partial X} \equiv G'$ and $G'' \equiv \frac{\partial^2 G}{\partial X^2}$ wherein the chemical potentials and molar free energy G will now include the gradient and elastic strain energies as described earlier.

The phenomenological description of the diffusion instability can now be formulated in terms of a modified Fick's laws writing the flux (First law) as

$$J_B = -M(G'' + 2\eta^2 Y) \frac{\partial C_B}{\partial y} + 2M\kappa \frac{\partial^3 C_B}{\partial y^3} \tag{197}$$

and the continuity equation (Fick's Second law) as

$$\frac{\partial C_B}{\partial t} = +M\left(G'' + 2\eta^2\gamma\right)\frac{\partial^2 C_B}{\partial y^2} - 2M\kappa\frac{\partial^4 C_B}{\partial y^4} \tag{198}$$

and for a cosinusoidal composition fluctuation $C_B - C_0 = A\cos(\beta y)$ this diffusion equation admits an analytical solution $C_B(y,t)$ as similar to that previously obtained with the simple approach. The new solution for the time evolution of the initially cosinusoidal fluctuation is

$$A(\beta, t) = A(\beta, 0)\exp(R(\beta)t) \tag{199}$$

where the amplification factor $R(\beta)$ is a function of the wavelength $\left(\beta = \frac{2\pi}{\lambda}\right)$ and given by

$$R(\beta) = -M\beta^2\left[G'' + 2\eta^2 Y + 2\beta^2\kappa\right] \tag{200}$$

as shown in **Figure 87**. The $A(\beta,0) = A_0$ is the initial amplitude of the fluctuation (Fourier component) in question. The maximum value of the amplification factor occurs at $\beta_{max} = \frac{\beta_C}{\sqrt{2}}$ where $\beta_C^2 = -\frac{(G'' + 2\eta^2 Y)}{2\kappa}$ recalling that only wavelengths greater than $\lambda_C = \frac{2\pi}{\beta_C}$ will be amplified and lead to a decrease in free energy; the supersaturated state is unstable with respect to such composition fluctuations while the shorter wavelength fluctuations will tend to be damped or decay as discussed earlier. Fourier components at and in the near vicinity of β_{max} will be the dominant concentration waves and essentially determine the scale of the microstructure during the early stages of decomposition and this length scale varies as $\left(\frac{\kappa}{(\Delta T)}\right)^{\frac{1}{2}}$ where $\Delta T = T_s - T$, in which T_S is the spinodal temperature and T is the aging temperature or temperature of the spinodal process.

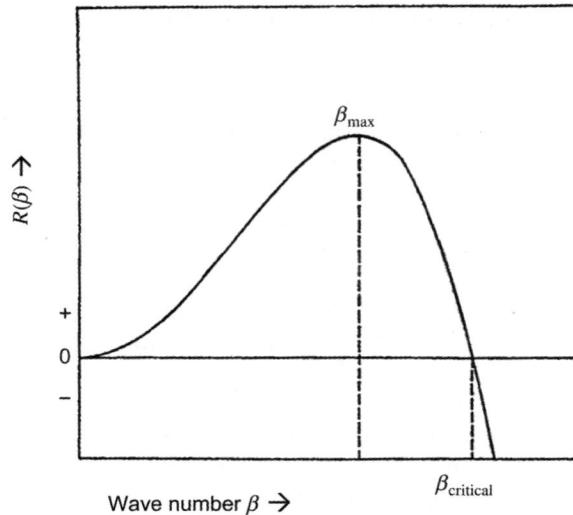

Figure 87 Variation of the amplification factor $R(\beta)$ with wave number, $\beta = \frac{2\pi}{\lambda}$ showing a maximum at $\beta = \beta_{max}$ and crossover at $\beta_{critical}$ where short wavelength fluctuations are suppressed.

The impact of Cahn's phenomenological or continuum theory based on an analytical solution of a modified diffusion equation was a major development in the theory of phase transformations bringing the spinodal concept and spinodal decomposition into the main stream of thinking about transformation behavior in metallic and nonmetallic systems. A new mode of phase separation distinct from classical nucleation based on a quantitative theory—albeit rigorously applicable to the early stages of decomposition—was recognized and the phenomenon began to be recognized to be relevant to a plethora of alloy systems of technological significance such as ferritic stainless steels (Fisher et al., 1953; Williams and Paxton, 1957; Imai et al., 1966; DeNys and Gielen, 1971; Vintaykin et al., 1966, 1970), Alnicos (Voss, 1969) and other permanent magnet alloys (Rossiter and Houghton, 1984) and numerous age hardening alloys (Ditchek and Swartz, 1979). Interestingly, Cahn's theory strongly impacted glass scientists almost immediately clarifying the structure and properties of phase-separated glasses, a phenomenon which is ubiquitous in glass technology. A new ingredient of major importance was introduced into the spinodal theory by Cahn, that being the elastic energy associated with the concentration waves and the influence of elastic anisotropy. Here we have a fundamental basis for the occurrence of modulated structures developing along certain crystallographic directions, for example $\langle 100 \rangle$, in numerous cubic systems as well as unidirectional modulations in decomposing TiO_2–SnO_2 tetragonal solid solutions (Stubican and Schultz, 1970). Thomas and coworkers (1970 and 1971) published the first comprehensive transmission electron microscopy (TEM) studies of the classic Cu–Ni–Fe sideband alloys revealing the emergence of modulated structures along the $\langle 100 \rangle$ matrix directions and the development of crystallographically aligned quasi-periodic two-phase mixtures. See **Figure 88**. The mechanism of decomposition was found to be homogeneous throughout the grains and uniform up to the grain boundaries as shown in **Figure 89**. The satellite reflections or sidebands and the emergence of a periodic and aligned microstructure were revealed prominently in the electron diffraction analysis of the spinodally decomposing alloys as displayed in **Figure 90**. It is important to note that Ardell and Nicholson (1966) in the mid-1960s pointed out in their studies of Ni–Al alloys that periodic microstructures mimicking the latter stages of spinodal decomposition could result from elastic interaction of coherent precipitates during stress-affected growth and coarsening of an initially random array of nucleated particles.

The Fe–Cr binary system exhibits a miscibility gap at low temperature giving rise to the formation of two bcc phases, one Fe rich and the other Cr rich and this miscibility gap including its metastable extension at elevated temperatures is significant in a number of technological contexts including ferritic stainless steels and tailored permanent magnet alloys based on the Fe–Cr–Co ternary system as cited earlier. Because of the similar atomic sizes these Fe–Cr spinodally decomposing alloys exhibit an isotropic spinodal morphology composed of interconnected veins of Cr-rich and Cr-depleted regions resulting from the amplification of concentration waves in three dimensions exhibiting no direction-ality because the elastic or misfit energy is small. The isotopic, sponge-like spinodal morphology is shown in **Figure 91** (Such a microstructure is expected in phase-separated glasses and indeed is found experimentally. In the glasses the absence of elastic anisotropy produces the isotropic behavior whereas in the metallic Fe–Cr–X metallic systems it is the small misfit.) This isotropic morphology in phase-separated Fe–Cr and Fe–Cr–Co was revealed by field-ion microscopy in the 1980s including atom probe studies of the time evolution of the concentration waves (Brenner et al., 1982, 1984). These early atom probe results showed the amplitude of the waves progressively increasing toward their equilibrium values indicated by the phase diagram. Later in the 1990s Miller et al. (1995) and Hyde et al. (1995b, 1995c) at Oxford in a series of papers carried out an impressive quantitative analysis of the system employing new atom probe techniques such as PoSAP and extensive computer simulations. **Figures 92** and **93** show reconstructions of the emerging microstructures stemming from the spinodal

Figure 88 Spinodal microstructure revealed by TEM in an aged CuNiFe alloy showing a periodic, crystallographically aligned two-phase mixture; the foil normal is approximately [001], and the particles of the second phase are aligned along the [100] and [010] matrix directions. After Butler and Thomas (1970).

reaction. Danoix and Auger (2000) have reviewed atom probe studies of the Fe–Cr system directly applicable to stainless steel technology. It should be noted that the ternary Cu–Ni–Cr system exhibits a ternary miscibility gap and behavior very similar to the Cu–Ni–Fe system. See **Figure 94**. Importantly, atom probe analysis of the phase separation process in the Cu–Ni–Cr alloys revealed a similar progressive change of extended composition fluctuations toward the equilibrium compositions of the resultant phases consistent with the concept of spinodal decomposition versus classical nucleation and growth (Abe and Soffa, 1991). Small-angle X-ray and neutron scattering have also been employed to effectively monitor the development of composition modulations during spinodal decomposition showing the selective growth and decay of Fourier components across the β spectrum. The early stages for which Cahn's theory is rigorously applicable are found to be difficult to access experimentally and most results show an early coarsening of the dominant wavelength of the emerging microstructure.

Classical nucleation and spinodal decomposition represent extremes in a decomposition spectrum within a miscibility gap characterized by a gradual change from metastability to instability. The Cahn–Hilliard nonclassical nucleation theory (Cahn and Hillliard, 1958) essentially provides the critical linkage for understanding the progressive change in the nature of the decomposition process as the supersaturation or undercooling of the parent phase is increased. Let us look briefly at this generalized nucleation theory. Cahn and Hilliard used their analysis of inhomogeneous systems and diffuse interfaces (Cahn and Hillliard, 1958) to reexamine the nucleation problem. They allowed the

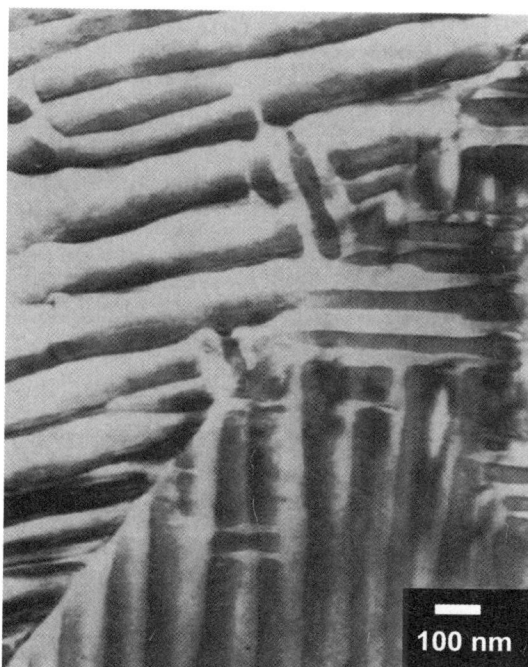

Figure 89   TEM micrograph of the decomposed CuNiFe alloy showing the phase separation being homogeneous up to the grain boundaries of the parent phase. After Butler and Thomas (1970).

composition profile of potential nuclei to vary and considered the interface between the parent and emerging phases to be generally diffuse rather than sharp as assumed in CNT. See **Figure 45**. At low supersaturations the critical nuclei look very much like classical nuclei but as the supersaturation increases the work to form the critical nucleus decreases continuously to zero at the spinodal. However, the effective radius or spatial extent of the critical fluctuation rapidly increases at high supersaturations approaching infinity in the vicinity of the spinodal. Furthermore, the interface becomes progressively more diffuse and the concentration difference between the center of the nucleus and the supersaturated solution decreases toward zero as well. What are we to make of this apparent singular behavior or discontinuity from the "nucleation side" to the "spinodal side?" As pointed out by Cahn (1962b), there is no discontinuity. As the supersaturation is increased toward the spinodal the critical nucleus or fluctuation becomes more diffuse and progressively exhibits characteristics very different from the "classical nucleus". The amplitude of the critical fluctuations begins to deviate significantly from the equilibrium composition of the precipitating phase and the interface between the incipient precipitate and parent phases becomes markedly extended. Furthermore, the "nucleation barrier" rapidly decreases approaching $k_B T$ and a range of finite-amplitude fluctuations against which the system is unstable becomes part of the spectrum of frequently occurring fluctuations and a well-defined critical nucleus loses its meaning. A variety of fluctuations of varying spatial extent and amplitude readily leads to decomposition of the supersaturated state. The system no longer follows an optimum path toward equilibrium. The transition regime is expected to extend into the "spinodal side" or inside the spinodal since true spinodal instabilities with their characteristic long wavelengths (large diffusion distances)

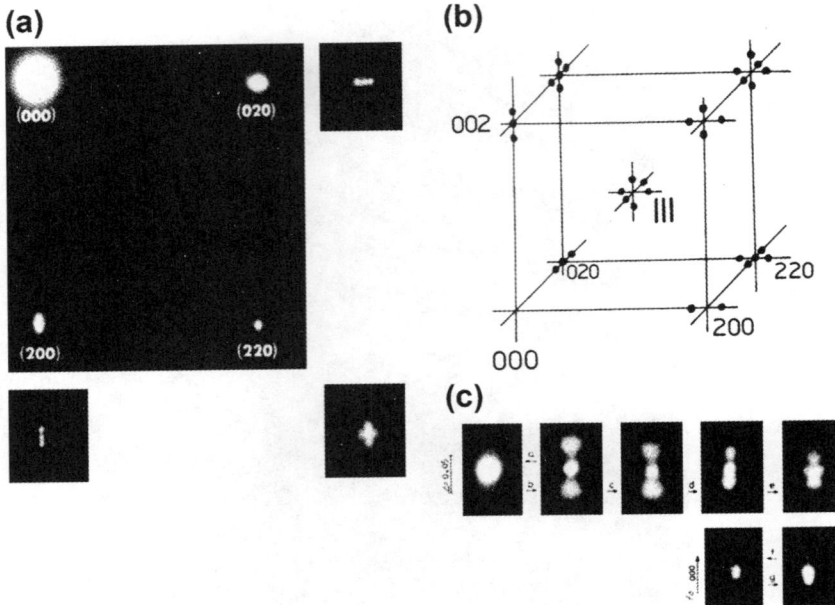

Figure 90   (a) An [001] electron diffraction pattern of a spinodally decomposed Cu–Ti alloy showing satellite configurations, consistent with periodic strain modulations along the [100] and [010] matrix directions. (b) Schematic of the positions of the satellites in reciprocal space. (c) Enlarged images of (200) reflections showing details of the satellites. After Hakkarainen (1971) and Soffa and Laughlin (1982).

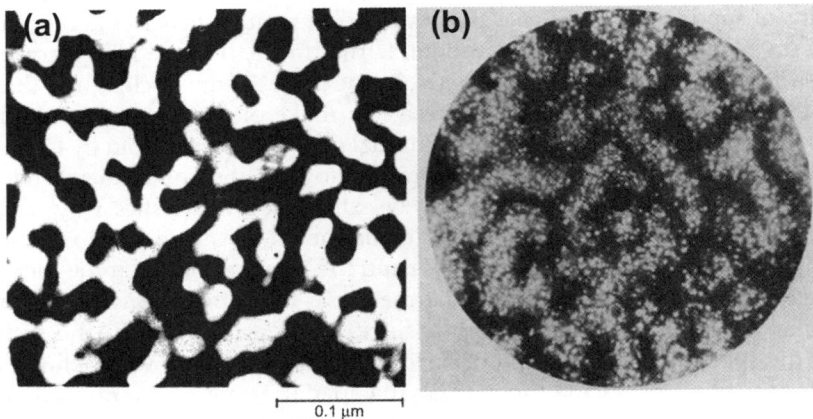

Figure 91   (a). TEM image of an isotropic spinodal morphology developed in an Fe–Cr–Co alloy (Zeltzer). (b) Field ion micrograph image of an isotropic spinodal morphology in an Fe–Cr–Co alloy. After Soffa et al. and Brenner et al. (1984).

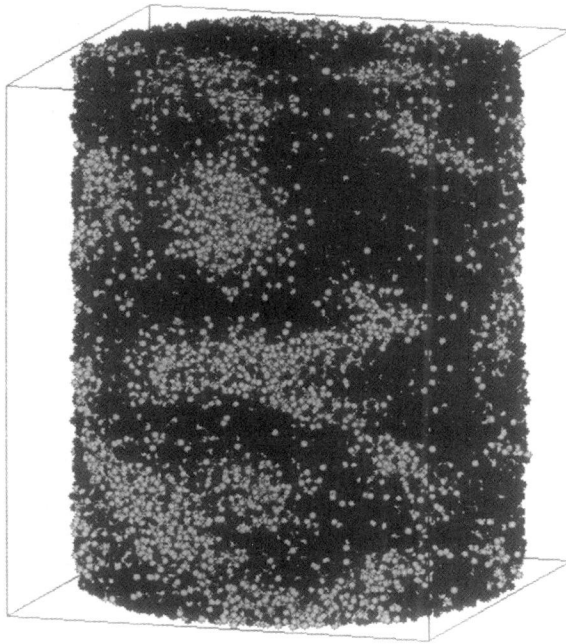

Figure 92 An atomic reconstruction of a spinodally decomposed Fe–Cr alloy aged at 773 K. The lighter spheres represent Fe atoms and the darker spheres Cr atoms. After Miller et al. (1995).

will be slow to evolve whereas finite-amplitude fluctuations with smaller spatial extent will develop more rapidly. Thus, the transition from metastability to instability is characterized by a hybrid process and the reaction path does not necessarily minimize the free energy of formation of the critical fluctuations but maximizes the rate of decomposition. A comprehensive pedagogical review article which this section closely followed is found in Hilliard (1970).

8.6.3 Spinodal Decomposition and Ordering

Clustering and ordering effects and instabilities during the decomposition of supersaturated solid solutions were essentially considered mutually exclusive behaviors until the 1960s and 1970s based on textbook treatments of these phenomena which discussed the energetics and stability of binary alloys in terms of first nearest-neighbor pairwise interactions within a zeroth approximation quasichemical theory or the classic Bragg–Williams description of ordering. In particular, spinodal decomposition and ordering tendencies were thought to be incompatible or unrelated behaviors in precipitation systems. However, it is now well established that an interplay of clustering and ordering tendencies can occur synergistically during the precipitation of ordered phases and influence the morphology and resultant microstructural scale. As early as 1963 Israel and Fine (1963) suggested that such behavior was involved in the formation of the metastable Ni_3Ti phase in the Ni–Ti system. Gentry and Fine (1972) in later studies of precipitation of Ni_3Al in Ni–Al alloys suggested that Cahn–Hilliard nucleation or spinodal decomposition was involved in the formation of the ordered phase. Corey et al. (1973) in their

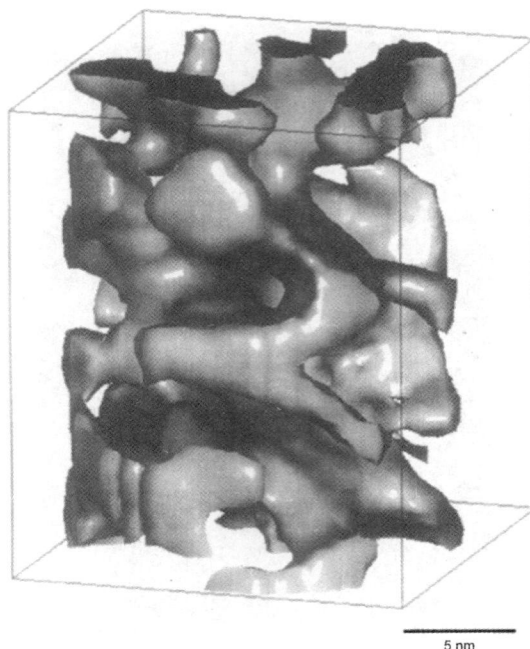

Figure 93   Isosurface reconstruction from PoSAP (position sensitive atom probe) analysis showing morphology of Cr-enriched regions in a spinodally decomposed Fe–Cr alloy aged at 773 K. After Miller et al. (1995).

contemporaneous work in the early 1970s on nonstoichiometric Ni_3Al-base alloys also concluded that these alloys undergo a two-step decomposition process consisting of an ordering transition followed by phase separation by spinodal decomposition. These authors also presented an attempt to rationalize the synergistics of the two-step process in terms of simple free energy–composition diagrams (graphical thermodynamics) relevant to the phase separation and ordering.

A thermodynamically sound basis for concomitant and synergistic ordering and phase separation behavior started to emerge in 1976 with the works of Allen and Cahn (1976), Ino (1978) and Kokorin and Chuistov (1976). Allen and Cahn (1976) addressed the mechanisms of phase transformations in the Fe–Al system revealed by electron microscopy (TEM) in association with a tricritical point which can occur when a line of critical points or a locus of higher order transitions ends uniquely on a miscibility gap. See **Figure 95**. The line of critical points represents the α (disordered; bcc; A2) → B2 disorder–order transition which is a higher order transition. Using free energy–composition curves associated with the phase diagram in the vicinity of the tricritical point they introduce explicitly the concept of spinodal phase separation contingent on prior ordering—the conditional spinodal reaction—and contrast a limit of metastability with a spinodal line. The limit of metastability is where the curvature of the free energy–composition curve changes sign (plus to minus) discontinuously on passing from a disordered state to an ordered state and is coincident with the extrapolated line of critical points associated with the A2 → B2 ordering in this case. Within this graphical thermodynamic scheme a disordered A2 phase which is metastable with respect phase separation can be rendered unstable upon continuously ordering and spinodally phase separate finally resulting in a disordered α phase and

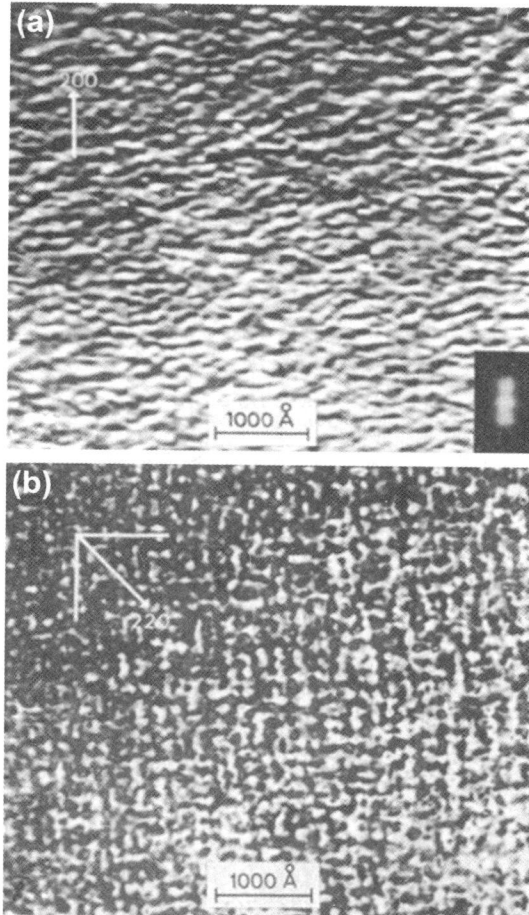

Figure 94 TEMs of Cu-31.6 Ni-1.7 Cr alloy aged at 650 °C for 1 h revealing matrix strain contrast striations along traces of the {1 0 0} matrix planes under different imaging conditions; the average wavelength of the modulated structure is about 170A. (a) Foil normal near (0 0 1); $g = [20\ 0]$; insert shows satellite flanking matrix reflection. (b) Foil normal near (001); $g = [2\ 2\ 0]$; interpenetrating modulations revealed indicative of triaxially modulated structure (Chou et al., 1978).

an ordered nonstoichiometric B2 phase. They also suggested that the continuous ordering of the initial disordered parent phase is expected to occur much more rapidly than any competing nucleation and growth mechanism. Ino (1978) using a straightforward quasichemical model but including first and second nearest-neighbor pairwise interactions of opposite sign was able to predict a similar behavior associated with the A2 → B2 ordering. The interplay between phase separation and ordering stemmed from writing the interchange energies as $V = \frac{1}{2}\left[2E_{AB}^1 - E_{AA}^1 - E_{BB}^1\right]$ and $U = \frac{1}{2}\left[2E_{AB}^2 - E_{AA}^2 - E_{BB}^2\right]$ for first and second nearest-neighbor interactions, respectively, and assuming random mixing on the α and β sublattices of the B2 structure. This is sometimes referred to as a generalized Bragg–Williams model. Kubo and Wayman (1980) later also described miscibility gaps intersecting the A2 → B2 transition in CuZn alloys. Kokorin and Chuistov (1976) addressed the possibility of spinodal decomposition in

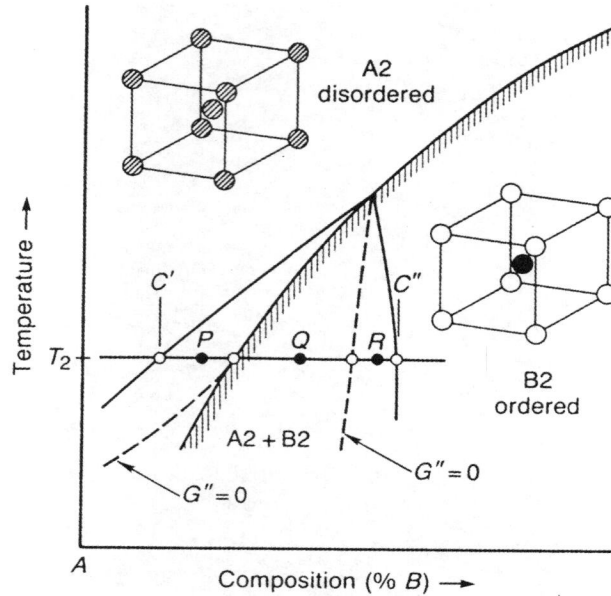

Figure 95 Region in the vicinity of a tricritical point showing the emergence of a spinodal line associated with the disordered solid solution at low temperatures ($G' = 0$). Alloys quenched to points P, Q and R exhibit different regions of thermodynamic stability with respect to clustering and ordering. After Laughlin and Soffa (1988).

conjunction with the formation of an ordered A_3B ($L1_2$) phase within a supersaturated binary fcc solid solution emphasizing the dependence of the free energy of the system on the composition and order parameter at a given temperature as well as incorporating second nearest-neighbor interactions in a generalized Bragg–Williams model. The $A1(fcc) \rightarrow L1_2$ ordering transformation is first order under equilibrium conditions. Khachaturyan et al. (1988) addressed the problem as well using a generalized Bragg–Williams model and SCW formalism to elucidate phase equilibria and precipitation of the $A_3B(L1_2$ phase) in Al–Li alloys. They described the possibility of spinodal decomposition of homogeneously ordered Al–Li solid solutions into a disordered phase and ordered phase mixture as reported experimentally by Radmilovic et al. (1989) in their TEM studies of the Al–Li system, similar to the discussion of Datta and Soffa (1973, 1976) in their studies of age hardening Cu–Ti alloys. Khachaturyan et al. (1988) defined a congruent ordering process whereby a disordered single-phase state transforms without composition change via nucleation and growth of ordered regions within a metastable solid solution (heterogeneous ordering).

Soffa and Laughlin (1989) extended this analysis to include a detailed graphical thermodynamic analysis of the interplay of the clustering and ordering tendencies in conjunction with a thermodynamically first-order disorder $\rightarrow$ order transformation in a precipitation system. Soffa et al. (2010, 2011) subsequently used the generalized Bragg–Williams approach in a computational thermodynamic analysis to generate free energy–composition diagrams emphasizing the role of second nearest-neighbor interactions on thermodynamic stability and the synergistics of ordering and spinodal decomposition. This graphical thermodynamics of Soffa and Laughlin (1989) represents a convolution of the usual free energy–composition diagram with the Landau graphical representation of ordering

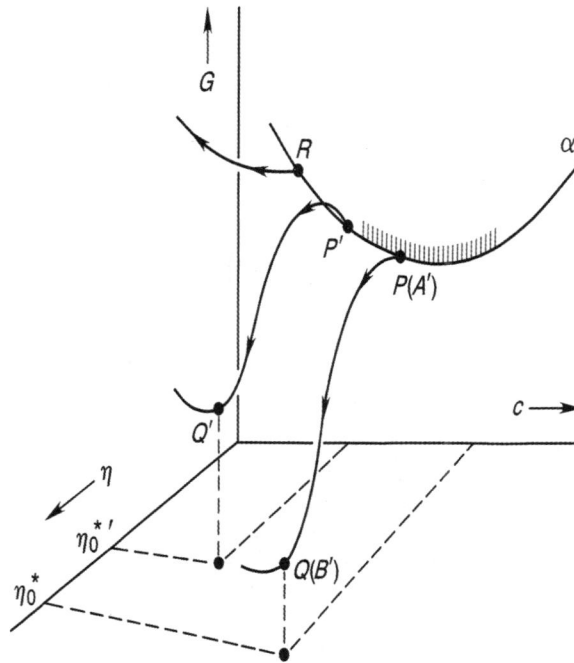

Figure 96 Schematic representation of G–c–η space showing different regions of stability and instability at points R, P' and P. A disordered alloy at R is stable with respect to atomic ordering, an alloy at P' is metastable with respect to ordering and one at P is unstable with respect to atomic ordering. The hatching along the free energy composition curve denotes thermodynamic instability with respect to ordering. After Soffa and Laughlin (1989).

transformations as shown in **Figure 96**. Ordering instabilities (T_i^-) and spinodal regions (T_S) can be delineated and an interesting array of reaction paths predicted in agreement with experimental evidence. **Figure 97** shows an alloy of composition C_0 quenched to temperature T_3 which might be expected to continuously order (below T_i^-) and then spinodally decompose (below T_S) producing an ordered precipitate within a disordered matrix. The continuous phase separation or spinodal reaction is contingent on the prior ordering and will be discussed subsequently.

8.6.4 Precipitation Sequences: Modes; Coherency and Metastable Phases

Let us now return to the classic Al–Cu age hardening system and examine, in detail, the reaction path during aging of supersaturated Al–Cu solid solutions giving rise to the strengthening which derives from the precipitation reaction. Alloys nominally containing 2–4 wt.% Cu (1.0–1.7 at.% Cu) when solution treated and quenched from the single-phase region ($\sim 500\,^\circ$C) and subsequently aged at $\sim 100\,^\circ$C undergo a multistage decomposition characterized by the formation of a series of metastable precipitate phases before the formation of the equilibrium precipitate $CuAl_2$ (θ). The reaction sequence can be summarized as: $\alpha_0 \rightarrow \alpha''' + $ G.P. I zones $\rightarrow \alpha'' + $ G.P. II zones (θ'') $\rightarrow \alpha' + \theta' \rightarrow \alpha_{eq} + CuAl_2$ (θ) where α_0 is the initial supersaturated FCC solid solution. It is useful to represent the multistage reaction path in this manner because the sets of phases ($\alpha'' + $ G.P. II zones (θ''), etc.) represent

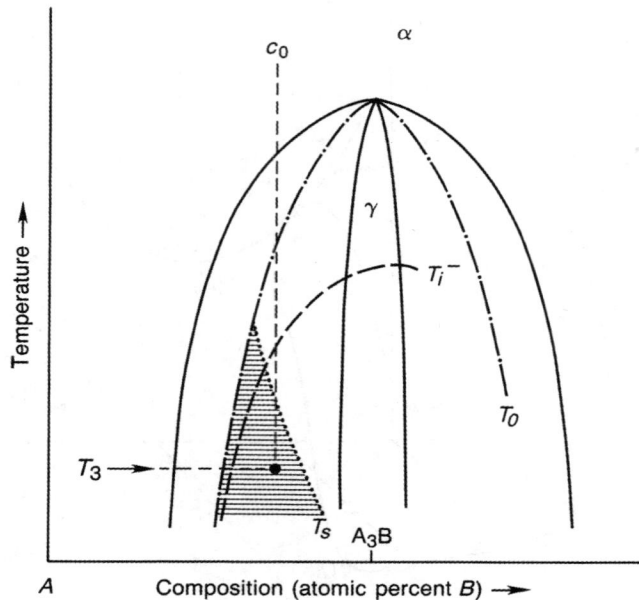

Figure 97 Phase diagram configuration associated with A_3B ordering (A1 → $L1_2$; first-order transition) showing ordering instability locus and spinodal regime (shaded) along with T_0 the temperature below which an ordered solution is thermodynamically favored (lower free energy) with respect to a disordered phase.

metastable two-phase equilibria which precede the formation of the stable $\alpha_{eq} + CuAl_2$ (θ) equilibrium phase mixture. The structure and morphology of these different phases are shown in **Figure 98**. The basic question to be addressed is: What fundamental principles underlie this multistage approach to stable equilibrium involving the appearance of a sequence of metastable precipitates? The rate of approach to equilibrium from the initial supersaturated state is controlled by the activation barriers along the reaction path generally associated with thermally activated processes involved in the construction of a new phase or phases which lower the free energy of the constituent assembly such as nucleation and atomic migration. The metastable transition precipitates during the early stages of decomposition of the supersaturated state are generally crystallographically similar to the matrix allowing the formation of low-energy coherent (semicoherent) interphase interfaces during the nucleation process. As discussed earlier, CNT shows that the nucleation barrier ΔG^* is proportional to $\dfrac{\sigma_{M-P}^3}{(\Delta G_V + \Delta G_S)^2}$ where σ_{M-P} is the interfacial free energy of the matrix–precipitate interphase interface, ΔG_v is the thermodynamic driving force per unit volume (the free energy released per unit volume of the new phase formed) and ΔG_s is the strain energy per unit volume attendant to the formation of the new phase (coherency strain energy), as defined earlier. Since the nucleation rate of a phase varies exponentially as $\exp\left(-\dfrac{\Delta G^*}{k_B T}\right)$ the nucleation of a coherent transition phase with $\sigma_{M-P} \sim 30$ erg/cm^2 will occur more easily compared with the equilibrium phase with $\sigma_{M-P} \sim 400 - 1000$ erg/cm^2 despite having a lower driving force and some associated strain energy expenditure, in general. The different stages can be depicted in a free energy–composition diagram and common tangent constructions can

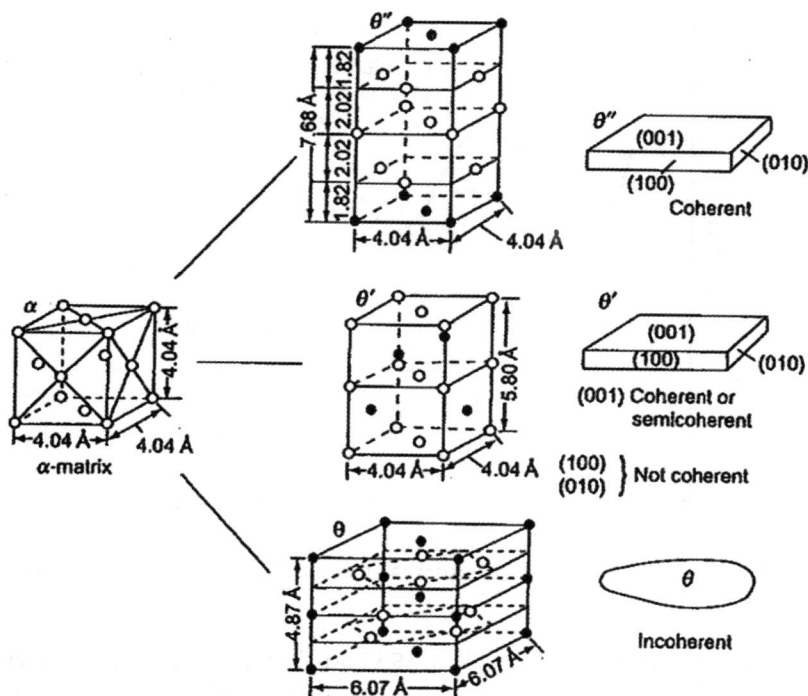

Figure 98   Schematic showing the structures of precipitate phases occurring in Al–Cu age hardening alloys including the morphologies and nature of their interphase interfaces. After Smallman (1963).

be employed at different temperatures to map out the loci of metastable solvi of precipitate phases. See **Figure 99**. Also, *C*-curves marking the start of precipitation for the different competing phases can be established with respect to the metastable solvi as discussed earlier and depicted in **Figure 100**. It should be mentioned that the thermodynamic validity of these metastable solvi, for example G.P. I solvus, can be categorically established by a *reversion* experiment wherein if the metastable two-phase mixture is rapidly reheated above the G.P. I solvus a temporary softening is often observed resulting from the re-solution of the G.P. I zones followed by further hardening on continued aging with the appearance of the θ'' phase. In the free energy–composition schemes depicted in **Figure 99**, G.P. I zones could form by spinodal decomposition at sufficiently high supersaturations. Indeed, Rioja and Laughlin (1977) have studied the early stages of G.P. I zone formation in aged Al-4 Wt.% Cu alloys using electron microscopy and diffraction. The observation of diffuse satellite reflections and the apparent formation of a modulated microstructure (the development of concentration waves along the ⟨100⟩ directions of the decomposing matrix) supports the notion of a spinodal mechanism. These conclusions have been supported by the small-angle X-ray scattering results reported by Kaskyap and Koppad (2011). The well-known plate-like G.P. I zones along the {100} matrix planes emerge during the later stages of the spinodal process and coarsening of the modulated structure under the influence of the strain energy. It is interesting to note that the heat of mixing of Al–Cu alloys in the solid state appears to be negative over most of the composition range except for perhaps an anomalous behavior

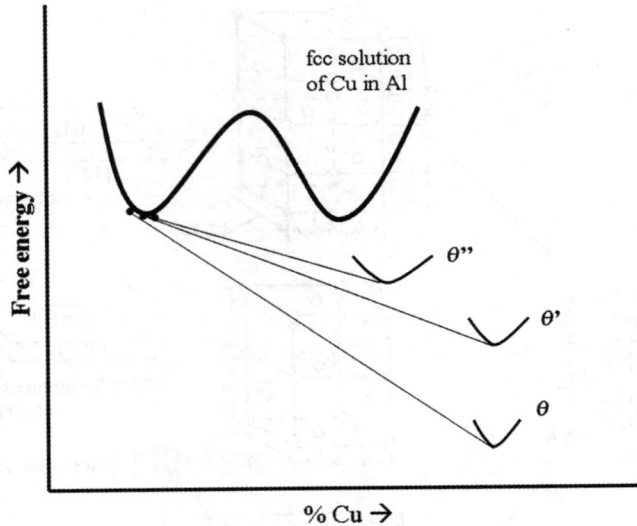

Figure 99 Hypothetical free energy–composition curves for the precipitate phases in Al–Cu alloys. After Fine (1964) and Hardy and Heal (1954).

near the Al-rich side of the alloy (Hardy and Heal, 1954; Meijering, 1952). However, even if the heat of mixing is negative over the entire composition range, an inflected G vs. X curve and a metastable miscibility gap can result if the heat of mixing curve is sufficiently inflected (Meijering, 1952); thus, a positive heat of solution or mixing *is not a prerequisite* for the occurrence of spinodal phase separation

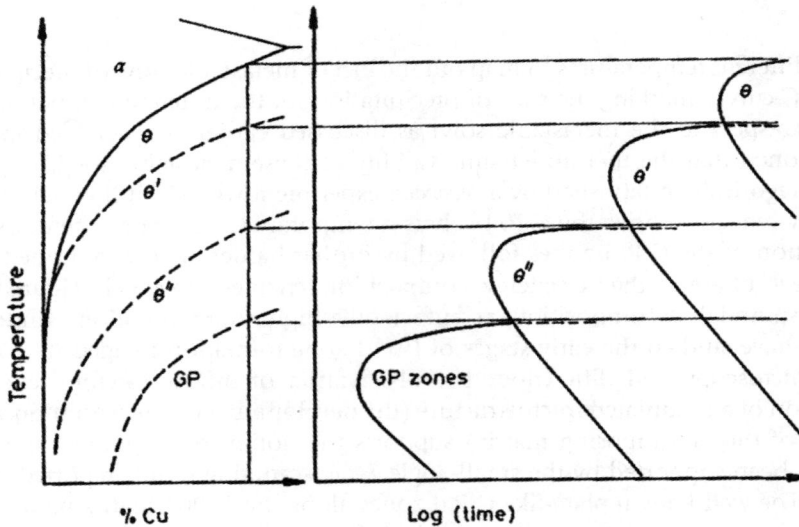

Figure 100 Portion of a schematic Al–Cu phase diagram depicting stable and metastable solvi as well as TTT curves (precipitation start) for the various precipitation reactions. After Smallman (1963).

in the system. Furthermore, it should be pointed out that the subsequent phases following G.P. I zone formation are, in fact, atomically ordered phases, namely, G.P. II (θ''), θ' and θ.

Clearly, there must be some synergistics between the formation of the n-th and (n+1)-th phases in the precipitate sequence. The copious formation of a coherent intragranular phase through homogeneous nucleation or spinodal decomposition creates a high density of interphase interfaces which can serve as effective nucleation sites for subsequent phase formation. Also, in some cases the next phase in a precipitation sequence may form through continuous transformation of the previous phase which may be the situation in the G.P. I zones → G.P. II zones (ordering) transition. The formation of the equilibrium precipitates are generally relegated to heterogeneous nucleation at high-angle grain boundaries after long aging times. It should be pointed out that the formation of a more stable phase in any precipitation sequence will lead to the dissolution of the less stable phase because of chemical potential/composition gradients which develop since the concentration in local equilibrium with the more stable phase is less than that in equilibrium with the less stable phase.

The Ni–Ti system is generally of the same genre as that of the Ni–Al alloys and both are of great importance in the metallurgy of modern superalloys with Ti along with Al playing a primary role in the formation of the Ni$_3$ (Al, Ti) γ' phase. The γ' phase exhibits the L1$_2$ superstructure in Ni–Al and Ni–Ti alloys and is essentially the major precipitating phase in numerous high-temperature high-strength alloys. (The γ' Ni$_3$Al and Ni$_3$Ti phases are isomorphous.) In the Ni–Al binary system the γ' is the equilibrium phase whereas in the Ni–Ti binary precipitation system the equilibrium phase is an Ni$_3$Ti (η) hexagonal phase which is preceded by the formation of a coherent γ' Ni$_3$Ti phase during aging. We will focus our attention on the formation of this metastable γ' in Ni–Ti alloys containing 10–15 atomic percent Ti because extensive experimental evidence is available indicating a complex interplay between clustering and ordering effects during the decomposition of supersaturated solid solutions in this system. There appears to be a synergism between ordering instabilities and spinodal decomposition during the precipitation of the γ'Ni$_3$Ti phase. These experimental studies include magnetic measurements (Israel and Fine, 1963), electron microscopy and diffraction (Saito and Watanabe, 1969; Ardell, 1970; Laughlin, 1976; Kompatscher et al., 2003), atom probe field ion microscopy (Sinclair et al., 1974; Grune, 1988) and small-angle neutron scattering (Kostorz et al., 1999; Kompascher et al., 2000). The results clearly point to a complex interplay between continuous ordering and phase separation/spinodal decomposition. The following analysis is based heavily on the thermodynamic analyses of Soffa and Laughlin (1989) and Khachaturyan et al. (1988).

In **Figure 101** a schematic phase diagram showing a metastable γ' solvus along with loci of thermodynamic instability with respect to ordering (T_i^-) and phase separation (T_S) is depicted. Also delineated is a region of congruent ordering between T_0 and T_i^- wherein a supersaturated fcc disordered solid solution can lower its free energy by ordering via a nucleation and growth process at constant composition and subsequently spinodally decompose. The T_S is a conditional spinodal mapping out a region between T_0 and T_S where a nonstoichiometric ordered state becomes unstable with respect to phase separation subsequent to ordering of an initially metastable disordered solid solution. For example, if an Ni–Ti alloy containing ∼ 12–14 atomic percent Ti (C$_0$ in **Figure 101**) is rapidly quenched to room temperature and then aged at 600 °C, the resulting supersaturated state α_0 (A) depicted in **Figure 102** may become unstable with respect to L1$_2$ ordering and is expected to continuously order to an imperfectly ordered, nonstoichiometric solid solution decreasing the free energy as shown. This ordered state at B is now phase separation and a spinodal process involving selective amplification of composition modulations within the ordered solution resulting in the continuous evolution of a metastable two-phase mixture composed of α' terminal solid solution and γ'(Ni$_3$Ti; L1$_2$). During the

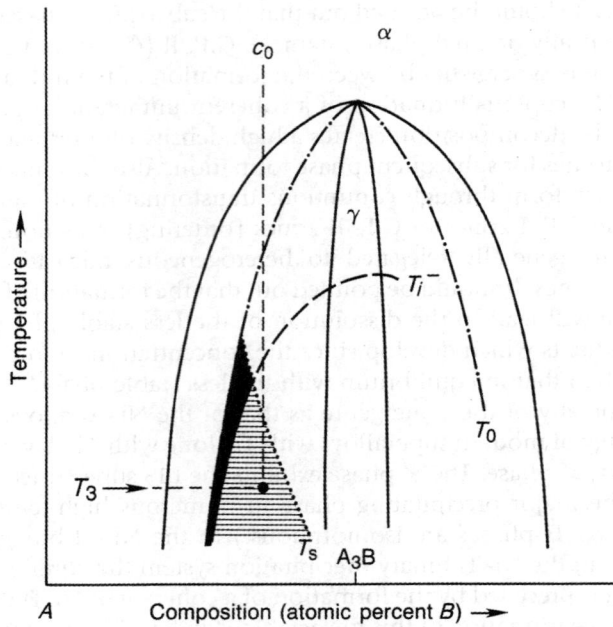

Figure 101 Schematic phase diagram showing a metastable γ' phase including a spinodal region (shaded) as well as a region of ordering instability below T_i^-. After Soffa and Laughlin (1989).

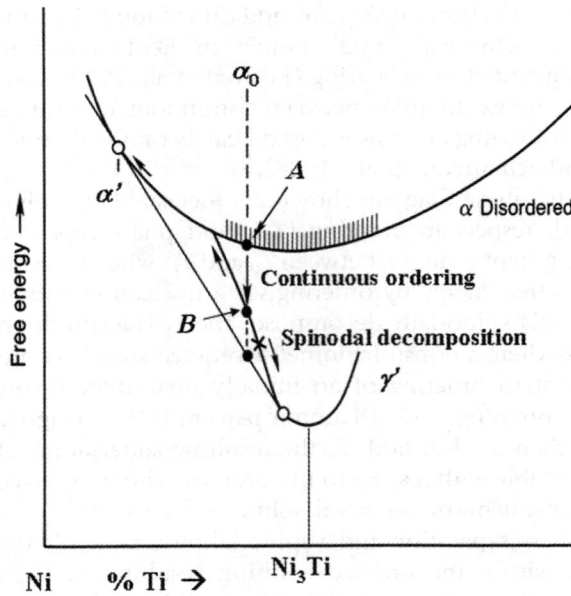

Figure 102 Hypothetical free energy–composition diagram that is consistent with the decomposition process in Ni–Ti alloys. A supersaturated α_0 solid solution will first order continuously and then decompose spinodally into the ordered γ' and disordered α' two-phase mixture.

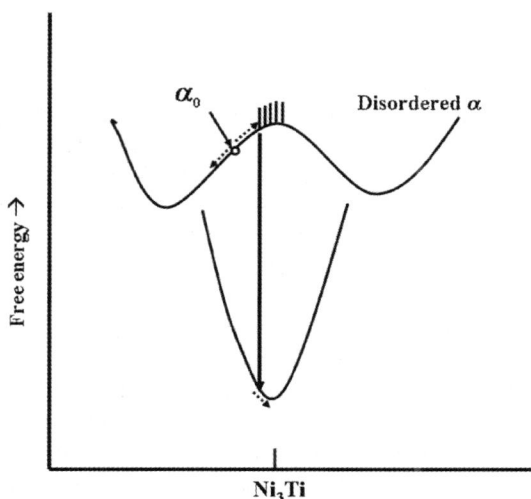

Figure 103 Precipitation reaction whereby the initial supersaturated solid solution α_0 begins to decompose spinodally and the solute-enriched regions become unstable with respect to atomic ordering (continuous ordering) leading to the formation of a two-phase mixture resulting from consecutive continuous transformations.

early stages of decomposition the alloy exhibits a so-called modulated structure and "sideband" state as composition waves develop preferentially along the $\langle 100 \rangle$ directions of the cubic matrix (~ 100A spacing) resulting in a periodic array of γ' particles aligned along the $\langle 100 \rangle$ directions. There is strong experimental evidence as well as computational thermodynamic analysis that an ordered state precedes the amplification of the concentration waves within this synergistic process involving ordering and clustering (Kompascher et al., 2000 and Soffa et al., 2010) consistent with the decomposition path shown in the free energy–composition scheme of **Figure 102**. The reaction path involving a conditional spinodal phase separation similar is to that discussed by Allen and Cahn (1976) and Ino (1978) cited earlier but in this case a first-order ordering transformation (A1 $\rightarrow$ L1$_2$) is involved rather than a second-order/higher order (A2 $\rightarrow$ B2) ordering. However, another decomposition path has been suggested (Laughlin, 1975) as depicted in the free energy–composition scheme in **Figure 103**. In this scenario the initial supersaturated disordered solid solution begins to phase separate spinodally into solute-enriched and solute-depleted regions with the solute-enriched regions becoming unstable with respect to ordering and continuously order to form the L1$_2$-based ordered phase. In both cases there is an interplay between ordering and clustering tendencies and continuous transformation involved in the precipitation of an ordered phase within a supersaturated solid solution. Generally, this behavior derives from solution energetics involving pairwise interactions beyond first nearest neighbors (Richards and Cahn, 1971; Ino, 1978; Soffa et al., 2010).

The Al–Li binary system is the basis for the development of a series of light-weight, high-strength precipitation-hardened alloys and the strengthening precipitate, δ', is a metastable Al$_3$Li (L1$_2$) phase characterized by a small misfit ($<0.1\%$) between the matrix and precipitate. The δ' phase can form coherently with the matrix with a very low interfacial free energy $\sigma_{M-P} \sim 20$ erg/cm^2 to produce fine homogeneous dispersions of quasi-spherical particles throughout the matrix after suitable heat treatment. (Sometimes so-called precipitate-free zones can develop near grain boundaries resulting from vacancy depletion or solute depletion near grain boundaries (Kelly and Nicholson, 1963).) The equilibrium δ phase has the B32 cubic structure and nucleates and grows primarily at the grain boundaries after prolonged aging. Baumann and Williams (1985) have applied CNT to describe the homogeneous nucleation of the metastable δ' phase and Wang and Shiflet (1998) have quantitatively described nucleation and growth of δ' on dislocations. Radmilovic, Fox and Thomas (1989) in their HREM and X-ray studies of Al–Li alloys containing ∼8–11 atomic percent Li called attention to an atomic ordering within the supersaturated state preceding precipitation of the δ' phase and explicitly suggested that the ordering was followed by spinodal decomposition similar to the conditional spinodal discussed earlier. Their observation is in agreement with the earlier report by Sato et al. (1988) that the early stages of decomposition in similar alloys exhibited imperfectly ordered domains that percolated throughout the microstructure before formation of a fully developed discrete $L1_2$ precipitate phase. These experimental studies were followed by a theoretical treatment by Khachaturyan et al. (1988) that indicated an interplay of ordering and spinodal phase separation in the precipitation of ordered intermetallic phases consistent with a generalized graphical thermodynamic description by Soffa and Laughlin (1989) as cited previously. Poduri and Chen (1997) subsequently reported a computer simulation of ordering and phase separation in the Al–Li alloys associated with δ' formation and were able to delineate the different regimes of behavior discussed earlier as well as revealing a regime of nonclassical nucleation involving fluctuations in composition and order parameter.

Copper–titanium alloys containing 1–6 atomic percent Ti can be age hardened to develop physical and mechanical properties comparable with the widely used high-strength Cu–Be alloy series. In the 1960s and 1970s the Cu–Ti alloys were recognized to be prototypical "sideband alloys" (see **Figure 90**) and it was suggested by various investigators that these alloys undergo spinodal decomposition during age hardening (Hakkarainen, 1971; Cornie et al., 1973; Laughlin and Cahn, 1976; Datta and Soffa, 1976; Soffa and Laughlin, 2004). The strengthening precipitate is a tetragonal $Cu_4Ti\left(D1_a; Ni_4Mo - type/I\frac{4}{m}\right)$ phase which forms below approximately 700–800 °C. At aging temperatures in the range 350–500 °C ultrahigh strengths can be achieved through the formation of fine-scale dispersions of coherent $D1_a$ precipitates aligned along the ⟨100⟩ matrix directions exhibiting a quasi-periodic microstructure. The precipitates are elongated along the c-axis of the tetragonal phase which is parallel to the cube directions of the matrix. See **Figure 104**. The equilibrium phase is a $Cu_4Ti(\beta)$ orthogonal (Pnma) phase above about 400–500 °C whereas at lower temperatures the $D1_a(\beta')$ phase is the stable phase. Relevant portions of the Cu–Ti binary phase diagram are shown in **Figure 105**. During prolonged aging at low and moderate temperatures (350–500 °C) a coarse cellular lamellar microconstituent composed of terminal solid solution and the equilibrium phase forms at the grain boundaries and grows out consuming the fine dispersion of coherent/semicoherent $D1_a$ particles. See **Figure 106**. At high aging temperatures ∼700 °C the equilibrium orthorhombic phase forms via classic Widmanstatten precipitation with platelets lying along the {111} matrix planes as shown in **Figure 106**. Ecob et al. (1980) have suggested that the Widmanstatten plates in the vicinity of the grain boundaries can catalyze the formation of the cellular colonies at the higher aging temperatures.

The decomposition of supersaturated Cu–Ti alloys containing 1–6 atomic percent Ti embodies a very complex synergy of ordering, clustering and precipitation behavior associated with the formation of the $D1_a$ phase. During decomposition there is a subtle interplay between SRO and LRO as well as phase separation and precipitation of an ordered intermetalic phase. The early stage ordering effects have been widely studied in Ni_4Mo-type systems for stoichiometric and off-stoichiometric compositions including effects of radiation (Bellon and Martin, 1988). The concentration wave approach shown in **Figure 107** has proven very useful for describing the SRO and LRO effects. Interestingly, the dilute

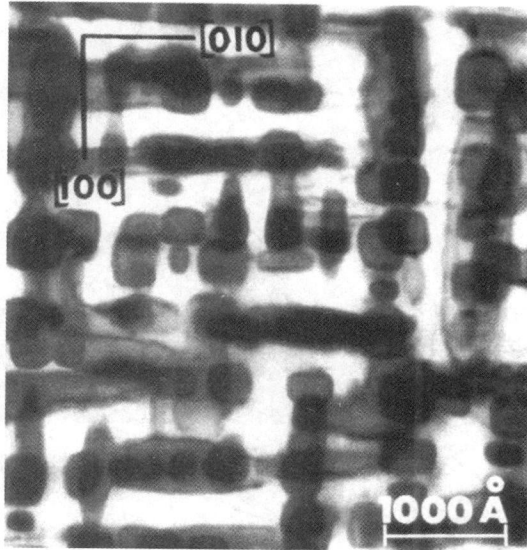

Figure 104   TEM microstructure of age-hardened Cu-4wt % Ti alloy aged at 500 °C for 2000 min. After Soffa and Laughlin (2004).

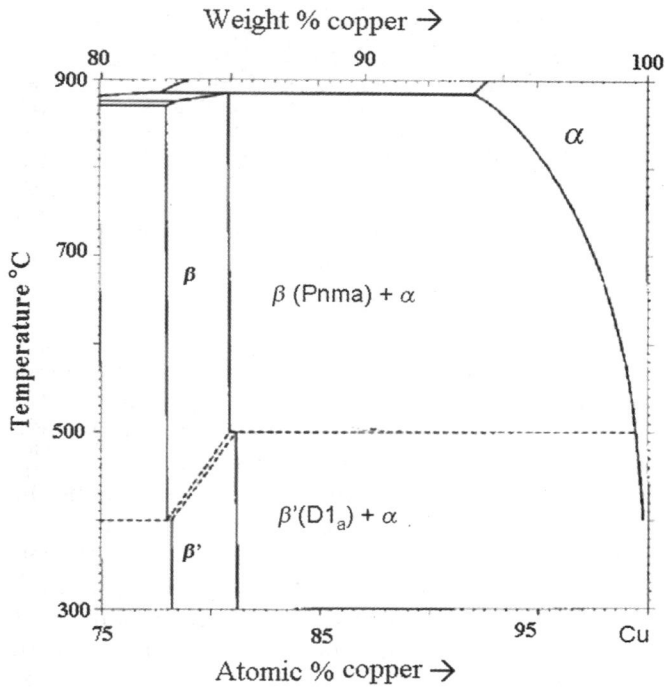

Figure 105   Detailed portion of Cu–Ti phase diagram showing the polymorphic transformation temperatures of Cu_4Ti phase. After Soffa and Laughlin (2004) and Brun et al. (1983).

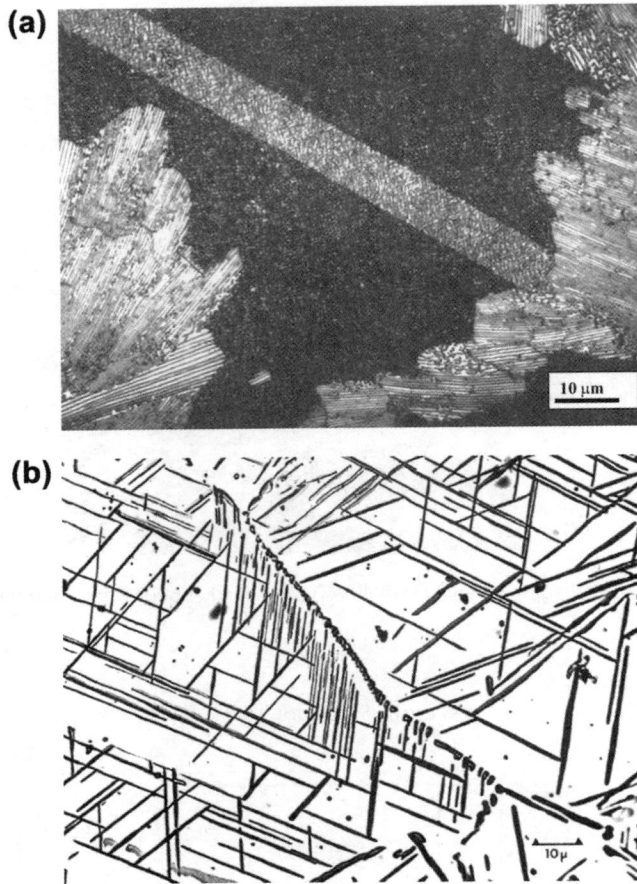

Figure 106   (a) Cellular microconstituent growing into the coherent/semicoherent fine-scale two-phase mixture ($\alpha' + \beta'$) near peak hardness of a Cu-4w/0 Ti alloy aged at 600 °C for 1000 min. (b) Widmanstätten precipitation in a Cu 3 wt.% Ti alloys held at 730 °C for 600 min. After Soffa and Laughlin (2004).

Cu–Ti alloys very early in the decomposition process during aging show diffuse diffracted intensity at reciprocal lattice locations $\langle 1\frac{1}{2}\, 0 \rangle$ deriving from the amplification of concentration waves along the {420} planes having a wave vector $\frac{1}{4} \langle 420 \rangle$ producing modulations of the type AABBAA.... This rapid amplification of "ordering waves" stems from an ordering instability associated with so-called special points in the k-space representation of the solution energetics (de Fontaine, 1975; Bellon and Martin, 1980). These early stage atomic rearrangements (SRO) are followed by the emergence of an imperfectly ordered $D1_a$ structure producing diffuse superlattice reflectons (LRO) at the positions $1/5\langle 420 \rangle$ (Hakkarainen, 1971; Laughlin and Cahn, 1975) which then gives rise to discrete coherent $Cu_4Ti/D1_a$ precipitates aligned along the $\langle 100 \rangle$ matrix directions. It is important to point out that the free energy–composition curve which is central to conventional thermodynamic discussions of metastability and instability in supersaturated solid solutions is subject to change as atomic rearrangements on various length scales within a nonequilibrium solid solution occur. The free energy curve evolves as the

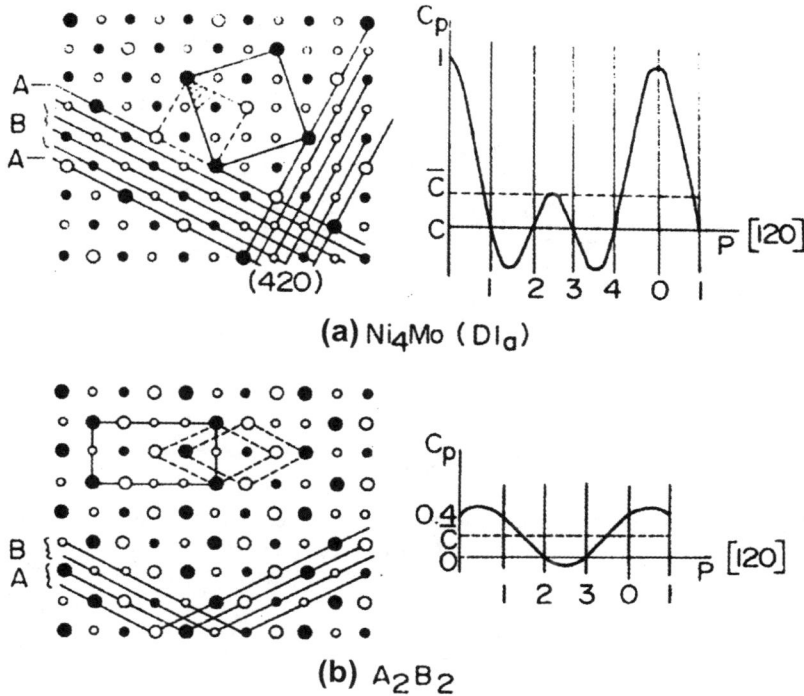

Figure 107 Concentration wave description of LRO (a) and SRO (b) in an Ni$_4$Mo-type system depicted as modulation in atomic arrangement of {420} planes. After Banerjee and Sundaraman (1992).

solution energetics are changed by local ordering (or clustering) and distort giving rise to regions of negative curvature and changes in thermodynamic stability with respect to ordering and phase separation (Liu and Loh, 1971). See **Figure 108**. Thus, a disordered solid solution which is initially metastable when passing through the SRO → LRO states described above can lower the free energy from P to Q and render the imperfectly ordered state unstable with respect to phase separation and lead to spinodal decomposition and continuous formation of the nearly stoichiometric D1$_a$ phase. This is a conditional spinodal reaction similar to the reaction paths discussed above and consistent with the apparent amplification of concentration waves within an imperfectly ordered state and the unambiguous appearance of a sideband state. It is concluded here that the decomposition of dilute Cu–Ti alloys at high supersaturations generally involves a conditional spinodal process resulting from the SRO → LRO atomic rearrangements occurring during the earliest stages of aging resulting in the precipitation of an intermetallic phase (Laughlin and Cahn, 1975; Datta and Soffa, 1973, 1976; Soffa and Laughlin 1982). However, Borchers (1999) in studies of a Cu-0.9 at.%Ti alloy using electron microscopy (TEM) and thermodynamic calculations analyzed the nucleation energetics and suggests that the copious formation of coherent ellipsoidal or oblong D1$_a$ particles from the onset of decomposition in this alloy was most likely the result of nonclassical nucleation or "big bang"/"catastrophic" nucleation. The analysis using classical nucleation indicates a nucleation barrier ΔG^* of approximately $k_B T$ for an estimated interfacial free energy $\sigma \sim 30$ erg cm^{-2} including an Eshelby estimate of the strain energy. Borchers concludes that the apparent modulated structures emerge through concomitant

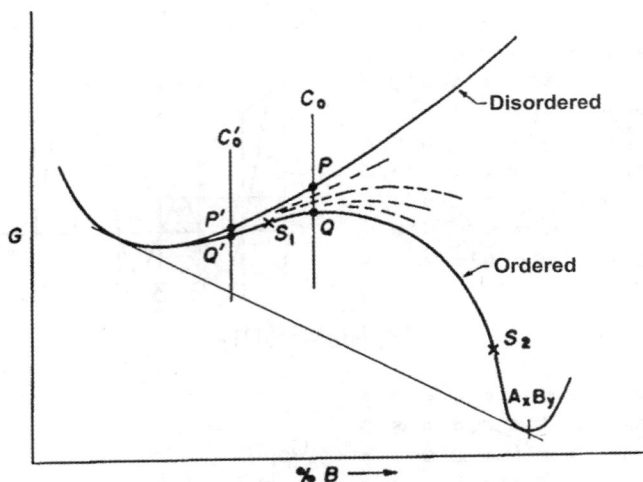

Figure 108 Hypothetical free energy–composition diagram for homogeneous disordered and ordered solid solution, showing a hierarchy of free energy curves. An alloy of composition C_0 first orders homogeneously (from P to Q), and then phase separates into two ordered phases until the solute lean phase disorders. After Soffa and Laughlin (1982).

growth and coarsening under the influence of elastic interaction from the earliest stages of decomposition.

Kahlweit (1970) and Kampmann and Kahlweit (1967, 1970) suggested that at high supersaturations separating a precipitation reaction into distinct stages of nucleation, growth and coarsening may not be applicable to describing the evolution of the particle density during aging. They analyzed a supersaturated system wherein nucleation, growth and coarsening were concomitant and competitive processes showing that the density of precipitate particles during decomposition rises to a maximum and then tends to decrease while the matrix is still markedly supersaturated as a result of competitive coarsening. Langer and Schwarz (1980) later approached the problem and found a similar behavior. Wendt (1981) and Wendt and Haasen (1983) also revealed such a trend in his studies of precipitation in an Ni-14 at.% Al alloy. Wagner and Kampmann (1991) proposed a modified Langer–Schwartz model and a detailed numerical formulation to predict the evolution of the size distribution of precipitate particles and applied it to a Cu-1.9 wt % alloy.

Spinodal decomposition has been identified in high-strength Cu–Ni–Sn alloys which also results in the precipitation of an ordered phase (Zhao and Notis, 1998). In these ternary alloys the initial disordered supersaturated state appears to undergo spinodal decomposition into two disordered phases with the solute-rich phase then lowering its free energy by ordering to form the ordered precipitate phase as discussed earlier.

8.7 Crystallography and Microstructure

8.7.1 Introduction

There are many ways that the symmetry of the parent and/or new phases comes into play in determining the microstructure of an alloy that has undergone a phase transformation. For example, it is well known that if the new phase has an arrangement of atoms on a plane that is the same as or similar to a plane in the parent phase, it is expected that the planes may be in contact with each other and determine the orientation relation between the new and parent phases. When an HCP Co alloy phase forms from an FCC Co alloy phase we find that the $(0001)_{HCP}$ planes are parallel to the $\{111\}_{FCC}$ planes. If the environmental fields (stress fields, magnetic fields, etc.) are isotropic, or non existent, we expect all the $\{111\}_{FCC}$ planes to have HCP particles of Co with their basal planes parallel to them. If however a field is applied in a specific direction one or more of the orientations of the new HCP phase may be missing.

The shape of the new phase is another way that symmetry controls microstructure. For example, a phase with one long direction and two unequal short directions will give rise to a distinctive microstructure.

Another effect of symmetry has to do with the relationship between the symmetry groups of the parent and new phases. When a phase undergoes a *disorder to order transition*, the phase with the lower symmetry (the ordered phase) can exist in two or more regions, called variants or *domains*, that are related to each other by one of the symmetry operations that was lost in the transition. In atomic ordering transitions, regions differing by a translation vector are called *antiphase domains* and regions differing by a rotation or reflection are called *orientational domains*.

Disorder to Order Transformations include

- Atomic Order
- Magnetic Order
- Displacive Order
- Ferroelectric Order

Subsequently, we will only discuss atomic ordering. See Dahmen (1987) for a full discussion of the role of symmetry on phase transformations.

8.7.2 Habit Planes and Orientation Relationships

The number of variants of a phase that exist in the microstructure of a transformed alloy can be determined by use of the symmetry of the parent phase if the orientation relationship is known. The example mentioned earlier of a plate-like HCP Co alloy phase precipitating from an FCC Co alloy phase will first be discussed. It was stated that the close-packed planes were parallel to each other. There are eight $\{111\}$ planes in the FCC structure, as the digits 1 or -1 can be placed in any of the three positions of the Miller indices. These are

$$(111), (\overline{11}1), (11\overline{1}), (\overline{1}11), (1\overline{1}1), (\overline{1}1\overline{1}), (\overline{1}11) \text{ and } (1\overline{11})$$

However, both the HCP and FCC phases contain a center of symmetry and so there are only four distinguishable orientations of the variants. For example, in this case, the $(0001)_{HCP}//(111)_{FCC}$ cannot be differentiated from the case of $(0001)_{HCP}//(\overline{111})_{FCC}$.

For the general case of an $\{hkl\}$ habit plane in a cubic phase with m$\overline{3}$m symmetry it can determined that there are 24 distinct orientation relationships that a plate-like new phase with that of the parent phase if the new phase has a center of symmetry. The $\pm h$ index could be placed in any of the three positions (six ways), the $\pm k$ index could then be placed in any of two positions (four ways) and the $\pm l$ index can only go in the remaining position (two ways). The product of these ways of indexing $\{hkl\}$ planes is 48 (the order of the cubic point group). However if the structures of the new phase contains a center of symmetry, only 24 of these is distinguishable. This method can also be used for parent phases of lower symmetry.

8.7.3 Shape of the Precipitate

The orientation relationship does not necessarily completely determine the number of ways that a new phase may appear in the parent matrix. If the shape of the new phase allows for additional degrees of freedom more orientations are possible. Consider a lath-shaped particle forming in a cubic matrix, where one dimension of the particle is much longer than the other two unequal dimensions. See **Figure 109**. The long dimension of the precipitate can be along any of the $\langle 100 \rangle$ directions of the cubic phase. However there are two ways for each $\langle 100 \rangle$ direction that the precipitate may align itself. See for example particles 5 and 6 of **Figure 109**. These particles have their larger flat surfaces perpendicular to the [010] and [100] directions, respectively.

In this case there are three possible habit planes and two orientations per habit producing six possible orientations of the particle. If the flat faces of the particles are along {hk0} of the cubic matrix, there are six possible planes and 2 orientations of the particle per habit yielding 12 possible orientations of the particles. For habit planes along {hkl} planes there are 24 possible orientations of the particles. In all these cases it was assumed that both the particle and matrix contained a center of symmetry. If either the particle or the matrix does not contain a center of symmetry (or if neither does) the above number of orientations should be multiplied by 2.

An interesting feature of the precipitations of particles with specific orientation relationships with the matrix can be seen in **Figure 110**. Here two hexagonal phases have precipitated in an Al alloy 6022. It can be seen that the overall symmetry of the selected area electron pattern (SAD) retains its 4 mm symmetry when all the variants of the new phases are present. This is quite general and shows that when all variants are present the symmetry of the parent phase is retained in the SAD if the diffracting region is large enough.

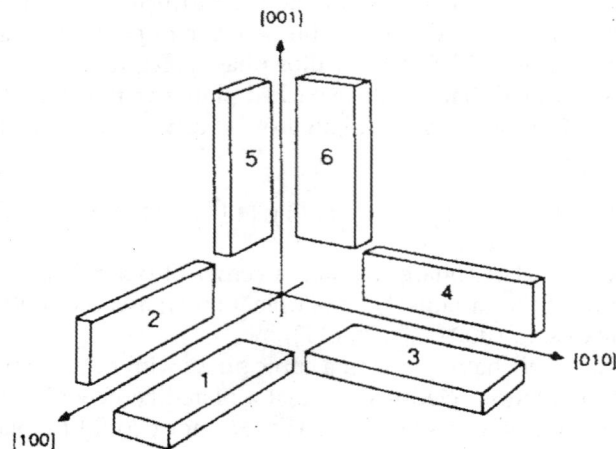

Figure 109   Possible arrangement of a lath with its face perpendicular to $\langle 100 \rangle$ directions of a cubic matrix. After Hugo and Muddle (1989).

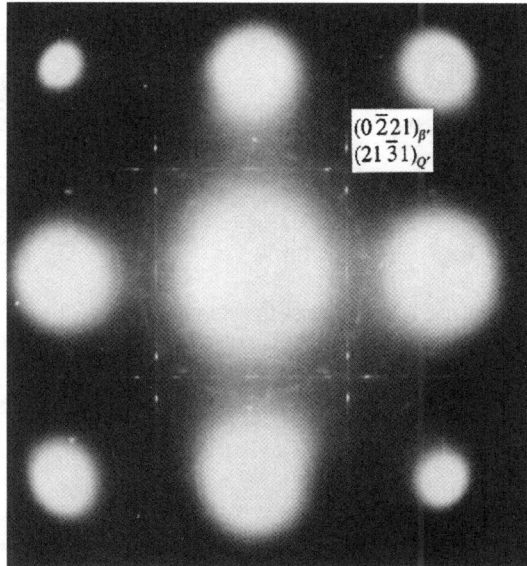

Figure 110 An [001] electron diffraction pattern from an Al–Mg–Si Cu alloy (6022) which has both Q' and b' precipitates. Both precipitates have their basal planes parallel to $\langle 001 \rangle$ of Al matrix. After Miao and Laughlin (2000).

8.7.4   Atomic Disorder to Order Transitions

All disorder to order phase transitions give rise to regions in the microstructure that are called domains. The domains exist as a necessary consequence of the lowering of the symmetry during the ordering process. For atomic ordering the usual cases involve a lowing of the translational symmetry, a lowing of the point symmetry or both translational and point symmetry being lowered. We will look at some examples.

8.7.4.1   Lowering of the Translational Symmetry

A common atomic ordering transformation in binary alloys is the FCC to $L1_2$ transformation. See **Figure 111**.

This transformation is an isostructural one in that the structure of both phases is cubic. What has changed during the atomic ordering is the translational symmetry: In the FCC structure, the smallest translation of the unit cell which leaves the crystal unchanged (i.e. to identical sites) is $\frac{1}{2} \langle 110 \rangle a$, whereas in the $L1_2$ structure the shortest translation to translation of the unit cell which leaves the crystal unchanged is $\langle 100 \rangle a$. This decrease in the translational symmetry gives rise to four possible domains, where the red atoms may be placed on any of the four equivalent sites of the FCC structure. See **Figure 112**.

These domains should arise with equal probabilities unless there is an outside influence on the transformation.

Another example of an isostructural transformation is the BCC to B2 transformation.

Because the ordering arrangements of the four domains displayed in **Figure 110** are out of phase with each other a defect called an antiphase domain boundary (APB) is produced when the domains impinge on each other. These APBs will influence many of the physical properties of the ordered phase.

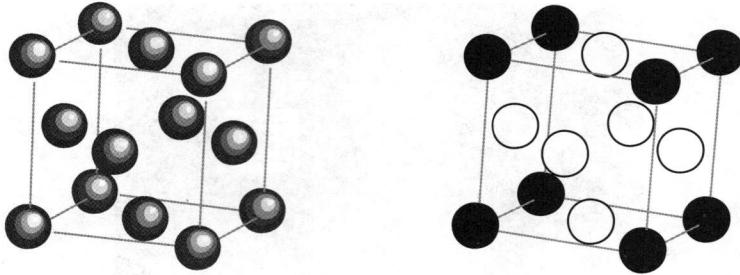

Figure 111   Unit cell of an FCC (Cu prototype, A1, cF4) phase transforming to a L1$_2$ (Cu$_3$Au prototype cP4). In the disordered cell, the probability of occupancy of all sites by a red atom is 25%, in other words the red atoms and blue atoms are assumed to be randomly arranged on the FCC Bravais lattice.

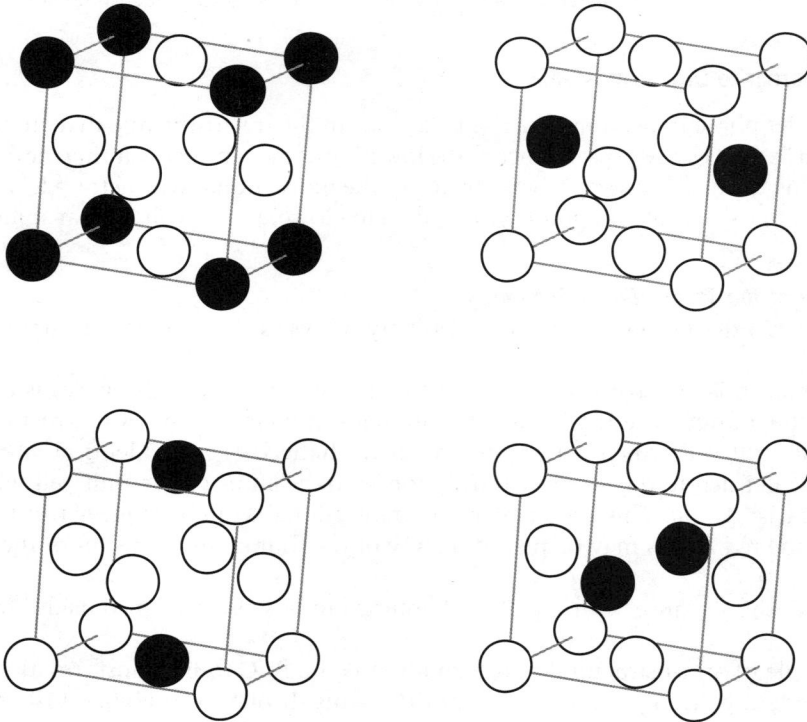

Figure 112   The four possible translational domains that may arise from the FCC to L1$_2$ disorder to order transformation.

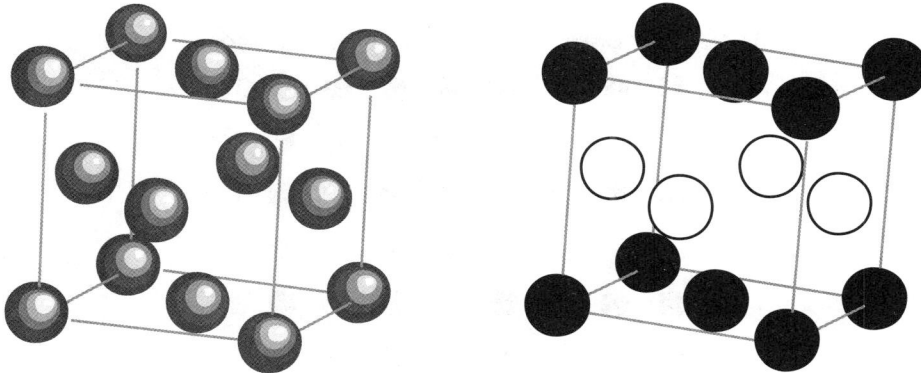

Figure 113 FCC structure to L10 structure. Note the $L1_0$ structure is no longer cubic (it is tetragonal) and can no longer be classified as a face centered cell.

8.7.4.2 *Lowering of the Point Group Symmetry*

In some disorder to order transitions the point group of the high symmetry disordered phase decreases on ordering. A well-known example of a disorder to order transformation is FCC to $L1_0$. See **Figure 113**.

This transformation has lowered the point group symmetry of the unit cell from $m\bar{3}m$ to $\frac{4}{m}mm$. It can be readily seen that the c-axis of the ordered phase could be parallel to any of the $\langle 100 \rangle$ directions of the parent FCC structure. Since both of the structures contain a center of symmetry, there are three ways to do this since $[001]_{L1_0}//[100]_{FCC}$ is the same as $[001]_{L1_0}//[\bar{1}00]_{FCC}$. More generally this can be determined by the ratio of the orders of the point groups of the two structures: 48 for the cubic structure and 16 for the tetragonal one.

This transformation has a new aspect to it. Adjacent domains may not have parallel c-axes. See **Figure 114**. In this case the domains are called *structural domains* or *variants* of the new phase. Sometimes they have been called twin domains, but that is not the best description, since the underlying disordered structure is not in a twin relationship. Clearly this microstructural feature will affect many of the physical properties of the ordered phase.

This transformation actually changes both the translational symmetry and the point group symmetry. The FCC Bravais lattice with four equivalent sites was changed to a simple tetragonal lattice (with two atoms in it) of one half the volume. This means that there would be not only the *structural domains* or *variants* of the ordered phase but also domains in antiphase with one another. Thus, there would be a total six domains of $L1_0$ arising from the disordered FCC structure. This also can be obtained by multiplying the ratio of the order of the points groups (which we found above to be 3) by the factor 2, which is the ratio (per unit volume) of the number of equivalent points in the disordered phase to that of the ordered phase.

8.8 Massive Transformation

The massive transformation is a distinct genre of diffusional solid-state phase transformation involving a compositionally invariant nucleation and growth process producing a change in crystal structure and/

(a)

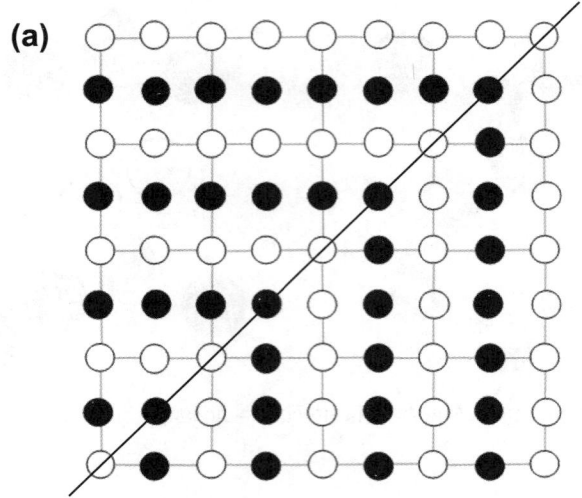

(b)

20 μm

(c)

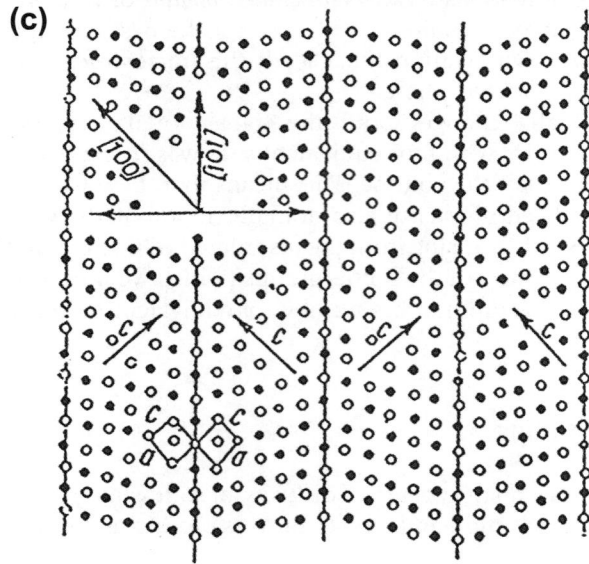

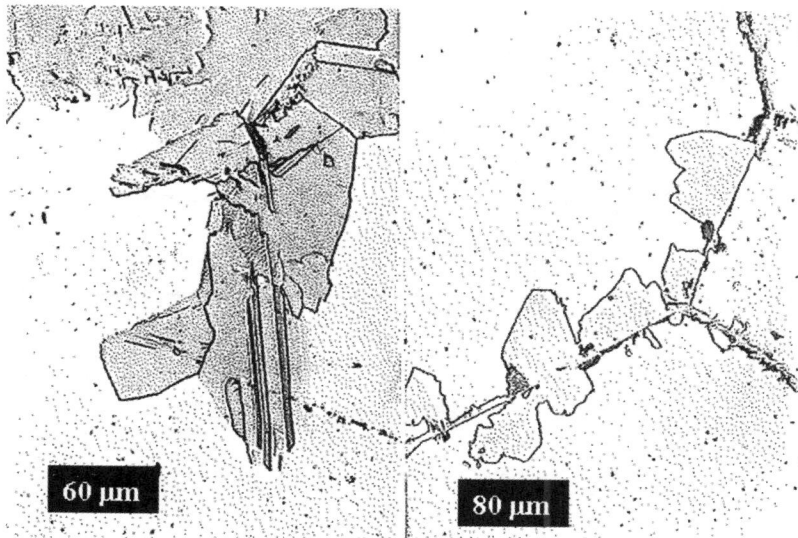

Figure 115   Massive transformation $\beta \to \alpha$ in β-brass. After Hull and Garwood (1956).

or degree of LRO. This partitionless change of phase is propagated by the migration of interphase interfaces controlled by interphase boundary diffusion processes and these interfaces are generally incoherent across which no systematic crystallographic orientation relationship is established and maintained between the parent and product phases. This mode of transformation occurs in pure metals, ferrous and nonferrous solid solutions and ceramic materials (Massalski, 1970; Fung et al., 1994; Aaronson et al., 2010). The transformation was first documented in studies of Cu–Zn and Cu–Al alloys during the 1930s (Phillips, 1930; Greninger, 1939), and later examined, in detail, by Hull and Garwood (1956) and Massalski (1958) in the 1950s. **Figure 115** shows the massive transformation product in the seminal β-brass alloys. A Symposium on the Mechanism of the Massive Transformaion was held in St. Louis, MO, at the Fall 2000 TMS/ASM Meeting as the transformation was recognized to occur in a growing number of technological contexts such as in TiAl-based alloys (Wang et al., 2002; Wittig, 2002).

Let us look briefly at the thermodynamic aspects of this compositionally invariant transformation in a binary system. We consider the classic $\beta(bcc) \to \alpha_m(fcc)$ massive transformation in a Cu- 38 at.% Zn solid solution occurring when the high-temperature β phase is quenched to room temperature or below with the $\alpha_m(fcc)$ massive product appearing at the grain boundaries of the parent phase. See **Figure 115**. A set of schematic free energy–composition curves are shown for the α and β phases in **Figure 116** exhibiting the important crossing of the free energy curves at the composition Cu-38 Zn at 700 °C. This

Figure 114   (a) Two domains of the $L1_0$ structure with perpendicular c-axes. These are called variants of the phase and the boundary is a structure domain boundary. For this case the boundary mirrors the structure of one domain into that of the other. (b) Optical micrograph showing polytwinned structure in Fe-Pd alloy resulting from stress-affected growth and coarsening of tetragonal L1o phase. (courtesy of H. Okumura). (c) Schematic of polytwinned microstructure with modulation of the c-axis across variant related c-domains. After Vlasova et al. (1969).

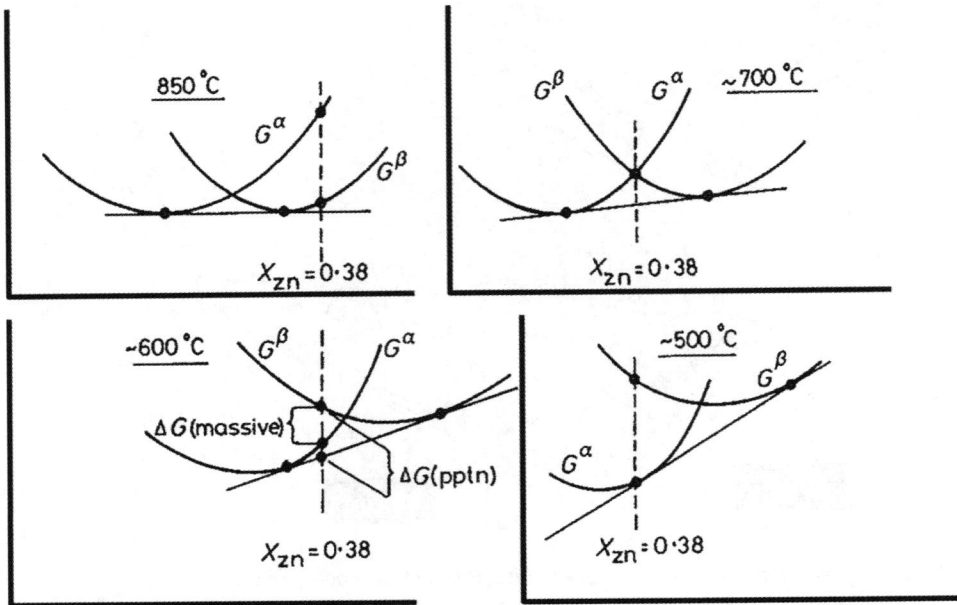

Figure 116   Free energy–composition diagrams showing the thermodynamics of the massive transformation $\beta \to \alpha$ in Cu–Zn (β-brass) delineating the temperature $T = T_0 \approx 700\,°C$ for the composition $X_{Zn} = 0.38$. After Porter and Easterling (1992).

intersection defines the temperature T_0 for this composition because it is clear that at any temperature below 700 °C this particular alloy can lower its free energy by transforming at constant composition by $\Delta G_{massive}$ as the free energy of the solution falls from one curve to another ($\beta \to \alpha$) with an attendant change in crystal structure. This massively transformed state α_m is metastable with respect to an $\alpha + \beta$ phase mixture that would lower the free energy to a point on the common tangent. A general set of free energy–composition curves are shown in **Figure 117** where regions between X_{eq}^{α} and X_0 allow for the

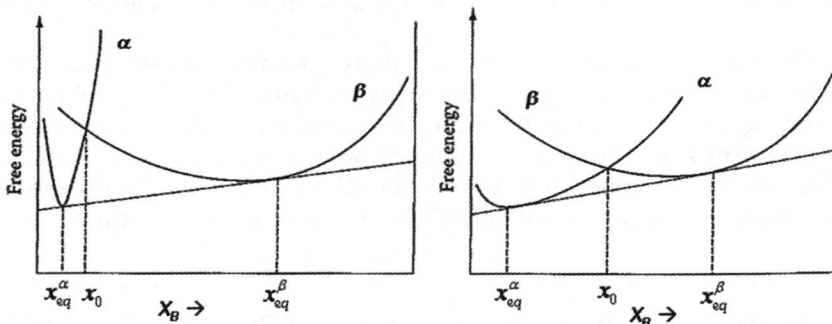

Figure 117   General free energy–composition schemes delineating regions of possible massive transformation of $\beta \to \alpha$ between compositions $X_{eq}(\alpha)$ and X_0. Note the importance of the crossover point at $X = X_0$ in both figures. After Aaronson et al. (2010).

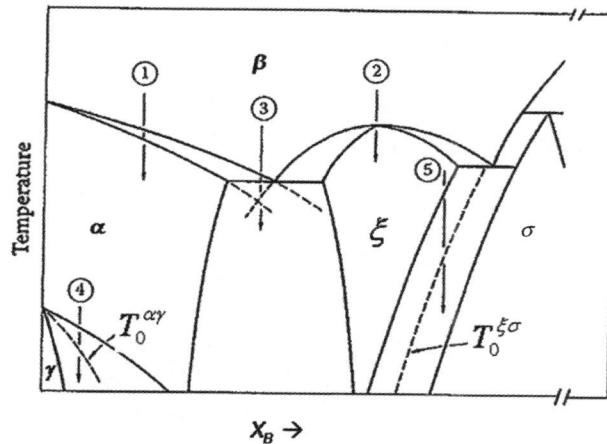

Figure 118 General phase diagram showing T_0 loci and regions of possible massive transformation. After Aaronson et al. (2010).

possibility of a compositionally invariant $\beta \rightarrow \alpha$ transformation (massive or martensitic as mentioned earlier). The T_0 concept is of paramount importance in mapping regions of possible compositionally invariant transformation below this temperature. The T_0 loci are depicted in a general hypothetical binary phase diagram in **Figure 118**.

The massive transformation product almost invariably nucleates heterogeneously at the grain boundaries of the high-temperature phase generally establishing an orientation relationship and a low-energy coherent or semicoherent interphase interface with one of the grains and an incoherent interface with the adjacent grain. Subsequent growth occurs virtually exclusively into the parent phase by migration of the disordered/incoherent boundary. However, it is sometimes found that the migrating interphase boundaries develop facets and serrated morphologies even in the absence of any orientation relationship with the parent phase and it appears that this results from the nature of the surface energetics and atomic attachment processes specific to the growing phase and not from constraints associated with crystallographic matching at the interface (Yanar et al., 2002).

The heterogeneous nucleation of the massive transformation product at grain boundaries has been treated quantitatively by several investigators over the years using CNT (Yanar et al., 2002; Veeraraghavan et al., 2003). The growth kinetics appear to follow a modified Burke–Turnbull description allowing for continuous atomic attachment or a ledge mechanism at the advancing interface. The activation energy for growth is typically found to be about 1/4 to 2/3 that for bulk diffusion in the system consistent with the notion of growth mediated by boundary or interface diffusion. Atomic attachment at the migrating interface can result in profuse twinning and faulting of the massive transformation product (Veeraraghavan et al., 1999; Yanar et al., 2002).

Buckley (1975) and Rajkovic and Buckley (1981) have revealed and analyzed a massive mode involved in the order–disorder transformation (A2 → B2) in Fe–Co and Fe–Co–X alloys. See **Figure 119**. At high transformation temperatures the ordering appears to occur via homogeneous or continuous ordering controlled by volume diffusion but at low temperatures where volume diffusion becomes sluggish the ordered phase nucleates at the grain boundaries of the parent-disordered bcc phase and grows behind an advancing incoherent interface utilizing the enhanced diffusivity at the

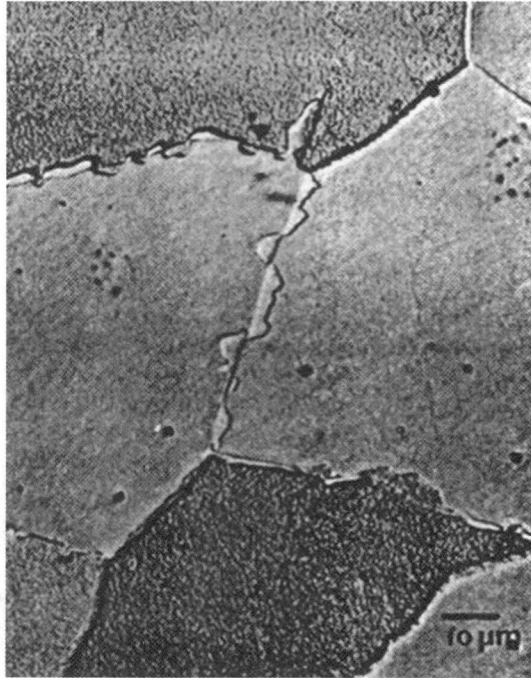

Figure 119 Massive transformation mode in the disorder → order transformation in Fe–Co alloy. After Rajkovic and Buckley (1981).

phase boundary. As mentioned earlier, the nuclei are coherent with respect to the adjacent grain into which growth of the ordered phase is negligible. It is important to note that the A2 → B2 is thermodynamically HIGHER ORDER making this a particularly interesting case.

8.9 Closure

Phase transformations in materials (metallic and nonmetallic) provide the metallurgist and materials scientist/engineer with one of the most effective tools for tailoring the structure and properties of engineering materials for application in modern technology. Understanding the fundamentals (thermodynamics, kinetics, crystallography and mechanistics, etc.) governing the evolution of material structure during synthesis, heat treatment and thermomechanical processing substantially removes the production and manufacture of materials from the realm of inefficient and unnecessarily expensive quasi-empiricism—albeit sometimes sophisticated empiricism—and allows for intelligent engineering design of structure–property relations in high-strength aluminum alloys, high-strength low-alloy steels and high-temperature nickel-base alloys. Ceramic materials are toughened by controlled precipitation from solid solution of a dispersed phase in an oxide matrix which inhibits crack propagation. It must be appreciated that the wide variety of structure–property relationships attainable with conventional steels more often than not involves controlling the distribution of carbides within an Fe-rich matrix or controlling the ferrite grain size resulting from the proeutectoid reaction in low-carbon steels. In quenched and tempered steels the important tempering step really involves precipitation from a supersaturated bcc or bct phase. If steels were the substance of the industrial revolution of the nineteenth and early twentieth century we might say that the Fe–C phase diagram with its eutectoid and associated phase transformations underpinned this cultural transformation. Extraordinary combinations of properties can now be achieved in age hardenable alloys through our understanding of metastability and the role of vacancies and trace elements in controlling the formation of phases during commercial heat treatments. Furthermore, our understanding and control of deleterious effects of fine-scale precipitation reactions during exposures of alloys to critical temperature ranges such as in the case ferritic and austenitic stainless steels have greatly enhanced our control over the properties of these materials in important engineering applications. The remarkable properties of Ni-base precipitation hardened superalloys (and they are indeed super) in turbine blades stems from a comprehensive understanding of the nucleation, growth and coarsening of the $\gamma'(Ni_3Ti, Al)$ phase during processing and in-service (as well as single-crystal growth during solidification). Over the past two decades or so a great deal of attention has focused on the nature of the disorder $\rightarrow$ order transformation in ferro-magnetic alloys because of the growing applications of ordered alloys, (for example, FePt) in a plethora of applications in magnetic thin films including futuristic spintronic materials. Furthermore, with the emergence of nanotechnology new fundamental challenges have arisen related to our lack of understanding of the thermodynamics and kinetics of transformations in small systems compared with bulk behavior and this realm will be a rich area for study well into the future.

Computational methodologies and computer simulations have had an enormous impact on the field of phase transformations and microstructural evolution in materials from both a scientific point of view and in the realm of practical application to the design of material structure and properties. CALPHAD and THERMOCALC and associated databases have been of tremendous value to systematizing phase equilibria in a myriad of alloy systems and Monte Carlo simulations and phase field approaches have provided insights into a wide range of issues related to transformation behavior. In this treatise we have not attempted to review this area because of space limitations, not because these contributions were deemed peripheral to the subject matter.

The field of phase transformations is a uniquely broad field of scientific endeavor and fundamental issues which have arisen in the analysis of pearlite growth and spinodal decomposition reach to the cutting edge of some of the most challenging problems in the thermodynamics and statistical mechanics of matter such as nonequilibrium thermodynamics and statistical mechanics, irreversible processes, self-assembly or self-organization, dissipative structures, critical phenomena, and so on. The Nobel Prize winner Prigogine (1980) has called attention to the important difference between BEING (thermodynamic equilibrium) and BECOMING (transformation) and THERMOSTATICS versus THERMODYNAMICS in the physical sciences and the domain of phase transitions is a marvelous example where these two dimensions come together to define the evolution of structure. Another eminent scientist and Nobel Prize recipient Steven Weinberg in his short book "The First Three Minutes" describing the "big bang" and the origins of the universe in virtually layman's terms relates the unfolding of this event essentially as a cascade of phase transitions.

References

Aaronson, H.I., Enomoto, M., Lee, J.K., 2010. Mechanisms of Diffusional Phase Transformations in Metals an Alloys. CRC Press, Baca Raton FL.

Abe, T., Brenner, S.S., Soffa, W.A., 1991. Decomposition of a Cu-Ni-Cr Ternary Alloy. Surf. Sci. 246 (1-3), 266–271.

Allen, S.M., Cahn, J.W., 1976. Mechanisms of Phase Transformations Within the Miscibility Gap of Fe-Rich Fe-Al Alloys. Acta. Met. 24, 425–437.

Ardell, A.J., 1988. Precipitate Coarsening in Solids: Modern Theories, Chronic Disagreement With Experiment. In Phase Transformations '87. Institute of Metals, Cambridge, 485–494.

Ardell, A.J., Nicholson, R.B., 1966. On modulated structure of aged Ni–Al alloys. Acta Metall. 14 (10), 1295–1309.

Ardell, A.J., 1970. The growth of gamma prime precipitates in aged Ni–T alloys. Metall. Trans. 1, 525–534.

Ardell, A.J., 1972. Effect of volume fraction on particle coarsening: theoretical considerations. Acta Metall. 20 (1), 61–71.

Ardell, A.J., Maheshwari, A., 1995. Coherent equilibrium in alloys containing spherical precipitates. Acta Metall. Mater. 43, 1825–1835.

Ardell, A.J., 1988. Precipitate Coarsening in Solids: Modern Theories, Chronic Disagreement With Experiment. Phase Transformations '87. Institute of Metals, Cambridge, 485–494.

Archer, R.S., Jeffries, Z., 1925. New developments in high-strength aluminum alloys. Trans. Am. Inst. Min. Metall. Eng. 71, 828–845.

Asimow, R., 1963. Clustering kinetics in binary alloys. Acta Metall. 11 (1), 72–73.

Avrami, M., 1939. Kinetics of phase change. I: general theory. J. Chem. Phys. 7, 1103–1112.

Avrami, M., 1940. Kinetics of phase change. II: transformation–time relations for random distribution of nuclei. J. Chem. Phys. 8, 212–224.

Avrami, M., 1941. Granulation, phase change and microstructure: kinetics of phase change. III. J. Chem. Phys. 9, 177–184.

Bain, E.C., 1923. Crystal structure of solid solutions. Trans. AIME 68, 625–639.

Banerjee, S., Sundaraman, M., 1992. Kinetics of order-disorder transformation under irradiation. In: Chen, H., Vasudevan, V.K. (Eds.), Kinetics of Ordering Transformations in Metals. The Minerals, Metals and Materials Society, pp. 227–253.

Barmak, K., 2010. A commentary on: Reaction kinetics in the processes of nucleation and growth. Met. Trans. A. 41, 2711–2712 by W. A. Johnson, R. F. Mehl, Trans. AIME 35, 416–458(1939).

Barrett, C.S., Massalski, T.B., 1966. The Structure of Metals. McGraw-Hill, NY, NY.

Baumann, S.F., Williams, D.B., 1985. Experimental observations on the nucleation and growth of Delta prime (Al_3Li) in dilute Al–Li alloys. Metall. Trans. A 16 (7), 1203–1211.

Becker, R., Doring, W., 1935. Kinetic treatment of grain formation in supersaturated vapours. Ann. Phys. 24, 719–752.

Becker, R., 1937. Über den Aufbau binarer Legierungen (On the constitution of binary alloys). Z. Metallk. 29, 245–249.

Becker, R., 1938. Die Keimbildung bei der Ausscheidung in metallischen Mischkristallen. Ann. Physik. 32, 128–140.

Bellon, P., Martin, G., 1988. Irradiation-induced formation of metastable phases-A master equation approach. Phys. Rev. B 38 (940), 2570–2582.

Bethe, H.A., 1935. Statistical theory of superlattices. Proc. Roy. Soc. London 150A, 552–575.

Borchers, C., 1999. Catastrophic nucleation during decomposition of Cu-0.9%Ti, Phil. Mag 79 (3), 537–547.

Borelius, G., 1937. Ann. Phys. 280, 507–??

Borchers, C., 1999. Catastrophic nucleation during decomposition of Cu-0.9%Ti, Phil. Mag 79 (3), 537–547.

Bradley, A.J., 1940. X- ray evidence of intermediate stages during precipitation from solid solution. Proc. Phys. Soc. vol. 52, 80–85.

Bradley, A.J., Taylor, A., 1940. An X-ray Investigation of Aluminum-rich iron nickel aluminum alloys after slow cooling. J. Institute of Metals 66, 53–65.

Bragg, W.L., Williams, E.J., 1934. The effect of thermal agitation on atomic arrangement in alloys. Proc. Roy. Soc. London 145A, 699–730.

Brenner, S.S., Miller, M.E., Soffa, W.A., 1982. Spinodal decomposition of iron-32 at. % chromium at 470 °C. Scr. Metall. 16, 831–836.

Brenner, S.S., Camus, P.P., Miller, M.E., Soffa, W.A., 1984. Phase separation and coarsening in Fe–Cr–Co alloys. Acta Metall. 32, 1217–1227.

Brun, J.Y., Hamar-Thilbault, S.J., Allibert, C.H., 1983. Cu-Ti and Cu-Ti-Al solid-state phase-equilibria in the cu-rich region. Zeitschrift Fur Metallkunde 74 (8), 525–529.

Buckley, R.A., 1975. Microstructure and kinetics of the ordering transformation in iron–cobalt alloys, FeCo, FeCo 0.4%Cr and FeCo 2.5% V. Metal Sci. 9, 243–247.

Buckley, R.A., Rajkovic, M., 1979. Order–Disorder Transformations, and the Metallography and Kinetics of Ordering in Fe-50% Co Based Alloys. Institution of Metallurgists, London.

Buerger, M.J., 1948. The role of temperature in minerology. Am. Mineral. 33 (3–4), 101–121.

Buerger, M.J., 1951. Crystallographic aspects of phase transformations. In: Phase Transformations in Solids. Wiley and Sons, New York, pp. 183–211.

Burgers, W.G., Groen, J., 1957. Mechanism and Kinetics of the Allotropic Transformation of Tin. Discuss. Faraday Soc. 23, 183–195.

Burke, J.E., Turnbull, D., 1952. Recrystallization and grain growth. Prog. Metals Phys. 3, 220–292.

Butler, E.P., Thomasp, G, 1970. Structure and properties of spinodally decomposed Cu-Ni-Fe alloys. Acta. Met. vol. 18, 347–365.

Cahn, J.W., 2006. Introduction to "A solid Solution Model for Inhomogeneous Systems". In: Agren, John, Brechet, Y., Hutchinson, C., . Philbert, J., Purdy, G. (Eds.), Thermodynamics and Phase Transformations: Selected works of mats Hillert. EDP Sciences.

Cahn, J.W., 1956. The kinetics of grain boundary nucleated reactions. Acta Metall. 4, 449–459.

Cahn, J.W., 1957. Nucleation on dislocations. Acta Metall. 5, 169–172.

Cahn, J.W., Hilliard, J.E., 1958. Free energy of a nonuniform system. 1. Interfacial free energy. J. Chem. Phys. 28 (2), 258–267.

Cahn, J.W., 1959. Free energy of a nonuniform system. 2. Thermodynamic basis. J. Chem. Phys. 30 (5), 1121–1124.

Cahn, J.W., Hilliard, J.E., 1959. Free energy of a nonuniform system. 3. Nucleation in a 2-component incompressible fluid. J. Chem. Phys. 31 (3), 688–699.

Cahn, J., 1961. On spinodal decomposition. Acta Metall. 9 (9), 795–801.

Cahn, J.W., 1962a. On spinodal decomposition in cubic crystals. Acta Metall. 10, 179–183.

Cahn, J.W., 1962b. Coherent fluctuations and nucleation in coherent solids. Acta Metall. 10, 907–913.

Cahn, J.W., 1966. The later stages of spinodal decomposition nad the beginnings of particle coarsening. Acta Metall. 14, 1685–1692.

Cahn, J.W., 1968. Spinodal decomposition. Trans. TMS AIME 242, 166–180.

Carslaw, H.S., Jaeger, J.C., 1946. Conduction of Heat in Solids, second ed. 1959. Oxford at the Claredon Press, Oxford.

Cheong, B., Laughlin, D.E., 1994. Thermodynamic consideration of the tetragonal lattice distortion of the $L1_0$ ordered phase. Acta Metall. Mater. 42 (6), 2123–2132.

Chou, A., Datta, A., Meier, G.H., Soffa, W.A., 1978. Microstructural behaviour and mechanical hardening in a Cu–Ni–Cr alloy. J. Mater. Sci. 13, 541–552.

Christian, J.W., 2002. The Theory of Transformation in Metals and Alloys. Pergamon Press.

Corey, C.L., Rosenblum, B.Z., Greene, G.M., 1973. Ordering transition in Ni_3Al alloys. Acta Metall. 21 (7), 837–844.

Cornie, J.A., Datta, A., Soffa, W.A., 1973. Electron microscopy study of precipitation in Cu–Ti sideband alloys. Metall. Trans. 4 (3), 727–733.

Dahmen, U., 1987. Phase Transformations, Crystallographic Aspects. In: Encyclopedia of Physical Science and Technology. Academic Press, San Diego, pp. 319–354.

Daniel, V., Lipson, H., 1943. An X-ray study of the dissociation of an alloy of copper, iron and nickel. Proc. Royal Soc. A 181, 368–378.

Daniel, V., Lipson, H., 1944. The dissociation of copper, iron and nickel: further X-ray work. Proc. Royal Soc. A 182, 378–387.

Danoix, F., Auger, P., 2000. Atom probe studies of the Fe–Cr system and stainless steels aged at intermediate temperature: a review. Mater. Charact. 44 (1–2), 177–201.

Datta, A., Soffa, W.A., 1976. Structure and properties of age hardened Cu–Ti alloys. Acta Metall. 24, 987–1001.

Dehlinger, U., 1937. Dei verschiedenen Arten der Ausscheidung. Z. Metallk 29, 401–403.

Dehlinger, U., 1939. Chemische Physik der Metalle und Legierungen. Akademische verlagsgesellschaft m.b.h.

DeNys, T., Gielen, P.M., 1971. Spinodal decompositionin the Fe–Cr system. Metall. Trans. 2, 1423–1428.

de Fontaine, D., 1973. In: Phase Transitions. Pergamon Press, New York.

de Fontaine, D., 1975. K-space symmetry rules for order-disorder reactions. Acta. Met. 23 (5), 553–571.

Ditchek, B., Schwartz, L., 1979. Applications of spinodal alloys. Ann. Rev. Mater. Sci. 9, 219–253.

Doi, M., Miyazaki, T., Wakatsuki, T., 1984. The effect of elastic interaction energy on the morphology of y′ precipitates in nickel-based alloys. Mater. Sci. Eng. 67, 247–253.

Dollins, C.C., 1970. Nucleation on dislocations. Acta Metall. 18, 1209–1215.

Ecob, R.C., Bee, J.V., Ralph, B., 1980. The cellular reaction in dilute copper-titanium alloys. Metallurgical Transactions A 11 (8), 1407–1414.

Ehrenfest, P., 1933. Phasenumwandlungen in ueblichen und erweiterten Sinn, classifiziert nach dem entsprechenden Singularitaeten des thermodynamischen Potentiales, vol. 36. Verhandlingen derKoninklijke Akademie van Wetenschappen, Amsterdam, pp. 153–157.

Eshelby, J.D., 1957. The determination of the elastic field of an ellipsoidal inclusion, and related problems. Proc. Royal Soc. A 241, 376–396.

Eyring, H., Glasstone, S., Laidler, K.J., 1939. Application of the theory of absolute reaction rates to overvoltage. J. Chem. Phys. 7 (12), 1053–1065.

Farkas, L., 1927. The speed of germinitive formation in over saturated vapours. Z. physikalische chemie–stochiometrie verwandtschaftslehre 125 (3/4), 236–242.

Finniston, H.M., 1971. Structural characteristics of materials. Elsevier materials science series, Amsterdam.

Fisher, R.M., Dulis, E.J., Carroll, K.G., 1953. Identification of the precipitate accompanying 885 °F embrittlement in Cr steels. Trans. AIME 197 (5), 690–695.

Fournelle, R.A., Clarke, J.B., 1972. The Genesis of the cellular precipitation reaction. Met. Trans. 3, 2757–2767.

Fung, K.Z., Vikar, A.V., Drobeck, D.L., 1994. Massive transformation in the Y_2O_3–Bi_2O_3 system. J. Amer. Ceram. Soc. 77, 1638–1648.

Gentry, W.O., Fine, M.E., 1972. Precipitation in Ni-11.1 at percent A1 and Ni-13.8 at percent A1 alloys. Acta Metall. 20 (2), 181–190.

Gibbs, J.W., 1873. Graphical methods in the thermodynamics of fluids. Trans. Conn. Acad. 2, 309–342. Also in Collected Works 1928, Vol. 1. Longmans, Green and Co., New York. 1–32.

Gibbs, J.W., 1875. On the Equilibrium of heterogeneous Substances. Trans. Conn. Acad. 3, 108–248, 1875–76; 343–524, 1877–78. Also in Collected Works, (1928) vol. 1. Longmans, Green and Co., New York. 55–349.

Ginzburg, V.L., Landau, L.D., 1950. J. Exptl. Theoret. Phys. (USSR) 20, 1064ff.

Ginzburg, V.L., Levanyuk, A.P., Sobyanin, A.A., 1987. Comments on the region of applicability of the Landau theory for structural phase transitions. Ferroelectrics 73, 171–182.

Girifalco, L.A., 1971. Statistical Physics of Materials. Wiley and Sons, NY, NY.

Glasstone, S., Laidler, K.J., Eyring, H., 1941. The Theory of Rate Processes; the Kinetics of Chemical Reactions, Viscosity, Diffusion and Electrochemical Phenomena. McGraw-Hill Book Company, New York; London.

Glicksman, M.E., 2000. Diffusion in Solids: Field Theory, Solid State Principals and Applications. Wiley, NY, NY.

Gomez-Ramirez, R., Pound, G.M., 1973. Nucleation of a second solid-phase along dislocations. Metall. Trans. 4, 1563–1570.

Greenwood, G.W., 1956. The growth of dispersed precipitates in solution. Acta Metall. 4, 243–248.

Greenwood, G.W., 1969. The mechanism of Phase Transformations in Crystalline solids. London UK, 103–108.

Greninger, A.B., 1939. The martensitic transformation in beta copper–aluminum alloys. Trans. AIME 133, 204–229.

Grune, R., 1988. Decomposition of Ni-12 at.% Ti by atom probe field-ion microscopy. Acta Metallurgica 36 (10), 2797–2809.

Guggenheim, E.A., 1952. Mixtures. Clarenden Press, Oxford.

Guinier, A., 1938a. Un nouveau type de diagrammes de rayons X, (A new type of X-ray diagram). Comptes Rendus Hebdomadaires Séances l'Académie Sci. 206, 1641–1643.

Guinier, A., 1938b. Structure of age-hardened Aluminium–Copper alloys. Nature 142, 569–570.

Gust, W., 1979. Discontinuous Precipitation in Binary Metallic Systems. In: Phase Transformations, vol. 1 II-27–68. Institution of Metallurgists, London.

Guy, A.G., Hren, J.L., 1974. Elements of Physical Metallurgy. Addison-Wesley series in metallurgy and materials.

Haasen, P., 2001. Physical Metallurgy. Cambridge University Press.

Hakkarainen, T., 1971. Formation of Coherent Cu_4Ti Precipitates in Copper-rich Copper–Titanium Alloys. Helsinki University of Technology, Helsinki.

Ham, F.S., 1959. Diffusion limited growth of precipitate particles. J. Appl. Phys. 30 (10), 1518–1525.

Hardy, H.K., Heal, T.J., 1954. Report on precipitation. Prog. Metals Phys. 5, 143–278.

Hermann, C., 1934. Tensors and crystal symmetry. Zeit. Kritallographie 89 (1), 32–48.

Hilliard, J.E., 1970. Spinodal decomposition. In: Phase Transformations. American Society for Metals, ASM, Park OH, pp. 497–560.

Hillert, M., 1956. A Theory of Nucleation for Solid Metallic Solutions, D. Sc. Thesis, Mass. Inst. Tech.

Hillert, M., 1957. The role of interfacial energy during solid state phase transformation. Jernkontorets Annaler 141, 757–789.

Hillert, M., 1961. A solid solution model for inhomogeneous systems. Acta Metall. 9, 525–536.

Hillert, M., Cohen, M., Averbach, 1961. Formation of modulated structures in copper--nickel--iron alloys. Acta Metall. 9 (6), 536–546.

Hillert, M., Jarl, M., 1978. Model for alloying effects in ferromagnetic metals. Calphad-Computer Coupling of Phase Diagrams and Thermochemistry 2 (3), 227–238.

Hillert, M., 2011. Discussion of A Commentary on Reaction Kinetics in Processes of Nucleation and Growth. Metall. Mats. Trans. 42A, 3241.

Howe, J.M., 1997. Interfaces in Materials: Atomic Structure, Thermodynamics and Kinetics of Solid–Vapor. In: Solid–Liquid and Solid–Solid Interfaces. John Wiley & Sons Inc, New York.

Hugo, G.R., Muddle, B.G., 1989. The role of symmetry in determining precipitate morphology. Mater. Forum 13, 147–152.

Hull, D., Garwood, R.D., 1956. The Diffusionless Transformations of metastable beta brass. In: The Mechanism of Phase Transformations in Metals. Institute of Metals, London, pp. 219–227.

Hyde, J.M., Miller, M.K., Hetherinton, M.G., Cerezo, A., Smith, G.D.W., Elliott, C.M., 1995a. Spinodal decomposition in fe-cr alloys - experimental-study at the atomic-level and comparison with computer-models .2. Development of domain size and composition amplitude. Act. Met. et. Mat. 43 (9), 3403–3413.

Hyde, J.M., Miller, M.K., Hetherinton, M.G., Cerezo, A., Smith, G.D.W., Elliott, C.M., 1995b. Spinodal decomposition in fe-cr alloys - experimental-study at the atomic-level and comparison with computer-models, 3. Development of morphology. Act Met. et. Mat. 43 (9), 3414–3426.

Imai, Y., Izumiyama, M., Masumoto, T., 1966. Phase Transformations of Fe–Cr Binary System at about 5000C, vol. 18. Sci. Rep. Res. Inst. Tohoku University, Ser. A, p. 56.

Ino, H., 1978. Pairwise interaction-model for decomposition and ordering processes in Bcc binary-alloys and its application to Fe–Be system. Acta Metall. 26 (5), 827–834.

Israel, D.H.B., Fine, M.E., 1963. Precipitation studies in Ni-10 at. Percent Ti. Acta Metall. 11 (9), 1051–1059.

Inden, G., 1976. Project Meeting CALPHAD V, 21–25 June 1976. Max-Planck –Inst. Eisenforschung, G.m.b.H, Dusseldorf/West Germany, pp. 111, 4-1.

Inden, G., 1981. The role of magnetism in the calculation of phase diagrams. Physica B & C 103 (1), 82–100.

Inden, G., 1982. The effect of continuous transformations on phase diagrams. Bull. Alloy Phase Diagrams 2, 412–422.

International Tables for Crystallography, 2006. Volume D, Physical Properties of Crystals. http://dx.doi.org/10.1107/97809553602060000104, International Union of Crystallography.

Jena, A.K., Chaturvedi, M.C., 1992. Phase Transformations in Materials. Prentice Hall, Englewood Cliffs, NJ.

Johansson, C.H., Linde, J.O., 1925. X-ray determination of the atomic structure of the Au–Cu and Pd–Cu mixed crystal series. Ann. Physik 78 (21), 439–460.

Johnson, C.A., 1965. Generalization of Gibbs–Thomson equation. Surf. Sci. 3 (5), 429–444.

Johnson, W.A., Mehl, R.F., 1939. Reaction kinetics in processes of nucleation and growth. Trans. AIME 135, 416–458.

Johnson, W.C., Voorhees, P.W., 1987. Phase equilibria in two-phase solids. Metall. Trans. 18A, 1213–1228.

Johnson, W.C., Mueller, W.H., 1991. Characteristics of phase equilibria in coherent solids. Acta Metall. Mater. 39, 89–103.

Kahlweit, M., 1970. Precipitation and Aging, in PHYSICAL CHEMISTRY: an Advanced Treatise. In: Jost, Wilhelm (Ed.), Solid State, vol. X. Academic Press, New York/London, pp. 719–759.

Kampmann, L., Kahlweit, M., 1967. Zur Theorie Von Fallungen. Berichte Der Bunsen-Gesellschaft Fur Physikalische Chemie 71 (1), 78.

Kampmann, L., Kahlweit, M., 1970. On theory of precipitations. 2. Ber. Bunsen Gesell Phys. Chem. 74 (5), 456–462.

Kashyap, K.T., Koppad, P.G., 2011. Small-angle scattering from GP zones in Al–Cu alloy. Bull. Mater. Sci. 34 (7), 1455–1458.

Khachaturyan, A.G., Airapetya, V.M., 1974. Spatially periodic distributions of new phase inclusions caused by elastic distortions. Physica Status Solidi (A) 28, 61–70.

Khachaturyan, A.G., 1978. Ordering in substitutional and interstitial solid solutions. Prog. Mater. Sci. 22 (1–2), 1–150.

Khachaturyan, A.G., 1983. Theory of Stuctural Transformations in Solids. John Wiley & Sons Inc, New York.

Khachaturyan, A.G., Semenovskaya, S.V., Morris, J.W., 1988. Theoretical analysis of strain-induced shape changes in cubic precipitates during coarsening. Acta Metallurgica 36 (6), 1563–1572.

Khachaturyan, A.G., Lindsey, T.F., Morris, J.W., 1988. Theoretical Investigation Of The Precipitation Of Delta. In Al–Li. Metallurgical Transactions A-physical Metallurgy and Materials Science 19 (2), 249–258.

Kashchiev, D., 2000. Nucleation: Basic Theory with Applications. Butterworths-Heinemann, Oxford.

Kaufman, L., Weiss, R.J., Clougherty, E.V., 1963. Lattice stability of metals-III- iron. Acta Metall. 11 (4), 323–328.

Kelly, A., Nicholson, R.B., 1963. Prog. Mater. Sci. 10, 151–391.

Kelton, K.F., Greer, A.L., 2010. Nucleation in Condensed Matter, Pergamon Materials Series. Elsevier, Netherlansds.

Kirkaldy, J.S., Young, D.J., 1987. Diffusion in the Condensed State. Institute of Metals, London.

Kirkwood, D.H., 1970. Precipitate number density in a Ni–Al alloy at early stages of ageing. Acta Metall. 18, 563–570.

Kokorin, V.V., Chuistov, K.V., 1976. Spinodal Decomposition in ordering Solid Solutions. Fizika Metallov Metallovedenie 42 (5), 1114–1117.

Kolmogorov, A.N., 1937. The statistical theory of metal crystallization. Izv. Akad. Nauk SSSR, Ser. Mater. 3, 355–360.

Kompascher, M., Schonfeld, B., Heinrich, H., Kostorz, G., 2000. Small-angle neutron scattering investigation of the early stages of decomposition in Ni-rich Ni–Ti alloys. J. Appl. Cryst. 33, 488–490.

Kompatscher, M., Schonfeld, B., Heinrich, H., Kostorz, G., 2003. Phase separation in Ni-rich Ni–Ti alloys; metastable states. Acta Mater. 51, 165–175.

Kostorz, G., Kompatscher, M., Schonfeld, B., 1999. Coherent precipitates in Ni-rich Ni–Ti single crystals. In: Koiwa, M., Otsuka, K., Miyasaki, T. (Eds.), Proceedings of the International Conference on Solid–Solid Phase Transformations'99 (JIMIC-3). JIM, Tokoyo, pp. 305–312.

Kubo, H., Wayman, C.M., 1980. Theoretical basis for spinodal decomposition in ordered alloys. Acta Metall. 28 (3), 395–404.

Kurnakow, N., Zemczuzny, S., Zasedatelev, M., 1916. The transformations in alloys of gold with copper. J. Inst. Metals 15, 305–331.

Laughlin, D.E., 1976a. Spinodal decomposition in Ni based Ni–Ti alloys. Acta Metall. 24, 53–58.

Laughlin, D.E., Cahn, J.W., 1975. Spinodal decomposition in age hardening Cippoer–Titanium alloys. Acta Metall. 23, 329–339.

Laughlin, D.E., Soffa, W.A., 1988. In: Salje, E.K.H. (Ed.), Exsolution, Ordering and Structural Transformations: Systematics and Synergistics. Physical Properties and Thermodynamic Behaviour of Minerals, NATO ASI Series C, vol. 225. D. Reidel Publishing Company, pp. 213–264.

Laughlin, D.E., Soffa, W.A., 1985. Spinodal Structures Metals Handbook. In: Metallography and Microstructures, Ninth ed., vol. 9. American Society for Metals. 652–654.

Landau, L.D., 1937. In: Collected Papers, D. ter Haar, (Ed.) Oxford.

Langer, J.S., Schwartz, A.J., 1980. Kinetics of Nucleation in near critical fluids. Phys. Rev. A21, 948–958.

Larche', F., Cahn, J.W., 1973. A linear theory of thermochemical equilibrium of solids under stress. Acta Metall. 21, 1051–1063.

Larche', F.C., 1979. In: Dislocations in Solids, edited by F.R.N. Nabarro. North-Holland Publishing Company, Amsterdam, Holland, vol. 4. Pergamon Press.

Larikov, L.N., Geichenko, V.V., Falchenko, V.M., 1975. Diffusional Processes in Ordered Alloys. Izdatel'stvo Naukova Dumka, Kiev.

Legoues, F.K., Aaronson, H.I., 1984. Influence of crystallography upon critical nucleus shapes and kinetics of homogeneous fcc–fcc nucleation .4. Comparisons between theory and experiment in Cu–Co alloys. Acta Metall. 32, 1855–1870.

Linde, A.D., 1982. A new inflationary universe scenario: a possible solution of the horizon, flatness, homogeneity, isotropy and primordial monopole problems. Phys. Lett. B108, 389–393.

Lifshitz, I.M., Slyozov, V.V., 1961. The kinetics of precipitation from supersaturated solid solutions. J. Phys. Chem. Solids 19, 35–50.

Liu, C.T., Loh, B.T.M., 1971. Solid Solution Theory and Spinodal Decomposition. Phil. Mag. 24, 367–380.

Livak, R.J., Thomas, G., 1971. Spinodally Decomposed Cu-Ni-Fe Alloys Of Asymmetrical Compositions Acta. Met. 19 (6), 497–505.

Livingston, J.D., Cahn, J.W., 1975. Discontinuous coarsening of aligned eutectoids. Acta Metall. 22 (4), 495–503.

Lukas, H.L., Fries, S.G., Sundman, Bo, 2007. Computational Thermodynamics (Calphad Method). Cambridge University Press.

Lyubov, B.Y., Solovyov, V.A., 1965. The possible existence of stable segregations of solute atoms and the formation of coherent nuclei of a new phase of the elastic field of edge dislocations. Fiz. Met. Metallov 19, 333–342.

Manna, I., Pabi, S.K., Gust, W., 2001. Discontinuous reactions in Solids. Int. Mater. Rev. 46 (2), 53–91.

Martin, J.W., Doherty, R.D., Cantor, B., 1997. Stability of microstructure in metallic systems. Cambridge University Press, NY, USA.

Martin, J.W., 1998. Precipitation Hardening: Theory and Applications. Butterworth-Heinemann, Oxford.

Massalski, T.B., 1958. The mode and morphology of massive transformations in Cu–Ga, Cu–Zn, Cu–Zn–Ga and Cu–Ga–Ge alloys. Acta Metall. 6 (4), 243–253.

Massalski, T.B., 1970. Massive transformations. In: Phase Transformations. American Society for Metals, ASM Park OH.

Massalski, T.B., Laughlin, D.E., 2009. The surprising role of magnetism on the phase stability of Fe (Ferro). CALPHAD: Comput. Coupling Phase Diagrams Thermochem. 33, 3–7.

Massalski, T.B., Soffa, W.A., Laughlin, D.E., 2006. The nature and role of incoherent interphase interfaces in diffusional solid-solid phase transformations. Metall. Mater. Trans. 37A, 825–831.

Mehl, R.F. and Jetter, L.K., (1940). American society for metals Symposium, Age Hardening of Metals, 342–317.

Meijering, J.L., 1952. Calculs thermodynamiques concernant la nature des zones Guinier-Preston dans les alliages aluminum-cuivre. Revue de Metallurgie. 49 (12), 906–910.

Meijering, J.L., 1963a. Miscibility gaps in ferromagnetic alloy systems. Philips Res. Rep. 18, 318–330.

Meijering, J.L., 1963b. Usefulness of the $1/\gamma$ plot in the theory of thermal faceting. Acta Metall. 11, 847–849.

Merica, P.D., Waltenberg, R.G., Scott, H., 1921. Heat treatment and constitution of duralumin. Trans. Am. Inst. Min. Metall. Eng. 64, 41–77.

Miao, W.F., Laughlin, D.E., 2000. Effects of Cu content and preaging on precipitation characteristics in aluminum alloy 6022. Metall. Mater. Trans. 31A, 361–371.

Miller, M.K., Hyde, J., Hetherington, M.G., Cerezo.A., Smith, G.D.W., Elliott, C.M., 1996. Spinodal decomposition in Fe-Cr alloys: Experimental study at the atomic level and comparison with computer models—I. Introduction and methodology. Act Met. et. Mat. 43 (9), 3385–3401.

Miyazaki, T., Imamura, H., Kozakai, T., 1982. The formation of g' precipitate doublets in Ni–A1 alloys and their energetic stability. Mater. Sci. Eng. 54, 9–15.

Mott, N.F., Nabarro, F.R.N., 1940. An attempt to estimate the degree of precipitation hardening, with a simple model. Proc. Phys. Soc. 52, 86–89.

Nabarro, F.R.N., 1940. The influence of elastic strain on the shape of particles segregating in an alloy. Proc. Phys. Soc. 52, 90–93 and The strains produced by precipitation in alloys. Proc. R. Soc. A, 175, 519–538.

Nembach, E., 1996. Particle Strengthening of Metals and Alloys. John Wiley and Sons, NY, NY.

Nicholson, R.B., Nutting, J., 1958. Direct observation of the strain field produced by coherent precipitated particles in an age-hardened alloy. Philos. Mag. 3 (29), 531–535.

Nicholson, R.B., Thomas, G., Nutting, J., 1959. Electron-microscopic studies of precipitation in aluminium alloys. J. Inst. Metals 87 (12), 429–438.

Nicholson, R.B., 1970. Nucleation at imperfections. In: Phase Transformations. American Society for Metals, ASM Park OH.

Nishizawa, T., Hasebe, M., Ko, M., 1979. Thermodynamic analysis of solubility and miscibility gap alpha iron alloys. Acta Metall. 21, 817–828.

Nix, F.C., Shockley, W., 1938. Order–disorder transformations in alloys. Rev. Mod. Phys. 10, 1–71.

Novotny, G.M., Ardell, A.J., 2001. Precipitation of Al3Sc in Al–Sc alloys. Mater. Sci. Eng. A318, 144–154.

Orowan, E., 1948. Theory of Yield Strength without Particle Shear. Institute of Metals Symposium on Internal Stress, London, p. 451.

Pfeifer, M.J., Voorhees, P.W., 1991. A graphical method for constructing coherent phase diagrams. Acta Metall. Mater. 39, 2001–2012.

Phillips, A.J., 1930. The alpha to beta transformation in brass. Trans. AIME 89, 194–200.

Pippard, A.B., 1966. Elements of Classical Thermodynamics. Cambridge University Press, Cambridge.

Poduri, R., Chen, L.-Q., 1997. Computer simulation of the kinetics of order-disorder and phase separation during precipitation of delta' (Al3Li) in Al-Li alloys. Acta. Mater. 45 (1), 245–255.

Porter, D.A., Easterling, K.E., 1992. Phase Transformations in Metals and Alloys. Chapman Hall, London.

Preston, G.D., 1938a. The diffraction of X-rays by age-hardening aluminium copper alloys. Proc. R. Soc. A167, 526–538.

Preston, G.D., 1938b. The diffraction of X-rays by an age-hardening alloy of aluminium and copper. Struct. Intermediate Phase Philos. Mag. 26, 855–871.

Preston, G.D., 1938c. Structure of age-hardened aluminium–copper alloys. Nature 142, 570.

Prigogine, I., 1980. From being to becoming: time and complexity in the physical sciences. W. H. Freeman, San Francisco.

Putnis, A., 1992. Introduction to Mineral Sciences. Cambridge University Press.

Radmilovic, V., Fox, A.G., Thomas, G., 1989. Spinodal decomposition of Al rich Al–Li alloys. Acta Metall. 37 (9), 2385–2394.

Rajkovic, M., Buckley, R.A., 1981. Ordering transformations in Fe 50Co based alloys. Metal Sci. 81 (1), 21–29.

Reed-Hill, R.E., 1972. Physical Metallurgy Principles, second ed. Van Nostrand.

Richards, M.J., Cahn, J.W., 1971. Pairwise interactions and ground state of orderd binary alloys. Acta Metall. 19 (11), 163–1277.

Rioja, R.J., Laughlin, D.E., 1977. The early stages of G. P. zone formation in Naturally aged Al-4w/o Cu alloys. Met. Trans. 8A, 1257–1261.

Robin, P.Y., 1974. Thermodynamic equilibrium across a coherent interface in a stressed crystal. Am. Mineral. 59, 1286–1298.

Rollins, C.C., 1970. Nucleation on dislocations. Acta. Met. 18, (11), 1209–1215.

Rossiter, P.I., Houghton, M.E., 1984. Contemporary Fe–Cr–Co and Mn–Al–C permanent magnet alloys: a review. Metals Forum 7 (3), 187–208.

Russell, K.C., 1970. Nucleation in solids. In: Phase Transformations. American Society for Metals, ASM Park OH, pp. 219–268.

Sato, T., Tanaka, N., Takahashi, T., 1988. High Resolution Lattice Images of Ordered Structures in Al–Li Alloys. Transactions of the Japan Institute of Metals. 29 (1), 17–25.

Saito, K., Watanabe, R., 1969. Precipitation in Ni-12at.% Ti. Jap. J. Appl. Phys. 83, 14–23.

Shewmon, P., 1963. Diffusion in Solids. Mcgraw-Hill, Ny. NY.

Sinclair, R., Ralph, B., Leake, J.A., 1974. Spinodal decomposition of a nickel–titanium alloy. Phys. Stat. Solidi (A) 26, 285–298.

Singh, A.K., Lele, S., 1991. Ground state structures of ordered binary HCP alloys. Philos. Mag. B 64 (3), 275–297.

Sirotin, Y.I., Shaskolskaya, M.P., 1982. Fundamentals of Crystal Physics. Mir Publishers, Moscow.

Smallman, R.E., 1963. Modern Physical Metallurgy. Butterworths.

Soffa, W.A., Laughlin, D.E., 1982. Recent experimental studies of continuous transformations in alloys: an overview. Proceedings of an International Conference on Solid–Solid Phase Transformations. AIME, 159–183.

Soffa, W.A., Laughlin, D.E., 1989. Decomposition and ordering processes involving thermodynamically first-order order->disorder transformations. Acta Metall. 37, 3019–3029.

Soffa, W.A., Laughlin, D.E., 2004. High-strength age hardening copper-titanium alloys: redivivus. Prog. Mater. Sci. 49, 347–366.

Soffa, W., Laughlin, D.E., Singh, N., 2010. Interplay of ordering and spinodal decomposition in the formation of ordered precipitates in binary fcc alloys: role of second nearest-neighbor interactions. Philos. Mag. 90 (1–4), 287–304.

Soffa, W., Laughlin, D.E., Singh, N., 2011. Re-examination of A1 $\rightarrow$ L1$_0$ ordering: generalized Bragg-Williams model with elastic relaxation. Solid State Phenomena 172, 608–616.

Soffa, W.A., 1985. Structures Resulting from Precipitation from Solid Solution, Metallography and Microstructures. In: ASM Handbook, ASM International, pp. 646–650.

Stanley, H.E., 1971. Introduction to Phase Transitions and Critical Phenomena. Oxford University Press, Oxford and New York.

Stowell, M.J., 2002. Precipitate nucleation: does capillarity theory work? Mater. Sci. Technol. 18 (2), 139–144.

Stubican, V.S., Schultz, A.H., 1970. Phase Separation by Spinodal Decomposition in the Tetragonal System 53 (4), 211–214.

Tammann, G., 1919. Die chemischen und galvanischen Eigenscaften von Mischkristallrein und ihre Atomverteilung. Zeits. F. Anorg. Allgem. Chemie 107 (1/3), 1–239.

Trivedi, R., 1970. The role of interfacial free energy and interface kinetics during the growth of precipitate plates and needles. Met. Trans 1 (4), 921–927.

Trivedi, R., 1975. The effects of crystallographic anisotropy on the growth kinetics of Widmanstatten precipitates. Acta Met 23, 713–722.

Trivedi, R., 1999. Theory of capillarity. In: Aaronson, H.I. (Ed.), Lectures on the Theory of Phase Transformations, second ed. TMS, Warrendale, PA.

Turnbull, D., 1956. Phase changes. In: Seitz, Turnbull (Eds.), Solid State Physics, vol. 3. Academic Press, New York, pp. 226–308.

Turnbull, D., Fischer, J.C., 1949. Rate of nucleation in condensed systems. J. Chem. Phys. 17 (1), 71–73.

Turnbull, D., 1950a. Correlation of liquid-solid interfacial energies calculated from supercooling of small droplets. J. Chem. Phys. 18 (5), 769.

Turnbull, D., 1950b. Isothermal rate of solidification of mercury droplets. J. Chem. Phys. 18 (5), 448, 448.

Turnbull, D., Tu, K.N., 1970. The cellular and pearlite reactions. In: Phase Transformations. American Society for Metals, ASM Park OH.

Van der Waals, J. D. and Kohnatamm, P. (1908). Lehrbuch der Thermodynamik. Maas und Suchtelen.

Veeraraghavan, D., Wang, P., Vasudevan, V.K., 1999. Kinetics and thermodynamics of the alpha->gamma(m), massive transformation in a Ti-47.5 at.% Al alloy. Acta Mater. 47 (11), 3313–3330.

Veeraraghavan, D., Wang, P., Vasudevan, V.K., 2003. Nucleation kinetics of the alpha->gamma(M) massive transformation in a Ti-47.5 at.% Al alloy. Acta Mater. 51 (6), 1721–1741.

Vineyard, G.H., 1957. Frequency factors and isotope effects in solid state rate processes. J. Phys. Chem. Solids 3, 121–127.

Vintaykin, Y.Z., Loshmanov, A.A., 1966. The character of embrittlement in Fe–Cr alloy at 475 °C. Fiz. Metallov. Metalloved 22, 473–476.

Vintaykin, Y.Z., Dmitiyov, V.N., Kolontsov, Y., 1970. Kinetic study of phase separation of Fe–Cr solid solutions by neutron diffraction analysis. Phys. Metals Metall. 29 (6), 1257–1267.

Vlasova, Y.Y., Vintaykin, Y.Z., 1969. Study of the fine structure of FePt alloys. Phys. Met. Metallogr. 27, 60–64.

Volmer, M., Weber, A., 1926. Nuclei-formation in supersaturated alloys. Z. Phys. Chem. 119, 227–301.

Vonnegut, B.J., 1948. Variation with temperature of the nucleation rate of supercooled liquid tin and water drops. Colloid Sci. 3, 563–569.

Voorhees, P.W., Johnson, W.C., 2004. Thermodynamics of elastically stressed crystals. Solid State Phys. 59, 1–201.

Voss, K.J., 1969. Alnico permanent magnet alloys. Chapter IX. In: Berkowitz, A., Kneller, E. (Eds.), Metallurgy and Magnetism, vol. 1. Academic Press, New York and London, pp. 473–512.

Wagner, C., 1961. Theorie der Alterung von Niederschlägen durch Umlösen. Z. Electrochem., 581–591.

Wagner, R., Kampmann, R., 1991. Homogeneous Second Phase Precipitation, in Phase Transformations. In: Haasen, Peter (Ed.), Materials/Materials Science And Technology, Chapter4, vol. 5. VCH Publishers, Weinheim and New York, p. 213.

Wagner, R., Kampmann, 1991. Homogeneous second phase precipitation, In: Cahn, R.W., Haasen, P., Kramer, E.J. (Eds.), Materials, Materials Science and Technology, vol. 5, Phase Transformations in Materials, P. Haasen, (Chapter 4), pp. 213–303.

Wang, P., Veeraraghavan, D., Vasudevan, V.K., 2002. Massive-parent interphase boundaries and their implications on the mechanisms of the alpha -> gamma (M) massive transformation in Ti–Al alloys. Metall. Mater. Trans. 33A (8), 2353–2371.

Wang, Z.M., Shiflet, G.J., 1998. Growth of Delta on dislocations in a dilute Al–Li alloy. Met. Maters. Trans. 29 (8), 2073–2085.

Weiss, R.J., Tauer, K.J., 1956. Components of the thermodynamic functions of iron. Phys. Rev. 102 (6), 1490–1495.

Weiss, P., 1907. L'Hypothèse du champ moléculaire et du propriété ferromagnétique. J. Physique 6, 661–690.

Wendt, H., 1981. Ph.D. Thesis, Univ.Goettingen.

Wendt, H., Haasen, P., 1983. Nucleation and growth of γ'-precipitates in Ni-14 at-percent Al. Acta Metall. 31, 1649–1659.

Wert, C., Zener, C., 1950. Interference of growing spherical precipitate particles. J. Appl. Phys. 21 (1), 5–8.

Wilm, A., 1911. Physikalisch-metallurgische Untersuchungen über magnesiumhaltige Aluminiumlegierungen. Metallurgie: Z. gesamte Hüttenkunde 8 (8), 225–227.

Wilkinson, D.S., 2000. Mass Transport in Solids and Fluids. In: Cambridge Solid State Science Series. Cambridge University Press.

Williams, D.B., Butler, E.P., 1981. Grain boundary discontinuous precipitation reactions. Int. Metals Rev. 26 (3), 153–183.

Williams, R.O., Paxton, H.W., 1957. The nature of aging of binary iron chromium alloys around 500C. J. Iron Steel Inst. 185, 358–374.

Williams, R.O., 1980. Long-period superlattices in the copper–gold system as two-phase systems. Metall. Trans. 11A, 247–253.

Williams, R.O., 1984. Calculation of coherent phase equilibria. CALPHAD 8 (1), 1–14.

Wittig, J.E., 2002. Mater. Trans. A (2002), vol. 33A, p. 2373. The massive transformation in titanium aluminides: Initial stages of nucleation and growth. Met. Mater. Trans. A 33 (8), 2373–2379.

Yanar, C., Wiezorek, J.M.K., Radmilovic, V., Soffa, W.A., 2002. Massive transformation and the formation of the ferromagnetic L1(0) phase in manganese–aluminum-based alloys. Met. Mater. Trans. A 33 (8), 2413–2423.

Zackay, V.F., Aaronson, H.I., 1962. The Decomposition of Austenite by Diffusional Processes. TMS of AIME, Interscience Publishers, NY.

Zeldovich, J.B., 1943. On the Theory of New Phase formation.

Zener, C., 1946. Kinetics of the decomposition of austenite. Trans. AIME 167, 550–595.

Zener, C., 1955. Impact of magnetism upon metallurgy. Trans. AIME 203, 619–630.

Zhao, J.C., Notis, M.R., 1998. Spinodal decomposition, ordering transformation and discontinuous precipitation in a Cu–15Ni–8Sn alloy. Acta Mater. 46 (12), 4203–4218.

Further Reading

Aaronson, H.I. (Ed.), 1999. Lectures on the Theory of Phase Transformations. TMS, Warrendale PA.

Aaronson, H.I., Enomoto, E., Lee, J.K., 2010. Mechanisms of Diffusional Phase Transformations in Metals and Alloys. CRC Press, Baca Raton.

Balluffi, R.W., Allen, S.M., Carter, W.C., 2005. Kinetics of Materials. Wiley, Hoboken, NJ.

Banerjee, S., Mukhopadhyay, P., 2007. Phase Transformations: Examples from Titanium and Zirconium Alloys. ELSEVIER, Amsterdam.

Burke, J., 1965. The Kinetics of Phase Transformations in Metals. Pergamon Press, Oxford.

Christian, J.W., 2002. The Theory of Transformation in Metals and Alloys. Pergamon Press.

Fine, M.E., 1964. Phase Transformations in Condensed Systems. Macmillian, NY.

Kelton, K.F., Greer, A.L., 2010. Nucleation in Condensed Matter. In: Pergamon Materials Series. Elsevier, Oxford.

Khachaturyan, A.G., 1983. Theory of Structural Transformations in Solids. John Wiley & Sons, NY, NY.

Kostorz, G. (Ed.), 2001. Phase Transformations in Materials. Wiley-VCH, Weinhem.

Machlin, E.S., 1991. An Introduction to Aspects of Thermodynamics and Kinetics Relevant to Materials Science. Giro press, Croton-on-Hudson, NY.

Martin, J.W., Doherty, R.D., Cantor, B., 1997. Stability of Microstructure in Metallic Systems. Cambridge University Press.

Rao, C.N.R., Rao, K.J., 1978. Phase Transitions in Solids. McGraw-Hill, NY.

Sharma, R.C., 2002. Phase Transformations in Materials. CBS Publishers, India.

Shewmon, P.G., 1969. Transformations in Metals. McGraw–Hill, NY, NY.

Wagner, R., Kampmann, L., 1991. Homogeneous second phase precipitation. In: Cahn, R.W., Haasen, P., Kramer, E.J. (Eds.), Phase Transformations in Materials, Materials Science and Technology, Phase Transformations in Materials (Volume Editor, P. Haasen), vol. 5, pp. 213–303 (Chapter 4).

Conferences/Symposia on Phase Transformations (in chronological order)

Age hardening of Metals, 1940. American Society for Metals. Cleveland, OH.

Smoluchowski, R., Mayer, J.E., Wehl, W.A. (Eds.), 1951. Phase Transformations in Solids. John Wiley and Sons, London.

The Mechanism of Phase Transformations in Metals, 1955. Institute of Metals, London.

Precipitation from Solid Solution, 1959. American Society for Metals. Cleveland, OH.

Zackay, V.F., Aaronson, H.I., 1962. The Decomposition of Austenite by Diffusional Processes. TMS of AIME, Interscience Publishers, NY.

Phase Transformations, 1969. American Society for Metals. ASM Park, OH.

The Mechanism of Phase Transformations in Crystalline Solids. Monograph and Report Series No. 33. 1969. Institute of Metals, London.

Russell, K.C., Aaronson, H.I., 1978. Precipitation Processes in Solids. Metallurgical Society of AIME, Warendale, PA.

Solid State Phase Transformations in Metals and Alloys: Ecole d'été d'Aussois, 3–15 septembre 1978, 1980. Orsay: Editions de Physique.

Aaronson, H.I., Laughlin, D.E., Sekerka, R.F., Wayman, C.M., 1982. Proceedings of an International Conference on Solid [to] Solid Phase Transformations. Metallurgical Society of AIME, Warrendale, Pa.

Hassen, P., Gerold, V., Wagner, R., Ashby, M.F., 1984. Decomposition of Alloys: Early Stages. Pergamon Press, Oxford.

Lorimer, G.W., 1988. Phase Transformations '87. Institute of Metals, London.

Johnson, W.C., Howe, J.M., Laughlin, D.E., Soffa, W.A. (Eds.), 1994. Proceedings of an International Conference on Solid [to] Solid Phase Transformations. Metallurgical Society of AIME, Warrendale, Pa.

Koiwa, M., Otsuka, K., Miyazaki, T., 1999. Solid [to] solid phase transformations. Jpn. Inst. Metals Proc. 12 (I & II).

Howe, J.M., Laughlin, D.E., Lee, J.K., Dahman, U., Soffa, W.A., 2005. Proceedings of an International Conference on Solid [to] Solid Phase Transformations in Inorganic Materials. Metallurgical Society of AIME, Warrendale, Pa.

Brechet, Y., Clouet, E., Deschamps, A., Finel, A., Soisson, F., 2011. Solid [to] Solid Phase Transformations in Inorganic Materials. Trans tech Publications, Zurich.

Biography

Professor Soffa received his B.S. in Metallurgy/Metallurgical Engineering from Carnegie Institute of Technology (CMU) in 1961 and his M.S. in Materials Engineering from Rensselaer Polytechnic Institute (RPI) in 1963. He completed his Ph.D. in Metallurgical Engineering/Physical Metallurgy at Ohio State University in 1967 after which he was granted a N.A.T.O Postdoctoral Fellowship to study at Oxford University in the Department of Metallurgy and Materials Science. Following his post-doctoral work at Oxford he joined the faculty in the Department of Materials Science and Engineering at the University of Pittsburgh in 1968 where he taught and did research for over thirty years. He was Department Chair at Pitt from 1995–2000. He was a Visiting Humboldt Scholar at the Institut fur Metallphysik at the University of Goettingen in 1979 and a Visiting Professor in the Department Materials Science at Berkeley (U of Cal) in 1987. While at Pitt he was also Director of the Engineering Physics Program. In 2000–2001 Professor Soffa was a Visiting Professor in the Department of Physics at the University of Vienna and at the Department of Applied Physics/ ETH-Zurich. In 2004 he joined the Department of Materials Science and Engineering at the University of Virginia becoming Professor Emeritus in 2012. Professor Soffa has published over 100 papers in archival journals and refereed conference proceedings in the areas of phase transformations, magnetic materials and materials physics.

David E. Laughlin is the ALCOA Professor of Physical Metallurgy in the Department of Materials Science and Engineering at Carnegie Mellon University, Pittsburgh, PA. He obtained his B.S. in Metallurgical Engineering from Drexel University in 1969 and his Ph.D. in Metallurgy and Materials Science from MIT in 1973. He has taught at CMU since 1974. He is Principal Editor of *Metallurgical and Materials Transactions* and has co-edited eight books. His research has centered on the structure of materials as observed by electron microscopy, phase transformations and magnetic materials. He has published more than 450 peer reviewed research papers and is co-inventor on eleven US patents. Laughlin is a Fellow of TMS and ASM International.

9 Phase Transformations: Nondiffusive

H.K.D.H. Bhadeshia, University of Cambridge, UK
C.M. Wayman, Deceased but originally from University of Illinois at Urbana-Champaign, USA

This chapter deals with phase changes where there is no long-range movement of atoms, with the change in crystal structure achieved by a deformation of the parent lattice into that of the product. There is no diffusional flow in this process, with atoms moving less than an interatomic distance and retaining their relationship with their neighbors during the phase change. The prototype of this behavior is a martensitic transformation, and the transformations of this kind are frequently referred to as "displacive", "shear-like" or "diffusionless." A vast amount of geometrical or crystallographic information on martensitic transformations has accrued over the years, and we shall see that all martensitic transformations have certain common crystallographic characteristics. These characteristics help distinguish martensitic reactions and have led to a successful phenomenological theory of martensite crystallography, which will be discussed in this chapter, along with a consideration of the thermodynamics and kinetics of the transformation. As will be seen, there are other phase changes that although not commonly recognized as martensitic are described well by the martensite crystallography theory. Finally, some other types of diffusionless, displacive solid-state transformations such as omega phase and charge-density wave formation are considered, both of which involve lattice distortions with correlated atomic displacements.

9.1 Martensitic Transformations

9.1.1 Introduction and General Characteristics

The name *martensite* is after the German scientist Adolf Martens. It was used originally to describe the hard constituent found in quenched steels. Many materials other than steel are now known to exhibit the same type of solid-state phase transformation, known as a *martensitic transformation*, frequently also called a *shear* or *displacive transformation*. Martensite occurs in, for example, nonferrous alloys, pure metals, ceramics, minerals, inorganic compounds, solidified gases and polymers (**Table 1**), although it is undoubtedly of greatest technological importance in metallic systems.

It has long been recognized that martensitic transformations are diffusionless; this is why martensite can grow at temperatures below 100 K where the mobility of atoms is negligible. It is not implied that all martensitic transformations occur at low temperatures; indeed, many of them occur at

Table 1   The temperature M_S at which martensite first forms on cooling, and the approximate Vickers hardness of the resulting martensite for a number of materials

Composition	M_S/K	Hardness HV
ZrO_2	1200	1000
Fe–31Ni–0.23C wt%	83	300
Fe–34Ni–0.22C wt%	<4	250
Fe–3Mn–2Si–0.4C wt%	493	600
Cu–15Al	253	200
Ar–40N$_2$	30	

comparatively high temperatures (**Table 1**), but these too are diffusionless. Martensite can also grow at speeds in excess of 1000 m s^{-1}, speeds that are inconsistent with the diffusion of atoms. In most cases, the amount of martensite that is obtained depends on the transformation temperature rather than the time at that temperature, in which case the transformation is called *athermal*. As pointed out by Christian (2003a, 2003b, 1979), martensitic transformations are thermodynamically of first order, i.e. they involve nucleation and growth so that the parent and product phases can be observed to coexist. It follows that the overall kinetics of a martensitic reaction depend on both the nucleation and growth stages and will largely be dominated by the slower of the two. For example, slow thermal nucleation may give rise to isothermal transformation characteristics. As will be seen later, the martensitic transformations do not obey classical nucleation theory associated with heterophase fluctuations.

The interface between the martensite and its parent has to have a structure that is consistent with the absence of diffusion and with the rapid rate of growth. A glissile interface like this does not require significant thermal activation in order to move, and must be coherent or semi-coherent, depending upon the crystallography of the particular material undergoing transformation. With a few exceptions, the interface is semi-coherent and the parent and product lattices are coherently accommodated only over local regions of the boundary. The accumulating misfit is then relieved periodically by other auxiliary deformation processes in such a way that the interface remains glissile. On the other hand, e.g. the fcc → hcp transformation in iron and cobalt alloys, the martensite/parent interface is fully coherent, and the deformation it causes when it moves (normal to itself) is sufficient to accomplish the required lattice change. In contrast, the macroscopic deformation caused by the movement of a semi-coherent martensite interface is related in a more complex way to the strain, which converts the parent to the product lattice. More discussion of interfaces will follow later, but for now it is emphasized that both semi-coherent and coherent martensite–parent interfaces must be glissile.

Martensitic transformations are most readily distinguished from other solid-state phase changes on the basis of their crystallographic characteristics, which imply a "military" or coordinated transfer of atoms from the parent to the product phase. In a "civilian" transformation the atoms move across the phase boundary in an uncoordinated manner by diffusion. Martensitic transformations feature a coordinated structural change involving a lattice correspondence and an ideally planar parent–product interface that during movement produces an invariant-plane strain shape deformation. A fine-scale inhomogeneity in the martensite, such as slip, twinning or faulting, is usually observed at the electron microscope scale. This secondary deformation, an intrinsic part of the transformation process, produces the invariant-plane condition at the macroscopic scale and provides a semi-coherent glissile interface between the martensite and the parent phase. Crystallographic features between the martensite and the parent phase such as the habit (invariant) plane and orientation relationship are

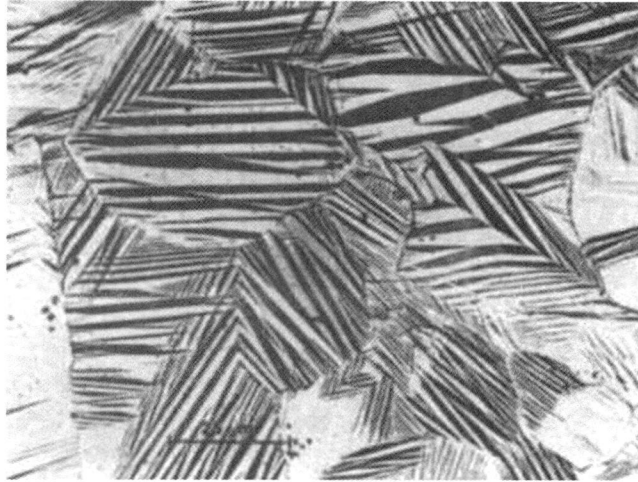

Figure 1 Optical micrograph showing surface relief due to a martensitic transformation in an Fe-24.5%Pt alloy. The specimen was polished to a flat condition at room temperature and then cooled to produce the martensite.

usually not expressible in terms of exact relations involving integral Miller indices. The various crystallographic features of martensitic transformations will be discussed and illustrated with representative experimental examples. After this, a brief account of the development of the phenomenological crystallographic theory will be presented, followed by an algebraic analysis.

9.1.2 Observations of Crystallographic Features

Figure 1 is an optical micrograph of a polycrystalline Fe–24.5Pt at.% alloy taken after transformation of the parent phase into martensite by cooling. The specimen was polished flat prior to transformation to martensite. The observed contrast results purely from deformation caused by the formation of martensite. The light and dark shades correspond to regions of martensite that have undergone distortion in different senses with respect to the initial surface. This surface tilting or macroscopic distortion is known as the *shape deformation* or *shape strain*. It can be demonstrated by serial sectioning that the martensite has a three-dimensional shape, which is that of lenticular plates. Crystallographic analysis would reveal that the plates showing different optical contrast also feature different lattice orientations with respect to the initial parent grain. X-ray analysis would also show that a structural change had occurred, which for this Fe-Pt alloy is an fcc → bcc transformation.[1] The differently oriented plates are crystallographic variants of the habit plane and orientation relationship.

Referring again to **Figure 1**, if straight scratches were purposely abraded on a flat specimen still in its parent phase, then the formation of martensite would displace the scratches in a characteristic manner. After analyzing several nonparallel scratches crossing a given plate and noting the initial position of the same scratches in the untransformed parent phase, a distortion matrix that describes the shape

---

[1] The terms fcc, bcc, bct and hcp are abbreviations for face-centered cubic, body-centered cubic, body-centered tetragonal and hexagonal close-packed, respectively.

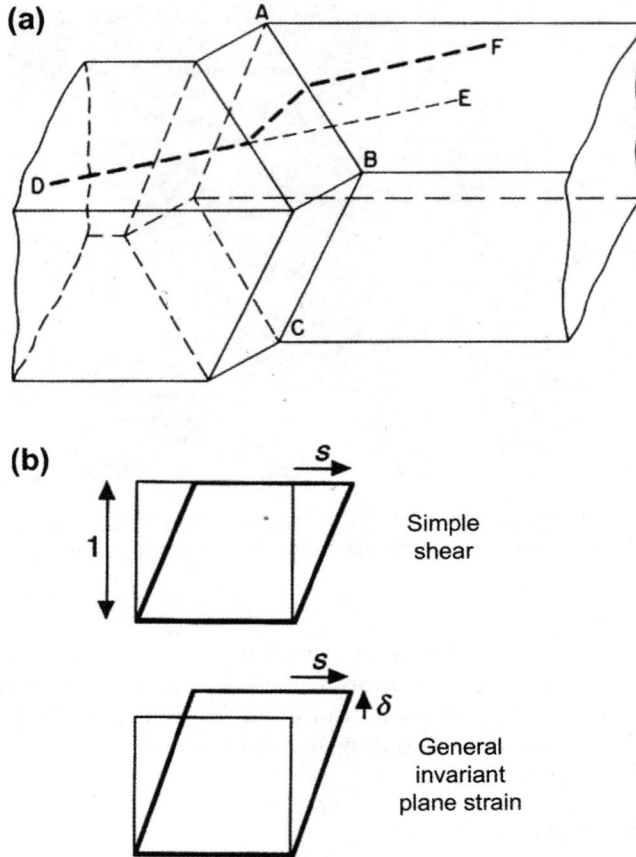

Figure 2 (a) An invariant-plane strain characteristic of martensitic transformation. The initially straight line DE is displaced to the position DF when the martensite plate with habit plane ABC is formed. The plane ABC is invariant (unrotated and undistorted) as a result of the formation of martensite. (b) An illustration of the difference between a simple shear and a general invariant-plane strain. s is the shear component of the strain and δ the dilatational component of strain. The dilatational strain defines the volume change of transformation.

deformation can be derived, which shows that a deformation that is homogeneous on an optical scale has taken place. This shape deformation resembles a simple shear, but in general is better described as an invariant-plane strain since the deformation may also include a volume change normal to the invariant plane (**Figure 2**). One plane is left undistorted and unrotated; this invariant plane is also known as the *habit plane*.

An invariant-plane strain is a homogeneous distortion such that the displacement of any point is in a common direction. The magnitude of the displacement is proportional to the distance from a fixed plane of reference, which is the invariant-plane (itself unaffected by the strain). In most martensitic transformations a volume change accompanies the structural change, which produces a normal component δ to the invariant-plane strain.

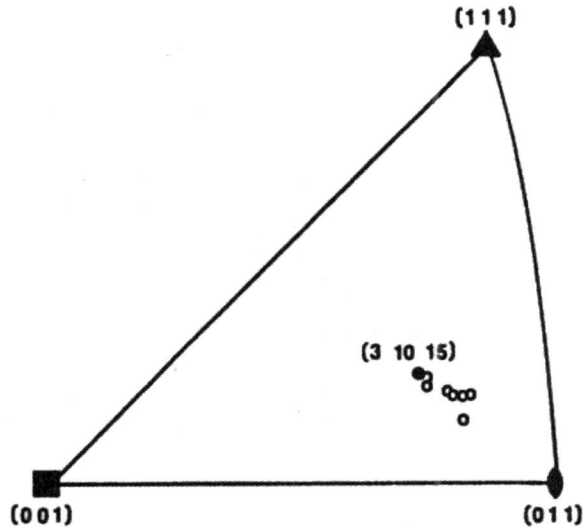

Figure 3 Unit triangle of stereographic projection showing experimental habit planes (open circles) for martensite formed in Fe–7Al–1.5C wt% alloy. The average habit plane cannot be expressed in terms of simple Miller indices, but is near the plane {3 10 15}$_P$.

It is well established that the shape deformation caused by the growth of martensite is an invariant-plane strain. Nevertheless, it is also true that for many martensitic transformations, an invariant-plane strain cannot convert the crystal structure of the parent into that of the product phase. In iron alloys the martensitic transformation converts an fcc parent into a bcc (or bct) product, but the measured shape deformation matrix when applied to the parent fcc will not produce a bcc structure. This apparent inconsistency will be explained later.

Martensite plates in a given alloy usually possess a unique habit plane, as shown for example in **Figure 3**. The stereographic projection shows the experimentally determined habit plane poles for different martensite plates, which, as can be seen, cluster near the plane in the parent phase with Miller indices (3 10 15)$_P$ (subscripts P and M, respectively, are used to designate the parent and martensite).

Early work suggested that, apart from a small relative rotation of corresponding unit cells in the parent and product phases, a homogeneous, lattice distortion would account for the known structural change in martensitic transformations. Bain (1924) suggested that the austenite (parent phase) → martensite transformation in steels could be explained by a homogeneous "upsetting" of the parent fcc lattice into the required bcc (or bct) lattice, **Figure 4**. There are in fact many ways to generate a bcc product from an fcc parent by means of a homogeneous distortion, and another possible correspondence is shown in **Figure 5**, but analysis shows that the Bain deformation (often called the *Bain strain*) involves the smallest principal strains and should therefore be favored. The lattice correspondence, a unique relationship between any lattice point in the initial lattice and the point it becomes in the final lattice, implied by the Bain distortion has also been verified experimentally using an ordered Fe$_3$Pt alloy, which undergoes nominally an fcc to bcc transformation. By observing corresponding super-lattice reflections in the parent and martensitic phases by means of transmission electron diffraction it was deduced that the Bain correspondence actually applies (Tadaki and Shimizu, 1970).

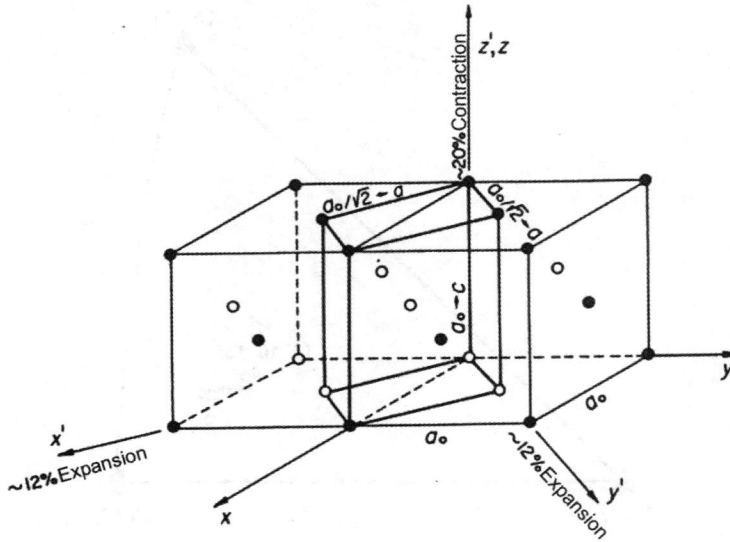

Figure 4 Lattice distortion and correspondence proposed by (Bain, 1924) for the fcc–bcc (bct) martensitic transformation in iron alloys. The correspondence related cell in the parent phase (indicated by bold lines) becomes a unit cell in the martensite as a consequence of a homogeneous "upsetting" with respect to the z' axis.

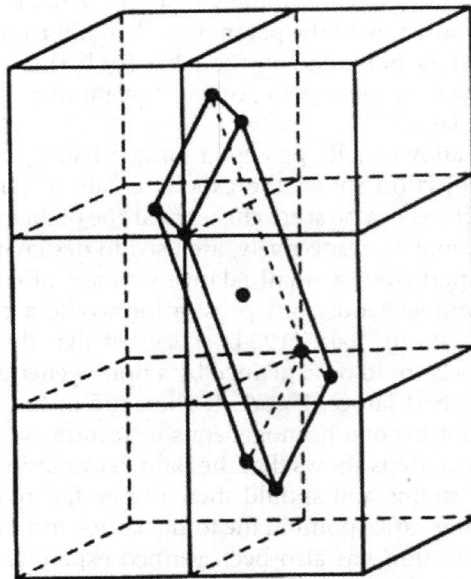

Figure 5 Alternative lattice correspondence for the fcc–bcc (bct) martensitic transformation, which involves larger principal distortions than the Bain correspondence.

From the lattice correspondence shown in **Figure 4** one would expect for example, $[001]_M \parallel [001]_P$, $[010]_M \parallel [110]_P$, $(112)_M \parallel (101)_P$, $(011)_M \parallel (111)_P$, etc. However, such exact parallelisms are not observed. For the Fe–Pt alloy shown in **Figure 1** the following orientation relationship was observed:

$$[001]_P - [001]_M : 9.10° \quad \text{apart},$$

$$[\overline{1}01]_P - [\overline{1}\,\overline{1}1]_M : 4.42° \quad \text{apart},$$

$$(111)_P - (011)_M : 0.86° \quad \text{apart},$$

which is typical of the orientation relationships found in iron alloys and steels. Further consideration of the above orientation relationship shows that the correspondence cell is not only distorted according to the Bain strain, but also is rotated (about 10° from $[001]_P$ toward $[110]_P$). This is termed a *rigid body rotation*. The combined distortion-rotation is known as the *lattice deformation*. Additional examination of the above orientation relationship shows that the close-packed planes $(111)_P$ and $(011)_M$ are nearly parallel to each other and the close-packed directions $[\overline{1}01]_P$ and $[\overline{1}\,\overline{1}1]_M$ are a few degrees apart. This is an example of the *Greninger–Troiano* orientation relationship. When the close-packed planes and directions in two coexisting structures are almost parallel, the *Kurdjumov–Sachs* orientation relationship is obtained.

In summary, the Bain strain correctly transforms the crystal structure of the austenite into that of martensite, and when combined with an appropriate rigid body rotation leads to the correct orientation relationship. In fact, the combined lattice deformation (Bain + rotation) is an *invariant-line strain*, which leaves a single line unrotated and undistorted. This line lies in the interface between the austenite and martensite, and permits the interfacial structure to be semi-coherent and glissile. Hence, two crystals can only transform into one another by a martensitic mechanism if they can be related by a lattice deformation that is at the very least an invariant-line strain.

9.2   Crystallographic Theory

We have seen that the Bain strain (**B**) can transform the parent into the product crystal structure with minimal atomic displacements. In itself the Bain strain alters every vector, but it can in combination with a rigid body rotation (**R**) be converted into a *lattice deformation* that is an invariant-line strain (**RB**), which also yields the experimentally observed orientation relationship. However, this invariant-line strain is inconsistent with the experimentally established invariant-plane strain shape deformation (**P₁**).

The reason for this is as follows. Referring to **Figure 4**, the correspondence cell is contracted ($\sim 20\%$) along the z' axis and expanded the same amount ($\sim 12\%$) along the x' and y' axes. Such a homogeneous distortion will leave no plane invariant (i.e. undistorted and unrotated). But suppose that the distortion along y' vanished. One of the principal distortions would then be less than unity (along z'), one greater than unity (along x') and the remaining one (along y') exactly unity. These conditions imply the distortion of an initial sphere (parent phase) into a triaxial ellipsoid (martensite) following which the sphere and ellipsoid can fit together along an undistorted plane of contact (habit plane) as shown in **Figure 6**. But this special set of conditions is not generally found in practice; the principal distortions are determined by the lattice correspondence and observed lattice parameters of the two phases.

The apparent inconsistency described above is resolved by envisioning a *lattice-invariant deformation* $\overline{P}$ involving no structural change, such as slip or twinning, to occur in conjunction with the lattice deformation **RB**. The additional deformation must be lattice-invariant because the necessary structural

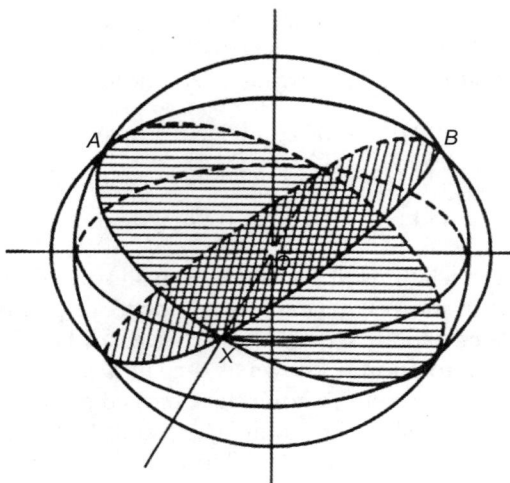

Figure 6 The Bain distortion has a mathematical analog wherein an initial unit sphere is distorted into an ellipsoid. If one of the principal distortions is unity and the other two are of opposite sign, then the sphere and the ellipsoid can fit together along the undistorted planes *AOX* and *BOX*.

change is effected by the Bain distortion alone. The role of the additional deformation is essentially to shear the ellipsoid resulting from the Bain distortion into tangency with the initial sphere; then one of the principal distortions (OX in **Figure 6**) becomes unity and an undistorted contact plane exists. This additional deformation (slip, twinning or faulting) being a shear is known as the *inhomogeneous shear* or *complementary shear* of the crystallographic theory.

These are the three phenomenological steps $\mathbf{B}$, $\mathbf{R}$, $\overline{\mathbf{P}}$ describing the total transformation crystallography. There is no time sequence implied as to which step occurs when. Of course the combined effect of these three operations must be equivalent to the shape deformation.

Working with the theory just described, different crystallographic features such as the habit plane and orientation relationship can be predicted by supposing the inhomogeneous shear to occur on different crystallographic planes and directions. For example, in most iron alloys an inhomogeneous shear on the $(112) - [\overline{1}\,\overline{1}1]_M$ twinning system will predict a habit plane near $(3\ 15\ 10)_P$, but assuming the inhomogeneous shear system to be $(011) - [\overline{1}\,\overline{1}1]_M$ predicts a habit plane near $(111)_P$. Since the lattice parameters and correspondence are usually known, the Bain distortion for a given transformation is specified, and the flexibility in the theory comes through different suppositions concerning the plane and direction of the inhomogeneous shear. Once the inhomogeneous shear system is assumed, a shear of a certain magnitude will produce an undistorted plane, which when rigidly rotated to its original position becomes the invariant habit plane.

The discussion and examples cited above have centered around iron alloys. Because of the importance of steels there has been substantial work on them and thus an abundance of experimental data exists. However, the principles presented are quite general and apply to all martensitic transformations.

9.2.1 Summary of Crystallographic Theory

The Bain strain converts the structure of the parent phase into that of the product phase. When combined with an appropriate rigid body rotation, the net homogeneous lattice deformation $\mathbf{RB}$ is an

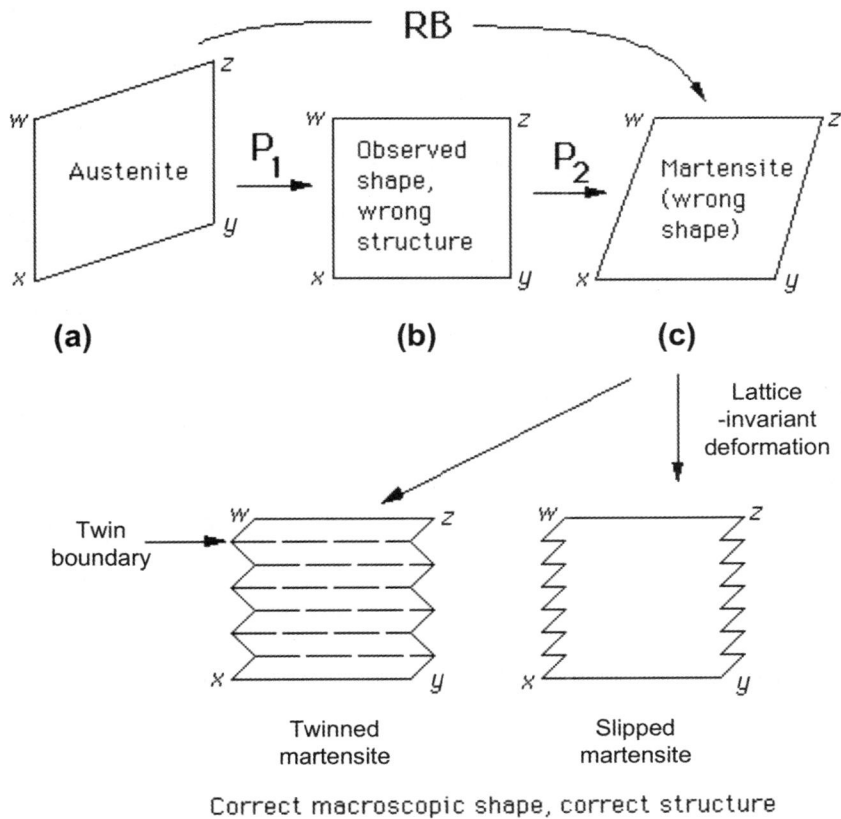

Figure 7   Schematic illustration of the phenomenological theory of martensite crystallography (Bhadeshia, 2006).

invariant-line strain (step *a* to *c* in **Figure 7**). However, the observed shape deformation is an invariant-plane strain P_1 (steps *a* to *b* in **Figure 7**), but this gives the wrong crystal structure. If, however, a second homogeneous shear P_2 is combined with P_1 (step *b* to *c*), then the correct structure is obtained but the wrong shape since

$$P_1 P_2 = RB$$

These discrepancies are all resolved if the shape changing effect of P_2 is canceled macroscopically by an inhomogeneous lattice-invariant deformation, which may be slip or twinning as illustrated in **Figure 7**.

The theory elegantly explains all the observed features of the martensite crystallography. It is easy to predict the orientation relationship, by deducing the Bain strain and adding a rigid body rotation that makes the net lattice deformation an invariant-line strain. The habit plane does not have rational indices because the amount of lattice-invariant deformation needed to recover the correct the macroscopic shape is not usually rational. The theory predicts a substructure in plates of martensite (either twins or slip steps) as is observed experimentally. The transformation goes to all the trouble of ensuring that the shape deformation is macroscopically an invariant-plane strain because this reduces the strain energy when compared with the case where the shape deformation might be an invariant-line strain. We now proceed to discuss the details of this theory.

9.2.2   The Inhomogeneous Shear and Martensite Substructure

The inhomogeneous shear was introduced in the crystallographic description of martensite transformations to ensure that the habit plane is macroscopically undistorted. **Figure 8** is a schematic representation of the appearance of internally twinned and internally slipped martensite plates. Although there are localized distortions at the interface, the "saw tooth" effect because of alternating twins (called transformation twins) or slip lamellae prevents the accumulation of any strain at the interface over large distances.

For internally twinned martensite the Bain distortion is envisioned to occur along different contraction axes in the two twin-related regions, and the twinning plane in the martensite is derived from a mirror plane in the parent. In the case of internally slipped martensite, the Bain distortion is the same in all regions of a plate. **Figure 9** shows that the same effective shear angle γ can be accomplished by slip or twinning. Note that the relative thickness of the two twin components determines the angle γ.

Because of the inhomogeneous shear mentioned above, one would expect to observe some kind of substructure in the martensite. Such observations have been made since the introduction of the theory. **Figure 10** is a transmission electron micrograph of a martensite plate in an Fe–Ni–C alloy. The fine striations crossing the plate are transformation twins ($\{112\}_M$ twinning plane). Regions adjacent to the martensite are retained austenite. If the twin plane is indexed specifically as $(112)_M$, the habit plane

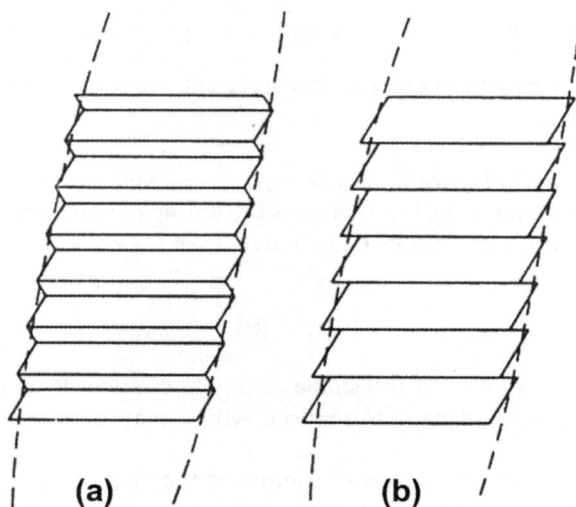

Figure 8   Schematic representations of the inhomogeneous shear in a martensitic transformation involving (a) internally twinned, and (b) internally slipped plates of martensite. The serrated effect at the interface (habit plane) in each case prevents the long-range accumulation of strain and consequently the interface remains macroscopically invariant though, in general, irrational.

(a) **(b)**

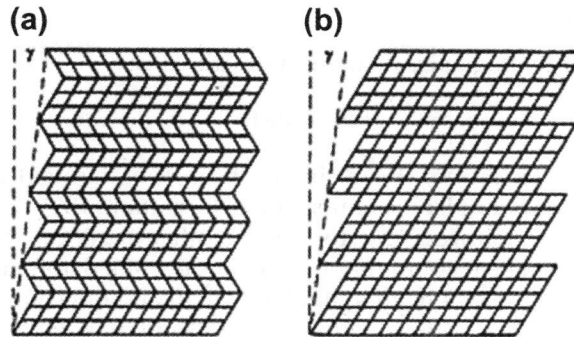

Figure 9 As shown for twinning (a) and slip (b) the same magnitude of the inhomogeneous shear angle can be accomplished by either mechanism of heterogeneous deformation.

becomes specifically $(3\ 15\ 10)_P$. If $(112) - [\bar{1}\ \bar{1}\ 1]_M$ twinning is used in the crystallographic calculations, the predicted habit plane also is $(3\ 15\ 10)_P$. Thus the particular variant of the twin plane is found to be consistent with the specific variant of the habit plane, and for the Fe–Ni–C alloy the experimental observations are in excellent agreement with those features that are predicted using the phenomenological crystallographic theory. This is also the case for many other martensitic transformations that have been studied in some detail using transmission electron microscopy and diffraction. Notable exceptions, however, are certain steels that transform to martensite with a $\{225\}_P$ habit plane. The martensite in these materials has a complex substructure, as seen in the electron microscope, and is not very well explained using the theory described above.

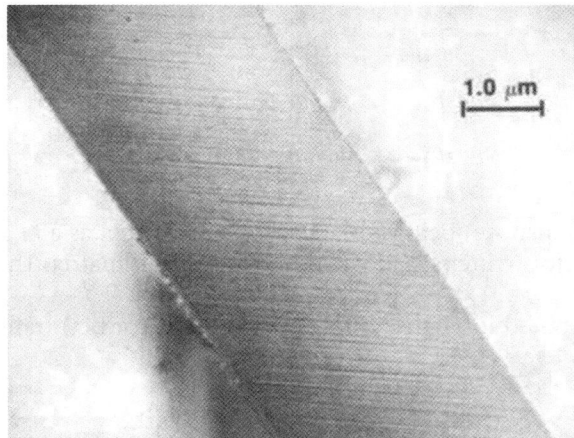

Figure 10 Transmission electron micrograph showing a martensite plate and adjacent retained austenite in an Fe–30Ni–0.4C wt% alloy. The striations within the plate running from left to right are transformation twins.

9.2.3   Mathematical Description

The basic equation of the crystallography theory just described is:

$$\mathbf{P}_1 = \mathbf{R}\overline{\mathbf{P}}\mathbf{B} \qquad (1)$$

where $\mathbf{B}$ represents the Bain distortion, $\overline{\mathbf{P}}$ is a simple shear, $\mathbf{R}$ is the rigid body rotation previously mentioned, and $\mathbf{P}_1$ is the invariant-plane strain shape deformation. $\mathbf{B}, \overline{\mathbf{P}}, \mathbf{R}$ and $\mathbf{P}_1$ can all be represented as (3×3) deformation matrices. A vector, on the other hand, can be represented by a 1×3 single-row matrix or a 3×1 single-column matrix. The product of a deformation matrix with a vector gives a new vector that may in general be of a different length and direction.

The matrix product $\mathbf{R}\overline{\mathbf{P}}\mathbf{B}$ is equivalent to the shape deformation $\mathbf{P}_1$ and the rotation matrix $\mathbf{R}$ rotates the plane left undistorted by $\overline{\mathbf{P}}\mathbf{B}$ to its original position. That is, $\mathbf{P}_1 = \mathbf{R}\overline{\mathbf{P}}\mathbf{B}$ is an invariant-plane strain.

Although Eqn (1) indicates that the inhomogeneous shear $\overline{\mathbf{P}}$ follows the Bain distortion, the same end result can be obtained by "allowing" the shear to occur in the parent phase prior to the Bain distortion. This alternative procedure simplifies computations. In this case the basic equation becomes:

$$\mathbf{P}_1 = \mathbf{R}\mathbf{B}\mathbf{P}$$

where $\mathbf{P}$ as before is a simple shear.

The matrix representation of the invariant-plane strain shape deformation is:

$$\mathbf{P}_1 = \mathbf{I} + m\mathbf{d}\mathbf{p}' \qquad (2)$$

$$= \begin{pmatrix} 1 & 0 & 0 \\ 0 & 1 & 0 \\ 0 & 0 & 1 \end{pmatrix} + m[d_1 \quad d_2 \quad d_3](p_1 \quad p_2 \quad p_3)$$

$$= \begin{pmatrix} 1 + md_1p_1 & md_1p_2 & md_1p_3 \\ md_2p_1 & 1 + md_2p_2 & md_2p_3 \\ md_3p_1 & md_3p_2 & 1 + md_3p_3 \end{pmatrix} \qquad (3)$$

where $\mathbf{p}'$ (prime meaning transpose) being a plane normal is written as a (1×3) single-row matrix, in contrast to $\mathbf{d}$, which is a vector written as a (3×1) single-column matrix. The magnitude of the shape deformation is m.

The Bain distortion illustrated in **Figure 4**, for the fcc $\rightarrow$ bcc (or bct) transformation, can be represented as:

$$\mathbf{B} = \begin{pmatrix} \sqrt{2}a/a_0 & 0 & 0 \\ 0 & \sqrt{2}a/a_0 & 0 \\ 0 & 0 & c/a_0 \end{pmatrix}. \qquad (4)$$

For typical values of the lattice parameters of the fcc lattice, a, and of the bct lattice, a_0 and c, the matrix **B** becomes

$$\mathbf{B} = \begin{pmatrix} 1.12 & 0 & 0 \\ 0 & 1.12 & 0 \\ 0 & 0 & 0.8 \end{pmatrix} \tag{5}$$

It is apparent from Eqn (5) that two of the principal distortions are greater than unity (i.e. 1.12) and the third is less than unity (i.e. 8). Therefore, the Bain distortion on its own cannot leave a plane invariant; as stated earlier, for a strain to be an invariant-plane strain, one of the principal distortions must be zero, one greater than unity and the other less than unity.

Referring back to Eqn (1), it is noted that **P** is a simple shear, and therefore can be represented as $(\mathbf{I} + m\mathbf{dp}')$. Also, the inverse of **P**, $\mathbf{P}^{-1} = (\mathbf{I} - m\mathbf{dp}')$ is a simple shear of the same magnitude, on the same plane, but in the opposite direction. That is, both **P** and $\mathbf{P}^{-1}$ are invariant-plane strains. It is now convenient to rewrite Eqn (1) as:

$$\mathbf{P}_1\mathbf{P}_2 = \mathbf{RB}, \tag{6}$$

where $\mathbf{P}_2 = \mathbf{P}^{-1}$. Since both $\mathbf{P}_1$ and $\mathbf{P}_2$ are invariant-plane strains, their product **RB** is and invariant-*line* strain **S**. the invariant-line is defined as the line of intersection of the *planes*, which are invariant to $\mathbf{P}_1$ and $\mathbf{P}_2$. If the invariant-line strain **S** is known, all the crystallographic features of a given martensitic transformation can be predicted. The Bain correspondence and distortion **B** are known from the lattice parameters of the parent and martensite phases. **R** is determined once the plane $\mathbf{p}_2'$ and the direction $\mathbf{d}_2$ of $\mathbf{P}_2$ are assumed. It should be noted that $\mathbf{RB} = \mathbf{S}$ constitutes the lattice deformation of the transformation.

Noting that the shape strain is $\mathbf{P}_1 = \mathbf{I} + m_1\mathbf{d}_1\mathbf{p}_1'$ and that the simple shear is $\mathbf{P}_2 = \mathbf{I} + m_2\mathbf{d}_2\mathbf{p}_2'$, it can be shown that:

$$\mathbf{d}_1 = [\mathbf{Sy}_2 - \mathbf{y}_2]/\mathbf{p}_1'\mathbf{y}_2, \tag{7}$$

$$\mathbf{p}_1' = (\mathbf{q}_2' - \mathbf{q}_2'\mathbf{S}^{-1})/\mathbf{q}_2'\mathbf{S}^{-1}\mathbf{d}_1, \tag{8}$$

where $\mathbf{y}_2$ is any vector lying in $\mathbf{p}_2'$ (except the invariant line **x**) and $\mathbf{q}_2'$ is any normal (other than $\mathbf{n}'$, the row eigenvector of $\mathbf{S}^{-1}$, i.e. $\mathbf{n}'\mathbf{S}^{-1} = \mathbf{n}'$) to a plane containing $\mathbf{d}_2$. The normalization factor for $\mathbf{d}_1$ in Eqn (7) is $1/m_1$ and therefore $\mathbf{P}_1, m_1, \mathbf{d}_1$ and $\mathbf{p}_1'$ are all determinable. The matrix **R** is determined from the requirement that **x** and $\mathbf{n}'$ which are displaced by the Bain distortion must be totally invariant. **R** defines the orientation relationship within any small region of the martensite plate not involving $\mathbf{P}_2$. Thus, the assumed correspondence and lattice parameters determine **B**, the assumption of $\mathbf{p}_2'$ and $\mathbf{d}_2$ allows **R** to be determined, and $\mathbf{RB} = \mathbf{S}$ defines the elements of $\mathbf{P}_1$.

The above description of the crystallographic theory parallels the analysis given by Bowles and Mackenzie (1954); Mackenzie and Bowles (1954a, 1954b) but the treatments of Wechsler et al. (1953) and Bullough and Bilby (1956) are equivalent.

Some variations in the basic theory just presented include the introduction of a dilatation parameter, which slightly relaxes the requirement for the habit plane to be undistorted, although such a uniform dilatation is not in fact observed. The second is the incorporation of two inhomogeneous shear systems such that

$$P_1 = RBS_2S_1, \tag{9}$$

where S_2 and S_1 are the two inhomogeneous shear systems involved. There are difficulties in this since as will be seen later, the interface must be glissile, which limits the multiple inhomogeneous shears that can be selected.

9.2.4 Other Crystallographic Observations

As pointed out earlier, the martensite habit plane is near $\{3\ 10\ 15\}_P$ for an Fe–24.5Pt at.% alloy. A similar habit plane is also found for martensite formed in Fe–Ni alloys containing approximately 30 wt% Ni. However, in some steels, e.g. Fe–8Cr–1.1C wt%, the habit plane is near $\{225\}_P$, and that of lath martensites in dilute iron alloys is near $\{111\}_P$. **Figure 11** shows some experimental habit plane determinations for a variety of iron alloys. However, the martensite–parent orientation relationship is almost the same in the $\{3\ 10\ 15\}$, $\{225\}$ and $\{111\}$ transformations. This is not surprising since the orientation relationship is determined only by the lattice parameters through **RB** and hence is insensitive to alloy composition, whereas the habit plane depends also on the magnitude, plane and direction of the lattice invariant deformation.

Techniques are improving for measuring the shape strain associated with martensitic transformations and the information so obtained is leading to a clearer understanding of the transformation crystallography. In addition, the use of transmission electron microscopy has greatly enhanced the understanding of the crystallography of martensitic transformations.

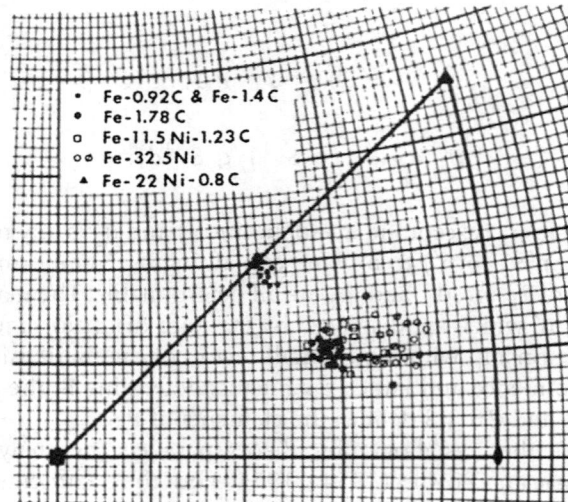

Figure 11 Unit triangle of stereographic projection showing the results of habit plane measurements on martensite formed in five different iron alloys.

9.3 Martensite Morphology and Substructure

Martensite in ferrous alloys generally takes the form of lenticular plates as seen in **Figure 1** for an Fe–Pt alloy. This morphology is readily understood because a lenticular plate, as a mechanical twin, is a low-strain energy shape in a "shear" transformation. However, other adaptations are found in ferrous alloys.

In low-carbon steels (up to 0.4 wt% C) and in some other ferrous alloys, the martensite takes the form of laths rather than plates. A typical lath may be about $0.3 \times 4 \times 200$ µm, the actual size being dependent on the austenite grain size. As mentioned above, the habit plane of such laths is near $\{111\}_P$. In Fe–Ni–C alloys, three distinct morphologies are found and are classified as the lenticular, thin plate, and butterfly morphologies. In one well-studied case it has been shown that butterfly martensite (**Figure 17**) forms at the highest temperatures, lenticular martensite forms at intermediate temperatures and thin-plate martensite forms at the lowest temperatures. In some materials, particularly ferrous alloys, a needlelike morphology known as surface martensite is formed under some conditions. Apparently, this morphology is easily nucleated at surfaces and can develop into a facetted form because matrix constraints are relaxed near a free surface.

Whereas the plate types of ferrous martensite frequently form in a self-accommodating manner involving several habit plane variants, lath martensites do not appear to do so. Instead, the laths tend to cluster together, organized into hierarchically into blocks and packets within a given austenite grain (Maki, 1990). A block consists of laths that are in virtually identical orientation in space, whereas a packet is the cluster of blocks that share the same austenite (111) close-packed plane to which the corresponding (011) martensite plane is almost parallel. The laths within a packet therefore have habit planes that make small angles with respect to each other (**Figure 12**), but have different crystallographic orientations. Those within the blocks also have similar crystallographic orientations. Given that there are 24 variants of the irrational orientation relationship between austenite and martensite, there are four different blocks possible in a given austenite grain. The importance of these structures is that they have a profound influence on the mechanical properties, in particular, the toughness of the steel.

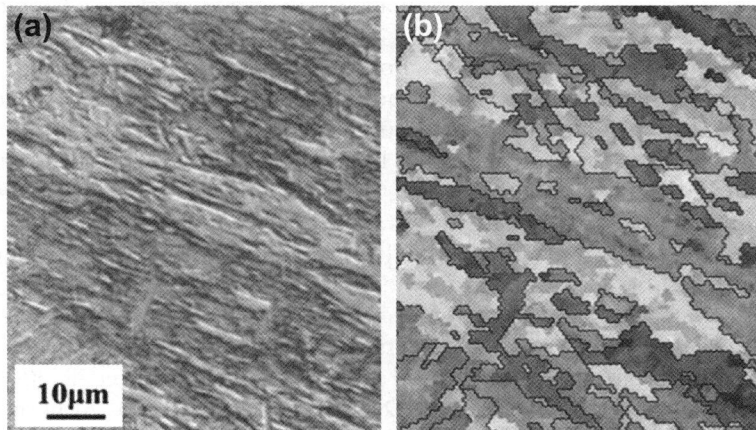

Figure 12 Lath martensite in a low-alloy steel (Wang et al., 2008). (a) Optical micrograph. (b) Corresponding orientation image showing block boundaries (black) connecting crystalline regions with misorientations in excess of 15°. A comparison of these images shows that the blocks are made up of many approximately parallel platelets.

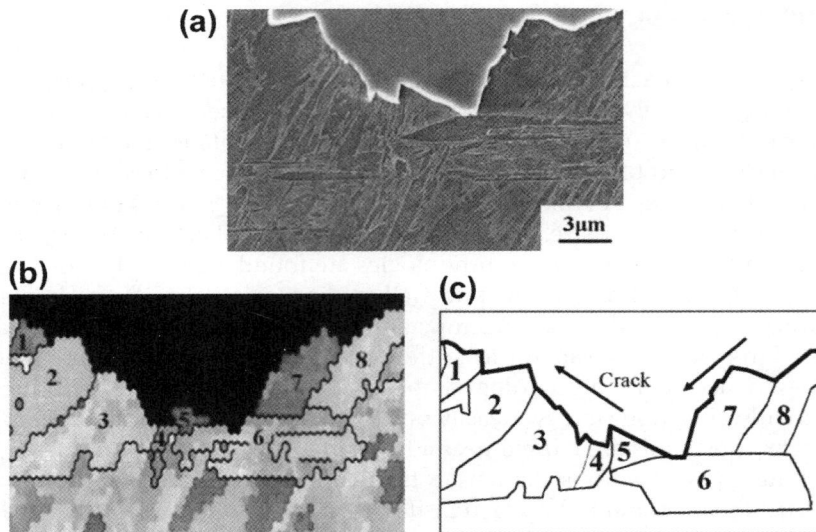

Figure 13 Lath martensite in a low-alloy steel (Wang et al., 2008). (a) Optical micrograph showing cleavage facets created by brittle fracture. (b) Corresponding orientation image showing cleavage crack deflected at block and packet boundaries. The misorientation between regions 6 & 7, and between 5 & 7, is the greatest and led to crack arrest, as shown in (c), whereas there is minimal deflection at the small misorientation between blocks 3 & 4. (For color version of this figure, the reader is referred to the online version of this book.)

The way in which they partition the austenite grain influences strength and toughness. Given that there is little crystallographic discontinuity within a block, it is the scale of the blocks rather than of the individual laths, which relates most closely to the strength (Morito et al., 2006). The crystallographic misorientations across packet boundaries are larger than between blocks within a packet, so it is the size of packet, which controls the deflection of propagating cleavage cracks, and hence the toughness, **Figure 13** (Wang et al., 2008).

The advent of techniques that permit the crystallographic orientation imaging at the same time as the underlying microstructure led to a plethora of studies concerning the nature of boundaries between adjacent platelets of martensite or bainite (Gourgues-Lorenzon, 2007). As we have seen, these boundaries influence the properties of materials. This in turn has led to the definition of a *crystallographic grain size*, which defines the scale of connected regions that have a similar orientation in space. The same regions may exhibit a substructure when observed using techniques such as optical microscopy, so the crystallographic grain size will in general be greater than or equal to the microstructural grain size. The difference is relevant when considering structure–property relationships, for example, the length scale over which cleavage cracks or other deformation phenomena are not significantly deviated. This size may include clusters of grains instead of the individual grains detected using ordinary microscopy on polished and etched samples (Bhattacharjee et al., 2004; Gourgues et al., 2000; Kim et al., 2007; Lambert-Perlade et al., 2004; Yan et al., 2010). Although there has been considerable work on the definitions of laths, blocks, and packets of martensite, there is no theoretical framework that permits the prediction of these features. The best that can be said is that there is a correlation between the features and the austenite grain size, as illustrated in **Figure 14**, although the dependence on alloy composition and heat-treatment parameters is unknown. We have noted

(a)

(b)

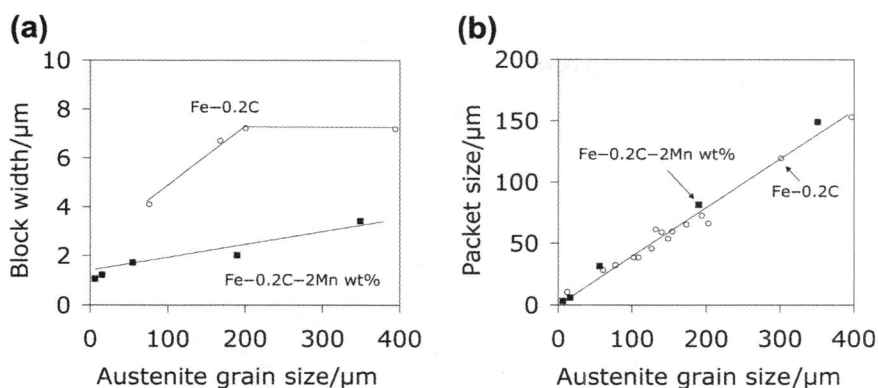

Figure 14 (a) Block width as a function of the austenite grain size. (b) Packet size as a function of austenite grain size. Data from Morito et al. (2006).

previously that each austenite grain can transform into four packets corresponding to the multiplicity of the $\{111\}_M$ planes. However, a comparison of the size scales on **Figure 14b** shows that each austenite grain is partitioned into only three packets, assuming that the packets form with uniform size.

Unlike the plate morphology, which typically contains transformation twins as shown in **Figure 10**, martensite laths are typically untwinned and contain a high density of internal dislocations as shown in **Figure 15a**. Since the inhomogeneous shear in ferrous lath martensite is not twinning, it is expected that slip is the mode involved and that the lattice-invariant deformation would be effected by the movement of interface dislocations. An example of such interface dislocations in an Fe–Ni–Mn martensite lath is shown in **Figure 15b**. These interface dislocations should be distinguished from those within the martensite; if the lath is elastically accommodate then it should in principle be free from dislocations other than those inherited from the austenite. Plastic accommodation of the shape deformation would lead to a greater density of dislocations in the austenite into which the martensite grows. Hence, the dislocations found inside laths of martensite are accommodation defects rather than an intrinsic feature of the transformation mechanism.

Plastic accommodation of the shape change occurs when the austenite is relatively weak; a high transformation temperature or a low solute content (i.e. reduced solid solution strengthening) would therefore promote the lath morphology, as is evident in **Figure 16a**. Thin-plate martensite (**Figure 17a**) is elastically accommodated to an extent that it exhibits a shape memory effect (SME); in contrast, lenticular martensite (**Figure 17b**) is not perfectly reversible because of the plastic deformation that its formation induces in the adjacent austenite. The term "butterfly martensite" refers simply to pairwise clusters of plates (**Figure 17c**), which are crystallographic variants with different specific habit planes and orientation relationships with the parent austenite grain. The detailed reasons for this particular pattern of plates is not understood, but it is believed that the accommodation defects produced by one plate stimulates the formation of the other one (Umemoto et al., 1984). Lath martensite is associated with significant plastic accommodation; the shear strain associated with lath martensite growth has been shown to be exceptionally large (Sandvik and Wayman, 1983).

The density of dislocations created due to plastic accommodation of a variety of displacive transformation products in iron is illustrated in **Figure 16b** as a function of the transformation temperature. There is a general increase as the transformation temperature is reduced, but the data exhibit considerable scatter so the actual trend is not well understood. The role of plastic accommodation of the shape

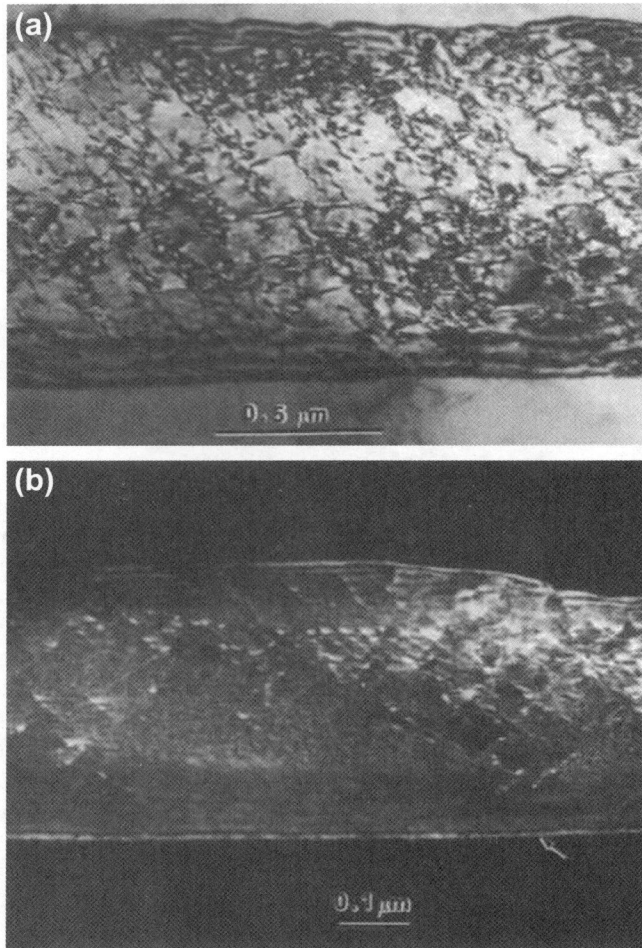

Figure 15 Martensite lath formed in an Fe–20Ni–5Mn wt% alloy. (a) Dislocations inside the lath. (b) Interfacial dislocations, parallel to the arrow.

deformation is vividly illustrated in **Figure 18**; the lack of orientation gradients in the austenite adjacent to the thin-plate martensite shows that the latter is elastically accommodated, consistent with the fact that the transformation front is reversible. On the other hand, there is a clear evidence of plastic relaxation in the austenite adjacent to the lenticular and lath forms of martensite. The plastic zones in the austenite extend a distance that scales with the martensite thickness. It is for this reason that thin plate martensite is almost free of dislocations whereas the other two often contain a large density of dislocations, inherited from the plastically deformed austenite as the transformation front progresses.

Another martensite morphology found in certain steels, notably austenitic stainless steels and Fe–Mn–C alloys, is the banded morphology, **Figure 17**d. However in these alloys, the transformation is not fcc → bcc, but fcc → hcp instead. The lattice deformation and the shape deformation for the fcc → hcp transformation are in fact identical ($P_1 = B$), an invariant-plane strain (**Figure 19**), so that

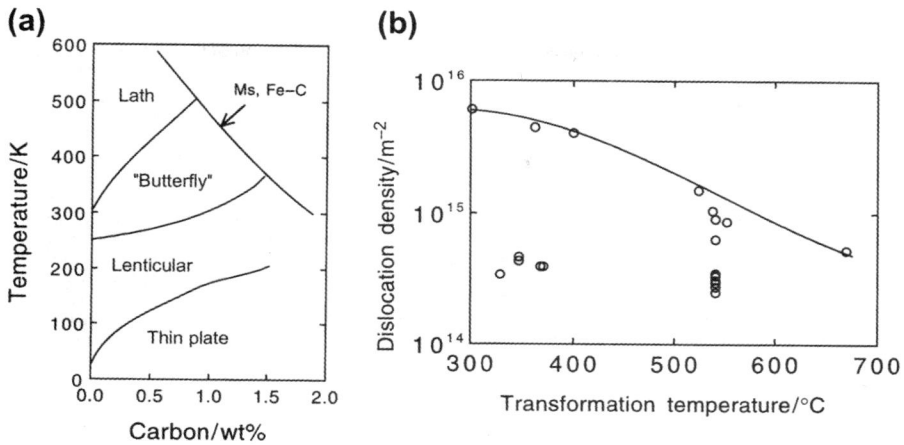

Figure 16   (a) Shape of martensite crystals as a function of the transformation temperature and carbon concentration of Fe–Ni–C alloys (Maki and Tamura, 1986). (b) Dislocation density of a variety of displacive transformation products in iron alloys (Bhadeshia, 2001a).

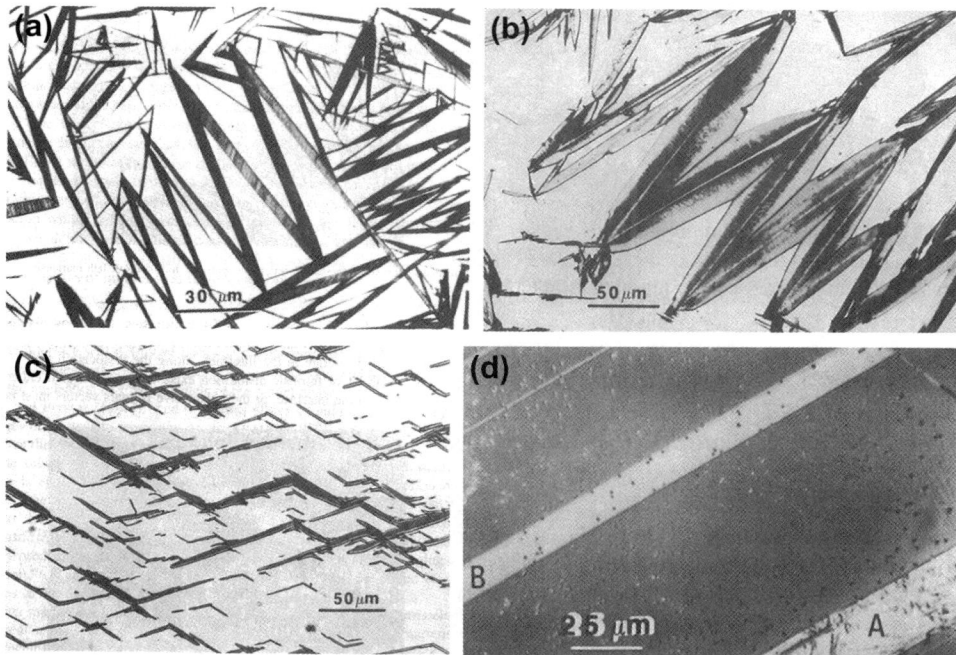

Figure 17   Optical micrographs. (a) Thin-plate morphology of martensite, observed in an Fe–31Ni–0.29C wt% alloy transformed at −160 °C. (b) lenticular martensite formed at −79 °C in an Fe–31Ni–0.23C wt% alloy. The more heavily etched linear central region of the plate is known as a midrib and has transformation twins that then give way to dislocated martensite away from the midrib. (c) Butterfly martensite morphology formed at 0 °C in an Fe–20Ni–0.7C wt% alloy. (d) Bands of hcp martensite (A and B) formed in fcc austenite matrix in an Fe–12Mn–0.48C wt% alloys.

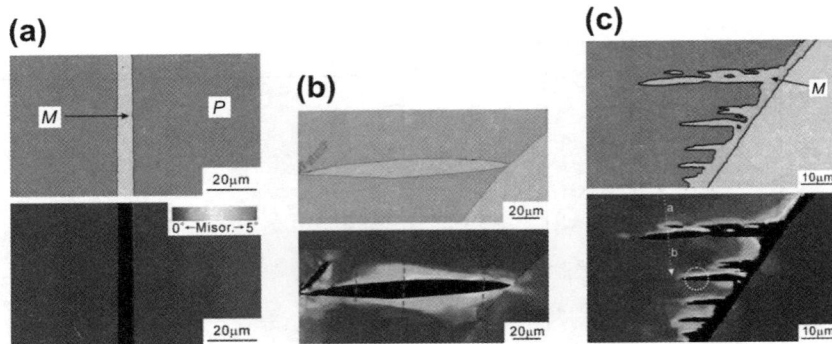

Figure 18 Plastic accommodation effects. The top image in each case is an orientation image to show the martensite plate in an austenitic matrix. The colors in the lower image show the crystallographic misorientation that exists in the austenite in the vicinity of the martensite. (a) Thin-plate martensite. (b) Lenticular martensite. (c) Lath martensite. After Miyamoto et al. (2009). (For interpretation of the references to color in this figure legend, the reader is referred to the online version of this book.)

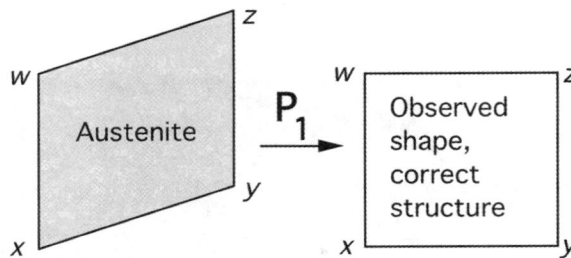

Figure 19 The deformation that accomplishes the crystal structure change from fcc $\rightarrow$ hcp is also the shape deformation P_1 for that transformation. Note that P_1 is an invariant-plane strain with both shear and dilatational components. *cf.* **Figure 7** for the more complex fcc $\rightarrow$ bcc martensitic transformation.

there is no necessity for a lattice invariant deformation as in the case of the fcc $\rightarrow$ hcp. The interface (habit plane) between the parent and product phases is therefore fully coherent and has rational indices, exactly $\{111\}_P$. This transformation is accomplished by distorting the fcc parent phase on $\{111\}_P$ planes by the passage of partial dislocations that generate the hcp martensite. As would be expected, the bands of martensite delineate $\{111\}_P$ planes. This type of fcc $\rightarrow$ hcp martensite transformation is similar to that which occurs in cobalt and cobalt alloys. It is important to note that the shape deformation is not simply a shear but also involves a volume change normal to the habit plane, since the hcp form is denser than the parent fcc (Brooks et al., 1979a, 1979b).

9.4 Martensite–Parent Interfaces

The phenomenological theory of martensitic transformations permits all the crystallographic features of the parent and product phases to be mathematically related, but does not yield detailed information about the mechanism of transformation. The latter depends critically on the structure of the transformation interface.

The fact that martensite can grow at low temperatures and high velocities means that the transformation interface must be very mobile and be able to move without any need for diffusion. The interface must be *glissile*. A fully coherent interface is glissile, but since the parent and martensite crystals cannot in many cases be converted into one another by a lattice deformation that is an invariant-plane strain, the interfacial structure is more likely to be semi-coherent.

A semi-coherent interface consists of coherent regions separated by discontinuities that prevent the misfit in the interface plane from accumulating over large distances. There are two kinds of semi-coherency (Christian and Crocker, 1980). If the discontinuities mentioned above are intrinsic dislocations with Burgers vectors in the interface plane, not parallel to the dislocation line, then the interface is said to be *epitaxially semi-coherent*. The normal displacement of such an interface requires the thermally activated climb of the intrinsic dislocations. A martensite interface cannot therefore be epitaxially semi-coherent.

In the second type of semi-coherency the discontinuities are screw dislocations, or alternatively dislocations whose Burgers vectors do not lie in the interface plane. This is the kind of semi-coherency associated with glissile martensite interfaces, whose motion is conservative (i.e. it does not lead to the creation or destruction of lattice sites). Two further conditions must be satisfied before even this interface can be said to be glissile (Christian and Crocker, 1980):

(1) A glissile interface requires that the glide planes of the intrinsic dislocations associated with the product lattice must meet the corresponding glide planes of the parent lattice edge to edge in the interface, along the dislocation lines.
(2) If more than one set of intrinsic dislocations exist, then these should either have the same line vector in the interface, or their respective Burgers vectors must be parallel. This ensures that the interface can move as an integral unit, and places severe restrictions on any theory involving more than one lattice-invariant deformation.

The intrinsic dislocations accomplish the lattice-invariant deformation as the interface glides. The interface also contains small, atomic height steps, which are coherent and whose motion leads to the transformation of the parent into the product phase. These have the character of transformation dislocations, and their Burgers vector can be defined in terms of the lattice deformation S (Christian, 1982). The density of such steps depends on the local curvature of the interface, but it must be emphasized that their motion is glissile and that it is their motion, which leads to transformation, whereas the intrinsic dislocations cause the lattice-invariant deformation.

When the transformation dislocations (the coherent steps) generate the final lattice, as in the fcc $\rightarrow$ hcp cobalt transformation, no other defects are required in the interface.

9.5   Energetics of Martensitic Transformations

9.5.1   Transformation Hysteresis and the Reverse Transformation

Two representative but widely different cases are considered here: the nonthermoelastic martensitic transformation in Fe–Ni ($\simeq 30$ wt%) alloys, and the thermoelastic martensite in typical β-phase alloys such as β-brass, often known as the "3/2 electrons per atom Hume-Rothery phases". In both cases the shear component of the macroscopic shape strain is about 0.2 (shear angle about $12°$) but the way in which this strain is accommodated by the parent phase is quite different for the two cases, as is the transformation hysteresis.

In the Fe–Ni alloys the martensite–parent interface appears to become immobilized after a plate of martensite has thickened to a certain extent, and when the martensite is heated, the interface will not move backward, apparently being pinned by its damaged environment. Instead, the reverse martensite → parent (austenite) transformation takes place by the nucleation of small platelets of the parent phase within the martensite plates (Kessler and Pitsch, 1967). And in ferrous alloys containing carbon, i.e. steels, the usual stages of martensite tempering occur (Krauss, 1980) within which the martensite decomposes through diffusional processes.

On the other hand, in alloys of the β-phase type, the shape deformation of the martensite is elastically accommodated by the matrix with no dislocation or "debris" generation; the interface therefore remains glissile, capable of "backward" motion leading to the shrinkage of martensite plates during heating. In addition, in such *thermoelastic transformations* (Kurdjumov and Khandros, 1949), stored elastic energy apparently predisposes the reverse martensite–parent transformation, allowing it to occur "prematurely" (Kaufman and Cohen, 1958). In such cases, the stored elastic energy contributes to the driving force for the reverse transformation.

Whether martensitic transformation in any particular alloy will be thermoelastic or not can be predicted by an analysis of the stress field around a plate of martensite. If the stress induced by the shape deformation causes the matrix to yield plastically, then the transformation is not thermoelastic (Ling and Owen, 1981; Olson and Owen, 1976). Austenite in iron–platinum alloys undergoes ordering, which greatly affects the elastic properties. The shear modulus decreases and the yield strength increases, so that it becomes possible to elastically accommodate the martensite. The ordered alloy therefore has thermoelastic martensite whereas the disordered alloy has nonthermoelastic martensite. The plastic deformation for the two cases is illustrated in **Figure 20** and has been confirmed experimentally to be the case (Vevecka et al., 1995).

When the parent phase is weakened by transformation at elevated temperatures, the plastic deformation around a plate can be so severe that it prematurely stops the growth of the plate. This is responsible for the subunit mechanism of bainite growth (Bhadeshia, 2001a; Chatterjee et al., 2006).

9.5.2   Thermoelastic and Nonthermoelastic Martensite

The differences between thermoelastic and nonthermoelastic martensitic transformations may be highlighted by comparing Au–Cd and Fe–Ni alloys, respectively (Otsuka and Wayman, 1977). **Figure 21** shows that there is a substantial difference in the transformation hysteresis for the two alloys. This hysteresis is the difference between the A_F (completion temperature for the reverse transformation) and M_S temperatures (Kaufman and Cohen, 1958). The thermoelastic transformation (Au–Cd) is characterized by a small hysteresis. Another difference is in the manner of the martensite–parent reverse transformation. In the thermoelastic case, the martensitic transformation on cooling proceeds by the continuous growth of martensite plates and the nucleation of new plates. When cooling is stopped, growth ceases, but resumes during further cooling, until the martensite plates impinge each other or on grain boundaries. The reverse transformation upon heating occurs by the backward movement of the martensite–parent interface, and the plates of martensite shrink and revert completely to the initial parent phase orientation.

On the other hand, a plate of nonthermoelastic martensite generated at a given temperature does not grow further upon subsequent cooling because the interface apparently becomes immobilized. Nor does the immobilized interface reverse its direction of motion during heating. Instead, the parent phase has to

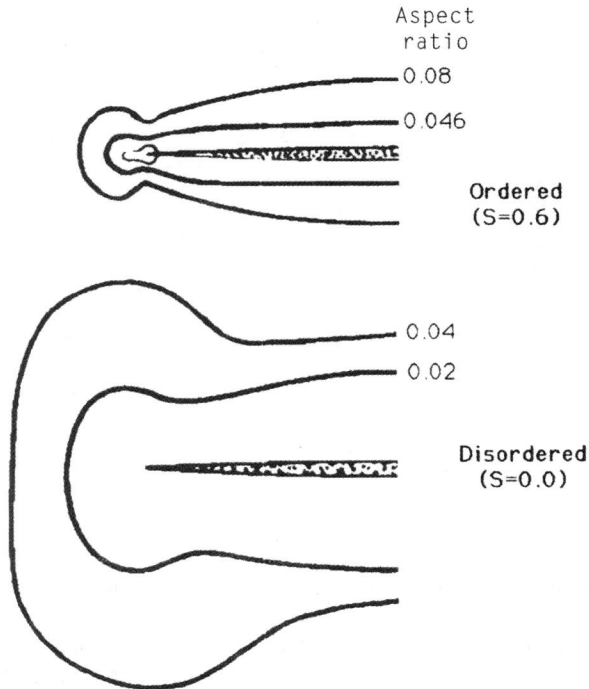

Figure 20   Ling and Owen's calculations showing the contours where the matrix satisfies a yield criterion, as a function of the aspect ratio of the martensite plate and the degree of ordering (S). The region between each contour and the martensite plate represents plastically deformed austenite. Only half the martensite plate is illustrated because the results are symmetrical about the vertical centerline.

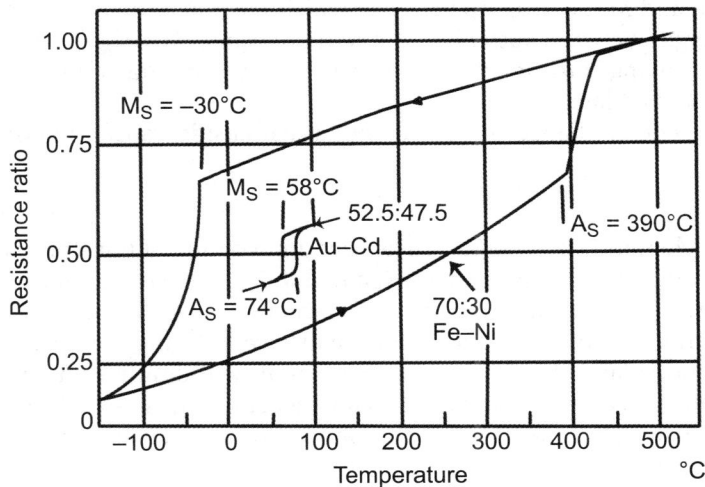

Figure 21   Graph showing electrical resistance change during heating and cooling for Fe–Ni and Au–Cd alloys, indicating the hysteresis for the martensitic transformation on heating for nonthermoelastic and thermoelastic transformation, respectively (Kaufman and Cohen, 1958).

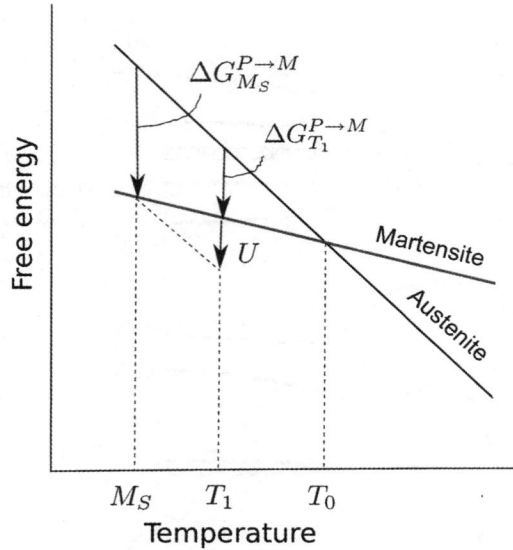

Figure 22 Schematic diagram showing the free energy change for martensitic transformation. See text for details.

nucleate within the immobilized martensite plates; since the parent phase can nucleate in many orientations, the original parent phase orientation is not recovered on heating (Kessler and Pitsch, 1967).

9.5.3 Free Energy Change Due to Martensitic Transformation

Figure 22 shows schematically the change in chemical free energies of martensite and austenite (parent phase) with temperature. T_0 is the temperature at which the austenite and martensite are in thermo-dynamic equilibrium and M_S is the temperature at which the transformation starts upon cooling. The difference in free energies between austenite (γ) and martensite (α'), $\Delta G_{M_S}^{\gamma \to \alpha'}$ at the M_S temperature, is the critical chemical driving force for the onset of the martensitic transformation (other features of **Figure 22** will be described later).

In general, the free-energy change associated with a martensitic transformation is given by:

$$\Delta G^{P \to M} = \Delta G_C^{P \to M} + \Delta G_{NC}^{P \to M}, \tag{10}$$

where $\Delta G_C^{P \to M}$ is the chemical free-energy change (per unit volume of transformation) associated with the transformation from the parent to martensite, and $\Delta G_{NC}^{P \to M}$ is the nonchemical energy opposing the transformation (consisting of elastic strain and surface energy). Since the martensite–parent interface is semi-coherent the surface energy term should be small except during the nucleation stage where the surface to volume ratio is large, but the elastic energy term will dominate at larger scales. The strain energy per unit volume was derived using the Eshelby theory by Christian (1958), for an isolated plate in the form of an oblate spheroid with length r much greater than the thickness c, located within elastically isotropic austenite, as:

$$G_{strain} = \frac{c}{r}\frac{\mu}{1-\nu}\left[\frac{\pi}{4}\delta^2 + \frac{\pi}{8}(2-\nu)s^2\right] \approx \frac{c}{r}\mu(s^2 + \delta^2), \tag{11}$$

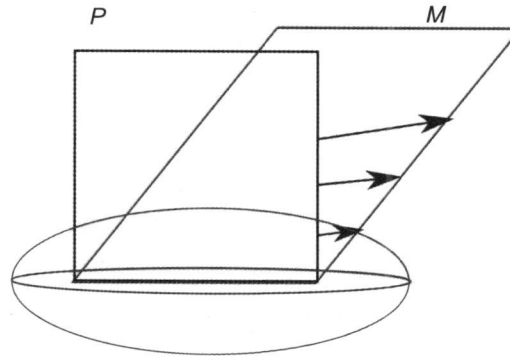

Figure 23 An illustration of why a thinner plate leads to less strain energy. The displacements, as indicated by the arrows, become larger with distance from the habit plane.

where μ and ν are the respective shear modulus and Poisson's ratio of austenite, and s and δ are respectively the shear and dilatational strains parallel and normal to the habit plane. Whilst the derivation of this equation is complicated, its form is easy to understand because the strain energy per unit volume for an elastically accommodated plate scales with the product of the modulus with the square of the strain, and the aspect ratio comes in because the displacements due to the shear scale with the distance from the central plane of the plate, as shown by the arrows in **Figure 23**. This is also the reason why martensite and mechanical twins have sharp peripheries (lenticular shape) where the displacement becomes minimal. For a constant length, a thinner plate is associated with a smaller strain energy term.

It follows that during thermoelastic equilibrium, the plate of martensite continues to increase its aspect ratio until the chemical and nonchemical terms balance. Consequently, a plate of constant length can be thickened or thinned when the transformation temperature is lowered or raised, respectively. This is the origin of the terminology *thermoelastic transformation*.

9.5.4 Nucleation of Martensite

Phase and chemical composition fluctuations can occur as random events due to the thermal vibration of atoms. An individual fluctuation may or may not be associated with a reduction in free energy, but it can only survive and grow if there is a reduction beyond its embryonic size. There is a cost associated with the creation of a new phase, the interface energy, a penalty that becomes smaller as the particle surface to volume ratio decreases. In a metastable system this leads to a critical size of fluctuation beyond which a growth is favored.

Consider the homogeneous nucleation of martensite (M) from the parent phase austenite (P) by the classical heterophase fluctuation mechanism. For a spherical particle of radius r with an isotropic interfacial energy σ_{MP}, the change in free energy as a function of radius is:

$$\Delta G = \frac{4}{3}\pi r^3 \Delta G_{\text{chemical}} + \frac{4}{3}\pi r^3 \Delta G_{\text{strain}} + 4\pi r^2 \sigma_{MP}, \tag{12}$$

where $\Delta G_{\text{chemical}} = G_V^M - G_V^P$, G_V is the Gibbs free energy per unit volume of M and G_{strain} is the strain energy per unit volume of M. The activation barrier and critical size obtained using Eqn (12) are given by:

$$G^* = \frac{16\pi\sigma_{MP}^3}{3(\Delta G_{\text{chemical}} + \Delta G_{\text{strain}})^2} \quad \text{and} \quad r^* = \frac{2\sigma_{MP}}{\Delta G_{\text{chemical}} + \Delta G_{\text{strain}}}. \tag{13}$$

The important outcome is that in classical nucleation the activation energy varies inversely with the square of the driving force. And the mechanism involves random phase fluctuations. It is questionable whether this applies to cases where thermal activation is in short supply. In particular, an activation barrier must be very small indeed if the transformation is to occur at a proper rate at low temperatures.

One mechanism in which the barrier becomes sufficiently small involves the spontaneous dissociation of specific dislocation defects, which are already present in the parent phase (Christian, 1951; Olson and Cohen, 1976). The dislocations are glissile so the mechanism does not require diffusion. The only barrier is the resistance to the glide of the dislocations. The nucleation event cannot occur until the undercooling is sufficient to support the faulting and strains associated with the dissociation process that leads to the creation of the new crystal structure, **Figure 24**.

The free energy per unit area of fault plane is:

$$G_F = n_P\rho_A(\Delta G_{\text{chemical}} + G_{\text{strain}}) + 2\sigma_{MP}\{n_P\}, \tag{14}$$

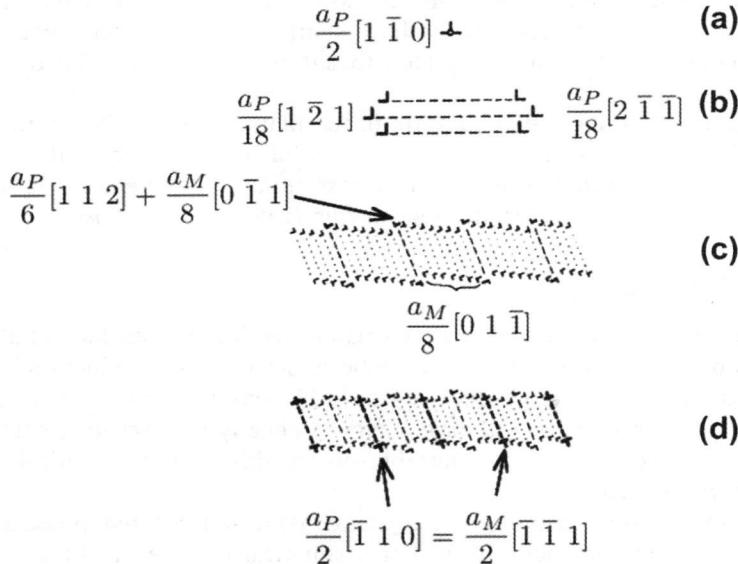

Figure 24 Olson and Cohen model for the nucleation of M martensite. (a) Perfect screw dislocation in austenite. (b) Three-dimensional dissociation over a set of three close-packed planes. The faulted structure is not yet that of M. (c) Relaxation of fault to a body-centered cubic structure with the introduction of partial dislocations in the interface. (d) Addition of perfect screw dislocations that cancel the long-range strain field of the partials introduced in (c).

where n_P is the number of close-packed planes participating in the faulting process, ρ_A is the spacing of the close-packed planes on which the faulting is assumed to occur. The fault energy can become negative when the austenite becomes metastable.

For a fault bounded by an array of n_P dislocations each with a Burgers vector of magnitude b, the force required to move a unit length of dislocation array is $n_P\tau_o b$. τ_o is the shear resistance of the lattice to the motion of the dislocations. G_F provides the opposing stress via the chemical free energy change $\Delta G_{chemical}$; the physical origin of this stress is the fault energy, which becomes negative so that the partial dislocations bounding the fault are repelled. The defect becomes unstable, i.e. nucleation occurs, when

$$G_F = -n_P\tau_o b. \tag{15}$$

Take the energy barrier between adjacent equilibrium positions of a dislocation to be G_o^*. An applied shear stress τ has the effect of reducing the height of this barrier (Conrad, 1964; Dorn, 1968):

$$G^* = G_o^* - (\tau - \tau_\mu)v^*, \tag{16}$$

where v^* is an activation volume and τ_μ is the temperature-independent resistance to a dislocation motion (**Figure 25**). In the context of nucleation, the stress τ is not externally applied but comes from the chemical driving force. On combining the last three equations we obtain

$$G^* = G_o^* + \left[\tau_\mu + \frac{\rho_A}{b}G_{strain} + \frac{2\sigma}{n_P b}\right]v^* + \frac{\rho_A v^*}{b}\Delta G_{chemical}. \tag{17}$$

It follows that with this model of nucleation the activation energy G^* will decrease *linearly* as the magnitude of the driving force $\Delta G_{chemical}$ increases. This direct proportionality contrasts with the inverse square relationship of classical theory.

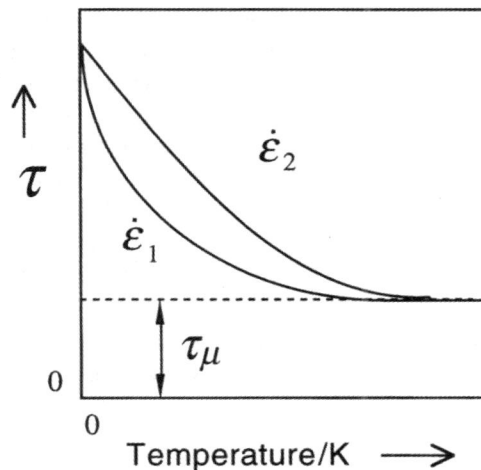

Figure 25 Temperature dependence of the applied stress necessary to move a dislocation at two different strain rates ($\dot{\varepsilon}_2 > \dot{\varepsilon}_1$). τ_μ is the athermal resistance, which never vanishes. After Conrad (1964).

Table 2   Characteristics of different modes of deformation/transformation

	Slip deformation	Mechanical twinning	Displacive transformation	Reconstructive transformation
Causes permanent change in shape	Yes	Yes	Yes	Yes
Invariant-plane strain shape change with a large shear component	Yes	Yes	Yes	No
Changes crystallographic orientation	No	Yes	Yes	Yes
Changes lattice type	No	No	Yes	Yes
Can lead to a density change	No	No	Yes	Yes

9.5.5   Mechanical Effects in Martensitic Transformations

9.5.5.1   *Introductory Comments*

The most familiar mechanisms of plastic deformation are slip, mechanical twinning and diffusion-induced creep. For very small plastic strains, the first two of these deformation modes are conservative—i.e. they preserve an atomic correspondence between the deformed and undeformed parts of the crystal so that the crystal contains a memory of its original shape. All of these deformation modes are *lattice-invariant* because although they cause a change in the shape, the crystal structure remains as it was prior to deformation.

Martensitic transformation can also be regarded as a mode of deformation, but a mode that at the same time causes a change in the crystal structure. The shape change is, of course, an invariant-plane strain with a large shear component and a relatively small dilatational strain. The distinguishing features of a variety of deformation modes are listed in **Table 2**.

Given that the growth of martensite causes a change in shape, it is natural to expect the transformation to be influenced by an externally applied stress. The permanent strain caused by martensitic transformation is called *transformation plasticity*. A phase change in a stress-free material is usually triggered by heat treatment, when the parent phase passes through an equilibrium transformation temperature. Alternatively, the application of a stress in isothermal conditions can trigger transformation in circumstances where it would not otherwise occur. Unusual effects can occur when stress and temperature work together. The transformation may occur at remarkably low stresses or at very small deviations from the equilibrium temperature. This is why even minute stresses can have a large influence on the development of microstructure. It is not surprising that transformation plasticity can be obtained at stresses that are much smaller than the conventional yield stress of the parent phase.

9.5.5.2   *Chemical and Mechanical Driving Forces*

The interaction of an externally applied stress (which is below the yield strength of the parent phase) can manifest itself in two ways:

- The stress can alter the driving force for the transformation.
- It can change the appearance of the microstructure by favoring the formation of those variants that best comply with the applied stress.

Both of these factors are illustrated in **Figure 26**, where a fine-grained polycrystalline sample of austenite was stressed at a temperature *above* its normal martensite-start temperature. The amount of martensite obtained is seen to vary directly with the magnitude of the applied stress. The stress manifests as a *mechanical driving force* whose contribution assists the chemical driving force, which on its

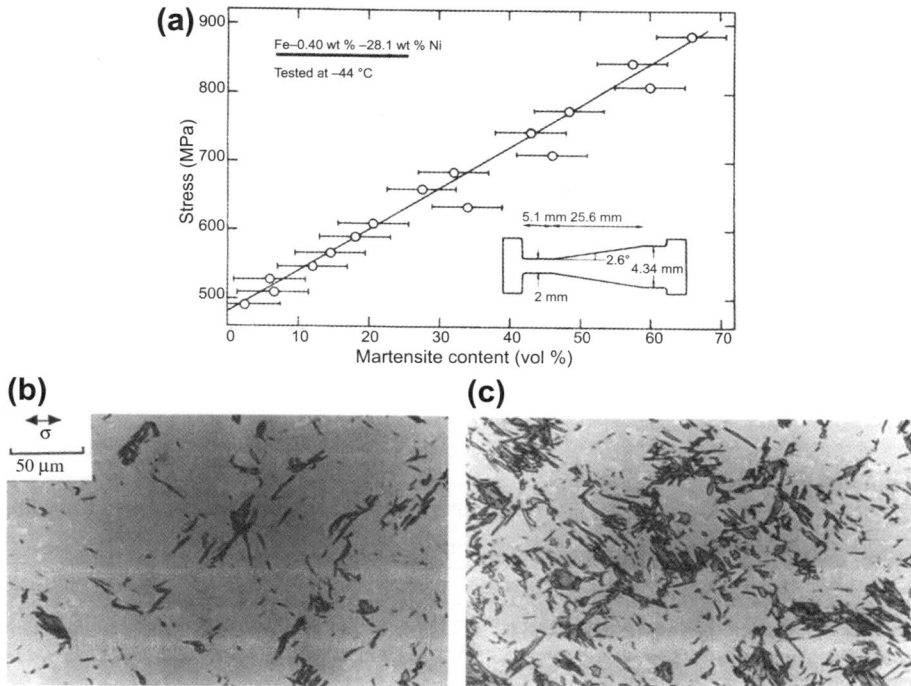

Figure 26 Stress-affected martensitic transformation in an Fe–28Ni–0.4C wt% alloy tested at a temperature above the martensite-start temperature. (a) Volume fraction of martensite as a function of stress. (b) Optical micrograph from a low-stress region. (c) Optical micrograph from a high-stress region. The arrow indicates the direction of the tensile stress.

own is inadequate to trigger transformation. Not only does the stress induce the formation of martensite, but only those variants that comply with the tensile stress grow in profusion (**Figure 26**b). Thus, most of the plates are tempted to grow on those planes that are close to the plane of maximum shear stress (45° to the tensile axis). The microstructure would have been much more chaotic in the absence of the stress. The stress caused the alignment of plates, and a more ordered, organized microstructure.

The chemical free energy change associated with a martensitic transformation was depicted in **Figure 22**. Referring to **Figure 22** again, when a stress is applied to the austenite at T_1 (between M_S and T_0), the mechanical driving force, U, due to the stress is added to the chemical driving force, $\Delta G_{T_1}^{\gamma \to \alpha'}$, and the martensitic transformation starts at the critical stress where the total driving force is equal to $\Delta G_{M_S}^{\gamma \to \alpha'}$. $U' = (\Delta G_{M_S}^{\gamma \to \alpha'} - \Delta G_{T_1}^{\gamma \to \alpha'})$ in **Figure 22** is the critical mechanical driving force necessary for the stress-induced martensitic transformation at T_1. The mechanical driving force is a function of stress and the orientation of a transforming martensite plate, and can be expressed (Patel and Cohen, 1953) as:

$$U = \tau s + \sigma_N \delta, \tag{18}$$

where τ is the resolved shear stress along the transformation shear direction in the martensite habit plane, σ_N is the normal stress resolved perpendicular to the habit plane, and s and δ are the shear and dilatational components of the shape deformation due to martensitic transformation, respectively. For

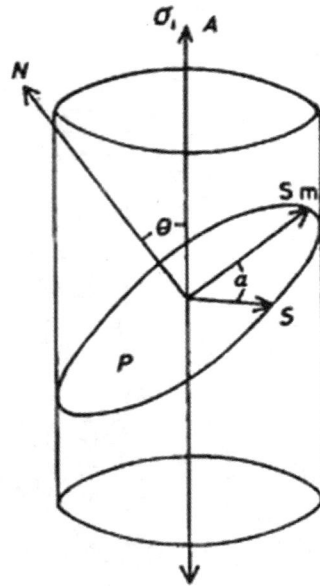

Figure 27 Schmid-factor diagram for the influence of an applied stress σ_1 along stress axis A in inducing a martensitic transformation on habit plane P with normal N. S is the direction of the shape strain for the martensite and S_m is the maximum shape strain elongation parallel to the habit plane.

a small applied stress σ_1, τ and σ (**Figure 27**) may be expressed as follows for any given orientation of a martensite plate (Onodera et al., 1976);

$$\tau = \frac{1}{2}\sigma_1 \sin 2\theta \ \cos \alpha, \tag{19}$$

$$\sigma_N = \pm\frac{1}{2}\sigma_1[1 + \cos 2\theta], \tag{20}$$

where σ_1 is the magnitude of the applied stress, θ is the angle between the applied stress and the normal to the habit plane, and α is the angle between the transformation shear direction and the maximum shear stress in the habit plane. The plus and minus signs in Eqn (20) correspond to uniaxial tension and compression, respectively. It follows that the mechanical driving force is

$$U = \frac{1}{2}\sigma_1[s \sin 2\theta \ \ \cos \alpha \pm \delta(1 + \cos 2\theta)]. \tag{21}$$

These concepts can be generalized to a system of stresses (Kundu et al., 2007). The applied system of stresses can be described by a 3×3 stress tensor σ_{lm}, which when multiplied by the unit normal to the habit plane gives the traction $\mathbf{t}$ describing the state of stress on that plane. The traction can then be resolved into σ_N and τ in the normal manner (Timoshenko and Goodier, 1982):

$$\sigma_N = |\mathbf{t}|\cos\{\theta\};$$

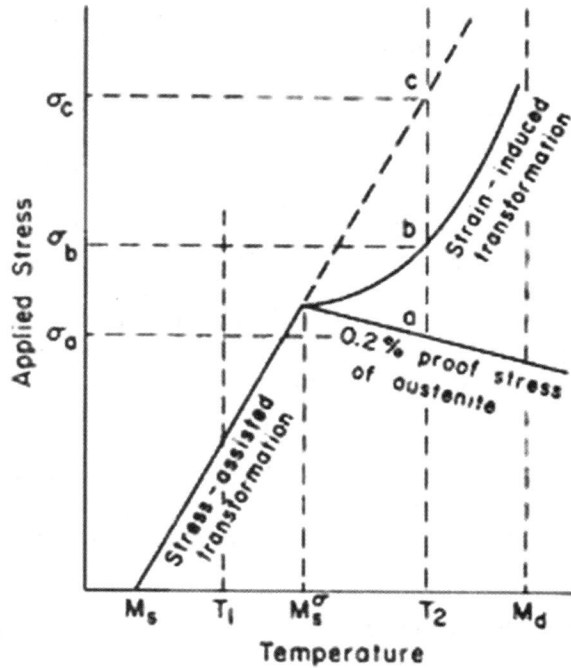

Figure 28 Schematic stress-temperature diagram showing the critical stress to initiate the formation of martensite as a function of temperature. The various regimes of behavior are described in the text.

$$\tau = |t|\cos\{\beta\}\cos\{\phi\}, \tag{22}$$

where $|t|$ is the magnitude of t, θ is the angle between the habit plane normal and t, β the angle between t and the direction of the maximum resolved shear stress, and ϕ the angle between the latter and the direction of shear for the plate concerned.

9.5.5.3 Critical Stress to Induce Martensitic Transformation

When martensite is induced in polycrystalline austenite (containing randomly oriented grains) by an external stress, the martensite plate whose orientation yields a maximum value of U, Eqn (21), will be formed first. If it is assumed that the angle α is zero, then the maximum value of U is obtained by setting $dU/d\theta = 0$. For Fe–Ni alloys ($s = 0.20$, $\delta = 0.04$), the value of θ, which would maximize U, is therefore found to be 39.5° for tension and 50.5° for compression. These angles become exactly 45° if $\delta = 0$.

If the chemical driving force ($\Delta G^{\gamma \to \alpha'}$) decreases linearly with an increase in temperature above M_S, then a proportionally greater mechanical driving force is required before martensite formation is triggered. The necessary stress might then be expected to increase linearly with temperature, and this is in fact observed experimentally (Olson and Cohen, 1972; Onodera et al., 1976) over the range $M_S \to M_S^\sigma$ (**Figure 28**). However, above M_S (e.g. at T_2), martensite is induced at a stress σ_b after a plastic deformation of the austenite occurs.

The microstructure also changes when martensite is generated under the influence of an applied stress. It has been observed (Otsuka and Wayman, 1977) that only a single variant, out of (up to) twenty four that might be possible, of martensite is induced when an elastic stress is applied. This is the variant, which gives rise to maximum elongation in the direction of the tensile axis.

When the austenite is deformed at temperatures above M_S^σ (e.g. at T_2 in **Figure 28**), it begins to deform plastically at a stress σ_a, and is strain-hardened up to σ_b. Then the martensitic transformation starts to take place. σ_b is considerably lower than σ_c, which is obtained by extrapolating the critical stress-temperature line between M_S and M_S^σ. This decrease (i.e. $\sigma_c - \sigma_b$) in the critical applied stress for martensite formation is due to plastic deformation of austenite. As to the role of plastic deformation of austenite on the deformation-induced martensitic transformation, two different views have been expressed. One is the strain-induced martensite nucleation hypothesis proposed by (Olson and Cohen, 1972, 1976, 1979), and the other holds that the stress is locally concentrated at obstacles (e.g. grain boundaries, twin boundaries, etc.) by plastic deformation of austenite, and thus the concentrated stress becomes equivalent to σ_c in **Figure 28**, the latter proposed by Onodera and Tamura (1979) and others.

In contrast to the M_S temperature where a martensitic transformation will begin as a consequence of the chemical driving force, reference is frequently made to the M_d temperature. M_d is the temperature above which the chemical driving force becomes so small that nucleation of martensite cannot be mechanically induced, even in the plastic strain regime.

9.5.5.4 *Transformation-Induced Plasticity* (TRIP)

As an example of TRIP behavior (Zackay et al., 1967), we consider an austenitic Fe–29Ni–0.26C wt% alloy ($M_S = -60\,^\circ C$, $M_d = 25\,^\circ C$) deformed in tension at various temperatures. The relationship between tensile properties and test temperature for this alloy is shown in the upper part of **Figure 29** (Tamura et al., 1970). Note that the elongation is nearly 100%. An inverse temperature dependence of the yield stress is observed between M_S^σ and M_S. This results from the shape strain of martensite formed before yielding of the austenite occurs. The tensile strength is increased with decreasing test temperature just above M_S^σ. Such an elongation enhancement is attributed mainly to the suppression of necking, due to the increase in work-hardening rate by the formation of martensite. Furthermore, it is thought that the initiation and propagation of cracks is suppressed by the formation of martensite during deformation, because the stress concentrations may be released by the formation of preferential variants of martensite at regions of stress concentration. Similar results are seen in **Figure 29** for two other alloys.

With an increase in carbon content, the martensite hardness is increased substantially, but the austenite hardness only slightly increases. Accordingly, in the case of high-carbon steels, since a larger work-hardening rate can be obtained even with a small amount of martensite formation, the uniform elongation by TRIP becomes much larger (Tamura et al., 1972, 1973).

9.5.5.5 *Magnetic Driving Force*

The magnetic properties of austenite and martensite can be different, in which case their respective free energies are affected in the presence of external magnetic fields. The magnetization M, the magnetic dipole moment per unit volume of sample, and magnetic field strength H have the same units ($A\,m^{-1}$) and are related by a dimensionless volume susceptibility χ so that $M = \chi H$. The susceptibility therefore measures the degree to which a material can be magnetized. This relationship assumes that the spontaneous magnetization of the phase is constant, but it may not be at large H, whence a high-field susceptibility χ_{HF} is defined.

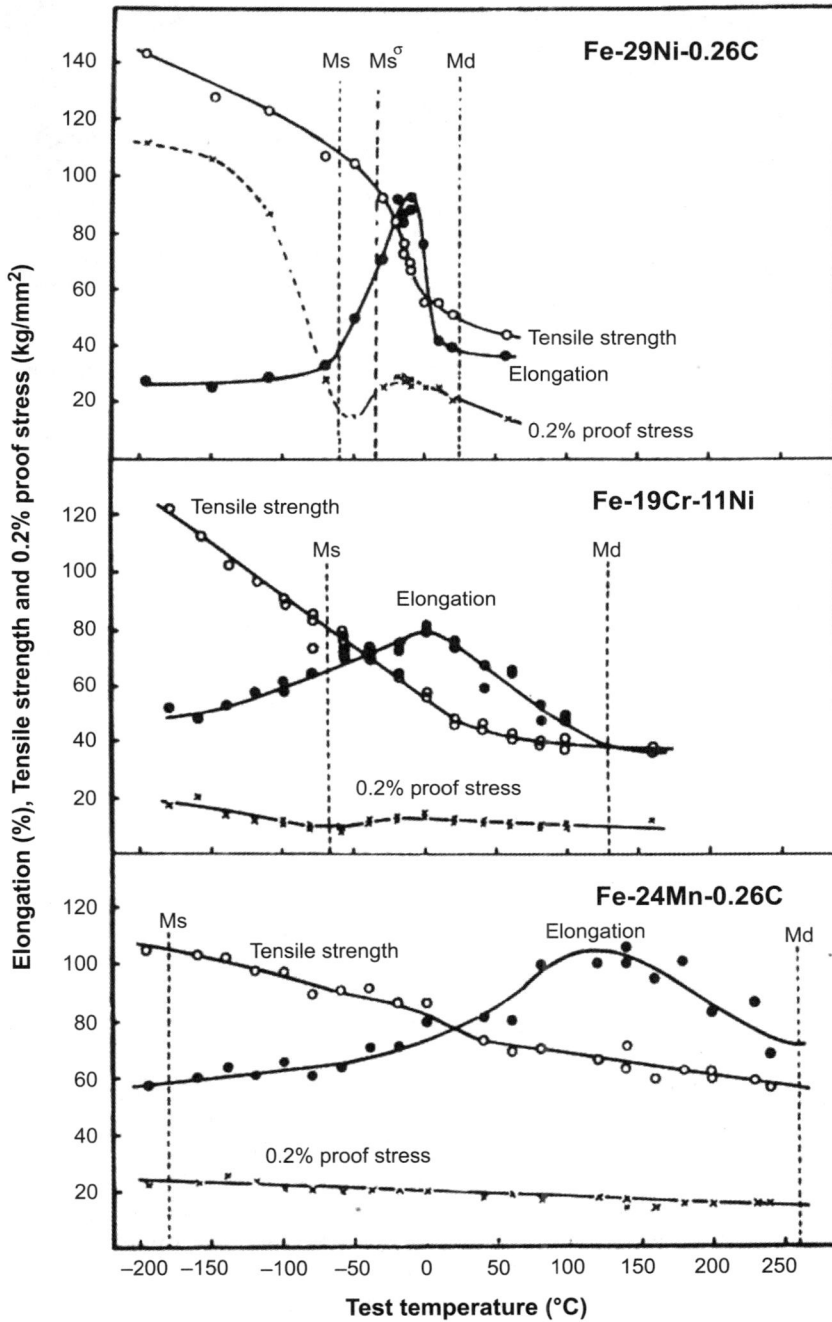

Figure 29 Experimental data showing the effect of test temperature on the tensile properties of metastable austenitic alloy strained at a rate $5.5 \times 10^{-4} \, \text{s}^{-1}$ (Tamura et al., 1970).

Figure 30 Magnetic field-induced martensite (large plates parallel to applied field) in an iron alloy. After Date et al. (1986).

The free-energy change accompanying a phase transformation is altered in a magnetic field if the spontaneous magnetizations of the parent and product phases are different. The magnetic energy contribution to the phase transformation, $\Delta G^{\gamma\alpha}_{mag}$ is then (Kakeshita et al., 1985; Krivoglaz and Sadovskiy, 1964):

$$\Delta G^{\gamma\alpha}_{mag} = \Delta M_T - \frac{1}{2}\chi_{HF}H^2, \tag{23}$$

where ΔM_T is the difference in magnetic moment between the parent and product phases $(M^\alpha - M^\gamma)$ at the temperature T. The second term on the right corrects for the fact that the magnetic susceptibility of the austenite is different at high field strengths. There may be other terms associated with the interaction of magnetostrictive effects in the parent phase with the volume change due to a field-free transformation (Kakeshita et al., 1985). A more complex theory would also take into account magnetic anisotropy. **Figure 30** shows the dramatic influence on the microstructure following martensitic transformation under the influence of an external magnetic field.

9.5.6 Mechanical Effects Specific to Thermoelastic Transformations

9.5.6.1 *General*

Thermoelastic martensitic transformations are accompanied by very unusual mechanical effects in both the parent and martensite phases. In addition to the now popular SME and the mechanical behavior of alloys exhibiting this type of behavior, closer inspection shows that SME alloys also show a variety of other kinds of interesting mechanical behavior. Examples are the superelasticity associated with the formation of a reversible stress-induced martensite (SIM), the rubberlike effect, "training" and the two-way SME, extensive deformation resulting from SIM-to-martensite transformations, unusual damping behavior, and finally, high stresses generated during the reverse martensite-to-parent transformation.

The number of materials exhibiting the SME is now extensive, including many Cu-based alloys, those of the noble metals based on Ag and Au, the classic Ni–Ti alloys (NITINOL), ternary variations of the same such as Ni–Ti–Cu and Ni–Ti–Fe, Ni–Al alloys, and Fe–Pt alloys, to mention a few. The martensite that forms in these alloys is thermoelastic and crystallographically reversible, and both the parent and product phases are in general atomically ordered (Wayman and Shimizu, 1972). The SME martensites are either internally twinned or internally faulted due to the lattice-invariant deformation.

In addition, the parent phase is usually of the B2 or DO3 type and the martensite crystal structure is given by the stacking sequence in the parent phase {110} planes, e.g. 2H, 3R, 9 and 18R.

9.5.6.2 The Shape Memory effect (SME)

A brief description of the SME is as follows. An object in the low temperature martensitic condition[2] when deformed and then unstressed regains its original shape on heating through the $A_S \rightarrow A_F$ temperature interval, as the parent phase grows at the expense of the martensite. The strains (6–8%) describing the original deformation of the martensite are completely recovered. The recovery of the original shape is entirely due to the reverse transformation of the deformed martensite.

The SME effect can be explained as follows (Saburi et al., 1979; Schroeder and Wayman, 1977a). A single crystal of the parent phase will usually transform into 24 orientations of martensite (variants of {hkl}). But when this mixture of many variants of martensite is deformed, the microstructure changes into a single orientation of martensite. This change occurs by the motion of martensite–martensite boundaries, during which the variant most compatible with the applied deformation grows at the expense of the others. The motion of the martensite–martensite boundaries involves some twinning/detwinning deformation, because the twins are in fact simply other orientations of martensite. It has been shown that twinning can convert one orientation (variant) of martensite to another (Schroeder and Wayman, 1977a). Consequently, when a fully martensitic mixture of many orientations of martensite is deformed, one plate orientation grows at the expense of the others. The favored martensite is that whose shear component of the shape deformation permits the maximum elongation of the specimen in the direction of the tensile axis (Saburi et al., 1979). The entire deformation is therefore due solely to martensite variant reorientation. Any other form of deformation (e.g. ordinary slip) tends to destroy the SME.

Although a single crystal of the parent phase transforms into many orientations of martensite, the reverse situation does not occur. Rather, the single crystal of martensite resulting from deformation below the M_F temperature transforms during heating into a single orientation of parent phase. This is a consequence of the martensite and parent lattice symmetries involved and the necessity to maintain atomic ordering during the reverse transformation. The highly symmetric parent phase (usually cubic) has many crystallographically equivalent ways in which the Bain distortion can occur, giving numerous variants of martensite. On the other hand, the relatively unsymmetric martensite (e.g. monoclinic in Cu–Zn–Al) does not have such a multiplicity of choices, so that only the original orientation of parent phase can form. The single crystal of martensite "unshears" to form a single crystal of the parent phase, thereby restoring the specimen to its original shape. A schematic representation of the shape memory process is shown in **Figure 31** (Saburi et al., 1979).

This description of the SME is quite general, irrespective of the particular alloy system or crystal structure of the martensite.

Figure 32 (Schroeder and Wayman, 1977a) shows a stress–strain curve for a Cu–Zn SME alloy single-crystal specimen deformed below the M_F temperature. It is seen that the martensite begins to deform (a') at a relatively low stress, 35 MN/m^2. The residual strain at point c was completely recovered during heating.

9.5.6.3 The Two-Way Shape Memory Effect

With the SME, as previously described, a specimen deformed by martensite variant conversion will undeform and regain its original shape during heating from A_S to A_F. In contrast the two-way shape

---

[2] usually below M_F, the temperature at which the martensitic transformation is completed.

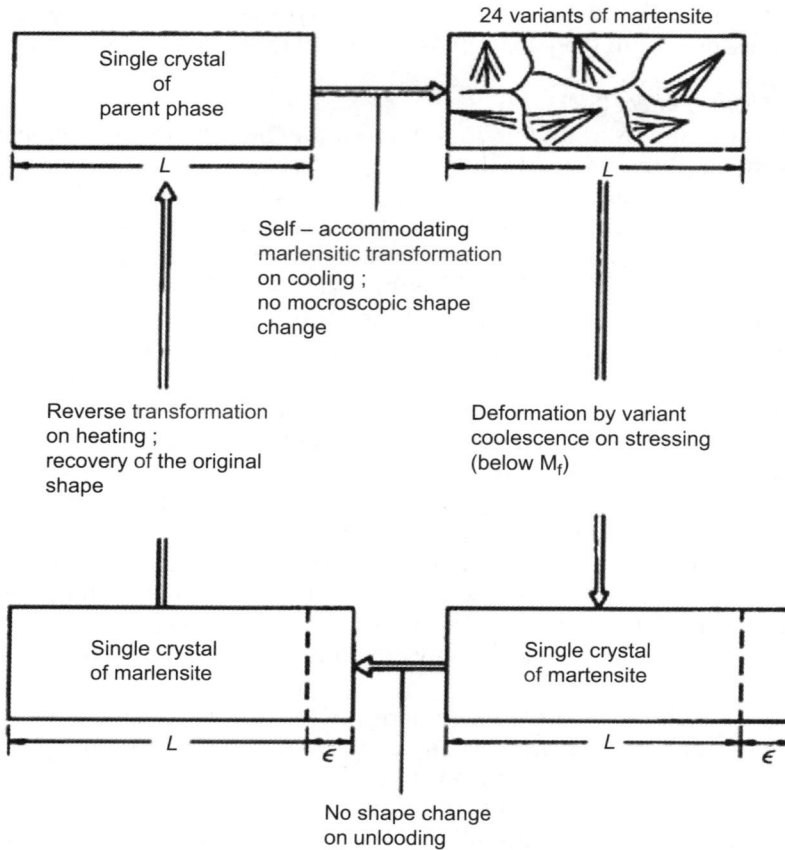

Figure 31 Schematic representation of deformation process associated with the shape memory effect.

memory (TWSM) involves a reversible deformation. A specimen will spontaneously deform during cooling from M_S to M_F and then undeform during heating from A_S to A_F. Such behavior occurs as a result of deformation in either the parent phase or in the martensite (Wasilewski, 1975). There are, in fact, two types of TWSM depending on the way in which the alloy is *trained* (Schroeder and Wayman, 1977b):

- *SME cycling* involves cooling a specimen below the M_F temperature, deforming it to produce a preferential martensite variant, as described earlier, and heating it to above the A_F temperature. This procedure is repeated several times, the manner of martensite deformation (e.g. tension, compression, bending) remaining unchanged.
- *SIM cycling* involves the deformation of a specimen above the M_S temperature in order to produce SIM, followed by reversal of the SIM when the applied load is released. This process is also repeated several times using the same means of stressing each time.

The two-way shape memory is observed after both SME and SIM cycling, and the terms *SME training* and *SIM training* have been suggested to describe the process of cycling in the desired manner

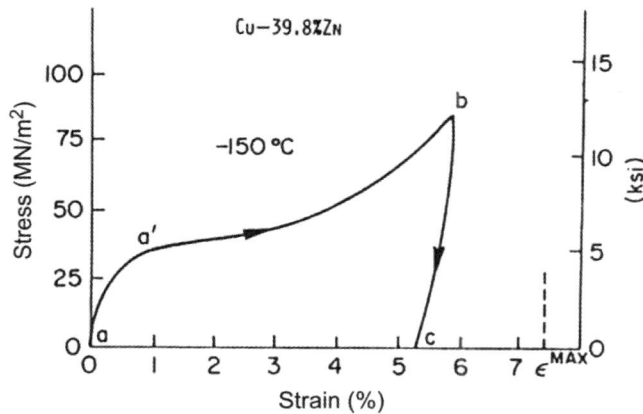

Figure 32 Stress–strain curve for a Cu–Zn shape-memory alloy deformed below its M_F temperature. The low flow stress a' associated with the martensite deformation is to be noted.

(Schroeder and Wayman, 1977b). In either case, the two-way behavior comes from the preferential formation and reversal of a trained variant of martensite formed after either SME or SIM cycling and cooling the specimen from M_S to M_F. That is, the training of a specimen to form a preferential variant of martensite upon cooling to M_F can result from prior deformation of the martensite formed by cooling followed by heating, or SIM cycling above the M_S temperature. In both cases the first part of the TWSM occurs upon cooling. In either case, the result is that a major portion of the martensite in a given specimen corresponds to a preferred variant, which produces a spontaneous strain during cooling, the amount of the strain corresponding to the shape strain of the preferred variant.

The effectiveness of SME training is found to be inferior to that of SIM training. Nevertheless, in both cases, the preferential formation of a selected variant of martensite is explained by the progressive introduction of a pattern of stresses or defects into the material during the training processes. These stresses (or defects) apparently become better defined and established with increased cycling.

9.5.6.4 The Engine Effect in Shape Memory Alloys

Although SME materials generally deform in the martensitic condition at comparatively low stresses, as seen in **Figure 32**, surprisingly large stresses are generated when the deformed martensite is heated from A_S to A_F. As a case in point (Jackson et al., 1972), a nearly equiatomic Ni–Ti alloy when martensitic will undergo deformation (as at point a' of **Figure 32**) at a stress of about 70 MN/m^2. Yet when it is constrained and heated to A_F it is found that a *thermomechanical recovery stress* of about 700 MN/m^2 is generated. In other words, heat can be used to create a mechanical force that can do work. This is the basis for the design of engines based on shape memory alloys (Goldstein and McNamara, 1978).

9.5.6.5 Pseudoelastic Effects

Pseudoelasticity refers to a situation where large strains, in excess of the elastic limit, are completely recovered upon unloading at a constant temperature. There are two categories, *superelasticity* and the *rubberlike* effect (Otsuka and Wayman, 1977). In superelasticity, the martensite is stress-induced and reverts to the parent phase on the removal of stress; the stress therefore contributes a mechanical driving

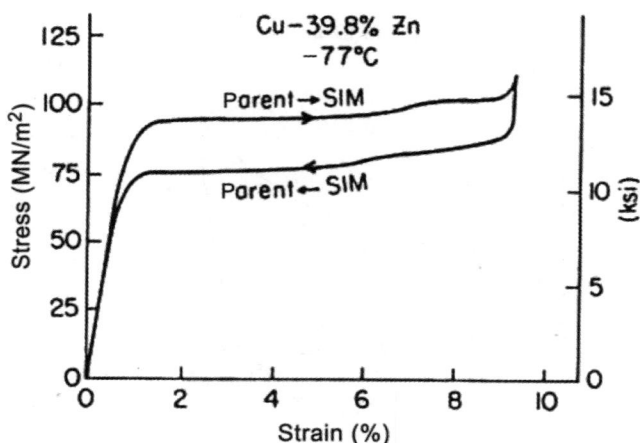

Figure 33 Stress–strain curve for a thermoelastic Cu–Zn shape memory alloy deformed above the M_S temperature, showing superelastic behavior as a consequence of the formation and reversion of a reversible, stress-induced martensite.

force for transformation. By contrast, rubberlike behavior involves the deformation of existing martensite with no phase transformation.

9.5.6.6   Pseudoelastic Effects: Superelasticity

The application of a suitable stress above the M_S temperature (but below M_d) induces the formation of martensite. In thermoelastic alloys, this SIM disappears when the stress is released, giving rise to superelasticity. A stress–strain curve showing typical superelastic behavior for a single-crystal sample is shown in **Figure 33**. Two plateau regions are seen. The upper plateau corresponds to the formation of a preferred variant of SIM plates. The preferred variant is that whose shape strain best complies with the applied stress (Schroeder and Wayman, 1977a). The parallel plates of martensite nucleate lengthen and coalesce, leading eventually to a sample that is a single crystal of martensite, at the end of the upper plateau region in **Figure 33**. Upon releasing the load, the stress–strain curve follows the lower plateau region, which involves the formation of parallel plates of only one variant of the parent phase. The levels of the plateau stresses shown in **Figure 33** depend on the test temperature. The upper plateau stress is naturally zero at the M_S temperature. The temperature dependence of the stress to produce SIM is shown in **Figure 34**. The slope of the M_S variation with temperature may be used to estimate the latent heat of transformation (Otsuka and Wayman, 1977). Stress–strain curves of the type shown in **Figure 34** are frequently referred to as *superelastic loops*.

9.5.6.7   Pseudoelastic Effects: Rubberlike Behavior

Rubberlike behavior occurs in the martensite phase of some alloys that undergo thermoelastic transformation. It does not involve a phase transformation, but is related to the reversible movement of transformation twin boundaries or martensite boundaries.

It is useful to compare the usual SME (**Figure 31**) with rubberlike behavior, since the latter often occurs in SME alloys that are aged in their martensitic state. In the single-crystal SME, applied deformation can convert a microstructure that consists of many variants of martensite into a single crystal of

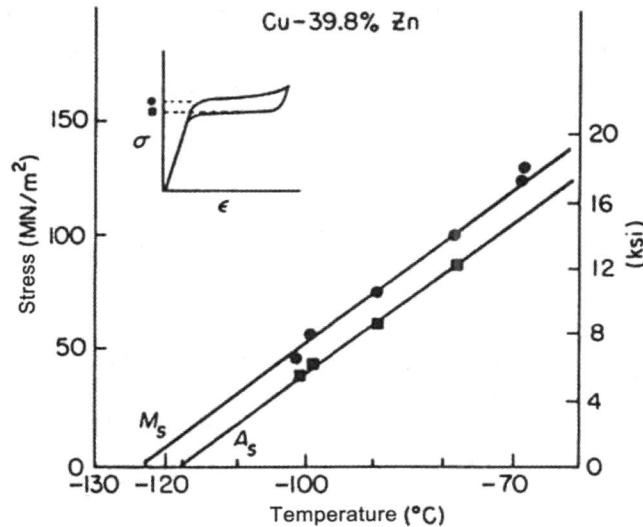

Figure 34 Stress–temperature plot of experimental data showing the temperature dependence of the applied stress required to produce stress-induced martensite in a Cu–39.8% Zn alloy.

martensite. Heating then displacively transforms the martensite into the parent phase single-crystal, thereby removing the applied deformation.

However, if the microstructure consisting of many variants of martensite is first allowed to age, and then deformed to a single crystal of martensite, the removal of the applied stress causes the single crystal of martensite to revert to the many variants, thereby reversing the applied deformation. This restoring force only arises after aging, and its origins lie in the changes that occur in the state of order on aging (Ahlers et al., 1978; Marukawas and Tsuchiya, 1995; Tsuchiya et al., 1995).

The long-range order in the parent structure is usually imperfect because the parent phase occurs at relatively high temperatures. On quenching, the martensite inherits this imperfect order. Aging at ambient temperature can, with the help of quenched-in excess vacancies, allow the state of order in the martensite to become more perfect. The martensite also has a lower crystallographic symmetry than the parent phase, so the *orientation* of the more perfect state of order is different for each martensite variant. This makes the conversion of one martensite variant into another more difficult, since the conversion would lead to wrongly oriented order in the growing variant. The resulting increase in energy provides the restoring force, which tends to make the single-crystal of martensite revert to the original many variants when the applied stress is removed.

9.5.6.8 Martensite-to-Martensite Transformations

Figure 35 shows the deformation behavior of a Cu–39.8Zn wt% single crystal strained at −88 °C, some 35 °C above the M_S temperature (Schroeder and Wayman, 1978) The first upper plateau corresponds to the formation of SIM as described previously. However, the second plateau, which starts at about 9% strain, corresponds to a second martensitic *transformation* that is stress-induced from the first martensite "mother". The two lower plateaux are a result of the reverse transformations, occurring in an inverse sequence. By means of these successive martensite-to-martensite transformations, completely recoverable strains as high at 17% can be realized. **Figure 35** shows a double superelastic loop. It should be

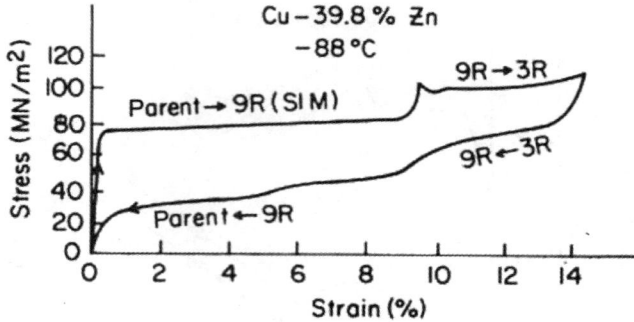

Figure 35 Stress–strain diagram showing the deformation behavior of a Cu–39.8%Zn alloy single crystal specimen deformed above the M_S temperature. A double superelastic loop is observed, which is the result of two successive stress-induced martensitic transformations. The first is that depicted by **Figure 33**, while the second originates from the first-formed martensite under the influence of stress.

noted that the second SIM in the Cu–Zn alloy can be formed from the first SIM, or from the normal thermally formed (upon cooling) martensite.

In the case of the Cu–Zn alloy discussed above, the initial martensite has a 9R structure, and additional deformation changes this into a 3R structure. This occurs by shearing on the basal (close-packed) plane of the 9R such that the structural stacking sequence is changed from ...ABCBCACAB... to ...ABCABCABC...

9.5.6.9 Magnetic Shape Memory

A material that in which the martensite is ferromagnetic has the capacity to respond to an externally applied magnetic field. **Figure 36** illustrates twin related variants of martensite. The application of a magnetic field causes the variant better aligned to H to grow by the translation of the twin boundary, resulting in a macroscopic change in shape, and vice versa. The changes can in principle occur much more rapidly than thermally driven shape memory alloy, and hence can be used in high-frequency

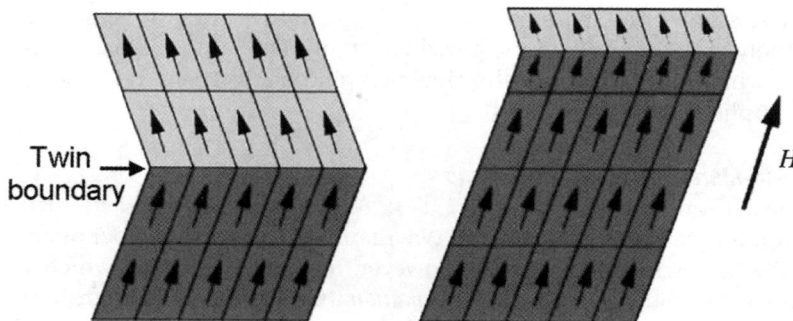

Figure 36 Twin related variants of ferromagnetic martensite. The arrows within the martensite represent magnetic moments. After Enkovaara et al. (2004).

applications. The conditions necessary for the material to exhibit such a memory effect are as follows (Enkovaara et al., 2004; Vassiliev, 2002):

- the material must obviously exhibit a transformation into ferromagnetic martensite;
- the magnetic anisotropy energy must be greater than that required to translate the twin boundary, so that the field does not result in a realignment of the magnetic moments;
- it is advantageous for the magnetic moment per atom to be large so that the applied field needed to effect the growth of the favored variant is relatively small.

Because the deformation is due to changes in the fractions of variants of martensite, the shape deformation can be large, much larger than magnetostrictive strains. Some 6% reversible strain has been observed in the Ni_2MnGa alloys, which have a Curie transition at 376 K and form thermoelastic martensite below 202 K. The transformation is from the cubic to tetragonal lattice, with a large increase in magnetic anisotropy, the easy axis being along the c-axis of the tetragonal form.

9.6 Crystallographically Similar Transformations

9.6.1 The Bainite Transformation in Steels

It is widely agreed that the bainite reaction in steels has certain martensitic characteristics but there may be some diffusion of carbon at some stage of the reaction. The surface relief accompanying the growth of bainite is an invariant-plane strain with a large shear component. There are also other martensitic-like features such as an irrational habit plane and transformation substructure, and various studies of the austenite–bainite orientation relationship imply that the Bain correspondence between the austenite and bainitic ferrite exists.

The detailed role of carbon has been clarified by Bhadeshia (2001a) and Caballero et al. (2012a, 2012b) that it does not partition during growth. Indeed, there is no diffusion of any solute during the growth of bainite, but carbon only partitions subsequent to transformation (see Chapter 21, Volume III on Steels). The atoms are transferred in a coordinated manner across the bainite–austenite interface. Consequently, it is reasonable to expect the phenomenological crystallographic theory to apply to the bainite transformation, and this is indeed found to be the case in practice.

9.6.2 Oxides and Hydrides

Various metallic oxides and hydrides form with a plate morphology and exhibit an invariant-plane strain shape deformation with a large shear component. Such plates are also internally twinned or internally slipped. An example of an internally twinned plate in a tantalum oxide is shown in **Figure 37**. In this case the observed crystallographic features of the oxide plates are accurately predicted by the phenomenological crystallographic theory developed for martensitic transformations (van Landuyt and Wayman, 1968a, 1968b). A lattice correspondence between therefore exists between the metallic atoms in the two phases and remains intact despite the occurrence of interstitial diffusion (of oxygen) during the phase change.

9.6.3 Diffusional–Displacive Transformations

The formation of the ordered orthorhombic CuAu II phase from the disordered cubic phase parent at about 390 °C generates an invariant-plane strain relief with a large shear component, and conforms well with the phenomenological crystallographic theory of martensitic transformations. This led

Figure 37 Transmission electron micrograph showing plates of the suboxide TaO_y following the oxidation of tantalum. The striations with the plates are twins and the general crystallography of the process is well described by the phenomenological theory developed for martensitic transformations.

Smith and Bowles (1960) to conclude that the mechanism by which the internally twinned plates of the CuAu II form is similar to that of a martensite, even though place changes between neighboring atoms are necessarily involved during the ordering reaction. The CuAu II ordering reaction was later considered by Bowles and Wayman (1979) who suggested that the place exchanges between neighboring atoms, required to achieve the ordering, are redundant in the sense that they do not contribute to the transformation shape change so that the theory still prescribes the displacements involved in producing the lattice change. They proposed that the growth of CuAu II plates occurs by the glide of transformation dislocations, i.e. the structure of the interface is glissile as required for displacive transformation.

Christian (1997) has analyzed this and other transformations in which there is apparently a change in the concentration of substitutional solutes and yet the crystallography is consistent with martensitic transformation. In a displacive transformation, there is a lattice correspondence between the parent and product phases, i.e. labeled vectors, planes and unit cells of the parent lattice become, as a result of a homogeneous deformation, corresponding vectors and planes in the new phase. This relationship can be expressed mathematically that the corresponding parent and product vectors $\mathbf{u}$ and $\mathbf{v}$ by the homogeneous deformation $\mathbf{B}$ as $\mathbf{v} = \mathbf{Bu}$. **Figure 38a** illustrates the corresponding unit cells of the parent and product crystals, and hence the shape change due to the transformation. In this case the interface is large in size so that free surfaces need not be taken into consideration—the only mass flow relevant, therefore, is that normal to the interface. It is evident that atomic jumps of the type illustrated at 1 and 2 cannot eliminate the shape change, but can result in a change in order or even composition. This illustrates a diffusional–displacive transformation and yet maintains a crystallography that is consistent with the theory of martensite.

It is important to note that since the shape change is not eliminated in this diffusional–displacive transformation, its consequences remain relevant, exactly in the same manner as a diffusionless martensitic transformation. Thus, the large strain energy due to the shape change must be accounted for in any thermodynamic treatment, and the shape of the transformation product will be overwhelmingly controlled by the need to minimize this strain energy.

(a)

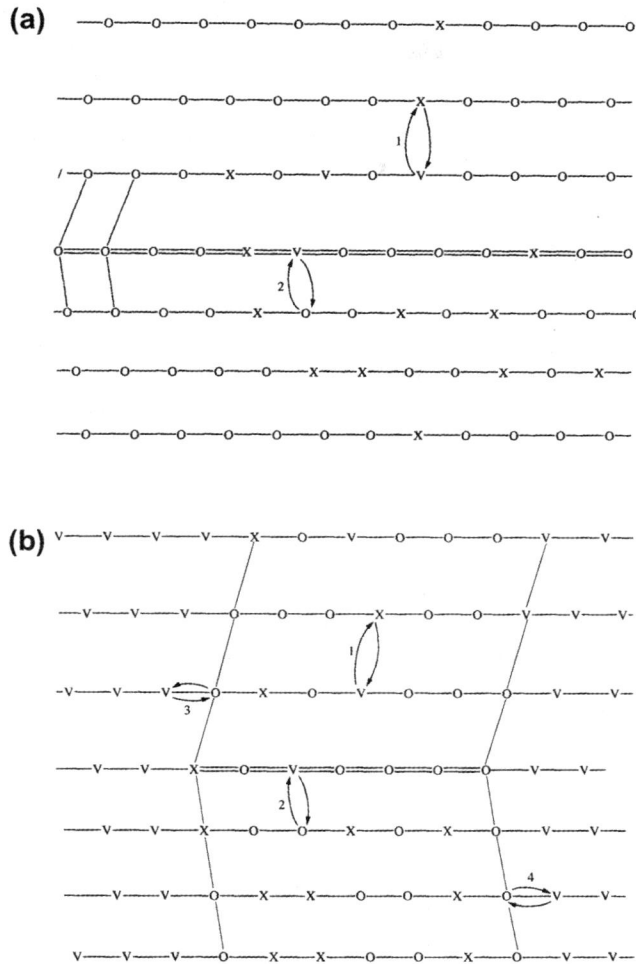

(b)

Figure 38 In these diagrams, O, X and V represent solvent atoms, solute atoms and vacancies respectively. (a) Coherent interface (double horizontal line) separating parent and product lattices. The change in shape of the unit cell is illustrated. (b) Finite coherent interface. Note that the delineated particle is surrounded by vacancies, i.e. free surfaces. After Christian (1997).

In contrast to the "diffusional–displacive" transformation where the crystallography, including the shape deformation, is preserved in spite of internal interchanges of atoms, **Figure** 38b illustrates a case where the sites predicted by the lattice correspondence are not conserved (Christian, 1997). If the vacancy at the free surface at the position marked 4 migrates horizontally through the entire crystal than that row of atoms is translated through one atomic distance. If sufficient vacancies flow from one free surface to the other, then the shear component of the shape change could be eliminated. In other words, diffusion would be required over a distance comparable to the dimensions of the product. Such flow has been described as reconstructive diffusion (Bhadeshia, 1982, 1985), the absence of which would preserve the shape deformation and the change amounts to a reconstructive transformation.

Finally, it is worth pointing out that some of the evidence for diffusional–displacive transformations where the crystallography is consistent with the theory of martensite but the composition of substitutional solutes changes bears more detailed study. The diffusional–displacive mechanism is claimed for the Ag–Cd system (Muddle et al., 1994), but the composition change could happen after diffusionless transformation once the system equilibrates over the heat-treatment periods implemented (Mujahid and Bhadeshia, 1999).

9.7 Omega Phase Formation

The $\beta \rightarrow \omega$ phase change (cubic–hexagonal) involves correlated atomic displacements and occurs in Ti, Zr and Hf alloys and in β-phase alloys of the noble metals. It has been extensively investigated. The reader is referred to two reviews: Sass (1972) and Williams et al. (1973). The $\beta \rightarrow \omega$ phase change occurs both athermally and isothermally, with the same mechanism for both paths (Williams et al., 1973).

Omega phase formation can occur reversibly and without diffusion at quite low temperatures, and the ω-phase cannot be suppressed by rapid quenching. These characteristics parallel those found in martensitic transformations, but unlike martensite, the ω-phase "particles" exhibit a cuboidal or ellipsoidal morphology, depending upon the relative parent and product lattice parameters. At least during the early stages of transformation, the ω particles are quite small (≈ 1.5 nm), and they are aligned along $\langle 111 \rangle$ directions. Their observed number-density is very high (10^{18}–10^{19} cm^{-3}). The smallness of the particles and their high number-density suggest that their nucleation is not as serious a barrier as their growth, in contrast to martensitic transformations where the opposite is the case. **Figure 39** shows an example of omega particles.

A number of alloys exhibiting the $\beta \rightarrow \omega$ transformation feature the following orientation relationship:

$$(1\,1\,1)_\beta \parallel (0\,0\,0\,1)_\omega; \tag{24}$$

$$[1\,\bar{1}\,0]_\beta \parallel [2\,\bar{1}\,1\,0]_\omega. \tag{25}$$

Interestingly, these same alloys (although at different compositions) also undergo a $\beta \rightarrow \alpha'$ martensitic transformation in which case the orientation relationship is of the Burger's type:

$$(1\,1\,0)_\beta \parallel (0\,0\,0\,1)'_{\alpha'}; \tag{26}$$

$$[1\,\bar{1}\,1]_\beta \parallel [1\,1\,\bar{2}\,0]_{\alpha'}. \tag{27}$$

It should be noted that the $\beta \rightarrow \omega$ orientation relationship shows a multiplicity of only four, which, when compared to that of martensitic transformations (where the multiplicity is much higher) suggests that the respective lattice correspondences are different.

Another comparison can be made between the diffusionless $\beta \rightarrow \omega$ and $\beta \rightarrow \alpha'$ (martensitic) transformations. When martensite forms, the atomic displacements are mostly homogeneous, as given by a Bain-type distortion. Each atom (apart from additional shuffles in some cases) is homogeneously

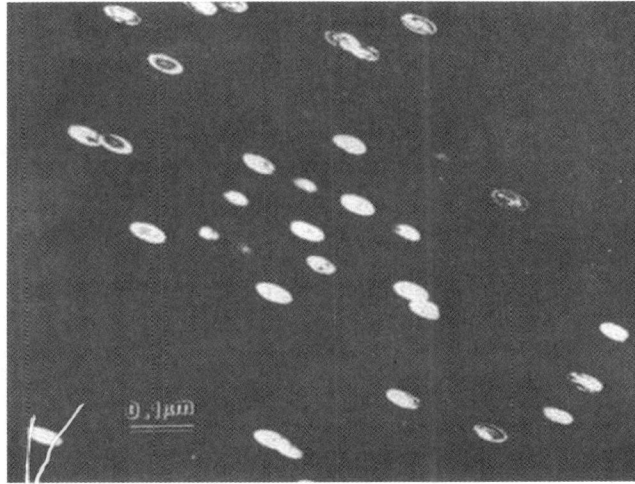

Figure 39 Transmission electron micrograph (dark field) showing ω-particles formed at 480 °C in a Ti–Mo–Sn–Zr alloy.

transported to its final position according to the Bain deformation and correspondence. But in the $\beta \to \omega$ transformation, some of the corresponding atoms undergo movements whereas others do not. **Figure 40** (Sass, 1972) shows a $(\bar{1}01)$ section through a β-unit cell. Atoms A, B, E and F remain in place during the $\beta \to \omega$ transformation, whereas atoms C and D undergo shuffle movements in opposite sense along $[\bar{1}\,\bar{1}\,\bar{1}]_{\beta}$, in the end reaching the position $\frac{1}{2}[111]$, initially having been situated at the $\frac{2}{3}$ and $\frac{1}{3}$ positions. Accordingly, the 1st, 4th, 7th, 10th, etc. planes remain unchanged while 2 and 3, 5 and 6, 8 and 9, etc., shuffle past each other in opposite senses. The $\beta \to \omega$ atomic movements are formally consistent with atom movements given by a $\frac{2}{3}\langle 111 \rangle$ displacement wave having nodes at $n\langle 111 \rangle_{\beta}$ positions ($n =$ any integer).

Advances in understanding the $\beta \to \omega$ transformation have come from the application of transmission electron microscopy and diffraction. Diffraction patterns frequently exhibit networks of diffuse intensity (sheets of intensity on $\langle 111 \rangle_{\beta}$ planes), which obscure the identification of diffraction maxima from the ω-phase. These diffuse diffraction effects are observed at temperatures above which those from the ω-phase are clearly identified, and the former have been ascribed to pretransformation "linear defects".

Impurities have a marked influence on the $\beta \to \omega$ transformation. For example, the presence of 1200 ppm of oxygen can lower the ω-start temperature by about 600 K, presumably because oxygen stiffens the matrix and depresses the transformation start temperature. Apparently the oxygen atoms somehow interact with the pretransformation linear defects along $\langle 111 \rangle_{\beta}$ and impede their ordering.

There are some outstanding problems with ω-phase formation, including the possible role of lattice vacancies in promoting linear defects and hence nucleation, the high nucleation frequency and low particle growth rate, and the extreme embrittlement of the matrix β-phase after the ω-phase forms.

9.8 Phase Changes and Charge Density Waves

A charge density wave (CDW) is a static modulation of conduction electrons and is a Fermi-surface-driven phenomenon usually accompanied by a periodic distortion of the lattice. The electronic

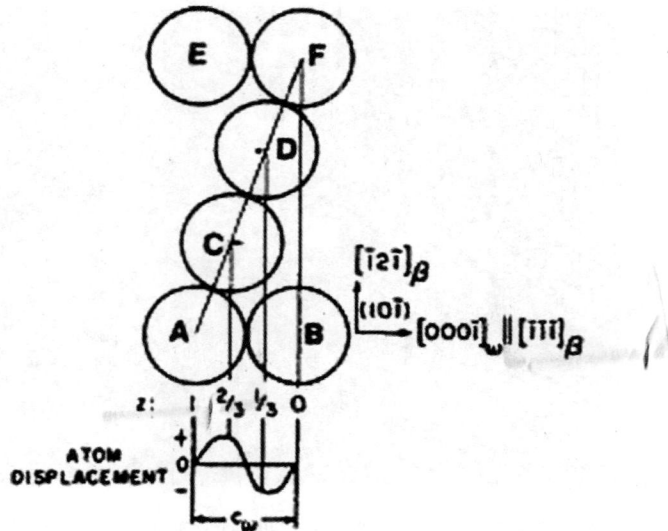

Figure 40 Schematic diagram showing atomic movements involved during ω-phase formation. This is a $(10\bar{1})$ section through the parent bcc unit cell. The $\frac{2}{3}\langle 111 \rangle$ longitudinal displacement waves cause atomic movements (in direction indicated by the two arrows) that are those needed to form the ω-structure (Sass, 1972).

energy of the solid is lowered by the distortion, the attendant strain energy of which is more than compensated by the reduction in electronic energy. The present understanding of CDWs in conductors follows from the pioneering work of Peierls (1955), Kohn (1959) and Overhauser (1968, 1971). A detailed description has been published by Wilson et al. (1975).

Peierls (1955) suggested that a simple one-dimensional solid is susceptible to a periodic lattice distortion that can lower the total energy. **Figure** 41a shows the familiar $E-k$ relationship for a one-dimensional monovalent metal. The first Brillouin zone with boundary at $\pm \pi/a$ is half occupied.

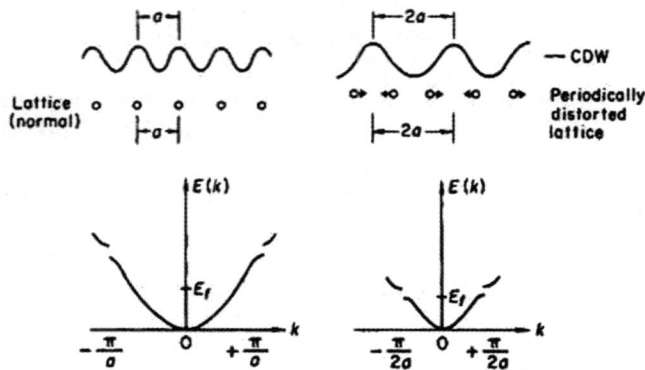

Figure 41 Showing a periodic lattice distortion, charge density waves, and energy band gaps before and after a periodic lattice distortion in a hypothetical one-dimensional metal. Note the overall lowering of Fermi level following the periodic lattice distortion.

The electronic charge density is of periodic form, being maximum in the vicinity of the ion cores. In **Figure 41b** the one-dimensional lattice has been distorted periodically, causing the formation of a CDW of period $2a$ associated with the new superlattice. Since the lattice has been doubled in real space, a new band gap will appear at $\pm\pi/2a$ in k-space, which permits some electrons to "spill down", thus lowering the Fermi energy. A favorable Fermi surface geometry is necessary for the formation of a CDW, which will most likely occur when the shape of the Fermi surface permits a connection by the same wavevector $\mathbf{Q}$, i.e. $\mathbf{Q} = 2\mathbf{k}_f$. This modulation with wavevector $\mathbf{Q}$ will modify the Fermi surface by creating gaps at these nested positions. It is possible that the "gain" by creating energy gaps may overcome the energy "cost" arising from the strain associated with the periodic lattice distortion, thus allowing the formation of a CDW. In other words, a structural change will occur when the CDW formation is accompanied by ion displacements that stabilize the charge perturbation. Another requirement for forming a CDW is a strong electron-phonon coupling, required to permit ionic displacements to reduce the otherwise prohibitive Coulomb energy, and precursor phenomena such as a soft phonon mode (Chan and Heine, 1973) might occur above the transition temperature to assist the CDW instability. The normal crystalline periodicity is altered in the presence of a CDW. However, the wavevector of a CDW is determined by the Fermi surface and is therefore not necessarily an integral fraction of a reciprocal lattice vector of the undistorted parent phase. Consequently, an "incommensurate" phase may result, which is considered to have lost its translational symmetry (Axe, 1977).

The incommensurate state described above may not actually correspond to the lowest possible energy state and, accordingly, the CDW or the lattice may undergo a further distortion that makes the two "commensurate" in which case the CDW wavevector is an integral fraction of the underlying lattice. The commensurate state is usually referred to as a "locked-in" state. Thus, there can be two phase changes associated with a CDW formation: the normal (parent)-to-incommensurate transition (usually second order) and the incommensurate-to-commensurate transformation (first order). Since a CDW is accompanied by a lattice distortion, diffraction techniques (electron, neutron, X-ray) can be used to reveal satellite reflections appearing near the Bragg reflections of the parent phase as a consequence of the new superlattice associated with the formation of the CDW. These satellites are separated from the associated Bragg reflections by a reciprocal lattice vector determined by the CDW wavevector.

Besides the incommensurate state mentioned above, (McMillan, 1976) introduced the concept of *discommensurations*, which are narrow out-of-phase regions (defects) between large in-phase (commensurate) regions. Accordingly, the incommensurate-to-commensurate change is viewed as a defect (discommensuration) "melting" transition.

A study (Hwang et al., 1983) of the so-called "premartensitic effects" in Ti–Ni alloys containing a few percent of Fe in order to suppress M_S while at the same time leaving the temperature regime of the "premartensite" phenomena essentially unchanged. The "premartensitic" resistivity anomaly (increase with decreasing temperature) in Ti–Ni type alloys has been ascribed (Hwang et al., 1983) to the formation of a three-dimensional CDW that appears to evolve in two stages. In the first, the B2 (CsCl) parent is gradually (second order) distorted into an incommensurate phase with decreasing temperature. The incommensurate phase is most simply described as one with "distorted cubic" symmetry. Its gradual transition involves the appearance of new $\frac{1}{3}\langle 110 \rangle_{B2}$ superlattice reflections, which intensify with decreasing temperature. At this stage, the $\frac{1}{3}\langle 110 \rangle$ superlattice reflections are in slightly irrational (incommensurate) positions. As the temperature is further decreased these superlattice reflections "lock in" to commensurate positions that are precise multiples of $\frac{1}{3}\langle 110 \rangle_{B2}$, and the resulting structure is rhombohedral (designated the R-phase) as a consequence of a homogeneous distortion that involves an expansion along $\langle 111 \rangle$ cube diagonals (and a small contraction in all directions in the plane normal to $\langle 111 \rangle$). This incommensurate-to-commensurate transformation is a first order in nature. Thus, the

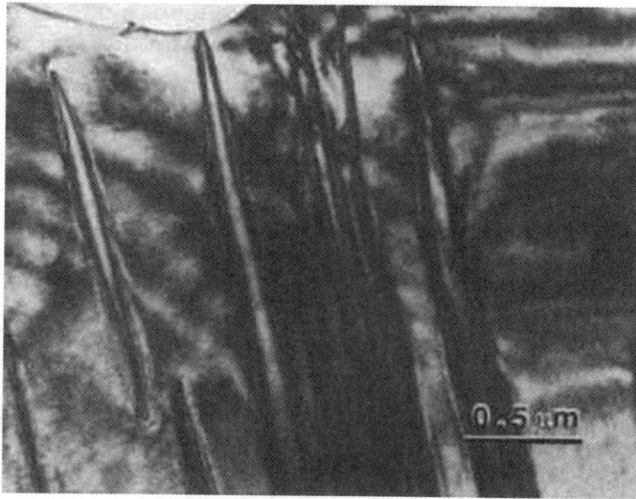

Figure 42 Twin-related domains as a consequence of an incommensurate-to-commensurate phase change in a $Ti_{50}Ni_{47}Fe_3$ at% alloy.

events observed during the so-called *premartensitic state* of Ti–Ni–X alloys identify with the sequence: B2(CsCl) → distorted cubic (incommensurate) → R, rhombohedral (commensurate). In addition, microdomains typical of antiphase domains in ordered alloys have been identified using dark field electron microscope images formed by using the new superlattice reflections.

Finally, it should be mentioned that the incommensurate-to-commensurate transformation in Ti–Ni–Fe alloys is also associated with the appearance of twin-like domains, as shown in **Figure** 42. The twin plane has been identified as a {110} plane. It would appear that the only difference between the "twins" and the "matrix" within which they exist is that the sense of the ⟨111⟩ expansion (Bain distortion) occurs along different c axes in each. In this respect, the formation of twins during the incommensurate-to-commensurate change appears to be a result of overall strain compensation, not unlike that occurring in a martensitic transformation. But in the case of the Ti–Ni–Fe alloys, when the actual martensite is formed at a lower temperature, the {110} twins and the commensurate R-phase are completely destroyed by the advancing martensite interface, and, seemingly, the "premartensitic" commensurate twinned structure presents no obstacle to the growth of the martensite plates.

Acknowledgments

We would like to acknowledge many helpful comments from numerous colleagues and members of his research group, and to thank particularly Professors G. Krauss, K. Shimizu, M. Umemoto and J.C. Williams for some of the figures.

References

Ahlers, M., Barceló, G., Rapacioli, R., 1978. A model for the rubber-like behaviour in Cu–Zn–Al martensites. Scr. Metall. 12, 1075–1078.
Axe, J.D., 1977. Electron–Phonon Interactions and Phase Transitions. Plenum Press, New York, USA.
Bain, E.C., 1924. The nature of martensite. Trans. AIME 70, 25–46.

Bhadeshia, H.K.D.H., 1982. Bainite: mobility of the transformation interface. J. Phys. Colloque C4 43, 449–454.

Bhadeshia, H.K.D.H., 1985. Diffusional formation of ferrite in iron and its alloys. Prog. Mater. Sci. 29, 321–386.

Bhadeshia, H.K.D.H., 2001a. Bainite in Steels, second ed. Institute of Materials, London.

Bhadeshia, H.K.D.H., Honeycombe, R.W.K., 2006. Steels: Microstructure and Properties, third ed. Butterworth-Heinemann, London.

Bhattacharjee, D., Knott, J.F., Davis, C.L., 2004. Charpy-impact-toughness prediction using an "effective" grain size for thermomechanically controlled rolled microalloyed steels. Metall. Mater. Trans. A 35, 121–130.

Bowles, J.S., Mackenzie, J.K., 1954. The crystallography of martensite transformations, part I. Acta Metall. 2, 129–137.

Bowles, J.S., Wayman, C.M., 1979. The growth mechanism of AuCu II plates. Acta Metall. 27, 833–839.

Brooks, J.W., Loretto, M.H., Smallman, R.E., 1979a. Direct observations of martensite nuclei in stainless steel. Acta Metall. 27, 1839–1847.

Brooks, J.W., Loretto, M.H., Smallman, R.E., 1979b. In situ observations of martensite formation in stainless steel. Acta Metall. 27, 1829–1838.

Bullough, R., Bilby, B.A., 1956. Continuous distributions of dislocations: surface dislocations and the crystallography of martensitic transformations. Proc. Phys. Soc. B 69, 1276–1286.

Caballero, F.G., Miller, M.K., Garcia-Mateo, C., Cornide, J., 2012a. New experimental evidence of the diffusionless transformation nature of bainite. J. Alloys Compd. http://dx.doi.org/10.1016/j.jallcom.2012.02.130.

Caballero, F.G., Miller, M.K., Garcia-Mateo, C., Cornide, J., Santofimia, M.J., 2012b. Temperature dependence of carbon supersaturation of ferrite in bainitic steels. Scr. Mater. 67, 846–849.

Chan, S.-K., Heine, V., 1973. Spin density wave and soft phonon mode from nesting Fermi surfaces. J. Phys. F 3, 795–809.

Chatterjee, S., Wang, H.S., Yang, J.R., Bhadeshia, H.K.D.H., 2006. Mechanical stabilisation of austenite. Mater. Sci. Technol. 22, 641–644.

Christian, J.W., 1951. A theory for the transformation in pure cobalt. Proc. Roy. Soc. Lond. A 206A, 51–64.

Christian, J.W., 1958. Accommodation strains in martensite formation, the use of the dilatation parameter. Acta Metall. 6, 377–379.

Christian, J.W., 1979. Thermodynamics and kinetics of martensite. In: Olson, G.B., Cohen, M. (Eds.), International Conference on Martensitic Transformations ICOMAT '79. Alpine Press, Massachusetts, USA, pp. 220–234.

Christian, J.W., 1982. Deformation by moving interfaces. Metall. Trans. A 13, 509–538.

Christian, J.W., 1997. Lattice correspondence atomic site correspondence and shape change in diffusional-displacive phase transformations. Prog. Mater. Sci. 42, 109–124.

Christian, J.W., 2003a. Theory of Transformations in Metal and Alloys, Part I, third ed. Pergamon Press, Oxford, U. K.

Christian, J.W., 2003b. Theory of Transformations in Metal and Alloys, Part II, third ed. Pergamon Press.

Christian, J.W., Crocker, A.G., 1980. In: Nabarro, F.R.N. (Ed.), Dislocations in Solids, vol. 3. North Holland, Amsterdam, Holland, Ch. 11, 165–252.

Conrad, H., 1964. Thermally activated deformation of metals. J. Metals 145, 582–588.

Date, M., Yamasishi, A., Yosida, T., Sugiyama, K., Kijima, S., 1986. Phase transitions in high magnetic field. J. Magn. Magn. Mater. 54–57, 627–631.

Dorn, J.E., 1968. Dislocation Dynamics. In: Rosenfield, A.R., Hahn, G.T., Bement, A.L., Jaffee, R.I. (Eds.). McGraw-Hill, New York, pp. 27–55.

Enkovaara, J., Ayuela, A., Zayak, A.T., Entel, P., Nordström, L., Dube, M., Jalkanen, J., Impola, J., Nieminen, R.M., 2004. Magnetically driven shape memory alloys. Mater. Sci. Eng. A 378, 52–60.

Goldstein, D.M., McNamara, L.J. (Eds.), 1978. Proceedings of the NITINOL Heat Engine Conference. Tech. Rep. NSWC MP 79–441. U.S. Naval Surface Weapons Centre, Maryland, U.S.A.

Gourgues, A.F., Flower, H.M., Lindley, T.C., 2000. Electron backscattering diffraction study of acicular ferrite, bainite, and martensite steel microstructures. Mater. Sci. Technol. 16, 26–40.

Gourgues-Lorenzon, A.F., 2007. Application of electron backscatter diffraction to the study of phase transformations. Int. Mater. Rev. 52, 65–128.

Hwang, C.M., Meichle, M.B., Salamon, M.B., Wayman, C.M., 1983. Transformation behaviour of a Ti50Ni47Fe3 alloy I. premartensitic phenomena and the incommensurate phase. Philos. Mag. A, 9–30.

Jackson, C.M., Wagner, H.J., Wasilewski, R.J., 1972. 55-Nitinol-the Alloy with a Memory: It's Physical Metallurgy Properties, and Applications. NASA, Washington, D.C., U.S.A. Tech. Rep. NASA Special Publication, 5110.

Kakeshita, T., Shimizu, K., Funada, F., Date, M., 1985. Composition dependence of magnetic field induced martensitic transformations in Fe–Ni alloys. Acta Metall. 33, 1381–1389.

Kaufman, L., Cohen, M., 1958. Thermodynamics and kinetics of martensitic transformation. Prog. Metal Phys. 7, 165–246.

Kessler, H., Pitsch, W., 1967. On the nature of the martensite to austenite reverse transformation. Acta Metall. 15, 401–405.

Kim, Y.M., Shin, S.Y., Lee, H., Wang, B., Lee, S., Kim, N.J., 2007. Effects of molybdenum and vanadium addition on tensile and charpy impact properties of API X70 linepipe steels. Metall. Mater. Trans. A 38, 1731–1742.

Kohn, W., 1959. Image of the Fermi surface in the vibration spectrum of a metal. Phys. Rev. Lett. 2, 393–394.

Krauss, G., 2005. Steels: Processing, Structure and Performance. ASM International, Metals Park, Ohio, USA.

Krivoglaz, M.A., Sadovskiy, V.D., 1964. Effect of strong magnetic fields on phase transformations. Fiz. Met. Metalloved. 18, 23–27.

Kundu, S., Hase, K., Bhadeshia, H.K.D.H., 2007. Crystallographic texture of stress–affected bainite. Proc. Roy. Soc. A 463, 2309–2328.

Kurdjumov, G., Khandros, L.G., 1949. First reports of the thermoelastic behaviour of the martensitic phase of Au–Cd alloys. Dokl. Akad. Nauk, SSSR 66, 211–213.

Lambert-Perlade, A., Gourgues, A.F., Pineau, A., 2004. Austenite to bainite phase transformation in the heat-affected zone of a high strength low alloy steel. Acta Mater. 52, 2337–2348.

Ling, H.C., Owen, W.S., 1981. A model of the thermoelastic growth of martensite. Acta Metall. 29 (10), 1721–1736. URL: www.scopus.com.

Mackenzie, J.K., Bowles, J.S., 1954a. The crystallography of martensite transformations II. Acta Metall. 2, 138–147.

Mackenzie, J.K., Bowles, J.S., 1954b. The crystallography of martensite transformations III FCC to BCT transformations. Acta Metall. 2, 224–234.

Maki, T., 1990. Microstructure and mechanical properties of ferrous martensites. Mater. Sci. Forum, 56–58, 157–168.

Maki, T., Tamura, I., 1986. Shape memory effect in ferrous alloys. In: ICOMAT 86 (International Conference on Martensitic Transformations). Japan Institute of Metals, Tokyo, Japan, pp. 963–970.

Marukawas, K., Tsuchiya, K., 1995. Short-range ordering as the cause of the rubber-like behaviour in alloy martensite. Scr. Metall. Mater. 32, 77–82.

McMillan, W., 1976. Theory of discommensurations and the commensurate-incommensurate charge-density-wave phase transition. Phys. Rev. 14, 1496–1502.

Miyamoto, G., Shibata, A., Maki, T., Furuhara, T., 2009. Precise measurement of strain accommodation in austenite matrix surrounding martensite in ferrous alloys by electron backscatter diffraction analysis. Acta Mater. 57, 1120–1131.

Morito, S., Huang, X., Furuhara, T., Maki, T., Hansen, N., 2006. The morphology and crystallography of lath martensite in alloy steels. Acta Metall. 54, 5323–5331.

Muddle, B.C., Nie, J.F., Hugo, G.R., 1994. Application of the theory of martensite crystallography to displacive phase transformations in substitutional nonferrous alloys. Metall. Mater. Trans. A 25, 1841–1856.

Mujahid, S.A., Bhadeshia, H.K.D.H., 1999. An analysis of compositional data on α plates in an Ag-44Cd at.% alloy. Model. Simul. Mater. Sci. Eng. 7, 1–13.

Olson, G.B., Cohen, M., 1972. Mechanism for strain-induced nucleation of martensite. J. Less Common Metals 28, 107–118.

Olson, G.B., Cohen, M., 1976. A general mechanism of martensitic nucleation, parts i–iii. Metall. Trans. A 7A, 1897–1923.

Olson, G.B., Cohen, M., 1979. US–Japan symposium. In: Mechanical Behaviour of Metals and Alloys Associated with Displacive Transformations. Rensselaer Polytechnic Institute, New York, U.S.A., p. 7.

Olson, G.B., Owen, W.S., 1976. Stress field of a martensitic particle and the conditions for thermoelastic behaviour. In: New Aspects of Martensitic Transformations. Japan Institute of Metals, Tokyo, Japan, pp. 105–110.

Onodera, H., Goto, H., Tamura, I., 1976. Effect of volume change on martensitic transformation induced by tensile or compressive stress in polycrystalline iron alloys. In: New Aspects of Martensitic Transformations. Japan Institute of Metals, Tokyo, Japan, pp. 327–332.

Onodera, H., Tamura, I., 1979. US–Japan symposium. In: Mechanical Behaviour of Metals and Alloys Associated with Displacive Transformations. Rensselaer Polytechnic Institute, New York, U.S.A., p. 24.

Otsuka, K., Wayman, C.M., 1977. Pseudoelasticity and stress induced martensitic transformations. Int. Q. Sci. Rev. J. 2, 81.

Overhauser, A.W., 1968. Exchange and correlation instabilities of simple metals. Phys. Rev. 167, 691–698.

Overhauser, A.W., 1971. Observability of charge-density waves by neutron diffraction. Phys. Rev. B 3, 3173–3182.

Patel, J.R., Cohen, M., 1953. Criterion for the action of applied stress in the martensitic transformation. Acta Metall. 1, 531–538.

Peierls, R.E., 1955. Quantum Theory of Solids. Oxford University Press, Oxford, U.K.

Saburi, T., Nenno, S., Wayman, C.M., 1979. Shape memory mechanisms in alloys. In: Olson, G.B., Cohen, M. (Eds.), International Conference on Martensitic Transformations ICOMAT '79. Alpine Press, Massachusetts, USA, pp. 619–627.

Sandvik, B.P.J., Wayman, C.M., 1983. Lath martensite: crystallographic and substructural features. Metall. Trans. A 14, 809–822.

Sass, S.L., 1972. The structure and decomposition of Zr and Ti bcc solid solutions. J. Less Common Metals 28, 157–173.

Schroeder, T.A., Wayman, C.M., 1977a. The formation of martensite and the mechanism of the shape memory effect in single crystals of Cu-Zn alloys. Acta Metall. 25, 1375–1381.

Schroeder, T.A., Wayman, C.M., 1977b. The two-way shape memory effect and other "training" phenomena in Cu-Zn single crystals. Scr. Metall. 11, 225–230.

Schroeder, T.A., Wayman, C.M., 1978. Martensite-to-martensite transformations in Cu–Zn alloys. Acta Metall. 26, 1745–1757.

Smith, R., Bowles, J.S., 1960. The crystallography of the cubic to orthorhombic transformation in the alloy AuCu. Acta Metall. 8, 405–415.

Tadaki, T., Shimizu, K., 1970. Electron microscope study of the martensitic transformation in ordered Fe_3Pt alloy. Trans. JIM 11, 44–50.

Tamura, I., Maki, T., Hoto, H., Tomota, Y., Okada, M., 1970. In: 2nd Int. Conf. on Strength of Metals and Alloys, vol. 3. ASM, Ohio, U.S.A., p. 894.

Tamura, I., Maki, T., Shimooka, S., Okada, M., Tomota, Y., 1972. In: 3rd Conf. on High Strength Martensitic Steels. Vice Ministry of Metallurgy and Engineering, Havirov, Czechoslovakia, p. 118.

Tamura, I., Tomota, Y., Ozawa, M., 1973. In: 3rd Int. Conf. on Strength of Metals and Alloys, vol. 1. Elsevier, Netherlands, p. 611 and vol. 2 p. 540.

Timoshenko, S.P., Goodier, J.N., 1982. Theory of Elasticity. McGraw Hill International Book Company, London.

Tsuchiya, K., Tateyama, K., Sugino, K., Marukawa, K., 1995. Effect of aging on the rubber-like behavior in Cu–Zn–Al martensites. Scr. Metall. Mater. 32, 259–264.

Umemoto, M., Hyodo, T., Maeda, T., Tamura, I., 1984. Electron microscopy studies of butterfly martensite. Acta Metall. 32, 1191–1203.

van Landuyt, J., Wayman, C.M., 1968a. A study of oxide plate formation in tantalum-I growth characteristics and morphology. Acta Metall. 16, 803–814.

van Landuyt, J., Wayman, C.M., 1968b. A study of oxide plate formation in tantalum-II. Crystallographic analysis. Acta Metall. 16, 815–822.

Vassiliev, A., 2002. Magnetically driven shape memory alloys. J. Magn. Magn. Mater. 242–245, 66–67.

Vevecka, A., Ohtsuka, H., Bhadeshia, H.K.D.H., 1995. Plastic accommodation of martensite in disordered and ordered iron–platinum alloys. Mater. Sci. Technol. 11, 109–112.

Wang, C., Wang, M., Shi, J., Hui, W., Dong, H., 2008. Effect of microstructural refinement on the toughness of low carbon martensitic steel. Scr. Mater. 58, 492–495.

Wasilewski, R.J., 1975. The shape memory effect in TiNi: one aspect of stress-assisted martensitic transformation. In: Perkins, J. (Ed.), Shape Memory Effects in Alloys. Plenum Press, New York, USA, pp. 245–272.

Wayman, C.M., Shimizu, K., 1972. The shape memory ('Marmem') effect in alloys. Metal Sci. 6, 175–181.

Wechsler, M.S., Lieberman, D.S., Read, T.A., 1953. On the theory of the formation of martensite. Trans. AIME J. Metals 197, 1503–1515.

Williams, J.C., Fontaine, D.D., Paton, N.E., 1973. The ω-phase as an example of an unusual shear transformation. Metall. Trans. 4, 2701–2708.

Wilson, P.C., Murty, Y.V., Kattamis, T.Z., Mehrabian, R., 1975. Effect of homogenization on sulphide morphology and mechanical properties of rolled AISI 4340 steel. Metals Technol. 2, 241–244.

Yan, P., Güngör, O.E., Thibaux, P., Bhadeshia, H.K.D.H., 2010. Crystallographic texture of induction–welded and heat–treated pipeline steel. Adv. Mater. Res. 89-91, 651–656.

Zackay, V.F., Parker, E.R., Farh, D., Busch, R., 1967. The enhancement of ductility in high-strength steels. Trans. ASM 60, 252–259.

Biography

Harry Bhadeshia is the Tata Steel Professor of Physical Metallurgy at the University of Cambridge and Professor of Computational Metallurgy at POSTECH. He graduated with a B.Sc. from the City of London Polytechnic, followed by a Ph.D. at the University of Cambridge. His research is concerned with the theory of solid-state transformations in metals, particularly multicomponent steels, with the goal of creating novel alloys and processes with the minimum use of resources. He is the author or coauthor of more than 600 research papers and six books on the subject.

ELSEVIER

国际材料前沿丛书
International Materials Frontier Series

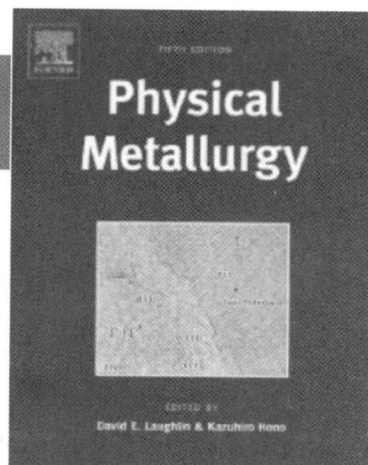

FIFTH EDITION

Physical
Metallurgy

EDITED BY
David E. Laughlin & Kazuhiro Hono

David E. Laughlin
Kazuhiro Hono

物理冶金学（第5版）
Physical Metallurgy (Fifth Edition)

（下）
影印版

中南大学出版社
www.csupress.com.cn
·长沙·

图字：18 - 2017 - 166 号

Physical Metallurgy（Fifth Edition）

David E. Laughlin，Kazuhiro Hono

ISBN：9780444537706

Elsevier（Singapore）Pte Ltd.

3 Killiney Road

#08 - 01 Winsland House I

Singapore 239519

Tel：（65）6349 - 0200

Fax：（65）6733 - 1817

First Published ＜2017＞

＜2017＞年初版

内容简介

　　该书全面系统地涵盖了材料科学领域中的相关知识，深刻地描述和解释了物理冶金学中的大多数方法，其中轻合金物理冶金、钛合金物理冶金、原子探针场离子显微镜、计算冶金和取向成像显微镜等都是当前科技发展的前沿。全书分为 3 册（上、中、下）：上册包含第 1 章 ~ 第 9 章的内容，中册包含第 10 章 ~ 第 19 章的内容，下册包含第 20 章 ~ 第 27 章的内容。

　　该书为材料类的经典书籍，第 5 版在第 4 版（1996 年）的基础上做了全面修改和扩充，且新增了几个主题以反映过去 18 年来物理冶金学的新进展。

　　本书可供相关学科领域（冶金、材料、物理、化学、生物医学等）的科研人员、工程技术人员使用，也可作为本科生、研究生等的参考用书。

作者简介

David E. Laughlin 美国匹兹堡卡内基梅隆大学教授，TMS 和 ASM International 的会士。他的主要研究方向为材料结构的电子显微镜观察、相变和磁性材料。1969 年获得美国德雷塞尔大学冶金工程专业学士学位，1973 年获得美国麻省理工学院冶金和材料科学专业博士学位。他从 1974 年起在美国卡内基梅隆大学任教。他发表了超过 450 篇同行评审的研究论文，并且共有 11 项美国专利。

Kazuhiro Hono 日本国立材料研究所研究员，磁性材料部主任，筑波大学教授。主要研究方向是金属材料，特别是磁性材料微观结构与性能的关系。1982 年获得日本东北大学学士学位，1988 年获得美国宾夕法尼亚州立大学材料科学与工程博士学位。

序

　　这套 3 卷本的《物理冶金学》，是剑桥大学 Robert W. Cahn 教授和哥廷根大学 Peter Haasen 教授的享有崇高声望的同类著作的第 5 版。《物理冶金学》的第 1 版于 1965 年以单卷本的形式出版。关于本系列出版物的历史请参见第 4 版《序》。本系列权威参考工具书提供了物理冶金学——材料科学与工程领域的最大学科的全面知识。本系列著作广泛而深刻地描述和解释了物理冶金学的大多数方法。书中的每篇文章或由新作者重写，或由第 4 版作者单独或联合新的合作者全面修改和扩充。

　　在《物理冶金学》第 1 版的《序》中，主编之一 R. W. Cahn 教授说："物理冶金学是现代材料科学赖以蓬勃发展的根源。"（R. W. Cahn，1965，《物理冶金学》，北荷兰出版公司，阿姆斯特丹 – 伦敦）。50 年过去了，这一说法仍然正确。本版的两位主编均是作为物理冶金学家培养的，但我们通常称自己为材料科学家。事实上，我们各自所在的部门并不使用"冶金学"一词。但材料科学的核心概念（有时称为"理论框架"），即材料的性能取决于其加工工艺和由此产生的微观结构，直接来源于物理冶金学和加工冶金学。若想了解材料科学的详尽历史，请查看 R. W. Cahn 主编的《材料科学的未来》，并与 R. F. Mehl 主编的《金属科学简史》进行对比。

　　在本系列著作的第 1 版和第 2 版中，R. F. Mehl 在《物理冶金学的发展历史》一文中写道："物理冶金学和提取冶金学交织在一起。从历史的角度看，在很长一段时间内，冶金学的两个分枝构成的一门统一的艺术，并由同样的艺术家实现。"早在 2014 年面世的 3 卷本《冶金学论丛》，由 Seshadri Seetharaman 主编，同样由 ELSEVIER 出版，是本系列著作的同类出版物。这两套共 6 卷的著作无疑地全面覆盖了冶金学。

　　《物理冶金学》第 5 版是继第 4 版（1996 年）大约 18 年之后出版的。2007 年，创始主编不幸辞世，延缓了第 5 版的面世。最后，由我们负责第 5 版的出版工作。

　　本版作者具有更多的国际元素。21 世纪通信的便捷确实使编辑工作变得容易。然而，这似乎并没有加快所有作者写作的进程！

　　本版增加了几个主题，以反映过去 18 年来物理冶金学的最新进展。部分章节由第 4 版的相同作者撰写，但均进行了更新，涵盖了新的论题和方法。我们感谢 45 名作者，他们各自勤奋地完成了有关章节的撰写和校对工作。这套著作获得的声誉当然应主要归功于作者而不是现任主编。当今，在我们的研究机构中，"精打细算的人"并不总是认可撰写类

似套书中的某一章节这样的工作。因此我们对每一位作者心存感激，他们将个人利益置之度外，详细报道他们精通的研究工作，而不在意其是否被 Web of Science 收录。

1970 年，本套著作主编之一（DEL）参加博士生入学资格考试，使用的教材就是《物理冶金学》的第 1 版。他从来不曾想过，近 50 年后他会亲自主编这套深受好评的《物理冶金学》著作。

我们愿将这 3 卷的著作献给我们的前任主编 Robert W. Cahn 教授和 Peter Haasen 教授。我们相信，我们在续写这套专著中付出的努力，可以达到他们曾经制定的高标准。

David E. Laughlin
Kazuhiro Hono

目　录

PHYSICAL METALLURGY

VOLUME III

FIFTH EDITION

PHYSICAL METALLURGY

VOLUME III

FIFTH EDITION

EDITED BY

DAVID E. LAUGHLIN

ALCOA Professor of Physical Metallurgy
Materials Science and Engineering
Carnegie Mellon University
Pittsburgh, PA, USA

KAZUHIRO HONO

Magnetic Materials Unit
National Institute for Materials Science
Tsukuba-city Ibaraki, Japan

ELSEVIER

AMSTERDAM • WALTHAM • HEIDELBERG • LONDON • NEW YORK
OXFORD • PARIS • SAN DIEGO • SAN FRANCISCO • SYDNEY • TOKYO

Elsevier
Radarweg 29, PO Box 211, 1000 AE Amsterdam, Netherlands
The Boulevard, Langford Lane, Kidlington, Oxford OX5 1GB, UK
225 Wyman Street, Waltham, MA 02451, USA

Fifth edition

British Library Cataloguing in Publication Data
A catalogue record for this book is available from the British Library

Library of Congress Cataloging in Publication Data
A catalog record for this book is available from the Library of Congress

Volume III ISBN: 978-0-444-59599-7
SET ISBN: 978-0-444-53770-6

For information on all Elsevier publications
visit our website at www.store.elsevier.com

Printed and bound in the UK

Working together
to grow libraries in
developing countries

www.elsevier.com • www.bookaid.org

CONTENTS

LIST OF CONTRIBUTORS TO VOLUME III

H.K.D.H. Bhadeshia
University of Cambridge, UK

Long-Qing Chen
*Department of Materials Science and Engineering,
The Pennsylvania State University, University Park,
PA, USA*

Russell Goodall
*Department of Materials Science and
Engineering, University of Sheffield, Sheffield,
United Kingdom*

Yijia Gu
*Department of Materials Science and Engineering,
The Pennsylvania State University, University Park,
PA, USA*

R. Kirchheim
*Institut für Materialphysik, Georg-August-
Universität Göttingen, Göttingen, Germany*

Andreas Mortensen
*Laboratory for Mechanical Metallurgy, Ecole
Polytechnique Fédérale de Lausanne (EPFL),
EPFL—STI—IMX—LMM, MXD 140 (Bâtiment
MX), Lausanne, Switzerland*

Jian-Feng Nie
*Department of Materials Engineering, Monash
University, Clayton, VIC, Australia*

A. Pundt
*Institut für Materialphysik, Georg-August-
Universität Göttingen, Göttingen, Germany*

Dierk Raabe
*Max-Planck-Institut für Eisenforschung, Düsseldorf,
Germany*

C.M.F. Rae
*Department of Materials Science and Metallurgy,
University of Cambridge, Cambridge, UK*

R.C. Reed
*Department of Materials, University of Oxford,
Oxford, UK*

Gerhard Wilde
*Institute of Materials Physics, University of
Münster, Münster, Germany*

LIST OF CONTRIBUTORS TO VOLUME III

PREFACE TO THE FIFTH EDITION

These three volumes represent the fifth edition of *Physical Metallurgy*, a prestigious and famous family formerly edited by Robert Cahn (University of Cambridge) and Peter Haasen (Universität Göttingen). Physical Metallurgy was first published as a single volume in 1965. See the preface to the fourth edition for a history of this series. It is an authoritative reference tool, providing a complete knowledge set in Physical Metallurgy, the largest discipline in the fields of Materials Science and Engineering. This series describes and explains most aspects of physical metallurgy across the full breadth and in considerable depth. Each article has been either rewritten by new authors, or thoroughly revised and expanded, either by the 4th edition authors alone or jointly with new co-authors.

In the preface to the first edition of Physical Metallurgy, the founding editor of this series stated that "Physical metallurgy is the root from which the modern science of materials has principally sprung." (R. W. Cahn (1965), Physical Metallurgy North Holland Publishing Company, Amsterdam-London). Over the next five decades this has continued to ring true. While both of the editors of this edition were educated as physical metallurgists, nowadays it is more common to call ourselves Materials Scientists, and indeed our respective departments do not utilize the word "metallurgy." But the core concept (or sometimes called its paradigm) of Materials Science, that the properties and performance of materials have their origin in their processing and resulting microstructure, is derived directly from Physical and Process Metallurgy. For an exhaustive history of Materials Science see "The Coming of Materials Science" by R. W. Cahn and compare this to "A Brief History of the Science of Metals," by R. F. Mehl.

In the article "The Historical Development of Physical Metallurgy" by R. F. Mehl, which appeared in the first two editions of this series, Mehl wrote: "*Physical Metallurgy* has been … interwoven with *Extractive Metallurgy*, and for a long time, historically, these two branches constituted a common art, practiced by the same artisans…" Early in 2014 there appeared the three volume *Treatise on Process Metallurgy*, edited by Seshadri Seetharaman and also published by ELSEVIER, which may be said to be the companion to this set. These six volumes certainly cover Metallurgy comprehensively.

This fifth edition of Physical Metallurgy is published some eighteen years after the 4th edition of this series, which was published in 1996. The lamented death of the founding editor in 2007 slowed down the appearance of this 5th edition. Finally we are ready to present the 5th edition.

This edition has a more international flavor to the listing of authors. Indeed in this 21st Century the ease with which correspondence can be sent makes the task of editing easier. It does not seem to speed up the writing and response of all authors however!

Several new subjects were added in this edition to update the progress in *physical metallurgy* in the last 18 years. Several of the chapters are written by the same authors as those in the fourth edition; but they have all been updated to include new topics and approaches.

We do thank the 45 authors for their hard work and diligence to get their chapters and proofs in. It is of course to authors, more than the current editors, that a series such as this gets its reputation. In a day when "bean counters" in our institutions do not always appreciate the work that goes into writing a chapter in a series such as this, we are grateful that each of the authors put that aside and wrote up work about which they are experts, whether or not it gets indexed in the Web of Science!

In 1970 one of the editors of these volumes (DEL) studied for his qualifying examinations for entrance into the Ph.D. program from the first edition of the series of *Physical Metallurgy*. Little did he suspect that nearly five decades later he would be editing the 5th edition to this well received series on *Physical Metallurgy*.

We wish to dedicate these volumes to our predecessor editors: Prof. Robert W. Cahn and Prof. Peter Haasen. We trust that our efforts in the continuation of this series will be up to the high standards which they have set.

David E. Laughlin
ALCOA Professor of Physical, Metallurgy,
Department of Materials Science and Engineering,
Carnegie Mellon University, Pittsburgh, PA USA

Kazuhiro Hono
ZNIMS Fellow, Naitonal Institute for
Materials Science, Tsukuba, 305-0047, Japan

PREFACE TO THE FOURTH EDITION

The first, single-volume edition of this Work was published in 1965 and the second in 1970; continued demand prompted a third edition in two volumes which appeared in 1983. The first two editions were edited by myself alone, but in preparing the third, which was much longer and more complex, I had the crucial help of Peter Haasen as co-editor. The third edition came out in 1983, and sold steadily, so that the publishers were motivated to propose the preparation of yet another version of the Work; we began the joint planning for this in early 1992. We agreed on the changes and additions we wished to make: the responsibility for commissioning chapters was divided equally between us, but the many policy decisions, made during a series of face-to-face discussions, were very much a joint enterprise. Peter Haasen was able to commission all the chapters which he had agreed to handle, and this task (which involved detailed discussions with a number of authors) was completed in early 1993. Thereupon, in May 1993, my friend of many years was suddenly taken ill; the illness worsened rapidly, and in October of the same year he died, at the early age of 66. When he was already suffering the ravages of his fatal illness, he yet found the resolve and energy to revise his own chapter and to send it to me for comments, and to modify it further in the light of those comments. He was also able to examine, edit and approve the revised chapter on dislocations, which came in early. These were the very last professional tasks he performed. Peter Haasen was in every sense co-editor of this new edition, even though fate decreed that I had to complete the editing and approval of most of the Chapter I am proud to share the title-page with such an eminent physicist.

The first edition had 22 chapters and the second, 23. There were 31 chapters in the third edition and the present edition has 32. The first two editions were single volumes, the third had to be divided into two volumes, and now the further expansion of the text has made it necessary to go to three volumes. This fourth edition is nearly three times the size of the first edition 30 years ago; this is due not only to the addition of new topics, but also to the fact that the treatment of existing topics has become much more substantial than it was in 1965. There are those who express the conviction that physical metallurgy has passed its apogee and is in steady decline; the experience of editing this edition, and the problems I have encountered in holding enthusiastic authors back from even more lengthy treatments (to avoid exceeding the agreed page limits by a wholly unacceptable margin), have shown me how mistaken this pessimistic assessment is! Physical metallurgy, the parent discipline of materials science, has maintained its central status undiminished.

The first three editions each opened with a historical overview. We decided to omit this in the fourth edition, for two main reasons: the original author had died and it would have fallen to others to revise his work, never an entirely satisfactory proceeding; it had also become plain (especially from the reaction of the translators of the earlier editions into Russian) that the overview was not well balanced

between different parts of the world. I am engaged in writing a history of materials science, as a separate venture, and this will incorporate proper attention to the history of physical metallurgy as a principal constituent. — It also proved necessary to leave out the chapter on superconducting alloys: the ceramic superconductor revolution has virtually removed this whole field from the purview of physical metallurgy. — Three entirely new topics are treated in this edition: one is oxidation, hot (dry) corrosion and protection of metallic materials, another is the dislocation theory of the mechanical behavior of intermetallic compounds. The third new topic is a leap into very unfamiliar territory: it is entitled "A Metallurgist's Guide to Polymers". Many metallurgists — including Alan Windle, the author of this chapter — have converted in the course of their careers to the study of the more physical aspects of polymers (regarded by many materials scientists as *the* "materials of the future"), and have had to come to terms with novel concepts (such as "semicrystallinity") which they had not encountered in metals: Windle's chapter is devoted to analysing in some depth the conceptual differences between metallurgy and polymer science, for instance, the quite different principles which govern alloy formation in the two classes of materials. I believe that this is the first treatment of this kind.

Six of the existing chapters (now numbered 1, 4, 21, 22, 27, 30) have been entrusted to new authors, while another five chapters have been revised by the previous authors with the collaboration of additional authors (8,13,16, 17, 19). Chapter 19, originally entitled "Alloys rapidly quenched from the melt" has been broadened and retitled "Metastable states of alloys". A treatment of quasicrystals has been introduced in the form of an appendix to Chapter 4, which is devoted to the solid-state chemistry of intermetallic compounds; this seemed appropriate since quasicrystallinity is generally found in such compounds. — Only three chapters still have the same authors they had in the first edition, written some 32 years ago.

27 of the 29 new versions of existing chapters have been substantially revised, and many have been entirely recast. Two Chapters (11 and 25) have been reprinted as they were in the third edition, except for corrected cross-references to other chapters, but revision has been incorporated in the form of an Addendum to each of these chapters; this procedure was necessary on grounds of timing.

This edition has been written by a total of 44 authors, working in nine countries. It is a truly international effort.

I have prepared the subject index and am thus responsible for any inadequacies that may be found in it. I have also inserted some cross-references between chapters (internal cross- references within chapters are the responsibility of the various authors), but the function of such cross-references is better achieved by liberal use of the subject index.

As always, the editors have been well served by the exceedingly competent staff of North—Holland Physics Publishing (which is now an imprint of Elsevier Science B.V. in Amsterdam; at the time of the first two editions, North—Holland was still an independent company). My particular thanks go to Nanning van der Hoop and Michiel Bom on the administrative side, to Ruud de Boer who is responsible for production and to Chris Ryan and Maurine Alma who are charged with marketing. Mr. de Boer's care and devotion in getting the proofs just right have been extremely impressive. My special thanks also go to Professor Colin Humphreys, head of the department of materials science and metallurgy in Cambridge University, whose warm welcome and support for me in my retirement made the creation of this edition feasible. Finally, my thanks go to all the authors, who put up with good grace with the numerous forceful, sometimes impatient, messages which I was obliged to send in order to "get the show on the road", and produced such outstanding chapters under pressure of time.

I am grateful to Dr. W.J. Boettinger, one of the authors, and his colleague Dr. James A. Warren, for kindly providing the computer-generated dendrite microstructure that features on the dust-cover.

The third edition was dedicated to the memory of Robert Franklin Mehl, the author of the historical chapter and a famed innovator in the early days of physical metallurgy in America. I would like to dedicate this fourth edition to the memory of two people: my late father-in- law, *Daniel Hanson* (1892—1953), professor of metallurgy at Birmingham University for many years, who did more than any other academic in Britain to foster the development and teaching of modern physical metallurgy; and the physical metallurgist and scientific publisher — and effective founder of Pergamon Press — *Paul Rosbaud* (1896—1963), who was retained by the then proprietor of the North—Holland Publishing Company as an adviser and in 1960, in the presence of the proprietor, eloquently urged upon me the need for a new, advanced, multiauthor text on physical metallurgy.

Robert W. Cahn
Cambridge
November 1995.

PREFACE TO THE THIRD EDITION

The first edition of this book was published in 1965 and the second in 1970. The book continued to sell well during the 1970s and, once it was out of print, pressure developed for a new edition to be prepared. The subject had grown greatly during the 1970s and R. W. C. hesitated to undertake the task alone. He is immensely grateful to P. H. for converting into a pleasure what would otherwise have been an intolerable burden!

The second edition contained 22 chapters. In the present edition, 8 of these 22 have been thoroughly revised by the same authors as before, while the others have been entrusted to new contributors, some being divided into pairs of chapters. In addition, seven chapters have been commissioned on new themes. The difficult decision was taken to leave out the chapter on superpure metals and to replace it by one focused on solute segregation to interfaces and surfaces—a topic that has made major strides during the past decade and which is of great practical significance. A name index has also been added.

Research in physical metallurgy has become worldwide and this is reflected in the fact that the contributors to this edition live in no fewer than seven countries. We are proud to have been able to edit a truly international text, both of us having worked in several countries ourselves. We would like here to express our thanks to all our contributors for their hard and effective work, their promptness and their angelic patience with editorial pressures!

The length of the book has inevitably increased, by 50% over the second edition, which was itself 20% longer than the first edition. Even to contain the increase within these numbers has entailed draconian limitations and difficult choices; these were unavoidable if the book was not to be priced out of its market. Everything possible has been done by the editors and the publisher to keep the price to a minimum (to enable readers to take the advice of G. Chr. Lichtenberg (1775): "He who has two pairs of trousers should pawn one and buy this book".).

Two kinds of chapters have been allowed priority in allocating space: those covering very active fields and those concerned with the most basic topics such as phase transformations, including solidification (a central theme of physical metallurgy), defects, and diffusion. Also, this time we have devoted more space to experimental methods and their underlying principles, microscopy in particular. Since there is a plethora of texts available on the standard aspects of X-ray diffraction, the chapter on X-ray and neutron scattering has been designed to emphasize less familiar aspects. Because of space limitations, we regretfully decided that we could not include a chapter on corrosion.

This revised and enlarged edition can properly be regarded as to all intents and purposes a new book.

Sometimes it was difficult to draw a sharp dividing line between physical metallurgy and process metallurgy, but we have done our best to observe the distinction and to restrict the book to its intended

theme. Again, reference is inevitably made occasionally to nonmetallics, especially when they serve as model materials for metallic systems.

As before, the book is designed primarily for graduate students beginning research or undertaking advanced courses, and as a basis for more experienced research workers who require an overview of fields comparatively new to them, or with which they wish to renew contact after a gap of some years.

We should like to thank Ir J. Soutberg and Dr A. P. de Ruiter of the North-Holland Publishing Company for their major editorial and administrative contributions to the production of this edition, and in particular we acknowledge the good-humored resolve of Dr W. H. Wimmers, former managing director of the Company, to bring this third edition to fruition. We are grateful to Dr Bormann for preparing the subject index. We thank the hundreds of research workers who kindly gave permission for reproduction of their published illustrations: all are acknowledged in the figure captions.

Of the authors who contributed to the first edition, one is no longer alive: Robert Franklin Mehl, who wrote the introductory historical chapter. What he wrote has been left untouched in the present edition, but one of us has written a short supplement to bring the treatment up to date, and has updated the bibliography. Robert Mehl was one of the founders of the modern science of physical metallurgy, both through his direct scientific contributions and through his leadership and encouragement of many eminent metallurgists who at one time worked with him. We dedicate this third edition to his memory.

Robert W. Cahn, Paris
Peter Haasen, Göttingen.
April 1983.

PREFACE TO THE FIRST AND SECOND EDITIONS

This book sets forth in detail the present state of physical metallurgy, which is the root from which the modern science of materials has principally sprung. That science has burgeoned to such a degree that no author can do justice to it at an advanced level; accordingly, a number of well-known specialists have consented to write on the various principal branches, and the editor has been responsible for preserving a basic unity among the expert contributions. This book is the first general text, as distinct from research symposium, which has been conceived in this manner. While principally directed at senior undergraduates at universities and colleges of technology, the book is therefore also appropriate for postgraduates and particularly as a base for experienced research workers entering fields of physical metallurgy new to them.

Certain topics have been left to one side or treated at modest length, so as to limit the size of the book, but special stress has been placed on others, which have rarely been accorded much space. For instance, a good deal of space is devoted to the history of physical metallurgy, and to point defects, structure and mechanical properties of solid solutions, theory of phase transformations, recrystallization, superpure metals, ferromagnetic properties, and mechanical properties of two-phase alloys. These are all active fields of research. Experimental techniques, in particular diffraction methods, have been omitted for lack of space; these have been ably surveyed in a number of recent texts. An exception has however been made in favor of metallographic techniques since, electron microscopy apart, recent innovations have not been sufficiently treated in texts.

Each chapter is provided with a select list of books and reviews, which will enable readers to delve further into a particular subject. Internal cross-references and the general index will help to tie the various contributions together.

I should like here to acknowledge the sustained helpfulness and courtesy of the publisher's staff, and in particular of Mr A. T. G. van der Leij, and also the help provided by Prof P. Haasen and Dr T. B. Massalski in harmonizing several contributions.

Brighton, June 1965 (and again 1970) R. W. Cahn

ABOUT THE EDITORS

David E. Laughlin is the ALCOA Professor of Physical Metallurgy in the Department of Materials Science and Engineering at Carnegie Mellon University, Pittsburgh, PA. He obtained his B.S. in Metallurgical Engineering from Drexel University in 1969 and his Ph.D. in Metallurgy and Materials Science from MIT in 1973. He has taught at CMU since 1974. He is Principal Editor of *Metallurgical and Materials Transactions* and has co-edited eight books. His research has centered on the structure of materials as observed by electron microscopy, phase transformations and magnetic materials. He has published more than 450 peer reviewed research papers and is co-inventor on eleven US patents. Laughlin is a Fellow of TMS and ASM International.

Kazuhiro Hono is NIMS Fellow and Director of Magnetic Materials Unit at National Institute for Materials Science and Professor of Materials Science at the University of Tsukuba. He obtained his B.S. from Tohoku University in 1982, and Ph.D. in Materials Science and Engineering from the Pennsylvania State University in 1988. His research interests are mirostructure-property relationships of metallic materials, in particular of magnetic materials.

20 Physical Metallurgy of Light Alloys

Jian-Feng Nie, Department of Materials Engineering, Monash University, Clayton, VIC, Australia

20.1 Introduction

Improvements in energy efficiency, reduction of greenhouse gas emissions and 3R (reduce, reuse and recycle) have been one of the central topics in recent years in environment and climate change. Reducing the overall weight of the planet transportation fleet is critical to achieving these goals. It is foreseeable that, driven by environmental legislation, lightweight products will gain much wider applications in the near future. Aluminum, magnesium and titanium are all lightweight metals. At room temperature, the density of aluminum is 2.70 g/cm^3, which is about one-third that of steel; the density of magnesium is 1.74 g cm^{-3}, one-fourth that of steel; and the density of titanium is 4.51 g cm^{-3}, approximately 55% that of steel. Light alloys based on aluminum, magnesium and titanium have become important classes of engineering materials. Among these materials, aluminum has found the widest applications in the transportation, construction, and packaging industries. The importance of aluminum alloys is reflected by the fact that the global aluminum production reached 45 million tons in 2011. Magnesium metal has received renewed and increasingly high interest since 2000 for potential applications in the automotive, aerospace and 3C (computer, communication and consumer electronic products) industries. The global magnesium production each year was about 250,000 tons before 2000. But over the past 5 years an average of more than 650,000 tons of magnesium metal was produced each year. Most of this was used to alloy aluminum, and only 30% being used to produce magnesium alloys. Because of the lack of innovative manufacturing processes and novel alloy compositions, the mechanical and chemical properties of magnesium alloys are still inferior to those of aluminum alloys, and therefore the market for magnesium is still small compared with that enjoyed by aluminum. The global titanium metal sponge production reached 186,000 tons in 2011. Titanium alloys are mostly used in the aerospace industry, with increasing applications in the chemical and biomedical (implants) industries.

Physical metallurgy, precipitation and age hardening in particular, forms the backbone of light alloy development. The age hardening phenomenon in aluminum alloys was discovered in 1906. In the following years, considerable progress has been made in the improvement of the age hardening response in aluminum alloys via the additions of microalloying elements. By the end of 1980s, the yield strength of precipitation-hardened aluminum alloys exceeded 700 MPa, which was already two to three times higher than that obtained in 1906. For magnesium alloys, remarkable progress was made between 2000 and 2010. In this decade, the yield strength of magnesium casting alloys was raised from less than 200 MPa to more than 300 MPa, and that of magnesium wrought alloys from typically

300 MPa to above 475 MPa. It is foreseeable that it will require a significantly shorter period, that is much less than 80 years, for magnesium alloys to reach the strength level currently enjoyed by ultrahigh-strength aluminum alloys. For titanium alloys a major breakthrough was made in the 1950s, with the advent of the $\alpha + \beta$ Ti-6Al-4V alloy that had an exceptionally good balance of strength, ductility, fatigue, and fracture properties. This alloy has been the dominant titanium alloy since then. The unique combination of mechanical properties is achieved through the careful control of diffusional and diffusionless phase transformations in this alloy.

This chapter provides a review of the fundamental principles of precipitation hardening and the microstructures of classical and novel alloys in aluminum, magnesium and titanium alloys. Apart from an emphasis on the processing–microstructure–property relationships in these alloys, the formation mechanisms of plate-shaped precipitates and their quantitative effects on strengthening, and microstructural factors that are important in controlling the nucleation of strengthening precipitates and the strength of light alloys, are also reviewed. The aim was to provide a basis for the rational design and development of light alloys with even higher performance.

20.2   Precipitation and Age Hardening

High-purity aluminum and magnesium have very low yield strength and therefore they are rarely used for structural applications. It is common to add some desirable alloying elements to aluminum or magnesium, and to process the resultant alloys with appropriate thermal or thermomechanical treatments, to increase their yield strength. One of the most effective processing treatments is precipitation hardening, which is also known as age hardening. The heat treatment scheme of precipitation-hardened alloys generally includes the following:

- Solution treatment at temperature T_0 in the single-phase region, **Figure 1**, to dissolve second-phase particles that have formed during solidification or postsolidification cooling to room temperature.
- Rapid quenching, typically water quench, to room temperature to obtain a supersaturated solid solution of the alloying elements in aluminum or magnesium.

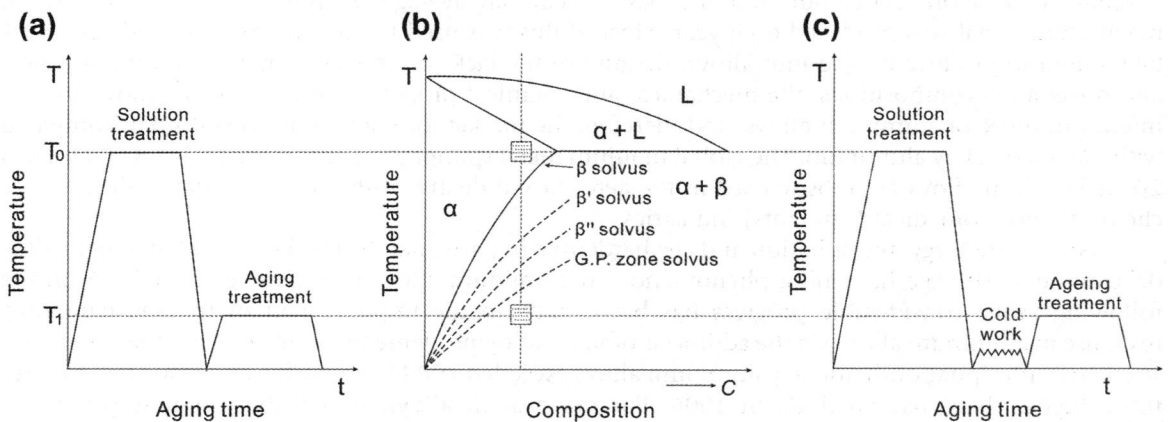

Figure 1   Schematic diagrams showing (a and c) heat treatment schemes for age hardening, and (b) part of a binary phase diagram for precipitation hardenable alloys.

● Aging by one or two heat treatments at a lower temperature T_1, or cold working the as-quenched product before aging, to allow controlled decomposition of the retained supersaturated solid solution to form a fine-scale dispersion of precipitates.

Precipitation generally involves nucleation, growth and coarsening of precipitates. While the three stages usually overlap and thus it is difficult to clearly separate them, each has distinguishable characteristics. The nucleation stage is where the precipitate number density N_v increases with aging time t, that is $\partial N_v/\partial t > 0$; the growth stage is where the precipitate size d increases with aging time, that is $\partial d/\partial t > 0$, and the coarsening stage is where precipitate number density decreases with aging time, that is $\partial N_v/\partial t < 0$.

20.2.1 Nucleation of Precipitates

Suppose that the aging condition is such that precipitation involves the exclusive formation of β. Since the structure and composition of β are different from those of the α matrix, the formation of the nucleus of β will create an α/β interface and a misfit strain energy. For nucleus with a spherical shape and a diameter of d, the total free energy change is given by

$$\Delta G = -\frac{\pi}{6}d^3(\Delta G_v - \Delta G_s) + \pi d^2 \gamma \tag{1}$$

where ΔG_v, ΔG_s and γ are the free energy change per unit volume, the misfit strain energy per unit volume, and the interfacial energy per unit area, respectively. Differentiation of Eqn (1) leads to

$$d^* = \frac{4\gamma}{\Delta G_v - \Delta G_s} \tag{2}$$

$$\Delta G^* = \frac{16\pi\gamma^3}{3(\Delta G_v - \Delta G_s)^2} \tag{3}$$

where d^* is the critical diameter of the nucleus, and ΔG^* is the activation energy barrier to the nucleation, **Figure 2**. If the nucleus adopts a shape of circular plate with diameter d and thickness t, then the activation energy barrier to nucleation is given by

$$\Delta G^* = \frac{8\pi\gamma_e^2\gamma_b}{(\Delta G_v - \Delta G_s)^2} \tag{4}$$

where γ_e and γ_f are the specific interfacial energy for the edge surface and the broad surface of the plate, respectively. The homogeneous nucleation rate of precipitates, $\dot{N}$, is of the form

$$\dot{N} = \omega N_0 e^{-\frac{\Delta G^*}{kT}} e^{-\frac{\Delta G_m}{kT}} \tag{5}$$

where ω is a constant, N_0 is the number of atoms per unit volume of the matrix, ΔG_m is the activation energy for atomic migration, k is the Boltzmann's constant, and T is the temperature in Kelvin.

In the above equations, ΔG_v is the driving force for precipitate nucleation, while the other two factors, ΔG_s and γ (or γ_e and γ_f), are the resistance to the precipitate nucleation. For dilute alloys

$$\Delta G_v \propto \Delta C \tag{6}$$

where $\Delta C = (C_0 - C_e)$ is the solute supersaturation, C_0 is the alloy composition, and C_e is the equilibrium solid solubility of the alloying element at the aging temperature. Therefore, the driving force for precipitate nucleation, ΔG_v, is strongly influenced by the *alloy composition* and the *aging temperature*. For given values of ΔG_s and γ, the larger the ΔG_v value is, the smaller the ΔG^* is, **Figure 2(b)**, and therefore the higher the nucleation rate $\dot{N}$. For a given driving force, any minimization in the value of ΔG_s, γ, or both can reduce ΔG^*, **Figure 2(c) and (d)**, and therefore promote precipitate nucleation. The misfit strain energy ΔG_s can be minimized by heterogeneous nucleation on lattice defects such as dislocations, aggregates of solute atoms and vacancies, twin boundaries and grain boundaries, second-phase particles; and minimization of interfacial energy is usually achieved by adopting an invariant plane strain, if any, as the habit plane of the precipitate.

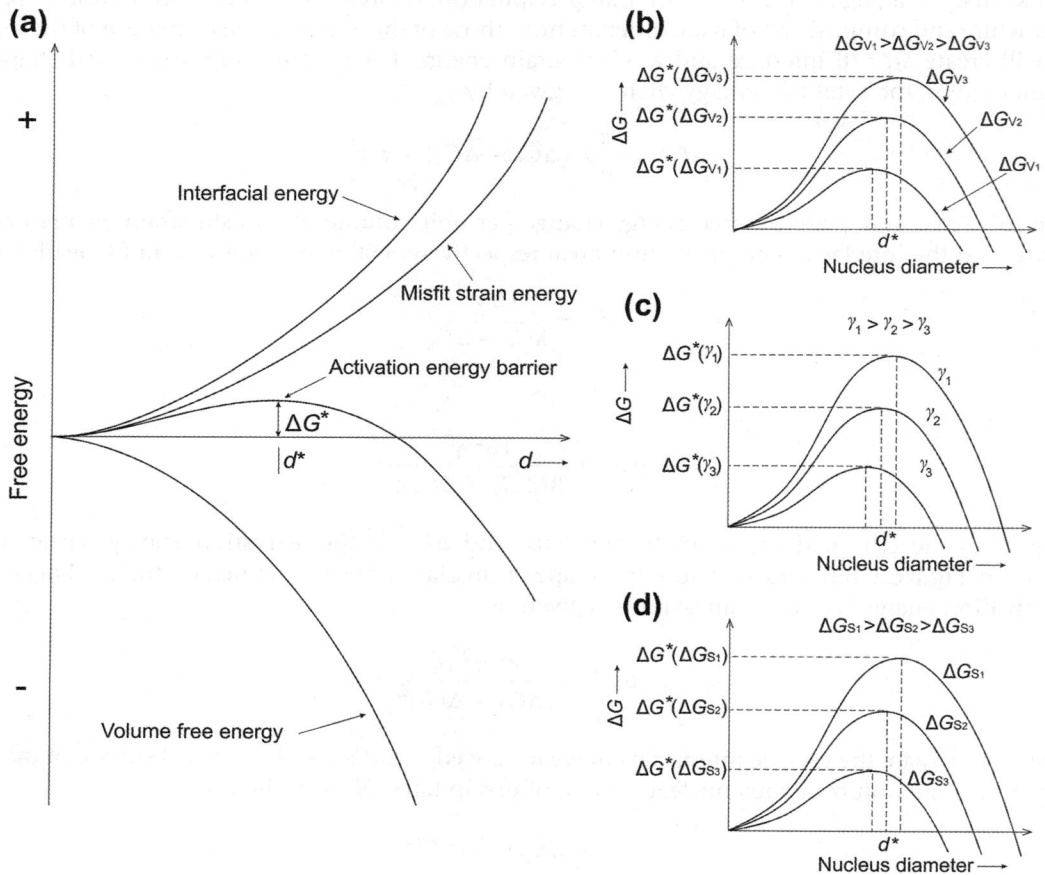

Figure 2 Schematic diagrams showing (a) the variation of ΔG as a function of particle size d for homogeneous nucleation and (b–d) the variation of the activation energy barrier to nucleation ΔG^* as a function of ΔG_v, ΔG_s and γ, respectively.

20.2.2 Growth and Coarsening of Precipitates

During precipitation the precipitate and the matrix have different compositions. Therefore, the growth of the precipitate phase requires long-range diffusion of solute atoms to the precipitate–matrix interface and the transfer of these atoms across the interface. Suppose the precipitate phase is β and rich in B element and the matrix phase is α and rich in A element, **Figure 3(a)**, the growth of the β phase into the α phase can only occur if diffusion is able to transport B atoms toward the advancing interface. If the B atoms can readily transfer across the interface, then the growth rate of the β phase will be governed by the diffusion rate of B atoms—a process that is known as *diffusion-controlled* growth. However, if the B atoms cannot readily cross the interface, the growth rate will then be governed by the interface kinetics—a process which is said to be *interface-controlled* growth. When the diffusion process and interface reaction occur at similar rates, then the migration of the interface will be *mixed controlled*.

For the growth of spherical precipitates, the precipitate radius R after time t is given by

$$R = \chi\sqrt{Dt} \tag{7}$$

where χ is a function of the solute supersaturation, and D is the diffusivity of solute atoms.

For plate-like precipitates with a partially coherent habit plane, their thickening occurs via the formation and lateral movement of ledges within the otherwise planar interface of the habit plane. For a precipitate plate that is thickening by the lateral movement of an array of linear ledges of constant spacing s and height h, **Figure 3(b)**, if u is the lateral migration rate of the plate and v is the thickening rate, then

$$v = uh/s \tag{8}$$

If the edges of the ledges are incoherent, then the matrix composition in contact with the edges is C_e, and growth is diffusion controlled. The necessary composition changes required for precipitate growth is achieved by long-range diffusion of B atoms to the ledges, **Figure 3(b)**. The migration of the edges is similar to that of plate lengthening, and its rate is given as

$$u = \frac{D\Delta C}{k(C_\beta - C_e)h} \tag{9}$$

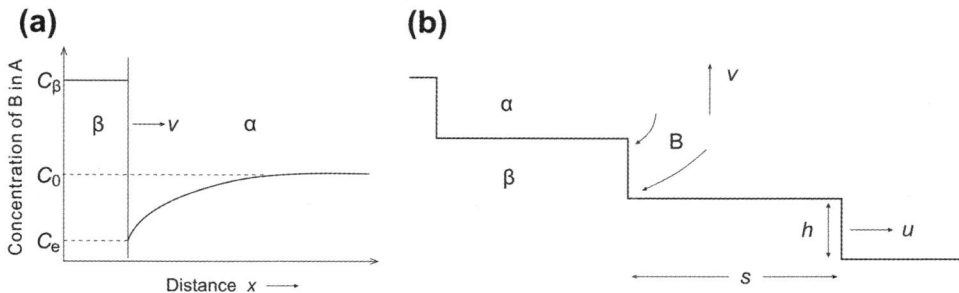

Figure 3 Schematic diagrams showing (a) thickening of a precipitate plate and (b) ledges on the habit plane of a precipitate plate (Porter and Easterling, 1992).

where $\Delta C = C_0 - C_e$, and C_β is the concentration of B atoms in the precipitate. The combination of the above two equations leads to a thickening rate that is independent of h and given by

$$v = \frac{D\Delta C}{k(C_\beta - C_e)s} \tag{10}$$

Equation (10) indicates that the thickening rate of the plate is inversely proportional to the inter-ledge spacing s. This equation is valid only when there is a constant supply of ledges, or for high-index interfaces containing an array of structural ledges. For precipitate plates that are free of ledges or with incoherent broad surfaces, their thickening rate is given by

$$v = \frac{\Delta C}{2(C_\beta - C_e)}\sqrt{\frac{D}{t}} \tag{11}$$

Therefore, there is a parabolic relationship between the half-thickness of the plate and time, which is similar to that made for growth of spherical particles.

For volume diffusion-controlled coarsening, the coarsening of spherical particles can be represented to a reasonable approximation by an equation of the form

$$r^3 - r_0^3 \propto D_v \gamma_i C_e t \tag{12}$$

where r and r_0 are the average radius of spherical particles at time t and zero, respectively, D_v is the volume diffusion coefficient of the solute element, γ_i is the specific interfacial energy between precipitate and matrix phases, and C_e is the equilibrium solid solubility of solute in the matrix phase in weight percentage. According to Eqn (12), the principal factors influencing the coarsening of precipitate particles are the equilibrium solid solubility and solid-state diffusivity of the solute element, and the interfacial energy between the particle and the matrix phase. To ensure low coarsening rates of strengthening particles at elevated temperatures, alloying elements should thus have low equilibrium solid solubility and low solid-state diffusivity in the matrix phase and the dispersed particles should preferably be coherent with the matrix. Furthermore, the dispersed particles should be intrinsically stable at elevated temperatures. A low interfacial energy between precipitate phases and the matrix phase is important not only for achieving low coarsening rates but also for homogeneous solid-state precipitation of the precipitate phases. If precipitate phases have a low interfacial energy with the matrix phase, there exists a low activation energy barrier to their nucleation, and consequently, they may nucleate more homo-geneously within the matrix phase, giving rise to a uniform distribution of precipitates.

At elevated temperatures $(T > 0.5T_m$, where T_m is the melting point in Kelvin of a given alloy), dislocation gliding is no longer the dominant deformation mode. Depending on the applied stress and the temperature, the plastic deformation of alloys occurs primarily by mechanisms of dislocation climb, grain boundary sliding or vacancy diffusion. However, a large volume fraction of thermally stable precipitate phase(s), in the form of fine-scale particles within the grains and at grain boundaries, can again contribute to the strength of the alloys by providing resistance to steady-state creep (i.e. dislocation creep, diffusional creep and grain boundary sliding) at elevated temperatures.

20.2.3 Morphology of Precipitates

The equilibrium shape of coherent precipitate is controlled by the balance between interfacial energy and elastic strain energy. For a circular-shaped precipitate plate with a coherent habit plane, its equilibrium aspect ratio, defined by the ratio of its diameter over its thickness, and hence its equilibrium shape are dictated by the ratio of $\gamma_b{:}\gamma_e$, where γ_b is the interfacial energy of the habit plane of the plate and γ_e is the interfacial energy of the edge plane of the plate. Precipitates formed during isothermal aging seldom exist in their equilibrium shapes.

20.2.3.1 Coherent Precipitates

The equilibrium shape of a coherent precipitate or zone can only be predicted from the γ-plot when the misfit between the precipitate and the matrix is small. γ is the interfacial free energy. When misfit is present, **Figure 4**, the formation of coherent interfaces raises the free energy of the system on account of the elastic strain fields that arise. If the total interface energy is $\Sigma S_i\gamma_i$ and the elastic strain energy is ΔG_s, then the condition for the equilibrium shape becomes (Porter and Easterling, 1992)

$$\sum S_i\gamma_i + \Delta G_s = \text{minimum}$$

The stresses maintaining coherency at the interfaces distort the precipitate lattice, and in the case of a spherical inclusion the distortion is purely hydrostatic, that is it is uniform in all directions. When the precipitate is a thin disc the in situ misfit is no longer equal in all directions, but instead it is perpendicular to the disc and almost zero in the plane of the broad faces, as shown in **Figure 4(b)**.

In general, the total elastic energy depends on the shape and elastic properties of both matrix and precipitate. However, if the matrix is elastically isotropic, the elastic moduli of the matrix and precipitate are equal, and Poisson's ratio is 1/3, then the total elastic strain energy ΔG_s is given by

$$\Delta G_s \approx 9G\varepsilon^2 V \tag{13}$$

where G is the shear modulus of the matrix (and precipitate), ε is the constrained misfit between the precipitate and the matrix, and V is the volume of the unconstrained hole in the matrix. Therefore, coherency strain produces an elastic strain energy that is proportional to the square of the lattice misfit (ε^2) and increases with the volume of the precipitate. If the precipitate and matrix have different elastic moduli, the elastic strain energy is no longer shape independent but is a minimum for a sphere, if the inclusion is hard, or a disc, if the inclusion is soft.

In general, most metals are elastically anisotropic. For example, most cubic metals are soft in $\langle 100\rangle$ directions and hard in $\langle 111\rangle$ directions. The shape with the minimum strain energy under these conditions is a disc parallel to $\{100\}$ since most of the misfit is then accommodated in the soft directions perpendicular to the disc. The influence of strain energy on the equilibrium shape of coherent

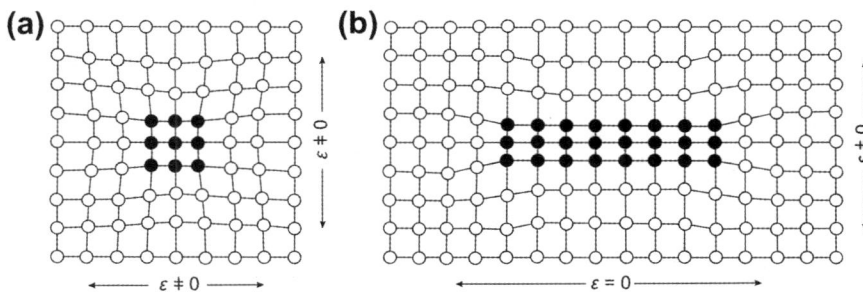

Figure 4   Misfit between coherent precipitate and surrounding matrix. (a) Spherical particle and (b) plate-shaped particle. (Revised from Porter and Easterling, 1992).

precipitates can be illustrated by reference to Guinier–Preston (GP) zones in precipitation hardenable aluminum alloys. In each case zones containing 50–100% solute can be produced. Assuming the zone is pure solute the misfit can be calculated directly from the atomic radii. When zone misfit is less than 5%, as in the case of GP zones in Al–Ag and Al–Zn alloys, strain energy effects are less important than interfacial energy effects and spherical zones minimize the total free energy. For zone misfit above 5%, as in the case of zones in Al–Cu alloys, the small increase in interfacial energy caused by choosing a disc shape is more than compensated by the reduction in coherency strain energy.

20.2.3.2 *Incoherent Precipitates*

There are no coherency strains when the precipitate is incoherent with the matrix. However, the misfit strain can still occur if the volume of the precipitate is different from that originally occupied by the matrix, that is before the precipitate is formed, **Figure 5(a)**. In this case the lattice misfit is replaced by the volume misfit Δ that is defined by

$$\Delta = \frac{V_\beta - V_\alpha}{V_\beta} \tag{14}$$

where V_α is the volume of the unconstrained hole in the matrix, and V_β is the volume of the unconstrained precipitate. When the matrix hole and precipitate are constrained to occupy the same volume the elastic strain fields again result as shown in **Figure 5(a)**. For a homogeneous incompressible ellipsoidal shaped precipitate in an isotropic matrix, the elastic strain energy is given (Nabarro, 1940) as

$$\Delta G_s = \frac{2\mu(V_\beta - V_\alpha)^2}{3V_\beta} E\left(\frac{y}{x}\right) = \frac{2}{3} G\Delta^2 V_\beta E\left(\frac{y}{x}\right) \tag{15}$$

where x and y are the semi-axes of the ellipsoidal precipitate, $E(y/x)$ is the strain energy of the ellipsoidal precipitate, and G is the shear modulus of the matrix. Therefore, the elastic strain energy is proportional to the square of the volume misfit Δ^2 and the volume of the precipitate. The value of $E(y/x)$ is a function of precipitate shape, **Figure 5(b)**. When $y/x = \infty$, the ellipsoid approximates a cylinder or needle and $E(y/x)$ is 3/4. In the situation where $y/x = 1$, a sphere is obtained and $E(y/x)$ is 1. When $y/x \ll 1$ or $y/x \to 0$, the ellipsoid becomes a thin plate or disc and $E(y/x)$ approaches zero. Therefore, for a given

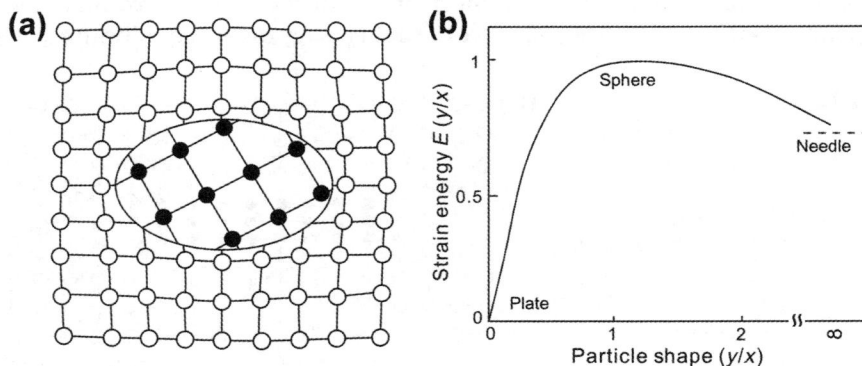

Figure 5   (a) Misfit strain associated with an incoherent precipitate (Revised from Porter and Easterling, 1992) and (b) variation of the misfit strain energy associated with an incoherent precipitate with its shape (Nabarro, 1940).

volume, a spherical shape has the highest strain energy while a thin plate has the lowest strain energy. The minimum strain energy is obtained when the precipitate adopts a plate shape with the minimum value of y/x, that is the largest plate aspect ratio $(A = x/y)$. The equilibrium shape of an incoherent precipitate will be an oblate spheroid with a y/x value that balances the opposing effects of elastic strain energy and interfacial energy.

20.2.4 Precipitation Hardening

For spherical particles of uniform diameter d, the relationship among the volume fraction f, size d and number density N_V of precipitates per unit volume of the matrix phase is given by

$$f = VN_v = \frac{1}{6}\pi d^3 N_v \tag{16}$$

If the spherical particles are assumed to have a triangular distribution in the slip plane of the matrix phase, as illustrated in **Figure 6**, then

$$d_p = \int_0^{\pi/2} \frac{d}{2} \sin\theta = \frac{\pi}{4}d \tag{17}$$

$$\frac{1}{2} = N_a S = N_a \frac{\sqrt{3}}{4}L_p^2 \tag{18}$$

where d_p is the planar diameter of the spherical particle in the slip plane, θ is the angle between lines oo' and oe in **Figure 6(b)**, N_a is the number density of precipitates per unit area of the matrix phase in the slip plane, S is the area of the triangle shown in **Figure 6(a)**, and L_p is the center-to-center interparticle spacing. Since $N_a = N_v d$, the combination of Eqns (16) and (18) yields

$$L_p = \sqrt{\frac{2}{\sqrt{3}N_a}} = \sqrt{\frac{2}{\sqrt{3}N_v d}} = d\sqrt{\frac{\pi}{3\sqrt{3}f}} = \sqrt[6]{\frac{4\pi}{9\sqrt{3}fN_v^2}} \tag{19}$$

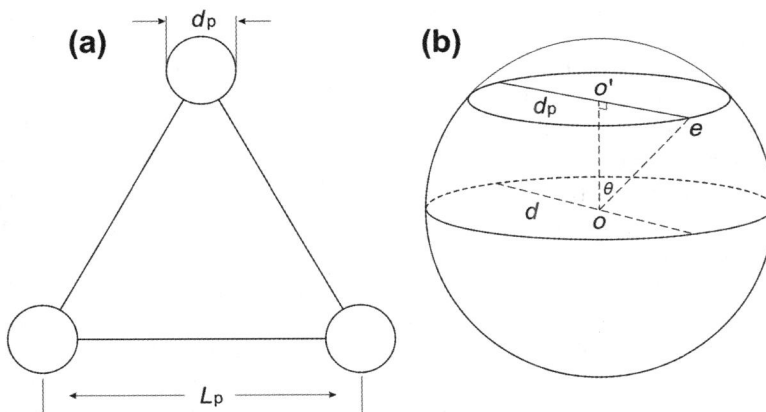

Figure 6 Schematic diagrams showing (a) triangular distribution of spherical particles in the slip plane and (b) the relationship between true diameter d and planar diameter d_p of a spherical particle. o is the center of the sphere, and o' is the center of the top circle after truncation.

The effective interparticle spacing, that is the surface-to-surface interparticle spacing, is then in the form

$$\lambda = L_p - d_p = \sqrt{\frac{2}{\sqrt{3}N_v d}} - \frac{\pi}{4}d = \left(\sqrt{\frac{\pi}{3\sqrt{3}f}} - \frac{\pi}{4}\right)d \tag{20}$$

Examination of Eqns (19) and (20) indicates that the effective interparticle spacing λ is a function of planar center-to-center interparticle spacing L_p and planar diameter of the spherical particle d_p, which is in turn a function of volume fraction, number density and diameter of the spherical particles. For a given volume fraction, the smaller the particle size, the smaller the effective interparticle spacing. Similarly, for the given particle size, the higher the volume fraction of precipitates, the smaller the effective interparticle spacing.

Assuming that the spherical particles are resistant to dislocation shearing, then the contribution of these spherical particles to the alloy strength is given by the Orowan strengthening equation (Orowan, 1948; Brown and Ham, 1971; Ardell, 1985; Reppich, 1993; Nembach, 1997)

$$\Delta\sigma = M\Delta\tau = \frac{MGb}{2\pi\sqrt{1-v}} \cdot \frac{1}{\lambda} \cdot \ln\frac{\pi d}{4b} \tag{21}$$

where $\Delta\sigma$ is the contribution of the particles to the alloy yield strength, M is the Taylor factor for a polycrystalline alloy, $\Delta\tau$ is the contribution of the particles to the critical resolved shear stress (CRSS) of the alloy, G is the shear modulus of the matrix phase, b is the magnitude of the Burgers vector of dislocations in the matrix phase, and v is Poisson's ratio.

From the above analysis, it is clear that the higher the volume fraction of precipitates, the higher the maximum strength achievable in the alloy. For a given volume fraction of precipitates, the higher the nucleation rate of precipitates, the higher the number density of precipitate one can get, and therefore the smaller the effective interparticles spacing is. This suggests that a higher yield strength can be achieved if a higher nucleation rate of intrinsically strong precipitates is obtainable. The precipitate volume fraction can be increased by increasing the concentration of alloying elements in the super-saturated solid solution, either by simply adding more alloying elements within the maximum equi-librium solid solubility or by reducing the equilibrium solid solubility at the aging temperature via appropriate alloying additions.

Figure 7 shows the relationship between aging temperature, alloy composition and maximum age hardening response of the alloy. In **Figure 7(a)**, the alloy composition is given as C_0, and the equilibrium precipitate phase is β. Three different aging temperatures, designated by T_1, T_2, and T_3, respectively, are shown in the diagram. The age hardening responses of the alloy at these three different aging temperatures are provided in **Figure 7(b)**. If the precipitation process involves the exclusive formation of β precipitates, then the equilibrium volume fraction of precipitates achievable at T_1 is the highest, that is $f_1 > f_2 > f_3$, according to the lever rule. Furthermore, according to Eqn (6), $\Delta G_{v1} > \Delta G_{v2} > \Delta G_{v3}$, that is the driving force for precipitation is the highest at aging temperature T_1. If the change of ΔG_s and γ because of the variation in aging temperature is negligible, then the activation energy barriers at these temperatures follow the sequence: $\Delta G_1^* < \Delta G_2^* < \Delta G_3^*$, and nucleation rate of precipitates: $\dot{N}_1 > \dot{N}_2 > \dot{N}_3$. Therefore, the thermodynamic consideration suggests that the maximum nucleation rate and therefore the maximum number density of precipitates can be achieved in the alloy when it is aged at T_1. This maximum number density of precipitates leads to minimum effective interparticle spacing, Eqn (20), and hence the maximum hardness or strength,

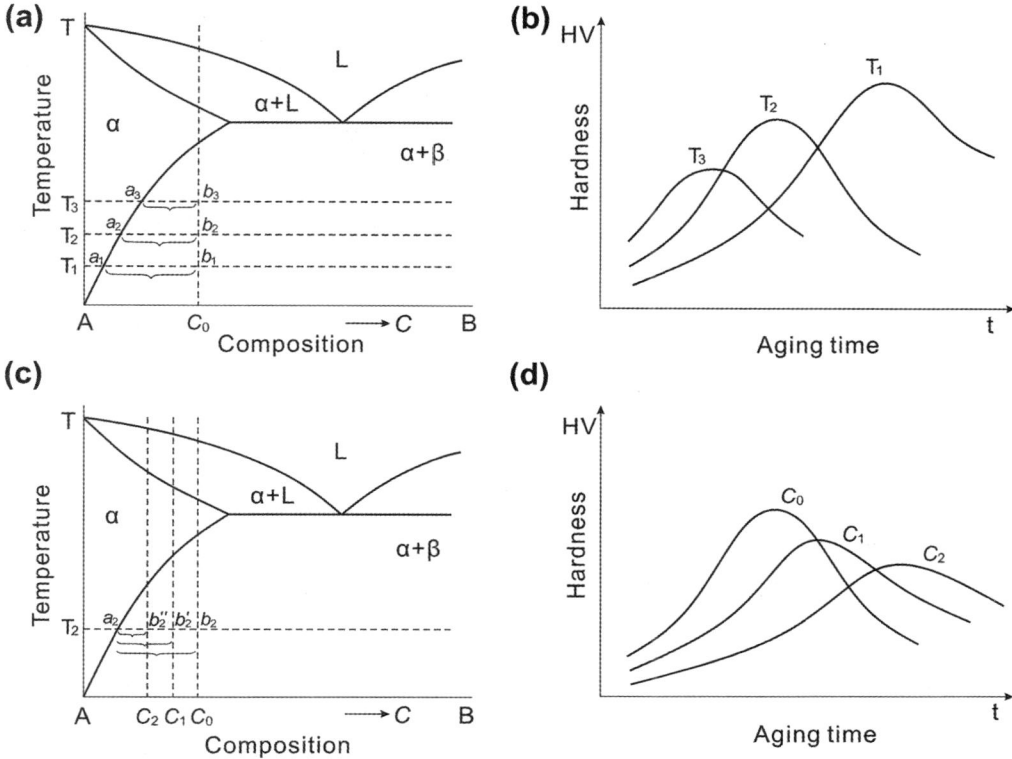

Figure 7 Schematic diagrams showing effects of aging temperature and alloy composition on age hardening response.

Eqn (21). Note that, while the maximum age hardening response is achieved at T_1, it takes much longer time to reach the peak hardness at this temperature, compared with the other two aging temperatures, **Figure 7(b)**. This is because of the lower diffusion rate of solute atoms in the matrix phase at lower temperatures.

The influence of alloy composition on the age hardening response is illustrated in **Figure 7(c) and (d)**. Three alloy compositions, designated C_0, C_1, and C_2, respectively, are given in **Figure 7(c)**, and the aging temperature is T_2. The equilibrium volume fractions of precipitates achievable in these three alloys have the following sequence: $f_0 > f_1 > f_2$. Furthermore, $\Delta G_{v0} > \Delta G_{v1} > \Delta G_{v2}$, that is the driving force for precipitation is the highest in alloy C_0. Therefore, the activation energy barriers to nucleation in these alloys follow the sequence: $\Delta G_0^* < \Delta G_1^* < \Delta G_2^*$, and nucleation rate of precipitates: $\dot{N}_0 > \dot{N}_1 > \dot{N}_2$. Consequently, $N_{v0} > N_{v1} > N_{v2}$, which leads to $\lambda_0 < \lambda_1 < \lambda_2$, and in turn leads to $\Delta\sigma_0 > \Delta\sigma_1 > \Delta\sigma_2$. Thermodynamically, the maximum nucleation rate, and therefore the maximum number density of precipitates, is expected to occur in alloy C_0, **Figure 7(d)**. This maximum number density of precipitates leads to minimum effective interparticle spacing, Eqn (11), and hence the maximum hardness or strength, Eqn (21). A higher concentration of solute atoms in the supersaturated matrix phase leads to accelerated kinetics of precipitation and thus a shorter time to reach peak hardness.

All the aging curves shown in **Figure 7** are typical for precipitation of a second phase from super-saturated solid solution matrix phase. The hardness of the alloy increases with aging time, gradually reaches the maximum and then declines. The segment of the aging curve before peak hardness is referred to as *under aged* and that beyond the peak hardness as *over aged*. The aging treatment is usually stopped at the peak-aged condition. Traditionally, the peak hardness condition is inferred to be the transition from dislocation looping or bypassing to particle shearing and therefore the concept of critical particle radius is often used in the literature. An alternative interpretation, which is based on a single mechanism of dislocation looping and will be described in detail in Section 20.3.2, suggests that the peak-aged condition would be expected to have, statistically, the minimum effective inter-particle spacing in the microstructure.

In summary, the major alloying element is often added close to the maximum solid solubility at the solution treatment temperature, and is combined with a relatively low aging temperature, to maximize the supersaturation, the driving force for precipitation, and thus the age hardening response. Furthermore, microalloying additions and cold work are also used to minimize the resistance to nucleation, either ΔG_s or γ or both.

20.2.5   Effects of Vacancies, Cold Work and Microalloying Elements

The equilibrium concentration of vacancies in solid metals and alloys decreases exponentially with decrease in temperature—it is relatively high at the solution treatment temperature and much lower at the aging temperature. When an alloy is rapidly quenched from high temperature, there is no time for the migration of vacancies to establish a new equilibrium concentration, and therefore the high vacancy concentration becomes quenched in. These vacancies in excess of the equilibrium concentration are metastable and they will, during the exposure at lower temperatures, be attracted together into vacancy clusters or other vacancy sinks. Some of these vacancy clusters collapse into dislocation loops which can grow by absorbing more vacancies. Both vacancy clusters and dislocation loops play an important role in the heterogeneous nucleation of precipitates, particularly the intermediate precipitate phases.

The quenched-in vacancies also significantly enhance the diffusion rate. If the alloy is quenched from different solution treatment temperatures and aged at the same temperature, the samples quenched from the highest temperature will have the highest rate of precipitation. Reducing the quenching rate will allow more time for vacancies to be lost during the quench. For large products the cooling rate varies greatly from the surface to the center when they are water quenched from the solution treatment temperature, which makes it difficult to achieve a uniform distribution of precipitates in such products.

Since one of the main sinks for excess vacancies are grain boundaries, the vacancies in their vicinity can be readily lost and their concentrations can be substantially lower than those in regions remote from the grain boundaries. The concentration of vacancies near a grain boundary is shown schematically in **Figure 8**. The vacancy concentration corresponds to the equilibrium value for the aging temperature in and immediately adjacent to a grain boundary, but in regions remote from the boundary it is for the solution treatment temperature. As a *critical vacancy supersaturation* is needed for precipitate nucleation to occur, the uneven distribution of vacancies influences the distribution of precipitates that form in the vicinity of grain boundaries on subsequent aging. A precipitation-free zone (PFZ) is often formed at each side of the grain boundary. These PFZs are formed for two reasons. First, a depletion of vacancies to levels below that needed for the nucleation of precipitates at the aging temperature. Second, the precipitation in the grain boundaries during cooling from the solution treatment temperature causes solute to be drained from the surrounding matrix. Similar PFZs can also

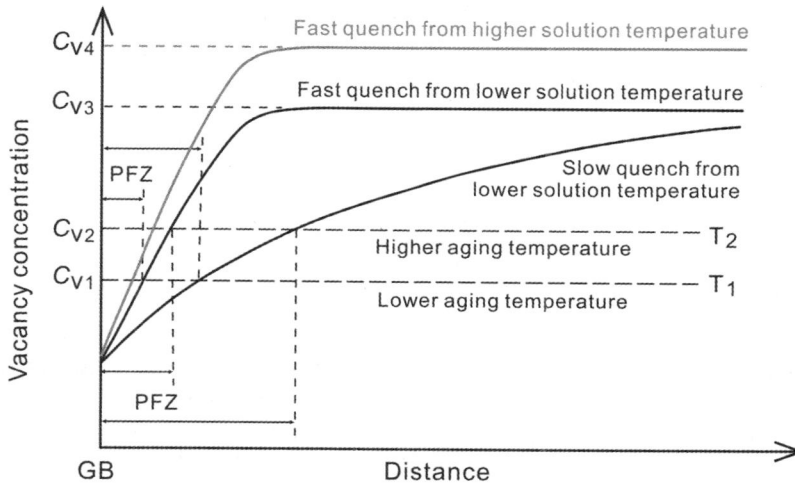

Figure 8 Schematic diagram showing PFZ adjacent to grain boundary and effects of quenching rate, solution treatment temperature, aging temperature on the width of PFZ. GB represents grain boundary.

form at other second-phase particles such as dispersoids, because vacancies can be lost at their interfaces with the matrix.

The width of the PFZs is determined mainly by the vacancy concentration, **Figure 8**, and can be altered by heat treatment conditions. Narrower PFZs are obtained when higher solution treatment temperatures, faster quenching rates, or lower aging temperatures are used. The higher solution treatment temperature and faster quenching rate can both increase the excess vacancy content or reduce the width of the vacancy concentration profile, **Figure 8**. At lower aging temperatures, the driving force for precipitation is higher, and the critical vacancy supersaturation for nucleation to occur is therefore lower and narrower PFZs can be formed.

Precipitation reactions may be greatly influenced by the presence of dislocations introduced by thermomechanical processing. Dislocations are often effective in promoting the formation of strengthening precipitate phases. For some precipitation-hardened aluminum alloys, it is common to strain them at room temperature (known as cold work) by a few percent after quenching from the solution treatment temperature, or extrusion temperature, and before aging to increase the density of dislocations on which heterogeneous precipitation occurs. The effect of dislocations is to reduce the ΔG_s contribution to ΔG^* by reducing the total strain energy of the embryo.

Equally, the precipitation reactions and hence the age hardening response can be strongly affected by the presence of trace or microalloying elements. Several mechanisms have been proposed to account for the effects of microalloying additions on precipitate nucleation (Martin, 1998):

(1) Preferential interaction with vacancies, leading to reduced diffusion rates and nucleation rate of GP zones.
(2) Raising the GP zones solvus, altering the temperature ranges over which precipitate phases are stable.
(3) Stimulating nucleation of an existing precipitate by reducing the interfacial energy or elastic strain between precipitate and matrix.

(4) Promoting formation of a different precipitate phase by changing the free energy relationships in an alloy system.
(5) Providing heterogeneous sites at which existing or new precipitates may nucleate. These sites include clusters of solute atoms.
(6) Increasing supersaturation so that the precipitation process is stimulated.

20.3   Aluminum Alloys

20.3.1   Clustering and Precipitation

20.3.1.1   2xxx Series Alloys

Sustained interest in improving the strength of precipitation hardenable aluminum alloys at both ambient and elevated temperatures has led to continued alloy development. Although phase equilibria restrict the total concentration of alloying elements to less than 8 wt% in most precipitation hardenable aluminum alloys, and thus impose limitations on the maximum volume fraction of strengthening precipitate phase achievable, remarkably high tensile yield strengths have been achieved in a range of alloys via the use of microalloying additions to promote formation of specific dispersed precipitate phases which, in appropriate combination, improve mechanical properties at both ambient and elevated temperatures. Most notable among these developments are the 2xxx series alloys based on the Al–Cu system. The precipitation processes in these alloys are relatively well studied, and several commercial alloys have been developed and are used in aircrafts. Although the major alloying element is Cu in these alloys and its content is typically less than 6 wt%, a substantial enhancement in age hardening response has been made over the period 1906–1989.

20.3.1.1.1   Al–Cu Binary Alloys

The Al-rich side of the equilibrium phase diagram of the Al–Cu system is shown in **Figure 9(a)**. The solvus lines for GP zones and metastable phases θ'' and θ' are also included. The equilibrium solid solubility of Cu in Al decreases substantially when temperature drops. The maximum solid solubility of Cu in Al is 2.40 at.% (or 5.65 wt%) at the eutectic temperature of 548.2 °C. The equilibrium inter-metallic phase θ has a complex body-centered tetragonal structure (space group I4/mcm, $a = 0.607$ nm, $c = 0.487$ nm), **Figure 9(b)**, a stoichiometric composition of Al_2Cu, and a melting point of ~ 590 °C. The θ phase does not have lattice planes that match well those of the matrix. A range of orientation relationships (ORs) and shapes has been observed between the θ precipitates and matrix (Vaughan and Silcock, 1967).

In binary Al–Cu alloys containing typically 2–4 wt% Cu, the precipitation sequence during isothermal aging may involve formation of GP zones, metastable θ'' and θ' phases, and eventually the equilibrium θ phase, that is

$$SSSS \rightarrow GP\ zones \rightarrow \theta'' \rightarrow \theta' \rightarrow \theta$$

The GP zones form as single atomic layer disks on $\{001\}_\alpha$ planes, **Figure 9(c)**. These zones were detected in 1938, independently by Guinier and Preston from streaks in X-ray diffraction patterns obtained from Al–Cu alloys. The results obtained from 1-dimensional atom probe (1DAP) in early studies suggested that the Cu composition in the GP zones is approximately 33 at.%, which is the same as that of the equilibrium θ phase. More recent work using near-atomic resolution 3-dimensional atom probe (3DAP) provided similar results. However, in this work, the diffuse character of the atom probe

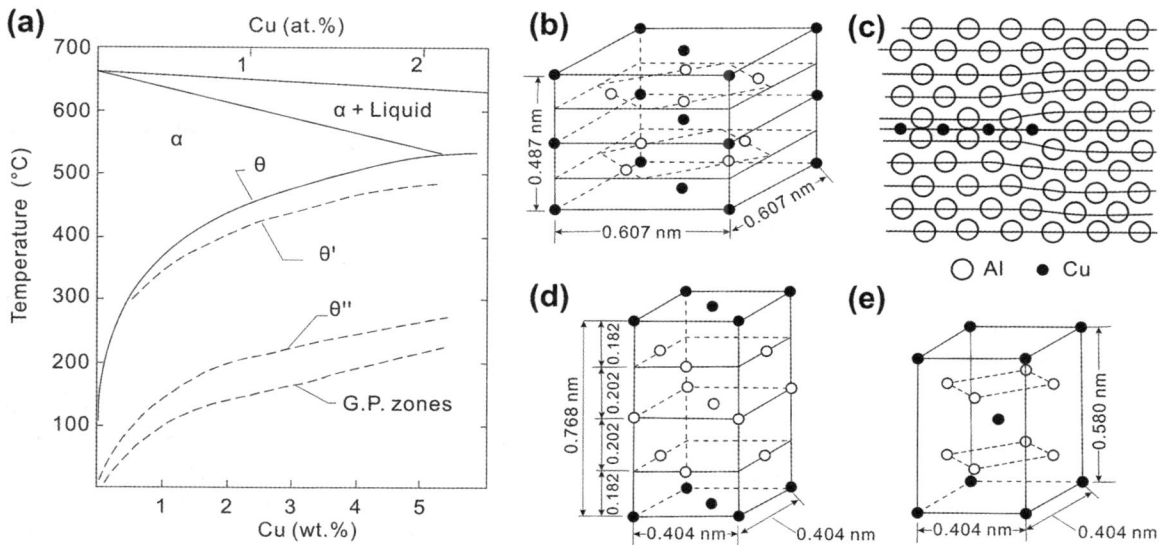

Figure 9   (a) Al-rich side of the Al–Cu binary phase diagram, (b) θ unit cell, (c) GP zone, (d) θ″ unit cell, and (e) θ′ unit cell (Porter and Easterling, 1992).

concentration–depth profiles was attributed to evaporation artifacts, and it was suggested that the actual Cu concentration in the GP zones was close to 100 at.% (Ringer and Hono, 2000).

For samples of an Al–4wt%Cu alloy that are water quenched and subsequently aged for a period of time at room temperature or a temperature below about 180 °C, the first precipitates to nucleate are metastable GP zones, rather than the equilibrium phase θ. This is because the activation energy barrier for the nucleation of GP zones is the smallest. GP zones are fully coherent with the matrix and therefore have a very low interfacial energy. Furthermore, these zones minimize their strain energy by choosing a disc shape perpendicular to the elastically soft $\langle 100 \rangle_\alpha$ directions in the aluminum matrix. Therefore, despite the fact that the driving force for precipitation of GP zones is less than for the equilibrium phase θ, the barrier to the nucleation is even lower, **Figure 10**, and the GP zones nucleate most readily.

The formation of GP zones is usually followed by the precipitation of θ″. θ″ has a tetragonal unit cell ($a = 0.404$ nm, $c = 0.768$ nm), **Figure 9(d)**, which is essentially a distorted face-centered cubic structure in which the Cu and Al atoms are arranged in an ordered manner on $(001)_\alpha$ planes. The atomic structure of the $(001)_{\theta''}$ plane is identical to that in the α-Al matrix. The $(010)_{\theta''}$ and $(100)_{\theta''}$ planes are very similar to the $(010)_\alpha$ and $(100)_\alpha$ planes, apart from a small contraction in the $[001]_{\theta''}$ direction. The θ″ precipitates form as fully coherent plates with a $\{001\}_\alpha$ habit plane and an OR of the following form:

$$(001)_{\theta''}//(001)_\alpha, [100]_{\theta''}//[100]_\alpha$$

θ′ also has a tetragonal structure (space group 14/mmm, $a = 0.404$ nm, $c = 0.580$ nm), **Figure 9(e)**, and a nominal composition of Al_2Cu. Its $(001)_{\theta'}$ plane is similar to $\{001\}_\alpha$, except for the absence of an atom in the face center. Its $(100)_{\theta'}$ and $(010)_{\theta'}$ planes are quite different from those of the matrix lattice in terms of arrangement and separation distance of atoms. Therefore, θ′ forms as rectangular or octagonal plates on $\{001\}_\alpha$ with the same OR as θ′, **Figure 11**. The broad faces of the θ′ plates are fully coherent at the early stage of growth but lose coherency as the plates grow, while the edges of the plates

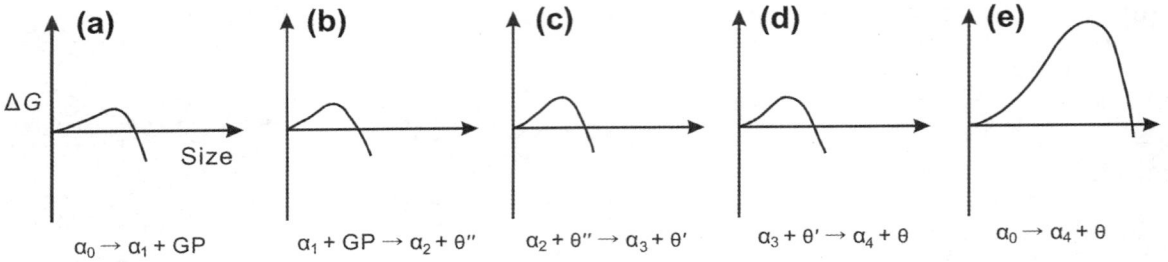

Figure 10 Activation energy barrier to the nucleation of individual precipitate phases in binary Al–Cu alloys. α_0 is the original supersaturated solid solution, and α_1, α_2, α_3 and α_4 are the compositions of the matrix in equilibrium with GP zones, θ'', θ' and θ, respectively (Porter and Easterling, 1992).

have a semi-coherent structure. Despite the fact that the θ' plates are nearly fully coherent across their broad faces and form with a volume change of less than 6%, it is commonly observed that nucleation of θ' requires the presence of dislocations or particles of a preexisting transition phase. The aspect ratio of θ' precipitate plates is typically less than 20:1, and the maximum volume fraction is below 5%. A very recent study of θ' plates using high-angle annular dark-field scanning transmission electron microscopy (HAADF-STEM) (Bourgeois et al., 2011) revealed the presence of excessive Cu atoms in the θ'/α-Al interface during early stages of the θ' precipitation.

Like GP zones, the intermediate metastable phases θ'' and θ' have lower activation energy barriers for nucleation than the equilibrium phase θ, **Figure 10**. The lower activation energy barriers are obtained because the crystal structures of these transition phases can achieve a higher degree of coherence and thus a lower interfacial energy contribution to ΔG^*. In addition, the θ'' and θ' phases adopt a plate shape that is perpendicular to the elastically soft $\langle 100 \rangle_\alpha$ directions in the aluminum matrix and hence a lower elastic strain contribution to ΔG^* is achieved. In contrast, the complex crystal structure of the equilibrium θ phase is incompatible with the matrix and results in high energy interfaces and high ΔG^*.

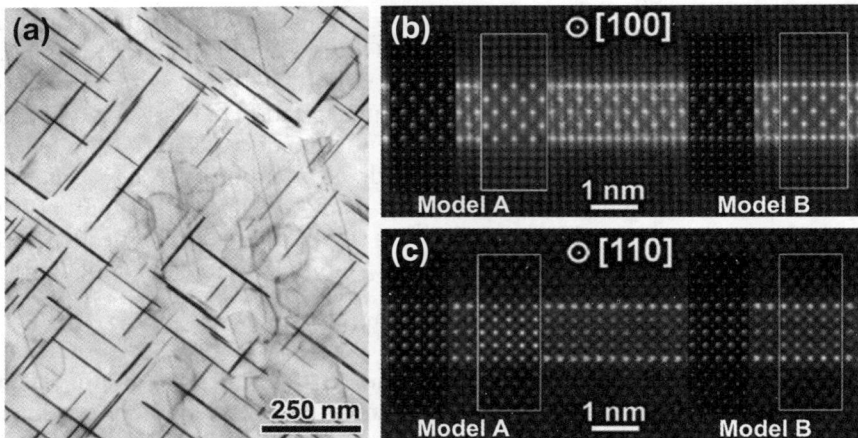

Figure 11 (a) Transmission electron microscopy image showing the distribution of θ' precipitates in peak-aged microstructure of an Al–4wt%Cu alloy and (b and c) HAADF-STEM images showing θ' precipitates and segregation of Cu atoms in the broad face of θ' precipitates (Bourgeois et al., 2011). (For color version of this figure, the reader is referred to the online version of this book.)

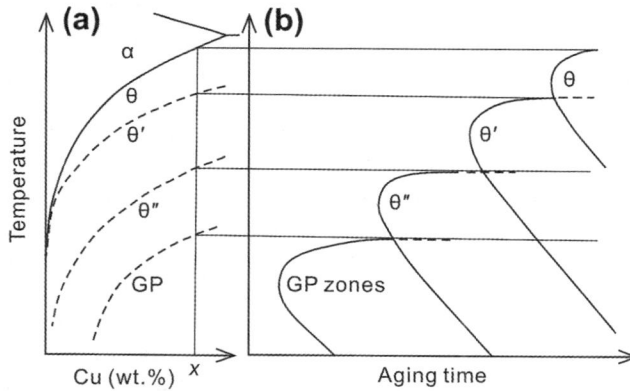

Figure 12 Schematic TTT diagrams showing the starting time of precipitation of various precipitate phases at different temperatures for a binary Al–Cu alloy with x wt% of Cu (Porter and Easterling, 1992).

The temperature–time–transformation (TTT) diagrams for individual precipitate phases in Al–Cu alloys are shown schematically in **Figure 12**. The θ'' phase forms in the early stage of aging by in situ transformation from the GP zones. After longer aging times the θ' phase starts to form by nucleating on preexisting dislocations in the matrix. This heterogeneous nucleation occurs because the strain field of a dislocation can reduce the misfit strain and hence decrease the activation energy barrier to nucleation. After still longer aging times the equilibrium θ phase nucleates on grain boundaries or at θ'/matrix interfaces. The full precipitation sequence is only possible when the alloy is aged for a long time at a temperature below the GP zones solvus. For example, if aging is carried out at a temperature above the θ'' solvus but below the θ' solvus, **Figure 9**, the first precipitate phase to occur will be θ'. If aging is carried out above the θ' solvus, then the only precipitate phase that is possible to form is θ. Also, if an alloy containing θ'' precipitates is heated to above the θ'' solvus the θ'' precipitates will dissolve. This phenomenon is known as *reversion* (Porter and Easterling, 1992).

The aging curves of Al–Cu alloys with various Cu concentrations are shown in **Figure 13**. The Al–Cu binary alloys are solution treated in the single-phase α region of the phase diagram, **Figure 9**, water

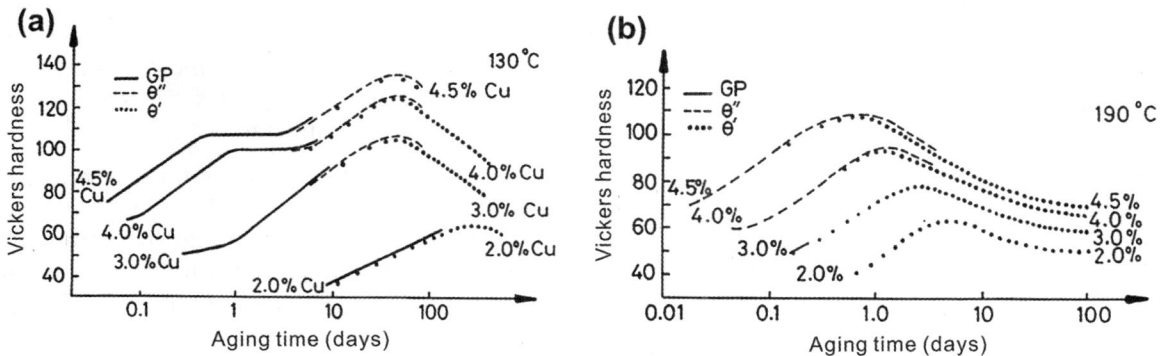

Figure 13 Effects of alloy composition and aging temperature on the maximum hardness achievable in Al–Cu binary alloys. (a) 130 °C and (b) 190 °C (Silcock et al., 1953–54).

quenched to room temperature, and then aged at 130 °C or 190 °C. The hardness or strength is relatively low in the as-quenched condition as the main resistance to dislocation movement is solid solution hardening. With the formation of GP zones and/or precipitates during isothermal aging, the hardness increases because of the extra stress required to force dislocations through the coherent zones or precipitates. The maximum hardness in each curve is associated with the existence of θ'' and θ' precipitates in the microstructure.

Both Al–4wt%Cu and Al–4.5wt%Cu alloys exhibit two stages of age hardening during isothermal aging at 130 °C, **Figure 13(a)**. Since 130 °C is below the GP zone solvus temperature, the first stage of hardening is associated with the formation of GP zones. After reaching a critical diameter of 5–10 nm, the GP zones become stable and a hardness plateau occurs. Further aging at 130 °C leads to θ'' precipitation and consequently another rise in hardness. Prolonged aging results in θ' precipitation and eventually the formation of the equilibrium θ phase. More than two precipitate phases can coexist at a given stage of the aging process. When the two alloys are aged at 190 °C, only a single stage hardening is observed, **Figure 13(b)**. This is because the aging temperature is now above the GP zones solvus but below the θ'' solvus and hence θ'' nucleates directly. However, the smaller driving force for precipitation, together with the lower volume fraction of θ'', results in a coarser dispersion and a lower peak hardness at 190 °C. At 190 °C, the hardness raises quickly to the maximum value as a result of high diffusivity. At lower aging temperature such as 130 °C, the hardness increases slowly as a result of low diffusion rates of Cu atoms. It can be seen that the peak hardness in the Al–4wt%Cu and Al–4.5wt%Cu alloys is not reached even after several tens of days aging at 130 °C. The temperatures that can be used in the heat treatment of commercial aluminum alloys are limited by economic considerations to those that produce desired mechanical properties within a reasonable period of time, usually up to 24 h.

The aging curves in **Figure 13** provide a comparison of the effects of Cu content in the alloy and aging temperature on the age hardening response. As shown clearly, the maximum hardness of the Al–Cu alloys is strongly influenced by the alloy composition and aging temperature:

- For a given aging temperature, a higher value of the maximum hardness is obtainable in the alloy with a higher concentration of Cu. This is because a higher concentration of Cu can lead to a higher volume fraction of precipitates. Furthermore, a higher concentration of Cu can lead to a larger driving force for precipitation to occur which in turn results in a smaller activation energy barrier to nucleation and hence a higher nucleation rate and thus more precipitates in the microstructure.
- For a given alloy composition, aging at a lower temperature leads to a larger value of maximum hardness. This is because of the fact that a lower aging temperature can lead to a larger driving force for precipitation, which in turn leads to a higher nucleation rate and therefore more precipitates. A lower aging temperature can also lead to a higher volume fraction of precipitates, which can again result in a higher value of the maximum hardness achievable at the temperature.

Although binary Al–Cu alloys have moderate tensile yield strength at ambient temperature, they exhibit a superior high strength at elevated temperatures compared with other precipitation-hardened aluminum alloys, **Figure 14**, particularly when they contain a higher Cu content and trace amounts of transition elements. The improved strength at elevated temperatures is attributable to the presence of thermally stable intermetallic precipitates at grain boundaries and θ' dispersed within the grains. Alloy 2219 [Al–6.3Cu–0.3Mn–0.18Zr–0.1V (wt%)] has served as the basis for several new alloy developments in which the addition of particular elements, individually or in combination, has stimulated changes to the precipitation process, which in turn have resulted in improved performance at both room and elevated temperatures.

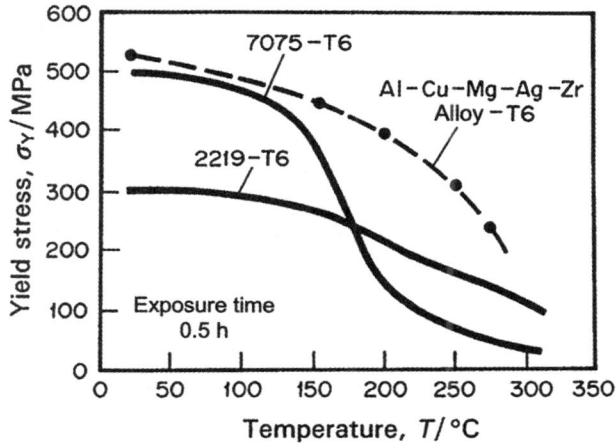

Figure 14 Variation of yield strength of peak-aged samples of Al–Cu–Mg–Ag–Zr, 2219 and 7075 alloys with testing temperature (Polmear and Couper, 1988).

20.3.1.1.2 Al–Cu–Sn Alloys

The maximum hardness value achievable in Al–Cu binary alloys during isothermal aging in the temperature range 130–200 °C is moderate, and this limited increment in hardness is associated with a relative coarse distribution of θ' precipitates, as illustrated in **Figure 11**. The nucleation of θ' precipitates is difficult compared with those precursor phases such as θ'', with the maximum number density typically less than 3000 mm^{-3} in an alloy containing 4 wt% Cu, even when trace amounts of Sn, Cd and In are added to provide heterogeneous nucleation sites. Therefore, it is common to add microalloying elements to enhance the age hardening response. Trace additions (typically less than 0.1 at.%) of Sn to Al–Cu alloys lead to an accelerated age hardening response and a substantial increase in the maximum hardness achievable, as is also illustrated in **Figure 15(a)**. Microalloying additions of

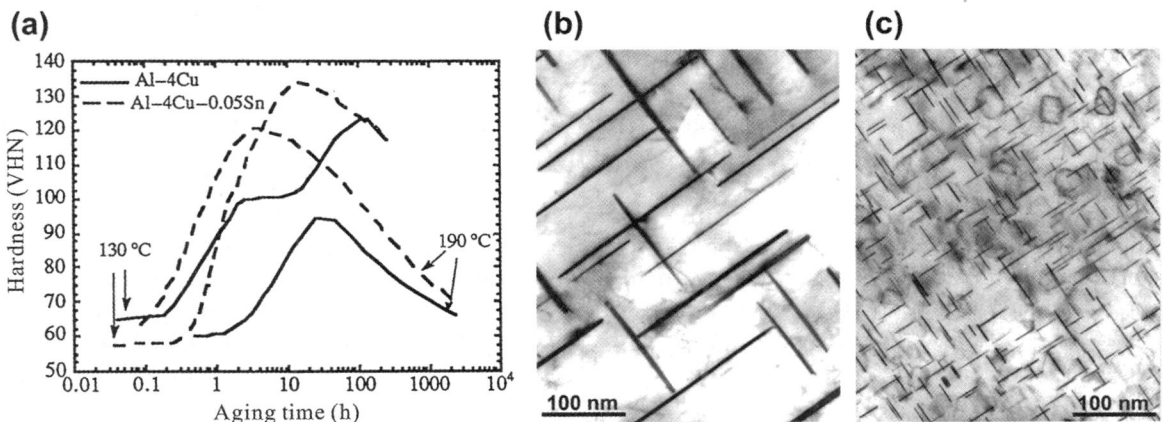

Figure 15 (a) Age hardening response of Al–4Cu and Al–4Cu–0.05Sn (wt%) alloys at 190 and 130 °C (Hardy, 1951–52). TEM images showing distribution of θ' precipitates in peak-aged samples of (b) Al–4Cu, and (c) Al–4Cu–0.05Sn alloys.

Cd and In to Al–Cu alloys produce similar effects. Trace additions of Sn, Cd and In to Al–Cu alloys increase the number density of θ' precipitates at the expense of GP zones and θ'', **Figure 15(b and c)**. Observations made with transmission electron microscopy (TEM) and atom probe field ion microscopy reveal that clusters of Sn atoms form before the nucleation of θ', and that, at later stages of aging, θ' precipitates are often associated with Sn particles, with one Sn particle almost invariably located at an end facet of individual θ' plates. Moreover, segregation of Sn atoms to the broad faces of the θ' plates has not been observed.

While it is well known that Sn has low equilibrium solubility (<0.02 at.%) and relatively high diffusivity in solid aluminum (the diffusion rate of Sn may exceed that of Cu by at least two orders of magnitude), the precise role of Sn (as well as Cd and In) in stimulating θ' nucleation is unclear and controversial. Three models have been proposed for the role of Sn in refining θ' dispersion. One model is that Sn atoms segregate to the θ'/α-Al interfaces, reducing the interfacial energy involved in θ' nucleation. This model was originally proposed by Silcock et al. (1955–56), and further refined by Silcock and Flower (2002), to account for the streaked diffraction maxima (so-called P diffraction spots) in their X-ray diffraction patterns obtained from samples in the early stages of aging. The P diffractions are also observed in Al–1.7Cu alloys with Cd and In additions. The model of Silcock et al. seems to be supported by the TEM work of Sankaran and Laird (1974) who claimed to detect what they referred to as Sn, Cd or In segregates and precipitates in association with the θ'/α-Al interface.

The second model is that the Sn atoms precipitate first during the early stages of aging, in the form of Sn-rich particles, which act as heterogeneous nucleation sites for θ' that forms subsequently. This model is essentially based on the minimization of interfacial energy or volumetric misfit strain energy. Hardy (1951–52) suggested that Sn atoms could precipitate first in the form of coherent Sn particles and subsequently act as heterogeneous nucleation sites for θ', and his model seems to be supported by the work of Suzuki et al. (1975), Kanno et al. (1980) and Ringer et al. (1995a, 1995b). Characterization of an Al–1.7Cu–0.01Sn alloy using TEM and 1DAP (Ringer et al., 1995a, 1995b) showed the occurrence of clustering of Sn atoms in the very early stages of aging, which was followed by precipitation of incoherent particles of the equilibrium phase β-Sn (space group I4$_1$/amd, $a = 0.583$ nm, $c = 0.318$ nm). The Cu atoms are not involved in the Sn clusters and Sn particles. The initial rapid clustering of Sn atoms is because of the strong interaction between Sn atoms and vacancies and enhanced diffusion of Sn atoms resulting from the initial excess concentrations of quenched-in vacancies. The θ' precipitates that have formed at the later stage of aging are frequently found to be associated with β-Sn particles, and this observation has long been used as the experimental evidence for heterogeneous nucleation of θ' on β-Sn particles.

The third model is that Sn atoms segregate to the θ' nucleus to reduce the shear strain energy involved in θ' nucleation (Nie and Muddle, 1999a; Gao et al., 1999), which will be discussed in the later section of this chapter. This model was proposed based on two critical observations: first, the formation of θ' precipitates generates a shear strain in the matrix phase; and second, the β-Sn particles are always attached to the end facet, rather than the broad face, of the θ' plates, **Figure 16**. The implication of this model is such that the β-Sn particles formed in the early stages of aging are not heterogeneous nucleation sites for θ', and that the β-Sn particle that is in contact with a θ' plate forms as the consequence of segregation of Sn atoms to the extension region of the matrix adjacent to the end facet of the θ' nucleus. Studies using atom probe and more recently HAADF-STEM have demonstrated unambiguously that Sn atoms do not segregate to the broad face of the θ' plate. The analysis of the ORs observed among θ', β-Sn and α-A phases (Gao et al., 1999) indicate that none of the observed crystallographic orientations provides a good match between the structures of θ' and β-Sn that might suggest the assistance of β-Sn in nucleating θ'.

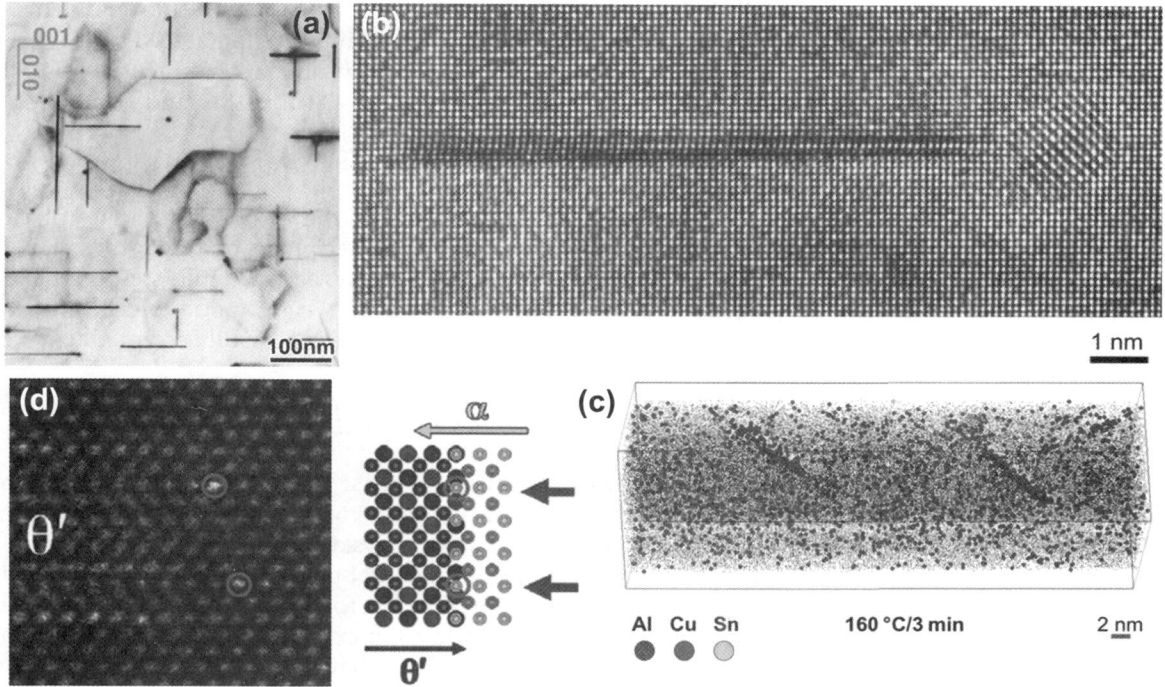

Figure 16 (a) Conventional TEM and (b) high-resolution TEM images and (c) 3DAP map showing β-Sn particles attached to one end facet of individual θ' plate in Al–4Cu–0.05Sn (wt%) alloy [Courtesy of L. Bourgeois for (b) and K. Hono for (c)]. (d) Atomic resolution HAADF-STEM image showing columns of Sn atoms located at one end facet of a single θ' plate (Bourgeois et al., 2012). (For color version of this figure, the reader is referred to the online version of this book.)

The Sn particles observed both in the very early stages of aging (Ringer et al., 1995a, 1995b) and in overaged samples (Gao et al., 1999) of Al–Cu–Sn alloys exhibit the following OR:

$$(100)_{Sn} // (111)_\alpha, [010]_{Sn} // [\overline{1}\overline{1}2]_\alpha$$

which is identical to that found in the binary Al–0.01at.%Sn alloy (Silcock et al., 1955–56). Another OR is also reported (Gao et al., 1999) to occur between β-Sn and α-Al in overaged samples of the Al–1.7at.%Cu–0.01at.%Sn alloy:

$$(100)_{Sn} // (111)_\alpha, [001]_{Sn} // [\overline{1}\overline{1}2]_\alpha$$

With the assistance of high-resolution TEM and electron microdiffraction, the β-Sn particles associated with θ' platelets, formed in the early stages of aging at 200 °C in the Al–1.7at.%Cu–0.01at.% Sn alloy, were found to exhibit an OR that was distinctly different from those reported previously for the Al–Cu–Sn and Al–Sn alloys (Bourgeois et al., 2005):

$$(001)_{Sn} // (001)_{\theta'} // (001)_\alpha, [110]_{Sn} // [100]_{\theta'} // [100]_\alpha$$

This observation implies that, while tin may assist θ' nucleation, the formation of θ' also influences the crystallography of β-Sn that forms in association with θ'. The continued aging at 200 °C leads to coarsening of those β-Sn particles that are in contact with θ' plates, and eventual replacement of this OR by $(100)_{Sn}//(111)_\alpha$, $[010]_{Sn}//[\bar{1}\,\bar{1}2]_\alpha$, which is observed in the Al–0.01at.%Sn and Al–1.7at.% Cu–0.01at.%Sn alloys.

20.3.1.1.3 *Al–Cu–Mg(–Ag) Alloys*

20.3.1.1.3.1 Alloy Compositions in the $\alpha + \theta + S$ Phase Field. The Al-rich side of the Al–Cu–Mg ternary phase diagram is shown in **Figure 17**. The identity of the strengthening precipitate phases in individual Al–Cu–Mg alloys is determined by the Cu:Mg weight ratio. Generally, the age hardening response and the maximum strength achievable increase with an increase in Cu:Mg weight ratio. For alloys with compositions lying in the $\alpha + \theta + S$ phase field, the precipitation reaction is generally accepted to have the following sequence:

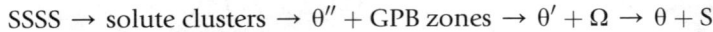

$$SSSS \rightarrow \text{solute clusters} \rightarrow \theta'' + \text{GPB zones} \rightarrow \theta' + \Omega \rightarrow \theta + S$$

The GPB zones (or Guinier–Preston–Bagaryatsky zones) and S phase will be described in detail in the next section. Apart from the above-mentioned precipitate phases, two other precipitate phases have also been reported in the literature, namely σ phase that has a cubic crystal structure (space group: $Pm\bar{3}$ $a = 0.831$ nm) (Samson, 1949) and an $Al_5Cu_6Mg_2$ composition and N phase that has a face-centered cubic structure with lattice parameters of approximately 0.580 nm. The formation and stability of these two phases seem to be affected by impurity elements in the alloy. For example, the presence of Si can promote the formation and stabilize the σ phase. The σ precipitates have a cuboidal shape and they usually have a cube-to-cube OR with the α-Al matrix phase. The N precipitates have a lath morphology, with their long axis parallel to $\langle 100 \rangle_\alpha$ directions. The OR between the N phase and the α-Al matrix is such that $(001)_N//(011)_\alpha$ and $[110]_N//[100]_\alpha$. The crystal structure of the N phase is closely similar to that of a ternary phase T_B (Al_7Cu_4Li) observed in Al–Cu–Li alloys (Hardy and Silcock, 1955–56), and

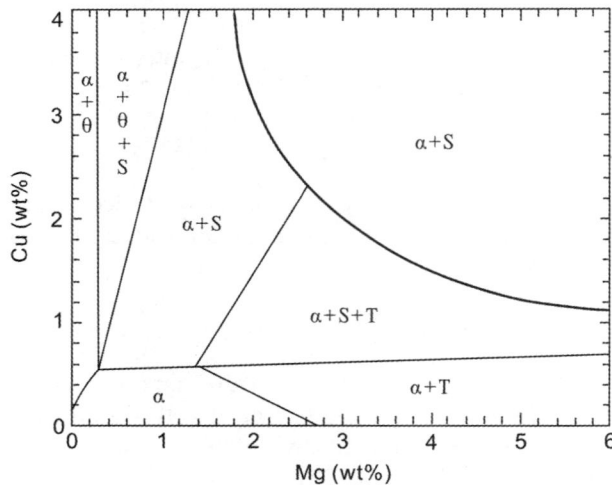

Figure 17 Al-rich side of the Al–Cu–Mg ternary phase diagram at 190 °C. The dark blue thicker line represents the solvus line at 500 °C (Brook, 1965).

the OR between N and matrix phases is identical to one of the several ORs observed between T_B and matrix phases. Therefore, the N phase could be regarded as an Mg isomorph of T_B. The T_B phase has a CaF_2-type structure (space group $Fm\bar{3}m$, $a = 0.583$ nm) with 12 atoms per unit cell. The structure of T_B is closely similar to that of θ', and as a result, with Al atoms partially replaced by Li atoms, θ' may transform continuously to T_B (Hardy and Silcock, 1955–56; Schneider and von Heimendahl, 1973). The tetragonal structure of θ' ($a = 0.404$ nm, $c = 0.580$ nm) can also be regarded as a tetragonal structure with $a = 0.571$ nm (0.404 nm $\times \sqrt{2}$) and $c = 0.580$ nm, and a slight expansion in parameter a can generate the cubic unit cell of T_B. Since the atomic radius of Mg atoms (0.160 nm) is slightly larger than that of Al atoms (0.143 nm), it is plausible that a partial replacement of Al atoms by Mg atoms in θ' results in a lattice expansion to form the cubic structure of the N phase. Interestingly, the N phase could also be regarded as a partially substituted θ' phase with another morphology and OR.

The addition of a small amount of Mg, typically in the range 0.3–0.5 wt%, to Al–Cu alloys results in an occasional replacement of θ' plates by a metastable Ω phase in the precipitation process at high temperatures. A quaternary addition of 0.4–0.5 wt% Ag can further facilitate the formation of the Ω phase and the development of a uniform and relatively dense distribution of the Ω precipitates in samples aged at lower temperatures, **Figure 18**. It is to be noted that additions of Ag alone to binary Al–Cu alloys do not facilitate nucleation of either θ' or Ω. The Ω precipitates form as thin hexagonal shaped plates on $\{111\}_\alpha$ planes. They are remarkably thermally stable at elevated temperatures. Associated with precipitation of such plates is a significant increase in the peak hardness, **Figure 18(a)**. The resultant Al-Cu-Mg-Ag alloys exhibit a tensile yield strength of above 500 MPa at ambient temperatures and improved creep resistance at temperatures up to 180–200 °C, **Figure 14**. Although the number density and volume fraction of $\{111\}_\alpha$ precipitate plates of Ω are similar to that of $\{100\}_\alpha$ plates of θ' in Al–Cu binary alloys, the aspect ratio of Ω plates is typically over 35:1, much larger than that of $\{100\}_\alpha$ plates of θ'. It appears that it is the formation of $\{111\}_\alpha$ plates and the increased aspect ratio of the precipitate plates reduce the effective interparticle spacing and thus increase the alloy tensile strength. The improved creep strength is attributable to the thermal stability of the Ω phase, which is associated with a low interfacial energy between Ω and matrix phases on the habit plane.

The crystal structure of the Ω precipitates was originally reported to be monoclinic ($a = 0.496$ nm, $b = 0.496$ nm, $c = 0.848$ nm, $\gamma = 120°$) or hexagonal ($a = 0.496$ nm, $c = 0.701$ nm). Subsequent studies using high-resolution TEM and electron microdiffraction (Knowles and Stobbs, 1988; Muddle and Polmear, 1989) revealed that the crystal structure of Ω had a face-centered orthorhombic (space group $Fmmm$, $a = 0.496$ nm, $b = 0.859$ nm, $c = 0.848$ nm), **Figure 18(c)**, which is now commonly accepted in the literature. The OR between the Ω precipitates and the α-Al matrix is such that $(001)_\Omega //(111)_\alpha$ and $[010]_\Omega//[10\bar{1}]_\alpha$. The structure of Ω is closely similar to that of θ phase in Al–Cu alloys. Based on convergent-beam electron diffraction patterns, Garg and Howe (1991) proposed a tetragonal structure (point group $4/mmm$, $a = b = 0.607$ nm, $c = 0.496$ nm) and redesignated Ω as θ_m. The difference in lattice parameter between θ and Ω is extremely small. Prolonged aging at temperatures above 250 °C results in the eventual replacement of the Ω phase by the equilibrium θ phase.

The precise roles of Mg and Ag in stimulating the nucleation and formation of the Ω phase are still not fully understood. Early studies using X-ray microanalysis (Muddle and Polmear, 1989) and 1DAP (Hono et al., 1993) revealed the strong segregation of Mg and Ag to the broad Ω/α-Al interface and the absence of Mg and Ag atoms in the interior part of the Ω phase. The segregation of the Mg and Ag atoms occurs intriguingly in two atomic layers. It has been now well established that co-clusters of Mg and Ag atoms form in the very early stages of decomposition of the Al–Cu–Mg–Ag alloys and that Ag modifies the precipitation process through a preferred Mg–Ag interaction. For example, the 3DAP work of Hono et al. (Murayama and Hono, 1998; Reich et al., 1998) shows that, in the Al–1.7Cu–0.3Mg–0.2Ag alloy

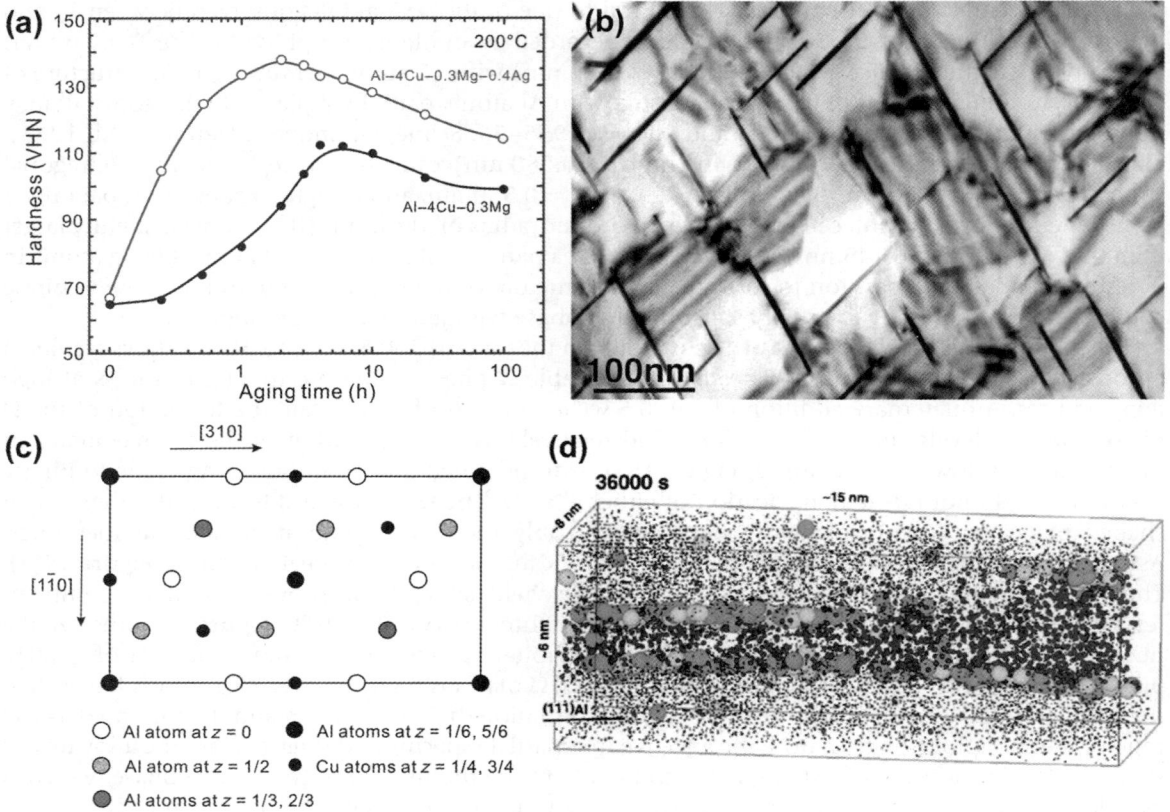

Figure 18 (a) Age hardening response of Al–4Cu–0.3 Mg and Al–4Cu–0.3Mg–0.4Ag (wt%) alloys. (b) TEM image showing the morphology and distribution of Ω precipitates in the Al–4Cu–0.3Mg–0.4Ag alloy. (c) Projection of Ω unit cell, and (d) 3DAP map showing segregation of Mg and Ag atoms in the Ω/α-Al interface. Mg and Ag atoms are represented by green and yellow color, respectively (Reich et al., 1998). (For color version of this figure, the reader is referred to the online version of this book.)

aged for 15 s at 180 °C, the Mg–Ag co-clusters containing 40–80 atoms have already formed. These clusters do not have a well-defined shape at this stage. Continued aging causes Cu atoms to segregate into the cluster and a change of the cluster shape to disc-like with its broad surface parallel to $\{111\}_\alpha$ planes, before the nucleation of Ω in association with these clusters. At later stages of aging, atoms of Mg and Ag partition to the broad faces of Ω plates, **Figure 18(d)**.

Suh and Park (1995) showed that Mg–Ag co-clusters prefer to form on the $\{111\}_\alpha$ planes as a result of elastic strain energy minimization. After 2.5 h aging at 180 °C, the microstructure contains predominantly a uniform distribution of Ω precipitates on $\{111\}_\alpha$, in addition to a small fraction of θ' plates on $\{001\}_\alpha$ planes. Therefore the role of Ag seems to effectively trap Mg and Cu atoms to form disc-like Mg–Ag–Cu co-clusters on $\{111\}_\alpha$ planes that subsequently act as heterogeneous nucleation sites for the Ω phase. Since the formation of Ω precipitates causes a larger contraction strain (~ 0.093) in the matrix in the direction normal to their habit plane, the larger atomic size of Mg can reduce the contraction strain in the direction normal to the habit plane when

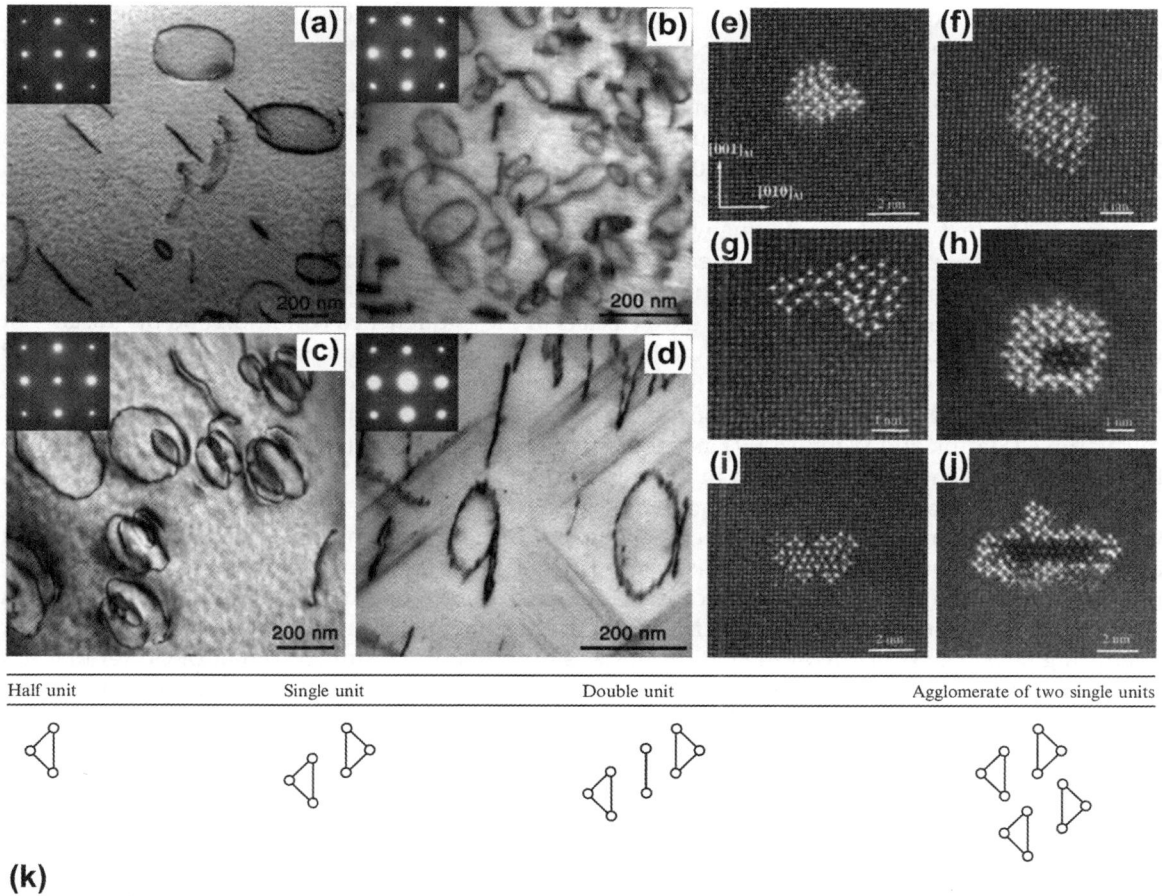

Figure 19 (a–d) TEM images showing dislocation loops and heterogeneous nucleation on these loops in an Al–1.1Cu–1.7 Mg (at.%) alloy (Marceau et al., 2010). (e–j) HAADF-STEM images showing cross-sections of $\langle 001 \rangle_\alpha$ rods of GPB zones, and (k) schematic diagrams showing structural units of GPB zones (Kovarik et al., 2008).

Mg atoms are present in the broad Ω/α-Al interface. The observation that Ω forms only as a minor phase in ternary Al–Cu–Mg alloys can be attributed to the presence of fewer Mg clusters.

20.3.1.1.3.2 Alloy Compositions in the $\alpha + S$ Phase Field. For Al–Cu–Mg alloy compositions lying in the $\alpha + S$ phase field of the Al–Cu–Mg phase diagram, the precipitation sequence is generally accepted to be the following for aging treatments below 218 °C:

$$\text{SSSS} \rightarrow \text{solute clusters} \rightarrow \text{GPB zones} \rightarrow \text{S}$$

Immediately following water quenching, a rapid coalescence and condensation of quenched-in vacancies occurs, resulting in the formation of a large number of dislocation loops and helices, **Figure 19(a–d)**. The loops lie on $\{110\}_\alpha$ and possess $a/2\langle 010 \rangle_\alpha$ Burgers vectors. The diffusion of

Mg and Cu atoms to these dislocation loops and helices creates the chemical environment for the precipitation of the S phase at these defects. In fact, S precipitates are often found to form heterogeneously on dislocation loops and helices.

Solute clusters were reported to occur in attempts to account for the rapid hardening phenomenon in an Al-2.5Cu-1.5 Mg (wt%) or Al-1.1Cu-1.7 Mg (at.%) alloy during isothermal aging at 150 and 200 °C, **Figure 20**. As shown in **Figure 20(a)**, hardening occurs very rapidly in the early stage of aging, and this process is followed by a hardness plateau over a lengthy period of time and a secondary stage hardening. The first stage of hardening was usually attributed to the formation of GPB zones in early studies, while the second stage of hardening was attributed to the formation of metastable or equilibrium precipitate phases. Characterizations using TEM and 1DAP (Ringer et al., 1997) indicated the formation of independent Mg clusters and Cu clusters in the as-quenched condition and the absence of precipitation just before and right after the hardening reaction. However, subsequent work by 3DAP revealed little or no evidence of a uniform distribution of Cu–Mg co-clusters in the α-Al matrix. Recently, dimers of Mg and Cu were proposed to occur in Al–Cu–Mg alloys based on thermodynamic calculations (Starink and Wang, 2009) and the formation of these solute dimers might be responsible for the rapid hardening phenomenon in **Figure 20(a)**.

GPB zones start to form toward the end of the hardness plateau shown in **Figure 20(a)**. The onset of GPB precipitation causes secondary hardening, during which the maximum hardness is achieved when a critical distribution of the GPB zones is reached. The GPB zones form as nanoscale rods with their long axis parallel to $\langle 001 \rangle_\alpha$. They are typically 1–2 nm in diameter, but their length may vary from 4 nm for samples aged after cold water quenching to more than 8 nm in samples subjected to lower quenching rates. The structure of the GPB zones was proposed to be similar to that of CuAuI that has a face-centered tetragonal structure (space group P4/mmm, $a = 0.55$ nm, $c = 0.404$ nm), but inconsistency was noted between the model and the positions of diffraction maxima in the X-ray diffraction

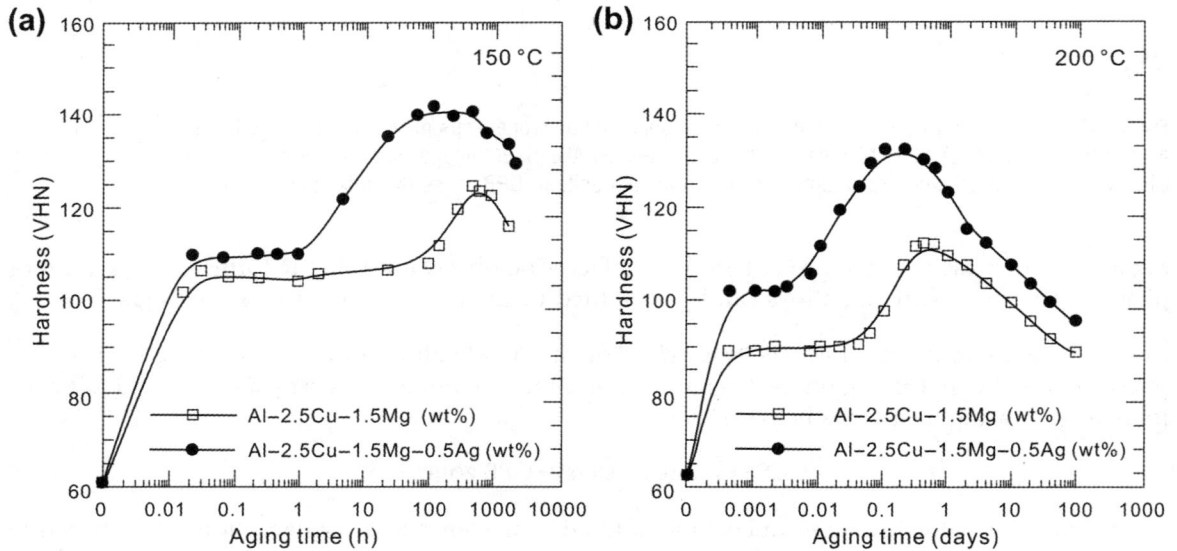

Figure 20 Aging curves of Al–Cu–Mg alloys at (a) 150 °C and (b) 200 °C (Vietz and Polmear, 1966).

patterns. A recent study of nanometer $\langle 001 \rangle_\alpha$ rods in an Al–1Cu–3Mg (wt%) alloy using atomic resolution HAADF-STEM and first-principles calculations (Kovarik et al., 2008) revealed that these rods have two distinctly different structures: one structure composed of an agglomeration of structural units with a translational periodicity along a single $\langle 001 \rangle_\alpha$ direction; while the other structure has a core having a hexagonal structure and a shell comprising structural units. The former was designated that of GPB zones, while the latter that of GPB II, **Figure 19(e–j)**. A single structural unit is made of a pair of triangles facing each other, **Figure 19(k)**. This pair of triangles is such oriented that their bases face each other and are separated by one atomic layer of $\{200\}_\alpha$ and mutually offset by $a/2\langle 010 \rangle_\alpha$.

Overaging occurs as the GPB zones are gradually replaced by the precipitation of S phase that persists throughout the subsequent aging process. Several structures have been reported for the equilibrium S phase. However, the accumulated experimental evidence indicates that the crystal structure of S is orthorhombic (space group Cmcm, $a = 0.400$ nm, $b = 0.923$ nm, $c = 0.714$ nm), with 16 atoms in the unit cell. Four distinguishably different ORs have been found between S and α phases:

$$\text{OR1}: \quad (001)_s//(021)_\alpha, [100]_s//[100]_\alpha$$
$$\text{OR2}: \quad (0\bar{1}1)_s//(012)_\alpha, [100]_s//[100]_\alpha$$
$$\text{OR3}: \quad (0\bar{1}2)_s//(011)_\alpha, [100]_s//[100]_\alpha$$
$$\text{OR4}: \quad (03\bar{1})_s//(011)_\alpha, [100]_s//[100]_\alpha$$

All four ORs are related to each other by a simple rotation about the $[100]_s//[100]_\alpha$ axis. For example, the OR2 can be obtained from the OR1 by $\sim 0.85°$ clockwise rotation of the S lattice; a clockwise rotation by $\sim 2.71°$ of the S lattice from the OR1 can lead to the OR3; and $\sim 4.88°$ clockwise rotation from the OR1 produces the OR4. Precipitates of the S phase have either lath or rod morphology, with their long axes parallel to $\langle 001 \rangle_\alpha$. While precipitates in both OR1 and OR4 have a lath shape, **Figure 21**, their habit planes are distinctly different; the former is atomically flat and is exactly parallel to $\{021\}_\alpha$, whereas the latter has an array of structural ledges with the macroscopic interface parallel to $\{043\}_{\alpha s}$. The cross-sections of the S precipitates in the OR2 and OR3 are faceted and nearly equiaxed. The precipitates associated with the OR1 often form in clusters and are commonly associated with heterogeneous nucleation on preexisting dislocation loops and helices in the matrix phase. The OR1 was originally reported by Bagaryatskii (1952), subsequently confirmed by Silcock (1960–61), and was commonly accepted in the literature as the only OR between S precipitates and the α-Al matrix. However, the existence of the other ORs was noted in the studies of Bagaryatskii (1952) and Ringer et al. (1996), and was studied in detail by Radmilovic et al. (1999), Winkelman et al. (2007), and Wang and Starink (2007).

Over a long period of time, the precipitation process in the Al–Cu–Mg alloys is said to involve the formation of S″ and S′ phases as metastable precursors to S. The structures of the S″ and S′ phases are essentially the same as that of S. They were originally reported (Bagaryatskii, 1952) to be distinguishable from S on the basis of small differences in lattice parameters, being coherent and partially coherent, respectively, with the α-Al matrix phase. The S′ phase was assumed in this early work to have lattice parameters ($a = 0.404$ nm, $b = 0.904$ nm, $c = 0.720$ nm), which are slightly different from those of the S phase. This assumption was not supported by the X-ray diffraction study made in the following years (Silcock, 1960–61). Subsequent studies made by X-ray diffraction or electron diffraction all indicated that the distinction between S′ and S was very slight and arbitrary (Wilson and Partridge, 1965; Gupta et al., 1987; Radmilovic et al., 1989). It is therefore not necessary to differentiate and to introduce the nomenclature of S′ and S because the difference between them is simply

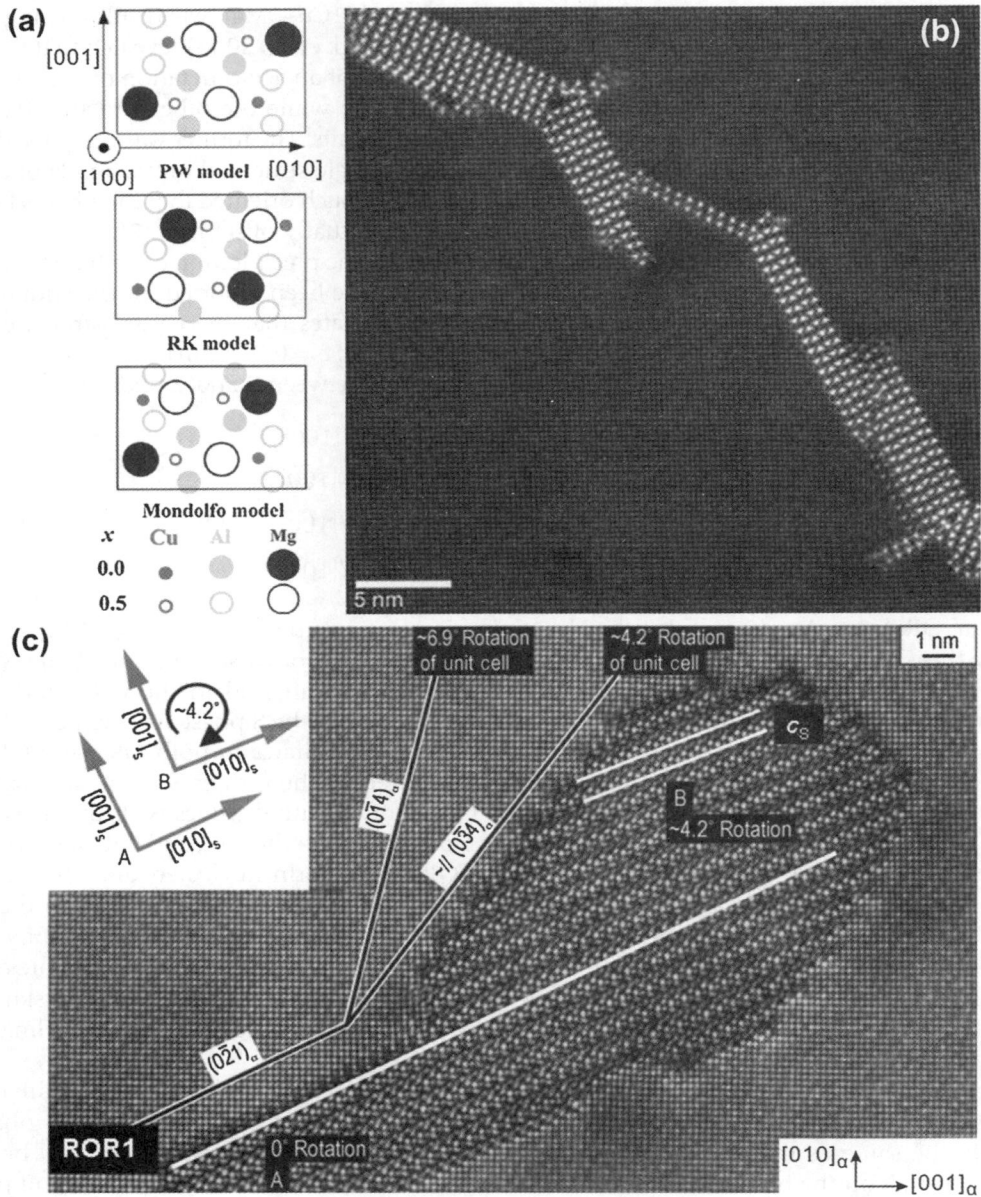

Figure 21 (a) Unit cell models of S phase (Liu et al., 2011). (b) HAADF-STEM (Ralston et al., 2010) and (c) high-resolution transmission electron microscopy (HRTEM) images showing S precipitates in two different ORs (Winkelman et al., 2007). (For color version of this figure, the reader is referred to the online version of this book.)

a subtle change in lattice parameters rather than a significant variation in structure or composition. Note that two other orthorhombic structures have also been reported for the S′ (or S) phase: an orthorhombic unit cell ($a = 0.400$ nm, $b = 0.925$ nm, $c = 0.718$ nm) (Mondolfo, 1976; Pérez-Landazábal et al., 1997) that has significantly different atomic positions compared with that proposed by Perlitz and Westgren, and a unit cell (space group Pmm2, four atoms in the unit cell) that has different lattice parameters ($a = 0.400$ nm, $b = 0.461$ nm, $c = 0.718$ nm) (Jin et al., 1990).

The addition of Ag to the Al–Cu–Mg alloys accelerates the aging process and increases significantly the maximum hardness achievable, even though it does not affect the characteristics of the hardness–time curve. Examination using 1DAP of the microstructure of an Al–1.1Cu–1.7Mg–0.1Ag alloy, aged for 5 min at 150 °C, revealed the existence of Mg clusters, Mg–Cu co-clusters, and more notable Mg–Ag co-clusters. The maximum hardness is associated with a uniform and dense distribution of nanoscale $\{111\}_\alpha$ platelets of X′ phase. Characterization of these platelets using electron microdiffraction (Chopra et al., 1995) indicated that they had a hexagonal structure ($a = 0.496$ nm, $c = 1.375$ nm) and an OR $(0001)_{X'}//(111)_\alpha$, $[10\bar{1}0]_{X'}//[1\bar{1}0]_\alpha$. The analysis of 1DAP data obtained from the X′ platelets (Ringer et al., 1997) revealed that they contained 50–65at.%Al, 20–25at.%Cu, 15–25at.%Mg, and up to 5at.%Ag.

20.3.1.1.3.3 Alloy Compositions in the $\alpha + S + T$ Phase Field.   For alloys with compositions within the $\alpha + S + T$ phase field, there has been a lack of detailed characterization of precipitate phases formed in them. The T phase has a body-centered cubic structure (space group Im$\bar{3}$, $a = 1.425$ nm) and an atomic composition of Al$_6$CuMg$_4$. The age hardening characteristics of Al–0.64Cu–4.7Mg and Al–0.64Cu–4.7Mg–0.1Ag (at.%) alloys are very similar to those for the alloys in the $\alpha + S$ phase field. The addition of Ag to the Al–Cu–Mg alloys significantly raises the level of the initial rapid hardening reaction, even though it has little effect on the character of the aging curve (Vietz and Polmear, 1966). The enhanced age hardening response is associated with a refined dispersion of spherical particles of the Z phase that has a cubic structure (point group m$\bar{3}$m, $a = 1.999$ nm) (Chopra et al., 1996) and contains 2–5at.%Ag, 20at.%Cu, 20–25at.%Mg, and 50–65at.%Al (Ringer and Hono, 2000). The Z precipitates have two ORs with the α-Al matrix: $(100)_Z//(100)_\alpha$, $[010]_Z//[010]_\alpha$ and $(011)_Z//(111)_\alpha$, $[01\bar{1}]_Z//[01\bar{1}]_\alpha$.

20.3.1.1.4 Al–Cu–Li(–Mg–Ag) Alloys
In Al–Cu–Li alloys, the predominant strengthening precipitate phase is T$_1$ (Al$_2$CuLi), which has a hexagonal structure ($a = 0.496$ nm, $c = 0.935$ nm) and forms as plates on $\{111\}_\alpha$ planes. The aspect ratio of the T$_1$ precipitate plates is remarkably large, typically above 50:1, **Figure 22(a)**. The melting point of the T$_1$ phase, despite some uncertainty, is above 635 °C, and is thus higher than that of θ' and Ω phases. T$_1$ phase was originally proposed (Hardy and Silcock, 1955–56) to have a hexagonal structure ($a = 0.497$ nm, $c = 0.935$ nm), and to share an OR with the matrix phase in the form $(0001)_{T_1}//(111)_\alpha$, $[10\bar{1}0]_{T_1}//[1\bar{1}0]_\alpha$. However, the detailed symmetry of the unit cell remains controversial. Many of the structural models proposed for the T$_1$ phase (Howe et al., 1988; Radmilovic and Thomas, 1987; Cassada et al., 1991a, 1991b; Huang, 1992; Donnadieu et al., 2011) do not appear to be fully consistent with experimental observations. A very recent study involving the use of atomic resolution HAADF-STEM and first-principles calculations (Dwyer et al., 2011) has shown that the T$_1$ precipitates have a hexagonal structure (space group P6/mmm, $a = 0.496$ nm, $c = 1.391$ nm), **Figure 22(b)**, that is isomorphous to that determined for the monolithic T$_1$ alloy (Vecchio and Williams, 1988) and that is closely similar to that proposed by van Smaalen et al.

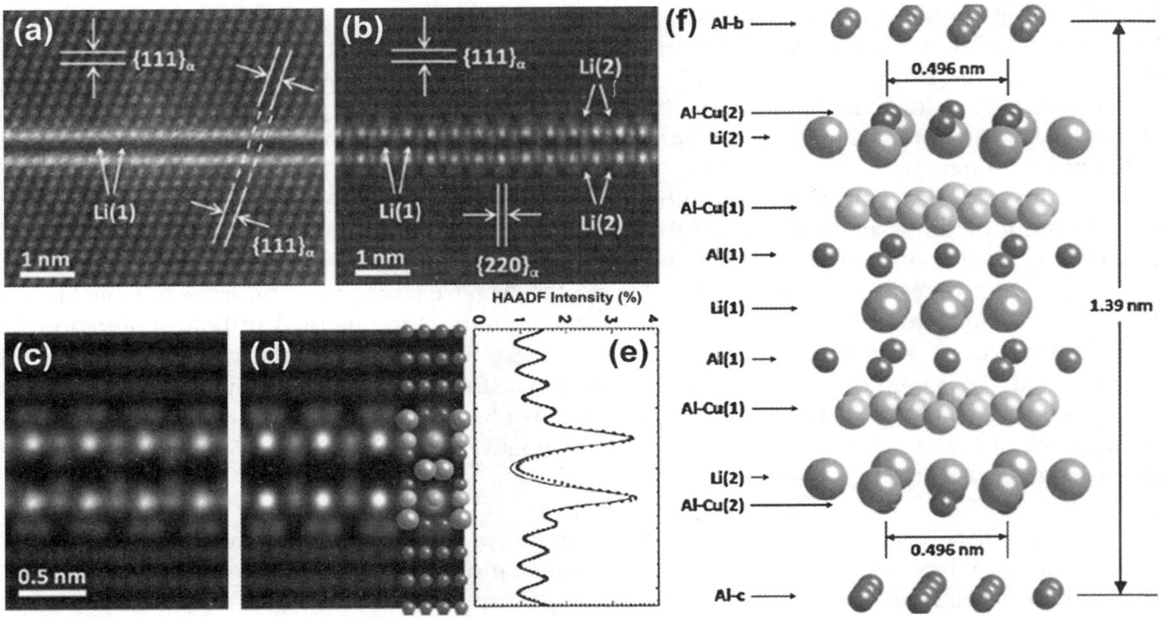

Figure 22   HAADF-STEM and simulated images showing the structure of T_1 (Dwyer et al., 2011). (For color version of this figure, the reader is referred to the online version of this book.)

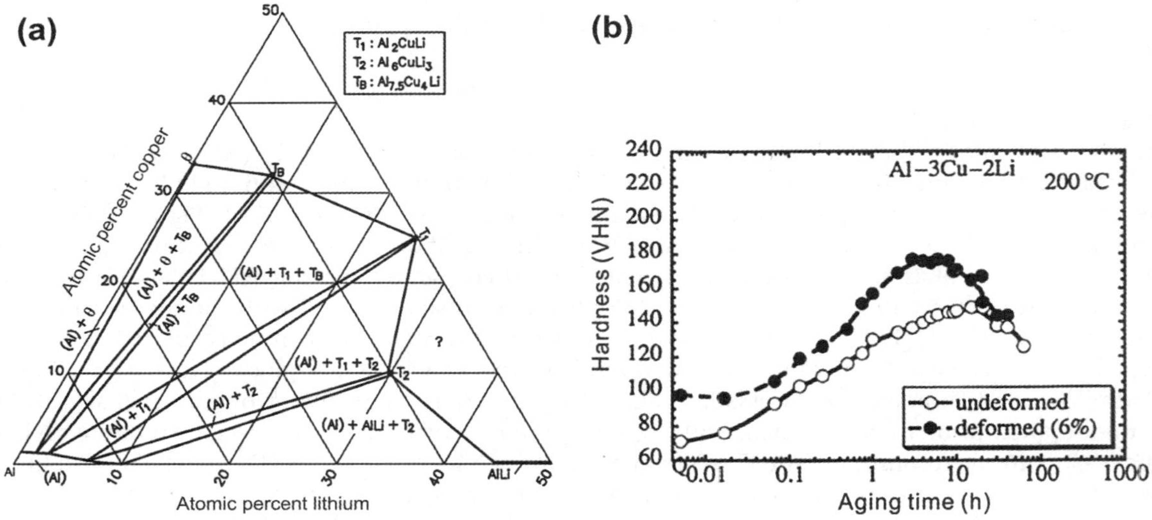

Figure 23   (a) Isothermal section of the ternary Al–Cu–Li phase diagram at 500 °C (Raghavan 2010). (b) Effect of cold work on age hardening response of an Al–3Cu–2Li (wt%) alloy at 200 °C (Ringer et al., 1995a).

(1990). While the isothermal section of the ternary Al–Cu–Li phase diagram indicates that T_1 is an equilibrium phase at 500 °C, **Figure 23(a)**, it has been reported that the T_1 precipitates are metastable and are replaced by T_B (Al_7Cu_4Li) and T_2 (Al_6CuLi_3) phases after prolonged aging in the temperature range 200–300 °C.

Although the T_1 phase is near to fully coherent with the matrix phase, its nucleation occurs with much difficulty in Al–Cu–Li ternary alloys during conventional isothermal aging treatments. Consequently, it is common practice to deform the matrix phase before aging, or to add microalloying additions of Mg and Ag to such alloys, to facilitate nucleation of the T_1 phase. Individual additions of Mg, In or Cd to Al–Cu–Li ternary alloys also promote the nucleation of the T_1 phase. The effect of cold work (T8 temper), after solution treatment and water quench and before aging, is shown in **Figure 23(b)**. It is apparent that the cold work leads to accelerated age hardening response and an increased maximum hardness. The dislocations introduced by the cold work act as heterogeneous nucleation sites for the T_1 phase and hence result in an increased number density of T_1 precipitate plates in the microstructure.

While the Al–Cu–Li alloys have an excellent combination of lower density and higher stiffness and strength, they suffers from toughness deterioration during low-temperature thermal exposure. It has been shown that the fracture toughness of aluminum alloy AA2090, peak aged at 150–160 °C, decreases significantly with secondary aging at 70–120 °C. This severe reduction in fracture toughness is accompanied by a simultaneous increase in yield strength. A systematic characterization of microstructures and age hardening behavior of duplex aged samples of the AA2090 alloy, aged initially to the maximum hardness at 200 °C and then subjected to long-term aging at 90–150 °C, indicates that secondary age hardening occurs at temperatures up to 90 °C and that this secondary hardening is because of the formation of GP zones and precipitation of very fine spherical particles of δ' phase (Al_3Li) in matrix regions adjacent to grain boundaries and between precipitates formed from the primary aging at 200 °C, **Figure 24**.

Combined additions of Mg and Ag to Al–Cu–Li alloys stimulate significantly the nucleation and formation of T_1 phase (Pickens et al., 1989). It is interesting to note that systematic additions of Li to Al–Cu–Mg–Ag alloys lead to a similar microstructure—the Ω precipitates are gradually replaced by the T_1 precipitate plates (Polmear and Chester, 1989). As shown in **Figure 25(a)**, the age hardening response and the maximum hardness increase with Li additions. At maximum strength, the Al–Cu–Li–Mg–Ag alloys commonly have a microstructure comprising an essentially continuous three-dimensional network of T_1 plates, with the addition of minor phases θ' and S, and exhibit exceptionally high strength. The tensile yield strength of the Al–Cu–Li–Mg–Ag alloys, such as the Weldalite 049™ series of ultrahigh-strength alloys X 2094 and X 2095, may exceed 700 MPa, which makes them competitive with high-strength low-alloy steels on the basis of strength alone. Although the number density and volume fraction of T_1 plates are both comparable with those of Ω in Li-free Al–Cu–Mg–Ag alloys, they are much thinner (typically less than 2 nm), and the aspect ratio is significantly larger than that of Ω, **Figure 25(b)**. The exceptionally high strength achieved in the Al–Cu–Li–Mg–Ag alloys is attributable to the uniquely larger aspect ratio of $\{111\}_\alpha$ plates on the primary slip plane of the matrix phase. The formation of T_1 phase also leads to increased creep strength. The improved creep resistance is attributable to higher thermal stability of T_1 phase, which appears to result from its lower interfacial energy with the matrix phase, compared with both the Ω and θ' phases.

Examination using a 3DAP of microstructures of the Al–Cu–Li–Mg–Ag alloys, aged isothermally at 180 °C, indicates the presence of pre-precipitate clusters of Mg atoms during early stages of aging, and segregation of Mg and Ag atoms to the broad faces of T_1 plates at later stages, **Figure 25(c)**.

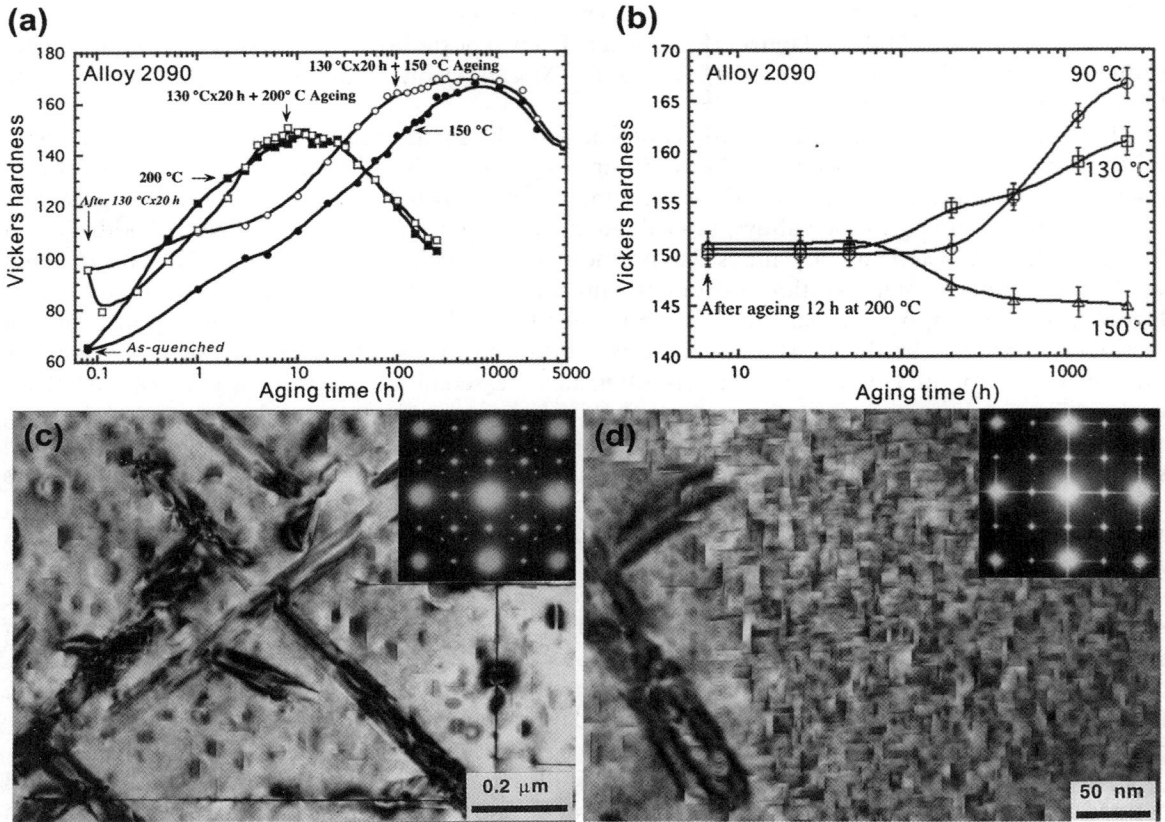

Figure 24   (a) Age hardening response of alloy AA2090 at 150 and 200 °C, with and without initial aging at 130 °C for 20 h. (b) Aging curves of same alloy during secondary aging at 90, 130 and 150 °C, after initial aging at 200 °C for 12 h. (c and d) TEM images showing distribution of precipitates in samples (c) aged to maximum hardness at 200 °C for 10 h with initial aging at 130 °C for 20 h and (d) after secondary aging at 90 °C for 150 days (Gao et al., 1998). (For color version of this figure, the reader is referred to the online version of this book.)

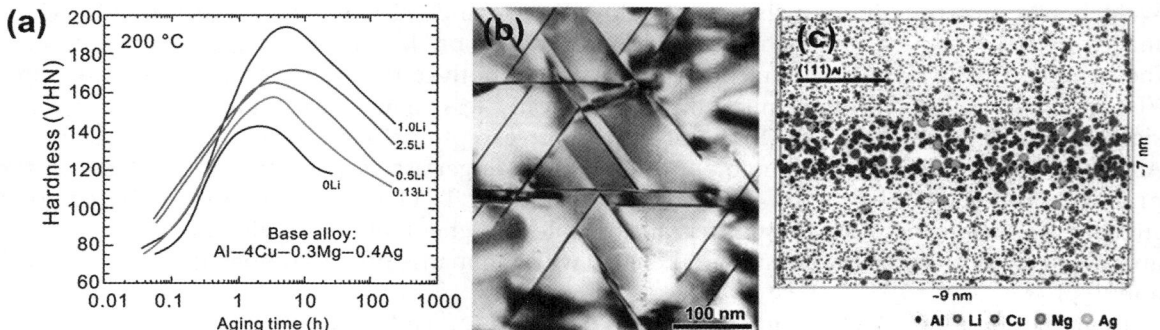

Figure 25   (a) Effects of Li additions on age hardening response of Al–Cu–Mg–Ag alloy (Polmear and Chester, 1989). (b) TEM image showing distribution of T_1 precipitates (Muddle et al., 1994). (c) 3DAP map showing Mg and Ag atoms in T_1/α-Al interface (Murayama and Hono, 2001). (For color version of this figure, the reader is referred to the online version of this book.)

20.3.1.1.5   Al–Cu–Mg–Si(–Ag) Alloys

Additions of small amounts of Si and/or Ge to Al–Cu–Mg(–Ag) alloys, with high Cu:Mg ratios, suppress formation of Ω, but promote the nucleation of θ'. The effects of the Si addition on the age hardening response of the Al–4Cu–0.3 Mg and Al–4Cu–0.3Mg–0.4Ag (wt%) alloys are shown in **Figure 26**. It is apparent that the Si additions accelerate the hardening kinetics and increase the maximum hardness achievable.

Figure 26(c) shows a dense distribution of $\{100\}_\alpha$ plates of θ' phase in the peak-aged microstructure of an Al–4Cu–0.3Mg–0.5Si (wt%) alloy. The peak hardness of this alloy is higher than those obtained in the Al–4Cu–0.3Mg and Al–4Cu (wt%) alloys. Characterization of the microstructural evolution during isothermal aging at 170 °C reveals the formation of co-clusters of Mg and Si (Ge) atoms during the very early stages of precipitation (Ringer et al., 1996). These Mg–Si (Ge) co-clusters adopt an elongated form parallel to $\langle 100 \rangle_\alpha$ directions before the formation of θ'. With prolonged aging these

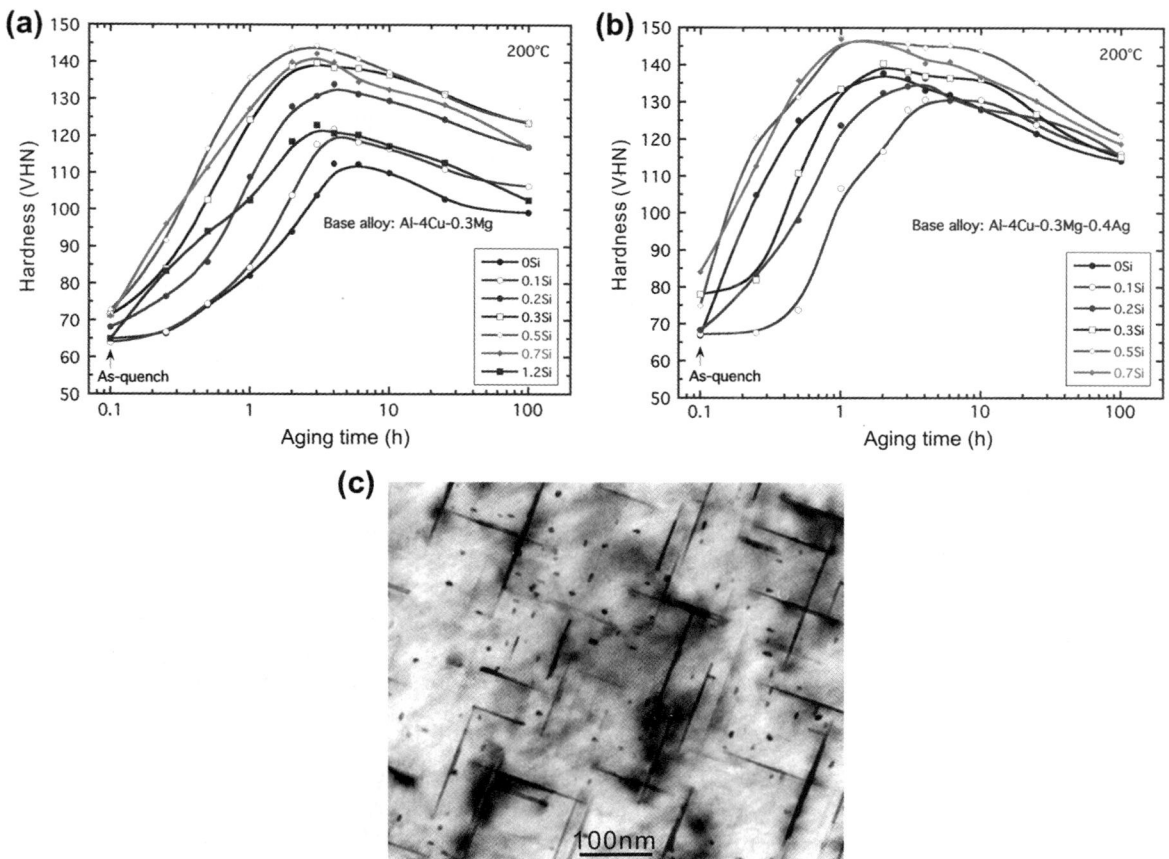

Figure 26   Effects of Si additions on age hardening response of (a) Al–Cu–Mg and (b) Al–Cu–Mg–Ag alloys. (c) TEM image showing distribution of θ' precipitates (Gao et al., 1996). (For color version of this figure, the reader is referred to the online version of this book.)

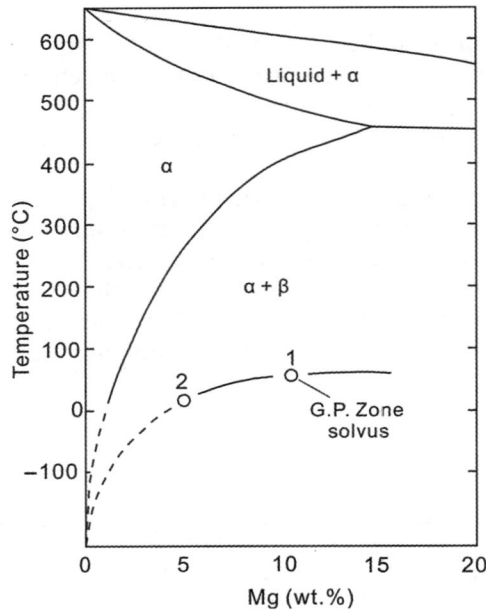

Figure 27   The Al-rich side of the Al–Mg phase diagram (Lorimer, 1978).

Mg–Si (Ge) clusters combine with Al and Cu atoms to evolve into $\langle 100 \rangle_\alpha$ rods of an equilibrium phase Q that has an $Al_4Cu_2Mg_8Si_7$ composition and a hexagonal structure ($a = 1.039$ nm, $c = 0.404$ nm), and θ' is eventually observed to form in association with the Q phase. One end facet of θ' plates is invariably attached to particles of the Q phase.

20.3.1.2   5xxx Series Alloys

The equilibrium intermetallic phase at the Al-rich side of the Al–Mg phase diagram, designated β, has a composition of Al_3Mg_2 or Al_8Mg_5 and a complex face-centered cubic structure (space group $Fd\bar{3}m$, $a = 2.82$ nm). The equilibrium solid solubility of Mg in Al is approximately 18.6at.% at the eutectic temperature 450 °C and it drops to 3.2–4.4at.% at 200 °C and about 2.1at.% at 100 °C, **Figure 27**. Therefore, there exists a great potential to develop precipitation hardenable alloys based on the Al–Mg system. Unfortunately, the age hardening of the binary Al-Mg alloys is abnormally low. During isothermal aging in the temperature range 160–240 °C, there is virtually no age hardening response when the Mg content in Al is less than 5 wt%. A moderate age hardening response starts to develop when the Mg content in the alloy is increased to 10 wt%, **Figure 28** (Kubota et al., 2004). The poor age hardening response is because of a coarse distribution of precipitates in the aged samples. While cold work can increase the hardness of binary Al–Mg alloys, it does not lead to any age hardening enhancement.

The precipitation sequence in binary Al–Mg alloys is not well studied. It is generally accepted that the precipitation involves the formation of GP zones, β' and β phases (Lorimer, 1978; Polmear, 2006), even though a β'' phase is also reported to form (Starink and Zahra, 1998). The GP zones have a spherical shape. Their solvus line lies below ambient temperature, for Mg contents less than 5 wt%, and close to ambient temperature when the Mg content in the alloy is in the range 5–10 wt%, **Figure 27**. The β'' phase

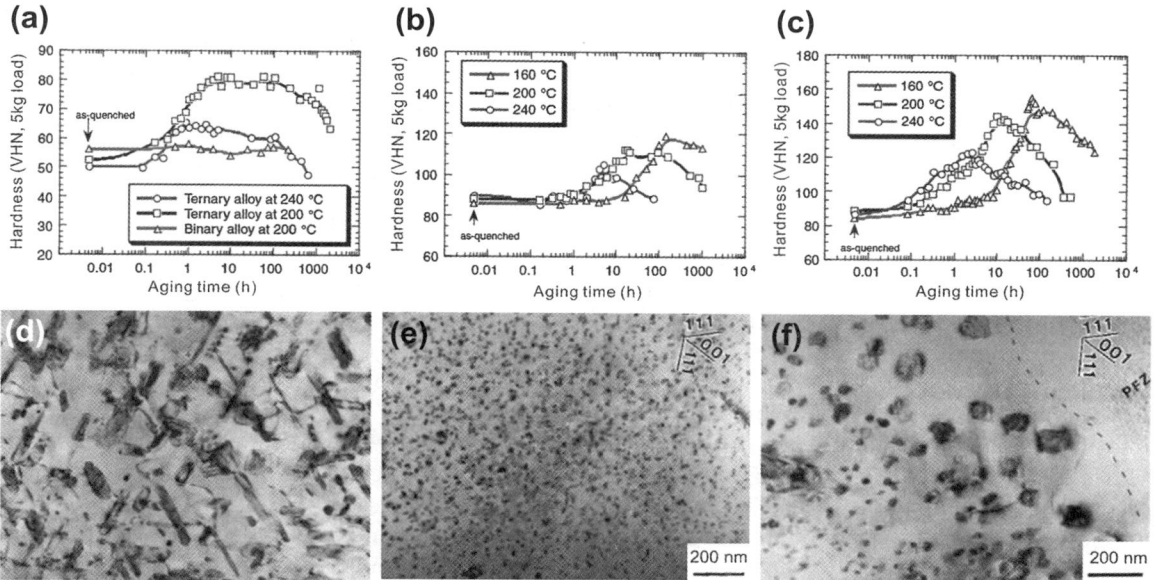

Figure 28 Aging curves of (a) Al–5Mg, (b) Al–10 Mg, and (c) Al–10Mg–0.5Ag (wt%) alloys, and TEM images showing distribution of precipitates in Al–10Mg–0.5Ag alloy aged for (d) 2 h at 240 °C and (e and f) 0.5 h at 240 °C. (e) From the core of grain and (f) adjacent to grain boundary (Kubota et al., 2004; Kubota et al., 2005a, 2005b).

reportedly has an ordered $L1_2$ structure (Al_3Mg) and a spherical morphology (Starink and Zahra, 1997). This phase has also been designated an ordered GP zone with an $L1_2$ structure (Lorimer, 1978). The β′ precipitates have a hexagonal structure, with $a = 1.002$ nm and $c = 1.636$ nm. They form as plate-shaped particles, with their broad surfaces parallel to $\{001\}_\alpha$. Their OR is such that $(0001)_{β′}//(001)_\alpha$ and $[01\bar{1}0]_{β′}//[110]_\alpha$. The equilibrium phase β forms as plates or laths in grain boundaries, with the following OR with the surrounding matrix: $(111)_β//(001)_\alpha$ and $[1\bar{1}0]_β//[010]_\alpha$ (Lorimer, 1978).

As indicated above, the age hardening response of binary Al–Mg alloys even with Mg levels as high as 10 wt% is still quite low. However, trace additions of Ag, in the range of 0.4–0.5 wt%, to Al–Mg alloys can lead to enhanced age hardening response, **Figure 28(c)**. The age hardening response in the Ag-modified Al–Mg alloys can be further enhanced if the Al–Mg–Ag alloys are cold worked by 7% after solution treatment and before aging. In the as-quenched condition, the number density of dislocations in the Ag-containing alloy is much lower than that in the Ag-free alloy. This is attributed to a strong interaction between Ag atoms and vacancies that reduces the vacancy population available to condense and collapse to form dislocation loops. The enhanced age hardening response in the Ag-containing Al–Mg alloys is associated with a much finer distribution of precipitates promoted by small concentrations of Ag. These finer scale precipitates were reported in the early studies to be T phase that has an $Mg_{32}(Al,Ag)_{49}$ composition and a body-centered cubic structure (space group $Im\bar{3}$, $a = 1.41$ nm) (Bergman et al., 1957). Using X-ray diffraction and electron diffraction techniques, the OR between T precipitates and the Al matrix is determined to be $(010)_T//(11\bar{1})_\alpha$ and $[001]_T//[1\bar{1}0]_\alpha$ (Auld, 1968; Kubota et al., 2005a). The T phase precipitates contain all three elements Al, Mg and Ag, and they adopt a rod-like shape with the rod axis parallel to $\langle 110 \rangle_\alpha$ directions, **Figure 28(d)**.

Recent TEM studies (Kubota et al., 2005b) indicate that T phase precipitates form predominantly in the peak-aged condition during isothermal aging at 240 °C of an Al–10Mg–0.5Ag (wt%) alloy. In the underaged condition at 240 °C, the majority of precipitates have an icosahedral quasicrystalline structure. These quasicrystalline precipitates have a rhombohedral shape. They are densely and uniformly distributed in the core part of individual Al grains, but are sparsely distributed and coarser in regions adjacent to grain boundaries, **Figure 28(e) and (f)**. The OR between the icosahedral quasi-crystalline phase and the matrix phase is such that i5//$\langle 011 \rangle_\alpha$ and i3//$\langle 111 \rangle_\alpha$.

Al–Mg alloys containing 0.8–5 wt% Mg are widely used in wrought forms. In the annealed condition, the strength of 5005 (Al–0.8 Mg) alloy is 40 MPa yield and 125 MPa tensile, and that of the strongest alloy 5456 is 160 MPa yield and 310 MPa tensile, together with an elongations exceeding 25% (Polmear, 2006). The Al–Mg alloys work harden rapidly, and the working hardening rate increases with the magnesium content in the alloy. Fully work hardened 5456 alloy (H19 temper) may develop a yield strength of ∼500 MPa. However, these work-hardened alloys may undergo age softening at room temperature—the tensile strength drops over time because of localized recovery within the deformed grains. An H3 temper, involving cold working to a slightly higher strain and then stabilizing by heating briefly in the temperature range 120–150 °C, is used to overcome this problem (Polmear, 2006). This temper stabilizes tensile properties at the slight expense of tensile strength. The Al–Mg wrought alloys with higher contents of Mg have received increasing interest for developing automotive sheet and marine plate. However, these alloys are prone to intergranular attack and stress corrosion cracking (SCC), because of the precipitation of β phase in slip bands and grain boundaries during service in corrosive environment or tropical conditions.

20.3.1.3 *6xxx Series Alloys*
20.3.1.3.1 Al–Mg–Si Alloys
The equilibrium precipitate phases in the ternary Al–Mg–Si system include Mg$_2$Si and Si, depending on the alloy composition, **Figure 29**. The Mg$_2$Si phase, often designated β, is also an equilibrium phase in

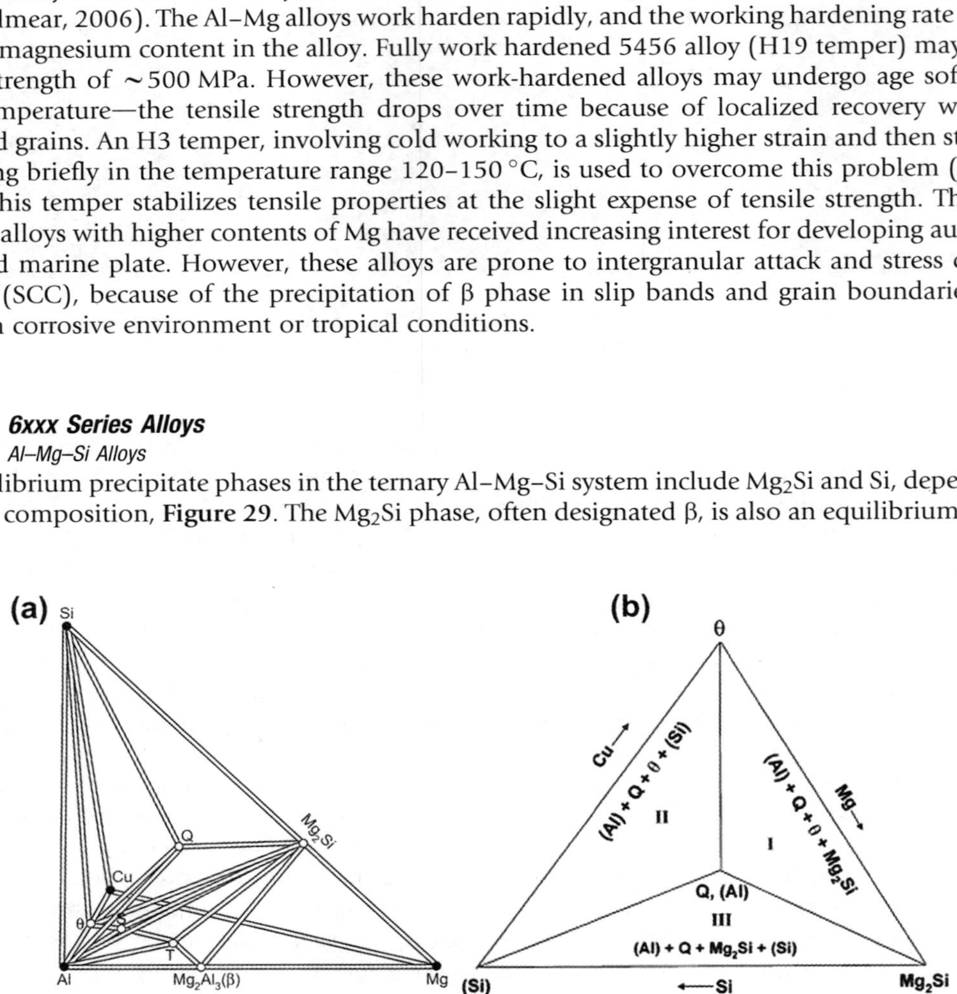

Figure 29 (a) Line diagram showing phase fields of the equilibrium phases in the Al–Mg–Si–Cu quaternary system at room temperature and (b) schematic projection of the Al corner onto the Q corner of the four-phase field tetrahedrons of the Al–Mg–Si–Cu system. The vertical line represents alloys with Mg/Si = 1 (Chakrabarti and Laughlin, 2004).

the binary Mg–Si system, and it has a face-centered cubic structure (space group $Fd\bar{3}m$, $a = 0.6325$ nm). The Si phase has a diamond cubic structure with a space group of $Fd\bar{3}m$ and a lattice parameter of 0.5431 nm.

Early studies suggested that, in the pseudo-binary Al–Mg$_2$Si system, the eutectic temperature is about 595 °C and the maximum equilibrium solid solubility of Mg$_2$Si in aluminum is approximately 1.85 wt% (Lorimer, 1978). The crystal structures of metastable precipitate phases and the precipitation sequence in the Al–Mg–Si alloys have proven to be the most complicated and controversial among precipitation hardenable aluminum alloys, even with the help of modern facilities such atomic resolution TEM. These subjects have been under intensive international research over the past 30 years. In early studies, the precipitation sequence reportedly involves the formation of GP zones, metastable β′ and equilibrium β phases (Lorimer, 1978). However, contemporary studies seem to suggest that the precipitation sequence involves the formation of, at least, solute clusters, GP zones, β″, β′ and β phases.

Pashley et al. (1966, 1967) hypothesized that spherical shaped clusters form during water quenching of their alloy. This hypothesis was supported by subsequent differential scanning calorimetry work made by Dutta and Allen (1991) and the 1DAP field ion microscopy study of an Al–0.80Mg–0.79-Si–0.18Cu (wt%) alloy made by Edwards et al. (1998). GP zones were originally reported to form as needles, with their long axis parallel to $\langle 001 \rangle_\alpha$, in Al–Mg–Si alloys aged at temperatures up to 200 °C. However, according to Dutta and Allen (1991), the GP zones have a spherical, rather than needle, shape. The 3DAP study of aged Al–Mg–Si alloys by Murayama and Hono (1999) revealed that (i) separate Mg and Si clusters are present in the solution-treated and water-quenched condition, and (ii) Mg and Si atoms aggregate to form Mg–Si co-clusters during natural aging, and (iii) spherical GP zones form during isothermal aging at 70 °C and the these GP zones have similar chemistry to those Mg–Si co-clusters. The GP zones are always observed after aging for 0.5 h at 175 °C, irrespective of pre-aging condition of the sample.

The β″ phase was reported to form, as $\langle 001 \rangle_\alpha$ rods, after GP zones and before the β′ phase. Structures reported in the early studies include monoclinic ($a = 0.616$ nm, $b = 0.71$ nm, $c = 0.616$ nm, $\beta = 82°$), monoclinic ($a = 0.30$ nm, $b = 0.33$ nm, $c = 0.40$ nm, $\gamma = 71°$), and hexagonal ($a = 0.705$ nm, $c = 0.405$ nm). Based on high-resolution electron microscopy and electron diffraction, Andersen et al. (1998) proposed that the β″ phase has a C-centered monoclinic structure (space group C2/m, $a = 1.516$ nm, $b = 0.405$ nm, $c = 0.674$ nm, $\beta = 105.3°$) and an Mg$_5$Si$_6$ atomic composition. This monoclinic structure is identical to the based-centered monoclinic structure determined by Edwards et al. (1998) for β″ phase in a Cu-containing Al–Mg–Si alloy. In the study of Edwards et al. (1998), the OR of the β″ phase with the matrix was proposed to be $(001)_{\beta''}//(310)_\alpha$ and $[010]_{\beta''}//[001]_\alpha$, and the Mg:Si atomic ratio in the β″ phase was about 1.2:1.

The metastable β′ phase was originally reported to have a face-centered cubic structure ($a = 0.64$ nm) that is similar to that of the equilibrium phase β. The β′ precipitates form as rods with their long axes parallel to $\langle 001 \rangle_\alpha$. The OR between the β′ precipitates and the aluminum matrix phase is such that $(100)_{\beta'}//(100)_\alpha$, $[011]_{\beta'}//[001]_\alpha$ (Kelly and Nicholson, 1963). A different crystal structure, which is hexagonal with $a = 0.705$ nm and $c = 0.405$ nm, was reported by Jacobs (1972). For this structure, its OR with the matrix phase is $(001)_{\beta'}//(100)_\alpha$, $[100]_{\beta'}//[011]_\alpha$. The crystal structure proposed by Jacobs (1972) was generally accepted for β′ between 1972 and 2007. Based on high-resolution TEM images and electron microdiffraction patterns, Cayron et al. (1999) further proposed that the space group of β′ was $P\bar{6}2m$. However, a more recent study (Vissers et al., 2007) using atomic resolution TEM, energy-dispersive X-ray spectroscopy and first-principles calculations indicates that the β′ phase has an atomic composition of Mg$_9$Si$_5$ and a hexagonal structure (space group $P6_3/m$, $a = 0.715$ nm, $c = 1.215$ nm) that is different from that proposed by Jacobs (1972) and Cayron et al. (1999).

The β precipitates have a platelet shape with their broad surface parallel to $\{001\}_\alpha$, and their OR with the aluminum matrix phase is such that $(001)_\beta//(001)_\alpha$, $[110]_\beta//[100]_\alpha$ (Jacobs, 1972), or $(001)_\beta//(001)_\alpha$, $[100]_\beta//[100]_\alpha$ (Kanno et al., 1980).

Apart from those mentioned above, some other precipitate phases have also been reported to form in Al–Mg–Si alloys, particularly in alloys containing an excessive amount of Si. Three types of precipitates (Matsuda et al., 2000) have been reported. The precipitates of all three types are rod shaped, with their long axes long axes parallel to the $\langle 001 \rangle_\alpha$.

The type A phase has a hexagonal structure (space group $P\bar{6}2m$, $a = 0.405$ nm, $c = 0.67$ nm), and its cross-section is faceted and octagonal. The type A phase is also designated U1 (FrØseth et al., 2003). But the structure of the U1 phase is refined to be trigonal (space group $P\bar{3}m1$, $a = 0.405$ nm, $c = 0.674$ nm) with an Al_2Si_2Mg composition. The type A precipitates have the largest size, in both cross-sectional area and length, among the metastable phases. The OR between this precipitate phase and the matrix phase is $[\bar{1}2\bar{1}0]_A//[001]_\alpha$ with $(0001)_A$ rotated clockwise by 20° with respect to $(100)_\alpha$. The type A precipitates contain all three elements in the alloy, namely Si, Mg and Al, but the Si content is much higher than that of Mg, with an Al_4Si_5Mg atomic composition. It should be noted that only $[\bar{1}2\bar{1}0]_A$ image and $[\bar{1}2\bar{1}0]_A$ electron microdiffraction pattern were provided for the above-proposed structure in the work of Matsuda et al. (2000). In their study, the β' phase was proposed to have a hexagonal unit cell with lattice parameters of $a = 0.407$ nm and $c = 0.405$ nm, which is different from that proposed by Jacobs (1972).

The type B precipitates are distinguishable from type A precipitates by their non-octagonal shape. They have an orthorhombic structure ($a = 0.683$ nm, $b = 0.794$ nm $c = 0.405$ nm) and an $Al_4Si_5Mg_2$ atomic composition (Matsuda et al., 2000; Matsuda et al., 1996). The OR between the type B phase and the matrix phase is $[001]_B//[001]_\alpha$ with $(010)_B$ rotated clockwise by 20° with respect to $(010)_\alpha$, that is $(010)_B//(\bar{1}30)_\alpha$ and $[001]_B//[001]_\alpha$. The lattice correspondence between type A and type B phases was discussed in the work of Matsuda et al. (2000). The type B phase is also known as U2 (FrØseth et al., 2003), who proposed that the structure of the U2 phase was orthorhombic (space group Pnma, $a = 0.675$ nm, $b = 0.405$ nm, $c = 0.794$ nm), and that the composition was $Al_4Si_4Mg_4$. In this notion, the orientation relation of U2 precipitates with the matrix phase becomes $(001)_{U2}//(130)_\alpha$ and $[010]_{U2}//[001]_\alpha$.

Type C precipitates have again a hexagonal structure. They have a rectangular cross-section. The lattice parameters of its unit cell are $a = 1.04$ nm and $c = 0.405$ nm. Its OR with the matrix phase is $(2\bar{1}\bar{1}0)_C//(3\bar{1}0)_\alpha$ and $[0001]_C//[001]_\alpha$. The structure of the type C phase is identical to that of B' phase that forms in Al–Mg–Si (Marioara et al., 2006) and Al–Mg–Si–Cu (Dumolt et al., 1984) alloys. Apart from the hexagonal structure, a base-centered orthorhombic structure ($a = 1.80$ nm, $b = 1.04$ nm, $c = 0.405$ nm) has also been reported for the B' phase (Dumolt et al., 1984). It is to be noted that the B' phase is remarkably similar to Q' and Q phases in Al–Mg–Si–Cu alloys, which will be described in the following section.

The relative fraction of each of the four metastable phases, types A, B and C and β', in an Al–1.0wt% Mg_2Si–0.4wt%Si alloy aged for different times at 250 °C was examined by Matsuda et al. (2000). For the selected aging temperature, the β' phase is the dominant one in the early stage of the aging, and is gradually replaced by the type B and type A phases with prolonged aging. The metastable precipitate phases in the well overaged condition are types A and C. The period of existence of the β' phase is relatively short compared with that in alloys that do not contain excessive amounts of Si. The equilibrium phases in the alloy are β-Mg_2Si and Si.

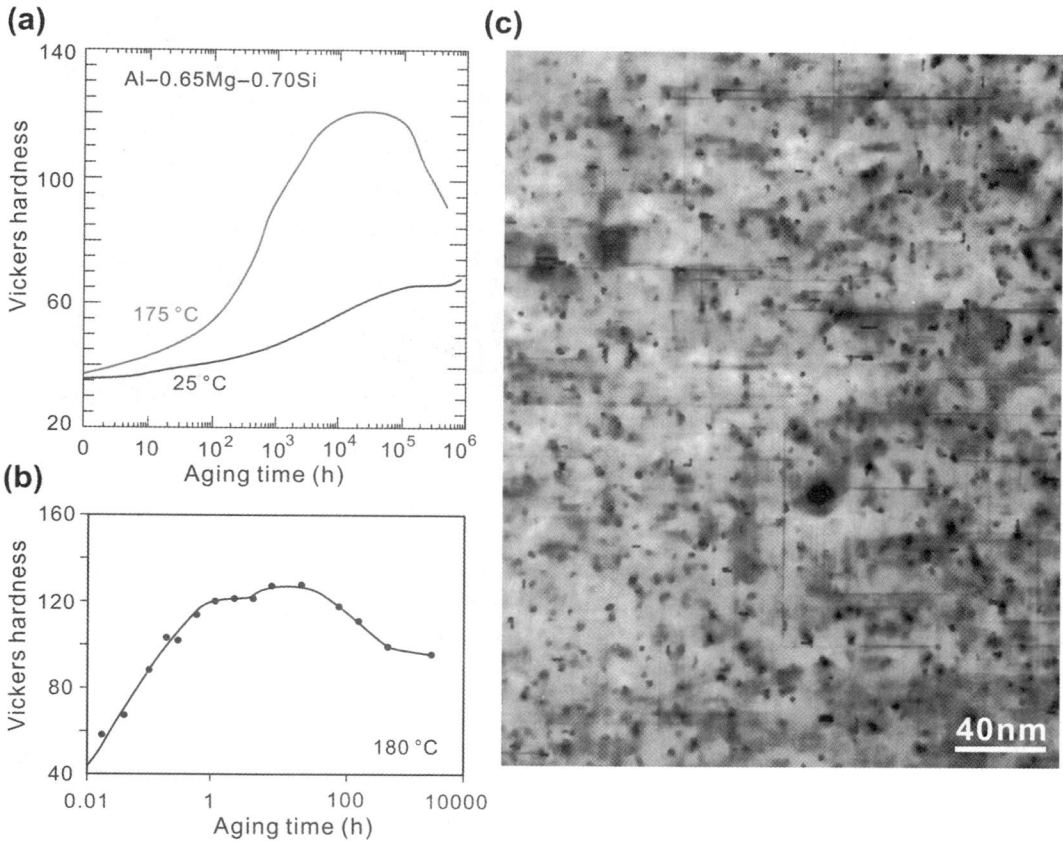

Figure 30 Age hardening response of (a) Al–Mg–Si and (b) Al–Mg–Si–Cu (6111) alloys and (c) TEM image showing the distribution of precipitate rods in 6111 aluminum alloy (Ringer and Hono, 2000; Goh et al., 2006). (For color version of this figure, the reader is referred to the online version of this book.)

A comparison of crystal structures, morphologies and stability of various precipitate phases in the Al–Mg–Si system is provided by van Huis et al. (2007). The generic precipitation sequence in the Al–Mg–Si alloys is suggested to be the following:

$$\text{SSSS} \rightarrow \text{solute clusters} \rightarrow \text{GP zones} \rightarrow \beta'' \rightarrow \beta', A(U1), B(U2), C \rightarrow \beta$$

The aging curves of the Al–Mg–Si and Al–Mg–Si–Cu alloys are shown in **Figure 30**. The 6111 alloy has been use for auto sheet. The current commercial processing routine involves pre-aging of the panel straight after solution treatment, followed by room temperature storage. The panel is then stamped and painted before the paint-baking cycle at 175 °C for 30 min. In commercial alloys Mg and Si are added either in what are called "balanced" amounts to form quasi-binary Al–Mg_2Si alloys (Mg:Si = 1.73:1) or with an excess of silicon above that needed to form Mg_2Si. The addition of excessive amounts of Si to the Al–Mg_2Si quasi-binary alloy leads to an accelerated age hardening response and can cause formation of additional precipitate phases and therefore a change in precipitation sequence. An atom

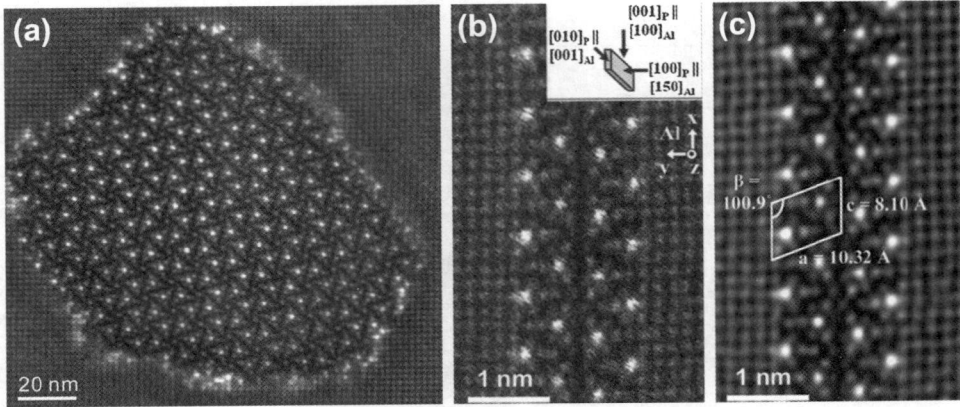

Figure 31   (a) HAADF-STEM image showing the cross-section of a Q precipitate in an Al–Mg–Si–Cu alloy (Fiawoo and Nie, unpublished research). (b) Original and (c) Fourier-filtered HAADF-STEM images showing the cross-section of a C phase precipitate in an Al–Mg–Si–Cu–Ag alloy. The unit cell is outlined in (c) (Torsæter et al., 2012).

probe study by Murayama and Hono (1999) revealed that the chemical composition of the GP zones changed with the alloy composition, and that the number density of the GP zones in the paint-bake condition increased with the number of Si atoms that were available to form Mg–Si co-clusters. The addition of excessive amount of Si is known to promote refined distribution of precipitates.

20.3.1.3.2 Al–Mg–Si–Cu Alloys

The additions of Cu to Al–Mg–Si alloys generally increase the precipitation kinetics during isothermal aging, and reduce the deterioration of the age hardening response arising from natural aging of the Al–Mg–Si alloys. The Cu level has a smaller but noticeable effect on the value of peak hardness. A comprehensive review of precipitate phases in Al–Mg–Si–Cu alloys is available in the literature (Chakrabarti and Laughlin, 2004). The equilibrium precipitate phases include β-Mg_2Si, Si and Q, **Figure 29**. The Q phase has reportedly a hexagonal structure ($a = 1.04$ nm, $c = 0.405$ nm), with a space group of either $P6_3/m$ or $P\bar{6}$, **Figure 31(a)**. This structure is very similar to that of Th_7S_{12}, and Arnberg and Aurivillius (1980) reported a detailed description of the positions of Al, Mg, Si and Cu atoms in the unit cell. The composition of the Q phase is not clearly established; the reported compositions include $Al_5Cu_2Mg_8Si_6$, $Al_4CuMg_5Si_4$, $Al_4Cu_2Mg_8Si_7$ and $Al_3Cu_2Mg_9Si_7$.

In Al–Mg–Si–Cu alloys, the precipitation sequence involves the formation of Q′, Q_c, Q_p, C, L and Q phases in addition to those precipitate phases formed in the Cu-free or Cu-lean alloys. The structural and compositional differences between Q′ and Q phases are both subtle—the lattice parameters of Q′ are slightly different from those of Q. In fact, the Q′ phase is similar to type C or B′ phase mentioned previously. The Q′ precipitates have a lath shape, with their long axis parallel to $\langle 001 \rangle_\alpha$. The broad surface of the Q′ laths is reportedly parallel to $\{510\}_\alpha$, even though the OR between the Q′ and α-Al matrix phases is such that $(21\bar{3}0)_{Q′}//(020)_\alpha$, $[0001]_{Q′}//[001]_\alpha$. The Q_c phase also has a hexagonal structure ($a = 0.667$ nm, $c = 0.405$ nm) (Cayron et al., 1999). Its OR with the matrix is such that $(2\bar{1}\bar{1}0)_{Q_c}//(110)_\alpha$ and $[0001]_{Q_c}//[001]_\alpha$. The Q_p phase has again a hexagonal structure ($a = 0.393$ nm, $c = 0.405$ nm). Its OR with the matrix is such that $(2\bar{1}\bar{1}0)_{Q_p}//(100)_\alpha$ and $[0001]_{Q_p}//[001]_\alpha$. The C phase has a monoclinic structure (space group $P2_1/m$, $a = 1.032$ nm, $b = 0.405$ nm, $c = 0.810$ nm, $\beta = 100.9°$) and an atomic composition of $AlCuMg_4Si_3$ or $AlCu_{0.7}Mg_4Si_{3.3}$ (Torsæter et al., 2012).

The C precipitates have a plate shape, with their broad surfaces parallel to $\{200\}_\alpha$, **Figure 31(b) and (c)**. The OR between this phase and the surrounding matrix is such that $[100]_C//[\bar{1}50]_\alpha$, $[010]_C//[001]_\alpha$ and $[001]_C//[100]_\alpha$. However, the accuracy of the relationship of $[100]_C//[\bar{1}50]_\alpha$ is questionable, because the angle between $[\bar{1}50]_\alpha$ and $[100]_\alpha$ is not identical to $100.9°$ that is proposed for the angle between $[100]_C$ and $[001]_C$. The structure of the L phase is not clearly established, even though it has been reported to have a needle shape with its long axis parallel to $\langle 100 \rangle_\alpha$ and an OR that is similar to that of the Q_p phase (Marioara et al., 2007). Two different ORs have been reported for Q precipitates, including $(21\bar{3}0)_Q//(020)_\alpha$ and $[0001]_Q//[001]_\alpha$, and $(2\bar{1}\bar{1}0)_Q//(200)_\alpha$ and $[0001]_Q//[001]_\alpha$. It remains unclear why the Q phase forms in two different ORs. The relationship between the Q', Q_c, Q_p, C, L and Q phases is not unambiguously defined either; further research is needed in this area.

20.3.1.3.3 Al–Mg–Si–Ag and Al–Mg–Ge Alloys

Ag additions to Al–Mg–Si alloys have been reported to produce some profound effects on the identity of precipitate phases (Marioara et al., 2012). Examination using HAADF-STEM of coarse $\langle 100 \rangle_\alpha$ precipitate rods in an Al–0.63Mg–0.37Si–0.12Ag (at.%) alloy aged for 33.3 h at 250 °C indicated that these rods, designated β'_{Ag} phase, had a hexagonal structure (space group $P\bar{6}2m$, $a = 0.690$ nm, $c = 0.405$ nm) and a composition of $Mg_3Al_3Si_2Ag$. The Ag atoms were found to replace some of those atomic sites of Si in the β'_{Ag} unit cell. These precipitate rods commonly contained ordered β'_{Ag} domains that were separated by bands resembling the features of anti-phase boundaries, **Figure 32**. It was further suggested that these bands were composed of full or half unit cells of the type B or U2 structure. Apart from segregation of Ag atoms into the precipitates, a strong segregation of Ag atoms in regions immediately adjacent to the precipitate/matrix interface was also observed.

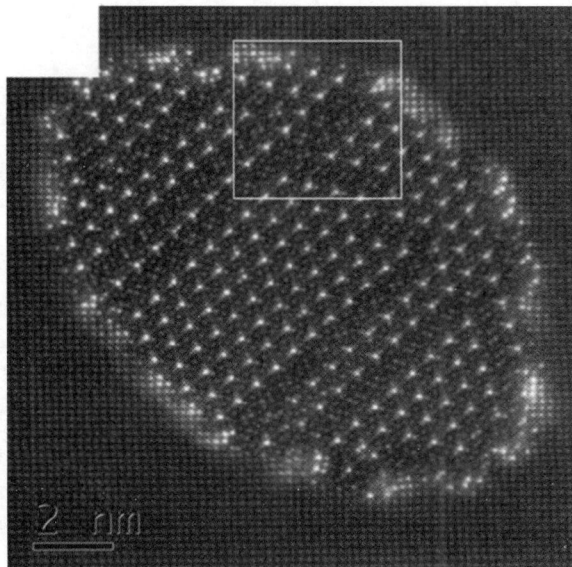

Figure 32   HAADF-STEM image showing the cross-section of a precipitate rod in Al–0.63Mg–0.37Si–0.12Ag (at.%) alloy (Marioara et al., 2012).

The precipitation in Al–Mg–Ge alloys is similar to that in Al–Mg–Si alloys, even though Ge has a larger atomic radius and is much heavier than Si. The precipitates formed in two Al–Mg–Ge alloys, Al–0.87Mg–0.43Ge and Al–0.59Mg–0.71Ge (at.%), were recently studied using conventional TEM and HAADF-STEM (Bjorge et al., 2010). These alloys were aged for 16 h at 200 °C, and their hardness values were similar to those of the counterpart Al–Mg–Si alloys. The microstructures of these two alloys contained a fine distribution of $\langle 100 \rangle_\alpha$ rods and laths, resembling to those of typical of the Al–Mg–Si alloys. However, the β'' phase was absent in these two alloys. The microstructure had a combination of β'_{Ge} and so-called disordered precipitate phase in the Mg-rich Al–0.87Mg–0.43Ge alloy, U1–Ge and disordered precipitates in the Ge-rich Al–0.59Mg–0.71Ge alloy. The β'_{Ge} and U1–Ge phases play a more important role in precipitation hardening in the Al–Mg–Ge alloys than in the Al–Mg–Si system.

The β'_{Ge} precipitates in the Al–0.87Mg–0.43Ge alloy have the same structure as the β' phase (space group P6$_3$/m, $a = 0.715$ nm, $c = 1.215$ nm) in the Al–Mg–Si alloys (Bjorge et al., 2012). The Ge atoms occupy the positions of Si atoms in the β' unit cell. However, measurements of the intensities of Ge columns in $\langle 001 \rangle_\beta$ HAADF-STEM images indicate that these Ge columns are not fully occupied by Ge—they have about either 30 at.% Al or a 20 at.% vacancy. The β'_{Ge} precipitates have a lath shape, with their long axis parallel to $\langle 100 \rangle_\alpha$ and broad face parallel to $\{001\}_\alpha$, Figure 33. They are coherent with the matrix phase in their broad faces, and this coherency implies that the lattice parameter a for the β'_{Ge} is 0.70 nm, rather than 0.715 nm of the β' phase. The β'_{Ge} precipitates had an elongated or square

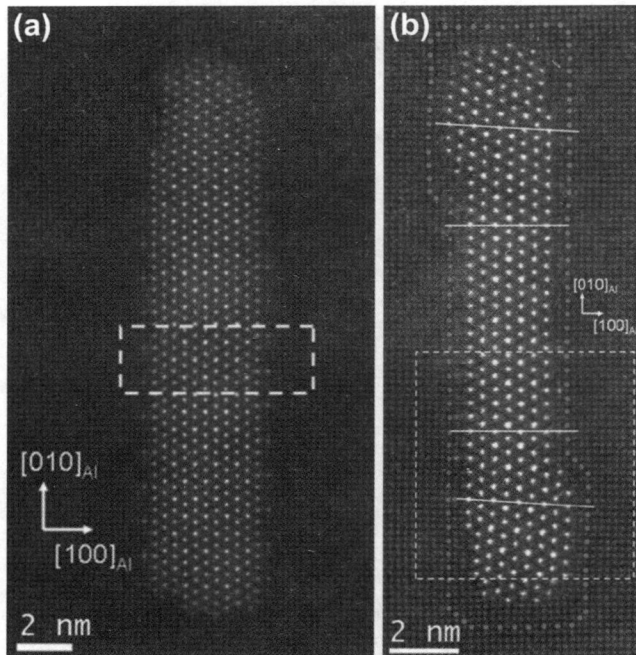

Figure 33 HAADF-STEM images showing the cross-section of β'-Ge precipitates in the Al–0.87Mg–0.43Ge alloy (Bjorge et al., 2012). The existence of a misfit dislocation at the ledge and the rotation of the precipitate lattice are illustrated by the help of white lines in (b). (For color version of this figure, the reader is referred to the online version of this book.)

cross-section. For precipitates with the rectangular cross-section, **Figure 33(a)**, a periodic segregation of Ge atoms was found in the β'_{Ge}/matrix interface along $\langle 010 \rangle_\alpha$. The periodicity of the segregation corresponds to the spacing of six $\{100\}_\alpha$ planes and the diagonal distance of the β'_{Ge} unit cell in this direction. The columns of the segregated Ge atoms are located at some specific sites that could become the Ge-rich column sites of the β'_{Ge} structure by a displacement of $a/2\langle 010 \rangle_\alpha$. The two ends of the cross-section of some elongated β'_{Ge} precipitates often contain ledges and are rotated clockwise by approximately 4° with respect to the central part of the particle, **Figure 33(b)**. This rotation leads to an elastic distortion in the matrix surrounding the particle and the formation of dislocations in regions near the ledges.

The precipitates in the Al–0.59Mg–0.71Ge alloy are mainly U1–Ge phase and an unknown phase. An HAADF-STEM image of a U1-like precipitate, viewed along the $[110]_{U1-Ge}//[001]_\alpha$ direction, is shown in **Figure 34(a) and (b)** (Bjorge et al., 2011). The brightest spots in the image are the Ge columns. Rows of Mg columns form a bright band in between the rows of Ge columns. Some of the Al columns are also highlighted in the image. This image is reportedly consistent with the U1 structure (space group P$\bar{3}$m1, $a = 0.405$ nm, $c = 0.674$ nm) in the Al–Mg–Si alloys, with Ge atoms replacing the Si atoms in the U1 unit cell. Apart from the OR displayed in the image shown in **Figure 34(a) and (b)**, another OR is also found to exist between U1–Ge precipitates and the matrix (Bjorge et al., 2011). Precipitates of the unknown phase have smaller cross-sections than the U1–Ge particles, and a hexagonal arrangement of Ge columns when viewed along their long axis, that is $\langle 001 \rangle_\alpha$. The planar faces of their cross-sections are parallel to $\{010\}_\alpha$, **Figure 34(c) and (d)**.

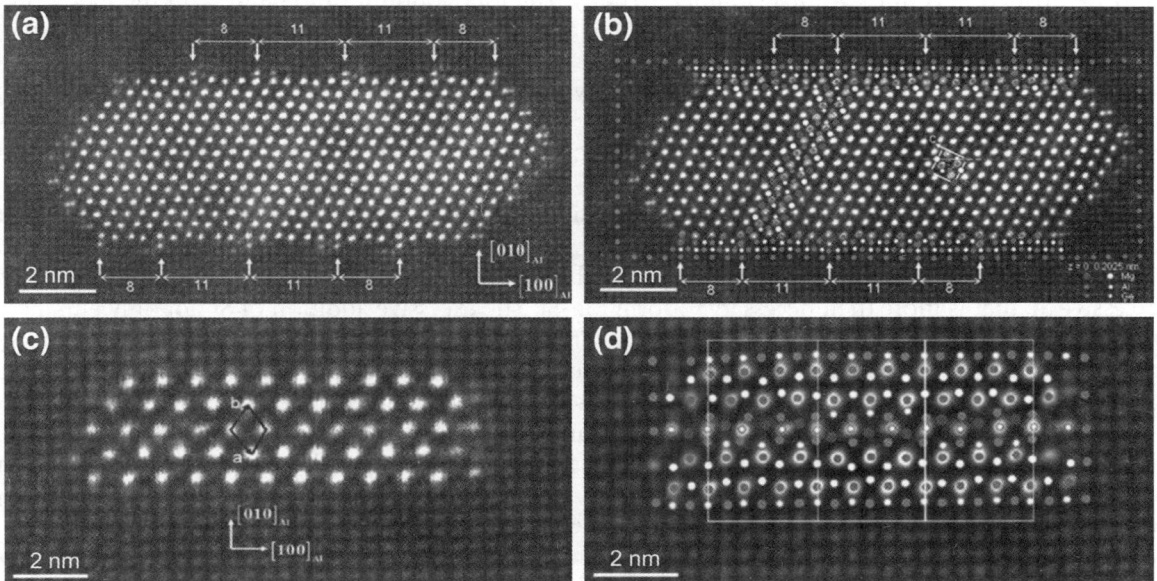

Figure 34 HAADF-STEM images showing cross-sections of (a and b) U1–Ge and (c and d) unknown precipitates in an Al–0.59Mg–0.71Ge alloy (Bjorge et al., 2011). Filled and unfilled circles in (b) correspond to two different layers of the structure, which are separated by 0.2025 nm along the viewing direction [001]$_\alpha$. (For color version of this figure, the reader is referred to the online version of this book.)

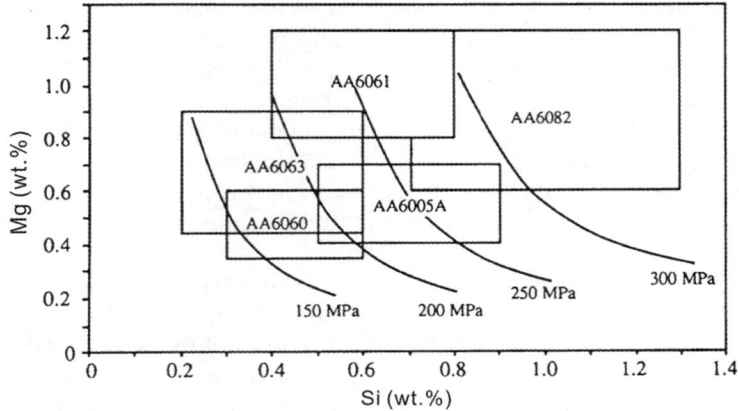

Figure 35 Compositional limits of some common 6xxx series alloys, together with contours representing common peak-aged (T6) values of yield strength (Court et al., 1994; Polmear, 2006).

20.3.1.3.4 Applications of 6xxx Alloys

The 6xxx alloys have medium strength, and the variation of their strength in the T6 condition as a function of Mg and Si contents is shown in **Figure 35**. Most 6xxx series alloys are used as extrusion products, even though a minor fraction of them are used in the form of sheet and plate. These alloys may be divided into three groups (Polmear, 2006). The total amount of Mg and Si in the first group of alloys is in the range 0.8 wt% to 1.2 wt%. These alloys can be readily hot extruded and quenched at the extrusion press. The quenching is usually done by spraying water onto the extrudate that has just exited the extrusion die. The strength is achieved by aging at 160–190 °C. These alloys respond well to color anodizing and other surface finishes. A high-purity version of this group of alloys, 6463 that has an Fe content below 0.15 wt%, responds well to chemical brightening and anodizing for use as automotive trim.

The total content of Mg and Si in the other two alloy groups is over 1.4 wt%, and therefore a higher age hardening response and higher strength can be achieved. However, it is usually necessary to solution treat these alloys after extrusion because they are more quench sensitive. One group contains excessive Si, that is more than necessary to form Mg_2Si. The excessive addition of Si not only enhances the age hardening response but also reduces ductility and causes intergranular embrittlement because of segregation of Si to grain boundaries. Small amounts of Cr, Mn or both are usually added to these alloys to counter intergranular embrittlement. This group of alloys is used as extrusions and forgings.

The other group has the addition of Cu to improve the age hardening response and therefore strength. Higher strengths may be achieved by increasing the Cu content in the alloy. For example, in the T6 condition, alloy 6013, Al–1.0Mg–0.8Si–0.8Cu–0.4Mn (wt%), has a yield strength of 330 MPa, which is much higher than that, 275 MPa, of alloy 6061 that has a maximum Cu content of 0.25 wt%. These alloys are intended for automotive and aircraft applications. They are prone to intergranular corrosion because of the higher Cu contents, but this problem can be reduced by a T78 temper.

There is strong experimental evidence that the actual atomic ratios Mg:Si in the intermediate precipitates that contribute maximum age hardening in Al–Mg–Si alloys are close to 1:1 rather than the expected 2:1 present in the equilibrium precipitate β-Mg_2Si. It has been suggested, therefore, that the balanced alloys actually have Mg contents in excess of that needed to promote the required age

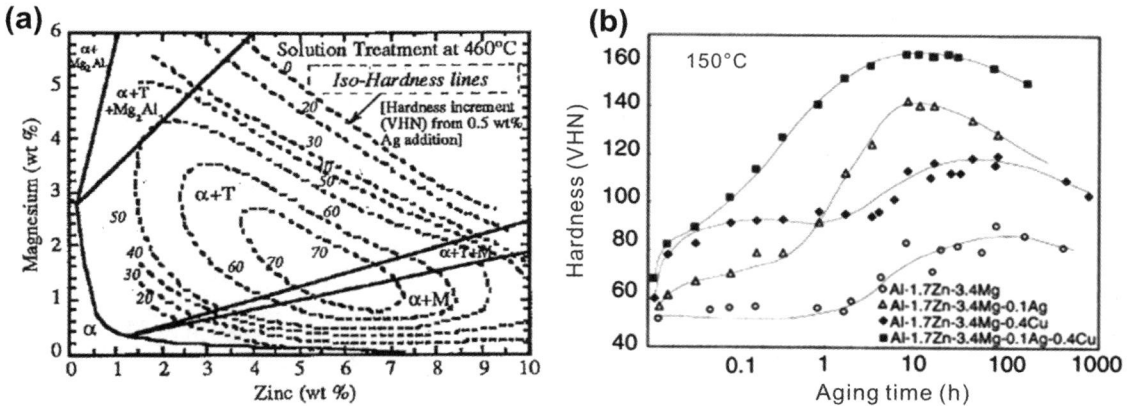

Figure 36 (a) Al-rich corner of the Al–Zn–Mg phase diagram at 200 °C (Vietz et al., 1963–64) and (b) age hardening response at 150 °C of Al–Zn–Mg alloys with various microalloying additions (Caraher et al., 1998).

hardening response. This has led to the design of a new series of alloys in which the Mg contents have been reduced to improve the hot working characteristics and increase productivity without compromising mechanical properties. As an example, the tensile properties of the readily extrudable alloy 6060, with a modified composition (Al-0.35Mg-0.5Si), are found to be comparable with those of the popular alloy 6063 (Al-0.5Mg-0.4Si) whereas extrusion speeds may be up to 20% greater.

20.3.1.4 7xxx Series Alloys

20.3.1.4.1 Precipitate Phases in Al–Zn–Mg and Al–Zn–Mg–Cu Alloys

The Al-rich corner of the ternary phase diagram at 200 °C is shown in **Figure 36(a)**. At 200 °C and above, there exists a relatively large $\alpha + T$ phase field that contains most of the alloy compositions of commercial significance. The T phase is a ternary intermetallic compound, with a stoichiometry of $Mg_{32}(Al,Zn)_{49}$. The other equilibrium intermetallic phases include η ($MgZn_2$) and β (Al_3Mg_2). The equilibrium solid solubilities of Zn and Mg in Al are relatively large at higher temperatures, and decrease with temperature. The age hardening curves of some typical Al–Zn–Mg alloys are provided in **Figure 36(b)**.

The precipitation sequence from the supersaturated solid solution for the Al–Zn–Mg and Al–Zn–Mg–Cu alloys is generally accepted to be

$$SSSS \rightarrow \text{clusters of solues} \rightarrow GP\ I\ \text{and}\ G\ P\ II\ \text{zones} \rightarrow \eta' \rightarrow \eta$$

The evolution of precipitates during isothermal aging is provided in **Figure 37**. The metastable and equilibrium precipitate phases are less well studied compared with those in the 2xxx series alloys, and the structure and chemistry of these precipitate phases are still controversial after more than 50 years study. For example, the final equilibrium precipitate phase is commonly accepted to be η rather than T. Clustering of solute atoms and possibly vacancies reportedly occurs before the formation of GP zones. A subsequent study using 3DAP of an Al–Zn–Mg–Cu (7050) alloy revealed the existence of solute clusters in samples that were water quenched and stored for 1.5 h at room temperature, and in samples that were aged for 24 h at 121 °C (Sha and Cerezo, 2004), but the size and shape of such clusters were such that they could be GP zones.

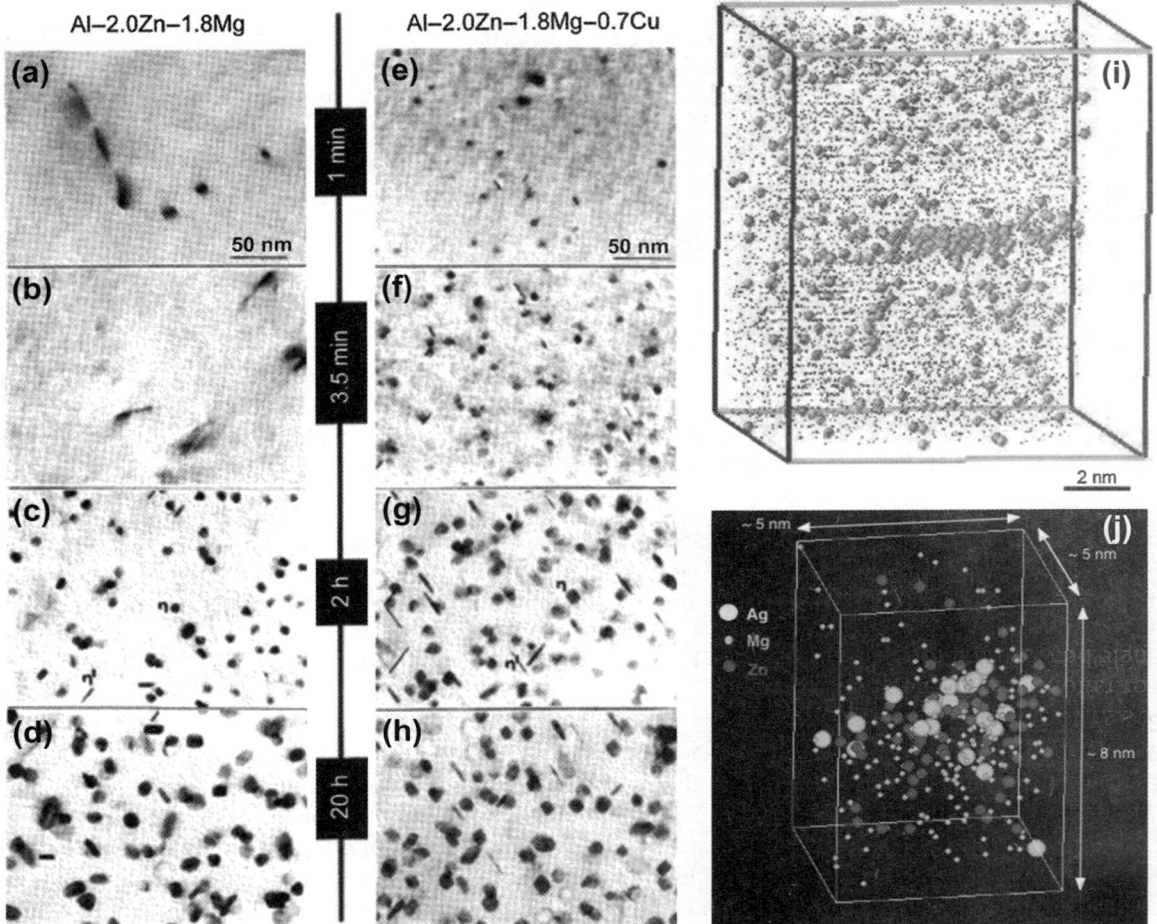

Figure 37 (a–h) TEM images showing distribution of precipitates in Al–Zn–Mg(–Cu) alloys aged for various times at 150 °C. 3DAP maps showing (i) a precipitate in 7050 Al alloy aged for 24 h at 121 °C, and (j) an Mg–Zn–Ag-rich aggregate in an Al–1.7Zn–3.4Mg–0.1Ag (at.%) alloy aged for 96 h at 150 °C. The aggregate in (j) comprises approximately 40 atoms and includes similar numbers of Mg, Ag, and Zn atoms (Ringer and Raviprasad, 2000; Sha and Cerezo, 2004; Nie et al., 2002). (For color version of this figure, the reader is referred to the online version of this book.)

Two types of GP zones, designated GP I and GP II, respectively, have been proposed to form in Al–Zn–Mg and Al–Zn–Mg–Cu alloys. The GP I zones have a spherical morphology. These zones are thought to be internally ordered with a structure similar to that of AuCuI, and they are coherent with the aluminum matrix. The GP I zones contain Al, Zn and Mg, with the Zn/Mg ratio close to unity. They form over a wide temperature range, from room temperature up to 140–150 °C, independent of quenching temperature. In the Al–Zn–Mg–Cu alloys, Cu atoms also segregate into GP I zones. The GP I zones formed at early stage of aging have significantly higher Cu content than those at later stages of aging. The GP II zones form in samples quenched from temperatures above 450 °C and aged at temperatures above 70 °C. The formation of these zones is sensitive to the level of vacancy supersaturation. The GP II zones are thin platelets formed on $\{111\}_\alpha$, typically one to two atomic layers thick and 3–6 nm wide. They are also more thermally stable than the GP I zones. These GP II zones may act as a precursor to or transform into η' phase during continued aging.

The metastable η' phase forms as thin platelets on $\{111\}_\alpha$, that are fully coherent with the aluminum matrix in their habit plane. It is generally accepted that the η' phase has a hexagonal crystal structure ($a = 0.496$ nm, $c = 1.402$ nm), although an orthorhombic structure with a stoichiometry of $MgZn_2$ (Gjonnes and Simensen, 1970) and a hexagonal structure with lattice parameters $a = 0.496$, $c = 0.868$ nm (Mondolfo, 1976) were also reported. The η' precipitates have a rational OR with the aluminum matrix:

$$(0001)_{\eta'}//\{111\}_\alpha, \langle11\overline{2}0\rangle_{\eta'}//\langle11\overline{2}\rangle_\alpha$$

The composition of the metastable η' phase is controversial. Using X-ray diffraction of single crystals, Auld and Cousland (1974) proposed a structural model with a composition of $Mg_4Zn_{11}Al$. A subsequent study using HRTEM reported a η' structural model with stoichiometry $Mg_2Zn_{5-x}Al_{2+x}$, which is distinct from the previous studies. First-principles calculations of energetics in the Zn-rich corner of the Al–Zn–Mg system (Wolverton, 2001) indicated that the structure models of Gjonnes and Simensen (1970) have higher energies than that of Auld and Cousland (1974). Consequently, a low-energy stoichiometry, $Mg_4Zn_{13}Al_2$, was proposed (Wolverton, 2001) by modifying the model of Auld and Cousland. However, the recent 3DAP work (Sha and Cerezo, 2004) indicated that the Zn:Mg ratio in the η' phase was 1.2–1.3, which is quite different from the much higher Zn:Mg ratio proposed by the first-principles calculations.

The equilibrium phase η has a hexagonal structure (space group $P6_3/mmc$, $a = 0.5221$ nm, $c = 0.8567$ nm). Nine different ORs have been reported to occur between the η phase and the aluminum matrix. A very recent study made by HRTEM (Liu et al., 2011) reported that an intermediate precipitate phase exists between η' and η. This intermediate phase has a hexagonal structure with lattice parameters $a = 0.496$ nm and $c = 0.935$ nm. Since its a is the same as that of the η' phase and its c is close to that of the η phase, this intermediate phase, termed η precursor, was suggested to act as a structural bridge to allow the a smooth transformation form the η' phase to the η phase.

20.3.1.4.2 *Microalloying effects*

As with the 6xxx series alloys the age hardening response of the 7xxx series alloys can be enhanced by additions of Cu and Ag (Caraher et al., 1998). The quaternary addition affects precipitation kinetics and precipitate distribution rather than introducing new metastable precipitate phases. Therefore, the age hardening enhancement is likely due to the preferential interaction between the microalloying element atoms and vacancies, which leads to an increased nucleation rate of precipitates.

The addition of Cu to Al–Zn–Mg alloys results in a significant improvement in strength and the resistance to stress corrosion cracking. For aging in the temperature range 100–235 °C, very rapid hardening occurs in the initial stages, **Figure 36(b)**, indicating the presence of an effect distinct from the basic aging process of the ternary alloys. This rapid hardening phenomenon was initially attributed to the formation of the S phase (Al_2CuMg). But subsequent studies using 3DAP (Sha and Cerezo, 2004) revealed that Cu atoms were present in the Mg/Zn clusters and the precipitate phases that formed later on. The participation of Cu atoms in the clustering and precipitation might reduce the activation energy barrier to aggregate nucleation and hence increase the nucleation rate that leads to the rapid hardening in the alloy. Once the first stage of hardening is completed, the Cu-bearing quaternary alloys age harden in the same way as the equivalent ternary alloys, that is Cu has little effect on the second stage of aging. The effect of Cu on peak hardness is similar to that obtained by increasing the Zn content in the ternary alloy. The improved resistance to stress corrosion cracking (SCC) has been attributed to the reduced difference in the electrode potential between grains and grain boundaries and the reduced susceptibility to intergranular weakness, both of which are due to the way Cu influences the precipitation reactions in these alloys.

Similar to the effects of Cu additions, microalloying additions of Ag to Al–Zn–Mg alloys can also enhance the age hardening response in the temperature range 120–220 °C. It is found that a critical concentration of Ag is required to stimulate the effect and that this critical concentration increases with increasing aging temperature. The reason for the increased hardening response is not well understood. One possibility is due to an increase in the upper temperature limit of stability of GP zones and η' phase. The experimental evidence indicates that Ag segregates into the η' phase or the precursor of this phase, **Figure 37(j)**, and promotes a finer dispersion of η' precipitates perhaps by increasing the nucleation rate. More quenched-in vacancies are retained from the solution treatment because of the strong interaction between vacancies, Mg and Ag atoms, and they accelerate the diffusion and precipitation kinetics. The addition of Ag also provides a unique opportunity to achieve high strengths in alloys having lower Zn and Mg contents, without having the SCC problem in these high-strength alloys. Not only are lower Zn- and Mg-containing alloys less susceptible to SCC, but the Ag addition significantly reduces the width of PFZs adjacent to grain boundaries, **Figure 38**, which is intimately linked to the onset of SCC.

Figure 36(a) shows contours representing constant increments in hardness that may be achieved with the addition of 0.5 wt% Ag. The hardness increment represents the maximum achievable at typical aging temperatures. The diagram shows that the response of Al–Zn–Mg alloys to Ag additions does not appear to be controlled by the alloy composition. The maximum hardening response has been suggested to occur when Ag is added to alloys in which [Zn] × [Mg] = 8.5 (wt%).

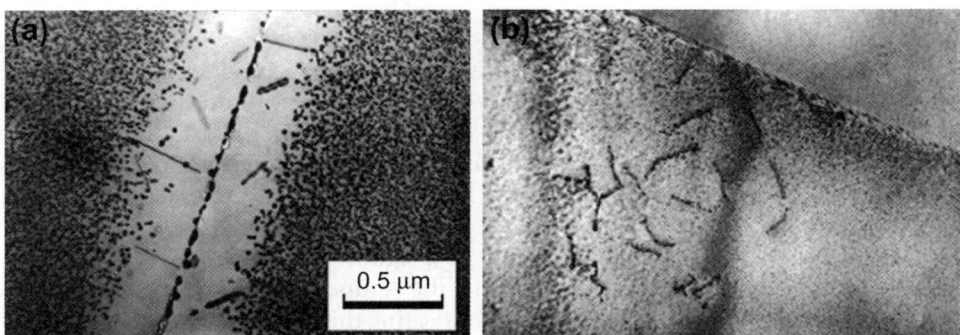

Figure 38   (a) Wide PFZ in an Al–4Zn–3Mg alloy aged 24 h at 150 °C and (b) effect of 0.3%Ag on PFZ width and precipitate distribution in the Al–4Zn–3Mg alloy aged 24 h at 150 °C (Polmear, 1972).

20.3.2 Precipitation Hardening and Microstructural Design

20.3.2.1 Effects of Precipitate Shape and Orientation on Orowan Strengthening
20.3.2.1.1 Spherical Particles
For a dispersion of spherical particles of uniform diameter d, distributed uniformly in the slip plane of the matrix phase, the mean planar diameter of the spherical particles in the slip plane is $d_p = \pi d/4$, and the increment in CRSS produced by the need for dislocations to bypass these particles is given by Eqn (21). For a triangular array of spherical particles of diameter d_s, volume fraction f and number density N_v, the effective interparticle spacing is given by Eqn (20).

20.3.2.1.2 $\{100\}_\alpha$ Precipitate Plates
If the shape of the particles is changed from spherical to that of thin parallel-sided plates, then both the effective planar interparticle spacing and the dislocation line tension will be changed. For plate-shaped particles formed on $\{100\}_\alpha$ planes, the angle between the particle habit plane and the $\{111\}_\alpha$ slip plane in the matrix phase is 54.74°. If it is assumed that the $\{100\}_\alpha$ plates have a uniform size and circular shape, and that the true diameter, d, is significantly larger than the true thickness, t, then $N_a = N_v d \sin 54.74°$, and

$$f = N_a S_a = N_v d S_a \sin 54.74° = N_v V = \frac{N_v \pi d^2 t}{4} \tag{22}$$

Therefore, $S_a = (\pi dt)/(4 \sin 54.74°) = 0.306\pi dt$. The cross-sectional area of the precipitate in the slip plane is approximately rectangular in shape, with the mean planar width $t_p = t/\sin 54.74° = 1.225t$ and the mean planar length $d_p = \pi d/4$. If it is assumed that the $\{100\}_\alpha$ precipitate plates are distributed ideally at the center of each surface of a cubic volume in the matrix phase, **Figure 39(a)**, then in a section on a $\{111\}_\alpha$ slip plane, the $\{100\}_\alpha$ precipitate plates will have a triangular array distribution shown in **Figure 39(b)**. In this case, the mean planar end-to-end spacing, λ, between the precipitate plates is given by

$$\lambda = L_p - \frac{d_p}{2} - \frac{\sqrt{3}}{2} t_p = \frac{1.030}{\sqrt{N_v d}} - \frac{\pi}{8} d - 1.061t \tag{23}$$

The parameter in the logarithmic term in the Orowan Eqn (21) is uncertain due to the plate shape of the particles. However, if we use the planar thickness of the particles as the particle size in the slip plane, then the appropriate version of the Orowan equation for aluminum alloys containing $\{100\}_\alpha$ precipitate plates is

$$\Delta\tau = \frac{Gb}{2\pi\sqrt{1-v}} \cdot \frac{1}{\left(\frac{1.030}{\sqrt{N_v d}} - \frac{\pi}{8}d - 1.061t\right)} \cdot \ln\frac{1.225t}{b} \tag{24}$$

or

$$\Delta\tau = \frac{Gb}{2\pi\sqrt{1-v}} \cdot \frac{1}{\left(0.931\sqrt{\frac{0.306\pi dt}{f}} - \frac{\pi}{8}d - 1.061t\right)} \cdot \ln\frac{1.225t}{b} \tag{25}$$

20.3.2.1.3 $\{111\}_\alpha$ Precipitate Plates
For $\{111\}_\alpha$ precipitate plates of uniform diameter d and thickness t, where $d \gg t$, the dihedral angle between the $\{111\}_\alpha$ habit plane and the $\{111\}_\alpha$ slip plane plates is 70.53°. If N_v and P represent the

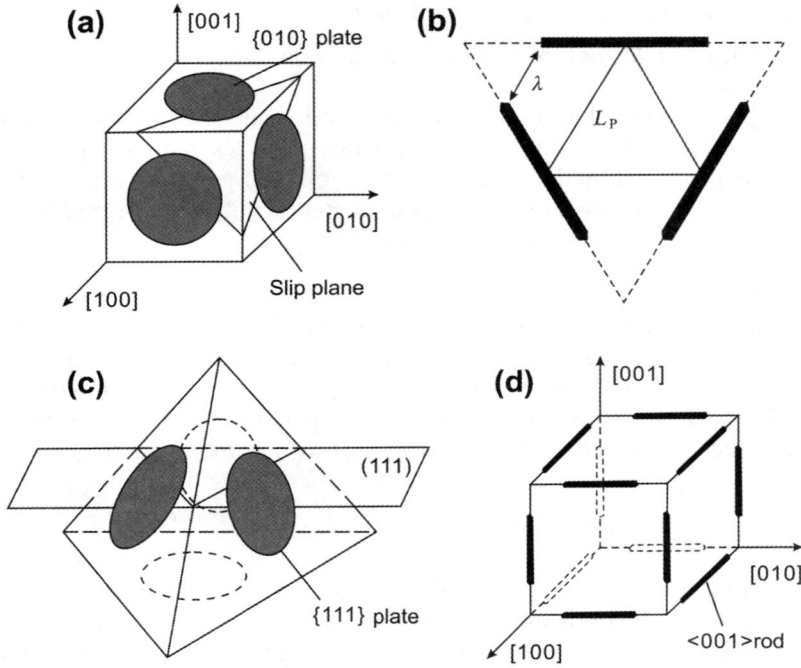

Figure 39 Schematic diagrams showing distribution of (a) circular $\{100\}_\alpha$ precipitate plates, (c) $\{111\}_\alpha$ plates, and (d) $\langle 001\rangle_\alpha$ rods in a cubic volume of the matrix phase, and (b) projection of the intersected $\{100\}_\alpha$ precipitate plates in a $\{111\}_\alpha$ plane of the matrix.

number density of precipitate plates per unit volume and the probability that a $\{111\}_\alpha$ precipitate plate is intersected by a $\{111\}_\alpha$ slip plane, respectively, then $N_a = N_v P = N_v d \sin 70.53°$, and

$$f = N_a S_a = N_v d S_a \sin 70.53° = N_v V = \frac{N_v \pi d^2 t}{4} \tag{26}$$

where V is the volume of a single precipitate plate, and S_a is the mean planar cross-sectional area of a precipitate plate intersected by a $\{111\}_\alpha$ plane. Therefore, $S_p = (\pi dt)/(4 \sin 70.53°) = 0.265\pi dt$. The cross-sectional area of the precipitate plate in the slip plane is approximately rectangular in shape with mean planar width $t_p = t/\sin 70.53° = 1.061t$ and mean planar length $d_p = \pi d/4$.

Assuming that the $\{111\}_\alpha$ precipitate plates are circular discs distributed uniformly at the center of each surface of a tetrahedral volume in the matrix phase, **Figure 39(c)**, then the $\{111\}_\alpha$ precipitate plates intersected by a $\{111\}_\alpha$ slip plane of the matrix phase will have a triangular distribution on that plane, **Figure 39(b)**, provided that the fourth $\{111\}_\alpha$ precipitate plate is not on the slip plane. The mean planar end-to-end distance λ, that is effective planar interparticle spacing, between the precipitate plates is given by Eqn (20):

$$\lambda = L_p - \frac{d_p}{2} - \frac{\sqrt{3}}{2} t_p = 0.931 \sqrt{\frac{0.265\pi dt}{f}} - \frac{\pi}{8} d - 0.919t \tag{27}$$

Assuming that the log term in Eqn (21) is $\ln(t_p/b)$ for plate-shaped particles, and the appropriate version of the Orowan equation for aluminum alloys containing $\{111\}_\alpha$ precipitate plates is

$$\Delta\tau = \frac{Gb}{2\pi\sqrt{1-\nu}} \cdot \frac{1}{0.931\sqrt{\frac{0.265\pi dt}{f}} - \frac{\pi}{8}d - 0.919t} \cdot \ln\frac{1.061t}{b} \tag{28}$$

20.3.2.1.4 $\langle 001 \rangle_\alpha$ Precipitate Rods
For rationally oriented $\langle 100 \rangle_\alpha$ precipitate rods of uniform diameter d and length l, where $l \gg d$, the angle between the rod axis and the normal to the $\{111\}_\alpha$ slip plane is $54.74°$.

$$N_a = N_v P = N_v l \cos 54.74° \tag{29}$$

$$f = N_a S_a = N_v l S_a \cos 54.74° = N_v V = \frac{\pi}{4}d^2 l N_v \tag{30}$$

Therefore, $S_a = 0.433\pi d^2$. The cross-section of the precipitate rod in the slip plane has an ellipsoidal shape, with the major axis of $d/\cos 54.74° = 1.732d$. Approximating this ellipsoidal area by a circle of diameter d_P, then $d_P = d\sqrt{1.732}$. Assuming that the $\langle 100 \rangle_\alpha$ precipitate rods have a distribution that defines a cubic volume within the matrix phase, then the $\langle 100 \rangle_\alpha$ precipitate rods intersected on a slip plane of the matrix phase will have a triangular distribution. The effective interparticle spacing between $\langle 100 \rangle_\alpha$ precipitate rods is given by

$$\lambda = L_p - d_p = \frac{1}{\sqrt{\frac{\sqrt{3}}{2}N_a}} - d_p = 1.075d\sqrt{\frac{0.433\pi}{f}} - \sqrt{1.732}d \tag{31}$$

Assuming $r_0 = b$, the appropriate version of the Orowan equation for aluminum alloys containing $\langle 100 \rangle_\alpha$ precipitate rods is thus given by

$$\Delta\tau = \frac{Gb}{2\pi\sqrt{1-\nu}} \cdot \frac{1}{\left(0.075\sqrt{\frac{0.433\pi}{f}} - \sqrt{1.732}\right)d} \cdot \ln\frac{\sqrt{1.732}d}{b} \tag{32}$$

20.3.2.1.5 Strengthening Effects
Assuming that the number density of precipitates per unit volume N_v and the volume of each precipitate remain constant when the precipitate shape varies from sphere to plate or rod, then a comparison of N_a(plate) and N_a(sphere), and N_a(rod) and N_a(sphere) gives

$$\frac{N_a(\text{plate})}{N_a(\text{sphere})} = 0.824A^{1/3} \quad \text{for} \quad \{111\}_\alpha \text{ precipitate plates} \tag{33}$$

$$\frac{N_a(\text{plate})}{N_a(\text{sphere})} = 0.714A^{1/3} \quad \text{for} \quad \{100\}_\alpha \text{ precipitate plates} \tag{34}$$

$$\frac{N_a(\text{rod})}{N_a(\text{sphere})} = 0.437A^{2/3} \quad \text{for} \quad \langle 100 \rangle_\alpha \text{ precipitate rods} \tag{35}$$

(a)

(b)

(c)

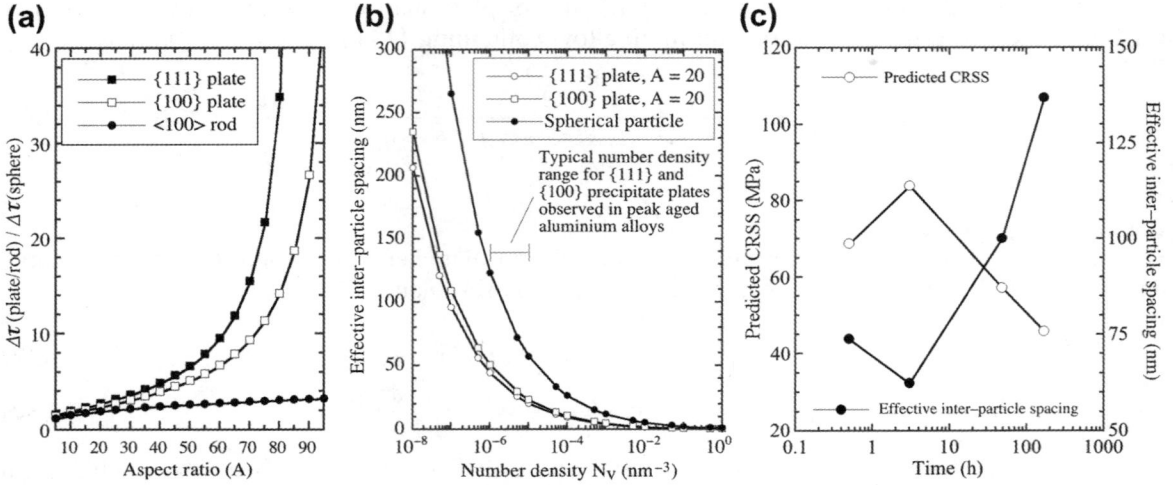

Figure 40 (a) Variation of ratio $\Delta\tau(\text{plate/rod})/\Delta\tau(\text{sphere})$ with aspect ratio for $\{111\}_\alpha$ and $\{100\}_\alpha$ plates and $\langle 100 \rangle_\alpha$ rods, calculated assuming Orowan strengthening at a precipitate volume fraction of 0.05. (b) Variation of effective interparticle spacing with number density of particles per unit volume, at a precipitate volume fraction of 0.05. (c) Variation with aging time of interparticle spacing and increment in CRSS arising from Orowan strengthening. Alloy Al–4Cu–0.05Sn (wt%) is aged isothermally at 200 °C (Nie and Muddle, 2008). (For color version of this figure, the reader is referred to the online version of this book.)

where A is again the aspect ratio (d/t for plates, and l/d for rods). As indicated by Eqns (33) and (34), $N_a(\text{plate})$ is invariably larger than $N_a(\text{sphere})$ when the plate aspect ratio is in the range in which $N_a(\text{plates}) = f/S_a$ is applicable. This implies that a change in precipitate shape from spheres to plates results in an increase in the number of precipitates intersecting a unit area of the slip plane and that this increment in N_a increases with an increase in plate aspect ratio. The precipitate rods formed in $\langle 100 \rangle_\alpha$ directions of the matrix phase are more effective in raising N_a than precipitate plates of the same aspect ratio, and $N_a(\text{rod})$ is larger than $N_a(\text{sphere})$ in the range of A for which $N_a(\text{rod}) = f/S_a$ is applicable. For a given f and N_v of precipitates, a change in precipitate shape from spheres to rods thus results in an increase in N_a, and the increment in N_a increases with an increase in rod aspect ratio.

For fixed volume fractions and number densities of precipitates, the variations in the ratio $\Delta\tau(\text{plate})/\Delta\tau(\text{sphere})$ with plate aspect ratio are shown in **Figure 40(a)**, for a precipitate volume fraction of 0.05. To simplify comparison, the logarithmic terms in the strengthening equations are assumed to be identical, and, in the case of spherical precipitates, the base microstructure is assumed to comprise precipitates in a regular triangular array distribution on the slip plane of the matrix phase. The Orowan increments in CRSS produced by $\{100\}_\alpha$ precipitate plates are invariably larger than those produced by equivalent volume fractions of spherical particles, and the difference increases substantially with increasing plate aspect ratio. There is a critical value of aspect ratio for each for which the effective interparticle spacing becomes zero, and the plates form a continuous three-dimensional network. If the plates are assumed to remain shear resistant, then the projected Orowan increment becomes infinitely large.

For a fixed volume fraction and a number density of precipitates, the variation in the ratio $\Delta\tau(\text{rod})/\Delta\tau(\text{sphere})$ with rod aspect ratio is also plotted in **Figure 40(a)**. Similar to those produced by the rationally oriented plates, the Orowan increment in CRSS produced by $\langle 100 \rangle_\alpha$ precipitate rods is invariably larger than that produced by spherical precipitates. The CRSS increment increases with an increase in rod aspect ratio. The CRSS increment is nearly independent of the volume fraction of precipitate rods, in the range of 2–8%.

The variation of effective interparticle spacing with precipitate number density per unit volume of the aluminum matrix phase is shown in **Figure 40(b)** for $\{111\}_\alpha$ and $\{100\}_\alpha$ plates and spherical particles. For identical volume fractions and number densities of precipitates, the $\{111\}_\alpha$ and $\{100\}_\alpha$ plates are much more effective, than spherical particles, in achieving a reduction in interparticle spacing particularly when N_v is less than 10^{-5} nm^{-3}. Typical values of number density of strengthening precipitate plates in high-strength and ultrahigh-strength aluminum alloys are in the range 10^{-6} to 10^{-5} nm^{-3}.

An Al–4Cu–0.05Sn (wt%) alloy is strengthened exclusively by θ' precipitate plates, when aged isothermally at 200 °C, **Figure 15**. Careful examination of the microstructures of fractured tensile samples (Nie and Muddle, 2008) has revealed no evidence that the θ' precipitates in this alloy are sheared during tensile plastic deformation, regardless of the aging condition of the alloy. The Orowan equation might thus be used to calculate the contribution of the θ' precipitates to the yield stress of the alloy in both underaged and overaged conditions. Experimental values of the increments in CRSS attributable to precipitation in such an alloy may be determined as the difference $(\sigma_y/M - \tau_m)$, where σ_y is the 0.2% yield strength of the alloy, $M = 3.06$ for polycrystalline face-centered cubic alloys, and τ_m is the contribution of the matrix phase to the CRSS of the alloy. Since samples were heat treated for 24 h at 525 °C to coarsen the grain size, and the concentration of solute Cu in Al is very low in overaged samples, contributions due to solid solution strengthening and grain boundary strengthening in overaged samples were assumed to be negligible.

Comparison of the predicted and measured increments in CRSS for the Al–4Cu–0.05Sn alloy indicates that they are generally in good agreement (Nie and Muddle, 2008). The CRSS predicted for underaged samples is slightly lower than the experimental value. However, this discrepancy can likely be attributed to solid solution strengthening, because the solute Cu concentration in the matrix in underaged samples is expected to be higher than in peak-aged and overaged samples. Another important observation is that the form of the age hardening response of the alloy can be accounted for exclusively using the Orowan equation. This again demonstrates that it is not necessary to invoke a transition from dislocation shearing to Orowan looping to account for the form of the hardening curve. The variation in the effective planar interparticle spacing with aging time is also shown in **Figure 40(c)**, and it is noted that the maximum value of yield strength corresponds approximately to the minimum value in the effective interparticle spacing.

20.3.2.2 Effects of Shearable Precipitates on Strengthening

Shearable precipitates can impede the movement of gliding dislocations through a variety of dislocation/particle interaction mechanisms, including those described as interfacial (chemical) strengthening, coherency strengthening, stacking fault strengthening, modulus strengthening and order strengthening (Reppich, 1993; Nembach, 1997). For each such proposed mechanism of strengthening, the contribution of the shearable precipitates to the CRSS of an alloy can be represented generally by an equation of the form (Friedel, 1963)

$$\Delta\tau = \left(\frac{2}{b\sqrt{\Gamma}}\right)\left(\frac{1}{L_p}\right)\left(\frac{F}{2}\right)^{3/2} \tag{36}$$

where $\Delta\tau$ is the increment in CRSS, b is the magnitude of the Burgers vector of the slip dislocations, Γ is the dislocation line tension in the matrix phase, L_P is the mean planar center-to-center interprecipitate spacing, and force F is a measure of the resistance of the precipitates to dislocation shearing. Since the effects of shearable precipitate plates or rods on strengthening remain largely unknown, it is convenient to restrict discussion to two of the more important mechanisms contributing to strength: interfacial strengthening and order strengthening.

20.3.2.2.1   Interfacial Strengthening

The interfacial strengthening, also known as chemical strengthening, results from the energy required to create an additional precipitate/matrix interface when a gilding dislocation shears a precipitate. In the following treatment of interfacial strengthening, the slip plane of precipitates is assumed to be parallel to that of the matrix phase, the Burgers vector of the dislocations is identical in the two phases, and the precipitates are assumed to be internally disordered. In such conditions shearing of precipitates will not lead to the formation of jogs and misfit dislocations at the precipitate/matrix interface, and the work done by moving a dislocation across a precipitate is identical to the energy of the created precipitate/matrix interface. These equations are based on an ideal regular triangular array of particles in the slip plane.

20.3.2.2.1.1   Spherical Particles.

For spherical particles, the projected shape of particles on the slip plane is spherically symmetrical. The force required to shear the particle is independent of the particle size and is given by

$$F = \frac{2\Delta A \gamma_i}{d_p} = \frac{\pi b \gamma_i}{2} \tag{37}$$

where γ_i is the specific interfacial energy between precipitate and matrix phases for the area created by dislocation shearing. However, the increment in CRSS is affected by the planar interparticle spacing, which in turn is a function of particle size. Assuming a triangular array distribution of spherical particles of uniform diameter d, Eqn (36) may be rewritten in the form

$$\Delta\tau_i = \frac{1.793\gamma_i^{3/2}}{d}\left(\frac{bf}{\Gamma}\right)^{1/2} \tag{38}$$

for interfacial strengthening. In this equation, f is the volume fraction of particles. As the interfacial energy for coherent particles is generally in the range $0.01-0.1\ \mathrm{J\ m^{-2}}$, and the volume fraction and diameter of coherent particles are typically $0.02-0.10$ and $2-20$ nm, respectively, in aluminum alloys of peak strength/hardness, the calculated CRSS increments are typically less than 5 MPa. Therefore, interfacial strengthening does not appear to be a major strengthening mechanism in aluminum alloys containing spherical particles.

20.3.2.2.1.2   {111}α and {100}α Precipitate Plates.

Due to the limited thickness of $\{111\}_\alpha$ plates of T_1 and Ω phases and to the strain contrast in the matrix phase that commonly surrounds the thin plates, it is often difficult to characterize the interaction between dislocations and T_1 precipitates using conventional TEM. These precipitates were therefore regarded for a long time as being resistant to shearing during tensile deformation (Huang and Ardell, 1987a, 1987b), although there have been some observations (Howe et al., 1988; Nie and Muddle, 2001) of T_1 plates apparently sheared by dislocations. There is now ample experimental evidence that the $\{111\}_\alpha$ precipitate plates in Al–Cu–Li alloys and Al–Cu–Mg–Ag alloys are sheared under tensile testing at room temperature, **Figure 41(a) and (b)**.

For $\{111\}_\alpha$ precipitate plates of uniform diameter d and thickness t, the dihedral angle between their habit plane and the $\{111\}_\alpha$ slip plane is $70.53°$. Assuming the slip plane is $(1\bar{1}1)_\alpha$, shearing of a $(111)_\alpha$ precipitate plate can occur in $[10\bar{1}]_\alpha$, $[110]_\alpha$, or $[011]_\alpha$ directions on the $(1\bar{1}1)_\alpha$ slip plane, **Figure 41(c)**. Since precipitate shearing in the $[011]_\alpha$ direction is crystallographically equivalent to shearing in the $[110]_\alpha$ direction, and shearing of $\{111\}_\alpha$ precipitate plates is experimentally observed only in these two directions, the precipitate shearing in the $[110]_\alpha$ direction is considered here. When a $(111)_\alpha$ plate is sheared in the $[110]_\alpha$ direction, **Figure 41(c)**, the displacement of two segments of the precipitate plate is $b \sin 60°$ in the $[121]_\alpha$ direction. The energy of the created particle/matrix interface is approximately $2\gamma_i d_p b \sin 60°$. Therefore the strength of the precipitate plate is

$$F = \frac{2d_p b \gamma_i \sin 60°}{t_p} = \frac{1.282 d b \gamma_i}{t} = 1.282 A b \gamma_i \qquad (39)$$

The mean planar center-to-center distance between precipitate plates, L_P, is given by Eqn (27). Combining L_P and Eqns (36) and (39), the interfacial strengthening increment in aluminum alloys containing $\{111\}_\alpha$ precipitate plates is given as

$$\Delta\tau_i = \frac{1.211 d \gamma^{3/2}}{t^2} \left(\frac{bf}{\Gamma} \right)^{1/2} \qquad (40)$$

For $\{100\}_\alpha$ precipitate plates, the dihedral angle between their habit plane and the $\{111\}_\alpha$ slip plane is $54.74°$. To shear the $\{100\}_\alpha$ plates with an additional particle/matrix interface area of $2d_p b \sin 60°$, the required shear force is

$$F = \frac{2d_p b \gamma_i \sin 60°}{t_p} = \frac{1.110 d b \gamma_i}{t} = 1.110 A b \gamma_i \qquad (41)$$

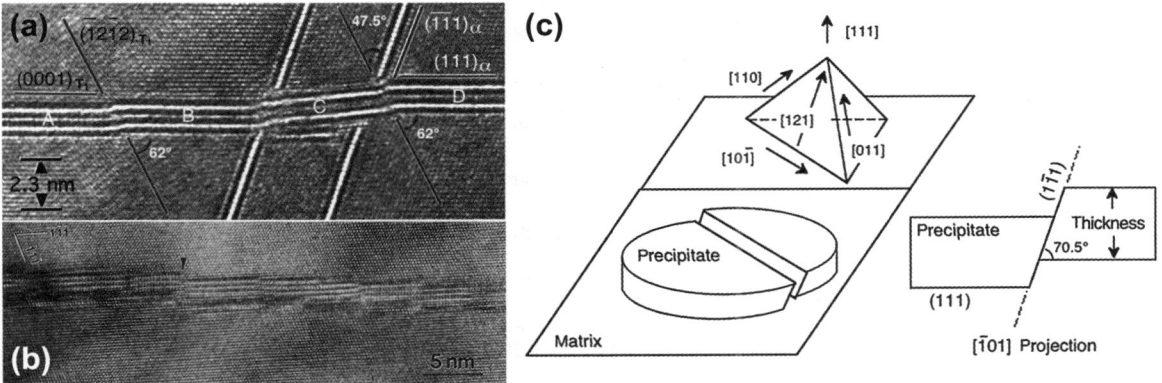

Figure 41 (a and b) HRTEM images showing sheared precipitate plates in Al–Cu–Li alloys (Nie and Muddle, 2001) and Al–Cu–Mg–Ag alloys (Li and Wawner, 1998). (c) Schematic diagrams showing shearing of a circular precipitate plate in the $[110]_\alpha$ direction on the $(1\bar{1}1)_\alpha$ plane of the aluminum matrix phase, and $[10\bar{1}]_\alpha$ projection of a sheared precipitate plate. Relative displacement of two segments of precipitate plate in the $[121]_\alpha$ direction is $b \sin 60°$. (For color version of this figure, the reader is referred to the online version of this book.)

The mean planar center-to-center distance between the $\{100\}_\alpha$ precipitate plates, L_P, is given by Eqn (23). Combining L_P and Eqns (36) and (41), the interfacial strengthening increment in CRSS in aluminum alloys containing $\{100\}_\alpha$ plates is

$$\Delta\tau_i = \frac{0.908 d\gamma^{3/2}}{t^2}\left(\frac{bf}{\Gamma}\right)^{1/2} \tag{42}$$

20.3.2.2.1.3 $\langle 001\rangle_\alpha$ *Precipitate Rods.* For rationally oriented $\langle 100\rangle_\alpha$ rod-shaped particles of uniform diameter d and length l, where $l >> d$, the cross-section of the particle in the slip plane has an ellipsoidal shape, with the major axis of $1.732d$ in the $\langle 121\rangle_\alpha$ directions. Since the projected shape of particles is not spherically symmetrical, the shear force will be a function of particle size. However, the shear force will be independent of the particle size if the projected shape of particles on the slip plane is regarded as a circle. Since the major axis of ellipsoids is less than $2d$, it is reasonable to approximate the ellipsoidal area by a circle of diameter d_P, with $d_P = d\sqrt{1.732}$. Therefore, the force required to shear precipitate rods is given as $\pi b\gamma_i/2$. Assuming that the $\langle 100\rangle_\alpha$ rods have a distribution that defines a cubic volume within the matrix phase, then the $\langle 100\rangle_\alpha$ rods intersected on a slip plane of the matrix phase will have a triangular array distribution, with a mean planar center-to-center interparticle spacing defined by L_P is Eqn (31). Using $F = \pi b\gamma_i/2$, the interfacial strengthening increment in CRSS in aluminum alloys containing $\langle 100\rangle_\alpha$ rods is given as

$$\Delta\tau_i = \frac{1.110 d\gamma^{3/2}}{d}\left(\frac{bf}{\Gamma}\right)^{1/2} \tag{43}$$

20.3.2.2.1.4   Strengthening Effects. For a given value of dislocation line tension, the ratio of $\Delta\tau_i(\text{rod})/\Delta\tau_i(\text{sphere})$ as a function of rod aspect ratio and volume fraction is plotted in **Figure 42(a)**. For a given

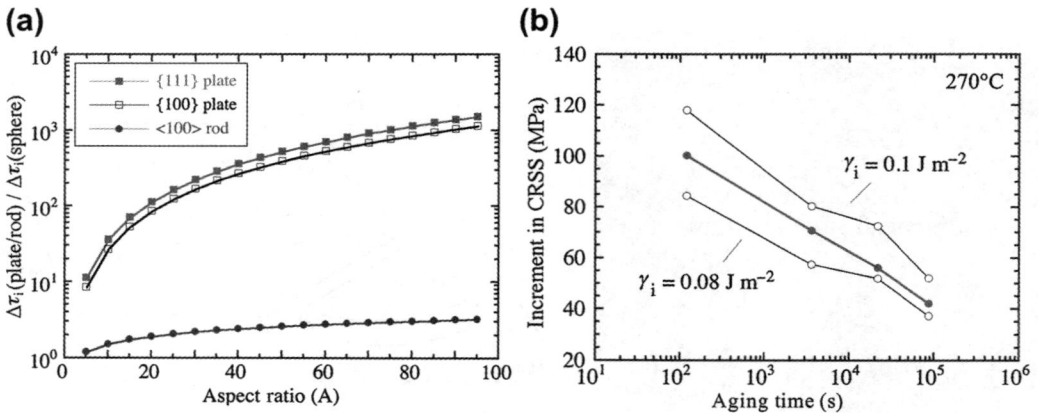

Figure 42   (a) Variation of ratio $\Delta\tau_i(\text{plate/rod})/\Delta\tau_i(\text{sphere})$ with aspect ratio for $\{111\}_\alpha$ and $\{100\}_\alpha$ plates and $\langle 100\rangle_\alpha$ precipitate rods, calculated assuming interfacial strengthening of sheared particles. (b) Variation with aging time of increment in CRSS arising from interfacial strengthening in an Al–2.86Cu–2.05Li–0.12Zr (wt%) alloy. Open and solid circles represent predicted and observed values from Nie and Muddle (2001). (For color version of this figure, the reader is referred to the online version of this book.)

volume fraction and number density of precipitates per unit volume, $\Delta\tau_i(\text{rod})$ is invariably larger than $\Delta\tau_i(\text{sphere})$ in the range of A for which $N_a(\text{rod}) = f/S_a$ is applicable ($A \geq 5$), and the ratio of $\Delta\tau_i(\text{rod})$ to $\Delta\tau_i(\text{sphere})$ increases gradually with an increase in A. However, the interfacial strengthening increment is limited, less than four times of that produced by spherical particles, when the rod aspect ratio is in the range 5:1–100:1. Since the interfacial strengthening contribution from spherical particles is trivially small, the interfacial strengthening contribution made by $\langle 100 \rangle_\alpha$ rods toward the alloy strength is negligible.

For a given value of dislocation line tension, the variations in the ratio $\Delta\tau_i(\text{plate})/\Delta\tau_i$ (sphere) with plate aspect ratio for various orientations of plate are also shown in **Figure 42(a)**. Unlike the results for spherical or rod particles, it is evident that the contribution due to interfacial strengthening may become significant when particles take, in particular, a plate-shaped form. For identical volume fractions and number densities of precipitates per unit volume, the yield stress increments produced by $\{111\}_\alpha$ and $\{100\}_\alpha$ precipitate plates are orders of magnitude larger than those produced by $\langle 100 \rangle_\alpha$ precipitate rods and by spherical particles. The increments in CRSS produced by $\{111\}_\alpha$ and $\{100\}_\alpha$ plates increase substantially with an increase in plate aspect ratio and are up to three orders of magnitude larger than that produced by spheres, when the plate aspect ratio is in the range of 5:1 to 95:1. Interfacial strengthening can thus potentially be a major strengthening mechanism in aluminum alloys containing rationally oriented shearable precipitate plates of large aspect ratio.

The capability of the interfacial strengthening model has been examined using experimental data by Huang and Ardell (1987b) for the scale and distribution of precipitates in an Al–2.85Cu–2.3Li–0.12Zr (wt%) alloy containing a uniform single distribution of $\{111\}_\alpha$ precipitate plates of T_1 phase. The alloy was aged isothermally at 190 °C, then reversion heat treated for 60 s at 265 °C to dissolve δ' (Al$_3$Li) and θ' precipitates and to isolate the contribution of T_1 precipitate plates to yield strength (Huang and Ardell, 1987b). It has been shown (Howe et al., 1988; Nie and Muddle, 2001) that T_1 precipitates are sheared in plastically deformed (3% strain) and fractured tensile samples in both under- and overaged conditions, and thus it is appropriate to apply the model for interfacial strengthening to predict the contribution of the T_1 precipitates to the yield strength of the alloy. Assuming a value of 0.08 J m^{-2} for the specific precipitate–matrix interfacial energy (currently accepted values for specific precipitate–matrix interfacial energy vary in the range 0.005–0.2 J m^{-2} for coherent precipitates), the predicted increments in CRSS produced by the T_1 precipitates were found to be in generally good agreement with values observed experimentally, **Figure 42(b)** (Nie and Muddle, 2001). It is important to emphasize, however, that the level of this agreement depends sensitively on the value selected for the energy of the precipitate–matrix interface created by the shearing dislocation and, at the present time, there is a lack of reliable data for the specific precipitate–matrix interfacial energy. With this qualification, the interfacial strengthening model does appear capable of accounting for the yield strength and, more generally, the form of the age hardening response of the alloy, when the effects of precipitate shape are included in the model.

The apparent success of this model, effectively over the entire aging regime, suggests that overaging is not necessarily accompanied by a transition from predominantly dislocation shearing to an Orowan looping mechanism. A comparison of predicted and observed increments in CRSS for overaged samples of an Al–2.86Cu–2.05Li–0.12Zr (wt%) alloy (AA2090) indicates that, for this independent set of measurements, the increments in CRSS predicted by the interfacial strengthening model are again in excellent agreement with observed values. Similar attempts to account for observed strength levels using alternative mechanisms of precipitation strengthening or Orowan hardening have proved unsuccessful.

20.3.2.2.2　Order Strengthening

The order strengthening occurs when a gliding dislocation shears an internally ordered precipitate and creates disorder within the precipitate in the form of an antiphase boundary (APB) on the slip plane of the precipitate phase. It is common that dislocations travel in pairs, with the second dislocation recovering the disorder created by the first dislocation. Since the calculated increment in CRSS for dislocation pairs does not differ largely from that for a single dislocation, and to simplify the analysis, only the interaction between particles and a dislocation is considered in the following analysis. Assuming the work done by moving the dislocation across the precipitate is identical to the energy of the created APB, the force required to shear the precipitate varies for precipitates of different shapes and orientations, and therefore the order strengthening increment is different in aluminum alloys containing precipitates of different shapes and orientations. Assuming a triangular array distribution of spherical particles of uniform diameter d, Eqn (36) may be rewritten in the form

$$\Delta\tau_o = 0.441 \left(\frac{df}{b^2\Gamma}\right)^{1/2} \gamma_{apb}^{3/2} \tag{44}$$

for order strengthening. In this equation, γ_{apb} is the specific APB energy on the slip plane of the precipitate phase.

When a dislocation shears through an internally ordered, coherent $\{111\}_\alpha$ and $\{100\}_\alpha$ precipitate plates, the area of APB within the particle is approximately

$$S = \frac{\pi}{4}d(d - b\sin 60°) \tag{45}$$

and the shear force to create this disordered area is

$$F = \frac{\pi}{4}d_p \sin 60° \left(1 - \frac{b\sin 60°}{t_p}\right)\gamma_{apb} \tag{46}$$

The mean planar center-to-center interparticle spacing for $\{111\}_\alpha$ and $\{100\}_\alpha$ precipitate plates is given in Eqns (27) and (23), respectively. Substituting F and L_P into Eqn (36), the order strengthening increment in CRSS in aluminum alloys containing a single type of precipitate plates is given as

$$\Delta\tau_o = 0.579 \left(\frac{f}{\Gamma}\right)^{1/2} \gamma_{apb}^{3/2}\frac{d}{b\sqrt{t}}\left[1 - \frac{0.816b}{t}\right]^{3/2} \tag{47}$$

for $\{111\}_\alpha$ precipitate plates, and

$$\Delta\tau_o = 0.539 \left(\frac{f}{\Gamma}\right)^{1/2} \gamma_{apb}^{3/2}\frac{d}{b\sqrt{t}}\left[1 - \frac{0.707b}{t}\right]^{3/2} \tag{48}$$

for $\{100\}_\alpha$ precipitate plates.

Examination of the variation of the ratio of $\Delta\tau_o$(plate)/$\Delta\tau_o$(sphere) as a function of plate aspect ratio indicates that, for a given volume fraction and number density of precipitates per unit volume, the order strengthening increment in CRSS produced by $\{111\}_\alpha$ plates is approximately 7% larger than that produced by $\{100\}_\alpha$ plates, and for both precipitate orientations, the CRSS increment increases with an increase in plate aspect ratio. The CRSS increment produced by both $\{111\}_\alpha$ and $\{100\}_\alpha$ plates are invariably larger than that produced by spheres.

For an internally ordered coherent $\langle 100 \rangle_\alpha$ precipitate rod, an APB area of approximately $0.433\pi d^2$ will be created if it is fully sheared by a dislocation. The shear force is given approximately by

$$F \approx \frac{\pi}{4}(d_p - 2b)\gamma_{apb} \tag{49}$$

Substituting Eqn (49) and the mean planar center-to-center interparticle spacing into Eqn (36), the order strengthening increment in CRSS in aluminum alloys containing $\langle 100 \rangle_\alpha$ precipitate rods is given as

$$\Delta\tau_o = 0.393 \left(\frac{f}{\Gamma}\right)^{1/2} \gamma_{apb}^{3/2} \left[\frac{\pi}{4}\left(\frac{d}{b^2}\right)^{1/3} - 2\left(\frac{b}{d^2}\right)^{1/3}\right]^{3/2} \tag{50}$$

For a given volume fraction and number density of precipitates per unit volume, $\Delta\tau_o$(rod) is invariably larger than $\Delta\tau_o$(sphere) when A is in the range of 5–95. However, the ratio of $\Delta\tau_o$(rod)/$\Delta\tau_o$(sphere) decreases gradually with an increase in A from 5 to 95.

Whether particles are sheared or shear resistant, the strengthening models confirm experimental observations that high tensile yield strength is associated with microstructures containing a high density of intrinsically strong plate-shaped precipitates with $\{111\}_\alpha$ or $\{100\}_\alpha$ habit planes and large aspect ratio. As shown in **Figure 40(b)**, the effective interparticle spacing between $\{111\}_\alpha$ precipitate plates is comparable with that between spherical precipitates, even when the number density of $\{111\}_\alpha$ plates is substantially lower than that of spherical particles. The ultrahigh tensile yield strength (≥ 700 MPa) observed in Al–Cu–Li–Mg–Ag (X2095) alloy (Pickens et al., 1989) and the high yield strength (550 MPa) of Al–Cu–Li (AA2090) alloys (Cassada et al., 1991a, 1991b) may, for example, be attributable to a uniform distribution of T_1 precipitate plates of large aspect ratio formed on $\{111\}_\alpha$ planes. It has also been shown that Al–Cu–Mg–Ag alloy, strengthened by the presence of thin plates of the Ω phase on $\{111\}_\alpha$ planes, may develop a tensile yield strength exceeding 500 MPa (Polmear and Couper, 1988). Furthermore, the well-known high strength of Al–Zn–Mg–Cu alloys (e.g. 7075) is associated with the combined effects of the number density and aspect ratio of precipitate plates of the phase η', which also form on $\{111\}_\alpha$ planes. A remarkably high value of hardness (180 VHN) has also been obtained in Al–Cu–Mg–Si alloys by refining the distribution of θ' precipitate plates formed on $\{100\}_\alpha$ planes, and complementing this by a fine-scale distribution of $\langle 100 \rangle_\alpha$ rods of the equilibrium quaternary Q phase.

Further improvements in the design of aluminum alloys for high strength will require an improved understanding of the metastable phase equilibria leading to those transition phases forming as plate-shaped products on low-index habit planes. In particular, there needs to be a concerted systematic effort to better understand the role of pre-precipitate clusters of microalloying additions in determining metastable phase equilibria. For existing alloys strengthened by plate-shaped precipitates, the strengthening models outlined above suggest that further improvements in strength might be achieved by increasing the number density and/or aspect ratios of the plates. One effective approach to achieving an increase in plate aspect ratio and number density may lie in the use of microalloying additions, which partition to either matrix or precipitate phase to improve coherency of the precipitate phase in the habit plane. This should, in turn, tend to decrease the formation kinetics of ledges on the plate broad faces and hence the thickening rate of these interfaces.

20.3.2.3   Critical Thickness of $\{111\}_\alpha$ Precipitate Plates

The classical form of isothermal age hardening response is conventionally attributed to competing modes of deformation involving either shearing or bypassing of dispersed particles by mobile dislocations and maximum strength is associated with a critical scale of strengthening precipitates, above which a transition from dislocation shearing to dislocation bypassing occurs. To the extent that alternative particle forms have been considered, the critical dimension for plate-like forms has been interpreted to correspond to plate thickness, but such analyses essentially ignore the full ramifications of an array of plate-like forms and the potential effects that the aspect ratio of such forms may have on any strengthening increment.

In their analysis of the sheared Ω precipitate plates in plastically deformed samples of an Al–Cu–Mg–Ag alloy, Li and Wawner (1998) reported that there existed a critical thickness of $\{111\}_\alpha$ plates, below which the plates were sheared during the plastic deformation and above which a transition from dislocation shearing to dislocation looping occurred. Based on high-resolution TEM observations that Ω phase has been sheared, they proposed a strengthening model for aluminum alloys containing $\{111\}_\alpha$ precipitate plates. That model is based essentially on the order strengthening mechanism that is accepted for spherical particles, and includes increments in strength arising from both order strengthening and interfacial (chemical) strengthening for sheared precipitates. However, it does not take into account the plate-like shape and orientation of the Ω precipitates. Despite this major deficiency, the model is claimed to be applicable not only to Ω precipitate plates in Al–Cu–Mg–Ag alloys but also to other $\{111\}_\alpha$ precipitate plates, such as T_1 phase in Al–Cu–Li alloys. They suggested that it was the reduction in the critical thickness of $\{111\}_\alpha$ plates that was a major factor in strengthening the alloy. This notion of a critical size for dispersed strengthening particles was originally proposed for spherical particles. If $\Delta\tau_s$ represents the increment in CRSS required for particle shearing and $\Delta\tau_o$ is the CRSS increment produced by Orowan looping, then, for a given volume fraction of spherical particles, it has been traditionally accepted that there is a critical particle diameter d_c such that $\Delta\tau_s/\Delta\tau_o < 1$, when particle diameter $d < d_c$, and $\Delta\tau_s/\Delta\tau_o > 1$, when $d > d_c$. A transition from dislocation shearing to dislocation looping thus occurs when the diameter of spherical particles increases above the critical size.

In the case of plate-shaped particles, Li and Wawner (1998) assumed that the critical particle size corresponded simply to the thickness of the precipitates. However, such an assumption ignores the plate-like form and the effect that the aspect ratio of the precipitate plates may have on any strengthening increment (Kelly, 1972; Oblak et al., 1974a, 1974b; Merle et al., 1981; Huang and Ardell, 1987b; Nie and Muddle, 2000a). As shown in Eqns (28), (40) and (47), the precipitate contribution to the CRSS increment is a function of the thickness and the aspect ratio (diameter) of the precipitate plates. It is thus clearly an oversimplification to propose (Li and Wawner, 1998) that there exists a simple critical thickness for precipitate plates, above which the ratio $\Delta\tau_s/\Delta\tau_o > 1$ and there is a transition from dislocation shearing to Orowan looping. If indeed such a transition occurs, then it will be at a value of resolved shear stress that is determined by both plate thickness and diameter (aspect ratio). It is perhaps to be emphasized that Li and Wawner (1998) produce no direct evidence that such a transition does occur for plates of Ω phase; their experimental evidence focuses on establishing that such plates are sheared. In a subsequent study of plastically deformed tensile samples of an Al–Cu–Li alloy (Nie and Muddle, 2001), examination of microstructures in fractured tensile samples revealed unambiguously that the T_1 precipitate plates were sheared in both peak-aged and overaged conditions (Nie and Muddle, 2001). Therefore, it is unlikely that there exists a simple critical thickness for these precipitate plates, above which there is a transition from predominantly dislocation shearing to Orowan looping.

It has been demonstrated (Nie and Muddle, 2008) that, at least for an Al–4Cu–0.05Sn alloy, it is not necessary to invoke a transition from precipitate shearing to Orowan looping to account for the form of the strengthening curve obtained from that alloy aged at 200 °C. The measured variations in the scale and distribution of θ' precipitates as a function of aging time indicate that the maximum increment in CRSS corresponds to the minimum value in the effective planar interparticle spacing, **Figure 40(c)**. This suggests that the peak strength of the alloy is not associated with a critical thickness, above which a transition from dislocation shearing of precipitates to dislocation looping occurs. Examinations using TEM of plastically deformed samples provided no evidence for the dislocation shearing of precipitates in the underaged samples. The quantitative analysis of the distribution of θ' precipitates indicated that the lower yield strength observed for the underaged samples is attributable to a larger value of the effective planar interprecipitate spacing, which is because of the relatively low number density and small diameters of θ' precipitates. The similarities in the stress–strain curves of the underaged and overaged samples further support the prevailing operation of a single deformation mechanism, that is Orowan strengthening, in both underaged and overaged samples during tensile deformation.

These observations highlight the need for further detailed consideration of the mechanism(s) of interaction between gliding dislocations and dispersed second-phase particles. The common strengthening phases, such as θ', in high-strength aluminum alloys are typically ordered intermetallic compounds and, while there is rarely direct evidence, they are likely to be intrinsically resistant to conventional plastic deformation. However, as dispersed plate-shaped particles, these phases commonly preserve a rational OR and coherent broad planar interfaces with face-centered cubic matrix. In this constrained configuration, their susceptibility to shear deformation arising from dislocations having their origin in the matrix phase may not necessarily be determined by the intrinsic properties of the precipitate phase, particularly when precipitate plates may be as little as 1–2 unit cells in thickness. If such plates were to be sheared locally, then the mechanism by which shear might be transferred from matrix to structurally distinct precipitate phase, and whether this mechanism is dependent on the scale of the precipitate crystal, remains to be clarified.

20.3.3   Formation Mechanisms of Plate-shaped Precipitates

20.3.3.1   Precipitation of γ' Plates in Al–Ag Alloys

The equilibrium solid solubility of Ag in aluminum is quite high: $\sim$23.5 at.% at the eutectic temperature 567 °C. It drops to less than 1.0 at.% at temperatures near 200 °C, **Figure 43**. For Al–Ag alloys containing up to $\sim$23.5 at.% Ag that are solution treated, quenched and subsequently aged isothermally at a low temperature, the supersaturation of Ag in α-Al solid solution is relieved by the following precipitation sequence:

$$SSSS \rightarrow GP \text{ zones} \rightarrow \gamma' \rightarrow \gamma$$

The GP zones are Ag-rich and they have a spherical shape. These zones are relatively thermally stable; they remain in the microstructure even after prolonged aging at low temperatures. The metastable γ' phase has a hexagonal structure (space group $P6_3/mmc$, $a = 0.2858$ nm, $c = 0.4607$ nm) and a composition of $AlAg_2$. It forms as hexagonal shaped plates on the $\{111\}_\alpha$ planes with large aspect ratio (typically from 10:1 to 100:1). The OR of the γ' phase with the surrounding matrix phase is

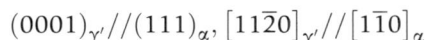

$$(0001)_{\gamma'} // (111)_\alpha, \, [11\bar{2}0]_{\gamma'} // [1\bar{1}0]_\alpha$$

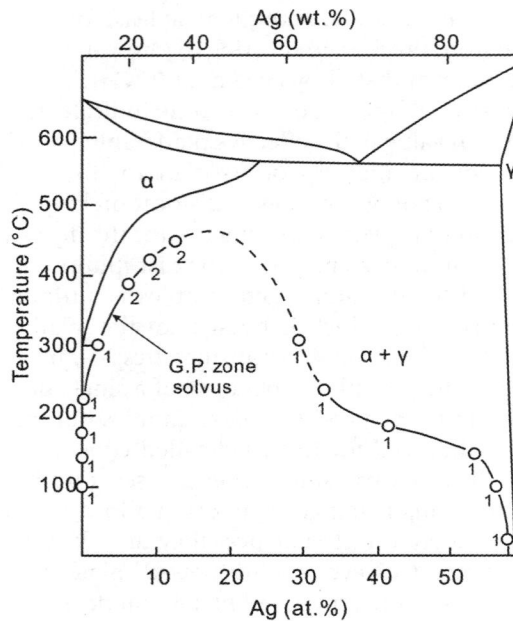

Figure 43 Al-rich side of the Al–Ag phase diagram (Lorimer, 1978).

Precipitates of γ' nucleate heterogeneously on stacking faults that are associated with dislocation loops and have the segregation of Ag atoms. The structure and composition of the equilibrium γ phase are similar to those of γ'. The only difference between γ and γ' phases seems to be a small variation in lattice parameters. The γ phase also forms as plate-like precipitates with $\{111\}_\alpha$ habit planes, and has the same OR with the matrix as γ'.

The γ' phase is near fully coherent with the matrix across the $(0001)_{\gamma'}//(111)_\alpha$ habit plane, and there is a contraction misfit of only $\sim 2\%$ in the spacing of the close-packed planes normal to the habit plane. The transformation is thus structurally similar to the martensitic face-centered cubic to hexagonal transformation in Co and Co-based alloys. However, there can be no question that there is a substantial composition change associated with the γ' plates, for they have a structure comprising alternating A and B layers of the composition of pure Ag and Al_2Ag, respectively, to give an average composition of $AlAg_2$.

Based on the observed OR, it is possible to define a lattice correspondence across the coherent broad faces of the γ' plates which is shown schematically in **Figure 44(a)**. As for the martensitic face-centered cubic to hexagonal transformation, the structural change associated with the formation of the γ' may be considered to be accomplished by the passage of transformation dislocations across the broad faces of the plates (Muddle et al., 1994). These dislocations define steps that are two close-packed planes in height and have a Burgers vector $\mathbf{b}_T$ of the form $(a/6)\langle 11\bar{2}\rangle_\alpha$ in the case of coherent transformation to an ideal hexagonal structure with zero volume change. If there is a finite misfit normal to the habit plane, as there is for the γ' phase, then the Burgers vector will contain a small additional component of magnitude ξ normal to the habit plane, that is $(2a\xi/3)\langle 111\rangle_\alpha$.

Figure 44(b) contains a schematic representation of an embryo γ' platelet within α-Al matrix. If, as depicted, there is no change in spacing of the close-packed planes across the habit plane, then the

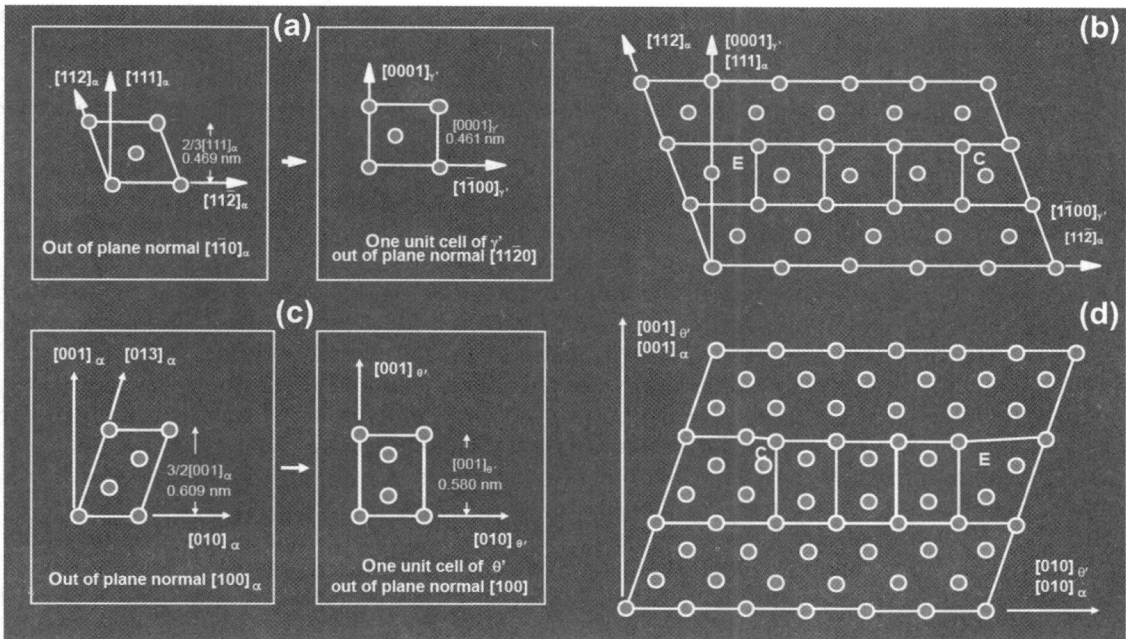

Figure 44 Schematic diagrams showing lattice correspondence between (a) γ' and α-Al lattices and (c) θ' and α-Al lattices. A precipitate plate of single unit cell height formed within aluminum for (b) γ' and (d) θ'. For clarity only atoms at zero layer are shown. (For color version of this figure, the reader is referred to the online version of this book.)

structural change can be considered to be accomplished by a shear of magnitude $8^{-1/2}$ on the $(111)_\alpha$ plane in the $[11\bar{2}]_f$ direction. In the formalism of the crystallographic theory, the total strain $\mathbf{S}_T$ is an invariant plane strain equivalent in this case to the shape strain $\mathbf{S}_R$, that is

$$\mathbf{S}_T = \mathbf{S}_R = \mathbf{I} + m\mathbf{d}\mathbf{p}' \tag{51}$$

where $\mathbf{I}$ is the identity matrix, m is the strain magnitude, $\mathbf{d}$ defines the direction of displacement, and $\mathbf{p}'$ is a unit vector normal to the habit plane. The displacement vector has the general form

$$m\mathbf{d} = \mathbf{b}_T = a/6[11\bar{2}] - 2a\xi/3[111] \tag{52}$$

where ξ is the misfit normal to the habit plane and, for the γ' phase, has a value of ~ 0.02. As indicated in **Figure 44(b)**, the transformation strain defined by the lattice correspondence has a substantial shear component, equivalent to the magnitude of an $(a/6)\langle 11\bar{2}\rangle_\alpha$ Shockley partial dislocation (~ 0.35).

There is strong experimental evidence (Howe et al., 1985; Howe et al., 1987) to support the proposal that the formation of γ' plates involves the lateral migration of steps across the otherwise coherent broad faces of the product phase, with the steps having the configuration of the transformation dislocations required by the theory. High-resolution electron microscope images, recorded parallel to close-packed $\langle 1\bar{1}0\rangle_\alpha//\langle 11\bar{2}0\rangle_{\gamma'}$ directions and equivalent to the projection shown in **Figure 44(b)**, have revealed that the interface commonly contains steps normal to the habit plane that are invariably an even multiple of close-packed layers in height and that may be interpreted as an accumulation of $(a/6)\langle 11\bar{2}\rangle_\alpha$ Shockley partial transformation dislocations on every second close-packed $(111)_\alpha$ plane.

In a given $\{111\}_\alpha$ habit plane, there are three equivalent $\langle 11\bar{2}\rangle_\alpha$ directions and the passage of a transformation dislocation, with a Burgers vector defined by Eqn (52) in any one of these directions, will generate the required hexagonal structure. If transformation involves migration of dislocations in a single direction on alternate $\{111\}_\alpha$ planes, then an appreciable shape change will accumulate over the transformed volume. If, alternatively, transformation dislocations are generated in all three $\langle 11\bar{2}\rangle_\alpha$ directions and the transformed volume contains equal volume fractions of all three variants of the hcp structure, then the variants will be self-accommodating and the shear component of the net shape deformation will be zero. The formation of a single variant within α matrix will, as indicated schematically in **Figure 44(b)**, give rise to regions of expansion (E) and contraction (C) in the matrix surrounding the precipitate plate and to the associated accumulation of significant shear strain energy (Nie and Muddle, 1999a). For transformation product forming within the bulk constraints of the parent phase, it is thus to be expected that all three transformation variants will operate within a given product plate to minimize the shape deformation, which is consistent with direct experimental observations for bulk γ' plates in Al–Ag alloys (Howe et al., 1987).

The accommodation of the shear component of the invariant plane strain transformation strain is thus seen to be an important factor controlling the growth of γ' plates. Symmetry within the habit plane allows potential for three equivalent variants of the product and partial, if not total, self-accommodation of the shear component of strain. Shear strain accommodation is also equally likely to be a key element in the nucleation of such plates. Although near to fully coherent with α-Al matrix (volumetric misfit $\sim 2.6\%$) and with a degree of coherency in the habit plane that is more than 99.5%, γ' plates are not readily nucleated from supersaturated solid solution and typically form in a coarse distribution during aging that is ineffective in precipitation hardening the matrix. It has been observed (Hren and Thomas, 1963; Nicholson and Nutting, 1961) that γ' plates nucleate preferentially on quenched-in $\{111\}_\alpha$ Frank dislocation loops, utilizing the existing adjustment in stacking effected by the associated stacking fault and thus accommodating the shear component of the transformation strain.

The above mechanism for the formation of γ' plates has been anticipated for many years (Christian, 1969) and the subsequent work, to provide experimental support for the model (Howe et al., 1987) and to confirm the applicability of the crystallographic theory, has meant that the proposed mechanism has gained broad acceptance. However, there have been few attempts to extend this form of analysis to other similar plate-shaped transformation products, despite the emergence of a number of important systems in which key strengthening precipitates form as coherent plates of large aspect ratio. The following discussion will consider selected examples.

20.3.3.2 Precipitation of θ' Plates in Al–Cu Alloys

The more general applicability of this approach to the formation of coherent plate-shaped transformation products was recognized by Dahmen and Westmacott (1983a) in their treatment of the precipitation of θ' phase in solution-treated, quenched and aged Al–Cu alloys. The θ' phase forms as thin rectangular plates parallel to $\{100\}_\alpha$ planes. The lattice correspondence is depicted schematically in **Figure 44(c)**, and **Figure 44(d)** shows a schematic representation of a θ' platelet that is one unit cell in thickness constrained within α matrix. The existence of this lattice correspondence across the near fully coherent habit plane again implies that the structural transformation can be described phenomenologically by an invariant plane strain. As with most invariant plane strain transformations, this strain has a small volumetric component of approximately -0.06 and a relatively large shear component of ~ 0.33. In the absence of a mechanism for accommodation of this shear strain, formation of a platelet of θ' such as that represented in **Figure 44(d)** will, as indicated schematically, lead to local strain fields of expansion and contraction in the matrix.

Dahmen and Westmacott (1983b) have produced compelling evidence of the importance of the accommodation of this shear component of the transformation strain for both nucleation and growth of θ'. For θ' phase forming directly from supersaturated solid solution, single unit cell platelets such as that depicted in **Figure 44(d)** are rarely observed experimentally and, when they are, they are invariably observed to be nucleated on $a/2\langle 100\rangle_\alpha$ dislocation shear loops formed through dissociation of $a/2$ $[110]_\alpha$ climb dislocations. A shear loop with displacement vector $a/2[010]_\alpha$ lying in the $(001)_\alpha$ plane provides a displacement field that will allow the single unit cell platelet of θ' in **Figure 44(d)** to form without accumulating shear strain.

For θ' plates nucleated directly from solid solution, plates of single unit cell thickness are never observed (Dahmen and Westmacott, 1983a). The minimum thickness observed is equivalent to two full unit cells and thicker plates have thicknesses that are discrete multiples of unit cells or half unit cells in height. Almost invariably, the smallest growth ledges (transformation dislocations) comprise pairs of structural units which form by $a/2\langle 100\rangle_\alpha$ shears of opposite sign and which are thus self-accommodating with respect to shear displacement. Dahmen and Westmacott (1983a) have proposed a credible model for the growth of θ' plates which then combine conservative and nonconservative structural units to minimize both the shear and dilatational components of the transformation strain, and the discrete plate thicknesses observed can be rationalized in terms of effective self-accommodation of these strain components.

20.3.3.3 *Precipitation of $\{111\}_\alpha$ Plates in Microalloyed Al–Cu and Al–Cu–Li Alloys*

The highest strength aluminum alloys produced by conventional precipitation hardening are those strengthened by intermediate precipitate phases forming as large aspect ratio plates on the primary slip planes of an α-Al matrix. These precipitate phases include the Ω and T_1 phases in microalloyed Al–Cu and Al–Cu–Li alloys, respectively, and the η' phase in Al–Zn–Mg alloys. There remains some controversy surrounding the crystal structures of all three of these phases and, such are the uncertainties associated with the form, structure and orientation of the η' phase, that it is not currently profitable to examine the formation of this phase in any detail.

Both Ω and T_1 form as thin plates of large aspect ratio ($>50:1$) on $\{111\}_\alpha$ planes and are near perfectly coherent with α-Al matrix across the habit plane. They differ predominantly in the magnitude of the dilation normal to the habit plane, which is a contraction of $\sim 9.6\%$ in the case of Ω phase and $\sim 0.32\%$ for T_1. For both phases there is a large shear component to the invariant plane strain that may be invoked to describe the structural change phenomenologically. This shear has a magnitude of ~ 0.18 in the case of Ω phase and 0.35 in the case of T_1 plates. The analysis, that is widely accepted for the formation of the γ' phase in Al–Ag alloys and that has also been effectively applied to the θ' phase in Al–Cu alloys, is thus also extended to these two further examples of coherent precipitate plates. However, the details of the crystallographic analysis require the knowledge of the precise structure that is adopted for each of the precipitate phases.

20.3.4 Effects of Microalloying Additions

Using an expression developed for calculating the strain energy associated with formation of a martensite plate (Christian, 1958), it is possible to demonstrate that, for a given precipitate volume

(a)

(b)

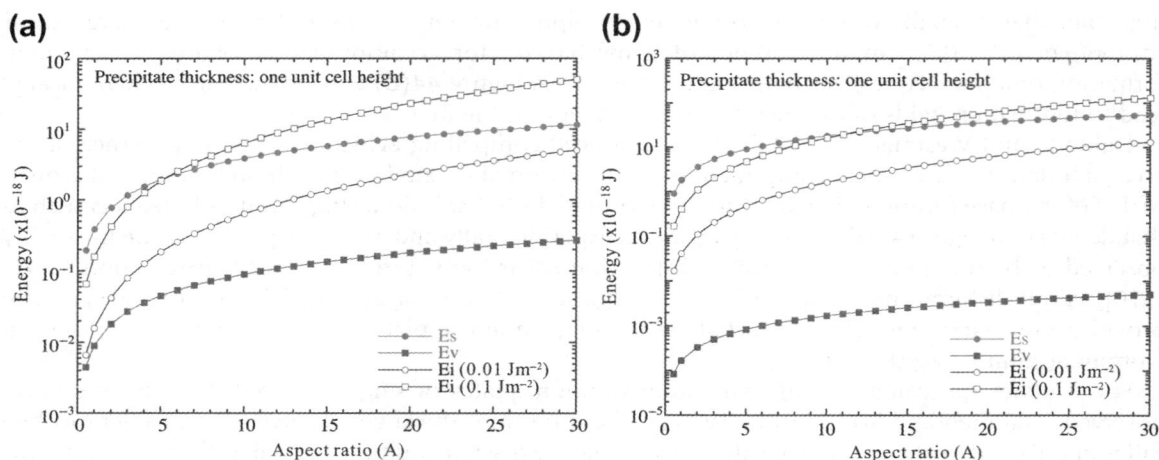

Figure 45 Variation of shear strain energy E_s, volumetric strain energy E_v and interfacial energy E_i as a function of plate aspect ratio for (a) θ', and (b) T_1. (For color version of this figure, the reader is referred to the online version of this book.)

and aspect ratio, the volume strain energy component associated with the formation of γ' and θ' plates is substantially smaller than the shear strain energy, **Figure 45**. For precipitate plates that form fully constrained by the matrix, this strongly suggests that it may be the manner in which the shear component of the transformation (shape) strain is accommodated that is a key factor in controlling nucleation behavior. It is to be recognized that the shear displacement associated with formation of the first planar element of a precipitate plate leads to local regions of extension and compression in the adjacent matrix, **Figure 44**. If there is an extrinsic supply of solute atoms larger in radius than Al (or clusters of such atoms) to the regions E, and a corresponding supply of vacancies or vacancy clusters to regions C, then the shear strain associated with plate formation may be reduced or even eliminated. Such an environment might be found in the vicinity of pre-precipitate clusters combing large solute atoms with a local excess of vacancies.

The Ω phase only forms in Al–Cu alloys microalloyed with Mg (in limited circumstances), and only forms readily in uniform distributions in Al–Cu alloys microalloyed with combinations of Mg and Ag. It has been observed (Murayama and Hono, 1998; Reich et al., 1998) that the Mg and Ag atoms co-cluster during the early stages of isothermal aging and that these clusters appear to develop in planar arrays parallel to $\{111\}_\alpha$ to provide preferred nucleation sites for Ω plates. The Mg and Ag atoms are observed to be segregated at the interfaces of the Ω plates and to remain so throughout plate growth.

Despite being near fully coherent with an α-Al matrix, the T_1 phase (Al$_2$CuLi) is also difficult to nucleate in the Al–Cu–Li solid solution. In ternary Al–Cu–Li alloys, T_1 plates nucleate preferentially on dislocations and commercial heat treatment schedules for such alloys invariably include a stage of plastic deformation, postsolution treatment and before aging (T8 treatment), to promote a more uniform distribution of this key strengthening precipitate. Microalloying additions of Mg and Ag to such Al–Cu–Li alloys have also proven effective in promoting a uniform fine-scale distribution of T_1 plates and the resulting commercial Weldalite™ series of alloys are the highest strength precipitation hardening alloys yet produced ($\sigma_y > 700$ MPa). The mechanism by which the Mg and Ag promote nucleation of T_1 is less well established, but it seems likely that it is associated with pre-precipitate clustering of the microalloying additions. The Mg and Ag atoms are once again observed to segregate to the broad faces of the precipitate plates from the earliest stages of precipitation detectable.

The presence of Mg and Ag at the precipitate–matrix interface may improve the (volumetric) fit between the two structures and/or favorably modify the chemical component of the interfacial energy to reduce the energy barrier to nucleation of both Ω and T_1 phases. However, in the case of T_1 phase, in particular, the volumetric misfit is already negligible and, in both cases, it is not altogether obvious why the presence of foreign atoms, at the interface between an Al-rich precipitate and an Al-based solid solution matrix, should reduce the interfacial energy. Such arguments ignore the substantial shear component to the invariant plane strain transformation strain that may be used to define the structural transformation. Although mechanistic models for the formation of these two precipitates are less well developed than those for the γ' and θ' phases, it should prove profitable to reexamine these transformation products using a similar approach. Once again it seems likely that accommodation of the shear–strain component of the transformation strain may prove a key factor controlling both nucleation and growth of these important precipitate phases.

The morphology and orientation of solute clusters may well determine the identity of the transition phase precipitated in a given trace-modified alloy. Since Sn, Cd and In atoms have relatively high binding energies with vacancies (Ozbilen and Flower, 1989) and are larger than Al atoms, **Figure 46**, clusters of such solute atoms would be expected to trap vacancies, and the shear strain associated with a critical nucleus might be reduced or eliminated if Sn, Cd or In atoms move into the extended region and their associated vacancies migrate into the compressed region. Presumably those vacancies associated with these solute atoms would be attracted away from them and into the compressed regions of embryos forming within the solute–vacancy clusters. This hypothesis is supported by evidence that clusters of Sn atoms form before the precipitation of θ' in Al–Cu–Sn alloys (Ringer et al., 1995a) and that a particle of Sn is ultimately observed in contact with an end facet, rather than the broad face of a θ' precipitate plate. Although Mg atoms are larger than Al atoms, they have a relatively low binding energy with vacancies (Ozbilen and Flower, 1989). Therefore, a high concentration of vacancies will not be expected to be associated with clusters of Mg atoms. However, the population of vacancies can be increased when Mg atoms co-cluster with those solute atoms (Ag, Si or Ge) that do bond strongly with

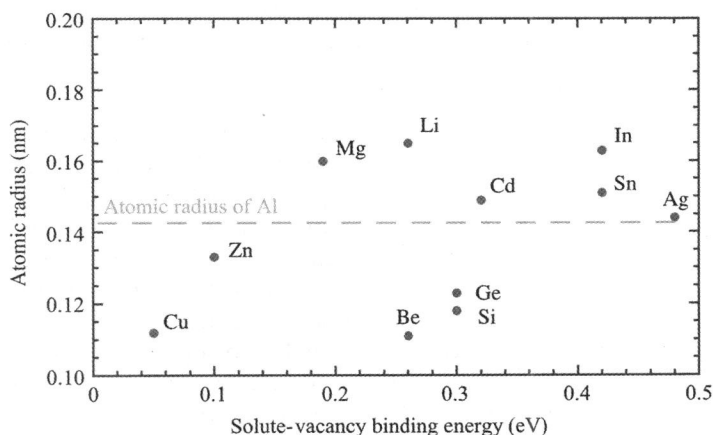

Figure 46   Atomic sizes of Al and various solutes as a function of solute–vacancy binding energy (Nie et al., 2002). (For color version of this figure, the reader is referred to the online version of this book.)

(a)

2324-T39 Type II ─────────
7055-T7751, 2524-T3 ─────── B-777 ER 2004
2324-T39, 7150-T651 ───── B-777, 1994
7075-T651, 7178-T651 ─── B-767, 1979
2024-T3 ──── B-707, 1954
2017-T4 ──── DC-3, 1935
Al-Cu casting ──── Junkers F 13, 1919
──── Wright Brothers, 1903

(b)

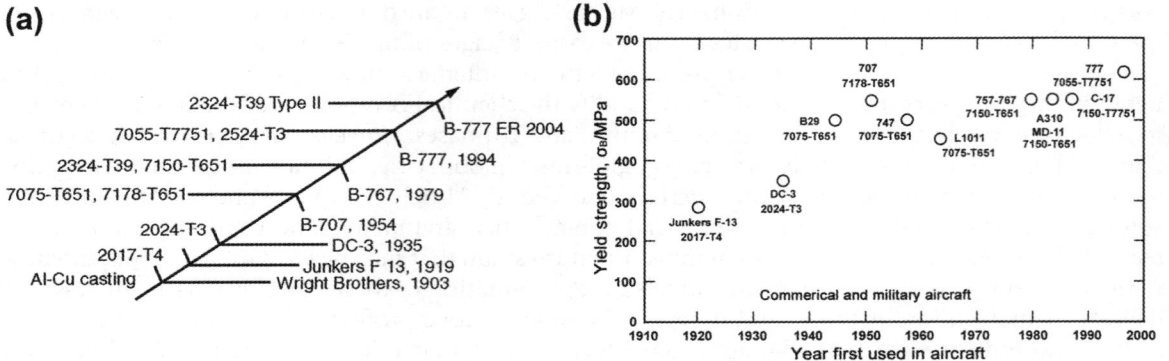

Figure 47 (a) Timeline highlighting the introduction of new aluminum alloys (Warren, 2004). (b) Historical record of wing upper skin alloys and tempers used for passenger aircraft (Lukasak and Hart, 1991).

vacancies (Ozbilen and Flower, 1989; Mondolfo, 1976; Raman et al., 1970). The presence of clusters or zones of Mg–Si(Ge) atoms and vacancies in $\langle 001 \rangle_\alpha$ directions can again provide a favorable environment for the nucleation of θ'. Given that pre-precipitate clusters of Mg and Ag atoms and vacancies form on $\{111\}_\alpha$ planes of the α-Al matrix solid solution (Reich et al., 1998; Suh and Park, 1995), the presence of such clusters will facilitate the nucleation of Ω and T_1, rather than θ'.

20.3.5 Aerospace Applications of 2xxx and 7xxx Series Alloys

The development of commercial aircraft has resulted in the development of a suite of high-performance aluminum alloys, **Figure 47**, including high-strength 7xxx series alloys (e.g. 7075, 7150, 7055, 7449, in chronological order of application), high damage tolerance 2xxx and 6xxx series alloys (e.g. 2024 and its variants, 2056, 6013 and 6056), and the workhorse high strength—good damage tolerance alloys for thick section applications (7010/7050, and more recently 7040/7140 and 7085) (Starke and Staley, 1996; Williams and Starke, 2003). The major structural design criteria for Boeing passenger aircrafts and for Airbus A380 wing, fuselage and empennage structures are shown in **Figure 48**.

20.3.5.1 Before Mid-1960s

The driver for aluminum alloy development has undergone significant evolution since the beginning of the aircraft industry. The first powered flight was made by Wright Brothers in 1903 in Kitty Hawk, North Carolina. The weight of that aircraft was less than 750 pounds ($\sim$340 kg) and the wingspan was about 40 feet ($\sim$12 m). The specific strength of aluminum alloys available to the Wright brothers was too low for structural applications, so they only used aluminum for the engine. For modern aircraft; the Airbus A380, its own weight is over 277 tons and can be more than 560 tons when fully loaded. The wing span of this plane is 80 m, the length is 73 m, and the height is 24 m. The giant leap in the aircraft development became possible with the advent of high-strength and high-toughness aluminum alloys. The wings and most of the fuselage of modern airplanes are made of precipitation-hardened aluminum alloys with highly specific properties. More than 61% of the structures of the airbus A380 is made of aluminum alloys.

The age hardening phenomenon in aluminum alloys was discovered accidentally by Alfred Wilm in Berlin in 1906, 3 years after the first successful flight of the Wright Brothers. This discovery led to the

(a)

Property requirements for jetliner and military transport applications

(b)

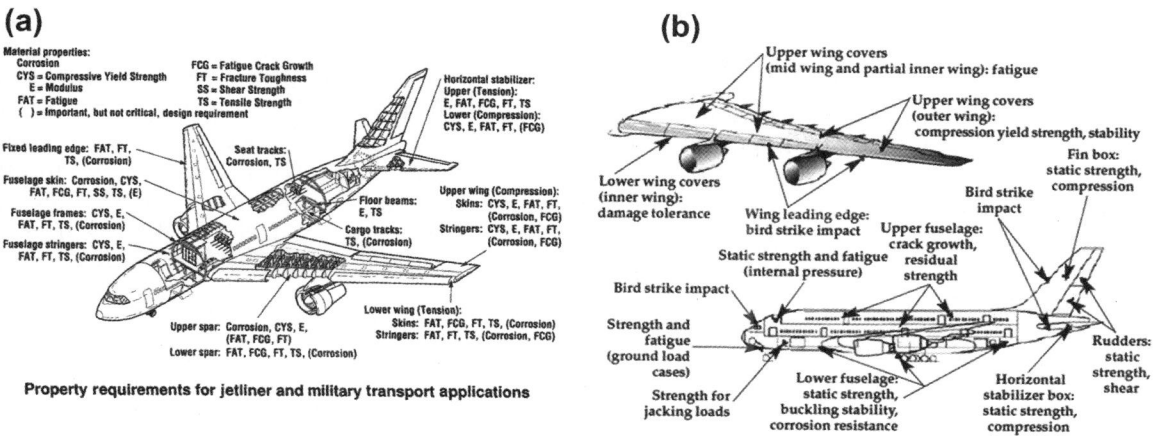

Figure 48 (a) Property requirements for Boeing passenger aircrafts (Staley and Lege, 1993). (b) Major structural design criteria and associated alloy characteristics of Airbus A380 wing, fuselage, and empennage (Lequeu et al., 2007).

development of an aluminum alloy, known as Duralumin (Al–3.5Cu–0.5Mg–0.5Mn; wt%), in the 1900s, that started the age of aluminum for aircraft. The Duralumin alloy was used for structural sections of Zeppelin airships and some early aircraft. The Duralumin was the forerunner of stronger Al–Cu–Mg alloy such as 2014 (Al–4.4Cu–0.5Mg–0.9Si–0.8Mn; wt%) and 2024 (Al–4.3Cu–1.5Mg–0.6Mn; wt%) that have tensile properties in the range 350–480 MPa, and which are still used today. Soon after the use of the Duralumin in dirigibles, the U.S. Navy contracted with Alcoa to further develop an aerospace sheet alloy, which is known as 2017-T4 nowadays. This alloy was used extensively on the first all aluminum airplane, the Junkers F13, that was manufactured in Germany in 1920. In 1921, Northrop built the Alpha 2, again with an extensive use of 2017-T4. Alcoa began developing higher strength alloys by increasing the alloy content and modifying the heat treatments (2024-T3 and 2014-T6) after it gained some insights into the mechanisms of precipitation hardening. The importance of corrosion was subsequently demonstrated by the development of alclad 2024-T3, the surface of which corrodes rather mildly and uniformly to protect the high-strength core from localized corrosion attack (Staley and Hunt, 2012).

During World War II, the military need for aircraft alloys having higher specific strength led to the development of several Al–Zn–Mg–Cu alloys, including the 7075-T6 alloy (Al–5.5Zn–2.5Mg–1.5-Cu–0.2Cr; wt%) that is well known nowadays. With specific strength as the main driver in that period, stronger alloys such as 7178-T6, with tensile strengths more than 600 MPa, were developed and 7178-T6 was used on the first commercially successful jetliner, the Boeing 707. However, stress–corrosion cracking problems began to appear when thicker products were used in larger planes. In an attempt to solve the problem, alloy 7079-T6 was introduced. This alloy had lower quench sensitivity and was developed particularly for large forgings—it had higher strength than 7075-T6 in thick sections. Although the 7079-T6 alloy performed well under accelerated SCC tests, the continuing problems with SCC in service and deficiencies in other properties led to being withdrawn from service.

A duplex aging treatment designated the T73 temper was then developed by Alcoa for alloy 7075 to solve the SCC problem. The T73 temper involved aging below (e.g. 110 °C) and then above (e.g. 160 °C) the GP zones solvus temperature—leading to uniform precipitation of the η' phase which reportedly formed at the sites of GP zones that were stabilized by the lower temperature age.

The distribution of η precipitates in grain boundaries is also changed and these modifications to the microstructure greatly reduce susceptibility to SCC. While tensile properties of 7075-T73 were approximately 15% lower than those of the T6 temper, the T73 temper did result in a significant increase in resistance to stress–corrosion cracking. For 7075 specimens tested in the short transverse direction, those in the T6 condition failed at stresses of only 50 MPa whereas those in the T73 temper were not cracked at stress levels of 300 MPa. The alloy 7075-T73 was used for fabricating larger aircraft components such as the die-forged integral of engine support and stabilizer spar. The DC10 was the first aircraft that used 7075-T73 forgings. Shortly afterward, the T76 temper was developed for achieving good resistance to exfoliation corrosion while regaining some of the strength loss, and it was used on the Lockheed aircraft L-1011.

20.3.5.2 Late 1960s to Late 1970s

A disadvantage of the T73 temper is that aging at the higher temperature lowers tensile properties by about 15%. The weight penalties resulted when replacing aircraft components aged to the T6 temper by the T73 temper. For this reason, much subsequent research was directed to achieve T6 strength levels combined with the T73 resistance to SCC. From late 1960s and early 1970s, the alloy development was influenced by the need for durability and damage tolerance. Premium strength 7175-T74 forgings were developed by Alcoa, which had higher resistance to SCC and matched the strength of 7075-T6 in thin sections. The need for an SCC-resistant alloy having higher strength in thick sections than 7075-T73 led the U.S. Navy and Air Force to fund Alcoa to make compositional modifications to develop alloy 7050 (Al–6.2Zn–2.3Cu–2.25Mg–0.1Zr; wt%). The higher level of Cu in 7050 helped to partially compensate for the loss of properties on overaging. The 7050 alloy had a desired combination of SCC resistance and high strength for thick structural sections in aircraft. Plates of alloy 7050-T74 were first used to retrofit 7079-T6 components in the A6 Intruder, and were chosen for the bulk of the aluminum structure in the F18. Shortly after the development of 7050 in the United States, alloy 7010-T74 was developed in Europe and, together with 7050-T74, was used for Airbus jetliners and military aircraft. In the following years, alloy 7475 was developed by Alcoa for higher fracture toughness. It was first used on the Panavia Tornado, and then used for the F16 (Staley and Hunt, 2012).

In the late 1970s, Alcoa and Boeing developed alloy 7150-T6, a higher purity version of 7050, for Boeing 757 and 767 jetliners that required the strength of 7178-T6 with improved toughness and fatigue resistance. Subsequently, Alcoa developed a new temper, T61, to improve the exfoliation corrosion resistance of 7150, and McDonnell–Douglas and Airbus are using 7150 for upper wing panels of their jetliners. For the lower wing structure of Boeing 757 and 767 jetliners, Alcoa and Boeing developed higher strength 2324-T39 plate and 2224-T3511 extrusions with sufficiently improved fracture toughness compared with 2024-T3.

20.3.5.3 During the 1980s

The Arab oil embargo in the 1980s led to projected high fuel cost and thus the development of nontraditional materials for aircraft. The perception at that time was that the next aircraft after the Boeing 757 and 767 would be built with polymer matrix composites unless radically different aluminum materials were developed. This perception pushed the established aluminum producers, as well as nonaluminum producers, to develop advanced aluminum alloys. A major effort was made over this period to develop the second generation of Al–Li alloys—Alcoa developed 2090; Pechiney developed 2091; and the Royal Aircraft Establishment developed 8090 and licensed the technology to Alcan. However, these Al–Li alloys have found only niche applications in aircraft, even though the alloy 8090 has found applications in space vehicles and helicopters.

Significant efforts were also made to develop nonconventional materials such as wrought powder metallurgy alloy X7093, and rapid solidified Al–Fe–V–Si alloys for elevated temperature applications, and laminated hybrids of aluminum sheet with aramid-fiber-reinforced (ARALL™) or glass-fiber-reinforced (GLARE™) composites. The structural laminates of ARALL™ and GLARE™ had high fatigue resistance and resistance to burn through in the event of a fire. They also offered the potential for adding extra functionality such as load monitoring and damage detection. The cargo door of the C-17 was fabricated from ARALL™. However, these materials were more expensive, typically 7–10 times that of monolithic aluminum sheet.

A major technical and commercial success in the 1980s was the development of the cost-effective T77 temper for alloy 7150. For the first time, the SCC resistance was improved without losing strength. The T77 temper was derived from the retrogression and re-aging (RRA) that involves three steps for 7075: (i) a T6 treatment, that is solution treatment at 465 °C, cold water quench, and age 24 h at 120 °C, (ii) a heat treatment for a short time (e.g. 5 min) in the temperature range 200–280 °C and cold water quench, and (iii) re-aging for 24 h at 120 °C. The RRA treatment resulted aluminum grains containing GP zones and finely dispersed phase η′, which was very similar to that obtained in a T6 temper, together with grain boundary regions characteristic of a T73 temper. Reduced susceptibility to SCC was attributed mainly to this latter change. Because of the short period (1–5 min) initially proposed for stage 2, an RRA treatment was difficult to apply to thick plates. Subsequently, the time and temperature conditions for this critical stage of the treatment were optimized (e.g. 1 h at 200 °C) and the RRA process was given the temper designation T77. The T77 temper was applied by Alcoa initially to alloy 7150, and later to the highly alloyed composition 7055 (Al–8Zn–2.05Mg–2.3Cu–0.16Zr–0.15Fe max-0.10Si max; wt%). The alloy 7055-T7751 offered a 25% advantage in tensile properties over the early alloy 7075-651 for similar levels of fracture toughness. When measured in the L–T direction, the 7055-T7751 plate had a typical yield strength of 635 MPa, a tensile strength of 650 MPa and an elongation to fracture of 11%.

Contrary to the earlier perception and predictions, the new Boeing 777 passenger aircraft came out with the use of many new aluminum alloys. Its upper wing panel was constructed with 7055-T77 skin and stringers along with 7150-T77 spar chords. Its low wing surface was built with 2324 skin and 2224 stringers. The fuselage skin was built with alclad 2524-T3 sheet. The body stiffeners were made of 7150-T77 extrusions. Seat tracks were made of 7150-T77. Many 7150-T77 forgings were also used. The weight saving associated with the use of alloy 7055 alone was estimated to be about 635 kg.

20.3.5.4 During the 1990s

The alloy development efforts in the 1990s recognized the strong need of airframers to lower their manufacturing costs while keeping inspection and maintenance costs down. The alloy development approach in this period was also characterized by the reintroduction of tailoring alloy properties for specific applications. The third generation of the Al–Li low-density alloys, including 2195, 2 × 96, 2 × 97, and 2098, were developed with high Cu contents, and were commercially used in the larger Super Light Weight fuel tank of the Space Shuttle and the F16. They replaced incumbent alloys 2x24 and 2219 in thin tanks and fuselage skins, bulkheads or spars. The third-generation Al–Li alloy 2195 was cost-effective because its high strength provided superior weight savings. Alloy 2097 also offered cost savings because it had high resistance to fatigue crack growth and therefore eliminated the need for regular replacement of 2124-T851 bulkheads because of fatigue cracks.

Al–Mg–Sc alloys with better corrosion resistance, lower density, and good weldability were also designed and evaluated. The Al–Mg–Sc sheet did not require cladding to provide high corrosion resistance and therefore offered opportunities for cost savings. Further cost savings became possible through design improvements using the improved weldability characteristics of Al–Mg–Sc. Alloy 7055-O sheet was also developed with higher resistance to stress corrosion cracking while being easier to manufacture. Age formable and superplastically formable alloys were evaluated for new generation large wing structure may require curvature.

20.3.5.5   After 2000

The airframe design of most recent passenger aircrafts such as Airbus A380, **Figure 48(b)**, required the use of more advanced high-strength and high damage-tolerant alloys. The alloy 2050-T84 was designed for this purpose, with a view to outperforming the property balance of 2xxx thin plate alloys and 7xxx thick plate alloys. It was the latest development by Alcan Aerospace in plate products. It had lower amounts of Li and reduced quench sensitivity that allowed it to be produced in thicker gauges than most other Al–Li alloys. This third-generation Al–Li alloy was approved for the lower wing structures of A380-800 and A380-800F. The evaluation results indicated 2050-T84 had better strength and toughness than 7050-T7451 for 5-inch thick plates, while reducing overall weight by 5%. Furthermore, 2050-T84 had improved thermal resistance and higher resistance to fatigue–crack growth.

20.4   Magnesium Alloys

Magnesium is the lightest of all commonly used structural metals, with a density of approximately two-thirds that of aluminum and a quarter that of steels. Magnesium is an abundant element, comprising 2.7% of the earth's crust, and available commercially with purity exceeding 99.8%. Magnesium has a relatively low melting temperature and high specific heat. Hence magnesium and its alloys may thus be readily cast to near-net shape by conventional casting methods. Because of such attractive features, magnesium alloys have received considerable research over the past decade for potentially wider and larger applications in the automotive, aerospace and 3C industries. Despite the considerable efforts made thus far, the adoption of magnesium alloys in engineering applications remains limited compared with that achieved for aluminum alloys. One important technical reason is that there are limited magnesium alloys for designers to select from for specific applications and, within these limited choices, the most cost-effective magnesium alloys have inadequate properties such as yield strength, creep resistance, formability and corrosion resistance. It has been the accumulated empirical experience, rather than basic understanding, that provides the tools for practical design and development of magnesium alloys with better mechanical and chemical properties.

Many of the magnesium casting and wrought alloys achieve their useful mechanical properties via age hardening, which involves solution treatment at a relatively high temperature within the α-Mg single-phase region, water quench to obtain a supersaturated solid solution of alloying elements in magnesium, and subsequent aging at a relatively low temperature to achieve a controlled decomposition of the supersaturated solid solution into a fine distribution of precipitates in the magnesium matrix. The decomposition of the supersaturated solid solution often involves the formation of a series of metastable or equilibrium precipitate phases that have quite different resistance to dislocation shearing. Therefore, the control of the precipitation is important if the maximum precipitation strengthening effect is to be achieved.

Attempts to improve the age hardening response of magnesium alloys inevitably requires an in-depth understanding of precipitation, precipitation hardening and microstructural factors that are most important in controlling the precipitation of strengthening phases and the strength of precipitation hardenable alloys. For precipitation-hardened magnesium alloys, their microstructures often contain a distribution of plate- or lath/rod-shaped precipitates of intermediate or equilibrium phases formed parallel or normal to the basal plane of the magnesium matrix phase. In the past century, the crystal structure, composition and OR of these precipitates have been characterized primarily using conventional TEM and electron diffraction. As a consequence of the resolution and limitation of these techniques, the characteristic features of some precipitate phases and the precipitation sequence in many alloys were not clearly established. In the first decade of this century, with the assistance of high-resolution TEM, particularly atomic resolution HAADF-STEM, and 3DAP, some puzzles on the structure and composition of precipitate phases in some existing magnesium alloys have been solved. These modern characterization facilities also greatly facilitate the identification of precipitates in magnesium alloys that were developed in recent years. Such knowledge on the crystallography of precipitate phases provides the basis for the understanding of the formation and strengthening mechanisms of the precipitate phases and, more importantly, for rational alloy design in practice.

It is the purpose of this part of the chapter to provide a comprehensive review of the literature on precipitation and hardening in most, if not all, age hardenable magnesium alloys. Since a few books on magnesium alloys (Raynor, 1959; Roberts, 1960; Emley, 1966; Rokhlin, 2003) and some review articles on precipitation in magnesium alloys (Lorimer, 1987; Polmear, 1994, 1995, 2006) and particle hardening (Kelly and Nicholson, 1963; Ardell, 1985; Reppich, 1993; Nembach, 1997) are already in the literature, the emphasis of this part will be focused on (1) the structure, morphology and orientation of precipitates, precipitation sequence and hardening response in each of the major alloy systems, (2) effects of precipitate shapes on strengthening, and (3) rational design of microstructures for larger age hardening response and therefore higher strength. Some unsolved issues that require further research are also highlighted and discussed.

20.4.1 Precipitation and Age Hardening

20.4.1.1 Mg–Al-Based Alloys

20.4.1.1.1 Precipitation

The magnesium-rich side of the Mg–Al binary phase diagram includes equilibrium solid phases α-Mg and β-$Mg_{17}Al_{12}$ and a eutectic temperature of 437 °C. The β phase has a body-centered cubic structure (space group $I\bar{4}3m$) with the lattice parameter $a \sim 1.06$ nm (Villars and Calvert, 1985). The equilibrium solid solubility of Al in α-Mg is 11.8 at.% (12.9 wt%) at the eutectic temperature, and decreases to about 3.3 at.% at 200 °C. The equilibrium volume fraction of precipitates achievable in the Mg–Al alloys aged at 200 °C can reach a substantially large value of 11.4%. This thermodynamic feature provides a unique opportunity for generating a large volume fraction of precipitates by using conventional aging treatments, that is solution treatment at about 420 °C, followed by water quench and subsequent aging at a temperature in the range 100–300 °C. Unfortunately, during the isothermal aging treatment in the temperature range 100–300 °C, the precipitation process appears to involve solely the formation of the equilibrium β phase, **Table 1**. While the β precipitates are resistant to dislocation shearing (Clark, 1968; Hutchinson et al., 2005), their distribution is relatively coarse, presumably because of the relatively high diffusion rate of Al atoms in the solid matrix of magnesium and a possibly high concentration of vacancies in the α-Mg matrix. Consequently, the age hardening response of Mg–Al alloys (Clark, 1968; Lagowski, 1971; Bettles et al., 1997; Celotto, 2000; Nie, 2002) is not as appreciable as expected, **Figure 49(a)**.

Earlier studies (Hutchinson et al., 2005; Porter and Edington, 1977) revealed that the precipitation of the equilibrium β phase occurs both discontinuously and continuously. The discontinuous precipitation is also known as cellular precipitation, and in this reaction the supersaturated solid

Table 1   Part of the whole precipitation sequence in individual magnesium alloy systems

System					
Mg-Al	SSSS → β (Mg₁₇Al₁₂) bcc (I4̄3m, a = 1.06 nm); (0001)α plate/lath				
Mg-Al -Ca	SSSS → ordered G.P. zones hcp a = 0.556 nm monolayer (0001)α disc	→ C15 (Al₂Ca) fcc, Fd3̄m a = 0.802 nm (0001)α plate			
Mg-Zn	SSSS → G.P. zones	→ β'₁ (Mg₄Zn₇) monoclinic, B/2m a = 2.60 nm b = 1.43 nm c = 0.52 nm γ = 102.5° [0001]α rod	→ β'₂ (MgZn₂) hcp, P6₃/mmc a = 0.52 nm c = 0.86 nm (0001)α plate	→ β (MgZn) monoclinic a = 1.61 nm b = 2.58 nm c = 0.88 nm β = 112.4°	
Mg-Zn-Al§	SSSS → G.P. zones	→ i icosahedral or approximant diamond shape	→ Φ and/or T Φ: orthorhombic, Pbcn T: bcc, Im3̄ a = 0.90 nm a = 1.40 nm b = 1.70 nm c = 1.97 nm (0001)α lath		
Mg-Sn(-Zn)	SSSS → β (Mg₂Sn) fcc (Fm3̄m, a = 0.68 nm), (0001)α and {11̄22}α plate/lath, [0001]α rod and polygon				
Mg-Ca-Zn	SSSS → ordered G.P. zones hcp a = 0.556 nm monolayer (0001)α disc	→ η' (MgCaZn) hcp, P6₃/mmc a = 0.56 nm c = 1.04 nm (0001)α plate	→ η (Mg₂(Ca,Zn)) hcp, P6₃/mmc a = 0.62 nm c = 1.01 nm (0001)α plate		
Mg-Nd	SSSS → ordered G.P. zones zig-zag shape d = 0.37 nm *	→ β'' (Mg₃Nd) hcp, DO₁₉ a = 0.64 nm c = 0.52 nm hexagonal prism	→ β' (Mg₇Nd) orthorhombic a = 0.74 nm b = 1.14 nm c = 0.52 nm lenticular shape	→ β₁ (Mg₃Nd) fcc, Fm3̄m a = 0.74 nm {101̄0}α plate	→ β (Mg₁₂Nd) tetragonal, I4/mmm a = 1.03 nm c = 0.59 nm [0001]α rod
Mg-Nd-Zn	SSSS → ordered G.P. zones hcp a = 0.556 nm monolayer (0001)α disc	→ γ'' (Mg₅(Nd,Zn)) hcp, P6̄2m a = 0.55 nm c = 0.52 nm (0001)α plate	→ γ (possibly Mg₃(Nd,Zn)) fcc, possibly Fm3̄m a = 0.70 nm plate on irrational plane		βe (Mg₄₁Nd₅) tetragonal, I4/m a = 1.47 nm c = 1.04 nm
Mg-Gd(-Y)	SSSS → ordered G.P. zones zig-zag shape d = 0.37 nm *	→ β'' (Mg₅Gd) hcp, DO₁₉ a = 0.64 nm c = 0.52 nm hexagonal prism	→ β' (Mg₇Gd) orthorhombic a = 0.65 nm b = 2.27 nm c = 0.52 nm lenticular shape	→ β₁ (Mg₃Gd) fcc, Fm3̄m a = 0.73 nm {101̄0}α plate	→ β (Mg₅Gd) fcc, Fm3̄m a = 2.23 nm {101̄0}α plate
Mg-Y	SSSS → β' (Mg₇Y) orthorhombic a = 0.65 nm b = 2.27 nm c = 0.52 nm globular shape	→ β (Mg₂₄Y₅) bcc, I4̄3m a = 1.13 nm {101̄0}α or {314̄0}α plate			
Mg-Y-Nd	SSSS → ordered G.P. zones zig-zag shape d = 0.37 nm *	→ β'' (Mg₃Nd) hcp, DO₁₉ a = 0.64 nm c = 0.52 nm hexagonal prism	→ β' (Mg₁₂YNd) orthorhombic a = 0.64 nm b = 2.24 nm c = 0.52 nm globular shape	→ β₁ (Mg₃(Nd,Y)) fcc, Fm3̄m a = 0.74 nm {101̄0}α plate	→ β (Mg₁₄Nd₂Y) fcc, Fm3̄m a = 2.20 nm {101̄0}α plate
Mg-Gd-Zn#	SSSS → γ'' (Mg₇₀Gd₁₅Zn₁₅) ordered hcp, P6̄2m a = 0.56 nm c = 0.44 nm (0001)α plate	→ γ' (MgGdZn) hcp, P3̄m1 a = 0.32 nm b = 0.78 nm (0001)α plate	→ γ (Mg₁₂GdZn) ordered hcp, 14H a = 1.11 nm c = 3.65 nm (0001)α plate		
Mg-Y-Zn	SSSS → I₂ stacking fault (0001)α plane	→ γ' (MgYZn) hcp, P3̄m1 a = 0.32 nm c = 0.78 nm (0001)α plate	→ γ (Mg₁₂YZn) ordered hcp, 14H a = 1.11 nm c = 3.65 nm (0001)α plate		
Mg-Y-Ag-Zn	SSSS → G.P. zones monolayer (0001)α disc	→ γ'' ordered hcp, P6̄2m a = 0.56 nm c = 0.45 nm (0001)α plate	→ γ' hcp, P3̄m1 a = 0.32 nm c = 0.78 nm (0001)α plate	→ γ+δ γ: ordered hcp, 14H δ: fcc, Fd3̄m a = 1.11 nm a = 1.59 nm c = 3.65 nm (0001)α plate	

§ precipitation process is not well studied; * d is separation distance of columns of RE atoms; # low Gd:Zn weight ratio and low Gd content

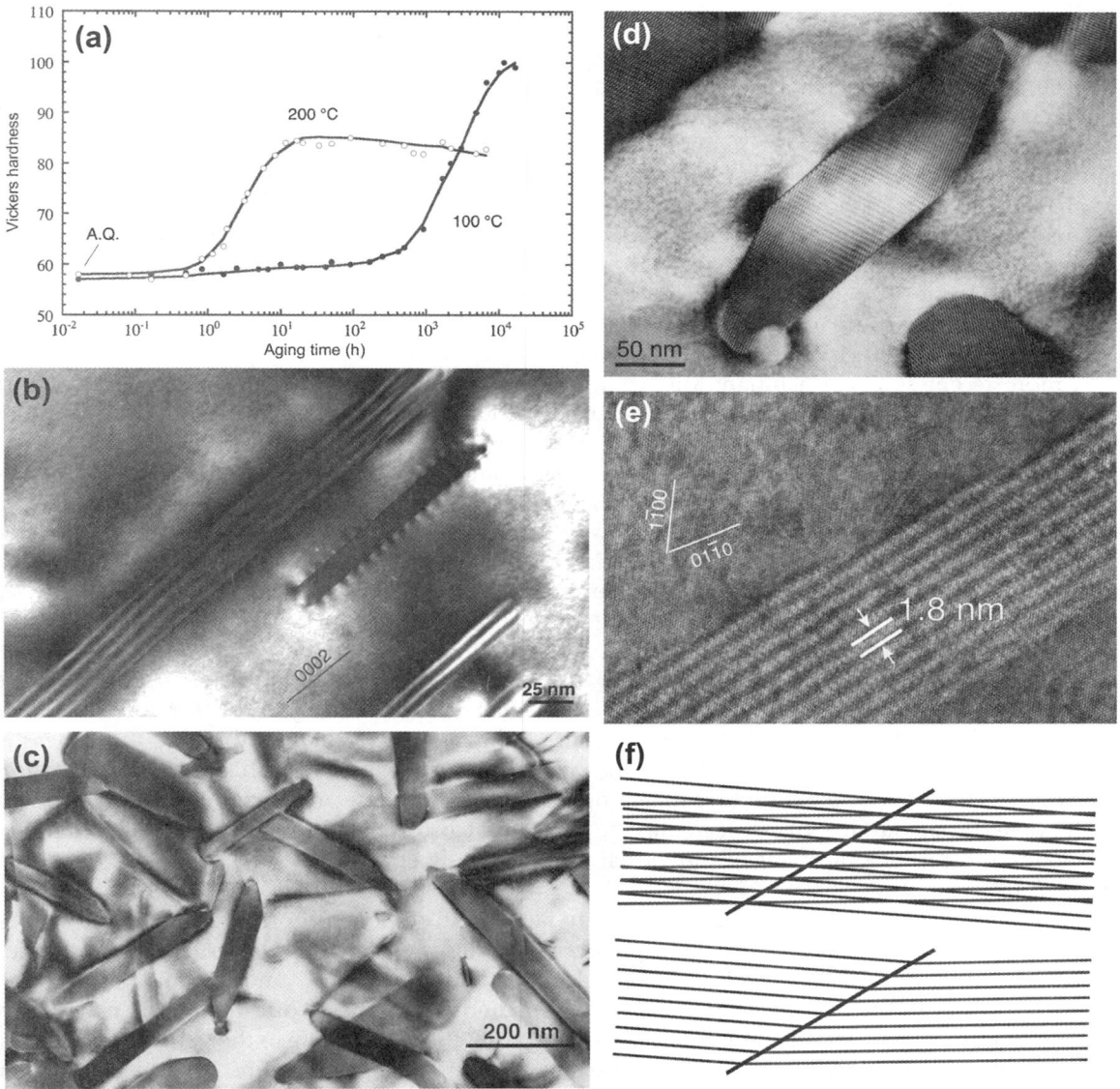

Figure 49   (a) Isothermal aging curves of magnesium alloy AZ91 at 100 and 200 °C. (b and c) TEM images showing distribution and morphology of β precipitates in samples aged for 8 h at 200 °C. Electron beam is parallel to $[2\bar{1}\bar{1}0]_\alpha$ in (b) and $[0001]_\alpha$ in (c). (d) $[0001]_\alpha$ TEM image showing the parallelogram shape of β precipitates. (e) HRTEM image showing the major planar interface of β precipitate such as that shown in (d); the planar interface is indicated by moiré planes. (f) Schematic diagram showing the relationship between two sets of lattice planes and their resultant moiré plane, which corresponds to that shown in (e), and demonstrating the coherent matching of the two sets of lattice planes within the moiré plane (Celotto, 2000; Nie et al., 2001; Nie, 2002; Nie and Muddle, 2002; Nie, 2006). (For color version of this figure, the reader is referred to the online version of this book.)

solution α' phase decomposes into the β phase and an α phase that is structurally identical to the α' phase but has a less saturated concentration of aluminum. The discontinuous precipitation initiates in grain boundaries and expands toward the grain center in a cellular form (Williams and Butler, 1981). The cell comprises a lamellar structure of β and α phases, and the cell interface separating α and α' is a high angle boundary. The continuous precipitation occurs inside grains. The continuous and discontinuous precipitations occur simultaneously and compete with each other during isothermal aging of Mg–Al alloys. Duly et al. (1994a, 1994b, 1995a) reported that the continuous precipitation is favored at both high and low aging temperatures, while that discontinuous precipitation dominates the microstructure at intermediate temperatures. They further proposed that the disappearance of discontinuous precipitation at high aging temperatures is because of the volume diffusion of solute that prevents the nucleation and growth of the cellular colonies, and that the absence of discontinuous precipitation at low aging temperatures is due to a lower driving force, which is caused by the occurrence of continuous precipitation in the early stage of aging treatment.

A more recent study of a binary Mg-9wt%Al alloy and alloy AZ91 (Braszczynska-Malik, 2009) indicates that, in the Mg-9wt%Al alloy samples aged at 150 °C or cooled from the solution temperature to room temperature, only discontinuous precipitates are observed, while that only continuous precipitates form when the binary and the AZ91 alloys are aged at 350 °C. It is also found that both discontinuous and continuous precipitates form when the alloys are aged at intermediate temperatures of 200 or 250 °C. It was proposed (Braszczynska-Malik, 2009) that whether discontinuous precipitation occurs or not also depends the concentration of vacancies in addition to the aging temperature.

For most continuous precipitates and discontinuous precipitates of the β phase in the lamellar structure, they were initially reported to adopt the exact Burgers OR, that is $(011)_\beta//(0001)_\alpha$, $[1\bar{1}1]_\beta//[2\bar{1}\bar{1}0]_\alpha$ (Crawley and Miliken, 1974). But subsequent TEM studies (Duly et al., 1995b; Nie et al., 2001) indicate that the OR is actually near the Burgers. The β precipitates in this OR have a plate shape, with their broad surfaces parallel to $(0001)_\alpha$, **Figure 49(b) and (c)**. While it is often to describe the β plates in this OR as incoherent (Polmear, 1995, 2006; Padfield, 2004), there is now ample experimental evidence to demonstrate that the equilibrium β phase is in fact *not incoherent*. Apart from the apparent lattice matching between the β phase and surrounding matrix phase in the plate broad surface, or habit plane, the lattice matching is also found in interfaces defining the major and minor side facets of individual β plates (Duly et al., 1995b; Nie et al., 2001; Zhang et al., 2004).

Despite the irrational orientation of these side facets with respect to both precipitate and matrix lattices, the major and minor side facets, **Figure 49(d)**, are invariably parallel to the moiré fringes defined by the intersection of $(1\bar{1}00)_\alpha$ and $(0\bar{3}3)_\beta$, and of $(10\bar{1}0)_\alpha$ and $(4\bar{1}1)_\beta$, respectively. **Figure 49(e)** shows the major side facet of a thin β plate that is embedded in the matrix phase. This major interface is parallel to the moiré fringes resulting from the overlapping of the $(1\bar{1}00)_\alpha$ and $(0\bar{3}3)_\beta$ planes, **Figure 49(f)**, and it contains some ledges whose unit height is defined by the interplanar spacing of the moiré fringes. The migration of this interface in its normal direction seems to involve the formation and lateral gliding of moiré ledges within the interface plane (Nie, 2004, 2006). These observations suggest the existence of commensurate matching of $(1\bar{1}00)_\alpha$ and $(0\bar{3}3)_\beta$ planes (Nie et al., 2001; Nie, 2006; Nie and Muddle, 2002) in the major facet interface, and of $(10\bar{1}0)_\alpha$ and $(4\bar{1}1)_\beta$ in the minor facet interface. This commensurate matching, together with the fact that $(0\bar{3}3)_\beta$ and $(4\bar{1}1)_\beta$ are the closest packed planes in the β lattice, and $\{1\bar{1}00\}_\alpha$ is the near-closest packed plane in the magnesium lattice, suggests that the major and minor side facets of each β plate have relatively low interfacial energies.

Apart from the near Burgers OR, two other ORs have also been reported for the β phase (Gjönnes and Östmoe, 1970; Crawley and Lagowski, 1974; Crawley and Miliken, 1974; Nie et al., 2001; Celotto, 2000), namely $(1\bar{1}0)_\beta//(1\bar{1}00)_\alpha$, $[111]_\beta//[0001]_\alpha$ and $(1\bar{1}0)_\beta \sim//(1\bar{1}00)_\alpha$, $[11\bar{5}]_\beta \sim//[0001]_\alpha$. For the former OR, the β precipitates have a rod shape with their long axes parallel to $[0001]_\alpha$. The $[0001]_\alpha$ rods have a hexagonal cross-section, with the bounding facets parallel to $\{1\bar{1}00\}_\alpha//(\bar{3}30)_\beta$. For the latter OR, the β precipitates develop a rod shape with their long axes inclined with respect to the $[0001]_\alpha$ direction. While these rods are more effective than the $(0001)_\alpha$ plates in impeding dislocation gliding on the basal plane, only a small fraction of them exists in the microstructure. It is currently unclear how to promote the rod-shaped precipitates at the expense of the $(0001)_\alpha$ plates. Such an effort inevitably requires an in-depth understanding of the transformation strains associated with each of the ORs and the corresponding activation energy barrier to nucleation.

20.4.1.1.2 *Effects of Cold Work and Alloying Additions*

Cold work after solution treatment and before aging, microalloying and macroalloying additions to Mg–Al alloys do not seem to produce a significant enhancement in age hardening response. Clark (1968) reported that cold work after solution treatment and before aging of an Mg–9wt%Al alloy could increase the age hardening response. The enhanced aging kinetics and the maximum hardness are attributable to a high number density of dislocations and twins in the cold worked samples before aging and heterogeneous nucleation of β precipitates on such lattice defects. It was noted that the density of precipitates formed at the twin interface and within twins is higher than that in the untwinned matrix regions. The preferential precipitation within twins may be attributed to the presence of stacking faults (Partridge, 1967) and a higher density of dislocations within the twins (Partridge and Roberts, 1964).

The effects of microalloying additions on precipitation and age hardening of magnesium alloy AZ91 were studied by Bettles et al. (1997). They added, separately, 0.1 at.% Li, B, Ca, Ti, Sr, Ag, Mo, Ba, Pb or 0.05 at.% Si to AZ91. Even though the additions of such microalloying elements could influence the kinetics of precipitation and hardening, they did not lead to any appreciable increase in the maximum hardness values of the alloys, **Figure 50**. Therefore, it was speculated that the individual additions of

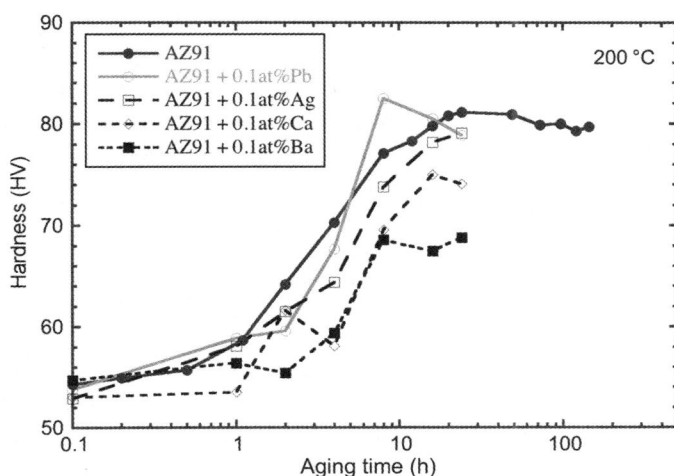

Figure 50 Effects of microalloying additions on the age hardening response of AZ91 at 200 °C. (For color version of this figure, the reader is referred to the online version of this book.)

these elements did not increase the nucleation rate of the β phase or result in the formation of any new precipitate phases that can further enhance the age hardening response.

The macroalloying addition of Ca to Mg–Al alloys has been shown to improve the creep resistance (Ninomiya et al., 1995; Luo et al., 2002). It was commonly reported in the early studies that the added Ca atoms react with Al atoms to form Al_2Ca that has a C15 structure (space group $Fd\bar{3}m$, $a = 0.802$ nm) (Ninomiya et al., 1995), or $(Mg,Al)_2Ca$ of a C14 structure (Luo et al., 2002), or a mixture of the two Laves phases during casting. In a subsequent study (Suzuki et al., 2004), it was reported that the intermetallic phase formed in the Mg–Al–Ca alloys has in fact a C36 structure ($a = 0.584$ nm, $c = 1.897$ nm). This C36 Laves phase has now been confirmed in many recent studies to exist as an equilibrium phase in the Mg–Al–Ca system, **Figure 51**. For a long time, it has been generally accepted that the Mg–Al–Ca alloys are not age hardenable. However, a paper published in 2005 (Suzuki et al., 2005) reported the precipitation of C15 when a high-pressure die-cast Mg-4.5Al-3.0Ca-0.14Sr-0.25Mn (wt%) alloy, designated AXJ530, is aged at 300 °C. The C15 precipitates form as plates on $(0001)_\alpha$ planes, with the following OR with the matrix phase: $(111)_{C15}//$ $(0001)_\alpha$ and $[10\bar{1}]_{C15}//[10\bar{1}0]_\alpha$. In a subsequent study (Suzuki et al., 2008), it was reported that the same alloy exhibits an age hardening phenomenon when it is aged at 175–250 °C, and that this age hardening response is due to the precipitation of the C15 plates. In a very recent study (Homma et al., 2011) of Mg-2Al-2Ca and Mg-2Al-2Ca-0.3Mn (wt%) alloys, produced by permanent mold casting

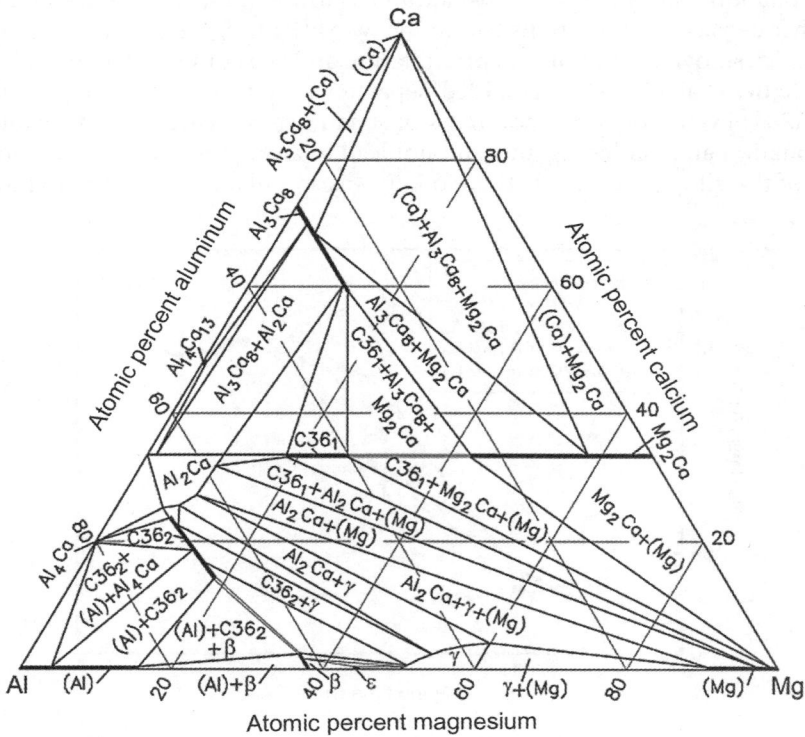

Figure 51   Isothermal section of the Mg–Al–Ca ternary phase diagram at 400 °C (Raghavan, 2011).

and creep tested at 175 and 200 °C under 50 MPa, the C15 plates were observed to form in the as-cast microstructures of the two alloys, and ordered GP zones form on the basal plane of the matrix phase in the crept samples. These ordered GP zones were inferred to be similar to those observed in Mg–RE–Zn and Mg–Ca–Al alloys, and their number density in the Mn-containing alloy is higher than that in the Mn-free alloy.

20.4.1.2 Mg–Zn-Based Alloys

20.4.1.2.1 Phase Equilibria and Precipitation

The magnesium-rich side of the Mg–Zn binary phase diagram is more complex than that of the Mg–Al binary phase diagram. The eutectic temperature is 340 °C, and the maximum solid solubility of Zn in magnesium is 6.2 wt% (or 2.4 at.%) at the eutectic temperature. The eutectic reaction is such that the liquid phase solidifies into a mixture of α-Mg and Mg_7Zn_3 phases. The Mg_7Zn_3 phase has an ortho-rhombic structure (space group Immm, $a = 1.4083$ nm, $b = 1.4486$ nm, and $c = 1.4025$ nm (Higashi et al., 1981)), and it is thermodynamically stable only at temperatures above 325 °C. At temperatures of and below 325 °C, the Mg_7Zn_3 phase decomposes, via an eutectoid reaction, into α-Mg and MgZn, the intermetallic phase that is in equilibrium with α-Mg at temperatures below 325 °C. The structure of this intermetallic phase was not unambiguously established before 2006, even though the Mg–Zn-based alloys have received considerable attention in the past 10 years. Based on the X-ray diffraction results, Khan (1989) proposed a rhombohedral structure for the MgZn phase and expressed the lattice parameters in a hexagonal version, with $a = 2.569$ nm and $c = 1.8104$ nm. However, this rhombohe-dral structure has so far not been confirmed by the others. A recent study made by TEM and electron microdiffraction (Gao and Nie, 2007a) suggests that the MgZn phase has a base-centered monoclinic structure ($a = 1.610$ nm, $b = 2.579$ nm, $c = 0.880$ nm, $\beta = 112.4°$).

During heat treatments of an Mg-8wt%Zn alloy at temperatures below 325 °C, it was observed (Gao and Nie, 2007a) that primary intermetallic particles of the Mg_7Zn_3 phase that had formed during the solidification process decompose into a divorced lamellar structure of α-Mg and Mg_4Zn_7. The Mg_4Zn_7 phase is metastable, and it is gradually replaced by the equilibrium phase MgZn after prolonged heat treatment. The Mg_4Zn_7 phase has a base-centered monoclinic structure and the following OR: $[001]_{Mg_4Zn_7} \sim //[0001]_\alpha$ and $(630)_{Mg_4Zn_7} \sim //(01\bar{1}0)_\alpha$. The structure of this metastable Mg_4Zn_7 phase is identical to that of the equilibrium Mg_4Zn_7 phase (space group B/2m, $a = 2.596$ nm, $b = 1.428$ nm, $c = 0.524$ nm, $\gamma = 102.5°$) in the Mg–Zn system (Yarmolyuk et al., 1975; Yang and Kuo, 1987), rather than the triclinic structure ($a = 1.724$ nm, $b = 1.445$ nm, $c = 0.520$ nm, $\alpha = 96°$, $\beta = 89°$, $\gamma = 138°$) assumed for the Mg_2Zn_3 phase by Gallot and Graf (1966) and adopted by the others (Polmear, 2006, 1994; Zhang et al., 2004). Despite the fact that it is still unclear whether the monoclinic phase Mg_4Zn_7 is actually identical to the Mg_2Zn_3 phase, it is commonly accepted (Clark et al., 1988) that these two phases are the same, as there exists only one intermetallic phase at compositions close to Mg-(60–63.6) at.% Zn in the Mg–Zn binary phase diagram.

The equilibrium solid solubility of Zn in magnesium decreases substantially with temperature and controlled decomposition of the supersaturated solid solution of Zn in magnesium can produce an age hardening effect (Clark et al., 1988; Clark, 1965; Mima and Tanaka, 1971a, 1971b; Wei et al., 1995). The aging curves of two binary Mg–Zn alloys are provided in **Figure 52(a)**. Depending on the alloy composition and aging temperatures, it is commonly accepted in the literature (Polmear, 2006; Padfield, 2004) that the decomposition of the supersaturated solid solution matrix phase reportedly involves the formation of GP zones, β_1' ($MgZn_2$), β_2' ($MgZn_2$) and β (Mg_2Zn_3). The formation of GP zones, which are described as coherent discs formed on $(0001)_\alpha$, has not been supported by direct experimental evidence so far. While GP zones, together with Zn clusters and GP 1 zones, were reported

(a)

(b)

(c)

(d)

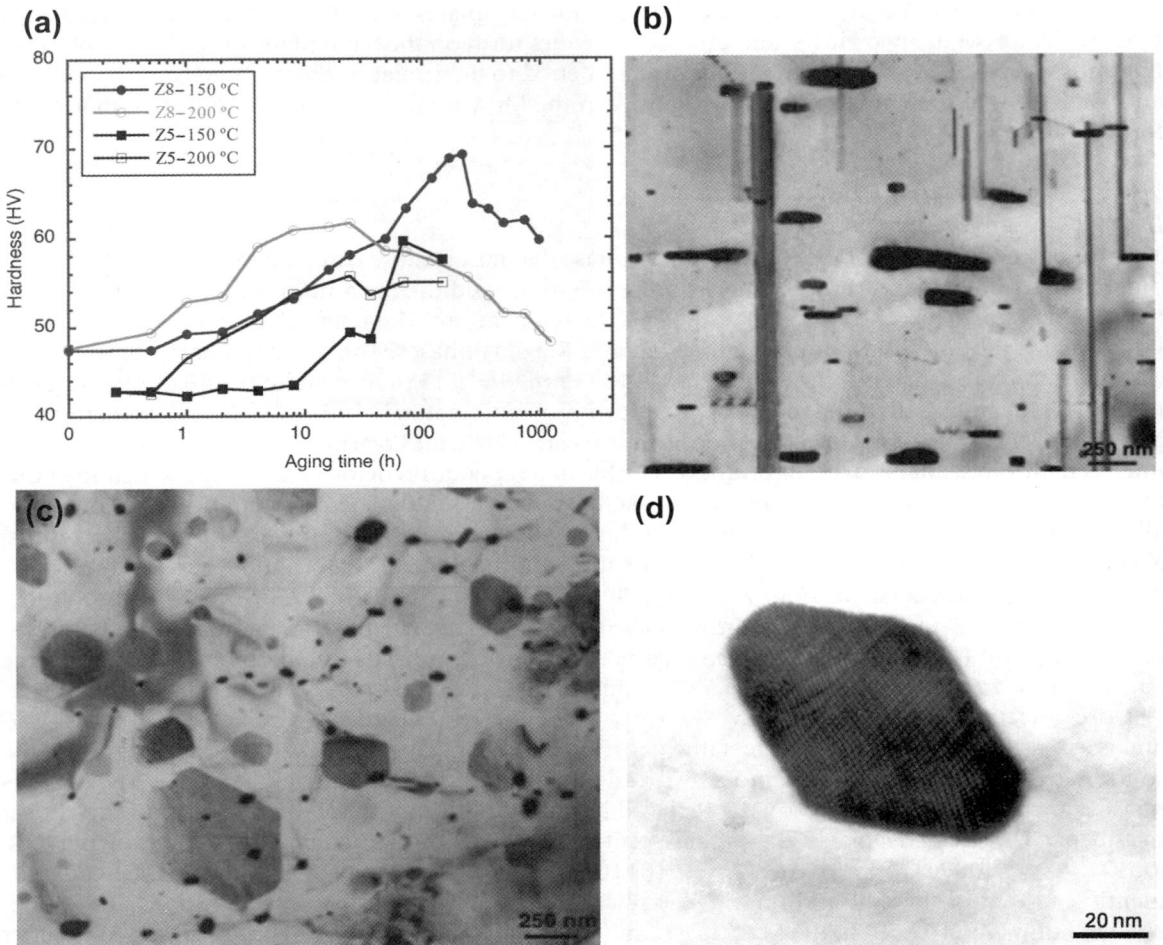

Figure 52   (a) Age hardening response of Mg–8wt%Zn and Mg–5wt%Zn alloys at 150 and 200 °C, and (b and c) TEM images showing $[0001]_\alpha$ rods and $(0001)_\alpha$ plates in Mg–8Zn alloy aged for 1000 h at 200 °C. (d) The cross-section of a $[0001]_\alpha$ rod of Mg_4Zn_7 and the substructure within this precipitate. Electron beam is parallel to $[2\bar{1}\bar{1}0]_\alpha$ in (b) and $[0001]_\alpha$ in (c and d) (Gao and Nie, 2007b). (For color version of this figure, the reader is referred to the online version of this book.)

in recent studies (Buha, 2008a; Buha and Ohkubo, 2008) to form in an Mg–2.8at.%Zn alloy aged at temperatures 22, 70, 98 and 160 °C, no compelling experimental evidence was provided to support the existence of such Zn clusters and GP zones. The hardness values reported in this work also seem too high compared with those reported by the others. Further careful characterization in the future using 3DAP and atomic resolution HAADF-STEM, and an in-depth analysis of these characterization results, are necessary before the notion of GP zones and solute clusters is formally accepted in the precipitation sequence of Mg–Zn alloys.

The metastable phase β_1', also described as MgZn' (Clark et al., 1988; Rokhlin and Oreshkina, 1988), forms as $[0001]_\alpha$ rods, while the metastable phase β_2' forms as $(0001)_\alpha$ plates. Based on observations from X-ray diffraction and selected area electron diffraction (SAED) patterns, both β_1' and β_2' phases were suggested to have a hexagonal structure ($a = 0.520$ nm, $c = 0.857$ nm) (Gallot and Graf, 1965; Gallot et al., 1964; Sturkey and Clark, 1959–60; Chun and Byrne, 1969; Wei et al., 1995) that is identical to that of MgZn$_2$ (space group P6$_3$/mmc, $a = 0.5221$ nm, $c = 0.8567$ nm) (Komura and Tokunaga, 1980). Another hexagonal structure ($a_{\beta_1'} = 0.556$ nm, $c_{\beta_1'} = 0.521$ nm) was also reported for β_1' (Rokhlin and Oreshkina, 1988), but it has not been confirmed so far. The ORs for these two precipitate phases are that $[0001]_{\beta_1'}//[11\bar{2}0]_\alpha$ and $(11\bar{2}0)_{\beta_1'}//(0001)_\alpha$ between β_1' and α-Mg, and $(0001)_{\beta_2'}//(0001)_\alpha$ and $[11\bar{2}0]_{\beta_2'}//[10\bar{1}0]_\alpha$ between β_2' and α-Mg.

A recent electron microscopy study of precipitate phases in an Mg–8wt%Zn alloy aged at 200 °C (Gao and Nie, 2007b) indicates that the precipitate structures and ORs are more complicated than those reported in early studies. **Figure 52(b–d)** shows precipitates typical of Mg–8wt%Zn samples aged for 1000 h at 200 °C. Most precipitates are β_1' rods/laths, while a fraction of β_2' plates is also visible, in the microstructure. Electron microdiffraction patterns obtained from the β_1' rods indicate that, contrary to the traditional view, they have a base-centered monoclinic structure ($a = 2.596$ nm, $b = 1.428$ nm, $c = 0.524$ nm, $\gamma = 102.5°$) that is similar to that of Mg$_4$Zn$_7$, and that the OR is such that $[001]_{\beta_1'} \sim //[0001]_\alpha$ and $(630)_{\beta_1'} \sim //(01\bar{1}0)_\alpha$. This OR is identical to that observed between Mg$_4$Zn$_7$ and α-Mg phases in the eutectoid reaction within primary particles of the Mg$_7$Zn$_3$ phase. The proposed crystal structure and the OR are subsequently confirmed in separate studies on precipitates in Mg–Zn–Y alloys (Singh and Tsai, 2007; Singh et al., 2010; Rosalie et al., 2010). In these latest electron microscopy studies, it is revealed that the Mg$_4$Zn$_7$ phase has a complex substructure and planar defects elongated along the long axis of β_1' rods, **Figure 52(d)**. In a very recent study (Rosalie et al., 2011), it was reported that the β_1' rods contained a mixture of Mg$_4$Zn$_7$ and MgZn$_2$ phases that have the following OR: $[010]_{Mg_4Zn_7}//[0001]_{MgZn_2}$ and $(20\bar{1})_{Mg_4Zn_7}//(0\bar{1}10)_{MgZn_2}$. Some domains of a face-centered cubic structure (C15) were also proposed to exist inside the β_1' rods. It is unclear at the present whether there is any in situ transformation from one structure to the other within the β_1' rods. It seems necessary to employ atomic resolution HAADF-STEM to resolve the complex substructure of the β_1' rods. A small fraction of the β_1' phase was also found (Gao and Nie, 2007b) to adopt a rarely reported blocky shape and a different OR with the α-Mg phase, for example $[001]_{\beta_1'} \sim //[10\bar{1}0]_\alpha$ and $(250)_{\beta_1'} \sim //(0001)_\alpha$.

All recent studies have confirmed that the β_2' phase has the MgZn$_2$ structure ($a = 0.523$ nm, $c = 0.858$ nm) and the OR reported in early studies. Similar to β_1' rods, the β_2' plates also have a complex substructure of domains and planar defects (Gao and Nie, 2007b). Most particles of the β_2' phase adopt a plate morphology, but a small fraction of the β_2' phase also exists as laths with their long axis parallel to $[0001]_\alpha$. These β_2' laths can be distinguished from the β_2' rods from their morphology because their cross-section has a larger aspect ratio and appears as a near parallelogram shape with the broad surface parallel to $\{10\bar{1}0\}_\alpha$. The OR between these β_2' laths and α-Mg is $[11\bar{2}0]_{\beta_2'}//[0001]_\alpha$ and $(0001)_{\beta_2'}//(11\bar{2}0)_\alpha$. This OR is clearly different from that associated β_2' plates, but identical to that reported for β_1' rods in early studies (Gallot and Graf, 1965; Gallot et al., 1964; Sturkey and Clark, 1959–60). A few β_2' laths with a rarely reported OR, $[11\bar{2}0]_{\beta_2'}//[0001]_\alpha$ and $(1\bar{1}06)_{\beta_2'}//(\bar{1}010)_\alpha$, were also found (Gao and Nie, 2007b). The broad surface of these laths is $\sim 6°$ from the nearest $\{10\bar{1}0\}_\alpha$ plane, instead of being parallel to $\{10\bar{1}0\}_\alpha$.

Surprisingly, the structure and composition of the equilibrium β phase have long been accepted as those of the Mg$_2$Zn$_3$ phase, that is the triclinic structure ($a = 1.724$ nm, $b = 1.445$ nm, $c = 0.520$ nm, $\alpha = 96°$, $\beta = 89°$, $\gamma = 138°$) and the Mg$_2$Zn$_3$ composition, even though the alloy composition lies in the (α-Mg + MgZn) two-phase field. Based on the Mg–Zn binary phase diagram and a more recent electron microscopy study, it seems appropriate to suggest that the equilibrium β phase has an MgZn composition and a base-centered monoclinic structure ($a = 1.610$ nm, $b = 2.579$ nm, $c = 0.880$ nm, $\beta = 112.4°$).

Recent studies indicate clearly that the precipitation sequence in Mg–Zn alloys containing 4–9 wt% Zn and aged isothermally at 120–260 °C is different from that accepted traditionally (Polmear, 2006). The probable precipitation sequence in the Mg–Zn alloys is provided in **Table 1**. While some detailed information has been gained in recent years, with the help of advanced characterization facilities, on the structure and morphology of precipitates in Mg–Zn-based alloys, there still is a lack of reports in the literature that elucidate the details of the full precipitation process and provide some insightful understanding of the nucleation and growth behaviors of the precipitate phases in this group of alloys. For example, why do the structures of the metastable β_1' and β_2' phases resemble closely to those of the Mg_4Zn_7 and $MgZn_2$ phases that exist as equilibrium phases in the Mg–Zn binary phase diagram? If there exists an in situ structural transformation from Mg_4Zn_7 to $MgZn_2$ within the β_1' rods, then why do some single $MgZn_2$ particles also exist in the microstructure?

20.4.1.2.2 Effects of Alloying Additions

Since the age hardening response of binary Mg–Zn alloys is limited, efforts have been made in the past to improve the age hardening response of Mg–Zn alloys via macro- and microalloying additions. Examples of macroalloying additions include Cu (Mose, 1983; Lorimer, 1987), Co (Geng et al., 2011) and Ba (Buha, 2008d), **Figure 53(a)**. In these alloys, the solution treatment temperature can be increased from 320–335 °C, typically used for Mg–Zn binary alloys, to 430–440 °C for Mg–Zn–X (X = Cu, Ba and Co) alloys without causing any local melting of the casting alloys,

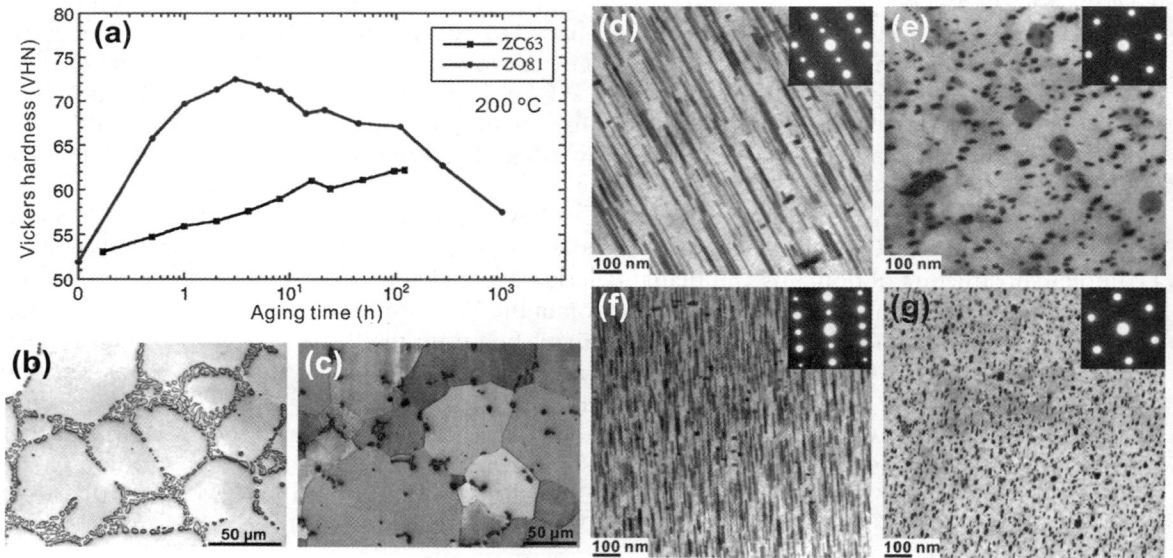

Figure 53   (a) Effects of macroalloying additions of Cu and Co on age hardening response of Mg–Zn alloys at 200 °C. (b and c) Reflected light micrographs showing retained intermetallic particles in solution treated Mg–6wt%Zn–3wt%Cu (ZC63) and Mg–8wt%Zn–1wt%Co (ZO81) alloys, respectively. (d–g) TEM images showing distribution of precipitates in (d and e) Mg–8Zn alloy and (f and g) Mg–8Zn–1Co alloy (Geng et al., 2011). (For color version of this figure, the reader is referred to the online version of this book.)

Figure 53(b) and (c). This might be the result of a substantially increased eutectic temperature in the Mg–Zn–X alloys. The use of a much higher temperature for the solution treatment allows more Zn atoms to be dissolved into the magnesium matrix after the solution treatment and possibly more vacancies to be achieved after the water quench. The higher concentrations of Zn atoms and vacancies can result in an enhanced age hardening response during the isothermal aging treatment. Comparison of the aging curves of Mg–8Zn and Mg–8Zn–1Co (wt%) alloys, **Figures 52(a) and 53(a)**, indicates that the maximum hardness value achievable at 200 °C is increased by about 18% and that the aging time needed to achieve the maximum hardness is reduced from ∼24 to ∼3 h. A further comparison of microstructures of peak-aged samples of these two alloys, **Figure 53(d–g)**, reveals that the increased maximum hardness is associated with a denser distribution of precipitates in the Mg–8Zn–1Co (wt%) alloy.

In contrast to macroalloying additions, microalloying additions to Mg–Zn alloys generally cannot raise the eutectic temperature and thus they do not permit higher temperatures to be used for the solution treatment (Buha, 2008b, 2008c; Mendis et al., 2009; Oh-ishi et al., 2009; Hono et al., 2010). However, the additions of appropriate alloying elements can equally result in a substantial enhancement in age hardening response, as demonstrated in **Figure 54(a)**. The additions of 0.1–0.35 wt% Ca to Mg–(4–6)Zn alloys (Bettles et al., 2004; Eom et al., 2003; Homma et al., 2010a), the addition of Ag or the combined addition of Ag and Ca to an Mg–6wt%Zn alloy (Park et al., 2003; Ben-Hamu et al., 2006; Mendis et al., 2007, 2010, 2011a, 2009; Oh-ishi et al., 2009), can result in a significant enhancement in age hardening response and tensile yield strength. The improved age hardening response is associated with a refined distribution of rod-shaped precipitates, **Figure 54(b)** and **Figure 55**. The analysis of 3DAP data suggests the co-segregation of Ca and Zn atoms in the pre-precipitation stage. The Ag atoms do not associate with the Ca/Zn clusters and are uniformly distributed in the magnesium matrix phase before they segregate to precipitates that form in the peak-aged condition. Since it is the combined addition of Ca and Ag that leads the largest increment in the maximum hardness, it is unclear at the

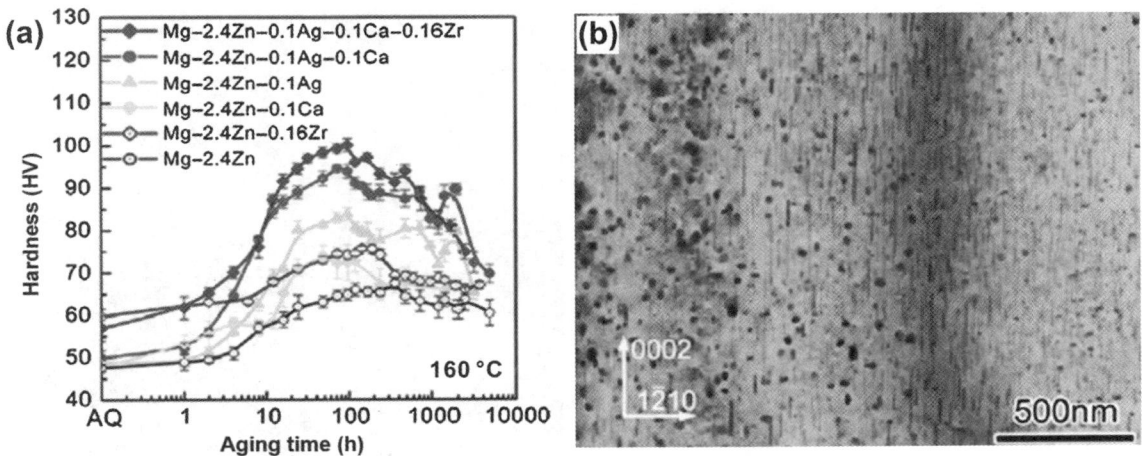

Figure 54 (a) Age hardening response of microalloyed Mg–2.4at.%Zn during isothermal aging at 160 °C. (b) TEM image showing distribution of precipitates in extruded and peak-aged samples of Mg–2.4Zn–0.1Ag–0.1Ca–0.16Zr (at.%) alloy (Mendis et al., 2009; Oh-ishi et al., 2009; Hono et al., 2010). (For color version of this figure, the reader is referred to the online version of this book.)

present how the Ca and Ag atoms facilitate the nucleation rate of the precipitates. A more thorough characterization of the distribution of the Ca and Ag atoms, and of the structure and ORs of the precipitates, and considerations of solute–vacancy binding energies in magnesium (Shin and Wolverton, 2010a, 2010b), are all required if the precise role of Ca and Ag in the nucleation is to be revealed.

Some other studies indicate that macroadditions of RE elements (Wei et al., 1995; Yang et al., 2008) or microalloying additions of Sn and In (Seremetis, 1999) to Mg–Zn alloys have little effects on age hardening response. A commercial magnesium alloy developed from the Mg–Zn–RE system is ZE41, Mg–4.2wt%Zn–1.3wt%RE–0.7wt%Zr (where RE represents rare-earth misch metal), and this alloy is often fabricated in the T5, instead of T6, condition for applications in helicopter transmission housings.

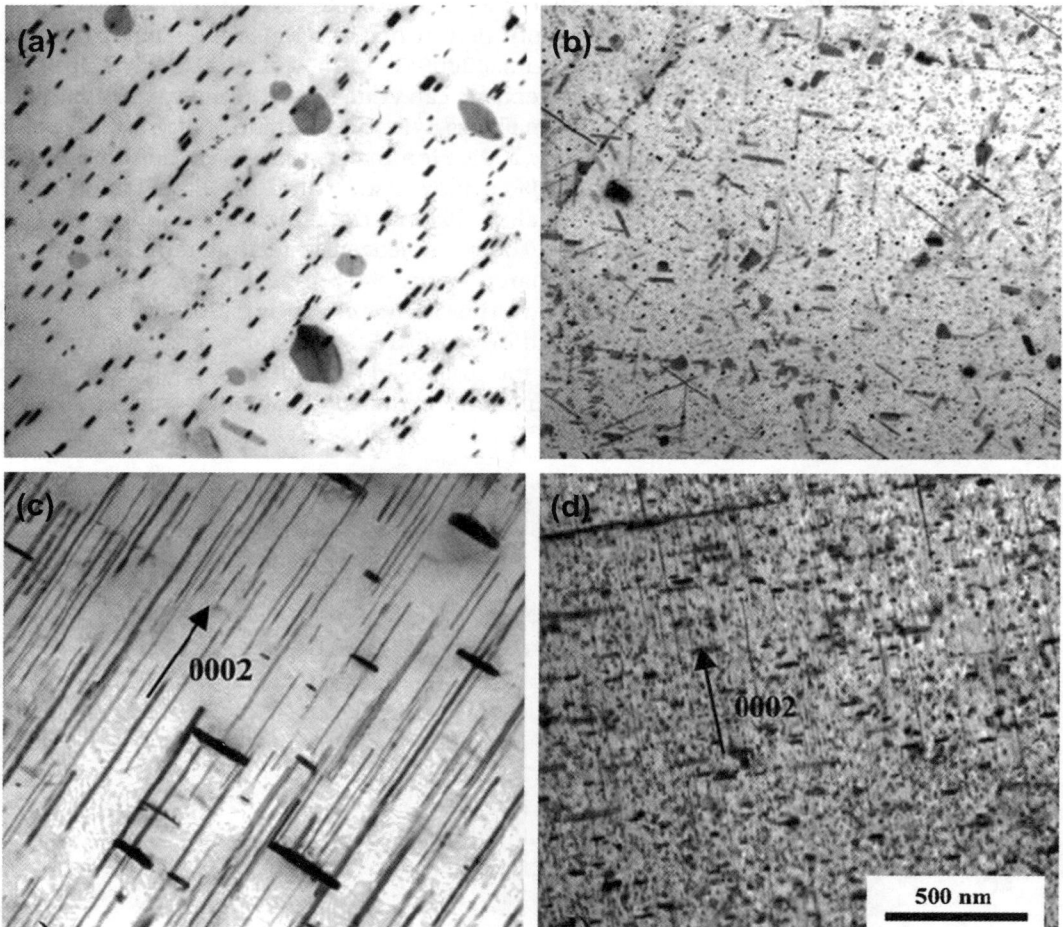

Figure 55 TEM images showing distribution of precipitates in (a and c) Mg–4wt%Zn, and (b and d) Mg–4wt%Zn–0.35wt% Ca alloys. Electron beam is parallel to $[0001]_\alpha$ in (a and b) and $[2\overline{1}\overline{1}0]_\alpha$ in (c and d). The samples are solution treated at 345 °C for 2 h, ramped to 530 °C in 2 h, and then at 530 °C for 12 h, water quenched and aged at 177 °C for 28.9 h (Bettles et al., 2004).

The most commonly studied alloying addition in recent years seems to be Y. The additions of Y to Mg–Zn alloys lead to the formation of relatively large particles of a quasicrystalline phase (Luo et al., 1993; Kim et al., 2003). The formation of such quasicrystalline particles does not contribute much to the alloy strength, and therefore the Mg–Zn–Y alloys are generally extruded to achieve finer magnesium grains for strengthening purpose (Singh et al., 2003; Somekawa et al., 2007; Xu et al., 2009).

20.4.1.3 Mg–Zn–Al-Based Alloys

Mg–Zn–Al-based alloys, with the Zn:Al weight ratio in the range 1:1–3:1, have received some interest in the past 15 years for developing casting alloys for elevated temperature applications. While the Mg–Zn–Al ternary phase diagram is relatively well established compared with other Mg-based ternary phase diagrams (Liang et al., 1997; Petrov et al., 2006; Ren et al., 2009; Raghavan, 2010), the identities of the equilibrium intermetallic phases in the Mg–Zn–Al alloys are still controversial. Based on X-ray diffraction observations, the equilibrium intermetallic phase in the Mg–Zn–Al alloys has been determined (Anyanwu et al., 2000; Zhang et al., 1998, 2000) to be the T phase which has an atomic composition of $Mg_{32}(Al,Zn)_{49}$ and a body-centered cubic structure (space group $Im\bar{3}$, $a \sim 1.4$ nm (Bergman et al., 1957)). In contrast, the equilibrium intermetallic phase in the Mg-8wt%Zn-(4–8)wt% Al alloys has been found (Mendis, 2000) to the ϕ phase. For the Mg–8wt%Zn–(4–8)wt%Al alloys, the latest version of the Mg–Zn–Al isothermal section at 320 °C, **Figure 56**, indicates that the equilibrium

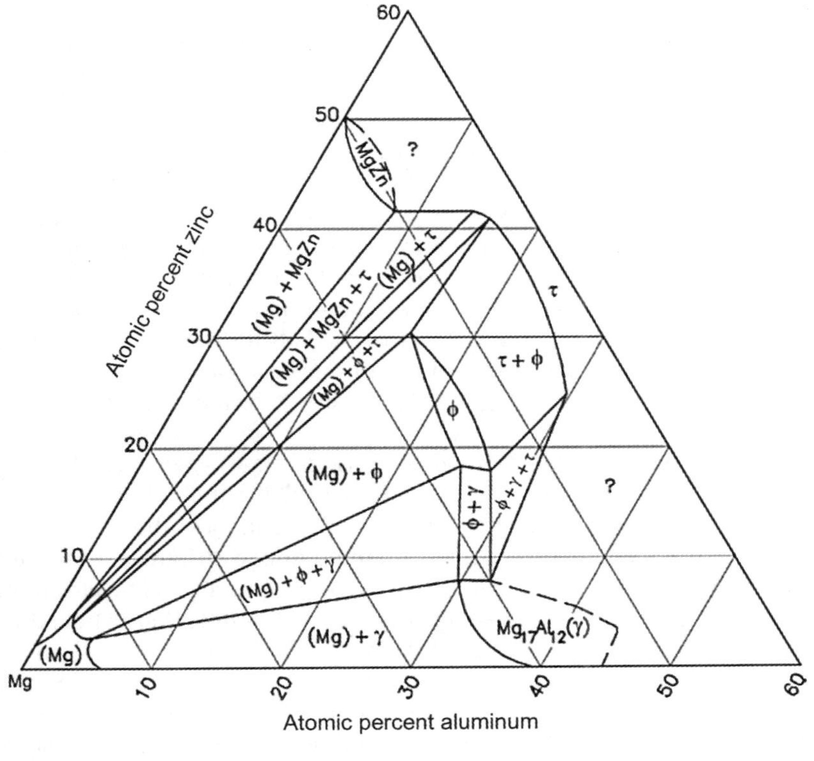

Figure 56 Isothermal section of the Mg–Zn–Al ternary phase diagram at 320 °C (Ren et al., 2009; Raghavan, 2010).

intermetallic phase is ϕ, instead of T. The ϕ phase was originally reported to have a composition of $Mg_5Al_2Zn_2$ (Clark and Rhines, 1959) and a primitive orthorhombic structure ($a = 0.8979$ nm, $b = 1.6988$ nm, and $c = 1.9340$ nm) (Donnadieu et al., 1997). A subsequent study using TEM and convergent-beam electron diffraction (Bourgeois et al., 2001a) confirms the primitive orthorhombic unit cell and lattice parameters proposed in the early studies, and further indicates that the ϕ phase has a space group of Pbcm and a composition of $Mg_{21}(Zn,Al)_{17}$.

For Mg–Zn–Al alloys with compositions lying in the α-Mg $+ \phi$ two-phase field and produced by high-pressure die casting or permanent mold casting, recent studies using TEM and convergent-beam electron diffraction patterns (Bourgeois et al., 2001b; Vogel et al., 2001) indicate that primary intermetallic particles in the as-cast condition have a quasicrystalline structure (point group of $m\overline{35}$, quasi-lattice parameter ~ 0.515 nm) and a composition of approximately $Mg_{55}Al_{19}Zn_{26}$. The quasi-lattice parameter is very close to those of icosahedral phases formed in rapidly solidified $Mg_{32}Al_{17}Zn_{32}$ (Rajasekharan et al., 1986) and $Mg_{32}(Al,Zn,Cu)_{49}$ (Mukhopadhyay et al., 1986) intermetallic alloys, but is smaller than the that (0.528 nm) reported for the icosahedral $Mg_{38.5}Al_{52}Zn_{9.5}$ phase in Al–Mg–Zn alloys (Cassada et al., 1986). Despite the fact that the equilibrium T phase is often regarded as the crystalline approximant of the icosahedral phase (Henley and Elser, 1986), the metastable quasicrystalline phase does not transform to the T phase after prolonged heating at elevated temperatures such as 325 °C. Instead, the quasicrystalline particles are gradually replaced by the ϕ phase, without any intermediate phases formed between them. Given that the composition of the Mg–8wt% Zn–4wt%Al alloy is in the α-Mg $+ \phi$ two-phase field, the replacement of the quasicrystalline phase by the equilibrium ϕ phase is not surprising at all. The icosahedral phase has a composition closer to that of the ϕ phase than to the T phase, and the ϕ unit cell has icosahedral clusters (Donnadieu et al., 1997; Bourgeois et al., 2001a). The transformation from a metastable icosahedral phase to a non-approximant crystalline phase has also been observed in an $Mg_{65}Zn_{25}Y_{10}$ intermetallic alloy (Abe and Tsai, 1999).

The aging curves of Mg–8Zn, Mg–8Zn–4Al and Mg–8Zn–8Al (wt%) alloys, which are solution treated for 4 h at 325 °C for Mg–8Zn and Mg–8Zn–4Al and 350 °C for Mg–8Zn–8Al, water quenched and aged at 200 and 150 °C, are provided in **Figure 57(a) and (b)**. The age hardening response is significantly enhanced by ternary addition of 4–8 wt% Al to the Mg–8Zn alloy. The maximum hardness values achieved in the Mg–8Zn–8Al alloy are considerably higher than those obtained in Mg–6wt% Zn–3wt%Cu, Mg–8wt%Zn–1wt%Co, and Mg–8wt%Zn–1.5wt%RE (Wei et al., 1995) alloys. The quaternary addition of 0.5 wt% Ca to Mg–8wt%Zn–4wt%Al alloy does not lead to any enhancement in age hardening response, **Figure 57(c)**, and an increase in the Ca content from 0.5 wt% to 1.0 wt% causes a reduced age hardening response, **Figure 57(c) and (d)**, even though the Ca addition retards the overaging of the alloys.

The effects of pre-aging on the age hardening response of an Mg–6Zn–3Al–1Mn (wt%) alloy were studied by Oh-ishi et al. (2008). Their alloy was aged initially at 70 °C for 48 h and followed by subsequent aging at 150 °C. It was found that the pre-aging enhances the age hardening response at 150 °C, and that the double-aged microstructure contains a finer distribution of precipitates. The sheet of this alloy, produced by twin-roll casting and hot rolling, exhibits a tensile yield strength of 319 MPa, after it is pre-aged at 70 °C for 24 h and then aged at 150 °C for 24 h (Park et al., 2007).

In a separate study of an Mg–2.4at.%Zn–2at.%Al (Mg–6.2wt%Zn–2.1wt%Al) alloy, it was reported (Mendis et al., 2010) that the combined addition of 0.1 at.% Ag and 0.1 at.% Ca to this alloy could lead to a significant enhancement of the aging response at 160 °C and that the maximum hardness achievable was raised by about 27%, from ~ 75 VHN in the Mg–2.4Zn–2Al alloy to ~ 95 VHN in the Mg–2.4Zn–2Al–0.1Ag–0.1Ca alloy. The maximum hardness value does not change when the

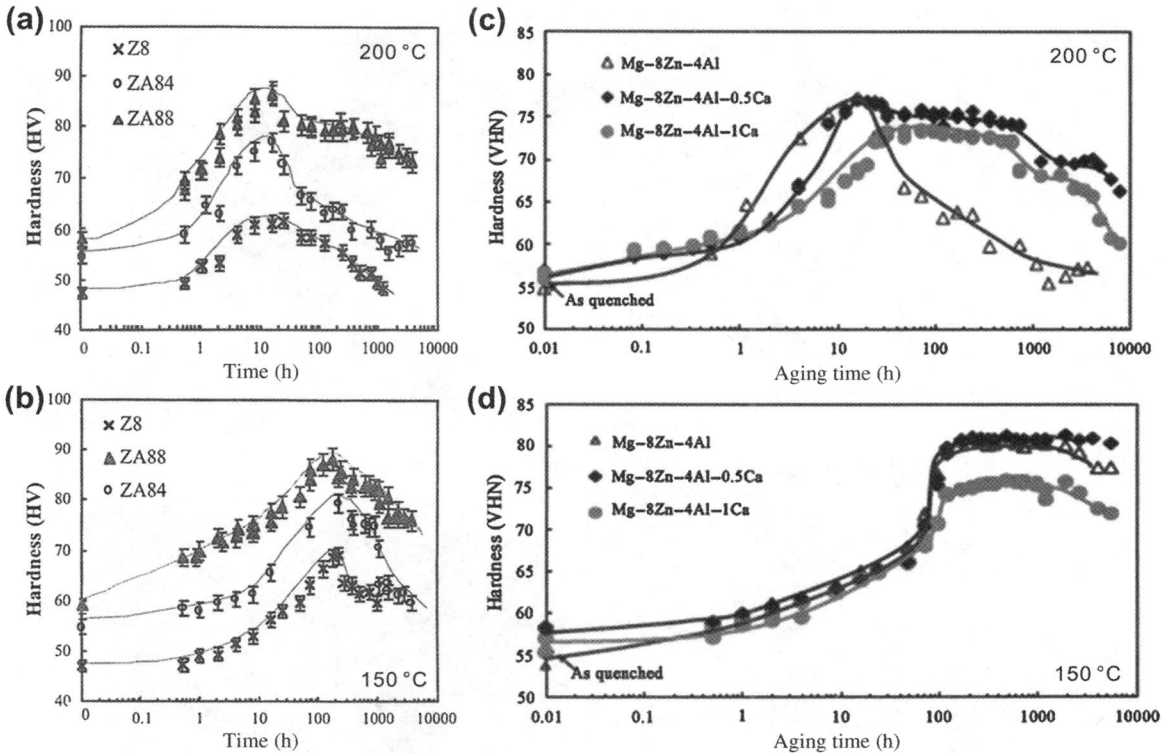

Figure 57 Age hardening response of Mg–8Zn, Mg–8Zn–4Al and Mg–8Zn–8Al (wt%) alloys during isothermal aging at (a) 200 °C and (b) 150 °C. Effects of Ca additions on age hardening response of Mg–8Al–4Zn (wt%) alloy at (c) 200 °C and (d) 150 °C (Mendis et al., 2001; Mendis, 2005). (For color version of this figure, the reader is referred to the online version of this book.)

Al content in the alloy is increased or decreased by 1 at.%, and remains unchanged even when the Al content is reduced to zero. These observations suggest that the age hardening response of the Mg–Zn–Al–Ag–Ca alloys is similar to that of the Mg–Zn–Ag–Ca alloys shown in **Figure 54(a)**.

While the Mg–Zn–Al alloys exhibit a substantial age hardening response during isothermal aging in the temperature range 100–200 °C, the solid-state precipitates formed in these alloys have not been characterized in detail. **Figure 58** shows microstructure typical of an Mg–8wt%Zn–8wt%Al alloy, homogenized and solution treated for 122 h at 325 °C, water quenched and then aged for 120 h at 200 °C. The microstructure contains predominantly a dispersion of rhombic precipitates. These precipitates appear to distribute heterogeneously throughout the magnesium matrix phase, distributing along lines that are approximately parallel to $\langle 10\bar{1}0\rangle_\alpha$ directions when the microstructure is viewed in the $[0001]_\alpha$ direction. Electron microdiffraction patterns obtained from such precipitates are indicative of an icosahedral structure with a quasi-lattice parameter of approximately 0.52 nm, instead of the T phase that has been reported for most solid-state precipitates formed in a high-pressure die cast Mg–8wt%Zn–5wt%Al alloy (Vogel et al., 2003). Currently, there is a lack of convergent-beam electron diffraction patterns to unambiguously establish whether these precipitates have a perfect icosahedral

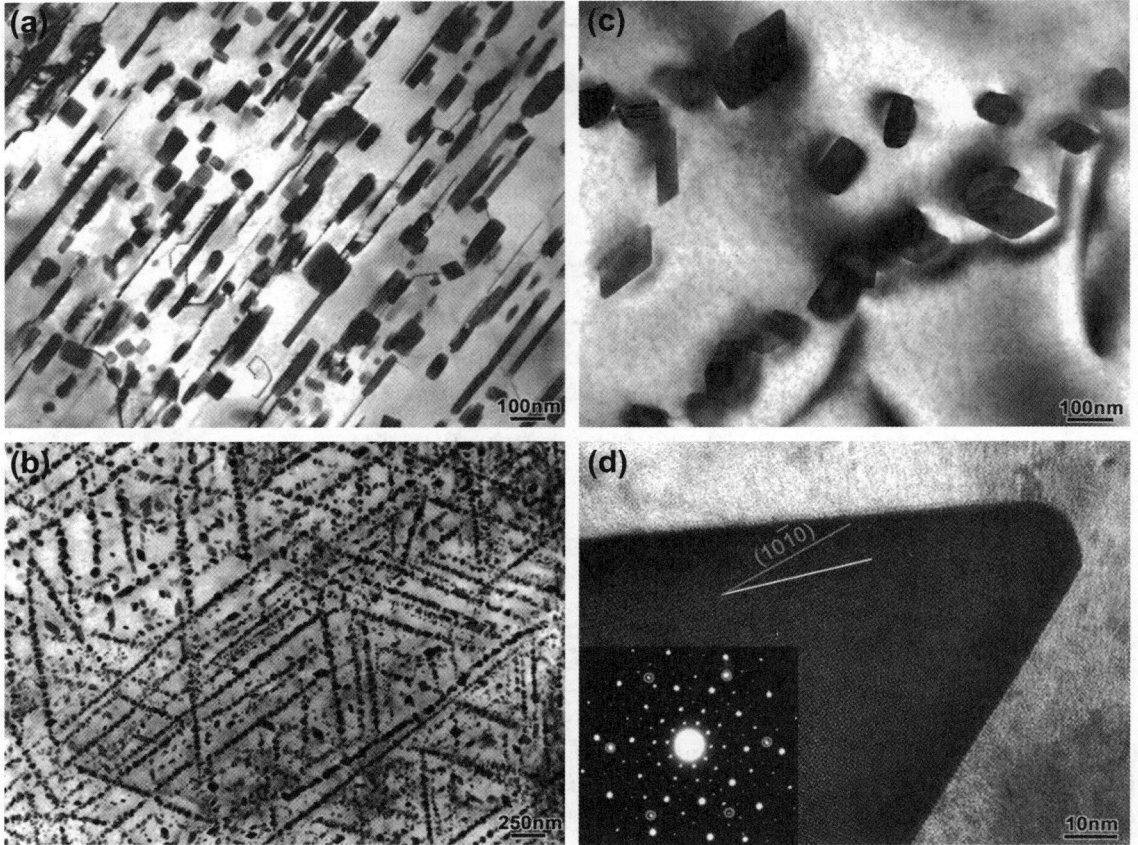

Figure 58 TEM images showing distribution and morphology of precipitates in an Mg–8wt%Zn–8wt%Al alloy aged at 200 °C for 120 h. Electron beam is parallel to $[2\bar{1}\bar{1}0]_\alpha$ in (a) and $[0001]_\alpha$ in (b–d).

symmetry or are a crystalline approximant with a very large unit cell parameter, but atomic resolution TEM does not reveal any periodic crystal structure or nanoscale twins existing within such particles.

The OR between the rhombic precipitates and the matrix phase is $(8/13, 5/8, \bar{3}/\bar{5})_i//(\bar{2}110)_\alpha$ and $[5\text{-fold}]_i//[0001]_\alpha$. The facets of the rhombic precipitate are not parallel, or close to parallel, to any of the close-packed planes of the two phases, **Figure 58(d)**. However, it is found (Nie et al., unpublished research) that the two rhombic interfaces are exactly parallel to the moiré planes defined by the intersection of $(3/5, 8/13, \bar{5}/\bar{8})_i$ and $(1\bar{1}00)_\alpha$ planes, and the $(\overline{10}/\bar{6}, 0/0, 0/0)_i$ and $(01\bar{1}0)_\alpha$ planes, respectively. This observation indicates that the $(3/5, 8/13, \bar{5}/\bar{8})_i$ and $(1\bar{1}00)_\alpha$ planes are fully coherent within one of the precipitate–matrix interfaces, while that the $(\overline{10}/\bar{6}, 0/0, 0/0)_i$ and $(01\bar{1}0)_\alpha$ planes are also fully coherent within another precipitate–matrix interface. Note that $(3/5, 8/13, \bar{5}/\bar{8})_i$ and $(\overline{10}/\bar{6}, 0/0, 0/0)_i$ are the closest packed planes in the icosahedral phase, and $\{10\bar{1}0\}_\alpha$ is a near closest packed plane in the matrix phase. Some of the icosahedral particles also have a truncated rectangular or square shape when viewed in the $[0001]_\alpha$ orientation, and these icosahedral particles have another OR with the matrix phase: $(8/13, 5/8, \overline{3}/\overline{5})_i // (1\overline{1}00)_\alpha$ and $[2\text{-fold}]_i // [0001]_\alpha$. Inspection of images of the icosahedral particles projected along the $[000l]_\alpha$ direction also reveals that some of them contain a high density of planar defects or polycrystalline aggregates. But the crystallography of such defects is unclear at the present.

The microstructure of Mg–8wt%Zn–8wt%Al alloy samples aged for 120 h at 200 °C also has a small fraction of relatively coarse, lath-shaped ϕ precipitates that have the following OR with the matrix: $(002)_\phi // (0002)_\alpha$, $[010]_\phi // [10\overline{1}0]_\alpha$. The broad surface of these ϕ laths is parallel to $(0001)_\alpha$. In addition to the precipitates of the equilibrium ϕ phase, some precipitates of the equilibrium phase β-$Mg_{17}Al_{12}$ are also found in samples aged for 120 h at 200 °C. The OR and morphology of these β precipitates are similar to those observed in Mg–Al binary alloys. While there is no doubt that the precipitation sequence in the Mg–8wt%Zn–8wt%Al alloys and alloys of similar compositions involves the formation of metastable icosahedral phase and the equilibrium ϕ and β phase, **Table 1**, the actual precipitation sequence in such alloys is more complex and therefore requires detailed characterization using advanced imaging and diffraction techniques in the future. For example, some relatively coarse particles that have a point group perfectly consistent with that of the T phase (space group $I\overline{3}m$, $a = 1.42$ nm) are occasionally observed in the sample aged for 120 h at 200 °C. The OR between these cubic precipitates and the matrix phase is such that $(002)_T // (10\overline{1}0)_\alpha$, $[100]_T // [0001]_\alpha$. The T phase precipitates were also reported to form during isothermal aging at 170 °C of high-pressure die cast Mg–8wt%Zn–4.8wt% Al–0.3wt%Mn (designated ZA85) (Vogel et al., 2003). While the T phase is an equilibrium phase in the ternary Mg–Zn–Al phase diagram, it is difficult to assess at the present whether this phase is an equilibrium phase in the Mg–8wt%Zn–8wt%Al ternary alloy, because three equilibrium phases, α-Mg, ϕ and β-$Mg_{17}Al_{12}$, have already been detected in this alloy.

In Mg–6.2Zn–2.1Al (wt%) alloy and those of similar compositions, it has been reported that the precipitation sequence is similar to that in Mg–Zn binary alloys (Mendis et al., 2010). However, it was reported (Oh-ishi et al., 2008) in a separate study that the addition of 3 wt% Al to an Mg–6wt%Zn–1wt %Mn alloy leads to a change in precipitate morphology. The basal plates formed in the Al-free alloy are replaced by cuboidal precipitates in the Al-containing alloy. In addition, spherical GP zones enriched in Zn also reportedly form in the Al-containing alloy after the alloy is aged at 70 °C. These GP zones reportedly act as heterogeneous nucleation sites for the metastable phases that precipitate during subsequent aging at 150 °C, resulting in a finer distribution of precipitates.

20.4.1.4 *Mg–Ca-Based Alloys*

The Mg–Ca system has some potential for developing precipitation hardenable alloys. The equilibrium solid solubility of Ca in magnesium is 0.82 at.% (1.35 wt%) at the eutectic temperature of 516.5 °C, and it is approximately zero at 200 °C. The equilibrium intermetallic phase at the Mg-rich end of the Mg–Ca phase diagram is Mg_2Ca that has a crystal structure (space group $P6_3/mmc$, $a = 0.623$ nm, $c = 1.012$ nm) with a close relationship to that of the magnesium matrix phase ($P6_3/mmc$, $a = 0.321$ nm, $c = 0.521$ nm) (Villars and Calvert, 1985). This similarity in crystal structure may result in a higher nucleation rate and hence higher number density of precipitates in Mg–Ca alloys. Assuming that the precipitates formed during isothermal aging at 200 °C have an Mg_2Ca composition, then the maximum volume fraction of precipitates achievable at 200 °C is calculated to be approximately 2.2% for an Mg–1wt%Ca alloy, which is adequate for precipitation strengthening. An example is the conventional precipitation hardenable Al–0.6wt%Si–1.0wt%Mg (6061) wrought alloy. The volume fraction of solid-state precipitates is about 2% in this alloy. However, a tensile yield strength of 275 MPa is achieved in the T6 condition. Since Ca has a low density (1.55 g cm^{-3}), Mg–Ca alloys have the added advantage of preserving the low density of magnesium, and the addition of Ca can also

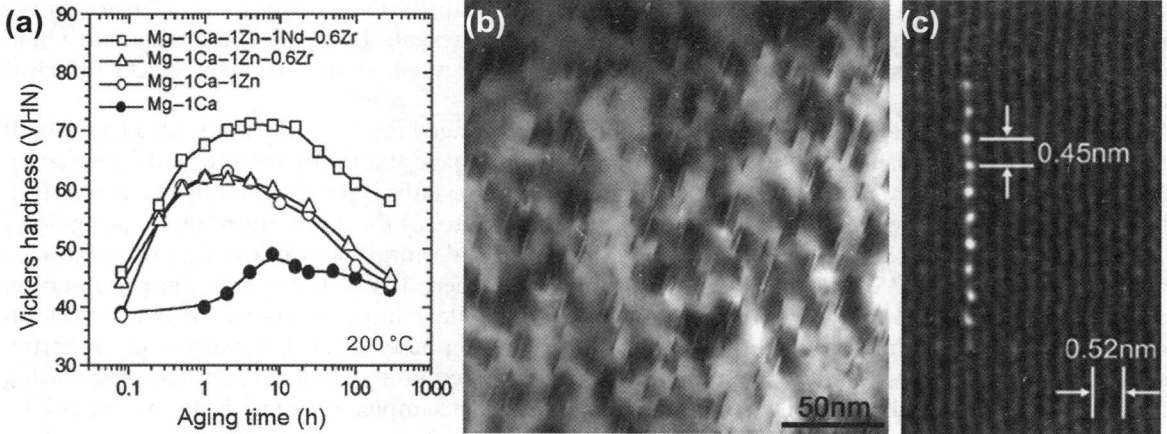

Figure 59 (a) Aging curves of Mg–Ca-based alloys at 200 °C. (b) TEM image showing the distribution of basal plates in Mg–1Ca–1Zn–1Nd–0.6Zr (wt%) alloy aged for 1000 h at 200 °C. (c) Atomic resolution HAADF-STEM image showing ordered GP zones in an Mg–0.3Ca–0.6Zn (at.%) alloy (Gao et al., 2005; Oh-ishi et al., 2009).

reduce flammability of molten magnesium and improve the oxidation and corrosion resistance of magnesium (Raynor, 1959). Therefore, efforts have been made in the past 15 years to develop precipitation hardenable alloys based on the Mg–Ca system (Nie and Muddle, 1997; Gao et al., 2005; Bamberger et al., 2006; Jayaraj et al., 2010; Oh et al., 2005; Oh-ishi et al., 2009).

A study made by Nie and Muddle (1997) indicates that the Mg–1wt%Ca alloy exhibits only a moderate age hardening response during isothermal aging at 200 °C. However, they noticed that the addition of 1 wt% of Zn to the binary alloy led to a substantial increase in peak hardness and an accelerated rate of aging. A subsequent study (Gao et al., 2005) indicates that the quaternary addition of 1 wt% Nd to the Mg–1Ca–1Zn–0.6Zr (wt%) alloy can lead to further increase in the maximum hardness and strength, **Figure 59(a)**. The resultant Mg–1Ca–1Zn–1Nd–0.6Zr (wt%) alloy exhibits a tensile yield strength of 153 MPa at room temperature, and 135 MPa at 150 °C. A comparison of the microstructures of these alloys indicates that the significant increase in maximum hardness and strength is associated with a refined distribution and improved thermal stability of basal precipitate plates, **Figure 59(b)**. The crystallographic features of these basal plates and their electron diffraction patterns resemble closely to those formed in Mg–Nd/Ce–Zn alloys. However, it remains to be unambiguously established whether the precipitates in the Mg–Ca–Zn and Mg–Nd/Ce–Zn alloys are structurally identical to each other. An improved understanding on this aspect can facilitate the optimization of alloy composition for improved thermal stability and creep resistance.

In the as-cast microstructure of the Mg–1Ca–1Zn (wt%) alloy, the primary intermetallic phase has a composition of $Mg_{69.4}Ca_{27}Zn_{3.6}$ and a hexagonal crystal structure (point group 6/mmm, $a \sim 0.61$ nm, $c \sim 1.02$ nm) that appears isomorphous with Mg_2Ca. Two types of solid-state precipitate phases are observed in the as-cast microstructure: a hexagonal phase with $a = 0.623$ nm, $c = 1.012$ nm, and a hexagonal phase with $a = 0.556$ nm, $c = 1.042$ nm (Nie and Muddle, 1997). Both precipitate phases form as thin plates on $(0001)_\alpha$, and they have identical ORs with respect to the matrix phase: $(0001)_p//(0001)_\alpha$, $[2\bar{1}\bar{1}0]_p//[10\bar{1}0]_\alpha$. Of these, the phase with $a = 0.556$ nm and $c = 1.042$ nm has a much more uniform distribution than the former hexagonal phase. It was argued

(Nie and Muddle, 1997) that, for the observed ORs between precipitate and matrix phases, a lattice parameter of $a = 0.556$ nm permits a perfect lattice matching between precipitate and matrix phases in the habit plane, and that this improved lattice matching gives rise to a higher nucleation rate and consequently an enhanced age hardening response in the Mg–Ca–Zn alloy. It was further suggested that an incorporation of Zn atoms into the Mg_2Ca unit cell can change the lattice parameters and therefore reduce the lattice misfit between the precipitate and magnesium matrix phases within the $(0001)_\alpha$ habit plane.

A subsequent study of microstructures of an Mg–0.3at.%Ca–0.3at.%Zn alloy aged for different times at 200 °C (Oh et al., 2005) indicates the following precipitation sequence: monolayers of GP zones on $(0001)_\alpha$, larger $(0001)_\alpha$ plates of an unidentified phase, and rectangular $Mg_2Ca(Zn)$ phase. Each GP zone was reported to contain approximately 18 at.% Ca and 8 at.% Zn. Two ORs were reported for the $Mg_2(Ca,Zn)$ phase: $(0001)_p//(0001)_\alpha$, $[1\bar{2}10]_p//[01\bar{1}0]_\alpha$, and $(0001)_p//(01\bar{1}0)_\alpha$, $[1\bar{2}10]_p//[2\bar{1}10]_\alpha$. In a separate but more recent study of precipitation and age hardening response at 200 °C of Mg–0.3at.%Ca–Zn alloys (Oh-ishi et al., 2009), it was reported that the maximum age hardening response is obtained when the Zn content is 0.6 at.%. Any increase or decrease in the Zn content in the alloy diminishes the age hardening response. In this recent study, HAADF-STEM was employed to characterize the precipitates in the Mg–0.3at.%Ca–0.6at.%Zn alloy aged for different times at 200 °C. It was found that monolayer GP zones, **Figure 59(c)**, were responsible for the age hardening and that these GP zones had an ordered structure that is identical to that in Mg–RE–Zn alloys (Ping et al., 2003). The total concentration of Ca and Zn atoms in each ordered GP zone is $\sim 33\%$. These ordered GP zones are thermally stable and still dominate the microstructure after over aging (16 h) at 200 °C.

An increase in the Zn content from 0.6 at.% to 1.6 at.% in the Mg–0.3at.%Ca alloy leads to a change in the precipitation sequence. It was reported (Oh-ishi et al., 2009) that the precipitation reaction in the Mg–0.3at.%Ca–1.6at.%Zn alloy at 200 °C included the formation of $[0001]_\alpha$ rods of metastable β'_1-Mg_4Zn_7 phase, $(0001)_\alpha$ plates of the equilibrium $Mg_6Ca_2Zn_3$ phase (space group $P\bar{3}1c$, $a = 0.97$ nm, $c = 1.0$ nm), and laths of an unknown phase. The broad surface of these laths is parallel to $(0001)_\alpha$ and their long axis is parallel to $[2\bar{1}\bar{1}0]_\alpha$ and $[10\bar{1}0]_\alpha$. The OR between the $Mg_6Ca_2Zn_3$ phase and the magnesium matrix is such that $(0001)_p//(0001)_\alpha$, $[10\bar{1}0]_p//[10\bar{1}0]_\alpha$. The $Mg_6Ca_2Zn_3$ phase was originally reported to form as cuboidal particles in melt-spun ribbons of an Mg–6wt%Zn–1.5wt%Ca alloy (Jardim et al., 2002). These cuboidal precipitates reportedly adopt two ORs: $(11\bar{2}0)_p//(0001)_\alpha$, $[0001]_p//[11\bar{2}0]_\alpha$ and $(11\bar{2}0)_p//(0001)_\alpha$, $[0001]_p//[21\bar{3}0]_\alpha$ (Jardim et al., 2004).

While it has been demonstrated that additions of Zn and Nd can enhance the age hardening response of Mg–Ca alloys, the effects of other alloying elements have received little attention. As discussed in the Section 20.4.1.1, the high-pressure die-cast Mg–Al–Ca–Sr (AXJ530) alloy exhibits some age hardening when they are aged at 175–250 °C (Suzuki et al., 2008) and this age hardening is associated with the formation of $(0001)_\alpha$ precipitate plates of Al_2Ca phase (space group $Fd\bar{3}m$, $a = 0.802$ nm). The total concentration of ternary and quaternary alloying elements such as Ca and Sr in this AXJ530 alloy is quite high for the purpose of castability and creep resistance, while the potential of developing dilute precipitation hardenable alloys based on the Mg–Al–Ca system was not explored. In a very recent study (Jayaraj et al., 2010), it was demonstrated that the ternary addition of an appropriate amount of Al (0.3 wt%) to an Mg–0.5wt%Ca alloy can remarkably enhance the age hardening response at 200 °C, **Figure 60(a)**. The maximum age hardening response achievable at the aging temperature is reduced if the Al content in the Mg–Ca–Al alloy is higher or lower than 0.3 wt%. The peak-aged microstructure contains a dense distribution of nanoscale precipitate plates on $(0001)_\alpha$.

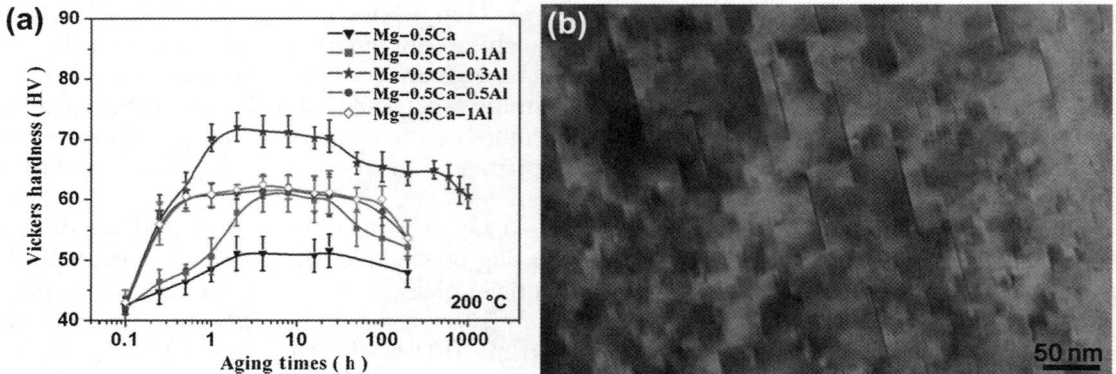

Figure 60   (a) Aging curves of Mg–Ca–Al alloys at 200 °C and (b) TEM image showing distribution of basal plates in Mg–0.5Ca–0.3Al alloy aged for 1000 h at 200 °C. The alloy compositions are in weight percentage (Jayaraj et al., 2010). (For color version of this figure, the reader is referred to the online version of this book.)

Based on the monolayer thickness of these precipitate plates, the enrichment of Ca in these particles, and the SAED patterns recorded from matrix regions containing such precipitates, these precipitate plates are inferred (Jayaraj et al., 2010) to be ordered GP zones such those observed in the Mg–Ca–Zn and Mg–RE–Zn alloys. The concentrations of Ca and Al atoms in the GP zone, measured from 3DAP, are approximately 6 at.% and 7 at.%, respectively.

The ordered GP zones are gradually replaced by $(0001)_\alpha$ plates of the equilibrium phase Al_2Ca with continued aging at 200 °C. The formation of the relatively larger basal plates of the Al_2Ca phase, **Figure 60(b)**, was reported to cause the overaging of the alloy. The OR between the Al_2Ca plates and the magnesium matrix phase is such that $(111)_p//(0001)_\alpha$ and $[01\overline{1}]_p//[0\overline{1}10]_\alpha$, which is identical to that associated with the Al_2Ca plates formed in AXJ530 alloy (Suzuki et al., 2005).

The occurrence of the age hardening phenomenon in the Mg–Ca–Al system offers the potential for developing a precipitation hardenable wrought alloy (Jayaraj et al., 2010). In 2011, an unusual Mg–3.5Al–3.3Ca–0.4Mn (wt%) extrusion alloy was developed. In the as-extruded condition, this alloy exhibits a tensile yield strength of 410 MPa, together with an elongation to fracture of 5.6% (Xu et al., 2011). Among the RE-free magnesium alloys, the strength level achieved in this alloy is exceptionally impressive, and this is attributed to the formation of plate-shaped and spherical-shaped precipitates, basal texture and refined grain size of magnesium. The identities of the two types of precipitates were not verified, but they were assumed (Xu et al., 2011) to be identical to those formed in Mg–0.5Ca–0.3Al (wt%) (Jayaraj et al., 2010) and Mg–6Al–3.2Ca–0.5Mn (wt%) (Homma et al., 2010b) alloys.

20.4.1.5 Mg–Sn-Based Alloys

Mg–Sn-based alloys have received some attention in recent years for developing casting and wrought alloy products (Hono et al., 2010; Gibson et al., 2010; Kang et al., 2005, 2007). The Mg–Sn binary system itself is ideal for developing precipitation hardenable alloys (van der Planken, 1969). The maximum equilibrium solid solubility of Sn in α-Mg is about 3.35 at.% (or 14.5 wt%) at the eutectic temperature 561 °C, and it decreases to about 0.1 at.% at 200 °C. The equilibrium volume fraction of precipitates obtainable at 200 °C is approximately 4.7% for an Mg–7wt%Sn alloy.

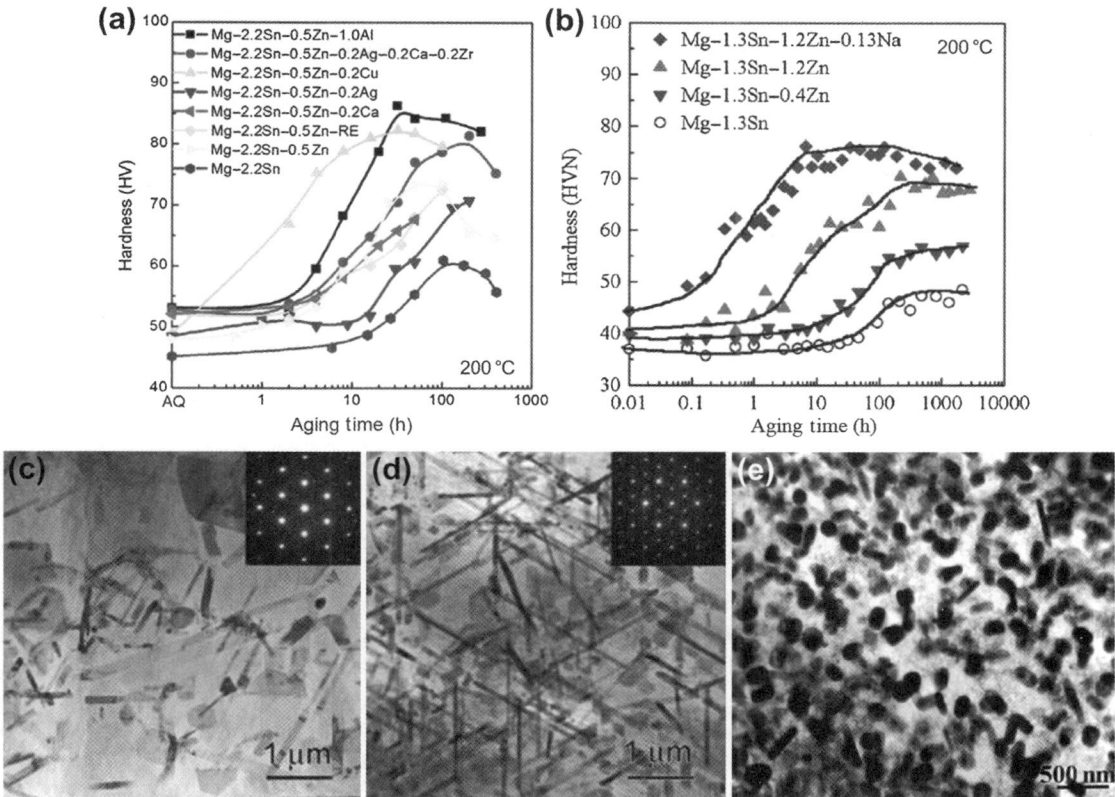

Figure 61 (a and b) Age hardening response at 200 °C of Mg–Sn alloys without and with additions of microalloying elements. Transmission electron micrographs showing distribution of precipitates in (c) Mg–2.2Sn, (d) Mg–2.2Sn–0.1Zn, (e) Mg–1.3Sn–1.2Zn–0.12Na alloys (Hono et al., 2010; Sasaki et al., 2006; Mendis et al., 2006a). All alloy compositions are in atomic percentage. (For color version of this figure, the reader is referred to the online version of this book.)

The equilibrium intermetallic phase in the Mg–Sn binary alloys is β-Mg_2Sn (space group $Fm\overline{3}m$, $a = 0.68$ nm). It is rather unfortunate that, during isothermal aging treatments in the temperature range 160–300 °C, the precipitation process in binary Mg–Sn alloys does not involve the formation of any metastable precipitate phases. The precipitates formed during the aging process are generally much coarser than the precipitates in other precipitation hardenable magnesium alloys. Therefore, attempts have been made in recent years (Mendis et al., 2006a, 2006b; Sasaki et al., 2006; Hono et al., 2010) to refine the distribution of precipitates in Mg–Sn alloys by microalloying additions. As shown in **Figure 61(a) and (b)**, ternary additions of Zn to Mg–Sn alloys can remarkably improve the age hardening response, and further microalloying additions of Cu, Na, Ag and Ca, and macroalloying additions of Al, to the resultant ternary Mg–Sn–Zn alloys can lead to an even greater age hardening response. A comparison of microstructures indicates that the Zn additions can increase the precipitate number density, even thorough the refined precipitates are still the β phase, and that quaternary additions of Na can lead to a much finer distribution of β phase, **Figure 61(c–e)**. While the microalloying elements such as Zn and Na can apparently enhance the nucleation rates and thus the number density of β precipitates and change the morphology of the precipitates, it remains to be established whether atoms of such microalloying elements segregate into the precipitate or the precipitate–matrix interface. It is unclear how microalloying elements can enhance the precipitate nucleation rate and the precipitate morphology, and whether the precipitate morphology change is associated with any change in OR. Further work is needed to gain an in-depth understanding of such fundamental issues.

Based on X-ray diffraction and crystallographic analysis in an early study, Derge et al. (1937) reported that three ORs exist between β and magnesium phases, namely:

$$OR1: \quad (111)_\beta // (0001)_\alpha \quad \text{and} \quad [1\bar{1}0]_\beta // [2\bar{1}\bar{1}0]_\alpha$$

$$OR2: \quad (111)_\beta // (0001)_\alpha \quad \text{and} \quad [2\bar{1}\bar{1}]_\beta // [2\bar{1}\bar{1}0]_\alpha$$

$$OR3: \quad (110)_\beta // (0001)_\alpha \quad \text{and} \quad [1\bar{1}1]_\beta // [2\bar{1}\bar{1}0]_\alpha$$

Twenty-five years later, these three ORs were confirmed by Henes and Gerold (1962) in their X-ray diffraction observations. They also reported an additional OR:

$$OR4: \quad (110)_\beta // (0001)_\alpha \quad \text{and} \quad [001]_\beta // [2\bar{1}\bar{1}0]_\alpha$$

and that these four ORs were associated with β precipitates that form during aging at 160–300 °C. In these early studies, it was not mentioned which one of the four ORs was most common and what the precise precipitate morphology was for each of the four ORs. Examination of the ORs reported in recent years (Mendis et al., 2006a, 2006b; Kang et al., 2007; Sasaki et al., 2011; Zhang et al., 2007, 2008) indicates that, even at the present, this information still remains unclear. While the most commonly reported OR in recent years appears to be OR-1, a variety of morphologies have been reported for this OR1, including *long laths*, formed with the broad surface parallel to $(0001)_\alpha$ and the long axis parallel to $\langle 2\bar{1}\bar{1}0 \rangle_\alpha$, in an Mg–1.9at.%Sn alloy aged for 240 h at 200 °C (Mendis et al., 2006b), *short laths*, formed with the broad surface parallel to $(0001)_\alpha$ and the long axis parallel to $\langle 2\bar{1}\bar{1}0 \rangle_\alpha$, in Mg–1.3Sn–1.2Zn (at.%) alloy aged for 211 h at 200 °C (Mendis et al., 2006a) and aged samples of Mg–2.2Sn–0.5Zn (at.%) alloy (Sasaki et al., 2006), *short* $[0001]_\alpha$ *laths* in Mg–1.3Sn–1.2Zn (at.%) alloy aged 211 h at 200 °C (Mendis et al., 2006a), *short rods* in Mg–7.8Sn–2.7Al–0.7Si–0.7Zn–0.2Mn (wt%) alloy produced by die casting (Kang et al., 2007), and *polygons* in Mg–1.3Sn–1.2Zn–0.12Na (at.%) alloy aged 6.7 h at 200 °C (Mendis et al., 2006a), Mg–2.2Sn–0.5Zn (at.%) alloy peak aged at 200 °C (Sasaki et al., 2011) and Mg–5.3Sn–0.3Mn–0.2Si (wt%) alloy aged for 10 h at 250 °C (Zhang et al., 2007).

In contrast, the *long laths*, formed on $(0001)_\alpha$ and elongated along $\langle 2\bar{1}\bar{1}0 \rangle_\alpha$ in Mg–5.3Sn–0.3Mn–0.2Si (wt%) alloy aged for 10 h at 250 °C, and the *short laths*, formed on $(0001)_\alpha$ in Mg–2.1Sn–1Zn–0.1Mn (at.%) alloy peak aged at 200 °C (Sasaki et al., 2009a), have also been reported to have the OR4; and the *polygons* in Mg–2.2Sn–0.5Zn (at.%) alloy peak aged at 200 °C (Sasaki et al., 2011) and aged samples of Mg–2.2Sn–0.5Zn (at.%) alloy (Sasaki et al., 2006) and *short laths* formed with broad surface parallel to the pyramidal plane of the matrix phase in Mg–1.3Sn–1.2Zn (at.%) alloy aged for 211 h at 200 °C (Mendis et al., 2006b) have been reported to have the OR3. Furthermore, Zhang et al. (2007) reported that the $(0001)_\alpha$ plates in their alloy adopt the following OR:

$$OR5: \quad (111)_\beta // (0001)_\alpha \quad \text{and} \quad [2\bar{1}\bar{1}0]_\alpha \text{ is} \sim 9° \text{away from} [1\bar{1}0]_\beta$$

The OR5 can be related to the OR-1 by a rotation of 9° about the $[111]_\beta//[0001]_\alpha$ axis. However, Sasaki et al. (2011) reported that the basal plates in their alloy have the OR2. Two other different ORs have also been reported, namely:

$$OR6: \quad (110)_\beta//(0001)_\alpha \quad \text{and} \quad [31\bar{1}]_\beta//[1\bar{1}00]_\alpha$$

$$OR7: \quad (111)_\beta//(01\bar{1}0)_\alpha \quad \text{and} \quad [1\bar{1}0]_\beta//[2\bar{1}\bar{1}0]_\alpha$$

The precipitate morphology is short lath elongated along $\langle 10\bar{1}0 \rangle_\alpha$ for the OR6 (Zhang et al., 2007), and rod elongated in the pyramidal plane of the matrix phase (Sasaki et al., 2006). It is currently unclear why there exist several ORs, and what microstructural factor dictates the formation of these ORs. It remains to be examined whether the appearance of different ORs, and morphologies associated with these ORs, is due to the variation of concentration of Sn atoms in the solid solution matrix. Any local segregation of Sn atoms may cause a change in the lattice parameter and in turn promote the formation of a particular OR.

20.4.1.6 Mg–Nd and Mg–Ce-Based Alloys
20.4.1.6.1 Mg–Nd and Mg–Ce Binary Alloys
The identity of the equilibrium intermetallic phase at the Mg-rich side of the Mg–Nd binary phase diagram has been controversial. The early version of the phase diagram (Nayeb-Hashemi and Clark, 1988) indicates that the equilibrium intermetallic phase is $Mg_{12}Nd$ that has a tetragonal structure (space group I4/mmm, $a = 1.031$ nm, $c = 0.593$ nm). However, more recent studies (Delfino et al., 1990; Gorsse et al., 2005; Okamoto, 2007) indicate that the equilibrium intermetallic phase is $Mg_{41}Nd_5$ (space group I4/m, $a = 1.474$ nm, $c = 1.040$ nm) instead of $Mg_{12}Nd$, **Figure 62**. Reported values of the maximum solid solubility of Nd in magnesium vary significantly, ranging from 0.10 at.%

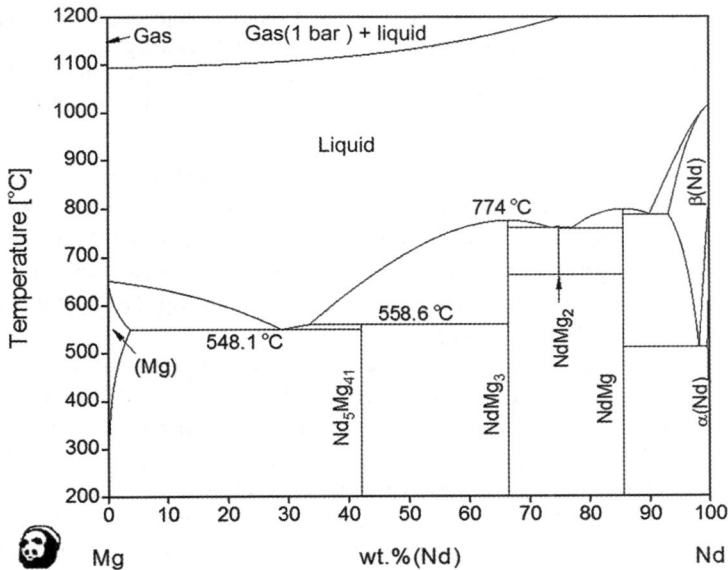

Figure 62 Mg–Nd binary phase diagram (Courtesy R. Schmid-Fetzer; Nie, 2012).

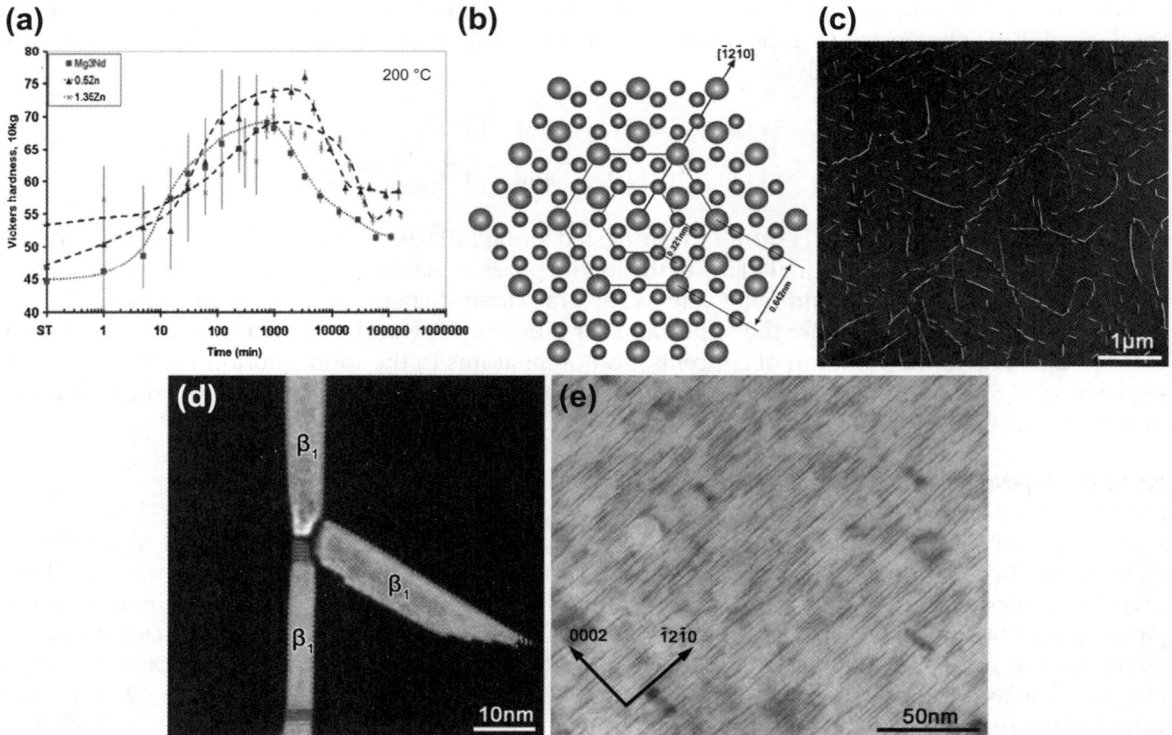

Figure 63 (a) Age hardening response at 200 °C of Mg–3wt%Nd alloys without and with Zn additions. (b) Schematic diagram showing DO$_{19}$ structure. The perspective viewing direction is [0001]. Large purple circles and small gray circles represent Nd and Mg atoms, respectively. TEM images showing distribution of precipitates in (c and d) Mg–3wt%Nd alloy aged for 10 h at 250 °C, and (e) Mg–3wt%Nd–1.35wt%Zn, aged for 52 h at 200 °C (Wilson et al., 2003; Zhu and Nie, unpublished research). (For color version of this figure, the reader is referred to the online version of this book.)

(0.59 wt%) at 548 °C to 0.63 at.% (3.62 wt%) at 552 °C (Nayeb-Hashemi and Clark, 1988; Rokhlin, 1995). A recent study using atom probe tomography (Kopp et al., 2011) indicates that the equilibrium solid solubility of Nd in magnesium is approximately 0.32 at.% (1.87 wt%) at 520 °C, and 0.11 at.% (0.65 wt%) at 400 °C. The equilibrium solid solubility of Nd decreases with temperature, down to nearly zero at 200 °C, implying a potential for precipitation hardening. The aging curve of an Mg–3wt% Nd alloy at 200 °C is shown in **Figure 63(a)**.

The decomposition of supersaturated solid solution of magnesium in the temperature range 60–350 °C is currently accepted as involving the formation of GP zones, β″, β′ and β phases. While the GP zones have been reported to form as needles along the [0001]$_\alpha$ direction (Pike and Noble, 1973) in an Mg–2.9wt%Nd (Mg–0.5at.%Nd) alloy, or platelets on {10$\bar{1}$0}$_\alpha$ in an Mg–2.5wt%Di (80%Nd–20% Pr)–0.6wt%Zr alloy (Gradwell, 1972), there was a lack of direct experimental evidence, or TEM images, to support the formation of such GP zones in these alloys at the early stage of aging. It was not clear whether GP zones did form in the Mg–Nd binary alloys and, if so, what morphology they might adopt.

In a very recent study, using HAADF-STEM, of precipitate phases formed in Mg–0.5at.%Nd samples aged at 170 °C (Saito and Hiraga, 2011), some peculiar zigzag arrays of Nd atoms, aligned approximately along $\langle 2\bar{1}\bar{1}0 \rangle_\alpha$, are observed and regarded as GP zones. Each zigzag array comprises several V-shaped, or N-shaped, units that are separated by an almost regular spacing. Each V-shaped and N-shaped unit is made of three and four columns of Nd atoms, respectively. The separation distance between two neighboring columns of Nd atoms within each unit is invariably ~ 0.37 nm, and the plane defined by the two neighboring columns of Nd atoms is invariably parallel to $\{10\bar{1}0\}_\alpha$.

Some hexagonal rings, defined by six columns of Nd atoms, are also observed in this latest work (Saito and Hiraga, 2011). The separation distance between two neighboring columns of Nd atoms within the hexagonal ring is again about 0.37 nm, and the prism plane of the hexagonal ring is parallel to $\{10\bar{1}0\}_\alpha$. Each of the hexagonal rings can be regarded as being constructed by linking three variants of the V-shaped unit mentioned in the paragraph above. This arrangement of Nd atoms resembles that in the DO_{19} structure (space group $P6_3/mmc$, $a = 0.641$ nm, $c = 0.521$ nm, Mg_3Nd composition) (Gradwell, 1972), even though the Saito and Hiraga did not mention it. Moreover, the area defined by the hexagonal ring is precisely that of the basal plane of the DO_{19} unit cell, **Figure 63(b)**. The β'' phase was originally reported to have a DO_{19} structure and an OR that is in the form $(0001)_{\beta''}//(0001)_\alpha$ and $[2\bar{1}\bar{1}0]_{\beta''}//[2\bar{1}\bar{1}0]_\alpha$. Therefore, these hexagonal rings may be regarded as the metastable β'' phase. The latest experimental observations made by HAADF-STEM indicate clearly that the β'' precipitates are not $\{2\bar{1}\bar{1}0\}_\alpha$ plates.

The HAADF-STEM work of Saito and Hiraga (2011) also reveals the existence of a precipitate phase not previously reported for the binary Mg–Nd alloys. This phase, designated β', has an orthorhombic structure ($a = 0.64$ nm, $b = 1.14$ nm, $c = 0.52$ nm), an Mg_7Nd composition, and a lenticular morphology. While not mentioned by the authors, the following OR can be deduced from their work: $(100)_{\beta'}//\{1\bar{2}10\}_\alpha$ and $[001]_{\beta'}//[0001]_\alpha$. Most of these features are similar to those of the β' phase in Mg–Gd, Mg–Y, Mg–Gd–Y and WE54 alloys, as will be discussed in Section 20.4.1.7.

The β' phase mentioned in earlier studies was reported to form as plates on $\{10\bar{1}0\}_\alpha$. Pike and Noble (1973) suggested that the β' phase had a hexagonal structure ($a = 0.52$ nm, $c = 1.30$ nm) and the following OR with respect to the matrix phases: $(1\bar{2}10)_{\beta'}//(10\bar{1}0)_\alpha$, $(\bar{1}014)_{\beta'}//(0001)_\alpha$. In contrast, Karimzadeh (1985) and Gradwell (1972) indicated that β' had a face-centered cubic structure ($a = 0.735$ nm) and a composition close to $Mg_{20}Nd_{17}$. Karimzadeh (1985) also suggested that the OR between β' and the matrix was such that $(\bar{1}12)_{\beta'}//(10\bar{1}0)_\alpha$, $[110]_{\beta'}//[0001]_\alpha$. The structure, OR and morphology proposed by Karimzadeh (1985) and Gradwell (1972) are essentially the same as those of β_1 phase (space group $Fm\bar{3}m$, $a = 0.74$ nm) in Mg–Gd, Mg–Gd–Y and WE54 alloys. To avoid any confusion and to be consistent with those symbols used for similar structures, it is appropriate to use β_1 to replace β' used in the previous studies. The β_1 precipitates often form heterogeneously on preexisting dislocations, leading to a nonuniform distribution of these precipitates, **Figure 63(c)**.

The β phase, which was originally regarded as the equilibrium phase but is in fact a metastable phase, has a body-centered tetragonal structure ($a = 1.031$ nm, $c = 0.593$ nm) and a composition of $Mg_{12}Nd$. It forms as rods with their long axis parallel to $[0001]_\alpha$. The cross-section of the β rods has a hexagonal shape, with the prism facets parallel to $\{10\bar{1}0\}_\alpha$. The OR between β and the matrix was reported (Wei et al., 1996) to be the following: $(002)_\beta//(10\bar{1}0)_\alpha$ and $[100]_\beta//[0001]_\alpha$.

Based on the above discussion of individual precipitate phases and the fact that the equilibrium precipitate phase is $Mg_{41}Nd_5$, instead of $Mg_{12}Nd$, the part of the whole precipitation sequence in Mg–Nd binary alloys is provided in **Table 1**. To preserve the symbols commonly used in the literature for similar structures and to avoid any confusion, in this newly proposed sequence the β' phase represents the newly reported orthorhombic phase, β_1 phase represents the β' phase reported in the previous studies, and β_e is used as the final precipitate phase, that is $Mg_{41}Nd_5$. It is to be noted that the whole precipitation sequence is far from well established; the early stage precipitation still remains to be unambiguously established, and even the later stage of the precipitation requires more characterization and analysis. For example, the gray contrast regions connecting β_1 precipitates or segments shown in **Figure 63(d)** cannot be attributable to any of the precipitate phases known in the Mg–Nd system. The TTT diagrams for various precipitate phases formed in the Mg–0.5at.%Nd alloy were studied and reported by Pike and Noble (1973). Since the identities of the precipitate phases were not unambiguously established in that work, some caution should be taken if such TTT diagrams are to be used.

The age hardening response and precipitation sequence of Mg–Ce binary and Mg–MM (MM represents Ce- or Nd-rich misch metal) alloys were also studied in the past (Wei et al., 1996; Hisa et al., 2002). Since the equilibrium solid solubility of Ce in magnesium is lower than that of Nd, a lower age hardening response is expected for the Mg–Ce and Mg–MM alloys. Wei et al. (1996) and Hisa et al. (2002) investigated the precipitation sequence of Mg–1.3wt%MM (Ce-rich MM) and Mg–1.3wt%Ce alloys. The identities and formation mechanisms of the intermediate precipitate phases are controversial in these two studies. Nevertheless, the precipitation in the Mg–Ce and Mg–MM alloys is expected to be similar to that in the Mg–Nd alloys.

20.4.1.6.2 *Mg–Nd–Zn and Mg–Ce–Zn Alloys*

The intermetallic particles formed in the as-cast microstructures of the Mg–3Nd–0.5Zn and Mg–3Nd–1.35Zn (wt%) alloys are similar to those in the Mg–Nd binary alloys. They have the tetragonal structure of the $Mg_{12}Nd$ phase, even though they contain some Zn. The intermetallic particles in the interdendritic regions of a sand cast Mg–2.5wt%RE–0.5wt%Zn alloy, designated MEZ, have an $Mg_{12}(La_{0.43}Ce_{0.57})$ composition and a high density of unidentified planar defects (Bettles et al., 2003). These intermetallic particles are more difficult to dissolve into the solid solution matrix phase during the solution treatment, and the volume fraction of the retained such particles increases with the Zn content in the alloy. The addition of Zn seems to lead to a reduction in eutectic temperature and therefore the use of lower temperature for solution treatment. For example, localized melting in grain boundaries was noted after the Mg–3Nd–1.35Zn alloy was solution treated for 24 h at 530 °C (Wilson, 2005), and therefore 510 °C was used for the solution treatment. While the retained intermetallic particles in the Mg–3Nd–1.35Zn alloy still have the structure of the $Mg_{12}Nd$ phase after 24 h solution treatment at 510 °C, they gradually decompose and are replaced by the γ phase (face-centered cubic, $a = 0.72$ nm) after prolonged heat treatments at 200 and 325 °C. The structure of this γ phase is similar to that of the Mg_3Nd phase, except the lattice parameter is a little smaller. The structures of the equilibrium intermetallic phases at the Mg-rich side of the Mg–Nd–Zn or Mg–Ce–Zn ternary phase diagram have not been unambiguously established (Zhang et al., 2011; Qi et al., 2011; Huang et al., 2010; Chiu et al., 2010). Therefore, it is difficult to know whether the Mg_3Nd is an equilibrium phase at the Mg-rich side of the Mg–Nd–Zn alloys and, if so, whether γ is the Mg_3Nd phase.

Ternary additions of Zn to Mg–Nd or Mg–Ce alloys do not lead to any appreciable change in the age hardening response, **Figures 63(a) and 64(a)**, but they lead to a significant improvement in creep resistance (Wilson, 2005). The age hardening response of the Mg–Nd–Zn alloys is greater than that of the Mg–Ce–Zn alloys due to a higher saturation of solutes in the as-quenched condition and a higher volume fraction of precipitates after aging. The tensile yield strength of an Mg–Nd–Zn-based alloy, designated NEZ, is much higher than that of an Mg–Ce–Zn-based alloy designated MEZ. Quaternary additions of Gd to the resultant Mg–Nd–Zn alloys can result in further improvement in age hardening response (Lyon et al., 2005), and this discovery leads to the development of a commercial alloy designated EV31, which is also known as Elektron 21. The EV31 alloy is age hardenable, and possesses adequate strength and excellent creep resistance at temperatures up to 200 °C. The addition of more than 1.35 wt% Zn to the Mg–3wt%Nd alloy results in decreased hardening response. The commercial alloy EZ33 (Mg–3.2wt%MM–2.7wt%Zn–0.7wt%Zr) has a large amount of insoluble intermetallic particles at grain boundaries after casting. This alloy does not have much age hardening response and therefore is often used in the T5 condition.

The precipitation sequence in an Mg–2.8wt%Nd–1.3wt%Zn alloy was reported to include a low-temperature reaction, and the formation of γ'' and γ phases in an early study (Nuttall et al., 1980). The low-temperature reaction was noted during isothermal resistivity measurements at temperatures between 50 and 150 °C. The nature of this low-temperature reaction was not fully characterized in that study, and it was speculated (Nuttall et al., 1980) that it was related to the formation of GP zones. In 2003, Ping et al. (2003) reported the first 3DAP result on magnesium alloys. In their study, the TEM and 3DAP were combined to characterize the structure, morphology and composition of nanoscale precipitates formed in an Mg–2.4RE–0.4Zn–0.6Zr (wt%, where 2.4 wt% RE includes 1.3 wt% Ce, 0.6 wt% La, 0.4 wt% Nd and 0.1 wt% Pr) casting alloy, solution treated for 16 h at 525 °C, cold water quenched and then aged for 48 h at 200 °C. Hitherto unreported ordered GP zones, that form as $(0001)_\alpha$ discs of a single atomic plane thickness, were observed in this alloy. The RE/Zn atoms have an ordered hexagonal distribution within individual GP zone planes, with $a = 0.556$ nm. The OR between the two-dimensionally ordered GP zones and α-Mg matrix phase is $[10\bar{1}0]_{GP}//[11\bar{2}0]_\alpha$. The 3DAP results indicate that the GP zones contain about 3.2 at.% Nd, 1.0 at.% Ce and 1.2 at.% Zn, **Figure 64(b)**. The effect of the presence of both RE and Zn atoms in the GP zones is to reduce the elastic strain associated with individual atoms of RE and Zn, because RE atoms are larger than Mg while Zn atoms are smaller than Mg.

In the study made by Nuttall et al. (1980), the γ'' phase was reported to have a hexagonal structure ($a = 0.556$ nm, $c = 1.563$ nm) and it formed as plates on the basal plane of α-Mg. The OR between γ'' and the matrix phase was reported to be such that $(0001)_{\gamma''}//(0001)_\alpha$, $[2\bar{1}\bar{1}0]_{\gamma''}//[10\bar{1}0]_\alpha$. In more recent studies (Wilson et al., 2003; Ping et al., 2003) of peak-aged samples of Mg–Nd–Zn and Mg–RE–Zn alloys, most precipitates form as $(0001)_\alpha$ plates, **Figure 63(e)**. They are similar to the γ'' phase, but were designated γ' (Ping et al., 2003). In contrary to the early work of Nuttall et al. (1980), the basal plates of γ'' were proposed to have a hexagonal structure ($a = 0.556$ nm, $c = 0.521$ nm) in these recent studies. This structure is similar to the Mg$_5$(Ce,Zn) phase (space group P$\bar{6}$2m, $a = 0.571$ nm, $c = 0.521$ nm) (Nishijima et al., 2007a). The precipitates formed in aged samples of an Mg–2.5RE–0.5Zn (wt%) alloy (MEZ), **Figure 64(c) and (d)**, are similar to those in the Mg–Nd–Zn alloys (Bettles et al., 2003).

The equilibrium phase γ was reported (Nuttall et al., 1980) to have a face-centered cubic structure ($a = 0.72$ nm) and an OR $(011)_\gamma//(0001)_\alpha$ and $[\bar{1}11]_\gamma//[2\bar{1}\bar{1}0]_{\alpha'}$ and to form as rods with their long axes parallel to $\langle 10\bar{1}0\rangle_\alpha$ and $\langle 2\bar{1}\bar{1}0\rangle_\alpha$ directions. A more recent study (Wilson, 2005) confirms the structure and OR proposed in the early study, but the morphology of the γ phase was proposed to be plate. The habit plane of the γ plate varies from plate to plate and is therefore irrational, but it is always parallel to the moiré plane defined by the intersection of $\{220\}_\gamma$ and $\{10\bar{1}0\}_\alpha$ planes, implying a coherent matching of these two sets of lattice planes within the habit plane. Note that the structure and the OR of the γ phase are similar to those of the β_1 phase in the Zn-free alloys, that is Mg–Nd binary alloys. The incorporation of Zn into the precipitates leads to the reduction in the lattice parameter from ~0.74 nm to ~0.72 nm, and the reduction in the lattice parameter leads to a change of the orientation of the precipitate plates to preserve the matching between the precipitate and matrix lattices.

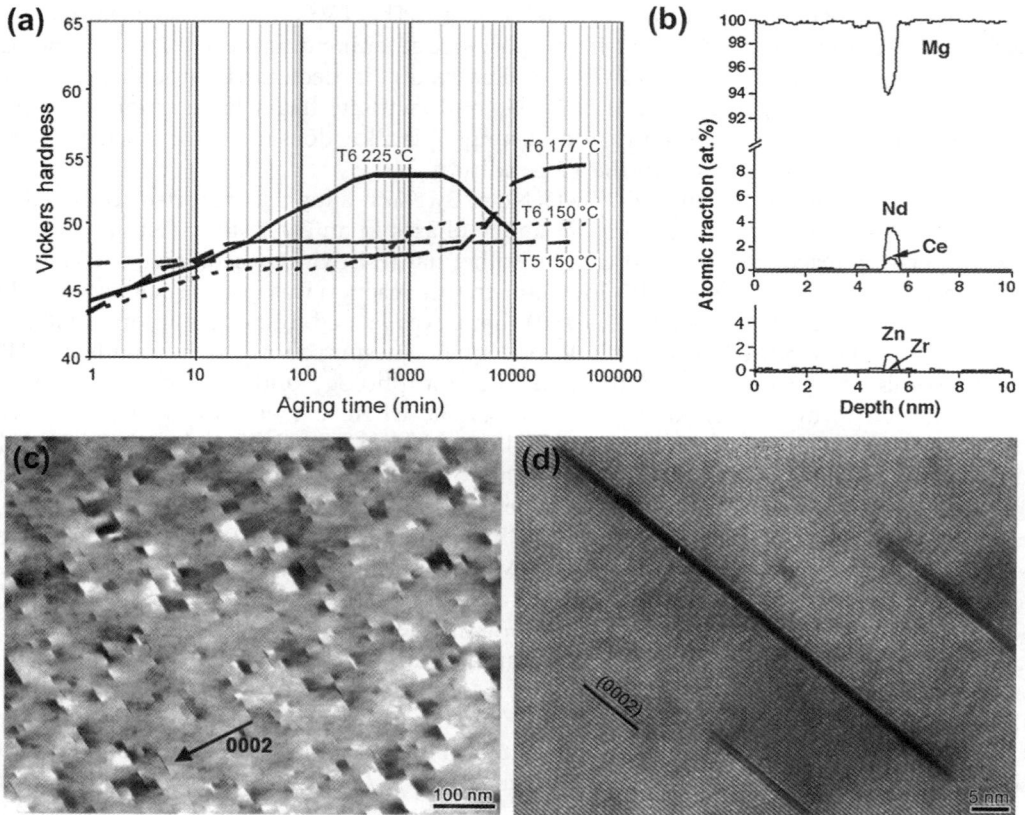

Figure 64   (a) Age hardening response of Mg–2.5wt%RE–0.5wt%Zn (MEZ) alloy at 150 °C, 177 °C and 225 °C. (b) 3DAP composition profile of ordered GP zones in an Mg–2.4wt%RE–0.4wt%Zn–0.6wt%Zr alloy aged for 48 h at 200 °C. (c and d) TEM images showing distribution and morphology of precipitates in the as-cast microstructure of MEZ alloy produced by sand-casting (Bettles et al., 2003; Ping et al., 2003).

20.4.1.6.3   Mg–Nd–Ag Alloys

The ternary phase diagram of the Mg–Nd–Ag system is rather incomplete (Berger and Weiss, 1988), and therefore the equilibrium solid solubility of Ag in Mg and the equilibrium intermetallic phases in the Mg-rich end of the Mg–Nd–Ag system are both unclear. According to the Mg–Ag binary phase diagram, the maximum equilibrium solid solubility of Ag in Mg is ~15 wt% at the eutectic temperature of 472 °C, and it falls to approximately 2 wt% at room temperature, and the intermetallic phase at the Mg-rich side of the phase diagram is Mg_3Ag (space group $P6_3/mmc$, $a = 0.488$ nm, $c = 0.779$ nm). More than 50 years ago, Payne and Bailey (1959–60) found that the relatively low tensile strength of Mg–Nd alloys could be considerably increased by Ag additions. This discovery subsequently led to the development of a commercial alloy designated QE22, Mg–2Nd–2.5Ag–0.7Zr (wt%). The alloy QE22 is age hardenable, and its aging curves at 150–300 °C are shown in **Figure 65**. Within the temperature and time selected, a maximum hardness value of ~85 VHN is obtainable when the alloy is aged at 150 °C. After the peak hardness is obtained at each temperature, the prolonged aging leads to only a slight reduction in hardness.

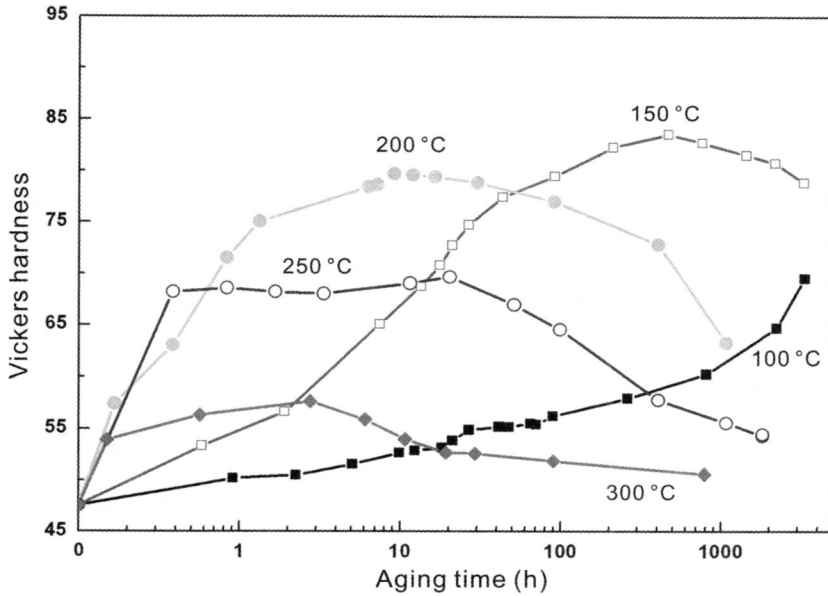

Figure 65   Age hardening response of Mg–2.1wt%Nd–2.5wt%Ag–0.7wt%Zr alloy (QE22) alloy at 100–300 °C (Kallisch, 1998). (For color version of this figure, the reader is referred to the online version of this book.)

In the temperature range 200–300 °C, the decomposition of the supersaturated solid solution phase of α-Mg in QE22 alloy was reported (Gradwell, 1972; Lorimer, 1987) to occur via two independent precipitation sequences. One sequence involves the formation of GP zones, in the form of $[0001]_\alpha$ rods, metastable γ phase that also forms as $[0001]_\alpha$ rods, and the equilibrium phase ($Mg_{12}Nd_2Ag$) that has a lath morphology and a yet to be determined hexagonal structure. The other sequence has the formation of GP zones of an ellipsoid shape, the metastable β phase of an equiaxed morphology, and the equilibrium phase $Mg_{12}Nd_2Ag$. Gradwell (1972) proposed that both types of GP zones formed simultaneously during aging at temperatures up to 250 °C. Without providing the transformation mechanisms, he further proposed that the rod-like GP zones transformed into the rod-shaped γ phase, while that the ellipsoidal GP zones transformed into the equiaxed β phase.

The metastable γ phase reportedly has a hexagonal structure ($a = 0.963$ nm, $c = 1.024$ nm), but the OR between γ and the magnesium matrix phase has not been reported. The metastable β phase also has a hexagonal structure ($a = 0.556$ nm, $c = 0.521$ nm), and its OR with α-Mg phase is such that $(0001)_\beta //$ $(0001)_\alpha$ and $[10\bar{1}0]_\beta // [11\bar{2}0]_\alpha$. The equilibrium phase $Mg_{12}Nd_2Ag$ was originally reported to have a complex hexagonal structure (Gradwell, 1972; Lorimer, 1987). But in two separate studies of alloy QE22 (Kiehn et al., 1997; Barucca et al., 2009), this equilibrium phase was inferred, without any strong supportive evidence, to be $(Mg,Ag)_{12}Nd$ that has a tetragonal structure ($a = 1.03$ nm and $c = 0.59$ nm), that is isomorphous to that of $Mg_{12}Nd$. To unambiguously establish the structures of all precipitate phases, including that of the equilibrium precipitate phase, in the alloy QE22, it seems necessary to employ modern facilities such as atomic resolution HAADF-STEM and electron microdiffraction in any efforts to be made in future studies.

Gradwell (1972) studied the precipitation hardening mechanism in QE22 by examining foils taken from specimens which had been strained 2% in tension after aging for various times at 200 °C. He concluded that peak hardness coincided with the transition from precipitate cutting to Orowan looping, and that maximum age hardening was associated with the presence of the γ and β precipitates. The alloy QE22 in the peak-aged condition has superior tensile properties and creep resistance over many other magnesium alloys. QE22 in its peak-aged condition exhibits a 0.2% proof strength of 205 MPa at room temperature, 195 MPa at 100 °C, and 165 MPa at 200 °C. The creep strength, the stress required to produce 0.2% creep strain in 500 h, of this alloy is 135 MPa at 150 °C and 65 MPa at 200 °C. However, this alloy is relatively expensive and thus has limited applications only in the aircraft and aerospace industries.

20.4.1.7 Mg–Gd- and Mg–Y-Based Alloys

20.4.1.7.1 Mg–Gd Binary Alloys

The equilibrium solid solubility of Gd in magnesium is relatively high (4.53 at.% or 23.49 wt%) at the eutectic temperature of 548 °C and decreases exponentially with temperature to approximately 0.81 at.% (5.0 wt%) at 250 °C, 0.61 at.% (3.82 wt%) at 200 °C, forming an ideal system for precipitation hardening. However, binary Mg–Gd alloys containing less than 10 wt% Gd show little or no precipitation hardening response during isothermal or isochronal aging of supersaturated solid solutions of these alloys (Vostry et al., 1999; Nie et al., 2005). It is often necessary to increase the Gd concentration to the range 10–20 wt% to enhance the precipitation hardening response (Nie et al., 2005; Kamado et al., 1992; Rokhlin and Nikitina, 1994; Li et al., 2011). The aging curve of an Mg–15Gd–0.5Zr (wt%) alloy is shown in **Figure 66(a)**. The hardness value in the as-quenched condition is more than 70 VHN, which is substantially higher than that of the magnesium alloys described in previous sections. This relatively high value of hardness is the result of solid solution strengthening in a supersaturated magnesium matrix. It was demonstrated very recently (Li et al., 2011) that an appreciably high 0.2% proof strength of 445 MPa can be achieved in an Mg–14wt%Gd–0.5wt%Zr alloy when this alloy is produced by the combined processes of hot extrusion, cold work and aging.

The precipitates in Mg–Gd binary alloys containing more than 10 wt% Gd were studied in recent years by conventional TEM (Gao et al., 2006) and HAADF-STEM (Nishijima and Hiraga, 2007; Nishijima et al., 2006). The precipitation sequence is now generally accepted to involve the formation of β″, β′, β_1 and β phases, which is similar to that originally proposed for Mg–Y–Nd-based alloys (Nie and Muddle, 2000). The metastable β″ phase has an ordered DO_{19} structure ($a = 0.641$ nm, $c = 0.521$ nm). The OR between β″ and α-Mg phases is such that $[0001]_{\beta''}//[0001]_{\alpha}$ and $\{2\bar{1}\bar{1}0\}_{\beta''}//\{2\bar{1}\bar{1}0\}_{\alpha}$. The β″ precipitates were originally reported to have a plate-like morphology, with their habit plane almost parallel to $\{2\bar{1}\bar{1}0\}_{\alpha}$. However, this morphology is not consistent with experimental observations made by HAADF-STEM images obtained in recent years, **Figure 66(b)**. The metastable β′ phase usually forms as lenticular particles with their broad surface parallel to $\{2\bar{1}\bar{1}0\}_{\alpha}$, **Figure 66(c)**. It has a base-centered orthorhombic Bravais lattice ($a = 0.650$ nm, $b = 2.272$ nm and $c = 0.521$ nm) and an OR with respect to the matrix phases: $(100)_{\beta'}//\{1\bar{2}10\}_{\alpha}$ and $[001]_{\beta'}//[0001]_{\alpha}$. Based on the analysis made from atomic resolution HAADF-STEM images, **Figure 67(a)**, an atomic structure of the β′ phase was proposed, **Figure 67(b) and (c)**. The zigzag arrays of Gd atoms in the unit cell, indicated by dotted line, are consistent with those of bright dots in atomic resolution HAADF-STEM images. The composition of the β′ phase inferred from this model is Mg_7Gd. The β_1 phase has a face-centered cubic structure that is isomorphous to that of the β_1 phase in WE54 alloy, which has the Mg_3Nd or Mg_3Gd structure (space group F$\bar{4}$3m, $a = 0.74$ nm). The OR is such that $\{\bar{1}12\}_{\beta_1}//\{10\bar{1}0\}_{\alpha}$ and $\langle110\rangle_{\beta_1}//[0001]_{\alpha}$, and its habit plane is parallel to $\{10\bar{1}0\}_{\alpha}$. Precipitates of the

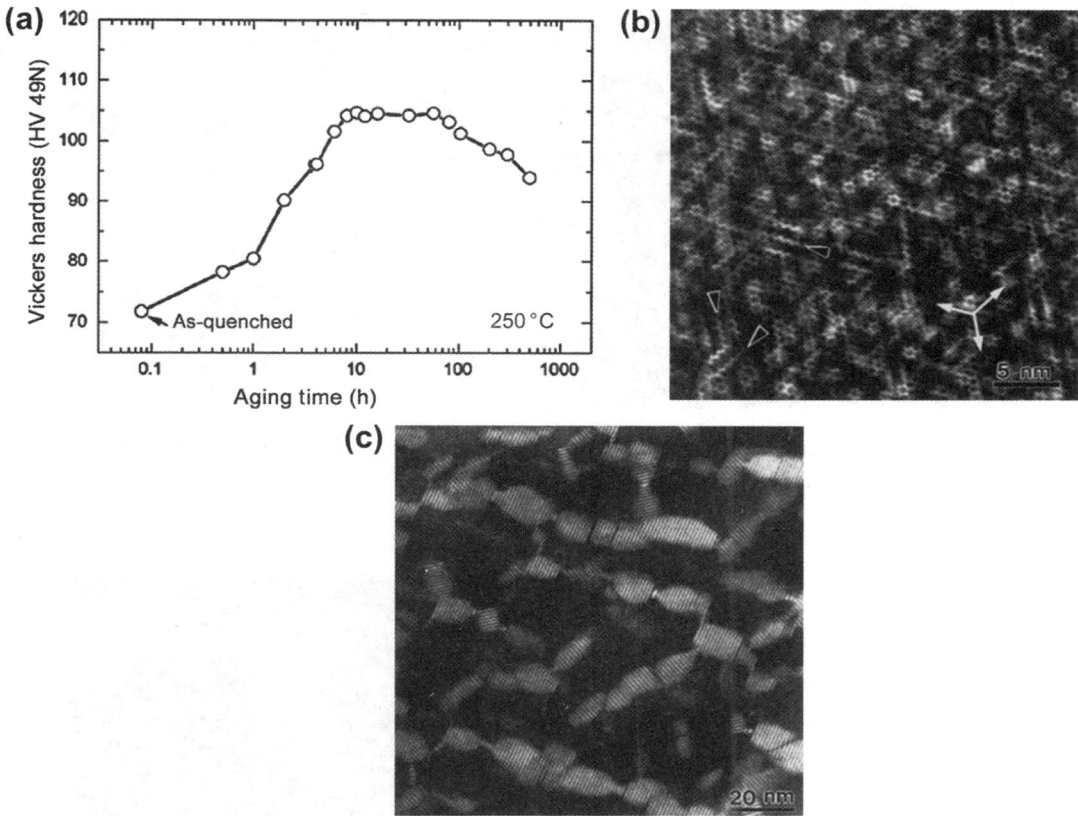

Figure 66 (a) Age hardening response of Mg–15Gd–0.5Zr (wt%) alloy at 250 °C. (b and c) HAADF-STEM images showing distribution and morphology of precipitates in Mg–5at.%Gd alloy aged at 200 °C for 5 and 10 h, respectively. Electron beam is parallel to $[0001]_\alpha$ in (b, c) (Gao et al., 2006; Nishijima and Hiraga, 2007).

equilibrium β phase have a face-centered cubic structure (space group F$\bar{4}$3m, $a = 2.2$ nm) and an Mg$_5$Gd composition. The OR of the β phase with α-Mg is $\langle 1\bar{1}1 \rangle_\beta // \langle 2\bar{1}\,\bar{1}0 \rangle_\alpha$ and $\{110\}_\beta // (0001)_\alpha$. The β phase forms as plates parallel to $\{10\bar{1}0\}_\alpha$ (Gao et al., 2006).

20.4.1.7.2 Mg–Y Binary Alloys

The Mg-rich part of the Mg–Y binary phase diagram indicates that the maximum solid solubility of Y in magnesium is 3.75 at.% (12.5 wt%) at 566 °C (the eutectic temperature) and it decreases to 0.75 at.% (2.69 wt%) Y at 200 °C. The equilibrium intermetallic phase β has a composition of Mg$_{24}$Y$_5$ and a body-centered cubic structure (space group I$\bar{4}$3m, $a = 1.128$ nm). The equilibrium volume fraction of precipitates achievable at 200 °C is approximately 4.5% for an Mg–5wt%Y alloy, and 10.6% for an Mg–8wt%Y composition. The age hardening response of binary Mg–Y alloys is remarkable, when the Y concentration in the alloy is at or above 8 wt% and the aging treatment is carried out at a temperature close to 200 °C (Mizer and Peters, 1972), **Figure 68(a)**, even though the hardening kinetics are quite sluggish during the first 100 h aging at 200 °C.

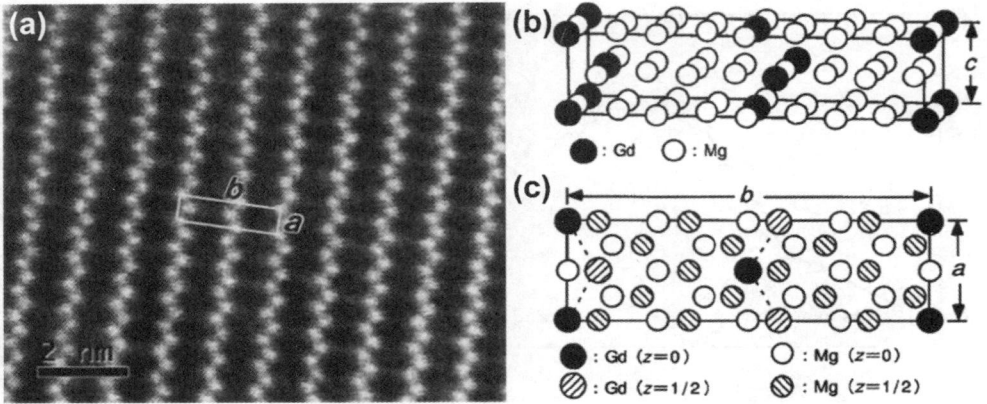

Figure 67 (a) [001] Atomic resolution HAADF-STEM image and (b and c) unit cell of β' phase in Mg–5at.%Gd alloy (Nishijima et al., 2006).

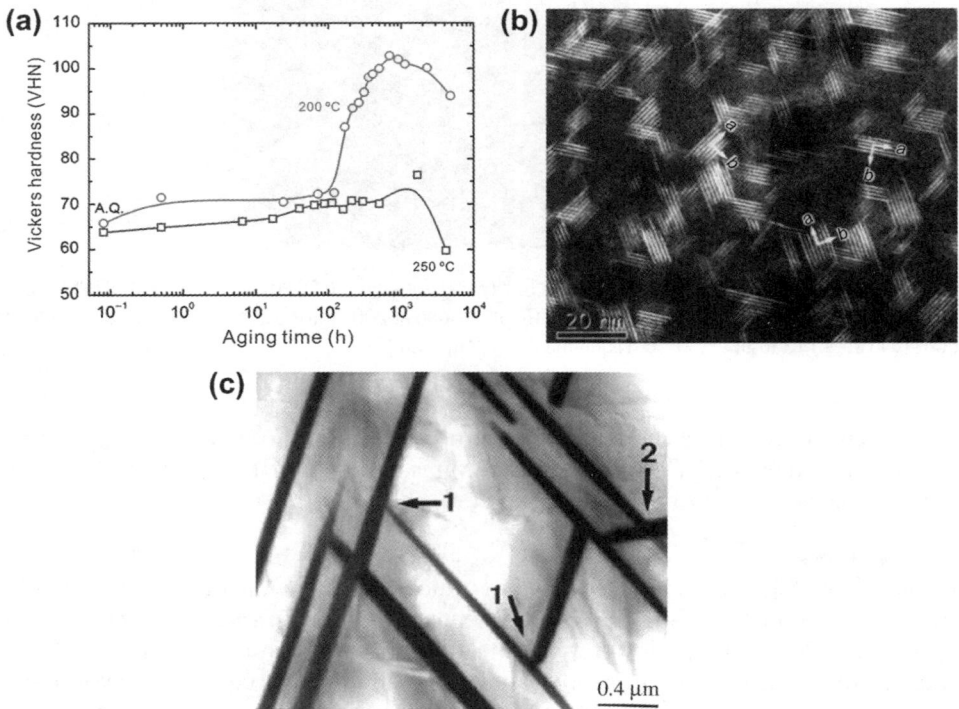

Figure 68 (a) Aging curve of Mg–8wt%Y alloy at 200 and 250 °C. (b) HAADF-STEM image of β' precipitates in Mg–2at.%Y alloy. (c) TEM image showing β precipitate plates in Mg–10.2wt%Y alloy, solution treated for 24 h at 550 °C, water quenched and then aged for 24 h at 350 °C (Nie, 2012; Nishijima et al., 2007b; Zhang and Kelly, 2003). (For color version of this figure, the reader is referred to the online version of this book.)

The precipitate phases formed during isothermal aging of Mg–Y binary alloys were reported (Lorimer, 1987) to include β'', β' and β. The β'' phase has a base-centered orthorhombic lattice ($a = 0.64$ nm, $b = 2.223$ nm and $c = 0.521$ nm) (Karimzadeh, 1985) and an OR $(001)_{\beta''}//(0001)_{\alpha}$, $[100]_{\beta''}//\{\bar{2}110\}_{\alpha}$. Since the structures of β' and β'' phases were reported to be the same (Lorimer, 1987), it is appropriate, at least at the present, to merge the β'' and β' phases into β' in the precipitation sequence. The morphology of the β' in an Mg–2at.%Y alloy, **Figure 68(b)**, is remarkably different from that in an Mg–5at.%Gd alloy, **Figure 66(c)**, even though the structure of β' (Mg$_7$Y) in the Mg–2at.%Y alloy is reportedly the same as that in the Mg–5at.%Gd alloy, **Figure 67(b) and (c)**. The difference was attributed to the difference between lattice parameters of Mg$_7$Gd ($a = 0.650$ nm, $b = 2.272$ nm and $c = 0.521$ nm) and Mg$_7$Y structures, which results in different lattice misfits with Mg matrix (Nishijima et al., 2007b).

Precipitates of the equilibrium phase β have a plate shape. The OR between β and magnesium matrix phase is exact Burgers, with habit plane of the β plates parallel to $\{10\bar{1}0\}_{\alpha}$, or near Burgers with the plate habit plane parallel to $\{31\bar{4}0\}_{\alpha}$, **Figure 68(c)**.

20.4.1.7.3 Mg–Y–Nd and Mg–Gd–Nd Alloys

The age hardening response of Mg–Y–Nd alloys is much higher than that of counterpart Mg–Y and Mg–Nd binary alloys (Vostry et al., 1988; Rokhlin et al., 2004). The most successful commercial magnesium alloys developed to date via precipitation hardening, in terms of strength and creep resistance, have been WE54 (Mg–5wt%Y–2wt%Nd–2wt%HRE) and WE43 alloys based on the Mg–Y–Nd system (Unsworth and King, 1987; Khosrhoshahi et al., 1997). The aging curves of magnesium alloy WE54 at 250 and 200 °C are shown in **Figure 69**. The hardness value in the as-quenched condition is 68 VHN, close to that of the Mg–15Gd–0.5Zr (wt%) alloy shown in **Figure 66(a)**, even though the total concentration of alloying additions is about 9 wt%. Cold work of this alloy by 6%, after solution treatment and before aging, can raise the hardness of as-quenched samples to 75 VHN. Only a slight increase in hardness is obtained when the level of plastic deformation is increased from 6 to 12%. During isothermal aging at 250 °C, the hardness of the undeformed sample increases gradually to a plateau hardness of 87 VHN after 24 h. In contrast, the sample with 6% deformation reaches a maximum hardness of ~97 VHN in 8 h, and that with 12% deformation achieves a maximum hardness of ~94 VHN in 4 h. Therefore, it is possible to obtain higher hardness values in much shorter aging time by cold work. With prolonged aging at 250 °C, the difference in hardness between undeformed and deformed samples diminishes gradually. Higher hardness values can be obtained for both undeformed and deformed samples at 200 °C, **Figure 69(b)**. Although the precipitation hardening response of deformed samples is accelerated, the maximum hardness achievable at 200 °C is not affected significantly by cold work. A maximum hardness between 106 and 110 VHN is achieved at 200 °C for both deformed and undeformed samples.

The microstructure of the undeformed sample aged for 24 h at 250 °C, which is close to the peak-aged condition, is shown in **Figure 69(c)**. It has a relative coarse distribution of $\{1\bar{1}00\}_{\alpha}$ plates of β_1 phase, with globular β' particles attached to the two ends of each of the prismatic plates. In contrast, the $\{1\bar{1}00\}_{\alpha}$ precipitate plates formed samples deformed by 6% before 4 h aging at 250 °C have much larger diameter and higher number density, and the number density of globular particles of β' is significantly reduced, **Figure 69(d)**. Apparently, the dislocations introduced by the cold work promote the nucleation and the growth $\{1\bar{1}00\}_{\alpha}$ precipitate plates at the expense of β' phase. Observations made by TEM (Hilditch et al., 1998) indicate that β_1 precipitates form heterogeneously on preexisting dislocations, similar to the intermediate precipitate phase Mg$_3$X (X = Nd, Ce or MM) in Mg–Nd, Mg–Ce and Mg–MM alloys that also nucleate preferentially on dislocations. The comparison of the microstructures and aging curves of the undeformed and deformed samples indicates that the remarkable increase in hardness, from 75 VHN in undeformed samples to 93 VHN in deformed samples, is associated with the formation of $\{1\bar{1}00\}_\alpha$ precipitate plates of larger diameter and higher number density. As discussed in Section 20.3, an increase in the prismatic plate diameter, number density, or a combination of both, can effectively reduce interparticle spacing and therefore increase hardness and strength.

The T6 condition of alloy WE54 typically involves a solution treatment of 8 h at 525 °C, a hot water quench and a subsequent aging treatment of 16 h at 250 °C. The aged microstructure at maximum hardness was initially reported (Lorimer, 1987; Khosrhoshahi et al., 1997) to contain metastable β′ and equilibrium β phases as dispersed precipitates, and both phases were described to form as plates on

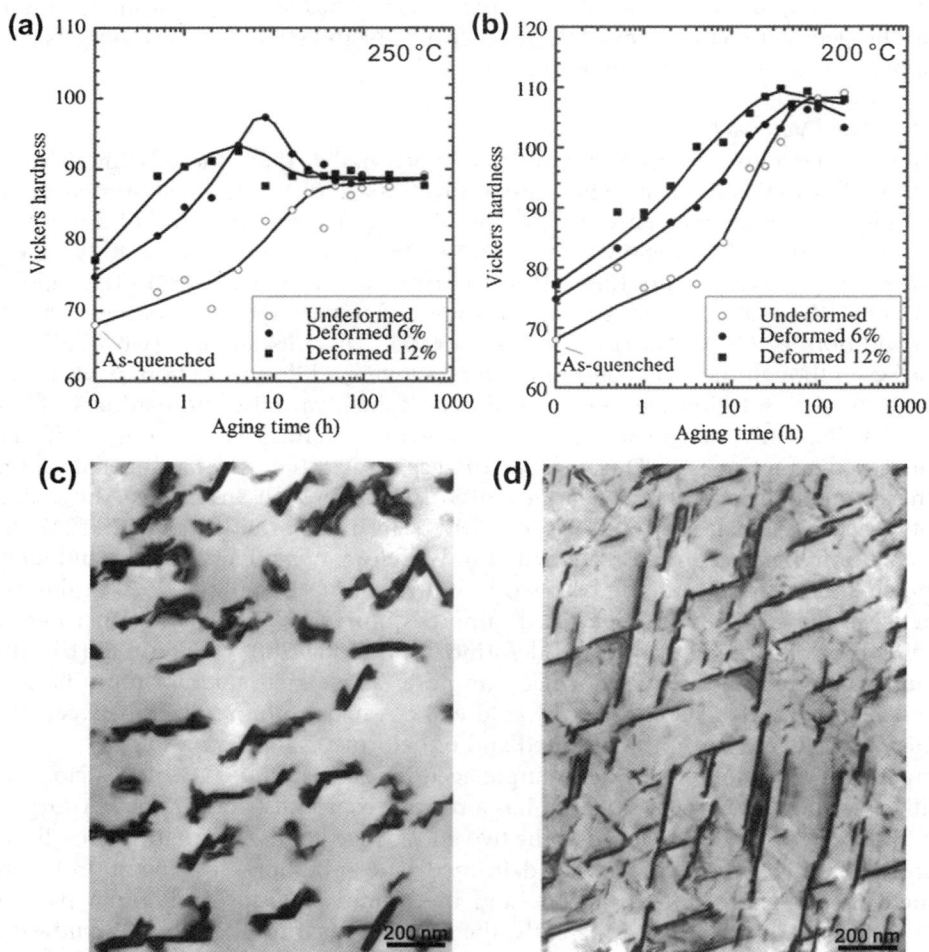

Figure 69   Age hardening response of WE54 alloy at (a) 250 °C and (b) 200 °C. (c and d) TEM images showing distribution of precipitates in (c) samples peak-aged at 250 °C and (d) cold worked 6% after water quench and then aged to peak hardness at 250 °C (Hilditch et al., 1998). (For color version of this figure, the reader is referred to the online version of this book.)

$\{1\overline{1}00\}_\alpha$ planes of the magnesium matrix phase. In early studies, the β' phase was reported to have an $Mg_{12}NdY$ composition and a base-centered orthorhombic structure ($a = 0.640$ nm, $b = 2.223$ nm, $c = 0.521$ nm) (Ahmed et al., 1992; Karimzadeh, 1985; Lorimer, 1987). The OR of β' was reported to have the form $(100)_{\beta'}//(1\overline{2}10)_\alpha$, $[001]_{\beta'}//[0001]_\alpha$. The proposed structure and orientation of the β' phase are similar to those of the β' formed in binary Mg–Y alloys. The β phase was reported to have an $Mg_{14}Nd_2Y$ composition (Ahmed et al., 1992) and a face-centered cubic structure ($a = 2.223$ nm) (Karimzadeh, 1985; Lorimer, 1987). The OR between β and α-Mg matrix is such that $(\overline{1}12)_\beta//(1\overline{1}00)_\alpha$, $[110]_\beta//[0001]_\alpha$, which is identical to that observed between β' and α-Mg in binary Mg–Nd alloys.

In the following years, a number of studies (Nie and Muddle, 2000; 1999b; Antion et al., 2003, 2006; Barucca et al., 2011) have been made to use modern facilities to characterize the precipitates in aged samples of WE54 and WE43. It was reported (Antion et al., 2003; Antion et al., 2006) that, during the early stage of aging of WE43 alloy, monolayer precipitates with a $D0_{19}$ ordering form with two possible habit planes: $\{11\overline{2}0\}_\alpha$ and $\{1\overline{1}00\}_\alpha$, and that the precursor $\{11\overline{2}0\}_\alpha$ monolayers form the β'' ($D0_{19}$) phase whereas the precursor $\{1\overline{1}00\}_\alpha$ monolayers lead to the formation of the β' phase. However, the similarity between these Mg–Y–Nd-based alloys and Mg–Gd alloys suggests that the precipitation in the early stage of the WE alloys is likely to be similar to that in Mg–Gd alloys, **Table 1**. **Figure 70(a)** shows the microstructure typical of samples aged to maximum hardness (48 h at 250 °C). The microstructure contains a dispersion of plate-shaped precipitates in contact with irregular globular

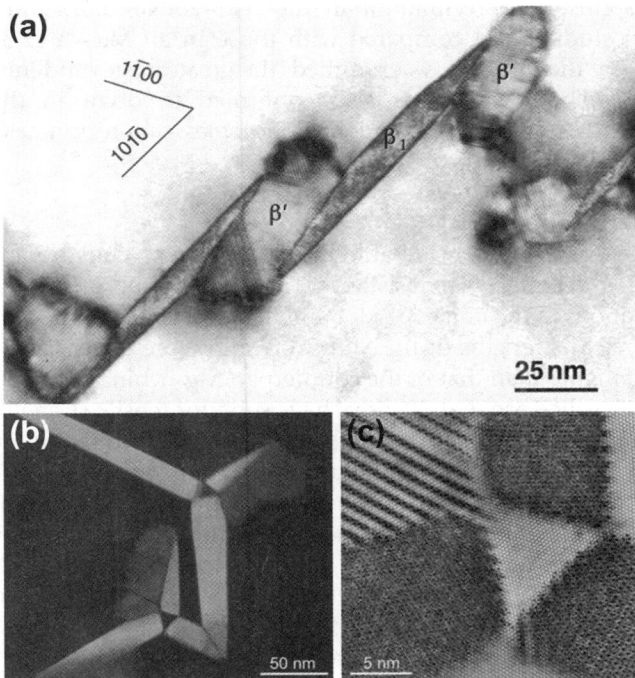

Figure 70 (a) Conventional TEM and (b) HAADF-STEM images showing arrangement of β' and β_1 precipitates in peak-aged samples of WE54 alloy. (c) HAADF-STEM image showing an enlargement of (b) (Nie and Muddle, 2000; Xu et al., 2012).

particles. Examination of electron microdiffraction patterns obtained from individual globular precipitate indicates that these particles are β' phase and that the structure and ORs proposed for the β' phase in early studies are correct. However, the observed morphology of the β' phase is clearly different from that from previous suggestions (Karimzadeh, 1985; Lorimer, 1987), that is β' forms as plates parallel to $\{1\bar{1}00\}_\alpha$.

Contrary to the early studies, Nie and Muddle (2000) reported that the $\{1\bar{1}00\}_\alpha$ plates have a face-centered cubic structure, with $a \sim 0.74$ nm, and an OR that is of the form $(\bar{1}12)_{\beta_1}//(1\bar{1}00)_\alpha$, $[110]_{\beta_1}//[0001]_\alpha$. This phase was designated β_1 because it had not been reported previously in WE alloys. The structure β_1 is similar to that of Mg_3X phase (X = Nd, La, Ce, Gd, Pr, Dy, and Sm), which has a face-centered cubic structure (space group $Fm\bar{3}m$, $a = 0.74 \pm 0.01$ nm) and forms exclusively as $\{1\bar{1}00\}_\alpha$ plates. The β_1 phase invariably forms jointly with other particles, either with β' precipitates such as that shown in **Figure 70(a)** or with themselves in the form of triads, **Figure 70(b)**. In the latter case, the magnesium matrix region isolated by the three β_1 variants is rotated by approximately 10.5°, **Figure 70(c)**. These observations indicate that the formation of β_1 phase involves a large shear strain (Nie and Muddle, 2000), which will be discussed in Section 20.4.3. The $\{1\bar{1}00\}_\alpha$ plates of β_1 transform in situ to the equilibrium phase β during prolonged aging at 250 °C. Examination of pattern symmetries in the higher order Laue zone and zero-order Laue zone patterns recorded from the β plates, together with recognition of absent reflections in the observed electron microdiffraction patterns, indicates a space group of $F\bar{4}3m$ ($a = 2.2 \pm 0.1$ nm). This structure is isomorphous with Mg_5Gd. The OR is similar to that observed between β_1 and the matrix phase.

The age hardening response and precipitation in Mg–7Gd–2.3Nd–0.6Zr and Mg–7Dy–2.3Nd–0.6Zr (wt%) alloys have been studied and compared with those in an Mg–4Y–2.3Nd–0.6Zr (wt%) alloy (Apps et al., 2003). Among the three alloys examined, the greatest age hardening response is found in the Gd-containing alloy. The β_1 phase is also confirmed to occur in the Mg–Gd–Nd–Zr and Mg–Dy–Nd–Zr alloys, and the structures and morphologies of precipitates and the precipitation sequences in the three alloys are identical.

20.4.1.7.4 Mg–Gd–Y-Based Alloys

The Mg–Gd–Y alloys have received considerable interest in recent years because of their potential in achieving higher strength and better creep resistance. The Gd:Y atomic ratio in these alloys is important. When the Gd:Y atomic ratio is in the range 3:1–1:1 and the total concentration of the alloying additions is 2.75 at.%, the tensile yield strengths of the Mg–Gd–Y alloys are lower than that of the counterpart Mg–Gd binary alloy and higher than that of the counterpart Mg–Y binary alloy (Anyanwu et al., 2001). The T5 samples of the hot-rolled Mg–Gd–Y alloys have superior tensile strengths to commercial WE54 alloy. The aging curves of two representative alloys in this category are shown in **Figure 71(a) and (b)**. The effects of Zn additions are also shown in **Figure 71(b)**. In general, the additions of Zn reduce the aging hardening response of these alloys. When the Zn content is more than 1.5–2.0 at.%, the age hardening response of the resultant Mg–Gd–Y–Zn alloys diminishes completely (Yamada et al., 2006). However, an impressively high value of 473 MPa has been achieved for the 0.2% proof strength in an extruded and aged Mg–1.8Gd–1.8Y–0.7Zn–0.2Zr (at.%) or Mg–10Gd–5.7Y–1.6Zn–0.7Zr (wt%) alloy (Homma et al., 2009).

The remarkable age hardening response achieved in the Mg–Gd–Y alloys is attributable to a dense distribution of precipitates in the microstructure. The precipitation sequence in this group of alloys appears to be similar to Mg–Gd, Mg–Gd–Nd and Mg–Y–Nd alloys. The β' precipitates formed in the peak-aged samples have a lenticular shape, **Figure 71(c)**, which is more effective in impeding dislocation slip than the globular β' precipitates formed in WE54 alloy.

(a)   **(b)**   **(c)**

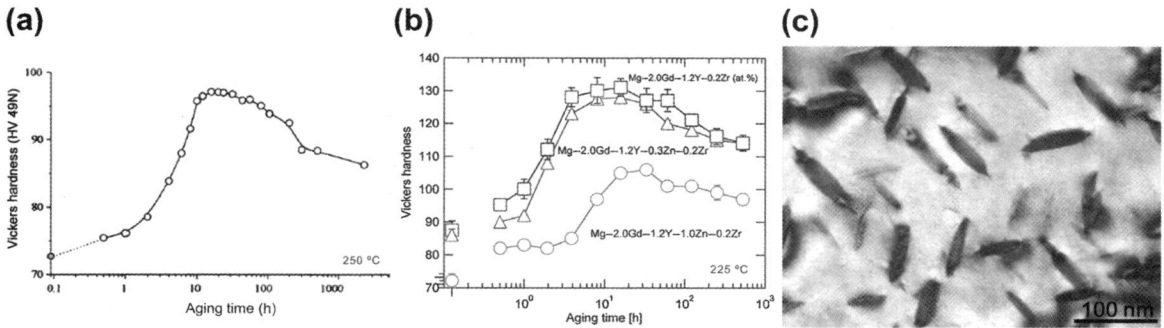

Figure 71   Aging curves of (a) Mg–10wt%Gd–3wt%Y–0.4wt%Zr alloy at 250 °C and (b) Mg–2at.%Gd–1.2at.%Y–(0–1)at.% Zn–0.2at.%Zr alloys at 225 °C. (c) TEM image showing β′ precipitates in peak-aged samples (16 h at 250 °C) of Mg–10wt% Gd–3wt%Y–0.4wt%Zr alloy (He et al., 2006; Homma et al., 2007). (For color version of this figure, the reader is referred to the online version of this book.)

20.4.1.7.5 Mg–Gd–Zn Alloys

Since the equilibrium solid solubility of Gd in magnesium is about 0.8 at.% at 250 °C and 0.6 at.% at 200 °C, Mg–Gd alloys containing less than 1.0 at.% (6.1 wt%) Gd have poor age hardening response at 200–250 °C because of low volume fractions of precipitates, **Figure 72(a)**. For example, the maximum volume fraction of precipitates achievable at 250 °C is approximately 1.1%, according to the lever rule calculation, if the precipitates are assumed to have a composition close to that of the equilibrium phase Mg$_5$Gd. Therefore, it has been of commercial interest to explore the possibility of enhancing the precipitation hardening response of the diluted Mg–Gd alloys via the use of microalloying additions of low-cost elements. A recent study (Nie et al., 2005) reported that additions of 1–2 wt% Zn to an Mg–6wt%Gd alloy could considerably enhance the solid solution strengthening effect and generate a relatively strong precipitation hardening response, **Figure 72(a)**.

The 3DAP work (Nie et al., 2008) did not reveal any detectable clusters of Gd and/or Zn atoms in the as-quenched samples of an Mg–6Gd–1Zn–0.6Zr (wt%) alloy. But the statistical analysis of the 3DAP data suggests that co-segregation of Gd and Zn atoms occurs in the as-quenched samples. The co-segregation phenomenon of Gd and Zn atoms is attributed to the atomic size difference between the solute and the solvent. The atomic radius is 0.180 nm for Gd, 0.133 for Zn, and 0.160 for Mg. Therefore, substituting an Mg atom by a Gd atom leads to a *compression* strain, while replacing an Mg atom by a Zn atom causes an *extension* strain. It is therefore energetically favorable for Gd and Zn atoms to segregate to each other to minimize the elastic strain associated with individual Gd or Zn atoms. It was further speculated (Nie et al., 2008) that the Gd and Zn atoms form Gd–Zn dimers in the α-Mg solid solution and that these dimers are more effective barriers, than those from individual atoms of Gd or Zn, to the motion of dislocations and therefore contribute to the large hardness increase in the as-quenched condition.

The experimental observations made by TEM indicate that the precipitation process during isothermal aging at 200 and 250 °C of the Mg–6Gd–1Zn–0.6Zr (wt%) alloy involves the formation of metastable γ″ and γ′ phases and that the peak hardness is associated with the γ″ precipitates, **Figure 72(b)**. Both precipitate phases form as plate-shaped particles on (0001)$_α$. Most precipitates in the peak-aged condition are γ″ phase that has an ordered hexagonal structure (space group P$\bar{6}$2m, $a = 0.556$ nm and $c = 0.444$ nm) with an ABA stacking sequence of the close-packed planes,

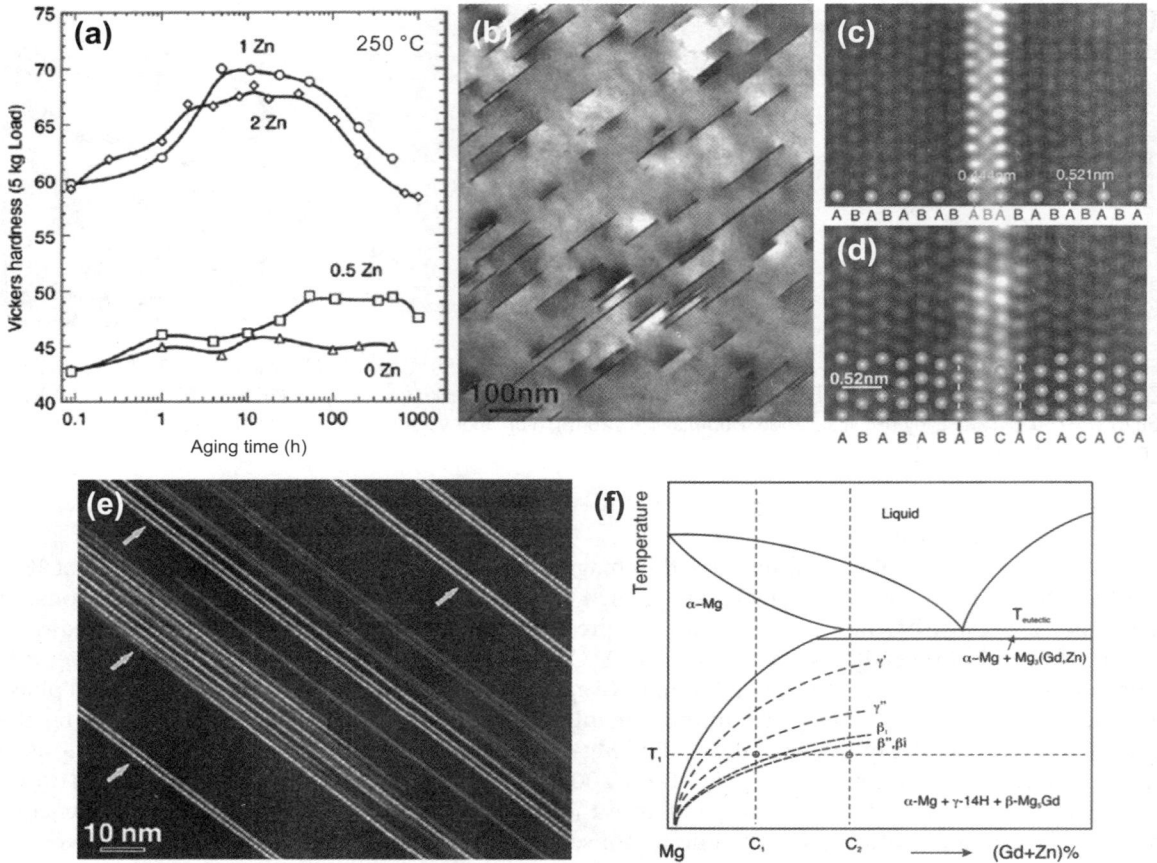

Figure 72 (a) 250 °C aging curves of Mg–6wt%Gd–Zn alloys. (b) Conventional TEM image showing the distribution of precipitates in peaked-aged samples of Mg–6Gd–1Zn–0.6Zr alloy. (c and d) Atomic resolution HAADF-STEM images showing structures of γ'' and γ' phases in Mg–6Gd–1Zn–0.6Zr alloy, aged for 2 h at 250 °C. (e) HAADF-STEM image showing precipitates in Mg–6Gd–1Zn–0.6Zr alloy, aged for 1000 h at 250 °C. (f) Schematic diagram showing probably solvus lines of metastable and equilibrium phases (Nie et al., 2005; Nie et al., 2008). (For color version of this figure, the reader is referred to the online version of this book.)

Figure 72(c), and a composition of $Mg_{70}Gd_{15}Zn_{15}$ (Nie et al., 2008). This structure is subtly different from that reported for the basal plates in an Mg–2Gd–1Zn (at.%) alloy (Nishijima et al., 2008) in terms of the distribution of Zn atoms and the c value of the unit cell. The OR between γ'' and α-Mg phases is such that $(0001)_{\gamma''}//(0001)_{\alpha}$ and $[10\bar{1}0]_{\gamma''}//[2\bar{1}\bar{1}0]_{\alpha}$. The thickness of the γ'' plates is often of a single unit cell height. This phase is fully coherent with the matrix in its habit plane, but with a relatively large misfit strain (-0.16) in the direction normal to the habit plane. It is to be noted that GP zones have been reported (Nishijima et al., 2008; Saito et al., 2011) to form in Mg–2Gd–1Zn (at.%) and Mg–1.5Gd–1Zn (at.%) alloys. However, inspection of the images of and electron diffraction patterns from these "GP zones" indicates that they are actually γ'' phase.

With continued aging at 250 °C, the γ″ precipitates are gradually replaced by γ′ precipitates. The γ′ phase has a disordered hexagonal structure (space group P3̄m1, $a = 0.321$ nm and $c = 0.780$ nm) and an MgGdZn composition. The OR is such that $(0001)_{\gamma'}//(0001)_\alpha$ and $[2\bar{1}\bar{1}0]_{\gamma'}//[2\bar{1}\bar{1}0]_\alpha$. The γ′ plates are perfectly coherent with the matrix phase in their habit plane and in the direction normal to their habit plane. Their thickness is again invariably of a single unit cell height, and their aspect ratio is considerably larger than that of γ″. In contrast to the γ″ plates, the γ′ plates have an ABCA stacking order of their close-packed planes, **Figure 72(d)**. Therefore, the formation of a γ′ precipitate of a single unit cell height generates a shear strain of about 0.35. This large shear strain can impose a large barrier to the nucleation of γ′ plates. While γ′ is a metastable phase, it is remarkably resistant to thickening during isothermal aging at 250 °C. It is still less than 1 nm in thickness, and remains the dominant precipitate phase in the microstructure of the Mg–6Gd–1Zn–0.6Zr (wt%) alloy after long-term aging (1000 h at 250 °C), **Figure 72(e)**.

Independently, Yamasaki et al. (2007) studied the precipitation process in an Mg–2Gd–1Zn (at.%) alloy. The Gd and Zn concentrations in their alloy are almost twice of those in the Mg–6wt%Gd–1wt%Zn–0.6wt%Zr (Mg–1at.%Gd–0.4at.%Zn–0.2at.%Zr) alloy. The precipitation in the Mg–2Gd–1Zn (at.%) alloy involves the formation of stacking faults and 14H phase on $(0001)_\alpha$ and β′, β₁ and β phases. The stacking faults and 14H precipitates reportedly form at intermediate and high temperatures (300–500 °C), while the β′, β₁ and β phases form at low temperatures (∼200 °C). It was suggested (Yamasaki et al., 2007) that the 1 at.% Zn addition to the Mg–2at.%Gd alloy reduced the stacking fault energy and therefore promoted the formation of stacking faults and the 14H phase. The stacking faults in the Mg–2at.%Gd–1at.%Zn alloy were reported to be I_1 and I_2 intrinsic stacking faults, and segregation of Gd and Zn atoms into the two atomic planes around each single stacking fault was observed. The appearance of these so-called stacking faults resembles that of the γ′ phase in the same alloy system and, to avoid confusion, it is appropriate to call them γ′ precipitates. Otherwise, it would be difficult to understand why these so-called stacking faults form preferentially only at intermediate and high temperatures, rather than at both high and low temperatures, that is in the temperature range 25–500 °C. Furthermore, the stacking fault energy would be unrealistically low when the extraordinarily long length (the separation distance between the two Shockley partials binding the stacking fault) is taken into account. The work by Nie et al. (2008) suggests that the I_1 and I_2 intrinsic stacking faults reported by Yamasaki et al. (2007) are in fact two twin-related variants of γ′ precipitate phase.

While some particles of the equilibrium intermetallic phase β-Mg₅Gd start to form in well overaged samples of the Mg–6Gd–1Zn–0.6Zr (wt%) alloy, the metastable phases β″, β′ and β₁, which usually form in Mg–Gd and Mg–Gd–Y alloys, and γ phase (14H) are not observed in the Mg–6Gd–1Zn–0.6Zr (wt%) or Mg–1Gd–0.4Zn–0.2Zr (at.%) alloy. In the Mg–2at.%Gd–1at.%Zn alloy, the 14H phase can form from the supersaturated solid solution phase of α-Mg, and the decomposed primary intermetallic particles Mg₃Gd (Fm3̄m, $a = 0.72$ nm) that have already formed in the as-cast microstructure. The 14H seems to be an equilibrium phase in this alloy and other alloys of similar compositions (Yamasaki et al., 2007; Yamasaki et al., 2005; Homma et al., 2007; Yamada et al., 2006). While Mg–1Gd–0.4Zn–0.2Zr, Mg–2Gd–1Zn and Mg–2.5Gd–1Zn (at.%) alloys have a similar Gd:Zn atomic ratio, the precipitate process in these alloys is quite different. It is currently difficult to rationalize the observations in different alloys due to a lack of isotherm sections of the ternary Mg–Gd–Zn phase diagram. Nevertheless, one plausible explanation is that the 200–250 °C aging temperature range is above the solvus lines of β″, β′ and β₁ for the Mg–1Gd–0.4Zn–0.2Zr (at.%) alloy that has lower contents of Gd and Zn, but is below the solvus lines of these metastable phase for the Mg–2Gd–1Zn (at.%) and Mg–2.5Gd–1Zn (at.%) compositions, **Figure 72(f)**.

20.4.1.7.6 Mg–Y–Zn Alloys

The experimental evidence accumulated thus far indicates that Zn additions to Mg–Y alloys significantly reduce the equilibrium solid solubility of Y in magnesium. Therefore, the volume fraction of solid-state precipitates is quite low in the resultant Mg–Y–Zn alloys, and the Mg–Y–Zn alloys exhibit little age hardening response. Consequently, Mg–Y–Zn alloys, for example Mg–2at.%Y–2at.%Zn (Mg–6.7wt%Y–4.9wt%Zn) (Yoshimoto et al., 2006), are usually hot extruded to achieve useful tensile properties.

It is perhaps for the lack of age hardening that the precipitation sequence in the Mg–Y–Zn alloys has not been well characterized in the past. In the early studies of Mg–Y–Zn alloys (Suzuki et al., 2003, 2004), some planar defects were observed to form on the basal plane of the magnesium matrix phase. Based on the invisibility analysis of the two-beam TEM images obtained from these planar defects, these defects were reported to be I_1 intrinsic stacking fault bounded by Frank partial dislocations ($b = \pm 1/6 \langle 20\bar{2}3 \rangle_\alpha$, $a + c$ type). This fault can be generated via the condensation of vacancies onto a single atomic plane, which leads to a dislocation loop bounded by Frank partials with $b = \pm 1/2 \langle 0001 \rangle_\alpha$, and subsequent formation of a Shockley partial loop ($b = \pm 1/3 \langle 10\bar{1}0 \rangle_\alpha$) within this atomic plane. Similar planar defects are also observed in an early study of Mg–Th–Zn alloys aged at 300–400 °C (Noble et al., 1971), but they were reported to be intrinsic stacking faults I_2 bounded by Shockley partial dislocations ($b = \pm 1/3 \langle 10\bar{1}0 \rangle_\alpha$). It is apparent that a wide range of planar features exists in the Mg–Y–Zn, Mg–Gd–Zn and Mg–Th–Zn alloys. Some of them have many of the characteristic features of stacking faults, but the analysis made did not draw any clear conclusions concerning their identity and relationships.

Stacking faults in magnesium alloys include intrinsic faults I_1 and I_2 and extrinsic fault E (Hirth and Lothe, 1968). An I_1 fault can be produced by removing the B plane, usually via the condensation of aggregates of vacancies, above an A plane and then shearing the remaining planes above the A plane by a displacement of $1/3 \langle 10\bar{1}0 \rangle_\alpha$. In this case, the stacking order of the closely packed planes changes from

$$\cdots\text{ABABABAB}\cdots\text{to}\cdots\text{ABABACAC}\cdots(I_1).$$

An I_2 fault is generated directly by the passage of a Shockley partial dislocation on $(0001)_\alpha$, or by directly shearing the hexagonal lattice by a displacement of $1/3 \langle 10\bar{1}0 \rangle_\alpha$. The passage of the Shockley partial, or shearing, changes the stacking sequence of closely packed planes from

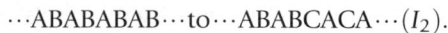

$$\cdots\text{ABABABAB}\cdots\text{to}\cdots\text{ABABCACA}\cdots(I_2).$$

These two types of intrinsic stacking faults can be distinguished by that fact that the I_1 fault is bound by a pair of Frank partials ($b = 1/6 \langle 20\bar{2}3 \rangle_\alpha$), while that the I_2 fault is bound by a pair of Shockley partial dislocations ($b = 1/3 \langle 10\bar{1}0 \rangle_\alpha$). The extrinsic fault E can be generated by inserting a C plane into the …ABAB… stacking sequence. The stacking of the closely packed planes is then changed from

$$\cdots\text{ABABABAB}\cdots\text{to}\cdots\text{ABABCABAB}\cdots(E).$$

Note that it is the I_2 fault, rather than I_1, that yields an ABCA stacking that is characteristic of the γ' structure in Mg–Gd–Zn and Mg–Y–Zn alloys. While the E fault can also generate the ABCA stacking, the packing order of closely packed planes outside the ABCA segment is distinguishably different from that associated with the I_2 fault. If the ABCA segment is taken as a precipitate plate, then the closely packed planes of the magnesium lattice at both sides of the plate are symmetrically arranged for the E fault and asymmetrically stacked for the I_2 fault.

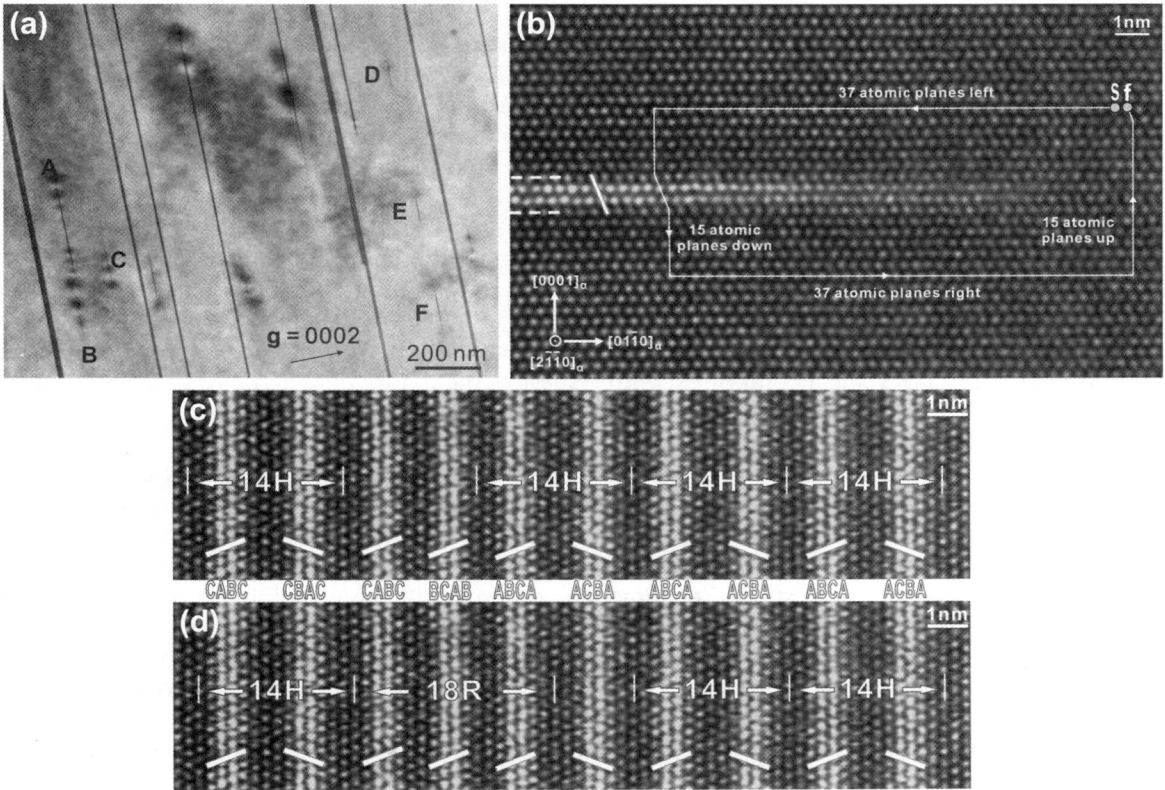

Figure 73   (a) Conventional TEM image showing the distribution of precipitates in Mg–8Y–2Zn–0.6Zr (wt%) alloy solution treated for 1 h at 500 °C and water quenched. (b and c) Atomic resolution HAADF-STEM images showing γ′, 18R and 14H phases in Mg–8Y–2Zn–0.6Zr (wt%) alloy solution treated for 1 h or 16 h at 500 °C (Zhu et al., 2010a; Zhu et al., 2009).

In more recent years, the planar defects in the microstructure of an Mg–8Y–2Zn–0.6Zr (wt.%) alloy, solution treated for 1 h at 500 °C and water quenched, were characterized in detail using conventional TEM imaging techniques and a modern Z-contrast imaging technique, that is atomic resolution HAADF-STEM (Zhu et al., 2010a), and were analyzed using both traditional $\mathbf{g \cdot b}$ and $\mathbf{g \cdot R}$ invisibility criteria and computer image simulation.

The alloy microstructure contains three types of planar features on $(0001)_\alpha$, **Figure 73(a)**. The first type of planar features are small ribbons. They are intrinsic stacking faults I_2, bounded by two Shockley partial dislocations, rather than I_1. The second type is a precipitate γ′ phase that has a single unit cell height and is always associated with Shockley partial dislocations, **Figure 73(b)** (Zhu et al., 2009). The structure of the γ′ phase appears similar to that in the Mg–Gd–Zn alloys, **Figure 72(d)**. The third type is the 14H precipitate phase that is again always associated with Shockley partial dislocations. In the literature this type of planar features is often inferred to be an intrinsic stacking fault I_2 bounded by a Frank partial dislocation (Suzuki et al., 2003). The 14H phase is an equilibrium phase in the Mg–Y–Zn system. The probable precipitation sequence in the Mg–Y–Zn alloys is given in **Table 1**.

The 14H phase was originally reported (Matsuda et al., 2005) to have a *disordered* hexagonal structure ($a = 0.321$ nm and $c = 3.694$ nm). The close-packed planes of this structure have a long-period stacking order and are arranged in an ACBCBABABABCBCA stacking sequence. But in a more recent study (Zhu et al., 2010c), it was proposed that 14H has in fact an *ordered* hexagonal structure ($a = 1.112$ nm, $c = 3.647$ nm), and an $Mg_{12}YZn$ composition that is identical to that of the equilibrium X phase in the Mg–Y–Zn system (Padezhnova et al., 1982; Shao et al., 2006). The stacking sequence of the close-packed planes is ABABCACACACBABA in the 14H lattice (Nie et al., 2008; Zhu et al., 2010c), and the OR between 14H and α-Mg is $(0001)_{14H}//(0001)_\alpha$ and $\langle 0\bar{1}10 \rangle_{14H}//\langle \bar{1}120 \rangle_\alpha$.

The 14H unit cell is made up of two structural units, or building blocks, that are separated by three $(0001)_\alpha$ planes of magnesium, **Figure 73(c)**. Each structural unit has an ABCA-type stacking sequence, and Y and Zn atoms have an *ordered* arrangement in the B and C layers, that is the two middle layers, of each structural unit. The stacking sequence of the close-packed planes in the structural unit is such that each structural unit has a shear component with respect to the matrix phase. The two structural units within the 14H unit cell are twin related. Therefore, the joint formation of these two blocks generates a zero net shear relative to the α-Mg matrix.

In the as-cast microstructures of Mg–Y–Zn alloys, produced by either conventional ingot casting or rapid solidification processing, it is often to observe primary intermetallic particles of a long-period stacking ordered (LPSO) structure. It seems that this LPSO structure can readily form from the melt, irrespective of the solidification rates in the casting. This LPSO phase can be produced in nanoscales and larger volume fractions when Mg–Y–Zn alloys are produced by rapid solidification process. For example, the size of the intermetallic particles of the LPSO phase ranges from 50 to 250 nm in an Mg-2at.%Y-1at.%Zn alloy produced by gas atomization, compaction and hot extrusion (Ping et al., 2002). This alloy can have a 0.2% proof strength exceeding 600 MPa and an elongation to fracture of 5% (Kawamura et al., 2001).

In the early studies (Abe et al., 2002; Ping et al., 2002) the intermetallic particles in the Mg–Y–Zn alloys were reported to have a 6H structure, which has a monoclinic unit cell ($a = 0.56$ nm, $b = 0.32$ nm, $c = 1.56$ nm and $\beta = 88°$) and an ABCBCB′ stacking sequence of the close-packed planes. In the proposed 6H structure, the A and B′ layers are significantly enriched by Zn and Y (Abe et al., 2002), with the Y and Zn content of approximately 10 at.% and 3 at.%, respectively, in each of these two layers (Ping et al., 2002). In subsequent studies (Itoi et al., 2004; Matsuda et al., 2005) the 6H structure was regarded as not correct and was superseded by an 18R structure (hexagonal unit cell, $a = 0.321$ nm, $c = 4.86$ nm) with an ACBCBCBACACACBABABA stacking sequence of the close-packed planes. This structure is identical to that of the $X–Mg_{12}YZn$ phase proposed by Luo and Zhang (2000). In all such studies, the term "order" refers to the ordered stacking of the close-packed planes, rather than ordered arrangement of Y and Zn atoms in the close-packed planes.

In a very recent study (Zhu et al., 2010c) the 18R unit cell was reported to be *ordered* monoclinic ($a = 1.112$ nm, $b = 1.926$ nm, $c = 4.722$ nm, and $\beta = 83.25°$), with Y and Zn atoms occupying some specific positions of the unit cell. The OR between the 18R and α-Mg phases is such that $(001)_{18R}//(0001)_\alpha$ and $[010]_{18R}//\langle 1\bar{2}10 \rangle_\alpha$. The 18R unit cell has an ACACBABABACBCBCBACA stacking sequence of its close-packed planes and is made up of three structural units, with two adjacent units separated by two $(0001)_\alpha$ planes of magnesium, **Figure 73(d)**. Similar to that in the 14H phase, each structural unit of the 18R phase has an ABCA-type stacking sequence, and Y and Zn atoms have an ordered arrangement in the B and C layers, that is the two middle layers of each structural unit. However, the three structural units within the 18R unit cell have the same stacking sequence, in contrary to the structural units within the 14H unit cell. The experimental measurements and this latter model of 18R indicate that its stoichiometric composition is $Mg_{10}YZn$, rather than $Mg_{12}YZn$ that has long been

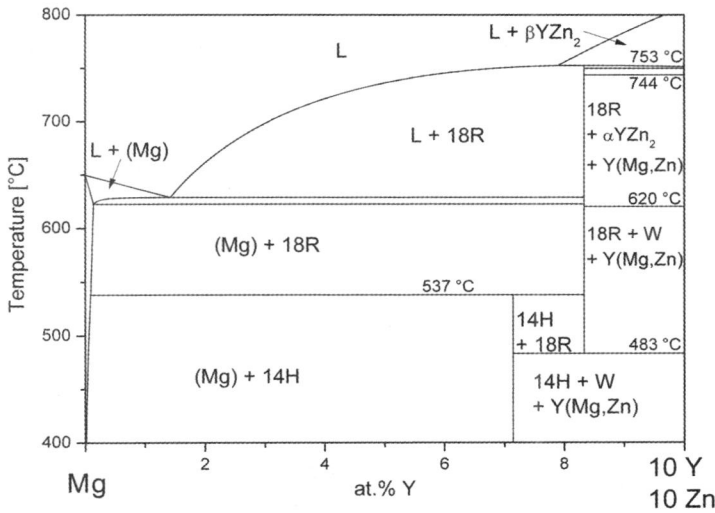

Figure 74   Vertical section of the Mg–Y–Zn ternary phase diagram from Mg-rich corner to $Mg_{80}Y_{10}Zn_{10}$ through 18R and 14H phases (Courtesy R. Schmid-Fetzer; Nie, 2012).

assumed and commonly accepted in the earlier studies. This error occurred presumably because the 18R structure was mistakenly taken as the structure of the equilibrium X–Mg_{12}YZn phase in the work of Luo and Zhang (2000). Luo and Zhang (2000) did not provide any information on their sample preparation conditions, and it is very likely that the intermetallic particles that they studied are the 18R phase, rather than the equilibrium X–Mg_{12}YZn phase. The 18R phase is observed predominantly in the as-cast microstructure of Mg–Y–Zn alloys. The accumulated experimental evidence, and the calculated Mg–Y–Zn phase diagram shown in **Figure 74** (Grobner et al., 2012), indicates that 18R is not thermodynamically stable at temperatures below 500 °C; it is gradually replaced by 14H after prolonged heat treatment at 350–500 °C.

It is to be noted that the stacking sequence of the close-packed planes of the structural units of the 14H and 18R phases is the same as that of the γ' phase, **Figure 72(d) and Figure 73(b)**, even though the γ' phase is disordered. In two very recent studies of LPSO structures in Mg–5at.%Gd–3.5at.%Al (Yokobayashi et al., 2011) and Mg–2at.%Y–1at.%Zn, Mg–9at.%Y–6at.%Zn and Mg–2at.%Er–1at.%Zn (Egusa and Abe, 2012) alloys, the structural unit of the LPSO structures was proposed to have an ordered enrichment of Y and Zn atoms in all four layers, instead of the middle two layers, of the ABCA units. In the Mg–5at.%Gd–3.5at.%Al alloy, the HAADF-STEM images obtained indicate that Gd atoms are arranged with an order in all four layers of the structural unit, with a greater enrichment in the inner two layers of the unit. However, it should be pointed that the SAED patterns recorded from the LPSO phase in this alloy, **Figure 2** in Yokobayashi et al. (2011), are not fully consistent with those obtained from the LPSO phase in the Mg–Y–Zn alloys, **Figure 11** in Yokobayashi et al. (2011), and therefore further efforts are needed in the future to establish whether the structure reported for the LPSO phase in the Mg–Gd–Al alloy is representative of that of the Mg–Y–Zn alloys.

In the Mg–2at.%Y–1at.%Zn, Mg–9at.%Y–6at.%Zn and Mg–2at.%Er–1at.%Zn alloys (Egusa and Abe, 2012), the HAADF-STEM images of the 18R and 14H phases do not show any strong evidence of a systematic ordering of Y and Zn atoms in the two outer layers, and the SAED patterns obtained from the LPSO phases in these three alloys are not fully self-consistent in terms of the intensity of some reflections. Again, a further systematic study is needed in the future to reconcile the HAADF-STEM images and SAED patterns obtained from different alloys and alloys prepared under different processing conditions.

While 18R and 14H are the most frequently observed LPSO structures in Mg–Y–Zn alloys and many other magnesium alloys such as Mg–Gd–Zn, Mg–Gd–Y–Zn, Mg–Dy–Zn, Mg–Ho–Zn, Mg–Er–Zn, Mg–Tm–Zn, Mg–Tb–Zn, Mg–Y–Cu(–Zn), and possibly Mg–Gd–Al (Kawamura and Yamasaki, 2007; Kawamura et al., 2006; Matsuura et al., 2006; Amiya et al., 2003; Hui et al., 2007; Homma et al., 2007; Yamada et al., 2006; Nie et al., 2008; Yamasaki et al., 2007; Yokobayashi et al., 2011), a few other long-period structures such as 10H and 24R have also been reported (Matsuda et al., 2005). However, whether these LPSO structures are also ordered or not, and their relationships with the ABCA-type building block and 18R and 14H structures, remain to be unambiguously established.

20.4.1.7.7   Mg–Gd–Ag Alloys

Recent studies (Gao and Nie, 2008; Yamada et al., 2009) indicate that the addition of 2 wt% Ag to the Mg–6Gd–0.6Zr (wt%) alloy can lead to a significant acceleration and increase in the age hardening response in the resultant Mg–6Gd–2Ag–0.6Zr alloy, **Figure 75(a)**. When combined with 1 wt% Zn, this

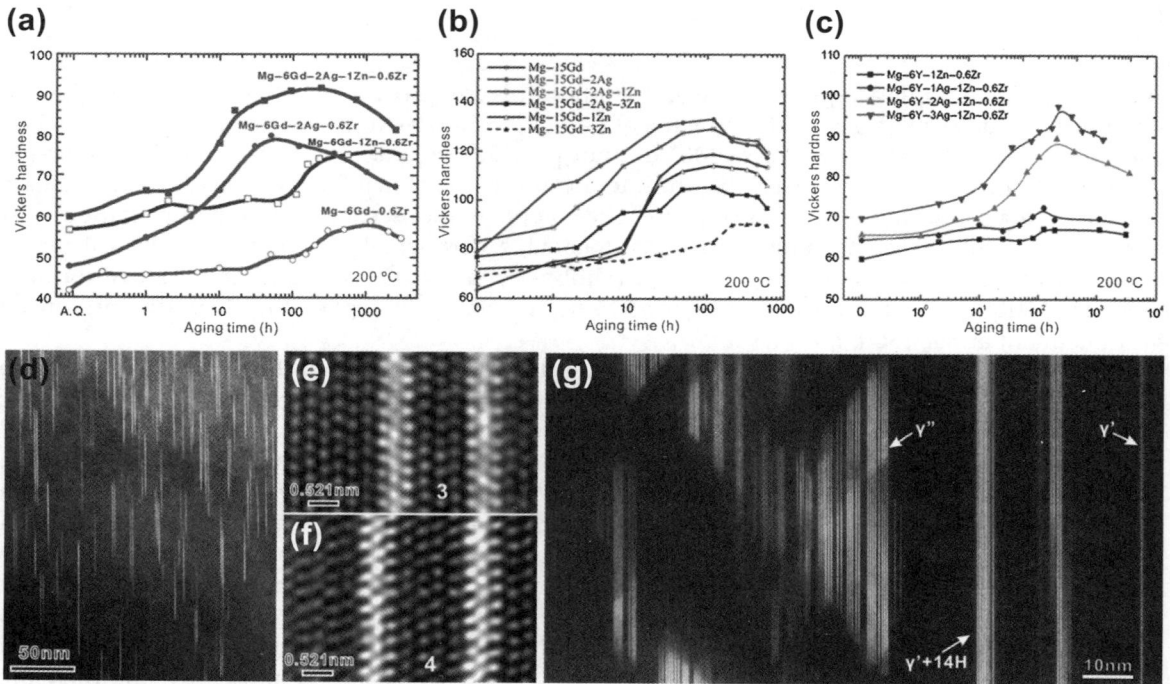

Figure 75 Aging curves at 200 °C of (a) Mg–6Gd–2Ag-based alloys, (b) Mg–15Gd–2Ag-based alloys, and (c) Mg–6Y–(1–3) Ag–1Zn-based alloys. HAADF-STEM images showing the distribution of precipitates in (d–f) peak-aged (224 h at 200 °C) and (g) overaged (5800 h at 200 °C) samples of Mg–6Y–2Ag–1Zn–0.6Zr alloy. All alloy compositions are in weight percentage (Gao and Nie, 2008; Ma, 2007; Zhu et al., 2008; Zhu, 2011). (For color version of this figure, the reader is referred to the online version of this book.)

addition can result in a further substantial increase in the peak hardness value, with a maximum hardness value of ~92 VHN obtainable at 200 °C. This improved age hardening response is associated with the formation of a dense distribution of nanoscale basal precipitate plates that are not available in the Mg–6Gd–0.6Zr alloy. The SAED patterns recorded from the nanoscale plates indicate that they may have a hexagonal structure with lattice parameters $a = 0.556$ nm, $c = 0$, 1, or 2 times of c_α (i.e. $c = 0$, 0.521, or 1.042 nm) and an OR of $(0001)_h//(0001)_\alpha$, $[10\bar{1}0]_h//[11\bar{2}0]_\alpha$. Based on what has been reported for similar plates formed in Mg–Nd–Ag (Gradwell, 1972; Lorimer, 1987) and Mg–Y(–Zn)–Ag alloys (Zhu, 2011), the lattice parameters of the structure of these plates are likely to be that $a = 0.556$ nm, $c = 0.521$ nm. An appreciably high maximum hardness value, more than 130 VHN, is obtained when the concentration of Gd is increased from 6 wt% to 15 wt%, **Figure 75(b)**. When tested at room temperature in the peak-aged condition, the Mg–15Gd–2Ag–0.6Zr alloy has a 0.2% proof strength of 320 MPa and an ultimate strength of 347 MPa (Ma, 2007).

20.4.1.7.8   Mg–Y(–Zn) –Ag Alloys

Figure 75(c) shows the aging curves at 200 °C of Mg–6Y–1Zn–0.6Zr alloys with systematic Ag additions. The Ag-free alloy has a rather poor age hardening response at 200 °C, which is attributable to little precipitation of fine-scale precipitates inside individual magnesium grains, the preferential formation of coarse precipitates of 14H in grain boundaries, and a relatively high fraction of retained intermetallic particles that have formed during solidification. Systematic additions of Ag to the Mg–6Y–1Zn–0.6Zr (wt%) alloy have been reported (Zhu et al., 2008) to reduce the volume fraction of retained intermetallic particles in the microstructure and to promote the formation of fine-scale precipitates at the expense of coarse precipitates of 14H. The alloy containing 2 wt% Ag exhibits a remarkable age hardening response during isothermal aging at 200 °C. An increase in the Ag content to 3 wt% leads to a further increase in the maximum hardness value achievable. In the peak-aged condition, the Mg–6Y–3Ag–1Zn–0.6Zr (wt%) alloy has a hardness value of approximately 100 VHN. It should be emphasized that a similar enhancement effect on the age hardening response is also expected in the counterpart Mg–Y–Ag alloys, that is alloys without any Zn additions. These observations are similar to those observed in Mg–Gd(–Zn) alloys containing Ag additions (Gao and Nie, 2008). The enhanced age hardening response is associated with a dense distribution of fine-scale basal precipitate plates of γ'', **Figure 75(d)**, that were not observed in the Ag-free alloy, and the number density of these fine-scale precipitates increases with an increase in the Ag content in the alloy. It is bit puzzling that the γ'' precipitate platelets are invariably of a single unit cell thickness, comprising three atomic planes, irrespective of the aging period that is used, **Figure 75(e) and (f)**. After prolonged aging (5800 h) at 200 °C, the γ'' platelets remain very thin, with a thickness of a single unit cell. While they are thermally stable, instead of thickening, they tend to form clusters in the direction normal to the platelet broad surface that are often separated by a few atomic planes, **Figure 75(g)**. It is currently unclear what factors cause the cluster distribution of the γ'' platelets.

Based on SAED patterns recorded from regions containing γ'' platelets and atomic resolution HAADF-STEM images of the γ'' platelets, the structure of the γ'' platelets is proposed to be hexagonal ($a = 0.556$ nm, $c = 0.450$ nm). The precise symmetry of the structure, that is the space group and arrangement of individual atoms in the unit cell, remain to be unambiguously established. The OR between γ'' and α-Mg phases is $(0001)_{\gamma''}//(0001)_\alpha$ and $[10\bar{1}0]_{\gamma''}//[\bar{2}100]_\alpha$. The observed structure, OR and the morphology of the γ'' phase are similar to those of platelets in Mg–Gd–Zn, Mg–Nd–Zn, Mg–Ce–Zn and Mg–Ca–Zn alloys.

Aging treatments of an Mg–6Y–2Ag–1Zn–0.6Zr (wt%) alloy in the temperature range 200–350 °C lead to the formation of GP zones, γ' and 14H phases. The GP zones have a monolayer atomic structure on $(0001)_\alpha$, and they form in the early stage of aging at 200 °C. These GP zones are replaced by γ'' precipitate plates during continued aging at 200 °C. Thus, the precipitation in the alloy seems to involve the formation of GP zone, γ'', γ', 14H and δ, **Table 1**, where δ is an equilibrium phase (space group Fd$\overline{3}$m, $a = 1.59$ nm) that forms in grain boundaries (Zhu et al., 2010b; Zhu, 2011).

20.4.1.8   Other Magnesium Alloy Systems

20.4.1.8.1   Mg–In–Ca Alloys

The equilibrium solid solubility of indium in Mg is approximately 19.4 at.% at the peritectic temperature of 484 °C, decreases only to 18.62 at.% at 327 °C and 13.95 at.% at 200 °C. Given the large solid solubility of indium in magnesium at temperatures close to 200 °C and the price of indium, the Mg–In system is clearly an unlikely candidate for developing alloys for engineering application. For this reason, the Mg–In alloys have so far received little attention. However, in a very recent study, Mendis et al. (2011b) reported that the addition of 0.3 at.% Ca to magnesium alloys containing a very dilute amount (0.6–1.0 at.%) of In could lead to a remarkable age hardening response, **Figure 76(a)**. This age hardening response is several times larger than that of the binary Mg–0.3at.%Ca alloy, and therefore it is not related with formation of precipitates intrinsic of binary Mg–In alloys or Mg–Ca alloys.

Inspection of peak-aged samples of an Mg–1In–0.3Ca (at.%) alloy using HAADF-STEM and 3DAP reveals that the this age hardening is associated with the formation of a dense distribution of precipitate platelets forming on $\{10\overline{1}0\}_\alpha$ planes of the magnesium matrix phase, **Figure 76(b)**. These prismatic platelets are typically three $\{10\overline{1}0\}_\alpha$ planes thick and 20 nm in diameter, and are fully coherent with the matrix phase. They are composed of approximately 7.0 at.% Ca and 3.5 at.% In, **Figure 76(c)**. The structure of these prismatic platelets is yet to be determined. The phase equilibria and the precipitation sequence in this alloy are also both unknown.

Nevertheless, the formation of prismatic plates in this alloy system is exciting. As will be described and discussed in the following section, prismatic precipitate plates are the most effective barriers to basal dislocations and twins that are operating in the plastic deformation process. Any detailed study in the future of the structure and formation mechanism of the $\{10\overline{1}0\}_\alpha$ precipitate plates will shed

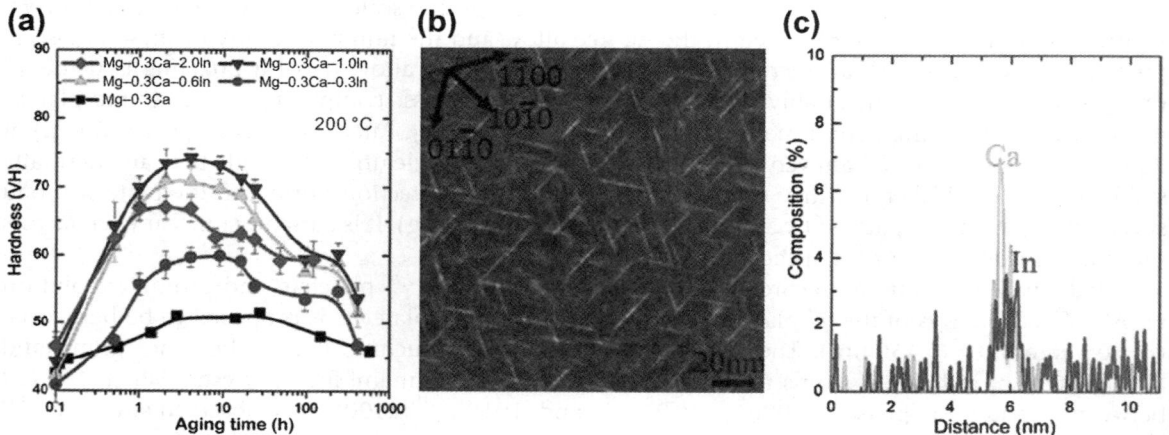

Figure 76   (a) Age hardening response of Mg–In–Ca alloys during isothermal aging at 200 °C, (b) HAADF-STEM image showing (c) 3DAP compositional profile of precipitate plates in Mg–1In–0.3Ca (at.%) alloy peak aged at 200 °C (Mendis et al., 2011b). (For color version of this figure, the reader is referred to the online version of this book.)

light on the exploration for the principles for generating prismatic precipitate plates in magnesium alloys.

20.4.1.8.2 Mg–Bi–Zn Alloys

The maximum equilibrium solid solubility of Bi in Mg is 1.12 at.% (or 8.87 wt%) at the eutectic temperature, 553 °C, and it decreases to approximately zero at 200 °C. The equilibrium intermetallic phase at the Mg-rich side of the phase diagram is α-Mg_3Bi_2 that has a hexagonal structure ($P\bar{3}ml$, $a = 0.4671$ nm, $c = 0.7403$ nm). The maximum volume fraction of the equilibrium precipitate phase α-Mg_3Bi_2 is about 3.38% for the alloy composition Mg–8.85 wt%Bi and 200 °C aging temperature. While the Mg–Bi system is an ideal one for designing precipitation hardenable alloys, the age hardening response of binary Mg–Bi alloys is too low to be considered for alloy development. It has been reported recently (Sasaki et al., 2009b) that the ternary addition of 0.5 at.% Zn to an Mg–0.8at.%Bi alloy can enhance the maximum hardness achievable by about 40% and that an even higher value of the maximum hardness can be obtained when the Zn concentration is increased from 0.5 at.% to 1.0 at.%, **Figure 77(a)**.

Microstructures of peak-aged samples of Mg–0.8Bi and Mg–0.8Bi–1.0Zn alloys are compared in **Figure 77(b) and (c)**. It is apparent that precipitates in the Zn-containing alloy have a denser distribution. These precipitates were reported to be the equilibrium phase α-Mg_3Bi_2, and they form as small platelets on $\{11\bar{2}0\}_\alpha$ planes of the magnesium matrix phase. The OR between the α-Mg_3Bi_2 platelets and the surrounding matrix phase is $(0001)_{\alpha-Mg_3Bi_2}//(11\bar{2}0)_{\alpha-Mg}$, $[11\bar{2}0]_{\alpha-Mg_3Bi_2}//[0001]_\alpha$. A fraction of $[0001]_\alpha$ rods also form in the Mg–0.8Bi–1.0Zn alloy and they are often in contact with the α-Mg_3Bi_2

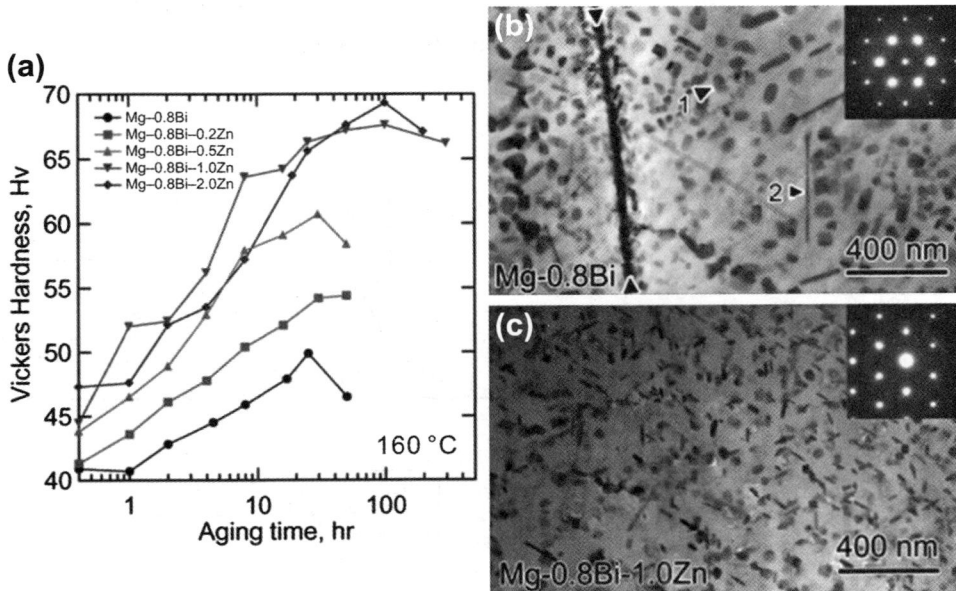

Figure 77 (a) Aging curves of Mg–Bi–Zn alloys at 160 °C. (b and c) TEM images showing distribution of precipitates in (b) Mg–0.8at.%Bi and (c) Mg–0.8at.%Bi–1at.%Zn alloys, both peak-aged at 160 °C (Sasaki et al., 2009b). (For color version of this figure, the reader is referred to the online version of this book.)

platelets. Instead of examining whether these rods are similar to those of Mg_4Zn_7 in binary Mg–Zn alloys, the $[0001]_\alpha$ rods were designated $MgZn_2$ with the following OR: $[0001]_{\beta_1'}//[11\bar{2}0]_\alpha$ and $(11\bar{2}0)_{\beta_1'}//(0001)_\alpha$. It was speculated (Sasaki et al., 2009b) that the $[0001]_\alpha$ rods form first during the aging process and then act as heterogeneous nucleation site for the Mg_3Bi_2 precipitates. If this speculation is accepted as representative, then any further increase in the nucleation rate of α-Mg_3Bi_2 precipitates would require a denser distribution of $[0001]_\alpha$ rods. Any detailed characterization in the future of precipitates in the early stages of aging using atomic resolution HAADF-STEM is expected to provide some insightful results on heterogeneous nucleation and the role of Zn in the precipitation of α-Mg_3Bi_2 in Mg–Bi–Zn alloys.

20.4.2   Effects of Precipitate Shape on Strengthening

20.4.2.1   Plastic Deformation by Basal Slip

20.4.2.1.1   Shear-Resistant Particles

The increment in CRSS produced by the need for basal slip dislocations to bypass point obstacles is given by Eqn (21). For spherical particles of uniform diameter d_t and distribution in a triangular array on the slip plane, the particle spacing is given by Eqn (20). If the particle shape is changed from sphere to plate, then both the effective planar interparticle spacing, **Figure 78(a) and (b)**, and the dislocation line tension will be changed. For magnesium alloys containing a triangular array of rationally oriented precipitate plates of uniform size, the strengthening effect produced by such particles can be estimated using the following equations (Nie, 2003):

$$\Delta\tau = \frac{Gb}{2\pi\sqrt{1-v}} \cdot \frac{1}{\left(0.825\sqrt{\frac{dt}{f}} - 0.393d - 0.886t\right)} \cdot \ln\frac{0.886\sqrt{dt}}{b} \qquad (53)$$

for prismatic precipitate plates,

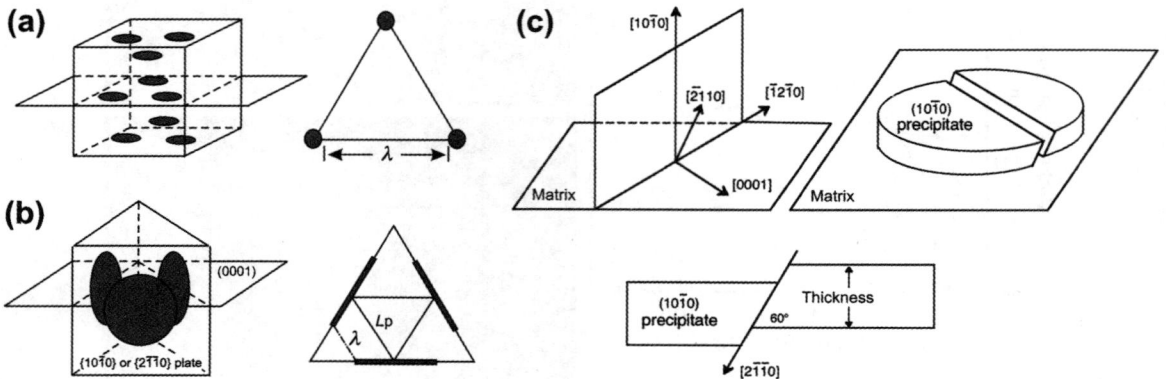

Figure 78 Schematic diagrams showing (a and b) effects of the orientation of precipitate plates on interparticle spacing for dislocation slip on basal plane of the magnesium matrix phase, and (c) a prismatic plate sheared by dislocations gliding in basal plane of the magnesium matrix phase (Nie, 2002, 2003). (For color version of this figure, the reader is referred to the online version of this book.)

$$\Delta\tau = \frac{Gb}{2\pi\sqrt{1-v}} \cdot \frac{1}{\left(\frac{0.953}{\sqrt{f}} - 1\right)d} \cdot \ln\frac{d}{b} \tag{54}$$

for basal precipitate plates or $[0001]_\alpha$ rods.

These equations are derived under the assumption of a uniform periodic triangular array of particles intersecting the slip plane of the magnesium matrix phase. For fixed volume fractions and number densities of precipitates, the variations in the ratio $\Delta\tau$(plate)/$\Delta\tau$(sphere) with plate aspect ratio are shown in **Figure 79(a)**, for a precipitate volume fraction of 0.03. To simplify comparison, the base microstructure is assumed to comprise precipitates in a regular triangular array distribution on the slip plane of the matrix phase. As shown in **Figure 79(a)**, the CRSS increments produced by prismatic plates are invariably larger than those produced by equivalent volume fractions of basal plates or spherical particles, and the difference increases substantially with increasing plate aspect ratio. For a given distribution of precipitate plates, the effective planar interplate spacing λ decreases with an increase in the plate aspect ratio. There is a critical value of aspect ratio for which the effective interparticle spacing becomes zero, and the prismatic plates form closed prismatic volumes, defined by prismatic plates. If the plates are assumed to remain resistant to dislocation shearing, then the dislocations generated within the prismatic volumes cannot escape and the theoretical Orowan increment becomes infinitely large. In practice, accumulation of dislocations may lead to local stress concentrations exceeding the yield strength of precipitates and precipitate shearing.

The effect of number density of precipitates on variation in the effective interparticle spacing with increase in the number density of precipitates is shown in **Figure 79(b)** for a volume fraction of 0.03. For identical volume fractions and number densities of precipitates, the prismatic precipitate plates are much more effective in reducing interparticle spacing than basal plates and spherical particles. For an aspect ratio of either 50:1 or 25:1, the smallest interparticle spacing is always associated with prismatic plates.

20.4.2.1.2 Shearable Particles

Shearable precipitates can impede the movement of gliding dislocations through a variety of dislocation/particle interaction mechanisms, including those described as interfacial or chemical strengthening, coherency strengthening, stacking fault strengthening, modulus strengthening and order strengthening. For each such proposed mechanism of strengthening, the contribution of the shearable precipitates to the CRSS of an alloy can be represented by Eqn (36). For a triangular array distribution of spherical particles of uniform diameter d_t, their contribution to the CRSS increment due to interfacial strengthening is given by Eqn (38).

For precipitates of morphologies other than sphere, it is currently difficult to assess their quantitative contributions to the CRSS increment because there is a lack of appropriate version of the strengthening equations for such precipitate shapes and orientations. Since it has been reported (Brown and Ham, 1971; Nie and Muddle, 1998b; Huang and Ardell, 1987a, 1987b; Kelly and Nicholson, 1963) that interfacial strengthening can become a major strengthening mechanism in alloys containing plate-shaped precipitates, it is convenient to restrict the present review to interfacial strengthening. The quantitative contribution of the interfacial strengthening to the CRSS for magnesium alloys containing rationally oriented precipitate plates is given by (Nie and Muddle, 1998a):

$$\Delta\tau_i = \frac{1.461 d\gamma^{3/2}}{d} \left(\frac{bf}{\Gamma}\right)^{1/2} \tag{55}$$

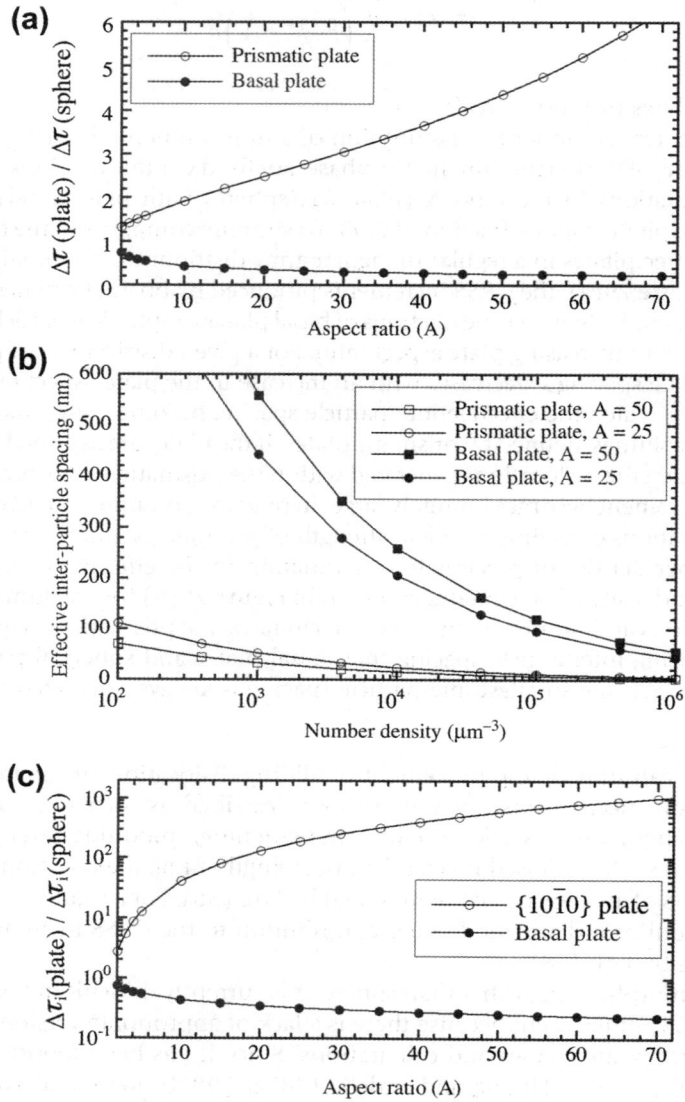

Figure 79 Variation of $\Delta\tau$(plate)/$\Delta\tau$(sphere) with (a) plate aspect ratio and (b) number density for prismatic and basal precipitate plates calculated assuming Orowan strengthening of shear-resistant particles with a volume fraction of 0.03. (c) Interfacial strengthening from sheared particles. (For color version of this figure, the reader is referred to the online version of this book.)

for $(0001)_\alpha$ plate,

$$\Delta\tau_i = \frac{1.359 d\gamma^{3/2}}{t^2} \left(\frac{bf}{\Gamma}\right)^{1/2} \tag{56}$$

for $\{10\overline{1}0\}_\alpha$ plate.

These equations are derived based on an ideal regular triangular array of particles in the slip plane. It is also assumed that there is continuity of the slip plane between magnesium matrix and precipitate, and that the magnitude of the Burgers vector of the dislocations in the magnesium matrix is identical to that in the precipitate phase, **Figure 78(c)**.

For a given value of dislocation line tension, the variations of $\Delta\tau_i$(plate)/$\Delta\tau_i$(sphere) with plate aspect ratio are shown in **Figure 79(c)**. It is evident that the contribution due to interfacial strengthening can become significant when particles take a plate shape on prismatic planes of the magnesium matrix phase. For identical volume fractions and number densities of precipitates per unit volume, the yield stress increments produced by prismatic precipitate plates are orders of magnitude larger than those produced by basal plates and spherical particles. The increments in CRSS produced by prismatic plates increase substantially with an increase in plate aspect ratio and are up to three orders of magnitude larger than that produced by spheres, when the plate aspect ratio is in the range of 10:1 to 70:1.

Many existing magnesium alloys are strengthened by $(0001)_\alpha$ precipitate plates. The precipitate plates formed in these alloys are often extraordinarily thin, less than 1 nm thick, and their aspect ratio is remarkably large. Further improvement in alloy strength might be achieved by increasing the number density and/or thickness of the $(0001)_\alpha$ plates. Such approaches seem, however, to have limitation since simultaneous increase in the number density and thickness of precipitate plates or increasing one factor while maintaining the other constant is practically difficult to achieve. Although a much higher number density of $(0001)_\alpha$ plates has been produced in many of the magnesium alloys, the age hardening response and the maximum hardness and strength values of these alloys are still much lower than those obtained in counterpart aluminum alloys. A substantial improvement in strength in these alloys would be difficult to achieve unless prismatic plates of large aspect ratio are introduced to replace, or coexist with, the $(0001)_\alpha$ precipitate plates.

20.4.2.2 Plastic Deformation by Twinning

20.4.2.2.1 Shear-Resistant Particles

It was reported by Clark (1968) that twinning becomes difficult to occur when the size and number density of $Mg_{17}Al_{12}$ precipitates increase. The precipitate–twin interactions in the Mg–Al alloys and effects of precipitate shape on twinning have been studied (Gharghouri et al., 1998; Robson et al., 2010, 2011). For simplicity, the propagation of twins is considered to involve the motion of twinning dislocations, and only $\{10\overline{1}2\}_\alpha$ twins are treated in the plastic deformation process. In addition, the lattice parameters of Mg are taken as $a = 0.5200$ nm, $c = 0.3203$ nm ($c/a = 1.6235$). For a $\{10\overline{1}2\}_\alpha$ twin to bypass spherical particles of uniform diameter and of a triangular array distribution, the increment in CRSS can be approximated by Eqn (21). For $(0001)_\alpha$ plates, of uniform aspect ratio A and thickness t that are ideally distributed in a triangular array in the $(10\overline{1}2)_\alpha$ twin plane, **Figure 80(a)**, the increment in CRSS for $(10\overline{1}2)_\alpha$ twinning dislocations to bypass $(0001)_\alpha$ precipitate plates can be approximated by (Nie, 2012):

$$\Delta\tau = \frac{Gb}{2\pi\sqrt{1-v}} \cdot \frac{1}{\left(\frac{1.327At}{1.152\sqrt{Af}+0.785Af-1.462t} - 1.462t\right)} \cdot \ln\frac{1.072t\sqrt{A}}{b} \tag{57}$$

For $\{10\bar{1}0\}_\alpha$ prismatic plates, with $d \gg t$, outlined in **Figure 80(b)**, if we assume that each of the three plates is ideally distributed at the center of each side of the isosceles triangle, then the increment in CRSS for $\{10\bar{1}2\}_\alpha$ twinning dislocations to bypass $\{10\bar{1}0\}_\alpha$ precipitate plates can be approximated by (Nie, 2012):

$$\Delta\tau = \frac{Gb}{2\pi\sqrt{1-v}} \cdot \frac{1}{\left(0.825t\sqrt{\frac{A}{f}} - 0.305At - 0.981t\right)} \cdot \ln\frac{0.914t\sqrt{A}}{b} \qquad (58)$$

Similarly, the increment in CRSS for $(10\bar{1}2)_\alpha$ twinning dislocations to bypass $\{11\bar{2}0\}_\alpha$ precipitate plates can be estimated by (Nie, 2012):

$$\Delta\tau = \frac{Gb}{2\pi\sqrt{1-v}} \cdot \frac{1}{\left\{t\sqrt{0.752\frac{A}{f} - \sqrt{\frac{A}{f}}(0.579A + 1.727) + 0.117A^2 + 0.690A + 1.021}\right\}} \cdot \ln\frac{0.886t\sqrt{A}}{b} \cdot$$

$$(59)$$

For the purpose of comparison, the logarithmic terms in above strengthening equations are assumed to be identical, and, in the case of spherical precipitates and basal plates, the base microstructure is

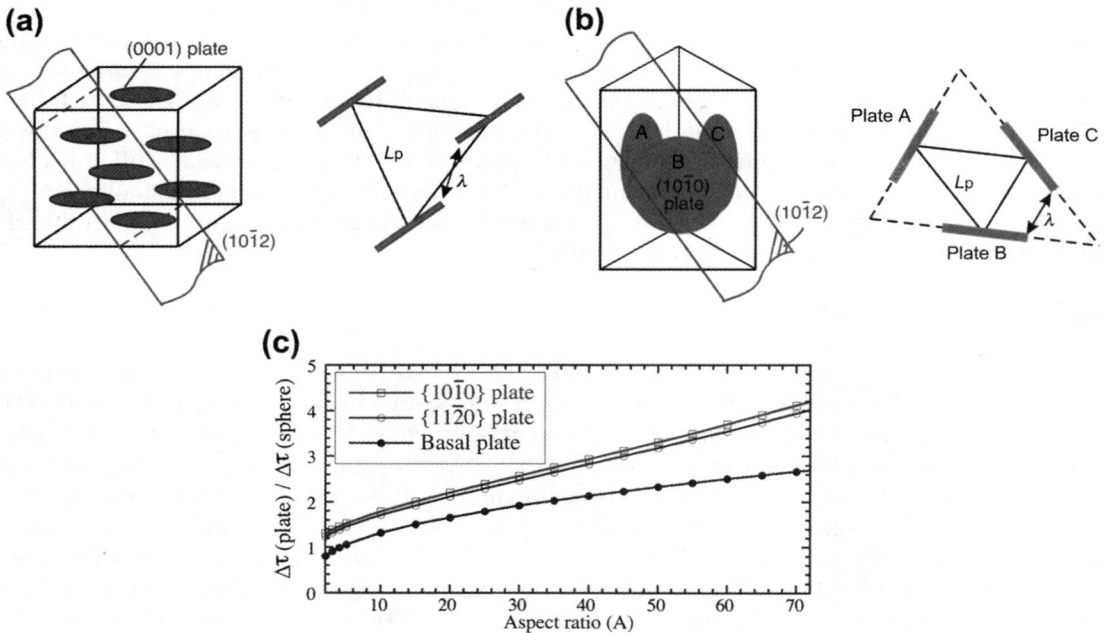

Figure 80 Schematic diagrams showing (a and b) effects of the orientation of precipitate plates on interparticle spacing for twinning on $\{10\bar{1}2\}_\alpha$ plane of the magnesium matrix phase, and (c) quantitative effects of the orientation and aspect ratio of precipitate plates on Orowan strengthening. The precipitate volume fraction is 3%. (For color version of this figure, the reader is referred to the online version of this book.)

assumed to comprise precipitates in a regular triangular array distribution in the $(10\bar{1}2)_\alpha$ twin plane. For identical volume fractions and number densities of precipitates per unit volume of the matrix phase, the ratio $\Delta\tau$(plate)/$\Delta\tau$(sphere) is invariably larger than unity, when the plate aspect ratio is in the range 5:1–70:1, which means that precipitate plates are more effective in impeding twinning dislocation propagation, **Figure 80(c)**. Furthermore, the ratio $\Delta\tau$(plate)/$\Delta\tau$(sphere) increases substantially with increasing plate aspect ratio. For a given distribution of prismatic plates, an increase in plate aspect ratio can lead to significant reduction in effective interparticle spacing. For either $\{10\bar{1}0\}_\alpha$ or $\{11\bar{2}0\}_\alpha$ plates, there is again a critical value of plate aspect ratio for which the effective interparticle spacing becomes zero. This implies that, at such a critical plate aspect ratio, the plates will divide a single magnesium grain into many enclosed prismatic volumes. If the plates are assumed to remain shear resistant, then the $(10\bar{1}2)_\alpha$ twins generated within a prismatic volume are constrained in this prismatic volume.

20.4.3 Microstructural Design for Higher Strength

For precipitation-hardened magnesium alloys, whether particles are sheared or shear resistant and whether the deformation mode is basal slip or twinning, the particle strengthening models indicate that precipitate plates formed on prismatic planes of the magnesium matrix phase provide the most effective barrier to gliding dislocations and propagating twins in the magnesium matrix. The models further suggest that a higher strength can be achieved if a high density of intrinsically strong plate-shaped precipitates with prismatic and basal habit planes and of large aspect ratio can be developed in the microstructure, **Figure 81**.

While it is now possible to generate a microstructure containing both prismatic and basal plates in some magnesium alloys such as those based on the Mg–Gd–Y–Zn system, and the basal plates in such a microstructure have a large aspect ratio, the aspect ratio and number density of the prismatic plates are much lower than those typical of precipitate plates formed in high-strength aluminum alloys. In high-strength precipitation-hardened aluminum alloys, the maximum hardness and yield strength are

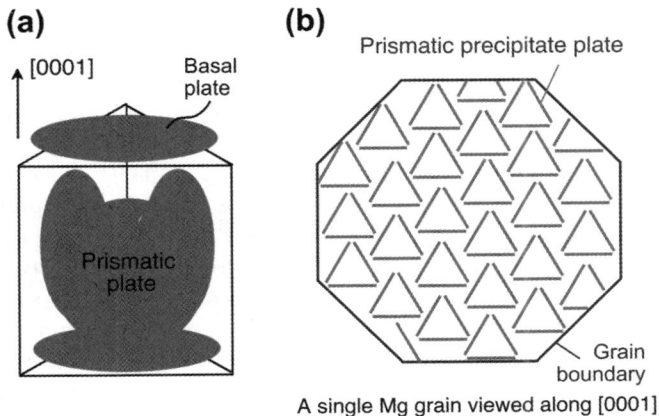

Figure 81 Microstructural design for precipitation hardenable magnesium alloys with higher strength. A near-continuous network of prismatic and basal precipitate plates in (a) can divide a single magnesium grain into many near-isolated blocks in (b) and therefore effectively impede propagation of dislocations and twins (Nie, 2012).

commonly associated with microstructures containing a high density of $\{100\}_\alpha$ and $\{111\}_\alpha$ precipitate plates of large aspect ratio (typically above 40:1) (Nie et al., 2001; Weakley-Bollin et al., 2004), and these strengthening precipitates are often intermediate or equilibrium phases that are intrinsically stronger than GP zones and metastable precipitates formed in the early stage of aging (Muddle et al., 1994). Further improvements in the alloy strength thus require an increase in the number density and/or aspect ratio of prismatic plates of intermediate or equilibrium precipitate phases (Nie et al., 1996). It is to be noted that the traditional approach to the strength improvement rarely considers the precipitate plate aspect ratio and, instead, puts an emphasis on precipitate number density. The particle strengthening models suggest that the precipitate plate aspect ratio is an important strengthening factor, as least as important as the precipitate number density, and that a remarkable enhancement in alloy strength could be achieved if the aspect ratio of the prismatic precipitate plates is substantially increased.

How to introduce intrinsically strong, prismatic precipitate plates into the magnesium matrix is currently a difficult question to answer. However, one possible approach may involve consideration of phase equilibria and lattice matching of magnesium and precipitate structures (Nie, 2006, 2008; Zhang and Kelly, 2005). The precipitate phase may be selected from the equilibrium or near-equilibrium phases that form via reactions among magnesium and/or added alloying elements. These phases are expected to be intrinsically strong and therefore more resistant to shearing and plastic deformation. Inspections of the structures of these equilibrium phases and that of magnesium can reveal whether it is possible to have the prismatic plane as an invariant strain plane (habit plane) between the precipitate and magnesium structures. One good example is the β_1 phase (space group $Fm\bar{3}m$, $a = 0.74$ nm) in Mg–Nd, Mg–Y–Nd, Mg–Gd, Mg–Gd–Y and Mg–Gd–Nd alloys. The structure of this phase can be generated from the magnesium lattice by an invariant plane strain transformation (Nie and Muddle, 2000). Therefore, it is fully predictable that the Mg_3Nd precipitates form as plates on $\{10\bar{1}0\}_\alpha$, even without examining the microstructures using TEM.

It is perhaps also constructive to compare precipitates of $Mg_{17}Al_{12}$ in Mg–Al alloys and $Mg_{24}Y_5$ in Mg–Y alloys (Zhang and Kelly, 2005). While the $Mg_{17}Al_{12}$ and $Mg_{24}Y_5$ phases have same crystal structure (space group $I\bar{4}3m$, 58 atoms per unit cell) and similar lattice parameters and ORs, they have drastically different habit planes. The $Mg_{17}Al_{12}$ phase has a lattice parameter of ~ 1.064 nm. For the observed OR between the $Mg_{17}Al_{12}$ phase and the matrix, $(1\bar{1}0)_p//(0001)_\alpha$, $[111]_p//[11\bar{2}0]_\alpha$, the lattice misfit between $(1\bar{1}0)_p$ and $(0001)_m$ is 2.7% if the lattice parameters of the Mg–9wt%Al alloy are taken as the following: $a = 0.31699$ nm, $c = 0.5155$ nm (Busk, 1950). The $Mg_{17}Al_{12}$ phase adopts a plate shape, with its habit plane parallel to $(1\bar{1}0)_p$ and $(0001)_m$. The equilibrium $Mg_{24}Y_5$ phase has a lattice parameter of ~ 1.125 nm (Nayeb-Hashemi and Clark, 1988), which is slightly larger than that of $Mg_{17}Al_{12}$. Given that its ORs is also $(1\bar{1}0)_p//(0001)_\alpha$, $[111]_p//[11\bar{2}0]_\alpha$, then the lattice misfit between $(1\bar{1}0)_p$ and $(0001)_m$ is −2.9%. The $Mg_{24}Y_5$ phase also adopts a plate shape, but its habit plane is parallel to $\{10\bar{1}0\}_\alpha$ instead of $(0001)_\alpha$. While both $(0001)_\alpha$ and $\{10\bar{1}0\}_\alpha$ habit planes are geometrically possible to occur for the Burger's OR (Nie, 2008), and that a small change in the lattice parameter of the precipitate phase, or the matrix phase, can cause either to form, it is currently unclear why the $Mg_{17}Al_{12}$ precipitates form predominantly on $(0001)_\alpha$. If Al atoms in the $Mg_{17}Al_{12}$ unit cell could be partially replaced by Y atoms, the lattice parameter of the $Mg_{17}Al_{12}$ phase would be modified to reduce the lattice misfit between precipitates and the matrix phase, and it becomes energetically possible to achieve higher nucleation rates and number density of precipitates. Unfortunately, Y and Al atoms react strongly in the molten metal to form Al_2Y particles before alloy casting, leaving few Y atoms in the magnesium solid solution matrix to form the $Mg_{17}(Al_{1-x}Y_x)_{12}$ phase during the aging process.

It is also to be noted that Mg–Th alloys are also strengthened by $\{10\bar{1}0\}_\alpha$ precipitate plates (Mushovic and Stoloff, 1969; Noble and Crook, 1970; Stratford, 1971; Stratford, 1972; Stratford and Beckley, 1972; Pike et al., 1972), even though these alloys were abandoned many years ago due to toxicity of Th. The maximum solid solubility of Th in magnesium is only about 0.52 at.% or 4.75 wt% at the eutectic temperature of 582 °C, and the maximum volume fraction of the equilibrium precipitate phase $Mg_{23}Th_6$ in an Mg–3.2wt%Th–0.7wt%Zr alloy (HK31) aged at 200 °C is estimated to be ∼1.8%. The prismatic precipitate plates formed in this alloy are quite thin. It is worth examining the crystal structure and formation mechanism of the key strengthening precipitate phase in peak-aged Mg–Th alloys. The knowledge to be gained from this alloy can be used to select precipitate phases that can form from nontoxic alloying elements but lead to similar or even better age hardening effects. Equally, an in-depth knowledge on the structure and formation mechanisms of prismatic precipitate plates formed in Mg–In–Ca alloys, **Figure 76(b)**, can also provide useful guidelines to the formation and manipulation of prismatic plates in magnesium alloys.

In magnesium alloys that are already strengthened by prismatic plates of intrinsically strong precipitate phases, any further enhancement in yield strength requires an increase in the precipitate plate aspect ratio and/or precipitate number density. Approaches to achieving an increase in plate aspect ratio and number density may lie in the use of (i) *microalloying additions*, which may partition to either matrix or precipitate phase to improve matching between the precipitate and matrix phase, (ii) *cold work* after solution treatment and before aging, which introduces dislocations to provide heterogeneous nucleation sites for intermediate and/or equilibrium precipitate phases, (iii) *duplex or multiple isothermal aging*, and (iv) *single or multiple non-isothermal aging*. Such approaches inevitably involve the consideration of the crystal structure, OR, formation mechanism of strengthening precipitate phases and the elastic strains associated their nucleation.

Currently, there is a lack of a selection rule for microalloying elements. The establishment of this selection rule inevitably requires an in-depth understanding of the formation mechanisms of the plate-shaped precipitates. It was proposed (Nie and Muddle, 2000) that the β_1 structure can be generated phenomenologically from the α-Mg lattice by a shear on the $(1\bar{1}00)_\alpha$ plane in the $[\bar{1}\bar{1}20]_\alpha$ direction, combined with an expansion in the direction normal to the shear plane, which was confirmed in a recent study using atomic resolution HAADF-STEM, **Figure 70(c)**. The transformation strain itself is an invariant plane strain, with the invariant plane parallel to the shear plane $(1\bar{1}00)_\alpha$. This formation mechanism is similar to that proposed for γ' plates in Mg–Gd–Zn (Nie et al., 2008) and Mg–Y–Zn (Zhu et al., 2009) alloys, pro-eutectoid α plates in a Ti–Cr alloy (Furuhara and Aaronson, 1991; Furuhara et al., 1991; Hirth et al., 1998) and for precipitate plates of γ' in Al–Ag alloys, θ' in Al–Cu alloys, T_1 in Al–Cu–Li(–Mg–Ag) alloys, and Ω in Al–Cu–Mg–Ag alloys (Nie and Muddle, 1999a; Nie et al., 1999). **Figure 82** shows schematically an initial β_1 platelet that has a thickness of one unit cell and is constrained within the α-Mg matrix phase. If the precipitate phase is much stronger than the matrix, then the formation of this platelet will result in regions of *expansion* (E) and *contraction* (C) in the matrix surrounding the platelet and to the associated accumulation of significant shear strain energy (Nie and Muddle, 2000). A mechanism is thus required to operate to minimize the shear strain energy during nucleation (and growth) of the precipitate phase. One potential approach to accommodating the shear strain energy involves an extrinsic supply of relatively larger solute atoms to the regions E, and of corresponding concentrations of vacancies to the regions C. Such an environment might be found in the vicinity of clusters combining larger solute atoms with a local excess of vacancies. Although there is currently no evidence for such solute-rich clusters in Mg–Nd, Mg–Gd, Mg–Gd–Y, Mg–Gd–Nd and Mg–Y–Nd alloys, it is to be noted that Nd, Gd and Y atoms have relatively high binding energies with

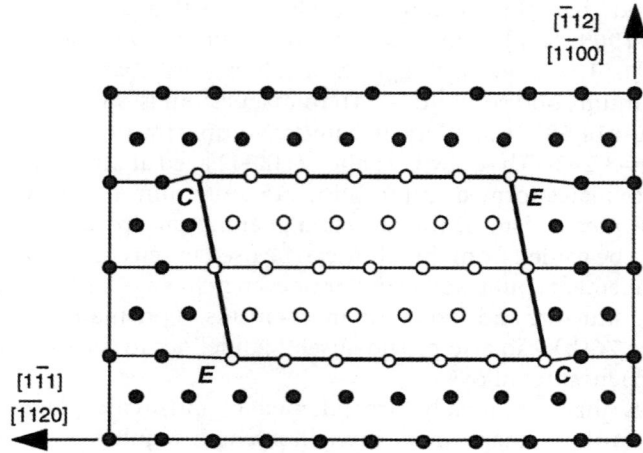

Figure 82 Schematic diagram showing a β_1 platelet constrained in matrix magnesium lattice. Open and filled circles represent atoms in β_1 and Mg lattices, respectively (Nie and Muddle, 2000).

vacancies (Nie and Muddle, 2000) and are all larger than Mg atoms. The shear strain associated with an embryo precipitate plate might be reduced or removed if Nd, Gd or Y atoms segregate to the extended regions and associated vacancies to the compressed regions during nucleation. If such a mechanism were to operate, then it is plausible that any particles evolved from such solute-rich clusters would be expected to form at the end facets, rather than the broad surfaces, of the β_1 plate. This is in perfect agreement with experimental observations that β_1 plates form invariably in association with β' particles which are in contact with the end facets of β_1 plates, **Figure 70(a)**. It is to be noted that similar phenomena have been documented in Al–Cu–Sn alloys, where the formation of θ' plates involves a large shear strain and Sn particles are invariably located at the end facet of θ' plates when they form in association with the θ' plates (Gao et al., 1999; Bourgeois et al., 2012).

The accumulated experimental evidence seems to suggest that the most effective microalloying elements in enhancing nucleation rates of plate-shaped precipitates are those having larger atomic sizes and strong binding energies with vacancies. While it is a straightforward exercise for the selection of elements with larger atomic sizes, it is currently difficult to judge which elements have large binding energies with vacancies in the matrix of magnesium. There is an urgent need for more theoretical (Shin and Wolverton, 2010a, 2010b) and experimental (Moia et al., 2010) work on this aspect.

An alternative way to accommodate the shear strain energy involved in the precipitate formation and therefore to facilitate the nucleation of β_1 precipitates is to introduce dislocations of appropriate Burgers vectors. As shown in **Figure 69**, the cold work before aging can promote the formation of β_1 phase and the maximum hardness. Although the interaction between preexisting dislocations and the β_1 phase has not been examined in detail using TEM, it is to be noted that, in Al–Cu and Al–Cu–Li alloys, a partial dislocation with a Burgers vector of $a/2\langle 001 \rangle_\alpha$ on $\{100\}_\alpha$ can cancel the shear strain associated with the formation of a plate of θ' of single unit cell thickness and thus facilitate nucleation of θ' (Dahmen and Westmacott, 1983a), and that the shear strain involved in the formation of T_1 precipitate plates of unit cell thickness may be eliminated if nucleation occurs on a Shockley partial dislocation (Cassada et al., 1991a, 1991b). It is interesting to note that some prismatic plates of

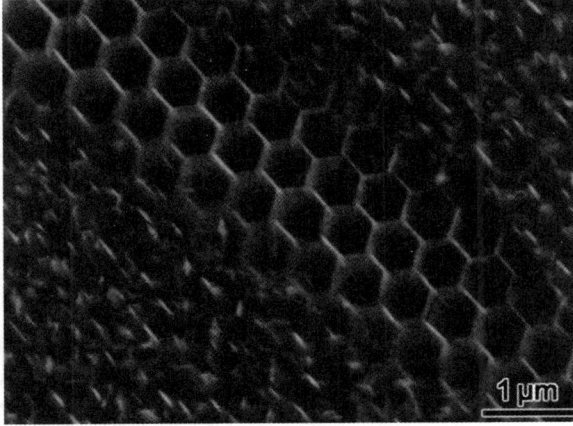

Figure 83 Scanning electron microscopy image showing the formation of a continuous network of prismatic plates of β_1 phase, that is honeycomb structure, in WE54 alloy (Xu et al., 2012).

β_1 phase can form a hexagonal shaped, continuous network in localized regions of the magnesium matrix, **Figure 83**. The formation of such a honeycomb structure of the β_1 plates is likely to be associated with a preexisting hexagonal network of dislocations inside magnesium grains. It is of significance to strengthening if this honeycomb structure could be extensively generated inside individual magnesium grains.

While the yield strength achieved thus far in magnesium alloys is still much lower than that obtained in counterpart aluminum alloys, **Figure 84**, it is to be noted that a significant progress has been made over the past 12 years. The yield strengths of precipitation hardenable magnesium alloys were generally

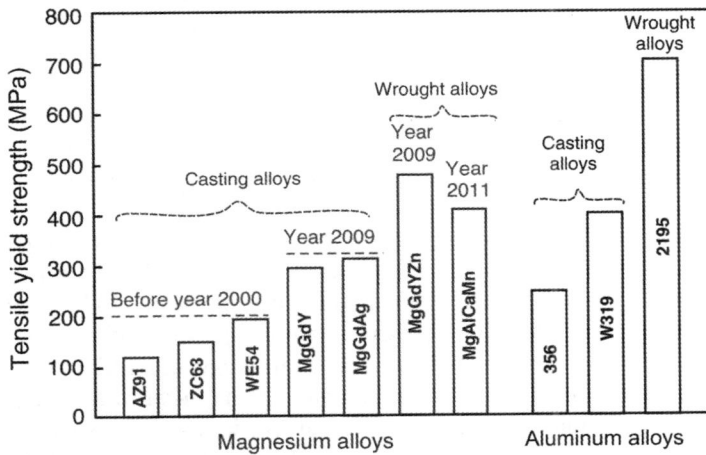

Figure 84 Development of precipitation hardenable casting and wrought magnesium alloys. Representative precipitation hardenable aluminum casting (356 and W319) and wrought (2195) alloys are also provided for the purpose of comparison of tensile yield strength. (For color version of this figure, the reader is referred to the online version of this book.)

less than 200 MPa, which are lower than those of aluminum alloys, before year 2000. But around years 2007–2009 the yield strength of magnesium alloys approached 320 MPa for casting alloys (Ma, 2007) and 475 MPa for extrusion alloys (Homma et al., 2009). A similar or even faster pace is expected in the development of high-strength magnesium alloys if the level of magnesium research activities enjoyed over the past 12 years can be maintained in the future. The precipitation hardening phenomenon in aluminum alloys was discovered accidentally by Alfred Wilm in 1906 (Wilm, 1906, 1911; Hardouin Duparc, 2005), and the strength of aluminum alloys made at that time was compatible to that of precipitation hardenable magnesium alloys in year 2000. In the following years the age hardening response and the strength of the alloys were gradually improved with the additions of microalloying elements, either individually or in combination. In 1989, it was discovered (Polmear and Chester, 1989) that additions of small amounts of lithium to Al–Cu–Mg–Ag alloys could stimulate an even greater age hardening response and an Al–Cu–Li–Mg–Ag alloy designated Weldalite 049™ was developed by Pickens et al. (1989). While the composition of this alloy is not too much different from that of the Al–Cu binary base alloy, this alloy has an extraordinary age hardening response and has an ultrahigh strength exceeding 700 MPa when tested at room temperature.

20.4.4 Applications of Magnesium Alloys

Driven by recent environmental legislation in Europe, the United States and Japan, where the need for improved fuel economy and reduced emissions in transport vehicles has given rise to a high demand for the use of lightweight alloys, magnesium alloys have potential widespread applications in the auto-motive industry, for example in such components as engine blocks, cylinder heads, transmission housings, seat frames, and wheels. In North America, the annual growth rate of magnesium alloys has averaged 14% in the period 1987 to 2001. In January 2001, three U.S. automotive manufacturers, General Motors, Ford and DaimlerChrysler, and the U.S. Department of Energy launched a 4-year USAMP project to evaluate the feasibility of using magnesium in powertrain components. In Aus-tralia, there was strong interest in the development of a magnesium metal industry in association with the large magnesite deposits discovered at Kunwarara in Queensland. In January 1997, Ford Motor Co. Australia announced that it would take 45,000 tons of magnesium products a year for ten years for its worldwide use in car manufacturing. In 2010, Magnesium Elektron Limited, a subsidiary of the Luxfer Group and the world's largest magnesium product producer, and POSCO, the world's third largest steel maker, signed an agreement on a joint project to combine their expertise and facilities to develop technologies for fabricating automotive magnesium sheet. In November 2010, Magnesium Elektron announced the installation of a new fully integrated magnesium extrusion facility at its manufacturing site in Manchester, UK.

Magnesium castings were used in the Volkswagen Beetle motor cars in 1930s. The car contained about 17 kg magnesium in its powertrain in the form of engine crankcase and transmission housing castings. The use of such magnesium products resulted in approximately 50 kg weight saving compared with those produced from cast iron. The higher magnesium price in the mid-1970s led to their replacement by aluminum alloy castings. Advances in manufacturing and processing technologies in recent years have stimulated renewed interest in lightweight magnesium products for applications in the automotive industry. Nowadays, Volkswagen Passat, Audi A4 and A6 use about 14 kg of AZ91 castings. GM Savana and Express vans use approximately 26 kg of magnesium alloys. Magnesium alloys WE43 and WE54 and a magnesium/aluminum hybrid engine block are also used in racing cars.

While it is true that magnesium alloys are yet to reach their potential in aerospace applications, magnesium has in fact been previously deployed in both civil and military aircrafts. Civil applications include intermediate casings for the Tay engines and gearboxes for the RB211, Tay and BR710 engines. Military aircrafts, including the F16, Tornado and Eurofighter Typhoon, capitalize on the lightweight characteristics of magnesium transmissions. The new Pratt and Whitney F119 gearbox, specified on the F22 aircraft, uses the creep-resistant magnesium alloy WE43. Magnesium alloys such as ZE41 and WE43 have been extensively used for a range of helicopter transmission casings because of their excellent castability and good mechanical properties at both ambient and elevated temperatures. The WE43 alloy has higher resistance to creep and corrosion, and is therefore being selected for many new helicopter programs including the MD500, Eurocopter EC120, NH90 and Sikorsky S92.

Flammability studies conducted in recent years indicate that wrought magnesium alloys can satisfy the aviation regulations on flammability. In very recent years, Airbus has been actively involved in several Research and Development projects on magnesium alloys, launching test campaigns to identify the best magnesium alloy for aircraft applications. There are ongoing global R&D activities that are focused on development of magnesium wrought products for window frame of A340 and doorstop for A380, and door panel and leading edge rib for Gulfstream G200 jets.

20.5 Titanium Alloys

Since there are several classical books and book chapters on titanium alloys (Williams, 1978; Lutjering and Williams, 2003; Polmear, 2006; Banerjee and Mukhopadhyay, 2007), the focus of this section will be placed on martensitic and massive transformations and precipitation in some titanium alloys.

20.5.1 Alloy Classifications

Pure titanium has a hexagonal structure (space group P6$_3$/*mmc*, $a = 0.295$ nm, $c = 0.468$ nm at room temperature) at lower temperatures but transforms to body-centered cubic structure (space group Im$\bar{3}$m, $a = 0.332$ nm at 900 °C) at lower temperatures above 882 °C. The hexagonal structure is denoted as the α phase, whereas the body-centered cubic structure is the β phase. The transformation temperature between α and β phases is strongly influenced by the presence of impurity or alloying elements. The substitutional element Al and the interstitial elements such as O, N and C dissolve preferentially in and expand the α phase. They can increase the α-to-β transformation temperature and therefore are strong α stabilizers. Other α stabilizers include B, Ga, Ge and RE (rare-earth elements). The β stabilizers are usually classified in two groups: those forming binary systems of the β isomorphous type and those favoring formation of a β eutectoid. The β isomorphous elements include V, Mo, W and Nb, and the additions of sufficient amounts of these elements can stabilize the β phase down to room temperature. The commonly used β eutectoid elements are Cr, Fe and Si. Alloying elements such as Zr, Hf and Sn do not cause much change to the α-to-β transformation temperature, and these elements have neutral effects on either α or β phase. On the basis of which phase is dominating the microstructure after processing, titanium alloys are classified into three main groups: α, $\alpha + \beta$ and β.

20.5.2 Martensitic Transformation

The transformation of the β phase to the α phase can occur by either a diffusionless martensitic reaction or a diffusional process. A well-defined OR, namely Burgers, exists between the two phases:

$$(110)_\beta // (0001)_\alpha, \langle 1\bar{1}1 \rangle_\beta // \langle 11\bar{2}0 \rangle_\alpha$$

This Burgers OR is also found to occur in other bcc/hcp transformations. The commonly observed martensite has a hexagonal structure and is denoted as α'. In the more dilute alloys, it forms as colonies of parallel-sided plates or laths. With increasing solute content, these colonies decrease in size and many degenerate into individual plates. These plates have a lenticular or acicular morphology and are internally twinned. The OR of α' martensite and β phase is near Burgers, and the habit planes for the untwinned and twinned martensites are close to $\{4\bar{3}\bar{3}\}_\beta$ and $\{3\bar{4}\bar{4}\}_\beta$, respectively. Another martensite, denoted as α'', has an orthorhombic structure ($a = 0.298$ nm, $b = 0.494$ nm, $c = 0.464$ nm). It is internally twinned on the $\{111\}_{\alpha''}$ planes, and its OR with the parent β phase is

$$(110)_{\alpha''} // (2\bar{1}\bar{1})_\beta, [001]_{\alpha''} // [0\bar{1}1]_\beta$$

The habit plane of α'' martensite is close to $\{7\bar{5}\bar{5}\}_\beta$ in binary Ti–Nb alloys, but can change to $\{7\bar{3}\bar{3}\}_\beta$ when the Nb concentration in the alloy is increased.

Another martensitic transformation in titanium is from α phase to ω phase by the application of pressure in the range 2–9 GPa. The $\alpha \rightarrow \omega$ transformation is associated with a large hysteresis and because of this the ω phase generated by the high pressure can be retained in a metastable state at ambient pressure and temperature. The metastable ω phase thus formed has two ORs with the α phase:

$$(01\bar{1}1)_\omega // (0001)_\alpha, [1\bar{1}01]_\omega // [11\bar{2}0]_\alpha$$

$$(11\bar{2}0)_\omega // (0001)_\alpha, [0001]_\omega // [11\bar{2}0]_\alpha$$

On the basis of the observed ORs, two distinct mechanisms have been proposed to account for the transformation from the α phase to the ω phase: one involves direct structural transformation from α to ω while the other involves the formation of β as an intermediate phase (i.e. $\alpha \rightarrow \beta \rightarrow \omega$).

20.5.3   Decomposition of β Phase

The high-temperature β phase in some titanium alloys can be retained upon quenching. It decomposes during heat treatments at elevated temperatures. The equilibrium α phase precipitates only at relatively high temperatures. For heat treatments at temperatures below 500 °C, the precipitation process typically involves the formation of metastable ω phase. The ω phase has a structure that is either hexagonal (space group P6/mmm) or trigonal (space group P$\bar{3}$2m1). The lattice parameters of the ω phase are $a \sim 0.460$–0.503 nm, $c \sim 0.282$–0.309 nm, depending on the actually alloy compositions. The OR between the ω and β phases is

$$(0001)_\omega // (111)_\beta, [11\bar{2}0]_\omega // [1\bar{1}0]_\beta$$

The ω phase can form during quenching from the β phase field above the β transus, or during isothermal aging at certain temperatures below the β transus. The ω phase formed in the former is termed athermal ω, while the ω phase formed in the latter is termed isothermal ω. The athermal $\beta \rightarrow \omega$ transformation is displacive and diffusionless and therefore the composition of the ω phase obtained in this transformation is very close to that of the β phase. In contrast, the formation of isothermal ω is accompanied by solute rejection by the ω to the β phase. The athermal ω precipitates are typically less than 5 nm, and they have a tendency to align in $\langle 111 \rangle_\beta$ directions. The isothermal ω precipitates have two types of morphologies: ellipsoidal and cuboidal. The ellipsoidal shape is controlled by interfacial energy minimization and is often observed in alloys stabilized by Nb or Mo. The cuboidal shape results from the minimization of elastic strains in the cubic matrix and is found in alloys stabilized by V, Cr, Mn or Fe.

The formation or the absence of the ω phase can directly affect the elastic modulus and the strength of β alloys. As a result, considerable efforts have been made to understand the formation mechanism of ω in β (and also in α) phase and to identify alloying elements that can suppress or facilitate ω formation. Currently, it is the crystallographic information obtained from TEM, rather than the first-principle predictions, that provides the essential clue to the understanding of ORs between ω and β and the transformation mechanism from β to ω. It has been challenging to establish a simple methodology that is capable of predicting the correspondence between the two lattices involved in the transformation and the transformation strains associated with the transformation. Under a given driving force, how easy it is for ω to nucleate inside β depends critically on the magnitudes of transformation strains and the type of lattice defects and heterogeneity in the β phase.

The addition of oxygen reportedly suppresses the formation of athermal ω, probably through the yet-to-be-confirmed interference with diffusionless atomic displacement in the structural transformation from β to ω. The oxygen addition can also inhibit isothermal ω formation during the aging of Ti–V alloys, but in this case the influence of oxygen on isothermal ω formation is not clear. It becomes even more puzzling after a recent report suggesting that oxygen can facilitate the formation of isothermal ω in a Ti–35Nb–7Zr–5Ta (wt%) alloy. It is unclear whether such contradictory reports are because of the formation or absence of clusters of oxygen atoms during the aging process. To resolve such issues, it is necessary to use the state-of-the-art facilities such as atomic resolution HAADF-STEM and 3DAP to characterize the microstructures of alloys of different compositions and having been subjected to different heat treatments. The precise role of oxygen and other alloying elements in the β to ω transformation has to be unambiguously established if β-titanium alloys exhibiting both high-strength and low-elastic modulus are to be obtained for biomedical applications.

20.5.4 Decomposition of α Phase

20.5.4.1 Precipitation of Hydrides

It has been known for a long time that hydrogen can be dissolved in titanium and the presence of hydrogen in titanium can lead to brittle fracture. However, it became known in recent years that the presence of hydrogen improves the workability and thus facilitates the plastic forming process at high temperatures. Therefore, hydrogen is added into some titanium alloys under a positive pressure before hot processing at high temperature, and then removed by heat treatments under vacuum after hot working, that is functioning as a "temporary alloying element".

The solid solubility of hydrogen in α titanium decreases from 6.7 at.% at 300 °C to almost zero at room temperature. The equilibrium hydride phase at the Ti-rich side of the Ti–H phase diagram, δ, has a face-centered cubic structure (space group Fm$\bar{3}$m, $a = 0.443$ nm) and a stoichiometry close to TiH$_2$. The metastable γ hydride phase has a face-centered tetragonal structure (space group P4$_2$/n, $a = 0.421$ nm, $c = 0.460$ nm), and a stoichiometry of TiH (Numakura and Koiwa, 1984). The H atoms have an ordered arrangement in the unit cell. The metastable γ hydride phase can coexist with the α and δ phases.

Since the temperature range in which γ hydride forms is too low for any significant self-diffusion of Ti atoms to occur, the $\alpha \rightarrow \gamma$ transformation process is therefore suggested to be a shear transformation of the hcp lattice of α with an accompanying interstitial diffusion and ordering of H atoms. Based on TEM observations, two ORs have been found to occur between γ hydride and the α matrix.

$$(111)_\gamma // (0001)_\alpha, [1\bar{1}0]_\gamma // [\bar{1}2\bar{1}0]_\alpha$$

$$(001)_\gamma // (0001)_\alpha, [1\bar{1}0]_\gamma // [\bar{1}2\bar{1}0]_\alpha$$

The γ hydride has a plate morphology for each of the two ORs. Its habit plane is nearly parallel to $\{10\bar{1}7\}_\alpha$ or $\{20\bar{2}5\}_\alpha$ in the former OR, and exactly parallel to $\{10\bar{1}0\}_\alpha$ in the latter. These habit planes and the associated ORs can all be predicted by the crystallographic theory.

20.5.4.2 *Precipitation in Ti–Cu Alloys*
There are quite a few alloy systems in which precipitation occurs as a result of decomposition of supersaturated solid solution of α phase. One of them is the Ti–Cu system. The equilibrium solid solubility of Cu in Ti decreases with temperature, from ∼2.1 wt% at the eutectoid temperature of 805 °C to 0.7 wt% at 600 °C, offering a potential for developing a precipitation hardenable alloy. The Ti–2.5Cu (IMI 230) sheet alloy is one of very few titanium alloys strengthened by precipitation and that has been commercially developed. It is typically solution treated at 805 °C, followed by air cooling or oil quenching to room temperature. A double aging treatment at 400 and 475 °C promotes precipitation of a fine dispersion of a metastable precipitate phase that forms as coherent thin plates on $\{10\bar{1}1\}_\alpha$ with a large aspect ratio. The crystal structure of this metastable phase has been largely unclear. A study made by TEM and SAED (Williams et al., 1971) suggested that it was either GP zones or a structure that was similar to that of the matrix phase. In the overaged condition, the microstructure contains a coarse distribution of the equilibrium phase Ti$_2$Cu (space group I4/mmm, $a = 0.2944$ nm, $c = 1.0786$ nm) which also forms as $\{10\bar{1}1\}_\alpha$ plates with a large aspect ratio. The OR between the equilibrium Ti$_2$Cu phase and the α matrix is

$$(0\bar{1}3)_{\text{Ti}_2\text{Cu}} // (0001)_\alpha, [100]_{\text{Ti}_2\text{Cu}} // [1\bar{2}10]_\alpha$$

A moderate increase in tensile strength by 150–170 MPa may be achieved in the IMI 230 alloy by precipitation hardening. The strength can be further enhanced if the alloy is cold formed after quenching and before aging. The IMI 230 alloy is weldable and its strength is recoverable if a duplex aging treatment is applied after the welding. The IMI 230 sheet has been used as vanes for gas turbine engines.

20.5.5 Massive Transformation

Several Ti alloys, including Ti–Au, Ti–Ag and Ti–Si alloys, also exhibit a β → α$_m$ massive transformation. Massive transformations, which involve diffusional nucleation and growth that leads to a crystal structure change without a change in bulk composition, have also been observed in other ferrous and nonferrous alloy systems. The effective levels of undercooling at which such transformations occur and thus the driving forces involved are usually sufficiently large that the product phase takes a massive, approximately equiaxed form with dimensions typically in the range of several microns. A confirmed feature characteristic of this class of phase transformation is the absence of a rational OR between the product and matrix phases, and indeed it is commonly suggested that the product phase adopts a random orientation with respect to the matrix phase. A fundamental issue that is associated with this observation, and which still remains most controversial after several decades of debate, concerns the structure and mobility of the massive–matrix interphase boundaries.

In the absence of a defined OR and the presence of high rates of growth, the structure of the massive–matrix interface has been suggested to be incoherent or disordered (Massalski, 1984; Perepezko, 1984). The implications of this view are that the crystallography of the matrix phase, and thus crystallographic relationships across the interphase boundary, play little role during the transformation, and that the massive transformation is similar in this regard to the vapor–solid or liquid–solid transformations. In contrast, it has been proposed (Aaronson et al., 1968), on the basis of nucleation theory and the common presence of planar facets on massive products, that a massive product phase nucleates with a low-energy OR with the parent matrix and is bound by partially coherent interfaces. In this view, it is suggested that planar, low-energy massive–matrix interphase boundaries are common during growth, and that these interfaces migrate in directions normal to the interface plane only by the formation and lateral movement of growth ledges.

One prominent example of the massive transformation is the formation of ordered tetragonal γ_m phase (L1$_0$, $a \sim 0.398$ nm, $c \sim 0.408$ nm) from disordered hexagonal α (P6$_3$/mmc, $a \sim 0.575$ nm, $c \sim 0.473$ nm) solid solution in Ti–Al alloys having compositions close to TiAl. When these alloys are quenched or rapidly cooled from the high-temperature α phase field, the α phase transforms massively into equiaxed grains of a reportedly disordered face-centered cubic phase, which subsequently orders to γ_m. The composition of the γ_m phase has been reported to be approximately equal to that of α_2. In contrast to the lamellar duplex structure of γ and α_2, which forms during slow cooling from the α phase field, there is no evidence of a rational OR detected between γ_m and α_2. The γ_m phase is invariably bound by a series of pronouncedly planar interface facets that are often oriented irrationally with respect to both the γ_m and α_2 phases. Examinations of the planar interfaces using TEM did not reveal any evidence of linear defects within the interface planes (Nie et al., 1998; Veeraraghavan et al., 1999), even though arrays of dislocations have been detected in some curved part of the γ_m/α_2 interface, **Figure 85(a)**. On the basis that there is no rational OR between γ_m and α_2, and that the planar interfaces are oriented irrationally with respect to both γ_m and α_2, the γ_m/α_2 interphase boundaries have often been inferred to be incoherent.

Systematic characterization of the orientation and interfacial structure of planar facets of the massive γ_m phase that forms in a Ti–46.5at.%Al alloy, using TEM, confirms that the planar γ_m/α_2 interfaces are irrationally oriented with respect to both the matrix phase α_2 and the γ_m phase, and that these phases do not share a rational OR across such facets. Examination of the planar interface segments under two-beam electron diffraction contrast in different orientations reveals, within the resolution of the microscope, no evidence of linear defects in such sections of the γ_m/α_2 boundary. However, when imaged parallel to a particular direction in the interface, these irrationally oriented planar interfaces are invariably parallel to the moiré plane that is defined by the intersection between two sets of most closely packed planes in the γ_m and α_2 phases, **Figure 85(b–d)**. The relationship is such that there is an effective continuity of these lattice planes across the interface and a *one-dimensional coherency* within the planar interface. The existence of even one-dimensional lattice plane matching demonstrates unambiguously that these interfaces are not incoherent. It has been suggested (Nie and Muddle, 2002) that these planar interface facets may migrate in a relatively glissile manner normal to the interface plane by nucleation and rapid lateral movement within the interface plane of interfacial defects that have the form of moiré ledges defined by the spacing of the moiré pattern that may be formed by interpenetrating the crystals across the interface.

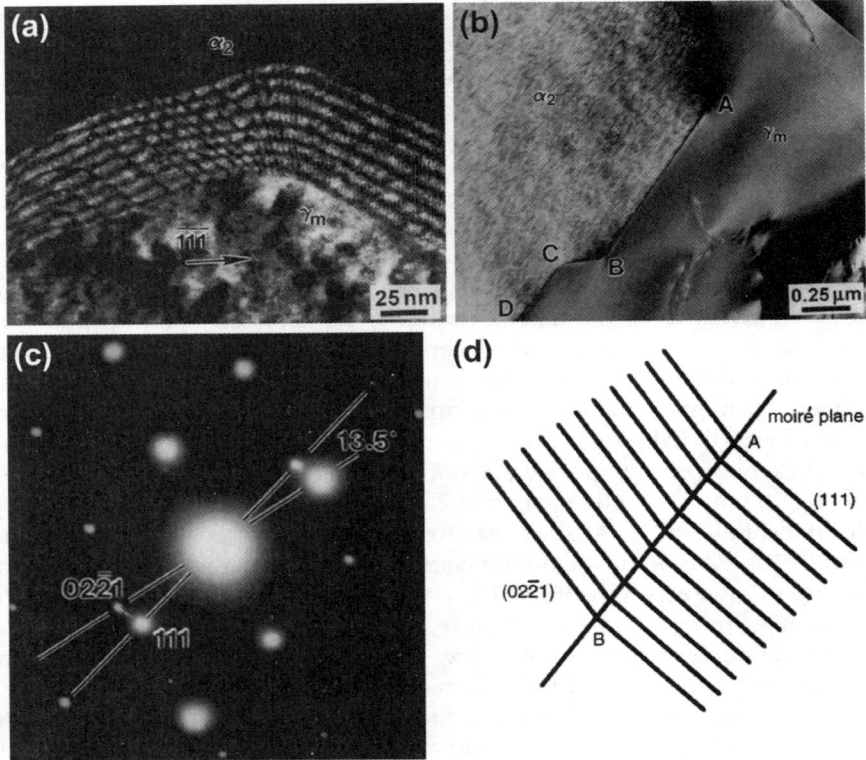

Figure 85   TEM images showing (a) arrays of misfit dislocations in a γ_m/α_2 interface, and (b) a planar γ_m/α_2 interface in a Ti–46.5at.%Al alloy. (c) SAED pattern recorded from the planar interface in (b), and (d) schematic diagram showing the relationship between $(111)_\gamma$ and $(02\bar{2}1)_{\alpha_2}$ reflections in (c) and their resultant moiré planes which is parallel to the planar interface in (b) (Nie and Muddle, 2002).

20.5.6   Alloy Developments and Applications

20.5.6.1   Aircraft Applications

Ti–6Al–4V-based $\alpha + \beta$ alloys have been used in both commercial and military aircraft for many years due to their sufficient specific strength, good damage tolerance and excellent corrosion resistance. The aircraft products made of such alloys include the landing gear beam in the Boeing 747, the bulkhead in fighter aircraft, and the wing box on the B1–B bomber. The use of a Ti landing gear beam permits less space in the airframe fuselage and therefore structural efficiency. A high-strength β alloy, Ti–10V–2-Fe–3Al and with a strength level of 1250 MPa, has been used for the landing gear of the Boeing 777. The usage of Ti alloys in aircraft is increasing relative to Al alloys. The original Boeing 747-100 contained about 2.6% Ti whereas the Boeing 777 contains about 8.3% Ti. It is the combination of weight reduction, lower maintenance cost and improved reliability that justify the use of costly Ti alloys.

The unique mechanical properties of titanium alloys make them excellent materials for high-temperature applications, for example the cooler portions of aircraft engines. However, there have been little research and development activities on new titanium alloys and little success in introducing new alloys into aircraft engines. The main reason for this situation is the low commercial benefit. In the 1980s, considerable effort was devoted to developing Ti-based intermetallic alloys for high-temperature applications. These alloys were based on the intermetallic phase of Ti_3Al, Ti_2AlNb and TiAl, respectively. Unfortunately, these intermetallic alloys suffer from low ductility, environmental sensitivity, and high cost and they have not found many commercial applications up to now.

20.5.6.2 Biomedical Applications

Biomaterials for implant applications need to have lightweight, complete inertness to the human body environment, sufficient strength, low modulus and high capacity to join with bones and other tissues. The metallic-based biomaterials currently used for orthopedic implants are 316L stainless steel, Co–Cr alloys and Ti–6Al–4V alloy. These materials have a number of shortcomings. Ni and Co in these biomedical materials are cytotoxic, and they can be gradually released into the human body due to corrosion in the human body environment. In addition, the stiffness of the 316L stainless steel and Co–Cr alloys is considerably higher than that of human bones, which can lead to gradual bone resorption and implant loosening over the years. While the Ti–6Al–4V alloy has a much lower elastic modulus than 316L stainless steel and Co–Cr alloys, and is therefore most commonly used for implants, its elastic modulus (about 110 GPa) is still much higher than that of human bones (4–30 GPa) (Geetha et al., 2009). Therefore, the development of biomedical titanium alloys with lower elastic moduli has been the focus of intensive international research in recent years.

As the Ti–6Al–4V alloy has a mixture of α and β, attempts have made to manipulate the alloy composition and thermomechanical processes to obtain microstructures comprising a single phase of β that can exhibit a lower elastic modulus. Since 2003 a group of novel β-titanium alloys have been developed for bone implant applications. Such alloys include Ti–29Nb–13Ta–4.6Zr (wt%), Ti–35.8Nb–2.1Ta–3.1Zr–0.3O (wt%) and Ti–24Nb–4Zr–8Sn (wt%). Nb is noncytotoxic, and these new Ti–Nb-based alloys have a low elastic modulus of 40–70 GPa, which is considerably lower than that of the Ti–6Al–4V alloy. These β-titanium alloys are generally solution treated in the β phase field followed by quenching. The solid solution alloys may also undergo a subsequent aging to form ω precipitates within the β phase to achieve higher strength. The elastic modulus of the β alloys depends critically on the alloy compositions and heat treatment conditions. For example, a very low elastic modulus can be obtained in a Ti–30Nb–10Ta–5Zr (wt%) alloy, while only a high elastic modulus is obtainable in Ti–30Nb–5Zr and Ti–30Nb–20Ta–5Zr alloys. The interpretation of these observations is that the presence of only β phase, and thus the absence of the ω phase, in the microstructure is responsible for the low elastic modulus of the Ti–30Nb–10Ta–5Zr alloy. However, this interpretation cannot explain why the Ti–30Nb–20Ta–5Zr alloy, where only β phase is present in the microstructure, has a high elastic modulus.

Apart from the influence of alloying elements and heat treatment conditions, the elastic modulus of the biomedical β-Ti alloys can also be strongly affected by cold work, that is plastic deformation at room temperature. In 2003, Saito et al. discovered a group of Ti–Nb–Ta–Zr–O alloys (called Gum Metal) that exhibit ultralow elastic moduli (Saito et al., 2003). They further discovered that if Gum Metal is severely cold worked, this significantly decreases the elastic modulus (the slope of the near-elastic segment of each curve), **Figure 86(a)**. It was further reported that the plastic deformation occurs via the formation and propagation of giant faults, instead of the conventional dislocation motion. While an apparent change in microstructure was noted after the severe cold work, **Figure 86(b) and (c)**, these investigators observed no dislocations in the cold worked sample. Despite the general acceptance that giant faults form and propagate in Gum Metal, these phenomena have not been experimentally confirmed and there is no fundamental understanding of the notion. Other

(a)

(b)

(c)

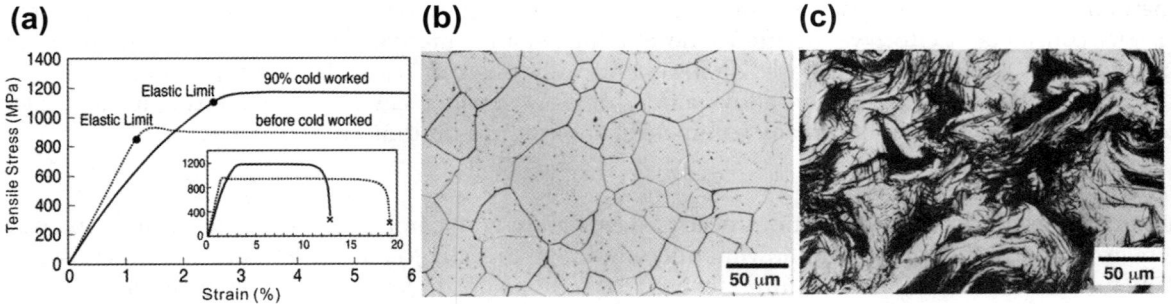

Figure 86 (a) Elastic softening of severe cold worked Gum Metal; (b) and (c) microstructures before and after severe cold work. The alloy is Ti–23 Nb–0.7Ta–2Zr–1.2O (Saito et al., 2003).

mechanisms can deliver such dislocation-free plastic deformation—for example the formation of a densely distributed, nanoscale domains of martensite in a cold worked sample and the reversible growth of these domains under the influence of external stress. However, the microstructures in the severe cold worked samples have not been thoroughly characterized to allow the clear identification of the deformation mechanisms.

Another new β-Ti alloy (Ti–24wt%Nb–4wt%Zr–8wt%Sn, designated Ti2448) was developed for biomedical applications. The bone plates made of this alloy have been tested by in vivo trials on fractured tibia of rabbits and beagle dogs. The results from the clinical trails indicate that Ti2448 is better than pure Ti in terms of bone reproduction in areas around the implant. This alloy is now being assessed for applications with bone healing and resorption in human bodies. Ti2448 has an extraordinarily low elastic modulus and intriguing elastic deformation behavior. Its elastic modulus is ~42 GPa when tested in the as-hot-rolled condition. However, it can be further decreased to ~20 GPa, a value that is now truly compatible with that of human bones, by cyclic loading and reloading tensile tests. As shown by the cyclic stress–strain curves in **Figure 87(a)**, the elastic modulus decreases with an

(a)

(b)

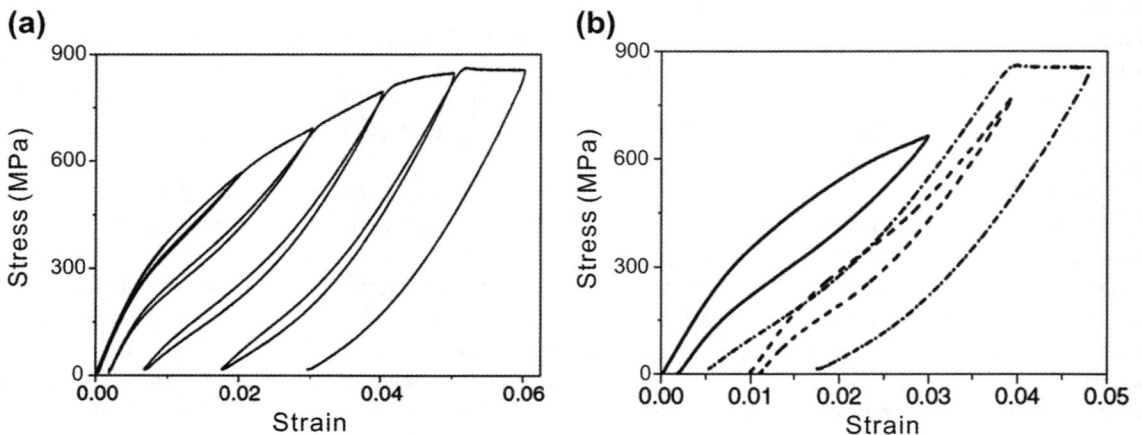

Figure 87 (a) Elastic softening phenomenon during a cyclic tensile test of Ti2448 alloy, and (b) recovery of elastic stiffness after natural aging; the dash–dot and dash curves represent the stress–strain curves without and with natural aging (Hao et al., 2007).

increasing tensile strain. The softened elastic modulus is unstable and can rise again if the unloaded sample is kept at room temperature for several days (naturally aging), **Figure 87(b)**. The cause of this phenomenon is unknown.

Based on observations made by X-ray diffraction, it was reported (Hao et al., 2007) that the gradual reduction in the elastic modulus during the cyclic loading–unloading tests is not because of any martensitic transformations, even though some stress-induced martensite was detected in the 6% tensile-strained sample. However, this does not exclude changes in nano strain domain structure in a strain-glass state. While three mechanisms—stress-induced martensitic transformations, incipient kink bands and giant faults—have been proposed for the explanation of the peculiar deformation behavior of Ti2448, none seems to be capable of describing all the features of the deformation. Four things remain to be firmly established: (i) what causes the gradual decrease in the elastic modulus during the cyclic tensile tests; (ii) whether martensite forms during the loading and transforms back to β after unloading; (iii) whether strain-glass transition leading to the formation of martensite nano-domains in β (Wang et al., 2010) can cause softening of the elastic modulus; and (iv) whether the rise of elastic modulus during aging at room temperature is due to the formation of ω (or other metastable phases).

Acknowledgments

The author wishes to acknowledge the generous and continuing support from the Australian Research Council in the past decade, the research work done by author's students, postdocs and colleagues, including Mr. Tim Hilditch, Miss Sabine Kallisch, Miss Olga Seremetis, Mr. Bryce Wood, Miss Zheng Ma, Dr. Yuman Zhu, Mr. Zhou Xu, Dr. Matthew Weyland, Dr. Laure Bourgeois, Dr. Xiaoling Xiao, Dr. Chamini Mendis, Dr. Robert Wilson, Mr. Sam Gao, and Dr. Suming Zhu, the access to the facilities in the Monash Centre for Electron Microscopy, the assistance from Drs. Yuman Zhu and Houwen Chen for formatting and drawing some figures in this paper, Dr. Xiya Fang for assistance in formatting references, and helpful discussions with Dr. Allan Morton, Prof. Barry Muddle, Prof. Hub Aaronson, Prof. Ian Polmear, Dr. Robert Sanders, and Dr. Kazuhiro Hono.

References

Aaronson, H.I., Laird, C., Kinsman, K.R., 1968. Scr. Metall. 2, 259–264.

Abe, E., Tsai, A.P., 1999. Phys. Rev. Lett. 83, 753–756.

Abe, E., Kawamura, Y., Hayashi, K., Inoue, A., 2002. Acta Mater. 50, 3845–3857.

Ahmed, M., Lorimer, G.W., Lyon, P., Pilkington, R., 1992. In: Mordike, B.L., Hehmann, F. (Eds.), Proceedings Magnesium Alloys and Their Applications. DGM Informationsgesellschaft, Germany, pp. 301–308.

Amiya, K., Ohsuna, T., Inoue, A., 2003. Mater. Trans. 44, 2151–2156.

Andersen, S.J., Zandbergen, H.W., Jansen, J., TrÆholt, C., Tundal, U., Reiso, O., 1998. Acta Mater. 46, 3283–3298.

Antion, C., Donnadieu, P., Perrard, F., Deschamps, A., Tassin, C., Pisch, A., 2003. Acta Mater. 51, 5335–5348.

Antion, C., Donnadieu, P., Tassin, C., Pisch, A., 2006. Philos. Mag. 86, 2797–2810.

Anyanwu, I.A., Kamado, S., Honda, T., Kojima, Y., Takeda, S., Ishida, T., 2000. Mater. Sci. Forum 350–151, 73–78.

Anyanwu, I.A., Kamado, S., Kojima, Y., 2001. Mater. Trans. 42, 1206–1211.

Apps, P.J., Karimzadeh, H., King, J.F., Lorimer, G.W., 2003. Scr. Mater. 48, 1023–1028.

Ardell, A.J., 1985. Metall. Trans. A 16A, 2131–2165.

Arnberg, L., Aurivillius, B., 1980. Acta Chem. Scand. A 34, 1–5.

Auld, J.H., 1968. Acta Metall. 16, 97–101.

Auld, J.H., Cousland, S.M., 1974. J. Aust. Inst. Metals 19, 194.

Bagaryatskii, Y.A., 1952. Dokl. Akad. Nauk SSSR 87, 397–400.

Bamberger, M., Levi, G., Vander Sande, J.B., 2006. Metall. Mater. Trans. A 37, 481–487.

Banerjee, S., Mukhopadhyay, P., 2007. Phase Transformations: Examples from Titanium and Zirconium Alloys. Elsevier, Amsterdam.

Barucca, G., Ferragut, R., Lussana, D., Mengucci, P., Moia, F., Riontino, G., 2009. Acta Mater. 57, 4416–4425.

Barucca, G., Ferragut, R., Fiori, F., Lussana, D., Mengucci, P., Moi, F., Riontino, G., 2011. Acta Mater. 59, 4151–4158.

Ben-Hamu, G., Eliezer, D., Kaya, A., Na, Y.G., Shin, K.S., 2006. Mater. Sci. Eng. A435–436, 579–587.

Berger, G., Weiss, A., 1988. J. Less Common Met. 142, 109–121.

Bergman, G., Waugh, J.L.T., Pauling, L., 1957. Acta Crystallogr. 10, 254–259.

Bettles, C.J., Humble, P., Nie, J.F., 1997. In: Lorimer, G.W. (Ed.), Proceedings 3rd International Magnesium Conference. The Institute of Materials, London, pp. 403–417.

Bettles, C.J., Venkatesan, K., Nie, J.F., 2003. Mater. Sci. Forum 419–422, 273–278.

Bettles, C.J., Gibson, M.A., Venkatesan, K., 2004. Scr. Mater. 51, 193–197.

Bjorge, R., Marioara, C.D., Andersen, S.J., Holmestad, R., 2010. Metall. Mater. Trans. A 41A, 1907–1916.

Bjorge, R., Nakashima, P.N.H., Marioara, C.D., Andersen, S.J., Muddle, B.C., Etheridge, J., Holmestad, R., 2011. Acta Mater. 59, 6103–6109.

Bjorge, R., Dwyer, C., Weyland, M., Nakashima, P.N.H., Marioara, C.D., Andersen, S.J., Etheridge, J., Holmestad, R., 2012. Acta Mater. 60, 3239–3246.

Bourgeois, L., Muddle, B.C., Nie, J.F., 2001a. Acta Mater. 49, 2701–2711.

Bourgeois, L., Mendis, C.L., Muddle, B.C., Nie, J.F., 2001b. Philos. Mag. Lett. 81, 709–718.

Bourgeois, L., Nie, J.F., Muddle, B.C., 2005. Philos. Mag. 85, 3487–3509.

Bourgeois, L., Dwyer, C., Weyland, M., Nie, J.F., Muddle, B.C., 2011. Acta Mater. 59, 7043–7050.

Bourgeois, L., Dwyer, C., Weyland, M., Nie, J.F., Muddle, B.C., 2012. Acta Mater. 60, 633–644.

Braszczynska-Malik, K.N., 2009. J. Alloys Compd. 477, 870–876.

Brook, G.B., 1965. Fulmer Research Institute Report No. 3.

Brown, L.M., Ham, R.K., 1971. In: Kelly, A., Nicholson, R.B. (Eds.), Strengthening Methods in Crystals. Elsevier Publishing Company Ltd, London, pp. 12–135.

Buha, J., 2008a. Mater. Sci. Eng. A 492, 11–19.

Buha, J., 2008b. Mater. Sci. Eng. A 492, 293–299.

Buha, J., 2008c. Acta Mater. 56, 3533–3542.

Buha, J., 2008d. Mater. Sci. Eng. A 491, 70–79.

Buha, J., Ohkubo, T., 2008. Metall. Mater. Trans. 39A, 2259–2273.

Busk, R.S., 1950. J. Met. 188, 1460–1464.

Caraher, S.K., Polmear, I.J., Ringer, S.P., 1998. In: Sato, T., Kumai, S., Kobayashi, T., Murakami, Y. (Eds.), Proceedings of 6th International Conference on Aluminum Alloys (ICAA6), vol. 2. Japan Institute for Light Metals, Tokyo, pp. 739–744.

Cassada, W.A., Shen, Y., Poon, S.J., Shiflet, G.J., 1986. Phys. Rev. B 34, 7413–7416.

Cassada, W.A., Shiflet, G.J., Starke, E.A., 1991a. Metall. Trans. A 22, 287–297.

Cassada, W.A., Shiflet, G.J., Starke, E.A., 1991b. Metall. Trans. A 22A, 299–306.

Cayron, C., Sagalowicz, L., Beffort, O., Buffat, P.A., 1999. Philos. Mag. A 79, 2833–2851.

Celotto, S., 2000. Acta Mater. 48, 1775–1787.

Chakrabarti, D.J., Laughlin, D.E., 2004. Prog. Mater. Sci. 49, 389–410.

Chiu, C.N., Grobner, J., Kozlov, A., Schmid-Fetzer, R., 2010. Intermetallics 18, 399–405.

Chopra, H.D., Liu, L.J., Muddle, B.C., Polmear, I.J., 1995. Philos. Mag. Lett. 71, 319–325.

Chopra, H.D., Muddle, B.C., Polmear, I.J., 1996. Philos. Mag. Lett. 73, 351–357.

Christian, J.W., 1958. Acta Metall. 6, 377–379.

Christian, J.W., 1969. In: Gifkins, R.C. (Ed.), Interfaces Conf. Butterworth's, London, p. 159.

Chun, J.S., Byrne, J.G., 1969. J. Mater. Sci. 4, 861–872.

Clark, J.B., 1965. Acta Metall. 13, 1281–1289.

Clark, J.B., 1968. Acta Metall. 16, 141–152.

Clark, J.B., Rhines, F.N., 1959. Trans. Am. Soc. Met. 51, 199–221.

Clark, J.B., Zabdyr, L., Moser, Z., 1988. In: Nayeb-Hashemi, A.A., Clark, J.B. (Eds.), Phase Diagrams of Binary Magnesium Alloys. ASM International, OH: Metals Park, pp. 353–364.

Court, S.A., Dudgeon, H.D., Ricks, R.A., 1994. In: Sanders, T.H., Starke, E.A. (Eds.), Proceedings of 4th International Conference on Aluminum Alloys, USA, vol. 1, p. 395.

Crawley, A.F., Lagowski, B., 1974. Metall. Trans. 5, 949–951.

Crawley, A.F., Miliken, K.S., 1974. Acta Metall. 22, 557–562.

Dahmen, D., Westmacott, K.H., 1983a. Physica Status Solidi (A) 80, 249–262.

Dahmen, D., Westmacott, K.H., 1983b. Scr. Metall. 17, 1241–1246.

Delfino, S., Saccone, A., Ferro, F., 1990. Metall. Trans. A 21, 2109–2114.

Derge, G., Kommell, A.R., Mehl, R.F., 1937. Trans. AIME 124, 367–378.

Donnadieu, P., Quivy, A., Tarfa, T., Ochin, P., Dezellus, A., Harmelin, M., Liang, P., Lukas, H.L., Seifert, H.J., Aldinger, F., Effenberg, G., 1997. Z. Metallkd. 88, 911–916.

Donnadieu, P., Shao, Y., Geuser, F.D., Botton, G.A., Lazar, S., Cheynet, M., de Boissieu, M., Deschamps, A., 2011. Acta Mater. 59, 462–472.

Duly, D., Cheynet, M.C., Brechet, Y., 1994a. Acta Metall. Mater. 42, 3843–3854.

Duly, D., Cheynet, M.C., Brechet, Y., 1994b. Acta Metall. Mater. 42, 3855–3863.

Duly, D., Simon, J.P., Brechet, Y., 1995a. Acta Metall. Mater. 43, 101–106.

Duly, D., Zhang, W.Z., Audier, M., 1995b. Philos. Mag. A 71, 187–204.

Dumolt, S.D., Laughlin, D.E., Williams, J.C., 1984. Scr. Metall. 18, 1347–1350.

Dutta, I., Allen, S.M., 1991. J. Mater. Sci. Lett. 10, 323–326.

Dwyer, C., Weyland, M., Chan, L.Y., Muddle, B.C., 2011. Appl. Phys. Lett. 98, 201909.

Edwards, G.A., Stiller, K., Dunlop, G.L., Couper, M.J., 1998. Acta Mater. 46, 3893–3904.

Egusa, D., Abe, E., 2012. Acta Mater. 60, 166–178.

Emley, E.F., 1966. Principles of Magnesium Technology. Pergamon Press, London.

Eom, J.P., Jin, Q.L., Lim, S.G., Hur, B.Y., Park, W.W., 2003. Mater. Sci. Forum 419–422, 307–312.

Fiawoo, M.F., Nie, J.F. Unpublished research.

Friedel, J., 1963. In: Thomas, G., Washburn, J. (Eds.), Electron Microscopy and Strength of Crystals. Interscience, New York, p. 605.

FrØseth, A.G., HØier, R., Derlet, P.M., Andersen, S.J., Marioara, C.D., 2003. Phys. Rev. B 67, 224106.

Furuhara, T., Aaronson, H.I., 1991. Acta Metall. Mater. 39, 2857–2872.

Furuhara, T., Howe, J.M., Aaronson, H.I., 1991. Acta Metall. Mater. 39, 2873–2886.

Gallot, J., Graf, R., 1965. C. R. Acad. Sci. 261B, 728–731.

Gallot, J., Graf, R., 1966. C. R. Acad. Sci. 262B, 1219.

Gallot, J., Lal, K., Graf, R., Guinier, A., 1964. C. R. Acad. Sci. 258B, 2818.

Gao, X., Nie, J.F., 2007a. Scr. Mater. 57, 655–658.

Gao, X., Nie, J.F., 2007b. Scr. Mater. 56, 645–648.

Gao, X., Nie, J.F., 2008. Scr. Mater. 48, 619–622.

Gao, X., Nie, J.F., Muddle, B.C., 1996. Mater. Sci. Forum 217–222, 1251–1256.

Gao, X., Nie, J.F., Muddle, B.C., 1998. In: Ferry, M. (Ed.), Proceedings of the Biennial Conference of the Institute of Materials Engineering, Australasia, pp. 573–578 (Melbourne).

Gao, X., Nie, J.F., Muddle, B.C., 1999. In: Koiwa, M., Otsuka, K., Miyazaki, T. (Eds.), Proceedings of the International Conference on Solid-Solid Phase Transformations. The Japan Institute of Metals, Tokyo, pp. 225–228.

Gao, X., Zhu, S.M., Muddle, B.C., Nie, J.F., 2005. Scr. Mater. 53, 1321–1326.

Gao, X., He, S.M., Zeng, X.Q., Peng, L.M., Ding, W.J., Nie, J.F., 2006. Mater. Sci. Eng. A 431, 322–327.

Garg, A., Howe, J.M., 1991. Acta Metall. Mater. 39, 1939–1946.

Geetha, M., Singh, A.K., Asokamani, R., Gogia, A.K., 2009. Prog. Mater. Sci. 54, 397–425.

Geng, J., Gao, X., Fang, X.Y., Nie, J.F., 2011. Scr. Mater. 64, 506–509.

Gharghouri, M.A., Weatherly, G.C., Embury, J.D., 1998. Philos. Mag. A 78, 1137–1149.

Gibson, M.A., Fang, X., Bettles, C.J., Hutchinson, C.R., 2010. Scr. Mater. 63, 899–902.

Gjönnes, J., Östmoe, T., 1970. Z. Metallkd. 61, 604–606.

Gjonnes, J., Simensen, C.J., 1970. Acta Met. 18, 881.

Goh, S.L., Nie, J.F., Bourgeois, L., Muddle, B.C., Embury, J.D., 2006. Mater. Sci. Forum 519–521, 1499–1504.

Gorsse, S., Hutchinson, C.R., Chevalier, B., Nie, J.F., 2005. J. Alloys Compd. 392, 253–262.

Gradwell, K.J., 1972. (PhD thesis), University of Manchester, U.K.

Grobner, J., Kozlov, A., Fang, X.Y., Geng, J., Nie, J.F., Schmid-Fetzer, R., 2012. Acta Mater. 60, 5948–5962.

Gupta, A.K., Gaunt, P., Chaturvedi, M.C., 1987. Philos. Mag. 55, 375–387.

Hao, Y.L., Li, S.J., Sun, S.Y., Zheng, C.Y., Yang, R., 2007. Acta Biomater. 3, 277–286.

Hardouin Duparc, O., 2005. Z. Metallkd. 96, 398–404.

Hardy, H.K., 1951–52. J. Inst. Met. 80, 483–492.

Hardy, H.K., Silcock, J.M., 1955–56. J. Inst. Met. 84, 423–428.

He, S.M., Zeng, X.Q., Peng, L.M., Gao, X., Nie, J.F., Ding, W.J., 2006. J. Alloys Compd. 421, 309–313.

Henes, V.S., Gerold, V., 1962. Z. Metallkd. 53, 743–748.

Henley, C.L., Elser, V., 1986. Philos. Mag. B 53, 59–66.

Higashi, I., Shiotani, N., Uda, M., 1981. J. Solid State Chem. 36, 225–233.

Hilditch, T., Nie, J.F., Muddle, B.C., 1998. In: Mordike, B.L., Kainer, K.U. (Eds.), Proc. Magnesium Alloys and Their Applications. Werkstoff–Informationsgesellschaft, Frankfurt, Germany, pp. 339–344.

Hirth, J.P., Lothe, J., 1968. Theory of Dislocations. McGraw–Hill Book Company, New York.
Hirth, J.P., Spanos, G., Hall, M.G., Aaronson, H.I., 1998. Acta Mater. 46, 857–868.
Hisa, M., Barry, J.C., Dunlop, G.L., 2002. Philos. Mag. A 82, 497–510.
Homma, T., Ohkubo, T., Kamado, S., Hono, K., 2007. Acta Mater. 55, 4137–4750.
Homma, T., Kunito, N., Kamado, S., 2009. Scr. Mater. 61, 644–647.
Homma, T., Mendis, C.L., Hono, K., Kamado, S., 2010a. Mater. Sci. Eng. A 527, 2356–2362.
Homma, T., Nakawaki, S., Kamado, S., 2010b. Scr. Mater. 63, 1173–1176.
Homma, T., Nakawaki, S., Oh-ishi, K., Hono, K., Kamado, S., 2011. Acta Mater. 59, 7662–7672.
Hono, K., Sano, N., Babu, S.S., Okano, R., Sakurai, T., 1993. Acta Metall. Mater. 41, 829–838.
Hono, K., Mendis, C.L., Sasaki, T.T., Oh-ishi, K., 2010. Scr. Mater. 63, 710–715.
Howe, J.M., Aaronson, H.I., Gronsky, R., 1985. Acta Metall. 33, 639–648, 649–658.
Howe, J.M., Dahmen, U., Gronsky, R., 1987. Philos. Mag. A 56, 31–61.
Howe, J.M., Lee, J., Vasudevan, A.K., 1988. Metall. Trans. A 19A, 2911–2920.
Hren, J.A., Thomas, G., 1963. Trans. Metall. Soc. AIME 227, 308–318.
Huang, J.C., 1992. Scr. Metall. Mater. 27, 755–760.
Huang, J.C., Ardell, A.J., 1987a. Mater. Sci. Technol. 3, 176–188.
Huang, J.C., Ardell, A.J., 1987b. J. Physique C3 (9), 373–383.
Huang, M.L., Li, H.X., Ding, H., Tang, Z.Y., Mei, R.B., Zhou, H.T., Ren, R.P., Hao, S.M., 2010. J. Alloys Compd. 489, 620–625.
Hui, X., Dong, W., Chen, G.L., Yao, K.F., 2007. Acta Mater. 55, 907–920.
Hutchinson, C.R., Nie, J.F., Gorsse, S., 2005. Metall. Mater. Trans. A 36, 2093–2105.
Itoi, T., Seimiya, T., Kawamura, Y., Hirohashi, M., 2004. Scr. Mater. 51, 107–111.
Jacobs, M.H., 1972. Philos. Mag. 26, 1–13.
Jardim, P.M., Solorzano, G., Vander Sande, J.B., 2002. Microsc. Microanal. 8, 487–496.
Jardim, P.M., Solorzano, G., Vander Sande, J.B., 2004. Mater. Sci. Eng. A 381, 196–205.
Jayaraj, J., Mendis, C.L., Ohkubo, T., Oh-ishi, K., Hono, K., 2010. Scr. Mater. 63, 831–834.
Jin, Y., Li, C.Z., Yan, M.G., 1990. J. Mater. Sci. 26, 421.
Kallisch, S., 1998. Undergraduate Final-Year Project Report. Monash University, Australia.
Kamado, S., Iwasawa, S., Ohuchi, K., Kojima, Y., Ninomiya, R., 1992. J. Jpn. Inst. Light Met. 42, 727–733.
Kang, D.H., Park, S.S., Kim, N.J., 2005. Mater. Sci. Eng. A 413–414, 555–560.
Kang, D.H., Park, S.S., Oh, Y.S., Kim, N.J., 2007. Mater. Sci. Eng. A 449–451, 318–321.
Kanno, M., Suzuki, H., Kanoh, O., 1980. J. Jpn. Inst. Met. 44, 1139–1145.
Karimzadeh, H., 1985. (PhD thesis), The University of Manchester, U.K.
Kawamura, Y., Yamasaki, M., 2007. Mater. Trans. 48, 2986–2992.
Kawamura, Y., Hayashi, K., Inoue, A., Masumoto, T., 2001. Mater. Trans. 42, 1172–1176.
Kawamura, Y., Kasahara, T., Izumi, S., Yamasaki, M., 2006. Scr. Mater. 55, 453–456.
Kelly, P.M., 1972. Scr. Metall. 6, 647–656.
Kelly, A., Nicholson, R.B., 1963. Prog. Mater. Sci. 10, 151–391.
Khan, Y., 1989. J. Mater. Sci. 24, 963–973.
Khosrhoshahi, R.A., Pilkington, R., Lorimer, G.W., Lyon, P., Karimzadeh, H., 1997. In: Lorimer, G.W. (Ed.), Proceedings of the 3rd International Magnesium Conference. The Institute of Materials, London, pp. 241–256.
Kiehn, J., Smola, B., Vostry, P., Stulikova, I., Kainer, K.U., 1997. Physica Status Solidi 164, 709–723.
Kim, I.J., Bae, D.H., Kim, D.H., 2003. Mater. Sci. Eng. A 359, 313–318.
Knowles, K.M., Stobbs, W.M., 1988. Acta Crystallogr. B44, 207–227.
Komura, Y., Tokunaga, K., 1980. Acta Crystallogr. 36B, 1548–1554.
Kopp, V., Lefebvre, W., Pareige, C., 2011. J. Phase Equilib. Diff. 32, 298–301.
Kovarik, L., Mills, M.J., 2012. Acta Mater. 60, 3861–3872.
Kovarik, L., Miller, M.K., Court, S.A., Mills, M.J., 2006. Acta Mater. 54, 1731–1740.
Kovarik, L., Court, S.A., Fraser, H.L., Mills, M.J., 2008. Acta Mater. 56, 4804–4815.
Kubota, M., Nie, J.F., Muddle, B.C., 2004. Mater. Trans. 45, 3256–3263.
Kubota, M., Nie, J.F., Muddle, B.C., 2005a. Mater. Trans. 46, 1288–1294.
Kubota, M., Nie, J.F., Muddle, B.C., 2005b. Mater. Trans. 46, 365–368.
Lagowski, B., 1971. Trans. Am. Foundrymen's Soc. 79, 115–120.
Laird, C., Aaronson, H.I., 1967. Acta Metall. 15, 73–103.
Lequeu, Ph., Lassince, Ph., Warner, T., July, 2007. Adv. Mater. Process., 41–44.
Li, B.Q., Wawner, F.E., 1998. Acta Mater. 46, 5483–5490.

Li, R.G., Nie, J.F., Huang, G.J., Xin, Y.C., Liu, Q., 2011. Scr. Mater. 64, 950–953.

Liang, H., Chen, S.L., Chang, Y.A., 1997. Metall. Mater. Trans. A 28, 1725–1734.

Liu, Y.C., Aaronson, H.I., 1970. Acta Metall. 18, 845–856.

Liu, Z.R., Chen, J.H., Wang, S.B., Yuan, D.W., Yin, M.J., Wu, C.L., 2011. Acta Mater. 59, 7396–7405.

Lorimer, G.W., 1978. In: Russell, K.C., Aaronson, H.I. (Eds.), Precipitation Processes in Solids. The Metallurgical Society of AIME, Warrendale, pp. 87–119.

Lorimer, G.W., 1987. In: Baker, C., Lorimer, G.W., Unsworth, W. (Eds.), Proceedings London Conference on Magnesium Technology. The Institute of Metals, London, pp. 47–53.

Lorimer, G.W., Khosroshahi, R.A., Ahmed, M., 1999. In: Koiwa, M., Otsuka, K., Miyazaki, T. (Eds.), Proceedings of the International Conference on Solid-Solid Phase Transformations. The Japan Institute of Metals, Tokyo, pp. 185–192.

Lukasak, D.A., Hart, R.M., 1991. Light Met. Age 49, Oct, 11.

Luo, Z.P., Zhang, S.Q., 2000. J. Mater. Sci. Lett. 19, 813–815.

Luo, Z.P., Zhang, S.Q., Tang, Y.L., Zhao, D.S., 1993. Scr. Metall. Mater. 28, 1513–1518.

Luo, A., Balogh, M., Powell, B.R., 2002. Metall. Mater. Trans. A 33, 567–574.

Lutjering, G., Williams, J.C., 2003. Titanium. Springer–Verlag, Berlin.

Lyon, P., Wilks, T., Syed, I., 2005. In: Neelameggham, N.R., Kaplan, H.I., Powell, B.R. (Eds.), Magnesium Technology. TMS, pp. 303–308.

Ma, Z., 2007. Undergraduate Final–Year Research Project Report. Monash University, Australia.

Marceau, R.K.W., Sha, G., Lumley, R.N., Ringer, S.P., 2010. Acta Mater. 58, 1975–1805.

Marioara, C.D., Nordmark, H., Andersen, S.J., Holmestad, R., 2006. J. Mater. Sci. 41, 471–478.

Marioara, C.D., Andersen, S.J., Stene, T.N., Hasting, H., Walmsley, J., van Helvoort, A.T.J., Holmestad, R., 2007. Philos. Mag. 87, 3385–3413.

Marioara, C.D., Nakamura, J., Matsuda, K., Andersen, S.J., Holmstad, R., Sato, T., Kawabata, T., Ikeno, S., 2012. Philos. Mag. 92, 1149–1158.

Martin, J.W., 1998. Precipitation Hardening, second ed. Butterworth–Heinemann, Oxford.

Massalski, T.B., 1984. Metall. Trans. A 15A, 421–425.

Matsuda, K., Ikeno, S., Sato, T., Kamio, A., 1996. Scr. Mater. 34, 1797–1802.

Matsuda, K., Sakaguchi, Y., Miyata, Y., Uetani, Y., Sato, T., Kamio, A., Ikeno, S., 2000. J. Mater. Sci. 35, 179–189.

Matsuda, M., Ii, S., Kawamura, Y., Ikuhara, Y., Nishida, M., 2005. Mater. Sci. Eng. A 393, 269–274.

Matsuura, M., Konno, K., Yoshida, M., Nishijima, M., Hiraga, K., 2006. Mater. Trans. 47, 1264–1267.

Mendis, C.L., 2000. (M. Eng. Sci. thesis), Monash University, Australia.

Mendis, C.L., 2005. (PhD thesis), Monash University, Australia.

Mendis, C.L., Muddle, B.C., Nie, J.F., 2001. In: Hanada, S., Zhong, Z., Nam, S.W., Wright, R.N. (Eds.), Proceedings of the 4th Pacific Rim International Conference Advanced Materials and Processing (PRICM-4). Japan Institute of Metals, pp. 1207–1210.

Mendis, C.L., Bettles, C.J., Gibson, M.A., Hutchinson, C.R., 2006a. Mater. Sci. Eng. A 435–436, 163–171.

Mendis, C.L., Bettles, C.J., Gibson, M.A., Gorsse, S., Hutchinson, C.R., 2006b. Philos. Mag. Lett. 86, 443–456.

Mendis, C.L., Oh-ishi, K., Hono, K., 2007. Scr. Mater. 57, 485–488.

Mendis, C.L., Oh-ishi, K., Kawamura, Y., Honma, T., Kamado, S., Hono, K., 2009. Acta Mater. 57, 749–760.

Mendis, C.L., Oh-ishi, K., Hono, K., 2010. Mater. Sci. Eng. A 527, 973–980.

Mendis, C.L., Bae, J.H., Kim, N.J., Hono, K., 2011a. Scr. Mater. 64, 335–338.

Mendis, C.L., Oh-ishi, K., Ohkubo, T., Hono, K., 2011b. Scr. Mater. 64, 137–140.

Merle, P., Fouquet, F., Merlin, J., 1981. Mater. Sci. Eng. 50, 215.

Mima, G., Tanaka, Y., 1971a. Jpn. Inst. Met. 12, 71–75.

Mima, G., Tanaka, Y., 1971b. Jpn. Inst. Met. 12, 76–81.

Mizer, D., Peters, B.C., 1972. Metall. Trans. 3, 3262–3264.

Moia, F., Ferragut, R., Calloni, A., Dupasquier, A., Macchi, C.E., Somoza, A., Nie, J.F., 2010. Philos. Mag. 90, 2135–2147.

Mondolfo, L.F., 1976. Aluminum Alloys: Structure and Properties. Butterworths, London, p. 501.

Mose, R.I., 1983. PhD Thesis, The University of Manchester, U.K.

Muddle, B.C., Polmear, I.J., 1989. Acta Metall. 37, 777–789.

Muddle, B.C., Ringer, S.P., Polmear, I.J., 1994. In: Somiya, S., Doyama, M., Roy, R. (Eds.), Advanced Materials '93, VI/Frontiers in Materials Science and Engineering, Transactions Materials Research Society of Japan, vol. 19B. Elsevier Science B.V., pp. 999–1023.

Mukhopadhyay, N.K., Subbanna, G.N., Ranganathan, S., Chattopadhyay, K., 1986. Scr. Metall. Mater. 20, 525–528.

Murayama, M., Hono, K., 1998. Scr. Mater. 38, 1315–1319.

Murayama, M., Hono, K., 1999. Acta Mater. 47, 1537–1548.

Murayama, M., Hono, K., 2001. Scr. Mater. 44, 701–706.

Murayama, M., Hono, K., Miao, W.F., Laughlin, D.E., 2001. Metall. Mater. Trans. A 32, 239–246.

Mushovic, J.N., Stoloff, N.S., 1969. Trans. Metall. Soc. AIME 245, 1449–1456.

Nabarro, F.R.N., 1940. Proc. R. Soc. Lond. A, Math. Phys. Sci. 175, 519–538.

Nayeb-Hashemi, A.A., Clark, J.B., 1988. Phase Diagrams of Binary Magnesium Alloys. ASM International, Ohio, USA.: Metals Park.
Nembach, E., 1997. Particle Strengthening in Metals and Alloys. John Wiley & Sons, New York.
Nicholson, R.B., Nutting, J., 1961. Acta Metall. 9, 332–343.
Nie, J.F. Unpublished research.
Nie, J.F., 2002. In: Kaplan, H. (Ed.), Proceedings of Magnesium Technology 2002. TMS, USA.: Warrendale, pp. 103–110.
Nie, J.F., 2003. Scr. Mater. 48, 1009–1015.
Nie, J.F., 2004. Acta Mater. 52, 795–807.
Nie, J.F., 2006. Metall. Mater. Trans. 37A, 841–849.
Nie, J.F., 2008. Acta Mater. 56, 3169–3176.
Nie, J.F., 2012. Metall. Mater. Trans. A 43A, 3891–3939.
Nie, J.F., Muddle, B.C., 1997. Scr. Mater. 37, 1475–1481.
Nie, J.F., Muddle, B.C., 1998a. In: Mordike, B.L., Kainer, K.U. (Eds.), Proc. Magnesium Alloys and Their Applications. Werkstoff-Informationsgesellschaft, Frankfurt, pp. 229–234.
Nie, J.F., Muddle, B.C., 1998b. J. Phase Equilib. 19, 543–551.
Nie, J.F., Muddle, B.C., 1999a. Mater. Forum 23, 23–40.
Nie, J.F., Muddle, B.C., 1999b. Scr. Mater. 40, 1089–1094.
Nie, J.F., Muddle, B.C., 2000. Acta Mater. 48, 1691–1703.
Nie, J.F., Muddle, B.C., 2001. Mater. Sci. Eng. A319–321, 448–451.
Nie, J.F., Muddle, B.C., 2002. Metall. Mater. Trans. 33A, 2381–2389.
Nie, J.F., Muddle, B.C., 2008. Acta Mater. 56, 3490–3501.
Nie, J.F., Xiao, X.L., Bourgeois, L. Unpublished research.
Nie, J.F., Muddle, B.C., Polmear, I.J., 1996. Mater. Sci. Forum 217–222, 1257–1262.
Nie, J.F., Muddle, B.C., Furuhara, T., Aaronson, H.I., 1998. Scr. Mater. 39, 637–645.
Nie, J.F., Aaronson, H.I., Muddle, B.C., 1999. In: Koiwa, M., Otsuka, K., Miyazaki, T. (Eds.), Proceedings of International Conference on Solid-solid Phase Transformations. The Japan Institute of Metals, Tokyo, pp. 157–160.
Nie, J.F., Xiao, X.L., Luo, C.P., Muddle, B.C., 2001. Micron 32, 857–863.
Nie, J.F., Muddle, B.C., Aaronson, H.I., Ringer, S.P., Hirth, J.P., 2002. Metall. Mater. Trans. 33A, 1649–1658.
Nie, J.F., Gao, X., Zhu, S.M., 2005. Scr. Mater. 53, 1049–1053.
Nie, J.F., Oh-ishi, K., Gao, X., Hono, K., 2008. Acta Mater. 56, 6061–6076.
Ninomiya, R., Ojiro, T., Kubota, K., 1995. Acta Metall. Mater. 43, 669–674.
Nishijima, M., Hiraga, K., 2007. Mater. Trans. 48, 10–15.
Nishijima, M., Hiraga, K., Yamasaki, M., Kawamura, Y., 2006. Mater. Trans. 47, 2109–2122.
Nishijima, M., Hiraga, K., Yamasaki, M., Kawamura, Y., 2007a. Mater. Trans. 48, 476–480.
Nishijima, M., Yubuta, K., Hiraga, K., 2007b. Mater. Trans. 48, 84–87.
Nishijima, M., Hiraga, K., Yamasaki, M., Kawamura, Y., 2008. Mater. Trans. 49, 227–229.
Noble, B., Crook, A., 1970. J. Inst. Met. 98, 375–380.
Noble, B., Pike, T.J., Crook, A., 1971. Philos. Mag. 23, 543–553.
Numakura, H., Koiwa, M., 1984. Acta Metall. 32, 1799–1807.
Nuttall, P.A., Pike, T.J., Noble, B., 1980. Metallography 13, 3–20.
Oblak, J.M., Duvall, D.S., Paulonis, D.F., 1974a. Mater. Sci. Eng. 13, 51.
Oblak, J.M., Paulonis, D.F., Duvall, D.S., 1974b. Metall. Trans. 5, 143–153.
Oh, J.C., Ohkubo, T., Mukai, T., Hono, K., 2005. Scr. Mater. 53, 675–679.
Oh-ishi, K., Hono, K., Shin, K.S., 2008. Mater. Sci. Eng. A 496, 425–433.
Oh-ishi, K., Watanabe, R., Mendis, C.L., Hono, K., 2009. Mater. Sci. Eng. A 526, 177–184.
Okamoto, H., 2007. J. Phase Equilib. Diff. 28, 405.
Orowan, E., 1948. In Symposium on Internal Stresses in Metals and Alloys. Institute of Metals, London, pp. 451–453.
Ozbilen, S., Flower, H.M., 1989. Acta Metall. 37, 2993–3000.
Padezhnova, E.M., Mel'nik, E.V., Miliyevskiy, R.A., Dobatkina, T.V., Kinzhibalo, V.V., 1982. Russ. Metall. 4, 185–188.
Padfield, P.V., 2004. ASM Handbook, vol. 9. ASM International, OH, USA.: Materials Park, p. 801.
Park, S.C., Lim, J.D., Eliezer, D., Shin, K.S., 2003. Mater. Sci. Forum 419–422, 159–164.
Park, S.S., Bae, G.T., Kang, D.H., Jung, I.H., Shin, K.S., Kim, N.J., 2007. Scr. Mater. 57, 793–796.
Partridge, P.G., 1967. Metall. Rev. 12, 169–194.
Partridge, P.G., Roberts, E., 1964. The 3rd European Regional Conference Electron Microscopy, pp. 213–214.
Pashley, D.W., Rhodes, J.W., Sendorek, A., 1966. J. Inst. Met. 94, 41–49.
Pashley, D.W., Jacobs, M.H., Vietz, J.T., 1967. Philos. Mag. 16, 51–76.

Payne, R.J.M., Baily, N., 1959–60. J. Inst. Met. 88, 417–427.

Perepezko, J.H., 1984. Metall. Trans. A 15A, 437–447.

Pérez-Landazábal, J.I., No, M.L., Madariaga, G., San Juan, J., 1997. J. Mater. Res. 12, 577–580.

Petrov, D., Watson, A., Grobner, J., Rogl, P., Tedenac, J.C., Bulanova, M., Turkevich, V., 2006. In: Effenberg, G., Ilyenko, S. (Eds.), Ternary Alloy Systems, vol. 11A3. Springer, Germany, pp. 191–209.

Pickens, J.R., Heubaum, F.H., Langan, T.J., Kramer, L.S., 1989. In: Starke, E.A., Sanders, T.H. (Eds.), Proceedings of the 5th International Conference on Aluminum–Lithium Alloys. Mater. Comp. Eng. Publications, Birmingham, U.K., pp. 1397–1414.

Pike, T.J., Noble, B., 1973. J. Less Common Met. 30, 63–74.

Pike, T.J., Nuttall, P.A., Noble, B., 1972. J. Inst. Met. 100, 249–254.

Ping, D.H., Hono, K., Kawamura, Y., Inoue, A., 2002. Philos. Mag. Lett. 82, 543–551.

Ping, D.H., Hono, K., Nie, J.F., 2003. Scr. Mater. 48, 1017–1022.

Polmear, I.J., 1972. J. Aust. Inst. Met. 17, 129–141.

Polmear, I.J., 1994. J. Mater. Sci. Technol. 10, 1–16.

Polmear, I.J., 1995. Light Alloys, third ed. Arnold, London.

Polmear, I.J., 2004. Mater. Forum 28, 1–14.

Polmear, I.J., 2006. Light Alloys, fourth ed. Elsevier/Butterworth-Heinemann, Oxford.

Polmear, I.J., Chester, R.J., 1989. Scr. Metall. Mater. 23, 1213–1217.

Polmear, I.J., Couper, M.J., 1988. Metall. Trans. A 19A, 1027–1035.

Porter, D.A., Easterling, K.E., 1992. Phase Transformations in Metals and Alloy, second ed. Chapman & Hall, London.

Porter, D.A., Edington, J.W., 1977. Proc. R. Soc. Lond. A 358, 335–350.

Qi, H.Y., Huang, G.X., Bo, H., Xu, G.L., Liu, L.B., Jin, Z.P., 2011. J. Alloys Compd. 509, 3274–3281.

Radmilovic, V., Thomas, G., 1987. J. Physique 48 (C3(9)), 385–396.

Radmilovic, V., Thomas, G., Shiflet, G.J., Starke, E.A., 1989. Scr. Metall. 27, 1141–1146.

Radmilovic, V., Kilaas, R., Dahmen, U., Shiflet, G.J., 1999. Acta Mater. 47, 3987–3997.

Raghavan, V., 2010. J. Phase Equilib. Diff. 31, 288–290.

Raghavan, V., 2011. J. Phase Equilib. Diff. 32, 52–53.

Rajasekharan, T., Akhtar, D., Gopalan, R., Muraleedharan, K., 1986. Nature 322, 528–530.

Ralston, K.D., Birbilis, N., Weyland, M., Hutchinson, C.R., 2010. Acta Mater. 58, 5941–5948.

Raman, K.S., Das, E.S.D., Vasu, K.I., 1970. Scr. Metall. 4, 291–293.

Raynor, G.V., 1959. The Physical Metallurgy of Magnesium and Its Alloys. Pergmon Press, London.

Reich, L., Murayama, M., Hono, K., 1998. Acta Mater. 46, 6053–6062.

Ren, Y.P., Qin, G.W., Pei, W.L., Guo, Y., Zhao, H.D., Li, H.X., Jiang, M., Hao, S.M., 2009. J. Alloys Compd. 481, 176–181.

Reppich, B., 1993. In: Cahn, R.W., Hassen, P., Kramer, E.J., Weinheim, VCH (Eds.), Materials Science and Technology: A Comprehensive Treatment, vol. 6, pp. 311–357 (Germany).

Ringer, S.P., Hono, K., 2000. Mater. Charact. 44, 101–131.

Ringer, S.P., Raviprasad, K., 2000. Mater. Forum 24, 59–94.

Ringer, S.P., Muddle, B.C., Polmear, I.J., 1995a. Metall. Mater. Trans. A 26A, 1659–1671.

Ringer, S.P., Hono, K., Sakurai, T., 1995b. Metall. Mater. Trans. A 26A, 2207–2217.

Ringer, S.P., Hono, K., Polmear, I.J., Sakurai, T., 1996. Appl. Surf. Sci. 94–95, 253–260.

Ringer, S.P., Sakurai, T., Polmear, I.J., 1997. Acta Mater. 45, 3731–3744.

Roberts, C.S., 1960. Magnesium and Its Alloys. Wiley, New York.

Robson, J.D., Stanford, N., Barrnet, M.R., 2010. Scr. Mater. 63, 823–826.

Robson, J.D., Stanford, N., Barrnet, M.R., 2011. Acta Mater. 59, 1945–1956.

Rokhlin, L.L., 1995. J. Phase Equilib. 16, 504–507.

Rokhlin, L.L., 2003. Magnesium Alloys Containing Rare Earth Metals. Taylor and Francis, London.

Rokhlin, L.L., Nikitina, N.I., 1994. Z. Metallkd. 85, 819–823.

Rokhlin, L.L., Oreshkina, A.A., 1988. Fizika Metallovi Metallovedenie 66, 559–562.

Rokhlin, L.L., Dobatkina, T.V., Tarytina, I.E., Timofeev, V.N., Balakhchi, E.E., 2004. J. Alloys Compd. 367, 17–19.

Rosalie, J.M., Somekawa, H., Singh, A., Mukai, T., 2010. Philos. Mag. 90, 3355–3374.

Rosalie, J.M., Somekawa, H., Singh, A., Mukai, T., 2011. Philos. Mag. 91, 2634–2644.

Saito, K., Hiraga, K., 2011. Mater. Trans. 52, 1860–1867.

Saito, T., Furuta, T., Hwang, J.H., Kuramoto, S., Nishino, K., Suzuki, N., Chen, R., Yamada, A., Ito, K., Seno, Y., Nonaka, T., Ikehata, H., Nagasako, N., Iwamoto, C., Ikuhara, Y., Sakuma, T., 2003. Science 300, 464–467.

Saito, K., Yasuhara, A., Hiraga, K., 2011. J. Alloys Compd. 509, 2031–2038.

Samson, S., 1949. Acta Chem. Scand. A 3, 809–834.

Sankaran, R., Laird, C., 1974. Mater. Sci. Eng. 14, 271–279.

Sasaki, T.T., Oh-ishi, K., Ohkubo, T., Hono, K., 2006. Scr. Mater. 55, 251–254.

Sasaki, T.T., Ju, J.D., Hono, K., Shin, K.S., 2009a. Scr. Mater. 61, 80–83.

Sasaki, T.T., Ohkubo, T., Hono, K., 2009b. Scr. Mater. 61, 72–75.

Sasaki, T.T., Oh-ishi, K., Ohkubo, T., Hono, K., 2011. Mater. Sci. Eng. A 530, 1–8.

Sato, T., Takahashi, I., Tezuka, H., Kamio, A., 1992. Kei-Kinzoku (J. Japan. Inst. Light Met.) 42, 804–809.

Schneider, K., von Heimendahl, M., 1973. Z. Metallkd. 64, 342.

Seremetis, O., 1999. Undergraduate Final-Year Research Project Report. Monash University, Australia.

Sha, G., Cerezo, A., 2004. Acta Mater. 52, 4503–4516.

Shao, G., Varsani, V., Fan, Z., 2006. Comput. Coupling of Phase Diagr. Thermochem. 30, 286–295.

Shin, D., Wolverton, C., 2010a. Scr. Mater. 63, 680–685.

Shin, D., Wolverton, C., 2010b. Acta Mater. 58, 531–540.

Silcock, J.M., 1960–61. J. Inst. Met. 89, 203–210.

Silcock, J.M., Flower, H.M., 2002. Scr. Mater. 46, 389–394.

Silcock, J.M., Heal, T.J., Hardy, H.K., 1953–54. J. Inst. Met. 82, 239–248.

Silcock, J.M., Heal, T.J., Hardy, H.K., 1955. J. Inst. Met. 84, 23–31.

Silcock, J.M., Heal, T.J., Hardy, H.K., 1955–56. J. Inst. Met. 84, 19.

Singh, A., Tsai, A.P., 2007. Scr. Mater. 57, 941–944.

Singh, A., Nakamura, M., Watanabe, M., Kato, A., Tai, A.P., 2003. Scr. Mater. 49, 417–422.

Singh, A., Rosalie, J.M., Somekawa, H., Mukai, T., 2010. Philos. Mag. Lett. 90, 641–651.

Somekawa, H., Singh, A., Mukai, T., 2007. Scr. Mater. 56, 1091–1094.

Staley, J.T., Hunt, W.H., 2012. Private communications.

Staley, J.T., Lege, D.J., 1993. J. Phys. IV 3 (C7), 179–190.

Starink, M., Wang, S.C., 2009. Acta Mater. 57, 2376–2389.

Starink, M.J., Zahra, A.M., 1997. Philos. Mag. 76, 701–714.

Starink, M.J., Zahra, A.M., 1998. Acta Mater. 46, 3381–3397.

Starke, E.A., 1989. In: Vasudevan, A.K., Doherty, R.D. (Eds.), Aluminum Alloys: Contemporary Research and Applications, Treatise on Materials Science and Technology, vol. 31. Academic Press, New York, p. 35.

Starke, E.A., Staley, J.T., 1996. Prog. Aerosp. Sci. 32, 131–172.

Stratford, D.J., 1971. J. Inst. Met. 99, 201–208.

Stratford, D.J., 1972. J. Inst. Met. 100, 381–384.

Stratford, D.J., Beckley, L., 1972. J. Met. Sci. 6, 83–89.

Sturkey, L., Clark, J.B., 1959–60. J. Inst. Met. 88, 177–181.

Suh, I.S., Park, J.K., 1995. Acta Metall. Mater. 43, 4495–4503.

Suzuki, H., Kanno, M., Araki, I., 1975. J. Jpn. Inst. Light Met. 25, 413–421.

Suzuki, M., Kimura, T., Koike, J., Maruyama, K., 2003. Scr. Mater. 48, 997–1002.

Suzuki, A., Saddock, N.D., Jones, J.W., Pollock, T.M., 2004. Scr. Mater. 51, 1005–1010.

Suzuki, A., Saddock, N.D., Jones, J.W., Pollock, T.M., 2005. Acta Mater. 53, 2823–2834.

Suzuki, A., Saddock, N.D., TerBush, J.R., Powellm, B.R., Jones, J.W., Pollock, T.M., 2008. Metall. Mater. Trans. A 39, 696–702.

Torsœter, M., Ehlers, F.J.H., Marioara, C.D., Andersen, S.J., Holmestad, R., 2012. Philos. Mag. http://dx.doi.org/10.1080/14786435.2012.693214.

Unsworth, W., King, J.F., 1987. In: Lorimer, G.W. (Ed.), Proceedings of London Conference on Magnesium Technology. The Institute of Metals, London, pp. 25–35.

Van der Planken, J., 1969. J. Mater. Sci. 4, 927–929.

van Huis, M.A., Chen, J.H., Sluiter, M.F.H., Zandbergen, H.W., 2007. Acta Mater. 55, 2183–2199.

Van Smaalen, S., Meetsma, A., Boer, J.D., Bronsveld, P., 1990. J. Solid State Chem. 85, 293–298.

Vaughan, D., Silcock, J.M., 1967. Physica Status Solidi 20, 725–736.

Vecchio, K.S., Williams, D.B., 1988. Metall. Trans. A 19, 2875–2884.

Veeraraghavan, D., Wang, P., Vasudevan, V.K., 1999. Acta Mater. 47, 3313–3330.

Vietz, J.T., Polmear, I.J., 1966. J. Inst. Met. 94, 410–419.

Vietz, J.T., Sargant, K.R., Polmear, I.J., 1963–64. J. Inst. Met. 92, 327–333.

Villars, P., Calvert, L.D., 1985. Pearson's Handbook of Crystallographic Data for Intermetallic Phases. ASM, Ohio, USA. (Metals Park).

Vissers, R., van Huis, M.A., Jansen, J., Zandbergen, H.W., Marioara, C.D., Andersen, S.J., 2007. Acta Mater. 55, 3815–3823.

Vogel, M., Kraft, O., Dehm, G., Arzt, E., 2001. Scr. Mater. 45, 517–524.

Vogel, M., Kraft, O., Arzt, E., 2003. Scr. Mater. 48, 985–990.

Vostry, P., Stulikova, I., Smola, B., Cieslar, M., Mordike, B.L., 1988. Z. Metallkd. 79, 340–344.
Wang, S.C., Starink, M.J., 2007. Acta Mater. 55, 933–941.
Wang, D., Wang, Y., Zhang, Z., Ren, X., 2010. Phys. Rev. Lett. 105, 205702.
Warren, A.S., 2004. Mater. Forum 28, 24–31.
Weakley-Bollin, S.C., Donlon, W., Wolverton, C., Jones, J.W., Allison, J.E., 2004. Metall. Mater. Trans. A 35, 2407–2418.
Wei, L.Y., Dunlop, G.L., Westengen, H., 1995. Metall. Mater. Trans. A 26, 1705–1716.
Wei, L.Y., Dunlop, G.L., Westengen, H., 1996. J. Mater. Sci. 31, 387–397.
Williams, J.C., 1978. In: Russell, K.C., Aaronson, H.I. (Eds.), Precipitation Processes in Solids. The Metallurgical Society of AIME, Warrendale, pp. 191–221.
Williams, D.B., Butler, E.P., 1981. Int. Met. Rev. 26, 153–183.
Williams, J.C., Starke, E.A., 2003. Acta Mater. 51, 5775–5799.
Williams, J.C., Taggart, R., Polonis, D.H., 1971. Metall. Trans. 2, 1139–1148.
Wilm, A., 1906. DRP 244554 (Germany patent).
Wilm, A., 1911. Metallurgie 8, 225–227.
Wilson, R., 2005. PhD Thesis, Monash University, Australia.
Wilson, R.N., Partridge, P.G., 1965. Acta Metall. 13, 1321.
Wilson, R., Bettles, C.J., Muddle, B.C., Nie, J.F., 2003. Mater. Sci. Forum 419–422, 267–272.
Winkelman, G.B., Raviprasad, K., Muddle, B.C., 2007. Acta Mater. 55, 3213–3228.
Wolverton, C., 2001. Acta Mater. 49, 3129.
Xu, D.K., Han, E.H., Liu, L., Xu, Y.B., 2009. Metall. Mater. Trans. A 40, 1727–1740.
Xu, S.W., Oh-ishi, K., Kamado, S., Uchida, F., Homma, T., Hono, K., 2011. Scr. Mater. 65, 269–272.
Xu, Z., Weyland, M., Nie, J.F., 2012. Unpublished work.
Yamada, K., Okubo, Y., Shino, M., Watanabe, H., Kamado, S., Kojima, Y., 2006. Mater. Trans. 47, 1066–1070.
Yamada, K., Hoshikawa, H., Maki, S., Ozaki, T., Kuroki, Y., Kamado, S., Kojima, Y., 2009. Scr. Mater. 61, 636–639.
Yamasaki, M., Anan, T., Yoshimoto, S., Kawamura, Y., 2005. Scr. Mater. 53, 799–803.
Yamasaki, M., Sasaki, M., Nishijima, M., Hiraga, K., Kawamura, Y., 2007. Acta Mater. 55, 6798–6805.
Yang, Q.B., Kuo, K.H., 1987. Acta Crystallogr. A43, 787–795.
Yang, J., Wang, L.D., Wang, L.M., Zhang, H.J., 2008. J. Alloys Compd. 459, 274–280.
Yarmolyuk, Y.P., Kripyakevich, P.I., Mel'nik, E.V., 1975. Sov. Phys. Crystallogr. 20, 329–331.
Yokobayashi, H., Kishida, K., Inui, H., Yamasaki, M., Kawamura, Y., 2011. Acta Mater. 59, 7287–7299.
Yoshimoto, S., Yamasaki, M., Kawamura, Y., 2006. Mater. Trans. 47, 959–965.
Zhang, M.X., Kelly, P.M., 2003. Scr. Mater. 48, 379–384.
Zhang, M.X., Kelly, P.M., 2005. Acta Mater. 53, 1085–1096.
Zhang, Z., Couture, A., Luo, A.A., 1998. Scr. Metall. Mater. 39, 45–53.
Zhang, Z., Themblay, R., Dube, D., Couture, A., 2000. Can. Metall. Q. 39, 503–512.
Zhang, M., Zhang, W.Z., Ye, F., 2004. Metall. Mater. Trans. A 36, 1681–1688.
Zhang, M., Zhang, W.Z., Zhu, G.Z., Yu, K., 2007. Trans. Nonferrous Met. Soc. China 17, 1428–1432.
Zhang, M., Zhang, W.Z., Zhu, G.Z., 2008. Scr. Mater. 59, 866–869.
Zhang, C., Luo, A.A., Peng, L., Stone, D., Chang, Y.A., 2011. Intermetallics 19, 1720–1726.
Zhu, Y.M., 2011. Ph.D. Thesis, Monash University, Australia.
Zhu, Y.M., Nie, J.F. Unpublished research.
Zhu, Y.M., Morton, A.J., Nie, J.F., 2008. Scr. Mater. 58, 525–528.
Zhu, Y.M., Wayland, M., Morton, A.J., Oh-ishi, K., Hono, K., Nie, J.F., 2009. Scr. Mater. 60, 980–983.
Zhu, Y.M., Morton, A.J., Weyland, M., Nie, J.F., 2010a. Acta Mater. 58, 464–475.
Zhu, Y.M., Morton, A.J., Nie, J.F., 2010b. Philos. Mag. Lett. 90, 173–181.
Zhu, Y.M., Morton, A.J., Nie, J.F., 2010c. Acta Mater. 58, 2936–2947.

Biography

Professor Jian-Feng Nie received his Ph.D. from Monash University in 1993. He was a postdoctoral research fellow in the University of Queensland in 1993, and CSIRO Materials Science and Technology in 1994. He was awarded a Logan Research Fellowship by Monash University in 1997, and was promoted to professor in 2009. His research interests include physical metallurgy of magnesium and aluminium alloys, crystallography of solid-solid phase transformations, applications of transmission electron microscopy, and processing-microstructure-property relationships in light alloys. He is editor of Metallurgical and Materials Transactions, and former Chair of Phase Transformations Committee of TMS.

21 Physical Metallurgy of Steels

H.K.D.H. Bhadeshia, University of Cambridge, UK

21.1 Introduction

Pure iron has a number of allotropic forms, the stabilities of which vary as a function of temperature, pressure, magnetic fields and size. It can assume many crystal structures, including the body-centered cubic, face-centered cubic, hexagonal close-packed, face-centered tetragonal, trigonal and, possibly, double hexagonal close-packed forms. The latter is sometimes postulated to exist deep within the earth's core where extreme temperatures and pressures create the appropriate conditions for this otherwise unstable form. There are also magnetic transitions defined by the Curie and Néel temperatures, together with complex magnetic states which lead to the greater thermal expansion coefficient of close-packed austenite relative to the less dense ferrite. When the size of iron is reduced sufficiently it can undergo a metal to insulator transition with the loss of the metallic state as the number of participating atoms decreases. The experimentally measured strength of pure iron ranges from 10 GPa to some 50 MPa but in principle can reach 22 GPa. Access to this vast range of properties is not limited to strength; parameters such as toughness and ductility exhibit similar versatility.

Iron is capable also of forming solutions with a very large number of other atomic species. There are countless precipitation and phase transformation reactions that can be induced with appropriate alloying. Indeed, a typical steel may contain at least 10 deliberately controlled additions with concentrations ranging from parts per million to a few weight percent. Alloying can be used to impose additional controls on the behavior of the steel, for example, to induce the formation of a coherent, electrically insulating and regenerating oxide film on the surface to render the steel stainless. In contrast, by alloying with copper, the steel can be induced to rust in an aesthetically pleasing manner.

Iron is incredibly cheap to manufacture on a large scale; about 1.3 billion tonnes of steel is produced annually, to be applied to improve the quality of life with such amazing reliability, that no ordinary person needs to know the technology involved. Indeed, many would not know that it is crystalline. It is primarily the allotropic transformations, precipitation and thermomechanical processing that dominate the performance of the majority of steels, and those are the topics that are described in the Chapter.

21.2   Martensite in Steels

The name *martensite* is after the German scientist Adolf Martens. It was used originally to describe the hard microscopic constituent found in quenched steels. Martensite remains of the greatest technological importance in steels where it can confer an outstanding combination of strength ($>3500\,\text{MPa}$) and toughness ($>200\,\text{MPa}\,\text{m}^{1/2}$). Many materials other than steel are now known to exhibit the same type of solid-state phase change, still generically known as a *martensitic transformation*, frequently also called a *shear* or a *displacive transformation*. Martensite occurs in, for example, nonferrous alloys, pure metals, ceramics, minerals, inorganic compounds, solidified gases and polymers (**Table 1**). We shall review first the experimental facts about martensite and then proceed to explain them.

21.2.1   Diffusionless Character

Martensitic transformations are diffusionless, but what evidence is there to support this? The phase can form at very low temperatures, where diffusion, even of interstitial atoms, is not conceivable over the time period of the experiment. **Table 1** gives values of the highest temperature at which martensite forms in a variety of materials; this temperature is known as the martensite start, or M_s temperature. It is obvious that although martensite *can* form at low temperatures, it need not do so. Therefore, a low transformation temperature is not sufficient evidence for diffusionless transformation.

Martensite plates can grow at speeds which approach that of sound in the metal. In steel this can be as high as $1100\,\text{m}\,\text{s}^{-1}$, which compares with the fastest recorded solidification front velocity of about $80\,\text{m}\,\text{s}^{-1}$ in pure nickel. Such large speeds are inconsistent with diffusion during transformation. Note that martensite need not grow so rapidly. For example, in shape-memory alloys or in single-interface transformations, the interface velocity is small enough to observe, and there are many confocal laser microscope studies showing relatively slow martensitic transformation.

The chemical composition of martensite can be measured and shown to be identical to that of the parent austenite. The totality of these observations demonstrates convincingly that martensitic transformations are diffusionless.

Table 1   The temperature M_s at which martensite first forms on cooling, and the approximate Vickers hardness of the resulting martensite for a number of materials

Composition	M_s/K	Hardness HV
ZrO_2	1200	1000
Fe–31Ni–0.23C wt%	83	300
Fe–34Ni–0.22C wt%	<4	250
Fe–3Mn–2Si–0.4C wt%	493	600
Cu–15Al	253	200
Ar–40N$_2$	30	

Unconstrained transformation

Constrained transformation

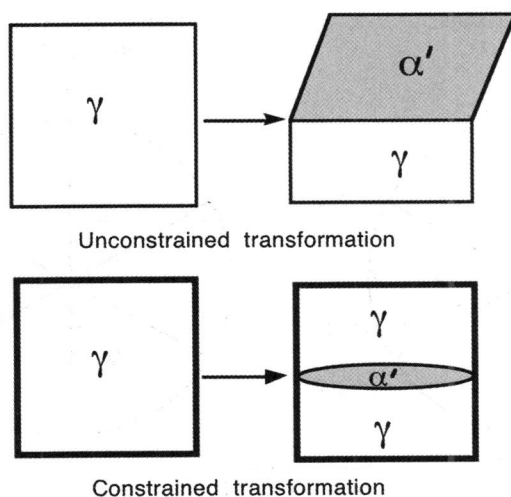

Figure 1 An illustration of the habit plane between austenite (γ) and martensite (α').

Table 2 Habit plane indices for martensite. With the exception of ε-martensite, the quoted indices are approximate because the habit planes are in general irrational

Composition/wt%	Approximate habit plane indices
Low-alloy steels, Fe–28Ni	$\{1\ 1\ 1\}_\gamma$
Plate martensite in Fe–1.8C	$\{2\ 9\ 5\}_\gamma$
Fe–30Ni–0.3C	$\{3\ 15\ 10\}_\gamma$
Fe–8Cr–1C	$\{2\ 5\ 2\}_\gamma$
ε-Martensite in 18/8 stainless steel	$\{1\ 1\ 1\}_\gamma$

21.2.2 The Habit Plane

This is the interface plane between austenite and martensite as measured on a macroscopic scale (**Figure 1**), for example, by using one- or two-surface crystallographic trace analysis on metallographic samples. For unconstrained transformations this interface plane is flat, but strain energy minimization introduces some curvature when the transformation is constrained by its surroundings. Nevertheless, the macroscopic habit plane is identical for both cases, as illustrated in **Figure 1**. Steels of vastly different chemical composition can have martensite with the same habit plane (**Table 2**), and indeed, other identical crystallographic characteristics.

21.2.3 Orientation Relationships

The formation of martensite involves the coordinated movement of atoms. It follows that the austenite and martensite lattices will be intimately related. All martensitic transformations therefore lead to

Kurdjumov-Sachs
$(011)_\alpha \| (111)_\gamma$

Nishiyama-Wasserman
$(011)_\alpha \| (111)_\gamma$

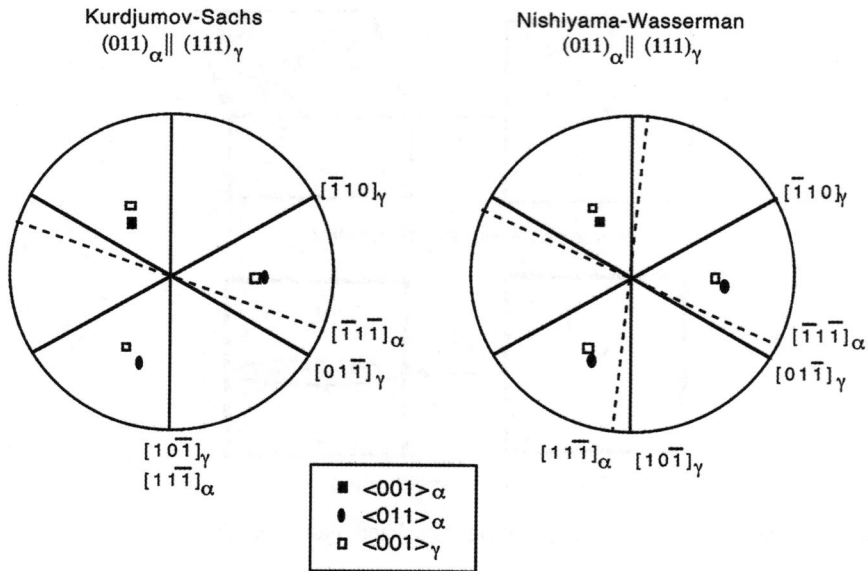

$[\bar{1}10]_\gamma$

$[\bar{1}\bar{1}1]_\alpha$
$[01\bar{1}]_\gamma$

$[10\bar{1}]_\gamma$
$[1\bar{1}\bar{1}]_\alpha$

$[\bar{1}10]_\gamma$

$[\bar{1}1\bar{1}]_\alpha$
$[01\bar{1}]_\gamma$

$[11\bar{1}]_\alpha$ $[10\bar{1}]_\gamma$

■	$<001>\alpha$
●	$<011>\alpha$
□	$<001>\gamma$

Figure 2 Stereographic representation of the Kurdjumov-Sachs (KS) and Nishiyama-Wasserman (NW) orientation relationships. The stereograms are both centered on $(1\,1\,1)_\gamma \| (0\,1\,1)_\alpha$. It is seen that the NW orientation can be generated from KS by an appropriate small rotation (5.25°) about $[0\,1\,1]_\alpha$. Only a few of the poles are marked to allow a comparison with the Bain orientation relationship. The neighboring pairs of poles would superpose exactly for the Bain orientation.

a reproducible orientation relationship between the parent and product lattices. It is frequently the case that a pair of corresponding close-packed[1] planes in the ferrite and austenite are parallel or nearly parallel, and it is usually the case that corresponding directions within these planes are roughly parallel (**Figure 2**):

Kurdjumov-Sachs

$$\{111\}_\gamma \| \{011\}_\alpha$$
$$\langle 10\bar{1}\rangle_\gamma \| \langle 11\bar{1}\rangle_\alpha$$

Nishiyama-Wasserman

$$\{111\}_\gamma \| \{011\}_\alpha$$
$$\langle 10\bar{1}\rangle_\gamma \text{ about 5.3° from } \langle 11\bar{1}\rangle_\alpha \text{ toward } \langle \bar{1}1\bar{1}\rangle_\alpha$$

Greninger-Troiano

$$\{111\}_\gamma \text{ about 0.2° from } \{011\}_\alpha$$
$$\langle 10\bar{1}\rangle_\gamma \text{ about 2.7° from } \langle 11\bar{1}\rangle_\alpha \text{ toward } \langle \bar{1}1\bar{1}\rangle_\alpha.$$

---

[1] The body-centered cubic lattice does not have a close-packed plane but $\{0\,1\,1\}_\alpha$ is the most densely packed plane.

Note that these have been stated approximately: the true relations are irrational, meaning that the indices of the parallel planes and directions cannot be expressed using rational numbers (the square root of 2 is not a rational number).

21.2.4 Athermal Nature of Transformation

In the vast majority of cases, the extent of reaction is found to be virtually independent of time:

$$1 - V_{\alpha'} = \exp\{\beta(M_s - T)\}, \quad \text{where} \quad \beta \simeq -0.011, \tag{1}$$

$V_{\alpha'}$ is the fraction of martensite and T is a temperature below M_s. This is the Koistinen and Marburger equation; note that time does not feature in this relation, so that the fraction of martensite depends only on the undercooling below the martensite start temperature. This athermal character is a consequence of very rapid nucleation and growth, so rapid that the time taken can in normal circumstances be neglected (**Figure 2**).

Isothermal martensite is possible when nucleation is hindered, although the growth rate of individual plates of martensite can still be rapid.

21.2.5 Structure of the Interface

Any process that contributes to the formation of martensite cannot rely on assistance from thermal activation. There must therefore exist a high level of continuity across the interface, which must be coherent or semi-coherent. A stress-free fully coherent interface is impossible for the $\gamma \rightarrow \alpha'$ transformation since the lattice deformation **BR** is an invariant-line strain (ILS). A semi-coherent interface must be such that the interfacial dislocations can glide as the interface moves (climb is not permitted). It follows that the Burgers vectors of the interface dislocations must not lie in the interface plane unless the dislocations are screw in character.

There is an additional condition for a semi-coherent interface to be glissile. The line vectors of the interfacial dislocations must lie along an *invariant-line*, that is, a line that joins the parent and product crystals without any rotation or distortion. Why is that? If there is any distortion along the dislocation line, then other dislocations are needed to accommodate that misfit. It will then be necessary to have more than one set of nonparallel dislocations in the interface. These nonparallel dislocations can intersect to form jogs which render the interface sessile.

It follows that for martensitic transformation to be possible, the deformation which changes the parent into the product must leave one or more lines invariant (unrotated, undistorted). A deformation which leaves one line invariant is called an "ILS".

21.2.6 The Shape Deformation

The passage of a slip dislocation through a crystal causes the formation of a step where the glide plane intersects the free surface (**Figure 3a,b**). The passage of many such dislocations on parallel slip planes causes macroscopic shear (**Figure 3c,d**). Slip causes a change in shape but not a change in the crystal structure, because the Burgers vectors of the dislocations are also lattice vectors.

During martensitic transformation, since the pattern in which the atoms in the parent crystal are arranged is *deformed* into that appropriate for martensite, there must be a corresponding change in the macroscopic shape of the crystal undergoing transformation. The dislocations responsible for the

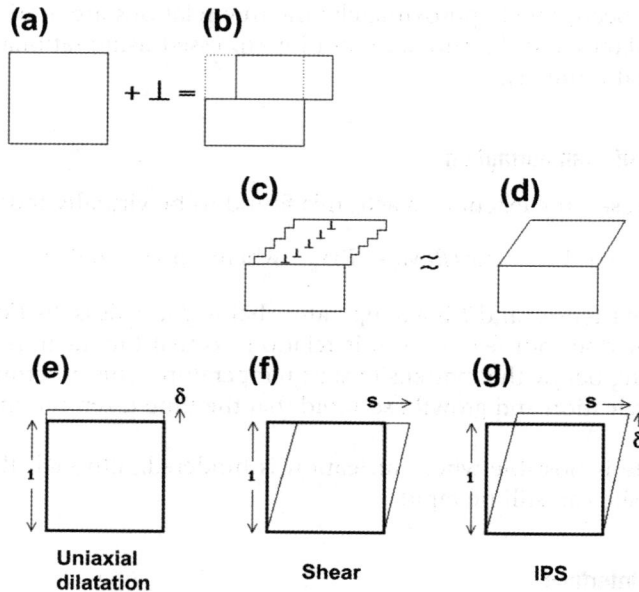

Figure 3 (a, b) Step caused by the passage of a slip dislocation. (c, d) Many slip dislocations, causing a macroscopic shear. (e) An IPS with a uniaxial dilatation. (f) An IPS which is a simple shear. (g) An IPS which is the combined effect of a uniaxial dilatation and a simple shear.

deformation are in the α'/γ interface, with Burgers vectors such that in addition to deformation they also cause the change in crystal structure. The deformation is such that an initially flat surface becomes uniformly tilted about the line formed by the intersection of the interface plane with the free surface. Any scratch traversing the transformed region is similarly deflected though the scratch remains connected at the α'/γ interface. These observations, and others, confirm that the measured shape deformation is an invariant-plane strain (IPS) (**Figure 3e–g**) with a large shear component (≈ 0.22) and a small dilatational strain (≈ 0.03) directed normal to the habit plane.

21.2.7 Bain Strain

We now consider the nature of the strain necessary to transform the face-centred cubic (f.c.c.) lattice of γ into the body centred-cubic (b.c.c.) lattice of α'. Such a strain was proposed by Bain in 1924 and hence is known as the "Bain Strain" (**Figure 4**). There is a compression along the z axis and a uniform expansion along the x and y axes.

The deformation describing the Bain Strain is given by

$$\mathbf{B} = \begin{pmatrix} \varepsilon_0 & 0 & 0 \\ 0 & \varepsilon_0 & 0 \\ 0 & 0 & \varepsilon_0' \end{pmatrix}, \tag{2}$$

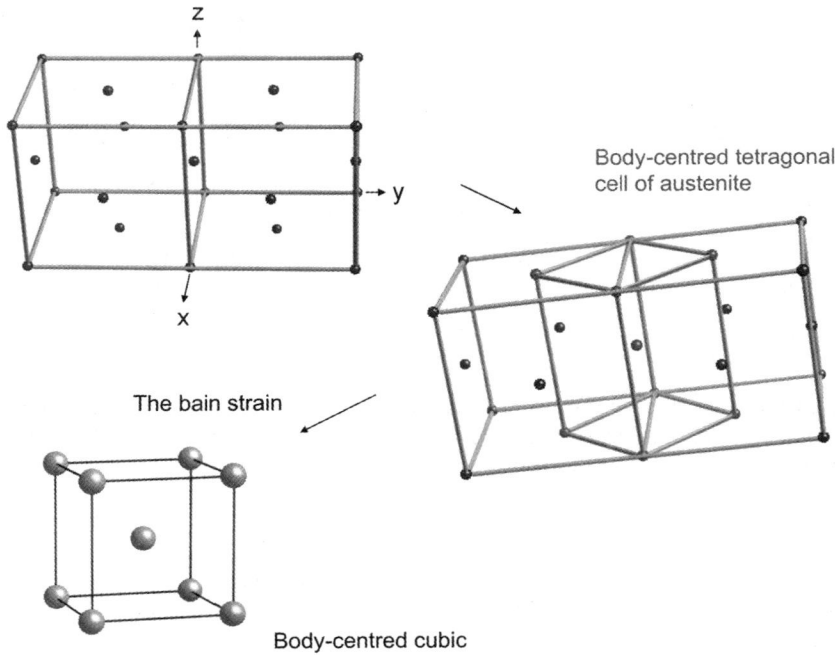

Body-centred tetragonal cell of austenite

The bain strain

Body-centred cubic

Figure 4 The Bain strain. (For color version of this figure, the reader is referred to the online version of this book.)

$$\varepsilon_0 = \frac{\sqrt{2}a_{\alpha'} - a_\gamma}{a_\gamma} \quad \varepsilon_0' = \frac{a_{\alpha'} - a_\gamma}{a_\gamma}, \tag{3}$$

where $a_{\alpha'}$ and a_γ are the lattice parameters of martensite and austenite, respectively. The contraction is therefore along the $[0\ 0\ 1]_\gamma$ axis and a uniform expansion on the $(0\ 0\ 1)_\gamma$ plane.

The Bain strain implies the following orientation relationship between the parent and product lattices:

$$[0\ 0\ 1]_{fcc} \| [0\ 0\ 1]_{bcc} \quad [1\ \bar{1}\ 0]_{fcc} \| [1\ 0\ 0]_{bcc} \quad [1\ 1\ 0]_{fcc} \| [0\ 1\ 0]_{bcc}$$

but in fact, the experimentally observed orientation relationships are irrational, as discussed earlier. We shall deal with this inconsistency later.

Temporarily neglecting the fact that the Bain orientation is inconsistent with experiments, we proceed to examine whether the Bain strain leaves at least one line invariant. After all, this is a necessary condition for martensitic transformation.

In **Figure 5a,b**, the austenite is represented as a sphere which, as a result of the Bain strain **B**, is deformed into an ellipsoid of revolution which represents the martensite. There are no lines which are left undistorted or unrotated by **B**. There are no lines in the $(0\ 0\ 1)_{fcc}$ plane which are undistorted. The lines ab and cd are undistorted but are rotated to the new positions $a'b'$ and $c'd'$. Such rotated lines are not invariant. However, the combined effect of the Bain strain **B** and the rigid body rotation **R** is indeed an ILS because it brings cd and $c'd'$ into coincidence (**Figure 5c**). This is the reason why

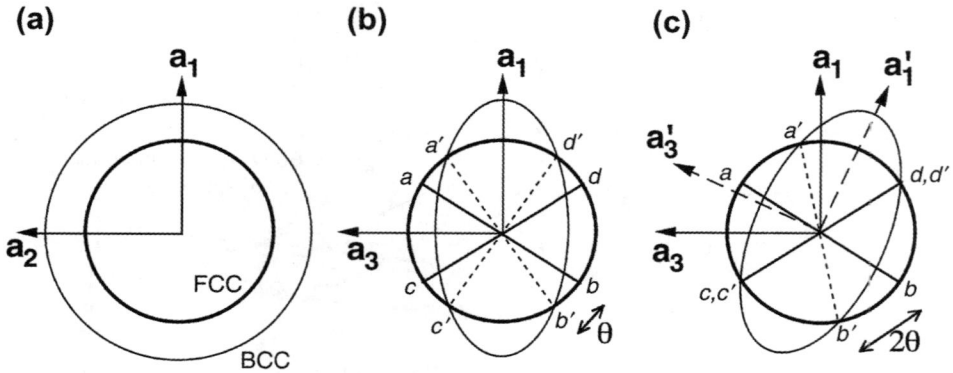

Figure 5 (a) and (b) The effect of the Bain strain on austenite, which when undeformed is represented as a sphere of diameter $ab = cd$ in three-dimensions. The strain transforms it to an ellipsoid of revolution. (c) The ILS obtained by combining the Bain strain with a rigid body rotation through an angle θ.

the observed irrational orientation relationship differs from that implied by the Bain strain. The rotation required to convert **B** into an ILS precisely corrects the Bain orientation into that which is observed experimentally.

As can be seen from **Figure 5c**, there is no rotation which can make **B** into an IPS since this would require two nonparallel invariant-lines. Thus, for the fcc → bcc transformation, austenite cannot be transformed into martensite by a homogeneous strain which is an IPS. And yet, the observed shape deformation leaves the habit plane undistorted and unrotated, that is, it is an IPS.

The phenomenological theory of martensite crystallography solves this remaining problem (**Figure 6**). The Bain strain converts the structure of the parent phase into that of the product phase. When combined with an appropriate rigid body rotation, the net homogeneous lattice deformation **RB** is an ILS (step a to c in **Figure 6**). However, the observed shape deformation is an IPS P_1 (step a to b in **Figure 6**), but this gives the wrong crystal structure. If a second homogeneous shear P_2 is combined with P_1 (step b to c), then the correct structure is obtained but the wrong shape since

$$P_1 P_2 = RB$$

These discrepancies are all resolved if the shape-changing effect of P_2 is canceled macroscopically by an inhomogeneous lattice-invariant deformation, which may be slip or twinning as illustrated in **Figure 6**.

The theory explains all the observed features of the martensite crystallography. The orientation relationship is predicted by deducing the rotation needed to change the Bain strain into an ILS. The habit plane does not have rational indices because the amount of lattice-invariant deformation needed to recover the correct the macroscopic shape is not usually rational. The theory predicts a substructure in plates of martensite (either twins or slip steps) as is observed experimentally. The transformation goes to all the trouble of ensuring that the shape deformation is macroscopically an IPS because this reduces the strain energy when compared with the case where the shape deformation might be an ILS.

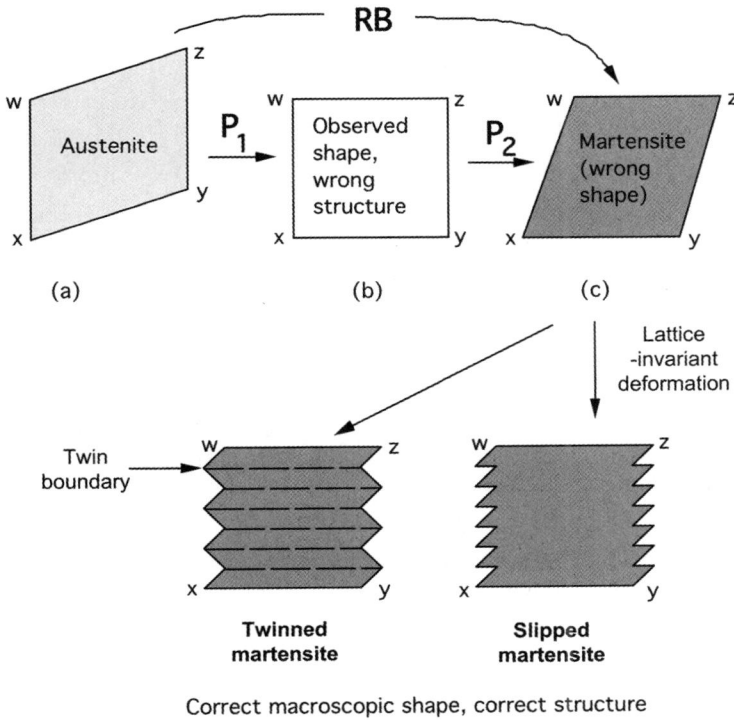

Figure 6 The phenomenological theory of martensite crystallography. (For color version of this figure, the reader is referred to the online version of this book.)

21.2.8 Thermodynamics of Martensitic Transformations

Martensite is not represented on phase diagrams because the latter deal with equilibrium. Martensite deviates from equilibrium in two important ways. It grows without diffusion, so it inherits the chemical composition of the parent austenite. In an equilibrium transformation the chemical elements partition into the parent and product phases in a manner which leads to a minimization of free energy. Secondly, the shape deformation associated with martensitic transformation causes strains; the resulting strain energy has to be accounted for in dealing with the thermodynamics of transformation. If the free energy change accompanying the transformation of austenite into ferrite without a composition change is written $\Delta G^{\gamma\alpha}$, then that for transformation into martensite has a smaller magnitude $\Delta G^{\gamma\alpha'}$ as illustrated in **Figure 7**. The difference is accounted for by the terms listed in **Table 3**.

In thermodynamic terms, the martensite start temperature is defined by that at which $\Delta G^{\gamma\alpha'}$ reaches a critical value $\Delta G_{M_s}^{\gamma\alpha'}$ (Bell and Owen, 1967; Imai et al., 1965; Kaufman and Cohen, 1958; Kaufman et al., 1962), with the entire effect of chemical composition the effect of solutes on the free energies of austenite and martensite. Others have introduced a dependence of $\Delta G_{M_s}^{\gamma\alpha'}$ on the strength of the austenite (Ghosh and Olson, 1994), a logical progression given that the transformation front has a glissile dislocation structure. Nevertheless, the difference $\Delta G^{\gamma\alpha'} - \Delta G_{M_s}^{\gamma\alpha'}$ is essentially treated as

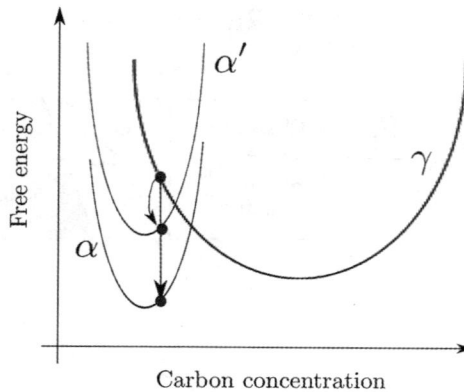

Figure 7 The free energy curves for ferrite (α), martensite (α') and austenite (γ) as a function of the carbon concentration. The driving forces for transformation to ferrite or martensite without a change in composition are $\Delta G^{\gamma\alpha}$ and $\Delta G^{\gamma\alpha'}$, represented by the long and short arrows, respectively.

Table 3   Typical values of stored energies associated with martensitic transformation

	J mol^{-1}
Strain energy	600
Twin interface energy	100
γ/α' Interface energy	1
Stored energy due to dislocations	20

empirical in all of these models, and we shall return to this issue with what should be a more fundamental thermodynamic definition of the start temperature.

The accumulation of data on the martensite start temperature has led to an acknowledgment that there are further issues that need to be addressed, some of which are dependencies of the measurement method. For example, during dilatometry, M_s changes from 419, 384, 382 to 373 K as the volume percent of martensite at which it is measured is changed from 0.08, 0.5, 0.54 and 1%, respectively. Acoustic emission can be much more sensitive than electrical resistivity measurements, so the M_s temperature measured using the latter technique is much lower than that associated with acoustic data. To avoid experimental artifacts, it clearly is necessary to specify the amount of martensite used to define the start temperature. This can be done using an offset method which is generally applicable but for dilatometric data defines M_s as the critical value of transformation strain (**Figure 8**). The critical strain is for 1 volume percent of martensitic transformation assuming that the latter occurs at room temperature, by using equations for the lattice parameters of austenite and martensite. This ensures that the method is reproducible and emphasizes that transformation start temperatures should be quoted in association with the sensitivity of the technique used.

It is known that M_s varies with the austenite grain size (**Figure 9**) and, indeed, that films of austenite are much more stable than larger blocks. These results can also be explained on the ability to detect transformation as a function of the austenite grain size. As martensite forms, it geometrically partitions

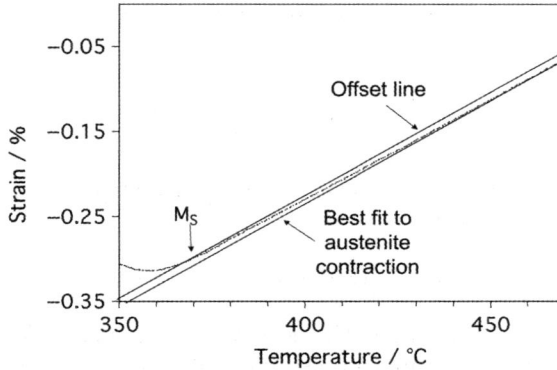

Figure 8 Offset method for defining M_s. The offset line is defined as representing a 1 vol.% strain relative to austenite at ambient temperature, with the strain calculated using standard lattice parameter equations (Yang and Bhadeshia, 2009).

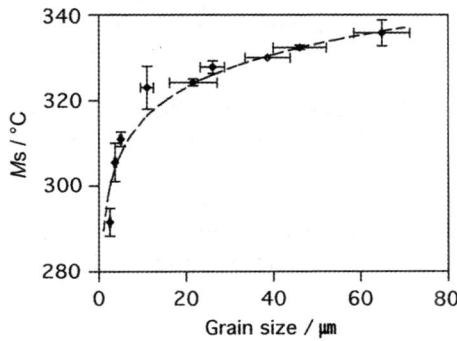

Figure 9 One example of the measured variation in the martensite start temperature as a function of the austenite grain size (Yang and Bhadeshia, 2009).

the austenite into ever smaller segments. The relationship between the number of martensite plates per unit volume, N_V related to the fraction $V_{\alpha'}$ of martensite is given by

$$\frac{dV_{\alpha'}}{dN_V} = m(1 - V_{\alpha'})\,\overline{V}_C \equiv \frac{m(1 - V_{\alpha'})}{N_V^C}, \tag{4}$$

where m is the aspect ratio of the martensite plate (assumed here to be 0.05 (Christian, 1979)) and $\overline{V}_C$ is the average size of the compartment resulting from the partitioning of an austenite grain by the presence of martensite plates. N_V^C is the number of austenite compartments per unit volume. Each new martensite plate produces one additional compartment so that

$$N_V^C = \frac{1}{V_\gamma} + N_V. \tag{5}$$

where V_γ is the average austenite grain volume. On combining eqns (29) and (33) we get

$$N_V = \frac{1}{V_\gamma}\left[\exp\left\{-\frac{\ln(1 - V_{\alpha'})}{m}\right\} - 1\right].$$ (6)

This equation can be used to convert the measured martensite fraction into N_V using an equation of the form (Yang and Bhadeshia, 2009)

$$N_V = a\left[\exp\{b(M_s^o - T)\} - 1\right],$$ (7)

where a and b are empirical fitting constants. The term M_s^o is defined as a fundamental martensite start temperature for an infinitely large austenite grain size; it will be shown later that it can be derived using thermodynamics. On combining eqns (6) and (7) we obtain

$$\begin{aligned}M_s^o - T &= \frac{1}{b}\ln\left[\frac{1}{aV_\gamma}\left\{\exp\left(-\frac{\ln(1 - V_{\alpha'})}{m}\right) - 1\right\} + 1\right],\\ &\equiv \frac{1}{b}\ln\left[\frac{1}{\bar{L}_\gamma^3}\left\{\exp\left(-\frac{\ln(1 - V_{\alpha'})}{m}\right) - 1\right\} + 1\right].\end{aligned}$$ (8)

In this equation, the term $M_s^o - T$ becomes $M_s^o - M_s$ when the fraction f is set to be the first detectable fraction of $(V_{\alpha'}^{M_s})$ martensite, and $M_s \to M_s^o$ as $V_\gamma \to \infty$. $\bar{L}_\gamma$ is the austenite grain size defined as a mean lineal intercept.

In eqn (8), the plate aspect ratio is $m = 0.05$ (Christian, 1979); $a = 1.57 \times 10^{-21}$ μm^3 and $b = 0.253$ (Yang and Bhadeshia, 2009). Given that M_s^o is the highest temperature at which martensite can form, it is calculated using thermodynamics alone, but accounting for the stored energy of martensite which is about $700\,J\,mol^{-1}$ (Christian, 1979).

M_s^o is therefore given by the temperature at which $G_\gamma - G_\alpha = 700\,J\,mol^{-1}$, where G_α and G_γ are the Gibbs free energies of ferrite and austenite of the same chemical composition. It represents the highest temperature at which martensite can in principle form.

21.2.8.1   More Complete Theory for Martensite Start

The ability to calculate M_s has acquired renewed importance in the context of a large variety of steels, both commercial and experimental, which rely on transformation-induced plasticity (TRIP) (Section 21.9), or which deal with ultrafine austenite and the influence of stresses and strains. The terms "stress-induced" and "strain-induced" martensite have been used rather loosely in the literature. Almost all the experiments where martensite has been stated to be strain-induced actually can be explained on the basis of the stress alone, which is used to produce the plastic strain. However, there are two special effects of large plastic strains that cannot be neglected. The scale of the austenite grain structure defined in terms of a mean lineal intercept is changed by the deformation. Secondly, the introduction of dislocation debris can mechanically stabilize the austenite. Mechanical stabilization occurs because the movement of glissile martensite–austenite interfaces is hindered by the dislocation debris, and can lead to a complete cessation of transformation. The phenomenon has recently been expressed quantitatively (Chatterjee et al., 2006).

The influence of stress per se is expressed via a mechanical driving force (Section 21.9) which supplements the chemical component, and has been estimated on the basis of the shape deformation of martensite as (Olson and Cohen, 1982)

$$U = -0.86\sigma \quad \text{for uniaxial tension,}$$
$$U = -0.58\sigma \quad \text{for uniaxial compression,} \tag{9}$$

where the driving force is in $J\,mol^{-1}$ and the stress in MPa. Such calculations can in fact be done for arbitrary stress tensors, but most experiments are done in uniaxial mode.

The grain size effect is expressed in eqn (8); plastic strain before transformation alters $\bar{L}_\gamma$ through a change in the shape of the austenite grains. This problem has been dealt with (Singh and Bhadeshia, 1998b) for a large variety of deformations in which an equiaxed austenite grain is strained, but for uniaxial tension or compression, the result is

$$\frac{\bar{L}_0}{\bar{L}} = \frac{(1 + 3\sqrt{3})S_{11}^{0.5} + 3(S_{11}^3 + 2)^{0.5}S_{11}^{-1} + (2 + 2S_{11}^3)^{0.5}S_{11}^{-1}}{3(2\sqrt{3} + 1)} \tag{10}$$

where $S_{11} = \exp\{\pm\varepsilon\}$ where $+\varepsilon$ and $-\varepsilon$ are the plastic strains in tension and compression, respectively, and $\bar{L}_0$ and $\bar{L}$ are the lineal intercepts in the undeformed and deformed states, respectively.

The second effect of plastic strain through mechanical stabilization is considered by calculating the additional driving force ΔG_{STA} needed in order for the interface to overcome the dislocation density (ρ) created by strain before transformation (Chatterjee et al., 2006):

$$\Delta G_{STA} = \frac{\mu b}{8\pi(1-v)}\left(\rho^{0.5} - \rho_0^{0.5}\right) \quad J\,m^{-3}$$
$$\text{with} \quad \rho = 2 \times 10^{13} + 2 \times 10^{14}\varepsilon \quad m^{-2}, \tag{11}$$

where μ is the shear modulus of austenite at 80 GPa, $b = 0.252$ nm is the magnitude of the Burgers vector of the dislocations, and the Poisson's ratio $v = 0.27$. To summarize, the martensite start temperature independent of grain size effect is obtained from and this result is then corrected for a grain size effect using eqn (8). Some indicative calculations are illustrated in **Figure 10** and more details are available in Yang et al. (2012).

$$\Delta G^{\gamma\alpha} + U < \Delta G_{M_s}^{\gamma\alpha} - \Delta G_{STA}. \tag{12}$$

21.3 Bainite in Steels

Bainite forms by the decomposition of austenite at a temperature which is above M_s but below that at which fine pearlite forms. All bainite forms below the T_0 temperature.

All time–temperature–transformation (TTT) diagrams consist essentially of two C-curves (**Figure 11**). If we focus first on the Fe–Mn–C steel with the higher hardenability (slower rates of transformation) then the two curves are separated. The upper C-curve represents the time required to initiate reconstructive transformations such as ferrite or pearlite, whereas the lower C-curve represents displacive transformations such as bainite or Widmanstätten ferrite. Note that as the hardenability of the steel decreases, the two curves tend to overlap so that in experiments it appears as if the TTT diagram contains just one curve with a complicated shape, describing all the reactions. This is not the case

(a)

(b)

(c)

(d)

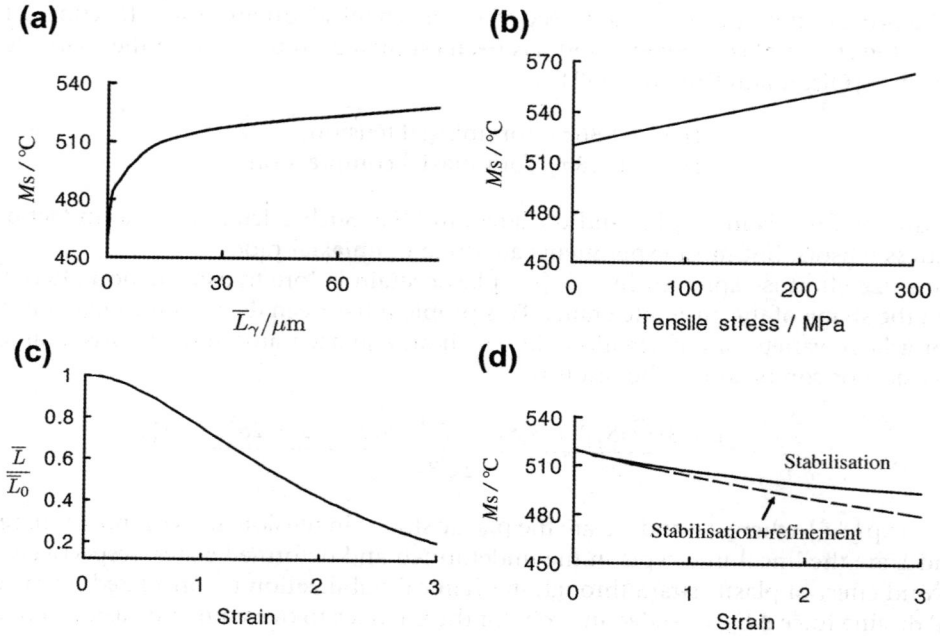

Figure 10   Calculations for Fe–0.1C wt% steel. (a) Influence of austenite grain size. (b) Influence of tensile stress. (c) Change in mean linear intercept for austenite grains as a function of plastic strain. (d) The effect of mechanical stabilization and grain refinement on the martensite start temperature.

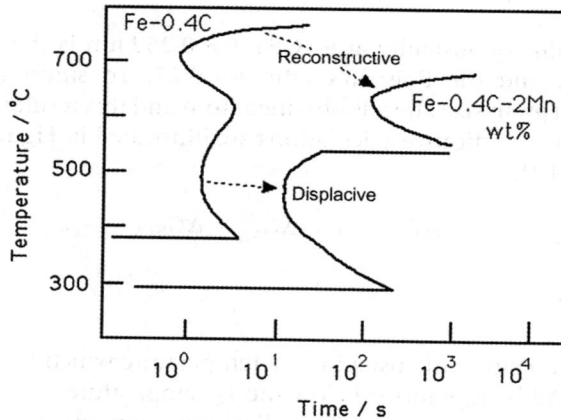

Figure 11   TTT diagrams for two steels, one of which has a high hardenability.

because it is possible to show that this is an experimental artifact caused by the inability to detect the two C-curves separately.

A further feature to note (**Figure 11**, Fe–Mn–C) is that the lower C-curve representing displacive transformations has a flat top. This represents the highest temperature T_h at which displacive

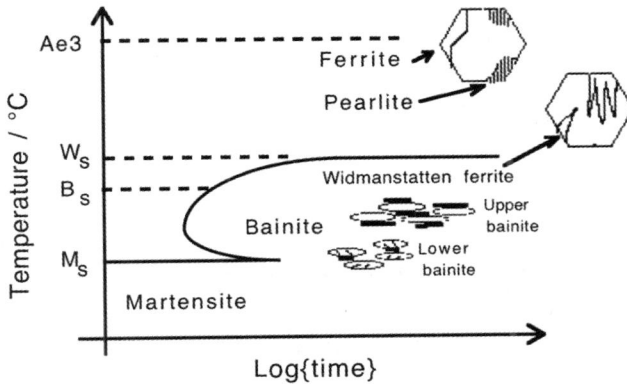

Figure 12 TTT diagrams showing the different domains of transformation.

transformations may occur. The temperature T_h may equal the bainite start temperature B_s if the hardenability is high enough, but otherwise, $T_h = W_s$ where W_s is the Widmanstätten ferrite start temperature (**Figure 12**). The latter does not form in high-hardenability steels and we shall discuss in this lecture the detailed differences between bainite and Widmanstätten ferrite.

Bainite is a nonlamellar aggregate of carbides and plate-shaped ferrite (**Figure 13**). As we shall see later, the carbide part of the microstructure is not essential; the carbides form as a secondary reaction, rather as in the tempering of martensite. The ferrite plates are each about 10 μm long and about 0.2 μm thick, making the individual plates invisible in the optical microscope.

Upper bainite consists of clusters of platelets of ferrite adjacent to each other and in almost identical crystallographic orientation, so that a low-angle boundary arises whenever the adjacent platelets touch. The ferrite always has a Kurdjumov-Sachs-type orientation relationship with the austenite in which it grows.

Elongated cementite particles usually decorate the boundaries of these platelets, the amount and continuity of the cementite layers depending on the carbon concentration of the steel.

The clusters of ferrite plates are known as "sheaves" (**Figure 14**); each sheaf is itself in the form of a wedge-shaped plate on a macroscopic scale. The sheaves inevitably nucleate heterogeneously at austenite grain surfaces. The cementite precipitates from the carbon-enriched austenite between the ferrite plates; the ferrite itself is free from carbides. Cementite precipitation from austenite can be prevented by increasing the silicon concentration to about 1.5 wt%; this works because silicon is

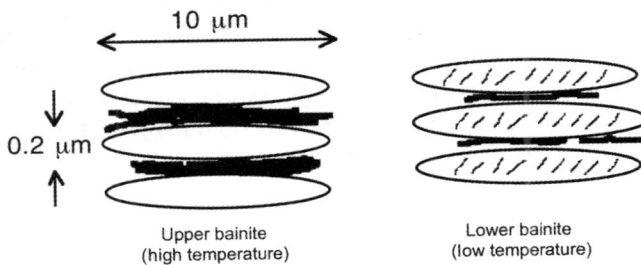

Figure 13 Schematic illustration of the microstructure of upper and lower bainite.

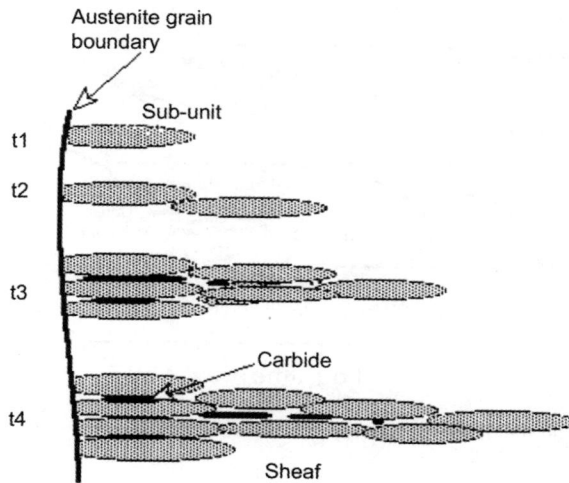

Figure 14 Evolution of a bainite sheaf as a function of time.

insoluble in cementite. Silicon-rich bainitic steels can have very good toughness because of the absence of brittle cementite.

21.3.1 Shape Deformation

The formation of bainite causes a deformation (**Figure 15**) which is an IPS with a shear component of about 0.26 and a dilatational strain normal to the habit plane of about 0.03. This is consistent with a displacive mechanism of transformation.

Bainite forms at a relatively high temperature when compared with martensite. The parent austenite is weaker at high temperatures and cannot accommodate the large shape deformation elastically. It therefore relaxes by plastic deformation in the region adjacent to the bainite. This is evident in **Figure 15**, but is also presented as a height scan in **Figure 16**. The effect of this plastic deformation is to stifle the growth of bainite plates before they hit any obstacle. This is why each bainite plate grows to a size which is often smaller than the austenite grain size and then comes to a halt. Further transformation happens by the formation of a new plate and this is why the sheaf morphology arises.

21.3.2 Substitutional Alloying Elements

These do not redistribute at all during transformation, even though equilibrium requires them to partition between the austenite and ferrite (**Figure 17**). The ratio of substitutional to iron atoms remains constant everywhere including across the interface. This is consistent with a displacive mechanism of transformation and the existence of an atomic correspondence between the austenite and bainitic ferrite. The results exclude any mechanism which involves local equilibrium at the interface, or solute drag effects associated with interfacial motion.

21.3.3 Interstitial Alloying Elements (C, N)

It appears from **Figure 17** that the carbon has partitioned into the austenite. It is simple to establish that martensitic transformation is diffusionless, by measuring the local compositions before and after

Figure 15 Atomic force microscope image of the displacements caused on a polished surface of austenite by the growth of bainite. Note the shear deformation (dark contrast) and indeed the plastic accommodation (light contrast tapering from the ridge of the region of dark contrast) of the shape change in the austenite adjacent to the bainite plates.

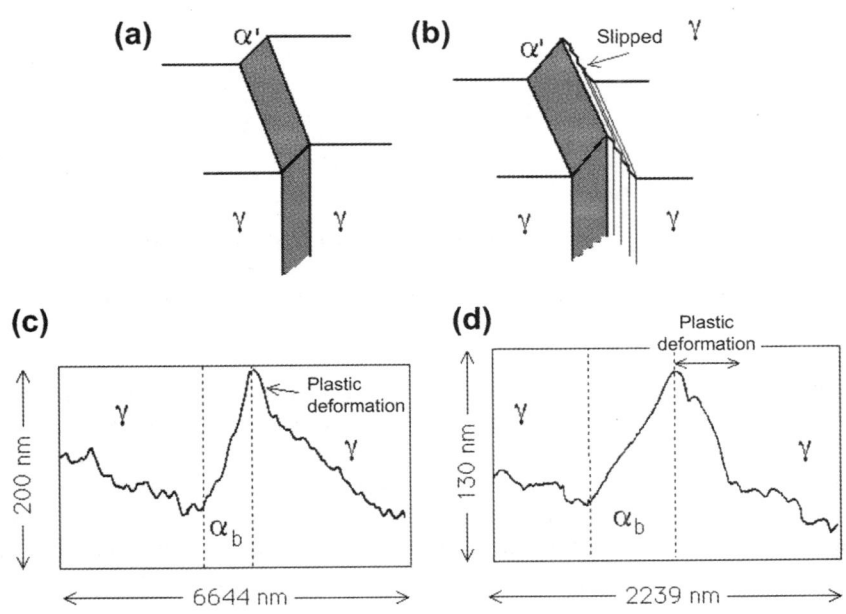

Figure 16   (a) A perfect IPS surface relief effect. (b) One where plastic relaxation of the shape change occurs in the adjacent matrix. (c and d) An actual atomic force microscope scan across the surface relief because of a bainite subunit (Swallow and Bhadeshia, 1996).

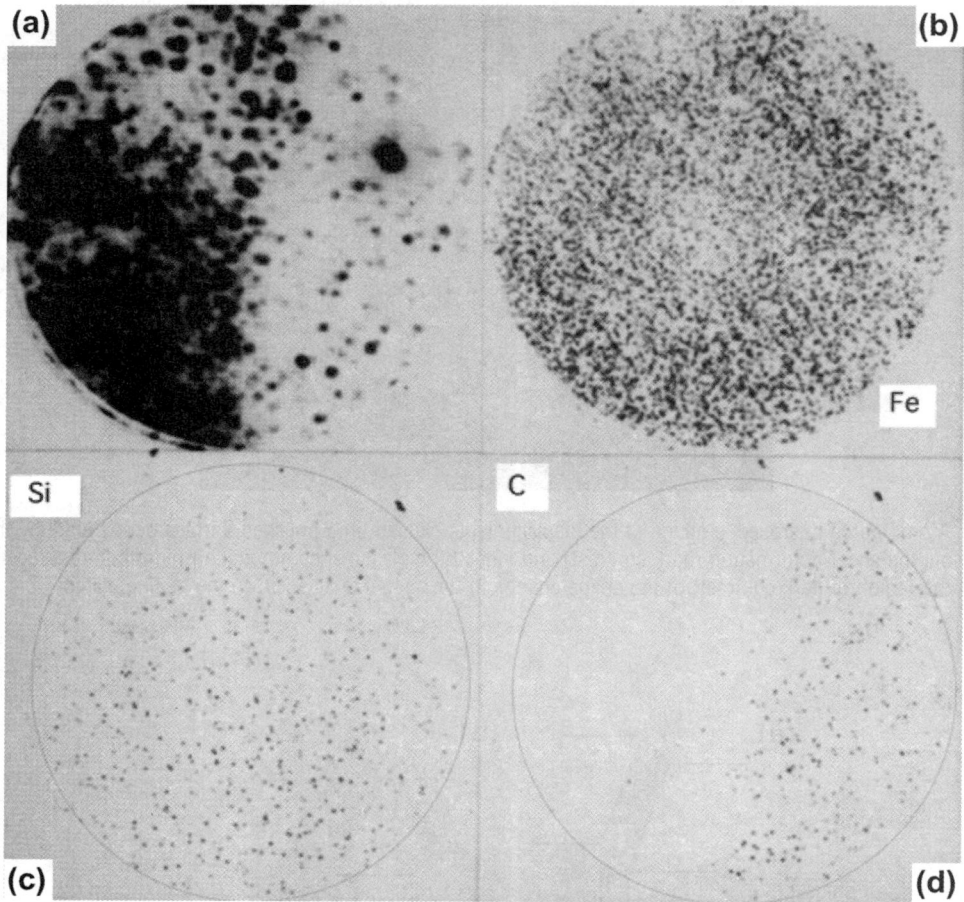

Figure 17 Imaging atom-probe micrographs, taken across an austenite–bainitic ferrite interface in an Fe–C–Si–Mn alloy. The images confirm quantitative data (Bhadeshia and Waugh, 1982) showing the absence of any substitutional atom diffusion during transformation. (a) Field-ion image; (b) corresponding silicon map; (c) corresponding carbon map; (d) corresponding iron map.

transformation. Bainite forms at somewhat higher temperatures where the carbon can escape out of the plate within a fraction of a second. Its original composition cannot therefore be measured directly.

There are three possibilities. The carbon may partition during growth so that the ferrite may never contain any excess carbon. The growth may on the other hand be diffusionless with carbon being trapped by the advancing interface. Finally, there is an intermediate case in which some carbon may diffuse with the remainder being trapped to leave the ferrite partially supersaturated. It is therefore much more difficult to determine the precise role of carbon during the growth of bainitic ferrite than in martensite.

Diffusionless growth requires that transformation occurs at a temperature below T_0, when the free energy of bainite becomes less than that of austenite of the same composition. A locus of the T_0 temperature as a function of the carbon concentration is called the T_0 curve, an example of which is

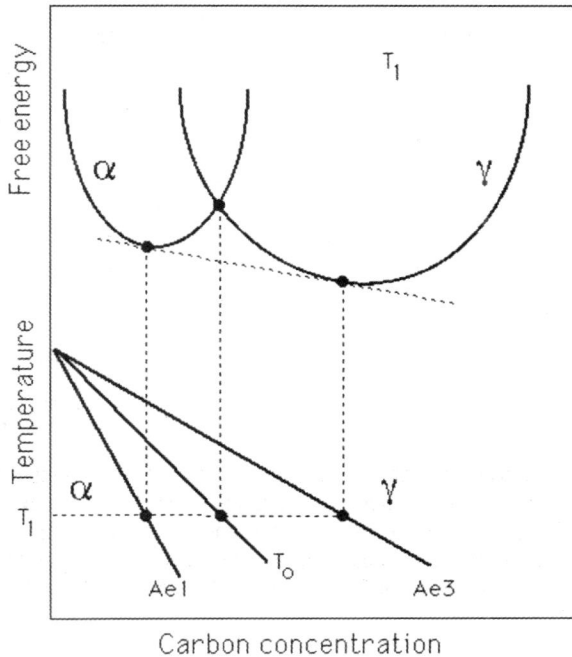

Figure 18 Schematic illustration of the origin of the T_0 construction on the Fe–C phase diagram. Austenite with a carbon concentration to the left of the T_0 boundary can in principle transform without any diffusion. Diffusionless transformation is thermodynamically impossible if the carbon concentration of the austenite exceeds the T_0 curve.

plotted on the Fe–C phase diagram in **Figure 18**. Growth without diffusion can only occur if the carbon concentration of the austenite lies to the left of the T_0 curve.

Suppose that the plate of bainite forms without diffusion, but that any excess carbon is soon afterward rejected into the residual austenite. The next plate of bainite then has to grow from carbon-enriched austenite (**Figure 19a**). This process must cease when the austenite carbon concentration reaches the T_0 curve. The reaction is said to be incomplete, since the austenite has not achieved its equilibrium composition (given by the Ae3 curve) at the point the reaction stops. If on the other hand, the ferrite grows with an equilibrium carbon concentration then the transformation should cease when the austenite carbon concentration reaches the Ae3 curve.

It is found experimentally that the transformation to bainite does indeed stop at the T_0 boundary (**Figure 19b**). The balance of the evidence is that the growth of bainite below the B_s temperature involves the successive nucleation and martensitic growth of subunits, followed in upper bainite by the diffusion of carbon into the surrounding austenite. The possibility that a small fraction of the carbon is nevertheless partitioned during growth cannot entirely be ruled out, but there is little doubt that the bainite is at first substantially supersaturated with carbon.

These conclusions are not significantly modified when the strain energy of transformation is included in the analysis.

There are two important features of bainite which can be shown by a variety of techniques, for example, dilatometry, electrical resistivity, magnetic measurements and by metallography. Firstly, there

(a)

(b)

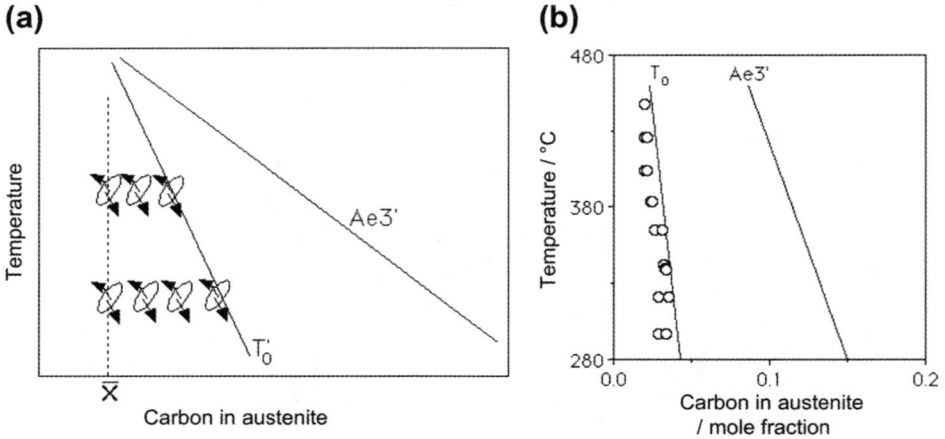

Figure 19   (a) Illustration of the incomplete reaction phenomenon. During isothermal transformation, a plate of bainite grows without diffusion, and then partitions its excess carbon into the residual austenite. The next plate therefore has to grow from carbon-enriched austenite. This process continues until diffusionless transformation becomes impossible when the austenite composition eventually reaches the T_0 boundary. (b) Experimental data showing that the growth of bainite stops when the austenite carbon concentration reaches the T_0 curve (Fe–0.43C–3Mn–2.12Si wt% alloy).

is a well-defined temperature B_s above which no bainite will form, which has been confirmed for a wide range of alloy steels. The amount of bainite that forms increases as the transformation temperature is reduced below the B_s temperature. The fraction increases during isothermal transformation as a sigmoidal function of time, reaching an asymptotic limit which does not change on prolonged heat treatment even when substantial quantities of austenite remain untransformed. Transformation in fact ceases before the austenite achieves its equilibrium composition, so that the effect is dubbed the "incomplete reaction phenomenon".

These observations are understood when it is realized that growth must cease if the carbon concentration in the austenite reaches the T_0 curve of the phase diagram. Since this condition is met at ever increasing carbon concentrations when the transformation temperature is reduced, more bainite can form with greater undercoolings below B_s. But the T_0 restriction means that equilibrium, when the austenite has a composition given by the Ae3 phase boundary, can never be reached, as observed experimentally. A bainite finish temperature B_F is sometimes defined, but this clearly cannot have any fundamental significance.

21.3.4   Nucleation and Growth Kinetics

Little is known about the nucleation of bainite except that the same nucleus develops into bainite or Widmanstätten ferrite (Section 21.5), depending on the prevailing magnitude of the thermodynamic driving force (Bhadeshia, 1981). The process involves the dissociation of dislocations (Olson and Cohen, 1976) which determines the activation energy Q, which is a linear function of the chemical driving force ΔG_m, so that the nucleation rate I becomes (Bhadeshia, 1982b)

$$I = C_1 \exp\left[-\frac{C_2 + C_3 \Delta G_m}{RT} \right],$$

(13)

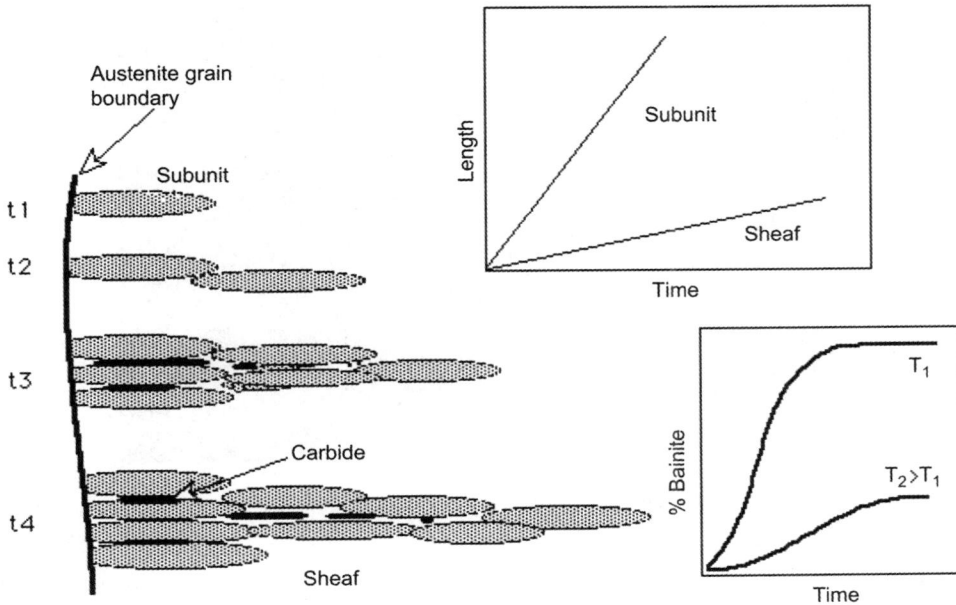

Figure 20 Evolution of a bainite sheaf by the repeated formation of rapidly growing platelets, each of which has a constant size limited by the accumulation of dislocation debris at the interface.

where C_i are constants and R is the gas constant. The dissociation process requires the partitioning of carbon but the nucleus evolves into bainite if the free energy change available is sufficient to sustain diffusionless growth, and into Widmanstätten ferrite if not. Bainite growth occurs by the propagation of subunits as illustrated in **Figure 20**. The growth rate of each of these platelets is far greater than that of the sheaf as a whole, and the precipitation of carbides is a secondary reaction. Detailed discussion and quantitative aspects can be found elsewhere (Bhadeshia, 2001).

The structure develops somewhat differently if nucleation occurs from intragranular sites such as inclusions. The platelets then radiate from point nucleation sites so that the microstructure becomes more chaotic, and is classically known as *acicular ferrite* (Bhadeshia, 2001).

21.3.5 Summary

Bainite grows by displacive transformation; the growth is accompanied by a shape deformation which is an IPS with a large shear component. The transformation is diffusionless but carbon atoms partition into the residual austenite (or precipitate as carbides), shortly after growth is arrested. The precipitation of carbides is therefore a secondary event.

21.4 Alloy Design: Strong Bainite

High-strength bainitic steels have not in practice been as successful as quenched and tempered martensitic steels, because the coarse cementite particles in bainite are detrimental for toughness.

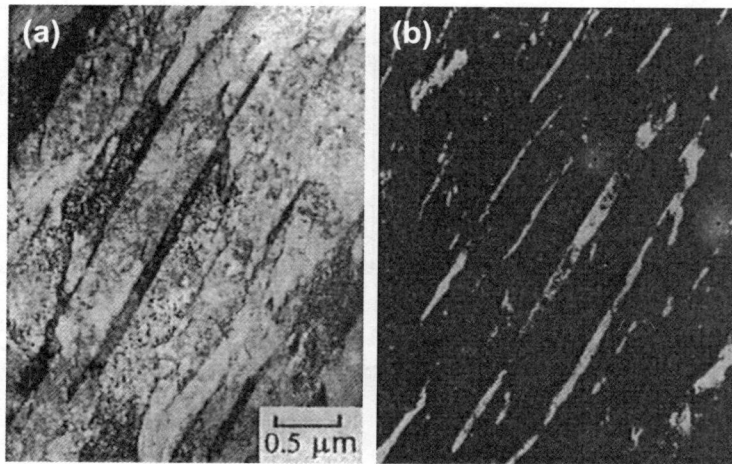

Figure 21 Transmission electron micrograph of a mixture of bainitic ferrite and stable austenite. (a) Bright field image. (b) Retained austenite dark field image.

However, it is now known that the precipitation of cementite during bainitic transformation can be suppressed. This is done by alloying the steel with about 1.5 wt% of silicon, which has a very low solubility in cementite and greatly retards its growth.

An interesting microstructure results when this silicon-alloyed steel is transformed into upper bainite. The carbon that is rejected into the residual austenite, instead of precipitating as cementite, remains in the austenite and stabilizes it down to ambient temperature. The resulting microstructure consists of fine plates of bainitic ferrite separated by carbon-enriched regions of austenite (**Figure 21**).

The potential advantages of the mixed microstructure of bainitic ferrite and austenite can be listed as follows:

(1) Cementite is responsible for initiating fracture in high-strength steels. Its absence is expected to make the microstructure more resistant to cleavage failure and void formation.
(2) The bainitic ferrite is almost free of carbon, which is known to embrittle ferritic microstructures.
(3) The microstructure derives its strength from the ultrafine grain size of the ferrite plates, which are less than 1 μm in thickness. It is the thickness of these plates which determines the mean free slip distance, so that the effective grain size is less than a micrometer. This cannot be achieved by any other commercially viable process. It should be borne in mind that grain refinement is the only method available for simultaneously improving the strength and toughness of steels.
(4) The ductile films of austenite which are intimately dispersed between the plates of ferrite have a crack blunting effect. They further add to toughness by increasing the work of fracture as the austenite is induced to transform to martensite under the influence of the stress field of a propagating crack. This is the TRIP effect.
(5) The diffusion of hydrogen in austenite is slower than in ferrite. The presence of austenite can, therefore, improve the stress corrosion resistance of the microstructure.
(6) Steels with the bainitic ferrite and austenite microstructure can be obtained without the use of any expensive alloying additions. All that is required is that the silicon concentration should be large enough to suppress cementite.

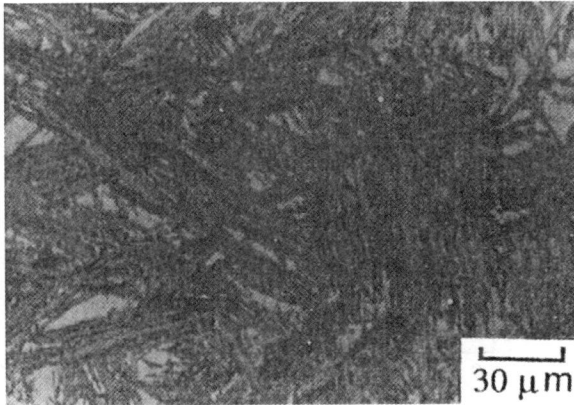

Figure 22 Optical micrograph of upper bainite in an Fe–0.43C–3Mn–2.02Si wt% showing the blocks of retained austenite between sheaves of bainite.

In spite of these appealing features, the bainitic ferrite/austenite microstructure does not always give the expected good combination of strength and toughness. This is because the relatively large "blocky" regions of austenite between the sheaves of bainite (**Figure 22**) readily transform into high-carbon martensite under the influence of stress. This untempered hard martensite embrittles the steel.

The blocks of austenite are detrimental to toughness, and anything that can be done to reduce their fraction, or increase their stability to martensitic transformation, would be beneficial. Both of these effects are controlled by the T_0' curve of the phase diagram. This curve determines the composition of the austenite at the point where the reaction to bainite stops. By displacing the curve to larger carbon concentrations, both the fraction of bainite that can form and the carbon concentration of the residual austenite can be increased. Modifications to the T_0' curve can be achieved by altering the alloy composition. It is therefore necessary to calculate the effect of substitutional solutes on the T_0' curve.

21.4.1 The Improvement in Toughness

An apparently ideal microstructure consisting of bainitic ferrite and ductile austenite in an Fe–3Mn–2.02Si–0.43C wt% exhibits poor toughness because of the presence of blocky unstable austenite (**Figure 23**). It is necessary to increase the amount of bainitic ferrite in the microstructure and to increase the stability of the austenite. Both of these aims can be achieved by changing the substitutional solute concentration such that the T_0' curve is shifted to higher carbon concentrations (i.e. T_0' is raised at any given carbon concentration).

Manganese has a large effect in depressing the T_0' temperature. An examination of thermodynamic data shows that one possibility is to replace all of the manganese with nickel (**Figure 23**). Thus, for an Fe–4Ni–2Si–0.4C wt% (3.69Ni, 3.85Si at%) alloy remarkable improvement in toughness achieved by doing this, without any sacrifice of strength, is illustrated in **Figure 23**, along with the T_0' curves as calculated above.

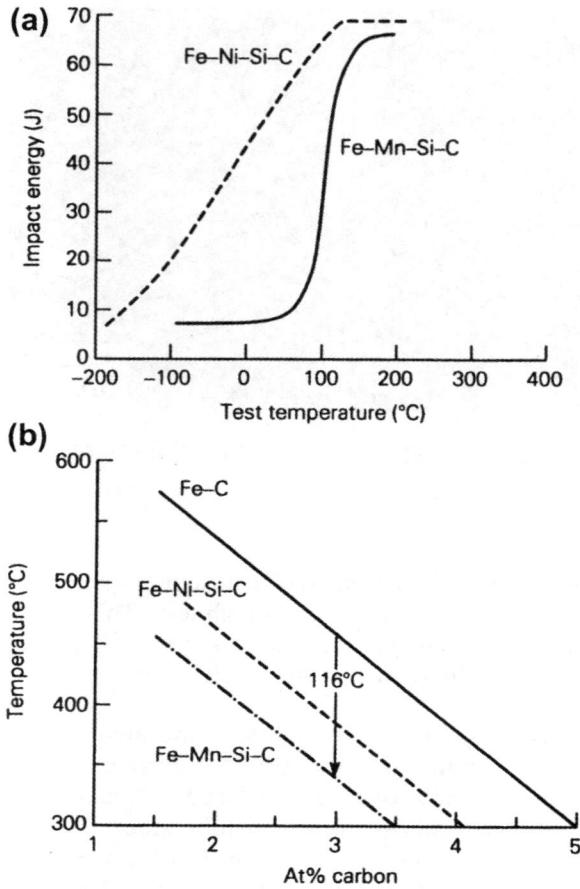

Figure 23   (a) Experimentally determined impact transition curves showing how the toughness improves as the amount of blocky austenite is reduced. (b) Calculated T_0' curves for the Fe–C, Fe–Mn–Si–C and Fe–Ni–Si–C steels.

21.5   Widmanstätten Ferrite

21.5.1   Morphology

Primary Widmanstätten ferrite either directly grows from the austenite grain surfaces, whereas secondary Widmanstätten ferrite develops from any allotriomorphic ferrite that may be present in the microstructure (**Figure 24**). Widmanstätten ferrite can form at temperatures close to the Ae3 temperature and hence can occur at very low driving forces; the undercooling needed amounts to a free energy change of only 50 J mol^{-1}. This is much less than that required to sustain diffusionless transformation.

21.5.2   Shape Change

The growth of a single plate of martensite is accompanied by an IPS of the type illustrated in **Figure 25a**. However, at the high temperatures (low undercoolings) at which Widmanstätten ferrite grows, the

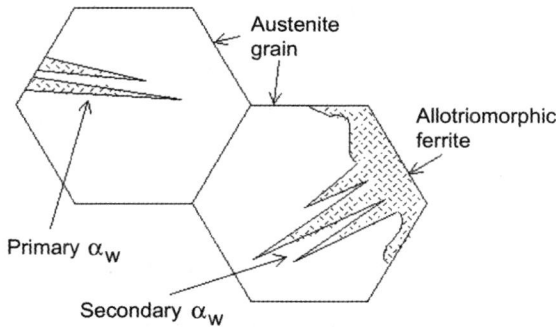

Figure 24 Morphology of primary and secondary Widmanstätten ferrite.

driving force is not sufficient to support the strain energy associated with a single plate. Widmanstätten ferrite formation therefore involves the simultaneous and adjacent cooperative growth of two plates, which are crystallographic variants such that their shape deformations mutually accommodate (**Figure 25b**). This has the effect of canceling much of the strain energy.

It follows that what is seen as a single plate in an optical microscope is actually a combination of two variants, usually separated by a low-misorientation boundary (**Figure 25c**). Widmanstätten ferrite has a habit plane which is close to $\{5\ 5\ 8\}_\gamma$. Hence, the two plates $\alpha_{\omega 1}$ and $\alpha_{\omega 2}$, which have different variants of this habit with the austenite, together form the thin-wedge shaped plate which is characteristic of Widmanstätten ferrite. A self-consistent set (Bhadeshia, 2011) describing the crystallography (Watson and McDougall, 1973) of a plate of Widmanstätten ferrite has the habit plane normal:

$$\mathbf{P} = (0.5057\ 0.4523\ 0.7346)_\gamma,$$

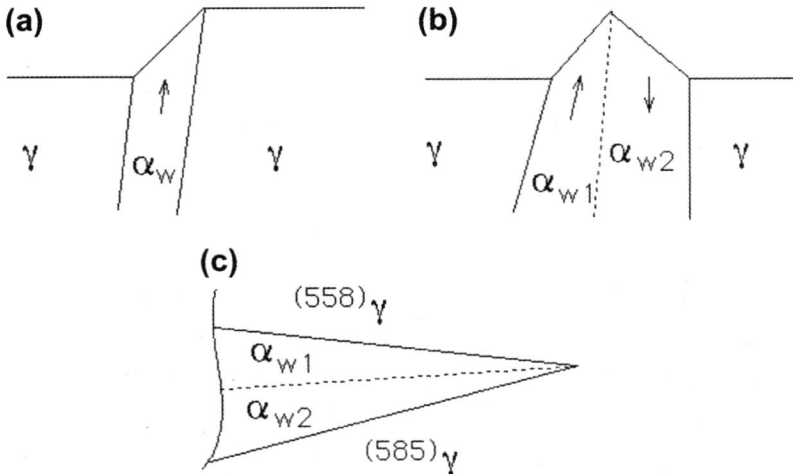

Figure 25 (a) A single IPS shape deformation. (b) The combined effect of two mutually accommodating back-to-back IPS deformations. (c) The morphology of two plates, with different habit plane variants, growing together in a mutually accommodating manner.

the orientation relationship being irrational but close to Kurdjumov-Sachs:

$$(1\ 0\ 1)_\alpha\ \|\ (0.5916\ 0.5772\ 0.5628)_\gamma$$
$$[1\ 1\ \overline{1}]_\alpha\ \|\ [0.6984\ \overline{0.7157}\ 0.0001]_\gamma$$

and the average magnitude of the shape deformation and direction

$$m = 0.36,$$
$$\mathbf{d} = [\overline{0.8670}\ 0.4143\ 0.2770]_\gamma.$$

The displacement vector **d** does not lie precisely within the habit plane **p** because it describes both the shear and dilatational strains, the latter being directed normal to the habit plane.

Because Widmanstätten ferrite forms at low undercoolings (and above the T_0 temperature), it is thermodynamically required that the carbon is redistributed during growth. α_ω therefore always has a paraequilibrium carbon content and grows at a rate which is controlled by the diffusion of carbon in the austenite ahead of the plate tip. For plates, diffusion-controlled growth can occur at a constant rate because solute is partitioned to the sides of the plate, whereas the growing tip can advance into fresh austenite. Since the transformation is nevertheless displacive, substitutional atoms do not partition and an atomic correspondence is maintained between the parent and product lattices.

21.5.3 Growth Kinetics

For isothermal transformation during growth in which there is no partitioning of substitutional solutes, the growth rate is governed by the rate at which carbon diffuses ahead of the Widmanstätten ferrite plate tip. In the first approximation, the concentrations at the interface are given by a tie-line of the phase diagram as shown in **Figure 26**. The diffusion flux of solute from the interface must equal the rate at which solute is incorporated in the precipitate so that

$$\underbrace{(c^{\gamma\alpha} - c^{\alpha\gamma})\frac{\partial z^*}{\partial t}}_{\text{rate solute partitioned}} = \underbrace{-D\frac{\partial c}{\partial z}}_{\text{diffusion flux from interface}} \simeq D\frac{c^{\gamma\alpha} - \bar{c}}{\Delta z}, \tag{14}$$

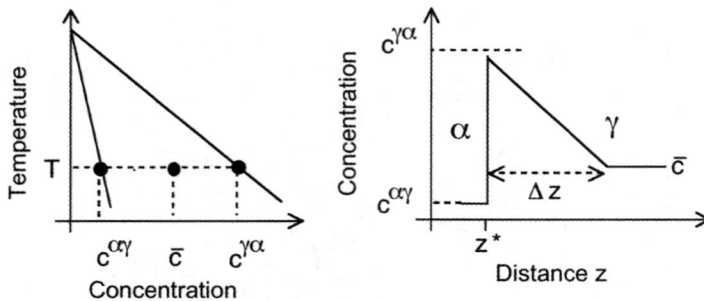

Figure 26 Phase diagram and its relationship to the concentration profile at the α/γ interface during diffusion-controlled growth.

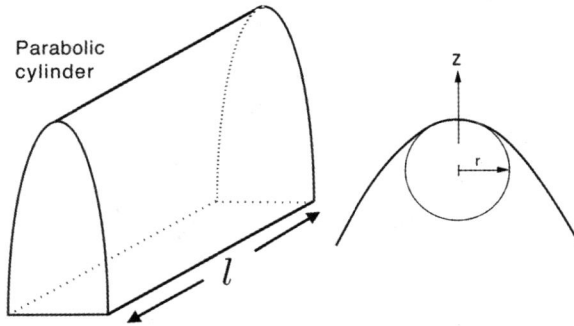

Figure 27 Widmanstätten ferrite plate represented as a parabolic cylinder, with tip radius r.

where z is a coordinate normal to the interface with a value z^* at the position of the interface. Note that concentration gradient here is assumed to be constant but it is in general evaluated at the position of the interface $(z = z^*)$.

The plate can be described as a parabolic cylinder in three dimensions (**Figure 27**), a shape which is preserved as the plate lengthens. If it is assumed that the diffusion distance Δz is equal to the plate tip radius r (**Figure 27**), then from eqn (14) it follows that the lengthening rate $v_1 = \partial z^*/\partial t$ is given by

$$v_1 \approx \frac{D}{r} \frac{c^{\gamma\alpha} - \bar{c}}{c^{\gamma\alpha} - c^{\alpha\gamma}}. \tag{15}$$

This leads to the obvious difficulty that $v_1 \to \infty$ as $r \to 0$, caused by the fact that the creation of additional interfacial area as the plate grows is neglected in the derivation.

Given that the change in surface area per atom added to the plate is v_a/r, the corresponding increase in the free energy per atom because of the creation of additional interface is $\sigma v_a/r$ where σ is the interfacial energy per unit area and v_a is the volume per atom. Therefore, the net free energy change per atom, Δg_r, as the plate grows is

$$\Delta g_r = \Delta g_\infty - \frac{\sigma v_a}{r}, \tag{16}$$

where Δg_∞ represents the free energy change per atom, driving the transformation in the absence of interface creation. At a critical radius r_c, $\Delta g_r = 0$ so that $\Delta g_\infty = \sigma v_a/r_c$ and $v_1 = 0$. Equation (16) can therefore be written as

$$\Delta g_r = \frac{\sigma v_a}{r_c} - \frac{\sigma v_a}{r} \quad \text{or} \quad \frac{\Delta g_r}{\Delta g_\infty} = 1 - \frac{r_c}{r}. \tag{17}$$

The velocity scales with the driving force when the latter is small, so eqn (15) can be rewritten to account for the interface creation as follows:

$$v_1 \approx \frac{D}{r} \left(\frac{c^{\gamma\alpha} - \bar{c}}{c^{\gamma\alpha} - c^{\alpha\gamma}} \right) \times \left(1 - \frac{r_c}{r} \right). \tag{18}$$

The accounting for interfacial energy in this manner is known as the *capillarity effect* (Christian, 1975) which governs the equilibrium between a curved particle and the matrix. **Figure 28** shows how the

(a)

(b)

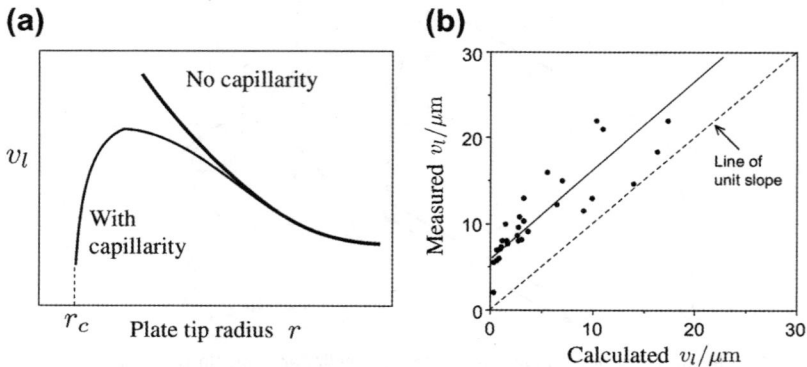

Figure 28   (a) Influence of interface curvature on plate lengthening rate. (b) Comparison of measured versus calculated lengthening rates.

lengthening rate now goes through a maximum, and it is often assumed that the plate picks a radius consistent with the maximum growth rate. This is approximately consistent with experimental measurements as shown in **Figure 28b** (Bhadeshia, 1985).

21.5.4   Summary

- An atomic correspondence is maintained for substitutional solutes, consistent with a displacive transformation mechanism.
- Ferrite has a paraequilibrium carbon content during growth which occurs at a constant rate, controlled by the diffusion of carbon in the austenite ahead of the plate tip.
- Growth involves the simultaneous and cooperative formation of a pair of adjacent self-accommodating plates of Widmanstätten ferrite.

21.6   Allotriomorphic Ferrite

An allotriomorph has a shape which does not reflect its internal crystalline symmetry. This is because it tends to nucleate at the austenite grain surfaces, forming layers which follow the grain boundary contours (**Figure 29**).

An idiomorph, on the other hand, has a shape which reflects the symmetry of the crystal as embedded in the austenite. Idiomorphs nucleate without contact with the austenite grain surfaces; they tend to nucleate heterogeneously on nonmetallic inclusions present in the steel.

These are both true diffusional transformations, that is, there is no atomic correspondence between the parent and product crystals, there is no IPS shape change accompanying transformation, the growth rate is diffusion controlled, interface controlled or mixed. Thermal activation is necessary for transformation, which can therefore only occur at high homologous temperatures.

The α/γ interface need not in this case be glissile; the motion of the interface involves diffusion and is not conservative.

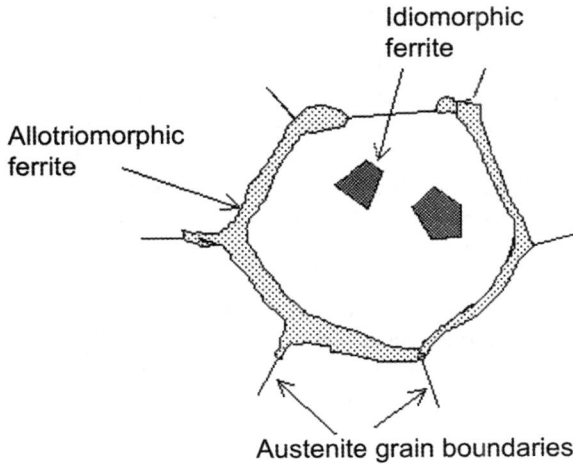

Figure 29 Allotriomorphic and idiomorphic ferrite.

21.6.1 Diffusion-Controlled Growth in Fe–C

The ferrite has a different chemical composition from the austenite in which it grows. We shall assume that the growth of ferrite (α) is controlled by the diffusion of carbon in the austenite (γ) ahead of the interface.

As the ferrite grows, so does the extent of its diffusion field. This retards growth because the solute then has to diffuse over ever larger distances. As we will prove, the thickness of the ferrite increases with the square root of time, that is, the growth rate slows down as time increases. We will assume in our derivation that the concentration gradient in the matrix is constant, and that the far-field concentration $\bar{c}$ never changes (i.e. the matrix is semi-infinite normal to the advancing interface). This is to simplify the mathematics without loosing the insight into the problem.

A theoretical treatment of the thickening of allotriomorphs follows the same procedure as outlined in eqn (14) and **Figure 26**. However, unlike the case for Widmanstätten ferrite, the solute accumulates in front of the interface so the diffusion distance Δz is no longer constant, so an additional condition is necessary to calculate the growth rate as a function of time. A second equation can be derived by considering the overall conservation of mass

$$(c^{\alpha\gamma} - \bar{c})z^* = \frac{1}{2}(\bar{c} - c^{\gamma\alpha})\Delta z. \tag{19}$$

On combining this with eqn (14) to eliminate Δz we get

$$\frac{\partial z^*}{\partial t} = \frac{D(\bar{c} - c^{\gamma\alpha})^2}{2z^*(c^{\alpha\gamma} - c^{\gamma\alpha})(c^{\alpha\gamma} - \bar{c})}. \tag{20}$$

It follows that

$$z^* \propto \sqrt{Dt}.$$

21.6.2   Thermodynamics of Irreversible Processes

Thermodynamics generally deals with measurable properties of materials, formulated on the basis of equilibrium. Thus, properties such as entropy and free energy are, on an appropriate scale, static and time invariant during equilibrium. There are other parameters not relevant to the discussion of equilibrium: thermal conductivity, diffusivity and viscosity, but which are interesting because they can describe a second kind of time independence, that of the steady state. Thus, the concentration profile does not change during steady-state diffusion, even though energy is being dissipated by the diffusion.

The thermodynamics of irreversible processes deals with systems which are not at equilibrium but are nevertheless *stationary*. The theory in effect uses thermodynamics to deal with *kinetic* phenomena. There is, nevertheless, a distinction between the thermodynamics of irreversible processes and kinetics. The former applies strictly to the steady state, whereas there is no such restriction on kinetic theory.

21.6.2.1   Reversibility

A process whose direction can be changed by an infinitesimal alteration in the external conditions is called reversible. Consider the example illustrated in **Figure 30**, which deals with the response of an ideal gas contained at uniform pressure within a cylinder, any change being achieved by the motion of the piston. For any starting point on the P/V curve, if the application of an infinitesimal force causes the piston to move slowly to an adjacent position still on the curve, then the process is reversible since energy has not been dissipated. The removal of the infinitesimal force will cause the system to revert to its original state.

On the other hand, if there is friction during the motion of the piston, then deviations occur from the P/V curve as illustrated by the cycle in **Figure 30**. An infinitesimal force cannot move the piston because energy is dissipated because of friction (as given by the area within the cycle). Such a process, *which involves the dissipation of energy*, is classified as irreversible with respect to an infinitesimal change in the external conditions.

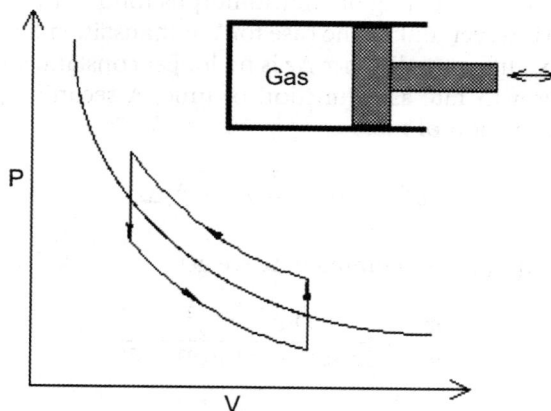

Figure 30   The curve represents the variation in pressure within the cylinder as the volume of the ideal gas is altered by positioning the frictionless piston. The cycle represents the dissipation of energy when the motion of the piston causes friction.

More generally, reversibility means that it is possible to pass from one state to another without appreciable deviation from equilibrium. Real processes are not reversible so equilibrium thermodynamics can only be used approximately, though the same thermodynamics defines whether a process can occur spontaneously without ambiguity.

For irreversible processes the *equations* of classical thermodynamics become *inequalities*. For example, at the equilibrium melting temperature, the free energies of the pure liquid and solid are identical ($G_{liquid} = G_{solid}$) but not so below that temperature ($G_{liquid} > G_{solid}$). Such inequalities are much more difficult to deal with though they indicate the natural direction of change. For steady-state processes, however, the thermodynamic framework for irreversible processes as developed by Onsager is particularly useful in approximating relationships even though the system is not at equilibrium.

21.6.2.2 *The Linear Laws*

At equilibrium there is no change in entropy or free energy. An irreversible process dissipates energy and entropy is created continuously. In the example illustrated in **Figure 30**, the dissipation was because of friction; diffusion ahead of a moving interface is dissipative. The rate at which energy is dissipated is the product of the temperature and the rate of entropy production (i.e. $T\sigma$) with

$$T\sigma = JX, \tag{21}$$

where J is a generalized flux of some kind, and X a generalized force. In the case of an electrical current, the heat dissipation is the product of the current (J) and the electromotive force (X).

As long as the flux–force sets can be expressed as in eqn (21), the flux must naturally depend in some way on the force. It may then be written as a function $J\{X\}$ of the force X. At equilibrium, the force is zero. If $J\{X\}$ is expanded in a Taylor series about equilibrium ($X = 0$), we get

$$J\{X\} = \sum_0^\infty a_n X^n$$
$$= J\{0\} + J'\{0\}\tfrac{X}{1!} + J''\{0\}\tfrac{X^2}{2!}\cdots \tag{22}$$

Note that $J\{0\} = 0$ since that represents equilibrium. If the high-order terms are neglected then we see that

$$J \propto X.$$

This is a key result from the theory, that the forces and their conjugate fluxes are linearly related ($J \propto X$) whenever the dissipation can be written as in eqn (22), at least when the deviations from equilibrium are not large. Some examples of forces and fluxes in the context of the present theory are given in **Table 4**.

21.6.2.3 *Multiple Irreversible Processes*

There are many circumstances in which a number of irreversible processes occur together. In a ternary Fe–Mn–C alloy, the diffusion flux of carbon depends not only on the gradient of carbon but also on that of manganese. Thus, a uniform distribution of carbon will tend to become inhomogeneous in the presence of a manganese concentration gradient. Similarly, the flux of heat may not depend on the temperature gradient alone; heat can be driven also by an

Table 4   Examples of forces and their conjugate fluxes. z is distance, ϕ is the electrical potential in Volts, and μ is a chemical potential. "e.m.f." stands for electromotive force

Force	Flux
e.m.f $= \dfrac{\partial \phi}{\partial z}$	Electrical current
$-\dfrac{1}{T}\dfrac{\partial T}{\partial z}$	Heat flux
$-\dfrac{\partial \mu_i}{\partial z}$	Diffusion flux
Stress	Strain rate

electromotive force (Peltier effect)[2]. Electromigration involves diffusion driven by an electromotive force. When there is more then one dissipative process, the total energy dissipation rate can still be written

$$T\sigma = \sum_i J_i X_i. \tag{23}$$

In general, if there is more than one irreversible process occurring, it is found *experimentally* that each flow J_i not only is related to its conjugate force X_i, but also is related linearly to all other forces present. Thus,

$$J_i = M_{ij} X_j \tag{24}$$

with $i, j = 1, 2, 3 \dots$. Therefore, a given flux depends on all the forces causing the dissipation of energy.

21.6.2.4 Onsager Reciprocal Relations

Equilibrium in real systems is always dynamic on a microscopic scale. It seems obvious that to maintain equilibrium under these dynamic conditions, a process and its reverse must occur at the same rate on the microscopic scale. The consequence is that provided the forces and fluxes are chosen from the dissipation equation and are independent, $M_{ij} = M_{ji}$. This is known as the Onsager theorem, or the Onsager reciprocal relations. It applies to systems near equilibrium when the properties of interest have even parity, and assuming that the fluxes and their corresponding forces are independent. An exception occurs with magnetic fields in which case there is a sign difference $M_{ij} = -M_{ji}$.

21.6.3 Ternary Steels

Consider now a ternary steel, say Fe–Mn–C. It would be necessary to satisfy two equations of the form of eqn (14), simultaneously, for each of the solutes:

$$\begin{aligned}
(c_1^{\gamma\alpha} - c_1^{\alpha\gamma})v &= -D_1 \nabla c_1, \\
(c_2^{\gamma\alpha} - c_2^{\alpha\gamma})v &= -D_2 \nabla c_2,
\end{aligned} \tag{25}$$

---

[2] In the Peltier effect, the two junctions of a thermocouple are kept at the same temperature but the passage of an electrical current causes one of the junctions to absorb heat and the other to liberate the same quantity of heat. This Peltier heat is found to be proportional to the current.

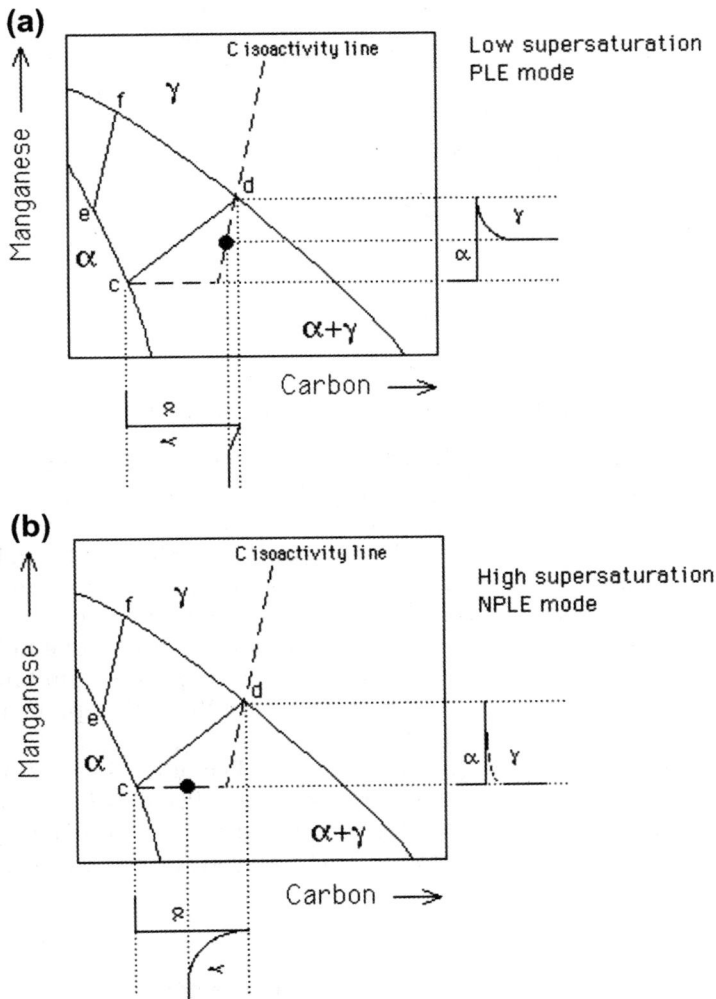

Figure 31 Schematic isothermal sections of the Fe–Mn–C system, illustrating ferrite growth occurring with local equilibrium at the α/γ interface. (a) Growth at low supersaturations (PLE) with bulk redistribution of manganese, (b) growth at high supersaturations (NPLE) with negligible partitioning of manganese during transformation. The bulk alloy compositions are designated by the symbol • in each case.

where the subscripts refer to the solutes (1 for carbon and 2 for Mn). The interface velocity v is dz^*/dt in eqn (20).

Because $D_1 \gg D_2$, these equations cannot in general be simultaneously satisfied for the tie-line passing through the alloy composition $\bar{c}_1, \bar{c}_2$. It is, however, possible to choose other tie-lines which satisfy eqn (25). If the tie-line is such that $c_1^{\gamma\alpha} = \bar{c}_1$ (e.g. line cd for alloy A in **Figure 31a**), then ∇c_1 will become very small, the driving force for carbon diffusion in effect being reduced, so that the flux of carbon atoms is forced to slow down to a rate consistent with the diffusion of manganese. Ferrite

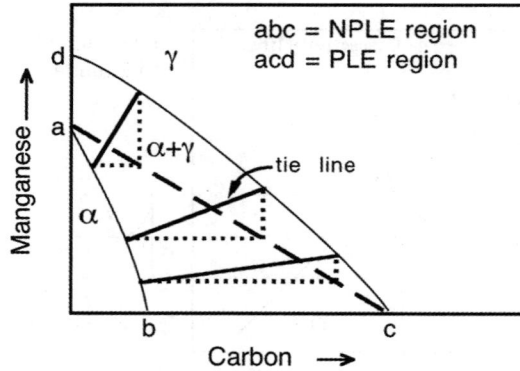

Figure 32   Regions of the two-phase field where either PLE or NPLE modes of transformation are possible.

forming by this mechanism is said to grow by a "partitioning, local equilibrium" (PLE) mechanism, in recognition of the fact that $c_2^{\gamma\alpha}$ can differ significantly from $\bar{c}_2$, giving considerable partitioning and long-range diffusion of manganese into the austenite.

An alternative choice of tie-line could allow $c_2^{\gamma\alpha} \rightarrow \bar{c}_2$ (e.g. line cd for the alloy in **Figure 31b**), so that ∇c_2 is drastically increased since only very small amounts of Mn are partitioned into the austenite. The flux of manganese atoms at the interface correspondingly increases and manganese diffusion can then keep pace with that of carbon, satisfying the mass conservation conditions of eqn (25). The growth of ferrite in this manner is said to occur by a "negligible partitioning, local equilibrium" (NPLE) mechanism, in recognition of the fact that the manganese content of the ferrite approximately equals $\bar{c}_2$, so that little if any manganese partitions into austenite.

What circumstances determine whether growth follows the PLE or NPLE mode? **Figure 32** shows the Fe–Mn–C phase diagram, now divided into domains where either PLE or NPLE is possible but not both. The domains are obtained by drawing right-handed triangles on each tie-line in the $\alpha + \gamma$ phase field and joining up all the vertices. For example, prove to yourself that if you attempt to define NPLE conditions in

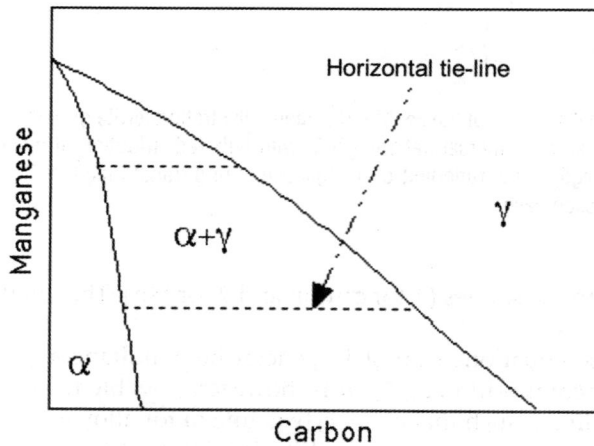

Figure 33   A paraequilibrium phase diagram.

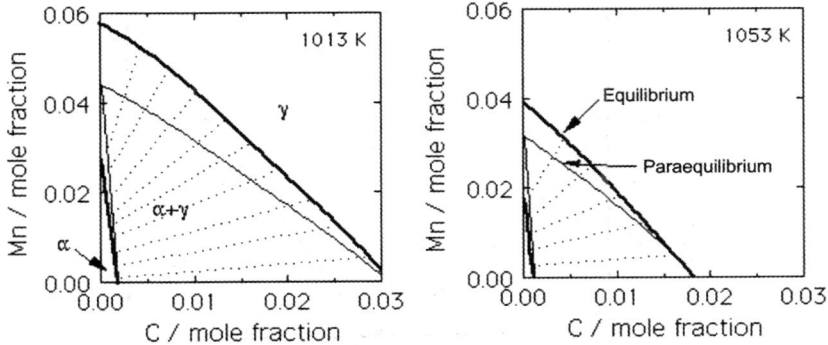

Figure 34 The paraequilibrium phase field lies within the equilibrium field. The tie-lines illustrated are for equilibrium.

the PLE domain, then the tie-line determining interface compositions will incorrectly show that both austenite and ferrite contain less carbon than $\bar{c}_1$, a circumstance which is physically impossible.

Paraequilibrium is a constrained equilibrium. It occurs at temperatures where the diffusion of substitutional solutes is not possible within the timescale of the experiment. Nevertheless, interstitials may remain highly mobile. Thus, in a steel, manganese does not partition between the ferrite and austenite, but subject to that constraint, the carbon redistributes until it has the same chemical potential in both phases.

Therefore, the tie-lines in the phase diagram (**Figure 33**) are all virtually parallel to the carbon axis, since Mn does not partition between ferrite and austenite. In an isothermal section of the ternary phase diagram, the paraequilibrium phase boundaries must lie within the equilibrium phase boundaries as illustrated in **Figure 34**.

21.7 Pearlite

A colony of pearlite when viewed in three dimensions consists of an interpenetrating bicrystal of ferrite and cementite (Hillert, 1962). In planar sections the phases appear as lamellae which grow at a common front with the austenite. Cementite (θ) is rich in carbon whereas ferrite (α) accommodates

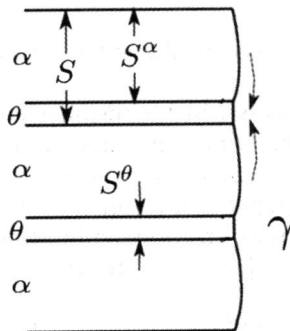

Figure 35 Diffusion flux parallel to the advancing interface. S is the interlamellar spacing.

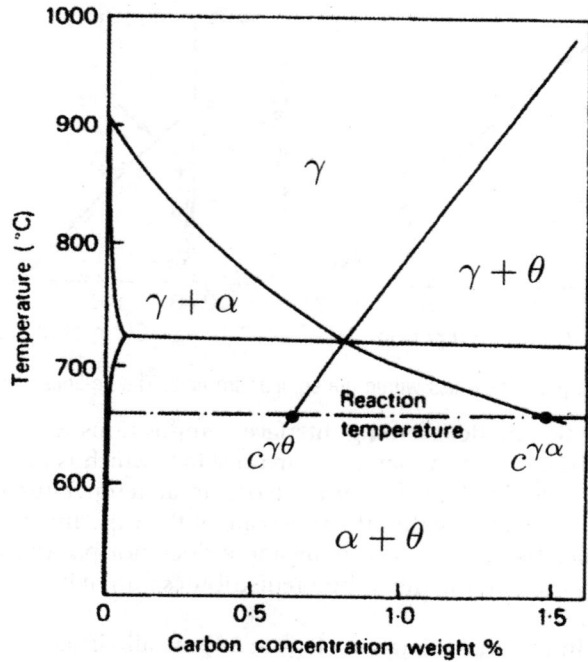

Figure 36 Phase diagram with extrapolated phase boundaries to identify the concentrations in the austenite which is in equilibrium with cementite or ferrite.

very little when it is in equilibrium with either cementite or austenite (γ). It is therefore necessary for carbon to be redistributed at the transformation front. This can happen either by diffusion in the austenite in a direction parallel to the transformation front (**Figures 35 and 36**).

The diffusion distance parallel to the interface can be approximated as aS where a is a constant and S is the interlamellar spacing. By analogy with eqn (14), it follows that the rate at which solute is absorbed by the cementite must equal the amount arriving there by diffusion, so that

$$v\left(c^{\theta} - c^{\gamma\theta}\right) = D\frac{\left(c^{\gamma\alpha} - c^{\gamma\theta}\right)}{aS},\tag{26}$$

where v is the speed of the growth front, D is the diffusivity of carbon in austenite and the concentration terms are self-explanatory.

However, there is an additional process which consumes energy, the creation of cementite/ferrite interfaces within the pearlite colony. The minimum value of interlamellar spacing possible is a critical spacing $S_C = 2\sigma^{\alpha\theta}/\Delta G$ where $\sigma^{\alpha\theta}$ is the interfacial energy per unit area and ΔG is the magnitude of the driving force for transformation in Joules per unit volume. Growth ceases when $S = S_C$. To allow for the energy consumed in the process of interface creation, and following the procedure outlined in Section 21.5, eqn (26) is modified by a term $1 - [S_C/S]$ as follows:

$$v = \frac{D}{aS}\frac{\left(c^{\gamma\alpha} - c^{\gamma\theta}\right)}{\left(c^{\theta} - c^{\gamma\theta}\right)}\left(1 - \frac{S_c}{S}\right).\tag{27}$$

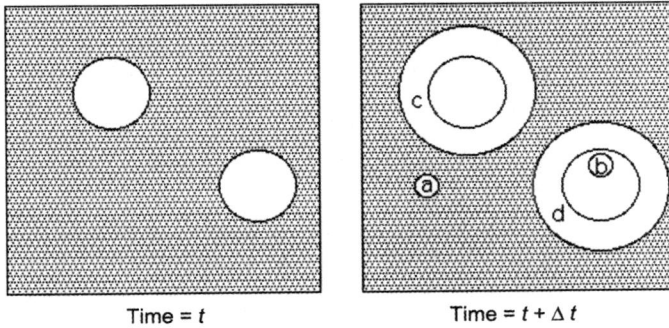

Time = t Time = $t + \Delta t$

Figure 37 An illustration of the concept of extended volume. Two precipitate particles have nucleated together and grown to a finite size in the time t. New regions c and d are formed as the original particles grow, but a and b are new particles, of which b has formed in a region which is already transformed.

We now need to specify the value that S will adopt during growth, and one assumption is that the spacing will correspond to that consistent with the maximum growth rate, that is, when $S = 2S_C$.

The theory presented here is necessarily oversimplified and uses a number of unnecessary assumptions; more rigorous models are available but can get complex as soon as interfacial diffusion and the influence of solutes other than carbon is taken into consideration (Christian, 1975; Pandit and Bhadeshia, 2011a,b; Puls and Kirkaldy, 1972).

21.8   Overall Transformation Kinetics

TTT diagrams represent the evolution of the volume fraction of phases by solid-state transformation of austenite as a function of time at a constant temperature. The change in volume fraction depends on nucleation, growth and impingement phenomena, all of which form the theoretical basis for the calculation of TTT diagrams, a theory commonly referred to as *overall transformation kinetics*, one which continues to have tremendous validity and success since the early foundations (Avrami, 1939, 1940, 1941; Johnson and Mehl, 1939).

21.8.1   Isothermal Transformation

To model transformation it is obviously necessary to calculate the nucleation and growth rates, but an estimation of the volume fraction requires *impingement* between particles to be taken into account.

This is done using the extended volume concept of Kolmogorov, Johnson, Mehl and Avrami. Referring to **Figure 37**, suppose that two particles exist at time t; a small interval δt later, new regions marked a, b, c and d are formed assuming that they are able to grow unrestricted in extended space whether the region into which they grow is already transformed. However, only those components of a, b, c and d which lie in previously untransformed matrix can contribute to a change in the real volume of the product phase (α):

$$\mathrm{d}V^\alpha = \left(1 - \frac{V^\alpha}{V}\right)\mathrm{d}V_e^\alpha, \tag{28}$$

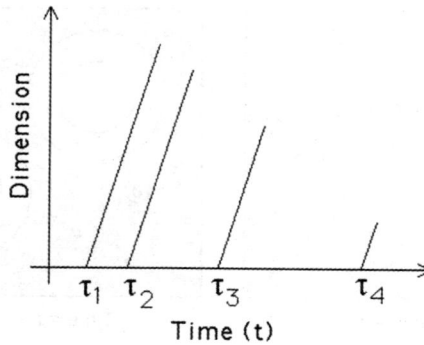

Figure 38 An illustration of the incubation time τ for each particle.

where it is assumed that the microstructure develops at random. The subscript e refers to extended volume, V^α is the volume of α and V is the total volume. Multiplying the change in extended volume by the probability of finding untransformed regions has the effect of excluding regions such as b, which clearly cannot contribute to the real change in volume of the product. For a random distribution of precipitated particles, this equation can easily be integrated to obtain the real volume fraction,

$$\frac{V^\alpha}{V} = 1 - \exp\left\{-\frac{V_e^\alpha}{V}\right\}. \tag{29}$$

The extended volume V_e^α is straightforward to calculate using nucleation and growth models and neglecting completely any impingement effects. Consider a simple case where α grows isotropically at a constant rate G and where the nucleation rate per unit volume, I_V. The volume of a particle nucleated at time $t = \tau$ (**Figure 38**) is given by

$$v_\tau = \frac{4}{3}\pi G^3 (t - \tau)^3. \tag{30}$$

The change in extended volume over the interval τ and $\tau + d\tau$ is

$$dV_e^\alpha = \frac{4}{3}\pi G^3 (t - \tau)^3 \times I_V \times V \times d\tau. \tag{31}$$

On substituting into eqn (29) and writing $\xi = V^\alpha/V$, we get

$$dV^\alpha = \left(1 - \frac{V^\alpha}{V}\right)\frac{4}{3}\pi G^3 (t - \tau)^3 I_V V \, d\tau$$

so that $\quad -\ln\{1 - \xi\} = \frac{4}{3}\pi G^3 I_V \int_0^t (t - \tau)^3 d\tau \tag{32}$

and $\quad \xi = 1 - \exp\{-\pi G^3 I_V t^4/3\}$.

This equation has been derived for the specific assumptions of random nucleation, a constant nucleation rate and a constant growth rate. There are different possibilities but they often reduce to the general form

$$\xi = 1 - \exp\{-k_A t^n\}, \tag{33}$$

where k_A and n characterize the reaction as a function of time, temperature and other variables. The values of k_A and n can be obtained from experimental data by plotting $\ln(-\ln\{1 - \xi\})$ versus $\ln\{t\}$. The specific values of k_A and n depend on the nature of nucleation and growth. Clearly, a constant nucleation and growth rate leads to a time exponent $n = 4$, but if it is assumed that the particles all begin growth instantaneously from a fixed number density of sites (i.e. nucleation is not needed) then $n = 3$ when the growth rate is constant. There are other scenarios and the values of the Avrami parameters are not necessarily unambiguous in the sense that the same exponent can represent two different mechanisms.

The form of eqn (33) is illustrated in **Figure 39**. Note that the effect of temperature is to alter the thermodynamic driving force for transformation, to alter diffusion coefficients and to influence any other thermally activated processes. The effect of manganese is via its influence on the stability of the parent and product phases.

The results of many isothermal transformation curves such as the ones illustrated in **Figure 39** can be plotted on at TTT diagram as illustrated in **Figure 40**. The curves typically have a C shape because the

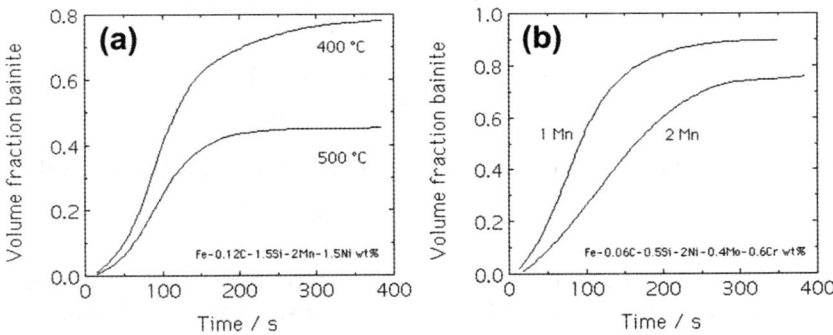

Figure 39 The calculated influence of (a) transformation temperature and (b) manganese concentration on the kinetics of the bainite reaction (Singh, 1998). Bainite is a particular kind of solid-state phase transformation that occurs in steels.

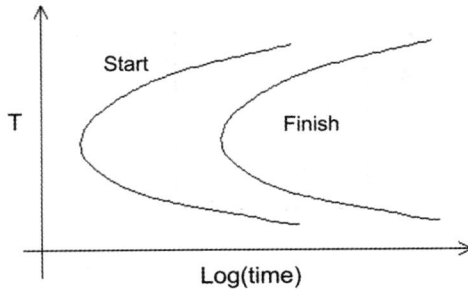

Figure 40 A TTT diagram.

driving force for transformation is small at high temperatures whereas the diffusion coefficient is small at low temperatures. There is an optimum combination of these two parameters at intermediate temperatures, giving a maximum in the rate of reaction. The curve marked *start* corresponds to a detectable limit of transformation (e.g. 5%), and that marked finish corresponds to say 95% transformation.

21.9   TRIP Steels

The steels developed to exploit the properties obtained when the martensite reaction occurs during plastic deformation are known as TRIP steels (Gerberich et al., 1970). The way in which they enhance the strength and uniform ductility of the steel is discussed later, but suffice it to say that a significant constituent of the microstructure must be austenite which is capable of transforming into martensite under the influence of an applied stress. Some of the original studies were conducted on alloys rich in solutes to preserve the austenite to ambient temperature; this can be expensive, but is a good starting point for the discussion of the TRIP effect.

As pointed out in Section 21.2, martensitic transformation is also a deformation, described accurately as an IPS with a shear s on the habit plane, and a dilatation δ normal to that plane (**Figure 41**). Given the orthonormal coordinate system Z defined by the unit vectors z_1 parallel to the direction of shear, and z_3 normal to the habit plane, the deformation matrix **P** becomes

$$(Z \quad \mathbf{P} \quad Z) = \begin{pmatrix} 1 & 0 & s \\ 0 & 1 & 0 \\ 0 & 0 & 1+\delta \end{pmatrix}. \tag{34}$$

The effect of the deformation on any vector **u** to produce a resultant vector **v** is then given by

$$(Z \quad \mathbf{P} \quad Z)[Z; \mathbf{u}] = [Z; \mathbf{v}]. \tag{35}$$

The two vectors will not in general be parallel, but a comparison of the magnitudes gives an impression of the strain expected because of martensitic transformation. This calculation of strain is

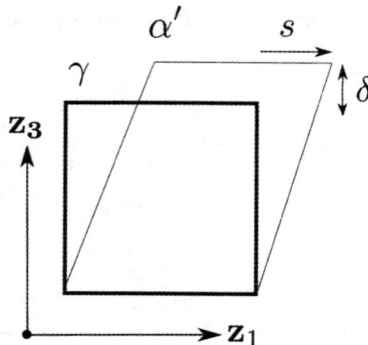

Figure 41   Coordinate axes for the derivation of the deformation matrix representing martensitic transformation ($\alpha\prime$) from austenite (γ).

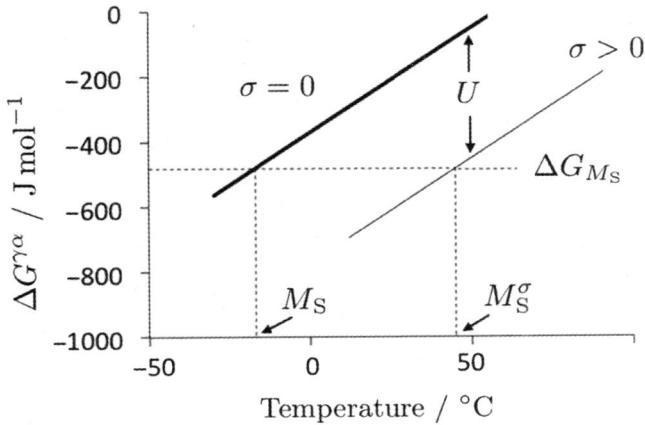

Figure 42 Plot of the chemical driving force for martensitic transformation against temperature, with σ representing the applied stress.

along a specific direction, **u**, because of the shape deformation associated with the formation of a plate of martensite. In dealing with **TRIP** steels, the problem needs to be posed somewhat differently, that is, what is the strain along a particular direction when a stress is applied to induce martensitic transformation in an otherwise stable austenite. It is necessary therefore to consider the thermodynamics of stress-affected martensitic transformation. **Figure 42** shows that in the absence of external stress ($\sigma = 0$), the martensite start temperature M_s is defined by the temperature at which the free energy change $\Delta G^{\gamma\alpha}$, when austenite decomposes into ferrite of the same composition, reaches a critical value ΔG_{M_s}.

In contrast, when transformation occurs under the influence of a stress, the latter interacts with the shape deformation and the interaction energy U is given by Patel and Cohen (1953) to be

$$U = \tau_0 s + \sigma_N \delta, \tag{36}$$

where τ is the shear stress on the habit plane and σ_N is the stress normal to that plane. Note that the strains involved are plastic, so the interaction energy is given simply by the product of the stress and strain, rather than half that value as is sometimes assumed on the basis of elastic strains. If the stress is such that it favors the formation of martensite then U supplements $\Delta G^{\gamma\alpha}$ and the martensite start temperature is raised to M_S^σ which is above ambient temperature, so that stress-induced martensitic transformation becomes feasible (**Figure 42**).

Each single crystal of austenite can in principle transform into 24 different crystallographic variants of martensite. Each of these variants is associated with a particular value of U depending on its orientation relative to the applied stress. Therefore, those variants with the largest interaction with the stress, that is, which transform in a manner that relieves the stress, are favored. This process is known as *variant selection* so that stress-induced martensite results in a biased microstructure with reduced variety. This is illustrated in **Figure 43**, where the martensite is generated by applying a tensile stress to polycrystalline metastable austenite, resulting in plates which are approximately at 45° to the tensile axis.

Assuming that a tensile stress σ_1 is applied, inclined at an angle θ to the habit plane normal, with the stress axis in the plane containing z_1 and z_3, then from eqn (36) and the Mohr's circle representation in **Figure 44**,

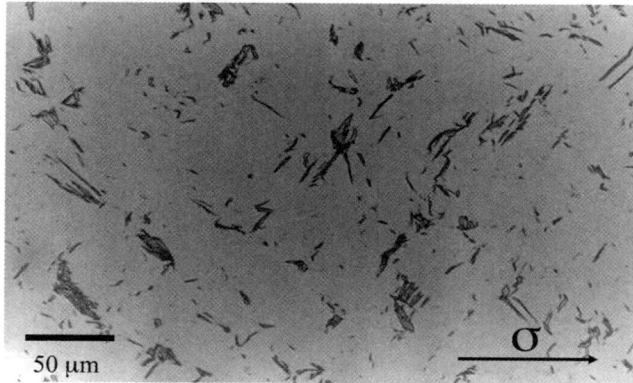

Figure 43 A nonrandom distribution of martensite habit plane orientations produced by stress-induced martensitic transformation at a temperature between M_s and M_s^σ. The sample is polycrystalline austenite.

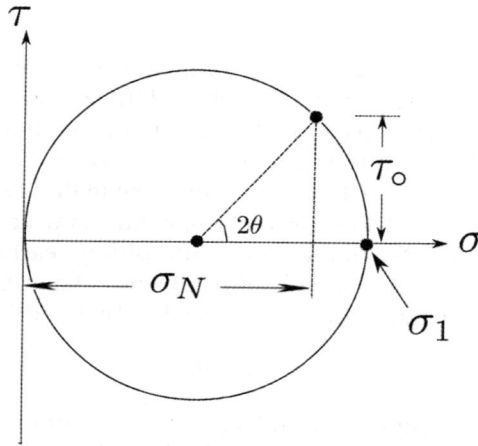

Figure 44 Mohr's circle representation of the shear and normal stresses on a habit plane normal inclined at θ to the tensile stress σ_1.

$$U = \underbrace{\frac{\sigma_1}{2}\sin 2\theta \times s}_{\tau_0} + \underbrace{\frac{\sigma_1}{2}(1+\cos 2\theta)\times \delta}_{\sigma_N},$$

$$\frac{dU}{d\theta} = \frac{\sigma_1}{2}[2s\cos 2\theta - 2\delta\sin 2\theta].$$

(37)

Setting the differential to zero gives the maximum value of U at $\tan 2\theta_{max} = s/\delta$, which for typical values of $s = 0.26$ and $\delta = 0.03$ gives $\theta_{max} = 46.3°$. Given that there are 24 habit plane orientations within a single austenite grain, it is likely that one close to this value will form first, hence explaining the observation of aligned plates in **Figure** 44 even though the sample has many orientations of austenite grains. The tensile axis can be represented as a unit vector

$$[Z; \mathbf{u}] = [\sin \theta_{max} \quad 0 \quad \cos \theta_{max}]$$

and using $\theta = 46°$, the elongation obtained along the tensile axis when a single crystal of austenite transforms completely into the most favored orientation of martensite is, using eqn (35), given by

$$[Z; \mathbf{v}] = [(\sin \theta_{max} + s \cos \theta_{max}) \quad 0 \quad (1 + \delta)\cos \theta_{max}],$$

$$\text{elongation} = 1 - \frac{|\mathbf{v}|}{|\mathbf{u}|}.$$

The elongation because of phase transformation is therefore calculated and is found to be 15% (Bhadeshia, 2002). This impressive value of elongation due to phase transformation alone supplements that due to ordinary dislocation plasticity, which can be a significant boon to the design of strong steels which usually suffer from early plastic instabilities. However, steels which are fully austenitic at ambient temperature can, in the context of iron-based alloys, be expensive.

21.9.1 TRIP-Assisted Steels

One way of producing cheap austenite is to stabilize it with carbon, but excessive carbon can harm other important engineering properties such as the ability to use the steel in a welded state. But an ingenious method involves a low-carbon steel (≈ 0.15 wt% C), which is first transformed into about 70% of allotriomorphic ferrite and the remaining austenite cooled to partly transform it into bainitic ferrite. Both of these transformations leave the residual austenite enriched in carbon to a concentration in excess of 1.2 wt%, thus leaving about 15% of retained austenite in the final microstructure. Since bainite is the last phase to form, the carbon concentration of the austenite is limited by the T_0 condition described in Section 21.3.

It is because the steel is not fully austenitic, that it is referred to as *TRIP-assisted* and typically has a lean composition Fe–0.15C–1.5Si–1.5Mn wt%. The silicon serves to prevent the precipitation of cementite from the carbon-enriched austenite, and the manganese enhances the hardenability required to implement heat treatments suitable for mass production. The microstructure and typical properties are illustrated in **Figure 45**, which also shows how the retained austenite transforms during the course of deformation.

The fraction of retained austenite in a TRIP-assisted steel is $V_\gamma = 0.20$ (**Figure 45**), in which case the maximum elongation to be expected if it all transforms into the favored crystallographic variant of martensite is, using eqn (38), scaled with V_γ given by $0.15 \times 0.2 = 0.03$. The transformation strain therefore makes a negligible contribution to the overall elongation which is about 25% as shown in **Figure 45b**. The major contribution from stress-affected transformation is not via the shape deformation of the martensite, but through a composite effect (Bhadeshia, 2002). The production of hard untempered martensite increases the work-hardening rate and hence delays the onset of plastic instabilities. This is why TRIP-assisted steels have a large uniform elongation.

21.9.2 δ – TRIP Steel

In this alloying concept, δ-ferrite which forms during solidification is stabilized by specific aluminum additions, and substitutes for the 70% of allotriomorphic ferrite in conventional TRIP-assisted steels

(a)

(b)

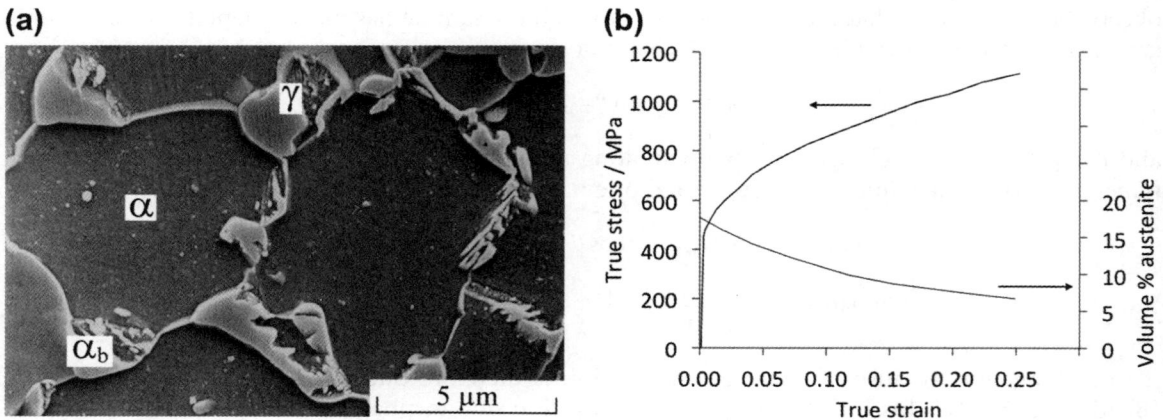

Figure 45 TRIP-assisted steel. (a) Typical final microstructure (micrograph courtesy of P. Jacques). (b) True stress versus true strain, and the stress-assisted decomposition of retained austenite. Data from Jacques (2004).

(Chatterjee et al., 2007). The advantage in doing this is that the δ-ferrite can never be fully removed from the microstructure so that a fully martensitic structure cannot be produced in the heat-affected zone of a resistance spot weld. Another benefit is that the aluminum substitutes for the role of silicon in suppressing cementite precipitation. The difficulty with silicon is that during hot processing, it forms a low-melting temperature oxide known as fayalite which adheres to the steel, making it difficult to remove other disfiguring oxides from the surface of the steel during descaling operations. Silicon can therefore lead to a reduction in the quality of the steel surface, and cause problems during coating processes. The chemical composition of the steel is typically Fe–0.4C–0.2Si–1Mn–3Al wt% and the microstructures obtained directly from casting and following processing are illustrated in **Figure 46**.

21.10 TWIP Steels

There are three essential modes by which steels can be permanently deformed at ambient temperature, without recourse to diffusion. Individual dislocations whose Burgers vectors correspond to lattice vectors can glide, leading to a change in shape without altering the crystal structure or volume. In contrast, a displacive transformation (e.g. martensite or bainite) results not only in a plastic strain but also a change of crystal structure and density; this is the phenomenon exploited in the TRIP steels.

The third mode of deformation is mechanical twinning, in which the crystal structure of the steel is preserved but the twinned region is reoriented in the process. Mechanical twinning results in a much larger shear strain, $s = 1/\sqrt{2}$, compared with displacive transformations where s is typically 0.25. There is a particular class of extraordinarily ductile alloys of iron, known as the TWIP steels, which exploit mechanical twinning to achieve their properties. TWIP stands for twinning-induced plasticity.

TWIP stands for twinning-induced plasticity. The alloys are austenitic and remain so during mechanical deformation, but the material is able to accommodate strain via both the glide of individual dislocations and mechanical twinning on the $\{1\ 1\ 1\}_\gamma\langle 1\ 1\ \bar{2}\rangle_\gamma$ system. The alloys typically contain a large amount of manganese, some aluminum and silicon (e.g. Fe–25Mn–3Si 3Al wt %) with carbon and nitrogen present essentially as impurities. Larger concentrations of

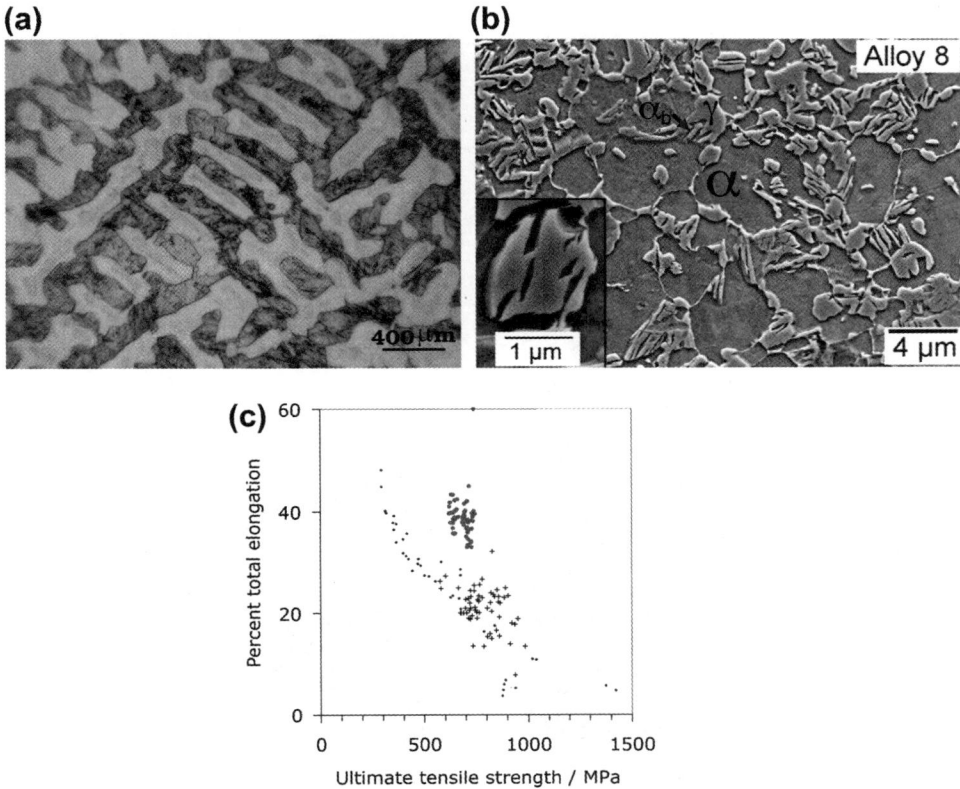

Figure 46 δ-TRIP steel. (a) Optical microstructure of the as-cast condition showing the δ-ferrite dendrites and martensite between the dendrite arms (Chatterjee et al., 2007). (b) The structure after processing to produce bainite and retained austenite (Yi et al., 2011). (c) The properties obtained (in red) compared against a range of established automotive steels. (For color version of this figure, the reader is referred to the online version of this book.)

carbon may be added to enhance strength. At high manganese concentrations, there is a tendency for the austenite to transform into ε-martensite (hexagonal close packed) during deformation. ε-martensite can form by the dissociation of a perfect $a/2\langle 0\ 1\ 1\rangle_\gamma$ dislocation into Shockley partials on a close packed $\{1\ 1\ \bar{1}\}_\gamma$ plane, with a fault between the partials. This faulted region represents a three-layer thick plate of ε-martensite. A reduction in the fault energy therefore favors the formation of this kind of martensite. The addition of aluminum counters this because it raises the stacking fault energy of the austenite. Silicon has the opposite effect of reducing the stacking fault energy, but like aluminum, it leads to a reduction in the density of the steel; the combination of Al and Si at the concentrations used typically reduces the overall density from some $7.8\ \mathrm{g\ cm}^{-3}$ to about $7.3\ \mathrm{g\ cm}^{-3}$.

The alloys have a rather low yield strength at 200–300 MN m^{-2} but the ultimate tensile strength (UTS) can be much higher, in excess of 1100 MN m^{-2}. This is because the strain-hardening coefficient is large, resulting in a great deal of uniform elongation, and a total elongation of some 60–95%. The effect of mechanical twinning is twofold. The twins add to plasticity, but they also have a powerful effect in increasing the work-hardening rate by subdividing the untwinned austenite into finer regions (**Figure 47**).

(a)

(b)

50 μm

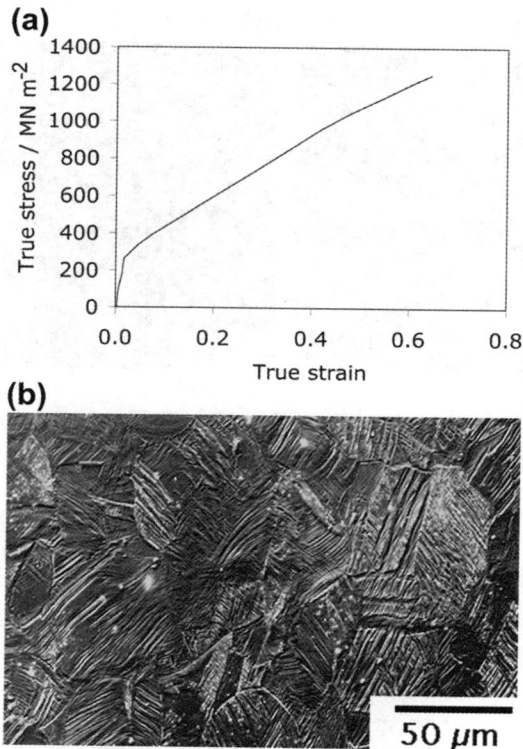

Figure 47   (a) Typical stress–strain curve for a TWIP steel. (b) Optical microstructure of a TWIP steel following deformation, showing profuse twinning (image and data courtesy of G Frommeyer, U Brüx, and P Neumann).

One major advantage of TWIP steels is that they are austenitic and they maintain attractive properties at cryogenic temperatures ($-150\,^\circ$C) and high strain rates, for example, $10^3\,\text{s}^{-1}$. They therefore have great potential in enhancing the safety of automobiles by absorbing energy during crashes.

21.11   Transformation Plasticity and Mitigation of Residual Stress

Residual stress is that which remains in a body which is stationary and at equilibrium with its surroundings. It can be very detrimental to the performance of a material or the life of a component.

Jones and Alberry conducted an elegant series of experiments to illustrate the role of transformations on the development of residual stress in steels. Using bainitic, martensitic and stable austenitic steels, they demonstrated that transformation plasticity during the cooling of a uniaxially constrained sample from the austenite phase field acts to relieve the buildup of thermal stress as the sample cools. By contrast, the nontransforming austenitic steel exhibited a monotonic increase in residual stress with decreasing temperature, as might be expected from the thermal contraction of a constrained sample. When the steels transformed to bainite or martensite, the transformation strain compensated for any thermal contraction strains that arose during cooling. Significant residual stresses were therefore found

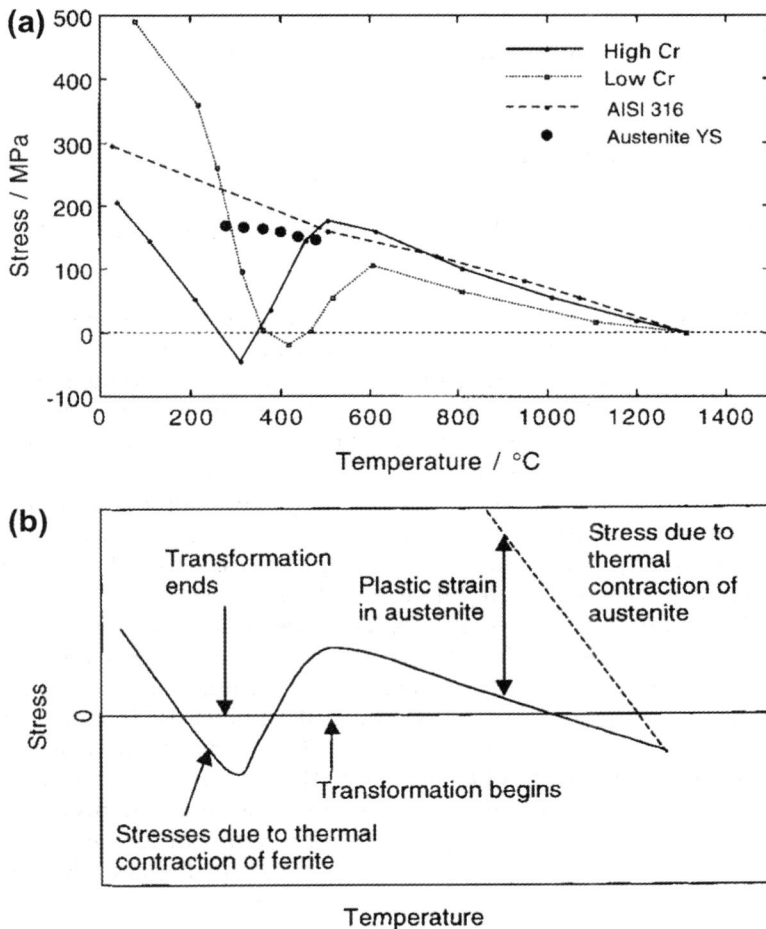

Figure 48　(a) The axial macrostress that develops in uniaxially constrained samples during cooling of a martensitic (9Cr1Mo), bainitic (2 ½ Cr1Mo) and austenitic steel (AISI 316). Also plotted are some experimental data for the yield strength of austenite in a low-alloy steel. (b) Interpretation of the Alberry and Jones experiments. The thermal expansion coefficient of austenite is much larger than that of ferrite.

to build up only after transformation was completed, and the specimens approached ambient temperature (**Figure 48**).

The experiments contain other revealing features. The thermal expansion coefficient of austenite $(1.8 \times 10^{-6}\,\mathrm{K}^{-1})$ is much larger than that of ferrite $(1.18 \times 10^{-6}\,\mathrm{K}^{-1})$, and yet, the slope of the line before transformation is smaller when compared with that after transformation is completed (**Figure 48**). This is because the austenite deforms plastically; its yield strength at high temperatures is reduced so much that the sample is unable to accommodate the contraction strains elastically. Thus, the high temperature austenite part of each curve is virtually a plot of the yield strength as a function of temperature, as is evident from the comparison versus the actual yield strength data also plotted on **Figure 48a**.

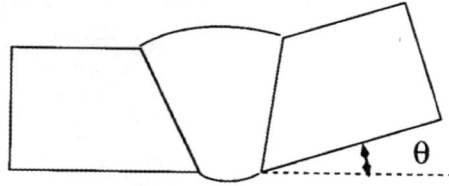

Figure 49 An illustration of the distortion caused when a pair of coplanar plates are welded together and the joint is then allowed to cool to ambient temperature.

In the region of the stress/temperature curve where transformation happens, the interpretation of experimental data of the kind illustrated in **Figure 48** is difficult. In the case of displacive transformations, the shape change because of transformation has a shear component which is much larger than the dilatational term. This will give rise to significant intergranular microstresses, part of which will be relaxed plastically. This shear component will on average cancel out in a fine-grained polycrystalline sample containing plates in many orientations so that the average type II microstress component will be zero. However, the very nature of the stress effect is to favor the formation of selected variants, in which case the shear component rapidly begins to dominate the transformation plasticity. **Figure 48a** shows that the stress can temporarily change sign as the sample cools. This is because the stress-selected variants continue to grow preferentially until transformation is exhausted.

Note that if transformation is completed at a higher temperature, then the ultimate level of stress at ambient temperature is larger, since the fully ferritic sample contracts over a larger temperature range. To reduce the residual stress level at ambient temperature requires the design of alloys with low transformation temperatures. The sort of high-strength welding alloys used for making submarine hulls tends to have very low transformation temperatures ($<250°C$). This fact may be fortuitous, but such alloys should be less susceptible to cracking induced by the development of macro residual stresses. **Figure 49** illustrates one kind of distortion found in welds, measured in terms of the angle θ through which the unconstrained plates rotate during the cooling to ambient temperature. It has now been demonstrated that the use of appropriate martensitic weld metal can dramatically reduce the distortion.

21.12 Bulk Nanostructured Steel

Nanostructured structural materials are the fashion of the day, but taking the concept to a point where it can be exploited commercially has until recently been impossible. The term "nanostructure" has unfortunately become a generic reference to a wide range of grain and precipitate structures. It is in fact possible to define it rigorously in terms of the interfacial area per unit volume, S_V, which must be large enough to make the governing length scale $\bar{L} = 2/S_V$ comparable with the diameter of carbon nanotubes, that is, of the order 20–50 nm. There is logic behind this definition, as illustrated by the data in **Table 5**. Thus, a coarse-grained structure containing nanoparticles, or a structure referred to as nanobainite but which has a scale closer to micrometers, does not have a large S_V and hence is not strong. **Figure 50** shows that the amount of surface per unit volume only becomes sensitive to the grain size when the latter is below about 50 nm.

The desire for such materials in the engineering context comes from the expectation of novel mechanical properties, particularly the stress that can safely be tolerated in service. It is difficult to invent such materials because any design must address three basic issues:

Table 5 The surface per unit volume (S_V) associated with a variety of steel-based structures. Symbols t, r, N_V refer to the true plate thickness, precipitate radius and number density per unit volume, respectively. The strength values (σ) are quoted for ambient temperature and are approximate, often estimated from hardness or microhardness

Structure		Parameters	S_V/nm^{-1}	σ/MPa
Nanostructured bainite	(Garcia-Mateo et al., 2003)	$t = 20$–40 nm	$2t^{-1} = 0.10 - 0.05$	>2100
Mechanical milling	(Kimura et al., 1999)	$\bar{L} = 200$ nm	$2\bar{L}^{-1} = 0.1$	2850
"Nanobainite"	(Timokhina et al., 2011a)	$\bar{L} = 200 - 400$ nm	$2\bar{L}^{-1} = 0.005 - 0.01$	<960 MPa
Nanoparticle strengthened	(Isheim et al., 2006)	$N_V = 1.1 \times 10^{24}$ m^{-3}, $r = 1.25$ nm	$4\pi r^2 N_V = 0.011$	800
Accumulative roll bonding	(Tsuji et al., 1999)	$\bar{L} = 420$ nm	$2\bar{L}^{-1} = 0.005$	870
Severe plastic deformation	(Valiev et al., 1996)	$\bar{L} = 100$ nm	$2\bar{L}^{-1} = 0.02$	1570
Angular processing	(Wang et al., 2004)	$\bar{L} = 200$ nm	$2\bar{L}^{-1} = 0.01$	1150

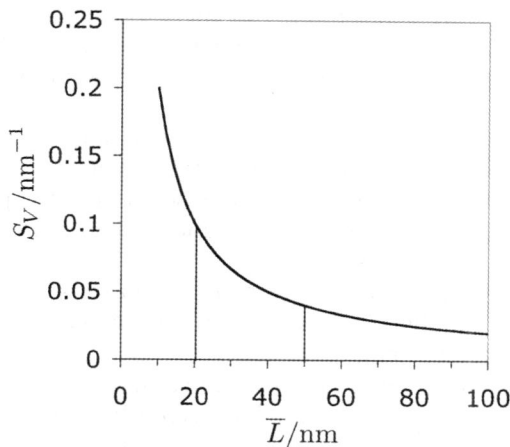

Figure 50 Grain surface per unit volume as a function of the mean lineal intercept, a measure of the grain size.

(i) It should ideally be possible to make samples which are large in all dimensions, not simply wires or thin sheets;

(ii) There are commercially available steels in which the distance between interfaces is of the order of 250–100 nm. The novelty is in approaching a structural scale in polycrystalline metals which is an order of magnitude smaller.

(iii) The material concerned must be cheap to produce if it is not to be limited to niche applications. A good standard for an affordable material is that its cost must be similar to that of bottled water when considering weight or volume.

These are formidable challenges and the process of design can begin with a consideration of how strength can be achieved. The long-range periodicity which is assumed to typify the crystalline state is in practice punctuated by defects, some of which make a significant contribution to configurational entropy and hence can exist in thermodynamic equilibrium. This kind of entropy scales with the number of entities (atoms) in the crystal and hence it is only possible to approximate perfection in small crystals. Such crystals can be strong because in the absence of defects, deformation must occur by the wholesale glide of planes over each other, rather than by the propagation of discontinuities such as dislocations. Micrometer-sized crystals of pure iron have achieved strength levels in excess of 10 GPa, although in principle the strength can exceed 20 GPa. The crystals become weaker as they are made larger, because of both the thermodynamically stabilized defects and also accidents of growth. This is fundamentally why the impressive mechanical properties of carbon *nanotubes* are not maintained (and indeed, should not be expected to be maintained) when the tubes become long.

An alternative method for manufacturing sizable strong materials is to introduce large numbers of defects such as interfaces or dislocations, which interfere with the ordinary mechanisms of slip or twinning. The defects can be introduced by deformation. Techniques which involve severe plastic deformation are limited in the shape of the final product and in the quantity that can be produced at reasonable cost. Examples include fine nanostructured wire with a strength in excess of 5 GPa; metals were subjected to equal-channel angular extrusion in which redundant work is used to achieve large plastic strains while maintaining the external shape of the object being deformed. Accumulative roll-bonding involves the repeated rolling and folding of sheet material to accomplish strain increments without thinning the sample entering the rolls; the process is suited for large-scale production but does not lead to particularly fine grains, which tend to be closer to micrometers than nanometers in size.

Thermomechanical processing is particularly suited to the large-scale production of fine-grained steels by phase transformation from the deformed parent austenite. However, the minimum ferrite grain size achieved in practice is about 1 µm, partly because the speed of production and the thickness of the steel lead to recalescence during transformation, and hence prevent the achievement of the large undercoolings needed to refine the grain size.

There are a couple of further difficulties. The ductility decreases sharply as the grain size in a poly-crystalline metal is reduced. Secondly, there is often a requirement for rapid heat treatment which becomes impractical for large components.

A recent development seems to avoid all of these difficulties, and meets the criteria outlined in the opening paragraph of this paper. A nanostructure has been achieved in large lumps of steel by phase transformation, with the design of the steel based on an understanding of the atomic mechanisms of crystal growth in the solid state.

21.12.1 Alloy Design

All of the conditions listed above can in principle be met by the phase transformation of austenite into bainite, partly because the reaction is particularly amenable to control by either isothermal or continuous cooling heat treatment. Furthermore, the transformation is displacive, that is, it leads to a shape deformation which is macroscopically an IPS with a large shear component, as illustrated in **Figure 51**. To minimize strain energy, the product is therefore shaped as thin plates, which, because they are lengthy, have a mean free slip distance of only twice the thickness. In other words, the grain size is related to the thickness rather than the length.

It has been known that the platelets of bainite become thinner as the transformation temperature is reduced, and modern analysis confirms this (Hu and Wu, 2011; Singh and Bhadeshia, 1998a). What then is the lowest temperature at which bainite can be generated? The lower limit must represent the martensite start temperature, so the answer lies in suppressing both the bainite and martensite start temperatures, and it is necessary to resort to detailed calculations. The atomic mechanism of the bainite

Figure 51 A sample of steel polished flat, austenitized and then transformed into bainitic ferrite, resulting in large upheavals of the surface, representing a shear strain of ≈ 0.46 and a dilatational strain normal to the habit plane of 0.03 (Peet and Bhadeshia, 2011).

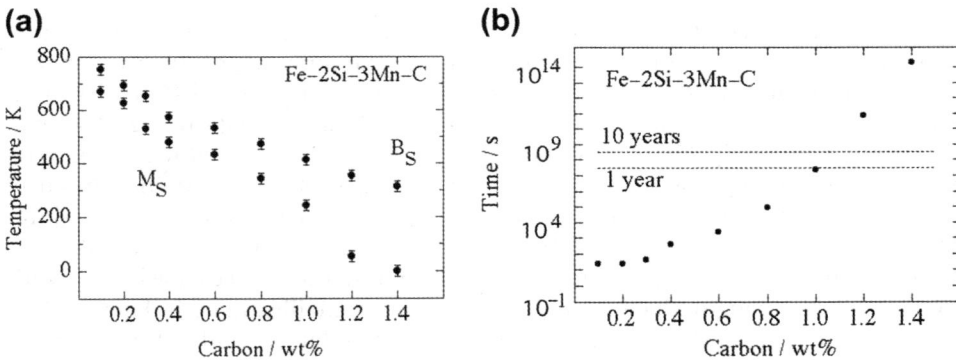

Figure 52 (a) Calculated transformation start temperatures in Fe–2Si–3Mn steel as a function of the carbon concentration. (b) The calculated time required to initiate bainite.

reaction is not yet fully understood or appreciated (Bhadeshia, 1999), but as demonstrated in a compendium[3], there has been sufficient progress to enable the quantitative design of useful steels.

Suppose that the theory is used to estimate the lowest temperature at which bainite can form. Such calculations are illustrated in **Figure 52a**, which shows for an example steel, how the B_s and M_s temperatures vary as a function of the carbon concentration. There appears to be no lower limit to the temperature at which bainite can be generated. On the other hand, the rate at which bainite forms slows down dramatically as the transformation temperature is reduced (**Figure 52b**). It may take hundreds or thousands of years to generate bainite at room temperature. For practical purposes, a transformation time of tens of days is reasonable, corresponding to a carbon concentration of about 1 wt%, in which

---

[3] Special issue of Current Opinion in Solid State and Materials Science, 8 (2004) 211–311.

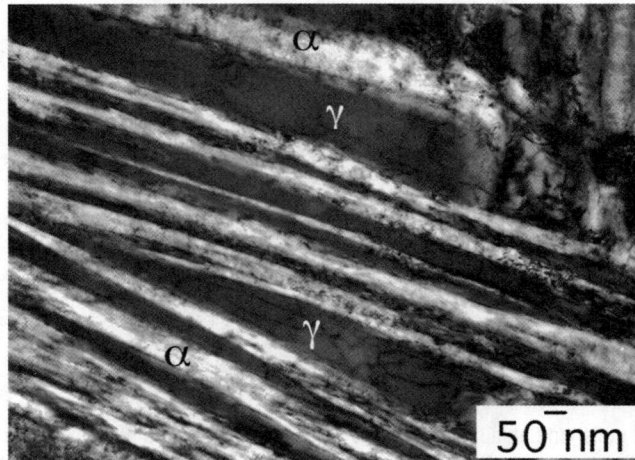

Figure 53   Fe–0.98C–1.46Si–1.89Mn–0.26Mo–1.26Cr–0.09V wt%, transformed at 200 °C for 15 days. Transmission electron micrograph.

case bainite can be generated at a temperature as low as 125 °C, which is so low that the diffusion distance of an iron atom is an inconceivable 10^{-17} m over the timescale of the experiment!

A steel designed on this basis has been manufactured and characterized (Caballero et al., 2002); **Figure 53** shows the structure obtained following isothermal transformation at 200 °C, consisting of platelets of bainitic ferrite only 200–400 Å thick, with intervening regions of the parent austenite (γ). This *retained* austenite is important because when it undergoes stress or strain-induced martensitic transformation, it enhances the work-hardening capacity of the material, thereby avoiding the usual problem of fine-grained metals where ductility diminishes as the grain size is reduced.

The bainite obtained by low-temperature transformation is harder than ever achieved, with values in excess of 700 HV. Some strength, ductility and toughness data are illustrated in **Figure 54**. The simple heat treatment involves the austenitization of a chunk of steel (at say 950 °C), followed by a gentle transfer into an oven at the low temperature (say 200 °C) to be held there for 10 days or so. There is no rapid cooling—residual stresses are avoided. The size of the sample can be large because the time taken to reach 200 °C from the austenitization temperature is much less than that required to initiate bainite. This is an important commercial advantage.

21.12.2   Excess Carbon in Bainitic Ferrite

The maximum solubility of carbon in ferrite that is in equilibrium with austenite is a little greater than 0.02 wt% at a temperature of about 600° C because of the retrograde shape of the $\alpha / \alpha + \gamma$ phase boundary (Aaronson et al., 1966; Bhadeshia, 1982a). It has been known in this context that bainitic ferrite is supersaturated with an excess of carbon (Bhadeshia and Waugh, 1982a,b; Stark et al., 1988, 1990). This carbon fails to partition into the residual austenite in spite of the fact that the process is not limited by atomic mobility (Bhadeshia, 1988). In fact, the accumulated evidence demonstrates that the carbon inherited by bainitic ferrite is reluctant to partition into the residual austenite in spite of prolonged heat treatment (Bhadeshia and Waugh, 1982b; Caballero et al., 2007, 2010a; Garcia-Mateo et al., 2004; Peet et al., 2004; Timokhina et al., 2011b). The early interpretations of these observations

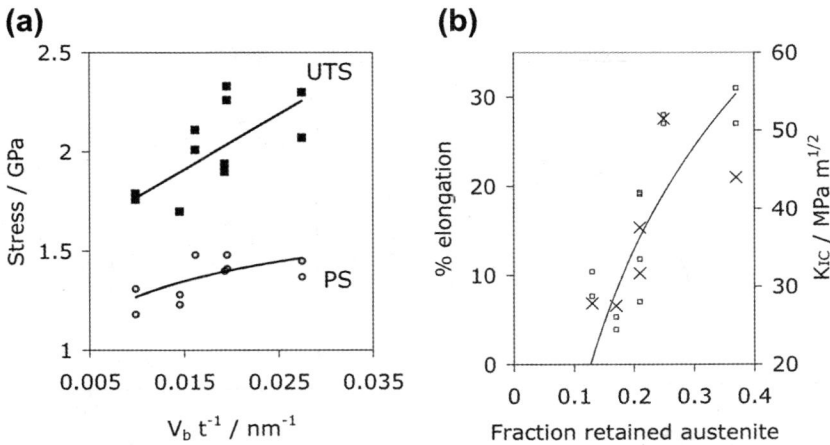

Figure 54   Some mechanical properties of two superbainitic steels. (a) The UTS and 0.2% proof strength as a function of the volume fraction of bainitic ferrite (V_b) divided by the ferrite platelet thickness t. (b) Ductility (points and curve) and toughness K_{Ic} represented as crosses.

attributed this reluctance to the presence of dislocations which trap the solute. However, recent work has conclusively shown, using the atom-probe technique, that large quantities of excess carbon remain in defect-free solid solution (Caballero et al., 2012a,b).

A possible explanation is that the tetragonality caused by the presence of carbon changes the equilibrium between austenite and body-centered cubic ferrite (Jang et al., 2012). Since the symmetry of the ferrite is changed, thermodynamic data were calculated using ab initio methods, and subsequently incorporated into phase diagram calculations. The results are shown in **Figure 55** where it is clear that the solubility of carbon in tetragonal ferrite in equilibrium with austenite is much larger than that for cubic ferrite.

It is *possible* therefore that the change in symmetry of the ferrite may explain the observed reluctance for the "excess" carbon present in bainitic ferrite to partition into the residual austenite despite prolonged heat treatment. The tetragonality may exist over a long range, but the possibility of a domain structure, such as that found in minerals which undergo cubic to tetragonal transitions, should not be ruled out.

21.12.3   Consequences on the Mechanism of Transformation

The discovery of the nanostructured bainite has in some cases revealed vital information about the choreography of atoms during the transformation:

(1) It is now clear that the crystallography of the low-temperature bainite platelets is such that there is an exceptionally large shear strain of ≈ 0.46 accompanying transformation (Peet and Bhadeshia, 2011), which compares with bainite formed at elevated temperatures where the strain is of ≈ 0.26 (Swallow and Bhadeshia, 1996). This partly explains why the plates are so thin, since the strain energy per unit volume scales with the square of the shear strain (Christian, 1958). The reasons why the crystallography changes for the lower transformation temperatures involved are not established.

(a)

(b)

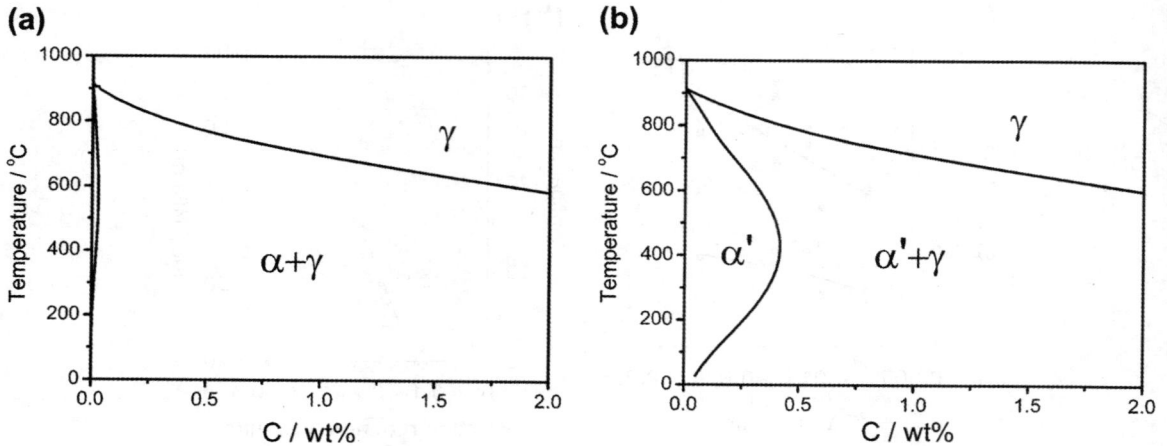

Figure 55 Binary phase diagrams of the Fe–C system allowing (a) equilibrium between body-centered cubic ferrite and austenite, (b) between body-centered tetragonal ferrite and austenite (Jang et al., 2012).

(2) Atomic resolution studies by Caballero and coworkers in particular (Caballero et al., 2010b, 2011, 2012b) and others (Timokhina et al., 2011b) have confirmed the presence of carbon at dislocations and boundaries, as postulated on the basis of early atom-probe studies (Bhadeshia and Waugh, 1982b). There is also some evidence of the presence of incredibly small carbides in a structure which is often thought to be a mixture of just bainitic ferrite and retained austenite. Other observations such as the incomplete reaction phenomenon and the configurational freezing of the substitutional solutes are all consistent with a vast body of previous research (Bhadeshia, 2001).

(3) The most exciting discovery from the atomic resolution studies is definitive proof that the excess carbon in bainitic ferrite is not just found at defects, but is present in solid solution (Caballero et al., 2012a,b). This is an insightful result which can only be explained by the displacive transformation mechanism.

(4) In contradiction to studies of the hardness of retained austenite and bainitic ferrite in high-temperature bainite (Furnemont et al., 2002), the nanohardness of the thin bainite platelets in the nanostructured steel exceeds that of retained austenite (Lan et al., 2011). The probable explanation is first the fine scale of the plates and second the excess carbon that they contain, when compared against bainite generated at higher temperatures.

(5) Since rapid cooling is unnecessary to generate the nanostructure, it has been suggested that residual stresses or the associated distortion because of heat treatment should be minimal (Bhadeshia, 2005). A recent study (Amey et al., 2012) confirms this for the nanostructured bainite, although the results are misinterpreted in terms of stress relaxation in the retained austenite, whereas the observations are macroscopic and better explained in terms of the homogeneous temperature achieved by the steel before transformation. As to the mechanism of the phase change, it has been demonstrated that the transformation plasticity, characteristic of displacive transformations and crystallographic variant selection, occurs when the nanostructure is generated under the influence of an externally applied stress (Kundu et al., 2007).

21.12.4 Weakness

The world's first bulk nanostructured metal is now in production; its essence lies in detailed solid-state phase transformation theory, meticulously researched and argued over many decades. There is now a framework of knowledge which allows such materials to be developed systematically, beginning with calculations that define experiments.

The weakness of the nanostructured bainite is that it cannot at the moment be satisfactorily welded. This limits its applications to objects which do not require joining, such as armor, shafts and bearings. It would be good to focus technological research in this area to develop practical and cheap methods which might turn out to be generically helpful to strong steels.

References

Aaronson, H.I., Domian, H.A., Pound, G.M., 1966. Thermodynamics of the austenite – proeutectoid ferrite transformation I, Fe–C alloys. Trans. Metall. Soc. AIME 236, 753–767.

Amey, C.M., Huang, H., del Castillo, P.E.J.R.-D., 2012. Distortion in 100Cr6 and nanostructured bainite. Mater. Des. 35, 66–71.

Avrami, M., 1939. Kinetics of phase change 1. J. Chem. Phys. 7, 1103–1112.

Avrami, M., 1940. Kinetics of phase change 2. J. Chem. Phys. 8, 212–224.

Avrami, M., 1941. Kinetics of phase change 3. J. Chem. Phys. 9, 177–184.

Bell, T., Owen, W.S., 1967. The thermodynamics of the martensite transformation in Fe–C and Fe–N. Trans. Metall. Soc. AIME. 239, 1940–1949.

Bhadeshia, H.K.D.H., Waugh, A.R., 1982a. An atom-probe strudy of bainite. In: Aaronson, H.I., Laughlin, D.E., Sekerka, R.F., Wayman, C.M. (Eds.), Solid–Solid Phase Transformations. Trans. Metall. Soc., Warrendale, Pennsylvania, USA, pp. 993–998.

Bhadeshia, H.K.D.H., Waugh, A.R., 1982b. Bainite: an atom probe study of the incomplete reaction phenomenon. Acta Metall. 30, 775–784.

Bhadeshia, H.K.D.H., 1981. Rationalisation of shear transformations in steels. Acta Metall. 29, 1117–1130.

Bhadeshia, H.K.D.H., 1982a. Application of first-order quasichemical theory to transformations in steels. Met. Sci. 16, 167–169.

Bhadeshia, H.K.D.H., 1982b. Bainite: overall transformation kinetics. J. Phys., Colloq. C4 43, 443–448.

Bhadeshia, H.K.D.H., 1985. Critical assessment: diffusion-controlled growth of ferrite plates in plain carbon steels. Mater. Sci. Technol. 1, 497–504.

Bhadeshia, H.K.D.H., 1988. Bainite in steels. In: Lorimer, G.W. (Ed.), Phase Transformations '87. Institute of Metals, London, UK, pp. 309–314.

Bhadeshia, H.K.D.H., 1999. The bainite transformation: unresolved issues. Mater. Sci. Eng., A 273–275, 58–66.

Bhadeshia, H.K.D.H., 2001. Bainite in Steels, second ed. Institute of Materials, London.

Bhadeshia, H.K.D.H., 2002. TRIP-assisted steels? ISIJ Int. 42, 1059–1060.

Bhadeshia, H.K.D.H., 2005. Large chunks of very strong steel. Mater. Sci. Technol. 21, 1293–1302.

Bhadeshia, H.K.D.H., 2011. Comments on "the mechanisms of the fcc–bcc martensitic transformation revealed by pole figures". Scr. Mater. 64, 101–102.

Caballero, F.G., Bhadeshia, H.K.D.H., Mawella, K.J.A., Jones, D.G., Brown, P., 2002. Very strong low temperature bainite. Mater. Sci. Technol. 18, 279–284.

Caballero, F.G., Miller, M.K., Babu, S.S., Garcia-Mateo, C., 2007. Atomic scale observations of bainite transformation in a high carbon high silicon steel. Acta Mater. 55, 381–390.

Caballero, F.G., Miller, M.K., Clarke, A.J., Garcia-Mateo, C., 2010a. Examination of carbon partitioning into austenite during tempering of bainite. Scr. Mater. 63, 442–445.

Caballero, F.G., Miller, M.K., Garcia-Mateo, C., 2010b. Tracking solute atoms during bainite reaction in a nanocrystalline steel. Mater. Sci. Technol. 26, 889–898.

Caballero, F.G., Miller, M.K., Garcia-Mateo, C., 2011. Slow bainite: an opportunity to determine the carbon content of the bainitic ferrite during growth. Solid State Phenom. 172–174, 111–116.

Caballero, F.G., Miller, M.K., Garcia-Mateo, C., Cornide, J., 2012a. New experimental evidence of the diffusionless transformation nature of bainite. J. Alloys Compd.. http://dx.doi.org/10.1016/j.jallcom.2012.02.130.

Caballero, F.G., Miller, M.K., Garcia-Mateo, C., Cornide, J., Santofimia, M.J., 2012b. Temperature dependence of carbon supersaturation of ferrite in bainitic steels. Scr. Mater.. http://dx.doi.org/10.1016/j.scriptamat.2012.08.007.

Chatterjee, S., Wang, H.S., Yang, J.R., Bhadeshia, H.K.D.H., 2006. Mechanical stabilisation of austenite. Mater. Sci. Technol. 22, 641–644.

Chatterjee, S., Murugananth, M., Bhadeshia, H.K.D.H., 2007. δ-TRIP steel. Mater. Sci. Technol. 23, 819–827.

Christian, J.W., 1958. Accommodation strains in martensite formation, the use of the dilatation parameter. Acta Metall. 6, 377–379.

Christian, J.W., 1975. Theory of Transformations in Metals and Alloys, Part I, second ed. Pergamon Press, Oxford, UK.

Christian, J.W., 1979. Thermodynamics and kinetics of martensite. In: Olson, G.B., Cohen, M. (Eds.), International Conference on Martensitic Transformations ICOMAT '79. pp. 220–234.

Furnemont, Q., Kempf, M., Jacques, P.J., Göken, M., Delannay, F., 2002. On the measurement of the nanohardness of the constitutive phases of TRIP-assisted multiphase steels. Mater. Sci. Eng., A 328A, 26–32.

Garcia-Mateo, C., Caballero, F.G., Bhadeshia, H.K.D.H., 2003. Development of hard bainite. ISIJ Int. 43, 1238–1243.

Garcia-Mateo, C., Peet, M., Caballero, F.G., Bhadeshia, H.K.D.H., 2004. Tempering of a hard mixture of bainitic ferrite and austenite. Mater. Sci. Technol. 20, 814–818.

Gerberich, W.W., Thomas, G., Parker, E.R., Zackay, V.F., 1970. Metastable austenites: decomposition and strength. In: Second International Conference on Strength of Metals and Alloys. ASM International, Ohio, USA, pp. 894–899.

Ghosh, G., Olson, G.B., 1994. Kinetics of FCC→BCC heterogeneous martensitic nucleation. Acta Metall. Mater. 42, 3361–3370.

Hillert, M., 1962. The formation of pearlite. In: Zackay, V.F., Aaronson, H.I. (Eds.), Decomposition of Austenite by Diffusional Processes. Interscience, New York, USA, pp. 197–237.

Hu, F., Wu, K., 2011. Isothermal transformation of low temperature super bainite. Adv. Mater. Res. 146–147, 1843–1848.

Imai, Y., Izumiyama, M., Tsuchiya, M., 1965. Thermodynamic Study on the Transformation of Austenite into Martensite in Fe–N and Fe–C Systems. In: Scientific Reports, A17. Research Institute of Tohoku University, 173–192.

Isheim, D., Gagliano, M.S., Fine, M.E., Seidman, D.N., 2006. Interfacial segregation at Cu-rich precipitates in a high-strength low-carbon steel studied on a sub-nanometer scale. Acta Mater. 54, 841–849.

Jacques, P.J., 2004. Transformation-induced plasticity for high strength formable steels. Curr. Opin. Solid State Mater. Sci. 8, 259–265.

Jang, J.H., Bhadeshia, H.K.D.H., Suh, D.W., 2012. Solubility of carbon in tetragonal ferrite in equilibrium with austenite. Scr. Mater. 68, 195–198.

Johnson, W.A., Mehl, R.F., 1939. Reaction kinetics in processes of nucleation and growth. Trans. Metall. Soc. AIME 135, 416–458.

Kaufman, L., Cohen, M., 1958. Thermodynamics and kinetics of martensitic transformation. Prog. Met. Phys. 7, 165–246.

Kaufman, L., Radcliffe, S.V., Cohen, M., 1962. Thermodynamics of the bainite reaction. In: Zackay, V.F., Aaronson, H.I. (Eds.), Decomposition of Austenite by Diffusional Processes. Interscience, New York, USA, pp. 313–352.

Kimura, Y., Hidaka, H., Takaki, S., 1999. Work-hardening mechanism during super-heavy plastic deformation in mechanically milled iron powder. Mater. Trans., JIM 40, 1149–1157.

Kundu, S., Hase, K., Bhadeshia, H.K.D.H., 2007. Crystallographic texture of stress-affected bainite. Proc. R. Soc., A 463, 2309–2328.

Lan, H.F., Liu, X.H., Du, L.X., 2011. Ultra-hard bainitic steels processed through low temperature heat treatment. Adv. Mater. Res. 156–157, 1708–1712.

Olson, G.B., Cohen, M., 1976. A general mechanism of martensitic nucleation, parts i–iii. Metall. Trans., A 7, 1897–1923.

Olson, G.B., Cohen, M., 1982. Stress-assisted isothermal martensitic transformation: application to TRIP steels. Metall. Trans., A 13, 1907–1914.

Pandit, A.S., Bhadeshia, H.K.D.H., 2011a. The growth of pearlite in ternary steels. Proc. R. Soc., A 467, 2948–2961.

Pandit, A.S., Bhadeshia, H.K.D.H., 2011b. Mixed diffusion-controlled growth of pearlite in binary steel. Proc. R. Soc., A 467, 508–521.

Patel, J.R., Cohen, M., 1953. Criterion for the action of applied stress in the martensitic transformation. Acta Metall., 1, 531–538.

Peet, M., Bhadeshia, H.K.D.H., 2011. Surface relief due to bainite transformation at 473 K. Metall. Mater. Trans., A 42, 3344–3348.

Peet, M., Babu, S.S., Miller, M.K., Bhadeshia, H.K.D.H., 2004. Three-dimensional atom probe analysis of carbon distribution in low-temperature bainite. Scr. Mater. 50, 1277–1281.

Puls, M.P., Kirkaldy, J.S., 1972. The pearlite reaction. Metall. Trans. 3, 2777–2796.

Singh, S.B., 1998, Ph.D. Thesis, University of Cambridge, U.K.

Singh, S.B., Bhadeshia, H.K.D.H., 1998a. Estimation of bainite plate-thickness in low-alloy steels. Mater. Sci. Eng., A A245, 72–79.

Singh, S.B., Bhadeshia, H.K.D.H., 1998b. Topology of grain deformation. Mater. Sci. Technol. 15, 832–834.

Stark, I., Smith, G.D.W., Bhadeshia, H.K.D.H., 1988. The element distribution associated with the incomplete reaction phenomenon in steels: an atom probe study. In: Lorimer, G.E. (Ed.), Phase Transformations '87. Institute of Metals, London, UK, pp. 211–215.

Stark, I., Smith, G.D.W., Bhadeshia, H.K.D.H., 1990. Distribution of substitutional alloying elements during the bainite transformation. Metall. Trans., A 21, 837–844.

Swallow, E., Bhadeshia, H.K.D.H., 1996. High resolution observations of displacements caused by bainitic transformation. Mater. Sci. Technol. 12, 121–125.

Timokhina, I., Beladi, H., Xiong, X.Y., Adachi, Y., Hodsgon, P., 2011a. Application of advanced experimental techniques for the microstructural characterization of nanobainitic steels. Solid State Phenom. 172–174, 1249–1254.

Timokhina, I.B., Xiong, X.Y., Beladi, H., Mukherjee, S., Hodgson, P.D., 2011b. Three-dimensional atomic scale analysis of microstructures formed in high strength steels. Mater. Sci. Technol. 27, 739–741.

Tsuji, N., Saito, Y., Utsunomiya, H., Tanigawa, S., 1999. Ultra-fine grained bulk steel produced by accumulative roll-bonding (arb) process. Scr. Mater. 40, 795–800.

Valiev, R.Z., Ivanisenko, Y.V., Rauch, E.F., Baudelet, B., 1996. Structure and deformation behaviour of Armco iron subjected to severe plastic deformation. Acta Mater. 44, 4705–4712.

Wang, Y., Ma, E., Valiev, R.Z., Zhu, Y., 2004. Tough nanostructured metals at cryogenic temperatures. Adv. Mater. 16, 328–331.

Watson, J.D., McDougall, P.G., 1973. The crystallography of Widmanstätten ferrite. Acta Metall. 21, 961–973.

Yang, H.S., Bhadeshia, H.K.D.H., 2009. Austenite grain size and the martensite-start temperature. Scr. Mater. 60, 493–495.

Yang, H.-S., Suh, D.W., Bhadeshia, H.K.D.H., 2012. More complete theory for the calculation of the martensite-start temperature in steels. ISIJ Int. 52, 162–164.

Yi, H.L., Lee, K.Y., Bhadeshia, H.K.D.H., 2011. Extraordinary ductility in Al-bearing δ-TRIP steel. Proc. R. Soc., A 467, 234–243.

Further Reading

Bhadeshia, H.K.D.H., Honeycombe, R.W.K., 2006. Steels: Microstructure and Properties, third ed. Butterworth-Heinemann, London.

Bhadeshia, H.K.D.H., 2001a. Bainite in Steels, second ed. Institute of Materials, London.

Bhadeshia, H.K.D.H., 2001b. Geometry of Crystals, second ed. Institute of Materials.

Christian, J.W., 2003a. Theory of Transformations in Metal and Alloys, Part I, third ed. Pergamon Press, Oxford, UK.

Christian, J.W., 2003b. Theory of Transformations in Metal and Alloys, Part II, third ed. Pergamon Press.

Davies, G., 2012. Materials for Automobile Bodies, second ed. Butterworth-Heinemann, London, UK.

DeCooman, B., 2004. Structure–properties relationship in TRIP steels containing carbide-free bainite. Curr. Opin. Solid State Mater. Sci. 8, 285–303.

Gladman, T., 1996. The Physical Metallurgy of Microalloyed Steels. IOM Communications, London.

Hillert, M., 1998. Phase Equilibria, Phase Diagrams and Phase Transformations. Cambridge University Press, Cambridge, UK.

Jacques, P.J., 2004. Transformation-induced plasticity for high strength formable steels. Curr. Opin. Solid State Mater. Sci. 8, 259–265.

Krauss, G., 2005. Steels: Processing, Structure and Performance. ASM International, Metals Park, Ohio, USA.

Llewellyn, D.T., Hudd, R.C., 1998. Steels: Metallurgy and Applications. Butterworth-Heinemann, London, UK.

Pickering, F.B., 1992. Constitution and Properties of Steels. VCH Publishers, London, UK. pp. 339–399.

Biography

Harry Bhadeshia is the Tata Steel Professor of Physical Metallurgy at the University of Cambridge and Professor of Computational Metallurgy at POSTECH. He graduated with a B.Sc. from the City of London Polytechnic, followed by a Ph.D. at the University of Cambridge. His research is concerned with the theory of solid-state transformations in metals, particularly multicomponent steels, with the goal of creating novel alloys and processes with the minimum use of resources. He is the author or coauthor of more than 600 research papers and six books on the subject.

22 Physical Metallurgy of the Nickel-Based Superalloys

R.C. Reed, Department of Materials, University of Oxford, Oxford, UK
C.M.F. Rae, Department of Materials Science and Metallurgy, University of Cambridge, Cambridge, UK

22.1 Introduction

The nickel-based superalloys are often the material of choice for high-temperature structural applications, particularly when resistance to creep and/or fatigue is needed and the risk of degradation due to oxidation and/or corrosion is severe. Their emergence can be traced to the development of the gas turbine engine, particularly those used for jet propulsion. Thus, at the time of writing they are approximately 75 years old; compared to other structural alloys based upon iron, aluminum or even titanium, they are relatively young. But superalloys are now being employed in an increasingly diverse range of applications: e.g. ultrasupercritical power plant (both nuclear and fossil fuel-fired), diesel engines and even fuel cells. Their use is particularly pronounced beyond 750 °C, since the properties of ferritic steels degrade markedly beyond this temperature. The urgency to improve the fuel economy—and associated CO_2 emissions—of such energy conversion systems is providing the technological incentive for this, underpinned by significant economic, societal and legislative pressures.

What are the physical factors that give rise to the usefulness of these alloys for elevated temperature applications? It is not the magnitude of the melting temperature of nickel (Ni) itself, for this is not particularly high and is in fact lower (at 1455 °C) than that of either iron or titanium. A first contributory factor is the face-centered cubic (FCC) crystal structure, for which the rates of the thermally activated processes controlling creep deformation are low; moreover, the FCC polymorph is stable from ambient conditions to the melting point, so that phase transformations are resisted. A second factor is the substantial solubility of alloying elements in the Ni matrix, denoted γ. Oxidation and corrosion/sulfidation is suppressed by Cr and Co, respectively. Additions of Al, Ti and Ta improve the flow stress and ultimate tensile strength, and Mo, Re and W—together with grain boundary strengtheners C and B where these are needed—confer the necessary time-dependent creep performance. This chemical complexity affords many possibilities for the design of new alloys. Third, through the use of judicious solidification processing, it has become possible to remove the grain boundaries, which are a source of weakness at elevated temperatures, and thus to deploy these alloys in single-crystal form, to advantageous effect. This situation is unique amongst structural materials. Finally and most crucially, Ni is able to support the precipitation of the $Ni_3(Al,Ti,Ta)$ phase, denoted γ', which exhibits the $L1_2$ crystal structure. In doing so, the so-called yield stress anomaly arises: the flow stress increases with temperature, an effect that can be exploited for high-temperature applications. The behavior of the planar defect structures—antiphase boundaries, stacking faults—in L1$_2$ is directly responsible for this effect. It is the influence of the L1$_2$ phase on the properties of the nickel-based superalloys, which stands out as the single, defining characteristic of the physical metallurgy of the superalloys.

In this review, it has been necessary to be somewhat selective in the choice of subject matter. Emphasis is placed on providing a summary of the microstructure/property relationships that determine the mechanical behavior of these materials. The role played by dislocations configurations— particularly in the precipitate phase—in promoting (and indeed limiting) plastic flow is considered, since the strength displayed under static and creep conditions is influenced markedly by them. Recently, use of high-resolution analytical methods (e.g. transmission electron microscopy, atom probe tomography) and advanced modeling methods has helped to elucidate the influence of such defects and the associated effects of alloy chemistry, and these findings are summarized. New time-dependent microtwinning mechanisms have been discovered in high-strength superalloys, and these are critiqued. Experimentation on single-crystal superalloys has clarified the conditions necessary for precipitate penetration and the configurations needed to cause it. When these are not satisfied, dislocation activity is restricted to the γ phase and the γ/γ' interfaces; nonconservative interfacial climb processes are then rate-controlling, and elements such as Re exert a profound influence, and a strong strain-rate dependence arises. The so-called Re-effect is now very much better understood. Fluxes of vacancies are associated with this process and the γ' phase morphology evolves via the so-called rafting effect, the mechanism of which has been clarified.

In what follows, no attempt is made to cover aspects of processing, which is nonetheless very relevant to these alloys. The reader is referred to Reed (2006) or the proceedings of the International Symposium for Superalloys—see Huron et al. (2012)—for coverage of this. Furthermore, the important topic of oxidation and corrosion behavior is not considered; this subject matter has been reviewed recently by Birks et al. (2006). The authors acknowledge the excellent reviews of Pope and Ezz (1984) and Pollock and Field (2002), which predate theirs.

22.2   Structure and Constitution of the Superalloys

Tables 1 and 2 list compositions of some important nickel-based superalloys used in the wrought and cast conditions respectively. The number of alloying elements is significant and can approach ten; consequently, at least in chemical terms, the superalloys are amongst the most complicated of structural alloys yet designed by man. Some broad trends become apparent from a consideration of the alloy chemistries. Significant amounts of Cr, Co, Al and Ti are present in most alloys. Small amounts of B, Zr and C are often included. Other elements that are added, but not to all alloys, include Re, W, Ta & Hf from the 5d block of transition metals, and Ru, Mo, Nb & Zr from the 4d block. Certain superalloys such as IN718 and IN706 contain significant proportions of Fe, and should be referred to as nickel–iron superalloys.

Thus many of the alloying elements are taken from the d-block of transition metals. Unsurprisingly, the behavior of each alloying element and its influence on the phase stability exhibited depends strongly upon its position within the periodic table. A first class of elements includes Ni, Co, Fe, Cr, Ru, Mo, Re and W, which prefer to partition to the austenitic γ and thereby stabilize it. These elements have an atomic radius very similar to Ni. A second group of elements Al, Ti, Nb and Ta have greater atomic radii and these promote the formation of ordered phases such as the compound Ni$_3$(Al,Ta,Ti), known as γ'. B, C and Zr constitute a third class, which tend to segregate to the grain boundaries of the γ phase, on account of their atomic sizes, which are very different from that of Ni. Carbide and boride phases can

Table 1 Compositions (in weight per cent) of some common superalloys used in the wrought condition

Alloy	Cr	Co	Mo	W	Nb	Al	Ti	Ta	Fe	Hf	C	B	Zr	Ni
Alloy 10	11.5	15	2.3	5.9	1.7	3.8	3.9	0.75	–	–	0.030	0.020	0.05	Bal
Astroloy	15.0	17.0	5.3	–	–	4.0	3.5	–	–	–	0.06	0.030	–	Bal
C-263	16	15	3	1.25	–	2.50	5.0	–	–	–	0.025	0.018	–	Bal
Hastelloy S	15.5	–	14.5	–	–	0.3	–	–	1.0	–	–	0.009	–	Bal
Hastelloy X	22.0	1.5	9.0	0.6	–	0.25	–	–	18.5	–	0.10	–	–	Bal
Haynes 230	22.0	–	2.0	14.0	–	0.3	–	–	–	–	0.10	–	–	Bal
Haynes 242	8.0	2.5	25.0	–	–	0.25	–	–	2.0	–	0.15	0.003	–	Bal
Haynes R-41	19.0	11.0	10.0	–	–	1.5	3.1	–	5.0	–	0.09	0.006	–	Bal
Incoloy 800	21.0	–	–	–	–	0.38	0.38	–	45.7	–	0.05	–	–	Bal
Incoloy 801	20.5	–	–	–	–	–	1.13	–	46.3	–	0.05	–	–	Bal
Incoloy 802	21.0	–	–	–	–	0.58	0.75	–	44.8	–	0.35	–	–	Bal
Incoloy 909	–	13.0	–	–	4.7	0.03	1.5	–	42.0	–	0.01	–	–	Bal
Incoloy 925	20.5	–	–	–	–	0.20	2.1	–	29.0	–	0.01	–	–	Bal
Inconel 600	15.5	–	–	–	–	–	–	–	8.0	–	0.08	–	–	Bal
Inconel 601	23.0	–	–	–	–	1.4	–	–	14.1	–	0.05	–	–	Bal
Inconel 617	22.0	12.5	9.0	–	–	1.0	0.3	–	–	–	0.07	–	–	Bal
Inconel 625	21.5	–	9.0	–	3.6	0.2	0.2	–	2.5	–	0.05	–	–	Bal
Inconel 690	29.0	–	–	–	–	–	–	–	9.0	–	0.025	–	–	Bal
Inconel 706	16.0	–	–	–	2.9	0.2	1.8	–	40.0	–	0.03	–	–	Bal
Inconel 718	19.0	–	3.0	–	5.1	0.5	0.9	–	18.5	–	0.04	–	–	Bal
Inconel 738	16.0	8.5	1.75	2.6	0.9	3.4	3.4	1.7	–	–	0.11	0.01	0.05	Bal
Inconel 740	25.0	20.0	0.5	–	2.0	0.9	1.8	–	0.7	–	0.03	–	–	Bal
Inconel X750	15.5	–	–	–	1.0	0.7	2.5	–	7.0	–	0.04	–	–	Bal
LSHR	13	21	2.7	4.3	1.5	3.5	3.5	1.6	–	–	0.030	0.030	0.050	Bal
ME3	13.1	18.2	3.8	1.9	1.4	3.5	3.5	2.7	–	–	0.030	0.030	0.050	Bal
MERL-76	12.4	18.6	3.3	–	1.4	0.2	4.3	–	–	0.35	0.050	0.03	0.06	Bal
Nimonic 75	19.5	–	–	–	–	–	0.4	–	3.0	–	0.10	–	–	Bal
Nimonic 80A	19.5	–	–	–	–	1.4	2.4	–	–	–	0.06	0.003	0.06	Bal
Nimonic 90	19.5	16.5	–	–	–	1.5	2.5	–	–	–	0.07	0.003	0.06	Bal
Nimonic 105	15.0	20.0	5.0	–	–	4.7	1.2	–	–	–	0.13	0.005	0.10	Bal
Nimonic 115	14.3	13.2	–	–	–	4.9	3.7	–	–	–	0.15	0.160	0.04	Bal
Nimonic 263	20.0	20.0	5.9	–	–	0.5	2.1	–	–	–	0.06	0.001	0.02	Bal
Nimonic 901	12.5	–	5.75	–	–	0.35	2.9	–	–	–	0.05	–	–	Bal
Nimonic PE16	16.5	1.0	1.1	–	–	1.2	1.2	–	33.0	–	0.05	0.020	–	Bal
Nimonic PK33	18.5	14.0	7.0	–	–	2.0	2.0	–	0.3	–	0.05	0.030	–	Bal

(Continued)

Table 1 Compositions (in weight per cent) of some common superalloys used in the wrought condition—cont'd

Alloy	Cr	Co	Mo	W	Nb	Al	Ti	Ta	Fe	Hf	C	B	Zr	Ni
N18	11.5	15.7	6.5	0.6	–	4.35	4.35	–	–	0.45	0.015	0.015	0.03	Bal
Pyromet 860	13.0	4.0	6.0	–	0.9	1.0	3.0	–	28.9	–	0.05	0.01	–	Bal
Pyromet 31	22.7	–	2.0	–	1.1	1.5	2.5	–	14.5	–	0.04	0.005	–	Bal
Rene 41	19.0	11.0	1.0	–	–	1.5	3.1	–	–	–	0.09	0.005	0.03	Bal
Rene 88DT	16.0	13.0	4.0	4.0	0.7	2.1	3.7	–	–	–	0.03	0.015	0.03	Bal
Rene 95	14.0	8.0	3.5	3.5	3.5	3.5	2.5	–	–	–	0.15	0.010	0.05	Bal
Rene 104	13.1	18.2	3.8	1.9	1.4	3.5	3.5	2.7	–	–	0.030	0.030	0.050	Bal
RR 1000	15.0	18.5	5.0	–	–	3.0	3.6	2.0	–	0.5	0.027	0.015	0.06	Bal
Udimet 500	18.0	18.5	4.0	–	–	2.9	2.9	–	–	–	0.08	0.006	0.05	Bal
Udimet 520	19.0	12.0	6.0	1.0	–	2.0	3.0	–	–	–	0.05	0.005	–	Bal
Udimet 630	18.0	–	3.0	3.0	6.5	0.5	1.0	–	18.0	–	0.03	–	–	Bal
Udimet 700	15.0	17.0	5.0	–	–	4.0	3.5	–	–	–	0.06	0.030	–	Bal
Udimet 710	18.0	15.0	3.0	1.5	–	2.5	5.0	–	–	–	0.07	0.020	–	Bal
Udimet 720	17.9	14.7	3.0	1.25	–	2.5	5.0	–	–	–	0.035	0.033	0.03	Bal
Udimet 720LI	16.0	15.0	3.0	1.25	–	2.5	5.0	–	–	–	0.025	0.018	0.05	Bal
Waspaloy	19.5	13.5	4.3	–	–	1.3	3.0	–	–	–	0.08	0.006	–	Bal

Table 2   Compositions (in weight per cent) of some common superalloys used in the cast condition

Alloy	Cr	Co	Mo	W	Al	Ti	Ta	Nb	Re	Ru	Hf	C	B	Zr	Ni
AM1	7.0	8.0	2.0	5.0	5.0	1.8	8.0	1.0	–	–	–	–	–	–	Bal
AM3	8.0	5.5	2.25	5.0	6.0	2.0	3.5	–	–	–	–	–	–	–	Bal
CM186LC	6.0	9.3	0.5	8.4	5.7	0.7	3.4	–	3.0	–	1.4	0.07	0.015	0.005	Bal
CM247LC	8.0	9.3	0.5	9.5	5.6	0.7	3.2	–	–	–	1.4	0.07	0.015	0.010	Bal
CMSX-2	8.0	5.0	0.6	8.0	5.6	1.0	6.0	–	–	–	–	–	–	–	Bal
CMSX-3	8.0	4.8	0.6	8.0	5.6	1.0	6.3	–	–	–	0.1	–	–	–	Bal
CMSX-4	6.5	9.6	0.6	6.4	5.6	1.0	6.5	–	3.0	–	0.1	–	–	–	Bal
CMSX-6	10.0	5.0	3.0	–	4.8	4.7	6.0	–	–	–	0.1	–	–	–	Bal
CMSX-10	2.0	3.0	0.4	5.0	5.7	0.2	8.0	–	6.0	–	0.03	0.03	–	–	Bal
EPM-102	2.0	16.5	2.0	6.0	5.55	–	8.25	0.07	5.95	3.0	0.15	0.10	0.014	0.007	Bal
GTD-111	14.0	9.5	1.5	3.8	3.0	5.0	3.15	–	–	–	–	0.10	0.004	0.02	Bal
GTD-222	22.5	19.1	–	2.0	1.2	2.3	0.94	0.8	–	–	–	0.08	0.014	0.06	Bal
IN100	10.0	15.0	3.0	–	5.5	4.7	–	–	–	–	–	0.18	0.01	0.10	Bal
IN-713LC	12.0	–	4.5	–	5.9	0.6	–	2.0	–	–	–	0.05	0.01	0.04	Bal
IN-738LC	16.0	8.5	1.75	2.6	3.4	3.4	1.75	0.9	–	–	–	0.11	0.016	0.018	Bal
IN-792	12.4	9.2	1.9	3.9	3.5	3.9	4.2	–	–	–	–	0.07	0.009	0.10	Bal
IN-939	22.4	19.0	–	2.0	1.9	3.7	–	1.0	–	–	–	0.15	0.015	0.03	Bal
Mar-M002	8.0	10.0	–	10.0	5.5	1.5	2.6	–	–	–	1.5	0.15	0.015	0.03	Bal
Mar-M246	9.0	10.0	2.5	10.0	5.5	1.5	1.5	–	–	–	1.5	0.15	0.015	0.05	Bal
Mar-M247	8.0	10.0	0.6	10.0	5.5	1.0	3.0	–	–	–	1.5	0.15	0.015	0.03	Bal
Mar-M200Hf	8.0	9.0	–	12.0	5.0	1.9	–	1.0	–	–	2.0	0.13	0.015	0.04	Bal
Mar-M421	15.0	10.8	1.8	3.3	4.5	1.6	–	2.3	–	–	–	0.18	0.019	–	Bal
MC2	8.0	5.0	2.0	8.0	5.0	1.5	6.0	–	–	–	0.1	–	–	–	Bal
MC-NG	4.0	–	1.0	5.0	6.0	0.5	5.0	–	4.0	4.0	0.1	–	–	–	Bal
MX4	2.0	16.5	2.0	6.0	5.55	–	8.25	–	5.95	3.0	0.15	0.03	–	–	Bal
Nasair 100	9.0	–	1.0	10.5	5.75	1.2	3.3	1.0	–	–	–	–	–	–	Bal
PWA1422	9.0	10.0	–	12.0	5.0	2.0	–	–	–	–	1.5	0.14	0.015	0.1	Bal
PWA1426	6.5	10.0	1.7	6.5	6.0	–	4.0	–	3.0	–	1.5	0.10	0.015	0.1	Bal
PWA1480	10.0	5.0	–	4.0	5.0	1.5	12.0	–	–	–	–	–	–	–	Bal

(Continued)

Table 2   Compositions (in weight per cent) of some common superalloys used in the cast condition—cont'd

Alloy	Cr	Co	Mo	W	Al	Ti	Ta	Nb	Re	Ru	Hf	C	B	Zr	Ni
PWA1483	12.2	9.2	1.9	3.8	3.6	4.2	5.0	–	–	–	–	0.07	–	–	Bal
PWA1484	5.0	10.0	2.0	6.0	5.6	–	9.0	–	3.0	–	0.1	–	–	–	Bal
PWA1487	5.0	10.0	1.9	5.9	5.6	–	8.4	–	3.0	–	0.25	–	–	–	Bal
PWA1497	2.0	16.5	2.0	6.0	5.55	–	8.25	–	5.95	3.0	0.15	0.03	–	–	Bal
Rene 80	14.0	9.0	4.0	4.0	3.0	4.7	–	–	–	–	–	0.16	0.015	0.01	Bal
Rene 125	9.0	10.0	2.0	7.0	1.4	2.5	3.8	–	–	–	0.8	0.11	0.017	0.05	Bal
Rene 142	6.8	12.0	1.5	4.9	6.15	–	6.35	–	2.8	–	0.05	0.12	0.015	0.02	Bal
Rene 220	18.0	12.0	3.0	–	0.5	1.0	3.0	5.0	–	–	1.5	0.02	0.010	–	Bal
Rene N4	9.0	8.0	2.0	6.0	3.7	4.2	4.0	0.5	–	–	–	–	–	–	Bal
Rene N5	7.0	8.0	2.0	5.0	6.2	–	7.0	–	3.0	–	0.2	–	–	–	Bal
Rene N6	4.2	12.5	1.4	6.0	5.75	–	7.2	–	5.4	–	0.15	0.05	0.004	–	Bal
RR2000	10.0	15.0	3.0	–	5.5	4.0	–	–	–	–	–	–	–	–	Bal
SRR99	8.0	5.0	–	10.0	5.5	2.2	3.0	–	–	–	–	–	–	–	Bal
TMS-75	3.0	12.0	2.0	6.0	6.0	–	6.0	–	5.0	–	0.1	–	–	–	Bal
TMS-138	2.9	5.9	2.9	5.9	5.9	–	5.6	–	4.9	2.0	0.1	–	–	–	Bal
TMS-162	2.9	5.8	3.9	5.8	5.8	–	5.6	–	4.9	6.0	0.09	–	–	–	Bal

also be promoted. Cr, Mo, W, Nb, Ta and titanium are particularly strong carbide formers; Cr and Mo promote the formation of borides.

The microstructure of a typical superalloy consists therefore of different phases, drawn from the following list, see (Reed, 2006):

(1) The gamma phase, denoted γ. This displays the FCC structure, and it forms a continuous, matrix phase in which other phases reside. It contains significant concentrations of elements such as Co, Cr, Mo, Ru and Re, where these are present.
(2) The gamma prime phase, denoted γ'. This usually forms as a precipitate and is often coherent with the γ-matrix; it is rich in elements such as Al, Ti and Ta. In nickel-iron superalloys and those rich in Nb, a related ordered phase γ'' is preferred instead of γ'.
(3) Carbides and borides. Carbon, often present at concentrations up to 0.2 wt%, combines with reactive elements such as Ti, Ta and Hf to form MC carbides. During processing or service, these can decompose to other species such as $M_{23}C_6$ and M_6C, which prefer to reside on the γ-grain boundaries, and which are rich in Cr, Mo and W. B can combine with elements such as Cr or Mo to form borides, which reside on the γ-grain boundaries.

Figure 1 illustrates the microstructure of a typical polycrystalline superalloy in the cast and heat-treated condition. Note the presence of the cuboidal precipitates of the γ' phase and the carbides at the grain boundaries.

Other phases can be found in certain superalloys particularly in the service-aged condition, for example, the topologically close-packed (TCP) phases such as μ, σ, Laves, etc. Excessive quantities of Cr, Mo, W and Re promote the precipitation of TCPs, which are rich in these elements, see Rae and Reed (2001). The resulting phases have a number of distinct characteristics: (1) a high and uniform packing density of atoms (2) a degree of nonmetallic, directional bonding and (3) complex crystal structures, each built up of distinct tessellated layers consisting of arrays of hexagons, pentagons and triangles, stacked into so-called Kasper coordination polyhedra. This last fact gives rise to the name *topologically close-packed* or TCP phases. Generally, the TCP phases have chemical formulae A_xB_y, where A and B are

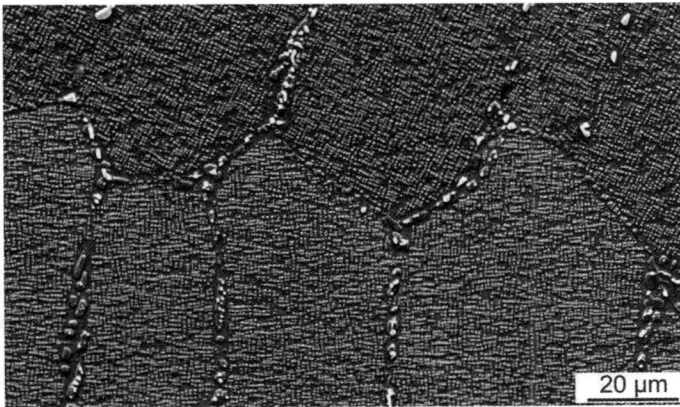

Figure 1 Microstructure of a polycrystalline nickel-based superalloy in the cast and heat-treated form. Note in particular the alignment of the γ' precipitates and the presence of carbide phase $M_{23}C_6$ at the grain boundaries. Courtesy of Paraskevas Kontis.

transition metals, such that A falls to one side of the group VIIA column defined by Mn, Tc & Re and the B atoms from the other. The μ phase is based on the ideal stoichiometry A_6B_7 and has a rhombohedral cell containing 13 atoms: examples include W_6Co_7 and Mo_6Co_7. The σ phase is based upon the stoichiometry A_2B and has a tetragonal cell containing 30 atoms: examples include Cr_2Ru, $Cr_{61}Co_{39}$ and $Re_{67}Mo_{33}$. The P phase, e.g. $Cr_{18}Mo_{42}Ni_{40}$, is primitive orthorhombic, containing 56 atoms per cell. Finally, the R phase, e.g. $Fe_{52}Mn_{16}Mo_{32}$, has a rhombohedral cell containing 53 atoms. It should be emphasized that the TCP phases can have very different compositions from those quoted here, and that the possible compositional ranges are often very wide. One should note that the compositions of the superalloys are chosen to avoid rather than to promote the formation of TCP compounds; where they form, this is evidence of microstructural instability so that properties are unlikely to be optimal.

The binary Ni–Al phase diagram is the most important one in superalloy metallurgy. It exhibits a number of ordered phases, which possess the following characteristics: (1) a significant degree of directional, covalent bonding such that precise stoichiometric relationships exist between the number of Ni and Al atoms in each unit cell, and (2) crystal structures in which Ni–Al rather than Ni–Ni or Al–Al bonds are preferred. The chemical formulae are Ni_3Al, $NiAl$, Ni_2Al_3, $NiAl_3$ and Ni_2Al_9. For each compound, the enthalpy of ordering is significant. This is demonstrated in **Figure 2**, in which the enthalpies of formation at 25 °C are plotted against the enthalpy of mixing at that temperature, see Blobaum et al. (2003). The enthalpy of formation is largest for the β-NiAl compound, which displays the CsCl crystal structure. However, in the present context, the greatest significance is the gamma prime (γ') phase, Ni_3Al, which displays the primitive cubic $L1_2$ crystal structure, see **Figure 3**, with Al atoms at the cube corners and Ni atoms at the centers of the faces. It is notable that each Ni atom has four Al and eight Ni as nearest neighbors, but that each Al atom is coordinated by twelve Ni atoms—thus Ni and

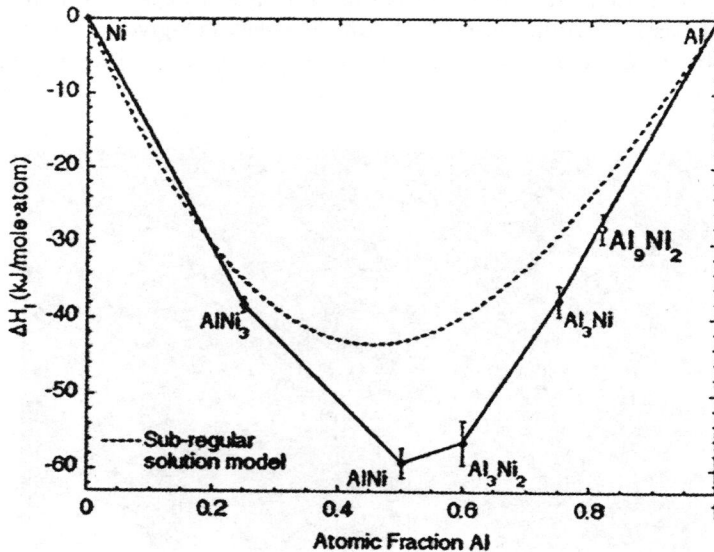

Figure 2 Variation of the enthalpy of formation ΔH_f for the various intermetallic compounds in the binary Ni–Al system. The broken line represents the enthalpy of mixing of the disordered FCC phase, with respect to the pure Ni and Al constituents. After Blobaum et al. (2003).

(a)

(b)

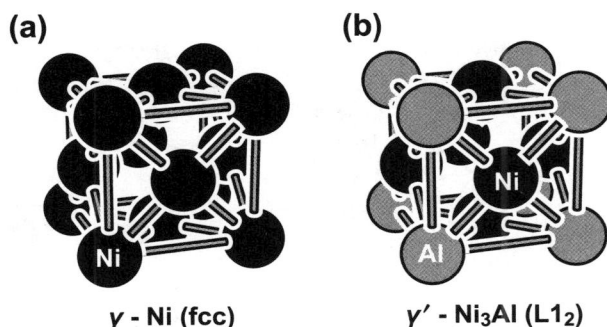

γ - Ni (fcc)

γ′ - Ni₃Al (L1₂)

Figure 3 Illustration of the critical crystal structures: (a) Ni and the face-centered cubic (disordered) lattice, and (b) Ni and Al in the ordered lattice of the Ni₃Al compound. Courtesy of Alessandro Mottura.

Al have distinct site occupancies. The ordering energy increases with decreasing temperatures; thermodynamic models indicate that it is ~ 3 kJ/mol at 1000 K.

Various pieces of evidence confirm the ordered nature of the γ' phase. Consider first the ternary phase diagrams Ni–Al–X where X = Co, Cr, Mo, W, etc, see **Figure 4**. In each case the γ' phase field is extended in a direction that depends upon the solubility of X in the γ' phase, see Ochiai et al. (1984). Elements such as Co and Pt promote γ' phase fields, which are parallel to the Ni-X axis on the ternary section, implying a constant Al fraction and thus providing confirmation that substitution for Ni on the first of the two sublattices is preferred. Elements such as Ti and Ta promote phase fields parallel to the Al-X axis, confirming that they replace Al on the second sublattice. Only rarely, e.g. for Cr, Fe and Mn, is mixed behavior observed. The behavior of the element X in this regard depends rather strongly on its size relative to Ni and Al. The lattice parameter a of stoichiometric γ' at room temperature is 0.3570 nm, which is equivalent to the Al–Al distance; this is only $\sim 1.5\%$ larger than the lattice parameter of pure Ni, which is 0.3517 nm. The Ni–Al distance is $a/\sqrt{2}$ or 0.2524 nm. Thus

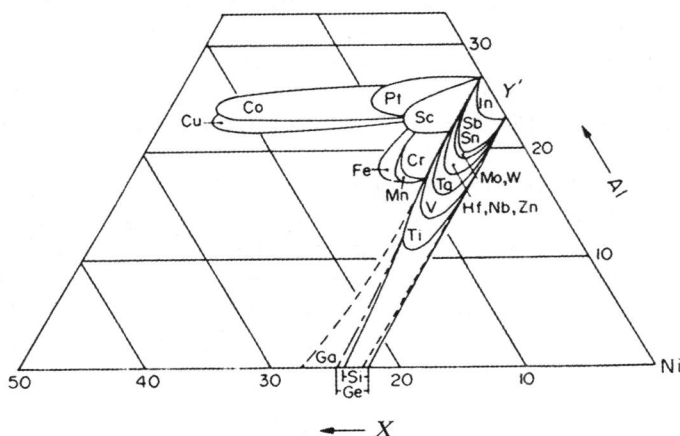

Figure 4 Superimposed ternary phase diagrams Ni–Al–X, illustrating the effect of X on the extent of the γ' phase field, which arises due to the site occupancy effect. After Ochiai et al. (1984).

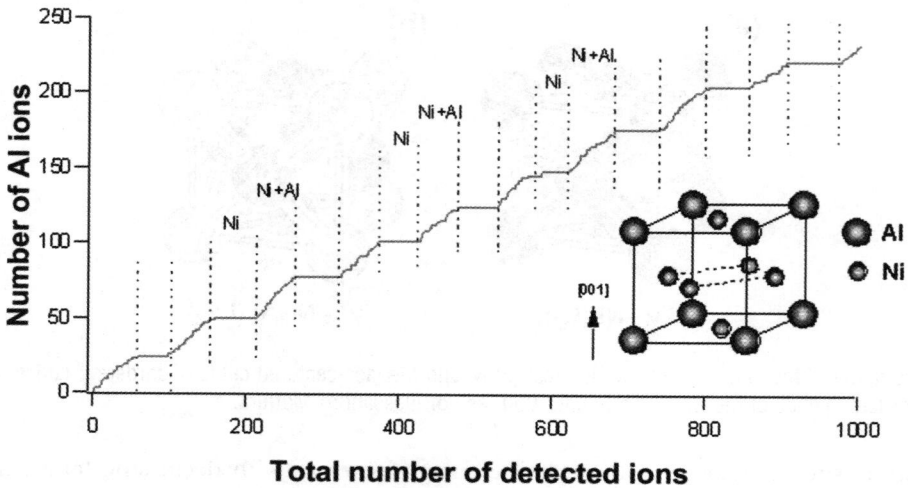

Figure 5 Ladder diagram taken from $\langle 001 \rangle$-orientated Ni_3Al single crystal, using atom probe field ion microscopy. Courtesy of Kazihiro Hono. (For color version of this figure, the reader is referred to the online version of this book.)

substitution for Al is favored by large elements such as Ta and Ti, whereas smaller atoms such as Co substitute for Ni.

Atom probe tomography has provided evidence of the ordered nature of the γ' phase. The sequential stripping of atoms from the $\{200\}$ planes allows a ladder diagram to be built up, see **Figure 5**, on which the total number of Al ions detected during analysis is plotted against the total number of ions collected. The mean gradient of the line is one quarter, but during the stripping of mixed Ni/Al and Ni planes the gradient changes to one half and zero respectively. These features of the ladder diagram provide very elegant and direct proof of the ordering of the γ' phase on the atomic scale. There is now unambiguous proof that Ni_3Al remains ordered until the melting temperature; this has been demonstrated, see Cahn et al. (1987), by doping Ni_3Al with varying amounts of Fe, which weakens the ordering; this lowers the critical ordering temperature, which can then be measured using dilatometry. The ordering temperature of pure stoichiometric Ni_3Al has then been estimated by extrapolation of the data to zero iron content; the result of $\sim 1375^\circ$ C must be termed "virtual" since it lies above the melting point.

Binary Ni–Al alloys of composition consistent with a two-phase γ/γ' microstructure exhibit γ' precipitates, which are often cuboidal in form—see Prikhodko and Ardell (2003). Transmission electron microscopy confirms that a cube–cube orientation relationship exists between the γ' precipitates and γ matrix in which they reside. Since the lattice parameters of the disordered γ and ordered γ' phases are very similar, the electron diffraction patterns exhibit some maxima that are common, e.g. from the $\{110\}$, $\{200\}$, $\{220\}$ reflections, but also others that are due only to γ', e.g. $\{100\}$, $\{210\}$ etc. The orientation relationship is described by

$$\{100\}_\gamma // \{100\}_{\gamma'}$$

$$\langle 010 \rangle_\gamma // \langle 010 \rangle_{\gamma'}, \tag{1}$$

i.e. a cube–cube orientation relationship is displayed. The γ/γ' interfaces have the $\langle 001 \rangle$ directions as their plane normals. Provided that the lattice misfit between the lattice parameters of the γ and γ' phases is not too large, the γ/γ' interface remains coherent and the interfacial energy remains low. The γ' precipitates often align along the elastically soft $\langle 100 \rangle$ direction. This preferential alignment can be seen in **Figure 1**.

22.3   Planar, Line and Point Defects in the Superalloys

The superior high-temperature performance of the superalloys—in particular the increase in yield strength with temperature—stems from the $L1_2$-ordered phase, which precipitates within the FCC matrix. To understand the origin of the considerable strengthening effect that arises, it is necessary to appreciate the possible dislocation configurations that can exist in the $L1_2$ phase. The mechanisms by which they can become locked and—just as importantly—the conditions needed to break them free are influenced strongly by the dissociation reactions by which certain 2D planar faults are formed. Much understanding has been gleaned from studies on single-phase ordered intermetallics, e.g. the Ni_3Al compound. Moreover, in the superalloys, point defects—and their interaction with solute atoms—are important at high temperatures (beyond $\sim 750\,°C$) when creep deformation is favored. Since plasticity is initiated in the matrix phase under such conditions and often restricted to it, attention is focused on vacancies and solute/vacancy interactions in the FCC phase. In this section, the important planar, line and point defects (2D, 1D and 0D defects respectively) are considered in turn.

22.3.1   Planar Faults in the $L1_2$ Phase

The replacement of 1/4 of the Ni atoms by Al to form the Ni_3Al compound reduces the symmetry of the FCC arrangement of the gamma matrix to primitive cubic. The shortest perfect Burgers vector is then $a\langle 100 \rangle$, rather than $a/2\langle 1\bar{1}0 \rangle$. An important consequence is that a perfect $a/2\langle 1\bar{1}0 \rangle\{111\}$ dislocation from the matrix phase cannot enter the $L1_2$ lattice without creating a planar defect such an antiphase boundary (APB) fault. A second $a/2\langle 1\bar{1}0 \rangle\{111\}$ dislocation is necessary to restore perfect order. Pairs of $a/2\langle 1\bar{1}0 \rangle$ dislocations can be glissile on the octahedral planes of γ', but they must be closely bound due to the high energy of the APB, which is formed between them.

In practice, the situation is considerably more complicated because further dissociation can occur into superlattice partial dislocations, which are analogous to the familiar $a/6\langle 112 \rangle$ Shockley partials of the FCC phase. A total of six distinct faults can be identified. The geometries of the faults are illustrated by projections onto the (111) or $(10\bar{1})$ planes, see **Figure 6** and **Figure 7** respectively (Vorontsov, 2011). **Figure 6** illustrates the three layers of the repeating $L1_2$ structure each plane containing 1/4 Al and 3/4 Ni atoms. These are arranged such that there are no Al–Al bonds in the perfect structure. The faults can be best visualized when viewed edge-on along the $[10\bar{1}]$ direction.

In **Figure 7**, the top left-hand panel (a) represents the perfect crystal; the panels below this illustrate incremental displacements of the lower three layers by $a/6[\bar{1}2\bar{1}]$ in the plane of the figure; the corresponding panels on the right-hand side show a displacement of $a/2[10\bar{1}]$ *perpendicular* to the plane of the figure. The APB fault (b) brings two Al atoms together as nearest neighbors and consequently, a large energy penalty is paid for its creation. **Table 3** lists experimental data for the different fault energies—including the APB fault—which can arise, as determined from the separation of partials using transmission electron microscopy or, in the case of the PE16 alloy, from the minimum size of stable dislocation loops surrounding the ordered particles. Whilst there is some uncertainty in the exact value

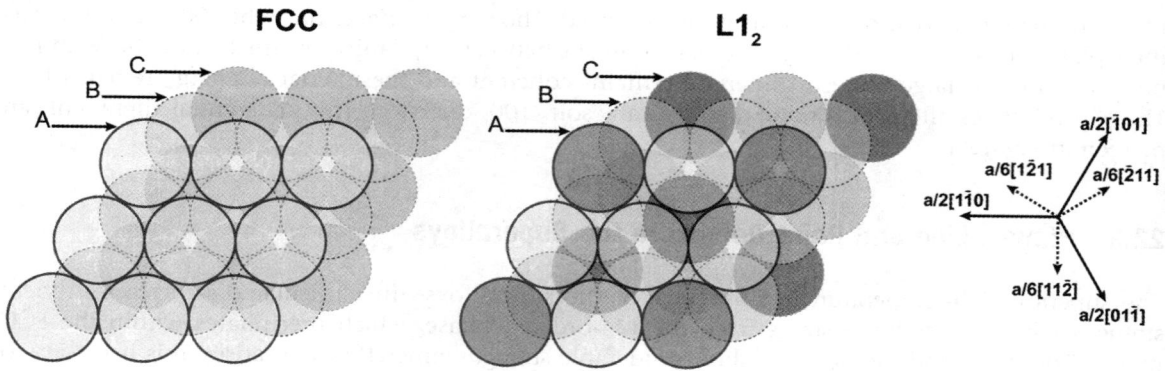

Figure 6 Illustration of the stacking of atoms in the disordered FCC and ordered crystal structures. (For color version of this figure, the reader is referred to the online version of this book.)

of the APB energy, it is large and probably in excess of 200 mJ m^{-2}. Note that a deviation in composition from the stoichiometric Ni$_3$Al toward increased aluminum would appear to increase the APB energy (Dimiduk et al., 1993).

The complex intrinsic stacking fault (CISF) (f) combines a stacking fault with Al–Al bonds to give a fault energy a little in excess of the APB energy, indicating that the chemical penalty from the Al–Al bonds is much larger than that of the structural stacking fault. This is confirmed by combining this with a displacement out of plane to give (e) the superlattice intrinsic stacking fault (SISF) for which the Al atoms maintain their separation; the fault energy is consequently much lower and comparable with stacking fault energies in FCC Ni. Panels (c) and (d) illustrate the metastable states of the extrinsic faults produced by shear on a single plane. To avoid the C–C stacking, the fourth atom layer alone is displaced to the left from C to B or B*. This allows two variants of the complex extrinsic stacking fault (CESF). The CESF-2 (h) consists of two CISFs on adjacent planes and would be expected to have exceptionally high energy. It can rearrange by shifting the fourth plane by $a/2[10\bar{1}]$ to give the much more stable super-lattice extrinsic stacking fault (SESF), see **Figure 7** (i). The shift of the fourth layer is essential for the movement of partial dislocations; nonetheless, it is not obvious how this can happen. This has been termed a shuffle by Kear (1968, 1969a, 1969b), or reordering by Kovaric et al. (2009) and will be discussed in more detail later in this review. The more stable form of the CESF, CESF-1 (g), is essentially a two-layer fault of a CISF over an APB, and, although it has a higher energy than the APB or CISF, it has been observed in superlattice structures and again plays a role in the movement of partial dislocations.

Values for the fault energies have been estimated using different modeling methods based upon for example the embedded atom method (EAM) or density functional theory (DFT). Values are given in **Table 4**. A comparison with **Table 3** indicates that the measurements and modeling are not in serious disagreement, although considerable uncertainty in the values exists. Calculations have usually been limited to the binary Ni$_3$Al compound; measurements have been made on both binary and ternary alloys. Despite the scatter in the data, the relative sizes of the fault energies are reasonably consistent. The SESF and SISF are lowest and exhibit quite similar values; the APB energy is higher and would appear to be sensitive to the Ni–Al ratio and alloying. The CESF and CISF energies are highest of all and are often not far from the sum of the APB and SISF/SESF energies, where these are known.

The APB energies are often given for the fault lying in the {111} close-packed plane. However, if the fault lies instead on the {001} plane, the nearest neighbor Al–Al bonds—which are energetically

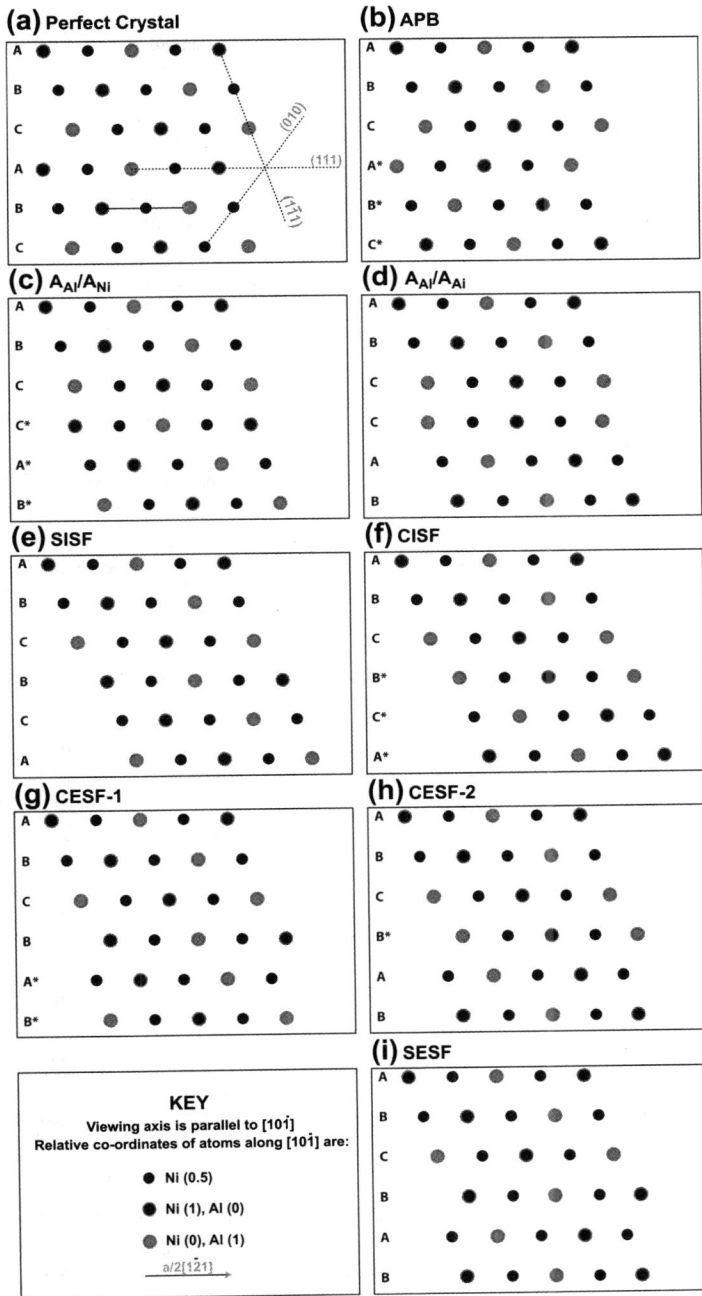

Figure 7 Possible faults in the L1$_2$ lattice, viewed along the [011] crystallographic direction. Courtesy of Vassili Vorontsov. (For color version of this figure, the reader is referred to the online version of this book.)

Table 3   Values reported for the APB, SISF and CISF fault energies (units mJ/m^2) as deduced using transmission electron microscopy

APBE {111}	APBE {100}	γ_{APB}{100}/γ_{APB}{111}	SISF	CISF	Material	Reference
300 ± 50	–		–	–	Ni$_3$Al	Taunt and Ralph (1974)
111 ± 15	90 ± 5	1.23	10 ± 5	–	Ni$_3$Al	Veyssiere et al. (1985)
169 ± 22	104 ± 11	1.62	–	–	Ni$_{77.1}$Al$_{22.9}$	Dimiduk et al. (1993)
163 ± 22	92 ± 10	1.77	–	–	Ni$_{75.8}$Al$_{24.2}$	Dimiduk et al. (1993)
190 ± 29	170 ± 25	1.12	–	–	Ni$_{74.1}$Al$_{25.9}$	Dimiduk et al. (1993)
180 ± 20	–		–	206 ± 30	Ni$_{76.0}$Al$_{24.0}$	Hemker and Mills (1993)
175 ± 15	–		–	–	Ni$_{78.0}$Al$_{22.0}$	Karnthaler et al. (1996)
166 ± 22	129 ± 13	1.29	–	–	Ni$_{75.0}$Al$_{24.1}$Sn$_{0.9}$	Dimiduk et al. (1993)
174 ± 27	146 ± 18	1.19	–	–	Ni$_{75.2}$Al$_{21.1}$Sn$_{3.7}$	Dimiduk et al. (1993)
192 ± 31	155 ± 19	1.24	–	–	Ni$_{75.6}$Al$_{23.4}$Sn$_{1.0}$	Dimiduk et al. (1993)
198 ± 37	201.36	0.99	–	–	Ni$_{75.1}$Al$_{20.8}$Sn$_{4.0}$	Dimiduk et al. (1993)
237 ± 32	200 ± 23	0.19	–	–	Ni$_{75.0}$Al$_{20.8}$Ta$_{1.0}$	Baluc et al. (1991)
250 ± 30	–		–	–	Ni$_{75.0}$Al$_{24.0}$Ta$_{1.0}$	Baluc et al. (1991)
150 ± 10	–		–	–	Ni$_{75.0}$Al$_{24.75}$Hf$_{0.25}$	Conforto et al. (2005)
173 ± 15	–		–	–	Ni$_{75.5}$Al$_{23.7}$B$_{0.7}$	Hemker and Mills (1993)
250 ± 30	–		–	–	Ni$_{76.44}$Al$_{17.40}$Ti$_{6.16}$	Korner (1988)
325^{+10}_{-104}					PE16	Baithery et al. (2001)

Table 4   Calculated estimates for the APB, SISF, CISF, SESF and CESF-1 fault energies (units mJ/m^2) made using density functional theory (DFT) and the embedded atom method (EAM)

APB	SISF	CISF	SESF	CESF-1	Method	Source
220	40	260	–		DFT	Fu and Yoo (1989)
142	13	–	–	–	EAM	Chen et al. (1986)
92	–				EAM	Foiles and Daw (1987)
220	79	267	–	–	DFT	Schoeck et al. (1999)
210	80	225	–	–	DFT	Mryasov et al. (2002)
240	147	308	–	–	DFT	Rosengaard and Skriver (1994)
126.9	248.5	392.8	–	–	EAM	Michelon and Antonelli (2008)
208.5	16.7	228.1	–	–	EAM	Michelon and Antonelli (2008)
226	11.4	198	–	–	EAM	Michelon and Antonelli (2008)
252	51	202	–	–	EAM	Mishin (2004)
188	–	–	–	–	DFT	Paxton and Sun (1998)
278	62	217	74	271	EAM	Voskoboinikov and Rae (2009)

unfavored—are avoided. This is because every other {001} layer in the L1$_2$ structure consists entirely of Ni atoms. Thus, the energy of the APB fault when it lies on {001} is much lower than when it resides on {111}. A reduction by as much as 30% has been estimated, see Fu and Yoo (1989). The lowering of the APB energy in this way—by cross-slip from {111} to {001}—is fundamental to the behavior of dislocations in the L1$_2$ phase. At high temperatures the movement of $a/2\langle 110 \rangle$ dislocations on the cube planes becomes possible despite the high Peierls stress, the advantage being that the APB thus created is of a much lower energy, see Flinn (1960). Note that the factor by which the APB energy on {111} exceeds that on {001}—as given in **Table 3**—has not been found to exceed the value of $\sqrt{3}$, which has been estimated to be necessary for spontaneous cross-slip from {111} to {001} (Paidar et al., 1984). However, as discussed in greater detail in Section 22.4, this cross-slip process is aided by the elastic anisotropy of the Ni_3Al compound (Yoo, 1986; Fu and Yoo, 1989).

22.3.2 Dislocations in the $L1_2$ Phase

The most important dislocation configuration in $L1_2$ intermetallic compounds is the pair of identical $a/2[110]$ dislocations separated by an APB. **Figure 8**—adapted from Kear (1969a)—illustrates this and other possible dissociations. Each individual dislocation also dissociates into two Shockley partials separated by a CISF; the separation of the configuration can be used to calculate both the APB and CISF energies (Vorontsov, 2011).

Another possibility is that the dislocation dissociates into two superpartials to give $a/3[211]$ with an SISF between them, see **Figure 8(b and c)**. Dissociation into two superpartials separated by an SESF fault is, in principle, also possible given the similarity in the energies of the two faults, see **Figure 8(d)**. The existence of $a\langle 110\rangle$ dislocations, which spontaneously dissociate into $a/3\langle 211\rangle$ superpartials is strongly dependent upon the composition of the alloy, and has been found to occur in some systems that display the $L1_2$ compound but not in others. For example, Pt_3Al is prone to the formation of stacking faults and phase instability with respect to the DO_{19} hexagonal structure, indicative of an extremely low SISF energy. This and similar $L1_2$ intermetallics do not exhibit anomalous yield effect;

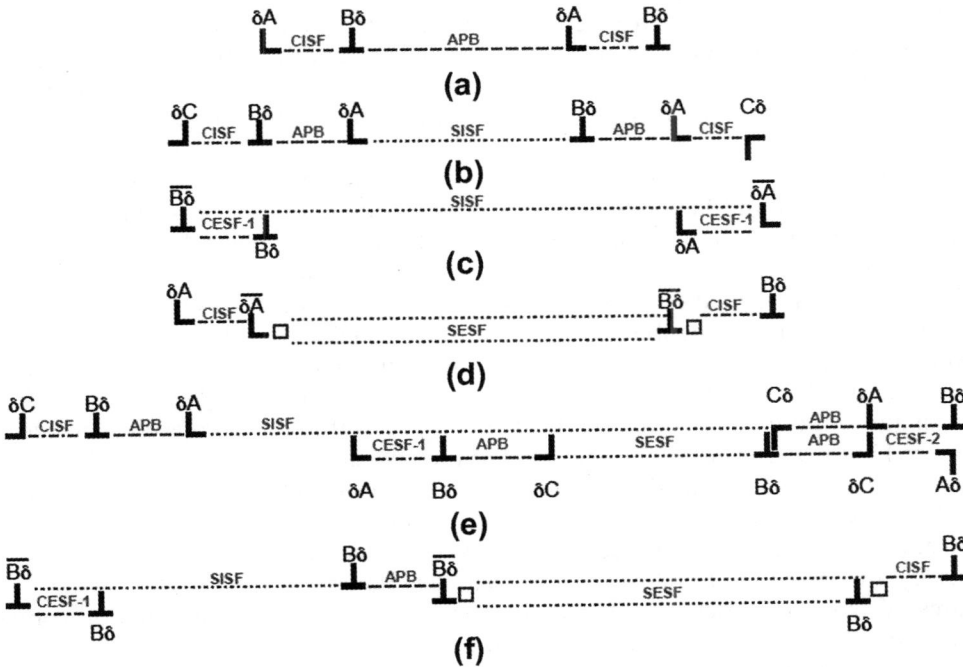

Figure 8 Configuration of possible planar faults in the $L1_2$ lattice, projected along [011]. After Kear (1969a).

instead the yield stress rises rapidly at low temperature due to the sessile nature of the superpartial dislocation, see Wee and Suzuki, 1979; Takasugi et al., 1987; Heredia et al., 1989; Yamaguchi and Umakoshi, 1990. But in the γ' phase of the nickel-based superalloys of technological relevance, there is strong evidence that this dissociation is not a spontaneous process (Rae et al., 2000).

Thus the $a/3\langle 112 \rangle$ superpartial is frequently observed in crept superalloys and forms part of the $a\langle 112 \rangle$ dislocation ribbon, which causes the primary creep effect in single-crystal superalloys, see Section 22.5. This combination of partials and faults—shown in **Figure 8(e and f)**—is able to cut through γ and γ' leaving no debris; it is an important source of deformation at temperatures close to the yield stress maximum in a number of different superalloys. In addition to its observation in the early stages of creep in single-crystal alloys, it has also been recorded recently in polycrystalline disc alloys, see Kovaric et al. (2009).

Two alternative, differing structures for the $a/3[211]$ dislocation bounding the SISF have been proposed by Kear (1969b): one consists of three partials coupled with APB and CISF faults, the other two partials separated by a CESF and spread over two planes—compare **Figures 8(b and c)**. Recent work imaging the terminating dislocations of SISFs in single-crystal alloys (Vorontsov et al., 2012a, 2012b) has shown conclusively that the former structure—**Figure 8(c)**—is correct. Similarly an SESF, itself already consisting of faults on two adjacent layers, is terminated by a CISF and two closely spaced partials, as illustrated in **Figure 8(d)**.

The implications of this are significant for the following reason. Although the SISF is itself confined to one plane, the termination is an extrinsic fault led by two partials on adjacent close-packed planes. The net Burgers vector of these is equivalent to the trailing partial but needs to be spread over the two planes to avoid the "C–C" stacking of **Figure 8e**, thus yielding the CESF-2 of **Figure 8g**. This is a very high-energy fault, but can be reduced if the single layer in the middle is displaced to remove the adjacent Al atoms in both layers. It has been recognized (Kear, 1969b) that displacement of a whole layer by shear solves this problem geometrically, but is energetically unfavorable; instead an equivalent change in the ordering of that layer by the diffusion-mediated exchange of adjacent Ni and Al atoms has been proposed. This process is termed a shuffle and is indicated on the diagrams by an empty square.

Recently, a related microtwinning mechanism has been observed in polycrystalline nickel-based superalloys such as ME3 and Rene 88DT in the intermediate temperature regime of 650 °C at low stresses and strain rates, see Kovaric et al. (2009). The reordering at the leading edge of both an SESF and an SISF needed for their advance—as outlined above—is equivalent in the envisaged microtwinning process. Activation energies have been calculated for each stage of the various diffusion pathways deemed possible, using ab initio modeling based upon a DFT. It has been concluded that a single vacancy in the fault plane can facilitate the reordering by moving along the dislocation, but not by the most highly correlated route originally suggested by Kolbe (2001); instead a slightly more convoluted path involving out-of-plane jumps is considered more probable—see **Figure 9**. The activation energies for the critical stage in this process are estimated to be similar to that for the self-diffusion coefficient of Ni. Hence, deformation by partial dislocations in the $L1_2$ lattice is an activated process with an activation energy similar to that needed for a climb process. In these alloys therefore, shearing by stacking faults—and by extension the formation of microtwins by the movements of partials on parallel planes—is therefore promoted by higher temperatures, so that a time-dependent creep-like deformation occurs. This situation contrasts with the widespread view that in FCC alloys, twinning and stacking fault-mediated shear are low-temperature mechanisms of deformation. Quantitative modeling is needed to place these ideas on a firm micromechanical basis.

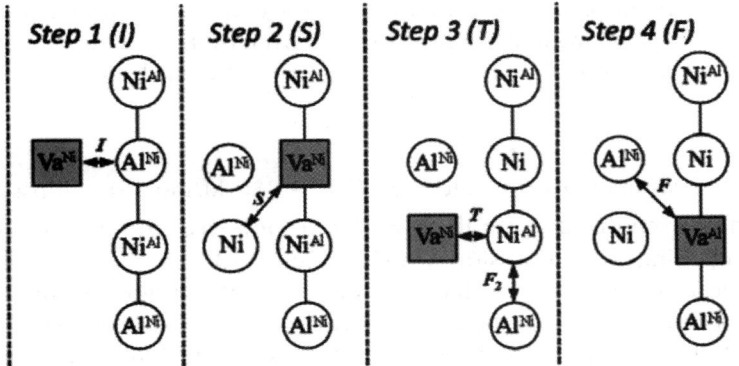

Figure 9 A four-step reordering sequence at an $a/2[211]$ dislocation that initiates with an exchange between a vacancy on a Ni site and an Al atom on a Ni site. After Kovaric et al. (2009).

22.3.3 Point Defects: Vacancies and Solute/Vacancy Interactions in the FCC Phase

The presence of vacancies in the nickel alloys is of great significance, since they mediate the diffusional flow required for (1) creep occurring at a rate dependent upon diffusional rearrangements, e.g. at dislocation cores or microtwins, (2) coarsening of the γ' precipitates on the scale of their periodicity; consequently the retardation of diffusional processes is important if the very best properties are to be attained.

Consider the self-diffusion of nickel. The diffusion coefficient D_{Ni}^{self} depends on two factors:

(1) the rate of successful atom-vacancy exchanges, denoted Γ_{Va}, which is given by

$$\Gamma_{Va} = v_o \exp\left\{ -\frac{\Delta G_{Ni-Va}^{mig}}{RT} \right\}, \qquad (2)$$

where v_o is an attempt frequency and ΔG_{Ni-Va}^{mig} is an activation energy associated with the migration of a nickel atom into a neighboring vacant site, through a distance $a/\sqrt{2}$ along the lattice vector $\langle 110 \rangle$, past the saddle point.

(2) the probability that the neighboring site is vacant. This is numerically equivalent to the fractional concentration of lattice vacancies in thermal equilibrium with the lattice f_{Va}^{eq}, and can be approximated by

$$f_{Va}^{eq} = \exp\left\{ -\frac{\Delta G_{Va}^{f}}{RT} \right\} = \exp\left\{ +\frac{\Delta S_{Va}^{f}}{R} \right\} \times \exp\left\{ -\frac{\Delta H_{Va}^{f}}{RT} \right\}, \qquad (3)$$

where ΔG_{Va}^{f}, ΔH_{Va}^{f} and ΔS_{Va}^{f} are the free energy, enthalpy and entropy of formation of a vacancy. For most metals, the preexponential term given by $\exp\{ +\Delta S_{Va}^{f}/R \}$ lies between 1 and 10, and is nearly invariant with temperature.

One has therefore to a first approximation:

$$D_{Ni}^{self} = \lambda^2 v_o \exp\left\{ +\frac{\Delta S_{Va}^{f} + \Delta S_{Ni-Va}^{mig}}{R} \right\} \times \exp\left\{ -\frac{\Delta H_{Va}^{f} + \Delta H_{Ni-Va}^{mig}}{RT} \right\}, \qquad (4)$$

where ΔG_{mig} has been partitioned into its enthalpy and entropy components, ΔH_{mig} and ΔS_{mig}, respectively. The term λ is the characteristic jump distance given by $a/\sqrt{2}$. One can see that the apparent activation energy for self-diffusion Q_{self} is composed of two terms, such that

$$Q_{self} = \Delta H_{Va}^{f} + \Delta H_{Ni-Va}^{mig}. \tag{5}$$

The experimental measurements indicate that $Q_{self} \sim 280$ kJ/mol—see Jonsson (1995). Hence $\Delta H_{Ni-Va}^{mig} \sim 120$ kJ/mol, given that ΔH_{Va}^{f} is about ~ 160 kJ/mol—see Janotti et al. (2004). This value is associated with the opening of the diffusion window along $\langle 110 \rangle$.

The presence of vacancies influences the rate of diffusion of substitutional solutes in nickel. The rates of interdiffusion of a number of important transition metals with Ni have been determined experimentally by Karunaratne et al. (2000) and Karunaratne and Reed (2003), and correlations between diffusion rates, activation energies and atomic numbers identified. Elements near the center of the d-block of the periodic table, e.g. Re & Ru, diffuse several orders of magnitude more slowly than those from the far west or far east of the d-block, e.g. Hf & Au—see **Figure 10**. These experimental results have been rationalized using quantum-mechanical modeling by Janotti et al. (2004) and Krcmar et al. (2005) using DFT; these findings disprove the traditional view that diffusion is least rapid when the lattice misfit between solute and solvent atom radii is the greatest. Such calculations have allowed an evaluation of the activation energies for solute/vacancy exchange, denoted E_b, and the vacancy formation energy for a solute/vacancy complex, denoted E_{X-Va}^{f}. The results indicate that E_{X-Va}^{f} does not vary strongly as one crosses the d-block and by no more than 40 kJ/mol; in fact the magnitudes of E_{Hf-Va}^{f} and E_{Au-Va}^{f} are close to 120 kJ/mol and thus the vacancy formation energy in pure Ni—implying that the solute-vacancy binding energy for elements such as Hf and Au is negligible. The Re–Va binding energy is about +40 kJ/mol. Thus the dependence of E_{X-Va}^{f} on atomic number is relatively weak. Instead, the major contribution to the variation of the interdiffusion coefficient with atomic number arises from the diffusion energy barrier E_b. This is significant for elements such as Re and Ru. Alloying with these elements, which are amongst the densest in the d-block owing to their electronic structure and many unpaired electrons, causes directional and incompressible Ni–Re and Ni–Ru bonds, which do not favor solute-vacancy exchanges. This effect dominates over any influences of atomic radius, misfit strain and differences in the solute-vacancy binding energy. These effects help to explain the influence of Re on the kinetics of thermally activated phenomena in these alloys. The so-called "Re effect" is considered further in Section 22.5.

22.4   Strengthening Mechanisms in Nickel-Based Superalloys

Traditionally, when considering the strengthening mechanisms that operate in superalloy metallurgy, emphasis has been placed on the analysis of pairs of dislocations, their geometrical interaction with the γ' precipitates and thus the role played by the APB fault. In what follows, this seminal approach is dealt with first. Next, the temperature dependence of the strengthening is addressed and in particular the role played by the γ' phase emphasized. It is this effect—and the so-called yield anomaly exhibited the L1$_2$ crystal structure—which is important in rationalizing the unique property of the superalloys: their maintenance of significant strength to a temperature of 700 °C or higher. Finally, consideration is given to the high-temperature microtwinning mode of deformation, which has been observed in the superalloys, which is considerably distinct from the traditional view of plasticity in these materials.

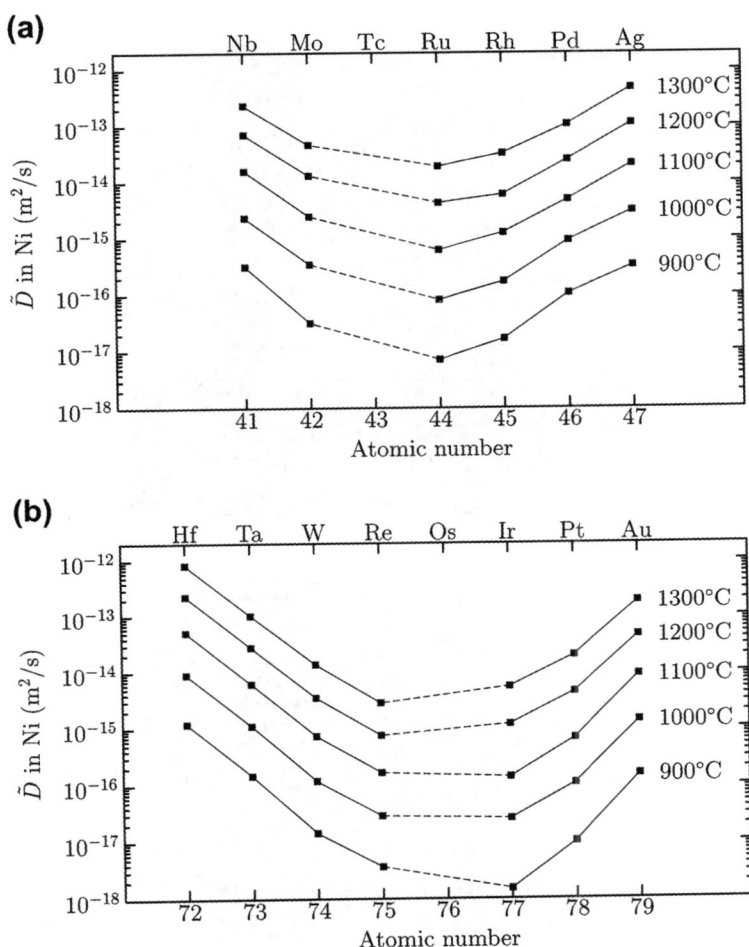

Figure 10 Interdiffusion coefficients measured from Ni/Ni–X diffusion couples for (a) X corresponding to the 4d block of transition metals and (b) the 5d block of transition metals. Note the very low rate of interdiffusion of Re. After Karunaratne and Reed (2003).

22.4.1 Strengthening by Particles of the Gamma Prime Phase

A large energy penalty is associated with a single $a/2\langle 1\bar{1}0\rangle\{111\}$ entering the γ' phase, due to the size of the APB energy. Therefore, in practice, dislocations travel through the γ/γ' structure in pairs, with a second $a/2\langle 1\bar{1}0\rangle\{111\}$ dislocation removing the APB introduced by the first. This is supported by the TEM evidence—see **Figure 11** (Nembach, 1997). Although detailed calculations are required for an exact estimate, the particle cutting stress is expected to be of the order γ_{APB}/b where γ_{APB} is the APB energy and b is the Burgers vector. With $\gamma_{APB} \sim 0.2$ Jm^{-2} and $b = 0.25$ nm, a first estimate of the cutting stress is $0.2/0.25 \times 10^{-9}/10^6$ or 800 MPa. Thus this effect is usually considerable, and contributes the majority of the strengthening effect in this class of alloy. More sophisticated estimates respect the

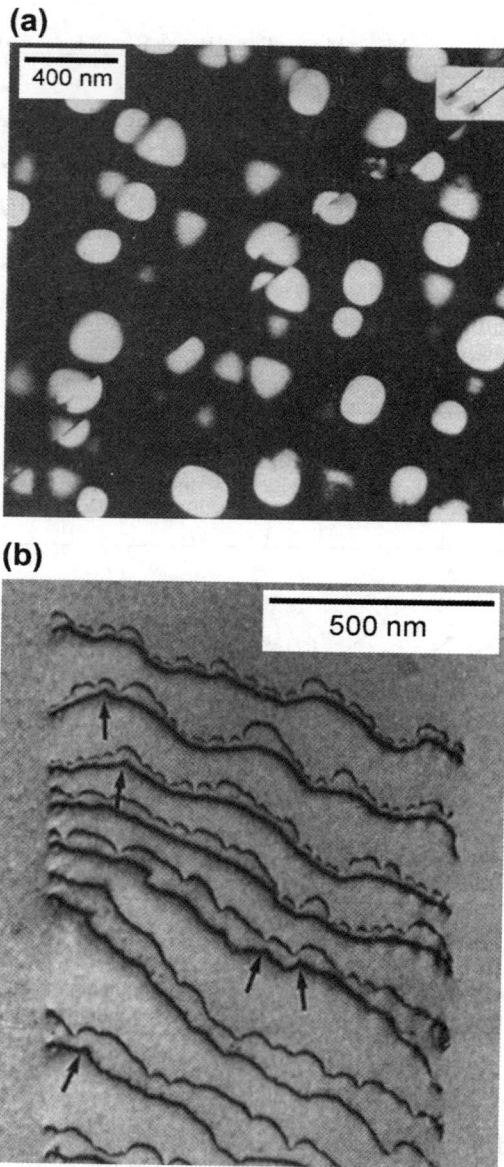

Figure 11 Transmission electron micrographs (Nembach et al., 1985; Nembach, 1997) illustrating dislocations traveling through the γ/γ' microstructure in pairs: (a) dark-field micrograph of sheared particles in Nimonic 105, of size about 140 nm; (b) passage of pair of edge dislocations in Nimonic PE16, $r = 8$ nm and $f = 0.09$.

balancing of forces and account properly for the geometry of the dislocation/particle interactions. The details now follow.

22.4.1.1 The Case of Weakly Coupled Dislocations

To develop expressions for the strengthening expected from γ' particles of volume fraction f, the case of "weakly coupled" dislocations—see (Ardell, 1985)—is considered first—see **Figure 12a**. By "weak" one means that the spacing of the two paired dislocations is large in comparison with the particle diameter; consequently, the second trailing dislocation is some way behind the first, so that faulted particles lie between the two. The γ' particles are modeled as spherical and the volume fraction of γ' is taken to be small. The particles intersecting the first and second dislocations are assumed to be spaced at intervals λ_1 and λ_2 respectively—close to the so-called "Friedel spacing". The radii of the particles along the first and second lines are r_1 and r_2 respectively.

The onset of particle shearing is controlled by the following forces, which act on the pair of dislocations: (1) the forces $\tau b \lambda_i$ ($i = 1,2$) driving the particle shearing process, due to the applied shear stress; (2) the elastic force per unit length R, repulsive in sign since the dislocations are of the same sense, acting to keep the pair separated and (3) the pinning forces F_i ($i = 1,2$), in magnitude equal to $2\gamma_{APB}r_i$, which are a consequence of the APB energy. Following Ardell (1985); Huther & Reppich (1978) the force balances are written

$$\tau b \lambda_1 + R\lambda_1 + F_1 = 0 \tag{6}$$

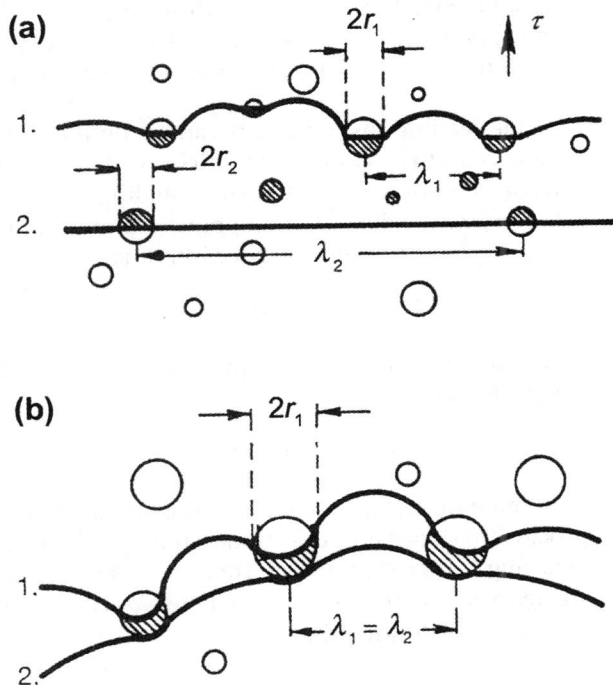

Figure 12 Particles of ordered γ' being sheared by pairs of dislocations: (a) weak pair-coupling, and (b) strong pair-coupling. Adapted from Huther and Reppich (1978).

for the first dislocation and

$$\tau b \lambda_2 - R \lambda_2 + F_2 = 0 \tag{7}$$

for the second, where $F_1 = -2\gamma_{APB}r_1$ and $F_2 = +2\gamma_{APB}r_2$. The terms F_1 and F_2 are of opposite signs since the first dislocation introduces the APB, whereas the second dislocation removes it. The net shear stress required for cutting, denoted τ_c, is found by elimination of R from Eqns (6) and (7), yielding (Huther and Reppich, 1978):

$$\tau_c = \frac{\gamma_{APB}}{2b}\left[\frac{2r_1}{\lambda_1} - \frac{2r_2}{\lambda_2}\right]. \tag{8}$$

Equation (8) indicates that τ_c depends critically upon the difference between the ratios $2r_1/\lambda_1$ and $2r_2/\lambda_2$, and therefore the way in which the dislocations interact with the γ' particles.

Thus, the onset of particle shearing is controlled by the behavior of the leading dislocation, which will in practice be held up by closely spaced, larger particles. It must bow in order to overcome the pinning forces due to the particles, i.e. to develop a component of the line tension in a direction opposite to that in which the pinning forces act. On the other hand, the trailing dislocation can be assumed to be straight; thus to a good approximation $2r_2/\lambda_2 = f$. An estimate of the ratio $2r_1/\lambda_1$ is necessary. This can be made by considering the conditions necessary for the leading dislocation to break away or "unzip" from a particle, see **Figure 13a**. The leading dislocation (initially pinned at X, Y and Z) breaks away from Y and becomes pinned at Y', bowing out to another configuration compatible with the stress field and sweeping out an area A on the slip plane. The new configuration has a radius of curvature denoted R.

In practice, the curvature of the leading dislocation is caused by the resistance imparted by the particles—therefore the force balance in the vicinity of a particle needs to be considered carefully—see **Figure 13b**. Each particle is considered to be a point obstacle providing a pinning force $F = 2\gamma_{APB}r$—the largest which can be exerted, consistent with an "unzipping" condition. The condition for static equilibrium is given by resolving forces in the vertical direction; it is

$$2T\cos\{\phi/2\} = F, \tag{9}$$

where $\phi = \pi - 2\theta$ and T is the line tension, which is approximately $\frac{1}{2}Gb^2$. The angles ϕ and θ are defined in **Figure 13b**. Also by geometry

$$\lambda_1 = 2R\sin\theta. \tag{10}$$

In order to proceed to an estimate of $2r_1/\lambda_1$, a relationship is required between λ_1 and the area A swept out by the unzipping process. To derive this, one models the random array of particles on the slip plane as a "square lattice" of spacing L, such that $A = L^2$. Then the number of particles per unit area N_s equals L^{-2} where $L = A^{-1/2}$; the term N_s is equal to $N_V \times 2r$ where N_V is the number of particles per unit volume, consistent with $N_V = f/\frac{4}{3}\pi r^3$. It follows that

$$L = \left(\frac{2\pi}{3f}\right)^{1/2} r. \tag{11}$$

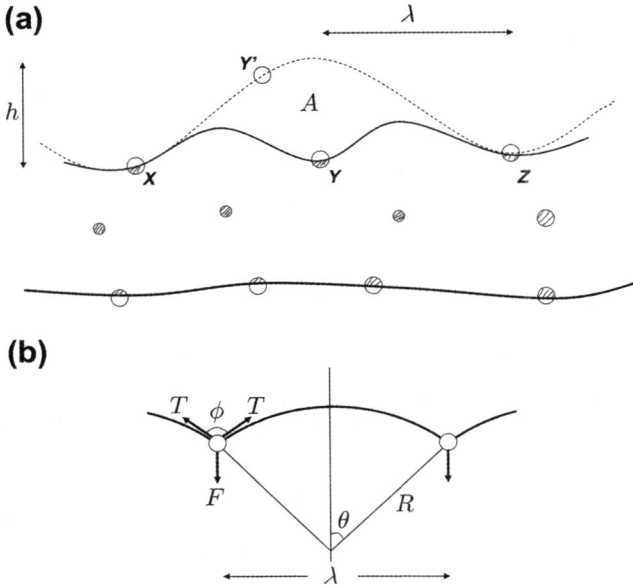

Figure 13 For weakly coupled dislocations, (a) illustration of the unzipping of the leading dislocation from the pinning point Y to Y'—an area A on the slip plane is swept out; (b) geometry for force balance calculation between the dislocation line tension and pinning force due to order hardening, in which the particles are assumed to be point sources.

The area A can be approximated by $h\lambda_1$, where h is defined by the semicircle passing through X, Y' and Z, see **Figure 13a**; this has radius R. From the property of a circle one has $R^2 = \lambda_1^2 + (R - h)^2 = \lambda_1^2 + R^2 - 2Rh + h^2$; if $h \ll R$ then $\lambda_1^2 = 2Rh$. Consequently

$$\lambda_1^3 = 2Rh \times \frac{L^2}{h} = 2L^2R = L^2\lambda_1/\cos\{\phi/2\} \tag{12}$$

and thus

$$\left(\frac{\lambda_1}{L}\right)^2 = \frac{1}{\cos\{\phi/2\}} = \frac{2T}{F} \tag{13}$$

so that

$$\lambda_1 = \left(\frac{2T}{F}\right)^{1/2} \times L \tag{14}$$

or making use of Eqn (11)

$$\lambda_1 = \left(\frac{2T}{F}\right)^{1/2} \times \left(\frac{2\pi}{3f}\right)^{1/2} \times r_1. \tag{15}$$

Equations (14) and (15) are the critical equations for the Friedel spacing λ_1; as expected it is proportional to the square lattice spacing L, which is in turn inversely proportional to the square root of the unzipped area A.

Finally, an estimate of the critical resolved shear stress (CRSS) τ_c can be derived. One inserts Eqn (15) into Eqn (8), and places F equal to $2\gamma_{APB}r_1$. The result is

$$\tau_c = \frac{\gamma_{APB}}{2b}\left[\left(\frac{6\gamma_{APB}fr}{\pi T}\right)^{1/2} - f\right], \tag{16}$$

where r_1 has been replaced by r since the result will be valid for the general case of a uniform distribution of particles.

One can see that the factor $\frac{1}{2}[(6\gamma_{APB}fr/(\pi T))^{1/2} - f]$ modifies the first estimate of γ_{APB}/b for the cutting stress. In practice, for large particles $(6\gamma_{APB}fr/(\pi T))^{1/2} \gg f$ so that to a good approximation $\tau_c \propto (fr)^{1/2}$. Thus significant fractions of larger particles are expected to promote hardening. The strong dependence upon γ_{APB} arises from the direct effect of the drag on the dislocations creating the fault, together with the increased number of precipitates intersected as the strength of pinning increases. Data (Ardell, 1985) for the age hardening of monocrystalline Ni–Al alloys tested in compression have confirmed that the $r^{1/2}$ dependence expected as $f \to 0$. However, as $r^{1/2}$ becomes large the linear dependence breaks down; there are a number of reasons for this behavior. First, the point obstacle approximation is no longer valid, since the volume fraction f of γ' is then large. Second, the dislocation pairs become more strongly coupled; an analysis of the kind given in the next section is required.

22.4.1.2 *The Case of Strongly Coupled Dislocations*
When the γ' particles are large—as will be the case for the overaged condition—the spacing of the dislocation pairs becomes comparable to the particle diameter. Thus any given particle may contain a pair of dislocations that are now "strongly coupled"—see **Figure 12b**. In this case, the behavior becomes critically dependent upon the elastic repulsive force per unit length R since this must be overcome if the trailing dislocation is to enter the γ' particle. The Friedel spacings λ_1 and λ_2 are now equal, and can be approximated by the square lattice spacing L; the terms r_1 and r_2 are different. The drag on the dislocation pair comes from the difference in the lengths of the two dislocation segments as they pass through the precipitate, and hence the forces on the dislocations are due to the effect of the APB energy.

Equation (8) remains valid for this situation. The force on the leading dislocation $F_1 = 2r_1\gamma_{APB} = \ell\{x\}\gamma_{APB}$ increases as it moves further into the precipitate until a maximum is reached as it passes through the diameter of the precipitate. The trailing dislocation experiences the opposite force $F_2 = -2r_2\gamma_{APB}$ leading to a maximum net force when the trailing dislocation just touches the precipitate. Thus that the maximum pinning effect occurs when x is half the effective cross-section of the precipitate; at this point one has $r_2 = 0$ and hence

$$\tau_c = \frac{\gamma_{APB}}{2b}\frac{2r_1}{\lambda_1} = \frac{\gamma_{APB}}{2b}\frac{2r_1}{L} \tag{17}$$

If the particles are spherical, the term $2r_1$ or $\ell\{x\}$ is given by $2(2rx - x^2)^{1/2}$ where r is the particle radius. The forces F_1 and F_2 are of opposite sign and have magnitudes that depend critically upon the line length $\ell\{x\}$ of the corresponding dislocation inside the particle, see **Figure 14**. The force–distance plot is as depicted in **Figure 15**. Substitution for $2r_1$ and L via Eqn (11) gives

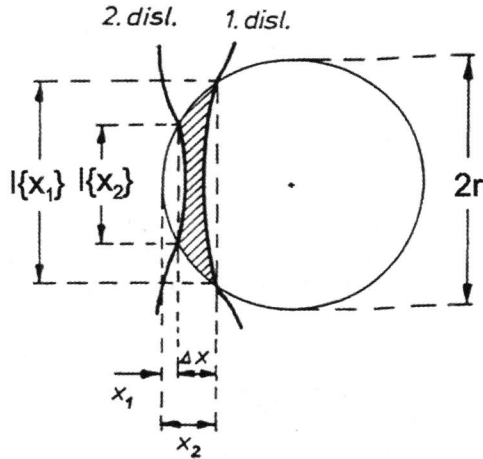

Figure 14 Representation of the strong interaction between a pair of dislocations and a γ' particle, assumed spherical. The hatched area denotes an APB. Adapted from Huther and Reppich (1978).

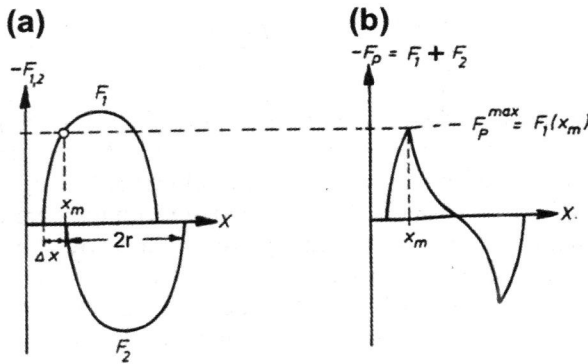

Figure 15 Force–distance profiles of dislocations cutting an ordered γ' particle, assumed spherical: (a) variation of the forces F_1 and F_2 acting on the first and second dislocations with distance x, measured from the point of penetration; (b) net force on the pair, their spacing assumed to be constant. Adapted from Huther and Reppich (1978).

$$\tau_c = \frac{\gamma_{\text{APB}}}{2br}\left(\frac{3f}{2\pi}\right)^{1/2}\ell\{x\} \tag{18}$$

An estimate of the spacing x is required at the point when the leading dislocation is inside the precipitate and the trailing one is just touching it. Combining Eqns (6) and (7) with $F_2 = 0$—since the trailing dislocation is outside the precipitate—and assuming $\lambda_1 = \lambda_2 = L$ gives

$$\tau b + R - \gamma_{\text{APB}} = 0 \tag{19}$$

for the leading dislocation and

$$\tau b - R = 0 \tag{20}$$

for the trailing one. Hence

$$2\tau b = \gamma_{APB} = 2R. \tag{21}$$

Note that $R = \gamma_{APB}/2$; the dislocation spacing halves and R doubles to γ_{APB} as soon as the second dislocation enters the precipitate—hence the curvature of the trailing dislocation in **Figure 14**.

Dislocation theory provides an estimate of the repulsive force between two similar parallel dislocations as

$$R = w \frac{Gb^2}{2\pi x} = \frac{\gamma_{APB}}{2}, \tag{22}$$

where is a dimensionless constant that is expected to be of order unity. This gives

$$x = w \frac{Gb^2}{\pi \gamma_{APB}} \tag{23}$$

and hence

$$\tau_c = \sqrt{\frac{3}{2}} \left(\frac{Gb}{r} \right) f^{1/2} \frac{w}{\pi^{3/2}} \left(\frac{2\pi \gamma_{APB} r}{w Gb^2} - 1 \right)^{1/2} \tag{24}$$

It is instructive to insert some estimates into these expressions. Taking $\gamma_{APB} \sim 0.2$ Jm^{-2}, $f = 0.3$, $r = 25$ nm, $G = 80$ GPa and $b = 0.25$ nm, one finds that $\tau_c = 280$ MPa. This is generally significantly lower than the Orowan bowing stress given by Gb/L—suggesting that Orowan mechanism is unlikely to be the major cause of strengthening in nickel alloys, at least at room temperature. Finally, as $r \to \infty$, an $r^{-1/2}$ dependence for τ_c is expected, so that the resistance provided by strongly coupled dislocations disappears at large particle sizes. This is in contrast to the $r^{1/2}$ dependence for weakly coupled dislocations, which is found provided that f is small; clearly in that case the strengthening increases monotonically as r increases. In fact, it can be shown that the transition from "weak" to "strong" coupling occurs when $r \sim 2T/\gamma_{APB}$ yielding a peak-strength of $\sim \frac{1}{2} \gamma_{APB} f^{1/2}/b$.

Many studies have confirmed that a maximum increment in hardening occurs in the superalloys for a γ' particle size, which lies at the transition from weak- to strong-coupling. For example, Reppich (1982) and Reppich et al. (1982) have measured the CRSS at room temperature for the nickel-based superalloys PE16 and Nimonic 105, aged such that the volume fraction f and mean particle diameter r varied substantially. These systems show a degree of solid-solution strengthening, estimates for which were subtracted from the total strengthening displayed. The results are given in **Figure 16**. For PE16, an optimum r was found to be in the range 26–30 nm. For Nimonic 105, the range was found to be 55–85 nm. Note in particular that strengthening by Orowan bowing is expected only at very large particle radii, greater than those found in practice.

These concepts have been used for practical applications to design heat treatment schedules for commercial superalloys, which places the γ' particles at a size that confers hardening at the transition from weak- to strong-coupling—and thus the maximum, which can be achieved, see Jackson & Reed (1999).

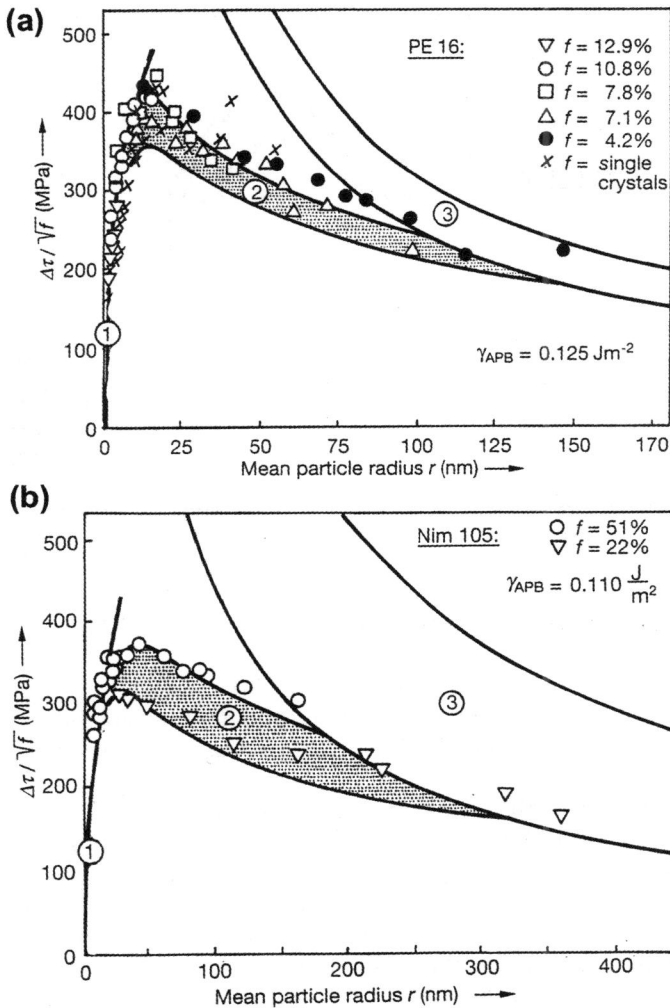

Figure 16 Data (Reppich, 1993) for the increase in critical resolved shear stress (CRSS) (divided by the square root of the volume fraction of the particles) as a function of the mean particle radius for two nickel-base superalloys: (a) Nimonic PE16, and (b) Nimonic 105. In regime 1, hardening is by weakly coupled pairs, regime 2 by strongly coupled pairs and regime 3 by the Orowan mechanism. Notice that the peak strength is associated with the transition from weak- to strong-coupling.

22.4.2 Temperature-Dependent Hardening: Role of Gamma Prime and the Anomalous Yield Effect

The nickel-based superalloys display a remarkable characteristic: between ambient conditions and about 700 °C, the yield stress does not drop markedly as the temperature is increased as it does for other engineering alloys; in fact, the flow stress can often *increase* with increasing temperature. **Figure 17** shows some typical data for a number of single-crystal alloys, tested along the $\langle 001 \rangle$ direction. Until a temperature of 800 °C is reached, the flow stress is maintained but thereafter it falls rather rapidly.

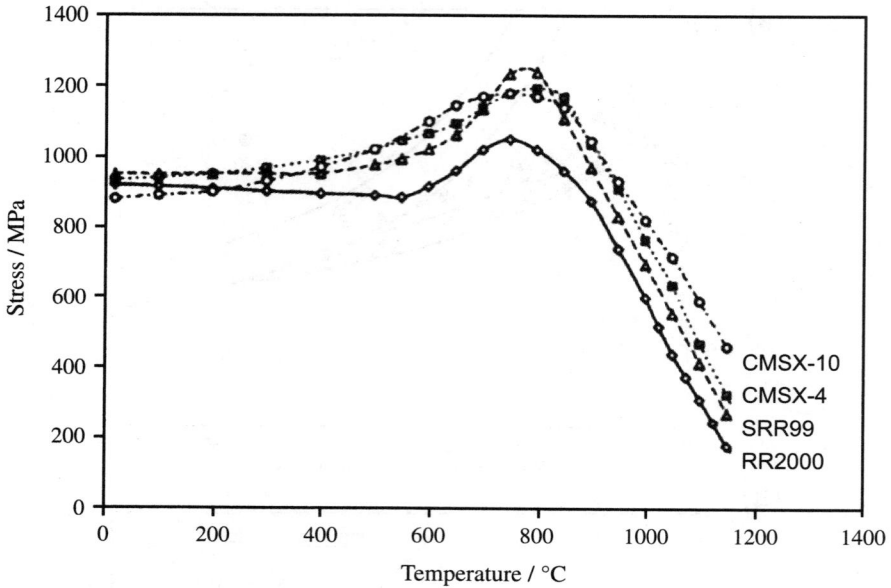

Figure 17 Variation of the yield stress of a number of single-crystal superalloys with temperature.

The theory of Section 22.4.1—with its great reliance on strengthening via the volume fraction of the γ' precipitates f and the APB fault energy γ_{APB} both of which *decrease* with increasing temperature—cannot explain this phenomenon.

Critical experiments of great significance that help to rationalize this effect are due to Piearcey et al. (1967) and Beardmore et al. (1969). Piearcey et al. tested the MarM200 single-crystal superalloy in the $\langle 001 \rangle$ direction, see **Figure 18**, with tensile behavior similar to that given in **Figure 17**—for the more modern single-crystal superalloys—being displayed. Also tested was cube-orientated Ni_3Al, alloyed such that its composition matched that found in MarM200. Comparison of the two curves (for the $\gamma + \gamma'$ and γ' alloys) demonstrates that the γ' imparts an increasing fraction of the strength as the temperature increases. Moreover, beyond 800 °C, the flow stress of single-phase γ' and the two-phase $\gamma + \gamma'$ alloys are not very different, indicating that at these high temperatures it is the γ', which controls properties. Related experiments were carried out by Beardmore et al., (1969); here, the temperature dependence of the yield stress was determined for a number of alloys containing varying amounts of γ'. Rather ingeniously, a series of alloys lying on the tie-line joining Ni_3Al and the Ni–25at%Cr alloy were considered, see **Figure 19a**. Once again, the alloy consisting of 100% γ' displayed a strong positive dependence on the temperature, until about 800 °C, see **Figure 19b**. Above the temperature corresponding to the peak stress, the yield stress of a two-phase $\gamma + \gamma'$ alloy obeys a rule of mixtures, corresponding to the weighted average of the values for γ and γ'. This contrasts strongly with the behavior at low temperatures, where two-phase alloys are very much stronger than the rule of mixtures would predict.

The above results confirm the critical role played by the $L1_2$ phase in conferring strength to the superalloys to elevated temperatures. They justify the considerable emphasis, which has been placed upon understanding and rationalizing the positive temperature-dependence of strengthening

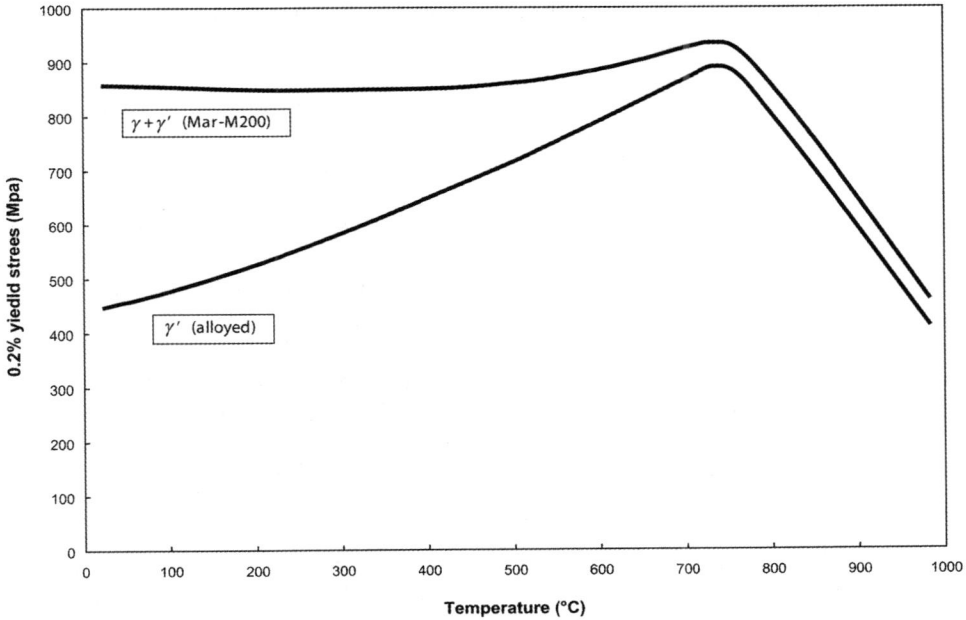

Figure 18 Variation with temperature of the yield stress of the MarM200 alloy in single-crystal form, and a monolithic Ni$_3$Al alloy of composition equivalent to the γ' particles in it. In both cases, testing was along $\langle 100 \rangle$. Adapted from Piearcey et al. (1967).

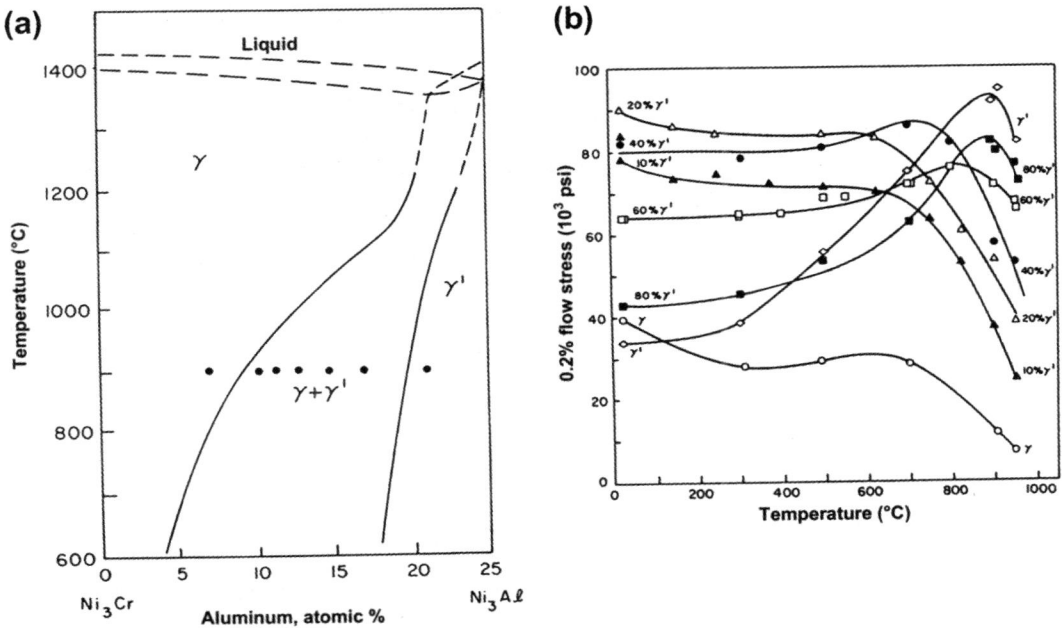

Figure 19 (a) Section through the Ni–Cr–Al ternary phase diagram at 75 at% Ni, showing the alloys used in Beardmore et al. (1969), and (b) variation of the yield stress with temperature for the alloys produced in this way.

particularly of the monolithic L1$_2$ intermetallic compound, which has become known as the *anomalous yield effect*.

In fact, an anomalous increase in the hardness of Ni$_3$Al with temperature was first demonstrated by Westbrook (1957). Flinn (1960) demonstrated the effect for the flow stress, and explained it by appealing to the cross-slip of $a/2\langle110\rangle$ partial screw dislocations from the {111} glide plane onto the {001} plane, thus rendering the full dislocation immobile. Thus, diffusive jumps of a short APB segment between the two displaced partials were envisaged. Later, Davies and Stoloff (1965) demonstrated that the yield stress rises steadily from $-196\,°C$, too low a temperature for a purely diffusion-based mechanism of cross-slip. Kear developed Flinn's original idea, proposing what has become known as the Kear–Wilsdorf lock, see **Figure 20** (Kear and Wilsdorf, 1962; Kear and Hornbecker, 1966); the basic idea is that the cross-slip from {111} to {001} is promoted by the anisotropy in the APB energy, so that hardening becomes more prevalent as the temperature rises due to a component of the cross-slipping process that is thermally activated. Many of these early studies were on ordered Cu$_3$Au, which exhibits the L1$_2$ structure but which disorders before the melting temperature is reached. In the disordered state, deformation is characterized by inhomogenous Luders bands and a low work-hardening rate. In the ordered material, in contrast, the rate of work-hardening is rapid and deformation homogeneous; TEM reveals long straight screw dislocations, a common feature in many ordered superalloys and Ni$_3$Al in particular—see **Figure 21**. Thus, the high work-hardening rate occurs as a result of the immobilization of dislocation loops as the screw components cross-slip onto lower energy APB configurations. Later, Thornton et al. (1970) examined Ni$_3$Al alloyed with Ti, Nb and Cr. Careful measurements of the plasticity on the microstrain level demonstrated that yield starts to occur at a stress roughly independent of temperature but that the rate of work-hardening rises very rapidly with temperature, consistent with initial flow of edge and screw dislocations and the increasingly rapid exhaustion of the mobile dislocations as they become immobilized by cross-slip. Hence one can argue that it is not so much a *yield anomaly* as a *work-hardening rate anomaly*, which is occurring. The active slip planes were also determined by slip trace analysis; a gradual transition from slip on (111) to slip on the cube plane (001) was noted.

Seminal experimentation was carried out by Takeuchi and Kuramoto (1973) and later Lall et al. (1979) who carried out mechanical testing of single-crystal L1$_2$-ordered Ni$_3$Ga and Ni$_3$(Al,Nb) respectively, to determine the dependence of the onset of plasticity on crystallographic orientation and temperature. Testing was carried out for orientations within the standard stereographic triangle, with

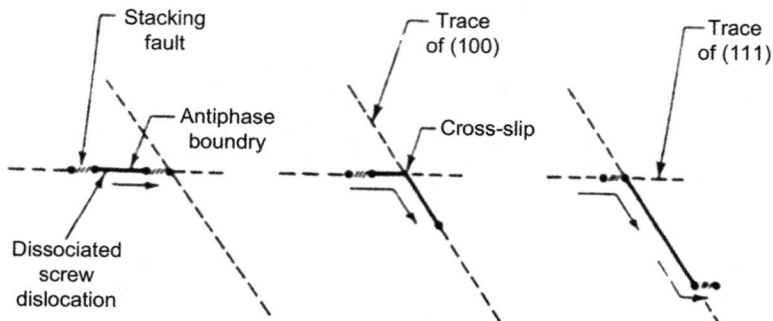

Figure 20   Adaptation of a diagram due to Kear (1964), which illustrates the concept of the Kear–Wilsdorf lock: cross-slip of a pair of screw dislocations from the octahedral {111} plane to the cube {100} plane, driven by a reduction in the APB energy.

(a) **(b)** **(c)**

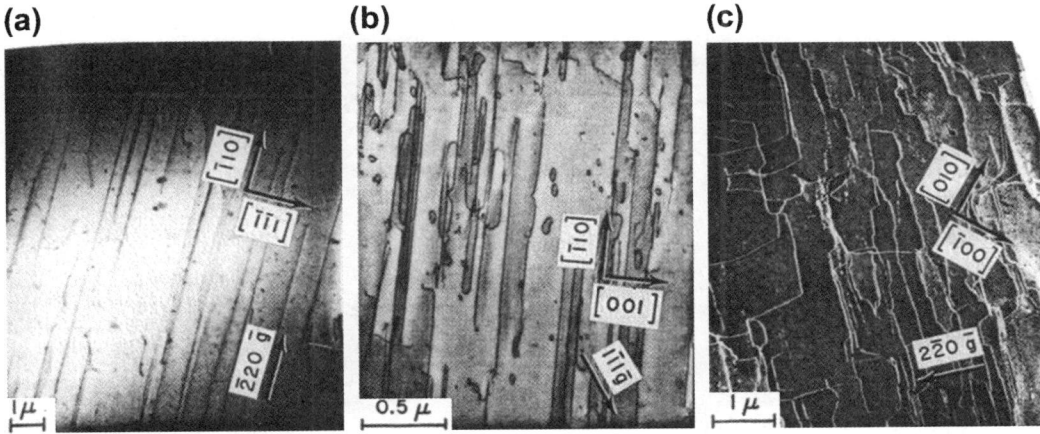

Figure 21 Dislocation structures on slip-plane sections of Ni$_3$(Al,Ti) single crystals after 4% deformation: (a) at 25 °C; (b) at 700 °C, indicative of octahedral slip and (c) 800 °C, indicative of octahedral slip. After Thornton et al. (1970).

use being made of slip trace analysis. Violations of the Schmid law were first identified by Takeuchi and Kuramoto (1973), which were rationalized by considering the effects of resolved shear stresses on the cube cross-slip plane; thus orientations with a higher resolved shear stress on the {001} plane show a notably higher resolved shear stress for slip on the {111} plane, indicating that the imposed stress assists cross-slip and hence increases the stress needed for yielding. Lall et al. (1979) and later Ezz et al. (1982) demonstrated the role of cube-slip in curtailing the increase in yield stress with temperature, making elegant use of slip trace analysis. But significantly, they identified further anomalies in relation to the Schmid law and in particular a strong tension/compression asymmetry—see **Figure 22**; this led to their consideration, not only of the resolved shear stress on the cube plane, but also of its effect on the dissociation of the Shockley partials, which make up the $a/2[110]$ dislocations. This dissociation introduces an edge component into the Burgers vector and the applied stress can then interact to contract or expand the dislocation core, assisting or hindering the cross-slip respectively. The tension/compression asymmetry was thus rationalized. Using these ideas, expressions for the increase in shear yield stress on the primary {111} plane over the value at −196 °C, denoted $\Delta\tau_{pb}$, were proposed (Pope & Ezz, 1984)

$$\Delta\tau_{pb} \propto \exp\left\{-\frac{H}{3kT}\right\} \tag{25}$$

with

$$H = H_0 - \tau_{cb}V_1 - \tau_{eb}V_2, \tag{26}$$

where H is the activation energy, τ_{cb} is the resolved stress on the {001} cube plane, τ_{eb} is the shear stress on the edge component of the Shockley partials and V_1 and V_2 are the two activation volumes, respectively. The above expression was derived on the basis of the dynamical breakaway of dislocations from pinning points—in the spirit of Friedel—with the difference being that these were assumed to

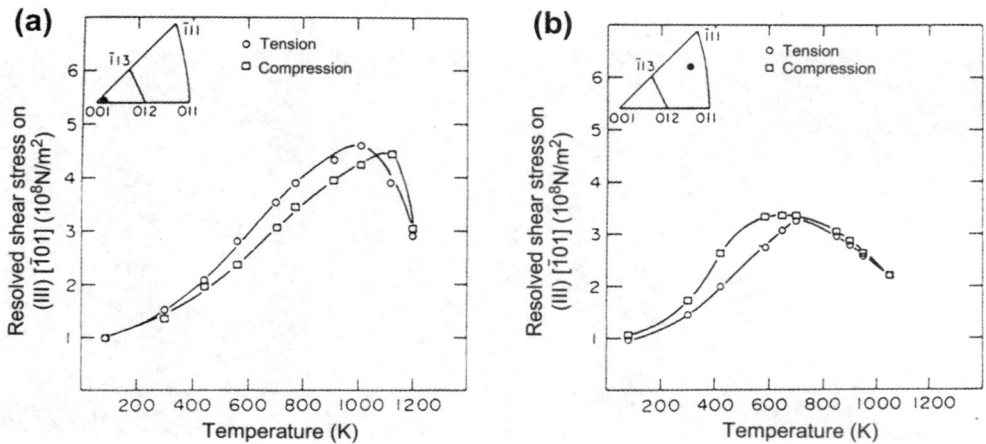

Figure 22   Variation with temperature of the critical resolved shear stress for $[\bar{1}01]\{111\}$ slip for $Ni_3(Al,Nb)$ single crystals, measured in tension and compression, for two orientations within the stereographic triangle (Ezz et al., 1982); note the considerable tension/compression asymmetry, which violates Schmid's law.

form continuously via activated cross-slip events. The factor of three in the denominator of Eqn (25) arises from assumptions made relating pinning site generation to the slow/linear viscous drag hypothesis, see Reed (2006). It leads to some confusion in quoting experimentally determined values of the activation energy; this is generally calculated from an equation such as that above without the factor of three. Liang and Pope (1977) have tabulated a comprehensive list of activation energies. The values for Ni₃Al-based compounds are generally rather less than for other L1₂-type intermetallics, for example in the range 0.03 to 0.05 eV for the data of Lall et al. (1979); the stress assistance of the cross-slip in the case of Ni₃Al is thus a much larger relative effect that leads to significant effects of crystallographic orientation. Thus, as the activation energy is rather small, the yield stress anomaly can become active at rather low temperatures, as originally found by Thornton et al. (1970).

The advent of computer modeling of dislocation core structures using empirical potentials has confirmed the basis of these ideas, by analysis of the dissociation planes and by estimation of the Peierls stress (Yamaguchi, 1982; Paidar, 1982). These calculations also extended the understanding from the purely geometrical analysis to include a consideration of the atomic scale arrangement of the atoms at the dislocation core, their response to stress and the specific movements necessary to execute the cross-slip needed for the anomaly. Paidar et al. (1984) recognized that cross-slip can occur by the formation of a double kink that moves in steps of $w = b/2$—where w is the distance cross-slipped on the (001) plane—and that for steps that are odd multiples of $b/2$ the dislocation dissociates on the $(1\bar{1}1)$ cross-slip plane, see **Figure 23**. For integral multiples of b however, the leading dislocation is mobile and can glide pulling the trailing dislocation on a parallel plane so that the pair can move forward by the separation of the dislocations into partials on the (111) primary glide plane. This introduces further complexity into the effects of the stress since it modifies the activation energy needed. Thus cross-slip is not an all or nothing process, but can occur incrementally, up or down, to produce a complex structure on a number of parallel planes—see Caillard (2001), **Figure 24**. The displacement by $w = b/2$ involving dissociation on a nonparallel {111} plane is known as a Paidar, Pope & Vitek (PPV) lock and is most probably unstable as demonstrated by (Schoek, 1994) using mechanical arguments. However the

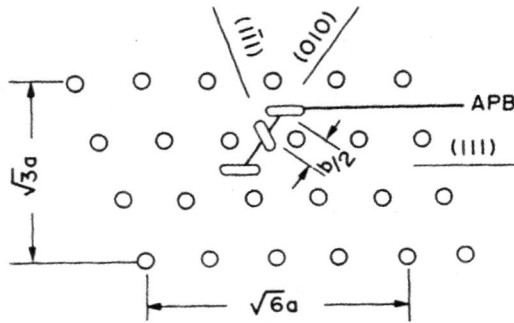

Figure 23 Possible core configurations of a $a/2[\bar{1}01]$ superpartial viewed along the $[\bar{1}01]$ direction. When on the (111) slip plane the core is on that plane, but when it cross-slips to the (010) plane it spreads onto the (111) or $(1\bar{1}1)$ alternating with position. After Paidar et al. (1984).

double jump ($w = nb$, where n is an integer) and dissociation on the parallel {111} plane is stable and becomes more so as the cross-slip distance increases. These are termed incomplete Kear–Wilsdorf locks and have been observed by a number of authors. Detailed analysis by (Caillard, 2001) showed that this is the case at low temperatures but that at higher temperatures a second peak arises where full Kear–Wilsdorf locks are formed. A direct link was demonstrated between the APB energy and the flow stress at maximum work-hardening rate for a number of systems with APB energies varying from 100 mJ/m^2 to 300 mJ/m^2. Caillard also argued that the CISF energy controls the onset of both locking and cube-slip—the greater the CISF energy, the lower the temperature for the onset of both because the separation of the partials is then smaller.

A further important contribution to the energy balance of cross-slip in Ni$_3$Al was made by Yoo (1986, 1987). It was realized that elastic anisotropy exerts a torque on a dislocation pair and, in the case of two parallel screw dislocations, acts to assist cross-slip from {111} to the {001} plane where the torque disappears due to symmetry, see **Figure 25**. The tangential force, F_t, rotating the pair off a plane at an angle θ from the {001} plane was estimated to be:

$$F_t = F_r\left(\frac{(A-1)\sin 2\theta}{2(A\sin^2\theta + \cos^2\theta)}\right), \tag{27}$$

where F_r is the repulsive force between the two dislocations and A is the anisotropy factor given by $A = 2C_{44}/(C_{11} - C_{12})$. This reaches a maximum of 0.62 F_r for a dislocation pair on a {111} plane assuming $A = 3.3$ for Ni$_3$Al. Hence the Yoo torque makes a very sizable contribution, tipping the balance—at least in the case of Ni$_3$Al—in favor of cross-slip. To balance the torque, Yoo calculated that the APB plane would need to rotate away from the {111} plane, the vertical displacement being about 0.2 nm for the average separation of 6.3 nm. These calculations confirm the important contributions made by elasticity to the process of cross-slip.

Although the work of Pope et al. (Paidar et al., 1984) is able to rationalize the sensitivity of the flow stress to crystallographic orientation, it does not deal adequately with the strain-rate dependence of the yield point. For such low-activation energies, one might anticipate a much higher strain rate dependence; in fact the flow stress is remarkably independent of the strain rate, although very sensitive to the strain via the dislocation exhaustion effect. It is important to distinguish between the activation energy

(a)

(b)

(c)

(d)

(e)

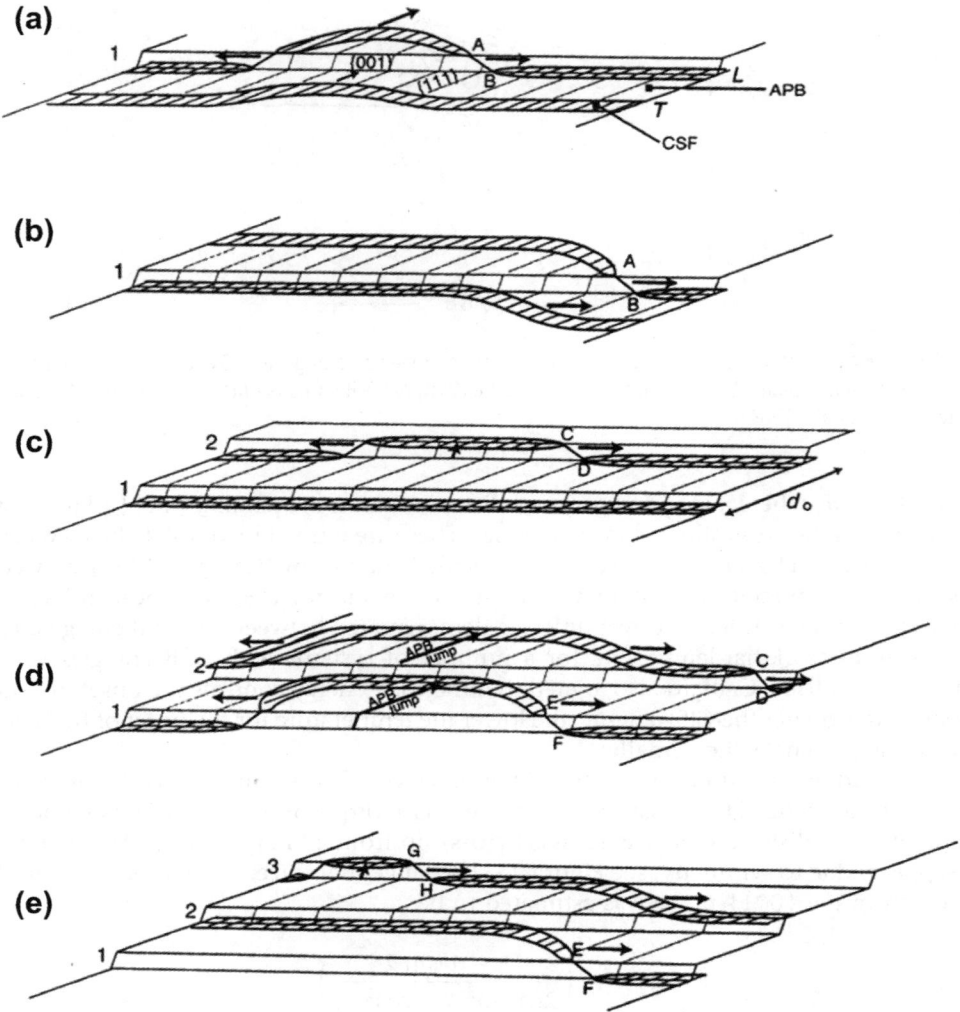

Figure 24 Schematic illustration of the APB jumps occurring in a cross-slipping dislocation pair, showing the mechanism by which the cross-slip needed to form the Kear–Wilsdorf lock can occur incrementally. After Caillard (2001).

needed for the formation of the locks—which provides the temperature dependence of the yield stress anomaly—and the activation of the unlocking mechanism, which is responsible for its strain rate sensitivity. The activation energy for unpinning the Kear–Wilsdorf locks would need to be of the order of 3–6 eV to give the observed insensitivity to strain rate, see Vitek and Sodani (1991).

An alternative approach has been built on the growing body of observations of the dislocation configurations and the debris they leave behind after flow. Examining the dislocation structures in Ni_3Al deformed in the anomalous regime, small elongated dipole loops have been reported—see for example Sun and Hazzledine (1988). Similarly, Chou and Hirsch (1982) and Hirsch and Sun (1993)

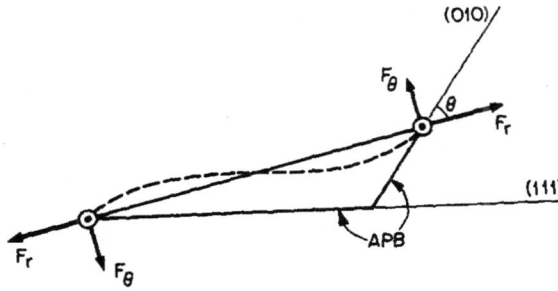

Figure 25 Illustration of the elastic interaction forces between a pair of superpartial screw dislocations that aid cross-slip from the {111} plane to {001}. The terms F_r and F_t are the radial and tangential forces acting on the superlattice dislocation pair.

identified faint contrast from prismatic "tubes" the walls of which are continuous APB faults attached to these dipoles. It is argued that a parallelepiped of the material is translated through $R = a/2[110]$ along the axis of the tube terminated by the dipole loop. Such features are typically tens of nanometers wide and many hundreds of nanometers long and are visible when $g \cdot R \neq$ integer, a condition achieved only when the imaging g vector is a superlattice vector. Faint contrast from the strain associated with the surface tension of the fault can also be seen in normal imaging conditions where g is normal to the axis of the tube. Explaining their formation led Chou & Hirsch to develop an alternative mechanism for the unpinning of a Kear–Wilsdorf lock, see Hirsch (1992).

Consider **Figure 26**, which illustrates a dislocation pair that has formed a Kear–Wilsdorf lock on the (001) plane, after a number of jumps on the (001) cross-slip plane and back onto the {111} glide plane as envisaged by Paidar et al. (1984). The upper loop is mobile and bows out over the top of the cross-slipped section; by nucleating an edge dipole, the trailing and leading dislocations are able to swap places. The upper segment bows out parallel to the lower—locked—dislocation pair to produce a two long screw dipoles (as a result of the switch), which can annihilate by cross-slip on the primary (111) plane and the low-energy (001) plane to give the APB tube lying along $[1\bar{1}0]$. It is argued that the critical step in this process is the formation of the edge dipole to facilitate the exchange of the leading and trailing dislocations. This involves bowing the dislocation loop out to a degree dependent upon whether only one or both ends of the looped segment is pinned. The loop then moves forward again until cross-slip re-pins it. The movement of the released segment forward creates an edge segment that not being prone to cross-slip is mobile; hence it can carry on expanding until halted by interaction with other dislocations. The key point is that the stress for unpinning is relatively insensitive to the strain rate, as it is effectively achieved by action of the stress field and not by an activated event; hence the agreement with the experimentally observed lack of strain-rate sensitivity. The most compelling evidences for this mechanism are the APB tubes and dipoles associated with deformation in the temperature range of the yield anomaly. No other explanations have been proposed for the observation of such configurations. It is also relevant that at the lower temperatures in the anomalous range the tubes are less apparent; this does however depend on the material type, and it is very likely that in this range the formation of smaller partial Kear–Wilsdorf (K–W) locks form, as envisaged by Paidar, Pope & Vitek.

In summary, a great amount of work has been carried out in an effort to rationalize the effects which cause the yield stress anomaly in single-phase Ni₃Al. Despite the uncertainties in explaining all the observed effects and in particular the absence of truly composition-dependent theory, which remains

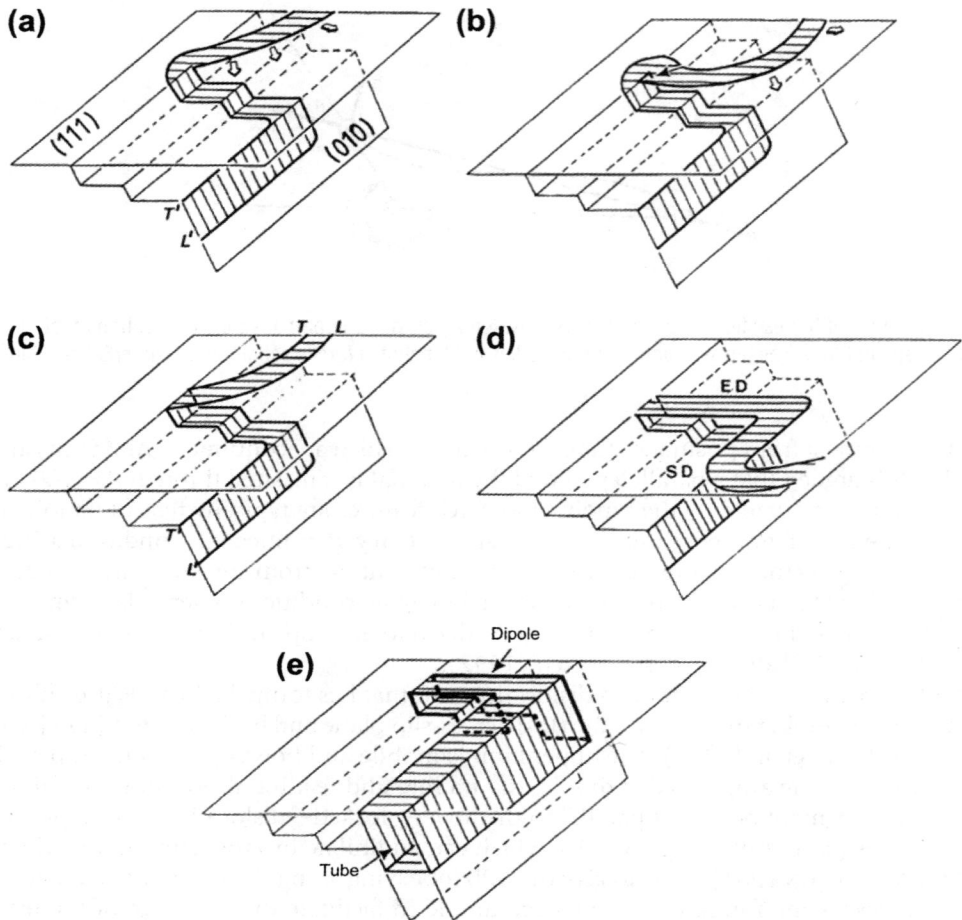

Figure 26 Illustration of the formation of APB tubes during the unlocking of a Kear–Wilsdorf lock. After Hirsch (1992).

a grand challenge, there is now good agreement on the basic principles governing the rise in yield stress with temperature. The following represents a summary of the state of knowledge:

(1) deformation in the anomalous regime occurs by the movement of pairs of $a/2\langle 110 \rangle$ dislocations on $\{111\}$ planes and at higher temperatures on $\{001\}$ planes when so-called cube-slip becomes activated;

(2) the increase in yield stress, which occurs from very low temperatures is due to the activated cross-slip of short segments of the dislocations from the $\{111\}$ glide plane to the $\{001\}$ plane, driven by a combination of the anisotropy of the APB energy and the significant Yoo torque on the dislocation pair resulting from the anisotropic elastic properties;

(3) the locks can be small so that they can consist of individual steps, the so-called Paidar–Pope–Vitek (PPV) locks, or larger full Kear–Wilsdorf (K–W) locks;

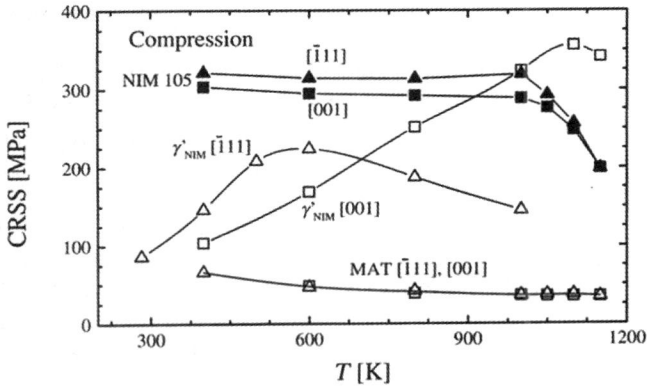

Figure 27 Experimental results from compression tests on the Nimonic 105 alloy: variation with temperature of the critical resolved shear stress (CRSS), on the octahedral system. Solid symbols: results from two-phase alloy. Open symbols: data from single-phase alloys corresponding to compositions of γ and γ', tested in single-crystal form. After Nitz et al. (1998).

(4) macroscopic yielding is governed by the stress to expand the loops between pinning points; unpinning occurs either by an athermal stress (consistent with the Hirsch model) for larger locks or an activated process for the smaller steps at lower temperatures;

(5) orientation effects do not follow the Schmid law; account needs to be taken of the effect of the applied stress both in assisting cross-slip directly or by altering the separation of partials so that the activation energy for cross-slip is affected indirectly;

(6) the common consensus is that anomalous yield effect ceases when the temperature is high enough to activate slip on the cube plane by either $a/2\langle 110 \rangle$ pairs or at higher temperatures by single perfect $a[100]$ dislocations. Both dissociate on the $\{111\}$ planes, resulting in an extremely high Peierls stress for flow on $\{001\}$, which is analogous to the situation for body-centered cubic (BCC) metals.

22.4.3 Hardening in Two-Phase Alloys: Role of Anomalous Hardening

The results of Section 22.4.2 concern in the main the anomalous yielding of monolithic Ni_3Al, but important questions remain that concern whether the effect is carried across to two-phase γ/γ' superalloys. The results so far are sparse, and more research is needed. Many studies have made use of TEM microscopy to identify the presence of dislocation pairs in γ' precipitates and confirm that these have indeed cross-slipped so that the APB plane is then lying on (001); however, these observations do not by themselves represent conclusive proof of anomalous hardening. However, revealing insights are gleaned from the seminal studies of Nitz et al. (1998) and Nembach (2000). The flow stress of single-crystal Nimonic 105 was determined as a function of temperature, with additional but similar experimentation for two single-phase alloys of compositions replicating the matrix and γ' precipitate at 800 °C, see **Figure 27**. Specimens of composition consistent with the γ matrix—which rely upon strengthening by solid solution hardening—exhibit a low flow stress that diminishes monotonically with temperature, as expected. The γ' phase shows all the same trends as the binary $L1_2$-ordered Ni_3Al, including a peak stress, orientation dependence and tension/compression anisotropy. But significantly, the two-phase γ/γ' alloy exhibits a flow stress consistently higher than the peak strength of the monolithic γ' alloy, with the strength maintained in this way above the temperature of the peak flow stress of the [111] orientation, to a temperature of nearly 800 °C.

Thus it is clear that the performance of such peak-aged two-phase alloys is superior to that of a weighted average of the constituent phases: i.e. it performs better than a sum of its parts. One could argue that the role of the γ' is merely to maintain a fine, coherent and hence stable array of precipitates that are able to resist the entry of single and paired dislocations, were it not for one crucial observation. The two-phase alloy retains substantial (10%) anisotropy, after the effects of the Schmid factor are eliminated. This mirrors the behavior of the γ' single-phase alloy. The implication of this is that not only are the ordered precipitates strengthening the alloy but that anomalous hardening is also operating in the precipitates, which in this system are less than 20 nm in diameter. The γ' alloy in the [111] orientation shows a maximum at about 600 K and switches to cube-slip at temperatures above that, but the two-phase alloy shows excellent flow stress in both the [111] and [001] orientations up to 1100 K. Note that the [001] orientation has zero-resolved shear stress on the cube-slip planes. The two-phase alloy in the [111] orientation thus suppresses the switch from octahedral glide to cube-slip between the temperatures 500 K and 1000 K, so that anomalous hardening is maintained in all orientations up to 1000 K. Between 1000 K and 1100 K the volume fraction of the γ' drops from 50 to 30% and it is most probably this, which causes the flow stress to drop rapidly. Thus we have the perfect storm: the small size of the precipitates inhibits cube-slip, which, although favored in the ordered material, does not occur in the disordered matrix. It would appear that the peak-aged precipitates are large enough to allow the cross-slip of the dislocation pairs to add to the flow stress and produce the anisotropy observed, but small enough for those sections of the dislocations in the matrix to constrain the flow to the octahedral planes.

So what happens at high temperature in the two-phase alloys? Does the belated transition to cube-slip occur? The occurrence of cube-slip will be discussed in more detail in the next section on single-crystal alloys. However, in the context of the work of Nitz et al. (1998), they were unable to identify the mechanism of flow above 1000 K when the flow stress drops rapidly. They noted that material cycled repeatedly through this transition during in situ deformation showed Shockley partial dislocations and hence presumably stacking faults. This suggests that the constraint of the matrix on the octahedral planes is enough to avoid cube-slip altogether and to produce flow by the creep-like stacking fault mechanisms; these are reviewed in Section 22.5.

22.4.4 Modeling of Strength in Two-Phase Alloys

It is now understood that high-strength polycrystalline nickel-based superalloys such as Rene 104, RR1000 and Udimet 720Li—used for turbine disc applications—exhibit multimodal distributions of γ' precipitates, each of which will have differing compositions and associated APB energies. Modeling the mechanical response under such situations is an immense challenge that has not yet been achieved with accuracy. The most comprehensive study to date, see Kozar et al. (2009), combined modeling and experimentation to quantify the contributions made to the flow-stress. Many of the equations described in the above sections were employed—together with the Hall–Petch equation and an empirical linear fit for the solid-solution hardening terms—to model the flow stress of the nickel-based superalloy alloy IN100. By giving the alloy a number of different heat treatments—supersolvus, subsolvus, and monomodal—it was possible to determine how the various expressions reproduce the effects of primary, secondary and tertiary distributions of γ' precipitates, and to refine the strengthening terms by deconvoluting the effects of each. The IN100 alloy has a relatively high fraction of 55–60% γ' and maintains a flow stress of approximately 1 GPa to temperatures in excess of 800 °C. As the alloy was not peak-aged for any one given γ' size in this study, both weak- and strong-coupling were considered together with solid-solution hardening of the matrix and anomalous hardening of the ordered phase. The high volume fraction of γ' makes the applicability of the weak and strong coupling equations somewhat contentious. Tertiary precipitates at ~ 10 nm diameter were much smaller than the separation of the dislocations in the γ channels of 30 nm causing weak-coupling at this level but strong coupling in the secondary precipitates of size in the range 200–400 nm and in the primary γ', which was around 1 μm.

The effect of the anomalous (or cross-slip-induced) hardening was tackled by looking directly for evidence of dislocation pairs trapped in cross-slipped configurations and quantifying this as a function of the precipitate size. The occurrence of cross-slipped dislocation pairs observed in γ' was found to be much more likely in precipitates of size 300 nm and greater, and this was assumed as a minimum size for anomalous hardening to have an effect. Note that this is somewhat at odds with the evidence of Nitz et al. (1998) who suggest that significant hardening can result from precipitates as small as 20 nm. Each contribution to the yield stress was combined by adding the individual calculated yield stresses, raised to a power k. In a detailed discussion Schilnzer and Nembach (1992) recommend a value of k in the range 1.13–1.19, but Kozar et al. chose a linear addition of the individual contributions to flow stress, since summations assuming otherwise significantly underestimated the flow stress. Thus, there remains some doubt about the best way to add up strengthening contributions from multimodal distributions of γ'. Better theory and modeling is needed here.

In Kozar et al. (2009), IN100 (between 55 and 60% total γ') was modeled in three microstructural conditions induced by subsolvus, supersolvus, and a monomodal γ' distribution. A modified IN100 composition was also studied of reduced γ' volume fraction (53% γ') but otherwise identical phase chemistry, consistent with a tie-line analysis. The super- and subsolvus heat treatments of the alloy gave very nearly the same yield stress but the Hall–Petch contribution dropped from being the dominant hardening mechanism in the subsolvus (34%) to 1/3 that value in the supersolvus condition, with a tenfold increase in grain size. This was more than compensated for by the increase in the γ' population of secondary and tertiary precipitates, which contribute nearly 80% of the strength to the supersolvus material as against 40% to the subsolvus through a combination of weak-pair coupling, strong-pair coupling and anomalous hardening (cross-slip). The contribution from the solid-solution hardening is fairly constant at about 10% but rises slightly with the increased volume fraction in the modified alloy. In this model the contribution from anomalous hardening is very dependent on the proportion of the precipitates that are above the 300 μm threshold; somewhat surprisingly it is highest for the supersolvus alloy contributing 22%. What emerges from such calculations is confirmation that it is the γ', which adds the strength, that the volume fraction of it matters critically and that γ' is utilized most effectively as small (peak-aged tertiary) precipitates.

22.4.5 Microtwinning as a High-Temperature Deformation Mechanism

Recently, the need for pairs of dislocations—either weakly- or strongly coupled—for the deformation of high-strength superalloys at temperatures of around 650 °C has been called into question by new observations made using high-resolution transmission electron microscopy (TEM). Particularly at lower stresses, deformation by partials and microtwins has been observed in alloys such as Rene 88DT and Rene 104, with the mechanism being found to be sensitive to the levels of applied load and the microstructural condition (Viswanathan et al., 2005, 2006; Kovaric et al., 2009). Fine and coarse bimodal precipitate distributions—each containing both secondary and tertiary precipitates—were tested under static constant load creep conditions at a temperature of 650 °C and a range of stresses.

The finer of the two distributions was found to display superior creep properties and different deformation characteristics. At higher stresses approaching the yield stress, the matrix was found to be sheared by $a/2\langle110\rangle$ dislocations, which loop around the tertiary precipitates; no sign of pairing or weak coupling was observed. At lower stresses and small precipitate size, microtwins some 3–12 atomic planes in thickness were the major features. It is reasonable to assume that these account for the observed strain.

The evidence confirms that the microtwins form by the passage of a succession of identical $a/6\langle112\rangle$ Shockley partials on adjacent slip planes. Hence their migration exhibits similarities with a cooperative transformation initiated at the grain boundaries, with growth extending across the grain with the next boundary acting as a termination site. However, for the coarser microstructure at the lower stress level, deformation is dominated by single stacking faults, which are distinct from microtwins. In this case faulting is found to be limited to the interiors of the larger secondary precipitates and does not develop into twins. It is argued that dislocations with $a/2\langle110\rangle$ Burgers vectors cut into the precipitates and subsequently stacking fault loops nucleate independently in the larger precipitates to replace the higher energy APB fault with the lower-energy SISF fault; this is the mechanism first proposed by Decamps and Morton (1991), which leaves faulted precipitates surrounded by Shockley partial dislocations. Orowan looping was found in both microstructures with increasing stress, but at a higher stress in the finer microstructure. The different mechanisms observed in these studies have been summarized on a deformation map of stress against temperature, see **Figure 28**.

Although the mechanism is similar to that observed in FCC single-phase metals, it is significant that twinning in such high-strength superalloys has been found to be a time-dependent process exhibited only during time-dependent creep deformation at lower stresses. This is because the required reordering

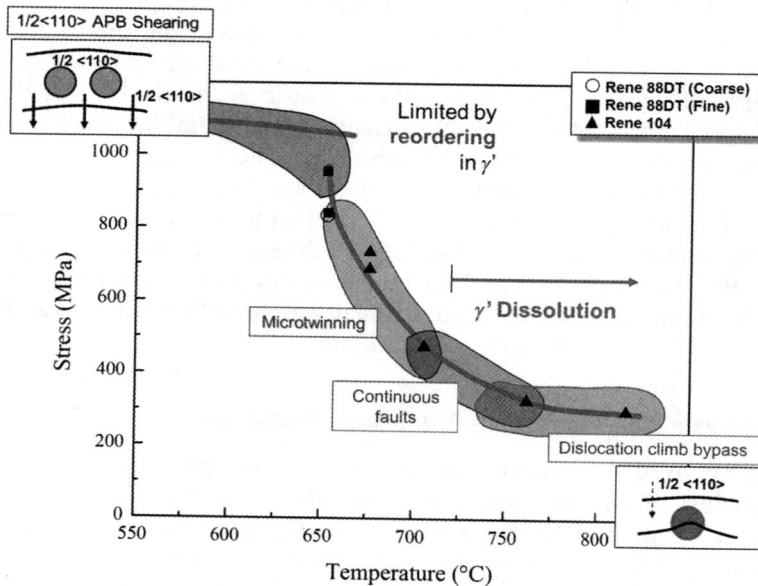

Figure 28 Deformation map for the Rene 88DT and Rene 104 high-strength superalloys, illustrating the dependence of the different mechanisms on temperature and applied stress. Courtesy of Mike Mills. (For color version of this figure, the reader is referred to the online version of this book.)

in the L1$_2$ crystal structure is as summarized in Section 22.3.2. As there is a strong dependence of the kinetics on the stress level, variations in behavior are observed between grains as the resolved shear stress changes, particularly when different deformation mechanisms are possible. Twinning, for example, is observed in single crystals of the [011] orientation in tensile creep and the [001] orientation in compression, but these are rarely seen in practice because these orientations and loading combinations give poor properties and are thus avoided. In polycrystalline superalloys a full range of orientations is inevitable.

22.5   Single-Crystal Superalloys

The single-crystal superalloys have been developed to display superior properties particularly under creep conditions; for the best performance, the γ' shearing of the previous section needs to be avoided. Thus, knowledge of the factors known to the important in promoting γ' shearing aids in the design of single-crystal superalloys, by enabling microstructures and alloy chemistries to be optimized (Reed, 2009). The removal of grain boundaries is carried out by judicious use of solidification processing (Carter et al., 2000; Elliott et al., 2007); this implies the elimination of the now unnecessary grain boundary strengthening elements such as C and B, which degrade properties at the very highest temperatures since they exacerbate incipient melting.

For the optimization of alloy chemistry, the following are needed:

(1) Strengthening of the γ' phase by alloying with Ti and Ta, both of which increase the APB energy; shearing of the γ' phase is then resisted (Caron et al., 1988);
(2) Alloying with γ' partitioning elements Al, Ti and Ta, such that the γ' fraction is beyond 0.5 and approaches 0.75; the γ channels between the γ' precipitates are then of reduced dimensions so that dislocation activity in them is minimized (Murakamo et al., 2004; van Sluytman and Pollock, 2012);
(3) Choice of compositions such that the lattice parameters of the γ and γ' phases are similar, so that resistance to precipitate coarsening is maximized (Harada and Murakami, 1999; MacKay et al., 2012);
(4) Alloying with refractory elements—particularly W and Re—to improve time-dependent properties, but not excessively so, because otherwise TCP phases (Rae and Reed, 2001) are promoted.

A typical microstructure of a single-crystal superalloy is depicted in **Figure 29**. Provided that the lattice misfit is small, the γ' precipitates are approximately cuboidal. A cube–cube orientation relationship is displayed, as proven readily by transmission electron microscopy (TEM). Elastic anisotropy plays a role in the development of the γ' morphology (Ricks et al., 1983): the preferred interface plane is {001} since its normal corresponds to the direction, which is elastically the softest. Atom probe tomography has proven an elegant way to study the partitioning characteristics of the alloying elements (Reed et al., 2004), see **Figure 30**. The lattice parameters a_γ and $a_{\gamma'}$ of the γ and γ' phases are altered by this partitioning. Note in particular the behavior of Re, which partitions strongly to the γ phase. In this way, the lattice misfit δ (Mughrabi, 2009) defined according to

$$\delta = 2 \times \left[\frac{a_{\gamma'} - a_\gamma}{a_{\gamma'} + a_\gamma} \right] \tag{28}$$

is made substantially more negative due to the increase of a_γ caused by the preference of Re to reside in γ rather than γ'. Thus, most commonly employed Re-containing superalloys have negative values of δ.

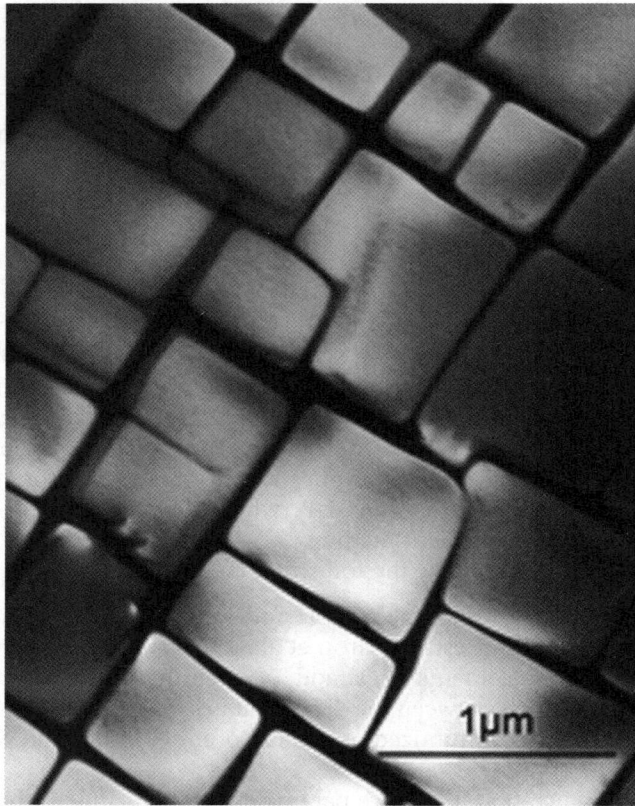

Figure 29 Microstructure of a typical single-crystal superalloy, showing presence of cuboidal γ' precipitates within a matrix of γ.

Moreover, with increasing temperature, δ becomes more negative, since a_γ increases less rapidly with temperature than does $a_{\gamma'}$ (Harada and Murakami, 1999).

Compositions of the single-crystal superalloys that are used widely for engine applications are given in **Table 2**; examples include Rene N4, Rene N5, PWA1480, PWA1484, SRR99, CMSX-4 and CMSX-10, amongst others.

22.5.1 The Yielding Behavior of Single-Crystal Superalloys

For the deployment of components fabricated from single-crystal superalloys in turbine systems, the yield stress is an important consideration. Adequate strength is needed in locations such as the root of the turbine blade, to complement the overarching requirement for creep resistance and oxidation performance. Thus yield stress is an important parameter in engineering design particularly along the technologically important [001] direction. Differences in behavior due to crystallographic anisotropy are also of concern.

Figure 31 shows the microstructure of [001]-oriented CMSX-4 single-crystal superalloy deformed to just above the yield point, at 750 °C. Dislocation pairs on several octahedral slip planes are seen

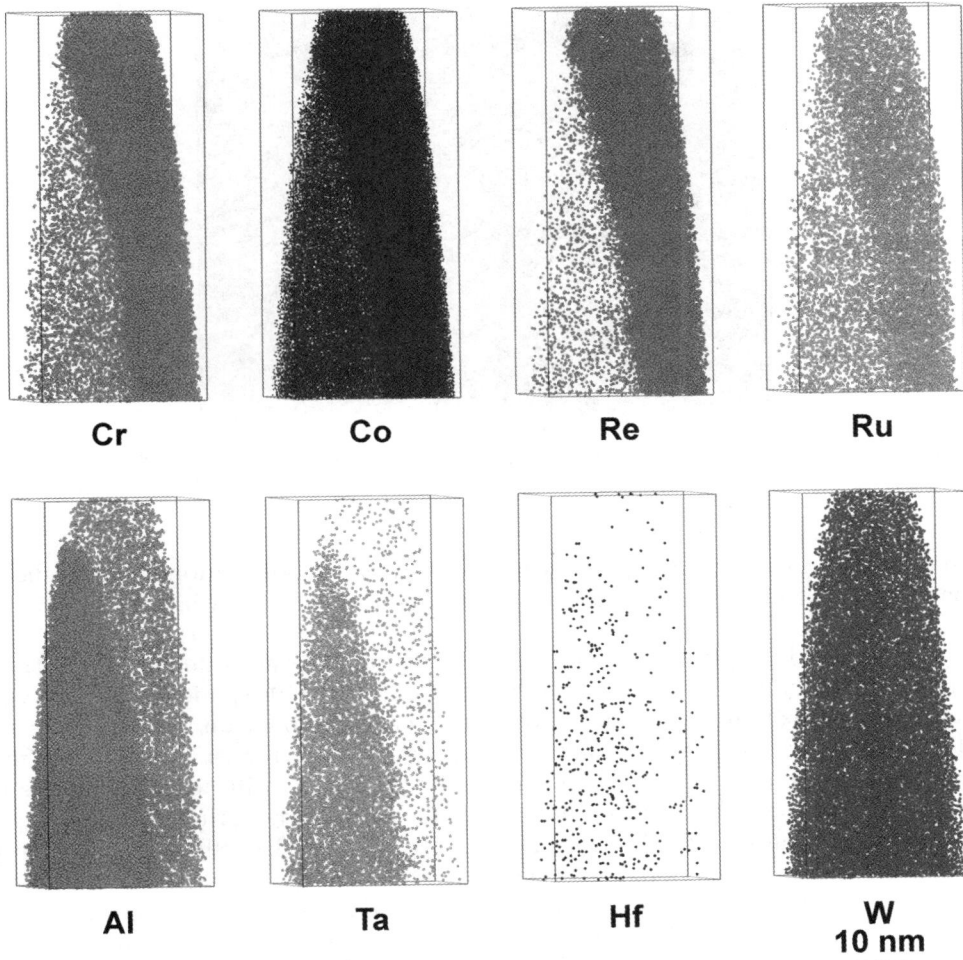

Figure 30 Atom probe tomography image of γ/γ' interface in a single-crystal superalloy, confirming preferred partitioning of elements to the γ and γ' phases. The γ phase is at top right, the γ' phase at bottom left. Courtesy of Mike Miller. (For color version of this figure, the reader is referred to the online version of this book.)

together with single dislocations in all the γ channels. Such postmortem TEM studies indicate that equations used for describing dislocation movement—based upon either weak- or strong-coupling—cannot be applied to the single-crystal superalloys without considerable caution, because the fraction of the γ' phase is very large, for some alloys exceeding 0.70. A key question to address is: to what extent do dislocations entering the γ' precipitates at peak yield temperatures—around 750 °C—replicate the known behavior of dislocations in the single-phase intermetallic Ni_3Al? There are two related issues: do dislocation pairs form Kear–Wilsdorf (KW) locks and do they glide on cube planes as they pass through the γ' precipitates? It is important not to confuse (1) the cross-slip needed to form KW locks (when the slip plane remains the octahedral plane but the APB fault plane becomes {001}), and (2) cube-slip

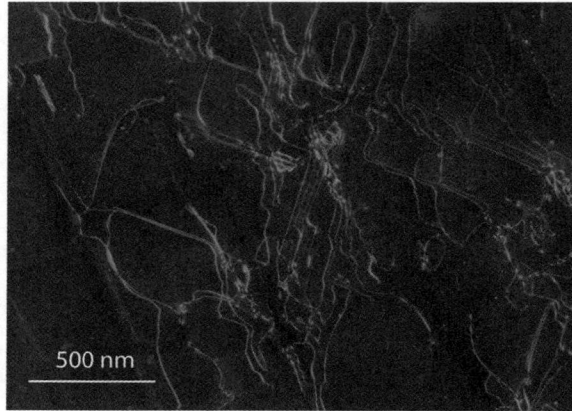

Figure 31 Weak beak beam image of the dislocation structure of CMSX-4 after tensile deformation to the yield point, at 750 °C.

for which the slip plane itself becomes {001}. Unfortunately, whilst observations of dislocation pairs in γ' precipitates have been commonly made, researchers rarely determine which plane that separates the pair.

A very systematic study of yielding in nickel superalloy single crystals is due to (Allen, 1995). Data are given in **Figure 32**, for the CMSX-4 superalloy. For the [001] and [011] orientations, the yield stress is found to display a maximum at 650 °C in both tension and compression, but this is not so readily apparent for the [111] orientation. In tension, the three orientations reverse their ranking from room temperature to the maximum at 700 °C. The rise in the yield stress with temperature, the tension–compression asymmetry and the noncompliance with the Schmid law for nickel-based single-crystal superalloys has been confirmed by the work of Nitz et al. (1998). The effect of γ' size on yielding behavior has also been studied by Shah and Duhl (1984) and Sengupta et al. (1994) but with conflicting findings. Shah and Duhl (1984) studied PWA1480 and found a demonstrable increase in yield stress with decreasing γ' size, whilst Sengupta showed that increasing γ' size in CMSX-4 caused a modest increase in flow properties. The data of Shah and Duhl (1984) are plotted on **Figure 32** to enable comparison with that of Allen (1995).

Both Allen and Shah & Duhl report a maximum in the yield stress at the same temperature irrespective of the orientation and a non-Schmid anisotropy consistent with that observed in the Ni_3Al intermetallic, which was considered in detail in Section 22.4. In such studies, the lengths of the dislocation pairs passing through the γ' precipitates are found to be shorter than the segments of cross-slipped pairs observed in the single-phase intermetallic; therefore it is unlikely that the mechanisms of dislocation escape and macroscopic yielding occurring in monolithic Ni_3Al are directly applicable to the two-phase alloys. All three orientations at the corners of the unit triangle show a sharp drop in yield stress at the same (and highest) temperature, although this temperature is not identical in all of the studies reported so far. Hence, it is clear that the dislocation pairs passing through the γ' in single-crystal alloys experience Kear–Wilsdorf locking, as in the single-phase intermetallic Ni_3Al. Moreover, despite the very significant fraction of γ' at the yield stress maximum, there is sufficient matrix γ phase to inhibit the onset of the cube slip, which would otherwise occur at a lower temperature for the {111} orientation. Clearly a dislocation segment gliding on a {001} plane inside the precipitate would be

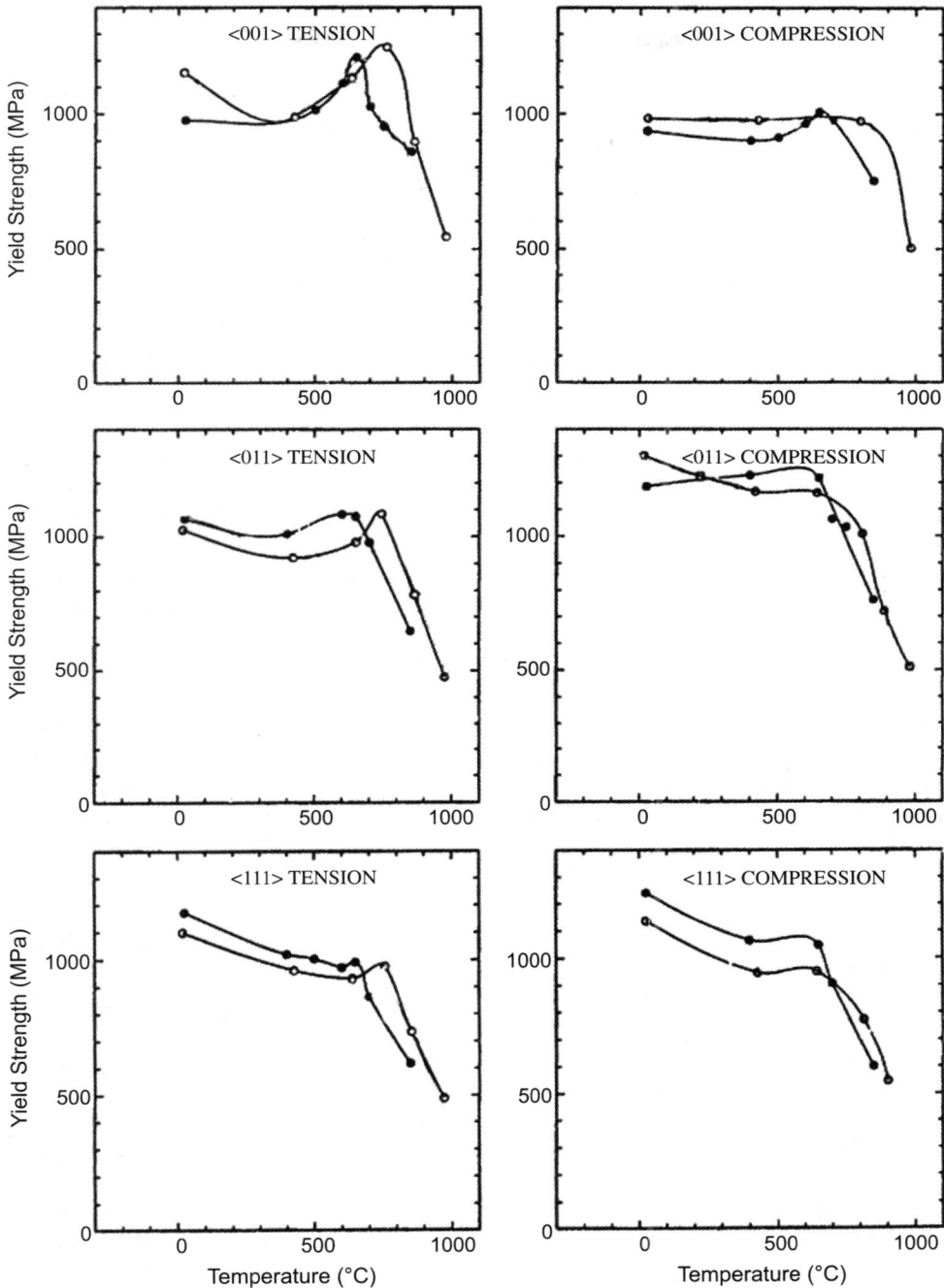

Figure 32 Variation of the 0.2% yield stress with temperature and crystallographic orientation for (i) the CMSX-4 single-crystal superalloy (Allen, 1995) and (ii) the PWA1480 single-crystal superalloy (Shah and Duhl, 1984). Solid symbols: Allen, 1995; open symbols: Shah and Duhl (1984).

incompatible with planar slip on the {111} plane in the matrix without climb at the γ/γ' interface, but at higher temperatures this becomes feasible.

Thus it is very clear that the two phases in the so-called single-crystal superalloys interact in a subtle way during yielding, with the matrix γ phase providing a reinforcing effect. Vattre et al. (2009) have used modeling methods to attempt to rationalize these effects. Dislocation dynamics calculations were used to estimate the critical stresses needed to propagate pairs of $a/2\langle110\rangle$ dislocations through a regular array of cubic γ' precipitates, with the volume fraction varying between 0.40 and 0.70. In these simulations, the dislocations are pulled into the precipitates as strongly paired couples by their bowing in the γ channels. For this modeling, the resistance of the dislocations pairs in the γ' is treated using equations for the anomalous flow stress for Ni$_3$Al developed by Demura et al. (2007) with dislocation glide being restricted to the octahedral planes. The precipitates are presumed to be perfectly coherent, elasticity is isotropic and a friction stress for solid-solution hardening is included. Assuming APB energies of between 100 and 350 mJ m^{-2}, the authors show a square root dependence on the APB energy, and a strong linear increase with γ' fraction at constant channel width. The dependence on APB energy is consistent with the analysis for the smaller volume fraction (polycrystalline) alloys of Section 22.4 but shows a greater sensitivity to the volume fraction f: linear rather than $f^{1/2}$. Vattre et al. find a linear decline in flow stress with channel width at constant volume fraction, i.e. the yield stress drops with increasing precipitate size: a trend that is in line with Shah and Duhl (1984) and alloys of lower volume fraction, but which contradicts Sengupta (1994).

An important point concerns the role of cube slip in the single-crystal superalloys, a topic that has proven controversial. Cube slip—the transition to favor glide on the {001} planes at high temperatures—is widely used in creep modeling because it successfully explains the observed orientation dependence of creep in single crystals, see Ghosh et al. (1990), Pan et al. (1995) and McLean (1995). In single-phase ordered L1$_2$ intermetallics cube slip undoubtedly occurs and is responsible for the end of the anomalous rise in the flow stress. Although macroscopic slip traces consistent with cube slip have also been seen on the surface of deformed two-phase single crystals, microscopically these have been shown to be caused by duplex slip involving the zigzag of dislocations of the same $a/2\langle110\rangle$ Burgers vector on two different {111} planes giving an average {001} slip plane. This has been demonstrated by Bettge and Osterle (1999) using tensile tests at 650 °C and 750 °C on the low-volume fraction, single-crystal alloy SC16, Sass and Feller-Kniepmeier (1998) via the creep of CMSX-4 at 850 °C; and Phillips and Mills (2010) in a detailed study of fatigue in the alloy Rene 104.

This effect has been simulated in a single-crystal microstructure by Vattre et al. (2009, 2010), who assumed boundary conditions to exclude the possibility of dislocations entering the precipitates and thus cutting them. A model was built that coupled dislocation dynamics with finite element estimates to evaluate the evolution of long-range stresses. The assumption was made that the dislocations are confined to the γ channels. A [111] orientation—for which cube slip might be expected—showed a yield drop and no work hardening as observed experimentally by Shah and Duhl (1984) and Bettge and Osterle (1999). Deformation was shown to occur by dislocations with a common Burgers vector but different {111} slip planes, which are able to combine in a zigzag configuration to give an apparent cube-slip on the macroscopic level. This mechanism eliminates most of the dislocation loops entering the channels through dipole annihilation and hence accumulates little dislocation density to inhibit further slip. Results from these simulations show a striking resemblance to the observations of (Bettge and Osterle, 1999) who studied a deformed single-crystal alloy of 40% γ' volume fraction—see **Figure 33**.

Such calculations provide insights into the performance of single-crystal superalloys under these conditions. The operation of coupled pairs of slip systems in each channel—the elimination of much of the dislocation activity, which might otherwise line the channels to produce a back-stress—serves to

(a)

(b)

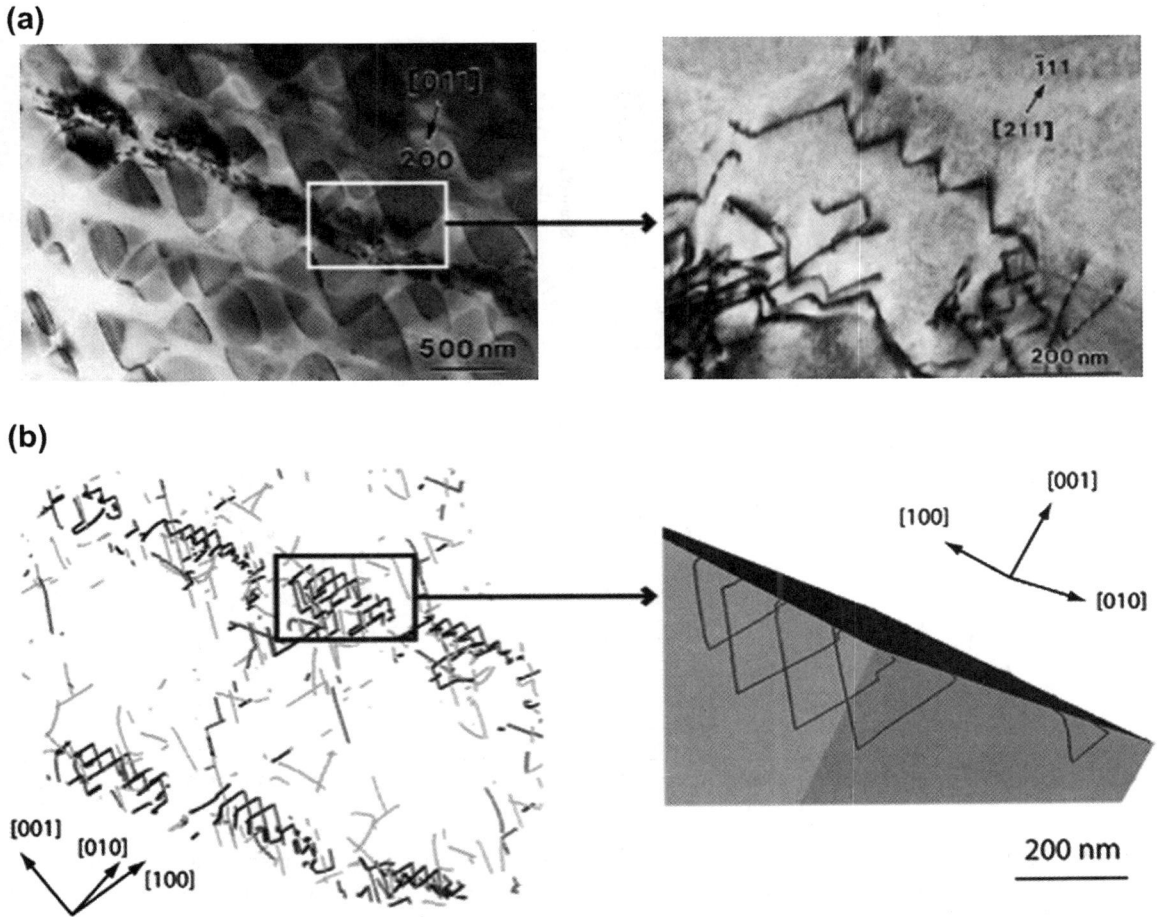

Figure 33 Comparison of observed and simulated dislocation structures due to yielding of a [111]-oriented single crystal at 650 °C: (a) TEM images due to Bettge and Osterle (1999); (b) simulations due to Vattre et al. (2010). (For color version of this figure, the reader is referred to the online version of this book.)

emphasize the role played by the microstructural architecture of the two-phase structure in determining the mechanical response. Although some penetration of the γ' by $a/2\langle 110\rangle$ dislocations may undoubtedly occur, a major role of the precipitates is to exclude and constrain dislocation activity to the γ channels.

22.5.2 Creep Deformation Behavior of Single-Crystal Superalloys

It is appropriate to consider the creep performance of single-crystal superalloys under static, uniaxial stress in the technologically important $\langle 001\rangle$ orientation. This corresponds to the direction of centrifugal loading of the turbine blades in engine applications, due to the preferred solidification direction during processing. **Figure 34** shows creep data (Matan et al., 1999a, 1999b) for the widely employed CMSX-4 superalloy (Harris et al., 1992) at temperatures between 850 °C and 1000 °C.

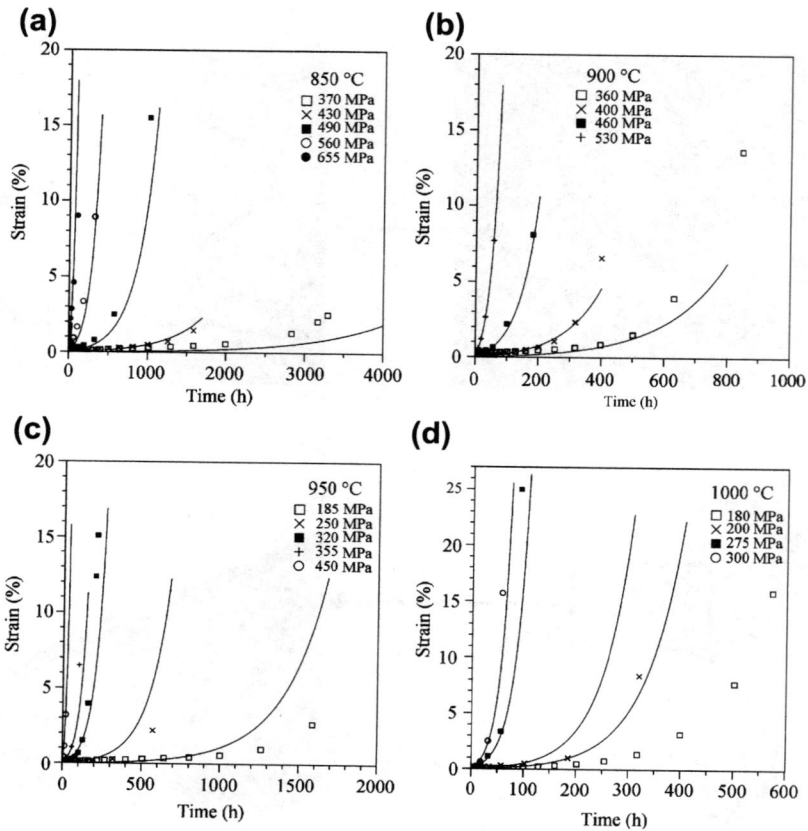

Figure 34 Comparison of creep data for CMSX-4 single-crystal superalloy and the tertiary creep model described by Eqns (29) and (31). Note that the agreement between model and experiment is generally good, except at high temperatures and low stresses, where the rafting effect becomes operative.

The points correspond to the raw creep data, with the lines to a fit to the data using the model described below. The following points are evident. Note that the applied stress does not exceed 500 MPa; values greater than this are not normally attained under engine conditions. First, there is no evidence of the primary creep, which requires the cutting of the γ' by APBs and the stacking faults, which requires a uniaxial stress of greater than about 500 MPa (Rae and Reed, 2007). The mechanism of this is considered in detail in Section 22.5.6. Second, the creep strain rate increases monotonically with increasing strain, in what has been termed a tertiary-type behavior (Ghosh et al., 1990; McLean, 1995). In fact, the creep strain is often found to increase exponentially with time (Matan et al., 1999a, 1999b). Third, at 1000 °C or higher, creep strain evolution is slower than that predicted by extrapolation of the data from lower temperatures, particularly for smaller values of stress when the creep strain is accumulated less rapidly. Compare for example the experimental data and predictions of **Figure 34d**.

Insight into the prevailing mechanism of deformation has been provided by characterization using electron microscopy (Pollock and Argon, 1992) and mathematical analysis of the measured creep data

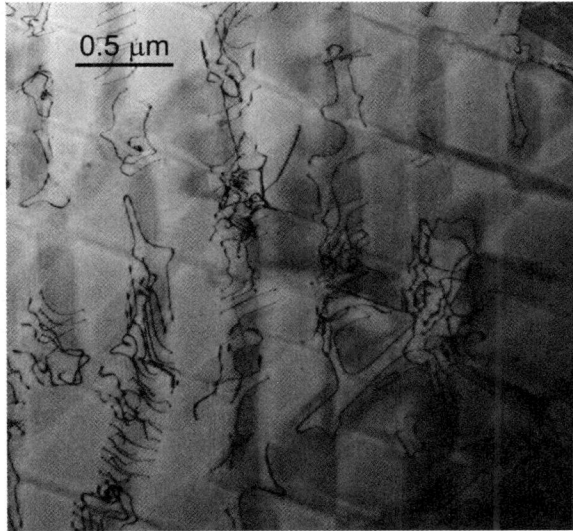

Figure 35 Transmission electron micrograph of the CMSX-4 single-crystal superalloy, deformed to 0.04% strain in 1890 h at 750 °C and 450 MPa. Note the localization of $\langle 1\bar{1}0 \rangle \{111\}$ dislocation activity in a limited number of γ channels. The foil normal is {111}.

(Ghosh et al., 1990; Matan et al., 1999a, 1999b); these confirm that the mode of deformation behavior is rather unique. Little if any evidence is found of the creep cavitation, which is so prevalent in poly-crystalline nickel-based superalloys (Betteridge and Shaw, 1987). Any γ' coarsening is very limited, provided that the temperature is not too high so that the rafting of Section 22.5.4 is avoided. Instead, the dominant, prevailing damage is a dislocation density that—starting from very low levels—increases steadily through life. In the tertiary regime, prestraining weakens the material, in contrast to the strengthening it causes in many polycrystalline engineering alloys. Dislocation activity is of the $a/2\langle 1\bar{1}0 \rangle \{111\}$ type (Pollock and Argon, 1992; Matan et al., 1999a, 1999b; Epishin and Link, 2004), and restricted to the γ channels between the γ' particles—see **Figure 35**. The γ' particles remain intact, i.e. they are not sheared by the creep dislocations; therefore, substantial cross-slip and/or climb is required for substantial deformation to occur, since the γ' fraction is high and consequently the γ channels too thin to allow the dislocations to percolate the γ/γ' structure, if this were not the case. The leading segments of the dislocation loops are screw in character, consistent with the requirement for cross-slip; as they expand through γ matrix channels, segments of mixed or edge character are deposited at the γ/γ' interfaces (Pollock and Argon, 1992). The macroscopic deformation is found to be relatively homogeneous, with usually two or more slip systems operative for crystal orientations near $\langle 001 \rangle$ (Matan et al., 1999a, 1999b).

Thus, for the single-crystal superalloys there is no evidence of a steady-state creep rate, consistent with an underlying constant dislocation density; traditional recovery-controlled creep equations and power-law representations, which have been used for other engineering alloys (Nabarro and de Villiers, 1995) are therefore inappropriate. Instead, the observed strain softening can be rationalized in the following way (McLean 1995; Reed et al., 1999). One appeals to the observation that, *provided one deformation is restricted to the tertiary creep regime so that γ' shearing is avoided*, the creep strain rate $\dot{\epsilon}$ is to a good approximation proportional to the accumulated macroscopic creep strain ϵ. The dislocation density ρ is the state variable characterizing the state of damage, which varies with ϵ according to

$$\rho = \rho_o + C\epsilon, \tag{29}$$

where ρ_o is the initial dislocation density and C is a constant describing the dislocation multiplication rate. The macroscopic creep strain rate $\dot{\epsilon}$ is then given by combining Eqn (29) with Orowan's equation

$$\dot{\epsilon} = \rho \mathbf{b} v, \tag{30}$$

where $\mathbf{b}$ is the Burgers' vector and v is the dislocation velocity. Provided v can be taken as constant, the proportionality of $\dot{\epsilon}$ and ϵ follows from the substitution of (29) into (30). One can then write

$$\dot{\epsilon} = \dot{\Gamma}_i + \Omega\epsilon, \tag{31}$$

where $\dot{\Gamma}_i$ is the initial strain rate. Calculations indicate (Matan et al., 1999a, 1999b) that each of Ω and $\dot{\Gamma}_i$ shows an Arrhenius-type dependence upon temperature, consistent with $\Omega = a \exp\left\{b\sigma - \frac{Q_1}{RT}\right\}$ and $\dot{\Gamma}_i = c \exp\left\{d\sigma - \frac{Q_2}{RT}\right\}$ where R is the gas constant. Six constants a, b, c & d and Q_1 & Q_2 are required to describe the temperature- and stress-dependence of the deformation behavior, and in practice the macroscopic creep curves are reproduced with reasonable accuracy. The curves given in **Figure 34** are produced using the simple model described, following the fitting procedures described in (Matan et al., 1999a, 1999b). Note that the creep performance at the highest temperatures is underpredicted; more will be said about this later in the context of the rafting effect. It follows from the model that at constant temperature and stress, the creep strain is proportional to $\exp\{kt\}$ where k is a constant and t is time. A criticism of the model is the relatively large number of fitting parameters needed. Moreover, it does not distinguish between the different components of the dislocation lines, and their capacities to be in trapped configurations—and thus needing to climb—or alternatively free to glide. A consequence is that the important composition-dependent terms are not identified. The computed activation energies Q_1 and Q_2 are also somewhat larger than expected (Matan et al., 1999a, 1999b), based upon the physical processes that are known to be occurring. These limitations are removed in the model of Section 22.5.4.

22.5.3   The Rhenium Effect

Additions of Re have been found to improve the creep resistance substantially (Reed, 2006). Thus whilst the so-called first generation single-crystal superalloys (PWA1480, Rene N4, SRR99) are Re-free, alloys such as CMSX-4, Rene N5 and PWA1484 (second generation alloys) contain Re at about 3 wt%; these have become widely employed in engine applications, due to their superior properties. Judicious alloying with Re improves the creep rupture life at temperatures of ~ 900 °C by an order of magnitude. The potency by which Re improves the high-temperature behavior is referred to in superalloy metallurgy as the Re-effect. Alloys exist with Re at greater concentrations—the so-called third generation alloys—of which CMSX-10 and Rene N6 are exemplars; however, due their increased density, cost and susceptibility to oxidation and precipitation of TCP phases such as σ (Rae and Reed, 2001), these have not been used as widely as was originally anticipated.

Historically, the choice of Re as the critical alloying addition to the single-crystal superalloys was made on the basis of empirical observations: the improved creep properties so conferred and a significant retardation in γ' coarsening kinetics (Giamei and Anton, 1985). The first attempt to provide a physical explanation for the Re-effect was made by Blavette et al. (1986). A one-dimensional atom probe was used to characterize Re-doped PWA1480. Datasets of order $\sim 10,000$ atoms were acquired, with emphasis placed on analyzing the γ matrix phase to which Re partitions strongly. On the basis of discontinuities in the gradient of the ladder diagram for Re, it was suggested that Re is prone to clustering in the γ phase, with a conjecture made that clusters of Re might act as efficient obstacles to dislocation motion (Blavette et al., 1988). However, more recent statistical analyses using friends-of-friends algorithms on three-dimensional atom probe tomography datasets of tens of millions of atoms confirm that Re is not prone to clustering in nickel-based superalloys (Mottura et al., 2010); the discontinuities observed on ladder diagrams for Re have been shown to be consistent with random statistical fluctuations in the distribution of Re. The absence of Re clustering is in agreement with calculations made using DFT, which indicate a large, repulsive pairwise Re-Re interaction energy of -0.42 eV (Mottura et al., 2012): sufficient to resist any clustering. Spectra from extended X-ray absorption spectroscopy experiments confirm that clustering does not occur (Mottura et al., 2008). Consistent with its known partitioning characteristics, the transformation $\gamma \rightarrow \gamma + \gamma'$ causes partitioning of Re ahead of the moving γ/γ' interface; the resulting concentration profile has been detected using atom probe tomography and rationalized using diffusion-controlled phase transformation theory (Mottura et al., 2010).

Whilst the Re-effect has yet to be rationalized completely, some clues have emerged from analysis of Ni–Re/Ni diffusion couples and interdiffusion coefficient data—and associated activation energies—extracted from them (Karunaratne and Reed, 2003). In comparison with other elements from the d-block, the diffusion of Re in Ni is very slow, see **Figure 10**, with only Os diffusing at such a low rate (Youssef et al., 2010). The measured values and experimental trends are consistent with calculations made using DFT methods; quantum-mechanical models for diffusion based upon the five-frequency model indicate that low rates of Re are due to a large Re/vacancy exchange energy of ~ 160 kJ/mol, see **Figure 36**, which is appreciably greater than for elements taken from the far-west and far-east of the d-block of transition metals—for example Hf and Au respectively. The correlation factor is close to unity for Re (Janotti et al., 2004). These results indicate that Re is expected to retard processes such as dislocation climb, which require vacancy-mediated diffusional flow, and thus dislocation creep in general. A quantitative treatment follows.

22.5.4 Model for Composition Dependence of Creep Deformation

It is reasonable to assume that the total content of dislocation segments will consist of (1) those free to glide through the γ channels, and (2) those trapped at the γ/γ' interfaces, which need to move nonconservatively to overcome the precipitates (Zhu et al., 2012). A postulate can then be made: at any instant, trapped dislocation configurations predominate; it is their rate of release—and the necessary climb processes implied—which is the origin of alloy composition-dependence of the macroscopic creep rate. On the other hand, it is the active glide of the recently released dislocations—before the next trapping event—which contributes most to the creep strain accumulation. Thus the mean free path through which the dislocation can travel becomes vital; this depends upon the scale of the microstructure but is independent of the alloy chemistry, once the latter's influence on the microstructural architecture is accounted for.

The situation is illustrated in **Figure 37**, where the configuration denoted "A" represents the instant at which the dislocation encounters the particle and becomes held up. The force available to support climb will in general be finite, due to the externally applied stress field. Thus a nonequilibrium vacancy concentration near the vicinity of the dislocation is set up, which induces a chemical potential

(a)

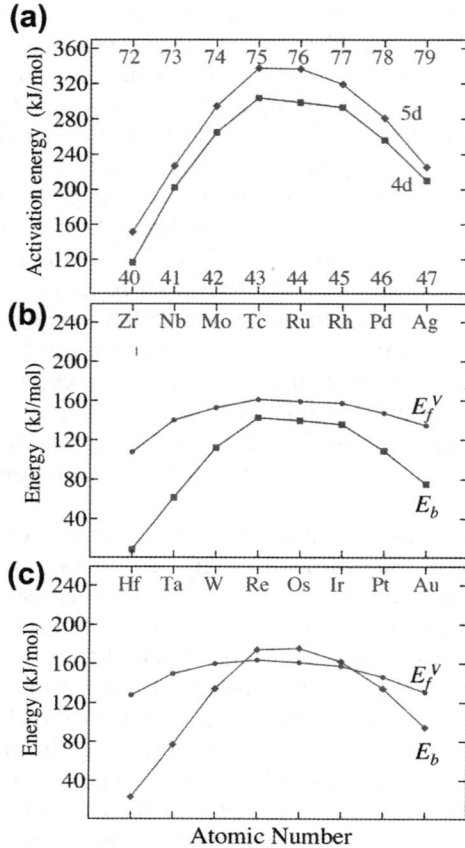

Figure 36 Predictions using density functional theory for the impurity diffusion of solutes in nickel: (a) variation of the activation energy for diffusion in Ni with atomic number, for d-block elements; (b) variation of barrier energy and solute-vacancy interaction energy for 4d elements; (c) variation of barrier energy and solute-vacancy interaction energy for 5d elements. After Janotti et al. (2004).

difference, and thus migration of vacancies. The absorption/emission of vacancies from dislocations results in the formation of jogs that propagate along the dislocation lines leading to climb of the dislocation along the surface of the particle, configuration "B". Eventually, the dislocation approaches a configuration such that thermal activation of a single jog is enough to release it from its trapped configuration. This is situation "C" in **Figure 37** and corresponds to the critical escape event. Clearly, any accurate model will require an estimate of the frequency of these escape events. The geometry of the situation implies $\phi_p = d^3/(w+d)^3$ so that $w = d(1/\phi_p^{1/3} - 1)$, where ϕ is the fraction of the γ' particles, which are assumed to be regularly spaced cubes of side d. The channels between the precipitates have width w.

Attention is focused on the microscopic creep shear strain rate $\dot{\gamma}$. This can be related to the mobile gliding dislocation density ρ_g via the Orowan equation $\dot{\gamma} = b\rho_g v_g$ where b is the Burgers vector and v_g the time-averaged glide velocity. It is assumed that dislocation activity is restricted to γ and the γ/γ'

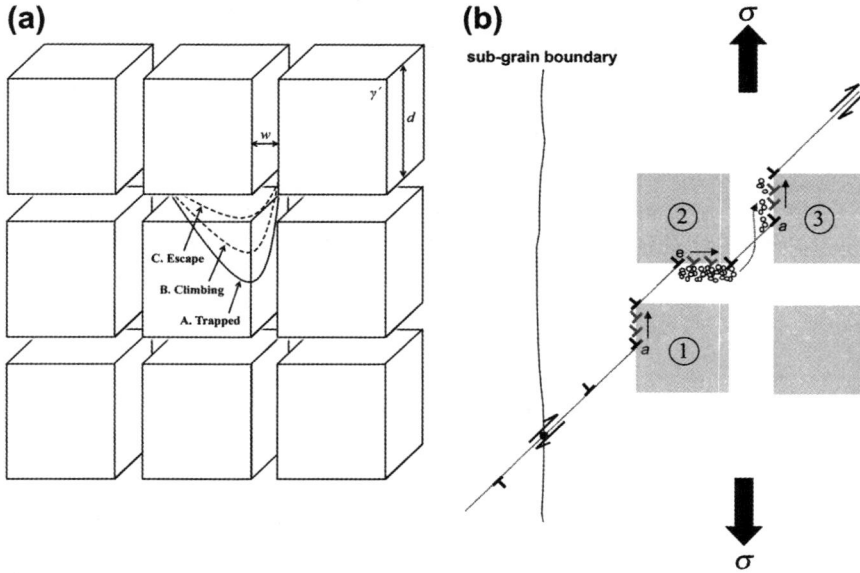

Figure 37   Schematic illustration of dislocation configurations during dislocation creep in a nickel-based single-crystal superalloy.

interfaces. To acknowledge that deformation is limited to the matrix channels, a simple correction is introduced so that

$$\dot{\gamma} = \left(1 - \phi_{\rm p}\right) b \rho_{\rm g} v_{\rm g}. \tag{32}$$

The key step is to partition the total mobile dislocation density $\rho_{\rm m}$ into that climbing at the interfaces $\rho_{\rm c}$—thus held up by the γ' particles—and that gliding within the γ channels $\rho_{\rm g}$. Hence

$$\rho_{\rm m} = \rho_{\rm g} + \rho_{\rm c}, \tag{33}$$

so that the net rate of change of gliding dislocations is the sum of the rates of release and arrest of gliding dislocations. The release rate can be related to the fraction of the total trapped dislocations in Configuration "C", which are released at a frequency $\Gamma_{\rm e}$, to be determined. The probability that a dislocation in this position will escape is dependent on the probability that the dislocation will encounter a particle. This is approximately equal to $\phi_{\rm p}$, and a geometrical factor taken to be $b/(d/2)$, which accounts for the dislocation being in the correct configuration for this to happen. Similarly, the rate at which a gliding dislocation becomes trapped depends on the time for the dislocation to move from one particle and then to be trapped again by the next particle. The rate is approximated by the glide velocity $v_{\rm g}$ divided by the interparticle spacing w. From these considerations the net rate of gliding dislocations is given by

$$\dot{\rho}_{\rm g} = \rho_{\rm c} \phi_{\rm p} \frac{2b}{d} \Gamma_{\rm e} - \rho_{\rm g} \frac{v_{\rm g}}{w}. \tag{34}$$

Substituting for ρ_c from Eqn (33) and factorizing for the gliding dislocation density ρ_g, one arrives at

$$\dot{\rho}_g + \rho_g\left(\phi_p\frac{2b}{d}\Gamma_e + \frac{v_g}{w}\right) = \rho_m\phi_p\frac{2b}{d}\Gamma_e. \tag{35}$$

It is assumed that $t_c \gg t_g$, where t_g is the time for a dislocation to glide from on particle to the next, and t_c is the time for a trapped dislocation to escape ($\Gamma_e = 1/t_c$). Since $v_g = w/t_g$, it follows that $v_g/w \gg \Gamma_e$. Thus Eqn (35) becomes

$$\dot{\rho}_g + \rho_g\frac{v_g}{w} = \rho_m\phi_p\frac{2b}{d}\Gamma_e. \tag{36}$$

Integrating and by application of the lower limit $\rho_g = 0$ at $t = 0$, one has

$$\rho_g = \rho_m\phi_p\frac{2b}{d}\Gamma_e\frac{w}{v_g}\left[1 - \exp\left\{-\frac{v_g}{w}t\right\}\right]. \tag{37}$$

Since one seeks an expression for the creep rate over a duration that is long compared to the escape time, one can assume $\left[1 - \exp\left\{-\frac{v_g}{w}t\right\}\right] \approx 1$. Thus

$$\rho_g v_g = \rho_m\phi_p b\frac{2w}{d}\Gamma_e. \tag{38}$$

Recalling the Orowan equation and substituting for w, the shear strain rate is then

$$\dot{\gamma} = 2\rho_m b^2\phi_p\left(1 - \phi_p\right)\left(1/\phi_p^{1/3} - 1\right)\Gamma_e. \tag{39}$$

It is necessary to estimate Γ_e, the frequency at which the jogged dislocation segment is expected to escape by a vacancy emitting process. Following (Friedel, 1964), one has

$$\Gamma_e = nv\exp\left\{-\frac{\Delta E - \Delta W}{kT}\right\}, \tag{40}$$

where n is the density of locations that can either emit or absorb vacancies and v is the Debye frequency. The term ΔE is the activation energy for the process and ΔW the work done by the dislocation moving from the pinned configuration to the released (gliding) state. During the release process the escape of climbing dislocation will allow localized glide in the matrix. The slipped area is bw so with the shear stress τ acting on the dislocation line, one has

$$\Delta W = \tau b^2 w. \tag{41}$$

Combining Eqns (40) and (41), the escape frequency can be written as

$$\Gamma_e = nv\exp\left\{-\frac{\Delta E}{kT}\right\}\exp\left\{\frac{\tau b^2 w}{kT}\right\}. \tag{42}$$

Since the necessary mass transport will be effected by vacancy flow, the effective diffusivity for the necessary climb has the form as

$$D_{\text{eff}} = nvb^2 \exp\left\{-\frac{\Delta E}{kT}\right\}. \tag{43}$$

Therefore, combining Eqns (42) and (43), the escape frequency of a vacancy emitting jogged segment on the horizontal channel is

$$\Gamma_{e,e} = \frac{D_{\text{eff}}}{b^2} \exp\left\{\frac{\tau b^2 w}{kT}\right\}, \tag{44}$$

where the second subscript 'e' refers to the vacancy *emission* occurring. Similarly, for a vacancy absorbing jogged segment on the vertical channel, the escape frequency is given by

$$\Gamma_{e,a} = \frac{D_{\text{eff}}}{b^2} \exp\left\{-\frac{\tau b^2 w}{kT}\right\}. \tag{45}$$

The number of vacancy emitting and absorbing jogs is expected to be approximately equal so that the overall escape frequency is given by the mean of Eqns (44) and (45). Thus

$$\Gamma_e = \frac{D_{\text{eff}}}{b^2} \sin h\left\{\frac{\tau b^2 w}{kT}\right\} \tag{46}$$

The expression for the shear strain rate on the active slip system is arrived at by combining Eqns (39) and (46) to give

$$\dot{\gamma} = 2\rho_m \phi_p D_{\text{eff}} \left(1 - \phi_p\right) \left(1/\phi_p^{1/3} - 1\right) \sinh\left\{\frac{\tau b^2 w}{kT}\right\}. \tag{47}$$

Notice that the microstructural parameters ϕ_p, w and the implied d appear in Eqn (47) in a way that make their contribution to $\dot{\gamma}$ unclear at first sight.

For practical applications one is interested in uniaxial deformation along the $\langle 001 \rangle$ axis, and thus the contribution of the shear strain rate $\dot{\gamma}$ of Eqn (47) to the macroscopic creep strain rate $\dot{\epsilon}_{\langle 001 \rangle}$. This can be estimated from the contributions from the active $a/2\langle 1\bar{1}0\rangle\{111\}$ slip systems. For the symmetrical loading considered here, there are eight octahedral slip system with nonzero-resolved shear stress and equal Schmid factors (Ghosh et al., 1990). Thus one has

$$\dot{\epsilon}_{\langle 001 \rangle} = \frac{8}{\sqrt{6}} \dot{\gamma}. \tag{48}$$

One then has the coupled equation set

$$\dot{\epsilon}_{\langle 001 \rangle} = \frac{16}{\sqrt{6}} \rho_m \phi_p D_{\text{eff}} \left(1 - \phi_p\right) \left(1/\phi_p^{1/3} - 1\right) \sinh\left\{\frac{\sigma b^2 w}{\sqrt{6} K_{\text{CF}} kT}\right\}; \tag{49}$$

$$\dot{\rho}_m = C\dot{\epsilon}_{\langle 001 \rangle}, \tag{50}$$

Table 5   Critical microstructural and kinetic parameters needed for the creep model described. After Zhu et al. (2012)

Initial dislocation density, $\rho_{m,0}$	10^{10} m^{-2}
Dislocation multiplication parameter, C	100 m^{-2}
Burgers vector, **b**	2.54×10^{-10} m
γ' particle size, d	400 nm
Volume fraction of γ' phase, ϕ_p	0.68

which can be integrated from time zero using standard finite difference methods. Note that in the case of Eqn (47) one needs an estimate of the dislocation multiplication parameter C and the initial dislocation density $\rho_{m,0}$. The term $K_{CF} = 1 + 2\phi_p^{1/3}/3\sqrt{3\pi}(1 - \phi_p^{1/3})$ is a constraint factor that accounts for the close proximity of the cuboidal particles in these alloys (Pollock and Argon, 1994).

The model described by Eqns (32)–(50) has advantages over the one presented in Section 22.5.2, despite its relationship to it. Just five parameters are needed, and all of these are physically based. Of these, three may be regarded as boundary conditions for the problem: the microstructural architecture defined by the size and spacing ϕ_p and w, and the initial dislocation density $\rho_{m,0}$. The remaining two parameters—the dislocation multiplication parameter C and the effective diffusion coefficient D_{eff}—are kinetic parameters that are needed to describe the dislocation flow problem. **Table 5** summarizes typical values that have been used by Zhu et al., (2012) for CMSX-4. **Figure 38** compares the predictions of this model with the time to 1% strain, for the CMSX-4 superalloy. Particularly at the lower temperatures the accuracy of the model is reasonable, although it creep strain is underpredicted at higher temperatures. The dashed lines are calculated using a modification to the basic model, the details of which are dealt with in the following section.

An estimate of the dependence of the creep strain evolution on alloy composition emerges from the model, via differences in ϕ_p and particularly D_{eff}. The major challenge is to work out how to estimate D_{eff} for alloys that contain a number of different alloying elements. By appealing to absolute reaction rate theory, Zhu et al. (2012) have employed a weighted average of the activation energies from the diffusion data of (Karunaratne and Reed, 2003); the influence of Re is then pivotal on the basis of its large activation energy for thermally activated processes. The influence of Re in this way is sufficient to explain the tenfold variation in the time to 1% strain, which corresponds to about one Larson–Miller parameter, see **Figure 39**. Predicted creep curves are given at 900 °C/400 MPa for SRR99, CMSX-4, TMS-75 and PW1497—which are representative of typical first, second, third and fourth generation alloys.

Thus, whilst better models are needed for the rate of climb of dislocations at γ/γ' interfaces—electron theory methods may become available for this—it would appear that the model described here is capable of accounting for the role played by Re, at least approximately.

22.5.5   The Rafting Effect

The underestimation of the creep resistance in the high-temperature/low-stress regime—as implied by the data in **Figure 34d**—can be rationalized at least in part by accounting for the so-called rafting effect (Pollock and Argon, 1994; Carroll et al., 2008). When rafting occurs, the γ'-precipitate size and spacing no longer remains invariant during deformation, which invalidates a major assumption of the model above. **Figure 40** illustrates the rafted morphology in the CMSX-4 superalloy—in this case after creep

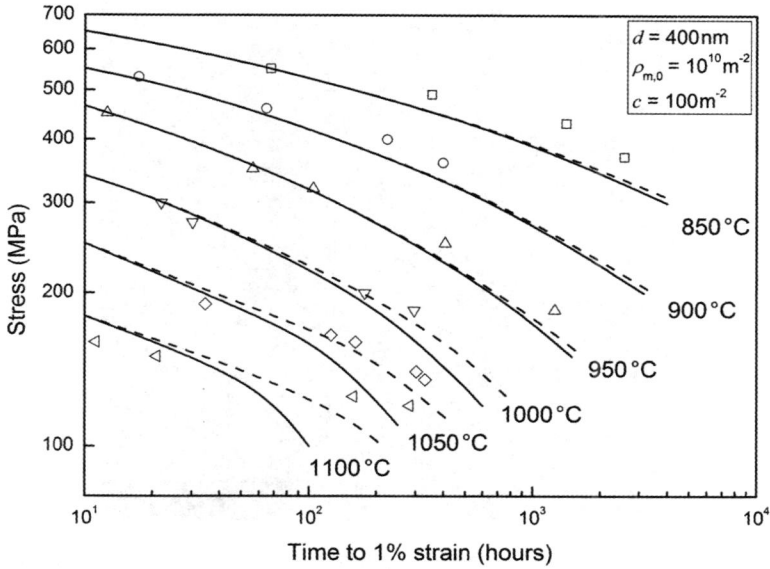

Figure 38 Comparison between simulated time to 1% strain and experimental creep data for CMSX-4. The dashed lines include the modifications to the model needed to account for the rafting effect.

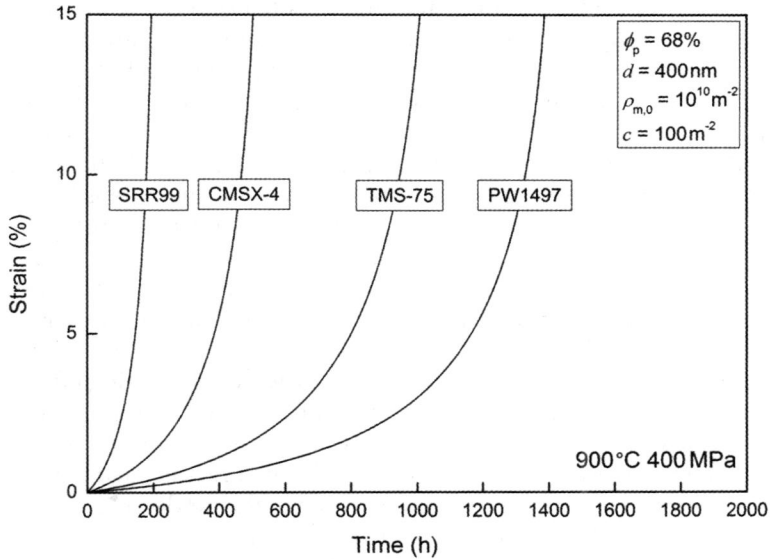

Figure 39 Predictions of creep curves for different generations of single-crystal superalloy at 900 °C and 400 MPa. After Zhu et al. (2012).

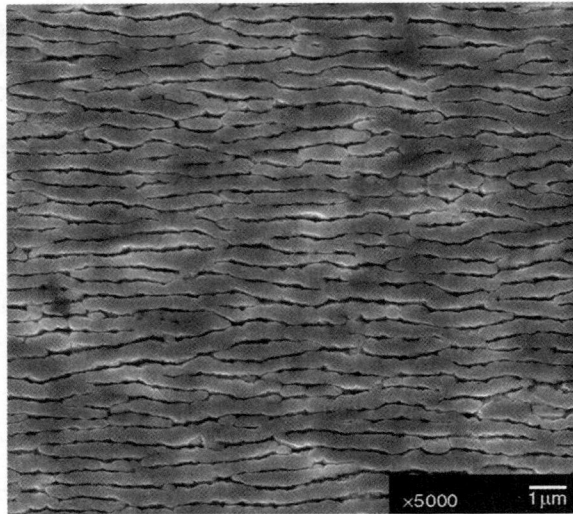

Figure 40   Rafted γ/γ' structure in the CMSX-4 superalloy, deformed at 1150 °C and 100 MPa to 0.39% strain in 10 h. After Reed et al. (2007).

deformation to 0.39% plastic strain in 10 h at 1150 °C and +100 MPa. The direction of applied stress lies in the plane of the micrograph in the vertical direction; thus the normals to the broad faces of the rafts—which are plate-shaped in form—lie perpendicular to the axis of loading. This is always found to be the case for negatively misfitting alloys such as CMSX-4, into which category most commercially developed alloys fall. Notice, see **Figure 40**, that as many as ten or more γ' precipitates coalesce in the transverse direction; thus a fully developed raft is formed from one hundred or more precipitates of total cross-sectional area equal to several hundreds of μm^2, since the rafts have a thickness of order one γ' particle. Under these circumstances for negatively misfitting alloys, rafting reduces the ability of the $a/2\langle 1\bar{1}0\rangle\{111\}$ dislocations to percolate γ since the vertical channels are gradually removed by it. Extensive rafting leads eventually to the inversion of the γ/γ' structure such that the continuous, matrix phase is γ' rather than γ. When the sign of the stress is reversed from tension to compression—or alternatively when the lattice misfit is positive—needle-shaped rafts are found whose long axes lie parallel to the direction of loading. This appears to be important for out-of-phase (OP) thermal–mechanical fatigue (TMF), see Section 22.5.5. The effects of lattice misfit and modulus mismatch on raft morphology are consistent with the elasticity arguments first made by Pineau (1976).

One can account for the effect of rafting on creep deformation by modifying the theory of above. However, it is necessary first to consider the mechanism by which the rafting effect occurs, so that the correct physics is introduced. The mechanism of the rafting effect has proven to be controversial (Nabarro and de Villiers, 1995). Whilst a driving force that is merely elastic in origin is available, in practice differential microplasticity on the scale of the γ/γ' structure is needed to initiate it (Buffiere and Ignat, 1995; Veron et al., 1996). The vacancy flow envisaged in the model of Section 22.5.4 is consistent with this picture. Nevertheless, the exact details of the necessary mass transport are not completely known, as pointed out by Pollock and Argon (1994). A first, potentially plausible, situation requires simple flow of material from the vertical γ channels into the horizontal ones, see **Figure 41**; the γ' precipitates are forged together, with their centroids separating consistent with an increase in the

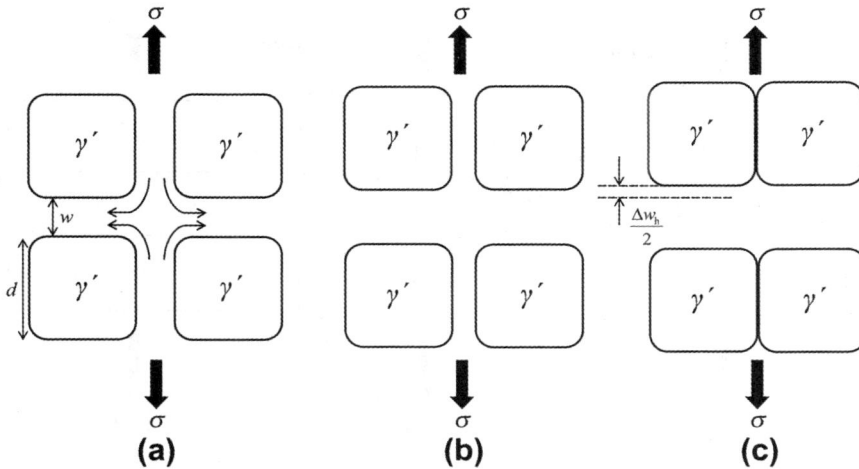

Figure 41 Schematic illustration of the changes in microstructural dimensions during rafting by a simple flow of matrix material from the vertical channels to the horizontal channels. From (a) to (c) the precipitates are assumed to retain their cuboidal shape and will coalesce along their vertical interfaces.

horizontal channel width Δw_h such that

$$\Delta w_h = w + \frac{w^2}{d}, \tag{51}$$

so that a net uniaxial strain ϵ_h will develop according to

$$\epsilon_h = \frac{\Delta w_h}{d + w}, \tag{52}$$

For a typical single-crystal microstructure, one has $d = 400$ nm and $\phi_p = 0.68$ so that $w = 54.9$ nm. In this first situation, the mass flow associated with a fully rafted structure is then consistent with $\Delta w_h = 62.4$ nm and $\epsilon_h = 10.4\%$. The latter value in particular in substantially greater than the strain of a few tenths of a percent, which is all that is required for rafting to be completed (Matan et al., 1999a, 1999b). On this basis, a second situation is considered more realistic, see **Figure 42**. Counter-related fluxes need to be invoked: (1) flow of γ-stabilizing elements (Co, Cr and Re, etc) from vertical to horizontal channels coupled with (2) flow of γ' forming elements (Al, Ti, Ta) into the vertical channels. In the second situation, the important point is that the centroids of the γ' particles do not separate, and instead one has

$$\Delta w_h = \frac{\Delta w_v}{1 + \Delta w_v/d} \tag{53}$$

so that $\Delta w_h = 48.3$ nm, a value very similar to that for the first scenario. However, in this case, the precipitates contract in the direction of the applied stress, so that no net strain is accumulated. The stereological data support the second situation (Matan et al., 1999a, 1999b).

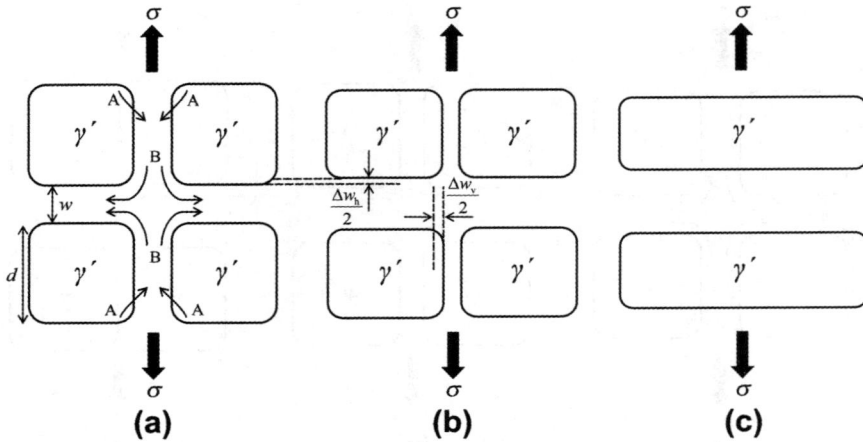

Figure 42 Schematic illustration of the changes in microstructural dimensions by a more complex preferential flow of γ- and γ'-forming elements: (a) initial state; (b) microstructure in a rafted state, no coalescence of γ' particles and (c) fully rafted structure. "A" indicates the flux of γ'-forming elements; "B" the flux of γ-stabilizing elements.

It follows from the above argument that the overarching influence of the rafting effect will be to cause a net decrease in the overall escape frequency Γ_e of Eqn (46), which will diminish to a vanishingly small value as w—more specifically the vertical channel width w_v—as rafting continues to completion. The rate of thinning $\dot{w}_v$ is expected to be proportional to the product ρv_{climb} on account of its anticipated dependence on the rate of vacancy absorption at the γ/γ' interfaces. It follows that (Zhu et al., 2012)

$$\dot{w}_v = A\rho d^2 v_{\text{climb}} = A\rho d^2 \frac{D_{\text{eff}}}{b}\sinh\left\{\frac{\tau b^2 w_v}{kT}\right\}, \tag{54}$$

where A is a parameter that is dimensionless on account of the introduction of the term d^2. Examination of the data in the literature, specifically the data for CMSX-4 in Epishin et al. (2008), suggests $A = 2.36 \times 10^{-2}$, which gives the optimum fit to the experimental data (Zhu et al., 2012). The width of vertical channels w_v is then a state variable that needs to be updated as the simulation of creep deformation proceeds, according to

$$w_v = w - \Delta w_v, \tag{55}$$

so that one employs in place of Eqn (49) (Zhu et al., 2012)

$$\dot{\epsilon} = \frac{16}{\sqrt{6}}\rho_m\phi_p D_{\text{eff}}\left(1 - \phi_p\right)\left(1/\phi_p^{1/3} - 1\right)\sinh\left\{\frac{\sigma b^2 w_v}{\sqrt{6}K_{\text{CF}}kT}\right\}. \tag{56}$$

Figure 38 shows the predictions of time to 1% strain for the case of CMSX-4, with rafting taken into account using Eqns (53)–(56). The effect of rafting is accounted for in the dash curves, but not in the solid curves. The influence of rafting is most prevalent at temperatures greater than 1000 °C, and lower stress levels that enable sufficient time for rafting to proceed. At lower temperatures the original predictions of the basic model prove sufficient. These numerical calculations help to clarify the difficult issue of whether rafting is beneficial or detrimental to creep. As shown, extrapolation to the rafting regime from the lower temperature/higher stress regime indicate that rafting is beneficial, when judged in this manner. But this finding must be placed in context. Rafting requires the introduction of dislocations which—consistent with the state variable approach of Section 22.5.1—will necessarily weaken the material. Thus, the performance of pre-rafted material is generally found to be poorer than virgin (uncrept) material (Pollock and Field, 2002). Moreover, the transverse properties of the rafted structure—particularly the yield stress—are particularly poor (Mughrabi, 2009). The effect of pre-rafting on properties is discussed in more detail in Pollock and Field (2002).

Further insight has emerged from studies of the dislocation activity, using transmission electron microscopy. When rafting occurs, it is always associated with interfacial dislocations at the γ/γ' interfaces that have the effect of relaxing the elastic misfit. With time, these reorganize themselves into equilibrium network configurations by processes involving climb. Consistent with the micro-mechanism of microplasticity in the tertiary regime of Section 22.5.1, the interfacial dislocation networks arise from $a/2\langle 1\bar{1}0 \rangle\{111\}$ creep dislocations originating in the γ phase; segments of these become captured by the γ/γ' interfaces, and the networks are formed by a series of dislocation reactions. The interfacial dislocations are primarily of edge character with the extra half planes in the γ' phase if $\delta \ll 0$. Thus by their introduction the elastic coherency of the interfaces is lost and misfit stresses are relieved. **Figure 43** is taken from the work of Field et al. (1992) who have analyzed the networks in detail using TEM microscopy—this specimen was interrupted after 20 h with 0.5% creep strain having been accumulated at 1093 °C. Note in particular the square configurations of four dislocations of Burgers vector of $a/2\langle 1\bar{1}0 \rangle$ type; they are of edge character and are thus efficient at relieving the misfit, but these segments have no shear stress resolved on them—thus they must have formed by reactions driven by reductions in misfit and dislocation line energies. A model has been proposed by these authors, which accounts for their formation, see **Figure 44**. Initially, on a γ/γ' interface, which has [001] as its normal, slip-deposited dislocations form two perpendicular sets, e.g. of $a/2[0\bar{1}1]$ dislocations, one resulting from slip on (111) and the other on $(\bar{1}11)$. These can react to form a single set with

Figure 43 a & b: TEM micrograph (a) and accompanying schematic interpretation (b) of an experimental single-crystal superalloy deformed to 0.5% creep strain in 20 h at 1093 °C, showing dislocation networks at the γ/γ' interfaces; arrows in (a) and (b) show equivalent positions. Foil normal {001}, equivalent to the direction of loading. After Field et al. (1992).

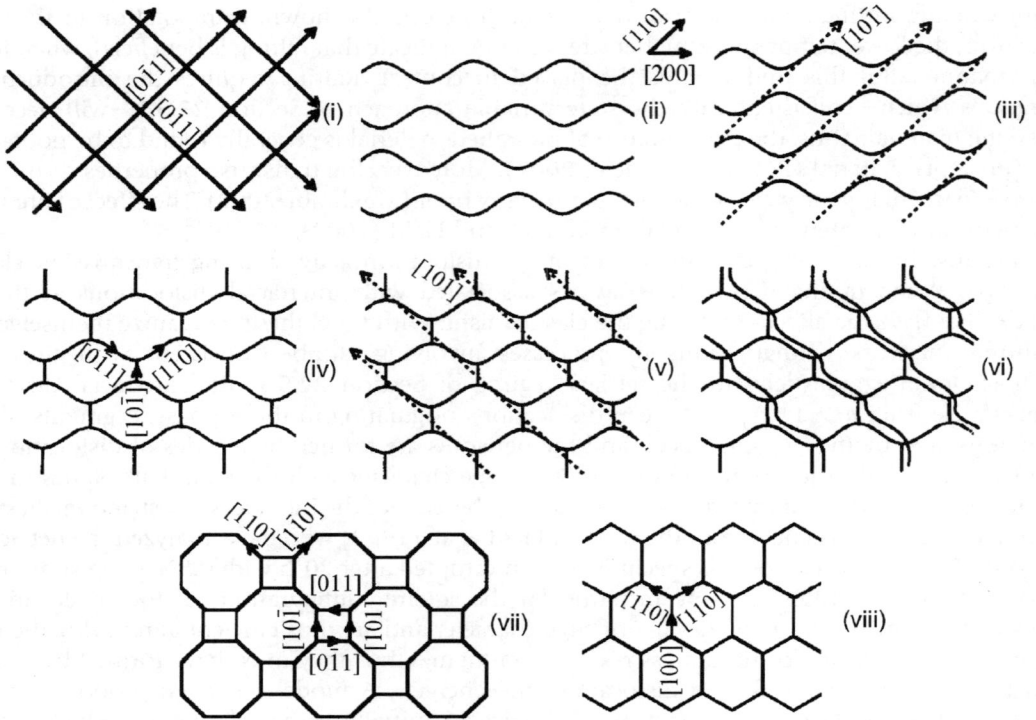

Figure 44 Schematic representation of dislocation reactions leading to the formation of mismatch accommodating nets on the (001) γ/γ' interfaces during creep. After Field et al. (1992).

average line vector [100]. Further $a/2\langle 1\overline{1}0\rangle$ dislocations resulting from slip on the other two octahedral planes $(1\overline{1}1)$ and $(\overline{1}\overline{1}1)$ can then be knitted into the net, consistent with reactions of the type (Field et al., 1992)

$$a/2[101] - a/2[0\overline{1}1] \rightarrow a/2[110] \tag{57a}$$

and

$$a/2[101] - a/2[1\overline{1}0] \rightarrow a/2[011] \tag{57b}$$

or alternatively

$$a/2[101] + a/2[10\overline{1}] \rightarrow a[100]. \tag{57c}$$

Thus it is apparent that the equilibrium network can arise only when sufficient creep dislocations have formed on all four of the {111} octahedral-type planes. These findings have been confirmed by more recent observations of Epishin et al. (2007), who have used novel etching techniques to remove the γ' rafts to reveal the dislocation activities.

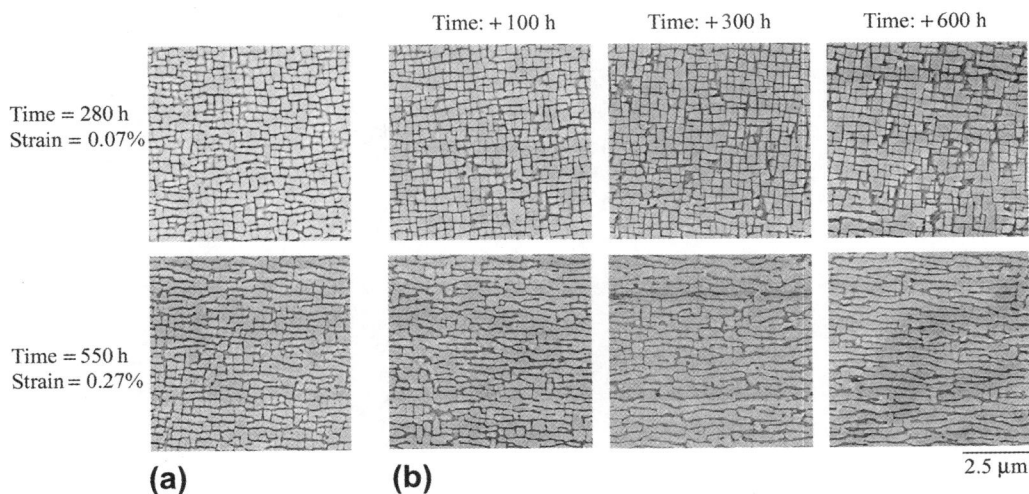

Figure 45 SEM micrographs of the CMSX-4 superalloy deformed at 950 °C. At left, γ/γ' microstructures after 280 h (0.07% strain) and 550 h (0.27% strain), with an applied stress of 185 MPa. At right, microstructures after further heat treatment of 100, 300 and 600 h, with zero applied stress. After Matan et al. (1999a, 1999b).

The relaxation of interfacial misfit by the above mechanisms is important because rafting occurs only once this has occurred (Pollock and Argon, 1994; Buffiere and Ignat, 1995; Veron et al., 1996; Matan et al., 1999a, 1999b). Clear evidence arises from observations on partially crept single crystals that continue to raft, even after the externally applied load is removed (Matan et al., 1999a, 1999b). The evidence suggests that the kinetics of rafting is not slowed by the removal of the stress field. **Figure 45** illustrates the results of experiments on CMSX-4 (Matan et al., 1999a, 1999b). For deformation at 950 °C and 185 MPa, an accumulated strain of about 0.1% is sufficient to relax the interfacial misfit and thereafter, rafting without the externally applied load continues at the same rate as with the load present.

The question of the direction of rafting is interesting. For negatively misfitting alloys, the rafts are found to lengthen normal to the direction of applied (positive) load. The traditional explanation (Mughrabi, 2009) invokes a consideration of the net stress in the vertical and horizontal channels, due to the superposition of misfit stresses and the applied uniaxial stress. Since the stress in the particles is positively hydrostatic and in the γ channels negatively biaxial, uniaxial loading leads to higher net positive stress in the horizontal channels. The relaxation of interfacial misfit needed for rafting is then able to occur first on the horizontal channels, so that the mass transport needed can be accommodated close to the γ/γ' interfaces. However, it has recently been pointed out (Zhu et al., 2012) that the climb processes assumed at the γ/γ' interfaces in the model of Section 22.5.3 give rise to a vacancy flux—and an opposing flux of atoms—from the horizontal to the vertical channels, which might be of a magnitude sufficient to support the mass transport needed for rafting. This arises from the opposite positive and negative senses of the climb needed at the vertical and horizontal interfaces. Reversing the sense of the externally applied load causes the direction of the vacancy flux to be reversed, from the vertical to horizontal channels. But more research is needed to elucidate the mass transport effects arising during the rafting effect.

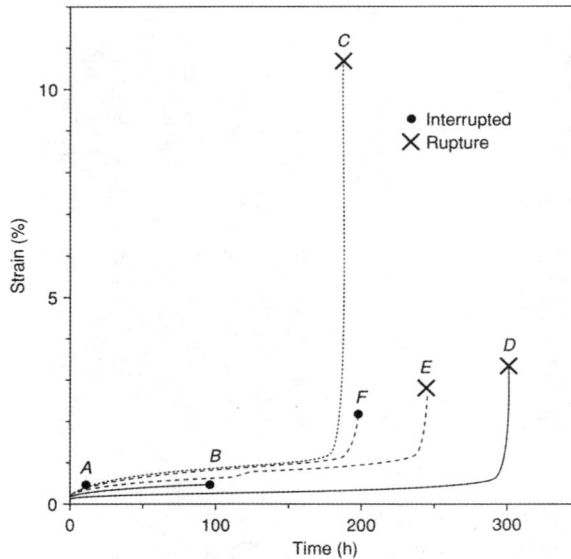

Figure 46 Constant-load creep data for the CMSX-4 superalloy, deformed at 1150 °C and 100 MPa. After Reed et al. (2007).

When single-crystal superalloys are tested in creep at temperatures beyond 1050 °C, rafting occurs very quickly (Reed et al., 2007). Creep curves for CMSX-4 at 1050 °C/100 MPa are given in **Figure 46**. The microstructure at point A after 10 h of testing is illustrated in **Figure 40**; one can see that rafts are fully formed despite a mere 5% of life being exhausted at this point. Some dissolution of γ' occurs due to the proximity to the γ' prime solvus temperature. The creep curves no longer conform to the exponential form of **Figure 34**; the plateau, which lasts for ∼200 h or longer is associated with a hardening during which the creep strain rate decreases with increasing strain. The factors influencing the onset of creep fracture at such extreme conditions are not well established, but damage observed includes TCP phase precipitation and casting porosity, there being evidence of the condensation of vacancies on these and also strain-controlled cavitation. The stochastic nature of these processes explains the scatter is observed in the final time to rupture. A correlation between the creep strain and the porosity levels has been found in this regime (Epishin and Link, 2004). It has been argued that the rearrangement of interfacial dislocation nets into equilibrium configurations provides a further source of vacancies.

22.5.6 Role of Stacking Fault Shear in Primary Creep Deformation

When a uniaxial stress greater than about 500 MPa is applied to single-crystal superalloys in the temperature range 600 to 900 °C, a mode of deformation known as stacking fault shear can operate (Rae and Reed, 2007). In Section 22.2 it was noted that although the complex stacking fault energy (SFE) is high, shear of the $L1_2$ lattice by a double Shockley partial results in intrinsic and extrinsic stacking faults that have extremely low energies—a fraction of the APB energy—so that shearing becomes feasible. However, to shear by partials in this way requires lattice reordering via a thermally activated mechanism, require shuffling and/or diffusional jumps. Characteristics of this mode of deformation include: (1) a strong dependence on the loading direction, with significant anisotropy

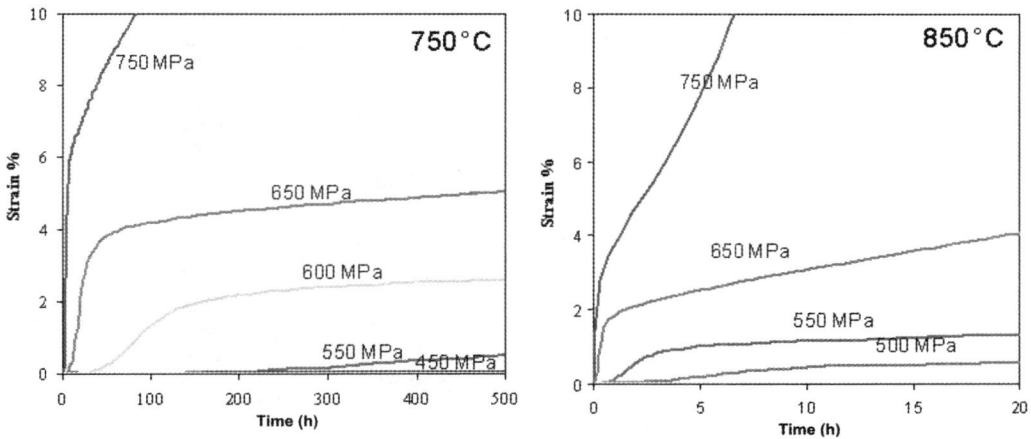

Figure 47 Constant-load creep data for CMSX-4 at low temperatures and high stresses, i.e. in the primary creep regime.

being displayed around the $\langle 001 \rangle$ direction (2) a macroscopic deformation of the form $a\langle 11\bar{2}\rangle\{111\}$, consistent with the underlying micromechanism, so that deformation is far from homogeneous and (3) a primary creep component characterized by rapid softening of several percent followed by hardening. A key characteristic is time-dependent plasticity occurring at or just below the yield point, so that strain-rate sensitive behavior is expected. For this reason, stacking fault shear can be observed during the low-cycle fatigue (LCF) process. Creep curves for the CMSX-4 superalloy that illustrate the primary creep effect are illustrated in **Figure** 47.

The micromechanism of deformation involves dislocation ribbons of overall Burgers vector $a\langle 11\bar{2}\rangle$. Dislocation ribbons form from the combination of $a/2\langle 110\rangle$ lattice dislocations in the γ matrix as described earlier. They require activated reordering to produce the SISF and on both sides of the SESF, but are active during primary creep (or LCF) moving as a self-contained ribbon often spread over several γ' precipitates. The ribbons are composed of four distinct dislocations separated by faults in the order (1) SISF (2) APB and (3) SESF; the leading and trailing dislocations have the Burgers vectors $a/3\langle 112\rangle$ and the intermediate dislocations bounding the APB are of $a/6\langle 112\rangle$. **Figure 48** illustrates the ribbon configuration imaged using transmission electron microscopy in the TMS-82 alloy. **Figure 49** is an interpretation of the ribbon configuration, identifying the location of SESF, APB and SISF. The observed density of the dislocation ribbons compared to the dislocations in the γ channels is relatively low, but ribbons in the early stages of creep—when the dislocation density is low—are able to cut through the precipitates leaving very little or no debris at γ/γ' interfaces, so that strain is accumulated rapidly. Formation of the $a\langle 11\bar{2}\rangle$ ribbons from matrix $a/2\langle 1\bar{1}0\rangle\{111\}$ dislocations requires geometrical compatibility. For instance, if the major Burgers vectors of the active slip systems in the lattice are, for example, $a/2[110]$ and $a/2[1\bar{1}0]$ then they are unable to combine to form ribbons, no stacking faults are observed, primary creep is totally absent and the creep rate is exceptionally low, see Rae et al. (2000). Thus in this way, the mechanism of primary creep can be nucleation-limited (ribbons cannot form) or propagation-limited (shear stresses are insufficient for ribbons to move). These ideas have been used by Rae and Reed (2007) to rationalize the orientation-dependence of primary creep deformation in single-crystal superalloys.

As the ribbons move from precipitate to matrix, significant changes in the fault energies arise. For example, the APB energy changes from ~ 200 mJ m^{-2} in the γ' to zero in the matrix and the SFE from

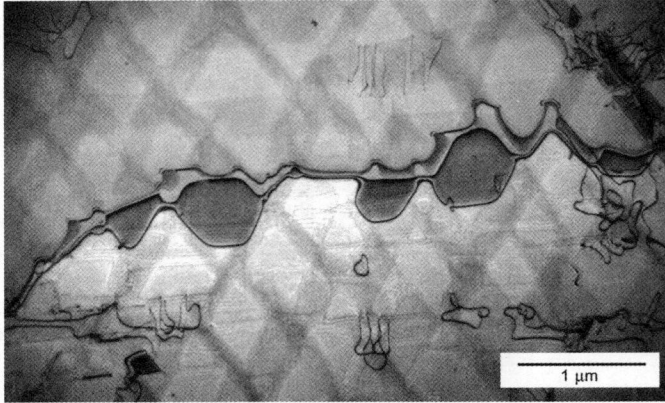

Figure 48   TEM micrograph of a $a\langle11\bar{2}\rangle$ ribbon in the single-crystal superalloy TMS-82, deformed in creep at 750 °C and 750 MPa to 11% strain. The foil normal is {111}.

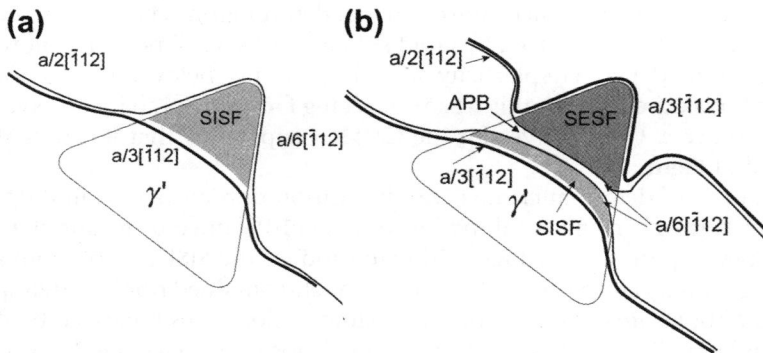

Figure 49   Illustration of the mechanism of shearing of the γ' by an $a\langle11\bar{2}\rangle$ ribbon during primary creep. An $a/3[\bar{1}12]$ dislocation enters the γ' leaving a superlattice intrinsic stacking fault (SISF) behind it, (a), leaving an $a/6[\bar{1}12]$ partial at the γ/γ' interface. In (b), the original $a/6[\bar{1}12]$ and a further $a/6[\bar{1}12]$ have entered the γ', leaving an $a/3[\bar{1}12]$ at the interface.

~ 150 mJ m^{-2} in the γ to ~ 10 mJ m^{-2} in γ'. Thus, in practice, large changes in the configuration of the fault are expected and are indeed observed. This characteristic gives rise to drag effects just as acknowledged for the expansion of stacking faults in alloys containing low SFE precipitates, see Hirsch and Kelly, 1965; Gerold and Hartmann, 1968; Reppich, 1993. Thus, the kinetics of ribbon movement is influenced in a complex way by the expansion and shrinkage of the different faults, the changes in the dislocation line energies and the size/spacing of the precipitates. Kovaric et al. (2009) have made phase field simulations of such dislocation dissociation at the γ/γ' interfaces to leave a stacking fault in the matrix and no planar defect in the precipitate; however, the volume fraction of γ' was rather low and therefore the calculations are probably more relevant to polycrystalline superalloys. Recently, Vorontsov et al. (2012a, 2012b) have relaxed this limitation to demonstrate the effects of fault energy and precipitate shape, for a γ' fraction typical of a single-crystal superalloy.

22.5.7 Thermal–Mechanical Fatigue of Single-Crystal Superalloys

Single-crystal superalloys have traditionally been designed to resist creep deformation. But TMF (Mughrabi, 2009) is of growing importance, owing to (1) lightweight component geometries of reduced wall thickness, so that stresses are enhanced (2) greater use of cooling air, which promotes thermal gradients and thus thermal stresses and (3) more arduous mission characteristics, e.g. shorter periods of peak, sustained operation with a greater number of engine start ups/shut downs.

TMF arises due to the superposition of mechanical and thermal loadings that are distinct. Thus in principle a TMF test involves cycling of the temperature T and the mechanical strain ϵ_{mech}—defined as the total strain with the thermal strain subtracted from it—with different phase shifts. An important example is OP (out of phase) cycling in which ϵ_{mech} and T are out of synchronization by $180°$: the maximum temperature coincides with a minimum in ϵ_{mech} and *vice versa*. Mechanical analysis indicates that the loading experienced by "hot spots" on the platforms of turbine blades are loaded in this way: localized temperature increases require thermal expansion that is constrained by cooler, surrounding material that places it in compression; upon cooling the situation is reversed and the material is pulled into tension, which can cause fatigue cracking. A further example is in-phase (IP) cycling, in which the maximum temperature coincides with a maximum in ϵ_{mech}; the cooled, trailing edges of turbine blade aerofoils are loaded in this way, particularly close to the base of the blade. A schematic illustration of the variation of stress and strain during OP-TMF cycling is given in **Figure 50**. Note in particular the locations of the cold and hot portions of the cycle, in which time-independent and time-dependent plasticity are expected.

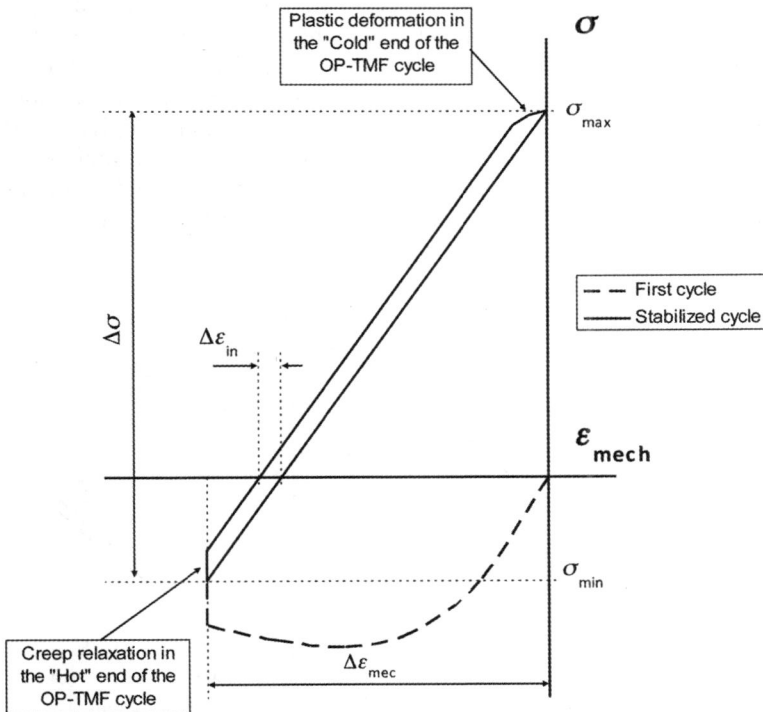

Figure 50 Schematic illustration of the variation of stress and strain during out-of-phase (OP) thermal-mechanical fatigue (TMF) testing.

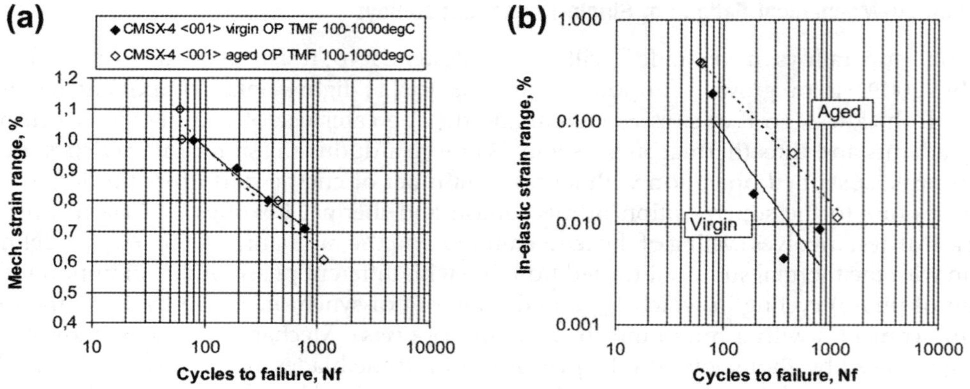
(a)
◆ CMSX-4 <001> virgin OP TMF 100-1000degC
◇ CMSX-4 <001> aged OP TMF 100-1000degC
Mech. strain range, %
Cycles to failure, Nf
(b)
In-elastic strain range, %
Aged
Virgin
Cycles to failure, Nf

Figure 51 Results from OP-TMF testing of CMSX-4 in the virgin and long-term aged condition. After Moverare et al. (2009).

Results for CMSX-4 tested by OP-TMF between 100 °C and 1000 °C, with a dwell period of 300 s at the highest temperature, are given in **Figure 51** (Moverare et al., 2009). Testing was carried out in both the (a) virgin and (b) aged conditions, the latter produced by a heat treatment of 4000 h at 1000 °C on specimens that had been subjected to 25 cycles. Note that the data of **Figure 51** imply very little influence of the aging on TMF life. The stabilized stress–strain hysteresis loops are given in **Figure 52** for the virgin and aged conditions. Note in particular the material is subjected to compression—consistent with OP-TMF testing—of maximum value −200 MPa at the hottest part of the cycle; upon cooling a tensile stress of about 600 MPa is reached. The hysteresis loop for the aged material is wider, due to some accumulation of creep strain at the highest temperature. Interestingly, and despite the fatigue performance not being influenced too strongly by it, the heat treatment influences the fracture surfaces in a remarkable way—see **Figure 53**. In the virgin state, failure is entirely crystallographic with the fracture surface parallel to the {111} plane, with very little ductility. Ageing promotes more homogeneous plastic deformation and pronounced necking.

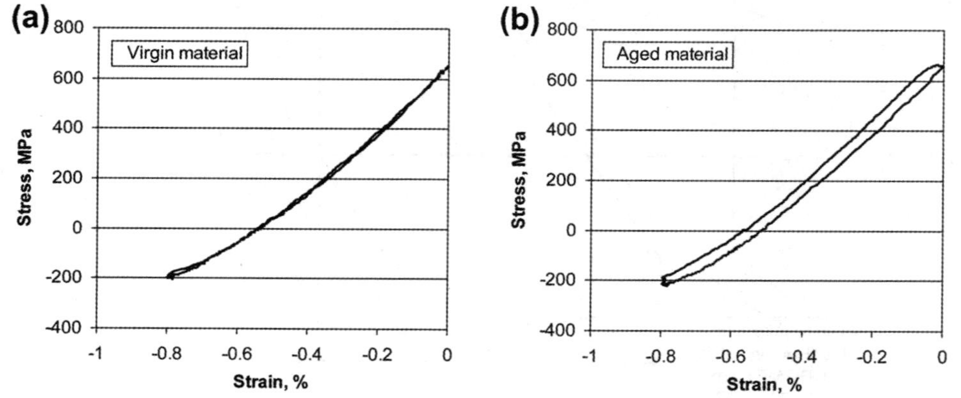
(a)
Virgin material
Stress, MPa
Strain, %
(b)
Aged material
Stress, MPa
Strain, %

Figure 52 Shapes of stabilized stress–strain hysteresis loops for CMSX-4 in (a) virgin and (b) long-term aged condition. After Moverare et al. (2009).

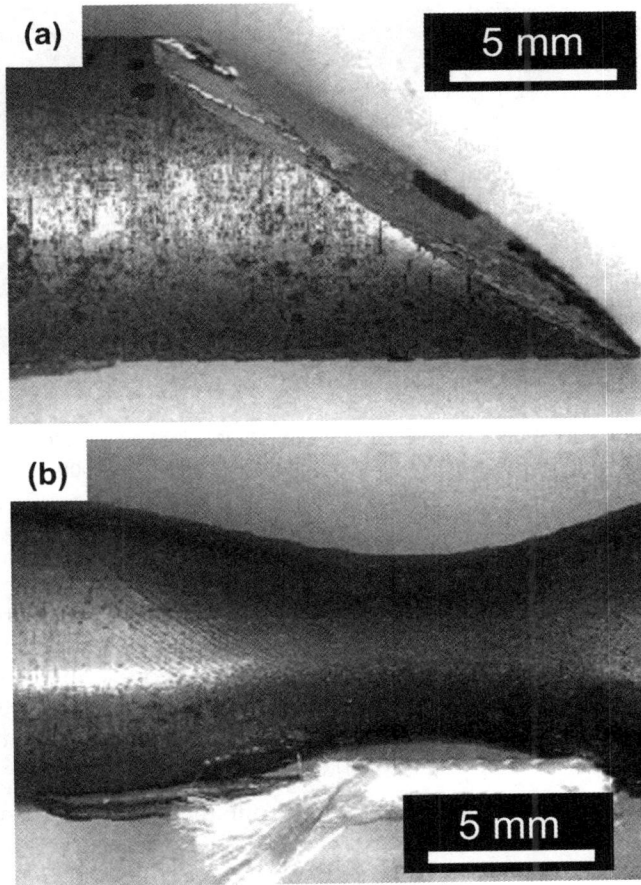

Figure 53 Fracture appearance for CMSX-4 subjected to OP-TMF testing (a) virgin and (b) long-term aged condition. After Moverare et al. (2009).

The micromechanisms of OP-TMF failure sheds light on the reasons for these effects. Consistent with the compressive stress experienced at the highest temperature, the rafts are now parallel to the direction of loading—see **Figure 54**. In the virgin condition, highly localized twins are present that extend across the entire cross-section of the testpiece; these are responsible for the lack of ductility observed since they are able to shear the rafted γ in an efficient manner. The twins are preferred sites for nucleation and growth of small, spherical TCP precipitates of diameter 100 nm. Moreover, extensive recrystallization (Cox et al., 2003; Zambaldi et al., 2007) occurs at the intersection of twin bands of different orientation, which leads to rapid crack propagation—because high-angle grain boundaries are formed, which are weak since grain boundary strengtheners C and B are absent. The greater ductility promoted by aging seems to be due to the precipitation of large plates of TCPs of length 5 μm or more. These seem to be efficient at arresting twin propagation; TCP precipitation in this way does not appear to be detrimental. But more research is needed to elucidate the factors leading to the nucleation and propagation of twins under such OP-TMF conditions and TMF conditions in general.

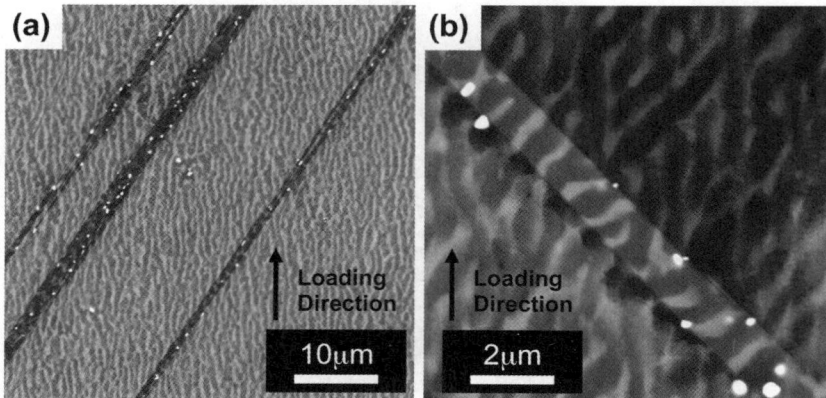

Figure 54 Backscatter electron micrographs of TMF damage showing localized deformation appearing as twin bands, decorated with TCP precipitates. Note the alignment of the γ' rafts in the loading direction.

22.6   Summary and Conclusions

The nickel-based superalloys remain unsurpassed as high-temperature materials, due to a remarkable capability to resist substantial loading at temperatures up to 1000 °C and beyond. This ability is sustained even in operating environments for which oxidation and corrosion become possible.

Substantial progress has been made in elucidating the mechanisms responsible for the prowess of these materials for structural applications, and in particular the factors that influence plastic deformation. The following represents a summary of the current state of knowledge:

(1) The major contribution to strengthening arises from the energy penalty associated with the shearing of the γ' precipitates: an $a/2\langle 1\bar{1}0\rangle$ dislocation, which is perfect in γ cannot enter γ' without creating a planar APB fault.

(2) The APB can however dissociate into configurations involving the SISF, superlattice extrinsic stacking fault (SESF) and complex stacking faults (CSFs) of different varieties.

(3) This dissociation can occur in different ways, giving rise to complicated effects of temperature- and strain-rate on deformation. For a given size and fraction of the γ' phase, the dependence of flow stress and rate of hardening on alloy chemistry arises from differences in the manner and kinetics of dissociation.

(4) The anomalous yield effect—an increase in flow stress with temperature—is the most striking corollary, arising due to contributions from the anisotropies of the APB energy and elastic properties; these promote the thermally activated cross-slip from {111} to {001} and thus a hardening increment that increases with temperature.

(5) Electron microscopy has been used to deduce values of the different fault energies particularly for the binary Ni–Al system. Some scatter exists in the reported values, which are in broad agreement with estimates made using modeling based upon modern electron theory. But further research is needed to clarify the influence of alloy chemistry on the fault values.

(6) At lower temperatures (below 650 °C) and higher stresses, shearing of γ' occurs by paired $a/2\langle 1\bar{1}0\rangle$ dislocations, in either weakly- or strongly coupled configurations; at higher temperatures of (above 800 °C), bypass of γ' by individual, unpaired $a/2\langle 110\rangle$ dislocations is predominant.

(7) At intermediate temperatures of around 725 °C, other mechanisms have been identified; the common feature is shearing of γ' by SISFs and/or SESFs. The dislocation configurations can be either of isolated character (only in γ') or can be extended thus traversing both γ and γ'. A viscous microtwinning mode has also been observed, which imparts very thin twins (microtwins) in γ' across approximately 2–50 atomic planes; this requires vacancy-mediated reordering in the wake of the gliding partial dislocations.

(8) Constant-load creep tests on $\langle 001 \rangle$-orientated single-crystal superalloys have elucidated the magnitude of stress needed to shear the γ' precipitates; at 750 °C a shear stress on $\{111\}$ of about 250 MPa is needed for shearing by $a\langle 112 \rangle$ ribbons, which consist of closely coupled SISF, APB and SESF faults. A strong dependence of deformation on crystallographic orientation is observed, due to the conditions needed for ribbon nucleation and propagation.

(9) When the conditions for penetration of the γ' precipitates are not met, deformation can still occur in the γ matrix but in a heavily time-dependent, viscous manner; for single crystals this leads to tertiary creep behavior, in which the dominant mode of damage is a dislocation content that increases steadily through life, from very low initial values.

(10) Under these circumstances, the rate-controlling step is nonconservative thermally activated climb of dislocation segments at γ/γ' interfaces. Quantitative modeling indicates that elements such as Re and W—which have high-activation energies for thermally activated processes such as solute/vacancy migration—have a profound effect on creep deformation, consistent with experimental observations.

(11) Many important unanswered questions remain concerning the plasticity of these two phase materials at elevated temperatures, particularly around 750 °C when the yield anomaly runs out.

(12) Little is known about the physical mechanisms causing degradation under TMF conditions, and further research is needed to elucidate the factors that are important in this case.

References

Allen, C.D., 1995. Plasticity of Nickel-based Single Crystal Superalloy, PhD thesis, Massachusetts Institute of Technology, USA.

Ardell, A.J., 1985. Precipitation hardening. Metall. Trans. A Phys. Metall. Mater. Sci. 16, 2131–2165 (Minerals Metals Materials Soc).

Baithery, D., Mohles, V., Nembach, E., 2001. Philos. Mag. Lett. 81 (12), 839–847.

Baluc, N., Schaublin, R., Hemker, K.J., 1991. Methods for determining precise values of antiphase boundary energies in Ni$_3$Al. Philos. Mag. Lett. 64, 327–334.

Beardmore, P., Davies, R.G., Johnston, T.L., 1969. On temperature dependence of flow stress of nickel-base alloys. Trans. Metall. Soc. AIME 245, 1537–1545.

Betteridge, W., Shaw, S.W.K., 1987. Development of superalloys. Mater. Sci. Technol. 3, 682–694 (Inst. Mater.).

Bettge, D., Osterle, W., 1999. "Cube slip" in near-[111] oriented specimens of a single-crystal nickel-base superalloy. Scr. Mater. 40, 389–395 (Pergamon-Elsevier Science Ltd).

Birks, N., Meier, G.H., Pettit, F.S., 2006. Introduction to the High Temperature Oxidation of Metals. Cambridge University Press.

Blavette, D., Caron, P., Khan, T., 1986. An atom probe investigation of the role of rhenium additions in improving creep resistance of Ni-base superalloys. Scr. Metall. 20, 1395–1400 (Pergamon-Elsevier Science Ltd).

Blavette, D., Grancher, G., Bostel, A., 1988. Statistical-analysis of atom-probe data-I: Derivation of some fine-scale features from frequency-distributions for finely dispersed systems. J. Phys. 49, 433–438 (Editions Physique).

Blobaum, K.J., Van Heerden, D., Gavens, A.J., Weihs, T.P., 2003. Al/Ni formation reactions: characterization of the metastable Al$_9$Ni$_2$ phase and analysis of its formation. Acta Mater. 51, 3871–3884.

Buffiere, J.Y., Ignat, M., 1995. A dislocation based criterion for the raft formation in nickel-based superalloys single-crystals. Acta Metall. Mater. 43, 1791–1797 (Pergamon-Elsevier Science Ltd).

Cahn, R.W., Siemers, P.A., Geiger, J.E., Bardhan, P., 1987. The order-disorder transformation in Ni$_3$Al and Ni$_3$Al–Fe alloys-I. Determination of the transition temperatures and their relation to ductility. Acta Metall. 35, 2737–2751.

Caillard, D., 2001. Yield stress anomalies and HT mech props of intermetallics and disordered alloys. Mater. Sci. Eng. A 319–321, 74–83.

Caron, P., Khan, T., Veyssiere, P., 1988. On precipitate shearing by superlattice stacking-faults in superalloys. Philos. Mag. A 57, 859–875 (Taylor & Francis Ltd).

Carroll, L.J., Feng, Q., Pollock, T.M., 2008. Interfacial dislocation networks and creep in directional coarsened Ru-containing nickel-base single-crystal superalloys. Metall. Mater. Trans. A Phys. Metall. Mater. Sci. 39A, 1290–1307 (Springer).

Carter, P., Cox, D.C., Gandin, C.A., Reed, R.C., 2000. Process modelling of grain selection during the solidification of single crystal superalloy castings. Mater. Sci. Eng. A Struct. Mater. Prop. Microstruct. Process. 280, 233–246 (Elsevier Science Sa).

Chen, S.P., Voter, A.F., Srolovitz, D.J., 1986. Computer-simulation of grain-boundaries in Ni$_3$Al—the effect of grain-boundary composition. Scr. Metall. 20, 1389–1394.

Chou, C.T., Hirsch, P.B., McLean, M., Hondros, E., 1982. Anti-phase domain boundary tubes in Ni$_3$Al. Nature 300, 621–623.

Conforto, E., Molenat, G., Caillard, D., 2005. Comparison of Ni-based alloys with extreme values of antiphase boundary energies: dislocation mechanisms and mechanical properties. Philos. Mag. 85, 117–137.

Cox, D.C., Roebuck, B., Rae, C.M.F., Reed, R.C., 2003. Recrystallisation of single crystal superalloy CMSX-4. Mater. Sci. Technol. 19, 440–446 (Maney Publishing).

Davies, R.G., Stoloff, N.S., 1965. Influence of long-range order upon strain hardening. Philos. Mag. 12, 297–304 (Taylor & Francis).

Decanps, B., Morton, A.J., Condat, M., 1991. On the mechanism of shear of gamma' precipitates by single (a/2)<111> dissociated matrix dislocations in Ni-based superalloys. Philos. Mag. A Phys. Condens. Matter Struct. Defects Mech. Prop. 64, 641–668.

Demura, M., Golberg, D., Hirano, T., 2007. An athermal deformation model of the yield stress anomaly in Ni$_3$Al. Intermetallics 15, 1322–1331 (Elsevier Sci Ltd).

Dimiduk, D.M., Thompson, A.W., Williams, J.C., 1993. The compositional dependence of antiphase-boundary energies and the mechanism of anomalous flow in Ni$_3$Al alloys. Philos. Mag. A 67, 675–698.

Elliott, A.J., Pollock, T.M., 2007. Thermal analysis of the bridgman and liquid-metal-cooled directional solidification investment casting processes. Metall. Mater. Trans. A Phys. Metall. Mater. Sci. 38A, 871–882 (Minerals Metals Materials Soc).

Epishin, A., Link, T., 2004. Mechanisms of high-temperature creep of nickel-based superalloys under low applied stresses. Philos. Mag. 84, 1979–2000.

Epishin, A., Link, T., Nazmy, M., Staubli, M., Klingelhoffer, H., Nolze, G., 2008. Microstructural degradation of CMSX-4: kinetics and effect on mechanical properties. In: Superalloys 2008. Minerals, Metals & Materials Society, Warrendale, PA, USA, pp. 725–731.

Epishin, A., Link, T., Nolze, G., 2007. SEM investigation of interfacial dislocations in nickel-base superalloys. J. Microsc. Oxford 228, 110–117 (Blackwell Publishing).

Ezz, S.S., Pope, D.P., Paidar, V., 1982. The tension compression flow-stress asymmetry in Ni$_3$(Al.Nb) single-crystals. Acta Metall. 30, 921–926 (Pergamon-Elsevier Science Ltd).

Field, R.D., Pollock, T.M., Murphy, W.H., 1992. The development of γ/γ' interfacial dislocation networks during creep in Ni-base superalloys. In: Antolovich, S., Stusrud, R., abd Mackay, R.W., Anton, D., Khan, T., Kissinger, R. (Eds.), Superalloys 1992. Minerals, Metals & Materials Soc, pp. 557–566.

Flinn, P.A., 1960. Theory of Deformation in Superlattices. In: Metallurgical Society of American Institute of Mining, Metallurgical and Petroleum Engineers – Transactions, vol. 218. American Institute of Mining, Metallurgical and Petroleum Engineers (AIME), New York, NY, United States, 145–154.

Foiles, S.M., Daw, M.S., 1987. Application of the embedded atom method to Ni$_3$Al. J. Mater. Res. 2, 5–15.

Friedel, J., 1964. Dislocations. Pergamon Press, New York.

Fu, C.L., Yoo, M.H., 1989. Stacking fault energies, crystal elasticity and their relation to the mechanical properties of L1$_2$ intermetallics. In: MRS Symposium, vol. 133, pp. 81–86.

Gerold, V., Hartmann, K., 1968. Theoretical and experimental investigations on stacking fault strengthening. Trans. Jpn. Inst. Metals 9, 509 (Japan Inst Metals).

Ghosh, R.N., Curtis, R.V., Mclean, M., 1990. Creep deformation of single-crystal superalloys—modeling the crystallographic anisotropy. Acta Metall. Mater. 38, 1977–1992 (Pergamon-Elsevier Science Ltd).

Giamei, A.F., Anton, D.L., 1985. Rhenium additions to a Ni-base superalloy—effects on microstructure. Metall. Trans. A 16, 1997–2005 (Minerals Metals Materials Soc).

Harada, H., Murakami, H., 1999. Design of Ni-base Superalloys. Springer-Verlag, Berlin, pp. 39–70.

Harris, K., Erickson, G.L., Brentnall, W.D., Aurrecoechea, J.M., Sikkenga, S.L., Kubarych, K.G., 1992. Development of the rhenium containing superalloys CMSX-4(R) and CM 186 LC(R) for single-crystal blade and directionally solidified vane applications in advanced turbine-engines. In: Antolovich, S., Stusrud, R., abd Mackay, R.W., Anton, D., Khan, T., Kissinger, R. (Eds.), Superalloys 1992, pp. 297–306.

Hemker, K.J., Mills, M.J., 1993. Measurements of antiphase boundary and complex stacking-fault energies in binary and B-doped Ni$_3$Al using TEM. Philos. Mag. A 68, 305–324.

Heredia, F.E., Tichy, G., Pope, D.P., Vitek, V., 1989. Temperature and orientation dependent plastic flow in Pt$_3$Al. Acta Metall. 37, 2755–2758 (USA).

Hirsch, P.B., 1992. A model of the anomalous yield stress for <111> slip in L1$_2$ alloys. Prog. Mater. Sci. 36, 63–88.

Hirsch, P.B., Kelly, A., 1965. Stacking-fault strengthening. Philos. Mag. 12, 881 (Taylor & Francis Ltd).

Hirsch, P.B., Sun, Y., 1993. Observations of dislocations relevant to the anomolous yield stress in L1$_2$ alloys. Mater. Sci. Eng. A 164, 395–400.

Huron, E.S., Reed, R.C., Hardy, M.C., Mills, M.J., Montero, R.E., Portella, P.D., Telesman, J. (Eds.), 2012. Superalloys 2012. TMS, Warrendale PA, USA.

Huther, W., Reppich, B., 1978. Interaction of dislocations with coherent, stress-free, ordered particles. Z. Metallkd. 69, 628–634 (West Germany).

Jackson, M.P., Reed, R.C., 1999. Heat treatment of UDIMET 720Li: the effect of microstructure on properties. Mater. Sci. Eng. A Struct. Mater. Prop. Microstruct. Process. 259, 85–97 (Elsevier Science Sa).

Janotti, A., Krcmar, M., Fu, C.L., Reed, R.C., 2004. Solute diffusion in metals: larger atoms can move faster. Phys. Rev. Lett. 92, 4.

Jonsson, B., 1995. Assessment of the mobilities of Cr, Fe and Ni in bcc Cr-Fe-Ni allays, ISIJ International, Iron Steel Inst Japan Keidanren Kaikan, vol. 35, pp. 1415–1421.

Karnthaler, H.P., Muhlbacher, E.T., Rentenberger, C., 1996. The influence of the fault energies on the anomalous mechanical behaviour of Ni$_3$Al alloys. Acta Mater. 44, 547–560.

Karunaratne, M.S.A., Carter, P., Reed, R.C., 2000. Interdiffusion in the face-centred cubic phase of the Ni–Re, Ni–Ta and Ni–W systems between 900 and 1300 degrees C. Mater. Sci. Eng. A Struct. Mater. Prop. Microstruct. Process. 281, 229–233 (Elsevier Science Sa).

Karunaratne, M.S.A., Reed, R.C., 2003. Interdiffusion of the platinum-group metals in nickel at elevated temperatures. Acta Mater. 51, 2905–2919.

Kear, B.H., 1964. Dislocation configurations and work-hardening in Cu$_3$Al crystals. Acta Metall. 12, 555–569.

Kear, B.H., Wilsdorf, H.G., 1962. Dislocation configurations in plastically deformed polycrystalline Cu$_3$Au alloys. Trans. Metall. Soc. AIME 224, 382.

Kear, B.H., Gaimei, A.F., Leverant, G.R., Oblak, J.M., 1969a. On Int/Ext SF pairs in the L1$_2$ lattice. Scr. Metall. 3, 123–130.

Kear, B.H., Gaimei, A.F., Leverant, G.R., Oblak, J.M., 1969b. Viscous slip in the L1$_2$ lattice. Scr. Metall. 3, 455–460.

Kear, B.H., Gaimei, A.F., Silcock, J.M., Ham, R.K., 1968. Slip and climb processes in γ′ ppt hardened Ni-base alloys. Scr. Metall. 2, 287–294.

Kear, B.H., Hornbecker, M.F., 1966. Deformation Structures in Polycrystalline Ni$_3$Al. In: ASM Transactions Quarterly, vol. 59. American Society of Metals (ASM), Cleveland, OH, United States, pp. 155–161.

Kolbe, M., 2001. The high temperature decrease of the critical resolved shear stress in nickel-base superalloys. In: Mater. Sci. Eng. A, Struct. Mater., Prop. Microstruct. Process. (Switzerland), vol. A319–321. Elsevier, Switzerland, pp. 383–387.

Korner, A., 1988. Weak-beam study of superlattice dislocations moving on cube planes in Ni$_3$(Al.Ti) deformed at room-temperature. Philos. Mag. A Phys. Condens. Matter Struct. Defects Mech. Prop. 58, 507–522.

Kovaric, L., Unocic, R.R., Li, J., Sarosi, P., Shen, C., Wang, Y., Mills, M.J., 2009. Microtwinning and other shearing mechanisms at intermediate temperatures in Ni-based superalloys. Prog. Mater. Sci. 54, 839–873.

Kozar, R.W., Suzuki, A., Milligan, W.W., Shirra, J.J., Savage, M.F., Pollock, T.M., 2009. Strengthening mechanisms in polycrystalline multimodal nickel-base superalloys. Metall. Trans. 40A, 1588–1603.

Krcmar, M., Fu, C.L., Janotti, A., Reed, R.C., 2005. Diffusion rates of 3d transition metal solutes in nickel by first-principles calculations. Acta Mater. 53, 2369–2376 (Pergamon-Elsevier Science Ltd).

Lall, C., Chin, S., Pope, D.P., 1979. Orientation and temperature-dependence of the yield stress of Ni$_3$(Al,Nb) single-crystals. Metall. Trans. A Phys. Metall. Mater. Sci. 10, 1323–1332 (Minerals Metals Materials Soc).

Liang, S., Pope, D., 1977. The yield stress of L1$_2$ ordered alloys. Acta Metall. 25, 485–493.

MacKay, R.A., Gabb, T.P., Garg, A., Rogers, R.B., Nathal, M.V., 2012. Influence of composition on microstructural parameters of single crystal nickel-base superalloys. Mater. Character. 70, 83–100 (Elsevier Science Inc).

Matan, N., Cox, D.C., Carter, P., Rist, M.A., Rae, C.M.F., Reed, R.C., 1999a. Creep of CMSX-4 superalloy single crystals: effects of misorientation and temperature. Acta Mater. 47, 1549–1563.

Matan, N., Cox, D.C., Rae, C.M.F., Reed, R.C., 1999b. On the kinetics of rafting in CMSX-4 superalloy single crystals. Acta Mater. 47, 2031–2045.

McLean, M., 1995. Nickel-base superalloys—current status and potential. Philos. Trans. Roy. Soc. Lond. A 351, 419–433.

Michelon, M.F., Antonelli, A., 2008. Nonphysical thermodynamical phases in L1$_2$ intermetallic alloys from semiempirical tight-binding potentials. Comput. Mater. Sci. 42, 68–73.

Mishin, Y., 2004. Atomistic modeling of the γ and γ′-phases of the Ni-Al system. Acta Mater. 52, 1451–1467.

Mottura, A., Finnis, M.W., Reed, R.C., 2012. On the possibility of rhenium clustering in nickel-based superalloys. Acta Mater. 60, 2866–2872 (Pergamon-Elsevier Science Ltd).

Mottura, A., Warnken, N., Miller, M.K., Finnis, M.W., Reed, R.C., 2010. Atom probe tomography analysis of the distribution of rhenium in nickel alloys. Acta Mater. 58, 931–942 (Pergamon-Elsevier Science Ltd).

Mottura, A., Wu, R.T., FinniS, M.W., Reed, R.C., 2008. A critique of rhenium clustering in Ni–Re alloys using extended X-ray absorption spectroscopy. Acta Mater. 56, 2669–2675 (Pergamon-Elsevier Science Ltd).

Moverare, J.J., Johansson, S., Reed, R.C., 2009. Deformation and damage mechanisms during thermal-mechanical fatigue of a single-crystal superalloy. Acta Mater. 57, 2266–2276 (Pergamon-Elsevier Science Ltd).

Mryasov, O.N., Gornostyrev, Y.N., van Schilfgaarde, A., Freeman, A.J., 2002. Superdislocation core structure in L1$_2$ Ni$_3$Al, Ni$_3$Ge and Fe$_3$Ge: Peierls–Nabarro analysis starting from ab-initio GSF energetics calculations. Acta Mater. 50, 4554.

Mughrabi, H., 2009. Microstructural aspects of high temperature deformation of monocrystalline nickel base superalloys: some open problems. Mater. Sci. Technol. 25, 191–204.

Murakumo, T., Kobayashi, T., Koizumi, Y., Harada, H., 2004. Creep behaviour of Ni-base single-crystal superalloys with various γ' volume fraction. Acta Mater. 52, 3737–3744.

Nabarro, F.R.N., de Villiers, H.L., 1995. The Physics of Creep. Taylor and Francis, London.

Nembach, E., 1997. Particle Strengthening of Metals and Alloys, John Wiley and Sons, New York.

Nembach, E., 2000. Order strengthening: recent developments with special reference to Al–Li alloys. Prog. Mater. Sci. 45, 275–338.

Nembach, E., Neite, G., 1985. Precipitation hardening of superalloys by ordered-γ' particles. Prog. Mater. Sci. 29, 177–319 (UK).

Nitz, A., Lagerpusch, U., Nembach, E., 1998. CRSS anisotropy and tension/compression asymmetry of a commercial superalloy. Acta Mater. 46, 4769–4779.

Ochiai, S., Oya, Y., Suzuki, T., 1984. Alloying behaviour of Ni_3Al, Ni_3Ga, Ni_3Si and Ni_3Ge. Acta Metall. 32, 289–298.

Paidar, V., Pope, D.P., Vitek, V., 1984. A theory of the anomolous yield behaviour in $L1_2$ ordered alloys. Acta Metall. 32, 435–448.

Paidar, V., Yamaguchi, M., Pope, D.P., Vitek, V., 1982. Dissociation and core structure of -$<110>$ screw dislocations in $L1_2$ ordered alloys II. effects of applied shear stress. Philos. Mag. A 45, 883–894.

Pan, L.M., I, S., Henderson, M.B., Mclean, M., 1995. Asymmetric creep deformation of a single-crystal superalloy. Acta Metall. Mater. 43, 1375–1384 (Pergamon-Elsevier Science Ltd).

Paxton, A.T., Sun, Y.Q., 1998. The role of planar fault energy in the yield anomaly in $L1_2$ intermetallics. Philos. Mag. A 78, 85–103.

Phillips, P.J., Unocic, R.R., Kovarik, L., Mourer, D., Wei, D., Mills, M.J., 2010. Low cycle fatigue of a Ni-based superalloy: non-planar deformation. Scr. Mater. 62, 790–793 (Pergamon-Elsevier Science Ltd).

Phillips, P., Unocic, R., Mills, M., 2013. Low cycle fatigue of a polycrystalline Ni-based Superalloy: deformation substructure analysis. Int. J. Fatigue 57, 50–57 (Elsevier Science Ltd).

Piearcey, B.J., Smashey, R.W., 1967. Carbide phases in Mar-M200. Trans. Metall. Soc. AIME 239, 451.

Pineau, A., 1976. Influence of uniaxial stress on morphology of coherent precipitates during coarsening—elastic energy considerations. Acta Metall. 24, 559–564 (Pergamon-Elsevier Science Ltd).

Pollock, T.M., Argon, A.S., 1992. Creep resistance of CMSX-3 nickel-base superalloy single-crystal. Acta Metall. Mater. 40, 1–30.

Pollock, T.M., Argon, A.S., 1994. Directional coarsening in nickel-base single-crystals with high-volume fractions of coherent precipitates. Acta Metall. Mater. 42, 1859–1874.

Pollock, T.M., Field, R.D., 2002. Dislocations and high temperature plastic deformation of superalloy single crystals. In: Nabarro, F.R.N., Duesbery, M.S. (Eds.), Dislocations in Solids, vol. 11, Elsevier, Amsterdam, pp. 593–595.

Pope, D.P., Ezz, S.S., 1984. Mechanical properties of Ni_3Al and nickel-base alloys with high volume fraction of γ'. Int. Metals Rev. 29, 136–167.

Prikhodko, S.V., Ardell, A.J., 2003. Coarsening of γ' in Ni–Al alloys aged under uniaxial compression-III. Characterization of the morphology. Acta Mater. 51, 5021–5036 (Pergamon-Elsevier Science Ltd).

Rae, C.M.F., Matan, N., Cox, D., Rist, M., Reed, R., 2000. On the primary creep of CMSX-4 superalloy single crystals. Metall. Trans. 31A, 2219–2228.

Rae, C.M.F., Reed, R.C., 2001. The precipitation of topologically close-packed phases in rhenium-containing superalloys. Acta Mater. 49, 4113–4125 (Pergamon-Elsevier Science Ltd).

Rae, C.M.F., Reed, R.C., 2007. Primary creep in single crystal superalloys: origins, mechanisms and effects. Acta Mater. 55, 1067–1081.

Reed, R.C., 2006. The Superalloys: Fundamentals and Applications. Cambridge University Press.

Reed, R.C., Cox, D.C., Rae, C.M.F., 2007. Damage accumulation during creep deformation of a single crystal superalloy at 1150 degrees C. Mater. Sci. Eng. A Struct. Mater. Prop. Microstruct. Process. 448, 88–96 (Elsevier Science Sa).

Reed, R.C., Matan, N., Cox, D.C., Rist, M.A., Rae, C.M.F., 1999. Creep of CMSX-4 superalloy single crystals: effects of rafting at high temperature. Acta Mater. 47, 3367–3381.

Reed, R.C., Tao, T., Warnken, N., 2009. Alloys-By-Design: application to nickel-based single crystal superalloys. Acta Mater. 57, 5898–5913.

Reed, R.C., Yeh, A.C., Tin, S., Babu, S.S., Miller, M.K., 2004. Identification of the partitioning characteristics of ruthenium in single crystal superalloys using atom probe tomography. Scr. Mater. 51, 327–331 (Pergamon-Elsevier Science Ltd).

Reppich, B., 1982. Some new aspects concerning particle hardening mechanisms in γ' precipitating ni-base alloys-I. Theoretical concept. Acta Metall. 30, 87–94 (Pergamon-Elsevier Science Ltd).

Reppich, B., 1993. Mater. Sci. Technol. 6, 312 (VCH).

Reppich, B., Schepp, P., Wehner, G., 1982. Some new aspects concerning particle hardening mechanisms in γ' precipitating nickel-base alloys-II. Experiments. Acta Metall. 30, 95–104 (Pergamon-Elsevier Science Ltd).

Ricks, R.A., Porter, A.J., Ecob, R.C., 1983. The growth of γ' -precipitates in nickel-base super-alloys. Acta Metall. 31, 43–53 (Pergamon-Elsevier Science Ltd).

Rosengaard, N.M., Skriver, H.L., 1994. Ab-initio study of antiphase boundaries and stacking-faults in $L1_2$ and Do_{22} compounds. Phys. Rev. B 50, 4848–4858.

Sass, V., Feller-Kniepmeier, M., 1998. Orientation dependence of dislocation structures and deformation mechanisms in creep deformed CMSX-4 single crystals. Mater. Sci. Eng. A Struct. Mater. Prop. Microstruct. Process. 245, 19–28 (Elsevier Science Sa).

Schilnzer, S., Nembach, E., 1992. The critical resolved shear-stress of γ' -strengthened nickel-based superalloys with γ' -volume fractions between 0.07 and 0.47. Acta Metall. Mater. 40, 803–813 (Pergamon-Elsevier Science Ltd).

Schoeck, G., Kohlhammer, S., Fahnle, M., 1999. Planar dissociations and recombination energy of [1-10] superdislocations in Ni$_3$Al: generalized Peierls model in combination with ab initio electron theory. Philos. Mag. Lett. 79, 849–857.

Schoek, G., 1994. The instability of Paidar–Pope–Vitek locks in L1$_2$ compounds. Philos. Mag. Lett. 70, 179–187.

Sengupta, A., Putatunda, S.K., Bartosiewicz, L., Hangas, J., Nailos, P.J., Peputapeck, M., Alberts, F.E., 1994. Tensile behavior of a new single-crystal nickel-based superalloy (CMSX-4) at room and elevated-temperatures. J. Mater. Eng. Perform. 3, 73–81 (ASM International).

Shah, D.M., Duhl, D., 1984. The effect of orientation, temperature and gamma prime size on the yield strength of a single crystal nickel base superalloy. In: Superalloys 1984, p. 105.

Sun, Y.Q., Hazzledine, P.M., 1988. Tem weak-beam study of dislocations in gamma prime in a deformed Ni-based superalloy. Philos. Mag. A Phys. Condens. Matter. Struct. Defects Mech. Prop. 58, 603–618 (Univ of Oxford, Oxford, Engl, Univ of Oxford, Oxford, Engl).

Takasugi, T., Hirakawa, S., Izumi, O., Ono, S., Watanabe, S., 1987. Plastic flow of Co$_3$Ti single crystals. Acta Metall. 35, 2015–2026 (USA).

Takeuchi, S., Kuramoto, E., 1973. Temperature and orientation dependence of the yield stress in Ni$_3$Ga single crystals. Acta Metall. 21, 415–425.

Taunt, R.J., Ralph, B., 1974. Observations of fine-structure of superdislocations in Ni$_3$Al by field-ion microscopy. Philos. Mag. 30, 1379–1394.

Thornton, P.H., Davies, R.G., Johnston, T.L., 1970. Temperature dependence of flow stress of γ' phase based upon Ni$_3$Al. Metall. Trans. 1, 207 (Metallurgical Soc. Amer. Inst.).

Van Sluytman, J.S., Pollock, T.M., 2012. Optimal precipitate shapes in nickel-base γ-γ' alloys. Acta Mater. 60, 1771–1783 (Pergamon-Elsevier Science Ltd).

Vattre, A., Devincre, B., Roos, A., 2009. Dislocation dynamics simulations of precipitation hardening in Ni-based superalloys with high γ' volume fraction. Intermetallics 17, 988–994 (Elsevier Sci Ltd).

Vattre, A., Devincre, B., Roos, A., 2010. Orientation dependence of plastic deformation in nickel-based single crystal superalloys: Discrete-continuous model simulations. Acta Mater. 58, 1938–1951 (Pergamon-Elsevier Science Ltd).

Veron, M., Brechet, Y., Louchet, F., 1996. Strain induced directional coarsening in ni based superalloys. Scr. Mater. 34, 1883–1886.

Veyssiere, P., Douin, J., Beauchamp, P., 1985. On the presence of superlattice intrinsic stacking-faults in plastically deformed Ni$_3$Al. Philos. Mag. A 51, 469–483.

Viswanathan, G.B., Karthikeyan, S., Sarosi, P.M., Unocic, R.R., Mills, M.J., 2006. Microtwinning during intermediate temperature creep of polycrystalline Ni-based superalloys: mechanisms and modelling. Philos. Mag. 86, 4823–4840.

Viswanathan, G., Sarosi, P., Henry, M., Whitis, D., Milligan, W., Mills, M., 2005. Investigation of creep deformation mechanisms at intermediate temperatures in Rene 88 DT. Acta Mater. 53, 3041–3057.

Vitek, V., Sodani, Y., 1991. A theory of the temperature and strain rate dependence of the yield stress of L1$_2$ compounds in the anomalous regime. Scr. Metall. 25, 939–944 (Pergamon-Elsevier Science Ltd).

Vorontsov, V., 2011. Thesis, PhD thesis, Cambridge University.

Vorontsov, V.A., Voskoboinikov, R.E., Rae, C.M.F., 2012a. Shearing of γ' precipitates in Ni-base superalloys: a phase field study incorporating the effective γ-surface. Philos. Mag. 92, 608–634 (Taylor & Francis Ltd).

Vorontsov, V., Kovarik, L., Mills, M., Rae, C.M.F., 2012b. HREM of dislocation ribbons in a CMSX-4 superalloy single crystal. Acta Mater. 60, 4866–4878.

Voskoboinikov, R.E., Rae, C.M.F., 2009. A new γ-surface in 111 plane in L1$_2$ Ni$_3$Al. Dislocations 3, 012009, 2008.

Wee, D.-M., Suzuki, T., 1979. The Temperature Dependence of Hardness of L1$_2$ Ordered Alloys, vol. 20. Transactions of the Japan Institute of Metals, Japan, pp. 634–46.

Westbrooke, J.H., 1957. Temperature dependence of the hardness of secondary phases common in turbine bucket alloys. Trans AIME 209, 898.

Yamaguchi, M., Paidar, V., Pope, D., Vitek, V., 1982. Dissociation and core structure of <110> screw dislocations in l1$_2$ ordered alloys I. core structure in an unstressed crystal. Philos. Mag. A 45, 867–882.

Yamaguchi, M., Umakoshi, Y., 1990. The deformation behaviour of intermetallic superlattice compounds. Prog. Mater. Sci. 34, 1–148.

Yoo, M.H., 1986. On the theory of anomolous yield behaviour of Ni$_3$Al - effect of elastic anisotropy. Scr. Metall. 20, 915–920.

Yoo, M.H., 1987. Stability of superdislocations and shear faults in L1$_2$ ordered alloys. Acta Metall. 35, 1559–1569 (Pergamon-Elsevier Science Ltd).

Youssef, Y.M., Lee, P.D., Mills, K.C., Reed, R.C., 2010. On the diffusion behaviour of Os in the binary Ni-Os system. Mater. Sci. Technol. 26, 1173–1176 (Maney Publishing).

Zambaldi, C., Roters, F., Raabe, D., Glatzel, U., 2007. Modeling and experiments on the indentation deformation and recrystallization of a single-crystal nickel-base superalloy. Mater. Sci. Eng. A Struct. Mater. Prop. Microstruct. Process. 454, 433–440 (Elsevier Science Sa).

Zhang, J.X., Murakumo, T., Harada, H., Koizumi, Y., Kobayashi, T., 2004. Creep deformation mechanisms in some modern single-crystal superalloys. In: Green, K.A., Pollock, T.M., Harada, H., Howson, T.E., Reed, R.C., Schirra, J.J., Walston, S. (Eds.), Superalloys 2004. Minerals, Metals & Materials Soc, pp. 189–195.

Zhu, Z., Basoalto, H., Warnken, N., Reed, R., 2012. A model for the creep deformation behaviour of nickel-based single crystal superalloys. Acta Mater. 60, 4888–4900.

Biography

Professor Roger Reed is an authority on the nickel-based superalloys used for high temperature applications. He has held teaching positions at Imperial College, Cambridge, The University of British Columbia, Birmingham and Oxford. His textbook 'The Superalloys: Fundamentals and Applications' was published by Cambridge University Press in 2006. He has been a member of the organising committee of the International Symposium for Superalloys since 2004.

Dr. Catherine Rae is a Reader in Physical Metallurgy at the University of Cambridge and has worked on Ni-based superalloys in collaboration with Rolls-Royce for more than 30 years. She is director of the EPSRC-Rolls-Royce Doctoral Training Centre and is a member of the organising committees of the International Symposium on Superalloys and the Eurosuperalloys Conference.

23 Recovery and Recrystallization: Phenomena, Physics, Models, Simulation

Dierk Raabe, Max-Planck-Institut für Eisenforschung, Düsseldorf, Germany

23.1 Phenomena, Terminology, and Methods: Recovery and Recrystallization

23.1.1 Basic Phenomena and Terminology

In engineering processing terms, severe plastic cold working is typically carried out for changing the shape of a metallic work piece. In physical metallurgy terms, cold deformation leads to the increase of the internal stored energy. This increase in the stored energy is microstructurally associated with the accumulation of lattice defects. More specific, the majority of the internal energy that is stored due to cold working of metals can be attributed to dislocations (Orowan, 1934; Taylor, 1934; Polanyi, 1934). Point defects such as vacancies and interstitials as well as interfaces arising from athermal processes such as deformation twinning or martensite formation contribute only a minor portion to the stored energy (Wever, 1924; Swan et al., 1963; Keh and Weissman, 1963).

During plastic straining only a relatively small fraction of the total deformation energy imposed through the tool actually remains stored in the material. The main part of this energy is dissipated as heat during cold working. This is also the reason why heavily cold-worked metallic alloys can become very hot upon plastic deformation with up to several 100 K of temperature increase when strained at high rates. Yet, the defect density that remains stored in heavily cold-worked metals is still high enough to serve as driving force for a number of nonequilibrium transformation phenomena that are usually summarized as recovery, recrystallization, and grain coarsening (Kalisher, 1881; Sorby, 1886, 1887; Stead, 1898; Rosenhain, 1914; Ewing and Rosenhain, 1899, 1900a, 1900b; Alterthum, 1922; Carpenter and Elam, 1920; Czochralski, 1927; Burgers and Louwerse, 1931). Typically, for these processes to occur, the total amount of cold working has to be above a certain critical threshold value, namely, typically 30–50% thickness reduction or above.

For many metals and alloys, the dislocation density increases from values of 10^{10}–10^{11} /m^2 in the annealed state to 10^{12}–10^{13} /m^2 after modest deformation and even up to 10^{16} /m^2 after very heavy deformation at low temperatures. When translating these values into an average spacing among the dislocations, we obtain 1 μm for weakly deformed materials and 10 nm for heavily deformed metals.

These high-lattice defect densities lead to changes in a number of properties, such as the electrical resistivity, hardness, strength, toughness, and ductility with the amount of cold working. Also, the corrosion behavior is altered since the potential changes around dislocations render deformed zones preferred attacking points to oxidation phenomena.

At the light optical scale, i.e. at a mesoscopic observation regime, the grains become elongated along the direction of cold working and their shapes are distorted. Often shear bands that are oblique mesoscopic inhomogeneities of high local deformation appear at higher stages of the plastic deformation. At an electron optical scale complex dislocation cell (DC) arrangements and twinning substructures evolve. The latter microstructure features are an essential characteristic when the stacking fault energy is low (of the order of 20–40 mJ/m^2). For materials with yet lower stacking fault energy deformation-induced martensitic transformations can take place. This is for instance the case for Cu-30 wt.% Zn brass (mechanical twinning), many highly alloyed austenitic Cr–Ni stainless steels and high-Mn containing TWIP (twinning-induced plasticity) and transformation-induced plasticity steels.

For the further downstream manufacturing of heavily cold-worked metals, the high strength and hardness that are also associated with a drop in ductility and toughness typically requires to first restore the original soft state of the material (Burke and Turnbull, 1952; Smith, 1948; Beck and Hu, 1966; Haessner, 1978; Humphreys and Hatherly, 1995, 2004; Doherty et al., 1997; Doherty, 2005).

In that context, recovery, recrystallization, and competitive grain coarsening are the most important and effective thermomechanical processing methods to render cold-worked metallic materials back into a softer and hence easily formable state (Kalisher, 1881; Sorby, 1886, 1887; Stead, 1898; Rosenhain, 1914; Ewing and Rosenhain, 1899, 1900a, 1900b; Alterthum, 1922; Carpenter and Elam, 1920; Czochralski, 1927; Burgers and Louwerse, 1931). Correspondingly, the microstructure as well as the mechanical and electrical properties of the crystalline materials observed after heavy plastic deformation can be restored back into their original states before conducting further manufacturing steps.

Static primary recrystallization starts when heating deformed metals to an elevated temperature (**Figure 1**). In practical terms this usually means holding the cold deformed materials at a temperature above about 1/3 of their respective absolute melting point for a certain period of time, typically 1 h. Higher recrystallization temperatures require in most cases much shorter annealing times. The relationship between plastic deformation, annealing temperature, and the resulting grain size is for specific alloys for practical applications often presented in terms of recrystallization diagrams (**Figure 2**) (Kalisher, 1881; Sorby, 1886, 1887; Stead, 1898; Rosenhain, 1914; Ewing and Rosenhain, 1899, 1900a, 1900b; Alterthum, 1922; Carpenter and Elam, 1920; Czochralski, 1927; Burgers and Louwerse, 1931). It should be noted though that in such diagrams the influence of chemical alloy variations and specifically also of the annealing time is often not taken into consideration. This means that even for practical purposes such recrystallization diagrams can only serve as a rough first guideline for assessing an appropriate softening heat treatment.

The term dynamic recrystallization refers to all recrystallization phenomena that occur during the plastic deformation (Doherty et al., 1997; Humphreys and Hatherly, 2004; Gifkins and Coe, 1951; Tamhankar et al., 1958; Rossard and Blain, 1960; Rossard, 1963; Sellars and Tegart, 1966; Drube and Stüwe, 1967; Stüwe, 1968; McQueen et al., 1976; Fritzmeier et al., 1979; Luton and Sellars, 1969; McQueen, 1968; Jonas et al., 1969; McQueen and Jonas, 1975; Sellars, 1978). Such phenomena hence typically occur when hot deforming the material at temperatures greater than about half of the absolute melting point.

The kinetics associated with the softening process during such an annealing treatment can be grouped into three main stages, namely, recovery, recrystallization, and competitive grain coarsening.

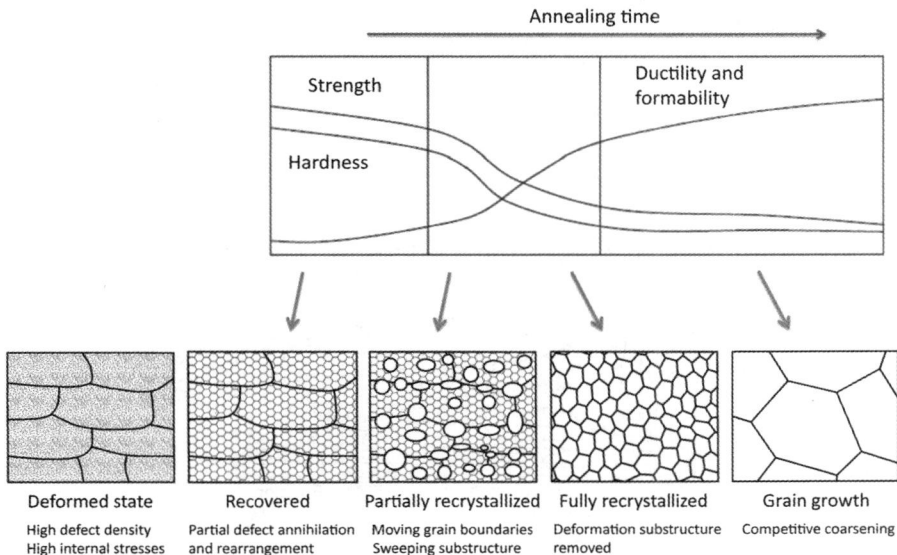

Figure 1 Schematical diagram illustrating the most generic steps during recovery, static primary recrystallization, and grain growth.

The latter mechanism is also referred to as grain growth (Carpenter and Elam, 1920; Burke and Turnbull, 1952; Doherty et al., 1997; Humphreys, 1997; Miodownik, 2002; Himmel, 1962; Cahn, 1965, 1966).

Recovery usually occurs at low temperatures and involves the thermally activated motion, condensation, and annihilation of point defects and specifically the annihilation and rearrangement of dislocations (Burgers and Louwerse, 1931; Doherty et al., 1997; Humphreys and Hatherly, 2004; Himmel, 1962). Owing to the high self-ordering tendency among dislocations recovery hence results in the formation of pronounced subgrain structures consisting of tilt and twist low-angle grain boundaries. With ongoing recovery, further dislocation annihilation and particularly competitive subgrain growth takes place (Humphreys, 1997). This mechanism overlaps with the onset of recrystallization, which is characterized by discontinuous nucleation phenomena (**Figure 3**). In some materials, recrystallization nucleation can result from discontinuous subgrain growth where some of the subgrains accumulate a sufficiently high misorientation relative to the as-deformed surrounding microstructure (Burgers, 1941; Doherty et al., 1997; Humphreys and Hatherly, 2004; Humphreys, 1997; Himmel, 1962).

Recovery leaves the grain topology of the deformed crystals usually unaffected as it does not involve the motion of high-angle grain boundaries. Instead the dislocation rearrangements described above occur within the as-deformed grains. Relatively small and continuous changes in hardness occurring during static recovery are essentially due to the gradual decrease in the elastic distortion of the material, which is attributed to the reduction in the dislocation density, to the formation of low-energy DC structures, and to the growth of these subgrains (Clarebrough et al., 1955) (**Figure 4**).

While static recovery thus describes the thermally activated formation and reorganization of dislocation substructures in a deformed material, recrystallization is concerned with those phenomena where the formation and subsequent motion of new high-angle grain boundaries is involved. The

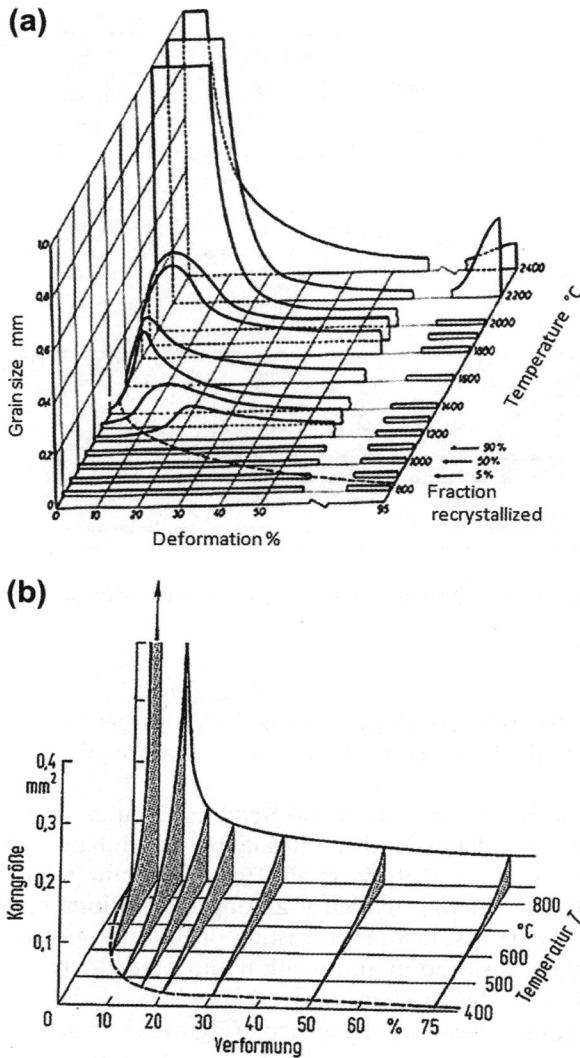

Figure 2   Recrystallization diagram of molybdenum (a) (grain size, deformation, temperature, recrystallized volume fraction) and (b) pure iron (Burgers, 1941; Doherty, 2005; Humphreys and Hatherly, 2004).

newly formed interfaces discontinuously sweep the inherited deformation substructure and lead to a softened polycrystal microstructure.

Recrystallization starts when sufficient thermal activation is available to lead to nucleation and growth of strain-free new grains in the plastically deformed matrix (Kolmogorov, 1937). Since newly formed grains sweep the deformation microstructure discontinuously, recrystallization nucleation can be described as a process leading to a local strain-free zone in the deformation matrix. This newly formed strain-free region is mainly surrounded by newly formed or bulged mobile high-angle grain

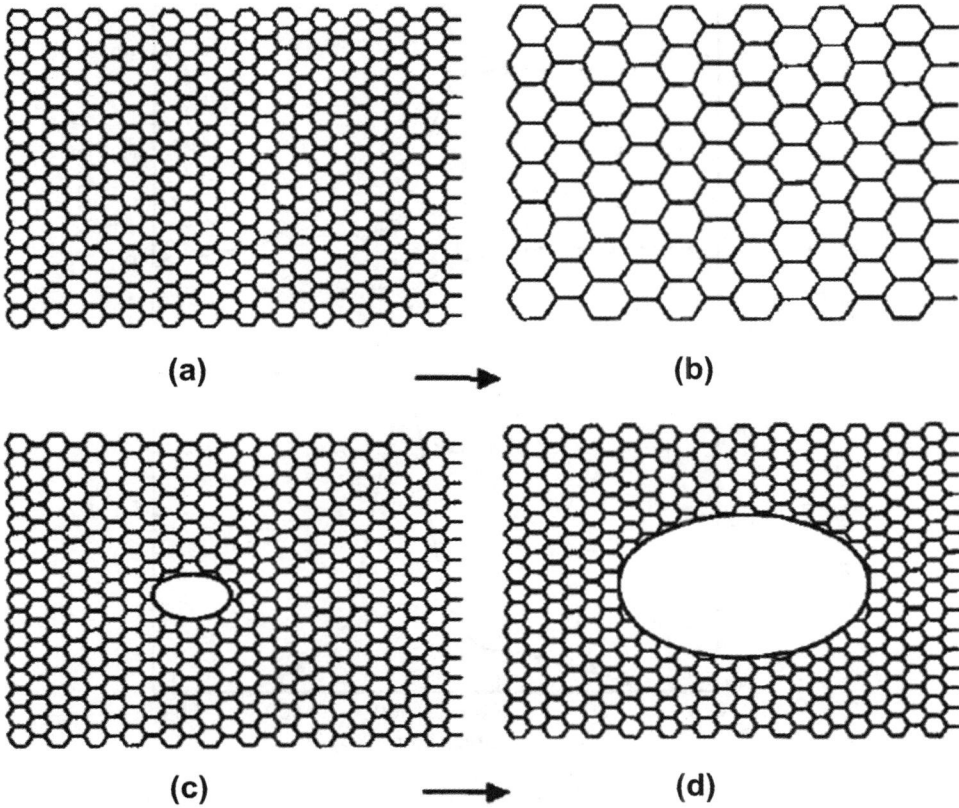

Figure 3 Schematic illustration of (a–b) continuous, and (c–d) discontinuous subgrain-coarsening phenomena during annealing (Humphreys, 1997).

boundaries (Beck and Hu, 1966; Humphreys and Hatherly, 1995; Humphreys, 1997; Himmel, 1962; Cahn, 1965, 1966; Kolmogorov, 1937; Johnson and Mehl, 1939; Avrami, 1939, 1940; Doherty and Cahn, 1972; Humphreys and Ferry, 1997; Faivre and Doherty, 1979; Bhatia and Cahn, 1978; Srolovitz et al., 1986a; Sebald and Gottstein, 2002; Engler et al., 1996; Engler, 1997; Humphreys, 1992a).

During recrystallization the strength and hardness both decrease considerably and often at much higher rates than during recovery. During the recrystallization process, the original ductility that the metal had prior to cold deformation is restored. The lowest temperature at which newly formed stress-free grains appear in the microstructure of a previously plastically deformed metal is termed the recrystallization temperature. This depends upon the grain size, the severity of plastic deformation, the strain path, and the presence of solute atoms or second-phase particles. The recrystallization temperature is usually 1/3–1/2 the absolute melting point of the material in Kelvin.

Recrystallization is finished when all originally deformed grains are swept by mobile grain boundaries. After that and locally also during the ongoing recrystallization grain growth, i.e. competitive capillary driven grain-coarsening begins. This phenomenon changes the strength of the material further owing to the increase in the average grain size and the associated reduction in Hall–Petch strengthening. Grain growth is thermodynamically driven by the reduction of the total grain-boundary area. While

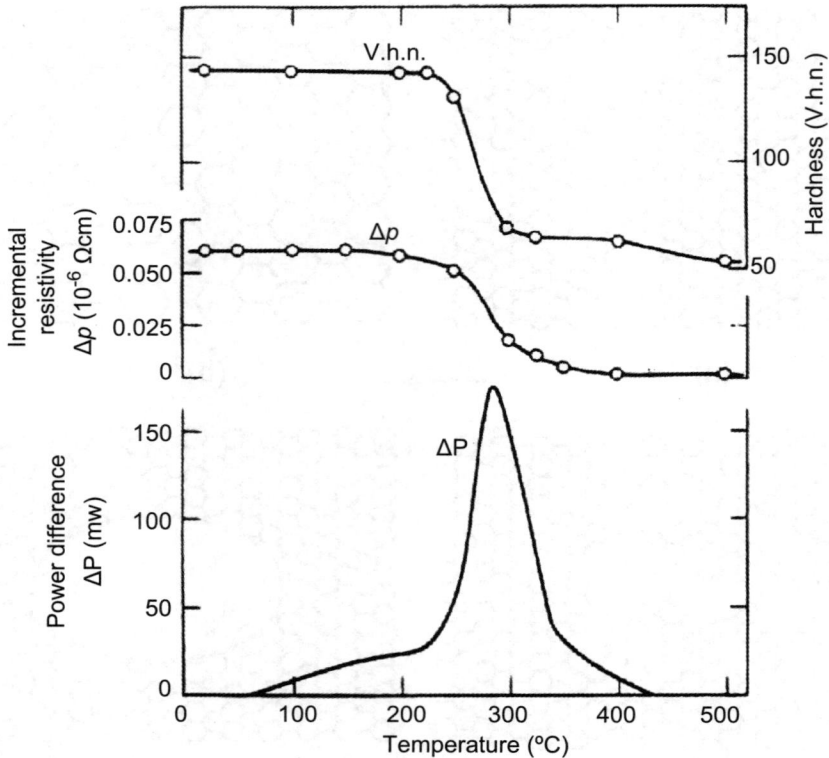

Figure 4 Changes in hardness, release in energy, and change in electrical resistivity upon isothermal heating of a copper rod (Clarebrough et al., 1955).

recovery and grain growth are relatively slow processes, recrystallization follows in most cases sigmoidal-type kinetics, i.e. it can proceed very fast after an initial more sluggish nucleation period.

23.1.2 Historical Notes

Many of the technical terms associated with phenomena such as recovery, recrystallization and grain growth are older than the knowledge that we have today about them.

This means that some of the terminology that is commonly used in this field is often a bit imprecise at first view. An example is the term "recrystallization". It describes the removal of the deformation substructure obtained from cold working via the formation and thermally activated motion of high-angle grain boundaries. The notion "recrystallization" seems odd since deformed metals remain, as a rule, crystalline upon plastic deformation because the elementary deformation carriers, the dislocations, proceed via translational shear increments. In other words the material is (usually) not rendered amorphous upon straining, hence, "recrystallization" is not a correct term as the crystalline state of the material is not altered during the heat treatment. It rather goes back to the times when structure observation was exclusively made by optical inspection of as-processed or etched metallic surfaces. The blurred appearance of many metals after strong plastic deformation led early metallurgists to the assumption that deformed metals were no longer crystalline. The advent of structure analysis via X-ray diffraction (XRD) by Max von Laue, when applied to cold-worked metals, revealed that deformed materials usually remain crystalline (Laue, 1913). Irrespective of this observation, the term "recrystallization" prevailed.

Also, the use of the term "nucleation" in conjunction with the early stages of recrystallization may be a bit misleading. Originally, it was assumed that the nucleation process by which new recrystallized grains are formed would be based on thermal fluctuations. This model assumption was adopted from classical phase transformation theory. In classical nucleation theory, pertaining to phase transformation, it is assumed that as a result of the permanent thermal motion and rearrangement of the atoms, tiny strain-free clusters, i.e. nuclei, would spontaneously form in the cold-worked matrix. Following the classical energy balance used to explain nucleation phenomena in association with phase transformations, these nuclei would be associated with an energy increase due to the formation of a new grain boundary and an energy reduction due to the formation of a new volume of strain free, hence lower energy material. If such nuclei, assumed to be spherical in shape, were larger than a critical radius then they would be thermodynamically stable and could start to grow and sweep the surrounding deformed matrix.

Later, it was suggested that this classical nucleation theory could not hold for recrystallization phenomena (Sellars, 1978; Humphreys, 1997; Miodownik, 2002; Himmel, 1962; Cahn, 1965, 1966). The main inconsistency with this type of nucleation model is that the stored deformation energy that could be released upon nucleation is very low (about $1-10$ M Jm^{-3} ($1-10$ MPa)) while the interface energy associated with a newly formed high-angle grain boundary is relatively high (about 1 Jm^{-2}). This means that in the context of the classical fluctuation theory nucleation would be impossible.

As will be discussed below in more detail, recrystallization nucleation must instead proceed via the growth of already preexisting subgrains or cells. The discontinuous and competitive coarsening of such subgrains during the nucleation or incubation time can then lead in certain cases to the formation of new mobile high-angle grain boundaries. A higher misorientation of such a gradually forming interface increases the mobility of that boundary and so the rate of growth of the subgrain increases until it finally acts as a nucleus. Since subgrains are typically surrounded by low-angle grain boundaries of usually not more than $1-2°$ misorientation each, the accumulation of further misorientation during subgrain coarsening requires local orientation gradients to be present in the vicinity of such a growing subgrain that finally prevails as a successful nucleus (Ferry and Humphreys, 1996; Holm et al., 2003; Rollett, 1997; Heidenreich, 1949; Beck, 1949; Cahn, 1950; Humphreys and Chan, 1996; Bailey, 1960; Cotterill and Mould, 1976; Bailey, 1963; Sandström et al., 1978; Varma and Willitis, 1984; Walter and Koch, 1963; Blum et al., 1995; Li, 1962; Doherty and Szpunar, 1984a). The details of these mechanisms will be discussed below.

Likewise, the term grain growth is not very precise as during grain growth most of the crystals actually shrink rather than grow. Only the *average* crystal size increases such as in many other capillary driven competitive ripening processes (e.g. such as for instance known from Ostwald ripening).

23.1.3 Experimental Methods to Study Recovery and Recrystallization

23.1.3.1 Introduction

While integral macroscopic property changes associated with recovery and recrystallization can be efficiently tracked in terms of hardness and electrical resistivity characterization, the associated microstructural changes are more challenging to measure. This applies particularly to recrystallization. This is due to a number of reasons:

Firstly, recrystallization nucleation takes place at very small dimensions, usually at the subgrain scale. Hence, it is difficult to predict where—in a large array of subgrains—nucleation will actually happen. Recrystallization nucleation is a relatively rare event and strongly dependent on the underlying type of nucleation mechanism (e.g. discontinuous subgrain coarsening; nucleation that shear bands; particle stimulated nucleation (PSN); grain-boundary bulging) (Humphreys and Hatherly, 2004; Humphreys, 1992a). When aiming at the observation of the early stages of nucleation hence adequate microscopy methods in conjunction with the capability to overlook a high field of view are required (Doherty et al., 1997; Cahn, 1965; Doherty and Cahn, 1972). Later stages of subgrain coarsening or cell growth could in principle already be the effect of a grain-growth competition phenomenon subsequent to the actual nucleation stage.

Secondly, recrystallization nucleation often starts in regions that contain microstructural inhomogeneities. These are for instance heavily distorted grain boundaries or heterointerfaces with strain gradients and high lattice curvatures before them; the vicinity of hard second-phase particles; microbands with high local dislocation densities; or shear bands that are characterized by both, locally high stored-dislocation densities and large lattice curvature around them. Probing and understanding nucleation events occurring at these sites requires to first understand the underlying deformation substructures and the specific characteristics that render these sites into suited nucleation sites (Ferry and Humphreys, 1996; Cahn, 1950). This means that recrystallization research is not a "one-mechanism" search but usually requires rather a systemic approach where a variety of possible phenomena and their interactions have to be taken into consideration (Humphreys, 1997; Miodownik, 2002; Holm et al., 2003; Rollett, 1997).

Thirdly, recrystallization can proceed very fast and sweep the deformation substructures inherited from cold working. This means that the microstructural mapping of recrystallization phenomena is often conducted using postmortem and ex situ experimental approaches rather than direct in situ observation.

Fourth, recovery and recrystallization phenomena can strongly depend on many subtle and hard-to-measure details such as material purity, the types of grain boundaries involved, the distribution of the stored energy, interfering ordering and/or phase transformation phenomena, and the complexity of the underlying deformation substructures inherited from preceding cold working. Many of these effects can have a drastic, namely, an exponential influence on recrystallization as they, in part, affect the thermal activation barriers. Thermal activation during recrystallization often follows an Arrhenius dependence, hence the exponential influence. A typical example is the dependence of the grain-boundary mobility and of the grain-boundary energy on the impurity content of the material (Molodov et al., 1984; Ibe and Lücke, 1966, 1972; Ibe et al., 1970; Babcock and Balluffi, 1989; Hashimoto and Baudelet, 1989; Gottstein and Shvindlerman, 1999; Aust and Rutter, 1959, 1960; Hu et al., 1990; Rath and Hu, 1969a; Molodov, 2001; Lücke and Stüwe, 1963; Gottstein et al., 1995, 1997, 1998; Shvindlerman et al., 1995, 1999; Upmanyu et al., 1999; Rollet et al., 2004; Liebman et al., 1956; Christian, 1965; Viswanathan and Bauer, 1973; Furtkamp et al., 1998; Li et al., 1953; Winning et al., 2001, 2002; Winning, 2003; Heinrich and Haider, 1996; Vandermeer and Juul Jensen, 1993; Beck et al., 1950; Molodov et al., in press).

These aspects show that recrystallization phenomena are among the most complex metallurgical topics and therefore require the use of advanced and combined experimental and modeling methods.

23.1.3.2 Indirect Experimental Methods

Indirect methods for analyzing the effects and kinetics of recovery and recrystallization as well as grain-growth phenomena typically probe macroscopic or mesoscopic mechanical, optical, calorimetric, or electromagnetic property changes that are associated with the underlying microstructural evolution during these processes. Examples are changes in resistivity, hardness, strength, or averaged values of the stored deformation energy. Electrical resistivity measurements are characterized by the fact that point defects and internal interfaces provide a rather high contribution to the resistivity while the dislocation cores and elastic long-range distortions contribute less (Burgers and Louwerse, 1931; Burgers, 1941; Burke and Turnbull, 1952; Smith, 1948; Beck and Hu, 1966; Haessner, 1978; Humphreys and Hatherly, 1995, 2004; Doherty et al., 1997; Doherty, 2005) (**Figure** 4).

A more quantitative approach to determine the changes in internal stresses associated with the annealing of cold work metals is provided by advanced XRD methods. Corresponding experiments can be conducted by using lab-scale XRD devices or the more brilliant high-energy synchrotron XRD methods (Poulsen et al., 2001; Margulies et al., 2001; Leslie et al., 1963a; Schmidt et al., 2004; Larson et al., 2002; Tamura et al., 2003, 2002; Bale et al., 2005; Chung and Ice, 1999; Yang et al., 2003; Chawla et al., 2004).

When exposing polycrystalline metals to a monochromatic XRD setup, the underlying assumption is that each scattering volume of the illuminated material contains a sufficiently large number of crystallites, such that there are always some of the grains that satisfy the Bragg–Brentano diffraction condition for any given orientation of the sample relative to the beam setup. A photon point, line, or areal detector then collects a pattern of diffracted counts versus angle corresponding to a section through the Debye–Scherrer cones. XRD methods typically provide excellent diffraction profiles with little instrumental broadening contribution that is particularly suitable for line profile and crystallographic texture analysis. While the former information (peak broadening) provides information about the elastic distortion of the material the latter information (crystallographic texture) provides information about crystallographic orientation changes. The latter are quite indicative of recrystallization since by definition it involves the motion of high-angle grain boundaries, and hence the orientation changes from deformation textures to recrystallization textures. Data collection in XRD measurement is typically relatively slow due to the low flux density of laboratory-scale monochromatic-beam setups.

Generally, when a material is plastically deformed, the reflections observed in constructive interference Bragg–Brentano of a monochromatic beam detected with XRD can be affected in two ways.

The first possibility is that due to compressive or dilatational stresses the average lattice parameter will systematically become smaller or larger, respectively, reflecting the presence of mean homogeneous strains.

The second possibility is that the breadth and the shape of the peaks can change due to the size of the diffracting elements such as the domain or crystal size or due to an inhomogeneous distribution of the local strains. The latter are also referred to as microstrains. Possible sources of inhomogeneous strain include lattice dislocations, grain-boundary defects, and intracrystalline gradients such as high-dislocation-density walls.

Hence, in conjunction with appropriate peak analysis models that allow the interpretation of XRD peaks on the basis of certain lattice defect distributions XRD methods can be used to provide an average information about the dislocation density of the bulk deformed material at a limited spatial resolution when compared to scanning electron microscope (SEM) or transmission electron microscopy (TEM)-based methods. Also, XRD analysis of defect structures requires the use of a well-justified underlying model that connects a certain dislocation density and distribution with a total displacement gradient field (Mughrabi et al., 1986; Straub et al., 1996).

An interesting advantage of XRD-based methods is that the line broadening can be different for different crystallographic textures. X-ray line broadening data from different such studies have shown that indeed a variation in the stored deformation energy exists. These results were confirmed by TEM, Electron Backscatter Diffraction (EBSD), and Electron Channeling Contrast Imaging (ECCI) measurements. For instance in cold rolled steels it was observed that crystallographic texture components with a {111} axis in normal direction are capable of storing twice as much deformation energy compared to grains that have a {001} axis in normal direction.

The second essential option provided by the use of XRD methods consists in the analysis of crystallographic textures, viz. orientation distributions (Beck and Hu, 1966; Ibe and Lücke, 1966; Poulsen et al., 2001; Chung and Ice, 1999; Bunge, 1969, 1982, 1986, 1987; Wassermann and Grewen, 1962; Bunge and Esling, 1982, 1991; Adams et al., 1993; Duggan et al., 1978; Dillamore et al., 1979; Hölscher et al., 1991; Ushioda et al., 1987; Raabe and Lücke, 1993). While the global orientation distributions do not change much during recovery owing to the very definition of this process that excludes the motion of high-angle grain boundaries, they can indeed change quite dramatically during primary recrystallization (**Figure 5**). This is due to the fact that primary static recrystallization is defined as the formation and motion of high-angle grain boundaries through the deformed material.

An essential advantage of using XRD methods for recrystallization research lies in the nondestructive nature of this approach. Many other methods to probe recrystallizing microstructures require cutting, grinding, or polishing of the samples under inspection. Using XRD beams with high energy, in contrast, allows for the in situ observation of samples during recrystallization without destroying them (Poulsen et al., 2001; Margulies et al., 2001).

Calorimetric methods to probe recovery and recrystallization driving forces and kinetics can be used to analyze enthalpy release rates upon recovery and recrystallization heat treatments (**Figure 6**)

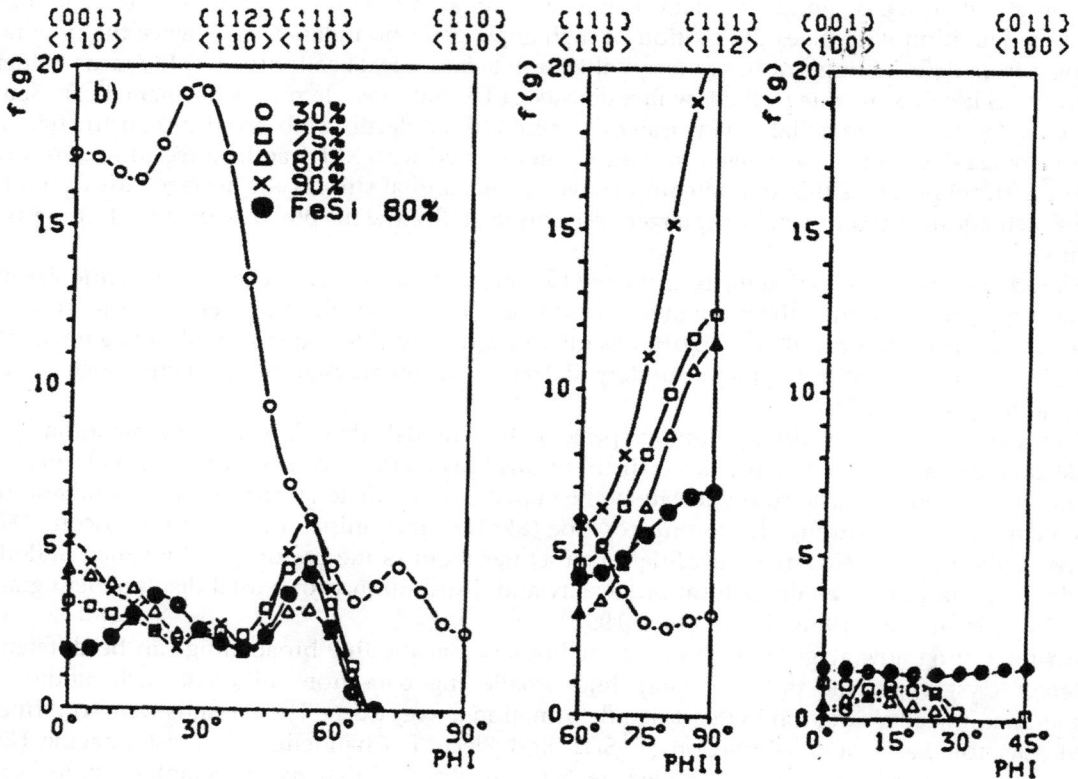

Figure 5 Texture of Fe-16%Cr and Fe-4%Si steels after annealing at 700 °C for different rolling reductions. Weak deformation (30% rolling) does not lead to a texture change indicating that recovery prevails (Hölscher et al., 1991).

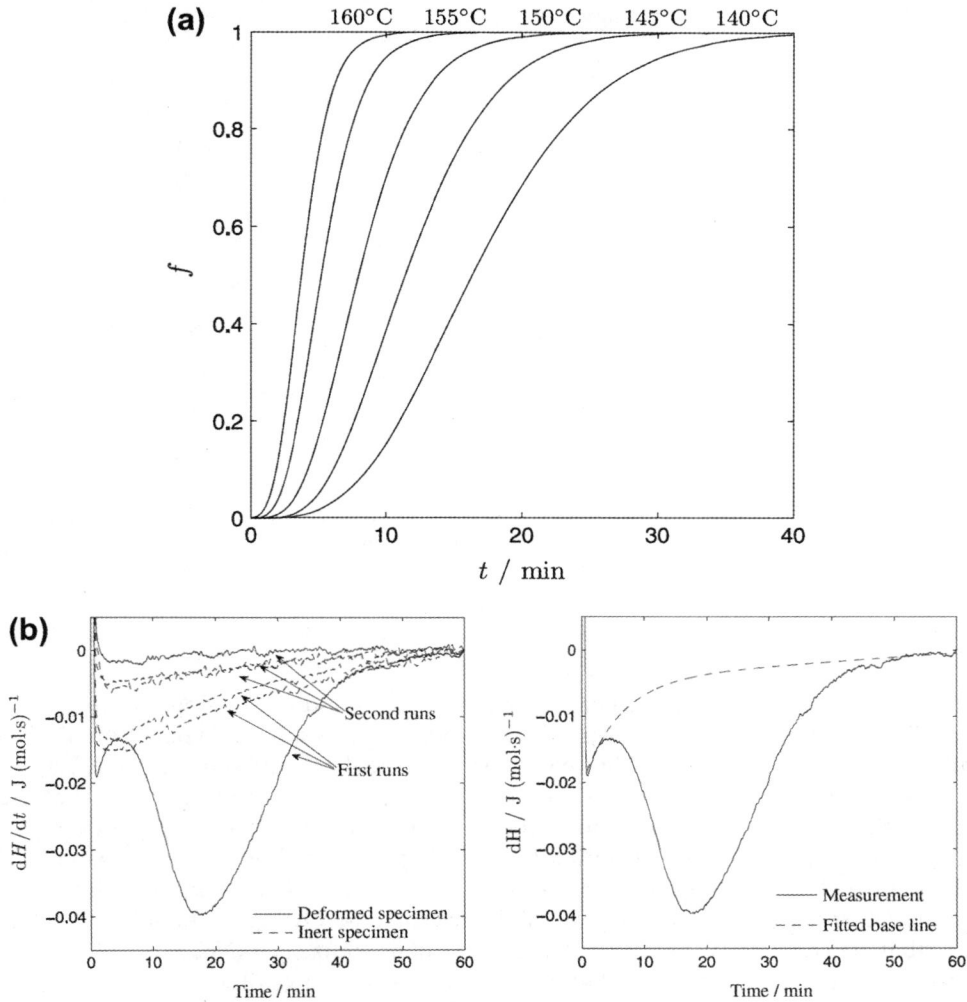

Figure 6   (a) The kinetics of the recrystallization of pure copper as investigated via DSC. The recrystallized fraction, f, determined by isothermal DSC, as a function of time at various temperatures. (b) Using DSC methods in recrystallization analysis: (left) the heat flux signal of a DSC device for the first and second runs of a deformed and an inert copper specimen at 408 K (135 °C). Right: the heat flux of the first run of the same deformed specimen as in the left-hand image together with a fitted base line. (c) Transformed fraction as determined from EBSD measurements for isothermal recrystallization at $T = 413$ K (140 °C) (data points), as a function of time. The error bars indicated for the data points result from choosing different image-quality thresholds in the analysis of the EBSD measurements. The full line represents the recrystallized fraction as determined by DSC for an identically conducted experiment. The DSC curve is shifted by 1 min to shorter times to obtain the best match with the EBSD results (Jägle and Mittemeijer, 2012).

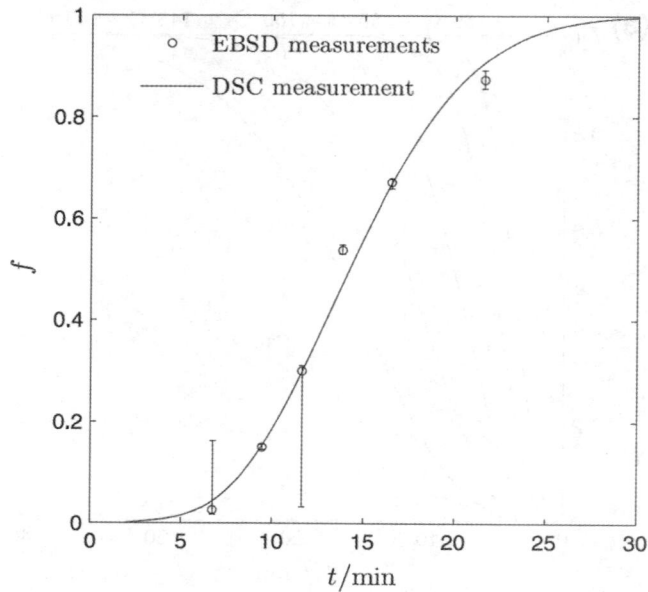

Figure 6   *(Continued).*

(Jägle and Mittemeijer, 2012). Such experiments are usually conducted using a differential scanning calorimetry (DSC) method. In such a setup, when applied to recrystallization phenomena, experiments are typically conducted isothermally (**Figure** 6a). Usually, the deformed samples to be probed have to be stored—like in most recrystallization experiments—in a refrigerator prior to the characterization. This is due to the fact that the activation barriers for initiating recovery and even recrystallization nucleation (for pure metals) can be of similar magnitude as provided by room temperature. Examples are heavily strained lead or high-purity aluminum polycrystals. Then specimens are heated to the probing temperature using the maximum possible heating rate. Owing to the relatively small energy release rates that are characteristic for recrystallization phenomena such DSC methods have to be very well calibrated as they operate close to their resolution limits (**Figure** 6b,c).

It must be underlined though that calorimetry that is conducted on samples during recovery, recrystallization, or grain growth provides an integral value of the energy release rates.

For all calorimetric methods, the global effective activation enthalpy associated with recovery and recrystallization can for isothermal measurements be calculated from a diagram that plots the logarithm of the time that has elapsed between two recrystallization states that are characterized by their respective recrystallized volume fractions and the inverse temperature. Precondition to such a classical analysis of the effective activation barrier is the assumption that the underlying processes are—when occurring simultaneously—described by one single value of the activation energy.

23.1.3.3 *Direct Experimental 2D Methods*

The most frequently used instruments for tracking recovery and recrystallization phenomena are optical microscopes. They typically allow one to map grain shape changes and also polygonization as well as subgrain coarsening effects when a corresponding sample is subjected to grain-boundary etching.

Polygonization, i.e. the regular alignment of dislocations that can be observed in heat-treated samples after preceding bending, was also made visible by using etch-pit methods.

The primary recrystallized portion of a previously cold-worked portion of material is usually quantified by determining the area fraction of recrystallized versus the remaining unrecrystallized zones. From stereological considerations, the equivalent recrystallized volume fraction can be determined.

In order to reveal crystallographic features, such as the grain-boundary misorientations or the grain orientations, EBSD methods, which are also referred to as orientation microscopy, are frequently used (Adams et al., 1993; Adam et al., 2001; Randle and Engler, 2000; Randle, 1992, 2004).

In EBSD, a stationary primary electron beam of an SEM is diffracted by atomic layers in crystalline materials. These diffracted electrons can be detected when they impinge on a fluorescent screen and generate visible lines. These are referred to as Kikuchi bands or electron backscatter patterns (EBSPs). These patterns are projections of the geometry of the lattice planes in the crystal, and they give direct information about the crystalline structure and crystallographic orientation of the grain from which they originate. When used in conjunction with a database that includes crystallographic structure information for phases of interest and with adequate analysis software for processing the EBSPs and indexing the lines, the data can be used to identify crystallographic textures and phases provided that they have a crystalline structure. EBSD provides information about the local phases, crystal orientations, and local misorientations down to 50 nm resolution—for instance when using an SEM that is equipped with a field emission gun. This renders it a powerful instrument for microstructural characterization in the context of recrystallization (**Figure 7**).

Another, more recently matured technique to track lattice defect ensembles at a wide field of view is ECCI (Gutierrez-Urrutia et al., 2009, 2010; Gutierrez-Urrutia and Raabe, 2011; Eisenlohr et al., 2012). ECCI is an SEM technique that is based on the effect that the backscattered electron intensity is strongly dependent on the orientation of the crystal lattice planes with respect to the incident electron beam due to electron channeling mechanism. Also minor local distortions in the crystal lattice due to dislocations or interfaces cause a modulation of the backscattered electron intensity enabling such linear or planar defects to be mapped. The ECCI method has been used to image dislocation structures in metals deformed during fatigue loading or in the vicinity of cracks, and even stacking faults. For quantifying the change in the dislocation structure during recovery or recrystallization, a special variant of the ECCI method is particularly suited. This approach consists of a combined EBSD and ECCI methods to achieve enhanced contrast by calculating the optimum channeling angle based on the information provided by the EBSD measurement (Gutierrez-Urrutia et al., 2013). In this specific combination ECCI is a powerful, versatile, fast, and experimentally robust method for determining dislocation and interface defect densities and cell-type dislocation arrangements at small preparation time and at a wide field of view (**Figure 8**) (Gutierrez-Urrutia et al., 2009, 2010; Gutierrez-Urrutia and Raabe, 2011).

When higher spatial resolutions are required than offered by conventional SEM, EBSD, or ECCI methods TEM has to be used in recrystallization research (Berger et al., 1988; Wilbrandt, 1980; Wilbrandt and Haasen, 1979, 1980a, 1980b; Weiland and Schwarzer, 1984; Ray et al., 1975a; Hartig and Feller-Kniepmeier, 1985; Klement and Haasen, 1993; Schwarzer and Weiland, 1986; Hu, 1962, 1963; Bailey and Hirsch, 1962; Zaefferer, 2005; Grewen and Huber, 1978). This is a microscopy technique whereby a beam of electrons is transmitted through an ultrathin specimen, interacting with the specimen as it passes through it. An image is formed from the interaction of the electrons transmitted through the specimen; the image is magnified and focused onto an imaging device, such as a fluorescent screen, on a layer of photographic film, or to be detected by a sensor such as a charge-coupled device

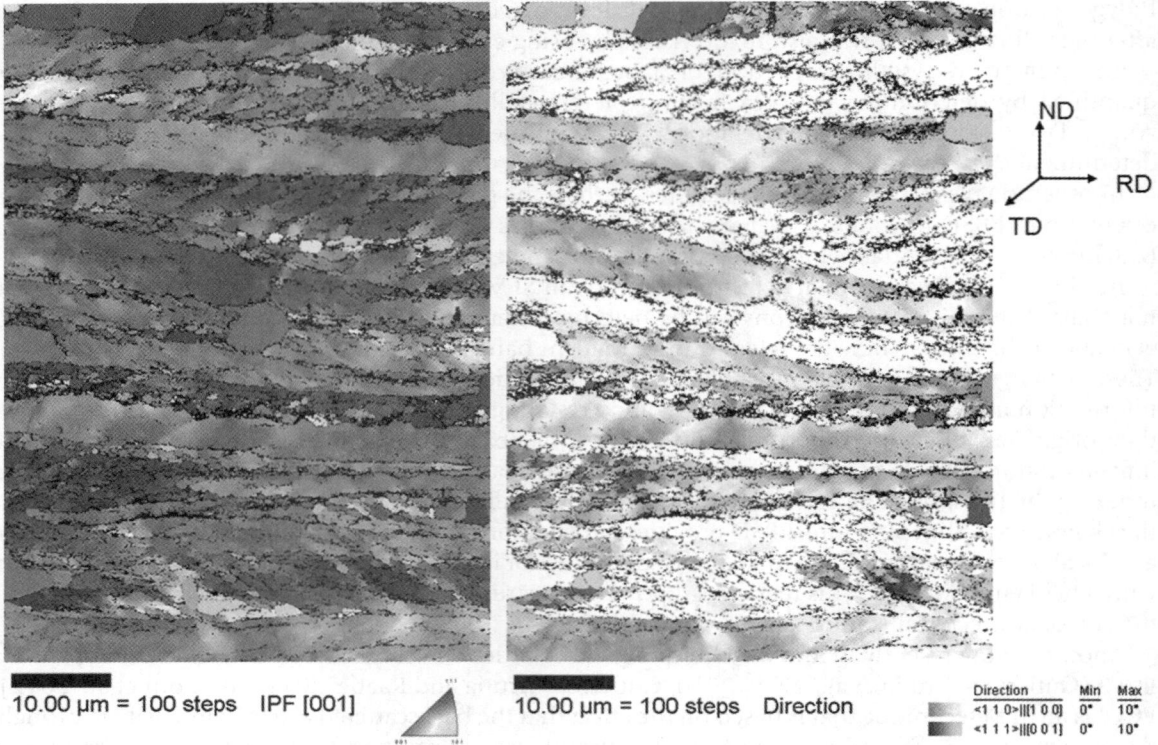

Figure 7 Formation of a new grain during the recrystallization of a 70% cold rolled low-C steel, measured by EBSD.

(CCD) camera. TEMs are capable of imaging microstructures of recrystallizing samples at a significantly higher resolution than light microscopes, owing to the small de Broglie wavelength of the electrons.

23.1.3.4 Tomographic Experimental Methods

By combining conventional EBSD microstructure imaging with sequential serial sectioning of series of subsequent microstructure slices, tomographic microstructure imaging can be performed. This method which is referred to as three-dimensional (3D) EBSD or EBSD tomography is a technique for the 3D high-resolution characterization of crystalline microstructures (Konrad et al., 2006; Zaafarani et al., 2006; Zaefferer et al., 2008). Typically, the technique is based on automated serial sectioning using a focused ion beam (FIB) and characterization of the sections by orientation microscopy based on EBSD in a combined FIB–SEM. These instruments are typically equipped with a field-emission electron gun for obtaining highest EBSD resolutions, a Ga^+ ion emitter unit (FIB), and secondary electron, and backscatter electron detectors. The FIB is usually operated at an accelerating voltage of 30 kV. The EBSD measurements are performed in a range between 15 and 20 kV (**Figure 9**).

On typical two-beam systems (electron beam for EBSD and ion beam for sectioning), the technique reaches a spatial resolution of about $50 \times 50 \times 50$ nm^3. The maximum observable volume depends on the optical setup and is typically of the order of $50 \times 50 \times 50$ μm^3. The technique provides all the main characterization features of two-dimensional (2D) EBSD-based orientation microscopy and extends

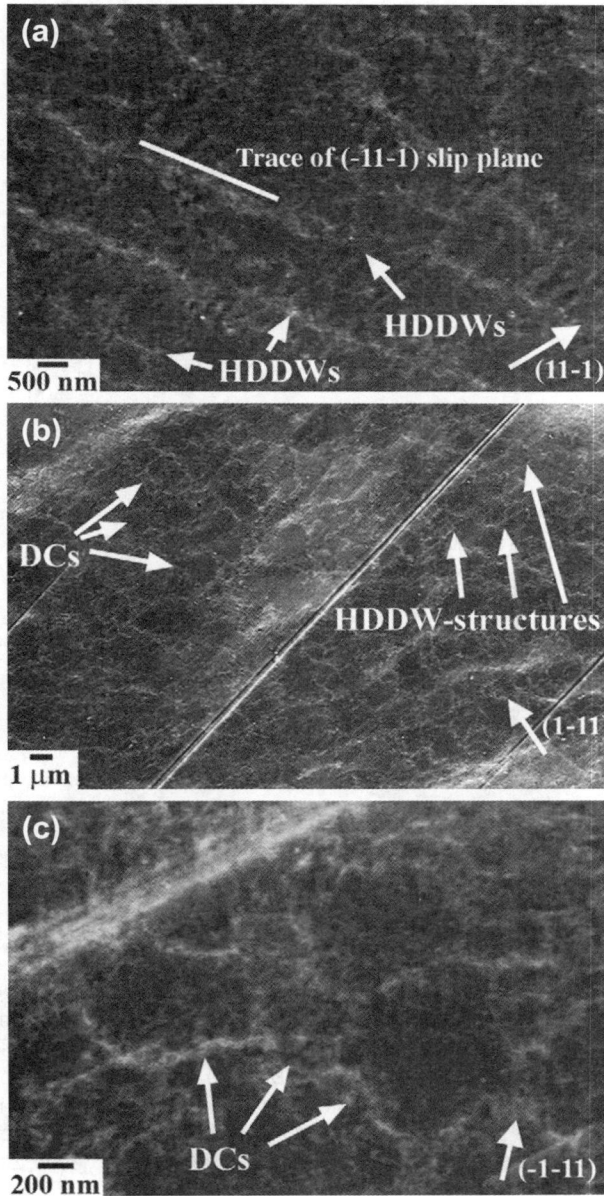

Figure 8 ECCI images of a deformed microstructure at the early stages of the deformation of an Fe-22%Mn-0.6C TWIP steel (wt.%) to a value below 0.1 true strain (ECCI). (a) High density dislocation walls (HDDWs) along the (-1 1 -1) slip plane on a sample tensile deformed to 0.05 true strain (HDDW). The ECCI image was obtained by orienting the grain into Bragg condition using the (1 1 -1)g vector (arrow). (b) DCs and HDDW structures in a sample tensile deformed to 0.1 true strain. The ECCI image was obtained by orienting the grain into Bragg condition using the (1 -1 1)g vector (arrow). (c) Details of the DC structure on a sample tensile deformed to 0.05 true strain. The ECCI image was obtained by orienting the grain into Bragg condition using the (-1 -1 1)g vector (arrow) (Gutierrez-Urrutia et al., 2009, 2010; Gutierrez-Urrutia and Raabe, 2011).

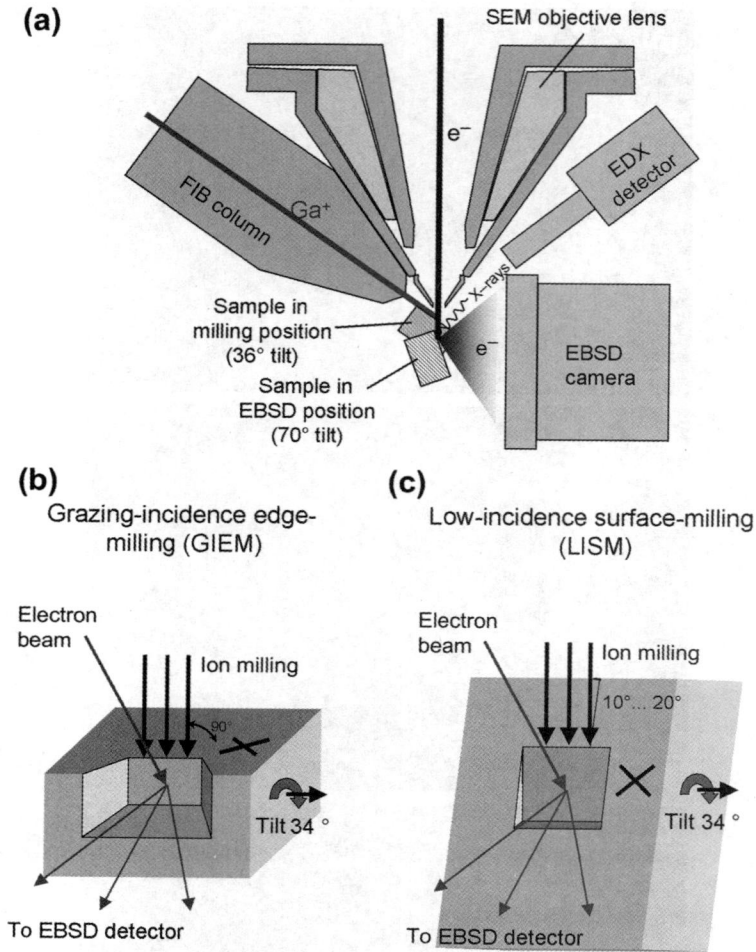

Figure 9 Schematics of the geometry of the 3D EBSD sectioning and mapping procedure. (a) Cross-section through the instrument chamber, showing the sample positions for milling and EBSD analysis in the tilt setup. To change between the two positions, the sample need only be tilted and moved in the y direction. During milling, the EBSD camera is retracted to the chamber wall. Working distance for milling is 8 mm; for EBSD, it is 13 mm. The camera is positioned at a distance of 22 mm from the sample. (b) And (c) Schematics of different variations of the tilt geometry: (b) the graizing incidence edge milling method and (c) the low-incidence surface milling method. For both schemes, the black cross indicates the marker for image alignment that is milled into the sample before the process begins (Konrad et al., 2006; Zaafarani et al., 2006; Zaefferer et al., 2008).

them into the third dimension of space. This enables new parameters of the microstructure to be obtained—for example, the full crystallographic characterization of all interface segments, including the morphology and the crystallographic indices of the interface planes and the crystallographic texture of all phases involved. Of specific interest is the high-resolution mapping of the five-dimensional grain-boundary segment orientation distribution function (**Figure 10**).

Figure 10 Presentation of the interface distribution (here the normal vectors of the grain boundaries are shown) obtained from a 3D EBSD method. The material is a Cu alloy after eight step Equal Channel Angular Pressing (ECAP) deformation.

An alternative method to track recrystallization in three dimensions consists in using tomographic synchrotron methods. Since such approaches are nondestructive they are particularly suited to study recrystallization as a function of time. Another advantage of synchrotron-based methods is the fact that they allow to probe grain topology and texture evolution also inside recrystallizing and recrystallized bulk materials so that the effects observed are not affected by free surfaces. Such free surface effects are particularly problematic in TEM samples and in EBSD in-situ experiments.

In Bragg-based 3D-XRD microscope setups, a monochromatic high-energy synchrotron beam is focused by using a Laue crystal in order to obtain a small full width at half maximum of the incident probing beam at the focal point. Every volume portion that fulfills the Bragg condition inside the illuminated region of the sample creates a diffraction spot on a detector. Coupling this method with sample rotation and a geometrical 3D path reconstruction setup allows one to map spot shapes and back trace them into the sample to achieve the shape of the grains. Such techniques are successful in achieving reflection from mm-deep grains due to the high energies. By using the high brilliance of third-generation synchrotron sources, microstructures in 4D were studied with a spatial resolution of micrometers and a time resolution of minutes (Poulsen et al., 2001; Margulies et al., 2001; Leslie et al., 1963a; Schmidt et al., 2004; Larson et al., 2002; Tamura et al., 2002, 2003; Bale et al., 2005; Chung and Ice, 1999) (**Figure 11**). **Figure 12** shows an example of some expansion snapshots of a growing grain as observed by XRD tomography (Poulsen et al., 2001).

An alternative to the monochromatic 4D XRD analysis approach involves an experimental procedure using polychromatic microbeam X-radiation (micro-Laue). The micro-Laue setup resolves individual

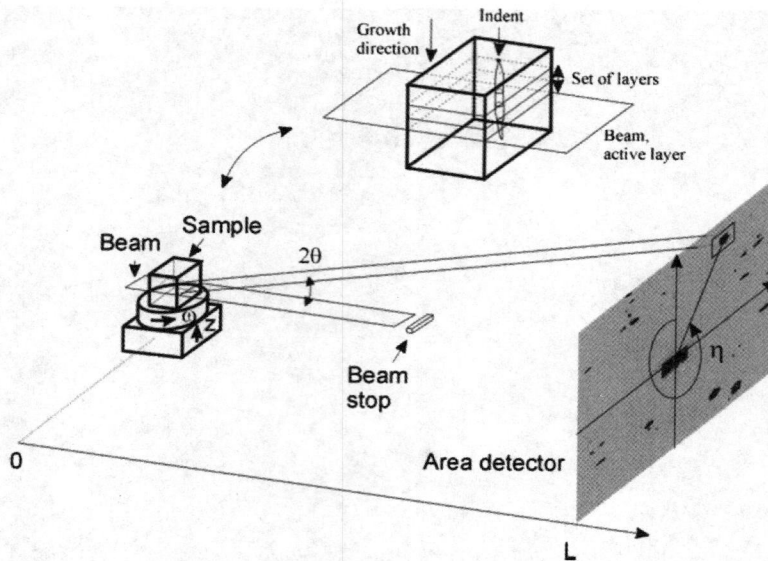

Figure 11   Sketch of an experimental 4D XRD setup (Margulies et al., 2001). Coordinate system (x,y,z) and angles are defined. The x-axis is along the beam direction, the y-axis is transverse to the beam direction, and the z-axis is normal to the beam plane. For the diffraction spot in question, the direction of the diffracted beam is parameterized. All grains within the stripe illuminated by the beam will give rise to diffracted spots during a scan. The inset shows the principle of obtaining a picture of the grain by repeatedly recording an oscillation photograph followed by a vertical translation of the sample stage. L denotes the distance between the sample and the CCD detector (Poulsen et al., 2001; Margulies et al., 2001; Leslie et al., 1963a; Schmidt et al., 2004; Larson et al., 2002; Tamura et al., 2003).

grains in the polycrystalline matrix. Results obtained from a list of grains sorted by crystallographic orientation depict the strain states within and among individual grains. Locating the grain positions in the plane perpendicular to the incident beam is trivial. However, determining the exact location of grains within a 3D space is quite challenging. Determining the depth of the grains within the matrix (along the beam direction) involved a triangulation method tracing individual rays that produce spots on the CCD back to the point of origin. Triangulation was experimentally implemented by simulating a 3D detector capturing multiple diffraction images while increasing the camera to sample distance. Hence, by observing the intersection of rays from multiple spots belonging to the corresponding grain, depth is calculated. Depth resolution is a function of the number of images collected, grain to beam size ratio, and the pixel resolution of the CCD. The 4D XRD method provides grain morphologies, strain behavior of each grain, and interactions of the matrix grains with each other and the centrally located single crystal fiber (Poulsen et al., 2001; Margulies et al., 2001).

23.2   Recovery

23.2.1   Introduction and Basic Phenomena

The early stages of the thermally activated restoration of the inner structure of cold-worked metals proceed at first without notable changes of the grain and phase topology. This stage is referred to as recovery (Vandermeer and Rath, 1990; Stüwe et al., 2002). It comprises a set of thermally activated

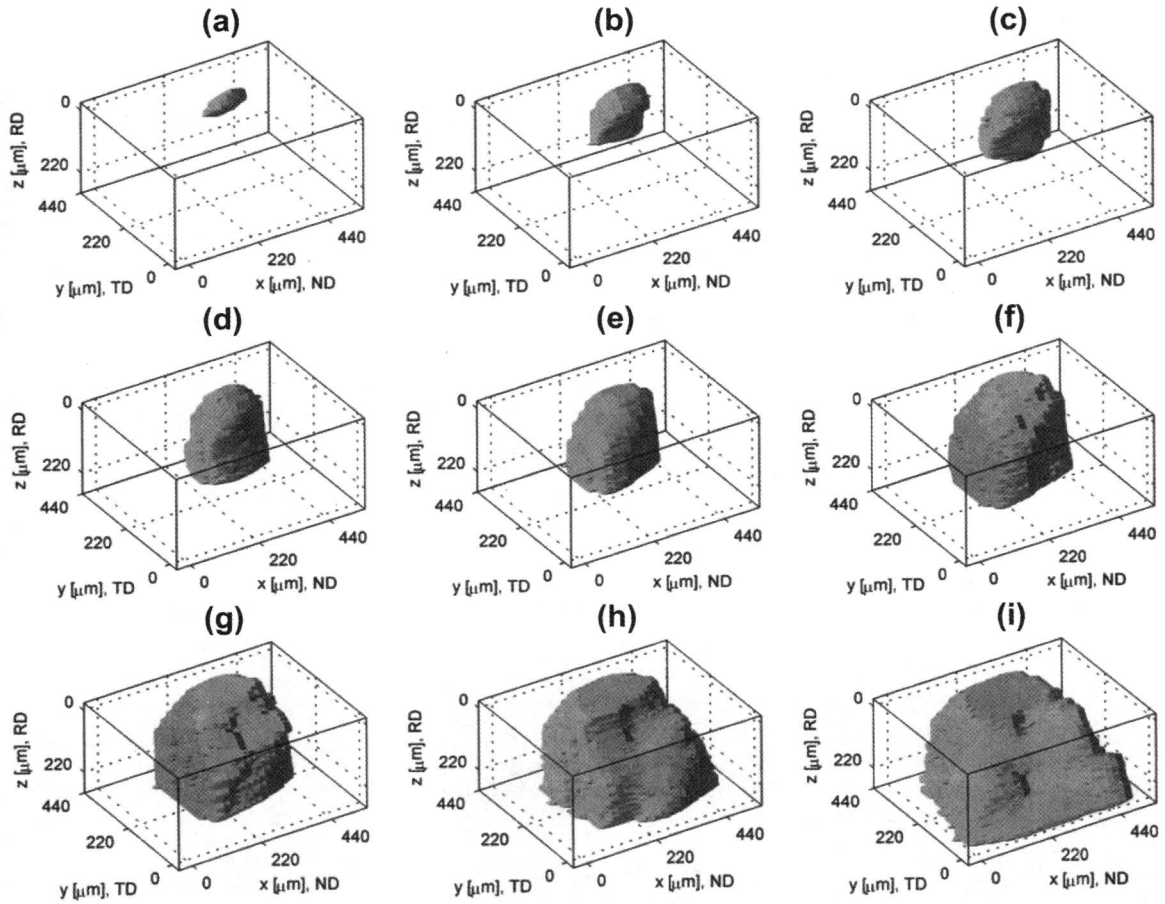

Figure 12 Expansion snapshots of a grain as observed by XRD tomography (Margulies et al., 2001).

gradual softening and microstructural reorganization phenomena in deformed metals to release some of the stored internal energy. The energy relieve is mainly due to point defect motion and dislocation relaxation and reorganization phenomena. When occurring during cold working, it is referred to as dynamic recovery. The properties that are mostly affected by recovery are those that change with the point defect density, such as the electrical resistivity (**Figures 4** and **13**).

Static recovery can be well observed at rather modest temperatures already far below the recrystallization temperature of about 1/3–1/2 of the absolute melting point. The microstructural mechanisms are essentially characterized by the reorganization of the dislocation substructure, driven by the elastic internal stresses that the material has inherited from the preceding plastic deformation and by point defect annihilation (**Figure 14**). The excess point defects that were created during plastic deformation are annihilated either by absorption at grain boundaries or dislocation climbing. This means that recovery proceeds by the annihilation of point defects and the annihilation and rearrangement of dislocations associated with the higher point-defect mobility. The overall dislocation density does not

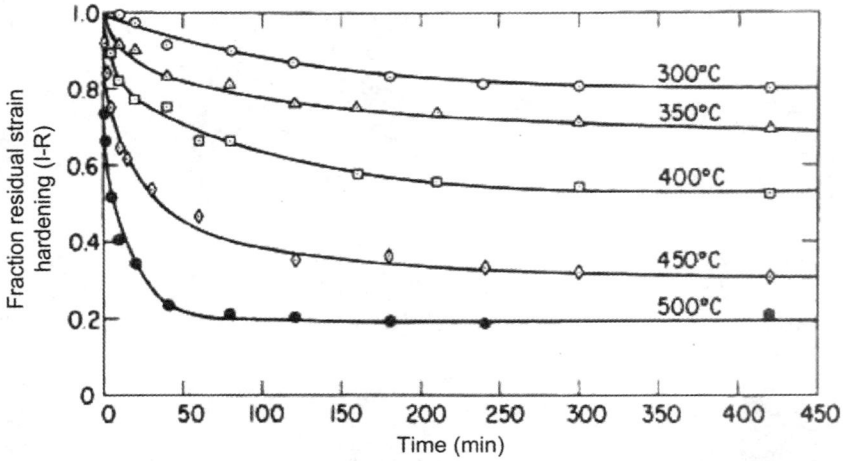

Figure 13   Recovery kinetics of deformed iron expressed in change in the relative flow stress (Leslie et al., 1963c).

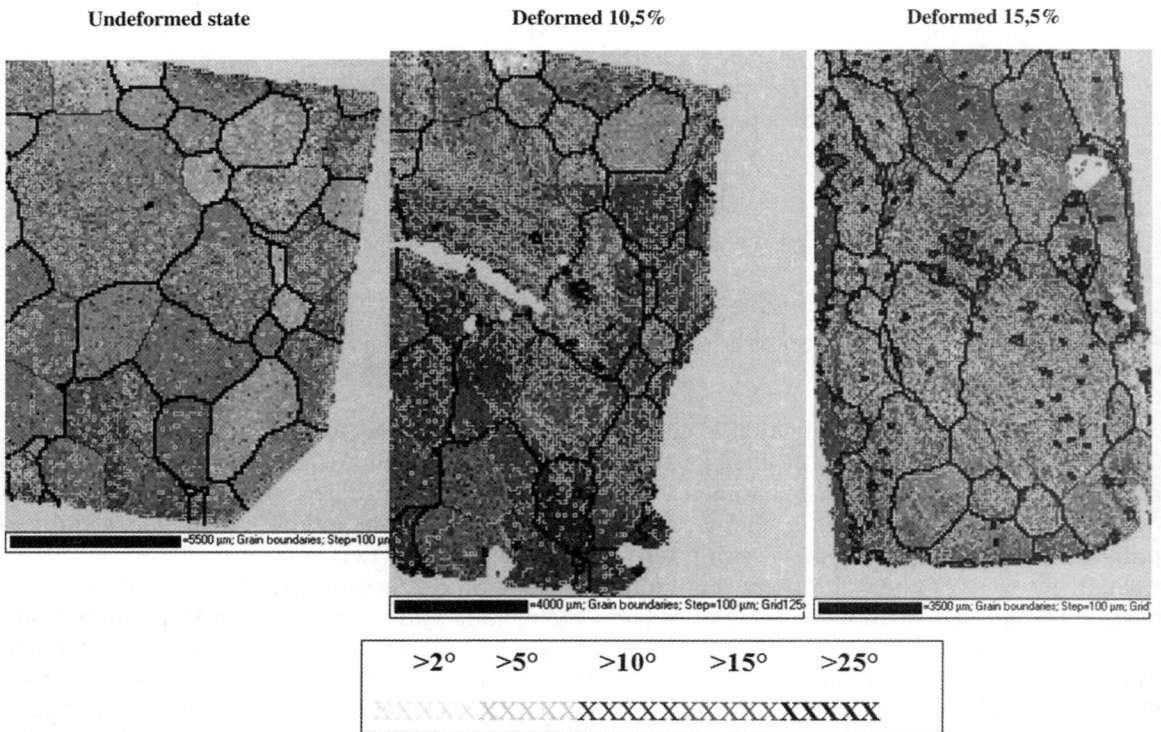

Figure 14   Dislocation substructure evolution during plastic deformation (characterized in terms of EBSD mapping and identification of low-angle grain boundaries) in an Al polycrystal. Images taken subsequently at the same positions.

drop strongly during recovery; however, the internal stresses associated with them, more specific with their configuration, do. The dislocation rearrangement processes lead to the gradual formation of subgrains and subgrain boundaries (e.g. tilt and/or twist low-angle boundaries). This means that recovery starts without incubation or respectively without any nucleation stage but instead starts immediately at temperatures where point defects and dislocations become sufficiently mobile to interact with each other or with other lattice defects (Smith, 1948; Beck and Hu, 1966).

In contrast to recrystallization, recovery does not involve any change in the grain structure of the cold-deformed material—instead, rather the main changes affect the dislocation arrangements inside the deformed grains.

These characteristic recovery phenomena, i.e. the decrease in the dislocation and point defect density and the gradual competitive coarsening of the subgrains, typically leads to the reduction in the internal stress fields and hence to a modest and gradual reduction in hardness (Humphreys and Chan, 1996) **(Figure 4)**.

23.2.2   Microstructure Changes during Recovery

23.2.2.1   Elementary Dislocation Mechanisms

The elementary mechanisms that are involved in recovery are the condensation and annihilation of point defects, particularly of quenched-in vacancies, and the gradual relaxation of internal elastic stress fields via cross-slip and reorganization of the dislocations that create these long-range distortions. Even when the density of the dislocations remains nearly the same during recovery, the total elastic energy inherited from cold working can efficiently be reduced just by relaxation of their local arrangement. This is due to the fact that dislocations can reduce their overall lattice distortion when assuming certain low-energy configurations (Vandermeer and Rath, 1990; Stüwe et al., 2002).

As quenched-in point defects can particularly annihilate at dislocations early stages of recovery also include dislocation climb. The reason for observing climb also at modest recovery temperatures, where it does usually not occur, is due to the osmotic pressure that results from a supersaturation of quenched-in point defects.

The second elementary dislocation effect associated with recovery is the cross-slip of dislocations. While dislocation climb is thermally activated owing to the required formation and motion of point defects, cross-slip of dislocations is thermally activated because the dislocation core, when it is spread or even dissociated into partial dislocations in the case of metals with low-stacking fault energy, has to be constricted before the dislocation can cross-slip. This constriction is thermally activated and hence the entire process of cross-slip depends on the stacking fault energy. The thermal activation of the cross-slip increases with the width of the stacking fault (Burgers and Louwerse, 1931; Doherty et al., 1997; Humphreys and Hatherly, 2004; Himmel, 1962).

23.2.2.2   Polygonization and Cell Formation

Structural recovery can be observed when a crystal is first bent in such a way that only one glide system operates, and then is subsequently annealed. The crystal dislocation substructure arising from single slip then breaks up into a number of strain-free subgrains, each preserving the local orientation of the original bent crystal, and separated by plane subboundaries that are normal to the glide vector of the active glide plane. This process is referred to as polygonization, because a continuously curved vector attached to the crystal lattice turns into a portion of a polygon (Beck and Hu, 1966; Haessner, 1978; Humphreys and Hatherly, 1995, 2004; Doherty et al., 1997; Doherty, 2005).

To better understand polygonization, the process must be interpreted in terms of the underlying dislocation distribution and their relaxation steps. When a crystal undergoes plastic deformation under single-slip and bending boundary conditions, then it is possible for all dislocations, both positive and negative, to pass right through the crystal and out at the surface such as in the easy-glide strain regime of a single crystal.

However, for accommodating the imposed curvature a certain density of excess dislocations of the same sign must remain inside the crystal. Such arrays of polarized dislocations are also referred to as geometrically necessary dislocations (GNDs).

The density of these excess dislocations is $1/R\,b$, where R is the mean radius of curvature and b is the magnitude of the Burgers vector. When the bent crystal is subjected to an annealing treatment, these excess dislocations relax into local low-energy configurations such as dislocation walls or tilt boundaries with their interfaces normal to the Burgers vector. When assuming such arrangements, the dislocations mutually compensate their distortion fields and hence reduce the total elastic energy.

This can be understood in terms of the Read–Shockley equation, which describes the elastic energy of an infinite low angle and boundary consisting of straight and infinite edge dislocations. In this low-energy configuration, the compressive field around the core of the dislocations overlaps within the dilatation field of the dislocation above it. This means that dislocations constituting a tilt boundary progressively relieve each other's energy as the tilt angle increases.

The process of polygonization involves individual glide and climb steps of the constituting dislocations. This process requires thermal activation and determines the rate of polygonization. Later stages of polygonization take place by the progressive merging of pairs of adjacent subboundaries. The driving force for this process comes from the progressive reduction of the boundary energy per dislocation in the boundary, as the misorientation of the interface angle increases. This means that the elastic energy associated with two independent low-angle grain boundaries, each characterized by the same misorientation, is higher than that of one single-merged low-angle grain boundary with twice that misorientation.

The rate of low-angle boundary merging is dependent by dislocation climb, since the dislocations in the two merged boundaries will not be uniformly spaced unless some dislocations climb.

In contrast to these well organized polygons that can from during recovery of microstructures resulting from single slip, the distribution of dislocations in a real plastically deformed metal is usually not uniform.

TEM, ECCI, EBSD, and XRD experiments of deformed polycrystals or single crystals subjected to more complex loading situations or strain-path changes that cannot be accommodated by single slip reveal that grains contain more complex subgrains or cells after plastic cold working.

While subgrains are characterized by a local accumulation of dislocations in a tangled in slightly blurred arrangement, cell walls (also referred to as cell boundaries) have a more sharp dislocation arrangement. The interiors of such substructures have a comparatively low-dislocation concentration while the small-angle cell boundaries have a high dislocation density.

Cell formation is characteristic of metals that glide on more than one glide system, so that dislocations with several different Burgers vectors are available to form cell walls. The sharpness of the cell walls formed during plastic deformation varies from metal to metal. There is a clear correlation between stacking-fault energy—and hence the degree of dissociation of dislocations and their ability to climb—and the sharpness of cell walls. Alloys with low stacking-fault energy such as Cu–Zn brass, Fe–Cr–Ni, Fe–Ni, or Fe–Mn austenitic steels have after deformation typically less pronounced, viz. less sharp DC structures. They rather reveal denser dislocation tangles with some regions of low dislocation density and multiple microbands. The values for the stacking fault energy for metals and alloys can vary substantially. For instance, chemically pure Aluminum (99.999 wt.%) has a very high stacking-fault energy of about 200 mJ/m^2. Commercial purity face-centered cubic-structured metals such as 99.95 wt.% nickel and 99.95 wt.% silver have intermediate stacking-fault energies of about 128 mJ/m^2 and 22 mJ/m^2, respectively. 70/30 copper–zinc alloys (α-brass) have very low stacking-fault energy below 15 mJ/m^2. Highly alloyed Fe–Mn, Fe–Mn–C, and Fe–Ni steels also have low stacking-fault energy in the range between 10 and 40 mJ/m^2 depending on the alloying content. A high aluminum content (2–8 wt.%) that is an attractive pathway to design weight-reduced face-centered cubic steels increases the stacking fault energies for instance of Fe-30 wt.% Mn–C steels from about 20 mJ/m^2 to about 80 mJ/m^2 (Reeh et al., 2012).

These examples of stacking fault energies are an important characteristic and play a significant role in the deformation of metals due to its influence on dislocation mobility and morphology. More specific, the stacking fault energy determines the distance between the partial dislocations and it hence has a direct influence on the ability of dislocations to undergo cross-slip during plastic straining. The lower the stacking fault energy the larger is the separation between the partial dislocations and thus cross-slip and dynamic and static recovery are inhibited.

Copper and silver have recognizable cells with thick tangled boundaries, and nickel and especially aluminum have better defined cells that rapidly sharpen and grow during annealing after deformation (Sandström et al., 1978; Varma and Willitis, 1984; Walter and Koch, 1963; Blum et al., 1995; Li, 1962).

During subsequent recovery, the subgrain boundaries sharpen and turn into well-defined thin-cell walls. These cells progressively grow larger while their interiors become further void of free dislocations. The DCs, both before and after recovery, are normally roughly equiaxed, i.e. their dimensions are nearly the same in all directions, although under some circumstances, depending on crystallographic orientation and crystal size, cells may be elongated and arranged in bands.

The mechanism by which the diffuse cells formed after deformation sharpen and grow is a complex form of polygonization. Since dislocations are present on several glide planes (or, if on a single glide plane, with different glide vectors), the cell walls can as a rule not be simple tilt boundaries. However, where the misorientation across cell walls has been examined, it turns out that they are effectively very similar to tilt boundaries. In zinc, each wall contains at least two families of mixed edge-cum-screw dislocations, but so arranged that the misorienting effects of the screw components cancel and the tilt axis is still parallel to the dislocation lines lying in the wall. There is evidence that cell walls in aluminum have a similar structure. Since a variety of dislocations are available, cell walls have a number of different orientations relative to the crystal axes and can thus completely enclose individual cells.

While the early steps of recovery lead to a gradual sharpening of the substructure walls, later stages of recovery are characterized by their competitive coarsening of existing substructure inside the grain. Kinematically, two extreme scenarios are conceivable in that context. Firstly, when the grain does not contain an overall in-grain net lattice curvature, gradual coarsening of the substructure does not lead to a higher local misorientation associated with each cell wall. Secondly, when the grain is curved, containing GND arrays, gradual subgrain coarsening leads to the accumulation of the content of polarized dislocations (dislocations of the same sign; GNDs) and hence to a higher average misorientation associated with each single-cell wall.

Competitive coarsening of the subgrain structure proceeds through the motion of small-angle subboundaries. The mobility of such very small-angle subboundaries (cell walls) under purely thermal activation (i.e. in the absence of stress) can be vanishingly small. However, as has been shown in the literature, the mobility of such small-angle subboundaries can be drastically increased under the presence of a small stress (Winning et al., 2001, 2002; Winning, 2003). This is true for pure tilt boundaries, but not for other kinds of small-angle boundaries.

23.2.3   Recovery during Hot Deformation

The individual processes associated with recovery are thermally activated. Therefore, recovery also occurs dynamically already during hot working. In that context, one has to consider the melting point of the alloy under investigation, because most of the activation barriers involved, such as climbing and cross-slip, scale with the absolute melting point (climb rate) and the stacking fault energy (cross-slip rate) of the material. Normalizing the actual deformation temperature, expressed in Kelvin, by the absolute melting point introduces the homologous temperature. This means that rolling pure tungsten at 400 °C can be regarded as a cold-rolling process and the deformation of pure Aluminum at 50 °C must be already regarded as a hot-working process (Haessner, 1978; Humphreys and Hatherly, 1995).

The occurrence of recovery phenomena, such as dislocation cross-slip, dislocation climb as well as substructure sharpening and competitive coarsening during deformation, is referred to as dynamic recovery. Under conditions of rapid deformation, this process is most pronounced in metals with high stacking-fault energy such as aluminum, iron, niobium, or tungsten. Owing to their high stacking-fault energy dislocations in these metals have high nonconservative mobility and a high cross-slip rate that enables dislocation reordering and hence subgrain formation.

It was shown that because of continuous growth of subgrains, or cells during hot working (the subgrain equivalent of normal grain growth), some subboundaries disappear while others grow. The consequence is that the mean subgrain misorientation remains constant at a few degrees, right up to high strains.

23.2.4   Property Changes during Recovery

23.2.4.1   *Mechanical Properties*

Recovery is characterized particularly by the rearrangement of the dislocation substructure associated with a partial and gradual reduction of the internally stored elastic energy that was inherited from the preceding cold working. This means that the degree and stage of those recovery phenomena that affect the dislocation substructure is to some extend reflected by the mechanical properties of the material that is being probed.

Since the rearrangement of the dislocation structure requires a certain short-range mobility of the dislocations involved, it can be expected that the kinetics of recovery depends to some extend on the stacking fault energy of the material. This is so because the stacking fault energy determines the cross-slip probability of dislocations. Cross-slip is one of the essential recovery steps that can lead to fast dislocation rearrangements, subgrain formation, and hence, to internal stress relaxation. A small value of the stacking fault energy produces a large stacking fault. Dislocation cross-slip requires that the stacking fault is constricted before the actual cross-slip event. This is a thermally activated process and the activation barrier is proportional to the size of the stacking fault.

In metals that have a high value of the stacking fault energy, the dislocation core remains nearly intact so that screw dislocations can undergo cross-slip events without the aid of thermal activation.

This is shown in **Figures** 4 and **13** for corresponding mechanical experiments conducted for copper and nickel. It can be observed that no drop in hardness accompanies stress relief or recovery in brass, copper, or nickel. These are all metals or respectively alloys with low or average stacking-fault energy and, owing to the reduced mobility of the underlying dislocations, only little climb and rearrangement of dislocations can take place, especially in copper–zinc brass and austenitic iron–nickel and iron–manganese steels. In that context, it is worth to mention that already slight dislocation rearrangements can account for a considerable reduction in the stored elastic energy.

From these considerations, it becomes apparent that two types of consequences are conceivable in that context. Firstly, recovery obviously reduces the remaining driving force for primary static recrystallization. Therefore, extended recovery may entail much slower recrystallization kinetics. Secondly, if recovery reduces the internal elastic energy to a very large extent, for instance because the heating rate is very low or in cases where the dislocations a very mobile and hence can rearrange very efficiently, recrystallization can be even entirely suppressed. Such a situation is referred to as "recrystallization in situ" (Burgers, 1941; Humphreys and Hatherly, 1995; Himmel, 1962). Typical materials where very strong recovery can occur, thus competing with primary recrystallization, are pure aluminum, iron, and other high-melting refractory metals with body-centered crystal structure (e.g. tungsten, molybdenum, and niobium). In both materials, the dislocations are very mobile enabling them to undergo easy cross-slip. Correspondingly, softening in these materials during recovery can be very fast so that the driving force that remains for recrystallization becomes very small.

The second important aspect that can shift the competition between recovery and recrystallization toward a higher recovery tendency is a low mobility of high-angle grain boundaries. Since recrystallization is described by the formation and motion of new high-angle grain boundaries, reduced mobility of these internal interfaces gives recovery enough time to also reduce the internal stored energy. This diminishes the driving force for recrystallization and promotes strong recovery.

Under such circumstances as described above, it is conceivable that the whole of the work hardening may be recovered without the occurrence of recrystallization.

For instance, it has been shown that a weakly deformed silicon–iron crystal, iron–chromium polycrystals or pure iron polycrystals and likewise a slightly deformed iron–aluminum polycrystals can recover completely (Hölscher et al., 1991; Raabe and Lücke, 1993, 1992). For iron and iron-based alloys, that can even happen up to high annealing temperatures of 700–800 °C, depending on the heating rate. Slightly deformed aluminum samples can recover almost completely at 400–600 °C while various studies reported that aluminum single or polycrystals will recover about half the work-hardening at 300–400 °C.

The general trend observed in that context is that the larger the preceding deformation was, the smaller the fraction of pure recovery of the work-hardening will be.

Some crystals of hexagonal metals such as zinc or cadmium are exceptions: these can recover completely even after very large tensile strains by easy glide.

The reason for this type of behavior that will be discussed in more detail below lies in the available lattice curvature. The formation of new and mobile high-angle grain boundaries during the nucleation stage of recrystallization requires the occurrence of local zones in the deformation substructure with very high local-lattice curvature, e.g. above 15°. In such substructures, subgrain nuclei can, during competitive subgrain growth, accumulate a sufficiently large curvature to finally form new and highly mobile high-angle grain boundaries. If subgrain growth takes place in an area where no net lattice curvature is available in the deformed substructure, no new high-angle grain boundaries can be formed (Cotterill and Mould, 1976; Bailey, 1963; Sandström et al., 1978; Varma and Willitis, 1984; Walter and Koch, 1963; Blum et al., 1995; Li, 1962; Doherty and Szpunar, 1984a; Raabe et al., 2002a; Raabe and Becker, 2000; Humphreys, 1992a; Bate, 1999).

A number of specific mechanisms determine the relationship between yield stress, dislocation density, and DC structure. While the early stages of recovery show a reduction in yield stress that can be related to the square root of the dislocation density, the later stages of recovery that are characterized by cell formation and gradual cell coarsening are characterized by an inverse relationship between the flow stress and the cell size. As the interiors of the cells become more and more devoid of dislocations during the later stages of recovery the average cell size (together with the grain size, which is, however, not changed during recovery) determines the strength of the material.

23.2.4.2 Electromagnetic Properties

Plastic deformation slightly increases the electrical resistivity. Various studies have been devoted to the stages by which the electrical resistivity returns to its fully annealed value. This is of interest both because it helps to disentangle the separate contributions made to the resistivity increase by dislocations and by deformation-induced vacancies, and because it helps to elucidate the complex mechanism of the damage caused by neutron irradiation in nuclear reactors; this damage also causes resistivity changes that anneal out in a different manner from those caused by plastic deformation (Beck and Hu, 1966; Haessner, 1978; Humphreys and Hatherly, 1995, 2004; Doherty et al., 1997; Doherty, 2005).

23.2.5 Recovery Textures

The mechanical properties of metals considerably depend on their microstructure and texture, which are a result both of composition and the thermomechanical processing to which they are exposed. While recrystallization is characterized by the formation and motion of new high-angle grain boundaries, recovery preserves the existing grain structure and also the original deformation texture that was created during cold working (Engler et al., 1996; Engler, 1997; Ibe and Lücke, 1966; Bunge, 1969, 1982, 1986, 1987; Wassermann and Grewen, 1962; Bunge and Esling, 1982; Beck and Sperry, 1950; Li, 1961; Dillamore et al., 1967; Hutchinson and Ryde, 1995; Doherty and Szpunar, 1984b; Doherty et al., 1988; Doherty, 1985; Samajdar and Doherty, 1994; Juul-Jensen, 1992, 1995; Juul-Jensen et al., 1988; Dillamore and Katoh, 1974; Raabe and Lücke, 1992; Klinkenberg et al., 1992; Raabe, 1995a; Juntunen et al., 2001; Ray et al., 1975b; Humphreys, 1977a; Ridha and Hutchinson, 1982a; Samajdar et al., 1992; Hjelen et al., 1991; Lee and Duggan, 1993; Doherty et al., 1993a; Duggan et al., 1993; Duggan and Chung, 1994; Samajdar and Doherty, 1995).

This means that the design of adequate recrystallization pathways can be used to create a desired crystallographic texture. Consequently, the proper understanding and use of recrystallization marks a necessary prerequisite for the design of the crystalline anisotropy of a polycrystalline sample (Raabe et al., 2002b; Roters et al., 2010).

In turn, as recovery affects only the in-grain dislocation and cell substructure it can lead to the inheritance of the original cold-working texture to the final texture of the heat-treated material. In this context, two recovery scenarios are conceivable:

One is that all grains undergo recovery instead of recrystallization—for instance when a high stacking-fault energy material is heat treated at low temperatures and after low strains. This would lead to a situation where all original grain shapes and the crystallographic texture remain the same as before the heat treatment and only the in-grain dislocation and cell structure have relaxed as described above (Raabe, 1995b, 1995a) (**Figure 15**).

The second possibility is that some grains undergo recrystallization while others only undergo recovery. Such a phenomenon leads either to the coexistence of recrystallized and recovered grains or to the gradual sweeping of recovered grains by recrystallizing ones—however with reduced driving forces (**Figure 16**).

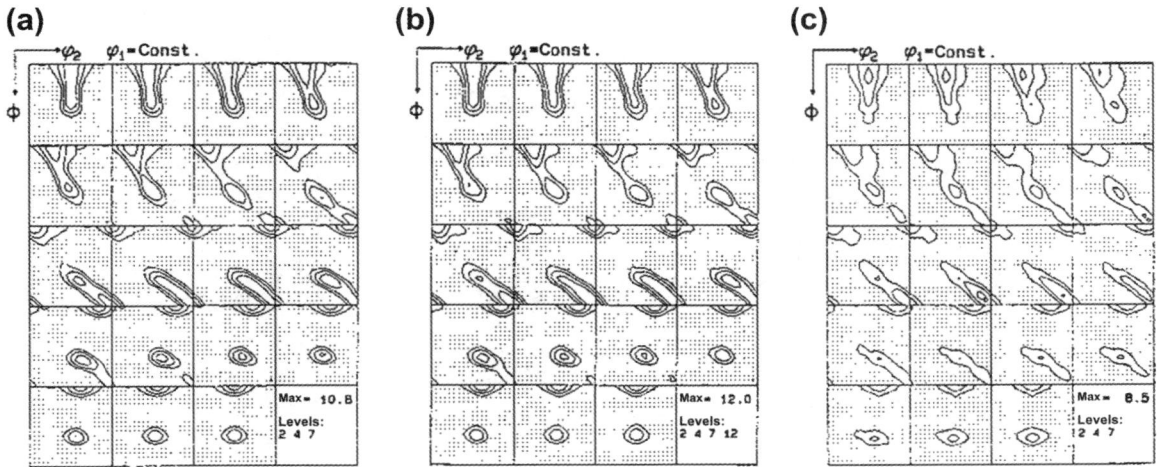

Figure 15 Texture of 90% cold-rolled and 1000 K annealed low-C steel measured by XRD. (a) 5; (b) 6s; (c) 7s (Raabe and Lücke, 1992).

Specifically regarding the second scenario, it is therefore important to understand, for an optimal design of the final microstructure and texture, why at the same heat-treatment temperature and for the same global strain imposed by a machine some grains in the same material recrystallize while others do not and recover instead (Raabe, 1995b, 1995a).

In certain aluminum alloys, this mixed approach involving joint recrystallization and recovery phenomena within the same microstructure is often used for the optimization of the material's anisotropy. In contrast to that, in some ferritic steels, the occurrence of elongated recovered grains within an otherwise-recrystallized microstructure is usually an undesired phenomenon as it leads to inhomogeneous plastic flow and surface roughness during manufacturing. This phenomenon is referred to as "ridging".

For most materials, including alloys with high stacking-fault energy, sufficient initial cold deformation in excess of 70% thickness reduction can be used to suppress strong recovery and the undesired inheritance of crystallographic deformation textures. These effects have been studied in detail on iron and low-carbon steels as well as on aluminum (Wassermann and Grewen, 1962; Bunge, 1982, , 1986, 1987; Bunge and Esling, 1982, 1991; Adams et al., 1993; Duggan et al., 1978; Dillamore et al., 1979; Hölscher et al., 1991; Ushioda et al., 1987; Raabe and Lücke, 1993; Jägle and Mittemeijer, 2012; Adam et al., 2001; Randle and Engler, 2000). It was observed that not only the total deformation or the heat-treatment temperature play an essential role for determining how much of the volume is recovered and how much is recrystallized, but also the orientations of the deformed grains, the grain size, and the accumulated shear strains are central parameters that determine the degree of recovery and the suppression of primary recrystallization after cold working at a microstructural scale.

The strong orientation dependence of recovery can be interpreted in two ways. The first approach is based on the assumption that the driving force is reduced by recovery so that the mechanical instability criterion (sufficient stored elastic energy) required for recrystallization is no longer fulfilled (Himmel, 1962). Recovery and recrystallization reveal different kinetic laws. Whereas in the first case (recovery) the properties change in a near-logarithmic manner, in the latter case (recrystallization) an incubation period followed by a sigmoidal law with a very sudden change in microstructure is commonly observed. It is thus conceivable that if recovery was strong enough to reduce the total driving force, i.e.

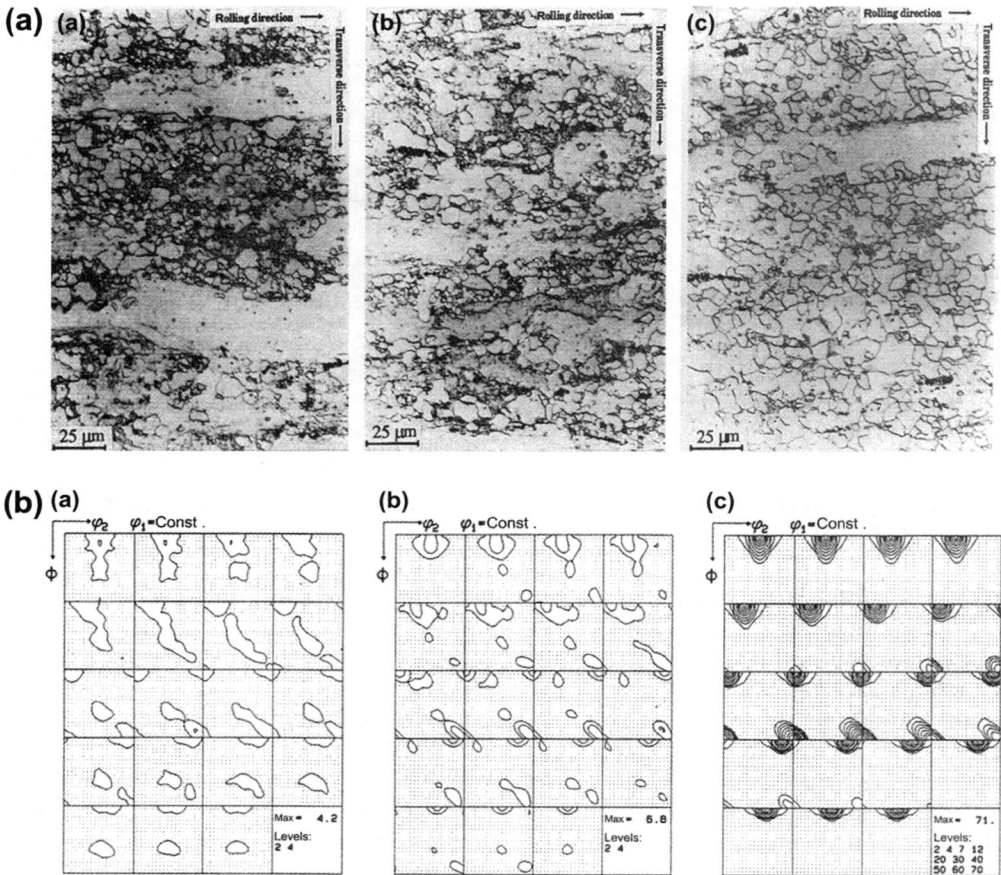

Figure 16 (a) Microstructure of a 90% cold rolled and 1000 K annealed low-C steel in flat sections. (a) 5; (b) 6s; (c) 7s. (b) Corresponding ODFs calculated only from the recovered grains. (a) 5; (b) 6s; (c) 7s (Raabe and Lücke, 1992).

the internal stress associated with the stored dislocations and cell walls, the movement of newly formed large-angle grain boundaries would be very slow or even suppressed. This assumption, however, contradicts three facts. First, the deformation energy stored in different crystals depends on their orientation. This is suggested by experimental results based on corresponding crystal plasticity finite element simulations (Roters et al., 2010) as well as the measurement of internal stresses, stored dislocation densities and cell sizes (Himmel, 1962; Cahn, 1965, 1966; Humphreys and Ferry, 1997; Engler et al., 1996; Bailey, 1960). Second, recovery is not an isolated process but a preprocess to recrystallization during which nuclei are formed. Strong recovery in the conventional sense should therefore provide a large number of nuclei rather than suppress them. Third, in case of a small grain size (polycrystals), it was observed that certain crystals are consumed by growing nuclei that proceed from the former grain boundary and or even from neighboring grains. This experience implies that grains with low stored deformation energy contain a sufficiently large net driving force for recrystallization, viz. a sufficiently high mechanic instability existed in such cases.

The second approach for explaining the suppression of recrystallization is based on the assumption that, albeit a mechanical instability potentially exists, the thermodynamic and kinetic instability criteria are not fulfilled in specific grains. In order to produce new large-angle grain boundaries during nucleation it is essential to provide areas with large local misorientations, quantified by certain densities of GNDs, already within the deformation microstructure. In most metallic alloys recrystallization nucleation then naturally results from discontinuous subgrain growth where some of the subgrains accumulate a sufficiently high misorientation relative to the surrounding microstructure. This accumulation of misorientation increases the mobility of the subgrain boundaries and so the rate of growth of the subgrain increases until it finally acts as a nucleus. Since subgrains are typically surrounded by low-angle boundaries of very low misorientation (usually below $2°$), the accumulation of further misorientation during subgrain coarsening requires local orientation gradients inside grains to be present in the deformation microstructure (**Figure 17**). However, such orientation gradients are sometimes not observed within some of the deformed crystals. Such grains can then not provide sufficient kinetic instability. **Figure 18** shows an extreme example of such a joint recrystallized and recovered microstructure. The sample is Fe with large grains, rolled to 90% thickness reduction. The micrograph shows a flat section, i.e. the deformation was practically the same for all areas shown. Upon heat treatment at 1000 K for 120 s some grains recrystallize entirely while others only recover.

23.3　Recrystallization

23.3.1　Introduction

Primary static recrystallization is defined by crystal nucleation and growth into the deformation substructure (**Figure 1**). More specifically, this process proceeds via the formation and motion of new high-angle grain boundaries. In some cases also low-angle grain boundaries can be mobile and, therefore, are able to contribute to the microstructural changes occurring during primary recrystallization. When recrystallization starts the deformation substructure is already in a recovered state because the recrystallization process requires an incubation period and recovery does not.

The level to which the preceding recovery reduces the stored energy from the cold-worked state depends on the stacking fault energy of the material and on the incubation time that has elapsed before the high-angle grain boundaries that have been newly formed during the recrystallization incubation period can sweep the deformation substructure (Stead, 1898; Rosenhain, 1914; Ewing and Rosenhain, 1899, 1900a, 1900b; Alterthum, 1922; Carpenter and Elam, 1920; Czochralski, 1927; Burgers and Louwerse, 1931; Burgers, 1941; Burke and Turnbull, 1952; Smith, 1948; Beck and Hu, 1966; Haessner, 1978; Humphreys and Hatherly, 1995; Doherty et al., 1997; Doherty, 2005).

While materials with a low stacking-fault energy can store a high portion of the internal elastic strains owing to the limited cross-slip and climb capabilities of the dislocations during recovery, alloys and metals and that have a high stacking fault energy such as aluminum and iron can substantially reduce the deformation energy and hence also the remaining driving force for primary recrystallization during the preceding recovery period. In extreme cases recovery can even suppress recrystallization and an elongated and softened grain structure prevails that preserves the crystallographic texture that was formed during cold working. Typically, such behavior is not desired owing to the resulting preserved deformation texture and the topological anisotropy that results from the elongated grain shapes. In some cases, however, partial recrystallization is a microstructural design goal. This applies for instance in cases where a texture that is composed of both recrystallized and deformation orientation components is desired for optimizing the overall crystallographic anisotropy of a polycrystalline product by

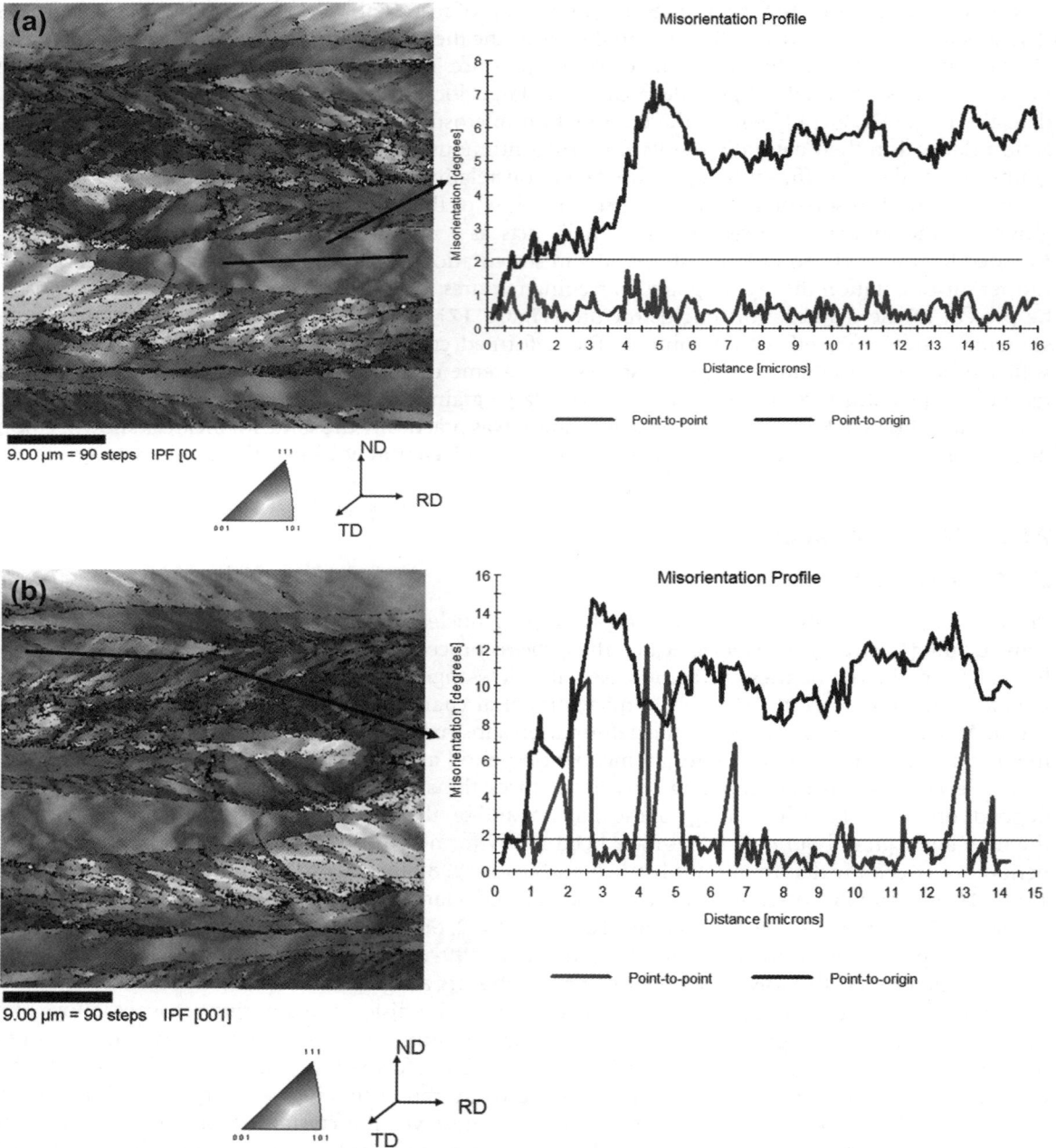

Figure 17 (a) Misorientation profile in a {001} oriented grain in 70% cold rolled low-C steel. (b) Misorientation profile in a {111} oriented grain in 70% cold rolled low-C steel. The misorientation profile reveals much higher jumps and higher overall values (Thomas et al., 2003).

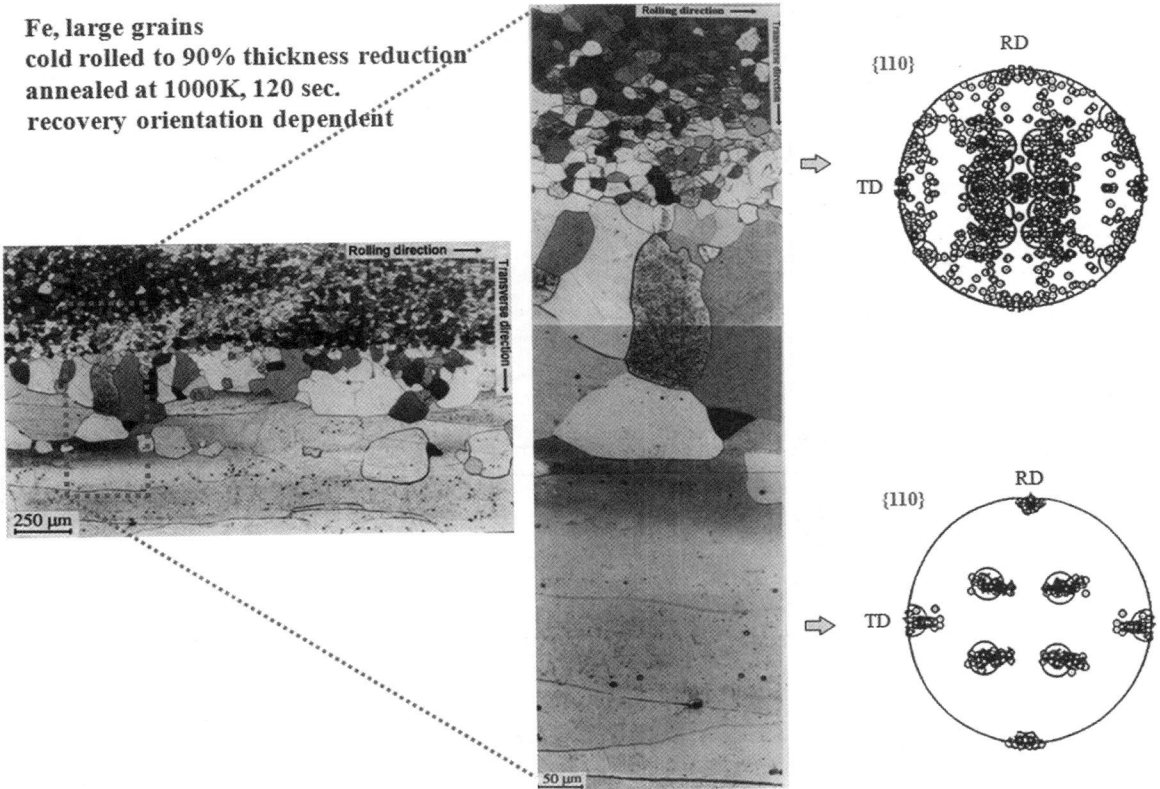

Figure 18 Extreme example of a joint recrystallized and recovered microstructure. The data were taken on a commercial purity iron sample with large grains (mm-sized), rolled to 90% thickness reduction. The micrograph shows a flat section, i.e. the deformation was practically the same for all areas shown. Upon heat treatment at 730 °C for 120 s some grains recrystallize entirely while others only recover (Raabe and Lücke, 1992).

balancing texture components with different individual anisotropy (Bunge, 1986, 1987; Bunge and Esling, 1991; Raabe et al., 2002b; Roters et al., 2010).

Primary static recrystallization, which follows the recovery stage and also competes with it (**Figures 16 and 18**), involves the replacement of the remaining recovered cold-worked structure by new strain-free and approximately equiaxed grains. This means that the analysis of a partially recrystallized grain structure can make use of the quantification of the crystallographic texture, internal elastic distortions, remaining dislocation substructures, grain shapes, and grain size distributions as characteristic microstructural features to differentiate between the as-recovered and the as-recrystallized portions of the material.

Recrystallization starts upon heating the cold-worked sample to temperatures in the range of 0.3–0.5 of the absolute melting point (**Figure 19**) (Burgers and Louwerse, 1931; Burgers, 1941). The temperature regime for recrystallization lies above that required for recovery since the former mechanism goes through a thermally activated nucleation process while the latter process requires only the thermal activation of cross-slip and climb under a state of high point-defect oversaturation and high elastic internal stresses. Although some cases occur such as in the case of heavily wire drawn Cu–Zr two phase

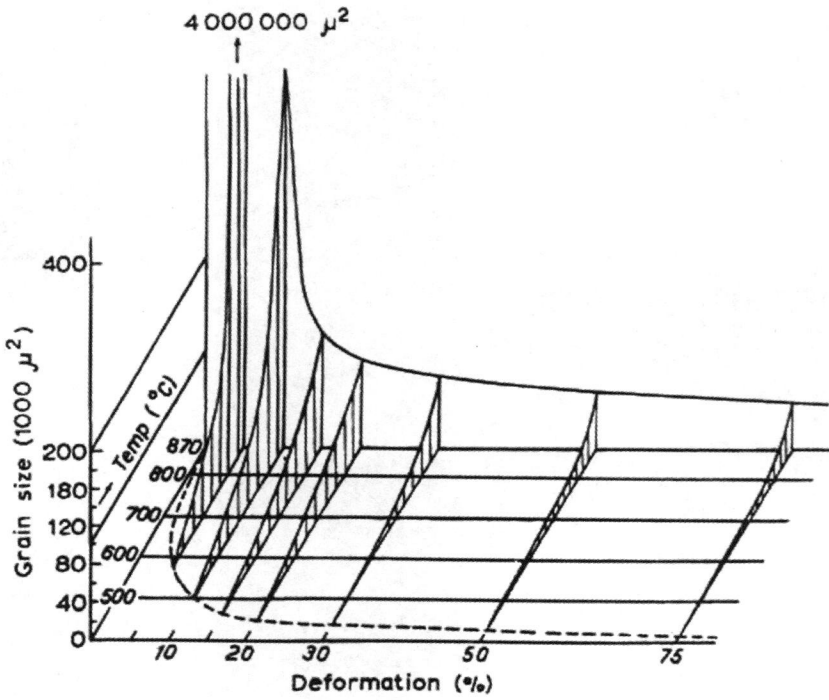

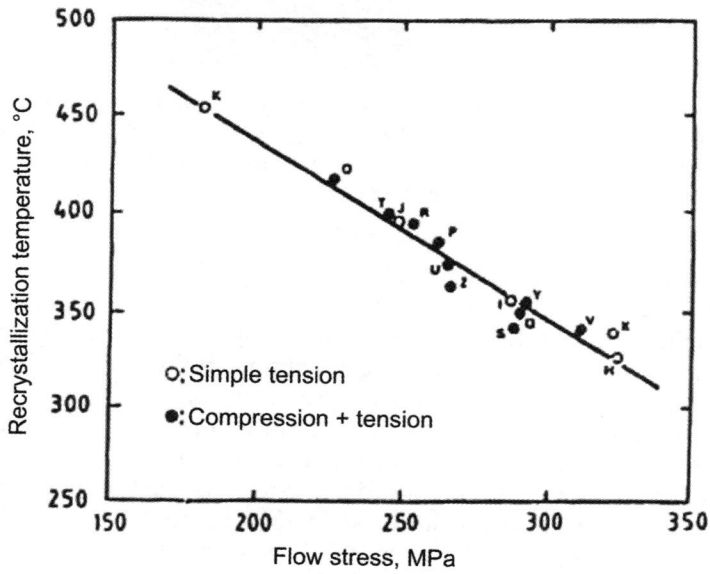

Figure 19 Recrystallization diagram of electrolytically refined iron (Burgers, 1941).

alloys where heavy cold working can even lead to the amorphization of the crystalline structure, there is usually no crystal structure change observed during recrystallization. For engineering estimates, it is often required to provide a simple measure under which circumstances recrystallization starts, namely, the recrystallization temperature. This measure is defined as the temperature at which 50 vol.% of cold-worked material has recrystallized after 1 h. As will be discussed in more detail below regarding the influence of the mobility of the high-angle grain boundaries involved during recrystallization, the actual recrystallization temperature as defined above is strongly dependent on the purity of a material. While high-purity materials can recrystallize already around 0.3 of the absolute melting point, impure materials may not recrystallize before a temperature of 0.5–0.7 of the melting point in Kelvin is reached. The influence of material purity on the recrystallization kinetics is so remarkable because solute atoms change the apparent activation energy of the grain-boundary mobility via an impurity drag mechanism (Lücke and Stüwe, 1963). This means that impurities can have a kinetic influence that acts on an Arrhenius function, namely, on the thermally activated mobility term associated with the motion of a high-angle grain boundary (Molodov, 2001; Lücke and Stüwe, 1963; Gottstein et al., 1995; Shvindlerman et al., 1995, 1999).

Besides this essential role of the interfaces involved a number of other metallurgical parameters are essential for the recrystallization microstructure and kinetics. Here specifically the amount of prior cold working, temperature, time, initial grain size, chemical composition, presence of second phases, crystallographic texture, and the amount of recovery prior to the start of recrystallization play important roles. Also, in a manufacturing environment, the heating rate has to be considered since many commercial heat-treatment cycles work with rather large slabs. This may lead to relatively low effective heating rates entailing substantial inhomogeneities of the recrystallization microstructure and kinetics across the macroscopic dimensions of the material. For instance, large steel coils, when processed in a commercial manufacturing environment, often reveal substantial and through-thickness gradients in the recrystallization microstructure and crystallographic texture (**Figure 20**) (Peranio et al., 2010). For instance, in the field of ferritic steels such inhomogeneities lead to a reduction in the forming properties (Huh et al., 2005; Raabe, 1996, 1997).

From the vast body of experimental data in this field, a small set of elementary empirical recrystallization rules can be filtered:

For recrystallization to start and proceed, typically a minimum amount of preceding plastic cold working is required. For most commercial alloys with a high stacking-fault energy such as body-centered cubic steels and certain aluminum alloys, this value amounts to about 40–50% cold reduction prior to the heat treatment. Cold reduction below a value of 35% is often not sufficient for primary static recrystallization to take place.

Rapid nucleation during the incubation period of recrystallization requires a certain amount of inhomogeneity present in the deformation microstructure. Such inhomogeneous deformation zones can be for instance crystallographically curved areas containing high amounts of GNDs close to the existing grain boundaries (**Figure 21**); shear bands (a crystallographic deformation inhomogeneity that can assume mesoscopic dimensions and penetrate multiple grains under an oblique angle to be dominant deformation axis) (**Figure 22**) (Dorner et al., 2007); microbands (zones of concentrated crystallographic shear, typically inside grains); transition bands (zones of divergent crystal reorientation inside grains, i.e. connected rotation zones in different orientation directions that preserve in part the original orientation in the middle between the branches of the transition band); large precipitates that create sufficient curvature and inhomogeneity in their vicinity during deformation (**Figure 23**); or in-grain areas that have a sufficiently high lattice curvature such as the distorted zones adjacent to grain boundaries. These microstructure features provide not only a high local deformation energy that is

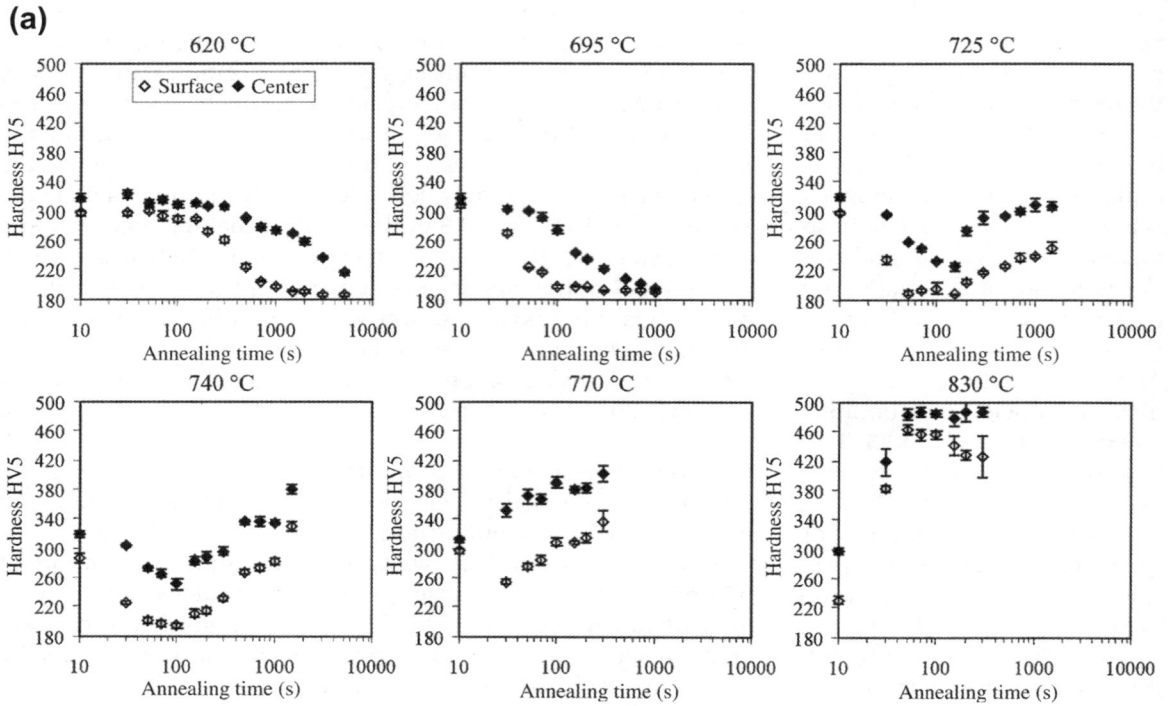

(a)

(b)

Annealed
in salt bath
at 740 °C for 100 s,
surface

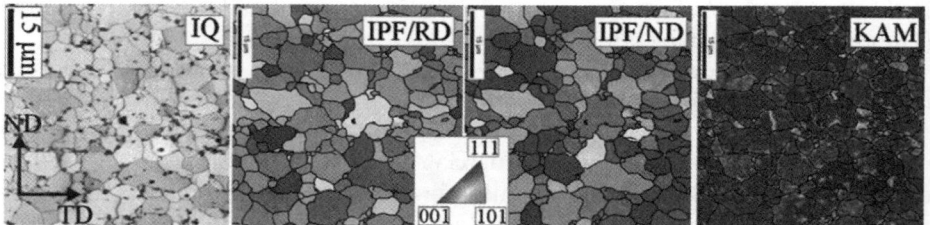

Annealed
in salt bath
at 740 °C for 100 s,
center

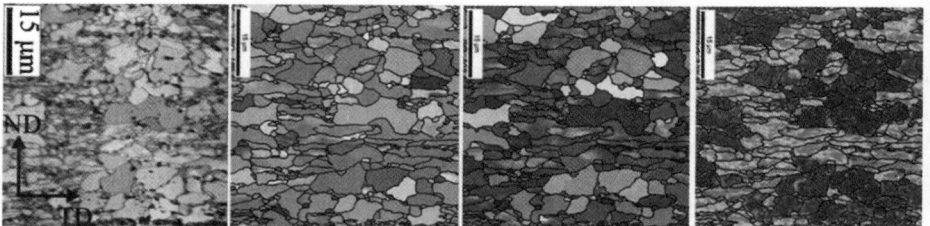

Figure 20   (a) Hardness measurement in dependence of the annealing time measured in the center and at the surface of C-steel sheets. The sheets were annealed in salt bath at ferritic and intercritical annealing temperatures. (b) EBSD microstructure maps: EBSD carried out in the center and at the surface of a transverse section of a sheet annealed in salt bath. The images show the Image Quality map (IQ), the Inverse Pole Figure maps (IPF/RD and IPF/ND), and the Kernel Average Misorientation map (KAM, blue: KAM = 0°, red: KAM = 3°) obtained at the same specimen area. Grain boundaries with angles larger than 3° are indicated by black lines in the IPF and KAM maps (Peranio et al., 2010).

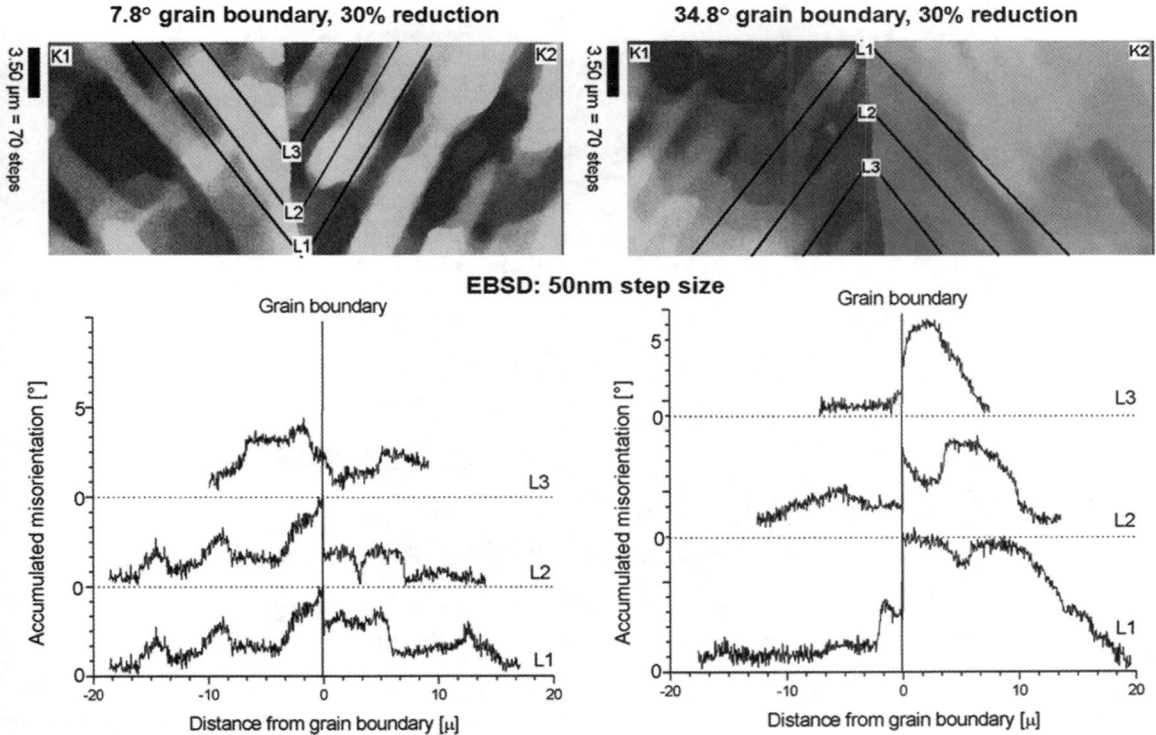

Figure 21 Misorientation before a small-angle grain boundary and before a high-angle grain boundary in Al bicrystals, strained 30%. Different misorientation profiles build up before the two different types of interfaces (Zaefferer et al., 2003).

required as a driving force during nucleation but they specifically promote the accumulation of a large local lattice curvature during discontinuous subgrain coarsening that finally leads to the formation of new high-angle grain boundaries.

Recrystallization phenomena are thermally activated. This means that the rate of the mechanisms that control the formation and motion of the newly formed high-angle grain boundaries depend on the annealing temperature. Usually the temperature dependence follows Arrhenius-type equations, establishing an exponential relationship. The recrystallization temperature of most materials lies between 0.4 and 0.6 of the absolute melting point. In engineering terms, the recrystallization temperature refers to that point where 50 volume % of the material is recrystallized after 1 h. The recrystallization temperature decreases with increasing plastic deformation. The smaller the degree of the preceding cold working is, the higher will be the recrystallization temperature. Also, it was observed that the finer the initial grain size is, the lower will be the recrystallization temperature (Haessner, 1978; Humphreys and Hatherly, 1995). The larger the initial grain size is, the greater is the degree of plastic deformation that is typically required to produce an equivalent recrystallization temperature. An increase in the degree of plastic cold working leads to lower required annealing temperatures and to a reduced grain size after recrystallization. Higher plastic deformation means greater lattice rotations (from dislocation slip or mechanical twinning) inside grains and also to higher stored energies (Raabe et al., 2002a). Therefore, the probability of generating new grains increases with strain. The higher the temperature of cold working, the less

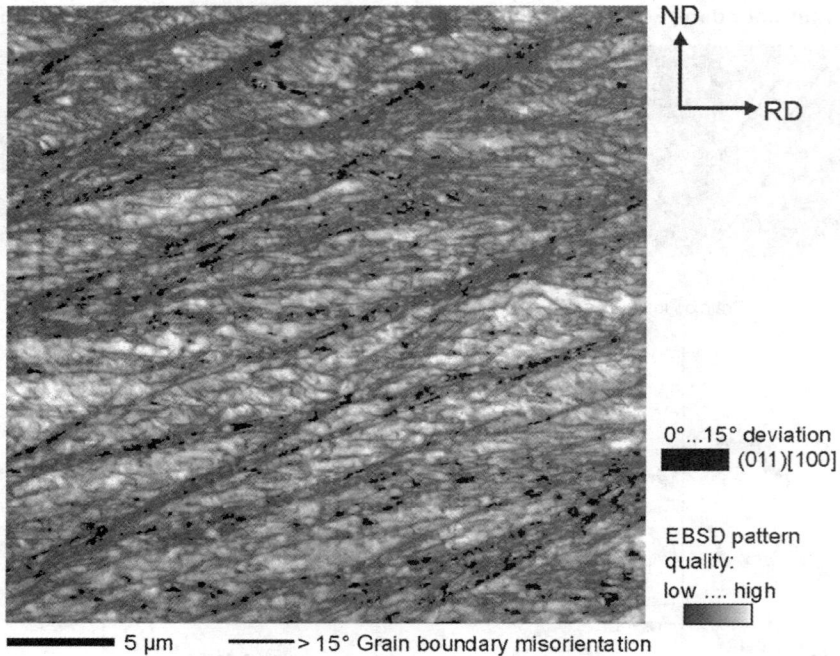

Figure 22   Goss-oriented regions inside of shear bands in a 89% deformed Fe-3 wt.% Si initial single crystal sample. The-diagram shows the EBSD pattern quality map (gray scale) revealing shear bands as regions with low EBSD pattern qualities. The shear bands are inclined by 29–36° to the RD. In addition, Goss-oriented regions are marked in black (Dorner et al., 2007).

is the strain energy stored and thus the recrystallization temperature is correspondingly higher. The overall recrystallization rate increases exponentially with the heat-treatment temperature. Higher recrystallization temperatures usually lead to an increasing probability to form nuclei, hence, to an increased density of nuclei and therefore to a smaller recrystallized grain size. This means that the grain size after recrystallization decreases with increasing prior strain, i.e. the nucleation density increases. In most materials, the grain size after recrystallization decreases as the strain increases.

Crystallographic recrystallization textures are typically profoundly different from those produced during cold deformation. Higher degrees of cold working often lead to more pronounced and sharp recrystallization textures. Prevalence of recovery leads to the inheritance of the cold deformation textures and hence to weaker recrystallization textures. Higher annealing temperatures often lead to more pronounced and sharp recrystallization textures.

During recrystallization, the mechanical properties that were changed during deformation due to strain hardening are restored to their respective values that they had before the cold-working procedure. During this stage of annealing, impurity atoms tend to segregate at grain boundaries, and retard their motion. This mechanism typically may obstruct the processes of nucleation and growth. This so-called solute drag effect can be used to retain the cold-worked strength at higher service temperatures. The presence of second-phase particles causes the retardation of recrystallization. This effect is referred to as particle pinning or Zener pinning. A special situation, where the precipitation of particles can very strongly accelerate (rather than slow down) primary recrystallization, occurs when recrystallization and precipitation of second phases from a supersaturated solid solution occur at the same time. The

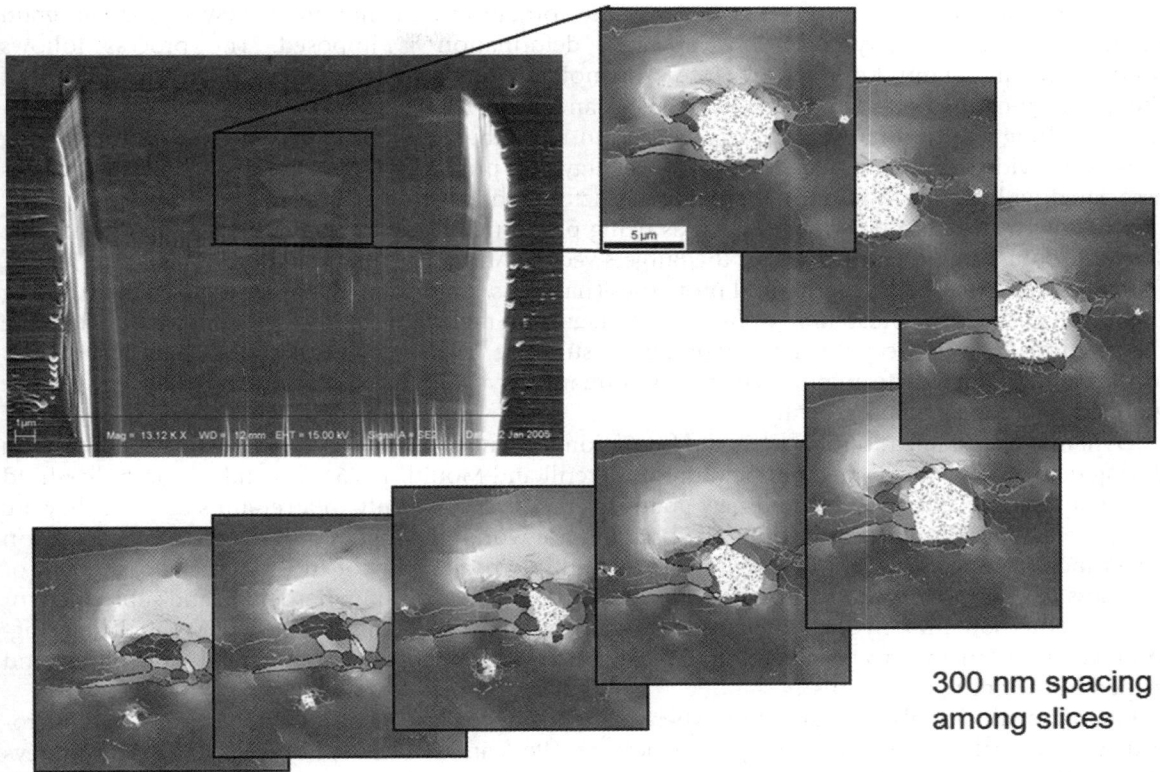

300 nm spacing
among slices

Figure 23 Orientation gradients around a hard Laves particles in a Fe₃Al matrix (Konrad et al., 2006).

acceleration is due to the fact that higher driving forces act when recrystallization (reduction in dislocation density) takes place together with phase transformation or respectively precipitation (release in transformation-free energy). Such situations are referred to as discontinuous precipitation phenomena. They are characterized by the fact that the driving force is not only coming from the stored deformation energy but also from the chemical driving force that is associated with the transformation.

The heating rate can also have a profound impact on the resulting recrystallization microstructure. This is so because a number of thermally activated processes compete during heat treatment. These are for instance dislocation climbing, dislocation cross-slip, motion of low-angle grain boundaries, motion of high and grain boundaries, formation or dissolution of second phases, and segregation. As these different mechanisms show and as will be discussed later in more detail, it is therefore useful to differentiate between chemical driving forces and mechanical driving forces.

23.3.2 Overview of Basic Recrystallization Phenomena

Primary static recrystallization describes the transformation of an as-deformed microstructure above the critical recrystallization temperature and above a critical threshold deformation. In scientific terms, the analysis of recrystallization typically addresses isothermal transformations. In manufacturing, however, recrystallization processes are usually not isothermal.

Primary static recrystallization proceeds by the formation and motion of new high-angle grain boundaries. During recrystallization, no new deformation is imposed. The process follows Johnson–Mehl–Avrami–Kolmogorov (JMAK) sigmoidal kinetics and typically leads to a refinement of the microstructure (Kolmogorov, 1937; Johnson and Mehl, 1939; Avrami, 1939, 1940). Grain structures resulting from primary static recrystallization typically consist of equiaxed crystals. The driving force is provided by the stored deformation energy, i.e. primarily by the long-range elastic stresses associated with the dislocation and subgrain structure that was formed during plastic straining. The driving force can hence be approximated as being proportional to the stored dislocation density, the shear modulus, and the magnitude of the Burgers vector (Aust and Rutter, 1959, 1960; Hu et al., 1990; Rath and Hu, 1969a). The mechanical properties (hardness, yield strength) decay at first slowly during the incipient recovery (incubation or nucleation stage) and then very rapidly, i.e. sigmoidally, when the newly formed grains sweep the deformation microstructure. Final impingement of the growing crystals leads to end of the transformation. The end of primary recrystallization is accompanied by competitive grain coarsening, i.e. grain growth.

Dynamic recrystallization describes the formation and motion of new high-angle grain boundaries during deformation at elevated temperatures (Cotterill and Mould, 1976). Beyond a certain threshold deformation, nucleation occurs and new grains from. These grow only to a certain size since they are continuously further deformed during the ongoing hot working. Kinetics of dynamic recrystallization phenomena are characterized by single peak or multiple peak behavior in the measured strength or hardness. This behavior reflects a locally heterogeneous sequence of strain hardening and subsequent or simultaneous softening by local dynamic recrystallization. In materials with a low or small stacking-fault energy, dynamic recrystallization typically occurs during most commercial hot-rolling and related hot-forming operations.

Grain growth describes the situation where competitive coarsening of the crystals leads to a microstructure where the average grain size slowly increases (Beck and Hu, 1966; Haessner, 1978; Humphreys and Hatherly, 1995; Doherty et al., 1997). The driving force of this phenomenon comes from the reduction in the total grain-boundary area. Like in other processes that undergo gradual structure coarsening under the effect of local interface curvature, such as originally described by the Gibbs–Thompson equation, typically the larger grains grow at the expense of smaller grains.

Secondary recrystallization refers to a specific grain-growth phenomenon where a very small number of grains grow to a dimension that exceeds the average grain size by one order of magnitude or more in terms of the grain diameter. However, secondary recrystallization is a slightly misleading term. When cast into a more appropriate term, it is sometimes also referred to as discontinuous grain coarsening or discontinuous grain growth. It does not describe the sweeping of the deformed microstructure such as encountered during primary static recrystallization but instead it refers to the extensive growth of a few large grains in an otherwise recrystallized grain structure. Hence, it has only a phenomenological similarity to primary static recrystallization because some literature describes the process in terms of a nucleation stage where some of the grains grow first extensively in the incipient stage of secondary recrystallization (pseudonucleation) and a subsequent "growth" stage where these huge crystals sweep the other regularly sized crystals (Rossard, 1963; Sellars and Tegart, 1966; Drube and Stüwe, 1967; Stüwe, 1968; McQueen et al., 1976). This latter growth stage does not require a specific additional driving force but the mere fact that some grains are much larger than others is topologically sufficient that the local curvature provides extensive further growth of these candidates. In physics terms, secondary recrystallization is a grain-growth process where a small number of grains grows extensively and can assume a size that is more than 100 times larger than that of the average size of the grains that surround it. The reason for this behavior can be local back-driving forces, inhomogeneous microstructures, and inhomogeneities in the grain-boundary properties in terms of energy and mobility. A typical example is the discontinuous grain growth of huge Goss-oriented grains in Fe-3 wt.% Si soft magnetic steels.

Another recrystallization type is referred to as recrystallization in situ (Beck and Hu, 1966; Haessner, 1978). Here again the classical terminology may appear a bit confusing because this phenomenon describes extensive recovery and not a nucleation and growth situation such as encountered in primary static recrystallization. In this situation, the very strong recovery of the as-deformed material proceeds to a level where the remaining driving force for static recrystallization becomes too low and where the subgrain growth has led in part to the formation of high-angle grain boundaries.

23.3.3 The Nucleation Stage of Primary Static Recrystallization

23.3.3.1 Introduction

Many structural transformations in metallic alloys take place via nucleation and growth. This is a heterogeneous two-stage transformation mechanism where those regions that are transformed first during the nucleation stage are separated by atomically sharp interfaces from the remaining untransformed regions that surround these nuclei (Humphreys and Hatherly, 2004; Cahn 1966; Johnson and Mehl, 1939; Shvindlerman and Gottstein, 1999). From a phenomenological viewpoint, recrystallization can also be formally described in terms of this conceptual framework. However, although some similarities indeed exists between conventional phase transformations and recrystallization, it must be emphasized that the latter type of microstructure transformation is a nonthermodynamic equilibrium process because the stored deformation energy (essentially the dislocations) is only in local mechanical equilibrium but not in thermodynamic equilibrium.

The nuclei in static primary recrystallization are defect-free, newly formed crystals that are—at least partially—surrounded by mobile high-angle grain boundaries. Hence, as a recrystallization nucleation, we jointly refer to those mechanisms that lead to the formation of new mobile high-angle grains boundaries or to the overcritical bulging of already existing inherited high-angle grains boundaries.

Although we formally define here two stages in recrystallization, namely, the nucleation stage and the growth stage, both phenomena are characterized by growth, i.e. by the motion of interfaces (Humphreys, 1997). More specific, the main difference between the two processes is that only the first stage, viz. recrystallization nucleation, requires some additional characteristics that determine how, why, where and in which direction the transformation starts.

Although in many cases the deformation substructure is in its function of providing a driving force for the transformation viewed as a homogeneous parameter, recrystallization nucleation cannot be described as a homogeneous phenomenon. In real microstructures, the driving force provided by the preceding cold working is of course not distributed homogeneously. For instance, near-grain boundary regions have in cold-worked metals usually both, higher lattice gradients and higher stored deformation energy compared to the grain interiors. However, the use of a scalar and homogeneous driving force works as a first approximation to understand some elementary kinetics and microstructural properties of recrystallization. Although this simplification works to understand the mechanical driving force acting during primary recrystallization, it does not work when addressing the nucleation stage itself. This means that recrystallization nucleation is principally a consequence of microstructure inhomogeneity. In order to understand nucleation in recrystallization one has to address three main questions: first, where are the highest local driving forces? Second, where does the microstructure reveal a sufficiently large local lattice curvature or already existing misorientations to allow the gradual or spontaneous formation of mobile high-angle grain boundaries that can later sweep the surrounding deformation substructure with an intrinsic high mobility? Third, where in the microstructure does a sufficiently high gradient exist in the stored energy across a high-angle grain boundary. The first criterion is referred to as thermodynamic instability criterion (existence of a driving force), the second one as kinetic instability criterion (formation of mobile interfaces), and the third on as mechanical instability criterion of recrystallization nucleation (gradient in driving force across interfaces) (Haessner, 1978; Humphreys and Hatherly, 1995, 2004; Doherty et al., 1997; Doherty, 2005).

Classical thermodynamic nucleation criteria developed for isothermal equilibrium phase transformations describe how the newly formed interface (energy loss) that surrounds a new and transformed region (energy gain) and separates it from the as-deformed hence untransformed area leads to a critical nucleation energy barrier that can be overcome by thermal fluctuation if the barrier is not too high.

For homogeneous nucleation, in which the nucleus can start to form in principle at any atomic site in a unit volume, the critical local increase of the free energy (nucleation barrier) must be supplied by thermal activation. In case of heterogeneous nucleation, some of the new interface energy (high-angle grain boundary energy) required can be contributed by an already existing portion of interface. This effect can lead to a notable reduction in the size of the nucleation barrier. This phenomenon is referred to as heterogeneous nucleation. It forms the theoretical basis for the understanding of most nucleation phenomena in materials science.

In recrystallization, however, classical nucleation theory has not proven to be a successful concept. Two specific reasons stand specifically against applying a classical nucleation model to recrystallization: the first one is the very low value of the stored energy associated with plastic deformation. It does usually not exceed several MPa in driving force. The second one is the high interfacial energy of the newly formed high-angle grain boundaries. They are usually of the order of 1 J/m^2 (Molodov, 2001; Lücke and Stüwe, 1963; Gottstein et al., 1995; Shvindlerman et al., 1995, 1999; Upmanyu et al., 1999; Rollet et al., 2004).

Hence, when using these values in homogeneous nucleation theory, an impossibly small density of new grains would be predicted. Even the most effective heterogeneous nucleation sites cannot reduce the barrier to recrystallization nucleation to any significant extent to reach the required nucleation energy.

An alternative theoretical approach how nucleation proceeds in primary static recrystallization was that a new grain does not necessarily develop via nucleation of a totally new crystal with an individual crystallographic orientation that did not exist before in the material but instead a recrystallized grain develops gradually from a recovered region of the existing deformed microstructure, a cell or a subgrain (Humphreys and Chan, 1996; Humphreys, 1992a). The new grain then has an orientation that will be essentially that of the deformed region from which it grew.

This means that according to this commonly accepted view the crystallographic orientations that are formed during the nucleation stage of primary recrystallization already exist in the deformed microstructure. From that we can conclude that in a system where nucleation proceeds exclusively by such a discontinuous subgrain coarsening mechanism all texture components of the recrystallization texture were already hidden at least in some very small portion of the deformation microstructure. This discussion reveals that recrystallization is not characterized by a true nucleation mechanism in the classical thermodynamic sense where new structure units are formed by thermal fluctuation that did not previously exist.

23.3.3.2 *Nucleation by Discontinuous Subgrain Coarsening*

It was described above that one typical process that can lead to recrystallization nucleation is the discontinuous subgrain-coarsening mechanism. It describes the competitive growth of some DCs at the expense of others that shrink and finally vanish during the recovery stage. It must be emphasized though that this mechanism can only act as a recrystallization nucleation mechanism in specific cases. For a gradual discontinuous subgrain coarsening to function as recrystallization nucleation mechanism with the result to produce new mobile high-angle grain boundaries that sweep the cold-worked substructure and less recovered cells, it has to occur inhomogeneously in the deformation micro-structure. If such competitive subgrain-coarsening phenomenon occurs everywhere in the material, the mechanical instability criterion would be lost and instead recrystallization in situ would take place.

This means that in order to fulfill the instability criteria outlined above, subgrain coarsening must proceed in a heterogeneous fashion where some areas reveal faster subgrain coarsening than others. This may lead first, to a sufficiently high accumulated misorientation to form a new high-angle grain boundary; and second to a directed gradient in driving force between the rapidly and the not-so-rapidly coarsened substructure. This is why the successful nucleation stage is often referred to as discontinuous subgrain coarsening and not just as subgrain coarsening (Haessner, 1978; Humphreys and Hatherly, 2004; Faivre and Doherty, 1979; Humphreys, 1992a; Ferry and Humphreys, 1996). In the classical literature, it is alternatively also called abnormal subgrain growth; however, the term discontinuous subgrain coarsening is more appropriate.

This means that recrystallization nucleation by discontinuous subgrain coarsening is characterized by the rapid growth of a very small minority of the recovered cells that then become the new grains that finally are surrounded by high-angle grain boundaries.

It has often been noted, that only a very small fraction of cells make the transition to a new grain. In a moderately cold-deformed aluminum sample, the subgrains are typically about 1 μm in size while after primary recrystallization a grain size of at least 100 μm is quite common in this material. This ratio suggests a volume increase by a factor of about 10^6. This means that only one subgrain out of a million reaches the transition to become a rapidly growing recrystallization nucleus capable of producing a recrystallized grain.

An essential feature of discontinuous subgrain growth leading to a recrystallization nucleus is the gradual accumulation of sufficient misorientation across the moving interfaces involved in subgrain coarsening. The mobility of grain boundaries, their velocity under a given driving pressure, is typically much lower for subgrain boundaries with a low angle of misorientation than for high-angle grain boundaries. As a result of this mobility difference, only subgrains that are highly misoriented, typically by more than about 15° with respect to at least part of their surrounding substructure, can grow sufficiently rapidly and hence become recrystallizing grains. This observation adds an essential crys-tallographic aspect to the topology kinetics described above. The requirement for a subgrain boundary to accumulate sufficient misorientation and, hence, achieve higher mobility means that successful nucleation takes place particularly in deformed areas that contain already a high lattice curvature from preceding cold working. Strong lattice curvature typically occurs at shear bands, around second-phase particles (Furu et al., 1993; Hansen and Bay, 1981; Hillert, 1988; Humphreys, 1977a, 1977b, 1979; Leslie et al., 1963b; Siqueira et al., 2011; Humphreys and Kalu, 1987; Humphreys and Ardakani, 1994; Hutchinson, 1989; Hutchinson et al., 1989; Juul Jensen et al., 1991, 1994; Leslie et al., 1963c; Liu et al., 1995; Rath and Hu, 1969b; Ridha and Hutchinson, 1982b; Russel and Ashby, 1970; Miyazaki et al., 2002; Fujita et al., 1996; Inagaki, 1987; Doherty and Martin, 1962–1963; Jones et al., 1979), at grain boundaries, and within instable crystallographic orientations (divergent texture components) (Raabe et al., 2002a).

At modest preceding plastic deformations of the order of 20–40% cold reduction, only few gradient zones are formed, particularly in high-purity metals. In such cases where deformation does not entail high in-grain orientation gradients, i.e. only few regions exist in the lattice with high misorientations relative to the neighboring substructure, capable nucleation sites do hence not occur frequently. At low reductions (less than 20%) in polycrystalline metals, the only high misorientations are found to occur at prior grain boundaries or at heterointerfaces in cases where a second phase occurs.

In investigations on heavily rolled copper and aluminum-containing particles larger than 1 μm, the density of sites with high local misorientations was shown to increase greatly with strain and, with the strength and size of the heterogeneities. In such cases, heavily strained, and also heavily curved, deformation zones formed around the coarse, second-phase particles (Haessner, 1978; Humphreys and Hatherly, 1995, 2004; Faivre and Doherty, 1979; Engler, 1997; Laue, 1913; Ferry, 2002; Huang et al., 2000).

In more severely strained alloys (e.g. heavy rolling reductions of 80% and more) a very high density, particularly in the normal direction, of high-angle misorientation regions is observed. Typically, the occurrence of a high local misorientation of some cells relative to their vicinity is only a necessary condition for a potential nucleation site to become active. It should be underlined that it is however not a sufficient condition since the vast majority of subgrains at high-angle grain boundaries, and other high misorientation sites, do not become new grains. It has long been recognized in addition to the mobility requirement for a subgrain to grow, a successful subgrain needs to have an energy advantage so that it grows rather than vanishes (Humphreys and Chan, 1996; Humphreys and Ardakani, 1994; Humphreys, 1977b). The usual form of this energy advantage is having a significantly larger subgrain (Humphreys, 1997, 1992a; Miodownik, 2002).

The size advantage may arise from the deformation process if a particular orientation on one side of a high-angle boundary has a larger subgrain size and thus a lower stored energy.

This effect was observed in body-centered cubic metals through the measurement of the mean subgrain size that was smallest in grains having crystallographic {111} oriented planes parallel to the rolling plane, and by measurement of the mean subgrain misorientation that was largest in grains with that orientation (Choi, 2003). In pure iron cold-rolled only to 50%, Inokuti and Doherty, (1977), (1978) found that nucleation occurred by invasion of {111}-oriented grains by neighboring crystals with a larger subgrain size.

This phenomenon was termed "strain-induced boundary migration". It had originally been identified by Beck et al. (1950). In compressed and also in rolled aluminum, cold deformed to only 40–50% reduction, there were no significant subgrain size differences observed in the cold-worked microstructure. However, the required size differences for nucleation for strain-induced boundary migration appeared by the process of subgrain coalescence in one of the grains, that can be considered as the parent grain during this process. The enlarged subgrain then grew into the adjacent grain by migration of the existing high-angle grain boundary between the deformed grains. Detailed analysis of this subgrain coalescence at grain boundaries was identified by Faivre and Doherty, (1979) as requiring the presence, in the parent grain, of an additional high misorientation at a transition band in order that coalescence could take place.

23.3.3.3 *Recrystallization Nucleation by Thermal Twinning*
The analysis of recrystallization nucleation by the discontinuous competitive coarsening of the subgrain structure has revealed that this mechanism does not lead to other orientations than those already present in the cold-worked microstructure. This means that as a rule no new DCs are spontaneously formed during the nucleation stage.

However, this mechanism does not explain certain recrystallization texture components that manifest a discontinuous deviation from all orientations that were originally present in the cold deformation texture. Such observations apply particularly to metals and alloys with a low stacking-fault energy such as brass or austenitic stainless steels. Such a more spontaneous change in the crystallographic orientation during recrystallization must be due to an alternative nucleation mechanism, namely, annealing twinning (Berger et al., 1988; Wilbrandt, 1980; Wilbrandt and Haasen, 1980a, 1980b). Indeed, it was observed that twins tend to form first as a thin lamella parallel to the progressing recrystallization front (**Figure** 24) (Berger et al., 1988).

The exact atomistic mechanisms associated with this phenomenon are not entirely understood. One assumption is that a growth defect that leads to fine twins at the recrystallization front while other mechanisms assumed that a dissociation of a moving grain boundary splits off a twin boundary. This mechanism was assumed to occur in order to reduce the overall energy associated with the grain boundary. The phenomenon was observed in a triple point configuration where several grains meet.

As later discussed by Wilbrandt and Haasen (Berger et al., 1988; Wilbrandt, 1980; Wilbrandt and Haasen, 1980a, 1980b) a random twinning sequence using randomly one out of the possible 12 $\{111\}<112>$ variants available in the face-centered cubic lattice would lead to nearly any orientation

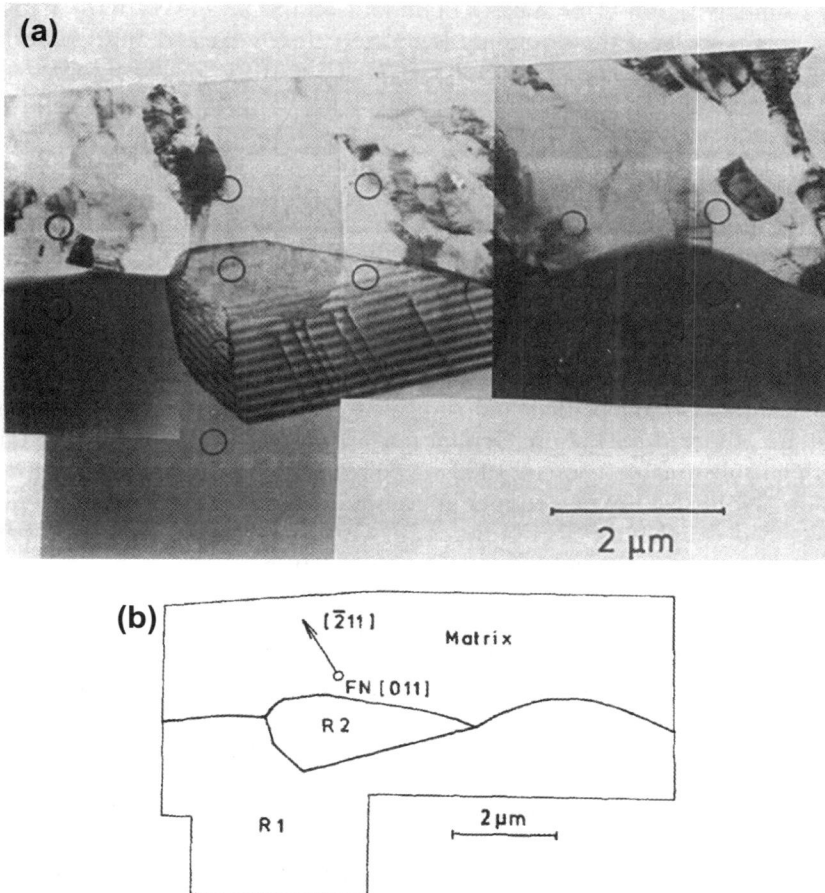

Figure 24 Al specimen annealed at 200 °C and 5 min at 300 °C showing twin formation (Berger et al., 1988).

possible and, hence, to a random crystallographic recrystallization texture. As this complete orientation randomization effect is typically not observed in crystallographic texture measurements (Engler et al., 1996; Ibe and Lücke, 1966; Bunge, 1982, 1986, 1987; Bunge and Esling, 1982, 1991; Adams et al., 1993), it was assumed that there must be a reason for the preference of certain twin sequences. More specific, the authors suggested that either a gain in grain-boundary energy or an advantage in growth kinetics might play the dominant role in thermal twinning. Wilbrandt also observed a number of strong hints regarding the first reason, namely, that the growing grain aims at lowering its grain-boundary energy by twinning until step by step a lower energy of the interface is attained (Berger et al., 1988; Wilbrandt, 1980; Wilbrandt and Haasen, 1980a, 1980b).

In that context, it has to be considered though that thermal twinning would typically lead to the formation of rather immobile boundaries. This applies specifically to the formation of coherent twin boundaries, which are the grain boundaries with the lowest possible mobility (Aust and Rutter, 1959, 1960; Hu et al., 1990). However, special high-angle boundaries, although having low grain-boundary energy, are among the most mobile ones in dilute alloys (Aust and Rutter, 1959, 1960; Hu et al., 1990). One imagines that a compact boundary structure produces low grain-boundary energy but not necessarily high boundary mobility. To move a boundary, atoms have to be transferred from one grain to the other and this is easier if the boundary is not too closely packed and has some free volume (Viswanathan and Bauer, 1973; Furtkamp et al., 1998; Li et al., 1953; Winning et al., 2001, 2002; Winning, 2003; Heinrich and Haider, 1996; Gottstein et al., 1995, 1997, 1998; Vandermeer and Juul Jensen, 1993; Beck et al., 1950; Molodov et al., in press). Thus, there is some competition between frequent formation of low-energy boundaries and the formation of a highly mobile one that covers a large volume of the specimen during its movement. In that sense, a selection mechanism for recrystallized orientations according to their growth rate may have an influence on the resulting recrystallization texture. This discussion underlines the importance the effects of both, energy and mobility of the high-angle grain boundaries, on the annealing texture and microstructure formed in metals during recrystallization in cases where twinning is involved.

23.3.3.4 *Recrystallization Nucleation at Prior Grain Boundaries*
Recrystallization textures of cold rolled and annealed metals often resemble the crystallographic orientations that are observed for the preferential nucleation of new grains from certain orientations within the deformed substructure (Bunge, 1982, 1986, 1987; Bunge and Esling, 1982, 1991). In that context, specifically the nucleation at a former grain boundary plays an important role. A number of works have shown evidence that the distribution of newly formed crystal orientations at an early stage of recrystallization bears a close resemblance to the final texture.

Additionally, it has to be considered that at a later stage of recrystallization also growth selection phenomena play a role as will be discussed later in a subsequent section.

Hutchinson, (1989) conducted experiments on the role of nucleation at former grain boundaries in iron and low-carbon steels. These studies have shown that new grains with characteristic crystallographic orientations were formed by preferred nucleation at certain original grain boundaries of the as-deformed crystals. In body-centered cubic steels, they observed that in {111}<uvw>-oriented texture components a zone of highly localized deformation exists adjacent to the grain boundary as a result of the constraint imposed by the neighboring crystal and that this heterogeneous structure is favored for the nucleation of new grains on annealing.

The model that had originally been put forward by Inagaki, (1987) describes a possible mechanism by which the local stress state at the boundary may cause a local lattice rotation around the normal direction in certain cases of adjacent {111}//normal oriented grains. According to this approach, such grains can reveal substantial local grain orientation changes at the grain boundaries between two abutting crystals that belong even to the same texture fiber. Typical texture components that are affected by such a mechanism in body-centered cubic steels are the $\{111\}<112>$ and $\{111\}<110>$ texture components (Dillamore et al., 1979; Hölscher et al., 1991; Ushioda et al., 1987; Raabe and Lücke, 1993, 1992; Klinkenberg et al., 1992; Raabe, 1995a; Juntunen et al., 2001; Ray et al., 1975b; Humphreys, 1977a; Ridha and Hutchinson, 1982a). The resulting recrystallization nuclei were found to differ substantially in orientation from the abutting interiors of the grains in which they were formed.

The reasoning behind this model is that the local incompatibility between abutting grains leads to more heavily distorted and more distinctly misoriented regions at the original grain boundaries.

Therefore, these highly distorted near-boundary regions also contain a higher local crystallographic orientation gradient relative to the surrounding deformation matrix and also a higher local stored energy compared to regions that are located inside of the deformed grains. Hence, it is plausible to assume that near-grain boundary regions can more rapidly form new mobile high-angle grain boundaries that then can grow fast owing to the higher local driving force.

In contrast to this model, experiments conducted on pure iron (Inokuti and Doherty, 1977, 1978) showed that after 40% cold reduction via cold-rolling nucleation at grain boundaries occurred primarily by strain-induced migration of the original grain boundary (strain-induced boundary migration). This process regenerated one of the originally existing grain orientations from the deformation texture, and favored those orientation components of low stored-deformation energy. Similar strain-induced boundary migration phenomena were also reported to occur in aluminum.

The mechanism of strain-induced boundary migration has to be clearly differentiated from the grain-boundary nucleation mechanism that has been explained above for body-centered cubic steels: strain-induced boundary migration is characterized by the fact that it does not involve the formation of new mobile high-grain boundaries but only the bulging of existing ones. Hence, strain-induced boundary migration occurs preferably at such former grain boundaries where a high difference in the stored energy occurs across the interface. Humphries could simulate and hence confirm such situations by assuming different substructures on either side of an already existing high-angle grain boundary (Humphreys, 1997). For this purpose, he used a vertex front-tracking model (Humphreys, 1997, 1992b; Miodownik, 2002; Bate, 1999).

23.3.3.5 Recrystallization Nucleation at Shear Bands

Shear bands are mesoscopic band-like deformation inhomogeneities in middle and heavily deformed metallic alloys. They appear under oblique angles relative to the main deformation axis and can penetrate multiple crystals in the deformation structure. They are characterized by a mesoscopic orientation that does not match a distinct crystallographic direction, for instance that of a specific slip or twinning system. Therefore, they are also referred to as noncrystallographic deformation zones (Haessner, 1978; Humphreys and Hatherly, 2004). In cold-rolled materials, they typically form under an angle of approximately 35–40° relative to the rolling direction (RD) (**Figure 22**).

Shear bands are found in many materials. Very frequently, they appear in materials with a low stacking-fault energy (**Figure 25**). However, they also observed in metallic alloys with high stacking-fault energy such as for instance in several aluminum alloys—for instance when these materials contain shearable particles or a high solute content of copper or magnesium.

The underlying micro- and nanostructures of such shear bands typically consist of very small DCs and contain very high dislocation densities. Shear bands often penetrate multiple grains and carry very high local shears. At the transition zones between the shear band interior and the surrounding matrix

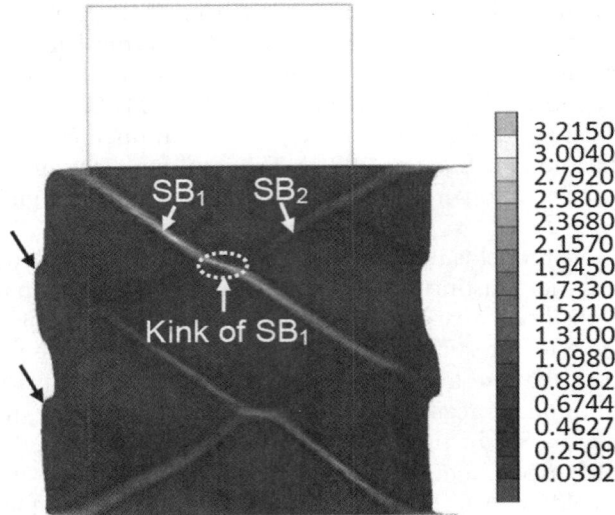

Figure 25   Shear band predicted in copper using a crystal plasticity finite element method. The color code represent the von Mises strain (Jia et al., 2012a,b).

material often very strong lattice curvature and very high densities of GNDs are found. These two properties, namely, the high driving forces and the high local misorientations at the rims of the shear bands are essential prerequisites for nucleation and subsequent growth of new grains. Consequently, such inhomogeneous deformation zones are highly potential sites for nucleation of recrystallization (Engler et al., 1996; Engler, 1997) (**Figures 26** and **27**).

When aiming at the study of the influence of recrystallization nucleation at such shear bands one is confronted with the difficulty that in many alloys nucleation at shear bands competes with nucleation events at other microstructural features. For example in Al-alloys, recrystallization nuclei at shear bands compete with nuclei forming at transition bands, cube-bands, grain boundaries, and second-phase particles (Engler and Vatne, 1998; Engler et al., 1989; Duckham et al., 2002; Korbel et al., 1986; Liu et al., 1989; Doherty and Baumann, 1993).

There is a considerable amount of experimental evidence that shear bands do indeed act as very successful nucleation sites during heat treatment of heavily deformed metallic alloys. Most prominent examples are such with alloys that have small stacking fault energy and high solute element content as for instance brass.

There is further evidence that nucleation at shear bands leads to a randomization of the recrystallization texture. This was first suggested by Ridha and Hutchinson, (1982a), who measured a weak, almost random recrystallization texture in copper for which grain nucleation at shear bands had been observed. This was in contrast to a sharp cube recrystallization texture corresponding to a condition about the occurrence of shear bands of a similar—yet presumably purer—batch of copper. It was suggested that the orientational randomization of the final texture was primarily caused by the nucleation of randomly oriented nuclei at shear bands and, in addition, by the destruction of the sites of cube-oriented nuclei by cutting through the cube transition bands.

More recently, local orientation measurements revealed that new grains formed in a partially recrystallized specimen of 90% cold-rolled polycrystalline Al-1.8 wt.% Cu (Engler and Vatne, 1998).

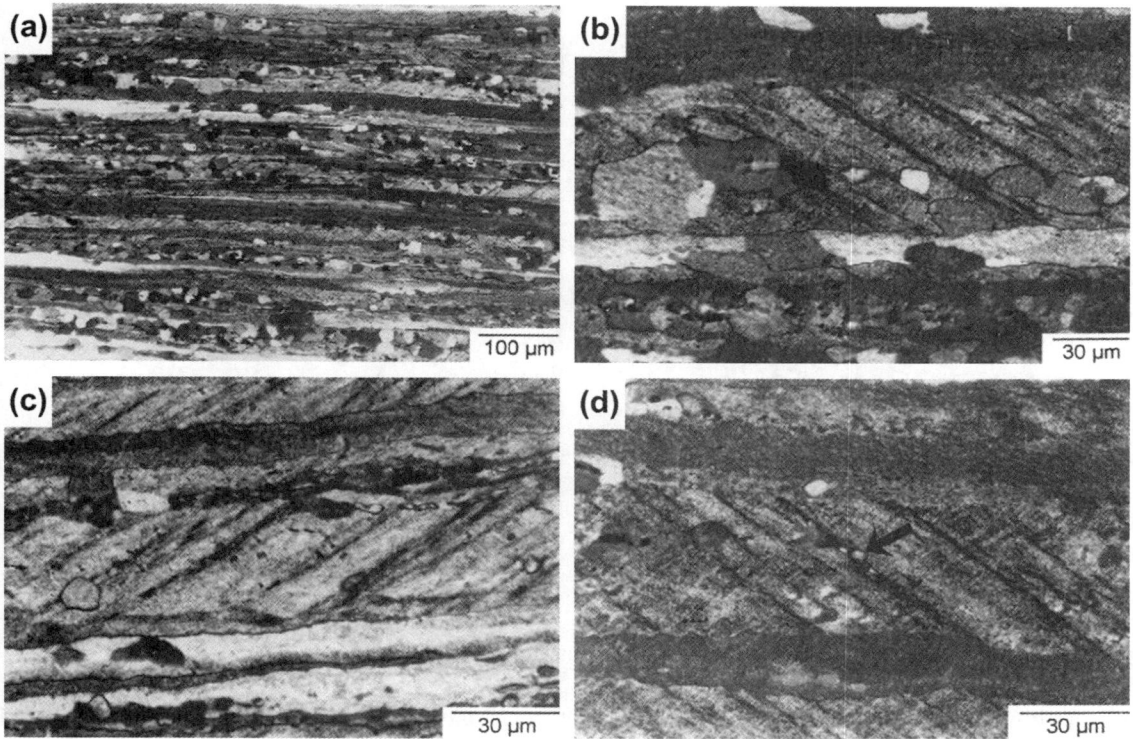

Figure 26   Optical micrographs showing evidence of grain nucleation at shear bands: (a) is a low magnitude view of a partially recrystallized structure; (b) shows grains that appear to have nucleated at shear bands; while (c) and (d) show grains at the beginning stages of nucleation at shear bands (e.g. grain indicated by arrow in (d)) (Duckham et al., 2002).

A distinction was made between grains nucleating at shear bands and grains nucleating at band-like structures parallel to the RD, presumed to be cube bands. The texture corresponding to nucleation at the presumed cube bands demonstrated a cube texture with only minor scatter. The texture corresponding to nucleation at the shear bands was less well-defined with notably less occupation of the exact cube orientation. There also was evidence that there may be preferred nucleation of some orientations at shear bands. Engler and Vatne, (1998) also investigated the influence of shear bands on recrystallization textures in binary Al-1.8 wt.% Cu and Al-3 wt.% Mg alloys by means of bulk and local orientation measurements. These results suggested that a recrystallization texture that is characterized by peaks around the Goss- and Q orientations, accompanied by a reduction in the strength of the cube orientation, provides a strong indication of recrystallization nucleation at shear bands. The Goss and Q orientations were suggested to result from positive and negative rotations around the transverse axis of the sheet associated with the formation of shear bands in copper-$\{112\}<111>$ and S $\{123\}<634>$ oriented grains. The Goss and Q orientations are metastable, in that they possess a low rate of further rotation, and are thus always present for possible nucleation during annealing.

It is the main feature of nucleation events occurring at shear bands that, like for most of the other nucleation mechanisms discussed above, the crystallographic orientation of the successful nucleus, after its competitive growth out of the surrounding subgrain structure, is already present in the

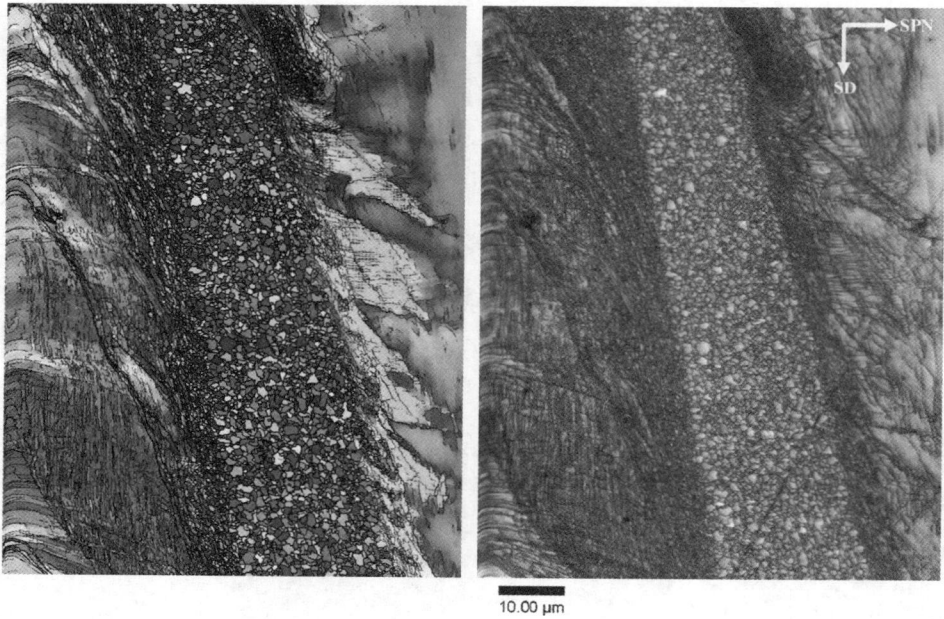

Figure 27 Nucleation at adiabatic shear band in IF steel (Lins et al., 2007).

deformation substructure at that shear band. Therefore, it makes sense to analyze the crystallographic rotations that occur at and in shear bands by simulations. This was recently done in a study using crystal plasticity finite element methods (Jia et al., 2012a, 2012b). It was observed that shear bands do not only lead to characteristic rotation zones, particularly at the transition between the band and the surrounding matrix, but also that this phenomenon is highly orientation-dependent.

More specific, the finite element simulations showed that Copper-oriented crystals, $(1\ 1\ 2)[1\ 1\ \bar{1}]$, and Brass-R oriented crystals, $(1\ 1\ 1)[1\ 1\ \bar{2}]$, had the largest tendency to form shear bands, associated with inhomogeneous texture distribution induced by the shear banding. To also understand the influence of the micromechanical boundary conditions on shear band formation, simulations on copper-oriented single crystals with varying sample geometry and loading conditions were performed. From that, it was found that shear banding can be understood in terms of a mesoscopic-softening mechanism. The predicted local textures and the shear banding patterns agreed well with experimental observations in face-centered cubic crystals with low stacking-fault energy. These observations also explain why in many materials recrystallization nucleation at shear bands does not necessarily lead to entirely random crystallographic textures. Instead, as only certain crystallographic texture components are prone to developing shear bands, for instance during cold rolling, the corresponding spectrum of nucleation orientations should recruit from those lattice rotations that occur in the transition zones between the shear bands and the matrix for these specific crystallographic orientations (**Figure 27**) (Lins et al., 2007).

23.3.3.6 Particle-Stimulated Nucleation

Particles can have two important effects on recrystallization (Haessner, 1978; Beck, 1949). The first one is that large, undeformable particles can generate potential sites for recrystallization nucleation in their vicinity. The second effect is that small particles exert a retarding force on the migration of sub-boundaries as well as of high-angle boundaries. This force is referred to as Zener pinning force. Such a pinning force can act twofold as it means that not only the growth of viable nuclei via the motion of high-angle grain boundaries will be retarded, but also the nucleation itself since this includes the migration of low-angle boundaries. In this section, only the effects of large particles on recrystallization nucleation is discussed (**Figure 28**) (Humphreys, 1977a, 1979; Leslie et al., 1963b; Siqueira et al., 2011; Humphreys and Kalu, 1987; Humphreys and Ardakani, 1994; Hutchinson, 1989; Hutchinson et al., 1989; Harun et al., 2006). Particle pinning forces will the discussed in a subsequent section.

The presence of large, hard particles during the plastic deformation of a softer metallic material will induce strong deformation gradients that are associated with both substructure refinement and lattice orientation gradients. PSN is the event by which a recrystallized grain is nucleated in such a deformation zone around a hard inclusion. Many industrial alloys, particularly those of iron and aluminum, often contain micrometer-sized, second-phase particles. This means that PSN most likely is an import nucleation mechanism in many engineering materials, and PSN has been observed in many alloy systems, including those of aluminum, iron, copper, and nickel (Humphreys and Hatherly, 2004) (**Figure 29**) (Siqueira et al., 2011). The work of Humphreys and other authors (Humphreys, 1977a, 1979; Leslie et al., 1963b; Siqueira et al., 2011; Humphreys and Kalu, 1987; Humphreys and Ardakani, 1994; Hutchinson, 1989; Hutchinson et al., 1989) revealed a number of characteristic features of PSN:

Firstly, it was found that recrystallized grains originate at preexisting subgrains within the deformation zone, but not necessarily directly at the particle surface. This observation is plausible since the large lattice curvature that is typically associated with the deformation zone around a hard inclusion (Calcagnotto et al., 2010). This type of microstructure promotes the rapid formation of mobile high-angle grain boundaries during subgrain coarsening inside a curved lattice area. When finally high-angle grain boundaries result from such a competitive coarsening mechanism, these can subsequently sweep the surrounding deformation matrix owing to their high mobility (Gottstein et al., 1995; Shvindlerman et al., 1995; Shvindlerman and Gottstein, 1999). This effect seems to prevail over the availability of an already existing interface (heterogeneous nucleation effect).

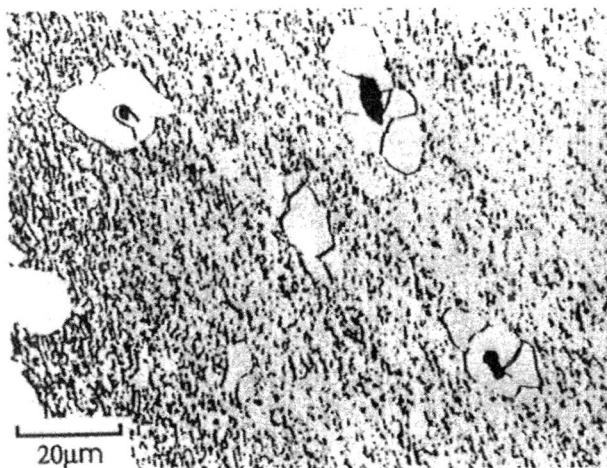

Figure 28 PSN in cold-rolled iron containing hard second-phase particles (Humphreys, 1977a).

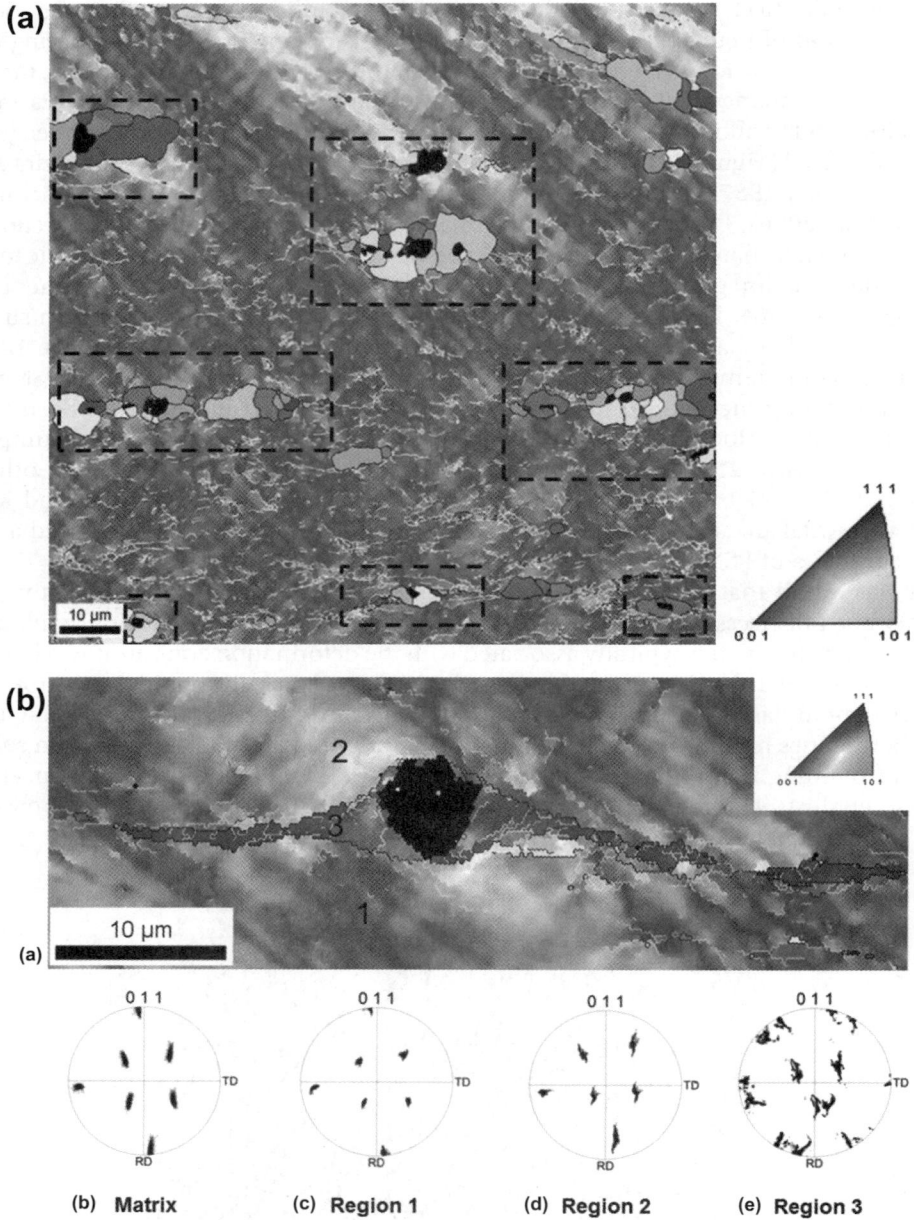

Figure 29   (a) EBSD map showing particle stimulated nucleation in cold rolled coarse-grained ferritic stainless steel. The IPF (inverse pole figure) map shows recrystallized grains (dash square) nucleated preferentially around the particles (black in the map) in a 45° rotated-cube large grain cold rolled and further annealed at 725°C for 15 min (De Siqueira et al., 2013). (b) EBSD results showing details in a 45° rotated-cube large grain: a) IPF map; b) {011} pole figure of the large grain; c–e) {011} pole figures corresponding to the regions 1, 2, and 3, respectively (De Siqueira et al., 2013).

Secondly, it was observed that recrystallization nucleation occurs by rapid subboundary migration. This observation is in line with the statement above, namely, that discontinuous subgrain coarsening in conjunction with a sufficiently high local lattice curvature leads to the rapid formation of mobile new grain boundaries during heat treatment.

Thirdly, it is a typical feature of recrystallization microstructures that are influenced by the PSN mechanism that the further growth of the newly formed grains may stagnate when the deformation zone has been consumed. This result can be understood, since the deformation zone that surrounds a sufficiently large hard particle in an otherwise soft metallic matrix extends only to a certain range into the vicinity of the particle. The material more remote from the interface to the hard particle is often much less deformed and has lower lattice curvature. This means that even if a new high-angle grain boundary was successfully formed in the vicinity of the inclusion, the driving force far away from the particle may no longer be sufficiently high to support further growth of such a newly formed nucleus. This means that microstructures where recrystallization via PSN plays a dominant role may reveal fine recrystallized areas around the particles and more coarse-grained or even recovered zones far away from them. In order to successfully optimize and homogenize the microstructure in such a material, it is hence required that the rigid second phase particles have a sufficiently high density so that the surrounding recrystallized grains can sweep the entire matrix.

Based on these observations, three main criteria for a successful nucleation according to the PSN mechanism can be stated (Humphreys, 1977a, 1979; Leslie et al., 1963b; Siqueira et al., 2011; Humphreys and Kalu, 1987; Humphreys and Ardakani, 1994; Hutchinson, 1989; Hutchinson et al., 1989):

(i) The presence of a deformation zone with sufficiently large lattice rotations. The formation of such a zone will depend on the deformation temperature because the effective recovery reactions in the case of high deformation temperatures will strongly retard the evolution of deformation zones.

(ii) The formation of a nucleus within the rotation zone surrounding the inclusion by competitive subgrain growth involving the motion of angle grain boundaries into a curved lattice region.

(iii) The issue of further grain growth beyond the deformation zone surrounding the hard particle. After the deformation zone is consumed, the potential nucleus must have reached a critical size and acquired a high-angle boundary misorientation (i.e. high mobility) in order to be able to grow into the surrounding matrix. Humphreys has shown that the latter is a restrictive criterion that in most cases determines the efficiency of the PSN mechanism for renewing the entire microstructure (Humphreys, 1977a, 1979; Leslie et al., 1963b; Siqueira et al., 2011; Humphreys and Kalu, 1987; Humphreys and Ardakani, 1994; Hutchinson, 1989; Hutchinson et al., 1989).

Besides these questions on the basic occurrence of new recrystallized grains in the vicinity of hard particles also the resulting orientation distribution of these new crystals is of relevance.

For this purpose Humphreys studied the orientations of PSN grains (Humphreys, 1977a, 1979; Leslie et al., 1963b; Siqueira et al., 2011; Humphreys and Kalu, 1987; Humphreys and Ardakani, 1994). On single crystals of Al–Si alloys deformed in tension, he observed that the orientations of the PSN nuclei are contained within the spread of the existing deformation zone. Annealing of weakly rolled crystals of the same alloys resulted in a sharp recrystallization texture that was rotated from the deformation texture by 30–40° about a crystallographic <112> axis. When assuming single slip, one can show that the orientations of the subgrains inside the deformation zones should, for face-centered cubic metals, rotate around a <112> axis lying in the slip plane and perpendicular to the slip direction. Such <112> rotations were indeed observed in Al–Cu alloys by Russel and Ashby, (1970).

In polycrystalline alloys, the analysis of PSN effects is more complicated (Engler et al., 1996; Engler, 1997; Humphreys, 1992a). Owing to the large number of initial grain orientations and the locally acting stress state—that can in a polycrystal substantially deviate from the global stress state—a more or less random-orientation distribution of the resulting subgrains in the deformation zones was observed. In total, it is therefore commonly observed that microstructures of deformed polycrystalline alloys that are affected by the PSN nucleation mechanism are either weakly textured or even randomly oriented. Both the initial orientations around the particles and also the size and shapes of the particles will have an influence on the orientations within the deformation zones.

The range of the potential crystal orientations resulting from PSN stimulated recrystallization nucleation is therefore enormous.

Some indications of preferred nuclei orientations appearing in the deformation zones of heavily strained materials around a hard inclusion have been reported in recent works by Engler et al. (1996), (1989), Engler (1997), Engler and Vatne (1998). At low levels of deformation (e.g. up to 50% cold-rolling reduction), they reported a randomization of the subgrain orientations within the deformation zones. At higher strains, however, a slight preferred occurrence of rotated cube subgrains appeared in the regions close to the large particles. Nuclei generated at the particles that were studied after a short-time heat treatment also revealed this slight preference of the rotated cube component. Weak recrystallization textures with some preference of a rotated cube texture have been reported in various aluminum alloys where PSN acted as one of the nucleation mechanisms.

A consequence of the general consensus, and success of the PSN mechanism, is that many workers regard PSN as the only possible nucleation mechanism for creating randomly oriented recrystallized grains (**Figure 30**).

23.3.3.7 *Suppression of Nucleation through Gradient-Free Deformation*
The process of primary recrystallization takes place by the formation and motion of mobile high-angle grain boundaries. The first process is referred to as nucleation.

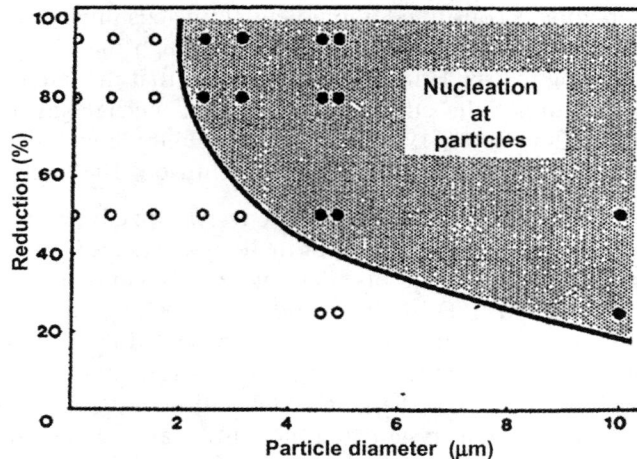

Figure 30 The conditions for particle size and strain for PSN to occur at Si-particles in cold rolled aluminum (Humphreys, 1977a,b).

As was explained above, for the onset of recrystallization a thermodynamic (transformation via nucleation), mechanical (net driving force vector), and kinetic (large-angle grain boundary with high mobility) instability is required. If one of these preconditions is not fulfilled, nonconservative dislocation rearrangement and mutual dislocation annihilation, viz. static recovery instead of recrystallization, can prevail during annealing (Raabe and Lücke, 1992).

Such behavior can be observed in cases of low cold-working prior to annealing. However, it is less well-established that even in case of sufficient initial cold-deformation in excess of 80% cold working, primary recrystallization can be suppressed if the initial nucleation stage is not taking place.

Owing to the mechanisms involved during recrystallization, the crystallographic texture is changed whereas during recovery it remains unchanged. Due to this principle difference, it is obvious that texture investigation is an appropriate diagnostic means of distinguishing between the two mechanisms.

In low-carbon steels, it was observed that samples that have a relatively coarse grain size and that were cold-rolled in a homogeneous form, without introducing substantial internal or external strain gradients, the nucleation stage and hence the entire recrystallization process can be suppressed (Raabe and Lücke, 1992; Raabe, 1995a, 1995b). It was found that certain rolling texture components, i.e. $\{001\}<1l0>$ grains of sufficient size, although 90% cold rolled, did not reveal any texture change upon heat treatment at 730 °C for 120 s. EBSD analysis revealed that these grains did not produce recrystallization nuclei. Neighboring grains that had an initially different crystallographic orientation and had undergone exactly the same cold-working procedure were in contrast swept entirely by recrystallization (**Figure 18**).

23.3.4 Role of Grain Boundaries in Recrystallization and Grain Growth

The preceding section was dealing with the formation of new mobile high-angle grain boundaries. This section is concerned with the properties of these interfaces and the great intrinsic influence that they have on the growth of the newly formed grains (Burgers, 1941; Burke and Turnbull, 1952; Smith, 1948; Beck and Hu, 1966; Haessner, 1978; Humphreys and Hatherly, 1995, 2004; Doherty et al., 1997; Doherty, 2005).

Grain boundaries have the essential effect of acting as the main kinetic carrier that enables a growing grain to sweep the surrounding deformation microstructure. This applies specifically to mobile high-angle grain boundaries that are formed during the nucleation process. The relevance of small-angle grain boundaries during the entire recrystallization process lies more in the early stages, namely, in their relevance for the discontinuous subgrain coarsening that was discussed in the preceding section as one of the leading nucleation mechanisms. Hence, the current section concentrates on high-angle grain boundaries and their role in recrystallization and grain growth.

High-angle grain boundaries act threefold on recrystallization:

Firstly, they can move under the influence of a sufficiently high driving force (Winning et al., 2001, 2002; Winning, 2003). However, depending on their crystallographic misorientation and plane orientation their kinetic properties, viz. their mobility, can vary substantially (**Figure 31**) (Winning et al., 2001, 2002; Winning, 2003). Secondly, high-angle grain boundaries have also a relatively high self-energy that is of the order of 1 J/m^2. Therefore, the overall reduction in the total grain-boundary density acts itself as a main driving force, namely for grain growth. Thirdly, both the mobility and the energy of high-angle grain boundaries can vary not only with their geometrical characteristics but also with chemical decoration effects (**Figures 32** and **33**) (Tytko et al., 2012). These features make the migration of grain boundaries the kinetically dominating process of microstructure formation during annealing of cold-worked materials.

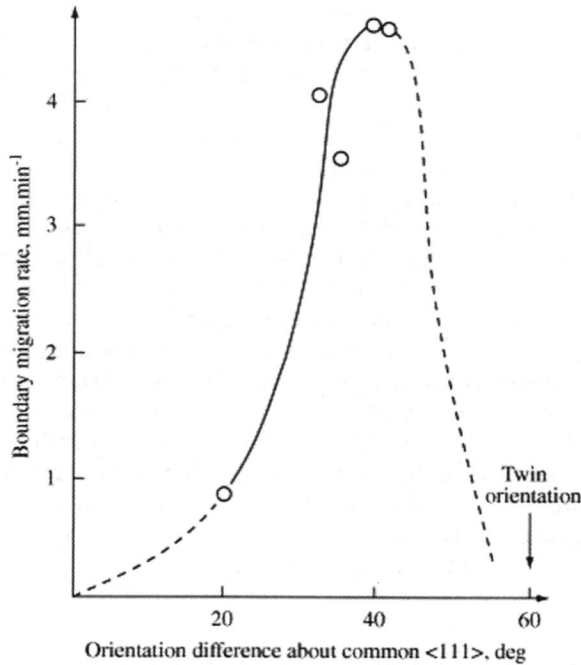

Figure 31 The migration velocity varies considerably with the misorientation angle (difference in orientation between two grains) about the <111> rotation axis, Al (Liebman et al., 1956).

For deriving an expression for the motion of high-angle grain boundaries, it is useful to start with Turnbull's classical rate equation of interface migration: according to Turnbull (Burke and Turnbull, 1952; Sandim et al., 2010) a phenomenological symmetric rate equation, which describes grain-boundary motion in terms of isotropic single-atom diffusion processes perpendicular through a homogeneous planar grain-boundary segment under the influence of free-energy gradients, can be written,

$$\dot{x} = n v_D \lambda_{gb} c \left\{ \exp\left(-\frac{\Delta G + \Delta G_t/2}{k_B T} \right) - \exp\left(-\frac{\Delta G - \Delta G_t/2}{k_B T} \right) \right\}$$

$$= n v_D \lambda_{gb} \exp\left(-\frac{\Delta H^f - \Delta S^f T}{k_B T} \right) \left\{ \exp\left(-\frac{\Delta H^m - T\Delta S^m - \frac{p\Omega}{2}}{k_B T} \right) - \exp\left(-\frac{\Delta H^m - T\Delta S^m + \frac{p\Omega}{2}}{k_B T} \right) \right\}$$

where $\dot{x}$ is the interface velocity (grain-boundary velocity in the current case), v_D the elementary Debye attack frequency, λ_{gb} the jump width through the interface (which is of the order of a Burgers vector), c the intrinsic concentration of in-plane self-diffusion carrier defects (e.g. grain-boundary vacancies or shuffle sources), n the normal of the grain-boundary segment (when written in vector notation), ΔG the activation energy of motion through in the interface, ΔG_t the Gibbs free enthalpy associated with the transformation (equivalent to the driving force for recrystallization; here written in general form), p the negative gradient in Gibbs free transformation enthalpy across the interface (driving force), Ω the

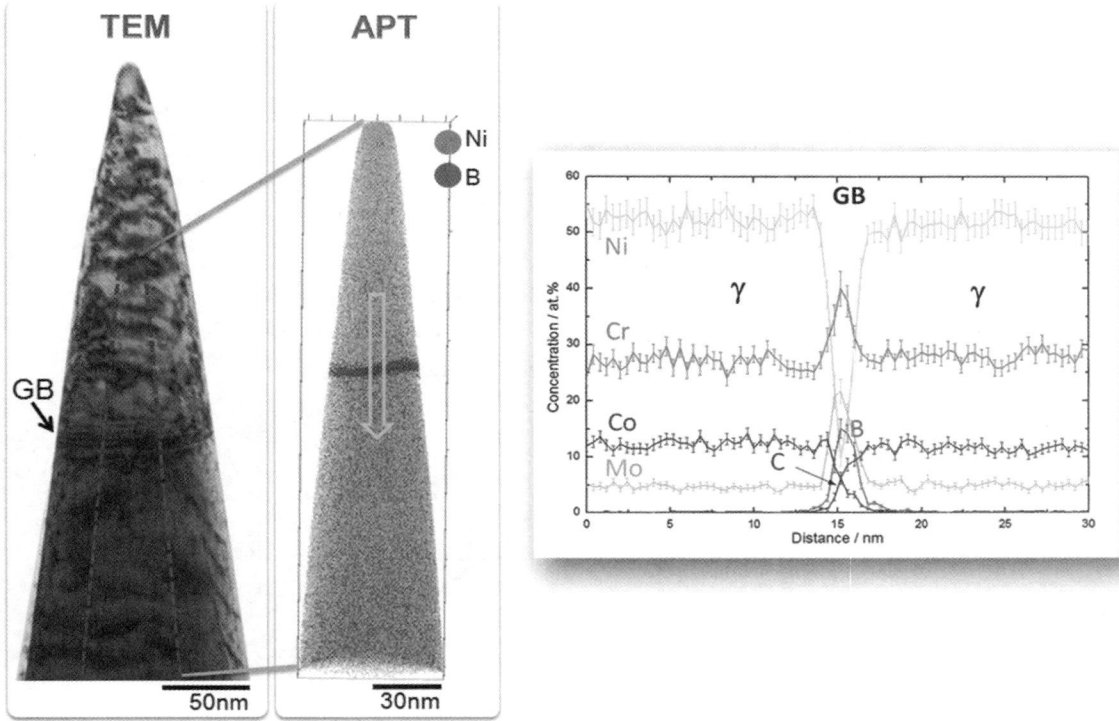

Figure 32   Atom probe characterization of B decoration of a grain boundary in a Ni-alloy (Tytko et al., 2012).

atomic volume, ΔS^f the entropy of formation, ΔH^f the enthalpy of formation, ΔS^m the entropy of motion, ΔH^m the enthalpy of motion, k_B the Boltzmann constant, and T the absolute temperature. The atomic volume is of the order of b^3, where b is the magnitude of the Burgers vector. Bold symbols indicate vector quantities. The Debye frequency is of the order of 10^{13}–10^{14}/s and the jump width of the order of the magnitude of the Burgers vector. Summarizing these terms leads to

$$\dot{x} = n\upsilon_D b \exp\left(-\frac{\Delta S^f + \Delta S^m}{k_B}\right) \sinh\left(\frac{p\Omega}{k_B T}\right) \exp\left(-\frac{\Delta H^f + \Delta H^m}{k_B T}\right)$$

$$\approx n\upsilon_D b \exp\left(-\frac{\Delta S^f + \Delta S^m}{k_B}\right)\left(\frac{p\Omega}{k_B T}\right)\exp\left(-\frac{\Delta H^f + \Delta H^m}{k_B T}\right)$$

which reproduces the well-known phenomenological Turnbull expression

$$\dot{x} = nmp = nm_0 \exp\left(-\frac{Q_{gb}}{k_B T}\right) p$$

where m is referred to as the mobility of the grain boundary and Q_{gb} the activation energy of boundary motion. The above equations provide a well-known phenomenological kinetic picture, where the

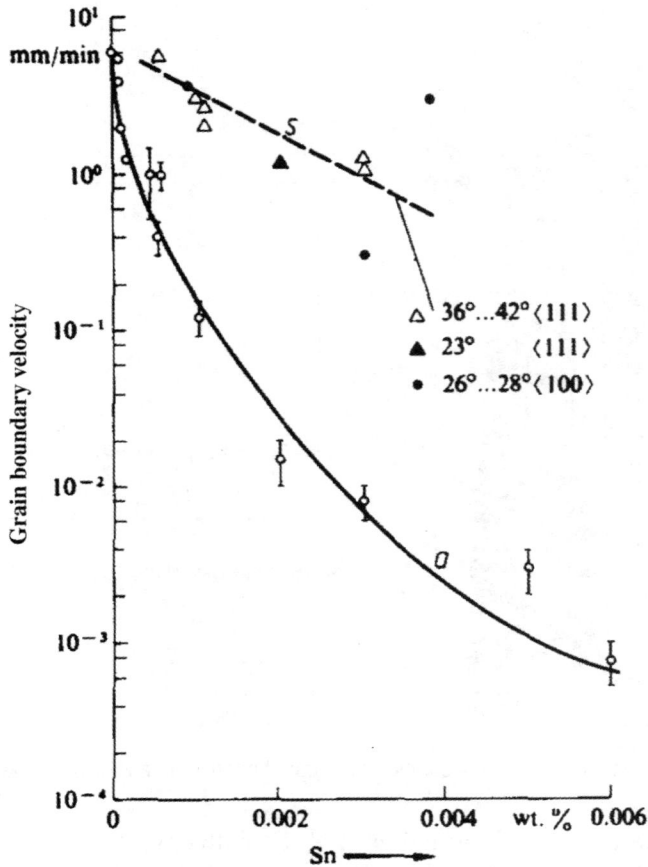

Figure 33 Velocity of special (s) and general (a) GBs in dilute Pb alloys vs. Sn content (Aust and Rutter, 1959).

atomistic processes associated with the grain-boundary motion are statistically described in terms of $m_0 = m_0(\Delta g, n)$ and $Q_{gb} = Q_{gb}(\Delta g, n)$.

When translating this kinetic picture into a feasible experiment, the challenge obviously lies firstly, in controlling the activation energy in terms of the chemical purity of the moving interface and secondly, in making sure that the driving force remains constant during an experiment. One can also learn from this analysis that retrieving intrinsic grain-boundary characteristics from the polycrystal experiment is a very difficult task.

The most systematic and precise way of measuring the grain-boundary mobility therefore is the use of a bicrystal experiment where the grain boundary between the crystals has a constant and self-reproducing curvature. This leads to a constant capillary driving pressure provided by the grain-boundary surface tension during the motion of the grain boundary. Gottstein et al. (1997) conducted multiple systematic experiments in the basis of this setup (**Figure** 34).

They studied the characteristics of grain-boundary motion by conducting successive high-temperature annealing treatments where the gradual change of the grain-boundary position, which moves according to its constant capillary driving force, is monitored as a function of time. The position-

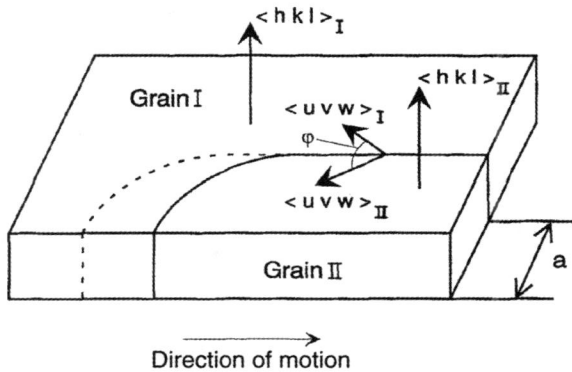

Figure 34 Setup for the measurement of grain-boundary mobility under constant driving force (Gottstein et al., 1997).

sensitive monitoring of the current interface position is conducted in such a way that under an incident X-ray beam one of the crystals is in Bragg-position while the other one is not. The reflected XRD intensity is used as the governing tracking signal. When the grain boundary has moved and the crystal that was originally in Bragg-position is swept by the other grain the reflected X-ray intensity drops. The sample is then automatically readjusted in such a way that the targeted crystal remains in Bragg-position. Hence, when the boundary moves, the specimen is shifted such that the reflected X-ray intensity remains constant during the measurement. The velocity of the moving grain boundary is then equal to the speed of sample movement. As the temperature(s), the grain-boundary velocity, and the driving pressure are known, the parameters of the mobility equation can be extracted from a number of data sets taken for different temperatures.

Several investigations have shown that the grain-boundary mobility depends on the misorientation between the abutting crystals. For instance Aust and Rutter, (1959), (1960) observed that certain low-coincidence grain boundaries move faster than noncoincidence grain boundaries. This result was later confirmed by Shvindlerman and Gottstein (Molodov, 2001; Lücke and Stüwe, 1963; Gottstein et al., 1995; Shvindlerman et al., 1995, Shvindlerman and Gottstein 1999) on Al, Zn, and Sn bicrystals. They observed the smallest enthalpy of activation for tilt boundaries between grains of exact coincidence orientation relationship (**Figure 35**).

On the other hand, numerous growth selection experiments on A1 single crystals, which were conducted by Ibe and Lücke (1966), (1972), Ibe et al. (1970), provided clear evidence that the maximum growth rate misorientation is actually close but yet distinctly different from the exact $\Sigma 7$ orientation relationship, which occurs at an angle of rotation 38.2°. The angular difference between both results, obtained by very different methods, is comparably small and has been attributed to the large scatter of results in growth selection experiments. However, as already pointed out by Ibe and Lücke (1966), (1972), Ibe et al. (1970), the overwhelming statistics of growth selection experiments substantiate that with progressing growth selection the fastest moving boundaries are observed for a <111> axis of rotation and for an angle of about 40°. Although the difference between growth selection and bicrystal experiments amounts to only 2° the difference is of substantial importance with regard to the interpretation of the misorientation dependence of the grain-boundary mobility. The finding of a high mobility for low Σ coincidence boundaries is commonly interpreted in terms of a decreased tendency to segregation and, therefore, less solute drag in long-range periodic boundary structures, like in CSL (coincidence site lattice) boundaries. This means that grain boundaries that are

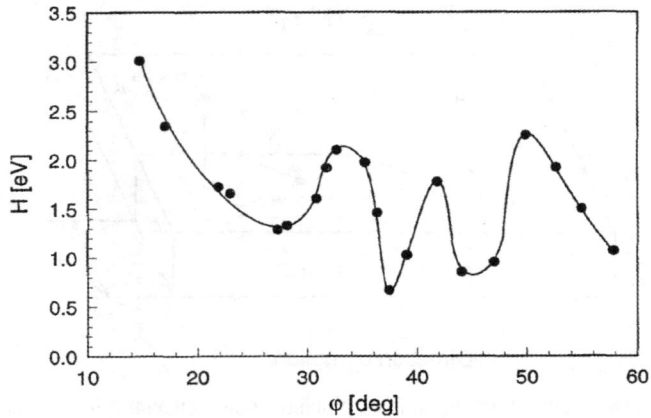

Figure 35   Misorientation dependence of activation enthalpy for grain-boundary motion of <111> tilt grain boundaries in A1 (Gottstein and Shvindlerman, 1999).

intrinsically more densely packed are less prone to be decorated by solute impurities, hence their higher boundary mobility. This however means, that for low CSL-grain boundaries the often observed high mobility is not a property of the grain boundary itself but rather of the orientation dependence of the interaction of solutes and grain boundaries. The finding of a maximum growth rate for off-coincidence (nonspecial) grain boundaries would be at variance with this interpretation.

In order to better understand these rather controversial observations Gottstein and Shvindlerman (Furtkamp et al., 1998; Winning et al., 2001; Gottstein et al., 1995, 1997, 1998) determined the grain boundary mobility of <111> tilt grain boundaries with angles of rotation between 35 and 42° using misorientation intervals of 0.4°. Interestingly, they observed that the activation enthalpy of the grain-boundary mobility reached a maximum for grain boundaries that had 40.5° misorientation. In contrast for the exact Σ7 coincidence misorientation, they observed a minimum of the activation enthalpy of the grain-boundary mobility.

Additionally, the authors observed that the mobility also depends on the preexponential factor of the mobility, which is not constant but again assumes a minimum for the exact Σ7 grain boundary and at maximum for the 40.5° interface. Therefore, they concluded that the temperature dependence of grain-boundary mobility varies with misorientation and that a temperature can be defined, namely, the so-called compensation temperature, where the mobilities of the differently oriented boundaries are the same. This result easily reconciles the seemingly contradictory results of previous bicrystal and growth selection experiments.

23.3.5 Effects of Solute Elements on Recrystallization

Solute atoms can interact twofold with grain boundaries. According to the Gibbs absorption theorem, solute atoms can reduce the self-energy of grain boundaries and hence enhance their thermodynamic stability. Such an effect can for instance be relevant in grain growth phenomena where the primary driving force stems from the grain-boundary energy. More specific, it has been suggested that different types of grain boundaries have different solubility for solute atoms (Gottstein and Shvindlerman, 1999) (**Figure 32**). This means that in grain growth, where many types of grain boundaries are involved, different types of interfaces can be stabilized differently due to the specific solute content they contain. Also, as discussed in the preceding section, solutes can even modify the intrinsic structure of the grain boundary (Gottstein and Shvindlerman, 1999). This means that at the grain boundary does not only act as a container for solute atoms without changing its own structure but it can rather interact with solutes in a way that a new joint structure can result. Such a situation leads to both, a change in mobility and change in the grain boundary energy.

A second strong effect of solute atoms exists on the mobility of grain boundaries. In this context, it has been observed that also tiny concentrations of impurities within an alloy can have extremely profound effects on the grain-boundary mobility (Stüwe, 1978; Lücke and Detert, 1957).

The equilibrium concentration of impurity atoms on a grain boundary can be greatly enhanced over that in the bulk. Hence, the effects of impurities on grain boundaries are much larger than can be expected based simply on the bulk impurity concentration.

The earliest model of the effects of impurities on grain-boundary motion was suggested by Lücke and Detert (1957). This model assumes a flat interface. The drag force, p_i, that the impurity atoms exert on that flat boundary portion can be written as

$$p_i = nC_0 f$$

where n is the number of impurity atoms per unit area of the boundary, C_0 is the equilibrium concentration (written as atomic fraction) of impurities on the boundary, and f is the force exerted on the boundary by one impurity atom. The concentration of interface impurities is in equilibrium related to the bulk concentration of the same atomic species far away from the boundary, C_∞, i.e. the mean bulk impurity concentration

$$C_0 = C_\infty \exp(- E/k_B T)$$

where $k_B T$ is the thermal energy and E the is the boundary impurity interaction energy.

In steady-state motion, the flat grain-boundary segment moves with a constant velocity v, which is related to the impurity drag force f via the Einstein relation as

$$v = \left(\frac{D}{k_B T}\right) f$$

where D is the diffusion coefficient of the impurity atoms in the abutting bulk matrix. Combining these equations above yields the expression

$$v = \left(\frac{D}{k_B T}\right) \frac{1}{nC_\infty \exp(- E/k_B T)} p.$$

Another model variant of impurity drag was put forward by Cahn (1962). In this approach, the grain boundary is also regarded as a flat interface portion much like in the Lücke–Detert approach outlined above, but in Cahn's model the impurity–boundary interaction energy, E, is a function of their separation distance, i.e. dE/dx. In the moving reference frame that is attached to the mobile grain boundary, the steady-state impurity concentration profile is derived from the diffusion equation according to

$$D\frac{dC}{dx} + \frac{DC}{k_B T}\frac{dE}{dx} + vC = vC_\infty$$

Cahn (1962) derived a formulation for the steady-state impurity concentration profile and solved it for the case of a triangular-shaped impurity–boundary interaction model. The resulting impurity drag force associated with any type of interaction profile can be calculated according to the integral formulation

$$p = -n \int\limits_{-\infty}^{+\infty} C(x) \frac{dE}{dx} dx$$

Under steady-state conditions, the driving force is equal to

$$p = p_0(v) + p(v, C_\infty)$$

where $p_0(v)$ describes the intrinsic force–velocity relation associated with grain-boundary migration in the material when it does not contain any foreign solute atoms. The resulting relationship between the solute force that acts in the grain boundary and its resulting velocity suggests the existence of two types of regimes. The first type of impurity drag regime is characterized by a monotonic, nonlinear relation. The second impurity drag regime is marked by a situation in which the grain boundary velocity is a multivalued function of the driving force. This can lead to a discontinuous transition in the grain-boundary velocity from the fully impurity-loaded slow configuration to the chemically unloaded fast configuration. This analysis results directly from the Cahn model in conjunction with the incorporation of realistic grain boundary parameters and realistic impurity profile. As outlined above, the analysis assumes that the intrinsic boundary velocity/driving force relation is linear (i.e. connected through the grain boundary mobility) and the impurity–boundary interaction potential is:

$$E(x) \begin{cases} E_0 + \dfrac{E_0}{a} x & -a \leq x \leq 0 \\[2mm] E_0 - \dfrac{E_0}{a} x & 0 \leq x \leq a \end{cases}$$

For the triangular potential employed, p depends only on the absolute value of E and not on its sign. This means that a repulsive impurity–boundary interaction leads to the same grain-boundary velocity–driving force relationship as for an attractive interaction.

Both, the Lücke and Cahn models consider the case of a dilute, ideal solution. Such an assumption is however commonly not quite appropriate since the impurity concentration on the boundary is often too high to be considered as being dilute.

These basic considerations show that impurity atoms affect grain-boundary mobility via solute drag. Usually, impurity atoms reduce the mobility of grain boundaries, but little is known how much solute drag depends on grain boundary structure, i.e. on misorientation across the boundary. To study this problem, at least for the most relevant grain boundaries in recrystallization, Gottstein and Shvindlerman (1999) investigated the orientation dependence of the grain boundary mobility in pure Al bicrystals of different chemical purity for <111> tilt boundaries within the angular interval 37–43°. The authors observed that the mobilities of both materials are different for the same type of boundary, but the difference in activation enthalpy is obviously largest for the exact Σ7 grain boundary and smallest for off-coincidence grain boundaries, in particular for the 40.5° <111> boundary. However, the different impurity content has a much deeper consequence, as it also affects the compensation temperature.

These results showed that the change between the ground state and activated state differs for even slightly differently pure material or, in other words, that the segregated impurity atoms also modify the structure of the grain boundary. This would support the frequent observation that the activation enthalpy of the grain-boundary motion can be very high, actually much higher than predicted by impurity drag theory.

23.3.6 Effect of Precipitates on Moving Grain Boundaries

One of the most important mechanisms for alloy design lies in the introduction of second phases. Specifically, many metallic alloys contain a high dispersion of small and second-phase precipitates that can occur in coherent, semicoherent for incoherent form. Such particles can exert a substantial back driving force on the moving grain boundaries (**Figure 36**) (Sandim et al., 2010). These interactions are also referred to as particle-pinning forces or Zener forces (Zener and Smith, 1948).

Grain-boundary-pinning forces arise when second-phase particles occur on the grain boundary. Their presence reduces the grain-boundary area and, hence, the grain-boundary energy. This energy saving, which must be replenished upon unpinning, is referred to as Zener pinning (Zener and Smith, 1948; Rios, 1987; Pimenta et al., 1986; Ashby et al., 1969; Köster, 1974; Martin and Doherty, 1976; Raabe and Hantcherli, 2005; Weygand et al., 1999; Holm et al., 2001; Miodownik et al., 2000). In the following treatment, we consider pinning effects imposed by a stable array of incoherent particles that reside on the grain boundaries of the deformed microstructure. In his first estimate, Zener and Smith, (1948) approximated the magnitude of the pinning force by assuming randomly distributed spherical particles. The boundary was assumed to move as a straight interface through the particle array and to experience a resistive force, F, from each particle. With the grain-boundary energy γ (in units of J/m^2), the force F due to one particle is given by $F = p\,r\,c$, where r is the particle radius. The surface A on which the force is applied amounts to $A = 2pr2/(3f)$, where f is the volume fraction of spherical particles. The Zener pressure then amounts to

$$P_Z = -\frac{3}{2}\,\gamma\frac{f}{r}$$

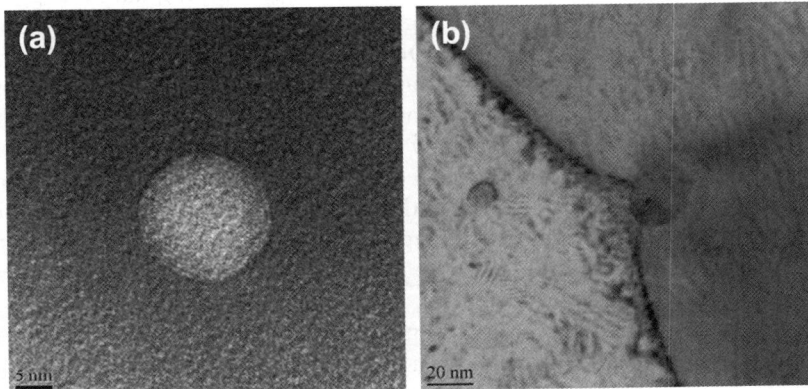

Figure 36 High-resolution transmission electron micrographs of a stainless steels sample deformed to 80% cold reduction and annealed at 800 °C for 48 h showing: (a) individual YCrO3 particle in the ferritic matrix; (b) dragging effect (particle–boundary interaction) during annealing. The RD is parallel to the scale bar (Sandim et al., 2010).

A shortcoming of this approach is that a grain boundary cannot be considered as a rigid interface, but it may have some flexible to bow out between particles when a driving force is applied (Rios, 1987; Pimenta et al., 1986; Ashby et al., 1969; Köster, 1974).

Diverse modifications have been proposed to correct for this flexibility (Rios, 1987; Pimenta et al., 1986; Ashby et al., 1969; Köster, 1974; Martin and Doherty, 1976). They give results that are of the same order of magnitude as the original formulation of Zener. One approach introduces a correction factor that depends on the volume fraction of the particles f:

$$P_{HZ} = -\Phi(f)\frac{3}{4}\gamma\frac{f}{r}$$

When assuming a Friedel-like particle behavior, one obtains the following equation for the drag force,

$$P_{FZ} = -2.6\gamma\frac{f^{0.92}}{r}.$$

When considering a stronger dependence of the corrector factor on the value of f than assumed in the original Friedel model, one obtains a modified expression for the Zener pressure according to

$$P_{FZ} = -0.33\,\gamma\frac{f^{0.87}}{r}; \quad f < 3 \text{ vol.\%}$$

For a reasonable choice of the grain-boundary energy (0.6 J/m^2), the precipitate volume fraction (1 vol.%), and the average particle radius (100 nm) the original Zener pinning force amounts to about 0.1 MPa. When considering the corrections discussed, the pinning force can rise to a maximum value of about 0.5 MPa.

23.3.7 Recrystallization In situ

Dislocation recovery proceeds particularly fast in alloys with high stacking-fault energy (Himmel, 1962; Cahn 1966). In alloys with low stacking-fault energy dislocation recovery is less pronounced (Burgers, 1941; Humphreys and Hatherly, 1995; Himmel, 1962). The reason for the connection of stacking-fault energy and recovery tendency of an alloy lies in the dependence of the cross-slip probability of screw dislocations on the recombination spacing of the underlying Shockley partial dislocations that open the stacking fault between them. In other words, a low stacking-fault energy leads to a wide-stacking fault and hence to a wide spacing among the associated Shockley partial dislocations. For cross-slip the partial dislocations must first recombine under the aid of phonons plus the applied local stress. Therefore, systems with higher stacking-fault energy have smaller stacking faults so that the partial dislocations involved can recombine and cross-slip at lower thermal activation (Mughrabi et al., 1986; Straub et al., 1996; Nes, 1995; Stüwe et al., 2002).

Frequent cross-slip leads to a more rapid reorganization of the stored dislocation arrangements that are formed during cold working and hence to a more rapid reduction in the stored internal energy.

In cases where extensive and rapid recovery leads to such a high reduction in the stored internal energy before the associated recrystallization nucleation leads to the formation of new mobile high-angle grain boundaries, primary recrystallization can be entirely suppressed. Sometimes, this effect occurs only in specific grains so that a mixed microstructure results, containing both recrystallized and recovered crystals (**Figure 18**). In such cases, the microstructure inherited from cold working is—at least in part (**Figure 16**)—transferred into a reorganized, softer microstructure without the motion of high-angle grain boundaries. Such a situation is referred to as recrystallization in situ. Typical materials where very strong recovery and hence recrystallization in situ can occur, competing with primary recrystallization, are pure aluminum and iron as well related alloys with high stacking-fault energy (Bailey, 1963; Sandström et al., 1978; Vandermeer and Rath, 1990; Nes, 1995; Stüwe et al., 2002).

Recrystallization in situ can be exploited also for technological applications: recrystallization phenomena that involve the motion of newly formed high-angle grain boundaries naturally transform the crystallographic texture that is inherited from cold working into a texture that is characterized by specific recrystallization texture components.

When recrystallization in situ takes place, however, the deformation texture is inherited. Thus, in order to design, optimize, and balance a certain annealing texture with specific anisotropy, the corresponding heat treatment can be conducted in a way that some areas of the cold-worked materials undergo primary recrystallization, discontinuously providing specific new texture components, while other texture components of the same microstructure undergo extensive recovery and even recrystallization in situ.

The reasons why in the same microstructure both, primary recrystallization and extensive recovery phenomena (recrystallization in situ) can occur, can be twofold. First, in many commercial products macroscopic gradients of the deformation energy occur within the same sample (**Figure 37**). Examples

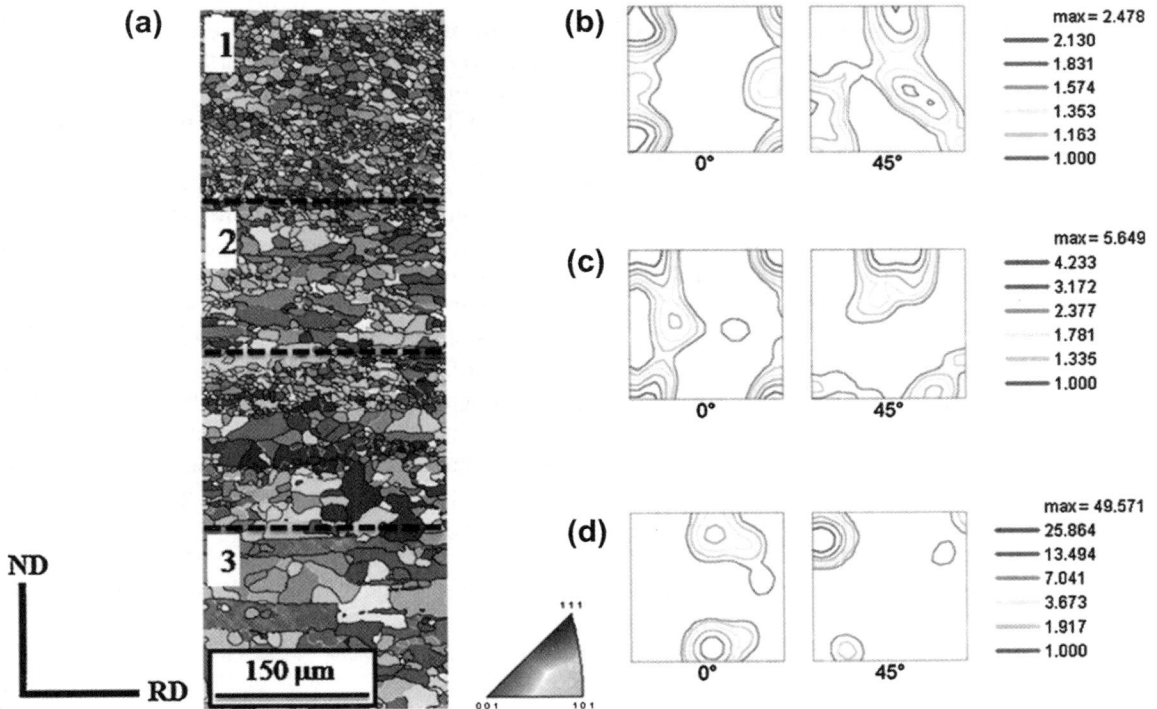

Figure 37 EBSD results showing inhomogeneous recrystallization behavior in a ferritic stainless steel after recrystallization annealing: (a) orientation map; (b–d) ODF (ϕ2-constant sections) corresponding to the specific recrystallized regions marked by 1, 2, and 3, respectively (Siqueira et al., 2011).

are macroscopic-rolling processes where typically higher plastic deformation rates occur close to the surface and subsurface areas of the sheet as compared to the center layers of the same product. A further difference is the applied deformation state: while the center layer during rolling is usually deformed by plane strain state, near-surface areas are subjected to shear deformation. Both types of deformation can lead to different texture components, different local orientation gradients and differences in the accumulated total deformation (Raabe and Becker, 2000; Bate, 1999; Roters et al., 2010). Also, microscopic differences in plastic deformation can occur. Examples are the deformation zones close to large particles and close to internal interfaces.

The second important reason for the inhomogeneity of recrystallization and recovery phenomena in the same sample lies in the dependence of these mechanisms on the host orientation in which they take place (**Figures 16** and **18**). For instance **Figure 16** clearly reveals that with progressing annealing treatment only specific grains in a body-centered cubic low-carbon steel sample are sluggish to undergo primary recrystallization. The orientation distribution function that is shown in **Figure 16**b summarizes only those orientations that did not recrystallize. For longer annealing times, it becomes apparent that only the 45°-rotated cube component, {001}<110>, prevails as a nonrecrystallized texture component (Raabe and Lücke, 1992).

Also, it must be considered that some crystallographic texture components reveal much higher plastic deformation and higher distortions than other texture components (Raabe et al., 2002a; Raabe and Becker, 2000; Bate, 1999; Roters et al., 2010). This must be explained in more detail: as characterized in terms of classical polycrystal homogenization theory (Taylor–Bishop–Hill theory) the Taylor factor—as a well-known measure of grain scale plasticity—provides information for each orientation how much shear is required (per unit deformation step) to render grain deformation compatible. In the classical Taylor theory, compatible deformation of all crystals is obtained when all grains are subjected to the same externally imposed strain state. In order to comply with this global boundary condition, however, each grain requires a certain amount of shear, distributed on the active slip systems, depending on its specific crystallographic orientation (Raabe, 1998; Raabe et al., 2004b). While some grains can follow the externally prescribed deformation state by activating a relatively small amount of internal shears, other grains require larger internal amounts of shear. Grains of the former type have a small Taylor factor (small amount of shear per unit strain) and grains of the latter type a high Taylor factor (high amount of shear per unit strain). Sometimes, the Taylor factor was therefore chosen as a criterion for high or low nucleation rates during primary recrystallization, respectively. In reality, this situation is more complex. Firstly, the deformation state of a grain is not only a function of the grain orientation but also of its grain neighborhood (Raabe et al., 2001; Kuo et al., 2003; Zaefferer et al., 2003). The interaction of the externally imposed load together with the influence of the neighboring grains adds up to a micro-mechanical boundary condition that determines the deformation state of a grain. Secondly, the total deformation that goes into a grain does not include any information whether the grain-scale deformation is heterogeneous or homogeneous. For relating the deformation state of a grain to its tendency to undergo either recrystallization or respectively recrystallization in situ, however, both the total deformation and the deformation inhomogeneity matter (**Figure 38**) (Raabe et al., 2002b; Roters et al., 2010).

23.3.8 Strain-Induced Grain-Boundary Migration

The mechanism of strain-induced grain-boundary migration (SIBM) is characterized by the bulging of a portion of an already existing high-angle grain boundary that sweeps the neighboring deformation microstructure and leaves a dislocation-free region behind the migrating boundary (Humphreys, 1997, 1992a; Li et al., 1953; Winning et al., 2001, 2002; Winning, 2003; Heinrich and Haider, 1996).

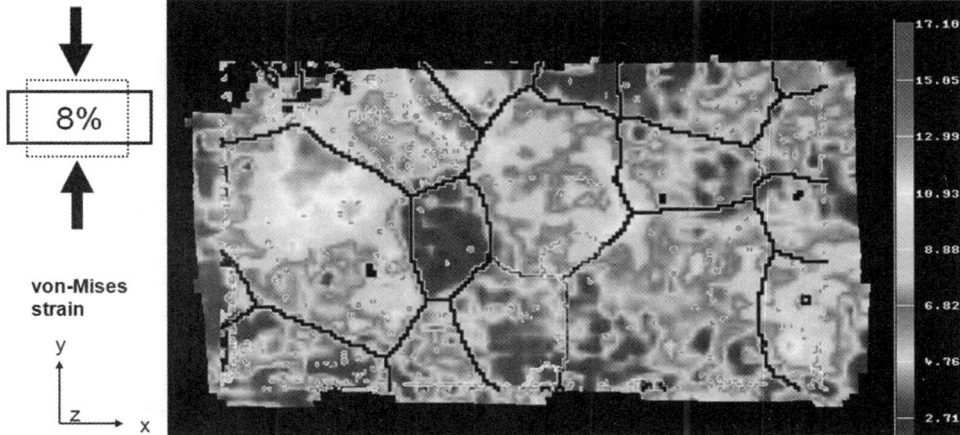

Figure 38 Experimental example of the heterogeneity of plastic deformation at the grain and subgrain scale using an aluminum polycrystal with large columnar grains. The image shows the distribution of the accumulated von Mises equivalent strain in a specimen after 8% plane strain thickness reduction (the deformation is given in % of $\Delta d = d$, where d is the sample extension along compression direction). The experiment was conducted in a lubricated channel–die setup. The strains were determined using digital image correlation. The high-angle grain boundaries indicated by black lines were taken from EBSD microtexture measurements. The equivalent strains differ across some of the grain boundaries by a factor of 4–5, giving evidence of the enormous orientation-dependent heterogeneity of plasticity even in pure metals (Raabe et al., 2002b; Roters et al., 2010).

It differs from primary static recrystallization as it does not involve the formation of new high-angle grain boundaries (nucleation stage in conventional primary recrystallization) but instead is based on the motion of an already existing internal interface (Humphreys, 1997). SIBM can nonetheless efficiently reduce the internal stored deformation energy.

The driving force for this bulging mechanism stems from the difference in stored dislocation (or DC) energy on opposite sides of the same grain boundary (**Figure 39**). Such situations occur between neighboring texture components of different crystallographic orientation in cases where they reveal

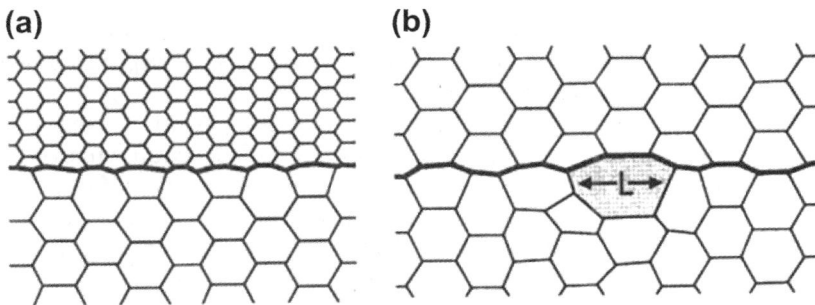

Figure 39 Models for recrystallization by SIBM after Humphreys (Humphreys, 1992a). (a) The conventional model in which SIBM is driven by differences in stored energy, (b) The possibility of SIBM occurring by discontinuous growth of a large subgrain adjacent to a high-angle boundary.

(a)

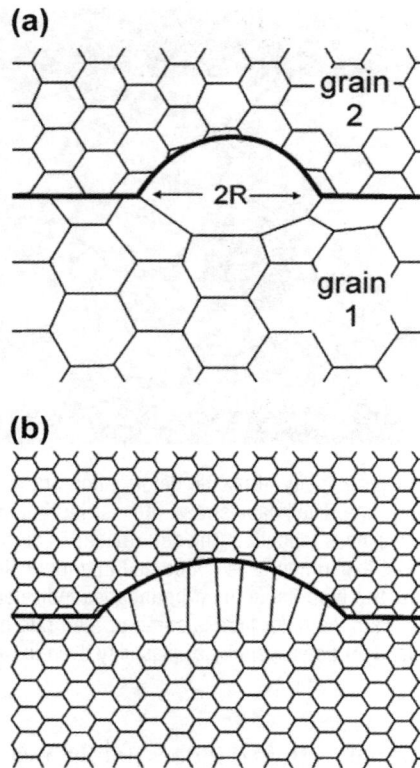

(b)

Figure 40   Schematic diagrams of: (a) single subgrain SIBM and (b) multiple subgrain SIBM. After Humphreys (Humphreys, 1992a).

a characteristic difference in the substructure evolution (**Figures 40** and **41**) (Humphreys, 1992a). For instance, in body-centered cubic steels that are deformed by cold rolling, the {001}<110> texture component reveals relatively weak strain hardening while the {111}<112> and {111}<110> texture components are characterized by higher stored dislocation densities and, correspondingly, by smaller DC sizes (**Figures 14** and **42**) (Hölscher et al., 1991; Ushioda et al., 1987; Raabe and Lücke, 1993; Hutchinson and Ryde, 1995; Samajdar and Doherty, 1994).

Consequently, when an existing high-angle grain boundary bulges into a neighboring grain with a lower stored energy and forms a new grain a pronounced global texture change does usually not occur. This means that the recrystallization texture is closely related to the deformation texture.

Strain-induced boundary migration is a frequently occurring recrystallization mechanism at relatively low strains. At deformations below 20–50% rolling reduction, the differences in stored energy among the different texture components are relatively strong (Hurley and Humphreys, 2003) (**Figure 43**). At higher strains, the internal deformation substructure and the gradual deformation texture evolution leads to a more homogeneous distribution of the stored internal deformation energy and the occurrence of conventional discontinuous nucleation phenomena via discontinuous subgrain coarsening as described above.

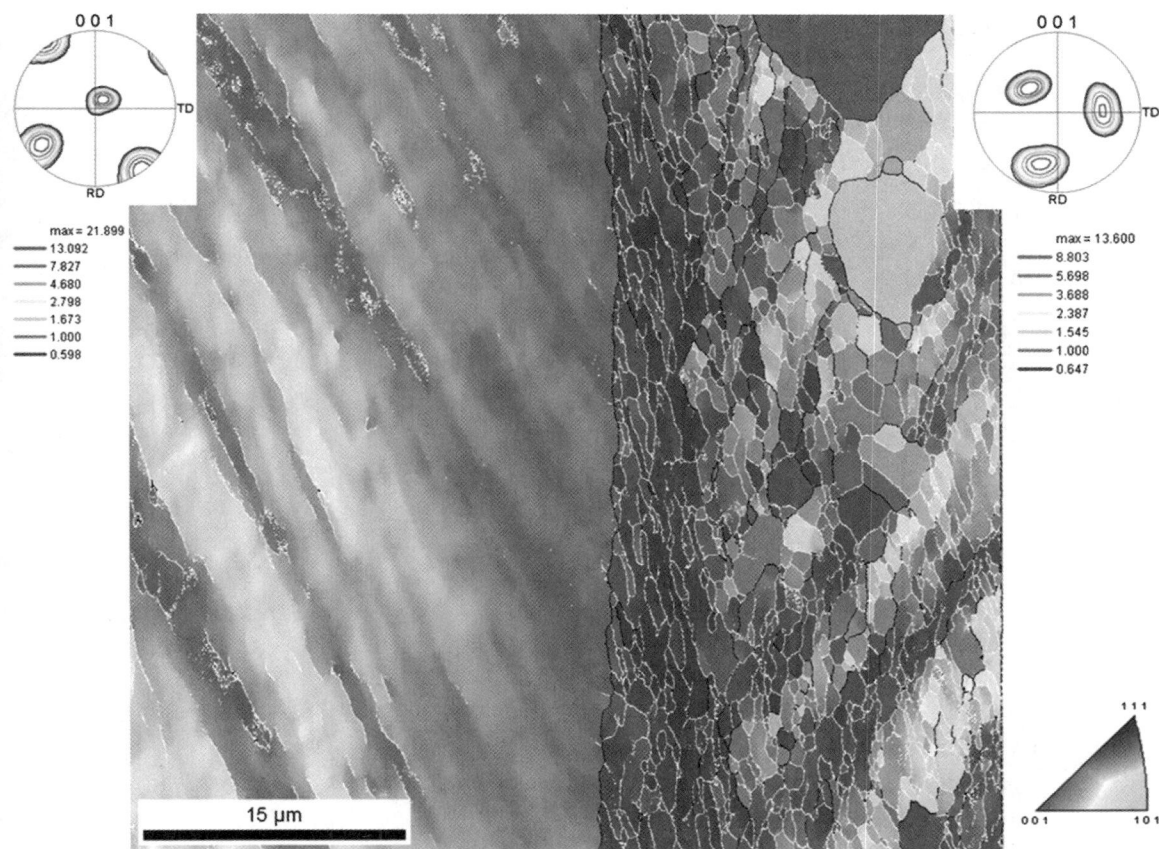

Figure 41   Recovery and recrystallization phenomena in neighboring grains in a 70% cold rolled and annealed (750 °C, 1 min) microalloyed low-C steel.

23.3.9   Recrystallization Textures

Recrystallization processes can lead to characteristic types of orientation changes, viz. to different types of crystallographic textures. One group of phenomena, such as recovery or recrystallization in situ, that are primarily characterized by the continuous rearrangement and relaxation of the internal stored dislocation substructure without involving the motion of high-angle grain boundaries, is naturally not associated with profound transformations of the initial deformation textures. This means that for instance cold-rolling textures remain essentially unchanged when only recovery or recrystallization in situ take place (**Figures 15, 16 and 18**).

This is different for primary recrystallization. As outlined above, the incipient stage of recrystallization is described by the occurrence of different types of nucleation phenomena. As nucleation is described as the formation of new mobile high-angle grain boundaries corresponding recrystallization changes are the consequence, particularly when assuming that the nuclei are randomly distributed and that all grain boundaries have the same velocity when sweeping the deformed microstructure. Typically,

{001}<110>component, BCC steel

{111}<110> component, BCC steel

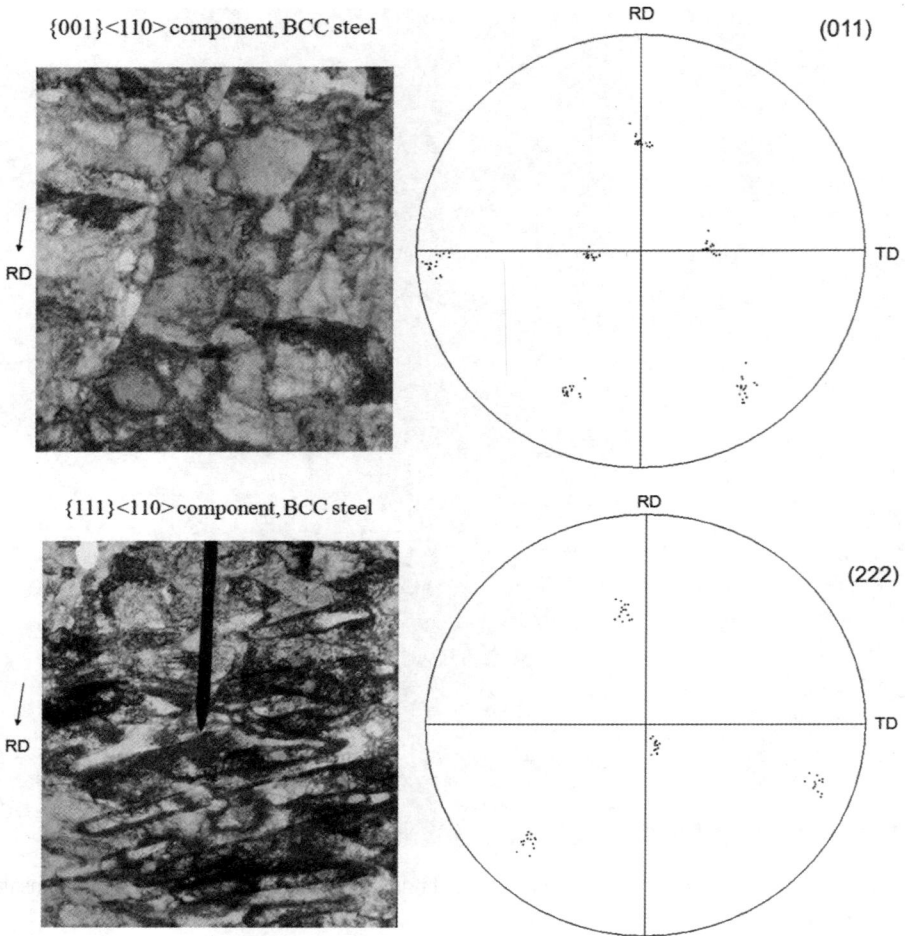

Figure 42   Example of dislocation substructure in 70% cold-rolled low-carbon steel as observed by TEM. The upper image shows DCs with a rather diffuse arrangement in a typical rolling-texture component {001}<110>. The bottom picture shows more sharper and elevated DCs in a texture component {111}<110> (Thomas et al., 2003).

this simplified picture is not correct but instead, usually both, the specific nucleation mechanism and also the growth selection that takes place through the growth competition among the moving grain boundaries together determine the transformation of the deformation texture into the final annealing texture. Corresponding modeling approaches that aim at capturing the multiple possible interactions that arise from these different nucleation phenomena and the different mobilities of the interfaces involved will be discussed in more detail in the next section.

New orientation components that are characteristic for certain nucleation phenomena are in most cases already existent within the deformation microstructure, however, only to a very small volume fraction. For instance, if a hard second phase particle is surrounded by local orientation gradients in the deformed matrix, it is likely that nuclei can form there since these are regions where a high

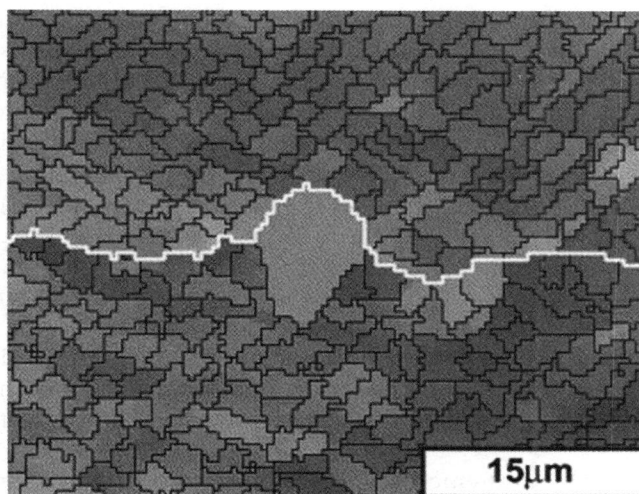

Figure 43 EBSD map (relative Euler contrast), showing an example of single subgrain SIBM during the early stages of annealing in an Al sample rolled to 50%. Bulging of the high angle grain boundary (white line) surrounding the large subgrain in the lower grain may be seen. Subgrain boundaries >0.5° are shown (Hurley and Humphreys, 2003).

misorientation can quickly accumulate to form high-angle grain boundaries. The original volume fraction from which such a nucleus emerges can be so small that it is hardly detectable in a statistical orientation distribution, such as for instance measured by using XRD, of the as-deformed material. In a case where such PSN-determined nuclei growing out of these curved lattice zones represent the dominant nucleation mechanism and where all grain boundaries have the same velocity (as a thought experiment), the final recrystallization texture will be very different from the original deformation texture **(Figures 23, 28 and 29)** (Doherty and Cahn, 1972; Humphreys and Ferry, 1997; Faivre and Doherty, 1979; Bhatia and Cahn, 1978).

A similar situation occurs for recrystallization nucleation at shear bands (Sebald and Gottstein, 2002; Engler et al., 1996; Engler, 1997). In this phenomenon, the nuclei are typically formed at the interface between the shear band interior and the surrounding matrix. This is the zone where the highest stored energy and also the highest lattice curvatures occur. This implies that rapid accumulation of high misorientations through discontinuous subgrain coarsening is possible during the nucleation stage. Hence, such nuclei are often formed rapidly and sweep the surrounding deformation matrix efficiently. The original volume from which they emerge and whose orientation they carry is—like for the PSN case described above—usually so small that it is also hardly visible in the original deformation orientation distribution. Yet, these originally tiny orientation zones are already present as a specific feature of the deformation microstructure and they can prevail in the final recrystallization texture. Typical sites where such mechanisms are assumed to prevail are deformation bands, shear bands, transition bands, PSN, and deformation twins (Berger et al., 1988). In all these cases, the inherited underlying substructures correspond to the most strongly deformed ones with often characteristic orientation deviations from the surrounding matrix orientation. Therefore, these sites have the potential to rapidly create highly mobile boundaries with respect to neighboring grains. This concept is referred to as "oriented nucleation" concept of the formation of recrystallization textures.

Besides these examples that show how nucleation phenomena can affect and initiate orientation changes also mobility differences among different types of grain boundaries can lead to essential and characteristic texture changes during recrystallization. Such mechanisms are referred to as "oriented growth". This concept assumes that nuclei of all orientations are present at the beginning of recrystallization but that the recrystallization texture depends on which orientations grow fastest into the deformed matrix. That the growth rate depends on misorientation has been early established for a long time by single crystal experiments by Liebman et al. (1956) and Aust and Rutter, (1959), (1960) (**Figure 31**). As was outlined above, depending on the purity of the material, certain types of grain boundaries, for instance the near $40°<111>$ misoriented grain boundary in face-centered cubic metals and alloys, can have a much higher mobility relative to other grain boundaries (Gottstein et al., 1995, 1998; Beck et al., 1950; Molodov et al., in press). This mobility advantage can also lead to specific orientation changes as for a randomly distributed spectrum of nucleation orientations only a few one will grow very fast at the expense of the deformation texture components, namely those that have the fastest orientation relationship to the main deformation texture orientations.

These two examples, describing extreme cases of a texture transformation during recrystallization, are referred to as "oriented nucleation model" and "growth selection model", respectively (Ibe et al., 1970; Ibe and Lücke, 1972; Bunge, 1969, 1982, 1986, 1987; Wassermann and Grewen, 1962; Bunge and Esling, 1982, 1991). They underline that one essential feature of primary recrystallization is the possibility of a complete change of the crystallographic texture from the original deformation texture resulting from heavy cold working to a completely modified recrystallization texture that can either consist of texture components that are due to a characteristic nucleation mechanism or due to the mobility advantage of a certain class of grain boundaries.

For this reason, the analysis of crystallographic textures has always been an important statistical (XRD) and local (EBSD, TEM) analysis vehicle for studying the underlying basic mechanisms of such recrystallization phenomena.

In this context, it is a striking observation that the formation of completely random orientation distributions through cold working and subsequent heat treatment involving recovery and recrystallization is not the rule but a rare exception in metal processing. In other words, the aim of producing polycrystalline alloys with random textures is usually difficult to achieve. Often only the smart mix of different acting mechanisms involving both, recovery and recrystallization or random nucleation phenomena, can help to obtain random textures.

In order to achieve minimum elastic and plastic anisotropy for sheet-metal forming applications such as those encountered in aerospace, food packaging, and automotive manufacturing, the textures of aluminum alloys and steels are often designed to assume minimum plastic anisotropy after recrystallization. **Figure 44** shows an example where an aluminum sheet with a high cube orientation has been produced by rolling and recrystallization. Such a preferred crystallographic recrystallization texture leads to undesired plastic anisotropy of the entire part. In such cases, a random texture is highly desired in the case of aluminum but difficult to produce. What is usually produced in such a case is a successful processing leading to a balanced recrystallization texture with a mixture of the two opposite anisotropies, namely retained and recovered rolling-texture components mixed with recrystallized cube-texture components in aluminum alloys (**Figure 45**) (Zhao et al., 2001, 2004).

In steels with a body-centered cubic structure, typically $\{111\}<uvw>$-dominated texture components are desired for sheet forming applications. Two main recrystallization components, namely, the $\{111\}<110>$ and the $\{111\}<112>$ orientations typically result from recrystallization annealing of heavily cold-rolled body-centered cubic steels (Bunge, 1986; Duggan et al., 1978; Ushioda et al., 1987;

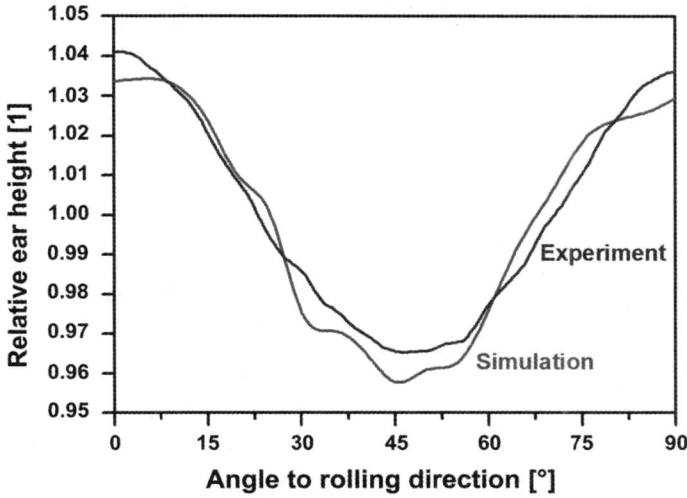

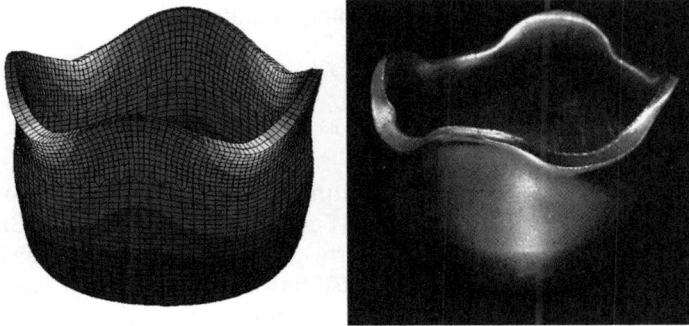

Figure 44 Rolled and recrystallized aluminum sheet with very strong cube texture and the resulting undesired plastic anisotropy. On the bottom left-hand side the corresponding crystal plasticity finite element simulation is shown. It considers the original crystallographic texture of the sheet before cup drawing (Zhao et al., 2004).

Raabe and Lücke, 1992, 1993; Randle, 1992; Klinkenberg et al., 1992; Raabe, 1995a; Juntunen et al., 2001; Ray et al., 1975b; Hutchinson, 1999).

As an example, **Figure 46** shows a fiber representation of the texture transition of cold-rolled tantalum (body-centered cubic structure) polycrystals during heat treatment. Annealing of the samples that were before cold rolled to 90 and 95%, respectively led to a typical recrystallization texture of body-centered cubic metals, namely, to a preferred {111}<uvw> texture (Raabe et al., 1994).

23.4   Driving Forces of Recrystallization and Grain-Growth Phenomena

From a thermodynamic standpoint, most recrystallization and grain-growth phenomena can be formally characterized as nonequilibrium transformations. In either type of process, a driving force acts on a grain-boundary segment. The free-enthalpy change of the system is then associated with the release of stored energy per volume that has been swept by the moving grain boundary (Rosenhain, 1914;

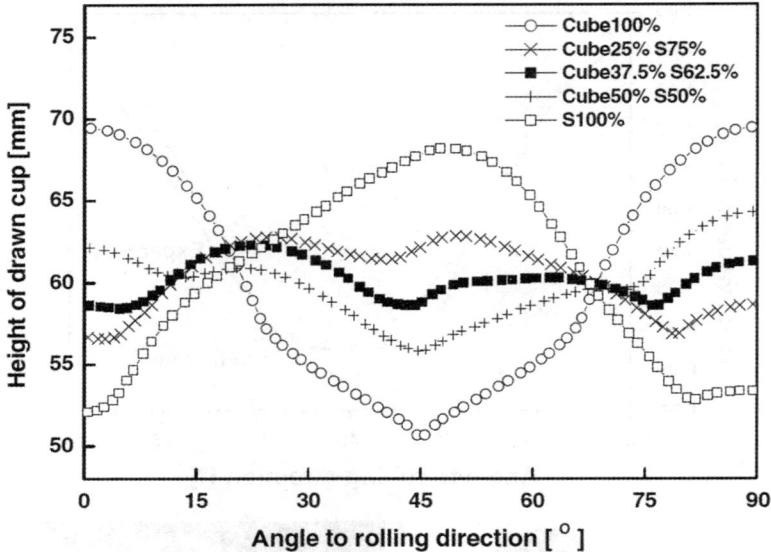

Figure 45 Optimization study by crystal plasticity finite element simulations regarding appropriate mixing of the cube orientation and the S orientation in aluminum alloys. The analysis reveals that appropriate mixing of the two texture components provide a strong reduction of the corresponding earing and thinning behavior (Zhao et al., 2004).

Ewing and Rosenhain, 1899, 1900a, 1900b; Alterthum, 1922; Carpenter and Elam, 1920; Czochralski, 1927; Burgers and Louwerse, 1931; Burgers, 1941; Burke and Turnbull, 1952; Smith, 1948; Beck and Hu, 1966; Haessner, 1978; Humphreys and Hatherly, 1995; Doherty et al., 1997).

In principle, two types of driving forces can be distinguished. The first one is a stored volume energy that acts equally in every portion of the affected grain that is being swept by a moving grain boundary. In such cases, the stored energy can often be simplified and written in scalar form. A most prominent (though simplified) example is the difference in the stored dislocation density across a moving interface. It should be emphasized though that for some experiments, i.e. in the case of differences in elastic or magnetic energy, the tensorial nature of these mechanisms must be taken into consideration. The second class of driving forces is of a configurational, i.e. a topological nature. This means that the driving force and thus also the release in the stored system energy depends on the exact local arrangement of the defects that are removed or rearranged by a moving grain boundary. A typical example is continuous grain growth, i.e. competitive grain coarsening, where the driving force depends on the local curvature and energy of the grain boundary but not on the size of the entire grain. In other words, in continuous grain growth, the local grain boundary portion that moves toward its center of curvature in order to reduce its total length does not "know" the size of the grain that it encompasses. This means that in this case only the local capillary driving force matters. It is however not a constant force that acts equally in each volume portion of the same crystal but it differs everywhere in the crystal and polycrystal depending on the local grain boundary configuration in terms of the curvature and grain-boundary intersection lines and points. During the grain-boundary motion when sweeping a volume dV, the change in the free-enthalpy dG can be written

$$dG = -p\,dV.$$

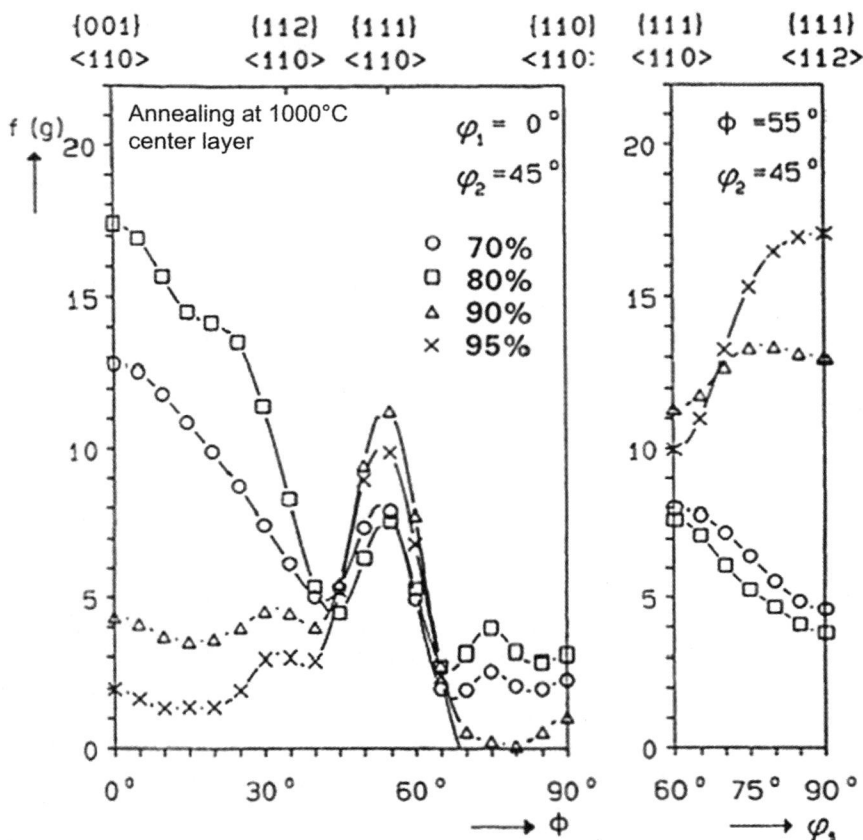

Figure 46 Fiber representation of the texture transition of cold rolled tantalum polycrystals during heat treatment. Annealing of the samples that were before cold rolled to 90 and 95%, respectively, lead to a typical recrystallization texture of BCC metals, namely, to a preferred {111} texture (Raabe et al., 1994).

The symbol p is referred to as driving force. For the case of primary static recrystallization the driving force stems from the stored dislocations and can hence be written

$$p = \rho\, E_{disl} = \frac{1}{2}\rho G b^2$$

where E_{disl} is the elastic energy of a dislocation per line segment, ρ is the dislocation density, G the shear modulus, and b the magnitude of the Burgers vector. Using typical values for the Burgers vector, the shear modulus and the dislocation density yields a maximum value for the driving force of primary static recrystallization of up to 10 MPa. When aiming at the prediction of recrystallization kinetics, it is recommended to use temperature-corrected values for the shear modulus and for the Burgers vector. Particularly, the magnitude of the shear modulus can be substantially reduced at high temperatures where recrystallization typically takes place.

The driving force for continuous capillary driven grain growth can be written as a Laplace pressure that acts on a curved portion of grain boundary. It reads

$$p = \frac{8\pi r\gamma \, dR}{4\pi r^2 \, dR} = \frac{2\gamma}{r}$$

where it must be emphasized that r is the local curvature of the affected portion of grain boundary and not the grain size. γ is the grain-boundary energy and dR the local growth increment (Raabe, 2000; Holm and Battaile, 2001; Rollett and Raabe, 2001; Humphreys, 1992b; Maurice and Humphries, 1998; Maurice, 2001; Weygand et al., 2001; Kinderlehrer et al., 2004, 2001; Anderson et al., 1984; Srolovitz et al., 1984; Glazier et al., 1990; Anderson and Rollett, 1990; Tavernier and Szpunar, 1991a, 1991b; Srolovitz et al., 1985, 1986b, 1988; Rollett et al., 1989b, 1989c, 1992; Peczak, 1995; Doherty et al., 1990; Miodownik et al., 1999).

For discontinuous grain coarsening where huge grains sweep much smaller grains the driving force reads

$$p = \frac{3d^2\gamma}{d^2} = \frac{3\gamma}{d}$$

where d is the grain size of the smaller grains that are being swept by much larger grains. This equation shows, since the average grain size d during discontinuous grain coarsening is much smaller than the local grain curvature r in the equation above describing continuous grain coarsening, discontinuous grain growth progresses usually much faster than continuous grain growth (Raabe, 2000; Holm and Battaile, 2001; Rollett and Raabe, 2001; Humphreys, 1992b; Maurice and Humphries, 1998; Maurice, 2001; Weygand et al., 2001; Kinderlehrer et al., 2004, 2001; Anderson et al., 1984; Srolovitz et al., 1984; Glazier et al., 1990; Anderson and Rollett, 1990; Tavernier and Szpunar, 1991a, 1991b; Srolovitz et al., 1985, 1986b, 1988; Rollett et al., 1989b, 1989c, 1992; Peczak, 1995; Doherty et al., 1990; Miodownik et al., 1999).

Using typical values yields a driving force of 0.01–0.05 MPa for discontinuous grain coarsening. A phenomenon that is sometimes referred to as tertiary recrystallization describes the grain-coarsening phenomena where the driving force stems from the difference in surface energy among neighboring crystals:

$$p = \frac{2\Delta\Omega}{h}$$

where $\Delta\Omega$ is the difference in the surface energy among two neighboring grains and h is the radius of curvature of the grain boundary that extends between the two parallel surfaces. Tertiary recrystallization is often observed in thin films. Typical driving forces for tertiary recrystallization are of a similar magnitude as those for grain coarsening, namely, of the order of 0.01–0.05 MPa.

Much higher driving forces can be obtained when recrystallization takes place together with a phase transformation. Such situations are referred to as discontinuous precipitation phenomena. They are characterized by the fact that the driving force for this process is not only coming from the stored deformation energy but additionally from a thermodynamic transformation energy contribution. A well-known example is the joint recrystallization and transformation of a plastically deformed and over saturated solution of Al–Ag that forms AlAg2 precipitates in an otherwise pure-aluminum matrix behind the moving grain boundary. The driving force for such a discontinuous precipitation can be written

$$p = R_g(T_1 - T_0)c_0 \ln c_0$$

where c_0 is the concentration of the over saturated solid solution, corresponding to the maximal solubility at T_0. The temperature T_1 is the chosen heat-treatment temperature that lies below T_0. Discontinuous precipitation phenomena can proceed extremely fast owing to the very high driving forces involved, namely, of the order of 500 MPa.

23.5 Dynamic and Metadynamic Recrystallization

Controlling the microstructure during hot working and heat treatment is of high relevance in the thermomechanical processing of metallic alloys. When crystalline metallic materials are deformed at temperatures above about half of the absolute melting point, the accumulated dislocations can continuously be removed by two different types of mechanisms, namely, by dynamic recovery and by dynamic recrystallization (Haessner, 1978; Humphreys and Hatherly, 1995, 2004; Doherty et al., 1997; Doherty, 2005).

The first type of process, termed dynamic recovery, is a gradually acting mechanism. It occurs in practically all materials in a continuous fashion and leads to the gradual mutual annihilation of dislocations by climbing and cross-slip and to the thermally activated formation of subgrains and subboundaries.

In materials with high stacking-fault energy, such as aluminum, dynamic recovery can continuously remove and balance the strain hardening imposed during the hot working process. Such a pronounced and continuous dynamic recovery process, therefore, leads to an overall steady-state plastic flow when the material is formed above half of its melting temperature.

In contrast to high stacking-fault energy materials, alloys with a moderate to low stacking-fault energy build up considerably higher dislocation densities during hot working when compared to metals that have high stacking-fault energy. Eventually in low stacking-fault energy materials, the accumulated dislocation density becomes sufficiently high to stimulate the nucleation of recrystallization during hot deformation (**Figure 47**) (McQueen et al., 1976; Fritzmeier et al., 1979; Luton and Sellars, 1969; McQueen, 1968; Jonas et al., 1969).

This mechanism is referred to as dynamic recrystallization. It consists in the formation of mobile high-angle boundaries that at least partially surround the nuclei and by the subsequent sweeping of the deformation substructure that was stored during hot working.

The study of dynamic recrystallization phenomena aims at two main directions. Firstly, it is of interest to understand the basic characteristics of high temperature flow curves and secondly, it is of high relevance for the optimization of metallurgical processing to understand the evolution of new grain structures under conditions of dynamic recrystallization. In the latter context specifically, the efficiency of dynamic recrystallization as a mechanism to refine and homogenize the grain size is of high interest.

In a given metallic alloy, the essential characteristics of dynamic recrystallization are determined by three main parameters: these are the initial grain size prior to plastic hot working; the temperature; and the strain rate (**Figure 48**) (Rossard and Blain, 1960; Rossard, 1963; Sakai et al., 1983; Sakai and Jonas, 1984). The initial grain size mainly determines the critical strain of dynamic recrystallization, the peak strain, and the kinetics of the process.

The finer the starting grain size is, the lower are the values for the critical and the peak strains. This effect is due to the fact that dislocations accumulate more rapidly in metallic alloys that have a smaller grain size. A similar effect occurs in those materials that have lower stacking-fault energy. The lower the stacking fault energy, the higher is the strain hardening rate and hence the stored deformation energy.

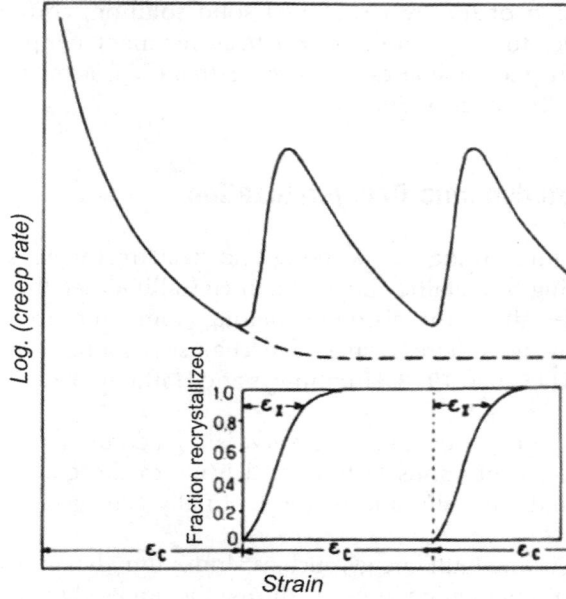

Figure 47 Schematical diagram of dynamic recrystallization (Luton and Sellars, 1969). The term ε_c refers to the critical strain.

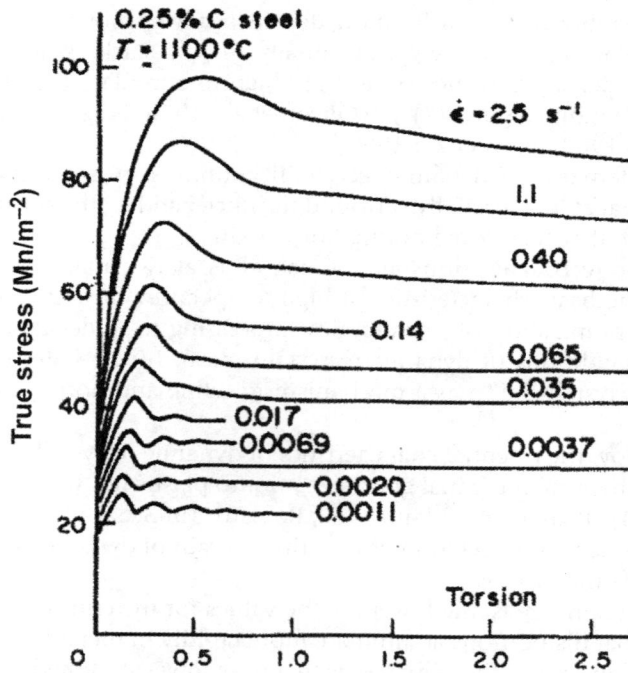

Figure 48 Influence of strain rate on the flow curves derived from hot torsion data at 1100 °C for a C-steel (Sakai and Jonas, 1984).

Peak stresses are also observed to be dependent on the initial grain size, however, the steady-state stress and the final grain size are independent of the initial grain size (McQueen and Jonas, 1975; Sellars, 1978; Sakai et al., 1983; Sakai and Jonas, 1984; Roberts, 1982; Ryan and McQueen, 1990; McQueen et al., 1990; Tsuji et al., 1997).

An important feature is that the flow curves of face-centered cubic metals can show characteristic undulations in the flow stress. This type of phenomenon is referred to as multiple-peak flow behavior (**Figure 49**). It is typical of dynamic recrystallization processes where subsequent waves of incomplete nucleation and growth processes sweep the deformation substructure during ongoing straining (McQueen et al., 1976; Fritzmeier et al., 1979; Luton and Sellars, 1969; McQueen, 1968; Jonas et al., 1969; McQueen and Jonas, 1975; Sellars, 1978).

The analysis and investigation of dynamic recovery processes dates back to the first observation of this phenomenon during the deformation of Pb in 1939 (McQueen et al., 1976; Fritzmeier et al., 1979; Luton and Sellars, 1969; McQueen, 1968; Jonas et al., 1969; McQueen and Jonas, 1975; Sellars, 1978). During the 20 years that followed, most of the observations were made on metals under creep

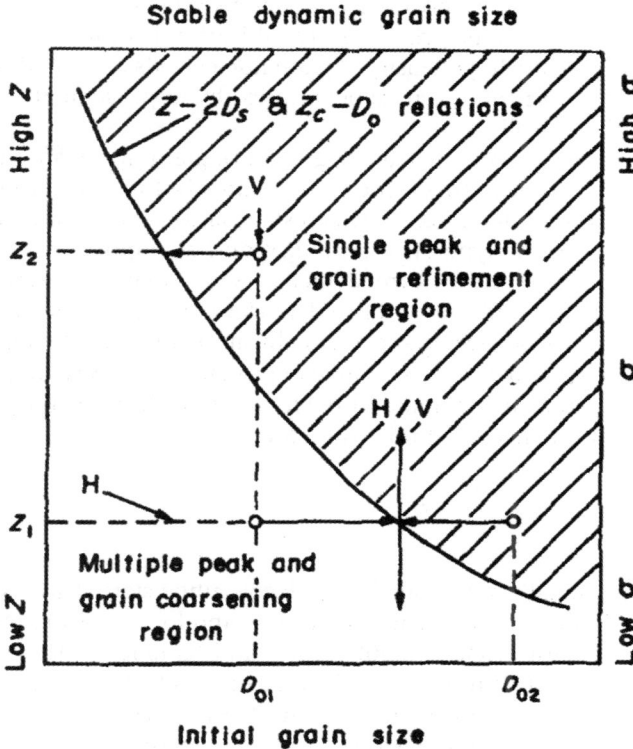

Figure 49 A microstructural mechanism map for distinguishing between the occurrence of two types of dynamic recrystallization. The central curve separates the single (grain refinement) from the multiple peak (grain-coarsening) region. Three distinct types of experiment are represented, namely, the so-called "vertical tests carried out over a range of strain rates and temperatures on material with a fixed initial grain size do: "horizontal" tests (H) carried out at a fixed temperature-compensated strain rate Zl with a series of initial grain sizes; and combined "horizontal/vertical" (H/V) tests involving changes in strain rate or temperature after a period of steady state deformation (Sakai and Jonas, 1984).

conditions. The first detailed investigation of metallic behavior under constant strain rate conditions was published by Rossard and Blain (1960), Rossard (1963).

The classical approach to dynamic recrystallization was primarily concerned with the aim to understand the transition from cyclic (i.e. multiple peak) to single-peak recrystallization (McQueen et al., 1976). As can be seen from **Figure 48**, the flow curves pass through this transition as the strain rate is increased, or the temperature decreased.

In some cases, the transition from multiple-peak to single-peak dynamic recrystallization behavior was characterized in terms of certain critical strain criterion.

It was observed though that this criterion is not valid for the high-temperature deformation regime of face-centered cubic metals in tension and compression.

An alternative criterion for the transition was suggested based on the grain size considerations. The latter indicate that *cyclic* flow curves are associated with grain *coarsening* and that *single peak* flow curves are associated with grain *refinement*.

It was observed that often a single peak behavior was associated with the formation of a necklace or cascade-type multilayer of sets of newly grown grains around the center of a nonrecrystallized larger inner part of a host grain. The growth process and hence also the grain size in this case appears to be deformation limited. By contrast, static primary recrystallization is nearly completely synchronized in fine-grained materials, because the high density of grain nuclei leads to a small spread in the nucleation strain. The grain size under these conditions is determined by impingement, and thus it is nucleation controlled and not growth controlled.

Metadynamic recrystallization is defined as the growth of grain nuclei after the plastic deformation stage with the specific property that these nuclei were already formed during the preceding plastic deformation. It is important to differentiate between static and metadynamic recrystallization, as they proceed at different rates. Also the type of recrystallization has important consequences on the final microstructure and properties of the deformed metals. Once dynamic recrystallization is initiated during deformation, the dynamically recrystallized nuclei continue to grow, even after straining is terminated. The main difference between static recrystallization and metadynamic recrystallization is in the nucleation mechanism of the new grains. Unlike in primary static recrystallization, metadynamic re recrystallization does not require an incubation time for grain nucleation, as dynamically recrystallized nuclei already occur within the deformed microstructure.

23.6   Grain Growth

Practicality all engineering metallic alloys are in a polycrystalline state. This means that each crystal is defined in terms of its crystallographic orientation, its size and local topological environment, and the properties of its grain boundaries (Raabe, 2000; Holm and Battaile, 2001; Rollett and Raabe, 2001; Humphreys, 1992b; Maurice and Humphries, 1998; Maurice, 2001; Weygand et al., 2001; Kinderlehrer et al., 2004, 2001; Anderson et al., 1984; Srolovitz et al., 1984, 1985, 1986b, 1988; Glazier et al., 1990; Anderson and Rollett, 1990; Tavernier and Szpunar, 1991a, 1991b; Rollett et al., 1989b, 1989c, 1992; Peczak, 1995; Doherty et al., 1990; Miodownik et al., 1999).

All such polycrystalline aggregates are subject to capillary-driven competitive grain-coarsening phenomena, provided that the temperature is high enough to overcome the activation barriers. In the metallurgical terminology, these phenomena are often referred to as grain growth.

It is described as a process by which the mean grain size of an aggregate of crystals increases. The driving force for this results from the decrease in free energy that accompanies reduction in total grain-boundary area. As the crystals become gradually larger, the curvature of the boundaries becomes smaller. This results in a tendency for larger grains to grow at the expense of smaller grains.

It should be emphasized that in many engineering applications grain growth is not desirable, as many mechanically beneficial properties of structural metallic alloys such as strength, strain-hardening, and ductility are improved inversely to the average grain size.

Given a sufficiently high temperature and no factors that impede grain-boundary migration such as second-phase particles or impurities on the grain boundaries, polycrystals will gradually evolve toward a single crystal. In real microstructures this goal is rarely achieved.

In two dimensions, it is in principle possible, provided that all grain boundaries have exactly the same grain-boundary energy and are connected with each other with 120° angles, that the microstructure is free of any grain-boundary curvature so that no capillary-driven grain-coarsening occurs.

It must be emphasized though that such a microstructure is only mechanically stable owing to the local mechanical equilibrium of the grain boundaries when abutting under 120°. In two dimensions such intersections are referred to as stable triple junctions.

From a thermodynamic viewpoint, the driving force of grain coarsening is the overall reduction in the total grain-boundary area per volume. Hence, even a microstructure that is under local mechanical equilibrium is not necessarily also in thermodynamic equilibrium.

In most cases, however, even in quasi 2D situations as sometimes encountered in thin-film grain structures, microstructures are not free of local curvatures of the grain boundaries involved. In three dimensions, curvature-free and at the same time compatible grain topologies are not possible in full mechanical equilibrium (Raabe, 2000; Holm and Battaile, 2001; Rollett and Raabe, 2001; Humphreys, 1992b; Maurice and Humphries, 1998; Maurice, 2001; Weygand et al., 2001; Kinderlehrer et al., 2004, 2001; Anderson et al., 1984; Srolovitz et al., 1984, 1985, 1986b, 1988; Glazier et al., 1990; Anderson and Rollett, 1990; Tavernier and Szpunar, 1991a, 1991b; Rollett et al., 1989b, 1989c, 1992; Peczak, 1995; Doherty et al., 1990; Miodownik et al., 1999). This means, that in three dimensions, all polycrystalline microstructures must necessarily contain some local grain-boundary curvature.

A classical kinetic approach to describe such a gradual coarsening phenomenon was suggested by Burke and Turnbull.

In this derivation, the velocity of a portion of grain boundary is related to the pressure gradient across it as described above in terms of the phenomenological Turnbull rate expression for grain-boundary motion as was derived above

$$\dot{x} = nmp = nm_0 \exp\left(-\frac{Q_{gb}}{k_B T}\right) p$$

where m is the mobility, n the normal vector of the interface, p the driving force, and Q_{gb} the activation energy of grain-boundary motion. Rewriting the growth rate $\dot{x}$ in terms of a scaling relation for the time-evolution of the average grain size D according to

$$\dot{x} = n\frac{dD}{dt} = nmp$$

The driving force p can be written

$$p = K\frac{\gamma}{D}$$

where K is a scaling constant, γ the grain-boundary energy, and D the average grain size. Integrating the resulting expression

$$n\frac{dD}{dt} = nK\frac{\gamma}{D}$$

with respect to time, we obtain that the average area of the grain evolves with time as

$$\langle D \rangle^2 - D_0^2 = K\gamma t$$

This shows that grains grow at a rate that is essentially proportional to the square root of time when considering that the initial grain size D_0 is much larger than the current average grain size $\langle D \rangle$.

It has to be emphasized at this point that in this derivation we have only considered the reduction in grain-boundary energy for the driving pressure. Possible additional sources that might enter into the total energy reduction are that the grain boundary might be separating grains consisting of different phases, or one grain may have a higher dislocation density than another. If neither of these effects apply, the pressure can arise simply as a product of the curvature of the grain boundary. This pressure acts toward the center of curvature.

It is interesting to note in that context that it is actually in a real microstructure not the average grain size that acts as a driving force for grain coarsening but it is the *local* curvature of the interface that matters. In other words, any local grain-boundary curvature aims at reducing the total grain-boundary area per volume by moving toward the center of the curvature radius. This elementary process of grain coarsening is explicitly guided by the local curvature and the grain does not "know" if actual grain size.

The actual connection between grain size and local curvature comes simply through the principle that for a large grains have either very low curvatures, and hence, slow growth rates of their grain boundaries, or the curvature of their grain-boundary facets point outward, toward the surrounding smaller grains.

As mentioned above, a neutral, curvature-free microstructure, irrespective of the actual grain size, occurs when all grains in a 2D grain structure have the same hexagonal grain shape and mechanically stable intersecting *triple junctions of* 120°. A triple junction is the point at which three grain boundaries meet. This local mechanical equilibrium among three intersecting grain boundaries in two dimensions does only apply when all interfaces have the same grain-boundary energy.

Mullins, (1956) pointed out the random grain structures are, however, inherently unstable. In 2D, grain boundaries that are associated with microstructures that are characterized by certain size distribution actually have to "curve" so that they can intersect at triple junctions under preservation of the local mechanical equilibrium. In two dimensions, this means that grains with less than six sides have centers of curvature lying inside their boundaries. Conversely, grains with more than six sides have centers of curvature lying outside their boundaries. This gives rise to the driving pleasure derived above, being proportional to both the grain-boundary energy and the local grain-boundary curvature. The resulting velocity of boundary migration is always directed toward the centers of the interface curvature.

For 2D grain structures, this observation was cast into the so-called N-6 rule. This rule states that an N-sided grain evolves under 2D grain growth according to

$$\frac{dD(N)}{dt} = m\gamma\frac{\pi}{3}(N-6)$$

where $D(N)$ is the 2D grain size as a function of its number of grain-boundary facets N and γ is the grain-boundary energy. The latter parameter is here assumed to be identical for all interfaces. This so-called Mullins-von Neumann rule (Mullins, 1956; von Neumann, 1952) shows that actually the number of grain-boundary sides determines the crystal growth and not the grain size itself. Topologically, it should be noted, though, that larger grains are very unlikely to contain a small number of boundary facets, and so the two formulations are approximately equivalent.

Although in three dimensions this problem is more complicated, the main kinetic and energetic characteristics of grain growth remain in principle unchanged, namely: (i) grain boundaries migrate toward their centers of their curvature; (ii) small grains have usually larger curvature values associated with their grain-boundary facets and hence reveal higher grain-boundary velocities so that these become smaller at the expense of larger grains; and (iii) the average grain size increases with time.

However, one main difference in the grain growth exists between two dimensions and three dimensions. In three dimensions, no grain topology exists (as in 2D), which satisfies both, zero curvature grain-boundary facets and local mechanical equilibrium at the grain-boundary interaction lines and points. The application of 3D surface evolver simulations has recently shown that in average grain shrinkage occurs for grain shapes that have a smaller number of facet areas than 14. Grains that have a larger number of grain facets will in average become larger during grain growth. The minimal-area tessellating grain shape in three dimensions is the 14-sided tetrakaidecahedra, a shape whose surface is made up of six hexagons and eight squares (**Figure 50**) (Krill and Chen, 2002).

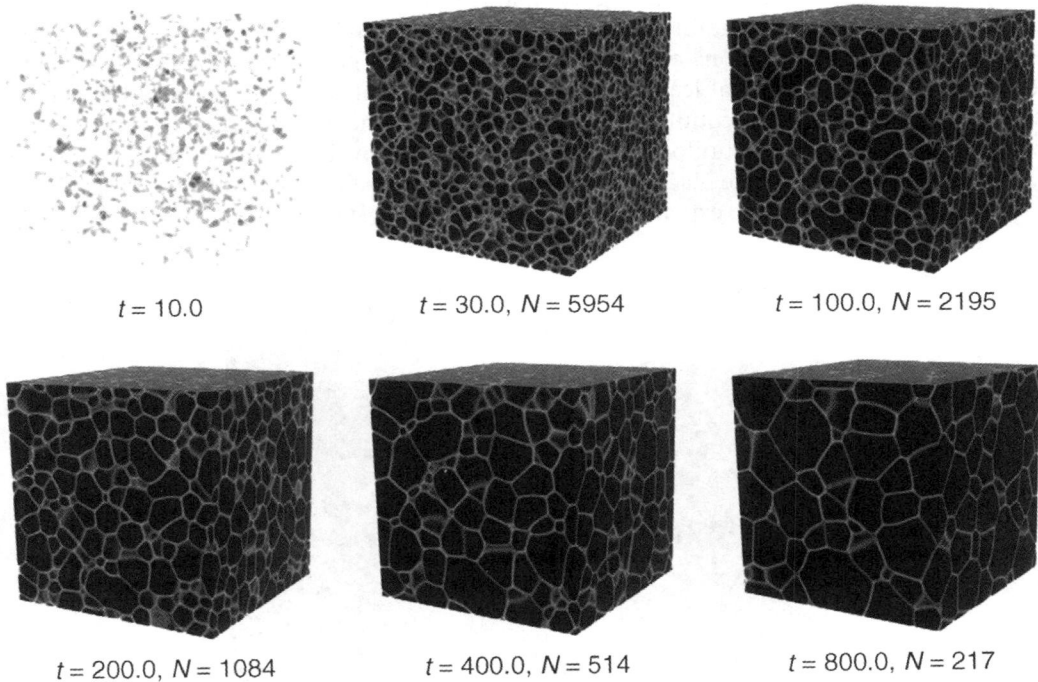

$t = 10.0$ $t = 30.0, N = 5954$ $t = 100.0, N = 2195$

$t = 200.0, N = 1084$ $t = 400.0, N = 514$ $t = 800.0, N = 217$

Figure 50 Phase-field simulation of microstructural evolution performed on a $180 \cdot 180 \cdot 180$ simple-cubic grid, visualized by mapping the interfaces. The elapsed time t and the number of grains N are specified under each image. The microstructure at $t = 10.0$ illustrates the homogeneous nucleation of crystallites from the supercooled liquid initial state (Krill and Chen, 2002).

23.7   Secondary Recrystallization: Discontinuous Grain Coarsening

In some cases during grain coarsening, a sudden and every rapid growth of a small number of grains can occur. The final dimensions of such rapidly growing grains can be of the order of multiple centimeters (**Figure 51**) (Dorner et al., 2006).

Since this process has a phenomenological similarity to a nucleation and growth mechanism involving only a few crystals, it is sometimes also referred to as secondary recrystallization. A more specific terminology refers to it as a discontinuous grain coarsening or to abnormal grain growth. The latter two terms are considered to be more adequate as no sweeping of cold-worked microstructures is involved during such discontinuous grain growth. The other crystals undergo continued grain growth.

A number of general features can be identified that characterize discontinuous grain coarsening:

During the nucleation stage of secondary recrystallization, only a small number of grains start to grow to very large dimensions. These few large grains are, however, not freshly nucleated in any way but they are particular crystals that already existed in the preceding microstructure. It is observed that the early stages of growth of these abnormally large grains are slow. This indicates that at the beginning of secondary recrystallization a certain incubation period occurs.

When those grains that show an extraordinary high growth rate exceed the average grain-size diameter by a factor of 4–5, they will prevail in the resulting microstructure and show further discontinuous growth even if the surrounding continuously coarsening grains are not impeded by impurities or particles. This means that a grain diameter advantage of a factor 4–5 is topologically sufficient to promote abnormal grain growth.

The crystallographic orientations of the abnormally growing grains typically deviate substantially from those that surround them, at least during the early stages of secondary recrystallization.

A common feature to all discontinuous grain coarsening phenomena, at least in their incipient stages, is the fact that normal uniform grain growth is selectively inhibited. This means that for secondary recrystallization to take place normal grain growth must be very slow so that large secondary grains can effectively grow at the expense of the surrounding crystals.

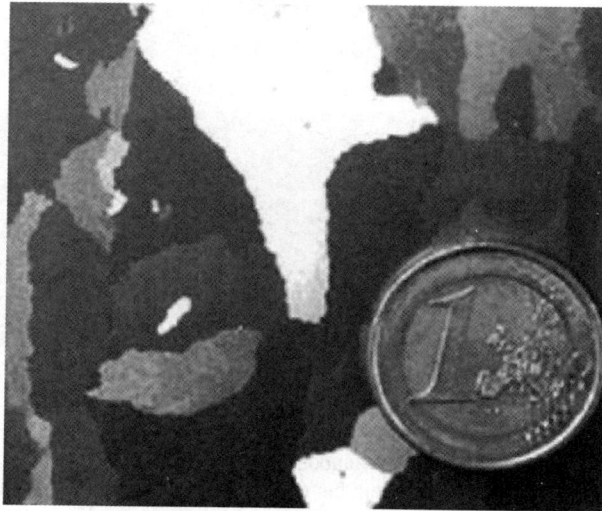

Figure 51   Typical size of Goss grain after abnormal grain growth (Dorner et al., 2006).

Such inhibition typically occurs by the presence of second dispersed phase particles, by impurities that decorate certain grain boundaries, by specific energetic, structural or kinetic grain-boundary characteristics, by an inherited topological advantage of certain grains, or by crystallographic texture effects. In some materials also the inhomogeneity of dissolving precipitates on grain boundaries seems to play a role for promoting secondary recrystallization. Microstructures resulting from discontinuous grain growth are often characterized by strong crystallographic textures.

Typically textures resulting from secondary recrystallization are different from those obtained after primary recrystallization and normal grain growth. Also it is observed that a well-defined minimum annealing temperature must be exceeded for initiating discontinuous grain coarsening. The largest grains are normally produced just above this temperature; at higher annealing temperatures smaller secondary grains result.

As outlined above, the driving forces for discontinuous grain coarsening result from the reduction in the total grain-boundary area per volume. Different to the driving force of continuous grain growth, which results from the local grain-boundary curvature, the driving force of secondary recrystallization, once the huge grains have started to grow abnormally, arises from reducing the grain-boundary energy in units of the small grains that surround the huge grains.

Owing to the fact that a number of interacting phenomena, such as described above, can give rise to abnormal grain growth, computer simulations under different intern boundary conditions can help to clarify which mechanism is decisive to produce a certain discontinuous grain growth phenomenon (Humphreys and Hatherly, 2004).

From such simulations, it was concluded that an inhomogeneous distribution of the grain-boundary energy alone is usually not sufficient to explain abnormal grain growth. In most cases, it was observed that a second mechanism has to assist promoting abnormal grain growth, such as certain grain boundaries with a much higher mobility compared to others (Anderson et al., 1984; Srolovitz et al., 1984, 1986b, 1988; Glazier et al., 1990; Anderson and Rollett, 1990; Tavernier and Szpunar, 1991a, 1991b; Rollett et al., 1989b).

Such a situation occurs naturally when discontinuous grain coarsening occurs in a strongly textured polycrystal, where only a few grains assume a different grain orientation relative to their grain neighborhood. Such a microstructure implies that most of the normally growing grains have small-angle grain boundaries between them while the abnormally growing ones might have high-angle grain boundaries around them. This means that normally growing grains, that undergo continuous grain coarsening, are surrounded by interfaces that have both low energy and low mobility. If a few grains with a high misorientation (hence surrounded by high-angle grain boundaries) are embedded in such a grain structure, they should have a growth advantage in terms of the higher grain-boundary energy and higher grain-boundary mobility.

Another effect that has been discussed as relevant in secondary recrystallization is the fact that the Zener force is proportional to the grain-boundary energy. Since the grain-boundary energy is a function of five crystallographic parameters, namely, the three misorientation angles and the interface plane normal inclination, the Zener force can act selectively, depending on the grain boundaries involved—more specific that grains that are surrounded by interfaces with a low grain-boundary energy are exposed to a small particle drag force. By way of contrast, grains with a high grain-boundary energy will experience a higher Zener drag force (Holm et al., 2001; Miodownik et al., 2000). Similar effects apply to impurity drag, where the solubility and the interaction force with the solutes is different for different types of grain boundaries. Hence, certain grain boundaries might have an advantage over others for instance in a situation where their solute content is below that of others.

Commercially highly important examples where abnormal grain growth plays an essential role for a product are the formation of secondary recrystallized grains with a strong Goss texture (110)[001] (i.e. the (110) plane is oriented parallel to the sheet surface and the [001] direction is oriented parallel to the RD) in silicon–iron electrical steels and the formation of cube oriented textures in nickel alloys that are used as substrates for growing ceramic superconductors.

One of these examples, namely, the Goss texture in soft magnetic FeSi alloys, is discussed in a bit more detail in the following:

The strong Goss texture, (110)(001), that develops in commercially produced Fe-3% Si soft magnetic steels is due to a secondary recrystallization process (Goss, 1935; Ruder, 1935; Burwell, 1940). The preferred abnormal growth of Goss-oriented grains is characterized by the presence of an inhibitor phase in the material. As inhibitor phase one refers to a second solid phase that is finely dispersed in the primary matrix and interacts with the grain boundaries. The nature of the interaction between these inhibitor particles and the FeSi grain boundaries is not exactly understood. It is assumed that they interact either through conventional Zener drag or via Zener drag plus solid solution when the particles start to dissolve in the intercritical-annealing regime. As intercritical annealing, we refer here to a heat treatment where partial phase transformation occurs (Goss, 1935; Ruder, 1935; Burwell, 1940).

In that context, it must be considered that the magnitude of the back driving forces associated with both Zener drag and impurity drag depends on the crystallographic character, energy, and misorientation of the grain boundary. In FeSi electrical steels, these small precipitates are usually nitrides and sulfides. It was observed that at an early stage (the "nucleation stage" of abnormal grain growth) of the secondary recrystallization process some fraction of the inhibitor phase starts to dissolve so that locally the back driving forces cease to pin the grain boundaries. This stage is characterized by a sudden size increase of very few Goss-oriented grains which—once they reach a size advantage of several grain diameters relative to their neighbor crystals—rapidly consume the entire polycrystalline matrix (**Figure 52**).

In an approach to understand this complex phenomenon different models were considered (Hu et al., 2008). One line of argumentation was concentrated on a possible initial size advantage of Goss grains in the microstructure after primary recrystallization. The rationale behind this assumption was that the largest driving force for grain growth occurs for the biggest grains in the matrix. It was hence discussed

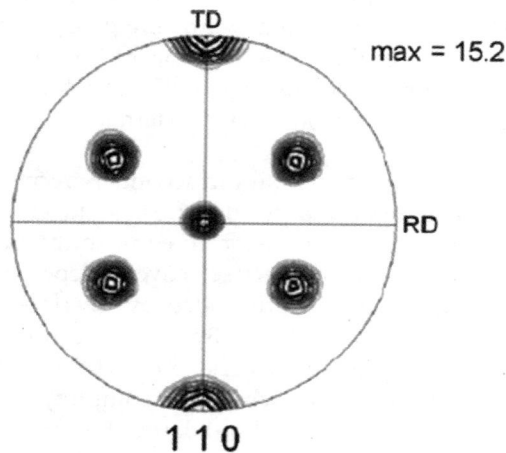

Figure 52   Goss texture of Fe-3 wt.% Si steel after abnormal grain growth (Dorner et al., 2006).

whether a faster growth of Goss-oriented nuclei during primary recrystallization might lead to a certain number of these grains with a larger initial size in the matrix prior to the onset of abnormal grain growth. However, by using large-scale EBSD measurements, it was found that Goss-oriented grains have the same average grain size as crystals with other crystallographic orientations that do not grow abnormally.

Hence, the argument of an initial size advantage of Goss-oriented grains did not seem plausible.

Another line of the discussion suggested the possibility that Goss grain might have a smaller grain neighborhood in the as-recrystallized state since they are during primary recrystallization formed in shear bands where the nucleation rate is high and hence many smaller grains are formed. It was discussed that this configuration might lead to a growth rate advantage during the early stages of grain growth (Burwell, 1940; Hillert, 1965; May and Turnbull, 1958; Shimizu and Harase, 1989; Lin et al., 1996; Hayakawa and Szpunar, 1997; Rajmohan et al., 1999; Morawiec, 2000; Chen et al., 2002). Another hypothesis assumed a preferential coalescence mechanism in clusters of Goss-oriented grains because of their very low relative misorientation (Matsuo, 1989; Inokuti et al., 1981). At last a selective process due to a special orientation relationship between Goss and matrix grains has been claimed to act as the main cause of the secondary recrystallization (Shimizu and Harase, 1989; Lin et al., 1996; Hayakawa and Szpunar, 1997; Rajmohan et al., 1999; Morawiec, 2000; Chen et al., 2002).

23.8 Phenomenological Kinetics of Recrystallization

Owing to their nature as nucleation and growth mechanisms, recrystallization phenomena can be kinetically described by a JMAK approach, which is characterized by a sigmoidal shape when plotting the transformed volume fraction as a function of time (**Figures 6a** and **53**). The topological model

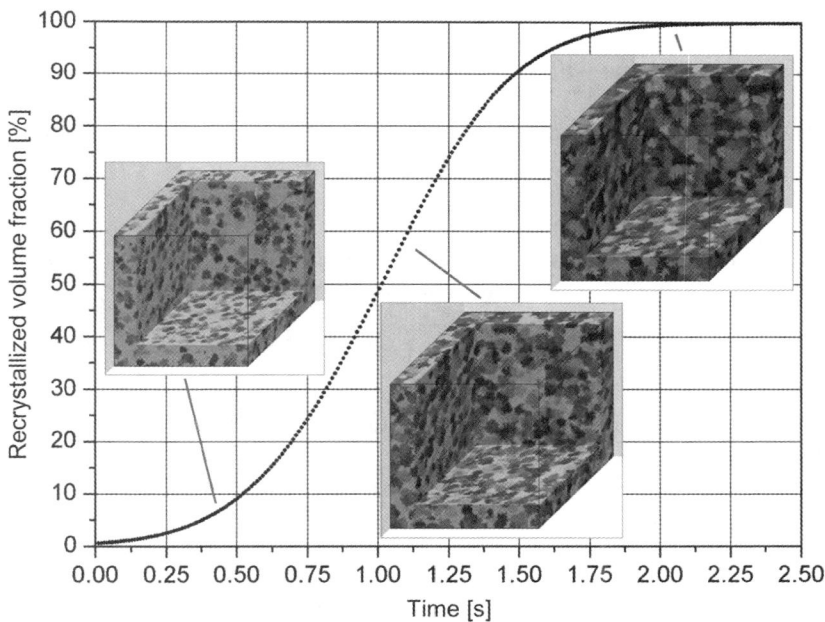

Figure 53 Sigmoidal recrystallization curve predicted by a cellular automaton model that includes inhomogeneity of the crystallographic texture and of the grain-boundary mobility (Raabe, 1999, 2002).

behind the Avrami kinetic formulation is the most common formulation for describing isothermal discontinuous transformation kinetics that are based on nucleation and growth into a homogeneous matrix (Kolmogorov, 1937; Johnson and Mehl, 1939; Avrami, 1939, 1940).

When applied to primary static isothermal recrystallization kinetics it describes the volume fraction that is transformed from the cold worked to the recrystallized state.

In the 1940s, various authors independently developed a similar kinetic formulation that is sometimes therefore referred to as the JMAK equation (Kolmogorov, 1937; Johnson and Mehl, 1939; Avrami, 1939, 1940).

More specific, the original Avrami model assumed site-saturated nucleation (all nuclei exist at $t = 0$s and no new nuclei are formed). statistically distributed nucleation (both in real and orientation space), and isotropic growth at a constant growth rate.

About the growth rate and the statistical nucleus distribution Kolmogorov, Johnson and Mehl made the same assumptions; however, they used not a site-saturated but alternatively also a time-dependent nucleation rate (continuous nucleation).

Interestingly, the theory was initially developed for low molecular-weight materials such as metals. Later, it was extended to the crystallization of high molecular-weight polymers (Raabe, 2004; Raabe et al., 2004a; Godara et al., 2006; Jia and Raabe, 2006).

Avrami based his approach on the topological consideration that a phase is nucleated by small nuclei that already exist in the preceding host phase in which the transformation spontaneously takes place and whose effective number is n_0 per unit volume. The number of nuclei per unit region at time t decreases from n_0 in two ways: (i) some of them become active growth nuclei in consequence of free energy fluctuations and with probability of occurrence p per unit time and (ii) some of them get swallowed by growing grains of the new phase. The number of growth nuclei can increase linearly with time (continuous nucleation) or the large majority of the growth nuclei can be formed near the beginning of the transformation (instantaneous nucleation or site-saturated nucleation). Further, the authors made the assumption that when one grain impinges upon another growth ceases. This means that now grain growth occurs.

According to the results of JMAK analysis, the dependence of the recrystallized volume fraction X on time can be written

$$X(t) = 1 - \exp(-k_0\, t^q)$$

where k_0 is a constant and q is the Avrami kinetic exponent. When using the underlying metallurgical quantities for site-saturated nucleation conditions we can write this equation as

$$X(t) = 1 - \exp\left(-n_0 \frac{4}{3}\pi v^3 t^3\right)$$

where v is the constant growth rate and n_0 is the effective number of nuclei per unit volume. The recrystallization time t_R is

$$t_R(n_0) = \left(n_0 \frac{4}{3}\pi\right)^{-1/3} \frac{1}{v}$$

and the grain size after recrystallization amounts to

$$d = 2vt_R = 2\left(\frac{1}{n_0}\frac{3}{4\pi}\right)^{1/3}$$

For the corresponding case of continuous nucleation the same equations read

$$X(t) = 1 - \exp\left(-\dot{n}\frac{1}{3}\pi v^3 t^4\right)$$

with the recrystallization time

$$t_R(\dot{n}) = \left(\dot{n}\frac{1}{3}\pi v^3\right)^{-1/4}$$

and the grain size after recrystallization amounts to

$$d = 2vt_R = 2\left(\frac{v}{\dot{n}}\frac{3}{\pi}\right)^{1/4}$$

For extracting the kinetic coefficient from of discrete simulations, one can rewrite this expression in the following form

$$\ln(-\ln(1 - X(t))) = \ln k_0 + q\ln(t).$$

From the analysis above, we learn that the kinetic coefficient for the site-saturated nucleation conditions amounts to 3 and that for continuous nucleation to 4 (Haessner, 1978; Humphreys and Hatherly, 1995).

Sometimes, kinetic coefficients far outside of the here presented regime of 3–4 are discussed in the literature (Sellars and Tegart, 1966; Humphreys, 1997; Himmel, 1962; Cahn, 1965, 1950; Ibe and Lücke, 1966; Hu et al., 1990; Rath and Hu, 1969a). However, such analysis should usually be treated with care since kinetic coefficients that substantially deviate from this regime are probably influenced by recovery and grain-growth phenomena. Then, however, the corresponding kinetic laws for recovery or grain growth rather than the conventional sigmoidal Avrami kinetics alone should be used for analyzing corresponding datasets. Alternatively, full-field numerical models must be used, which consider the various mechanisms in the same simulation run (Humphreys, 1997, 1992a; Miodownik, 2002; Raabe and Becker, 2000; Raabe, 2000; Holm and Battaile, 2001; Rollett and Raabe, 2001; Humphreys, 1992b; Maurice and Humphries, 1998; Maurice, 2001; Weygand et al., 2001; Kinderlehrer et al., 2004, 2001; Anderson et al., 1984; Srolovitz et al., 1984; Glazier et al., 1990; Anderson and Rollett, 1990; Tavernier and Szpunar, 1991a, 1991b; Srolovitz et al., 1985, 1986b, 1988; Rollett et al., 1989b, 1989c; 1992; Peczak, 1995; Doherty et al., 1990; Miodownik et al., 1999; Raabe, 1998).

Another main result can be retrieved from the statistical analysis presented here: as pointed out before, however, real recrystallization phenomena are characterized by more complex and spatially not statistically distributed nucleation mechanisms, by a nonhomogeneous deformation substructure (hence an inhomogeneous driving force), and by a nonconstant growth rate of the grain boundaries involved. Understanding this high degree of inherent inhomogeneity of recrystallization phenomena naturally advocates the use of more complex and spatially and/or orientationally discrete simulation methods involving also the preceding deformation history of the material (Raabe and Becker, 2000; Bate, 1999; Roters et al., 2010; Raabe, 1998, 2007; Zambaldi et al., 2007; Radhakrishnan et al., 2000, 1998; Hurley and Humphreys, 2003; Hallberg, 2011; Carel et al., 1996; Nagai et al., 1990; Bernacki et al., 2007, 2008, 2009; Merriman et al., 1994; Zhao et al., 1996; Bernacki and Coupez, 2011; Logé et al., 2008; Janssens et al., 2007). Such simulation approaches will be presented in the next sections.

23.9   Modeling Recrystallization and Grain-Growth Phenomena

23.9.1   Introduction

The use of models for describing recrystallization phenomena with the aim of predicting crystallographic texture, kinetics, microstructure, and mechanical properties in the context of materials processing is a challenging task (Raabe, 1998, 2007; Zambaldi et al., 2007; Radhakrishnan et al., 2000, 1998; Hurley and Humphreys, 2003; Hallberg, 2011; Carel et al., 1996; Nagai et al., 1990; Bernacki et al., 2009; Merriman et al., 1994; Zhao et al., 1996; Bernacki et al., 2007, 2008; Bernacki and Coupez, 2011; Logé et al., 2008; Janssens et al., 2007; Crumbach et al., 2004; Rollett et al., 1989a; Raabe et al., 2004b). This judgment is motivated by the common observation that substantial changes in recrystallization phenomena can be stimulated by rather small modifications in the metallurgical state (e.g. chemical purity, thermodynamic state, crystallographic texture, microstructure inhomogeneity, and microstructure inheritance from preceding process steps) or in the external boundary conditions (time, temperature fields, strain fields, and joint thermal and mechanical constraints) (McQueen and Jonas, 1975; Miodownik, 2002; Ferry, 2002). This sensitivity of recrystallization is due to the fact that most of the metallurgical mechanisms involved during texture formation in the course of recrystallization such as grain nucleation, grain-boundary mobility, or impurity drag effects are thermally activated. Typically, these mechanisms follow Arrhenius functions the arguments of which can strongly depend on some of the parameters listed above and interact with each other in a nonlinear fashion. Similarly, the microstructural state of the deformed material from which recrystallization proceeds is often not well known or not so well reproduced in models. The requirement to establish a better connection between the inherited deformation microstructure and the onset of recrystallization has been reflected in a number of investigations that aimed at using for instance crystal plasticity finite element simulation results as starting configurations for recrystallization simulations (Raabe and Becker, 2000; Bate, 1999; Roters et al., 2010; Zambaldi et al., 2007; Radhakrishnan et al., 2000; Bernacki et al., 2007; Bernacki and Coupez, 2011). Although various types of homogenization models of crystal deformation are nowadays capable of providing information about the average behavior of the material in the course of a thermomechanical process, it are often the details and inhomogeneities, i.e. the singularities in the deformed structure that strongly affect recrystallization. This means that even a good knowledge of average plasticity parameters does not generally solve the open questions pending in the field of recrystallization modeling (Raabe and Becker, 2000; Bate, 1999).

Moreover, in commercial metal-manufacturing processes typically encountered in physical metallurgy many of the influencing factors, be they of a metallurgical or of a processing nature, are usually neither exactly known nor sufficiently well defined to apply models that require a high degree of precision with regard to the input parameters.

These remarks underline that the selection of an appropriate recrystallization multiscale model for the prediction of crystallographic texture, microstructure, and properties for a given process must follow a clear concept as to what exactly is expected from such a model and what *cannot* be predicted by it in view of the points made above (Rollett, 1997). This applies in particular to cases where a model for the simulation of recrystallization textures is to be used in conjunction with real manufacturing processes.

The main challenge of using multiscale process models for recrystallization, therefore, lies in selecting the right model for a well-defined task, i.e. it must be agreed that microstructural property is to be simulated and what kind of properties should be subsequently calculated from these microstructure data. From that, it is obvious that no model exists that could satisfy all questions that may arise in the context of recrystallization textures (Humphreys, 1992a; Rollett, 1997; Mahin, 1980; Saetre, 1986). Also one has to clearly separate between the aim of predicting recrystallization textures and microstructures (Miodownik, 2002;

Sebald and Gottstein, 2002; Humphreys, 1992a; Raabe and Becker, 2000) on the one hand and the materials properties (functional or mechanical) on the other. While the first task may by pursued by formulating an appropriate recrystallization model within the limits addressed above, the second challenge falls into the wide realm of microstructure–property theory. This means that one should generally separate between the prediction of the microstructure and the prediction of some property from that particular microstructure. Typically through-process modelers are interested in the final materials properties in the first place rather than in the details of the microstructure of a recrystallized material. The modern attitude toward this discrepancy is the commonly accepted understanding that a decent description of materials properties requires the use of internal (i.e. of microstructurally motivated) parameters that can be coupled to suited microstructure–property laws. A typical example along that philosophy would be the grain-size prediction via a recrystallization model and the subsequent application of the Hall–Petch law for the estimation of the yield strength (Janssens et al., 2007; Crumbach et al., 2004).

This section addresses exclusively the first question, more precisely, only models for primary static recrystallization will be tackled placing particular attention on the prediction of microstructure and crystallographic texture.

23.9.2   Statistical Models for Predicting Recrystallization Textures

The analytical formulation of physically based statistical models with a simple mathematical structure and yet at the same time equipped with a solid metallurgical basis for fast applications in the area of process simulation remains an important challenge in materials science (Janssens et al., 2007; Crumbach et al., 2004). This applies in particular for statistical recrystallization models that pursue the aim of predicting crystallographic texture, certain microstructure parameters (e.g. grain size), and even certain mechanical properties during processing with a precision that is sufficiently robust for industrial applications.

In the past, various statistical variants of the original JMAK approaches were suggested for the prediction of recrystallization textures (Kolmogorov, 1937; Johnson and Mehl, 1939; Avrami, 1939, 1940).

These models typically combine JMAK-type kinetic evolution equations with the texture dependence of grain nucleation and the misorientation dependence of the motion of grain boundaries (Humphreys, 1992a; Rollett, 1997; Gottstein and Shvindlerman, 1999). Examples of such orientation-dependent JMAK approaches for crystallographic texture prediction are the Bunge transformation model (Bunge and Köhler, 1992) and the Gottstein kinetic model variants (Crumbach et al., 2006).

This section on statistical texture models in the field of recrystallization is essentially inspired by these two formulations, particularly by the microgrowth selection model of Sebald and Gottstein, which is consistently formulated on the basis of nucleation and crystal growth kinetics while the Bunge–Köhler model essentially uses crystallographic transformations of the orientation distribution functions of the deformed samples without considering kinetics (Bunge and Köhler, 1992).

In the Sebald–Gottstein model, the crystallographic texture of the deformed material is discretized in terms of a large set of discrete single orientations that approximate a given orientation distribution function of a plastically deformed specimen using typically cold-rolled sheet material.

Sebald and Gottstein investigated several nucleation mechanisms with respect to the orientation and misorientation distribution they create at the incipient stages of recrystallization. The information that was provided by the submodels for nucleation are the orientation-dependent density of nuclei within the deformed crystals with a given crystallographic orientation. This means that the nucleation submodels initiate the orientation and misorientation distributions of the nuclei for primary recrystallization. The different types of nucleation processes investigated so far were random nucleation creating both, a random nucleus texture and a random misorientation distribution, nucleation at shear bands forming a random nucleus texture with a nonrandom misorientation distribution; and nucleation due to preexisting nuclei creating a nucleus texture similar to the deformation texture in conjunction with a narrow misorientation distribution.

The growth rate of the nuclei corresponds to their grain-boundary velocity, which is given by the product of the grain-boundary mobility and the driving force. The mobility of a grain boundary depends on the misorientation between the growing and the deformed grain. The model can distinguish three categories of grain boundaries, namely, small angle, high-angle, and special boundaries. Small-angle grain boundaries are assumed to be essentially immobile. In the case of aluminum, it is generally accepted that grain boundaries with near $40°<111>$ misorientation may show particularly high mobilities so that such interfaces are treated as special boundaries (Ibe and Lücke, 1966, 1972; Ibe et al., 1970; Gottstein and Shvindlerman, 1999; Aust and Rutter, 1959, 1960; Molodov, 2001; Lücke and Stüwe, 1963; Gottstein et al., 1995; Shvindlerman et al., 1995, Shvindlerman and Gottstein, 1999). This is realized in the simulations of Gottstein and Sebald by assigning higher mobilities to such grain boundaries. The mobility of an average high-angle grain boundary is typically set to 20% of the maximum occurring mobility. The driving force for primary static recrystallization is the difference of the stored energy density between the deformed matrix and the nucleus, which in such models typically approximated in terms of the Taylor factor. Growth of the newly formed nuclei is assumed to be isotropic, but it ceases when the nuclei impinge. This means that a growing nucleus can only grow into the nonrecrystallized volume fraction. This portion of the material can be calculated according to the JMAK theory, which provides a relation between the increase in the recrystallized volume fraction for unconstrained growth (the so-called expanded volume fraction) and the true or constrained increase under consideration of grain impingement for a random spatial distribution.

When compared to the Potts Monte Carlo (Anderson et al., 1984; Srolovitz et al., 1984, 1985, 1986b, 1988; Glazier et al., 1990; Anderson and Rollett, 1990; Tavernier and Szpunar, 1991a, 1991b; Rollett et al., 1989b, 1989c, 1992; Peczak, 1995), cellular automaton (Raabe et al., 2004b; Hesselbarth and Göbel, 1991; Pezzee and Dunand, 1994; Sheldon and Dunand, 1996; Davies, 1995, 1997; Marx et al., 1997, 1998; Davies and Hong, 1999; Raabe, 1999, 2002, 2001; Janssens, 2003), or vertex-type front-tracking models (Humphreys, 1992b; Maurice and Humphries, 1998; Maurice, 2001; Weygand et al., 2001; Kinderlehrer et al., 2004, 2001), statistical JMAK-type approaches such as exemplarily presented here are more efficient for physically based recrystallization texture predictions in the field of materials processing owing to their semistatistical formulation. On the other hand, statistical models neglect important local features of the microstructures such as grain topology, grain neighborhood, and the local curvature of an interface. This can be a disadvantage when applying statistical models to heterogeneous microstructures.

23.9.3 Spatially Discrete Models for Simulating Recrystallization

23.9.3.1 Introduction

The design of time and space discretized recrystallization models for predicting texture and microstructure in the course of materials processing, which predict kinetics and energies in a local fashion are of interest for two reasons. First, from a fundamental point of view, it is desirable to understand better the dynamics and the topology of microstructures that arise from the interaction of large numbers of lattice defects that are characterized by a wide spectrum of intrinsic properties and interactions in spatially heterogeneous materials under complex engineering boundary conditions. For instance, in the field of recrystallization (and grain growth), the influence of local grain-boundary characteristics (mobility, energy), local driving forces, and local crystallographic textures on the final microstructure is of particular interest (Rollett, 1997; Humphreys, 1997). An important point of interest in that context, however, is the question how local such a models should be in its spatial discretization in order to really provide microstructural input that cannot be equivalently provided by statistical methods. In the worst case, a problem in that field may be that spatially discrete recrystallization models may have the tendency to pretend a high degree of precision without actually providing it. In other words, even in highly discretized recrystallization models the physics always lies in the details of the constitutive description of the kinetics and thermodynamics of the deformation structure and of the interfaces involved. The mere fact that a model is formulated in a discrete fashion does not, as a rule, automatically render it a sophisticated model per se. Second, from a practical point of view, it makes sense to predict microstructure parameters such as the crystal size or the crystallographic texture, which determines the mechanical and physical properties of materials subjected to industrial processes on a sound phenomenological basis (Raabe, 1998).

In this section on spatially discrete models particular attention is laid on cellular automata (Raabe et al., 2004b; Hesselbarth and Göbel, 1991; Pezzee and Dunand, 1994; Sheldon and Dunand, 1996; Davies, 1995, 1997; Marx et al., 1997, 1998; Davies and Hong, 1999; Raabe, 1999, 2002, 2001; Janssens, 2003), Potts-type Monte Carlo multispin models (Anderson et al., 1984; Srolovitz et al., 1984, 1985, 1986b, 1988; Glazier et al., 1990; Anderson and Rollett, 1990; Tavernier and Szpunar, 1991a, 1991b; Rollett et al., 1989b, 1989c, 1992; Peczak, 1995), and vertex (front-tracking) models (Humphreys, 1992b; Maurice and Humphries, 1998; Maurice, 2001; Weygand et al., 2001; Kinderlehrer et al., 2004, 2001) because those three have been used in the past successfully for the simulation of recrystallization textures and microstructures. Related spatially discrete models of recrystallization phenomena such as the phase-field model as metallurgically motivated derivative of the Ginzburg–Landau kinetic theory also represent elegant ways to predict the evolution of topology during recrystallization but they are up to now less well-established for predicting recrystallization textures (Raabe, 1998).

23.9.3.2 *Cellular Automaton Models of Recrystallization*
Cellular automata are algorithms that describe the discrete spatial and temporal evolution of complex systems by applying local transformation rules to lattice cells that typically represent volume portions. The state of each lattice site is characterized in terms of a set of internal state variables. For recrystallization models these can be lattice defect quantities (stored energy), crystal orientation, or precipitation density. Each site assumes one out of a finite set of possible discrete states. The opening state of the automaton is defined by mapping the initial distribution of the values of the chosen state variables onto the lattice (Hesselbarth and Göbel, 1991; Pezzee and Dunand, 1994; Sheldon and Dunand, 1996; Davies, 1995, 1997; Marx et al., 1997, 1998; Davies and Hong, 1999; Raabe, 1999, 2002, 2001; Janssens, 2003).

The dynamical evolution of the automaton takes place through the application of deterministic or probabilistic transformation rules (switching rules) that act on the state of each lattice point. These rules determine the state of a lattice point as a function of its previous state and the state of the neighboring sites. The number, arrangement, and range of the neighbor sites used by the transformation rule for calculating a state switch determines the range of the interaction and the local shape of the areas that evolve. Cellular automata work in discrete time steps. After each time-interval, the values of the state variables are updated for all points in synchrony mapping the new (or unchanged) values assigned to them through the transformation rule. Owing to these features, cellular automata provide a discrete method of simulating the evolution of complex dynamical systems that contain large numbers of similar components on the basis of their local interactions.

Cellular automata are—like all other continuum models that work above the discrete atomic scale—not intrinsically calibrated by a characteristic physical length or time scale. This means that a cellular automaton simulation of continuum systems requires the definition of elementary units and transformation rules that adequately reflect the kinetics at the level addressed. If some of the transformation rules refer to different real-time scales (e.g. recrystallization and recovery, bulk diffusion, and grain-boundary diffusion) it is essential to achieve a correct common scaling of the entire system. The requirement for an adjustment of time scaling among various rules is due to the fact that the transformation behavior of a cellular automaton is sometimes determined by noncoupled Boolean routines rather than by local solutions of coupled differential equations.

The following examples on the use of cellular automata for predicting recrystallization textures are designed as automata with a probabilistic transformation rule (Raabe, 1999). Independent variables are time and space. The latter is discretized into equally shaped cells each of which is characterized in terms of the mechanical driving force (stored deformation energy) and the crystal orientation (texture). The starting data of such automata are usually derived from experiment (for instance from a microtexture map) or from plasticity theory (for instance from crystal plasticity finite element simulations). The initial state is typically defined in terms of the distribution of the crystal orientation and of the driving force. Grains or subgrains are mapped as regions of identical crystal orientation, but the driving force may vary inside these areas.

The kinetics of the automaton gradually evolve from changes in the state of the cells (cell switches). They occur in accord with a switching rule (transformation rule) that determines the individual switching probability of each cell as a function of its previous state and the state of its neighbor cells. The switching rule is designed to map the phenomenology of primary static recrystallization. It reflects that the state of a nonrecrystallized cell belonging to a deformed grain may change due to the expansion of a recrystallizing neighbor grain that grows according to the local driving force and boundary mobility. If such an expanding grain sweeps a nonrecrystallized cell the stored dislocation energy of that cell drops to zero and a new orientation is assigned to it, namely that of the expanding neighbor grain. The mathematical formulation of the automaton used in this report can be found in (Raabe, 2002). It is derived from a probabilistic form of a linearized symmetric rate equation, which describes grain-boundary motion in terms of isotropic single-atom diffusion processes perpendicular through a homogeneous planar grain-boundary segment under the influence of a decrease in Gibbs energy. This means that the local progress in recrystallization can be formulated as a function of the local driving forces (stored deformation energy) and interface properties (grain-boundary mobility). The most intricate point in such simulations consists in identifying an appropriate phenomenological rule for nucleation events.

Figure 54 shows an example of a coupling of a cellular automaton with a crystal plasticity finite element model for predicting recrystallization textures in aluminum (Raabe and Becker, 2000). The major advantage of such an approach is that it considers the inherited material deformation heterogeneity as opposed to material homogeneity.

This type of coupling the two models, therefore, seems more appropriate when aiming at the simulation of textures formed during materials processing. Nucleation in this coupled simulation works above the subgrain scale, i.e. it does not explicitly describe cell walls and subgrain coarsening phenomena. Instead, it incorporates nucleation on a more phenomenological basis using the kinetic and thermodynamic instability criteria known from classical recrystallization theory. The kinetic instability criterion means that a successful nucleation process leads to the formation of a mobile high-

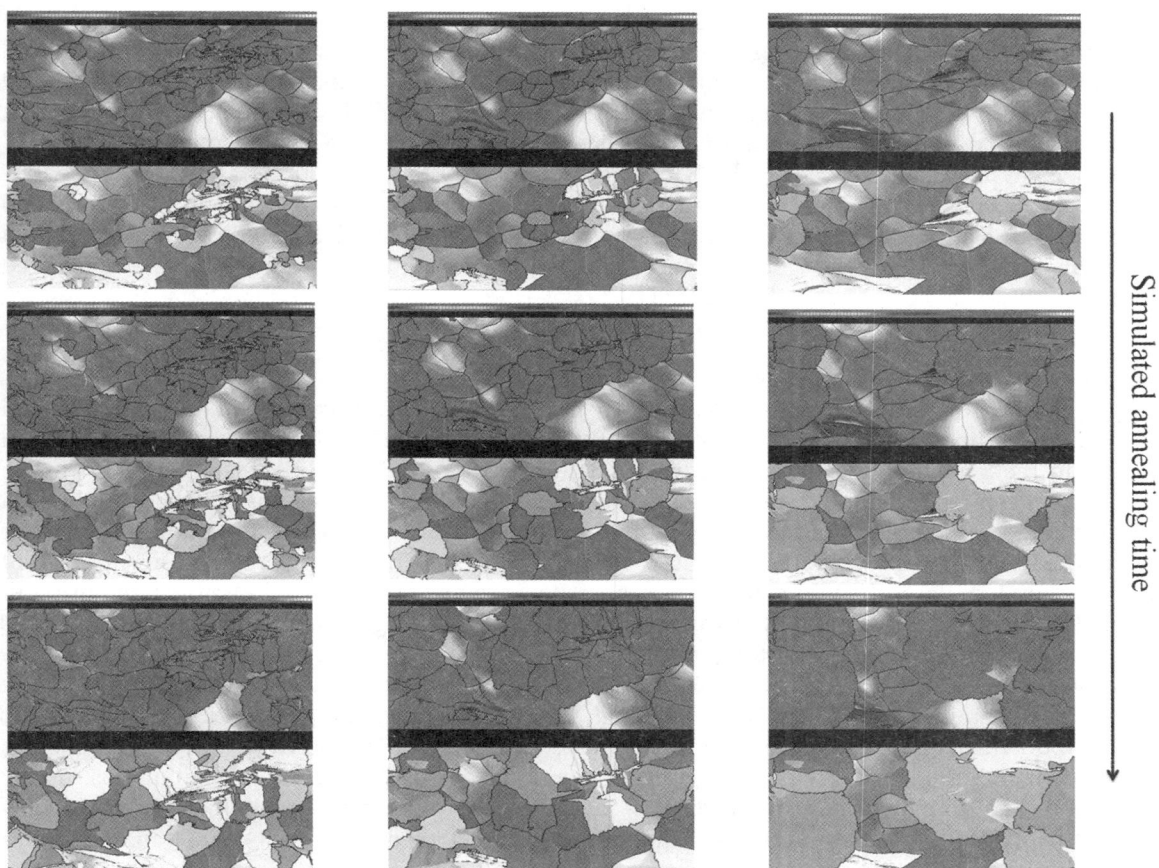

Simulated annealing time

Figure 54 Series of subsequent stages of a 2D cellular automaton simulation of primary static recrystallization in a deformed aluminum polycrystal on the basis of crystal plasticity finite element data. The figure shows the change both in dislocation density (upper figures) and in microtexture (lower figures), as a function of the annealing time during isothermal recrystallization. The gray areas in the upper figures indicate a stored dislocation density of zero, i.e. these areas are recrystallized. The simulation parameters are: 800 K; thermodynamic instability criterion: site-saturated spontaneous nucleation in cells with at least 50% (left-hand series), 60% (middle series), or 70% (right-hand series) of the maximum occurring dislocation density (threshold value); kinetic instability criterion for further growth of such spontaneous nuclei: misorientation above 15°; activation energy of the grain-boundary mobility: 1.46 eV; preexponential factor of the grain-boundary mobility: $m_0 = 8.3.10 \pm 3$ m^3/(Ns); mesh size of the cellular automaton grid (scaling length): 61.9 μm per grid point (Raabe and Becker, 2000).

angle grain boundary that can sweep the surrounding deformed matrix. The thermodynamic instability criterion means that the stored energy changes across the newly formed high-angle grain boundary providing a net driving force pushing it forward into the deformed matter. Nucleation in this simulation is performed in accord with these two aspects, i.e. potential nucleation sites must fulfill both, the kinetic and the thermodynamic instability criterion. The used nucleation model does not create any new orientations. At the beginning of the simulation the thermodynamic criterion, i.e. the local value of the dislocation density was first checked for all lattice points. If the dislocation density was larger than some critical value of its maximum value in the sample, the cell was spontaneously recrystallized without any orientation change, i.e. a dislocation density of zero was assigned to it and the original crystal orientation was preserved. In the next step, the conventional cellular growth algorithm was used, i.e. the kinetic conditions for nucleation were checked by calculating the misorientations among all spontaneously recrystallized cells (preserving their original crystal orientation) and their immediate neighborhood considering the first, second, and third neighbor shells. If any such pair of cells revealed a misorientation above 15°, the cell flip of the unrecrystallized cell was calculated according to its actual transformation probability. In case of a successful cell flip the orientation of the first recrystallized neighbor cell was assigned to the flipped cell.

23.9.3.3 *Potts-type Monte Carlo Multispin Models of Recrystallization*
The application of the Metropolis Monte Carlo method in microstructure simulation has gained momentum particularly through the extension of the Ising lattice model for modeling magnetic spin systems to the kinetic multistate Potts lattice model (Potts, 1952). The original Ising model is in the form of an $^1/_2$ spin lattice model where the internal energy of a magnetic system is calculated as the sum of pair-interaction energies between the continuum units that are attached to the nodes of a regular lattice. The Potts model deviates from the Ising model by generalizing the spin and by using a different Hamiltonian. It replaces the Boolean spin variable where only two states are admissible (spin up, spin down) by a generalized variable that can assume one out of a larger spectrum of discrete possible ground states, and accounts only for the interaction between *dissimilar* neighbors (Anderson et al., 1984; Srolovitz et al., 1984, 1985, 1986b, 1988; Glazier et al., 1990; Anderson and Rollett, 1990; Tavernier and Szpunar, 1991a, 1991b; Rollett et al., 1989b, 1989c, 1992; Peczak, 1995; Doherty et al., 1990; Miodownik et al., 1999). The introduction of such a spectrum of different possible spins enables one to represent domains discretely by regions of identical state (spin). For instance, in microstructure simulation such domains can be interpreted as areas of similarly oriented crystalline matter. Each of these spin orientation variables can be equipped with a set of characteristic state variable values quantifying the lattice energy, the dislocation density, the Taylor factor, or any other orientation-dependent constitutive quantity of interest. Lattice regions that consist of domains with identical spin or state are in such models translated as crystal grains. The values of the state variable enter the Hamiltonian of the Potts model. The most characteristic property of the energy operator when used for coarsening models is that it defines the interaction energy between nodes with like spins to be zero, and between nodes with unlike spins to be one. This rule makes it possible to identify interfaces and to quantify their energy as a function of the abutting domains.

According to Srolovitz et al. (Anderson et al., 1984; Srolovitz et al., 1984; Glazier et al., 1990; Anderson and Rollett, 1990; Glauber, 1963; Metropolis et al., 1953; Sahni et al., 1983; Hassold and Holm, 1993; Safran et al., 1983; Holm et al., 1998; Rollett et al., 1998) a typical energy operator for Potts-type grain growth and recrystallization simulations can be written

$$E = E^{GG} + E^{El} = \sum_{i=1}^{N} \left(\frac{J}{2} \sum_{j=1}^{nnn} (1 - \delta_{S_i S_j}) + H^{El} f(Q_u - S_i) \right)$$

where E is a scaled energy proportional to the total excess energy associated with the presence of lattice defects, E^{GG} a scaled energy proportional to the excess energy associated with grain-boundary energy, E^{El} a scaled energy proportional to the excess energy associated with elastically stored energy, N the number of discrete lattice sites, *nnn* the geometrically weighted number of neighbor sites in the first, second, and third neighbor shell, S the orientational state variable, $\delta_{S_i S_j}$ the Kronecker symbol, which assumes a value of 1 if $S_i = S_j$ and a value of 0 if $S_i \neq S_j$, J an energy proportional to the grain-boundary energy, and H^{El} an energy proportional to the stored elastic energy. J and H^{El} have a positive sign. Their respective proportionality factors relating them to realistic energies scale the simulation with respect to temperature. The factor $^1/_2$ in eqn (1) corrects that each interface segment is counted twice. The function $f(Q_u - S_i)$ describes whether a site is recrystallized or not. The variable Q_u is the number of distinct crystal orientations of unrecrystallized grains. The recrystallized sites are given orientation variables larger than Q_u. The step function $f(Q_u - S_i)$ assumes a value of 1 for arguments equal to or larger than 0 $(Q_u \geq S_i > 0)$ and a value of 0 if the site is recrystallized, i.e. for arguments below 0 $(S_i > Q_u)$.

The kinetic evolution of the Potts model, which occurs in the form of domain growth, is usually simulated using Metropolis or Glauber dynamics (Glauber, 1963; Metropolis et al., 1953), i.e. it proceeds by randomly selecting lattice sites, switching their orientational state randomly to a new one, and weighting the resulting energy change in terms of Metropolis Monte Carlo sampling. This means that an orientation flip is generally accepted if it leads to a state of lower or equal energy and is accepted with thermal probability if it leads to an energy increase. The switching probability W can be

$$W = \begin{cases} \exp(-\Delta E / k_B T) & \text{if} \quad \Delta E > 0 \\ 1 & \text{if} \quad \Delta E \leq 0 \end{cases}.$$

The evaluation of the thermal fluctuation, which leads to an energy increase, is conducted by generating a pseudorandom number between 0 and 1 and comparing it to the actual switching probability $\exp(-\Delta E / k_B T)$. If the random number is equal or lower than the thermal probability the switch is accepted (Anderson et al., 1984; Srolovitz et al., 1984, 1985, 1986b, 1988, Glazier et al., 1990; Anderson and Rollett, 1990; Tavernier and Szpunar, 1991a, 1991b; Rollett et al., 1989b, 1989c, 1992; Peczak, 1995). If it is larger the switch is rejected. If the new configuration is rejected, one counts the original position as a new one and repeats the process by switching another site. The microstructural evolution of the system is reflected by the development of the domain size and shape.

The Potts model is very versatile for describing coarsening phenomena. It takes a quasimicroscopic metallurgical view of grain growth or ripening, where the crystal interior is composed of lattice points (e.g. atom clusters) with identical energy (e.g. orientation) and the grain boundaries are the interfaces between different types of such domains (Holm et al., 2001; Miodownik et al., 2000). As in a real ripening scenario, interface curvature leads to increased wall energy on the convex side and thus to wall migration entailing local shrinkage. The discrete simulation steps in the Potts model, by which the system proceeds toward thermodynamic equilibrium, are typically calculated by randomly switching lattice sites and weighting the resulting interfacial energy changes in terms of Metropolis Monte Carlo sampling (Anderson et al., 1984; Srolovitz et al., 1984, 1985, 1986b, 1988; Glazier et al., 1990; Anderson and Rollett, 1990; Tavernier and Szpunar, 1991a, 1991b; Rollett et al., 1989b, 1989c, 1992; Peczak, 1995).

23.9.3.4 *Vertex Models of Recrystallization*
Vertex and related front tracking simulations are another alternative for engineering process models with respect to recrystallization phenomena (Humphreys, 1992b; Maurice and Humphries, 1998;

Maurice, 2001; Weygand et al., 2001; Kinderlehrer et al., 2004, 2001). Their use is currently less common when compared to the widespread application of Monte Carlo and cellular automaton models owing to their geometrical complexity and the required small integration time steps. Despite these differences to lattice-based recrystallization models they have an enormous potential for predicting interface dynamics at small scales also in the context of process simulations (Raabe et al., 2004b; Humphreys, 1997).

Topological network and vertex models idealize solid materials or soap-like structures as homogeneous continua that contain interconnected boundary segments that meet at vertices, i.e. boundary junctions (Humphreys, 1997; Humphreys, 1992b; Maurice and Humphries, 1998; Maurice, 2001; Weygand et al., 2001; Kinderlehrer et al., 2004, 2001). Depending on whether the system dynamics lies in the motion of the junctions or of the boundary segments, they are sometimes also referred to as boundary dynamics or, more generalized, as front-tracking models. The grain boundaries appear as lines or line segments in 2D and as planes or planar segments in 3D simulations. The dynamics of these coupled interfaces or interface portions and of the vertices determine the evolution of the entire network.

The dynamical equations of the boundary and node (vertex) motion can be described in terms of a damped Newtonian equation of motion that contains a large frictional portion or, mathematically equivalent, in terms of a linearized first-order rate equation. Using the frictional form of the classical equation of motion with a strong damping term results in a steady-state motion where the velocity of the defect depends only on the local force but not on its previous velocity. The overdamped steady-state description is similar to choosing a linearized rate equation, where the defect velocity is described in terms of a temperature-dependent mobility term and the local driving pressure.

The calculation of the local forces in most vertex models is based on equilibrating the line energies of subgrain walls and high-angle grain boundaries at the interface junctions according to Herring's equation. The enforcement of the local mechanical equilibrium at these nodes is for obvious topological reasons usually only possible by allowing the abutting interfaces to curve.

These curvatures in turn act through their capillary force, which is directed toward the center of curvature, on the junctions. In sum, this may lead to their displacement. In order to avoid the artificial enforcement of a constant boundary curvature between two neighboring nodes, the interfaces are usually decomposed into sequences of piecewise straight boundary segments.

Most vertex and network models use switching rules that describe the topological recombination of approaching vertices when such neighboring nodes are closer than some critical spontaneous recombination spacing. This is an analogy to the use of phenomenological annihilation and lock-formation rules that appear in dislocation dynamics. As in all continuum models, the use of such empirical recombination laws replaces a more exact atomistic treatment.

It is worth noting in this context that the recombination rules, particularly the various values for the critical recombination spacing of specific configurations, can affect the topological results of a simulation. Depending on the underlying constitutive continuum description, vertex simulations can consider crystal orientation and, hence, misorientations across the boundaries, interface mobility, and the difference in elastic energy between adjacent grains. Due to the stereological complexity of grain boundary arrays and the large number of degrees of freedom encountered in such approaches, most network simulations are currently confined to the 2D regime. Topological boundary dynamics models are different from kinetic cellular automaton or Potts Monte Carlo models in that they are not based on minimizing the total energy but directly calculate the motion of the lattice defects, usually on the basis of capillary and elastic forces.

References

Adams, B.L., Wright, S.I., Kunze, K., 1993. Metall. Mater. Trans. 24A, 819.

Alterthum, H., 1922. Zur Theorie der Rekristallisation. Z. Metallkd. 14, 417–424.

Anderson, M.P., Rollett, A.D., 1990. Simulation and Theory of Evolving Microstructures. The Minerals, Metals and Materials Society. TMS Publication, Warrendale, PA.

Anderson, M.P., Srolovitz, D.J., Grest, G.S., Sahni, P.S., 1984. Acta Metall. 32, 783.

Ashby, M.F., Harper, J., Lewis, J., 1969. The interaction of crystal boundaries with second-phase particles. Trans. Metall. Soc. AIME 245 (8), 413–420.

Aust, K.T., Rutter, J.W., 1959. Grain boundary migration in high-purity lead and dilute lead–tin alloys. Trans. Metall. Soc. AIME 215 (1), 119–127.

Aust, K.T., Rutter, J.W., 1960. Kinetics of grain boundary migration in high-purity lead containing very small additions of silver and of gold. Trans. Metall. Soc. AIME 218 (4), 682–688.

Avrami, M., 1939. Kinetics of phase change I, general theory. J. Chem. Phys. 7 (12), 1103–1112.

Avrami, M., 1940. Kinetics of phase change. II. Transformation time relations for random distribution of nuclei. J. Chem. Phys. 8, 212–224.

Babcock, S.E., Balluffi, R.W., 1989. Acta Metall. 37, 2357–2367.

Bailey, J.E., Hirsch, P.B., 1962. The recrystallization process in some polycrystalline metals. Proc. R. Soc. London 267 (1328), 11–30.

Bailey, J.E., 1960. Electron microscope observations on the annealing processes occurring in cold worked silver. Philos. Mag. 5 (53), 485–497.

Bailey, J.E., 1963. In: Thomas, G., Washburn, J. (Eds.), Electron Microscope Observations on Recovery and Recrystallization Processes in Cold Worked Metals. Electron Microscopy and Strength of Crystals. Interscience, New York, pp. 535–564.

Bale, H.A., Hanan, J.C., Tamura, N., 2005. Average and grain specific strain of a composite under stress using polychromatic microbeam X-rays. Adv. X-Ray Anal., 49.

Bate, P., 1999. Modelling deformation microstructure with the crystal plasticity finite-element method. Philos. Trans. R. Soc., A 357, 1589–1601.

Beck, P., Hu, H., 1966. ASM Seminar on Recrystallization. Grain Growth and Texture, Met.Park, Ohio, USA.

Beck, P.A., Sperry, P.R., 1950. J. Appl. Phys. 21, 150.

Beck, P.A., Sperry, P.R., Hu, H., 1950. The orientation dependence of the rate of boundary migration. J. Appl. Phys. 21, 420–425.

Beck, P.A., 1949. The formation of recrystallization nuclei. J. Appl. Phys. 20 (6), 633–634.

Berger, A., Wilbrandt, P.-J., Ernst, F., Klement, U., Haasen, P., 1988. On the generation of new orientations during recrystallization: recent results on the recrystallization of tensile-deformed fcc single crystals. Prog. Mater. Sci. 32, 1–95.

Bernacki, M., Coupez, R.L.T., 2011. Level set framework for the finite-element modelling of recrystallization and grain growth in polycrystalline materials. Scr. Mater. 64, 525–528.

Bernacki, M., Chastel, Y., Digonnet, H., Resk, H., Coupez, T., Logé, R., 2007. Development of numerical tools for the multiscale modelling of the recrystallization in metals, based on a digital material framework. Comput. Methods Catal. Mater. Sci. 7, 142–149.

Bernacki, M., Chastel, Y., Coupez, T., Logé, R.E., 2008. Level set framework for the numerical modelling of primary recrystallization in polycrystalline materials. Scr. Mater. 58, 1129–1132.

Bernacki, M., Resk, H., Coupez, T., Logé, R.E., 2009. Finite element model of primary recrystallization in polycrystalline aggregates using a level set framework. Modell. Simul. Mater. Sci. Eng. 17, 1–22.

Bhatia, M.L., Cahn, R.W., 1978. Proc. R. Soc. London, Ser. A 302, 341.

Blum, W., Schlögl, C., Meier, M., 1995. Subgrain formation and subgrain boundary migration in Al–5Mg during high temperature deformation in the range of class – a behavior in comparison with pure aluminium. Z. Metallkd. 86 (9), 631–637.

Bunge, H.-J., Esling, C., 1982. Quantitative Texture Analysis. DGM, Oberursel.

Bunge, H.-J., Esling, C., 1991. Advances and Applications of Quantitative Texture Analysis. DGM, Oberursel.

Bunge, H.J., Köhler, U., 1992. Model calculations of primary recrystallization textures. Scr. Metall. Mater. 27, 1539–1543.

Bunge, H.-J., 1969. Mathematische Methoden der Texturanalyse. Akademie Verlag, Berlin.

Bunge, H.-J., 1982. Texture Analysis in Materials Science. Butterworths, London. (Reprint: Cuvillier Verlag, Göttingen 1993.).

Bunge, H.-J., 1986. Experimental Techniques of Texture Analysis. DGM, Oberursel.

Bunge, H.-J., 1987. Theoretical Methods of Texture Analysis. DGM, Oberursel.

Burgers, W.G., Louwerse, P.C., 1931. Über den Zusammenhang zwischen Deformationsvorgang und Rekristallisationstextur bei Aluminium. Z. Phys. 67, 605–678.

Burgers, W.G., 1941. Rekristallisation, verformter Zustand und Erholung. Akademischer Verlagsgesellschaft, Leipzig.

Burke, J.E., Turnbull, D., 1952. Recrystallization and grain growth. Prog. Met. Phys. 3, 220–292. (London: Pergamon Press).

Burwell, J.T., 1940. Trans. Metall. Soc. AIME 140, 353.

Cahn, R.W., 1950. A new theory of recrystallization nuclei. Proc. Phys. Soc. London, Sect. A 63 (364), 323–336.

Cahn, J.W., 1962. Acta Metall. 10, 789.

Cahn, R.W. (Ed.), 1965. Physical Metallurgy. North-Holland Publishing Co., Amsterdam. Also 2nd, 3rd and 4th editions.

Cahn, I.R.W., 1966. In: Margohn, H. (Ed.), In Recovery, Recrystallization and Grain Growth. ASM, Metals Park, OH, pp. 99.

Calcagnotto, M., Ponge, D., Raabe, D., 2010. Mater. Sci. Eng., A 527, 2738–2746.

Carel, R., Thompson, C., Frost, H., 1996. Acta Mater. 44, 2419–2494.

Carpenter, H.C.H., Elam, C.F., 1920. Crystal growth and recrystallization in metals. J. Inst. Met. 24, 83–131.

Chawla, N., Ganesh, V.V., Wunsch, B., 2004. Three-dimensional (3D) microstructure visualization and finite element modeling of the mechanical behavior of SiC particle reinforced aluminum composites. Scr. Mater. 51, 161–165.

Chen, N., Zaefferer, S., Lahn, L., Günther, K., Raabe, D., 2002. In: Lee, S. (Ed.), Proceedings of the 13th International Conference on Textures of Materials (ICOTOM 13). Seoul, South Korea, pp. 949.

Choi, S.-H., 2003. Simulation of stored energy and orientation gradients in cold-rolled interstitial free steels. Acta Mater. 51, 1775–1788.

Christian, J.W., 1965. The theory of transformations in metals and alloys. In: Raynor, G.V. (Ed.), 1965. International Series of Monographs in Metal Physics and Physical Metallurgy, vol. 7. Pergamon Press, London, pp. 710–742.

Chung, J.-S., Ice, G.E., 1999. Automated indexing for texture and strain measurement with broad-band-pass X-ray microbeams. J. Appl. Phys. 86, 5249–5255.

Clarebrough, L.M., Haregraves, M.E., West, G.W., 1955. The release of energy during annealing of deformed metals. In: Proceedings of the Royal Society, A, vol. 232.

Cotterill, P., Mould, P.R., 1976. Recrystallization and Grain Growth in Metals. Surrey University, London. p. 85.

Crumbach, M., Goerdeler, M., Gottstein, G., Neumann, L., Aretz, H., Kopp, R., 2004. Through-process texture modelling of aluminium alloys. Modell. Simul. Mater. Sci. Eng. 12. S1.

Crumbach, M., Goerdeler, M., Gottstein, G., 2006. Modelling of recrystallisation textures in aluminium alloys: I. Model set-up and integration. Acta Mater. 54 (12), 3275–3289.

Czochralski, J., 1927. Geschichtlicher Beitrag zur Frage der Rekristallisation. Z. Metallkd. 19, 316–320.

Davies, C.H.J., Hong, L., 1999. Cellular automaton simulation of static recrystallization in cold-rolled AA1050. Scr. Mater. 40, 1145–1152.

Davies, C.H.J., 1995. The effect of neighbourhood on the kinetics of a cellular automaton recrystallisation model. Scr. Metall. Mater. 33, 1139–1154.

Davies, C.H.J., 1997. Growth of nuclei in a cellular automaton simulation of recrystallisation. Scr. Mater. 36, 35–46.

De Siqueira, R.P., Sandim, H.R.Z., Raabe, D., 2013. Particle stimulated nucleation in coarse-grained ferritic stainless steel. Metall. Mater. Trans. A 44A, 469–478.

Dillamore, I.L., Katoh, H., 1974. Met. Sci. 8, 73.

Dillamore, I.L., Smith, C.J.E., Watson, T.W., 1967. Met. Sci. J. 1, 49.

Dillamore, I.L., Roberts, J.G., Bush, A.C., 1979. Met. Sci. 13, 73.

Doherty, R.D., Baumann, S.F., 1993. In: Morris, J.G., et al. (Eds.), Aluminum Alloys for Packaging. TMS, Warrendale, PA, pp. 369.

Doherty, R.D., Cahn, R.W., 1972. Nucleation of new grains in recrystallization of cold worked metals. J. Less Common Met. 28 (2), 279–296.

Doherty, R.D., Martin, J.W., 1962–1963. The effect of a dispersed second phase on the recrystallization of aluminium–copper alloys. J. Inst. Met. 91, 332–338.

Doherty, R.D., Szpunar, J.A., 1984a. Kinetics of subgrain coalescence-a reconsideration of the theory. Acta Metall. 32 (10), 1789–1798.

Doherty, R.D., Szpunar, J.A., 1984b. Acta Metall. 32, 1789.

Doherty, R.D., Gottstein, G., Hirsch, J.R., Hutchinson, W.B., Lucke, K., Nes, E., Wilbrandt, P.J., 1988. In: Kallend, J.S., Gottstein, G. (Eds.), ICOTOM 8. TMS, Warrendale, PA, pp. 563.

Doherty, R.D., Li, K., Anderson, M.P., Rollett, A.D., Srolovitz, D.J., 1990. Proceedings of the International Conference on Recrystallization in Metallic Materials, Recrystallization. In: Chandra, T. (Ed.), The Minerals, Metals and Materials Society. TMS Publication, Warrendale, PA, pp. 129.

Doherty, R.D., Kashyap, K., Panchanadeeswaran, S., 1993. Acta Metall. Mater. 41, 3029.

Doherty, R.D., Hughes, D.A., Humphreys, F.J., Jonas, J.J., Juul Jensen, D., Kassner, M.E., King, W.E., McNelley, T.R., McQueen, H.J., Rollett, A.D., 1997. Current issues in recrystallization: a review. Mater. Sci. Eng., A 238, 219–274.

Doherty, R.D., 1985. Scr. Metall. 19, 927.

Doherty, R.D., 2005. Primary recrystallization. In: Cahn, R.W., et al. (Eds.), Encyclopedia of Materials: Science and Technology. Elsevier, pp. 7847–7850.

Dorner, D., Zaefferer, S., -Lahn, L., Raabe, D., 2006. Overview of microstructure and microtexture development in grain-oriented silicon steel. J. Magn. Magn. Mater. 304, 183–186.

Dorner, D., Zaefferer, S., Raabe, D., 2007. Retention of the Goss orientation between microbands during cold rolling of an Fe3%Si single crystal. Acta Mater. 55, 2519–2530.

Drube, B., Stüwe, H.P., 1967. Z. Metallkd. 58, 799–804.

Duckham, A., Engler, O., Knutsen, R.D., June 28, 2002. Moderation of the recrystallization texture by nucleation at copper-type shear bands in Al–1Mg. Acta Mater. 50 (11), 2881–2893.

Duggan, B.J., Chung, C.Y., 1994. Mater. Sci. Forum 113–115, 1765.

Duggan, B.J., Hatherley, M., Hutchinson, W.B., Wakefield, P.T., 1978. Met. Sci. 12, 343.

Duggan, B.J., Lucke, K., Kohlhoff, G.D., Lee, C.S., 1993. Acta Metall. Mater. 41, 1921.

Eisenlohr, A., Gutierrez-Urrutia, I., Raabe, D., May 2012. Adiabatic temperature increase associated with deformation twinning and dislocation plasticity. Acta Mater. 60 (9), 3994–4004.

Engler, O., Vatne, H.E., 1998. Modeling the recrystallization textures of aluminum alloys after hot deformation. JOM 50, 23–27.

Engler, O., Hirsch, J., Lücke, K., 1989. Acta Metall. 37, 2743.

Engler, O., Vatne, H.E., Nes, E., 1996. The roles of oriented nucleation and oriented growth on recrystallization textures in commercial purity aluminium. Mater. Sci. Eng., A 205, 187–198.

Engler, O., 1997. Influence of particle stimulated nucleation on the recrystallization textures in cold deformed Al-alloys Part II—Modeling of recrystallization textures. Scr. Mater. 37, 1675–1683.

Ewing, J.A., Rosenhain, W., 1899. The crystalline structure of metals. Proc. R. Soc. 65, 85–90.

Ewing, J.A., Rosenhain, W., 1900a. The crystalline structure of metals. Philos. Trans. R. Soc., A 193, 353–372.

Ewing, J.A., Rosenhain, W., 1900b. The crystalline structure of metals. Philos. Trans. R. Soc., A 195, 279–301.

Faivre, P., Doherty, R.D., 1979. Nucleation of recrystallization in compressed aluminium – studies by electron microscopy and Kikuchi diffraction. J. Mater. Sci. 14 (4), 897–919.

Ferry, M., Humphreys, F.J., 1996. Discontinuous subgrain growth in deformed and annealed {110} (001) aluminium single crystals. Acta Mater. 44, 1293–1308.

Ferry, M., 2002. Mechanism of discontinuous subgrain growth in as-deformed aluminum single crystals. Mater. Sci. Forum 408–412, 979–984.

Fritzmeier, L., Luton, M.J., McQueen, H.J., 1979. Strength of Metals and Alloys. In: ICSMA 5, vol. 1. Pergamon Press, Frankfurt. pp. 95.

Fujita, N., Ohmura, K., Kikuchi, M., Suzuki, T., Funaki, S., Hiroshige, I., 1996. Scr. Metall. Mater. 35, 705–710.

Furtkamp, M., Gottstein, G., Molodov, D.A., Semenov, V.N., Shvindlerman, L.S., 1998. Grain boundary migration in Fe3.5%Si bicrystals with <001> tilt boundaries. Acta Mater. 46, 4103–4110.

Furu, T., Marthinsen, K., Nes, E., 1993. Mater. Sci. Forum 113–115, 41.

Gifkins, R.C., Coe, H.C., 1951. Metallurgia 43, 47.

Glauber, R.J., 1963. J. Math. Phys. 4, 294.

Glazier, J.A., Anderson, M.P., Grest, G.S., 1990. Philos. Mag. B 62, 615.

Godara, A., Raabe, D., Van Puyvelde, P., Moldenaers, P., 2006. Polym. Test. 25, 460–469.

Goss, N.P., 1935. Trans. Am. Soc. Met. 23, 511.

Gottstein, G., Shvindlerman, L.S., 1999. Grain Boundary Migration in Metals – Thermodynamics, Kinetics, Applications. CRC Press, Boca Raton.

Gottstein, G., Molodov, D.A., Czubayko, U., Shvindlerman, L.S., 1995. J. Phys. IV, colloque C3, supplément au Journal de Physique III 5, 9.

Gottstein, G., Shvindlerman, L.S., Molodov, D.A., Czubayko, U., 1997. In: Duxbury, P.M., Pence, T.J. (Eds.), Dynamics of Crystal Surfaces and Interfaces. Plenum Press, New York, pp. 109.

Gottstein, G., Molodov, D.A., Shvindlerman, L.S., 1998. Interface Sci. 6, 7.

Grewen, J., Huber, J., 1978. In: Haessner, F. (Ed.), Recrystallization of Metallic Materials, second ed. Dr. Riederer-Verlag, Stuttgart, pp. 111.

Gutierrez-Urrutia, I., Raabe, D., 2011. Dislocation and twin substructure evolution during strain hardening of an Fe-22 wt.% Mn-0.6 wt.% C TWIP steel observed by electron channeling contrast imaging. Acta Mater. 59, 6449–6462.

Gutierrez-Urrutia, I., Zaefferer, S., Raabe, D., 2009. Electron channeling contrast imaging of twins and dislocations in twinning-induced plasticity steels under controlled diffraction conditions in a scanning electron microscope. Scr. Mater. 61, 737–740.

Gutierrez-Urrutia, I., Zaefferer, S., Raabe, D., 2010. The effect of grain size and grain orientation on deformation twinning in a Fe-22 wt.% Mn-0.6 wt.% C TWIP steel. Mater. Sci. Eng., A 527, 3552–3560.

Gutierrez-Urrutia, I., Zaefferer, S., Raabe, D., 2013. Coupling of Electron Channeling with EBSD: toward the quantitative characterization of deformation structures in the SEM. JOM 65, 1229–1236.

Haessner, F. (Ed.), 1978. Recrystallization. Dr. Riederer Verlag, Stuttgart.

Hallberg, H., 2011. Metals 1, 16–48.

Hansen, N., Bay, B., 1981. Acta Metall. 29, 65.

Hartig, Ch., Feller-Kniepmeier, M., May 1985. Electron microscopic investigation of the microstructure of rolled and annealed Ni-single crystals. Acta Metall. 33 (5), 743–752.

Harun, A., Holm, E.A., Clode, M.P., Miodownik, M.A., Jul 2006. On computer simulation methods to model Zener pinning. Acta Mater. 54 (12), 3261–3273.

Hashimoto, S., Baudelet, B., 1989. Scr. Metall. 23, 1855.

Hassold, G.N., Holm, E.H., 1993. Comput. Phys. 7, 7.

Hayakawa, Y., Szpunar, J.A., 1997. Acta Mater. 45, 1285.

Heidenreich, R.D., 1949. Electron microscope and diffraction study of metal crystal textures by means of thin sections. J. Appl. Phys. 20 (10), 993–1010.

Heinrich, M., Haider, F., 1996. Primary recrystallization in slightly tensile deformed aluminium single crystals. Philos. Mag. 74, 1047–1057.

Hesselbarth, H.W., Göbel, I.R., 1991. Simulation of recrystallization by cellular automata. Acta Metall. 39, 2135–2144.

Hillert, M., 1965. Acta Metall. 13, 227.

Hillert, M., 1988. Acta Metall. 36, 3177.

Himmel, L. (Ed.), 1962. Recovery and Recrystallization of Metals. AIME. published by Interscience Publishers (1963).

Hjelen, J., Orsund, R., Nes, E., 1991. Acta Metall. Mater. 39, 1377.

Holm, E.A., Battaile, C.C., 2001. The computer simulation of microstructural evolution. JOM 9, 20.

Holm, E.A., Zacharopoulos, N., Srolovitz, D.J., 1998. Acta Mater. 46, 953.

Holm, E.A., Miodownik, M.A., Rollett, A.D., 2003. On abnormal subgrain growth and the origin of recrystallization nuclei. Acta Mater. 51, 2701–2716.

Holm, E.A., Hassold, G.N., Miodownik, M.A., Sep 3, 2001. On misorientation distribution evolution during anisotropic grain growth. Acta Mater. 49 (15), 2981–2991.

Hölscher, M., Raabe, D., Lücke, K., 1991. Rolling and recrystallization textures of bcc steels. Steel Res. 62, 567–575.

Hu, H., Rath, B.B., Vandermeer, R.A., 1990. An historical perspective and overview of the annealing studies of cold worked metals. In: Chandra, T. (Ed.), Recrystallization '90. TMS, Warrendale, pp. 3–16.

Hu, Y., Miodownik, M.A., Randle, V., Jun 2008. Experimental and computer model investigations of microtexture evolution of non-oriented silicon steel. Mater. Sci. Technol. 24 (6), 705–710.

Hu, H., 1962. Direct observations on the annealing of Si–Fe crystals in the electron microscope. Trans. Metall. Soc. AIME 224 (1), 75–84.

Hu, H., 1963. Recrystallization by subgrain coalescence. In: Thomas, G., Washburn, J. (Eds.), Electron Microscopy and Strength of Crystals. Interscience, New York, pp. 564–573.

Huang, Y., Humphreys, F.J., Ferry, M., 2000. The annealing behavior of deformed cube-oriented aluminium single crystals. Acta Mater. 48, 2543–2556.

Huh, M.-Y., Lee, J.-H., Park, S.H., Engler, O., Raabe, D., 2005. Effect of through-thickness macro and micro-texture gradients on ridging of 17%Cr ferritic stainless steel sheet. Steel Res. 76, 797–806.

Humphreys, F.J., Ardakani, M.G., 1994. Acta Metall. 42, 749.

Humphreys, F.J., Chan, H.M., 1996. Discontinuous and continuous annealing phenomena in aluminium–nickel alloy. Mater. Sci. Technol. 12 (2), 143–148.

Humphreys, F.J., Ferry, M., 1997. Applications of electron backscattered diffraction to studies of annealing of deformed metals. Mater. Sci. Technol. 13 (1), 85–90.

Humphreys, F.J., Hatherly, M., 1995. Recrystallization and Related Annealing Phenomena, vol. 8. Pergamon Press, Oxford, 235.

Humphreys, F.J., Hatherly, M., 2004. Recrystallisation and Related Annealing Phenomena. Elsevier.

Humphreys, F.J., Kalu, P.N., 1987. Acta Metall. 35, 2815.

Humphreys, F.J., 1977a. Acta Metall. 25, 1323.

Humphreys, F.J., 1977b. Acta Metall. 25, 1323–1344.

Humphreys, F.J., 1979. Acta Metall. 27, 1801.

Humphreys, F.J., 1992a. Modelling mechanisms and microstructures of recrystallisation. Mater. Sci. Technol. 8, 135–143.

Humphreys, F.J., 1992b. A network model for recovery and recrystallisation. Scr. Metall. 27, 1557–1562.

Humphreys, F.J., 1997. A unified theory of recovery, recrystallisation and grain growth, based on the stability and growth of cellular microstructures-i the basic model. Acta Meter. 45, 4231–4240.

Hurley, P.J., Humphreys, F.J., 2003. Acta Mater. 51, 3779–3793.

Hutchinson, W.B., Ryde, L., 1995. Microstructural and crystallographic aspects of recrystallization. In: Hansen, N., et al. (Eds.), Proc. 16th Riso Symposium. Riso National Lab, Roskilde, Denmark, pp. 105.

Hutchinson, W.B., Oscarsson, A., Karlsson, 1989. Mater. Sci. Technol. 5, 1118.

Hutchinson, W.B., 1989. Acta Metall. 37, 1047.

Hutchinson, B., 1999. Philos. Trans. R. Soc., A 357, 1471–1485.

Ibe, G., Lücke, K., 1966. Recrystallization, Grain Growth and Textures. American Society for Metals, Metals Park, Ohio. pp. 434.

Ibe, G., Lücke, K., 1972. Texture 1, 87.

Ibe, G., Dietz, W., Fraker, A.-C., Liicke, K., 1970. Z. Metallkd. 6, 498.

Inagaki, H., 1987. Z. Metallkd. 78, 630–638.

Inokuti, Y., Doherty, R.D., 1977. Texture 2, 143.

Inokuti, Y., Doherty, R.D., 1978. Acta Metall. 26, 61.

Inokuti, Maeda, C., Itoh, Y., Shimanaka, H., 1981. ICOTOM6. Tokyo 948.

Jägle, E.A., Mittemeijer, E.J., April 2012. Metall. Mater. Trans. 43A, 1117.

Janssens, K.G.F., Raabe, D., Kozeschnik, E., Miodownik, M.A., Nestler, B., 2007. Computational Materials Engineering. Elsevier, London.

Janssens, K.G.F., 2003. Random grid, three-dimensional, space-time coupled cellular automata for the simulation of recrystallization and grain growth. Modell. Simul. Mater. Sci. Eng. 11, 157.

Jia, J., Raabe, D., 2006. Eur. Polym. J. 42, 1755–1766.

Jia, N., Roters, F., Eisenlohr, P., Kords, C., Raabe, D., 2012a. Acta Mater. 60, 1099–1115.

Jia, N., Eisenlohr, P., Roters, F., Raabe, D., Zhao, X., 2012b. Orientation dependence of shear banding in face-centered-cubic single crystals. Acta Mater. 60, 3415–3434.

Johnson, W.A., Mehl, R.F., 1939. Reaction kinetics in the process of nucleation and growth. Trans. Metall. Soc. AIME 135, 416–458.

Jonas, J.J., Sellars, C.M., Tegart, W.J.McG., 1969. Metall. Rev. 14, 1.

Jones, A.R., Ralph, B., Hansen, N., 1979. Nucleation of recrystallization in aluminium containing dispersions of alumina. Met. Sci. 13, 149–154.

Juntunen, P., Raabe, D., Karjalainen, P., Kopio, T., Bolle, G., 2001. Optimizing continuous annealing of if steels for improving their deep drawability. Metall. Mater. Trans. A 32, 1989.

Juul Jensen, D., Hansen, N., Liu, Y.L., 1991. Mater. Sci. Technol. 2, 369.

Juul Jensen, D., Bolingbroke, R.K., Shi, H., Shahani, R., Furu, T., 1994. Mater. Sci. Forum 157–162, 1991.

Juul-Jensen, D., Hansen, N., Humphreys, F.J., 1988. Acta Metall. 33, 2155.

Juul-Jensen, D., 1992. Scr. Metall. Mater. 27, 533.

Juul-Jensen, D., 1995. Acta Metall. Mater. 43, 4117.

Kalisher, S., 1881. Über den Einfluss der Wärme auf die Molekularstruktur des Zinks. Ber. Dtsch. Chem. Ges. XIV, 2727–2753.

Keh, A.S., Weissman, S., 1963. In: Thomas, G., Washburn, J. (Eds.), Deformation Structure in Body-centered Cubic Metals. Electron Microscopy and Strength of Crystals. Interscience, New York, pp. 231–300.

Kinderlehrer, D., Livshits, I., Manolache, F., Rollett, A.D., Ta'asan, S., 2001. An approach to the mesoscale simulation of grain growth, influences of interface and dislocation behavior on microstructure evolution. In: Aindow, M., et al. (Eds.). Mat. Res. Soc. Symp. Proc., 652, p. Y1.5.

Kinderlehrer, D., Livshits, I., Rohrer, G.S., Ta'asan, S., Yu, P., 2004. Mesoscale evolution of the grain boundary character distribution, recrystallization and grain growth. Mater. Sci. Forum 467–470, 1063–1068.

Klement, U., Haasen, P., April 1993. In situ HVEM-investigations of the early stages of recrystallization in Cu-0.2 at.% Mn-single crystals. Acta Metall. Mater. 41 (4), 1075–1087.

Klinkenberg, C., Raabe, D., Lücke, K., 1992. Influence of volume fraction and dispersion rate of grain boundary cementite on the cold rolling textures of low carbon steels. Steel Res. 63, 227.

Kolmogorov, A.N., 1937. Statistical theory of crystallization of metals. Izv. Akad. Nauk SSSR, Met. 1, 355–359.

Konrad, J., Zaefferer, S., Raabe, D., 2006. Investigation of orientation gradients around a hard Laves particle in a warm rolled Fe3Al-based alloy by a 3D EBSD-FIB technique. Acta Mater. 54, 1369–1380.

Korbel, A., Embury, J.D., Hatherley, M., Martin, P.L., Erblöh, H.W., 1986. Acta Metall. 34, 1999.

Köster, U., 1974. Recrystallization involving a second phase. Metal Sci. 8, 151–160.

Krill III, C.E., Chen, L.-Q., 2002. Computer simulation of 3-D grain growth using a phase field model. Acta Mater. 50, 3057–3073.

Kuo, J.-C., Zaefferer, S., Zhao, Z., Winning, M., Raabe, D., 2003. Deformation behaviour of aluminium-bicrystals. Adv. Eng. Mater. 5 (8), 563–566.

Larson, B.C., Yang, W., Ice, G.E., Budai, J.D., Tischler, J.Z., 2002. Three-dimensional X-ray structural microscopy with submicrometre resolution. Nature 415, 887–890.

Laue, M.V., 1913. Röntgenstrahlinterferenzen. Z. Phys. 14, 1075–1079.

Lee, C.S., Duggan, B.J., 1993. Acta Metall. Mater. 41, 2691.

Leslie, W.C., Michalak, J.T., Aul, F.W., 1963a. In: Spencer, et al. (Eds.), Iron and Its Dilute Solutions. Publ. Interscience, NY.

Leslie, W.C., Plecity, F.J., Michalak, J.T., 1963b. Acta Metall. 11, 561.

Leslie, W.C., Michalak, J.T., Aul, F.W., 1963c. In: Spencer, Werner, (Eds.), 1963c. Iron and Its Dilute Solid Solutions, 119. Interscience, New York, pp. 1470.

Li, C.H., Edwards, E.H., Washburn, J., Parker, E.R., 1953. Stress-induced movement of crystal boundaries. Acta Metall. 1, 223–229.

Li, J.C.M., 1961. J. Appl. Phys. 32, 525.

Li, J.C.M., 1962. Possibility of subgrain rotation during recrystallization. J. Appl. Phys. 33 (10), 2958–2965.

Liebman, B., Lücke, K., Masing, G., 1956. Untersuchung über die Orientierungsabhängigkeit der Wachstumsgeschwindigkeit bei der primären Rekristallisation von Aluminium-Einkristallen. Z. Metallkd. 47 (2), 57–63.

Lin, P., Palumbo, G., Harase, J., Aust, K.T., 1996. Acta Mater. 44, 4677.

Lins, J.F.C., Sandim, H.R.Z., Kestenbach, H.-J., Raabe, D., Vecchio, K.S., 2007. A microstructural investigation of adiabatic shear bands in an interstitial free steel. Mater. Sci. Eng., A 457, 205–218.

Liu, J., Mato, M., Doherty, R.D., 1989. Scr. Metall. 23, 1811.

Liu, X., Solberg, J.K., Gjengedal, R., Kluken, A.O., 1995. Mater. Sci. Technol. 11, 469.

Logé, R., Bernacki, M., Resk, H., Delannay, L., Digonnet, H., Chastel, Y., Coupez, T., 2008. Linking plastic deformation to recrystallization in metals using digital microstructures. Philos. Mag. 88, 3691–3712.

Lücke, K., Detert, K., 1957. A quantitative theory of grain boundary motion and recrystallization in metals in the presence of impurities. Acta Metall. 5 (11), 628–637.

Lücke, K., Stüwe, H.P., 1963. In: Himmel, L. (Ed.), On the Theory of Grain Boundary Motion. Recovery and Recrystallization of Metals. Interscience, New York, pp. 171–210.

Luton, M.J., Sellars, S.M., 1969. Acta Metall. 17, S. 1033.

Mahin, K., Hanson, K., Morris, J., 1980. Acta Metall. 28, 443.

Margulies, L., Winther, G., Poulsen, H.F., 2001. In situ measurement of grain rotation during deformation of polycrystals. Science 291 (5512), 2392–2394.

Martin, J.W., Doherty, R.D., 1976. Stability of Microstructure in Metallic Systems. Cambridge University Press, Cambridge. pp. 7–10.

Marx, V., Raabe, D., Engler, O., Gottstein, G., 1997. Simulation of the texture evolution during annealing of cold rolled bcc and fcc metals using a cellular automaton approach. Textures Microstruct. 28, 211–218.

Marx, V., Reher, F.R., Gottstein, G., 1998. Stimulation of primary recrystallization using a modified three-dimensional cellular automaton. Acta Mater. 47, 1219–1230.

Matsuo, M.I., 1989. ISIJ Int. 29, 809.

Maurice, C., Humphries, J., 1998. 2- and 3-d curvature driven vertex simulations of grain growth. In: Grain Growth in Polycrystalline Materials III, vol. 1. The Minerals, Metals and Materials Society, Warrnedale, pp. 81–90.

Maurice, C., 2001. 2- and 3-d curvature driven vertex simulations of grain growth. In: Proceedings of the First Joint International Conference on Recrystallization and Grain Growth, vol. 1. Springer-Verlag, Berlin, pp. 123–134.

May, J.E., Turnbull, D., 1958. Trans. Metall. Soc. AIME 769, 781.

McQueen, H.J., Jonas, J.J., 1975. Plastic Deformation of Materials. Acadamic Press, New York. pp. 393–493.

McQueen, H.J., Petkovic, R.A., Weiss, H., Hinton, L.G., 1976. In: Balance, J.B. (Ed.), The Hot Deformation of Austenite. AIME, New York, pp. 113–139.

McQueen, H.J., Evangelist, E., Ryan, N.D., 1990. In: Chandra, T. (Ed.), Recrystallization ('90) in Metals and Materials. TMS-AIME, Warrendale, PA, pp. 89.

McQueen, H.J., 1968. J. Met. 20 (4), 31.

Merriman, B., Bence, J., Osher, S., 1994. Motion of multiple junctions: a level set approach. J. Comput. Phys. 112, 334–363.

Metropolis, N., Rosenbluth, A.W., Rosenbluth, M.N., Teller, A.T., Teller, E., 1953. J. Chem. Phys. 21, 1087.

Miodownik, M.A., Martin, J.W., Cerezo, A., 1999. Philos. Mag. 79, 203.

Miodownik, M.A., 2002. A review of microstructural computer models used to simulate grain growth and recrystallisation in aluminium alloys. J. Light Met. 2, 125–135.

Miodownik, M., Holm, E.A., Hassold, G.N., Jun 13, 2000. Highly parallel computer simulations of particle pinning: Zener vindicated. Scr. Mater 42 (12), 1173–1177.

Miyazaki, A., Takao, K., Furukimi, O., 2002. ISIJ Int. 42, 916.

Molodov, D.A., Straumal, B.B., Shvindlerman, L.S., 1984. Scr. Metall. 18, 207.

Molodov, D.A., Gottstein, G., Shvindlerman, L.S., 1998. Grain Growth in Polycrystalline Materials III: Proceedings of [the] Third International Conference on Grain Growth, ICGG-3, June 14-19, 1998, Carnegie Mellon University, Pittsburgh, PA, USA.

Molodov, D.A., 2001. Grain Grain boundary boundary character – a key factor for grain boundary control. In: Gottstein, G., Molodov, D.A. (Eds.), 2001. Recrystallization and Grain Growth, vol. 1. Springer Verlag, Aachen, pp. 21–38.

Morawiec, A., 2000. Scr. Metall. 43, 275.

Mughrabi, H., Ungár, T., Kienle, W., Wilkens, M., 1986. Long-range internal stresses and asymmetric X-ray line-broadening in tensile-deformed [001]-orientated copper single crystals. Philos. Mag. A 53 (6), 793–813.

Mullins, W.W., 1956. J. Appl. Phys. 27, 900.

Nagai, T., Ohta, S., Kawasaki, K., Okuzono, T., 1990. Computer simulation of cellular pattern growth in two and three dimensions. Phase Trans. 28, 177–211.

Nes, E., 1995. Recovery revisited. Acta Metall. Mater. 43 (6), 2189–2207.

Orowan, E., 1934. Zur Kristallplastizität: iii. Über den Mechanismus des Gleitvorganges. Z. Phys. 89, 634–659.

Peczak, P., 1995. Acta Metall. 43, 1297.

Peranio, N., Li, Y.J., Roters, F., Raabe, D., 2010. Microstructure and texture evolution in dual-phase steels: competition between recovery, recrystallization, and phase transformation. Mater. Sci. Eng., A 527, 4161–4168.

Pezzee, C.E., Dunand, D.C., 1994. The impingement effect of an inert, immobile second phase on the recrystallization of a matrix. Acta Metall. 42, 1509–1522.

Pimenta Jr., F.C., Arruda, A.C.F., Padilha, A.F., 1986. Resistance to recrystallization in Al-1%Mn alloys. Z. Metallkd. 77 (8), 522–528.

Polanyi, M., 1934. Über eine Art Gitterstörung die einen Kristall plastich machen könnte. Z. Phys. 89, 660–664.

Potts, R.B., 1952. Proc. Cambridge Philos. Soc. 48, 106.

Poulsen, H.F., Nielsen, S.F., Lauridsen, E.M., Schmidt, S., Suter, R.M., Lienert, U., Margulies, L., Lorentzen, T.D., Juul Jensen, D., 2001. Three-dimensional maps of grain boundaries and the stress state of individual grains in polycrystals and powders. J. Appl. Crystallogr. 34, 751–756.

Raabe, D., Becker, R., 2000. Coupling of a crystal plasticity finite element model with a probabilistic cellular automaton for simulating primary static recrystallization in aluminum. Modell. Simul. Mater. Sci. Eng. 8, 445–462.

Raabe, D., Hantcherli, L., 2005. 2D cellular automaton simulation of the recrystallization texture of an IF sheet steel under consideration of Zener pinning. Comput. Mater. Sci. 34, 299–313.

Raabe, D., Lücke, K., 1992. Annealing textures of bcc metals. Scr. Metall. 27, 1533.

Raabe, D., Lücke, K., 1993. Textures of ferritic stainless steels. Mater. Sci. Technol. 9, 302–312.

Raabe, D., Schlenkert, G., Weisshaupt, H., Lücke, K., 1994. Texture and microstructure of rolled and annealed tantalum. Mater. Sci. Technol. 10, 229–305.

Raabe, D., Sachtleber, M., Zhao, Z., Roters, F., Zaefferer, S., 2001. Micromechanical and macromechanical effects in grain scale polycrystal plasticity experimentation and simulation. Acta Mater. 49, 3433.

Raabe, D., Zhao, Z., Park, S.–J., Roters, F., 2002a. Theory of orientation gradients in plastically strained crystals. Acta Mater. 50, 421–440.

Raabe, D., Klose, P., Engl, B., Imlau, K.-P., Friedel, F., Roters, F., 2002b. Concepts for integrating plastic anisotropy into metal forming simulations. Adv. Eng. Mater. 4, 169–180.

Raabe, D., Chen, N., Chen, L., 2004a. Polymer 45, 8265–8277.

Raabe, D., Roters, F., Barlat, F., Chen, L.-Q. (Eds.), 2004b. Continuum Scale Simulation of Engineering Materials. WILEY-VCH, Weinheim.

Raabe, D., 1995a. On the orientation dependence of static recovery in low-carbon steels. Scr. Metall. 33, 735–740.

Raabe, D., 1995b. Investigation of the orientation dependence of recovery in low-carbon steel by use of single orientation determination. Steel Res. 66, 222–229.

Raabe, D., 1996. On the influence of the chromium content on the evolution of rolling textures in ferritic stainless steels. J. Mater. Sci. 31, 3839–3845.

Raabe, D., 1997. Texture and microstructure evolution during cold rolling of a strip cast and of a hot rolled austenitic stainless steel. Acta Mater. 45, 1137–1151.

Raabe, D., 1998. Computational Materials Science. Wiley-VCH, Weinheim.

Raabe, D., 1999. Introduction of a scaleable 3D cellular automaton with a probabilistic switching rule for the discrete mesoscale simulation of recrystallization phenomena. Philos. Mag. A 79, 2339–2358.

Raabe, D., 2000. Scaling Monte Carlo kinetics of the Potts model using rate theory. Acta Mater. 48, 1617.

Raabe, D., 2001. Mesoscale simulation of recrystallization textures and microstructures. Adv. Eng. Mater. 3, 745–752.

Raabe, D., 2002. Cellular automata in materials science with particular reference to recrystallization simulation. Annu. Rev. Mater. Res. 32, 53.

Raabe, D., 2004. Acta Mater. 52, 2653–2664.

Raabe, D., 2007. J. Strain Anal. Eng. Des. 42, 253–268.

Radhakrishnan, B., Sarma, G.B., Zacharia, T., 1998. Acta Mater. 46 (12), 4415–4433.

Radhakrishnan, B., Sarma, G.B., Weiland, H., Baggethun, P., 2000. Modell. Simul. Mater. Sci. Eng. 8 (5), 737–750.

Rajmohan, N., Szpunar, J.A., Hayakawa, Y., 1999. Acta Mater. 47, 2999.

Randle, V., 1992. Microtexture Determination and Its Applications. The Institute of Materials, London.

Randle, V., 2004. Application of electron backscatter diffraction to grain boundary characterisation. Int. Mater. Rev. 49 (1), 1–11.

Randle, V., Engler, O., Aug 7, 2000. Introduction to Texture Analysis: Macrotexture, Microtexture and Orientation Mapping. Crc Pr Inc.

Rath, B.B., Hu, H., 1969a. Effect of driving force on the migration of high-angle tilt grain boundaries in aluminum bicrystals. Trans. Metall. Soc. AIME 245 (7), 1577–1585.

Rath, B.B., Hu, H., 1969b. Trans. Metall. Soc. AIME 245, 1243–1577.

Ray, R.K., Hutchinson, W.B., Duggan, B.J., 1975a. A study of the nucleation of recrystallization using HVEM. Acta Metall. 23 (7), 831–840.

Ray, R.K., Hutchinson, W.B., Duggan, B., 1975b. Acta Metall. 23, 831.

Reeh, S., Music, D., Gebhardt, T., Kasprzak, M., Japel, T., Zaefferer, S., Raabe, D., Richter, S., Schwedt, A., Mayer, J., Wietbrock, B., Hirt, G., Schneider, J.M., 2012. Elastic properties of face-centred cubic Fe-Mn-C studied by nanoindentation and ab initio calculations. Acta Materialia 60, 6025–6032.

Ridha, A.A., Hutchinson, W.B., 1982a. Acta Metall. 30, 1925.

Ridha, A.A., Hutchinson, W.B., 1982b. Acta Metall. 30, 1929.

Rios, P.R., 1987. A theory for grain-boundary pinning by particles. Acta Metall. 35 (12), 2805–2814.

Roberts, W., 1982. In: Krauss, G. (Ed.), Deformation, Processing and Structure. ASM, Metals Park, OH, pp. 109.

Rollet, A.D., Gottstein, G., Shvindlerman, L., Molodov, D., 2004. Grain boundary mobility – a brief review. Z. Metallkd. 95 (4), 226–229.

Rollett, A.D., Raabe, D., 2001. A hybrid model for mesoscopic simulation of recrystallization. Comput. Mater. Sci. 21, 69.

Rollett, A.D., Srolovitz, D.J., Doherty, R.D., Anderson, M.P., 1989a. Computer simulation of recrystallization in non-uniformly deformed metals. Acta Metall. 37 (2), 627–639.

Rollett, A.D., Srolovitz, D.J., Doherty, R.D., Anderson, M.P., 1989b. Acta Metall. 37, 627.

Rollett, A.D., Srolovitz, D.J., Anderson, M.P., 1989c. Acta Metall. 37, 1227.

Rollett, A.D., Luton, M.J., Srolovitz, D.J., 1992. Acta Metall. 40, 43.

Rollett, A.D., 1998. In: Weiland, H. (Ed.), Proceedings of the Third International Conference on Grain Growth. The Minerals, Metals and Materials Society, TMS Publication, Warrendale, PA.

Rollett, A.D., 1997. Overview of modeling and simulation of recrystallization. Prog. Mater. Sci. 42, 79–99.

Rosenhain, W., 1914. An Introduction to Physical Metallurgy. Constable, London.

Rossard, C., Blain, P., 1960. Mem. Sci. Rev. Metall. 57, 173.

Rossard, C., 1963. Écrouissage, Restauration, Recrystalisation. Presses Universitaires de France, Paris. p. 111.

Roters, F., Eisenlohr, P., Hantcherli, L., Tjahjanto, D.D., Bieler, T.R., Raabe, D., 2010. Overview of constitutive laws, kinematics, homogenization and multiscale methods in crystal plasticity finite-element modeling: theory, experiments, applications. Acta Mater. 58, 1152–1211.

Ruder, W.E., 1935. Am. Soc. Microbiol. 23, 534.

Russel, K.C., Ashby, M.F., 1970. Acta Metall. 18, 891.

Ryan, N.D., McQueen, H.J., 1990. High Temp. Technol. 8, 185.

Saetre, T., Hunderi, O., Nes, E., 1986. Acta Metall. 34, 981.

Safran, B.A., Sahni, P.S., Grest, G.S., 1983. Phys. Rev. B28, 2693.

Sahni, P.S., Srolovitz, D.J., Grest, G.S., Anderson, M.P., Safran, B.A., 1983. Phys. Rev. B28, 2705.

Sakai, T., Jonas, J.J., 1984. Acta Metall. 32 (2), 189.

Sakai, T., Akben, M.G., Jonas, J.J., 1983. Acta Metall. 31, 631.

Samajdar, I., Doherty, R.D., 1994. Scr. Metall. Mater. 31, 527.

Samajdar, I., Doherty, R.D., 1995. Scr. Metall. Mater. 32, 845.

Samajdar, I., Doherty, R.D., Kunze, K., 1992. Scr. Metall. 27, 1459.

Sandim, H.R.Z., Renzetti, R.A., Padilha, A.F., Raabe, D., Klimenkov, M., Lindau, R., Möslang, A., 2010. Annealing behavior of ferritic–martensitic 9%Cr–ODS–eurofer steel. Mater. Sci. Eng., A 527, 3602–3608.

Sandström, R., Lehtinen, B., Hedman, E., Groza, I., Karlsson, S., 1978. Subgrain growth in Al and Al-1% Mn during annealing. J. Mater. Sci. 13 (6), 1229–1242.

Schmidt, S., Nielsen, S.F., Gundlach, C., Margulies, L., Huang, X., Juul Jensen, D., July 2004. Science 305 (9), 229–232.

Schwartz, A.J., Kumar, M., Field, D.P., Adams, B.L. (Eds.), June 30 2009. Electron Backscatter Diffraction in Materials Science, second ed, Springer, Berlin.

Schwarzer, R., Weiland, H., 1986. In: Bunge, H.J. (Ed.), Experimental Techniques of Texture Analysis. DGM Informationsgesellschaft, Oberursel.

Sebald, R., Gottstein, G., 2002. Modeling of recrystallization textures: interaction of nucleation and growth. Acta Mater. 50, 1587–1598.

Sellars, C.M., Tegart, W.J.McG., 1966. Mem. Sci. Rev. Metall. 63, 731–746.

Sellars, C.M., 1978. Philos. Trans. R. Soc., A 288, 147–158.

Sheldon, R.K., Dunand, D.C., 1996. Computer modeling of particle pushing and clustering during matrix crystallization. Acta Mater. 44, 4571–4582.

Shimizu, R., Harase, J., 1989. Acta Metall. 37, 1241.

Shvindlerman, L.S., Gottstein, G., 1999. Recrystallization and related phenomena. In: Sakai, T., Suzuki, H.G. (Eds.), The Japan Inst. of Metals, Proc, Grain Boundary and Triple Junction Migration – the Latest Advances, vol. 13, pp. 431–438.

Shvindlerman, L.S., Czubayko, U., Gottstein, G., Molodov, D.A., 1995. In: Hansen, N., Juul Jensen, D., Liu, Y.L., Ralph, B. (Eds.), Proceedings 16th RISØ International Symposium on Materials Science: Microstructural and Crystallographic Aspects of Recrystallization, pp. 545.

Siqueira, R.P., Sandim, H.R.Z., Oliveira, T.R., Raabe, D., 2011. Composition and orientation effects on the final recrystallization texture of coarse-grained Nb-containing AISI 430 ferritic stainless steels. Mater. Sci. Eng., A 528, 3513–3519.

Smith, C.S., 1948. Grains, phases and interfaces, an interpretation of microstructure. Trans. Metall. Soc. AIME 175, 15.

Sorby, H.C., 1886. The application of very high powers to the study of the microscopical structure of steel. J. Iron Steel Inst. 30 (1), 140–145.

Sorby, H.C., 1887. On the microscopical structure of iron and steel. J. Iron Steel Inst. 31 (1), 255–288.

Srolovitz, D.J., Anderson, M.P., Sahni, P.S., Grest, G.S., 1984. Acta Metall. 32, 793.

Srolovitz, D.J., Grest, G.S., Anderson, M.P., 1985. Acta Metall. 33, 2233.

Srolovitz, D.J., Grest, G.S., Anderson, M.P., 1986a. Computer simulation of recrystallization—I. Homogeneous nucleation and growth. Acta Metall. 34, 1833–1845.

Srolovitz, D.J., Grest, G.S., Anderson, M.P., 1986b. Acta Metall. 34, 1833.

Srolovitz, D.J., Grest, G.S., Anderson, M.P., Rollett, A.D., 1988. Acta Metall. 36, 2115.

Stead, J.E., 1898. The crystalline structure of iron and steel. J. Iron Steel Inst. 53 (1), 145–205.

Straub, S., Blum, W., Maier, H.J., Ungar, T., Borbély, A., Renner, H., November 1996. Long-range internal stresses in cell and subgrain structures of copper during deformation at constant stress. Acta Mater. 44 (11), 4337–4350.

Stüwe, H.P., Padilha, A.F., Siciliano Jr, F., 2002. Competition between recovery and recrystallization. Mater. Sci. Eng., A A233, 361–367.

Stüwe, H.P., 1968. Deformation Under Hot Working Conditions. Iron Steel Inst., London. pp. 1–6.

Stüwe, H.P., 1978. In: Haessner, F. (Ed.), Driving and Dragging Forces in Recrystallization. Recrystallization of Metallic Materials. Dr. Riederer Verlag, Stuttgart, pp. 11–21.

Swan, P.R., 1963. In: Thomas, G., Washburn, J. (Eds.), Dislocations Arrangements in Face Centered Cubic Metals. Electron Microscopy and Strength of Crystals. Interscience, New York, pp. 131–181.

Tamhankar, R., Plateau, J., Crussard, C., 1958. Rev. Metall. 55, 383.

Tamura, N., Spolenak, R., Valek, B.C., Manceau, A., Meier Chang, M., Celestre, R.S., MacDowell, A.A., Padmore, H.A., Patel, J.R., 2002. Submicron X-ray diffraction and its applications to problems in materials and environmental science. Rev. Sci. Instrum. 73, 1369–1372.

Tamura, N., MacDowell, A.A., Spolenak, R., Valek, B.C., Bravman, J.C., Brown, W.L., Celestre, R.S., Padmore, H.A., Batterman, B.W., Patel, J.R., 2003. Scanning X-ray microdiffraction with sub-micrometer white beam for strain/stress and orientation mapping in thin films. J. Synchrotron Radiat. 10, 137–143.

Tavernier, P., Szpunar, J.A., 1991a. Acta Metall. 39, 549.

Tavernier, P., Szpunar, J.A., 1991b. Acta Metall. 39, 557.

Taylor, G.I., 1934. The mechanism of plastic deformation of crystals. Part I-Theoretical. Proc. R. Soc. London 145 (A), 312–387.

Thomas, I., Zaefferer, S., Friedel, F., Raabe, D., 2003. High resolution EBSD investigation of deformed and partially recrystallized IF steel. Adv. Eng. Mater. 5, 566–570.

Tsuji, N., Matsubara, Y., Sakai, T., Saito, Y., 1997. ISIJ Int. 37, 797.

Turnbull, D., Fisher, J.C., 1949. Rate of nucleation in condensed systems. J. Chem. Phys. 17, 71–73.

Tytko, D., Choi, P., Klöwer, J., Kostka, A., Inden, G., Raabe, D., 2012. Microstructural evolution of a Ni-based superalloy (617B) at 700 °C studied by electron microscopy and atom probe tomography. Acta Mater. 60, 1731–1740.

Upmanyu, M., Srolovitz, D.J., Shvindlerman, L.S., Gottstein, G., 1999. Misorientation dependence of intrinsic grain boundary mobility: simulation and experiment. Acta Mater. 47, 3901–3914.

Ushioda, K., Schlippenbach, U. v., Hutchinson, W.B., 1987. The effect of carbon content on recrystallisation and texture development in steel. Textures Microstruct. 7, 11–28.

Vandermeer, R.A., Juul Jensen, D., 1993. The migration of high angle grain boundaries during recrystallization. Interface Sci. 6 (1–2), 95–104.

Vandermeer, R.A., Rath, B.B., 1990. Interface migration during recrystallization: the role of recovery and stored energy gradients. Metall. Trans. A 21 (A), 1143–1149.

Varma, K., Willitis, B.L., 1984. Subgrain growth in aluminum during static annealing. Metall. Trans. A 15, 1502–1503.

Viswanathan, R., Bauer, C.L., 1973. Kinetics of grain boundary migration in copper bicrystals with [001] rotation axes. Acta Metall. 21 (8), 1099–1109.

von Neumann, J., 1952. Metal Interface. ASM, Cleveland, Ohio. pp. 108.

Walter, J.L., Koch, E.F., 1963. Substructures and recrystallization of deformed (100) [001]-oriented crystals of high purity silicon–iron. Acta Metall. 11 (8), 923–938.

Wassermann, G., Grewen, J., 1962. Texturen metallischer Werkstoffe. Springer-Verlag, Berlin.

Weiland, H., Schwarzer, R., 1984. Proc. ICOTOM 7, 857.

Wever, F., 1924. Über die Walzstruktur kubisch kristallisierender Metalle. Z. Phys. 28, 69–90.

Weygand, D., Brechet, Y., Lepinoux, J., 1999. Zener pinning and grain growth: a two-dimensional vertex computer simulation. Acta Mater. 47, 961–970.

Weygand, D., Brechet, Y., Lepinoux, J., 2001. A vertex simulation of grain growth in 2d and 3d. Adv. Eng. Mater. 3, 67–71.

Wilbrandt, P.-J., Haasen, P., 1979. HVEM study of the development of the recrystallization texture in deformed copper single crystals. Krist. Tech. 14 (11), 1379–1384.

Wilbrandt, P.-J., Haasen, P., 1980a. Z. Metallkd. 71, 273.

Wilbrandt, P.-J., Haasen, P., 1980b. Z. Metallkd. 71, 385.

Wilbrandt, P.-J., 1980. Phys. Status Solidi 61, 411.

Winning, M., Gottstein, G., Shvindlerman, L.S., 2001. Stress induced grain boundary motion. Acta Mater. 49, 211–219.

Winning, M., Gottstein, G., Shvindlerman, L.S., 2002. On the mechanisms of grain boundary migration. Acta Mater. 50, 353–363.

Winning, M., 2003. Motion of <100> tilt grain boundaries. Acta Mater. 50, 6465–6475.

Yang, W., Larson, B.C., Ice, G.E., Tischler, J.Z., Budai, J.D., Chung, K.S., Lowe, W.P., 2003. Spatially resolved Poisson strain and anticlastic curvature measurements in Si under large deflection bending. Appl. Phys. Lett. 82, 3856–3858.

Zaafarani, N., Raabe, D., Singh, R.N., Roters, F., Zaefferer, S., 2006. Three dimensional investigation of the texture and microstructure below a nanoindent in a Cu single crystal using 3D EBSD and crystal plasticity finite element simulations. Acta Mater. 54, 1863–1876.

Zaefferer, S., Kuo, J.-C., Zhao, Z., Winning, M., Raabe, D., 2003. On the influence of the grain boundary misorientation on the plastic deformation of aluminum bicrystals. Acta Mater. 51, 4719–4735.

Zaefferer, S., Wright, S.I., Raabe, D., 2008. Three-dimensional orientation microscopy in a focused ion beam-scanning electron microscope: a new dimension of microstructure characterization. Metall. Mater. Trans. A 39, 374–389.

Zaefferer, S., 2005. In: Proceedings of the 14th International Conference on textures of materials ICOTOM 14, 2005, Leuven, Belgium, Materials Science Forum. Application of Orientation Microscopy in SEM and TEM for the Study of Texture Formation during Recrystallisation Processes, vol. 495–497. Trans Tech Publications, Switzerland, pp. 3–12.

Zambaldi, C., Roters, F., Raabe, D., Glatzel, U., 2007. Mater. Sci. Eng., A 454–455, 433–440.

Zener, C., Smith, C.S., 1948. Grains, phases and interfaces: an interpretation of microstructure. Trans. Metall. Soc. AIME 175, 11–51.

Zhao, H., Chan, T., Merriman, B., Osher, S., 1996. A variational level set approach to multiphase motion. J. Comput. Phys. 127, 179–195.

Zhao, Z., Roters, F., Mao, W., Raabe, D., 2001. Introduction of a texture component crystal plasticity finite element method for industry-scale anisotropy simulations. Adv. Eng. Mater. 3, 984–990.

Zhao, Z., Mao, W., Roters, F., Raabe, D., 2004. A texture optimization study for minimum earing in aluminium by use of a texture component crystal plasticity finite element method. Acta Mater. 52, 1003–1012.

Biography

Dierk Raabe holds a Ph.D. degrees in physical metallurgy anm etal physics from RWTH Aachen. Currently he is Chief Executive of the Max-Planck Institut für Eisenforschung in Düsseldorf and Professor at RWTH Aachen. His focus is in physical metallurgy and materials physics. Specifically he works on the simulation and mechanical properties of metallic alloys. The aim is the design of materials with superior properties (strength, elongation, damage tolerance) for the fields of energy, mobility and health from the atomic to the macro-scale under consideration of synthesis and processing. He received the Leibniz-Award and an ERC advanced grant. He is a member of the Science Advisory Board of the German Government and chairs the Governors Board of RWTH Aachen University. He is a member of the National Academy Leopoldina.

24   Porous Metals

Russell Goodall, Department of Materials Science and Engineering, University of Sheffield, Sheffield, United Kingdom.
Andreas Mortensen, Laboratory for Mechanical Metallurgy, Ecole Polytechnique Fédérale de Lausanne (EPFL), EPFL–STI–IMX–LMM, MXD 140 (Bâtiment MX), Lausanne, Switzerland.

Nomenclature

a crack length (eqn (97))

A_{PB} cross-sectional area Plateau borders (eqn (2))

b number of beams (eqn (61))

B Bulk modulus (eqn (43))

c_I inertia coefficient (eqn (30))

c_p heat capacity (massic) (eqns (34) and (35))

c_Δ ratio between the radius of the concave channel and the area of a single Plateau border (normally ≈ 0.4) (eqn (2))

c flow stress constant in foam flow curve (eqns (84) and (88))

c_m flow stress constant in metal flow curve (eqn (83))

C concentration (eqn (85))

C_{GL} curvature of the gas–liquid interface (eqn (1))

C_1, C_2, C_3 constants in scaling expressions for foam elastic moduluses, eqns (54) and (55)

C_4, C_5 constants in Andrews–Gibson–Ashby creep equation, eqn (91)

C_6, C_7 constants in fracture toughness scaling laws, eqns (93) and (94)

C_1', C_2' parameters in scaling expressions for foam flow stress, eqns (77) and (78)

C_E constant in scaling expressions for foam elastic moduli, eqns (56), (85) and (92)

C_P Paris law crack growth rate constant (eqn (97))

d_o parameter defined in eqn (69)

D dimension ($= 2$ or 3) (eqns (14), (16), (45) and (61))

$D_{f,p}$ diffusion coefficient through stagnant liquid within the pores of the porous metal (eqn (41))

D_D dispersion diffusion coefficient (eqn (41))

Da Darcy number (eqn (33))

D_m (single-valued) diameter of spherical solid elements making the porous medium (eqns (24) and (27))

D_p average pore radius (eqn (29))

e exponent in percolation-based predictions of permability (eqn (24))

E Young's modulus (eqns (10), (11) and (43))

E_1, E_2, E_3 parameters entering the elliptical yield criterion in (eqn (67)) *f* parameter that describes the viscous drag in a channel with a concave triangular shape (normally ≈ 49) (eqn (2))

F form coefficient (eqns (21), (26), (27), (30) and (31))

F_E fractional reduction in Young's modulus of the metal caused by pores (eqns (80), (85) and (92))

g gravitational acceleration (eqn (2))

G Shear modulus (eqn (43))

G_1, G_2, G_3, G_4 functions of V_m entering the modified Gurson (GTN) yield criterion (eqn (65))

h heat transfer coefficient between a substrate and a far-field fluid (eqn (42))

H parameter defined in eqn (47)

$\dot{H}$ heat flux from a solid substrate or other site (eqn (42))

g constant of proportionality on the order of unity (eqn (95))

G_P pressure gradient along the direction of flow (eqn (32))

G_{Ic} toughness of the porous metal (eqns (93)–(96))

$G_{Ic,m}$ toughness of the dense metal (eqn (95))

h strip thickness (eqns (10) and (11))

h_{sf} average (volumetric) heat transfer coefficient between solid and fluid phases in a small volume of material containing the porous metal and a fluid (eqns (34) and (35)),

j number of joints (eqn (61))

J_{Ic} J-integral for initiation of crack propagation in the porous metal (eqn (96))

k thermal or electrical conductivity; subscripts m for dense metal, f for porous metal, subscript c for conduction through foam, subscript r for radiation through foam (eqns (12)–(20)), subscript mf denoting metal foam (needed in eqn (35)), subscript D for effective thermal conductivity expressing the role of dispersion within the fluid (eqns (34 and (35))

K Darcian permeability (eqns (21), (23)–(26), (28), (29), (31))

K_{Ic} fracture toughness of the porous metal (eqns (93)–(96))

l characteristic microscopic distance (as opposed to L which is macroscopic there) (text after eqn (40); pore size in text of Section 24.5.6)

L various lengths (eqns (5), (6), (8), (19), (22), (26), (33), (40) and (42))

m_f parameter in ductile fracture ≈ 1.7 (eqn (96))

M geometrical optimization parameter defined in eqn (61)

M steady-state creep rate stress exponent, eqn (89)

n exponent in metal or foam flow curve ($n < 1$ in eqns (83) and (84), $n > 1$ in eqn (88))

N number of counts (eqn (6)), parameter taking integer or fractional values (eqn (15)); exponent in Young's modulus scaling law (eqns (56) and (85)), number of cycles (eqn (97))

N_p Paris law exponent (eqn (97))

r radius of curvature of Plateau border cross-section (eqn (1)), radius of capillary (eqn (4))

p exponent in fracture toughness scaling law, eqn (93)

P pressure (in fluid, eqn (21))

P_1, P_2, P_3 parameters entering the parabolic yield criterion in (eqn (66))

Pe Peclet number (eqn (40))

q exponent in fracture toughness scaling law, eqn (94)

q_1, q_2 parameters entering the modified Gurson yield criterion (eqn (65))

Q activation energy for creep, eqn (89)

$\dot{Q}$ volumetric heat generation rate within each phase (eqns (34) and (35))

R particle radius (eqn (3)), true 3D pore radius (eqn (7)), strip radius of curvature (eqns (10), (11) and (14)), gas constant (eqn (89))

Re Reynold's number (eqn (22))

S generic surface area (eqn (5))

S_V surface area per unit volume (eqns (7) and (8))

t time (many equations)

T temperature (many equations)

T_w wall temperature (eqn (42))

U potential (eqn (12))

v_D "superficial" (or Darcian or seepage) fluid velocity (eqns (21) and (31))

V generic volume (eqn (5))

V_m volume fraction metal = relative density = $(1 - V_p)$ (eqn (9) and many others)

V_o the volume fraction porosity in the initial loose-packed compact of leachable particles before densification (eqn (29))

V_p volume fraction pores = porosity = $(1 - V_m)$ (eqn (31))

z altitude (eqn (2))

Z average coordination number of packed particles or average connectivity of beams (text)

ΔE_{Capill} binding energy of a spherical particle to a liquid/vapor interface (eqn (3))

ΔP pressure differential driving fluid flow (eqns (24) and (26))

ΔP_{GL} pressure difference between gas in bubble and liquid (eqn (1))

ΔT change in temperature (eqn (10))

α coefficient of thermal expansion (eqn (10)); parameter defined in eqn (69)

α_{DF} parameter entering the Deshpande–Fleck yield criterion (eqns (72) and (73))

β uniaxial yield asymmetry ratio of the microcellular metal (eqn (71))

β_1, β_2 parameters defined in eqn (52)

χ effective tortuosity parameter (eqn (28))

δ_t crack tip opening displacement at the onset of crack propagation (eqn (96))

ε emissivity in radiative heat transfer (eqn (19)), strain (eqns (84) and (85))

ε_{impact} deformation strain at a shock front (eqns (86) and (87))

ε_m metal strain (eqn (83))

ε_D densification strain (compressive) (eqn (79))

$\dot{\varepsilon}$ steady-state creep rate (eqn (89))

$\dot{\varepsilon}_o$ reference steady-state creep rate (eqn (89))

Φ fraction solid contained in cell walls (as opposed to struts, eqns (54), (77) and (91))

γ liquid/gas surface tension (eqns (1)–(3)); parameter defined in eqn (68)

γ_1, γ_2 parameters defined in eqn (52)

κ curvature (eqn (11))

μ_{eff} "effective" fluid dynamic viscosity (eqn (31))

μ_f fluid dynamic viscosity (eqns (2), (21), (22) and (31))

μ_p exponent in conductivity scaling law (eqn (18))

ν Poisson ratio, eqns (43) and (52)

ν_{pl} experimental uniaxial "apparent plastic Poisson's ratio" (eqns (68), (69) and (73))

θ contact angle of liquid on solid in vapor (eqns (3) and (4)); generic angle (eqns (10) and (11)), Lode angle (eqn (64))

ρ_f fluid density (eqns (2), (21) and (31))

ρ porous metal density (eqn (9))

ρ_m dense metal density (eqn (9))

σ Stefan's constant (eqn (19))

σ stress (eqn (62) and following, eqn (89))

σ' deviatoric stress (eqn (63))

$\sigma_1, \sigma_2, \sigma_3$ principal stresses (eqn (62) and following)

σ_e equivalent stress (eqn (63))

σ_f ductile (microcellular) metal flow stress near the tip of a crack (eqn (96))

σ_{impact} stress on the porous materials on its impacted side (eqn (86))

σ_m mean stress (eqn (62))

σ_o porous metal stress constant in steady-state creep rate (Norton's law), eqns (89)–(91)

$\sigma_{o,m}$ dense metal stress constant in steady-state creep rate (Norton's law), eqns (89) and (91)

σ_y yield stress (eqns (66), (67) and (86))

$\sigma_{y,c}$ microcellular metal uniaxial compressive yield stress (eqn (71))

$\sigma_{y,m}$ dense metal yield stress (eqn (65))

$\sigma_{y,t}$ microcellular metal uniaxial tensile yield stress (eqn (71))

V_{front} shock front velocity (eqn (87))

V_{impact} impacted microcellular metal velocity in shock front deformation (eqns (86) and (87))

Recurrent subscripts:

m: metal

p: pore

f: fluid

c: properties characteristic of the fluid-filled porous metal composite (eqns (36) and (37))

24.1 Introduction

Highly porous cellular materials are interesting in many ways. The combined physical and mechanical properties of these materials are unique: notable examples include their low density, their high mechanical energy absorption efficiency, their damping capacity, and their permeability to flowing fluids, which gives them the capacity to interact with their environment. Cellular materials are thus widely used in engineering and also represent a high proportion of natural structural "materials" such as wood or bone.

Highly porous polymers are well known and quite usual, as are of course the many natural porous materials; comparatively, highly porous metals are newcomers to engineering. Metallic cellular materials have the general attractive features of metals, such as high flow stress and toughness, solid state mechanical formability, resistance to thermal exposure and to many environments, coupled with high thermal and electrical conductivity, combined with the attributes of cellular materials. Microcellular metals are now commercially available; as a result these are nowadays an area of intense current research and development, in both academia and industry.

As with all materials, there is in highly porous metals an intimate link between the nature and microstructure of the material on one hand, and its properties on the other. As with all materials again, this link never was simple (witness this book's thickness…); in porous metals it has an added level of complexity because the pores give the material an extra level of structure situated between the global structure of the metallic part (the part shape and macrostructure) and the inner structure of the metal or alloy it is composed of (the microstructure). This extra level of structure describes the shape and scale of the pores. It is often called the "mesostructure" or the "architecture" of the porous material, and greatly influences its properties. To give a basic example: two main classes of cellular materials exist, namely "open-cell" materials in which porosity is interconnected, and "closed-cell" materials in which each cell is enclosed by the metal. The former are weaker than the latter, but they can "breathe". Indeed, open-cell materials can be infiltrated with gas or a liquid, which can circulate through them past the metal. This in turn renders open-pore microcellular metals suitable for a whole new spectrum of applications, for example, in catalysis, energy conversion or thermal management. A second example: closed-cell materials include honeycombs; the considerable variation in strength and stiffness of honeycombs with the orientation of applied stress illustrates the importance of their mesostructure.

We present in what follows the making, structure and properties of highly porous metals, describing how they are made and how they behave, all the while emphasizing the many interesting questions raised by the fact that highly porous metals and alloys have an "architecture". Given the importance of porous metals in a broader sense (they, for example, underlie the whole field of ductile fracture), and also the wide range of research effort that has been devoted to porous metallic materials, we place certain (loosely defined) boundaries on what we cover. First, we restrict coverage to porous metals that can be viewed as a material, in the sense used by Anthony Kelly to define a composite material (Kelly and Mortensen, 2001). In brief, to count as a material a porous metal must be something that is treated by the designer as a material, with a structure and well-defined properties of its own. Secondly, although we have not restricted ourselves to a fixed range of values, we tend to focus on metals that contain at least 10 or so percent porosity, and generally much more.

We thus here describe metals and alloys that contain a good amount of porosity divided into many pores. This still is a vast subject. It has conferences of its own (notably the "Metfoam" conferences (Banhart et al., 2003; Nakajima and Kanetake, 2006; Lefebvre et al., 2008a)), it is a regular focus of journals in metallurgy and materials science (e.g. in *Advanced Engineering Materials* or *Scripta Materialia*), it is the subject matter of three books (Ashby et al., 2000; Degischer and Krizt, 2002; Dukhan, 2013a) and of several review articles (Evans et al., 1999; Gibson, 2000; Gergely et al., 2000; Banhart, 2001; Lu, 2002; Lefebvre et al., 2008b; Zhao, 2012; Han et al., 2012). We describe here how highly porous metals are made, what their structure is (emphasizing the mesostructure), and then examine their properties. We look first at their nonmechanical properties, such as the permeability to fluids or their conductivity, giving these the attention we believe they deserve given the fact that many applications of highly porous metals will be driven by their transport properties. We then turn to principal mechanical properties of highly porous metals, covering much but not all of the subject (e.g. damping and acoustic properties of these materials are not covered here, interesting though these questions might be; a recent pertinent review is in Perrot et al. (2013)).

Before we begin, a few words concerning semantics are in order. When we say "porous metal," the word "metal" refers to all materials based on metal, including pure metals, metal alloys, and also metal matrix composites. We have also used the word "foam" quite loosely: "metal foam," and also "microcellular metal," designates similarly any highly porous metallic material having a regular mesostructure and a microstructure, be it made by foaming or otherwise. We do so simply because it is now customary to do so in the literature, despite the fact that most of these metal foams are not really "foams" in the strict sense of being made by bubbling.

24.2 Processing

Microcellular metal, also called metal foam, can be produced using a very wide range of processes. These have been reviewed in whole or in part at various dates by Thiele (1972), Davies and Zhen (1983), Shapovalov (1994), Baumeister (1997), Banhart and Baumeister (1998b), Huschka (1998), Evans et al. (1988), Aluminium-Zentrale (1999), Körner and Singer (2000), Ashby et al. (2000), Gergely et al. (2000), Banhart (2000), (2001), (2006), Wadley (2002), Degischer and Krizt (2002), Dunand (2004), Conde et al. (2006a), Colombo and Degischer (2010), which collectively give a broad and quite comprehensive view of many aspects of the processing technology.

Most of these reviews classify the diverse production methods using different process-centered criteria, for example, the state of the metal (solid, liquid, gas or vapor) during the pore creation process. Here, due to the importance of the mesostructure in governing the properties displayed by the foams, we separate the methods according to what defines the topology of the porosity present in the material produced; this develops natural "families" of porous metals, which tend to share many characteristics and general properties. This way of classifying porous metals leads to a division of the processes into six natural and distinct groups:

(i) "Isolated Porosity": where there is dilute porosity, such that pores can be considered as being isolated;

(ii) "True Metal Foam": where a gas phase creates a swarm of contacting bubbles separated by thin metal membranes (these foams hence tend to be closed celled). These we term true metal foams;

(iii) "Foam Precursor": where a preexisting foam (generally a polymer such as polyurethane) is used to create the structure of the metal foam;

(iv) "Porosity Created by Packing": where an assembly of individual elements (which either make up the porous solid themselves once combined or act to form the pores) leaves open spaces that can be exploited to create a foam;

(v) "Porosity Created by a Phase Change": where phase transitions (generally invariant) of a single phase to multiple phases, one of which is a gas, result in the formation of a porous structure.

(vi) "Regular Lattices": where a uniform porous structure, often made up of regular beam elements, is repeated many times to form a material (a variety of techniques can be employed, but all need special treatment to form lattices).

In this section we consider the methods that have been used for the creation of metallic parts with controlled levels of porosity. We cover briefly the basic physics that govern some of these processes. In our selection from the very large number of proposed methods, we concentrate on those that have most frequently been used, or those that have industrial significance, while also discussing some of the methods that are more peripheral but demonstrate different concepts in foam production. **Table 1** gives a broad overview of the main microcellular metals produced by the methods described here.

24.2.1   Isolated Porosity

Generally, the presence of isolated pores (less than 20–30% by volume) in a metal would be considered a defect. This can arise at several stages during processing, for example, as microshrinkage during casting, or by the release of dissolved gas. It can also be created as internal damage by processes such as irradiation or ductile deformation under tensile mean stress. This last mechanism is most dramatically exemplified by the structure of polycrystalline samples having undergone extensive tertiary high-temperature creep (e.g. (Ashby et al., 1979; Courtney, 1990)).

Some specific effects are known to generate isolated pores. One is the Kirkendall effect, which arises when there is a diffusion couple with unequal diffusion rates, resulting in one part of the couple moving faster than the other (Smigelskas and Kirkendall, 1947; Shewmon, 1989). This can result in Kirkendall porosity within the faster moving part since the net flow of atoms in one direction must be balanced by a net flow of vacancies in the other. Although some of these vacancies may be consumed at free surfaces or by processes such as dislocation climb, a significant number can coalesce and nucleate internal voids that will grow as long as there is a net flow of atoms away from their location (on the side of the lower melting point metal). Porosity levels can be enhanced by stresses developed across the interface, as these tend to be compressive on the side where material flows to, and tensile on the side

Table 1   Summary of the approximate pore size and density range of foams produced by each of the methods discussed. Note that the values given are indicative of the general range, not extreme limits

Method	Closed or open cell	Volume fraction solid, V_s	Pore size, μm	Commercial foam example
Gas entrapment	Closed/partially open	0.5–0.8	10–100	
Direct gas injection	Closed	0.05–0.5	1,000–10,000	Cymat
Foaming agent in melt	Closed	0.1–0.35	1,000–5,000	Alporas
Foaming agent in precursor	Closed	0.1–0.35	500–5,000	Alulight
Investment casting	Open	0.05–0.15	1,000–10,000	Duocel
Deposition onto template	Open	0.02–0.05	1,000–10,000	INCO Foam
Sintered particles	Open/partially closed	0.1–0.45	10–1,000	
Sintered fibers	Open	0.05–0.3	100–5,000	
Syntactic foams	Closed	0.1–0.7	100–5,000	
Spaceholder method	Open	0.1–0.35	10–10,000	Versarien Foam
Replication	Open	0.1–0.35	10–10,000	Corevo Foam, Constellium
Gasar/lotus structure	Closed (highly directional)	0.25–0.95	5–10,000	
Dealloying	Open	0.15–0.5	0.5–100	
Regular lattices	Open	0.05–0.2	1,000–10,000	

where it flows from. Kirkendall porosity tends to be on a small scale (microporosity) and at a relatively low volume fraction. The process has been observed in thin films (Jiang et al., 2009), and in powder composites (Pezzee and Dunand, 1994; He et al., 2007; Dong et al., 2009; Wen et al., 2010). It has also been employed as a means of increasing the pore fraction in sintered powder compacts (Weber and Knüwer, 1997; Wang et al., 1998).

Another metallurgical effect that generates porosity is specific to relatively noble metals, notably copper. When copper containing oxygen not fixed by oxide-forming alloying elements is exposed at high temperature to gaseous hydrogen, the hydrogen will dissolve and diffuse into the metal and reduce cuprous oxide Cu_2O to form water vapor, H_2O. This water then remains within the metal, where it forms small bubbles situated where Cu_2O was present (Brick et al., 1977).

In each of these situations the pores that are formed will mostly be created in an uncontrolled manner, for example, along grain boundaries. There are a small number of cases where regularly spaced isolated pores may be considered an advantage, and it may be a requirement to deliberately engineer them in a material; other processes are then used. In the production of tungsten filaments, a small volume fraction of microscopic pores strengthens the metal, by a process akin to the Orowan mechanism (Schade, 2010, 2011). To this end, a "dopant" is added; during further processing and use, the volatilization of this dopant causes pore growth by plastic deformation of surrounding metal through the pressure induced, leading to the formation of fine and relatively well-distributed porosity within the metal (Brett and Friedman, 1972; Schade, 2010, 2011, 2002).

Somewhat higher levels (though still <50%) of porosity can be created with methods based around the principle of gas entrapment. Here, in contrast to the gas foaming of liquid or semiliquid metal described in the next section, gas creates porosity within a solid metal. The structure produced is not subject to the same constraints as a liquid metal foam, and can take various forms.

Such methods are based on the trapping of gas within a metal, with the material being then heated under vacuum to a temperature where creep or superplastic behavior is seen; the pressure of the gas then causes pores to expand. Various methods can be used to entrap gas in a material. One is to ion implant a species in solid metal that, during a subsequent anneal, will precipitate in the form of gas bubbles (Shapovalov, 1994; Kim and Welsch, 1990; Ozgur et al., 1996). Resulting materials feature very fine pores, and generally low levels of porosity. Another is to sputter deposit the metal in the presence of pressurized gas which is then expanded, again by heating (Davies and Zhen, 1983). It has also been found that milling of some powders under argon at cryogenic temperatures will result in sufficient pickup of gas for the process to work (VanLeeuwen et al., 2011). A more direct method is to compact a powder under an environment of inert gas (e.g. argon using hot isostatic pressing, [HIP]), such that gas becomes trapped within the particle network (Schwartz et al., 1998). When this structure is heated under vacuum, the gas may develop enough pressure to expand the structure. This type of process is best carried out in metals that can elongate to high strain below their melting point, by creep (Kearns et al., 1988) or superplasticity (Dunand and Teisen, 1998; Oppenheimer and Dunand, 2010), or alternatively in a partly liquid state (Chino et al., 2002, 2004). For this reason, titanium and its alloys have often been selected (Schwartz et al., 1998; Dunand and Teisen, 1998; Evans et al., 1999; Davis et al., 2001; Murray and Dunand, 2003, 2004; Murray et al., 2003; Oppenheimer and Dunand, 2010). In general, maximum porosity values around 50% are the limit (Elzey and Wadley, 2001). If the vacuum heat treatment is continued after this point is reached (often accompanied by thermal cycling), then rather than increasing porosity, there is a progression from closed pores to an open porous network, as the walls separating the cells thin and break (Murray and Dunand, 2003). By replacing the powder particles with fine aligned wires, a structure with aligned porosity can be created (Spoerke et al., 2008). Aligned pores have also been produced in powder compacts by the application of a uniaxial stress to the sample during the heat treatment stage (Davis et al., 2001).

There have been attempts (Ricceri and Matteazzi, 2003) to increase the volume fraction porosity in titanium by the addition of TiH_2 powder within the metal powder mix; this material breaks down at high temperature releasing H_2, and in theory this larger quantity of gas should allow higher levels of porosity. However, the effectiveness of this approach is limited by the high solubility of hydrogen in solid titanium, which allows it to rapidly escape through the structure by diffusion (Dunand and Derby, 1993).

24.2.2 True Metal Foam

24.2.2.1 *Elementary Foam Physics*
A true foam is made up of many tightly packed bubbles of gas in a liquid. But more than this, to qualify a foam as a material the structure must have some degree of stability, both mechanically and over time (otherwise it is unlikely to be accessible in an engineering sense).

Although this is not easy, one can foam a metal; once solidified, the resulting highly porous metal is a material in its own right, which can be useful in various applications; **Figure 1**a gives an example. Metal foams as such are, in most ways, no different than other foams that one finds in daily life; however, the underlying physics of their formation and stability have some specificity, which explains to a large extent how these are made and why metal foam is more difficult to produce and has a less regular structure than, say, widely used polyurethane foam.

"The secret" behind the formation and stability of most conventional nonmetallic foams is in the use (or at times the unwanted presence) of surfactants (Pugh, 1996; Weaire and Hutzler, 1999). Effective foam-producing surfactants (i) reduce the energy of liquid/vapor interfaces to which they segregate, in turn decreasing the excess energy associated with all the free surface that foams contain, and (ii) cause two liquid/vapor interfaces that approach one another through the liquid to repel one

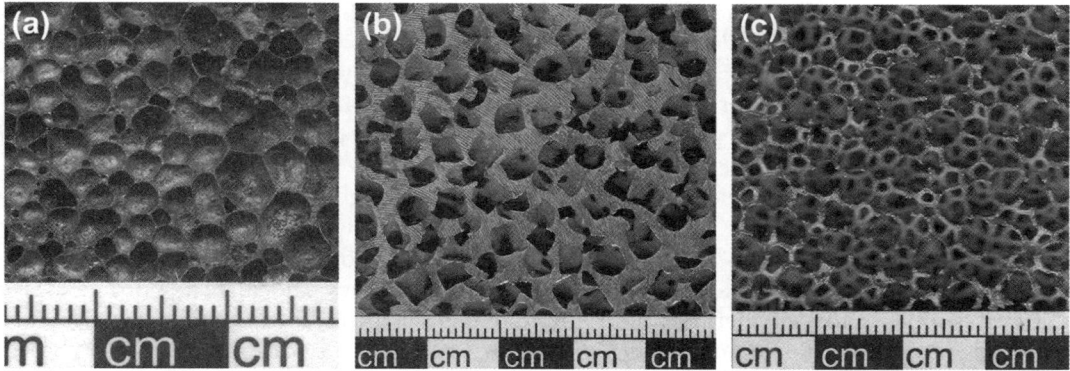

Figure 1   Examples of foamed metal: (a) closed-cell aluminum foam produced by the precursor route with TiH$_2$ (FORMGRIP), (b) a high-density open-celled aluminum foam produced by replication (Corevo foam), and (c) a low-density open-celled nickel foam produced by deposition onto a polymer template.

another. Foam-producing surfactants are generally asymmetric long-chain molecules (e.g. fatty acids, alcohols, polymers or proteins) that segregate to one side of the interface and repel one another by a combination of steric, Coulomb and long-range forces. Their influence translates physically into what is known as a "disjoining pressure," defined as the repulsive force exerted per unit area of paired liquid/vapor interfaces when these are brought near one another. The disjoining pressure increases, at least over certain ranges, as the two interfaces approach one another; since such surfaces also attract one another (through van der Waals forces notably) the thickness of such a liquid film is stabilized at one or at times several energy minima (depending on force profiles). A sufficiently high disjoining pressure has the consequence that bubbles can be pressed one against the other without rupturing the thin film of liquid that remains to separate the two. Emulsions are stabilized by essentially the same mechanisms, acting this time at two liquid/liquid interfaces.

Once a foam is formed, the bubbles tend to float upward, or in other words the liquid falls or "drains" from the foam (the two descriptions being indistinguishable in the physics of the process). This in turn presses the bubbles against one another, causing these to pack closely, forming flat facets where these meet. Where bubble surfaces do not touch, these are curved and separated by thicker regions of liquid; most frequently, and increasingly so as the bubbles are more closely packed, such regions are linear junctions between three touching bubbles. These tri-bubble junctions are relatively straight, and have a perimeter defined by three circular arcs that meet with a common tangent; these are known as "Plateau borders" (commemorating JAF Plateau). Plateau borders in turn meet four at a time at "nodes". Between the nodes, the Plateau borders thus define thicker struts, which surround thin flat facets across which two liquid/vapor interfaces are pressed against one another. Along these facets, the disjoining pressure opposes the pressure exerted by gas within the bubbles, whereas in curved regions of the interface the same gas pressure is opposed by the pressure in the liquid to which is added the effect of surface tension of curved liquid surfaces, which creates a pressure difference between the liquid and the gas, ΔP_{GL} required to balance the surface tension acting through the curvature in the Plateau borders. This is determined by Laplace's law, which for a straight Plateau border is

$$\Delta P_{GL} = C_{GL}\gamma \approx \frac{\gamma}{r},$$ (1)

where γ is the liquid/gas surface tension, C_{GL} is the curvature of the gas–liquid interface and r is the radius of curvature of the (roughly straight) Plateau border cross-section. The pressure in each phase is dictated by global force equilibrium; along a free surface it is negative in the liquid, positive in the gas. Forces in presence increase in inverse proportion with the average bubble diameter (eqn (1). They also increase as the fraction liquid decreases since this causes the bubbles to be more tightly packed together, this in turn decreasing r (eqn (1) again).

Now, gravity will cause the liquid pressure to vary along the height of a foam. Hence, the value of r, and with it the relative fractions of liquid and gas, will vary with altitude at local mechanical equilibrium. More precisely, a lower liquid pressure at the foam top will imply that r must be more strongly curved to equilibrate the gas pressure within the bubbles; the Plateau borders are hence thinner, corresponding to a smaller fraction liquid than below. The bubbles are thus increasingly close packed as one moves from the foam bottom, which with a sufficient proportion of liquid will be bubble free, to the top of the foam. All this is visible in a head of beer foam, in which the bubble packing fraction can be seen through the glass to decrease from the top down. Such gravity-induced gradients also affect most other usual foams, such as polyurethane foam and as a result any metal foam produced therefrom (see Section 24.2.3).

In most foams all bubbles are not exactly of the same size. As a result the gas pressure in different cells will be unequal, and this drives diffusion through the relatively thin cell walls, since under the Laplace equation (eqn (1)), smaller cells have a higher gas pressure (Weaire, 2002). This leads to the elimination of some of the cells, causing in turn a gradual increase in the average cell size; this bubble coarsening phenomenon is called "disproportionation" in the foam literature and closely resembles grain growth in metals, as Cyril Stanley Smith pointed out long ago (Stanley Smith, 1964; Smith, 1981; Pugh, 1996; Weaire and Hutzler, 1999). Foams thus coarsen in time; this too is visible in beer foam. Cell walls may also rupture occasionally, causing two bubbles to become one; this second coarsening mechanism, called "coalescence" (and quite visible in soap froth), will also lead to an increase in the mean bubble size. If coalescence is extensive enough, it will lead to "collapse" of the foam when most of the cell walls have failed (Pugh, 1996; Weaire and Hutzler, 1999; Weaire, 2002). **Figure 2** summarizes these basic effects underlying the physics of foams (or emulsions, which as mentioned already are essentially similar except that the "bubbles" are liquid). Detailed coverage of basic foaming physics is in Refs. Pugh, (1996), Weaire and Hutzler (1999).

A description of the physics of foam formation, structure and stability normally includes two limiting cases: that of a highly wet foam and a dry foam. In a highly wet foam, the bubbles in the fluid are dilute and either do not interact or do so weakly; the limit of this structure is evidently when the pore fraction has reached that for random close-packed hard spheres (64%). Cell walls have then reduced to a tiny circle or a point and the bubbles are essentially spherical. Past this limit, bubbles are disconnected free-floating spheres surrounded by liquid or solid.

A dry foam on the other hand is nearly all bubble, with a liquid phase volume fraction that approaches zero. The liquid then exists only along the boundaries of pores that take the form of variously shaped, close-packed flat-faced polyhedra (Weaire and Phelan, 1996). A dry foam is a useful limiting case, as it will obey a series of simple rules defined by Plateau in 1873 and by Euler (Weaire and Hutzler, 1999). These rules treat the structure essentially as grain boundaries in a polycrystal having a perfectly isotropic grain boundary energy, in which no more than three cell faces meet at a cell edge, and no more than four cell edges meet at a node. The liquid will tend to concentrate where cell edges meet (where three or four pores come together) to form a network of channels with concave triangular cross-section and nodes where these meet (Weaire and Phelan, 1996).

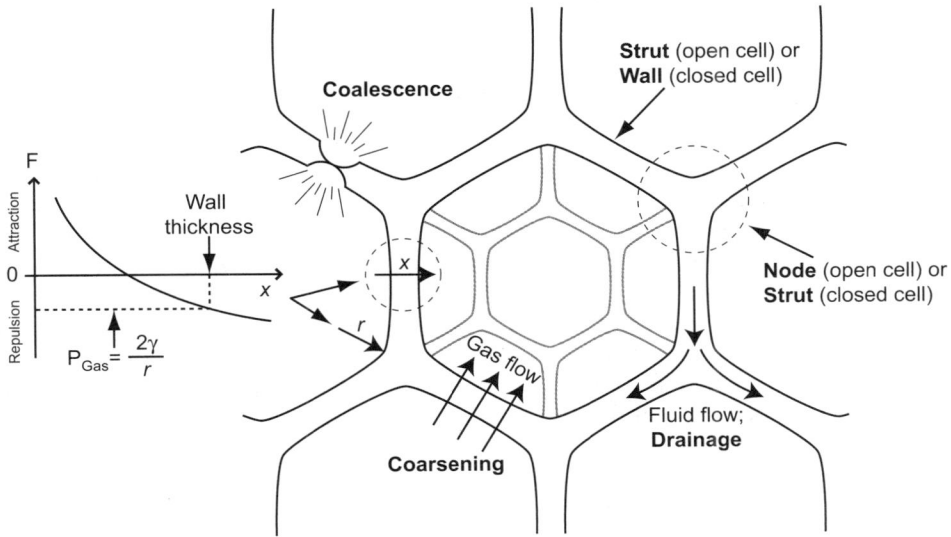

Figure 2   A schematic diagram of a foam structure, indicating some of the important structural elements and processes.

Different bubble packing arrangements and ideal pore shapes have been proposed in the dry limit. Widely used is the Kelvin cell (or *tetrakaidecahedron*), **Figure 3**, which is a space-filling periodic structure where this same cell is repeated in a bcc arrangement. This was long assumed to be the most stable dry foam form (Weaire, 2009). Examination of the question using surface energy minimization modeling

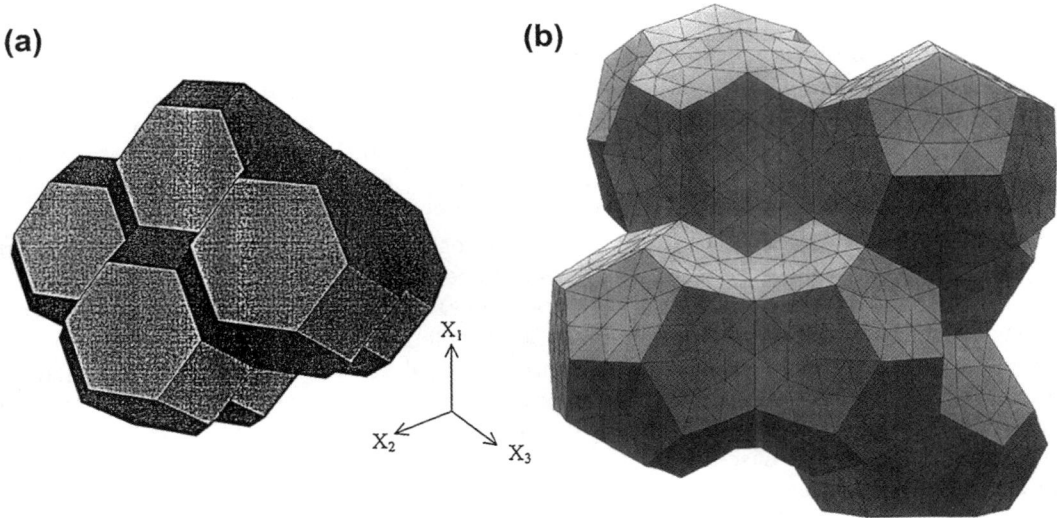

Figure 3   Ideal foam structures. (a) A unit cell of the Kelvin structure, reprinted from *Acta Materialia,* vol. 46, Simone, A. E. & Gibson, L. J. "Effects of Solid Distribution on the Stiffness and Strength of Metallic Foams" pp. 2139-2150 (Simone and Gibson, 1998c), with permission from Elsevier and Prof. L.J. Gibson, MIT and (b) that of the Weaire-Phelan structure, reprinted from "The Physics of Foam", *Journal of Physics: Condensed Matter,* **8**, pp. 9519–9524 (Weaire and Phelan, 1996) with kind permission from Prof D. Weaire, Trinity College Dublin, and IOP Publishing.

techniques has led to the conclusion that a different periodic form (called the Weaire–Phelan structure after the discoverers (Weaire and Phelan, 1994)) with eight bubbles per repeating unit cell, **Figure 3**, is actually slightly more stable.

Real foams have a structure in between the two limits of a highly wet and a dry foam, although the tendency will be for the overall structure to go toward that of a dry foam having slightly fatter (wet) cell edges and nodes well before the fraction liquid approaches zero (Weaire, 2002). Real foams also have a degree of disorder, the shape and size of bubbles being irregular and obeying rules only in terms of average characteristics (e.g. the average number of edges bordering cell walls is near five but this number varies from facet to facet, while the number of facets varies from one bubble to another) (Weaire and Phelan, 1996).

In structures that are not mechanically equilibrated, liquid will flow along Plateau borders and through nodes; this process is known as "drainage" and is caused by gravity or, in some cases, by centrifugal force. In practice, the transition of a foam from a wet foam to a dry foam, and the evolution that leads to the cell walls becoming thin enough to break, is generally driven by the downward flow of liquid out of the foam. When at or close to the limit of a dry foam, the regular structure permits models of drainage to be more easily formulated. In the usual treatment, drainage is assumed to occur by flow along cell edges only (Verbist et al., 1996), and this leads to the "foam drainage equation"

$$\frac{\partial A_{PB}}{\partial t} + \frac{1}{3f\mu_f}\frac{\partial}{\partial z}\left(\rho_f g A_{PB}^2 - \frac{c_\Delta \gamma}{2}\sqrt{A_{PB}}\ \frac{\partial A_{PB}}{\partial z}\right) = 0, \tag{2}$$

where A_{PB} is the cross-sectional area of the cell edges (also known as Plateau borders), which is linked to the proportion of liquid present, t is the time, z is the vertical height, μ_f is the fluid viscosity, γ is the fluid surface tension, ρ_f is the fluid density, g is the acceleration due to gravity, f is a parameter that describes the viscous drag in a channel with a concave triangular shape (normally ≈ 49) and c_Δ is the ratio between the radius of the concave channel and the area of a single Plateau border (normally ≈ 0.4) (this particular form of the equation is from (Brunke et al., 2005)).

This is a nonlinear partial differential equation, the solution of which for the relevant boundary conditions at the top and bottom of the foam (e.g. is the liquid replenished, does it flow away) gives the amount of liquid present in a foam at a particular height and time. The effect of this process on real foams is that there will be an increasing gradient of porosity (increasing toward the top of the foam) which evolves over time (Verbist et al., 1996).

This makes the solidification time critical in determining the structure of a foam and its uniformity (Gergely and Clyne, 2004); this process for metals has been numerically modeled in a 1D form (Cox et al., 2001). The analysis considers the competition between the speed of drainage (as determined by the drainage equation) and the speed of solidification (as determined by the rate of heat extraction) and also the effect of temperature on the melt viscosity. The analysis shows that, for a particular metal, a thicker foam section or a higher initial volume fraction liquid is more likely to lead to the final structure being coalesced at the top with a fully dense drainage pool at the bottom, a situation likely to lead to foam collapse. Models based on the standard foam drainage equation have been compared with experimental results and reasonable agreement found (Brunke et al., 2005). Numerical modeling has also been used to examine drainage in both films and Plateau borders (Gergely and Clyne, 2004; Anderson et al., 2010), and reasonable agreement has also been found (Gergely and Clyne, 2004). This work suggests that drainage can be reduced by high initial porosity (in agreement with the modeling of flow only through cell edges) and small pore size, and also by having a higher proportion of liquid on cell faces, rather than edges, this ratio in turn being dictated by the nature of the disjoining pressure.

24.2.2.2 Metal Foams

With metals, several difficulties arise in making foam. The first is that molecular surfactants, such as the proteins or polymers that are used to stabilize aqueous or organic foams, are essentially excluded in metals. The physics of metallic foams are thus, by nature, different from those of most foams. The second is that metals have comparatively high densities and low viscosities: drainage is therefore very rapid eqn (2). And finally metals also have high surface energies: as a result the thermodynamic driving force of coarsening and cell wall rupture processes is much higher.

Making stable metal foam is thus much more difficult than with, say, aqueous solutions or polymers. Still, metal foams exist (**Figure 1**) and have been around for many years (Meller, 1926; Banhart, 2001). This is because molecular surfactants are not the only way to stabilize a foam: an alternative stabilization agent, which is generally used together with surfactants but has been shown to work alone, is provided by small particles that segregate to liquid/vapor interfaces (or, in emulsions, to the interface between the two liquids) (Binks, 2002; Hunter et al., 2008; Leal-Calderon and Schmitt, 2008; Horozov, 2008). That small particles with the appropriate wetting characteristics can stabilize foams or emulsions have been known for some time (such emulsions are known as "Pickering emulsions"); however, the underlying physical laws and mechanisms are complex, interesting, and still in the process of being unraveled. Research on the topic has in fact been particularly active over the past decade, in part because particle stabilization gives greater (ultra) stability to foams and emulsions than does more classical surfactant-based stabilization alone.

In essence, to stabilize a foam, particles must bind tightly to the interface while residing mostly within the liquid. To this end, the contact angle θ that the liquid makes with the solid particle material must be in a narrow range, situated roughly from 65 to 80° if the particles are smooth, spherical and homogeneous. There are two main opposing reasons for this. The first is that the capillary binding energy of a spherical particle to an interface

$$\Delta E_{\text{Capill}} = \pi R^2 \gamma (1 \pm \cos \theta)^2, \tag{3}$$

where γ is the liquid/gas surface tension, R the particle radius, and the sign inside the bracket is positive for removal of the particle into the liquid, negative for removal into the gas phase. This binding energy, which attracts and holds the particles along the interface, peaks at $\theta = 90°$ and falls rapidly on either side of this value. The contact angle θ must therefore be near 90° or else the particles will not segregate to the interface. The second factor defining the useful range for θ is that particles bend the interface toward the side opposite to that at which they segregate; this is illustrated in **Figure 4**. Hence, slightly wetted solid particles (with θ a bit below 90°) will bind strongly to the interface while, if these are tight packed or repel one another from a finite distance, (i) bending the interface so that it delineates round bubbles and (ii) giving it some degree of mechanical stiffness. Both factors tend to stabilize the bubbles against collapse, and can do so over long time periods because particle-lined bubbles cannot shrink indefinitely, which hinders disproportionation (coarsening akin to grain growth, see above). These are the main mechanisms that underlie particle stabilization of foams or emulsions; the rest is complex and, as mentioned, not yet fully understood. Reviews, relevant examples and recent research can be found in Refs. Binks (2002), Sun and Gao, (2002), Kaptay (2004), (2006), Somosvari et al. (2007), Hunter et al. (2008), Leal-Calderon and Schmitt (2008), Cervantes Martinez et al. (2008), Horozov (2008), Stocco et al. (2011).

Banhart gives in Ref. Banhart (2006) a comprehensive and reasoned review of the literature up to 2006 on the physics and processes of metal foaming, and has also recently summarized the state of the art in metal foam engineering up to today (Banhart, 2013). It is now recognized that solid

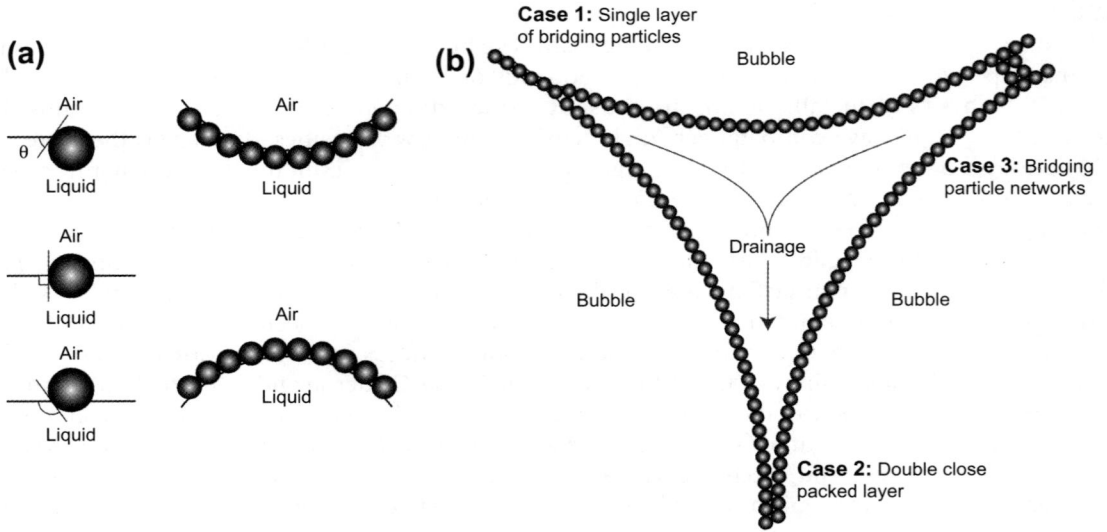

Figure 4 Schematic diagram of mechanisms underlying foam stabilization by solid particles: (a) influence of contact angle θ on particle situation and interface curvature; (b) general structure of particle-stabilized foam, detailing three mechanisms that have been proposed for how solid particles may stabilize a liquid film at its narrowest section.

inclusions in the liquid play a key role in the stabilization of metal foams (Ip et al., 1999; Babscan et al., 2004; Banhart, 2006; Körner, 2008): when metals are foamed successfully in the liquid form, there is always a solid particulate component present. The solid phase may be generated by the metal being in a semisolid state (e.g. (Körner et al., 2004; Wubben and Odenbach, 2005; Ravi Kumar et al., 2010; Helwig et al., 2011)), it may consist in deliberately added ceramic particles (e.g. (Jin and Kenny, 1990; Leitlmeier et al., 2002; Yang and Nakae, 2003; Kennedy and Asavavisitchai, 2004a, 2004b; Babscan et al., 2004, 2005, 2012; Deqing and Ziyuan, 2003; Deqing et al., 2006; Haesche et al., 2008)), or fibers (e.g. (Cao et al., 2008; Wei and Sun, 2011)), ceramic particles produced by reaction in the melt (e.g. (Kumar et al., 2011b)), or oxide generated at an earlier processing stage (e.g. (Korner et al., 2005; Ha et al., 2005; Wubben and Odenbach, 2005; Babcsán and Banhart, 2006; Babcsan et al., 2006b; Asavavisithchai and Kennedy, 2006d, 2006a, 2006c; Dudka et al., 2008; Chethan et al., 2011)) or formed during foaming by reaction with the atmosphere (Banhart et al., 2001b; Gergely et al., 2003; Babscan et al., 2004, 2005; Haesche et al., 2010; Mukherjee et al., 2010a). It is further observed that when more solid particulate matter is present in the melt, the cell wall thickness increases (Deqing and Ziyuan, 2003), and that the permeability of the liquid cell walls to the gas makes a difference, a change in this permeability causing a secondary expansion upon metal solidification (Mukherjee et al., 2010c). In particular, it has been shown that the wettability of the particles by the metal is an important parameter (Asavavisithchai and Kennedy, 2006b; Klinter et al., 2010; Kumar et al., 2011b)—as is the case in particle-stabilized foaming more generally. Direct in situ observations of foaming and liquid foam structure have also been conducted, shedding light on the structure of the foam before solidification (which pushes solid particles and hence alters their distribution) (Haibel et al., 2006) or on the kinetics of pore formation or cell wall rupture (Banhart et al., 2001a, 2001b; Bellmann et al., 2002; Helfen et al., 2002, 2005; Stanzick et al., 2002a, 2002b; Garcia-Moreno et al., 2004, 2005, 2008, 2009, 2011; Babcsan et al., 2007; Rack et al., 2009b, 2009a;

Mukherjee et al., 2010b). From these observations, there are a number of possible explanations for how the presence of solid particles (which in practical metal foam production can be highly diverse in size, form and composition) can improve foam stability.

One possibility for the stability of metal foams is that the solid particles act as points between which the liquid can bridge, so stabilizing a thin film in a pore wall (Kaptay, 2006); such an arrangement will only be stable for a particular wetting angle of the metal on the particles, and this may explain why certain particles (e.g. SiC) work better than others at stabilizing foams. However, this scheme only works if the particles and the pore walls have similar thickness, something that is most often not the case (Banhart, 2006). Still, there is some experimental evidence in relation to the influence of wetting angle (Klinter et al., 2010) and in the form of cell walls bridged by one or a few particles (e.g. Figure 16a in Ref. Gergely et al. (2000)). A further development of the idea is that small particles, if they reside at the surface of the pore, will change the form of the surface from flat to corrugated, which affects the pressure difference between the center of the films and the edges of the pores (as described by the Laplace equation, eqn (1)), reducing the pressure difference by introducing curvature and so reducing the rate at which liquid moves to the cell edges and then drains. However, this simple picture also falls down when confronted with experimental evidence, as the cell walls observed in metal foams do not have this shape (Banhart, 2006).

Another possibility was considered in detail by Kaptay (2004), this being the idea that networks of particles might connect across the pore surface and so provide a solid bridge to support the film. Experimental evidence for this mechanism in foams produced by powder metallurgy has been found (Korner et al., 2005), with particles assembling into clusters within the metal. An alternative suggestion is that the action of solid particles is less to change the stability of the pore walls by acting at the surface, but rather to slow down drainage by the increase in viscosity that results from having small particles present within the liquid metal (Gergely and Clyne, 2004); however, this too is opposed by experiment (showing, e.g., that the absence of gravity brings little improvement in foam stability (Wubben and Odenbach, 2005; Garcia-Moreno et al., 2009)). Tomography of aluminum foams made by a liquid metal route with SiC particles using TiH_2 as the gas release agent has found that particles segregate to the surface but do not interact across the cell faces, leading to the suggestion of a mechanism where the interactions between particles at the surface give the surface some additional strength, possibly in addition to an increase in viscosity through those particles that are situated in the film interior (Haibel et al., 2006). Microstructural investigations of oxides in aluminum foams produced from aluminum powder, stabilized by oxides from the aluminum particles, have found that the oxide particles are distributed in hierarchical clusters, in a way comparable with the particle network theories (Dudka et al., 2008). A related effect to the presence of particles at the surface could arise where the interior of the pore is oxidized to create a solid "crust" lining the pore surfaces (rather than a distribution of fine oxide particles through the liquid). In this case it may also be that the oxide layer is imparting greater strength to the pore walls, and helping to resist rupture, without necessarily having an effect on drainage or creating a disjoining pressure (though the effect of oxide layers and particles is often combined (Babscan et al., 2004; Babcsan et al., 2005)). Along the same lines, oxide layers can also act to damp surface waves that might contribute to instabilities leading to rupture (Babcsan et al., 2006a). It has also been shown in zinc foams that increased oxygen levels in the precursor lead to higher expansions and expansion rates being achieved on foaming (Chethan et al., 2011). Here the oxides are seen as clusters distributed on the pore walls.

In conclusion the exact mechanism(s) by which metal foams are produced is complex and the debate continues, concerning for example, whether the same mechanism operates to stabilize all foams or whether, given the large diversity in foam production routes and structures, different mechanisms occur in different situations (e.g. (Banhart, 2006; Haibel et al., 2006; Körner, 2008)), or on the importance of drainage in cell stabilization (e.g. (Banhart, 2006; Lehmhus and Busse, 2004; Körner, 2008; Garcia-Moreno et al., 2009)). Still, metal foams of varying levels of regularity can be produced; several different methods have been used in practice to this end.

24.2.2.3 Production of True Metal Foams

These methods are sometimes called *direct foaming*, and they all involve gas somehow being introduced into a liquid metal.

24.2.2.3.1 Bubbling Gas in Liquid

One of the simplest ways of performing the foaming operation is by "bubbling," that is, by injecting gas within the melt. The gas is selected so as to be unreactive (nitrogen, argon or other noble gases) or mildly reactive (oxygen-containing gas), and a system must be provided to remove the froth formed at the top of the liquid metal before it solidifies. As already mentioned, the nature of liquid metals (low viscosity, high surface tension and no surfactant molecules to create a disjoining force) is such that a close-packed swarm of bubbles in molten metal will not be held for very long: once at the surface the bubbles will quickly burst. Therefore this method requires the prior incorporation in the metal of solid (generally ceramic) particles which can stabilize a froth structure longer than in unreinforced molten metals (Jin and Kenny, 1990; Åsholt, 1997; Prakash et al., 1995, 1997; Simone and Gibson, 1998a; Babcsan et al., 2012). The effect of these particles is to increase the apparent viscosity of the melt, which acts to reduce the rate of drainage under eqn (2) (and thus bubble bursting), and also to provide for capillary stabilization of the cell walls, as mentioned above. These particles are of an inert material that resists the required melt temperature, for example, aluminum oxide, silicon carbide, titanium diboride, coupled with the required wetting characteristics (although this is seldom measured). Their average size varies over a wide range of possible values, from a few tens of nanometers (oxide skins formed on the metal surface) to a few tens of micrometers (comminuted SiC particles, for example) (e.g. (Babcsan et al., 2012; Jin and Kenny, 1990)). Of course, these particles have an effect throughout the foam life cycle, and will tend to make the foam produced stiffer, more brittle and harder to machine than it would be if it were made of unreinforced metal. The fact that such particles need to be present for the majority of the liquid-state metal processing routes to function has led to the suggestion that metal foam would be a good route for the use of recycled aluminum metal matrix composites (as similar ceramic particles can be left in from the previous stage) (Gergely et al., 2000), and indeed attempts have used recycled aluminum, with the attendant oxide, to form foams (Park et al., 2005; Amsterdam et al., 2006a). There has been some attempt to avoid ceramic particles and control the viscosity of the molten metal through control of the temperature (Weber, 1985), with bubbles being frozen in place, but this requires a very good degree of temperature control.

Foams are made by gas bubbling and sold commercially by Cymat Technologies Ltd, Ontario, Canada (http://www.cymat.com/); the technology was formerly developed by Hydro Aluminum and by Alcan. In this process, an inert gas such as argon is bubbled through molten aluminum-containing ceramic particles (e.g. SiC) to increase the viscosity (Jin and Kenny, 1990; Asholt, 1999) (see **Figure 5**). The bubbles rise to the top of the liquid, and the foam formed is skimmed off and solidified on a conveyor belt to retain the porous structure. In some configurations this is done vertically to reduce disturbance to the foam structure (Simone, 1997), and this appears to be similar to the process used to make Metcomb® (Leitlmeier et al., 2002) foams. These materials are made by allowing the bubbles to rise and fill an inverted mold (which starts the process already filled with

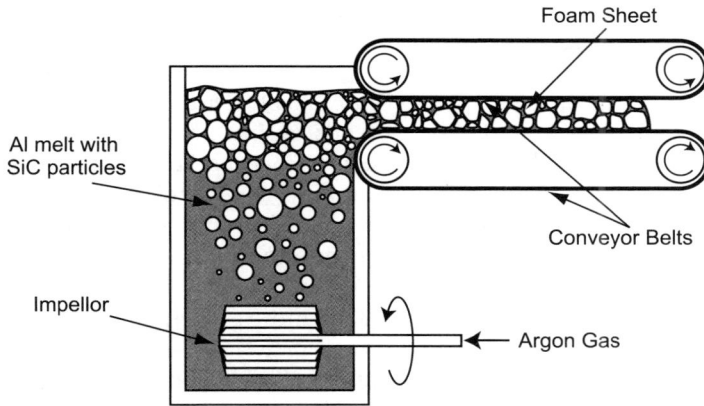

Figure 5   Schematic diagram of the production process for direct gas injection (Cymat-style foams).

liquid metal). This produces a more regular structure as it avoids the need to mechanically displace the foam. The cooling of the mold is controlled so that once the foam is formed it can be solidified relatively rapidly.

In getting uniform and regular pores, the injection of gas bubbles of the correct size is obviously very important. This can be controlled by features such as the propeller arrangement of **Figure 5**, which is intended to break up the gas stream into smaller monosized bubbles with the intention of forming many uniform pores in the final product. This can also be achieved by vibrating (Banhart, 2001) or using an ultrasonic (Babcsan et al., 2012) nozzle. A higher propeller speed gives a smaller pore size for a given air injection rate, while a higher rate of injection produces larger pores (Deqing et al., 2006). The use of ultrasonic vibration on the nozzle has been shown to enable precise and narrow control of bubble size (Babcsan et al., 2012). Other critical parameters to the success of this method are the selection of the correct ceramic addition, its distribution throughout the metal, and removal of the foam for solidification in a manner that does not deform the pore structure excessively. Without a high degree of control in these areas, the foam that is produced can have an irregular structure: good process control is thus important (Banhart, 2001).

24.2.2.3.2   *Gas Release Agent*

The injection of gas directly into the molten metal often gives, particularly in earlier versions of the process, a large distribution of pore sizes and frequently irregular pore shapes. To overcome this, methods that introduce gas with a more homogenous distribution throughout the metal have been developed. One approach, patented in 1956 (Elliott, 1956), is to employ a gas release agent to generate bubbles more homogeneously throughout a melt; here too a ceramic addition is used to stabilize the pores. Gas release agents are chemicals that break down at a particular temperature, close to (ideally slightly higher than) the melting point of the metal or alloy, to release a gas. Examples are

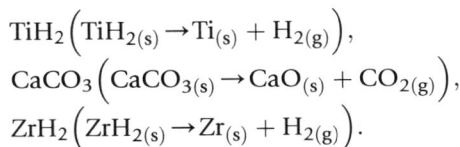

$$TiH_2 \left(TiH_{2(s)} \rightarrow Ti_{(s)} + H_{2(g)} \right),$$
$$CaCO_3 \left(CaCO_{3(s)} \rightarrow CaO_{(s)} + CO_{2(g)} \right),$$
$$ZrH_2 \left(ZrH_{2(s)} \rightarrow Zr_{(s)} + H_{2(g)} \right).$$

The gas release kinetics of these agents has been characterized by several authors (Kennedy, 2002; Kennedy and Lopez, 2003; Yang and Nakae, 2000; Gergely et al., 2003; von Zeppelin et al., 2003; Lemhus and Busse, 2004; Yang et al., 2007a; Ibrahim et al., 2008; Malachevsky and D'Ovidio, 2009; Lemhus, 2010; Haesche et al., 2010; Jimenez et al., 2011, 2012; Golestanipour et al., 2011). The temperature range over which gas is released depends on both the nature of the compound and its pretreatment, which can be used to tailor the process (Gergely and Clyne, 1998; 2000; Frei et al., 2002; Gergely et al., 2003; Lemhus and Busse, 2004; Matijasevic and Banhart, 2006; Matijasevic-Lux et al., 2006). Simultaneously, gas release agents may generate by-products in the metal, in the form of aluminides or, for $CaCO_3$ notably, in the form of oxides, the latter often coating and hence possibly stabilizing bubbles in the metal (Gergely et al., 2003; Bryant et al., 2008; Gnyloskurenko et al., 2005) (see the discussion of liquid metal foams above). Many more gas release agents have been used or tried (Banhart, submitted for publication), including related materials such as MgH_2 (Tuncer et al., 2011b). More complex chemicals which give off a larger volume of hydrogen for a given mass of foaming agent have been suggested, including Li/Na/K BH_4, and $LiAlH_4$ (Banhart et al., 2011). Yet other options are hydrated materials, such as hydroxides, sulfates or chlorides (Hardy and Peisker, 1967), natural materials such as dolomite, which releases CO_2 (Papadopoulos et al., 2011), and vermiculite, which releases water vapor. Water in turn may further react with the metal or with oxide-forming alloying elements to form hydrogen gas and a metal oxide, which may assist in raising the viscosity and stabilizing the cell walls. Although many of these systems are optimized for use with aluminum, $CaCO_3$ has been used also for magnesium (Yang et al., 2008a). Chemicals such as $MgCO_3$ or $SrCO_3$ with higher decomposition temperatures have been used to produce steel foams (Park and Nutt, 2000; 2001b), lead carbonate has been used for lower melting temperature alloys of lead (Irretier and Banhart, 2005), while TiH_2 was used to foam zinc (Chethan et al., 2011). Some consideration has also been given to economic ways to produce the foaming agent (Hur et al., 2003), and to the importance of powder compaction process parameters (Bonacorsi and Proverbio, 2006).

If the gas release agents are obtained in powder form, they may be distributed throughout the metal uniformly (through power mixing and pressing or by stirring into the melt at a temperature below the decomposition temperature). Then the breakdown of the chemical has been theorized to allow a more uniform gas generation than is found with direct gas injection (Gergely et al., 2000); however, there is evidence, from tomographic observations of the foaming of Al–Si-based alloys, that the nucleation of pores does not correlate with the positions of gas-generating particles (Rack et al., 2009b). There is also a difference in the behavior observed with different gas release agents. When $CaCO_3$ is used, the pore size is smaller and tends to be more uniform (Gergely et al., 2003). This is attributed to the oxidizing character of the CO_2 gas evolved (as opposed to the reducing action of H_2), which will interact with the liquid metal at gas–metal interfaces and form oxides, in turn increasing the bubble stability.

For the process to work correctly, it may be necessary to delay the breakdown of the gas release agent, for example, by carrying out fine-tuning of the decomposition point to allow the powder to be dispersed, or by controlling the amount of hydrogen released (Gergely and Clyne, 1998; Gergely et al., 2000). This can be achieved by using a heat treatment to generate an oxide layer at the surface of the powdered gas release agent, which then retards its decomposition (Speed, 1976; Gergely and Clyne, 1998; Matijasevic and Banhart, 2006; Matijasevic-Lux et al., 2006). Other pretreatments have been used, such as coating $CaCO_3$ particles with calcium fluoride to enhance their wettability by aluminum melts (Nakamura et al., 2002; Gnyloskurenko et al., 2005). For further control of decomposition and foaming, high external gas pressure can be applied before foaming is desired (e.g. during heating), and then released. This method gives rapid foaming, which can help to avoid problems associated with drainage or cell wall rupture (Gergely and Clyne, 2004). Another approach that has been used to increase control is to use an alloy with a melting temperature well below the foaming agent decomposition temperature (Lehmhus and Busse, 2004).

These methods may be further classified based on how the gas-generating agent is dispersed throughout the metal.

24.2.2.3.2.1 Gas Release Agent in Liquid. A process for the production of closed-cell aluminum foams by liquid foaming using gas release agents was developed by the Shinko Wire Company in 1987, and was marketed until recently as Alporas foam (Akiyama et al., 1987), although production has been discontinued. The process, see **Figure 6**, involves the addition of calcium, which reacts with melt impurities to form ceramic particles, which perform the function of melt viscosity increase (Miyoshi et al., 1998). After mixing of titanium hydride, the melt is poured into a mold for foaming to take place. Once this is complete, forced cooling through air jets may be used to freeze and preserve the foam structure (Miyoshi et al., 2000).

A similar process has been investigated using a combination of foaming and injection molding with magnesium (Körner et al., 2004; Hirschmann et al., 2007) and aluminum (Tuncer et al., 2011b; Hartmann et al., 2011). This process, called integral foam molding, requires the decomposition of the gas release agent to be ongoing during the injection of the liquid into the mold cavity. The high cooling rate at the mold wall limits the extent of bubbling in this region, and results in the outer surface of the component being effectively a dense skin. The internal part of the foam can however foam normally (no pressure is applied on the injected metal during solidification, as it would be for the casting of a dense part). No stabilization agent is added, instead relying on the presence of nuclei produced in the solidifying metal to increase viscosity and stabilize pores.

24.2.2.3.2.2 Gas Release Agent in a Precursor. With many similarities to the process of adding gas releasing agents to the liquid, processes exist to make foams by the production of a (virtually) dense precursor that can be heated when desired to cause foaming. The main technological challenge to be solved is the distribution of the gas release agent throughout the preform without its decomposition; that is, the avoidance of high temperatures before foaming. Ways of achieving this were developed in the 1960s using powder metallurgical techniques, one using carbonate gas release agents (Pashak, 1960), and another also considering the use of hydrides (Allen et al., 1963). In this method, the metal powder and foaming agent were first blended, then cold pressed to form a compact. To aid breakup of the oxide on the particle surface and so help consolidation, the compact was taken to close to full density by extrusion; as demonstrated later (Kennedy and Asavavaisthchai, 2002) this also avoids channels down which the generated gas can escape. If the compaction of the powder is carried out under vacuum the results are better, thought to be because the consolidation is improved and the loss of hydrogen before melting of the metal is so reduced (Jiménez et al., 2009). Whichever process is used, the product is a precursor which can be heated, in a mold if desired, to cause foaming. This step is sometimes referred to as "baking," as it results in the precursor growing in size in a manner reminiscent of bread or soufflé rising in an oven.

Several variants of this process have since been developed, and some implemented on an industrial scale, known as Alulight (http://www.alulight.com) and Foaminal. The Fraunhofer Institute for Applied Materials Research in Bremen have done extensive work on the process, showing that a wide range of compaction techniques may be used, including uniaxial and isostatic pressing, or powder rolling (Baumeister, 1990; Baumeister and Schrader, 1991; Kunze et al., 1993) as well as extrusion (Baumgärtner et al., 2000). They also developed a precursor which can be roll bonded between sheets

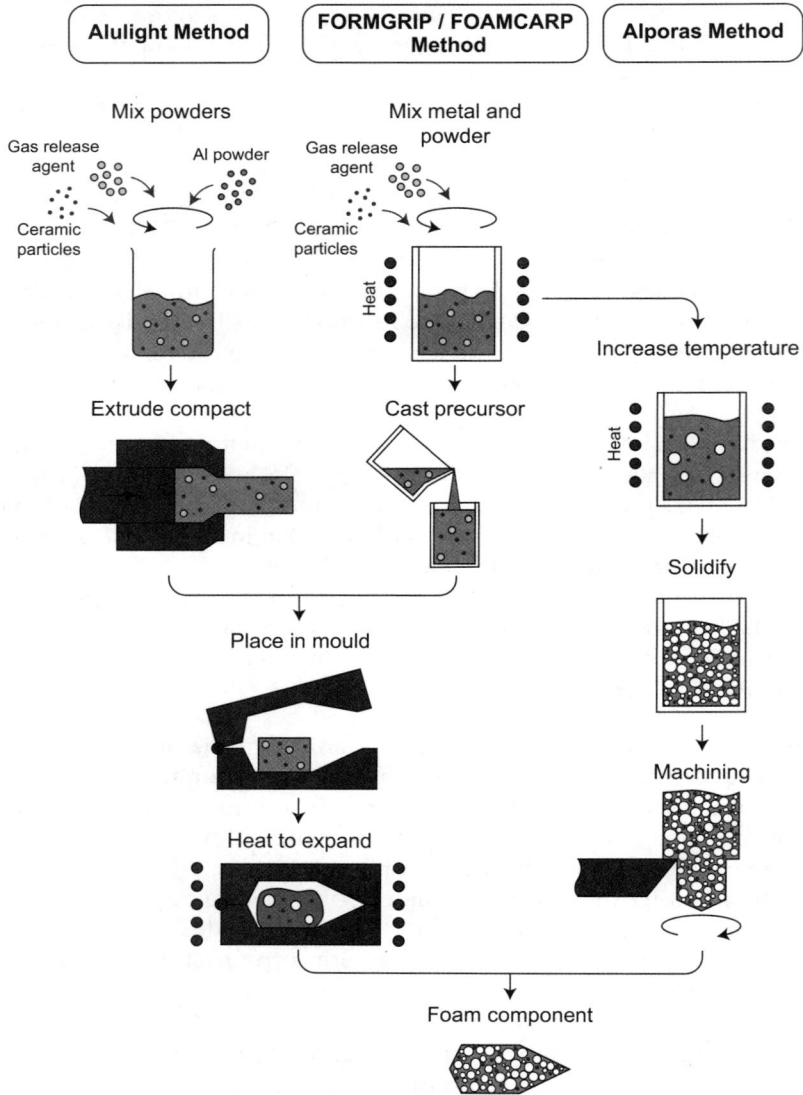

Figure 6 Schematic diagram of the methods for foam production that utilize a gas release agent. (For color version of this figure, the reader is referred to the online version of this book.)

and then foamed to directly produce a sandwich panel with a foamed core (Banhart and Baumeister, 1998b; Baumgärtner et al., 2000; Banhart and Seeliger, 2012). Metal foams of this type are produced commercially by Pohltec Metal foam GmbH (www.metalfoam.de). Similar processes to produce parts with dense skins and a porous core have been proposed (Yu et al., 1998a; Zare and Manesh, 2011), all of which share the advantage that complex shapes may be produced without the need for shaping or machining once the foam has been made (Simančik and Schoerghuber, 1998). Other variants allow the production of an extruded foam with the desired cross-section without an enclosing dense metal shell (Shiomi et al., 2010).

The key aim in processing is to produce a well compacted part with very low levels of porosity, as pre-existing pores or cracks will provide paths for gas escape (Banhart, 2000). Controlling the baking stage is also critical to the product formed, as after a certain amount of time the foam will begin to collapse. It has been shown that the composition of the metal has an influence on the nucleation and growth of pores (Kim et al., 2004a). In an attempt to better understand the whole foaming process, modeling has also been carried out (Korner et al., 2002; Belkessam and Fritsching, 2003; Zhang et al., 2009a). The solidification behavior of foams is also complex; notably a solidification expansion is often observed, as was already mentioned above. This has been investigated experimentally, and is found to be due to competition between continuing gas production (by the gas release agent or precipitation of gas dissolved in the metal) and gas loss by diffusion (which becomes more difficult as the metal solidifies) (Mukherjee et al., 2010c).

Another solid state method to distribute particles of a foaming agent throughout a metal for later foaming is based on Accumulative Roll Bonding. This process was originally developed to form highly strained samples of metal without resulting in very small sample thicknesses (Saito et al., 1999). To produce a precursor the gas release agent powder is placed between dense sheets of metal, which are deformed in rolling to cause them to bond and entrain the foaming agent. The resulting sheet can be cut, stacked and rerolled a number of times, with each step acting to further distribute the foaming agent; a uniform distribution is seen after six cycles (Kitazono et al., 2004b). The final product can be heat treated to produce a foamed sheet, with the potential advantage that the need to start from a metal powder is avoided. Yet another variant uses friction stir processing to produce the precursor (Hangai et al., 2011).

Liquid metal routes to precursor formation have been explored by Gergely, Clyne and coworkers. This resulted in two protocols for the creation of foamable precursors, the Foaming of Reinforced Metals by Gas Release in Precursors (FORMGRIP) (Gergely and Clyne, 2000) and Foaming Of Aluminum MMCs by the Chalk Aluminum Reaction in Precursors (FOAMCARP) (Gergely et al., 2003) methods. The processes proceed as shown schematically in **Figure 6**. The gas release agent powder is mixed into molten aluminum alloy containing ceramic addition (the powder may be blended with aluminum powder to aid dispersal in the melt). To achieve this with the TiH_2 powder used in the FORMGRIP process, a pretreatment is needed to delay decomposition (Gergely and Clyne, 1998; 2000). The FOAMCARP process uses $CaCO_3$. In both cases, the melt is solidified and if required machined to size. Then, as for the powder compaction processes, in the baking step the precursor is heated to above the decomposition temperature of the foaming agent.

Foams produced with the FOAMCARP method are observed to have smaller, more uniform pores (around 1 mm or less in FOAMCARP). This is attributed to the oxidizing nature of the gas released, leading to generation of a stabilizing oxide film on the pore interiors (Gergely et al., 2003). Another advantage of this process is that the foaming agent is considerably cheaper and needs no pretreatment.

24.2.2.4 Defects and their formation

A variety of defects can arise in foam production; some examples for a closed-cell foam are indicated in **Figure 7**. In methods where gas is used to foam a liquid, drainage processes can result in nonuniform density along what was the vertical direction during the process (Beals and Thompson, 1997; Simone and Gibson, 1998c), leading to the foam having nonuniform properties. The processes that occur during foam generation can also lead to defects, such as the rupture of cell walls producing cells of a much larger size and significantly different coordination than the rest of the distribution and/or being of irregular shape. Such bubbles will frequently be a site for failure to initiate

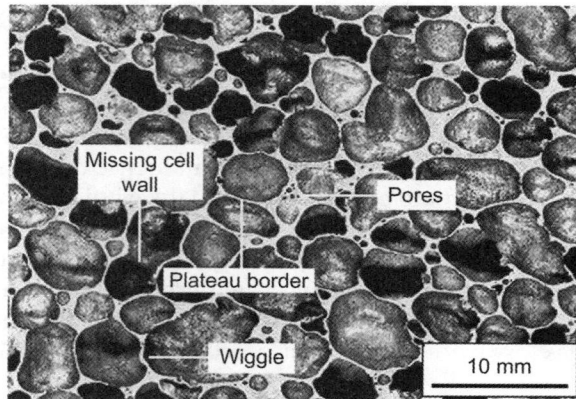

Figure 7   A micrograph showing some defects that occur in closed cell gas blown foams. Reproduced from Acta Materialia, vol. 52, Ramamurty, U. & Paul, A. "Variability in mechanical properties of a metal foam" pp. 869–876 (Ramamurty and Paul, 2004) with permission from Elsevier and Prof Ramamurty, Indian Institute of Science, Bangalore.

(Bastawros et al., 2000; Markaki and Clyne, 2001; Ramamurty and Paul, 2004; Jeon and Asahina, 2005). Other features that have been identified as defects include elliptical cells and nonplanar (i.e. curved, kinked or wrinkled) cell membranes (Bastawros et al., 2000; Ramamurty and Paul, 2004), coarse eutectic, fine oxide films and large brittle particles in cell walls (Markaki and Clyne, 2001; San Marchi et al., 2004) and partially joined or fully collapsed cells (Jeon and Asahina, 2005). It has also been suggested that the "secondary" expansion sometimes observed during the solidification of metal foams could give rise to a large number of the defects observed (Mukherjee et al., 2010b).

In processes where there is some directionality, such as the injection of gas into a melt (where the pores may be elongated and sheared during collection and solidification of foam on the conveyor belt), or when there is directionality in the structure of powder preforms (e.g. after extrusion (Kim et al., 2004a)), the pores produced may be nonuniform in shape. Anisotropic cells with a preferred orientation will obviously lead to foam properties being different along different directions (e.g. for mechanical properties (Beals and Thompson, 1997)).

24.2.3   Processes Based on a PU Foam Precursor

Since polymeric foams are far easier to produce than metallic ones and because their structure is far more regular, several methods use a polymer foam precursor to form a metal foam. This precursor is usually a polyurethane foam which is then reticulated to cause cell walls to burst, leaving the cell edge network in an open-celled foam.

24.2.3.1   Investment Casting

Some processes begin with such foams and use them as a pattern for producing the metallic foam structure by conventional precision casting, such as the investment casting method, a process closely related to the lost wax technique. Yet, despite the fact that this is the method used to make some of the most widely used and most extensively characterized open-cell metal foam, it remains one of the least well covered in the literature. The review papers that do cover it, such as (Davies and Zhen, 1983; Shapovalov, 1994; Baumeister, 1997; Huschka, 1998; Aluminium-Zentrale, 1999; Banhart, 2001), normally have to be content with inferences based on (generally transient) web pages, a small number of (mostly conference) research papers, or other sources of limited accessibility.

In the basic method, a polymeric open-celled foam is infiltrated with a ceramic investment compound, which, after removal of the polymer by pyrolysis, is then used as a mold for the metal, **Figure 8**. After infiltration and (preferably directional) solidification, the mold must be removed from the resulting metallic network. Once this is done by shaking or high pressure water jets, the metal is exposed and will reproduce the initial structure of the polymer foam, although metal shrinkage may cause a significant reduction in size. This process necessarily produces open-celled foams, generally with rather coarse cell sizes on the order of a millimeter in diameter. More specific details of the process as applied to magnesium foams are given in papers by Yamada et al. (1999), (2000).

Although the producer of "Duocel" foams (ERG Materials and Aerospace Corporation, Oakland CA, USA, www.ergaerospace.com) remains vague about their processing method, it is likely that these foams are produced by such an investment casting process. Materials produced by ERG are open-celled foams of high structural quality, **Figure 9**, with densities spanning 3–12% by volume, homogeneous pore sizes between half a millimeter up to about 5 mm, and slightly anisotropic

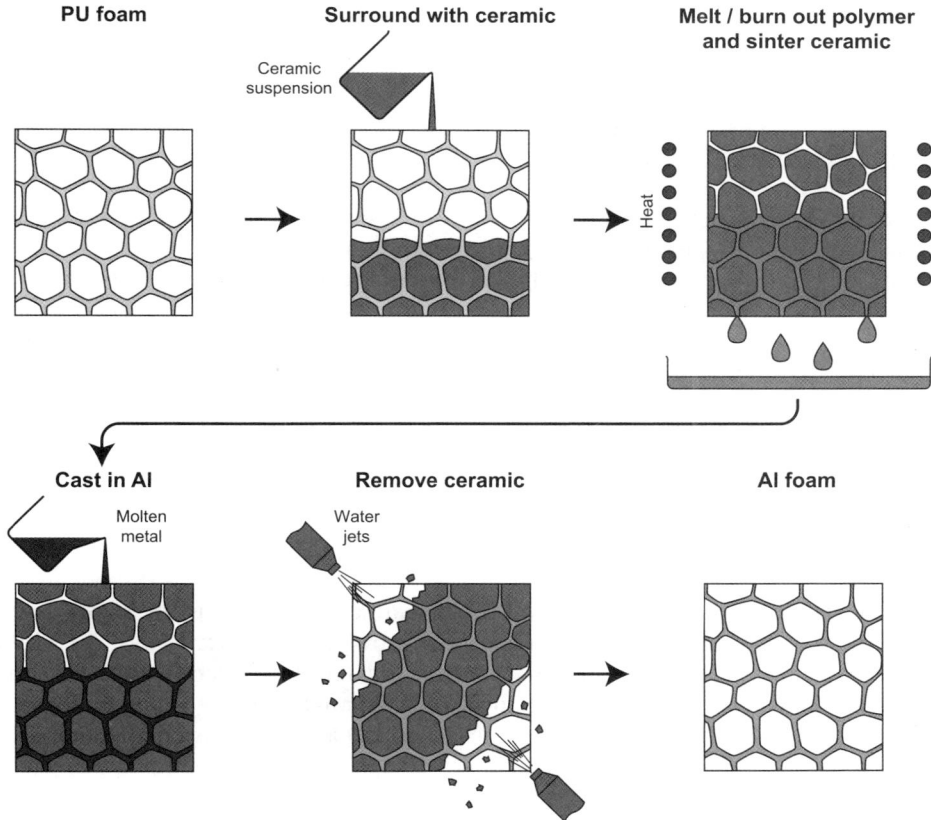

Figure 8 Schematic diagram of the production of foams using an investment casting process. (For color version of this figure, the reader is referred to the online version of this book.)

Figure 9   Image of aluminum alloy foams with different pore sizes, produced by the ERG materials and Aerospace Corporation and marketed under the trade name Duocel®. (For color version of this figure, the reader is referred to the online version of this book.)

properties (Nieh et al., 1998; ERG, 2005; Margevicius et al., 1998; Wagner et al., 1999). A newer company, m-pore GmbH of Dresden, produces similar foams using an investment casting method (http://www.m-pore.de/).

Although many workers have used Duocel foam in research, comparatively few have processed material themselves using the technique. Indications are that, where the necessary equipment exists, the method can be highly versatile. A wide range of metals appear possible; ERG claim aluminum, copper, tin, zinc, nickel, Inconel, silicon, silver and gold (http://www.ergaerospace.com/), though their most commonly examined product is made from 6101 aluminum alloy (Nieh et al., 2000). Other groups have, as mentioned, produced magnesium foams (Yamada et al., 1999, 2000). The process has also been used to produce density-graded metal foams, by taking a conical polymer precursor and constraining this inside a cylinder before the addition of the refractory material, such that at the wider base of the cone the polymer precursor is squeezed to a higher density (Brothers and Dunand, 2006b).

24.2.3.2 *Deposition on Polymer Foam*

Another way that uses an existing polymer foam to create a microcellular metal consists of coating metal onto an open-celled polymer foam, using an electrolytic, chemical vapor, or physical vapor deposition process, or using a powder slurry that is deposited and subsequently dried and sintered, with in all cases deposition generally followed by removal of the polymer via leaching or pyrolysis (Davies and Zhen, 1983; Shapovalov, 1994; Baumeister, 1997; Banhart and Baumeister, 1998b; Huschka, 1998; Aluminium-Zentrale, 1999; Banhart, 2001; Quadbeck et al., 2011), **Figure 10**. Resulting structures are typically open-celled with pore sizes and porosities roughly similar to the investment cast foams; like these foams they replicate the polymer foam structure, but since they do so by coating, instead of replacing, the polymer they will have voids within the metallic struts left by the polymer after it is removed.

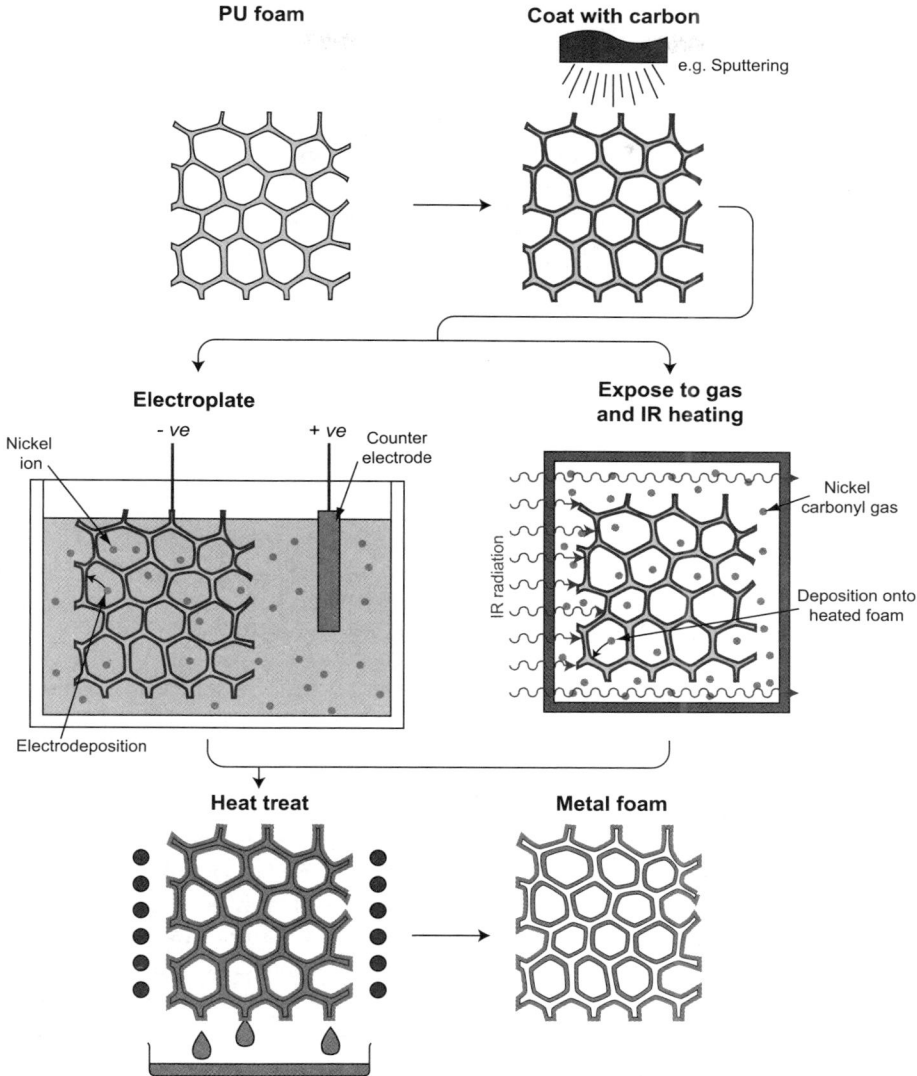

Figure 10   Schematic diagram of the production of foam using a deposition onto a precursor route. (For color version of this figure, the reader is referred to the online version of this book.)

In the electrolytic form of the technique, the original polymer foam may need to be first rendered electrically conductive. This can be achieved by precoating the foam with carbon or metal, either in the form of a slurry or by another thin film deposition technique, such as electroless plating or sputtering (Bugnet and Doniat, 1989; Brannan et al., 1992; Badiche et al., 2000; Liu and Liang, 2000; Liu et al., 2001; Liu, 2004). The method should, in principle, function for any metal or alloy that can be electrodeposited. Examples have been reported for copper (Bugnet and Doniat, 1989) and zinc (Tian and Guo, 2011), but by far the most common use is for nickel, **Figure 11**

Figure 11   Images of nickel foams with various pore sizes produced by the deposition onto a template method. (For color version of this figure, the reader is referred to the online version of this book.)

(Bugnet and Doniat, 1989; Brannan et al., 1992; Badiche et al., 2000; Liu and Liang, 2000; Liu et al., 2001; Liu, 2004; Wang et al., 2003). Nickel–chromium alloy foams of this type have also been created by alternate deposition of elemental layers and thermal treatment to cause interdiffusion (Banhart, 2001).

The metal may also be deposited onto the polymer foam by vapor deposition. One method is arc vapor deposition, which has been used to make nickel foams (Pinkhasov, 1991). An alternative uses chemical vapor deposition (CVD) of nickel tetracarbonyl ($Ni(CO)_4$), which decomposes at 150–200 °C to form nickel, allowing processing at relatively low temperatures (Babjack et al., 1990; Sherman et al., 1991; Paserin et al., 2004). In this process the polymer foam may be precoated in advance with carbon, with the reaction being carried out in a chamber with infrared radiation (Babjack et al., 1990). This is absorbed by the carbon coating on the foam, and causes localized heating, such that the gas breaks down. Heating being localized, this yields metal only in the regions near the foam surface, at a rate that is governed by local deposition kinetics (rather than transport). This attempts to combat the tendency for the foams processed by these deposition techniques (both gas phase and electrochemical) to have a larger density near the surface than in the center. Nevertheless, this variation is not eliminated, and there is a limiting upper thickness above which pieces of these foams cannot be made, of roughly 1–2 cm depending on pore size.

It is also possible to process similar material by first converting an open-pore polymer foam to carbon, and then using a slurry to coat the carbon foam with powder (of superalloy and a binder), which can then be sintered together to form a solid metal coating much thicker than the carbon. This method has been used to produce open-pore microcellular nickel-base alloy/carbon core composites (Queheillalt et al., 2004).

Such foams have found a number of commercial producers. Electrodeposition and powder coating methods are used by Sumitomo Electric Industries, Japan, to produce Celmet® foam (http://www.sei. co.jp/). Retimet® foam is produced by electrodeposition by Dunlop Equipment, Coventry, UK (http:// www.dunlop-equipment.com/prod_retimet.htm), as is Recemat® by Recemat International, the Netherlands (http://www.recemat.com/). The most well-known user of CVD techniques is Inco foam, produced by Inco Special Products (www.novametcorp.com/products/incofoam/).

24.2.4 Porosity Created by Packing

It is also possible to create a porous solid by making use of the fact that the majority of solid shapes will pack (either randomly or in an ordered manner) leaving gaps between them.

24.2.4.1 *Elementary Physics of Particle Packing*

The packing of powders is a phenomenon that governs a wide range of engineering situations, including soils, particle beds, powder processing of ceramics and metals, as well as in the structure of some porous materials. As such, considerable effort has been made to understand the basic behavior of packed particles and develop predictive tools. While it will not be attempted to cover all of this here, basic ideas useful in the present context will be presented.

Often the powder particles are taken to be spheres, to simplify the analysis. In describing a random distribution of powder particles, the main parameters of interest are normally the average coordination number, Z, and the proportion of space filled. The two will be linked and can be predicted for monomodal spherical powders (Arzt, 1982; Fischmeister and Arzt, 1983; German, 1989). From simple geometry, the maximum packing efficiency, obtained with face-centered cubic and hexagonal close-packed structures, can be calculated to fill 0.74 of the available volume if the spheres are undeformed. These highly organized arrangements are unlikely to form in a stochastically packed loose hard powder however, even after extensive vibration (or "tapping"). The random defects and disorder that would occur in a real structure result in random packing of monosized spheres filling only 0.64 of the available space (Arzt, 1982; Fischmeister and Arzt, 1983; German, 1989).

This value is often used as the typical packing fraction of a powder, even though it will be (strongly) affected by a distribution in particle sizes, and by departures in their shape from a sphere. Increased distributions of particle sizes generally give higher packing efficiencies (Cumberland and Crawford, 1987; Ma and Lim, 2002; Bierwagen and Sanders, 1974; Suzuki et al., 2001; Yu and Standish, 1987b; 1987a; German, 1989) (the reason can be seen by considering the extreme where a bimodal distribution reaches higher packing fractions by having smaller particles fill the voids between close-packed larger ones; packing fractions up to 0.8 have been achieved experimentally in this way (Molina et al., 2002)). Where particles depart from spherical form the packing efficiency is generally reduced (Cumberland and Crawford, 1987; Yu and Standish, 1993; German, 1989), due to particles contacting each other at, on average, greater distances of separation than would be the case for the same volume of material arranged in a sphere. There are however instances where the reverse is observed, for example (Donev et al., 2004; Marmottant et al., 2008); this tends to occur with nonspherical particles that are still regular in shape and have been well tapped. With nonspherical particles the number of interparticle contact points per particle (particle coordination) also differs, this being generally higher than for spheres (Donev et al., 2004; Chaikin et al., 2006; Hopkins, 2004; Marmottant et al., 2008).

Loose packing of particles is used in relatively few processing methods for porous metals; normally particles are compacted or sintered to bond and produce (generally) higher powder packing densities. The physics of powders has been extensively studied to understand such powder densification processes; results of simulations of particles compacted to different densities are shown in **Figure 12**. In this case the spheres deform at the contact points, increasing the volume occupied, the coordination number, and the average particle–particle contact area, A. The links between the space filled, Z and A can be predicted for monodisperse spheres (Arzt, 1982; Fischmeister and Arzt, 1983; Helle et al., 1985), and these relationships have been investigated experimentally (Fischmeister and Arzt, 1983; Uri et al., 2006; Aste et al., 2005; Georgalli and Reuter, 2006; Atwood et al., 2004). These is also evidence that the same relationships can be used for A (but not Z) in compacts of roughly equiaxed irregular granular

(a)   **(b)**   **(c)**

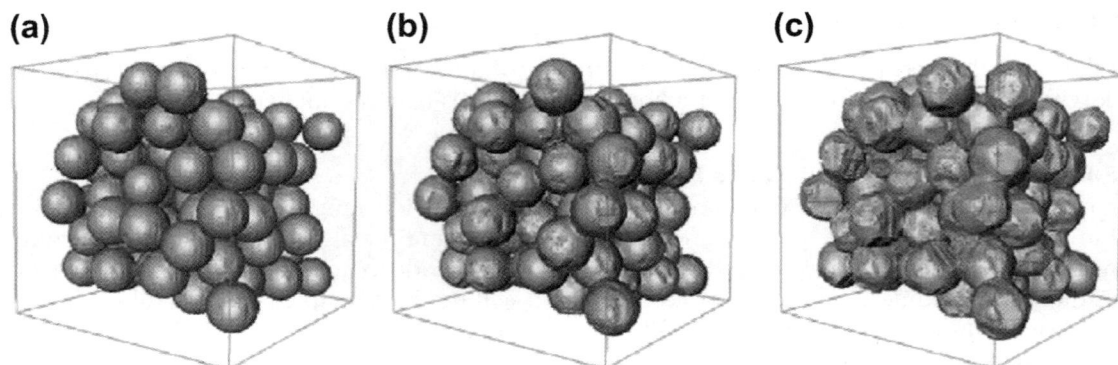

Figure 12 3D renderings of simulated packings of spherical particles to different volume fractions solid; (a) $V_s = 0.64$, (b) $V_s = 0.76$, and (c) $V_s = 0.90$. Reproduced from *Journal of the European Ceramic Society*, vol. 28, Marmottant, A., Salvo, L., Martin, C. L. & Mortensen, A. "Coordination measurements in compacted NaCl irregular powders using X-ray microtomography" pp. 2441–2449 (Marmottant et al., 2008) with permission from Elsevier. (For color version of this figure, the reader is referred to the online version of this book.)

particles (Marmottant et al., 2008). More drastically nonspherical particles have also been specifically investigated (Cumberland and Crawford, 1987; Peronius and Sweeting, 1985), modeled (Cumberland et al., 1989) and simulated (Donev et al., 2004; Chaikin et al., 2006).

To transform a loose array of particles into a coherent material, bonds must be made between them. This can be achieved by the addition of another binding phase, such as an adhesive (as is done to bind small foamed elements in advanced pore morphology foams (Stöbener et al., 2008; Stobener and Rausch, 2009)), but will more normally be achieved by a thermal treatment (sintering) or a mechanical operation (pressing). These two procedures can be combined in HIP, but as this process is normally used to ensure full density is achieved in high-specification metallic components, it is more seldom used for the creation of porous metals.

Sintering is a process driven by the surface energy of small particles. When placed in contact and supplied with sufficient thermal energy to promote the motion of atoms by diffusion or through evaporation–condensation, fine powder particles can lower their energy by reducing their surface area through bonding. Several mechanisms can operate (Frost and Ashby, 1982), all of which lead to bonding while only some of them lead to densification. As different conditions will favor different mechanisms, it may be possible to select conditions that will permit the particles to form good bonds, but without a significant reduction in sample volume, preserving the majority of the porosity present from the random packing of the particles. Depending on sintering conditions, sodium chloride conveniently falls in this category (which may be convenient in replication processing; see Section 24.2.4.3) (Goodall et al., 2006b).

Pressing (or cold pressing if it is carried out at ambient temperature) is a process whereby bonds are created by pressing particles against one another using simple mechanical force. Mechanical compaction of the particles has the advantage that it breaks any surface oxide skin, particularly if the powder compact is simultaneously deformed in shear (as in extrusion–compaction). The simplest arrangement is to use a well-fitting plunger and die, but this will introduce significant anisotropy between the directions aligned with and normal to the pressing direction. By the use of a fluid surrounding the powder compact held in a flexible die to transfer pressure, uniform compaction can be achieved; in this case it is described as cold isostatic pressing (CIP). The method will always result in some densification, and is therefore not usually preferred for the fabrication of porous materials; however, it is used in methods that involve the production of a perform, as described later.

24.2.4.2 Assembled Elements
One of the oldest and most conventional routes for the production of porous metal bodies is by partial densification of packed metal powder or fibers (Thiele, 1972; Fedorchenko, 1979; Davies and Zhen, 1983; Shapovalov, 1994; Vityaz et al., 1989; Lenel, 1980; Huschka, 1998; Aluminium-Zentrale, 1999). This is particularly effective for high melting point materials, such as titanium and its alloys (e.g. (Oh et al., 2003)) or steels (e.g. (Park and Nutt, 2000)).

24.2.4.2.1 Sintered Particles
When powders are used, this tends to produce material of relatively high density since the pore fraction cannot fall below that of initial packing of the powder; about 40–60% for equiaxed metal powders; with fibers on the other hand much lower solid fractions are feasible (see below). Nevertheless, in the low porosity range accessible by powders, control of the pore fraction is possible by varying the sintering time or conditions to affect the densification mechanism, as well as the powder size distribution (Oh et al., 2003). These pore fractions can also be enlarged by inclusion of second phases which are subsequently removed by pyrolysis or by other means, as will be described in a later section. Metal foams produced by sintering are marketed by Schunk Sinter Metals (http://www.sintermetalltechnik.com) for use as filters and silencers.

Sintering may also be applied to powders held in a foamed liquid suspension (Jee et al., 2000; Guo et al., 2000). For this purpose, a foamable polymer would normally be chosen, such as a multicomponent polyurethane foaming system, with which it is very easy to generate a porous structure. The initial foams produced are closed cell, but on heat treatment and sintering it is observed that the cell faces open up (Jee et al., 2000). This is possibly due to the cell faces becoming denuded of powder particles as drainage processes take place. Although relatively low strength, the final foams obtained are highly porous with porosity above 90% (Guo et al., 2000). This process has been applied to produce copper foams (Xie and Evans, 2004), where the copper particles were oxidized during heating in air to remove the polymer, and then reduced during sintering; this may improve the strength of the foam at the intermediate stage. If the metal powder particles are very fine (of the order of 60 µm or less), they may be placed into aqueous suspension and will act as stabilizers, in a very similar manner to the theorized action of ceramic particles in liquid metal foams. It may then be possible to foam the liquid, and to dry and vacuum sinter the foams produced to give a metallic foam (Studart et al., 2012). A related approach for the production of iron foam starts with the formation of a ceramic foam of iron oxide that is later reduced at 1240 °C in Ar-4%H_2 (Verdooren et al., 2004). Sintering in a temperature gradient created during spark plasma sintering (SPS) was also demonstrated to be a method for the production of (slightly) porous metals containing a gradient in relative density (Kwon et al., 2003a).

A specific variant of sintering, namely exothermic reactive sintering also referred to as combustion synthesis, or self-propagating high temperature synthesis (SHS), does not require heating to elevated temperature and frequently produces highly porous material (Munir and Anselmi-Tamburini, 1989). This starts with the production of powder compacts of two materials that will react exothermically with one another when ignited to form a third material; the compact can be produced by hot extrusion in a more industrially scalable manner than cold compaction (Kobashi and Kanetake, 2010). Systems used include Ni–Ti (Kobashi and Kanetake, 2002; Kim et al., 2004b; Biswas, 2005), Ti–B_4C (Kobashi et al., 2006a), Al–Ni (Kobashi et al., 2006b) and Al–CuO (Nabavi and Khaki, 2011). In this last case, the process has been carried out between two dense metal skins to produce a sandwich panel. Generally metallic materials produced in this way are (or at least contain a substantial amount of) an intermetallic

(Biswas, 2005); alternatively, metal/ceramic composites may be produced, by thermitic reactions, for example. In these methods, there is some possible control over the cell size and porosity by varying the powder size, the proportion of the different species, and the amount of initial porosity in the powder compact before reaction (Kobashi and Kanetake, 2002; Kobashi et al., 2006a, 2006b); however, these highly exothermic reactions being difficult to harness, structures obtained tend to be somewhat irregular. The final porosity in these materials appears to arise from the structure not having time to sinter or collapse, coupled in some cases with the internal evolution of heated gas (Kanetake and Kobashi, 2006).

24.2.4.2.2 *Sintering Other Forms*

Hollow spheres of metal may be themselves sintered or brazed. The structure produced is then somewhat similar to syntactic metallic foams (see Section 24.2.4.2.3), but without the matrix between the hollow spheres, and so tend to be of lower density. Rather, it is made of (relatively compliant) spherical shells partially bonded one to the other at discrete circular necks (Nagel et al., 1997; Uslu et al., 1997; Baxter et al., 1998; Hurysz et al., 1998; Sypeck et al., 1998; Evans et al., 1999; Weise et al., 2007; Lim et al., 2002; Roy et al., 2011).

Very low-density metal structures can also be made using metal wires; cleaning pads of steel wool are probably one of the most commonly encountered highly porous metals in daily life. Stronger wire-based structures can be made by joining together arrays of metallic fibers (e.g. by sintering or brazing) (Fedorchenko, 1979). Sintered random or semi-random coiled carbon steel, titanium or aluminum wire structures were produced and tested by Liu et al. (2009b), (2008), (2009c), (2010), Tan et al. (2009), varying the porosity by precompacting the wire arrays before sintering. Bonded wire structures of stainless steel fibers of low diameter (down to 50–100 µm) have been produced (Ducheyne et al., 1987; Delannay, 2005; Clyne et al., 2006; Tan et al., 2006a) to give solids with $V_m \approx 0.4$ for applications in biomedical implant surface coatings. The fact that an order of magnitude difference was found between the tensile and compressive Young's modulus (higher in tension, explained by fiber stretching in tension and fiber buckling in compression) highlights the possibility of unusual mechanical behavior being designed into such systems. Anisotropic behavior has been seen in lower density fiber arrays (Delannay and Clyne, 1999), where 12 µm diameter stainless steel fibers were sintered to form a solid with $V_m \approx 0.2$, with fibers oriented in a plane, similar to the material imaged in **Figure 13**. Other work in this vein is reported in (Markaki et al., 2003).

24.2.4.2.3 *Syntactic Foam*

An alternative processing route uses hollow ceramic or glass spheres (Cochran, 1988), or other highly porous ceramic forms. These are loosely packed together and then infiltrated with metal (Thiele, 1972; Davies and Zhen, 1983; Drury et al., 1989; Microcell, 1992; Hartmann and Singer, 1997; Hartmann et al., 1998; Balch et al., 2005; Brothers and Dunand, 2004; Rabiei and O'Neill, 2005; Balch and Dunand, 2006; Wu et al., 2007; Colombo and Degischer, 2010; Murasawa et al., 2011), producing a three-phase (metal, ceramic, void) composite otherwise known as a "syntactic foam," **Figure 14**. Such materials are evidently dependent on the initial packing of the spheres to determine the quantity of capsules included and also on the sphere shell thickness to determine the density; from the discussion of particle packing earlier it will be seen that it will be difficult to achieve greater than 64% porosity in these materials. Any metal suitable for casting could be used, while difficult to cast metals, such as titanium, have rather been used with a powder metallurgy route (Xue and Zhao, 2011). To date aluminum and magnesium as the infiltrant are most common (Zhao, 2011), though the use of other materials, such as zinc alloy, have also been reported (Daoud, 2008). There can be reactions between the ceramic and the metal (Daoud et al., 2007), although selection of the correct alloy for a particular type of sphere can avoid or minimize this (Orbulov et al., 2009). Similar materials can be created using

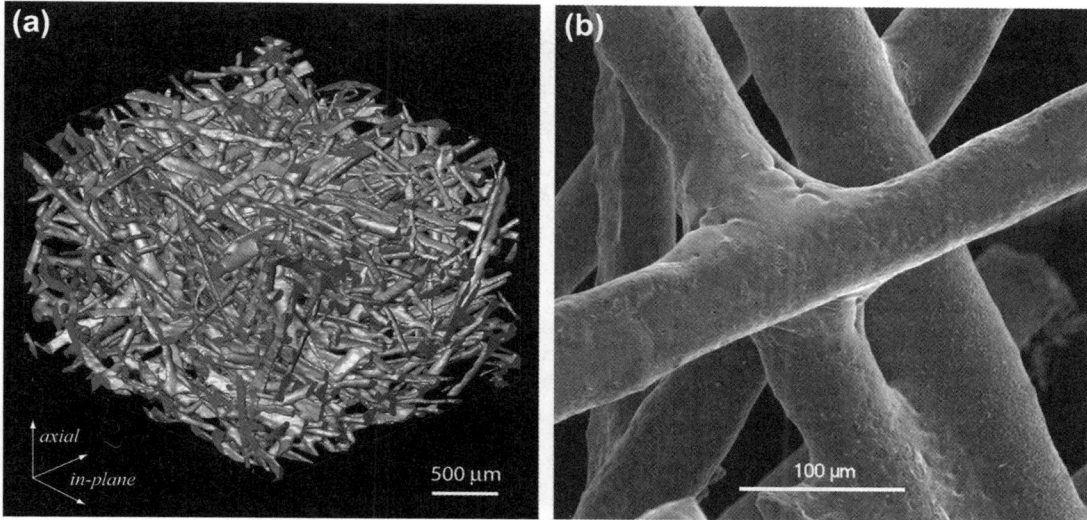

Figure 13   Porous stainless steel made by sintering fibres. (a) a 3D reconstruction of a tomographic image of a sintered fibre network (Tan et al., 2006a) and (b) a close up image of the join formed between two fibres (Markaki et al., 2003). Reproduced respectively from (a) *Advanced Engineering Materials* vol. 8, Tan, J. C., Elliott, J. A. & Clyne, T. W. "Analysis of Tomography Images of Bonded Fibre Networks to Measure Distributions of Fibre Segment Length and Fibre Orientation" pp. 495-500 (Tan et al., 2006a) with permission from Professor T.W. Clyne, Cambridge University, and Wiley, and (b) *Composites Science and Technology* vol. 63, Markaki, A. E., Gergely, V., Cockburn, A. & Clyne, T. W., "Production of a highly porous material by liquid phase sintering of short ferritic stainless steel fibres and a preliminary study of its mechanical behaviour", pp. 2345–2351 (Markaki et al., 2003), with permission of Prof T W Clyne, Cambridge University, and Elsevier.

metal spheres if there is a difference in melting point between the spheres and the infiltrant; one example is infiltrating the spaces between steel spheres with aluminum (Rabiei and O'Neill, 2005; Rabiei et al., 2006; Vendra and Rabiei, 2007).

The resulting "syntactic" materials generally have closed cells and, in the early examples, relatively coarse microstructures, the spheres being several millimeters in diameter. Current research is carried out more on materials with smaller cell sizes, often fabricated with microcapsules (Orbulov et al., 2009; Murasawa et al., 2011). Bimodal distributions have also been created (Tao et al., 2009). The materials produced often show comparatively high strengths, which may be attributed to the low fractions of porosity, coupled with the presence of the load-bearing ceramic phase (Daoud, 2008).

24.2.4.2.4   *Thermal Spraying*
Thermal spraying is, like SHS, another metallurgical process that has a tendency to produce porous materials. It consists of the deposition followed by solidification of a stream of atomized metal, forming layers that can be built up to form bulk shapes such as ingots. It is often advantageous for forming high melting point materials, as only a very small amount of material needs to be molten at any one time. More specifically, the process consists of passing a powder through a heat source (a flame or plasma at high temperature) which both heats and melts the powder and provides it with an impulse toward a target (which may be an object to be coated, or the surface of a built-up ingot cooled at the far end). The molten droplets impinge on the surface and solidify, gradually building up material in layer or bulk form, depending on the duration of the process.

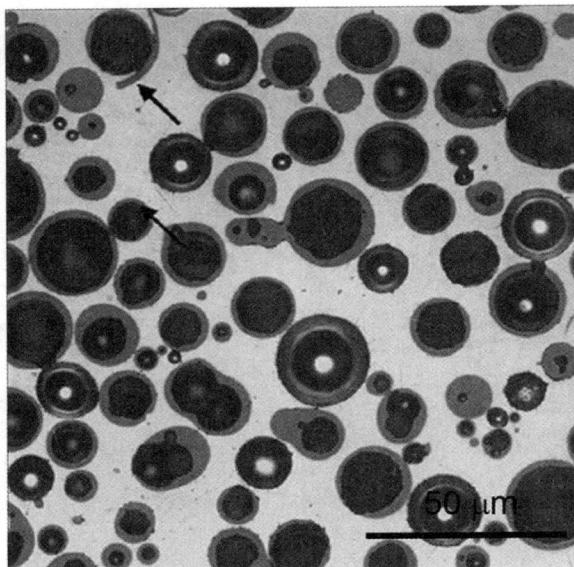

Figure 14 A cross section through a syntactic foam produced by infiltrating hollow carbon spheres with a $Zr_{57}Nb_5Cu_{15.4}Ni_{12.6}Al_{10}$ (Vit106) metallic glass (Brothers and Dunand, 2004). Highlighted in the image are some fragments of broken spheres, while the variable shape and wall thickness of the carbon spheres is also apparent. Image reprinted from *Applied Physics Letters* vol. 84, Brothers, A. H. & Dunand, D. C. "Syntactic bulk metallic glass foam", pp. 1108–1110 (Brothers and Dunand, 2004) with kind permission of Professor Dunand, Northwestern University, and the American Institute of Physics. Copyright 2004 American Institute of Physics.

It is part of the nature of the process that the layers produced contain (usually) small amounts of porosity. This may, for example, contribute to the low through-thickness thermal conductivity for a plasma sprayed sample. Some workers have attempted to increase these porosity levels by, for example, incorporating chemicals in the powder to be sprayed that will break down and release gas. Gas release agents used include $BaCO_3$ (Kelley et al., 1993) and SiO_2 or MnO_2 (Banhart and Knuwer, 1998). The former breaks down to give CO_2 gas in a similar way to the gas release agents used in direct foaming of liquid metal. The latter were used with steel, and reacted with carbon to liberate CO.

24.2.4.3 *Porosity Created by Removable Phase (replication)*
24.2.4.3.1 *Powder Processing with Space Holder*
While powder metallurgical processes can produce near fully dense parts from metal powder, porosity can be deliberately introduced by including in the powder mix an additional phase to retain spaces in the material (this phase is often called the space holder). If this phase can be removed, for example, by chemical or thermal means after or during sintering, a porous metal is produced (Fedorchenko, 1979; Weber and Knüwer, 1997), **Figure 15**.

The process is often used for the higher melting point metals, which can be challenging to process in the liquid phase. Notable examples include titanium (Bram et al., 2000; Wen et al., 2001; Laptev et al., 2004; Dunand, 2004; Esen and Bor, 2007), stainless steel (Bram et al., 2000; Bakan, 2006) and superalloys (Bram et al., 2000), although there has also been extensive use for the manufacture of aluminum foams in this way (Zhao and Sun, 2001; Wen et al., 2003; Sun and Zhao, 2003; Zhao et al.,

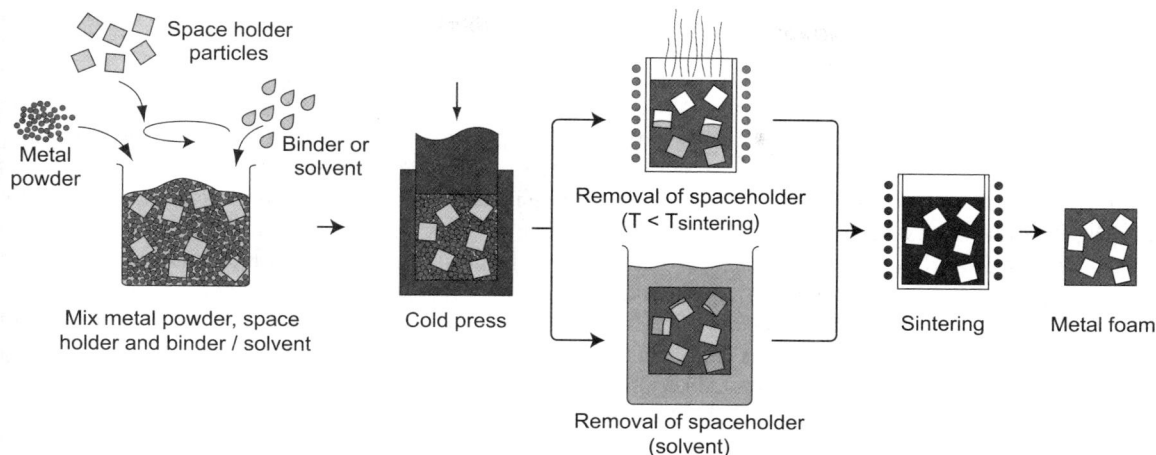

Figure 15 A schematic of the space holder method with metal powders. Note that here the space holder particles are shown as isolated cubes, which may be the case if they were salt crystals, for example: In three dimensions the space holder must, however, in all cases form an interconnected network (or it cannot be removed). Other space holder geometries are also possible. (For color version of this figure, the reader is referred to the online version of this book.)

2004d; Jiang et al., 2005b; Sun and Zhao, 2005a; Hakamada et al., 2005b; Bin et al., 2007). In the case of titanium, TiH_2 is sometimes added to promote sintering (Rak et al., 2003); it does not contribute to porosity generation as it does in foams processed by liquid metal routes, but the hydrogen released acts as a sintering aid by helping to remove the oxide layer formed at the particle surface.

The earliest powders to be used were carbamide (another name for urea $(NH_2)_2CO$) and ammonium hydrogen carbonate, which were removed by a thermal treatment (Bram et al., 2000). Although the second addition is often described as ammonium bicarbonate, examination of the chemical formula given reveals that it is in fact ammonium hydrogen carbonate $((NH_4)HCO_3)$ that is used. These materials were blended with a solvent and metal powder to coat the space holder particles with powder. These coated particles were then pressed together and the space holder burnt out at below 200 °C before sintering at elevated temperatures to bind the powder particles. The result was titanium, 316L stainless steel or various Ni-based superalloy foams (reported V_m in the range 0.2–0.4) with pores of variable shape and pore sizes in the range 0.1–2.5 mm (Bram et al., 2000). Following this, carbamide and ammonium hydrogen carbonate have frequently been used in the same process to make similar foams from titanium (Laptev et al., 2004; Niu et al., 2009), magnesium (Wen et al., 2001, 2004) or aluminum (Jiang et al., 2005b, 2005a; Bin et al., 2007; Barletta et al., 2009), and other metals such as stainless steel (Bakan, 2006; Gulsoy and German, 2008). Expanded polystyrene granules have also been used with stainless steel powder, with the powder held together with a PVA binder to avoid the need for compaction, and allow relatively low-density foams to be created (Shimizu et al., 2012). In the case of aluminum, dissolution rather than a thermal treatment is used to remove the carbamide before sintering. It has been shown that machining at the green body stage allows near-net shape production of parts (Laptev et al., 2004, 2005). Titanium has also been processed using magnesium particles as the space holder (Esen and Bor, 2007). The magnesium is removed by a thermal treatment that causes it to evaporate before the temperature is increased to allow sintering. Other workers have produced closed-cell foams, plating Ni–P alloy on the surface of polystyrene spheres which are then pressed and sintered, the organic residue of the polystyrene remaining trapped in the pores (Kishimoto et al., 2003).

At almost the same time as the initial work on the process with carbamide and other thermally degradable space holders, workers in the UK developed a similar process with NaCl as the space holder in the production of aluminum foams, the space-holding phase being removed by dissolution in water (Zhao and Sun, 2001; Zhao et al., 2004d; Hakamada et al., 2005b; Surace et al., 2009) (this process has been modeled to aid minimization of the amount of residual NaCl in the foam after treatment (Zhao, 2003)). Similar steps were followed, blending space holder grains and metal powders before compaction, but in this case sintering was carried out with the space holder in place. The melting point of salt ($T_m = 801$ °C) being higher than that of aluminum ($T_m = 660$ °C) means that the upper temperature limit was the melting point of the metal; liquid Al does not wet salt and if sufficiently fluid it would not remain blended with the salt particles. It has however been shown that sintering temperatures 10–20 °C above the melting point can be used to improve the sintering of such foams with only minor dewetting (Zhao et al., 2004d). After sintering the NaCl was dissolved in water to produce foams with V_m in the range 0.2–0.7, although it was noted that residual salt remained trapped for $V_m > 0.4$ (Zhao and Sun, 2001); this correlates with the volume fraction at which the salt grains would be expected to cease to percolate, and is in agreement with analytical models (Zhao, 2003). Modeling heat flow in the compacted salt bodies shows that the effect of the free volume on reducing the thermal diffusivity, and therefore the sintering rate, is dominant (Sun and Zhao, 2005b). The method has also been used with blended Al and Mg elemental powders, the intention being that a small quantity of magnesium will improve the sintering performance by reducing the oxide on the aluminum (Sun and Zhao, 2003); this Mg addition has been reported on occasions as having little effect on the sintering (Zhao et al., 2004d), and on others as being effective (Sun and Zhao, 2005a). Slightly higher sintering temperatures were achieved using potassium carbonate (K_2CO_3, $T_m = 891$ °C) allowing the production of Cu foams (Zhao et al., 2005c; Thewsey and Zhao, 2008), with the process being called Lost Carbonate Sintering (commercial production of foams has recently started using this method by Versarien Ltd, http://www.versarien.com/, with grades such as Versarien Cu63, a copper based foam with 63% porosity). In a similar method the use of NaF has allowed the processing of NiTi at 1250 °C (Bansiddhi and Dunand, 2007). NiTi has also been processed from prealloyed powder with NaCl space holders, with the use of Nb as a trace addition to create a in situ eutectic that allows full densification (Bansiddhi and Dunand, 2011a) (a variant replaces the NaCl with Nb wires as the space holder (Bansiddhi and Dunand, 2011b)). An alternative process uses elemental powders to form NiTi, requiring a heat treatment at only 950 °C (Zhao et al., 2009b). Here an NaCl space holder is used, which is removed by dissolution after pressing of the powders. A similar process uses NaCl with elemental powders to produce a stainless steel foam (Scott and Dunand, 2010), or porous Ni_3Al (Liu et al., 2011) using NaCl as a space holder. The process is reported to occur by evaporation of the salt, not requiring a dissolution step. Sugar space holders removed before sintering have also been used to produce microcellular silver (Asavavisithchai and Nisaratanaporn, 2008) or aluminum (Michailidis and Stergioudi, 2011; Michailidis et al., 2011).

A variant of the basic process uses SPS, where a pulsed electric current is passed through the powder to assist sintering and reduce sintering times. Pressure is often applied simultaneously, with the punches also serving as the electrodes to which the supply is connected. This has been performed on aluminum with salt grains as a space holder (Wen et al., 2003), and nearly closed-cell aluminum foams have also been produced in this way (Hakamada et al., 2005c). The method has been applied to a Ni–P alloy using polymer space holder granules (Song et al., 2003, 2004; Song and Kishimoto, 2006), though as this method produced closed-cell foams, the polymer was unable to escape and remained in the foam in a thermally degraded form. In yet another variant, preforms of metal powder blended with salt were prepared by friction stir processing (Hangai et al., 2012).

Another variant uses metal as the place holder: steel wires can be cocompacted with titanium, for example, and then leached, by electrochemical means, for example; a main challenge in this route is to limit interdiffusion of the two metals during sintering (Kwok et al., 2008; Jorgensen and Dunand, 2010; Neurohr and Dunand, 2011b; 2011a). Yet another variation, with considerable scope in the near-net shape processing of complex parts, is when a binder is added as well as a space holder to the metal particles. One route to obtaining a part of complex shape is machining at the green body stage (Laptev et al., 2004). An alternative is possible if the binder is chosen correctly, allowing an injectable mix to be produced. This can then be processed by injection molding to complex shapes (Guoxin et al., 2008; Köhl et al., 2009, 2011), following the procedure for metal injection molding of dense parts. This is followed by a binder removal step, before space holder removal and sintering can take place. For some systems, the binder phase may be enough to preserve free space, without the need for a specific space holder, particularly if sintering is rapid (Ismail et al., 2011).

24.2.4.3.2 Infiltration of a Preform

Instead of blending metal powder with a space holder and sintering, if the space holder has a higher melting point than the metal, then one can infiltrate the open-pore space left within a removable compacted or sintered powder body of that space holder material, **Figure 16**. This technique is reviewed

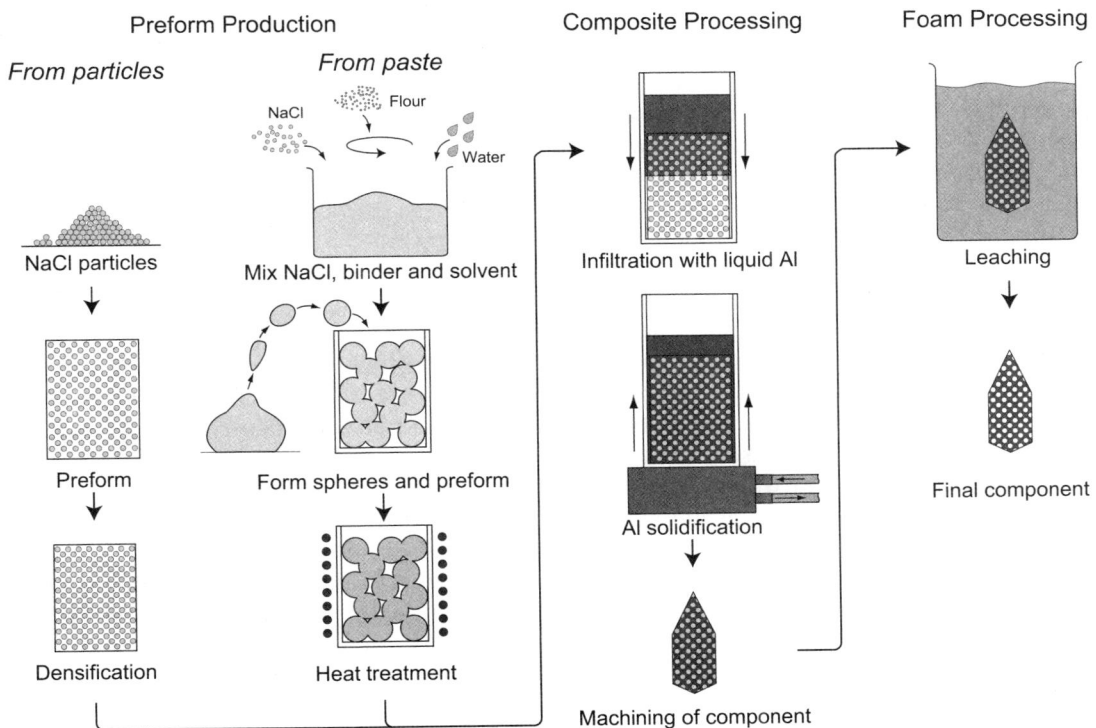

Figure 16 Schematic diagram of the replication process for open-cell foam manufacture. (For color version of this figure, the reader is referred to the online version of this book.)

in Refs. San Marchi and Mortensen (2002), Conde et al. (2006a). After solidification of the metal the second phase is removed as before by, for example, leaching with a solvent, chemical attack, or oxidative pyrolysis. With this approach one obtains a foam having the pore structure of the powder compact; the fact that such powder compacts generally feature low pore contents means that resulting foams are highly porous. A very common use of this method employs sodium chloride as the preform material to make aluminum (Polonsky et al., 1961; Seliger and Deuther, 1965; Kuchek, 1966; San Marchi et al., 1999; San Marchi and Mortensen, 2001, 2002; Despois et al., 2004, 2007; Gaillard et al., 2004; Despois and Mortensen, 2005; Goodall et al., 2006a, 2007) or aluminum alloy (San Marchi et al., 2004; Goodall et al., 2006d; Kadar et al., 2007; Goodall and Mortensen, 2007) foams. This has the advantages of ready availability and being easy to handle combined with nontoxicity and ease of removal in water after processing. With a melting point of 801 °C, NaCl is suitable for use with other metals having a melting point near that of aluminum or lower, and has, in addition been used in the fabrication of bulk metallic glass (BMG) foams, by quenching a mixture of salt grains and liquid Pd-based alloy (Wada and Inoue, 2003).

This salt-based "replication" process (LeMay et al., 1990) was first developed in the 1960s, casting metal by conventional means on to uncompressed grains of salt (Polonsky et al., 1961; Seliger and Deuther, 1965; Kuchek, 1966). The fact that in this case the preform could be leached in finite time even though the contacts between the pores were in theory just the point contacts existing between the particles was explained by invoking the movement of hydrated NaCl due to adsorbed atmospheric water to the wetted high angle contact points. Nevertheless, although technically open-celled foam, these structures had a low degree of pore interconnectivity. Further advances came once methods to densify the preform were employed, although this type of coarse salt process has continued to be used for the production of aluminum and aluminum alloy foams (Han et al., 2004; Cao et al., 2006; Abdulla et al., 2011); with very careful control of the pressure during introduction of the aluminum, it is possible to produce foams with larger interpore windows than those governed by the interparticle contacts.

Sintering was the first method employed to densify the NaCl particles, allowing preform relative densities of up to 0.8 (corresponding to a foam V_m of 0.2) (San Marchi et al., 1999; San Marchi and Mortensen, 2001), and this has continued to be used in several studies since (Kadar et al., 2007). The sintering behavior of NaCl in the range 500–790 °C is examined in Ref. Goodall et al. (2006b), indicating the regions of temperature, grain size and density space where sintering results in densification or only in morphological changes at the particle contact points. Further work (San Marchi and Mortensen, 2002; Despois et al., 2004) then developed CIP as a more rapid method to achieve a wider range of homogeneous densities (preform relative densities of 0.91 have been achieved with spherical pores, corresponding to foam V_m of 0.09 (Goodall et al., 2007), while with angular pores $V_m = 0.05$ has been produced by this route (Amsterdam et al., 2008d)). Although the net effect is similar, the two procedures produce topological differences in the preform, namely a more rounded structure in the case of sintering and more angular as a result of CIP treatment, which are transferred to the resulting foams and can affect the foam properties (Goodall et al., 2006a).

To further control the foam produced workers have examined modification of the starting salt grains, either by producing these grains by antisolvent precipitation, using additives that control the growth habit of the crystals and thus produce exotic forms (Gaillard et al., 2004), or by a remelting process to produce roughly spherical crystals (Goodall et al., 2006c, 2007). Preforms have also been made combining multiple sizes of salt grains to produce microcellular aluminum with bimodal pore size distributions (Li et al., 2003a; Despois et al., 2006a; Yu et al., 2006; Chmielus et al., 2010). The need to produce a preform has also been avoided by mixing into molten metal salt particles and using a permeable piston to compact the particles and remove excess aluminum (Jamshidi-Alashti and Roudini, 2012).

A recent development allows large pores of complex shape to be produced by blending small salt grains and a binder into a mouldable paste which is later fired to remove the binder phase (Goodall and Mortensen, 2007; Mortensen and Goodall, 2012), **Figure 16**. The resulting internally microporous structure of this preform sharply decreases the dissolution time in the final step of the process. In procedures of this type, relying on a dough containing salt, the mixing of the paste is a critical factor in determining the mechanical properties of the dough (Angioloni and Rosa, 2005). Aluminum alloy foams are being produced commercially using this technique. They are marketed under the name Corevo®, and are produced by Constellium (http://www.constellium.com/); see **Figure 1b**. A variant of this process that produces spherical salt agglomerated grains without the need to shape the particles by a mechanical method is the dispersal of a salt–flour–water mix in warm oil, before heat treatment to remove the flour and sinter (Jinnapat and Kennedy, 2010). Particles made by this method were used in a space holder route to produce aluminum foams (Jinnapat and Kennedy, 2011).

When small salt grains are used, pressure infiltration of the preform is required because aluminum does not naturally wet table salt; this is often carried out using argon gas pressure. Resulting open-cell foams with very fine and controlled pore sizes (down to the range 1–10 μm) can be produced in this way, **Figure 17**. The general process is well understood, having as it does many similarities to the production of metal matrix composites by infiltration (see e.g. (Mortensen and Jin, 1992; Mortensen, 2000)). It has been noted that the infiltration pressure can influence the foam relative density and thus its properties; as there is a range of sizes of features to be infiltrated, there will be a pressure range over which infiltration takes place. Some workers have looked at the influence of capillarity and infiltration pressure on the structure of foams made using this type of process (Despois et al., 2007; Chen and He, 1999; Berchem et al., 2002).

One way to view the replication process is as a casting process incorporating a very complicated core: the method is clearly suitable for the production of parts where a foam is intimately bonded to dense metal. Examples of such structures have been produced in the form of variously shaped castings (Despois et al., 2006a), of sandwich beams that also displayed graded levels of porosity across the beam

Figure 17 An image of aluminum-replicated foam samples, processed using NaCl as a preform material. Both samples have roughly 70% porosity, and the cylinder on the left-hand side has a mean pore size of 400 μm, and the cylinder on the right has a mean pore size of 75 μm. (For color version of this figure, the reader is referred to the online version of this book.)

thickness (Pollien et al., 2005)—a structure that may save weight in some configurations (Conde et al., 2006b). Graded porosity was also produced in magnesium using infiltration–replication (Bach et al., 2003). Other refractory phases may also be incorporated at the casting stage, as shown by the production of composites by prepacking the reinforcement phase between the salt particles; the process is hence also suited for the production of highly porous metal matrix composites (San Marchi et al., 1999).

Several other materials than NaCl have been employed in the replication process; some extend the process to higher melting point metals. These include infiltration processes close to conventional casting with sand or other refractory cores, the core material being removed by high pressure water jets (Lu and Ong, 2001; Berchem et al., 2002) or by other forms of washing (Chou and Song, 2002) or shaking (Dairon et al., 2011), but also a number of works using alternative powdered salts with higher melting points. Examples include SrF_2 ($T_m = 1477\,^\circ C$), BaF_2 ($T_m = 1368\,^\circ C$), used for BMG foams and removed with HNO_3 and HCl, the BMG remaining unaffected due to its high resistance to chemical attack (Brothers et al., 2005, 2006; Brothers and Dunand, 2005a, 2005b, 2006a), and also for NiTiCu superelastic alloy (Young et al., 2012). At lower temperatures, water-soluble $MgSO_4$ was shown to be suitable for the production of microcellular Ag–Cu eutectic (Diologent et al., 2009a). In a method that is potentially suitable for some high melting point and acid-resistant materials, foams have been made using silica gel beads, which are removed after casting with a treatment in HF. This process was first demonstrated with brass (Castrodeza and Mapelli, 2009), and has since been used for shape memory Cu–Zn–Al alloy (Bertolino et al., 2010). The highest melting point processing of replicated foams so far carried out has been at $1450\,^\circ C$ to process superalloy material. Different preform materials have been successfully employed, sodium aluminate, $NaAlO_2$ ($T_m = 1477\,^\circ C$), being used to produce J5 alloy (Ni-22.5Mo-12.5Cr-1Ti-0.5Mn-0.1Al-0.1Y) (Boonyongmaneerat and Dunand, 2008), and SrF_2 for IN792(Ni–12.6Cr–9.0Co–1.9Mo–4.3W–4.3Ta–3.4Al–4.0Ti–1.0Hf–0.09C–0.02B–0.06Zr) (DeFouw and C.Dunand, 2012); all compostions given in wt%. Sodium aluminate has also been used to process Ni–Mn–Ga martensitic alloys (yielding foams with interesting magnetic shape memory and magnetocaloric properties) (Boonyongmaneerat et al., 2007; Chmielus et al., 2010, 2009; Zhang et al., 2011; Sasso et al., 2011; Dunand and Müllner, 2011).

In one alternative approach, the removable porous structure is removed by burning after infiltration (Ma and He, 1994). This has been further developed by a system that uses the rapid infiltration of a polymer bead bed to ensure that the thermal energy of the metal does not significantly degrade the preform, which is later removed by a burning treatment. Resin-bonded polystyrene beads have been used in this way with aluminum (Ma et al., 1999).

24.2.5   Porosity Created through Phase Change

24.2.5.1   Solidification with Gas Evolution

In the processing of metals, invariant transformations are frequently observed, where there is a transition between a single phase and multiple phases at particular conditions of composition and temperature. One example would be a eutectic transformation

$$L \to \alpha + \beta.$$

In conventional metallurgy, this may produce a characteristic fine-scale, two-phase eutectic microstructure, the two phases produced organizing themselves according to their mode of growth (faceted/nonfaceted), the temperature gradient and their volume fraction ratio, structures obtained tending to minimize interfacial energy between the two phases. The outcome is governed by both minimization of energy and kinetics, and also by the time taken to separate the constituents of each out of the liquid. With roughly isotropic interfacial energy, the most commonly observed form is a lamellar eutectic when both phases occupy roughly equal volume fractions (e.g. as seen in Al–Cu alloys), while with a large difference in phase fractions rod eutectics are often observed. Here, the minority phase forms cylindrical rods, largely aligned along the growth direction, in a matrix of the other phase (the Al–Ni system is an example).

It is possible to create porous metals in this way, by seeking systems where a gas can be dissolved in a liquid metal in much higher quantities than when it is solid. The key ideas and process were developed in Ukraine and called the "Gasar" process (Shapovalov, 1993, 1998; Drenchev et al., 2006; Shapovalov and Withers, 2008) from a contraction of *gas armirovat* (Russian for gas reinforced). They have also been termed "lotus" structured metals (due to the similarity of the structure to that of a lotus root) by researchers in Japan who subsequently have contributed much research on this process (Nakajima et al., 2004; Nakajima, 2007). The process comprises the directional solidification of (normally) a hydrogen-saturated metallic melt having the composition of a eutectic between the metal and gaseous hydrogen (Shapovalov, 1992, 1994, 1998; Apprill et al., 1998). To keep the gas in solution, the process will have to be carried out under gas pressure, and control of the cooling will be required to impose a solidification direction, entailing somewhat complex equipment, **Figure 18**. Examples of equipment with different configurations to give different structures are given in (Shapovalov and Withers, 2008). Materials produced in this way feature relatively large and generally elongated and parallel closed hydrogen-filled bubbles, which can be created in a wide range of configurations and over a relatively wide range of volume fractions (Shapovalov, 1994, 1992, 1998; Simone and Gibson, 1996;

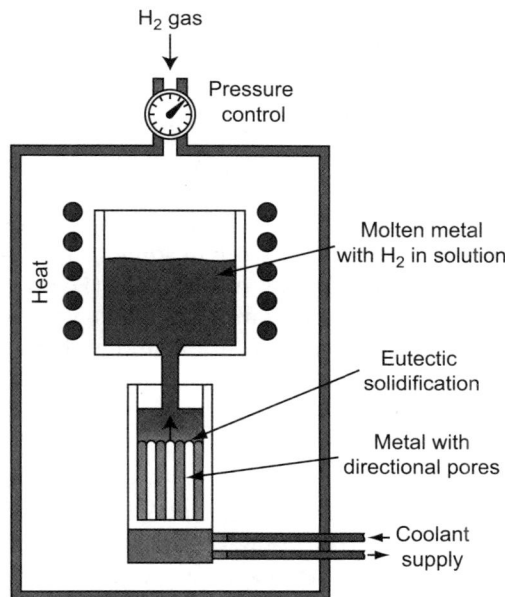

Figure 18 Schematic diagram of the apparatus used for Gasar production. (For color version of this figure, the reader is referred to the online version of this book.)

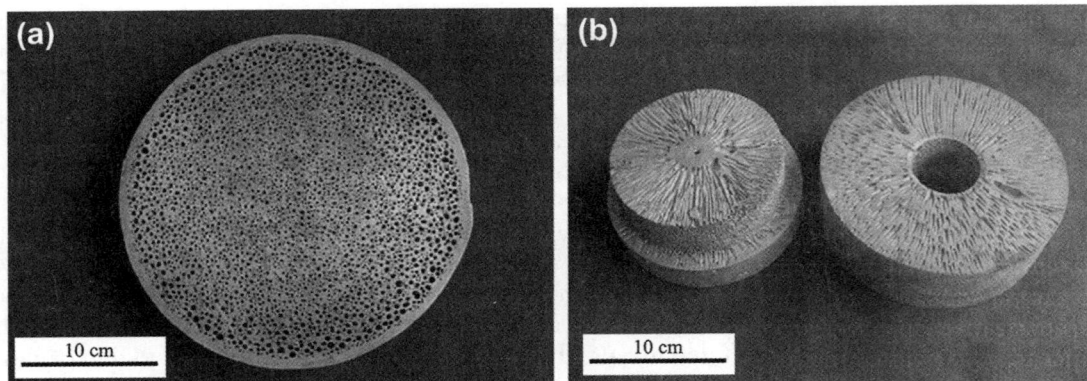

Figure 19 Magnesium foams processed by the Gasar method (Shapovalov and Boyko, 2004), solidified to produce (a) axially and (b) radially aligned pores. Reproduced from *Advanced Engineering Materials*, vol. 6, Shapovalov, V. & Boyko, L. "Gasar - a new class of porous materials", pp. 407–410 (Shapovalov and Boyko, 2004) with permission of Professor Shapovalov, Materials and Electrochemical Research Corp, Arizona, and Wiley.

1997; Kee et al., 1998; Shapovalov and Boyko, 2004; Hyun et al., 2001; 2004; Hyun and Nakajima, 2003; Kashihara et al., 2006; Kujime et al., 2006; Yuan et al., 2005; Xue et al., 2007; Nakajima et al., 2001; Xie et al., 2004; Fiedler et al., 2012).

Control of the structure may be achieved through the partial pressure of hydrogen (often in excess of 50 bar), the direction of cooling (which determines the pore orientation, **Figure 19**) and the speed of the solidification front (which determines the microstructural scale). It has been noted that, as samples are solidified and grow in length, there can be a progressive alteration in pore structure as the extraction of heat through the sample alters the cooling rate; the effect is particularly pronounced with low conductivity metals, such as stainless steels (Nakajima et al., 2004). It has been found that the structure displays short-range order (Wan et al., 2007), which would be expected from its nature as a eutectic microstructure. There have been studies of how the gas bubbles nucleate and so form the structure during metal/gas directional solidification (Wei et al., 2003), as well as the development of a related process of monotectic alloy directional solidification followed by selective dissolution (Yasuda et al., 2004, 2006). The process has also been shown to work in some cases with gases other than hydrogen; for example, porous iron has been produced using nitrogen gas (Hyun and Nakajima, 2002).

The Gasar process appears to work for a wide range of metals, with reports of aluminum, magnesium, copper, nickel, titanium, iron, steel, cobalt, chromium and molybdenum (Banhart, 2001; Simone and Gibson, 1996; Hyun and Nakajima, 2002; Drenchev et al., 2006) appearing in the literature.

24.2.5.2 *Dealloying*

Dealloying is a process for the production of nanoscale porous metals and metallic glasses (Erlebacher et al., 2001; Erlebacher and Sieradzki, 2003; Jayaraj et al., 2006; Lee and Sordelet, 2006a; Lu et al., 2006a, 2007a, 2007b; Hodge et al., 2006; Parida et al., 2006; Lee et al., 2007a; Chen-Wiegart et al., 2012). The basic principles of the process are not new; they have been exploited for hundreds of years. An early application of the dealloying process was to obtain an outer layer of pure noble metal on the surface of a lower cost alloy made with copper; the process was used by pre-Colombian peoples in Peru and Ecuador in many of the artifacts that are now known (Smith, 1981). An Au–Cu alloy, often with less than a third gold, was used to make the artifact shape, and this was then treated with a corrosive agent to remove the copper from the top 50–200 μm of metal. This was then polished or heat treated to form a dense gold surface to the artifact. In the nineteenth century, the same process was encountered in a particular form of wet corrosion, when under certain conditions zinc migrates out of Cu–Zn alloys (brasses), leaving behind a layer of porous copper (Smith, 1993; Al-Kharafi et al., 2004). Similarly, Ni–Al alloys can also be dealloyed to produce nanoporous nickel catalysts known as Raney nickel; other nanoporous or "skeletal" metal catalysts are also produced in this way (Smith and Trimm, 2005). At this point the phenomenon was called "dezincification" for brass or "dealloying" more generally. Often useful but at times also a nuisance, it can be prevented with suitable third additions to the alloy (e.g. As to brass).

Although these and similar processes were well known at different points in history, it was only much more recently that the process has been investigated and properly explained. One of the main contributors to this understanding, Erlebacher, defines the key requirements for the process to be (Erlebacher and Seshadri, 2009):

- An alloy with a solid solution over a wide composition range.
- There must be a significant difference in the reduction potentials of the two elements, and one must be soluble in its oxidized state in a suitable electrolyte.
- The element which is not dissolved must be able to diffuse along the surface of the alloy.

It is important to note that the dealloying process is not simply a selective dissolution process; there is simultaneously rearrangement of the more noble component. This behavior determines the structure produced, and is the reason why third alloying elements can "poison" the process (by blocking movement of the noble component).

The mechanism developed to describe the process is roughly as follows (Dietterle et al., 1995; Erlebacher, 2004). Initially, the surface will present a random mixture of atoms of both elements in the solid solution. Once this is placed into a suitable solvent, the most easily reduced of these starts to be removed, leaving the more noble behind, see **Figure 20**. Surface diffusion, if sufficiently rapid, allows these atoms to move and cluster together on the surface, and particularly at the edges of the steps produced as material is removed. There will hence be a competition between the rate of surface diffusion and reorganization of the noble atoms, and the rate of dissolution of the more easily reduced atoms. This can lead to the development of regions where the surface is passivated by remaining noble atoms and regions that are etched away, which may undercut passivated regions as the quantity of more noble atoms becomes limited. As the undercutting proceeds, different etched regions will interconnect. Thus a porous network structure is produced, which can continue to grow and consists of regions of alloy where the surface is passivated by the more noble element. In some systems, a homogenous composition can be formed where coarsening of the structure takes place a small distance behind the etch front. This coarsening process brings new material to the surface until virtually all of the more easily reduced element is removed (Snyder et al., 2008). There is evidence from X-ray nanotomography experiments that this coarsening process does not lead to self-similar structures, with the interfacial shape changing with the structural scale (Chen-Wiegart et al., 2012).

Kinetically controlled processes thus play key roles in determining whether dealloyed structures will be produced as well as the final mesostructural scale. Using this as a method for control, there has been progress in the understanding of how to control the pore structure formed via the temperature and chemistry of the process, and how to incorporate certain additional elements in the surface of the porous structure (Erlebacher and Seshadri, 2009). For example, this can be by thermal annealing to increase the pore size as the structure reorganizes, or by adding additional elements that segregate to the surface and act to stabilize it (Chen et al., 2010).

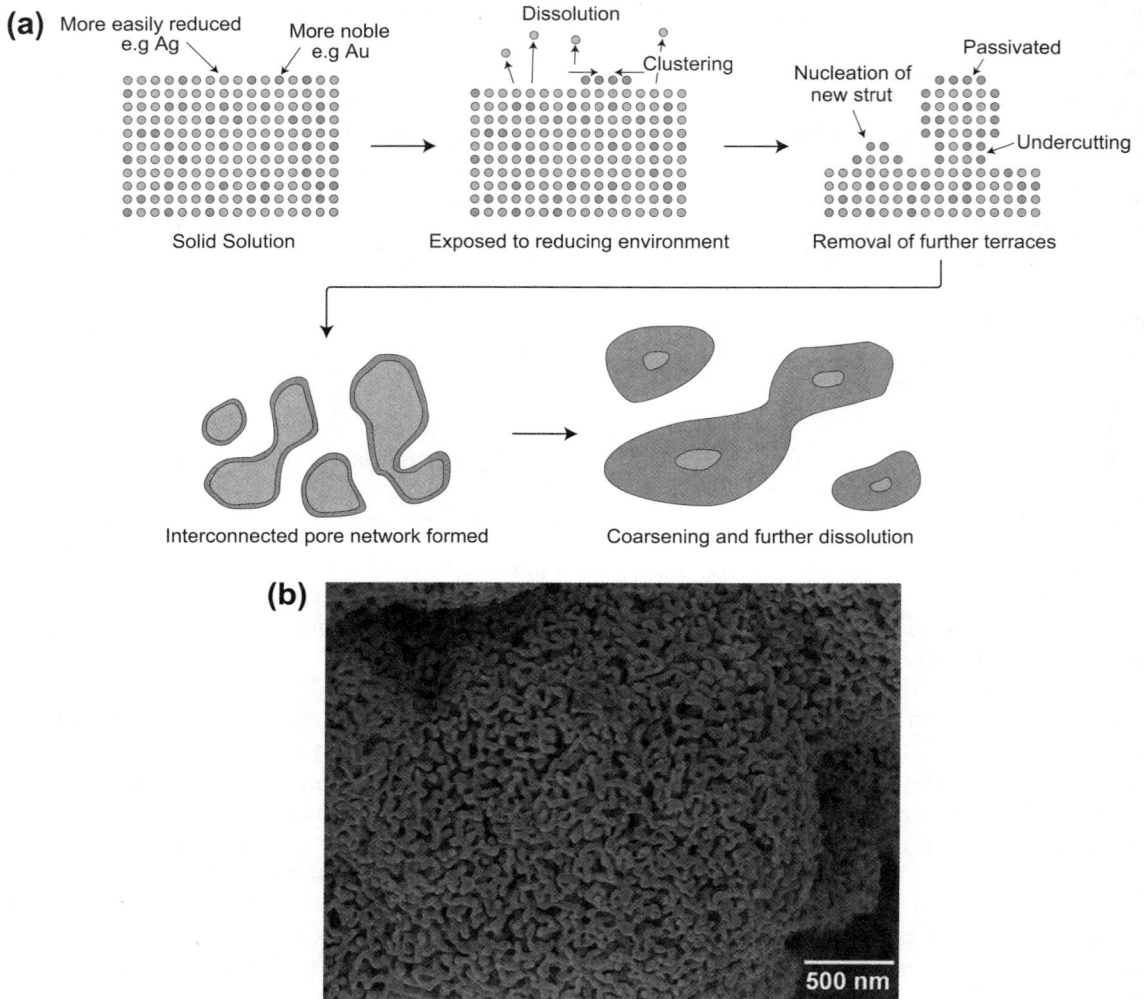

Figure 20 The dealloying process showing (a) a schematic diagram of the operation of the process, and (b) a micrograph of nonporous gold structures produced. The micrograph is reproduced from Materials Science and Engineering: A, vol. 528, Cox, M. E. & Dunand, D. C. "Bulk gold with hierarchical macro-, micro- and nano-porosity", pp. 2401–2406. with permission from Elsevier and Prof Dunand, Northwestern University. (For color version of this figure, the reader is referred to the online version of this book.)

The structures formed being at very small scale and of limited dimensions in at least one direction, dealloyed metals are frequently identified for different applications than other types of metal foam. Common examples would be as highly sensitive detectors (making use of the large surface area), or applications for liquid storage, such as drug delivery in the biomedical field and, of course, catalysis (with Raney nickel as a classical example). Similar structures have been made out of magnesium, aluminum, iron and titanium. In some cases, a very different mechanism is used: if thin layers of the metal are briefly exposed to intense ultrasound (60 s at a maximum intensity of 57 W cm^{-2}) in aqueous solution, the combination of physical and chemical attack can produce a porous layer of up to 200 nm thickness (Gensel et al., 2012).

Pores created by dealloying are so fine that, if the process is combined with other porous metal processing routes, multiscale hierarchical porosity samples can be produced; this was recently demonstrated in gold (Cox and Dunand, 2011). Finally, we cite a recent and ingenious extension of the process that uses liquid metal (magnesium) instead of an aqueous medium as the host for the "departing" metal species. After solidification the "selective solvent" metal is removed by selective etching, creating fine-scale porous structures similar to those produced by conventional dealloying but of nonnoble titanium metal (Wada et al., 2011b, 2011a).

24.2.5.3 Freeze Casting

The freeze casting process is another method for developing porosity-containing structures from powders, with the process being reviewed in (Li et al., 2012b). The process was originally developed for ceramics, but has been demonstrated for metal powders, initially titanium (Chino and Dunand, 2008; Fife et al., 2009; Li and Dunand, 2011) and more recently stainless steel (Driscoll et al., 2011) or copper (Ramos and Dunand, 2012), **Figure 21**.

The principle is based on the phenomenon of particle pushing, commonly encountered in the solidification of particulate composites (Mortensen and Jin, 1992). Operationally, it is relatively simple (Fukasawa et al., 2001, 2002; Deville et al., 2006; Zhang and Cooper, 2007; Chino and Dunand, 2008); it

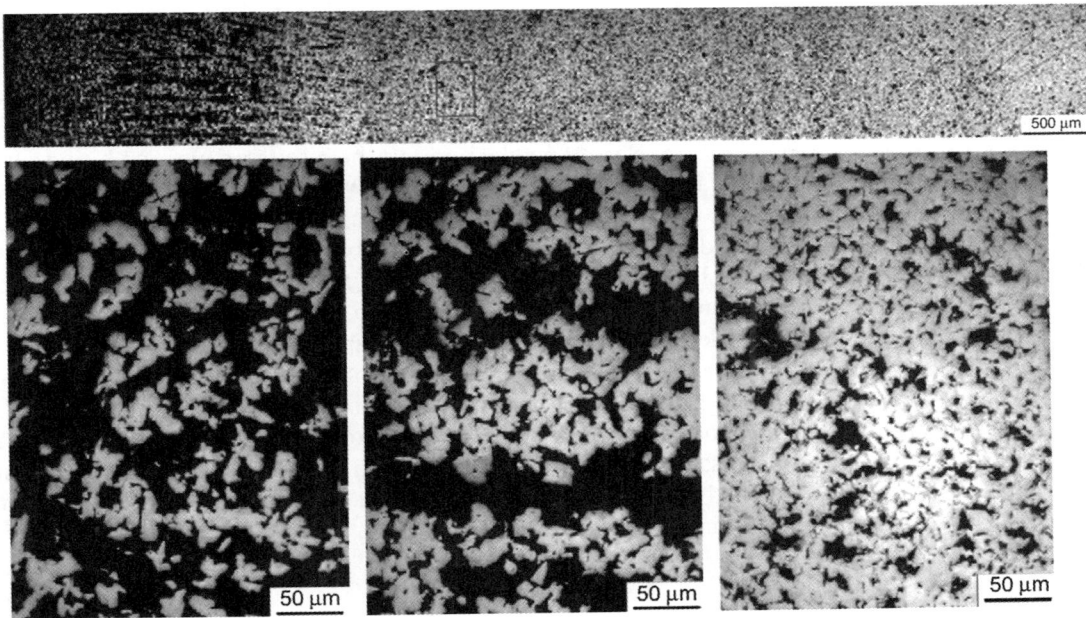

Figure 21 Micrographs of cross-sections through a billet of titanium produced by the freeze casting process (Li and Dunand, 2011). The top image shows the structure from the edge of the billet on the left-hand side (the first part to freeze) to the center on the right (the last part to freeze). The lower images represent higher magnification views of the edge, intermediate and center regions, respectively, going from left to right. Reproduced from Acta Materialia, vol. 59, Li, J. C. & Dunand, D. C. "Mechanical properties of directionally freeze-cast titanium foams", pp. 146–158 (Li and Dunand, 2011), with permission of Professor Dunand, Northwestern University, and Elsevier.

starts with a water-based slurry, which is frozen. As the ice dendrites grow, they push solid particles into the remaining liquid, creating regions occupied by the ice (on the scale of perhaps hundreds of microns) that are free of powder particles, and at the same time bringing the powder particles closer together and removing water from their environment. The effect of this is to pack the particles close together, leading to the formation of a network with some degree of strength (which can be improved using a water-soluble binder). The ice crystals are then sublimated to leave open spaces in the powder compact that become pores in a foam after the metal powder is sintered. Aligned pores with a high aspect ratio are made by carrying out the initial solidification in a directional manner (Fukasawa et al., 2001, 2002; Deville et al., 2006; Zhang and Cooper, 2007; Chino and Dunand, 2008). By doing this, the ice crystals tend to line up along the direction of heat extraction, although they also tend to form with a somewhat planar shape (Fukasawa et al., 2001, 2002; Deville et al., 2006; Zhang and Cooper, 2007; Fife et al., 2009).

The method requires some degree of control. It has mostly been used with ceramics (Fukasawa et al., 2001, 2002; Deville et al., 2006; Zhang and Cooper, 2007; Deville et al., 2007); metal powder particles are typically much larger than the submicron particles used with ceramics, yet the powder particles cannot be too large, or they are not pushed by the growing ice crystals (Chino and Dunand, 2008). Also, smaller powder particles sinter more completely (Fife et al., 2009), and this increases mechanical properties (Li and Dunand, 2011). There may also be several pore types and scales of porosity that are formed (Ramos and Dunand, 2012). In a two-step variant of the process, freeze cast porous ceramic was used as a template for growth of gold nanowire brushes, creating a porous metal/ceramic composite (Olevsky et al., 2007).

24.2.6   Regular Lattices

The porous materials that have been explained in previous sections have, to a greater or lesser extent, stochastic structures. There is an important class of regular periodic metallic structures exhibiting a large amount of porosity, namely periodic truss assemblies (Sypeck and Wadley, 2001; Wadley et al., 2003; Tian et al., 2004; Kooistra et al., 2004; Queheillalt and Wadley, 2005a; 2005b; Wadley, 2006; Bouwhuis and Hibbard, 2006; Kooistra and Wadley, 2007). Often based on cubic lattices, such as simple cubic or the diamond structure, or an ancient basket weaving pattern known as Kagome, these structures prove very efficient, and can be optimized to give the best performance at given mass or porosity for given loading conditions (see below) (Wicks and Hutchinson, 2001; Deshpande and Fleck, 2001a). There is also some evidence that they are more defect tolerant than conventional open-cell foams (Wallach and Gibson, 2001a). Although one could debate whether such truss assemblies are really a material as opposed to a structure (this will depend on the ratio of truss to component lengths, although there is no precisely defined limit), recent processing advances allow material to be created with many lattice unit cells in all directions, in excess of the 7–10 typically required for foam samples to be considered representative of a material, rather than a structure (see below).

24.2.6.1   *Casting*

Polymer processing methods can be used to create a regular structure that can then be converted into a metal foam using the techniques discussed above for a random polyurethane (PU) precursor foam. This has included polymer injection molding (Deshpande and Fleck, 2001a), wire cutting of a polymer block (Ho et al., 2010) and rapid prototyping techniques (Chiras et al., 2002). Following creation of the polymer form, investment casting may be used to create this shape in metal; this has been done with tetrahedral lattices in aluminum alloy and brass (Deshpande and Fleck, 2001a) and also in beryllium copper alloy (Chiras et al., 2002). Casting defects are in some cases found near the nodes, which limit the performance if the alloy is not ductile.

24.2.6.2 Weaving

The weaving or cross-assembly of metal wires to form a regular network can be an effective method to produce lattice structures, and one that has a lower overall cost than methods requiring a bespoke precursor to be produced. However, to form a part with full integrity, a bonding method (such as brazing) would be required to join the woven fibers, or layers of metal fiber textile (Sypeck and Wadley, 2001; Wadley, 2002; Wadley et al., 2003). Woven wire Kagome truss structures have been produced and tested by assembly of continuous helical steel wires, which were subsequently brazed together to form relatively large 3D periodic lattice structures (Lee et al., 2007b; Kang, 2009). Appropriately placing the brazing material at nodes to prevent buckling yielded a material with outstanding specific compressive strength values (Kang, 2009). Somewhat similar structures of high load-bearing efficiency can also be produced by placing regularly crossing stiff rods or cylinders and bonding these, again by brazing, sintering or welding, a process that can produce structures large enough to qualify as a material if repeated a sufficient number of times (Wadley et al., 2003; Queheillalt and Wadley, 2005a; Wadley, 2006; Moongkhamklang et al., 2008, 2010).

24.2.6.3 Sheet Metal Processes

Another relatively simple method is to use sheet forming processes to produce perforated and folded metal sheets that can be stacked and bonded to form a lattice. For example, if a sheet is perforated with almost tessellating hexagonal holes, and then deformed at the nodes by simple sheet forming methods, a truss structure based around tetrahedra is formed (Wadley, 2002; Kooistra et al., 2004; Wadley, 2006). Other methods exist to make a wide range of shapes, including honeycomb-like forms (Wadley, 2006). The structure can then be bonded to faceplates to make a sandwich panel. This method has also been used to generate structures from Ti–6Al–4V alloy (Queheillalt and Wadley, 2009), where it was observed that failure is more likely to occur in the region of the nodes. This could be because this is where the metal has been most deformed, or because these form the sites where it is bonded to the face sheets.

24.2.6.4 Additive Layer Manufacturing (3D Printing)

Rapid prototyping technology is well established for the formation of small number of complex 3D parts from polymers. These often will contain internal voids or have a regular, repeating lattice structure (they are first designed using Computer Aided Design (CAD) software, and the repetition of a simple unit to produce a lattice is easily achieved). Polymer parts produced in this way have been the basis for structures that are claimed to be the lightest materials ever created, with a density less than even that of aerogels (down to 0.9 kg m^{-3} (Schaedler et al., 2011)). To form these materials, lattices were produced and used as a template, being coated with several hundred microns of electroless nickel before etching away the polymer. Interestingly, the Gibson–Ashby scaling law for the variation in Young's modulus with density (see below) is found to be an adequate predictor of the performance of these materials, even at the very low density range.

Three-dimensional printing has been used to make porous parts from a Ti–6Al–4V alloy powder (Li et al., 2006b). The metal powder is blended with a water-based carrier of high viscosity, which is deposited by extrusion through a computer-controlled nozzle mounted on a moving x–y stage. After printing the required shape, which is built up layer by layer, the lattice is dried and the metal powder sintered under vacuum. The method is successful in producing simple square-based lattices, indicating that there is no significant collapse of the overhang either during deposition or sintering. However, in common with many of the layer manufacturing methods, there is a limitation on the finest feature size that can be created (around 500 µm in Ref. Li et al. (2006b), somewhat lower values in more recent references (Hong et al., 2011; Luyten et al., 2011)).

Further developments of rapid prototyping, the processing methods that are variously termed "additive (layer) manufacture", "rapid manufacturing" or "powder bed processing" are extremely versatile in creating metallic parts of complex shapes, and porous metals are no exception. These methods, which have developed from the 3D printing and Rapid Prototyping techniques used for polymers, usually involve a thin layer of metallic powder being spread on a build bed. Following a 3D design uploaded to the system, a highly directional heat source, such as a laser or an electron beam, is used to melt the powder in a pattern corresponding to the first layer of the desired part. A second layer of powder is deposited, and the process repeated, with the region melted being slightly different. With a large number of repeated layers, a 3D part can be created, with the final step being the removal of any unmelted powder (which can be recycled for future processing), **Figure** 22. The method lends itself to the production of parts with complex geometry, and this includes regular lattice materials. It is relatively

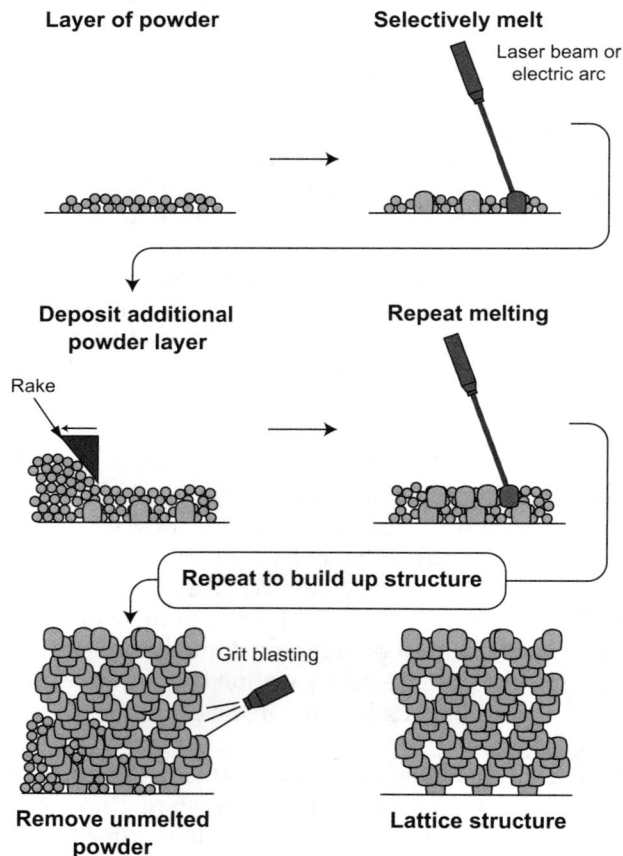

Figure 22   Schematic diagram of the production of a lattice via the additive layer manufacturing process. (For color version of this figure, the reader is referred to the online version of this book.)

easy to design a regular lattice as the CAD packages employed (such as Netfabb GmbH, http://www.netfabb.com/), can take a simple unit cell and repeat it many times in a defined space.

Of the many variants of this process that exist, the most commonly exploited for metallic lattices uses an electron beam source, and is often termed Electron Beam Melting (EBM) (Heinl et al., 2007, 2008b; Cansizoglu et al., 2008; Murr et al., 2010, 2011; Ramirez et al., 2011; Li et al., 2012a). An example of this is the equipment produced by Arcam AB (www.arcam.com). **Figure 23** shows an example of a periodic lattice structure produced in this way from titanium.

The process has most often been used for titanium and Ti–6Al–4V alloy (having more generally been developed to overcome the high cost of net shape fabrication of titanium parts), producing lattices with the struts in a square arrangement, or following the pattern of bonds in diamond (Heinl et al., 2007, 2008b; Cansizoglu et al., 2008; Murr et al., 2010; Li et al., 2012a). Lattices have also been created from Co–Cr-based alloys (Murr et al., 2011) as prototypes for biomedical implant applications, and from copper (Ramirez et al., 2011) for thermal and electrical management, although the presence of Cu_2O precipitates, which were found to form during the process and contribute to strengthening, may indicate that their thermal and electrical performance is not as high as would be expected for a pure copper lattice.

The advantage of the EBM technique, along with all similar processes of this type, is in the complexity of structures which can be made in a highly reproducible manner; in many cases these forms could simply not be accessed by conventional foam production methods. For example, EBM has been used to create lattices that show highly auxetic behavior (of negative Poisson's ratio; see below) (Schwerdtfeger et al., 2010; Yang et al., 2012b). In contrast, stochastic foam parts have also been built

Figure 23 A titanium lattice, based on the diamond structure, manufactured by EBM. Image kindly provided by Mr E Hernandez Nava, the University of Sheffield. (For color version of this figure, the reader is referred to the online version of this book.)

(Murr et al., 2010, 2011; Ramirez et al., 2011) utilizing tomographic scans of real open-cell foams as the input into the machine. The flexibility is even sufficient to make foams with a choice of dense or hollow pore struts (Murr et al., 2010).

There is, however, a lower limit to the size of features that can be created, and hence the pore size. This arises through the interplay between many factors, including the dimensions and power of the heat source beam, the size of the powder particles used, and how closed the structure is (which limits the ease with which unsintered powder can be removed). There will also be differences in the quality of the structure formed (e.g. step size on sloped surfaces, additional powder particles adhering to the surface, internal porosity) with different machine conditions (e.g. beam power, beam speed, layer thickness). It has been demonstrated that the angle of the struts relative to the direction of the build up of layers affects the mechanical performance (Cansizoglu et al., 2008), and in one case it has been found that different energy inputs affect the lattice density, though this does not cause apparent differences in mechanical properties outside those expected for varying density (Heinl et al., 2008a). It is, however, likely that differences would be seen over a wider range of process conditions and properties.

Some work on more advanced properties has been carried out, with tests on the fatigue behavior in compression finding a relatively low fatigue resistance, although this is coupled with the parent metal being in a fatigue-prone form (Li et al., 2012a) and it is possible that improvements in microstructure could show a benefit. Recent work has also comprised the development of a more involved, but highly versatile, process combining additive layer manufacturing and replication processing: a positive polymer template of the microcellular metal structure is produced, then infiltrated with a salt paste which is dried and sintered, and then infiltrated with molten magnesium. Leaching the NaCl after metal solidification leaves a relatively coarse structure and rough metal surface, but with great freedom in architectural design (Nguyen et al., 2011; Kirkland et al., 2011).

24.2.7 Metals and Alloys

Most metal foams, in both industry and research, are based on aluminum, nickel or titanium; however, a growing number of metallic systems have now been produced and characterized. **Table 2** collects some of the work on the processing of less commonly used microcellular metals.

Frequently, to improve final properties or as an aid to processing through depression of the melting point, alloys are used to produce metal foams. Here there is the possibility of more complex processing–microstructure relationships, as the scale of the foam structure and that of the underlying microstructure become similar, and therefore prone to interact.

24.2.8 Secondary Processing

There is a significant body of work on further processing of metal foams, involving many additional operations that can be done to increase the utility of the materials produced by the methods described above. The conversion of materials to other forms has been investigated; for example, the creation of nickel aluminide foams from nickel foams by a pack aluminization process, where aluminide is deposited on a metal surface, followed by thermal homogenization (Hodge and Dunand, 2001). A similar coating and reaction treatment has alloyed nickel foams with chromium (Choe and Dunand, 2004b).

Also of potential importance in real processing, the effect of treatments easily adapted from the processing of dense metal components, such as heat treatments and recovery, have been considered (Lehmhus and Banhart, 2003; vanSchaik et al., 2003; Yang et al., 2007c; Park and Nutt, 2000; Conde

Table 2 Examples of the many different metallic systems where the processing in porous form has been investigated, and an indication of some of the processing routes that have been used. Note that as aluminum, nickel and titanium are relatively common examples processed by the methods discussed in detail in the text, they have not been included in the table

Metal	Principal process routes	References
Copper	Sintering powder; space holder; investment casting; Gasar	(Goodstein et al., 1966; Hakamada et al., 2007a; Xie and Evans, 2004; Zhao et al., 2005c; Simone and Gibson, 1996; Nakajima et al., 2001; Xie et al., 2004; Hyun et al., 2001)
Magnesium	Gas release agent; investment casting; space holder; replication; Gasar	(Körner et al., 2003, 2004; Wen et al., 2004; Yamada et al., 1999, 2000) (Wen et al., 2004; Witte et al., 2007c; Yuan et al., 2005)
Zinc	Gas release agent; deposition onto template	(Thornton and Magee, 1975a; Kitazono et al., 2004a; Kitazono and Takiguchi, 2006; Chethan et al., 2011) (Tian and Guo, 2011)
Iron-based alloys	Replication; space holder; Gasar	(Yu et al., 1998b; Park and Nutt, 2000; 2001b; Stephani et al., 2006; Hyun and Nakajima, 2002; Kashihara et al., 2006) (Berchem et al., 2002; Wang et al., 2005; Bakan, 2006; Gulsoy and German, 2008; Shimizu et al., 2012)
Gold	Dealloying; plating onto template; gas release agent	(Erlebacher et al., 2001; Parida et al., 2006; Nagai et al., 2006; Hakamada and Mabuchi, 2007; Banhart, 2008)
Tantalum	Deposition onto template	(Zardiackas et al., 2001; Sevilla et al., 2007; Arciniegas et al., 2007)
Lead and Lead–Tin alloy	Gas release agent; replication	(Irretier and Banhart, 2005; Belhadi et al., 2006)
Brass	Replication	(Castrodeza and Mapelli, 2009)
Ag–Cu eutectic	Replication	(Diologent et al., 2009a)
Metallic glasses	Replication; trapped gas expansion	(Wada and Inoue, 2003, 2004; Brothers and Dunand, 2004, 2005a, 2005b, 2006a; Brothers et al., 2005, 2006; Inoue et al., 2006; Xie et al., 2006; Jayaraj et al., 2006; Lee and Sordelet, 2006a; Demetriou et al., 2007d, 2007e) (Demetriou et al., 2007a)
Shape memory alloys (e.g. NiTi)	Trapped gas expansion; sintering powder; space holder; replication	(Greiner et al., 2005; Bansiddhi et al., 2008; Sevilla et al., 2007; Arciniegas et al., 2007; Yuan et al., 2004, 2006; Oppenheimer et al., 2004; Kim et al., 2004b; Biswas, 2005; Bansiddhi and Dunand, 2007) (Zhao et al., 2009b; Bertolino et al., 2010) (Köhl et al., 2011; Young et al., 2012)
Ni–Mn–Ga martensitic alloys	Replication	(Boonyongmaneerat et al., 2007; Chmielus et al., 2010, 2009; Zhang et al., 2011; Sasso et al., 2011; Dunand and Müllner, 2011)

and Mortensen, 2008), as have more specific processes, such as the controlled nonuniform chemical milling of a foam to produce density-graded structures (Matsumoto et al., 2007). In this section we will focus on some of the areas that are likely to be important to foams in a wider range of applications, and where a significant body of research has been performed.

24.2.8.1 Machining

Many foams can be machined easily using standard workshop tools; in some cases, workability is compared to wood (a natural porous material). With low-density foams, the low yield strength can cause a problem, as the material around the machined region is significantly deformed and damaged. For precise machining, load-free techniques, such as electro-discharge machining, are therefore often employed.

For foams produced by the melt route, where ceramic particles are added to increase the viscosity and improve the foam stability, the presence of this hard phase can result in machining being difficult (Banhart, 2000). At the opposite extreme, foam made of soft metal such as pure aluminum with large grains, can also be difficult to machine due to ductility: the metal then simply smears instead of cutting under the tool. Foams produced by the replication process are on the other hand very easy to machine if this is done before dissolution, while the material is a metal/salt composite, with short chips and good precision (but needing care to prevent salt from corroding equipment); the smooth cylindrical samples in **Figure 17** were produced in this way (Despois et al., 2006a; Conde et al., 2006a).

As an alternative to machining to final shape, forming processes could be considered (Hahn et al., 2003; Ito and Kobayashi, 2006; Jackson et al., 2008), though these can lead to problems as foams tend to be weak and can fail, or deform nonuniformly (see below). One approach has been to use local laser heating to control deformation spatially and prevent problems associated with cold working (Quadrini et al., 2010). By progressive treatments, bend angles of up to 90° were achieved, and the process has also been applied to sandwich panels with metal foam cores (Guglielmotti et al., 2009).

24.2.8.2 Bonding

To be useful in most applications, metal foams will have to be attached to other materials (dense or foam). Some methods can produce seamless dense metal/porous metal structures during the initial foam processing. Space holder methods, such as replication processing, allow this with considerable flexibility, as does sometimes metal foaming if dense metal is collected on the underside by drainage, or if part of the metal is solidified before bubbles appear. Precursor methods may also create structures of dense metal near foamed metal; friction-stir welding was thus used to bond an aluminum-based precursor to steel before foaming (Hangai et al., 2010).

Many metal bonding techniques are available, and processes such as mechanical bonding (i.e. the use of fasteners), adhesive bonding (which often makes sense given the low inherent strength of microcellular metals and the fact that glue can anchor well onto the material) or forming a metallurgical joint (e.g. welding, soldering or brazing) have been examined (Olurin et al., 2000c; Bernard et al., 2002; Kitazono et al., 2002; Shirzadi et al., 2004; Born et al., 2006; Longerich et al., 2007; Shirzadi et al., 2008; Jarvis et al., 2011).

The most difficult of these cases is when there is a requirement to produce a joint that is electrically or thermally conducting to a high degree, or that must resist elevated temperatures (i.e. when a metallic joint is necessary). There are a number of issues that are present for foams that do not affect regular metals. One is simply the small amount of material present at the interface, which provides little material for bonding (even less if a foam-to-foam joint is being attempted). Another potential difficulty is in ensuring that any third material (e.g. a filler or braze) that is used is retained in the joint region and does not penetrate too far into the foam (Longerich et al., 2007; Jarvis et al., 2011). A problem that is reduced in the bonding of foams is generally residual stresses brought about by thermal expansion mismatch. While this can be a problem for dense materials, the low stiffness of foams typically allows differential strains to be accomodated with little residual stress (Jarvis et al., 2011). Foams have in fact been used as stress-relieving interlayers in the bonding of dissimilar materials, for example, stainless steel foams in the joining of stainless steel and alumina plates (Shirzadi et al., 2008). Finally, like regular materials, the properties of a foam may be affected by heating carried out as part of a bonding process (Bernard et al., 2002).

Welding of foams is challenging, as the melting of the foam material present at the joint will result in it retreating away from the joint; significant amounts of filler metal would be needed. It has been found that welding is ineffective at joining two foams (Shirzadi et al., 2004), though welding has been successfully applied to join metal foams to dense metal: laser welding with a filler metal for nickel foam, and capacitor discharge welding for iron-based foam (Longerich et al., 2007). The best results were obtained with a nickel mesh welded at the interface, to add a degree of mechanical keying.

A preferable process to welding might be soldering (as carried out by Huang et al. (2012)) or, preferably, brazing (from the point of view of control and scalability), which has been investigated in different systems by various authors (Shirzadi et al., 2004; Jarvis et al., 2011). It is normally observed that the joints formed are stronger than the foams themselves, normally a good result in bonding. Yet it has also been observed that when testing brazed joints in shear, a high strength interface (with low contamination and low porosity within the joint region) results in poorer overall properties than a weak joint (with higher levels of contamination and noticeable porosity in the braze layer) (Jarvis et al., 2011). This is because a strong joint supports a boundary layer of cells (Chen and Fleck, 2002) and so concentrates the deformation over a smaller foam region, which may fail more rapidly as a result. This shows that the low strength and stiffness of foams may mean that the normal desire of creating the strongest interface possible may not be appropriate in all cases. As noted by Han et al. (2012), joining of foam to dense metal by brazing could be considered essential to get the best properties from a metal foam heat exchanger. It is however reported that some brazing trials have led to problems with creep of the foam at the braze temperature; as discussed in more detail later, the creep of foams can be significant under conditions where the dense metal would not show any problems.

24.2.8.3 *Coating*

In many situations the performance of a metallic component can be improved by the addition of a coating. This is normally to obtain the desired color or surface finish, impart a resistance to a corrosive environment, or to improve the wear characteristics, although it can also be to add a functional layer. Where microcellular metals are used in fluid control (e.g. filters, silencers) or as electrodes, a high degree of corrosion resistance may be required. It may also be necessary to have a functional surface added where a foam is to serve as a catalyst support, or as a filter (e.g. (Seo et al., 2007)). However, there is another area where coatings can be beneficial, namely by making use of the high specific surface area that can be accessed for an open-cell foam.

If a material with superior mechanical properties is added as a surface layer on a dense component, it may improve the wear characteristics, but it is unlikely to have a significant influence on the mechanical behavior of the whole part. With a metal foam however, the amount of coating that can be potentially deposited could allow for significant effects. It has been observed that an increase in the thickness of the (nanometric) oxide covering a fine-scale replicated aluminum foam can have a measurable effect on its mechanical behavior, doubling its flow stress at small (75 μm) pore sizes, apparently by hindering dislocation escape through the surface (Diologent et al., 2009d). Most microcellular metals deform

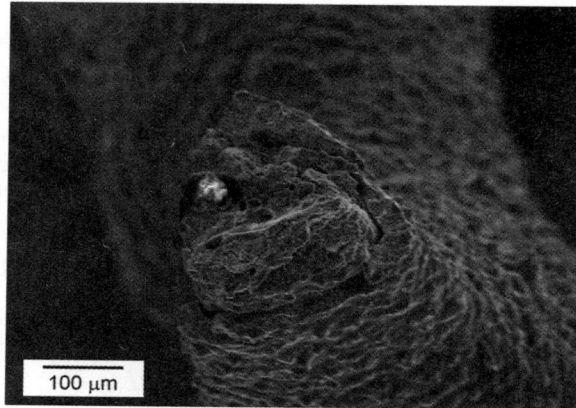

Figure 24 An SEM micrograph of a fractured surface of on open-celled Duocel aluminum foam with a plasma electrolytic oxidation coating. The coating and the strut have very different mechanical behavior, and are clearly distinguished in the image. Image kindly provided by Mr T Abdulla, the University of Sheffield.

predominantly by bending: since a thin outer layer of strong material remote from the neutral axis strongly enhances the bending resistance of beams, load-bearing coatings on microcellular metals may strengthen these appreciably. Coatings of different alloy systems have been applied by electrodeposition, including Ni, Ni–Fe and Ni–W coatings of 25–400 μm on aluminum foams (Boonyongmaneerat et al., 2008; Bouwhuis et al., 2009; Lausic et al., 2012) and regular lattices (Suralvo et al., 2008). Oxide coatings have been generated by a plasma-assisted electrochemical surface treatment known as Plasma Electrolytic Oxidation (PEO, described in Ref. Yerokhin et al. (1999)). Workers have examined PEO layers of 2–100 μm created on investment cast foams (Duocel) (Dunleavy et al., 2011; Abdulla et al., 2011) and on replicated foams (Abdulla et al., 2011). An example image of a broken strut from an investment cast foam treated in this way is shown in **Figure 24**.

In these studies, it is often found that the coating is not uniform throughout the structure, with thicker deposits near the foam surface (Boonyongmaneerat et al., 2008; Suralvo et al., 2008; Bouwhuis et al., 2009; Abdulla et al., 2011; Lausic et al., 2012). This effect, which can lead to substantial variations in foam structure and hence properties with location, is generally agreed to be due to a reduced potential in the center of the foam, and a limited passage of electrolyte into the structure (Abdulla et al., 2011). It is also pointed out that if the foam is to be subjected to bending, then having a larger amount of coating near the surface may allow it to be more mechanically efficient (Bouwhuis et al., 2009).

Closed-cell foams have been coated with ceramics by plasma spraying and high-velocity-oxyfuel-spraying, and these have been found to increase the mechanical properties of the whole artifact (Maurer et al., 2002). For open-cell foams where the coating can penetrate into this structure, the material properties would be expected to be more intimately combined. In low-density versions of these foams, where bending of the struts is an important deformation mechanism, it would be expected that, as it is at the surface, far from the strut neutral axis, a thin layer of a stiff, strong coating would have a significant effect (provided it adheres well to the underlying material) (Boonyongmaneerat et al., 2008). This has been found to be the case, with absolute and specific strength (Suralvo et al., 2008; Boonyongmaneerat et al., 2008), energy absorption (Boonyongmaneerat et al., 2008), and modulus (Bouwhuis et al., 2009) being improved. Refractory coatings formed by combustion synthesis over a nickel foam have been shown to improve the foam's elevated temperature corrosion resistance

(Smorygo et al., 2008). Where the coating is ceramic (as in the PEO process), it has been observed that the ceramic–metal composite produced has a relatively high toughness (Dunleavy et al., 2011), and improved corrosion resistance (Liu et al., 2012)]. Even though the effectiveness of these ceramic PEO coatings decreases with increasing coating thickness (due to an accumulation of defects) the specific strength of a foam can be raised by the addition of a coating (Abdulla et al., 2011); a similar result was obtained on aluminum microtruss materials using anodization (Bele et al., 2011). Attempts have been made to account for the strength increase with simple models of the additional strength imparted by a sleeve of coating material, a procedure that appears to work in at least some cases (Bouwhuis et al., 2009).

24.3 Structure

The structure of porous metals poses an interesting problem of multiplicity in relevant scales: unlike bulk metals which have a microstructure of scale (roughly from 100 μm downwards) generally far below other macroscopic dimensions of the component it composes (roughly from 1 mm upward), porous metals have an additional element of structure, namely the scale(s) and shape (or "architecture") of the pores. Porous materials have for this reason often been called "microarchitectured" materials (Fleck et al., 2010; Ashby, 2011). Along with the properties of the (dense) metal itself, this architecture of porous metals has the main role in determining the properties displayed. For some properties, such as the permeability, the foam structure is the sole determinant of the property, and even for others, such as the electrical conductivity (where the metal from which the foam is made has a strong role) the foam structure will have a major effect.

We thus make here a distinction between (i) the structure of the material relating to the pores, which we call the "mesostructure"; this is described with parameters such as the density and the pore size and shape, and (ii) the structure of the constituent metal which we call the "microstructure"; this comprises, for example, the orientation, nature, size and distribution of precipitates or grains within the metal. There is normally, but not always, a distinction of scale: the pores are most often an order of magnitude or more larger than the size of microstructural features in the metal; however, this is not always so. For example, in porous alloys produced by the replication process, metal grains encompass many pores and second phases formed during solidification can be nearly as large as the pores (San Marchi et al., 2004; Conde and Mortensen, 2010). It is also important to note from the onset that, because of the different aspects of the methods used to process foams described in the previous section and sometimes also because pores can influence the microstructural development of metals, it cannot be taken for granted that the microstructural state of the constituent metal in a foam is the same as in a dense sample of the same material having the same global thermal history.

24.3.1 The Structure of a Foam

As will be discussed in later sections, the structure of a foam is a key determinant of the properties it will display. To truly understand what this means we need to be precise in the definition of structure to be employed. Usually, the structure can be taken to encompass the distribution of the solid phase (and/or the pore phase) in space in three dimensions. As this is typically a complex arrangement, we normally use certain measures to describe it in quantitative terms. These can include the density (or volume fraction solid) and the mean pore diameter, although depending on the type of porosity (open or closed cell) and the density, it may be possible to picture the porous metal as being more like an arrangement of thin, regular strut-like elements, rather than isolated or semi-isolated pores. In this case the most important parameters are not those of the pores, but those relating to the struts: the strut length (the distance between contact points or nodes) the strut diameter and the cross-sectional shape.

Many porous metals are stochastic structures, and are therefore subject to statistical variations in parameters associated with their structure. Provided samples are sufficiently large, these variations will be smoothed out in the measurements and calculations to give an average value, and analyses such as the pore size determination allow standard statistical measures of distribution and confidence to be extracted to characterize the extent of these irregularities. Nevertheless, individual, localized differences can have a large effect on some foam properties, particularly mechanical behavior. It is therefore in some situations important to identify outliers such as individual pores of particularly large size (such as can appear in a gas blown foam when the wall between two pores breaks) or areas of particularly high or low density.

24.3.2   Characterization Methods

There are a wide range of techniques that can be applied to determine parameters of interest in porous materials.

24.3.2.1   *Determination of Density*

For a regular shape, determination of the density is relatively trivial: one simply measures the mass and calculates the volume from measurements of the dimensions (from which density = mass/volume follows). In many situations, however, a specimen for which the density has to be measured is not a regular shape. In this case one can make use of Archimedes' principle, that the volume of fluid displaced by an object is equal to the volume of the object itself. If we weigh the sample in air, and then make the same measurement with the sample suspended in fluid of known density (which, for obvious reasons, cannot be placed directly on the weighing pan and must not be included in the weight measurement; the arrangement is easy if a spring balance or other system where the sample is hung from the balance is used but requires more complicated equipment for other configurations), the difference in the values gives the mass of fluid displaced, which we can convert to a volume of fluid (and therefore of the sample) using the fluid density. This procedure is effective for dense materials and where the pores are isolated (closed porosity), but for open-cell foams when there are no trapped bubbles it will obviously give us the density of the parent metal and not the foam. To overcome this, for foams with sufficiently small pores, it may be possible to seal the surface with a material such a lacquer, wax or grease to prevent liquid entry. Weighing the sample before and after lacquering allows its mass to be taken into account; provided the lacquer enters the pores and does not build up the sample surface, which would increase the volume artificially, then no correction for its volume is required.

The simple measuring and weighing technique can be very effective for samples with dimensions of the order of tens of millimeters or higher, as a simple calculation can show. Typical Vernier or digital micrometers are able to measure to 0.05 mm resolution or higher, so the error in measuring the side length of a sample of aluminum foam of 20 mm would be ±0.25%. Weighing scales vary in accuracy, but equipment capable of making measurements to a resolution of 0.01 g is not difficult to locate (mass measurements are among the most precise extant). If our foam has a porosity of 80%, then the density would be 540 kg m^{-3}, and we would expect the weight to be 4.3 g, leading to an accuracy of ±0.23%. Combining these two errors with a linear superposition method and using the data for this assumed case gives an error of 6.3 kg m^{-3}, or 1.2%. This is an accuracy comparable with that of many other techniques that would be applied in the laboratory; however, since many porous metal properties depend strongly on relative density (see below), experimental points on charts of porous metal properties have generally a significant degree of scatter—something that is apparent throughout the literature on the subject.

For smaller samples, or where we wish to obtain more information on the porosity (e.g. pore size and distribution, open/closed ratio), other methods exist. One such technique is pycnometry, where gas is used to measure the volume of a sample (or at least the volume not accessible to gas), and so allows the amount of closed porosity present to be found provided the density of the parent metal is known. In a pycnometer, the sample is placed in an evacuated chamber. Another, identical but empty chamber is filled with gas (often helium due to its high mobility and low reactivity) at a known pressure. A valve separating the two chambers is then opened, and the drop in pressure in the empty chamber is recorded. If no sample had been placed inside, then the volume would have doubled and the pressure of the gas halved under Boyle's law. The volume will not fall so much, due to the presence of the sample, and from the pressure measurement this volume difference can be found. This, in conjunction with the weight of the sample, gives the density. As any (even slightly) open porosity will be accessed by the gas it will be ignored, so the method allows truly closed porosity to be distinguished and measured. In conjunction with a method such as the simple weighing and measuring, or assessment by Archimedes principle with grease or lacquer sealing the surface, this can allow calculation of the ratio between closed and open porosity in a sample of mixed character (Murray and Dunand, 2003).

Another process that is capable of giving some information about both types of porosity is porosimetry, which uses the infiltration of a nonwetting liquid that is forced into open pores (Giesche, 2002). This is normally carried out using mercury, as it is relatively uniform in its surface tension and wetting behavior with respect to other materials. The process involves the sample being placed in a sealed chamber and being infiltrated with mercury, the volume of mercury intruding into the sample (as a decrease in the apparent volume of sample plus mercury measured from when no overpressure is applied) being measured for various applied pressures. The relationship between the radius of the smallest pores infiltrated (implicitly assumed to be circular), r, and the pressure applied at any given point in the measurement, p, is given by

$$p = \frac{2\gamma_{Hg}\cos\theta}{r}, \tag{4}$$

where γ_{Hg} is the surface tension of mercury ($0.48\ \text{N m}^{-1}$) and θ is the wetting angle (generally taken as $140°$). Thus, as the pressure is raised, smaller and smaller pores are infiltrated, and the volume of each size can be determined. The method allows the distribution of pore sizes to be obtained in one measurement, although there are certain limitations. Evidently, only open pores can be accessed by the mercury, and eqn (4) gives the size of the smallest opening yet encountered by the liquid. Where there are large pores separated by small windows, the large pores will not be filled until a pressure sufficient to force the liquid through the small window is reached. The recording of wetting and dewetting pressure/volume characteristics can help to reveal if this has occurred, as a significant amount of mercury will remain trapped inside the sample. We note in passing that mercury porosimetry and its high-temperature equivalent can also be useful in predicting the infiltration behavior of metals into preforms, as required notably in the replication process (Bahraini et al., 2005; Molina et al., 2007; Bahraini et al., 2008; Jinnapat and Kennedy, 2011).

Another method which can be used for porosity characterization is BET. The method is so called as the procedure is based on Brunauer–Emmett–Teller theory, which predicts the way gases absorb on the surface of solids with increasing pressure. The theory suggests that absorption will initially be a monolayer of gas atoms, which then starts to fill pores, and by measuring the pressure changes as a known volume of gas is released into a chamber containing the specimen relatively accurate results can be obtained. However, though the method is precise for the determination of surface area on the atomic scale (taking into account surface steps, etc), it is experimentally difficult to apply for a material with pores larger than a few microns in size, and is mostly used in the characterization of materials that have submicron pore structures.

24.3.2.2 *Optical Microscopy and Scanning Electron Microscopy*

Optical microscopy and Scanning Electron Microscopy (SEM) are important tools in many aspects of metallurgy, and details relating to successful preparation and imaging of different metals can be found in many standard reference works, such as (Brandes, 1983; Brandes and Brook, 1992). For optical microscopy, preparation of foam samples for imaging follows the same steps as for dense materials; sectioning, mounting, grinding, polishing and etching. For some of these, specific steps are advisable due to certain features of foams. In sectioning, care must be taken not to damage the foam structure, which may be relatively weak (see the machining discussion in Section 24.2.8.1), also, mounting the sample may not be easy if the resin does not enter all pores. Cold mounting resins are more likely to be effective, given a greater time to flow into the structure without the risk of crushing it. Some suppliers offer low viscosity mounting resins, and degassing of the resin before use along with the employment of vacuum mounting systems can help to avoid trapped air bubbles or uninfiltrated regions. Care is required when polishing, particularly if there may be unfilled pores revealed by the grinding process; such pores can become traps for polishing media and lead to contamination of polishing wheels. Careful cleaning of samples between stages (e.g. using an ultrasonic bath) is advisable.

For SEM, the preparation is much easier; as the foam will already be conductive there is usually no need to apply a gold or carbon coating. Also, the benefit in using SEM for foams is normally not to take advantage of the high magnifying power, but rather the larger depth of field at moderate magnifications. This allows the depth of the structure to be seen in a single image, even if the magnification employed could be easily achieved with an optical system (this is the case, e.g. with the images in (Miwa and Revankar, 2011b)). Although usually used to access planar information, with geometric relationships used to convert this to true 3D sizes (Underwood, 1970; Russ and Dehoff, 2000; Higginson and Sellars, 2003), if a series of sections are taken (e.g. by successive grinding of a surface) and imaged, it may be possible to build up a 3D image of the structure. This process would usually be automated for best results (Spowart, 2006).

Images of foams can be analyzed for structural information following the same principles used for quantification of features in micrographs of dense materials (Underwood, 1970; Russ and Dehoff, 2000; Higginson and Sellars, 2003), either using manual procedures or using image analysis software. Due to the random nature of foams, it will be necessary to analyze a large number of features to obtain a high degree of confidence in the answer. For the analysis of features of several millimeters size, such as some pores, it may therefore be appropriate to capture an image using another method, for example, a flat bed scanner, to which the usual techniques can then be applied.

Optical microscopy has been used to reveal many aspects of foam structure. At low magnifications it has been employed to examine pores to identify form, for example (Hakamada et al., 2005b, 2007a), and deformation mechanisms, for example (Bastawros et al., 2000; Hakamada et al., 2005b, 2007a) after testing. At a smaller scale, it is possible to observe features in the metal, including the presence of defects, for example (Mukherjee et al., 2010d), or oxides, for example (Dudka et al., 2008), to examine or quantify the microstructure, for example (Park and Nutt, 2000; Conde and Mortensen, 2010), and to characterize surface coatings, for example (Dunleavy et al., 2011). Importantly, optical microscopy has been used to help deduce possible mechanisms of stabilization of liquid foams, for example (Gergely and Clyne, 2004).

SEM has been used because of its large depth of field, appropriate for viewing highly topographic specimens (once again to look at structure, e.g. (Goodall et al., 2007) or deformation mechanisms, for example (Abdulla et al., 2011)), and also because of its chemical analysis capability, through methods such as energy dispersive X-ray spectrometry, for example (Conde and Mortensen, 2010). Backscattered Electron imaging can give contrast due to the atomic number of the elemental species present. This has been used, for example, to identify second phases within the structure (Markaki and Clyne, 2001; Conde and Mortensen, 2008; Amsterdam et al., 2008d), and to observe the coating on a foam (Dunleavy et al., 2011).

24.3.2.3   X-ray Imaging and X-ray Computer Tomography

Foams can be imaged by simple X-ray radioscopy, the high contrast between pores and metal giving a general idea of the distribution of material (and so uniformity of density) and any large scale defects. Such methods have been used to look at the effects of processing variables, such as cooling rate or solidification conditions, on the structure and expansion dynamically during during foaming (Dudka et al., 2008; Mukherjee et al., 2010b,, 2010d, 2010c, 2010a), and are particularly appropriate for process control (Stanzick et al., 2002b) or where images in rapid succession are required to capture a sudden change such as bubble coalescence (Garcia-Moreno et al., 2004, 2008, 2011; Rack et al., 2009b). The method has been applied to the investigation of foaming of metals by gas injection (Banhart et al., 2001b; Garcia-Moreno et al., 2004, 2005), and a system to allow this to be done under microgravity has been built (Babcsan et al., 2006b; Garcia-Moreno et al., 2009).

Diffraction techniques, based on X-rays or neutron beams, have also been used to probe the physics of metal foaming. For the foaming of lead foams, a system based on neutrons has been used (Stanzick et al., 2002a). Titanium hydride decomposition kinetics have been probed by diffraction (Jimenez et al., 2011, 2012), and small-angle neutron scattering has been used to obtain information about the pore size distribution (Bellmann et al., 2002, 2003).

A more advanced use of X-ray imaging is X-ray computer tomography (XCT, sometimes called X-ray computer microtomography, or micro-CT, when applied to the investigation of small volumes of sample at high resolution). In this process, a large series of X-ray images of a structure are taken with a rotation of a few degrees between each. Computer algorithms are then applied which reconstruct the 3D shape that would give rise to that series of 2D images when viewed at those angles. This produces a 3D file showing the position in space of a particular phase, or phases, of interest. Metal foams are particularly suited to imaging in this manner, as there is a strong absorption contrast between the two phases (metal and air); as a result porous metals have been extensively examined using this technique.

Once the image file has been obtained, it can be treated in a number of different ways. The file can be sliced to produce multiple 2D sections along any plane, which can be investigated using the some procedures as for optical microscopy. It may also be analyzed directly to extract structural information in three dimensions. It is relatively easy to interrogate the database of 3D position information, and determine what fraction of points are in the solid phase, thus determining the volume fraction. More complex analysis also permits values for other parameters to be found, such as the pore size and distribution, connectivity and cell-to-cell window size, surface area, strut size (Olurin et al., 2002; Maire et al., 2003b; Elmoutaouakkil et al., 2002; Liebscher and Redenbach, 2012). Some features, such as the mean strut length, can only be accessed with accuracy from 3D measurements (Dillard et al., 2005). 3D correlation techniques also allow strain to be plotted if images are taken after different deformations (Marmottant et al., 2005).

The process can be carried out using benchtop systems, or much larger facilities such as the European Synchrotron Radiation Facility in Grenoble, France (www.esrf.eu/). A synchrotron can provide X-rays at much higher energy and intensity, and is therefore capable of examining larger volumes of material simultaneously, making measurements more rapidly. It can also reach high magnifications; voxels (the smallest discrete three dimensional units discerned by the process) down to below 1 µm resolution are achievable (Toda et al., 2006a) (for a small imaged volume).

XCT has been used to study the foaming process. In the "ex situ" approach samples are foamed at different times, quenched and observed (Mukherjee et al., 2010b, 2010d) (Helfen et al., 2002, 2005; Rack et al., 2009b). Alternatively, in a few experiments researchers have managed to observe foam structures "in situ," meaning while the metal is still liquid. This has the significant advantage of elucidating solid particle distributions free of redistribution caused during solidification by particle pushing, thus giving some of the most informative data to date on the physics of metal foaming (Haibel et al., 2006). High powered versions of the system are used to get the smallest scale resolution (down to submicron level, sometimes called microtomography), for example, to observe the changes in phase in different regions of TiH_2 particles on heating (Jimenez et al., 2012).

Tomography has been used to visualize the internal structure of foams (Haibel et al., 2006; Fife et al., 2009; Hartmann et al., 2011; Dunleavy et al., 2011) and to investigate structural changes in foams with different processing parameters, such as the variation pore size and pore wall thickness in titanium foams produced by sintering with space holders of different size (Tuncer et al., 2011b). Where foams are open celled, as in this last case, a watershed algorithm can be used to distinguish separate pores (see, e.g. (Benouali et al., 2005a)). Tomographic investigations can also identify different solid phases, and have been used to investigate the penetration of biological tissue into foam structures and its growth in the cellular environment (Baril et al., 2011). The technique is particularly useful here as the large difference in mechanical properties between the metallic implants and the biological tissue mean that the histological techniques (thin sectioning for observation in transmitted light microscopy) conventionally used in medical research are experimentally very challenging.

Structural changes on deformation have been investigated by imaging deformed samples, confirming mechanisms of strut bending and buckling (Maire et al., 2003a), and revealing, for example, the presence of narrow deformation bands in compression of closed-cell foams (Bart-Smith et al., 1998; Bastawros et al., 2000; Saadatfar et al., 2012) and syntactic foams (Zhang et al., 2009b), or their absence in open-cell replicated foams (Marmottant et al., 2005), and the role of defects in initiating deformation (Benouali et al., 2002, 2005a; Mukherjee et al., 2010d). Tomography has allowed the crushing of foam below an indenting body to be imaged (Kadar et al., 2004b); the surrounding foam remaining largely unaffected. Fatigue tests have also been performed with periodic interruptions for tomographic assessment, revealing that with respect to fatigue the defect tolerance of closed-cell aluminum foams is higher than for uniaxial deformation (Kolluri et al., 2008). Mechanical tests have also been carried out in situ in tomography equipment, allowing the deformation process to be imaged at different stages, both in compression (Salvo et al., 2004; Marmottant et al., 2005; Toda et al., 2006a) and tension (Dillard et al., 2003). Such tests have revealed that for a closed-pore foam made by a powder route with a blowing agent, the cell walls first buckled, and then proceeded to rupture (Salvo et al., 2004), whereas for an open-cell nickel foam struts bend and then buckle, with cracks in fracture initiating at nodes (Dillard et al., 2005). Detailed probing of the microstructure within deforming cell walls has shown that, in some cases at least, micropores within cell walls in closed-cell foams initiate failure (Toda et al., 2006a).

Tomography methods have also been used to obtain foam structural information for the validation of, or input into, simplified geometrical models of foams based on repeating unit cells (Perrot et al.,

2007; De Jaeger et al., 2011), and to investigate the homogeneity of real foam structures (Benouali et al., 2005a) (Solorzano et al., 2008a). Such measurements have been further used to develop a model relating this to the scatter in properties measured macroscopically (Liebscher et al., 2012), and for the validation of models of foam structure generated by random sphere packings (Redenbach, 2009).

Three-dimensional images of real foam structures obtained by tomography can also be used as input in finite element modeling simulations of foam properties (Maire et al., 2003b). The reported use of this includes modeling the mechanical response (Singh et al., 2010; Michailidis, 2011; Betts, 2012; Saadatfar et al., 2012), conduction through the solid phase (Veyhl et al., 2011b; Fiedler et al., 2009, 2012; Laschet et al., 2009), or by radiative heat transfer (Coquard et al., 2012). Such methods can obtain high accuracy as they reproduce in detail the structure of the foam, although their evident limitation is that predictive power is only obtained where the structure does not vary significantly from that analyzed.

24.3.3   The Mesostructure

By the mesostructure of a foam we identify the structure associated with the pores, including pore size, shape and interconnectivity. This sits between the scale of features within the metal (the microstructure, discussed below) and the structure related to the component (the macrostructure, which is not dealt with here).

24.3.3.1   Density and Pore Size
The two most important and widely measured parameters in assessing the structure of metal foams are the relative density and the pore size, which are relatively simple measures (perhaps deceptively so).

The relative density or the porosity is one of the most significant parameters in determining the majority of porous metal physical and mechanical properties. It can be recorded in many ways, either as relative parameters, such as the volume fraction solid or the porosity, or by an absolute measurement such as the density. In this work we have elected to use the volume fraction solid, V_m, as our standard parameter, and have generally given this in equations and figures, except where a simplification is possible or it is more appropriate to use one of the other terms. **Table 3** gives several of these measures and where relevant how they are calculated.

There is potential ambiguity in a pore size measurement. Manual techniques tend to give the simple diameter of a pore (perhaps in a random direction), whereas image analysis techniques will often give the diameter of a circle (or sphere) with the same area (or volume) as the pore being measured, for example (Tuncer et al., 2011b). For nonequiaxed or nonspherical pores, there evidently could be a difference. To illustrate the possibilities, we can imagine a pore with a cubic shape of side length a, see **Figure 25**. In a 2D section, we would measure the diameter somewhere between a and $1.73a$, depending on the intercept angle, while the diameter of a circle of equivalent area would be between $1.13a$ and $1.34a$. In 3D the cube would give a diameter of $1.24a$ if assimilated to a sphere of equal volume. As an alternative, we could consider a pore with a prolate spheroid shape and an aspect ratio of 1.5, see **Figure 25**. If we measured the lateral diameter as b, the longitudinal diameter would therefore be $1.5b$. Assimilating this to a sphere of the same volume would give a single diameter of $1.14b$. There is therefore the possibility of significant error where the measurement method is not reported, or the choice is inappropriate for the pore shape. An alternative where nonequiaxed pores are present is to use an algorithm to find the best fit between a grain in 2D or 3D and a regular shape that is easily defined, such as an oval or an ellipsoid, for example (Benouali et al., 2002, 2005a).

Table 3   Summary of common measures for the solid/pore ratio and feature sizes in foams, and how they are defined

Parameter	Symbol	Units	Description/calculation
Volume fraction solid	V_m	–	The proportion of volume occupied by metal (0–1)
Volume fraction pores or porosity	V_p	–	The proportion of volume occupied by free space (0–1), $= 1 - V_s$
Density	P	kg m^{-3}	The mass per unit volume, $= M/V$ (of dense or porous material)
Pore/cell size	D	m (mm may be most appropriate)	The diameter across a 3D region free of metal, taken in a random direction and normally projected onto a random 2D section. The diameter of a circle/sphere with the same area/volume as the pore
Window size	D_w	m (mm may be most appropriate)	The diameter across a 2D region free of metal within a wall separating two pores, taken in a random direction and normally projected onto a random 2D section
Cell wall thickness	l_w	m (μm may be most appropriate)	The distance across an area of metal separating two pores, measured at right angles to the pore surface
Strut diameter	D_s	m (μm may be most appropriate)	The distance across a strut in the direction normal to the strut axis, measured in a random position
Strut length	l_s	m (mm or μm may be most appropriate)	The distance along a strut between the center of two nodes

Pore size can be measured using image analysis software (which are often based on counting the number of pixels within a pore and relating this to a dimension of a regular shape of the same area), or through manual methods such as comparing pores with circles of different size to find the best match, to produce a distribution. In this case it must be remembered that, as on average pores will be intercepted on some plane away from their diameter, the true diameter of pores will be larger than recorded along 2D cuts. This can be handled statistically (Underwood, 1970), and the apparent size of a 3D feature when measured through the intersection with a random 2D plane (L) is given by Tomkeieff (1945)

$$L = \frac{4V}{S},\tag{5}$$

where V is the volume and S is the surface area. If pores are assumed to be spheres this means that features will be on average a factor of 3/2 larger than measured. This method is used to obtain a true feature size in (Simone and Gibson, 1996), for example.

An alternative measurement method is linear intercept, where a line is drawn on the surface of the section, and the number of pore boundaries it crosses is counted. A mean pore size can then be extracted using the equation below. Note that it is necessary to know the porosity of the material to use this method (which can also be extracted from image analysis, for example, by point counting, see (Higginson and Sellars, 2003)):

$$L_{Pore} = \frac{2(1 - V_m)L_{Total}}{N_{Boundaries}},\tag{6}$$

where L_{Pore} is the average 2D diameter of the pores as they intersect with the plane (i.e. a correction factor as above would have to be applied), L_{Total} is the total line length examined (the true line length,

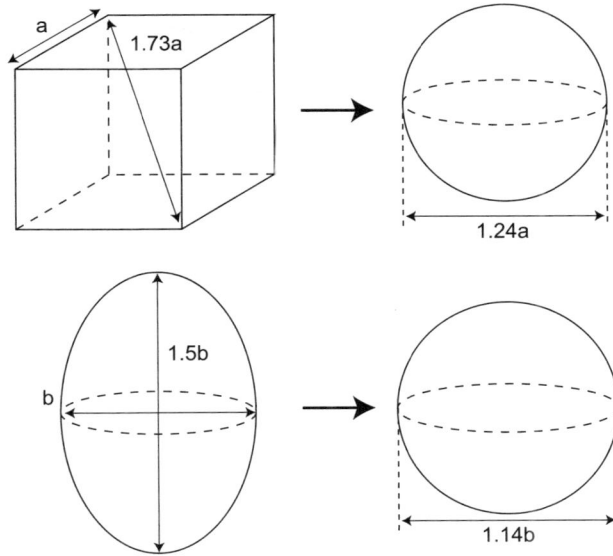

Figure 25   A simple example of the possible effect of different types of pore size measurement on values obtained if the pores are not the assumed shape (spherical).

accounting for the magnification of the image), and $N_{Boundaries}$ is the number of pore–metal boundaries crossed.

One common way of representing the pore size, particularly with foams made by methods based on a polyurethane precursor, is the pores per (linear) inch (PPI). This is chiefly because foams are sold under this measure, which has come about as it is the standard measure of pore size used to classify the polymer foams that are used as a template. However, caution is required with this measurement, as often for polymer foams the value relates to pores per *square* inch, not linear inch, so the conversion to an equivalent average diameter must be made with care. As an example, common PPI values are 5, 10, 20 and 40. If these were in number of pores encountered in a linear inch they would convert to pore diameters of 5.1 mm, 2.5 mm, 1.3 mm and 0.6 mm. However, taking the square root and converting, as is necessary for pores per square inch, gives 11.4 mm, 8.0 mm, 5.7 mm and 4.0 mm, giving a much smaller difference across the range. Observation of metal foams produced and sold under these classifications reveals that the latter conversion is in many cases much more realistic, but even this may not be the whole story. For many of the multistep processing methods, we might expect there to be dimensional changes, for example, a reduction in size during polymer burnout and ceramic sintering in the investment casting process, and this is often in fact the case. It is therefore good practice to make some independent assessment of the pore size of material available before carrying out an investigation where this parameter may be of importance.

Another issue to be aware of is the precise definition of a pore. Most authors define the "pore size" as the diameter of the dominant, roughly equiaxed, free spaces in the structure (the cells), and the "window size" as the diameter of the openings between these cells. However, some authors define the pores to be the openings, and the interior distance to be the cell diameter (Miwa and Revankar, 2011b), so care must be taken when using and reporting results. Other dimensions that may be important are the "wall thickness" and the "strut diameter" and "strut length," but these are much more self

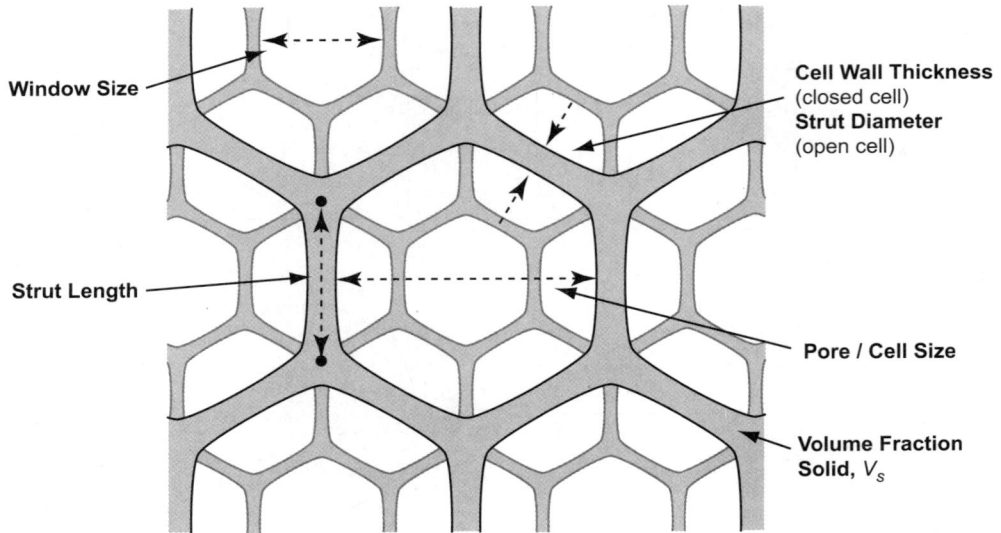

Figure 26   Schematic diagram of different structural features and measurements in the mesostructure of foams.

explanatory and less prone to confusion (the strut length would usually be taken as the mean node-to-node distance). **Figure 26** shows these on a generic foam mesostructure.

As foams are stochastic structures, the pore size distribution (or the distribution in any of the size measurements) is also something that may be important, and simply giving the average value for any parameter may be misleading. Certainly, bimodal pore size distributions have been observed (Zhao et al., 2009b), and more complex distributions could be imagined.

24.3.3.2   Pore Shape
Beyond the cell size, the actual shape of cells may be important. Assessing and quantifying this is a challenge because of the often complex shapes involved. A technique frequently employed to simplify the description of shape is to liken a pore to a circle or ellipse (or a sphere and prolate/oblate spheroid for 3D measurements). In making this assessment, the largest diameter that can be recorded at any angle is measured, along with the diameter in an orthogonal direction, and the ratio found between these. If the value is close to 1, the pores may be assumed to be equiaxed, with a greater degree of nonequiaxed character as the value found gets further away from unity.

By assessing the angle relative to the sample axis at which the largest diameter of each pore is found, it is possible to see if there is a preferred orientation (if there is a tendency for the angle to be around a particular value) or if the pores are random, albeit nonequiaxed (if the probability of any angle is roughly the same). Similarly, the strut shape, either along the profile or in the cross-section may be important (particularly for deformation dominated by bending (Goodall et al., 2006a)), and this can be measured in a similar way.

24.3.3.3   Surface Area
There are various ways to characterize the surface area of a foam. It can be measured directly with techniques such as XCT or BET (see above for information on these methods). It can also be estimated,

by assuming a pore structure and calculating the surface area of that structure for the particular volume fraction solid and pore size required (Duan et al., 2006a; Liu, 2010). The effectiveness of such an estimate is likely to be strongly related to how well the particular structure considered matches up to the true structure of the foam. For example, the pentagonal dodecahedreon model of Duan (Duan et al., 2006a) is effective for foams made by deposition onto polymer precursors, where the structure is similar, but is obviously expected to be much less effective for a Gasar foam, for example.

Image analysis methods can also be used to compute the surface area from 2D sections, through simple geometrical analysis. Similar methods are used to calculate grain boundary areas per unit volume in metallographic analysis of dense materials (Higginson and Sellars, 2003). A spherical pore shape is assumed, which means the surface area per unit sample volume is in this case

$$S_V = \frac{3}{2R}(1 - V_m),$$ (7)

where S_V is the surface area per unit volume, R is the true pore radius and V_m is the volume fraction solid. As the analysis on a 2D section will give a reduced value for the pore diameter, L (see above), this becomes

$$S_V = \frac{2}{L}(1 - V_m).$$ (8)

As with other approaches, the effectiveness of estimates obtained by this method will depend on how accurate the description of the foam structure is. Note that the method ignores area lost to cell windows, and therefore would be expected to function best for closed-cell foams, or open-cell foams with high density.

24.3.3.4 *Defects*

A defect in the foam structure can have a large effect on the properties displayed, particularly if there are not a large number of pores across the specimen's smallest dimension. As a result, the investigation of defects in the foam structure is of great importance and this has been the subject of a large amount of research effort, using some of the techniques outlined above; for example, X-ray tomography has been used to investigate and identify cell rupture and other defects in gas blown foams (Mukherjee et al., 2010b). Many kinds of defects can be identified, and those particular ones that are seen in gas blown foams (where arguably the control over individual pore structure is less precise) are discussed more fully elsewhere, in the processing and mechanical properties sections.

24.3.3.5 *Anisotropy*

A general and important feature of many foams that depend at some stage of their processing on using a gas to foam a liquid (this includes all closed-cell metal foams processed with the metal in the liquid state, and also foams made from a reticulated polymer template) is that the structure will not be perfectly isotropic: the pores will tend to be elongated along the direction of gravity. This anisotropy of structure is likely to be mirrored by an anisotropy in the properties displayed (Huber and Gibson, 1988).

24.3.4 **The Microstructure**

The overwhelming majority of metal foams will have grains (the exception being BMG foams (Brothers and Dunand, 2005b)), and many foams are multiphase structures, being made from alloys or having ceramic additions, often used to aid foaming (see above). As such, they will show a microstructure, which will generally affect the properties displayed. Microstructure can be investigated in the same way as for dense metals, through microscopy techniques (Brandes, 1983; ASTM, 2000; Kim et al., 2012), with some care required when preparing samples.

As expected, all the microstructural features encountered in dense metals may be observed in foams, including dendritic, eutectic and oxide structures (Markaki and Clyne, 2001; Conde and Mortensen, 2010). Investigations of the microstructure within the metal itself are relatively few, with most authors concentrating on the pore structure as the defining feature of a foam, and arguably the most significant factor determining its properties. However, particularly in liquid state methods where solidification largely determines the microstructure, it is likely that the pores will provide a physical constraint and may influence microstructure formation. This effect has been investigated specifically by Conde and Mortensen (2010), for the alloy Al-4.5 wt %Cu. It was found that the extent of microsegregation and the dendritic character of the microstructure diminished as the cooling rate was reduced, and also as the pore size was reduced, being almost absent for all cooling rates at a pore size of 75 μm. This was attributed to the physical constraint of the narrower channels between smaller pores. Models developed for microstructure prediction in metal matrix composites (Mortensen, 1990; Mortensen and Jin, 1992) were shown to be effective at predicting these changes (Conde and Mortensen, 2010).

Al-4.5wt% Cu is an age hardening alloy: this behavior was also seen in the foams produced and was shown to lead to significant enhancements of metal foam strength (Conde and Mortensen, 2008). High carbon steel foams processed by powder pressing with a carbonate gas release agent to cause expansion in the solid state during the sintering stage (1300 °C) were examined microstructurally (Park and Nutt, 2000), as formed foams showed a pearlitic structure with hypereutectoid cementite. It was reported that annealing reduced the amount of pearlite, presumably through a coarsening mechanism.

24.4 Physical Properties

One of the more interesting features of porous metals, and of porous or microcellular materials in general, is that these represent an extreme in the theory of composite materials: porous metals are composites of metal combined with a second phase of, essentially, "nothing". In most respects, therefore, these are composites with an infinite phase contrast. This in turn causes imprecision in predictions to be greatly amplified compared with other composite materials. Thus bounds tend to be far apart and when a physical property is exactly predicted, then the result is most often either a highly simplified, or a highly specific, prediction. This said, the presence of pores in metal can on the other hand create new functionalities, and can open new ranges of variation for certain physical properties. Examples of each will be found in what follows.

24.4.1 Density

Some physical properties of porous metals are trivial. Their density is the most obvious example: it is that of the constituent metal times its volume fraction $V_m = (1 - V_p)$ where V_p is the volume fraction of all pores present within the material:

$$\rho = V_m \rho_m, \tag{9}$$

which is why V_m is generally called the "relative density". The same relation holds for the heat capacity of porous metals, heat stored by gas within the pores being in general negligible. When pores are filled with a denser phase, generally a liquid, then the heat capacity is equally simple to predict, being given by a simple rule of mixtures.

24.4.2 Thermal Expansion

Another a priori trivial property is the thermal expansion; however, its apparent simplicity is deceiving. The thermal expansion of a uniform material is a homothetic transformation: voids in a homogeneous material therefore expand with the rest and do not affect the material's thermal expansivity. Put more physically, the relative expansion of all dimensions of whatever shaped solid is the same as that of the uniformly expanding material from which the solid is made. Pores then make no difference.

This simple picture however breaks down as soon as a second phase is considered. Gas trapped in closed pores or in closed foam cells may, on heating, exert an increased pressure tending to increase artificially the global thermal expansion of a porous metal (remember that this is sometimes used for foaming, Section 24.2). Unless temperatures come close to the melting point of the metal, however, the high stiffness and yield stress of metals will render this influence of trapped gas negligible. To see this, consider a metal foam of relative density $V_m = 0.1$, containing gas within closed cells, the gas pressure being near atmospheric (as will generally be the case given micro-cellular material fabrication processes, see Section 24.2). Cooling or heating the foam by a few 100 K from a temperature of a few 100 K will cause a change in gas pressure that is of the same order of magnitude as the initial gas pressure, that is, a few atmospheres (1 atm = 0.1 MPa). This, in turn, will cause a buildup of tensile stress within the metal reaching at most 10 times that, that is, a few MPa (from force equilibrium). If the metal is very near its melting point, then a few MPa may cause it to deform irreversibly (which is why die cast aluminum, which can contain closed pores filled with highly pressurized gas, is often not solutionized); however, in most circumstances, the metal will remain elastic. Metal stiffnesses being of the order of magnitude of 100 GPa, corresponding strains are on the order of a few times 10^{-6}, corresponding to a change in foam coefficient of thermal expansion (CTE) of a few times $10^{-8}\,K^{-1}$; this is negligible compared to the CTE of metals, which is of order of magnitude $10^{-5}\,K^{-1}$. Trapped gas thus exerts no discernable influence on the thermal expansion of porous metals except at very high temperatures. The thermal expansion of homogeneous porous metals, whether open or closed cell, therefore remains that of the corresponding dense metal.

What precedes is only true, however, if the metal CTE is constant on the scale of the average pore-to-pore distance: such will be the case with single-phase metals of course, and also with alloys having a very fine-scale microstructure, such as a precipitation or dispersion hardened metal, in which second phases are homogeneously distributed at a scale much finer than that of the pores. If, on the other hand, the porous metallic material is made of several phases that are distributed at a scale commensurate with that of the pores, then the question becomes both far more complex—and rich in possibilities. Consider the following example.

Bimetallic strips combine two materials having widely different CTEs, which are intimately bonded in fixed and generally uniform proportion along relatively long bands. These are a well-known means of creating large displacements in response to temperature change, and are thus extensively used in thermally driven sensing or actuation (Freund and Suresh, 2003). Assuming a strip sufficiently long for engineering beam theory to be applicable with neglect of end effects, the change in curvature that results from a change in temperature ΔT is given by Timoshenko (1925), Lakes (1996)

$$\Delta\kappa = \frac{6(\alpha_2 - \alpha_1)\Delta T}{h_1 + h_2}\frac{\left(1 + \dfrac{h_1}{h_2}\right)^2}{3\left(1 + \dfrac{h_1}{h_2}\right)^2 + \left(1 + \dfrac{E_1 h_1}{E_2 h_2}\right)\left(\left(\dfrac{h_1}{h_2}\right)^2 + \dfrac{E_2 h_2}{E_1 h_1}\right)}, \tag{10}$$

where α is the CTE of each of the two phases (subscript 1 and 2 with Phase 2 situated on the outer, convex, side of the strip) composing the bilayer strip, with h and E their thickness and Young's modulus, respectively. If two similar curved strips are assembled end-on-end and heated or cooled by ΔT, then the end separation distance will change (one strip would give the same result but the physics are more intuitively obvious with two because then the basic element remains symmetric and hence straight as it deforms). The resulting longitudinal thermal expansion of the bi-strip translates into a longitudinal CTE equal to (Lakes, 1996)

$$\alpha = \frac{6(\alpha_2 - \alpha_1)\theta R}{h_1 + h_2}\frac{\left(1 + \dfrac{h_1}{h_2}\right)^2}{3\left(1 + \dfrac{h_1}{h_2}\right)^2 + \left(1 + \dfrac{E_1 h_1}{E_2 h_2}\right)\left(\left(\dfrac{h_1}{h_2}\right)^2 + \dfrac{E_2 h_2}{E_1 h_1}\right)}\left(\frac{\cot(\theta/2)}{2} - \frac{1}{\theta}\right), \tag{11}$$

where R is the radius of curvature of the strip and θ the included angle. The result is proportional to $(\alpha_1 - \alpha_2)$, depends only weakly on E_1/E_2 (Timoshenko, 1925) and, most interestingly, is proportional to the bilayer aspect ratio, $[\theta R/(h_1 + h_2)]$. Therefore, (i) its sign can be freely chosen by placing the most expansive phase on the inner or outer surface of the bilayer, and (ii) there is no theoretical bound on the longitudinal CTE of the bistrip, since there is no theoretical bound on its aspect ratio. Such strips can be linked at their ends and assembled into a wide variety of 2D or 3D open-cell strut/node structure (e.g. along edges of a square or cubic lattice, respectively). Hence, they can serve as the building block of a porous metallic material, the CTE of which can be tailored at will to have, for example, a zero, negative, or very large thermal expansion—all of course within limits set by other considerations, such as the elastic stiffness of the material (see Section 24.5.1), or when it is very flimsy the simple need for the resulting material to hold its own weight.

The periodic bimetallic strip design was proposed in a short and elegant 1996 paper by Lakes, (1996), (2007); other design schemes have since been proposed for multiphase porous materials featuring highly tailorable CTEs. Several among these are based on assemblies of a different, triangular, beam assembly depicted in **Figure 27** and analyzed in detail by Miller et al. (2008). Here, two beams of low CTE material are connected to a third beam of higher CTE. Dimensional changes of this element are (like the bimetallic strip) anisotropic, with expansion in one direction, contraction in the perpendicular direction, and a global area change that can be positive or negative for given ΔT, the value of areal expansion being a function of the relative length and CTE of each of the two beam types. This basic triangular element can also be assembled in a number of arrangements to form 2D or 3D microcellular materials having tailored thermal expansion, including notably zero or negative CTE values (Miller et al., 2008; Palumbo et al., 2011; Grima et al., 2007; Lim, 2012). For example, the structure to the right of **Figure 28** having fourfold symmetry, it has an isotropic thermal expansion coefficient (Miller et al., 2008).

(a)

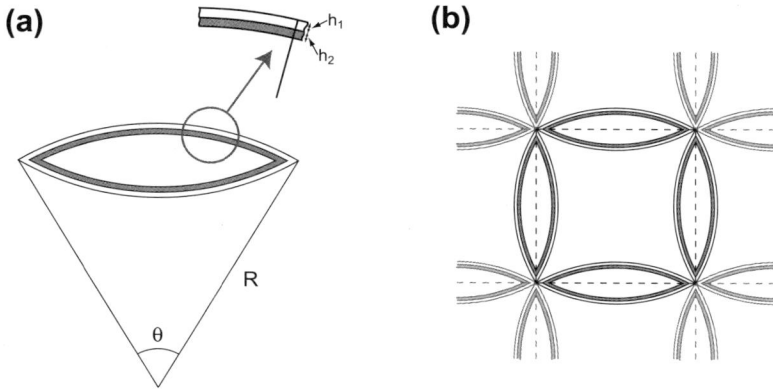

(b)

Figure 27 (a) The basic bilamellar strip of Ref. Lakes (1996) combining two materials of different thermal expansion (duplicated here to create a symmetric element); (b) assembled into a square or cubic (or other) lattice this element creates a periodic microcellular material of controlled thermal expansion.

Other, more complex, tailorable CTE porous structures have been proposed, often with a view of optimizing a combination of the CTE with another property (e.g. elastic modulus or global expansion stress). These are often variations of the triangular design in **Figure 28**, featuring solid segments instead of one of the beams, or optimized structures, often with reentrant angles designed to amplify deflections caused by differential thermal contraction (Sigmund and Torquato, 1996; Chen et al., 2001a; Aboudi and Gilat, 2005; Kelly et al., 2005, 2006; Steeves et al., 2007, 2009; Faure and Doyoyo, 2007).

"Real" porous metallic structures of this kind have been produced and tested using various processing methods, such as direct metal deposition, or extrusion and assembly. Although at times of coarse scale (raising the question of whether these are materials or structures) or structurally imperfect, the high potential held by porous metallic materials for applications requiring "exotic" CTE values is now demonstrated (Chen et al., 2001a; Palumbo et al., 2011; Steeves et al., 2009; Qi and Halloran, 2004; Oruganti et al., 2004; Rhein et al., 2011).

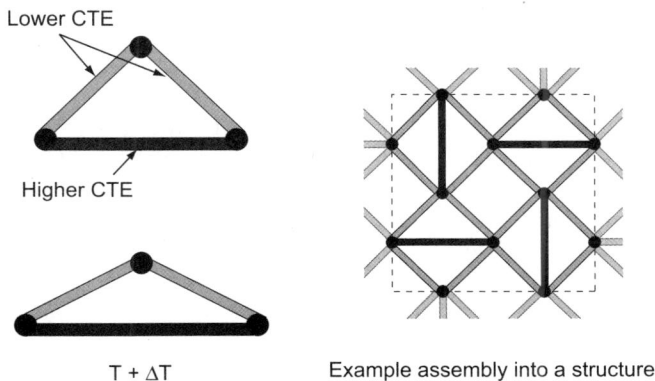

Figure 28 Two-material triangle structure described in Ref. Miller et al. (2008). (For color version of this figure, the reader is referred to the online version of this book.)

24.4.3 Conductivity (Electrical & Thermal)

The conduction of electricity (respectively heat) through solids is governed by Ohm's (resp. Fourier's) law, which states that the flux j of electrical charge (resp. heat) is proportional to the gradient of electrical potential U (resp. temperature T) times an electrical (resp. thermal) conductivity, k_m, characteristic of the solid under the relevant conditions (of temperature, pressure, etc):

$$j = -k_m \, \text{grad}(U), \qquad (12)$$

where k_m is a second-rank symmetric tensor in anisotropic media; this is a scalar when the medium is isotropic. At steady state and with no source terms, conservation of charge (resp. energy) dictates that U obeys the Laplace equation: $\Delta U = 0$.

Pores are electrical insulators; hence, within a porous solid, pore surfaces are everywhere parallel to current lines and normal to equipotentials. Being obstacles to the flow of current, pores (i) reduce the volume of conducting solid available per unit volume of (porous) material while also (ii) generally causing detours in the path of current flow. The second effect is why the conductivity k_f of a porous solid is nearly always lower than the conductivity k_m of the dense solid of which it is made times the relative density V_m of the porous solid:

$$k \leq V_m k_m. \qquad (13)$$

Equality holds when the pores are infinite parallel cylinders oriented along the direction of current flow (i.e. normal to equipotentials). This is the most general upper bound on k_f (Torquato, 2002a); it corresponds to the simple "slab" model for composites (i.e. the "rule of mixture" geometry).

Near the two limits, of dilute pores ($V_m \approx 1$) or dilute metal ($V_m \approx 0$), simple expressions hold. Isolated spherical pores in a dense isotropic metal lower its electrical conductivity to second order in the pore volume fraction $V_p = 1 - V_m$ according to (Torquato, 2002a; McLachlan et al., 1990)

$$k = k_m \left[1 - \frac{3}{D-1} V_p + \frac{3}{(D-1)^2} V_p^2 + O\left(V_p^3\right) \right], \qquad (14)$$

where D ($=2$ or 3) is the dimension of the porous material. Similar expressions for solids containing ellipsoidal pores, or for 2D porous solids, can be found in Refs. Torquato (2002a), Hyun and Torquato (2000).

Equally simple is the opposite extreme of highly porous microcellular solids containing only a few percent solid arranged in regular sheets or rods of even thickness. Current flow lines are then everywhere roughly parallel; however, these generally do not run along the (average macroscopic) electric field applied on the porous material. The ensuing detour in the path of current flow can be calculated by integration across angles (Lemlich, 1978; Torquato, 2002a). For random structures this gives

$$k = \frac{k_m V_m}{N} \qquad (15)$$

with $N = 1$ and 2 for flow, respectively, along, or across, a 2D honeycomb of randomly oriented walls. In 3D highly porous solids with straight walls or struts, $N = 3/2$ for a closed-cell microcellular metal with randomly oriented walls, while $N = 3$ for an open-cell microcellular metal with randomly oriented struts (Lemlich, 1978; Datye and Lemlich, 1983; Torquato, 2002a). Note that:

(i) structures that contain a parallel axis and plane of symmetry have isotropic thermal conductivity across all directions perpendicular to that plane (Nye, 1985) p. 206. Equation (15) is therefore also valid for regular nonrandom thin straight wall or strut structures of high symmetry. These include hexagonal, square, triangular or Kagome honeycombs in 2D, or any cubic 3D structure (since averaging their conductivity across all directions should give the same result as for randomly oriented walls or struts).

(ii) These results come out trivially from consideration of square/cubic structures if one aligns one side of the square or cube parallel to the macroscopic applied potential gradient (this makes the relation easy to remember).

If cell walls or struts are not straight, uniform and slender, then the situation becomes more complex, as it also does for V_m anywhere between these two extremes. Phase contrast being extreme in a porous material, bounds are of limited help without further specification of phase topology since the lower bound, also for isotropic structures, is zero (simply because structures containing continuous walls of zero flux void space can be proposed). The lower Hashin–Shtrikman bound is thus zero. The upper Hashin–Shtrikman bound is simple:

$$k = k_m V_m \frac{D - 1}{D - V_m}, \tag{16}$$

where D (=2 or 3) is the dimension of the porous material. This two-point bound is the highest conductivity that can be reached in isotropic porous metals. It is also "best possible" since it can be realized: the conductivity in eqn (16) is that of a uniform layer of solid coating a cylindrical or spherical pore. Note how it also equals, at very low V_m, expressions given in eqn (15) for isotropic straight wall structures: these are hence optimal. Far narrower bounds are obtained if further "three-point" information on the distribution of porosity is known; see Refs. Hyun and Torquato (2000), Torquato (2002a).

Mean-field (or effective medium) approximations extend predictions for the dilute pore limit ($V_m \approx 1$) to V_m values well below unity, at which one can no longer assume that each pore disturbs the flow of electricity as if it were embedded in an infinite solid. The most classical among these is the Maxwell approximation: this assumes that pore-to-pore interactions do not disturb significantly the flow of current around each pore, at given applied electric field, compared to what would be seen with dilute pores. On the other hand, the Maxwell approximation accounts for the fact that a finite volume fraction of porosity will alter the average electric field in the solid alone compared to the average electric field applied to the porous material as a whole. The resulting expression is simple: it coincides with the Hashin–Shtrikman upper bound, namely eqn (16) (Chang and Lemlich, 1980; Torquato, 2002a; Milton, 2002).

Mean-field predictions for porous metals are reviewed in Refs. Torquato (2002a), Milton (2002), Weber et al. (2003b), Goodall et al. (2006d). Noteworthy among these is the "differential effective medium" (DEM) approximation, also known as Bruggeman's approximation. In the DEM scheme, one progressively adds an infinitely small amount of porosity to the material, treating the already porous solid each time as a homogeneous porous continuum of uniform conductivity to which a few dilute pores are added—an assumption that can only, in a strict sense, be legitimate if each newly added set of pores is much larger than all pores considered previously. The resulting expression is obtained by integration of a differential equation (hence the name) expressing the influence of an incremental addition of dilute nonconducting inclusions to a material having already a given finite porosity. With a porous metal, the end result is again simple:

$$k = k_m \left[V_m^{\left(\frac{D}{D-1}\right)} \right] \tag{17}$$

with D ($=2$ or 3) again the dimension of the porous material.

Other expressions from mean-field theory together with their extension for randomly oriented or aligned spheroidal pores can be found in Refs. McLachlan et al. (1990), Clyne (2000), Torquato (2002a), Milton (2002), Weber et al. (2003b), Weber (2005), Goodall et al. (2006d), Tane et al. (2005), Ordonez-Miranda and Alvarado-Gil (2012). In the DEM scheme, with randomly oriented nonspherical pores the power law is maintained while the exponent changes, increasing somewhat above 1.5 with oblate spheroidal pores, for example.

It is interesting that a power law relation between conductivity and relative density

$$k = k_m V_m^{\mu_p} \tag{18}$$

is also obtained if k_f is governed, at low V_m, by the onset of percolation within the conducting solid phase (Balberg, 1987; McLachlan et al., 1990; Stauffer and Aharony, 1994; Sahimi, 1993, 1994; Hunt and Ewing, 2009; Torquato, 2002a; Milton, 2002; Kovacik, 1998a; Kovacik and Simancik, 1998; Simančik and Kovacik, 2002). In percolation models, the conductivity evolves with V_m because the number of connected conducting cell walls or struts increases as V_m increases past a certain (percolation) threshold. Near the percolation threshold, the conductivity varies as a power law of the difference between V_m and its threshold value; to obtain eqn (18) this threshold must be assumed to be met at $V_m = 0$. Classical fixed network percolation models, in which the conducting cell walls or struts are randomly distributed on a fixed grid and have a fixed resistance, yield the (universal) value $\mu_p = 2$. A "Swiss Cheese" continuum percolation model of randomly disposed spherical holes in a conducting metal (where resistances hence vary), gives $\mu_p \approx 2.5$ (Stauffer and Aharony, 1994; Balberg, 1987; Sahimi, 1993, 1994; Hunt and Ewing, 2009; Torquato, 2002a; Halperin et al., 1985; Feng et al., 1987; Kovacik, 1998a; Kovacik and Simancik, 1998; Simančik and Kovacik, 2002).

Power law relations yet again emerge from certain geometric models of specific microcellular metals, in which their structure is decomposed into a series of struts interconnected at nodes. Models in which a fixed geometry of connected struts thickens uniformly as V_m increases (as would occur in regular lattices) yield $\mu_p = 1$, that is, a linear relation. An exponent $\mu_p = 1.8$ is obtained for structures that fill the open space between compacted solid spheres (Roberts and Schwartz, 1985), this being the geometry of microcellular metals produced by replication or other space holder processes.

Other predictions of microcellular metal conductivity (as governed by Ohm's or Fourier's law) have been proposed in the literature by analysis of various specific geometries. These are often regular periodic assemblies of simply shaped objects representing nodes and struts in 2D (Lu and Chen, 1999; Gu et al., 2001) or 3D (Dharmasena and Wadley, 2002; Duan et al., 2007; Liu et al., 1999; Schmierer and Razani, 2006a, 2006b; Coquard et al., 2008; Dai et al., 2010; Boomsma and Poulikakos, 2011; 2001; Ashby et al., 2000; Paek et al., 2000; Calmidi and Mahajan, 1999, Bhattacharya et al., 2002; Zhao, 2012; Nakayama, 2013), essentially treated as elements assembled into equivalent thermal or electrical circuits. More complex unit cells of specific geometrical shape have also been proposed (Krishnan et al., 2006, 2008; Ochsner et al., 2006; Fiedler et al., 2008a, 2008b; Solorzano et al., 2009b; Belova et al., 2011), as have unit cells resulting from a calculation of low-density ($V_m \leq 0.1$) Plateau border structures dictated by capillary equilibrium (Phelan et al., 1996; Garcia-Gonzales et al., 1999; Ahern et al., 2005), or representative volume elements (RVEs) collected from actual foam samples by computed X-ray microtomography that were then simulated numerically (Fiedler et al., 2009; Laschet et al., 2009; Veyhl et al., 2011b; Fiedler et al., 2012). All of these predictions are, of course, specific to the structures they assume. Many empirical relations have also been proposed; for powder compacts these are reviewed in Ref. Montes et al. (2012).

So far, we have discussed the electrical conductivity of porous metals: their thermal conductivity can be somewhat more complex, because in heat transfer the pores are not perfect insulators. Indeed, (i) any gas or liquid the pores may contain can also transport heat by conduction, and sometimes also by convection; and (ii) heat can also be transferred across pores by radiation.

While the transport of heat by conduction through a stagnant (i.e. nonconvecting) gas is theoretically possible, the conductivity of gas will be many times lower than that of essentially any metal or alloy. Therefore, this contribution is unlikely to be significant (the situation is different with foams made of insulating solid, such that high-performance insulators are sometimes produced to have evacuated closed pores). Conduction through stagnant gas or liquid contained within the pores is thus generally negligible in porous metals, excepting a few cases of liquid-filled foam having very low relative density (Singh and Kasana, 2004). The contribution of fluid convection within closed pores is also negligible when pores are a few millimetres or less in diameter: the critical Grashof number needed to initiate density-driven convection within single cells is then not reached (Collishaw and Evans, 1994; Clyne et al., 2006; Gibson and Ashby, 1997) p. 287; this has been validated experimentally in polymer foams (Skochdopole, 1961). With open pores, on the other hand, a fluid (gas or liquid) can traverse the material: convective heat transport across cells then becomes a distinct (and technologically interesting) possibility. This is examined in the next section.

Heat transport by radiation through highly porous metals merits consideration because some metallic foams are liable to be used at elevated temperature. The problem of radiative heat transport across an open-pore microcellular metal is complex, given the intrinsically nontrivial physics of radiative heat transport and the complex geometry of "real" pore surfaces (Bird et al., 1960; Poirier and Geiger, 1994; Clyne et al., 2006). Metals are opaque to photons; hence, there is no radiative heat transport through the solid (this is on the other hand is an important consideration in foams of partly transparent material such as polyurethane or glass (Glicksman, 1994; Collishaw and Evans, 1994)). Rather, radiative heat is carried by "bouncing" electromagnetic radiation that is reflected, emitted, and absorbed at the surface of pores within the microcellular metal.

Consider purely radiative heat transport along a thermal gradient that is oriented perpendicularly to a series of identical parallel plates. Each plate is a gray body of equal and constant emissivity ε. The net heat flux from one plate to the next is

$$q = \sigma \frac{(T + \Delta T)^4 - (T)^4}{\frac{2}{\varepsilon} - 1} \approx \frac{4\sigma\varepsilon LT^3}{2 - \varepsilon}[-\mathrm{grad}(T)], \qquad (19)$$

where L is the plate separation, σ is Stefan's constant ($5.67 \times 10^{-8}\,\mathrm{W\,m^{-1}\,K^{-4}}$), $(T + \Delta T)$ is the temperature on the upstream plate, T that on the downstream plate, and $\mathrm{grad}(T)$ is the average temperature gradient if we neglect the temperature drop across each metal plate. In a real porous material the geometry will be different: distance L is to be replaced by the average mean free path of photons from one cell strut or wall surface to the next, and the more complex average solid radiation view factors will result in a multiplicative prefactor different from four; still, the underlying physical situation is the same. In particular, when the thermal gradient is sufficiently shallow for the approximation in the right hand of eqn (19) to be valid, one finds that radiative heat transport through the pores is tantamount to an increase in the apparent thermal conductivity of the porous metal k_f; this is known as the Rosseland approximation (Glicksman, 1994). The effective foam thermal conductivity is then expressed as the sum of a term k_c corresponding to heat transfer by conduction through the pore walls, augmented by k_r, which accounts for heat transport by radiation through the pores

$$k = k_c + k_r \qquad (20)$$

with $k_{f,c}$ equal to the dense solid metal conductivity times a function of V_m, and $k_{f,r}$ proportional to the cube of the local temperature times the average pore size, as in eqn (19). Note that, although this approach of combining heat transfer by different mechanisms into an effective foam thermal conductivity is often used, caution is required with its predictions. Thermal conduction in the pore walls is essentially driven by the temperature gradient alone, whereas the radiative contribution is strongly dependent on the absolute temperature, as outlined above (eqn (19). Thus, changes in the relevant temperature range can have a significant impact on the accuracy of the combined value obtained.

The importance of radiative heat transfer through porous metals thus increases with temperature; less intuitively, it increases with the average pore size; again this is seen in eqn (19)); this is because with increasing pore size, the absorption/reemission of energy becomes less frequent. Working numbers, one finds from eqn (19) that, for parallel plates of somewhat oxidized metal with, say, $\varepsilon \approx 0.5$ (Poirier and Geiger, 1994) held 100 μm apart, $k_{f,r} \approx 7.5 \ 10^{-12} \ T^3$. With $T = 273$ K, this comes to $k_{f,r} = 1.5 \ 10^{-4}$ W $(\text{m K})^{-1}$; at 1273 K, on the other hand, $k_r \approx 1.5 \ 10^{-2}$ W $(\text{m K})^{-1}$. For comparison, commercial metal foam conductivities are in the range of 0.5–20 W/(m K) (Ashby et al., 2000) p.52. At room temperature, it is thus a relatively safe assumption that radiative heat transfer contributes little to the thermal conductivity of porous metals: conduction through the solid dominates. Near room-temperature, therefore, the electrical and thermal conductivities of porous metals obey the same laws, and are linked by the Wiedemann-Franz relation. At elevated temperature, on the other hand, depending on temperature, pore size, pore shape, relative density and intrinsic properties of the metal (including its conductivity and emissivity), there may be a noticeable contribution of radiative heat transport, which increases roughly with the third power of (absolute) temperature. Radiative heat transfer through porous solids is additionally characterized by several constants; these are discussed and modeled in the context of high-temperature microcellular solids in Refs. (Tien and Vafai, 1990; Loretz et al., 2008b; Loretz et al., 2008a; Haussener et al., 2010; Coquard et al., 2009; Coquard et al., 2012; Zhao, 2012).

There is a relatively ample supply of experimental data in the literature on the electrical or thermal conductivity of metal foams. **Table 4** lists sources of data in the research literature; these are supplemented by data from commercial suppliers, much of which is accessible via http://www.metalfoam.net/. The electrical conductivity is easiest to measure and has the advantage of giving data at a single, well-defined, temperature free of the possible influence of radiative transport; electrical conductivity data therefore tend to be the more trustworthy. Data from several sources the literature are plotted in **Figure 29**.

Given the greater practical importance of heat transfer properties, many measurements also exist of the room-temperature thermal conductivity of porous metals. At room temperature, radiative transport of heat is negligible as seen above: as expected, data obey the same laws as the electrical conductivity. At higher temperatures, there are experimental data in the literature that show a contribution from heat transport through the foam by radiation, over and above transport by conduction through the metal (Zhao et al., 2004b, 2004c; Loretz et al., 2008a; Coquard et al., 2009). An analysis of such data is in Ref. Zhao et al. (2008).

Table 4   Sources of data on the conductivity of highly porous metals

Material type	Data on electrical conductivity	Data on thermal conductivity without visible influence of radiative transport
All (reviews)	(Ashby et al., 2000; Simançik and Kovacik, 2002; Goodall et al., 2006d; Cuevas et al., 2009)	(Ashby et al., 2000; Simançik and Kovacik, 2002; Solorzano et al., 2009b; Fiedler et al., 2009; Zhao, 2012; Han et al., 2012; Nakayama, 2013)
Open-pore microcellular metal	(Langlois and Coeuret, 1989; Dharmasena and Wadley, 2002; Huang et al., 2009; Li and Zhu, 2005; Liu et al., 2000b, 1999; Goodall et al., 2006d)	(Paek et al., 2000; Kim et al., 2001; Calmidi and Mahajan, 1999, 2000; Bhattacharya et al., 2002; Coquard et al., 2008; Solorzano et al., 2008b; Thewsey and Zhao, 2008; Sadeghi et al., 2011)
Closed-pore microcellular metal	(Babscan et al., 2003; Feng et al., 2002; Kovacik and Simancik, 1998; Kim et al., 2005a)	(Laschet et al., 2009; Solorzano et al., 2008a, 2008b; Babscan et al., 2003; Reutter et al., 2008a)
Syntactic foam		(Shabde et al., 2006; Fiedler et al., 2008b; Solorzano et al., 2009a)
Metal with 2D unidirectional pores	(Nakajima, 2007)	(Nakajima, 2007; Chiba et al., 2009)

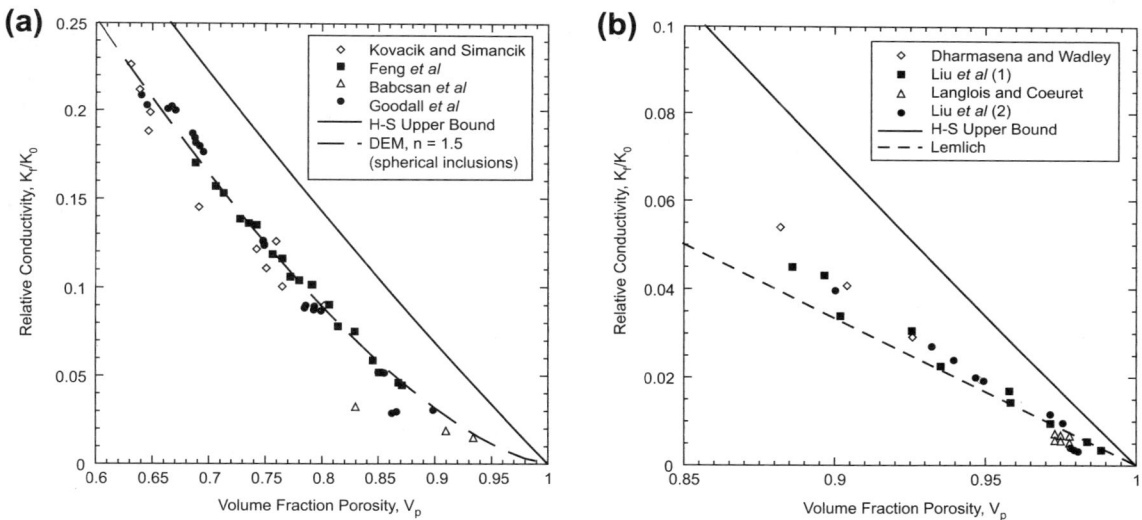

Figure 29 Comparison of experimental data for the electrical conductivity of microcellular metals with mean-field expressions; (a) data for open-cell replicated aluminum foam with spherical pores, Goodall et al. (Goodall et al., 2006d), and various closed-cell foams Kovacik and Simancik (Kovacik and Simancik, 1998), Feng et al. (Feng et al., 2002) and Babcsan et al. (Babscan et al., 2003), and the prediction of the DEM model for spherical pores (DEM); (b) data for foams made using a polyurethane precursor from Dharmasena and Wadley (Dharmasena and Wadley, 2002), Liu et al. (1) (Liu et al., 1999), Liu et al. (2) (Liu et al., 2000b) and Langolis and Coeuret (Langlois and Coeuret, 1989), and the predictions of the Lemlich model (Lemlich). The Hashin–Shtrickman upper bounds are shown in both graphs for comparison (H–S Upper bound).

Data on the electrical or room-temperature thermal conductivity of metal foams can be supplemented by literature data for analogous constants in other systems, such as the electrical conductivity of metals containing nonconductive ceramic inclusions or the electrical conductivity of rocks containing an electrolytic liquid caught between nonconducting sandstone particles. Taken together, this ensemble of data show clearly that, in the cleanest systems (e.g. ones for which there are no small-scale pores or oxide films within the metal making the foam):

(i) at very high porosities, metal foams made by methods where a polyurethane foam precursor is used, which have relatively straight struts linked at small nodes, the conductivity roughly follows the Lemlich rule (eqn (15) with $N = 3$), while

(ii) foams of higher relative density follow the DEM expression (eqn (17)) or, equivalently, the scaling relation in eqn (18), with a V_m exponent μ_p globally in the range 1.5–1.8. Metals containing ceramic particles (Weber et al., 2003a, 2003b), porous concrete (Bouvard et al., 2007), and wet sandstones (Sahimi, 1993, 1994; Hunt and Ewing, 2009) follow exactly the same law (known for rocks as "Archie's law").

Why the DEM predicts with such robustness the conductivity of microcellular metals is not yet clear: their structure of monomodal tight-packed pores are a far cry from the widely distributed pore sizes implicitly assumed in the DEM scheme. Different theoretical explanations have therefore been proposed for highly porous metals. These include percolation-based models cited above (which need relatively arbitrary assumptions to be made to display the proper exponent, e.g. (Wong et al., 1984)), and models that take as their starting point a description of the pore space defined between randomly packed spheres (which yield the same scaling law as eqn (17) or (18), with the proper exponent $\mu_p \approx 1.8$ (Roberts and Schwartz, 1985; Hunt and Ewing, 2009)).

Note that liquid (nonmetallic) foams, which can conduct electricity through the liquid if it is an electrolyte, behave differently: data (Phelan et al., 1996; Weaire and Hutzler, 1999) show that the conductivity is again well predicted by the Lemlich rule (eqn (15) with $N = 3$) at low fractions liquid (something that is to be expected given their straight shape and the fact that films hardly contribute at all to conduction), and then veers at higher fractions liquid toward the Maxwell prediction (eqn (16) with $D = 3$), that is, the optimal upper bound for isotropic 3D solids. A simple and easy-to-remember interpolation through the data proposed by Curtayne and cited in (Weaire and Hutzler, 1999) is $k = (k_m/3)(V_p + V_p^{3/2} + V_p^2)$. This interpolation should a priori apply to structures, including of metal, that are geometrically similar to the interconnected liquid phase found in liquid foams and emulsions. It differs from that derived for structures such as replicated foam because the strut shape is, in liquid foams and emulsions, entirely dictated by capillary forces and therefore differs from the space defined between compacted solid spheres that have densified by, say, CIP or sintering, or architectures produced by bubbling in "real" metal foams.

In closing this section, we note that the same laws and hence the same equations describe the magnetic permeability or the diffusion coefficient of porous metals knowing that of their (homogeneous) base constituent (Torquato, 2002a).

24.4.4   Fluid Flow

24.4.4.1   *Governing Laws and Parameters*

If their pores are open, microcellular metals can "breathe" in the sense that they can be filled with a fluid that may flow into or out of the material. This "breathing" function is one of the main reasons why highly porous microcellular materials exist (particularly in living systems); the same can in fact be said of highly porous metals.

Microcellular metals traversed by a fluid can perform many functions, as filters (Antsiferov et al., 2005; Lee et al., 2006a), in catalysis (Pestryakov et al., 1994, 1995; de Wild and Verhaak, 2000; Antsiferov et al., 2005; Sirijaruphan et al., 2005a, 2005b; Giani et al., 2005b; Ferrouillat et al., 2006), as battery electrodes (Ashby et al., 2000; Banhart, 2002), in chemical reactors (Hutter et al., 2010), in heat exchangers (Lage et al., 1996; Antohe et al., 1996b, 1996a; Lu et al., 1998; Kim et al., 2001, 2000; Rachedi and Chikh, 2001; Girlich and Franzke, 2001; Boomsma et al., 2003b; Hsieh et al., 2004; Phanikumar and Mahajan, 2002, Bhattacharya and Mahajan, 2006, Mahjoob and Vafai, 2008), as scaffolds for tissue or bone regeneration (Dunand, 2004; Shimko et al., 2005; Witte et al., 2007a, 2007b; Arciniegas et al., 2007; Imwinkelried, 2007; Spoerke et al., 2008; Zhuang et al., 2008; Sargeant et al., 2008; Kwok et al., 2008; Bansiddhi et al., 2008; Bansiddhi and Dunand, 2008; Levine, 2008; Asavavisithchai et al., 2010; Murr et al., 2009; Gaytan et al., 2010; Zheng and Gu, 2011; Luyten et al., 2011; Torres et al., 2012; Lefebvre, 2013), in heat pipes (Faghri, 1995)(Nouri-Borujerdi and Layeghi, 2005), as part of a phase-change material (Khateeb et al., 2004, 2005; Hong and Herling, 2006b, 2006a; Lafdi et al., 2007; Siahpush et al., 2008; Yang and Garimella, 2010; Zhou and Zhao, 2011; Zhao and Wu, 2011; Wu and Zhao, 2011; Hasse et al., 2011; Qu et al., 2012; Li et al., 2012c; Zhao, 2012), or in solid oxide fuel cells (Scott and Dunand, 2010), to give a noncomprehensive list. Microcellular metals are, compared to other microcellular materials, particularly interesting in several of the above applications because they are good conductors of heat and electricity, and are resistant to a variety of environments over a relatively wide range of temperatures. The resistance to flow of a fluid through open-pore microcellular metals is thus an important engineering characteristic of these materials.

Resistance to flow of a fluid through a porous medium is measured by the excess pressure ΔP that must be applied on the fluid, where it enters the porous medium, to cause it to flow, at steady state at a given unidirectional average velocity, through a unit length of the porous medium. For velocity and pore size ranges generally encountered in engineering, the law governing such unidirectional fluid flow is the Dupuit–Forchheimer modification of Darcy's law (Bear, 1972; Lage, 1995; Nield and Bejan, 2006, Auriault et al., 2007). For unidirectional flow through a homogeneous isotropic material this is written

$$-\frac{\partial P}{\partial x} = \frac{\mu_f}{K} v_D + \rho_f F v_D^2 \quad (1-D);$$

or

$$-\nabla P = \left(\mu_f K^{-1} + \rho_f F \| v_D \| \right) v_D \quad (3-D),$$

(21)

where P is the local pressure in the fluid, x is distance along the direction of macroscopic fluid flow, and v_D is the "superficial" (or "Darcian" or "seepage") fluid velocity, defined as the number of cubic meters of fluid seeping in one second through one square meter cut through the porous medium perpendicularly to the average direction of flow.

Characteristics of the fluid are its dynamic viscosity μ_f and its density ρ_f. Two parameters characterize the porous material: the Darcian permeability K (units: m^2) and the form coefficient F (units: m^{-1}). With an isotropic solid, or for flow along one of its principal directions if it is anisotropic, these are scalars. Note that not all porous metals have isotropic pore structures: honeycombs are an obvious example; lattice strut structures, some of which have two or three "easy flow" directions (see **Figure 23**), can also show measurably anisotropic through-flow resistance (Wadley, 2006). Then, when the equation is written in full 3D vectorial form, the permeability becomes a tensor and the second, Forchheimer, term of eqn (21) becomes a homogeneous vectorial function of degree 2 in the average fluid velocity (Auriault et al., 2007).

The first (Darcy regime) term on the right-hand side of eqn (21) accounts for viscous friction losses within the fluid; these are often high within a Newtonian fluid that must flow with a no-slip condition along the many solid pore surfaces present. The second (Forchheimer regime) term is taken to correspond to a gradual increase in the contribution of form drag to resistance of flow past the solid phase, this in turn being either linked to an increasing contribution of inertial losses associated with the irregular flow path of microscopic fluid volumes, or to the appearance of pore-scale eddies within the fluid (Lage, 1995; Nield and Bejan, 2006). For a given porous medium, the transition from the linear to the quadratic regime takes place around a certain value of a Reynold's number Re

$$Re = \frac{\rho_f \, v_D L}{\mu_f}, \tag{22}$$

the definition of which is more ambiguous than with a straight pipe, given the often complex shape of the pores and the ensuing lack of a clear definition for the relevant characteristic length L. Various definitions have thus been proposed for L: a mean pore size, a mean "solid element" size, or some other measurable internal characteristic length of the microcellular metal. Alternatively, flow-related characteristics of the porous medium have been used to define L, notably $L = \sqrt{K}$ (giving the "permeability-based Reynolds number") or $L = KF$ (Lage, 1995; Nield and Bejan, 2006). The transition Re value, at which flow changes from a dominance of viscous friction (first, linear, Darcy term dominant) to one of form loss (second, quadratic, Forchheimer term dominant), depends to some extent on the choice made in defining L, and also on the nature of the porous medium. In practice, the critical Re_c values vary with the medium and the author by as much as three orders of magnitude, from roughly from 0.1 to 100, while being most often situated between 1 and 10 if the definition of L is based on K (Bear, 1972; Scheidegger, 1974; Lage, 1995; Lage et al., 1997; Nield and Bejan, 2006)). At Re roughly one hundred to one thousand times higher than Re_c, F has been found to become velocity dependent within a transition region (an effect at times expressed by a cubic term) and then to settle at a different value characteristic of very high flow rates, at which strong turbulence is taken to dominate. Underlying fundamentals are discussed in Refs. (Bear, 1972; Bear and Bachmat, 1990; Lage, 1995; Lage et al., 1997; Hsu, 2005; Nield and Bejan, 2006). Note that an additional $v_D^{3/2}$ term has at times been added to eqn (21) and that it is strictly only valid for steady-state flow (Hsu, 2005).

When viscous friction dominates (Re < Re_c), eqn (1) simplifies to Darcy's law

$$v_D = -\frac{K}{\mu_f} \frac{\partial P}{\partial x} \tag{23}$$

and flow becomes linear. The Darcian permeability K to a Newtonian fluid is a basic linear physical property of the porous metal. It comes from a solution of the Stokes and continuity equations in the fluid, with an imposed average flow velocity and the condition that, everywhere along the fluid/metal interface, the fluid velocity be zero (Torquato, 2002a). In three dimensions, K is as mentioned a scalar if the porous metal is isotropic; when it is anisotropic K is a symmetric positive definite second-rank tensor (Bear, 1972; Torquato, 2002a; Nield and Bejan, 2006). Dimensional analysis dictates that K be proportional, all else constant, to the square of the microscopic scale of the porous medium L, whatever its definition (average pore radius, average diameter of solid elements making the porous metal,...).

Being a basic linear property of importance in many other contexts (the processing of composite materials, soil science, reservoir engineering, hydrogeology, chemical engineering, tissue engineering, …), there is an extensive literature on the permeability of porous materials. There are several approaches toward predicting K; these can be separated into a few broad classes.

The first is made of "tube bundle" models. These assimilate pores in the material to a swarm of equal straight cylindrical tubes. The resulting permeability is then calculated using the Hagen–Poiseuille equation, according to which the average velocity of a fluid flowing at steady state in the laminar regime through a straight cylindrical pipe of radius R equals the pressure gradient times $R^2/(8\mu)$ (Bear, 1972; Bird et al., 1960; Poirier and Geiger, 1994). The resulting prediction for K takes the form

$$\frac{K}{D_m{}^2} = \frac{(1 - V_m)^3}{\alpha(V_m)^2},$$

(24)

where D_m is the (single-valued) diameter of spherical solid elements making the porous medium. If one assumes in the derivation that the specific surface of the tubes is the same as that of the porous medium, an upper bound estimate is obviously obtained (since an absence of constrictions or detours along the path of the liquid will obviously ease its flow against viscous friction). For packed beds of similar particles having a uniform diameter D_m, $\alpha = 150$ reproduces data relatively well (this is twice the value 72 obtained by assuming that the tubes have the same specific surface as the spherical particles) (Bird et al., 1960; Poirier and Geiger, 1994). The resulting prediction is known as the Blake–Kozeny equation and is often used to calculate the permeability of granular porous media (Bird et al., 1960; Poirier and Geiger, 1994). Adjustments exist that adapt α to other solid phase morphologies (Nield and Bejan, 2006); for example, one can take $\alpha \approx 28$ for flow parallel to packed equal-sized cylindrical fibers (Michaud, 2011). More exotic tube bundle models have also been proposed, in which constrictions or bends are, for example, inserted within the tubes (Bear, 1972; Scheidegger, 1974); however, new parameters thus introduced make the approach useless from a practical standpoint. Although it is (too) often used, the tube bundle approach suffers a fundamental weakness, namely that pores in most porous materials are just not straight tubes. This causes strong discrepancies between the model and the physics of flow through porous media; for example, the origin of deviations from linearity is completely missed (which in steady flow within pipes correspond to a single transition from laminar to turbulent flow, while in porous media it does not).

Another class of models toward the prediction of K, and of the resistance to through flow of porous media more generally (F can often be calculated as well), is based on known expressions for the drag of a flowing fluid against a solid body of known shape (Bear, 1972; Scheidegger, 1974; Happel and Brenner, 1973; Torquato, 2002a). Force equilibrium across the solid/fluid system then gives the pressure gradient in the fluid from the force exerted on each solid element making the porous metal, knowing their number density per unit volume of porous medium. The main weakness of the approach is that it implicitly assumes an infinitely dilute solid, since it neglects perturbations imparted by each solid element (particle, fiber, foam strut, ….) on the pattern of flow around its neighbors (also problematic is the fact that a simple solution seemingly does not exist for one of the shapes most pertinent to microcellular metals, namely the cylinder (Torquato, 2002a)). In this sense, drag force models are, for permeability, the equivalent of dilute models used to predict the conductivity of slightly porous solids (see Section 24.4.3). Similarly to conductivity, mean-field approximations exist to account for interaction between flow field disturbances around neighboring solid elements in highly porous materials, by which finite volume effects can be accounted for in slightly more dense solids (V_m on the order of 10%) (Sangani and Yao, 1988; Sahimi, 1993). Again similarly to other linear composite properties such as conductivity, bounds can also be derived for the permeability of porous media (Torquato, 2002a).

A third approach is to assume a unit cell. This can then be used to simulate realistically the internal pore structure of specific porous media, and then to calculate their resistance to through flow of a fluid, in the linear or quadratic regimes. Such models start by defining a representative elementary volume (REV) for the porous material, and calculate the pressure drop for flow of a fluid through the pore space of the REV, generally assuming periodic boundary conditions. Many models have been proposed on this basis for a diversity of porous media, using either simplified analytical approaches or numerical methods to predict pressure gradients within the flowing fluid. Reviews pertinent to fibrous media are in Refs. Jackson and James (1986), Michaud (2011), while Ref. Mahjoob and Vafai (2008) gives a review focused on metal foam.

An interesting opportunity offered by numerical models is the search for optimal structures, for example, structures giving maximum permeability in all three directions at fixed V_m (Jung and Torquato, 2005) or "best possible" combinations of permeability and another property, such as bulk elastic modulus (Guest and Prévost, 2006).

Two other approaches can be used to predict K in porous media having comparatively low volume fractions of interconnected porosity, that is, at higher V_m. One is based on lubrication theory, the other on percolation theory.

Lubrication theory models are in a sense the high-V_m counterpoint to low V_m drag force models. Here, the starting point is the fact that the physics of laminar fluid flow (specifically the no-slip condition along the pore surface) are such that the narrowest constrictions exert a dominant influence on the rate at which a fluid can flow through the pores. In simple terms this is called the "bottleneck effect": we all know that the rate at which a bottle empties is entirely determined by the width of its neck, pretty much regardless of the dimensions or shape of the rest of the bottle. Hence, if "bottleneck" constrictions can be realistically described in terms of size, shape and density, it suffices to calculate the rate of flow through the bottlenecks using known expressions for creeping flow of fluids past solids, of which there are many (Happel and Brenner, 1973). Assuming negligible losses elsewhere within the fluid, one then obtains a realistic estimate of the porous medium permeability if its bottleneck(s) are well described. Where applicable, such predictions have proven both simple and reliable when compared against experimental data: for flow perpendicular to parallel cylinders notably (Happel, 1959; Sangani and Acrivos, 1982; Sangani and Yao, 1988; Jackson and James, 1986; Michaud, 2011) and for flow through isotropic replicated foam (Despois and Mortensen, 2005).

A second approach that has been proposed for porous media of high V_m is to view the pore network geometry as a disordered percolation-dominated network just past the percolation threshold; this yields a power law dependence of the permeability on porosity $V_p = (1 - V_m)$

$$K \alpha V_p^e \tag{25}$$

with a "universal" exponent e equal to 2 if the pore constrictions are of constant diameter, and different from two if such is not the case (*i.e.*, in pore networks determined by "continuum percolation" processes) (Sahimi, 1993, Hunt and Ewing, 2009, Sahimi, 2011, Thompson et al., 1987). This approach is most interesting for non-saturated flow, a subject not addressed here.

We note in closing that, in some cases, cross-property relations from continuum theory can be used to estimate K knowing other linear physical properties of the porous solid (Sahimi, 1993; Torquato, 2002a).

24.4.4.2 *Flow Through Porous Metals*

There is a relatively extensive literature on the resistance to fluid through flow of open-pore metals. In part this is because of their many potential applications in situations involving fluid through flow (see above); however, a second, more trivial reason is that highly porous metals are much more convenient to study than many other highly porous materials, such as polymer foams or living tissue: being much softer, the latter deform when traversed by a flowing fluid (*e.g.*, (Beavers and Wilson, 1975; Beavers et al., 1981b; Beavers et al., 1981a; Sommer and Mortensen, 1996)). Also, unlike natural media (such as tissue or plant-based porous media), metal foams do not interact chemically or physically with water, and hence, for example, do not swell with time, again making porous metals convenient for experimentation.

There are thus many reports giving data for the resistance to through flow of porous metals. These are conducted using either gas (air) or liquid (most often water) as the fluid, generally around room temperature. In principle, such measurements are comparatively simple: a fluid is made to flow across a straight fluid-filled sample of porous metal, and the pressure differential across the sample, ΔP, together with the volume or mass of fluid that seeps per second through the material, are alternatively either imposed or measured (depending on the setup). Standards furthermore exist for the measurement (Innocentini et al., 2010). Other than the obvious requirement that the sample to pore size ratio be sufficiently high (greater than roughly 20), the main (and nontrivial) experimental difficulty is in avoiding entrance effects and wall effects, that is, in ensuring that the measured average flow rate is representative of the material (Baril et al., 2008; Dukhan and Patel, 2011; Dukhan and Ali, 2012b, 2012a; Dukhan, 2013b). Wall effects can appear either because the walls alter the shape of pores in their vicinity (as is well known for packed particle beds) (Poirier and Geiger, 1994), or because the walls alter the flow path of the fluid in their vicinity, or simply because there is a gap between the sample and the container wall. Wall effects can artificially increase or decrease the rate of flow; to be avoided, these require (i) tightly fitting container walls (compliant elastic or shrinkable tubes are convenient in this regard) and (ii) that the samples show at least 20 pores across their diameter (Poirier and Geiger, 1994) and that this diameter of course be entirely sampled by the liquid in straight-line unidirectional flow (Innocentini et al., 2010). Entrance and exit effects are avoided by testing a sufficiently long sample, and at times by ducting the fluid immediately before and after the test section.

When using gas as the fluid, an added difficulty is that the fluid properties (viscosity, density) are pressure dependent. Density variations (which are necessarily present, eqn (21)) then enter the continuity equation, such that v_D is no longer constant along the bed (Scheidegger, 1974; Bird et al., 1960; Bonnet et al., 2008). These complications are often dealt with by taking the gas pressure as constant thoughout the porous medium, and equal to the average of inlet and outlet pressures; however, this approximation is only valid if the pressure drop ΔP is small compared with the average pressure P. When using a liquid, a required precaution is to ensure complete pore filling, that is, a complete absence of remanent bubbles within the sample, and if flow is not horizontal to take into account the role of gravity (by adding $g\rho_f$ times the altitude to the local pressure P).

Data are most conveniently reported by recasting eqn (21) in the form

$$\frac{\Delta P}{v_D L} = \frac{\mu}{K} + \rho_f F v_D, \tag{26}$$

where L is the sample length along the direction of flow and ΔP the pressure differential. Plotting the left-hand term as a function of v_D will then give a two-part curve if both flow regimes are covered: a horizontal line at low v_D where Darcy's law holds, eqn (23), from which the permeability K can be deduced knowing μ. A linear ascending portion will obtain at higher v_D, the slope of which gives F, with the transition between the two regimes visible and thus giving the transition Reynolds number Re_c. Note that the value for K that is deduced by extrapolating, to zero v_D, data gathered only in the (inertial)

Forchheimer regime will in general not give the same K as the Darcian permeability that obtains in the low Re, viscosity-dominated flow regime (Dukhan and Minjeur, 2011; Dukhan, 2013b).

Reports of measured porous metal through-flow characteristics in the literature are listed in **Table 5**; reviews of data or correlations in the literature (giving additional data from harder to find references) can be found in Refs. Dukhan (2006), Despois and Mortensen (2005), Bonnet et al. (2008), Mahjoob and Vafai (2008), Gerbaux et al. (2009), Dukhan (2013b). A few studies exist addressing multiphase flow through porous metals (Topin et al., 2006; Gerbaux et al., 2009). There also exists an extensive literature on through-flow characteristics of ceramic foams, which is relevant since the structure of microcellular ceramics often resembles that of microcellular metals (e.g. when these are produced by coating or replicating a polyurethane foam) and since there is no difference between (inert) porous metals or ceramics with regard to fluid through-flow. References (Richardson et al., 2000; Moreira et al., 2004; Lacroix et al., 2007; Edouard et al., 2008; Dimopoulos Eggenschwiler et al., 2009) are examples.

Measured values of K span many orders of magnitude, with reported values ranging very roughly from 10^{-12} to 10^{-6} m^2. There are several reasons for this wide range of variation of the data. The first is the size dependence of both K, which varies as the square of the average pore diameter, and F, which roughly varies linearly with the inverse of the pore diameter. The second is its strong dependence on relative density V_m, or in other words on the pore volume fraction V_p; this is immediately visible on examination of the Blake–Kozeny equation (eqn (24)). Values for F vary somewhat less; however, data span more than one order of magnitude. From the wealth of data, no single empirical correlation emerges to provide a reasonable prediction of K or F knowing only the two main parameters, namely the average pore diameter and the relative density: the resistance to through flow of highly porous metals is a strong function of the pore architecture. Appropriate expressions for prediction of K or F thus depend strongly on how the porous metal was made, or in other words on the porous metal mesostructure.

Most expressions that have been proposed are therefore specific to a specific class of porous metal. For flow through packed beds of metal powder the Ergun equation is frequently used; here K is given by the Blake–Kozeny expression (eqn (24) with $\alpha = 150$), while F is (Bird et al., 1960; Poirier and Geiger, 1994)

$$F = \frac{1.75V_m}{D_m(1 - V_m)^3}. \tag{27}$$

This expression has been adapted for porous metal structures other than packed particles; to this end the two numerical constants (150 and 1.75) are adapted to fit the data (Langlois and Coeuret, 1989; Alazmi and Vafai, 2000; Tadrist et al., 2004; Dukhan and Patel, 2008; Smorygo et al., 2011; Dukhan, 2013b) or adjusted on the basis of a geometrical model (Huu et al., 2009), while D_m is written by adaptation of a dimensional characteristic of the porous metal (such as the number of pores per inch, commonly reported by manufacturers).

The most extensively characterized microcellular metals in the context of fluid flow are those produced, by casting or coating, from a polyurethane foam precursor (see Section 24.2.3). For those foams, a comparatively extensive databank exists for both K and F (**Table 5**). On this basis, several models have been proposed toward predicting these quantities knowing the foam relative density V_m and a measure of its microstructural scale, such as the average pore diameter, the average metal ligament length, or the number of pores per inch (PPI; typical values range from 5 to 40 PPI). Models of this class were proposed by Diedericks and DuPlessis (1997), DuPlessis et al. (1994), Smit and du Plessis (1999), Fourie and DuPlessis (2002), Smit et al. (2005); Mahajan and collaborators (Bhattacharya et al., 2002), and several other authors (Huu et al., 2009; Dukhan,

Table 5   Literature references reporting experimental measurements of Darcian permeability K (units: m^2) and form coefficient F (units: m^{-1}) for porous metal samples

Porous metallic material	Fluid	Measurement of K	Measurement of F
Cast aluminum from polyurethane template	Air	(Antohe et al., 1996a, 1996b; Lage et al., 1997; Azzi et al., 2007; Leong and Jin, 2006a; Kim et al., 2001, 2000; Paek et al., 2000; Hwang et al., 2002; Bhattacharya and Mahajan, 2002; Bhattacharya et al., 2002; Calmidi and Mahajan, 2000, Liu et al., 2006; Dukhan et al., 2006a; Dukhan, 2006, 2013b; Dukhan and Ali, 2012b; 2012a; Garrity et al., 2010; Mancin et al., 2010b, 2011, 2012b; Kamath et al., 2011)	(Antohe et al., 1996a, 1996b; Lage et al., 1997; Leong and Jin, 2006a; Kim et al., 2001, 2000; Paek et al., 2000; Hwang et al., 2002; Bhattacharya and Mahajan, 2002, Bhattacharya et al., 2002; Calmidi and Mahajan, 2000, Liu et al., 2006; Dukhan et al., 2006a; Dukhan, 2006, 2013b; Dukhan and Patel, 2008, 2011; Dukhan and Ali, 2012b; 2012a; Garrity et al., 2010; Mancin et al., 2010b, 2011, 2012b; Kamath et al., 2011)
(ibidem)	Water	(Hunt and Tien, 1988; Boomsma and Poulikakos, 2002; Boomsma et al., 2003a; Tadrist et al., 2004; Noh et al., 2006; Hutter et al., 2011a, 2011b)	(Hunt and Tien, 1988; Boomsma and Poulikakos, 2002; Boomsma et al., 2003a; Hetsroni et al., 2005; Tadrist et al., 2004; Noh et al., 2006; Hutter et al., 2011a, 2011b)
(ibidem)	Other liquid	(Antohe et al., 1996a; Boomsma et al., 2003a)	(Antohe et al., 1996a; Boomsma et al., 2003a)
Nickel or nickel-based alloys coated onto polyurethane foam	Air or exhaust gas (Lee et al., 2006a)	(Albanakis et al., 2009; Topin et al., 2006; Bonnet et al., 2008; Lee et al., 2006a; Medraj et al., 2007; Smorygo et al., 2011; Gerbaux et al., 2009)	(Albanakis et al., 2009; Topin et al., 2006; Bonnet et al., 2008; Lee et al., 2006a; Medraj et al., 2007; Gerbaux et al., 2009)
(ibidem)	Water or water-based solutions	(Beavers and Sparrow, 1969; Vafai and Tien, 1982; Hunt and Tien, 1988; Montillet et al., 1992; DuPlessis et al., 1994; Topin et al., 2006; Bonnet et al., 2008; Langlois and Coeuret, 1989; Innocentini et al., 2010; Miwa and Revankar, 2009, 2011a)	(Beavers and Sparrow, 1969; Vafai and Tien, 1982; Hunt and Tien, 1988; Montillet et al., 1992; DuPlessis et al., 1994; Topin et al., 2006; Bonnet et al., 2008; Langlois and Coeuret, 1989; Innocentini et al., 2010)
(ibidem)	Non-Newtonian liquid	Sabiri cited in (Smit and du Plessis, 1999)	Sabiri cited in (Smit and du Plessis, 1999)
FeCrAlY coated onto polyurethane foam	Air	(Giani et al., 2005b)	(Giani et al., 2005b)

(Continued)

Table 5 Literature references reporting experimental measurements of Darcian permeability K (units: m^2) and form coefficient F (units: m^{-1}) for porous metal samples—cont'd

Porous metallic material	Fluid	Measurement of K	Measurement of F
Tantalum coated onto polyurethane foam	Water	(Shimko et al., 2005)	
Inconel 625 coating nickel coated onto polyurethane foam	Air		(Jazi et al., 2009)
Copper open-pore foam	Air	(Topin et al., 2006; Mancin et al., 2012a)	(Topin et al., 2006; Zhao et al., 2004a; Mancin et al., 2012a)
(ibidem)	Water	(Topin et al., 2006)	(Topin et al., 2006; Zhang et al., 2007a)
Reaction-sintered iron	Air	(Laschet et al., 2008, 2009)	(Laschet et al., 2009)
Slip reaction foam sintered HastelloyB, Inconel 625 and Fe-based (NC) alloys	Air	(Reutter et al., 2008b, 2008a)	(Reutter et al., 2008b, 2008a)
Porous stainless steel	Gas (air or N$_2$ with CO and CO$_2$)	(Incera Garrido et al., 2008)	(Incera Garrido et al., 2008)
Replicated aluminum (cast)	Water	(Despois and Mortensen, 2005; Despois et al., 2007; Goodall et al., 2007)	
Replicated aluminum (powder based)	Air		(Banhart, 2001)
Replicated aluminum (powder based)	Water	(Hakamada et al., 2006)	
Replicated titanium (powder based)	Water	(Imwinkelried, 2007)	
Tetrakaidecahedra produced by selective e-beam melting	Air	(Inayat et al., 2011)	(Inayat et al., 2011)
Regular truss or frame structures	Water	(Kim et al., 2006; Xu et al., 2007; Hutter et al., 2011a, 2011b)	(Kim et al., 2006; Xu et al., 2007; Hutter et al., 2011a, 2011b)
Wire structures	Air	(Tian et al., 2004, 2007)	(Tian et al., 2004, 2007)
(ibidem)	Water	(Tian et al., 2007)	(Tian et al., 2007)

2013b). To give an example, the correlation proposed by Fourie and DuPlessis (2002) (eqns (35), A.7 and last line of Section 24.2.1) for foams of this class is

$$K = \frac{(1-V_m)^2 (0.57D_p)^2}{36(\chi-1)\chi},$$

(28)

where χ is an effective tortuosity obeying $(1 - V_m) = \chi\left(\frac{3-\chi}{2}\right)^2$.

In replicated microcellular metals the pore space is defined by the architecture of packed powder compacts. For these, a lubrication theory analysis leads to the following relation, shown to predict accurately data for this class of microcellular metal (Despois and Mortensen, 2005; Goodall et al., 2007)

$$K = \frac{(1 - V_m)D_p^2}{4\pi}\left(\frac{V_o - V_m}{3V_o}\right)^{3/2},$$

(29)

where D_p is the average pore radius, equal to the average radius of (leachable) space holder particles used in making the microcellular metal, with V_o the volume fraction porosity in the initial loose-packed compact of leachable particles before its densification $(1 - V_o = 0.64$ with random dense packed spherical particles).

In the nonlinear Forchheimer flow regime, it is often found that F is adequately predicted as

$$F = \frac{c_I}{\sqrt{K}},$$

(30)

where c_I is an inertia coefficient that varies somewhat with the porous metal, but is situated between roughly 0.03 and 0.11 by confrontation with experimental data for flow through investment cast microcellular aluminum (Beavers and Sparrow, 1969; Vafai and Tien, 1982; Lage, 1995; Nield and Bejan, 2006; Paek et al., 2000; Bhattacharya et al., 2002; Noh et al., 2006). Data exist, however, in which c_I was found to vary with the Reynolds number (Fourie and DuPlessis, 2002). In Ref. Ashby et al. (2000) (eqn (13.13), c_I is taken to vary strongly, being proportional to the Reynolds number (defined on the basis of metal ligament diameter) raised to the power (−0.4) with the Darcy term absent; the equation must thus most likely be viewed as a single term expression aiming to cover both regimes.

Numerical models have, finally, also been developed toward simulation of fluid flow through porous metals, using architectures that are either constructed by simplification of the architecture of highly porous metals (Boomsma et al., 2003a; Lu et al., 2005; Kim et al., 2005b; Krishnan et al., 2006, 2008; Bai and Chung, 2011), or starting from tomography files of actual metal foam samples (Laschet et al., 2008, 2009; Kopanidis et al., 2010; Brun et al., 2009; Hétu et al., 2013) (such simulations incidentally often show nicely the "bottleneck" effect across pores). Results are of course specific to geometries assumed in each case.

24.4.4.3 *Wall Effects*

When a fluid-filled porous metal is used it will nearly always adjoin a dense solid container or substrate. Along the interface with this dense solid the pattern and laws of fluid flow through the porous metal may be altered, for several reasons.

A first reason was already mentioned in the previous section: the pore shape might be disrupted near solid walls, be it because solid packing is modified, or more mundanely because there is a small gap between the dense and the porous solids, or yet again due to effects of gluing, brazing or machining. Such effects are generally hard to describe quantitatively. Often, the affected region will be restricted to a layer only a few pore diameters wide, making it easier to ignore (there are exceptions, however, one example being with metal particles packed against a wall, where the wall can disturb particle packing in a zone many particle diameters wide).

A second reason why dense solid surfaces may affect flow in porous metals has to do with boundary layer effects. The no-slip condition dictates that, along a dense solid surface, the fluid velocity be that of the solid, that is, zero if the solid surface is stationary. The Darcy–Forchheimer equation (eqn (20)) does not allow for this—as should be, since it is an averaged equation for flow through a maze of many no-slip solid surfaces. Two extreme situations may be envisaged.

The first extreme situation is when the porous metal is relatively dense: a perfect interface between such a material and a dense wall affects the pore wall geometry only along a layer roughly one or two average pore diameters thick. This should not perturb flow significantly, so that the above equations can be used. The solid wall is then simply a boundary across which flow is not permitted, but along which tangential flow occurs with at most a small variation in K and/or F, the no-slip condition being respected as it is embedded in the law itself.

The second extreme is that where the porous medium is highly permeable: a solid wall may then cause a wide-ranging perturbation in the pattern of flow within the porous medium. As it happens, low-density microcellular metals of the type derived from polyurethane foams (Section 24.2.3) and often used in experimentation, represent a textbook example of this. These metal foams are at the same time very ($\geq 90\%$) porous and sufficiently rigid not to collapse in the presence of a flowing fluid—microcellular metals have thus been called "hyperporous" materials (Nield and Bejan, 2006). These materials have as a consequence been used as a standard test-case for a good many studies of flow and transport through porous media (as detailed above and below).

For such "hyperporous" metals, fluid flow has been described using a hybrid of the Navier–Stokes equation for laminar flow of a pure "clear" fluid, and the Darcy–Forchheimer equation. Many different variants of this equation, which in its simpler forms is called the "Brinkman equation," have been proposed; a general expression derived by averaging flow of a Newtonian fluid through a limited number of solid obstacles is (Nield and Bejan, 2006, Hsu, 2005)

$$\rho_f \left[\frac{\partial v_D}{\partial t} + \nabla \left(\frac{v_D \cdot v_D}{V_p} \right) \right] = -\nabla(V_p P) + \mu_{eff} \nabla^2 v_D - \frac{V_p \mu_f}{K} v_D - V_p \rho_f F \|v_D\| v_D, \qquad (31)$$

where $V_p = 1 - V_m$ is the pore volume fraction, μ_{eff} is an "effective" fluid dynamic viscosity generally taken to be the same as the fluid viscosity μ_f and P is the pressure in the fluid (Hsu and Cheng, 1990; Nield and Bejan, 2006). This complete combination of the Navier–Stokes and Darcy–Forchheimer equations is generally used in simplified form, by removing the inertial term (second term on the left-hand side), by neglecting one or the other of the two terms in the Darcy–Forchheimer equation (last two terms on the right-hand side), and/or by taking $V_p \approx 1$ and writing the equation in the form appropriate for 2D boundary layer flow. With these simplifications, for fully developed flow through a porous metal along a plane solid surface oriented normal to the y direction, the following equation results

$$G_P = \mu_f \frac{\partial^2 v_D}{\partial y^2} - \frac{\mu_f}{K} v_D - \rho_f C \|v_D\| v_D, \qquad (32)$$

where G_P is the pressure gradient along the direction of flow (x direction, parallel to v_D) (Vafai and Tien, 1982; Kaviany, 1985; Hunt and Tien, 1988; Tien and Hunt, 1987; Haji-Sheikh, 2004; Haji-Sheikh and Vafai, 2004; Nield and Bejan, 2006). This version of the Brinkman equation is often used to model fluid flow in porous metals along solid surfaces across which heat is exchanged between solid and fluid, a situation of interest in many possible applications of porous metals. The respective importance of the two (Navier–Stokes vs. Darcy–Forchheimer) terms is a governed by the dimensionless Darcy number

$$\mathrm{Da} = \frac{K}{L^2},\tag{33}$$

where L is a characteristic dimension of the system at hand (e.g. the length or width of the porous metal sample). At high Da, meaning when the porous medium is highly permeable, flow is dominated by boundary layer effects created by the solid wall. At low Da, flow is dominated by the porous medium (first term of the right-hand side of eqn (32) omitted; it then reduces to eqn (21) or (26)) and described with sufficient precision by neglecting the presence of the wall, other than by the zero-flux boundary condition it creates. In other words, within the porous medium, boundary layer effects will affect flow only within a layer of thickness on the order of $\sqrt{K}$, which will generally be rather small for usual values of K yet can, for the upper end of the range measured (for coarse low-density aluminum foam), reach a millimeter. With high permeabilities, also, turbulence (otherwise damped in fine-scale pores) may play a role; Antohe and Lage offer a treatment of this case (Antohe and Lage, 1997).

24.4.5   Convective Transport

24.4.5.1   *Heat and Mass Transport Through Porous Media: Basics*
If we neglect radiative heat transport and assume the solid phase to be stationary, the equation of energy conservation within a fully saturated porous metal writes (Bear, 1972; Hsu, 2005; Nield and Bejan, 2006)

$$(\rho c_p)_f \left(V_p \frac{\partial T_f}{\partial t} + v_D \cdot \nabla T_f \right) = V_p \nabla \cdot (k_f \nabla T_f) + V_p \nabla \cdot (k_D \nabla T_f) + h_{sf}(T_f - T_m) + V_p \dot{Q}_f \tag{34}$$

for the fluid, and

$$(\rho c_p)_m V_m \frac{\partial T_m}{\partial t} = V_m \nabla \cdot (k_{mf} \nabla T_m) - h_{sf}(T_f - T_m) + V_m \dot{Q}_m \tag{35}$$

for the solid phase, where T is the average phase temperature, (ρc_p) is volumetric heat capacity, h_{sf} is the average (volumetric) heat transfer coefficient between solid and fluid phases, $\dot{Q}$ is the volumetric heat generation rate within each phase, k is the thermal conductivity with subscripts f and mf denoting the fluid and metal foam, respectively, $V_p = 1 - V_m$ is the pore (and hence fluid) volume fraction, and k_D is an effective thermal conductivity expressing the role of dispersion within the fluid. Both k_{mf} and k_D can be tensors, the former with its principal directions oriented with the porous metal structure, the latter with a principal direction aligned with the average direction of fluid flow (lateral dispersion differing in general from longitudinal dispersion). In eqn (34), from left to right the terms represent the local temperature change, heat flow by convection, heat flow by conduction, heat flow by dispersion, heat exchange between fluid and solid, and local heat generation (by chemical reaction or viscous dissipation).

Heat exchange between the solid and the liquid has been simulated by several authors, in full 3D models of fluid flow and heat transfer conducted for idealized geometries, considering the problem as one of heat transfer between a Navier–Stokes fluid and discrete solid elements such as spheres or cylinders. Several authors have proposed such models for low-density metallic foams, fibrous metals, honeycomb structure or lattice truss structures; examples are in (Lu et al., 1998; Gu et al., 2001; Hayes et al., 2004; Kim et al.,

2005b; Kamiuto and Yee, 2005; Krishnan et al., 2008; Ghosh, 2008, 2009b; 2009a; deLemos and Saito, 2009; Kopanidis et al., 2010; Clyne et al., 2006; Zhao et al., 2010; Bai and Chung, 2011; Wang et al., 2011; Tamayol and Hooman, 2011). A recent review is offered by Nakayama (2013).

In many practical cases, the microstructural scale of the porous metal is sufficiently small that the time for heat equilibration between the solid and the liquid phases within a small RVE can be neglected in comparison with the time scale of the overall heat transfer process one seeks to quantify. With metals of course this equilibration time is made short by the high metal conductivity. For this reason, one often takes $T_f = T_m = T$. If we furthermore neglect any local heat generation (i.e. assume negligible viscous dissipation and no chemical interaction between solid and liquid, or no phase change within the solid or liquid), then eqns (34) and (35) merge into a single equation

$$(\rho c_p)_c \frac{\partial T}{\partial t} + (\rho c_p)_f v_D \cdot \nabla T = \nabla \cdot (k_c \nabla T) + V_p \nabla \cdot (k_D \nabla T) \tag{36}$$

in which subscript c denotes properties characteristic of the fluid-filled porous metal composite (Bear, 1972; Hsu, 2005; Nield and Bejan, 2006, Hsiao and Advani, 1999). Obviously, as noted above (Section 24.4.1)

$$(\rho c_p)_c = V_p (\rho c_p)_f + V_m (\rho c_p)_m \tag{37}$$

and, given the high thermal conductivity of metals (exception made for the rare possible case where the fluid is a metal), one can without introducing much error take the stagnant metal/fluid composite thermal conductivity, k_c, equal to that of the empty porous metal (Section 24.4.3).

As with the momentum balance equation, the full 3D equation is often simplified for the specific problem at hand. For steady-state fully developed unidirectional flow along the x direction over a flat plate normal to the y direction, one arrives at the heat transfer companion of eqn (32) (Nield and Bejan, 2006, Vafai and Tien, 1982; Hunt and Tien, 1988; Kaviany, 1985; Alazmi and Vafai, 2000)

$$(\rho c_p)_f v_D \cdot \frac{\partial T}{\partial x} = (k_c + V_p k_D) \frac{\partial^2 T}{\partial y^2}. \tag{38}$$

Dispersion effects, summarized in the second term to the right of eqn (34) and the last term of eqns (36) and (38), have their origin in the complexity and diversity of individual "fluid particle" paths within the fluid as it flows through the porous metal (Bear, 1972; Bear and Bachmat, 1990; Hsu and Cheng, 1990; Sahimi, 1993; Liu and Masliyah, 2005; Nield and Bejan, 2006). Mathematically, dispersion emerges during the pore-level spatial averaging procedure that is used to arrive at these equations, from the fact that the average of a product is not the product of averages. There is, as a result, an added term linked with the average value of the product of fluctuations; this is seen if one writes local temperature and fluid particle velocities as the sum of their average and a fluctuating term

$$v = \bar{v} + v', \quad T = \bar{T} + T', \quad \int_V v \nabla T dV - v_D \cdot \nabla T_f = \int_V v' \nabla T' dV \equiv -V_p \nabla \cdot (k_D \nabla T). \tag{39}$$

Physically, dispersion takes a variety of forms, two of which are illustrated in **Figure 30**, showing

(i) how flow separation can carry two neighboring elements along widely different paths, causing in turn a new mechanism of advective energy transport that goes beyond what the pore-wide average flow velocity would suggest, and

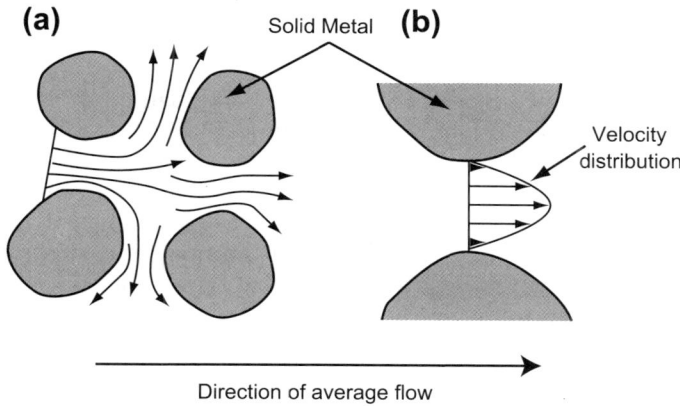

Figure 30 Physical origins of dispersion: (a) dispersion caused by bifurcations in flow across pores; (b) dispersion caused by differences in fluid velocity within a pore.

(ii) how the no-slip condition at the solid/fluid interface causes strong gradients in flow rate, with the consequence that two neighboring fluid particles again will often flow along widely different flow paths (Bear, 1972; Bear and Bachmat, 1990).

Seen from the standpoint of an individual flowing fluid particle, these phenomena imply a constant change in its neighboring particles, with which heat can be rapidly exchanged given the proximity. The result is an added mechanism for heat flow which is a direct result of flow path complexity within the porous medium intermeshed with accelerated heat exchange by conduction. A recent publication presents a superb visualization of dispersion mechanisms in a supersized replica of microcellular metal, showing also the importance of dispersion-driven transport in such structures (Onstad et al., 2011).

Translating this into an effective additional conductivity term k_D it is then evident that

(i) k_D will be a strong function of the average flow rate v_D,
(ii) k_D is likely to depend on thermal properties of the fluid, and
(iii) k_D is a priori also dependent on pore size and pore volume fraction, something that is indeed apparent in experimental data (Bear, 1972; Bear and Bachmat, 1990; Hsu and Cheng, 1990; Liu and Masliyah, 2005; Nield and Bejan, 2006).

Translating dispersion into an effective additional conductivity term k_D is, however, not always appropriate (Sahimi, 1993). Also, k_D is not necessarily proportional to v_D: this is mainly observed at high flow rates (roughly Re above 10) (Hsu and Cheng, 1990; Liu and Masliyah, 2005); several models predict that k_D be proportional to v_D squared (Hsu and Cheng, 1990; Hsiao and Advani, 1999). It is evident that the rate of dispersive heat transfer will differ between the average direction of flow and directions transverse to that direction. One thus distinguishes between longitudinal and transverse dispersion, the former being higher than the latter (a factof of 20 is noted for packed beds); k_D is hence anisotropic (Nield and Bejan, 2006, Nakayama, 2013).

Dispersion tends to be increasingly important as

(i) the fluid thermal thermal diffusivity increases (making it in practice important in the context of heat flow for flowing water and unimportant for flowing air) (Nakayama, 2013) and
(ii) the flow rate, or more appropriately as the Peclet number

$$\mathrm{Pe} = \frac{(\rho c_p)_f \nu_D L}{k_c} \tag{40}$$

increases, where L is a characteristic length of the system at hand (Sahimi, 1993; Hsu and Cheng, 1990; Nield and Bejan, 2006). The subject is highly complex, however, such that there are unfortunately no established generally valid equations for the prediction of k_D. At high Re, k_D is often taken to vary as α $(\rho c_p)_f$ l ν_D where l is a characteristic microscopic distance (e.g. the pore diameter of $\sqrt{K}$), and α a constant very roughly of order of magnitude 0.1 (see below).

With a few usual assumptions (of no cross-flow effects and no solid/fluid interaction or chemical reaction) the equation for mass transport is the same as that for the transport of heat (Sahimi, 1993):

$$\frac{\partial C}{\partial t} + \nu_D \cdot \nabla C = \nabla \cdot (D_{f,p} \nabla C) + V_p \nabla \cdot (D_D \nabla C) \tag{41}$$

with C the concentration (in moles or mass per unit volume) of the species in solution within the liquid, $D_{f,p}$ the diffusion coefficient through the stagnant liquid within the pores of the porous metal, and D_D a dispersion diffusion coefficient, also anisotropic with different values along and transverse to the direction of fluid flow. The greatest difference with heat flow is that intrinsic diffusion coefficients $D_{f,p}$ are in general so low that purely diffusive transport can be ignored altogether in comparison with convective or dispersive solute transport. Similar explanations, restrictions and complexity as for dispersive heat transport exist for dispersive solute transport, this being a question that has been extensively investigated given its considerable importance, in hydrogeology for example (Bear, 1972; Bear and Bachmat, 1990; Sahimi, 1993).

24.4.5.2 *Heat Transport through Filled Porous Metals*

The technological potential of microcellular metals in heat transfer applications, coupled with the convenience of using "hyperporous" microcellular metals in experimental research on convective heat transfer through porous media, have resulted in an extensive and rapidly growing literature on their use in the context of flow-driven transport of heat.

Many different configurations have been examined, in both analysis and experimentation. Typically these comprise a rectangular or cylindrical metal foam piece glued or soldered onto a metal plate or pipe equipped with a heater and thermocouples, and a fluid (generally air or water) circulating either freely (natural convection) or pushed by a pump or fan (forced or mixed convection). Geometries are generic configurations amenable to numerical analysis using a limited set of geometrical parameters; alternatively applications-specific configurations are used.

The result is generally a predicted or measured Nusselt number, defined as:

$$Nu \equiv \frac{hL}{k} = \frac{\dot{H}x}{\left(T_{w(x)} - T\right)k} \text{ (local) or } = \frac{\dot{H}L}{\left(\overline{T_w} - T\right)k} \text{ (global).} \tag{42}$$

Nu is the general dimensionless measure of a given structure's performance with regard to heat transfer. In eqn (42), h is the relevant heat transfer coefficient between the substrate and the far-field fluid, L a length scale for the object or device and k a relevant heat conductivity (generally that of the fluid, sometimes that of the stagnant fluid-filled porous metal). Since either the heat flux $\dot{H}$ from the solid substrate (or other site from which heat is to be moved) or the wall temperature T_w are in practice specified along the solid/porous metal contact surface, and since h is generally position dependent, the two right-hand local or global definitions are both relevant in reporting results from simulation or experiments.

The literature on convective heat transport through metallic porous media is part of a much wider body of work on heat transfer through porous media, a subject that has considerable importance in chemical engineering or electronics. A far wider body of relevant information and data than what pertains to porous metals therefore exists on the subject; an extensive and authoritative review of this is in Ref. Nield and Bejan (2006), (Tien and Vafai, 1990) gives coverage up to 1990. Recent reviews specifically pertinent to porous metals are in Refs. (Zhao, 2012; Han et al., 2012; Nakayama, 2013).

Given the high potential of microcellular metals in heat exchange applications and given the experimental convenience that these "hyperporous" materials offer, the body of work on heat exchange aided by metal foam is considerable. Forced convection configurations generally comprise situations where a porous metal sample in a fluid-containing duct either fills the entire cross-section offered to fluid flow, or only a portion thereof. In both situations the porous metal is fixed to a solid surface, which in models is generally taken to be either isothermal, or a constant heat flux surface. In experiments this is a dense solid metal heat spreader (generally of copper) attached to, or containing, a resistive heater or a flowing water cooling coil. In forced convection, fluid flow and heat transfer can be treated independently and in succession, by solving first for fluid flow then for heat transport. Natural convection is more complicated, the two transport phenomena being mutually linked.

A typical 2D situation that has been extensively tackled in modeling is that of a porous medium along a flat plate, of (i) semi-infinite extent (Vafai and Tien, 1981), or (ii) finite extent ending in the fluid (Vafai and Sozen, 1990; Jeng and Tzeng, 2005) or (iii) finite extent ending against another solid boundary (Hunt and Tien, 1988; Antohe et al., 1996b; Lu et al., 1998, 2005; Lu, 1999; Calmidi and Mahajan, 2000, Alazmi and Vafai, 2000; Angirasa, 2002b; Dukhan et al., 2006b, 2005; Xu et al., 2007; Leong and Jin, 2006b; 2006a; Tzeng and Jeng, 2007). This last situation is the most extensively treated in both modeling and experiment.

Porous metals in forced convection have been investigated experimentally for flow through samples extending across the fluid duct in Refs. Hunt and Tien (1988), Calmidi and Mahajan (2000), Kim et al. (2000), (2001), Hwang et al. (2002), Boomsma et al. (2003a), Zhao et al. (2004a), Tian et al. (2004), Dempsey et al. (2005), Lu et al. (2005), Kim et al. (2005b), Shih et al. (2006), Tzeng and Jeng (2006), Xu et al. (2007), Giani et al. (2005a), Dukhan et al. (2005), (2006b), Dukhan and Chen (2007), Jazi et al. (2009), Garrity et al. (2010), Mancin et al. (2010a), (2011), (2012a), (2012b), Kamath et al. (2011), Dai et al. (2011) and in Refs. Angirasa (2002a), Kim et al. (2003), Ejlali et al. (2009), Jeng et al. (2010) for configurations where there is also a section of clear fluid offered to fluid flow next to the porous metal sample. Reviews of data and correlations can be found in Refs. Mahjoob and Vafai (2008), Kurtbas and Celik (2009).

Corresponding geometries having cylindrical symmetry (whether square or circular; here heating is along the entire setup periphery) have also been addressed, of a cooling pipe within an infinite or finite sample of porous metal, or of a cylindrical tube lined either fully or partly with porous metal, in both analysis and experiment (Cookson et al., 2006; Lu et al., 2006b; Zhao et al., 2006a; Alazmi and Vafai, 2000; Boger and Heibel, 2005; Noh et al., 2006; Du et al., 2010; Xu et al., 2011; Hutter et al., 2011a; Odabaee et al., 2011; Yang et al., 2012a). Relatively similar conditions were studied in Ref. Kurtbas and Celik (2009), who characterized cooling by air in forced or mixed conditions flowing through a horizontal metal foam filling a rectangular duct heated from all sides; at lower velocities natural convection also intervened. Other more complex geometries have also been studied in the context of convective heat transport through porous metals in forced convection; examples are in

(i) simulations or experimental investigations of realistic cooling systems employing porous metals (Lage et al., 1996; Rachedi and Chikh, 2001; Zhang et al., 2007a; T'Joen et al., 2010; Ding et al., 2011; Odabaee and Hooman, 2012b, 2012a; Ribeiro et al., 2012), at times in combination with solid metal cooling elements (Bhattacharya and Mahajan, 2002; Seyf and Layeghi, 2010),
(ii) an exploration of metal foams in automotive exhaust gas recirculation systems (Muley et al., 2011),
(iii) an investigation of metal foams as heat sinks for transmission windows in accelerators (Hetsroni et al., 2005),
(iv) or as a rotating heat sink (Jeng et al., 2008),
(v) in heat exchange with various cylinders embedded in microcellular aluminum (Tzeng, 2007) and
(vi) in forced convection patterns other than parallel through flow along a solid heat spreader (Hsieh et al., 2004).

Natural convection has also been investigated, in both simulation (Hong et al., 1985; Rachedi and Chikh, 2001; Phanikumar and Mahajan, 2002; Zhao et al., 2005a; Jeng and Tzeng, 2008; Mahdi et al., 2006) and experiment (Angirasa, 2002a; Zhao et al., 2005a; Bhattacharya and Mahajan, 2006; Kathare et al., 2008; Kurtbas and Celik, 2009) in a variety of configurations. Depending on the material and configuration, natural air flow can enhance the rate of heat flow above that delivered by metal conduction alone (e.g. (Zhao et al., 2005a)). Mixed convection conditions, combining natural and forced air flow, were also recently explored (Kamath et al., 2011). Boiling heat transfer through metal foam, both forced and natural, was examined in Refs. Tadrist et al. (2004), Topin et al. (2006), Choon et al. (2006), Xu et al. (2008), Lu and Zhao (2009), Carbajal et al. (2006), Zhao et al. (2009a), Yang et al. (2010b), Zhu et al. (2011b), Du et al. (2012), Zhao (2012). In one study, natural convective heat transfer was studied in assessing the performance of microcellular metals as a direct resistive heating strip in air (Hetsroni et al., 2008).

A practical issue also apparent in experimental work—and hence in engineering practice—is the interface between the porous metal and the substrate across which heat is to be transported: a high quality bond of low thermal resistance can lead to significant improvements in transport. To this end accumulative layer processing methods have been shown to have particular potential (Hutter et al., 2011a); alternatively, brazing and other metal-to-metal bonding processes have been found to be attractive (Muley et al., 2011; Han et al., 2012).

Often, it is observed that metal foams outperform, at times significantly, other designs tested; examples can be found in (Kim et al., 2000, 2003; Boomsma et al., 2003a; Bhattacharya and Mahajan, 2006, Leong and Jin, 2006b; Wadley, 2006; Dai et al., 2011). The reasons why are, however, not entirely clear; the literature gives a few pointers.

One pointer that comes across in various studies is that, in forced laminar flow over a constant temperature or constant heat-flux surface, all else constant, including all thermal properties and the average fluid flow rate, it does not make a tremendous difference whether there is a porous material present or not. This is seen by comparing, to take a simple example, the Nusselt number for flow within a confined, flat or circular, conduit. The two following situations are compared in **Table 6**: (i) fully developed steady-state laminar flow (parabolic velocity profile in an essentially clear fluid corresponding to high Da in eqn (33)); and (ii) "slug flow" (constant velocity across the porous medium corresponding to low Da in eqn (33)). These widely different flow patterns give different results in the various configurations; however, the difference is at most a factor of two. Between these two extremes, the transition (as modeled by means of the Brinkman equation, eqn (31)) is smooth, taking place around the range Da $\approx 10^{-4}$ to 10^{-2} (Kaviany, 1985). The general change in flow of the fluid created by

Table 6 Nusselt numbers for confined flow in two extreme situations (clear fluid or fine porous material) (Nield and Bejan, 2006, Poirier and Geiger, 1994)

Geometry	Fluid velocity distribution	Condition at wall	Nusselt number
Parallel plates	Parabolic (clear fluid)	Uniform temperature	7.60
	Flat (slug flow)	Uniform temperature	4.93
	Parabolic (clear fluid)	Uniform heat flux	8.23
	Flat (slug flow)	Uniform heat flux	6
Circular tube	Parabolic (clear fluid)	Uniform temperature	3.66
	Flat (slug flow)	Uniform temperature	5.78
	Parabolic (clear fluid)	Uniform heat flux	4.36
	Flat (slug flow)	Uniform heat flux	8

insertion of a porous metal thus makes a difference that creates an enhancement of heat transfer in through-flow; however, by a factor only around two.

This in turn implies that other factors must cause the high rates of heat transfer observed with metal foams. The first factor is obvious, namely the comparatively high thermal conductivity of porous metals (Section 24.4.3). Given the wide range of variation of this parameter as a function (i) of the purity of the metal (which determines its intrinsic conductivity k_m), (ii) of the architecture of the porous metals, particularly at low V_m and (iii) of the value of V_m, there is ample room for optimization of the performance of highly porous metals in the transport of heat. A second factor must be dispersive transport. There is indeed clear evidence that dispersive heat transfer plays a tangible role in liquid-mediated heat flow through porous metals. Hunt and Tien (Hunt and Tien, 1988) showed in forced flow experiments conducted with water through microcellular nickel, carbon or aluminum having V_m between 3 and 6% that measured Nusselt numbers, and hence measured heat transfer rates, far exceed predictions conducted without account of heat transfer by dispersion (i.e., assuming $k_D = 0$). This led to conclude that dispersion is important in highly porous metal convective heat transfer. Their data were collected at comparatively high Reynolds number (Re $\approx$ 8–80) and gave $k_D = 0.025 \, (\rho c_p)_f \, K^{1/2} \, v_D$, that is, a dispersion-related contribution to thermal conduction that is proportional to the pore size and to the flow rate. This finding was confirmed by Calmidi and Mahajan (2000), who showed in similar experiments, also conducted at high Reynolds number (Re $\approx$ 10–140) for water flowing through similar microcellular aluminum samples, that dispersion is the dominant heat transport mechanism (with k_D well above k_c). Calmidi and Mahajan (Calmidi and Mahajan, 2000) also found that the effect is seen with water but not air: flow of air through the same samples showed no effect of dispersion in heat transfer.

Both heat conduction through the metal and dispersive heat transfer are strong functions of the architecture of highly porous materials. The overall performance of fluid-filled porous metals in heat transfer applications is also a strong function of fluid flow parameters (parameters K and F of the Darcy–Forchheimer equation), which themselves are also strong functions of the architecture and scale of pores within the metal. These factors, combined with the fact that the subject is to date barely explored, lead to conclude that there is an exciting and rich body of research ahead of us on the architectural optimization of microcellular metals for energy transfer (Nield and Bejan, 2006; Bejan, 2004; Tian et al., 2004; Evans et al., 2001; Wadley, 2006; Hutter et al., 2011a).

24.4.5.3 Mass Transport through Filled Porous Metals

If we leave aside classical nanoporous metal catalysts such as Raney nickel, less has been written on mass transport through open-pore microcellular metals than on heat transfer. In part this is because porous metals are slightly less unique with regard to mass transfer: as heat or corrosion resistant supports for catalysts, ceramic foams are equally attractive. There have nonetheless been several interesting investigations of porous metal foams in the context of mass transport dominated applications.

As far as 30 years back, Vafai and Tien published a study parallel to their heat transfer studies, with theoretical and experimental results on the 2D transfer of mass through a fluid in a porous medium, to or from a planar boundary, in forced flow along the boundary. Predictions and measurements were given of the Sherwood number, the mass transfer equivalent of the Nusselt number in heat transfer (Vafai and Tien, 1982). The study shows that boundary layer effects cause a far smaller Sherwood number than is predicted without account of these and consequently that the Brinkman equation (Eq 4-24 in the present configuration) must be used instead of the Darcy-Forchheimer equation alone when solving for fluid flow.

Studies of transport effects in porous metals have been published that feature a cylindrical sample in through flow, with reactants at an inlet that often does not comprise the entire sample cross-section (Giani et al., 2005b; Hutter et al., 2010). Ref. Giani et al. (2005b) presents an experimental investigation of the mass transfer rate (as measured by computing the relevant Sherwood number) and fluid flow through Pd/Al_2O_3 coated FeCrAlY foam substrates, comparing these with other designs. It is shown, on the basis of dimensionless merit index measurements, that metal foams outperform packed beds significantly, and compare either unfavorably or favorably with honeycomb designs depending on design goals. These authors also propose a single correlation for the Sherwood number that fits their data as well as data from a subsequent study of ceramic foam catalysts (Giani et al., 2005b; Groppi et al., 2007). This correlation was later refined in a study of ceramic and metal foam catalysts (Incera Garrido et al., 2008). (Hutter et al., 2010) report an investigation of mixing induced by insertion of microcellular metal samples within a plug flow reactor, showing enhanced mixing performance as a result. Two interesting findings are to be noted: (i) that there is an optimum pore size, and (ii) that dispersion seems to play a role in the process.

Dispersion in mass transport through metal foams was investigated by Montillet et al. (1993), who conducted experiments showing that, of the various porous materials characterized, nickel foams showed the least longitudinal dispersion (transverse dispersion was on the other hand not measured). Hutter et al. (2011b) present data on axial dispersion through a foam-filled cylindrical reactor, giving a correlation that incorporates other data from the literature. Ferrouillat et al. (2006) present a study of combined heat and mass transfer, in which metallic foams were found to be efficient as substrates in heat exchanger reactors, where combined reactant mixing and heat transfer is used to drive reactions in a fluid while preventing runaway due to exothermicity, giving good reaction characteristics - but at greater pumping power cost, however. Similar combined roles of porous metal as chemical reaction substrate and heat transfer medium are shown in the context of a metal-supported catalyst for CO cleanup from fuel in Ref. van Dijk et al. (2010), and hydride support in Ref. Ahluwalia (2007). Ref. Chin et al. (2006) presents data from a comparison of CO oxidation through metal foam supported catalysts, varying the pore size and relative density, and comparing the metal foam with a ceramic foam support. There are also a few studies on metal foams that are more directly oriented at catalysis (in which the porous metal serves mainly as a substrate and not a mass transport or mixing medium); here mass transport other than at the very surface of the foam is not a major focus (Pestryakov et al., 1995; Sirijaruphan et al., 2005a, 2005b). In the context of catalysis, it is also worth noting that the literature for ceramic foams is equally interesting; it is also extensive given its relevance to automotive catalysts (Richardson et al., 2003; Dimopoulos Eggenschwiler et al., 2009).

A technical application, in which porous metal has considerable relevance as a permeable convective mass transport medium, is with fuel cells. (Brandon and Brett, 2006) provides a didactic overview of the subject. The cathode of a fuel cell is subjected to elevated temperature within a medium of high oxygen chemical activity, leaving only expensive noble metals as viable metallic candidates. The anode serves (i) as a vector for gases comprising hydrogen, hydrocarbons, water and the carbon oxides, (ii) as a catalyst for oxidation reactions comprising ionic oxygen carried through the solid electrolyte, and (iii) as a collector of the electrical current that results from this reaction.

The most classical solid oxide fuel cell anode material is a porous composite of nickel combined with yttria-stabilized zirconia, produced by reduction of NiO in a mixture of NiO and yttria-stabilized zirconia; another is nickel combined with rare-earth-doped ceria (Fergus, 2007; Goodenough and Huang, 2007; Sun and Stimming, 2007; Brandon and Brett, 2006; Zhu and Deevi, 2003). Other applications of porous metals in fuel cells include upstream or downstream components that serve to duct gases, often while collecting current, transferring heat, carrying mechanical load, or altering the gas composition by reaction. Highly porous metal has been investigated as a material for bipolar/end plates in polymer electrolyte membrane (Kumar and Reddy, 2003; 2004), proton exchange membrane (Tang et al., 2010) or direct methanol fuel cells (Jiang et al., 2004; Arisetty et al., 2007). Here its role is mainly in ducting the gas and as a current conductor. Other applications are in fuel preprocessing, notably CO removal from gases upstream of nickel-based anodes in solid oxide fuel cells (Chin et al., 2006; Smorygo et al., 2009; van Dijk et al., 2010; Bae et al., 2010), as a gas diffusion layer in polymer electrolyte membrane fuel cells (Brandon and Brett, 2006) or in direct methanol fuel cells (Chen and Zhao, 2007b, 2007a; Arisetty et al., 2007), or as porous support materials in solid oxide fuel cells (Brandon and Brett, 2006).

Attractive attributes of metal foams in the context of chemical reactors and catalysis are (i) that their complex internal architecture promotes lateral mixing of flowing gases or fluids, (ii) that their low relative density implies a low heat capacity which, in turn, enables a rapid response to transients or changes in operating conditions, and (iii) that their comparatively high thermal conductivity reduces the formation of "hot spots" and eases evacuation of heat through metal tubes that can be brazed to the metal foam (Chin et al., 2006). In summary, "breathing" porous metals also hold tangible promise for mass-transport related applications.

24.5 Mechanical Behavior

As a general class of materials, metals and alloys have comparatively attractive mechanical properties. Metals are strong, relatively stiff, and yet can deform significantly before breaking, by irreversible plastic deformation mechanisms that make metallic materials tough and comparatively easy to shape. These features alone make porous metals quite different from other porous materials: porous elastomers can also deform to great strain but generally do so reversibly and at far lower stress than porous metals while many ceramic-, carbon- and many polymer-based porous materials are brittle.

Many of the mechanical characteristics of porous metals are thus specific to porous metals: their behaviour in room or elevated temperature plastic deformation, fracture or fatigue, bears a strong signature of their metallic nature. In elastic deformation, on the other hand, nearly all materials are initially linear: the low-strain, linear elastic, deformation of porous metals is therefore a sub-branch of the wider domain defined as linear elastic composite mechanics applied to porous materials.

24.5.1 Elastic Constants

At small macroscopic strain, a porous linear elastic material is also linear elastic. Understanding the elastic behavior of porous metals is thus mostly a matter of predicting its elastic constants knowing

(i) those of the material of which it is made and (ii) its internal architecture, specifically the shape and distribution of the pores it contains. The subject is vast because its relevance is huge: nearly all natural materials are porous, be they mineral or organic (sand, rock, wood, tissue, bone are examples), as are many man-made materials of technological importance such as foamed polystyrene, glasswool, concrete and many other ceramics, plus other common "man-made" organic materials such as bread or biscuit. There is hence a considerable literature devoted to the linear elastic constants of porous materials. Experimental data are on the other hand relatively scarce, particularly for simple baseline materials that may serve as clean tests of theory.

Predicting the elastic stiffness of porous materials is not only relevant; it is also interesting. This is because we have here an extreme in composite mechanics: the elastic phase contrast between a metal and its pores is infinite, as it was for all practical considerations where conductivity was concerned. The subject thus largely parallels that of conduction through porous metals (Section 24.4.3); however, it is inherently more complicated given the higher dimensionality of the laws governing elastic deformation compared with those that govern the conduction of heat or current. At a basic, intuitive, level this is easily seen if one considers a highly porous 3D open-pore microcellular material such at that sketched in **Figure 26**: its many struts will conduct heat or current mainly along their length, opposing a local resistance that can, in the simplest cases, be summed to estimate that of the material (e.g. (Lemlich, 1978)). Under stress, however, the struts will elongate, but they will also bend and twist, in varying and unknown proportions. Back-of-the-envelope formulation of estimates for the elastic constants of even the most elementary porous structures is therefore not a trivial matter.

24.5.1.1 *Bounds*

As with conductivity, the elastic constants of porous materials can be bounded. For simplicity we restrict in what follows attention to homogeneous isotropic porous metals. We thus ignore the influence of second phases within the porous metal and ignore the potential influence of crystalline anisotropy (unless its grains are much smaller than the pores and randomly oriented, or unless the metal has, like aluminum, nearly isotropic elastic constants, at the scale of their pores most metals are elastically anisotropic).

Isotropic materials have two independent moduli; generally theory focuses on the two "scalar" deformation moduli, namely the bulk (B) and shear (G) moduli (these are scalar in the sense that they relate a single scalar deformation to a single scalar stress). Young's modulus E and Poisson's ratio ν are, in isotropic solids, linked to these by:

$$E = \frac{9BG}{G + 3B} \quad \text{and} \quad \nu = \frac{3B - 2G}{6B + 2G}. \tag{43}$$

In the most general case, the bulk and shear modulus, B and G, respectively, of porous metals are bounded by

$$0 \leq B \leq V_m B_m \quad \text{and} \quad 0 \leq G \leq V_m G_m, \tag{44}$$

where B_m and G_m are the bulk and shear modulus of the dense metal, and V_m the relative density (or fraction metal). The same bound applies to Young's modulus E, given the absence of stress transfer (and hence Poisson ratio effects) in porous metals (Torquato, 2002a). If the porous metal itself is isotropic, with no other information on the porosity than its volume fraction $V_p = 1 - V_m$, the upper Hashin–Shtrikman bound reads

$$B_{HS+} = \frac{2(D-1)V_m G_m B_m}{D(1-V_m)B_m + 2(D-1)G_m} \tag{45}$$

and

$$G_{HS+} = \frac{V_m G_m H}{(1-V_m)G_m + H}, \tag{46}$$

where

$$H = G_m \frac{D^2 B_m + 2(D+1)(D-2)G_m}{2D(B_m + 2G_m)} \tag{47}$$

with D (=2 or 3) the dimension of the porous material (Torquato, 2002a). The lower bound is zero, also when the porous metal is isotropic (as for the conductivity and for the same reason: the pores could be the only continuous phase).

24.5.1.2 Estimates at Low to Medium Porosity

When the pore volume fraction is low, classical "dilute" micromechanical solutions for a low volume fraction of one or a few aligned or randomly oriented ellipsoidal inclusions yield relatively sound and workable estimates. The resulting decrease in stiffness of the metal is calculated knowing the (low) volume fraction of pores, V_p, and Eshelby's solution of the ellipsoidal inclusion in an elastic continuum (Eshelby, 1957; Mura, 1982; Torquato, 2002a; Nemat-Nasser and Hori, 1999).

For spherical pores, older, relatively simple and explicit expressions, such as those proposed by Mackenzie (1950), Wachtman (1996)

$$\frac{1}{B} = \frac{1}{B_m V_m} + \frac{3V_p}{4G_m V_m} + O\left(V_p^3\right) \tag{48}$$

and

$$G = G_m \left[1 - \frac{5V_p(3B_m + 4G_m)}{9B_m + 8G_m}\right] + O\left(V_p^2\right) \tag{49}$$

can be used, at porosity levels of a few percent, for rapid estimations of their effect on the metal moduli. As V_m increases past a few percent, as long as the pores remain convex and isolated one of many effective medium schemes can be used to obtain estimates; in spirit these often parallel the schemes that are used for the conductivity of composites (Section 24.4.3). Reviews are given by several authors (Torquato, 2002a; Nemat-Nasser and Hori, 1999; Milton, 2002; Berryman, 1980a, 1980b; Berryman and Berge, 1996; Sevastianov et al., 2002; Hentschel and Page, 2006; Despois et al., 2006b).

The Maxwell approximation for conduction is, for elastic constants, generally called the Mori–Tanaka approximation. Its logic is the same: in essence, it takes the pores to behave as if these were surrounded by an infinite metal, but then balances overall stress assuming that the metal has a finite volume fraction V_m (Mori and Tanaka, 1973; Torquato, 2002a; Nemat-Nasser and Hori, 1999; Milton, 2002; Benveniste, 1987; Berryman, 1980a, 1980b; Berryman and Berge, 1996). For spherical pores the result coincides with the Hashin–Shtrikman upper bound, eqns (45)–(47). It also coincides for spherical voids with another approximate (Kuster and Toksöz) scheme used predominantly in geophysics; however, more generally the two approximations do not give the same result and both can violate rigorous bounds (and are hence not be realizable) (Berryman and Berge, 1996). Expressions for application of the Mori–Tanaka scheme to aligned or randomly oriented ellipsoidal inclusions can be found in Refs. Torquato (2002a), Berryman (1980a), (1980b), Berryman and Berge (1996), Zhao et al. (1989), Kitazono et al. (2003).

Another approach is the self-consistent approximation (Budiansky, 1965; Nemat-Nasser and Hori, 1999; Torquato, 2002a; Milton, 2002; Berryman, 1980a, 1980b; Berryman and Berge, 1996). This comes in several variants but in essence takes the pore to be surrounded by the porous metal, the modulus of which is unknown; as such this scheme is implicit. The self-consistent approximation tends to yield very low estimates of the stiffness of porous materials; in its classical form, it also predicts $G = E = 0$ below a certain finite relative density, which equals $V_m = 1/2$ for spherical pores (Torquato, 2002a). This percolation-like threshold is of necessity artificial, if only because this same approximation also yields a percolation threshold for the conductivity—but at a different relative density, namely $V_m = 2/3$ with spherical pores; the difference of course makes little sense.

Equations for several other schemes that can be applied to porous 3D metals containing spherical pores are listed in Ref. (Despois et al., 2006b) (and thus not repeated here). These include notably:

(i) the generalized self-consistent scheme of Christensen (Christensen, 1990; Nemat-Nasser and Hori, 1999; Torquato, 2002a; Milton, 2002), which takes the spherical pores to be embedded in a concentric spherical annulus of the metal to create a core–shell composite having the relative density of the porous material, this in turn being embedded in an infinite medium having the unknown effective properties of the composite,

(ii) the DEM approximation, which estimates the stiffness of finite porosity solids by integration of the differential effect of successive dilute pore additions; this model was proven to be realizable, and is for this reason interesting for highly porous solids, given that it therefore respects bounds (McLaughlin, 1977; Norris, 1985; Christensen, 1990; Zimmerman, 1991; Torquato, 2002a; Milton, 2002),

(iii) the Torquato identical spheres estimations, which incorporate three-point microstructural parameters, meaning microstructural information beyond the pore volume fraction and shape. For spherical inclusions, such microstructural information includes whether inclusions overlap or not, or whether the inclusions lie along random or regular periodic patterns. Resulting estimates match quite accurately predictions from numerical simulation of periodic cells, also at high inclusion volume fraction (Torquato, 2002a; Torquato et al., 1998, 1999).

(iv) Ju and Chen's formula, based on an estimation of pore-to-pore interactions in a metal containing ellipsoidal pores (Ju and Chen, 1994b, 1994a), which for a uniform random distribution of equal-sized spherical pores up to roughly 50% porosity reads (Ju and Chen, 1994a; Kakavas and Anifantis, 2003)

$$\frac{B}{B_m} = 1 + \frac{30V_m(1 - \nu_m)(3\gamma_1 + 2\gamma_2)}{3\beta_1 + 2\beta_2 - 10V_m(1 + \nu_m)(3\gamma_1 + 2\gamma_2)}, \tag{50}$$

$$\frac{G}{G_m} = 1 + \frac{30V_m(1 - \nu_m)\gamma_2}{\beta_2 - 4V_m(4 - 5\nu_m)\gamma_2} \tag{51}$$

with

$$\beta_1 = 2(5\nu_m - 1), \beta_2 = 5\nu_m - 7,$$

$$\gamma_1 = V_m \frac{-8.75\nu_m^2 + 7.63\nu_m - 0.5}{\beta^2}, \quad \gamma_2 = 0.5 + V_m \frac{6.87\nu_m^2 - 9.69\nu_m + 7.63}{\beta^2}. \tag{52}$$

A derivation in the same vein leading also to explicit expressions for spherical pores is given in Ref. Wang and Tseng (2003). Mean-field expressions have also been proposed for syntactic foams, in which hollow spheres of one phase are embedded in a matrix of another phase (Bardella and Genna, 2001; Balch and Dunand, 2006).

All of these expressions are applicable to metals containing pores that can be treated as roughly convex, separate "bubble-like" inclusions surrounded with metal. Data exist for such structures, gathered using mostly porous ceramics (Walsh et al., 1965; Berryman and Berge, 1996; Rice, 1996a; Boccaccini, 1994; Boccaccini and Fan, 1997; O'Rourke et al., 1997; Ramakrishnan and Arunacha-lam, 1990; 1993) but also porous metals (Li and Zhu, 2006). Comparison of predictions with data shows variegated agreement, largely because of scatter in data, and also because "real" pores are often neither spherical nor realistically represented by ellipsoids (Walsh et al., 1965; Nielsen, 1984; Ferrari and Filipponi, 1991; Zimmerman, 1991; Berryman and Berge, 1996; Ju and Chen, 1994a; Rice, 1996a; Boccaccini, 1994; Boccaccini and Fan, 1997; Wang and Tseng, 2003). For this reason, other expressions have been proposed as a pragmatic approach in the ceramics community (Rice, 1996b, 1996a; Wachtman, 1996; Arnold et al., 1996; Boccaccini and Fan, 1997; Ji et al., 2006); because these expressions incorporate ad hoc adjustable parameters, they unfortunately give little predictive power.

Pore shapes that deviate significantly from that of isolated rounded bubbles, including in particular interconnected pores, have been tackled using periodic unit-cell structures that are solved numerically, generally using finite element methods. Given the relevance of the question to many situations, there have been a good many studies in this vein. These include simulations of the linear elastic deformation of consolidated sphere packings (Chapman and Higdon, 1994; Bouhlel et al., 2010), or of numerically generated structures of varying complexity (Poutet et al., 1996b; Roberts and Garboczi, 2000, 2001, 2002b, 2002a; Spoerke et al., 2005; Shen et al., 2006; Konstantinidis et al., 2005; Kujime et al., 2007; Nachtrab et al., 2011), syntactic foam structures (Bardella and Genna, 2001) or microtomography-based (Maire et al., 2003b; Poutet et al., 1996a, 1996b; Guessasma et al., 2008; Guessasma and Bassir, 2010; Babin et al., 2005; Dillard et al., 2005; Caty et al., 2008; Singh et al., 2010; Saadatfar et al., 2012) high-porosity structures.

Among studies highly relevant to porous metals are the several finite element unit-cell simulations by Roberts and Garboczi (2000), (2001), (2002a), (2002b). These explore realistic numerically generated high-porosity structures containing either closed or open pores of convex or mixed convex/concave geometry; structures explored thus include ones resembling those of mean-field theory as well as more convoluted interpenetrating metal/pore structures. The authors show in particular, from their rich databank of simulated structure rigidities, that the Young's modulus E of the various structures does not depend much on the Poisson ratio of the dense material (something that also comes out of the Hashin–Shtrikman equation for G, eqn (46)). This makes prediction of E possible using correlations involving only the relative density as a parameter, which the authors propose from their simulations for the various structures examined. Another finding is that for all but the stiffest (closed-cell) structure, the porous material Poisson ratio converges to a constant, structure-dependent, value in the vicinity of 0.2 as V_m tends toward the percolation threshold characteristic of the structure (which is not always situated at $V_m = 0$). Hence, the lateral expansion of uniaxially stressed highly porous materials is generally dominated, not by that of the solid making the material, but by the architecture of the solid within the highly porous material. Also interesting is the finding that the DEM mean-field scheme, already quite successful as concerns conductivity (Section 24.4.3), gives good agreement with these finite element results.

A different but equally interesting study is that of Poutet et al. (Poutet et al., 1996b) who simulated a range of structures, including periodic and random sphere packings, as well as several fractal structures, both regular and random. Predicted modulus values are all below the Hashin–Shtrikman bound, and display a wide range of variation. Among the structures examined, that constructed by a random percolation model was the least rigid, reaching zero stiffness at a percolation threshold situated at $V_m = 0.322$. For $V_m > 0.6$ predictions for this structure essentially lie along a line similar to what is predicted by the classical self-consistent model (shown to be realizable by another fractal construction (Milton, 1984; Torquato, 2002a)). Since such structures are expected to be among the most compliant architectures of an interconnected porous solid (given the low level of solid phase continuity), the observation suggests that the classical self-consistent model can perhaps be viewed as a pragmatic lower bound for the modulus of porous materials containing less than 50% porosity.

24.5.1.3 *Microcellular Metals*

As the relative density V_m starts approaching zero, the material becomes microcellular: this changes the picture in two fundamental ways. The first is that predictions from the various models cited above, which already tended to cover a wide range of values, start drifting apart (again, porous materials are composites of infinite phase contrast): predicted values differ by more than one order of magnitude as V_m falls below 0.1, see **Figure 31**. In other words, the stiffness of microcellular metals becomes very hard to estimate when the fraction solid approaches zero, even if one has a good general idea of their mesostructure. The reason is that the stiffness of highly porous metals depends very strongly on the geometry of the pores, or more appropriately stated, on the architecture of the solid metal between the pores.

The pore architecture can in most low-V_m microcellular materials be viewed as an assemblage of up to three main elements, **Figure 26**; this is the second fundamental change as V_m approaches zero. These elements are (i) relatively straight beam-like struts, situated at the junction of three pores when these retain an identity in the material (as is the case with "real" foams and their derivatives, and also with replicated microcellular metals); the struts join at (ii) relatively equiaxed nodes and are the location where (iii) plate-like cell walls join, generally three at a time. Cell walls are present in closed-cell microcellular metals, in which relatively flat walls separate two pores and are bordered by struts and nodes where three or four cells meet, respectively (Gibson and Ashby, 1997; Weaire and Hutzler, 1999; Kraynik, 2003, 2006). Nearly all low-density microcellular metals can be viewed as regular or stochastic assemblies of these three basic elements. Open-cell microcellular metals (open-cell microcellular metals and lattice structures) combine the former two elements, struts and nodes, while all three elements are found in closed-cell foams; see Section 24.3.3 of this review. The implication is that the rigidity of microcellular metals can be analyzed using models that assemble and analyze diverse configurations of variously shaped beams, nodes and plates. There are many such models in the literature, particularly dealing with linear elastic deformation.

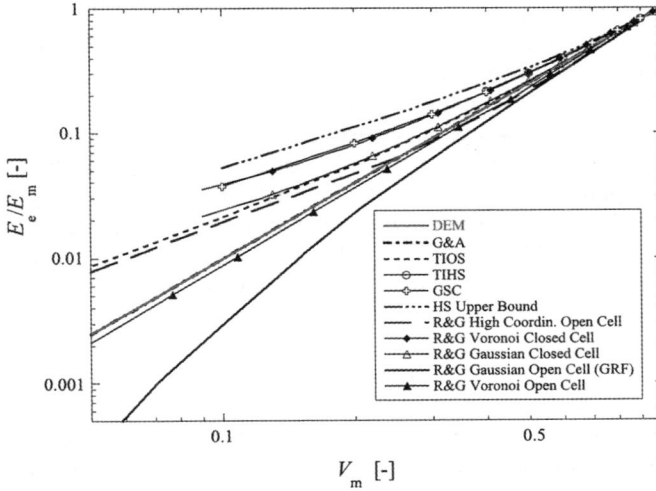

Figure 31 Different model predictions for the stiffness E_e of a porous material as a function of its relative density V_m. Reproduced from *Acta Materialia*, vol. 54, Despois, J. F., Mueller, R. & Mortensen, A. "Uniaxial deformation of microcellular metals", pp. 4129–4142 (Despois et al., 2006b) with permission from Elsevier. Curve denominations:

- HS: Hashin–Shtrikman upper bound (eqns (45)–(47) into eqn (43));
- GSC: Generalized self-consistent approximation (eqns (5) and (6) of (Despois et al., 2006b));
- DEM: Differential effective medium model (eqns (12) and (13) of (Despois et al., 2006b));
- TIOS: Torquato identical overlapping sphere estimation (eqns (14) and (15) of (Despois et al., 2006b));
- THIS: Torquato identical hard sphere estimation (same equations as TIOS with other values of parameters ζ_2 and η_2 defined in (Despois et al., 2006b));
- G&A: Gibson–Ashby estimate for open-cell foam (eqn (56) with $C = 1$ and $N = 2$));
- R&G: Gaussian Open Cell: finite element simulation results by Roberts and Garboczi of the structure in **Figure 4**(d) of Ref. Roberts and Garboczi (2002b)("Gaussian Random Field" model);
- R&G Voronoi Open Cell: finite element simulation results by Roberts and Garboczi of the structure in **Figure 4**(a) of Ref. Roberts and Garboczi (2002b) ("Open-cell Voronoi Tessellation" model);
- R&G High Coordin. Open Cell: finite element simulation results by Roberts and Garboczi of the structure in **Figure 4**(c) of Ref. Roberts and Garboczi (2002b) ("High-coordination number foam" model);
- R&G Gaussian Closed Cell: finite element simulation results by Roberts and Garboczi of the structure in **Figure 9** of Ref. Roberts and Garboczi (2001) ("Closed-cell Gaussian Random Field" model);
- R&G Voronoi Closed Cell: finite element simulation results by Roberts and Garboczi of the structure in **Figure 2** of Ref. Roberts and Garboczi (2001) ("Closed-cell Voronoi Tessellation" model).

Among the simplest and earliest contributions in this vein are estimations that parallel Lemlich's derivation (Lemlich, 1978) of the thermal conductivity of microcellular materials (Section 24.4.3): these similarly apply basic orientational averaging to beams or plates stressed uniformly (Gent and Thomas, 1959; 1963; Kuipers, 1964; Lederman, 1971; Christensen, 1986). The result is, at very low relative density, a linear relation between the foam Young's modulus, E, and the relative density V_m, with a constant of proportionality that depends on the geometry assumed and the Poisson ratio of the metal, for both two- and 3D microcellular structures.

Such predictions are seldom obeyed: the modulus of highly porous microcellular materials, both 2D and 3D, scales with the relative density raised to a power that is generally (but not always, as we shall see below) well above unity; it is nearer two in 3D microcellular solids. The reason is that the deformation of microcellular materials is most often dominated by modes of strut deformation other than stretching, namely bending and twisting. This observation is expressed in analyses of the deformation of open-pore microcellular solids that use engineering beam theory to calculate the stiffness of more or less elaborate, 2D or 3D, periodic assemblies of straight beams.

Early contributions pertaining to 3D open-pore microcellular materials (Ko, 1965; Menges and Knipschild, 1975; Gibson and Ashby, 1997) (together with a few later and more complex models (Zhu et al., 1997a, 2000; Gong et al., 2005a; Jang et al., 2008)) yield predictions of the form

$$\frac{E}{E_{\mathrm{m}}} = \frac{\alpha V_{\mathrm{m}}^2}{V_{\mathrm{m}} + \beta}, \tag{53}$$

thus giving at low relative density ($V_{\mathrm{m}} \approx 0$) a different exponent (2 instead of 1) in the scaling law between E and V_{m} for 3D assemblies of slender straight beams. We note in passing that the DEM model for a solid containing spherical pores also returns two as the scaling exponent between E and V_{m} (Despois et al., 2006b).

In microcellular solids, this scaling exponent was shown, by means of elegant dimensional arguments coupled with a simple cube-like model and an analogy to 2D regular microcellular solids (honeycombs), to be a consequence of the fact that the linear elastic deformation of microcellular materials is dominated by bending (Thornton and Magee, 1975b; 1975a; Gibson and Ashby, 1982, 1997; Ashby, 1983; Gibson, 1989). More specifically (i) struts mostly bend and (ii) cell walls mostly stretch in the low-strain deformation of microcellular materials (Gibson and Ashby, 1982, 1997; Gibson, 1989). This, coupled with the observation that refinements in analysis lead to corrections that mostly cancel one another, led Gibson and Ashby to propose that the Young and shear moduli of 3D microcellular materials can be predicted over a wide range of relative densities by the simple expression

$$\frac{E}{E_{\mathrm{m}}} \quad \text{or} \quad \frac{G}{G_{\mathrm{m}}} = C_1 \Phi^2 V_{\mathrm{m}}^2 + C_2(1 - \Phi)V_{\mathrm{m}}, \tag{54}$$

where C_1 and C_2 are constants, and Φ is the fraction of the solid that is contained in cell edges (the struts, as opposed to the cell walls; the influence of nodes can generally be neglected). The first term is associated with bending of the struts (in 2D honeycomb structures the corresponding exponent is three instead of two), while the second is associated with stretching of the cell walls (bending of the cell walls generally contributes little to stiffening closed-cell microcellular materials; in 3D microcellular solids the corresponding exponent would be three (Grenestedt, 1999a)).

The bulk modulus of microcellular materials, on the other hand, is in principle proportional to the relative density

$$\frac{B}{E_{\mathrm{m}}} = C_3 V_{\mathrm{m}} \tag{55}$$

with $C_3 \approx 1/9$ (1/4 in 2D). The reason is that small-strain hydrostatic deformation, which should at first glance expand or contract the porous material homothetically, must therefore cause a uniform uniaxial stretching of all struts and cell walls (Walsh et al., 1965; Christensen, 1986, 2000; Kraynik and

Warren, 1994; Gibson and Ashby, 1997; Kraynik et al., 1999)—unless there are defects, such as curved struts and walls that will bend also under hydrostatic stress. In practice, imperfections in the walls or struts of foamed metals are nearly always present to the extent that K and E are essentially equal for most microcellular metals (Ashby et al., 2000).

Comparison of these expressions with experimental data for microcellular materials of all classes shows that eqn (54) is, overall, relatively well obeyed although the scaling exponents observed can deviate somewhat from 2 (Gibson and Ashby, 1982, 1997; Gibson et al., 1982; Ashby, 1983; Gibson, 1989; Lemay, 1991; Kovacik and Simancik, 1998; Andrews et al., 1999a; Ashby et al., 2000; Gui et al., 2000; Badiche et al., 2000; Roberts and Garboczi, 2001, 2002b; Kwon et al., 2003b; Benouali et al., 2005b; Despois et al., 2006b; Li and Zhu, 2006; Brothers and Dunand, 2005b, 2007; Imwinkelried, 2007; Boonyongmaneerat and Dunand, 2008; Erk et al., 2008; Amsterdam et al., 2008b; Wada et al., 2008; Scott and Dunand, 2010; Oppenheimer and Dunand, 2010; Roy et al., 2011; Zhou et al., 2011; Tuncer et al., 2011b; Schaedler et al., 2011). As concerns Young's modulus, C_1 and C_2 are both near unity (as extrapolation to zero porosity would suggest). For G, the constants are somewhat smaller, being nearer 3/8; however, there are fewer data for G (Gibson and Ashby, 1997). In practice, and this holds largely true for microcellular metals in particular, the contribution of cell wall stretching is generally negligible, corresponding to $\Phi \approx 1$. Data specific to microcellular metals are quite scattered, however, and the appropriate values for constants C_1 and C_2 can be well below unity for certain materials (Ashby et al., 2000; Despois et al., 2006b).

Predictions for the Poisson ratio, ν, of microcellular metals vary significantly (more on this below; see Section 24.5.1.5). Data suggest taking, as for dense metals, $\nu_e = 1/3$ also for microcellular metals (Ashby et al., 2000).

Detailed beam/node/plate-based models of microcellular material elastic deformation have, since, been proposed to address more precisely both the deformation of the beams and also the relation between their architecture and the resulting porous material stiffness. In two-dimensional structures, engineering beam theory has been used by a number of authors to model the linear elastic stiffness of various well-defined structures, such as the hexagonal honeycomb (Gibson and Ashby, 1997; Gibson et al., 1982; Zhu et al., 2001; Torquato, 2002a; Torquato et al., 1998; Warren and Kraynik, 1987; Papka and Kyriakides, 1994; Gong et al., 2005a; Jang et al., 2008). For three-dimensional structures, such models generally start with the relatively well-known topology of foams and emulsions, in which struts are mostly connected to three cell walls each, while being linked four at a time to a node at either end (Weaire and Fortes, 1994; Gibson and Ashby, 1997; Weaire and Hutzler, 1999; Weaire et al., 1999).

Capillary equilibrium, if it is obeyed, dictates that the angle between cell walls joining along a strut be 120°, and that the angle between four tetrahedrally connected struts meeting at a node be 109° (see Section 24.3). As noted above (Section 24.2.2.1), the most basic regular space-filling unit cell satisfying these criteria is a distorted version of the tetrakaidecahedron, or Kelvin cell, modified to have appropriately curved faces and edges, a more recent alternative being the Weaire–Phelan cell (Weaire and Fortes, 1994; Weaire and Hutzler, 1999), **Figure 3**. Since many metal foams are derived from polyurethane foam (Section 24.2.3), and since the same topological relations are also obeyed in microcellar metals produced by other processes (replication, notably, Section 24.2.4.3), a good-many theoretical analyses of the linear elastic behavior of microcellular materials are based on analysis of the deformation of a regular, flat-faced and straight edged, tetrakaidecahedron (in a regular, flat face tetrahedron, requisite angle relations are however not strictly obeyed, some beams meeting, for example, at 90° instead of 109°). If dealing with open-cell structures, the beams follow face edges; if closed-cell structures are modeled solid material lines the cell faces as well.

Analyses are based either on engineering beam theory and shell mechanics, or use finite element analysis. Unit cells variously combine various numbers of struts/nodes/walls. The end result is generally expressed as a simple power law scaling of stiffness and relative density

$$\frac{E}{E_m} \quad \text{or} \quad \frac{G}{G_m} = C_E V_m^N. \tag{56}$$

The linear elastic deformation of the regular tetrahedral four-beam element per se was analyzed by Warren and Kraynik (1988), Kraynik and Warren (1994), using beam theory and taking into account microbeam bending, stretching, and twisting, and by Sihn and Roy (2004) by giving the struts and node the shape dictated by the space between four spheres. It is found in both studies that N remains equal to 2 in eqn (56). Assuming triangular beam cross-sections, the authors concluded that $C_E = 11/10$ for E and $11/30$ for G at small V_m (for higher V_m C_E is given as a polynomial fraction), while in eqn (55) $C_3 = 1/9$. Regular packings of tetrahedrally connected beams (as in the regular Kelvin cell) were subsequently modeled by several teams, using either beam theory (in both engineering and Timoshenko variants) or finite element analysis. In such models, N is found to vary over a range going from roughly 1.7 to 2 (Chen and Lakes, 1995; Choi and Lakes, 1995; Zhu et al., 1997a, 2000; Warren and Kraynik, 1997; Li et al., 2003b; Gong et al., 2005a; Gan et al., 2005; Jang et al., 2008), in good agreement with earlier calculations for regular packings of simple tetrahedrally connected regular beam elements (Ko, 1965; Menges and Knipschild, 1975) (see eqn (53) above, which has furthermore been shown to fit simulation results rather well; for example (Zhu et al., 1997a; Gan et al., 2005)). Varying the cross-sectional shape of the strut of course leaves N unaffected but increases C_E significantly as the beam moment of inertia increases for a constant cross-sectional area and hence constant V_m (in that order, C_E varies from 0.9 to 1.1 to 1.53 for E going from circular to triangular to Plateau border beam cross-sections) (Warren and Kraynik, 1988; Kraynik and Warren, 1994; Warren and Kraynik, 1997; Li et al., 2003b; Jang et al., 2008).

For closed-cell structures, higher modulus values, coupled with lower N, are obtained for tetrakaidecadron-based structures, with $N \approx 1$ and $C_E \approx 1/3$ when cell walls represent the majority of the material, and N somewhat above unity otherwise (Grenestedt, 1999a; Mills and Zhu, 1999a, 1999b; Kraynik et al., 1999; Simone and Gibson, 1998c; Nammi et al., 2010; Song et al., 2010b). Unit-cell finite element models have also been proposed for other highly porous metal architectures, including bonded hollow sphere assemblies (Grenestedt, 1999a; Sanders and Gibson, 2003a, 2003b; Gasser et al., 2003, 2004a; Fiedler et al., 2006), syntactic foam (Vendra and Rabiei, 2010; Nguyen and Gupta, 2010), cubically connected beams (Warren and Kraynik, 1988; Grenestedt, 1999a; Kaoua et al., 2009) or plates (Grenestedt, 1999a), together with the various slightly less porous numerically simulated structures mentioned above.

Expressions above suppose that there is some regularity in the structure and deformation of porous metals; a different view is taken in Ref. Kovacik (1998b), Kovacik and Simancik (1998). Here, it is proposed that, in highly porous microcellular metals, Young's modulus is governed by percolation, E scaling in the same way with V_m as conduction. A scaling exponent of 1.65 was thus proposed for foamed aluminum (Kovacik and Simancik, 1998).

24.5.1.4 Influence of the Architecture of Microcellular Metals on their Stiffness
Strut/node/wall-based unit-cell models also enable exploration of the influence of variations in the architecture of the solid on the stiffness of microcellular linear elastic materials. Two factors can be

altered, (i) the arrangement of the beams (i.e. the angles at which these meet at nodes, and their lengths between nodes), and (ii) their shape. Both have been examined.

Hollowing the struts is one way to improve their structural performance (Andrews, 2006; Wadley, 2006). Another interesting venue to improve the stiffness of microcellular materials is to tailor the distribution of solid along the struts: reducing the cross-section of beams where the moment is smaller to increase their cross-section where it is larger will stiffen the foam at constant relative density. The influence of strut cross-section nonuniformity (i.e. tapered struts) in open-cell microcellular structures was therefore explored by several authors, to conclude that significant stiffness improvements can be obtained for both open-cell (Warren and Kraynik, 1988; Kim and Al-Hassani, 2001a; Gong et al., 2005a; Jang et al., 2008) and closed-cell microcellular materials (Simone and Gibson, 1998c) as well as honeycombs (Warren and Kraynik, 1987; Simone and Gibson, 1998c; Kim and Al-Hassani, 2001a; Chuang and Huang, 2002; Yang and Huang, 2004; Huang and Chang, 2005; Yang et al., 2008b; Harders et al., 2005; Gong et al., 2005a). The exponent N remains at, or slightly below, two if the overall strut or cell wall shape is kept constant with changing relative density (Warren and Kraynik, 1988; Gong et al., 2005a; Jang et al., 2008; Simone and Gibson, 1998c) (Warren and Kraynik, 1987; Simone and Gibson, 1998c; Kim and Al-Hassani, 2001a; Harders et al., 2005), while it was found to be slightly above 2 at very low density, falling at higher density, in capillarity-governed anisotropic open-cell foam structures featuring tapered struts (Gong et al., 2005a; Jang et al., 2008).

Noteworthy in this regard is the work of Gong, Jang and Kyriakides, who measured and simulated the distribution of matter along the struts of a polymer as well as a commercial (Duocel®) open-cell aluminum foam, to show that the redistribution of solid matter under the action of capillary forces in the original liquid (polymer) foam structure that defines the structure of the microcellullar material tends to improve the material stiffness by as much as 70% compared to simpler "straight-beam" configurations (Gong et al., 2005a). Another interesting exploration of strut shape was by Markaki and Clyne, who simulated the deformation of isotropic random fiber structures as a single, randomly oriented, fiber rigidly connected to its neighbors, viewing the fiber aspect ratio as an additional structural parameter that may vary at fixed relative density (given that no explicit assumption was made as to the arrangement and connectivity of the fibers). The final result is a simple expression for modulus which, for fixed aspect ratio, varies somewhat more slowly with relative density than does power two scaling in eqn (56) (Clyne et al., 2005; Markaki and Clyne, 2005, 2004).

The influence of variations in the dimensions, geometrical distribution and thickness of the struts or cell walls has been explored by generating large unit cells containing many tetrahedrally connected beams. Such structures have generally been produced by disturbing randomly a starting regular distribution of cells, such as the body centered cubic packing characteristic of a regular periodic Kelvin cell structure (Van der Burg et al., 1997; Grenestedt, 1999b, 2005; Zhu et al., 2000; Li et al., 2005, 2006a, Jang et al., 2008; Daxner et al., 1999b, 2002; Luxner et al., 2005, 2007, 2009a, Babaee et al., 2012). An interesting result of such simulations is that irregularity in the cell geometry or size can lead, in 3D open-pore microcellular solids, to an increase in E or G, by as much as 50%, while the bulk stiffness is reduced by as much as 20% (leading in turn to reductions in Poisson's ratio; see eqn (58) below) (Van der Burg et al., 1997; Zhu et al., 2000; Li et al., 2006a, Jang et al., 2008, 2010; Babaee et al., 2012). Analogous results were also obtained for "disturbed" irregular honeycombs, and also for honeycombs containing bimodal cell sizes (Chen et al., 1999; Daxner et al., 1999b, 2002; Zhu et al., 2001; Fazekas et al., 2002; Li et al., 2005; Alkhader and Vural, 2008) (more drastic deviations from regular cell geometries do, however, produce a general reduction in the stiffness of honeycombs (Chen et al., 1999)). Analysis of detailed stress distributions in Ref. Van der Burg et al. (1997) showed that this effect of disorder is linked with a relative increase in tensile versus bending or torsion components of strut deformation with increasing disorder; this in turn has the effect that N decreases slightly below 2 as the level of disorder increases. Effects of more drastic alterations in the geometry and distribution of struts or cell walls within foams can be found in fully numerical simulations mentioned above (Roberts and Garboczi, 2001, 2002a, 2002b).

Just as the redistribution of mass within cells can improve the stiffness of microcellular materials, their overall structural performance in elastic structures can be improved by redistributing mass across cells, that is, over distances much larger than the cell diameter, on par with those of the component made of the material. Such functional grading of microcellular materials for improved component stiffness has been explored, both theoretically and experimentally, by Pollien et al. (2005), Ajdari et al. (2009), Taylor et al. (2011).

Alternatively, the role of defects can be explored: beams or cell walls might be more compliant because these are bent, contain wiggles, are of uneven thickness or altogether absent. The effect of corrugations or curvature in cell faces of closed-cell honeycombs or in Kelvin-cell 3D microcellular solids has thus been explored. Strut corrugations sharply reduce the stiffness of open-cell foams. So do corrugations in closed-cell foams, but to a lesser extent because cell walls always deform to some extent by stretching - still, stiffness decreases can reach 50% also in this case (Simone and Gibson, 1998b; Grenestedt, 1998; Chen et al., 1999; Daxner et al., 1999b; Grenestedt, 2005). A wall-to-wall distribution in the thickness of straight cell walls also gives, at equal relative density (and hence equal average thickness), a more compliant microcellular material (Grenestedt and Bassinet, 2000) (Li et al., 2005, 2006a). Reentrant angle beams sharply lower the elastic stiffness (but yield interesting Poisson's ratios; see below) (Choi and Lakes, 1995).

In both 2D (Silva and Gibson, 1997; Albuquerque et al., 1999; Ajdari et al., 2008; Symons and Fleck, 2008; Cui et al., 2011) and 3D (Roberts and Garboczi, 2001; Wallach and Gibson, 2001a; Gan et al., 2005; Luxner et al., 2009a) microcellular solids, missing struts or missing walls can cause sharp reductions in stiffness. One interesting observation in this regard is that the consequences of missing or otherwise degraded struts depend on the architecture of the porous material. For example, the shear modulus of honeycombs is little affected by missing or altered struts in both hexagonal, bending-dominated, or high (redundant) connectivity triangular structures, while in the Kagome lattice, missing or degraded cell walls lead to strong changes in stiffness because they cause a transition from stretch- to bending-dominated deformation (more on this below) (Symons and Fleck, 2008). This in turn has led to suggest that, in Kagome structures (see Section 24.5.1.7 below), actuated struts can be used for sensing or morphing applications (Wicks and Guest, 2004; Symons et al., 2005a, 2005b; Symons and Fleck, 2008).

Several contributions have also accounted for the fact that many microcellular materials are not isotropic (e.g. (Beals and Thompson, 1997; Nieh et al., 2000; Mu and Yao, 2010)). This concerns notably polyurethane foam and hence all microcellular metals derived therefrom: cells are elongated in the "rise" direction, with the result that the foam is elastically anisotropic (generally, it is orthotropic). Simple unit-cell approaches have been proposed to account for anisotropic porous metal microstructures (Huber and Gibson, 1988; Gibson and Ashby, 1997; Gong et al., 2005a; Dillard et al., 2005; Rösler and Näth, 2010), as have mean-field models (Ichitsubo et al., 2002; Kitazono et al., 2003; Tane and Ichitsubo, 2004; Tane et al., 2004a, 2004b, 2006, 2007) and models based on elongated periodic unit cells (Silva et al., 1995; Gong et al., 2005a; Sullivan et al., 2008; Jang et al., 2008) or based on tomographic images of actual open-cell foams as well as structures dictated by capillary equilibrium of Plateau border networks (Jang et al., 2008; Singh et al., 2010).

An interesting feature of the deformation of highly porous solids is that, even though the material these are made from remains linear elastic, the porous material itself can, beyond a certain strain, deform in highly nonlinear fashion: this is regularly observed in the deformation of polymer foams, for example. Bending, rotation, elastic buckling and cooperative movement of slender beams that locally sustain only small strains can lead to large nonlinear end point displacements that translate, in turn, into a globally nonlinear deformation law for the microcellular materials—even though it is made of strictly linear elastic solid. This subject, of the high-strain deformation of linear elastic microcellular materials, has been tackled by many authors (e.g. (Warren and Kraynik, 1991; Kraynik and Warren, 1994; Zhu et al., 2006, 1997b; Zhu and Windle, 2002; Mills and Zhu, 1999b; Gan et al., 2005; Brydon et al., 2005)). This said, the subject is of reduced relevance in microcellular metallic materials, given that most metals yield plastically and do so at relatively low elastic strain. With a few exceptions (see Section 24.5.3.5) localized plastic yielding tends to be the major cause for nonlinearity in the deformation of microcellular metals.

Plastic deformation brings irreversible changes in the local metal architecture of metal foams. Uniaxial compression into the plastic regime will generally cause a lowering of the Young's modulus along the compression direction (Bastawros et al., 2000; San Marchi and Mortensen, 2001; San Marchi et al., 2004; Lefebvre et al., 2006; Friedl et al., 2008; Aly et al., 2009; Papadopoulos et al., 2010; Ochiai et al., 2010a; Li and Dunand, 2011). A very simple model based on the tetrahedrally connected upright four-beam element captures the effect surprisingly well for replicated aluminum (San Marchi and Mortensen, 2001). Similarly, uniaxial tensile deformation accompanied with local plastic yielding in the metal will align struts parallel to the stress axis, causing in principle the material's Young's modulus to increase along the tensile direction. This has been observed; however, a decrease may be found instead. The reason is that a buildup of internal damage can override this geometrical evolution of the material architecture (San Marchi et al., 2004; Aly et al., 2009; Ochiai et al., 2010a). As a result of those effects, when monotonic mechanical (e.g. uniaxial) testing is used to measure an elastic (e.g. Young's) modulus of microcellular metals, it is advisable to measure the porous metal stiffness at various strains using repeated unloading/reload cycles and to then deduce the initial value of the elastic modulus by extrapolation to zero strain, because the initial slope of the stress-strain curve is often affected by microplasticity and/or misalignement effects (San Marchi and Mortensen, 2001; San Marchi et al., 2004; Lefebvre et al., 2006; Despois et al., 2006b; Friedl et al., 2008; Ochiai et al., 2010a). In more spectacular fashion, large-strain hydrostatic compression of copper foam has been shown to cause strong reductions in (isotropic) stiffness of the material (despite the increasing relative density) (Choi and Lakes, 1992a, 1995). Such compressed microcellular metals have unique elastic properties, as we see next.

24.5.1.5 Poisson's Ratio and Auxetic Metals

The Poisson ratio, ν, of microcellular materials is an interesting property. This is the ratio of lateral contration to longitudinal extension of the material under longitudinal tensile stress

$$\nu_{12} = -\varepsilon_2/\varepsilon_1. \tag{57}$$

In dense solids, ν is generally slightly below $1/2$ (which corresponds to no volume change in a 3D isotropic solid): pulling on a solid material in its linear elastic range of deformation generally causes its volume to shrink slightly.

In most microcellular materials ν is found to vary roughly within the same range as for dense solid materials, that is, roughly between 0.2 and 0.5. This is reflected in models, which predict various values in this range for most porous materials (Gibson and Ashby, 1997; Ashby et al., 2000; Roberts and Garboczi, 2000, 2001, 2002a, 2002b; Sihn and Roy, 2004); however, the presence of pores make it

(a)   **(b)**

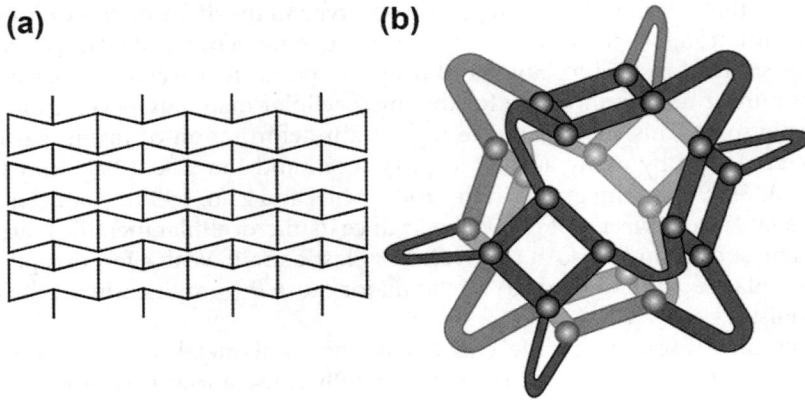

Figure 32   A 2D (top) and a 3D (bottom) auxetic structure, after R. Lakes et al. (Choi and Lakes, 1995, 1996; Prall and Lakes, 1997).

possible to vary ν much more widely. In fact, if one engineers the material's architecture appropriately, ν can be made to vary essentially over the entire physically allowable range, which stretches from -1 to $1/(D-1)$ in isotropic materials (where $D = 2$ or 3 is the material's dimension) —in essence for the same reason that porous materials can display a much wider range of CTE values (see Section 24.4.2) (Weaire and Fortes, 1994; Choi and Lakes, 1995). Materials with a negative Poisson ratio are thus possible: these contract laterally when compressed and expand when pulled upon; such a material is called "auxetic".

Mechanisms that can be used to this end are found in a few usual objects, such as an accordeon (zero apparent Poisson ratio), or puppets that extend their arms and legs when a string is pulled (negative apparent Poisson ratio). Microcell architectures have been proposed to vary ν over essentially the entire allowable range (Sigmund, 1994, 1995; Choi and Lakes, 1995; Prall and Lakes, 1997; Milton, 1992; Weaire and Fortes, 1994; Gaspar et al., 2005; Smith et al., 2000; Grima et al., 2006; Doyoyo and Wan Hu, 2006; Hughes et al., 2010; Taylor et al., 2011; Schwerdtfeger et al., 2010). One simple design to this end is a microcellular structure containing concave pores, or in other words solid struts or walls that have reentrant angles, **Figure 32**. These unfold as the structure is pulled in one direction, causing expansion in transverse directions (much, as Choi and Lakes note (Choi and Lakes, 1995), as with a piece of crumpled paper). The material hence has a negative Poisson's ratio.

In isotropic 3D elastic solids, the relation

$$\nu = \frac{1}{2} - \frac{E}{6B} \tag{58}$$

also gives hints as to how a low Poisson ratio microcellular material can be designed: it should have a bulk modulus K that is a low fraction of its Young's modulus E. This can be achieved if the microcellular metal has many bent or folded beams or cell walls. This in turn suggests an economical way to produce low or negative Poisson ratio microcellular materials, namely to create the rentrant angle struts or cell walls by buckling, or in other words by precompression of "normal" microcellular materials. Auxetic porous metals have thus been produced by predeformation well into the plastic regime, so as to create many bent or wiggled struts within the structure. This is a particularly inexpensive method that is made easy, and has been demonstrated, with metals, given their ability for plastic deformation (Friis et al., 1988; Choi and Lakes, 1992a, 1995) (unlike polymers, which need to be predeformed at elevated temperature (Lakes, 1987; Friis et al., 1988; Choi and Lakes, 1992b)). Other methods and designs producing auxetic microcellular metals, such as solid free-form fabrication, were demonstrated in Refs. (Hassan et al., 2008; Bianchi et al., 2011; Schwerdtfeger et al., 2010a; Schwerdtfeger et al., 2010b; Yang et al., 2012c).

24.5.1.6 Size Effects in Elastic Deformation

A salient feature peculiar to microcellular materials is the existence of size effects, meaning of variations in materials properties with the scale of the material, also in linear elastic deformation (Tekoglu et al., 2011). In one sense, this is relatively easy to understand: within a band roughly one cell diameter wide stretching along a free surface, a certain number of foam struts or cell walls either are left "dangling" (meaning unconnected at one end and therefore essentially unloaded) if the foam was cut, or alternatively are differently connected or in some cases covered by a thin skin of dense metal, all depending on how the part was produced (see Section 24.2). It has thus been documented that, if the diameter or width of a cut sample of microcellular metal is less than a certain number of cell diameters (roughly seven in Refs. Andrews et al. (2001), Onck (2003b), Jeon and Asahina (2005), roughly 20 in Refs. Brezny and Green (1990), Rakow and Waas (2004), (2005)), the measured stiffness becomes dependent on the width of the sample and more scattered.

In finite element simulations of honeycombs, similar effects are found:

(i) in uniaxial deformation or bending, straight honeycomb samples less than 10–20 cells wide see their Young's modulus decrease with decreasing width (Tekoglu and Onck, 2005; Tekoglu et al., 2011),

(ii) in shear, honeycombs bonded to a stiff plate see their shear modulus increase as the number of cells across the sandwich decreases (unlike a free surface, bonding to a plate creates a stiffened boundary layer) (Tekoglu and Onck, 2005; Tekoglu et al., 2011) such that

(iii) boundary conditions on the honeycomb area exert a strong influence on simulations results (Chen et al., 1999);

(iv) the simulated area must be at least roughly 20 cells wide in all directions for the results to be accurate within a few percent (Chen et al., 1999; Onck et al., 2001; Fazekas et al., 2002; Mangipudi and Onck, 2011a; Tekoglu et al., 2011) and

(v) these boundary layer effects also depend on the honeycomb structure, being far more pronounced for Kagome than for triangular or hexagonal honeycomb structures (Phani and Fleck, 2008; Cui et al., 2011).

In recent 3D simulations of capillarity-equilibrated, random open-cell structures, the average Young's modulus obtained was also found to increase with cell size (Jang et al., 2010).

For meaningful measurement of the elastic constants of microcellular or porous materials, therefore, test samples must exceed a certain size. As a practical guide, it is worth recalling a simple calculation given by Kocks on the testing of polycristalline metals: to get meaningful data at least 20 grains must run across the gage section of a tensile sample (Kocks, 1970). Below this a large proportion of grains touch the free surface (with 10 grains or pores across the sample diameter, roughly one-third of all grains or pores lie along the specimen surface). Similarly, samples of microcellular material should a *priori* be at least 20 cells across in all directions if meaningful data characteristic of the material are to be measured. In some cases, this may mean that test bars must be rather large (since some metal foams have pores that are on average several millimeters wide).

This size effect can also complicate significantly the mechanical analysis of structures made of porous metal, in all regimes of deformation including linear elastic. Linear continuum elasticity is in principle size independent; however, with porous materials the pores, with their finite diameter, introduce an intrinsic dimensional scale to the material that may affect its mechanical behavior, particularly in the presence of stress concentration sites. Take the simple case of a plate of porous metal through which a hole is bored. If the hole is, say, one hundred times larger than all pores, the problem can be reasonably tackled using classical (Cauchy) linear elasticity knowing the elastic constants of the porous metallic material. If, on the other hand, the hole is no bigger than a pore, it becomes just another pore and the material is, for all practical purposes, stressed homogeneously and does not "see" the hole. Between these two extremes there is a range of hole sizes for which neither of these two solutions holds. The same of course applies to any other site of stress concentration that a porous material may contain (another hollow, a notch, an inclusion or the interface with another material, a triple line, ...).

To enable the mechanical analysis of such intermediate-scale inhomogeneities, generalized continuum approaches have been formulated. Here, additional kinematic or constitutive variables are introduced besides the stress and (small) strain characteristic of classical Cauchy continua. Those extra variables in turn give, through gradient terms and associated modifications of the material constitutive laws, a size scale to the problem. Then, if the medium represents successfully the material, its analysis enables prediction of the mechanical behavior of porous metals containing elements of shape (holes, notches, inclusions, etc) that have any size scale above that of the pores. Generalized continuum models have been proposed by several authors to account for size effects in the elastic deformation of microcellular materials in general (Lakes, 1983, 1986; Chen and Huang, 1998) and microcellular metals in particular (Onck, 2003b; Tekoglu and Onck, 2003; 2005; Dillard et al., 2006; Neff and Forest, 2007; Forest and Trinh, 2011); the two last references give recent reviews of the subject.

When the pore size falls roughly below 10 nm, a different, more intrinsic, size effect can become manifest in the elastic deformation of porous metals. This is because, at the very high internal surface to volume ratio characteristic of nanoporous metals, surface stresses may contribute noticeably to the internal stress equilibrium within the solid (or in more elementary terms because a significant proportion of atoms in the solid become situated along pore surfaces and are thus differently bonded). As a result, the response to (reversible elastic) strain can be altered, changing in turn the elastic modulus of the material (Weissmüller et al., 2009; Wang et al., 2006a; Duan et al., 2006b; Re et al., 2011). Experimental data exist to suggest several-fold increases in Young's modulus as the average diameter of ligaments decreases below a few tens of nanometers. Interpretation of the data is however complicated, for example, because it is difficult to ascertain that there was no simultaneous change in the architecture of the ligaments with their size (Weissmüller et al., 2009; Cuenot et al., 2004; Mathur and Erlebacher, 2007). Surface stress effects coupled with elastic deformation in open-pore nanoporous metals and their ability to "breathe" fluids also bring other interesting phenomena: finite strain can, for example, appear with a change in chemical, or electrochemical, environment, this in turn having obvious relevance to actuation or sensing (Jin et al., 2008; Jin and Weissmüller, 2010; Detsi et al., 2011). This is currently a rich area for research.

24.5.1.7 *Architectural Optimization*

How do elastic properties of microcellular materials compare with what is best possible, that is, with upper bounds from theory, for example, Hashin–Shtrikman bounds when dealing with isotropic materials? Written for isotropic microcellular solids in the limit of vanishing relative density V_m, these (eqns (45)–(47)) reduce to (Grenestedt, 1999a; Torquato, 2002a)

$$\frac{B}{E_m} = \frac{2V_m}{9(1 - \nu_m)} \text{ in 3D;} \quad = \frac{V_m}{4} \text{ in 2D,} \tag{59}$$

$$\frac{G}{E_m} = \frac{(7 - 5\nu_m)V_m}{30(1 - \nu_m^2)} \text{ in 3D;} \quad = \frac{V_m}{8} \text{ in 2D.} \tag{60}$$

These bounds are realized, or nearly so, in some theoretical structures other than the hole-in-a-sphere structure that were used to show that the expressions are best possible (Torquato, 2002a).

In honeycombs, the above limits are reached by regular triangular cell structures, and also by Kagomé honeycombs (Gibson and Ashby, 1997; Torquato et al., 1998; Christensen, 2000; Hyun and Torquato, 2000; Torquato, 2002b, 2002a) —in fact for the triangular honeycomb, the Hashin–Shtrikman bound is nearly reached at all V_m values (Hyun and Torquato, 2000). Those relatively simple structures are, thus, optimal—which is not the case for several other designs commonly used, such as the classical hexagonal honeycomb, where bending dominates the in-plane deformation of cell walls such that G is proportional to V_m^3 and takes much lower values.

In 3D microcellular solids, these upper bounds are nearly reached by regular closed-cell structures in which essentially all solid material is confined to the cell walls and uniformly or optimally distributed—geometries that in fact approach the hole-in-a-sphere geometry shown to realize the bound. One example is the Kelvin cell structure which gives nearly isotropic elastic modulus values not far from those in eqns (59) and (60) (Simone and Gibson, 1998c; Grenestedt, 1999a; Mills and Zhu, 1999a, 1999b; Kraynik et al., 1999).

Other, often more complicated, structures have also been found numerically, using various optimization algorithms applied to periodic unit cells in which the distribution of solid matter is incrementally iterated (often along 2D square or 3D cubic symmetry patterns) (Bendsøe and Sigmund, 2003; Torquato, 2010). Final structures maximize a chosen performance function, such as the elastic rigidity in one or several directions (Sigmund, 1994, 1995; Neves et al., 2000; Guedes et al., 2003; Bruck et al., 2007; Qiu et al., 2009; Huang et al., 2011), or combinations of properties, such as stiffness under multiload conditions, or combinations of (bulk) elastic rigidity with a physical property such as conductivity (Torquato and Donev, 2004) or fluid permeability (Guest and Prévost, 2006), all at given relative density and often reaching theoretical upper bounds for the relevant property or property combination.

Looking at structures that realize or approach optimality from the standpoint of elastic rigidity, one finds that these are all structures in which the dominant stress state within the solid phase making the microcellular material is predominantly one of tension or of compression, also when the microcellular solid is subjected to stress states other than hydrostatic pressure, such as uniaxial tension or pure shear. In 3D microcellular structures, this naturally occurs in regularly shaped closed-cell structures, provided of course the walls are straight, well shaped, properly oriented and interconnected.

An important contribution to the question was the realization by Deshpande et al. (2001a), Ashby (2006) that the same predominance of tensile or alternatively of compressive stresses within the solid of microcellular solids can also be achieved in open-cell structures—provided the struts are properly arranged and connected. This, in turn, can be achieved only if (but not always if; this is a necessary but not sufficient condition) Maxwell's stability criterion is satisfied by the architecture of the microcellular open-pore solid viewed as a structure of beams connected at joints. Here, the beams and joints are the struts and nodes of the microcellular solid, respectively. Specifically, optimal structures must statisfy, if there are b beams and j joints in the structure

$$M = b - Dj + 3(D - 1) \geq 0 \qquad (61)$$

(with D the microcellular solid dimension). If all joints (nodes) are made loose, meaning if the connection is free to rotate and the angle between the beams or struts can change without work or stress, then if M is less than zero the structure will collapse under any stress: it is then a "mechanism". If instead of being loose the joints or nodes are made rigid, such a mechanism will be able to resist deformation, but will only do so in proportion to the bending stiffness of the beams or struts. This is what is generally seen in open-cell microcellular structures. The reason is that nearly all regular space-filling structures, including in particular the Kelvin cell in 3D or the hexagonal honeycomb in 2D (and of course distorted versions thereof, and similar structures such as replicated open-cell metals), have M < 0. If M ≥ 0, on the other hand, depending on specifics of the structure, it might carry load even if its joints are loose: this is a "stretch-dominated structure," in which beams or struts are subjected to uniaxial loading if the structure is deformed.

To achieve M ≥ 0 in microcellular materials, therefore, Maxwell's rule translates into a requirement that the average connectivity (average number of struts linking at a node) Z of the structure be at least 4 in 2D, and 6 in 3D (because with many cells, $b \approx Zj/2$). If, in addition, the structure is such that it is made of similar beams similarly connected everywhere, such that the truss is invariant as seen from any joint, then it can be shown that Z must be 6 in 2D (as in the regular triangular honeycomb), and 12 in 3D (Deshpande et al., 2001a). The former is achieved with a 2D triangular cell honeycomb; the latter can, for example, be produced with a 3D structure made of trusses joining sites of a face centered cubic lattice (this is called the "octet truss lattice" (Deshpande et al., 2001b)).

Such stretch-dominated space-filling open-cell structures thus exist and come close to, or even realize, upper bounds for elastic stiffness. Two-dimensional structures of this kind include the triangular cell honeycomb, which was already mentioned above; viewed in 3D, such 2D materials are, like any other honeycomb, anisotropic microcellular solids that are closed-cell in two directions and open-cell along the third direction (Wadley, 2002, 2006). Truly 3D and truly open-cell stretch-dominated microcellular structures include tetrahedral (octet truss), **Figure 33** (Deshpande et al., 2001b),

10 mm

Figure 33   Photograph of a three-dimensional aluminium octet-truss lattice produced by precision casting. Reproduced from *Journal of the Mechanics and Physics of Solids*, vol. 49, Deshpande, V. S., Fleck, N. A. & Ashby, M. F. "Effective Properties of the Octet-truss Lattice Material", pp. 1747–1769 (Deshpande et al., 2001b), with permission from Professor Fleck of Cambridge University and Elsevier.

pyramidal ("lattice block") and 3D Kagome periodic beam assemblies, (Evans et al., 2001; Wallach and Gibson, 2001b; Wadley, 2002; Chiras et al., 2002a; Zhou et al., 2004c; Aboudi and Gilat, 2005; Wadley, 2006; Hutchinson and Fleck, 2005; Hutchinson and Fleck, 2006; Faure and Doyoyo, 2007; Queheillalt and Wadley, 2011). These are repeatable, space-filling, strut-based structures that can be used to build high-performance 3D microcellular solids.

The in-plane and normal moduli of such highly connected lattice truss structures have been calculated on the basis of beam models (which hence apply only to very low relative densities): predicted values are indeed proportional to V_m. The constant of proportionality is (of necessity) somewhat below unity and depends on the structure type and its geometrical parameters, (Deshpande and Fleck, 2001a; Queheillalt and Wadley, 2005a; Hutchinson and Fleck, 2005; Hutchinson and Fleck, 2006; Wadley, 2006; Kooistra and Wadley, 2007; Kooistra et al., 2008); this is illustrated for the octet-truss lattice in **Figure 34** (Deshpande et al., 2001b). As seen, the different scaling with V_m causes the octet-truss (or "FCC" because nodes of the octet-truss structure lie along a face-centered cubic lattice) lattice to be far stiffer at low V_m along its <111> direction than stochastic, bending-dominated, metal foams. Note also that its modulus lies only a factor 2.5 below the (isotropic solid) Hashin-Shtrikhman upper bound. To give another example, the in-plane shear modulus G of tetrahedral lattice truss sandwich core structures was calculated and measured to be $G = 0.11 E_m V_m$ (Kooistra et al., 2008), while the shear modulus of the three-dimensional Kagome lattice gave $G = 0.125 E_m V_m$ (Hutchinson and Fleck, 2006). These values are roughly half the Hashin-Shtrikman upper bound for the stiffness of an isotropic porous solid in the limit of low density (Eq. 5-18). If one considers the wide range of stiffness values potentially displayed by porous materials at very low V_m (**Figure 31**), one realizes that this is a very high-performance structure from the standpoint of stiffness.

Processing methods exist for such structures (see Section 24.2.6); these tend to produce mostly coarse structures with cell sizes well above the millimeter. As a result, relatively little work exists on

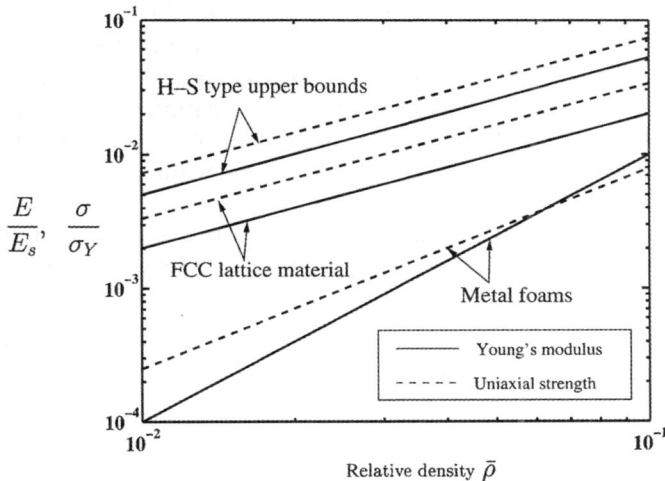

Figure 34   Structural performance as measured by calculated values of Young's modulus and uniaxial yield stress plotted versus relative density V_m for the three-dimensional octet-truss (or "FCC") lattice along the FCC lattice <111> direction, contrasted with corresponding values for the Hashin-Shtrikman upper bound (Eqs. 43 to 45, 59 and 60) and the Gibson-Ashby scaling law for bending-dominated open-cell microcellular solids (Eq. 14 with $N = 2$ and $C_E = 1$). Reproduced from *Journal of the Mechanics and Physics of Solids*, vol. 49, Deshpande, V. S., Fleck, N. A. & Ashby, M. F. "Effective Properties of the Octet-truss Lattice Material", pp. 1747–1769 (Deshpande et al., 2001b), with permission from Professor Fleck of Cambridge University and Elsevier.

stretch-dominated cellular structures that can, in truth, be viewed as a material; such an investigation is reported in Ref. (Deshpande et al., 2001b). Most contributions, both theoretical and experimental, on high-connectivity open-cell structures have, instead, focused on their use as a one- or two-layer core within sandwich beams: in this configuration they perform well but function, not so much as a material, but rather as a discrete set of connected ribs (e.g. (Wicks and Hutchinson, 2001; Evans et al., 2001; Deshpande and Fleck, 2001a; Wallach and Gibson, 2001b; Chiras et al., 2002; Hutchinson et al., 2003; McShane et al., 2006; Wadley, 2002, 2006; Rathbun et al., 2005; Zok et al., 2004; Li et al., 2008; Bouwhuis et al., 2008; Ho et al., 2010; Queheillalt and Wadley, 2011; Pingle et al., 2011)).

24.5.2 Yield

Departure from linear elastic deformation is, in porous metals, generally observed because plastic deformation initiates somewhere within the metal. In this respect porous metals differ qualitatively from porous polymers or ceramics, where buckling or microfracture generally bring purely elastic deformation to an end (Gibson and Ashby, 1997); yield criteria for porous metals are, hence specific to this class of materials. A few exceptions exist, however, in which yield is governed by buckling: these include (i) honeycombs, in which microbuckling and folding of long, regular and slender walls is observed also in metal (e.g. (Gibson and Ashby, 1997; Mohr and Doyoyo, 2004b; 2004a)), and (ii) high yield strain or brittle metals, an example being bulk metallic glass (see Section 24.5.3.5).

Plasticity sets in early during the deformation of porous metals: it is only in the very first stages—if at all—of monotonic deformation that porous metals behave in strictly linear elastic fashion. The reason for this is that nearly all porous metals contain a high density of stress concentration sites, causing localized plastic deformation to be manifest at very low applied strain. For this reason the intrinsic linear elastic modulus of a porous metal is, as already mentioned, best measured using load/unload cycles (e.g. (San Marchi and Mortensen, 2001; Despois et al., 2006b)), alternatively very low stress (acoustic) methods can be used. Microplasticity then spreads until one notices, at a relatively well-defined stress, that the porous metal stress–strain curve deviates in clearly noticeable fashion from linearity.

Yield surfaces quantify the onset of plasticity in a material: these delineate, in 3D (principal) stress space, the range of stress combinations below which deformation remains elastic. If the material is isotropic (which unless otherwise indicated is assumed for simplicity in what follows), this surface does not depend on the spatial orientation of the principal stress directions: it is then a single surface characteristic of the material.

A convenient way of expressing or drawing yield surfaces is to use, instead of the three principal stress values σ_1, σ_2 and σ_3, the hydrostatic (mean) stress σ_m, the Von Mises equivalent stress σ_e and the Lode angle θ, respectively, defined as

$$\sigma_m = \frac{\sigma_1 + \sigma_2 + \sigma_3}{3}, \tag{62}$$

$$\sigma_{eq} = \sqrt{\frac{(\sigma_1 - \sigma_2)^2 + (\sigma_2 - \sigma_3)^2 + (\sigma_3 - \sigma_1)^2}{2}} = \sqrt{\frac{3\left(\sigma_1'^2 + \sigma_2'^2 + \sigma_3'^2\right)}{2}} \quad \text{with} \quad \sigma_i' = \sigma_i - \sigma_m, \tag{63}$$

$$-\sqrt{3}\tan(\theta) = \frac{2\sigma_3 - \sigma_1 - \sigma_2}{\sigma_1 - \sigma_2} \quad \text{or} \quad \cos(3\theta) = \frac{27}{2}\frac{(\sigma_1 - \sigma_m)(\sigma_2 - \sigma_m)(\sigma_3 - \sigma_m)}{(\sigma_{eq})^3}, \tag{64}$$

where σ' is the deviatoric part of the stress tensor, (Hill, 1950). These coordinates are convenient because in stress space, the radius r, the polar angle θ and the altitude z in a cylindrical coordinate system centered at the origin of the $(\sigma_1, \sigma_2, \sigma_3)$ axes and having its axis oriented along their bissectrix $(\sigma_1 = \sigma_2 = \sigma_3)$, are, respectively, $r = (2/3)^{1/2} \sigma_e$, θ, and $z = \sqrt{3}\,\sigma_m$. With an isotropic solid, symmetry dictates that this yield surface be unchanged as σ_1, σ_2 and σ_3 are permutated; this in turn dictates that the yield surface be of threefold symmetry about the bissectrix, and symmetric about each of the three planes defined by the bissectrix and one of the three $(\sigma_1, \sigma_2, \sigma_3)$ axes (Hill, 1950; Bai and Wierzbicki, 2008).

Plastic yield in dense metals is most often insensitive to hydrostatic stress because no significant change of volume accompanies dislocation creation, dislocation glide, or twinning. Their yield criterion is therefore independent of σ_m, and their yield surface is a (general) cylinder aligned along the bissectrix $(\sigma_1 = \sigma_2 = \sigma_3)$. If yield is also independent of the Lode angle θ, then this cylinder is circular and the corresponding yield criterion is the von Mises yield criterion $(\sigma_e = \sigma_y)$ (Hill, 1950).

When a metal contains pores, on the other hand, it may yield under hydrostatic stress, even though the dense metal itself does not. The yield surface of porous metals is, therefore, not a cylinder; rather, it is a closed surface, capped at either end by the yield stress under purely tensile or compressive hydrostatic stress. An example of such a yield surface, drawn in the most general case but assuming that yield is symmetric between tension and compression (or in other words that yield is indifferent to the sign of σ_m) is given in **Figure 35**. If the yield surface is independent of the Lode angle θ, or in other words if it is independent of the third invariant of the stress tensor, then the three "lobes" visible on the surface drawn in **Figure 35** disappear. The yield surface is then a surface of revolution about the bissectrix, and the yield criterion is expressed using only the equivalent and mean stress values (σ_e, σ_m).

Plastic yield in porous metals is a problem of considerable practical importance: it governs, among other questions, the ductile fracture of metals by internal voiding, and also the cold pressing of metal powder compacts. Many expressions have thus been proposed to express the onset of yield, or in other words to draw the yield surface, of porous metals (Shahbeyk, 2013). For values of V_m down to roughly 50%, the most commonly used yield criteria are variations of the Gurson criterion, proposed on the basis of a simplified analysis of plastic flow around a spherical void (Gurson, 1977). Such yield criteria can often be written as (Pardoen and Besson, 2004)

$$\Phi = G_1 \left(\frac{\sigma_{eq}}{\sigma_{y,m}}\right)^2 + G_2 \cosh\left(G_3(V_m, S)\frac{\sigma_m}{\sigma_{y,m}}\right) + G_4 = 0, \tag{65}$$

where $\sigma_{y,m}$ is the dense metal yield stress, and the G_i are functions of the pore fraction $(V_p = 1 - V_m)$ and of the void geometry, which is quantified using a shape parameter. Gurson's original criterion takes $G_1 = 1$, $G_2 = 2(1 - V_m)$, $G_3 = 3/2$, $G_4 = -[1 + (1 - V_m)^2]$. To improve agreement with finite element simulations of periodic structures containing single voids, Tvergård and Needleman proposed instead $G_2 = 2q_1(1 - V_m)$, $G_3 = 3q_2/2$, $G_4 = -[1 + \{q_1(1 - V_m)\}^2]$, where q_1 and q_2 vary somewhat with the yield stress and strain hardening exponent of the metal. Values for those parameters can be found in Refs. Faleskog et al. (1998), Pardoen and Besson (2004) for configurations of interest to ductile fracture (round voids surrounded by solid); on average these remain near $q_1 = 1.5$ and $q_2 = 1$. To account for void-to-void coalescence, above a threshold porosity $V_p = (1 - V_m) \approx 15\%$, $(1 - V_m)$ is replaced in the above expression by a different, linear function of V_m, to account for the ensuing material softening. This is known as the "Gurson–Tvergaard–Needleman" or "GTN" criterion.

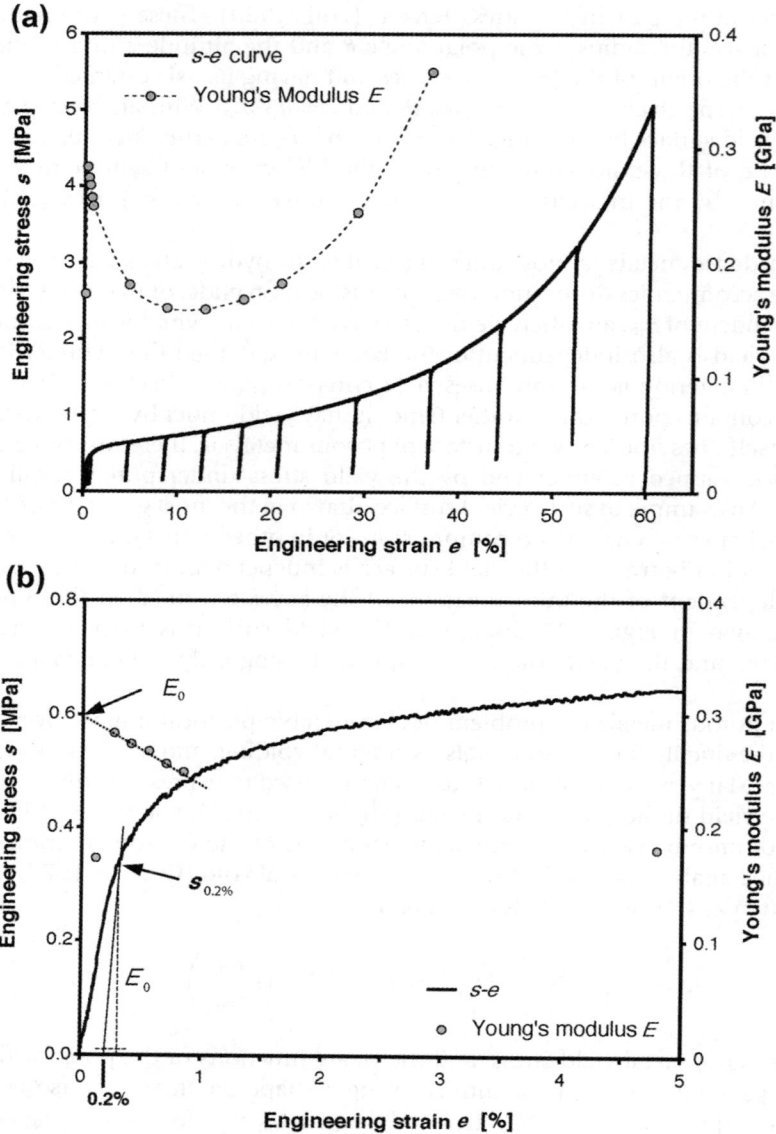

Figure 35   (a) Typical engineering stress (s) —engineering strain (e) curves for replicated open-pore microcellular pure aluminum loaded in compression (for this particular foam, the relative density was 14%). Unload–reload cycles have been performed at various deformations and the Young's modulus for each deformation step was evaluated from the slope of s–e during the corresponding cycles. (b) Details for the onset of deformation along the s–e curve represented in (a) and method of determination of the initial Young's modulus E_0 and the flow stress at 0.2% permanent deformation $s_{0.2\%}$—the unload–reload cycles have, for clarity, been omitted from this latter figure. Reproduced from *Acta Materialia*, vol. 54, Despois, J. F., Mueller, R. & Mortensen, A. "Uniaxial deformation of microcellular metals", pp. 4129–4142 (Despois et al., 2006b) with permission from Elsevier.

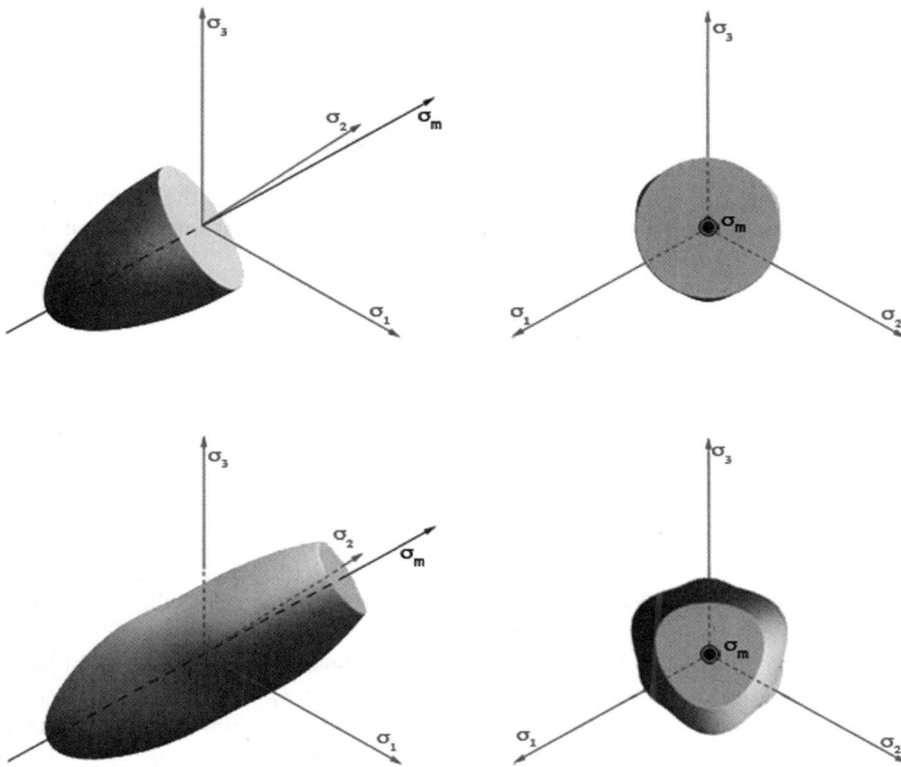

Figure 36 General shape of the yield surface of a porous metal, for a material that is insensitive to the sign of σ_m, showing two intersections with planes normal to the bissectrix (Π-planes). This particular yield surface corresponds to the yield criterion of McElwain et al. (McElwain et al., 2006a, 2006b) and was drawn using Mathematica™ by Dr. Etienne Combaz (EPFL, now with Novelis S.A., Sierre, Switzerland). Reproduced from *Acta Materialia*, vol. 58, Combaz, E., Bacciarini, C., Charvet, R., et al. "Yield surface of polyurethane and aluminium replicated foam", pp. 5168–5183 (Combaz et al., 2010) with permission from Elsevier. (For color version of this figure, the reader is referred to the online version of this book.)

Many other yield criteria have been proposed for metals containing a limited volume fraction of closed pores. These have been derived either on the basis of mean-field theory or from finite element simulations; relevant references include Refs. Duva and Hutchinson (1984), Rousselier et al. (1989), Tvergaard (1990), Thomason (1990), Qiu and Weng (1993), Leblond et al. (1995), Gologanu et al. (1997), Suquet (1997), Ponte-Castañeda and Suquet, (1998), Garajeu et al. (2000), Pardoen and Hutchinson (2000), Pardoen and Besson (2004), Huang and Wang (2006), Bilger et al. (2005), (2007), Garajeu and Suquet (2007), Nahson and Hutchinson (2008) and **Table 3** of (Lassance et al., 2007).

Yield criteria have been proposed specifically for metal powder compacts (Fleck et al., 1992; Fleck, 1995b, 1995a; Storåkers et al., 1999; Brown and Abou-Chedid, 1993, 1994; Olevsky, 1998; Redanz and Tvergaard, 2003). Powder compacts differ from dense metals containing rounded pores in that the porosity is far from round and generally higher. Also, powder particles have contacts of relatively well defined shape, which can show varying degrees of cohesion (see Section 24.2.4.1). If the powder particles are not fully bonded at these contacts, the compact will show a difference in deformation behavior between tension and compression, as is (to a greater degree) the case in yield of loose powder compacts, such as soil or sand (Watson and Wert, 1993; Brown and Abou-Chedid, 1994; Akisanya et al., 1997; Sridhar and Fleck, 2000; Aubertin and Li, 2004). With appropriately adated values for q_1 and q_2, the GTN expression has also been shown to describe adequately data from finite element simulations and experiment for the yield of compacted spherical metal powder (Kushch et al., 2008).

At high levels of porosity, characteristic of microcellular metals, other expressions have been proposed (Shahbeyk, 2013). Most among these depend on the two first invariants of the stress tensor only, and take the form of a parabolic expression

$$P_1 \left(\frac{\sigma_{eq}}{\sigma_y} \right) + P_2 \left(\frac{\sigma_m}{\sigma_y} \right)^2 - P_3 = 0 \tag{66}$$

or an elliptic expression

$$E_1 \left(\frac{\sigma_{eq}}{\sigma_y} \right)^2 + E_2 \left(\frac{\sigma_m}{\sigma_y} \right)^2 - E_3 = 0, \tag{67}$$

where P_i or E_i are functions of the relative density V_m, potentially of other parameters such as the pore shape, and at times also account for effects such as tension/compression asymmetry in the foam behavior. The yield stress in the denominator, σ_y, is either the flow stress of the microcellular metal or the yield stress of the dense metal making the microcellular material ($\sigma_{y,m}$).

Among the former, parabolic, yield criteria is the model developed on the basis of beam theory by Gibson et al. (1989), Triantafillou (1990), Gibson and Ashby (1997); here $P_1 = P_3 = 1$, $P_2 = 0.81 V_m$ and σ_y is the uniaxial compressive plastic collapse strength of the microcellular material. This yield criterion gives a surface that stretches far along the axis of hydrostatic loading ($\sigma_1 = \sigma_2 = \sigma_3$). It was therefore capped by its authors with another surface, along which beams buckle elastically before yielding.

A yield criterion that is near-parabolic but allows for a marked difference between tension and compression is that of Miller and Hutchinson (Miller, 2000; Miller and Hutchinson, 1998). This was derived by general consideration of the von Mises and Drucker–Prager yield criteria, coupled with the introduction of parameters designed to allow for specifics of the deformation of microcellular metals, such as tension/compression asymmetry. Here $P_1 = 1$, $P_2 = (\alpha/d_0)$, $P_3 = d_0 - \gamma \ (\sigma_m/\sigma_y)$ (with σ_m positive in compression; note the presence of a term linear in σ_m, which gives the yield surface a conical Drucker–Prager component); σ_y is the microcellular metal uniaxial compressive yield stress, $\sigma_{y,c}$, and

$$\gamma = \frac{6\beta^2 - 12\beta + 6 + 9\dfrac{\beta^2 - 1}{1 + \nu_{pl}}}{2(\beta + 1)^2}, \tag{68}$$

$$\alpha = \frac{45 + 24\gamma - 4\gamma^2 + 4\nu_{pl}(2 + \nu_{pl})(-9 + 6\gamma - \gamma^2)}{16(1 + \nu_{pl})^2}, \tag{69}$$

$$d_0 = \frac{1}{2} \left(1 - \frac{\gamma}{3} + \sqrt{\left(1 - \frac{\gamma}{3} \right)^2 + 4\frac{\alpha}{9}} \right), \tag{70}$$

where v_{pl} is the apparent "plastic" Poisson ratio (or "plastic expansion ratio") and β is the uniaxial yield asymmetry ratio of the microcellular metal

$$\beta = \frac{\sigma_{y,c}}{\sigma_{y,t}} \tag{71}$$

with $\sigma_{y,t}$ defined as the microcellular metal uniaxial tensile yield stress. An expression proposed by Doyoyo and Wierzbicki (2003) shows similar features: it contains both linear and quadratic σ_m terms, and accounts for tension/compression asymmetry.

Note that parabolic criteria define surfaces ending with a cusp where these intersect the hydrostatic stress axis. Elliptic yield surfaces are, on the other hand, smoothly convex everywhere. Among the elliptical expressions one finds the most frequently used criterion of all, namely that proposed by Deshpande and Fleck (2000b), (2001b), Ashby et al. (2000). Here σ_y is again the uniaxial compressive yield stress of the microcellular material, $\sigma_{y,c}$, while

$$E_1 = \frac{1}{1 + \left(\dfrac{\alpha_{DF}}{3}\right)^2}, \quad E_2 = \frac{\alpha_{DF}^2}{1 + \left(\dfrac{\alpha_{DF}}{3}\right)^2}, \quad E_3 = 1 \tag{72}$$

and α_{DF} is linked to the experimental uniaxial "apparent plastic Poisson's ratio" v_{pl} (defined as the ratio of lateral contraction to uniaxial elongation in uniaxial tensile deformation, or lateral expansion to uniaxial compression in uniaxial compression) by

$$\alpha_{DF} = 3\sqrt{\frac{0.5 - v_{pl}}{1 + v_{pl}}}. \tag{73}$$

Both quantities, $\sigma_{y,c}$ and α_{DF}, can be measured in a simple compression test of the foam itself, making this model clear and relatively easy to implement. A further attraction of this model is that it is available for direct implementation in finite element simulation software, such as *Abaqus* (Combaz et al., 2011b) or *LS-DYNA* (Hanssen et al., 2002b; Reyes et al., 2003, 2004).

Elliptical expressions for the yield surface also emerge from nonlinear (elastic) composite micro-mechanics (Ponte-Castañeda, 1991, 1992; Willis, 1991; Suquet, 1992, Suquet, 1997; Ponte-Castañeda and Suquet, 1998; Qiu and Weng, 1992, 1993). One of the simplest such expressions is that obtained from the variational principle of Ponte-Castañeda (Ponte-Castañeda, 1991) for a nonlinear incompressible metal containing pores that give it, in linear elastic deformation, a modulus equal to the Hashin–Shtrikhman upper bound (eqns (45)–(47)). The yield surface is then (Ponte-Castañeda, 1991; Qiu and Weng, 1992; Suquet, 1992, 1997; Ponte-Castañeda and Suquet, 1998)

$$\frac{9}{4}(1 - V_m)(\sigma_m)^2 + \left(1 + \frac{2(1 - V_m)}{3}\right)(\sigma_e)^2 - (\sigma_{y,m}V_m)^2 = 0, \tag{74}$$

where $\sigma_{y,m}$ is the yield stress of the dense metal making the foam. This reduces to eqn (67) with

$$E_1 = E_3 = 1, \quad E_2 = \frac{27(1 - V_m)}{4[3 + 2(1 - V_m)]} \tag{75}$$

together with the added implicit result that

$$\sigma_y = \sigma_{y,m} \frac{V_m}{\sqrt{1 + \dfrac{2(1 - V_m)}{3}}}. \tag{76}$$

In other words, the model comes also with a statement of the microcellular metal yield stress knowing that of the dense metal from which it is made. This prediction, in turn, can be adapted to account for specific features of the porous metal microstructure, such as the pore shape, orientation, distribution and, of course, volume fraction—several models exist to this end (Michel and Suquet, 1992; Suquet, 1992, 1993, 1995; Leblond et al., 1994; Suquet, 1993, 1995, 1997; Ponte-Castañeda, 1996, 2002a, 2002b; Ponte-Castañeda and Suquet, 1998; Garajeu et al., 2000; Pastor and Ponte Castaneda, 2002).

This greater predictive power is an important distinction between predictions of composite micromechanical theory and, for example, the yield criterion of Deshpande and Fleck, who use more pragmatically the measured porous metal uniaxial yield properties. Composite micromechanics criteria have, thus, in a sense more predictive power; however, it is found that mean-field models seldom predict the flow stress of porous metals accurately, for a variety of reasons (see Section 24.5.3). We note in passing that mean-field yield expressions have also been proposed for anisotropic structures containing spheroidal pores (Kitazono et al., 2003; Tane et al., 2007).

All of the above expressions are functions only of the first two stress tensor invariants, expressed in terms of (σ_e, σ_m). In stress space these describe closed surfaces of revolution about the $(\sigma_1 = \sigma_2 = \sigma_3)$ bissectrix, roughly shaped like a circular cross-section jelly bean or rice grain, with a pointed tip if the corresponding yield criterion is parabolic, or a smooth rounded tip if it is elliptic. Recent theory has, however, come to the conclusion that yield of highly porous metals is likely to depend on all three stress tensor invariants; this result has emerged from computer simulations of McElwain et al. (McElwain et al., 2006a, 2006b) (noting, though, that these were conducted for cubic structures, which are not isotropic), from more recent mean-field composite theory (Danas et al., 2008a), and also from analysis of polymer foams (Wang and Pan, 2006). Such θ dependence breaks circular symmetry in the yield surface, which now shows three lobes on either side of the $\sigma_m = 0$ plane, the lobes expressing its dependence on the Lode angle. **Figure 35** illustrates this with a 3D drawing of the yield surface predicted by McElwain et al. (2006a), (2006b). In Ref. Huang (2003) a yield surface made entirely of flat surfaces is proposed, based on capping the Tresca criterion with two triangular pyramids; this too obviously depends on θ.

Turning to numerical models of the yield surface of periodic microcellular metals, these give results that are strong functions of the assumed structures (Daxner et al., 2000a; Kim and Al-Hassani, 2001b, 2002; Daxner et al., 2002; Sanders and Gibson, 2003b, 2003a; Zhang and Lee, 2003; Ströhla et al., 2003; Ochsner et al., 2003; Gasser et al., 2004b; Fiedler et al., 2005; Wicklein and Thoma, 2005; Florence and Sab, 2005; Aboudi and Gilat, 2005; Xie and Chan, 2006; Demiray et al., 2007). Such numerical structures are generally also anisotropic, a case that has also been envisaged in a few of the proposed analytical models in the literature (e.g. (Theocaris, 1991)).

Honeycombs parallel 3D microcellular metals as far as in-plane yield is concerned. With in-plane isotropy, yield is expressible in terms of the two in-plane principal stresses, or alternatively in terms of the in-plane mean and maximum shear stresses: this makes the yield criterion a 2D curve that is conveniently plotted on a graph. Both parabolic and elliptical yield criteria have been proposed, which take the form of narrow loops elongated along the bissectrix of the two principal stress axes

(Gibson and Ashby, 1997; Chen et al., 1999; Daxner et al., 1999b, 2000a, 2002; Chen and Fleck, 2002; Alkhader and Vural, 2009a, 2009b, 2010). Given the greater simplicity of honeycombs, the relation between mesoscopic features of the material and its yield surface have been explored to greater depth than for 3D microstructural materials, leading to interesting correlations between mesostructure and yield. Cell wall defects such as corrugated, curved or missing cell walls, together with irregularity in the length and orientation of cell walls tend to lower the yield stress in all directions; however, the effect is significantly more marked as concerns the hydrostatic yield stress of honeycombs than their yield stress in pure shear (Chen et al., 1999; Daxner et al., 1999b, 2000a, 2002). In other words, defects reduce the aspect ratio of the yield loop, shortening it mainly along the bissectrix while leaving its width normal to this direction (i.e. the yield stress in shear) far less affected—or nearly constant as concerns corrugated or curved walls. The reason for the strong effect on hydrostatic yield is that defects break the symmetry which causes cell walls to deform uniaxially when the honeycomb is subjected to equal in-plane stresses; with defects, cell walls always bend and hence yield much more easily, also under in-plane hydrostatic stress. Facing general 3D stress states, honeycombs become orthotropic solids; out-of-plane yield of honeycombs has also been studied extensively (Gibson and Ashby, 1997; Mohr and Doyoyo, 2004b).

In three-dimensional microcellular metals, bend-dominated strut deformation is expected in nearly all realistic open-cell structures. Exceptions include certain ideally regular periodic structures, notably stretch-dominated lattice truss materials, the topology of which is specifically designed such that struts deform under near-uniaxial tension or compression (see Section 24.5.1.7). Here, yield of the microcellular material is determined either by general yield within a family of struts, or by the onset of strut buckling (Deshpande et al., 2001b, Wadley, 2006). Resulting yield surfaces are complex (if only because these materials are often anisotropic) and experimental data for the octet-truss lattice have compared relatively well with theory in Ref. (Deshpande et al., 2001b). Properties of these structures have often been explored with particular focus on collapse within sandwich structures (Deshpande and Fleck, 2001a; Wicks and Hutchinson, 2001; Wallach and Gibson, 2001b; Chiras et al., 2002; Hutchinson et al., 2003; Hutchinson and Fleck, 2005; 2006; Wadley, 2006; Doyoyo and Wan Hu, 2006; Doyoyo and Hu, 2006; Faure and Doyoyo, 2007).

Several authors have conducted experimental investigations of the yield of highly porous metals, using either a series of different tests (shear, compression, lateral uniform compression plus axial tension or compression, Arcan butterfly sample testing…) (Triantafilou et al., 1989; Ehlers et al., 1999; Gioux et al., 2000; Deshpande and Fleck, 2000b; Doyoyo and Wierzbicki, 2003; Sun et al., 2003; Daxner et al., 2003; Blazy et al., 2004; Sridhar and Fleck, 2005; Ochsner et al., 2005; Ruan et al., 2007; Peroni et al., 2008; Hou et al., 2011b; Zhou et al., 2012), or alternatively using a specific experimental setup designed to explore a wider spectrum of triaxial stress states (Combaz et al., 2010, Combaz et al., 2011a). Few sets of data are sufficiently comprehensive to provide a critical test of the many models proposed, something made all the more difficult by the fact that most models contain, as seen in what precedes, many adjustable parameters.

One clear conclusion from confrontation with experiment is that the parabolic expression proposed by Gibson et al. (eqn (66) with $P_1 = P_3 = 1$, $P_2 = 0.81V_m$) agrees overall poorly with data because the corresponding yield surface extends too far and is too pointed in the vicinity of purely hydrostatic stress states. The main elliptical yield surface, that by Deshpande and Fleck, and the parabolic expression of Miller and Hutchinson so far agree relatively well with data if the appropriate yield stress and plastic expansion ratio values are used. A recent study of yield in replicated microcellular aluminum (Combaz et al., 2010, 2011a), in which a more extensive set of data was generated, has shown that the yield stress of microcellular metals can indeed depend on all three invariants of the stress tensor, the yield surface showing "lobes" characteristic of Lode-angle dependence. Another finding of this work is that if one scales the foam flow stress by its uniaxial yield stress (to account for the influence of relative density), then the measured yield surfaces become independent of relative density. In other words its shape does not change, meaning that an increase in V_m only causes a homothetic expansion of the yield surface. This feature is shown by the Miller–Hutchinson and Deshpande–Fleck expressions if other parameters (such as v_p) do not vary with V_m, but not by several other expressions that have been proposed in the literature for the flow of porous metals (see above). Additional conclusions of that study are (i) that there can be, depending on the scale of the cells, an asymmetry between yield in tension or compression, (ii) that the measured yield surfaces are marginally better expressed by a parabolic (eqn (66)) than by an elliptical (eqn (67)) expression, and (iii) that the direction of flow past yield is overall compatible with flow according to the normality rule.

24.5.3 Plastic Flow

24.5.3.1 General Features of Plastic Flow in Microcellular Metals

Past yield, which in smooth samples generally initiates at pore-level sites of stress concentration, under uniaxial compression most microcellular metals and honeycombs (deformed in-plane) see a concentration of deformation along one or a few narrow bands that traverse the specimen, with the remainder of the specimen still deforming elastically (e.g. (Papka and Kyriakides, 1994; Prakash et al., 1996; Sugimura et al., 1997; Gibson and Ashby, 1997; Simone and Gibson, 1998a; Bart-Smith et al., 1998; McCullough et al., 1999b; Kriszt et al., 1999; Gradinger and Rammerstorfer, 1999; Bastawros et al., 2000; Bastawros and Evans, 2000; Olurin et al., 2000a; Park and Nutt, 2001a; Markaki and Clyne, 2001; Foroughi et al., 2002; Ruan et al., 2002; Meguid et al., 2002; Maire et al., 2003a; Zhou et al., 2004a; Song and Nutt, 2005; Issen et al., 2005; Jeon and Asahina, 2005; Tan et al., 2005a; Guden and Yüksel, 2006; Werther et al., 2006; Krishna et al., 2007; Ruan et al., 2007; Hakamada et al., 2007b, 2007a; Kolluri et al., 2007; Amsterdam et al., 2008d; Kadar et al., 2008; Mukherjee et al., 2009; Jang and Kyriakides, 2009a; Cady et al., 2009; Mukherjee et al., 2010d; Mu et al., 2010; Nosko et al., 2010; Jang et al., 2010; Michailidis et al., 2011; Goglio et al., 2011; Ravi Kumar et al., 2010; Saadatfar et al., 2012)). The same holds true of bonded hollow sphere foams (Friedl et al., 2008; Roy et al., 2011), and of syntactic foams, with the difference that syntactic foams being more brittle, these might fracture in the band (e.g. (Hartmann et al., 1998; Balch et al., 2005; Weise et al., 2007; Zhang and Zhao, 2007; Wu et al., 2007; Palmer et al., 2007; Daoud, 2008; Tao and Zhao, 2009; Zhang et al., 2009b; Mondal and Das, 2010)). Plastic deformation then generally remains concentrated there, leading eventually to "densification" of the band, meaning that opposite sides of cells within the band have met. This then displaces deformation elsewhere—often to another discrete band, where deformation remains similarly concentrated until it has in turn densified, and so forth. Such "Tetris-esque" gradual collapse of discrete pore bands causes the uniaxial compressive stress–strain curve of microcellular metals to display an abrupt onset of yield, often followed by a slight decrease of stress, the stress–strain curve tracing thereafter an irregular and wavy stress plateau. This continues over a significant compressive strain until contact of opposing cell walls has occurred along at least one column traversing the entire sample along the compression direction. "Densification" of the sample has then started: the stress rises much faster with increasing strain.

The overall result is a three-stage curve featuring a (frequently wavy) plateau between two rising portions (Papka and Kyriakides, 1994; Prakash et al., 1996; Sugimura et al., 1997; Gibson and Ashby, 1997; Banhart and Baumeister, 1998a; Simone and Gibson, 1998a; Bart-Smith et al., 1998; McCullough et al., 1999b; Gradinger and Rammerstorfer, 1999; Bastawros et al., 2000; Bastawros and Evans,

2000; Olurin et al., 2000a; Park and Nutt, 2001a; Markaki and Clyne, 2001; Foroughi et al., 2002; Ruan et al., 2002; Meguid et al., 2002; Lehmhus and Banhart, 2003; Maire et al., 2003a; Zhou et al., 2004a; Song and Nutt, 2005; Issen et al., 2005; Jeon and Asahina, 2005; Tan et al., 2005a; Guden and Yüksel, 2006; Werther et al., 2006; Krishna et al., 2007; Ruan et al., 2007; Hakamada et al., 2007b, 2007a; Kolluri et al., 2007; Friedl et al., 2008; Chino and Dunand, 2008; Kadar et al., 2008; Mukherjee et al., 2009; Cady et al., 2009; Hao et al., 2009; Mu et al., 2010; Jang and Kyriakides, 2009a; Nosko et al., 2010; Mukherjee et al., 2010d; Jang et al., 2010; Ye and Dunand, 2010; Yang et al., 2010a; Song et al., 2010a; Jeon et al., 2010; Michailidis et al., 2011; Goglio et al., 2011). Similar features, of nonuniform deformation and a three-stage curve, are also observed in polymer foams, which deviate from linear elasticity by microbuckling rather than by initiation of plastic deformation in the struts (Gibson and Ashby, 1997; Gong et al., 2005a, 2005b; Gong and Kyriakides, 2005). Compressed metal wire also shows a similar three-stage compression curve (Liu et al., 2008, 2010), with some complications linked with the influence of fiber contact points (the number of which is not constant) (Bouaziz et al., 2011).

In tension, foams deform as most metals do, but then break rather rapidly, often at a strain below 10 percent: Microcellular metals can (but do not always) have a relatively low tensile ductility (e.g. (Sugimura et al., 1997; Andrews et al., 1999a; Motz and Pippan, 2001; Hanssen et al., 2002b; Sun et al., 2003; San Marchi et al., 2004; Hakamada et al., 2005c; Ruan et al., 2007; Friedl et al., 2008; Ochiai et al., 2010b; Rösler and Näth, 2010; Combaz et al., 2011a)). A difference between the uniaxial tensile and compressive yield stress has also been noted in some microcellular metals; however, such a difference is not universal, nor is its sign (e.g. (Ruan et al., 2007; Combaz et al., 2011a)).

With aligned elongated pores, as found in samples produced by directional solidification of gas-evolving metals (Section 24.2.5.1), somewhat different, highly anisotropic, behavior is obtained. Stress–strain curves measured along the pores give roughly rule of mixture behavior until buckling or tearing sets in, while transverse flow stresses are much lower (Simone and Gibson, 1996, 1997; Hyun and Nakajima, 2002; Nakajima, 2007, 2013).

Reasons for the localization of deformation that is observed in the deformation of microcellular metals are twofold. The first is extrinsic: many microcellular metals have highly inhomogeneous structures (e.g. (Koza et al., 2004; Doyoyo and Mohr, 2006; Nosko et al., 2010)); this holds particularly true of earlier work, conducted on highly imperfect closed-cell foams. Foaming is a stochastic process of bubble nucleation and growth followed by coarsening or collapse, as exposed above (Section 24.2.2). The result is a highly irregular structure which, furthermore, can present numerous internal defects such as punctured or wavy cell walls, abnormally large pores (meaning locally lowered V_m), or "T"-shaped cell wall junctions. Such defects cause a significant weakening of the material at a local level, something that is already visible in their elastic modulus (Section 24.5.1.4). As a result, the strength of microcellular metals is often well below what one might a priori expect from their average characteristics. Also, structural inhomogeneity on a macroscopic scale creates weaker regions, which deform first, causing deformation to be highly inhomogeneous. The effect this has on deformation of the material is, in turn, often amplified by the fact that test samples frequently contain only a few pores across their width, this being a simple result of the fact that pores in commercial foams tend to be large (several millimeters, meaning that samples should in principle have widths of several centimeters at least). With a limited number of pores traversing the sample, averaging does little to reduce the influence of statistical variations in pore characteristics. The general result is strong inhomogeneity in deformation coupled with visible fluctuations in the flow stress plateau along the stress–strain curve.

The second reason for the appearance of a plateau in compression testing is intrinsic: compression weakens most porous materials along the compression direction. Pores flatten and struts align into the plane normal to the applied compressive stress direction. This lowers the load-bearing capacity of the material along the stress axis (this is visible in the evolution of Young's modulus with deformation). This also makes compressive deformation intrinsically unstable: even in the absence of structural inhomogeneity, once initiated, plastic deformation in highly porous metals remains localized unless another mechanism, such as densification or work hardening of the metal, causes the hardening that is needed to initiate or drive deformation elsewhere along the sample. In tension, the reverse holds true: deformation aligns pores and struts along the stress axis, which per se hardens the material. If a stress plateau is observed after tensile yield, then, the cause is in initial structural heterogeneity, coupled with internal damage evolution.

Internal damage in microcellular metals has been identified and studied by several authors, using detailed examination of the evolution of individual struts or cell walls in deformed microcellular samples and also by tracking the evolution of Young's modulus with deformation. Damage in microcellular metal takes various forms, including notably fracture of brittle second phases or at micropores in the metal within the struts, the tearing of cell walls, as well as cell wall or strut buckling after a certain amount of compressive deformation (McCullough et al., 1999b; Maire et al., 1999, 2003a; Kriszt et al., 2000; Badiche et al., 2000; Motz and Pippan, 2001; Markaki and Clyne, 2001; Zhou et al., 2002, 2004b, 2004c; Salvo et al., 2004; Kadar et al., 2004a; San Marchi et al., 2004; Blazy et al., 2004; Marmottant et al., 2005; Onck et al., 2005; Amsterdam et al., 2005, 2006a, 2006b, 2008a, 2008c, 2008d; Ohgaki et al., 2006a, 2006b; Toda et al., 2006a, 2006b; Rammerstorfer et al., 2006; Nakajima, 2007; Aly, 2007; Friedl et al., 2008; Kadar et al., 2008; Fallet et al., 2008; Aly et al., 2009; Jang and Kyriakides, 2009a; Brown et al., 2010; Ochiai et al., 2010b; Lhuissier et al., 2010; Mu et al., 2010; Mukherjee et al., 2010d; Jang et al., 2010; Ravi Kumar et al., 2010; Nadella et al., 2010; Tuncer et al., 2011a; Vendra et al., 2011). A characteristic of internal damage in microcellular metals is that it sets in early, at low values of the overall strain. This is a dual result of (i) the presence of strong stress/strain gradients within the material even if uniform load is applied, and (ii) the frequent presence of brittle second phases within the metal making the material.

Smooth ascending stress–strain curves devoid of a plateau in either tension or compression can, however, be observed if the sample (i) contains many pores of roughly identical size and morphology and (ii) is made of a ductile metal with an intrinsic capacity for work hardening and a low content of brittle second phases or other sources of internal damage accumulation. Replicated microcellular pure aluminum produced by infiltration thus displays smooth stress–strain curves, in both tension and compression, as do other microcellular metals of uniform mesostructure produced from ductile metals (Badiche et al., 2000; Dillard et al., 2003; Goussery et al., 2004; Despois et al., 2006b, 2004; Goodall et al., 2006a, 2007; Yang et al., 2007d; Friedl et al., 2008; Neville and Rabiei, 2008; Rabiei and Vendra, 2009; Jin et al., 2009; Diologent et al., 2009d, 2011; Wang et al., 2010b; Wada et al., 2011a; Dou and Derby, 2011; Jamshidi-Alashti and Roudini, 2012); Deformation is then relatively uniform across the sample if the material is initially of relatively uniform relative density (Marmottant et al., 2005). In some instances, though, deformation bands can be seen during compressive deformation if geometrical softening has a greater effect than metal work hardening (Friedl et al., 2008). Cast open-pore micro-cellular metal samples can also show smooth stress–strain curves devoid of local maxima, for example (Zhou et al., 2002).

The uniaxial plastic flow of microcellular metals is also characterized by an unusually small amount of lateral contraction: unlike dense metals, volume is not conserved. Rather, "apparent plastic Poisson's ratios," ν_{pl} of eqn (73), tend to be well below the value ($\nu_{pl} = 1/2$) characteristic of constant volume deformation of isotropic solids. Measured values vary, according to the material, between roughly 0.2 and zero; for example (Gioux et al., 2000; Deshpande and Fleck, 2000b; Hanssen et al., 2002b; Sridhar and Fleck, 2005; Ochsner et al., 2005; Ruan et al., 2007; Combaz et al., 2010, 2011a). This has a number of consequences; notably, when a microcellular metal is indented, there is little or none of the lateral flow characteristic of the indentation of dense solids. If the indenter radius is much larger than the cell size, then the indentation hardness of micro-cellular metals will tend to be well below three times their uniaxial compressive flow stress (the approximate value characteristic of dense metals). Instead, the hardness approaches the uniaxial compressive flow stress as ν_{pl} approaches zero (augmented with flat-punch indenters by the shear force required for the indent to cut through the material at the indent periphery) (Olurin et al., 2000b; Andrews et al., 2001; Kumar et al., 2003; Yan and Pun, 2010; Vodenitcharova et al., 2010).

24.5.3.2 Analytical Expressions for the Uniaxial Flow Stress

Predictions of the flow stress, or of the flow curve, of highly porous metals abound in the literature. Approaches used broadly parallel those used to predict elastic constants. For simplicity we keep coverage focused on highly porous (i.e. microcellular) isotropic porous metals.

The most basic and most frequently used prediction is, as for elastic stiffness, that of Gibson and Ashby who consider the deformation of an elementary unit cell containing beams in bending and cell walls in uniaxial stress (Gibson and Ashby, 1997). The metal is assumed ideally plastic. Yield of the microcellular metal then occurs upon formation of a plastic hinge within bending beams, coupled with yield of cell faces subjected to tensile stress, if these are present. A simple derivation yields an expression for the uniaxial yield stress of isotropic microcellular metals that parallels that for their modulus (eqn (54)), but has different exponents

$$\frac{\sigma}{\sigma_{y,m}} = C_1' \Phi^{3/2} V_m^{3/2} + C_2'(1 - \Phi)V_m, \tag{77}$$

where $\sigma_{y,m}$ is the yield stress of the dense metal, C_1' and C_2' are constants, Φ is the fraction solid that is contained in cell edges (as in eqn (54). The first term is associated with bending of the struts while the second is associated with stretching of the cell walls, as for Young's modulus. At higher relative densities, an added term can be added to adapt the material volume, so as to account for the actual shape of cell corners in open-cell foams (Gibson and Ashby, 1997); this gives

$$\frac{\sigma}{\sigma_{y,met}} = C_1' V_m^{3/2}\left(1 + V_m^2\right). \tag{78}$$

Open-cell microcellular metals are thus predicted to have a yield stress proportional to their relative density raised to the power (3/2). We note in passing that the same law governs yield by brittle fracture of struts or cell walls in microcellular materials made of brittle solids such as ceramics (Gibson and Ashby, 1997); the observed scaling exponent is thus not a discriminant of yield mechanisms. In honeycombs, the transverse yield stress is proportional to V_m^2 (Gibson and Ashby, 1997). In dual-porosity structures that combine pores on two very different scales, analytical predictions of the flow stress have also been combined, treating each pore level in turn as being present in a homogeneous single-phase solid (Lakes, 1993; Esen and Bor, 2007; Cox and Dunand, 2011; Li and Dunand, 2011).

In stretch-dominated lattice truss materials eqn (77) is not obeyed; rather, the yield stress varies linearly with the relative density V_m—if yield occurs before buckling or cooperative deformation mechanisms. For the octet-truss material tested the constant of proportionality is on the order of one-third along the <111> direction of the underlying FCC lattice (Deshpande et al., 2001b; Wadley, 2006; Pingle et al., 2011), **Figure 34**. Given the slenderness of struts in such (typically low V_m) structures, however, buckling, either elastic or plastic, often governs the departure from linear elastic behavior. This makes usage of beam cross-sections that delay buckling (e.g. hollow beams) an interesting strategy for high strength in such structures (Queheillalt and Wadley, 2005a, 2005b, 2011; Wadley, 2006). A linear dependence of the flow stress on V_m is of course also expected with unidirectional pores when these are stressed along the pore axis; directionally solidified metal/gas structures or honeycombs are examples (note, however, that the latter will generally buckle in compression before generalized yield is observed) (Gibson and Ashby, 1997; Nakajima, 2007).

Comparison with experimental data gives overall a relatively good confirmation to eqn (77); however, data are more limited and scatter is greater than for elastic modulus values; also, exponents noticeably different from 1.5 have been noted with open-pore microcellular metals (Gibson and Ashby, 1997; Sugimura et al., 1997; Banhart and Baumeister, 1998a; Santosa and Wierzbicki, 1998; McCullough et al., 1999b; Mukai et al., 1999a; Gui et al., 2000; Badiche et al., 2000; Kanahashi et al., 2000; Foroughi et al., 2002; Ramamurty and Paul, 2004; Kim et al., 2005a; Zhou et al., 2005b; Rakow and Waas, 2005; Tan et al., 2005a; Despois et al., 2006b; Wang et al., 2006b; Hakamada et al., 2007a; Imwinkelried, 2007; Kolluri et al., 2007; Boonyongmaneerat and Dunand, 2008; Amsterdam et al., 2008b; Idris et al., 2009; Hao et al., 2009; Tuncer et al., 2011a, 2011b; Jamshidi-Alashti and Roudini, 2012). Scaling exponents higher than 3/2 have in some cases been interpreted as a sign that strut or cell wall buckling intervenes noticeably in the deformation process (Hakamada et al., 2007a; Tuncer et al., 2011a).

Comparison between theory and experiment as concerns yield or flow stress values is hampered by the fact that, unlike the modulus, which remains the same for the metal whatever its form (meaning whether the metal is in bulk form or is situated within struts or cell walls of a foam) $\sigma_{y,m}$ can for a variety of reasons differ significantly in the foam from the value measured for the same metal in dense form. Overall, experimental data suggest $C_1' \approx 0.3$ and $C_2' \approx 0.4$. At the same time, the high variability of $\sigma_{y,m}$ implies that conventional metal and alloy hardening strategies, such as age hardening, will equally strengthen metal foams. There are thus two strategies to increase the strength of porous metals: (i) to improve their architecture (which will also increase their elastic stiffness) (e.g. (Miyoshi et al., 1999; Goodall et al., 2006a, 2007)), and (ii) to harden the metal itself (Thornton and Magee, 1975b, 1975a; Kanahashi et al., 2001a, 2001b; Conde and Mortensen, 2008; Lefebvre and Baril, 2008).

The analysis leading to eqn (77) assumes ideal plasticity in the metal; hence, the foam does not work harden initially, deforming as already mentioned to display a long yield plateau that ends, in compression, with densification. Densification is predicted by Gibson and Ashby (1997) to occur at a compressive strain ε_D roughly equal to

$$\varepsilon_D \approx 1 - 1.5\, V_m, \tag{79}$$

which marks the end of the plateau in uniaxial compression. Obviously, the densification strain will depend on the strain path (Reyes et al., 2003). If the pore shape deviates significantly from a sphere, ε_D may obey a different law (Chan and Xie, 2003) or might be significantly smaller than is predicted by eqn (79) (Gaillard et al., 2004).

Simple analytical expressions can also be derived from mean-field theory of nonlinear composite deformation. Specifically, variational estimates of Ponte Castañeda and Suquet, which correspond to a secant modulus formulation of nonlinear composite deformation (Ponte-Castañeda, 1991; Qiu and Weng, 1992; Suquet, 1993, 1995, 1997; Buryachenko, 1996; Hu, 1996; Ponte-Castañeda and Suquet, 1998) can be adapted to yield simple predictions of the uniaxial deformation of isotropic nonlinear porous metals knowing the stress–strain curve of the dense metal of which these are made (Mueller and Mortensen, 2006; Despois et al., 2006b; Mueller et al.,

2007). In essence, the approach uses as its starting point the ratio of the elastic modulus of the porous metal and that of the dense metal in the foam. This is a quantity that can be either predicted (see above) or measured:

$$E = F_E E_m, \tag{80}$$

where F_E is a function of V_m, whose value is less than unity. In simple terms F_E is the fractional reduction caused by the pores on the metal's load-bearing capacity during linear elastic deformation. This function can be measured without ambiguity because E_m will be the same whether the metal is dense or porous (assuming isotropy). Adopting F_E as a measure of the load-bearing capacity characteristic of the porous metal architecture at hand, it can be shown by adapting the variational estimate of Ponte Castañeda and Suquet with a few simplifying assumptions (e.g. taking the metal within the foam to deform without changing volume) that, if the dense metal flow curve is given as

$$\sigma_m = f(\varepsilon_m), \tag{81}$$

then that of the porous metal can be estimated as

$$\sigma = \sqrt{F_E V_m} f\left(\sqrt{\frac{F_E}{V_m}} \varepsilon\right). \tag{82}$$

In particular, if the dense metal displays a power law uniaxial stress–strain curve

$$\sigma = c_m \varepsilon^n, \tag{83}$$

then the porous metal follows the same power law

$$\sigma = c\, \varepsilon^n, \tag{84}$$

with

$$c = \left[F_E^{\left(\frac{1+n}{2}\right)} (V_m)^{\left(\frac{1-n}{2}\right)} \right] c_m = \left[C_E^{\left(\frac{1+n}{2}\right)} (V_m)^{\left[\frac{1-n}{2} + N\left(\frac{1+n}{2}\right)\right]} \right] c_m, \tag{85}$$

where constants C_E and N are from the scaling relation $F_E = C_E V_m^N$ often used to express the ratio of the porous metal modulus to that of the metal of which it is made (eqn (56)).

This estimate recovers the Gibson–Ashby expression for yield of an open-cell microcellular ideally plastic metal ($n = 0$ and $N = 2$), namely eqn (77) with $\Phi = 1$, further specifying that $C_1' = \sqrt{C_1}$. It also recovers the linear relation between flow stress and relative density that obtains for stretch-dominated truss structures (noting, however, that these are often nonisotropic, in which case the variational approximation does not hold). The result of a beam-based analysis for an open-cell microcellular power law metal (San Marchi and Mortensen, 2001) is also recovered. These mean-field predictions thus have a similar agreement with data as do the Gibson–Ashby expressions; however, by calibration of the model using the elastic modulus, any isotropic pore architecture can be tackled so long as its load-bearing capacity is known by measurement (or derivation) of F_E. The approach furthermore forces consistency between the flow stress constant for linear elastic and plastic flow (e.g. C_1 and C_1' in the Gibson–Ashby law).

This last point, in turn, raises an interesting question, pointed out also by Gibson and Ashby in their confrontation of theory with data: experimentally measured flow stresses of microcellular metals are roughly one-half to one-third of what consistency within theory would dictate (Gibson and Ashby, 1997; Mukai et al., 1999a; Zhou et al., 2005b; Despois et al., 2006b; Hakamada et al., 2007a; Diologent et al., 2011). Reasons for this "knock-down" factor, of roughly one-half to one-third, in the flow stress of highly porous metals are not well known at present. One explanation could be uncertainty in the in situ flow stress of the metal within the foam ($\sigma_{y,m}$ in eqns (77) and (78) or c_m in eqn (83)): this is nearly always poorly known since the flow stress of the metal depends strongly on its condition, which is likely to differ according to whether the metal is in dense or microcellular foam (Despois et al., 2006b). Another reason might be an intrinsic deficiency of mean-field models (note that predictions of the yield stress in open-cell aluminum from finite element simulations have shown overall good agreement with data (Jang and Kyriakides, 2009b; Jang et al., 2010)). A third possibility is that this discrepancy between theory and experiment in the value of the yield stress of highly porous metals, and also perhaps the paradox whereby cell shape anisotropy can raise the modulus while lowering the yield stress of numerically simulated open-cell structures (see above), might be a percolation effect. Indeed, in many microcellular materials, the yield stress of individual elements making the material (struts, cell walls) is statistically distributed, either by virtue of shape (short beams yield later than long ones at given cross-section) or by virtue of microstructural variations within the material (causing spatial variations in the metal yield stress from strut to strut, and/or within struts). Structures akin to open-pore microcellular materials have been modeled as a percolation theory network model in a few publications dealing with elastic deformation (Feng et al., 1984; Sahimi and Arbabi, 1989a, 1989b); however, percolation effects in the deformation of microcellular materials remain relatively unexplored, so the question is at present purely speculative.

Expressions such as those above are limited to small-strain deformation, as this is implicitly assumed in their derivation. At large strain, these will gradually lose validity because the architecture of the metal within the foam (or the foam mesostructure) evolves: struts align along the tensile axis in unaxial tension, for example, and this will increase their load-bearing capacity. This is notably visible in a measurable evolution of Young's modulus with plastic strain (see **Figure 35**). The issue of a strain-induced evolution in the load-bearing capacity of porous materials has been addressed by several authors in the context of the deformation of metals containing isolated ellipsoidal pores (this problem having importance in ductile fracture) (Ponte-Castañeda and Zaidman, 1994; Ponte-Castañeda, 1997; Kailasam et al., 1997a, 1997b; Aravas and Ponte Castaneda, 2004; Pardoen and Besson, 2004; Danas et al., 2008a, 2008b; Danas and Ponte Castañeda, 2009a, 2009b; Bilger et al., 2005; Garajeu et al., 2000). With microcellular metals, Onck et al. address the issue of large-strain deformation and resulting topological changes to the foam structure in 2D foams using numerical methods (Mangipudi et al., 2010). A treatment for 3D open-cell microcellular metals that is simplistic but agrees well with experiment is offered in Ref. San Marchi and Mortensen (2001).

24.5.3.3 *Meso- and Macromechanics of Microcellular Metal Deformation*

Several authors have proposed numerical analyses of the plastic deformation of microcellular metals, both two dimensional and three dimensional. These often constitute extensions of models proposed for linear elastic deformation. Numerical analyses tackle similarly the influence of metal mesostructure, including the shape and length distribution of beams or cell walls (e.g. (Chen et al., 1999; Chuang and Huang, 2002; Yang and Huang, 2004; Huang and Chang, 2005; Yang et al., 2008b) for honeycombs (Clyne et al., 2005; Markaki and Clyne, 2005), for fiber networks (Simone and Gibson, 1998b, 1998c;

Daxner et al., 2002; Santosa and Wierzbicki, 1998; Michailidis et al., 2008, 2010a, 2010b; Jang and Kyriakides, 2009b; Jang et al., 2010; Song et al., 2010b; Takahashi et al., 2010; Babaee et al., 2012), for microcellular metals (Lim et al., 2002; Caty et al., 2008), for sintered hollow spheres (Kujime et al., 2007), for directionally solidified structures or (Deshpande and Fleck, 2001a; Wadley, 2006; Luxner et al., 2007, 2009a, 2009b) for truss structures). Calculations have also been conducted to examine the influence of how beams or cell walls are assembled (e.g. (Silva and Gibson, 1997; Chen et al., 1999; Zhu et al., 2001; Fazekas et al., 2002; Hardenacke and Hohe, 2009; Mangipudi et al., 2010) for honeycombs or (Deshpande et al., 2001b; Zhu and Windle, 2002; Jang and Kyriakides, 2009b; Jang et al., 2010; Babaee et al., 2012) for 3D microcellular solids), and the role of topological defects within the structure (e.g. (Prakash et al., 1996; Chen et al., 1999; Ajdari et al., 2008) for honeycombs and (Simone and Gibson, 1998c, 1998b; Andrews et al., 1999a; Daxner et al., 2002; Luxner et al., 2009a, Singh et al., 2010; Saadatfar et al., 2012) for microcellar metal).

By and large, mesostructural factors have similar effects on load bearing in plastic deformation as they do in linear elastic deformation. Comparatively, however, a difference is created by the fact that bifurcations toward localized deformation are more prone to be observed in plastic deformation than in the linear elastic regime: this is amply documented by experiment, as indicated above, but also comes out of numerical simulations. Luxner et al. investigated the influence of cell-level structural disorder on strain localization during uniaxial compression of a 3D simple cubic lattice truss structure unit cell (in Kelvin cell structures, such strain localization was not observed, on the other hand). It was shown that the tendency for strain localization is reduced and then eliminated as the degree of cell local structural disorder increases (Luxner et al., 2007, 2009b). Their results showed a transition not unlike that seen in the longitudinal tensile failure of ductile/brittle two-phase materials, where a transition is noted from localized to delocalized damage evolution as the brittle phase strength distribution becomes wider (Phoenix and Beyerlein, 2000; Hauert et al., 2009). Similarly, increased local disorder reduces the sensitivity of the flow stress to missing struts in these 3D lattice truss structures. On the other hand, disorder also comes with a strong reduction in the microcellular material's flow stress (Luxner et al., 2007, 2009a, Babaee et al., 2012).

Large-cell simulations of post-yield deformation in elastoplastic open-cell structure by Jang and Kyriakides (2009b), Jang et al. (2010) shed interesting light on the progress of deformation inhomogeneity and its dependence on sample size and mesostructural homogeneity. These authors show in particular that increasing disorder causes increased homogeneity in post-yield plastic deformation. Other large-cell simulations have also been constructed on the basis of tomographic files collected on actual samples of microcellular metal (Youssef et al., 2005; Jeon et al., 2010, 2009; Takano et al., 2010; Veyhl et al., 2011a), thus opening the door to comparison of simulation with local measurements of deformation if such simulation is coupled with in situ deformation experiments conducted within the tomograph (Marmottant et al., 2005).

A disordered honeycomb numerical model incorporating damage was proposed by Mangipudi and Onck, to show how the gradual progression of damage in highly deformed metal foam, originating notably at brittle second phases (see Section 24.5.3.1), can influence the progression of both deformation and uniaxial failure of microcellular metals. Trends observed are summarized in **Figure 37** (Mangipudi et al., 2010; Mangipudi and Onck, 2011a). These authors also showed that, at large strains, the work hardening law of metal foams deviates from that of the corresponding dense metal, this being effect of the evolving architecture of the highly deformed material (an effect also visible in the evolution of the unloading modulus with strain; see above) (Mangipudi et al., 2010). In tension this is manifest as a slight increase in foam strain hardening exponent over that of the dense metal (as one would expect given that tensile deformation stiffens the material).

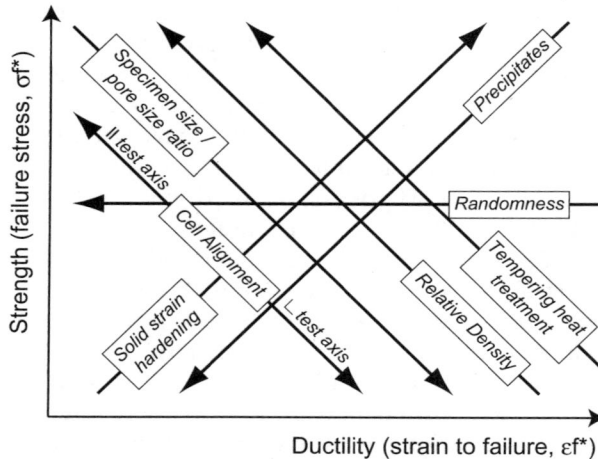

Figure 37 General trends for the dependence of strength and ductility of porous metals, as deduced from finite element simulations of 2D foam structures by Mangipudi et Onck. (Mangipudi et al., 2010, Mangipudi and Onck, 2011a).

The potential of functional porosity gradients, introduced to optimize the overall response of a porous metal structure, has also been explored for porous metals by a few authors (Daxner et al., 2000b; Pollien et al., 2005; Conde et al., 2006b; Brothers and Dunand, 2006b, 2008; Matsumoto et al., 2007; Ajdari et al., 2009) (we note in passing that this is something that is regularly practiced in natural microcellular structures such as plants or bone). Uncontrolled gradients in porosity can exist in microcellular metals: random spatial variations in relative density were introduced in a numerical model of closed-cell aluminum foam to show that these can account for the wavy shape of compression curves collected in compression testing of this class of material (Meguid et al., 2002). The influence of longitudinal density gradients on the compression stress–strain curve was also studied by a few authors (Daxner et al., 1999a; Gradinger and Rammerstorfer, 1999; Kenesei et al., 2004); their effect is obviously to raise the apparent (average) work-hardening rate of the material. Density gradients have also been measured by tomographic means, and coupled with finite element continuum modeling of macroscopic porous metal structures to simulate their deformation with account for the existence of density gradients in their midst (Degischer and Kottar, 2002; Degischer et al., 2013).

Numerical modeling has also addressed a few specific features of the plastic deformation of highly porous metals. Given the highly inhomogeneous stress distribution and the frequent preponderance of bending in the deformation of microcellular metals, one would expect to observe a significant Bauschinger effect after stress reversal, or in other words that work hardening be in large part kinematic. This was shown in 3D simulations of the deformation of open-cell elastic ideally plastic material with a Kelvin unit cell (Demiray et al., 2007) and was also demonstrated experimentally (Motz and Pippan, 2002; Krupp et al., 2006).

24.5.3.4 *Size Effects in Plastic Deformation*
24.5.3.4.1 *Pore Size Effects*
Porous metals show size effects also in plastic deformation. To start with, the same size effects that hold for elastic deformation also obtain for plastic deformation: the ratio between the pore size on one hand,

and the scale of features defining the shape of porous metal structures on the other (such as the diameter of a hole, the depth of a notch, the radius of an indentation punch, the width of a sandwich core, or simply the width of uniaxial specimens), manifests itself above a certain value, at which the deformation or fracture of edge or boundary pores contributes significantly to the overall sample response. This, in turn, makes the plastic deformation of porous structures scale dependent.

Test samples must therefore have a width that exceeds a certain number of cell diameters in all dimensions for data to be characteristic of bulk material deformation, for example (Andrews et al., 2001; Idris et al., 2009). Samples containing holes or notches flow or fail at a net section stress that can depend on the ratio between pore and neck size. Similarly, (i) metal foam sandwich cores flow at a stress that depends on the ratio of pore diameter to core width (and on how the sandwich is assembled), and (ii) the hardness of microcellular metals can depend on the radius of the indentation punch.

Experimental studies exist of the effect for uniaxial test samples (e.g. (Bastawros et al., 2000; Andrews et al., 2001; Jeon and Asahina, 2005; Chino et al., 2003)), structures containing holes and notches (e.g. (Sugimura et al., 1997; Paul et al., 1999; Motz and Pippan, 2001; Onck and Bastawros, 2000; Onck, 2003b; Olurin et al., 2000a, 2001a; Fleck et al., 2001; Andrews and Gibson, 2001b, 2002; McCullough et al., 1999b; Antoniou et al., 2004; Combaz et al., 2011b)), for sandwich structures (e.g. (Kesler and Gibson, 2002; Chen and Fleck, 2002)), or in indentation (e.g. (Olurin et al., 2000b; Andrews et al., 2001; Ramamurty and Kumaran, 2004; Kadar et al., 2004b)). General trends in experiment are that smooth samples tend to show a flow stress that decreases with decreasing ratio of width to cell diameter or, in other words, as the number of cells across the sample decreases. The underlying mechanism for this is that cells are weaker along free surfaces, which in turn eases shear band formation; the effect has, however, been found to be less pronounced in plastic than in elastic deformation (Chen et al., 1999; Chen and Fleck, 2002; Bastawros et al., 2000; Onck et al., 2001; Andrews et al., 2001; Onck, 2003b; Tekoglu and Onck, 2005; Jeon and Asahina, 2005; Mangipudi and Onck, 2011a, 2011b; Mangipudi et al., 2010; Tekoglu et al., 2011). Where cells are bonded to dense metal the inverse effect is observed: bonding to solid metal constrains cell edge deformation; this in turn will often hinder the formation of shear bands (Chen et al., 1999; Chen and Fleck, 2002; Onck et al., 2001; Andrews et al., 2001; Onck, 2003b; Tekoglu and Onck, 2005; Mukherjee et al., 2009; Mangipudi and Onck, 2011a; Jarvis et al., 2011).

Double-edge notched flat samples and circular-notched round samples show an increased net section flow or tensile fracture stress compared to smooth samples (Sugimura et al., 1997; Motz and Pippan, 2001; Onck and Bastawros, 2000; Onck, 2003b; Olurin et al., 2000a; Andrews and Gibson, 2001b, 2002; McCullough et al., 1999b; Antoniou et al., 2004; Combaz et al., 2011b) (although there are exceptions (Paul et al., 1999)). To explain the effect, several mechanisms have been proposed: (i) triaxiality in the middle of the ligament, an effect apparently more marked in the presence of work hardening (Sugimura et al., 1997; Motz and Pippan, 2001; Onck, 2001a; Combaz et al., 2011b), (ii) a defect sampling effect (narrower ligaments sample fewer defects) (Olurin et al., 2000a), and finally (iii) alterations (compared to continuum predictions) in deformation patterns around notch tips over dimensions on the scale of the cell size, making the effect one that disappears as the notch to cell size ratio increases (Onck and Bastawros, 2000; Onck, 2001b, 2003a, 2003b; Onck et al., 2001; Onck, 2003b, 2003a; Andrews and Gibson, 2001b, 2001c, 2002; Antoniou et al., 2004; Mangipudi and Onck, 2011b). With one notch only, the effect is absent (Onck, 2003b; Antoniou et al., 2004).

In samples containing a central hole or crack the effect is weak: these flow or fail at a net section stress that is not far from that seen with smooth uniaxial samples (McCullough et al., 1999a; Olurin et al., 2001a; Fleck et al., 2001; Combaz et al., 2011b) provided the central defect is not so large that it develops into a growing crack (Fleck et al., 2001). Overall, these trends are captured by numerical simulation (Andrews and Gibson, 2001c; Fleck et al., 2001; Onck, 2003b; Jang et al., 2010; Mangipudi and Onck, 2011b; Combaz et al., 2011b).

Yet again in parallel with linear elastic deformation, higher order continuum approaches to the plasticity of microcellular metals have also been explored to account for such size effects related to the ratio between specimen or feature size and cell size, and also to handle the occurrence of instabilities in the plastic deformation of cellular elastoplastic materials (Chen and Fleck, 2002; Forest et al., 2005; Dillard et al., 2006).

24.5.3.4.2 Plasticity Size Effects

With plastic deformation a second, more intrinsic, size effect can appear: the plastic flow stress of porous metals can also depend on the intrinsic size of the pores, also for test samples much larger than the pores. This occurs when the size of pores or struts within the metal becomes commensurate with that of substructural features that govern the plastic deformation of metals, such as the mean dislocation separation distance, the cell or the subgrain size, or the distance between precipitates. Substructural features in metals and alloys can easily span several micrometers. Therefore, when the pore or strut diameter falls below a few micrometers, free surfaces start interfering with dislocation glide. This in turn alters the local in situ flow stress of the metal within the porous material. Thus, unlike in elasticity, where surface stress induced intrinsic size effects only become visible when pores become smaller than a few tens of nanometer, in plastic deformation intrinsic (metal flow stress) size effects become visible when pore diametres fall below a few tens of micrometer; **Figure 38** gives an illustration for replicated microcellular pure aluminum.

This is one of many manifestations of what is more generally known, in metallurgy, as the crystal plasticity size effect. In microcellular metals the effect closely resembles that found in testing small-scale metal specimens (such as micrometric test samples produced by focused ion beam milling (Uchic et al.,

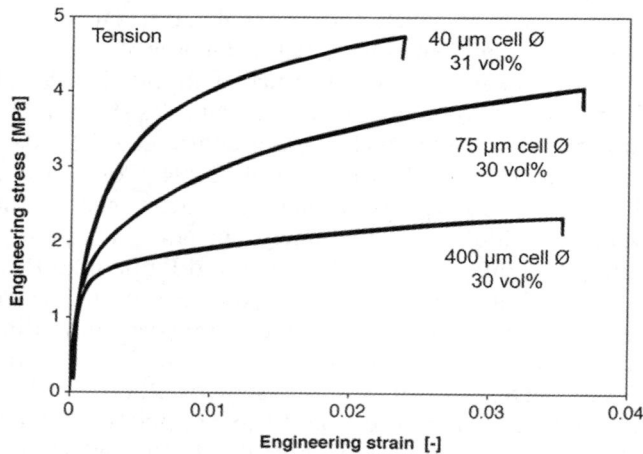

Figure 38 Tensile curves of replicated 99.99% pure aluminum, showing at constant relative density ($V_m = 30\%$) the influence of pore, and hence metal strut, dimensions. The finer the foam the higher the flow stress: this is a manifestation of plasticity size effects. Reprinted from *Advanced Engineering Materials* vol. 6, Despois, J. F., Conde, Y., Marchi, C. S. & Mortensen, A., "Tensile Behaviour of Replicated Aluminium Foams", pp. 444–447 (Despois et al., 2004) with permission from Wiley.

2009b; Dehm, 2009)). Measurements of the pore size-dependent flow stress of highly porous metals containing fine-scale structures are thus interesting, not only per se, but also for the light these shed on the deformation of small metallic structures more generally (such as those found in microelectronics or microsystems).

Studies on the plasticity size effect in highly porous metals have been conducted on two types of microcellular metal. The first is fine (μm scale) replicated open-pore microcellular metals, in which the flow stress increases markedly with decreasing pore size below roughly 100 μm (Despois et al., 2004; Goodall et al., 2006a, 2007), **Figure 38**. Since samples of such microporous metals can be made to a scale amenable to bulk metal testing methods, their modulus or relaxation response can be measured without ambiguity. The (strong) influence of the surface structure on the flow stress of fine-scale microcellular aluminum (Diologent et al., 2009d), and the influence of surfaces on the thermal activation of slip near surfaces (which were found to halve the activation area) (Diologent et al., 2011) were thus brought to light.

Nanoporous metals produced either by selective dissolution of a two-phase alloy (Lu et al., 2007a; Rösler et al., 2005; Rösler and Näth, 2010) or by dealloying (Erlebacher and Seshadri, 2009; Weissmüller et al., 2009) have nanometric pores, one to two orders of magnitude finer in scale than those in replicated metal foam. Here, the size effects on flow stress can be spectacular—but are more challenging to measure because samples are generally too small for bulk material characterization and are often brittle. In nanoporous metals (chiefly gold), it is now well documented that the in situ metal flow stress increases with decreasing strut size, and can reach values (given the very small strut sizes that can be attained by dealloying) estimated to approach the theoretical shear strength, to the point where at times the nanoporous metal is harder than the same metal in bulk form (Biener et al., 2005, 2006; Hodge et al., 2006, 2007; Volkert et al., 2006; Jin et al., 2007; Lee et al., 2007a; Hakamada and Mabuchi, 2007; Sun et al., 2007; Weissmüller et al., 2009; Jin et al., 2009; Fan and Fang, 2009; Li and Misra, 2010; Dou et al., 2010; Dou and Derby, 2011).

Data suggest, albeit not very clearly, a Hall-Petch-like linear dependence of nanoporous metal flow stress versus the inverse square root of the ligament size (Biener et al., 2006; Volkert et al., 2006; Hodge et al., 2007; Weissmüller et al., 2009; Jin et al., 2009; Fan and Fang, 2009; Dou and Derby, 2010, 2011; Dou et al., 2010). Trends and values in data are not all that clear, however (e.g. (Jin et al., 2009; Balk et al., 2009)), for several reasons. One has to do with the material itself: there may be variations of the metal structure or composition with specifics of the sample production process. Also, the volume fraction and morphology of the metal ligaments may vary simultaneously with size such that, as the ligament size changes, other parameters (mesostructure, composition) change as well. This makes deconvolution of effects difficult. Another difficulty in data interpretation is that local in situ ligament strength values obtained from the various test configuration employed (nanoindentation, compression of nanopillars, compression of larger specimens, tensile testing of thin sample, bend testing…) depend on assumptions made on two fronts: (i) how to back calculate a uniaxial flow stress for the porous material from, say, nanoindentation test data (is the hardness then one or three times the uniaxial flow stress?) and (ii) as with microporous metals, how to back calculate the in situ flow law of the metal in the ligament knowing that of the nanoporous material (to this end the Gibson–Ashby relations, eqn (77) with $\Phi = 1$ or eqn (78), are generally used).

The question is scientifically interesting because it raises intriguing questions that are likely to be a focus of intense research in coming years. Examples of such questions are (i) how slip proceeds through the material knowing that ligaments are coherent (grains are much larger than ligaments and hence a single dislocation can run across a good many ligaments and pores) (Parida et al., 2006; Jin et al., 2009; Sun et al., 2009), (ii) the role of nanoscale strain gradients and associated crystal defects

(Dou and Derby, 2010, 2011; Dou et al., 2010), (iii) the influence of the pore surface state on flow within the metal (e.g. these materials provide extreme test cases for Rehbinder-type surface plasticity effects) (Rehbinder and Shchukin, 1972; Jin and Weissmüller, 2011), (iv) whether surface stresses interact with vectors of plastic deformation (Crowson et al., 2009), or (v) what the role of (potentially very high) internal stresses might be (Li and Misra, 2010). On several of these questions, the plasticity of fine-scale porous metals is intimately linked with the relatively intense current research effort on the plastic deformation of small-scale metal samples; recent reviews of this subject can be found in Refs. Uchic et al. (2009b), (2009a), Dehm (2009), Kraft et al. (2010), Greer and De Hosson (2011), Legros (2011), Godet et al. (2011), Weinberger and Cai (2012).

Intrinsic plasticity size effects are also manifest in damping characteristics of microcellular metals, where the level of (generally dislocational) internal friction has in several studies been shown to depend on the average pore size (Han et al., 1997; Liu et al., 2000a, 1998a, 1998b; Wei et al., 2002a; Golovin and Sinning, 2003, 2004; Hakamada et al., 2009) besides other parameters such as the metal structure, the temperature or the relative density (Banhart et al., 1996; Han et al., 1999; Wei et al., 2002b; Hao et al., 2007; Yang et al., 2007b). Size-dependent plasticity is also reflected in the relaxation response of fine-scale replicated microcellular aluminum, which as mentioned above was shown to be pore size dependent (Diologent et al., 2011). Acoustic emission signals were also found to depend on pore size at fixed foam composition and fabrication process (Kadar et al., 2008).

24.5.3.5 *Shape Memory Alloys and Bulk Metallic Glass*

Microcellular metals can deform by mechanisms other than dislocational plasticity: twinning and transformational plasticity have been demonstrated in highly porous thermal and magnetic shape memory alloys. These materials are interesting in several respects. One is that strains achieved upon twinning or martensitic transformation can, depending on the architecture of the porous material, be amplified when the metal is in microcellular form (by simple virtue of lever-type geometrical effects), while deformation-induced stresses are, on the other hand, strongly reduced (Elzey et al., 2005; John et al., 2007; Hassan et al., 2008; Dunand and Müllner, 2011).

A second interesting feature of highly porous metals deforming by twinning or transformational plasticity is in the fact that micropores provide sites for relief of microstrain incompatibility. Specifically, the performance of bulk polycristalline Ni–Mn–Ga magnetic shape memory alloys suffers from such incompatibilities: grain boundaries hinder the progress of magnetically induced twinning. Additionally, Ni–Mn–Ga alloys and their grain boundaries being brittle, twinning strain incompatibilities along grain boundaries cause these to fracture upon magnetic straining of the material. With a high fraction of pores having a size larger or commensurate with that of grains (leading to struts with a bamboo structure), such strain incompatibilities are relieved along the pore surfaces; microcracks then either do not appear, or remain geometrically confined. This, in turn, enables increased magnetic field-induced strains in Ni–Mn–Ga alloys while at the same time providing for greater material integrity through cycling compared with the dense polycrystalline material (Boonyongmaneerat et al., 2007; Chmielus et al., 2009, 2010; Dunand and Müllner, 2011; Zhang et al., 2011). The interaction between pores and magnetically induced twinning was also shown to beneficial to functional characteristics of the martensitic phase transition: pores improve the magnetocaloric effect in a Ni2MnGa-based Heusler alloy optimized for coincidence of the Curie and martensitic transformation temperatures (Sasso et al., 2011).

More conventional, thermal, shape memory alloys in highly porous form have been produced and characterized by many authors (also if we do not consider springs as highly porous metal), with an eye to applications in the biomedical field (Bansiddhi et al., 2008), in actuation (Elzey et al., 2005; Hassan et al., 2008), or in energy absorption, damping and impedance matching (Zhao et al., 2006c; Hassan et al., 2009). Such highly porous shape memory alloys have been shown to display attractive characteristics of elasticity, superelasticity and thermal shape memory effect, which have been measured and, at times, compared with predictions of micromechanical theory (Yuan et al., 2004, 2006; Greiner et al., 2005; Biswas, 2005; Zhang et al., 2007b; John et al., 2007; Hassan et al., 2008, 2009; Bansiddhi and Dunand, 2007, 2008, 2009; Oppenheimer and Dunand, 2009; Aydogmus and Bor, 2009; Neurohr and Dunand, 2011b, 2011a).

A beneficial interaction of porosity with microscopic agents of deformation is also found in BMGs. These materials are generally brittle, failing at low strain by the sudden propagation of discrete shear bands, the spacing of which scales as the size of the sample (Conner et al., 2003). Unlike most ceramics, BMGs are capable of some plastic flow above a generally very high yield stress (Ashby and Greer, 2006; Schuh et al., 2007). Crack tips are therefore surrounded by a plastic zone, which imparts to these materials a high fracture toughness despite their low ductility. When sample dimensions fall below a few times the radius of this plastic zone, fracture is prevented because shear displacements at which cracks nucleate are no longer attained (Conner et al., 2003). BMGs then become more ductile. The corresponding critical size depends on the alloy and varies, roughly, from somewhat below 1 μm to a few millimeters (Inoue et al., 1982; Conner et al., 2003; Ashby and Greer, 2006; Kumar et al., 2011a). This makes division of the material into many interconnected samples of small size, or in other words making microcellular BMG, a strategy to produce relatively ductile materials out of BMG (Conner et al., 2003; Schuh et al., 2007; Kumar et al., 2011a). Additionally, pores have the advantage of providing stress concentration sites that can nucleate shear bands. Pores hence multiply their number, reducing in turn associated displacements and hence the probability for shear band induced crack nucleation. Pores also provide free surfaces at which shear bands can be arrested, reducing their length and, again, associated displacements. Note that, since the role of pores is largely one of aiding nucleation, even low pore volume fractions should be beneficial: these too will cause a sharp increase in the density of shear bands within the BMG, in turn reducing the probability that these turn into cracks.

Several processes have been used to produce porous BMG, ending in open or closed-cell materials, with pores 10 μm in diameter and above occupying from a few percent to a majority of the material volume. Some processes take advantage of specificities of BMG deformation at elevated temperature, while all are designed to ensure sufficiently rapid postprocess cooling to prevent crystallization (Schroers et al., 2004, 2003; Wada and Inoue, 2003, 2004; Wada et al., 2005b, 2005a, 2006; Inoue et al., 2006, 2007; Brothers and Dunand, 2005a, 2005b, 2006a, 2007; Brothers et al., 2005; Lee and Sordelet, 2006b; Xie et al., 2006; Demetriou et al., 2007a, 2007c, 2007d, 2007e; Cox et al., 2010; Kumar et al., 2011a). Alternative routes based on selective dissolution have also been shown to produce nanoporous metallic glass structures (Gebert et al., 2004; Lee and Sordelet, 2006a; Jayaraj et al., 2006). In some cases porous BMG/crystalline metal composites were produced (Brothers and Dunand, 2004; Brothers et al., 2007, 2011).

Test data show that even a few percent porosity improves the compressive "ductility" of BMGs dramatically ($\approx 20\%$ uniaxial compressive strain to failure), by the mechanism previously mentioned, namely the formation of a high density of homogeneously distributed shear bands (Wada et al., 2005b, 2005a; Inoue et al., 2006, 2007; Schuh et al., 2007; Brothers and Dunand, 2007). In more highly porous microcellular BMGs, much lower flow stresses and stiffnesses are found, while compressive deformation can proceed up to densification of the material, showing that the BMG size effect noted above is indeed operational (Wada et al., 2005b, 2006; Inoue et al., 2006; Brothers and Dunand, 2007). The effect is, however, not universally observed, being, for example, a function of the pore geometry or orientation (Wada et al., 2006; Brothers and Dunand, 2007) —and, of course, of the size of struts in relation to plastic crack tip zone sizes in the BMG at hand.

Yield stresses of BMGs being far higher, and their elastic modulus values somewhat lower, than those of crystalline metals, struts and cell walls in microcellular (high porosity) BMGs tend to buckle before they yield. Their compressive yield is therefore generally governed by elastic buckling of the struts or cell walls; in this sense these materials resemble microcellular polymers and elastomers. As a result (Gibson and Ashby, 1997), the compressive yield stress of open-cell microcellular BMGs often scales with V_m^2 instead of $V_m^{3/2}$ (Brothers and Dunand, 2005b, 2007; Inoue et al., 2006; Wada et al., 2008; Schramm et al., 2010; Wang et al., 2010a) (one BMG metal foam however showed a $V_m^{3/2}$ scaling of yield stress (Demetriou et al., 2008), suggesting that it deforms predominantly by strut yield or strut fracture instead of buckling, both mechanisms giving, as already noted, the same scaling law (Gibson and Ashby, 1997)). Hence, the flow stress and Young's modulus scale similarly with relative density (Gibson and Ashby, 1997), a fact that was noted in Refs. Inoue et al. (2006), Wada et al. (2008), Demetriou et al. (2007b). Despite this difference with crystalline microcellular metals, yield in microcellular BMGs has been documented to have many features in common with yield in other highly porous metals, such as the formation of discrete deformation bands (Wada and Inoue, 2004; Demetriou et al., 2007b, 2008) and a wavy or serrated stress–strain curve past yield at a peak in the curve (Wada and Inoue, 2004; Inoue et al., 2006; Brothers and Dunand, 2007; Brothers et al., 2007; Demetriou et al., 2007c, 2007b, 2008; Cox et al., 2010). Another parallel with the deformation of highly porous crystalline metals is the fact that inhomogeneity in the mesostructure hastens yield, and thus reduces structural performance (Demetriou et al., 2008).

24.5.4 Dynamic Deformation

One of the main envisaged applications of highly porous microcellular metals is in the absorption of mechanical energy, notably in crash or other impact situations (e.g. aluminum honeycombs are placed in French TGV trains to protect the driver in the event of a collision (Zhao and Gary, 1998)). There have thus been many studies of their behavior under high strain rate deformation conditions.

Several test configurations have been used to measure this, ranging from the use of conventional uniaxial testing machines at the top range of their cross-head displacement rate capabilities (which does not deliver very high strain rates), to dedicated testing methods. Among these, by far the most usual is the Split Hopkinson Pressure Bar (SHPB) method, in which a compression sample is sandwiched between two long bars equipped with strain gages. One bar is static while the other is struck with a projectile to start the test. Analysis of SPHB data is not trivial but methods are well established and output is often of high quality; however, because highly porous metals are weak, and also because millimetric cell sizes in porous metals may require that samples have a wide diameter (see Sections 24.5.1.6 and 24.5.3.4.1), modifications have been made to the apparatus to test microcellular metals. For instance steel bars that are most frequently used in the SHPB apparatus have been replaced with bars of wider diameter and/or bars made of lower stiffness materials (Zhao et al., 2005b, 2006b; Lee et al., 2006c; Elnasri et al., 2007; Cady et al., 2009; Fang et al., 2010; Chakravarty, 2010; Hou et al., 2011a, 2011b). This in turn may necessitate refinement in data recovery procedures (e.g. because polymeric bars are viscoelastic). Other high strain-rate testing methods have also been used (Ma et al., 2009; Zhu et al., 2011a), including notably (i) impact tests conducted by sending a striker bar onto an anvil bar with the sample attached either to the striker bar (direct impact tests) or to the anvil bar (reverse impact test) (Hanssen et al., 2002b; Tan et al., 2002, 2005a; Lopatnikov et al., 2003; Radford et al., 2005; Lee et al., 2006c; Nemat-Nasser et al., 2007; Bourne et al., 2008), (ii) bipendulum tests in which the sample is impacted by a suspended pendulum bar (Montanini, 2005), or (iii) various more complex tests designed to gage the performance of microcellular metals in "real" impact energy absorption situations (Kumar et al., 2003; Ramachandra et al., 2003; Hanssen et al., 2002a; Zhao et al., 2007; da Cunda et al., 2011; Zhu et al., 2011a).

The global result of these many tests is an inconsistent picture. Some datasets show a clear stress enhancement effect of high strain rate on the deformation stress of highly porous microcellular metals (Mukai et al., 1999a, 1999b; Paul and Ramamurty, 2000; Kanahashi et al., 2000, 2001a, 2001b; Park and Nutt, 2002; Tan et al., 2002, 2005a; Hayes et al., 2004; Montanini, 2005; Balch et al., 2005; Han et al., 2005; Zhao et al., 2005b, 2006b; Elnasri et al., 2007; Yu et al., 2006, 2008; Cady et al., 2009; Liu et al., 2009a; Daoud, 2009; Fang et al., 2010; Hou et al., 2011a, 2011b; Mu et al., 2011), while others do not, or barely do (Kenny, 1996; Lankford and Dannemann, 1998; Böllinghaus et al., 1999; Dannemann and Lankford, 2000; Deshpande and Fleck, 2000a; Hall et al., 2000; Yi et al., 2001; Kanahashi et al., 2001b; Ruan et al., 2002; Hanssen et al., 2002b; Miyoshi et al., 2002; Feng et al., 2003; Hayes et al., 2004; Montanini, 2005; Mukai et al., 2006; Wang et al., 2006b; Lee et al., 2006c; Nemat-Nasser et al., 2007) (dual references in these two lists indicate alloy-dependent conclusions by the same authors; data up to 2009 are also tabulated in Refs. Ma et al. (2009), Liu et al. (2009d)). Contradictory conclusions were at times also reached for the same material. Some sources for inconsistency in data are sought in Ref. Ma et al. (2009): these include variations, from one study or test to the other, in parameters such as the cell-to-diameter ratio or the relative density of samples tested, or variations in testing procedure, notably in the method used to measure stress (direct and reverse impact tests can, for example, lead to conflicting conclusions (Ma et al., 2009)).

Overall, the rather extensive body of experimental work probing the dynamic deformation of highly porous metals counts, very roughly, a proportion near one-half of datasets that show a strong effect of strain rate on the flow stress of highly porous metals. The effect is hence not universally observed, but it is real. Four main mechanisms have been proposed to explain this enhancement of the flow stress of highly porous metals under rapid deformation (Tan et al., 2002; Zhao et al., 2005b, 2006b).

One is generally discounted: resistance to the flow of air out of, or into, pores that see a sudden change in their volume will oppose a certain resistance to deformation of the porous material, manifest as an increase in its flow stress. Back-of-the-envelope estimations of this increase in flow stress generally lead to the conclusion that it is negligible in comparison to the yield stress of engineering microcellular metals (Deshpande and Fleck, 2000a; Tan et al., 2005a; Zhao et al., 2005b; Liu et al., 2009d). Note, however, that depending (i) on the viscosity of the fluid within the pores of microcellular metals, (ii) on the average pore diameter and (iii) on the deformation rate, this effect might not be negligible (Gibson and Ashby, 1997; Coussy, 2004; Benke and Weichert, 2005; Dormieux et al., 2006; Dawson et al., 2008).

The second source of strain-rate dependence in the flow stress of microcellular metals is obvious: strain-rate effects in the metal itself will be manifest in the behavior of foams made therefrom. Barring other effects, a microcellular metal is expected to deform with roughly the same strain-rate dependence as the dense metal from which it is made—this is seen either by back-of-the-envelope calculations (Deshpande and Fleck, 2000a), or if one considers the fact that the steady-state creep stress exponent of a highly porous metal is, at small strains, that of the same metal in dense form (see the discussion of creep, in Section 24.5.5). The majority of microcellular metals or honeycombs tested to date are aluminum based: aluminum does not have a very strong strain-rate dependence of its flow stress at room temperature (San Marchi et al., 2002; Zhao et al., 2005b). So with aluminum-based materials this is not expected to be a major factor. With other metals, such as carbon steel, magnesium or zinc-based alloys, the base metal room-temperature strain-rate dependence being noticeable it is not surprising that test data betray a clear influence of strain rate on the corresponding porous metal flow stress (Mukai et al., 1999b; Kanahashi et al., 2001a, 2001b; Park and Nutt, 2002; Liu et al., 2009a; Daoud, 2009; Ma et al., 2009; Tane et al., 2010; Song et al., 2011).

Simulations confirm that strain-rate dependence in the dense metal flow law enhances the strain-rate dependence of the same metal in porous form. Still, this effect is not alone in causing dynamic stress enhancement in porous metals since these can exhibit strain-rate dependent deformation even though the metal of which they are made is relatively rate independent (Han et al., 2005; da Cunda et al., 2011). This, in turn, shows the importance of inertia in the dynamic deformation of porous metals.

The "microinertia" effect is a third cause for a strain-rate dependence of the flow stress of microcellular metals (Deshpande and Fleck, 2000a; Tan et al., 2002; Zhao et al., 2005b, 2006b). This is an extension to microcellular structures of an effect known to influence the dynamic behavior of structures in general: inertia can retard, in rapid deformation, the onset of buckling in loaded structures made of beams or shells. The effect is intuitively accessible; however, mechanically it is complex and depends strongly on details of the structure geometry and its buckling bifurcation.

Specifically it is known that if, upon buckling, the yielding load of the structure under quasistatic deformation decreases sharply from a peak value (indicating a clear branching bifurcation; these are called Type II structures), then inertia effects will have a strong influence on the value of the peak load attained. If, on the other hand, past the onset of buckling the quasistatic load versus deformation curve of the structure remains flat (Type I structures), then inertia effects will not change the buckling load strongly. Experimental observations coupled with finite element calculations have, for example, shown this effect with clarity in the flow stress and buckling bifurcation of a single-layer pyramidal lattice core structure of 304 stainless steel: at high deformation rates, inertia effects alter the truss deformation pattern, this leading in turn to a potentially strong increase in the peak stress (Lee et al., 2006b). Aluminum honeycomb data are also consistent with this effect, in that the peak flow stress of the honeycomb along the axis of the hexagonal cells (out of plane direction) shows a strong dependence of its value with strain rate, whereas the (flat) "in-plane" stress–strain curve of the honeycomb is little affected by rapid straining (Zhao and Gary, 1998; Deshpande and Fleck, 2000a).

Other evidence for the importance of microinertia effects has been found in the fact that metal foams with brittle cell walls (Cymat aluminum foam), which hence do not deform much past a buckling instability, have shown little effect of strain rate on their flow stress (Zhao et al., 2005b; Elnasri et al., 2007; Fang et al., 2010) (note however that other experiments show a strain-rate-dependent peak stress for this foam (Montanini, 2005; Ma et al., 2009)...). Tuncer et al. also proposed that microinertia explains the dependence of stress enhancement at high strain rate ($600\ \mathrm{s}^{-1}$) on the cell aspect ratio in a titanium foam (Tuncer et al., 2011a). Data from a comparison of dynamic and quasistatic compression tests conducted on samples of foamed Pd-based BMG suggest a transition in the scaling law from $\sigma_y \propto V_m^2$, typical of microbuckling-dominated yield of highly porous BMG, to $\sigma_y \propto V_m^{3/2}$ (Schramm et al., 2010), that is, the same law as found for open-cell microcellular ductile metals or brittle solids (Gibson and Ashby, 1997): this too might be a manifestation of microinertia effects, since it suggests a change in the point of the microbuckling bifurcation at high strain rate.

The importance of microinertia remains, however, unsettled. As Deshpande and Fleck point out (Deshpande and Fleck, 2000a), most microcellular structures are bending-dominated Type I structures that do not have a clear buckling bifurcation; also, it is known that this effect is reduced by the presence of imperfections, which are numerous in "real" microcellular metals (Zhao et al., 2005b; Lee et al., 2006b). It has been argued on the other hand that microinertia arises, not due to a Type II strut architecture, but rather due to the presence of large nodes where struts are interlinked (Tan et al., 2002). Tan et al. give a quantitative account of their initial yield stress data with a simple model that assumes plastic wave propagation through nonbuckling cell walls at high strain rates (Tan et al., 2005a). Also, examination of detailed strut deformation modes in experiment do show microbuckling events in quasistatic deformation that, at a local level, should be of Type II (Marmottant et al., 2005; Zhao et al., 2005b, 2006b). These events hence might cause microinertia-induced local hardening at high rates of deformation.

A fourth mechanism that can lead to an enhanced flow stress in microcellular metals at high rates of deformation is also related to inertia, but differs from microinertia in that it operates at the macroscale. At moderately high strain rates, highly porous metals deform as they do quasistatically: the stress is uniform and plastic deformation often concentrates in random bands (e.g. (Deshpande and Fleck, 2000a; Tan et al., 2005a)). When, on the other hand, the surface of a porous material is impacted at very high rates (typically on the order of 100 m s^{-1} and higher), then plastic deformation shock fronts can develop in materials that have a stress–strain curve that is upward concave (Tan et al., 2005b; Elnasri et al., 2007)—a feature of many microcellular metals, both two and three dimensional (Section 24.5.2). There is then a transition in the deformation mode of highly porous metals when these are struck at high velocity on one side, the other side resting on an immobile and stiff anvil. Namely, at higher impact velocities, instead of deforming (more or less) homogeneously along the sample, deformation occurs almost entirely at a narrow moving "shock front," where the material is suddenly crushed from nearly zero strain to a high strain, the value of which is situated well within the densification regime of rapidly rising stress. This front appears where the porous material is impacted, and then travels toward the far end of the sample. Across the front, there is a drop in stress, dictated by conservation of mass and momentum conservation. This is given as

$$\sigma_{\text{impact}} - \sigma_{\text{y}} = \frac{\rho V_{\text{impact}}^2}{\varepsilon_{\text{impact}}}, \tag{86}$$

where σ_{impact} is the stress on the porous materials on its impacted side, σ_{y} its yield stress, ρ its density before impact, V_{impact} the velocity of its impacted side (in the referential of the undeformed porous metal, the anvil side of the front thus being at rest), and $\varepsilon_{\text{impact}}$ the strain in the impact-deformed region to which the porous metal is deformed as the shock front passes it by (Tan et al., 2005a, 2005b, 2006b; Radford et al., 2005; Lee et al., 2006c; Elnasri et al., 2007; Pattofatto et al., 2007; Harrigan et al., 2010; Zhu et al., 2011a). The shock front velocity V_{front} is simply given by mass conservation (if we assume the undeformed porous metal to be at rest on the anvil) as

$$V_{\text{front}} = \frac{V_{\text{impact}}}{\varepsilon_{\text{impact}}}. \tag{87}$$

To bring the problem to a solution, several simplified models have been proposed that describe in tractable terms the shock-wave compaction behavior of the porous metal. In the rigid perfectly plastic locking model the porous metal is assumed to deform at constant stress until it "locks" at a fixed densification strain $\varepsilon_{\text{lock}}$; at that point its stress–strain curve is taken to go from horizontal (at σ_{y}) to vertical (Ashby et al., 2000; Tan et al., 2002, 2005a, 2005b, 2006b; Radford et al., 2005; Lee et al., 2006c; Harrigan et al., 2010). Various other simplifications of the stress–strain curve, and both analytical and numerical methods, have been used in analysis of the locally concentrated high strain rate deformation of highly porous metals including both honeycombs and microcellular metals (Lopatnikov et al., 2003, 2004, 2007; Karagiozova et al., 2010; Harrigan et al., 2010; Zhu et al., 2011a). An alternative assumption that does away with having to assume a value for $\varepsilon_{\text{lock}}$, was proposed by H.

Zhao and coworkers. This consists in simply assuming that the intrinsic deformation law of the impacted material remains the same despite the high rate of deformation; one then uses the quasistatic stress–strain curve to link σ_{impact} and ε_{impact} (Elnasri et al., 2007; Pattofatto et al., 2007). For a concave upward stress–strain curve typical of many porous metals, this can be represented as

$$\sigma_{impact} = \sigma_y + c\left(\varepsilon_{impact}\right)^n,\qquad(88)$$

where c is a constant, and exponent n exceeds unity (because the curve is concave upward). Note, however, that this approach implicitly assumes that microinertia effects are absent, or alternatively that medium-rate flow curve data can be used to account for these.

The existence and consequences of shock front formation in microcellular metals and honeycombs has been evidenced with clarity, both in simulation (Ma et al., 1997; Li and Meng, 2002; Hönig and Stronge, 2002a, 2002b; Ruan et al., 2003; Zheng et al., 2005; Li et al., 2007; Liu et al., 2009d; Song et al., 2010b), and experimentally by means of high-speed filming, interrupted tests and strain measurements in shock deformation of highly porous metals (Tan et al., 2005a; Radford et al., 2005; Lee et al., 2006c; Elnasri et al., 2007; Pattofatto et al., 2007; Bourne et al., 2008; Fang et al., 2010). An interesting further result of simulation is that randomness in the structure of porous metals "dilutes" shock fronts and reduces their occurrence (Zheng et al., 2005; Li et al., 2007; Song et al., 2010b).

The stress enhancement that is brought at high strain rates by this mechanism differs from that which results from other effects in that it is inhomogeneous: it is only on the impact side of the porous material that the flow stress is increased; downstream the porous metal remains at, or near, its (quasistatic) yield stress, σ_y. This fact has been evidenced (Pattofatto et al., 2007), and explains some of the disprepancies in data for the flow stress of porous metals at high rates of deformation: the stress–strain curve depends on which side of the sample the stress was measured during the dynamic experiment. The importance of this phenomenon in blast or impact mitigation applications is obvious; a discussion is given in Ref. Ashby et al. (2000). Refs. Hanssen et al. (2002a), Zhao et al. (2007) give examples of experimental trial tests of porous metals in this class of application.

24.5.5 Creep

When their temperature exceeds roughly 40% of their (absolute) melting point, metals and alloys show a strongly increased sensitivity to strain rate, and start creeping (i.e. deform gradually in time under steady load). The same of course holds also for porous metals. Beyond its importance in the creep fracture of structural metals (most creeping metals are, shortly before they fracture, often full of pores), creep has practical importance for many potential applications of highly porous metals: in catalysis, gas filtration, fuel cells, or heat exchangers, for example.

Pores accelerate creep: this is seen with both closed cell (Andrews et al., 1999c; Zhang et al., 2002; Haag et al., 2003; Hakamada et al., 2005a; Carcel et al., 2010), and open cell (Andrews et al., 1999b; Hodge and Dunand, 2003; Choe and Dunand, 2004b, 2004a; Hakamada et al., 2005b; Zurob and Brechet, 2005; Boonyongmaneerat and Dunand, 2009; Scott and Dunand, 2010; Diologent et al., 2009c, 2009b; Soubielle et al., 2011a) microcellular metals. Creep curves of microcellular metals generally have classical shapes, with a primary stage of gradually lowering strain rate followed by an often clearly visible zone of steady-state creep ending, if the sample is subjected to tensile stress but not in compression, with tertiary creep and fracture (Andrews et al., 1999b, 1999c; Choe and Dunand, 2004b; Couteau and Dunand, 2008; Diologent et al., 2009c, 2009b; Scott and Dunand, 2010; Soubielle et al., 2011a), **Figure 39**.

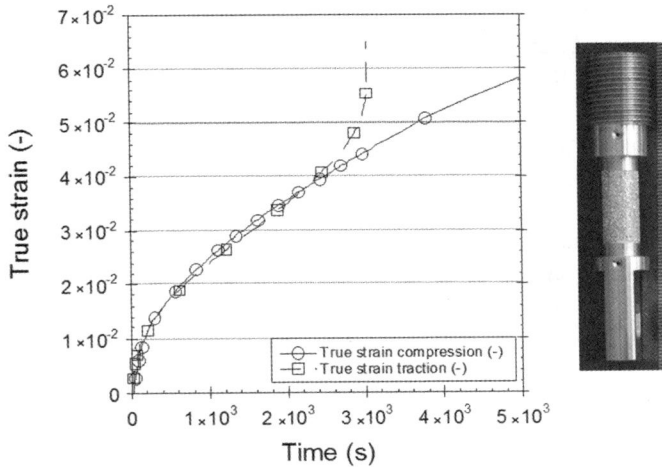

Figure 39 Left: two superimposed creep curves for replicated 99.99% pure aluminum, tested once in tension, once in compression with the same test parameters ($T = 250$ °C, $s = 0.5$ MPa, $V_m = 0.203$). Right: a typical test sample (ø 10 mm, length 20 mm, held by gluing on aluminum heads with ceramic glue). From EPFL research published in Ref. Soubielle et al. (2011a). (For color version of this figure, the reader is referred to the online version of this book.)

Measured values of the steady-state creep rate stress exponent M and activation energy Q are often the same as those of the (dense) metal or alloy making the foam: this was found in open-cell microcellular metals, of aluminum alloy (Andrews et al., 1999b; Diologent et al., 2009c, 2009b), Ni–Al and Ni–Cr–Al (Choe and Dunand, 2004b) or Fe–Cr–Mo alloy (Scott and Dunand, 2010). Sometimes, however, M or Q differ significantly from values for the same metal in dense form: in Alporas™ aluminum alloy closed-cell foam (Andrews et al., 1999c; Zhang et al., 2002; Haag et al., 2003; Carcel et al., 2010), in NiAl open-cell hollow strut foams at the upper end of applied stresses explored (while at lower stress the exponent was that of the bulk metal) (Hodge and Dunand, 2003) and in microcellular 99.99% pure aluminum (Soubielle et al., 2011a).

In principle, at the simplest level of analysis such deviations should not occur: provided strains remain small (such that the pore shape and size do not change), pores in a metal that creeps according to Norton's law should accelerate its rate of creep without changing its stress exponent or activation energy. Suppose that a metal creeps according to Norton's law:

$$\dot{\varepsilon} = \dot{\varepsilon}_0 \cdot \left(\frac{\sigma}{\sigma_{0,m}}\right)^M \quad \text{with} \quad \dot{\varepsilon}_0 = K \exp(-Q/RT), \tag{89}$$

where $\dot{\varepsilon}$ is the steady-state creep rate, σ the uniaxial applied stress, K, $\dot{\varepsilon}_0$ is a reference strain rate and $\sigma_{0,m}$ is a stress constant characteristic of the dense metal, M is the steady-state creep rate stress exponent, Q the activation energy for creep (often equal to that for self-diffusion within the creeping metal), R the gas constant and T temperature. Then, when it contains pores, at small strain the porous metal creep rate writes

$$\dot{\varepsilon} = \dot{\varepsilon}_0 \cdot \left(\frac{\sigma}{\sigma_0}\right)^M, \tag{90}$$

where $\sigma_o (< \sigma_{o,\,met})$ is a stress constant characteristic of the porous metal. Using engineering beam analysis, Andrews, Gibson and Ashby (Gibson and Ashby, 1997; Andrews et al., 1999c, 1999b) predict that for highly porous microcellular metals, σ_o is given by

$$\sigma_o = \sigma_{o,m} \left(\frac{M+2}{C_4} \right)^{\left(\frac{1}{M} \right)} \frac{M}{C_5 (2M+1)} (\Phi V_m)^{\left(\frac{1+3M}{2M} \right)} + \frac{2}{3} (1 - \Phi) V_m, \qquad (91)$$

where Φ is the fraction solid that is contained in cell walls (as in eqn (54) or (77)). Constants $C_4 = 0.6$ and $C_5 = 1.7$ are estimated by requiring that the equation agree with elastic and perfectly plastic foam behavior, respectively, in the two limits where M tends toward infinity or unity.

Hodge et al. (Hodge and Dunand, 2003) assume that the rate of creep of microcellular metals is controlled not by strut bending, but by the uniaxial deformation of struts aligned along the stress axis. This assumption gives a creep rate that scales linearly with V_m, while M and Q remain the same as those of the material making the foam.

Simplified variational estimates of Ponte Castañeda and Suquet for nonlinear elastic composite deformation can also be adapted to predict the steady state power-law creep rate of porous solids (Despois et al., 2006b, Mueller et al., 2007, Diologent et al., 2009c). The starting point of this calculation is that, provided certain frequently obeyed conditions are respected, equations for the elastic deformation and viscous flow of isotropic materials are interchangeable (Hoff, 1954; Ponte-Castañeda and Suquet, 1998). The approach presented in Section 24.5.3.2 for the monotonic nonlinear deformation of porous metals can thus be paralleled, using again the elastic modulus of the porous metal for calibration of its load-bearing capacity. This yields

$$\sigma_o = \sigma_{o,m} \, F_E^{\left(\frac{1+M}{2M} \right)} V_m^{\left(\frac{M-1}{2M} \right)} = \sigma_{o,m} \, C_E^{\left(\frac{1+M}{2M} \right)} V_m^{\left(\frac{M-1+N(1+M)}{2M} \right)}, \qquad (92)$$

where F_E is the ratio of the Young's modulus of the foam, E, over that of the dense metal, E_m (eqn (80)), and the last right-hand term is obtained after insertion of eqn (56), which defines the modulus scaling constant C_E and exponent N. Dense pure metals have high creep exponents: typically, M exceeds three, being nearer five in pure aluminum or copper, for example.

Predictions of the preceding equations have been confirmed by the (relatively few) experimental studies that have included systematic variations of the relative density V_m (Andrews et al., 1999b; Diologent et al., 2009b, 2009c; Boonyongmaneerat and Dunand, 2009; Scott and Dunand, 2010; Soubielle et al., 2011a). Note that the creep rate of the porous metal scales according to analysis with relative density as $V_m^{-\left(\frac{1+3M}{2} \right)}$ according to the model of Andrews et al. (eqn (90)), or as $V_m^{-\left(\frac{M-1+N(1+M)}{2} \right)}$ according to the variational estimate, eqn (92). With M equal to three, five or more in metals and alloys and with N near two, it transpires that porous metals must see their creep rate increase steeply with porosity; this too was confirmed by experiment.

Alternative models have also been proposed (i) to account for a hollow core in the struts making open-pore microcellular metals (Hodge and Dunand, 2003; Zurob and Brechet, 2005; Andrews, 2006), (ii) to assume that deformation is dominated by other modes of deformation than strut bending and also to introduce a difference in width between struts or nodes (Boonyongmaneerat and Dunand, 2009), (iii) to compare analytical or experimental results with finite element simulations of periodic open-pore microcellular structures (Hodge and Dunand, 2003; Oppenheimer and Dunand, 2007),

(iv) to predict creep rates in sintered hollow sphere structures (Marcadon, 2011), or (v) to expand the Andrews–Gibson–Ashby derivation (Lin and Huang, 2005b; Chen and Huang, 2008b).

Several experimental investigations of creep in microcellular metals have, as already mentioned, confirmed that M and Q are the same as those of the same metal or alloy in dense form. Deviations found in other data have been explained on the basis of mechanical or metallurgical arguments; explanations depend on the study. One potential cause for stress exponents in excess of that describing creep of the dense metal is that, when pores are present, the metal may be deforming locally beyond the range of validity of Norton's law for the dense alloy. This may notably occur if, in regions of high stress, it has entered the range of power law breakdown (Andrews et al., 1999c; Haag et al., 2003; Hodge and Dunand, 2003; Carcel et al., 2010). This explanation was supported in one study by finding fine subgrains—less than 1 μm wide—in more heavily deformed regions of crept closed-pore aluminum foam (Zhang et al., 2002): such subgrains suggest that stresses in those locations were more than one order of magnitude higher than elsewhere. Variations of the stress exponent M with test parameters were also found in Refs. Boonyongmaneerat and Dunand (2009), Diologent et al. (2009c), (2009b); these are explained by the variational model knowing how the creep exponent varies in the dense metal.

Other reasons for M to deviate from its value in the dense metal are metallurgical. Choe et al. found the stress exponent of a low-V_m reticulated Ni–Cr foam to fall below that of the bulk alloy. This was attributed to a shift toward diffusional creep, in turn caused by a small grain size of the Ni–Cr in foam struts (50–100 μm) (Choe and Dunand, 2004a). In replicated pure aluminum it was shown that the creep exponent increases from M to $M+3$ when the pores are finer than the subgrains in the metal at the relevant applied stress (Soubielle et al., 2011a). This transition, which is absent if the dislocation substructure is of finer scale than the strut width (as is the case with Al–Mg or Al–Ni alloys (Diologent et al., 2009c, 2009b)) is in accordance with the substructure-invariant model of Sherby et al. (Young et al., 1975; Sherby et al., 1977): pores can pin subgrain boundaries and this changes the metal creep law.

Plasticity size effects are thus manifest also in the high-temperature deformation of porous metals. This in turn implies that pore scale refinement can strengthen porous metals against creep—a fact that is well known from the behavior of nanoporous tungsten (Schade, 2010, 2011). Another cause for lowered creep rates is oxidation: at high temperature in long-duration creep experiments, if a scale of oxide grows and has the cohesion, adhesion and strength to carry load, it will lower the rate of creep of the porous metal in comparison to what is expected free of oxidation (Scott and Dunand, 2010; Soubielle et al., 2011a). This effect is in fact engineered in syntactic foams: the presence of a ceramic shell around each pore lends these structures far greater creep resistance than unreinforced porous metals, to the point where at low stress these composite structures can be more creep resistant than the dense metal. Load transfer between matrix and ceramic in these materials can also cause transitions in creep regime (Couteau and Dunand, 2008).

Honeycombs behave similarly to microcellular metals, except that their behavior is strongly anisotropic. Analytical expressions have been derived (Andrews et al., 1999b; Lin and Huang, 2005a; Chen and Huang, 2008a), and finite element simulations have also been conducted (Andrews and Gibson, 2001a; Oruganti and Ghosh, 2008), to predict their rate of transverse creep knowing that of the (generally power law) metal of which they are made. Expressions and creep rates are globally similar to those for microcellular metals, M and Q being also preserved. The more regular structure of honeycombs makes it easier to predict the evolution of their creep rate when higher strain is reached, or when microstructural irregularities are present; both have been shown to affect the creep rate significantly (Andrews and Gibson, 2001a; Oruganti and Ghosh, 2008).

In tension, creeping microcellular metals fail in rather classical fashion (**Figure 39**). With both open and closed-pore microcellular aluminum alloys, the Monkman-Grant relation has been found to hold (this states that the time to tensile failure is roughly proportional to the inverse of the steady-state creep rate) (Andrews et al., 1999c; Andrews et al., 1999b; Chen and Huang, 2008b; Diologent et al., 2009b; Soubielle et al., 2011a).

In compression, creep failure corresponds to the onset of densification preceded by accelerated creep. In the creep regime this is generally driven by the buckling of cell walls or struts (Gibson and Ashby, 1997; Andrews et al., 1999b; Cocks and Ashby, 2000); examples of buckled cell walls in crept closed-pore aluminum alloy foam can be found in Ref. Carcel et al. (2010). Analytical and numerical predictions of the gradual, creep-induced buckling of honeycomb walls, or of 3D microcellular metal struts and cell walls, give the time after which a small imperfection has grown to the point where it induces buckling and ensuing cell collapse (Gibson and Ashby, 1997; Cocks and Ashby, 2000; Lin and Huang, 2006; Chen and Huang, 2009b, 2009a). This time depends on the imperfection size, the strut or wall geometry and all other creep parameters including the applied stress, but is in all cases finite if the solid is within the range of stresses and temperatures at which it creeps. Hence, all highly porous metals will fail at some time if they creep, this time depending strongly on their structure and service environment (given the strong dependence of creep rates on stress and temperature). An interesting observation in this regard is to be found in the pioneering work of Andrews, Gibson and Ashby, who measured similar failure times with open-pore microcellular aluminum in both tension and compression, both obeying the Monkman–Grant law (Andrews et al., 1999b). Given the difference in failure mechanism between tension and compression, that the same values were found is a priori unexpected—and hence interesting.

24.5.6 Fracture

Strictly speaking, fracture of porous metals includes the subject of ductile fracture in its near-entirety: a few exceptions set aside (such as highly pure metal single crystals, which rupture by continuous slip to a line or a point), ductile metals become porous before they fracture. Coverage of ductile fracture exceeds by far the scope of this chapter; the reader is hence referred to dedicated reviews or books (Ref. Pineau and Pardoen (2007) notably). Here, we will restrict attention to the fracture of microcellular metals.

Like most of their properties, the toughness and the fracture toughness of microcellular materials vary strongly with the relative density V_m. This variation is generally expressed by means of the usual scaling law (taking for simplicity the Poisson's ratio as close to zero and thus making no difference between plane stress or plane strain fracture)

$$G_{Ic} = C_6 V_m^p \tag{93}$$

or

$$K_{Ic} = \sqrt{EG_{Ic}} = C_7 V_m^q, \tag{94}$$

where C_6, C_7, p and q are constants characteristic of the porous material. A few values of p or q can be derived with relative ease.

Suppose first that the dense material making the microcellular solid is brittle and breaks in deterministic fashion wherever the stress in a certain direction reaches a critical value. A large unstable crack will then cut its way through the material by progressively breaking struts or cell walls at peak stress locations near the crack tip; this can be by bending or in tension, the precise crack path depending on the cell architecture and the loading configuration. The material being brittle, this will consume roughly $G_{Ic,m}$ J/m^2 of fractured solid surface (neglecting, e.g. the extra energy dissipated by elastic unloading within loose fractured struts), if $G_{Ic,m}$ is the toughness of the (dense) solid. If the architecture of the microcellular metal and the microfracture site and mechanism remain the same as the relative density V_m varies, it is then reasonable to assume that the areal fraction of solid traversed by the crack be proportional to the solid volume fraction, V_m. One then has

$$G_{Ic} = g G_{Ic,m} V_m, \qquad (95)$$

where g is a constant of proportionality on the order of unity (which it equals for a random straight crack plane). This simple reasoning predicts, for brittle microcellular materials, that $p = 1$, while q equals 2, 3/2 or 1 for honeycombs stressed in-plane, open-pore microcellular solids, or closed-pore microcellular solids, respectively (since N, eqn (56), then equals, respectively, 3, 2 or 1). The same, or only slightly different, results are obtained after stress-based derivations that map continuum crack tip stress fields onto the porous solid, to then calculate forces and moments on struts or cell walls when they fracture (Ashby, 1983; Maiti et al., 1984; Choi and Lakes, 1996; Gibson and Ashby, 1997). All else constant, also, G_{Ic} and K_{Ic} of the porous metal will vary with the only microstructural length scale of the struture, namely the cell size l, raised to the power 1 and 1/2, respectively—provided intrinsic properties of the material making the microcellular solid are not scale dependent. Note that such scale dependence is, however, often extant, for example, in very brittle materials, the strength of which is governed by the statistics of defects they contain. The toughness of such microcellular materials is then to be treated using Weibull statistics coupled with beam or plate theory; this is to be found in Refs. Huang and Gibson (1991b), (1991a), Gibson and Ashby (1997).

Metals being often ductile, crack separation and fracture in microcellular metals will frequently occur by ductile rupture, after necking and sometimes (depending on the metal) after appreciable deformation, of struts or cell walls along the crack path. These features can be seen in the fractograph of a compact tension sample of replicated pure aluminum, **Figures 40** and **41**. Crack tips in ductile microcellular metals are generally surrounded by a plastic zone, such that one can link the fracture initiation toughness to the crack tip opening displacement δ_t at the onset of crack propagation by the well-known relation

$$G_{Ic} = J_{Ic} = \frac{(K_{Ic})^2}{E} = m_f \sigma_f \delta_t, \qquad (96)$$

where J_{Ic} is the critical J-integral value (which is sometimes easier to measure than K_{Ic} in ductile porous metals), $m_f \approx 1.7$ and σ_f is the ductile (microcellular) metal flow stress (Barsom and Rolfe, 1999). The solid being ductile it is likely to fail in tension; all else constant δ_t is then proportional to the average cell diameter, l. If the architecture of the porous material does not change as V_m varies, then l and (to a first approximation at least) δ_t remain constant while σ_f is proportional to V_m^2, $V_m^{3/2}$ or V_m^1 for bending-dominated honeycombs stressed in-plane, bending-dominated open-pore microcellular metals, or closed-pore microcellular metals, respectively (eqn (77)). This in turn gives $p = 2$, 1.5 or 1, and $q = 2.5$, 1.75 or 1, for bending-dominated honeycombs stressed in-plane, bending-dominated open-pore microcellular metals, or closed-pore microcellular metals, respectively (Olurin et al., 2000a; Combaz and Mortensen, 2010).

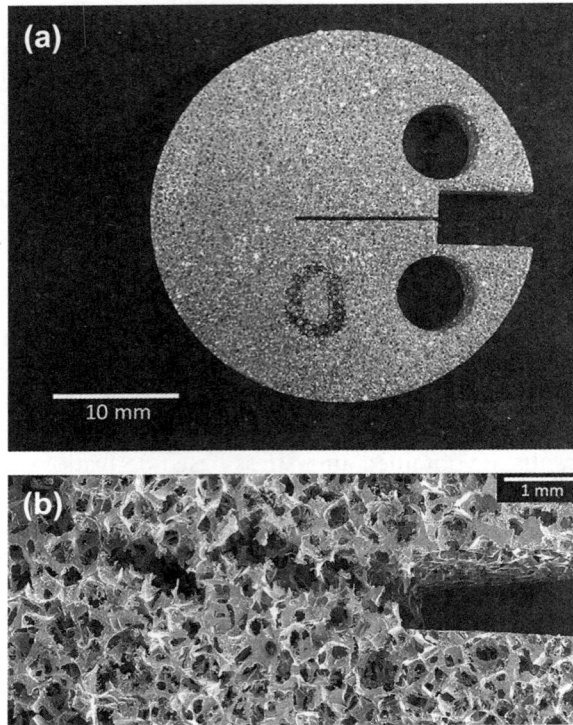

Figure 40 (a) Disc-shaped compact tension fracture specimen of replicated microcellular pure aluminum (mean pore size 400 μm), (b) scanning electron microscopy close-up of crack path in a similar sample after testing. Reproduced from *Acta Materialia*, vol. 58, Combaz & Mortensen, A., "Fracture toughness of Al replicated foam" pp. 4590–4603, (Combaz and Mortensen, 2010) with permission from Elsevier.

Stretch-dominated 2D or 3D open-pore structures will give different exponents. These are easily derived knowing scaling relations for σ_f and E; however, more complex behavior is found with 2D Kagome lattices. The reason is that these will switch locally, where struts are broken, from stretch-dominated to bending-dominated deformation. This, in turn, causes the formation of long bands of bent cell walls emanating from the tip of a crack, leading to toughness that scales as $V_m^{1/2}$ (Fleck and Qiu, 2007).

If the porous metal architecture varies with V_m, as is notably the case in powder compacts, foams, emulsions and microcellular materials derived therefrom (e.g. by replication processing), then l and δ_t will also vary with V_m. This may result in p and q values that are again different from those derived for a constant architecture. For instance, in open-pore microcellular replicated pure aluminum (**Figures 39 and 40**), p and q are nearer three; this was shown to be in accord with the evolution of topology (caused by the changing pore coordination with changing V_m) and with the higher N values that obtain in microcellular metals of this class (Conde, 2009; Combaz and Mortensen, 2010).

There are relatively few studies reporting measurements of fracture toughness in microcellular metals. The crack must be significantly wider than the cell size for meaningful measurement; with millimetric pores this can be a limiting factor. On the other hand the crack tip cannot physically be

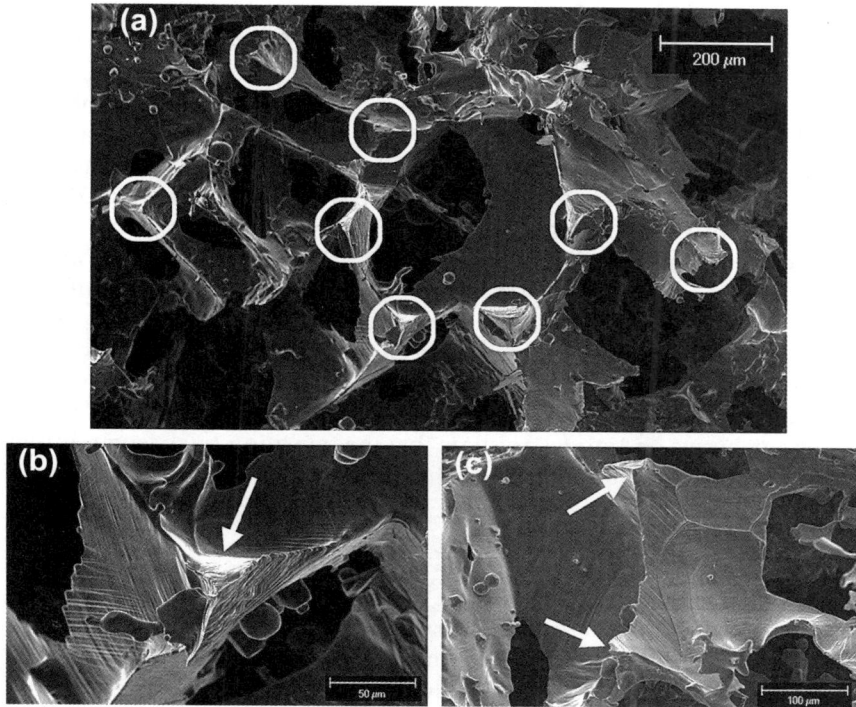

Figure 41 Scanning electron fractographs along the crack path in a disc-shaped compact tension specimen of relative density $V_m = 22\%$ showing tensile strut fracture by ductile necking to a point (highlighted with circles in (a) and with arrows in (b) and (c)) coupled with slip line markings betraying slip activity within the struts (the grain size of the pure aluminum making the foam is significantly larger than the pore diameter; the struts are thus individually single crystalline). Reproduced from *Acta Materialia*, vol. 58, Combaz & Mortensen, A., "Fracture toughness of Al replicated foam" pp. 4590–4603, (Combaz and Mortensen, 2010) with permission from Elsevier.

much finer than the pore size: precracking by fatigue is thus often unnecessary, particularly in open-pore microcellular materials. Hence, testing for toughness is in fact not particularly difficult with highly porous metals if the sample to pore size ratio is appropriate; **Figure 40** shows a compact tension fracture sample of replicated pure aluminum before and after testing.

Measured values of the fracture toughness of microcellular metals tend to be low, J_{Ic} taking values in the range 0.01–1 kJ/m^2. This corresponds to K_{Ic} values which can be as low as a few times 0.1 MPa$\sqrt{m}$ (Choi and Lakes, 1996; Sugimura et al., 1997; McCullough et al., 1999a; Olurin et al., 2000a; Fleck et al., 2001; Motz and Pippan, 2002; Combaz and Mortensen, 2010; Kashef et al., 2010). Still, despite their low toughness, crack growth tends to occur under large-scale yielding in these materials. Reasons for this are the low yield stress of microcellular metals, coupled with the fact that their compressibility increases the plastic zone size (Chen et al., 2001b). Where the relative density was varied systematically, scaling relations between measured J_{Ic} or K_{Ic} values and V_m in porous metals confirm those derived above for the relevant class of material (McCullough et al., 1999a; Olurin et al., 2000a; Combaz and Mortensen, 2010).

There is in general a significant degree of stable crack growth, indicating R-curve behavior in microcellular materials. This is linked with the formation of a fracture process zone along the crack front, several cell diameters wide within the plane of the crack (Sugimura et al., 1997; McCullough et al., 1999a; Olurin et al., 2000a; Fleck et al., 2001; Motz and Pippan, 2002; Combaz and Mortensen, 2010; Kashef et al., 2010). The mechanics of growing cracks in microcellular metals have been addressed also in more sophisticated analyses (Schmidt and Fleck, 2001; Chen et al., 2001b; Guo et al., 2004). Of particular interest in this regard is the question of what is identified as a crack within microcellular metals: clearly a few neighboring fractured cell walls or struts will not produce the same stress field as will a crack of similar length in a continuum. There is thus a transition, as cracks increase in length relative to the cell size, between cracks being a blunt defect that mainly increases the net section stress (small ratio of crack length a to cell size l) and "real" cracks that propagate once a critical stress intensity is exceeded in the surrounding microcellular material viewed as a continuum; roughly, this transition occurs when the ligament size exceeds the plastic zone radius. This question, which was already mentioned above in the context of elastic and plastic size effects, was addressed in Refs. Chen et al. (2001b) using a cohesive law model, to show that the transition crack length is roughly one third of the square of the ratio of steady-state toughness divided by the peak cohesive traction stress (eqn (28) of Ref. Chen et al. (2001b)).

24.5.7 Fatigue

The fatigue of porous metals is, as with fracture, a vast subject if it is treated in its entirety: even a single micropore can reduce the fatigue life of an entire metallic component. We will therefore again restrict attention to coverage of highly porous, microcellular metals. These have been studied using both classical approaches, namely by cyclic testing of smooth bars, and using fracture mechanics, by means of crack growth experiments in notched samples.

"Classical" fatigue testing of smooth samples has been conducted on nearly all types of highly porous microcellular metals, in both tension or compression: closed-pore foams (Banhart and Brinkers, 1999; Harte et al., 1999; Sugimura et al., 1999; Rabiei et al., 2000; Zettl et al., 2000, 2001; Amsterdam et al., 2006b; McCullough et al., 2000; Beck et al., 2002; Harders et al., 2003; Motz et al., 2005; Hakamada et al., 2007b; Kim and Kim, 2008; Ingraham et al., 2009), open-pore microcellular metals (Harte et al., 1999; Zhou and Soboyejo, 2004; Zhou et al., 2005a; Krupp et al., 2006; Lin et al., 2010; Soubielle et al., 2011b), bonded hollow spheres (Motz et al., 2005; Caty et al., 2009) and syntactic metal foam (Vendra et al., 2009). With honeycombs, studies have mostly placed focus on the fatigue of sandwich structures comprising honeycomb, e.g. (Sharma et al., 2006; Jen and Chang, 2008; Bianchi et al., 2012). A few open-pore microcellular metals designed for bone repair applications have been subjected to compression–compression fatigue endurance limit measurements within simulated body fluid (Imwinkelried, 2007; Arciniegas et al., 2007).

A recurrent observation in such studies is that repeated cyclic loading of highly porous metals causes progressive straining by cyclic creep (also called ratchetting): the material accumulates, cycle by cycle, a small finite strain increment in the direction of the applied mean stress. This leads to a total accumulation of strain which can be high, depending on the material and test parameters (Banhart and Brinkers, 1999; Harte et al., 1999; Sugimura et al., 1999; Rabiei et al., 2000; McCullough et al., 2000; Schulz et al., 2000; Harders et al., 2003; Zhou and Soboyejo, 2004; Zhou et al., 2005a; Motz et al., 2005; Amsterdam et al., 2006b; Hakamada et al., 2007b; Lin et al., 2010; Kolluri et al., 2008; Kim and Kim, 2008; Vendra et al., 2009; Caty et al., 2009). Cyclic creep is then superseded, past a certain number of cycles (and associated cumulative strain) by a stage of rapidly accelerating strain accumulation, ending in rupture or collapse of the foam. The fatigue life of foams is thus generally defined by the onset of such accelerated deformation (which in tension corresponds to cracking). In microcellular BMG the same observations, of cyclic creep followed by an acceleration of deformation and damage leading to fracture, are observed although underlying mechanisms differ somewhat (plastic deformation being mediated by shear bands) (Wang et al., 2010a).

Fatigued porous metals typically fail in compression–compression by the multiplication or by the broadening of discrete crush bands, often after significant damage accumulation (Sugimura et al., 1999; Banhart and Brinkers, 1999; Harte et al., 1999; McCullough et al., 2000; Zhou and Soboyejo, 2004; Zhou et al., 2005a; Motz et al., 2005; Ashby et al., 2000; Kim and Kim, 2008). Alternatively, under tension-tension or tension-compression fatigue they can fail by the formation of a single dominant crack (Harte et al., 1999; McCullough et al., 2000; Schulz et al., 2000; Zettl et al., 2000, 2001; Motz et al., 2005; Amsterdam et al., 2006b; Krupp et al., 2006; Ingraham et al., 2009; Soubielle et al., 2011b). As a result of these different failure mechanisms, the fatigue life of microcellular materials can be quite different between tensile and compressive stress states (e.g. (McCullough et al., 2000; Caty et al., 2009)).

The fatigue strength of both closed-cell (Harte et al., 1999; Sugimura et al., 1999; Hakamada et al., 2007b; Kolluri et al., 2008; Banhart and Brinkers, 1999; McCullough et al., 2000; Zettl et al., 2000, 2001; Kim and Kim, 2008) or open-cell (Harte et al., 1999; Zhou and Soboyejo, 2004; Zhou et al., 2005a; Soubielle et al., 2011b) metal foams, as well as bonded metal hollow sphere structures (Caty et al., 2009), increases strongly with the relative density, decreases with increasing stress amplitude, and can also be orientation dependent (e.g. (Harte et al., 1999; McCullough et al., 2000)). Closed-cell aluminum foams have been found to obey the Coffin–Manson law (Zettl et al., 2001; Ingraham et al., 2009). The lifetime in stress-controlled tension–tension R = 0.1 fatigue of replicated open-pore microcellular aluminum was found to obey the cyclic creep equivalent of the Monkman–Grant relation: the number of cycles to failure is roughly proportional to the inverse of the (stabilized steady) cyclic creep rate, measured here in terms of strain per cycle (Soubielle et al., 2011b). This, in turn, suggests a parallel between cyclic and monotonous creep deformation and rupture in these materials.

As in other materials, the fatigue life of microcellular materials can be also influenced by internal defects that are somewhat larger than the average pore size; examples are precracks, larger cells or holes, and nonuniform cell wall thicknesses, all of which can trigger crack or crush band or crack initiation (e.g. (Beck et al., 2002; Kolluri et al., 2008)). In the presence of large stress concentrators such as a hole, metal foams tend to fail roughly as they would under the highest net section stress present (Ashby et al., 2000; McCullough et al., 2000; Rabiei et al., 2000).

The growth of cracks in fatigued prenotched samples has been shown to obey the Paris law (Olurin et al., 2001b; Motz et al., 2005; Kashef et al., 2011)

$$\frac{da}{dN} = C_P (\Delta K)^{N_P}. \tag{97}$$

The crack growth rate constant C_P was found to vary with relative density in similar fashion to the fracture toughness: $C_P \propto V_m^{1.7}$ (Olurin et al., 2001b)(see previous subsection). The Paris law exponent N_p is variable but tends to be high (values on the order of 10–40 have been found for metal foams, 6 for bonded hollow spheres) and dependent on the R-ratio (of minimum to maximum stress or stress intensity factor) and relative density. These effects have been attributed to crack bridging and closure (Olurin et al., 2001b; Motz et al., 2005). Similar results are found for fatigue crack growth in nonmetallic foams (Gibson and Ashby, 1997). Micromechanical models have also been proposed to describe the fatigue behavior of honeycombs (Schaffner et al., 2000; Huang and Liu, 2001a) or microcellular metal (Gibson and Ashby, 1997; Huang and Liu, 2001b; Harders et al., 2003; Demiray et al., 2009) on the basis of fatigue properties of their dense constituent metal.

24.6   Conclusion

To us, the main conclusion that emerges from the panorama that precedes is how rich it is: porous metals are not just metals with holes (an interesting and important subject already), but also metals that can do things other metals cannot. Microcellular metals can be permeable to gas and liquid. They then can "breathe" and so perform, in their midst, a variety of functions (drive chemical reactions, store and transport heat or electricity, ...) that are likely to become important in years to come, for example in energy conversion or biomedical applications. Their mechanical behavior is also unique: porous metals can change their volume, absorb mechanical energy, and also show interesting and variegated size effects. Despite many interesting and innovative contributions to date, the subject holds much promise for discovery and impact, in both science and engineering.

References

Abdulla, T., Yerokhin, A., Goodall, R., 2011. Effect of plasma electrolytic oxidation coating on the specific strength of open-cell aluminium foams. Mater. Des. 32, 3742–3749.

Aboudi, J., Gilat, R., 2005. Micromechanical analysis of lattice blocks. Int. J. Solids Struct. 42, 4372–4392.

Ahern, A., Verbist, G., Weaire, D., et al., 2005. The conductivity of foams: a generalisation of the electrical to the thermal case. Colloids Surf. A. physicochem. Eng. Asp. 263, 275.

Ahluwalia, R.K., 2007. Sodium alanate hydrogen storage system for automotive fuel cells. Int. J. Hydrogen Energy 32, 1251–1261.

Ajdari, A., Nayeb-Hashemi, H., Canavan, P., Warner, G., 2008. Effect of defects on elastic-plastic behavior of cellular materials. Mater. Sci. Eng. A. 487, 558–567.

Ajdari, A., Canavan, P., Nayeb-Hashemi, H., Warner, G., 2009. Mechanical properties of functionally graded 2-D cellular structures: a finite element simulation. Mater. Sci. Eng. A 499, 434–439.

Akisanya, A.R., Cocks, A.C.F., Fleck, N.A., 1997. The yield behaviour of metal powders. Int. J. Mech. Sci. 39, 1315–1324.

Akiyama, S., Ueno, H., Imagawa, K., et al., 1987.

Alazmi, B., Vafai, K., 2000. Analysis of variants within the porous media transport models. J. Heat Transfer 122, 303–326.

Albanakis, C., Missirlis, D., Michailidis, N., et al., 2009. Experimental analysis of the pressure drop and heat transfer through metal foams used as volumetric receivers under concentrated solar radiation. Exp. Therm. Fluid Sci. 33, 246–252.

Albuquerque, J.M., Vaz, M.F., Fortes, M.A., 1999. Effect of missing walls on the compression behaviour of honeycombs. Scr. Mater. 41, 167–174.

Alkhader, M., Vural, M., 2008. Mechanical response of cellular solids: role of cellular topology and microstructural irregularity. Ind. Eng. Chem. Res. 46, 1035–1051.

Alkhader, M., Vural, M., 2009a. An energy-based anisotropic yield criterion for cellular solids and validation by biaxial FE simulations. J. Mech. Phys. Solids 57, 871–890.

Alkhader, M., Vural, M., 2009b. The partition of elastic strain energy in solid foams and lattice structures. Acta Mater. 57, 2429–2439.

Alkhader, M., Vural, M., 2010. A plasticity model for pressure-dependent anisotropic cellular solids. Int. J. Plast. 26, 1591–1605.

Al-Kharafi, F.M., Ateya, B.G., Allah, R.M.A., 2004. Selective dissolution of brass in salt water. J. Appl. Electrochem. 34, 47–53.

Allen, B.C., Mote, M.W., Sabroff, A.M., 1963. US Patent, US 3,087,807.

Aluminium-Zentrale, 1999. Aluminiumschaum, Düsseldorf, Aluminium-Merkblätter, Aluminium-Zentrale e.V., Blatt Nr. W17.

Aly, M.S., Almajid, A., Nakano, S., Ochiai, S., 2009. Fracture of open cell copper foams under tension. Mater. Sci. Eng. A 519, 211–213.

Aly, M.S., 2007. Behavior of closed cell aluminium foams upon compressive testing at elevated temperatures: experimental results. Mater. Lett. 61, 3138–3141.

Amsterdam, E., Onck, P.R., De Hosson, J.T.M., 2005. Fracture and microstructure of open cell aluminum foam. J. Mater. Sci. 40, 5813.

Amsterdam, E., Babcsan, N., De Hosson, J.T.M., et al., 2006a. Fracture behavior of metal foam made of recycled MMC by the melt route. Mater. Trans. 47, 2219–2222.

Amsterdam, E., De Hosson, J.T.M., Onck, P.R., 2006b. Failure mechanisms of closed-cell aluminum foam under monotonic and cyclic loading. Acta Mater. 54, 4465.

Amsterdam, E., De Hosson, J.T.M., Onck, P.R., 2008a. The influence of cell shape anisotropy on the tensile behavior of open cell aluminium foam. Adv. Eng. Mater. 10, 877–881.

Amsterdam, E., De Hosson, J.T.M., Onck, P.R., 2008b. On the plastic collapse stress of open-cell aluminum foam. Scr. Mater. 59, 653–656.

Amsterdam, E., de Vries, J.H.B., De Hosson, J.T.M., Onck, P.R., 2008c. The influence of strain-induced damage on the mechanical response of open-cell aluminum foam. Acta Mater. 56, 609–618.

Amsterdam, E., Goodall, R., Mortensen, A., et al., 2008d. Fracture behavior of low-density replicated aluminum alloy foams. Mater. Sci. Eng. A 496, 376–382.

Anderson, A.M., Brush, L.N., Davis, S.H., 2010. Foam mechanics: spontaneous rupture of thinning liquid films with Plateau borders. J. Fluid Mech. 658, 63–88.

Andrews, E., Gibson, L.J., 2001a. The role of cellular structure in creep of two-dimensional cellular solids. Mater. Sci. Eng. A 303, 120–126.

Andrews, E.W., Gibson, L.J., 2001b. The influence of crack-like defects on the tensile strength of an open-cell aluminum foam. Scr. Mater. 44, 1005–1010.

Andrews, E.W., Gibson, L.J., 2001c. The influence of cracks, notches and holes on the tensile strength of cellular solids. Acta Mater. 49, 2975–2979.

Andrews, E.W., Gibson, L.J., 2002. On notch-strengthening and crack tip deformation in cellular metals. Mater. Lett. 57, 532.

Andrews, E., Sanders, W., Gibson, L.J., 1999a. Compressive and tensile behaviour of aluminum foams. Mater. Sci. Eng. A 270, 113–124.

Andrews, E.W., Gibson, L.J., Ashby, M.F., 1999b. The creep of cellular solids. Acta Mater. 47, 2853–2863.

Andrews, E.W., Huang, J.S., Gibson, L.J., 1999c. Creep behavior of a closed-cell aluminum foam. Acta Mater. 47, 2927–2935.

Andrews, E.W., Gioux, G., Onck, P., Gibson, L.J., 2001. Size effects in ductile cellular solids. Part II: experimental results. Int. J. Mech. Sci. 43, 701–713.

Andrews, E.W., 2006. Open-cell foams with hollow struts: mechanical property enhancements. Mater. Lett. 60, 618.

Angioloni, A., Rosa, M.D., 2005. J. Cereal Sci. 41, 327–331.

Angirasa, D., 2002a. Experimental investigation of forced convection heat transfer augmentation with metallic fibrous materials. Int. J. Heat Mass Transfer 45, 919–922.

Angirasa, D., 2002b. Forced convective heat transfer in metallic fibrous materials. J. Heat Transfer 124, 739–745.

Antohe, B.V., Lage, J.L., 1997. A general two-equation macroscopic turbulence model for incompressible flow in porous media. Int. J. Heat Mass Transfer 40, 3013–3024.

Antohe, B.V., Lage, J.L., Price, D.C., Weber, R.M., 1996a. Experimental determination of permeability and inertia coefficients of mechanically compressed aluminum porous matrices. Trans. ASME – J. Fluids Eng. 119, 404–412.

Antohe, B.V., Lage, J.L., Price, D.C., Weber, R.M., 1996b. Numerical characterization of micro heat exchangers using experimentally tested porous aluminum layers. Int. J. Heat Fluid Flow 17, 594–603.

Antoniou, A., Onck, P.R., Bastawros, A.F., 2004. Experimental analysis of compressive notch strengthening in closed-cell aluminum alloy foam. Acta Mater. 52, 2377.

Antsiferov, V.N., Makarov, A.M., Khramtsov, V.D., 2005. High-porosity permeable cellular metals and alloys in catalytic processes of gas cleaning. Adv. Eng. Mater. 7, 77–91.

Apprill, J.M., Poirier, D.R., Maguire, M.C., Gutsch, T.C., 1998. Gasar porous metals process control. In: Schwartz, D.S., Shih, D.H., Evans, A.G., Wadley, H.G. (Eds.), Porous and Cellular Materials for Structural Applications. San Francisco, CA, USA, Materials Research Society Symposium Proceedings, vol. 521. (Warrendale PA, USA).

Aravas, N., Ponte Castaneda, P., 2004. Numerical methods for porous metals with deformation-induced anisotropy. Comput. Methods Appl. Mech. Eng. 193, 3767.

Arciniegas, M., Aparicio, C., Manero, J.M., Gil, F.J., 2007. Low elastic modulus metals for joint prosthesis: tantalum and nickel–titanium foams. J. Eur. Ceram. Soc. 27, 3391–3398.

Arisetty, S., Prasad, A.K., Advani, S.G., 2007. Metal foams as flow field and gas diffusion layer in direct methanol fuel cells. J. Power Sources 165, 49–57.

Arnold, M., Boccaccini, A.R., Ondracek, G., 1996. Prediction of the Poisson's ratio of porous materials. J. Mater. Sci. 31, 1643–1646.

Arzt, E., 1982. The influence of an increasing particle coordination on the densification of spherical powders. Acta Metall. 30, 1883–1890.

Asavavisithchai, S., Kennedy, A.R., 2006a. The effect of compaction method on the expansion and stability of aluminium foams. Adv. Eng. Mater. 8, 811–815.

Asavavisithchai, S., Kennedy, A.R., 2006b. The effect of Mg addition on the stability of Al–Al2O3 foams made by a powder metallurgy route. Scr. Mater. 54, 1331.

Asavavisithchai, S., Kennedy, A.R., 2006c. Effect of powder oxide content on the expansion and stability of PM-route Al foams. J. Colloid Interface Sci. 297, 715.

Asavavisithchai, S., Kennedy, A.R., 2006d. The role of oxidation during compaction on the expansion and stability of Al foams made via a PM route. Adv. Eng. Mater. 8, 568–572.

Asavavisithchai, S., Nisaratanaporn, A., 2008. Fabrication of open-cell silver foams using a replication process. In: Lefebvre, L.P., Banhart, J., Dunand, D.C. (Eds.), Metfoam 2007-Porous Metals and Metallic Foams, Proc. 2007 Metfoam Conference. Montreal, Sept. 5–7, 2007. DESTech Publications, Inc, Lancaster, PA.

Asavavisithchai, S., Oonpraderm, A., Ruktanonchai, U., 2010. The antimicrobial effect of open-cell silver foams. J. Mater. Sci.: Mater. Med. 21, 1329–1334.

Ashby, M.F., Greer, A.L., 2006. Metallic glasses as structural materials. Scr. Mater. 54, 321–326.

Ashby, M.F., Ghandhi, C., Taplin, D.M.R., 1979. Fracture-mechanism maps and their construction for F.C.C. metals and alloys. Acta Metall. 27, 699–729.

Ashby, M.F., Evans, A., Fleck, N.A., et al., 2000. Metal Foams: A Design Guide. Butterworth Heinemann, Boston USA.

Ashby, M.F., 1983. The mechanical properties of cellular solids. Metall. Trans. 14A, 1755–1769.

Ashby, M.F., 2006. The properties of foams and lattices. Philos. Trans. R. Soc., A 364, 15.

Ashby, M., 2011. Hybrid materials to expand the boundaries of material–property space. J. Am. Ceram. Soc. 94, s3–s14.

Åsholt, P., 1997. Manufacturing of aluminium foams from PMMC melts material characteristics and typical properties. In: Banhart, J. (Ed.), Metallschäume. Bremen, March 1997. Verlag MIT, Bremen.

Asholt, P., 1999. Aluminium foam produced by the melt foaming route process. In: Banhart, J., Ashby, M.F., Fleck, N.A. (Eds.), Metal Foams and Porous Metal Structures. MIT Press-Verlag, Bremen, Germany.

Aste, T., Saadatfar, M., Senden, T.J., 2005. Geometrical structure of disordered sphere packings. Phys. Rev. E 71, 061302.

ASTM, 2000. Section 3, Metals Test Methods and Analytical Procedures. Annual Book of ASTM Standards. American Society for Testing and Materials, West Conshohocken, Pa.

Atwood, R.C., Jones, J.R., Lee, P.D., Hench, L.L., 2004. Analysis of pore interconnectivity in bioactive glass foams using X-ray microtomography. Scr. Mater. 51, 1029–1033.

Aubertin, M., Li, L., 2004. A porosity-dependent inelastic criterion for engineering materials. Int. J. Plast. 20, 2179–2208.

Auriault, J.-L., Geindreau, C., Orgéas, L., 2007. Upscaling Forchheimer law. Transp. Porous Media 70, 213–229.

Aydogmus, T., Bor, S.J., 2009. Processing of porous TiNi alloys using magnesium as space holder. J. Alloys Compd. 478, 705–710.

Azzi, W., Roberts, W.L., Rabiei, A., 2007. A study on pressure drop and heat transfer in open cell metal foams for jet engine applications. Mater. Des. 28, 569–574.

Babaee, S., Jahromi, B.H., Ajdari, A., et al., 2012. Mechanical properties of open-cell rhombic dodecahedron cellular structures. Acta Mater. 60, 2873–2885.

Babcsán, N., Banhart, J., 2006. Metal foams towards high-temperature colloid chemistry. In: Binks, B.P., Horozov, T.S. (Eds.), Colloidal Particles at Liquid Interfaces. Cambridge University Press, Cambridge.

Babcsan, N., Leitlmeier, D., Banhart, J., 2005. Metal foams–high temperature colloids: part I. Ex situ analysis of metal foams. Colloids Surf., A 261, 123–130.

Babcsan, N., Garcia-Moreno, F., Banhart, J., 2006a. Role of oxidation during blowing of aluminium foams by external gas injection. In: Nakajima, H., Kanetake, N. (Eds.), MetFoam 2005 Porous Metals and Metal Foaming Technology. Japan Institute of Metals, Kyoto, Japan.

Babcsan, N., Garcia-Moreno, F., Leitlmeier, D., Banhart, J., 2006b. Liquid metal foams – feasible in-situ experiments under low gravity. Mater. Sci. Forum 508, 275–280.

Babcsan, N., Garcia-Moreno, F., Banhart, J., 2007. Metal foams–high temperature colloids: part II. In situ analysis of metal foams. Colloids Surf., A 309, 254–263.

Babcsan, N., Beke, S., Makk, P., et al., 2012. Aluhab – the superior aluminium foam. In: Weiland, H., Rollett, A.D., Cassada, W.A. (Eds.), 13th International Conference on Aluminum Alloys (ICAA13). TMS (The Minerals, Metals & Materials Society), Pittsburgh, PA.

Babin, P., Valle, G., Dendievel, R., et al., 2005. Mechanical properties of bread crumbs from tomography based finite element simulations. J. Mater. Sci. 40, 5867–5873.

Babjack, J., Ettel, V.A., Paserin, V., 1990. Method of forming nickel foam

Babscan, N., Meszaros, I., Hegman, N., 2003. Thermal and electrical conductivity measurements on aluminum foams. Materialwiss. Werkstofftech. 34, 391–394.

Babscan, N., Leitlmeier, D., Degischer, H.P., Banhart, J., 2004. The role of oxidation in blowing particle-stabilised aluminium foams. Adv. Eng. Mater. 6, 421–428.

Bach, F.W., Bormann, D., Wilk, P., 2003. Cellular magnesium. In: Banhart, J., Fleck, N., Mortensen, A. (Eds.), Cellular Metals: Manufacture, Properties, Applications, Proc. Conf. Metfoam 2003. Berlin, 23–25 June 2003. Verlag MIT Publishing, Berlin, Germany.

Badiche, X., Forest, S., Guibert, T., et al., 2000. Mechanical properties and non-homogeneous deformation of open-cell nickel foams: application of the mechanics of cellular solids and of porous materials. Mater. Sci. Eng. A 289, 276–288.

Bae, G., Bae, J., Kim-Lohsoontorn, P., Jeong, J., 2010. Performance of SOFC coupled with n-C4H10 autothermal reformer: carbon deposition and development of anode structure. Int. J. Hydrogen Energy 35, 12346–12358.

Bahraini, M., Molina, J.M., Kida, M., et al., 2005. Measuring and tailoring capillary forces during liquid metal infiltration. Curr. Opin. Solid State Mater. Sci. 9, 196.

Bahraini, M., Molina, J.M., Weber, L., Mortensen, A., 2008. Direct measurement of drainage curves in infiltration of SiC particle preforms. Mater. Sci. Eng. A 495, 203–212.

Bai, M., Chung, J.N., 2011. Analytical and numerical prediction of heat transfer and pressure drop in open-cell metal foams. Int. J. Therm. Sci. 50, 869–880.

Bai, Y., Wierzbicki, T., 2008. A new model of metal plasticity and fracture with pressure and lode dependence. Int. J. Plast. 24, 1071–1096.

Bakan, H.I., 2006. A novel water leaching and sintering process for manufacturing highly porous stainless steel. Scr. Mater. 55, 203.

Balberg, I., 1987. Recent developments in continuum percolation. Philos. Mag. B 56, 991–1003.

Balch, D.K., Dunand, D.C., 2006. Load partitioning in aluminum syntactic foams containing ceramic microspheres. Acta Mater. 54, 1501.

Balch, D.K., O'Dwyer, J.G., Davis, G.R., et al., 2005. Plasticity and damage in aluminum syntactic foams deformed under dynamic and quasi-static conditions. Mater. Sci. Eng. A 391, 408.

Balk, T., Eberl, C., Sun, Y., et al., 2009. Tensile and compressive microspecimen testing of bulk nanoporous gold. JOM 61, 26–31.

Banhart, J., Baumeister, J., 1998a. Deformation characteristics of metal foams. J. Mater. Sci. 33, 1431–1440.

Banhart, J., Baumeister, J., 1998b. Production methods for metallic foams. In: Schwartz, D.S., Shih, D.H., Evans, A.G., Wadley, H.G. (Eds.), Porous and Cellular Materials for Structural Applications. San Francisco, CA, USA, Materials Research Society Symposium Proceedings, vol. 521. (Warrendale PA, USA).

Banhart, J., Brinkers, W., 1999. Fatigue behavior of aluminum foams. J. Mater. Sci. Lett. 18, 617–619.

Banhart, J., Knuwer, M., 1998. Proceedings of the 1998 Congress, European Powder Metallurgy Association, 5, 265.

Banhart, J., Seeliger, H.-W., 2012. Recent trends in aluminum foam sandwich technology. Adv. Eng. Mater. 14, 1082–1087.

Banhart, J., Baumeister, J., Weber, M., 1996. Damping properties of aluminium foams. Mater. Sci. Eng. A 205, 221–228.

Banhart, J., Bellmann, D., Clemens, H., 2001a. Investigation of metal foam formation by microscopy and ultra small-angle neutron scattering. Acta Mater. 49, 3409–3420.

Banhart, J., Stanzick, H., Helfen, L., Baumbach, T., 2001b. Metal foam evolution studied by synchrotron radioscopy. Appl. Phys. Lett. 78, 1152–1154.

Banhart, J., Fleck, N., Mortensen, A. (Eds.), 2003. Cellular Metals: Manufacture, Properties, Applications, Proc. Metfoam 2003 Conf. Berlin, 23–25 June 2003. Verlag MIT Publishing, Berlin, Germany.

Banhart, J., Garcia-Moreno, F., Seeliger, H.W., 2011. Aluminium Foams – Scientific Challenges and Industrial Applications. Presented at Euromat 2011. Montpellier, France.

Banhart, J., 2000. Manufacturing routes for metallic foams. JOM 52, 22–27.

Banhart, J., 2001. Manufacture, characterization and application of cellular metals and metal foams. Prog. Mater. Sci. 46, 559–632.

Banhart, J., 2002. Chapter 7.2: functional applications. In: Degischer, H.P., Kriszt, B. (Eds.), Handbook of Cellular Metals. Wiley-VCH Verlag, Weinheim Germany.

Banhart, J., 2006. Metal foams: production and stability. Adv. Eng. Mater. 8, 781–794.

Banhart, J., 2008. Gold and gold alloy foams. Gold Bull. 41, 251–256.

Banhart, J., 2013. Light-Metal Foams – History of Innovation and Technological Challenges. Advanced Engineering Materials. 15 (3), 82–111.

Bansiddhi, A., Dunand, D.C., 2007. Shape-memory NiTi foams produced by solid-state replication with NaF. Intermetallics 15, 1612–1622.

Bansiddhi, A., Dunand, D.C., 2008. Shape-memory NiTi foams produced by replication of NaCl space-holders. Acta Biomater. 4, 1996–2007.

Bansiddhi, A., Dunand, D.C., 2009. Shape-memory NiTi–Nb foams. J. Mater. Res. 24, 2107–2117.

Bansiddhi, A., Dunand, D., 2011a. Processing of NiTi foams by transient liquid phase sintering. J. Mater. Eng. Perform. 20, 511–516.

Bansiddhi, A., Dunand, D.C., 2011b. Niobium wires as space holder and sintering aid for porous NiTi. Adv. Eng. Mater. 13, 301–305.

Bansiddhi, A., Sargeant, T.D., Stupp, S.I., Dunand, D.C., 2008. Porous NiTi for bone implants: a review. Acta Biomater. 4, 773–782.

Bardella, L., Genna, F., 2001. On the elastic behavior of syntactic foams. Int. J. Solid. Struct. 38, 7235–7260.

Baril, E., Mostafid, A., Lefebvre, L.P., Medraj, M., 2008. Experimental demonstration of entrance/exit effects on the permeability measurements of porous materials. Adv. Eng. Mater. 10, 889–894.

Baril, E., Lefebvre, L.P., Hacking, S.A., 2011. Direct visualization and quantification of bone growth into porous titanium implants using micro computed tomography. J. Mater. Sci.: Mater. Med. 22, 1321–1332.

Barletta, M., Gisario, A., Guarino, S., Rubino, G., 2009. Production of open cell aluminum foams by using the dissolution and sintering process. ASME J. Manuf. Sci. Eng. 131, 041009-1–041009-10.

Barsom, J., Rolfe, S., 1999. Fracture and Fatigue Control in Structures, third ed. Butterworth Heinemann/ASTM, Woburn (MA, USA).

Bart-Smith, H., Bastawros, A.F., Mumm, D.R., et al., 1998. Compressive deformation and yielding mechanisms in cellular Al alloys determined using X-ray tomography and surface strain mapping. Acta Mater. 46, 3583–3592.

Bastawros, A.-F., Evans, A.G., 2000. Deformation heterogeneity in cellular Al-alloys. Adv. Eng. Mater. 2, 210–214.

Bastawros, A.-F., Bart-Smith, H., Evans, A.G., 2000. Experimental analysis of deformation mechanisms in a closed-cell aluminum alloy foam. J. Mech. Phys. Solids 48, 301–322.

Baumeister, J., Schrader, H., 1991. German Patent DE 4,101,630.

Baumeister, J., 1990. German Patent, DE 4,018,360.

Baumeister, J., 1997. Uberblick: Verfahren zur Herstellung von Metallschäumen. In: Banhart, J. (Ed.), Metallschäume. Bremen, March 1997. Verlag MIT, Bremen.

Baumgärtner, F., Duarte, I., Banhart, J., 2000. Industrialization of powder compact foaming processes. Adv. Eng. Mater. 2, 168–174.

Baxter, N.E., Sanders, T.H., Nagel, A.R., et al., 1998. Metallic foams from alloy hollow spheres. In: Schwartz, D.S., Shih, D.H., Evans, A.G., Wadley, H.G. (Eds.), Porous and Cellular Materials for Structural Applications. San Francisco, CA, USA, Mater. Res. Soc. Symp. Proc. vol. 521. (Warrendale PA, USA).

Beals, J.T., Thompson, M.S., 1997. Density gradient effects on aluminium foam compression behaviour. J. Mater. Sci. 32, 3595–3600.

Bear, J., Bachmat, Y., 1990. Introduction to Modeling of Transport Phenomena in Porous Media. Kluwer Academic Publishers, Dordrecht, The Netherlands.

Bear, J., 1972. Dynamics of Fluids in Porous Media. American Elsevier, New York.

Beavers, G.S., Sparrow, E.M., 1969. Non-darcy flow through fibrous porous media. J. Appl. Mech. 36, 711–714.

Beavers, G.S., Wilson, T.A., 1975. Flow through a deformable porous material. J. Appl. Mech. 42, 598–602.

Beavers, G.S., Hajii, A., Sparrow, E.M., 1981a. Fluid flow through a class of highly deformable porous media part I: experiments with air. J. Fluids Eng. 103, 432–439.

Beavers, G.S., Wittenberg, K., Sparrow, E.M., 1981b. Fluid flow through a class of highly deformable porous media part II: experiments with water. J. Fluids Eng. 103, 440–444.

Beck, T., Löfe, D., Baumgärtner, F., 2002. The fatigue behavior of an aluminium foam sandwich beam under alternating bending. Adv. Eng. Mater. 4, 787–790.

Bejan, A., 2004. Designed porous media: maximal heat transfer density at decreasing length scales. Int. J. Heat Mass Transfer 47, 3073–3083.

Bele, E., Bouwhuis, B.A., Codd, C., Hibbard, G.D., 2011. Structural ceramic coatings in composite microtruss cellular materials. Acta Mater. 59, 6145–6154.

Belhadi, A.E., Dahmoun, S.D., Azzaz, M., et al., 2006. Elaboration et characterisation de mousses metalliques a base d'etain-plomb. Materiaux 2006. Dijon, France.

Belkessam, O., Fritsching, U., 2003. Modelling and simulation of continuous metal foaming process. Modell. Simul. Mater. Sci. Eng. 11, 823–837.

Bellmann, D., Clemens, H., Banhart, J., 2002. USANS investigation of early stages of metal foam formation. Appl. Phys. A 74, S1136–S1138.

Bellmann, D., Clemens, H., Matijasevic, B., et al., 2003. Investigation of early stages of metal foam formation by small-angle neutron scattering techniques. In: Banhart, J., Fleck, N., Mortensen, A. (Eds.), Cellular Metals: Manufacture, Properties, Applications, Proc. Conf. Metfoam 2003. Berlin, 23–25 June 2003. Verlag MIT Publishing, Berlin, Germany.

Belova, I.V., Veyhl, C., Fiedler, T., Murch, G.E., 2011. Analysis of anisotropic behaviour of thermal conductivity in cellular metals. Scr. Mater. 65, 436–439.

Bendsøe, M.P., Sigmund, O., 2003. Topology Optimization – Theory, Methods and Applications. Springer Verlag, Berlin, Germany.

Benke, S., Weichert, D., 2005. Thermo-plasticity of metal foams. Comput. Mater. Sci. 32, 268.

Benouali, A.H., Froyen, L., Delerue, J.F., Wevers, M., 2002. Mechanical analysis and microstructural characterization of metal foams. Mater. Sci. Technol. 18, 489–494.

Benouali, A.H., Froyen, L., Dillard, T., et al., 2005a. Investigation on the influence of cell shape anisotropy on the mechanical performance of closed cell aluminium foams using micro-computed tomography. J. Mater. Sci. 40, 5801–5811.

Benouali, A.H., Froyen, L., Dillard, T., et al., 2005b. Investigation on the influence of cell shape anisotropy on the mechanical performance of closed cell aluminium foams using micro-computed tomography. J. Mater. Sci. 40, 5801.

Benveniste, Y., 1987. A new approach to the application of more-Tanaka's theory in composite materials. Mech. Mater. 6, 147–157.

Berchem, K., Mohr, U., Bleck, W., 2002. Controlling the degree of pore opening of metal sponges, prepared by the infiltration preparation method. Mater. Sci. Eng. A 323, 52–57.

Bernard, T., Bergmann, H.W., Haberling, C., Haldenwanger, H.G., 2002. Joining technologies for Al foam-Al sheet compound structures. Adv. Eng. Mater. 4, 798–802.

Berryman, J.G., Berge, P.A., 1996. Critique of two explicit schemes for estimating elastic properties of multiphase composites. Mech. Mater. 22, 149–164.

Berryman, J.G., 1980a. Long-wavelength propagation in composite elastic media I. Spherical inclusions. J. Acoust. Soc. Am. 68, 1809–1819.

Berryman, J.G., 1980b. Long-wavelength propagation in composite elastic media II. Ellipsoidal inclusions. J. Acoust. Soc. Am. 68, 1820–1831.

Bertolino, G., Larochette, P.A., Castrodeza, E.M., et al., 2010. Mechanical properties of martensitic Cu–Zn–Al foams in the pseudoelastic regime. Mater. Lett. 64, 1448–1450.

Betts, C., 2012. Benefits of metal foams and developments in modelling techniques to assess their materials behaviour: a review. Mater. Sci. Technol. 28, 129–143.

Bhattacharya, A., Mahajan, R.L., 2002. Finned metal foam heat sinks for electronics cooling in forced convection. ASME: J. Electron. Packag. 124, 155–163.

Bhattacharya, A., Mahajan, R.L., 2006. Metal foam and finned metal foam heat sinks for electronics cooling in buoyancy-induced convection. ASME: J. Electron. Packag. 128, 259–266.

Bhattacharya, A., Calmidi, V.V., Mahajan, R.L., 2002. Thermophysical properties of high porosity metal foams. Int. J. Heat Mass Transfer 45, 1017–1031.

Bianchi, M., Scarpa, F., Banse, M., Smith, C.W., 2011. Novel generation of auxetic open cell foams for curved and arbitrary shapes. Acta Mater. 59, 686–691.

Bianchi, G., Aglietti, G., Richardson, G., 2012. Static and fatigue behaviour of hexagonal honeycomb cores under in-plane shear loads. Appl. Compos. Mater. 19, 97–115.

Biener, J., Hodge, A.M., Hamza, A.V., et al., 2005. Nanoporous Au: a high yield strength material. J. Appl. Phys. 97, 024301.

Biener, J., Hodge, A.M., Hayes, J.R., et al., 2006. Size effects on the mechanical behavior of nanoporous Au. Nano Lett. 6, 2379.

Bierwagen, G.P., Sanders, T.E., 1974. Studies of the effects of particle size distribution on the packing efficiency of particles. Powder Technol. 10, 111–119.

Bilger, N., Auslender, F., Bornert, M., et al., 2005. Effect of a nonuniform distribution of voids on the plastic response of voided materials: a computational and statistical analysis. Int. J. Solids Struct. 42, 517.

Bilger, N., Auslender, F., Bornert, M., et al., 2007. Bounds and estimates for the effective yield surface of porous media with a uniform or a nonuniform distribution of voids. Eur. J. Mech. – A/Solids 26, 810–836.

Bin, J., Zejuna, W., Naiqin, Z., 2007. Effect of pore size and relative density on the mechanical properties of open cell aluminum foams. Scr. Mater. 56, 169–172.

Binks, B.P., 2002. Particles as surfactants'similarities and differences. Curr. Opin. Colloid Interface Sci. 7, 21–41.

Bird, R.B., Stewart, W.E., Lightfoot, E.N., 1960. Transport Phenomena. John Wiley & Sons, New York, N.Y.

Biswas, A., 2005. Porous NiTi by thermal explosion mode of SHS: processing, mechanism and generation of single phase microstructure. Acta Mater. 53, 1415–1425.

Blazy, J.S., Marie-Louise, A., Forest, S., et al., 2004. Deformation and fracture of aluminium foams under proportional and non proportional multi-axial loading: statistical analysis and size effect. Int. J. Mech. Sci. 46, 217–244.

Boccaccini, A.R., Fan, Z., 1997. A new approach for the Young's modulus–porosity correlation of ceramic materials. Ceram. Int. 23, 239–245.

Boccaccini, A.R., 1994. Comment on "effective elastic moduli of porous ceramic materials". J. Am. Ceram. Soc. 77, 2779–2781.

Boger, T., Heibel, A.K., 2005. Heat transfer in conductive monolith structures. Chem. Eng. Sci. 60, 1823–1835.

Böllinghaus, T., vonHagen, H., Bleck, W., 1999. Dynamic behavior of melt-foamed aluminum under compressive and tensile loads. In: Clyne, T.W., Simancik, F. (Eds.), Euromat 99. 27–30 Sept. 1999, Munich, Germany. DGM/Wiley-VCH, Weinheim, Germany.

Bonacorsi, L., Proverbio, E., 2006. Powder compaction effect on foaming behavior of uni-axial pressed PM precursors. Adv. Eng. Mater. 8, 864–869.

Bonnet, J.-P., Topin, F., Tadrist, L., 2008. Flow laws in metal foams: compressibility and pore size effects. Transp. Porous Media 73, 233–254.

Boomsma, K., Poulikakos, D., 2001. On the effective thermal conductivity of a three-dimensionally structured fluid-saturated metal foam. Int. J. Heat Mass Transfer 44, 827–836.

Boomsma, K., Poulikakos, D., 2002. The effects of compression and pore size variations on the liquid flow characteristics in metal foams. J. Fluids Eng. 124, 263–272.

Boomsma, K., Poulikakos, D., 2011. Corrigendum for the paper: Boomsma, K., Poulikakos, D. "On the effective thermal conductivity of a three-dimensionally structured fluid-saturated metal foam". Int. J. Heat Mass Transfer 44 (2001), 827–836. Int. J. Heat Mass Transfer, 54, 746–748.

Boomsma, K., Poulikakos, D., Ventikos, Y., 2003a. Simulations of flow through open cell metal foams using an idealized periodic cell structure. Int. J. Heat Fluid Flow 24, 825–834.

Boomsma, K., Poulikakos, D., Zwick, F., 2003b. Metal foams as compact high performance heat exchangers. Mech. Mater. 35, 1161–1176.

Boonyongmaneerat, Y., Dunand, D.C., 2008. Ni–Mo–Cr foams processed by casting replication of sodium aluminate preforms. Adv. Eng. Mater. 10, 379–383.

Boonyongmaneerat, Y., Dunand, D.C., 2009. Effects of strut geometry and pore fraction on creep properties of cellular materials. Acta Mater. 57, 1373–1384.

Boonyongmaneerat, Y., Chmielus, M., Dunand, D.C., M√ollner, P., 2007. Increasing magnetoplasticity in polycrystalline Ni–Mn–Ga by reducing internal constraints through porosity. Phys. Rev. Lett. 99, 247201.

Boonyongmaneerat, Y., Schuh, C.A., Dunand, D.C., 2008. Mechanical properties of reticulated aluminium foams with electrodeposited coatings. Scr. Mater. 59, 336–339.

Born, C., Wagner, G., Eifler, D., 2006. Ultrasonically welded aluminium foams/sheet metal – joints. Adv. Eng. Mater. 8, 816–820.

Bouaziz, O., Masse, J.P., Bréchet, Y., 2011. An analytical description of the mechanical hysteresis of entangled materials during loading'unloading in uniaxial compression. Scr. Mater. 64, 107–109.

Bouhlel, M., et al., 2010. Microstructural effects on the overall poroelastic properties of saturated porous media. Modell. Simul. Mater. Sci. Eng. 18, 045009.

Bourne, N.K., Bennett, K., Milne, A.M., et al., 2008. The shock response of aluminium foams. Scr. Mater. 58, 154–157.

Bouvard, D., Chaix, J.M., Dendievel, R., et al., 2007. Characterization and simulation of microstructure and properties of EPS lightweight concrete. Cem. Concr. Res. 37, 1666–1673.

Bouwhuis, B., Hibbard, G., 2006. Compression testing of periodic cellular sandwich cores. Metall. Mater. Trans. B 37, 919–927.

Bouwhuis, B., Bele, E., Hibbard, G., 2008. Plastic hinging collapse of periodic cellular truss cores. Metall. Mater. Trans. A 39, 2329–2339.

Bouwhuis, B.A., McCrea, J.L., Palumbo, G., Hibbard, G.D., 2009. Mechanical properties of hybrid nanocrystalline metal foams. Acta Mater. 57, 4046–4053.

Bram, M., Stiller, C., Buchkremer, H.P., et al., 2000. High-porosity titanium, stainless steel and superalloy parts. Adv. Eng. Mater. 2, 196–199.

Brandes, E.A., Brook, G.B., 1992. Smithells Metals Reference Book, Seventh ed. Butterworth-Heinemann, Woburn MA.

Brandes, E.A. (Ed.), 1983. Smithells Metals Reference Book. Butterworths, London.

Brandon, N.P., Brett, D.J., 2006. Engineering porous materials for fuel cell applications. Philos. Trans. R. Soc., A 364, 147–159.

Brannan, J.R., Bean, A.S., Vaccaro, A.J., Stewart, J.J., 1992. Continuous electroplating of conductive foams.

Brett, J., Friedman, S., 1972. High-temperature porosity in tungsten. Metall. Trans. 3, 769–778.

Brezny, R., Green, D.J., 1990. Characterization of edge effects in cellular materials. J. Mater. Sci. 25, 4571–4578.

Brick, R.M., Pense, A.W., Gordon, R.B., 1977. Structure and Properties of Engineering Materials. McGraw-Hill, New York, USA.

Brothers, A.H., Dunand, D.C., 2004. Syntactic bulk metallic glass foam. Appl. Phys. Lett. 84, 1108–1110.

Brothers, A.H., Dunand, D.C., 2005a. Ductile bulk metallic glass foams. Adv. Mater. 17, 484–486.

Brothers, A.H., Dunand, D.C., 2005b. Plasticity and damage in cellular amorphous metals. Acta Mater. 53, 4427.

Brothers, A.H., Dunand, D.C., 2006a. Amorphous metal foams. Scr. Mater. 54, 513.

Brothers, A.H., Dunand, D.C., 2006b. Density-graded cellular aluminum. Adv. Eng. Mater. 8, 805–809.

Brothers, A.H., Dunand, D.C., 2007. Porous and foamed amorphous metals. MRS Bull. 32, 639–643.

Brothers, A.H., Dunand, D.C., 2008. Mechanical properties of a density-graded replicated aluminum foam. Mater. Sci. Eng. A 489, 439–443.

Brothers, A.H., Scheunemann, R., DeFouw, J.D., Dunand, D.C., 2005. Processing and structure of open-celled amorphous metal foams. Scr. Mater. 52, 335.

Brothers, A.H., Prine, D.W., Dunand, D.C., 2006. Acoustic emissions analysis of damage in amorphous and crystalline metal foams. Intermetallics 14, 857–865.

Brothers, A.H., Dunand, D.C., Zheng, Q., Xu, J., 2007. Amorphous Mg-based metal foams with ductile hollow spheres. J. Appl. Phys. 102, 023508-6.

Brothers, A.H., Mangrich, B., Cox, M., Dunand, D.C., 2011. Effect of crystalline metallic particles on the compressive behavior of a cellular amorphous metal. Scr. Mater. 64, 1031–1034.

Brown, S., Abou-Chedid, G., 1993. Appropriate yield functions for metal powder compaction. Scr. Metall. Mater. 28, 11–16.

Brown, S., Abou-Chedid, G., 1994. Yield behavior of metal powder assemblages. J. Mech. Phys. Solids 42, 383–399.

Brown, J., Vendra, L., Rabiei, A., 2010. Bending properties of Al–steel and steel–steel composite metal foams. Metall. Mater. Trans. A 41, 2784–2793.

Bruck, H.A., Gilat, R., Aboudi, J., Gershon, A.L., 2007. A new approach for optimizing the mechanical behavior of porous microstructures for porous materials by design. Modell. Simul. Mater. Sci. Eng. 15, 653–674.

Brun, E., Vicente, J., Topin, F., et al., 2009. Microstructure and transport properties of cellular materials: representative volume element. Adv. Eng. Mater. 11, 805–810.

Brunke, O., Hamann, A., Cox, S.J., Odenbach, S., 2005. Experimental and numerical analysis of the drainage of aluminium foams. J. Phys.: Condens. Matter 17, 6353–6362.

Bryant, J., Crowley, M., Wang, W., et al., 2008. Development of Alcoa aluminum foam products. In: Lefebvre, L.P., Banhart, J., Dunand, D.C. (Eds.), Metfoam 2007-Porous Metals and Metallic Foams, Proc. 2007 Metfoam Conference. Montreal, Sept. 5–7, 2007. DESTech Publications, Inc, Lancaster, PA.

Brydon, A.D., Bardenhagen, S.G., Miller, E.A., Seidler, G.T., 2005. Simulation of the densification of real open-celled foam microstructures. J. Mech. Phys. Solids 53, 2638–2660.

Budiansky, B., 1965. On the elastic moduli of some heterogeneous materials. J. Mech. Phys. Solids 13, 223.

Bugnet, B., Doniat, D., 1989. Porous metal structure and method of manufacturing of said structure.

Buryachenko, V.A., 1996. The overall elastoplastic behavior of multiphase materials with isotropic components. Acta Mech. 119, 93–117.

Cady, C.M., Gray Iii, G.T., Liu, C., et al., 2009. Compressive properties of a closed-cell aluminum foam as a function of strain rate and temperature. Mater. Sci. Eng. A 525, 1–6.

Calmidi, V.V., Mahajan, R.L., 1999. The effective thermal conductivity of high porosity fibrous metal foams. Trans. ASME – J. Heat Transfer 121, 466–471.

Calmidi, V.V., Mahajan, R.L., 2000. Forced convection in high porosity metal foams. Trans. ASME – J. Heat Transfer 122, 557–565.

Cansizoglu, O., Harrysson, O., Cormier, D., et al., 2008. Properties of Ti–6Al–4V non-stochastic lattice structures fabricated via electron beam melting. Mater. Sci. Eng. A 492, 468–474.

Cao, X.-q., Wang, Z.-h., Ma, H.-w., et al., 2006. Effects of cell size on compressive properties of aluminum foam. Trans. Nonferrous Met. Soc. China 16, 351.

Cao, Z.-K., Li, B., Yao, G.-C., Wang, Y., 2008. Fabrication of aluminum foam stabilized by copper-coated carbon fibers. Mater. Sci. Eng. A 486, 350–356.

Carbajal, G., Sobhan, C.B., Peterson, G.P., et al., 2006. Thermal response of a flat heat pipe sandwich structure to a localized heat flux. Int. J. Heat Mass Transfer 49, 4070–4081.

Carcel, B., Carcel, A.C., Arrué, P., 2010. Creep behaviour of closed cell aluminium foams from stress relaxation. Key Eng. Mater. 423, 131–136.

Castrodeza, E.M., Mapelli, C., 2009. Processing of brass open-cell foam by silica-gel beads replication. J. Mater. Process. Technol. 209, 4958–4962.

Caty, O., Maire, E., Youssef, S., Bouchet, R., 2008. Modeling the properties of closed-cell cellular materials from tomography images using finite shell elements. Acta Mater. 56, 5524–5534.

Caty, O., Maire, E., Douillard, T., et al., 2009. Experimental determination of the macroscopic fatigue properties of metal hollow sphere structures. Mater. Lett. 63, 1131–1134.

Cervantes Martinez, A., Rio, E., Delon, G., et al., 2008. On the origin of the remarkable stability of aqueous foams stabilised by nanoparticles: link with microscopic surface properties. Soft Matter 4, 1531–1535.

Chaikin, P.M., Donev, A., Man, W., et al., 2006. Some observations on the random packing of hard ellipsoids. Ind. Eng. Chem. Res. 45, 6960–6965.

Chakravarty, U.K., 2010. An investigation on the dynamic response of polymeric, metallic, and biomaterial foams. Compos. Struct. 92, 2339–2344.

Chan, K.C., Xie, L.S., 2003. Dependency of densification properties on cell topology of metal foams. Scr. Mater. 48, 1147.

Chang, K.S., Lemlich, R., 1980. A study of the electrical conductivity of foam. J. Colloid Interface Sci. 73, 224–232.

Chapman, A.M., Higdon, J.J.L., 1994. Effective elastic properties for a periodic bicontinuous porous medium. J. Mech. Phys. Solids 42, 283–305.

Chen, C., Fleck, N.A., 2002. Size-effects in the constrained deformation of metallic foams. J. Mech. Phys. Solids 50, 955–977.

Chen, F., He, D., 1999. Preparation, structure control and acoustic properties of porous aluminium with open cells. In: Banhart, J., Ashby, M.F., Fleck, N.A. (Eds.), Metal Foams and Porous Metal Structures. Bremen, Germany. Verlag MIT, Bremen, Germany.

Chen, J.Y., Huang, Y., 1998. Fracture analysis of cellular materials: a strain gradient model. J. Mech. Phys. Solids 46, 789–828.

Chen, T.-J., Huang, J.-S., 2008a. Creep-rupturing of cellular materials: regular hexagonal honeycombs with dual imperfections. Compos. Sci. Technol. 68, 1562–1569.

Chen, T.-J., Huang, J.-S., 2008b. Creep-rupturing of open-cell foams. Acta Mater. 56, 2283–2289.

Chen, T.-J., Huang, J.-S., 2009a. Creep-buckling of hexagonal honeycombs with dual imperfections. Compos. Struct. 89, 143–150.

Chen, T.-J., Huang, J.-S., 2009b. Creep-buckling of open-cell foams. Acta Mater. 57, 1497–1503.

Chen, C., Lakes, R., 1995. Analysis of the structure–property relations of porous foam materials. Cell. Polym. 14, 186–202.

Chen, R., Zhao, T.S., 2007a. A novel electrode architecture for passive direct methanol fuel cells. Electrochem. Commun. 9, 718–724.

Chen, R., Zhao, T.S., 2007b. Porous current collectors for passive direct methanol fuel cells. Electrochim. Acta 52, 4317–4324.

Chen, C., Lu, T.J., Fleck, N.A., 1999. Effect of imperfections on the yielding of two-dimensional foams. J. Mech. Phys. Solids 47, 2235–2272.

Chen, B.-C., Silva, E.C.N., Kikuchi, N., 2001a. Advances in computational design and optimization with application to MEMS. Int. J. Numer. Methods Eng. 52, 23–62.

Chen, C., Fleck, N.A., Lu, T.J., 2001b. The mode I crack growth resistance of metallic foams. J. Mech. Phys. Solids 49, 231.

Chen, Y.-c. K., Chu, Y.S., Yi, J., et al., 2010. Morphological and topological analysis of coarsened nanoporous gold by X-ray nanotomography. Appl. Phys. Lett. 96, 043122–043123.

Chen-Wiegart, Y.-c. K., Wang, S., Chu, Y.S., et al., 2012. Structural evolution of nanoporous gold during thermal coarsening. Acta Mater. 60, 4972–4981.

Chethan, A., Garcia-Moreno, F., Wanderka, N., et al., 2011. Influence of oxides on the stability of zinc foam. J. Mater. Sci. 46, 7806–7814.

Chiba, H., Ogushi, T., Nakajima, H., et al., 2009. The uncertainty in SCHF-DT thermal conductivity measurements of lotus-type porous copper. Adv. Eng. Mater. 11, 848–851.

Chin, P., Sun, X., Roberts, G.W., Spivey, J.J., 2006. Preferential oxidation of carbon monoxide with iron-promoted platinum catalysts supported on metal foams. Appl. Catal. Gen. 302, 22–31.

Chino, Y., Dunand, D.C., 2008. Directionally freeze-cast titanium foam with aligned, elongated pores. Acta Mater. 56, 105–113.

Chino, Y., Nakanishi, H., Kobata, M., et al., 2002. Processing of a porous 7075 Al alloy by bubble expansion in a semi-solid state. Scr. Mater. 47, 769.

Chino, Y., Mabuchi, M., Yamada, Y., et al., 2003. An experimental investigation of effects of specimen size parameters on compressive and tensile properties in a closed cell Al foam. Mater. Trans. 44, 633–636.

Chino, Y., Mabuchi, M., Nakanishi, H., et al., 2004. Effect of metal powder size on the gas expansion behavior of 7075 Al alloy in a semisolid state. Mater. Sci. Eng. A 382, 35.

Chiras, S., Mumm, D.R., Evans, A.G., et al., 2002. The structural performance of near-optimized truss core panels. Int. J. Solids Struct. 39, 4093–4115.

Chmielus, M., Zhang, X.X., Witherspoon, C., et al., 2009. Giant magnetic-field-induced strains in polycrystalline Ni–Mn–Ga foams. Nat. Mater. 8, 863–866.

Chmielus, M., Witherspoon, C., Wimpory, R.C., et al., 2010. Magnetic-field-induced recovery strain in polycrystalline Ni–Mn–Ga foam. J. Appl. Phys. 108, 123526–123527.

Choe, H., Dunand, D.C., 2004a. Mechanical properties of oxidation-resistant Ni–Cr foams. Mater. Sci. Eng. A 384, 184–193.

Choe, H., Dunand, D.C., 2004b. Synthesis, structure, and mechanical properties of Ni–Al and Ni–Cr–Al superalloy foams. Acta Mater. 52, 1283–1295.

Choi, J.B., Lakes, R.S., 1992a. Non-linear properties of metallic cellular materials with a negative Poisson's ratio. J. Mater. Sci. 27, 5375–5381.

Choi, J.B., Lakes, R.S., 1992b. Non-linear properties of polymer cellular materials with a negative Poisson's ratio. J. Mater. Sci. 27, 4678–4684.

Choi, J.B., Lakes, R.S., 1995. Analysis of elastic modulus of conventional foams and of re-entrant foam materials with a negative Poisson's ratio. Int. J. Mech. Sci. 37, 51–59.

Choi, J.B., Lakes, R.S., 1996. Fracture toughness of re-entrant foam materials with a negative Poisson's ratio: experiment and analysis. Int. J. Fract. 80, 73–83.

Choon, N.K., Chakraborty, A., Aye, S.M., Xiaolin, W., 2006. New pool boiling data for water with copper-foam metal at sub-atmospheric pressures: experiments and correlation. Appl. Therm. Eng. 26, 1286.

Chou, K.S., Song, M.A., 2002. A novel method for making open-cell aluminum foams with soft ceramic balls. Scr. Mater. 46, 379–382.

Christensen, R.M., 1986. Mechanics of low density materials. J. Mech. Phys. Solids 34, 563–578.

Christensen, R.M., 1990. A critical evaluation for a class of micro-mechanics models. J. Mech. Phys. Solids 38, 379.

Christensen, R.M., 2000. Mechanics of cellular and other low-density materials. Int. J. Solids Struct. 37, 93–104.

Chuang, C.-H., Huang, J.-S., 2002. Elastic moduli and plastic collapse strength of hexagonal honeycombs with Plateau borders. Int. J. Mech. Sci. 44, 1827–1844.

Clyne, T.W., Markaki, A.E., Tan, J.C., 2005. Mechanical and magnetic properties of metal fibre networks, with and without a polymeric matrix. Compos. Sci. Technol. 65, 2492–2499.

Clyne, T.W., Golosnoy, I.O., Tan, J.C., Markaki, A.E., 2006. Porous materials for thermal management under extreme conditions. Philos. Trans. R. Soc., A 364, 125.

Clyne, T.W., 2000. Chapter 3.16-thermal and electrical conduction in MMCs. In: Clyne, T.W. (Ed.), Comprehensive Composite Materials, Metal Matrix Composites, vol. 3. Pergamon, Oxford UK.

Cochran, J.K., 1988. Ceramic hollow spheres and their applications. Curr. Opin. Solid State Mater. Sci. 3, 474–479.

Cocks, A.C.F., Ashby, M.F., 2000. Creep-buckling of cellular solids. Acta Mater. 48, 3395–3400.

Collishaw, P.G., Evans, J.R.G., 1994. An assessment of expressions for the apparent thermal conductivity of cellular materials. J. Mater. Sci. 29, 486–498.

Colombo, P., Degischer, H.P., 2010. Highly porous metals and ceramics. Mater. Sci. Technol. 26, 1145–1158.

Combaz, E., Mortensen, A., 2010. Fracture toughness of Al replicated foam. Acta Mater. 58, 4590–4603.

Combaz, E., Bacciarini, C., Charvet, R., et al., 2010. Yield surface of polyurethane and aluminium replicated foam. Acta Mater. 58, 5168–5183.

Combaz, E., Bacciarini, C., Charvet, R., et al., 2011a. Multiaxial yield behaviour of Al replicated foam. J. Mech. Phys. Solids 59, 1777–1793.

Combaz, E., Rossoll, A., Mortensen, A., 2011b. Hole and notch sensitivity of aluminium replicated foam. Acta Mater. 59, 572–581.

Conde, Y., Mortensen, A., 2008. Age-hardening response of replicated microcellular Al-4.5%Cu. Adv. Eng. Mater. 10, 849–852.

Conde, Y., Mortensen, A., 2010. Solidification of Al-4.5 wt pct Cu-replicated foams. Metall. Mater. Trans. A 41, 2048–2055.

Conde, Y., Despois, J.F., Goodall, R., et al., 2006a. Replication processing of highly porous materials. Adv. Eng. Mater. 8, 795–803.

Conde, Y., Pollien, A., Mortensen, A., 2006b. Functional grading of metal foam cores for yield-limited lightweight sandwich beams. Scr. Mater. 54, 539.

Conde, Y., 2009. Micro-, Meso- and Macrostructures in Replicated Microcellular Aluminium. Thèse No. 4309. Institute of Materials, Lausanne. (Switzerland, Ecole Polytechnique Fédérale de Lausanne).

Conner, R.D., Johnson, W.L., Paton, N.E., Nix, W.D., 2003. Shear bands and cracking of metallic glass plates in bending. J. Appl. Phys. 94, 904–911.

Cookson, E.J., Floyd, D.E., Shih, A.J., 2006. Design, manufacture, and analysis of metal foam electrical resistance heater. Int. J. Mech. Sci. 48, 1314.

Coquard, R., Loretz, M., Baillis, D., 2008. Conductive heat transfer in metallic/ceramic open-cell foams. Adv. Eng. Mater. 10, 323–337.

Coquard, R., Rochais, D., Baillis, D., 2009. Experimental investigations of the coupled conductive and radiative heat transfer in metallic/ceramic foams. Int. J. Heat Mass Transfer 52, 4907–4918.

Coquard, R., Rousseau, B., Echegut, P., et al., 2012. Investigations of the radiative properties of Al–NiP foams using tomographic images and stereoscopic micrographs. Int. J. Heat Mass Transfer 55, 1606–1619.

Courtney, T.H., 1990. Mechanical Behavior of Materials. Mc Graw-Hill, New York, NY.

Coussy, O., 2004. Poromechanics. John Wiley & Sons, Chichester, West Sussex, UK.

Couteau, O., Dunand, D.C., 2008. Creep of aluminum syntactic foams. Mater. Sci. Eng. A 488, 573–579.

Cox, M.E., Dunand, D.C., 2011. Bulk gold with hierarchical macro-, micro- and nano-porosity. Mater. Sci. Eng. A 528, 2401–2406.

Cox, S.J., Bradley, G., Weaire, D., 2001. Metallic foam processing from the liquid state. Eur. Phys. J.: Appl. Phys. 14, 87–96.

Cox, M., Mathaudhu, S., Hartwig, K., Dunand, D., 2010. Amorphous Zr-based foams with aligned, elongated pores. Metall. Mater. Trans. A 41, 1706–1713.

Crowson, D.A., Farkas, D., Corcoran, S.G., 2009. Mechanical stability of nanoporous metals with small ligament sizes. Scr. Mater. 61, 497–499.

Cuenot, S., Fretigny, C., Demoustier-Champagne, S., Nysten, B., 2004. Surface tension effect on the mechanical properties of nanomaterials measured by atomic force microscopy. Phys. Rev., B 69, 165410.

Cuevas, F., Montes, J., Cintas, J., Urban, P., 2009. Electrical conductivity and porosity relationship in metal foams. J. Porous Mater. 16, 675–681.

Cui, X., Zhang, Y., Zhao, H., et al., 2011. Stress concentration in two-dimensional lattices with imperfections. Acta Mech. 216, 105–122.

Cumberland, D.J., Crawford, R.J., 1987. The Packing of Particles. Elsevier, Amsterdam.

Cumberland, D.J., Crawford, R.J., Sprevak, D., 1989. A statistical model for the random packing of real powder particles. Eur. Polym. J. 25, 1173.

da Cunda, L.A.B., Oliveira, B.F., Creus, G.J., 2011. Plasticity and damage analysis of metal foams under dynamic loading. Analyse des Plastizitäts- und Schädigungsverhaltens von Metallschäumen bei dynamischer Belastung. Materialwiss. Werkstofftech. 42, 356–364.

Dai, Z., Nawaz, K., Park, Y.G., et al., 2010. Correcting and extending the Boomsma-Poulikakos effective thermal conductivity model for three-dimensional, fluid-saturated metal foams. Int. Commun. Heat Mass Transfer 37, 575–580.

Dai, Z., Nawaz, K., Park, Y., et al., 2011. A comparison of metal-foam heat exchangers to compact multi-louver designs for air-side heat transfer applications. Heat Transfer Eng. 33, 21–30.

Dairon, J., Gaillard, Y., Tissier, J.C., et al., 2011. Parts containing open-celled metal foam manufactured by the foundry route: processes, performances, and applications. Adv. Eng. Mater. 13, 1066–1071.

Danas, K., Ponte Castañeda, P., 2009a. A finite-strain model for anisotropic viscoplastic porous media: I, theory. Eur. J. Mech. – A/Solids 28, 387–401.

Danas, K., Ponte Castañeda, P., 2009b. A finite-strain model for anisotropic viscoplastic porous media: II, applications. Eur. J. Mech. – A/Solids 28, 402–416.

Danas, K., Idiart, M.I., Castañeda, P.P., 2008a. A homogenization-based constitutive model for isotropic viscoplastic porous media. Int. J. Solids Struct. 45, 3392–3409.

Danas, K., Idiart, M.I., Ponte Castañeda, P., 2008b. A homogenization-based constitutive model for two-dimensional viscoplastic porous media. C. R. Mec. 336, 79–90.

Dannemann, K.A., Lankford, J., 2000. High strain rate compression of closed-cell aluminium foams. Mater. Sci. Eng. A 293, 157–164.

Daoud, A., El-khair, M.T.A., Abdel-Aziz, M., Rohatgi, P., 2007. Fabrication, microstructure and compressive behavior of ZC63 Mg–microballoon foam composites. Compos. Sci. Technol. 67, 1842–1853.

Daoud, A., 2008. Synthesis and characterization of novel ZnAl22 syntactic foam composites via casting. Mater. Sci. Eng. A 488, 281–295.

Daoud, A., 2009. Effect of strain rate on compressive properties of novel Zn12Al based composite foams containing hybrid pores. Mater. Sci. Eng. A 525, 7–17.

Datye, A.K., Lemlich, R., 1983. Liquid distribution and electrical conductivity in foam. Int. J. Multiphase Flow 9, 627–636.

Davies, G.J., Zhen, S., 1983. Metallic foams: their production, properties and applications. J. Mater. Sci. 18, 1899–1911.

Davis, N.G., Teisen, J., Schuh, C., Dunand, D.C., 2001. Solid state foaming of titanium by superplastic expansion of argon filled pores. J. Mater. Res. 16, 1508–1518.

Dawson, M.A., McKinley, G.H., Gibson, L.J., 2008. The dynamic compressive response of open-cell foam impregnated with a Newtonian fluid. J. Appl. Mech. 75, 041015-11.

Daxner, T., Böhm, H.J., Rammerstorfer, F.G., 1999a. Mesoscopic simulation of inhomogeneous metallic foams with respect to energy absorption. Comput. Mater. Sci. 16, 61–69.

Daxner, T., Böhm, H.J., Rammerstorfer, F.G., 1999b. Influence of micro- and meso-topological properties on the crash-worthiness of aluminium foams. In: Banhart, J., Ashby, M.F., Fleck, N.A. (Eds.), Metal Foams and Porous Metal Structures. Bremen, Germany. Verlag MIT, Bremen, Germany.

Daxner, T., Böhm, H.J., Ramerstorfer, F.G., et al., 2000a. Simulation des elasto-plastischen Verhaltens von Metallschaum mit Hilfe von 2D und 3D Einheitszellen-Modellen. Materialwiss. Werkstofftech. 31, 447–450.

Daxner, T., Rammerstorfer, F.G., Böhm, H.J., 2000b. Adaptation of density distributions for optimising aluminium foam structures. Mater. Sci. Technol. 16, 935–939.

Daxner, T., Böhm, H.J., Seitzberger, M., Rammerstorfer, F.G., 2002. Chapter 6.1-modeling of cellular materials. In: Degischer, H.P., Kriszt, B. (Eds.), Handbook of Cellular Metals. Wiley-VCH Verlag, Weinheim Germany.

Daxner, T., Böhm, H.J., Rammerstorfer, F.G., 2003. Numerical investigation of local yielding in metallic foams. In: Banhart, J., Fleck, N., Mortensen, A. (Eds.), Cellular Metals: Manufacture, Properties, Applications, Proc. Conf. Metfoam 2003. Berlin, 23–25 June 2003. Verlag MIT Publishing, Berlin, Germany.

De Jaeger, P., T'Joen, C., Huisseune, H., et al., 2011. An experimentally validated and parameterized periodic unit-cell reconstruction of open-cell foams. J. Appl. Phys. 109, 103519-10.

de Wild, P.J., Verhaak, M.J.F.M., 2000. Catalytic production of hydrogen from methanol. Catal. Today 60, 3–10.

DeFouw, J.D., C.Dunand, D., 2012. Processing and compressive creep of cast replicated IN792 Ni-base superalloy foams. Mater. Sci. Eng. A 558, 129–133.

Degischer, H.P., Kottar, A., 2002. Chapter 4.3 Consideration on Quality Features. In: Degischer, H.P., Kriszt, B. (Eds.), Handbook of Cellular Metals. Wiley-VCH Verlag, Weinheim Germany.

Degischer, H.P., Krizt, B. (Eds.), 2002. Handbook of Cellular Metals. Wiley-VCH Verlag, Weinheim Germany.

Degischer, H.P., Foroughi, B., Kottar, A., 2013. Tomography based material model and simulation of deformation. In: Dukhan, N. (Ed.), Metal Foams – Fundamentals and Applications. DEStech Publications Inc, Lancaster PA, USA.

Dehm, G., 2009. Miniaturized single-crystalline fcc metals deformed in tension: new insights in size-dependent plasticity. Prog. Mater. Sci. 54, 664–688.

Delannay, F., Clyne, T.W., 1999. Elastic properties of cellular metals processed by sintering mats of fibres. In: Banhart, J., Ashby, M.F., Fleck, N.A. (Eds.), Metal Foams and Porous Metal Structures. MIT Verlag.

Delannay, F., 2005. Elastic model of an entangled network of interconnected fibres accounting for negative Poisson ratio behaviour and random triangulation. Int. J. Solids Struct. 42, 2265.

deLemos, M.J., Saito, M., 2009. Heat-transfer coefficient for cellular materials modeled as an array of elliptic rods. Adv. Eng. Mater. 11, 837–842.

Demetriou, M.D., Duan, G., Veazey, C., et al., 2007a. Amorphous Fe-based metal foam. Scr. Mater. 57, 9–12.

Demetriou, M.D., Hanan, J.C., Veazey, C., et al., 2007b. Yielding of metallic glass foam by percolation of an elastic buckling instability. Adv. Mater. 19, 1957–1962.

Demetriou, M.D., Schramm, J.P., Veazey, C., et al., 2007c. High porosity metallic glass foam: a powder metallurgy route. Appl. Phys. Lett. 91, 161903–3.

Demetriou, M.D., Veazey, C., Schroers, J., et al., 2007d. Expansion evolution during foaming of amorphous metals. Mater. Sci. Eng. A 449–451, 863–867.

Demetriou, M.D., Veazey, C., Schroers, J., et al., 2007e. Thermo-plastic expansion of amorphous metallic foam. J. Alloys Compd. 434–435, 92–96.

Demetriou, M.D., Veazey, C., Harmon, J.S., et al., 2008. Stochastic metallic-glass cellular structures exhibiting benchmark strength. Phys. Rev. Lett. 101, 145702.

Demiray, S., Becker, W., Hohe, J., 2007. Numerical determination of initial and subsequent yield surfaces of open-celled model foams. Int. J. Solids Struct. 44, 2093–2108.

Demiray, S., Becker, W., Hohe, J., 2009. Investigation of the fatigue behavior of open cell foams by a micromechanical 3-D model. Mater. Sci. Eng. A 504, 141–149.

Dempsey, B.M., Eisele, S., McDowell, D.L., 2005. Heat sink applications of extruded metal honeycombs. Int. J. Heat Mass Transfer 48, 527–535.

Deqing, W., Ziyuan, S., 2003. Effect of ceramic particles on cell size and wall thickness of aluminum foam. Mater. Sci. Eng. A 361, 45.

Deqing, W., Xiangjun, M., Weiwei, X., Ziyuan, S., 2006. Effect of processing parameters on cell structure of an aluminum foam. Mater. Sci. Eng. A 420, 235.

Deshpande, V.S., Fleck, N.A., 2000a. High strain rate compressive behaviour of aluminium alloy foams. Int. J. Impact Eng. 24, 277–298.

Deshpande, V.S., Fleck, N.A., 2000b. Isotropic constitutive models for metallic foams. J. Mech. Phys. Solids 48, 1253–1283.

Deshpande, V.S., Fleck, N.A., 2001a. Collapse of truss core sandwich beams in 3-point bending. Int. J. Solids Struct. 38, 6275.

Deshpande, V.S., Fleck, N.A., 2001b. Multiaxial yield behaviour of polymer foams. Acta Mater. 49, 1859–1866.

Deshpande, V.S., Ashby, M.F., Fleck, N.A., 2001a. Foam topology bending versus stretching dominated architectures. Acta Mater. 49, 1035–1040.

Deshpande, V.S., Fleck, N.A., Ashby, M.F., 2001b. Effective properties of the octet-truss lattice material. J. Mech. Phys. Solids 49, 1747–1769.

Despois, J.F., Mortensen, A., 2005. Permeability of open-pore microcellular materials. Acta Mater. 53, 1381.

Despois, J.F., Conde, Y., Marchi, C.S., Mortensen, A., 2003. Tensile behaviour of replicated aluminium foams. In: Banhart, J., Fleck, N., Mortensen, A. (Eds.), Cellular Metals: Manufacture, Properties, Applications, Proc. Conf. Metfoam 2003. Berlin, 23–25 June 2003. Verlag MIT Publishing, Berlin, Germany.

Despois, J.F., Conde, Y., Marchi, C.S., Mortensen, A., 2004. Tensile behaviour of replicated aluminium foams. Adv. Eng. Mater. 6, 444–447.

Despois, J.F., Marmottant, A., Conde, Y., et al., 2006a. Microstructural tailoring of open-pore microcellular aluminium by replication processing. Mater. Sci. Forum 512, 281–288.

Despois, J.F., Mueller, R., Mortensen, A., 2006b. Uniaxial deformation of microcellular metals. Acta Mater. 54, 4129–4142.

Despois, J.F., Marmottant, A., Salvo, L., Mortensen, A., 2007. Influence of the infiltration pressure on the structure and properties of replicated aluminium foams. Mater. Sci. Eng. A 462, 68–75.

Detsi, E., Chen, Z.G., Vellinga, W.P., et al., 2011. Reversible strain by physisorption in nanoporous gold. Appl. Phys. Lett. 99, 083104-3.

Deville, S., Saiz, E., Nalla, R.K., Tomsia, A.P., 2006. Freezing as a path to build complex composites. Science 311, 515–518.

Deville, S., Saiz, E., Tomsia, A.P., 2007. Ice-templated porous alumina structures. Acta Mater. 55, 1965–1974.

Dharmasena, K.P., Wadley, H.N.G., 2002. Electrical conductivity of open-cell metal foams. J. Mater. Res. 17, 625–631.

Diedericks, G.P.J., DuPlessis, J.P., 1997. Modelling of flow through homogeneous foams. Math. Eng. Ind. 6, 133–154.

Dietterle, M., Will, T., Kolb, D.M., 1995. Step dynamics at the Ag(111)-electrolyte interface. Surf. Sci. 327, L495–L500.

Dillard, T., Nguyen, F.N., Forest, S., et al., 2003. In-situ observation of tensile deformation of open-cell nickel foams by means of X-ray microtomography. In: Banhart, J., Fleck, N., Mortensen, A. (Eds.), Cellular Metals: Manufacture, Properties, Applications, Proc. Conf. Metfoam 2003. Berlin, 23–25 June 2003. Verlag MIT Publishing, Berlin, Germany.

Dillard, T., N'Guyen, F.N., Maire, E., et al., 2005. 3D quantitative image analysis of open-cell nickel foams under tension and compression loading using X-ray microtomography. Philos. Mag. 85, 2147–2175.

Dillard, T., Forest, S., Ienny, P., 2006. Micromorphic continuum modelling of the deformation and fracture behaviour of nickel foams. Eur. J. Mech. – A/Solids 25, 526–549.

Dimopoulos Eggenschwiler, P., Tsinoglou, D., Seyfert, J., et al., 2009. Ceramic foam substrates for automotive catalyst applications: fluid mechanic analysis. Exp. Fluids 47, 209–222.

Ding, X.R., Lu, L.S., Chen, C., et al., 2011. Heat transfer enhancement by using four kinds of porous structures in a heat exchanger. Appl. Mech. Mater. 52–54, 1632–1637.

Diologent, F., Combaz, E., Laporte, V., et al., 2009a. Processing of Ag–Cu alloy foam by the replication process. Scr. Mater. 61, 351–354.

Diologent, F., Conde, Y., Goodall, R., Mortensen, A., 2009b. Microstructure, strength and creep of aluminium–nickel open cell foam. Philos. Mag. 89, 1121–1139.

Diologent, F., Goodall, R., Mortensen, A., 2009c. Creep of aluminium–magnesium open cell foam. Acta Mater. 57, 830.

Diologent, F., Goodall, R., Mortensen, A., 2009d. Surface oxide in replicated microcellular aluminium and its influence on the plasticity size effect. Acta Mater. 57, 286–294.

Diologent, F., Goodall, R., Mortensen, A., 2011. Activation volume in microcellular aluminium: size effects in thermally activated plastic flow. Acta Mater. 59, 6869–6879.

Donev, A., Cisse, I., Sachs, D., et al., 2004. Improving the density of jammed disordered packings using ellipsoids. Science 303, 990–993.

Dong, H.X., Jiang, Y., He, Y.H., et al., 2009. J. Alloys Compd. 484, 907.

Dormieux, L., Kondo, D., Ulm, F.-J., 2006. Microporomechanics. John Wiley & Sons, Chichester, West Sussex, UK.

Dou, R., Derby, B., 2010. Strain gradients and the strength of nanoporous gold. J. Mater. Res. 25, 746–753.

Dou, R., Derby, B., 2011. Deformation mechanisms in gold nanowires and nanoporous gold. Philos. Mag. 91, 1070–1083.

Dou, R., Xu, B., Derby, B., 2010. High-strength nanoporous silver produced by inkjet printing. Scr. Mater. 63, 308–311.

Doyoyo, M., Hu, J.W., 2006. Multi-axial failure of metallic strut-lattice materials composed of short and slender struts. Int. J. Solids Struct. 43, 6115–6139.

Doyoyo, M., Mohr, D., 2006. Experimental determination of the mechanical effects of mass density gradient in metallic foams under large multiaxial inelastic deformation. Mech. Mater. 38, 325–339.

Doyoyo, M., Wan Hu, J., 2006. Plastic failure analysis of an auxetic foam or inverted strut lattice under longitudinal and shear loads. J. Mech. Phys. Solids 54, 1479–1492.

Doyoyo, M., Wierzbicki, T., 2003. Experimental studies on the yield behavior of ductile and brittle aluminum foams. Int. J. Plast. 19, 1195–1214.

Drenchev, L., Sobczak, J., Malinov, S., Sha, W., 2006. Gasars: a class of metallic materials with ordered porosity. Mater. Sci. Technol. 22, 1135–1147.

Driscoll, D., Weisenstein, A.J., Sofie, S.W., 2011. Electrical and flexural anisotropy in freeze tape cast stainless steel porous substrates. Mater. Lett. 65, 3433–3435.

Drury, W.J., Rickles, S.A., Sanders, T.H., Cochran, J.K., 1989. Deformation energy absorption characteristics of a metal/ceramic cellular solid. In: Lee, E.W., Chia, E.H., Kim, N.J. (Eds.), Lightweight Alloys for Aerospace Applications. The Minerals, Metals & Materials Soc, Warrendale, PA, USA.

Du, Y.P., Qu, Z.G., Zhao, C.Y., Tao, W.Q., 2010. Numerical study of conjugated heat transfer in metal foam filled double-pipe. Int. J. Heat Mass Transfer 53, 4899–4907.

Du, Y.P., Zhao, C.Y., Tian, Y., Qu, Z.G., 2012. Analytical considerations of flow boiling heat transfer in metal-foam filled tubes. Heat and Mass Transfer 48, 165–173.

Duan, D.L., Zhang, R.L., Ding, X.J., Li, S., 2006a. Calculation of specific surface area of foam metals using dodecahedron model. Mater. Sci. Technol. 22, 1364–1367.

Duan, H.L., Wang, J., Karihaloo, B.L., Huang, Z.P., 2006b. Nanoporous materials can be made stiffer than non-porous counterparts by surface modification. Acta Mater. 54, 2983–2990.

Duan, D.L., Li, S., Zhang, R.L., Jiang, S.L., 2007. Calculation of apparent resistivity of metallic open cell foams by dodecahedron model. Mater. Sci. Technol. 23, 661–664.

Ducheyne, P., Aernoudt, E., Meester, P.D., 1987. The mechanical behaviour of porous austenitic stainless steel fibre structures. J. Mater. Sci. 13, 2650–2658.

Dudka, A., Garcia-Moreno, F., Wanderka, N., Banhart, J., 2008. Structure and distribution of oxides in aluminium foam. Acta Mater. 56, 3990–4001.

Dukhan, N., Ali, M., 2012a. Effect of confining wall on properties of gas flow through metal foam: an experimental study. Transp. Porous Media 91, 225–237.

Dukhan, N., Ali, M., 2012b. Strong wall and transverse size effects on pressure drop of flow through open-cell metal foam. Int. J. Therm. Sci. 57, 85–91.

Dukhan, N., Chen, K.-C., 2007. Heat transfer measurements in metal foam subjected to constant heat flux. Exp. Therm. Fluid Sci. 32, 624–631.

Dukhan, N., Minjeur II, C., 2011. A two-permeability approach for assessing flow properties in metal foam. J. Porous Mater. 18, 417–424.

Dukhan, N., Patel, P., 2008. Equivalent particle diameter and length scale for pressure drop in porous metals. Exp. Therm. Fluid Sci. 32, 1059–1067.

Dukhan, N., Patel, K., 2011. Effect of sample-length on flow properties of open-cell metal foam and pressure–drop correlations. J. Porous Mater. 18, 655–665.

Dukhan, N., Quinones-Ramos, P.D., Cruz-Ruiz, E., et al., 2005. One-dimensional heat transfer analysis in open-cell 10-ppi metal foam. Int. J. Heat Mass Transfer 48, 5112–5120.

Dukhan, N., Picon-Feliciano, R., Alvarez-Hernandez, A.R., 2006a. Air flow through compressed and uncompressed aluminum foam: measurements and correlations. J. Fluids Eng. 128, 1004–1012.

Dukhan, N., Picon-Feliciano, R., Alvarez-Hernandez, A.R., 2006b. Heat transfer analysis in metal foams with low-conductivity fluids. J. Heat Transfer 128, 784–792.

Dukhan, N., 2006. Correlations for the pressure drop for flow through metal foam. Exp. Fluids 41, 665–672.

Dukhan, N. (Ed.), 2013a. Metal Foams – Fundamentals and Applications. DEStech Publications Inc, Lancaster PA, USA.

Dukhan, N., 2013b. Principles of fluid flow through open-cell metal foam. In: Dukhan, N. (Ed.), Metal Foams – Fundamentals and Applications. DEStech Publications Inc, Lancaster PA, USA.

Dunand, D.C., Derby, B., 1993. Creep and thermal cycling. In: Suresh, S., Mortensen, A., Needleman, A. (Eds.), Fundamentals of Metal Matrix Composites. Butterworth-Heinemann, Boston.

Dunand, D.C., Müllner, P., 2011. Size effects on magnetic actuation in Ni–Mn–Ga shape-memory alloys. Adv. Mater. 23, 216–232.

Dunand, D.C., Teisen, J., 1998. Superplastic foaming of titanium and Ti–6Al–4V. In: Schwartz, D.S., Shih, D.H., Evans, A.G., Wadley, H.G. (Eds.), Porous and Cellular Materials for Structural Applications. San Francisco, CA, USA, Materials Research Society Symposium Proceedings, vol. 521. (Warrendale PA, USA).

Dunand, D.C., 2004. Processing of titanium foams. Adv. Eng. Mater. 6, 369–376.

Dunleavy, C.S., Curran, J.A., Clyne, T.W., 2011. Plasma electrolytic oxidation of aluminium networks to form a metal-cored ceramic composite hybrid material. Compos. Sci. Technol. 71, 908–915.

DuPlessis, P., Montillet, A., Comiti, J., Legrand, J., 1994. Pressure drop prediction for flow through high porosity metallic foams. Chem. Eng. Sci. 49, 3545–3553.

Duva, J.M., Hutchinson, J.W., 1984. Constitutive potentials for dilutely voided nonlinear materials. Mech. Mater. 3, 41–54.

Edouard, D., Lacroix, M., Huu, C.P., Luck, F., 2008. Pressure drop modeling on solid foam: state-of-the art correlation. Chem. Eng. J. 144, 299–311.

Ehlers, W., Müllerschön, H., Klar, O., 1999. On the behaviour of aluminium foams under uniaxial and multiaxial loading. In: Banhart, J., Ashby, M.F., Fleck, N.A. (Eds.), Metal Foams and Porous Metal Structures. Bremen, Germany. Verlag MIT, Bremen, Germany.

Ejlali, A., Ejlali, A., Hooman, K., Gurgenci, H., 2009. Application of high porosity metal foams as air-cooled heat exchangers to high heat load removal systems. Int. Commun. Heat. Mass Transf. 36, 674–679.

Elliott, J.C., 1956. Method of Producing Metal Foam. US Patent. U.S.

Elmoutaouakkil, A., Salvo, L., Maire, E., Peix, G., 2002. 2D and 3D characterisation of metal foams using X-ray tomography. Adv. Eng. Mater. 4, 803–807.

Elnasri, I., Pattofatto, S., Zhao, H., et al., 2007. Shock enhancement of cellular structures under impact loading: part I experiments. J. Mech. Phys. Solids 55, 2652.

Elzey, D.M., Wadley, H.N.G., 2001. The limits of solid state foaming. Acta Mater. 49, 849–859.

Elzey, D.M., Sofla, A.Y.N., Wadley, H.N.G., 2005. A shape memory-based multifunctional structural actuator panel. Int. J. Solids Struct. 42, 1943–1955.

ERG, 2005. ERG Materials and Aerospace Corporation, Duocell Aluminum Foam, Materials and Aerospace Corporation, Oakland, CA 94608 (web site http://www.ergaerospace.com/).

Erk, K.A., Dunand, D.C., Shull, K.R., 2008. Titanium with controllable pore fractions by thermoreversible gelcasting of TiH2. Acta Mater. 56, 5147–5157.

Erlebacher, J., Seshadri, R., 2009. Hard materials with tunable porosity. MRS Bull. 34, 561–566.

Erlebacher, J., Sieradzki, K., 2003. Pattern formation during dealloying. Scr. Mater. 49, 991.

Erlebacher, J., Aziz, M.J., Karma, A., et al., 2001. Evolution of nanoporosity during dealloying. Nature 410, 450–453.

Erlebacher, J., 2004. An atomistic description of dealloying – porosity evolution, the critical potential, and rate-limiting behavior. J. Electrochem. Soc. 151, C614–C626.

Esen, Z., Bor, S., 2007. Processing of titanium foams using magnesium spacer particles. Scr. Mater. 56, 341.

Eshelby, J.D., 1957. The determination of the elastic field of an ellipsoidal inclusion, and related problems. Proc. R. Soc. London, Ser. A 241, 376–396.

Evans, A.G., Hutchinson, J.W., Ashby, M.F., 1988. Cellular metals. Curr. Opin. Solid State Mater. Sci. 3, 288–303.

Evans, A.G., Hutchinson, J.W., Ashby, M.F., 1999. Multifunctionality of cellular metal systems. Prog. Mater. Sci. 43, 171–221.

Evans, A.G., Hutchinson, J.W., Fleck, N.A., et al., 2001. The topological design of multifunctional cellular metals. Prog. Mater. Sci. 46, 309–327.

Faghri, A., 1995. Heat Pipe Science and Technology. Taylor and Francis, Washington DC, USA. pp. 765–760.

Faleskog, J., Gao, X., Shih, C., 1998. Cell model for nonlinear fracture analysis' I. Micromechanics calibration. Int. J. Fract. 89, 355–373.

Fallet, A., Lhuissier, P., Salvo, L., Bréchet, Y., 2008. Mechanical behaviour of metallic hollow spheres foam. Adv. Eng. Mater. 10, 858–862.

Fan, H.L., Fang, D.N., 2009. Modeling and limits of strength of nanoporous foams. Mater. Des. 30, 1441–1444.

Fang, D.-N., Li, Y.-L., Zhao, H., 2010. On the behaviour characterization of metallic cellular materials under impact loading. Acta Mech. Sin. 26, 837–846.

Faure, N., Doyoyo, M., 2007. Thermomechanical properties of strut-lattices. J. Mech. Phys. Solids 55, 803–818.

Fazekas, A., Dendievel, R., Salvo, L., Brechet, Y., 2002. Effect of microstructural topology upon the stiffness and strength of 2D cellular structures. Int. J. Mech. Sci. 44, 2047.

Fedorchenko, I.M., 1979. Progress in work in the field of high-porosity materials from powders and fibers. Soviet Powder Metall. Met. Ceram. 18, 615–622.

Feng, S., Sen, P.N., Halperin, B.I., Lobb, C.J., 1984. Percolation on two-dimensional elastic networks with rotationally invariant bond-bending forces. Phys. Rev., B 30, 5386–5390.

Feng, S., Halperin, B.I., Sen, P.N., 1987. Transport properties of continuum systems near the percolation threshold. Phys. Rev., B 35, 197.

Feng, Y., Zheng, H., Zhu, Z., Zu, F., 2002. The microstructure and electrical conductivity of aluminum alloy foams. Mater. Chem. Phys. 78, 196.

Feng, Y., Tao, N., Zhu, Z., et al., 2003. Effect of aging treatment on the quasi-static and dynamic compressive properties of aluminum alloy foams. Mater. Lett. 57, 4058.

Fergus, J., 2007. Materials challenges for solid-oxide fuel cells. JOM 59, 56–62.

Ferrari, M., Filipponi, M., 1991. Appraisal of current homogenizing techniques for the elastic response of porous and reinforced glass. J. Am. Ceram. Soc. 74, 229–231.

Ferrouillat, S., Tochon, P., Peerhossaini, H., 2006. Micromixing enhancement by turbulence: application to multifunctional heat exchangers. Chem. Eng. Process. 45, 633–640.

Fiedler, T., Ochsner, A., Gracio, J., Kuhn, G., 2005. Structural modeling of the mechanical behavior of periodic cellular solids: open-cell structures. Mech. Compos. Mater. 41, 277–290.

Fiedler, T., Sturm, B., Oechsner, A., et al., 2006. Modelling the mechanical behaviour of adhesively bonded and sintered hollow-sphere structures. Mech. Compos. Mater. 42, 559–570.

Fiedler, T., Oschner, A., Belova, I.V., Murch, G.E., 2008a. Recent advances in the prediction of the thermal properties of syntactic metallic hollow sphere structures. Adv. Eng. Mater. 10, 361–365.

Fiedler, T., Solorzano, E., Oechsner, A., 2008b. Numerical and experimental analysis of the thermal conductivity of metallic hollow sphere structures. Mater. Lett. 62, 1204–1207.

Fiedler, T., Solorzano, E., Garcia-Moreno, F., et al., 2009. Lattice Monte Carlo and experimental analyses of the thermal conductivity of random-shaped cellular aluminum. Adv. Eng. Mater. 11, 843–847.

Fiedler, T., Veyhl, C., Belova, I.V., et al., 2012. On the anisotropy of lotus-type copper. Adv. Eng. Mater. 14, 144–152.

Fife, J.L., Li, J.C., Dunand, D.C., Voorhees, P.W., 2009. Morphological analysis of pores in directionally freeze-cast titanium foams. J. Mater. Res. 24, 117–124.

Fischmeister, H.F., Arzt, E., 1983. Densification of powders by particle deformation. Powder Metall. 26, 82–88.

Fleck, N.A., Qiu, X., 2007. The damage tolerance of elastic-brittle, two-dimensional isotropic lattices. J. Mech. Phys. Solids 55, 562–588.

Fleck, N.A., Kuhn, L.T., McMeeking, R.M., 1992. Yielding of metal powder bonded by isolated contacts. J. Mech. Phys. Solids 40, 1139–1162.

Fleck, N.A., Olurin, O.B., Chen, C., Ashby, M.F., 2001. The effect of hole size upon the strength of metallic and polymeric foams. J. Mech. Phys. Solids 49, 2015–2030.

Fleck, N.A., Deshpande, V.S., Ashby, M.F., 2010. Micro-architectured materials: past, present and future. Proc. R. Soc. A.

Fleck, N.A., 1995a. A crystal plasticity view of powder compaction. Acta Metall. Mater. 43, 3177–3184.

Fleck, N.A., 1995b. On the cold compaction of powders. J. Mech. Phys. Solids 43, 1409–1431.

Florence, C., Sab, K., 2005. Overall ultimate yield surface of periodic tetrakaidecahedral lattice with non-symmetric material distribution. J. Mater. Sci. 40, 5883–5892.

Forest, S., Trinh, D.K., 2011. Generalized continua and non-homogeneous boundary conditions in homogenisation methods. ZAMM – J. Appl. Math. Mech./Z. Angew. Math. Mech. 91, 90–109.

Forest, S., Blazy, J.S., Chastel, Y., Moussy, F., 2005. Continuum modeling of strain localization phenomena in metallic foams. J. Mater. Sci. 40, 5903.

Foroughi, B., Kriszt, B., Degischer, H.P., 2002. Chapter 6.2-Mesomodel of real cellular structures. In: Degischer, H.P., Kriszt, B. (Eds.), Handbook of Cellular Metals. Wiley-VCH Verlag, Weinheim Germany.

Fourie, J.G., DuPlessis, J.P., 2002. Pressure drop modelling in cellular metallic foams. Chem. Eng. Sci. 57, 2781–2789.

Frei, J., Gergely, V., Mortensen, A., Clyne, T.W., 2002. The effect of prior deformation on the foaming behavior of "FORMGRIP" precursor material. Adv. Eng. Mater. 4, 749–752.

Freund, L.B., Suresh, S., 2003. Thin Film Materials, Stress, Defect Formation and Surface Evolution. Cambridge University Press, Cambridge, UK. pp. 102–104.

Friedl, O., Motz, C., Peterlik, H., et al., 2008. Experimental investigation of mechanical properties of metallic hollow sphere structures. Metall. Mater. Trans. B 39, 135–146.

Friis, E.A., Lakes, R.S., Park, J.B., 1988. Negative Poisson's ratio polymeric and metallic foams. J. Mater. Sci. 23, 4406–4414.

Frost, H.J., Ashby, M.F., 1982. Deformation Mechanism Maps. Pergamon Press, Oxford, U.K.

Fukasawa, T., Ando, M., Ohji, T., Kanzaki, S., 2001. Synthesis of porous ceramics with complex pore structure by freeze-dry processing. J. Am. Ceram. Soc. 84, 230–232.

Fukasawa, T., Deng, Z.-Y., Ando, M., et al., 2002. Synthesis of porous silicon nitride with unidirectionally aligned channels using freeze-drying process. J. Am. Ceram. Soc. 85, 2151–2155.

Gaillard, C., Despois, J.F., Mortensen, A., 2004. Processing of NaCl powders of controlled size and shape for the microstructural tailoring of aluminium foams. Mater. Sci. Eng. A 374, 250.

Gan, Y.X., Chen, C., Shen, Y.P., 2005. Three-dimensional modeling of the mechanical property of linearly elastic open cell foams. Int. J. Solids Struct. 42, 6628–6642.

Garajeu, M., Suquet, P., 2007. On the influence of local fluctuations in volume fraction of constituents on the effective properties of nonlinear composites. Application to porous materials. J. Mech. Phys. Solids 55, 842–878.

Garajeu, M., Michel, J.C., Suquet, P., 2000. A micromechanical approach of damage in viscoplastic materials by evolution in size, shape and distribution of voids. Comput. Methods Appl. Mech. Eng. 183, 223–246.

Garcia-Gonzales, R., Monnereau, C., Thovert, J.F., et al., 1999. Conductivity of real foams. Colloids Surf., A 151, 497–503.

Garcia-Moreno, F., Fromme, M., Banhart, J., 2004. Real-time X-ray radioscopy on metallic foams using a compact micro-focus source. Adv. Eng. Mater. 6, 416–420.

Garcia-Moreno, F., Babcsan, N., Banhart, J., 2005. X-ray radioscopy of liquid metalfoams: influence of heating profile, atmosphere and pressure. Colloids Surf., A 263, 290–294.

Garcia-Moreno, F., Rack, A., Helfen, L., et al., 2008. Fast processes in liquid metal foams investigated by high-speed synchrotron X-ray microradioscopy. Appl. Phys. Lett. 92, 134104-3.

Garcia-Moreno, F., Mukherjee, M., Jimenez, C., Banhart, J., 2009. X-ray radioscopy of liquid metal foams under microgravity. Trans. Indian Inst. Met. 62, 451–454.

Garcia-Moreno, F., Solorzano, E., Banhart, J., 2011. Kinetics of coalescence in liquid aluminium foams. Soft Matter 7.

Garrity, P.T., Klausner, J.F., Mei, R., 2010. Performance of aluminum and carbon foams for air side heat transfer augmentation. J. Heat Transfer 132, 121901–121909.

Gaspar, N., Ren, X.J., Smith, C.W., et al., 2005. Novel honeycombs with auxetic behaviour. Acta Mater. 53, 2439–2445.

Gasser, S., Paun, F., Cayzeele, A., Brechet, Y., 2003. Uniaxial tensile elastic properties of a regular stacking of brazed hollow spheres. Scr. Mater. 48, 1617.

Gasser, S., Paun, F., Brechet, Y., 2004a. Finite elements computation for the elastic properties of a regular stacking of hollow spheres. Mater. Sci. Eng. A 379, 240.

Gasser, S., Paun, F., Riffard, L., Brechet, Y., 2004b. Microplastic yield condition for a periodic stacking of hollow spheres. Scr. Mater. 50, 401–405.

Gaytan, S.M., Murr, L.E., Martinez, E., et al., 2010. Comparison of microstructures and mechanical properties for solid and mesh cobalt-base alloy prototypes fabricated by electron beam melting. Metall. Mater. Trans. A 41, 3216–3227.

Gebert, A., Kündig, A.A., Schultz, L., Hono, K., 2004. Selective electrochemical dissolution in two-phase La–Zr–Al–Cu–Ni metallic glass. Scr. Mater. 51, 961–965.

Gensel, J., Borke, T., Pérez, N.P., et al., 2012. Cavitation engineered 3D Sponge networks and their application in active surface construction. Adv. Mater. 24, 985–989.

Gent, A.N., Thomas, A.G., 1959. The deformation of foamed elastic materials. J. Appl. Polym. Sci. 1, 107–113.

Gent, A.N., Thomas, A.G., 1963. Mechanics of foamed elastic materials. Rubber Chem. Technol. 36, 597–610.

Georgalli, G.A., Reuter, M.A., 2006. Modelling the co-ordination number of a packed bed of spheres with distributed sizes using a CT scanner. Miner. Eng. 19, 246–255.

Gerbaux, O., Vercueil, T., Memponteil, A., Bador, B., 2009. Experimental characterization of single and two-phase flow through nickel foams. Chem. Eng. Sci. 64, 4186–4195.

Gergely, V., Clyne, T.W., 1998. The effect of oxide layers on gas-generating hydride particles during production of aluminium foams. In: Schwartz, D.S., Shih, D.H., Evans, A.G., Wadley, H.G. (Eds.), Porous and Cellular Materials for Structural Applications. San Francisco, CA, USA, Materials Research Society Symposium Proceedings, vol. 521. (Warrendale PA, USA).

Gergely, V., Clyne, T.W., 2000. The FORMGRIP process: foaming of reinforced metals by gas release in precursors. Adv. Eng. Mater. 2, 175–178.

Gergely, V., Clyne, T.W., 2004. Drainage in standing liquid metal foams: modelling and experimental observations. Acta Mater. 52, 3047.

Gergely, V., Degischer, H.P., Clyne, T.W., 2000. Recycling of MMCs and production of metallic foams (Chapter 3).30. In: Clyne, T.W. (Ed.), Comprehensive Composite Materials, Metal Matrix Composites, vol. 3. Pergamon, Oxford UK.

Gergely, V., Curran, D.C., Clyne, T.W., 2003. The FOAMCARP process: foaming of aluminium MMCs by the chalk–aluminium reaction in precursors. Compos. Sci. Technol. 63, 2301.

German, R.M., 1989. Particle Packing Characteristics. Metal Powder Industries Federation, Princeton.

Ghosh, I., 2008. Heat-transfer analysis of high porosity open-cell metal foam. J. Heat Transfer 130, 034501–034506.

Ghosh, I., 2009a. Heat transfer correlation for high-porosity open-cell foam. Int. J. Heat Mass Transfer 52, 1488–1494.

Ghosh, I., 2009b. How good is open-cell metal foam as heat transfer surface? J. Heat Transfer 131, 101004–101008.

Giani, L., Groppi, G., Tronconi, E., 2005a. Heat transfer characterization of metallic foams. Ind. Eng. Chem. Res. 44, 9078–9085.

Giani, L., Groppi, G., Tronconi, E., 2005b. Mass-transfer characterization of metallic foams as supports for structured catalysts. Ind. Eng. Chem. Res. 44, 4993–5002.

Gibson, L.J., Ashby, M.F., 1982. The mechanics of three-dimensional cellular materials. Proc. R. Soc. London, Ser. A 382, 43–59.

Gibson, L.J., Ashby, M.F., 1997. Cellular Solids – Structure and Properties, second ed. Cambridge University Press, Cambridge, U.K.

Gibson, L.J., Ashby, M.F., Schajer, G.S., Rbertson, C.I., 1982. The mechanics of two-dimensional cellular materials. Proc. R. Soc. London, Ser. A 382, 25–42.

Gibson, L.J., Ashby, M.F., Zhang, J., Triantafilou, T.C., 1989. Failure surfaces for cellular materials under multiaxial loads - I. Modelling. Int. J. Mech. Sci. 31, 635–663.

Gibson, L.J., 1989. Modelling the mechanical behavior of cellular materials. Mater. Sci. Eng. A 110, 1–36.

Gibson, L.J., 2000. Mechanical behavior of metallic foams. Annu. Rev. Mater. Sci. 30, 191–227.

Giesche, H., 2002. Chapter 2.7-Mercury porosimetry. In: Schüth, F., Sing, K.S.W., Weitkamp, J. (Eds.), Handbook of Porous Solids. Wiley-VCH, Weinheim, Germany.

Gioux, G., McCormack, T.M., Gibson, L.J., 2000. Failure of aluminum foams under multiaxial loads. Int. J. Mech. Sci. 42, 1097–1117.

Girlich, D., Franzke, U., 2001. Open pore metal foams for heat exchange in ventilation and refrigeration. Adv. Eng. Mater. 3, 351–352.

Glicksman, L., 1994. Heat transfer in foams. In: Hilyard, N.C., Cunningham, A. (Eds.), Low Density Cellular Plastics – Physical Basis of Behaviour. Chapman & Hall, London.

Gnyloskurenko, S., Nakamura, T., Byakova, A., et al., 2005. Development of lightweight Al alloy and technique. Can. Metall. Q. 44, 7–12.

Godet, J., Guenolé, J., Brochard, S., Pizzagalli, L., 2011. Mechanical properties of nanowires investigated by simulation. In: Thomas, O., Ponchet, A., Forest, S. (Eds.), Mechanics of Nano-objects. Transvalor - Presses des Mines, Paris, France.

Goglio, L., Manfredini Vassoler, J., Peroni, M., 2011. Measurement of longitudinal and transverse strain in an aluminium foam; Messung der longitudinalen und transversalen Dehnung in Aluminium-Schäumen. Materialwiss. Werkstofftech. 42, 342–349.

Golestanipour, M., Amini-Mashhadi, H., Abravi, M.S., et al., 2011. Manufacturing Al/SiCp composite foams using calcium carbonate as foaming agent. Mater. Sci. Technol. 27, 923–927.

Gologanu, M., Leblond, J., Perrin, G., Deveaux, J., 1997. Recent extensions of Gurson's model for porous ductile metals. In: Suquet, P. (Ed.), Continuum Micromechanics. Springer-Verlag, Wien New York.

Golovin, I.S., Sinning, H.R., 2003. Damping in some cellular metallic materials. J. Alloys Compd. 355, 2.

Golovin, I.S., Sinning, H.R., 2004. Internal friction in metallic foams and some related cellular structures. Mater. Sci. Eng. A 370, 504.

Gong, L., Kyriakides, S., 2005. Compressive response of open cell foams part II: initiation and evolution of crushing. Int. J. Solids Struct. 42, 1381.

Gong, L., Kyriakides, S., Jang, W.Y., 2005a. Compressive response of open-cell foams. Part I: morphology and elastic properties. Int. J. Solids Struct. 42, 1355.

Gong, L., Kyriakides, S., Triantafyllidis, N., 2005b. On the stability of Kelvin cell foams under compressive loads. J. Mech. Phys. Solids 53, 771–794.

Goodall, R., Mortensen, A., 2007. Microcellular aluminium? – Child's play! Adv. Eng. Mater. 9, 951–954.

Goodall, R., Despois, J.F., Marmottant, A., et al., 2006a. The effect of preform processing on replicated aluminium foam structure and mechanical properties. Scr. Mater. 54, 2069.

Goodall, R., Despois, J.F., Mortensen, A., 2006b. Sintering of NaCl powder: mechanisms and first stage kinetics. J. Eur. Ceram. Soc. 26, 3487.

Goodall, R., Marmottant, A., Despois, J.F., et al., 2006c. Replicated microcellular aluminium with spherical pores. In: Nakajima, H., Kanetake, N. (Eds.), 4th Int. Conf. on Porous Metals and Metal Foaming Technology (Metfoam 2005). Japan Institute of Metals, Kyoto, Japan. (Sendai, Japan).

Goodall, R., Weber, L., Mortensen, A., 2006d. The electrical conductivity of microcellular metals. J. Appl. Phys. 100, 044912.

Goodall, R., Marmottant, A., Salvo, L., Mortensen, A., 2007. Spherical pore replicated microcellular aluminium: processing and influence on properties. Mater. Sci. Eng. A 465, 124–135.

Goodenough, J.B., Huang, Y.-H., 2007. Alternative anode materials for solid oxide fuel cells. J. Power Sources 173, 1–10.

Goodstein, D.L., McCormick, W.D., Dash, J.G., 1966. Sintered copper for use at low temperature. Cryogenics 6, 167–168.

Goussery, V., Bienvenu, Y., Forest, S., et al., 2004. Grain size effects on the mechanical behavior of open-cell nickel foams. Adv. Eng. Mater. 6, 432–439.

Gradinger, R., Rammerstorfer, F.G., 1999. On the influence of meso-inhomogeneities on the crush worthiness of metal foams. Acta Mater. 47, 143–148.

Greer, J.R., De Hosson, J.T.M., 2011. Plasticity in small-sized metallic systems: intrinsic versus extrinsic size effect. Prog. Mater. Sci. 56, 654–724.

Greiner, C., Oppenheimer, S.M., Dunand, D.C., 2005. High strength, low stiffness, porous NiTi with superelastic properties. Acta Biomater. 1, 705.

Grenestedt, J.L., Bassinet, F., 2000. Influence of cell wall thickness variations on elastic stiffness of closed-cell cellular solids. Int. J. Mech. Sci. 42, 1327–1338.

Grenestedt, J.L., 1998. Influence of wavy imperfections in cell walls on elastic stiffness of cellular solids. J. Mech. Phys. Solids 46, 29–50.

Grenestedt, J.L., 1999a. Effective elastic behavior of some models for perfect cellular solids. Int. J. Solids Struct. 36, 1471–1501.

Grenestedt, J.L., 1999b. Influence of cell shape variations on elastic stiffness of closed cell cellular solids. Scr. Mater. 40, 71–77.

Grenestedt, J., 2005. On interactions between imperfections in cellular solids. J. Mater. Sci. 40, 5853.

Grima, J.N., Gatt, R., Ravirala, N., et al., 2006. Negative Poisson's ratios in cellular foam materials. Mater. Sci. Eng. A 423, 214–218.

Grima, J.N., Farrugia, P.S., Gatt, R., Zammit, V., 2007. A system with adjustable positive or negative thermal expansion. Proc. R. Soc. A 463, 1585–1596.

Groppi, G., Giani, L., Tronconi, E., 2007. Generalized correlation for gas/solid mass-transfer coefficients in metallic and ceramic foams. Ind. Eng. Chem. Res. 46, 3955–3958.

Gu, S., Lu, T.J., Evans, A.G., 2001. On the design of two-dimensional cellular metals for combined heat dissipation and structural load capacity. Int. J. Heat Mass Transfer 44, 2163.

Guden, M., Yüksel, S., 2006. SiC-particulate aluminum composite foams produced from powder compacts: foaming and compression behavior. J. Mater. Sci. 41, 4075–4084.

Guedes, J.M., Rodrigues, H.C., Bendsøe, M.P., 2003. A material optimization model to approximate energy bounds for cellular materials under multiload conditions. Struct. Multidiscip Optim. 25, 446–452.

Guessasma, S., Bassir, D., 2010. Optimization of the mechanical properties of virtual porous solids using a hybrid approach. Acta Mater. 58, 716–725.

Guessasma, S., Babin, P., Valle, G.D., Dendievel, R., 2008. Relating cellular structure of open solid food foams to their Young's modulus: finite element calculation. Int. J. Solids Struct. 45, 2881–2896.

Guest, J.K., Prévost, J.H., 2006. Optimizing multifunctional materials: design of microstructures for maximized stiffness and fluid permeability. Int. J. Solids Struct. 43, 7028–7047.

Guglielmotti, A., Quadrini, F., Squeo, E.A., Tagliaferri, V., 2009. Laser bending of aluminum foam sandwich panels. Adv. Eng. Mater. 11, 902–906.

Gui, M.C., Wang, D.B., Wu, J.J., et al., 2000. Deformation and damping behaviors of foamed Al–Si–SiCp composite. Mater. Sci. Eng. A 286, 282–288.

Gulsoy, H.O., German, R.M., 2008. Sintered foams from precipitation hardened stainless steel powder. Powder Metall. 51, 350–353.

Guo, Z.X., Jee, C.S.Y., Ozgüven, N., Evans, J.R.G., 2000. Novel polymer-metal based method for open-cell metal foam production. Mater. Sci. Technol. 16, 776–780.

Guo, R., Mai, Y.W., Fan, T., et al., 2004. Plane stress crack growing steadily in metal foams. Mater. Sci. Eng. A 381, 292.

Guoxin, H., Lixiang, Z., Yunliang, F., Yanhong, L., 2008. Fabrication of high porous NiTi shape memory alloy by metal injection molding. J. Mater. Process. Technol. 206, 395–399.

Gurson, A., 1977. Continuum theory of ductile rupture by void nucleation and growth: part I – yield criteria and flow rules for porous ductile media. J. Eng. Mater. Technol. 99, 2–15.

Ha, W., Kim, S.K., Jo, H.H., Kim, Y.J., 2005. Optimisation of process variables for manufacturing aluminium foam materials using aluminium scrap. Mater. Sci. Technol. 21, 495–499.

Haag, M., Wanner, A., Clemens, H., et al., 2003. Creep of aluminum-based closed-cell foams. Metall. Mater. Trans. 34A, 2809–2817.

Haesche, M., Weise, J.R., Garcia-Moreno, F., Banhart, J., 2008. Influence of particle additions on the foaming behaviour of AlSi11/TiH2 composites made by semi-solid processing. Mater. Sci. Eng. A 480, 283–288.

Haesche, M., Lehmhus, D., Weise, J., et al., 2010. Carbonates as foaming agent in chip-based aluminium foam precursor. J. Mater. Sci. Technol. 26, 845–850.

Hahn, M.C., Otto, A., Geiger, M., 2003. High-temperature processing of aluminium foam. In: Banhart, J., Fleck, N., Mortensen, A. (Eds.), Cellular Metals: Manufacture, Properties, Applications, Proc. Conf. Metfoam 2003. Berlin, 23–25 June 2003. Verlag MIT Publishing, Berlin, Germany.

Haibel, A., Rack, A., Banhart, J., 2006. Why are metal foams stable? Appl. Phys. Lett. 89.

Haji-Sheikh, A., Vafai, K., 2004. Analysis of flow and heat transfer in porous media imbedded inside various-shaped ducts. Int. J. Heat Mass Transfer 47, 1889–1905.

Haji-Sheikh, A., 2004. Estimation of average and local heat transfer in parallel plates and circular ducts filled with porous materials. J. Heat Transfer 126, 400–409.

Hakamada, M., Mabuchi, M., 2007. Mechanical strength of nanoporous gold fabricated by dealloying. Scr. Mater. 56, 1003–1006.

Hakamada, M., Nomura, T., Yamada, Y., et al., 2005a. Compressive deformation behavior at elevated temperatures in a closed-cell aluminum foam. Mater. Trans. 46, 1677–1680.

Hakamada, M., Nomura, T., Yamada, Y., et al., 2005b. Compressive properties at elevated temperatures of porous aluminum processed by the spacer method. J. Mater. Res. 20, 3385–3390.

Hakamada, M., Yamada, Y., Nomura, T., et al., 2005c. Effect of sintering temperature on compressive properties of porous aluminum produced by spark plasma sintering. Mater. Trans. 46, 186–188.

Hakamada, M., Wajima, T., Ikegami, Y., et al., 2006. Fluid conductivity of porous aluminum fabricated by powder-metallurgical spacer method. Jpn. J. Appl. Phys. 45, L575–L577.

Hakamada, M., Asao, Y., Kuromura, T., et al., 2007a. Density dependence of the compressive properties of porous copper over a wide density range. Acta Mater. 55, 2291–2299.

Hakamada, M., Kuromura, T., Chino, Y., et al., 2007b. Monotonic and cyclic compressive properties of porous aluminum fabricated by spacer method. Mater. Sci. Eng. A 459, 286–293.

Hakamada, M., Watanabe, H., Kuromura, T., et al., 2009. Effects of pore characteristics finely-controlled by spacer method on damping capacity of porous aluminum. Mater. Trans. 50, 427–429.

Hall, I.W., Guden, M., Yu, C.J., 2000. Chrushing of aluminum closed cell foams: density and strain-rate effects. Scr. Mater. 43, 515–521.

Halperin, B.I., Feng, S., Sen, P.N., 1985. Differences between lattice and continuum percolation transport exponents. Phys. Rev. Lett. 54, 2391.

Han, F.-S., Zhu, Z.-G., Liu, C.-S., 1997. Nonlinear internal friction character of foamed aluminum. Scr. Mater. 37, 1441–1447.

Han, F., Zhu, Z., Liu, C., Gao, J., 1999. Damping behavior of foamed aluminium. Metall. Trans. 30A, 771–776.

Han, F., Cheng, H., Wang, J., Wang, Q., 2004. Effect of pore combination on the mechanical properties of an open cell aluminum foam. Scr. Mater. 50, 13.

Han, F.S., Cheng, H.F., Li, Z., Wang, Q., 2005. The strain rate effect of an open cell aluminum foam. Metall. Trans. 36A, 645–650.

Han, X.-H., Wang, Q., Park, Y.-G., et al., 2012. A review of metal foam and metal matrix composites for heat exchangers and heat sinks. Heat Transfer Eng. 33, 991–1009.

Hangai, Y., Koyama, S., Hasegawa, M., Utsunomiya, T., 2010. Fabrication of aluminum foam/dense steel composite by friction stir welding. Metall. Mater. Trans. A 41, 2184–2186.

Hangai, Y., Oba, Y., Koyama, S., Utsunomiya, T., 2011. Fabrication of A-1050-A6061 functionally graded aluminum foam by friction stir processing route. Metall. Mater. Trans. A 42, 3585–3589.

Hangai, Y., Yoshida, H., Yoshikawa, N., 2012. Friction powder compaction for fabrication of open-cell aluminum foam by the sintering and dissolution process route. Metall. Mater. Trans. A 43, 802–805.

Hanssen, A.G., Enstock, L., Langseth, M., 2002a. Close-range blast loading of aluminium foam panels. Int. J. Impact Eng. 27, 593.

Hanssen, A.G., Hopperstad, O.S., Langseth, M., Ilstad, H., 2002b. Validation of constitutive models applicable to metal foams. Int. J. Mech. Sci. 44, 359–406.

Hao, G.L., Han, F.S., Wu, J., Wang, X.F., 2007. Damping properties of porous AZ91 magnesium alloy reinforced with copper particles. Mater. Sci. Technol. 23, 492–496.

Hao, G.L., Han, F.S., Li, W.D., 2009. Processing and mechanical properties of magnesium foams. J. Porous Mater. 16, 251–256.

Happel, J., Brenner, H., 1973. Low Reynolds Number Hydrodynamics with Special Applications to Particulate Media, Seconds edition. Noordhoff International Publishing, Leyden.

Happel, J., 1959. Viscous flow relative to arrays of cylinders. AIChE J. 5, 174–177.

Hardenacke, V., Hohe, J., 2009. Local probabilistic homogenization of two-dimensional model foams accounting for micro structural disorder. Int. J. Solids Struct. 46, 989–1006.

Harders, H., Rösler, J., Hupfer, K., 2003. Fatigue behaviour of metal foams: simulation and experiment. In: Banhart, J., Fleck, N., Mortensen, A. (Eds.), Cellular Metals: Manufacture, Properties, Applications, Proc. Conf. Metfoam 2003. Berlin, 23–25 June 2003. Verlag MIT Publishing, Berlin, Germany.

Harders, H., Hupfer, K., Rosler, J., 2005. Influence of cell wall shape and density on the mechanical behaviour of 2D foam structures. Acta Mater. 53, 1335–1345.

Hardy, P.W., Peisker, G.W., 1967. Method of Producing a Lightweight Foamed Metal. US Patent 3,300,296.

Harrigan, J.J., Reid, S.R., Seyed Yaghoubi, A., 2010. The correct analysis of shocks in a cellular material. Int. J. Impact Eng. 37, 918–927.

Harte, A.-M., Fleck, N.A., Ashby, M.F., 1999. Fatigue failure of an open cell and a closed cell aluminium alloy foam. Acta Mater. 47, 2511–2524.

Hartmann, M., Singer, R.F., 1997. Herstellung und Eigenschaften Syntakticher Magnesiumschäume. In: Banhart, J. (Ed.), Metallschäume. Bremen, March 1997. Verlag MIT, Bremen.

Hartmann, M., Reindel, K., Singer, R.F., 1998. Fabrication and properties of syntactic magnesium foams. In: Schwartz, D.S., Shih, D.H., Evans, A.G., Wadley, H.G. (Eds.), Porous and Cellular Materials for Structural Applications. San Francisco, CA, USA, Materials Research Society Symposium Proceedings, vol. 521. (Warrendale PA, USA).

Hartmann, J., Trepper, A., Körner, C., 2011. Aluminum integral foams with near-microcellular structure. Adv. Eng. Mater. 13, 1050–1055.

Hassan, M.R., Scarpa, F., Ruzzene, M., Mohammed, N.A., 2008. Smart shape memory alloy chiral honeycomb. Mater. Sci. Eng. A 481–482, 654–657.

Hassan, M.R., Scarpa, F., Mohamed, N.A., 2009. In-plane tensile behavior of shape memory alloy honeycombs with positive and negative Poisson's ratio. J. Intell. Mater. Syst. Struct. 20, 897–905.

Hasse, C., Grenet, M., Bontemps, A., et al., 2011. Realization, test and modelling of honeycomb wallboards containing a phase change material. Energ. Build. 43, 232–238.

Hauert, A., Rossoll, A., Mortensen, A., 2009. Ductile-to-brittle transition in tensile failure of particle-reinforced metals. J. Mech. Phys. Solids 57, 473–499.

Haussener, S., Coray, P., Lipinski, W., et al., 2010. Tomography based heat and mass transfer characterization of reticulate porous ceramics for high-temperature processing. J. Heat Transfer 132, 023305.

Hayes, A.M., Wang, A., Dempsey, B.M., McDowell, D.L., 2004. Mechanics of linear cellular alloys. Mech. Mater. 36, 691–713.

He, Y.H., Jiang, Y., Xu, N.P., et al., 2007. Adv. Mater. 19, 2102.

Heinl, P., Rottmair, A., Körner, C., Singer, R.F., 2007. Cellular titanium by selective electron beam melting. Adv. Eng. Mater. 9, 360–364.

Heinl, P., Körner, C., Singer, R.F., 2008a. Selective electron beam melting of cellular titanium: mechanical properties. Adv. Eng. Mater. 10, 882–888.

Heinl, P., Muller, L., Korner, C., et al., 2008b. Cellular Ti–6Al–4V structures with interconnected macro porosity for bone implants fabricated by selective electron beam melting. Acta Biomater. 4, 1536–1544.

Helfen, L., Baumbach, T., Stanzick, H., et al., 2002. Viewing the early stage of metal foam formation by computed tomography using synchrotron radiation. Adv. Eng. Mater. 4, 808–813.

Helfen, L., Baumbach, T., Pernot, P., et al., 2005. Investigation of pore initiation in metal foams by synchrotron-radiation tomography. Appl. Phys. Lett. 86.

Helle, A.S., Easterling, K.E., Ashby, M.F., 1985. Hot-isostatic pressing diagrams: new developments. Acta Metall. 33, 2163–2174.

Helwig, H.M., Garcia-Moreno, F., Banhart, J., 2011. A study of Mg and Cu additions on the foaming behaviour of Al–Si alloys. J. Mater. Sci. 46, 5227–5236.

Hentschel, M., Page, N., 2006. Elastic properties of powders during compaction. Part 3: evaluation of models. J. Mater. Sci. 41, 7902–7925.

Hetsroni, G., Gurevich, M., Rozenblit, R., 2005. Metal foam heat sink for transmission window. Int. J. Heat Mass Transfer 48, 3793–3803.

Hetsroni, G., Gurevich, M., Rozenblit, R., 2008. Natural convection in metal foam strips with internal heat generation. Exp. Therm. Fluid Sci. 32, 1740–1747.

Hétu, J.F., Ilinca, F., Marcotte, J.P., et al., 2013. Numerical simulation of the flow through metallic foams: multi-scale modeling and experimental validation. In: Dukhan, N. (Ed.), Metal Foams – Fundamentals and Applications. DEStech Publications Inc, Lancaster PA, USA.

Higginson, R., Sellars, M., 2003. Quantitative Metallography. Maney, London.

Hill, R., 1950. The Mathematical Theory of Plasticity. Oxford University Press, Oxford, UK.

Hirschmann, M., Körner, C., Singer, R.F., 2007. Integral foam molding – a new process for foamed magnesium castings. Mater. Sci. Forum 539–543, 1827–1832.

Ho, S., Ravindran, C., Hibbard, G.D., 2010. Magnesium alloy micro-truss materials. Scr. Mater. 62, 21–24.

Hodge, A.M., Dunand, D.C., 2001. Synthesis of nickel–aluminide foams by pack-aluminization of nickel foams. Intermetallics 9, 581.

Hodge, A.M., Dunand, D.C., 2003. Measurement and modeling of creep in open-cell NiAl foams. Metall. Mater. Trans. A 34, 2353–2363.

Hodge, A.M., Hayes, J.R., Caro, J.A., et al., 2006. Characterization and mechanical behaviour of nanoporous gold. Adv. Eng. Mater. 8, 853–857.

Hodge, A.M., Biener, J., Hayes, J.R., et al., 2007. Scaling equation for yield strength of nanoporous open-cell foams. Acta Mater. 55, 1343.

Hoff, N.J., 1954. Approximate analysis of structures in the presence of moderately large creep deformations. Q. Appl. Math. 12, 49–55.

Hong, S.-T., Herling, D.R., 2006a. Effects of surface area density of aluminum foams on thermal conductivity of aluminum phase change material composites. Adv. Eng. Mater. 9, 554–557.

Hong, S.-T., Herling, D.R., 2006b. Open-cell aluminum foams filled with phase change materials as compact heat sinks. Scr. Mater. 55, 887.

Hong, J.T., Tien, C.L., Kaviany, M., 1985. Non-Darcian effects on vertical-plate natural convection in porous media with high porosities. Int. J. Heat Mass Transfer 28, 2149–2157.

Hong, E., Ahn, B.Y., Shoji, D., et al., 2011. Microstructure and mechanical properties of reticulated titanium scrolls. Adv. Eng. Mater. 13, 1122–1127.

Hönig, A., Stronge, W.J., 2002a. In-plane dynamic crushing of honeycomb. Part I: crush band initiation and wave trapping. Int. J. Mech. Sci. 44, 1665–1696.

Hönig, A., Stronge, W.J., 2002b. In-plane dynamic crushing of honeycomb. Part II: application to impact. Int. J. Mech. Sci. 44, 1697–1714.

Hopkins, M.A., 2004. Discrete element modeling with dilated particles. Eng. Computation 21, 422–430.

Horozov, T.S., 2008. Foams and foam films stabilised by solid particles. Curr. Opin. Colloid Interface Sci. 13, 134–140.

Hou, B., Ono, A., Abdennadher, S., et al., 2011a. Impact behavior of honeycombs under combined shear-compression. Part I: experiments. Int. J. Solids Struct. 48, 687–697.

Hou, B., Pattofatto, S., Li, Y.L., Zhao, H., 2011b. Impact behavior of honeycombs under combined shear-compression. Part II: analysis. Int. J. Solids Struct. 48, 698–705.

Hsiao, K.-T., Advani, S.G., 1999. Modified effective thermal conductivity due to heat dispersion in fibrous porous media. Int. J. Heat Mass Transfer 42, 1237–1254.

Hsieh, W.H., Wu, J.Y., Shih, W.H., Chiu, W.C., 2004. Experimental investigation of heat-transfer characteristics of aluminum-foam heat sinks. Int. J. Heat Mass Transfer 47, 5149–5157.

Hsu, C.T., Cheng, P., 1990. Thermal dispersion in a porous medium. Int. J. Heat Mass Transfer 33, 1587–1597.

Hsu, C.T., 2005. Dynamic Modeling of Convective Heat Transfer in Porous Media, Second ed. In: Handbook of Porous Media. CRC Press.

Hu, G.K., 1996. A method of plasticity for general aligned spheroidal void or fiber-reinforced composites. Int. J. Plast. 12, 439–449.

Huang, J.-S., Chang, F.-M., 2005. Effects of curved cell edges on the stiffness and strength of two-dimensional cellular solids. Compos. Struct. 69, 183–191.

Huang, J.S., Gibson, L.J., 1991a. Fracture toughness of brittle foams. Acta Metall. Mater. 39, 1617–1626.

Huang, J.S., Gibson, L.J., 1991b. Fracture toughness of brittle honeycombs. Acta Metall. Mater. 39, 1627–1636.

Huang, J.-S., Liu, S.-Y., 2001a. Fatigue of honeycombs under in-plane multiaxial loads. Mater. Sci. Eng. A 308, 45–52.

Huang, J.-S., Liu, S.-Y., 2001b. Fatigue of isotropic open-cell foams under multiaxial loads. Int. J. Fatigue 23, 233.

Huang, Z.P., Wang, J., 2006. Nonlinear mechanics of solids containing isolated voids. Appl. Mech. Rev. 59, 210–229.

Huang, X.L., Wu, G.H., Lv, Z., et al., 2009. Electrical conductivity of open-cell Fe–Ni alloy foams. J. Alloys Compd. 479, 898–901.

Huang, X., Radman, A., Xie, Y.M., 2011. Topological design of microstructures of cellular materials for maximum bulk or shear modulus. Comput. Mater. Sci. 50, 1861–1870.

Huang, Y., Gong, J., Lv, S., et al., 2012. Fluxless soldering with surface abrasion for joining metal foams. Mater. Sci. Eng. A 552, 283–287.

Huang, W.M., 2003. A simple approach to estimate failure surface of polymer and aluminum foams under multiaxial loads. Int. J. Mech. Sci. 45, 1531.

Huber, A.T., Gibson, L.J., 1988. Anisotropy of foams. J. Mater. Sci. 23, 3031–3040.

Hughes, T.P., Marmier, A., Evans, K.E., 2010. Auxetic frameworks inspired by cubic crystals. Int. J. Solids Struct. 47, 1469–1476.

Hunt, A., Ewing, R., 2009. Percolation Theory for Flow in Porous Media. Springer, Heidelberg, Germany.

Hunt, M.L., Tien, C.L., 1988. Effects of thermal dispersion on forced convection in fibrous media. Int. J. Heat Mass Transfer 31, 301–309.

Hunter, T.N., Pugh, R.J., Franks, G.V., Jameson, G.J., 2008. The role of particles in stabilising foams and emulsions. Adv. Colloid Interface Sci. 137, 57–81.

Hur, B.Y., Ahn, D.K., Kim, S.Y., et al., 2003. Hydrogen treatment of Ti scrap. Mater. Sci. Forum 439, 143–148.

Hurysz, K.M., Clark, J.L., Nagel, A.R., et al., 1998. Steel and titanium hollow sphere foams. In: Schwartz, D.S., Shih, D.H., Evans, A.G., Wadley, H.G. (Eds.), Porous and Cellular Materials for Structural Applications. San Francisco, CA, USA, Materials Research Society Symposium Proceedings, vol. 521. (Warrendale PA, USA).

Huschka, S., 1998. Modellierung eines Materialgesetzes Zur Beschreibung des Mechanischen Eigenschaften von Aluminiumschaum. Reihe 5, Nr. 525. VDI Verlag GmbH, Düsseldorf, Germany.

Hutchinson, R.G., Fleck, N.A., 2005. Microarchitectured cellular solids – the hunt for statically determinate periodic trusses. ZAMM – J. Appl. Math. Mech./Z. Angew. Math. Mech. 85, 607–617.

Hutchinson, R.G., Fleck, N.A., 2006. The structural performance of the periodic truss. J. Mech. Phys. Solids 54, 756–782.

Hutchinson, R.G., Wicks, N., Evans, A.G., et al., 2003. Kagome plate structures for actuation. Int. J. Solids Struct. 40, 6969–6980.

Hutter, C., Allemann, C., Kuhn, S., Rudolf von Rohr, P., 2010. Scalar transport in a milli-scale metal foam reactor. Chem. Eng. Sci. 65, 3169–3178.

Hutter, C., Büchi, D., Zuber, V., Rudolf von Rohr, P., 2011a. Heat transfer in metal foams and designed porous media. Chem. Eng. Sci. 66, 3806–3814.

Hutter, C., Zenklusen, A., Lang, R., Rudolf von Rohr, P., 2011b. Axial dispersion in metal foams and streamwise-periodic porous media. Chem. Eng. Sci. 66, 1132–1141.

Huu, T.T., Lacroix, M., Pham Huu, C., et al., 2009. Towards a more realistic modeling of solid foam: use of the pentagonal dodecahedron geometry. Chem. Eng. Sci. 64, 5131–5142.

Hwang, J.J., Hwang, G.J., Yeh, R.H., Chao, C.H., 2002. Measurement of interstitial convective heat transfer and frictional drag for flow across metal foams. J. Heat Transfer 124, 120–129.

Hyun, S.K., Nakajima, H., 2002. Fabrication of lotus-structured porous iron by unidirectional solidification under nitrogen gas. Adv. Eng. Mater. 4, 741–744.

Hyun, S.K., Nakajima, H., 2003. Anisotropic compressive properties of porous copper produced by unidirectional solidification. Mater. Sci. Eng. A 340, 258.

Hyun, S., Torquato, S., 2000. Effective elastic and transport properties of regular honeycombs for all densities. J. Mater. Res. 15, 1985–1993.

Hyun, S.K., Murakami, K., Nakajima, H., 2001. Anisotropic mechanical properties of porous copper fabricated by unidirectional solidification. Mater. Sci. Eng. A 299, 241.

Hyun, S.-K., Ikeda, T., Nakajima, H., 2004. Fabrication of lotus-type porous iron and its mechanical properties. Sci. Technol. Adv. Mater. 5, 201.

Ibrahim, A., Körner, C., Singer, R.F., 2008. The effect of TiH2 particle size on the morphology of Al-foam produced by PM process. Adv. Eng. Mater. 10, 845–848.

Ichitsubo, T., Tane, M., Ogi, H., et al., 2002. Anisotropic elastic constants of lotus-type porous copper: measurements and micromechanics modeling. Acta Mater. 50, 4105.

Idris, M.I., Vodenitcharova, T., Hoffman, M., 2009. Mechanical behaviour and energy absorption of closed-cell aluminium foam panels in uniaxial compression. Mater. Sci. Eng. A 517, 37–45.

Imwinkelried, T., 2007. Mechanical properties of open-pore titanium foam. J. Biomed. Mater. Res. 81A, 964–970.

Inayat, A., Schwerdtfeger, J., Freund, H., et al., 2011. Periodic open-cell foams: pressure drop measurements and modeling of an ideal tetra-kaidecahedra packing. Chem. Eng. Sci. 66, 2758–2763.

Incera Garrido, G., Patcas, F.C., Lang, S., Kraushaar-Czarnetzki, B., 2008. Mass transfer and pressure drop in ceramic foams: a description for different pore sizes and porosities. Chem. Eng. Sci. 63, 5202–5217.

Ingraham, M.D., DeMaria, C.J., Issen, K.A., Morrison, D.J., 2009. Low cycle fatigue of aluminum foam. Mater. Sci. Eng. A 504, 150–156.

Innocentini, M., Lefebvre, L., Meloni, R., Baril, E., 2010. Influence of sample thickness and measurement set-up on the experimental evaluation of permeability of metallic foams. J. Porous Mater. 17, 491–499.

Inoue, A., Hagiwara, M., Masumoto, T., 1982. Production of Fe–P–C amorphous wires by in-rotating-water spinning method and mechanical properties of the wires. J. Mater. Sci. 17, 580–588.

Inoue, A., Wada, T., Wang, X.M., Greer, A.L., 2006. Bulk non-equilibrium alloys and porous glassy alloys with unique mechanical characteristics. Mater. Sci. Eng. A 442, 233.

Inoue, A., Wada, T., Louzguine-Luzgin, D.V., 2007. Improved mechanical properties of bulk glassy alloys containing spherical pores. Mater. Sci. Eng. A 471, 144–150.

Ip, S.W., Wang, Y., Toguri, J.M., 1999. Aluminum foam stabilization by solid particles. Can. Metall. Q. 38, 81–92.

Irretier, A., Banhart, J., 2005. Lead and lead alloy foams. Acta Mater. 53, 4903.

Ismail, M.H., Goodall, R., Davies, H.A., Todd, I., 2011. Porous NiTi alloy by metal injection moulding/sintering of elemental powders: effect of sintering temperature. Mater. Lett. 70, 142–145.

Issen, K.A., Casey, T.P., Dixon, D.M., et al., 2005. Characterization and modeling of localized compaction in aluminum foam. Scr. Mater. 52, 911–915.

Ito, K., Kobayashi, H., 2006. Production and fabrication technology development of aluminum useful for automobile leightweighting. Adv. Eng. Mater. 8, 828–835.

Jackson, G.W., James, D.F., 1986. The permeability of fibrous porous media. Can. J. Chem. Eng. 64, 364–374.

Jackson, K.P., Allwood, J.M., Landert, M., 2008. Incremental forming of sandwich panels. J. Mater. Process. Technol. 204, 290–303.

Jamshidi-Alashti, R., Roudini, G., 2012. Producing replicated open-cell aluminum foams by a novel method of melt squeezing procedure. Mater. Lett. 76, 233–236.

Jang, W.-Y., Kyriakides, S., 2009a. On the crushing of aluminum open-cell foams: part I experiments. Int. J. Solids Struct. 46, 617–634.

Jang, W.-Y., Kyriakides, S., 2009b. On the crushing of aluminum open-cell foams: part II analysis. Int. J. Solids Struct. 46, 635–650.

Jang, W.Y., Kraynik, A.M., Kyriakides, S., 2008. On the microstructure of open-cell foams and its effect on elastic properties. Int. J. Solids Struct. 45, 1845–1875.

Jang, W.-Y., Kyriakides, S., Kraynik, A.M., 2010. On the compressive strength of open-cell metal foams with Kelvin and random cell structures. Int. J. Solids Struct. 47, 2872–2883.

Jarvis, T., Voice, W., Goodall, R., 2011. The bonding of nickel foam to Ti–6Al–4V using Ti–Cu–Ni braze alloy. Mater. Sci. Eng. A 528, 2592–2601.

Jayaraj, J., Park, B.J., Kim, D.H., et al., 2006. Nanometer-sized porous Ti-based metallic glass. Scr. Mater. 55, 1063–1066.

Jazi, H.R.S., Mostaghimi, J., Chandra, S., et al., 2009. Spray-formed, metal-foam heat exchangers for high temperature applications. J. Therm. Sci. Eng. Appl. 1, 031008-7.

Jee, C.S.Y., Ozguve, N., Guo, Z.X., Evans, J.R.G., 2000. Preparation of high porosity metal foams. Metall. Mater. Trans. B 31, 1345–1352.

Jen, Y.-M., Chang, L.-Y., 2008. Evaluating bending fatigue strength of aluminum honeycomb sandwich beams using local parameters. Int. J. Fatigue 30, 1103–1114.

Jeng, T.-M., Tzeng, S.-C., 2005. Numerical study of confined slot jet impinging on porous metallic foam heat sink. Int. J. Heat Mass Transfer 48, 4685–4694.

Jeng, T.-M., Tzeng, S.-C., 2008. Heat transfer in a lid-driven enclosure filled with water-saturated aluminum foams. Numer. Heat Transfer, Part A 54, 178–196.

Jeng, T.-M., Tzeng, S.-C., Liu, T.-C., 2008. Heat transfer behavior in a rotating aluminum foam heat sink with a circular impinging jet. Int. J. Heat Mass Transfer 51, 1205–1215.

Jeng, T.-M., Tzeng, S.-C., Tang, F.-Z., 2010. Fluid flow and heat transfer characteristics of the porous metallic heat sink with a conductive cylinder partially filled in a rectangular channel. Int. J. Heat Mass Transfer 53, 4216–4227.

Jeon, I., Asahina, T., 2005. The effect of structural defects on the compressive behavior of closed-cell Al foam. Acta Mater. 53, 3415.

Jeon, I., Katou, K., Sonoda, T., et al., 2009. Cell wall mechanical properties of closed-cell Al foam. Mech. Mater. 41, 60–73.

Jeon, I., Asahina, T., Kang, K.-J., et al., 2010. Finite element simulation of the plastic collapse of closed-cell aluminum foams with X-ray computed tomography. Mech. Mater. 42, 227–236.

Ji, S., Gu, Q., Xia, B., 2006. Porosity dependence of mechanical properties of solid materials. J. Mater. Sci. 41, 1757–1768.

Jiang, R., Rong, C., Chu, D., 2004. Determination of energy efficiency for a direct methanol fuel cell stack by a fuel circulation method. J. Power Sources 126, 119–124.

Jiang, B., Zhao, N.Q., Shi, C.S., et al., 2005a. A novel method for making open cell aluminum foams by powder sintering process. Mater. Lett. 59, 3333.

Jiang, B., Zhao, N.Q., Shi, C.S., Li, J.J., 2005b. Processing of open cell aluminum foams with tailored porous morphology. Scr. Mater. 53, 781.

Jiang, Y., Deng, C., He, Y., et al., 2009. Mater. Lett. 63, 22–24.

Jiménez, C., Garcia-Moreno, F., Mukherjee, M., et al., 2009. Improvement of aluminium foaming by powder consolidation under vacuum. Scr. Mater. 61, 552–555.

Jimenez, C., Garcia-Moreno, F., Pfretzschner, B., et al., 2011. Decomposition of TiH2 studied in situ by synchrotron X-ray and neutron diffraction. Acta Mater. 59, 6318–6330.

Jimenez, C., Garcia-Moreno, F., Rack, A., et al., 2012. Partial decomposition of TiH2 studied in situ by energy-dispersive diffraction and ex situ by diffraction microtomography of hard X-ray synchrotron radiation. Scr. Mater. 66, 757–760.

Jin, I., Kenny, L.D., 1990. Method of Producing Lightweight Foamed Metal.

Jin, H.-J., Weissmüller, J., 2010. Bulk nanoporous metal for actuation. Adv. Eng. Mater. 12, 714–723.

Jin, H.-J., Weissmüller, J., 2011. A material with electrically tunable strength and flow stress. Science 332, 1179–1182.

Jin, H.-J., Kramer, D., Ivanisenko, Y., Weissmüller, J., 2007. Macroscopically strong nanoporous Pt prepared by dealloying. Adv. Eng. Mater. 9, 849–854.

Jin, H.-J., Parida, S., Kramer, D., Weissmüller, J., 2008. Sign-inverted surface stress-charge response in nanoporous gold. Surf. Sci. 602, 3588–3594.

Jin, H.-J., Kurmanaeva, L., Schmauch, J., et al., 2009. Deforming nanoporous metal: role of lattice coherency. Acta Mater. 57, 2665–2672.

Jinnapat, A., Kennedy, A., 2010. The manufacture of spherical salt beads and their use as dissolvable templates for the production of cellular solids via a powder metallurgy route. J. Alloys Compd. 499, 43–47.

Jinnapat, A., Kennedy, A., 2011. The manufacture and characterisation of aluminium foams made by investment casting using dissolvable spherical sodium chloride bead preforms. Metals 1, 49–64.

John, A.S., David, S.G., John, F., 2007. Superelastic NiTi honeycombs: fabrication and experiments. Smart Mater. Struct. 16, S170.

Jorgensen, D.J., Dunand, D.C., 2010. Ti–6Al–4V with micro- and macropores produced by powder sintering and electrochemical dissolution of steel wires. Mater. Sci. Eng. A 527, 849–853.

Ju, J., Chen, T., 1994a. Effective elastic moduli of two-phase composites containing randomly dispersed spherical inhomogeneities. Acta Mech. 103, 123–144.

Ju, J., Chen, T., 1994b. Micromechanics and effective moduli of elastic composites containing randomly dispersed ellipsoidal inhomogeneities. Acta Mech. 103, 103–121.

Jung, Y., Torquato, S., 2005. Fluid permeabilities of triply periodic minimal surfaces. Phys. Rev., E 72, 056319.

Kadar, C., Chmelik, F., Rajkovits, Z., Lendvai, J., 2004a. Acoustic emission measurements on metal foams. J. Alloys Compd. 378, 145.

Kadar, C., Maire, E., Borbely, A., et al., 2004b. X-ray tomography and finite element simulation of the indentation behavior of metal foams. Mater. Sci. Eng. A 387–389, 321.

Kadar, C., Chmelik, F., Kendvai, J., et al., 2007. Acoustic emission of metal foams during tension. Mater. Sci. Eng. A 462, 316–319.

Kadar, C., Chmelik, F., Cieslar, M., Lendvai, J., 2008. Acoustic emission of salt-replicated foams during compression. Scr. Mater. 59, 987–990.

Kailasam, M., Ponte-Castañeda, P., Willis, J.R., 1997a. The effect of particle size, shape, distribution and their evolution on the constitutive response of nonlinearly viscous composites I. Theory. Philosophical Transactions of the Royal Society, A 355, 1835–1852 (errata on p. 2520).

Kailasam, M., Ponte-Castañeda, P., Willis, J.R., 1997b. The effect of particle size, shape, distribution and their evolution on the constitutive response of nonlinearly viscous composites II. Examples. Philosophical Transactions of the Royal Society, A 355, 1853–1872.

Kakavas, P.A., Anifantis, N.K., 2003. Effective moduli of hyperelastic porous media at large deformation. Acta Mech. 160, 127–147.

Kamath, P.M., Balaji, C., Venkateshan, S.P., 2011. Experimental investigation of flow assisted mixed convection in high porosity foams in vertical channels. Int. J. Heat Mass Transfer 54, 5231–5241.

Kamiuto, K., Yee, S.S., 2005. Heat transfer correlations for open-cellular porous materials. Int. Commun. Heat. Mass Transf. 32, 947–953.

Kanahashi, H., Mukai, T., Yamada, Y., et al., 2000. Dynamic compression of an ultra-low density aluminium foam. Mater. Sci. Eng. A 280, 349–353.

Kanahashi, H., Mukai, T., Yamada, Y., et al., 2001a. Experimental study for the improvement of the crashworthiness in AZ91 magnesium foam controlling its microstructure. Mater. Sci. Eng. A 308, 283–287.

Kanahashi, H., Mukai, T., Yamada, Y., et al., 2001b. Improvement of the crashworthiness of ultralight metallic foam by heat-treatment for microstructural modification of base material. Mater. Trans. 42, 2087–2092.

Kanetake, N., Kobashi, M., 2006. Innovative processing of porous and cellular materials by chemical reaction. Scr. Mater. 54, 521.

Kang, K.-J., 2009. A wire-woven cellular metal of ultrahigh strength. Acta Mater. 57, 1865–1874.

Kaoua, S.-A., Dahmoun, D., Belhadj, A.-E., Azzaz, M., 2009. Finite element simulation of mechanical behaviour of nickel-based metallic foam structures. J. Alloys Compd. 471, 147–152.

Kaptay, G., 2004. Interfacial criteria for stabilization of liquid foams by solid particles. Colloids Surf., A 230, 67–80.

Kaptay, G., 2006. On the equation of the maximum capillary pressure induced by solid particles to stabilize emulsions and foams and on the emulsion stability diagram. Colloids Surf., A 282–283, 387.

Karagiozova, D., Langdon, G.S., Nurick, G.N., 2010. Blast attenuation in Cymat foam core sacrificial claddings. Int. J. Mech. Sci. 52, 758–776.

Kashef, S., Asgari, A., Hilditch, T.B., et al., 2010. Fracture toughness of titanium foams for medical applications. Mater. Sci. Eng. A 527, 7689–7693.

Kashef, S., Asgari, A., Hilditch, T.B., et al., 2011. Fatigue crack growth behavior of titanium foams for medical applications. Mater. Sci. Eng. A 528, 1602–1607.

Kashihara, M., Hyun, S.K., Yonetani, H., et al., 2006. Fabrication of lotus-type porous carbon steel by unidirectional solidification in nitrogen atmosphere. Scr. Mater. 54, 509–512.

Kathare, V., Davidson, J.H., Kulacki, F.A., 2008. Natural convection in water-saturated metal foam. Int. J. Heat Mass Transfer 51, 3794.

Kaviany, M., 1985. Laminar flow through a porous channel bounded by isothermal parallel plates. Int. J. Heat Mass Transfer 28, 851–858.

Kearns, M.W., Blenkinsop, P.A., Barber, A.C., Farthing, T.W., 1988. Manufacture of a novel porous metal. Int. J. Powder Metall. 24, 59–64.

Kee, A., Matic, P., Everett, R.K., 1998. A mesoscale computer simulation of multiaxial yield in Gasar porous copper. Mater. Sci. Eng. A 249, 30–39.

Kelley, P., Wong, C.R., Moran, A., 1993. Controlled porosity in spray-formed phosphor bronze. Int. J. Powder Metall. 29, 161–170.

Kelly, A., Mortensen, A., 2001. Composite materials: overview. In: Buschow, K.H.J., Cahn, R.W., Flemings, M.C., Ilschner, B., Kramer, E.J., Mahajan, S. (Eds.), Encyclopedia of Materials: Science and Technology. Elsevier Science Ltd, Oxford.

Kelly, A., McCartney, L.N., Clegg, W.J., Stearn, R.J., 2005. Controlling thermal expansion to obtain negative expansivity using laminated composites. Compos. Sci. Technol. 65, 47–59.

Kelly, A., Stearn, R.J., McCartney, L.N., 2006. Composite materials of controlled thermal expansion. Compos. Sci. Technol. 66, 154–159.

Kenesei, P., Kadar, C., Rajkovits, Z., Lendvai, J., 2004. The influence of cell-size distribution on the plastic deformation in metal foams. Scr. Mater. 50, 295.

Kennedy, A.R., Asavavaisthchai, S., 2002. Foaming of compacted Al–TiH2 powder mixture. Mater. Sci. Forum 396–402, 251–256.

Kennedy, A.R., Asavavisitchai, S., 2004a. Effects of TiB2 particle addition on the expansion, structure and mechanical properties of PM Al foams. Scr. Mater. 50, 115.

Kennedy, A.R., Asavavisithchai, S., 2004b. Effect of ceramic particle additions on foam expansion and stability in compacted Al–TiH2 powder precursors. Adv. Eng. Mater. 6, 400–402.

Kennedy, A.R., Lopez, V.H., 2003. The decomposition behavior of as-received and oxidized TiH2 foaming-agent powder. Mater. Sci. Eng. A 357, 258.

Kennedy, A.R., 2002. The effect of TiH2 heat treatment on gas release and foaming in Al–TiH2 preforms. Scr. Mater. 47, 763–767.

Kenny, L.D., 1996. Mechanical properties of particle stabilized aluminum foam. Mater. Sci. Forum 217–222, 1883–1890.

Kesler, O., Gibson, L.J., 2002. Size effects in metallic foam core sandwich beams. Mater. Sci. Eng. A 236, 228–234.

Khateeb, S.A., Farid, M.M., Selman, J.R., Al-Hallaj, S., 2004. Design and simulation of a lithium-ion battery with a phase change material thermal management system for an electric scooter. J. Power Sources 128, 292.

Khateeb, S.A., Amiruddin, S., Farid, M., et al., 2005. Thermal management of Li-ion battery with phase change material for electric scooters: experimental validation. J. Power Sources 142, 345.

Kim, H.S., Al-Hassani, S.T.S., 2001a. A morphological elastic model of general hexagonal columnar structures. Int. J. Mech. Sci. 43, 1027–1060.

Kim, H.S., Al-Hassani, S.T.S., 2001b. Plastic collapse of cellular structures comprised of doubly tapered struts. Int. J. Mech. Sci. 43, 2453.

Kim, H.S., Al-Hassani, S.T.S., 2002. The effect of doubly tapered strut morphology on the plastic yield surface of cellular materials. Int. J. Mech. Sci. 44, 1559.

Kim, A., Kim, I., 2008. Effect of specimen aspect ratio on fatigue life of closed cell Al–Si–Ca alloy foam. Acta Mech. Sin. 21, 354–358.

Kim, K.T., Welsch, G., 1990. Mater. Lett. 9, 295.

Kim, S.Y., Paek, J.W., Kang, B.H., 2000. Flow and heat transfer correlations for porous Fin in a plate-Fin heat exchanger. J. Heat Transfer 122, 572–578.

Kim, S.Y., Kang, B.H., Kim, J.-H., 2001. Forced convection from aluminum foam materials in an asymmetrically heated channel. Int. J. Heat Mass Transfer 44, 1451–1454.

Kim, S.Y., Paek, J.W., Kang, B.H., 2003. Thermal performance of aluminum-foam heat sinks by forced air cooling. IEEE Trans. Compon. Packag. Technol. 26, 262–267.

Kim, A., Cho, S.S., Lee, H.J., 2004a. Foaming behaviour of Al–Si–Cu–Mg alloys. Mater. Sci. Technol. 20, 1615–1620.

Kim, J.S., Kang, J.H., Kang, S.B., et al., 2004b. Porous TiNi biomaterial produced by self-propagating high-temperature synthesis. Adv. Eng. Mater. 6, 403–406.

Kim, A., Hasan, M.A., Nahm, S.H., Cho, S.S., 2005a. Evaluation of compressive mechanical properties of Al-foams using electrical conductivity. Compos. Struct. 71, 191–198.

Kim, T., Hodson, H.P., Lu, T.J., 2005b. Contribution of vortex structures and flow separation to local and overall pressure and heat transfer characteristics in an ultralightweight lattice material. Int. J. Heat Mass Transfer 48, 4243–4264.

Kim, T., Hodson, H.P., Lu, T.J., 2006. On the prediction of pressure drop across banks of inclined cylinders. Int. J. Heat Fluid Flow 27, 311–318.

Kim, H.G., Lee, T.W., Lee, J.Y., et al., 2012. Microstructural characterization of Ni–22Fe–22Cr–6Al metallic foam by transmission electron microscopy. J. Electron Microsc. 61, 299–304.

Kirkland, N.T., Kolbeinsson, I., Woodfield, T., et al., 2011. Synthesis and properties of topologically ordered porous magnesium. Mater. Sci. Eng. B 176, 1666–1672.

Kishimoto, S., Song, Z.-L., Shinya, N., 2003. Development of metallic closed cellular materials containing organic materials. J. Alloys Compd. 355, 161.

Kitazono, K., Takiguchi, Y., 2006. Strain rate sensitivity and energy absorption of Zn–22Al foams. Scr. Mater. 55, 501.

Kitazono, K., Kitajima, A., Sato, E., et al., 2002. Solid-state diffusion bonding of closed-cell aluminum foams. Mater. Sci. Eng. A 327, 128.

Kitazono, K., Sato, E., Kuribayashi, K., 2003. Application of mean-field approximation to elastic-plastic behavior for closed-cell metal foams. Acta Mater. 51, 4823.

Kitazono, K., Sato, E., Kuribayashi, K., 2004a. Enhanced foaming of cellular metals by internal stress superplasticity. Mater. Sci. Forum 447–448, 541–545.

Kitazono, K., Sato, E., Kuribayashi, K., 2004b. Novel manufacturing process of closed-cell aluminum foam by accumulative roll-bonding. Scr. Mater. 50, 495.

Klinter, A., Leon, C., Drew, R., 2010. The optimum contact angle range for metal foam stabilization: an experimental comparison with the theory. J. Mater. Sci. 45, 2174–2180.

Ko, W.L., 1965. Deformations of foamed elastomers. J. Cell. Plast. 1, 45–50.

Kobashi, M., Kanetake, N., 2002. Processing of intermetallic foam by combustion reaction. Adv. Eng. Mater. 4, 745–747.

Kobashi, M., Kanetake, N., 2010. Foaming Technique of Porous Aluminium/intermetallics Composites by Precursor Method. Proceedings of the 12th international conference on aluminium alloys. Japan Institute of Light Metals, Kyoto, Japan.

Kobashi, M., Kuze, K., Kanetake, N., 2006a. Cell structure control of porous titanium composite synthesised by combustion reaction. Adv. Eng. Mater. 8, 836–840.

Kobashi, M., Wang, R.X., Inagaki, Y., Kanetak, N., 2006b. Effects of processing parameters on pore morphology of combustion synthesized Al–Ni foams. Mater. Trans. 47, 2172–2177.

Kocks, U.F., 1970. The relation between polycrystal deformation and single-crystal deformation. Metall. Trans. 1, 1121–1143.

Köhl, M., Habijan, T., Bram, M., et al., 2009. Powder metallurgical near-net-shape fabrication of porous NiTi shape memory alloys for use as long-term implants by the combination of the metal injection molding process with the space-holder technique. Adv. Eng. Mater. 11, 959–968.

Köhl, M., Bram, M., Moser, A., et al., 2011. Characterization of porous, net-shaped NiTi alloy regarding its damping and energy-absorbing capacity. Mater. Sci. Eng. A 528, 2454–2462.

Kolluri, M., Karthikeyan, S., Ramamurty, U., 2007. Effect of lateral constraint on the mechanical properties of a closed-cell Al foam: I. Experiments. Metall. Mater. Trans. A 38, 2006–2013.

Kolluri, M., Mukheriee, M., Garcia-Moreno, F., et al., 2008. Fatigue of a laterally constrained closed cell aluminum foam. Acta Mater. 56, 1114–1125.

Konstantinidis, I.C., Papadopoulos, D.P., Lefakis, H., Tsipas, D.N., 2005. Model for determining mechanical properties of aluminum closed-cell foams. Theor. Appl. Fract. Mech. 43, 157–167.

Kooistra, G.W., Wadley, H.N.G., 2007. Lattice truss structures from expanded metal sheet. Mater. Des. 28, 507.

Kooistra, G.W., Deshpande, V.S., Wadley, H.N.G., 2004. Compressive behavior of age hardenable tetrahedral lattice truss structures made from aluminium. Acta Mater. 52, 4229.

Kooistra, G.W., Queheillalt, D.T., Wadley, H.N.G., 2008. Shear behavior of aluminum lattice truss sandwich panel structures. Mater. Sci. Eng. A 472, 242–250.

Kopanidis, A., Theodorakakos, A., Gavaises, E., Bouris, D., 2010. 3D numerical simulation of flow and conjugate heat transfer through a pore scale model of high porosity open cell metal foam. Int. J. Heat Mass Transfer 53, 2539–2550.

Körner, C., Singer, R.F., 2000. Processing of metal foams – challenges and opportunities. Adv. Eng. Mater. 2, 159–165.

Körner, C., Thies, M., Singer, R.F., 2002. Modeling of metal foaming with lattice Boltzmann automata. Adv. Eng. Mater. 4, 765–769.

Körner, C., Hirschmann, M., Lamm, M., Singer, R.F., 2003. Magnesium integral foams. In: Banhart, J., Fleck, N., Mortensen, A. (Eds.), Cellular Metals: Manufacture, Properties, Applications, Proc. Conf. Metfoam 2003. Berlin, 23–25 June 2003. Verlag MIT Publishing, Berlin, Germany.

Körner, C., Hirschmann, M., Bräutigam, V., Singer, R.F., 2004. Endogenous particle stabilization during magnesium integral foam production. Adv. Eng. Mater. 6, 385–390.

Korner, C., Arnold, M., Singer, R.F., 2005. Metal foam stabilization by oxide network particles. Mater. Sci. Eng. A 396, 28.

Körner, C., 2008. Foam formation mechanisms in particle suspensions applied to metal foams. Mater. Sci. Eng. A 495, 227–235.

Kovacik, J., Simancik, F., 1998. Aluminium foam-modulus of elasticity and electrical conductivity according to percolation theory. Scr. Mater. 39, 239–246.

Kovacik, J., 1998a. Electrical conductivity of two-phase composite material. Scr. Mater. 39, 153–157.

Kovacik, J., 1998b. The tensile behaviour of porous metals made by Gasar process. Acta Mater. 46, 5413–5422.

Koza, E., Leonowicz, M., Wojciechowski, S., Simancik, F., 2004. Compressive strength of aluminium foams. Mater. Lett. 58, 132.

Kraft, O., Gruber, P.A., Munig, R., Weygand, D., 2010. Plasticity in confined dimensions. Annu. Rev. Mater. Res. 40, 293–317.

Kraynik, A.M., Warren, W.E., 1994. The elastic behavior of low-density cellular plastics. In: Hilyard, N.C., Cunningham, A. (Eds.), Low Density Cellular Plastics – Physical Basis of Behavior. Chapman-Hall, London, U.K.

Kraynik, A.M., Neilsen, M.K., Reinelt, D.A., Warren, W.E., 1999. Foam micromechanics. In: Sadoc, J.F., Rivier, N. (Eds.), Foams and Emulsions, Proc. NATO Advanced Study Institute on Foams, Emulsions and Cellular Materials. Cargese, Corsica, France. Kluwer Academic Publishers, Dordrecht, the Netherlands.

Kraynik, A.M., 2003. Foam structure: from soap froth to solid foams. MRS Bull. 28, 275–278.

Kraynik, A., 2006. The structure of random foam. Adv. Eng. Mater. 8, 900–906.

Krishna, B.V., Bose, S., Bandyopadhyay, A., 2007. Strength of open-cell 6101 aluminum foams under free and constrained compression. Mater. Sci. Eng. A 452–453, 178–188.

Krishnan, S., Murthy, J.Y., Garimella, S.V., 2006. Direct simulation of transport in open-cell foam. J. Heat Transfer (ASME) 128, 793–799.

Krishnan, S., Garimella, S.V., Murthy, J.Y., 2008. Simulation of thermal transport in open-cell metal foams: effect of periodic unit-cell structure. J. Heat Transfer 130, 024503–024505.

Kriszt, B., Foroughi, B., Kottar, A., Degischer, H.P., 1999. Mechanical behaviour of aluminium foams under uniaxial compression. In: Clyne, T.W., Simancik, F. (Eds.), Euromat 99. 27–30 Sept. 1999, Munich, Germany. DGM/Wiley-VCH, Weinheim, Germany.

Kriszt, B., Foroughi, B., Faure, K., Degischer, H.P., 2000. Behaviour of aluminium foam under uniaxial compression. Mater. Sci. Technol. 16, 792–796.

Krupp, U., Ohrndorf, A., Guillèn, T., et al., 2006. Isothermal and thermomechanical fatigue behavior of open-cell metal sponges. Adv. Eng. Mater. 8, 821–827.

Kuchek, H.A., 1966. Method of Making Porous Metallic Article. US Patent 3,236,706.

Kuipers, M., 1964. Note on the macroscopic elastic constants of materials with a foam-like structure. Appl. Sci. Res. 13, 138–143.

Kujime, T., Hyun, S.K., Nakajima, H., 2006. Fabrication of lotus-type porous carbon steel by the continuous zone melting method and its mechanical properties. Metall. Mater. Trans. A 37, 393–398.

Kujime, T., Tane, M., Hyun, S.K., Nakajima, H., 2007. Three-dimensional image-based modeling of lotus-type porous carbon steel and simulation of its mechanical behavior by finite element method. Mater. Sci. Eng. A 460–461, 220–226.

Kumar, A., Reddy, R.G., 2003. Modeling of polymer electrolyte membrane fuel cell with metal foam in the flow-field of the bipolar/end plates. J. Power Sources 114, 54–62.

Kumar, A., Reddy, R.G., 2004. Materials and design development for bipolar/end plates in fuel cells. J. Power Sources 129, 62.

Kumar, P.S., Ramachandra, S., Ramamurty, U., 2003. Effect of displacement-rate on the indentation behavior of an aluminum foam. Mater. Sci. Eng. A 347, 330.

Kumar, G., Desai, A., Schroers, J., 2011a. Bulk metallic glass: the smaller the better. Adv. Mater. 23, 461–476.

Kumar, G.S.V., Chakraborty, M., Moreno, F.G., Banhart, J., 2011b. Foamability of MgAl2O4 (Spinel)-Reinforced aluminum alloy composites. Metall. Mater. Trans. A 42, 2898–2908.

Kunze, H.D., Baumeister, J., Banhart, J., Weber, M., 1993. Powder Metall. Int. 25, 182–185.

Kurtbas, I., Celik, N., 2009. Experimental investigation of forced and mixed convection heat transfer in a foam-filled horizontal rectangular channel. Int. J. Heat Mass Transfer 52, 1313–1325.

Kushch, V.I., Podoba, Y.O., Shtern, M.B., 2008. Effect of micro-structure on yield strength of porous solid: a comparative study of two simple cell models. Comput. Mater. Sci. 42, 113–121.

Kwok, P.J., Oppenheimer, S.M., Dunand, D.C., 2008. Porous titanium by electro-chemical dissolution of steel space-holders. Adv. Eng. Mater. 10, 820–825.

Kwon, Y.S., Suk, M.J., Kim, J.S., Shin, C.G., 2003a. Fabrication of porous material with porosity gradient by pulsed electric current sintering method. In: Banhart, J., Fleck, N., Mortensen, A. (Eds.), Cellular Metals: Manufacture, Properties, Applications, Proc. Conf. Metfoam 2003. Berlin, 23–25 June 2003. Verlag MIT Publishing, Berlin, Germany.

Kwon, Y.W., Cooke, R.E., Park, C., 2003b. Representative unit-cell models for open-cell metal foams with or without elastic filler. Mater. Sci. Eng. A 343, 63–70.

Lacroix, M., Nguyen, P., Schweich, D., et al., 2007. Pressure drop measurements and modeling on SiC foams. Chem. Eng. Sci. 62, 3259–3267.

Lafdi, K., Mesalhy, O., Shaikh, S., 2007. Experimental study on the influence of foam porosity and pore size on the melting of phase change materials. J. Appl. Phys. 102, 083549.

Lage, J.L., Weinert, A.K., Price, D.C., Weber, R.M., 1996. Numerical study of a low permeability microporous heat sink for cooling phased-array radar systems. Int. J. Heat Mass Transfer 39, 3633–3647.

Lage, J.L., Antohe, B.V., Nield, D.A., 1997. Two types of nonlinear pressure-drop versus flow-rate relation observed for saturated porous media. Trans. ASME – J. Fluids Eng. 119, 700–706.

Lage, J.L., 1995. The fundamental theory of flow through permeable media from Darcy to turbulence. In: Ingham, D.B., Pop, I. (Eds.), Transport Phenomena in Porous Media. Elsevier Science, Oxford.

Lakes, R.S., 1983. Size effects and micromechanics of a porous solid. J. Mater. Sci. 18, 2572–2580.

Lakes, R.S., 1986. Experimental microelasticity of two porous solids. Int. J. Solids Struct. 22, 55–63.

Lakes, R., 1987. Foam structures with a negative Poisson's ratio. Science 235, 1038–1040.

Lakes, R., 1993. Materials with structural hierarchy. Nature 361, 511–515.

Lakes, R., 1996. Cellular solid structures with unbounded thermal expansion. J. Mater. Sci. Lett. 15, 475–477.

Lakes, R., 2007. Cellular solids with tunable positive or negative thermal expansion of unbounded magnitude. Appl. Phys. Lett. 90, 221905-3.

Langlois, S., Coeuret, F., 1989. Flow-through and flow-by porous electrodes of nickel foam. I. Material characterization. J. Appl. Electrochem. 19, 43–50.

Lankford, J., Dannemann, K.A., 1998. Strain rate effects in porous materials. In: Schwartz, D.S., Shih, D.H., Evans, A.G., Wadley, H.G. (Eds.), Porous and Cellular Materials for Structural Applications. San Francisco, CA, USA, Materials Research Society Symposium Proceedings, vol. 521. (Warrendale PA, USA).

Laptev, A., Bram, M., Buchkremer, H.P., Stover, D., 2004. Study of production route for titanium parts combining very high porosity and complex shape. Powder Metall. 47, 85–92.

Laptev, A., Vyal, O., Bram, M., et al., 2005. Green strength of powder compacts provided for production of highly porous titanium parts. Powder Metall. 48, 358–364.

Laschet, G., Kashko, T., Angel, S., et al., 2008. Microstructure based model for permeability predictions of open-cell metallic foams via homogenization. Mater. Sci. Eng. A 472, 214–226.

Laschet, G., Sauerhering, J., Reutter, O., et al., 2009. Effective permeability and thermal conductivity of open-cell metallic foams via homogenization on a microstructure model. Comput. Mater. Sci. 45, 597–603.

Lassance, D., Fabregue, D., Delannay, F., Pardoen, T., 2007. Micromechanics of room and high temperature fracture in 6xxx Al alloys. Prog. Mater. Sci. 52, 62–129.

Lausic, A.T., Bouwhuis, B.A., McCrea, J.L., et al., 2012. Mechanical anisotropy in electrodeposited nanocrystalline metal/metal composite foams. Mater. Sci. Eng. A 552, 157–163.

Leal-Calderon, F., Schmitt, V., 2008. Solid-stabilized emulsions. Curr. Opin. Colloid Interface Sci. 13, 217–227.

Leblond, J., Perrin, G., Suquet, P., 1994. Exact results and approximate models for porous viscoplastic solids. Int. J. Plast. 10, 213–235.

Leblond, J., Perrin, G., Devaux, J., 1995. An improved Gurson-type model for hardenable ductile metals. Eur. J. Mech. – A/Solids 14, 499–527.

Lederman, J.M., 1971. The prediction of the tensile properties of flexible foams. J. Appl. Polym. Sci. 15, 693–703.

Lee, M.H., Sordelet, D.J., 2006a. Nanoporous metallic glass with high surface area. Scr. Mater. 55, 947.

Lee, M.H., Sordelet, D.J., 2006b. Synthesis of bulk metallic glass foam by powder extrusion with a fugitive second phase. Appl. Phys. Lett. 89, 021921–021923.

Lee, J., Sung, N., Cho, G., Oh, K., 2006a. Modeling of filtration for a metal foam diesel particulate filter. Key Eng. Mater. 326–328, 1153–1156.

Lee, S., Barthelat, F., Hutchinson, J.W., Espinosa, H.D., 2006b. Dynamic failure of metallic pyramidal truss core materials – experiments and modeling. Int. J. Plast. 22, 2118–2145.

Lee, S., Barthelat, F., Moldovan, N., et al., 2006c. Deformation rate effects on failure modes of open-cell Al foams and textile cellular materials. Int. J. Solids Struct. 43, 53–73.

Lee, D., Wei, X., Chen, X., et al., 2007a. Microfabrication and mechanical properties of nanoporous gold at the nanoscale. Scr. Mater. 56, 437.

Lee, Y.-H., Lee, B.-K., Jeon, I., Kang, K.-J., 2007b. Wire-woven bulk Kagome truss cores. Acta Mater. 55, 6084–6094.

Lefebvre, L.P., Baril, E., 2008. Efffect of oxygen concentration and distribution on the compression properties of titanium foams. Adv. Eng. Mater. 10, 868–876.

Lefebvre, L.P., Blouin, A., Rochon, S.M., Bureau, M.N., 2006. Elastic response of titanium foams during compression tests and using laser-ultrasonic probing. Adv. Eng. Mater. 8, 841–846.

Lefebvre, L.P., Banhart, J., Dunand, D.C. (Eds.), 2008a. Metfoam 2007-Porous Metals and Metallic Foams, Proc. 2007 Metfoam Conf., Montreal, Sept. 5–7, 2007. DESTech Publications, Inc, Lancaster, PA.

Lefebvre, L.P., Banhart, J., Dunand, D.C., 2008b. Porous metals and metallic foams: current status and recent developments. Adv. Eng. Mater. 10, 775–787.

Lefebvre, L.-P., 2013. Porous metals and metallic foams in orthopedic applications. In: Dukhan, N. (Ed.), Metal Foams – Fundamentals and Applications. DEStech Publications Inc, Lancaster PA, USA.

Legros, M., 2011. Small-scale plasticity – a review. In: Thomas, O., Ponchet, A., Forest, S. (Eds.), Mechanics of Nano-objects. Transvalor – Presses des Mines, Paris, France.

Lehmhus, D., Banhart, J., 2003. Properties of heat-treated aluminium foams. Mater. Sci. Eng. A 349, 98.

Lehmhus, D., Busse, M., 2004. Potential new matrix alloys for production of PM aluminium foams. Adv. Eng. Mater. 6, 391–396.

Lehmhus, D., 2010. Dynamic collapse mechanisms in foam expansion. Adv. Eng. Mater. 12, 465–471.

Leitlmeier, D., Degischer, H.P., Frankl, H.J., 2002. Development of a foaming process for particulate reinforced aluminium melts. Adv. Eng. Mater. 4, 735–740.

LeMay, J.D., Hopper, R.W., Hrubesh, L.W., Pekala, R.W., 1990. Low-density microcellular materials. Mater. Res. Soc. Bull. 15, 19–45.

Lemay, J.D., 1991. Mechanical structure–property relationships of microcellular, low density foams. In: Sieradzki, K., Green, D.J., Gibson, L.J. (Eds.), Nov. 1990 Symposium on Mechanical Properties of Porous and Cellular Materials. Materials Research Society, Boston.

Lemlich, R., 1978. A theory for the limiting conductivity of polyhedral foam at low density. J. Colloid Interface Sci. 64, 107–110.

Lenel, F.V., 1980. Powder Metallurgy. Metal Powder Industries Federation, Princeton, NJ.

Leong, K.C., Jin, L.W., 2006a. Characteristics of oscillating flow through a channel filled with open-cell metal foam. Int. J. Heat Fluid Flow 27, 144–153.

Leong, K.C., Jin, L.W., 2006b. Effect of oscillatory frequency on heat transfer in metal foam heat sinks of various pore densities. Int. J. Heat Mass Transfer 49, 671–681.

Levine, B., 2008. A new era in porous materials: applications in orthopaedics. Adv. Eng. Mater. 10, 788–792.

Lhuissier, P., Salvo, L., Brechet, Y., 2010. Sintered hollow spheres: random stacking behaviour under uniaxial tensile loading. Scr. Mater. 63, 277–280.

Li, J.C., Dunand, D.C., 2011. Mechanical properties of directionally freeze-cast titanium foams. Acta Mater. 59, 146–158.

Li, Q.M., Meng, H., 2002. Attenuation or enhancement' a one-dimensional analysis on shock transmission in the solid phase of a cellular material. Int. J. Impact Eng. 27, 1049–1065.

Li, H., Misra, A., 2010. A dramatic increase in the strength of a nanoporous Pt'Ni alloy induced by annealing. Scr. Mater. 63, 1169–1172.

Li, C.-F., Zhu, Z.-G., 2005. Apparent electrical conductivity of porous titanium prepared by the powder metallurgy method. Chin. Phys. Lett. 22, 2647–2650.

Li, C., Zhu, Z., 2006. Dynamic Young's modulus of open-porosity titanium measured by the electromagnetic acoustic resonance method. J. Porous Mater. 13, 21–26.

Li, J.R., Cheng, H.F., Yu, J.L., Han, F.S., 2003a. Effect of dual-size cell mix on the stiffness and strength of open-cell aluminum foams. Mater. Sci. Eng. A 362, 240–248.

Li, K., Gao, X.L., Roy, A.K., 2003b. Micromechanics model for three-dimensional open-cell foams using a tetrakaidecahedral unit cell and Castigliano's second theorem. Compos. Sci. Technol. 63, 1769–1781.

Li, K., Gao, X.L., Subhash, G., 2005. Effects of cell shape and cell wall thickness variations on the elastic properties of two-dimensional cellular solids. Int. J. Solids Struct. 42, 1777–1795.

Li, K., Gao, X.L., Subhash, G., 2006a. Effects of cell shape and strut cross-sectional area variations on the elastic properties of three-dimensional open-cell foams. J. Mech. Phys. Solids 54, 783–806.

Li, L.P., Wijn, J.R. d., Blitterswijk, C.A. v., Groot, K. d., 2006b. Porous Ti6Al4V scaffold directly fabricating by rapid prototyping: preparation and in vitro experiment. Biomaterials 27, 1223–1235.

Li, K., Gao, X.L., Wang, J., 2007. Dynamic crushing behavior of honeycomb structures with irregular cell shapes and non-uniform cell wall thickness. Int. J. Solids Struct. 44, 5003–5026.

Li, Q., Chen, E., Bice, D., Dunand, D., 2008. Mechanical properties of cast Ti–6Al–4V lattice block structures. Metall. Mater. Trans. A 39, 441–449.

Li, S.J., Murr, L.E., Cheng, X.Y., et al., 2012a. Compression fatigue behavior of Ti–6Al–4V mesh arrays fabricated by electron beam melting. Acta Mater. 60, 793–802.

Li, W.L., Lu, K., Walz, J.Y., 2012b. Freeze casting of porous materials: review of critical factors in microstructure evolution. Int. Mater. Rev. 57, 37–60.

Li, W.Q., Qu, Z.G., He, Y.L., Tao, W.Q., 2012c. Experimental and numerical studies on melting phase change heat transfer in open-cell metallic foams filled with paraffin. Appl. Therm. Eng. 37, 1–9.

Liebscher, A., Redenbach, C., 2012. 3D image analysis and stochastic modelling of open foams. Int. J. Mater. Res. formerly Z. Metallkd. 103, 155–161.

Liebscher, A., Proppe, C., Redenbach, C., Schwarzer, D., 2012. Uncertainty quantification for metal foam structures by means of image analysis. Probabilist. Eng. Mech. 28, 143–151.

Lim, T.J., Smith, B., McDowell, D.L., 2002. Behavior of a random hollow sphere metal foam. Acta Mater. 50, 2867–2879.

Lim, T.-C., 2012. Negative thermal expansion structures constructed from positive thermal expansion trusses. J. Mater. Sci. 47, 368–373.

Lin, J.-Y., Huang, J.-S., 2005a. Creep of hexagonal honeycombs with Plateau borders. Compos. Struct. 67, 477–484.

Lin, J.-Y., Huang, J.-S., 2005b. Stress relaxation of cellular materials. J. Compos. Mater. 39, 233–245.

Lin, J.-Y., Huang, J.-S., 2006. Creep-buckling of hexagonal honeycombs with Plateau borders. Compos. Sci. Technol. 66, 51–60.

Lin, J.-g., Zhang, Y.-f., Ma, M., 2010. Preparation of porous Ti35Nb alloy and its mechanical properties under monotonic and cyclic loading. Trans. Nonferrous Met. Soc. China 20, 390–394.

Liu, P.S., Liang, K.M., 2000. Preparation and corresponding structure of nickel foam. Mater. Sci. Technol. 16, 575–578.

Liu, S., Masliyah, J., 2005. Dispersion in Porous Media. Handbook of Porous Media, Second ed. CRC Press.

Liu, C.S., Zhu, Z.G., Han, F.S., Banhart, J., 1998a. Internal friction of foamed aluminium in the range of acoustic frequencies. J. Mater. Sci. 33, 1769–1775.

Liu, C.S., Zhu, Z.G., Han, F.S., Banhart, J., 1998b. Low-frequency internal friction of foamed Al. Philos. Mag. A 78, 1329–1337.

Liu, P.S., Li, T.F., Fu, C., 1999. Relationship between electrical resistivity and porosity for porous metals. Mater. Sci. Eng. A 268, 208–215.

Liu, C.S., Zhu, Z.G., Han, F.S., Banhart, J., 2000a. Study on nonlinear damping properties of foamed Al. Philos. Mag. A 80, 1085–1092.

Liu, P., Chen, H., Liang, K., et al., 2000b. Relationship between apparent electrical conductivity and preparation conditions for nickel foam. J. Appl. Electrochem. 30, 1183–1186.

Liu, P.S., Liang, K.M., Tu, S.W., et al., 2001. Relationship between tensile strength and preparation conditions for nickel foam. Mater. Sci. Technol. 17, 1069–1072.

Liu, J.F., Wu, W.T., Chiu, W.C., Hsieh, W.H., 2006. Measurement and correlation of friction characteristic of flow through foam matrixes. Exp. Therm. Fluid Sci. 30, 329.

Liu, P., He, G., Wu, L.H., 2008. Fabrication of sintered steel wire mesh and its compressive properties. Mater. Sci. Eng. A 489, 21–28.

Liu, J., Yu, S., Song, Y., et al., 2009a. Dynamic compressive strength of Zn–22Al foams. J. Alloys Compd. 476, 466–469.

Liu, P., He, G., Wu, L., 2009b. Uniaxial tensile stress-strain behavior of entangled steel wire material. Mater. Sci. Eng. A 509, 69–75.

Liu, P., He, G., Wu, L.H., 2009c. Impact behavior of entangled steel wire material. Mater. Charact. 60, 900–906.

Liu, Y.D., Yu, J.L., Zheng, Z.J., Li, J.R., 2009d. A numerical study on the rate sensitivity of cellular metals. Int. J. Solids Struct. 46, 3988–3998.

Liu, P., Tan, Q., Wu, L., He, G., 2010. Compressive and pseudo-elastic hysteresis behavior of entangled titanium wire materials. Mater. Sci. Eng. A 527, 3301–3309.

Liu, Y., He, X., Tang, H., Huang, B., 2011. Synthesis of carbon nanotubes by fine Ni particles in Ni3Al foam. Int. J. Mater. Res. formerly Z. Metallkd. 102, 1174–1179.

Liu, J., Zhu, X., Huang, Z., et al., 2012. Characterization and property of microarc oxidation coatings on open-cell aluminum foams. Journal of Coatings Technology and Research 9, 357–363.

Liu, P.S., 2004. Tensile fracture behavior of foamed metallic materials. Mater. Sci. Eng. A 384, 352.

Liu, P.S., 2010. A new method for calculating the specific surface area of porous metal foams. Philos. Mag. Lett. 90, 447–453.

Longerich, S., Piontek, D.P., Ohse, A., et al., 2007. Joining strategies for open porous metallic foams on iron and nickel base materials. Adv. Eng. Mater. 9, 670–678.

Lopatnikov, S.L., Gama, B.A., Jahirul Haque, M., et al., 2003. Dynamics of metal foam deformation during Taylor cylinder‘Hopkinson bar impact experiment. Compos. Struct. 61, 61–71.

Lopatnikov, S.L., Gama, B.A., Haque, M.J., et al., 2004. High-velocity plate impact of metal foams. Int. J. Impact Eng. 30, 421.

Lopatnikov, S.L., Gama, B.A., Gillespie Jr, J.W., 2007. Modeling the progressive collapse behavior of metal foams. Int. J. Impact Eng. 34, 587–595.

Loretz, M., Coquard, R., Baillis, D., Maire, E., 2008a. Metallic foams: radiative properties/comparison between different models. J. Quant. Spectrosc. Radiat. Transfer 109, 16–27.

Loretz, M., Maire, E., Baillis, D., 2008b. Analytical modeling of the radiative properties of metallic foams: contributions of X-ray tomography. Adv. Eng. Mater. 10, 352–360.

Lu, T.J., Chen, C., 1999. Thermal transport and fire retardance properties of cellular aluminium alloys. Acta Mater. 47, 1469–1485.

Lu, T.J., Ong, J.M., 2001. Characterization of close-celled cellular aluminum alloys. J. Mater. Sci. 36, 2773–2786.

Lu, W., Zhao, C.-Y., 2009. Numerical modelling of flow boiling heat transfer in horizontal metal-foam tubes. Adv. Eng. Mater. 11, 832–836.

Lu, T.J., Stone, H.A., Ashby, M.F., 1998. Heat transfer in open-cell metal foams. Acta Mater. 46, 3619–3635.

Lu, T.J., Valdevit, L., Evans, A.G., 2005. Active cooling by metallic sandwich structures with periodic cores. Prog. Mater. Sci. 50, 789.

Lu, H.-B., Li, Y., Wang, F.-H., 2006a. Dealloying behaviour of Cu–20Zr alloy in hydrochloric acid solution. Corros. Sci. 48, 2106.

Lu, W., Zhao, C.Y., Tassou, S.A., 2006b. Thermal analysis on metal-foam filled heat exchangers. Part I: metal-foam filled pipes. Int. J. Heat Mass Transfer 49, 2751–2761.

Lu, H.-B., Li, Y., Wang, F.-H., 2007a. Synthesis of porous copper from nanocrystalline two-phase Cu–Zr film by dealloying. Scr. Mater. 56, 165.

Lu, X., Bischoff, E., Spolenak, R., Balk, T.J., 2007b. Investigation of dealloying in Au–Ag thin films by quantitative electron probe microanalysis. Scr. Mater. 56, 557–560.

Lu, T.J., 1999. Heat transfer efficiency of metal honeycombs. Int. J. Heat Mass Transfer 42, 2031–2040.

Lu, T., 2002. Ultralight porous metals: from fundamentals to applications. Acta Mech. Sin. 18, 457–479.

Luxner, M., Stampfl, J., Pettermann, H., 2005. Finite element modeling concepts and linear analyses of 3D regular open cell structures. J. Mater. Sci. 40, 5859.

Luxner, M.H., Stampfl, J., Pettermann, H.E., 2007. Numerical simulations of 3D open cell structures – influence of structural irregularities on elasto-plasticity and deformation localization. Int. J. Solids Struct. 44, 2990–3003.

Luxner, M.H., Stampfl, J. r., Pettermann, H.E., 2009a. Nonlinear simulations on the interaction of disorder and defects in open cell structures. Comput. Mater. Sci. 47, 418–428.

Luxner, M.H., Woesz, A., Stampfl, J., et al., 2009b. A finite element study on the effects of disorder in cellular structures. Acta Biomater. 5, 381–390.

Luyten, J., Thijs, I., Ravelingien, M., Mullens, S., 2011. Bone engineering with porous ceramics and metals. Adv. Eng. Mater. 13, 1002–1007.

Ma, L., He, D., 1994. Fabrication and pore structure control of new type aluminium foams (English abstract). Chin. J. Mater. 8, 11–17.

Ma, J., Lim, L.C., 2002. Effect of particle size distribution on sintering of agglomerate-free submicron alumina powder compacts. J. Eur. Ceram. Soc. 22, 2197.

Ma, A.B., Gan, H., Imura, T., et al., 1997. Wear properties of aluminum alloy reinforced by short alumina fibers with gradient distribution. Trans. Jpn. Inst. Met. 38, 812–816.

Ma, L., Song, Z., He, D., 1999. Cellular structure controllable aluminum foams produced by high pressure infiltration process. Scr. Mater. 41, 785–789.

Ma, G.W., Ye, Z.Q., Shao, Z.S., 2009. Modeling loading rate effect on crushing stress of metallic cellular materials. Int. J. Impact Eng. 36, 775–782.

Mackenzie, J.K., 1950. The elastic constants of a solid containing spherical holes. Proc. Phys. Soc. London B63, 2–11.

Mahdi, H., Lopez, P., Fuentes, A., Jones, R., 2006. Thermal performance of aluminium-foam CPU heat exchangers. Int. J. Energy Res. 30, 851–860.

Mahjoob, S., Vafai, K., 2008. A synthesis of fluid and thermal transport models for metal foam heat exchangers. Int. J. Heat Mass Transfer 51, 3701–3711.

Maire, E., Wattebled, F., Buffière, J.Y., Peix, G., 1999. Deformation of a metallic foam studied by X-ray computed tomography and finite element calculations. In: Clyne, T.W., Simancik, F. (Eds.), Euromat 99. 27–30 Sept. 1999, Munich, Germany. DGM/Wiley-VCH, Weinheim, Germany.

Maire, E., Elmoutaouakkil, A., Fazekas, A., Salvo, L., 2003a. In situ X-ray tomography measurements of deformation in cellular solids. MRS Bull. 28, 284–289.

Maire, E., Fazekas, A., Salvo, L., et al., 2003b. X-ray tomography applied to the characterization of cellular materials. Related finite element modeling problems. Compos. Sci. Technol. 63, 2431.

Maiti, S.K., Ashby, M.F., Gibson, L.J., 1984. Fracture toughness of brittle cellular solids. Scr. Metall. 18, 213–217.

Malachevsky, M.T., D'Ovidio, C.A., 2009. Thermal evolution of titanium hydride optimized for aluminium foam fabrication. Scr. Mater. 61, 1–4.

Mancin, S., Zilio, C., Cavallini, A., Rossetto, L., 2010a. Heat transfer during air flow in aluminum foams. Int. J. Heat Mass Transfer 53, 4976–4984.

Mancin, S., Zilio, C., Cavallini, A., Rossetto, L., 2010b. Pressure drop during air flow in aluminum foams. Int. J. Heat Mass Transfer 53, 3121–3130.

Mancin, S., Zilio, C., Rossetto, L., Cavallini, A., 2011. Heat transfer performance of aluminum foams. J. Heat Transfer 133, 060904–060909.

Mancin, S., Zilio, C., Diani, A., Rossetto, L., 2012a. Experimental air heat transfer and pressure drop through copper foams. Exp. Therm. Fluid Sci. 36, 224–232.

Mancin, S., Zilio, C., Rossetto, L., Cavallini, A., 2012b. Foam height effects on heat transfer performance of 20-ppi aluminum foams. Appl. Therm. Eng. 49, 55–60.

Mangipudi, K.R., Onck, P.R., 2011a. Multiscale modelling of damage and failure in two-dimensional metallic foams. J. Mech. Phys. Solids 59, 1437–1461.

Mangipudi, K.R., Onck, P.R., 2011b. Notch sensitivity of ductile metallic foams: a computational study. Acta Mater. 59, 7356–7367.

Mangipudi, K.R., van Buuren, S.W., Onck, P.R., 2010. The microstructural origin of strain hardening in two-dimensional open-cell metal foams. Int. J. Solids Struct. 47, 2081–2096.

Marcadon, V., 2011. Mechanical modelling of the creep behaviour of hollow-sphere structures. Comput. Mater. Sci. 50, 3005–3015.

Margevicius, R.W., Stanek, P.W., Jacobson, L.A., 1998. Effects of thermomechanical processing on the resulting mechanical properties of 6101 aluminum foam. In: Schwartz, D.S., Shih, D.H., Evans, A.G., Wadley, H.G. (Eds.), Porous and Cellular Materials for Structural Applications. San Francisco, CA, USA, Materials Research Society Symposium Proceedings, vol. 521. (Warrendale PA, USA).

Markaki, A.E., Clyne, T.W., 2001. The effect of cell wall microstructure on the deformation and fracture of aluminium-based foams. Acta Mater. 49, 1677–1686.

Markaki, A.E., Clyne, T.W., 2004. Magneto-mechanical stimulation of bone growth in a bonded array of ferromagnetic fibres. Biomaterials 25, 4805–4815.

Markaki, A.E., Clyne, T.W., 2005. Magneto-mechanical actuation of bonded ferromagnetic fibre arrays. Acta Mater. 53, 877–889.

Markaki, A.E., Gergely, V., Cockburn, A., Clyne, T.W., 2003. Production of a highly porous material by liquid phase sintering of short ferritic stainless steel fibres and a preliminary study of its mechanical behaviour. Compos. Sci. Technol. 63, 2345–2351.

Marmottant, A., Despois, J.F., Salvo, L., et al., 2005. In-situ tomography investigation of replicated aluminium foams: quantitative analysis of localization using 3D strain mapping. In: Nakajima, H., Kanetake, M. (Eds.), 4th Int. Conf. on Porous Metals and Metal Foaming Technology (Metfoam 2005). Kyoto, Japan. Japan Institute of Metals, Sendai, Japan.

Marmottant, A., Salvo, L., Martin, C.L., Mortensen, A., 2008. Coordination measurements in compacted NaCl irregular powders using X-ray microtomography. J. Eur. Ceram. Soc. 28, 2441–2449.

Mathur, A., Erlebacher, J., 2007. Size dependence of effective Young's modulus of nanoporous gold. Appl. Phys. Lett. 90, 061910–061913.

Matijasevic, B., Banhart, J., 2006. Improvement of aluminium foam technology by tailoring of blowing agent. Scr. Mater. 54, 503.

Matijasevic-Lux, B., Banhart, J., Fiechter, S., et al., 2006. Modification of titanium hydride for improved aluminium foam manufacture. Acta Mater. 54, 1887.

Matsumoto, Y., Brothers, A.H., Stock, S.R., Dunand, D.C., 2007. Uniform and graded chemical milling of aluminum foams. Mater. Sci. Eng. A 447, 150.

Maurer, M., Zhao, L., Lugscheider, E., 2002. Surface refinement of metal foams. Adv. Eng. Mater. 4, 791–797.

McCullough, K.Y.G., Fleck, N.A., Ashby, M.F., 1999a. Toughness of aluminium alloy foams. Acta Mater. 47, 2331–2343.

McCullough, K.Y.G., Fleck, N.A., Ashby, M.F., 1999b. Uniaxial stress–strain behaviour of aluminium alloy foams. Acta Mater. 47, 2323–2330.

McCullough, K.Y.G., Fleck, N.A., Ashby, M.F., 2000. Stress-life fatigue behaviour of aluminum alloy foams. Fatigue and Fracture of Engineering Materials and Structures 199.

McElwain, D.L.S., Roberts, A.P., Wilkins, A.H., 2006a. Yield criterion of porous materials subjected to complex stress states. Acta Mater. 54, 1995–2002.

McElwain, D.L.S., Roberts, A.P., Wilkins, A.H., 2006b. Yield functions for porous materials with cubic symmetry using different definitions of yield. Adv. Eng. Mater. 8, 871–876.

McLachlan, D.S., Blaszkiewicz, M., Newnham, R.E., 1990. Electrical resistivity of composites. J. Am. Ceram. Soc. 73, 2187.

McLaughlin, R., 1977. A study of the differential scheme for composite materials. Int. J. Eng. Sci. 15, 237–244.

McShane, G.J., Radford, D.D., Deshpande, V.S., Fleck, N.A., 2006. The response of clamped sandwich plates with lattice cores subjected to shock loading. European Journal of Mechanics – A/Solids 25, 215.

Medraj, M., Baril, E., Loya, V., Lefebvre, L.-P., 2007. The effect of microstructure on the permeability of metallic foams. J. Mater. Sci. 42, 4372.

Meguid, S.A., Cheon, S.S., El-Abbasi, N., 2002. FE modelling of deformation localization in metallic foams. Finite Elem. Anal. Des. 38, 631–643.

Meller, A.D., 1926. Produit métallique pour l'obtention d'objets laminés, moulés ou autres, et procédés pour sa fabrication. In: Direction de la Propriété Industrielle, R.F. Ed. France.

Menges, G., Knipschild, F., 1975. Estimation of the mechanical properties for rigid polyurethane foams. Polym. Eng. Sci. 15, 623–627.

Michailidis, N., Stergioudi, F., 2011. Establishment of process parameters for producing Al-foam by dissolution and powder sintering method. Mater. Des. 32, 1559–1564.

Michailidis, N., Stergioudi, F., Omar, H., Tsipas, D.N., 2008. Investigation of the mechanical behaviour of open-cell Ni foams by experimental and FEM procedures. Adv. Eng. Mater. 10, 1122–1126.

Michailidis, N., Stergioudi, F., Omar, H., Tsipas, D., 2010a. FEM modeling of the response of porous Al in compression. Comput. Mater. Sci. 48, 282–286.

Michailidis, N., Stergioudi, F., Omar, H., Tsipas, D.N., 2010b. An image-based reconstruction of the 3D geometry of an Al open-cell foam and FEM modeling of the material response. Mech. Mater. 42, 142–147.

Michailidis, N., Stergioudi, F., Tsouknidas, A., 2011. Deformation and energy absorption properties of powder-metallurgy produced Al foams. Mater. Sci. Eng. A 528, 7222–7227.

Michailidis, N., 2011. Strain rate dependent compression response of Ni-foam investigated by experimental and FEM simulation methods. Mater. Sci. Eng. A 528, 4204–4208.

Michaud, V.J., 2011. Chapter 14: permeability properties of reinforcements in composites. In: Boisse, P. (Ed.), Composite Reinforcements for Optimum Performance. Woodhead Publishing, Oxford UK.

Michel, J.C., Suquet, P.M., 1992. The constitutive law of nonlinear viscous and porous materials. J. Mech. Phys. Solids 40, 783–812.

Microcell, 1992. Microcell Technology. Commercial Literature, Edison, N.J., USA.

Miller, R.E., Hutchinson, J.W., 1998. A continuum plasticity model for the constitutive behaviour of foamed metals. Mater. Res. Soc. Symp. Proc. 521.

Miller, W., Mackenzie, D.S., Smith, C.W., Evans, K.E., 2008. A generalised scale-independent mechanism for tailoring of thermal expansivity: positive and negative. Mech. Mater. 40, 351–361.

Miller, R.E., 2000. A continuum plasticity model for the constitutive and indentation behaviour of foamed metals. Int. J. Mech. Sci. 42, 729–754.

Mills, N.J., Zhu, H.X., 1999a. The compression of closed-cell polymer foams. In: Sadoc, J.F., Rivier, N. (Eds.), Foams and Emulsions, Proc. NATO Advanced Study Institute on Foams, Emulsions and Cellular Materials. Cargese, Corsica, France. Kluwer Academic Publishers, Dordrecht, the Netherlands.

Mills, N.J., Zhu, H.X., 1999b. The high strain compression of closed-cell polymer foams. J. Mech. Phys. Solids 47, 669–695.

Milton, G.W., 1984. Correlation of the electromagnetic and elastic properties of composites and microgeometries corresponding with effective medium approximations. In: Johnson, D., Sen, P. (Eds.), Physics and Chemistry of Porous Media. American Institute of Physics, New York.

Milton, G.W., 1992. Composite materials with poisson's ratios close to − 1. J. Mech. Phys. Solids 40, 1105–1137.

Milton, G.W., 2002. The Theory of Composites. Cambridge University Press, Cambridge, UK.

Miwa, S., Revankar, S., 2009. Hydrodynamic characterization of nickel metal foam, part 1: single-phase permeability. Transp. Porous Media 80, 269–279.

Miwa, S., Revankar, S., 2011a. Hydrodynamic characterization of nickel metal foam, part 2: effects of pore structure and permeability. Transp. Porous Media, 1–14.

Miwa, S., Revankar, S.T., 2011b. Hydrodynamic characterization of nickel metal foam, part 2: effects of pore structure and permeability. Transp. Porous Media 89, 323–336.

Miyoshi, T., Itoh, M., Akiyama, S., Kitahara, A., 1998. Aluminum foam "Alporas": the production process, properties and applications. In: Schwartz, D.S., Shih, D.H., Evans, A.G., Wadley, H.G. (Eds.), Porous and Cellular Materials for Structural Applications. San Francisco, CA, USA, Materials Research Society Symposium Proceedings, vol. 521. (Warrendale PA, USA).

Miyoshi, T., Itoh, M., Mukai, T., et al., 1999. Enhancement of energy absorption in a closed cell aluminium by the modification of cellular structures. Scr. Mater. 41, 1055–1060.

Miyoshi, T., Itoh, M., Akiyama, S., Kitahara, A., 2000. Alporas aluminum foam: production process, properties and applications. Adv. Eng. Mater. 2, 179–183.

Miyoshi, T., Mukai, T., Higashi, K., 2002. Energy absorption in a closed-cell Al–Zn–Mg–Ca–Ti foam. Mater. Trans. 43, 1778–1781.

Mohr, D., Doyoyo, M., 2004a. Deformation-induced folding systems in thin-walled monolithic hexagonal metallic honeycomb. Int. J. Solids Struct. 41, 3353–3377.

Mohr, D., Doyoyo, M., 2004b. Large plastic deformation of metallic honeycomb: orthotropic rate-independent constitutive model. Int. J. Solids Struct. 41, 4435–4456.

Molina, J.M., Saravanan, R.A., Arpon, R., et al., 2002. Pressure infiltration of liquid aluminium into packed SiC particulate with a bimodal particle size distribution. Acta Mater. 50, 247–257.

Molina, J.M., Rodrìguez-Guerrero, A., Bahraini, M., et al., 2007. Infiltration of graphite preforms with Al–Si eutectic alloy and mercury. Scr. Mater. 56, 991–994.

Mondal, D.P., Das, S., 2010. Effect of thickening agent and foaming agent on the micro-architecture and deformation response of closed cell aluminum foam. Einfluss des Dickungs- und Treibmittels auf die Mikrostruktur und das Deformationsverhalten von geschlossenzelligem Aluminiumschaum. Materialwiss. Werkstofftech. 41, 276–282.

Montanini, R., 2005. Measurement of strain rate sensitivity of aluminium foams for energy dissipation. Int. J. Mech. Sci. 47, 26.

Montes, J.M., Cuevas, F.G., Cintas, J., Munoz, S., 2012. Thermal conductivity of powder aggregates and porous compacts. Metall. Mater. Trans. A 43, 4532–4538.

Montillet, A., Comiti, J., Legrand, J., 1992. Determination of structural parameters of metallic foams from permeametry measurements. J. Mater. Sci. 27, 4460–4464.

Montillet, A., Comiti, J., Legrand, J., 1993. Axial dispersion in liquid flow through packed reticulated metallic foams and fixed beds of different structures. Chem. Eng. J. 52, 63–71.

Moongkhamklang, P., Elzey, D.M., Wadley, H.N.G., 2008. Titanium matrix composite lattice structures. Compos. Appl. Sci. Manuf. 39, 176–187.

Moongkhamklang, P., Deshpande, V.S., Wadley, H.N.G., 2010. The compressive and shear response of titanium matrix composite lattice structures. Acta Mater. 58, 2822–2835.

Moreira, E.A., Innocentini, M.D.M., Coury, J.R., 2004. Permeability of ceramic foams to compressible and incompressible flow. J. Eur. Ceram. Soc. 24, 3209–3218.

Mori, T., Tanaka, K., 1973. Average stress in matrix and average elastic energy of materials with misfitting inclusions. Acta Metall. 21, 571–574.

Mortensen, A., Goodall, R., 2012. Porous metal article and method of producing a porous metal article. US Patent US 8,151,860 B2.

Mortensen, A., Jin, I., 1992. Solidification processing of metal matrix composites. Int. Mater. Rev. 37, 101–128.

Mortensen, A., 1990. Solidification of reinforced metals. In: Rohatgi, P.K. (Ed.), Solidification of Metal Matrix Composites. The Minerals, Metals & Materials Society, Indianapolis, IN.

Mortensen, A., 2000. Melt infiltration of metal matrix composite (Chapter 3).20. In: Clyne, T.W. (Ed.), Comprehensive Composite Materials, Metal Matrix Composites, vol. 3. Pergamon, Oxford UK.

Motz, C., Pippan, R., 2001. Deformation behaviour of closed-cell aluminium foams in tension. Acta Mater. 49, 2463–2470.

Motz, C., Pippan, R., 2002. Fracture behaviour and fracture toughness of ductile closed-cell metallic foams. Acta Mater. 50, 2013–2033.

Motz, C., Friedl, O., Pippan, R., 2005. Fatigue crack propagation in cellular metals. Int. J. Fatigue 27, 1571–1581.

Mu, Y., Yao, G., 2010. Anisotropic compressive behavior of closed-cell Al–Si alloy foams. Mater. Sci. Eng. A 527, 1117–1119.

Mu, Y., Yao, G., Liang, L., et al., 2010. Deformation mechanisms of closed-cell aluminum foam in compression. Scr. Mater. 63, 629–632.

Mu, Y., Yao, G., Cao, Z., et al., 2011. Strain-rate effects on the compressive response of closed-cell copper-coated carbon fiber/aluminum composite foam. Scr. Mater. 64, 61–64.

Mueller, R., Mortensen, A., 2006. Simplified prediction of the monotonic uniaxial stress–strain curve of non-linear particulate composites. Acta Mater. 54, 2145.

Mueller, R., Soubielle, S., Goodall, R., et al., 2007. On the steady-state creep of microcellular metals. Scr. Mater. 57, 33.

Mukai, T., Kanahashi, H., Miyoshi, T., et al., 1999a. Experimental study of energy absorption in a close-celled aluminum foam under dynamic loading. Scr. Mater. 40, 921–927.

Mukai, T., Kanahashi, H., Yamada, Y., et al., 1999b. Dynamic compressive behavior of an ultra-lightweight magnesium foam. Scr. Mater. 41, 365–371.

Mukai, T., Miyoshi, T., Nakano, S., et al., 2006. Compressive response of a closed-cell aluminum foam at high strain rate. Scr. Mater. 54, 533.

Mukherjee, M., Kolluri, M., Garcia-Moreno, F., et al., 2009. Strain hardening during constrained deformation of metal foams' effect of shear displacement. Scr. Mater. 61, 752–755.

Mukherjee, M., Garcia-Moreno, F., Banhart, J., 2010a. Collapse of aluminum foam in two different atmospheres. Metall. Mater. Trans. B 41, 500–504.

Mukherjee, M., Garcia-Moreno, F., Banhart, J., 2010b. Defect generation during solidification of aluminium foams. Scr. Mater. 63, 235–238.

Mukherjee, M., Garcia-Moreno, F., Banhart, J., 2010c. Solidification of metal foams. Acta Mater. 58, 6358–6370.

Mukherjee, M., Ramamurty, U., Garcia-Moreno, F., Banhart, J., 2010d. The effect of cooling rate on the structure and properties of closed-cell aluminium foams. Acta Mater. 58, 5031–5042.

Muley, A., Kiser, C., Sundén, B., Shah, R.K., 2011. Foam heat exchangers: a technology assessment. Heat Transfer Eng. 33, 42–51.

Munir, Z.A., Anselmi-Tamburini, U., 1989. Self-propagating exothermic reactions: the synthesis of high-temperature materials by combustion. Mater. Sci. Rep. 3, 277–365.

Mura, T., 1982. Micromechanics of Solids. Martinus Nijhoff, The Hague.

Murasawa, G., Makuta, T., Cho, H., 2011. Fabrication of salami-type porous metal and its high attenuation characteristic. Scr. Mater. 65, 827–829.

Murr, L.E., Quinones, S.A., Gaytan, S.M., et al., 2009. Microstructure and mechanical behavior of Ti–6Al–4V produced by rapid-layer manufacturing, for biomedical applications. J. Mech. Behav. Biomed. Mater. 2, 20–32.

Murr, L.E., Gaytan, S.M., Medina, F., et al., 2010. Characterization of Ti–6Al–4V open cellular foams fabricated by additive manufacturing using electron beam melting. Mater. Sci. Eng. A 527, 1861–1868.

Murr, L.E., Amato, K.N., Li, S.J., et al., 2011. Microstructure and mechanical properties of open-cellular biomaterials prototypes for total knee replacement implants fabricated by electron beam melting. J. Mech. Behav. Biomed. Mater. 4, 1396–1411.

Murray, N.G.D., Dunand, D.C., 2003. Microstructure and solid state foaming of titanium. Compos. Sci. Technol. 63, 2311–2316.

Murray, N.G.D., Dunand, D.C., 2004. Effect of thermal history on the superplastic expansion of argon-filled pores in titanium: part I kinetics and microstructure. Acta Mater. 52, 2269.

Murray, N.G.D., Schuh, C.A., Dunand, D.C., 2003. Solid state foaming of titanium by hydrogen-induced internal stress superplasticity. Scr. Mater. 49, 879–883.

Nabavi, A., Khaki, J.V., 2011. Manufacturing of aluminum foam sandwich panels: comparison of a novel method with two different conventional methods. J. Sandwich Struct. Mater. 13, 177–187.

Nachtrab, S., Kapfer, S.C., Arns, C.H., et al., 2011. Morphology and linear-elastic moduli of random network solids. Adv. Mater. 23, 2633–2637.

Nadella, R., Sahu, S.N., Gokhale, A.A., 2010. Foaming characteristics of Al–Si–Mg (LM25) alloy prepared by liquid metal processing. Mater. Sci. Technol. 26, 908–913.

Nagai, K., Wada, D., Nakai, M., Norimatsu, T., 2006. Electrochemical fabrication of low density metal foam with mono dispersed sized micro- and sub-micrometer pore. Fusion Sci. Technol. 49, 686–690.

Nagel, A.R., Uslu, C., Lee, K.J., et al., 1997. Steel closed cell foams from direct oxide reduction. In: Ward-Close, C.M., Froes, F.H., Chellman, D.J., Cho, S.S. (Eds.), Synthesis/Processing of Lightweight Metallic Materials II, The Minerals. Metals & Materials Soc, Warrendale, PA, USA.

Nahson, K., Hutchinson, J., 2008. Modification of the Gurson model for shear failure. Eur. J. Mech. – A/Solids 27, 1–17.

Nakajima, H., Kanetake, N. (Eds.), 2006. Porous Metals and Metal Foaming Technology, Proc. of the 4th Int. Conf. on Metals and Metal Foaming Technology. The Japan Institute of Metals, Sendai, Japan.

Nakajima, H., Hyun, S.K., Ohashi, K., et al., 2001. Fabrication of porous copper by unidirectional solidification under hydrogen and its properties. Colloids Surf., A 179, 209–214.

Nakajima, H., Ikeda, T., Hyun, S.K., 2004. Fabrication of Lotus-type porous metals and their physical properties. Adv. Eng. Mater. 6, 377–384.

Nakajima, H., 2007. Fabrication, properties and application of porous metals with directional pores. Prog. Mater. Sci. 52, 1091–1173.

Nakajima, H., 2013. Fabrication, properties and applications of lotus-type porous metals. In: Dukhan, N. (Ed.), Metal Foams – Fundamentals and Applications. DEStech Publications Inc, Lancaster PA, USA.

Nakamura, T., Gnyloskurenko, S., Sakamoto, K., et al., 2002. Development of a new foaming agent for metal foam. Mater. Trans. 43, 1191–1196.

Nakayama, A., 2013. Heat transfer in metal foams. In: Dukhan, N. (Ed.), Metal Foams – Fundamentals and Applications. DEStech Publications Inc, Lancaster PA, USA.

Nammi, S.K., Myler, P., Edwards, G., 2010. Finite element analysis of closed-cell aluminium foam under quasi-static loading. Mater. Des. 31, 712–722.

Neff, P., Forest, S., 2007. A geometrically exact micromorphic model for elastic metallic foams accounting for affine microstructure. Modelling, existence of minimizers, identification of moduli and computational results. J. Elast. 87, 239–276.

Nemat-Nasser, S., Hori, M., 1999. Micromechanics: Overall Properties of Heterogeneous Materials, Second ed. North-Holland Elsevier, Amsterdam NL.

Nemat-Nasser, S., Kang, W.J., McGee, J.D., et al., 2007. Experimental investigation of energy-absorption characteristics of components of sandwich structures. Int. J. Impact Eng. 34, 1119–1146.

Neurohr, A.J., Dunand, D.C., 2011a. Mechanical anisotropy of shape-memory NiTi with two-dimensional networks of micro-channels. Acta Mater. 59, 4616–4630.

Neurohr, A.J., Dunand, D.C., 2011b. Shape-memory NiTi with two-dimensional networks of micro-channels. Acta Biomater. 7, 1862–1872.

Neves, M.M., Rodrigues, H., Gedes, J.M., 2000. Optimal design of periodic linear elastic microstructures. Comput. Struct. 76, 421–429.

Neville, B.P., Rabiei, A., 2008. Composite metal foams processed through powder metallurgy. Mater. Des. 29, 388–396.

Nguyen, N.Q., Gupta, N., 2010. Analyzing the effect of fiber reinforcement on properties of syntactic foams. Mater. Sci. Eng. A 527, 6422–6428.

Nguyen, T.L., Staiger, M.P., Dias, G.J., Woodfield, T.B.F., 2011. A novel manufacturing route for fabrication of topologically-ordered porous magnesium scaffolds. Adv. Eng. Mater. 13, 872–881.

Nieh, T.G., Kinney, J.H., Wadsworth, J., Ladd, A.J.C., 1998. Morphology and elastic properties of aluminum foams produced by a casting technique. Scr. Mater. 38, 1487–1494.

Nieh, T.G., Higashi, K., Wadsworth, J., 2000. Effect of cell morphology on the compressive properties of open-cell aluminum foams. Mater. Sci. Eng. A 283, 105–110.

Nield, D.A., Bejan, A., 2006. Convection in Porous Media, third ed. Springer Science, New York.

Nielsen, L.F., 1984. Elasticity and damping of porous materials and impregnated materials. J. Am. Ceram. Soc. 67, 93–98.

Niu, W., Bai, C., Qiu, G., et al., 2009. Preparation and characterization of porous titanium using space-holder technique. Rare Metals 28, 338–342.

Noh, J.-S., Lee, K.B., Lee, C.G., 2006. Pressure loss and forced convective heat transfer in an annulus filled with aluminum foam. Int. Commun. Heat Mass Transfer 33, 434–444.

Norris, A.N., 1985. A differential scheme for the effective moduli of composites. Mech. Mater. 4, 1–16.

Nosko, M., Simancik, F., Florek, R., 2010. Reproducibility of aluminum foam properties: effect of precursor distribution on the structural anisotropy and the collapse stress and its dispersion. Mater. Sci. Eng. A 527, 5900–5908.

Nouri-Borujerdi, A., Layeghi, M., 2005. A review of concentric annular heat pipes. Heat Transfer Eng. 26, 45–58.

Nye, J.F., 1985. Physical Properties of Crystals. Oxford Science Publishers, Oxford, UK.

Ochiai, S., Nakano, S., Fukazawa, Y., et al., 2010a. Change of Young's modulus with increasing applied tensile strain in open cell nickel and copper foams. Mater. Trans. 51, 925–932.

Ochiai, S., Nakano, S., Fukazawa, Y., et al., 2010b. Tensile deformation and failure behavior of open cell nickel and copper foams. Mater. Trans. 51, 699–706.

Ochsner, A., Winter, W., Kuhn, G., 2003. On an elastic-plastic transition zone in cellular metals. Arch. Appl. Mech. 73, 261–269.

Ochsner, A., Kuhn, G., Gracio, J., 2005. Investigation of cellular solids under biaxial constraint. Exp. Mech. 45, 325–330.

Ochsner, A., Tane, M., Nakajima, H., 2006. Prediction of the thermal properties of lotus-type and quasi-isotropic porous metals: numerical and analytical methods. Mater. Lett. 60, 2690.

Odabaee, M., Hooman, K., 2012a. Metal foam heat exchangers for heat transfer augmentation from a tube bank. Appl. Therm. Eng. 36, 456–463.

Odabaee, M., Hooman, K., 2012b. Metal foam heat exchangers for thermal management of fuel cell systems. AIP Conf. Proc. 1453, 237–242.

Odabaee, M., Hooman, K., Gurgenci, H., 2011. Metal foam heat exchangers for heat transfer augmentation from a cylinder in cross-flow. Transp. Porous Media 86, 911–923.

Oh, I.-H., Nomura, N., Masahashi, N., Hanada, S., 2003. Mechanical properties of porous titanium compacts prepared by powder sintering. Scr. Mater. 49, 1197–1202.

Ohgaki, T., Toda, H., Kobayashi, M., et al., 2006a. In-situ high-resolution X-ray CT observation of compressive and damage behaviour of aluminium foams by local tomography technique. Adv. Eng. Mater. 8, 473–475.

Ohgaki, T., Toda, H., Kobayashi, M., et al., 2006b. In situ observations of compressive behaviour of aluminium foams by local tomography using high-resolution X-rays. Philos. Mag. 86, 4417.

Olevsky, E.A., Wang, X., Bruce, E., et al., 2007. Synthesis of gold micro- and nano-wires by infiltration and thermolysis. Scr. Mater. 56, 867–869.

Olevsky, E.A., 1998. Theory of sintering: from discrete to continuum. Mater. Sci. Eng. R 23, 41–100.

Olurin, O.B., Fleck, N.A., Ashby, M.F., 2000a. Deformation and fracture of aluminium foams. Mater. Sci. Eng. A 291, 136–146.

Olurin, O.B., Fleck, N.A., Ashby, M.F., 2000b. Indentation resistance of an aluminium foam. Scr. Mater. 43, 983–989.

Olurin, O.B., Fleck, N.A., Ashby, M.F., 2000c. Joining of aluminium foams with fasteners and adhesives. J. Mater. Sci. 35, 1079–1085.

Olurin, O.B., Fleck, N.A., Ashby, M.F., 2001a. Tensile and compressive failure of notched cellular foams. Adv. Eng. Mater. 3, 55–58.

Olurin, O.B., McCullough, K.Y.G., Fleck, N.A., Ashby, M.F., 2001b. Fatigue crack propagation in aluminium alloy foams. Int. J. Fatigue 23, 375.

Olurin, O.B., Arnold, M., Körner, C., Singer, R.F., 2002. The investigation of morphometric parameters of aluminium foams using micro-computed tomography. Mater. Sci. Eng. A 328, 334–343.

Onck, P.R., Bastawros, A.F., 2000. Notch effects in metal foams. In: Miannay, D., Costa, P., François, D., Pineau, A. (Eds.), Euromat 2000: Advances in Mechanical Behaviour, Plasticity and Damage. Elsevier, Tours.

Onck, P., Andrews, E.W., Gibson, L.J., 2001. Size effects in ductile cellular solids. Part I: modeling. Int. J. Mech. Sci. 43, 681–699.

Onck, P., Merkerk, R., Raaijmakers, A., Hosson, J., 2005. Fracture of open- and closed-cell metal foams. J. Mater. Sci. 40, 5821.

Onck, P.R., 2001a. Application of a continuum constitutive model to metallic foam DEN-specimens in compression. Int. J. Mech. Sci. 43, 2947–2959.

Onck, P.R., 2001b. Notch-strengthening in two-dimensional foams. J. Phys. IV 11, Pr5-211–Pr5217.

Onck, P.R., 2003a. The role of strain pattern formation in understanding scale effects in metal foams. In: Banhart, J., Fleck, N., Mortensen, A. (Eds.), Cellular Metals: Manufacture, Properties, Applications, Proc. Conf. Metfoam 2003. Berlin, 23–25 June 2003. Verlag MIT Publishing, Berlin, Germany.

Onck, P.R., 2003b. Scale effects in cellular metals. MRS Bull. 28, 279–283.

Onstad, A., Elkins, C., Medina, F., et al., 2011. Full-field measurements of flow through a scaled metal foam replica. Exp. Fluids 50, 1571–1585.

Oppenheimer, S.M., Dunand, D.C., 2007. Finite element modeling of creep deformation in cellular metals. Acta Mater. 55, 3825–3834.

Oppenheimer, S.M., Dunand, D.C., 2009. Porous NiTi by creep expansion of argon-filled pores. Mater. Sci. Eng. A 523, 70–76.

Oppenheimer, S., Dunand, D.C., 2010. Solid-state foaming of Ti–6Al–4V by creep or superplastic expansion of argon-filled pores. Acta Mater. 58, 4387–4397.

Oppenheimer, S.M., O'Dwyer, J.G., Dunand, D.C., 2004. Porous, superelastic NiTi produced by powder metallurgy. TMS Letters 1, 93–94.

Orbulov, I.N., Dobranszky, J., Nemeth, A., 2009. Microstructural characterisation of syntactic foams. J. Mater. Sci. 44, 4013–4019.

Ordonez-Miranda, J., Alvarado-Gil, J., 2012. Effect of the pore shape on the thermal conductivity of porous media. J. Mater. Sci. 47, 6733–6740.

O'Rourke, J.P., Ingber, M.S., Weiser, M.W., 1997. The effective elastic constants of solids containing spherical inclusions. J. Compos. Mater. 31, 910–934.

Oruganti, R.K., Ghosh, A.K., 2008. FEM analysis of transverse creep in honeycomb structures. Acta Mater. 56, 726–735.

Oruganti, R.K., Ghosh, A.K., Mazumder, J., 2004. Thermal expansion behavior in fabricated cellular structures. Mater. Sci. Eng. A 371, 24.

Ozgur, M., Mullen, R.L., Welsch, G., 1996. Analysis of closed cell metal matrix composites. Acta Mater. 44, 2115–2126.

Paek, J.W., Kang, B.H., Kim, S.Y., Hyun, J.M., 2000. Effective thermal conductivity and permeability of aluminium foam material. Int. J. Thermophys. 21, 453–464.

Palmer, R.A., Gao, K., Doan, T.M., et al., 2007. Pressure infiltrated syntactic foams process development and mechanical properties. Mater. Sci. Eng. A 464, 85–92.

Palumbo, N.M.A., Smith, C.W., Miller, W., Evans, K.E., 2011. Near-zero thermal expansivity 2-D lattice structures: performance in terms of mass and mechanical properties. Acta Mater. 59, 2392–2403.

Papadopoulos, D., Omar, H., Stergioudi, F., et al., 2010. A novel method for producing Al-foams and evaluation of their compression behavior. J. Porous Mater. 17, 773–777.

Papadopoulos, D.P., Omar, H., Stergioudi, F., et al., 2011. The use of dolomite as foaming agent and its effect on the microstructure of aluminium metal foams-comparison to titanium hydride. Colloids Surf., A 382, 118–123.

Papka, S.D., Kyriakides, S., 1994. In-plane compressive response and crushing of honeycomb. J. Mech. Phys. Solids 42, 1499–1532.

Pardoen, T., Besson, J., 2004. Micromechanics-based constitutive models of ductile fracture. In: Besson, J. (Ed.), Local Approach to Fracture. Les Presses de l'Ecole des Mines, Paris, France.

Pardoen, T., Hutchinson, J., 2000. An extended model for void growth and coalescence. J. Mech. Phys. Solids 48, 2467–2512.

Parida, S., Kramer, D., Volkert, C.A., et al., 2006. Volume change during the formation of nanoporous gold by dealloying. Phys. Rev. Lett. 97, 035504.

Park, C., Nutt, S.R., 2000. Pm synthesis and properties of steel foams. Mater. Sci. Eng. A 288, 111–118.

Park, C., Nutt, S.R., 2001a. Anisotropy and strain localization in steel foam. Mater. Sci. Eng. A 299, 68–74.

Park, C., Nutt, S.R., 2001b. Effect of process parameters on steel foam synthesis. Mater. Sci. Eng. A 297, 62–68.

Park, C., Nutt, S.R., 2002. Strain rate sensitivity and defects in steel foam. Mater. Sci. Eng. A 323, 358–366.

Park, S.H., Hur, B.Y., Song, K.H., 2005. Mater. Sci. Forum 475–479, 2683–2686.

Paserin, V., Marcuson, S., Shu, J., Wilkinson, D.S., 2004. CVD technique for inco nickel foam production. Adv. Eng. Mater. 6, 454–459.

Pashak, J., 1960. US Patent US 2,935,396.

Pastor, J., Ponte Castaneda, P., 2002. Yield criteria for porous media in plane strain: second-order estimates versus numerical results. C. R. Mec. 330, 741.

Pattofatto, S., Elnasri, I., Zhao, H., et al., 2007. Shock enhancement of cellular structures under impact loading: part II analysis. J. Mech. Phys. Solids 55, 2672.

Paul, A., Ramamurty, U., 2000. Strain rate sensitivity of a closed-cell aluminum foam. Mater. Sci. Eng. A 281, 1–7.

Paul, A., Seshacharyulu, T., Ramamurty, U., 1999. Tensile strength of a closed-cell Al foam in the presence of notches and holes. Scr. Mater. 40, 809–814.

Peroni, L., Avalle, M., Peroni, M., 2008. The mechanical behaviour of aluminium foam structures in different loading conditions. Int. J. Impact Eng. 35, 644–658.

Peronius, N., Sweeting, T.J., 1985. On the correlation of minimum porosity with particle size distribution. Powder Technol. 42, 113–121.

Perrot, C., Panneton, R., Olny, X., 2007. Periodic unit cell reconstruction of porous media: application to open-cell aluminum foams. J. Appl. Phys. 101, 113538.

Perrot, C., Chevillotte, F., Jaouen, L., Hoang, M.T., 2013. Acoustic properties and applications. In: Dukhan, N. (Ed.), Metal Foams – Fundamentals and Applications. DEStech Publications Inc, Lancaster PA, USA.

Pestryakov, A.N., Fyodorov, A.A., Shurov, V., et al., 1994. Foam metal catalysts with intermediate support for deep oxidation of hydrocarbons. React. Kinet. Catal. Lett. 53, 347–352. Akademiai Kiado, Budapest.

Pestryakov, A.N., Fyodorov, A.A., Gaisinovich, M.S., et al., 1995. Metal-foam metal catalysts with supported active phase for deep oxidation of hydrocarbons. React. Kinet. Catal. Lett. 54, 167–172. Akademiai Kiado, Budapest.

Pezzee, C.F., Dunand, D.C., 1994. The impingement effect of an inert, immobile second phase on the recrystallization of a matrix. Acta Metall. Mater. 42, 1509–1524.

Phani, A.S., Fleck, N.A., 2008. Elastic boundary layers in two-dimensional isotropic lattices. J. Appl. Mech. 75, 021020–021028.

Phanikumar, M.S., Mahajan, R.L., 2002. Non-darcy natural convection in high porosity metal foams. Int. J. Heat Mass Transfer 45, 3781–3793.

Phelan, R., Weaire, D., Peters, E.A.J.F., Verbist, G., 1996. The conductivity of a foam. J. Phys.: Condens. Matter 8, L475–L482.

Phoenix, S.L., Beyerlein, Y., 2000. Chapter 1.19-Statistical strength theory for fibrous composite materials. In: Kelly, A., Zweben, C. (Eds.), Comprehensive Composite Materials, vol. 1. Pergamon, Oxford UK.

Pineau, A., Pardoen, T., 2007. 2.06-Failure of metals. In: I. M., Ritchie, R.O., Karihaloo, B. (Eds.), Comprehensive Structural Integrity. Pergamon, Oxford.

Pingle, S.M., Fleck, N.A., Deshpande, V.S., Wadley, H.N.G., 2011. Collapse mechanism maps for a hollow pyramidal lattice. Proc. R. Soc. A 467, 985–1011.

Pinkhasov, E., 1991. Method of making open-pore structures. US Patent US 5,011,638.

Poirier, D.R., Geiger, G.H., 1994. Transport Phenomena in Materials Processing. TMS, Warrendale, PA.

Pollien, A., Conde, Y., Pambaguian, L., Mortensen, A., 2005. Graded open-cell aluminium foam core sandwich beams. Mater. Sci. Eng. A 404, 9.

Polonsky, L., Lipson, S., Markus, H., 1961. Lightweight cellular metal. Mod. Cast. 39, 57–71.

Ponte-Castañeda, P., Suquet, P., 1998. Nonlinear composites. Adv. Appl. Mech. 34, 171–302.

Ponte-Castañeda, P., Zaidman, M., 1994. Constitutive models for porous materials with evolving microstructure. J. Mech. Phys. Solids 42, 1459.

Ponte-Castañeda, P., 1991. The effective mechanical properties of nonlinear isotropic composites. J. Mech. Phys. Solids 39, 45–71.

Ponte-Castañeda, P., 1992. New variational principles in plasticity and their application to composite materials. J. Mech. Phys. Solids 40, 1757.

Ponte-Castañeda, P., 1996. Exact second-order estimates for the effective mechanical properties of nonlinear composite materials. J. Mech. Phys. Solids 44, 827–862.

Ponte-Castañeda, P., 1997. Nonlinear composite materials: effective constitutive behavior and microstructure evolution. In: Suquet, P. (Ed.), Continuum Micromechanics. Springer-Verlag, Wien, New York.

Ponte-Castañeda, P., 2002a. Second-order homogenization estimates for nonlinear composites incorporating field fluctuations: I–theory. J. Mech. Phys. Solids 50, 737.

Ponte-Castañeda, P., 2002b. Second-order homogenization estimates for nonlinear composites incorporating field fluctuations: II–applications. J. Mech. Phys. Solids 50, 759.

Poutet, J., Manzoni, D., Hage-Chehade, F., et al., 1996a. The effective mechanical properties of reconstructed porous media. Int. J. Rock Mech. Min. Sci. 33, 409–415.

Poutet, J., Manzoni, D., Hage-Chehade, F., et al., 1996b. The effective mechanical properties of random porous media. J. Mech. Phys. Solids 44, 1587–1620.

Prakash, O., Sang, H., Embury, J.D., 1995. Structure and properties of Al–SiC foam. Mater. Sci. Eng. A 199, 195–203.

Prakash, O., Richebois, P., Bréchet, Y., et al., 1996. A note on the deformation behaviour of two-dimensional model cellular structures. Philos. Mag. A73, 739–751.

Prakash, O., Embury, J.D., Sang, H., et al., 1997. Light weight cellular structures based on aluminium composites. In: Ward-Close, C.M., Froes, F.H., Chellman, D.J., Cho, S.S. (Eds.), Synthesis/Processing of Lightweight Metallic Materials II. The Minerals, Metals & Materials Soc, Warrendale, PA, USA.

Prall, D., Lakes, R.S., 1997. Properties of a chiral honeycomb with a Poisson's ratio of − 1. Int. J. Mech. Sci. 39, 305–314.

Pugh, R.J., 1996. Foaming, foam films, antifoaming and defoaming. Adv. Colloid Interface Sci. 64, 67–142.

Qi, J., Halloran, J.W., 2004. Negative thermal expansion artificial material from iron–nickel alloys by oxide co-extrusion with reductive sintering. J. Mater. Sci. 39, 4113–4118.

Qiu, Y.P., Weng, G.J., 1992. A theory of plasticity for porous materials and particle-reinforced composites. Trans. ASME − J. Appl. Mech. 59, 261–268.

Qiu, Y.P., Weng, G.J., 1993. Plastic potential and yield function of porous materials with aligned and randomly oriented spheroidal voids. Int. J. Plast. 9, 271–290.

Qiu, K., Zhang, W., Domaszewski, M., Chamoret, D., 2009. Topology optimization of periodic cellular solids based on a superelement method. Eng. Optim. 41, 225–239.

Qu, Z.G., Li, W.Q., Wang, J.L., Tao, W.Q., 2012. Passive thermal management using metal foam saturated with phase change material in a heat sink. Int. Commun. Heat Mass Transfer 39, 1546–1549.

Quadbeck, P., Kümmel, K., Hauser, R., et al., 2011. Structural and material design of open-cell powder metallurgical foams. Adv. Eng. Mater. 13, 1024–1030.

Quadrini, F., Guglielmotti, A., Squeo, E.A., Tagliaferri, V., 2010. Laser forming of open-cell aluminium foams. J. Mater. Process. Technol. 210, 1517–1522.

Queheillalt, D.T., Wadley, H.N.G., 2005a. Cellular metal lattices with hollow trusses. Acta Mater. 53, 303.

Queheillalt, D.T., Wadley, H.N.G., 2005b. Pyramidal lattice truss structures with hollow trusses. Mater. Sci. Eng. A 397, 132.

Queheillalt, D.T., Wadley, H.N., 2009. Titanium alloy lattice truss structures. Mater. Des. 30, 1966–1975.

Queheillalt, D.T., Wadley, H.N.G., 2011. Hollow pyramidal lattice truss structures. Int. J. Mater. Res. 102, 389–400.

Queheillalt, D.T., Katsumura, Y., Wadley, H.N.G., 2004. Synthesis of stochastic open cell Ni-based foams. Scr. Mater. 50, 313.

Rabiei, A., O'Neill, A.T., 2005. A study on processing of a composite metal foam via casting. Mater. Sci. Eng. A 404, 159.

Rabiei, A., Vendra, L.J., 2009. A comparison of composite metal foam's properties and other comparable metal foams. Mater. Lett. 63, 533–536.

Rabiei, A., Evans, A.G., Hutchinson, J.W., 2000. Heat generation during the fatigue of a cellular Al alloy. Metall. Mater. Trans. 31A, 1129–1136.

Rabiei, A., Vendra, L., Reese, N., et al., 2006. Processing and characterisation of a new composite metal foam. Mater. Trans. 47, 2148–2153.

Rachedi, R., Chikh, S., 2001. Enhancement of electronic cooling by insertion of foam materials. Heat Mass Transfer 37, 371–378.

Rack, A., García-Moreno, F., Baumbach, T., Banhart, J., 2009a. Synchrotron-based radioscopy employing spatio-temporal micro-resolution for studying fast phenomena in liquid metal foams. J. Synchrotron Radiat. 16, 432–434.

Rack, A., Helwig, H.-M., Butow, A., et al., 2009b. Early pore formation in aluminium foams studied by synchrotron-based microtomography and 3-D image analysis. Acta Mater. 57, 4809–4821.

Radford, D.D., Deshpande, V.S., Fleck, N.A., 2005. The use of metal foam projectiles to simulate shock loading on a structure. Int. J. Impact Eng. 31, 1152–1171.

Rak, Z., Berkeveld, L.D., Snijders, G, 2003. Method for producing a porous titanium material article. In Patent, W. Ed.

Rakow, J.F., Waas, A.M., 2004. Size effects in metal foam cores for sandwich structures. AIAA J. 42, 1331–1337.

Rakow, J.F., Waas, A.M., 2005. Size effects and the shear response of aluminum foam. Mech. Mater. 37, 69–82.

Ramachandra, S., Sudheer Kumar, P., Ramamurty, U., 2003. Impact energy absorption in an Al foam at low velocities. Scr. Mater. 49, 741.

Ramakrishnan, N., Arunachalam, V.S., 1990. Effective elastic moduli of porous solids. J. Mater. Sci. 25, 3930–3937.

Ramakrishnan, N., Arunachalam, V.S., 1993. Effective elastic moduli of porous ceramic materials. J. Am. Ceram. Soc. 76, 2745–2752.

Ramamurty, U., Kumaran, M.C., 2004. Mechanical property extraction through conical indentation of a closed-cell aluminum foam. Acta Mater. 52, 181.

Ramamurty, U., Paul, A., 2004. Variability in mechanical properties of a metal foam. Acta Mater. 52, 869.

Ramirez, D.A., Murr, L.E., Li, S.J., et al., 2011. Open-cellular copper structures fabricated by additive manufacturing using electron beam melting. Mater. Sci. Eng. A 528, 5379–5386.

Rammerstorfer, F.G., Pahr, D.H., Daxner, T., Vonach, W.K., 2006. Buckling in thin walled micro and meso structures of lightweight materials and material compounds. Comput. Mech. 37, 470–478.

Ramos, A.I.C., Dunand, D.C., 2012. Preparation and characterization of directionally freeze-cast copper foams. Metals 2, 265–273.

Rathbun, H.J., Zok, F.W., Evans, A.G., 2005. Strength optimization of metallic sandwich panels subject to bending. Int. J. Solids Struct. 42, 6643–6661.

Ravi Kumar, N.V., Ramachandra Rao, N., Sudhakar, B., Gokhale, A.A., 2010. Foaming experiments on LM25 alloy reinforced with SiC particulates. Mater. Sci. Eng. A 527, 6082–6090.

Re, X., Xide, L., Qinghua, Q., et al., 2011. Surface effects on the mechanical properties of nanoporous materials. Nanotechnology 22, 265714.

Redanz, P., Tvergaard, V., 2003. Analysis of shear band instabilities in compaction of powders. Int. J. Solids Struct. 40, 1853–1864.

Redenbach, C., 2009. Microstructure models for cellular materials. Comput. Mater. Sci. 44, 1397–1407.

Rehbinder, P.A., Shchukin, E.D., 1972. Surface phenomena in solids during deformation and fracture processes. Progress in Surf. Sci. 3, 97–188.

Reutter, O., Sauerhering, J., Fend, T., et al., 2008a. Characterization of heat and momentum transfer in sintered metal foams. Adv. Eng. Mater. 10, 812–815.

Reutter, O., Smirnova, E., Sauerhering, J., et al., 2008b. Characterization of air flow through sintered metal foams. J. Fluids Eng. 130, 051201–051205.

Reyes, A., Hopperstad, O.S., Berstad, T., et al., 2003. Constitutive modeling of aluminum foam including fracture and statistical variation of density. Eur. J. Mech. – A/Solids 22, 815–835.

Reyes, A., Hopperstad, O.S., Hanssen, A.G., Langseth, M., 2004. Modeling of material failure in foam-based components. Int. J. Impact Eng. 30, 805–834.

Rhein, R.K., Novak, M.D., Levi, C.G., Pollock, T.M., 2011. Bimetallic low thermal-expansion panels of Co-base and silicide-coated Nb-base alloys for high-temperature structural applications. Mater. Sci. Eng. A 528, 3973–3980.

Ribeiro, G.B., Barbosa Jr, J.R., Prata, A.T., 2012. Performance of microchannel condensers with metal foams on the air-side: application in small-scale refrigeration systems. Appl. Therm. Eng. 36, 152–160.

Ricceri, R., Matteazzi, P., 2003. P/M processing of cellular titanium. Int. J. Powder Metall. 39, 53–61.

Rice, R.W., 1996a. Comparison of physical property–porosity behaviour with minimum solid area models. J. Mater. Sci. 31, 1509–1528.

Rice, R.W., 1996b. Evaluation and extension of physical property–porosity models based on minimum solid area. J. Mater. Sci. 31, 102–118.

Richardson, J.T., Peng, Y., Remue, D., 2000. Properties of ceramic foam catalyst supports: pressure drop. Appl. Catal. Gen. 204, 19.

Richardson, J.T., Remue, D., Hung, J.K., 2003. Properties of ceramic foam catalyst supports: mass and heat transfer. Appl. Catal. Gen. 250, 319.

Roberts, A.P., Garboczi, E.J., 2000. Elastic properties of model porous ceramics. J. Am. Ceram. Soc. 83, 3041–3048.

Roberts, A.P., Garboczi, E.J., 2001. Elastic moduli of model random three-dimensional closed-cell solids. Acta Mater. 49, 189–197.

Roberts, A.P., Garboczi, E.J., 2002a. Computation of the linear elastic properties of random porous materials with a wide variety of microstructure. Proc. R. Soc. London, Ser. A 58, 1033–1054.

Roberts, A.P., Garboczi, E.J., 2002b. Elastic properties of model random three-dimensional open-cell solids. J. Mech. Phys. Solids 50, 33–55.

Roberts, J.N., Schwartz, L.M., 1985. Grain consolidation and electrical conductivity in porous media. Phys. Rev., B 31, 5990–5997.

Rösler, J., Näth, O., 2010. Mechanical behaviour of nanoporous superalloy membranes. Acta Mater. 58, 1815–1828.

Rösler, J., Näth, O., Jäger, S., et al., 2005. Fabrication of nanoporous Ni-based superalloy membranes. Acta Mater. 53, 1397–1406.

Rousselier, G., Devaux, J.C., Mottet, G., Devesa, G., 1989. A methodology for ductile failure analysis based on damage mechanics: an illustration of a local approach of fracture. ASTM STP 995. In: Landes, J.D., Saxena, A., Merckle, J.G. (Eds.), Non-linear Fracture Mechanics, Elastic-plastic Fracture, Volume II. ASTM, Philadelphia, USA.

Roy, S., Wanner, A., Beck, T., et al., 2011. Mechanical properties of cellular solids produced from hollow stainless steel spheres. J. Mater. Sci. 46, 5519–5526.

Ruan, D., Lu, G., Chen, F.L., Siores, E., 2002. Compressive behaviour of aluminum foams at low and medium strain rates. Compos. Struct. 57, 331–336.

Ruan, D., Lu, G., Wang, B., Yu, T.X., 2003. In-plane dynamic crushing of honeycombs – a finite element study. Int. J. Impact Eng. 28, 161–182.

Ruan, D., Lu, G., Ong, L.S., Wang, B., 2007. Triaxial compression of aluminium foams. Compos. Sci. Technol. 67, 1218–1234.

Russ, J.C., Dehoff, R.T., 2000. Practical Stereology, Second ed. Kluwer Academic/Plenum Publishers, New York.

Saadatfar, M., Mukherjee, M., Madadi, M., et al., 2012. Structure and deformation correlation of closed-cell aluminium foam subject to uniaxial compression. Acta Mater. 60, 3604–3615.

Sadeghi, E., Hsieh, S., Bahrami, M., 2011. Thermal conductivity and contact resistance of metal foams. J. Phys. D: Appl. Phys. 44 (125406), 7pp.

Sahimi, M., Arbabi, S., 1989a. Force distribution, multiscaling, and fluctuations in disordered elastic media. Phys. Rev., B 40, 4975–4980.

Sahimi, M., Arbabi, S., 1989b. Force distribution, multiscaling, and fluctuations in disordered elastic media. Phys. Rev., B 40, 4975.

Sahimi, M., 1993. Flow phenomena in rocks: from continuum models to fractals, percolation, cellular automata, and simulated annealing. Rev. Mod. Phys. 65, 1393–1534.

Sahimi, M., 1994. Applications of Percolation Theory. Taylor and Francis, London UK.

Sahimi, M., 2011. Flow and Transport in Porous Media and Fractured Rock – From Classical Methods to Modern Approaches, Second, Revised and Enlarged Edition. Wiley-VCH Verlag, Weinheim, Germany.

Saito, Y., Utsunomiya, H., Tsuji, N., Sakai, T., 1999. Novel ultra-high straining process for bulk materials—development of the accumulative roll-bonding (ARB) process. Acta Mater. 47, 579–583.

Salvo, L., Belestin, P., Maire, E., et al., 2004. Structure and mechanical properties of AFS sandwiches studied by in-situ compression tests in X-ray microtomography. Adv. Eng. Mater. 6, 411–415.

San Marchi, C., Mortensen, A., 2001. Deformation of open-cell aluminum foam. Acta Mater. 49, 3959–3969.

San Marchi, C., Mortensen, A., 2002. Chapter 2.06: infiltration and the replication process for producing metal sponges. In: Degischer, H.P., Kriszt, B. (Eds.), Handbook of Cellular Metals. Wiley-VCH Verlag, Weinheim Germany.

San Marchi, C., Despois, J.F., Mortensen, A., 1999. Fabrication and compressive response of open-cell aluminum foams with sub-millimeter pores. In: Clyne, T.W., Simancik, F. (Eds.), Euromat 99. 27–30 Sept. 1999, Munich, Germany. DGM/Wiley-VCH, Weinheim, Germany.

San Marchi, C., Cao, F., Kouzeli, M., Mortensen, A., 2002. Quasistatic and dynamic compression of aluminum-oxide particle reinforced pure aluminum. Mater. Sci. Eng. A 337, 202–211.

San Marchi, C., Despois, J.F., Mortensen, A., 2004. Uniaxial deformation of open-cell aluminum foam: the role of internal damage. Acta Mater. 52, 2895.

Sanders, W.S., Gibson, L.J., 2003a. Mechanics of BCC and FCC hollow-sphere foams. Mater. Sci. Eng. A 352, 150–161.

Sanders, W.S., Gibson, L.J., 2003b. Mechanics of hollow sphere foams. Mater. Sci. Eng. A 347, 70–85.

Sangani, A.S., Acrivos, A., 1982. Slow flow past periodic arrays of cylinders with application to heat transfer. Int. J. Multiphase Flow 8, 193–206.

Sangani, A.S., Yao, C., 1988. Transport processes in random arrays of cylinders. II. Viscous flow. Phys. Fluids 31, 2435–2444.

Santosa, S., Wierzbicki, T., 1998. On the modeling of crush behavior of a closed-cell aluminum foam structure. J. Mech. Phys. Solids 46, 645–669.

Sargeant, T.D., Guler, M.O., Oppenheimer, S.M., et al., 2008. Hybrid bone implants: self-assembly of peptide amphiphile nanofibers within porous titanium. Biomaterials 29, 161–171.

Sasso, C.P., Zheng, P., Basso, V., et al., 2011. Enhanced field induced martensitic phase transition and magnetocaloric effect in Ni55Mn20Ga25 metallic foams. Intermetallics 19, 952–956.

Schade, P., 2002. Bubble evolution and effects during tungsten processing. Int. J. Refract. Met. Hard Mater. 20, 301–309.

Schade, P., 2010. 100 years of doped tungsten wire. Int. J. Refract. Met. Hard Mater. 28, 648–660.

Schade, P., 2011. Corrigendum to 100 years of doped tungsten wire. Int. J. Refract. Met. Hard Mater. 28, 647–734. Int. J. Refract. Met. Hard Mater., 2010, 29, 559.

Schaedler, T.A., Jacobsen, A.J., Torrents, A., et al., 2011. Ultralight metallic microlattices. Science 334, 962–965.

Schaffner, G., Guo, X.-D.E., Silva, M.J., Gibson, L.J., 2000. Modelling fatigue damage accumulation in two-dimensional Voronoi honeycombs. Int. J. Mech. Sci. 42, 645–656.

Scheidegger, A.E., 1974. The Physics of Flow through Porous Media. University of Toronto Press, Toronto.

Schmidt, I., Fleck, N.A., 2001. Ductile fracture of two-dimensional cellular structures. Int. J. Fract. 327.

Schmierer, E.N., Razani, A., 2006a. Direct simulation of transport in open-cell foam. J. Heat Transfer (ASME) 128, 1194–1203.

Schmierer, E.N., Razani, A., 2006b. Self-consistent open-celled metal foam model for thermal applications. J. Heat Transfer 128, 1194–1203.

Schramm, J.P., Demetriou, M.D., Johnson, W.L., et al., 2010. Effect of strain rate on the yielding mechanism of amorphous metal foam. Appl. Phys. Lett. 96, 021906-3.

Schroers, J., Veazey, C., Johnson, W.L., 2003. Amorphous metallic foam. Appl. Phys. Lett. 82, 370–372.

Schroers, J., Veazey, C., Demetriou, M.D., Johnson, W.L., 2004. Synthesis method for amorphous metallic foam. J. Appl. Phys. 96, 7723–7730.

Schuh, C.A., Hufnagel, T.C., Ramamurty, U., 2007. Mechanical behavior of amorphous alloys. Acta Mater. 55, 4067–4109.

Schulz, O., desLigneris, A., Haider, O., Starke, P., 2000. Fatigue behavior, strength and failure of aluminum foam. Adv. Eng. Mater. 2, 188–191.

Schwartz, D.S., Shih, D.S., Lederich, R.J., et al., 1998. Development and scale-up of the low density core process for Ti-64. In: Schwartz, D.S., Shih, D.H., Evans, A.G., Wadley, H.G. (Eds.), Porous and Cellular Materials for Structural Applications. San Francisco, CA, USA, Materials Research Society Symposium Proceedings, vol. 521. (Warrendale PA, USA).

Schwerdtfeger, J., Heinl, P., Singer, R.F., Körner, C., 2010. Auxetic cellular structures through selective electron-beam melting. Phys. Status Solidi B 247, 269–272.

Scott, J.A., Dunand, D.C., 2010. Processing and mechanical properties of porous Fe–26Cr–1Mo for solid oxide fuel cell interconnects. Acta Mater. 58, 6125–6133.

Seliger, H., Deuther, U., 1965. Die Herstellung von Schaum-und Zellaluminium. Feiburg. Forschungsh. 103, 129–158.

Seo, Y., Kim, Y., Lee, Y., et al., 2007. Hybridized porous structure of Al alloy by surface modification of pores. J. Porous Mater. http://dx.doi.org/10.1007/s10934-008-9253-4.

Sevastianov, I., Kovacik, J., Simancik, F., 2002. Correlation between elastic and electric properties for metal foams: theory and experiment. Int. J. Fract. 114, L23–L28.

Sevilla, P., Aparicio, C., Planell, J.A., Gil, F.J., 2007. Comparison of the mechanical properties between tantalum and nickel–titanium foams implant materials for bone ingrowth applications. J. Alloys Compd. 439, 67–73.

Seyf, H.R., Layeghi, M., 2010. Numerical analysis of convective heat transfer from an elliptic pin Fin heat sink with and without metal foam insert. J. Heat Transfer 132, 071401–071409.

Shabde, V., Hoo, K., Gladysz, G., 2006. Experimental determination of the thermal conductivity of three-phase syntactic foams. J. Mater. Sci. 41, 4061–4073.

Shahbeyk, S., 2013. Yield/failure criteria, constitutive models, and crashworthiness applications of metal foams. In: Dukhan, N. (Ed.), Metal Foams – Fundamentals and Applications. DEStech Publications Inc, Lancaster PA, USA.

Shapovalov, V., Boyko, L., 2004. Gasar – a new class of porous materials. Adv. Eng. Mater. 6, 407–410.

Shapovalov, V.I., Withers, J.C., 2008. Hydrogen technology for porous metals (Gasars) production. In: Baranowski, B., Zaginaichenko, S.Y., Schur, D.V., Skorokhod, V.V., Veziroglu, A. (Eds.), Proceedings of the NATO Advanced Research Workshop on Using Carbon Nanomaterials in Clean-energy Hydrogen System. Ukraine. Springer, Dordrecht NL.

Shapovalov, V.I., 1992. Structure formation behaviour of alloys during gas-eutectic transformation and prospects of the use of hydrogen in alloying. In: Lavernia, E.J., Gungor, M.N. (Eds.), Microstructural Design by Solidification Processing. The Minerals, Metals & Materials Society, Warrendale, PA.

Shapovalov, V.I., 1993. Method for manufacturing porous articles.

Shapovalov, V., 1994. Porous metals. Mater. Res. Soc. Bull. 19, 24–28.

Shapovalov, V.I., 1998. Formation of ordered gas-solid structures via solidification in metal-hydrogen systems. In: Schwartz, D.S., Shih, D.H., Evans, A.G., Wadley, H.G. (Eds.), Porous and Cellular Materials for Structural Applications. San Francisco, CA, USA, Materials Research Society Symposium Proceedings, vol. 521. (Warrendale PA, USA).

Sharma, N., Gibson, R.F., Ayorinde, E.O., 2006. Fatigue of foam and honeycomb core composite sandwich structures: a tutorial. J. Sandwich Struct. Mater. 8, 263–319.

Shen, H., Oppenheimer, S.M., Dunand, D.C., Brinson, L.C., 2006. Numerical modeling of pore size and distribution in foamed titanium. Mech. Mater. 38, 933–944.

Sherby, O.D., Klundt, R.H., Miller, A.K., 1977. Flow stress, subgrain size, and subgrain stability at elevated temperature. Metall. Trans. A 8A, 843–850.

Sherman, A.J., Williams, B.E., Delarosa, M.J., Laferia, R., 1991. Characterization of porous cellular materials fabricated by chemical vapor deposition. In: Sieradzki, K., Green, D.J., Gibson, L.J. (Eds.), Mechanical Properties of Porous and Cellular Materials. Boston, MA, USA, Materials Research Society Symposium Proceedings, vol. 207. (Warrendale PA, USA).

Shewmon, P., 1989. Diffusion in Solids, Second ed. The Minerals, Metals and Materials Society, Warrendale PA.

Shih, W.H., Chiu, W.C., Hsieh, W.H., 2006. Height effect on heat-transfer characteristics of aluminum-foam heat sinks. J. Heat Transfer 128, 530–537.

Shimizu, T., Matsuzaki, K., Nagai, H., Kanetake, N., 2012. Production of high porosity metal foams using EPS beads as space holders. Mater. Sci. Eng. A 558, 343–348.

Shimko, D.A., Shimko, V.F., Sander, E.A., et al., 2005. Effect of porosity on the fluid flow characteristics and mechanical properties of tantalum scaffolds. J. Biomed. Mater. Res., Part B 73B, 315–324.

Shiomi, M., Imagama, S., Osakada, K., Matsumoto, R., 2010. Fabrication of aluminium foams from powder by hot extrusion and foaming. J. Mater. Process. Technol. 210, 1203–1208.

Shirzadi, A.A., Kocak, M., Wallach, E.R., 2004. Joining stainless steel metal foams. Sci. Technol. Weld. Joining 9, 277–279.

Shirzadi, A.A., Zhu, Y., Bhadeshia, H.K.D.H., 2008. Joining ceramics to metals using metal foam. Mater. Sci. Eng. A 496, 501–506.

Siahpush, A., O'Brien, J., Crepeau, J., 2008. Phase change heat transfer enhancement using copper porous foam. ASME Trans. – J. Heat Transfer 130, 082301.

Sigmund, O., Torquato, S., 1996. Composites with extremal thermal expansion coefficients. Appl. Phys. Lett. 69, 3203–3205.

Sigmund, O., 1994. Materials with prescribed constitutive parameters: an inverse homogenization problem. Int. J. Solids Struct. 31, 2313–2329.

Sigmund, O., 1995. Tailoring materials with prescribed elastic properties. Mech. Mater. 20, 351–368.

Sihn, S., Roy, A.K., 2004. Modeling and prediction of bulk properties of open-cell carbon foam. J. Mech. Phys. Solids 52, 167–191.

Silva, M.J., Gibson, L.J., 1997. The effects of non-periodic microstructure and defects on the compressive strength of two-dimensional cellular solids. Int. J. Mech. Sci. 39, 549.

Silva, M.J., Hayes, W.C., Gibson, L.J., 1995. The effects of non-periodic microstructure on the elastic properties of two-dimensional cellular solids. Int. J. Mech. Sci. 37, 1161–1177.

Simançik, F., Kovacik, J., 2002. Chapter 5.3: electrical, thermal and acoustic properties of cellular metals. In: Degischer, H.P., Kriszt, B. (Eds.), Handbook of Cellular Metals. Wiley-VCH Verlag, Weinheim Germany.

Simançik, F., Schoerghuber, F., 1998. Complex foamed aluminium parts as permanent cores in aluminium castings. In: Schwartz, D.S., Shih, D.H., Evans, A.G., Wadley, H.G. (Eds.), Porous and Cellular Materials for Structural Applications. San Francisco, CA, USA, Materials Research Society Symposium Proceedings, vol. 521. (Warrendale PA, USA).

Simone, A.E., Gibson, L.J., 1996. The tensile strength of porous copper made by the Gasar process. Acta Mater. 44, 1437–1447.

Simone, A.E., Gibson, L.J., 1997. The compressive behavior of porous copper made by the Gasar process. J. Mater. Sci. 32, 451–457.

Simone, A.E., Gibson, L.J., 1998a. Aluminum foams produced by liquid-state processes. Acta Mater. 46, 3109–3123.

Simone, A.E., Gibson, L.J., 1998b. The effects of cell face curvature and corrugations on the stiffness and strength of metallic foams. Acta Mater. 46, 3929–3935.

Simone, A.E., Gibson, L.J., 1998c. Effects of solid distribution on the stiffness and strength of metallic foams. Acta Mater. 46, 2139–2150.

Simone, A.E., 1997. PhD Thesis, MIT, Cambridge, MA.

Singh, R., Kasana, H.S., 2004. Computational aspects of effective thermal conductivity of highly porous metal foams. Appl. Therm. Eng. 24, 1841–1849.

Singh, R., Lee, P.D., Lindley, T.C., et al., 2010. Characterization of the deformation behavior of intermediate porosity interconnected Ti foams using micro-computed tomography and direct finite element modelling. Acta Biomater. 6, 2342–2351.

Sirijaruphan, A., Goodwin, J.J.G., Rice, R.W., et al., 2005a. Metal foam supported Pt catalysts for the selective oxidation of CO in hydrogen. Appl. Catal. Gen. 281, 1.

Sirijaruphan, A., Goodwin Jr, J.G., Rice, R.W., et al., 2005b. Effect of metal foam supports on the selective oxidation of CO on Fe-promoted Pt/[gamma]-Al2O3. Appl. Catal. Gen. 281, 11.

Skochdopole, R.E., 1961. The thermal conductivity of foamed plastics. Chem. Eng. Prog. 57, 55–59.

Smigelskas, A.D., Kirkendall, E.O., 1947. Trans. AIME 171, 130.

Smit, G.J.F., du Plessis, J.P., 1999. Modelling of non-Newtonian purely viscous flow through isotropic high porosity synthetic foams. Chem. Eng. Sci. 54, 645–654.

Smit, G.J.F., du Plessis, J.P., Wilms, J.M., 2005. On the modeling of non-Newtonian purely viscous flow through high porosity synthetic foams. Chem. Eng. Sci. 60, 2815.

Smith, A.J., Trimm, D.L., 2005. The preparation of skelettal catalysts. Annu. Rev. Mater. Res. 35, 127–142.

Smith, C.W., Grima, J.N., Evans, K.E., 2000. A novel mechanism for generating auxetic behaviour in reticulated foams: missing rib foam model. Acta Mater. 48, 4349–4356.

Smith, C.S., 1981. A Search for Structure. MIT Press, Cambridge MA.

Smith, W.F., 1993. Structure and Properties of Engineering Alloys, Second ed. McGraw-Hill, New York, USA.

Smorygo, O., Mikutski, V., Leonov, A., et al., 2008. Nickel foams with oxidation-resistant coatings formed by combustion synthesis. Scr. Mater. 58, 910–913.

Smorygo, O., Mikutski, V., Marukovich, A., et al., 2009. Structured catalyst supports and catalysts for the methane indirect internal steam reforming in the intermediate temperature SOFC. Int. J. Hydrogen Energy 34, 9505–9514.

Smorygo, O., Mikutski, V., Marukovich, A., et al., 2011. An inverted spherical model of an open-cell foam structure. Acta Mater. 59, 2669–2678.

Snyder, J., Livi, K., Erlebacher, J., 2008. Dealloying silver/gold alloys in neutral silver nitrate solution: porosity evolution, surface composition, and surface oxides. J. Electrochem. Soc. 155, C464–C473.

Solorzano, E., Reglero, J.A., Rodrlguez-Pèrez, M.A., et al., 2008a. An experimental study on the thermal conductivity of aluminium foams by using the transient plane source method. Int. J. Heat Mass Transfer 51, 6259–6267.

Solorzano, E., Rodriguez-Perez, M.A., Saja, J.A. d., 2008b. Thermal conductivity of cellular metals measured by the transient plane source method. Adv. Eng. Mater. 10, 371–377.

Solorzano, E., Rodrlguez-Perez, M.A., de Saja, J.A., 2009a. Thermal conductivity of metallic hollow sphere structures: an experimental, analytical and comparative study. Mater. Lett. 63, 1128–1130.

Solorzano, E., Rodriguez-Perez, M.A., Lazaro, J., Saja, J.A. d., 2009b. Influence of solid phase conductivity and cellular structure on the heat transfer mechanisms of cellular materials: diverse case studies. Adv. Eng. Mater. 11, 818–824.

Sommer, J., Mortensen, A., 1996. Forced unidirectional infiltration of deformable porous media. J. Fluid Mech. 311, 193–217.

Somosvari, B.M., Babcsan, N., Barczy, P., Berthold, A., 2007. PVC particles stabilized water'ethanol compound foams. Colloids Surf., A 309, 240–245.

Song, Z., Kishimoto, S., 2006. The cell size effect of closed cellular materials fabricated by pulse current assisted hot isostatic pressing on the compressive behavior. Scr. Mater. 54, 1531.

Song, Z., Nutt, S., 2005. Energy of compressed aluminum foam. Adv. Eng. Mater. 7, 73–77.

Song, Z., Kishimoto, S., Shinya, N., 2003. Fabrication of closed cellular nickel alloy containing polymer by sintering method. J. Alloys Compd. 355, 166.

Song, Z., Kishimoto, S., Shinya, N., 2004. A novel pulse current assisted sintering method for fabrication of metallic cellular structures. Adv. Eng. Mater. 6, 211–214.

Song, Y., Tane, M., Ide, T., et al., 2010a. Fabrication of Al-3.7 pct Si-0.18 pct Mg foam strengthened by AlN particle dispersion and its compressive properties. Metall. Mater. Trans. A 41, 2104–2111.

Song, Y., Wang, Z., Zhao, L., Luo, J., 2010b. Dynamic crushing behavior of 3D closed-cell foams based on Voronoi random model. Mater. Des. 31, 4281–4289.

Song, Y.H., Tane, M., Nakajima, H., 2011. Appearance of a plateau stress region during dynamic compressive deformation of porous carbon steel with directional pores. Scr. Mater. 64, 797–800.

Soubielle, S., Diologent, F., Salvo, L., Mortensen, A., 2011a. Creep of replicated microcellular aluminium. Acta Mater. 59, 440–450.

Soubielle, S., Salvo, L., Diologent, F., Mortensen, A., 2011b. Fatigue and cyclic creep of replicated microcellular aluminium. Mater. Sci. Eng. A 528, 2657–2663.

Speed, S.E., 1976. Foaming of Metal by the Catalyzed and Controlled Decomposition of Zirconium Hydride and Titanium Hydride.

Spoerke, E.D., Murray, N.G., Li, H., et al., 2005. A bioactive titanium foam scaffold for bone repair. Acta Biomater. 1, 523.

Spoerke, E.D., Murray, N.G.D., Li, H., et al., 2008. Titanium with aligned, elongated pores for orthopedic tissue engineering applications. J. Biomed. Mater. Res. 84A, 402–412.

Spowart, J.E., 2006. Automated serial sectioning for 3-D analysis of microstructures. Scr. Mater. 55, 5–10.

Sridhar, I., Fleck, N.A., 2000. Yield behaviour of cold compacted composite powders. Acta Mater. 48, 3341–3352.

Sridhar, I., Fleck, N.A., 2005. The multiaxial yield behaviour of an aluminium alloy foam. J. Mater. Sci. 40, 4005–4008.

Stanley Smith, C., 1964. Some elementary principles of polycrystalline microstructure. Metall. Rev. 9, 1–48.

Stanzick, H., Klenke, J., Danilkin, S., Banhart, J., 2002a. Material flow in metal foams studied by neutron radioscopy. Appl. Phys. A 74, S1118–S1120.

Stanzick, H., Wichmann, M., Wiese, J., et al., 2002b. Process control in aluminium foam production using real-time X-ray radioscopy. Adv. Eng. Mater. 4, 814–823.

Stauffer, D., Aharony, A., 1994. Introduction to Percolation Theory – Revised, second ed. CRC Press, Boca Raton USA.

Steeves, C.A., dos Santos e Lucato, S.L., He, M., et al., 2007. Concepts for structurally robust materials that combine low thermal expansion with high stiffness. J. Mech. Phys. Solids 55, 1803–1822.

Steeves, C., Mercer, C., Antinucci, E., et al., 2009. Experimental investigation of the thermal properties of tailored expansion lattices. Int. J. Mech. Mater. Des. 5, 195–202.

Stephani, G., Andersen, O., Göhler, H., et al., 2006. Iron based cellular structures – status and prospects. Adv. Eng. Mater. 8, 847–852.

Stobener, K., Rausch, G., 2009. Aluminium foam–polymer composites: processing and characteristics. J. Mater. Sci. 44, 1506–1511.

Stöbener, K., Lehmhus, D., Avalle, M., et al., 2008. Aluminum foam-polymer hybrid structures (APM aluminum foam) in compression testing. Int. J. Solids Struct. 45, 5627–5641.

Stocco, A., Garcia-Moreno, F., Manke, I., et al., 2011. Particle-stabilised foams: structure and aging. Soft Matter 7.

Storåkers, B., Fleck, N.A., McMeeking, R.M., 1999. The viscoplastic compaction of composite powders. J. Mech. Phys. Solids 47, 785–815.

Ströhla, S., Winter, W., Kuhn, G., 2003. Yield behaviour of metal foams in the elastic-plastic transition region. In: Banhart, J., Fleck, N., Mortensen, A. (Eds.), CEllular Metals: Manufacture, Properties, Applications, Proc. Conf. Metfoam 2003. Berlin, 23–25 June 2003. Verlag MIT Publishing, Berlin, Germany.

Studart, A.R., Nelson, A., Iwanovsky, B., et al., 2012. Metallic foams from nanoparticle-stabilized wet foams and emulsions. J. Mater. Chem. 22.

Sugimura, Y., Meyer, J., He, M.Y., et al., 1997. On the mechanical response of closed cell Al alloy foams. Acta Mater. 45, 5245–5259.

Sugimura, Y., Rabiei, A., Evans, A.G., et al., 1999. Compression fatigue of a cellular Al alloy. Mater. Sci. Eng. A 269, 38–48.

Sullivan, R.M., Ghosn, L.J., Lerch, B.A., 2008. A general tetrakaidecahedron model for open-celled foams. Int. J. Solids Struct. 45, 1754–1765.

Sun, Y.Q., Gao, T., 2002. The optimum wetting angle for the stabilization of liquid-metal foams by ceramic particles: experimental simulations. Metall. Mater. Trans. A 33, 3285–3292.

Sun, C., Stimming, U., 2007. Recent anode advances in solid oxide fuel cells. J. Power Sources 171, 247–260.

Sun, D.X., Zhao, Y.Y., 2003. Static and dynamic energy absorption of Al foams produced by the sintering and dissolution process. Metall. Mater. Trans. B 34, 69–74.

Sun, D.X., Zhao, Y.Y., 2005a. Phase changes in sintering of Al/Mg/NaCl compacts for manufacturing Al foams by the sintering and dissolution process. Mater. Lett. 59, 6.

Sun, D.X., Zhao, Y.Y., 2005b. Simulation of thermal diffusivity of Al/NaCl powder compacts in producing Al foams by the sintering and dissolution process. J. Mater. Process. Technol. 169, 83.

Sun, D.Z., Andrieux, F., Haberling, S., 2003. Characterization and simulation of mechanical behaviour of aluminium foams. In: Banhart, J., Fleck, N., Mortensen, A. (Eds.), Cellular Metals: Manufacture, Properties, Applications, Proc. Conf. Metfoam 2003. Berlin, 23–25 June 2003. Verlag MIT Publishing, Berlin, Germany.

Sun, Y., Ye, J., Shan, Z., et al., 2007. The mechanical behavior of nanoporous gold thin films. JOM 59, 54–58.

Sun, Y., Ye, J., Minor, A.M., Balk, T.J., 2009. In situ indentation of nanoporous gold thin films in the transmission electron microscope. Microsc. Res. Tech. 72, 232–241.

Suquet, P., 1992. On bounds for the overall potential of power law materials containing voids with an arbitrary shape. Mech. Res. Commun. 19, 51.

Suquet, P.M., 1993. Overall potentials and extremal surfaces of power law or ideally plastic composites. J. Mech. Phys. Solids 41, 981–1002.

Suquet, P.M., 1995. Overall properties of nonlinear composites: a modified secant moduli theory and its link with Ponte Castaneda's nonlinear variational procedure. C. R. Acad. Sci. 320, 563–571.

Suquet, P., 1997. Effective properties of nonlinear composites. In: Suquet, P. (Ed.), Continuum Micromechanics. Springer-Verlag, Wien, New York.

Surace, R., De Filippis, L.A.C., Ludovico, A.D., Boghetich, G., 2009. Influence of processing parameters on aluminium foam produced by space holder technique. Mater. Des. 30, 1878–1885.

Suralvo, M., Bouwhuis, B.A., McCrea, J.L., et al., 2008. Hybrid nanocrystalline periodic cellular materials. Scr. Mater. 58, 247–250.

Suzuki, M., Sato, H., Hasegawa, M., Hirota, M., 2001. Effect of size distribution on tapping properties of fine powder. Powder Technol. 118, 53–57.

Symons, D.D., Fleck, N.A., 2008. The imperfection sensitivity of isotropic two-dimensional elastic lattices. J. Appl. Mech. 75, 051011–051018.

Symons, D.D., Hutchinson, R.G., Fleck, N.A., 2005a. Actuation of the Kagome double-layer grid. Part 1: prediction of performance of the perfect structure. J. Mech. Phys. Solids 53, 1855–1874.

Symons, D.D., Shieh, J., Fleck, N.A., 2005b. Actuation of the Kagome double-layer grid. Part 2: effect of imperfections on the measured and predicted actuation stiffness. J. Mech. Phys. Solids 53, 1875–1891.

Sypeck, D.J., Wadley, H.N.G., 2001. Multifunctional microtruss laminates: textile synthesis and properties. J. Mater. Res. 16, 890–897.

Sypeck, D.J., Parish, P.A., Wadley, H.N.G., 1998. Novel hollow powder porous structures. In: Schwartz, D.S., Shih, D.H., Evans, A.G., Wadley, H.G. (Eds.), Porous and Cellular Materials for Structural Applications. San Francisco, CA, USA, Materials Research Society Symposium Proceedings, vol. 521. (Warrendale PA, USA).

Tadrist, L., Miscevic, M., Rahli, O., Topin, F., 2004. About the use of fibrous materials in compact heat exchangers. Exp. Therm. Fluid Sci. 28, 193–199.

Takahashi, Y., Okumura, D., Ohno, N., 2010. Yield and buckling behavior of Kelvin open-cell foams subjected to uniaxial compression. Int. J. Mech. Sci. 52, 377–385.

Takano, N., Fukasawa, K., Nishiyabu, K., 2010. Structural strength prediction for porous titanium based on micro-stress concentration by micro-CT image-based multiscale simulation. Int. J. Mech. Sci. 52, 229–235.

Tamayol, A., Hooman, K., 2011. Thermal assessment of forced convection through metal foam heat exchangers. J. Heat Transfer 133, 111801–111807.

Tan, P.J., Harrigan, J.J., Reid, S.R., 2002. Inertia effects in uniaxial dynamic compression of a closed cell aluminium alloy foam. Mater. Sci. Technol. 18, 480–488.

Tan, P.J., Reid, S.R., Harrigan, J.J., et al., 2005a. Dynamic compressive strength properties of aluminium foams. Part I–experimental data and observations. J. Mech. Phys. Solids 53, 2174.

Tan, P.J., Reid, S.R., Harrigan, J.J., et al., 2005b. Dynamic compressive strength properties of aluminium foams. Part II–'shock' theory and comparison with experimental data and numerical models. J. Mech. Phys. Solids 53, 2206.

Tan, J.C., Elliott, J.A., Clyne, T.W., 2006a. Analysis of tomography images of bonded fibre networks to measure distributions of fibre segment length and fibre orientation. Adv. Eng. Mater. 8, 495–500.

Tan, P.J., Reid, S.R., Harrigan, J.J., et al., 2006b. Corrigendum to "Dynamic compressive strength properties of aluminium foams. Part II–'shock' theory and comparison with experimental data and numerical models", 2005. J. Mech. Phys. Solids 53, 2206–2230. J. Mech. Phys. Solids, 54, 445.

Tan, Q., Liu, P., Du, C., et al., 2009. Mechanical behaviors of quasi-ordered entangled aluminum alloy wire material. Mater. Sci. Eng. A 527, 38–44.

Tane, M., Ichitsubo, T., 2004. Effective-mean-field approach for macroscopic elastic constants of composites. Appl. Phys. Lett. 85, 197.

Tane, M., Ichitsubo, T., Hirao, M., et al., 2004a. Elastic constants of lotus-type porous magnesium: comparison with effective-mean-field theory. J. Appl. Phys. 96, 3696.

Tane, M., Ichitsubo, T., Nakajima, H., et al., 2004b. Elastic properties of lotus-type porous iron: acoustic measurement and extended effective-mean-field theory. Acta Mater. 52, 5195.

Tane, M., Hyun, S.K., Nakajima, H., 2005. Anisotropic electrical conductivity of lotus-type porous nickel. J. Appl. Phys. 97, 103701.

Tane, M., Hyun, S.-K., Nakajima, H., 2006. Evaluation of elastic and thermoelastic properties of lotus-type porous metals via effective-mean-field theory. Scr. Mater. 54, 545.

Tane, M., Ichitsubo, T., Hirao, M., Nakajima, H., 2007. Extended mean-field method for predicting yield behaviors of porous materials. Mech. Mater. 39, 53.

Tane, M., Kawashima, T., Yamada, H., et al., 2010. Strain rate dependence of anisotropic compression behavior in porous iron with unidirectional pores. J. Mater. Res. 25, 1179–1190.

Tang, Y., Yuan, W., Pan, M., Wan, Z., 2010. Feasibility study of porous copper fiber sintered felt: a novel porous flow field in proton exchange membrane fuel cells. Int. J. Hydrogen Energy 35, 9661–9677.

Tao, X.F., Zhao, Y.Y., 2009. Compressive behavior of Al matrix syntactic foams toughened with Al particles. Scr. Mater. 61, 461–464.

Tao, X.F., Zhang, L.P., Zhao, Y.Y., 2009. Al matrix syntactic foam fabricated with bimodal ceramic microspheres. Mater. Des. 30, 2732–2736.

Taylor, C.M., Smith, C.W., Miller, W., Evans, K.E., 2011. The effects of hierarchy on the in-plane elastic properties of honeycombs. Int. J. Solids Struct. 48, 1330–1339.

Tekoglu, C., Onck, P.R., 2003. A comparison of discrete and Cosserat continuum analyses for cellular materials. In: Banhart, J., Fleck, N., Mortensen, A. (Eds.), Cellular Metals: Manufacture, Properties, Applications, Proc. Conf. Metfoam 2003. Berlin, 23–25 June 2003. Verlag MIT Publishing, Berlin, Germany.

Tekoglu, C., Onck, P., 2005. Size effects in the mechanical behavior of cellular materials. J. Mater. Sci. 40, 5911.

Tekoglu, C., Gibson, L.J., Pardoen, T., Onck, P.R., 2011. Size effects in foams: experiments and modeling. Prog. Mater. Sci. 56, 109–138.

Theocaris, P., 1991. The elliptic paraboloid failure criterion for cellular solids and brittle foams. Acta Mech. 89, 93–121.

Thewsey, D.J., Zhao, Y.Y., 2008. Thermal conductivity of porous copper manufactured by the lost carbonate sintering process. Phys. Status Solidi A 205, 1126–1131.

Thiele, W., 1972. Aluminium used as an impact energy absorbing material. Met. Mater. 6, 349–352.

Thomason, P.F., 1990. Ductile Fracture of Metals. Pergamon Press, Oxford UK.

Thompson, A.H., Katz, A.J., Kronin, C.E., 1987. The microgeometry and transport properties of sedimentary rock. Adv. Phys. 36, 625–694.

Thornton, P.H., Magee, C.L., 1975a. Deformation characteristics of zinc foams. Metall. Trans. 6A, 1801–1807.

Thornton, P.H., Magee, C.L., 1975b. The deformation of aluminum foams. Metall. Trans. 6A, 1253–1263.

Tian, Q., Guo, X., 2011. Manufacturing microporous foam zinc materials with high porosity by electrodeposition. J. Wuhan Univ. Technol., Mater. Sci. Ed. 26, 843–846.

Tian, J., Kim, T., Lu, T.J., et al., 2004. The effects of topology upon fluid-flow and heat-transfer within cellular copper structures. Int. J. Heat Mass Transfer 47, 3171–3186.

Tian, J., Lu, T.J., Hodson, H.P., et al., 2007. Cross flow heat exchange of textile cellular metal core sandwich panels. Int. J. Heat Mass Transfer 50, 2521–2536.

Tien, C.L., Hunt, M.L., 1987. Boundary-layer flow and heat transfer in porous beds. Chem. Eng. Process. 21, 53–63.

Tien, C.-L., Vafai, K., 1990. Convective and radiative heat transfer in porous media. Adv. Appl. Mech. 27, 225–281.

Timoshenko, S., 1925. Analysis of bi-metal thermostats. J. Opt. Soc. Am. 11, 233–255.

T'Joen, C., De Jaeger, P., Huisseune, H., et al., 2010. Thermo-hydraulic study of a single row heat exchanger consisting of metal foam covered round tubes. Int. J. Heat Mass Transfer 53, 3262–3274.

Toda, H., Ohgaki, T., Uesugi, K., et al., 2006a. Quantitative assessment of microstructure and its effects on compression behavior of aluminum foams via high resolution X-ray tomography. Metall. Mater. Trans. 37A, 1211–1219.

Toda, H., Takata, M., Ohgaki, T., et al., 2006b. 3-D image-based mechanical simulation of aluminium foams: effects of internal microstructure. Adv. Eng. Mater. 8, 459–467.

Tomkeieff, S.I., 1945. Nature 155, 24.

Topin, F., Bonnet, J.P., Madani, B., Tadrist, L., 2006. Experimental analysis of multiphase flow in metallic foams: flow laws, heat transfer and convective boiling. Adv. Eng. Mater. 8, 890–899.

Torquato, S., Donev, A., 2004. Minimal surfaces and multifunctionality. Proc. R. Soc. London, Ser. A 460, 1849–1856.

Torquato, S., Gibiansky, L.V., Silva, M.J., Gibson, L.J., 1998. Effective mechanical and transport properties of cellular solids. Int. J. Mech. Sci. 40, 71–82.

Torquato, S., Yeong, C.L.Y., Rintoul, M.D., et al., 1999. Elastic properties and structure of interpenetrating boron carbide/aluminium multiphase composites. J. Am. Ceram. Soc. 82, 1263–1268.

Torquato, S., 2002a. Random Heterogeneous Materials – Microstructure and Macroscopic Properties. Springer-Verlag, New York.

Torquato, S.H. a. S., 2002b. Optimal and manufacturable two-dimensional, Kagomé-like cellular solids. J. Mater. Res. 17, 137–144.

Torquato, S., 2010. Optimal design of heterogeneous materials. Annu. Rev. Mater. Res. 40, 101–129.

Torres, Y., Rodríguez, J., Arias, S., et al., 2012. Processing, characterization and biological testing of porous titanium obtained by space-holder technique. J. Mater. Sci. 47, 6565–6576.

Triantafillou, T., 1990. Constitutive modeling of elastic-plastic open-cell foams. J. Eng. Mech. 116, 2772–2778.

Triantafilou, T.C., Zhang, J., Shercliff, T.L., et al., 1989. Failure surfaces for cellular materials under multiaxial loads - II. Comparison of models with experiment. Int. J. Mech. Sci. 31, 665–678.

Tuncer, N., Arslan, G.R., Maire, E., Salvo, L., 2011a. Influence of cell aspect ratio on architecture and compressive strength of titanium foams. Mater. Sci. Eng. A 528, 7368–7374.

Tuncer, N., Arslan, G.R., Maire, E., Salvo, L., 2011b. Investigation of spacer size effect on architecture and mechanical properties of porous titanium. Mater. Sci. Eng. A 530, 633–642.

Tvergaard, V., 1990. Material failure by void growth to coalescence. Adv. Appl. Mech. 27, 83–151.

Tzeng, S.-C., Jeng, T.-M., 2006. Convective heat transfer in porous channels with 90-deg turned flow. Int. J. Heat Mass Transfer 49, 1452–1461.

Tzeng, S.-C., Jeng, T.-M., 2007. Interstitial heat transfer coefficient and dispersion conductivity in compressed metal foam heat sinks. J. Electron. Packag. 129, 113–119.

Tzeng, S.-C., 2007. Spatial thermal regulation of aluminum foam heat sink using a sintered porous conductive pipe. Int. J. Heat Mass Transfer 50, 117–126.

Uchic, M.D., Shade, P.A., Dimiduk, D.D., 2009a. Microcompression testin of fcc metals: a selected overview of experiments and simulations. JOM 61, 36–41.

Uchic, M.D., Shade, P.A., Dimiduk, D.D., 2009b. Plasticity of micrometer-scale single crystals in compression. Annu. Rev. Mater. Res. 39, 361–386.

Underwood, E.E., 1970. Quantitative Stereology. Addison-Wesley Publishing Company, Reading., MA.

Uri, L., Walmann, T., Alberts, L., et al., 2006. Structure of plastically compacting granular packings. Phys. Rev., E 73, 051301.

Uslu, C., Lee, K.J., Sanders, T.H., Cochran, J.K., 1997. Ti–6Al–4V hollow sphere foams. In: Ward-Close, C.M., Froes, F.H., Chellman, D.J., Cho, S.S. (Eds.), Synthesis/Processing of Lightweight Metallic Materials II. The Minerals, Metals & Materials Soc., Warrendale, PA, USA.

Vafai, K., Sozen, M., 1990. Analysis of energy and momentum transport for fluid flow through a porous bed. J. Heat Transfer 112, 690.

Vafai, K., Tien, C.L., 1981. Boundary and inertia effects on flow and heat transfer in porous media. Int. J. Heat Mass Transfer 24, 195–203.

Vafai, K., Tien, C.L., 1982. Boundary and inertia effects on convective mass transfer in porous media. Int. J. Heat Mass Transfer 25, 1183–1190.

Van der Burg, M.W.D., Shulmeister, V., Van der Giessen, E., Marissen, R., 1997. A numerical study of large deformations of low-density elastomeric open-cell foams. J. Cell. Plast. 33, 31–54.

van Dijk, H.A.J., Boon, J., Nyqvist, R.N., van den Brink, R.W., 2010. Development of a single stage heat integrated water-gas shift reactor for fuel processing. Chem. Eng. J. 159, 182–189.

VanLeeuwen, B.K., Darling, K.A., Koch, C.C., Scattergood, R.O., 2011. Novel technique for the synthesis of ultra-fine porosity metal foam via the inclusion of condensed argon through cryogenic mechanical alloying. Mater. Sci. Eng. A 528, 2192–2195.

vanSchaik, A., Reuter, M.A., vanNieuwkoop, P., 2003. Secondary recovery of aluminium from aluminium foams. In: Banhart, J., Fleck, N., Mortensen, A. (Eds.), Cellular Metals: Manufacture, Properties, Applications, Proc. Conf. Metfoam 2003. Berlin, 23–25 June 2003. Verlag MIT Publishing, Berlin, Germany.

Vendra, L.J., Rabiei, A., 2007. A study on aluminum–steel composite metal foam processed by casting. Mater. Sci. Eng. A 465, 59–67.

Vendra, L., Rabiei, A., 2010. Evaluation of modulus of elasticity of composite metal foams by experimental and numerical techniques. Mater. Sci. Eng. A 527, 1784–1790.

Vendra, L., Neville, B., Rabiei, A., 2009. Fatigue in aluminum–steel and steel–steel composite foams. Mater. Sci. Eng. A 517, 146–153.

Vendra, L., Brown, J., Rabiei, A., 2011. Effect of processing parameters on the microstructure and mechanical properties of Al–steel composite foam. J. Mater. Sci. 46, 4574–4581.

Verbist, G., Weaire, D., Kraynik, A.M., 1996. The foam drainage equation. J. Phys.: Condens. Matter 8, 3715–3731.

Verdooren, A., Chan, H.M., Grenestedt, J.L., et al., 2004. Production of metallic foams from ceramic foam precursors. Adv. Eng. Mater. 6, 397–399.

Veyhl, C., Belova, I.V., Murch, G.E., Fiedler, T., 2011a. Finite element analysis of the mechanical properties of cellular aluminium based on microcomputed tomography. Mater. Sci. Eng. A 528, 4550–4555.

Veyhl, C., Belova, I.V., Murch, G.E., et al., 2011b. Thermal analysis of aluminium foam based on microcomputed tomography. Materialwiss. Werkstofftech. 42, 350–355.

Vityaz, P.A., Kostornov, A.G., Kaptsevich, V.M., Sheleg, V.G., 1989. Porous materials. In: Arunachalam, V.S., Roman, O.V. (Eds.), Powder Metallurgy-recent Advances. Oxford and IBH Publishing Company, New Delhi.

Vodenitcharova, T., Idris, M., Hoffman, M., 2010. Experimental and analytical study on the deformation response of closed-cell Al foam panels to local contact damage – mechanical properties extraction. Mater. Sci. Eng. A 527, 6033–6045.

Volkert, C.A., Lilleodden, E.T., Kramer, D., Weissmuller, J., 2006. Approaching the theoretical strength in nanoporous Au. Appl. Phys. Lett. 89, 061920.

von Zeppelin, F., Hirscher, M., Stanzick, H., Banhart, J., 2003. Desorption of hydrogen from blowing agents used for foaming metals. Compos. Sci. Technol. 63, 2293.

Wachtman, J.B., 1996. Mechanical Properties of Ceramics. John Wiley & Sons, New York.

Wada, T., Inoue, A., 2003. Fabrication, thermal stability and mechanical properties of porous bulk glassy Pd–Cu–Ni–P alloy. Mater. Trans. 44, 2228–2231.

Wada, T., Inoue, A., 2004. Formation of porous Pd-based bulk glassy alloys by a high hydrogen pressure melting-water quenching method and their mechanical properties. Mater. Trans. 45, 2761–2765.

Wada, T., Inoue, A., Greer, A.L., 2005a. Enhancement of room-temperature plasticity in a bulk metallic glass by finely dispersed porosity. Appl. Phys. Lett. 86, 251907-3.

Wada, T., Takenaka, K., Nishiyama, N., Inoue, A., 2005b. Formation and mechanical properties of porous Pd–Pt–Cu–P bulk glassy alloys. Mater. Trans. 46, 2777–2780.

Wada, T., Kinaka, M., Inoue, A., 2006. Effect of volume fraction and geometry of pores on mechanical properties of porous bulk glassy Pd42.5Cu30Ni7.5P20 alloys. J. Mater. Res. 21, 1041–1047.

Wada, T., Wang, X., Kimura, H., Inoue, A., 2008. Preparation of a Zr-based bulk glassy alloy foam. Scr. Mater. 59, 1071–1074.

Wada, T., Setyawan, A.D., Yubuta, K., Kato, H., 2011a. Nano- to submicro-porous beta-Ti alloy prepared from dealloying in a metallic melt. Scr. Mater. 65, 532–535.

Wada, T., Yubuta, K., Inoue, A., Kato, H., 2011b. Dealloying by metallic melt. Mater. Lett. 65, 1076–1078.

Wadley, H.N.G., Fleck, N.A., Evans, A.G., 2003. Fabrication and structural performance of periodic cellular metal sandwich structures. Compos. Sci. Technol. 63, 2331.

Wadley, H., 2002. Cellular metals manufacturing. Adv. Eng. Mater. 4, 726–733.

Wadley, H.N.G., 2006. Multifunctional periodic cellular metals. Philos. Trans. R. Soc., A 364, 31.

Wagner, I., Hintz, C., Sahm, P.R., 1999. Precision-cast near net shape components based on cellular metal materials. In: Clyne, T.W., Simancik, F. (Eds.), Euromat 99. 27–30 Sept. 1999, Munich, Germany. DGM/Wiley-VCH, Weinheim, Germany.

Wallach, J.C., Gibson, L.J., 2001a. Defect sensitivity of a 3D truss material. Scr. Mater. 45, 639.

Wallach, J.C., Gibson, L.J., 2001b. Mechanical behavior of a three-dimensional truss material. Int. J. Solids Struct. 38, 7181–7196.

Walsh, J.B., Brace, W.F., England, A.W., 1965. Effect of porosity on compressibility of glass. J. Am. Ceram. Soc. 48, 605–608.

Wan, J., Li, Y., Liu, Y., 2007. Spatial distribution of pores in lotus-type porous metal. J. Mater. Sci. 42, 6446–6452.

Wang, D.A., Pan, J., 2006. A non-quadratic yield function for polymeric foams. Int. J. Plast. 22, 434–458.

Wang, L., Tseng, K.K., 2003. A multi-scale framework for effective elastic properties of porous materials. J. Mater. Sci. 38, 3019–3027.

Wang, N., Starke, E.A., Wadley, H.N.G., 1998. Porous Al alloys by local melting and diffusion of metal powders. In: Schwartz, D.S., Shih, D.H., Evans, A.G., Wadley, H.G. (Eds.), Porous and Cellular Materials for Structural Applications. San Francisco, CA, USA, Materials Research Society Symposium Proceedings, vol. 521. (Warrendale PA, USA).

Wang, D.L., Dai, C.S., Wu, N., et al., 2003. Multiple-step constant current density electrodeposition of continuous metal material with high porosity. J. Appl. Electrochem. 33, 725–732.

Wang, Y., Liang, Z., Yuan, X., Xu, Y., 2005. Preparation of cellular iron using wastes and its application in dyeing wastewater treatment. J. Porous Mater. 12, 225–232.

Wang, J., Duan, H.L., Huang, Z.P., Karihaloo, B.L., 2006a. A scaling law for properties of nano-structured materials. Proc. R. Soc. London, Ser A 462, 1355–1363.

Wang, Z., Ma, H., Zhao, L., Yang, G., 2006b. Studies on the dynamic compressive properties of open-cell aluminum alloy foams. Scr. Mater. 54, 83.

Wang, G., Demetriou, M.D., Schramm, J.P., et al., 2010a. Compression-compression fatigue of Pd[sub 43]Ni[sub 10]Cu[sub 27]P[sub 20] metallic glass foam. J. Appl. Phys. 108, 023505–023507.

Wang, Q.Z., Cui, C.X., Liu, S.J., Zhao, L.C., 2010b. Open-celled porous Cu prepared by replication of NaCl space-holders. Mater. Sci. Eng. A 527, 1275–1278.

Wang, Y., Wang, J., Jia, P., 2011. Performance of forced convection heat transfer in porous media based on Gibson–Ashby constitutive model. Heat Transfer Eng. 32, 1093–1098.

Warren, W.E., Kraynik, A.M., 1987. Foam Mechanics: the linear elastic response of two-dimensional spatially periodic cellular materials. Mech. Mater. 6, 27–37.

Warren, W.E., Kraynik, A.M., 1988. The linear elastic properties of open-cell foams. J. Appl. Mech. 55, 341–346.

Warren, W.E., Kraynik, A.M., 1991. The nonlinear elastic behavior of open-cell foams. J. Appl. Mech. 58, 376–381.

Warren, W.E., Kraynik, A.M., 1997. Linear elastic behavior of a low-density Kelvin foam with open-cells. J. Appl. Mech. 64, 787–794.

Watson, T.J., Wert, J.A., 1993. On the development of constitutive relations for metallic powders. Metall. Trans. 24A, 2071–2981.

Weaire, D., Fortes, M.A., 1994. Stress and strain in liquid and solid foams. Adv. Phys. 43, 685–738.

Weaire, D., Hutzler, S., 1999. The Physics of Foam. Oxford University Press, Oxford, UK.

Weaire, D., Phelan, R., 1994. Philos. Mag. Lett. 69, 107.

Weaire, D., Phelan, R., 1996. The physics of foam. J. Phys.: Condens. Matter 8, 9519–9524.

Weaire, D., Phelan, R., Verbist, G., 1999. The structure and geometry of foams. In: Sadoc, J.F., Rivier, N. (Eds.), Foams and Emulsions. Cargese, Corsica, France. Kluwer Academic Publishers, Dordrecht, the Netherlands.

Weaire, D., 2002. Foam physics. Adv. Eng. Mater. 4, 723–724.

Weaire, D., 2009. Kelvin's ideal foam structure. J. Phys.: Conf. Ser. 158, 012005.

Weber, M., Knüver, M., 1997. Evaluierung Verschiedener Herstellungs- und Anwendungs-möglichkeiten für Hochporöse Stahlwerkstoffe. In: Banhart, J. (Ed.), Metallschäume. Bremen, March 1997. Verlag MIT, Bremen.

Weber, L., Dorn, J., Mortensen, A., 2003a. On the electrical conductivity of metal matrix composites containing high volume fractions of non-conducting inclusions. Acta Mater. 51, 3199.

Weber, L., Fischer, C., Mortensen, A., 2003b. On the influence of tne shape of randomly oriented, non-conducting inclusions in a conducting matrix on the effective electrical conductivity. Acta Mater. 51, 495.

Weber, J., 1985. German Patent DE 3,516,737.

Weber, L., 2005. Non-conducting inclusions in a conducting matrix: influence of inclusion size on electrical conductivity. Acta Mater. 53, 1945.

Wei, L., Sun, Y., 2011. Study on bubble's stability in process of preparing foam aluminum by powder metallurgy method. Adv. Mater. Res. 146–147, 370–373.

Wei, J., Cheng, H., Gong, C., et al., 2002a. Effects of macroscopic pores on the damping behavior of foamed commercially pure aluminum. Metall. Mater. Trans. A 33, 3565–3568.

Wei, J.N., Gong, C.L., Cheng, H.F., et al., 2002b. Low-frequency damping behavior of foamed commercially pure aluminum. Mater. Sci. Eng. A 332, 375–381.

Wei, P.S., Huang, C.C., Lee, K.W., 2003. Nucleation of bubbles on a solidification front – experiment and analysis. Metall. Mater. Trans. B 34, 321–332.

Weinberger, C.R., Cai, W., 2012. Plasticity of metal nanowires. J. Mater. Chem. 22, 3277–3292.

Weise, J., Zanetti-Bueckmann, V., Yezerska, O., et al., 2007. Processing, properties and coating of micro-porous syntactic foams. Adv. Eng. Mater. 9, 52–56.

Weissmüller, J., Newman, R.C., Jin, H.-J., et al., 2009. Nanoporous metals by alloy corrosion: formation and mechanical properties. MRS Bull. 34, 577–586.

Wen, C.E., Mabuchi, M., Yamada, Y., et al., 2001. Processing of biocompatible porous Ti and Mg. Scr. Mater. 45, 1147.

Wen, C.E., Mabuchi, M., Yamada, Y., et al., 2003. Processing of fine-grained aluminum foam by spark plasma sintering. J. Mater. Sci. Lett. 22, 1407.

Wen, C.E., Yamada, Y., Shimojima, K., et al., 2004. Compressibility of porous magnesium foam: dependency on porosity and pore size. Mater. Lett. 58, 357.

Wen, C.E., Xiong, J.Y., Li, Y.C., Hodgson, P.D., 2010. Porous shape memory alloy scaffolds for biomedical applications: a review. Phys. Scr. T139, 014070.

Werther, D.J., Howard, A.J., Ingraham, J.P., Issen, K.A., 2006. Characterization and modeling of strain localization in aluminum foam using multiple face analysis. Scr. Mater. 54, 783.

Wicklein, M., Thoma, K., 2005. Numerical investigations of the elastic and plastic behaviour of an open-cell aluminium foam. Mater. Sci. Eng. A 397, 391.

Wicks, N., Guest, S.D., 2004. Single member actuation in large repetitive truss structures. Int. J. Solids Struct. 41, 965–978.

Wicks, N., Hutchinson, J.W., 2001. Optimal truss plates. Int. J. Solids Struct. 38, 5165–5183.

Willis, J.R., 1991. On methods for bounding the overall properties of nonlinear composites. J. Mech. Phys. Solids 39, 73–86.

Witte, F., Ulrich, H., Palm, C., Willbold, E., 2007a. Biodegradable magnesium scaffolds: part II: peri-implant bone remodeling. J. Biomed. Mater. Res. 81A, 757–765.

Witte, F., Ulrich, H., Rudert, M., Willbold, E., 2007b. Biodegradable magnesium scaffolds: part 1: appropriate inflammatory response. J. Biomed. Mater. Res. 81A, 748–756.

Witte, F., Ulrich, H., Rudert, M., Willbold, E., 2007c. Biodegradable magnesium scaffolds; part I: appropriate inflammatory response. J. Biomed. Mater. Res. 81, 748–756.

Wong, P.-Z., Koplik, J., Tomanic, J.P., 1984. Conductivity and permeability of rocks. Phys. Rev., B 30, 6606–6614.

Wu, Z.G., Zhao, C.Y., 2011. Experimental investigations of porous materials in high temperature thermal energy storage systems. Sol. Energy 85, 1371–1380.

Wu, G.H., Dou, Z.Y., Sun, D.L., et al., 2007. Compression behaviors of cenosphere-pure aluminum syntactic foams. Scr. Mater. 56, 221.

Wubben, T., Odenbach, S., 2005. Stabilisation of liquid metallic foams by solid particles. Colloids Surf., A 266, 207.

Xie, L.S., Chan, K.C., 2006. The effect of strut geometry on the yielding behaviour of open-cell foams. Int. J. Mech. Sci. 48, 249.

Xie, S., Evans, J.R.G., 2004. High porosity copper foam. J. Mater. Sci. 39, 5877–5880.

Xie, Z., Ikeda, T., Okuda, Y., Nakajima, H., 2004. Sound absorption characteristics of lotus-type porous copper fabricated by unidirectional solidification. Mater. Sci. Eng. A 386, 390–395.

Xie, G., Zhang, W., Louzguine-Luzgin, D.V., et al., 2006. Fabrication of porous Zr–Cu–Al–Ni bulk metallic glass by spark plasma sintering process. Scr. Mater. 55, 687.

Xu, J., Tian, J., Lu, T.J., Hodson, H.P., 2007. On the thermal performance of wire-screen meshes as heat exchanger material. Int. J. Heat Mass Transfer 50, 1141–1154.

Xu, J., Ji, X., Zhang, W., Liu, G., 2008. Pool boiling heat transfer of ultra-light copper foam with open cells. Int. J. Multiphase Flow 34, 1008–1022.

Xu, H.J., Qu, Z.G., Tao, W.Q., 2011. Analytical solution of forced convective heat transfer in tubes partially filled with metallic foam using the two-equation model. Int. J. Heat Mass Transfer 54, 3846–3855.

Xue, X., Zhao, Y.Y., 2011. Ti matrix syntactic foam fabricated by powder metallurgy: particle breakage and elastic modulus. JOM 63, 43–47.

Xue, W., Li, Y., Yuan, L., 2007. Structural features in radial-type porous magnesium fabricated by radial solidification. Mater. Sci. Eng. A 444, 306.

Yamada, Y., Shimojima, K., Sakaguchi, Y., et al., 1999. Processing of an open-cellular AZ91 magnesium alloy with a low density of 0.05 g cm^{-3}. J. Mater. Sci. Lett. 18, 1477–1480.

Yamada, Y., Shimojima, K., Sakaguchi, Y., et al., 2000. Processing of cellular magnesium materials. Adv. Eng. Mater. 2, 184–187.

Yan, W., Pun, C.L., 2010. Spherical indentation of metallic foams. Mater. Sci. Eng. A 527, 3166–3175.

Yang, Z., Garimella, S.V., 2010. Melting of phase change materials with volume change in metal foams. J. Heat Transfer 132, 062301–062311.

Yang, M.-Y., Huang, J.-S., 2004. Numerical analysis of the stiffness and strength of regular hexagonal honeycombs with Plateau borders. Compos. Struct. 64, 107–114.

Yang, C.C., Nakae, H., 2000. Foaming characteristics control during production of aluminum alloy foam. J. Alloys Compd. 313, 188.

Yang, C.C., Nakae, H., 2003. The effects of viscosity and cooling conditions on the foamability of aluminum alloy. J. Mater. Process. Technol. 141, 202.

Yang, D.H., Hur, B.Y., He, D.P., Yang, S.R., 2007a. Effect of decomposition properties of titanium hydride on the foaming process and pore structures of Al alloy melt foam. Mater. Sci. Eng. A 445–446, 415.

Yang, Y.J., Han, F.S., Wen, C.E., Shu, Y.F., 2007b. Damping properties of open cell microcellular pure Al foams. Mater. Sci. Technol. 23, 1336–1340.

Yang, Y.J., Han, F.S., Yang, D.K., et al., 2007c. Effect of heat treatment on compressive properties of open cell Al/Al2O3 composite foams. Powder Metall. 50, 50–53.

Yang, Y.J., Han, F.S., Yang, D.K., Zheng, K., 2007d. Compressive behaviour of open cell Al Al2O3 composite foams fabricated by sintering and dissolution process. Mater. Sci. Technol. 23, 50–53.

Yang, D.H., Hur, B.Y., Yang, S.R., 2008a. Study on fabrication and foaming mechanism of Mg foam using CaCO3 as blowing agent. J. Alloys Compd. 461, 221–227.

Yang, M.-Y., Huang, J.-S., Sam, C.-P., 2008b. In-plane elastic moduli and plastic collapse strength of regular hexagonal honeycombs with dual imperfections. Int. J. Mech. Sci. 50, 43–54.

Yang, D.-H., Shang-Run, Y., Hui, W., et al., 2010a. Compressive properties of cellular Mg foams fabricated by melt-foaming method. Mater. Sci. Eng. A 527, 5405–5409.

Yang, Y., Ji, X., Xu, J., 2010b. Pool boiling heat transfer on copper foam covers with water as working fluid. Int. J. Therm. Sci. 49, 1227–1237.

Yang, C., Nakayama, A., Liu, W., 2012a. Heat transfer performance assessment for forced convection in a tube partially filled with a porous medium. Int. J. Therm. Sci. 54, 98–108.

Yang, L., Cormier, D., West, H., et al., 2012b. Non-stochastic Ti–6Al–4V foam structures with negative Poisson's ratio. Mater. Sci. Eng. A 558, 579–585.

Yang, L., Harrysson, O., West, H., Cormier, D., 2012c. Compressive properties of Ti–6Al–4V auxetic mesh structures made by electron beam melting. Acta Mater. 60, 3370–3379.

Yasuda, H., Ohnaka, I., Fujimoto, S., et al., 2004. Fabrication of porous aluminum with deep pores by using Al–In monotectic solidification and electrochemical etching. Mater. Lett. 58, 911.

Yasuda, H., Ohnaka, I., Fujimoto, S., et al., 2006. Fabrication of aligned pores in aluminum by electrochemical dissolution of monotectic alloys solidified under a magnetic field. Scr. Mater. 54, 527.

Ye, B., Dunand, D.C., 2010. Titanium foams produced by solid-state replication of NaCl powders. Mater. Sci. Eng. A 528, 691–697.

Yerokhin, A., Nie, X., Leyland, A., et al., 1999. Plasma electrolysis for surface engineering. Surf. Coat. Technol. 122, 73–93.

Yi, F., Zhu, Z., Zu, F., et al., 2001. Strain rate effects on the compressive property and the energy-absorbing capacity of aluminum alloy foams. Mater. Character. 47, 417.

Young, C.M., Robinson, S.L., Sherby, O.D., 1975. Effect of subgrain size on the high temperature strength of polycrystalline aluminium as determined by constant strain rate tests. Acta Metall. 23, 633–639.

Young, M., DeFouw, J., Frenzel, J., Dunand, D., 2012. Cast-replicated NiTiCu foams with superelastic properties. Metall. Mater. Trans. A 43, 2939–2944.

Youssef, S., Maire, E., Gaertner, R., 2005. Finite element modelling of the actual structure of cellular materials determined by X-ray tomography. Acta Mater. 53, 719–730.

Yu, A.B., Standish, N., 1987a. Porosity calculations of multi-component mixtures of spherical particles. Powder Technol. 52, 233–241.

Yu, A.B., Standish, N., 1987b. Porosity calculations of ternary mixtures of spherical particles. Powder Technol. 52, 249–253.

Yu, A.B., Standish, N., 1993. Characterisation of non-spherical particles from their packing behaviour. Powder Technol. 74, 205–213.

Yu, C.J., Eifert, H.H., Banhart, J., Baumeister, J., 1998a. Mater. Res. Innovations 2, 181–188.

Yu, C.J., Eifert, H.H., Knuewer, M., et al., 1998b. Investigation for the selection of foaming agents to produce steel foams. In: Schwartz, D.S., Shih, D.H., Evans, A.G., Wadley, H.G. (Eds.), Porous and Cellular Materials for Structural Applications. San Francisco, CA, USA, Materials Research Society Symposium Proceedings, vol. 521. (Warrendale PA, USA).

Yu, J.L., Li, J.R., Hu, S.S., 2006. Strain-rate effect and micro-structural optimization of cellular metals. Mech. Mater. 38, 160–170.

Yu, S., Luo, Y., Liu, J., 2008. Effects of strain rate and SiC particle on the compressive property of SiCp/AlSi9Mg composite foams. Mater. Sci. Eng. A 487, 394–399.

Yuan, B., Chung, C.Y., Zhu, M., 2004. Microstructure and martensitic transformation behavior of porous NiTi shape memory alloy prepared by hot isostatic pressing processing. Mater. Sci. Eng. A 382, 181.

Yuan, L., Yanxiang, L., Jiang, W., Huawei, Z., 2005. Evaluation of porosity in lotus-type porous magnesium fabricated by metal/gas eutectic unidirectional solidification. Mater. Sci. Eng. A 402, 47.

Yuan, B., Zhang, Y.P., Chung, C.Y., et al., 2006. A comparative study of the porous TiNi shape-memory alloys fabricated by three different processes. Metall. Mater. Trans. 37A, 755–761.

Zardiackas, L.D., Parsell, D.E., Dillon, L.D., et al., 2001. Structure, metallurgy, and mechanical properties of a porous tantalum foam. J. Biomed. Mater. Res., Part B 58, 180–187.

Zare, J., Manesh, H.D., 2011. A novel method for producing of steel tubes with Al foam core. Mater. Des. 32, 1325–1330.

Zettl, B., Mayer, H., Stanzl-Tschegg, S.E., Degischer, H.P., 2000. Fatigue properties of alumnium foams at high numbers of cycles. Mater. Sci. Eng. A 292, 1–7.

Zettl, B., Mayer, H., Stanzl-Tschegg, S.E., 2001. Fatigue properties of Al–1Mg–0.6Si foam at low and ultrasonic frequencies. Int. J. Fatigue 23, 565–573.

Zhang, H., Cooper, A.I., 2007. Aligned porous structures by directional freezing. Adv. Mater. 19, 1529–1533.

Zhang, T., Lee, J., 2003. A plasticity model for cellular materials with open-celled structure. Int. J. Plast. 19, 749–770.

Zhang, L.P., Zhao, Y.Y., 2007. Mechanical response of Al matrix syntactic foams produced by pressure infiltration casting. J. Compos. Mater. 41, 2105–2117.

Zhang, P., Kraft, O., Arzt, E., et al., 2002. Microstructural changes in the cell walls of a closed-cell aluminium foam during creep. Philos. Mag. A 2895.

Zhang, H.Y., Pinjala, D., Joshi, Y.K., et al., 2007a. Development and characterization of thermal enhancement structures for single-phase liquid cooling in microelectronics systems. Heat Transfer Eng. 28, 997.

Zhang, Y.P., Li, D.S., Zhang, X.P., 2007b. Gradient porosity and large pore size NiTi shape memory alloys. Scr. Mater. 57, 1020–1023.

Zhang, B., Kim, T., Lu, T.J., 2009a. Analytical solution for solidification of close-celled metal foams. Int. J. Heat Mass Transfer 52, 133–141.

Zhang, Q., Lee, P.D., Singh, R., et al., 2009b. Micro-CT characterization of structural features and deformation behavior of fly ash/aluminum syntactic foam. Acta Mater. 57, 3003–3011.

Zhang, X.X., Witherspoon, C., Mullner, P., Dunand, D.C., 2011. Effect of pore architecture on magnetic-field-induced strain in polycrystalline Ni–Mn–Ga. Acta Mater. 59, 2229–2239.

Zhao, H., Gary, G., 1998. Crushing behaviour of aluminium honeycombs under impact loading. Int. J. Impact Eng. 21, 827–836.

Zhao, Y.Y., Sun, D.X., 2001. A novel sintering-dissolution process for manufacturing Al foams. Scr. Mater. 44, 105–110.

Zhao, C.Y., Wu, Z.G., 2011. Heat transfer enhancement of high temperature thermal energy storage using metal foams and expanded graphite. Sol. Energy Mater. Sol. Cells 95, 636–643.

Zhao, Y., Tandon, G., Weng, G., 1989. Elastic moduli for a class of porous materials. Acta Mech. 76, 105–131.

Zhao, C.Y., Kim, T., Lu, T.J., Hodson, H.P., 2004a. Thermal transport in high porosity cellular metal foams. J. Thermophys. Heat Transfer 18, 309–317.

Zhao, C.Y., Lu, T.J., Hodson, H.P., 2004b. Thermal radiation in ultralight metal foams with open cells. Int. J. Heat Mass Transfer 47, 2927–2939.

Zhao, C.Y., Lu, T.J., Hodson, H.P., Jackson, J.D., 2004c. The temperature dependence of effective thermal conductivity of open-celled steel alloy foams. Mater. Sci. Eng. A 367, 123–131.

Zhao, Y., Han, F., Fung, T., 2004d. Optimisation of compaction and liquid-state sintering in sintering and dissolution process for manufacturing Al foams. Mater. Sci. Eng. A 364, 117.

Zhao, C.Y., Lu, T.J., Hodson, H.P., 2005a. Natural convection in metal foams with open cells. Int. J. Heat Mass Transfer 48, 2452–2463.

Zhao, H., Elnasri, I., Abdennadher, S., 2005b. An experimental study on the behaviour under impact loading of metallic cellular materials. Int. J. Mech. Sci. 47, 757–774.

Zhao, Y.Y., Fung, T., Zhang, L.P., Zhang, F.L., 2005c. Lost carbonate sintering process for manufacturing metal foams. Scr. Mater. 52, 295.

Zhao, C.Y., Lu, W., Tassou, S.A., 2006a. Thermal analysis on metal-foam filled heat exchangers. Part II: tube heat exchangers. Int. J. Heat Mass Transfer 49, 2762–2770.

Zhao, H., Elnasri, I., Li, H., 2006b. The mechanism of strength enhancement under impact loading of cellular materials. Adv. Eng. Mater. 8, 877–883.

Zhao, Y., Taya, M., Izui, H., 2006c. Study on energy absorbing composite structure made of concentric NiTi spring and porous NiTi. Int. J. Solids Struct. 43, 2497–2512.

Zhao, H., Elnasri, I., Girard, Y., 2007. Perforation of aluminium foam core sandwich panels under impact loading–An experimental study. Int. J. Impact Eng. 34, 1246.

Zhao, C.Y., Tassou, S.A., Lu, T.J., 2008. Analytical considerations of thermal radiation in cellular metal foams with open cells. Int. J. Heat Mass Transfer 51, 929–940.

Zhao, C.Y., Lu, W., Tassou, S.A., 2009a. Flow boiling heat transfer in horizontal metal-foam tubes. J. Heat Transfer 131, 121002–121008.

Zhao, X., Sun, H., Lan, L., et al., 2009b. Pore structures of high-porosity NiTi alloys made from elemental powders with NaCl temporary space-holders. Mater. Lett. 63, 2402–2404.

Zhao, C.Y., Dai, L.N., Tang, G.H., et al., 2010. Numerical study of natural convection in porous media (metals) using Lattice Boltzmann Method (LBM). Int. J. Heat Fluid Flow 31, 925–934.

Zhao, Y.Y., 2003. Stochastic modelling of removability of NaCl in sintering and dissolution process to produce Al foams. J. Porous Mater. 10, 105–111.

Zhao, Y.Y., 2011. Metal matrix syntactic foams: manufacture, matrix material, microstructure, modulus and more. JOM 63, 35.

Zhao, C.Y., 2012. Review on thermal transport in high porosity cellular metal foams with open cells. Int. J. Heat Mass Transfer 55, 3618–3632.

Zheng, Y., Gu, X., 2011. Research activities of biomedical magnesium alloys in China. JOM 63, 105–108.

Zheng, Z., Yu, J., Li, J., 2005. Dynamic crushing of 2D cellular structures: a finite element study. Int. J. Impact Eng. 32, 650–664.

Zhou, J., Soboyejo, W.O., 2004. Compression-compression fatigue of open cell aluminum foams: macro-/micro- mechanisms and the effects of heat treatment. Mater. Sci. Eng. A 369, 23–35.

Zhou, D., Zhao, C.Y., 2011. Experimental investigations on heat transfer in phase change materials (PCMs) embedded in porous materials. Appl. Therm. Eng. 31, 970–977.

Zhou, J., Mercer, C., Soboyejo, W.O., 2002. An investigation of the microstructure and strength of open-cell 6101 aluminum foams. Metall. Mater. Trans. 33A, 1413–1427.

Zhou, J., Gao, Z., Cuitino, A.M., Soboyejo, W.O., 2004a. Effects of heat treatment on the compressive deformation behavior of open cell aluminum foams. Mater. Sci. Eng. A 386, 118–128.

Zhou, J., Shrotriya, P., Soboyejo, W.O., 2004b. Mechanisms and mechanics of compressive deformation in open-cell Al foams. Mech. Mater. 36, 781.

Zhou, J., Shrotriya, P., Soboyejo, W.O., 2004c. On the deformation of aluminum lattice block structures: from struts to structures. Mech. Mater. 36, 723–737.

Zhou, J., Gao, Z., Cuitino, A.M., Soboyejo, W.O., 2005a. Fatigue of As-fabricated open cell aluminum foams. J. Eng. Mater. Technol.-Trans. ASME 127, 40–45.

Zhou, J., Allameh, S., Soboyejo, W.O., 2005b. Microscale testing of the strut in open-cell aluminum foams. J. Mater. Sci. 40, 429–439.

Zhou, G.-Y., Tu, S.-T., Xuan, F.-Z., Wang, Z., 2011. Viscoelastic model to describe mechanical response of compact heat exchangers with plate-foam structure. Int. J. Mech. Sci. 53, 1069–1076.

Zhou, Z., Wang, Z., Zhao, L., Shu, X., 2012. Loading rate effect on yield surface of aluminum alloy foams. Mater. Sci. Eng. A 543, 193–199.

Zhu, W.Z., Deevi, S.C., 2003. A review on the status of anode materials for solid oxide fuel cells. Mater. Sci. Eng. A 362, 228.

Zhu, H.X., Windle, A.H., 2002. Effects of cell irregularity on the high strain compression of open-cell foams. Acta Mater. 50, 1041–1052.

Zhu, H.X., Knott, J.F., Mills, N.J., 1997a. Analysis of the elastic properties of open-cell foams with tetrakaidecahedral cells. J. Mech. Phys. Solids 45, 319–343.

Zhu, H.X., Mills, N.J., Knott, J.F., 1997b. Analysis of the high strain compression of open-cell foams. J. Mech. Phys. Solids 45, 1875.

Zhu, H.X., Hobdell, J.R., Windle, A.H., 2000. Effects of cell irregularity on the elastic properties of open-cell foams. Acta Mater. 48, 4893–4900.

Zhu, H.X., Hobdell, J.R., Windle, A.H., 2001. Effects of cell irregularity on the elastic properties of 2D Voronoi honeycombs. J. Mech. Phys. Solids 49, 857–870.

Zhu, H.X., Thorpe, S.M., Windle, A.H., 2006. The effect of cell irregularity on the high strain compression of 2D Voronoi honeycombs. Int. J. Solids Struct. 43, 1061–1078.

Zhu, F., Chou, C.C., Yang, K.H., 2011a. Shock enhancement effect of lightweight composite structures and materials. Composites, Part B 42, 1202–1211.

Zhu, Y., Hu, H., Ding, G., et al., 2011b. Influence of oil on nucleate pool boiling heat transfer of refrigerant on metal foam covers. Int. J. Refrig. 34, 509–517.

Zhuang, H., Han, Y., Feng, A., 2008. Preparation, mechanical properties and in vitro biodegradation of porous magnesium scaffolds. Mater. Sci. Eng. C 28, 1462–1466.

Zimmerman, R.W., 1991. Elastic moduli of a solid containing spherical inclusions. Mech. Mater. 12, 17.

Zok, F.W., Waltner, S.A., Wei, Z., et al., 2004. A protocol for characterizing the structural performance of metallic sandwich panels: application to pyramidal truss cores. Int. J. Solids Struct. 41, 6249–6271.

Zurob, H., Brechet, Y., 2005. Effect of structure on the creep of open-cell nickel foams. J. Mater. Sci. 40, 5893.

Biography

Dr Goodall is Senior Lecturer in Metallurgy at the Department of Materials Science and Engineering at the University of Sheffield. He joined the university in 2008 from the Swiss Federal Institute of Technology (EPFL) in Lausanne, Switzerland. He obtained his MEng degree from Oxford University and his PhD from the University of Cambridge, before carrying out postdoctoral research at EPFL; this was where he first undertook research on metal foams, in particular those processed using the replication process. He currently leads a research group looking at the processing and properties of metal sponges and additively manufactured lattices, and also novel alloy development and mechanical behaviour.

Andreas Mortensen is professor and head of the Laboratory for Mechanical Metallurgy at EPFL. He was dean of doctoral studies for EPFL from 2000 to 2005 and director of EPFL's Institute of Materials from 2006 to 2012. Prior to joining the faculty of EPFL in 1997, he was postdoctoral researcher at Nippon Steel in 1986 and then from 1986 to 1996, he was an assistant professor, associate professor and then full professor in the Department of Materials Science and Engineering of the Massachusetts Institute of Technology. His research covers the spectrum from processing to mechanical properties of metallic materials with focus on advanced materials such as metallic composites or microcellular metals.

25 Hydrogen in Metals

R. Kirchheim and A. Pundt, Institut für Materialphysik, Georg-August-Universität Göttingen, Göttingen, Germany.

25.1 Introduction

Hydrogen in metals attracts interest from scientists since many decades. Most of the interesting properties are related to the small size of hydrogen: its interstitial diffusion accompanied by quantum mechanical tunnel transport results in an extraordinary high mobility of hydrogen atoms in materials. For metals, H diffusivity may reach values as known for ions in aqueous solutions. Thus, thermodynamic equilibrium is reached within comparably short times even at room temperature. Therefore, metal–hydrogen systems are often used as model systems to study physical or chemical properties and their change with concentration (see, for example Oates and Flanagan, 1981). In 1937, Lacher (1937) already used Pd–H (Oates and Flanagan, 1981; Flanagan and Oates, 1991) to study solute–solute interactions and interpreted it in the framework of a quasi-chemical approach (Lacher, 1937). The quantum mechanical tunneling as a diffusion mechanism also for atoms in solids was first discovered and discussed for hydrogen tunneling in metals (Flynn and Stoneham, 1970; Völk and Alefeld, 1975; Birnbaum and Flynn, 1976). Völk and Alefeld (1978), Zabel and Peisl (1979, 1980), and Steyrer and Peisl (1986) studied hydrogen density modulations that are related to the sample geometry; and Zabel and his colleagues, as published by Miceli et al. (1985), Uher et al. (1987), Song et al. (1996, 2000), and Uher et al. (1987), firstly used metal–hydrogen systems to study the behavior of systems with reduced dimensions and modulated hydrogen affinity. Kirchheim (1988) and colleagues extensively studied metal–hydrogen systems as representative for solute/solvent systems. The high mobility of hydrogen further allows studying the impact of defects that usually annihilate at elevated temperatures, see Gottstein (2001). It was, therefore, suggested to use hydrogen as a probe for defects (Cahn, 1990; Flanagan et al., 2001a, 2001b; Kirchheim, 2004) and perform site energy spectroscopy by gradually increasing the hydrogen chemical potential.

Hydrogen as the smallest of all atoms allows a dense packing in metal hosts often accompanied with a large negative heat of solution for hydrogen. The hydrogen density in metal hydrides can be even larger than that in liquid hydrogen (Latroche, 2004; Akiba, 1999). This stimulates the use of metal hydrides for energy storage (Züttel et al., 2004). The search for new energy storage metal hydrides was strongly directing the research activities of many groups, especially during the past decade. Since, nowadays, mobile energy carriers become more and more important in human society, volume and weight of the hosts materials are of major concern (Züttel and Schlapbach, 2001, 2002, 2004; Conte et al., 2004; Zaluska et al., 2000; Reiser et al., 2000; Ichikawa et al., 2004; Hirscher et al., 2003; Orimo et al., 2004; Klassen et al., 2001). Actually, for portable applications energy storage alloys like FeTi, LaNi$_5$ and multicompound alloys are successfully used working under ambient pressures and temperatures. They reach hydrogen weight capacities of 1.5–2.2% (Hong, 2001; Macdonald et al., 1981). Züttel and Schlapbach (2001) and Züttel et al. (2008) have recently overviewed issues of current hydrogen storage. In the handbook edited by Hirscher (2010), different actual storage principles are collected, including metal hydrides but also physisorption storage in porous materials and storage in pressurized gas tanks. Metal hydrides with sufficient weight capacities are suitable for stationary storage, especially when long-term storage times are required, because of their durability.

Hydrogen embrittlement (HE) is another subject of research also driven by its technological importance (Moody et al., 1990; Thompson and Moody, 1994; Somerday et al., 2008). Here, hydrogen plays a negative role resulting in metal failure even at low temperatures and very low concentrations. For iron-based alloys, namely high-strength steel, embrittlement effects are observed at H concentrations as low as a few ppm. The interaction of H atoms with lattice defects, the effect of hydrogen on plasticity, on the mobility of dislocations and crack tips has to be studied in detail. Again, the high mobility of hydrogen plays a crucial role. High mobilities occur especially in body-centered cubic (bcc) metals, where adjacent interstitial sites with very short distances are present. Thus, ferritic steels with bcc lattice are more susceptible to HE than austenitic steels possessing face-centered cubic (fcc) lattice structure.

Because of the large relative mass change of its isotopes, pronounced isotope effects occur in most of hydrogen's properties (Wicke and Brodowski, 1978a, 1978b). The isotope's different scattering length for neutrons, even changing sign from hydrogen to deuterium, gave rise to an extensive use of neutron scattering and diffraction methods in the field of metal–hydrogen systems (Ross, 1997).

Hydrogen may remarkably change the physical properties of a material. Examples range from metal–insulator transitions in some rare earths, like yttrium, when transforming from the dihydride into the trihydride (Schlapbach et al., 1987; Vajda, 1995; Huiberts et al., 1996), to changes in the magnetic ordering (cf. Hémon et al., 2000; Vajda, 2000) and interlayer coupling of multilayers (Zabel and Weidinger, 1995). In addition, microstructural changes have been observed during hydrogen absorption. Examples are the generation of superabundant vacancies (SAV) (Fukai and Okuma, 1994) and dislocations (Schober and Wenzl, 1978; Jamieson et al., 1976; Flangan and Lynch, 1976), the decomposition of miscible alloys (Noh et al., 1996; Lücke et al., 2002), and the change of mechanical properties of Ti alloys by decreasing the grain size (Nakahigashi and Yoshimura, 2002) or by changing the volume fraction of solid solution to hydride phase (Eliezer et al., 2000).

In the following sections some general aspects regarding hydrogen in metals will be reviewed and impacts of defects and nanostructuring on metals' properties will be presented.

25.2   Fundamentals

25.2.1   Hydrogen Adsorption at Metal Surfaces

The transfer of hydrogen to the lattice interior usually occurs via surface adsorption, namely physi- and chemisorption of hydrogen at the metal surface (see, e.g. Zangwill, 1996). While physisorption accounts for adsorption of the hydrogen molecule via van der Waals forces, chemisorption implies chemical bonds between the adsorbate and the metal. Chemisorption itself can be classified as non-dissociative and dissociative chemisorptions. Only the latter is important for subsequent hydrogen uptake in metals (Schlapbach, 1992). If thereby the hydrogen molecule has to overcome an activation barrier, chemisorption and dissociation are called activated.

The potential energy curve of hydrogen in the vicinity of a metal surface is illustrated in **Figure 1** (Züttel et al., 2008). The exemplary curves are derived using Lennard–Jones potentials. At the left side of

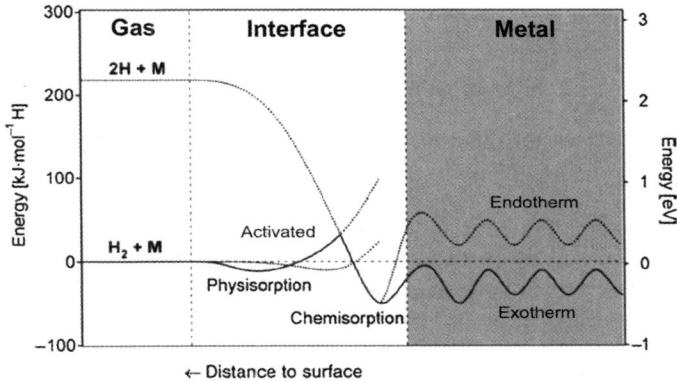

Figure 1 Lennard–Jones potential of hydrogen approaching a metallic surface. Far from the metal surface the potentials of a hydrogen molecule and of two hydrogen atoms are separated by the dissociation energy. The first attractive interaction of the hydrogen molecule is the van der Waals force leading to the physisorbed state. Closer to the surface the hydrogen has to overcome an activation barrier for dissociation and formation of the hydrogen metal bond. Hydrogen atoms sharing their electron with the metal atoms at the surface are then in the chemisorbed state. In the next step the chemisorbed hydrogen atom can jump in the subsurface layer and finally diffuse on the interstitial sites through the host metal lattice. From Züttel et al. (2008) and ref. therein.

Figure 1, the potential energy of the hydrogen gas phase and the metal ($H_2 + M$) is shown and set to zero, while potential energies of hydrogen dissolved in the metal are plotted on the right-hand side. With regard to the left side, the potential energy of the system can be lowered (exothermal reaction) or increased (endothermal reaction) by hydrogen dissolution in the metal. Metal–surface interactions with hydrogen are visible in the interface region. Hydrogen absorption occurs, when the reaction is exothermal, that is the system's potential energy level in the metal lays below that of the hydrogen gas molecule and the metal.

The fundamental processes at surfaces will be explained by taking Pd–H as a model system. At the Pd surface hydrogen adsorbs as an atom, as demonstrated by Conrad et al. (1974) for Pd (110) and (111) surfaces and by Behm et al. (1980) for Pd (100) surfaces. Hydrogen atoms adhere to different positions of the metal lattice depending on the temperature and the surface reconstruction. Thereby, different surface phases exist with different surface coverages (Θ) (Schlapbach, 1992; Christmann, 1988). Θ is the local adsorbate hydrogen concentration, given by the number of hydrogen atoms per lattice atom. The existence of surface phases is plotted in so-called surface phase diagrams (see, e.g. Foiles and Daw, 1985), where the surface coverage Θ replaces the concentration axis of conventional bulk phase diagrams.

The equilibrium position of the surface metal atoms can be different from that given by the lattice periodicity of the bulk (Schlapbach, 1992). This so-called surface relaxation occurs mainly in vertical direction, where atom positions often show 5–10% contraction between the first and second layers, whereas a slight expansion occurs between the second and third layers as well as the third and fourth atomic layers. Surface atom positions can also be different in lateral direction due to surface reconstruction.

Adsorbates like hydrogen themselves can change the relaxation of metal surface atoms and induce reconstruction. Hydrogen on Pd (110) forms a $(1 \times 2)H$ phase at low temperatures with an extraordinary high coverage of $\Theta = 1.5$ H/Pd (Rieder et al., 1983). At higher temperatures, the surface transforms into the $(2 \times 1)H$ phase by releasing about 0.5 H/Pd. At room temperature, this phase is still quite stable (Conrad et al., 1974; Behm et al., 1983; Dong et al., 1998; Cattina et al., 1983) with about $\Theta \approx 1H/Pd$, found for Pd (100) and Pd (111) surfaces (Okuyama et al., 1998).

Different hydrogen surface phases on Pd(111) were recently visualized by Mitsui et al. (2003a, 2003b) using scanning tunneling microscopy (STM) at 37–90 K. The reason for this visualization of surface hydrogen is still under debate. Obviously, hydrogen surface diffusion is too fast for today's STM techniques (Nikitin et al., 2003; Rick et al., 1993).

Additional hydrogen sorption sites were detected that are located below the surface, so-called subsurface sites. The existence of subsurface hydrogen was proven by Behm et al. (1983) and Rieder et al. (1983) according to separated desorption peaks in thermal desorption spectra. Rieder et al. (1983) allocated subsurface hydrogen in a depth of 1–2 atomic layers for Pd (110), and Muschiol et al. (1998) mentioned even 5–10 atomic layers for Pd (210). Thus, deep reaching subsurface hydrogen is found for open surface planes like (210) in the fcc lattice. Theoretical calculations predict and confirm the presence of hydrogen subsurface sites for many different surfaces (Wilke et al., 1994; Okuyama et al., 1998, 1999; Daw and Foiles, 1987; Dong et al., 1998).

Recently, hydrogen molecule physisorption was suggested for hydrogen storage at cryogenic temperatures (as reviewed by Panella and Hirscher, 2010). Low temperatures are required because heats of adsorption are in the range of 1–10 kJ/mol. Furthermore, only materials with a large surface to volume ratio are of interest for applications since the storage capacity scales with the surface area. Materials of interest are, actually, metal–organic frameworks and carbon-based nanostructures.

25.2.2 Hydrogen Absorption in Metals

25.2.2.1 Site Occupation

Because of its small size, hydrogen is absorbed on interstitial sites of the host lattice. For the simple lattice structures fcc, hexagonal close packed (hcp) and bcc, interstitial sites with octahedral (O-site) and tetrahedral (T-site) symmetry are commonly occupied (Fukai, 2005) as shown in **Figure 2**.

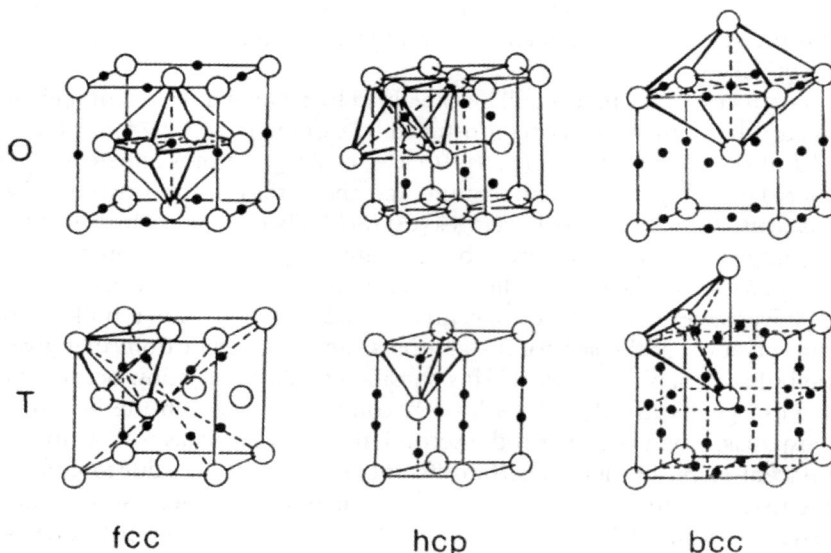

fcc   hcp   bcc

Figure 2 Interstitial sites (octahedral (O) sites and tetrahedral (T) sites) in fcc, hcp and bcc lattices. From Fukai (2005).

Occupation is accompanied by a host lattice distortion that raises the total energy of the system by a strain energy contribution. Therefore, the size of the interstitial site is an important argument for its occupation by hydrogen. Simple geometrical calculations assuming ball-like atoms give r_i/r_h, the relation of the radius of the interstitial site relative to that of the host lattice atom. For fcc and hcp host lattice structures the relation is 0.414 (O) and 0.225 (T); for bcc host lattice structures it is 0.155 (O) and 0.291 (T). Thus, for fcc and hcp host lattice structures the O-sites are predominately filled while for bcc structures the T-sites are often preferred. However, these simple arguments only explain trends of hydrogen solubility.

The number of available interstitial sites per host lattice atom, $r = n_i/n_M$, also differs. There exist 1 O-site and 2 T-sites for hcp and fcc host lattices, and 3 O-sites and 6 T-sites for bcc–host lattices. This becomes important when configurational entropy and maximum solubility are discussed in Section 25.2.5.

25.2.2.2 Lattice Expansion

Hydrogen is mostly dissolved in interstitial sites with a volume smaller than that of hydrogen in the metal. Therefore, solving hydrogen atoms in interstitial lattice sites causes host metal atom displacements and a crystal lattice distortion. This can be expressed by either a strain or a stress field which depends on $(\Delta v/\Omega)$, the volume change per hydrogen atom Δv relative to the metal atom mean volume Ω (Peisl, 1978). This quantity can be measured by several methods: the crystal distortion changes the total volume of the metal by

$$\Delta V = n_H \Delta v \tag{1}$$

If the total volume V of the metal, containing N metal atoms of volume Ω each, changes by ΔV, this scales with the concentration $c_H = n_H/N$.

$$\Delta V/V = n_H \Delta v/N\Omega = c_H(\Delta v/\Omega) \tag{2}$$

This expression holds when volume changes are only due to hydrogen atom uptake, and the defect density is small and constant.

For a freestanding metal the volume expansion can be measured just by total length changes when hydrogen randomly occupies interstitial lattice sites and the crystal expands isotropically. For small changes $\Delta V/V = 3\Delta L/L + O(3(\Delta L/L)^2) + ...$ second-order terms can be neglected and the volume change can be transferred into a length change

$$\Delta L/L = 1/3 \cdot c_H(\Delta v/\Omega) \tag{3}$$

or, by the same argumentation, a lattice parameter change $\Delta a/a$.

$$\left(a_H^3 - a_0^3\right)/a_0^3 = c(\Delta v/\Omega) = 3\Delta a/a + O(3(\Delta a/a)^2) + ... \tag{4}$$

$\Delta a/a$ can be precisely determined by X-ray or neutron diffraction. Collections of experimental results on volume, length and lattice expansion measured for different metals during hydrogen absorption, also including data of relaxation methods (Gorsky effect and Snoek effect) and other methods are given by Peisl (1978) and by Fukai (2005). Mean values for $(\Delta v/\Omega)$ of some metals are summarized in **Table 1**.

Table 1   Collection of relative volume changes per hydrogen atom ($\Delta v/\Omega$), compared with the host metal atom volume Ω (Peisl, 1978)

Metal	($\Delta v/\Omega$)	Literature
Nb–H	0.174 ± 0.005	Peisl (1978)
Ta–H	0.155 ± 0.005	Peisl (1978)
V	0.19 ± 0.01	Peisl (1978)
Pd	0.19 ± 0.01	Peisl (1978)
Ni	0.28	Bauer et al. (1968)
Er	0.13	Beaudry and Spedding (1975)
Tm	0.12	Beaudry and Spedding (1975)
Lu	0.11	Beaudry and Spedding (1975)
Y	0.1	Beaudry and Spedding (1975)

At the first glance, the volume change ΔV appears linear over a wide concentration range. Baranowski et al. (1971) and Krukowski and Baranowski (1976) have discovered that the volume change is, up to concentrations of 0.7 H/M, independent of the host metal initial volume. They have measured and collected the data for many fcc host metals, presented in **Figure 3**, where the increase is plotted in terms of the volume change of the unit cell. The data mainly refer to interstitial hydrogen existing within a single phase.

As deduced from the dashed line in **Figure 3**, the slope is $\Delta V/c = 11.5$ Å^3 for different metals. This value is associated with the addition of four hydrogen atoms to the fcc unit cell (Baranowski et al., 1971). As Peisl concluded, the addition of one hydrogen atom, therefore, increases the crystal volume

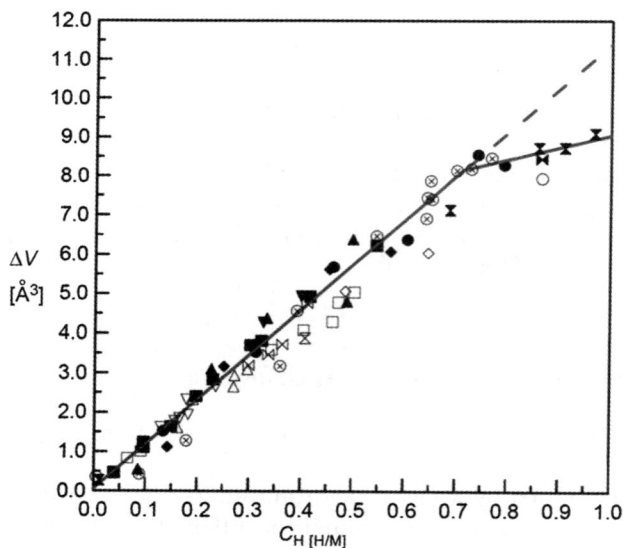

Figure 3   The volume change ΔV appears linear over a wide concentration range. Data from Baranowski et al. (1971) and Krukowski and Baranowski (1976).

by $v_H = 2.9$ Å^3. On closer examination, Fukai found differences depending on the type of the host metal, host metal structure and the site occupancy. He generalized his findings to the following: (a) v_H is larger for lanthanides than for d-band metals, (b) v_H is larger for T-site occupancy ($v_H = 2.9 \pm 0.3$ Å^3) than for O-site occupancy ($v_H = 2.2 \pm 0.3$ Å^3) (Fukai, 2005, p. 107). Fukai argues that the origin for these differences is the different type of bonding, which for d-band metals is the formation of local bonding states with surrounding metal atoms with a corresponding depletion of antibonding states. For lanthanide metals, the large v_H is believed to result from charge transfer from the metal to the hydrogen atoms. Hydrogen atoms bear some negative charges and become larger in size (Fukai, 2005).

Interestingly, there are also exceptions from these general rules: for the dihydrides of the lanthanides hydrogen absorption might also result in a lattice contraction. This occurs simultaneously to a change in the bonding character from metallic to covalent (Fukai, 2005, see 3.67).

For concentrations above 0.7 H/M a downward trend in the volume increase is found for many metals. Fukai (2005), p. 108 suggested that the formation of SAV of up to 10% accounts for this effect. Since vacancy formation generally leads to lattice contraction it certainly explains a reduced volume change—SAV further offer extra sites for H occupation leading to an increased H concentration and a further reduction of the curve's slope.

When nanomaterials are regarded, lattice parameter changes can be different from the results presented so far. Here, the need of stabilizers changes the sample environment and often hinders free sample expansion in certain directions. Complicated strain conditions result that change the volume expansion of the nano-metal compared with the free bulk metal because of the Poisson effect. Furthermore, plastic deformation, for example the formation of dislocations, has to be considered that also reduces the measured lattice expansion (see Section 25.4.4.4).

25.2.3 H Absorption at Low Concentrations: Sieverts' Law

For low H concentrations, the concentration of hydrogen c_H in a defect-free metal scales with the square root of the environmental gas pressure, $c_H \sim \sqrt{p_{H_2}}$. This relation was discovered in 1929 by Sieverts (1929) who studied many gases dissolving in metals and is, therefore, called Sieverts' law. It implies the thermodynamic equilibrium between a molecular gas consisting of two atoms and a single atom in a host lattice and applies for hydrogen solution in many metals at low concentrations. The proportionality, called solubility, can be formulated as

$$c_H = \sqrt{\frac{p_{H_2}}{p_0}} \exp\left(-\frac{\Delta H_H^{M-gas} - T\Delta S_H^{M-gas}}{k_B T}\right) \tag{5}$$

Experimental values of measured enthalpies and entropies of solution are summarized in **Table 2**. They apply for infinite dissolution.

A theoretical derivation of Sieverts' law will be given in the following. A more comprehensive treatment can be found, for example in the textbook of Fukai (2005).

As will be seen later, hydrogen molecules dissociate at the metal surface and dissolve as atoms within the metal according to the reaction:

$$H_2(gas) \rightarrow 2H(metal).$$

Thermodynamic equilibrium requires that all chemical potentials of all phases are equal. Here, we have to consider hydrogen in the gas phase with the chemical potential $\mu_{H_2}^{gas}$, and hydrogen in the metal

Table 2   Enthalpy and entropy of solution of hydrogen in metals, in the low-concentration limit

Metal	ΔH_s (eV per atom)	$\Delta S_s^0/k$	T (°C)	Reference
Li	−0.54	−7	200–700	Wenzl (1982)
Mg	+0.22	−4	500	Fromm and Gebhardt (1976a, 1976b); Fromm and Hörz (1980)
Al	+0.70	−6	500	Wenzl (1982)
Sc	−0.94	−7	−	Lieberman and Wahlbeck (1965)
Y	−0.85	−6	−	Fromm and Gebhardt (1976a, 1976b); Fromm and Hörz (1980)
La (fcc)	−0.83	−8	−	Fromm and Gebhardt (1976a, 1976b); Fromm and Hörz (1980)
Ce (fcc)	−0.77	−7	−	Wenzl (1982)
Ti (hcp)	−0.55	−7	500–800	Wenzl (1982)
Ti (bcc)	−0.62	−6	900–1100	Wenzl (1982)
Zr (hcp)	−0.66	−6	500–800	Wenzl (1982)
Zr (bcc)	−0.67	−6	860–950	Wenzl (1982)
Hf (bcc)	−0.38	−5	300–800	Fromm and Gebhardt (1976a, 1976b); Fromm and Hörz (1980)
V	−0.28	−8	150–500	Fromm and Gebhardt (1976a, 1976b); Fromm and Hörz (1980) and Wenzl (1982)
Nb	−0.35	−8	>0	Lieberman and Wahlbeck (1965), Veleckis and Edwards (1969), Fromm and Gebhardt (1976a, 1976b); Fromm and Hörz (1980) and Wenzl (1982)
Ta	−0.39	−8	>0	Wenzl (1982)
Cr	+0.60	−5	730–1130	Fromm and Gebhardt (1976a, 1976b); Fromm and Hörz (1980)
Mo	+0.54	−5	900–1500	Fromm and Gebhardt (1976a, 1976b); Fromm and Hörz (1980)
W	+1.1	−5	900–1750	Fromm and Gebhardt (1976a, 1976b); Fromm and Hörz (1980)
Fe (bcc)	+0.25	−6	<900	Fromm and Gebhardt (1976a, 1976b); Fromm and Hörz (1980)
Ru	+0.56	−5	1000–1500	Fromm and Gebhardt (1976a, 1976b); Fromm and Hörz (1980)
Co (fcc)	+0.33	−6	1000–1492	Fromm and Gebhardt (1976a, 1976b); Fromm and Hörz (1980)
Rh	+0.28	−6	800–1600	Fromm and Gebhardt (1976a, 1976b); Fromm and Hörz (1980)
Ir	+0.76	−5	1400–1600	Fromm and Gebhardt (1976a, 1976b); Fromm and Hörz (1980)
Ni	+0.17	−6	350–1400	Fromm and Gebhardt (1976a, 1976b); Fromm and Hörz (1980)
Pd	+0.10	−7	−78–75	Fromm and Gebhardt (1976a, 1976b); Fromm and Hörz (1980)
Pt	+0.48	−7	−	Fromm and Gebhardt (1976a, 1976b); Fromm and Hörz (1980)
Cu	+0.44	−6	<1080	Wenzl (1982)
Ag	+0.71	−5	550–961	Fromm and Gebhardt (1976a, 1976b); Fromm and Hörz (1980)
Au	+0.37	−9	700–900	Fromm and Gebhardt (1976a, 1976b); Fromm and Hörz (1980)
U (α)	+0.10	−6	<668	Fromm and Gebhardt (1976a, 1976b); Fromm and Hörz (1980)

Adapted from Fukai (2005).

with the chemical potential μ_H^M. The presence of two atoms counting for the gas phase and just one for the metal phase has to be considered in the equilibrium condition:

$$\tfrac{1}{2}\mu_{H_2}^{gas} = \mu_H^M \tag{6}$$

As known from basic thermodynamics, the chemical potential of the ideal gas phase is Becker (1985)

$$\mu_{H_2}^{gas} = \mu_0 + k_B T \ln\left(\frac{p_{H_2}}{p_0}\right) \tag{7}$$

with Boltzmann's constant $k_B = 1.381 \cdot 10^{-23}$ J/K, standard pressure $p_0 = 1.013 \cdot 10^5$ Pa and the standard chemical potential μ_0 of the gas phase. It mainly contains the energy of dissociation of the molecule as well as vibrational contributions to the entropy (Becker, 1985).

The chemical potential $\mu_H^M = \frac{\partial G}{\partial n_H}\big|_{T,p}$ of hydrogen in a metal (M) can be derived by determining the Gibbs free energy G

$$G_H^M = H_H^M + TS_H^M = H_H^M + T\left(S_H^{M\text{ conf}} + S_H^{M\text{ exc}}\right) \tag{8}$$

with the enthalpy H^M and the entropy S^M, which can be split into a configurational part, $S^{M,\text{conf}}$, and an excess part, $S^{M,\text{exc}}$, that mainly contains vibrational and electronic contributions to the entropy. The configurational part of the entropy depends on the number of hydrogen n_H atoms distributed over the number of available interstitial sites n_i by

$$S_H^{M\text{ conf}} = k_B \ln\left[\binom{n_i}{n_H}\right] = k_B \ln\left(\frac{n_i!}{n_H!(n_i - n_H)!}\right) = k_B[(n_i!) - \ln(n_H!) - \ln((n_i - n_H)!)] \tag{9}$$

where the binomial coefficient $\binom{n_i}{n_H}$ gives the number of possible configurations. Stirling's formula $\ln(x!) = x \ln(x) - x$ (Bronstein and Semendjajew, 1985) can be applied when large numbers n_i and n_H are regarded, yielding

$$S_H^{M\text{ conf}} = k_B n_i \ln\left(\frac{n_i}{n_i - n_H}\right) - k_B n_H \ln\left(\frac{n_H}{n_i - n_H}\right) \tag{10}$$

The derivative of this equation is

$$\frac{\partial S_H^{M\text{ conf}}}{\partial n_H} = k_B \ln\left(\frac{n_i - n_H}{n_H}\right) \tag{11}$$

In total, the chemical potential for hydrogen in a metal is obtained to be

$$\mu_H^M = h_H^M - Ts_H^{M\text{ exc}} + kT \ln\left(\frac{c_H}{r - c_H}\right) \tag{12}$$

using the partial enthalpy $h_H^M = \frac{\partial H_H^M}{\partial n_H}$ and the partial entropy $s_H^{M\text{ vibr}} = \frac{\partial S_H^{M\text{ exc}}}{\partial n_H}$. This equation usually is simplified to

$$\mu_H^M = \mu_H^{M0} + kT \ln\left(\frac{c_H}{r - c_H}\right) \tag{13}$$

with the standard chemical potential $\mu_H^{M0} = h_H^M - Ts_H^{M\text{ exc}}$.

The chemical potential of hydrogen in a metal can also be derived by defining a density of site energies (DOSE) or an energy landscape, respectively, and filling the sites according to the Fermi–Dirac statistics (FD statistics). This statistics takes a limited occupation per site into account and is well known for calculations of the occupation of electrons in the electronic density of states (DOS) (Ashcroft and Mermin, 1976). However, it can also be used for hydrogen's occupation of sites in the energetic landscape of the sample. For both cases, an exclusion law holds: for electrons the Pauli principle holds that results in a limitation to just two electrons (with opposite spin) in one electron energy level. For hydrogen, the spatial restriction limits the number of hydrogen atoms that can be placed in one interstitial site. Usually, one hydrogen atom only fits in one interstitial site of a metal. The DOSE does not contain any information about the localization of the sites. This concept will be further discussed and applied in Section 25.3.1.2.1. Thus, for site occupation the Fermi function

$$f(\varepsilon_i) = \frac{1}{e^{\beta(\varepsilon_i - \mu)} + 1} \tag{14}$$

can be used, with $\beta = (k_B T)^{-1}$ and the chemical potential of hydrogen in the metal μ_H^M, which is abbreviated with μ for simplicity (Becker, 1985). It gives the minimum Gibbs free energy for the thermal occupancy of energy level ε_i. The concentration of hydrogen can, then, be calculated by counting hydrogen atoms in all available energy levels

$$c = r \int_0^\infty n(\varepsilon) f(\varepsilon) d\varepsilon \tag{15}$$

with the DOSE, $n(\varepsilon)$. Using the ideal periodic lattice potential landscape with just one site energy gives a delta function like DOSE $n(\varepsilon) = \delta(\varepsilon - \varepsilon_0)$. Implementing this in the above equation yields

$$c = r \int_0^\infty \delta(\varepsilon - \varepsilon_0) \frac{1}{e^{\beta(\varepsilon - \mu)} + 1} d\varepsilon = \frac{T}{e^{\beta(\varepsilon_0 - \mu)} + 1} \tag{16}$$

and thus

$$\mu_H^M = \varepsilon_0 + k_B T \ln\left(\frac{c_H}{r - c_H}\right) \tag{17}$$

similar to Eqn (13). This treatment directly implements the dominance of the statistical occupation of sites, and thus the dominance of the entropy in the solid solution regime.

Sieverts' law can be obtained by regarding thermodynamic equilibrium between a metal and the two-atomic hydrogen gas according to Eqn (6). It follows

$$\frac{1}{2}\mu_0 + \frac{1}{2}k_B T \ln\left(\frac{p_{H_2}}{p_0}\right) = h_H^M - Ts_H^{M\ exc} + k_B T \ln\left(\frac{c_H}{r - c_H}\right)$$

$$\frac{1}{2}\ln\left(\frac{p_{H_2}}{p_0}\right) - \frac{h_H^M - Ts_H^{M\ exc} - \frac{1}{2}\mu_0}{k_B T} = \ln\left(\frac{c_H}{r - c_H}\right) \tag{18}$$

and finally

$$\frac{c_H}{r - c_H} = \sqrt{\frac{p_{H_2}}{p_0}} \exp\left(-\frac{h_H^M - Ts_H^{M\ exc} - \frac{1}{2}\mu_0}{k_B T}\right) \tag{19}$$

For low concentrations $c_H << 1$ H/M and $r = 1$ like for Pd, this results in Sieverts' law

$$c_H = \sqrt{\frac{p_{H_2}}{p_0}} \exp\left(-\frac{h_H^M - Ts_H^{M\ exc} - \frac{1}{2}\mu_0}{k_B T}\right) \tag{20}$$

The treatment does not only apply to hydrogen, but is of general nature. It also holds for other two-atomic gases like oxygen and nitrogen: as demonstrated, the square root dependency in Sieverts' law follows from the two-atomic hydrogen gas molecules being dissolved as single atoms in the metal. If the solution in a material occurs by gas molecules (like in polymers) or if the initial gas is a one-atomic gas like He or Ne, a linear relation between concentration and gas pressure would result.

The enthalpy change during hydrogen absorption, which is also called the heat of solution, is $\Delta h_H^{M-gas} = h^M - \frac{1}{2}h^{gas}$. It can be measured by calorimetry. The entropy changes are expressed as $\Delta s_H^{M-gas} = s_H^{M\ exc} + k_B \ln c_H - \frac{1}{2}s^{gas}$. The thermodynamic quantities can be determined from the pressure dependency of the solubility.

For $c << r$, comparison of Eqn (18) with Eqn (5) yields

$$\Delta H_H^{M-gas} - T\Delta S_H^{M-gas} = h_H^M - \frac{1}{2}\mu_0 - T(s_H^{M\ exc} + k_B \ln r) \tag{21}$$

As long as the temperature dependency of the other terms can be neglected, the Arrhenius plot of the equilibrium pressure gives enthalpy and entropy of formation.

The temperature dependency of the solubility gives the enthalpy of solution and the entropy of solution more accurately. At constant pressure, the enthalpy

$$\Delta h_H^M = \frac{\partial \ln c_H}{\partial \left(1/k_B T\right)}\bigg|_p \tag{22}$$

and, from this, the entropy of solution

$$\frac{\Delta s_H^M}{k_B} = \frac{\Delta h_H^M}{k_B T} + \ln c_H \tag{23}$$

can be derived from the slope and the intercept of the tangent of the solubility at the requested temperature when plotted as an Arrhenius plot (plotting $\sqrt{p}(1 - c_H)/c_H$ in decadal scale as a function of the inverse temperature $1/(k_B T)$). Temperature dependencies of both Δh_H^M and Δs_H^M become detectable (Fukai, 2005).

25.2.4 Interaction between Hydrogen Atoms

At higher H concentrations, the interaction between the dissolved hydrogen atoms becomes visible leading to a concentration dependency of the heat of solution. For Pd, V, Nb, Ta, Ti, Y and Sc, the heat of solution initially decreases but raises at high concentrations. The initial decrease hints at an attractive H–H interaction. A pair interaction of dissolved hydrogen atoms was first treated in statistical thermodynamics by Lacher (1937), but the origin of the interaction was unclear at that time. Alefeld (1972) suggested that an average elastic interaction contributes to the heat of solution. A mean field theory of the elastic interaction was finally developed by Horner and Wagner (1974).

For high concentrations, Griessen (1986) have exemplarily shown for the Pd–H system that the concentration dependency of the enthalpy can be explained by electronic band-filling models. In these models, the extra electron of hydrogen enters the rigid electronic DOS of the host lattice. Increasing the hydrogen concentration thus leads to a raise in the metals' Fermi level E_F. Enthalpy changes due to band filling are small when the DOS is high (e.g. for D-bands) and narrow. Enthalpy changes are large when low DOS bands (like s-bands) are needed to be filled. This happens for $c_H > 0.6$ H/Pd when the 4-d band of the Pd–host lattice is completely filled and energetically higher laying levels with low number density in the DOS have to be occupied. For such cases, E_F strongly raises and, thus, increases the enthalpy of solution. These contributions commonly are abbreviated as electronic contributions. Not regarded here are changes in the host lattice DOS due to alloying with hydrogen.

An additional short-range repulsive interaction resulting from interactions of many H atoms is also under discussion (Fukai, 2005). Its presence is based on different experimental findings:

(1) Westlake criterion (Westlake, 1980): H atoms do not come closer than 2.1 Å, resulting in a minimum hole size of 0.38 Å (only known exception to this rule: LnNiIn (Ln = La, Ce, Pr, Nd) with H–H distances of 1.5 Å (Oates et al., 1969; Boureau, 1981).
(2) Mutual blocking: Filling of interstitial sites by other hydrogen atoms results in a reduced configurational entropy. For bcc alloys this blocking extends to second nearest neighbors (Boureau, 1984; Clapp and Moss, 1966).
(3) Ordered structures: In practically all M–H systems ordered structures are found at low temperatures, indicating a repulsive H–H interaction.

25.2.5 Hydride Formation

After reaching the terminal concentration c_α (the solubility limit) of the solid solution MHc_α at a given temperature, a hydride phase MHc_β with a higher concentration c_β can form by the reaction

$$\frac{1}{c_\beta - c_\alpha} MH_{c_\alpha} + \frac{1}{2} H_2 \rightarrow \frac{1}{c_\beta - c_\alpha} MH_{c_\beta} \tag{24}$$

where the difference $c_\beta - c_\alpha$ accounts for the width of the miscibility gap between the two phases. The hydride can be formed with the host lattice structure, just with a different lattice parameter (as found for Pd–H), or it can differ from the structure of the solid solution. The denotation "hydride" is used for the high concentration phases after passing a two-phase region. Unfortunately, denotations are often inconsistent, especially when hydrides originating from a double minimum of the Gibbs free energy of one phase (fcc phase for Pd–H system) are regarded. Solid solution and hydride phase are commonly named α- and β-phases, and not as correctly, α and α' phases.

In equilibrium the transition is described by changes of entropy and enthalpy of formation:

$$\Delta g_H^{\alpha-\beta} = \Delta h_H^{\alpha-\beta} + \Delta T \cdot s_H^{\alpha-\beta} \tag{25}$$

Alternatively, by separating the pressure term

$$\Delta G_H^{\alpha-\beta} = \Delta H_H^{\alpha-\beta} + T\left(\Delta S_H^{\alpha-\beta} + \ln p/p_0\right) \tag{26}$$

The enthalpy and entropy changes evolving during the phase transition can be derived from a van't Hoff plot of the plateau pressure (Fukai, 2005). But, in experiments the plateau pressure differs between loading and unloading: a hysteresis is present (compare Section 25.2.6). This can be seen in **Figure 4**, where measured isotherms of Pd–D are shown under loading and unloading conditions (Frieske and Wicke, 1973). For data evaluation, unloading plateau pressures are often used.

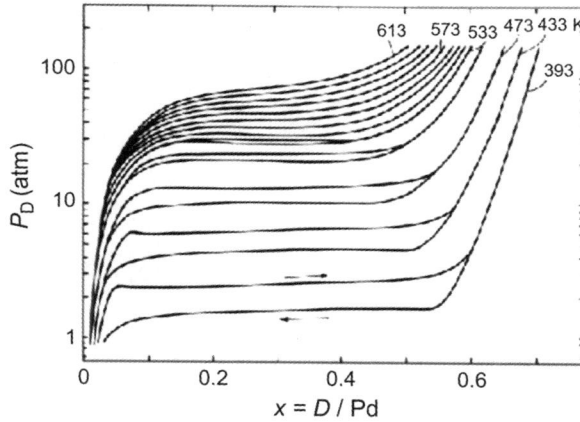

Figure 4 Isotherms of Pd–D. A hysteresis (Section 25.2.6) is detected between loading and unloading curve. From Frieske and Wicke (1973).

System parameters concerning phase transitions are listed for many elements and alloys, see, for example collections by Fukai (2005) or, recently, by Sandrock (2012). The IEA/DOE/SNL Hydride (HYDPARK) database collection of Sandrock includes AB5, AB2, AB, A2B alloys as well as complex materials. In **Table 3** some selected values are given. The thermodynamic quantities are given per hydrogen molecule.

Only in case of binary M–H alloys the experimentally determined plateau pressure is mostly constant, which is in accordance with Gibbs' phase rule. For ternary alloys the plateau is sloped according to the different chemical environments of each interstitial site. In the latter cases, for van't Hoff plots usually the pressure at concentrations in the middle of the miscibility gap is applied.

Table 3   Enthalpies and entropies of formation of selected elements and alloys, plateau pressure at 25 °C, hydrogen content and lattice structures

	P, atm (@ 25 °C)	H/M	wt.%	Initial	Hydride	ΔH, kJ/mol H_2	ΔS, kJ/K-mol H_2	Source
$Pd-Pd_{0.6}$	$8.2 \cdot 10^{-3}$	0.77	0.72	A1	fcc	−41.0	−0.0976	Wicke and Brodowski (1978a, 1978b)
$Nb-Nb_{0.65}$								
$Mg-MgH_2$	$1 \cdot 10^{-6}$	2.0	7.66	A3	C4	−74.5	−0.135	Stampfer et al. (1960)
$Zr-ZrH_2$	$6.4 \cdot 10^{-28}$	2.0	2.16	A3	C1	−217	−0.188	Beck and Mueller (1968)
$Ti-TiH_2$	$4 \cdot 10^{-20}$	1.97	3.98	A3	C1	−164	−0.179	Mueller (1968)
$V-VH_2$	2.1	2.0	3.81	A2		−40.1	−0.1407	Reilly and Wiswall (1970)
$TiFe-TiFeH$	4.1	0.975	1.86	B2	(P2221)	−28.1	−0.106	Reilly and Wiswall (1988)
$LaNi_5-LaNi_5H$	1.8	1.08	1.49	D2d	See Yvon, 1988	−30.8	−0.108	Lundin and Lynch, (1975)
Mg_2Ni-Mg_2NiH	$1 \cdot 10^{-5}$	1.33	3.6	C16	C1	−64.5	−0.122	Noreus (1989)
$Pd_{0.7}Ag_{0.3}-$ $(Pd_{0.7}Ag_{0.3})H_{0.3}$	$3.3 \cdot 10^{-4}$	0.34	0.32	A1		−50.0	−0.101	Brodowski et al. (1965)

From IEA/DOE/SNL Hydride (HYDPARK) database, data provided by Sandrock (2012).

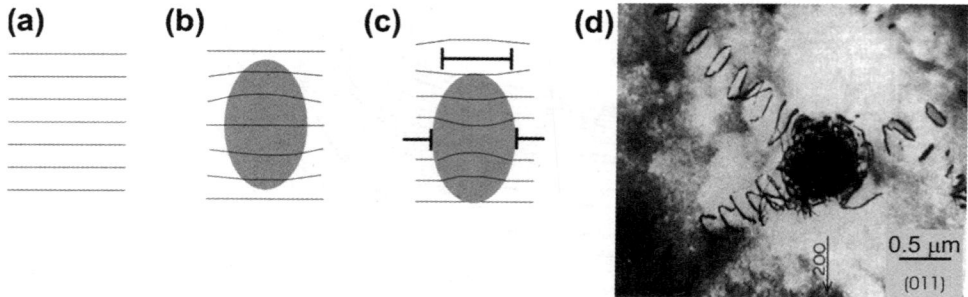

Figure 5 Scheme of the formation of a misfitting hydride in (a) the metal matrix, (b) coherent hydride and (c) semicoherent hydride. An intrinsic dislocation ring forms around the hydride while an extrinsic dislocation loop is emitted into the metal matrix. (d) Such loops have been revealed by Transmission Electron Microscopy (TEM) at hydride locations in Nb by Schober, (1973) or Makenas and Birnbaum (1980). Extrinsic dislocation loops (visible as black rings) are emitted in $a/2 \langle 111 \rangle$ directions into the metal matrix. The dark center is the position of the former hydride. From Makenas and Birnbaum (1980).

25.2.6 Hysteresis

Hysteresis (compare **Figure 4**) arising in a physical property of the system between hydrogen loading and unloading verifies the presence of energy dissipating processes. The hysteresis is commonly related to the formation and migration of dislocations emerging at the newly formed phase in the two-phase region of the phase diagram (Schober, 1973; Makenas and Birnbaum, 1980; Flanagan et al., 1980, 1982, 1991). Because of the local volume change, dislocation production is required during both hydride formation and hydride decomposition. Dislocations can be formed when the elastic strain energy between the two phases exceeds the formation energy of a dislocation, as shown in **Figure 5**. Thereby, dislocation emission results in strain energy release.

Recently, Schwarz and Katchaturyan (1995, 2006) showed that, for an open system, a hysteresis also appears when the elastic strain energy at the coherent metal/hydride boundaries is considered. A coherent phase boundary describes the situation where the atomic lattice of the hydride and that of the adjacent solid solution are tightly linked and no dislocation is present. Schwarz and Katchaturyan showed that coherency stress between the hydride precipitate and the matrix generates a macroscopic thermodynamic energy barrier that increases the chemical potential during hydride formation, or lowers the chemical potential for the back transformation into the α-phase. The hysteresis is a direct result of the thermodynamic barrier. The verification of this theory is a matter of actual research.

25.2.7 Phase Diagrams

Hydrogen interaction with the host metal lattice and among the hydrogen atoms themselves can lead to the formation of different phases, depending on temperature, concentration and pressure.

Phase diagrams of metal–hydrogen systems can be, for example, found in Massalski's collection of phase diagrams, or in the data compilation of Fromm and Gebhard (1976a, 1976b). Some metal–hydrogen systems have been intensively studied, among them that of palladium–hydrogen (Pd–H), niobium–hydrogen (Nb–H), vanadium–hydrogen (V–H) and tantalum–hydrogen (Ta–H). Especially Pd–H is often used for model studies because of its simplicity and the easy way of changing hydrogen

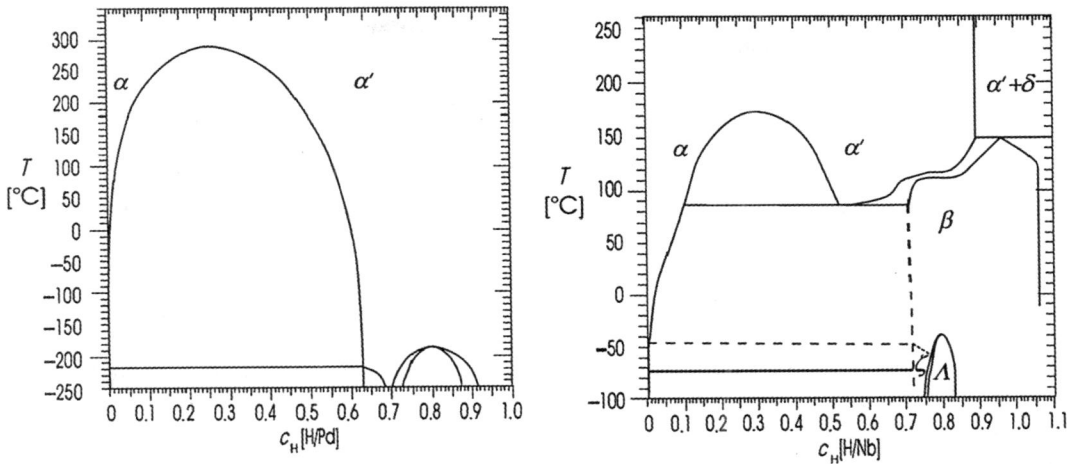

Figure 6 Phase diagrams of Pd–H and Nb–H, as adapted by Pundt (2005) from Fukai (2005); Frieske and Wicke (1973); Schober and Wenzl (1978).

concentrations via changes of the gas pressure or via cathodic polarization. The phase diagrams of Pd–H and Nb–H are plotted in **Figure 6**.

Hydrogen atom occupation of interstitial host lattice sites implies that the host atoms themselves do not change in number. However, the total number of atoms changes during hydrogen ab- and desorption. Therefore, the concentration of hydrogen in a material c_H is commonly given as the number of H atoms (n_H) per number of host lattice atoms (n_M) or per formula unit (n_{chem}). Concentrations can, thereby, easily reach 300%. Seldom, concentrations are given in atomic-%, recalculating concentrations to a constant number of atoms or as mole-H per volume.

25.2.8 Controlled Alloying with Hydrogen

Many metals form a protecting and stable surface oxide under ambient conditions, kinetically hindering the hydrogen uptake. Natural oxides can be removed by surface sputtering under ultrahigh vacuum (UHV) conditions and subsequent deposition of the protective layer which consist of, for example, palladium or its alloys. This layer allows for hydrogen chemisorption and protects the metal from oxidation.

Metals usually contain an initial and mostly unknown amount of hydrogen. For concentration control it is crucial to remove this initial hydrogen. Techniques are outgassing under UHV conditions and elevated temperatures or electrochemically by anodic polarization. Both techniques need to be conducted with care. At elevated temperatures, alloying can occur with protecting layers or substrates, especially when low-temperature eutectics are present in the binary phase diagram of the adjacent layers. Under reversed electrochemical conditions oxidation of the sample surface can appear. This is visible by a strongly negative electrochemical potential.

25.2.8.1 Gas Phase Loading

Gas volumetry: The direct relation of the hydrogen content of a metal and the environmental gas pressure allows controlled alloying from the hydrogen gas phase just by changing the pressure. For massive samples in a closed system, a pressure drop Δp accompanies the hydrogen uptake. Thereby, each change of the hydrogen content Δc_H^M can be evaluated as

$$\Delta c_H^M = \frac{\Delta n_H}{n_M} = \frac{2\Delta p \cdot V}{n_M RT} \tag{27}$$

with the gas constant $R = 8314$ J/(K·mole), absolute temperature T and recipient volume V. The factor 2 accounts for the 2 atoms in the H_2 molecule. The total concentration is derived from $c_H^M = \sum \Delta c_H^M$. By plotting the equilibrium pressure p as a function of the hydrogen content, the isotherm is obtained. Loading can be performed by increasing, unloading by reducing the partial pressure of hydrogen, respectively. In case of nano-metals like thin films, pressure drops can only be detected when a sufficient amount of sample mass (n_M) and small recipient volumes V are chosen (Feenstra et al., 1986). The method is, thus, hardly applied for concentration determination on nano-metals. Indirect measurements of the H concentration via the sample's resistance or light transmittance are commonly used.

Gas gravimetry: The hydrogen content change Δc_H^M can be calculated from the sample's weight change during hydrogen uptake. For this method, constant pressure conditions can be realized by using large volumes. The method is limited by requiring large samples and sensitivity of the used balance.

25.2.8.2 Electrochemical Loading

When the sample is immersed in an electrolyte and a current is passed to the sample from a counter electrode, the hydrogen content change can be calculated from Faradays Law.

$$\Delta c_H^M = \frac{\Delta n_H}{n_M} = \frac{Q}{F} \cdot \frac{V_M}{V_S} \tag{28}$$

With Faradays' Constant $F = 96485$ C/mole, the flown charge Q measured, the sample volume V_S and its molar volume V_M.

The chemical potential can be obtained by measuring the electromotive force U between the sample and a reference electrode. Nernst equation links the hydrogen pressure to the voltage difference $U–U_0$.

$$U - U_0 = -\frac{RT}{2F} \ln \frac{p_{H_2}}{p_0} \tag{29}$$

yielding

$$\mu_H(U) = -(U - U_0)F \tag{30}$$

The reference voltage U_0 depends on the choice of the reference electrode. When equilibrium values μ_H are taken and plotted as a function of the concentration, they give the composition pressure isotherm.

The main advantage of this method is its applicability at low hydrogen concentrations down to a few ppm, and providing the electro-chemical potentials. Furthermore, also the hydrogen content of nano-metals like thin films can be easily determined because every electron is measured. Special care has to be taken when high loading current densities (for Pd more than 0.3 mA/cm^2) are used: Hydrogen can be lost at the sample surface by gas formation. Hydrogen can also be desorbed from the sample by reversing the direction of the current. Here, high negative voltages (for Pd < -0.8 V) should be avoided because side reactions occurring in the electrolyte can adulterate the calculated hydrogen concentration.

The choice of a suitable electrolyte is also crucial for these experiments. For Pd-covered Nb, Ti and V, $PO_3H_4(85\%)$:glycerine(85%) in 1:2 volume ratios is successfully used. For ignoble metals, attack by the electrolyte needs to be considered and avoided by appropriate shielding. Further, oxygen has to be removed from the electrolyte to avoid hydrogen loss by water formation. This can be done by persistent argon gas bubbling through the electrolyte.

Electrochemical hydrogen loading is rarely applied to thin films because of undesired wet chemical conditions. However, when used under proper conditions, the hydrogen concentration can be accurately determined. Then, in situ measurement of the hydrogen concentration and another physical quantity is independently possible.

25.2.9 Measuring Hydrogen Concentrations

Controlled hydrogen loading (and unloading) directly results in the knowledge of the mean hydrogen concentration, providing that the initial concentration was negligible. The three main methods, namely gas volumetry, gas gravimetry (Section 25.2.8.1) and electrochemical loading (Section 25.2.8.2) have been already described. Gas volumetry and gas gravimetry commonly need large sample quantities because the measured signal (pressure change or sample weight change, respectively) scales with the chamber volume or the sample amount (Feenstra et al., 1986). For successful gas volumetry measurements on thin films, Munter and Heuser (1998) reduced the chamber volume to that of two directly connected blind flanges.

The hydrogen content in a sample can be determined by destructive measurements like hot extraction. This method allows measuring the mean hydrogen concentration with high accuracy. For hot extraction, the sample is heated in an UHV chamber or in an inert gas stream. Exhausted elements are quantitatively measured by mass spectrometry or by measuring heat conduction of the inert gas. This method is often used for steels, where small hydrogen contents of 0.1 $\mu g_H/g_{steel}$ − 1000 $\mu g_H/g_{steel}$ are needed to be detected.

When microstructural defects are regarded, the local concentration can deviate strongly from the mean concentration. This is of special importance for HE, where hydrogen enrichment in the dilatation field below the dislocation line occurs even at very low hydrogen concentrations and chemical potentials: While the mean concentration is about 0.001 H/M, it is about 0.5 H/M in the local vicinity of a dislocation line (see Section 25.3.2.2, or Maxelon et al., 2001a, 2001b). Special methods are required to detect these local concentrations.

Thermal desorption spectroscopy (TDS) is used to sensitively detect hydrogen surface coverages. From surface sites of different adsorption energies, hydrogen desorption depends on temperature and results in desorption peaks when the sample is heated. As long as hydrogen desorbs from surface sites, clear TDS peaks develop. Absorbed hydrogen undergoing long-range diffusion results in a broad TDS signal. Quantitative measurements need careful calibration, ultraclean systems, mass-resolved gas analysis and knowledge about contaminations inside the samples and their side reactions. Especially, when lightweight materials are studied, metal contaminations can lead to momentous misinterpretations, as deduced, for example, for the case of hydrogen interacting with carbon nanotubes by Haluska et al. (2001).

For samples with layered defect occurrence or layered metal structures, the ^{15}N technique is often used. The intensity of the 6.385 MeV resonance of the reaction $^1H(^{15}N, \alpha\gamma)^{12}C$ is used to measure the hydrogen concentration directly (Hjörvarsson et al., 1989, 1991; Hjörvarsson and Rydén, 1990). The depth resolution of this method scales with the detection depth, starting from about 1 nm depth resolution to larger values. The method is conventionally applied ex situ, after hydrogen loading, for selected samples. Hydrogen loss has to be suppressed by protecting layers.

It was further shown that neutron reflectometry can be applied in situ during gas loading to determine hydrogen/deuterium concentrations in thin films (Munter and Heuser, 1998 and Rehm et al., 1999).

Secondary ion mass spectrometry (SIMS, see, e.g. Benninghoven et al., 1987) has been used for local hydrogen distribution measurements in thin films and multilayers, as well (Scholz et al., 1997; Kesten et al., 2002). Hydrogen enrichments in certain layers can be detected with high accuracy. However, SIMS measurements do not give absolute values for elemental concentrations, and the method usually implies a further issue to consider: The true hydrogen distribution is not measured since hydrogen strongly diffuses toward the material surface during the measurement. The fresh metal surface acts as a hydrogen trap which is permanently formed during the SIMS sputtering process (Kesten et al., 2002). In principle, this issue might be suppressed by using parameters, as deduced by Gemma et al. (2011) for another surface destructive method.

Another method to measure hydrogen depth profiles is glow discharge optical emission spectrometry providing high erosion rates and erosion depths as well as high sensitivity and hydrogen quantification (Wilke et al., 2011).

Coincident elastic proton–proton scattering analysis (pp scattering) allows to perform 3D hydrogen microscopy by scanning a focused proton beam across the sample. A lateral resolution of less than 1 µm can be obtained; the depth resolution is on the order of 3 µm. The yield of the hydrogen signal directly gives the hydrogen density in H-at/cm^2 (Moser et al., 2011). Sensitivity in the atomic ppm range is obtained; even 0.08 ppm detection limit has been demonstrated in diamond (Reichart et al., 2004). Quantification of hydrogen enrichments at defects requires low defect densities because of the lateral resolution limit of 1 µm (Wagner et al., 2013). pp scattering has the lowest damage potential of all ion beam analysis methods for hydrogen detection (Reichart et al., 2002).

On a local scale, also hydrogen enrichments at line or point defects can be measured. Recently, Gemma et al. succeeded in accurately measuring the local deuterium concentration and distribution by using 3D atom probe tomography (3D-APT) (Gemma et al., 2009, 2011, 2012). By this technique the sample is field evaporated atom by atom from the actual sample surface. It allows to simultaneously measuring the local position and the mass-to-charge ratio by time of flight (Miller et al., 1996; Gault et al., 2012). Thereby, the elemental distribution of a material can be gained with atomic resolution. In detail, the local concentration at a defect is accessible with a resolution of one atomic layer in depth and, laterally, with about 1 nm. While the method was restricted to materials with high electronic conductance in the past, recent APT systems allow to also measuring insulating materials like oxides. These recent systems evaporate sample surface atoms by using an UV or a light laser pulses with nano- to femtosecond duration. The determination of real concentrations is still a challenging task for these systems and parameters have to be checked and adjusted for each new system measured.

Gemma et al. (2009, 2011, 2012) determined three important steps for a reliable hydrogen concentration determination using 3D-APT: (a) use of deuterium to drastically reduce quantum mechanical tunneling processes, (b) use of low temperatures to reduce deuterium diffusion, (c) in situ loading to avoid deuterium loss from the nanosized sample. Deuterium depletion at Fe/V interfaces and deuterium enrichment at oxide particles were detected, whereby the mean concentration matched the expected concentration.

A method for determining hydrogen concentrations located at defects is positron annihilation (see, e.g. Dupasquier and Mills, 1995). Positron annihilation spectroscopy (PAS), combining positron lifetime spectroscopy (Bečvář et al., 2000) and coincidence measurement of Doppler broadening (Lynn et al., 1977), is used to study defect concentrations in hydrogenated metals (Shirai et al., 2002; Čížek et al., 2004). The lifetime of the positron delocalized in the lattice or trapped at various defects can be separated. It also depends on the concentration of hydrogen located at the defect. A hydrogen atom bound to a vacancy results in an increase of the local electron density which is reflected by a decrease of the lifetime of a trapped positron. As theoretical calculations reveal, the lifetime of a positron decreases with increasing local hydrogen content at the defect. Thereby, the local hydrogen concentration can be determined. For Nb, about four hydrogen atoms are found to be bound to one vacancy (Čížek et al., 2004). Since the mean implantation depth of fast positrons emitted by a radioisotope source is several tens or even hundreds of microns conventional PAS gives information from bulk of a sample (Dupasquier and Mills, 1995). Slow positron implantation spectroscopy (SPIS) is a technique that uses moderated (slow) positrons and provides depth-resolved information from surface up to several microns (Schultz and Lynn, 1988). SPIS is employed for characterization of defects in hydrogenated thin films (Checchetto et al., 2004; Čížek et al., 2007; Eijt et al., 2009).

Indirect methods are often used when mean concentration measurements of every sample are difficult because of a limited sample quantity, for example in the case of thin films. Then, a physical property is calibrated by the known hydrogen concentration, for one *typical* sample. Further on, the physical property is used to measure the hydrogen content in other samples. However, especially nanosamples differ strongly in their physical properties. This method, therefore, needs to be done with care and knowledge about the microstructural development of the sample (Pundt and Kirchheim, 2006; Pundt, 2004). Physical properties that are commonly used are the electrical resistivity (Huang et al., 1991) or the optical light transmittance and reflectance (Pasturel et al., 2006; Gremaud et al., 2006). Furthermore, for thin films, Eigen frequency measurements of a quartz crystal microbalance are used to determine the weight change during hydrogen uptake. However, these measurements are affected by mechanical stress that arises during hydrogen absorption between the thin film and the quartz crystal. Thus, they are difficult to interpret and often give conflicting results (Feenstra et al., 1986; Bucur and Flanagan, 1974). Lattice strain of nanosized systems, as measured by X-ray diffraction (XRD), also, is not linearly linked to the hydrogen concentration (Yang et al., 1996; Laudahn et al., 1999a). However, in situ gas loading is a widely used technique for nanomaterials because the sample surface can be kept clean. For nanosystems of small sample amount, this method requires indirect hydrogen concentration determination.

25.3 Hydrogen in Defective Metals

25.3.1 Diffusion and Trapping—General Theory

Most of the exciting properties of hydrogen metal systems are related to the small size of the H atom which leads to a high mobility in materials. Namely in metals its diffusivity is very high at room temperature and may reach values which are the same as for ions in aqueous solutions. The physical reasons for the high H mobility are twofold. On the one hand H atoms are dissolved interstitially and migrate via a direct interstitial mechanism which at dilute concentrations does not require the presence of vacancies. On the other hand migrating from one site to an adjacent one may occur via quantum mechanical tunneling (Fukai, 2005). The consequences of the high H mobility are manifold:

(i) Thermal equilibrium is established in rather short times at room temperature between the H dissolved in the metal and either hydrogen gas or protons in aqueous solutions. Thus thermo-dynamic properties, especially the chemical potential of hydrogen, can be obtained simply by measuring the partial pressure or the electrochemical potential.

(ii) Hydrogen storage in metals and its use as an energy carrier become possible at room temperature.

(iii) Thermodynamic equilibrium can be obtained between hydrogen and defects at temperatures where defects do not annihilate.

(iv) Hydrogen can easily redistribute and segregate at defects produced during processing or during plastic deformation, that is grain boundaries, dislocations, crack tips. This interaction gives rise to hydrogen being trapped by defects and to hydrogen reducing the formation energy of defects. These effects are important for understanding HE.

25.3.1.1   One Kind of Trap (Oriani Model)
25.3.1.1.1   Solubility for One Kind of Trap

The simplest approach to trapping is a two-site (two energy levels) model assuming that the material is containing two types of interstitial sites (normal ones and traps) of different binding energy, E_o and E_t with $E_t < E_o$. The following quantities are labeled with the suffix t for traps and the suffix f for normal sites:

N_o = Total number of all interstitial sites

N_{to} = Number of traps

$\beta = N_o/N_a$ with N_a being the number of metal atoms

Ω = Atomic volume of the matrix (metal)

$n_{to} = N_{to}/N_o$ = Molar ratio of traps

$N_{fo} = N_o - N_{to}$ = Number of normal sites containing the freely moving hydrogen atoms

$n_{fo} = N_{fo}/N_o$ = Molar ratio of normal sites

$E_o - E_t > 0$ = Binding energy to a trap

N_H = Total number of H atoms

$n_H = N_H/N_o$ = Fraction of interstices occupied by hydrogen

N_t = Number of H atoms in traps

$n_t = N_t/N_{to}$ = Fractional occupancy of traps

$N_f = N_H - N_t$ = Number of H atoms in normal sites

$n_f = N_f/N_{fo}$ = Fractional occupancy of normal sites

c_t = Concentration of H atoms in traps = $c_t = (\beta/\Omega)n_t$

c_f = Concentration of H atoms in normal sites = $c_f = (\beta/\Omega)n_f$

$c = c_f + c_t$ = Total concentration of H atoms

Note that the definition of concentration c is in agreement with the mostly used one as number of atoms or moles per volume; whereas in Refs. 3.2 and Kirchheim (1982) c_f or c_t is equivalent with the fractional occupancies n_f or n_t.

Applying Statistical Mechanics (Lacher, 1937) analogously to the ideal solution model yields for the chemical potential μ (assuming equality of the chemical potentials in the subsystems)

$$\mu = \mu_f = E_o + RT \ln \frac{n_f}{1 - n_f} \quad \text{and} \quad \mu = \mu_t = E_t + RT \ln \frac{n_t}{1 - n_t}. \tag{31}$$

For a saturation of traps $n_t \to 1$ the chemical potential goes to infinity. The last equation may be also written as

$$N_f = \frac{N_o - N_{to}}{1 + \exp\left(\frac{E_o - \mu}{RT}\right)} \quad \text{and} \quad N_t = \frac{N_{to}}{1 + \exp\left(\frac{E_t - \mu}{RT}\right)} \qquad (32)$$

which in terms of concentration c is

$$c_f = \frac{\beta/\Omega - c_{to}}{1 + \exp\left(\frac{E_o - \mu}{RT}\right)} \quad \text{and} \quad c_t = \frac{c_{to}}{1 + \exp\left(\frac{E_t - \mu}{RT}\right)} \qquad (33)$$

This equation is a special case of the more general Eqn (46) valid for a distribution of binding energies. During the derivation a random distribution of H atoms is assumed among either normal or trap sites neglecting H–H interaction and vibrational entropy. In equilibrium the chemical potentials of trapped and free hydrogen atoms have to be equal. By eliminating μ Eqn (33) gives

$$\frac{n_f}{n_t} = \exp\left(\frac{E_t - E_o}{RT}\right) < 1 \quad \text{or} \quad \frac{N_f}{N_{fo} - N_f} = \frac{N_t}{N_{to} - N_t} \exp\left(\frac{E_t - E_o}{RT}\right) < 1 \qquad (34)$$

which is in agreement with Oriani's model (Oriani, 1970) for $N_f \ll N_{fo}$, that is for very small hydrogen concentrations in normal sites which is mostly the case in ferritic steels.

Solving the last equations for the chemical potential yields (Pfeiffer and Wipf, 1976)

$$\mu = E_o - RT \ln\left[-\left(\frac{1+\gamma}{2} - \frac{N_o - N_{to}(1-\gamma)}{2N_H}\right) + \sqrt{\left(\frac{1+\gamma}{2} - \frac{N_o - N_{to}(1-\gamma)}{2N_H}\right)^2 - \gamma\frac{N_H - N_o}{N_H}}\right]. \qquad (35)$$

The amount of hydrogen N_H a one-trap material picks up being in equilibrium with gaseous hydrogen of pressure p_H is obtained from the last equation by inserting $\mu = 0.5\,RT \ln p_H$. In this case E_o corresponds to the energy of normal sites with gaseous hydrogen at 295 K and 1 bar as a reference state. The relation between hydrogen content and pressure is called solubility or a pressure composition isotherm, respectively. This has to be distinguished from the terminal solubility, which is the maximum concentration of hydrogen before hydride formation occurs.

For the case of no traps and/or zero binding energy $E_o - E_t = 0$ ($\Rightarrow \gamma = 1$) the last equation reduces to the well-known relation

$$\mu = E_o + RT \ln\frac{N_H}{N_o - N_H}. \qquad (36)$$

A comparison with experimental data is limited, because most materials, especially steels, contain more than just one type of traps. An exception is nitrogen in niobium which traps hydrogen. Pfeiffer and Wipf (1976) not only derived the above equation but also showed by measuring resistivity changes that the solubility of hydrogen in Nb–N is in excellent agreement with these relations.

25.3.1.1.2 Diffusivity for One Kind of Trap

Following Oriani (1970) in his trap model he assumed that free H atoms only contribute to the overall flux (the validity of this assumption is proven in 3.1.2.2 a). Then Fick's first law becomes

$$J = -D_f \frac{\partial c_f}{\partial x},$$
(37)

where D_f is the diffusion coefficient of hydrogen in a lattice containing normal sites only. With the continuity equation

$$\frac{\partial c}{\partial t} = -\frac{\partial J}{\partial x}$$
(38)

we obtain

$$\frac{\partial c_f}{\partial t} + \frac{\partial c_t}{\partial t} = D_f \frac{\partial^2 c_f}{\partial x^2}$$
(39)

For the sake of simplicity and for a comparison with Oriani's result we further on assume $c_f \ll \beta/\Omega$ yielding the following relation from Eqn (33)

$$c_t = \frac{c_f c_{to}}{c_f + (\beta/\Omega - c_{to}) \exp\left(\frac{E_t - E_o}{RT}\right)}$$
(40)

with the following derivative

$$\frac{\partial c_t}{\partial t} = \frac{\partial c_f}{\partial t} \frac{c_{to}(\beta/\Omega - c_{to}) \exp\left(\frac{E_t - E_o}{RT}\right)}{\left[c_f + (\beta/\Omega - c_{to}) \exp\left(\frac{E_t - E_o}{RT}\right)\right]^2}.$$
(41)

Inserting the last equation into Eqn (39) yields

$$\frac{\partial c_f}{\partial t} = \frac{D_f}{\left\{1 + \frac{c_{to}(\beta/\Omega - c_{to}) \exp\left(\frac{E_t - E_o}{RT}\right)}{\left[c_f + (\beta/\Omega - c_{to}) \exp\left(\frac{E_t - E_o}{RT}\right)\right]^2}\right\}} \frac{\partial^2 c_f}{\partial x^2}.$$
(42)

This is a nonlinear partial differential equation with no known analytical solution. However for the very special case of traps being mostly empty corresponding to $c_f \ll (\beta/\Omega - c_{to})\exp[(E_t - E_o)/RT]$ the last equation becomes linear

$$\frac{\partial c_f}{\partial t} = \frac{D_f}{1 + \frac{c_{to}}{[(\beta/\Omega - c_{to})\exp\left(\frac{E_t - E_o}{RT}\right)]}} \frac{\partial^2 c_f}{\partial x^2} \equiv D_{eff} \frac{\partial^2 c_f}{\partial x^2}.$$
(43)

Thus all known solutions of diffusion problems can be used immediately by replacing D by the effective coefficient D_{eff} defined by Eqn (43). This is a well-known procedure for evaluating electrochemical and thermal desorption experiments. The results are of limited validity, because the assumption of mostly empty traps is not fulfilled in many cases.

25.3.1.2 *Distribution of Traps of Different Binding Energy*

25.3.1.2.1 *Solubility for a Distribution of Binding Energies*

From a system containing sites of one energy only like in a single crystal we come to a two-level system by introducing isolated point defects like foreign atoms or vacancies, by restricting the interaction with hydrogen to the nearest interstices. As spatial correlation does not play a role all the foreign atoms may be combined to a layer in between the host metal and by neglecting the interface this becomes a two-level system, too.

Going from zero- to one- and two-dimensional defects of the lattice, it is more realistic to deal with a manifold distribution of site energies instead of just two. This will be discussed in more detail in the following sections dealing with dislocations and grain boundaries. Finally, in a perfect amorphous structure all sites may be considered to be different from each other representing a system without degeneracy in terms of Statistical Thermodynamics. A DOSE is defined under these circumstances like in solid-state physics as the normalized number of sites $n(E) = \delta N_0 / N_0$ in a given energy window E, $E + dE$ with (Kirchheim, 1988)

$$\int_{-\infty}^{\infty} n(E) dE = 1 \qquad (44)$$

This definition is equivalent with the DOS function used for electrons or other fermions in quantum mechanics. Examples of a DOSE for defected lattices are shown in **Figure 7**.

The various sites of a DOSE compete for the occupancy with hydrogen. By occupying the sites of lowest energy the system reduces its total energy whereas the configurational entropy increases by filling sites being present in large numbers; both effects reduce the Gibbs free energy. If we allow a site to be occupied by one H atom only, the corresponding minimization can be treated in the framework of the FD statistics (Kirchheim, 1988). Here the configurational entropy in each energy window of the DOSE

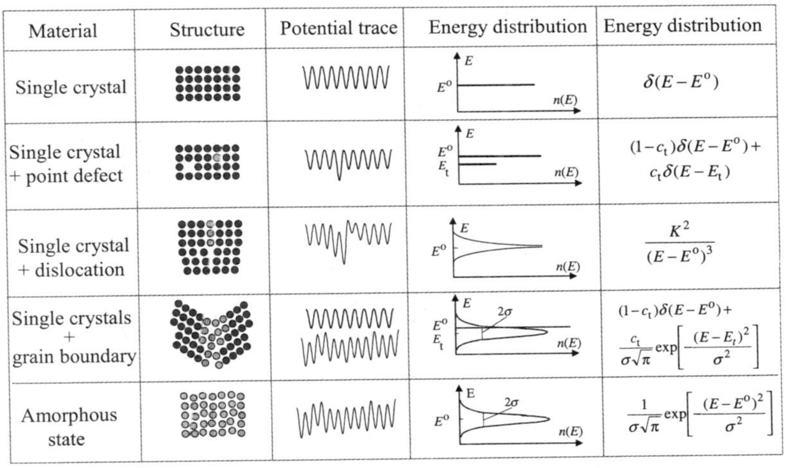

Material	Structure	Potential trace	Energy distribution	Energy distribution
Single crystal			E, E^0, $n(E)$	$\delta(E - E^0)$
Single crystal + point defect			E, E^0, E_t, $n(E)$	$(1 - c_t)\delta(E - E^0) +$ $c_t \delta(E - E_t)$
Single crystal + dislocation			E, E^0, $n(E)$	$\dfrac{K^2}{(E - E^0)^3}$
Single crystals + grain boundary			E, E^0, E_t, 2σ, $n(E)$	$(1 - c_t)\delta(E - E^0) +$ $\dfrac{c_t}{\sigma\sqrt{\pi}} \exp\left[-\dfrac{(E - E_t)^2}{\sigma^2}\right]$
Amorphous state			E, E^0, 2σ, $n(E)$	$\dfrac{1}{\sigma\sqrt{\pi}} \exp\left[-\dfrac{(E - E^0)^2}{\sigma^2}\right]$

Figure 7 Schematic presentation of a distribution of site energies which starts with the most degenerate one, i.e. the single crystal in the top line. In the following lines increasing structural disorder leads to a decreasing degeneracy by increasing the dimensions of lattice defects. For diffusion potential traces or energy landscapes, respectively, including the defect. (For color version of this figure, the reader is referred to the online version of this book.)

is calculated under the assumption of single occupancy. During this procedure it does not matter whether electrons are distributed among energy states in reciprocal space or particles among sites in real space. The minimum of Gibbs free energy yields the following result for the thermal occupancy of energy level E_i or the corresponding site, respectively,

$$o(E_i) = \frac{1}{1 + \exp[(E_i - \mu)/RT]} \tag{45}$$

where μ is the chemical potential of the particles also being called Fermi energy in this context. Despite the fact that the previous equations for the two-level system and the following ones are valid for all kinds of interstitial particles (Kirchheim, 2002) we restrict ourselves to hydrogen in metals. Other interstitials like small molecules in polymers are treated in Kirchheim (2002).

Integrating over all sites of a DOSE yields the total hydrogen content, that is

$$n_H = \int_{-\infty}^{\infty} \frac{n(E)\mathrm{d}E}{1 + \exp[(E - \mu)/RT]}. \tag{46}$$

The amount of hydrogen being in equilibrium with a gas phase of hydrogen partial pressure p_H is obtained by inserting $\mu = 0.5\,RT \ln p_H$. In this case the energies E correspond to the energy of sites with the reference state being gaseous hydrogen at 295 K and 1 bar.

A closed solution of the integral in Eqn (46) will be possible for simple forms of $n(E)$ only. To understand the aforementioned competition between energy decrease and entropy gain the following approximation based on the step or $T = 0$ approximation of the Fermi-Dirac function (i.e. $o(E) \approx 1$ for $E < \mu$ and $o(E) \approx \exp[(\mu - E)/kT]$ for $E > \mu$) will be applied

$$n_H = \int_{-\infty}^{\mu} n(E)\mathrm{d}E + \int_{\mu}^{\infty} \frac{n(E)\mathrm{d}E}{\exp[(E - \mu)/RT]} \equiv n_{H1} + n_{H2} \tag{47}$$

The first term n_{H1} on the right-hand side of the equation is that part of the total concentration c arising from particles below the Fermi level and n_{H2} corresponds to sites above μ. The system decreases its energy by occupying sites below the Fermi level whereas it increases its configurational entropy for the sites above μ.

For $N_H/N_o << 1$ or $\mu << 0$, respectively, and $n_{H1} << n_{H2} \approx n_H$ we obtain

$$n_H \approx \int_{-\infty}^{\infty} \frac{n(E)\mathrm{d}E}{\exp[(E - \mu)/kT]} = \exp\left[\frac{\mu}{kT}\right] \int_{-\infty}^{\infty} n(E)\exp\left[-\frac{E}{kT}\right]\mathrm{d}E = \frac{a}{\gamma} \tag{48}$$

Then Henry's Law $a = \gamma n_H$ is fulfilled with the thermodynamic activity $a = \exp(\mu/kT)$ being proportional to n_H. An interesting exception is an exponential DOSE (Kirchheim, 2002).

H–H interaction has to be taken into account for those cases, where segregation at extended defects takes place and, therefore, locally high concentrations n_{loc} occur. Then the chemical potential may be written in a first-order approximation as (Kirchheim, 1988; Lacher, 1937)

$$\mu = \mu_{id} + W n_{loc}, \tag{49}$$

where W is an interaction parameter. With the assumption that the sites associated with defects are a minority the DOSE is separated

$$n(E) = v_f \delta(E - E_o) + n_{\text{def}}(E) \tag{50}$$

with v_f being the fraction of normal sites, δ the Dirac-Delta function and n_{def} the fraction corresponding to the defect like a grain boundary. Then Eqn (46) gives

$$n_{\text{loc}} = \int_{-\infty}^{\infty} \frac{n_{\text{def}}(E)dE}{1 + \exp\left(\frac{E - \mu - W n_{\text{loc}}}{RT}\right)} \quad \text{and} \quad n_f = v_f \exp\left(\frac{\mu - E^o}{RT}\right). \tag{51}$$

With the trivial equation $n = n_f + n_{\text{loc}}$ we can solve analytically or numerically depending on the form of $n_{\text{def}}(E)$, Eqn (51) to get the relation $\mu(c)$.

25.3.1.2.2 Diffusivity for a Distribution of Binding Energies

Analytical modeling of diffusion with the presence of a DOSE requires some restricting assumptions regarding the transition rates of H atoms to adjacent sites. For H atoms to perform a random walk the transition rates to all next nearest-neighboring sites have to be equal. For thermally activated jumps as well as for quantum mechanical tunneling to the next site the activation barrier and the jump distance have to be the same. If not, correlation effects have to be taken into account. The analytical relations derived in the following are based on the assumption of equal transition rates, that is by assuming constant saddle point energies for thermally activated hopping over a saddle point. The validity of this assumption is checked by comparing analytical results with both experimental data and kinetic Monte Carlo simulations (Kirchheim, 1988). Examples are shown in **Figure 8** for a broad Gaussian DOSE.

Further on we have to distinguish between H atoms migrating in a concentration gradient yielding the chemical diffusion coefficient or migrating in a lattice of constant concentration yielding the tracer diffusion coefficient. A more detailed discussion and derivation of the following results is provided in Kirchheim (2002).

25.3.1.2.2.1 Diffusion in a Concentration Gradient.

We consider two adjacent lattice planes of distance l (jump distance). The x-axis is parallel to the normal of the planes and the average concentrations n_{H1} and n_{H2} within the planes are different. Thus we have a concentration gradient from plane 1 at x to plane 2 at $x + l$. The flux J is obtained by averaging over the jumps in between the planes from sites of energy E_1 in plane 1 to sites of energy E_2 in plane 2 and vice versa.

$$J = \frac{\beta l}{6\Omega} \int\int \left[\frac{\Gamma_1(x)n(E_1)}{1 + \exp\left(\frac{E_1 - \mu(x)}{RT}\right)} - \frac{\Gamma_2(x + l)n(E_2)}{1 + \exp\left(\frac{E_2 - \mu(x+l)}{RT}\right)} \right] dE_1 dE_2 \tag{52}$$

where the jumps occur in all six directions and only 1/6 of them to the plane under consideration. The ratio β/Ω on the right-hand side of Eqn (52) has to be included because n was defined as the fraction of occupied interstices whereas the flux J is defined as particles or moles per area and time. The first term in brackets corresponds to the jumps of particles in sites of energy E_1 out of plane 1 at x (cf. **Figure 9**) and the second term accounts for particles in sites of energy E_2 in plane 2 at $x + l$, where the jump

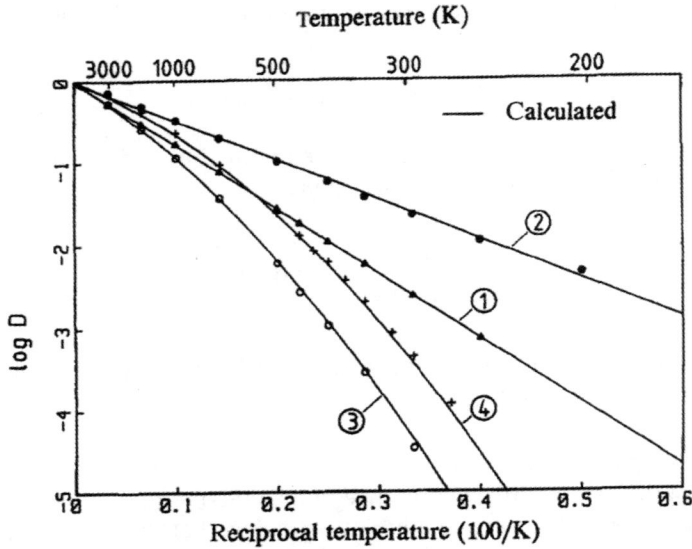

Figure 8   Diffusion coefficient from dynamic Monte Carlo simulations for $n_H = 10^{-3}$ and for a Gaussian distribution of site and saddle point energies with a width of 0 or 10 kJ/Mol for sites and 0 and 7.1 kJ/Mol for saddle points. Case 1: no distribution in site and saddle point energies. Case 2: no distribution in sites but in saddle point energies. Case 3: distribution in sites but not in saddle point energies. Case 4: distribution in both site and saddle point energies. The curves for cases 3 and 4 were calculated from the approximate equations for the ideal dilute case (cf. text). The curves for cases 1 and 2 are nearly straight lines, as no trapping in low-energy sites occurs. A detailed description of the diffusion behavior for the various cases is provided in Kirchheim (1988).

frequencies Γ_i have to be weighted with the fraction of occupied sites of a given energy $n(E_i)o(E_i)$. Jump frequencies are calculated for thermally activated processes over saddle points of constant energy $E^o + Q^o$ with an attempt frequency ν_o (cf. **Figure 9**) including blocking of sites by a factor of $n(E_i)$ $[1 - o(E_i, \mu)]$

$$\Gamma_1 = \nu_o n(E_2)[1 - o(E_2, \mu(x+l))]\exp\left(-\frac{Q^o + E^o - E_1}{RT}\right) \tag{53}$$

$$\Gamma_2 = \nu_o n(E_1)[1 - o(E_1, \mu(x))]\exp\left(-\frac{Q^o + E^o - E_2}{RT}\right) \tag{54}$$

   In Ref. 3.2 similar calculation was presented for the first time; but the fact that the occupancy depends on the chemical potential (cf. Eqn (45)) that depends on position x was not taken into account properly leading to a slightly different expression for J. Inserting Eqns (53) and (54) in Eqn (52) and partly integrating yields

$$J = \frac{\beta l \nu_o (1 - n_H)}{6\Omega}\exp\left(-\frac{Q^o + E^o}{RT}\right)\left[\int\frac{\exp\left(\frac{E_1}{RT}\right)n(E_1)dE_1}{1 + \exp\left(\frac{E_1 - \mu(x)}{RT}\right)} - \int\frac{\exp\left(\frac{E_2}{RT}\right)n(E_2)dE_2}{1 + \exp\left(\frac{E_2 - \mu(x+l)}{RT}\right)}\right] \tag{55}$$

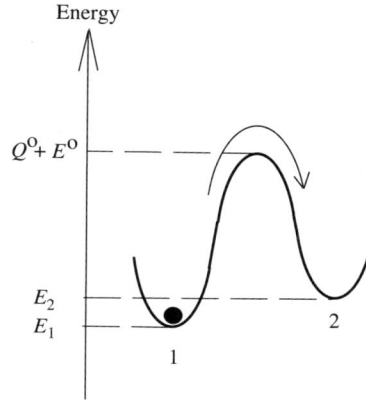

Figure 9 Potential trace for an interstitial jumping from site 1 with energy E_1 to site 2 with energy E_2. E^o is an average energy of the DOSE and Q^o the activation energy with respect to this reference. Thus particles in a material containing sites of energy E^o only have a jump frequency of $\Gamma = \Gamma^o \exp[-Q/RT]$.

The first term in brackets is equal to $1 - n_H$. Expanding the second term in brackets in a Taylor series around x gives (Kirchheim, 2002)

$$J = -\frac{\beta l v_o (1 - n_H)^2}{6\Omega} \exp\left(-\frac{Q^o}{RT}\right) \exp\left(\frac{\mu - E^o}{RT}\right) \frac{l}{RT} \frac{\partial \mu}{\partial x}. \tag{56}$$

Thus it is shown in the framework of Statistical Thermodynamics that the flux is proportional to the gradient of the chemical potential which is independently obtained from Irreversible Thermodynamics. The derivative in Eqn (56) is modified

$$J = -\frac{\beta l v_o (1 - n_H)^2}{6\Omega} \exp\left(-\frac{Q^o}{RT}\right) \exp\left(\frac{\mu - E^o}{RT}\right) \frac{l}{RT} \frac{\partial \mu}{\partial n_H} \frac{\partial n_H}{\partial x} \tag{57}$$

and the resulting relation is compared with Fick's first law yielding for the chemical diffusion coefficient

$$D = \frac{l^2 v_o (1 - n_H)^2}{6} \exp\left(-\frac{Q^o}{RT}\right) \exp\left(\frac{\mu - E^o}{RT}\right) \frac{1}{RT} \frac{\partial \mu}{\partial n_H} \tag{58}$$

For a two-level system with a small concentration of free hydrogen (ideal dilute case) and $n_H \ll 1$ Eqn (57) becomes

$$J = -\frac{\beta}{\Omega} D_f \frac{n_f}{RT} \frac{\partial \mu}{\partial x} = -D_f \frac{\partial c_f}{\partial x}, \tag{59}$$

which is equivalent with Oriani's assumption (cf. Eqn (37)).

25.3.1.2.2.2 Random Walk with Constant H Concentration. It will be shown in the following that the assumption of thermally activated hopping can be replaced by less restrictive conditions which includes tunneling through activation barriers.

We consider P particles in N sites of a lattice with a given DOSE. The equilibrium distribution of the particles according to FD statistics shall be stationary in space and one tagged particle shall do a random walk on the partially occupied lattice. It might not be really necessary to assume a stationary distribution for the remaining particles. What might be the important effect only is that the moving particle will not be able to occupy all lattice sites as they are differently blocked according to Eqn (45). One might call this simplification the one-particle approximation resembling the one-electron approximation of solid-state physics. Further on the walk is uncorrelated besides the few blocking events and, therefore, the mean distance R after Z jumps is given according to simple random walk theory (Heumann, 1992; Shewmon, 1963)

$$R^2 = Zl^2, \tag{60}$$

where l is the jump distance. Then a tracer diffusion coefficient can be defined by Heumann (1992) and Shewmon (1963)

$$D^* \equiv \lim_{t \to \infty} \frac{R^2}{6t} = \frac{l^2}{6} \lim_{t \to \infty} \frac{Z}{t}, \tag{61}$$

where t is the total time required for the walk. The time t is the sum of all the times of residence τ_m the particle stayed in the various sites visited during the walk, if the time interval required for the site exchange is small compared with τ_m. Thus we get

$$t = \sum_{m=1}^{Z} \tau_m. \tag{62}$$

This way the jump rate as the reciprocal of the residence time has not to be described by a special mechanism like thermally activated hoping for instance. Among the various sites we combine those having the same energy E_i and average over the distribution of empty sites, that is sum over all $N–P$ sites. Hereby it is assumed that the Z sites which have to belong to the empty category are as representative for empty sites as are the $N–P$ ones. Thus we obtain

$$t = \sum_{m=1}^{Z} \tau_m = \sum_i \sum_{E_m=E_i} \tau_m \equiv \frac{Z}{N-P} \sum_{i=1}^{N-P} t_i, \tag{63}$$

where t_i is the average time the particle spends in sites of energy E_i. It is tacitly assumed that the number of jumps Z and the number of lattice sites N are large enough to apply the laws of random walk and statistical mechanics. According to the ergodic hypothesis the fraction of time a particle spends in sites of type i is equal to the fraction of particles in this type of site yielding

$$\frac{t_i}{t} = \frac{n_i' o(E_i)}{P} = \frac{n_i[1 - o(E_i)]o(E_i)}{P}, \tag{64}$$

where n_i' is the number of free sites of energy E_i which is obtained from the site energy distribution by multiplying with $[1 - o(E_i)]$. For the dilute solution most of the free sites have a low occupancy $[o(E_i) << 1]$ and Eqn (45) gives

$$o(E_i) \approx \exp\left[\frac{\mu - E_i}{RT}\right]. \tag{65}$$

This way FD statistics is replaced for the empty sites by Boltzmann statistics. For convenience we choose an arbitrary reference site having energy E^o and a low value of $o(E^o)$. Then the following relation is derived from Eqns (64) and (65)

$$\frac{t_i}{t^o} = \frac{n_i[1 - o(E_i)]}{n^o[1 - o(E^o)]} \exp\left[\frac{E^o - E_i}{RT}\right],$$ (66)

where t^o is the fraction of the total time t particles reside within sites of energy E^o. Inserting Eqn (66) in Eqn (63) with $[1 - o(E^o)] \approx 1$ and Eqn (45) gives

$$t = \frac{Zt^o}{(N-P)n^o} \sum_{i=1}^{N-P} n_i \exp\left[\frac{E^o - E_i}{RT}\right] \left[\frac{\exp[(E_i - \mu)/RT]}{1 + \exp[(E_i - \mu)/RT]}\right]$$ (67)

or

$$t = \frac{Zt^o \exp[(E^o - \mu)/RT]}{N(1-c)n^o} \sum_{i=1}^{N-P} \frac{n_i}{1 + \exp[(E_i - \mu)/RT]}$$ (68)

where the last sum on the right-hand side is nothing else than the discrete form of Eqn (46) and, therefore, it can be expressed by the concentration n_H. Then the following simple result is obtained

$$t = \frac{cZt^o \exp[(E^o - \mu)/RT]}{(1 - n_H)n^o} = \frac{Zt^o}{\gamma^o(1 - n_H)n^o} = \frac{Z\tau}{(1 - n_H)\gamma^o}$$ (69)

where τ is the mean residence time as defined by

$$\tau = \frac{1}{n^o} \sum_{s=1}^{n^o} \tau_s \quad \text{for} \quad E_s = E^o$$ (70)

Inserting Eqn (69) into Eqn (61) yields

$$D^* = \frac{l^2(1 - n_H)}{6\tau} \frac{\exp[(\mu - E^o)/RT]}{n_H}$$ (71)

If we consider the hypothetical material with sites of energy E^o only the residence time τ as defined by Eqn (70) is enlarged compared with the empty lattice due to blocking of sites. The dilute residence time is then $\tau^o = \tau(1 - n_H)$ and the diffusion coefficient in the dilute regime of the reference lattice is given by

$$D^o = \frac{l^2}{6\tau^o} = \frac{l^2}{6\tau(1 - n_H)}.$$ (72)

Comparing the chemical diffusion coefficient with the tracer one (cf. Eqn (58)) and the trivial relation $\nu_o \exp(-Q^o/RT) = 1/\tau$ yields

$$D = D^* \frac{(1 - n_H)}{RT} \frac{\partial \mu}{\partial \ln n_H} = D^*(1 - n_H) \frac{\partial \ln a_H}{\partial \ln n_H}.$$ (73)

This again is a result which is also obtained in Irreversible Thermodynamics with the derivative on the very right-hand side of Eqn (73) called thermodynamic factor. The term $(1 - n_H)$ vanishes for $n_H \rightarrow 0$ because the thermodynamic factor $\frac{\partial \ln a_H}{\partial \ln n_H} \rightarrow \frac{1}{1-n_H}$ for $n_H \rightarrow 0$. This is expected because D has to become D^* for $n_H \rightarrow 0$.

The concentration dependence of both the chemical and tracer diffusion coefficients is very pronounced because both depend exponentially on the effective activation energy $Q^o + E^o - \mu$. The chemical potential μ or Fermi energy, respectively, increases with increasing concentration because of the second derivative of Gibbs free energy being always positive in equilibrium. Thus an increasing concentration always leads to decreasing activation energy and an increasing diffusivity, if we neglect the effect of n_H in the denominator of Eqn (73). The atomistic interpretation of this dependence is very simple. With increasing concentration sites of higher energy have to be occupied and, therefore, the Fermi energy rises and comes closer to the saddle point energy. Thus an increasing number of particles experience a smaller activation barrier for jumping.

25.3.1.3   *Diffusion in a Stress Field*

Assuming an isotropic strain field around a hydrogen atom the corresponding elastic energy E_{el} is the product of hydrostatic stress and partial molar volume (cf. Section 25.3.4.1). In a field of changing hydrostatic stress σ_h the elastic energy changes with position giving rise to a corresponding gradient of energy or a driving force, respectively,

$$\vec{F}_{el} = \nabla E_{el} = \nabla V_H \sigma_h = V_H \nabla \sigma_h. \tag{74}$$

where V_H is the partial molar volume of hydrogen.

The driving force may be superimposed on the gradient of the chemical potential leading to a total flux (mobility times driving force) of

$$J = \frac{c_f D_f}{RT}(-\nabla \mu + V_H \nabla \sigma_h). \tag{75}$$

Using the values of the free hydrogen atoms for the mobility $c_f D_f/(RT)$ the last equation includes effects of trapping. With tensile stresses being positive the hydrogen is driven into regions of larger tensile stress, that is from compressed into expanded regions of a lattice. Thus a counteracting concentration gradient is build up leading to a steady state with $J = 0$ and no changes of H concentration. For the one-dimensional case this corresponds to

$$RT \, d \ln c_f = V_H d\sigma_h \quad \text{or} \quad c_f = c_f^o \exp\left(\frac{V_H \sigma_h}{RT}\right), \tag{76}$$

with c_f^o being the concentration for $\sigma_h = 0$. The attainment of the steady state may be a complicated diffusion problem depending on the diffusion coefficient, its dependence on stress (see below), initial and boundary conditions. For the simplest case of a single crystalline lattice the steady state is approached by an exponential time constant leading to an elegant determination of the diffusion coefficient via the Gorsky effect (Peisl, 1978).

Besides stress gradients there may be additional driving forces like electric fields or temperature gradients giving rise to additional contributions to the flux called electro- and thermomigration (Wipf, 1978).

For the case of a constant stress field there is no driving force for an H flux; but there may be a difference of the elastic energies between equilibrium and saddle point position, if the strain induced by hydrogen in these two positions is different. In other words, if the partial molar volume in the equilibrium position V_H is different from the one in the saddle point position V_{Hs} the activation energy for diffusion Q is changed by the corresponding difference of elastic energies

$$Q = Q^o - \sigma_h(V_{Hs} - V_h). \tag{77}$$

Anticipating that H atoms in the saddle point configuration cause a larger distortion than in the equilibrium position leads to a positive activation volume $V^* = V_{Hs} - V_H > 0$ with a concurrent increase of the diffusion coefficient for tensile stresses.

25.3.2 Diffusion and Trapping—Special Cases and Experimental Results

In the preceding chapter the effect of a distribution of site energies (DOSE) on solubility and diffusivity of interstitial particles has been treated rather generally. Most of the statements apply for other interstitials besides hydrogen as well and the form of the DOSE was not specified and, therefore, can be used for all kind of defects. In this chapter we focus on hydrogen and the most important defects, foreign atoms, vacancies, dislocations and grain boundaries up to a totally defected lattice of an amorphous metal. For each case the appropriate expression for the DOSE is proposed and experimental results regarding trapping and diffusion are presented. Besides the fundamental interest in the interaction between solutes and defects there is the very much applied interest of understanding HE, most importantly in steels. In these materials hydrogen solubility is very low which makes it difficult to measure hydrogen-related property changes. In addition, many of the reported trapping energies to defects in steels are obtained by using Oriani's model, that is simplifying trapping by assuming two energy levels only and more severe by assuming mostly empty traps. A list of trap energies in hydrogen in a carbon steel is presented in **Table 4**.

25.3.2.1 Interaction with Solute Atoms and Vacancies

An appropriate DOSE for the interaction of hydrogen with point defects of dilute concentration is a two-level system

$$n(E) = n_f(E) + n_t(E) = (1 - n_{to})\delta(E - E^o) + n_{to}\delta(E - E_t). \tag{78}$$

Table 4 Binding energies of hydrogen to various defects relative to the standard state of gaseous hydrogen as compiled in Stroe (2006)

Traps	Binding energy (kJ/mol)	Degassing temperature (° C)	Material	Reference
Matrix	6.9	Room temperature	Fe	Bosson et al. (1999)
Grain boundaries	17	112	Fe	Choo and Lee (1982)
Dislocations	20–26	215	Fe	Choo and Lee (1982)
		200	Fe	Lee and Lee (1984)
		272	Steel	Lee et al. (1982)
Micro voids	35–48	338	Steel	Lee et al. (1982)
		305	Fe	Choo and Lee (1982)
		480	Steel	Otsubo et al. (1982)
Carbide interfaces	97	723	Fe	Lee and Lee (1984)

Inserting this in Eqn (46) immediately leads to Eqn (35) and the Oriani trapping model. Thus the effect of a two-level system on solubility and diffusivity is already treated sufficiently in Section 25.3.1.1.

The interaction of hydrogen with foreign atoms both substitutional and interstitial is rather weak with a binding energy of about 10 kJ/mol in iron (Hirth, 1980). This has been determined by a large variety of experimental techniques measuring permeation, internal friction, resistivity and neutron scattering. Because of the small value of the interaction energy it is difficult to decide how much of it is elastic or electronic interaction. However, interaction with vacancies is very strong with measured binding energies between 30 and 100 kJ/Mol. They are in good agreement with calculated values using the effective medium theory (Myers et al., 1992).

In a binary concentrated alloy $A_x B_{1-x}$ and for tetrahedral sites hydrogen atoms can occupy five different types of tetrahedra, that is with 4A, 3A1B, 2A2B, 1A3B and 4B matrix atoms on the corners of the tetrahedra. For a random distribution of A and B the fraction of the various tetrahedral is a binominal one and the DOSE will be

$$n(E) = f \sum_{i=0}^{4} \binom{4}{i} x^i (1-x)^{4-i} \delta(E - E_i), \tag{79}$$

where E_i is the site energy of the corresponding tetrahedron. This DOSE proposed for a concentrated alloy has been used by Wagner (1973b) to model the thermodynamic activity of interstitial oxygen in liquid iron alloys. Measurements of hydrogen solubility were performed by Feenstra et al. (1988) in niobium–vanadium alloys over a wide range of alloy composition and hydrogen concentration. Because of the large data set they were able to show unambiguously that the fraction of the various tetrahedra follows the binominal distribution of a random alloy. Hydrogen prefers tetrahedra having a higher number of V atoms at their corners in agreement with the vanadium hydride being a stronger hydride former when compared with niobium, that is $E_4 < E_3 < \ldots < E_0$ where the subscript refers to the number of V atoms. Besides the expected behavior the site energies with respect to i they also depend on alloy composition, x. This is explained (Feenstra et al., 1988) by an overall change of the size of tetrahedral sites following the changes of the lattice parameter. According to the positive partial molar volume of hydrogen the site energy is lowered, when the site volume increases. Therefore, a V_4 site has a lower site energy in a niobium-rich alloy in comparison with a vanadium-rich alloy. All these results are compiled in **Figure 10** as the DOSE of the Nb–V alloy.

25.3.2.2 Interaction with Dislocations

Theoretical models describing the interaction of hydrogen with dislocations (Hirth and Lothe, 1968) are based on an elastic interaction between the stress field around dislocations and the strain caused by a solute atom. Hydrogen in metals causes volume expansion and, therefore, interacts strongly with the stress field of edge-type dislocations. As stresses of dislocations are calculated by applying continuum theory they are less reliable within the dislocation core and, therefore, trapping of hydrogen in dislocation cores has to be treated separately. In the following the DOSE is calculated based on these considerations for edge dislocations.

The hydrostatic part of the stress field of an edge dislocation is given by Hirth and Lothe (1968)

$$p = \frac{\sigma_{ii}}{3} = \frac{Gb(1+\nu)}{3\pi(1-\nu)} \frac{\sin \theta}{r}, \tag{80}$$

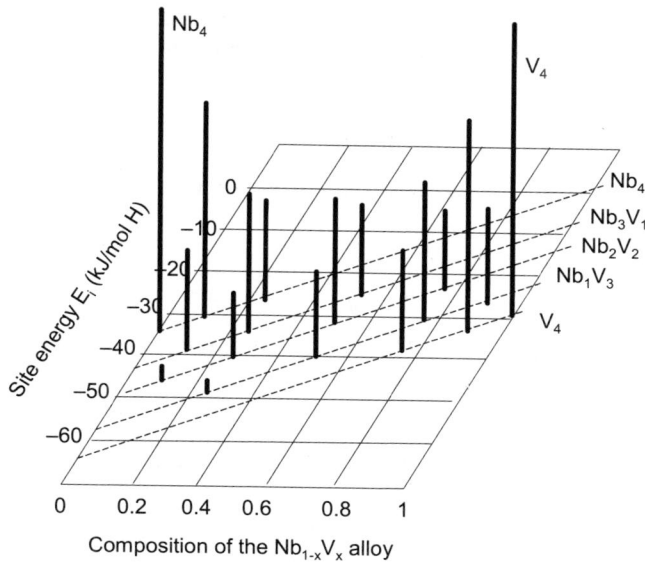

Figure 10   DOSE for the various $Nb_{1-x}V_x$ tetrahedra in crystalline Nb–V alloys as a function of composition (Feenstra et al., 1988).

where G is the shear modulus, ν Poisson's ratio, b the magnitude of the Burger's vector, θ and r are cylindrical coordinates as defined in **Figure 11** with the z-axis along the dislocation line.

Then the interaction energy with H atoms on a circle of constant pressure is obtained from

$$pV_H = \frac{Gb(1+\nu)}{6\pi(1-\nu)R} V_H \equiv \frac{AV_H}{R}, \qquad (81)$$

where R is the diameter of the cylinder and A is defined by the last equation.

For H in Pd the number of octahedral sites that are chosen by hydrogen is the same as the number of Pd atoms. We assume that like in the β-phase of Pd only the fraction x (being ca. 0.6 at room

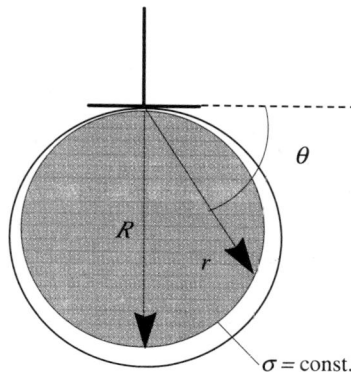

Figure 11   Circles of constant hydrostatic stress at an edge dislocation as calculated in continuum mechanics (cf. Eqn (80)). The elastic interaction energy with hydrogen is constant at these lines and the corresponding DOSE is the number of sites within the two circles.

temperature) is occupied. Then in a material containing ρ dislocations per unit area the number of sites, n, in a cylinder of radius R and unit length is $\rho\pi R^2$ and the DOSE becomes

$$n(E) = x\rho\frac{dn}{dE} = x\rho\frac{dn}{dR}\frac{dR}{dE} = x\rho\pi R\frac{AV_H}{E^2} = \frac{x\rho\pi AV_H}{E^3}. \tag{82}$$

Inserting this in Eqn (46), using the step approximation, that is the first part on the right-hand side of Eqn (47) and solving for μ gives

$$\mu = \sqrt{\frac{x\rho\pi AV_H}{c}}. \tag{83}$$

The last equation is checked by plotting in **Figure 12** measured values of μ for H in cold-rolled Pd (Kirchheim, 1981) versus the reciprocal square root of concentration.

Different to the theoretical prediction the interaction energy (difference of chemical potentials between deformed and annealed sample) becomes constant at very low concentrations which is explained in terms of a direct interaction with the dislocation core which was not included in Eqn (83). The value of about -50 kJ/Mol-H has been also determined for H in Fe (Kummnick and Johnson, 1980). The linear dependence on $1/\sqrt{c}$ in **Figure 12** yields a slope which corresponds to reasonable dislocation densities for heavily deformed metals ($2\cdot10^{11}$ cm^{-2}). Contrary to Eqn (83), the straight line corresponding to the interaction with the long-range stress field of the dislocation does not intercept the ordinate at 0 but at a value of about -20 kJ/Mol-H. This is attributed to a direct H–H interaction. It can be calculated from the values obtained for this interaction in well-annealed Pd at high H concentrations (Wicke and Blaurock, 1980). As a consequence of this H–H interaction segregation of hydrogen at edge dislocation leads to the formation of hydride cylinders below the glide plane. Direct evidence can be provided by small-angle neutron scattering (SANS). Scattering by randomly oriented cylinders has to be described by the following macroscopic cross-section (Maxelon et al., 2001a)

$$\frac{d\Sigma}{d\Omega} = \frac{2\pi^3\rho R_0^4\Delta g^2}{Q}\exp\left[-\frac{1}{4}Q^2R_0^2\right], \tag{84}$$

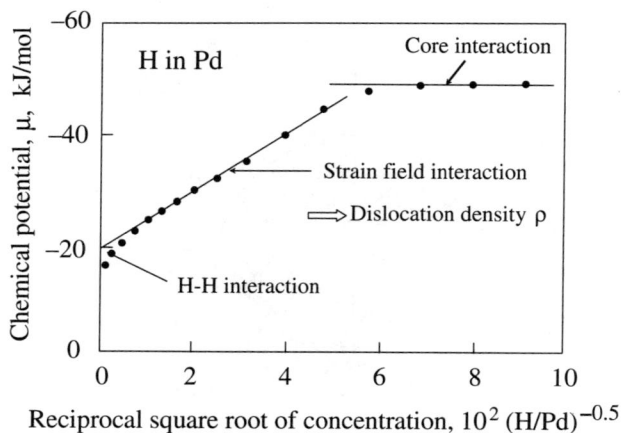

Figure 12 Electrochemically measured chemical potentials of hydrogen in heavily deformed palladium plotted according to Eqn (83) versus $1/\sqrt{c}$. The various regions of the data points represent regions of a predominant interaction mechanism.

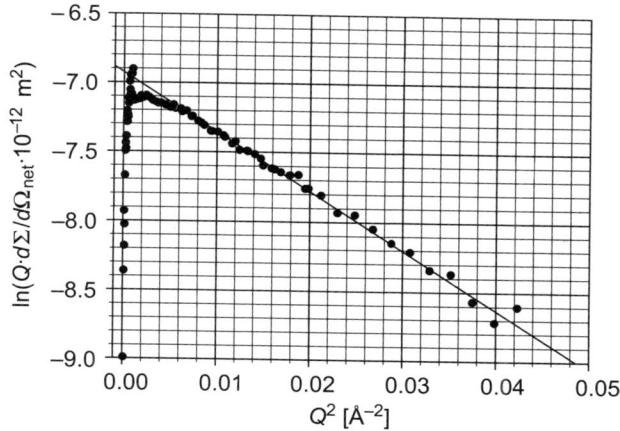

Figure 13 A modified Guinier Plot of the macroscopic scattering cross-section $d\Sigma/d\Omega$ (cf. Eqn (84)) measured by SANS for deformed Pd with 1 at.-% H. The slope of the linear part yields the radius of hydrogen-enriched cylinders formed by segregation at the dislocation line.

where ρ is the dislocation density, R_0 the radius of the cylinders, Δg is the difference of scattering length densities and Q is the magnitude of the scattering vector. By plotting the logarithm of the product of Q and measured values of the macroscopic cross-section (after appropriate subtraction of background and incoherent scattering) versus Q^2 straight lines are expected according to Eqn (84). This is in agreement with experimental findings as shown in **Figure 13**. The slope of the straight lines yields the radius of the cylinders and the intercept with the ordinate yields the dislocation density.

SANS experiments with a deformed iron-based alloy yield (Malard et al., 2012) less pronounced effects because the H concentration is much lower when compared with Pd and cannot be varied in a well-defined way.

For the Ni–H system it has been also shown by combining ab initio electronic structure calculations, semiempirical Embedded Atom Method (EAM) potentials and a lattice gas Hamiltonian (von Pezold et al., 2011) that H atoms at edge dislocation occupy all octahedral sites near the dislocation core, that is form a kind of cylindrical hydride, if and only if H–H interaction is taken into consideration. Otherwise a Cottrell type of H cloud is formed in the dilated region of the edge dislocation.

In agreement with theory it can be shown that only free hydrogen atoms contribute to the diffusion coefficient. H atoms trapped at the hydride cylinders do also not contribute to the electrical resistivity when compared with the free atoms. Thus the ratio of diffusivity in deformed and annealed metals and the corresponding one for the resistivity increment is equal to the ratio of free to total hydrogen concentration. The latter is determined from measured chemical potentials via electromotoric force (EMF) results. Thus we obtain

$$\frac{\rho_H}{\rho_H^0} = \frac{D^*}{D^o} = \frac{c_f}{c} = \exp\left(\frac{\mu}{k_B T}\right) \tag{85}$$

It is shown in **Figure 14** that this simple relation holds for hydrogen in strongly deformed palladium. At very low H concentrations the fraction of free hydrogen is negligible and, therefore, all the hydrogen is trapped at dislocations. At intermediate concentrations hydrogen atoms are partitioned between sites far away from dislocations and those close to them. But sites at dislocations never become saturated

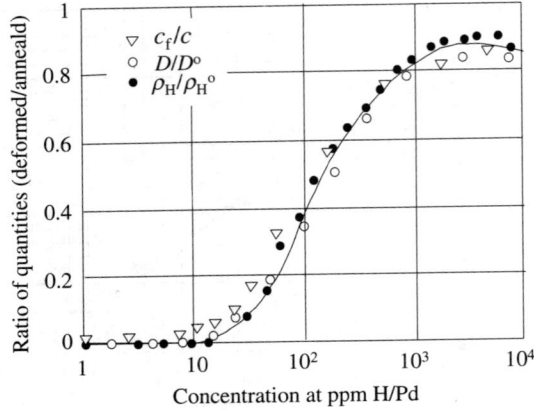

Figure 14 Hydrogen resistivity increment, hydrogen diffusivity and the concentration of free hydrogen in heavily deformed palladium divided by the corresponding values in well-annealed palladium. These ratios are plotted versus H concentration and in agreement with Eqn (85) they are about the same over the whole range of concentrations within the α-phase of Pd.

because of both the long-range elastic interaction and an attractive H–H interaction. Thus the corresponding cylinder being enriched in hydrogen is steadily growing.

The pronounced changes of H activity, diffusivity and resistivity in crystalline Pd which are caused by plastic deformation and arise from the presence of dislocations do not occur after cold rolling of amorphous PdSi alloys (Kirchheim et al., 1985). The absence of a detectable H trapping in deformed amorphous alloys is considered to reveal the absence of edge dislocation-like defects.

The experimental results presented in this section are in qualitative agreement with a variety of other studies (Kirchheim et al., 1985; Flanagan and Lynch, 1976; Züchner, 1970; Heuser et al., 1991; Heuser and King, 1997; Ross and Stefanopoulus, 1994). Often the interaction with dislocations is approximated by a two-level system. Depending on how much of the sites near the core are saturated the evaluated binding energies vary between −50 and −20 kJ/mol.

25.3.2.3 Interaction with Grain Boundaries

Again the concept of a site energy distribution is appropriate to study H segregation at grain boundaries. In this case the DOSE is bimodal containing a volume fraction f of sites belonging to the grain boundaries which may be obtained for nearly spherical grains from the grain diameter d and by assuming a certain thickness Δ of the boundary

$$f = \frac{A\Delta}{V} = \frac{3\Delta}{d}, \tag{86}$$

where V is the sample volume. For symmetric boundaries with a long-range periodic arrangement of atoms there will be a discrete spectrum of site energies. For a polycrystalline sample with a large number of different grain boundaries a Gaussian distribution may be appropriate. Thus the DOSE becomes

$$n(E) = (1-f)\delta(E) + f\frac{1}{\sigma\sqrt{\pi}}\exp\left[-\frac{(E-E_s)^2}{\sigma^2}\right], \tag{87}$$

where the first term on the right-hand side represents the sites in the grain of energy $E = 0$, E_s is the average site energy of all grain boundaries with respect to the sites within the grains and σ is the width of the Gaussian. Sites with $E < 0$ are traps for hydrogen.

Similar to dislocations the number of traps provided by the grain boundaries is rather small and, to study them by gradually filling, a large density of them is required, that is a small grain size. This will be the case for nanocrystalline metals. Electrochemical measurements (Mütschele and Kirchheim, 1987a, 1987b) of the chemical potential μ which were converted into partial pressures are presented in **Figure 15** for a nanocrystalline sample and a single crystal of Pd. In the latter case Sieverts' Law is fulfilled in the solid solution range.

For the sake of simplicity the sites within the grains are assumed to have the same energy as sites in a single crystal. This is still a good approximation for higher concentrations, where the interfacial stress affects the site energy within the grains (Weissmüller and Lemier, 1999). Thus at a given chemical potential or partial pressure, respectively, the concentration in the grains has to be the same as in the single crystal and, therefore, its contribution to the total concentration can be subtracted yielding the amount segregated at the boundaries. Fitting Eqn (46) to the experimental results presented in **Figure 15** yields values for E_{seg} and σ. Although two parameters (E_{seg} and σ) are available, experimental results at large H concentrations cannot be fitted (dashed curve in **Figure 15**). The discrepancy arises from neglecting H–H interaction (cf. Mütschele and Kirchheim, 1987a, 1987b). A more detailed description of the procedure and the results is given in Kirchheim (1988) and Mütschele and Kirchheim (1987a, 1987b). For H in nanocrystalline Ni experimental data are not available over the same large range of H concentration but it can be described within the same framework of a distribution of segregation energies as well (Arantes et al., 1993). For iron and steel experimental data are evaluated under the assumption of one trap energy instead of a Gaussian distribution. Values obtained this way range from 20 to 60 kJ/mol (cf. **Table 4** and Hirth, 1980).

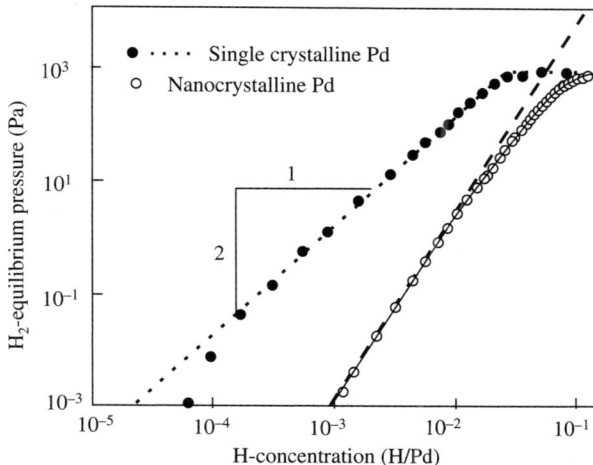

Figure 15 H$_2$ equilibrium pressure versus H concentration for single crystalline (•) and nanocrystalline Pd (○) (Mütschele and Kirchheim, 1987a) at 295 K. The dotted line has a slope of 2 within the α-phase according to Sieverts' Law for $c < 0.01$ and it becomes a plateau within the $\alpha + \beta$ two-phase region for $c > 0.015$. The solid and dashed lines for the nanocrystalline sample are calculated assuming a distribution of site energies and including or excluding H–H interaction as explained in the text.

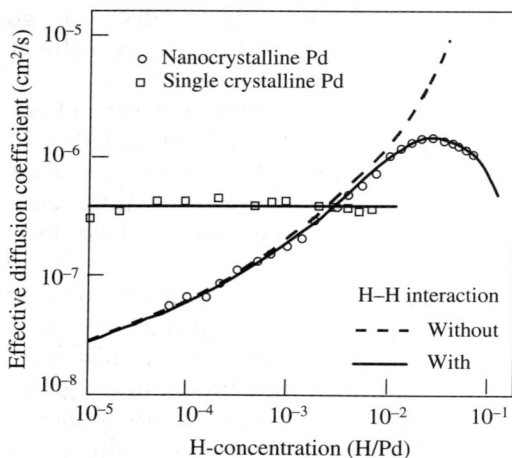

Figure 16 Effective diffusion coefficient of hydrogen at 295 K in nanocrystalline Pd ($\bigcirc$) and single crystalline Pd ($\square$) as a function of the total H concentration. The horizontal line through the single crystalline data corresponds to H diffusion of noninteracting H atoms. The curves through the nanocrystalline data are calculated using Eqns (71) and (73) and assuming that the effective diffusion coefficient corresponds to the grain boundary diffusion coefficient (cf. text). H–H interaction is included by adding a term Wc_{gb} to the chemical potential, where W equals -30 kJ/Mol as in polycrystalline Pd (Wicke and Blaurock, 1980) and cgb is the local concentration in the grain boundaries. Note that grain boundary diffusion of interstitials at low concentrations is slower than in single crystals.

H diffsuion in nanocrystalline Pd was measured via a time lag method (Kirchheim, 1988 and Mütschele and Kirchheim, 1987a, 1987b), where the samples were cathodically charged with hydrogen from one side and the delayed response of the electrochemical potential at the adjacent side was monitored (Kirchheim, 1988). As the transport of H through the sample is a mixture of grain boundary and bulk diffusion, the numbers evaluated from the time lag were called effective diffusion coefficients (see discussion below). The results are presented in **Figure 16**. Different to substitutional solutes there is no short circuit diffusion along grain boundaries for the interstitial hydrogen at low concentrations, because the effective diffusion coefficient is smaller than the value for single crystalline Pd. Substitutional solutes migrate along grain boundaries via a vacancy mechanism and their diffusion coefficient is enlarged because vacancy concentrations in grain boundaries are larger compared with the grain interior. Hydrogen migrating via an interstitial mechanism has sufficient vacancies on the interstitial lattice and, therefore, trapping by the low-energy sites in the DOSE reduces its diffusion coefficient at low H concentrations below the value within the grains. With increasing concentration traps are saturated continuously and the diffusion coefficient finally becomes larger than in the grains or the singly crystal, respectively. Local H concentrations in the boundaries may finally reach high values where H–H interaction and blocking of sites become important. For Pd the attractive H–H interaction leads to a decreasing diffusion constant at high hydrogen contents. Similar effects were observed for the H in nickel grain boundaries (Arantes et al., 1993; Kirchheim et al., 1993).

25.3.2.4 *Behavior in Amorphous Alloys*
Metal/nonmetal glasses usually have a concentration of about 20 at.-% nonmetal and palladium–silicon alloys were mostly used to measure hydrogen solubility and diffusivity (Berry and Pritchet, 1981; Finocchiaro et al., 1984; Szökefalvi-Nagy et al., 1987; Richter et al., 1986). In some studies the

DOSE of the amorphous alloys was considered to be a two-level system (Finocchiaro et al., 1984; Richter et al., 1986) whereas others preferred the following Gaussian distribution

$$n(E) = \frac{1}{\sigma\sqrt{\pi}}\exp\left[-\left(\frac{E - E^0}{\sigma}\right)^2\right] \tag{88}$$

where σ is the width and E^0 the average value of this function. Inserting this DOSE in Eqn (46) and application of the step approximation yields

$$c = \frac{1}{2}erfc\left(\frac{E^0 - \mu}{\sigma}\right). \tag{89}$$

Solving the last equation for μ gives

$$\mu = E^0 - \sigma erf^{-1}(1 - 2c). \tag{90}$$

where the inverse error function erf^{-1} was used. For two amorphous Pd–Si alloys the chemical potential as measured electrochemically and with a high-pressure equipment is presented in **Figure 17**. Thus it was possible to cover a range of hydrogen pressures extending over 18 orders of magnitude (Szökefalvi-Nagy et al., 1987). There is no pressure plateau visible in **Figure 17** indicating that there is no hydride formation as in crystalline alloys. Griessen (1983) explained that as a competition between occupation

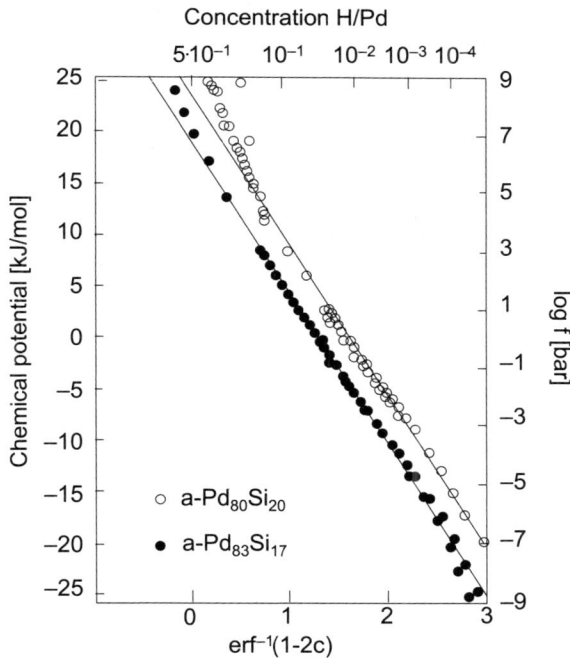

Figure 17 Pressure–concentration isotherms for hydrogen in two amorphous Pd–Si alloys obtained by various experimental techniques at 295 K (Szökefalvi-Nagy et al., 1987). The results are plotted in accordance with Eqn (90) and the fugacity of hydrogen, f, is used at high chemical potentials instead of the partial pressure. The slope of the straight lines yields a value for the width σ of the Gaussian DOSE.

of low-energy sites and H–H interaction also revealing an interesting analogy with ferromagnetism and the Stoner criterion.

In a first-order approximation (Kirchheim, 1988) the DOSE of amorphous PdSi alloys can be derived from the first peak in the radial distribution function. This reflects the distribution of atomic distances which may be considered as a distribution of stress. Multiplying this stress distribution with the partial molar volume of hydrogen yields the DOSE. In a more rigorous treatment Richards (1983) showed that the assumption of an interaction potential between hydrogen and metal of radial symmetry is sufficient to come to the same conclusion. Then the width σ of the DOSE can be calculated from measurable and known quantities yielding a value which is only about 30% larger than the experimental one.

Chemical diffusion coefficients in amorphous Pd–Si alloys of the same composition but prepared by different techniques are presented in **Figure 18**. They differ by up to 2 orders of magnitude, although the radial distribution function as measured with X-rays revealed no differences. The different diffusion coefficient may arise from different distributions of saddle point energies (Kirchheim, 1988). However the concentration dependence of the diffusion constant is well described by the general Eqns (71) and (73). Experimental values are always in good agreement with the calculated ones (Kirchheim, 1988; Berry and Pritchet, 1981; Hirscher and Kronmüller, 1991) as shown for instance in **Figure 18**. Only one fitting parameter D^o is required to describe the measured concentration dependence. With a logarithmic D-axis this parameter moves the calculated curves up or down but does not change slope and/or curvature.

For an amorphous metal–metal alloy $A_{1-x}B_x$ the same considerations as for a crystalline-disordered AB alloy have to be applied, to obtain the fractions of the various A_iB_{4-i} tetrahedra. But different to the crystalline case each type of tetrahedral site has a broad distribution of site energies. Assuming a Gaussian one for each of them with an average energy E_i and a width σ_i DOSE becomes

$$n(E) = \frac{f}{\sigma_i\sqrt{\pi}} \sum_{i=1}^{4} \binom{4}{i} x^i (1-x)^{4-i} \exp\left(-\frac{(E-E_i)^2}{\sigma_i^2}\right). \tag{91}$$

Binary metallic glasses are often alloys of an early and a late transition metal having different hydride formation energies. A compilation of these values (Fromm and Gebhardt, 1976a, 1976b) shows that they in almost all cases increase within the transition series from left to right (Pd being an exception).

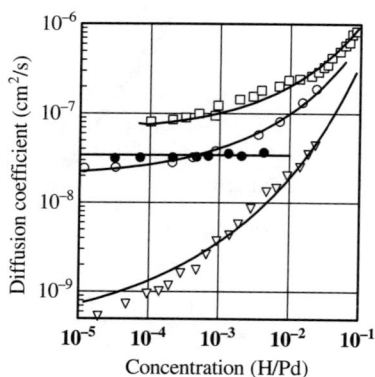

Figure 18   Hydrogen diffusion coefficient as a function of H concentration in amorphous $Pd_{80}Si_{20}$ alloys prepared by the double piston technique (squares), melt spinning (open circles) and sputtering (triangles). The concentration dependence vanishes after crystallization of the melt spun alloy (closed circles) (Kirchheim, 1988; Lee and Stevenson, 1985).

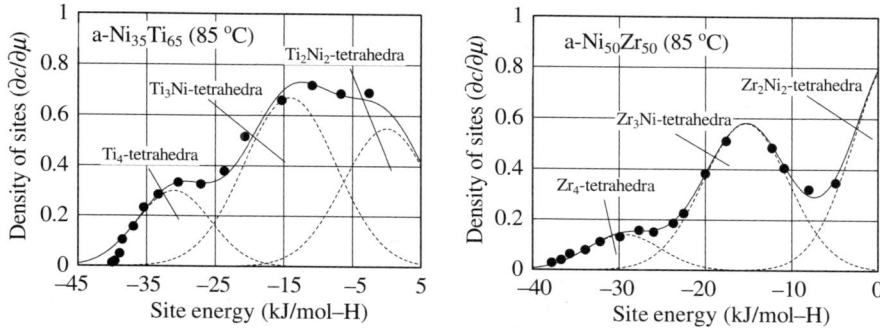

Figure 19 Measured DOSE $n(E) \approx \partial c/\partial \mu$ (cf. Eqn (91)) for an amorphous Ni–Ti alloy (left figure, filled circles) and the contributions from different tetrahedral sites (dashed Gaussian curves and their sum as a solid line). The Gaussian curves have about the same width and the ratio of their areas corresponds to the binomial distribution (cf. Eqn (91)). The right figure is the same but for an amorphous Ni–Zr alloy of different composition.

Thus the average site energy E_i for the A_iB_{4-i} tetrahedron is expected to increases with increasing i for A being the late transition component. This expectation is in agreement with the experimental findings (Harris et al., 1987; Jaggy et al., 1989). By using the step approximation of the FD statistics, that is first term in Eqn (47) the DOSE is obtained directly from measurements of $\mu(c)$ and compared with Eqn (91) as shown in **Figure 19**. For the sake of simplicity the same width σ was assumed for each type of tetrahedron. The position and the area of the various Gaussian functions yield values for f and E_i. In a first-order approximation the average energies are linear combinations of the site energies for A_4 and B_4 and the site energies for the latter are about the same as the hydride formation energies of the crystalline metals A and B.

The chemical diffusion coefficient of hydrogen in early transition/late transition metallic glasses is increasing with increasing concentration (Jaggy et al., 1989). However, the interpretation is not straightforward as for the case of amorphous metal/nonmetal alloys.

25.3.3   Hydrogen as a Defactant

25.3.3.1   Theoretical Background

So far the distribution of trap energies (DOSE) or the microstructure, respectively, remained unchanged during changes of either H concentration or chemical potential. Under these conditions H atoms distribute among all sites minimizing Gibbs free energy according to FD statistics. In the following we allow defects to be produced and annihilated in the presence of hydrogen for either a closed system (no changes of both the amount of metal- and H atoms) or a partially open system (constant amount of metal atoms and varying H concentration but constant chemical potential of hydrogen). Following Wagner (1973a) we define for this partially open system a thermodynamic state function Φ

$$\Phi = F - \mu_H n_H. \tag{92}$$

with F being Helmholtz free energy. For constant volume and temperature the corresponding differential is

$$d\Phi = dW + \mu_M dn_M - n_H d\mu_H, \tag{93}$$

where μ_M and n_M are the chemical potential and molar fraction of the metal atoms. Kirchheim (2007a) and (2007b) suggested to introduce a general work term dW_i which accounts for the creation and annihilation of defect i with a free energy of formation γ_i (dimensions: energy per area, per length or per number depending on the dimensionality of the defect)

$$dW_i = V\gamma_i d\rho_i, \tag{94}$$

where V is the volume and ρ_i is the defect density, that is area per volume for grain and phase boundaries, length per volume for dislocations, number per volume for vacancies and other point defects. For planar defects the term reduces to the well-known product γda, where da is the change of area. If defects are defined in a more general way as discontinuities of a lattice, the term includes surfaces as well. Inserting Eqn (94) in Eqn (93) yields the following derivatives

$$\left.\frac{\partial\Phi}{\partial\rho_i}\right|_{V,T,\mu_H,n_M} = V\gamma_i$$

$$\left.\frac{\partial\Phi}{\partial n_M}\right|_{V,T,\mu_H,\rho_i} = \mu_M \tag{95}$$

$$\left.\frac{\partial\Phi}{\partial\mu_H}\right|_{V,T,n_M,\rho_i} = -n_H$$

With Φ being a thermodynamic state function its second derivatives are independent of the sequence of differentiation leading to

$$\frac{\partial^2\Phi}{\partial\mu_H\partial\rho_i} = -\left.\frac{\partial n_H}{\partial\rho_i}\right|_{V,T,\mu_H,n_M} = \frac{\partial^2\Phi}{\partial\rho_i\partial\mu_H} = V\left.\frac{\partial\gamma_i}{\partial\mu_H}\right|_{V,T,\rho_i,n_M} \tag{96}$$

Defining excess hydrogen Γ_H by

$$\Gamma_H \equiv \frac{1}{V}\left.\frac{\partial n_H}{\partial\rho_i}\right|_{V,T,\mu_H,n_M} \tag{97}$$

leads to the Gibbs adsorption isotherm well known to surface scientist

$$\left.\frac{\partial\gamma}{\partial\mu_H}\right|_{V,T,\rho_i,n_M} = -\Gamma_H. \tag{98}$$

In aqueous liquids polar molecules with a hydrophobic tale and a hydrophilic head are known to give rise to an excess at surfaces and thus reducing the surface energy via the Gibbs adsorption isotherm. These molecules are called surfactants (SURFace ACTing AgeNTS). As the last equation is also valid for other defects the new term defactants (DEFect ACTing AgeNTS) was introduced by Kirchheim (2009). Defactants are components segregating at defects in solids reducing the defect formation energy like surfactants do with surfaces.

Before the generalized Gibbs adsorption isotherm and the role of defactants will be discussed, we take a closer look at the excess hydrogen defined by Eqn (97).

Equation (97) states that an incremental change of the defect density $\delta\rho_i$ causes an incremental change of the number of H atoms δn_H in a partially closed system, that is for constant volume, constant

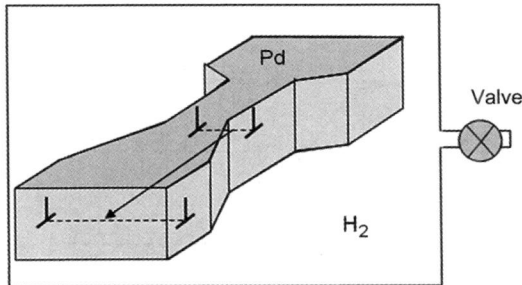

Figure 20 Wagner's Gedanken experiment (Wagner, 1973a) modified for dislocations. A vessel with a closed valve contains a dog bone-shaped single crystal of palladium and gaseous H_2. By an external force (e.g., shear stress) the dislocation is moved from the thin region to the thick region. Thus the dislocation length or ρ_i, respectively, is increased and hydrogen from the newly generated part $\delta\rho_i$ of the defect is absorbed from the gas phase and/or from the solid solution in Pd. Thereby the partial pressure of H_2 is decreased and the number of moles of hydrogen δn_H per additional dislocation length, which have to be added via the valve, to establish the initial H_2 pressure (chemical potential), is defined as excess hydrogen Γ_H.

temperature, constant number of metal atoms and constant chemical potential of hydrogen. The excess hydrogen Γ_H is positive, if n_H increases with increasing ρ_i. **Figure 20** shows a Gedanken experiment exemplifying the meaning and measurement of excess hydrogen for dislocations. The figure shows how the length l of an edge dislocation in palladium can be extended leading to an increased defect density $\delta\rho_i = \delta l/V^{-1}$. The palladium is in equilibrium with gaseous hydrogen and both metal and gas are in a container of constant volume at constant temperature. Thus the number of metal atoms n_B does not change. However the number of H atoms n_H may be changed by opening a valve and removing or adding hydrogen, to maintain a constant partial pressure of hydrogen or a constant chemical potential μ_H respectively. Thus increasing the length of the edge dislocation by δl leads to H segregation at this additional segment of the line defect. The required additional hydrogen is supplied from both the gas phase and the solid solution causing a decrease of the partial pressure of hydrogen in the gas phase. Then a certain amount δn_H is added from a reservoir (reservoir not shown in **Figure 20**) via the valve for establishing the original pressure of gaseous hydrogen. Through this process H atoms are partitioned between dislocation, solid solution and gas phase to attain a new equilibrium with the extended dislocation. The same procedure can be applied to a grain boundary located in the thinner part of the palladium and moved to the thicker region. Or for vacancies their density may be increased by bombarding the metal with high-energy electrons.

In the following we discuss the generalized Gibbs adsorption isotherm given by Eqn (98). Its validity can be checked by measuring defect formation energies as a function of the chemical potential of the solute. Very often an ideal dilute solution is assumed and the logarithm of solute concentration, $\ln c_s$, is used instead of the chemical potential. With surfactant molecules on liquid surfaces the defect formation energy or the surface energy, respectively, can be measured by a variety of different techniques. For solids measuring techniques for defect energies are sparse and/or complicated, but there are a variety of methods to measure excess solute at defects. Knowing the excess as a function of the chemical potential $\Gamma_H(\mu_H)$ allows an integration of the generalized Gibbs adsorption isotherm yielding curves like the ones shown in **Figure 21**. These three curves can be constructed by some reasonable and qualitative assumptions concerning the function $\Gamma_H(\mu_H)$. At very low chemical potentials ($\mu_H \rightarrow -\infty$) the H concentration and, therefore, the excess Γ_H become negligible and then the slope of the γ versus

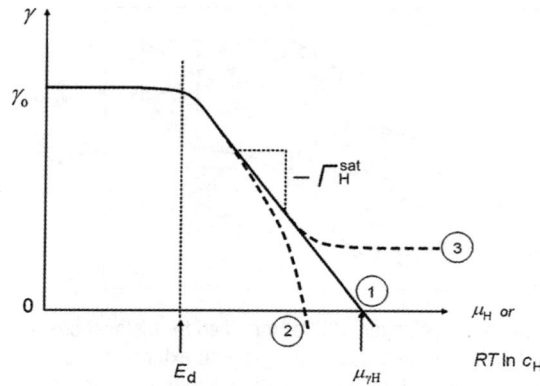

Figure 21   Three possible ways of the dependence of the defect energy, γ, on the chemical potential, μ_H of hydrogen according to the generalized Gibbs adsorption isotherm (Eqn (98)) (cf. text).

μ_H curve is zero (cf. Eqn (98)). If the chemical potential reaches the interaction energy, E_d, of hydrogen with the defect, the excess increases and the slope becomes negative. Saturating the defect with Γ_H^{sat} leads to a line with constant slope (line 1) which intercepts the abscissa at $\mu_{\gamma H}$. If the excess does not saturate but increases further on, line 2 describes the expected dependency. Line 3 describes the behavior, if a new H-rich phase is formed and the logarithm of concentration, c_H, instead of the chemical potential is used as the variable for the abscissa. In the latter case the chemical potential and, therefore, the excess and γ remain constant during the phase separation despite increasing concentration. So far and in the following hydrogen is a solute in a metal; but all of the statements about thermodynamic quantities remain valid for other solutes in other solvents.

In the following we are treating a simple special case of the generalized Gibbs adsorption isotherm. We assume an ideal dilute solution for hydrogen in normal lattice sites leading to the following equation within the framework of the two-level system (cf. Section 25.3.1.1)

$$\mu_H = E_o + RT \ln \frac{n_f}{1 - n_f} \tag{99}$$

Assuming further on that the H concentration in normal sites is small $n_f \ll 1$ the last equation reduces to

$$\mu_H = E_o + RT \ln n_f = E_t \tag{100}$$

for strong trapping that is $RT \ln 0.9 > E_t > RT \ln 0.1$. In **Figure 21** this corresponds to the point, where the line bends down at E_d (being equivalent with E_t). For higher values of the chemical potential the defect or trap sites, respectively, are saturated ($\Gamma_H = \Gamma_H^{sat}$) and line 1 becomes a straight one with the slope $-\Gamma_H^{sat}$. Any point on this straight part fulfills the relation

$$\frac{\gamma_o - \gamma}{\mu_H - E_t} = \Gamma_H^{sat}. \tag{101}$$

Rearranging the last equation and inserting Eqn (100) gives

$$\gamma = \gamma_o - \Gamma_H^{sat}(E_o - E_t + RT \ln n_H). \tag{102}$$

The last expression is equivalent with the one derived by Weissmüller (1993) with the limiting assumptions given before and for grain boundaries with $E_o - E_t$ being the positive value of the segregation energy.

Looking at **Figure 21** the intriguing question arises whether defect formation energies may become negative. Independent of a system being closed or partially open negative or positive values of the defect formation energies correspond to nonequilibrium situations. In both cases the Helmholtz free energy or the state function Φ can be lowered by defect formation for $\gamma < 0$ or defect annihilation for $\gamma > 0$. Only if $\gamma = 0$, a stable situation is possible, where the defect density remains constant. Point defects like vacancies are special because their formation is accompanied with the generation of sufficient configurational entropy leading to zero Helmholtz free energy γ and a corresponding equilibrium concentration of vacancies. In the presence of defactants γ becomes negative and abundant vacancies are formed until the configurational entropy balances the decrease caused by defactants and γ becomes zero again. As the generalized Gibbs adsorption isotherm is valid for all kinds of materials besides metals and all kinds of solutes besides hydrogen a further detailed discussion will be peripheral for the topic of this book chapter. The interested reader is referred to Kirchheim (2009). For the following we keep in mind that the defect formation energies will be lowered for excess hydrogen and may even become zero.

25.3.3.2   *Special Cases and Supporting Experimental Evidence*

25.3.3.2.1   *Vacancies*

There is evidence provided by a variety of experimental techniques that the concentration of vacancies in metals is increased in the presence of hydrogen (Sugimoto and Fukai, 1992; Fukai, 2003; Čížek et al., 2004; Myers et al., 1992; Sakaki et al., 2006) which is interpreted as a decrease of their formation energy. The magnitude of this decrease is the larger the larger the chemical potential of hydrogen is and the larger the excess hydrogen is. The excess hydrogen Γ_H is equivalent with the number of H atoms surrounding the vacancy. Usually only the ones in the next-nearest interstitial positions, for example six for tetrahedral positions in a bcc lattice contribute to the excess. Thus applying high hydrogen pressures up to 3 GPa Fukai (2003) discovered in several metals large vacancies concentrations of up to 30 at.-%. At these hydrogen pressures the corresponding fugacity is orders of magnitude larger due to a repulsive interaction among the hydrogen molecules in the gas phase (Sugimoto and Fukai, 1992). Then values of the chemical potential may be reached where the formation energy of vacancies, E_V, becomes negative (cf. **Figure 21**) and a large amount of vacancies called SAV are formed. The process of vacancy formation is stopped when the defects start to interact repulsively giving rise to positive contribution to γ.

Negative vacancy formation energies have been calculated from first principles in the Fe–H system (Tateyama and Ohno, 2003), where E_V was obtained as a function of the chemical potential of molecular hydrogen, μ_{H_2}, via the following equation

$$E_V = E_V^o - \frac{Z}{2}\mu_{H_2} = E_V^o - Z\mu_H. \tag{103}$$

E_V^o is the vacancy formation energy in the absence of hydrogen. Eqn (103) is the integrated version of Eqn (98) for constant excess solute Γ_H, or coordination number Z, respectively. For an excess of $Z = 5$ and $Z = 6$ H atoms per vacancy and a partial H_2 pressure of 2 GPa negative formation energies of vacancies have been calculated (Tateyama and Ohno, 2003). Under these circumstances the vacancy formation decreases the free energy of the system and the vacancy concentration should go to infinity, unless the formation energy increases with increasing concentration finally reaching a value of zero. Then a metastable equilibrium can be attained (Kirchheim, 2009). The presence of SAV has been indirectly confirmed by a much faster interdiffusion of copper and nickel subject to a hydrogen pressure of 5 GPa (Hayashi et al., 1998).

25.3.3.2.2 *Dislocations*

In the introduction to this chapter the decrease of the line energy of a dislocation by hydrogen has been discussed as an example of hydrogen acting as a defactant. Direct experimental evidence is provided by nanoindentation experiments (Katz et al., 2001; Nibur et al., 2006; Barnoush and Vehoff, 2006; Barnoush and Vehoff, 2008; Barnoush et al., 2009; Wen et al., 2009; Tal-Gutelmacher et al., 2010). During these experiments a nanosized volume is deformed by an indenter tip and the probability of a dislocation being present in this volume is negligible. Thus plastic deformation has to be accompanied by the generation of dislocation loops which requires shear stresses of the order of the theoretical strength of a metal. By increasing the load steadily during the indentation the occurrence of the first dislocation loop is observed as a step-like increase of the indenter depth. This event is called "pop-in" and the corresponding shear stress is proportional to the line energy of the dislocation loop. For a large variety of metals including nickel (Barnoush and Vehoff, 2006), aluminum (Barnoush and Vehoff, 2008), iron aluminide (Barnoush et al., 2009), vanadium (Tal-Gutelmacher et al., 2010) and stainless steel (Katz et al., 2001; Nibur et al., 2006; Zhang et al., 2009) it has been observed that in the presence of hydrogen pop-in events occur at much lower loads than without hydrogen. Thus hydrogen decreases the line energy of a dislocation which is a direct piece of evidence that hydrogen acts as a defactant for dislocations.

The same can be concluded from the results of a simple experiment. By cold rolling a sheet of palladium with increasing amounts of hydrogen from 0 to 1 at.-% (terminal solubility at room temperature) with the same degree of plastic deformation (50% reduction of cross-section) it was shown by XRD and TEM that the dislocation density increases with increasing H concentration (Chen et al., 2013).

For H in Pd the defactant concept can be used to evaluate the decrease of the line energy of dislocations in the following way. From measurements of the chemical potential μ_H the excess hydrogen was determined as the difference of H contents between a deformed (large value of the total length l of dislocations) and an annealed sample at the same value of μ_H (Kirchheim, 1981). The result is shown in **Figure 22**.

The difference of H concentrations (cf. **Figure 22**) which are given as ratios of the number of hydrogen n_H and palladium atoms n_{Pd} is converted to the excess Γ_H via the following relation

$$\left[\left(\frac{n_H}{n_{Pd}} \right)_{def} - \left(\frac{n_H}{n_{Pd}} \right)_{an} \right] = \frac{l\Gamma_H \Omega_{Pd}}{V} = \rho \Gamma_H \Omega_{Pd} \tag{104}$$

where V are the sample volume, ρ the dislocation density and Ω_{Pd} the molar volume of Pd. Subscripts an and def refer to the annealed and deformed samples. Thus in agreement with Eqn (98) an excess in units of mole-H/m is defined. This excess hydrogen and its dependence on the EMF or chemical potential, respectively, are assumed to be independent of the dislocation density, ρ. Therefore, the concentration differences for the various degrees of deformation will all fall on a master curve, if these differences are multiplied with a constant factor or a relative dislocation density, respectively. The EMF values have to be multiplied with Faraday's constant $F = 96\ 500$ J/(V mol) to obtain the relative chemical potentials of hydrogen. With a dislocation density of

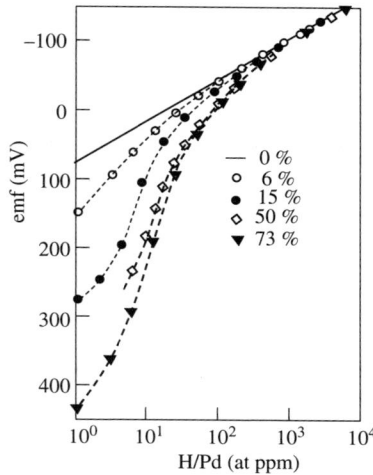

Figure 22 Measured EMF at 295 K for Pd samples deformed by cold rolling as a function of the total hydrogen concentration for different degrees of reduction of cross-section (Kirchheim, 1981).

$2 \cdot 10^{15}$ m^{-2} for the 73% deformed Pd sample (Kirchheim, 1981) the data in **Figure 22** were integrated between -400 and 150 mV. According to Eqn (98) the result corresponds to a reduction of the line energy of $\Delta \gamma = -4.7 \cdot 10^{-10}$ J/m. The line energy, γ_0, of a naked dislocation with Burgers vector, $\mathbf{b} = 0.275$ nm, in Pd with a shear modulus $G = 48$ GPa is estimated from the relation $\gamma_0 = 0.5\, Gb^2$ (Hirth (1996)) to be $1.8 \cdot 10^{-9}$ J/m. Thus the calculated reduction induced by H segregation is about 26%.

The EMF results discussed before yield the total H excess for all dislocations and the dislocation density has to be determined independently to calculate the excess per unit length of the dislocation which leads to the reduction of its line energy. However, this is not the case for SANS of hydrogenated and deuterated Pd samples, because it yields the H excess per unit length directly (Maxelon et al., 2001a; Heuser et al., 1991). In Maxelon et al. (2001a) this was expressed in terms of a hydride of composition PdH$_\alpha$ ($\alpha = 0.6$) segregating as a cylinder of radius r below the glide plane of an edge dislocation. Thus the H excess is given by

$$\Gamma_{\mathrm{H}} = \frac{\alpha \pi r^2}{\Omega_{\mathrm{Pd}}}. \tag{105}$$

Experimental results in terms of Γ_{H} are presented in **Figure 23** as a function of the total H concentration.

To obtain the chemical potential of hydrogen, we take advantage of the experimental fact that within the concentration range of the SANS results hydrogen activity a_{H} is proportional to the total H concentration in terms of H/Pd (Maxelon et al., 2001a; Flanagan et al., 1976). Then we have

$$\mathrm{d}\mu_{\mathrm{H}} = RT\mathrm{d}\ln(a_{\mathrm{H}}) = RT\mathrm{d}\ln\left(\frac{\mathrm{H}}{\mathrm{Pd}}\right) \tag{106}$$

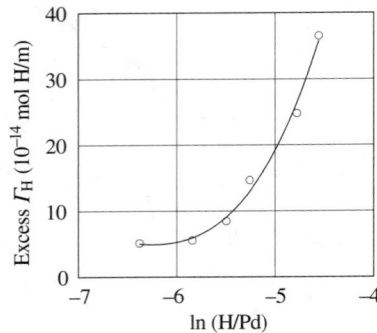

Figure 23   Excess hydrogen Γ_H segregated at a dislocation as obtained by SANS (Maxelon et al., 2001a) plotted versus the logarithm of total hydrogen concentration.

and Eqns (98), (105) and (106) yield the change of line energy between the minimum and maximum values of the H content

$$\Delta\gamma = -\int \Gamma_H d\mu_H = -RT \int\limits_{-6.4}^{-4.6} \Gamma_H d\left[\ln\left(\frac{H}{Pd}\right)\right] \qquad (107)$$

Using the SANS data presented in **Figure 23** gives $\Delta\gamma = -5.9 \cdot 10^{-10}$ J/m which corresponds to about 33% of the line energy of a naked dislocation. Thus the formation energy of a dislocation is remarkably reduced within the concentration range given in **Figure 23**.

25.3.3.2.3   Kink Pairs
During plastic deformation dislocations have to be generated and they have to move through the lattice. As discussed in the previous section the first process of dislocation formation is facilitated by hydrogen decreasing the formation or line energy, respectively. In the following we will show that the motion of dislocations as well may become easier in the presence of hydrogen.

On the atomic scale movement of dislocations starts with the creation of a pair of kinks. A shear stress drives the kinks in opposite directions and thus moves the dislocation from one Peierls valley to the next one (Hirth and Lothe, 1968). Thus the velocity of the dislocation may be determined by either the formation rate of kink pairs or the motion of these kinks. The formation of kinks and their motion are experimentally observed as a relaxation peak in mechanical spectroscopy (Seeger, 1979). For the bcc metals Fe and Nb the corresponding peak for kink creation on screw dislocations is labeled γ peak and is positioned at about 300 K. At temperatures below 100 K and the α peak arises from 71° dislocations (Seeger, 1979). Heavier solutes like carbon and oxygen give rise to a new peak at temperatures well above 300 K which is interpreted as solute drag on kinks moving along screw dislocations (Snoek, 1941; Köster et al., 1954). Hydrogen also gives rise to a Snoek–Köster peak between the α and γ relaxation temperatures. In analogy to carbon and oxygen this is interpreted as hydrogen drag on kinks moving along 71° dislocations (Seeger, 1979). An alternative interpretation is given by Hirth (1980) where the new hydrogen peak is assumed to arise from a decreasing kink formation energy of screw dislocations lowering the position of the γ peak below 300 K. This assumption can now be substantiated within the framework of the defactant concept. Hydrogen acts as a defactant on kinks, which in the case of screw dislocations have edge character and, therefore, should have larger excess hydrogen than screw dislocations. According to Eqn (98) the large excess gives rise to a larger decrease of the kink formation energy. As H atoms are very mobile in bcc metals one would expect a small dragging force on the moving kinks reducing only partly the effect of a decreased kink formation energy. Then dislocation motion could be faster when compared with a hydrogen-free metal. The reasoning is picked up again in the following chapter discussing softening and hardening by hydrogen.

25.3.3.2.4 Grain Boundaries

It is generally accepted that solute atoms segregating to grain boundaries reduce the boundary energy. The exceptional thermal stability of nanocrystalline alloys regarding grain growth has been interpreted by several groups (Weissmüller, 1993; Weissmüller, 1994; Gao and Fultz, 1994; Hong et al., 1994; Kirchheim, 2004; Krill et al., 2006) in terms of a reduced or even zero grain boundary energy as a consequence of excess solute at the boundaries. However for hydrogen there is not much evidence. Mechanical properties of Ti alloys were improved by decreasing the grain size (Yoshimura and Naka-higashi, 2002) where hydrogen was used as a temporary alloying element. Thus hydrogen stabilized as a defactant a larger grain boundary area (or smaller grain size) during the thermal treatment and to avoid H embrittlement it was removed at lower temperatures where no grain growth occurred.

The experimental results for hydrogen in nanocrystalline palladium were not evaluated in the framework of the defactant concept in the past. But in agreement with the widely accepted procedure the data presented in **Figure 15** have been interpreted in the framework of a trapping model (Mütschele and Kirchheim, 1987a, 1987b). This atomistic approach is easier to comprehend and allows interpreting the pronounced concentration dependence of H diffusivity in a straightforward way (cf. **Figure 16**). The thermodynamic approach of the defactant concept provides additional information about the grain boundary energy. Thus the following data evaluation is another example of the duality of the solute defect interaction; depending on which approach is used different perspectives and different insights become accessible.

Analogously to deformed Pd EMF measurements were performed at $T = 298$ K for nanocrystalline Pd as a function of H concentration (Mütschele and Kirchheim, 1987a, 1987b). In **Figure 15** the corresponding partial pressures of hydrogen are compared with the ones for a Pd single crystal. The concentration difference Δc between the two samples at a given pressure is a measure of the H excess like discussed for deformed Pd in Section 25.3.3.2.2.

$$\Delta c = \left(\frac{n_H}{n_{Pd}}\right)_{nano} - \left(\frac{n_H}{n_{Pd}}\right)_{single} = \frac{a\Gamma_H}{n_{Pd}} = \frac{a\Gamma_H\Omega_{Pd}}{V} = \frac{3\Gamma_H\Omega_{Pd}}{g}, \tag{108}$$

where a is the total grain boundary area, V is the sample volume, Ω_{Pd} is the atomic volume of Pd and g is the grain size. The last equation is exact for cubic- and spherically shaped grains and it has been assumed that the excess is independent of the total grain boundary area. Numerical integration of the Gibbs adsorption equation (Eqn (98)) for the data in **Figure 15** yields the change of the grain boundary energy

$$\Delta\gamma = \frac{2.3k_BT}{2}\int_{-3}^{3}\Gamma_H d\log p = -0.89\,J/m^2. \tag{109}$$

This value is about the same as the experimentally determined energy for grain boundaries in nanocrystalline Pd (Birringer et al., 2003) meaning that H segregation may finally lead to zero formation energy.

25.3.4 Hydrogen Effects on Mechanical Properties

Dissolved H atoms may change both the elastic and the plastic behavior of metals. To change the elastic constants of an alloy rather large H concentrations are required which usually lead to the decomposition into a solid solution and a hydride phase.

Contrary to elasticity the plasticity can be affected by a few weight ppm of hydrogen. This is due to an attractive interaction of H atoms with those defects which are responsible for a certain type of plastic deformation, that is vacancies for Nabarro-Herring creep and climb of edge dislocations, kinks for the motion of dislocations, stacking faults for cross-slip of dislocations. In this context three fundamental question arise:

  (i) How much hydrogen is segregating to a certain defect?
 (ii) How much is the generation and annihilation of defects changed in the presence of hydrogen?
(iii) How much is the mobility of defects (strain rate) affected by hydrogen?

Answers to question (i) were provided in a general way in Section 25.3.1 by using the DOSE concept. Question (ii) is treated in the framework of the defactant theory in a general way. The third question will be addressed in the following.

25.3.4.1 Hydrogen Interacting with Stress Fields

In Section 25.3.2.2 the DOSE for a dislocation was derived by considering contributions from the elastic interaction only and by assuming an isotropic strain around a hydrogen atom. This is valid for hydrogen atoms in octahedral sites of the fcc lattice (H in Pd) due to the symmetry of the lattice; it should not be valid for H atoms in tetrahedral sites of the bcc lattice of Fe, V, Nb, Ta. However, experimental evidence by measuring the local strain around H atoms or from a missing Snoek relaxation indicates that the expected tetragonal distortion is not present or negligible (Peisl, 1978). Then the elastic interaction energy is obtained from

$$E_{el} = -V_H \frac{(\sigma_{11} + \sigma_{22} + \sigma_{33})}{3} = -V_H \sigma_h, \tag{110}$$

with V_H being one-third of the trace of the strain tensor induced by a dissolved hydrogen atom or the partial molar volume, respectively, and σ_{ii} being the principal values of the stress tensor. Thus the elastic energy (Eqn (110)) is simply the product of partial molar volume and hydrostatic stress σ_h and the minus sign reflects a decrease of the elastic energy of the system for positive hydrostatic stresses. Then the stress gives rise to a corresponding change of the chemical potential μ

$$\Delta\mu = -V_H \sigma_h. \tag{111}$$

This has been proven experimentally (Kirchheim, 1986) by applying a tensile load σ_{11} to a palladium wire containing hydrogen in solid solution and measuring changes of the electromotive force $\Delta E = \Delta\mu/F$ (F is Faraday's constant). The stress was imposed on a wire as shown schematically in **Figure 24**. In agreement with Eqn (111) ΔE increases proportional with σ_{11} (see **Figure 25**), but no changes were observed by applying a moment of torque to the wire, that is for $\sigma_h = 0$ (cf. **Figure 24**).

From an atomistic point of view an externally applied tensile stress increases the lattice constant with a concomitant increase in the size of the octahedral site. Thus incorporation of a hydrogen atom into these sites requires less elastic distortion and consequently the site energy will be lowered giving rise to a corresponding decrease of the chemical potential.

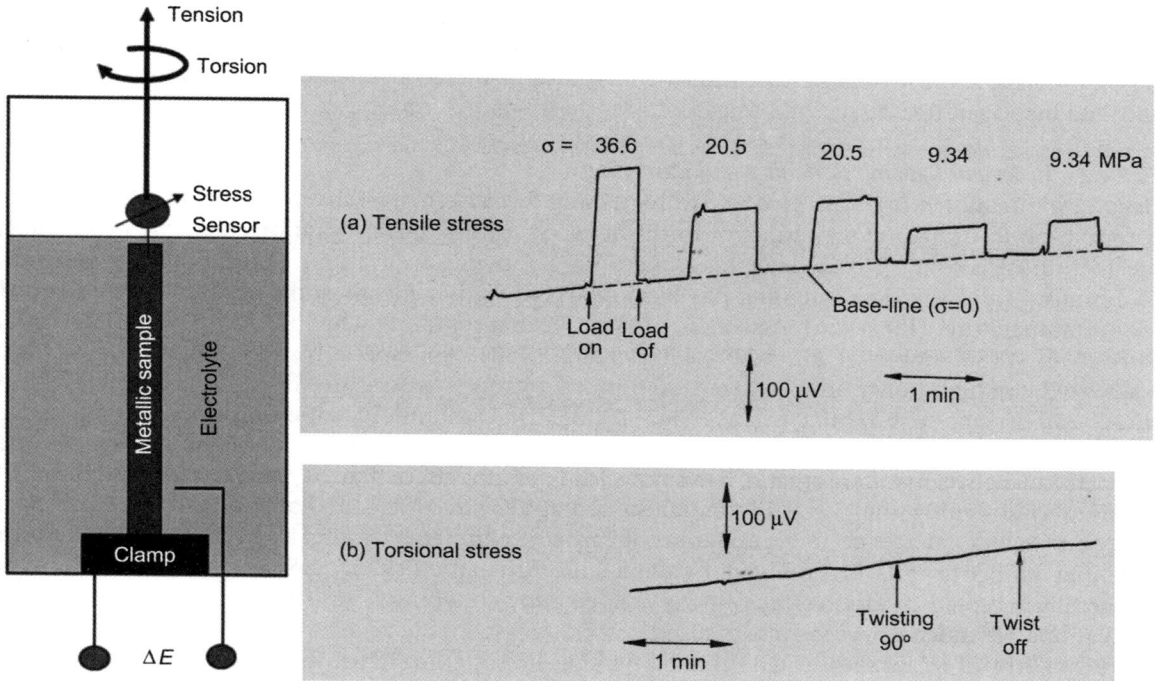

Figure 24 Schematic setup (left picture) for measuring EMF changes as a function of stress (Kirchheim, 1986). The EMF is the potential difference between the sample and a reference electrode. EMF traces for intermittently applying tension and torsion (right two pictures).

In a gradient of hydrostatic stress H atoms move from regions of lower hydrostatic stress to regions of higher stress (cf. Section 25.3.1.3) building up a gradient of H concentration leading to a counteracting hydrogen flux. A steady state is reached when the stress gradient-driven H flux and the concentration gradient one are equal. Then the stationary concentration is given by Eqn (76) which is exponentially approached. For a plate the corresponding time constant will be

$$\tau = \frac{l^2}{\pi^2 D}, \tag{112}$$

where l is the plate thickness and D the hydrogen diffusion coefficient. Stress gradients are imposed during Gorsky effect measurements and they are present at a crack tip under load. In the latter case H atoms move to the crack tip increasing the local concentration at the tip. This may even give rise to hydride formation creating a brittle phase in front of the tip and accelerating crack propagation. For a typical partial molar volume of hydrogen around 2 cm^3/mol-H Eqn (76) predicts an increase of the H concentration by a factor of 2 at room temperature for a hydrostatic stress of about 1 GPa. This is usually much higher than the yield strength of metals and, therefore, increases of H concentration at crack tips are expected to be small. In addition, it is not an increased H concentration triggering defect formation but an increased chemical potential. This quantity will decrease compared with the bulk value right after load is applied and it will approach the bulk value with increasing time. Thus the defactant concept predicts no change of defect formation by changing the H concentration and keeping the chemical potential constant. However, for stress-assisted defect formation it will be the combined action of stress and the defactant hydrogen which lead to the accelerated defect formation compared with the hydrogen-free case.

25.3.4.2 Hydrogen Causing Softening and Hardening

Usually solute atoms in alloys give rise to hardening, because higher stresses are required for plastic deformation to overcome the attractive forces between solutes and dislocations and/or to counteract the friction of a solute cloud dragging along with a moving dislocation. Contrary to these effects, pronounced solid solution softening has been observed for bcc metals and interstitial solutes below room temperature (Pink and Arsenault, 1979; Neuhäuser and Schwink, 1993; Kimura, 1985; Sato et al., 1980; Kubin et al., 1979; Bang et al., 1980). An especially large softening has been discovered for high-purity iron, where 200 ppm of nitrogen reduce the flow stress at 190 K from about 200 to about 100 MPa (Sato et al., 1980). The amount of softening depends on solute concentration, temperature, and crystal orientation and it is counteracted by solute hardening. The competition between softening and hardening leads to the effect that oxygen in a single crystal of niobium is softening the metal in one crystallographic direction and hardening it in a different direction (Bang et al., 1980). A compilation of earlier experimental results and of various models for softening in bcc metals is provided by Pink and Arsenault (1979). Because of the absence of a generally accepted model research in the area of alloy softening is in progress by using computer simulations (Trinkle and Woodward, 2005).

The majority of recent studies agree that hardening and softening in bcc metals at low temperatures is based on the formation and motion of kink pairs at screw dislocations (Seeger, 1984). In the framework of this model hardening is caused by the attractive interaction between solute atoms and moving kinks, whereas softening is explained by an elastic interaction between solute atoms and kink pairs during their formation and separation stage (Neuhäuser and Schwink, 1993; Sato et al., 1980; Bang et al., 1980). The latter explanation is rather vague and qualitative only. Within the defactant concept softening can be interpreted in a general way in the framework of thermodynamics. Within the new concept the nucleation energy of kink pairs at screw dislocations is reduced. If double kink nucleation is the rate-controlling step for plastic deformation solid solution softening will occur. In a quantitative treatment Eqn (98) has to be used, to account for the decrease of the formation energy of a kink pair. Softening is pronounced for screw dislocations as their kinks are edge type. Edge-type dislocation is expected to have a higher solute excess as besides the core their long-range hydrostatic stress field interacts attractively with solute atoms. Interstitial solutes with their large strain tensor, especially the large trace of this tensor, interact strongly with edge-type kinks.

Besides having high chemical potential for facilitating kink formation the solute atoms should be mobile enough within the lattice or dislocation core, to transport the necessary excess of solute atoms to a region where double kink creation occurs. Thus it is comprehensible that softening is not observed at very low temperatures, that is below 100 K (Pink and Arsenault, 1979). In molecular dynamic simulations of nanoindentation (Wen et al., 2009; Mao and Li, 2012) softening is not observed to the same extent as in experiments (cf. Section 25.3.3.2.2), because the timescale of the simulation does not allow H diffusion to supply sufficient excess hydrogen to the region where the dislocation loop is created. Nevertheless, on the average pop-in loads in computer simulations are smaller in samples containing hydrogen, because due to fluctuations in H concentration there may be regions providing larger excess hydrogen for the creation of a dislocation loop in the simulated volume.

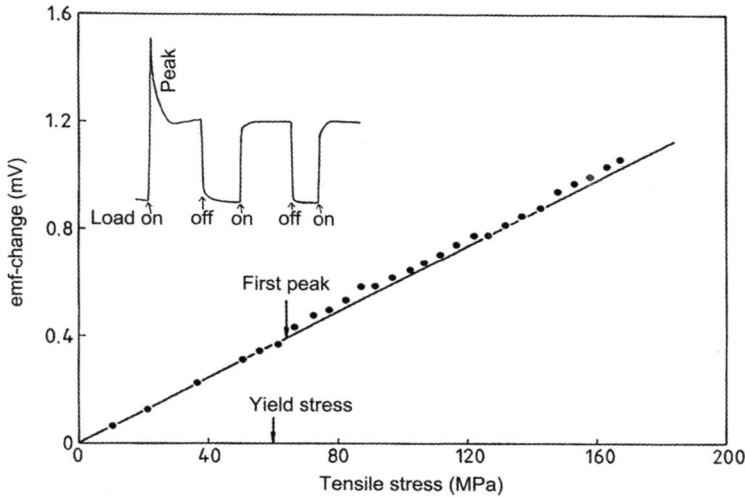

Figure 25 EMF changes as a function of the applied tensile stress (Kirchheim, 1986). After the tensile stress exceeded the yield strength of the annealed Pd wire a peak appeared with the first load cycle (see inset). This may be due to dislocations generated on the surface of the Pd wire decreasing the chemical potential of hydrogen by trapping of H atoms. The following cycles do not show this peak because the flow stress is increased by work hardening.

Softening by hydrogen was observed in pure iron (Kimura et al., 1977; Matsui et al., 1979a, 1979b, 1979c; Moriya et al., 1979); but effects are small compared with nanoindentation experiments. Softening has also been detected in α-Ti (Senkov and Jonas, 1996). However, the measuring temperatures were in the range of 500–800 °C. Compared with bcc metals like iron the H diffusivity in hcp–Ti is much lower and, therefore, higher temperatures may be necessary to reach a sufficient mobility for the solute atoms to cause softening in a dynamic strain rate test.

As mentioned earlier solutes like hydrogen usually give rise to hardening. To understand the interplay between softening and hardening, the strain rate is expressed by the Orowan equation (Argon, 2006; Kirchheim, 2012):

$$\frac{d\varepsilon}{dt} = \mathbf{b}\rho v = \mathbf{b}\rho a \left[\tau_g + \tau_m\right]^{-1} \tag{113}$$

where ε is the plastic strain, $\mathbf{b}$ is the Burgers vector, ρ is the dislocation density, a is the distance a double kink moves the dislocation ahead (distance between two Peierls valleys), τ_g is the time constant for double kink generation and τ_m is the time necessary to move the kinks to the ends of a dislocation line. In pure metals kinks move very fast (Seeger, 1979) and the condition $\tau_m \ll \tau_g$ holds leading to

$$\frac{d\varepsilon}{dt} \approx b\rho a / \tau_g \tag{114}$$

Defactants segregating to kinks reduce the kink formation energy and, therefore, decrease τ_g which via Eqn (114) leads to an increased strain rate or softening, respectively. On the other hand increasing the solute content leads to an increasing solute drag on the moving kinks which increases τ_m. With decreasing τ_g and increasing τ_m the condition $\tau_m \ll \tau_g$ yielding Eqn (114) no longer is valid but with increasing hydrogen concentration reverses to $\tau_m \gg \tau_g$. Then hydrogen drag on kinks instead of hydrogen enhancing kink formation dominates and Eqn (113) becomes

$$\frac{d\varepsilon}{dt} \approx b\rho a / \tau_m. \tag{115}$$

With increasing solute concentration τ_m increases further on, because excess solute at the kinks increases as well increasing the dragging forces. Then Eqn (115) predicts a decreasing stain rate, that is hardening. Both softening and hardening are given by the following relations

$$\frac{d\varepsilon}{dt} = b\rho a \left[\tau_g + \tau_m\right]^{-1} = \begin{cases} \tau_g > \tau_m \Rightarrow b\rho a / \tau_g & \text{and} \quad \tau_g \downarrow \text{ (solute softening)} \\ \tau_m > \tau_g \Rightarrow b\rho a / \tau_m & \text{and} \quad \tau_m \uparrow \text{ (solute hardening)} \end{cases}, \tag{116}$$

where the vertical arrows indicate decreasing or increasing time constants with increasing solute concentration.

In a fatigue experiment Murakami et al. (2010) observed with increasing hydrogen concentration in stainless steel an increasing crack velocity followed by a decreasing velocity when compared with the hydrogen-free sample. If the crack velocity is determined by plastic flow the faster crack growth at low H content is due to an increase of dislocation velocity by a decreased double kink formation energy with the condition $\tau_m < \tau_g$ still being valid. For larger H contents the chemical potential is larger and the kink formation should be even easier, that is τ_g decreases. In addition, with increasing H content the excess hydrogen at the kinks becomes larger, too, and so does the characteristic time τ_m for kink motion. Thus the condition $\tau_m > \tau_g$ is reached with increasing H content and the behavior changes from softening to hardening.

25.3.4.3 H Embrittlement

Hydrogen is an omnipresent element which may enter metals during their production and/or thermomechanical treatment in humid air via the decomposition of water leading to hydrogen and metal oxide. But during service metals may be exposed to friction and wear which leads to the formation of bare metal surfaces which may react with (i) humid air, (ii) water as an impurity in lubricants or (iii) may even decompose organic substances. All reactions produce hydrogen in "statu nascendi", that is in atomic form which accelerates hydrogen entry into the metal when compared with the exposure to gaseous hydrogen, that is hydrogen molecules. The natural oxide which is present on most metal surface may be semipermeable for water molecules or may be destroyed during corrosion leading to hydrogen generation as well.

Even small amounts of hydrogen like a few weight ppm may lead to a pronounced loss of ductility or fracture strength during tensile testing. During fatigue testing external and internal hydrogen leads to accelerated crack growth. This phenomenon is of concern for high-strength steels (yield stress above 600 N/mm^2). In the presence of hydrogen, their mechanical resistance can decrease considerably, below the yield strength. For many steels, there is a threshold stress below which HE does not occur. However, this threshold stress is a function of the strength of the material and of the environment. Generally, the higher the yield or tensile strength, the lower the threshold stresses. HE is definitely linked to absorption of H, and there is often an incubation time for charging and transport of H and/or defect generation, resulting in delayed fracture. The corresponding degradation of mechanical strength is called HE. Atomistic or micromechanical mechanisms leading to this embrittlement are discussed controversial at present. It may also be that several mechanisms are valid and one or the other prevails depending on external or internal conditions (Kirchheim, 2010). Frequently proposed models for HE are

I. HEDE (Hydrogen-enhanced decohesion, Oriani, 1970) explains H embrittlement by a decreasing strength of the metal/metal bond by H atoms in between the metal atoms. Then breaking of bonds at a crack tip is facilitated. In this model one H atom is sufficient to cause enhanced crack propagation, because it could jump from one just broken bond to an adjacent intact one.

II. HELP (Hydrogen-enhanced local plasticity, Birnbaum and Sofronis, 1994) is based on the experimental observation of an enhanced dislocation mobility in the presence of internal hydrogen. In combination with localized slip the mechanism explains the enhanced velocity of cracks due to enhanced plasticity at the tip. The localization of slip leads to flat fracture surfaces which are common for brittle fracture.

III. AIDE (Adsorption-induced dislocation emission, Lynch, 1988) similar to HELP assumes enhanced plasticity and localized slip at the crack tip. However, plasticity is enhanced by external hydrogen giving rise to a faster rate of dislocation generation at the crack surface.

IV. HESIV (Hydrogen-enhanced stress-induced vacancy mechanism, Nagumo, 2004) assumes that void growth in front of the crack tip is enhanced in the presence of hydrogen. Thus coalescence of voids occurs at an earlier time compared with H-free sample leading to accelerated crack growth. Enhanced void growth is due to higher vacancy concentrations stabilized by H atoms (cf. Section 25.3.3.2.1).

V. HYFO (Hydride formation, Gahr and Birnbaum, 1978) explains H embrittlement by the formation of a hydride in the tensile stress region in front of a crack tip during mode one fracture. This has been observed in a TEM for so-called easy hydride forming metals like IVB and VB group metals. Once the hydride has formed the crack propagates through the brittle hydride immediately, the tensile stress field moves forward and H atoms follow forming a hydride again in front of the advanced crack tip.

Among the various models for HE the so-called HELP mechanism has attracted considerable attention. It is based on in situ observations in an environmental TEM where dislocations started moving in the presence of hydrogen gas (Robertson and Birnbaum, 1986). This result is independent of the type of dislocations (edge, screw, mixed or partial) and the nature of the metal. The generality of the effect has been explained in terms of a hydrogen shielding model in which the presence of hydrogen atmospheres around dislocations and elastic obstacles decreases the interaction energy between them (Birnbaum and Sofronis, 1994). Applying the experimental findings to dislocations produced at a crack tip during ductile fracture of a hydrogen-containing sample corresponds to an enhanced local plasticity. The crack propagates faster and the fracture surface appears rather flat like the ones after brittle fracture.

Again a more general explanation for parts of the HELP mechanism can be provided by applying the defactant concept of the reduced formation energy of defects. Then dislocation sources become more active if the line energy of dislocations is reduced in the presence of hydrogen. The newly generated dislocations move through the field of view in a TEM or push former dislocations within this field leaving the impression as if the dislocation mobility has been enhanced. It may also very well be that the mobility is increased directly. This could be a consequence of the ease of kink pair formation by hydrogen segregation (cf. Section 25.3.3.2.3). If kink pair generation determines dislocation mobility, the arguments developed in the previous chapter apply leading to enhanced dislocation mobility. All the effects are especially pronounced in an environmental TEM, because the high-energy electron beam cracks the hydrogen molecule leading to a very high hydrogen fugacity or chemical potential, respectively. Then, as a consequence of Eqn (98), the dislocation line energy is very low (see **Figure 21**).

For very high hydrogen concentrations or for the formation of a new hydride phase other effects like the ones discussed in the following may play the decisive role. Hydrides are usually very brittle despite the fact that in bcc metals the H atoms are very mobile and, therefore, are not expected to hinder dislocation generation and motion. In a hydride all of the appropriate interstices are occupied with hydrogen except some sites above the glide plane of a dislocation within the compressed region of the lattice which have higher site energies for atoms expanding the lattice (cf. **Figure 7**). Therefore the excess hydrogen associated with an edge dislocation line becomes negative. Then during the generation of a dislocation or its extension hydrogen above the glide plane has to be repelled into a reservoir (gas phase) corresponding to a negative excess in agreement with Wagner's definition (Wagner, 1973a). Thus in the framework of the defactant concept, the creation of new dislocations in hydrides requires more energy than in the pure metal. If double kink formation for screw dislocations is considered to be the rate-determining step for plastic deformation, the reasoning is analogous regarding the edge character of the generated kinks.

25.4   Hydrogen in Nanosized Systems: Thin Films, Multilayers and Clusters

25.4.1   Hydrogen in Nanosized Metals

Nanosized metals and their interaction with hydrogen are of interest for two main reasons: first, hydrogen loading times become short because of the small penetration lengths. Loading becomes possible even for metals with low hydrogen diffusivity. This is of special interest when such systems are applied for storage. Second, metal–hydrogen systems can be used for model to study fundamental physical aspects of the nano-size limitation. These can be, for example, the so-called size effects or the impact of mechanical stress.

Two major consequences appear when systems are reduced to the nanoscale: (1) contributions of microstructure become visible or dominant; these are surface and subsurface site contributions, grain boundary or interface contributions or those of other defects like dislocations or vacancies. (2) Nano-metals, per se, are unstable and need to be stabilized by a special environment; usually this is done by substrates or scaffolding materials. The role of these stabilizers cannot be ignored.

25.4.2   Nanosized Metals: Thin Films, Multilayers and Clusters

Nano-metals can be classified by their extension in different room directions, as shown in **Figure 26**. Thin films or plates (a) are restricted to the nanoscale in one room direction, wires or rods, (2) are restricted in two room directions and clusters and (c) or nanoparticles in all three room directions.

25.4.2.1   *Metal Films*

Metal films are usually prepared by deposition on substrates, namely flat surfaces of single crystals or polycrystals or substrates with amorphous finish. Basic information about films can be obtained, for example, from the textbook of Ohring (1992). Usually, film/substrate partners are chosen with strong interatomic bonds at the interface, namely large adhesion energy. These strongly bond films remain at the substrate, even when mechanical stress is present. Mechanical stress often arises during film preparation, for example, during film cooling when the thermal expansion of the film and the substrate differ. If the adhesion energy is too small to bind the film to the substrate, cracks form at the film/substrate interface. These cracks result in film buckles since the stress can be locally released by film bending. Buckle extension is in the order of micrometers.

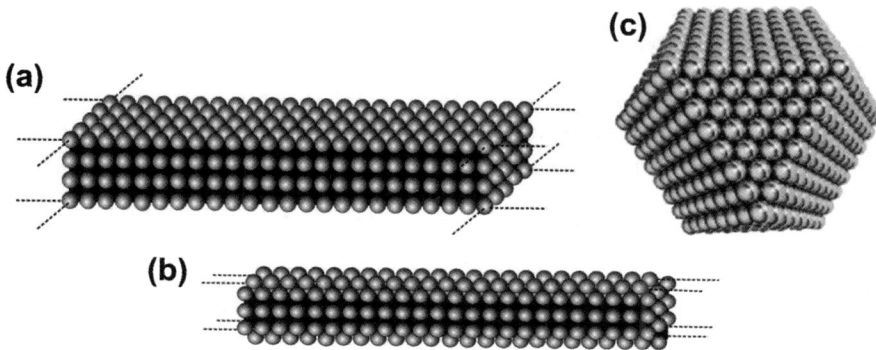

Figure 26 Classification of nano-metals by restriction of their size into the nanoscale in one dimension (a, films or sheets), two (b, nanowires or rods) or in all three room direction (c, clusters or nanoparticles). Dashed lines illustrate that the nano-metal further extends in this direction.

The interatomic bonds also influence the growth mode. In combination with the lattices misfit, the deposition conditions and the temperature, the interfacial energy between the substrate and the film determines if film growth occurs (1) layer by layer ("van-der-Merwe" type), and the film is atomically flat, (2) by island growth, "Volmer–Weber" growth, or (3) with a mixture of both, "Stransky–Krastanov" growth (Ohring, 1992).

The lattice match of the film and the chosen substrate determines whether the film grows oriented or with random in-plane orientation. If the adjacent lattices match considerably, the films grow with orientation relation on the substrate. Slight differences in lattice parameters can be adjusted by implementing misfit dislocations. These misfit dislocations only appear for films above a critical thickness. Models as developed by van der Merwe (1963a, 1963b), Matthews (1975), Matthews and Blakeslee (1974, 1975), Nix (1989) or Jain et al. (1997) base on energy minimization or force arguments: While the self-energy of a dislocation in a film scales with the logarithm of the film thickness $E \sim \ln d$, the elastic energy stored in a film scales linearly with the film thickness $E \sim d$. This concurrence results in a critical thickness of the growing film where the presence of a dislocation becomes energetically favorable.

For a growing film with certain misfit $\delta = (a_{\text{film}} - a_{\text{substrate}})/a_{\text{substrate}}$ to the substrate, dislocation formation sets in above a critical thickness d_{crit} (Matthews and Blakeslee, 1974).

$$d_{\text{crit}} = \frac{b}{\delta} \frac{1}{8\pi(1+v)} \left(\ln \frac{\xi d_{\text{crit}}}{b} \right) \qquad (117)$$

with Poissons ratio v and Burgers vector $\mathbf{b}$. The dislocation core radius was, here, estimated to be $\mathbf{b}$. The factor ξ adds an energy contribution of the dislocation core. For nanocrystalline films, the grain size becomes important.

Grain boundaries and the presence of differently oriented glide planes hinder the migration of dislocations and increase the (hydrogen-related) yield stress $\sigma_{\text{H},y}$ (Arzt, 1998).

According to Venkatraman and Bravman (1992), the yield stress in a film of thickness d increases by

$$\sigma_{\text{H},y}^{\text{film}} = \sigma_{\text{H,bulk}} + \frac{C1}{d^n} + \frac{C2}{b^m} \qquad (118)$$

C1 and C2 are constants that depend on the film orientation. As discussed before, this calculation results in a lower limit for the yield stress of films. Thereby, $\sigma_{H,bulk}$ is the yield stress of the bulk metal. The exponent n usually ranges from 0.5 to 1. With $n = 0.5$ it is the Hall–Petch relation. But, the Hall–Petch relation seems not to apply for small nano-metals any more. According to Arzt this appears when a dislocation ring does not fit into one grain (Arzt, 1998). The case $n = 1$ represents the model described before (Nix, 1989). For nanocrystalline samples, Thompson considered the grain size with an inverse dependency and $m = 1$ (Thompson, 1993a, 1993b). The Nix-Thompson model, that added a factor of 2, describes the general trend of experimental data, but does not give the exact values (Leung and Nix, 2000).

Usually, matching between the film and the substrate lattice appears in several directions. A region grown in one direction is called a domain. The growth mode influences the lateral extension of the domains.

Deposition conditions (technique, temperature, pressure, kinetic energy of the atoms, etc.) and the lattice misfit between the film and the substrate influence the film's microstructure which ranges from the epitaxial film with large in-plane domains to the nanocrystal with random in-plane grain orientation.

To reduce the film surface energy, films mainly grow fiber textured, in vertical direction. They usually possess low-indexed surface planes: in vertical direction, films with bcc lattice are $\langle 110 \rangle$- oriented, films with fcc lattice are $\langle 111 \rangle$-oriented and films with hcp lattice structure are $\langle 0001 \rangle$-oriented. Via in-plane epitaxial relations to the substrate, the films can be forced to vertically grow in other orientations.

Films usually contain mechanical stress right after deposition, the so-called intrinsic stress (e.g. Ohring, 1992; Koch, 1994, 1997). Intrinsic stress usually is in the order of some ± 100 MPa for evaporation and sputtering methods (Koch, 1994, 1997) and ranges up to the GPa range for laser-deposited films (Krebs et al., 2003; Scharf et al., 2003).

Hydrogen molecules enter the film via dissociative chemisorption at the film surface (compare Section 25.2.1). For films, the atomic number of adsorption sites at the surface of metals is in the order of $10^{16}/cm^2$, according to about 10^{19} internal lattice sites for a cm^2 film of 100 nm thickness.

25.4.2.2 Nanowires

Nanowires are restricted in one more direction compared with films. Metal nanowires can be prepared, for example, by (a) metal deposition on vicinal substrates, by growth along step edges, (b) by treating metal films with lithographical techniques by etching away the intermediate regions, and (c) by film deposition on substrates covered with masks. The number of adsorption sites at nanowire surfaces is, again, $10^{16}/cm^2 = 10^9/(cm \cdot nm)$. Adsorption sites exist at the nanowire surfaces. A nanowire of 1 cm length and 100 nm perimeter possesses about 10^{11} sorption sites and about 10^{13} inner sites.

25.4.2.3 Clusters and Nanoparticles

As shown in **Figure 26c**, clusters or nanoparticles are chunks of metals with nanoscale extensions in all three room directions. Basic information about clusters and nanoparticles can be obtained, for example, from the textbook of Sugano and Koizumi (1998) or that edited by Schmid (2005). Here, the two names are used by distinguishing their size: while clusters have diameters up to 10 nm, nano-particles can be larger, that is <1 μm. To reduce the surface area, clusters or nanoparticles tend to adjust with spherical morphology. But, when atoms are arranged in a lattice, the surface energies of different planes usually differ, and low-indexed planes finish the cluster surface. Kinks at surfaces increase the surface energy (Lang et al., 1992), therefore flat and complete surface planes yield the most stable clusters. These clusters of special stability are called 'magic clusters'. For magic clusters of cuboctahedral morphology that results from a cubic lattice structure and similar surface energies for the {111} and the {100} planes, a shell-like surface arrangement of i shells yields N_i atoms (Sugano and Koizumi, 1998):

$$N_i = 1 + \sum_{j=1}^{i} (10j^2 + 2) \tag{119}$$

This formula is also applicable for icosahedral clusters. Each shell contains a special number of atoms: the j-th shell contains n_j atoms

$$n_j = 10j^2 + 2 \tag{120}$$

Table 5 summarizes for clusters with cuboctahedral structure and morphology, the number of atoms located in surface N_S and one subsurface layer N_{SS}, as well as the number of inner bulk-like sites N_B. For Pd, the apparent cluster diameter d_{WS} is given by using the Wigner–Seitz radius of Pd, $r_{WS} = 1.55 \times 10^{-1}$ nm. The number of surface and subsurface sites is larger than the number of bulk-like inner sites, for clusters of up to eight full shells, or, in case of Pd, for clusters of up to 3.9 nm in diameter. For about 4.4 nm clusters, the number of surface and subsurface sites equals that of the bulk-like inner sites (shaded gray in **Table 5**). For larger clusters, the bulk-like inner sites dominate.

In clusters, atoms also arrange in crystal structures that are unknown from bulk metals. They have fivefold symmetries and are not room filling. In the simplest case, icosahedral structures emerge. Here, every inner atom is surrounded by four atoms arranged in an elastically strained tetrahedra. The surface morphology of a resulting icosahedron is closer to the sphere, compared with the cuboctahedron. This surface area reduction stabilizes the icosahedron compared with any other shape. But, tetraheder straining becomes stronger for larger clusters, thereby increasing the strain energy of larger icosahedrons. Therefore, above a critical size, the bulk-like atomic arrangement (fcc, bcc or hcp) is energetically

Table 5   Summary of the trap content of cuboctahedral clusters with closed shell(s): the number of atoms located in surface N_S and one subsurface layer N_{SS}, as well as the number of inner bulk-like sites N_B. The apparent cluster diameter d_{WS} is calculated for Pd, by using the Wigner–Seitz radius

S	N	N_S	N_{SS}	N_B	d_{WS} (nm)
1	13	12	1	–	0.73
2	55	42	12	1	1.2
3	147	92	42	13	1.6
4	309	162	92	55	2.1
5	561	252	162	147	2.6
6	923	362	252	309	3.0
7	1415	492	362	561	3.5
8	2057	642	492	923	3.9
9	2869	812	642	1415	4.4
10	3871	1002	812	2057	4.9
11	5083	1212	1002	2869	5.3
12	6525	1442	1212	3871	5.8

Courtesy of N. Jisrawi.

favorable. For Pd clusters, Jisrawi calculated a phase transition to happen at 4.7 nm in diameter (Pundt et al., 2004).

The number of adsorption sites for hydrogen on cluster surfaces scale with the shell number, according to Eq. 119. But, atoms located at edges and corners act different because they expose a larger contact area to the gas phase.

Since the surface of nanoparticles is close to spheres, the influence of surface curvature on the chemical potential of clusters is often discussed similar to that of a liquid droplet. For droplets surface curvature changes the chemical potential μ by

$$\Delta\mu = \frac{4\gamma\overline{V}_H}{d} \tag{121}$$

with cluster diameter d, and surface tension γ. Typical surface tensions are 73 mN/m for water and 471 mN/m for mercury, at 18 °C. But, for solid clusters and nanoparticles this model is not really applicable. The nanoscale surfaces are not spherical at all: clusters are bordered by atomically flat surfaces. For solids, surface energies depend on the crystallographic orientation. Surface energies of a solid are generally about 20–30% higher than that of a liquid (Tyson and Miller, 1977).

For clusters the situations becomes even more difficult, since structural fluctuations, permutations and morphological changes become more important when the size is small (Sugano and Koizumi, 1998, **Figure 2.1**). For very small clusters below 2 nm in diameter, the atomic arrangement changes by adding or removing one single atom. Here, molecular chemistry comes into play and general trends are not present any more. In this range, physical properties, and also the hydrogen uptake, change with the number of atoms (Schmid, 2005).

25.4.3 Microstructural Contributions in Nanosized Metals and H Solubility

Nanosized metals usually contain a large volume fraction of defects. As discussed in the previous chapter, such defects have a different energetic landscape for hydrogen solution, thereby affecting the local and global solubility of the sample. Energetic landscapes of selected defects are presented in **Figure 7**. For nano-metals, especially surfaces (b,c), interfaces and grain boundaries (d) and dislocations (e) have to be considered. A scheme of these contributions is depicted in **Figure 27**, exemplarily

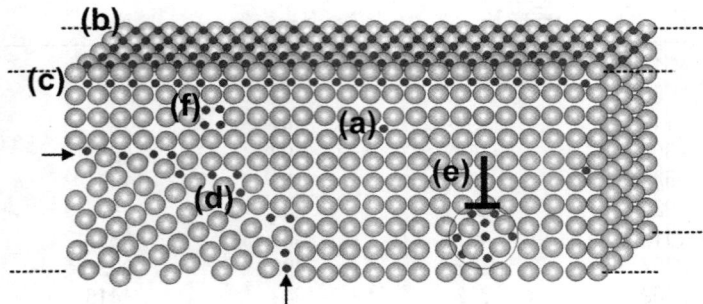

Figure 27 Scheme of the typical microstructure of a nano-metal and its local hydrogen solubility, demonstrated for a part of a thin film. Beside conventional interstitial lattice sites (a), surface (b) and subsurface (c) sites exist. Furthermore, grain boundary sites (d), sites in the strain field of dislocations or in metal vacancies need to be considered. After Pundt and Kirchheim (2006). (For color version of this figure, the reader is referred to the online version of this book.)

drawn for a cross-section of a thin film. For some cases, also vacancies have to be taken into account. For quantification of these contributions on the nano-metal's hydrogen solubility, the impact of these defects on the local hydrogen solubility are be weighted with their volume fraction to the total nano-metal volume and then added to the bulk part contribution of concentration c_0. This simple approach we name "nano-metal solubilty" model. It implies a size independency of the defects local energy environment.

For thin films, surface and subsurface contributions increase the solubility limit c_α by

$$c_{\alpha,S} = \frac{(d - d_{\text{surf}})}{d} c_0 + \frac{d_s}{d} c_s \qquad (122)$$

by approximating the total surface and subsurface thickness with d_{surf} and its mean concentration with c_{surf}. With regard to experimental data from Christmann (1988), or Muschiol et al. (1998) it seems reasonable to approximate these contributions for Pd–H, by $d_s = 0.7$ nm with $c_s = 1$ H/Pd-atom, at room temperature.

Grain boundary contributions, also, increase the solubility limit by

$$c_{\alpha,GB} = \frac{(d - d_{\text{GB}})^2}{d^2} c_0 + \frac{2 d_{\text{GB}} d}{d^2} c_{\text{GB}} \qquad (123)$$

when assumed that the grain size equals the film thickness d (Dornheim, 2002). The local grain boundary concentration c_{GB} can be averaged to be between hydride and α-phase (Mütschele and Kirchheim, 1987a, 1987b). For Pd we estimate $c_{\text{GB}} = 0.3$ H/Pd. For Pd–H, a grain boundary width d_{GB} of 0.9 nm was suggested (Mütschele and Kirchheim, 1987a, 1987b).

Misfit dislocations occur at the interface between a film and the substrate, for films growing with orientation relation and with lattice mismatch on a substrate. Thus, for films on substrates, dislocations usually exist at the film/substrate interface, with dislocation distance a_x and a_y differing in x- and y-directions, respectively. Their contribution shifts the solubility limit by

$$c_{\alpha,L} = c_0 + \frac{1}{d} \cdot \left(\frac{1}{a_x} + \frac{1}{a_y} \right) \cdot \pi r^2 (c_D - c_0) \qquad (124)$$

with the mean local concentration at the dislocation line c_D and the cylindrical enrichment zone with radius r (Dornheim, 2002). For Pd–H, hydrogen-enriched cylindrical zones were measured to grow up to $r = 1.4$ nm in maximum (Maxelon et al., 2001b). They are aligned in the dilatation field at the dislocation line. The local concentration in the dilatation zone is estimated to be that of the hydride. But as shown in **Figure 28**, the orientation of the misfit dislocation needs to be considered. It controls whether the dilatation field appears (a) in the film or (b) in the substrate. For a dilatation field appearing in the film, the solubility should be increased. For a compression zone present in the film the local concentration c_D is reduced compared with c_0 thereby reducing the film's α solubility limit. The dilatation fields in the substrate should not affect the film solubility limit. But, if the dislocation line localizes above the film/substrate interface, the dilatation field might come into play, again.

To estimate the microstructural impact on the film solubility limit, the different contributions can be plotted separately or in combination (Pundt, 2004). This is exemplarily shown in **Figure 29**, for a Pd–H film with bulk solubility limit $c_0 = 0.01$ H/Pd. The dashed lines give the combined impact of surfaces and grain boundaries for $d = d_G$ (red dashed line), or that of surfaces and misfit dislocations (with

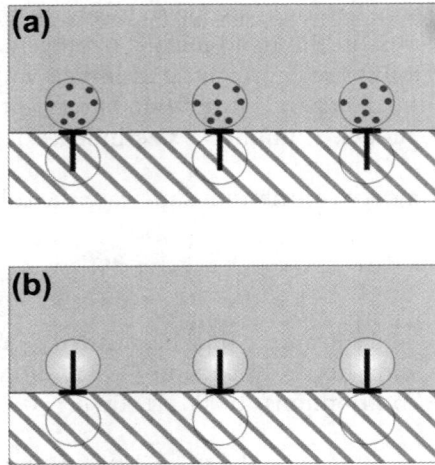

Figure 28 Sketch of two different arrangements of dislocations adjusting the misfit between a film and its substrate. (a) A compressed film can be relaxed by a dilatation field in the film. (b) A tensile strained film can be relaxed by the compression field in the film.

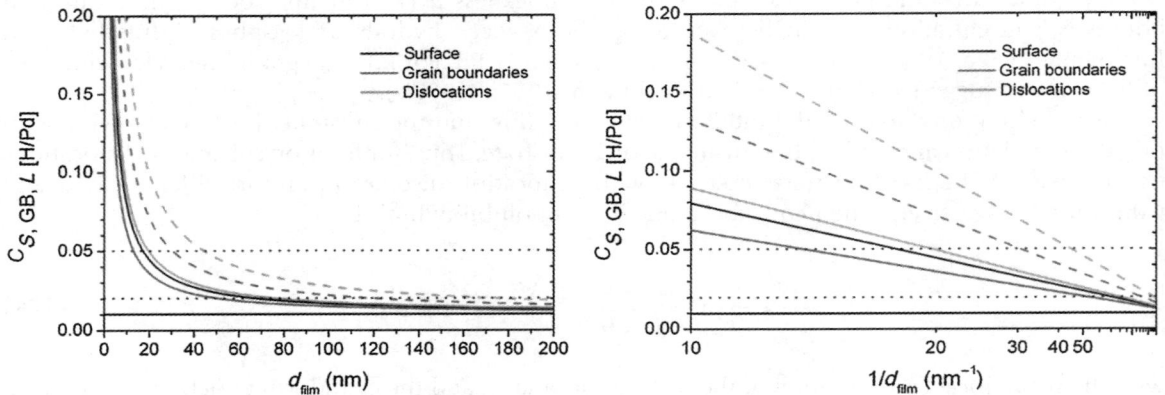

Figure 29 Pd films: Maximum change of the solubility limit by microstructural defects (surface, grain boundaries for grains with $d_{grain} = d_{film}$, dislocations with 10 nm distance) for different film thickness d_{film}, as achieved by using the "nano-metal solubilty" model.

$a_x = a_y = 10$ nm) for a formerly compressed film (green dashed line). This approximates the microstructure of a nanocrystalline film or an epitaxial Pd film, respectively. The solubility limit is increased by a factor of 2, for a 120 nm nanocrystalline film or a 180 nm epitaxial film. Under the same assumptions, it is increased by a factor of 5 for a 30 nm nanocrystalline film or a 50 nm epitaxial film when only microstructural contributions are considered.

25.4.3.1 Experimental Results on Solubility Limits

The solubility limits reported on M–H thin films usually differ from those measured in bulk M–H systems (Feenstra et al., 1988; Salomons, 1987; Lee and Glosser, 1985). Steiger et al. (1993) measured the hydrogen concentration in epitaxial Nb and Nb/Ti/Nb films by ^{15}N-technique, and found α-phase-like behavior up to concentrations of 0.14 H/Nb. This is clearly above the solubility limit of 0.06 H/Nb for bulk Nb–H. But, Steiger et al. also claimed that they usually find a reduced hydrogen concentration when compared with bulk at the same hydrogen pressure, meaning a reduced solubility for thin films—but not a reduced solubility limit. Feentra et al. (1986) and Salomons et al. (1987) intensively studied phase boundaries of Pd–H thin films by gas gravimetry. They found that isotherms of 300 nm thick films are similar to those measured for bulk whereas those of 50 nm films have narrower plateau regions. Salomons et al. (1987) proposed that this is due to a lower critical point of the miscibility gap originating from the abundant presence of grain boundaries in the nanocrystalline thin films. Lee and Glosser (1985) obtained similar results by gas volumetry on a 49 nm film. They detected that the critical point of the miscibility gap was by about 26 K reduced for the thinner film compared with the bulk system (Feenstra et al., 1986; Salomons et al., 1987). For many of these results, microstructure or stress state or the real hydrogen concentration in the film is unknown.

During the past years, phase boundaries are intensively studied for Nb–H and Y–H films by performing in situ XRD during stepwise electrochemical loading with special regard on microstructural aspects, mechanical stress and actual hydrogen concentration (Laudahn et al., 1999a; Pundt et al., 2000a, 2000b, 2000c; Dornheim, 2002). By varying the film thickness between 25 and 300 nm changing the different grain sizes from about 10 nm to epitaxial films general trends were found for films below 200 nm thickness, namely

(i) an increased α-phase solubility limit for thinner films and
(ii) a reduced solubility limit of the hydride phase.

This is in accordance with the "nano-metal solubility" model, presented above. Theoretically, microstructural contributions in Nb thin films change the solubility limit by a factor of 2, when the film thickness is reduced to 30 nm, as shown in **Figure 30**. For the calculations on Nb films, estimates are used to obtain curves of microstructural impacts on solubility limit. In **Figure 30**, $c_0 = 0.06$ H/Nb is taken from the bulk phase diagram at 300 K. For the surface $c_S = 1$ and $d_s = 0.7$ nm are estimated. At grain boundaries, $c_{GB} = 0.4$ H/Nb is taken as the concentration in the middle of the miscibility gap and $d_{GB} = 0.9$ nm. For the dislocation core, $r_D = 0.4$ nm was used and the local concentration $c_D = 0.7$ H/Nb as that of the hydride. For the dislocation distance a_x and a_y experimental data were used.

For epitaxial films above 200 nm, the solubility increase is small. But, solubility limits measured for different films attached to hard substrates are often higher (Dornheim, 2002; Pundt, 2004; Pundt and Kirchheim, 2006). This demonstrates that other factors also affect the solubility limits of thin films.

25.4.3.2 *Hydrogen Solubility Limits for Clusters and Nanoparticles*

For metal nanoparticles and, especially for clusters, the impact of surface and subsurface contributions is expected to become important, because their volume content with regard to regular lattice sites becomes large.

For clusters, the presence of dislocations is energetically not favorable below a certain size. For larger nanoparticles dislocations need to be considered. Also, grain boundaries might be present for larger nanoparticles (Sugano and Koizumi, 1998). Twin boundaries can also exist in clusters (Kishore et al., 2005).

For very small Pt, Ni and Rh clusters, Cox et al. (1990) found a strongly increased H solubility (cf. **Figure 31**). Isotherms reported on Pd clusters differ much from those of bulk materials, and also, from each other. This can be seen in **Figure 32** which summarizes absorption isotherms of about 3 nm Pd clusters, at 37 °C or slightly above room temperature (Aben, 1986; Pundt et al., 1999; Sachs et al.,

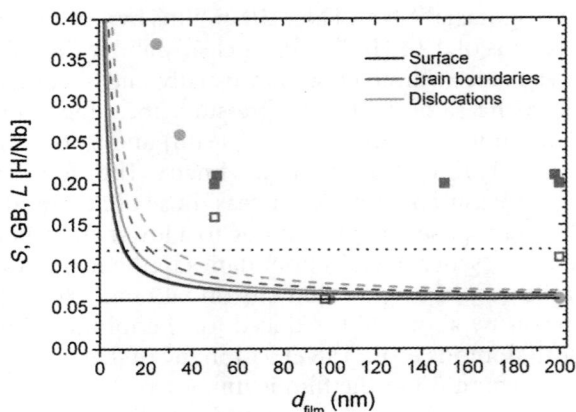

Figure 30 Microstructural impact on the solubility limit of Nb–H films of different thickness d_{film}, as derived by using the "nano-metal solubility" model. The calculation is done for impact of surface (S), grain boundaries (GB) and dislocations (L), the sum of surface and grain boundaries (dashed red) as well as surface and dislocations (dashed green). Experimental data are also implemented. For details see text. (For interpretation of the references to color in this figure legend, the reader is referred to the online version of this book.)

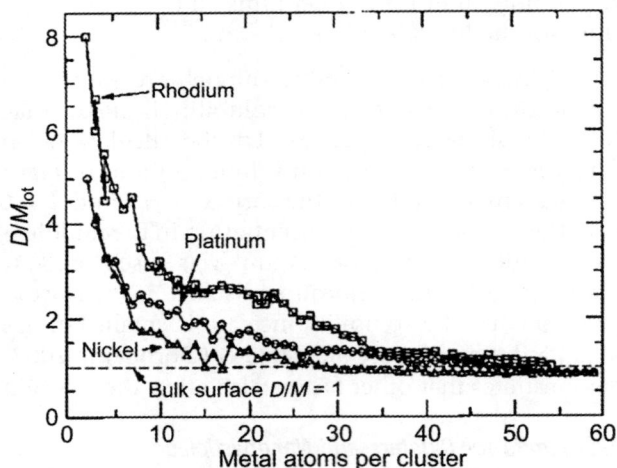

Figure 31 Increase of deuterium uptake as function of atoms per cluster, for Rh, Pt and Ni. Cox et al. (1990).

2001a, 2001b; Züttel et al., 2000; Kishore et al., 2005; Suleiman et al., 2003; Yamauchi et al., 2009). Each group uses different stabilizers; some also uses different precharging situations. But, for all cluster isotherms, the solubility limit of the solid solution is enhanced.

For oxide-free Pd clusters, furthermore, the hydride phase solubility limit is reduced. Typical experimental results on isotherms (● loading, ○ unloading) of 3 nm Pd–H clusters are shown exemplarily in **Figure 33** (data from Sachs et al., 2001a, 2001b). For comparison the isotherms for polycrystalline Pd–H are plotted as a straight line. The changes in solubility limits between a polycrystal and a cluster can be clearly seen.

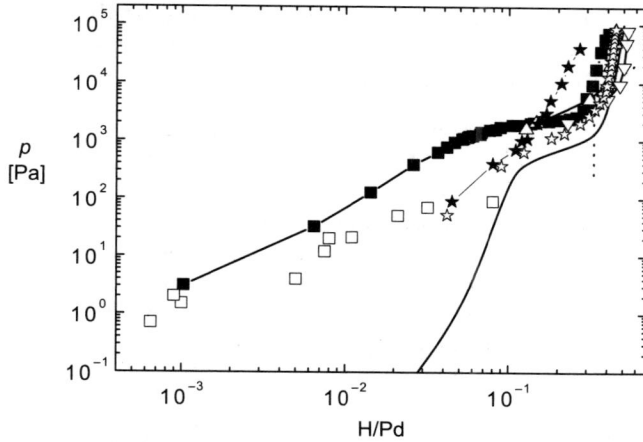

Figure 32 Isotherms of different, about 3 nm in diameter clusters at about 300 K, for comparison (-■- Pd (3 nm) tenside, 37 °C; □ Pd (3,4 nm) polyamidimide (PAI) (Sachs et al., 2001a, 2001b), 37 °C; ••• Pd (2,5 nm) on SiO$_2$, 70 °C (Aben, 1986); — Pd (3,9 nm) in Cu, 20 °C (Züttel et al., 2000); △ Pd (4 nm) reverse micelle, 50 °C; ▽ Pd (5 nm) reverse micelle, 50 °C (Kishore et al., 2005); ★ Pd (3 nm) tenside, 20 °C, icosahedral (Suleiman et al., 2003), ★ (Yamauchi et al., 2009).

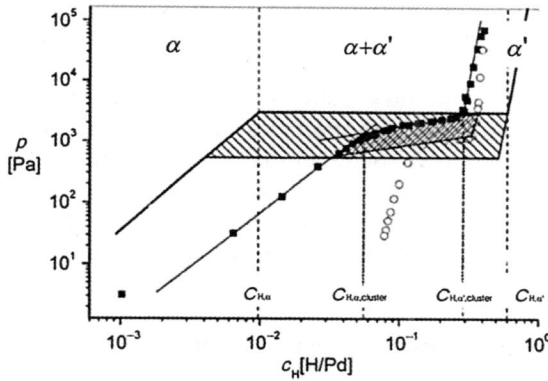

Figure 33 Typical loading and unloading isotherms of a 3 nm Pd cluster (310 K), opening up a hysteresis (data taken from Sachs et al., 2001a, 2001b). Typical isotherms and hysteresis for a Pd polycrystal are also sketched, for comparison. The solubility limits of the two types of Pd samples differ strongly.

From measurements on flat surfaces it is known that H is bound to the surface at room temperature and cannot leave the surface. However, it can leave the subsurface sites even at room temperature (Sachs et al., 2001a, 2001b). The presence of these low-energy sorption sites shifts the solubility limit of the α-phase of clusters from $c_\alpha = 0.01$ H/Pd to about $c_{\alpha,\max} = 0.06$ H/Pd (Frieske and Wicke, 1973). The lower solubility limit of the hydride phase is $c_{\alpha',\min} = 0.3$ H/Pd, compared with $c_{\alpha'} = 0.6$ for bulk Pd. Thus, similar to nanocrystalline samples, a narrowed two-phase region is measured, when compared with bulk. Using the model of a nontransforming shell-like surface region, here the subsurface region, and an inner core that behaves bulk like, explains reasonably well the width reduction of the two-phase region. This results just from reducing the amount of cluster volume that will transform into hydride phase. Sachs et al. (2001a, 2001b) suggested calculating the fraction of nontransforming sites f_n by using the measured solubility limits of the cluster $c_\alpha^{cluster}$ and $c_{\alpha'}^{cluster}$:

$$f_n = \frac{N_{SS}}{N} = 1 - \frac{c_{\alpha'}^{cluster} - c_\alpha^{cluster}}{c_{\alpha'} - c_\alpha} = 1 - \frac{0.3 - 0.06}{0.6 - 0.01} = 0.59 \tag{125}$$

giving 59% of nontransforming sites for the cluster sample in **Figure 33**. For 2 (± 0.5) nm and 3 (± 0.5) nm clusters, stabilized in (Oct)$_4$NBr–Pd and (Bu)$_4$NBr–Pd, respectively, Sachs et al. (2001a, 2001b) give a value of $f = 63\%$, that is 37% of the sites are bulk like. When simplifying the number of hydrogen sites with the number of metal sites, for a five-shell cluster of 561 atoms, at least two subsurface layers (See **Table 5**) are needed to completely relate the reduced width of the miscibility gap to microstructural defects (Sachs et al., 2001a, 2001b; Pundt, 2004).

$$f_n = \frac{N_{SS,\,1} + N_{SS,\,2}}{N} = \frac{162 + 92}{561} = 45\% \tag{126}$$

It should be taken under consideration that, for clusters, surface sites might be also available for hydrogen loading. This further increases the calculated percentage of the nontransforming sites content. Yamauchi et al. (2009) alternatively discussed the hydrogen absorption behavior of clusters by their changed electronic properties.

The influence of the cluster shape on the isotherm was recently studied by Kishore et al. (2005). Clusters with spherical morphology were compared with samples of platelet and mixed morphologies. Kishore et al. explained their results by using the core shell model. For 70 nm × 2 nm platelets the ratio between subsurface sites and bulk-like sites is smaller than that of clusters. This leads to an increase in the width of the plateau region. Actually, nanoparticles with layered stackings or of artificial shape are of interest. Hydrogenation behavior of core shell nanoparticles consisting of two different materials where one is located in the interior and the other at the outside is reported (Kusada et al., 2012). Also artificial particle shapes like cubes (Li et al., 2012) are actually prepared and studied for their hydrogen storage properties. Langhammer et al. 2010 use an indirect nanoplasmonic sensor to measure isotherms of individual clusters.

25.4.4   Mechanical Stress and Stabilization of Nanosized Metals

One important factor to further consider is the mechanical stress arising during hydrogen loading between the nanosystem and its stabilizer. For an Nb film stabilized on a 25 µm thin and soft polymer substrate, much more bulk-like solubility limits were detected (Pundt and Kirchheim, 2006). For free Pd films that were removed from their substrates, plateau pressures are bulk like (Wagner and Pundt, 2008). Such an effect can also be seen in other Pd film results of Gremaud et al. (2009) and Pivak et al. (2011) or, for films on different substrates (Pivak et al., 2009), as taken by hydrogenography. This qualitatively verifies a strong impact of mechanical stress, arising between the film and the substrate, on solubility limits of thin films.

The mechanical stress relates to the need of stabilizers to maintain nano-metals. Nano-metals usually are thermodynamically unstable because of their large surface that, weighted by the surface energy density γ_{surf}, increases the systems total energy F_{tot}.

$$\Delta F_{tot}(A, V) = (-\Delta F + \Delta F_{stress}) \cdot V + \gamma_{surf} \cdot A \tag{127}$$

Thus, when nano-metals touch they agglomerate to increase their bulk volume V and to decrease their surface area, which in sum decreases the systems' energy. To maintain the nano-metals morphology and size, nano-metals are usually stabilized by fixing them on substrates or embedding them in scaffolds or stabilizers (Sachs et al., 2001a, 2001b; Suleiman et al., 2003; Zlotea, 2010). This stabilization acts as a kinetic barrier by reducing the probability of nano-metals' getting in touch.

The need of stabilizers implements mechanical stress that, because of the metals' large volume expansion during hydrogen absorption, can reach tremendous values. Thus, when hydrogen is absorbed in interstitial lattice sites the stress state of the film changes. For Nb–H thin films deposited on substrates like Si or Al_2O_3, compressive stress of about -3 GPa is commonly measured during hydrogen absorption. For comparison, intrinsic mechanical stress occurring after film preparation usually is in the order of 0.1 GPa.

Depending on the stabilization method different mechanical stress fields arise. For thin films, multilayers or clusters deposited on flat substrates, mechanical stress appears mainly in in-plane direction. This stress can be determined by substrate–curvature measurements using Stoneys formula (Stoney, 1909) or by XRD. For small clusters embedded in surfactants, mechanical stress occurs in all three room directions. Zlotea and Latroche (2013) discuss the role of nanoconfinement.

25.4.4.1   Hydrogen-Induced Mechanical Stress and Strain in Metal Films
25.4.4.1.1   The Linear Elastic Regime: Theory of Linear Elasticity

Hydrogen-induced elastic lattice expansion and mechanical stress can be calculated by using the theory of linear elasticity (Landau and Lifshitz, 1989). It depends on the nano-metal morphology, the clamping condition and the mechanical stiffness of the stabilizer. Since the volume usually expands upon hydrogen absorption (compare Section 25.2.2.2), only compressive stress will arise during hydrogen absorption, when the theory of linear elasticity is used for calculation. Thus, when tensile stress components are measured, different explanations are needed like stress release, phase transformations, reorganization of hydrogen in lattice sites, or the seldom case of volume decrease upon hydrogen loading.

In the elastic regime, Hookes law holds (cf. e.g. Landau and Lifshitz, 1989):

$$\sigma_{ij} = C_{ijkl} \cdot \varepsilon_{kl} \tag{128}$$

with the fourth rank tensor of elasticity C_{ijkl}, also named elastic stiffness tensor, that generally couples the strain matrix ε_{kl} with the stress matrix σ_{ij}. The subscripts i,j,k,l run from 1 to 3. Fourth rank tensors usually contain $3^4 = 81$ components.

The elastic energy density is given by (Landau and Lifshitz, 1989):

$$f = \frac{1}{2} \cdot C_{ijkl} \cdot \varepsilon_{ij} \cdot \varepsilon_{kl} \tag{129}$$

The right-hand side is summed over all subscripts (Einstein convention).

Because of the symmetry of ε_{kl} and σ_{ij}, the tensor of elasticity just contains $6^2 = 36$ independent components in maximum (Weißmantel and Hamann, 1989; Nye, 1985). This allows the, often more practical, presentation of Hooks law in Voigt notation:

$$\boldsymbol{\sigma}_\alpha = C_{\alpha\beta} \cdot \boldsymbol{\varepsilon}_\beta \tag{130}$$

including the six-component stress vectors $\boldsymbol{\sigma}_\alpha$ and $\boldsymbol{\varepsilon}_\beta$ and the quadratic matrix of elasticity $C_{\alpha\beta}$.

$$C_{\alpha\beta} = \begin{pmatrix} C_{11} & C_{12} & C_{13} & C_{14} & C_{15} & C_{16} \\ C_{21} & C_{22} & C_{23} & C_{24} & C_{25} & C_{26} \\ C_{31} & C_{32} & C_{33} & C_{34} & C_{35} & C_{36} \\ C_{41} & C_{42} & C_{43} & C_{44} & C_{45} & C_{46} \\ C_{51} & C_{52} & C_{53} & C_{54} & C_{55} & C_{56} \\ C_{61} & C_{62} & C_{63} & C_{64} & C_{65} & C_{66} \end{pmatrix} \tag{131}$$

by Voigt substitution $11 \equiv 1$, $22 \equiv 2$, $33 \equiv 3$, $(yz)23 = 32 \equiv 4$, $(xz)13 = 31 \equiv 5$, $(xy)12 = 21 \equiv 6$. For further symmetry reasons this tensor simplifies. Its components can be found for the different lattice structures, for example, in Landolt-Börnstein (1988). For triclinic system, 21 components are independent, for cubic systems the number of independent components reduces to 3 and for isotropic systems it reduces to 2.

In detail, for cubic systems $C_{ij} = 0$ for i, $j = 4,5,6$ and $i \neq j$, $C_{12} = C_{21} = C_{13} = C_{13} = C_{32} = C_{23}$, $C_{ii} = C_{11}$ for $i = 1,2,3$ and $C_{ii} = C_{44}$ for $i = 4,5,6$. Therefore $C_{\alpha\beta}$ simplifies to

$$C_{\alpha\beta}^{\text{cub}} = \begin{pmatrix} C_{11} & C_{12} & C_{13} & 0 & 0 & 0 \\ C_{21} & C_{22} & C_{23} & 0 & 0 & 0 \\ C_{31} & C_{32} & C_{33} & 0 & 0 & 0 \\ 0 & 0 & 0 & C_{44} & 0 & 0 \\ 0 & 0 & 0 & 0 & C_{55} & 0 \\ 0 & 0 & 0 & 0 & 0 & C_{66} \end{pmatrix} = \begin{pmatrix} C_{11} & C_{12} & C_{12} & 0 & 0 & 0 \\ C_{12} & C_{11} & C_{12} & 0 & 0 & 0 \\ C_{12} & C_{12} & C_{11} & 0 & 0 & 0 \\ 0 & 0 & 0 & C_{44} & 0 & 0 \\ 0 & 0 & 0 & 0 & C_{44} & 0 \\ 0 & 0 & 0 & 0 & 0 & C_{44} \end{pmatrix} \tag{132}$$

For isotrop systems, additionally $C_{44} = 1/2\,(C_{11} - C_{12})$ holds.

For hexagonal systems the stiffness tensor is given by

$$C_{\alpha\beta}^{\text{hex}} = \begin{pmatrix} C_{11} & C_{12} & C_{13} & 0 & 0 & 0 \\ C_{21} & C_{22} & C_{23} & 0 & 0 & 0 \\ C_{31} & C_{32} & C_{33} & 0 & 0 & 0 \\ 0 & 0 & 0 & C_{44} & 0 & 0 \\ 0 & 0 & 0 & 0 & C_{55} & 0 \\ 0 & 0 & 0 & 0 & 0 & C_{66} \end{pmatrix} = \begin{pmatrix} C_{11} & C_{12} & C_{13} & 0 & 0 & 0 \\ C_{12} & C_{11} & C_{13} & 0 & 0 & 0 \\ C_{13} & C_{13} & C_{33} & 0 & 0 & 0 \\ 0 & 0 & 0 & C_{44} & 0 & 0 \\ 0 & 0 & 0 & 0 & C_{44} & 0 \\ 0 & 0 & 0 & 0 & 0 & \frac{1}{2}(C_{11} - C_{12}) \end{pmatrix} \tag{133}$$

Other systems can be found, for example; in Ref. Weißmantel and Hamann (1989). The coordinates of the stiffness tensor $C_{\alpha\beta}$ are given with respect to the internal crystal coordinate system (**Tables 6 and 7**).

The stress and strain calculation will be demonstrated, here, for the condition of a film fixed to a hard substrate.

25.4.4.1.2 *Hydrogen-Absorbing Metal Film Ideally Fixed to a Hard Substrate*

Stress and strain of a film in the ideally fixed state can be calculated by doing the Gedanken experiment of compressing a free metal film (expanded in all three room dimensions by hydrogen with

Table 6 Elastic constants of some cubic crystals at 300 K in Voigt notation

Material	C_{11} [GPa]	C_{44} [GPa]	C_{12} [GPa]	References
Pd	244	71.6	173	Landolt-Börnstein (1988)
Nb	245.6	29.3	138.7	Carroll (1965)
Ta	267	82.5	161	Bolef (1961)
V	228	42.6	119	Bolef (1961)
W	501	151.4	189	Huntington (1958)

Table 7 Elastic constants of some hexagonal crystals at 300 K in Voigt notation

Material	C_{11} [GPa]	C_{33} [GPa]	C_{44} [GPa]	C_{12} [GPa]	C_{13} [GPa]	References
Y	77.9	76.9		28.5	76.9	Smith and Gjevre (1960)
Mg	59.7	61.7	16.4	26.2	21.7	Huntington (1958).
Ti	162.4	180.7	46.7	92.0	69.0	Fisher and Renken (1964)

ε_0, **Figure 34**) back to the initial in-plane area of the clamped metal film, with $-\varepsilon_0$. This leads to a compressive biaxial stress state, in lateral directions (Laudahn et al., 1999a).

Due to the Poisson reaction the in-plane compression results in an additional out-of-plane expansion $\Delta\varepsilon_{zz}$. It is assumed here that the substrate is ideally hard and does not expand when the film is stressed. The relation between the additional out-of-plane expansion $\Delta\varepsilon_{zz}$ and the in-plane biaxial stress σ is given by biaxial module M.

$$\sigma = -M \cdot \varepsilon_0 \tag{134}$$

In vertical direction, the stress is zero.

$$\sigma_z = 0 \tag{135}$$

Stress and strain depend on the film orientation and texture with regard to the substrate and have to be calculated using the direction-dependent elastic constants appropriate for each case (Pundt, 2005).

The internal coordinate system of the crystal needs to be correlated to the external coordinate system of the film and the substrate. The external coordinate system, as used in **Figure 35**, describes the

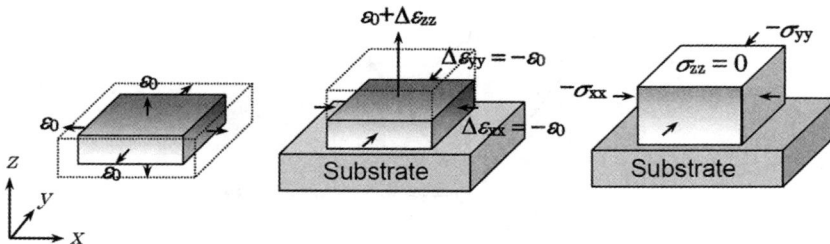

Figure 34 Sketch on calculation of stress and strain using theory of linear elasticity. (a) The free sample expands in all three room directions by ε_0 upon hydrogen absorption. (b) For the fixed film, in lateral directions this expansion has to be inverted, $\Delta\varepsilon_{xx} = \Delta\varepsilon_{yy} = -\varepsilon_0$. Via converse contraction, the lattice additionally expands in vertical direction by $\Delta\varepsilon_{zz}$. (c) Monitors the stress state.

experimental situation. Thus, for each film orientation the external strain has to be linked to the internal strain tensor, before the stress reaction can be calculated by Hookes law.

For linkage often rotation matrixes are used, with angles φ, ψ, and χ around the x-, y- and z-axis:

$$(a_{im}) = \begin{pmatrix} 1 & 0 & 0 \\ 0 & \cos\varphi & -\sin\varphi \\ 0 & \sin\varphi & \cos\varphi \end{pmatrix}, \begin{pmatrix} \cos\psi & 0 & -\sin\psi \\ 0 & 1 & 0 \\ \sin\psi & 0 & \cos\psi \end{pmatrix}, \begin{pmatrix} \cos\chi & -\sin\chi & 0 \\ \sin\chi & \cos\chi & 0 \\ 0 & 0 & 1 \end{pmatrix} \quad (136)$$

For rotation matrixes $(a_{ij})^{-1} = {}^T(a_{ij})$ holds (Kowalsky, 1974). When rotations around more than one axis are needed, the change of the axis by each subrotation also needs to be considered. The second rotation, then, considers a new axis resulting from the first rotation (Fukai, 2005). The determination of this transformation matrix is described exemplarily in the paragraph related to (110)-oriented films.

The determination of the needed rotation matrixes and angles is only simple for very special cases. A general way to directly get the transformation matrix is presented here, by using direction cosines, $\cos(<(r_i', r_j))$, that link the internal coordinate system r_i to the external coordinate system with r_j'.

The matrix elements are given by forming the scalar product of unit vectors

$$a_{ij} = \cos(\langle (r_i' r_j) \rangle) \quad (137)$$

They offer the possibility of getting the transformation matrix for any rotation of the two coordinates systems.

$$(a_{ij}) = \begin{pmatrix} \cos(<(x',x)) & \cos(<(x',y)) & \cos(<(x',z)) \\ \cos(<(y',x)) & \cos(<(y',y)) & \cos(<(y',z)) \\ \cos(<(z',x)) & \cos(<(z',y)) & \cos(<(z',z)) \end{pmatrix} \quad (138)$$

For thin films, the z-orientation of the internal coordinate system is determined by the film orientation. The two other internal axes just have to be chosen under the condition that the scalar products of all three unit vectors are zero:

$$e_1 \cdot e_2 = e_1 \cdot e_3 = e_2 \cdot e_3 = 0 \quad (139)$$

Thus, just one angle is needed to describe the in-plane orientation of the internal coordinate system.

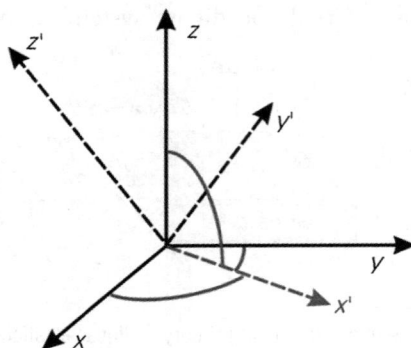

Figure 35   The internal coordinate system (x, y, z) of the crystal needs to be correlated to the external coordinate system of the film and substrate.

The directional dependency of the elastic properties can be taken into account by three ways:

(i) **The transformation of the strain matrix (ε_{ij}):** a film coordinate system is setup (marked with prime signs, directions as shown exemplarily in **Figure 35**). A transformation matrix (a_{ij}) is derived by expressing the internal crystal strains as a function of the external film directions (marked with prime sign). Coordinate transformation can, then, be done by (Kowalsky, 1974):

$$(\varepsilon_{ij}) = {}^T(a_{im}) \cdot (\varepsilon'_{mn}) \cdot (a_{jn}) \tag{140}$$

The transformation cannot be done in Voigt notation, since the coefficients of the Voigt vectors are not the components themselves, and therefore do not transform like the components of a first-rank tensor.

(ii) **The transformation of the elastic stiffness tensor:** the fourth rank tensor is transformed with an appropriate transformation matrix a. The elastic constants of the crystal, then, can be expressed as function of external film coordinates (marked with prime sign).

$$C_{ijkl} = C'_{mnop} \cdot {}^T a_{mi} \cdot a_{nj} \cdot {}^T a_{ok} \cdot a_{pl} \tag{141}$$

with subscripts $i,j,...,p = 1,2,3$. Also, this transformation cannot be done in Voigt notation, since the coefficients of the Voigt matrix are not the components themselves, and therefore do not transform like the components of a second-rank tensor.

(iii) When just the in-plane biaxial stress and the vertical strain are needed, the transformation of the elastic stiffness tensor can be avoided by using the invariance principle of the free energy density f regarding such transformations. In Voigt notation it is determined by

$$f_{el}^V = \frac{1}{2} \cdot C_{ij} \cdot \varepsilon_i \cdot \varepsilon_j \tag{142}$$

Again, the right-hand side of the equation is summed over all subscripts. The strain derivatives in desired directions give the related stresses. With the following additional assumptions and conditions, the vertical strain can be calculated.

Assumption 1: In the following paragraphs, the elastic stress is determined for different lattice orientations under the assumption that, for external coordinates, shear stress can be neglected:

$$\varepsilon'_4 = \varepsilon'_5 = \varepsilon'_6 = 0 \tag{143}$$

Condition 1: The Ansatz of Eqn (135) is used:

$$\sigma_3^{V'} = \sigma'_{33} = 0 \tag{144}$$

Assumption 2: Furthermore, it is assumed that the hydrogen-induced bulk lattice expansion ε_0 is direction independent. Since the film is laterally clamped to the substrate this yields for the in-plane strain condition:

$$\varepsilon'_1 = \varepsilon'_2 = -\varepsilon_0 \tag{145}$$

In vertical direction this gives an additional term that is, for simplicity, named ε_3'. In detail, it is $\Delta\varepsilon_3'$, and the total strain in z-direction is $\varepsilon_0 + \Delta\varepsilon_3'$.

25.4.4.1.3 *(100)-Oriented Cubic Films*

For this simple case the internal coordinate system coincides with the external sample system. Thus, $\varepsilon_i = \varepsilon_i'$ in Voigt notation, Eqn (130) and Eqn (144) gives

$$\sigma_1' = C_{11}\cdot\varepsilon_1' + C_{12}\cdot\varepsilon_2' + C_{12}\cdot\varepsilon_3' \quad \text{(a)}$$
$$\sigma_2' = C_{12}\cdot\varepsilon_1' + C_{11}\cdot\varepsilon_2' + C_{12}\cdot\varepsilon_3' \quad \text{(b)} \tag{146}$$
$$\sigma_3' = C_{12}\cdot\varepsilon_1' + C_{12}\cdot\varepsilon_2' + C_{11}\cdot\varepsilon_3' = 0 \quad \text{(c)}$$

by using the compliance tensor describing the cubic lattice Eqn (132). With Eqn (145), rearrangement of Eqn (146) (c) yields the total vertical expansion $\varepsilon_{3,\text{tot}}'$

$$\varepsilon_3' = \Delta\varepsilon_{zz} = -\frac{C_{12}}{C_{11}}\cdot(\varepsilon_1' + \varepsilon_2') = 2\frac{C_{12}}{C_{11}}\cdot\varepsilon_0 \quad \text{and} \quad \varepsilon_{3,\text{tot}}' = \varepsilon_0 + \Delta\varepsilon_{zz} = \left(1 + 2\frac{C_{12}}{C_{11}}\right)\cdot\varepsilon_0 \tag{147}$$

This can be implemented in Eqn (146) (a) and (b) to get the in-plane stress

$$\sigma_1' = \left(C_{11} - \frac{C_{12}^2}{C_{11}}\right)\cdot\varepsilon_1' + \left(C_{12} - \frac{C_{12}^2}{C_{11}}\right)\cdot\varepsilon_2' \quad \text{(a)}$$
$$\sigma_2' = \left(C_{12} - \frac{C_{12}^2}{C_{11}}\right)\cdot\varepsilon_1' + \left(C_{11} - \frac{C_{12}^2}{C_{11}}\right)\cdot\varepsilon_2' \quad \text{(b)} \tag{148}$$

With Eqn (145) follows

$$\sigma_1' = \sigma_2' = -\left(C_{11} + C_{12} - 2\frac{C_{12}^2}{C_{11}}\right)\cdot\varepsilon_0, \tag{149}$$

for the in-plane biaxial stress of the (100)-oriented film. All other directions can be obtained by taking the direction cosines.

25.4.4.1.4 *Isotropic Films*

Since the films behave isotropic, the inner crystal coordinate system can be chosen to be that of the external film coordinate system, and results obtained in the last paragraph can be used. The additional relation holding for isotropical materials, $C_{44} = 1/2\ (C_{11} - C_{12})$, does not change the final equations obtained.

For isotropic films it is common to use an alternative description where the biaxial modulus $\frac{E}{1-v}$ and Poisson's ratio v is used, namely

$$\frac{E}{1-v} = C_{12} + C_{11} - \frac{2C_{12}^2}{C_{11}} \tag{150}$$

and

$$\frac{v}{1-v} = \frac{C_{12}}{C_{11}} \tag{151}$$

Table 8 Results for isotropical films, as derived from Eqns (152) and (153)

Metal	E [GPa]	ν	α_H	ε_{zz}/c_H	σ_{xx}/c_H [GPa]	References
Pd	121	0.39	0.063	0.144	−12.6	Brandes (1983)
Nb	103	0.387	0.058	0.131	−9.75	Bolef (1961)
V	127.6	0.365	0.063	0.136	−12.7	Brandes (1983)
Gd	56.1	0.257	0.33	0.051	−2.27	Scott (1978)
Y	60.9	0.296	0.33	0.052	−2.50	Scott (1978)

This results in

$$\varepsilon'_{3,\text{tot}} = \varepsilon_0 + \Delta\varepsilon_{zz} = \left(1 + 2\frac{v}{(1-v)}\right) \cdot \varepsilon_0 \tag{152}$$

$$\sigma'_1 = \sigma'_2 = -\left(\frac{E}{1-v}\right) \cdot \varepsilon_0 \tag{153}$$

Mean values of E and v are given, for example, by Brandes (1983) or, for the rare earths by Scott (1978). **Table 8** summarizes some results for isotropical films, as they result by film deposition on amorphous substrates.

25.4.4.1.5 (110)-Oriented Cubic Films
Bcc films usually grow (110) oriented, the external z-direction correlates with the [110] internal crystallographic direction. Rotations by $\varphi = -90°$ around the x-axis and afterward by $\chi = -135°$ around the reoriented z-axis give the new external coordinates x', y', z' in [1–10], [001] and [110] crystallographic and orthogonal directions (see **Figure 36**).

(A) By matrix multiplications the total rotation matrix can be obtained. The composition of the subrotation is when the vector basis also transforms (Fischer, 2012), like

$$(a_{ij}) = \begin{pmatrix} 1 & 0 & 0 \\ 0 & 0 & 1 \\ 0 & -1 & 0 \end{pmatrix} \cdot \begin{pmatrix} -\frac{1}{\sqrt{2}} & \frac{1}{\sqrt{2}} & 0 \\ -\frac{1}{\sqrt{2}} & -\frac{1}{\sqrt{2}} & 0 \\ 0 & 0 & 1 \end{pmatrix} = \begin{pmatrix} -\frac{1}{\sqrt{2}} & \frac{1}{\sqrt{2}} & 0 \\ 0 & 0 & 1 \\ \frac{1}{\sqrt{2}} & \frac{1}{\sqrt{2}} & 0 \end{pmatrix} \tag{154}$$

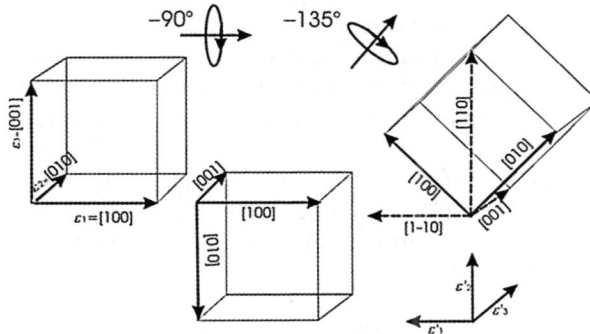

Figure 36 Subrotations required for (110)-oriented films: $\varphi = -90°$ around the x-axis and afterward by $\chi = -135°$ around the reoriented z-axis gives the new external coordinates x', y', z' in [1–10], [001] and [110] crystallographic directions.

and

$$
{}^{T}(a_{ij}) = \begin{pmatrix} -\frac{1}{\sqrt{2}} & 0 & \frac{1}{\sqrt{2}} \\ \frac{1}{\sqrt{2}} & 0 & \frac{1}{\sqrt{2}} \\ 0 & 1 & 0 \end{pmatrix}
\tag{155}
$$

(B) Another way to obtain the transformation matrix is by expressing the external film directions x', y', z' as functions of the internal coordinates x, y, z. This can be arranged in the following transformation scheme, expressing, for example, $x' = -\frac{1}{\sqrt{2}}x + \frac{1}{\sqrt{2}}y + 0z$ (**Figure 37**)

$$
\begin{array}{c|ccc}
 & x & y & z \\ \hline
x' & -\dfrac{1}{\sqrt{2}} & \dfrac{1}{\sqrt{2}} & 0 \\
y' & 0 & 0 & 1 \\
z' & \dfrac{1}{\sqrt{2}} & \dfrac{1}{\sqrt{2}} & 0
\end{array}
\tag{156}
$$

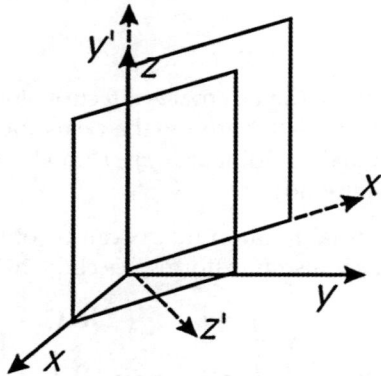

Figure 37 Orientation relation between the inner crystal lattice coordinate system and the outer film coordinate system.

The new coordinate's vectors are unit vectors, again. The transformation scheme directly represents the elements of the transformation matrix.

(C) A general way of getting the transformation matrix is the use of direction cosines, $\cos(<(r'_i, r_j))$, that link the internal coordinate system r_i to the external coordinate system with r'_j, compare Eqns (137) and (139) on page 70. For example, the coordinate systems relations for the (110)-oriented film in **Figure 37** give $\cos(<(x',x)) = \cos(-135°) = -1/\sqrt{2}$, $\cos(<(y',y)) = \cos(-90°) = 0$, $\cos(<(z',z)) = \cos(90°) = 0$, $\cos(<(y',x)) = \cos(90°) = 0$, $\cos(<(z',x)) = \cos(-45°) = 1/\sqrt{2}$, $\cos(<(x',y)) = \cos(-45°) = 1/\sqrt{2}$ and $\cos(<(z',y)) = \cos(-45°) = 1/\sqrt{2}$, and so on, the coordinates of transformation matrix Eqn (138). Even more easily this can be done by using the scalar product. For unit vectors $\mathbf{r}_i$, $\mathbf{r}'_j$ this simplifies to

$$
a_{ij} = \cos(<(r'_i, r_j)) = r'_i \cdot r_j
\tag{157}
$$

With the direction cosines, any transformation matrix can be easily obtained.

The transformation of the fourth rank tensor is described by Eqn (141). In analogy to the last section the vertical strain $\Delta\varepsilon_{zz}$ and the in-plane stress could be determined.

Here, we will present an alternative way to determine film stress and strain by avoiding the lengthy treatment involved in Eqn (141). This way follows the description used by Sander (1999) for magnetoelastic coupling in thin films.

With the elastic energy Eqn (142) and the elastic stiffness tensor of the cubic system (132), both in Voigt notation, one obtains:

$$f_{el} = \frac{1}{2}C_{11}(\varepsilon_1^2 + \varepsilon_2^2 + \varepsilon_3^2) + C_{12}(\varepsilon_1\varepsilon_2 + \varepsilon_2\varepsilon_3 + \varepsilon_1\varepsilon_3) + \frac{1}{2}C_{44}(\varepsilon_4^2 + \varepsilon_5^2 + \varepsilon_6^2) \tag{158}$$

The internal crystal coordinates now have to be expressed by the external coordinate system. This can be done by coordinate transformation Eqn (140) in matrix notation by use of Eqns (154) and (155) and the simplification Eqn (143), that shear strains $\varepsilon'_{12} = \varepsilon'_{13} = \varepsilon'_{23} = 0$ in the external coordinates. For the strain tensor this gives, by regarding the transformation factors 2

$$(\varepsilon_{ij}) = \begin{pmatrix} \frac{1}{2}(\varepsilon'_{11} + \varepsilon'_{33}) & \frac{1}{2}(-\varepsilon'_{11} + \varepsilon'_{33}) & 0 \\ \frac{1}{2}(-\varepsilon'_{11} + \varepsilon'_{33}) & \frac{1}{2}(\varepsilon'_{11} + \varepsilon'_{33}) & 0 \\ 0 & 0 & \varepsilon'_{22} \end{pmatrix}, \quad \text{Voigt notation } (\varepsilon_i^V) = \begin{pmatrix} \frac{1}{2}(\varepsilon'_1 + \varepsilon'_3) \\ \frac{1}{2}(\varepsilon'_1 + \varepsilon'_3) \\ \varepsilon'_2 \\ (-\varepsilon'_1 + \varepsilon'_3) \\ 0 \\ 0 \end{pmatrix} \tag{159}$$

The free energy density, in Voigt notation, is now expressed as function of the external film coordinates (Sander, 1999):

$$f_{el}^V = \frac{1}{4}C_{11}(\varepsilon_1'^2 + 2\varepsilon_2'^2 + 2\varepsilon_1'\varepsilon_3' + \varepsilon_3'^2) + \frac{1}{4}C_{12}(\varepsilon_1'^2 + 2\varepsilon_1'\varepsilon_3' + 4\varepsilon_1'\varepsilon_2' + 4\varepsilon_3'\varepsilon_2' + \varepsilon_3'^2) + \frac{1}{2}C_{44}(\varepsilon_3' - \varepsilon_1')^2 \tag{160}$$

The stress σ'_i in external coordinates in a certain direction is then given by the strain derivative of the free energy density

$$\sigma'_i = \frac{\partial f_{el}^V}{\partial \varepsilon'_i} \tag{161}$$

For the derivative in z- or, namely, ε'_3-direction, $\sigma'_3 = 0$ holds. Thus

$$\frac{\partial f_{el}^V}{\partial \varepsilon'_3} = \frac{1}{4}C_{11}(2\varepsilon'_1 + 2\varepsilon'_3) + \frac{1}{4}C_{12}(2\varepsilon'_1 + 4\varepsilon'_2 + 2\varepsilon'_3) + C_{44}(\varepsilon'_3 - \varepsilon'_1)$$

$$= \frac{1}{2}(C_{11} + C_{12} - 2C_{44})\varepsilon'_1 + C_{12}\varepsilon'_2 + \left(\frac{1}{2}C_{12} + \frac{1}{2}C_{11} + C_{44}\right)\varepsilon'_3 = 0 \tag{162}$$

This gives the additional vertical strain

$$\varepsilon_3' = -\frac{(C_{11} + C_{12} - 2C_{44})\varepsilon_1' + 2C_{12}\varepsilon_2'}{(C_{12} + C_{11} + 2C_{44})} \tag{163}$$

With $\varepsilon_1' = \varepsilon_2' = -\varepsilon_0$ this gives a total vertical strain of

$$\varepsilon_{3,\text{tot}}' = \varepsilon_3' + \varepsilon_0 = \left(1 + \frac{(C_{11} + 3C_{12} - 2C_{44})}{(C_{12} + C_{11} + 2C_{44})}\right)\varepsilon_0 \tag{164}$$

This can be implemented in Eqn (160). The in-plane stress can be gained from the strain derivative of the free energy density in the appropriate directions:

$$\frac{\partial f_{el}^V}{\partial \varepsilon_1'} = \sigma_1' = \frac{4(C_{11} + C_{12})C_{44}\varepsilon_1' + 4C_{12}C_{44}\varepsilon_2'}{C_{12} + C_{11} + 2C_{44}} = -4 \cdot \frac{(C_{11} + 2C_{12})C_{44}}{C_{12} + C_{11} + 2C_{44}}\varepsilon_0 \tag{165}$$

$$\frac{\partial f_{el}^V}{\partial \varepsilon_2'} = \sigma_2' = \frac{4C_{12}C_{44}\varepsilon_1' + \left(C_{11}^2 - 2C_{12}^2 + C_{11}(C_{12} + 2C_{44})\right)\varepsilon_2'}{C_{12} + C_{11} + 2C_{44}}$$

$$= -\frac{4C_{12}C_{44} + C_{11}^2 - 2C_{12}^2 + C_{11}(C_{12} + 2C_{44})}{C_{12} + C_{11} + 2C_{44}}\varepsilon_0 \tag{166}$$

When, again, $\varepsilon_1' = \varepsilon_2' = -\varepsilon_0$ is considered, as can be seen, for (110)-oriented films the stress differs in the two perpendicular in-plane directions $x' = [1{-}10]$ and $y' = [001]$.

Stress values for the other in-plane directions can be obtained by matrix multiplications.

25.4.4.1.6   (111)-Oriented Cubic Films

Fcc films usually grow (111) oriented, meaning the [111] direction of the crystal correlates to the external z-direction of the film. For the orthogonal system, the two other in-plane axes are found by

$$\begin{pmatrix} 1 \\ 1 \\ 1 \end{pmatrix} \cdot \begin{pmatrix} x'_1 \\ x'_2 \\ x'_3 \end{pmatrix} = 0, \quad \begin{pmatrix} 1 \\ 1 \\ 1 \end{pmatrix} \cdot \begin{pmatrix} y'_1 \\ y'_2 \\ y'_3 \end{pmatrix} = 0 \quad \text{and} \quad \begin{pmatrix} x'_1 \\ x'_2 \\ x'_3 \end{pmatrix} \cdot \begin{pmatrix} y'_1 \\ y'_2 \\ y'_3 \end{pmatrix} = 0 \tag{167}$$

One possible set is (111), (1−10) and (−1−12) for the external coordinates x', y', z'.

The related transformation matrix (a_{ij}) can be obtained by using the method of direction cosines C. Here, only the scalar products of [100], [010], [001] and the external coordinates $1/\sqrt{2}[-110]$, $1/\sqrt{6}[-1-12]$, $1/\sqrt{3}[111]$ have to be determined, as shown in **Figure 38**.

The transformation scheme and the resulting matrix are given by

$$
\begin{array}{c|ccc}
 & x & y & z \\
\hline
x' & -\dfrac{1}{\sqrt{2}} & \dfrac{1}{\sqrt{2}} & 0 \\
y' & -\dfrac{1}{\sqrt{6}} & -\dfrac{1}{\sqrt{6}} & \sqrt{\dfrac{2}{3}} \\
z' & \dfrac{1}{\sqrt{3}} & \dfrac{1}{\sqrt{3}} & \dfrac{1}{\sqrt{3}}
\end{array}
\qquad
a_{ij} = \begin{pmatrix} -\dfrac{1}{\sqrt{2}} & \dfrac{1}{\sqrt{2}} & 0 \\[8pt] -\dfrac{1}{\sqrt{6}} & -\dfrac{1}{\sqrt{6}} & \sqrt{\dfrac{2}{3}} \\[8pt] \dfrac{1}{\sqrt{3}} & \dfrac{1}{\sqrt{3}} & \dfrac{1}{\sqrt{3}} \end{pmatrix} \tag{168}
$$

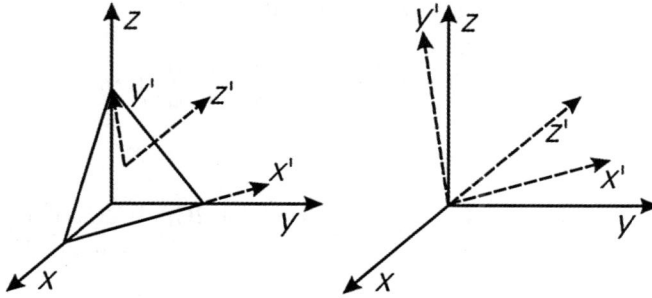

Figure 38 Orientation relation between the internal and the external coordinate systems for (111)-oriented films, for using the method of direction cosines.

With this transformation matrix the crystal strain can be expressed as function of the external strains (marked with prime sign).

$$(\varepsilon_{ij}) = \begin{pmatrix} \frac{1}{2}\varepsilon'_{11} + \frac{1}{6}\varepsilon'_{22} + \frac{1}{3}\varepsilon'_{33} & -\frac{1}{2}\varepsilon'_{11} + \frac{1}{6}\varepsilon'_{22} + \frac{1}{3}\varepsilon'_{33} & \frac{1}{3}\left(-\varepsilon'_{22} + \varepsilon'_{33}\right) \\ -\frac{1}{2}\varepsilon'_{11} + \frac{1}{6}\varepsilon'_{22} + \frac{1}{3}\varepsilon'_{33} & \frac{1}{2}\varepsilon'_{11} + \frac{1}{6}\varepsilon'_{22} + \frac{1}{3}\varepsilon'_{33} & \frac{1}{3}\left(-\varepsilon'_{22} + \varepsilon'_{33}\right) \\ \frac{1}{3}\left(-\varepsilon'_{22} + \varepsilon'_{33}\right) & \frac{1}{3}\left(-\varepsilon'_{22} + \varepsilon'_{33}\right) & \frac{2}{3}\varepsilon'_{22} + \frac{1}{3}\varepsilon'_{33} \end{pmatrix} \tag{169}$$

or in Voigt notation

$$(\varepsilon_i^V) = \begin{pmatrix} \frac{1}{2}\varepsilon'_1 + \frac{1}{6}\varepsilon'_2 + \frac{1}{3}\varepsilon'_3 \\ \frac{1}{2}\varepsilon'_1 + \frac{1}{6}\varepsilon'_2 + \frac{1}{3}\varepsilon'_3 \\ \frac{2}{3}\varepsilon'_2 + \frac{1}{3}\varepsilon'_3 \\ -\varepsilon'_1 + \frac{1}{3}\varepsilon'_2 + \frac{2}{3}\varepsilon'_3 \\ \frac{2}{3}\left(-\varepsilon'_2 + \varepsilon'_3\right) \\ \frac{2}{3}\left(-\varepsilon'_2 + \varepsilon'_3\right) \end{pmatrix} \tag{170}$$

Again, by using the assumption $\varepsilon'_4 = \varepsilon'_5 = \varepsilon'_6 = 0$. The elastic free energy density (Sander, 1999) is

$$f_{el}^V = C_{11}\left(\left(\frac{1}{2}\varepsilon'_1 + \frac{1}{6}\varepsilon'_2 + \frac{1}{3}\varepsilon'_3\right)^2 + \frac{1}{2}\left(\frac{2}{3}\varepsilon'_2 + \frac{1}{3}\varepsilon'_3\right)^2\right) + C_{12}\left(\left(\frac{1}{2}\varepsilon'_1 + \frac{1}{6}\varepsilon'_2 + \frac{1}{3}\varepsilon'_3\right)^2\right.$$
$$\left. + 2\left(\frac{1}{2}\varepsilon'_1 + \frac{1}{6}\varepsilon'_2 + \frac{1}{3}\varepsilon'_3\right)\left(\frac{2}{3}\varepsilon'_2 + \frac{1}{3}\varepsilon'_3\right)\right) + 2C_{44}\left(\left(-\frac{1}{2}\varepsilon'_1 + \frac{1}{6}\varepsilon'_2 + \frac{1}{3}\varepsilon'_3\right)^2 + 2\left(-\frac{1}{3}\varepsilon'_2 + \frac{1}{3}\varepsilon'_3\right)^2\right) \tag{171}$$

From the derivative $\frac{\partial f_{el}^V}{\partial \varepsilon'_3}$ and $\sigma'_3 = 0$, the additional expansion in z'-direction can be gained, analogous to the last paragraph.

$$\varepsilon'_3 = -\frac{(C_{11} + 2C_{12} - 2C_{44})}{(C_{11} + 2C_{12} + 4C_{44})}(\varepsilon'_1 + \varepsilon'_2) = 2\frac{(C_{11} + 2C_{12} - 2C_{44})}{(C_{11} + 2C_{12} + 4C_{44})} \cdot \varepsilon_0 \tag{172}$$

For the total film expansion $\varepsilon'_{3,\text{tot}} = \varepsilon'_3 + \varepsilon_0$ this results in

$$\varepsilon'_{3,\text{tot}} = \left(1 + 2\frac{(C_{11} + 2C_{12} - 2C_{44})}{(C_{11} + 2C_{12} + 4C_{44})}\right) \cdot \varepsilon_0 \tag{173}$$

The in-plane stress is given by

$$\begin{aligned}
\frac{\partial f^V_{el}}{\partial \varepsilon'_1} = \sigma'_1 = {} & \left(\frac{1}{2}C_{11} + \frac{1}{2}C_{12} + C_{44} - \frac{1}{3}\frac{(C_{11} + 2C_{12} - 2C_{44})^2}{C_{11} + 2C_{12} + 4C_{44}}\right)\varepsilon'_1 \\
& + \left(\frac{1}{6}C_{11} + \frac{5}{6}C_{12} - \frac{1}{3}C_{44} - \frac{1}{3}\frac{(C_{11} + 2C_{12} - 2C_{44})^2}{C_{11} + 2C_{12} + 4C_{44}}\right)\varepsilon'_2
\end{aligned} \tag{174}$$

$$\begin{aligned}
\frac{\partial f^V_{el}}{\partial \varepsilon'_2} = \sigma'_2 = {} & \left(\frac{1}{2}C_{11} + \frac{1}{2}C_{12} + C_{44} - \frac{1}{3}\frac{(C_{11} + 2C_{12} - 2C_{44})^2}{C_{11} + 2C_{12} + 4C_{44}}\right)\varepsilon'_2 \\
& + \left(\frac{1}{6}C_{11} + \frac{5}{6}C_{12} - \frac{1}{3}C_{44} - \frac{1}{3}\frac{(C_{11} + 2C_{12} - 2C_{44})^2}{C_{11} + 2C_{12} + 4C_{44}}\right)\varepsilon'_1
\end{aligned} \tag{175}$$

And for the case of hydrogen-induced lattice expansion by $\varepsilon'_1 = \varepsilon'_2 = -\varepsilon_0$ this results in the lateral stress

$$\sigma'_1 = \sigma'_2 = -\left(\frac{6C_{44}(C_{11} + 2C_{12})}{C_{11} + 2C_{12} + 4C_{44}}\right)\varepsilon'_0 \tag{176}$$

For the (111)-oriented cubic film this holds for all in-plane directions.

25.4.4.1.7   *(0001)-Oriented Hexagonal Films*

Hexagonal films often grow (0001) oriented, and the z-axis of the crystal correlates with the z'-direction of the film. For this orientation, the internal and the external coordinate systems coincide and, therefore, no matrix transformation is needed.

The free energy density can be determined by using the stiffness tensor of the hexagonal system Eqn (133), and the assumption $\varepsilon'_4 = \varepsilon'_5 = \varepsilon'_6 = 0$. (Sander, 1999)

$$f^V_{el} = \frac{1}{2}C_{11}\left(\varepsilon'^2_1 + \varepsilon'^2_2\right) + \frac{1}{2}C_{33}\varepsilon'^2_3 + C_{12}\varepsilon'_1\varepsilon'_2 + C_{13}\left(\varepsilon'_1\varepsilon'_3 + \varepsilon'_2\varepsilon'_3\right) \tag{177}$$

The additional vertical strain follows by using $\sigma'_3 = 0$, from the related derivative

$$\frac{\partial f^V_{el}}{\partial \varepsilon'_3} = \sigma'_3 = C_{33}\varepsilon'_3 + C_{13}\left(\varepsilon'_1 + \varepsilon'_2\right) = 0 \tag{178}$$

$$\varepsilon'_3 = -\frac{C_{13}\left(\varepsilon'_1 + \varepsilon'_2\right)}{C_{33}} \tag{179}$$

For the total vertical strain of the hydrogen-loaded film $(\varepsilon_1' = \varepsilon_2' = -\varepsilon_0)$ this results in

$$\varepsilon_{3,\text{tot}}' = \left(1 + 2\frac{C_{13}}{C_{33}}\right)\varepsilon_0 \tag{180}$$

Implementation of Eqn (179) simplifies the free energy density Eqn (177) yields

$$f_{\text{el}}^V = \frac{1}{2}C_{11}\left(\varepsilon_1'^2 + \varepsilon_2'^2\right) + C_{12}\varepsilon_1'\varepsilon_2' - \frac{C_{13}^2}{2C_{33}}\left(\varepsilon_1' + \varepsilon_2'\right)^2 \tag{181}$$

The derivatives give the in-plane stress

$$\frac{\partial f_{\text{el}}^V}{\partial \varepsilon_1} = \sigma_1' = C_{11}\varepsilon_1' + C_{12}\varepsilon_2' - \frac{C_{13}^2}{C_{33}}\left(\varepsilon_1' + \varepsilon_2'\right) = \left(C_{11} - \frac{C_{13}^2}{C_{33}}\right)\varepsilon_1' + \left(C_{12} - \frac{C_{13}^2}{C_{33}}\right)\varepsilon_2' = -\left(C_{11} + C_{12} - 2\frac{C_{13}^2}{C_{33}}\right)\varepsilon_0$$

$$\frac{\partial f_{\text{el}}^V}{\partial \varepsilon_2} = \sigma_2' = C_{11}\varepsilon_2' + C_{12}\varepsilon_1' - \frac{C_{13}^2}{C_{33}}\left(\varepsilon_1' + \varepsilon_2'\right) = \left(C_{12} - \frac{C_{13}^2}{C_{33}}\right)\varepsilon_1' + \left(C_{11} - \frac{C_{13}^2}{C_{33}}\right)\varepsilon_2' = -\left(C_{11} + C_{12} - 2\frac{C_{13}^2}{C_{33}}\right)\varepsilon_0 \tag{182}$$

for the hexagonal (0001)-oriented film.

25.4.4.1.8 Examples

First, the hydrogen-induced expansion factors have to be determined. Upon hydrogen loading the lattice expands homogeneously by Peisl (1978)

$$\frac{(V - V_0)}{V_0} = \left(\frac{\Delta v}{\Omega}\right) \cdot c_H \tag{183}$$

For different M–H systems, expansion factors are summarized by Peisl, as given in **Table 9**. For homogeneous hydrogen distribution this transforms into uniaxial lattice expansion

$$\frac{a^3 - a_0^3}{a_0^3} = \left(\frac{\Delta v}{\Omega}\right) \cdot c_H \approx \frac{3(a - a_0)}{a_0} = 3 \cdot \frac{\Delta a}{a_0}. \tag{184}$$

For Nb–H, the hydrogen-induced lattice expansion scales by $\varepsilon_0 = 0.058 \cdot c_H$. For a (110)-oriented epitaxial Nb–H film, the total vertical strain of a hydrogen-loaded Nb film Eqn (164) results in

$$\varepsilon_{3,\text{tot}}' = \left(1 + \frac{(245.6 + 3 \cdot 138.7 - 2 \cdot 29.3)}{(138.7 + 245.6 + 2 \cdot 29.3)}\right)0.0580 \cdot c_H = 0.137 \cdot c_H \tag{185}$$

For the in-plane stress of the Nb–H film, different stress values are obtained according to Eqns (165) and (166)

$$\sigma_1' = -4 \cdot \frac{(245.6 + 2 \cdot 138.7)29.3}{138.7 + 245.6 + 2 \cdot 29.3} \cdot [\text{GPa}]0.058 \cdot c_H = -8.03[\text{GPa}] \cdot c_H \tag{186}$$

$$\sigma_2' = -\frac{4 \cdot 138.7 \cdot 29.3 + 245.6^2 - 2 \cdot 138.7^2 + 245.6(138.7 + 2 \cdot 29.3)}{138.7 + 245.6 + 2 \cdot 29.3}[\text{GPa}] \cdot 0.058 \cdot c_H$$

$$= -11.3[\text{GPa}] \cdot c_H \tag{187}$$

Table 9   Theoretical lattice expansion of cubic bulk metals upon hydrogen (Peisl, 1978), resulting in vertical total strain $\varepsilon_{3,\text{tot}}$ and in-plane stresses σ_1, σ_2 and mean stress $\langle\sigma\rangle$, as calculated by the theory of linear elasticity on ideally clamped films (by Eqns (165) and (166))

Metal	$\frac{\Delta v}{\Omega}\big\|_{bulk}$	$\varepsilon_0 = \frac{\Delta a}{a_0}\big\|_{bulk}$	$\varepsilon_{3,tot}\big\|_{film}$	$\sigma_1[1{-}10]$ [GPa]$\cdot c_H$	$\sigma_2[001]$ [GPa]$\cdot c_H$	$\langle\sigma\rangle$ [GPa]$\cdot c_H$
Nb	0.174 ± 0.005	0.0580	0.137	-8.03	-11.3	-9.60
Ta	0.155 ± 0.005	0.0516	0.103	-16.9	-13.9	-15.4
V	0.19 ± 0.05	0.0633	0.136	-11.6	-13.3	-12.4
Pd	0.19 ± 0.05	0.0633	0.144	-18.0	-18.0	-18.0

The mean stress $\langle\sigma\rangle$ is, since these are the two extreme values and the stress curve is sinusoidal, the average of both. This mean stress is expected for a random in-plane orientation of a fiber-textured, (110)-oriented film.

$$\langle\sigma\rangle = -9.68[\text{GPa}]\cdot c_{\text{H}} \tag{188}$$

For different M–H systems, the calculated total vertical expansion and the mean in-plane stress are given in **Table 9**.

25.4.4.2 Experimental Data on Elastic Stress in Thin Films

Lattice expansion of films during hydrogen gas absorption has intensively been studied. Miceli et al. (1991) found that epitaxial Nb and Nb/Ta films expand strongly and only in the vertical direction, when loaded at low hydrogen gas pressures. At higher gas pressures they found both lattice expansion and lattice contraction, depending on the film direction. Some experimental data are exemplarily summarized in **Table 10**. For epitaxial Nb films and H concentrations below 0.02 H/Nb (determined by N[15] method, see Section 25.2.9), Reimer et al. (1993a, 1993b) reported on an unusual large vertical expansion factor ($\alpha_{\text{H}} = 0.53$). Up to now, there is no explanation for this unexpected large expansion. Laudahn et al. (1999a) determined the lattice expansion of 200 nm Nb films with microstructure ranging from the epitxial film to the nanocrystal. They experimentally obtained expansion factors of $\alpha_{\text{H}} = 0.10$ to 0.14. Nanocrystalline films showed, with $\varepsilon_{\text{H}} = 0.14$, the best agreement with the theoretical expansion factor $\alpha_{3,\text{tot}}\big|_{\text{film}} = 0.136$. Thereby, Laudahn et al. (1999a) confirmed the elastic behavior of these films in the low concentration regime. The measured strain evolution upon hydrogen uptake on a nanocrystalline Nb film is shown in **Figure 39**. Dornheim (2002) determined, for a 100 nm epitaxial Nb film, also a reduced expansion factor of $\alpha_{\text{H}} = 0.11$. The reduced expansion factors can, for this case, be explained by an open film morphology allowing lateral expansions of the upper film regions.

In contrast to this, Dornheim (2002) determined for a 27 nm Nb film deposited on a mechanically soft 20 μm polycarbonate substrate, bulk-like film expansion with an expansion factor of only $\alpha_{\text{H}} = 0.055$. This verifies the strong impact of the substrate mechanical properties on the film expansion.

For low concentrations, the linear compressive stress increase was also verified experimentally, in good agreement with the expectations of the theoretical calculations (Laudahn, 1998). Some experimental results on Nb–H, Y–H and Pd–H are summarized in **Figure 40**, for the low concentration regime. The linearity of the initial stress increase is visualized by straight lines.

Often, the experimentally measured slope of the stress increase is slightly reduced compared with the predictions resulting from theory of linear elasticity. This might be explained with a different hydrogen solubility of the metal due to the mechanical stress (Weissmüller and Lemier, 1999; Weissmüller, 2001), but also by an enlarged free volume compared with bulk metal, caused by additional grain boundaries

Table 10 Comparison between measured and calculated stress and strain data in a 200 nm thick polycrystalline, fiber-textured[1] M–H films (M = Pd, Nb, V, Y) following hydrogen loading in the low concentration regime. calculation database on theory of linear elasticity on a film ideally fixed to an elastically hard substrate

	$E/(1-\nu)$ [GPa]	$f = \varepsilon_{zz}/c_H\|_{theor}$ [1]	$f = \varepsilon_{zz}/c_H\|_{exp}$ [1]	References	$\sigma_{xx}/c_H\|_{theor}$ [GPa]	$\sigma_{xx}/c_H\|_{exp}$ [GPa]	References
Pd(111)	286	0.126	–	–	–18.0	–13.7	Pundt et al. (2000a, 2000b, 2000c)
Nb(110)	133	0.136	0.10–0.14 $c < 0.08$ H/Nb 0.53 $c < 0.02$ H/Nb	Laudahn et al. (1999a); Reimer et al. (1993a, 1993b)	–9.6 –	–6.9 to –8.7 –	Pundt (2005); Laudahn et al. (1999a)
V(110)	203	0.137	0.058	Gemma (2011)	–12.5	–7.6 to –12.2	Gemma (2011)
Y(0001)	187	0.061(3)	0.045	Dornheim et al. (2003)	–1.6	–1.3	Dornheim et al. (2003)

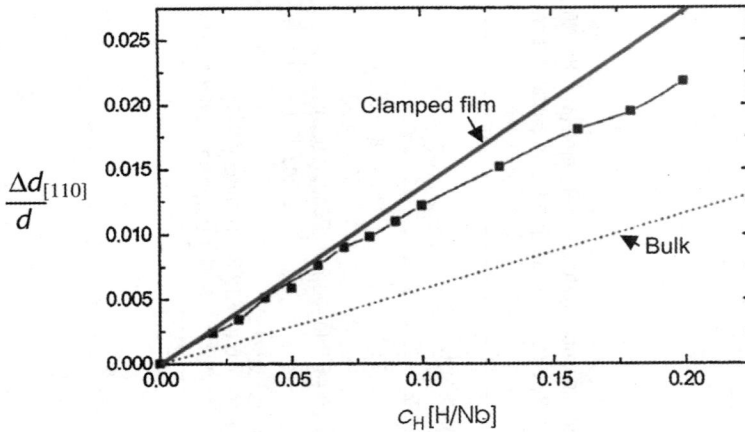

Figure 39 Lattice expansion of a nanocrystalline 200 nm Nb film, depending on the hydrogen concentration. Data after Laudahn et al. (1999a). (For color version of this figure, the reader is referred to the online version of this book.)

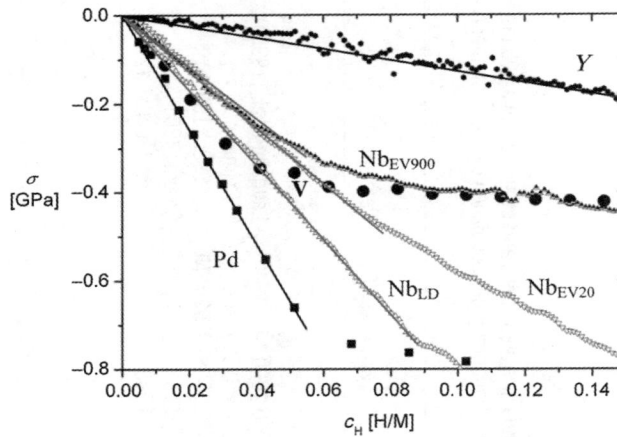

Figure 40 Compressive stress change upon hydrogen uptake, as measured for 200 nm Niobium Nb, Yttrium Y and Palladium Pd thin films (Laudahn, 1998; Pundt et al., 2000a, 2000b, 2000c; Dornheim et al., 2003; Wagner, 2014). The initial linearity of the measured compressive stress σ on the concentration c_H confirms linear elastic behavior of the films (abbreviations used: LD laser deposited at 20 °C; EV20, electron evaporated at 20 °C; EV900, electron evaporated at 900 °C, Pd, V and Y: sputter deposited at 400 °C, 20 °C and 20 °C, respectively).

and dislocations. The theoretical value of the ideal elastic stress increase, therefore, can be regarded as an upper stress value (Huang and Spaepen, 2000). Also the film topography has to be considered. Only for smooth film surfaces the theoretical predictions are expected to be experimentally measured.

For films of about 50 nm thickness or less, the linear elastic behavior can range up to high concentrations. For thinner films, the elastic energy W_{el} stored in the film, $W_{el} = \dfrac{E}{2(1-v^2)} \cdot \varepsilon^2 dA = \dfrac{\sigma^2}{2E} \cdot dA(1-v^2)$, is smaller compared with that of thicker films, as it scales with the film thickness d (A is

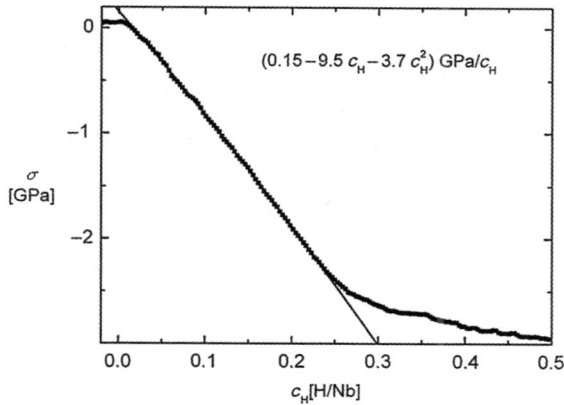

Figure 41 Stress evolving during hydrogen absorption in a 50 nm Nb film that is partly surface oxidized. Dislocation formation occurs at 0.25 H/Nb. The stress dependency in the linear elastic range reflects the H concentration dependency of the elastic constants. After Pundt et al. (2000a, 2000b, 2000c).

the film area, ν is the Poisson ratio). Thus, to supply the energy needed for dislocation formation, a higher stress σ is required. This goes hand in hand with a higher hydrogen concentration (see, e.g. Eqn (188)). The critical stress for dislocation formation can be shifted even upward when the film surface is oxidized (Nix, 1989; Jain et al., 1997). For these high concentrations the change of elastic constants upon the hydrogen concentration is already visible in a curve bending, in the elastic range. This can be seen in **Figure 41** where measured stress evolving during hydrogen loading of a 50 nm Nb film is shown. A fitting line is also implemented in the graph.

The line reproduces the concentration dependency published for bulk Nb–H for concentrations up to 0.25 H/Nb (Pundt et al., 2000a, 2000b, 2000c). On the one hand, the measurement shows that elastic properties of Nb–H do not change compared with bulk, even when the film thicknesses is reduced down to 50 nm. On the other hand it demonstrates that stress measurements on thin M–H films can be used to measure elastic constants of the material (Laudahn et al., 1999b).

For hydrogenated V films, the stress change and the lattice expansion is found to behave more complicated. Even though the calculated elastic linear compressive stress increase (with $E/(1 - \nu)$ from Bradfielck, 1964) is found in certain low concentration regions, also tensile stress increase is detected in the experiments (Gemma, 2011). Especially for low concentrations, this is attributed to the hydrogen-enhanced formation of vacancies as expected also by the defectant model (compare with Section 25.3.3.2.1). But, a strong thickness dependency is reported. For multilayered V/Fe film packages this even leads to tensile stress increase at higher concentrations (Gemma, 2011). According to Hjörvarsson this might be attributed to a preferential O_z-site occupation in the Fe/V superlattice (Olsson et al., 2001). This topic actually is still under discussion.

25.4.4.3 Stress in Clusters

For hydrogen-loaded clusters mechanical stress results from two different origins: First, from fixing to the stiff stabilizer or the surrounding scaffold and, second, from the strained conditions in surface and subsurface shells, especially when hydrogen is absorbed. The surface region absorbs large amounts of

---

[1] Grains of textured films are oriented along the vertical (out-of-plane) film direction, but of random orientation in plane. The out-of-plane alignment is driven by surface energy minimization.

hydrogen at low chemical potentials, when the concentration in the core is still small. Thereby, the surface shells expand and elastically strain the inner core of the cluster (compare with the model of Weissmüller, 2001). Like in films and nanocrystalline material this might also affect the clusters' hydrogen solubility. Volume expansion, furthermore, changes the electronic DOS. Expansion here results in band narrowing, or, for smaller clusters, in reduced niveau splittings. From this simple argument, band filling with electrons might cost less energy.

The presence of mechanical stress becomes visible for clusters embedded in different matrices (Suleiman et al., 2005). During phase transformation, the plateau pressure of M–H clusters increases with the concentration. The increase depends on the stabilizer (Suleiman et al., 2005). The arising stress in clusters, as originating from the surrounding stabilizer that is attached to the cluster surface, is difficult to determine from the experiments.

25.4.4.4   *Stress-Release Mechanisms*
25.4.4.4.1   *Film Plasticity*

The linear elastic stress increase holds up to the "yield hydrogen concentration" of the films (Laudahn et al., 1999a; Pundt, 2004; Pundt, 2013). For many different films, stress release occurs via dislocation formation. Therefore, the yield hydrogen concentration can be linked to the yield stress $\sigma_{y,tot}$ of the film when the intrinsic stress σ_i and the yield stress σ_y is considered.

$$\sigma_{y,tot} = \sigma_y + \sigma_i \qquad (189)$$

Often, the intrinsic stress in films is small. **Figure 42** summarizes stress–concentration curves as measured for 200 nm Nb films of different microstructures, curves (a) to (c). For all three films, at about $c_y \approx 0.08$ H/Nb marked by a vertical line, the linear compressive stress increase ends with an abrupt

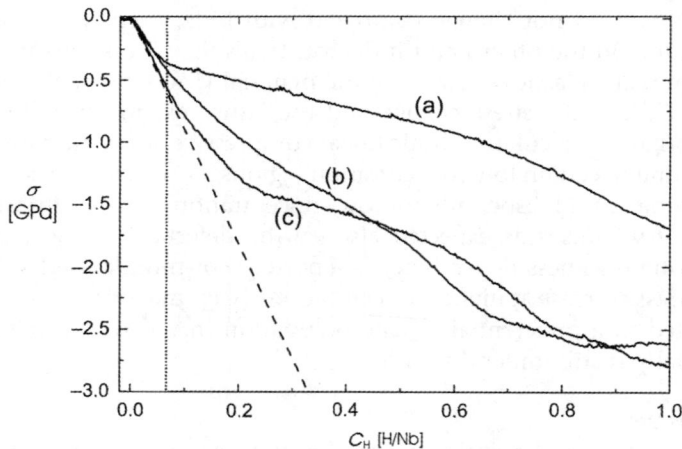

Figure 42   Stress–concentration curves for 200 nm Nb films with different microstructures due to different preparation methods: (a) molecular beam epitaxy at 900 °C (large epitaxial domains); (b) electron evaporation at 20 °C (grain size about 50 nm), and (c) laser deposition at 20 °C (grain size less than 20 nm). The stress curves of all films leave the ideal elastic behavior (dotted line) at approximately 0.08 H/Nb (*shaded area*), marking the onset of stress-releasing processes. Data from Laudahn et al. (1999a).

slope change. Above the yield concentration, the compressive stress increases with a reduced slope. Thus, the yield concentration marks the onset of stress release in the film.

For the epitaxial Nb film (a) $\sigma_i \approx +0.2$ GPa. With $\sigma_y = -0.3$ GPa at 0.06 H/Nb, this results in a film's yield stress $\sigma_{i,tot} \approx -0.1$ GPa. For the nanocrystalline Nb film (b) $\sigma_i \approx -0.3$ GPa. With $\sigma_y = -0.4$ GPa at 0.08 H/Nb, this results in a larger yield stress of $\sigma_{y,tot} \approx -0.7$ GPa. For film (c) with $\sigma_i \approx -0.3$ GPa and $\sigma_y = -0.7$ GPa at 0.10 H/Nb, this results in a larger yield stress of $\sigma_{y,tot} = -1.0$ GPa. Thus, the yield stress in Nb films is smallest for epitaxial films, and largest for nanocrystalline films.

The result is in accordance with results of tensile-tested bulk samples, when the grain size dependency is regarded. Single crystals have the smallest yield stress, while nanocrystals yield the highest stress, as described by the Hall–Petch relation (see, e.g. Haasen, 1994). For nanocrystals, dislocation glide is hindered by the presence of grain boundaries. This is also applicable for film samples, as studied by Venkatraman and Bravman (1992) or Arzt (1998).

Furthermore, stress release is most efficient for epitaxial films (curve a) in **Figure 42**). The slope change is strongest for this type of films (a) and the final stress is the smallest. It is -1.7 GPa while it is -2.8 GPa for the nanocrystalline film. This result is also in accordance with mechanical properties measured in a tensile test on Nb-bulk samples (Ravi and Gibala, 1971) (see **Figure 43**). Because of the high mobility of dislocations in single crystals, initial stress release is best, here.

Similarities to further stages in bulk stress–strain curves are not discussed, here, because similarities between tensile-tested bulk samples (with no hydrogen involved) and films compressed by hydrogen absorption should not be stressed over the solid solution limit of the M–H system. Above that limit, the presence of hydride precipitates will result in additional dislocations, since there is a large misfit between the matrix solid solution phase and the expanded hydride phase. This contribution of hydrides on the stress release can be seen in the additional slope change of film (c) in **Figure 42**, at $c_H = 0.2$ H/Nb. This is the α-phase solubility limit for this type of film. For the epitaxial film (a), the yield concentration coincides with the phase boundary which is at $c_H = 0.08$ H/Nb.

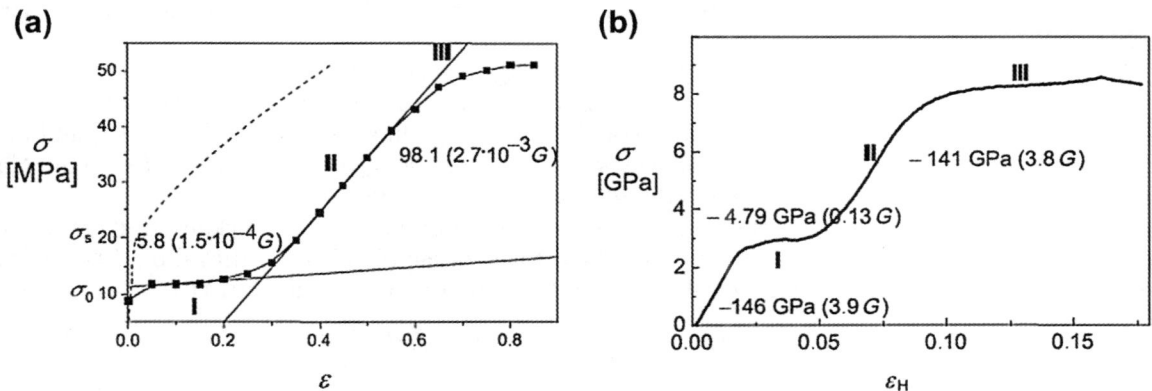

Figure 43 (a) Tensile test stress–strain curves of an Nb single crystal samples (after Ravi and Gibala, 1971). Included is further a typical stress–strain curve for a polycrystalline crystal (dotted line). (b) Hydrogen-induced stress–strain curve for a nanocrystalline Nb film (200 nm). The "strain" ε_H is calculated by assuming elastic lattice expansion with $\varepsilon_H = 0.058 \, c_H$. The curves resemble by shape, but the stress in thin films is much larger than that obtained in the single crystal. After Pundt (2005).

25.4.4.4.2 Determination of the Yield Concentration in Films

For films, stress release via misfit dislocations is intensively studied for films growing on substrates with a certain lattice misfit, as discussed in Section 25.4.1. For a growing film, dislocation formation sets in above a critical thickness d_{crit}, as described by Eqn (117) for equilibrium conditions.

The concept of dislocation formation during film growth can be transferred to hydrogen-loaded thin films as suggested by Pundt et al. (2000a, 2000b, 2000c, 2006). Upon hydrogen loading, the misfit between the film and the substrate changes because only the film lattice is strained by hydrogen atoms. Thereby, the elastic energy stored in the film changes and, finally, exceeds the self-energy of a dislocation. At this "yield concentration" $c_{H,y}$, the presence of a dislocation becomes energetically favorable. Above that value, plastic deformation sets in. In experiments, the yield stress is usually higher than the theoretical value determined by Eqn (117). When dislocations also need to nucleate, a nucleation barrier has to be considered. Furthermore, dislocation motion can be hindered by obstacles or other dislocations. Both terms increase the critical film thickness.

For hydrogen-loaded films, the total film thickness is constant and hydrogen loading results in an increase of the misfit $\delta = \varepsilon_0 = \alpha_H \cdot c_H$. According to Eqn (117), dislocation presence, therefore, is linked to a certain hydrogen concentration $c_{H,y}$.

$$c_{H,y} = \frac{\mathbf{b}}{\alpha_H \cdot d} \frac{1}{8\pi(1+\nu)} \ln\left(\frac{\xi d}{\mathbf{b}}\right) \tag{190}$$

where d is the film thickness, $\mathbf{b}$ is Burgers vector, α_H is the bulk lattice expansion factor upon hydrogen loading. Up to the yield concentration the film strains elastically. Therefore, the related mechanical stress can be determined by

$$\sigma_{H,y} = -\frac{E}{1-\nu}\alpha_H c_{H,y} = -\left(\frac{E}{1-\nu}\right)\cdot\frac{\mathbf{b}}{d}\frac{1}{8\pi(1+\nu)}\ln\left(\frac{\xi d}{\mathbf{b}}\right) \tag{191}$$

According to the formula, thinner films yield larger hydrogen concentrations $\sigma_{H,y}$ and thereby, mechanical stress $\sigma_{H,y}$, before stress release sets in (in accordance with the discussion in Section 25.4.4.2). In Section 25.4.4.1.1 we also showed that thin films yield much higher elastic stress than thick films when fixed to a hard substrate.

When grain sizes become small, the yield stress is strongly influenced by the grain size. Reisfeld et al. (1996) verified this strong influence of the grain size on the physical properties of Nb–H, by measuring Nb films between 20 and 200 nm thickness. For all films, the grain size was about 7–17 nm, and the isotherms were identical. Furthermore, oxide surface layers increase $c_{H,y}$ or $\sigma_{H,y}$ as shown for 50 nm Nb covered with an oxide layer in **Figure 41** (Pundt et al., 2000a, 2000b, 2000c).

Up to now, the inverse dependency of the yield concentration on the film thickness and grain size expected from Eqn (118) is qualitatively confirmed for hydrogen-loaded Nb thin films (data from Pundt et al., 2000a, 2000b, 2000c, Dornheim, 2002) and Hamm and Pundt (2014).

Nb films	d [nm]	Lateral grain size [nm]	$\sigma_{H,y}$ [GPa]
Epitaxial	200	Large	−0.1
Nanocrystalline (EV20)	200	200	−0.3
Nanocrystalline (LD20)	200	20	−0.7
Nb	90		
Epitaxial	50	Large	−2.4(±ca. 0.2)
Epitaxial	10	Large	−6.0(±ca. 0.5)

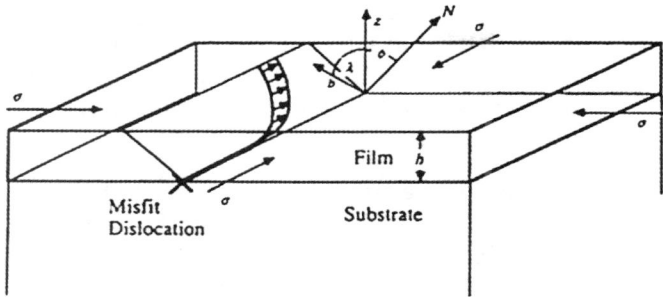

Figure 44 Dislocation glide in a thin film. The threading part links the stress-releasing part with the film surface. Hirsch (1991).

The yield stress can also be seen in measurements of Reimer et al. (1993a, 1993b) and Song et al. (1996): below the yield concentration the rocking curve is small and it even sharpens (Reimer et al., 1993a, 1993b); above the yield concentration the rocking curve broadens (Song et al., 1996). This can be attributed to the presence of dislocations.

Dislocations in thin films affect the film surface. During dislocation formation and migration, dislocation lines elongate at the interface between the film and the substrate thereby releasing stress between the film and the substrate (Jain et al., 1997; Nix, 1989). These propagating dislocations are linked to the film surface as shown in the sketch of **Figure 44**, taken from Hirsch (1991). Therefore, in the plastic regime, glide steps appear during hydrogen loading at the film surface.

Recently, hydrogen-induced glide steps were studied by STM on different epitaxial films (Gd–H, Nb–H, Pd–H) (Pundt et al., 2000a, 2000b, 2000c; Nörthemann et al., 2003, Nörthemann and Pundt, 2008; Nörthemann and Pundt, 2011; Wagner et al. 2011). Number density and length of glide steps increase when increasing the hydrogen content. STM micrographs of a 12 nm epitaxial Gd film on a W

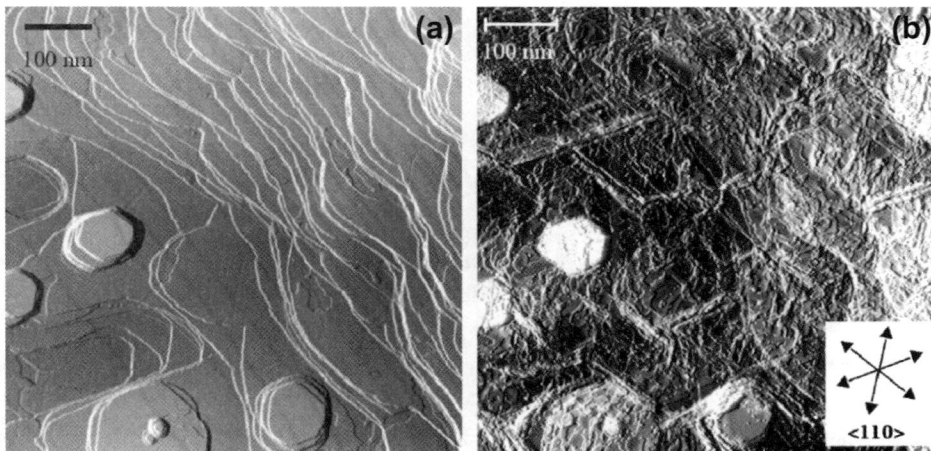

Figure 45 (a) STM image of a Gadolinium film (12 nm) epitaxial grown on a Tungsten substrate. Surface steps are visible. (b) The same position after pronounced hydrogen offer. Glide lines occur in ⟨110⟩ directions, marking plastic deformation of the film. After Pundt et al. (2000a).

substrate, before and after pronounced hydrogen loading, are shown in **Figure 45** (after Pundt et al., 2000a, 2000b, 2000c). In **Figure 45**(b) the film surface is completely covered with glide steps, visible as straight lines. For the {111}-oriented fcc Gd–H_2 film, gliding occurs on {111}-glide planes leaving straight surface steps in ⟨110⟩-directions (Pundt et al., 2000a, 2000b, 2000c).

25.4.4.4.3 Twinning and Film Detachment

Another mechanism of stress release is stress-induced twinning. Twins have been found in high-resolution electron microscopy (HREM) studies of Grier et al. (2000) for hydrogenated Ho films and Kerssemakers et al. (2000a, 2000b) for hydrogenated Y films. But, these samples are thinned for HREM study and the film conditions differ to the clamped film. Up to now, the question remains how the interface between a twining film and the substrate can remain intact upon stress release by twinning.

When the adhesion between the film and the substrate is small, the films tend to detach at certain local positions from the substrate (Pundt and Pekarski, 2003, 2004a, 2004b, 2004c, 2007). Thereby, the films locally form buckles that can have circular or elongated morphologies (Colin et al., 2000; Yu et al., 1991; Gioia and Ortiz, 1997; Song et al., 1999; Crosby and Bradley, 1999; Audoly, 1999). For the elongated morphologies, straight-sided buckles (Euler mode), telephone cord or undulated blisters are reported with a length in the mm range and a width of μm size (**Figure 46**).

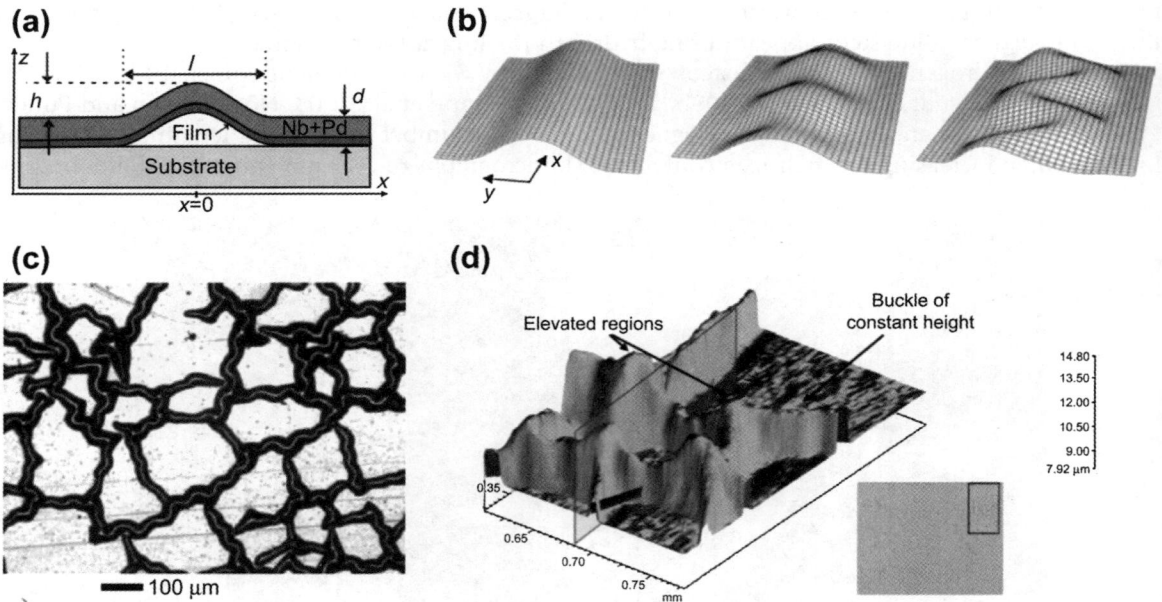

Figure 46   (a) Side view of a local buckle (after Pundt et al., 2007). (b) Buckle shapes resulting from finite element model calculations, after Moon et al. (2004). The film is detached from the substrate by the length l forming a buckle of height h. Hydrogen-induced telefone cord buckles as measured (c) by light microscopy (Pundt and Pekarski, 2003) and (d) by white light interferometry (Pundt et al., 2007). The buckled film is still intact. A modulation of the buckle height is further detected. (For color version of this figure, the reader is referred to the online version of this book.)

As the adhesion between the film and its substrate fails as soon as the elastic energy accumulated in the film exceeds the adhesion energy, hydrogen loading was suggested to be used as a measure for adhesion energies (Pundt and Pekarski, 2003; Pundt et al., 2004a, 2004b, 2004c). Hydrogenography measurements (**Figure 51**) and coincident elastic proton–proton scattering analysis (pp scattering, see Section 25.2.9) have recently shown that the film detachment strongly affects the local hydrogen distribution in the two-phase region. Buckled film regions are in the hydride phase while attached film regions contain α-phase (Kürschner et al., 2014; Wagner et al., 2013).

25.4.4.4.4 Stress Release in Clusters

Clusters are expected to strain elastically upon hydrogen loading. From Pressure Composition Temperature (PCT) curves or XRD, no hint on stress release in clusters is detected. The curves monotonously change upon hydrogen loading and step-like behavior is not visible. However, such measurements average over a large ensemble of clusters and XRD peaks are very broad. This, maybe, covers interesting properties of cluster. Thus, future in situ measurements especially with environmental TEM on individual clusters should clarify the behavior of clusters.

25.4.4.4.5 Stress Impact on Free Energy Curves and Chemical Potentials

Isotherms of thin films fixed to hard substrates show increasing chemical potentials in the plateau region. Here, a linear increase is expected when mechanical stress is considered. For hydrogen-loaded films, it is expected that the strain energy increases with the square dependency on the hydrogen concentration (with Dornheim, 2002). This shifts the Gibbs free energy F of the hydride more strongly compared with the solid solution. The chemical potential at the phase transition, thereby, is increased compared with the stress-free state. When the free energy curve of the hydride phase is lifted by square dependency on the concentration with regard to the free energy curve of the solid solution, the slope of the forming double tangents scales linearly on the hydrogen concentration. This is schematically shown in **Figure 47**.

Experimentally an increase of about $+3$ kJ/mol has been reached by hydrogen-induced stress for the palladium hydrogen system (Wagner and Pundt, 2008; Pundt, 2013).

Because of the same argument, plateau regions for films stressed by hydrogen are expected to be sloped. Measurements on hydrogen gas-loaded Nb films have confirmed that in this case the phase transition even stops under open conditions for pressures in the phase transition region

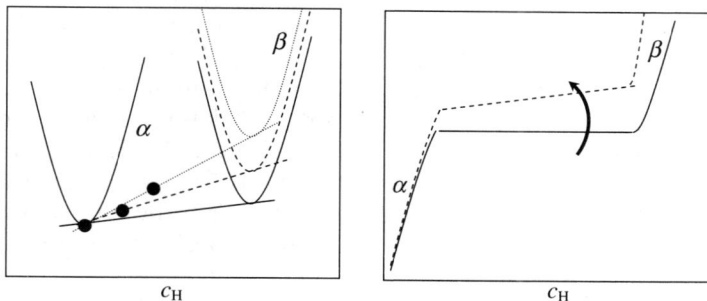

Figure 47 Change of the free energy F upon increasing hydrogen concentration c_H, when the system is fixed to a substrate. The strain energy contribution gradually increases the free energy of the final phase β. This results in an increasing chemical potential μ in the two-phase region.

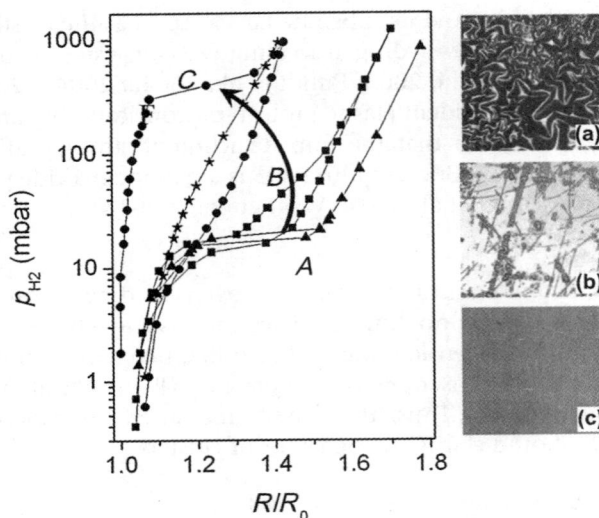

Figure 48 Mechanical stress impact on plateau pressures for about 200 nm Pd (2%Fe) films. The plateau region is visible by a strong increase in the relative resistivity R/R_0. For a highly delaminated film (A) the plateau pressure is close to the bulk value of 18 mbar. For the strongly bond film (C), it increases to 400 mbar during first loading. For further details see text. After Wagner and Pundt (2008). (For color version of this figure, the reader is referred to the online version of this book.)

(Nörthemann and Pundt, 2011). This is not known from (stress-free) bulk samples, where for an open system the sample always completely transfers into the hydride phase.

Consequently, the chemical potential increase in the plateau region can be reduced by reducing the mechanical stress. If the stress completely released, like for the case of a film completely delaminated from the substrate, it results in a flat plateau pressure. Such a flat plateau was verified for completely delaminated Pd film, reproducing the plateau pressure of Pd–H bulk of 18 mbar, **Figure 48**(A). Wagner and Pundt (2008) determined an increased plateau pressure of 400 mbar for clamped Pd–H films, shown in **Figure 48**(C), when loaded with hydrogen for the first time. Stress release by plastic deformation results in a reduced plateau pressure, but it is still larger than that of bulk Pd–H. For the second loading, the plateau region is strongly sloped. Its mean value of about 100 mbar is closer to that of bulk. For a partly detached film (B), two different plateau regions were detected. Their contents are related to the volume fractions of detached and attached film.

For in situ delamination, a drop in the chemical potential should result. This was confirmed for Pd-capped Nb films deposited on polycarbonate PC substrate by Nikitin (2008) (**Figure 49**). The adhesion energy between the three-layered film and PC is small. Therefore, only a small elastic energy is required to detach the three-layered film from the PC (compare Section 25.4.4.4.3). The measurement was done by stepwise electrochemical hydrogen loading. Above 0.13 H/Nb the film starts to detach from the substrate by forming buckles. This was verified by simultaneously monitoring the sample by light microscopy.

For M–H clusters, the plateau pressure is always increasing in the two-phase region (compare **Figure 32**, **Figure 33**). The origin of this increase (slope) is still under debate. The discussion includes the impact of the broad size distribution of clusters on the slope because of the size dependency of the surface tension (Salomons et al., 1987). But, since similar slopes were found in isotherms of

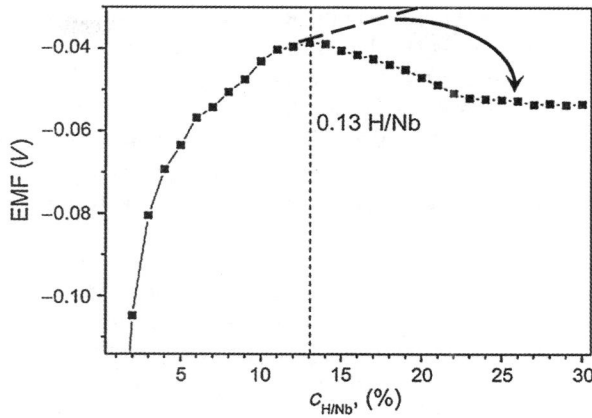

Figure 49 Mechanical stress impact on chemical potential as measured by the EMF, for a Pd/Nb/Pd//PC stack by stepwise hydrogen loading (EMF relative to an Ag–AgCl standard electrode). Above 0.13 H/Nb the metal stack starts to detach from the PC substrate by forming buckles. This was verified by simultaneously monitoring the sample by light microscopy. Above this concentration the EMF drops because of stress release, as marked by the arrow. The typical EMF development for an attached stack is sketched with the dashed line, for comparison. Adopted from Nikitin (2008).

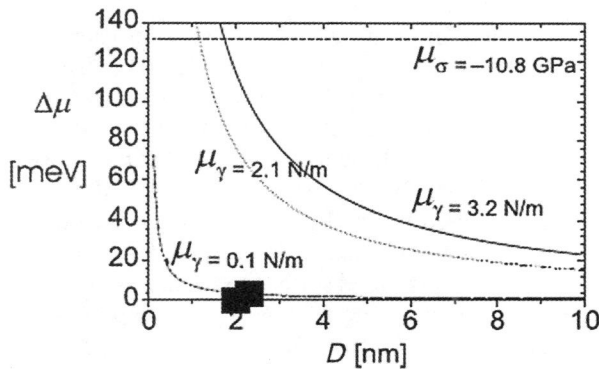

Figure 50 Size dependency of chemical potentials for clusters, after surface tension models by Salomons et al. (1987) with $\gamma = 3.2$ N/m, and Miedema using $\gamma = 2.1$ N/m. Experimental data on single-sized clusters yield much smaller shifts (bold squares, data from Sachs et al., 2001a, 2001b), hinting on $\gamma = 0.1$ N/m $= 100$ mN/m for such models. After Pundt (2005).

single-sized clusters (Sachs et al., 2001a, 2001b) this contribution was evaluated to be small. Surface tension is discussed to change plateau pressures. For 7.8 nm clusters, Salomons et al. (1987) measured an energy change $\mu = 31.7$ meV. According to Eqn (121) this results in a surface tension of 3.4 N/m (Pundt, 2005)—which is a rather large value. Curves, considering surface tension, are plotted in **Figure 50** (Pundt, 2005). Experimental data on single-sized clusters are also implemented. The results suggest that surface tension models with surface tension of about 0.1 N/m result in good description of the experimental values. This value of 100 mN/m further meets the expectations for surface tension data. As can be further seen, only for diameters below 2 nm, the surface tension noticeably contributes to the chemical potential of clusters. For larger clusters of ellipsoidal geometry studied by Kishore et al. (2005), the plateau pressure increase is smaller than that found for clusters.

However, the isotherm slope in the plateau region changes when clusters of comparable size were embedded different matrixes (Suleiman et al., 2005). This verifies the impact of the stabilizer on the isotherm. The origin of this impact can be electronically and mechanically (Ramos, 2002; Tannenbaum et al., 2004). Systematic studies on Pd–H clusters differentiating between the two contributions and the study of the impact of the stabilizers on metal properties is a topic of current research.

25.4.4.5 *Hydride Morphologies*

Different hydride morphologies are reported for thin films. For Nb films, hydrides with columnar morphology are reported that range through the complete film. Plate-like precipitates with aspect ratios of 5–10 were also reported for epitaxial Nb films (Heuser et al., 2002). For Y–H and Ho–H, triangular surface pattern are reported that might result from hydrides of even more complex morphology. Their borders lay in ⟨110⟩-directions of the cubic lattice (Grier et al. (2000); Keersemakers et al., 2000). Y–H and Ho–H both form dihydrides with an fcc lattice structure, and trihydrides with hexagonal lattice structure. For Pd films, also columnar hydrides ranging through the complete film were found. For Mg_2Ni–H, Lohstroh et al. (2004) reported on a layered hydrogenation surprisingly starting at the substrate–film interface. Actual studies point toward a strong impact of mechanical stress also on the hydride morphology (Nörthemann and Pundt, 2008; Waninger and Pundt, 2013; Kürschner, 2014). For buckled thin Nb films, preferential hydride formation was found in the stress-released buckled film volumes (Wagner et al., in preparation).

The hydride formation in clamped film can be studied by the film surface topography which changes when hydrides form inside of the film (Nörthemann and Pundt, 2011). During hydrogen absorption and hydride formation, the lattice expands predominately in vertical direction. Thus, local hydride formation inside of the film becomes visible as a local height increase at the film surface. For a concentration in the two-phase region of the phase diagram, hydride formation and surface modification occur localized in certain film regions. For epitaxial Nb films, Nörthemann and Pundt (2011) detected two different types of surface topographies that were explained by the hydrides' coherency to the surrounding matrix. In this work, finite element model calculations were used to link the surface topographies to internal hydrides of different morphologies. Small sphere cap-shaped surface topographies were found that relate to coherent hydrides of cylindrical morphology. Furthermore, large surface topographies were reported. They are extended in the films' elastically soft directions. This surface topography related to large incoherent hydrides (with lateral diameters of several tens of nanometers) scales with the film thickness (Nörthemann et al., 2003; Nörthemann and Pundt, 2008). This is in good agreement with the theory of linear elasticity when columnar hydrides are present that extend the film thickness (Nörthemann et al., 2003). By comparing the height changes for Y–H (Dornheim, 2002) and for Pd films (Wagner et al., 2011), the same argument holds.

Mooij and Dam (2013) exploited the change of the optical transmission upon hydrogen loading (by "hydrogenography") for 3 and 10 nm Mg films, to monitor the hydride extension. Large circle areas are reported representing the hydride film volume, as shown in **Figure 51**(a). This corresponds well to earlier reports on small number of hydrides forming in the Mg–H bulk system (Vigeholm et al., 1987). By using "hydrogenography", Kürschner et al. (2014) showed for hydrogenated Pd films that the hydride precipitate size strongly depends on the local adhesion, **Figure 51**(b)–(d). When the film is attached to the substrate, the extensions of hydride precipitates are too small to be detectable in a light microscope, **Figure 51**(d). But, as soon as the film locally detaches and forms buckles (compare Section 25.4.4.4.3), large hydrides with about 10 μm in diameter appear that preferentially show

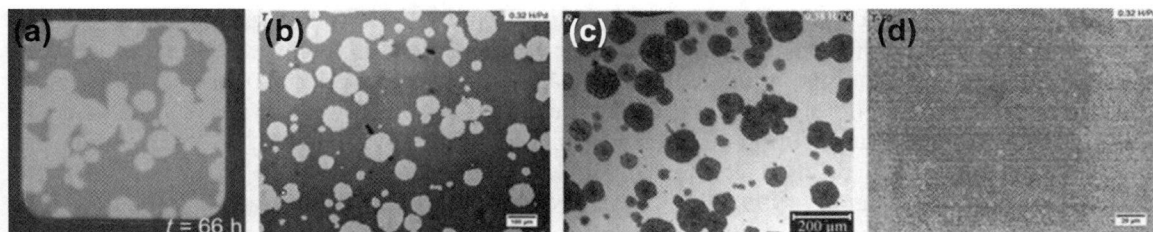

Figure 51 Hydride extensions as determined by, hydrogenography, in optical transmission and in reflectance. (a) For a 10 nm Mg film, after 66 h at 70 Pa and 90 °C. Frame size is 9 × 9 mm (from Mooij and Dam, 2013). (b)–(d) micrographs for a 30 nm Pd film by electrochemical hydrogen loading at 0.32 H/Pd were shown: (b) for the locally detached film in transmission and (c) the same position in reflectance. Part (d) shows the situation for the attached film at the same hydrogen concentration in transmission. Here, the extensions of hydride precipitates are too small to be detectable in a light microscope. For the Pd films, the strong impact of the adhesion condition on the lateral distribution and extension of hydrides can be directly seen (Kürschner et al., 2014).

circular growth, **Figure** 51(b). In this stage, many hydrides have already reached a diameter of about 100 μm. In reflectance (**Figure** 51(c)), the local detachment at the very positions of the hydride precipitates can be clearly seen. The extension of these local detachments is typically in the μm range (compare Section 25.4.4.4.3). Any nucleation and growth kinetics of hydrides is strongly influenced by the stress condition. Nonconsideration of detachment effects can result in unrealistic nucleation barriers. If this also applies for the results on Mg films is a topic of actual research. A low number of hydride precipitates expanding from the Mg surface can, however, explain the blocking effect.

Hydrogen cycling results in surface morphology changes by local expansion or contraction. This is due to formation and dissolution of hydride precipitates that can be controlled by the hydrogen gas pressure. For Y films, Keersemakers et al. (2000) reported on transformation of single domains when concentration differences are kept small. Interestingly, inner domains transfer at different chemical potentials compared with the border parts. For thin Pd films, reversible changes of surface morphologies were recently detected by Wagner et al. (2011). For Nb–H films, Nörthemann and Pundt (2008) calculated a critical film thickness of about 27 nm, below which only coherent hydride precipitation might occur. Below this film thickness, only reversible changes of the surface topography are expected.

For clusters, the morphologies of hydrides are still an open question. For small clusters, it is expected that they switch by elastical transformation in the hydride phase. The surface shell with high local hydrogen concentration might act as nucleus, here. XRD studies of Suleiman et al. (2003, 2005) on different Pd–H clusters undoubtedly prove the existence of hydride formation for clusters down to 4.7 nm, at 300 K. XRD pattern confirm the cubic lattice structure of clusters down to that size.

25.4.4.6 *Hydride Formation Kinetics*

Slow kinetics can be a barrier to practical application when loading time becomes an important issue. Since the loading time t also depends on the diffusion length x, nanomaterials attract special attention (Pundt and Kirchheim, 2006; de Jongh and Adelhelm, 2010). Considering one-dimensional diffusion into a plate yields (compare Section 25.3.1), with $t = \frac{x^2}{2D}$, for hydrogen in Pd ($D = 3 \cdot 10^{-8}$ m^2/s at 300 K, compare **Figure 18**), a loading time $t = 17$ s is required for a diffusion length $x = 1$ mm and it is just $t = 17$ μs needed for $x = 1$ μm. Often, nanosized systems need to be protected to avoid oxidation. This inevitably results in a bi- or multilayered structure. For multilayered film systems, for example Schmitz et al. (1998) or Baldi et al. (2009) developed models to predict the total hydrogen diffusivity. One key issue for storage applications is to optimize storage capacity and hydrogenation kinetics, simultaneously.

Hydride formation kinetics, for instance, can be measured gravimetrically or gas-volumetrically by monitoring the data-evolution upon time (compare Sections 25.2.8 and 25.2.9, or de Jongh and Adelhelm, 2010). Such methods have been intensively applied for the study of ball-milled nano-crystalline powders (Huot et al., 1999; Huot, 2010). For thin films and multilayers, hydride formation kinetics can be measured by time-resolved permeation techniques either by using gas-phase loading or the electrochemical double cell (Boes and Züchner, 1976, Schmitz et al., 1998). Three sophisticated methods for studying hydride formation kinetics will be addressed, in the following.

The hydride formation kinetics on nanomaterials, namely thin films, can be studied by measuring the film surface topography changes upon hydrogenation by STM (compare Section 25.4.4.5). Nörthemann and Pundt (2011) determined for epitaxial Nb films hydride nucleation and growth kinetics, by using the theory of Johnson, Mehl, Avrami and Kolmogoroff (Christian, 1975). The studies revealed a good match between the theory parameters interpretation and the experimental results (Nörthemann and Pundt, 2011). This surprises since the theory was originally developed for bulk materials and not for thin films.

For 3 and 10 nm Mg films, Mooij and Dam (2013) exploited the change of the optical transmission (by "hydrogenography") during hydride formation for studying the phase transformation kinetics. Large circle areas are reported representing the hydrided film volume. The hydride precipitate diameters and also the distances among the hydride precipitates, surprisingly, are in the range of millimeters. Mooij and Dam analyzed the growth kinetics and explained their results by an extremely large edge boundary energy between the matrix and the by about 30% expanded thickness of the MgH_2 (Mooij and Dam, 2013; Fujii et al., 2002). These results are still under debate. As presented above, film detachment can change nucleation and growth strongly and result in similar findings. Film detachment is not visible in the optical transmission. Improvements on Mg hydrogenation kinetics are of special interest for hydrogen storage applications, since Mg has a very large gravimetric hydrogen storage capacity but very slow hydrogenation kinetics.

Also, XRD studies are performed for studying the hydrogenation kinetics of nanomaterials (Dornheim, 2002; Baldi et al., 2009; Uchida et al., 2011). Here, peak shifts as well as vanishing and developing peaks monitor the hydrogenation kinetics. Short loading times, as they appear for many nanomaterials require in situ studies at Synchrotron facilities to catch the fast kinetics of the nano-system with sufficient intensity.

25.4.4.7 *Hysteresis in Nanosized Metals*

For M–H systems, hysteretic behavior is commonly present between film loading and unloading in the region of the phase transition (compare Section 25.2.6). The origin of hysteresis in M–H systems was attributed to the generation and motion of dislocations (Birnbaum and Flynn, 1976; Flanagan et al., 1980). As shown above, dislocations also form in thin films upon hydrogen loading, when the elastic strain energy suffices to exceed the dislocation's self-energy (Section 25.4.4.4.2, Pundt et al., 2000a, 2000b, 2000c or Nörthemann and Pundt, 2008).

For some M–H films, exceptionally large hysteresis areas were reported. An origin for the enlargement can be a structural change in the film lattice, which needs to stay fixed to the substrate. For such a case, the phase transition occurs after adding an additional strain energy contribution needed to couple the new lattice of the hydride to the matrix and to the substrate. The final phase can form with complete orientation relation with regard to the source phase.

Kinetic barriers can also generate a large nano-metal hysteresis—especially when nucleation barriers are large and isotherms are measured under open conditions, for example under constant gas flow. This condition is often applied for "hydrogenography" measurements on thin films. Especially from these measurements, large hystereses areas were recently reported. As Pundt argued at Delft University in 2012, such barriers lead to an overestimation of the loading plateau pressure—and an underestimation of the unloading plateau pressure. As soon as the barrier is passed, the system switches into the final state by the given external gas pressure. Mooij and Dam (2013) studied the presence of such barriers for the Mg–H system in detail (compare also **Figure 51**).

Cluster isotherms also show a hysteresis between loading and unloading (Züttel et al., 2000; Sachs et al., 2001a, 2001b; Suleiman et al., 2003, 2005, 2009; Yamauchi et al. 2009). But, for small clusters, the self-energy of a dislocation is larger than energy of a strained cluster—it is reasonable to assume that dislocations are not formed in clusters and that the lattice is just strained. Also, dislocations are not observed inside small clusters by HREM (Sachs et al., 2001a, 2001b). So, it seems reasonable to assume that dislocations are not present in small clusters. Consequently, a dislocation-based mechanism cannot explain the hysteresis measured for M–H clusters. The origin of the hysteresis may be of completely different nature. Since clusters strain elastically upon hydrogen loading, the thermodynamic of the (open) coherent system needs to be considered, as described by Schwarz and Katchaturyan (compare Section 25.2.6). For gravimetric measurements, the gas volume has been sufficiently large to call the system also "open" (Sachs et al., 2001a, 2001b). For the hysteresis between absorption and desorption plateaus, p_{ab} and p_{des}, the theory of the open coherent system predicts (Schwarz and Khachaturyan, 1995, 2006) for Pd–H

$$\ln\left(\frac{p_{ab}}{p_{des}}\right) = \frac{4\Omega G_s \frac{1+\nu}{1-\nu}\varepsilon_0^2(c_{\alpha'} - c_\alpha)}{kT} = \ln(1.86), \tag{192}$$

by using the bulk data of Pd, the volume of one hydrogen atom in Pd, $\Omega = 2.607$ Å^3 (Peisl, 1978), the shear modulus of Pd, $G_s = 47.7 \times 10^9$ Pa, the Poisson number $\nu = 0.385$ (Landolt-Börnstein, 1984), the change in the lattice constant, $\varepsilon_0 = 0.063$ (Peisl, 1978), and the boundaries of the two-phase region $c_\alpha = 0.008$, and $c_{\alpha'} = 0.607$. It is, thereby, supposed that elastic constants do not differ from the bulk system. Experimental determined ratios p_{ab}/p_{des} of the measured plateau pressures were 1.1 for 2 nm clusters, 1.8 for 3 nm clusters and 2.0 for 5 nm clusters. Thus, by taking the thermodynamics of the open coherent system into account, the existence of hysteresis of Pd–H clusters isotherms can be explained. Even the size of the hysteresis effect of 2, 3 and 5 nm Pd–H clusters fits reasonably well (Sachs et al., 2001a, 2001b). Sachs et al. found these similarities, even though the cluster structure is expected to differ for the small and larger clusters (Suleiman et al., 2009).

25.4.4.8 *Finite Size Effects: Strain Effects and the Drop of the Critical Temperature*

One topic of fundamental interest is the temperature and concentration of the critical point for phase separation as a function of the nanoscale. The critical point for phase separation is the maximum point of the miscibility gap between the solid solution and the hydride phase. For M–H systems, hydrogen can be regarded in the frame of a lattice gas model (Lee and Yang, 1952; Hill, 1960; Wagner and Horner, 1974; Wagner, 1978). For such systems, the sample shape or boundary condition should affect the phase transition (Zabel and Weidinger, 1995; Song et al., 1996). The hydrogen solubility is determined by an interplay between the attractive hydrogen–metal interaction and the repulsive hydrogen–hydrogen interaction. While the first is mainly electronic in nature (Griessen and Driessen, 1994), the second is proportional to the square of the relative volume change $\Delta V/V$. As Alefeld (1969) already

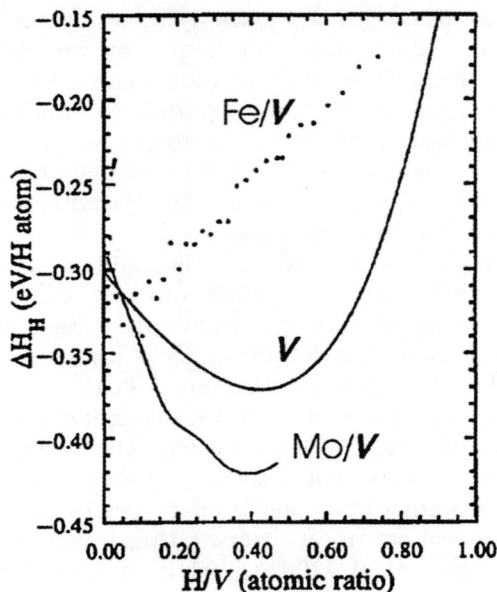

Figure 52 Change in the enthalpy of hydrogen solution in bulk V and in Fe/V and Mo/V multilayers. Hydrogen absorption leads to an increase for Fe/V but a decrease for Mo/V, verifying the impact of the strain state. From Hjörvarrson et al. (1997).

showed, for an elastically free sample, the interstitial hydrogen atom acts as a dilatation center and, therefore, indirectly attracts other hydrogen atoms independent of the distance. But for a sample fixed at all sides, the elastic interaction has to be repulsive. Thus, the thin film clamped to a substrate is something in between (Zabel and Weidinger, 1995). The changed hydrogen–hydrogen interaction results in a change of the critical point temperature.

For the experimental verification of these models epitaxial films are needed to exclude trap contributions, like those of grain boundaries. Hjörvarsson et al. showed for epitaxial multilayers that the hydrogen–hydrogen interaction energy can be changed in sign by changing the strain condition of the film (Stillesjo et al., 1995; Hjörvarsson et al., 1997; Andersson et al., 1997a, 1997b). Epitaxial V layers yield compressive stress between layers of Fe or tensile stress between layers of Mo. Hydrogen absorption leads to an increase or a decrease in the hydrogen–hydrogen interaction energy, respectively. This is shown in **Figure 52**. Hjörvarsson et al. (1997) claimed that phase separation can be prevented by choosing the appropriate boundary conditions.

Hjörvasson et al. also measured a different volume expansion of V upon hydrogen loading, depending on the strain state. This is shown in **Figure 53**). The expansion of V adjacent to Fe layers is almost 10 times larger than in V/Mo (Hjörvasson et al., 1997; Andersson et al., 1997a, 1997b; Stillesjö et al., 1996).

For 70 and 140 nm epitaxial Nb films Zabel and Weidinger derived a hydrogen–hydrogen interaction energy of 0.1 eV, which is 50% lower than the bulk value (Zabel and Weidinger, 1995). By reducing the film thickness of the Nb–H films, Song et al. (1996) reported a reduced critical point temperature of epitaxial Nb–H films. Song et al. (1996) measured pressure–expansion isotherms on epitaxial Nb(110) films deposited on sapphire substrates at $T = 473-573$ K, for thickness between 32 and 520 nm. They

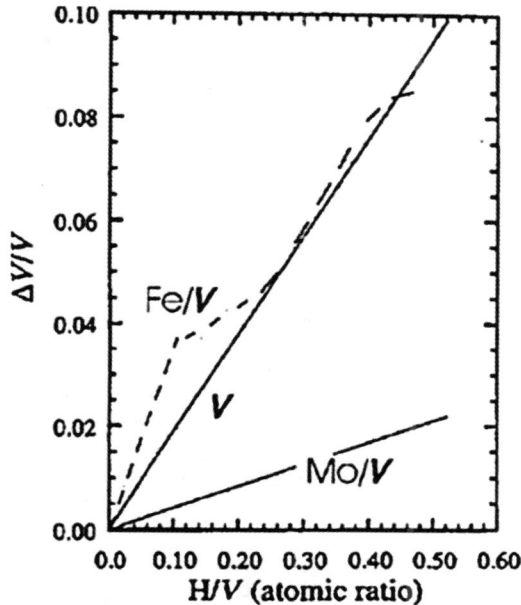

Figure 53 Hydrogen-induced lattice expansion for bulk V and for Mo/V and Fe/V multilayers. The ultrathin vanadium layers are under biaxial compression (Fe/V) and biaxial tension (Fe/Mo). From Hjörvarsson et al. (1997).

determined a decrease of the critical temperature T_c, with $T_c = 300$ K for a 50 nm film, and extrapolate to $T_c = 0$ K for a 4 nm Nb film. They argued that the clamping of the epitaxial film to the substrate prevents the appearance of macroscopic critical fluctuations in the hydrogen concentration from developing, thereby lowering the critical temperature for the phase transition (Song et al., 1996). But, Song et al. used the slope at the point of deflection of these curves to determine the hydrogen–hydrogen interaction energy, from which the critical temperature is evaluated (Song et al., 1996). With regard to Section 25.4.4.4, it follows that the lattice expansion is nonlinearly coupled to the hydrogen–pressure as soon as plastic processes are involved, especially in the two-phase region. In addition, because of clamping, isotherms have a different slope that also contains stress contributions. Thus, the pressure–expansion isotherm will have a different slope when compared with a fictive stress-free thin film needed for the study of pure finite size effects. Stress contributions act in the same direction as critical fluctuation-related finite size effects; they are thickness dependent and result in an overestimation of any size-dependencies (Pundt, 2005).

This topic was experimentally addressed by Dornheim (2002). For 35 and 25 nm epitaxial Nb–H films he verified phase transitions at 300 K. Thus, contrary to the argumentation by Song et al. (1996), the critical point of these films obviously is still well above 300 K. Furthermore, as Weissmüller pointed out, that elastic interactions suppress concentration fluctuations in regions of the phase diagram outside the coherent spinodal, in particular at the critical point of the incoherent miscibility gap (Weissmüller and Lemier, 2000; Cahn, 1979). Thus, critical fluctuations are expected at strongly reduced temperatures. The agreement of the critical point temperatures of the thick film (see **Figure 54**) with that of the incoherent miscibility gap is, in contradiction with this, indicating other precipitation mechanisms.

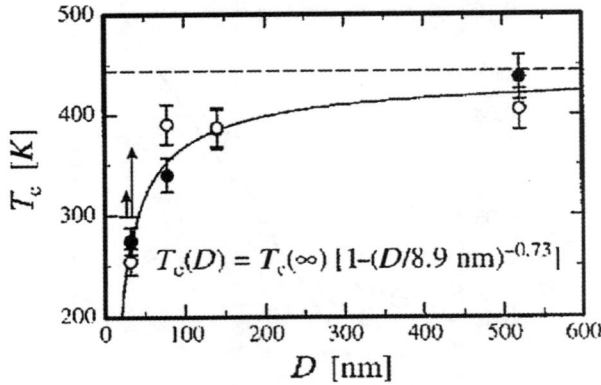

$$T_c(D) = T_c(\infty) \, |1-(D/8.9 \text{ nm})^{-0.73}|$$

Figure 54 Dependence of critical point temperature on film thickness of epitaxial Nb films, estimated by taking pressure–expansion isotherms (from Song et al., 1996). Data obtained from electrochemically loaded epitaxial Nb films (*arrows*) (M. Dornheim, A. Pundt, and R. Kirchheim, unpublished data) for epitaxial films are implemented. The interpretation of the results is very different.

(a)

(b)

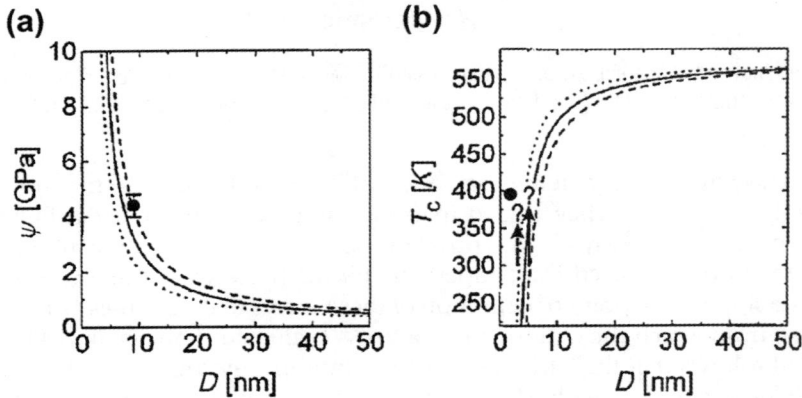

Figure 55 Results of Weismüller and Lemier (134), predicting a pressure-related reduction of the critical temperature. (a) Concentration–pressure coefficient $\psi = \delta\sigma/\delta cH$ and (b) critical temperature Tc versus grain size D for n-Pd–H. Lines indicate grain boundary layer thicknesses of 0.7 nm ($\cdots$), 0.9 nm (—), and 1.1 nm (— —). Solid black dot denotes experimental value for n-Pd–H. Experimental data for approximately 3 and 5 nm Pd clusters show hysteresis at room temperature—therefore suggesting a critical temperature well above 310 K (arrows) (Sachs et al., 2001a, 2001b; Pundt et al., 2004a, 2004b, 2004c, 2005). Yamauchi et al. (2009) measured a critical temperature of 390 K for 2.6 nm Pd clusters ($\bullet$).

Weissmüller and Lemier (2000) predicted a significant reduction of the critical point temperature for small nanoparticles due to interfacial stress between the shell and the core. They demonstrated the impact of interfacial stress for nanocrystalline Pd between grain boundaries and their coherently matching interior, on the critical point temperature. These results are plotted in **Figure 55**. Isotherms of Pd–H particles subjected to interfacial stress show a reduced critical point temperature T_c.

The theory of Weissmüller and Lemier (2000) is also applicable for clusters, since (i) their core–shell substructure might be comparable to the core interface structure of nanocrystals and (ii) additional stress belonging to the stabilizer or the matrix is present. According to their results, the critical temperature of nanocrystalline Pd with grain diameters of 3 nm should be below room temperature.

Suleiman et al. (2003, 2005, 2009) measured, for different sized clusters between 2, 7 and 6 nm, always a hysteresis, at 300 K (compare Section 25.4.4.7). For bulk metals the hysteresis can be taken as a fingerprint for the phase transition: Above the critical point no hysteresis is measured (Fukai, 2005, p. 4). If the hysteresis can be taken as a fingerprint for a phase transition also for clusters, phase transition is also present for clusters smaller than 3 nm (Pundt et al., 2004a, 2004b, 2004c, 2006). Also, the good agreement between the measured equilibrium pressures of clusters and the theory of Schwarz and Katchaturyan (1996) suggests that the Pd–H system decomposes even when the particle size is about 2 nm in diameter (Suleiman et al., 2005; Pundt, 2005). The isotherms of Züttel et al. (2000) show hysteresis even for $Pd_{55}phen_{36}O_{30}$–H compressed in copper matrix. These Pd clusters are of about 1 nm in diameter. Thus, phase transition seems to be present for these small clusters, hinting on a critical temperature T_c above 300 K. Yamauchi et al. (2009) directly measured the critical temperature of $T_c = 390$ K for 2.6 ± 0.4 nm-sized Pd clusters, which is strongly lowered compared with $T_c = 565$ K for bulk Pd. Yamauchi et al. derived T_c by strong changes of the isotherms shape, as measured for four different temperatures below 390 K.

However, the structure of small clusters is often not cubic any more (Suleiman et al., 2009). And, it is not yet clarified, to what extent also elastic strain between the surface and subsurface sites and the inner sites of the clusters might result in a hysteresis effect.

25.4.4.9 New Phases for Nano-Metals

For small clusters, new phases result. Pd–H clusters (stabilized in TetraOctylAmmoniumBromide TAOB) above 4.7 nm in diameter have cubic lattice structures, as verified by HREM and XRD. For Pd–H clusters (stabilized in TAOB) of less than 4.7 nm in diameter, fivefold symmetry structures were detected for clusters in HREM and XRD pattern hinting on icosahedral structures was detected by Pundt et al. (2002, 2004a, 2004b, 2004c) and Suleiman et al. (2003, 2009). Thus, for these TOAB-stabilized Pd clusters, 4.7 nm can be regarded as the transition size.

For icosahedral clusters, isotherms were reported to be strongly sloped upward over the complete pressure range. This results from the broad DOSE expected for icosahedral clusters: With increasing shell number the local lattice strain increases (Sugano and Koizumi, 1998). Larger Pd clusters with cubic lattice structure show less sloped isotherms. The solubility of cubic clusters is always increased compared with icosahedral clusters. The transition size strongly depends on the stabilizer. Suleiman et al. (2009) also reported on 3.1 nm cubic clusters.

Hydrogen absorption might even change the lattice structure reversible, during loading and unloading. A structural transition from fcc to the icosahedral phase was suggested to explain reversible changes found in XRD pattern of Pd clusters by Pundt et al. (2002). Recently, Zlotea et al. (2010a, 2010b) also found this hydrogen-induced structural transition for 2.5 nm Pd clusters embedded in mesoporous carbon.

25.4.5 Nanosized Metals for Hydrogen Storage

Hydrogen can be regarded as the energy carrier of the future. The number of books dedicated to this field is, naturally, increasing strongly, see, for example the book edited by Züttel et al. (2008) or that of Walker (2008), or the book edited by Hirscher (2010). Hydrogen can be prepared from water by electrolysis, by using renewable energy sources like wind or solar. Züttel et al. (2008) demonstrated the power of the "hydrogen cycle" that he recently extended. The storage of hydrogen has to fulfill requirements depending strongly on the application. For stationary storage, the weight or the volume of the storage is of minor importance. Actual projects demonstrate the applicability of stationary hydrogen storage in private houses, larger buildings or even towns. One actual example is the "Fukuoka Hydrogen Town" model project of Japan started in 2008. For mobile applications weight and volume become important factors. During the past decade, a lot of research was dedicated to the finding of novel hydrogen storage materials that fulfill the requirements for mobile applications. Methods like "hydrogenography" (Gremaud et al., 2009) or other high-throughput methods (Ludwig et al., 2007a, 2007b) were developed to screen thermodynamic properties of concentration gradient samples.

The book edited by Hirscher (2010) gives a good overview on many different types of solid-state hydrogen storage materials: (a) open-volume hydrogen adsorption materials that can be applied under cryogenic temperatures. (b) Conventional metallic hydrides. (c) Metal alloys, consisting of different metals to achieve optimized parameters, also by tailoring reaction enthalpies. (d) Amides and imides, ammonia borane, or (e) covalent alane as well as (f) clathrates. The field of solid-state storage materials is very wide and we, therefore, like to stimulate to further reading one of the relevant books.

However, loading and unloading time is a key issue for many storage application independent of the storage material used. de Jongh and Adelhelm (2010) collected data for nanosized metals, nanosized chemical hydrides (ammonia borane) with many decomposition substeps, and nanosized complex hydrides (sodium alanate). Mg–H is one of the most promising hydrides because of its high gravimetric hydrogen storage density of 7.7 wt%. Its application is hampered by its slow kinetics. Nanostructuring can be a powerful tool to overcome this hurdle. Significant progresses in hydrogenation kinetics were achieved by nanocrystallization performing mechanical alloying (ball milling) or Equal-Channel Angular Pressing (ECAP) treatments (Huot, 2010 and references therein, Skripnyuk et al., 2004).

The results gained by studying hydrogen behavior of different nanomaterials like thin films, film packages and clusters can be now used to tailor properties of nanomaterials for storage applications. For the perspective of hydrogen storage, the following aspects are of importance (mainly taken from Pundt, 2013):

(1) Microstructural defects usually act as strong hydrogen trapping sites. Their content needs to be minimized for maximum storage capacity.
(2) Mechanical stress, evolving between the nano-metal and the stabilizer, reduces the storage capacity of the metal. Soft and thin stabilizers are suitable for maximum storage capacity.
(3) When hydride destabilization (increase of chemical potentials) is desired, mechanical stress can be used. For this, hard and thick substrates with strong adhesion are preferred as well as metals with small dimensions in the 20 nm range.
(4) Metal/metal interfaces can lead to hydrogen depletion zones by chemical intermixing, electron transfer or elastic lattice strain. The number of metal/metal interfaces should be kept small.
(5) For very small metals, like clusters, new (fivefold symmetry) structures have to be considered that do not show up in the bulk phase diagram. Their storage properties are expected to differ from conventional structures.

Acknowledgments

AP likes to thank Stefan Wagner, Magnus Hamm, and Jantje Schommartz for proofreading and valuable comments, and Marc Warninger for extensive editing help. Furthermore, Helmut Takahiro Uchida is thanked for preparing raw data files from literature, for different figures. Jakub Cizek is thanked for comments on positron annihilation, Ryota Gemma for additions on Ti, and Najeh Jisrawi for providing his unpublished calculations on clusters.

References

Aben, P.C., 1986. J. Catal. 10, 224–226.

Akiba, E., 1999. Curr. Opin. Solid State Mater. Sci. 4, 267–272.

Alefeld, G., 1969. Phys. Status Solid 32, 67–77.

Alefeld, G., 1972. Ber. Bunsenges. Phys. Chem. 76, 335 bzw. 746.

Andersson, G., Hjörvarsson, B., Zabel, H., 1997a. Phys. Rev. B 55, 15905–15911.

Andersson, G., Hjörvarsson, B., Isberg, P., 1997b. Phys. Rev. B 55, 1774–1781.

Arantes, D.R., Huang, X.Y., Marte, C., Kirchheim, R., 1993. Acta Metall. Mater. 41, 3215.

Argon, A.S., 2006. Strengthening Mechanisms in Crystal Plasticity. Oxford University Press, Oxford.

Arzt, E., 1998. Acta Mater. 46 (16), 5611–5626.

Ashcroft, N.W., Mermin, D.N., 1976. Solid State Phys. (Cengage Learning Emea).

Audoly, B., 1999. Phys. Rev. Lett. 83, 4134.

Baldi, A., Pálsson, G.K., Gonzalez-Silveira, M., Schreuders, H., Slaman, M., Rector, J.H., Krishnan, G., Kooi, B.J., Walker, G.S., Fay, M.W., Hjörvarsson, B., Wijngaarden, R.J., Dam, B., Griessen, R., 2009. Cond. Mat. Mat. Sci..

Bang, G.W., Nagakawa, J., Meshii, M., 1980. Scr. Metall. 14, 289.

Baranowski, B., Majchrzak, S., Flanagan, T.B., 1971. J. Phys. F Metal Phys. 1, 258.

Barnoush, A., Vehoff, H., 2006. Int. J. Mater. Sci. 97, 1224.

Barnoush, A., Vehoff, H., 2008. Corr. Sci. 50, 259.

Barnoush, A., Biess, Ch., Vehoff, H., 2009. J. Mater. Res. 24, 1105.

Bauer, H.J., Berninger, G., Zimmermann, G., 1968. Z. Naturforsch. 23a, 2023.

Beaudry, B.J., Spedding, F.H., 1975. Metall. Trans. 6B, 419.

Beck, R.L., Mueller, W.M., 1968. Metal Hydrides, 241. Academic Press, 306.

Becker, R., 1985. Theorie der Wärme. In: Heidelberger Taschenbücher Bd, vol. 10. Springer.

Bečvář, F., Čížek, J., Lešták, L., Novotný, I., Procházka, I., Šebesta, F., 2000. Nucl. Instrum. Methods Phys. Res. A 443, 557.

Behm, R.J., Christmann, K., Ertl, G., 1980. Surf. Sci. 99, 320–340.

Behm, R.J., Penka, V., Cattina, M.G., Christmann, K., Ertl, G., 1983. J. Chem. Phys. 78, 7486–7490.

Benninghoven, A., Rudenauer, F.G., Werner, H.W., 1987. Secondary Ion Mass Spectrometry: Basic Concepts, Instrumental Aspects, Applications and Trends. John Wiley & Sons.

Berry, B.S., Pritchet, W.C., 1981. Scr. Metall. 15, 637.

Birnbaum, H.K., Flynn, C.P., 1976. Phys. Rev. Lett. 37, 25–28.

Birnbaum, H.K., Sofronis, P., 1994. Mat. Sci. Eng. A176, 191.

Birringer, R., Hoffmann, M., Zimmer, P., 2003. Z. Metallkd. 94, 1052.

Boes, N., Züchner, H., 1976. J. Less Comm. Metals 49, 223–240.

Bolef, D.F., 1961. J. Appl. Phys. 32, 100–105.

Bosson, J.C., Coet, J., Charles, J., 1999. Etude du dosage de l'hydrogène piégé dans les aciers (Recherche technique acier, Communautés Européennes, Rapport EUR 18801FR Chen, Y.Z., B).

Boureau, G., 1981. J. Phys. Chem. Solids 42, 743.

Boureau, G., 1984. J. Phys. Chem. Solids 45, 973.

Bradfielck, G., 1964. Use in Industry of Elasticity Measurements in Metals with the Help of Mechanical Vibrations'. National Physical laboratory. Notes on Applied Science No, 30. HMSO.

Brandes, E.A., 1983. Smithells Metals Reference Book. Butterworth & Co.

Brodowsky, H., Poeschel, E., 1965. Z. Phys. Chem. (N.F.) 44, 143(527).

Bronstein, N., Semendjajew, K.A., 1985. Taschenbuch der Mathematik. Harri Deutsch, Thun, Frankfurt/Main.

Bucur, R.V., Flanagan, T.B., 1974. Z. Phys. Chem. (N.F.) 88, 225–241.

Cahn, J.W., 1979. Acta Metall. 9, 795–801.

Cahn, R.W. (Ed.), 1990. Encyclopedia of Materials Science and Engineering, vol. 2. Pergamon Press, Oxford, UK (Suppl.).

Carroll, K.J., 1965. J. Appl. Phys. 33, 3689–3690.

Cattina, M.G., Penka, V., Behm, R.J., Christmann, K., Ertl, G., 1983. Surf. Sci. Lett. 126, A111–A112.

Checchetto, R., Bazzanella, N., Miotello, A., Brusa, R.S., Zecca, A., Mengucci, A., 2004. J. Appl. Phys. 95, 1989.

Chen, Y.Z., Barth, H.P., Deutges, M., Borchers, C., Liu, F., Kirchheim, R., 2013. Scripta Mater. 68, 743–746.

Choo, W.Y., Lee, J.Y., 1982. Metall. Trans. 13A, 135.

Christian, J.W., 1975. The Theory of Phase Transformation in Metals and Alloys. Pergamon Press, Oxford.

Christmann, K., 1988. Surf. Sci. Rep. 9, 1–163.

Čížek, J., Procházka, I., Kužel, R., Bečvář, F., Cieslar, M., Brauer, G., Anwand, W., Kirchheim, R., Pundt, A., 2004. Phys. Rev. B 69, 224106.

Čížek, J., Procházka, I., Daniš, S., Melikhova, O., Vlach, M., Žaludová, N., Brauer, G., Anwand, W., Mücklich, A., Gemma, R., Nikitin, E., Kirchheim, R., Pundt, A., 2007. J. Alloys Compd. 446-447, 484.

Clapp, P.C., Moss, S.C., 1966. Phys. Rev. 142, 418.

Colin, J., Cleymand, C., Coupeau, J., Grilhe, J., 2000. Philos. Mag. A80, 2559.

Conrad, H., Ertl, G., Latta, E.E., 1974. Surf. Sci. 41, 435–446.

Conte, M., Prosini, P.P., Passerini, S., 2004. Mater. Sci. Eng. B 108, 2–8.

Cox, D.M., Fayet, P., Brickman, R., Hahn, M.Y., Kaldor, A., 1990. Catal. Lett. 4, 271–278.

Crosby, K.M., Bradley, R.M., 1999. Phys. Rev. E 59, R2542–R2545.

Daw, M.S., Foiles, S.M., 1987. Phys. Rev. B 35, 2128–2136.

de Jongh, P., Adelhelm, P., 2010. In: Hirscher, M. (Ed.), Handbook of Hydrogen Storage. Wiley-VCH (Chapter 10).

Deutges, M., 2012. Master thesis, University of Goettingen.

Dong, W., Ledentu, V., Sautet, P., Eichler, A., Hafner, J., 1998. Surf. Sci. 411, 123–136.

Dornheim, M., 2002. Doctoral thesis, Universität Göttingen.

Dornheim, M., Pundt, A., Kirchheim, R., van der Molen, S.J., Geyer, U., 2003. J. Appl. Phys. 93, 8958–8964.

Dupasquier, A., Mills Jr., A.P. (Eds.), 1995. Positron Spectroscopy of Solids. IOS Press, Amsterdam, Oxford, Tokyo, Washington DC.

Eijt, S.W.H., Kind, R., Singh, S., Schut, H., Legerstee, W.J., Hendrikx, R.W.A., Svetchnikov, V.L., Westerwaal, R.J., Dam, B., 2009. J. Appl. Phys. 105, 043514.

Eliezer, D., Eliaz, N., Senkov, O.N., Froes, F.H., 2000. Mater. Sci. Eng. A Struct. 280, 220.

Feenstra, R., de Groot, D.G., Rector, J.H., Salomons, E., Griessen, R., 1986. J. Phys. F 16, 1953–1963.

Feenstra, R., Brower, R., Griessen, R., 1988. Europhys. Lett. 7, 425.

Finocchiaro, R.S., Tsai, C.L., Giesen, B.C., 1984. J. Non-Cryst. Solids 61+62, 661.

Fisher, E.S., Renken, C.J., 1964. Phys. Rev. 135, A482–A494.

Fischer, G., 2012. Lehrbuch Lineare Algebra und Analytische Geometrie. Springer Spektrum. Chap. 5.3.6.

Flanagan, T., Lynch, J., 1976. J. Less Comm. Metals 49, 25.

Flanagan, T.B., Oates, W.A., 1991. The palladium hydrogen system. Annu. Rev. Mater. Sci. 21, 269–304.

Flanagan, T.B., Bowerman, B.S., Biehl, G.E., 1980. Scr. Metall. 14, 443–447.

Flanagan, T.B., Lynch, J.F., Clewley, J.D., von Turkovich, B., 1976. J. Less Comm. Metals 49, 13.

Flanagan, T.B., Balasubramaniam, R., Kirchheim, R., 2001a. Platin. Met. Rev. 45 (Pt. 1), 114–121.

Flanagan, T.B., Balasubramaniam, R., Kirchheim, R., 2001b. Platin. Met. Rev. 45 (Pt. 2), 166–174.

Flynn, C.P., Stoneham, A.M., 1970. Phys. Rev. B 1, 3966–3978.

Foiles, S.M., Daw, M.S., 1985. J. Vac. Sci. Technol. A 3, 1565–1566 (Abstr.).

Frieske, H., Wicke, E., 1973. Ber. Bunsenges. Phys. Chem. 77, 48.

Fromm, E., Gebhardt, E. (Eds.) 1976a. Gase and Kohlenstoff in Metallen, Reine und angewandte Metallkunde in Einzeldarstellungen, 26. Band (Springer, Berlin, Heidelberg).

Fromm, E., Gebhardt, E., 1976b. Gase und Kohlenstoff in Metallen. Springer, Berlin.

Fromm, E., Hörz, G., 1980. Int. Met. Rev. 25, 269.

Fujii, H., Higuchi, K., Yamamoto, K., Kajioka, H., Orimo, S., Toiyama, K., 2002. Mater. Trans. Jpn. Inst. Met. 43, 2721–2727.

Fukai, Y., 2003. Phys. Scr. T103, 11.

Fukai, Y., 2005. The Metal-Hydrogen System—Basic Bulk Properties. Springer, Berlin.

Fukai, Y., Okuma, N., 1994. Phys. Rev. Lett. 73, 1640.

Gahr, S., Birnbaum, H.K., 1978. Acta Metall. 26, 1781.

Gao, Z.Q., Fultz, B., 1994. Nanostruct. Mater. 9, 939.

Gault, B., Moody, M.P., Cairney, J.M., Ringer, S.P., 2012. Atom probe microscopy. Springer Series in Materials Science 160. New York.

Gemma, R., 2011. Dissertation thesis Göttingen.

Gemma, R., Al-Kassab, T., Kirchheim, R., Pundt, A., 2009. Ultramiocroscopy 109, 631–636.

Gemma, R., Al-Kassab, T., Kirchheim, R., Pundt, A., 2011. J. Alloys Compd. 509S, 872–876.

Gemma, R., Al-Kassab, T., Kirchheim, R., Pundt, A., 2012. Scr. Mater. 67, 903–906.

Gioia, G., Ortiz, M., 1997. Adv. Appl. Mech. 33, 129.

Gottstein, G., 2001. Physikalische Grundlagen der Materialkunde. Springer, Berlin.

Gremaud, R., Borgschulte, A., Chacon, C., van Mechelen, J.L.M., Schreuders, H., Züttel, A., Hjörvarsson, B., Dam, B., Griessen, R., 2006. Appl. Phys. A 84, 77.

Gremaud, R., Gonzalez-Silveira, M., Pivak, Y., de Man, S., Slaman, M., Schreuders, H., Dam, B., Griessen, R., 2009. Acta Mater. 57, 1209.

Grier, E.J., Kolosov, O., Petford-Long, A.K., Ward, R.C.C., Wells, M.R., Hjörvarsson, B., 2000. J. Phys. D 33, 894–900.

Griessen, R., 1983. Phys. Rev. B 27, 7575.

Griessen, R., Driessen, A., 1994. Phys. Rev. B 30, 4372–4381.

Griessen, R.P., 1986. In: Bambakidis, G., Bowman Jr., R.C. (Eds.), Hydrogen in Disordered and Amorphous Solids. Plenum, New York, pp. 153–172.

Haasen, P., 1994. Physikalische Metallkunde. Springer, Berlin.

Haluska, M., Hulman, M., Hirscher, M., Becher, M., Roth, S., Stepanek, I., Bernier, P., 2001. AIP Conf. Proc. 591, 603.

Hamm, M., Pundt, A., 2014. Actual reseach data.

Harris, J.H., Curtin, W.A., Tenhover, M.A., 1987. Phys. Rev. B36, 5784.

Hayashi, E., Kurokawa, Y., Fukai, Y., 1998. Phys. Rev. Lett. 80, 5588.

Hémon, S., Cowley, R.A., Ward, R.C.C., Wells, M.R., Douysset, L., Ronnow, H., 2000. J. Phys. Condens. Matter 12, 5011.

Heumann, Th., 1992. Diffusion in Metallen. Springer, Berlin, London, New York.

Heuser, B.J., King, J.S., Summerfield, G.S., Boué, F., Epperson, J.E., 1991. Acta Metall. Mater. 39, 2815.

Heuser, B., Allain, M.M.C., Chen, W.C., 2002. Phys. Rev. B 66, 155419.

Heuser, B.J., King, J.S., 1997. J. Alloys Compd. 261, 225.

Hill, T.L., 1960. An Introduction to Statistical Mechanics. Addison-Wesley, Reading, pp. 523.

Hirsch, P.B., 1991. Nucleation and propagation of misfit dislocations in strained epitaxial layer systems. In: Werner, J.H., Strunk, H.P. (Eds.), Springer Proceedings in Physics, Polycrystalline Semiconductors II, vol. 54. Springer, Berlin, Heidelberg, pp. 470–481.

Hirscher, M., 2010. Handbook of Hydrogen Storage. Wiley-VCH, Weinheim.

Hirscher, M., Kronmüller, H.J., 1991. J. Less Comm. Metals 17, 658.

Hirscher, M., Becher, M., Haluska, M., von Zeppelin, F., Roth, S., et al., 2003. J. Alloys Compd. 356–357, 433–437.

Hirth, J.P., Lothe, J., 1968. Dislocations in Solids. McGraw Hill, New York.

Hirth, J.P., 1980. Metall. Trans. A 11A, 861.

Hirth, J.P., 1996. Dislocations. In: Cahn, R.W., Haasen, P. (Eds.), Physical Metallurgy, vol. 3. North Holland, Amsterdam, p. 1832.

Hjörvarsson, B., Rydén, J., Karlsson, E., Birch, Sundgren, J.E., 1991. Phys. Rev. B 43, 6440–6445.

Hjörvarsson, B., Rydén, J., 1990. Nucl. Instrum. Methods Phys. Res. B 45, 36–40.

Hjörvarsson, B., Andersson, G., Karlsson, E., 1997. J. Alloys Compd. 253–254, 51–57.

Hjörvarsson, B., Rydén, J., Ericsson, T., Karlsson, E., 1989. Nucl. Instrum. Methods Phys. Res. B 42, 257–263.

Hong, H., 2001. J. Alloys Compd. 321, 307–313.

Hong, L.B., Bansal, C., Fultz, B., 1994. Nanostruct. Mater. 4, 949.

Horner, H., Wagner, H., 1974. J. Phys. C 7, 3305–3325.

Huang, H., Spaepen, F., 2000. Acta Mater. 48, 3261–3269.

Huang, X., Mader, W., Kirchheim, R., 1991. Acta Metall. Mater. 39, 893.

Huiberts, J.N., Griessen, R., Rector, J.H., Wijngaarden, R.J., Koeman, N.J., et al., 1996. Yttrium and Lanthanum hydride films with switchable optical properties. Nature 380, 231–234.

Huntington, H.B., 1958. Solid State Phys. 7, 213–251.

Huot, J., 2010. In: Hirscher, M. (Ed.), Handbook of Hydrogen Storage. Wiley-VCH (Chapter 4).

Huot, J., Liang, G., Boily, S., 1999. J. Alloys Compd. 293–295, 495.

Ichikawa, T., Isobe, S., Hanada, N., Fujii, H., 2004. J. Alloys Compd. 365, 271–276.

Jaggy, F., Kieninger, W., Kirchheim, R., 1989. Z. Phys. Chem. (N.F.) 163, 431.

Jain, S.C., Harker, A.K., Cowley, R.A., 1997. Philos. Mag. A 75, 1461–1515.

Jamieson, H.C., Weatherly, G.C., Manchester, F.D., 1976. J. Less Comm. Metals 50, 85.

Kanamori, J., Kakehashi, Y., 1977. J. Phys. (Paris) Colloq. C7 (38), 274.

Katz, Y., Tymiak, N., Gerberich, W.W., 2001. Eng. Fract. Mech. 68, 619.

Kerssemakers, J.W.J., van der Molen, S.J., Koeman, N.J., Günther, R., Griessen, R., 2000a. Nature 406, 489.

Kerssemakers, J.W.J., van der Molen, S.J., Koeman, N.J., Günther, R., Griessen, R., 2000b. Nature 406, 489–491.

Kesten, P., Pundt, A., Schmitz, G., Weisheit, M., Krebs, H.-U., Kirchheim, R., 2002. J. Alloys Compd. 330–332, 225–228.

Kimura, H., 1985. Trans. Jpn. Inst. Met. Mater. 8, 527.

Kimura, H., Matsui, H., Moriya, S., 1977. Scr. Metall. 11, 473–474.

Kirchheim, R., 1981. Acta Metall. 29, 835–845.

Kirchheim, R., 1982. Acta Metall. 30, 1069.

Kirchheim, R., 1986. Acta Metall. 34, 34.

Kirchheim, R., 1988. Prog. Mat. Sci. 32, 262.

Kirchheim, R., 2002. Acta Mater. 50, 413.

Kirchheim, R., 2004. Solid State Phys. 59, 203–305.

Kirchheim, R., 2007a. Acta Mater. 55, 5129–5138.

Kirchheim, R., 2007b. Acta Mater. 55, 5139–5148.

Kirchheim, R., 2009. Int. J. Mater. Res. 100, 483–487.

Kirchheim, R., 2010. Scr. Mater. 62, 67–70.

Kirchheim, R., 2012. Scr. Mater. 67, 767.

Kirchheim, R., Szökefalvi-Nagy, A., Solz, U., Speitling, A., 1985. Scr. Metall. 19, 843.

Kirchheim, R., Kownacka, I., Filipek, S.M., 1993. Scr. Metall. Mater. 28, 1229.

Kishore, S., Nelson, J.A., Adair, J.H., Eklund, P.C., 2005. J. Alloys Compd. 389, 234–242.

Klassen, T., Oelerich, W., Bormann, R., 2001. Mater 10, 603–608.

Kobayashi, H., Kitagawa, H., et al., 2010. J. Am. Chem. Soc. 132, 15896–15898.

Koch, R., 1994. J. Phys. Condens. Matter 6, 9519–9550.

Koch, R., 1997. Intrinsic stress of epitaxial thin films and surface layers. In: King, D.A., Woodruff, D.P. (Eds.), Growth and Properties of Ultrathin Epitaxial Layers. Elsevier, Amsterdam, pp. 1–37.

Köster, W., Bangert, L., Hahn, R., 1954. Eisenhüttenwesen 25, 569.

Kowalsky, H.-J., 1974. Einführung in die lineare Algebra. deGruyter.

Krebs, H.-U., Weisheit, M., Faupel, J., Süske, E., Buback, M., et al., 2003. Adv. Solid State Phys. 43, 505–518.

Krill III, C.E., Ehrhardt, H., Birringer, R., 2006. Z. Metallkd. 96, 1134.

Krukowski, M., Baranowski, B., 1976. J. Less Comm. Metals 49, 385.

Kubin, L.P., Louchet, F., Peyrade, J.P., Groh, P., Cottu, J.P., 1979. Acta Metall. 27, 343.

Kummnick, A.J., Johnson, H.H., 1980. Acta Metall. 28, 33.

Kürschner, J., Wagner, S., Pundt, A., 2014. Journal of Alloys and Compounds.

Kusada, K., Yamauchi, M., Kobayashi, H., Kitagawa, H., Kubota, Y., 2012. J. Am. Chem. Soc. 132, 15896–15898.

Lacher, J.R., 1937. J. Proc. R. Soc. Lond. Ser. A 161, 525–545.

Landau, L.D., Lifschitz, E.M., 1989. Lehrbuch der theoretischen Physik: Elastizitätstheorie. Akademie-Verlag, Berlin.

Landolt-Börnstein, 1984. In: Hellwege, K.H., Madelung, O. (Eds.), Neue Serie III, Bd. 18. Springer-Verlag.

Lang, B., Joyner, R.W., Somorjai, G.A., 1992. Surf. Sci. 50, 454.

Langhammer, C., Zhadanov, V.P., Zoric, I., Kasemo, B., 2010. Chem. Phys. Lett. 488, 62–66.

Latroche, M., 2004. J. Phys. Chem. Solids 65, 517–522.

Laudahn, U., 1998. Dissertation thesis, University of Göttingen (Cuvillier Verlag).

Laudahn, U., Pundt, A., Bicker, M., von Hülsen, U., Kirchheim, R., et al., 1999a. J. Alloys Compd. 293, 490–494.

Laudahn, U., Fähler, S., Krebs, H.U., Pundt, A., Bicker, M., Hülsen, U.V., Geyer, U., Kirchheim, R., 1999b. Appl. Phys. Lett. 74, 647.

Lee, H.G., Lee, J.Y., 1984. Acta Metall. 32, 131.

Lee, J.Y., Lee, J.L., Choo, W.Y., 1982. Thermal analysis of trapped hydrogen in AISI 4340 steel. In: 1st Conf. Int. Hydrogen Problems in Steel.

Lee, M.W., Glosser, R., 1985. J. Appl. Phys. 57, 5236–5239.

Lee, T.D., Yang, C.N., 1952. Phys. Rev. 87, 410–419.

Lee, Y.S., Stevenson, D.A., 1985. J. Non-Cryst. Solids 72, 249.

Leung, O.S., Nix, W.D., 2000. Mater. Res. Soc. Symp. Proc. 594, 51.

Li, G., Kobayashi, H., Kubota, Y., Kitagawa, H., 2012, Internationakl Symposium on Metal-Hydrogen Systems—Fundamentals and Application, Program and Abstratcs, October 21–26, Kyoto, Japan. TuP-15.

Lieberman, M.L., Wahlbeck, P.G., 1965. J. Phys. Chem. 69, 3514.

Lohstroh, W., Westerwaal, R.J., Noheda, B., Enache, S., Griessen, R., et al., 2004. Phys. Rev. Lett. 93, 197404.

Lücke, R., Schmitz, G., Flanagan, T.B., Kirchheim, R., 2002. J. Alloys Compd. 330–332, 219.

Ludwig, A., Cao, J., Savan, A., Ehmann, M., 2007a. J. Alloys Compd. 446–447, 516–521.

Ludwig, A., Cao, J., Dam, B., Gremaud, A., 2007b. Appl. Surf. Sci. 254, 682–686.

Lundin, C.E., Lynch, F.E., 1975. Solid State Hydrogen Storage Materials for Application to Energy Needs. Rept. AFSOR, F44620-74-C0020. Denver Research Institute (260).

Lynch, S.P., 1988. Acta Metall. 44, 2639.

Lynn, K.G., MacDonald, J.R., Boie, R.A., Feldman, L.C., Gabbe, J.D., Robbins, M.F., Bonderup, E., Golovchenko, J., 1977. Phys. Rev. Lett. 38, 241.

Macdonald, D., Mckubre, M.C.H., Scott, A.C., Wentrcek, P.R., 1981. Ind. Eng. Chem. Fundam. 20, 290–297.

Makenas, B.J., Birnbaum, H.K., 1980. Acta Metall. 28, 979.

Malard, B., Remy, B., Scott, C., Deschamps, A., Chêne, J., Dieudonné, T., Mathon, M.H., 2012. Mater. Sci. Eng. A 536, 110–116.

Mao, W., Li, Z., 2012. Comput. Mater. Sci. 54, 28–31.

Matsui, H., Kimura, A., Kimura, H., 1979a. In: Haasen, P., Gerold, V., Kostorz, G. (Eds.), Strength of Metals and Alloys, vol. 2. Pergamon, Oxford, pp. 977–982.

Matsui, H., Kimura, H., Moriya, S., 1979b. Mater. Sci. Eng. 40, 207–216.

Matsui, H., Kimura, H., Kimura, A., 1979c. Mater. Sci. Eng. 40, 227–234.

Matthews, J.W., 1975. J. Vac. Sci. Technol. 12, 126–133.

Matthews, J.W., Blakeslee, A.E., 1974. J. Cryst. Growth 27, 118–125.

Matthews, J.W., Blakeslee, A.E., 1975. J. Cryst. Growth 29, 273–280.

Maxelon, M., Pundt, A., Pyckhout-Hintzen, W., Kirchheim, R., 2001a. Scr. Mater. 44, 817–822.

Maxelon, M., Pundt, A., Pyckhout-Hintzen, W., Barker, J., Kirchheim, R., 2001b. Acta Mater. 49, 2625.

Miceli, P.F., Zabel, H., Cunningham, J.E., 1985. Phys. Rev. Lett. 54, 917–919.

Miceli, P.F., Zabel, H., Dura, J.A., Flynn, C.P., 1991. J. Mater. Res. 6, 964–968.

Miller, M.K., Cerezo, A., Hetherington, M.G., Smith, G.D.W., 1996. Atom Probe Field Ion Microscopy. Oxford University Press.

Mitsui, T., Rose, M.K., Fomin, E., Ogletree, D.F., Salmeron, M., 2003a. Nature 422, 705–707.

Mitsui, T., Rose, M.K., Fomin, E., Ogletree, D.F., Salmeron, M., 2003b. Surf. Sci. 540, 5–11.

Moody, N.R., Thompson, A.W., Warrendal, T.M.S. (Ed.), 1990. Hydrogen effects on materials behavior Pennsylvania.

Mooij, L., Dam, B., 2013. Phys. Chem. Chem. Phys. 15, 2782–2792.

Moon, M.W., et al., 2004. Acta Mater. 52, 3151–3159.

Moriya, S., Matsui, H., Kimura, H., 1979. Mat. Sci. Eng. 40, 217–225.

Moser, M., et al., 2011. Nucl. Instrum. Methods Phys. Res. B 219, 2217–2228.

Mueller, W.M., 1968. Titanium Hydrides. Chapt. 8 in Metal Hydrides. Academic Press, p. 336(308).

Munter, A., Heuser, B., 1998. Phys. Rev. B 58, 678–684.

Murakami, Y., Kanezaki, T., Mine, Y., 2010. Met. Mater. Trans. A 41A, 2548.

Muschiol, U., Schmidt, P.K., Christmann, K., 1998. Surf. Sci. 395, 182.

Mütschele, T., Kirchheim, R., 1987a. Scr. Metall. 21, 135–140.

Mütschele, T., Kirchheim, R., 1987b. Scr. Metall. 21, 1101–1104.

Myers, S.M., Baskes, M.I., Birnbaum, H.K., Corbett, J.W., Deleo, G.G., Estreicher, S.K., Haller, E.E., Jena, P., Johnson, N.M., Kirchheim, R., Pearton, S.J., Stavola, M.J., 1992. Rev. Mod. Phys. 64, 559.

Nagumo, M., 2004. Mater. Sci. Technol. 20, 940.

Nakahigashi, J., Yoshimura, H., 2002. J. Alloys Compd. 330–332, 384.

Neuhäuser, H., Schwink, C., 1993. Solid solution strengthening. In: Cahn, R.W., Haasen, P., Kramer, E.J. (Eds.), Materials Science and Technology, vol. 6. VCH, Weinheim, p. 234.

Nibur, K.A., Bahr, D.F., Somerday, B.P., 2006. Acta Mater. 54, 2677.

Nikitin, I., Dong, W., Busnengo, H.F., Salin, A., 2003. Surf. Sci. 547, 149–156.

Nix, W.D., 1989. Metall. Trans. A 20, 2217–2245.

Noh, H., Clewley, J.D., Flanagan, T.B., 1996. Scr. Mater. 34, 665.

Noreus, D., 1989. Z. Phys. Chem. (N.F.) 163, 575(331).

Nörthemann, K., Pundt, A., 2008. Phys. Rev. B 78, 014105/1–014105/6.

Nörthemann, K., Pundt, A., 2011. Phys. Rev. B 83, 155420–155430.

Nörthemann, K., Kirchheim, R., Pundt, A., 2003. J. Alloys Compd. 356–357, 541–544.

Nye, J.F., 1985. Physical Properties of Crystals. Clarendon Press, Oxford.

Oates, W.A., Flanagan, T.B., 1981. Proc. Solid State Chem. 13, 193–272.

Oates, W.A., Lambert, J.A., Gallagher, P.T., 1969. Trans. AIME 245, 47.

Ohring, M., 1992. The Materials Science of Thin Films. Academic Press, San Diego.

Okuyama, H., Siga, W., Takagi, N., Nishijima, M., Aruga, T., 1998. Surf. Sci. 401, 344–354.

Okuyama, H., Siga, W., Takagi, N., Nishijima, M., Aruga, T., 1999. Surf. Sci. 427, 277–281.

Olsson, S., Blomquist, P., Hjörvarsson, B., 2001. J. Phys.: Condens. Matter 13, 1685–1698.

Oriani, R.A., 1970. Acta Metall. 18, 147.

Orimo, S., Nakamori, Y., Züttel, A., 2004. Mater. Sci. Eng. B 108, 51–53.

Otsubo, T., Goto, S., Sato, H., 1982. Proceedings JIMIS 2 Hydrogen in metals, Traduction IRSID T 4258.

Panella, B., Hirscher, M., 2010. In: Hirscher, M. (Ed.), Handbook of Hydrogen storage, Chap. 2 'physisorption in porous materials'. Wiley-VCH.

Pasturel, M., Slaman, M., Schreuders, H., Rector, J.H., Borsa, D.M., Dam, B., Griessen, R., 2006. J. Appl. Phys. 100, 023 515.

Peisl, H., 1978. In: Völkl, J., Alefeld, G. (Eds.), Hydrogen in metals I. Springer, Berlin, pp. 53–73.

von Pezold, J., Lymperakis, L., Neugebauer, J., 2011. Acta Mater. 59, 2969.

Pfeiffer, G., Wipf, H., 1976. J. Phys. F Metal Phys. 6, 167.

Pink, E., Arsenault, R.J., 1979. Prog. Mat. Sci. 24, 1.

Pivak, Y., Schreuders, H., Slaman, M., Griessen, R., Dam, B., 2011. Int. J. Hydrogen Energy 36, 4056.

Pivak, Y., Gremaud, R., Gross, K., Gonzalez-Silveira, M., Walton, A., Book, D., Schreuders, H., Dam, B., Griessen, R., 2009. Scr. Mater. 60, 348–351.

Pundt, A., 2004. Adv. Eng. Mater. 6, 11–21.

Pundt, A., 2005. Nanoskalige Metall-Wasserstoff-Systeme. Universitätsverlag, Göttingen.

Pundt, A., 2013. Materials Challenges in Alternative and Renewable Energy: MCARE 2012.

Pundt, A., Pekarsky, P., 2003. Scr. Mater. 48, 419.

Pundt, A., Kirchheim, R., 2006. Annu. Rev. Mater. Res. 36, 555–608.

Pundt, A., Sachs, C., Winter, M., Reetz, M.T., Kirchheim, R., et al., 1999. J. Alloys Compd. 193–195, 480–483.

Pundt, A., Nikitin, E., Kirchheim, R., Pekarski, P., 2004a. Acta Mater. 52, 1579–1587.

Pundt, A., Suleiman, M., Baehtz, C., Reetz, M.T., Kirchheim, R., Jisrawi, N., 2004b. Mat. Sci. Eng. B 108, 19–23.

Pundt, A., Suleiman, M., Reetz, M.T., Baehtz, C., Jisrawi, N.M., 2004c. Mat. Sci. Eng. B 108, 19–23.

Pundt, A., Brekerboom, L., Niehues, J., Wilbrand, P., Nikitin, E., 2007. Scr. Mater. 57, 889–892.

Pundt, A., Dornheim, M., Guerdane, M., Teichler, H., Jisrawi, N.M., et al., 2002. Eur. Phys. J. D19, 333–337.

Pundt, A., Getzlaff, M., Bode, M., Kirchheim, R., Wiesendanger, R., 2000a. Phys. Rev. B 61, 9964–9967.

Pundt, A., Laudahn, U., von Hülsen, U., Geyer, U., Getzlaff, M., Bode, M., Wiesendanger, R., Kirchheim, R., 2000b. Mat. Res. Soc. Symp. Proc. 594, 75.

Pundt, A., Laudahn, U., von Hülsen, U., Geyer, U., Kirchheim, R., et al., 2000c. Mat. Res. Soc. Symp. Proc. 594, 75–87.

Ramos, M.M.D., 2002. Vacuum 64, 225–260.

Ravi, K.V., Gibala, R., 1971. Met. Trans. 2, 1219.

Rehm, C., Fritzsche, H., Maletta, H., Klose, F., 1999. Phys. Rev. B 59, 3142–3150.

Reichart, P., Dollinger, G., Bergmaier, A., Datzmann, G., Hauptner, A., Körner, H.-J., 2002. Nucl. Instrum. Methods Phys. Res. B 197, 134–149.

Reichart, P., Datzmann, G., Hauptner, A., Hertenberger, R., Wild, C., Dollinger, G., 2004. Science 306, 1537–1540.

Reilly, J.J., Wiswall, R.H., 1970. Inorg. Chem. 9, 1678(313).

Reilly, J.J., Wiswall, R.H., 1988. Top. Appl. Phys. 63, 87(320).

Reimer, P.M., Zabel, H., Flynn, C.P., Matheny, A., Weidinger, A., et al., 1993a. Z. Phys. Chem. (N.F.) 181, 367–373.

Reimer, P.M., Zabel, H., Flynn, C.P., Dura, J.A., Ritley, K., 1993b. Z. Phys. Chem. (N.F.) 181, 375–380.

Reiser, A., Bogdanowic, B., Schlichte, K., 2000. Int. J. Hydrogen Energy 25, 425–430.

Reisfeld, G., Jisrawi, N.M., Ruckman, M.W., Strongin, M., 1996. Phys. Rev. B 53, 4974–4979.

Richards, P.M., 1983. Phys. Rev. B27, 2095.

Richter, D., Driesen, G., Hempelmann, R., Anderson, I.S., 1986. Phys. Rev. Lett. 57, 731.

Rick, S.W., Lynch, D.L., Doll, J.D., 1993. J. Chem. Phys. 99, 8183–8193.

Rieder, K.H., Baumberger, M., Stocker, W., 1983. Phys. Rev. Lett. 51, 1799–1802.

Robertson, I.M., Birnbaum, H.K., 1986. Acta Metall. 34, 353.

Ross, D.K., Stefanopoulus, K.L., 1994. Z. Phys. Chem. (N.F.) 183, 29.

Ross, D.K., 1997. In: Wipf, H. (Ed.), Hydrogen in metals III. Springer, Berlin.

Sachs, C., Pundt, A., Kirchheim, R., Winter, M., Reetz, T., Fritsch, D., 2001a. Phys. Rev. B 64, 075408-1-10.

Sachs, C., Pundt, A., Kirchheim, R., Winter, M., Fritsch, D., et al., 2001b. Phys. Rev. B 64, 075408.

Sakaki, K., Date, P., Mizuno, M., Araki, H., Shirai, Y., 2006. Acta Mater. 54, 4641.

Salomons, E., Griessen, R., de Groot, D.E., Magerl, A., 1988. Europhys. Lett. 5, 449–454.

Salomons, E.M., Feenstra, R., de Groot, D.G., Hector, J.H., Griessen, R., 1987. J. Less Comm. Metals 130, 415–420.

Sander, D., 1999. Rep. Prog. Phys. 62, 809.

Sandrock, G., 2012, IEA-DOE-SNL Hydride DBs.xls, offline (can be provided by request from A. Pundt with kind permission of G. Sandrock). Alternative: The U.S. Department of Energy's (DOE) Office of Energy Efficiency and Renewable Energy (EERE) has launched a comprehensive hydrogen storage materials database <http://hydrogenmaterialssearch.govtools.us/> that includes information from the DOE/IEA Hydpark databases, Hydrogen Storage Material Centers of Excellence, and the Fuel Cell Technologies Program.

Sato, A., Nakamura, Y., Mori, T., 1980. Acta Metall. 28, 1077.

Scharf, T., Faupel, J., Sturm, K., Krebs, H.-U., 2003. J. Appl. Phys. 94, 4273–4278.

Schlapbach, L., Burger, J.P., Bonnet, J.E., Thiry, P., Petroff, Y., Surface Science 189/190 (1987) 747–750.

Schlapbach, L., 1992. In: Schlapbach, L. (Ed.), Hydrogen in intermetallic compounds II. Springer.

Schlapbach, L., 1992. Surface properties and activation. In: Schlapbach, L. (Ed.), Hydrogen in Intermetallic Compounds II, 65. Springer, Berlin, Heidelberg, New York, pp. 15–85.

Schmid, G., 2005. Nanoparticles, from theory to application. Wiley-VCH.

Schmitz, G., Kesten, Ph., Kircheim, R., Yang, Q.M., 1998. Phys. Rev. B 58, 7333.

Schober, T., 1973. Scr. Metall. 7, 1119.

Schober, T., Wenzl, H., 1978. In: Alefeld, G., Völkl, J. (Eds.), Hydrogen in Metals II, Topics Appl. Phys., vol. 29. Springer, Berlin, Heidelberg, p. 11.

Scholz, J., Züchner, H., Paulus, H., Müller, K.-H., 1997. J. Alloys Comp. 253/254, 459.

Schultz, P.J., Lynn, K.G., 1988. Rev. Mod. Phys. 60, 701.

Schwarz, R.B., Khachaturyan, A.G., 1995. Phys. Rev. Lett. 74, 2523–2526.

Schwarz, R.B., Khachaturyan, A.G., 2006. Acta Mater. 54, 313–323.

Scott, T.E., 1978. Bd. 1, Kap. 8, Publ. In: Gschneidner, K.A., Eyring, L. (Eds.), Handbook on the Physics and Chemistry of Rare Earths. North-Holland, Amsterdam.

Seeger, A., 1979. Phys. Stat. Sol.(a) 55, 457.

Seeger, A., 1984. Structure and diffusion of kinks in monoatomic crystals. In: Veyssiere, P., Kubin, L., Castaing, J. (Eds.), Dislocations. CNRS, Paris, pp. 141–178.

Senkov, O.N., Jonas, J.J., 1996. Met. Mater. Trans. A 27A, 1869.

Shewmon, P.G., 1963. Diffusion in Solids. McGraw Hill, New York.

Shirai, Y., Araki, H., Mori, T., Nakamura, W., Sakaki, K., 2002. J. Alloys Compd. 330, 125.

Sieverts, A., 1929. Die Aufnahme von Gasen durch Metalle. Z. Metallkd. 21.

Skripnyuk, V., Rabkin, E., Estrin, Y., et al., 2004. Acta Mater. 52, 405–414.

Smith, J.F., Gjevre, J.A., 1960. J. Appl. Phys. 31, 645.

Snoek, D.L., 1941. Physica (Utrecht) 8, 711.

Somerday, B., Sofronis, P., Jones, R., 2008. Effects of Hydrogen on Materials, ASM International. Materials Park, Ohio.

Song, G., Remhof, A., Labergerie, D., Zabel, H., 1999. Ann. Rep. Ruhr-Universität Bochum 43.

Song, G., Remhof, A., Labergerie, D., Sutter, C., Zabel, H., 2000. J. Alloys Compd. (293–295), 476–479.

Song, G., Geitz, M., Abromeit, A., Zabel, H., 1996. Phys. Rev. B 54, 14093–14101.

Stampfer, J.F., Holley, C.E., Suttle, J.F., 1960. J. Am. Chem. Soc. 82 (299), 3504.

Steiger, J., Blässer, S., Boebel, O., Erxmeyer, J., Weidinger, A., et al., 1993. Z. Phys. Chem. (N.F.) 181, 381–386.

Steyrer, G., Peisl, J., 1986. Europhys. Lett. 2, 835–841.

Stillesjö, F., Hjörvarsson, B., Zabel, H., 1996. Phys. Rev. B 54, 3079–3083.

Stillesjö, F., Ólafsson, S., Isberg, P., Hjörvarsson, B., 1995. J. Phys. Condens. Matter 7, 8139–8150.

Stoney, G.G., 1909. Proc. Roy. Soc. London A82, 172.

Stroe, M.E., 2006, PhD-thesis, Université Libre de Bruxelles, Faculty of Applied Sciences.

Sugano, S., Koizumi, H., 1998. Microcluster Physics. Springer, Berlin, Heidelberg.

Sugimoto, H., Fukai, Y., 1992. Acta Metall. Mater. 40, 2327.

Suleiman, M., 2003. Doctoral thesis. Universität Göttingen.

Suleiman, M., Faupel, J., Borchers, C., Krebs, H.-U., Kirchheim, R., Pundt, A., 2005. J. Alloys Compd. 404–406, 523–528.

Suleiman, M., Jisrawi, N.M., Dankert, O., Reetz, M.T., Pundt, A., et al., 2003. J. Alloys Compd. 356–357, 644–648.

Suleiman, M., Borchers, C., Guerdane, M., Jisrawi, N.M., Fritsch, D., Kirchheim, R., Pundt, A., 2009. Z. Phys. Chem. 223, 169–181.

Szökefalvi-Nagy, A., Filipek, S., Kirchheim, R., 1987. J. Phys. Chem. Solids 48, 613.

Tal-Gutelmacher, E., Gemma, R., Volkert, C.A., Kirchheim, R., 2010. Scr. Mater. 63, 1032–1035.

Tannenbaum, R., King, S., Lecy, J., Tirrell, M., Potts, L., 2004. Langmuir 20, 4507–4514.

Tateyama, Y., Ohno, T., 2003. Phys. Rev. B 67, 174105.

Thompson, A.W., Moody, N.R., 1994. Hydrogen Effects in materials. TMS Warrendale, PA.

Thompson, C.V., 1993a. J. Mater Res. 8, 237.

Thompson, C.V., 1993b. Scr. Mater. 28, 168.

Trinkle, D.R., Woodward, C., 2005. Science 310, 1665.

Tyson, W.R., Miller, W.A., 1977. Surf. Sci. 62, 267–276.

Uchida, H.T., Kirchheim, R., Pundt, A., 2011. Scr. Mater. 64, 935–937.

Uher, C., Cohn, J., Miceli, P., Zabel, H., 1987. Phys. Rev. B 36, 815–818.

Vajda, P., 1995. In: Gschneidner, K.A., Eyring, L. (Eds.), Handbook of the Physics and Chemistry of Rare Earths, vol. 20. Elsevier.

Vajda, P., 2000. Physica B: Condensed Matter, 289–290, 435–442.

van der Merwe, J.H., 1963a. J. Appl. Phys. 34, 117–122.

van der Merwe, J.H., 1963b. J. Appl. Phys. 34, 123–127.

Veleckis, E., Edwards, R.K., 1969. J. Phys. Chem. 73, 683.

Venkatraman, R., Bravman, J.C., 1992. J. Mater. Res. 7, 2040.

Vigeholm, B., Jensen, K., Larsen, B., Schrøder Pedersen, A., 1987. J. Less Common Metals 131, 133–141.

Völkl, J., Alefeld, G., 1975. Hydrogen diffusion in metals. In: Nowick, A.S., Burton, J.J. (Eds.), Diffusion in Solids. Academic, New York, pp. 232–295.

Völkl, J., Alefeld, G., 1978. Diffusion of hydrogen in metals. In: Alefeld, G., Völkl, J. (Eds.), Topics in Metals I. Springer-Verlag, Berlin/Heidelberg/New York, pp. 321–344.

Wagner, S., Moser, M., Greupel, C., Peeper, K., Reichart, P., Pundt, A., Dollinger, G., 2013. Int. J. Hydrogen Energy 38, 13822–13830.

Wagner, C., 1973a. Nachrichten der Akademie der Wissenschaften in Göttingen, II. Mathematisch-Physikalische Klasse 1 (3).

Wagner, C., 1973b. Acta Metall. 21, 1297.

Wagner, H., 1978. Elastic interaction and phase transition in coherent metalhydrogen alloys. See Ref 3, pp. 5–50.

Wagner, H., Horner, H., 1974. Adv. Phys. 23, 587–627.

Wagner, S., Pundt, A., 2008. Appl. Phys. Lett. 92, 051914–051918.

Wagner, S., 2014. Dissertation Thesis, Göttingen, in preparation.

Wagner, S., Uchida, H.T., Baehtz, C., Lukac, F., Vlceck, M., Vlach, M., Burlaka, V., Bell, A., Pundt, A., 2011. Scr. Mater. 64, 978–981.

Walker, G., 2008. Solid-state Hydrogen Storage: Materials and Chemistry. Woodhead Publishing Ltd.

Waninger, M., Pundt, A., 2013. Actual Research.

Weißmantel, C., Hamann, C., 1989. Grundlagen der Festkörperphysik. VEB Deutscher Verlag der Wissenschaften, Berlin.

Weissmüller, J., 1993. Nanostruct. Mater. 3, 261.

Weissmüller, J., 1994. J. Mater. Res. 9, 4.

Weissmüller, J., 2001. Thermodynamics of nanocrystalline solids. In: Knauth, P., Schoonman, J. (Eds.), Nanocrystalline Metals and Oxides: Selected Properties and Applications. Kluwer Acad. Publ., Boston, pp. 1–39.

Weissmüller, J., Lemier, C., 1999. Phys. Rev. Lett. 82, 213–216.

Weissmüller, J., Lemier, C., 2000. Philos. Mag. Lett. 80, 411–418.

Wen, M., Zhang, L., An, B., Fukuyama, S., Yokogawa, K., 2009. Phys. Rev. B 80, 094113.

Wenzl, H., 1982. Intern. Metals Rev. 27, 140.

Westlake, D.G., 1980. J. Less Comm. Metals 75, 177.

Wicke, E., Brodowski, H., 1978a. Top. Appl. Phys. 29, 73(302).

Wicke, E., Brodowsky, H., 1978b. Hydrogen in Palladium and Palladium Alloys. See Ref. 4, pp. 73–151.

Wicke, E., Blaurock, J., 1980. Ber. Bunsenges. Phys. Chem. 85, 1091.

Wilke, M., Teichert, G., Gemma, R., Pundt, A., Kirchheim, R., Romanus, H., Schaaf, P., 2011. Thin Solid Films 520, 1660–1667.

Wilke, S., Hennig, D., Löber, R., 1994. Phys. Rev. B 50, 2548–2560.

Wipf, H., 1978. In: Alefeld, G., Völkl, J. (Eds.), Hydrogen in Metals II, Topics in Applied Physics, vol. 29. Springer Verlag, Berlin, pp. 273–304.

Yamauchi, M., Kobayashi, H., Kitagawa, H., 2009. Chem. Phys. Chem. 10, 2566–2576.

Yang, Q.M., Schmitz, G., Fähler, S., Kirchheim, R., 1996. Phys. Rev. B 54, 9131–9140.

Yoshimura, H., Nakahigashi, J., 2002. Int. J. Hydrogen Energy 27, 769.

Yu, Y., Kim, C., Sanday, S.C., 1991. Thin Solid Films 196, 229.

Zabel, H., Weidinger, A., 1995. Hydrogen in thin metal films and superlattices. Comm. Condens. Mater. Phys. 17, 239–262.

Zabel, H., Peisl, J., 1979. Phys. Rev. Lett. 42, 511–514.

Zabel, H., Peisl, J., 1980. Acta Metall. 28, 589–599.

Zaluska, A., Zaluski, L., Ström-Olsen, J.O., 2000. J. Alloys Compd. 307, 157–166.

Zangwill, A., 1996. Physics at Surfaces. Cambridge University Press.

Zhang, L., An, B., Fukuyama, S., Yokokawa, K., 2009. Jap. J. Appl. Phys. 48, 09JB08–1.

Zlotea, C., Cuevas, F., Paul-Boncour, V., Leroy, E., Dibandjo, P., Gadiou, R., Vix-Guterl, C., Latroche, M., 2010a. J. Am. Chem. Soc. 132, 7720–7729.

Zlotea, C., Campesi, R., Cuevas, F., Leroy, E., Dibandjo, P., Volkringer, C., Loiseau, T., Férey, G., Latroche, M., 2010b. J. Am. Chem. Soc. 132, 2991–2997.

Zlotea, C., Lahoche, M., 2013. Golloids and Surfaces A: Physiochem. Eng. Aspects 439, 117–130.

Züchner, H., 1970. Z. Naturforsch. A 25, 1490–1496.

Züttel, A., Borgschulte, A., Schlapbach, L., 2008. Hydrogen as a future energy carrier. Wiley-VCH, Weinheim.

Züttel, A., Schlapbach, L., 2001. Nature 414, 353–358.

Züttel, A., Nützenagel, C., Schmid, G., Emmenegger, C., Schlapbach, L., et al., 2000. Appl. Surf. Sci. 162, 571–575.

Züttel, A., Sudan, P., Mauron, Ph., Kiyobayashi, T., Schlapbach, L., 2002. Int. J. Hydrogen Energy 27, 203–212.

Züttel, A., Wenger, P., Sudan, P., Mauron, Ph., Orimo, S., 2004. Mat. Sci. Eng. B 108, 9–18.

Biography

Reiner Kirchheim received his Ph.D. in Physics from the University of Stuttgart in 1973. Until 1993 he worked as a senior research scientist at the Max-Plank-Institute for Metals Research in Stuttgart. In 1993 he became a full professor in Materials Physics at the University of Göttingen. Since 2009 he has been a Distinguished Professor of the state of Lower Saxony continuing his work at the University of Göttingen. In 2010 he was elected an external Member of the Max-Planck-Institute for Iron Research in Düsseldorf. Since 2011 he is also Principal Investigator at the World Premier International Institute for Carbon-Neutral Energy Research, Kyushu University, Japan.

Astrid Pundt studied Solid State Physics at the TU Braunschweig. In 1991 she moved to the Göttingen Institute of Metal Physics in the group of Peter Haasen. She achieved the doctoral degree in 1994. Since 1996, her own research group focuses on thermodynamics and kinetics of nano-materials like thin films, wires and clusters by using hydrogen-metal systems as platform. She finished her habilitation treatise in 2001 and works as a Professor (apl) at Göttingen University since 2008. In 2013 she was invited to Uppsala University for lectures and research. Astrid Pundt received the Heisenberg grant, is head of the Peter-Haasen price board, acts in several committees and has organized international conferences and symposia. She published more than 90 papers.

26 Physical Metallurgy of Nanocrystalline Metals

Gerhard Wilde, Institute of Materials Physics, University of Münster, Münster, Germany

26.1 Introduction

Nanostructured materials and composites present a promising class of engineering materials with structural length scales well below 1 μm. This microstructure characteristics result in properties and property combinations that are often enhanced or even unique compared to the polycrystalline counterparts. It has attracted enormous interest in basic research as well as in the application-oriented engineering sciences or—more recently—even in public media. In the historical perspective, and even disregarding the obvious coincidence of structure formation on a nanometer scale and basic synthetic chemistry, nanostructured materials have been around for a very long time. Natural examples can be found in abundance, such as the surface of butterfly wings, shark teeth or the Abalone shell—both latter examples utilize a nanocomposite approach to achieve high hardness or strength combined with a large toughness. Early man-made examples include damascene sword making (or equivalently, the Japanese art of sword making), where layers of two different steel grades with largely different mechanical properties are welded and repeatedly forged in over a hundred layers, yielding a nanocomposite that combines high hardness and a large elastic limit. A different—and probably earlier—technique of damascene sword making was based on a banding phenomenon of nanostructured Fe_3C precipitates in the legendary Indian Wootz steel (Srinivasan and Ranganathan, 2004). Even the size dependence of materials properties, which represents the very foundation of today's interest in nanostructured materials, has been exploited already several hundred years ago, although without addressing or understanding the underlying mechanisms. A prominent example is given by church windows from the Middle Ages: colloidal Gold- or Silver-nanocrystals are dispersed in window glass to obtain the special bright red or green colors. In this case, the size-dependent shifts of the surface plasmon wavelength of Gold or Silver are responsible for the color of these materials. Thus, utilizing nanoparticles, the color could be tuned without hampering the transparency of the glass (**Figure 1**).

Despite these early examples for man-made nanostructured materials, the basic scientific idea concerning the synthesis of nanostructured materials and concerning their modified properties was presented by Gleiter and Marquardt (1984). Conceptually, nanostructured materials were described as consisting of initially isolated single crystallites of nanometer dimensions that form

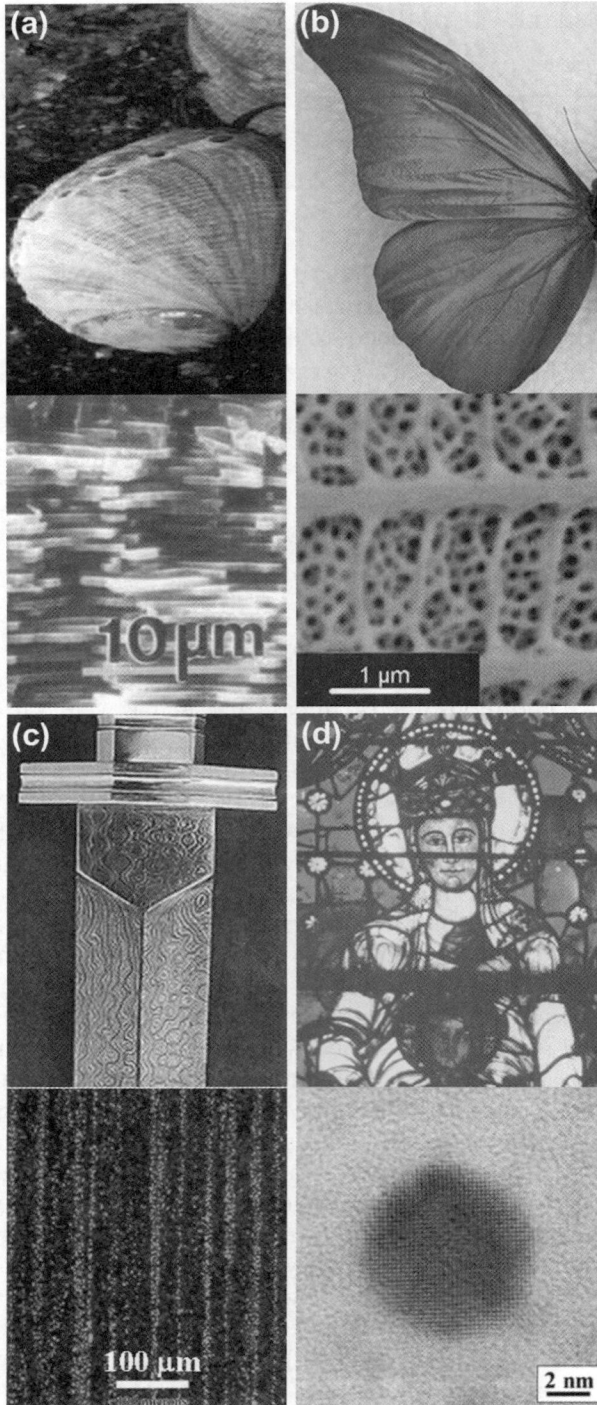

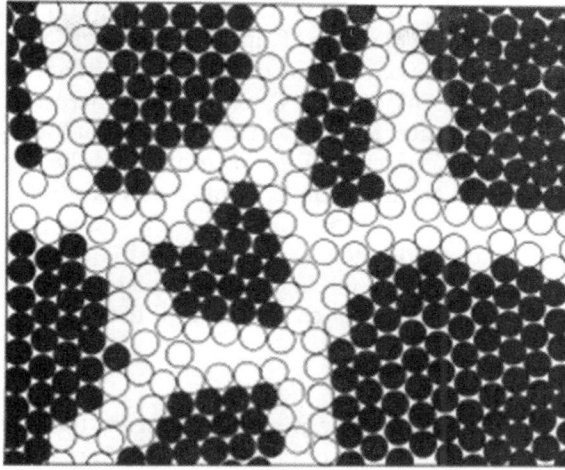

Figure 2 Two-dimensional schematic representation of a nanocrystalline material that consists of a grain boundary- and a grain core-component, as indicated by the black and white disks. The color coding represents the different local configurations of atoms in the core region of nanoscale grains or in the grain boundary region that is related to differences, e.g. in the number of nearest neighbors, the local specific volume etc. Reproduced with permission from Elsevier (Gleiter, 1996).

a three-dimensional (macroscopic) nanocrystalline (NC) solid upon consolidation. Due to disregistry at the interfaces, the grain-boundary (GB) regions were postulated to be characterized by low packing efficiency and low degree of order. While this concept has led to the coming of nanomaterials, recent experimental evidence indicates that the interfaces between the regions with a coherent lattice are atomically sharp in pure, single-phase materials, indicating fast relaxation of the atomic positions that are energetically most unfavorable into energetically as well as spatially near-by positions within the disordered core region of random high-angle grain boundaries (high-angle GBs) (**Figure 2**).

Metallic Materials have been addressed rather early on in the course of investigating nanostructured materials and have presented model systems for studying the impact of size confinement effects and of

Figure 1 Examples of natural and man-made nanostructures in normal view and at higher resolution revealing the nanoscale microstructure, respectively. (a) Abalone shell with the typical appearance of the outer shell. The microstructure reveals thin alternating layers of aragonite and layers of the protein matrix that form a composite with high strength and high toughness (Reprinted with permission from the Materials Research Society, MRS (DiMasi and Sarikaya, 2004)). (b) Butterfly wing of the butterfly Morpho rhetenor (Reprinted with permission from Macmillan Publishers Ltd: Nature (Vukusic and Sambles, 2003), copyright 2003). The scanning electron microscopy image displays the photonic structure of the wing surface that causes the iridescence (Reprinted figure with permission from Biro et al. (2003). Copyright (2003) by the American Physical Society). (c) Celtic sword that was forged in the so-called "damascene" tradition, using the so-called wootz-steel. The surface pattern results from a carbide-banding phenomenon produced by the microsegregation of minor amounts of carbide-forming elements present in the wootz ingots (Reprinted with permission from The Minerals, Metals and Materials Society, TMS (Verhoeven et al., 1998)). (d) Window of Notre Dame de la Belle Verrière. The red color in medieval church windows is often generated by Gold nanoparticles, and green is due to Silver nanoparticles that are dispersed within the glass. The micrograph shows a silver nanoparticle in high resolution. Reprinted with permission from the Materials Research Society, (Hofmeister et al., 2005). (For interpretation of the references to color in this figure legend, the reader is referred to the online version of this book.)

the impact of internal interfaces, i.e. the GBs and particularly random high-angle GBs, on macroscopic materials properties. After almost three decades of intensive research, a wealth of insight has been gained on all aspects of materials science of this new class of materials and research on nanostructured metallic materials has also produced enhanced insight into the microstructure–property relation of nonmetallic materials or into the structures and properties of defects of the crystal lattice, e.g. the GBs or the triple junctions (TJs) between GBs, because of the abundance of these defects at small grain sizes. Due to the continuing high interest in NC materials, it is inherently impossible to cover the full range of research on these materials or to present a full coverage even of a selected range of subtopics. Thus, the present chapter is meant to cover a range of interconnected topics in the area of NC materials that, to the author's belief, represent best the aspects that fall under the chapter topic of "physical metallurgy of NC materials" and that encompass published research results until the first half of the year 2012.

26.1.1   Scope

In the present chapter, we will focus on those aspects that concern the physical metallurgy of nano-structured metallic materials directly, i.e. the science and technology of nanostructured metals and alloys and the effects of processing, composition and environment on their properties. Aspects related to the specific nature of GBs or triple lines as well as the basic issues concerning diffusion, deformation or phase transformations are covered in other chapters of the present volume. Additionally, the term "nanostructured" material will be used as synonymous to "NC" materials. Thus, the scope will include materials with macroscopic dimensions and nanoscaled microstructures. Concerning isolated nano-particles, one-dimensional nanostructures or thin films and surfaces, we also refer to other chapters in this volume or to the numerous reviews and textbooks available on these particular aspects. Within this chapter, we will first introduce current synthesis aspects and the basic elements of the microstructure before the characteristics of diffusion, deformation, thermodynamics and phase transformations are addressed, followed by reviewing selected examples of microstructure–property relations. In the last chapter, open issues and future perspectives will be discussed as an outlook on ongoing and future work in this field.

Nanostructured or NC materials are often defined in terms of a characteristic length scale of a distinct feature of their microstructure. Most often, the average distance of incoherent interfaces in the microstructure, i.e. the grain size, is utilized for this purpose, but there are examples where the spacing of coherent interfaces on the scale of a few nanometers controls the properties. As concerning the critical length scale that defines the demarcation between nanostructured and conventional materials, different views are suggested that span a size range between a few tens to a few hundreds nanometers. Yet, the important issue about nanostructured materials is the dependence of their properties on the size scale of their microstructure. That dependence varies for different properties and, even more strongly, for different classes of materials. Thus, we will address in this chapter metallic materials with micro-structures that are sufficiently fine-grained as defined by the average spacing of coherent or incoherent internal interfaces, so that the resulting property under observation is significantly modified as compared to the respective behavior of coarse-grained (cg) material of identical average chemical composition.

26.2   Basic Concepts

Nanostructured metallic materials are characterized by high densities of internal interfaces and regions of coherent lattice symmetries that are significantly smaller than 1 μm in each spatial dimension.

In many cases of real materials, a great variety of grain-to-grain disorientations will be present in one material. Thus, GBs are treated as random high-angle GBs with properties that are thought of as averaged over the five-dimensional configuration space of the degrees of freedom of the GB disorientation.

In any case, the GBs separating the individual crystallites have a different atomic structure (e.g. the interatomic distance and the coordination number) than the regions of the crystalline cores. Thus, generally two different mechanisms contribute to the modification of properties or physical effects in nanostructured materials. The first mechanism is based on size confinement that occurs if the characteristic dimensions of a material (e.g. the diameter of the region with coherent crystal lattice) are of the same length as certain characteristic length scales such as a mean free path, the electron wave length, a correlation length etc.

In addition, the GBs present regions of increased (local) specific volume, increased free energy and reduced order. Thus, we can expect that all properties of a material are affected by nanostructuring, not only properties that are characterized by an intrinsic length scale of the order of a few nanometers.

With the GB width of relaxed random high-angle GBs in a typical fcc metal of the order of about 0.5 nm and taking space-filling average grain shapes into account, it can easily be shown that more than 10% of all atoms that constitute the material are situated at GBs if the average grain size is less than 20–30 nm (**Figure 3**; Palumbo et al., 1990).

At even smaller grain sizes, additional types of defects such as TJs that connect three—of three adjacent grains (**Figure 3**) or quadruple points that connect four TJs need to be taken into account. Thus, high densities of defect cores, specifically GBs, define nanostructured materials and constitute the main difference between nanostructured and other classes of materials (e.g. quasicrystals, glasses, poly- or single crystals).

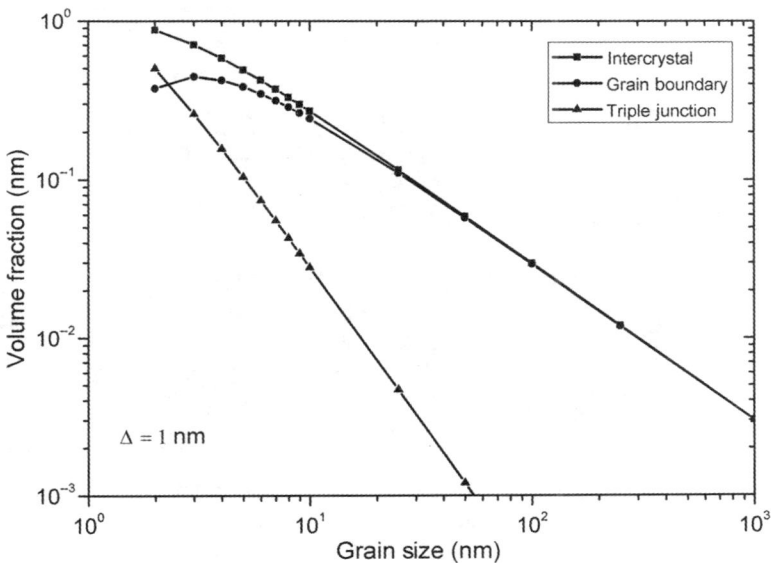

Figure 3 The effect of grain size on calculated volume fractions for intercrystalline regions, GBs and TJs, assuming a grain boundary thickness of 1 nm. Reprinted with permission from Elsevier (Palumbo et al., 1990).

According to an overwhelming majority of recent results from experiments and from atomistic simulations, the GBs in NC materials have similar atomic structures as GBs in cg polycrystalline materials. Naturally, that structure might vary with synthesis or processing pathway due to the interaction of lattice defects such as dislocations or vacancies that are created during processing with the GBs. Yet, this dependence is not unique for GBs in nanostructured materials. Thus, the average grain size presents a proper metric for scaling of nanostructure-induced property variations. It also is clear that the excess free energy associated with the density of internal interfaces constitutes a thermodynamic driving force for coarsening that poses severe limitations concerning the kinetic stability of NC single-phase materials. In this respect, multiphase materials offer additional options for enhancing the microstructure stability kinetically or even thermodynamically, as discussed in the following.

A variant of the NC material indicated schematically in **Figure 2** is given by nanostructured composite materials where the density of heterophase interfaces (or phase boundaries) constitutes the microstructure element that defines the characteristic length scale of the material. A classification scheme for nanostructured materials has been suggested (Gleiter, 2000), according to the dimensionality of crystalline regions and the distribution of chemically dissimilar phases (**Figure 4**).

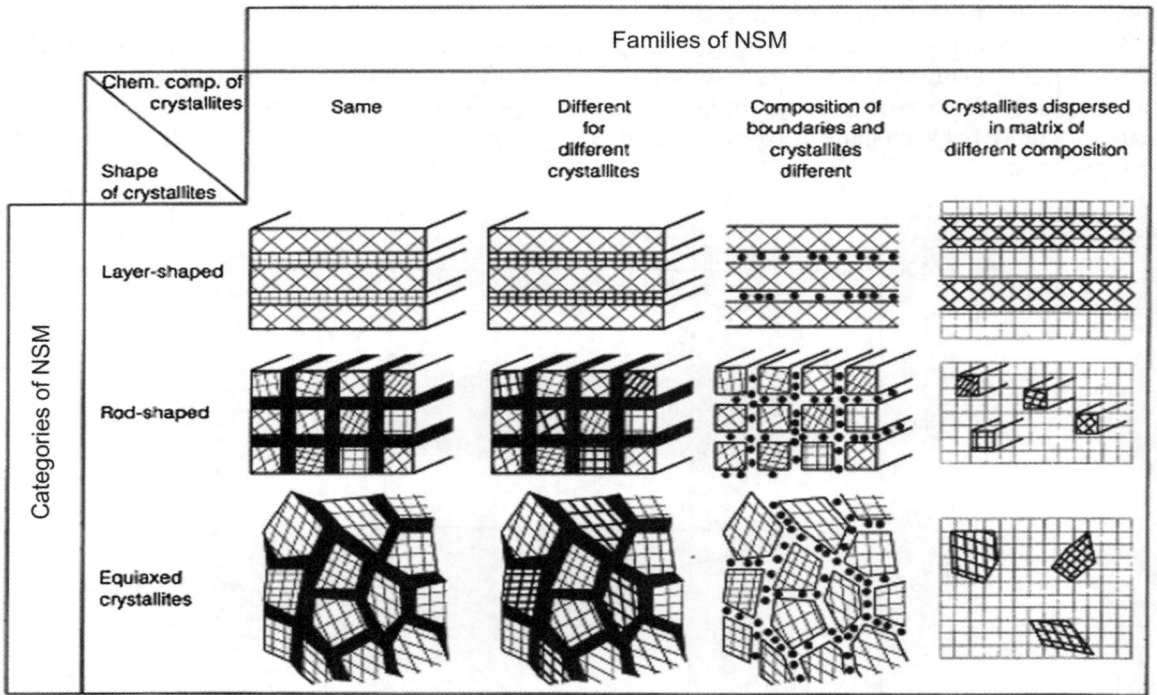

Figure 4   Classification scheme for nanostructured materials according to their chemical composition and the dimensionality (shape) of the crystallites (structural elements) forming the material. The boundary regions of the first and second family of nanostructured materials are indicated in black to emphasize the different atomic arrangements in the crystallites and in the boundaries. The chemical composition of the (black) boundary regions and the crystallites is identical in the first family. In the second family, the (black) boundaries are the regions where two crystals of different chemical composition are joined together causing a steep concentration gradient. Reprinted with permission from Elsevier (Gleiter, 2000).

26.3 Synthesis Options

The numerous options for generating metallic materials with high densities of internal interfaces can be subdivided into three main routes, starting from an amorphous, condensed phase, from vapor or from a polycrystalline solid. The NC structures that can be synthesized by these different types of processing pathways as well as the most important peculiarities are summarized below. Due to the excellent coverage of these aspects in the literature, the summary is not meant to be complete but to provide a basic overview.

26.3.1 Nanocrystallization of Amorphous Phases

Obviously, the method of achieving nanostructured metallic materials through the nanocrystallization of amorphous phases is restricted to alloys and compositions that permit initial vitrification or to processes that allow either for extreme rates of heat extraction of a liquid phase, such as melt spinning or splat quenching or to processes that can drive the material far away from thermodynamic equilibrium at low homologous temperatures, such as ball milling or irradiation. The key to obtaining nanostructured materials through this approach is to control the rate of nucleation (that should be high) and the rate of subsequent crystal growth (which should be low). Due to the different temperature dependencies of these two rates, temperature regimes can be found that allow for copious nucleation at low growth rates (**Figure 5**). This approach has been applied to various alloys, yet mostly to the class of the so-called marginally glass-forming systems, since in these glasses the crystallization upon heating starts at unusually high nucleation densities and low temperatures, i.e. at low growth rates (Lu, 1996; Wilde et al., 2003). Due to the kinetic restrictions for initial vitrification, these materials are obtained in the form of ribbons or powders. Massive NC/amorphous composites have been synthesized via applying dedicated thermal treatments to selected bulk metallic glasses (Xing et al., 1999).

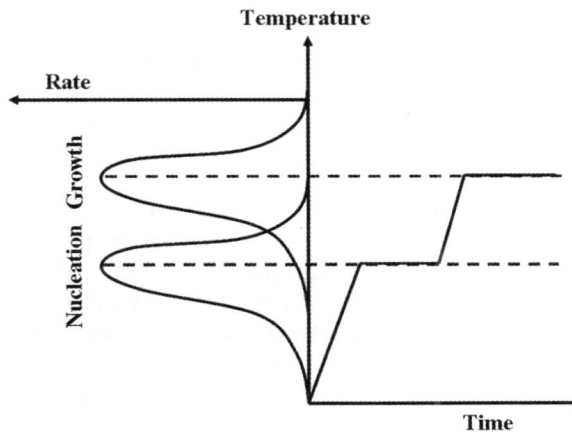

Figure 5 Schematic representation of the dependence of nucleation rate and growth rate on temperature. On the right side of the diagram, a temperature–time profile is indicated that is commonly applied for synthesizing glass ceramics from silicate glasses. Holding near temperatures where the nucleation rate attains its maximum yields a high number density of small crystallites. Rapid heating to higher temperatures and annealing at a temperature where the growth rate is high serves to fully crystallize the material by rather uniform growth of all crystals. The final product microstructure is thus characterized by a narrow size distribution of rather small grains.

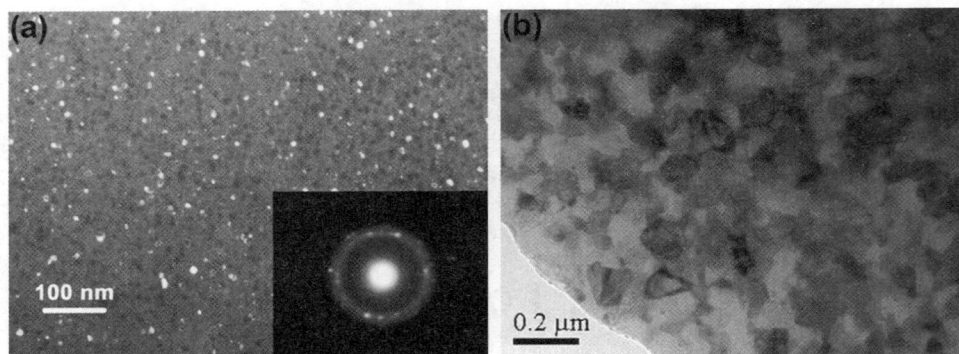

Figure 6   (a) TEM micrograph of an aluminum-rich Al–Y–Fe alloy that had been initially vitrified by melt spinning. Afterward, the metallic glass has been deformed at room temperature by torsion under high applied pressure to yield a very high number density of Al-nanocrystals. (b) After further thermal treatment, the nanocrystals grow and impinge, yielding a completely nanocrystallized Al-rich alloy with a rather uniform and small grain size (Boucharat et al., 2005).

An alternative route to obtain completely NC materials from an amorphous precursor is given by severe deformation processing of metallic glasses (Boucharat et al., 2005). Due to the enhanced atomic mobility inside the shear bands (see Chapter 4, Volume I), and due to the initiation of multiple shear bands during severe deformation, high number densities of nanocrystals can be formed at the deformation temperature. Upon subsequent thermal treatment, the nanocrystals impinge early during growth, rendering completely NC microstructures. **Figure 6** shows the example of a NC Al-rich Al–Y–Fe alloy with an average grain size of only 75 nm after high-pressure torsion (HPT) and subsequent annealing at low temperature (Boucharat et al., 2005).

More details on nanostructure synthesis by metallic glass devitrification can be found in the excellent review by Lu (1996). Methodologically, similar processes are based on rapid melt quenching of alloys with a miscibility gap in the liquid state that can form nanocomposite microstructures upon crystallization (Goswami et al., 1999).

26.3.2   Deposition or Precipitation Routes

The second route consists of synthesis methods that utilize a liquid or gaseous precursor state from which the nanostructured material forms through deposition or precipitation and that are often followed by a consolidation process. Aerosol reactor processes, chemical vapor deposition (CVD), inert gas condensation (IGC), physical vapor deposition (PVD), chemical reaction precipitation and electrochemical deposition (ED) constitute frequently applied examples of this route. A schematic representation, showing the principle of operation of the IGC method is indicated in **Figure 7** (Gleiter, 1989).

IGC has the advantage that, after additional consolidation, mostly one-domain nanocrystals with chemically unmodified surfaces are statistically assembled to form three-dimensional NC materials with random texture and mostly general high-angle GBs connecting mostly defect-free NC grains. Due to the large grain-boundary excess free energy, coarsening via grain growth occurs even at ambient temperature for single-phase material, i.e. at homologous temperatures as low as $T/T_m = 0.17$ (Ames et al., 2008), with T_m, the melting temperature, thus posing severe restrictions on the options for thermomechanical consolidation. Yet, relative densities higher that 95% have been reported after

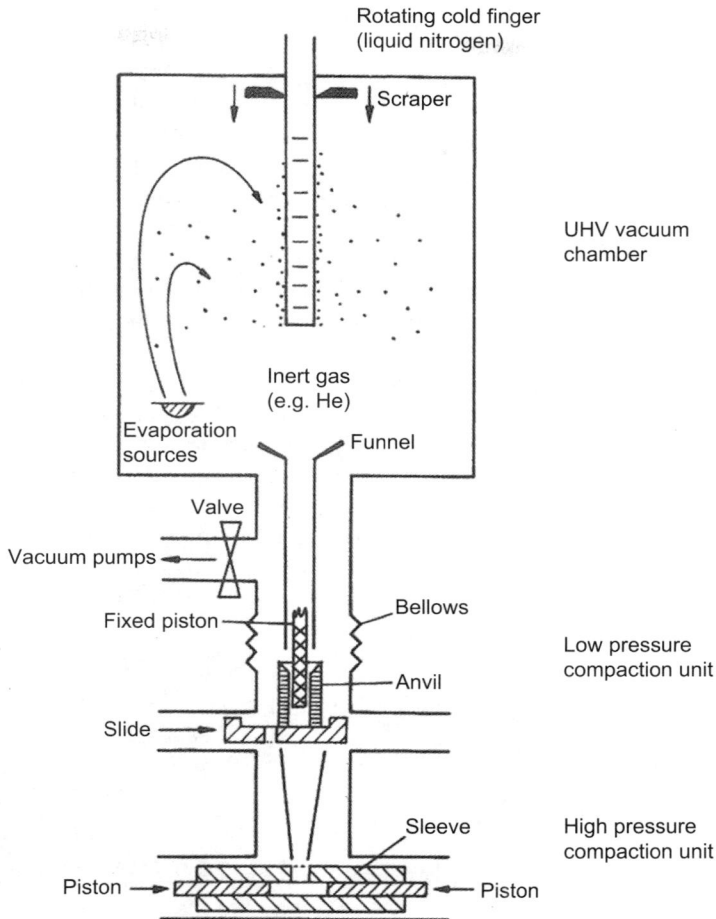

Figure 7 Schematic representation of the IGC process. The metallic material evaporates and condenses in the inert gas in the form of clusters or small nanocrystallites. Convection and thermophoresis transports the particles to the cold finger where it forms a nanocrystalline coating. The coating is scraped off, collected and in situ consolidated to render compact nanocrystalline specimens. Reproduced with permission from Elsevier (Gleiter, 1989).

room-temperature consolidation of metals with relatively high melting temperatures such as Pd. One way to overcome the restrictions imposed by the rapid coarsening kinetic is given by utilizing alloys instead of pure metals (Schäfer et al., 2011) or by doping with GB-active elements (see Section 26.4.2 of the current chapter). An overview on the IGC process can be found in (Gleiter, 2000).

Chemical reaction–precipitation processes are widely used for generating nanoparticle dispersions, utilizing a wide range of solvents and different types of chemical reactions (hydrolysis, chemical reduction of complex precursors etc.). The resulting metallic nanoparticles are usually capped by ligand molecules with charged end groups that prevent agglomeration within the solvent. Subsequent processing steps that involve filtration, debinding of the ligand molecules and thermomechanical consolidation steps yield macroscopically extended materials with nanoscale grain structures. These methods have been addressed in reviews such as (Burda et al., 2005) to which the interested reader is referred.

Deposition processes such as CVD or aerosol processes (Hahn, 1997) mostly serve for synthesizing nonmetallic materials due to the large negative formation enthalpy of metal-oxides or metal-nitrides and the fast reaction kinetics at metal nanoparticle surfaces. PVD and ED methods can either be used in conjunction with nanoporous substrates or nanostructured host phases where the metallic nanophase forms isolated dot- or tube-like structures. A wide range of nanostructured substrates or host materials is used, based on e.g. diblock copolymers, porous metal oxides, mesoporous Si, zeolithes etc. These fascinating groups of methods for synthesizing heterophase nanocomposites allow controlling the shape, spacing and geometrical arrangement of the metallic nanostructures and thus provide exciting opportunities for tailoring functionalities of metal–nonmetal composite nanostructures (Lei et al., 2007, 2011).

Depending on the deposition parameters (deposition rate, substrate temperature, and addition of chemical agents) PVD and ED processes can also be used for direct deposition of NC layers. Lu et al. (2004) have recently used electrodeposition to synthesize thick layers of fine-grained pure Cu. Due to the specific ED conditions, domains with high densities of twin GBs were formed, with average twin lamellae thicknesses down to less than 20 nm (see **Figure 8**). This hierarchical microstructure consisting of random high-angle GBs with average spacing of about 2 μm and twin boundaries within these domains with an average spacing of 20 nm showed exceptional mechanical properties with increased strength and increased ductility at the same time (see: Section 26.6.1.2).

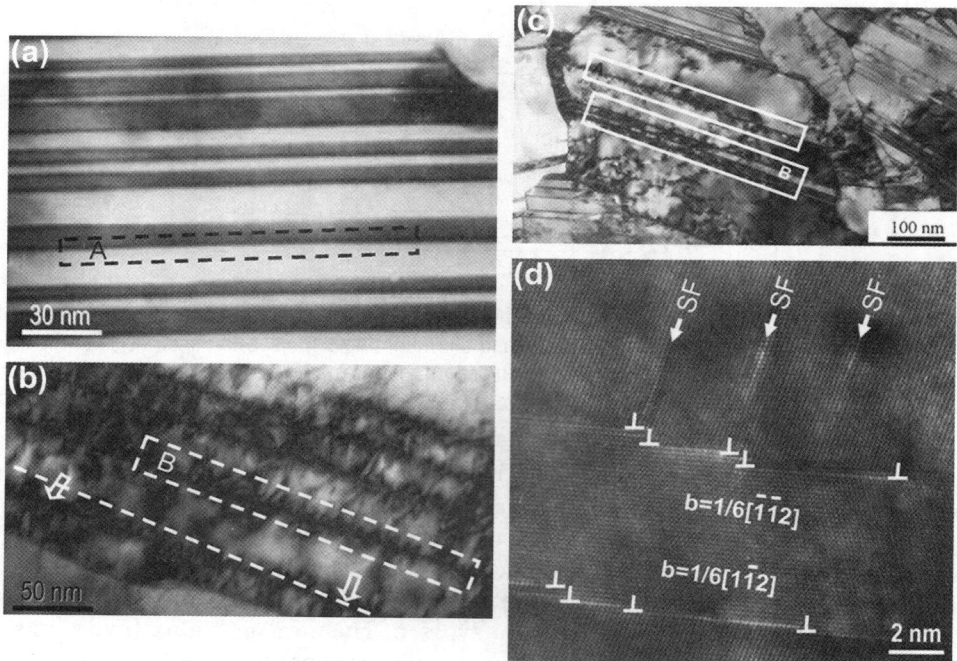

Figure 8   TEM micrograph showing the microstructure of nanotwinned copper after electrodeposition. The microstructure consists of grains with random high-angle GBs that have an average spacing of a few micrometers and nanoscale twin GBs with spacing of less than 20 nm. Reproduced with permission from Elsevier (Dao et al., 2006).

26.3.3 Defect Accumulation

The third general route for synthesizing nanostructured metallic materials is based on a cg initial state. The process of nanostructuring involves either ion-implantation of elements that are immiscible in the matrix or the introduction of large densities of lattice defects, specifically dislocations, which can rearrange to form new GBs and thus serve to successively break up the original cg microstructure. This group of methods that consists of several variants such as ball milling, high-energy impact, high strain rate deformation, sliding wear and friction or irradiation with high-energy particles has been used frequently for obtaining metastable or even thermodynamically unstable states of matter such as metallic glasses, supersaturated solid solutions or NC materials, by increasing the free energy of the material under conditions (such as sufficiently low processing temperatures) that prevent significant relaxation of the energized state during the synthesis process. The microstructures formed during these processes, particularly the resulting average grain size, can be controlled by adjusting the processing conditions such as the strain imparted, the strain rate, the temperature during processing etc. On the level of microstructure formation, these processing conditions affect the creation- and annihilation rate of lattice defects and thus the total accumulated/retained defect density. A clear indication for the occurrence of dynamic recovery processes is given by the dependence of the minimum grain size on the melting temperature of different materials after ball milling (**Figure 3.3**; Koch, 1997). The fact that for all methods a saturation grain size is reported that to some extend can be shifted by choosing proper processing conditions (e.g. the processing temperature) is intriguing and indicates that dynamic recovery processes are active during processing that effectively limit the lowest attainable grain size.

With ball milling, two remarkable methodological developments are cryogenic milling, which has recently been reviewed by Witkin and Lavernia (2006) and mechanical milling assisted by electrical discharge (**Figure 9**; Calka and Wexler, 2002). While this group of methods has already been upscaled to industrial-scale processing, consolidation of the resulting powder agglomerates is required for obtaining massive NC materials, rendering the obtained materials susceptible for contamination and porosity retention.

Recently, spark plasma sintering has been utilized for consolidating mechanically milled powder to full density without significant coarsening. Song et al. have developed a closed process including powder synthesis by ball milling or IGC and subsequent consolidation by spark plasma sintering (**Figure 10**; Song et al., 2006), without exposing the material to air or other reactive gases. Applied on pure Cu, massive samples with a narrow grain size distribution and a low average grain size of about 30 nm were obtained. So far, it is still an open question that why this material shows a high stability against rapid coarsening given the high atomic mobility of Cu even at room temperature.

26.3.3.1 Severe Plastic Deformation

A separate group of methods that has been developed in recent years for synthesizing materials with high densities of internal interfaces is summarized under the term "severe plastic deformation, (SPD)" (Valiev, 2004). These methods generally are defined by applying large strains normally at low or moderate strain rates to massive, cg materials without changing the macroscopic (geometrical) shape of the processed samples. Thus, repeated application of the straining process is possible so that very high strain levels can be reached. Similar to mechanical milling, a material-dependent and processing-dependent saturation grain size is obtained, which indicates a dynamic steady state between defect production, defect interaction and defect annihilation.

Out of the various variants of SPD processing methods, equal-channel angular pressing (ECAP), HPT and accumulative roll bonding (ARB) are the most widely used methods. Several of the SPD processing

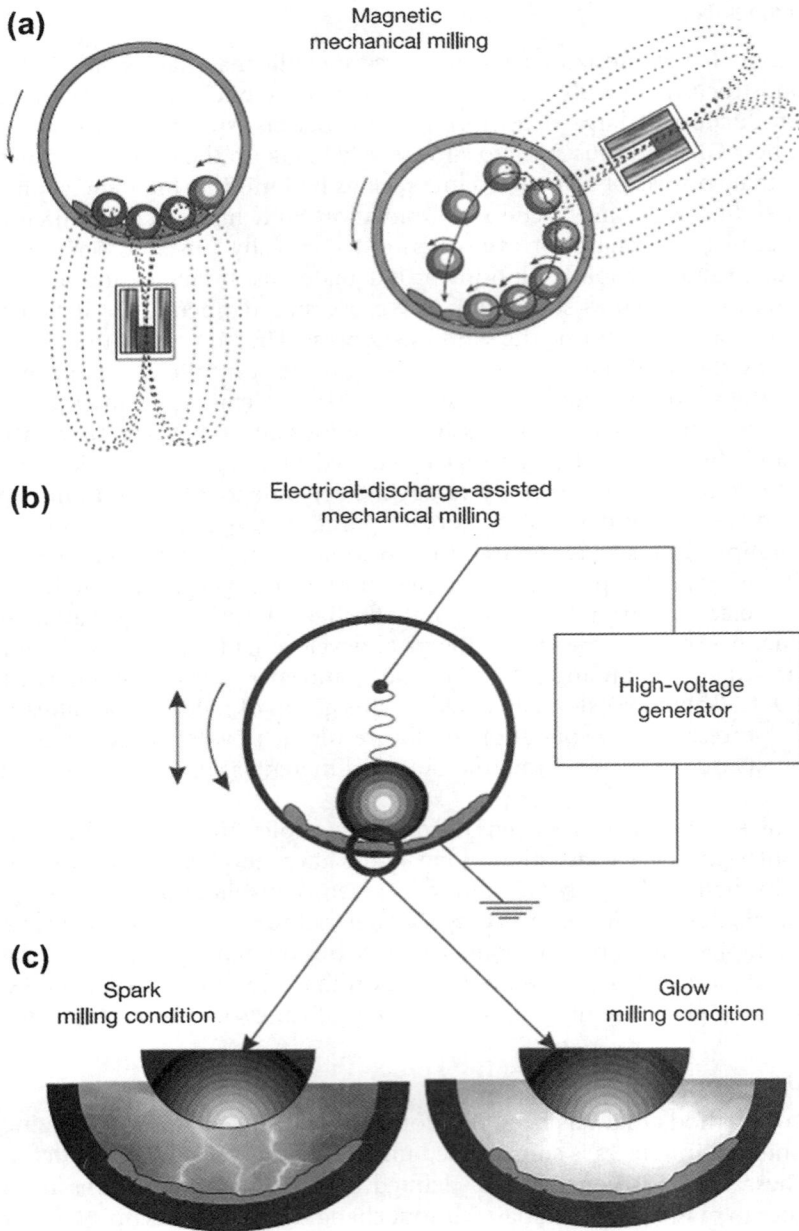

Figure 9 Schematic illustrations of magneto-ball milling and electric-discharge-assisted mills. (a) Conventional milling under external magnetic field. External magnets control ball motion producing milling modes with various combinations of impact and shear. (b) Electric-discharge-assisted mechanical milling method (either rotational or vibrational type mill). (c) Type of milling is characterized by type of electric discharge, either hot milling under spark discharge or cold milling under glow discharge (Calka and Wexler, 2002). (For color version of the figure, the reader is referred to the online version of this book.)

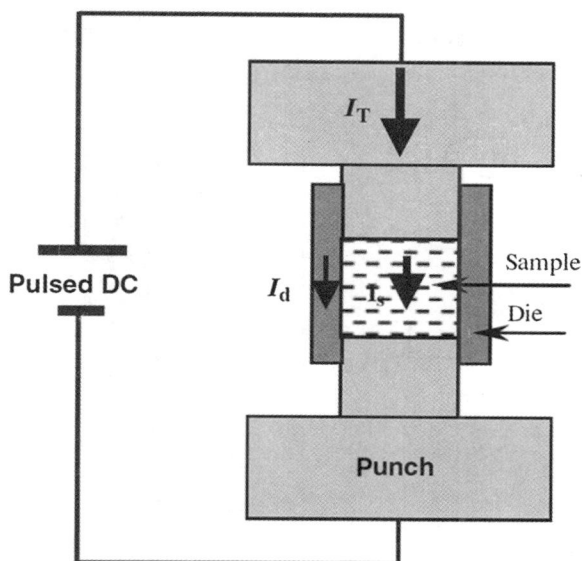

Figure 10 Diagram illustrating the parallel connection of the resistances between the sintering sample and the graphite die in spark plasma sintering. Reproduced with permission from John Wiley and Sons (Song et al. 2006).

routes, particularly ECAP and HPT (**Figure 11**; Valiev et al., 2000), apply a high quasi-hydrostatic pressure in the shear zone where the major part of the deformation occurs.

It has been argued that the high applied pressure, that is of the order of several gigapascal, increases the vacancy migration enthalpy significantly, thus hindering relaxation and dynamic recovery processes during straining, which then leads to the achievement and retention of ultrafine-grained (UFG) or even NC microstructures (Zehetbauer et al., 2003). However, Dinda et al. (2005) obtained NC pure Ni with an average grain size of slightly less than 10 nm by repeated cold rolling and folding, which is a variant of the ARB process. The estimated hydrostatic pressure component of this deformation process was about 0.8 GPa, which is too small to significantly alter the activation barrier for vacancy migration in Ni. Comparing the resulting microstructures that have been obtained through processing at different temperatures, including cryogenic temperatures, and using different deformation methods, i.e. ECAP, HPT or ARB, the number of activated slip systems and the minimization of dynamic recovery seem to be most important parameters for obtaining microstructures with ultrafine grain sizes.

Several SPD techniques that are based on the ECAP principle allow for upscaling to produce UFG microstructures on an industrial scale. In fact, these materials have already reached the market in the area of biomedical applications. More recently, SPD methods, particularly HPT and ECAP, have been used in combination with other methods that can drive materials away from thermodynamic equilibrium or they have been applied sequentially with other SPD methods (Boucharat et al., 2005; Wilde et al., 2006) to obtain nanostructured metals or alloys with extremely small grain sizes. Additionally, HPT has been used for consolidating powder samples including materials with NC microstructures that were initially produced by ball milling or IGC. One limiting factor in this context is given by deformation-induced grain coarsening that has been reported to be significant, specifically for NC and UFG materials. With the observation of more pronounced coarsening under mechanical loading

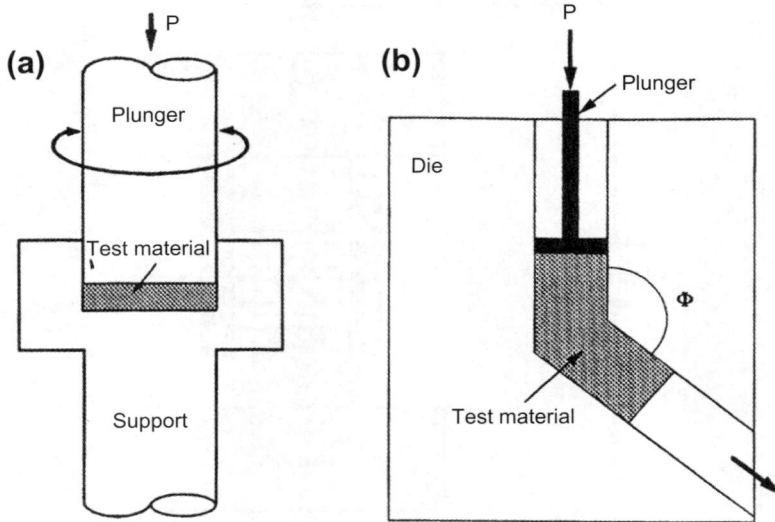

Figure 11 Schematic drawings that indicate the operating principles of the two most common methods of severe plastic deformation (SPD): (a) High Pressure Torsion and (b) Equal Channel Angular Pressing. Reproduced with permission from Elsevier (Valiev et al., 2000).

conditions at lower temperatures within the cryogenic regime, (Zhang et al., 2005) it seems clear that this type of coarsening is stress-driven.

With equal ECAP and variants thereof, large volumes can be processed and transformed into an UFG microstructure that is characterized by high densities of high-angle GBs. The microstructure evolution, particularly the final average grain size at saturation, the distributions of grain orientations and grain shapes and the distribution of GB disorientations between adjacent grains depends on the details of the processing pathway (**Figure 12**; Valiev and Langdon, 2006).

Specifically the application of defined rotations of the samples with respect to the extrusion axis between repeated ECAP steps affects the final microstructure to a large extent, since different numbers of different slip systems are activated (**Figure 13**; Iwahashi et al., 1998). Further details concerning this rapidly growing area of research can be found in some of the reviews on SPD (e.g. Valiev et al., 2000; Valiev and Langdon, 2006).

26.3.3.2 *Methods to Obtain Graded Nanostructures*

In recent years, a number of methods have been developed that apply large strains only to some part of the entire volume of a material. In the case of friction stir welding e.g. only the joining region of two work pieces is so severely deformed that the melting temperature is locally exceeded. Quite generally, nanostructuring methods such as friction stir processing or surface mechanical attrition produce graded microstructures that are often beneficial for applications where wear or corrosion protections are involved. Several methods have been proposed to develop such microstructures that are also based on applying large strains. Common to these methods is the creation of microstructures that are often nanostructured near the surface layer and have successively increasing grain sizes and decreasing defect densities with increasing depth into the material. While methods such as machining at high

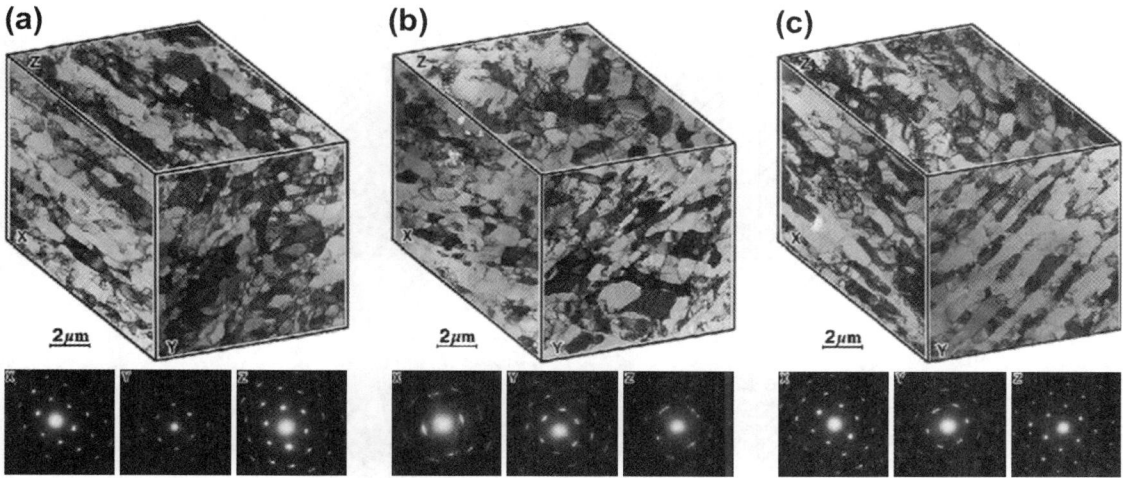

Figure 12 Microstructures on the *X*, *Y* and *Z* planes for polycrystalline aluminum after ECAP through two passes using (a) route A, (b) route B and (c) route C together with the associated SAED patterns. Reproduced with permission from Elsevier (Valiev and Langdon, 2006).

deformation rates or surface mechanical attrition affect micrometer-thick layers, friction stir processing and related methods affect thick layers of several millimeters in thickness. While these methods are expected to have significant commercial impact (cf. Section 26.8.3 of the current chapter), it seems that the microstructure evolution and the range of property modification achievable by these and related methods are similar to the SPD methods discussed above and thus do not require a separate discussion. More extensive reviews on the synthesis of graded nanostructures can be found in Li et al. (2008) and Zhang and Hansen (2007).

26.4 Microstructure Aspects

NC materials do not possess new or unique microstructure elements as such. Therefore, the reader is referred to earlier chapters in this volume that focus specifically on the description of the structure and properties of lattice defects (c.f. Chapters 10, 12 and 14, Volume II). However, due to basic scaling laws, two- and one-dimensional defects will become increasingly important at decreasing grain size, i.e. at decreasing volume of the regions with a coherent lattice (see **Figure 3**). Additionally, it has long been argued that due to the short distances between efficient sinks for defects of the crystal lattice, i.e. the GBs, NC grains would be expected to be mostly free of defects of the grain interiors, such as full or partial dislocations. However, recent results obtained by transmission electron microscopy, local strain analyses by geometric phase analysis (GPA) or through line profile analyses based on X-ray diffraction clearly indicate that NC materials that have been obtained via defect accumulation or severe straining and also deposited NC materials that have been mechanically strained often contain large densities of defects such as vacancies, full or partial dislocations or twin boundaries within the nanosized grains that are bordered by random high-angle GBs. It should however be noted at this point that utmost care is required to extract useful information on defects and defect densities from TEM or HRTEM images, as advocated by Rentenberger et al. (2004), due to the propensity of misinterpretation that is given by the

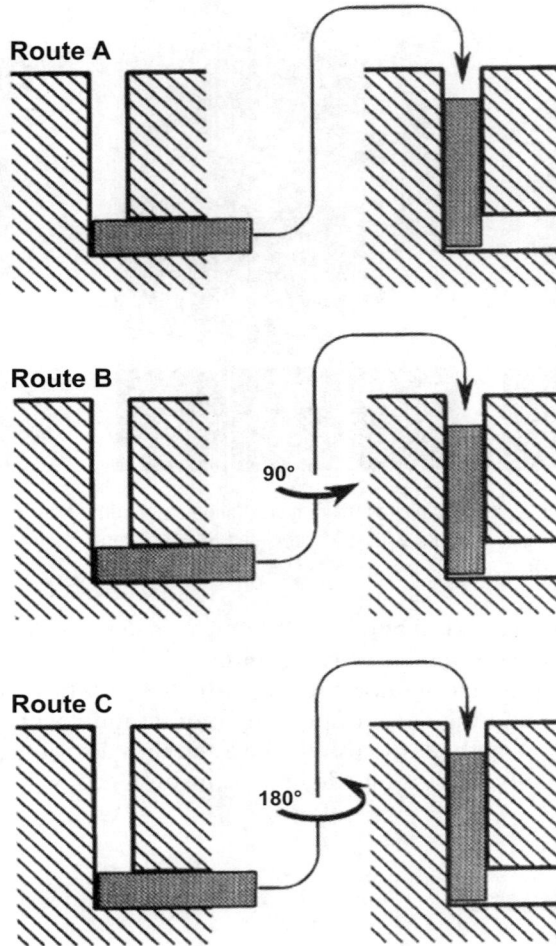

Figure 13   (a) Schematic illustration of the three routes used for ECAP. (b) Shearing patterns associated with ECAP through routes A, B and C, respectively. Reproduced with permission from Elsevier (Iwahashi et al., 1998).

Figure 14   (a) High-resolution TEM micrograph of a Pd grain (nanocrystalline) oriented along the [011]-direction exhibiting several cases of deformation twinning as indicated by the white lines. Note that the grain boundaries on top and bottom showing the transition to the neighbouring grains are imaged. The $\langle 111 \rangle$-planes are bending at an angle of about 14° in both cases (top and bottom) (Rösner et al., 2010a). (b) The in-plane rigid-body rotation w_{xy}. Rotations on a scale from −50° to +50° (anticlockwise positive) are seen. The individual maps of matrix and twins were stitched together to display the complete strain distribution for the whole grain interior. Twins and matrix are significantly distorted with respect to each other and show a strong rotation gradient from top to bottom. The boxes (left: twin, right: matrix) mark the areas used for the strain profile analysis. (c) The misorientation between matrix and twin is plotted as the difference of the two individual profiles indicated by the black frames in **Figure 14b** showing on average a misorientation angle of 3° (d) A rotation gradient from the top to the bottom of the Pd grain is shown by the sum of the two individual profiles indicated by the black boxes in **Figure 14b**. This analysis clearly reveals that the whole grain is bent (Rösner et al., 2010a).

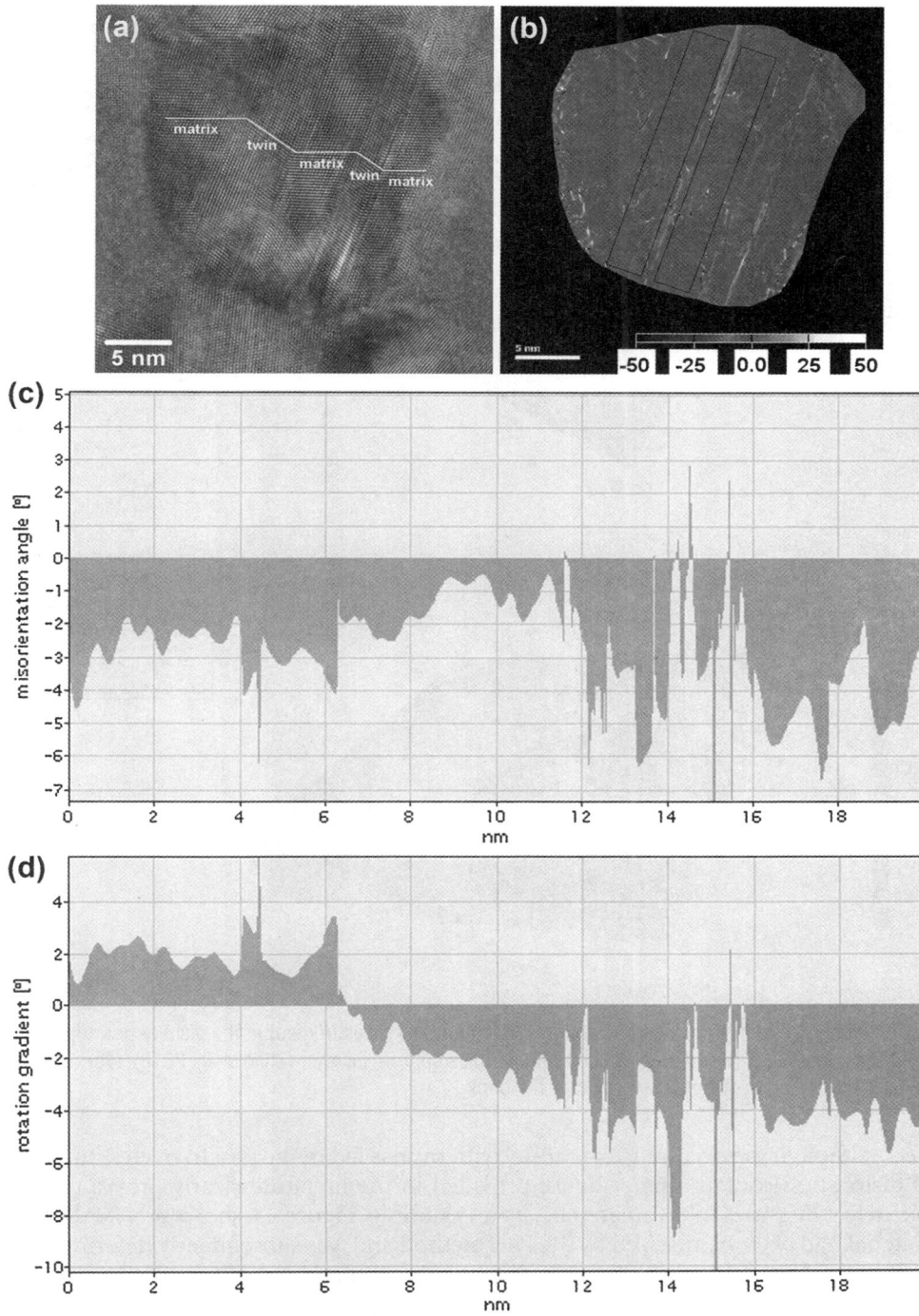

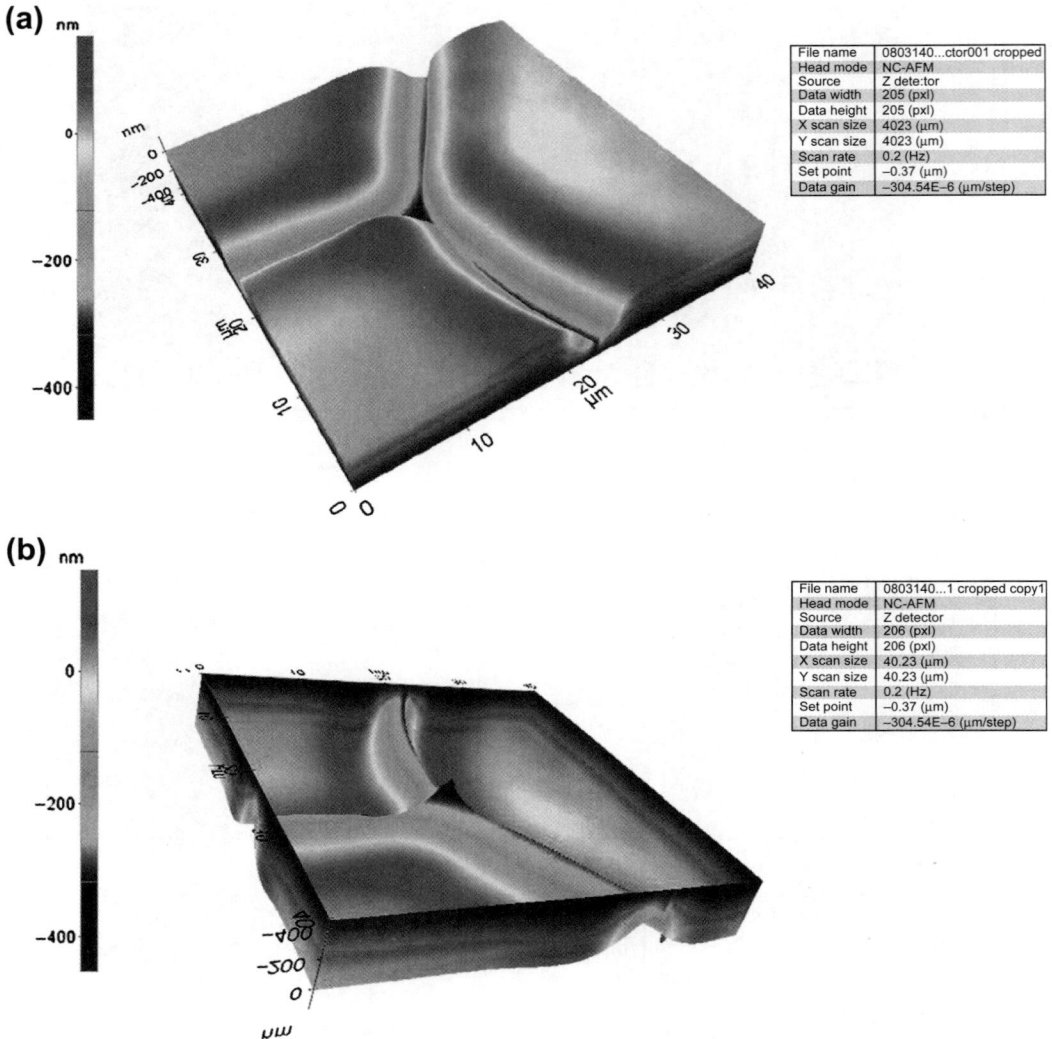

File name	0803140...ctor001 cropped
Head mode	NC-AFM
Source	Z dete:tor
Data width	205 (pxl)
Data height	205 (pxl)
X scan size	4023 (μm)
Y scan size	4023 (μm)
Scan rate	0.2 (Hz)
Set point	−0.37 (μm)
Data gain	−304.54E−6 (μm/step)

File name	0803140...1 cropped copy1
Head mode	NC-AFM
Source	Z detector
Data width	206 (pxl)
Data height	206 (pxl)
X scan size	40.23 (μm)
Y scan size	40.23 (μm)
Scan rate	0.2 (Hz)
Set point	−0.37 (μm)
Data gain	−304.54E−6 (μm/step)

Figure 15 Top view (a) and bottom view (b) of an AFM topography measurement in the vicinity of a tripe junction. AFM image 40 × 40 μm, step size 0.19 μm. Reproduced with permission from Elsevier (Zhao et al., 2010). (For color version of the figure, the reader is referred to the online version of the book.)

Moire' effect. Although the occurrence of Moire' patterns has led in the past to overestimations of strain fields and dislocation content, the evidence provided in the literature clearly proves the existence of remarkable defect densities in NC materials. As an example, **Figure 14** shows a NC Pd grain.

That material had been synthesized by the IGC method and was subsequently deformed by rolling at room temperature. **Figure 14a** displays the high-resolution transmission electron microscopy (HRTEM) image of the crystal lattice. Clearly, twin and matrix lamellae are observed in this grain that has a diameter of about 25 nm. The color-coded image in **Figure 14b** shows a strain map that was calculated by applying the so-called "Geometrical Phase Analysis" (Hytch et al., 1998; Takeda and Suzuki, 1996) on the HRTEM image shown in **Figure 14a**. This method allows calculating the in-plane components of the strain tensor of a TEM specimen quantitatively with almost atomic-level spatial resolution from the HRTEM images. The abrupt color changes in the strain map shown in **Figure 14b**, particularly near the twin/matrix boundaries, indicate a high density of dislocations and dislocation debris inside this grain in addition to the twinning faults (Rösner et al., 2010a). The fact that the defects within the nanocrystal interiors do not annihilate rapidly at the GBs has not been sufficiently clarified up to now. However, some indications exist that strain fields near the GBs that are caused by the Laplace pressure, which occurs at any curved interface, coupled to sufficiently high defect densities that can cause locked configurations of dislocations due to the interaction of their stress fields might sufficiently increase the energy barrier for the dislocations to glide into the GB, so that pile-ups or more complex dislocation reactions can occur. In fact, the results of the strain field analysis as indicated in **Figures 14c and 14d** in the form of rotational gradients along the twin boundaries as well as recent results obtained on UFG Nickel after severe straining by ECAP processing indicate that regions in the material where due to shear localization the defect density was highest exhibited enhanced thermal stability against recrystallization and grain growth compared to neighboring regions with significantly reduced initial defect density (Divinski et al., submitted for publication).

If (mostly) defect-free grain interiors and space-filling average grain shapes are assumed and typical values for the GB excess free energy density are taken into account, then thermodynamic excess contributions of high-angle GBs start to become significant at average grain sizes below about 30–50 nm (**Figure 3**). For TJs, i.e. linear defects that connect three GBs (**Figure 15**), similar estimates are more ambiguous due to the fact that the excess free energy density associated with those defects is not well known and suggestions in literature in fact range from positive to negative values with respect to the excess energy of the same volume of GB that is replaced by the triple junction.

If recent experimental results obtained through different methods on different pure fcc materials are taken into account, then TJs carry an enhanced excess free energy that translates to an excess free energy per unit length of this linear defect of about $1-2\ 10^{-9}\ \mathrm{Jm}^{-1}$ (Gottstein et al., 2010). This magnitude agrees also well with recent measurements of the strain distribution at a triple junction by the so-called Geometric Phase Analysis method (see **Figure 16**). These results indicate that the core of a triple junction has dislocation character (Rösner et al., 2011).

At this magnitude of the line energy, the total excess free energy contribution of the triple junction defects in an NC microstructure would start to become significant at average grain sizes less than about 5–10 nm. Thus, for the vast majority of NC materials, the excess free energy contributions of triple junction defects do not need to be considered. However, these defects are of importance, especially concerning plastic deformation (c.f. Section 26.6), since the TJs present obstacles for processes such as GB sliding, due to incompatibilities that arise from the different orientations of the grains that meet at the junction. As discussed in Section 26.5, TJs might also play an important role concerning diffusion along short-circuit pathways, particularly in severely deformed materials where the incompatibilities at the TJs for slip transfer might have served to create diffusion channels of extremely high diffusivity.

In three-dimensional microstructures, it is necessary and obvious that the TJs, that represent linear defects, join to form zero-dimensional defects, i.e. quadruple points. At these points, TJs of four adjacent grains end. While only few experimental studies on the properties of TJs exist, next to no information is available on quadruple points. However, in view of the above discussion on the significance of TJs, it becomes clear that even smaller grain sizes should be achieved in order to expect significant contributions from these point-like defects, since they are even more sparsely distributed within the microstructure. Thus, unrealistically small average grain sizes that are close to the cluster

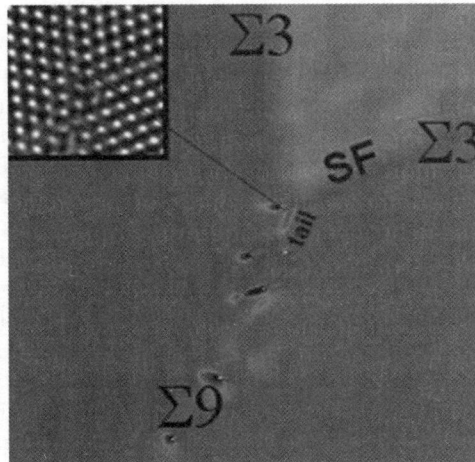

Figure 16   Strain map of the region in the vicinity of a triple junction in Pd after IGC and subsequent HPT deformation. The atomic-level strains have been obtained by applying the GPA method on an aberration-corrected HRTEM image. The "dipole" configuration with tensile and compressive regions at the triple junction is similar to the strain distribution at an edge dislocation in the fcc lattice. The insert shows the HRTEM image of the immediate vicinity of the triple junction (Rösner et al., 2011). (For color version of the figure, the reader is referred to the online version of the book.)

limit, which amounts to 2–3 nm for typical 3d-transition metals, must be reached for the contributions of quadruple points to become measurable.

26.4.1   Grain Boundaries in NC Materials

From the discussion given above, it becomes clear that the GB network represents the characteristic and distinguishing microstructure element that determines to a large extent the properties and behavior that is specific for NC materials. Since the earliest works on NC materials, there have been speculations that the structure (particularly the width) of random high-angle GBs in NC materials is different than in cg polycrystalline materials of identical distribution of the constituents. However, for pure, elemental solids, aberration-corrected HRTEM analyses, which became possible within the past few years, clearly indicate the presence of atomically sharp interfaces also for NC materials. Based on these results and applying to grain sizes down to about 20 nm, no increase of the GB width was observed. Similar results were also reported, based on GB diffusion measurements. From such measurements, if conducted at different temperatures, i.e. within different diffusion regimes, the average diffusional width, δ, of the GBs can be obtained. It was shown that $\delta = 0.5$ nm holds as a good approximation for pure transition metals, irrespective of the grain size (Prokoshkina et al., 2013). Diffusion measurements represent at the same time a sensitive probe for structure modifications, since according to transition state theory, the interatomic interaction potential enters the exponent of the rate equation for the thermally activated jump directly. Thus, small modifications of the local atomic structure are reflected in significant changes of the atomic diffusivity. Yet, in the case of pure, elemental NC materials, where any artifacts due to connected porosity etc. could be eliminated, rather similar GB diffusivities have been measured as for the cg counterparts.

Recently, based on measurements by nanoindentation of the local mechanical properties of a series of NC W–Ni alloys with different nominal compositions and, coupled to the composition, with grain sizes in the range between 5 and 50 nm, Trelewicz and Schuh (2007) suggested that the GBs at the smallest attainable grain sizes bear similarities to an amorphous phase, with the result that an NC solid with such small grain sizes should be described as an amorphous/crystalline nanocomposite material. While the core of random high-angle GBs represents a volume of reduced order, it is at present unclear how to separate effects due to the potentially heterogeneous distribution of alloying elements from purely structural gradients that would occur in a pure, elemental solid. Thus, according to the available experimental results, and also in accordance with careful evaluations of the results obtained by atomistic simulations (Van Swygenhoven et al., 2000), the width of GBs in NC elemental materials as well as the structure of the GB cores can be treated as for cg polycrystals (c.f. Chapter 14, Volume II).

The presence of interfaces, as mentioned already above, inherently causes interface stresses, as described by the Young–Laplace equation. Mechanical equilibrium thus requires the presence of volume stresses to compensate these interface stresses. While the occurrence of interface stresses due to the presence of grain, or phase boundaries, is not limited to any range of grain sizes, it is important to note that the bulk stresses that are induced by the interfaces are much larger for NC materials since they are of the order of the interface stress divided by the grain size. Thus, for an average grain size of the order of a few tens nanometers, the volume stress can exceed values above 1 GPa. Through quantitative measurements of the lattice parameter and the total interface area, Birringer and Zimmer (2009) showed the existence of linear relations between the GB energy and the interface stress with the absolute melting temperature of the material (**Figure 17**). At this high magnitude, the stresses need to be taken into account, e.g. for analyzing thermodynamic equilibria (c.f. Section 26.7 of the current chapter) or for describing the evolution of fine-grained microstructures as a function of time [c.f. Chapter 10, Volume II and Chapter 25 of this volume].

Recently, the importance of externally applied shear stresses on GB migration in NC materials has been elucidated. At the beginning of this development was the observation that NC materials, if compacted through deformation-assisted methods, showed significantly increased grain sizes even though the processing had been performed at temperatures that were too low to allow for measurable grain growth under conditions, where no external stress was applied. By comparing otherwise analogous deformation measurements, Gianola et al. (2006) showed that the effect of a unit applied shear strain on grain growth was strongly enhanced at cryogenic temperatures compared to room temperature. This result clearly indicates that the process of deformation-assisted grain growth is stress controlled. One explanation for the observation of stress-induced GB migration is indicated by Cahn and Taylor (2004) who developed a unified approach to GB motion that illustrates how normal motion of a GB (in this case even a high-angle GB) can result from a shear stress applied tangential to its boundary plane. This normal motion naturally results in a tangential displacement of the two grains, i.e. grain growth. This concept agrees well with experimental measurements by Winning et al. (2001, 2002) and with results of atomistic simulations by Suzuki and Mishin, 2005.

26.4.2 Segregation and Stability of Nanocrystalline Microstructures

As each type of defect of the crystal lattice contributes a specific excess energy density multiplied by its abundance to the total free enthalpy of the material, it is obvious that pure, elemental NC materials are far from thermodynamic equilibrium, rendering them susceptible to relaxation of the excess free enthalpy through recovery, recrystallization or grain growth. It has been shown (Gottstein

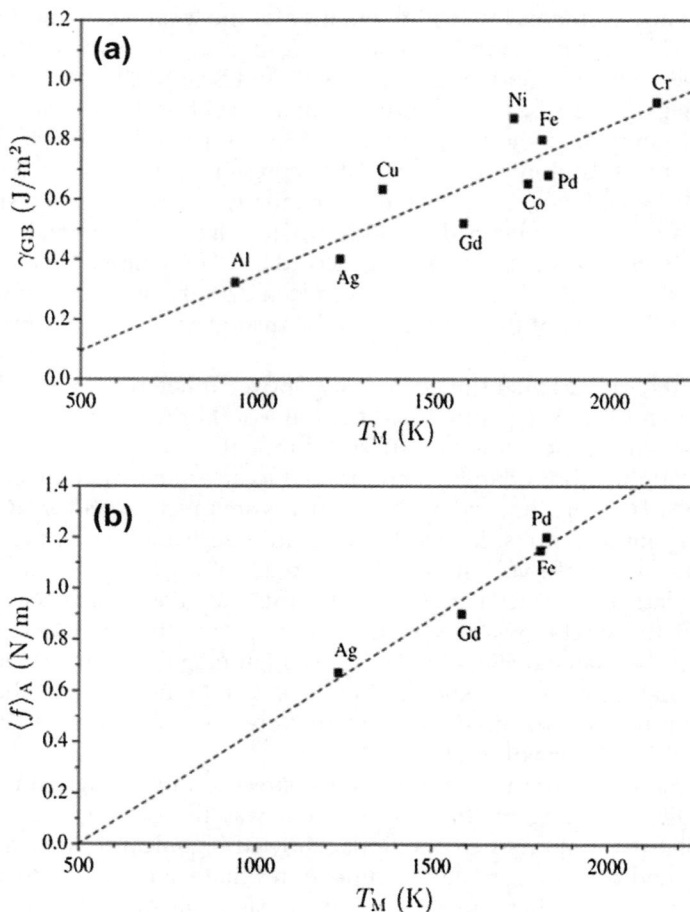

Figure 17 Linear scaling relation between the melting temperature, T_m and (a) the grain boundary energy γ_{GB} and (b) the grain boundary stress $\langle f \rangle_A$. Reproduced with permission from Elsevier (Birringer and Zimmer, 2009).

et al., 2010; Chokshi, 2008) that the thermodynamic contribution of TJs to the driving force for grain growth as well as their kinetic contribution, the so-called triple junction drag, need to be taken into account for describing the grain growth kinetics of materials that have grain sizes below about 50 nm. **Figure 18** shows the comparison between experimental grain growth data of pure NC-Pd at room temperature obtained by IGC with the predictions of different grain growth models. It is clear that neglecting the triple junction drag leads to a large overestimation of the grain growth rate. The anomalous upward curvature of the grain size evolution after about 40 000 s has also been observed in other NC materials and has been explained by anomalous grain growth that is most likely occurring due to complex annihilation processes of lattice defects that take place concomitantly (Paul, Krill, 2011).

Concerning the stability of NC materials, it is clear that for pure elemental solids the high density of internal defects create a significant driving force for coarsening that might only kinetically be hindered,

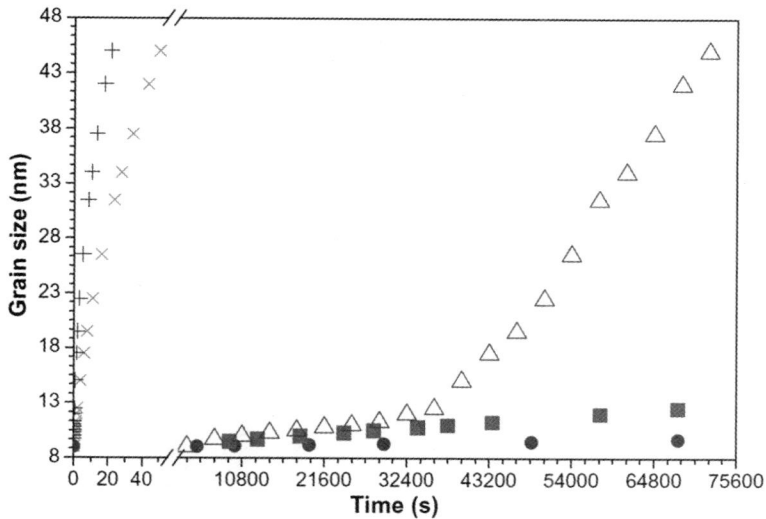

Figure 18 Grain growth in nanocrystalline Pd. The triangles represent the experimental results. The full symbols represent different approximations that take the influence of the triple junctions and the quadruple points into account, in contrast to the crosses (+, ×), where the drag effect of the junctions was neglected. Reproduced with permission from Elsevier (Gottstein et al., 2010). (For color version of the figure, the reader is referred to the online version of the book.)

as e.g. by triple junction drag or by interactions of GBs with complex dislocation arrangements. However, that situation changes when instead of pure, elemental solids, NC alloys are considered. Due to capillary effects, finite driving forces for constituent enrichment or depletion at GBs exist, causing the occurrence of segregation layers or segregation zones at the internal interfaces. As described by the Gibbs adsorption theorem, the GB energy is reduced by solutes with a tendency for GB segregation. Thus, adding solutes with a high negative segregation enthalpy to an NC material can enhance its stability in a two-fold way: through a reduction of the mobility of the GBs by Zener-Pinning and, additionally, by reducing the driving force for grain growth, i.e. the grain-boundary excess free energy (Liu and Kirchheim, 2004). Theoretically, even a vanishing grain-boundary excess free-energy density could be obtained, if the segregation enthalpy would amount to a sufficiently large (negative) value and if the GB excess of the solute would be sufficiently high. This (hypothetical) situation would in fact lead to a thermodynamically stable, finite grain size.

Figure 19 shows the comparison of the grain size of different Pd–Zr alloys, with Zr being a strongly segregating solute in Pd, as a function of temperature obtained from experiment with the predictions of Kirchheim's model that is based on the adsorption theorem. Later on, measurements by TEM indicated that the grain sizes that had been determined originally from XRD line broadening had to be corrected to significantly larger values for temperatures above 600 K, indicating either a strong kinetic contribution to the microstructure stabilization or a pronounced temperature dependence of the grain-boundary excess free enthalpy (Van Leeuwen et al., 2010). Yet, the high stability of the grain size up to temperatures of about 600 K compared to pure NC Pd that shows significant grain growth even at room temperature (Ames et al., 2008) indicates clearly the effectiveness of GB segregation for the stabilization of the nanostructure.

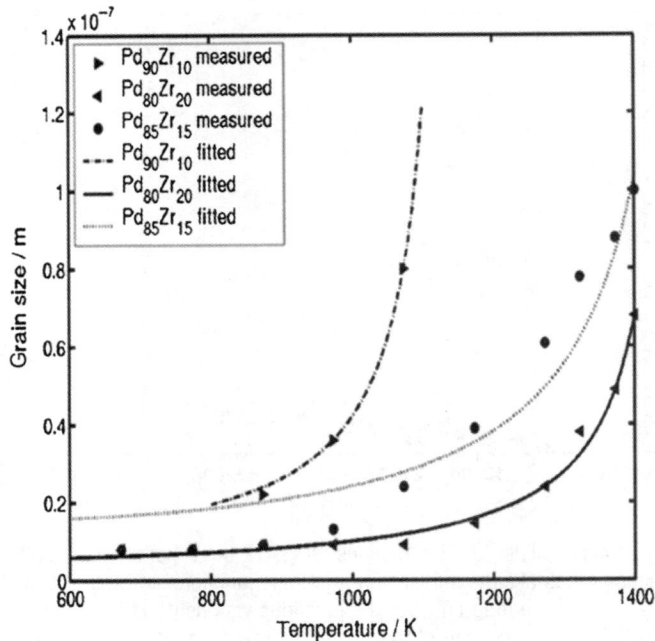

Figure 19 Comparison of experimental grain growth data obtained on different nanocrystalline solid solutions of $Pd_{100-x}Zr_x$ upon thermal annealing with the predictions of Kirchheims model that is based on the adsorption theorem. Reproduced with permission from Elsevier (Liu and Kirchheim, 2004).

26.4.3   Grain Boundaries in Materials Processed by Severe Plastic Deformation

SPD processing creates large densities of dislocations that self-organize first to form cell walls and that lead to microstructure refinement through the formation of new high-angle GBs. Yet, either through starting the processing from an already polycrystalline material or due to the interaction of newly formed GBs with dislocations that form during continued processing, copious GB-dislocation reactions are bound to occur. Due to possible incompatibilities of the adjoining interfaces, strain fields might be formed, leading also to the modification of the GB structure. Somewhat unfortunately, such modified GBs that are characterized by enhanced elastic strain fields and, accordingly, increased excess energy densities have been termed "nonequilibrium GBs" (Grabski, 1985; Valiev et al., 1986) in order to distinguish them from relaxed high-angle GBs. According to (Valiev et al., 1986), the formation of a "nonequilibrium GB" is characterized by three main features, namely, excess grain-boundary energy (at the specified crystallographic parameters of the boundary), the presence of long range elastic stresses (**Figure 20**) and enhanced free volume. However, in a strict sense, each GB is a nonequilibrium defect if segregation effects (see previous section) are not to be considered. Thus, in order to avoid complications with nomenclature, the term "high-energy GB" will be used here.

A model for these high-energy GBs has been developed by Nazarov, Romanov and Valiev in a series of papers (Nazarov et al., 1990, 1993) describing their formation. Lattice dislocations that are created during the plastic straining move toward high-angle GBs on their respective glide planes during continued straining and then, when reaching a high-angle GB, transform into the so-called "extrinsic GB

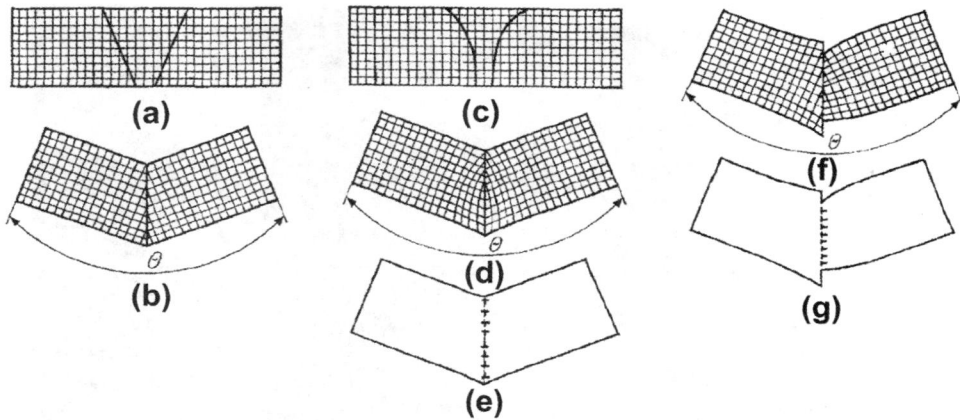

Figure 20 Schematic representation of the formation of grain boundaries by imaginary cuts. From (a) to (b), equilibrium grain boundaries would result. From (c) to (d) and from (a) to (f), nonequilibrium grain boundaries with long-range stresses would result. In (e) and (g), the arrangements of grain boundary dislocations are shown that would create similar elastic strain fields as indicated in (d) and (f). Reproduced with permission from John Wiley and Sons (Valiev et al., 1986).

dislocations", i.e. dislocations that do not contribute toward the disorientation of the two adjacent grains. As a net effect, high-angle GBs with high densities of such extrinsic GB dislocations would also contain increased energy and free volume and considerable microstrain associated with the GB region (Nazarov et al., 1993). Although quantitative estimates for these above-mentioned excess quantities as well as for the characteristic and temperature-dependent relaxation time of the high-energy GBs can be obtained, does the notion of "extrinsic GB dislocations" in the context of high-angle GBs stay somewhat undefined. In fact, measurements of the relaxation behavior of such GBs have revealed that the relaxation of high-energy GBs does not take place in a single step, as assumed in the above-mentioned model, but involves different states with distinct stability and behavior (Divinski et al., 2011). Thus, due to the absence of a complete model, we will summarize the experimental facts concerning these deformation-induced defects. The numerous experiments done concerning the severe deformation of metals and alloys clearly evidence that mostly high-angle GBs are formed after SPD processing and a fraction of these GBs possess a specific high-energy structure. Structure-sensitive probes have been applied that are sensitive to modifications of the atomic structure, such as grain-boundary diffusion measurements (see Section 26.5) or HRTEM analyses, in order to identify and characterize transformations of the GB structure due to the severe deformation processing. **Figure 21a** shows the HRTEM image of a GB with both adjacent grains oriented with their $\langle 110 \rangle$ zone axis parallel to the electron beam in a $Pd_{90}Ag_{10}$ alloy that had been severely deformed by repeated rolling and folding (Wilde et al., 2010). The nonequilibrium character of this GB is manifested in the nonuniform-faceted form (**Figure 21a**) of the two joining grains. GBs with similar features are commonly observed for materials after SPD processing. Yet, it should be noted that not all GBs in severely deformed materials present morphologies as in **Figure 21a**. In fact, only a minority of GBs with an average spacing of a few grain diameters display nonuniform faceting, implying that also during SPD processing, the localization of deformation controls the evolution of the microstructure.

In order to analyze whether the nonuniformly faceted GBs might correspond to nonequilibrium GBs, the residual microstrain present at the GB shown in **Figure 21a** was analyzed by the method of GPA. One result of this analysis is displayed in **Figure 21b** as the rigid body rotation (color coded) as a function of the position in real space. In addition to the disorientation between the two neighboring grains, a clear

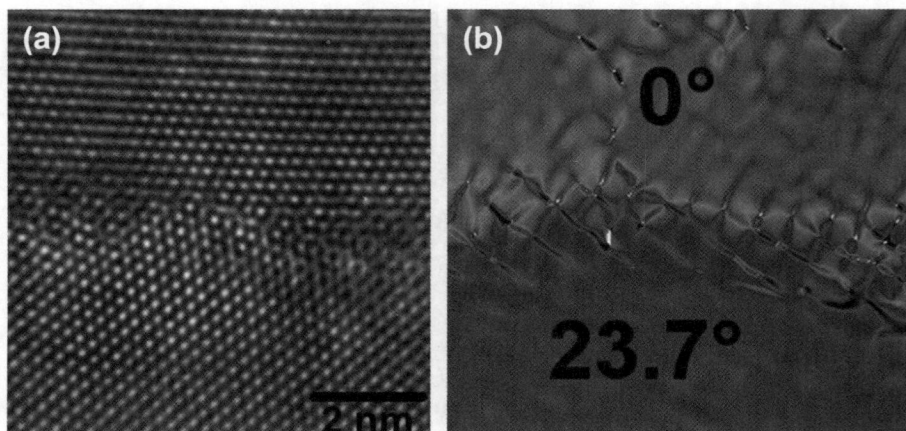

Figure 21   (a) HRTEM image of a random high-angle grain boundary in a $Pd_{90}Ag_{10}$ alloy that was severely deformed by cold rolling and folding. Since both adjacent grains are in Bragg orientation, the development of a high density of nanoscaled ledges at the boundary can be observed. (b) Rigid-body rotation as a function of the position, as determined from the aberration-corrected HRTEM image via GPA (Wilde et al., 2010). (For color version of the figure, the reader is referred to the online version of the book.)

and significant variation of the color representing the variation of the relative rotations of the lattice is observed in the near-boundary region. It should be noted that the bright spots ("hot spots") in **Figure 21b** represent regions where a discontinuity in the transmitted phase of the electron wave occurs, i.e. these spots mark the positions of the core region of full or partial dislocations. Additionally, local structures within the distorted GB region that show an abrupt change of the local orientation of the crystal lattice with respect to the orientation of the lattice of the parent grain could also contribute such bright features in the strain maps, since the linear density of the hot spots that is estimated to be about $10^9 \, m^{-1}$ is much too high to be associated with dislocations only. By integrating the strain over rectangular regions with the long axis perpendicular to the estimated boundary plane, the width of the boundary in terms of the strain distribution was found to be in a range of 1.5–2 nm, which is about 2–3 times the value for relaxed GBs analyzed by the same method (the GB width was directly determined by following the rotational component of the averaged strain field around the GB). The observed topology of the strain distribution at this GB in the severely strained Pd–Ag alloy clearly serves to contribute an enhanced excess free-energy density to the GB energy, supporting the existence of high-energy GBs after SPD processing. This first result of strain mapping at such a GB is also direct support for the interpretations of GB segregations and the diffusion studies discussed in the following chapter.

Thus, with respect to the GB structure in SPD-processed materials with ultrafine grain size, recent studies enable to conclude that:

- High-energy GBs exist in UFG materials and these specific GBs possess an increased free energy density, increased width, high density of dislocations (full or partial) associated with the near-boundary region and correspondingly large residual microstrain.
- The structural width of high-energy GBs is significantly smaller than 10 nm, if the rotational component of the strain gradient across the interface is used as a measure of the width. It reaches a value of 1.5–2.0 nm being at least twice as large as the width of relaxed high-angle GBs in annealed cg materials. The shear components of the strain field reveal similar values of the GB width.

The density of "hot spots" in GPA of a high-energy GB is remarkably large, about 10^9 m^{-1} (**Figure 21b**) and, thus, these features cannot be directly interpreted in terms of, e.g. extrinsic GB dislocations. In fact, these features in the strain map indicate the presence of distinct structural units with an abrupt change of the local orientation of the crystal lattice (in addition to full or partial dislocations), which might result from severe dislocation accumulation and/or dislocation dissociation at the GB. Note that the density of "extrinsic GB dislocations" was estimated based on diffusion measurements to be about 5×10^7 m^{-1} in Ni processed by ECAP.

These results provide a strong evidence that SPD-processing synthesizes material with a significant fraction of high-angle GBs that possess higher excess free-energy density, enhanced atomic mobility along the boundary plane, significant residual strain fields located at the near-boundary region and strongly increased segregation at the boundary and in the near-boundary region. The results also indicate that the final state of GBs created by SPD depends either on the dynamic recovery processes occurring in the vicinity of the boundaries, but also on possible interactions between lattice defects and impurities and/or solute elements. This could lead to a large variety of GBs, exhibiting various roughness, strain distribution, disorientation, local defect density, and these GB features can play a significant role in the properties of UFG materials. In addition, these features are closely related to the SPD processing regimes (temperature, strain rate and degree, applied pressure). For more details on high-energy GBs in severely deformed materials, the reader is referred to the original work by Valiev et al. (1986, 2000). A recent overview on the current state of the art concerning high-energy GBs is given at Sauvage et al. (2012).

26.5 Diffusion Characteristics

Atomic diffusion in NC metallic materials is of high importance, since most of the processes that are involved in coarsening depend directly on the diffusivity. In addition, the diffusion coefficient is extremely sensitive concerning structural changes, since such changes directly modify the interatomic interaction potential and since the interatomic interaction potential enters the exponent of the rate equation that describes the diffusion coefficient within transition state theory. Thus, atomic diffusion is an important quantity concerning the synthesis and stability of nanostructured materials, since NC materials by default possess a high density of potential short-circuit pathways for diffusion—the GBs. At the same time, it is also a sensitive probe concerning structural modifications on the atomic scale. Since these materials are most often synthesized through routes that are far from thermodynamic equilibrium, it is anticipated that the densities of other types of defects of the crystal lattice, mainly excess vacancies or dislocations, are also enhanced. Thus, it is of importance to analyze the diffusion characteristics in NC materials systematically and in detail.

26.5.1 Atomic Mobility in Nanocrystalline Materials

The atomic diffusion in polycrystalline materials represents a special case of solid-state diffusion, since the internal interfaces (i.e. the GBs) and their structure affect the kinetics and even the mechanisms of atomic diffusion. Generally, it is necessary to differentiate between grain interfaces—GBs—and interfaces between different phases—the phase boundaries or heterophase interfaces. NC materials represent a state of solid matter with a high density of GBs and TJs, i.e. a high volume density of potential short-circuit diffusion paths. A nearly ideal state of nanomaterials is represented by materials synthesized by IGC, if the residual porosity, which is inherent at the percent level to materials consolidated from powder at low homologous temperatures, is disregarded. If the materials are produced by a top-down method, e.g. by SPD, a high density of other defects—vacancies, dislocations, dislocation walls and/or loops, etc.—exists. Therefore, diffusion kinetics in such materials is typically enhanced, since the atomic and mesoscale defects present fast diffusion pathways with respect to diffusion within the crystalline lattice. The high values of the diffusion coefficient along the GBs and TJs (Kaur et al., 1995), combined with the high volume fraction of atoms positioned inside/close to these defects (up to 30–50%), lead to an unusually high effective diffusion permeability of nanostructured materials at low temperatures.

One important and long-standing issue in this regard is whether structure and diffusion properties of interfaces in nanomaterials are different from those in the well-annealed polycrystalline counterparts, or more specifically: does the mechanism of GB diffusion exhibit size dependence? In spite of a considerable progress in the understanding of diffusion in NC materials, a number of other fundamental problems also remain still unresolved: what is the relation between the GB structure and the corresponding energetic and kinetic properties? Could one apply the concept of an "average GB" in a nanostructured material? What is the effect of the production route? Which are the main short-circuit paths in nanostructured materials?

First studies of GB diffusion in NC materials reported significantly enhanced values of the specific GB diffusivity with respect to the cg counterparts (Gleiter, 1992). Later on, similar or much slower values were also observed. In order to analyze and understand the diffusion properties of NC materials, the diffusion measurements have to be combined with microstructure studies since the evolution of the microstructure, its stability, impurity content, solute segregation, density and arrangement of point and line defects in the material affect the diffusion kinetics considerably. While the detailed interdependencies are complex and in several instances not yet known, it seems clear that the atomic transport within the crystalline lattice of the grains remains unaffected by the microstructure being NC or ultrafine grained. Thus, in order to describe atomic diffusion in NC materials, it is sufficient to have a closer look at the case of GB diffusion, since the aspects of lattice diffusion have already been described in the chapter on Diffusion in this volume.

The present section is organized as follows. After a very brief introduction of the Fisher model of GB diffusion (Fisher, 1951) and the basic kinetic regimes of diffusion, the effect of variable microstructures (due to recrystallization or grain growth) is discussed. Finally, recently obtained results on diffusion in nanostructured and UFG materials are reviewed.

26.5.1.1 Grain-Boundary Diffusion

Grain-boundary diffusion measurements are very sensitive to the structural state of GBs. Small changes in atomic positions can result in corresponding changes of energy barriers for diffusion jumps. Since the diffusion rate depends exponentially on the energy barriers, a significant change in the GB diffusivities, both for solute and matrix atoms, can be expected.

The commonly accepted Fisher model describes the fast GB diffusion with a leakage of diffusing atoms into the bulk by atomic migration normal to the GBs (**Figure 22**).

Fisher postulated that a GB can be represented as a homogeneous slab of width δ and GB diffusion coefficient $D_{gb} \gg D_v$. In a general case when bulk and GB diffusion are concurrent processes and $\delta \ll \sqrt{D_v t}$ (D_v is a bulk diffusion coefficient and t is the annealing time), the boundary condition:

$$c_v\left(\pm\frac{\delta}{2}, y, t\right) = c_{gb}(y, t),$$
(1)

(a)

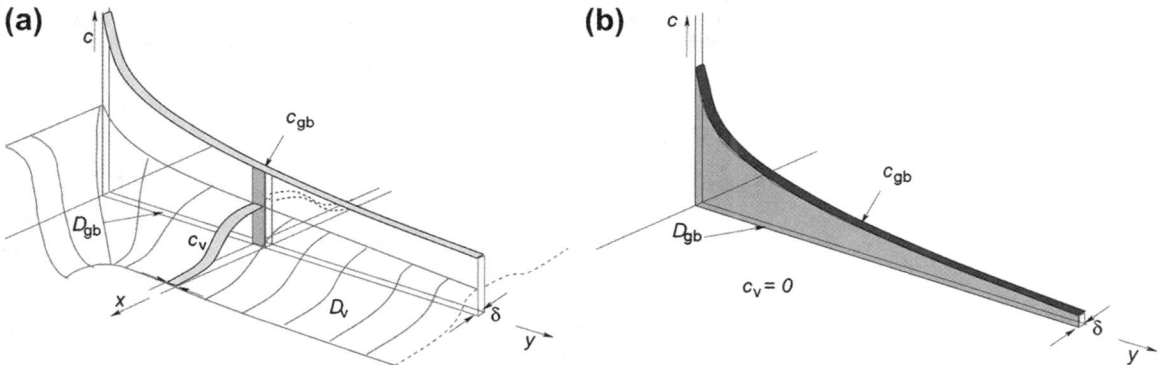

(b)

Figure 22 Schematic drawing of tracer distribution around a single grain boundary under B (a)—and C (b)—type kinetic conditions, after Harrison (1961). The GB is represented as a homogeneous slab of the thickness δ and the diffusivity D_{gb}, embedded in the crystalline bulk of the diffusivity D_v. Tracer concentrations in GB and bulk are given by c_{gb} and c_v, respectively, and their ratio, $s = c_{gb}/c_v$, is the solute segregation factor ($s = 1$ in the case of self-diffusion). (For color version of the figure, the reader is referred to the online version of the book.)

defines the continuity of the tracer concentration at the GB/bulk interface, i.e. the lack of GB segregation (c_v and c_{gb} are the diffusant concentrations in the bulk and in the GB). Therefore, the model is applied for self-diffusion. Later Gibbs took GB segregation in a linear form (by the so-called Henry isotherm) into account:

$$c_{gb}(y,t) = s \cdot c_v \left(\pm \frac{\delta}{2}, y, t \right), \tag{2}$$

where s is the segregation coefficient.

Following Harrison, the GB diffusion measurements can be classified into three types: A, B and C:

- C regime: At low temperatures (or short times of the diffusion annealing treatment), the bulk diffusion length is small with respect to the GB width δ and the tracer atoms concentrate exclusively in the GBs (**Figure 22b**). In this C-type kinetics regime, the GB diffusivity D_{gb} can directly be determined from an experimental diffusion penetration profile and the resulting diffusion coefficient is independent of the actual width of the GB.
- B regime: With increasing temperature, the bulk diffusion length becomes much larger than the GB width and diffusion out of the GB into the bulk cannot be neglected. If the bulk diffusion fluxes from different GBs do not overlap, such conditions correspond to the B-type kinetics and the only parameter, which can be determined from such a GB diffusion experiment, is the so-called triple product P of the segregation factor s, the GB width δ, and the GB diffusivity D_{gb}: $P = s \cdot \delta \cdot D_{gb}$.
- A regime: At even higher temperatures (correspondingly for very long diffusion times), the bulk diffusion fluxes from different GBs overlap and diffusion proceeds in an effectively homogeneous medium characterized by an effective diffusion coefficient $D_{eff} = gD_{gb} + (1 - g)D_v$, where $g = sf/(1 - f + sf)$ with f being the fraction of material present within or adjacent to GBs in the polycrystal.

The dominant majority of GB diffusion investigations were carried out in the B-type kinetics regime of GB diffusion (Kaur et al., 1989, 1995). The approximate solution for the quasi-stationary state ($\partial c_{gb}/\partial t = 0$) is given as:

$$c_{gb}(y) = c_0 \exp\left(-\frac{y}{L}\right),\qquad(3)$$

where c_0 is the surface concentration and

$$L^2 = \frac{s\delta D_{gb}\sqrt{\pi t}}{2\sqrt{D_v}}.\qquad(4)$$

L can be obtained from GB diffusion measurements by analyzing the concentration profile, which gives a straight line in the coordinates $\ln(c_{gb})$ vs. y with the slope L^{-1}. The values of L and D_v permit to calculate the triple product of GB diffusion $P = s\delta D_{gb}$. The defined effective activation enthalpy of GB diffusion, Q_{gb}, with:

$$P = P_0 \exp\left(-\frac{Q_{gb}}{RT}\right),\qquad(5)$$

is a sum of the GB diffusion enthalpy, ΔH_{gb}, and the enthalpy of segregation, $H_{seg} < 0$:

$$Q_{gb} = \Delta H_{gb} + H_{seg}\qquad(6)$$

The exact solution of the Fisher model of GB diffusion was later derived by Whipple (1954) and Suzuoka (1964) and has been generalized by LeClaire (1963).

Although the Fisher model seems to be oversimplified, it showed its robustness and applicability to a wide variety of diffusion problems. The parameter δ is introduced as an effective GB thickness, which is defined as the thickness of an interface region with a significantly enhanced value of the diffusion coefficient. The value of δ can be measured, if GB self-diffusion experiments are performed both at low temperatures (within the C-type regime), determining D_{gb}, and at higher temperatures (the B-type regime), measuring the double product P, $P = \delta\, D_{gb}$. Then $\delta = P/D_{gb}$. The results available in the literature are summarized in **Table 1**. The tabulated results confirm that the commonly accepted value $\delta = 0.5$ nm can safely be used for the data analysis.

A note is due here. Applying the above result and performing B- and C-type measurements of solute diffusion in a given matrix, the corresponding segregation factor $s = P/\delta\cdot D_{gb}$ can be determined. Moreover, a number of approaches were undertaken to extend the analysis beyond the applicability of the Fisher model. Klinger and Bokstein (2009) introduced the concept of a "wide" GB with a position-dependent diffusion coefficient $D_{eff}(x)$. Nonlinear segregation was included in several theoretical models and even a segregation isotherm was determined in specially designed bicrystal measurements.

If one performs diffusion measurements within a new system, especially with a solute of which the GB segregation is not known, some experiments cannot be classified strictly to belong to one of the introduced regimes and may fall into a transition area between the given Harrison regimes. The detailed analysis indicated that two intermediate (transition) regimes can be specified, AB and BC. These regimes can be specified by their specific exponential dependencies of the tracer distribution. The mathematical analysis revealed that the power of 3/2 characterizes the depth dependence of the logarithm of the tracer concentration in the AB regime and 5/4 was established for the BC regime. Penetration profiles specific for the AB regimes were indeed measured. The existence of the BC regime is still not proven.

Table 1   GB width determined from diffusion experiments

Material	Tracer	δ (nm)	Ref
NiO	^{63}Ni	0.7	
Ag	^{110m}Ag	0.43 ± 0.27	
Ag	^{110m}Ag	0.5	
γ-FeNi	^{59}Fe	0.5	
γ-FeNi	^{63}Ni	0.55 ± 0.43	
Fe	^{59}Fe	0.5	
Ni	^{63}Ni	0.54 ± 0.1	
Ti	^{57}Co	0.5	
Cu*	–	0.3	
Al/Pb*	–	0.7	
HPT-Pd*	–	$1.5 \div 2.0$	
Pd*	–	0.6	

Denotes results of high-resolution TEM measurements performed using the GPA method.

26.5.2   Special Issues for Nanocrystalline Materials

26.5.2.1   Hierarchical Microstructures

NC materials were found to often reveal a hierarchy of internal interfaces. Such a microstructure feature is expected for materials that underwent initial powder processing, such as for example a sintered Fe–Ni alloy that had initially been subjected to high-energy ball milling. The NC grains ($d \sim 100$ nm) turned out to be clustered in agglomerates with an average size dag from 30 to 50 µm (Divinski et al., 2002b, 2002c). A similar effect was reported for IGC materials, too (Baladin et al., 1997), and even for UFG materials produced by SPD (Divinski et al., 2009), however for completely different reasons. The effect of a hierarchical microstructure on diffusion in a nanostructured material has partially been considered by Baladin et al. (1997) (see also **Figure 23**).

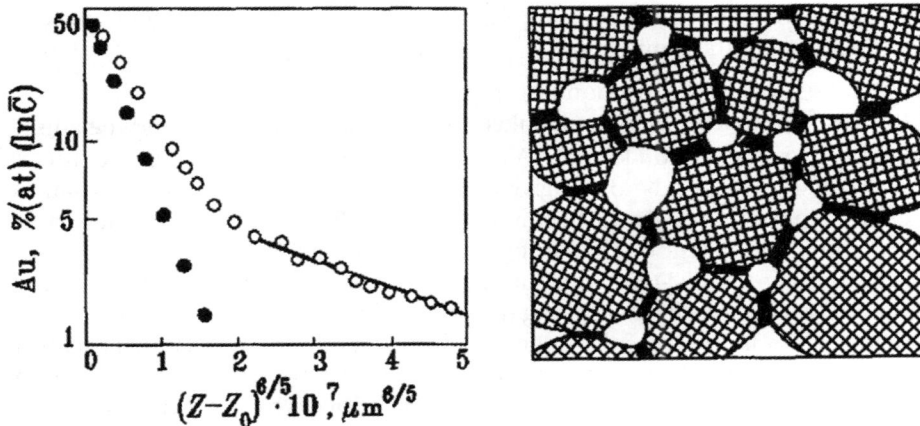

Figure 23   Concentration profile for Au interdiffusion in nanocrystalline Ni before (solid circles) and after (open circles) annealing at 448 K. Reproduced with permission from Elsevier (Bokstein et al., 1995).

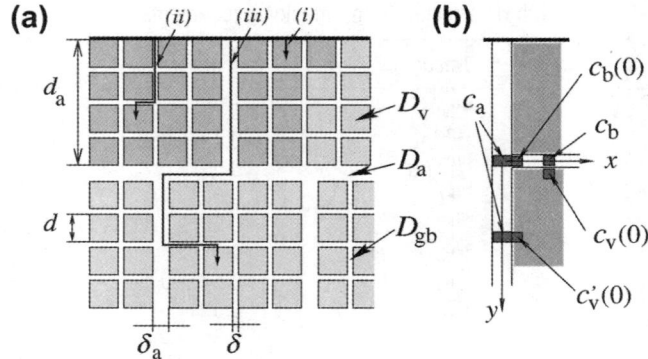

Figure 24 Schematic representation of diffusion in a nanocrystalline material with a hierarchic microstructure (a). Small nanograins (of the size d) are clustered in agglomerates of the size d_a. δ and δ_a are the widths of GBs between the nanocrystalline grains and interagglomerate boundaries, respectively; D_v, D_{gb}, and D_a are the bulk, GB, and interagglomerate boundary diffusivities, respectively. Possible diffusion paths of individual tracer atoms, which contribute to the fluxes (I), (II) and (III) (see text), are illustrated. In (b) the concentrations, which are relevant to the definitions of the segregation factors s and s_a, are indicated. (For color version of the figure, the reader is referred to the online version of the book.)

The complete analysis of interface diffusion in different possible kinetic regimes has been carried out by Divinski et al. (2002b, 2002c, 2004b). In the case of NC materials with a complex hierarchic micro-structure (**Figure 24a**), the kinetics of diffusion transport deviates from Harrison's scheme and demands a detailed consideration of the different diffusion pathways and the succession of pathways that are open for the diffusing species. The basic assumption is that $D_a \gg D_{gb} \gg D_v$ (the diffusion coefficients are explained in **Figure 24**). Taking this relation into account, three different diffusion fluxes can generally be introduced (counted according to the values of the relevant penetration depths, see **Figure 24a**):

- diffusion along interagglomerate boundaries with subsequent out diffusion to nanoboundaries and then into the grain interiors,
- GB diffusion along NC boundaries with subsequent out diffusion into the grain interiors and
- direct bulk diffusion from the sample surface into the nanoscaled grains (similar to Harrisons type-A diffusion).

Thus, the regimes of interface diffusion for the processes (*i*) and (*ii*) have to be specified separately in order to describe the overall diffusion kinetics. A two-letter designation was suggested, for example the C–B regime. Each letter corresponds to the given Harrison kinetic regime, which is satisfied for the particular interface type: at first for diffusion along the GBs of the nanocrystals and then for interagglomerate-boundary diffusion. Therefore, e.g. the C–B regime describes the case when the C regime of diffusion along the NC-GBs (no out-diffusion into bulk) is satisfied and when simulta-neously the (quasi) B regime of diffusion along the interagglomerate boundaries (fast diffusion along the interagglomerate boundaries with subsequent out diffusion into the adjacent nano-GBs) is fulfilled. Such a classification has a general character and has already been applied for triple junction diffusion (Edelhoff et al., 2011) (e.g. fast diffusion along triple lines followed by out diffusion to GBs and then into bulk), dislocation or sub-boundary diffusion (fast transport along a GB followed by out diffusion to the dislocation network, which intersects the GB and which is followed by leakage of the diffusant into the bulk) and other similar configurations.

In the case of solute diffusion, the segregation of the solute to both types of internal interfaces has to be taken into account. More than one segregation coefficient thus has to be introduced. The solute atoms can generally be in excess in (see **Figure 24b**):

(1) NC-GBs with respect to the adjacent bulk,
(2) interagglomerate interfaces with respect to the adjacent bulk and
(3) the interagglomerate interfaces with respect to the adjacent positions in the GBs between the NC grains, which intersect this interagglomerate interface.

Two segregation factors are required to describe the diffusion problem under consideration: s, which characterizes the solute excess in a nano-GB with respect to an adjacent bulk plane, and s_a, which corresponds to an excess of the solute in the interagglomerate boundary with respect to the adjacent position in the NC-GB:

$$s = \frac{c_b}{c_v(0)} \tag{7}$$

And

$$s_a = \frac{c_a}{c_b(0)}. \tag{8}$$

Here $c_v(0)$, c_b, $c_b(0)$, and c_a are the corresponding solute concentrations in the bulk just near a NC-GB, in a NC-GB, in a nano-GB just near an interagglomerate boundary, and in an interagglomerate boundary, respectively. The definition of these concentrations is illustrated in **Figure 24b**.

The segregation factor $s_{av} = c_a/c'_v(0)$ (the case (b) above) is not important in the present consideration, since the direct out diffusion from the interagglomerate boundaries into the bulk has been neglected (here $c'_v(0)$ is the bulk solute concentration just near the interagglomerate boundary, **Figure 24b**). Note that the excess of solute atoms in the interagglomerate boundaries with respect to the bulk, s_{av}, may be presented as $s_{av} = s_a s$, if the segregation behavior corresponds to dilute limit conditions (linear segregation). The segregation factor s is important for the flux (*ii*) and both factors, s and s_a, affect the flux (*iii*).

In dependence on the given kinetic conditions, five regimes (C–C, C–B, B–B, A–B and A) and one subregime (AB–B) were introduced to describe diffusion in a material with a bimodal distribution of internal interfaces. The relevant parameters along with the diffusion characteristics, which can experimentally be determined, and the typical concentration dependencies of the penetration profiles are given in **Table 2**. Examples of penetration profiles measured for Fe, Ni, or Ag diffusion in the NC γ-Fe–Ni alloy are shown in **Figure 25a–f**.

● The C–C Regime: This regime corresponds to very low temperatures and short diffusion times which suppress any out diffusion from the internal interfaces in the material. The condition $\alpha > 1$ means that bulk diffusion is negligible, i.e. $\sqrt{D_v t} \ll s \cdot \delta$. If the tracer enters the NC-GBs, it remains there. The diffusion length along the NC-GBs is also very small in this regime, since the next condition, $\alpha_a > 1$, can approximately be rewritten as $\sqrt{D_{gb} t} < s_a \delta_a / 2\lambda \cong s_a \cdot d$. Therefore, the diffusion length along the NC-GBs, $\sqrt{D_{gb} t}$, is very small and this flux cannot be detected by conventional sectioning methods. The factor λ in the expression for α_a takes into account the fact that out-diffusion from an interagglomerate boundary only occurs through regions where the NC boundaries (of the width δ) intersect with the interagglomerate boundary. For cubic grains $\lambda = 2\delta/d$ holds.

Table 2   Parameters of the kinetic regimes of GB diffusion in a nanocrystalline material with a bimodal structure. c is the layer concentration, y is the penetration depth, q_1 and q_2 are the numerical factors, and $\lambda = 2\delta/d$ is the density of nano-GBs intersecting an interagglomerate boundary. Other parameters are defined in the text

Regime	Conditions	Measured parameters	Typical concentration dependence
C–C	$\alpha = \dfrac{s\delta}{\sqrt[2]{D_v t}} > 1$; $\alpha_a = \dfrac{s_a\delta_a}{\sqrt[2]{D_{gb} t}} > 1$	$D_a = \dfrac{1}{4t} q_2^{-1}$	$c \sim \exp\left(-q_2 y^2\right)$ (Figure 2a)
C–B	$\alpha > 1; \alpha_\alpha < 0.1$, $\beta_a = \dfrac{P_a}{2D_{gb}\sqrt{D_{gb}}} \geq 2\sqrt{D_{gb} t} < d_a/4$	$D_{gb} = \dfrac{1}{4t} q_1^{-1}$ $P_a = \dfrac{s_a\delta_a D_a}{\lambda} = 1.31\sqrt{\dfrac{D_{gb}}{t}} q_2^{-5/3}$	$c \sim \exp\left(-q_1 y^2\right) + \exp\left(-q_2 y^{6/5}\right)$ (Figure 2b)
B–B	$\alpha < 0.1, \beta \geq 2, \sqrt{D_v t} < d/4$; $\alpha'_\alpha < 0.1, \beta'_\alpha \geq 2, \sqrt{P\left(\dfrac{\pi t}{4D_v}\right)^{1/4}} < d_a/2$	$P = s\delta D_{gb} = 1.31\sqrt{\dfrac{D_v}{t}} q_1^{-5/3}$ $P_a = \dfrac{s_a\delta_a D_a}{\lambda} = 1.80\sqrt{\dfrac{D_{gb}}{s\delta}}\left(\dfrac{D_v}{t}\right)^{1/4} q_2^{-2}$	$c \sim \exp\left(q_1 y^{6/5}\right) + \exp\left(-q_2 y\right)$ (Figure 2c)
AB–B	$d/4 < \sqrt{D_v t} < 3d$; $\alpha''_\alpha < 0.1, \beta''_\alpha \geq 2, \sqrt{D^a_{eff} t} < d_a/4$	$sD_{gb} = 16.48 \dfrac{D_v^{0.1}}{\delta^{0.2}\rho^{0.9}} q_1^{-4/3}$ $P'_a = s_a\delta_a D_a = 1.31\sqrt{\dfrac{D^a_{eff}}{t}} q_2^{-5/3}$	$c \sim \exp\left(-q_1 y^{3/2}\right) + \exp\left(-q_2 y^{6/5}\right)$ (Figure 2d)
A–B	$\sqrt{D_v t} > 3d, \sqrt{D^a_{eff} t} < d_a/4$; $\alpha''_\alpha < 0.1, \beta''_\alpha \geq 2$	$D^a_{eff} \cong \dfrac{1}{2}\dfrac{sf_{gb}}{1+sf_{gb}} \; D_{gb} = \dfrac{1}{4t} q_1^{-1}$ $P'_a = s_a\delta_a D_a = 1.31\sqrt{\dfrac{D^a_{eff}}{t}} q_2^{-5/3}$	$c \sim \exp\left(-q_1 y^2\right) + \exp\left(-q_2 y^{6/5}\right)$ (Figure 2e)
A	$\sqrt{D^a_{eff} t} > 3d_a$	$D^M_{eff} \cong \dfrac{sf_{gb}/2}{1+sf_{gb}} \; D_{gb} = \dfrac{ss_a f_a/2}{1+ss_a f_a} \; D_a = \dfrac{1}{4tq_2}$	$c \sim \exp\left(-q_2 y^2\right)$ (Figure 2f)

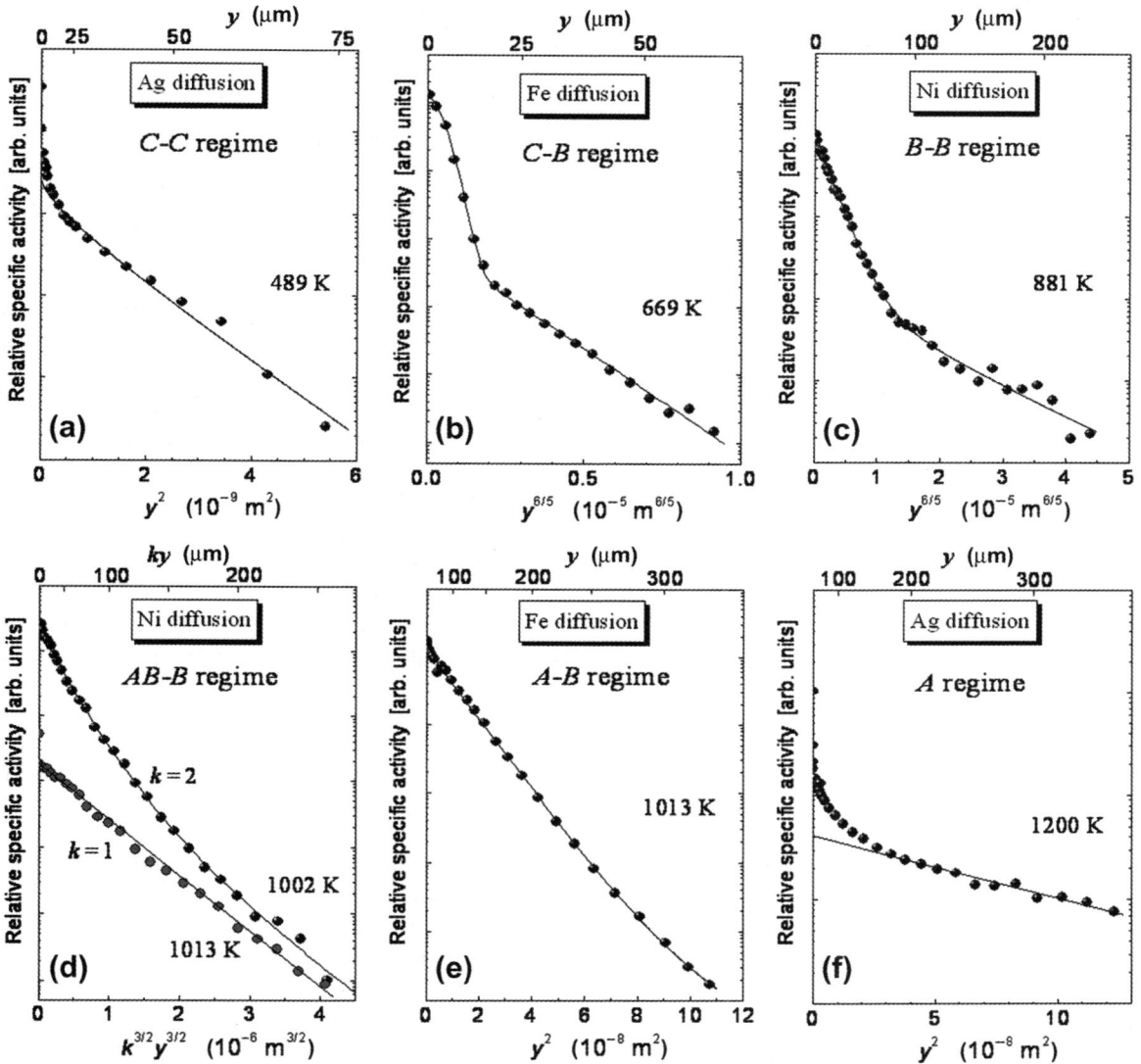

Figure 25 Examples of penetration profiles measured for Ag, Fe or Ni diffusion in nanocrystalline γ-Fe-40 wt. % Ni alloy in the C–C (a), C–B (b), B–B (c), AB–B (d), A–B (e), and A (f) kinetic regimes. k is a numerical factor; y is the penetration depth. The solid lines represent the relevant fits used to determine interface diffusion parameters (see **Table 2**).

Since nano-GB diffusion is almost "frozen out", the tracer is dominantly located in the interagglomerate boundaries. The diffusion profile corresponds to the error function or Gaussian solution of the diffusion equation in dependence of the given initial conditions. An example of such a profile, experimentally measured for Ag solute diffusion in the NC γ-FeNi alloy, is shown in **Figure 25a**. Due to the small solid solubility of Ag in the FeNi alloy, the initial conditions corresponded to the thick layer solution and the error function fitting was applied to extract the diffusivity D_a of interagglomerate boundaries in that case. Due to the applied mechanical sectioning method, the NC-GB diffusivity D_{gb} could not be analyzed in that experiment.

The segregation factors s and s_a should be known to evaluate α and α_a. Since s and s_a are not known a priori, the estimates $s = s_a = 1$ can initially be used. The self-consistent evaluation of the whole data set in all diffusion regimes, however, allows to calculate the segregation factor s and to estimate s_a, see below.

● The C–B Regime: With increasing temperature of diffusion anneals, the diffusion length along the nano-GBs increases and out diffusion from the interagglomerate boundaries is becoming important, $\sqrt{D_{gb}t} \gg s_a \cdot d$. This introduces the formal B regime conditions for diffusion along the interagglomerate boundaries. However, since bulk diffusion is still suppressed, $\sqrt{D_v t} \ll s \cdot \delta$, the C kinetic conditions are valid for the diffusion along the nanocrystal GBs. These conditions define the C–B diffusion regime in the material with a hierarchic interface structure (**Table 2**).

In this diffusion regime, it is not possible to determine independent values of s_a, δ_a or D_a, but only their product $P_a = s_a \delta_a D_a / \lambda$ can be determined from the penetration profiles. The specific feature of the interagglomerate boundaries is that out diffusion does not proceed uniformly but only at positions, where the GBs intersect the interagglomerate boundaries. The density of such positions, λ, enters explicitly into the expression for the product P_a.

The diffusion length for diffusion along the NC-GBs, $\sqrt{D_{gb}t}$, should be smaller than the size d_a of the agglomerates in this regime. Otherwise, the diffusion fluxes from different interagglomerate boundaries will overlap and the formal A-kinetic regime might become important. In the C–B diffusion regime, the parameter α is larger than unity and the tracer does not penetrate into the bulk. The diffusion process is thus confined to nano-GBs and interagglomerate boundaries only. The tracer is mainly located in the nano-GBs.

An example of such penetration profile measured for Fe diffusion in nano-γ-FeNi is presented in **Figure 25b**. A two-stage shape of the penetration profile is clearly seen. The first part, which is characterized by the $\ln c \sim y^2$ depth dependence of the concentration profile (c is the layer tracer concentration and y the penetration depth), corresponds to nano-GB diffusion in the C regime. As a result, the nano-GB diffusivity D_{gb} can directly be determined from this part.

The second part of the penetration profile in **Figure 25b** corresponds to the faster diffusion mode from the surface into the interagglomerate boundaries with subsequent out diffusion into adjacent NC-GBs. Since formal B-type conditions are fulfilled in this diffusion mode, the Suzuoka solution of the interface diffusion problem has to be applied, see **Table 2**. The solid line in **Figure 25b** represents the relevant fit.

● The B–B Regime: If temperature and/or time of diffusion anneal increases further, the bulk diffusion flux becomes more significant and cannot be neglected. Then the diffusion process will be dominated by two fluxes:
  – GB diffusion along the NC-GBs with subsequent out diffusion into the grain interior;
  – faster diffusion along interagglomerate boundaries with subsequent out diffusion to nano-boundaries and then into the grains.

Since the bulk diffusion length, $\sqrt{D_v t}$, has to be smaller than the grain size to satisfy the conditions of this B–B regime (**Table 2**), the total contribution of direct volume diffusion from the sample surface into the nanograins can be neglected. Correspondingly, two-stage penetration profiles should be observed. An example of such a profile, which was measured for Ni diffusion in compacted nano-γ-FeNi, is shown in **Figure 25c**.

The B regime conditions are satisfied for the flux (I). Therefore, the Suzuoka solution of the GB diffusion problem is applied to analyze this term. As a result, the first part of the penetration profile should be linear in the coordinates of $\ln c$ vs. $y^{6/5}$ and only the triple product $P = s\delta D_{gb}$ can be determined, but not the nano-GB diffusivity D_{gb} itself, see **Table 2**. The parameter $\beta = P/2D_v\sqrt{D_v t}$ has to be large enough in order to observe a distinct GB diffusion-related tail. The effect of β on the accuracy in the determination of P was analyzed by Monte Carlo simulation of GB diffusion and $\beta \geq 2$ can be used as a lower limit of the B regime.

The diffusion flux (*II*) represents a fundamentally new situation, which was analyzed in Divinski et al. (2002c, 2004b) for self- and solute diffusion, respectively. The Fisher model of interface diffusion was elaborated for the case of a hierarchic structure. The analysis has shown that the logarithm of concentration c should linearly decrease with the penetration depth y. A more careful numerical solution of this diffusion problem resulted in a power-law dependence of the logarithm of concentration, $\ln c \sim y^n$, with $n \approx 1.05$. The relevant fitting of the experimental penetration profile allows determining the triple product P_a for interagglomerate-boundary diffusion in the B–B regime: $P_a = s_a\delta_a D_a/\lambda$. It was shown that the approximate solution for P_a, is precise enough to be applied in a diffusion experiment. The parameters α'_a and β'_a, which determine the conditions of the B–B regime, are defined as follows:

$$\alpha'_a = \frac{s_a\delta_a}{\lambda\sqrt{s\delta D_{gb}}}\left(\frac{4D_v}{\pi t}\right)^{1/4} \tag{9}$$

and

$$\beta'_a = \frac{P_a}{D_{gb}\sqrt{s\delta D_{gb}}}\left(\frac{4D_v}{\pi t}\right)^{1/4}. \tag{10}$$

The fit in **Figure 25c** describes the experimental points over almost four decades of decrease in concentration, supplying reliable data on both nano-GB and interagglomerate-boundary diffusion.

● The AB–B Regime: As it was stated above, the temperature interval of the B–B regime has an upper limit by the condition that an overlap of the bulk diffusion fluxes from different NC-GBs must be avoided. With increasing temperature, this condition will be violated. For interface diffusion in a unimodal structure, it is known that the A kinetic regime is realized when the bulk diffusion length is much larger than the grain size d. In between, a transition from the B-like to A-like kinetics is expected. As it was stated above, in a relatively narrow interval of values of the bulk diffusion length, $d/4 < \sqrt{D_v t} < 3d$, a new subregime can be identified with an own kinetic of interface diffusion: the AB regime. The key feature of this subregime is the unusual depth dependence of the concentration profile: $\ln c \sim y^{3/2}$. Such profiles were indeed measured in the experiments on Fe diffusion in nano-γ-FeNi. The measured decrease of the logarithm of concentration with the penetration depth over more than three decades definitely allowed to rule out the B ($\ln c \sim y^{6/5}$) or A ($\ln c \sim y^2$) kinetics in the relevant cases.

In the case of a nanomaterial with a hierarchy of interface characteristics, the subregime AB–B can be introduced. Examples of such profiles, which were measured for Ni diffusion in nano-γ-FeNi, are shown in **Figure 25d**. The first part of the two-stage profile measured at $T = 1002$ K is linear in the coordinates of $\ln c$ vs. $y^{3/2}$ (nano-GB diffusion) and the second one is linear in the coordinates of $\ln c$ vs.

$\gamma^{6/5}$ (interagglomerate-boundary diffusion). The solid line, which represents the relevant fit in **Figure 25d**, describes the experimental data over about five orders of magnitude in concentration. The profile, which was measured at $T = 1013$ K (**Figure 25d**), reveals only a part related to nano-GB diffusion and is almost perfectly linear over three decades in the coordinates of $\ln c$ vs. $y^{3/2}$. The product sD_{gb} can be determined from the first part of the profile (see **Table 2**), whereas the triple product $P_a'' = s_a \delta_a D_a$ can be evaluated from the deeper part of the concentration profile. The interagglomerate-boundary diffusion problem in such conditions is still not solved exactly. In Divinski et al. (2002c), it was suggested to describe out diffusion into the agglomerates approximately by an effective diffusivity D_{eff}^a, which in the case of self-diffusion is:

$$D_{eff}^a = f_b D_{gb} + (1 - f_b)D_v \cong f_b D_{gb}. \tag{11}$$

Here f_b is the volume fraction of the nano-GBs in an agglomerate. This fraction can be determined as $f_b = q\delta/d$ (q is a numerical factor). In Eqn (10), the contribution of bulk diffusion was neglected, since both $D_v << D_{gb}$ and $f_b << 1$ hold for the nanomaterial under consideration. Then, interagglomerate-boundary diffusion can be described by the formal Suzuoka solution with the effective diffusivity D_{eff}^a instead of the GB diffusivity D_{gb}. The following parameters are important in the AB–B diffusion regime:

$$\alpha_a'' = \frac{s_a \delta_a}{2\sqrt{D_{eff}^a t}} \tag{12}$$

and

$$\beta_a'' = \frac{P_a''}{2D_{eff}^a \sqrt{D_{eff}^a t}}. \tag{13}$$

- The A–B Regime: If the bulk diffusion length becomes remarkably larger than the grain size, $\sqrt{D_v t} > 3d$, the A kinetic is valid for tracer diffusion along NC-GBs. Then the agglomerates are characterized by an effective diffusivity D_{eff}^a, which is described by a modified Hart–Mottlock equation in the case of solute diffusion:

$$D_{eff}^a = \tau D_{gb} + (1 - \tau)D_v. \tag{14}$$

Here τ is the fraction of time, which a tracer atom spends in the nano-GBs:

$$\tau = \frac{sf_b}{1 + sf_b}. \tag{15}$$

Since in the present case $D_v << D_{gb}$ and $f_b << 1$, only the nano-GB diffusion contribution becomes important in Eqn (13) and it is reduced to the form presented in **Table 2**. The additional factor $^1/_2$ originates from a recent Monte Carlo study of GB diffusion in the A kinetics regime (Divinski et al., 2002a) and it corresponds to the case of cubic grains. We suppose that this model gives a better description of GB diffusion in real nanomaterial with respect to a system of parallel GB slabs, which was adopted in the original Hart–Mottlock description.

If the resulting diffusion length inside of the agglomerates, $\sqrt{D_{eff}^a t}$, is smaller than the agglomerate size, $\sqrt{D_{eff}^a t} < d_a/4$, the fluxes from different interagglomerate boundaries do not overlap and diffusion occurs in the A–B kinetic regime in the bimodal interface microstructure under consideration. The penetration profiles generally should be composed of two parts. The first part corresponds to bulk diffusion in a homogeneous material (inner part of agglomerates) characterized by an effective diffusion coefficient D_{eff}^a. An example of such penetration profile is presented in **Figure 25e** for the case of Fe diffusion in nano-γ-FeNi. The penetration profile turns out to be slightly curved at large depths, see **Figure 25e**. However, the number of relevant experimental points is too small to extract reliably the interagglomerate-boundary diffusivity. The deviation from linearity in this profile could only be detected due to the extremely high sensitivity of the applied radionuclide counting facilities, which allowed an accurate detection of the penetration profile over five decades of the decrease of concentration.

- The A Regime: With increasing temperature, the effective diffusivity of agglomerates, D_{eff}^a in Eqn (14), becomes larger and the relevant diffusion length can be remarkably larger than the agglomerate size, $\sqrt{D_{eff}^a t} > 3d_a$. Then, the A kinetics regime is valid for tracer diffusion in this material. The NC alloy as a whole can be considered as a homogeneous material with the effective diffusivity D_{eff}^M. The penetration profile should follow the thin layer solution of the diffusion equation and the slope of the fitting line in the coordinates of ln c vs. y^2 gives the effective diffusivity D_{eff}^M.

An example of such a profile measured for Ag diffusion in NC-FeNi alloy at $T = 1200$ K (Divinski et al. 2004b) is shown in **Figure 25f**. The contribution of the interagglomerate-boundary diffusivity D_a to the effective diffusivity D_{eff}^M was found to be less than 10% and the relevant relation in **Table 2** allows a consistent determination of the nano-GB diffusivity D_{gb}. The knowledge of the segregation factor s is imperative for such calculations and it can be determined by an iterative approach.

26.5.2.2 Effect of Grain-Boundary Migration on Diffusion

Since NC materials and also materials with UFG microstructures are materials that are far from thermodynamic equilibrium, coarsening and grain growth are often encountered processes, specifically if the materials are analyzed or processed at elevated temperatures. Thus, GB motion, at least to some extent has to be expected to occur during diffusion annealing treatments. However, GB motion during a diffusion annealing treatment also changes the pertinent kinetic conditions. As a result, the penetration profiles become linear against depth instead of being linear against depth squared in the formal C (Glaeser and Evans, 1986) or B kinetics regimes (Mishin and Razumovskii, 1992). To analyze the diffusion profiles, a simplifying assumption is adopted that the migrating boundaries move with the same constant velocity V during the time t of the diffusion annealing treatment. It is additionally assumed that the GB migration distance Vt and the volume diffusion length $\sqrt{D_v t}$ remain smaller than the average grain size d. In such a model, GB motion perpendicular to the diffusion direction is considered that is generally acceptable for relatively large grain sizes. The robustness and applicability of the method was carefully investigated for high-purity polycrystalline copper and systematic results were derived using different tracers (Divinski et al., 2001b). The key point is that only trace amounts of solutes are applied so that the tracer atoms serve as a marker of GB motion and do not modify the GB or its kinetic characteristics.

Existence or absence of the GB motion in a particular experiment can (e.g.) be proven by careful measurement of the penetration profiles over four to six orders of magnitude in the concentration. Correspondingly, one can differentiate kinetic conditions even by the shape of the penetration profiles, see **Figure 26**.

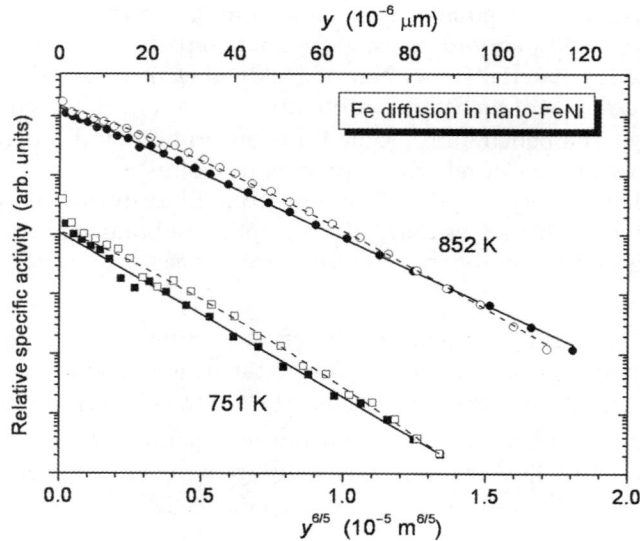

Figure 26 Penetration profiles plotted against depth y (open symbols, upper X-axis) and depth to the power 6/5 (full symbols, bottom X-axis). It is obvious that the profiles are systematically curved when plotted against depth and become almost linear against $y^{6/5}$ (as it is expected for the formal B-type conditions). The particular measurements of Fe diffusion were performed in the nanocrystalline FeNi alloy with hierarchic microstructure in the C–B regime.

In the particular case of Fe diffusion in the NC-FeNi alloy, the GB motion was effectively suppressed and undisturbed interface diffusion has been measured at elevated temperatures. The penetration profiles such as presented in **Figure 26** provide a strong support toward the correctness of the given diffusion measurements.

In the case of NC materials, the main assumption of the models by Glaeser and Evans or Mishin and Razumovskii—perpendicular motion of GBs to the diffusion front—is violated. If recrystallization/grain growth occurs during the diffusion annealing treatment and the diffusion length is less than the final grain size, then the analysis has to be modified (Amouyal et al., 2008; Klinger and Rabkin, 2009). The corresponding model was recently analyzed in Amouyal et al. (2008) for recrystallization and in Klinger and Rabkin (2009) for grain growth.

GB diffusion from an instantaneous surface source into an initially diffusant-free polycrystal, in which all GBs are characterized by the same GB diffusion coefficient D_{gb}, and the C regime conditions are assumed. The recrystallization process that is considered in the model is related to GB (or recrystallization front) migration. It is taken into account that the kinetics of recrystallization front migration:

$$\lambda = \exp(- Bt^n) \tag{16}$$

is not a steady state and that the total area of GBs substantially decreases with time. In the following a recrystallization process occurring simultaneously with diffusion is considered. In the case of an NC or UFG matrix, the following assumptions can be made:

- The bulk is separated into two interpenetrating, but continuous microstructural components. They comprise the NC/UFG matrix and the cg recrystallized material. The isolated islands of the NC/UFG matrix within the recrystallized regions will be disregarded.

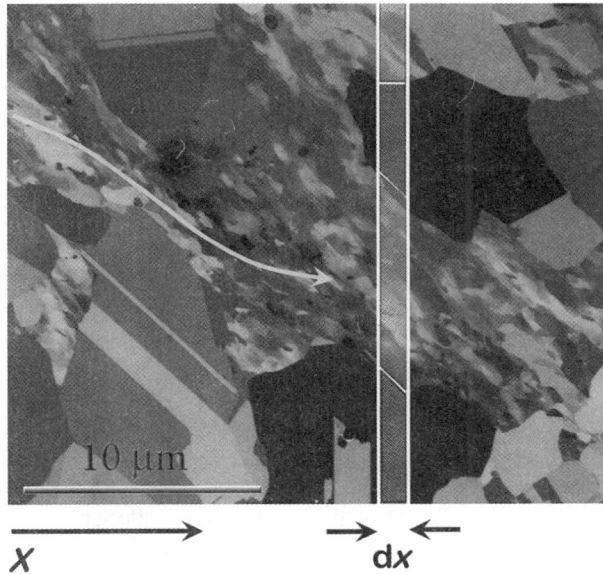

Figure 27 FIB micrograph showing a typical microstructure of the partially recystallized specimen. The diffusion flux is marked by the arrow within the UFG matrix. An infinitesimal slice of the thickness dx consists of both recystallized regions (blue) and untransformed UFG matrix (red). (For interpretation of the references to color in this figure legend, the reader is referred to the online version of this book.)

- The size of the newly formed grains is much larger than the grain size in the NC/UFG matrix. Thus, GB diffusion in the recrystallized regions can be disregarded and GB diffusion occurs in the NC/UFG matrix only. The newly formed coarse grains "freeze" the local concentration of solute that existed prior to recrystallization.
- No grain growth occurs in the NC/UFG matrix during recrystallization.
- The model is illustrated in **Figure 27**. Following the assumption of homogeneity, any internal slice of the infinitesimal thickness dx "cut" from the plane normal to the average diffusion flux should incorporate both, recrystallized and nonrecrystallized regions (**Figure 27**).

The resulting distribution of tracer atoms is described by

$$\tilde{C} = \frac{A}{\sqrt{t}} \frac{1}{X} \int\limits_{X}^{\infty} (2s^2 - 1) \exp\left[-s^2 + (\ln \lambda) \cdot (X/s)^{2n}\right] ds, \tag{17}$$

where $X = \frac{x}{2\sqrt{Dt}}$. Equation (17) gives a full description of the diffusion concentration profile that develops in recrystallizing material in terms of parameters λ and n of the Johnson, Mehl, Avrami, Kolmogorov (JMAK) equation. These parameters are determined in a set of independent experiments in which the volume fraction of the original UFG matrix in partly recrystallized microstructure, λ, is determined as a function of annealing time. For the short annealing times ($t \rightarrow 0$, $\lambda \rightarrow 1$) the radiotracer penetration profile can be linearized in the standard Gaussian coordinates $\ln \tilde{C}$ vs. x^2.

For $n = 1$, the integral in Eqn (17) can be calculated analytically, giving $\tilde{C} = \text{const} \cdot \exp(-2Z)$, which means a linear dependence of $\ln \tilde{C}$ on x, similarly to the result of Glaeser and Evans for GBs migrating

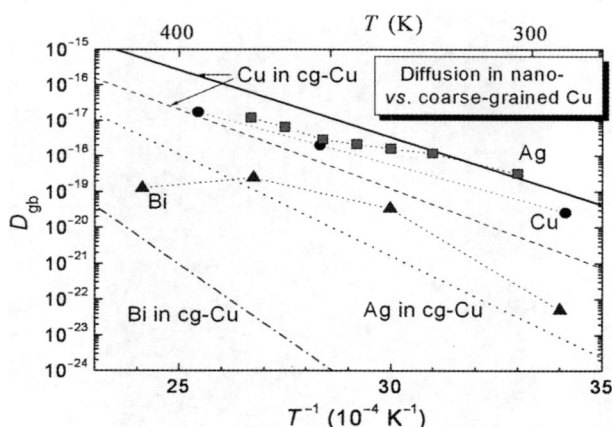

Figure 28   Self-diffusion of Cu (Horvath et al., 1987) and solute diffusion of Ag (Schumacher et al., 1989) and Bi (Höfler et al., 1993) in nanocrystalline Cu (symbols) in comparison with their diffusivities in coarse-grained material (cg-Cu) (Surholt and Herzig, 1997; Divinski et al. 2001a, 2004a, 2004c) (straight lines). GB self-diffusion of Cu in coarse-grained samples was measured in the materials of the purities 5N (dashed line) and 5N8 (solid line) (Surholt and Herzig, 1997).

with a constant velocity. However, for $n > 1$ the dependence of $\ln\tilde{C}$ on x obtained by numerical integration of Eqn (17) becomes nonlinear. Still, it can be approximated with a high accuracy by the following expression:

$$\tilde{C}\big|_{t \to \infty} = \text{const} \cdot \exp(-\varphi_0 Z^{m_0}), \tag{18}$$

where m_0 is an increasing function of n ($m_0(n = 1) = 1$).

26.5.3   Experimental Results of Diffusion Studies

Since the pioneering work (Birringer et al., 1984), diffusion in NC materials was the subject of numerous researches. A number of excellent overviews of on the diffusion properties of nanostructured materials have been published, see e.g. the articles of Gleiter (1992, 1995) or Würschum et al. (2003). Theoretical models of diffusion enhancement associated with transformations of GB defects in NC materials were reviewed in Ovidko (2005).

26.5.3.1   Materials from Inert Gas Condensation

Inert gas condensation is capable of producing a microstructure that is usually texture-free and consists of equiaxed grains. However, as indicated above, it has several limitations including specimen volume and yield, incomplete densification and difficulties associated with retaining the fine grain size during consolidation. The state of the art of the research on the diffusion behavior of interfaces in pure NC copper produced by IGC is illustrated in **Figure 28**, where the data for self- (Cu (Horvath et al., 1987)) and solute (Ag (Schumacher et al., 1989)), Bi (Höfler et al., 1993) GB diffusion in nano-Cu are compared with the corresponding diffusivities determined for the cg material. Note that different methods were applied for the diffusion study: the radiotracer technique (Cu diffusion (Horvath et al.,

1987)), electron-beam microanalysis (Ag diffusion) (Schumacher et al., 1989), and Rutherford back-scattering (Bi diffusion) (Höfler et al., 1993).

One recognizes that GB diffusion of Cu in both nano- and cg-materials (both measured with the radiotracer technique) proceeds with similar rates. Due to the high volume fraction of GBs in a nano-material, their impurity content is relatively low and must be compared with that in a cg high-purity material.

GB diffusivities of solutes in nano-Cu (especially of Bi) appear to be significantly enhanced with respect to the corresponding values in cg-Cu. This enhancement becomes more pronounced with decreasing temperature and approaches about seven orders of magnitude for Bi at $T = 350$ K, **Figure 28**. However, the reliability of this comparison is somewhat questionable, since it is well known that the radiotracer technique is superior with respect to electron-beam microanalysis and to Rutherford backscattering method regarding sensitivity and dynamical range of the analyzed concentration penetration profiles. It would thus be beneficial to apply the radiotracer method as the most sensitive method for a systematic investigation of diffusion of a solute in copper as a function of the grain size.

The coefficients of GB self-diffusion in NC-Fe (with a relative density higher than 91%) have been determined by Würschum et al. (Herth et al., 2001). The self-diffusivities were found to be similar to those extrapolated from high-temperature data of conventional GBs. These results suggest that the GBs in nano-Fe prepared by the compaction of inert gas condensed powder are similar to those in conventional polycrystalline Fe (see **Figure 29**). A similar result has also been obtained for Cu, where

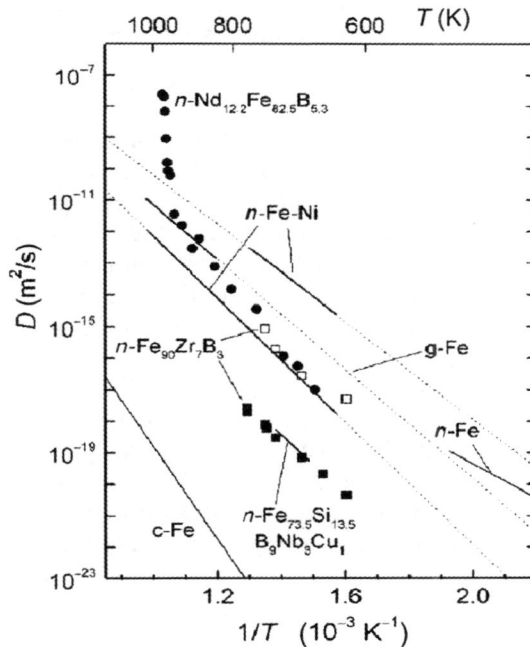

Figure 29 Arrhenius plots of ^{59}Fe tracer diffusivities in nanocrystalline Fe and in nanocrystalline Fe-rich alloys. For $Fe_{90}Zr_7B_3$ the diffusivities in two types of interfaces (solid and open squares) are shown. The data of n-Fe refer to relaxed grain boundaries. Reproduced with permission from John Wiley and Sons (Würschum et al., 2003).

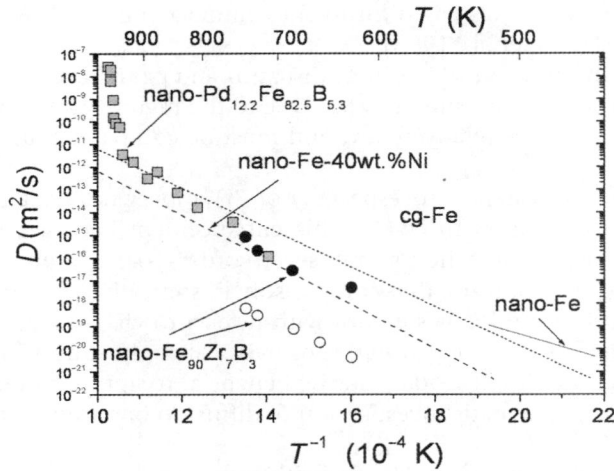

Figure 30   Interfacial diffusion of Fe in nanocrystalline Fe-based alloys produced by crystallization of amorphous phases: $Pd_{12.2}Fe_{82.5}B_{5.3}$ (Eggersmann et al., 2001), $Fe_{90}Zr_7B_3$ (Herth et al., 2004). The data on GB diffusion in coarse-grained Fe (Divinski et al., 2004a), nano-Fe (produced by inert gas condensation) (Würschum et al., 1999), and nano-Fe-40 wt. % Ni (Divinski et al., 2002b) are also presented.

self-diffusion measured in cg and in NC material by the same method (radiotracer diffusion) also indicated similar GB self diffusion rates. These results also indicate that HAGBs in NC materials have a similar structure as in polycrystalline materials.

26.5.3.2   NC Materials Obtained via Controlled Crystallization of Amorphous Precursors

As described above, the controlled crystallization of amorphous precursors can also yield NC materials. In such a case, the experimental measurements of GB diffusion are highly complicated by the presence of residual amorphous phase fractions (Würschum et al., 1999). Selected diffusion data, which were mainly obtained by Würschum et al. in Fe-based alloys, are summarized in **Figure 30**.

   In this case, interfacial diffusion is observed to be similar or even slower than in the respective cg materials. In $Fe_{90}Zr_7B_3$ and $Fe_{90}Zr_{10}$ alloys, a second, faster diffusion process was also observed in addition to the slow diffusion through the amorphous GB phase (Herth et al. 2004). Adopting the general treatment of diffusion in hierarchic structures, the diffusivities of the faster diffusion paths were determined and it was found that their diffusivity is similar to diffusion along conventional GBs. It thus seems that these faster diffusion paths are related to intercrystalline regions in the alloys (see **Figure 31**). Integranular melting of UFG $Nd_2Fe_{14}B$ alloy was also observed (Eggersmann et al., 2001). Similar results were also found for Mo tracer diffusion in a $Fe_{76}Mo_8Cu_1B_{15}$ alloy, which was studied using a serial sectioning method in the temperature range 548–648 K (Cermak and Stloukal, 2007).

26.5.3.3   Materials Obtained by Sintering

Pressureless sintering of ball-milled powders represents an attractive route for producing a bulk NC material with almost theoretical density. The NC-Fe–Ni alloy was produced during hydrogen reduction of ball-milled oxide powders. The detailed preparation scheme is described in Refs. Knorr et al. (2000),

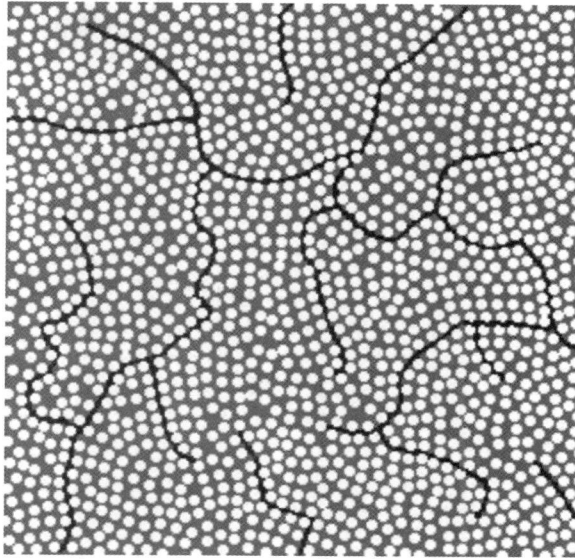

Figure 31 Two interface-type model of nanocrystalline $Fe_{90}Zr_7B_3$. Open circles: nanocrystallites, gray background: intergranular amorphous phase (interface type I), black lines: interfaces without amorphous phase (interface type II) with properties similar to grain boundaries. Source: American Institute of Physics (Herth et al., 2004).

Divinski et al. (2002a). A density of about 98% of the theoretical density was obtained. During sintering, the grain size increased from 30 to about 100 nm.

Diffusion of Fe, Ni and Ag in NC γ-FeNi alloy was measured in the temperature interval from about 500 to 1200 K. The penetration profiles were analyzed according to the strict mathematical conditions of the given kinetic regimes. For the analysis of Fe and Ni diffusion, the relevant segregation factors s_{Fe} and s_{Ni} are close to unity. Fe and Ni show complete mutual miscibility in the γ-phase and one may expect only slight (if any) segregation of both Fe and Ni to internal interfaces in the γ-Fe–40wt.%Ni alloy. On the other hand, Ag reveals very small solubility in FeNi and a strong Ag segregation to internal interfaces was established (Divinski et al., 2004b).

Since $s = s_a = 1$ for diffusion of Fe and Ni, it is relatively easy to identify the kinetic regime for self-diffusion. Ni diffusion in the NC γ-FeNi alloy was measured in the kinetics regimes C–B, B–B, AB–B, whereas Fe diffusion was analyzed in the regimes C–B, B–B, AB–B, and A–B, **Figure 32a and b**.

The nano-GB diffusivity D_{gb} was directly determined in the C–B and AB–B regimes. In absence of segregation, the triple product P is reduced to the double product $P = \delta D_{gb}$. Having determined the double product P in the B–B regime and having measured D_{gb} directly, the GB width δ can be determined as $\delta = P/D_{gb}$.

In **Figure 32a and b**, the resulting values for Ni and Fe diffusion along NC-GBs are plotted as functions of the inverse temperature. It is obvious that the P values are systematically below the values of D_{gb}. It is the ratio P/D_b, which is equal to the diffusion width of the GBs. The direct estimates give $\delta = (5.5 \pm 4.3) \times 10^{-10}$ m and $(7.3 \pm 5.2) \times 10^{-10}$ m for Ni and Fe diffusion, respectively. These values fit reasonably the commonly accepted value of the GB width, $\delta = 0.5$ nm. This value of δ is used for the comparison of diffusion data derived in different kinetic regimes. The effective diffusivity D_{eff}^a

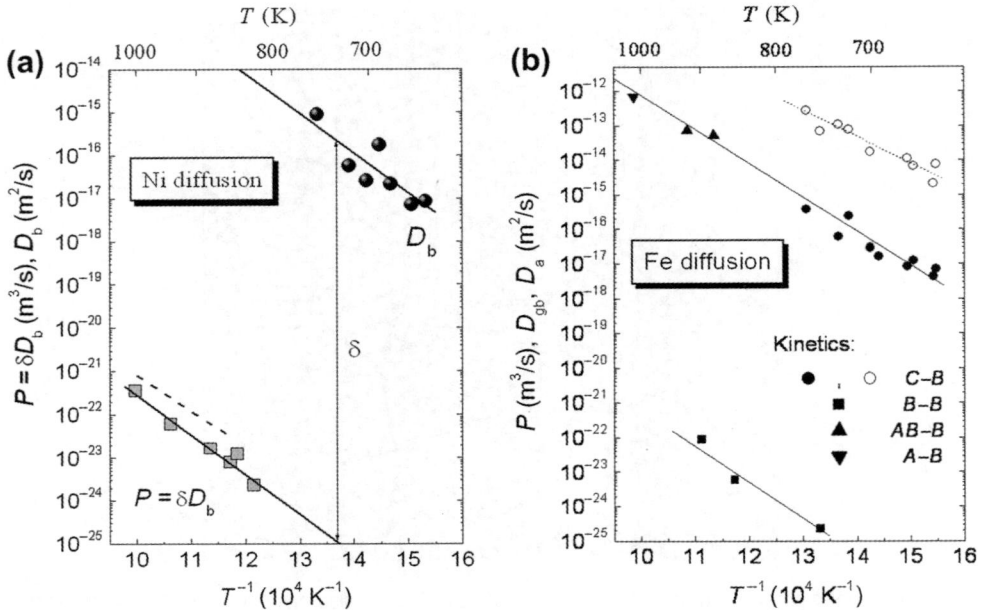

Figure 32 Arrhenius diagram for Ni (Divinski et al., 2002c) (a) and Fe (Divinski et al., 2002a, 2002b) (b) diffusion in nanocrystalline γ-Fe-40 wt. % Ni alloy with a hierarchic microstructure. The filled symbols represent the diffusivity of nanocrystalline GBs and open symbols correspond to interagglomerate boundary diffusion. The comparison of the directly measured double product $P = \delta D_{gb}$ and the nano-GB diffusivity D_{gb} for Ni (a) and Fe (b) diffusion gives an estimate for the GB width, $\delta \approx 0.5$ nm. The dashed line in (a) represents the GB diffusivity of Ni in the coarse-grained FeNi alloy (Kang et al., 2004). The estimate $\delta_a = 1$ nm was used to recalculate the measured P_a values to the relevant diffusivities D_a for Ni (a) and Fe (b) diffusion.

determined in the A–B kinetics regime gives the value of the nano-GB diffusivity $D_{gb} = 2D_{eff}^a/f_b$. Estimating the volume fraction of nano-GBs f_b as $f_b = \delta/d$ and using the same estimate of the GB width $\delta = 0.5$ nm, the diffusivity D_{gb} can be determined as: $D_{gb} = 2dD_{eff}^a/\delta$. The temperature dependencies of Ni and Fe diffusivities D_{gb} along NC-GBs are presented in **Figure 32a and b**, respectively (filled symbols; the type of symbol depends on the given kinetic regime). Although Fe and Ni diffusion were measured in very different kinetic regimes and different mathematical treatments were applied according to the different regimes, the results show a systematic behavior and the nano-GB diffusivity follows Arrhenius-type temperature dependence over a rather extended interval.

In **Figure 32a** the Ni-GB diffusivity in cg γ-FeNi, which is of similar composition as the nano-γ-FeNi alloy, is plotted in comparison with diffusion in nano-γ-FeNi. Ni diffusion in both materials was found to be similar. In the NC material (the grain size $d \sim 100$ nm), the GB structure is well relaxed and, according to the diffusion data, similar to that in the cg material.

Diffusion in interagglomerate boundaries was measured in the C–B kinetics for Fe tracer and in the C–B, B–B and AB–B kinetics for Ni tracer. The diffusivity of the interagglomerate boundaries is much faster than that of the NC boundaries and the activation enthalpy Q_a is remarkably smaller. The value of Q_a (especially for Ni diffusion) approaches values of activation enthalpies that are typical for surface diffusion. This behavior is most probably related to an increased free volume at such boundaries.

Ag diffusion in the NC γ-FeNi was measured in various diffusion regimes. A succession of the kinetics C–C, C–B, AB–B and A was observed and the diffusivities of both nano-GBs and interagglomerate boundaries were determined. The diffusivity D_{gb} of the NC-GBs was directly determined in the C–B kinetics regime. In order to establish the limits of the relevant kinetic regimes and to analyze the experimental data, the knowledge of the segregation factor s is imperative. Due to the strong segregation, it was impossible to measure GB diffusion of Ag under formal B-type kinetic conditions with the aim to estimate the pertinent segregation factor s.

Ag segregates strongly at GBs in the γ-Fe-40wt.%Ni alloy, e.g. $s = 1000$ at $T \sim 700$ K. Having determined the segregation factor s, a special GB diffusion experiment in the NC material was designed in which the conditions of the AB–B kinetics were satisfied (the B–B kinetics cannot be fulfilled for formal reasons). For this purpose, the time and temperature of the diffusion anneal have to be chosen very carefully. The value determined for sD_{gb} is multiplied by the GB width δ and is plotted in **Figure 33** (triangle up). Almost perfect agreement with the data of $P = s\delta D_{gb}$ measured in cg material (diamonds) is obtained, **Figure 33**. This fact supports the conclusion that the diffusivities of the NC ($d \sim 100$ nm) and cg ($d \sim 0.5$ mm) materials are very similar.

Ag diffusion along the interagglomerate boundaries proceeds much faster than along the nano-GBs, **Figure 33** (open symbols). Having determined $P_a = s_a \delta_a D_a / \lambda$ and D_a separately in the C–B and C–C regimes, respectively, the factor $s_a \delta_a / \lambda$ can be estimated. Taking λ as $\lambda = 2\delta/d$ with $\delta = 0.5$ nm

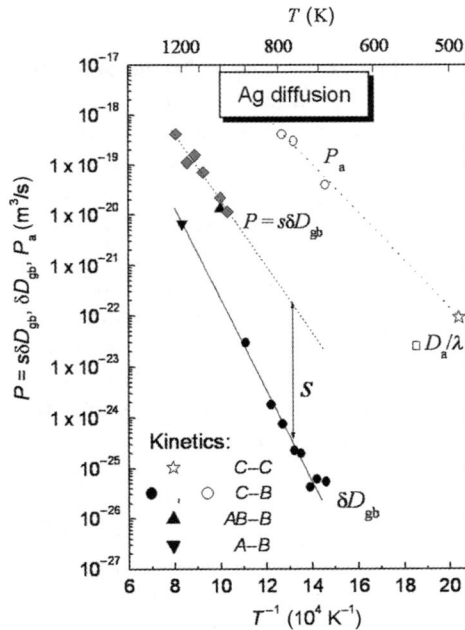

Figure 33 Arrhenius diagram for Ag (Divinski et al., 2004b) interface diffusion in nanocrystalline γ-Fe-40 wt. % Ni alloy. Filled symbols represent the diffusivity of nanocrystalline GBs and open symbols correspond to interagglomerate boundary diffusion. The estimate $\delta_a = 1$ nm was used to recalculate measured P_a values to the relevant diffusivities D_a. The triple product P for Ag diffusion in coarse-grained FeNi alloy is also shown (diamonds). The method of calculation of the segregation factor s for Ag in FeNi is illustrated.

and $d = 100$ nm, an upper estimate of the product $s_a \delta_a$ is obtained, $s_a \delta_a \sim 1$ nm. Since the interagglomerate boundaries present a more open structure with respect to the nano-GBs, the value of $\delta_a \approx 1$ nm seems to be a good estimate (Divinski et al., 2004b). Thus, the segregation factor s_a for Ag seems to be about unity. This means that there is practically no excess of Ag atoms in the interagglomerate boundaries with respect to the NC-GBs and the segregation behavior of these two internal interfaces with respect to the bulk is similar. In **Figure 33**, the value of D_a measured in the C–C regime is multiplied by the factor δ_a / λ (using the estimates $\delta_a = 1$ nm and $\lambda = 0.01$) for comparison with the P_a values measured in the C–B regime (open star and circles, respectively). Assuming $s_a = 1$, the temperature dependence of P_a can be presented by a linear Arrhenius relationship as indicated by the dotted line in **Figure 33**.

The entire amount of diffusion data supports the conclusion that the diffusivity of nano-GBs in the present NC material (grain size $d \sim 100$ nm) is similar to that of conventional cg material. One important reason for the apparently relaxed state of the GBs in the NC state is given by grain growth, which occurred in the present nanomaterial during the sintering process (from 30 to ~ 100 nm).

The presence of interagglomerate boundaries in the FeNi-material affected considerably the diffusion processes and altered the kinetic regimes of interface diffusion in the material. These interagglomerate boundaries present the fastest short-circuit diffusion paths in the material and therefore have to be taken into account in the analysis of diffusion and sintering processes in nanoalloys produced by the powder metallurgy route. Moreover, establishing of an optimum size of agglomerates provides a unique route to optimize the functional properties of the materials and to decrease significantly the sintering temperature (Lee et al., 2005).

26.5.3.4   Ultrafine-Grained Materials Obtained by Severe Plastic Deformation

Severe plastic deformation (SPD) processes have been studied extensively due to their potential to produce fully dense, fine-grained structures with attractive and rather unusual mechanical properties, as indicated above. Strictly speaking, the grain sizes achieved so far for pure and rather ductile metals such as Cu, Ni or Ti are actually outside the NC regime, in the range of about 150–350 nm.

It has been claimed in the literature that because of high dislocation activity during SPD the GBs in the final ultrafine microstructure exhibit higher energy, higher density of extrinsic GB dislocations, higher excess volume and higher microstrain than their counterparts in cg material of identical composition. Most phenomenological and structural models of modified GBs in SPD-processed materials (here: high-energy GBs) are based on dislocation-GB interaction. According to Nazarov et al. (1994), the high-energy GBs evolve from cell boundaries by absorbing lattice dislocations during plastic deformation. These dislocations are stored in nonperiodic, disordered arrays that results in long-range stress fields associated with such newly formed GBs. In spite of the fact that the hypothesis of high-energy GBs is quite plausible, its unequivocal experimental proof is still lacking.

Lian et al. (1995) have estimated the parameters of GB self-diffusion in severely plastically deformed Cu from grain growth kinetics. The corresponding activation enthalpy was about 71 and 107 kJ/mol at low and higher temperatures, respectively. The increase of the activation enthalpy was considered to result from relaxation of the so-called "nonequilibrium" structure of GBs that was suggested to be characteristic for materials processed by SPD (see above). However, these data can also be interpreted in terms of a purity effect on GB self-diffusion in copper: the newly created GBs after SPD are relatively pure from residual impurities and their segregation can be induced by subsequent heat treatment at higher temperatures. This alternative explanation is based on the fact that the activation enthalpy of 72 kJ/mol corresponds to GB self-diffusion in high-purity copper, whereas it increases to values of about 80–90 kJ/mol for less pure material (Surholt and Herzig, 1997).

Creep measurements in UFG Cu of technical purity yielded also higher values of the creep rate than expected from the measurements on cg materials in conditions of GB sliding-controlled creep (Valiev et al., 1994). This indicates an enhanced diffusivity of the GBs. Model-based estimates of the GB diffusivity from the creep data resulted in activation enthalpies of GB diffusion of about 70–78 kJ/mol. These values correspond to GB self-diffusion in high-purity cg copper, leaving some ambiguity concerning the verification of enhanced diffusivities in highly deformed materials.

There exist several additional investigations of GB diffusion in materials produced by SPD (Kolobov et al., 2001, Grabovetskaya et al., 1997, Würschum et al., 1997, Amouyal et al., 2007). Using secondary ion mass spectroscopy (SIMS), diffusion of Cu was exemplarily studied in nanostructured Ni produced by ECAP (Grabovetskaya et al., 1997). It was concluded that copper diffuses by several orders of magnitude faster along GBs of the UFG than in cg material. Annealing treatments before the diffusion experiment on the other hand, allowed GB relaxation and, as a result, similar diffusion rates were observed in both materials (Kolobov et al., 2001; Grabovetskaya et al., 1997). However, these conclusions should be treated with caution, since only shallow penetration profiles were recorded in the corresponding measurements. In contrast with the above reports, Würschum et al. (1997) have found that the GB diffusivity of ^{59}Fe in UFG Pd was comparable with that for cg Pd. Similarly, Fujita et al. (2002) observed no enhanced diffusivity in ECAP-processed UFG Al and Al-3 wt.% Mg alloy containing small precipitates that stabilized the UFG microstructure against grain growth. These results indicate that high-energy GBs, if existing in the as-deformed state, recover rapidly during the initial stages of diffusion annealing in the considered cases of UFG Pd and Al.

Recently, GB diffusion has been investigated by radiotracer diffusion analyses for a series of metals and alloys that had been processed by ECAP. The main finding is a clear bimodality of the GB diffusivities in the severely deformed materials: while the majority of GBs exhibits the diffusivities that are very close to those of relaxed high-angle GBs in the respective high-purity cg counterparts, there is a fraction of GBs that exhibit unusually high diffusivities. The detailed analysis of diffusion data allowed proposing a model of the GB distribution in a severely plastically deformed material. It was concluded that the "fast" GBs are well-separated from each other and form a network with mesh size on the micron scale embedded in the network of GBs with conventional diffusivity. The mesh size of the latter network corresponds to the grain size of the UFG material, namely a few hundred nanometers. This hierarchical model resembles the situation observed in the NC-FeNi alloy produced by pressureless sintering considered above, although the nature of the "fast" diffusion paths is basically different in these two cases.

The diffusion along the "fast" GBs in SPD-processed material was found to be enhanced by more than two orders of magnitude, **Figure 34**. Although the short-circuit diffusion paths with lower diffusivity are indicated as "slow" paths, they are not "slow" at all in absolute terms—their diffusivity is similar to that of general high-angle GBs in cg high-purity materials, where these are the fastest short-circuit diffusion paths, see **Figure 34**.

Whereas agglomeration can naturally occur in ball-milled and sintered powder materials, thus leading to the presence of agglomerate boundaries in addition to GBs, the bimodality of the GB diffusivities after SPD is most likely due to the interaction of the high density of lattice defects (dislocations and vacancies) that are created during the deformation with the GBs, leading to GBs with higher excess energy density and a more open structure with a higher amount of excess volume. Such a "high-energy" state of some of the GBs after ECAP was recently confirmed by scanning force microscopy measurements of the relative GB energy in severely plastically deformed Cu (Amouyal and

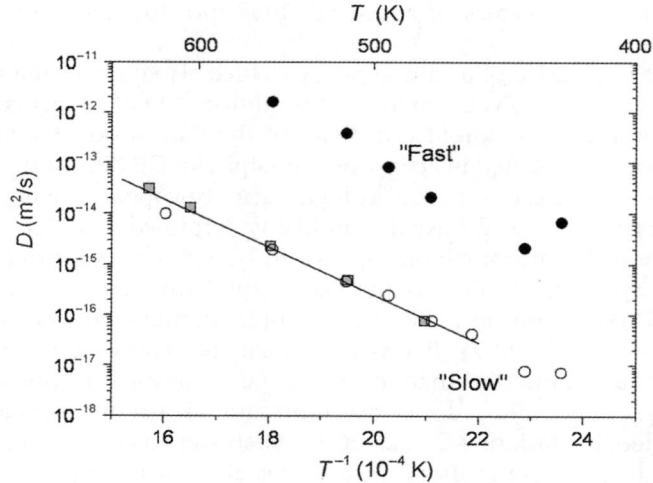

Figure 34 GB diffusion of Ni in UFG Cu–0.17wt.%Zr alloy deformed by ECAP (Amouyal et al., 2007) (circles) in comparison to the Ni diffusivity in high-purity coarse-grained Cu (Divinski et al., 2007) (squares). The diffusivities of "slow" (open circles) and "fast" (full circles) short-circuit diffusion paths in UFG-Cu alloy are shown.

Rabkin, 2007; Divinski et al., 2010b). The diffusion measurements suggest that "slow" short-circuit diffusion paths in the UFG materials have diffusivities that are very similar to those of conventional high-angle GBs in the cg counter parts, **Figure 34**.

The investigations indicate a strong heterogeneity of GB diffusivities in heavily deformed materials. A similar conclusion was made for hydro-extruded Al (Beke et al., 1987). Although this fact complicates the direct comparison of the diffusion data in UFG and cg materials, reliable conclusions can be drawn after detailed investigation of the microstructure and the diffusion behavior.

Since the above-mentioned "high-energy" GBs present defects with a high specific excess-energy density, it is expected that the relaxation toward a state of lower total Gibbs free energy is rather quick, even at low homologous temperatures. Thus, it is an important question, if it is theoretically possible to measure the diffusion contribution of high-energy GBs. According to Nazarov (2000), the relaxation time τ of the "nonequilibrium" state of GBs can be represented as:

$$\tau = \frac{kTd^3}{A\delta G\Omega D_{gb}},$$
(19)

where d and δ are the grain size and the GB width, G the shear modulus, Ω the atomic volume and a numerical factor that depends on the specific model of the relaxation and amounts to about $A = 150$. Estimations according to that model indicate that for example for pure Ni, the relaxation times are longer than typical annealing times for GB diffusion measurements for temperatures below about 600–700 K. For pure Cu, relaxation would be too fast even at room temperature ($\tau = 2.3 \times 10^4$ s at 300 K) according to the model by Nazarov. Certainly, other effects such as dislocation–boundary interactions or the elastic interactions between dislocations need to be taken into account for a more careful evaluation of the relaxation times, but the estimates according to Eqn (19) as well as the data presented above indicate that SPD results in modified structures of a fraction of the GBs that are sufficiently stable to significantly affect the diffusion behavior of the material.

The investigation of GB diffusion in a pure metal with an UFG microstructure can be significantly complicated by simultaneous recrystallization. This problem has recently been considered for Ni diffusion in UFG-Cu produced by ECAP (Amouyal et al., 2008). The detailed analysis of the microstructures obtained after annealing treatments at different times and temperatures allowed a quantitative description of the recrystallization kinetics in terms of the JMAK formalism. Using the JMAK-type expression for the volume fraction of the recrystallized area, a model that considers simultaneous diffusion and recrystallization was developed. This model enables the quantitative derivation of the diffusion parameters from experimentally measured penetration profiles under the boundary condition of simultaneously occurring recrystallization.

The experimentally measured penetration profiles exhibited two distinct branches with different slopes and were quite similar to those observed earlier in the thermally stable Cu–0.17 wt.%Zr alloy (Amouyal et al., 2007). The similarity of diffusion penetration profiles measured in the structurally stable Cu–Zr alloy (Amouyal et al. 2007) and in pure Cu (Amouyal et al., 2008) implies that hierarchical microstructures are typical for UFG materials produced by ECAP, and that the "fast" GBs can be associated with the high-energy GB state formed after ECAP. The activation enthalpy of Ni-GB diffusion in UFG Cu (the "slow" short-circuit diffusion paths) turned out to be very close to the corresponding activation enthalpy of Ni-GB diffusion in UFG Cu–0.17 wt.% Zr alloy and in cg Cu.

As discussed above, first direct measurements of GB diffusion in severely deformed materials yielded ambiguous results—both similar and enhanced rates of atomic transport were deduced with respect to the GB diffusivities in reference cg materials. Systematic measurements by the radiotracer technique discovered a hierarchic nature of internal interfaces that are developing as a result of strong dislocation activity during SPD processing and, presumably, of a localization of plastic flow (Divinski and Wilde, 2008). Both "conventionally fast" as well as "ultrafast" short-circuit diffusion paths were observed in SPD processed materials, with the latter being embedded in a network of GBs akin relaxed high-angle GBs as they exist in annealed cg materials (where these boundaries constitute the fastest short-circuit diffusion paths) (Divinski and Wilde, 2008). This is an important discovery of the radiotracer method, which provides sample-averaged information. The existence of a hierarchy of interfaces in plastically deformed metals has been pointed out by Hansen (Liu et al., 2002; Lu and Hansen, 2009) by introducing the so-called extended or geometrically necessary boundaries (GNBs) and incidental dislocation boundaries (IDBs). However, the diffusion studies indicate another type of hierarchy, which corresponds to different types of internal interfaces, since the diffusivity of dislocations or low-angle dislocation GBs is definitely lower than that of general high-angle GBs. Generalizing these findings, the following hierarchy of interfaces in SPD materials can be proposed (in the order of decreasing diffusivities):

- High-energy interfaces (probably of different types and representing a certain spectrum of diffusivities and structures);
- General high-angle GBs (with diffusivities and, probably, structure being similar to those of relaxed high-angle GBs);
- Highly defected twin boundaries with diffusivities similar to those of the previous level (Wang et al., 2010). Note that diffusion along relaxed twin boundaries is hardly measurable;
- low-angle boundaries, dislocation walls, single dislocations.

A comprehensive theory of SPD processing and grain refinement has to include all these levels of the hierarchy, which on the other hand may critically depend on processing routes and regimes (temperature, strain rate, applied pressure and so on). It is important that these levels correspond to different scales with the mesh size ranging from several micrometers (the nonequilibrium boundaries) down to hundred (dislocation walls) or even tens of nanometers (nanotwins). The appearance of general high-angle GBs (with properties similar to those of relaxed interfaces in the cg counterparts) in SPD materials depends obviously on the processing temperature because this fact might be a clear indication of dynamic recovery processes during SPD.

In **Figure 35a**, the results of radiotracer measurements of GB self-diffusion in UFG Ni (3N8 purity) after ECAP-processing (circles and dashed lines, (Divinski et al., 2011)) are compared to the grain-

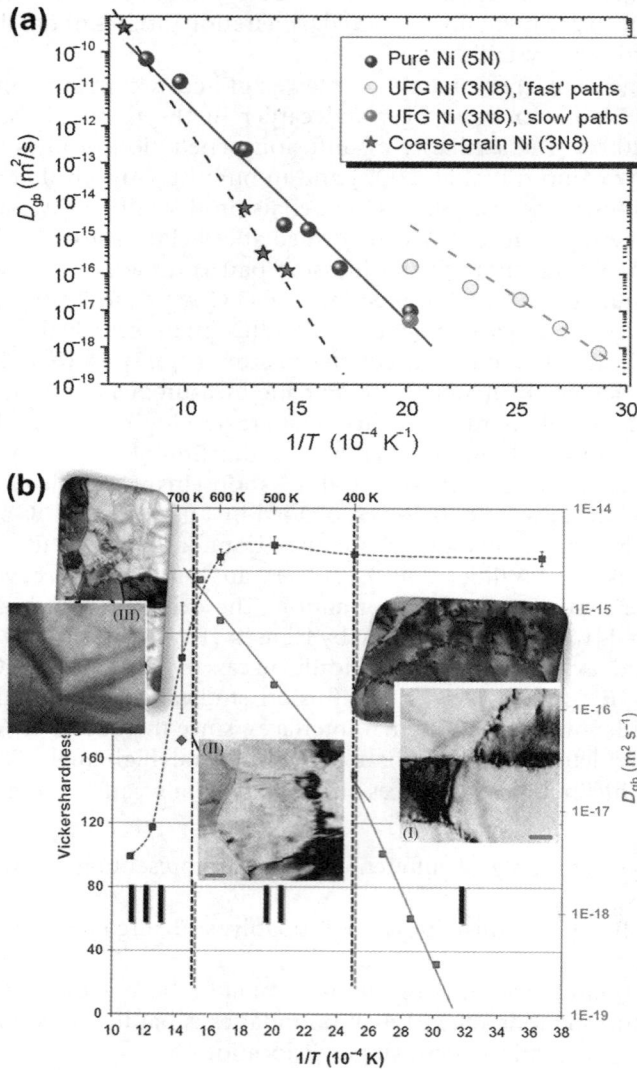

Figure 35 (a) Arrhenius plot of the grain boundary self diffusion in Ni of different purities and in different microstructure states, i.e. after recovery and relaxation and after ECAP processing. (b) Comparison of different regimes (I to III) of grain boundary diffusion of ECAP-deformed Ni (99.8% purity) with the hardness and with the characteristic appearance of the contrast at grain boundaries in bright-field TEM images. (For the color version of the figure, the reader is referred to the online version of the book.)

boundary diffusivity values that were obtained on a cg, polycrystalline material with a relaxed (annealed) GB structure and a grain size of about 100 µm (the solid line, (Divinski et al., 2010a)). Clearly, short-circuit diffusion is significantly faster in UFG Ni than in cg Ni and this diffusion enhancement depends critically on the temperature interval.

At lower temperatures, below about 400 K (region I), an almost linear Arrhenius behavior is measured, with the corresponding activation enthalpy being roughly half of that, which characterizes the cg Ni material. The experimental data at the temperatures above 400 K show a cross-over to a distinctly different, yet consistent Arrhenius dependence (region II). This fact points to the attainment of a partially relaxed state of the high-energy GBs with a different metastable structural configuration as compared to that produced during SPD at room temperature and which is kinetically stable over a significant interval of annealing temperatures. The interval III corresponds to relaxation of the high-energy state of GBs and overlapping processes of recrystallization/grain growth.

The regions II and III appear to be well separated in ECAP Ni due to the relatively low purity of the material used (99.6wt%). As a result of GB segregation of residual impurities, the UFG microstructure turned out to be relatively stable and no significant recrystallization/grain growth was detected below 600 K (Divinski et al., 2011).

The extremely fast tracer penetration in severely deformed pure Ni is consistent with the previous results of radiotracer diffusion measurements on ultrafast transport in SPD-processed pure Cu and Cu-based alloys. On the other hand, such a clear kink in the Arrhenius dependence was not observed for GB diffusion in UFG pure Cu and Cu-based alloy. This fact correlates with a lower homologous temperature of SPD processing for nickel ($T/T_m = 0.17$) in comparison to that for copper ($T/T_m = 0.22$), since the SPD process was carried out at room temperature in all cases under consideration. The low homologous temperature of deformation reduces the dynamic recovery processes that affect the concentrations of point defects, impurity segregation and atomic transport along interfaces and result in modified GB structures.

Such an observation substantiates the complexity of high-energy GBs. Specifically, it follows from these results that in addition to common parameters required specifying a relaxed high-angle GB (i.e. the disorientation, inclination and the translation vector), extra parameters have to be introduced to characterize the high-energy state. In a simplest approximation, one may think of the defect density and/or the free-volume density. Kinetic parameters, e.g. relaxation time(s), and thermodynamic parameters, e.g. segregation, compound formation or chemical ordering/disordering effects at interfaces might also be involved in view of an inherent metastability of these interface states.

The region II substantiates a specific structure state of GBs in SPD processed materials. It is important to note at this point, that a direct comparison of the diffusion measurements, which represent a macroscopically averaging method, with the highly local microstructure analyses results obtained by TEM-based techniques is not feasible. Yet, in all cases where SPD-processed material was studied, similar contrast features as in **Figure 14a** were observed by TEM and similar fast and ultrafast contributions to GB diffusion were found. This correlation is sketched in **Figure 35b** for ECAP Ni. The serrated contrast at the majority of GBs and strain contours around these GBs in the as-prepared state correlates with significant enhancement of interface diffusivity (region I). In the temperature interval II, the kink in the diffusion rate correlates with partial relaxation of the serrated contrast while the bulk strain/stress state is conserved to a large extent. Only in the interval III the recrystallization/grain growth processes trigger interface relaxation and recovery of the GB diffusivity. We conclude that the combined results on kinetics and structure of interfaces indicate that the high-energy GBs, revealing an increased width, a high extrinsic GB dislocation density and high residual strain levels associated with them, possess a significantly enhanced diffusivity. Analyses by GPA on a GB that displayed the serrated contrast features confirmed the presence of high-defect densities and large microstrains associated with such GBs (**Figure 14b**).

Figure 35 indicates clearly that the term "high-energy GB" encompasses a wide range of different states of interfaces with basically different kinetic/structure properties, cf. regions I and II. Remarkably enhanced diffusivities (although with significantly different effective activation enthalpies) and specific GB structures observed in regions I and II substantiate a high-energy state of interfaces in the corresponding temperature intervals. However, the pertinent interfaces reveal different TEM contrasts with presumably different strain/stress levels and defect populations. In the particular case of ECAP Ni, deformed at room temperature, we may talk about at least two distinct states of the high-energy interfaces. The fundamental questions arise. What is common between the states I and II? Which properties have to be used for an unambiguous definition of the high-energy state? The atomistic/structure reasons of the diffusivity enhancement also have to be understood.

The high diffusion rates are believed to be related to higher excess free energies of these defect-modified GBs in severely deformed Ni. Adopting the semi-empirical Borisov formalism (Borisov et al., 1964) the excess free energy of high-energy interfaces in ECAP Ni was found to be about 30% larger than in the annealed cg material, whereas about 10% increase was reported for SPD Cu.

We propose to generalize this formalism and to use the above-mentioned approach as a measure of the excess free energy of interfaces irrespective of their state. This phenomenological model gives rise to a definition of the high-energy state of an interface with respect to relaxed general high-angle GBs and furthermore introduces a convenient measure for the specific high-energy state. In a series of papers by Nazarov et al. (1990, 1993) the "nonequilibrium" state of interfaces in SPD materials was related to the content of the extrinsic GB dislocations (see Section 26.4.3 of this chapter). It is interesting that both structure and kinetic approaches can be combined providing an extensive characterization of ECAP Ni. The cross-over in the diffusion behavior at 400 K, **Figure 35**, correlates with a characteristic change of the relaxation time of the array of extrinsic GB dislocations calculated according to Ref. Nazarov et al. (1990) (and with the change of a typical HRTEM contrast at the interfaces) (Divinski et al., 2011). Whereas the agreement is encouraging for the region I in **Figure 35**, the diffusion approach allows characterizing the state of interfaces in other regions where the dislocation approach of Nazarov et al. fails since it predicts fully relaxed GBs.

At this point, it should be noted that along with arrays of extrinsic GB dislocations, as suggested by Nazarov et al., other defects should be considered too. The following processes/phenomena contribute to the high-energy state of GBs in SPD-processed materials:

- abundant vacancies and vacancy-like defects in interfaces produced by severe deformation;
- redistribution of the related excess free volume, release of local strains/stresses
- chemical effects (ordering) may be important in alloys and compounds affecting the atomic redistribution and retarding e.g. the stress/strain relaxation;
- segregation can be especially important in alloys involving even 2D compound formation along interfaces (Rodin et al., 2009).

The effective activation enthalpy of interface diffusion in ECAP Ni in the region I is similar to the effective activation enthalpy, which was found for recovery of vacancies in the material by DSC, suggesting that interface self-diffusion in as-prepared UFG Ni is governed by deformation-induced vacancies or vacancy-like defects. Basically, redistribution of these defects along with the stress relaxation in grains determined the transition from state I to state II with increasing temperature.

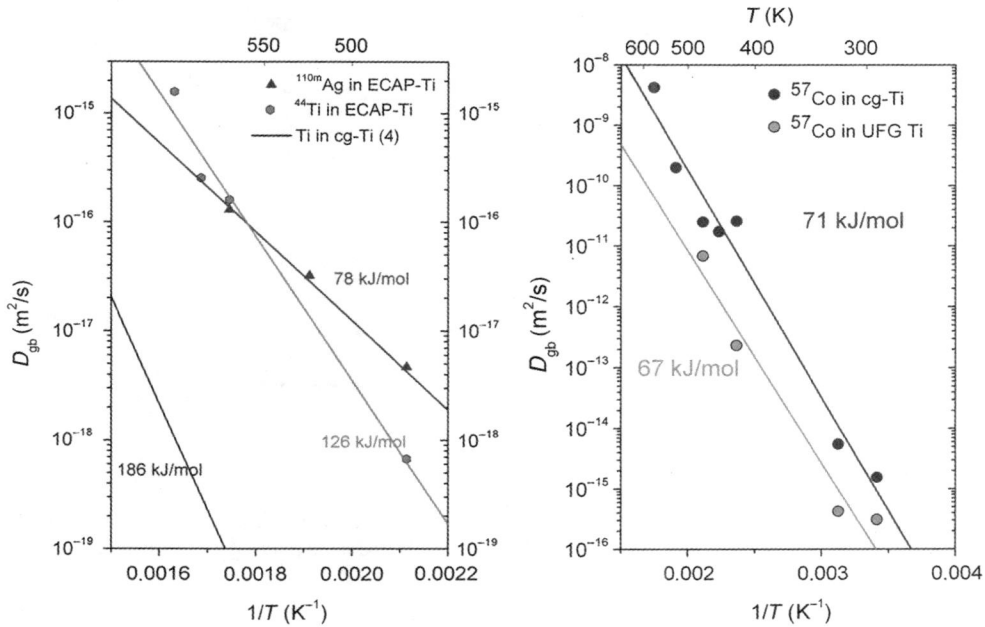

Figure 36 Arrhenius plot of the tracer diffusivities of Co, Ti and Ag in α-Ti before and after severe plastic deformation (Fiebig et al., 2011).

In order to give further insights into GB structures affected by SPD processing, GB diffusion of substitutionally (Ag) and interstitially (Co) diffusing solutes was investigated in cg as well as in UFG α-Ti produced by ECAP (Fiebig et al., 2011).

Co is known to be a so-called "ultrafast diffuser" in the crystalline bulk of CG α-Ti (Perez et al., 2003). It occurred that Co is an ultrafast diffuser in GBs of CG α-Ti, too. Astonishingly, at least at a first sight, GB diffusion of Co in UFG α-Ti is slower than the interface diffusion in CG α-Ti while for Ag as diffusing species, GB diffusion in the UFG material is significantly faster (see **Figure 36**). Due to SPD processing, a high concentration of defects, including those at GBs, is created. The associated excess free volume offers effective (substitutional) traps for interstitially diffusing Co atoms.

On the other hand, the Ag diffusivity is dramatically increased in UFG α-Ti as a result of SPD. It is assumed that the physical origin of the increased diffusivity of Ag in UFG α-Ti is the formation of high-energy GBs during the deformation process and the increase of the excess free volume of the interfaces.

In general the diffusion along high-energy GBs depends on the diffusion mechanism of the tracer. Interstitially diffusing atoms are trapped or scattered due to the high concentration of lattice defects in GBs, which were induced by the SPD. Accordingly, the interstitial diffusivity of such kind of elements can be slowed down.

A recent study by Schafler (2011) using X-ray line profile analysis showed that only half of the deformation-induced vacancies remain after unloading material from the high-pressure conditions maintained during the deformation. Thus, a significantly higher vacancy concentration is present in the material during deformation, resulting also in a potentially much higher atomic mobility during the

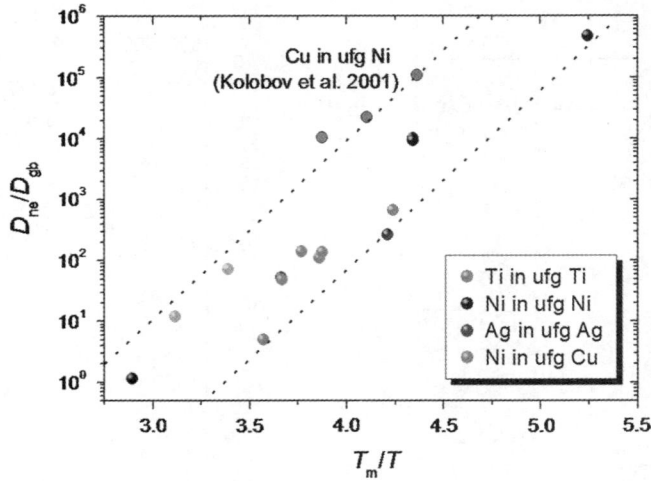

Figure 37 Enhancement of the grain boundary diffusivity after severe plastic deformation as a function of the reduced temperatures of the measurements. (For the color version of the figure, the reader is referred to the online version of the book.)

SPD treatment, since the GB diffusion measurements, which indicate GB diffusivities that are increased by orders of magnitude after severe deformation, were performed ex-situ, after the deformation and the high pressure has ceased.

It remains unclear if due to the increased value of the vacancy migration enthalpy under high hydrostatic pressure (Zehetbauer et al., 2005) the distribution of vacancies is rather homogeneous during the severe deformation. Yet, due to the presence of a thermodynamic driving force, preferential diffusion of vacancies to the most potent sinks that are within diffusion distance, i.e. toward the high-angle GBs, can be safely assumed. This scenario would lead to a composite structure with potentially more compliant GB regions and strong grain interiors. However, this scenario does not agree with the experimental observation showing that only a fraction of the GBs accumulates extra excess energy. Yet, whether this selectivity is present during SPD processing or only results as a residual effect after unloading and/partial (dynamic) relaxation remains to be analyzed.

The existing hierarchy of internal interfaces in UFG materials substantiates the fact that not all GBs are modified by SPD-processing to become high-energy GBs. According to the diffusion data, a fraction of boundaries that form an interconnected network with a "mesh size" of about 5–10 grain diameters transforms to become high-energy GBs. This feature has to be taken into account for adequate analyses of the interface structure in SPD materials by intrinsically local TEM investigations. Additionally, this result is also in line with TEM observations that indicated a nonuniform distribution of contrast variations near GBs in SPD-processed materials.

In summary, GB diffusion in UFG and in NC materials processed by SPD indicate the presence of a distinct hierarchy of interconnected networks of GB types with significantly different diffusivities, independent of the specific material or the specific deformation path that had been applied (see **Figure 37**).

The comparison between interstitial and substitutional GB diffusion furthermore indicates that the redistribution of free volume due to lattice dislocations and due to vacancies moving to GBs presents a likely basis for the observed modifications of a fraction of the GBs. At the same time, the GB diffusion measurements on NC materials that were processed without involving severe deformation have shown unchanged GB diffusivities. This result clearly indicates that the GB structure per se is not affected by reducing the size of the grains. It thus appears that complex defect interactions are required for modifying the structures of GBs in extent to the variations given by crystallography and that a small grain size is presenting a boundary condition for obtaining such interactions, since the small grain size effectively prevents self organization of lattice dislocations into cell walls or GBs.

26.5.4 Diffusion along Triple Junctions

The enhanced diffusivity of NC materials was also proposed to be explained by the contribution of TJs (Chen and Schuh, 2007; Ovidko and Scheinerman, 2004). Quantitative estimates (Chen and Schuh, 2007) indicate that triple junction diffusion can formally explain anomalies of several orders of magnitude in materials with a very small grain size, of the order of 10 nm or less. However, there are presently no direct experimental evidences of the importance of the triple junction contribution to the diffusion phenomena in nanomaterials, especially, since diffusion measurements that are unaffected by porosity are hardly available for materials with grain sizes below 50 nm. Within the grain size range that has become experimentally available (i.e. average grain sizes of the order of 30–50 nm or larger), the fraction of the material's volume directly associated with TJs is still negligible. A comprehensive study of the grain size dependence of the diffusivity of nanomaterials could provide further insight into this problem.

26.6 Plastic Deformation of Ultrafine-Grained and Nanocrystalline Metallic Materials

The mechanisms of deformation in NC materials differ from those of conventional, cg materials, which have been described in detail in earlier chapters in this volume. Due to the size-dependent mechanism changes and due to the importance of the mechanical properties for synthesis and processing but also for a wide range of "functional" applications, where minimum mechanical performance forms the basis of the applicability and the service life time, the response of NC and UFG materials to external mechanical stress fields has developed almost into a research field in its own right. Thus, we will focus here on the most important and most specific alterations of plasticity for materials with very small grain sizes. Again, as mentioned in Section 26.1.1, modifications of the characteristic mechanisms brought about by reducing the size of the entire testing specimen (e.g. as in compression tests of nanopillars) will not be included, since the characteristics of fine-grained microstructures form the central topic of this chapter.

Molecular dynamics (MD) simulations have contributed strongly for delivering new insight into deformation processes in NC materials (Van Swygenhoven et al., 2004; Derlet et al., 2003; Schiotz and Jacobsen, 2003; Yamakov et al., 2003). A transition in the mechanical behavior from dislocation-based deformation mechanisms to GB-mediated ones (Argon and Yip, 2006), which manifests in a change of the slope or even a change of the sign of the slope of the Hall–Petch relationship (Schiotz et al., 1998), has been found for decreasing grain sizes. A detailed review on the results of MD-simulations of NC materials is given by Wolf et al. (2003). Concerning experimental results, the picture is somewhat blurred by the multitude of observations on a rather wide range of chemically different materials that were processed by different methods and by the additional convolution due to possible artifacts such as impurities, pores, microcracks etc. One result that has become clear is that there is no sharp demarcation between different regimes of plastic deformation. Even at the smallest grain sizes that were obtained for real materials (as opposed to MD simulations), still the remains of dislocation processes were observed. Thus, grain size reduction in reality changes the relative importance of different mechanisms, which also means that several different mechanisms are active at the same time at small grain sizes, leading to a rather complex behavior and, by trend, to complex interactions of different types of defects. The mechanical properties of NC materials have been reviewed by Meyers et al. (2006) and the mechanical properties of UFG materials have been summarized in Valiev et al. (2000).

26.6.1   Nanocrystalline Metals and Alloys

One of the earliest and most discussed observations of a size-dependent change of the dominant deformation mechanism is presented by the so-called "negative Hall–Petch" behavior that is characterized by decreasing yield strength with decreasing grain size. According to the traditional explanation for the "normal" Hall–Petch behavior that is based on dislocation pile-up near the GBs, the yield stress σ_y varies with the grain size, d, as follows:

$$\sigma_y = \sigma_0 + k \cdot d^{1/2}, \tag{20}$$

where σ_0 denotes the friction stress and k is a constant. A more general expression is obtained by using a power law with an exponent $-n$ ($0.3 \leq n \leq 0.7$).

Thus, it is straightforward to expect the change to "negative" Hall–Petch behavior at grain sizes that are too small to accommodate dislocation pileups, i.e. at sizes that are comparable to the spacing between dislocations. The critical grain size, d_c, for the change in the Hall–Petch behavior would thus be expected when only one dislocation fits into the grain (Nieh and Wadsworth, 1991). From the calculations by Nieh and Wadsworth (1991), d_c is obtained as:

$$d_c = \frac{3 \cdot G \cdot b}{\pi (1 - \nu) \cdot H} \tag{21}$$

with the shear modulus, G, the magnitude of the burgers vector, b, Poisson's ratio, ν and the hardness, H. Thus, the transition from classical Hall–Petch behavior is expected at grain sizes of the order of 20–30 nm for 3d-transition metals. After first results obtained on electrodeposited thin films that were controversially discussed due to impurity segregation at the internal interfaces, Schiotz and Jacobsen (2003) reactivated the discussions on the occurrence of a "negative" Hall–Petch behavior by a MD study concerning the size dependence of the flow stress of Cu in the grain size range between 5 and 50 nm, showing a clear maximum at about $\langle d \rangle = 15$ nm (see **Figure 38**).

Several quantitative explanations for the complex deformation behavior are available. These include GB sliding, diffusion, motion of TJs, presence of pores and impurities etc. At grain sizes below about 20–30 nm, deformation may occur by the emission of partial dislocations or formation of deformation twins and at even finer grain sizes deformation could occur by GB sliding. These mechanisms, and the grain size ranges of their operation, are influenced by temperature, lattice spacing and stacking fault energy.

26.6.1.1   Inverse Hall–Petch Effect

Observations of an "inverse Hall–Petch" behavior were explained in terms of diffusion creep by fast transport along the numerous "disordered" GBs. For lower strain rates, a mechanism based on grain-boundary sliding and on coplanar alignment of GBs to form the so-called "mesoscopic glide planes"

(a)

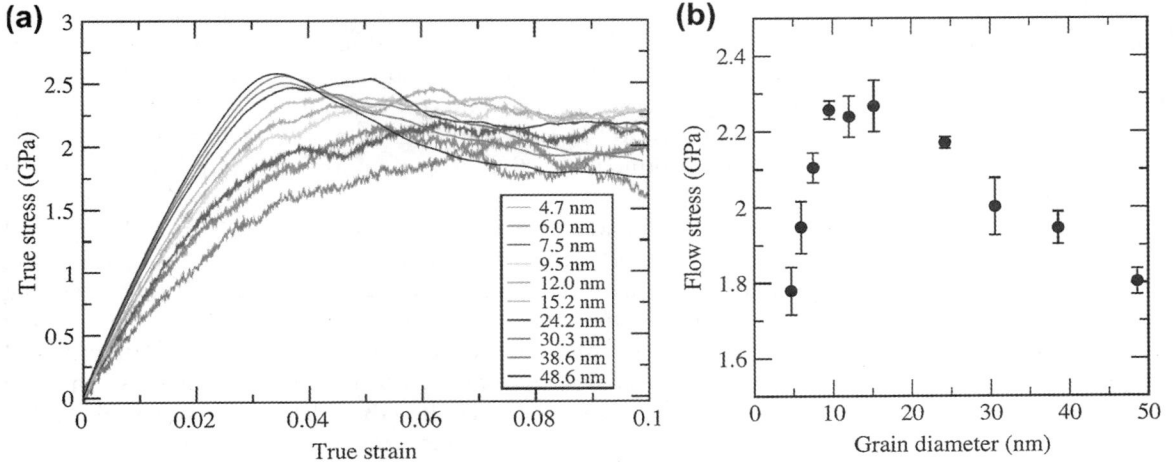

(b)

Figure 38 (a) Simulation results concerning the grain size dependence of the flow stress in Cu. The image shows the results of 10 simulations with varying grain sizes. (b) Flow stress as a function of the grain diameter as obtained from the simulations indicated in **Figure 38a**. Source: Schiotz and Jacobsen (2003). (For the color version of the figure, the reader is referred to the online version of the book.)

has been suggested by Hahn and Padmanabhan (1997); this provides explanations for the occurrence of the "inverse Hall–Petch" behavior and for a moderate work hardening, respectively (**Figure 39**).

The creep resistance of NC materials is more than what is warranted by coble creep and the creep rate also does not vary as D^{-3}, where D is the (average) grain size. Diffusion models such as that underlying the so-called "Coble" creep assume that GB sliding is always faster than diffusion. Thus, GB sliding could only generate very small creep strains because of the many densely spaced steric hindrances (i.e. blocking grains, TJs or quadruple points). Thus, classical models assume that the stoppage of flow at TJs

(a) **(b)**

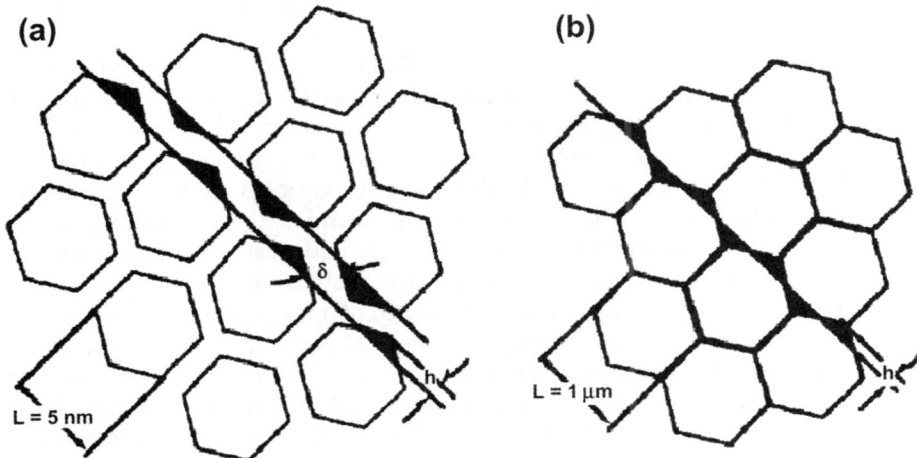

Figure 39 2D-schematic of grain arrangement in a nanocrystalline material. A mesoscopic plane interface can be formed by grain boundary migration. Reproduced with permission from Elsevier (Padmanabhan, 2009).

is overcome by diffusion processes or dislocation motion. As a result, the internal stresses that build up during sliding are reduced and plastic deformation can proceed (Padmanabhan et al., 2007; Padmanabhan and Gleiter, 2004). The dominance of GB sliding during superplastic flow was noted even in the early work of Pearson (1934). Many latter workers reinforced this finding. The boundary sliding-controlled flow model assumes that rate controlling flow is confined to a three-dimensional continuous network of internal interfaces that surround grains that do not deform except for what is required to ensure strain and geometric compatibility.

Within the approach of Hahn et al. (1997) and Hahn and Padmanabhan (1997), a high-angle GB is divided into a number of atomic scale ensembles that surround free-volume sites present at discrete locations characteristic of a boundary. Each of these ensembles constitutes a basic unit of sliding (where due to the presence of free volume the shear resistance is less than in the rest of the boundary). Sliding occurs by the movement of similar boundary volume elements, of oblate spheroid shape with ground area $\pi\delta2$ and height δ located symmetrically on the boundary plane so that the height on either side of the boundary is $\delta/2$. Here δ denotes the GB width. To produce substantial sliding on a mesoscopic scale, two or more grains must cooperate to form a plane interface by boundary migration that by further interconnection with other plane interfaces will lead to long-range sliding (Padmanabhan and Schlipf, 1996). The driving force for plane interface formation is the minimization of the total free energy of the system caused by the much larger anisotropic work done by the applied stress, compared with the surface free energies associated with the concerned GBs. Once a plane interface is formed, the localized sliding shears can lead to mesoscopic sliding over dimensions of many grains and eventually lead to large strains and superplastic deformation.

In physical terms, in metallic systems the deviation from the Hall–Petch relation, which eventually leads to the inverse Hall–Petch behavior, could be attributed to competition between the GB sliding controlled process and dislocation dominated deformation. These are two independent mechanisms and the one that requires less stress (energy) will be the favored mode of deformation under a given set of experimental conditions. Evidently, when GB sliding is dominant, grain refinement will weaken the material (inverse Hall–Petch), but when crystallographic deformation is dominant (conventional Hall–Petch), grain refinement will strengthen the material. This transition region from inverse Hall–Petch to Hall–Petch relationship is yet to be placed in a strict, quantitative framework.

Rather recently, the so-called "coupled motion" of GBs under shear has been proposed and already confirmed by MD simulations and experimental analyses as an alternative mechanism for stress accommodation that might also lead to stress-induced grain coarsening, which has been observed frequently and which has also been suggested as a mechanism that accommodates (at least partially) externally applied stresses in NC materials. In this context, "coupled motion", as discussed earlier, denotes the process that GBs move in the direction of the normal to the GB plane under an applied shear stress. The magnitude of normal motion of the GBs depends on the relative disorientation of the two grains that are adjacent to the boundary. That disorientation defines the geometrical "coupling factor" of the respective GB. With respect to the model of mesoscopic slide plane formation, as introduced above, coupled motion of GBs obviously represents a competing mechanism, since along a mesoscopic slide plane that extends by definition over several grains with GBs of different type the coupling factor varies. As a consequence, the different magnitude of normal motion of the different GBs that constitute the segments of the mesoscopic glide plane would destroy the alignment of these GBs (Schäfer and Albe, 2012a) (see **Figure 40**). Recent MD simulations indicate the importance of intergranular solutes on the competition between coupled motion and GB sliding (Schäfer and Albe, 2012b).

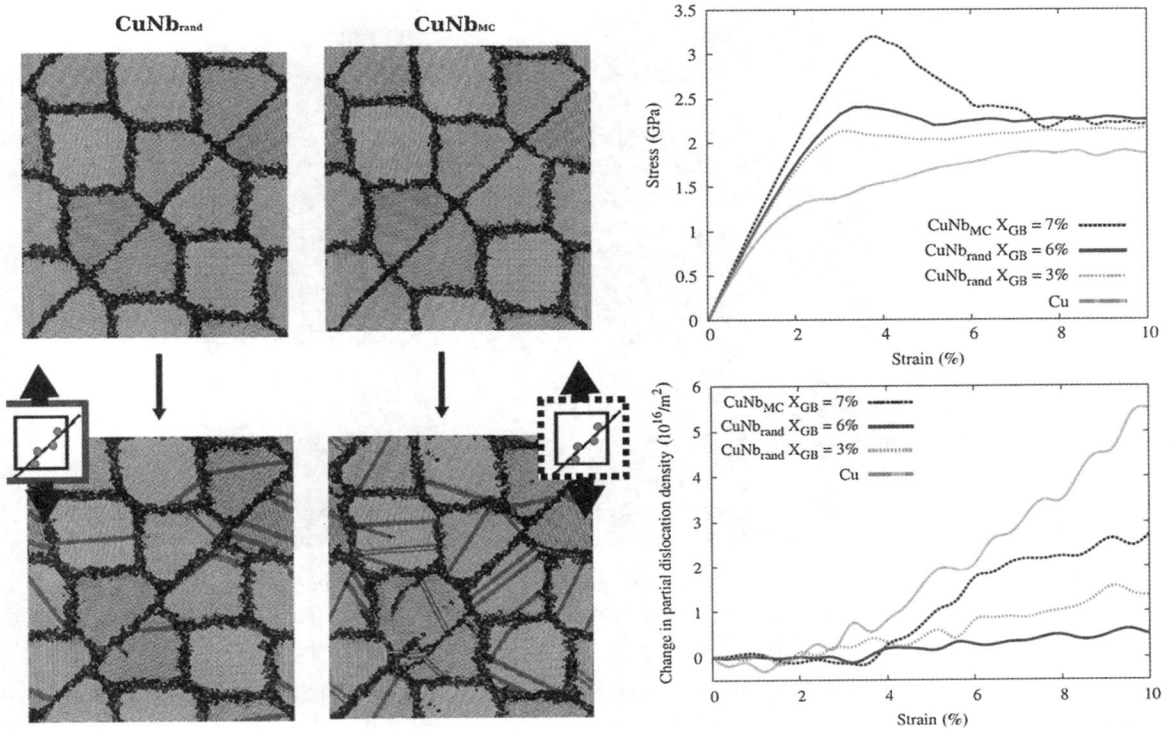

Figure 40 Snapshot of Molecular Dynamics simulations concerning the competition between grain boundary sliding along the mesoscopic plane interface and coupling leading to grain boundary motion parallel to the grain boundary normal of the individual grain boundary segments. Reproduced with permission from Elsevier (Schäfer and Albe, 2012). (For the color version of the figure, the reader is referred to the online version of the book.)

26.6.1.2 Twinning at Small Grain Sizes

Traditionally, strengthening through grain size decrease has been discussed with respect to the interaction between dislocations and random high-angle GBs. Rather recently, it has been demonstrated that also nanoscale growth twins, if present at high spatial density, can act as to efficiently enhance the mechanical strength, since they present effective barriers to dislocation motion while also helping to retain ductility. MD simulations (Jin et al., 2008) on the purely stress-driven interaction between a 60-degree nonscrew lattice dislocation and a coherent twin boundary in an fcc metal have shown that dislocations can efficiently be stopped at twin boundaries (see **Figure 41**). Firstly, the dislocation dissociates into different partial dislocations gliding into the twin as well as along the twin boundary. A sessile dislocation lock may be generated at the coherent twin boundary if the transited slip is incomplete. The details of the interaction are controlled by the material-dependent energy barriers for the formation of Shockley partial dislocations from the site where the lattice dislocation impinges upon the boundary.

Lu et al. demonstrated for electrodeposited films of polycrystalline Cu that consisted of a microstructure with nanospaced twin lamellae within the micron-sized grains a significant size-dependence of the mechanical properties on the twin lamellar spacing (Lu et al., 2004). In fact, the dependence on the lamellae spacing was found to be comparable to the traditional dependence of the yield strength on

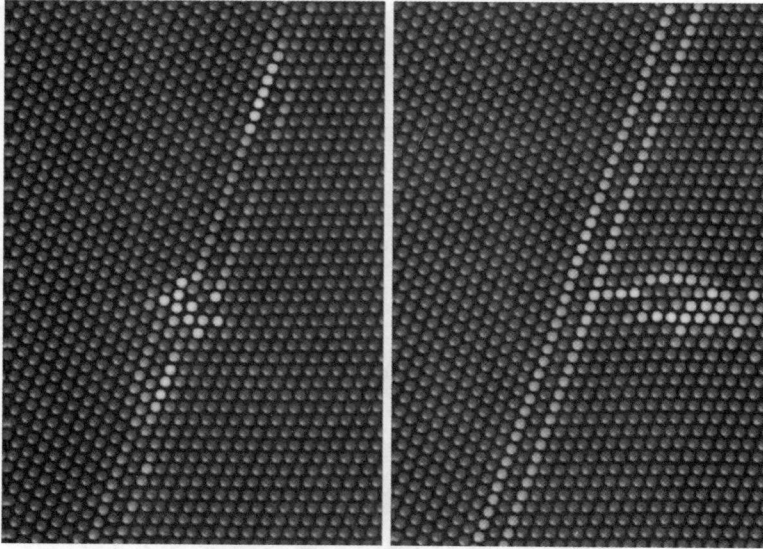

Figure 41 Snapshot of a Molecular Dynamics simulation on the interaction of a traveling dislocation with a coherent twin boundary in an fcc lattice. Reproduced with permission from Elsevier (Jin et al., 2008). (For the color version of the figure, the reader is referred to the online version of the book.)

grain size. As shown in **Figure 42**, the stress–strain curve of the material with the smallest lamellae spacing shows the highest tensile strength. Additionally, it was found that the sample with the smallest twin spacing showed the largest ductility of all samples with multiple twinning. In fact, for the nanotwinned sample with average twin lamellae spacing of about 15 nm, the ultimate tensile strength reaches about 1070 MPa. This value is even larger than the respective value for Cu of similar purity that consists of submicron-sized grains (without massive twinning).

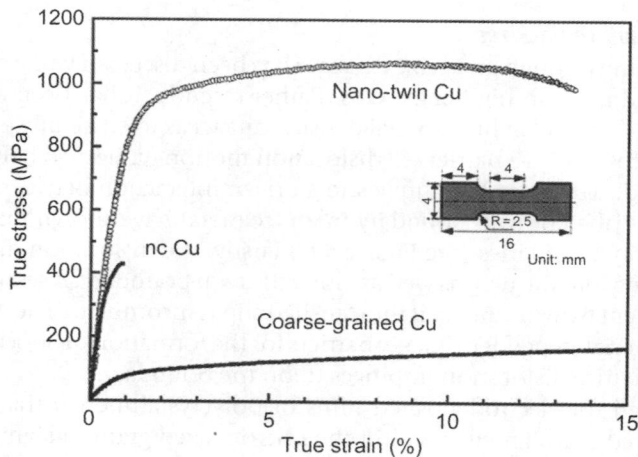

Figure 42 Effect of nanotwin density on the mechanical response of copper. Reproduced with permission from (Lu et al., 2004).

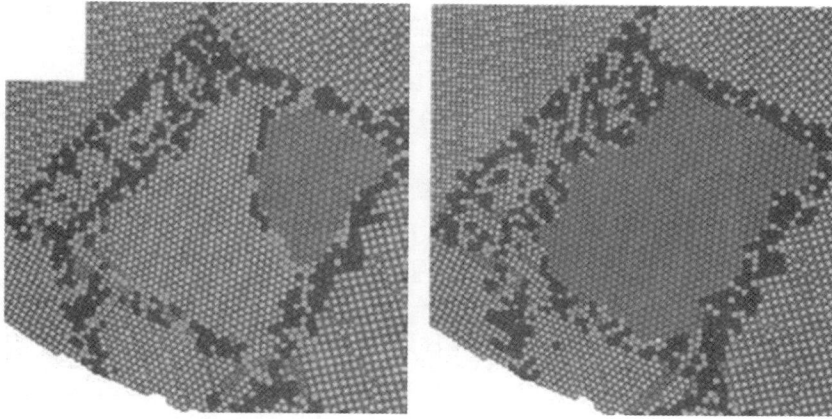

Figure 43 Snapshot of a Molecular Dynamics simulation of the emission of a partial dislocation from a grain boundary region near a triple junction in Cu. Reproduced with permission from (Van Swygenhoven et al., 2004). (For the color version of the figure, the reader is referred to the online version of the book.)

In contrast to growth twins, as present in the electrodeposited Cu, the observation of mechanical twins in materials (such as Al) with high stacking-fault energy has roused immense interest, since such observation would indicate a clear size-induced mechanism change. MD simulations of the deformation of NC materials with high stacking-fault energies, such as Al, indicated that twinning occurs as an alternate mode of plastic deformation at rather large strains and high stress levels, with twins originating at GBs as well as in the grain interior, the latter resulting from the interactions of stacking faults (see **Figure 43**).

While it is conceivable that sufficiently large stress concentrations can develop at nanoscaled GBs under shear (and slip), as elucidated by Asaro and Suresh (see also Section 26.6.1.3 below and (Asaro and Suresh, 2005)), it must be kept in mind that the strain rate in MD simulations is many orders of magnitude larger than in controlled experiments. As known from experiments, e.g. with shock loading, high strain-rate deformation (particularly when occurring at low homologous temperatures) favors deformation twinning even for materials with high stacking-fault energies. In addition, experimental observations of deformation-twin formation in NC materials with high stacking-fault energies have shown sparse distributions of grains that are twinned (often, multiple twins with nanoscale twin lamellae thicknesses are then observed). Thus, it seems unclear how the sparse occurrence of twinning should accommodate a macroscopic stress field that is applied on the entire microstructure.

26.6.1.3 *Ex-situ Studies of the Dominant Deformation Mechanisms*
Markmann et al. (2003) have shown that, similar to what is known for conventional materials, the dominant deformation mechanism in NC materials is a function of the strain rate. Diffusion creep is dominant in the limit of very low strain rate, and in NC materials it becomes noticeable at much lower temperatures than in cg materials. At higher strain rates, partial dislocations must be active as evidenced by the creation of stacking faults. Results of MD simulations indicate that the GBs (and most probably the GB areas adjacent to TJs and quadruple points) are active sources for partial dislocation emission. These observations are also in line with the observed strong decrease in strain rate sensitivity (see **Figure 44**), or alternatively the increase in activation volume, with increasing grain size, which can be explained by the presence of stress concentrations at sliding GB facets (Asaro and Suresh, 2005).

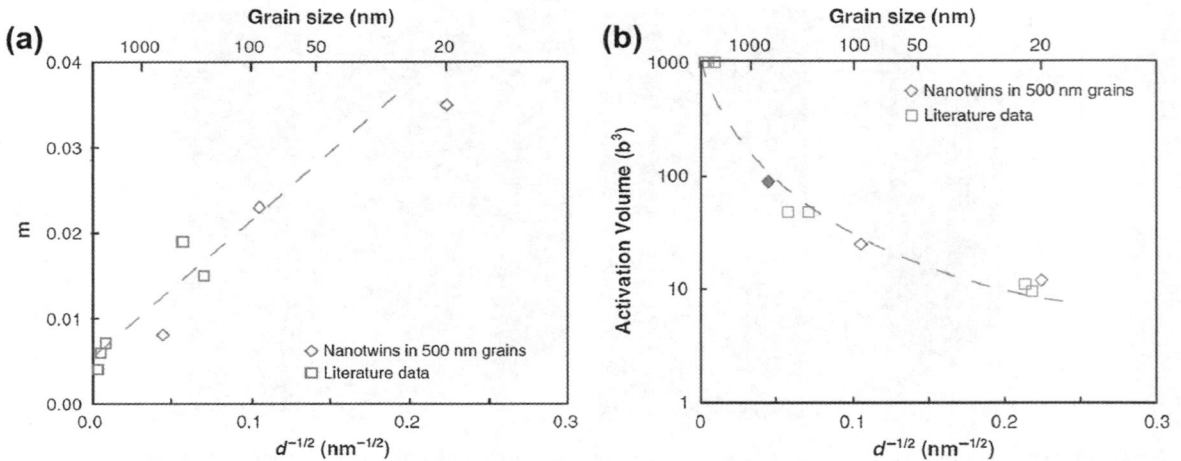

Figure 44 (a) A plot of the effect of grain size on the loading rate sensitivity index, m, of pure Cu and Ni at room temperature. Also indicated are points, denoted by open diamonds, for pure Cu where twins with a width of 20 or 90 nm were introduced by pulsed electrodeposition inside grains with an average size of approximately 500 nm. For these cases, the twin width is plotted instead of the grain size. (b) A plot of the effect of grain size on the activation volume, measured in units of b^3, for pure Cu and Ni. Also indicated are two data points, denoted by open diamonds (corresponding to the same set of experiments for which m values were shown in **Figure 44a**), for pure Cu where twins with a width of 20 or 90 nm were introduced via pulsed electrodeposition inside grains with an average size of 500 nm. For these cases, the twin width is plotted instead of the grain size. Reproduced with permission from Elsevier (Asaro and Suresh, 2005).

In addition, the absence of a deformation texture after straining of NC materials (Markmann et al., 2003) indicates that grain-boundary sliding (and possibly also grain rotation, see **Figure 45**) take place along with dislocation-based plasticity. The experimental findings at large strain rates in NC materials agree with predictions from MD simulations, where even much higher strain rates are necessarily imposed: dislocation activity, i.e. the emission of partial dislocations from GBs, as well as grain-boundary sliding were predicted based on these studies (Yamakov et al., 2003; Yamakov et al., 2001; Yamakov et al., 2002; Yamakov et al., 2002; Derlet et al., 2003; Van Swygenhoven and Derlet, 2001; Van Swygenhoven and Hasnaoui (2002); Van Swygenhoven and Hasnaoui, 2003).

Yet, in experiments, mostly either macroscopically averaging measurements (such as creep or tensile tests) are conducted or measurements with high spatial resolution (such as TEM) are used for analyzing the deformed state after the deformation has ceased. In any case, even if deformation studies are undertaken in situ, e.g. by testing thin foils inside the TEM, the time resolution of the methods available today does not allow true in situ observations of the active mechanisms that accommodate external stresses in nanostructured materials. Additionally, and in contrast to many simulated microstructures, NC metals consist in all real cases of distributions of grain sizes, grain shapes and possibly also of nonuniform distributions of chemical elements as well as nonrandom distributions of the lattice orientation in addition to artifacts such as pores or microcracks. Thus, it is expected to observe multiple stress-accommodation mechanisms acting in a given sample concomitantly. This complicating effect needs to be taken into account when analyzing experimental results and particularly when comparing them to the results of computer simulations.

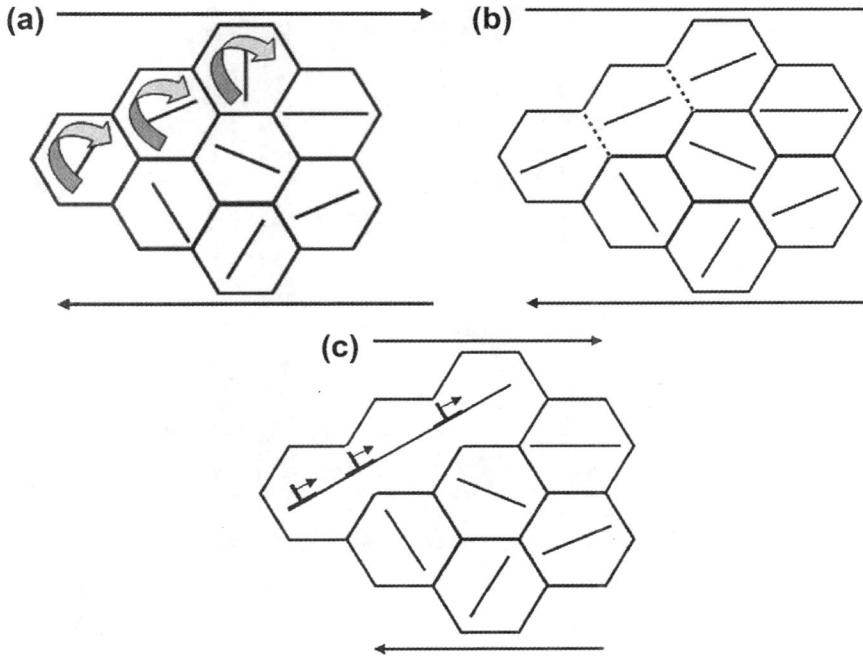

Figure 45 Rotation of neighboring nanograins during plastic deformation and creation of elongated grains by annihilation of grain boundary area. Reproduced with permission from Elsevier (Meyers et al., 2006).

Defect structures in plastically deformed NC-Pd that was initially almost free of dislocations have been investigated by TEM methods. Material with an average grain size of about 15 nm was prepared by inert-gas condensation, and was subsequently plastically deformed by cold rolling up to a true strain of 0.32 at a strain rate of about 0.3 s^{-1}. Abundant deformation twinning on $\{111\}$ planes was found and Shockley partial dislocations were identified (Rösner et al., 2004). Remarkably, in each grain, twinning occurs only on a single set of parallel planes, as has been shown in **Figure 14**.

This implies that only one out of the five independent slip systems required for the general deformation of a grain is active, which suggests that rigid-body grain rotation and GB sliding must have been active along with twinning. Until recently, only hints of the mechanisms at play have been obtained through changes in contrast that indicated that dislocations (Hugo et al., 2003; Mitra et al., 2004; Kumar et al., 2003) as well as GB rotation (Shan et al., 2004) were activated in the nanometer-sized grains. However, in situ tensile tests performed in a transmission electron microscope (TEM) in combination with high-resolution TEM are feasible. The in situ TEM tensile tests revealed that cracks were formed while the sample was elongated. The regions along the crack as well as the crack tip itself mark the starting points for a comprehensive TEM study of deformation processes in NC materials while the TEM sample is still under full load. The investigation revealed that the NC-Pd ruptured along GBs. Twins were formed in the grains next to the crack as exhibited in **Figure 46** indicating that the deformation processes must have emerged from the GBs. The observation of deformation twinning confirms furthermore the results of former TEM studies of plastically deformed NC-Pd (Markmann et al., 2003; Rösner et al., 2004).

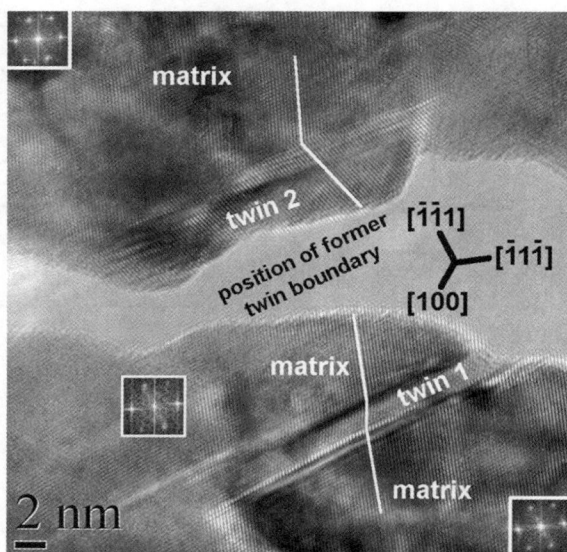

Figure 46 High-resolution micrograph taken under full load during an in situ TEM tensile test. The crack has propagated along the grain boundaries. A twin has been formed in a grain next to the crack (Rösner et al., 2009).

The experiments revealed that deformation twins provided intragranular propagation paths for the propagating crack tip. Such twins are believed to have formed at the head of the stress field of the advancing crack tip during the in situ tensile test because they were present only in a few grains (about 1–2% of the total number of grains) next to the intergranular crack. These features were observed independent of grain size or alloying addition. The sparse occurrence of twinning supports the interpretation that crack formation has to be regarded as a deformation feature in NC materials due to the limited availability of dislocation-based deformation processes, in conjunction with a near-total absence, at low homologous temperatures, of GB mediated deformation for stress relief prior to fracture. Thus, while growth twins have been shown to improve the mechanical properties and can give rise to high strength and high ductility of a material at the same time, mechanical twinning seems to introduce easy propagation pathways for intergranular cracks. With the propensity for mechanical twinning through the emission of partial dislocations from GBs at small grain sizes, as indicated by the results of Van Swygenhoven et al., and with the absence of lattice dislocation-based processes and the absence of diffusive processes at low homologous temperatures it seems that additional microstructural features, such as precipitates or additional phases need to be introduced to prevent the embrittlement of NC materials. Suggestions for several different routes to potentially enhance the ductility have been summarized by Ma (2006).

26.6.2 Ultrafine-Grained Materials

UFG materials obtained by SPD of initially cg microstructures have gained enormous interest in the past decade mainly due to the early observation of the unusual and highly beneficial combination of high yield strength and high tensile ductility of Cu and Ti after ECAP processing (see **Figure 47**). Since

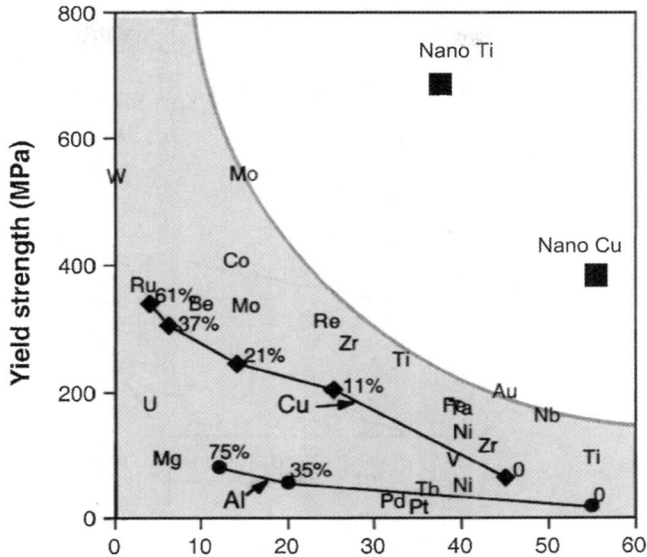

Figure 47 The combination of high strength and high ductility in ultrafine-grained Cu and Ti processed by SPD sets these materials apart from coarse-grained metals. Reproduced with permission from Materials Research Society (Valiev et al., 2002).

then, many different metals and alloys have been processed similarly, with mixed results concerning their mechanical response, specifically concerning their ductility and their capacity for homogeneous elongation without early necking: while for some materials, equal property improvements were reported, a large number of studies failed in finding significant tensile ductility in UFG samples after SPD processing.

From a basic point of view, the early observations on Cu and Ti are striking: the combination of enhanced yield strength and large (or even enhanced) ductility points to an underlying change of the mechanism that accommodates the external mechanical stress. While a change of the mechanisms governing plastic deformation is expected for materials that are characterized by nanoscale grain sizes or nanoscale twin separations (see above), it is less expectable for UFG materials that have grain sizes of a few hundreds of nanometers. One important fact that distinguishes UFG materials obtained by SPD methods from the NC materials obtained, e.g. by powder consolidation, is the large density of stored lattice defects in the former. Thus, large dislocation densities already exist in these materials and additionally, due to their large density and the small grain size, the formation of more complex defect structures such as dislocations in locked configurations or the presence of rotational defects need to be taken into account. As a consequence of the suggested interaction between GBs and lattice dislocations during the severe straining under a high hydrostatic pressure, the formation of the so-called "nonequilibrium GBs" that carry excess densities of extrinsic GB dislocations have also been proposed as a mechanism to enhance ductility (Valiev et al., 2002). It has been argued that such boundaries provide a large number of excess dislocations for slip (Hasnaoui et al., 2002) and can even enable grains to slide or rotate at room temperature, leading to a significant increase in the strain hardening exponent. Yet, as described in Section 26.5.3.4, the concept of nonequilibrium GBs based on extrinsic GB dislocations is not well-suited to explain the complete evolution of GB characteristics

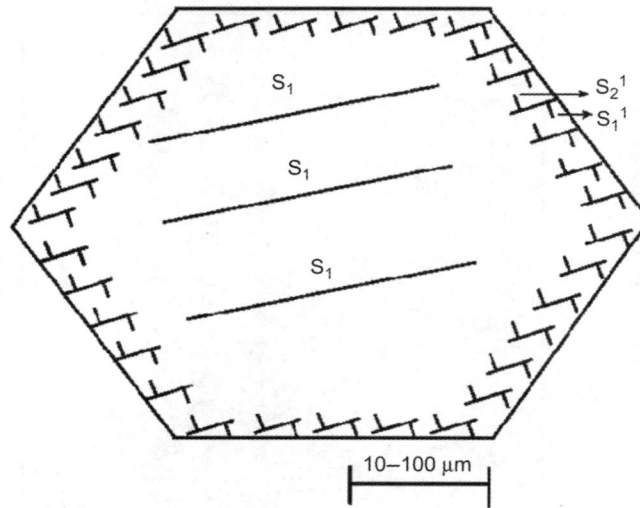

Figure 48 Schematic depiction of a deformed grain showing the more intense cross-slip and work hardening along the grain boundary region. Reproduced with permission from Elsevier (Meyers et al., 2006).

as a function of temperature and time. Additionally, one has to be aware that the idea of "extrinsic GB dislocations" encounters obvious difficulties with an atomistic description of such defects, e.g. in terms of a Displacement Shift Coincide (DSC) lattice and its physical meaning when the DSC lattice is very fine or infinitely small. It is also not clear why the concentration of the extrinsic GB dislocations has to be only such that in totality they do not change the disorientation of these "nonequilibrium" boundaries. Thus, while it has been shown that a fraction of high-angle GBs exist after SPD processing that possess higher excess free-energy densities and that allow for enhanced atomic diffusion along these GBs, their structure has not been properly analyzed and up to the present, it is unclear whether the presence of these microstructural features enhances or deteriorates the mechanical properties.

Taking the experimental observations into account, core-and-mantle models might be able to describe the mechanical response, as outlined by Meyers et al. (2006). This type of model is generally based on the formation of dislocations at/near GBs and on the formation of work-hardened GB layers (see **Figure 48**).

Different specific forms of such models have been proposed (e.g. Ashby and Verall, 1973; Ashby, 1970; Hirth and Lothe, 1982; Thompson, 1977; Margolin, 1998), but the general approach is similar as in the model according to Li (1963). It is assumed that the dislocations are formed from GB ledges (in the original work, these structures are called: "GB jogs"). If these ledges are present with a spacing of about 10–100 nm, then sufficient supply of these dislocation formation sites is present so that the dislocations, after emission into the grain cores, can cross-slip and multiply. These interactions would then form a work-hardened layer adjacent to the GBs. For smaller grain sizes, naturally the volume fractions of this "mantle" with respect to the unaffected "core" of the grains increases, which causes an increased yield stress. Thus, in the grain interior easy glide operates, resulting in a low work-hardening rate. In the grain-boundary mantle region, cross-slip leads to increased work hardening, which might rationalize the observed mechanical response from UFG materials that were processed by methods of SPD.

26.7   Phase Transformations in Nanocrystalline Metals

Nanostructured materials, as already mentioned above, are characterized by extraordinary large specific interface areas, such that the properties of nanostructured materials are significantly affected by the specific properties of the interfaces. As indicated in Section 26.5 of this chapter, in many cases the experimental results obtained so far indicate that the specific properties of the interfaces are unaffected by the specific interface area of a given material. However, at very large specific interface areas, i.e. at very small sizes of the respective building blocks (grains or particles) and due to the discrete nature of matter, effects due to local "curvature" might become significant. Yet, such effects are in reality mostly not of significance if bulk nanostructured materials (i.e. continuous and massive materials that consist of NC grains) are concerned.

Previous work concerning the thermodynamic properties of interfaces (or of materials with a large specific interface area) is mainly based on Gibbs' classical treatment (Gibbs, 1906, 1957). Further reading on this subject may be found in Howe (1997), Lojkowski and Fecht (2000). Additionally, tantamount of articles are devoted to the case of extremely small nanoparticles, since a different thermodynamic approach (based on microcanonical and canonical ensembles, see e.g. Baletto and Ferrando, 2005) is required for describing such particles with sizes below the cluster limit. At the same time, many computer simulations exist concerning such extreme cases of nanostructured matter, particularly since this size range allows for realistic simulations even by ab initio methods. Yet, as indicted in Section 26.1.1, isolated particles are not within the scope of this chapter, since the microstructure-property relations of materials (in contrast to clusters) are forming the focus of this volume.

26.7.1   Thermodynamic Functions

In general, the minimum of the Free Enthalpy, G, of a thermodynamic system defines equilibrium. Naturally, nanostructured materials do not represent the equilibrium state with respect to the infinite single crystal. However, under conditions where either coarsening is prevented (as in nanostructured composites with negligible mutual solubility of the matrix and the pore phase) or effectively hindered (e.g. by low homologous temperatures), nanostructured materials can be metastable or kinetically stabilized against coarsening on long time scales. Under such conditions, it is important to analyze how the thermodynamic potentials of the available phases are affected by the size confinement. In fact, that analysis is also important concerning the metastability of the material, since the thermodynamic driving force for coarsening is determined by the thermodynamic excess potentials of the nano-structured materials.

In the following, independent phases will be denoted by their individual subscripts and in all cases except where explicitly mentioned, we shall assume the individual phases to be homogeneous. Thus, for a heterogeneous system consisting of two different bulk phases, necessarily an additional interface term (denoted by the subscript Σ) needs to be taken into account. Following Kondepudi and Prigogine (1998) and Faghri and Zhang (2006) and neglecting the difference between Free Enthalpy and Free Energy (that is permissible for condensed matter with low compressibility), the total Free Energy of a system consisting of a heterogeneous mixture of two homogeneous bulk phases might be written as follows:

$$F = F_1 + F_2 + F_\Sigma$$

With

$$dF_1 = -S_1 dT - p_1 dV_1 + \mu_1 dN_1$$
$$dF_2 = -S_2 dT - p_2 dV_2 + \mu_2 dN_2 \qquad (22)$$
$$dF_\Sigma = -S_\Sigma dT + \gamma dA + \mu_\Sigma dN_\Sigma$$

Here, S, denotes the entropy, T, the temperature, p, the pressure, V, the volume, μ, the chemical potential, N, the number of particles (i.e. atoms) and γ, the interface excess free-energy density.

Note: for small particles the interface area, A, is not an independent state variable, but is a function of T, N and x (x: concentration of alloying elements in the case of multicomponent systems) and is determined by the Wulff construction. In fact, even for a real one-phase system, the surface contributions should be taken into account. However, for bulk systems, the contribution of the surface excess is rather negligible and additionally, the associated excess terms can be renormalized.

From the above Eqn (22), it becomes clear that the specific interface excess free-energy density, γ (that will be abbreviated as "interface energy") and correspondingly the interface stresses, σ, lead to additional and finite terms in the Free Energy of nanostructured materials. Thus, since the specific interface area and thus the contribution of the interface to the total values of the thermodynamic potentials is large for nanostructured materials, one needs to consider their impact separately.

From the above description, it directly follows that the chemical potential may be changed at external or internal interfaces, leading e.g. to preferred segregation at GBs (due to the local variation of the Free Energy). Naturally, the amount of segregation is directly connected with the interface energy as well as the interface stress. Moreover, for sufficiently strong segregations, phase transformations might occur localized at the interfaces (Divinski et al., 2005; Straumal et al., 2008). Therefore, in the following the effects of the interface energy, γ, the interface stress, σ, and segregation are discussed in more detail.

26.7.2 Interface Energy, Interface Stress and Segregation at Interfaces

At the outset, it seems appropriate to address the issue of definitions and terminology due to a confusion about the terminology of the "interface energy" and the "interface tension". The interface energy, γ, denotes an excess energy term connected with the energy (in units of $[J/m^2]$) that is necessary to create a new segment of interface. Thus, the interface energy is related e.g. to the number of broken bonds for cleavage. The interface tension on the other hand is defined, according to Gibbs, as a work term, $\gamma \, dA$, with dA denoting the change of the interface area when adding one atom into an existing interface. Thus, the interface tension as a quantity is introduced as a mechanical force term in the units $[N/m]$—equal to the interface tension of a liquid. Fischer et al. (2008) have shown that it is sufficient to use the independent parameters interface energy, γ, and interface stress, σ, to fully characterize interfaces/surfaces of any material.

The interface stress, σ, is defined as the mechanical work that is required to stretch a segment of an interface. As such, σ is defined as a tensorial quantity, σ_{ij}, since the work required to enlarge an interface in the plane of the original interface depends on the relative orientations of the interface and the mechanical force acting on it. In solids, the correlation of the stress tensor, σ_{ij}, with the interface energy γ is given as:

$$\sigma_{ij} = \gamma \delta_{ij} + \frac{\partial \gamma}{\partial u_{ij}}, \qquad (23)$$

where u_{ij} denotes the strain tensor. For liquids γ and σ are identical, but for solids γ and σ are different, as seen from Eqn (23). It is important to note that the quantity $\partial\gamma/\partial u$ can also be negative, as first observed by Couchman and Jesser (1977) and later also by Zhao et al. (2001).

It is well known that mechanical stresses influence intensive thermodynamic potentials such as the pressure. The Young–Laplace equation relates the mechanical stress with the pressure of a liquid or amorphous particle of radius R as:

$$p_{\text{particle}} - p_0 = \frac{2\sigma}{R}, \tag{24}$$

where p_{particle} and p_0 denote the pressure in a particle and the ambient pressure outside the particle, respectively. A similar treatment leads to the Gibbs–Thomson effect, i.e. the impact of the curvature on the pressure inside small particles that leads—in a generalized form—to a description of phenomena controlled by the interface curvature, such as e.g. Ostwald ripening or nucleation.

As indicated above, considering a thermodynamic stability, the single crystal as the state of lowest free energy would be favored. However, that statement is true only for single-phase materials. Naturally, with two phases that are both stable (such as e.g. in a eutectic alloy), an interface between the two phases must also exist. In that case, the thermodynamically stable state would be defined by minimizing the total interface excess free energy. Yet, with binary or higher component alloys, the formation of a defect such as an interface is accompanied by a more or less pronounced redistribution of the component atoms around the defect. If solute atoms segregate at the defect, the total free energy is reduced. Thus coarsening, driven by a decrease in the overall interface area, is retarded in the presence of solute atoms. Based on statistical thermodynamics, Kirchheim developed a generalized form of Gibbs' adsorption equation for solutes segregating at defects in solids, indicating that metastable equilibrium concentrations also of extended defects such as dislocations or GBs could result if the segregation enthalpy is sufficiently large (see also Chapter 27 of this volume).

This treatment has direct consequence concerning the explanation of the exceptional thermal stability of NC alloys with regard to grain growth, which has been interpreted in terms of a reduced or even zero GB energy as a consequence of excess solute at the boundaries. As an example: grains of NC-nickel coarsen at around 380 K whereas NC Ni–P alloys with a few at.% P are stable up to 620 K (Farber et al., 2000) (see **Figure 49**).

Above this temperature, rapid grain growth occurs, accompanied by the precipitation of the equilibrium phase Ni_3P. Using a tomographic atom probe, it could be shown that GBs are saturated with an excess phosphorous of 1.5×10^5 mol m^{-2}, whereas within the grains the P concentration was reduced to 1 at.%. This value is much higher than the solubility limit in the temperature range between room temperature and 620 K and, therefore, the alloy is highly supersaturated, corresponding to a high chemical potential of phosphorous. This state of large values of the GB excess and the chemical potential of the solute at the GB corresponds to low GB energies. In order to suppress grain growth, the latter has to be zero. However, this is a metastable rather than stable equilibrium because precipitation of the equilibrium Ni_3P phase (which would correspond to the stable equilibrium) is kinetically hindered. This treatment is also in line with observations of kinetic changes, i.e. enhanced activation energies for GB migration (Lojkowski and Fecht, 2000).

In addition to segregation at GBs, segregation to triple lines has also been shown to have important consequences, e.g. on the stability of NC multilayer materials. Stender et al. (2011) analyzed the impact of thermal treatments on a NC Fe–Cr multilayer by atom probe tomography (see **Figure 50**). The results revealed a strong modification of the structure (and diffusional width) of the triple line due to massive

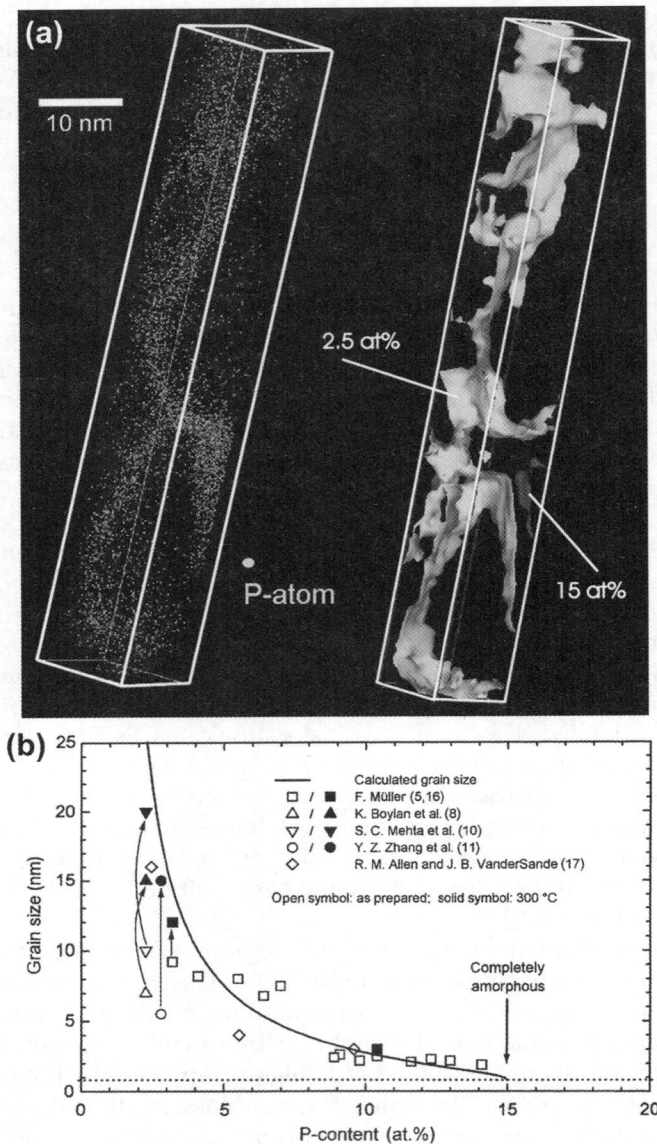

Figure 49 (a) left-3D reconstruction of the P-distribution of the Ni–P layer after heating up to 400 °C at 5 °C/min revealed by TAP analysis. Right-Isoconcentration surfaces representing 15 at.% P (red) and 2.5 at.% P (blue/yellow). Blue indicates the direction of lower and yellow indicates the direction of higher P-contents. The P-segregation in grain boundaries is directly observable. (b) Relationship between grain size and P-content. The experimentally determined grain sizes of NiP-alloys after preparation and after heat treatment at low temperatures before Ni_3P precipitation is observed are depicted as open and solid symbols, respectively. Reproduced with permission from Elsevier (Färber et al., 2000). (For interpretation of the references to color in this figure legend, the reader is referred to the online version of this book.)

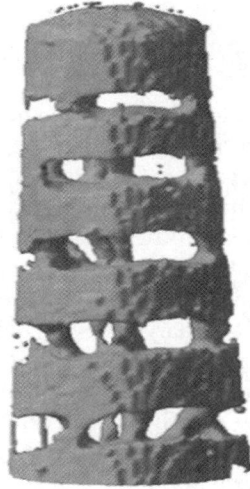

Figure 50 Isoconcentration surface at 60 at. % Cr of an annealed Fe–Cr multilayer obtained by atom probe tomography, revealing triple junction spreading across several consecutive layers. Source: American Physical Society (Stender et al., 2011). (For the color version of the figure, the reader is referred to the online version of the book.)

segregation, which changed the atomic transport characteristics along that defect drastically, in turn affecting also severely the stability of the multilayer structure.

26.7.3 Phase Diagrams of Nanostructured Materials

Constitutional phase diagrams represent crucial information concerning the choice of processing pathways and concerning the relative stabilities of different phases and phase mixtures in composition–temperature space. Naturally, and well known e.g. from early experiments that utilized rapid quenching methods, excess terms of the thermodynamic potentials modify the thermodynamic equilibria and thus the topology of the respective phase diagrams. Well-known effects encompass size-dependent solubility (Shirinyan et al., 2005), size-dependent dissolution (Swaminathan et al., 2009), and phase transitions and phase diagrams at GBs (Straumal, 2001) and at TJs (Straumal et al., 2008). For nanostructured materials, the importance of the excess interface contributions classically have been taken into account for binary equilibria by accounting for the "surface" contribution to the total enthalpy and the total entropy. These approaches are based on treating nanostructured materials as effective core–shell systems with a homogeneous core (that is characterized by bulk properties) and an interface shell with modified properties. Following e.g. the approach by Wautelet et al. (2000), the specific Gibbs free energy per volume, g, of a mechanical mixture is given as:

$$g_m = x_1 h_1 + x_2 h_2 - T(x_1 s_1 + x_2 s_2), \tag{25}$$

where x_i is the atomic fraction, h_i the enthalpy and s_i the entropy of each component. The configurational entropy of mixing is given as:

$$\Delta s_m = -k(x_1 \ln(x_1) + x_2 \ln(x_2)) \tag{26}$$

in a system with N atoms and accounting for the interface, the total Free Enthalpy is given as:

$$Ng_{particle} = N(g_m - T\Delta s_m) + fN^{2/3}(x_1\gamma_1 + x_2\gamma_2), \tag{27}$$

where f is a geometrical factor and γ_i denotes the interface energy. $fN^{2/3}$ is the number of atoms at the interface. This approach has been shown to include also the case of nonspherical systems (Wautelet et al., 2003) and taking into account segregation effects lead to changes of the compositions in the core and at the interfaces (Vallée et al., 2001). In that case, also an additional influence of the shape of small particles on the resulting phase equilibrium in the presence of segregation has to be expected.

Recently, Tanaka proposed the prediction of phase diagrams for nanostructured binary alloy systems based on thermodynamic databases (Tanaka, 2010). However, at present it seems that the complex interrelation of capillary-driven effects in binary or even higher component systems has not been sufficiently well explored to allow for precise predictions. Additionally, for nanostructured systems some of the approximations that are conventionally used for constructing and calculating phase equilibria might not longer apply, as detailed below, which poses severe constraints to such approaches.

As nanostructured materials are structures far away from thermodynamic equilibrium and since they have short transport pathways, fast diffusion and rapid transformation kinetics often lead to coarsening and to the deterioration of the microstructure and the associated properties. Thus, ensuring the stability of the nanoscale structures is a key issue. Aside from restricting the range of candidate materials to the class of refractories such as ceramics or high-melting point metals that are kinetically stabilized at or near-ambient conditions, a composite approach involving either two nanosized phases or an extended polycrystalline or amorphous matrix and a NC pore phase are obvious solutions for the latter issue since the atomic transport required for coarsening is severely hampered by a composite structure with limited mutual solubility. This route also includes surface-functionalized nanoparticles as e.g. presented by metallic nanoparticles with a shell consisting of organic ligands or of a natural oxide of the metal (Stahl et al., 2002). However, it is inherent to NC materials that the analysis of microstructure-property relations needs to consider internal interfaces rather than the surface of the nanoscaled structural units. Especially with two-phase nanocomposites, heterophase interfaces with the additional degree of freedom given by the position-dependent composition and possible concentration gradients need to be regarded. An important and basic aspect concerning the functionality of a given material is presented by the respective phase equilibrium that determines the stable structure and the phase distribution and thus the related materials properties. In fact, modifying the phase equilibrium by alloying to improve the performance of a material has been the first and most successful step to modern materials science. However, the phase diagrams are mostly unknown for nanostructured materials. In fact, some observations on ligand-capped magnetic nanoparticles indicate that the energetic contribution due to the bonds at the interface effectively shift the underlying phase stability ranges such that the equilibrium phase is different for the cg or the NC material Stahl et al., 2002. Yet, as will be shown below, already the presence of internal heterophase interfaces contributing an excess free energy is sufficient to severely modify the phase equilibrium and the associated phase transformations in nanosize alloy systems. Even the accepted rules to construct phase diagrams need to be modified if nanoscaled alloy systems are considered (Weissmüller et al. 2004).

In addition to the energetic contribution of an external surface, a qualitatively similar contribution arises due to the excess energy associated with internal heterophase interfaces in multiphase, multi-component nanosystems that are necessarily formed due to temperature- or composition-dependent variations of the relative amount of matter per phase. Therefore, the free energy balance must contain terms of the form $\gamma \Delta A$ (specific excess interface free-energy density of the heterophase interface multiplied by the change in area of that interface), on top of the term $A \Delta\gamma$, which is dominant in elemental systems, as indicated in the previous section.

Much less work has been devoted to phase equilibria of nanoscale alloys, despite their importance for future nanotechnology devices, which will require the extra degrees of freedom in materials design provided by the use of alloys as opposed to elemental solids. Alloys differ from elemental materials in the fact that constitutional alloy phase diagrams exhibit intervals of temperature and composition in which two (or more) phases coexist at equilibrium. In the following and without loss of generality, attention will be restricted to binary alloys where at constant pressure at maximum three phases can coexist in defined points of the phase diagram (zero degrees of freedom for three-phase coexistence, according to Gibbs' phase rule).

The central questions in modeling size-dependent alloy phase diagrams are therefore: can two phases coexist in a small particle and how are the conditions of equilibrium defined? As compared to elemental particles, this question raises a new issue related to the energetics of the internal interface separating the phases within the particle, since varying the relative amount of matter in the phases requires the creation or removal of internal interface area. While it is established that interfacial enrichment or depletion in solute (interfacial segregation) (Weissmüller, 1993, Kirchheim, 2007a, 2007b) and elastic interactions between the interfaces and the bulk (interface stress) (Weissmüller and Lemier, 1999) can significantly affect the relative stability of single-phase states in nanoscale alloys at constant interfacial area, the consequences of capillarity for the two-phase coexistence within a particle remain widely unexplored. Yet, the capillary energy of the interface between coexisting phases can lead to significant changes in the constitutional phase diagram, i.e. of the composition-temperature fields in which the different phases represent the thermodynamically stable state, and which may be observable even for sizes as large as 100 nm, i.e. well above the structure size of next-generation microelectronics devices. These changes are not mere shifts of temperatures or of compositions at equilibrium; instead, several qualitative rules, which are universally obeyed in the conventional alloy-phase diagrams of macroscopic systems, are no longer applicable at the nanometer scale. It has to be emphasized here that these considerations inherently imply that the pathway to global thermodynamic equilibrium, which would involve the formation of coarse phases with minimized interface areas, is at least kinetically prevented.

In order to analyze the impact of this internal interface to the thermodynamic equilibrium at different particle sizes, an idealized particle embedded in a solid matrix is regarded where the matrix, as in an experiment, serves to prevent coarsening; this implies that the particle shape and consequently (when volume changes during the phase transition can be neglected) the particle–matrix interface area are fixed. The excess free energy due to the outer surface of the particle is then a constant, which can be ignored altogether since it does not affect the phase equilibrium. Thus, in the following discussion in this section, the notion of an interfacial area A refers exclusively to the internal interfaces between coexisting phases.

The free enthalpy per particle, G, can be related to the molar free enthalpy, g_0, by:

$$G(T, N, x) = N g_0(T, x) + \sum_i \gamma_i(T) A_i(T, N, x), \tag{28}$$

where the subscripts label the possible interfaces. N is the total amount of matter, i.e. the sum of the amounts of solvent, N_1 and solute, N_2 and x denotes the solute fraction, $x = N_2/N$. At equilibrium, as already mentioned, the A_i are not independent state variables, but internal thermodynamic parameters that are functions of T, N, x, determined by the Wulff construction. Generally, the functional dependence of the A_i on N is not linear; this leads to the size-dependence of the chemical potentials of single-phase particles embodied in Gibbs–Thompson–Freundlich-type equations.

In cases where the particle contains two phases, φ and β, (e.g. solid, S and liquid, L) that coexist, the Gibbs Free Energy, $\tilde{G}$, of the two-phase state with arbitrary compositions (that are not necessarily the compositions at equilibrium that minimize the Free Enthalpy) is given as:

$$\tilde{G} = (N^\varphi/N) \cdot G^\varphi + (1 - (N^\varphi/N)) \cdot G^\beta. \tag{29}$$

The phase fractions for the macroscopic case are then given by the lever rule.

However, the formation of a new phase necessarily entails changes of the area of interfaces and the creation of new interfaces. Thus, the dependence of $\tilde{G}$ on the phase fraction will cease to be linear, contrary to Eqn (29) and consequently the Gibbs Free Energy of two-phase states (in a nanoparticle) are expressed as (Weissmüller et al., 2004):

$$\tilde{G} = (N^\varphi/N) \cdot G^\varphi + (1 - (N^\varphi/N)) \cdot G^\beta + \Delta G_c. \tag{30}$$

The term ΔG_c represents the deviation from linearity and becomes equal to zero for single-phase particles. Between two single-phase states, a curved graph as indicated in **Figure 51a** must result. Thus, the capillary term, $\partial A/\partial V\beta$, in general removes the coincidence of the tangent lines at equilibrium, as indicated in **Figure 51a**.

It should be noted that Eqn (28) holds in general, also for the macroscopic case. However, when the capillary term is negligible, as for macroscopic bulk systems, the condition that the tangents coincide at two-phase coexistence holds and deviations are negligible. The compositions of the coexisting phases in the bulk case are then given by the points of tangency of the common tangent to the free enthalpy functions G_φ and G_β; at each temperature these compositions are constants in the two-phase state, independent of the overall composition and the tangent line represents the total free enthalpy.

For small particles, however, the energetic contribution of the internal interface becomes significant and leads to a convex Free Enthalpy curve as indicated in **Figure 51a**. Different geometric assemblies of the two phases are possible, depending on the relative values of the γ_i, but, in general, ΔG_c is a nonlinear function of the phase fraction. The most important consequence of the loss of linearity is that the tangent rule ceases to apply. Instead, the compositions of the two coexisting phases at equilibrium is not a priori known and is determined by energy minimization: the entire set of functions $\tilde{G}^{SL}$ with the solute fraction of the liquid as a parameter needs to be calculated (**Figure 51b**). The set of curves for $\tilde{G}^{SL}$ also indicates that the composition of the coexisting phases cannot be read from the phase boundaries as in the case of the bulk material. The respective stable states are given by the lower enveloping curve of the Gibbs Free Energy of all possible phase states and the transition between different phase states that define the minimum of the total Gibbs Free Energy mark the boundaries of the stability ranges of the different phases.

For the bulk case, the well-known phase diagrams result from an equivalent treatment that minimizes the total Gibbs Free Energy (**Figure 52a**). However, it is important to note that for nanoscale systems not only the topology of the phase boundaries are changed, but that the way in which the resulting "phase diagrams" are to be used is completely different: the compositions of the coexisting phases are not longer invariant upon isothermal variations of the solute fraction and the composition of the majority phase is not longer continuous across phase-boundary lines (Wilde et al., 2007). Thus, at first sight, a more appropriate term for the resulting phase diagrams would be "stability" diagrams

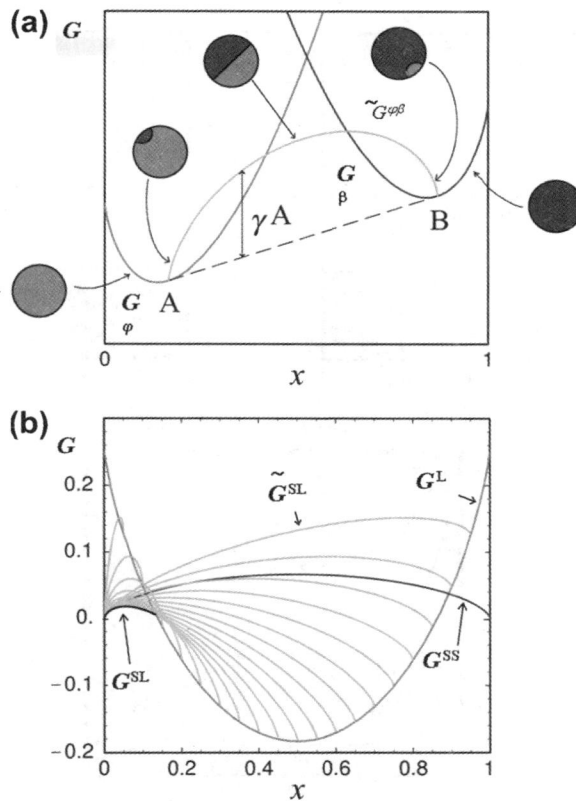

Figure 51 (a) Schematic diagram of molar Gibbs Free Energies $G\varphi$ (red) and $G\beta$ (blue) versus solute fraction x for two phases φ and β. Construction of the Gibbs Free Energy curve for the two-phase coexistence of alloys represented by points A and B. Black dashed: macroscopic system; solid green: finite size system. Inserts represent cross sections through the particle, illustrating the geometric arrangement of the phases (red: phase φ, blue: phase β) and the maximum of the interfacial area A for equal amounts of phase φ and β. (b) Example of the molar Gibbs Free Energies (in units of the enthalpy of melting) with various values for x_L used in the computation. Parameters are $T/T_M = 0.75$ and $D = 5$ nm. G_L represents the Gibbs Free Energy of the single-phase liquid state, G_{SS} represents the two-phase solid state and $G_{SL}(T,x)$ is the lower envelope of the set of functions $\hat{G}^{SL}(T, x, x_L)$ (Weissmüller et al., 2004). (For interpretation of the references to color in this figure legend, the reader is referred to the online version of this book.)

since the properties conventionally associated with phase diagrams are not longer applicable for nanoscaled alloys. It should be emphasized however, that these properties merely result from the applicability of linear approximations in the macroscopic world—in principle the results derived for nanoscale systems are generally applicable—they just become significant at small system size and—this is important to note—the calculated diagrams as well as the construction rules extrapolate to the accepted behavior for large system sizes.

In order to compute phase diagrams (or stability diagrams) for different particle sizes, assumptions for the equations of state need to be done. A simple case is given by an alloy with no solid solubility and an ideal liquid solution, which for the bulk case results in a simple eutectic phase diagram that is

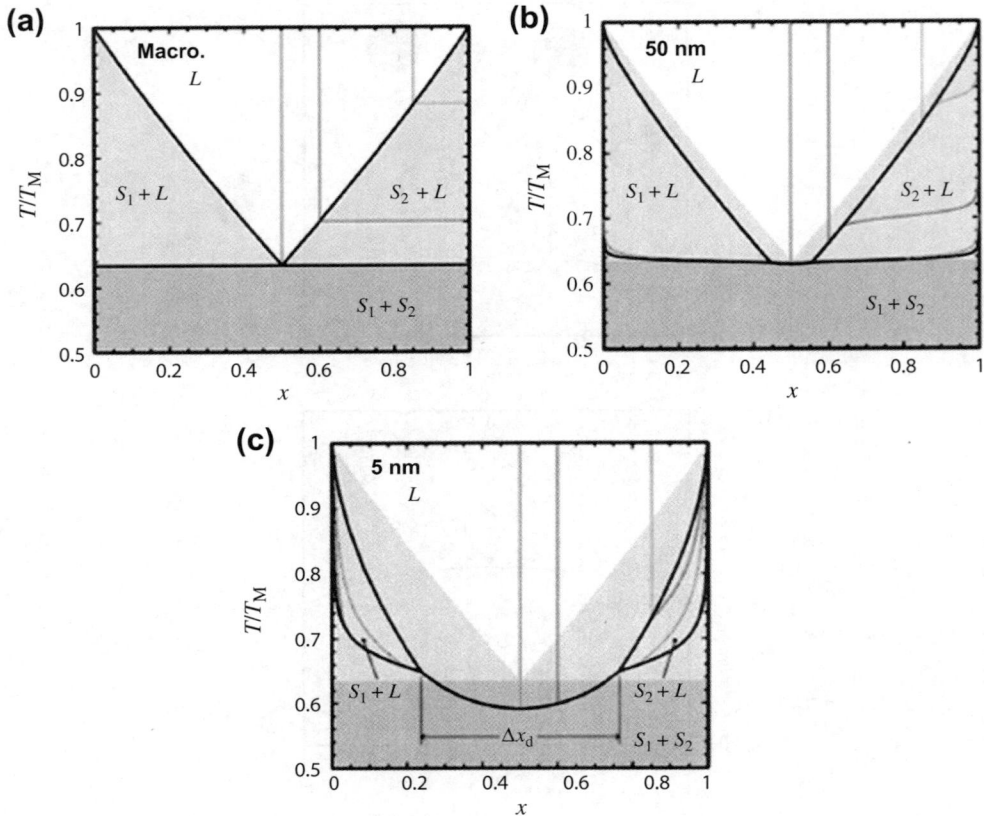

Figure 52 (a) Phase diagram of a bulk eutectic alloy with no solid solubility. Black: phase coexistence lines; colored: lines of equal solute fraction x_L in the liquid phase for three arbitrarily chosen values of x_L. (b) and (c): as in (a), but finite size systems with particle diameters $D = 50$ nm and $D = 5$ nm respectively. Δx_d: discontinuous melting interval where a direct transition from a two-phase solid to a single-phase liquid without three-phase coexistence occurs. Capital letters indicate the phases that are stable in the respective regions of temperature/composition space. $S_{1,2}$: solid phases; L: liquid. The gray shades represent the topologic features of the bulk phase diagram for easier comparison (Weissmüller et al., 2004). (For interpretation of the references to color in this figure legend, the reader is referred to the online version of this book.)

symmetric concerning the equiatomic composition. In a reduced representation, the three material constants that need to be specified (basically the atomic volume, melting entropy and interfacial free energy in scaled representations) are similar for most metals. Details of the computation as well as concerning the analytical model are given elsewhere (Weissmüller et al., 2004). The resulting phase diagrams for the bulk system and for two-alloy particles with different sizes are summarized in **Figure 52**. It is seen that, as the particle size is reduced, the phase diagram undergoes several qualitative changes, each of which breaks one of the rules that apply universally to the construction of the phase diagram for macroscopic systems. First, it is observed that the invariance of the solidus temperature is lost in favor of a significant composition-dependence. Second, as illustrated by the colored lines representing states of identical composition x_L of the liquid phase at equilibrium, the compositions of the constituent phases in two-phase equilibria are no longer invariant at constant temperature. Thirdly, the equicomposition lines lose their continuity at the intersection with the liquidus line (Wilde, 2006). This implies that there is a discrete jump in liquid fraction across the liquidus of the small alloy particles, consistent with the result of numerical modeling matched to Sn–Bi nanoparticles (Jesser et al., 1999), where the ends of the tie lines were found to detach from the phase-boundary lines. It should be stressed that relaxing the stringent boundary conditions that were used for constructing a simple model system does not qualitatively change the resulting phase equilibria. In fact, recent calculations based on a model eutectic with finite solubility of the terminal phases have shown that the stability fields of the different one- and two-phase states are shifted and that the two-phase solid–liquid stability fields are detached from the terminal phases (Gährken and Wilde, 2008) as also found by numerical calculations (Jesser et al., 1999). However, the most fundamental consequence of the finite system size is a topological change in the phase diagram, the degeneration of the eutectic point of the macroscopic system into a line representing an interval of compositions Δx_d (defined in **Figure 52(c)**) for which the particle undergoes a discontinuous transition between the two-phase solid–solid state and the single-phase liquid state. In the macroscopic system, three phases can coexist at equilibrium at the eutectic point; by contrast, discontinuous melting in this model is a transition between a two-phase equilibrium (solid–solid) and a single-phase state, without three-phase equilibrium (clearly, three phases will coexist during melting, but this situation resembles a transient, nonequilibrium configuration). It is because of this loss of three-phase equilibrium in the finite-size system that the transition from a eutectic point to a discontinuous melting line can be reconciled with the phase rule.

In fact, recent experimental studies of isothermal composition variation within the electron microscope (Lee and Mori, 2004) as well as calorimetric investigations on a Bi–Cd eutectic that closely resembles the assumptions of the simple model eutectic (Bunzel et al., 2004) are in complete agreement with the model results.

In order to verify the results of the theoretical study, experimental analyses were performed on a series of $Al_{98}(Bi_x–Cd_{1-x})_2$ alloys that had been synthesized via melt-spinning. The Bi–Cd system presents similar conditions concerning the constitutive behavior as assumed in the simplified theoretical model system, especially concerning the negligible mutual solubility of Bi and Cd in the solid state and the negligible solubility of both components in solid Al. Thus, eutectic Bi–Cd nanoparticles embedded in an Al matrix were obtained after rapid quenching, as indicated in the TEM bright field image in **Figure 53a**. However, the melt-spinning process resulted in a bimodal size distribution of the Bi–Cd particles with larger particles located at GBs of the Al matrix and small particles within the Al-grains, as already observed for the Al–Pb alloys. In situ melting experiments within the TEM have served to associate the calorimetric melting signals with the respective particle fractions. Thus, the melting signal of the different size fractions could be deconvoluted. Quantitative analyses of the size distribution from TEM bright field images have shown that the average sizes of the smaller and the larger particles did not depend on the alloy composition (**Figure 53b**).

Figure 54a shows the experimental results of calorimetric melting experiments on a series of $Al_{98}(Bi_x–Cd_{1-x})_2$ alloys with $x \in [0, 1]$. The two different peaks that are labeled as peak 1 and peak 2, respectively, refer to the melting signals of the two size fractions. It is clear from the calorimetric results in conjunction with the in situ TEM melting experiments that peak 2 is associated with the melting process of the smaller particles that are located within the Al-grains. This peak shows an onset temperature about 3 K above the signal maximum due to the larger particles (peak 1) at small Cd concentration but decreases significantly with increasing Cd concentration. At about 60 at.% Cd, the onset of melting off the smaller particles is about 6 K below the onset for the larger particles. Thus, the total variation of the onset temperature is about 9 K. Since the standard deviation for the onset of

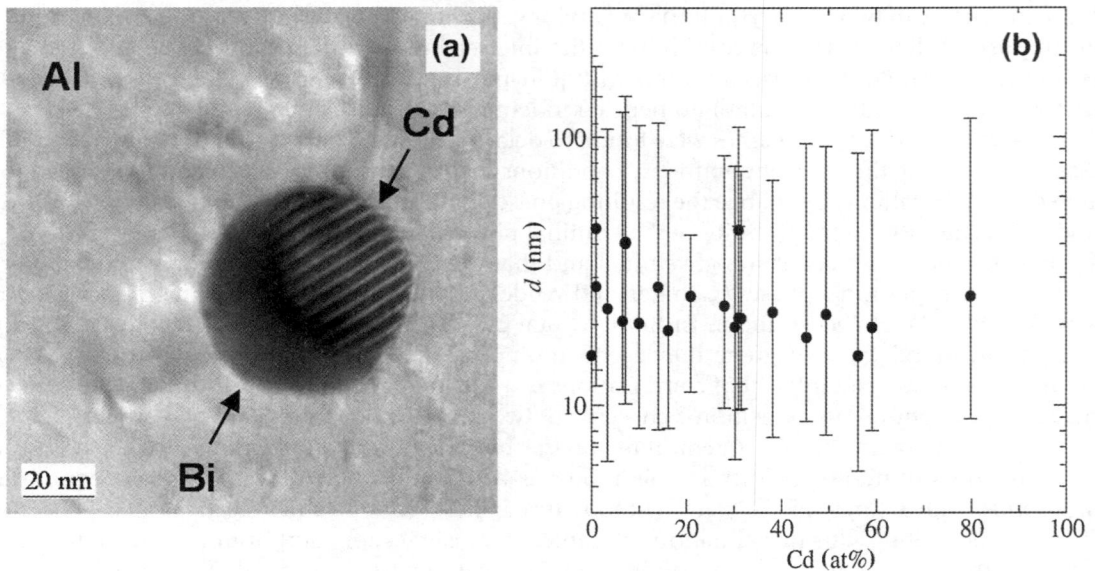

Figure 53 (a) TEM bright field image of a Bi–Cd particle embedded in an Al grain. The particle belongs to the "smaller" fraction that shows pronounced size dependence of the melting behavior. The two differently appearing parts of the particle (i.e. with or without Moire' effect) are due to the eutectic nature of the particle with one side consisting of Bi and the other of Cd. (b) Average particle diameter as measured from TEM bright field images. The particle sizes refer to the "smaller" particles that are located within the Al grains. The error bars indicate the 95% confidence range. The results indicate clearly that the average size of the particles is independent of the alloy composition (Bunzel et al., 2004).

melting (determined on the same macroscopic alloy system) is only about ± 0.8 K, these results indicate the validity of the theoretical approach. In fact, calculating calorimetric curves for the theoretical model system for the same particle size as observed experimentally results in a similar variation of the onset temperature for melting in dependence of the alloy composition (**Figure 54b**). It should be emphasized that observing a dependence of the solidus temperature on the alloy composition is a clear evidence for the importance of the interface contributions since for macroscopic systems and according to standard phase diagram construction, the solidus temperature in a bulk eutectic system such as Bi–Cd would strictly remain constant.

26.7.4   Reversible Phase Transformations

It is one of the earliest findings concerning finite-size effects on materials properties that a decrease of the diameter, D, of a particle leads to a shift of the melting temperature, $T_{m,D}$, compared to the bulk melting temperature, $T_{m,0}$ (Buffat and Borel, 1976). In fact, a similar question has been addressed already by Thompson, the later Lord Kelvin, in the context of premelting of surfaces. In that historical case, the question was, whether a liquid film is present on the surface of ice at temperatures below 0 °C, which would account for the low friction coefficient experienced during ice skating. While it has been shown that for this special case, the pressure dependence of the phase transformation between solid and liquid H_2O is decisive, the size- (or curvature-) dependent melting phenomenon has been

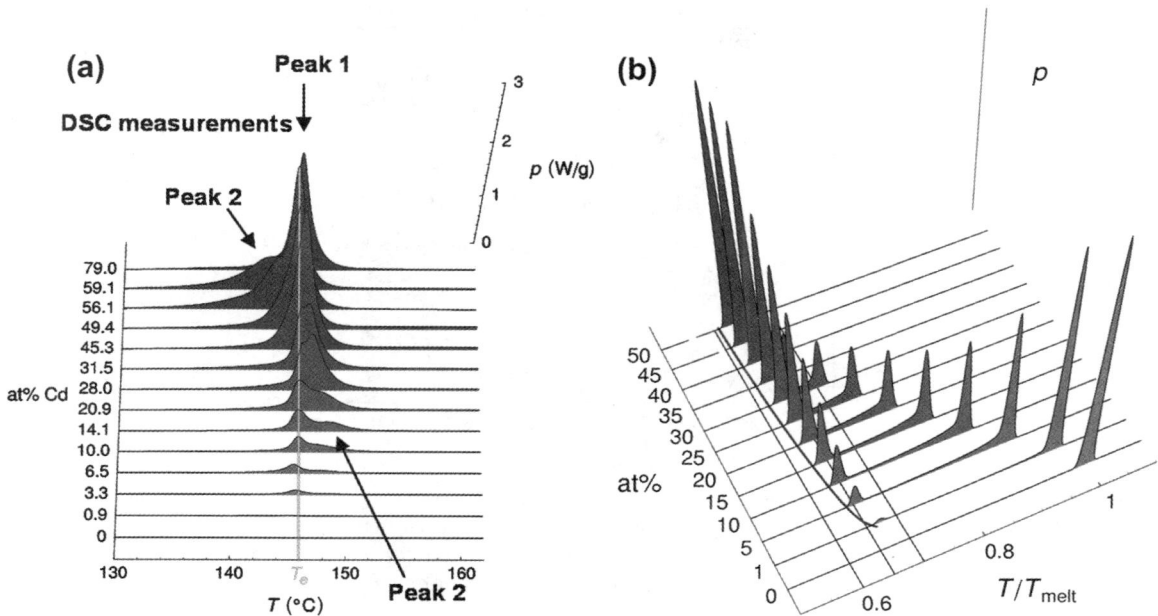

Figure 54 (a) Calorimetric results showing the onset of melting of a series of $Al_{98}(Bi_x\text{–}Cd_{1-x})_2$ alloys. P denoted the calorimetric signal in [W/g]. The two peaks that are indicated in the figure refer to the fraction of larger (peak 1) and smaller (peak 2) particles. The results indicate that the onset of melting (the "eutectic" temperature) for the smaller particles depends on the alloy composition and is not longer constant as for the larger particles or the bulk alloy. (b) Calculated calorimetry curves based on the theoretical model eutectic for a particle diameter of 20 nm (i.e. similar to the average particle size observed experimentally). The blue curve outlines the onset of melting for the different alloy compositions. The variation of the melting onset is comparable to the experimental results shown in **Figure 54a** (Bunzel et al., 2004). (For interpretation of the references to color in this figure legend, the reader is referred to the online version of this book.)

addressed in many theoretical and experimental works. A thorough overview on this subject can be found in the reviews by Chattopadhyay and Goswami (1997), Mei and Lu (2007).

26.7.4.1 *Melting and Freezing*

When the size of a particle is reduced, then the excess free energy—the product of the surface area A and of an interfacial free energy density γ—diminishes more slowly than the free energies of the bulk phases and capillary effects will therefore increasingly affect the thermodynamic equilibrium. In the last decades, the melting of nanoscale Pb particles embedded in Al has been of interest since Pb–Al nanocomposites serve as model systems to investigate the size-dependent melting phenomenon (Johnson et al., 2004; Chattopadhyay and Goswami, 1997; Mei and Lu, 2007).

Nanometer-sized Pb particles embedded in an Al host were produced by different techniques as for instance melt spinning, ball milling or ion implantation. The melting point of nanometer-sized Pb particles was found to deviate strongly from the bulk melting point of Pb. For instance, an elevated melting point of the Pb inclusions was found in DSC, TEM and XRD in situ heating experiments for a melt-spun material (Moore et al., 1987; Zhang and Cantor, 1991; Sheng et al., 1996; Moore et al., 1987b; Sheng et al., 1998; Gabrisch et al., 2001).

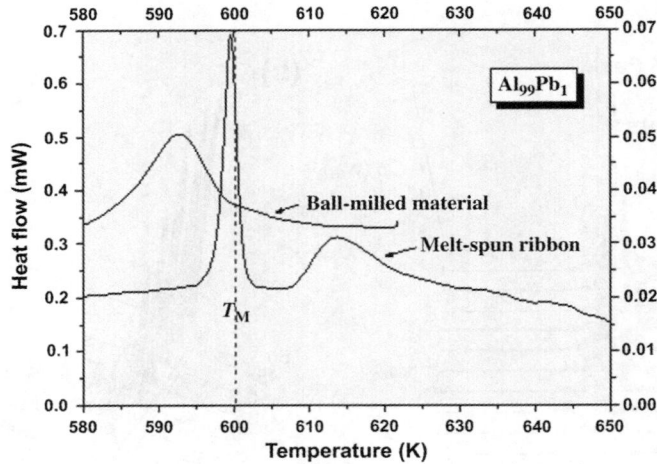

Figure 55 Comparison of melting signals of nanometer-sized Pb inclusions embedded in Al matrix fabricated by melt-spinning and ball-milling, respectively. The first peak of the melt-spun material, close to the nominal melting point of bulk Pb, is related to the melting signal of larger Pb inclusions located at the grain boundaries of the Al matrix. The smaller peak represents the melting of the smaller faceted Pb particles in the grain interior of the polycrystalline matrix grains (Rösner and Wilde, 2006).

On the other hand, material of identical composition and similar average particle size that was fabricated by ball milling revealed a significant depression of the melting point. **Figure 55** displays the different melting behavior of ball-milled and melt-spun material, respectively.

The observation of a melting point shift was always linked to the size and morphology of the Pb inclusions as shown in **Figure 56**. It should be noted that for both cases the nanometer-sized Pb

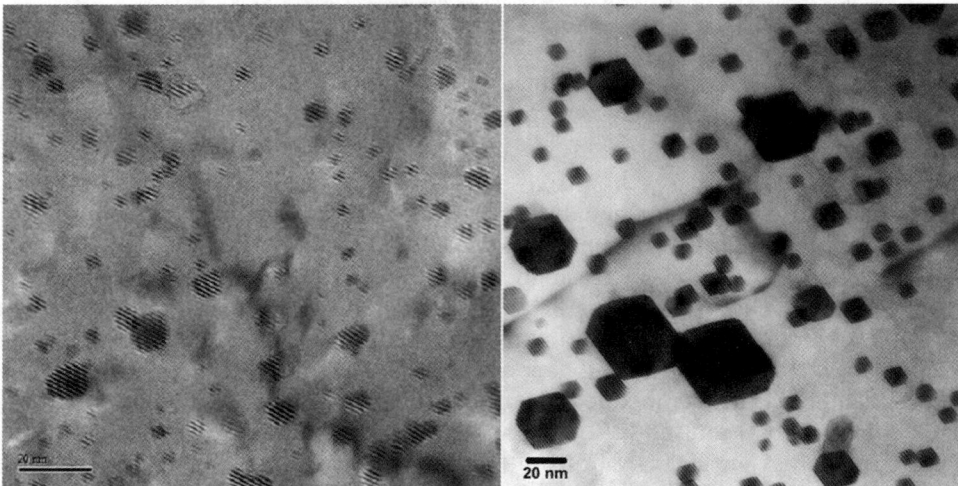

Figure 56 Comparison of the morphology of Pb inclusions located in the grain interior of the Al matrix. Left: ball-milled material. Right: melt-spun material.

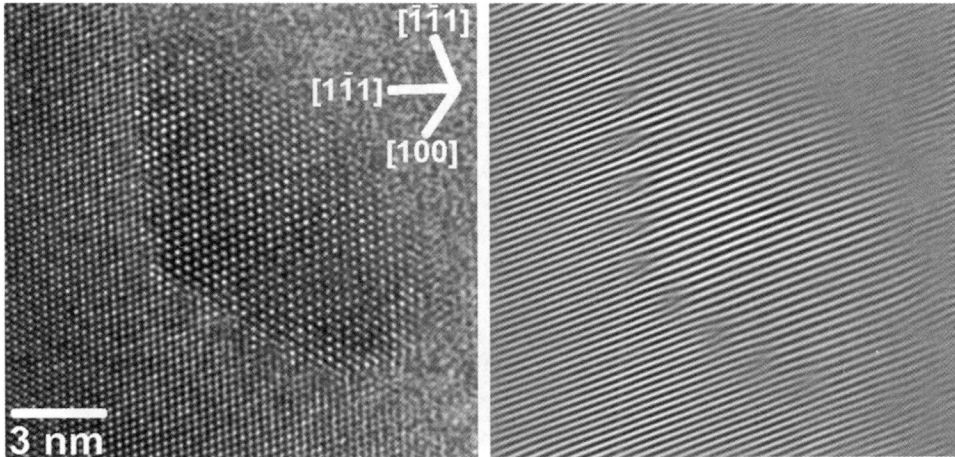

Figure 57 Left: High-resolution TEM micrograph of an uncovered Pb inclusion (ball-milled material) at Scherzer focus ($\Delta f = -68$ nm) showing a heterointerface with the Al matrix remaining on two sides. Such Pb particles were located in the amorphous edge area of the TEM specimen. Right: Bragg-filtered image using the $[-1,-1,1]$ reflection only for a better visibility of the misfit dislocations at the Al–Pb interfaces.

inclusions do exhibit a cube-on-cube orientation relationship. Pb inclusions in ball-milled material show a spherical morphology whereas they appear faceted in melt-spun material.

Remarkably, it has been shown that an increase in the melting point of ball-milled Al–Pb composites can be achieved by a heat treatment at high temperatures leading to an increased amount of faceted Pb particles (Rösner et al., 2003). From these findings attained so far, the melting process at small system sizes seems to be determined by the interface energy and the related interface topology rather than merely by the size of the particle. Thus, a more detailed understanding of the interface morphology is required. One point is to elucidate the reason for a strict maintenance of a cube-on-cube orientation relationship of Pb inclusions, which has been observed after rapid melt quenching as well as for ball-milled material (**Figure 55**). Several interface studies based on high-resolution TEM were undertaken (Rösner et al., 2004; Rösner et al., 2006; Rösner et al., 2007) revealing that the total excess free energy can be minimized (for a given particle size) by an efficient accommodation of misfit in the form of interfacial dislocations that is energetically particularly favorable if the "classical" cube-on-cube orientation relationship is maintained, since then the lowest number of misfit dislocations are needed to accommodate the misfit. As a result for both morphologies, e.g. faceted or spherical, the misfit was found to be accommodated via misfit dislocations (**Figure 57**) on about every fifth Al-plane.

This result is also corroborated by in situ observations of melting inside the TEM that showed for Pb particles situated at GBs of the Al matrix that the transformation front started at the rounded interfaces (Johnson et al., 2004). In addition, it was shown that upon altering the topology of the particle–matrix interfaces between a faceted and a curved morphology, the melting behavior could reversibly be changed from melting at higher to melting at lower temperatures with respect to the bulk melting temperature (Rösner and Wilde, 2006; Rösner et al., 2009). Moreover, experiments on the melting behavior of thin, matrix-confined thin films (Zhang et al., 2000) and also on matrix-encased nanocrystals (Moros et al., 2011) showed that also the melting enthalpy is affected (reduced) by small

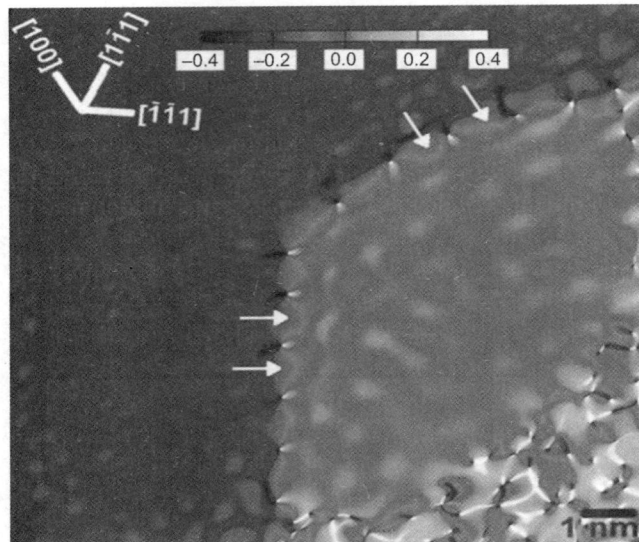

Figure 58 Experimental mean dilatation (trace of the strain tensor) using Pb as the reference lattice. Strains ranging from −40% to +40% are displayed. The Al matrix is displayed in blue color, indicating a strain state of −22%. The Pb particle is depicted in a mixture of green/red, representing the (approximately) zero strain level (reference lattice). A diffuse greenish zone along the interface region is discerned, indicating the presence of considerable strains. Note the offset position of the individual dislocation cores at the interfaces and the curvature (bulging nature) of the mean dilatation between them marked by arrows (Rösner et al., 2010). (For interpretation of the references to color in this figure legend, the reader is referred to the online version of this book.)

system sizes. These observations clearly point out the importance of the interface structure on the thermodynamics of the phase transformations. It should also be noted that the observation of melting occurring at temperatures above the bulk melting temperature does not substantiate the occurrence of the so-called *superheating*. Superheating, in a similar way as undercooling, describes a kinetic bypassing of the equilibrium phase-transformation temperature in such a way that the retained phase is in a metastable state. The results described above on the melting transformation of nanoscaled inclusions with faceted interfaces to the matrix indicate however that the shift of the phase transformation temperature is (at least partially) independent of kinetics and thus represents a shift of the thermo-dynamic equilibrium.

For small particle sizes, most models predict a shift of the melting temperature, T_m, following a linear dependence of T_m on the inverse particle diameter, D, of the form:

$$T_m = T_m, 0 - c/D, \tag{31}$$

where c is a constant. This linear dependence is obtained from any scaling approach that is based on the relative fractions of atoms situated at interfaces and in the grain or particle interior, irrespective of the specific model assumptions. In fact, models that are based on thermodynamics and that arrive at equivalents of the Kelvin equation (Couchman and Jesser, 1977), liquid-skin models (Wronski, 1967) or homogeneous melting models (Pawlow, 1909) all yield a similar dependence on the mean radius of the nanocrystals that are approximately described as spheres.

In fact, for small system sizes (i.e. large curvature), the Gibbs–Thomson effect leads directly to a shift of the thermodynamic melting temperature, given as:

$$T_{\mathrm{m}}^{\mathrm{bulk}} - T_{\mathrm{m}}(d) = \frac{4 \cdot \gamma \cdot T_{\mathrm{m}}^{\mathrm{bulk}}}{d \cdot \Delta H_{\mathrm{f}} \cdot \rho_{\mathrm{s}}} \tag{32}$$

with the bulk melting temperature $T_{\mathrm{m}}^{\mathrm{bulk}}$. $T_{\mathrm{m}}(d)$ denotes the melting temperature of a solid particle with the diameter d, ΔH_{f} is the enthalpy of fusion and ρ_{s} the density of the solid. A comprehensive overview on the existing groups of models is given in Mei and Lu (2007).

One problem associated with most models is that they predict only melting point reductions, i.e. these models consider the case of free (isolated) particles rather than the more common case of nanocrystals that are embedded within a solid matrix. In contrast, the theory of size-dependent melting by Couchman and Jesser (1977) considers both, melting temperature increase or decrease, depending on the relative magnitude of the interface energy between the solid particle and the matrix and the liquid particle and the matrix. Specifically, the equation relating the melting temperature to the particle size as derived by Couchman and Jesser is given as:

$$T_{\mathrm{m}} = T_{\mathrm{m,b}}\left\{1 - \frac{3}{L}\left(\frac{\nu_{\mathrm{s}}\gamma_{\mathrm{s}}}{r_{\mathrm{s}}} - \frac{\nu_{\mathrm{l}}\gamma_{\mathrm{l}}}{r_{\mathrm{l}}}\right)\right\}, \tag{33}$$

where L denotes the latent heat, r the radius of the particle, ν the atomic volume, $T_{\mathrm{m,b}}$ is the melting temperature of the bulk. The subscripts l and s denote the liquid and solid phase, respectively.

Pressure effects due to differential thermal expansion of the matrix and the embedded phase or strains are not included in this expression. However, a strain analysis of an embedded Pb particle in an Al matrix by the GPA method (Rösner et al., 2010b) unambiguously revealed considerable strain at the interface (see **Figure 58**), which has to be taken into account.

An adequate way to describe the melting behavior (first-order transition) considering strain/pressure is given by the Clapeyron equation:

$$(\Delta V/\Delta S)\Delta p = \Delta T_{\mathrm{M}}; \tag{34}$$

ΔV, ΔS, Δp, ΔT_{M} are the changes in volume, entropy, pressure and melting temperature, respectively. Therefore any increase in pressure would shift the melting point to higher temperatures. Since the outer regions of the Pb particle are under compressive strain (negative) of large magnitude, a higher melting temperature is expected according to Eqn (34). Similarly, strain contributions have been included empirically into Eqn (33) by Zhao et al. (2001), resulting in a modified Couchman–Jesser equation:

$$T_{\mathrm{m}} = T_{\mathrm{m,b}}\left\{1 - \frac{3}{L}\left(\frac{\nu_{\mathrm{s}}\gamma_{\mathrm{s}}}{r_{\mathrm{s}}} - \frac{\nu_{\mathrm{l}}\gamma_{\mathrm{l}}}{r_{\mathrm{l}}} - \Delta E\right)\right\}. \tag{35}$$

With the same meaning of the parameters in Eqn (35) as for Eqn (33) and with ΔE as the strain energy difference between an embedded solid or liquid particle, a description of the experimentally observed shifts of the melting temperature can be obtained.

It should be noted in this context that the strain peaks (hot spots) in the strain maps in **Figure 58** originating from the misfit dislocation cores at the interfaces do not have an influence on the melting behavior (decrease or increase of the melting temperature), since the misfit is accommodated in the same manner for both species of Pb particles (Rösner et al., 2004; Rösner and Wilde, 2006), i.e.

spherical or faceted. However, it has been reported that steps/ledges at particle–matrix interfaces lead to melting point depression (Goswami et al., 1998). Thus, two facts could account for an increase of the melting point, namely the faceted shape or the compressive strain, whereas the occurrence of steps could account for a melting point decrease. Currently, it cannot be said how these three contributions add up. More experimental facts in terms of strain mapping of curved particle–matrix interfaces as well as a direct observation of the melting of such excavated Pb particles have to be awaited before any further conclusions can be arrived at. In summary, the results obtained on different model systems concerning the size dependence of the melting transformation point out the importance of the interface contributions to the thermodynamic potentials of nanostructured materials and they highlight the strong impact atomic details of the interface can have on the macroscopic properties of these materials. A rather recent review on this subject (Nanda, 2009) provides an extensive overview on the different models and on some of the experimental results concerning the melting of nanoparticles and embedded nanocrystals.

26.7.4.2 *Martensitic Transformations*

In contrast to the melting/freezing transformation that is strongly dependent on the kinetics of atomic diffusion, martensitic transformations proceed diffusionless by atomic displacements of less than one interatomic distance (see also Chapter 9, Volume I). Thus, the ordered structure stays ordered during the transformation, giving rise to only negligible changes in the configurational entropy, but significant modifications of the thermodynamic potentials are caused by changes of the vibrational entropy or, as in some Fe alloys, from changes of the magnetic order (Pfeiler, 2007). Martensitic transformations can be triggered by changes of temperature, by applied stresses or external magnetic fields and occur by shearing of the crystalline lattice. Thus, a strict relation between the austenitic phase and the martensitic phase (the two phases that participate in the transformation) is maintained. Additionally, twins are typically formed in the martensitic phase during the transformation. This introduces an additional interface term due to twinning, leading to the dependence of the transformation on the grain- or particle size (Lange, 1982). Moreover, the volume change and the shear transformation within a given microstructure (i.e. within a constrained volume) leads to a strained state of the martensite, which, in turn, leads to transformation toughening (Lange, 1982). Therefore, the strain (and stress) aspects occurring in bulk NC materials are most important for analyzing the size-dependence of martensitic transformations.

Experimentally, the size effects concerning martensitic transformations were first studied in Fe–Ni alloys (Cech and Turnbull, 1956). Later, Kuhrt and Schultz (1993) showed for NC Fe–Ni alloys that the transformation stress is accompanied by modifications of the kinetics of the martensitic transformation. It was suggested that the abundant GBs hinder the growth of the martensitic phase. In line with this interpretation, it was shown that a decreasing grain size increases the hysteresis between the martensite finish (Mf) and the austenite finish (Af) temperatures (for definitions see Chapter 9, Volume I) upon thermal cycling and decreases the martensite transformation temperature (Fukuda et al., 1998; Frommen et al., 2004).

Twins show a self accommodating way to adapt for the shape and the volume change during the martensitic transformation (Otsuka and Ren, 2005). Nanograins of NiTi formed a herring-bone morphology of twin variants caused by an autocatalytic process of mutual strain accommodation (Waitz, 2005), rendering the twinning transformation size dependent. Moreover, it was shown that the transformation path from austenite to martensite is size-dependent (Waitz et al., 2005); small NiTi grains include an additional R-phase besides the austenite and the martensite phases. The self accommodation of twins in nanograins was more recently analyzed by the finite element method

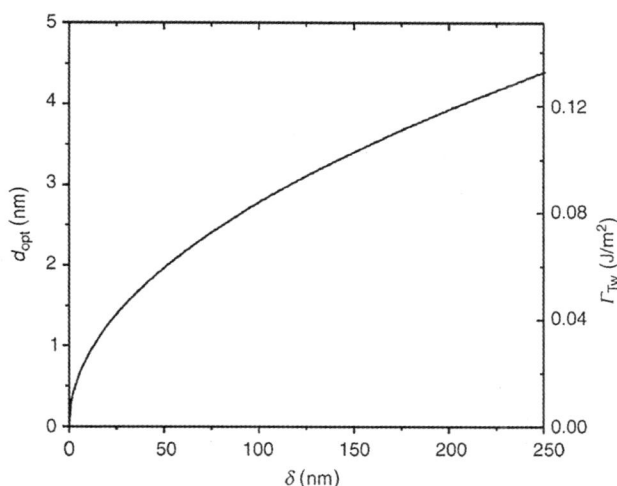

Figure 59 The optimum twin bandwidth d_{opt} as a function of the grain size. Reproduced with permission from Elsevier (Waitz et al., 2007).

(FEM) (Waitz et al., 2007). The twin width obtained from these calculations (**Figure 59**) clearly displays a strong dependence on the grain size.

The twin-boundary energy Γ_{TW} and the optimum twin width d_{opt} correlate. As seen in **Figure 59**, at small grain sizes the twin width decreases down to almost atomic dimensions, i.e. to widths of a few lattice constants. These theoretical results show good agreement with the experiments (Waitz et al., 2007). As a consequence of the strain accomodation, the total energy barrier for the martensitic transformation increases with decreasing grain size. This size effect is rather pronounced: NiTi with a grain size below 15 nm does not martensitically transform at all. Recently, the systematic suppression of the martensitic transformation with decreasing grain size was also experimentally shown (Peterlechner et al., 2011). In line with the increased energy barrier for the martensitic transformation, NC-NiTi after deformation and heating was shown to have enhanced properties: higher recovery stresses and strain, increased hardness, and a wider temperature window for superelastic applications (Tsuchiya et al., 2009).

26.8 Selected Examples of Application-Related Microstructure–Property Relations

Nanostructured materials are already applied in a multitude of products. Most of these are however related to nonmetallic materials, e.g. semiconductor nanostructures for microelectronics or oxidic nanoparticles in catalysis, optical applications or pharmaceutical products. With these applications, it is often the reduced system size that serves the purpose due to size confinement or even quantum confinement effects or due to reduced time scales for information transport.

Other technological applications of nanostructured materials that are related to thin films (e.g. GMR-Sensors, magnetic recording media or soft magnets for transformation cores) or particles (catalysts, luminescent surfaces, optical sensors) are treated in the respective chapters of this volume or in text books of chemistry. Naturally, the most important target of physical metallurgy, structural materials, needs to compete with traditional materials and processing routes concerning performance and price. The latter aspect naturally is more critical for nanostructured materials where more elaborate processing steps are involved in the initial synthesis. Thus applications of nanostructured or UFG materials are expected in areas, where the specific and often enhanced properties and property combinations are required and necessary, irrespective of the prize.

26.8.1   Ultrafine-Grained Ti for Medical Applications

Raab et al. (2004) have shown that Ti of commercial purity could be processed in kg-amounts by ECAP and also by variants of this process that utilizes wire-shaped material, so that equiaxed microstructures with average grain sizes of the order of 200 nm exhibiting hardness values larger than 3 GPa and a tensile strength of 1420 MPa were obtained (Semenova et al., 2008).

Additionally, Estrin et al. (2011) found that the small grain size of the materials also enhanced the growth rate of human cells onto the materials, making it highly suitable for dental or bone implants, particularly in situations where a minimum strength is required at small dimensions and sizes of the implanted structure. These beneficial property combinations served to implement this material already into the market.

26.8.2   Superplastic Deformation and Near-Net-Shape Forming at Low Temperatures

Superplastic forming as a near-net-shape production process most importantly requires fine grain sizes and—for materials with conventional grain sizes in the micrometer range—high homologous temperatures. It is a favorable shape-forming method, since complex shapes can be obtained in a cost-effective manner, i.e. without involving multistep processing routes. Thus, superplastic forming of NC materials at low homologous temperatures or high strain rates has been among the first examples for applying NC materials.

Naturally, since structural superplasticity depends on GB sliding as stress-accommodating mechanism, diffusion, GB migration and atomic displacements at the shear plane (the so-called "mesoscopic glide plane") are involved, that hamper the microstructure stability, especially at small grain sizes.

However, the utilization of structural superplasticity in NC materials has successfully been proven on NC-TiO_2 (Karch and Birringer, 1990) and on NC baron-doped intermetallic Ni_3Al (Mishara et al., 1998) as well as on a light weight aluminum alloy (Valiev et al., 1998) both processed by HPT prior to superplastic forming. The results reproduced in **Figure 60** indicate that the Ni_3Al specimen experienced very high tensile elongation without visible macroscopic necking. It should be noted that the deformations temperature of 650 °C for the NC material is about 400 °C lower that the superplastic forming temperature for microcrystalline Ni_3Al.

26.8.3   Hard Coatings

The application of hard coatings consisting of metal nitrides or oxides by plasma-assisted surface reactions or deposition methods based on CVD or sputtering methods for wear minimization purpose is well known and described in detail in recent reviews (Choy, 2003; Kelly and Arnell, 2000; Sun and Bell, 1991). Recently, Lu et al. (Tong et al., 2003) have shown that graded nanomaterials with a nanostructured surface and cg interior obtained by surface mechanical attrition allow nitriding Fe at 300 °C, i.e. at more than 200 °C lower temperatures than for conventional, cg Fe. During this treatment, a nitride surface layer of approximately 10 μm thickness was formed that presented a hardness of about

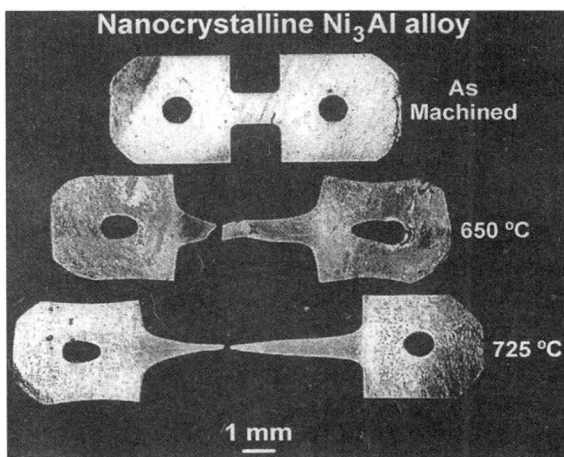

Figure 60 Appearance of nanostructured Ni$_3$Al samples prior and after tension at a temperature of 650 °C at 1×10^{-3} s^{-1} by 390% elongation and at a temperature of 725 °C at 1×10^{-3} s^{-1} by 560%, respectively. Reproduced with permission from Elsevier (Valiev et al., 2000).

5 GPa (**Figure 61**). One might also speculate that the graded initial microstructure that can yield compositionally graded structures upon adjusting proper processing conditions serves to enhance the life cycle time and damage tolerance of the coating due to the gradually decreasing lattice mismatch between the nitride and the parent phase.

Other examples concerning the application of NC metallic materials are found in the areas of hydrogen storage and nanostructured steel. For recent examples concerning these two areas, the reader is referred to the respective chapters (21, 27) in this volume.

26.9 Open Issues and Future Perspectives

Three decades after the first conceptual approach, the area of nanostructured metallic materials has been developed to a research direction in its own right and new synthesis and processing routes as well as new analysis methods for investigating microstructural aspects and microstructure–property relations have been established. However, despite the international "rush" for nanosciences and nanomaterials, there are still open questions and unresolved issues as well as unexplored directions with promising potential. In the following, we will discuss selected examples of these open issues that, according to a strictly personal view of the author, are either of special importance for the understanding of nanostructured materials or offer a high prospect for important future developments. It also needs to be stated that the field is still far from establishing a "nanomaterials science" as present for conventional materials. So far, processing-microstructure-property correlations are established only for selected cases and a unified view that takes into account the elements of the microstructure with their thermodynamic contributions and their impact on stability and reliability together with the functional property under investigation is hardly available. Yet, some of the future directions discussed below, such as e.g. GB engineering, could promote and utilize such a more complete understanding.

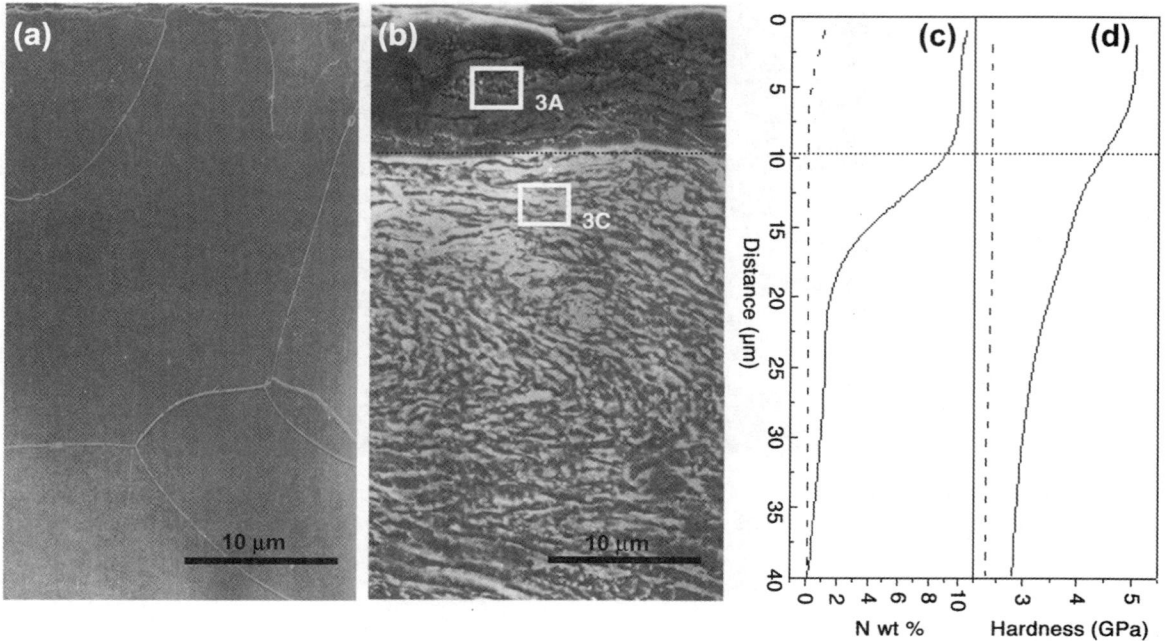

Figure 61 Cross-sectional observation of (A) an original coarse-grained Fe sample and (B) an SMAT Fe sample after nitriding at 300 °C for 9 h. (C) nitrogen concentration and (D) microhardness along the depth from the top surface layer in the original Fe sample (dashed) and in the nitrided one (solid lines), respectively. Source: Tong et al., 2003.

26.9.1 Bulk Nanostructured Materials

Synthesizing bulk quantities of nanostructured materials in massive shapes still poses an important task. In particular, the synthesis of bulk nanostructured materials without internal porosity or contaminations at the interfaces is difficult to achieve in multistep processing schemes due to the high reactivity of "fresh" metal surfaces and due to stability restrictions that limit any consolidation procedure due to the occurrence of thermally induced or stress-induced GB migration.

Methods that are based on grain fragmentation of a cg initial state such as the "SPD" techniques on the other hand hardly achieve microstructures for bulk specimens that are genuinely in the nanoscale regime, except for specific alloy compositions. For a selected range of materials and for specific applications, nanocrystallization of bulk glasses can lead to fully dense bulk NC materials, e.g. for nanostructured magnetic materials with low coercivity and high saturation polarization. On the other hand, for technical applications of bulk nanostructured materials, it would often be sufficient to achieve a nanoscale grain size on and near the surface of materials that are extended in the lateral dimensions. Methods such as surface mechanical attrition are candidate methods for synthesizing such nano-structurally graded surface layers. Another interesting option is given by rolling deformation of chemically dissimilar surface layers. As indicated by severe cold rolling of heterophase multilayers, surface bonding, intermixing and nanostructure formation can be obtained in one processing step. Additionally, this method is directly applicable for industrial-scale processing, since it presents a derivative of well-known roll-cladding processes, which are established processing routes. More extensively focused on academic purposes, advanced synthesis routes that target the preparation of bulk NC samples that are free from porosity can be based on sequentially combining the initial synthesis of NC powders and the consolidation by extrusion or SPD. The feasibility of this approach has already been demonstrated, although deformation-induced coarsening needs to be taken into account. Advanced microstructure stabilization methods that are based on chemically modifying the GB structure could be combined with such processing pathway to obtain fully dense bulk nanostructured materials.

26.9.2 Multifunctional Properties

The large fraction of atoms situated at or near-defect cores allow for a larger variability of materials properties and, through defect engineering, provide the opportunity for property tuning. One inventive way to adjust new properties of nanomaterials proposed by Gleiter et al. (2001) is based on modifying the effective number of electrons per atom in space charge layers. Gleiter suggested using nano-structured porous metal electrodes that present a large interface area with a liquid electrolyte that infiltrates into the open porosity. Thus, using an electrochemical setup with the nanostructured porous metal as one electrode in contact with a liquid electrolyte results in a large volume fraction of the electrode in contact with a liquid electrolyte resulting in a large volume fraction of the electrode material being effectively inside the space charge layer, i.e. in a region of the material where charge neutrality is not maintained and the electron-per-atom ratio does not correspond to the respective ratio for the material in the charge-neutral state. The number of electrons per atom determines many properties such as magnetic or electronic properties and also the interatomic interaction potential and thus the equilibrium interatomic spacing. Therefore, since almost all properties of a specific crystalline material depend on the interatomic spacing, a wide range of materials properties are prone to tuning via altering the effective number of electrons per atom. Since this type of modification is restricted to space charge layers, it is also restricted to materials with extremely large specific surface areas, i.e. nano-structured materials.

Recently, the validity of this concept has been verified by experimental work that indicated the tune-ability of the lattice constant of nanostructured Pt (Weissmüller et al., 2003) or the modification of the magnetization and the electrical conductivity of nanostructured Fe and Au, respectively (Sagmeister et al., 2006; Traussnig et al., 2011). The application of this approach by all solid-state devices might yield tuneable materials with completely new functionalities e.g. in the areas of sensors or transducers or as active components in nanoscale electro-optical or electromechanical systems.

A different approach toward structured materials with multifunctional properties is based on applying SPD for achieving fine grain sizes and high defect densities. Recently, Estrin et al. showed that slightly Ni-rich NiTi after severe deformation via HPT showed increased strength, increased corrosion resistance and enhanced cell adhesion (Estrin et al., 2011). This combination of properties is suitable for materials for medical applications such as bone—or orthodontic implants.

26.9.3 Grain-Boundary Engineering

Watanabe proposed in 1980 adjusting the relative fractions of low-angle GBs or special GBs with specific symmetries of the respective CSL lattice such as $\Sigma3$ twin boundaries by suitable thermo-mechanical processing routes to adjust the resulting macroscopic properties of polycrystalline materials (Watanabe et al., 1980). This concept might be extendable to encompass materials with high densities of internal interfaces, i.e. NC materials. In addition, it has been suggested that not only interfaces with a specific structure, i.e. with specific values of the disorientations of the adjacent grains but also random high-angle GBs might be modified such that they possess specific properties such as high atomic mobility along the interface, high solubility for GB-active solutes, high stability against GB motion, low scattering cross-section for electron waves or a combination of several of the properties.

Progress in this respect has recently been achieved in the area of severely deformed materials. Valiev et al. suggested that the large amount of lattice dislocations created during SPD led to the interaction of the dislocations with the high-angle GBs which, in the course of their interaction, should result in the creation of the so-called "extrinsic GB dislocations" that form a nonperiodic array in the GB plane. These GBs were then termed "nonequilibrium GBs" to distinguish them from random high-angle GBs and to point out that these boundaries were expected to have enhanced excess free-energy densities.

Recently, the existence of a fraction of high-angle GBs with significantly enhanced strain fields has been shown by applying GPA to materials after severe deformation. This method that is applied to HRTEM images measures the components of the strain tensor in the plane of the thin film specimen. Similarly, treated samples have also found to possess a fraction of GBs with largely enhanced diffusion rates. Measurements by atom probe spectroscopy have also shown that materials after SPD possess a fraction of GBs with strongly enhanced GB segregation (Sauvage et al., 2012). These results, although not obtained on a single material or a single GB indicate the opportunities for modifying the properties of random high-angle GBs by applying large plastic strains. So far, it remains an open question how these interfaces couple with external shear stresses and how their presence affects the stability of the UFG state. However, the possibility for utilizing segregation to stabilize the microstructure against coarsening and to utilize the enhanced atomic mobility along the GBs, e.g. for accommodating externally applied shear stresses and i.e. for enhancing the ductility, might lead to materials for applications in medical applications.

Recently, the idea of advanced GB engineering through modifying the structure of high-angle GBs has received support from an experimental investigation of the microstructure of a NC material that had been severely deformed by HPT after synthesizing by consolidation of inert-gas condensed powder (Rösner et al., 2011). As indicated in the HRTEM image in **Figure 62**, a triple junction constituting of $\sum 3 : \sum 3 : \sum 9$ special GBs has formed by the severe deformation, which is an energetically most favorable configuration.

In addition, strong rotational gradients of opposite sign have been observed along the $\sum 9$- GB, which are indicative of a disclination dipole between the triple junction and the quadruple point. Certainly, the presence of rotational strain gradients (that correspond to position-dependent bending of lattice planes) reduces the mobility of dislocations and thus stabilizes the GB position. In addition to this kinetic stabilization, the $\sum 3 : \sum 3 : \sum 9$ configuration (as a configuration of low excess free energy) also stabilizes the given structure since all transient structures that would necessarily be sampled during GB motion would energetically be less favorable. The fact that this configuration has been adopted without special annealing treatments, but just after the severe deformation, might indicate a propensity for adjusting an energetically favorable configuration once the system has been exited far enough out of equilibrium.

Self-healing materials are another group of materials that might benefit from such advanced grain-boundary engineering concepts. It has already been shown that the high density of internal interfaces provides a large effective sink for radiation-induced defects. Thus, applications in power reactors such as e.g. material for the so-called "first wall" in fusion reactors might be possible. However, in conjunction with the creation of lattice defects such as vacancies that inherently enhance the atomic mobility and thus the relaxation kinetics, also thermal energy is dissipated, which causes coarsening and creep. Thus,

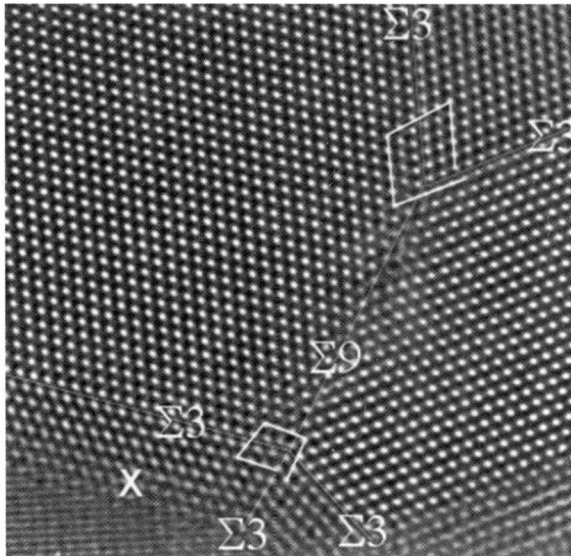

Figure 62   HRTEM image of the region in the vicinity of a triple junction in Pd after IGC and subsequent HPT deformation. The triple junction connects two $\Sigma 3$ and one $\Sigma 9$ special grain boundaries (Rösner et al., 2011).

stabilization, e.g. by controlled doping of GBs with stabilizing segregants and/or elements or particles that pin GB motion, might offer a now route for obtaining radiation-resistant materials. One recent example in this area is presented by the so-called "nanocluster" steel, which consists of nanoscale oxide particles that are dispersed in a steel matrix. It was shown that this material, which has an average grain size of only 20 nm, still shows a very high creep strength even at temperatures as high as 1400 K. The absence of unusual GB diffusivity highlights the effectiveness of the nanocluster dispersion for preventing GB motion and coarsening.

Acknowledgments

Over the years, research had not been possible without the support by the Deutsche Forschungsgemeinschaft, the Alexander von Humboldt-Foundation, the German Academic Exchange Service, the German-Israel-Foundation and the Volkswagen-Foundation. Support by these associations and foundations is most gratefully acknowledged. Valuable input for this chapter by Dr S. Divinski, Dr M. Peterlechner and Dr H. Rösner is also gratefully acknowledged. Special thanks also goes to Ms. S. Gurnik, who has provided invaluable support with the typewriting and the organization of the figures.

References

Ames, M., Markmann, J., Karos, R., Michels, A., Tschöpe, A., Birringer, R., 2008. Acta Mater. 56, 4255–4266.
Amouyal, Y., Divinski, S.V., Estrin, Y., Rabkin, E., 2007. Acta Mater. 55, 5968–5979.
Amouyal, Y., Divinski, S.V., Klinger, L., Rabkin, E., 2008. Acta Mater.
Amouyal, Y., Rabkin, E., 2007. Acta Mater. 55, 6681.

Argon, A.S., Yip, S., 2006. Philos. Mag. Lett. 86, 713.
Asaro, R.J., Suresh, S., 2005. Acta Mater. 53, 3369–3382.
Ashby, M.F., 1970. Philos. Mag. 21, 399.
Ashby, M.F., Verall, R.A., 1973. Acta Metall. Mater. 21, 149.
Balandin, I.L., Bokstein, B.S., Egorov, V.K., Kurkin, P.V., 1997. Nanostruct. Mater. 8, 37.
Baletto, F., Ferrando, R., 2005. Rev. Mod. Phys. 77, 371–423.
Beke, D.L., Godeny, I., Erdelyi, G., Kedves, F.J., 1987. Philos. Mag. A 56, 659–671.
Biro, L.P., Balint, Z., Kertesz, K., Vertesy, Z., Mark, G.I., Horvath, Z.E., Balazs, J., Mehn, D., Kiricsi, I., Lousse, V., 2003. Phys. Rev. E 67, 0219071.
Birringer, R., Gleiter, H., Klein, H.P., Marquardt, P., 1984. Phys. Lett. A 102, 365.
Birringer, R., Zimmer, P., 2009. Acta Mater. 57, 1703–1716.
Borisov, V.T., Golikov, V.T., Shcherbedinsky, G.V., 1964. Phys. Metall. Metallogr. 17, 881.
Bokstein, B.S., Brose, H.D., Trusov, L.I., Khvostantseva, T.P., 1995. Nanostruct. Mater. 6, 873–876.
Boucharat, N., Hebert, R.J., Rösner, H., Valiev, R.Z., Wilde, G., 2005. Scr. Mater. 53, 823–828.
Buffat, P., Borel, J., 1976. Phys. Rev. A 13, 2287–2298.
Bunzel, P., Wilde, G., Rösner, H., Weissmüller, J., 2004. In: Herlach, D.M. (Ed.), Solidification and Crystallization. Wiley, Weinheim, pp. 157–165.
Burda, C., Chen, X.B., Narayanan, R., El-Sayed, M.A., 2005. Chem. Rev. 105, 1025–1102.
Cahn, J., Taylor, J.E., 2004. Acta Mater. 52, 4887–4898.
Calka, A., Wexler, D., 2002. Nature 419, 147–151.
Cech, R.E., Turnbull, D., 1956. Trans. Am. Inst. Metals 206, 124–132.
Cermak, J., Stloukal, I., 2007. J. Phys. Chem. Solids 68, 1249–1254.
Chattopadhyay, K., Goswami, R., 1997. Prog. Mater. Sci. 42, 287–300.
Chen, Y., Schuh, C.A., 2007. Scr. Mater. 57, 253–256.
Chokshi, A.H., 2008. Scripta Mater. 59, 726–729.
Choy, K.L., 2003. Prog. Mater. Sci. 48, 57–170.
Couchman, P.R., Jesser, W.A., 1977. Nature 269, 481–483.
Dao, M., Lu, L., Shen, Y.F., Suresh, S., 2006. Acta Mater. 54, 5421–5432.
Derlet, P.M., Hasnaoui, A., Van Swygenhoven, H., 2003a. Scr. Mater. 49, 629.
Derlet, P.M., Van Swygenhoven, H., Hasnaoui, A., 2003b. Philos. Mag. 83, 3569–3575.
DiMasi, E., Sarikaya, M., 2004. J. Mater. Res. 19, 1471–1476.
Dinda, G.P., Rösner, H., Wilde, G., 2005. Scr. Mater. 52, 577–582.
Divinski, S., Hisker, F., Kang, Y.S., Lee, J.S., Herzig, C., 2002a. Z. Metallkd. 93, 256.
Divinski, S., Hisker, F., Kang, Y.S., Lee, J.S., Herzig, C., 2002b. Z. Metallkd. 93, 265.
Divinski, S., Hisker, F., Kang, Y.S., Lee, J.S., Herzig, C., 2002c. Interface Sci. 11, 67.
Divinski, S.V., Geise, J., Rabkin, E., Herzig, C., 2004a. Z. Metallkd. 95, 945–952.
Divinski, S.V., Hisker, F., Kang, Y.S., Lee, J.S., Herzig, C., 2004b. Acta Mater. 52, 631–645.
Divinski, S., Lohmann, M., Herzig, C., 2004c. Acta Mater. 52, 3973.
Divinski, S., Lohmann, M., Herzig, C., 2001a. Acta Mater. 49, 249.
Divinski, S.V., Lohmann, M., Surholt, T., Herzig, C., 2001b. Interface Sci. 9, 357–363.
Divinski, S., Lohmann, M., Herzig, C., Straumal, B., Baretzky, B., Gust, W., 2005. Phys. Rev. B 71, 104104.
Divinski, S.V., Reglitz, G., Rösner, H., Estrin, Y., Wilde, G., 2011. Acta Mater. 59, 1974.
Divinski, S.V., Reglitz, G., Wilde, G., 2010a. Acta Mater. 58, 386.
Divinski, S.V., Wilde, G., Rabkin, E., Estrin, Y., 2010b. Adv. Eng. Mater. 12, 779–785.
Divinski, S.V., Reglitz, R., Peterlechner, M., Wilde, G., submitted to J. Appl. Phys.
Divinski, S.V., Ribbe, J., Reglitz, G., Estrin, Y., Wilde, G., 2009. J. Appl. Phys. 106, 063502.
Divinski, S., Ribbe, J., Schmitz, G., Herzig, C., 2007. Acta Mater. 55, 3337–3346.
Divinski, S.V., Wilde, G., 2008. Mater. Sci. Forum 584–586, 1012.
Edelhoff, H., Prokofjev, S.I., Divinski, S.V., 2011. Scr. Mater. 64, 374–377.
Eggersmann, M., Ye, A., Herth, S., Gutfleisch, O., Würschum, R., 2001. Interface Sci. 9, 337.
Estrin, Y., Ivanova, E.P., Michalska, A., Truong, V.K., Lapovok, R., Boyd, R., 2011. Acta Biomater. 7, 900–906.
Faghri, A., Zhang, Y., 2006. Transport Phenomena in Multiphase Systems. Academic Press, Amsterdam.
Farber, B., Cadel, B., Menand, A., Schmitz, G., Kirchheim, R., 2000. Acta Mater. 48, 789.
Fiebig, J., Divinski, S.V., Rösner, H., Estrin, Y., Wilde, G., 2011. J. Appl. Phys. 110, 083514.
Fischer, F.D., Waitz, T., Vollath, D., Simha, N.K., 2008. Prog. Mater. Sci. 53, 481–527.
Fisher, J.C., 1951. J. Appl. Phys. 22, 74.

Frommen, C., Wilde, G., Rösner, H., 2004. J. Alloys Compd. 377, 232–242.

Fujita, T., Horita, Z., Langdon, T.G., 2002. Philos. Mag. A 82, 2249.

Fukuda, K., Iizuka, E., Taguchi, H., Ito, S., 1998. J. Am. Ceram. Soc. 81, 2729–2731.

Gabrisch, H., Kjeldgaard, L., Johnson, E., Dahmen, U., 2001. Acta Mater. 49, 4259–4269.

Gianola, D.S., Van Petegem, S., Legros, M., Brandstetter, S., Van Swygenhoven, H., Hemker, K.J., 2006. Acta Mater. 54, 2253–2263.

Gibbs, J.W., 1906. The Scientific Papers of J. Willard Gibbs. In: Thermodynamics, vol. 1. Longmans, Green, and Co, New York and Bombay.

Gibbs, J.W., 1957. Collected Works. Yale University Press, New Haven.

Glaeser, A.M., Evans, J.W., 1986. Acta Metall. 34, 1545–1552.

Gleiter, H., 1989. Prog. Mater. Sci. 33, 223.

Gleiter, H., 1992. Phys. Status Solidi B 172, 41.

Gleiter, H., 1995. Z. Metallkd. 86, 78.

Gleiter, H., 1996. In: Cahn, R., Haasen, P. (Eds.), Physial Metallurgy. Elsevier, Amsterdam, pp. 843–942.

Gleiter, H., 2000. Acta Mater. 48, 1.

Gleiter, H., Marquardt, P., 1984. Z. Metallkd. 75, 263–267.

Gleiter, H., Weissmüller, J., Wollersheim, O., Würschum, R., 2001. Acta Mater. 49, 737–745.

Goswami, R., Chattopadhyay, K., Ryder, P.L., 1998. Acta Mater. 46, 4257–4271.

Goswami, R., Ryder, P., Chattopadhyay, K., 1999. Philos. Mag. Lett. 79, 481–489.

Gottstein, G., Shvindlerman, L.S., Zhao, B., 2010. Scr. Mater. 62, 914–917.

Grabovetskaya, G.P., Ratotska, I.V., Kolobov, Y.R., Puchkareva, L.N., 1997. Fiz. Met. Metallovedenie 83, 112.

Grabski, M.W., 1985. J. Phys. 46, 567–579.

Gährken, M., Wilde, G., 2008. Unpublished results.

Hahn, H., 1997. Nanostruct. Mater. 9, 3–12.

Hahn, H., Mondal, P., Padmanabhan, K.A., 1997. Nanostruct. Mater. 9, 603–609.

Hahn, H., Padmanabhan, K.A., 1997. Philos. Mag. B 76, 559–571.

Harrison, L.G., 1961. Trans. Faraday Soc. A 8, 1191.

Hasnaoui, A., Van Swygenhoven, H., Derlet, P.M., 2002. Acta Mater. 50, 3927.

Herth, S., Eggersmann, M., Eversheim, P.D., Würschum, R., 2004. J. Appl. Phys. 95, 5075–5080.

Herth, S., Michel, T., Tanimoto, H., Eggersmann, M., Dittmar, R., Schaefer, H.E., Frank, W., Würschum, R., 2001. In: Limoge, Y., Bocquet, J.L. (Eds.), Diiffusion in Materials: DIMAT2000. PTS 1&2, pp. 1199–1204.

Hirth, J.P., Lothe, J., 1982. Theory of Dislocations. McGraw-Hill, New York.

Hofmeister, H., Tan, G.L., Dubiel, M., 2005. J. Mater. Res. 20, 1551–1562.

Horváth, J., Birringer, R., Gleiter, H., 1987. Solid State Commun. 62, 319.

Howe, J.M., 1997. Interfaces in Materials: Atomic Structure, Thermodynamics and Kinetics of Solid–Vapor, Solid–Liquid and Solid–Solid Interfaces. John Wiley & Sons. Inc., New York.

Hugo, R.C., Kung, H., Weertman, J.R., Mitra, R., Knapp, J.A., Follstaedt, D.M., 2003. Acta Mater. 51, 1937–1943.

Hytch, M.J., Snoeck, E., Kilaas, R., 1998. Ultramicroscopy 74, 131–146.

Höfler, H.J., Averback, R.S., Hahn, H., Gleiter, H., 1993. J. Appl. Phys. 74, 3832.

Iwahashi, Y., Horita, Z., Nemoto, M., Langdon, T.G., 1998. Acta Mater. 46, 3317–3331.

Jesser, W.A., Shiflet, G.J., Allen, G.L., Crawford, J.L., 1999. Mater. Res. Innov. 2, 211–216.

Jin, Z.H., Gumbsch, P., Albe, K., Ma, E., Lu, K., Gleiter, H., Hahn, H., 2008. Acta Mater. 56, 1126–1135.

Johnson, E., Anderson, H.H., Dahmen, U., 2004. Mater. Sci. Eng. A 375–377, 951–955.

Kang, Y.S., Lee, J.S., Divinski, S.V., Hisker, F., Herzig, C., 2004. Z. Metallkd. 95, 76.

Karch, J., Birringer, R., 1990. Ceram. Int. 16, 291–294.

Kaur, I., Gust, W., Kozma, L., 1989. Handbook of Grain and Interphase Boundary Diffusion Data. Ziegler Press, Stuttgart.

Kaur, I., Mishin, Y., Gust, W., 1995. Fundamentals of Grain and Interphase Boundary Diffusion. Wiley, Chichester.

Kelly, P.J., Arnell, R.D., 2000. Vacuum 56, 159–172.

Kirchheim, R., 2007a. Acta Mater. 55, 5129–5138.

Kirchheim, R., 2007b. Acta Mater. 55, 5139–5148.

Klinger, L.M., Bokstein, B.S., 2009. Defect Diffusion Forum 289–292, 711–718.

Klinger, L., Rabkin, E., 2009. Int. J. Mater. Res. 100, 530–535.

Knorr, P., Nam, J.G., Lee, J.S., 2000. Metall. Mater. Trans. A 31, 503–510.

Koch, C.C., 1997. Nanostruct. Mater. 9, 13–22.

Kolobov, Y.R., Grabovetskaya, G.P., Ivanov, M.B., Zhilyaev, A.P., Valiev, R.Z., 2001. Scr. Mater. 44, 873.

Kondepudi, D., Prigogine, I., 1998. Modern Thermodynamics. John Wiley & Sons, New York.

Kuhrt, C., Schultz, L., 1993. J. Appl. Phys. 73, 1975–1980.

Kumar, K.S., Suresh, S., Chisholm, M.F., Horton, J.A., Wang, P., 2003. Acta Mater. 51, 387–405.
Lange, F.F., 1982. J. Mater. Sci. 17, 225–234.
LeClaire, A.D., 1963. Br. J. Appl. Phys. 14, 351.
Lee, J.-S., Cha, B.-H., Kang, Y.-S., 2005. Adv. Eng. Mater. 7, 467–473.
Lee, J.G., Mori, H., 2004. Philos. Mag. 84, 2675.
Lei, Y., Cai, W.P., Wilde, G., 2007. Prog. Mater. Sci. 52, 465–539.
Lei, Y., Yang, S., Wu, M., Wilde, G., 2011. Chem. Soc. Rev. 40, 1247–1258.
Li, J.C.M., 1963. Trans. Metals Soc. 227, 239.
Li, W.L., Tao, N.R., Lu, K., 2008. Scr. Mater. 59, 546–549.
Lian, J., Valiev, R.Z., Baudelet, B., 1995. Acta Metall. Mater. 43, 4165–4170.
Liu, Q., Huang, X., Lloyd, D.J., Hansen, N., 2002. Acta Mater. 50, 3789.
Liu, F., Kirchheim, R., 2004. Thin Solid Films 466, 108–113.
Lojkowski, W., Fecht, H.-J., 2000. Prog. Mater. Sci. 45, 339–568.
Lu, K., 1996. Mater. Sci. Eng. R Rep. 16, 161–221.
Lu, K., Hansen, N., 2009. Scr. Mater. 60, 1033.
Lu, L., Shen, Y.F., Chen, X.H., Qian, L.H., Lu, K., 2004. Science 304, 422–426.
Ma, E., 2006. J. Metals 4, 49.
Margolin, H., 1998. Acta Mater. 46, 6305.
Markmann, J., Bunzel, P., Rösner, H., Liu, K.W., Padmanabhan, K.A., Birringer, R., Gleiter, H., Weissmüller, J., 2003. Scr. Mater. 49, 637–644.
Mei, Q., Lu, K., 2007. Prog. Mater. Sci. 52, 1175–1262.
Meyers, M.A., Mishra, A., Benson, D.J., 2006. Prog. Mater. Sci. 51, 427–556.
Mishara, R.S., Valiev, R.Z., McFadden, S.X., Mukherjee, A.K., 1998. Mater. Sci. Eng. A 252, 174.
Mishin, Y.M., Razumovskii, I.M., 1992. Acta Metall. Mater. 40, 839–845.
Mitra, R., Chiou, W.A., Weertman, J.R., 2004. J. Mater. Res. 19, 1029–1037.
Moore, K.I., Chattopadhyay, K., Cantor, B., 1987. Proc. Roy. Soc. Lond. A 414, 499.
Moore, K.I., Zhang, D.L., Cantor, B., 1987. Acta Metall. Mater. 38, 1327.
Moros, A., Rösner, H., Wilde, G., 2011. Scr. Mater. 65, 883–886.
Nanda, K.K., 2009. Pramana 72, 617–628.
Nazarov, A.A., 2000. Interface Sci. 8, 315–322.
Nazarov, A.A., Romanov, A.E., Valiev, R.Z., 1990. Scr. Metall. Mater. 24, 1929.
Nazarov, A.A., Romanov, A.E., Valiev, R.Z., 1990. Physica. Status Solidi. A 122, 495–502.
Nazarov, A.A., Romanov, A.E., Valiev, R.Z., 1993. Acta Metall. Mater. 41, 1033.
Nazarov, A.A., Romanov, A.E., Valiev, R.Z., 1994. Nanostruct. Mater. 4, 93.
Nieh, T.G., Wadsworth, J., 1991. Acta Metall. Mater. 39, 3037–3045.
Otsuka, K., Ren, X., 2005. Prog. Mater. Sci. 50, 511–678.
Ovidko, I.A., 2005. Int. Mater. Rev. 50, 65.
Ovidko, I.A., Sheinerman, A.G., 2004. Rev. Adv. Mater. Sci. 6, 41–47.
Padmanabhan, K.A., 2009. In: Wilde, G. (Ed.), Nanostructured Materials. Elsevier, Amsterdam, pp. 51–126.
Padmanabhan, K.A., Dinda, G.P., Hahn, H., Gleiter, H., 2007. Mater. Sci. Eng. A 452, 462–468.
Padmanabhan, K.A., Gleiter, H., 2004. Mater. Sci. Eng. A 381, 28–38.
Padmanabhan, K.A., Schlipf, J., 1996. Mater. Sci. Technol. 12, 391–399.
Palumbo, G., Thorpe, S.J., Aust, K.T., 1990. Scr. Metall. Mater. 24, 1347–1350.
Paul, H., Krill, C.E., 2011. Scr. Mater. 65, 5–8.
Pawlow, P., 1909. Z. Phys. Chem. 65, 37.
Pearson, C.E., 1934. J. Inst. Metals 54, 111–116.
Perez, R.A., Nakajima, H., Dyment, F., 2003. Mater. Trans. 44, 2.
Peterlechner, M., Waitz, T., Gammer, G., Antretter, T., 2011. Int. J. Mater. 102, 634–642.
Pfeiler, W., 2007. Alloy Physics. Wiley-VCH.
Prokoshkina, D., Esin, V.A., Wilde, G., Divinski, S.V., 2013. Acta Mater. 61, 5477–5486.
Rentenberger, C., Waitz, T., Karnthaler, H.P., 2004. Scr. Mater. 51, 789–794.
Rodin, A.O., Klinger, L.M., Bokstein, B.S., 2009. Defect Diffusion Forum 289–292, 711–718.
Rösner, H., Boucharat, N., Markmann, J., Padmanabhan, K.A., Wilde, G., 2009. Mater. Sci. Eng. A 525, 102–106.
Rösner, H., Boucharat, N., Padmanabhan, K.A., Markmann, J., Wilde, G., 2010. Acta Mater. 58, 2610–2620.
Rösner, H., Freitag, B., Wilde, G., 2007. Philos. Mag. Letter 87, 341–347.
Rösner, H., Koch, C.T., Wilde, G., 2010. Acta Mater. 58, 162.

Rösner, H., Kübel, C., Ivanisenko, Y., Kurmanaeva, L., Divinski, S.V., Peterlechner, M., Wilde, G., 2011. Acta Mater. 59, 7380–7387.

Rösner, H., Markmann, J., Weißmüller, J., 2004. Philos. Mag. Lett. 84, 321–334.

Rösner, H., Scheer, P., Weissmüller, J., Wilde, G., 2003. Philos. Mag. Lett. 83, 511–523.

Rösner, H., Scherer, T., Wilde, G., 2009. Scr. Mater. 60, 168–171.

Rösner, H., Weissmüller, J., Wilde, G., 2006. Philos. Mag. Lett. 86, 623–632.

Rösner, H., Wilde, G., 2006. Scr. Mater. 55, 119–122.

Raab, G.I., Soshnikova, E.P., Valiev, R.Z., 2004. Mater. Sci. Eng. A 387, 674–677.

Sagmeister, M., Brossmann, U., Landgraf, S., Würschum, R., 2006. Phys. Rev. Lett. 96, 156601.

Sauvage, X., Wilde, G., Divinski, S.V., Horita, Z., Valiev, R.Z., 2012. Mater. Sci. Eng. A 540, 1–12.

Schafler, E., 2011. Scr. Mater. 64, 130.

Schiotz, J., Di Tolla, F.D., Jacobsen, K.W., 1998. Nature 391, 561–563.

Schiotz, J., Jacobsen, K.W., 2003. Science 301, 1357–1359.

Schumacher, S., Birringer, R., Strauss, R., Gleiter, H., 1989. Acta Metall. 37, 2485.

Schäfer, J., Albe, K., 2012a. Scr. Mater. 66, 315.

Schäfer, J., Albe, K., 2012b. Acta Mater. 60, 6076–6085.

Schäfer, J., Stukowski, A., Albe, K., 2011. Acta Mater. 59, 2957–2968.

Semenova, I.P., Valiev, R.Z., Yakushina, E.B., Sahimgareeva, G.H., Lowe, T.V., 2008. J. Mater. Sci. 43, 7354.

Shan, Z.W., Stach, E.A., Wiezorek, J.M.K., Knapp, J.A., Follstaedt, D.M., Mao, S.X., 2004. Science 305, 654–657.

Sheng, H.W., Ren, G., Peng, L.M., Hu, Z.Q., Lu, K., 1996. Philos. Mag. Lett. 73, 179–186.

Sheng, H.W., Lu, K., Ma, E., 1998. Acta Mater. 46, 5195–5205.

Shirinyan, A., Gusak, A., Wautelet, M., 2005. Acta Mater. 53, 5025–5032.

Song, X.J., Liu, X.M., Zhang, J.X., 2006. J. Am. Ceram. Soc. 89, 494–500.

Srinivasan, S., Ranganathan, S., 2004. India's Legendary Wootz Steel—an Advanced Material of the Ancient World. Tata Steel, India.

Stahl, B., Gajbhiye, N.S., Wilde, G., Kramer, D., Ellrich, J., Ghafari, M., Hahn, H., Gleiter, H., Weißmüller, J., Würschum, R., Schloßmacher, P., 2002. Adv. Mater. 14, 24–27.

Stender, P., Balogh, Z., Schmitz, G., 2011. Phys. Rev. B 83, 1–4.

Straumal, B., 2001. Int. J. Inorg. Mater. 3, 1113–1115.

Straumal, B.B., Kogtenkova, O., Zieba, P., 2008. Acta Mater. 56, 925–933.

Sun, Y., Bell, T., 1991. Mater. Sci. Eng. A 140, 419–434.

Surholt, T., Herzig, C., 1997. Acta Mater. 45, 3817.

Suzuki, A., Mishin, Y., 2005. Mater. Sci. Forum 502, 157–162.

Suzuoka, T., 1964. J. Phys. Soc. Jpn. 19, 839.

Swaminathan, P., Sivaramakrishnan, S., Palmer, J.S., Weaver, J.H., 2009. Phys. Rev. B 79, 144113.

Takeda, M., Suzuki, J., 1996. J. Opt. Soc. Am. 13, 1495–1500.

Tanaka, T., 2010. Mater. Sci. Forum 653, 55–75.

Thompson, A.W., 1977. In: Thompson, A.W. (Ed.), Work Hardening in Tension and Fatigue.

Tong, W.P., Tao, N.R., Wang, Z.B., Lu, J., Lu, K., 2003. Science 299, 686–688.

Traussnig, T., Topolovec, S., Nadeem, K., Szabo, D.V., Krenn, H., Würschum, R., 2011. Phys. Status Solidi Rapid Res. Lett. 5, 150–152.

Trelewicz, J.R., Schuh, C.A., 2007. Acta Mater. 55, 5948–5958.

Tsuchiya, K., Hada, Y., Koyano, T., Nakajima, K., Ohnuma, M., Koike, T., Todaka, Y., Umemoto, M., 2009. Scr. Mater. 60, 749–752.

Valiev, R.Z., 2004. Nat. Mater. 3, 511–516.

Valiev, R.Z., Alexandrov, I.V., Zhu, Y.T., Lowe, T.C., 2002. J. Mater. Res. 17, 5–8.

Valiev, R.Z., Gertsman, V.Y., Kaibyshev, O.A., 1986. Phys. Status Solidi 97, 11–56.

Valiev, R.Z., Islamgaliev, R.K., Alexandrov, I.V., 2000. Prog. Mater. Sci. 45, 103–189.

Valiev, R.Z., Islamgaliev, R.K., Stolyarov, V.V., Mishra, R.S., Mukherjee, A.K., 1998. Mater. Sci. Forum 269–273, 969.

Valiev, R.Z., Kozlov, E.V., Ivanov, Y.F., Lian, J., Nazarov, A.A., Baudelet, B., 1994. Acta Metall. Mater. 42, 2467–2475.

Valiev, R.Z., Langdon, T.G., 2006. Prog. Mater. Sci. 51, 881–981.

Vallée, R., Wautelet, M., Dauchot, J.P., Hecq, M., 2001. Nanotechnology 12, 68–74.

Van Leeuwen, B.K., Darling, K.A., Koch, C.C., Scattergood, R.O., Butler, B.G., 2010. Acta Mater. 58, 4292–4297.

Van Swygenhoven, H., Derlet, P.M., 2001. Phys. Rev. B 64, 224105.

Van Swygenhoven, H., Derlet, P.M., Hasnaoui, A., 2002. Phys. Rev. B 66, 024101.

Van Swygenhoven, H., Derlet, P.M., Hasnaoui, A., 2003. Adv. Eng. Mater. 5, 345.

Van Swygenhoven, H., Derlet, P.M., Hasnaoui, A., 2004. Acta Mater. 52, 2251–2258.

Van Swygenhoven, H., Farkas, D., Caro, A., 2000. Phys. Rev. B 62, 831–838.

Verhoeven, J.D., Pendray, A.H., Dauksch, W.e, 1998. J. Metals 50, 58–64.

Vukusic, P., Sambles, J.R., 2003. Nature 424, 852–855.

Wahl, P., Traussig, T., Landgraf, S., Jin, H.J., Weissmüller, J., Würschum, R., 2010. J. Appl. Phys. 108, 073706.

Waitz, T., 2005. Acta Mater. 53, 2273–2283.

Waitz, T., Antretter, T., Fischer, F.D., Simha, N.K., 2007. J. Mech. Phys. Solids 55, 419–444.

Waitz, T., Spisak, D., Hafner, J., Karnthaler, H.P., 2005. Europhys. Lett. 71, 98–103.

Wang, Z.B., Lu, K., Wilde, G., Divinski, S.V., 2010. Acta Mater. 58, 2376.

Watanabe, T., Kitamura, S., Karashima, S., 1980. Acta Metall. 28, 455–463.

Wautelet, M., Dauchot, J., Hecq, M., 2000. Nanotechnology 11, 6.

Wautelet, M., Dauchot, J.P., Hecq, M., 2003. J. Phys. Condens. Matter 15, 3651–3655.

Weissmüller, J., 1993. Nanostruct. Mater. 3, 261–272.

Weissmüller, J., Bunzel, P., Wilde, G., 2004. Scr. Mater. 51, 813–818.

Weissmüller, J., Lemier, C., 1999. Phys. Rev. Lett. 82, 213–216.

Weissmüller, J., Viswanath, R.N., Kramer, D., Zimmer, P., Würschum, R., Gleiter, H., 2003. Science 300, 312–315.

Whipple, R.T.P., 1954. Philos. Mag. 45, 1225.

Wilde, G., 2006. Surf. Interface Anal. (SIA) 38, 1047–1062.

Wilde, G., Boucharat, N., Dinda, G.P., Rösner, H., Valiev, R.Z., 2006. Mater. Sci. Forum 503–504, 425–432.

Wilde, G., Boucharat, N., Hebert, R.J., Rösner, H., Tong, S., Perepezko, J.H., 2003. Adv. Eng. Mater. 5, 125–130.

Wilde, G., Bunzel, P., Rösner, H., Weissmüller, J., 2007. J. Alloys Compd. 434–435, 286–289.

Wilde, G., Ribbe, J., Reglitz, G., Wegner, M., Rösner, H., Estrin, Y., Zehetbauer, M., Setman, D., Divinski, S., 2010. Adv. Eng. Mater. 12, 758–764.

Winning, M., Gottstein, G., Shvindlerman, L.S., 2001. Acta Mater. 49, 211–219.

Winning, M., Gottstein, G., Shvindlerman, L.S., 2002. Acta Mater. 50, 353–363.

Witkin, D.B., Lavernia, E.J., 2006. Prog. Mater. Sci. 51, 1–60.

Wolf, D., Yamakov, V., Phillpot, S.R., Mukherjee, A.K., Gleiter, H., 2003. Acta Mater. 53, 1.

Wronski, C.R.M., 1967. Br. J. Appl. Phys. 18, 1731.

Würschum, R., Herth, S., Brossmann, U., 2003. Adv. Eng. Mater. 5, 365.

Würschum, R., Michel, T., Scharwächter, P., Frank, W., Schaefer, H.E., 1999. Nanostruct. Mater. 12, 555–558.

Würschum, R., Reimann, K., Gruss, S., Kubler, A., Scharwaechter, P., Frank, W., Kruse, O., Carstanjen, H.D., Schaefer, H.E., 1997. Philos. Mag. B 76, 407.

Xing, L.Q., Eckert, J., Löser, W., Schultz, L., Herlach, D.M., 1999. Philos. Mag. A 79, 1095–1108.

Yamakov, V., Wolf, D., Phillpot, S.R., Gleiter, H., 2002. Acta Mater. 50, 5005.

Yamakov, V., Wolf, D., Phillpot, S.R., Gleiter, H., 2003. Acta Mater. 51, 4135.

Yamakov, V., Wolf, D., Salazar, M., Phillpot, S.R., Gleiter, H., 2001. Acta Mater. 49, 2713.

Yamakov, V., Wolf, D., Salazar, M., Phillpot, S.R., Gleiter, H., 2002. Acta Mater. 50, 61.

Zehetbauer, M., Schafler, E., Ungar, T., 2005. Z. Metallkd. 96, 1044.

Zehetbauer, M., Stüwe, H.P., Vorhauer, A., Schafler, E., Kohaut, J., 2003. Adv. Eng. Mater. 5, 330–337.

Zhang, D.L., Cantor, B., 1991. Acta Metall. Mater. 39, 1595.

Zhang, M., Efremov, M.Y., Schiettekatte, F., Olson, E.A., Kwan, A.T., Lai, S.L., Wisleder, T., Greene, J.E., Allen, L.H., 2000. Phys. Rev. B 62, 10548–10557.

Zhang, H.W., Hansen, N., 2007. J. Mater. Sci. 42, 1682–1693.

Zhang, K., Weertman, J.R., Eastman, J.A., 2005. Appl. Phys. Lett. 87, 061921.

Zhao, B., Verhasselt, J.C., Shvindlerman, L.S., Gottstein, G., 2010. Acta Mater. 58, 5646–5653.

Zhao, M., Zhou, X.H., Jiang, Q., 2001. J. Mater. Res. 16, 3304–3308.

Biography

Prof. Gerhard Wilde obtained his Ph.D. degree at the Technical University of Berlin in 1997. From 1997 to 1999 he worked as an Alexander von Humboldt Fellow at the University of Wisconsin–Madison. From 1999 until 2006 he worked as a group leader at the Institute of Nanotechnology of the Karlsruhe Research Center in Germany. In 2004 he was appointed as interim professor at Saarland University and in 2006 he was appointed as full professor at the Department of Physics of the University of Muenster in Germany. He is a scientific member of Center for Nanotechnology. He is holding several honorary Professorships at Hyderabad and Chennai in India and at Shenyang and Shanghai in China. His major research interests include phase transformations, metallic glasses and nanostructured and nanocrystalline materials and interfaces. He has authored more than 230 scientific publications and delivered more than 140 plenary, keynote and invited lectures.

27 Computational Metallurgy

Long-Qing Chen and Yijia Gu, Department of Materials Science and Engineering, The Pennsylvania State University, University Park, PA, USA

27.1 Introduction

Computational materials science involves the application of computers to understanding and predicting the structures and properties of materials and their relationships to processing conditions, based on fundamental physics, thermodynamics, kinetics, mechanics, and numerical algorithms. It has experienced remarkable growth over the past 30 years across the disciplines of materials science and engineering, physics, chemistry, and mechanics.

Computer simulation and modeling allows one to probe the fundamental electronic/atomic interactions that determine the relative stability and physical properties of various thermodynamic states and crystal structures of a material. It provides insights on the thermodynamic driving forces and kinetic mechanisms leading to the observed microstructures under a given processing condition and measured responses under external loads. Computer simulations and modeling is playing an increasingly important role not only in materials research but also in materials design.

The discussion of this article will be focused on computational materials science and engineering approaches applied to understanding the process–microstructure–property relationships of metallic material systems, or computational metallurgy (Eberhart, 1994). Different processing conditions, specified by temperature, composition, cooling rate, and external loads such as stress, may lead to dramatically different microstructures and thus significant variations in properties. A microstructure is referred to an inhomogeneous distribution of structural features which include phases of different compositions and/or crystal structures, grains of different orientations, domains of different structural variants or magnetizations, as well as structural defects such as interphase boundaries, grain boundaries, domain walls, cracks, surfaces, and dislocations. The length scales of these structural features range from angstroms (crack tips) or nanometers (interfacial width, dislocation core, nuclei, small domains and grains) to hundreds of microns (large grains in polycrystalline solids). The emphasis of this chapter will be on modeling and predicting the stability (thermodynamics) and the temporal/spatial evolution (kinetics) of microstructures, or computational microstructural science, in metallic solids.

The thermodynamic stability of a microstructure within the classical thermodynamic description is determined by the total Gibbs free energy of a system, G, which may include the bulk chemical free energy density of each individual phases, the energies of defects such as grain boundaries, surfaces, interfaces, and dislocations, as well as the electrostatic, elastic, and magnetic interaction energies,

$$G = \sum_{i=1}^{p} V_i g_i + \sum_{i=1}^{q} S_i \gamma_i + \sum_{i=1}^{m} L_i E_i^d + E_{\text{elec}} + E_{\text{elast}} + E_{\text{mag}} \tag{1}$$

where g_i is bulk chemical free energy density of ith phase with volume V_i, γ_i is the specific interfacial energy of ith interface (grain boundary, surface or interphase boundary) with interfacial area S_i, E_i^d is the dislocation core energy per unit length of ith dislocation with length L_i, and E_{elec}, E_{elast}, and E_{mag} are the total electrostatic, elastic and magnetic energies, respectively.

A microstructure evolves to minimize its total free energy during materials processing or in service at sufficiently high temperatures at which atoms can migrate or change positions from one crystal lattice site to another. Microstructure evolution may take place as a result of phase changes or transformations driven by the reduction in the bulk chemical free energy (the first term in Eqn (1)). A phase transformation may lead to a change in the number of phases, for example a phase separation or decomposition reaction, or to a simple change of crystal structure for one of the phases to another with lower bulk chemical free energy density. A phase separation or decomposition reaction requires compositional redistribution within a microstructure while a simple change of crystal structure for one of the existing phases takes place at a fixed composition and therefore does not require long-range atomic diffusion.

Microstructures may coarsen, that is the overall spatial scale of domains, particles, or grains in a microstructure increases as a function of time, through interface motion to reduce the total interfacial energy (the second term in Eqn (1)). Similar to phase transformations, interface motion during microstructure coarsening can also be accomplished by long-range atomic diffusion over phase domains, or short-range diffusion across an interface, or even a small atomic displacement without atomic diffusion jumps.

Microstructure may undergo changes driven by the reduction of elastic, electrostatic or magnetostatic interaction energies or applied fields such as stress, electric, and magnetic fields. For example, a ferroelastic domain wall, or a twin wall, can be moved by an applied stress. Under a mechanical load, dislocations may also be generated and moved, that is a system may undergo plastic deformation, to release the strain or strain energy in a crystal. An applied electric field may lead to mass transfer and plastic deformation due to electromigration. Since elastic, electrostatic, and magnetic self-interactions are long-range interactions in space, they may lead to self-organized microstructure patterns such as the alignment of particles or structural variants along certain crystallographic directions or the formation of interfaces along particular lattice planes called habit planes (Christian, 2002; Khachaturian, 1983).

The primary goal for computational metallurgy and design is to understand the fundamental thermodynamic forces and kinetic mechanisms leading to the observed microstructures, identify the optimum microstructures possessing the most desirable properties, and thus predict desired processing conditions to optimize microstructures. To predict the microstructure evolution of a materials system, it requires not only the information on the thermodynamic driving forces (Eqn (1)) but also kinetic parameters or the rate constants such as atomic diffusion mobility, interfacial mobility, rate constants for chemical reactions or phase transformations (Christian, 2002; Balluffi et al., 2005). Therefore, one of the major computational materials science efforts is on computing the quantities that entered the total Gibbs free energy expression (1), for example the various enthalpic and entropic contributions to the bulk free energy density for an individual phase, the interfacial (grain boundary, surface, and interphase boundary) energies, and lattice parameters and elastic constants of individual phases for obtaining the elastic energy density. Computing the kinetic parameters is typically more challenging than obtaining thermodynamic information since it usually involves thermodynamically unstable or transition states. As a result, the amount of literature on kinetic parameter determination is significantly less than thermodynamic calculations. Nevertheless, significant progresses have been made in obtaining atomic diffusion mobility using first-principles calculations (see e.g. Van der Ven and Ceder, 2005; Mantina, 2008; Mantina et al., 2009).

Another major effort in computational metallurgy is to design and develop computational approaches to predict the microstructure evolution of an alloy under a given processing condition. Several computational methods ranging from direct atomistic/electronic structure level simulation techniques to continuum approaches exist. However, due to the complexity of microstructures and their evolution kinetics as well as the temporal and spatial scales that are involved, it is unrealistic to expect a single computational method to directly predict the microstructure evolution of a system and its overall responses under load starting with arbitrarily specified initial state, temperature, composition and time. This calls for multiscale modeling. One approach is to integrate electronic structure calculations, atomistic simulations and finite-element methods in a single simulation (Broughton et al., 1999; Curtin and Miller, 2003; Tadmor et al., 1996). However, the type of microstructure problems that can be handled is usually very limited, for example to systems with a single defect such as a crack or a grain boundary. A more practical multiscale approach is to use a combination of different computational techniques at different length and timescales through information passing from one scale to another. For example, one employs electronic/atomistic calculations to obtain the thermodynamic and kinetic information and then use it as input to microstructure evolution models (Vaithyanathan et al., 2002; Hoyt et al., 1999). This approach will be the focus in the discussion of this chapter.

Since there are several different computational techniques at each length scale, this chapter will be presented as an overview rather than an in-depth discussion of a particular method. The rest of the chapter is organized as follows. Section 27.2 discusses the possibility of using electronic/atomistic level first-principles calculations, in combination with experimental measurements, to determine the structures and thermodynamic properties of individual structural features in a microstructure and the kinetic properties of atoms and interfaces, namely atomic and interface mobility. Computational microstructure evolution models will be presented in Section 27.3. Section 27.4 briefly outlines the computational methods to obtain the responses of a microstructure under an applied load including domain switching and the effective properties of a microstructure such as effective elastic modulus, transport properties, and plastic deformation.

27.2   Structures and Properties of Single Crystals and Interfaces

A microstructure may be considered as a composite of individual single crystallites of different crystal structures/orientations/compositions and the interfaces between them. To understand and predict the stability of a microstructure requires information on the structures and properties of single crystals and interfaces, which include the lattice parameter dependence on temperature and composition, the bulk chemical free energy density of each phase as a function of temperature, pressure, and composition, the interfacial energy and mobility, elastic constants, the diffusional mobility of atomic species, as well as electrical and magnetic properties if electrostatics and magnetic materials are involved.

27.2.1   0 K Thermodynamic Properties of Elemental Crystals or Stoichiometric Compounds

The total energy of a perfect crystal at 0 K represents the ground-state stability and can be used as a reference value for obtaining the temperature dependence of thermodynamic properties. First-principles methods based on the density functional theory (DFT) (Hohenberg and Kohn, 1964; Kohn and Sham, 1965) have now become a mature computational materials technique that allows one to determine the ground-state phase stability of perfect crystals, such as elemental metals and metallic compounds. It is particularly appealing since it, in principle, requires only the knowledge of the atomic species and their positions as input. According to DFT, the total energy of a collection of atoms, E, is a functional of the spatially dependent electron density distribution, $n(r)$, that is

$$E_{\text{tot}} = E[n(r)] = E_k[n(r)] + E_e[n(r)] + E_{xc}[n(r)] \tag{2}$$

where E_k is the noninteracting kinetic energy, E_e is the electrostatic energy which arises from the Coulombic interactions among electrons and atomic nuclei, and E_{xc} is the exchange–correlation energy which includes all other quantum mechanical contributions that are not included in the first two terms. The exact form of the exchange–correlation functional form is not known, and hence a number of functionals have been proposed to approximate it. Two well-known approximations that implemented in essentially all DFT codes are the local density approximation (LDA) in which the exchange–correlation energy depends only on the local electronic density (Kohn and Sham, 1965) and the generalized gradient approximation (GGA) in which the exchange–correlation energy depends not only on the local electron density but also its gradient (Perdew and Yue, 1986; Perdew et al., 1996a).

For periodic systems like a perfect crystal, the plane wave basis set is usually used to represent electron wave functions, and the pseudopotential method (Schwerdtfeger, 2011; Pickett, 1989) is employed to treat the core electrons to dramatically enhance the computational efficiency. The accuracy of pseudopotential method can be determined by comparing the results from the full-potential line-arized augmented plane wave method (Wimmer et al., 1981; Blaha et al., 1990; Jansen and Freeman, 1984). The total energy at the ground state, E_0, of a given crystal with specified atomic positions can then be obtained by minimizing the total energy functional with respect to the electron density distribution. The electronic density which produces the minimum energy is the ground-state electron density, $n_0(r)$.

With the availability of several software packages (Hoyt et al., 1999), computation of the energies and properties of elemental and stoichiometric compounds at 0 K has now become almost routine (Curtarolo et al., 2003; Wang et al., 2004b; Curtarolo et al., 2005). The quantities that can be calculated include the total energies as a function of volume for different crystal structures of an element or a stoichiometric compound from which the equilibrium lattice parameters and the relative stabilities can be determined. The curvature of the total energy vs volume at the equilibrium volume can be related to the bulk modulus of a single crystal. The full set of elastic constants can be obtained by computing the total energy as a function of a given strain component and then measuring the curvature of the energy vs strain curve at zero strain or simply by applying a small strain and obtaining the corresponding stress component assuming the linear Hooke's law (Wang et al., 2010). The total energy as a function of strain, that is the deformation energy, is useful for modeling microstructure evolution process such as deformation twinning (Heo et al., 2011). A similar energy landscape, the gamma surface, can be generated for dislocation microstructure evolution modeling (see e.g. Woodward, 2005).

27.2.2   Finite-Temperature Thermodynamic Properties of Elemental Solids and Stoichiometric Compounds

While obtaining total energy and properties of ground state using DFT calculations is relatively straightforward for many cases, accurate determination of finite-temperature structures and thermodynamic properties is more challenging. At finite temperatures, the thermodynamic quantity of interest is the Helmholtz (F) or Gibbs (G) free energy rather than the total energy itself, that is

$$F(T, V) = E - ST \quad \text{or} \quad G(T, p) = E - ST + pV \tag{3}$$

where E is the internal energy, S is the entropy, p is pressure, T is temperature, and V is volume. Since we are mostly dealing with ambient pressure and solid state in practice, the contribution of the pV term is usually not significant, so theoretically, we can approximate the Helmholtz free energy as the Gibbs free energy.

In principle, there are two approaches that one can obtain the free energy of a pure element or a stoichiometric compound as a function of temperature. The first is to perform molecular dynamics (MD) (Alder and Wainwright, 1959; Rahman, 1964; Ryckaert et al., 1977) simulations of a crystal at finite temperatures by solving the Newton's equations of motion for each atom. The directly available structure and properties from an MD simulation include the internal energy or enthalpy and volume as a function of temperature, from which the heat capacity and thermal expansion coefficient can be derived based on basic thermodynamic relations. One of the disadvantages is the fact that it requires reliable interatomic potentials for the case of classical/semiclassical MD simulations and large computational memory and time for ab initio MD simulations. More importantly, the free energy as a function of temperature is not directly available and can only be obtained through a thermodynamic integration which requires a reference crystal, such as a harmonic crystal, whose free energy is known. An interesting recent application of ab initio MD was to obtain the free energy as a function of a structural order parameter describing a phase transition at a given temperature, for example the fcc to bcc transition in W (Ozolins, 2009).

The second approach to obtaining the free energy as a function of temperature is to combine first-principles DFT calculations and lattice dynamics or phonon theory in the harmonic approximation (HA) or quasi-harmonic approximation (QHA). Using either the supercell method or linear response theory (Baroni et al., 2001; Baroni and Moroni, 2005; Van De Walle and Ceder, 2002; Van De Walle et al., 2002), which are incorporated in most of the DFT codes, one can obtain vibrational frequencies, $\omega_j(\mathbf{q}, V)$ where $j = 1, \ldots, 3N$ with N the total number of atoms in a unit cell at wave vector $\mathbf{q}$. In the linear-response method, the normal frequencies (i.e., phonon frequencies) associated with microscopic displacements of atoms in a crystal are calculated by means of dynamical matrix obtained through the DFT calculations. The supercell method adopts the frozen phonon approximation through which the changes in forces are calculated in the real space by displacing the atoms from their equilibrium positions. According to basic lattice dynamics theory (Van De Walle and Ceder, 2002), the vibrational contribution to the free energy of the lattice ions is given by

$$F_{\text{vib}}(T, V) = k_B T \sum_{\mathbf{q}} \sum_j \ln\left\{ 2 \sinh\left[\frac{\hbar \omega_j(\mathbf{q}, V)}{2 k_B T} \right] \right\} \tag{4}$$

Therefore, the total free energy at a given temperature is given by

$$F(T, V) = E(v) + F_{\text{vib}}(T, V) \tag{5}$$

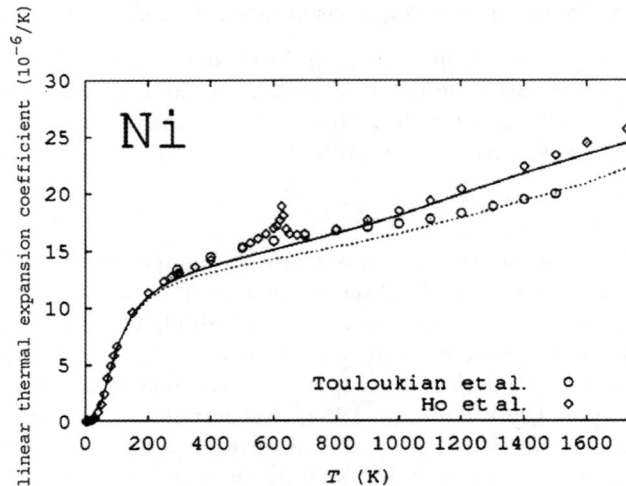

Figure 1 Calculated linear thermal expansion coefficient of Ni with and without considering the thermal electronic contribution is plotted using solid and dash lines. The open circles and open diamonds are values from Touloukian and Ho, respectively (Wang et al., 2004c).

where $E(V)$ is the static total energy of a crystal with volume V. Since V is a function of temperature, the static energy E is also an implicit function of temperature. In cases where the thermal electron contribution to the free energy is important, one may also need to add the electron free energy.

The heat capacity as a function of temperature can be directly evaluated from the free energy as a function of temperature using normal thermodynamic relations. To calculate the thermal expansion as a function of T, one approach is to obtain the free energy $F(T,V)$ as a function of lattice parameter, and then identify the volume that minimizes the free energy as the equilibrium volume at a given temperature (Wang et al., 2004c). **Figure 1** shows an example of the linear thermal expansion coefficient of Ni obtained using this approach.

27.2.3   Finite-Temperature Thermodynamic Properties of Binary Solid Solutions

Going from pure elements or stoichiometric ordered compounds to solid solutions is a giant step and poses another significant challenge to DFT calculations. Even though a crystalline solution is homogeneous macroscopically, displaying the same symmetry as the underlying crystalline lattice, an instantaneous atomic configuration within a solution is microscopically inhomogeneous, and thus has a lower symmetry.

One strategy to model such a microscopically inhomogeneous but macroscopically homogeneous atomic configuration is to employ a sufficiently large supercell so that the periodicity artificially imposed by the periodic boundary conditions has minimal effect on the computed atomic structures and properties. Each lattice site within the supercell is randomly assigned either A or B atoms according to the desired composition. To reduce the number of atoms in a supercell, the so-called special quasirandom structures (SQS) have been proposed (Zunger et al., 1990; Wei et al., 1990). The SQS are specially designed small-unit-cell periodic lattice structures with the nearest-neighbor pair and multisite correlation functions corresponding to a random solid solution. The properties such as the volume and total energy of SQS can be used to approximate those of a random solution at the same composition. For example, SQS have been generated for various crystal structures including fcc (Zunger et al., 1990), bcc (Jiang et al., 2004), hcp (Shin et al., 2006), and $L1_2$ (Wang et al., 2005b), using the Alloy Theoretic Automated Toolkit (ATAT) code (Van De Walle et al., 2002). However, it should be pointed out that correlations beyond a certain distance of small supercell SQS structures will be different from those of random alloys. Therefore, only a few specific compositions are possible. Furthermore, the information about the vibrational and configuration disorder to the total free energy is not directly available. Although it is possible to carry out the phonon calculations of SQS structures to obtain the vibrational free energy (Van De Walle et al., 1998), evaluation of the configuration entropy of a nonideal solid solution is not trivial. Finally, real solutions always contain short-range order, and hence SQS structures may not be good approximations.

Another approach to dealing with solid solutions is the Korringa–Kohn–Rostoker coherent potential approximation (Stocks et al., 1978) which assumes that a lattice site is occupied by the compositional average of A and B atoms for a binary system. It allows one to compute the thermodynamic properties of a solution throughout the composition range rather than a discrete few in SQS. However, the mean-field nature of treating the atoms at each lattice site makes it difficult to treat the local environments within the solution (Kissavos et al., 2005). The free energy of a solid solution, similar to SQS, is not directly available.

Finally, the cluster expansion method (Sanchez et al., 1984; Fontaine, 1994) in combination with Monte Carlo (MC) simulations (Binder, 1979) and thermodynamic integration is a common approach for obtaining the free energy as a function of composition (Wolverton and Zunger, 1995; Ozolins et al., 1998). In this approach, the total energies of many small-unit-cell ordered structures are first obtained from first-principles calculations. The Hamiltonian for an atomic configuration is written in terms of atomic clusters defined on an Ising lattice. The coefficients in the cluster expansion are then fitted to the total energies of ordered structures. With the cluster expansion coefficients, one can perform MC simulations to obtain the energy or enthalpy of formation of a solid solution with a given composition. The entropy and free energy of the solution can be obtained by thermodynamic integration using the ideal solution at infinite temperature as the reference (Wolverton and Zunger, 1995).

27.2.4 Finite-Temperature Thermodynamic Properties of Multicomponent Solution Phases

Obtaining the structure and properties of multicomponent solutions directly from first-principles calculation is currently not possible. For a general multicomponent solution, the free energy as a function of composition can only be obtained through empirical approaches such as the calculation of phase diagrams (CALPHAD) approach (Kaufman and Bernstein, 1998; Saunders and Miodownik, 1998). The CALPHAD approach starts with the evaluation of the thermodynamic descriptions of unary and binary systems based on thermochemical data either from experiments or first-principles calculations or now more commonly a combination of both (Chen et al., 2001; Wolverton et al., 2002; Liu et al., 2004). It follows the classical thermodynamic treatment which typically writes the molar free energy of a solution in terms of three contributions, that is

$$G_m = G_m^{\text{ref}} + G_m^{\text{id}} + G_m^{\text{xs}} \tag{9}$$

where G_m^{ref} represents the reference free energy of a phase. For example, for the simple case of a random solution, G_m^{ref} is the sum of the chemical potentials of pure species weighted by their compositions at a given temperature. G_m^{id} is the free energy of an ideal solution, and G_m^{xs} is the excess free energy due to nonideality of a solution. Clearly, the first two terms are fixed by composition and the chemical potentials of pure species or the molar free energies of compounds, and hence the main effort involved in a database development is to describe the G_m^{xs} term using a particular mathematical and thermodynamic model and to determine the corresponding parameters (Saunders and Miodownik, 1998). For metallic solid solutions, a commonly used model is the sublattice model (Sundman and Agren, 1981; Andersson et al., 1986) which has been employed in the majority of thermodynamic assessments of existing databases. It should be mentioned that when evaluating model parameters, it is often very difficult to find a set of model parameters for all individual phases to represent all the information in the system with high accuracy, particularly for systems with several solution phases. Furthermore, it should be cautioned that the nonequilibrium segments of the CALPHAD free energies are not necessarily unique or even correct.

27.2.5 Lattice Parameters and Molar Volumes of Binary Solutions

Lattice parameter as a function of temperature and composition is an important piece of information for the modeling of microstructure evolution. Different regions in a solid microstructure impose mechanical constraint on each other, and thus when the lattice parameters change in any region due to temperature or composition variations or due to a structural phase transformation, local elastic strains and stresses will develop, which can profoundly affect the local morphologies and spatial distributions of neighboring domains or particles and their temporal evolution (Khachaturian, 1983).

Similar to the temperature dependence of lattice parameter of a material characterized by its thermal expansion coefficient, the composition dependence of lattice parameter is described by a composition expansion coefficient, $e_0 = 1/a(da/dc)$. A constant or composition-independent e_o is typically referred to as the Vegard's law (Vegard, 1921) which is a reasonable approximation for very dilute solutions. The composition expansion coefficient can be estimated using first-principles calculations by comparing the lattice parameters of a supercell with a solute atom and one without (Wang et al., 2004a). For example, the lattice parameters of binary fcc-Ni solutions with solute atoms Al, Co, Cr, Hf, Mo, Nb, Re, Ru, Ta, Ti and W were first obtained (Wang et al., 2007a; Wang, 2006). By comparing the lattice parameters of pure fcc-Ni supercell from first-principles calculations, the linear composition coefficients of fcc-Ni lattice parameter for these solutes can be determined.

The lattice parameter measurements in practical alloys always show some deviations from Vegard's law, particularly for concentrated solutions, that is because the composition expansion coefficient is also a function of composition itself. Modeling of lattice parameters for practical alloys requires a more phenomenological approach such as CAPHPAD (Sundman and Agren, 1981). For example, one may add an excess contribution to describe the deviation of the lattice parameter from the Vegard's law,

$$a = \sum_i x_i^0 a_i + a^{ex} \tag{10}$$

where x_i^0 is the composition of species i in the solution, a_i is the lattice parameter of pure i assuming all species having the same crystal structures, and a^{ex} is the excess due to solute interactions. **Figure 2** shows an example of assessed lattice parameters of γ' and γ phases in the Ni–Al system (Wang et al., 2004a).

(a)

(b)

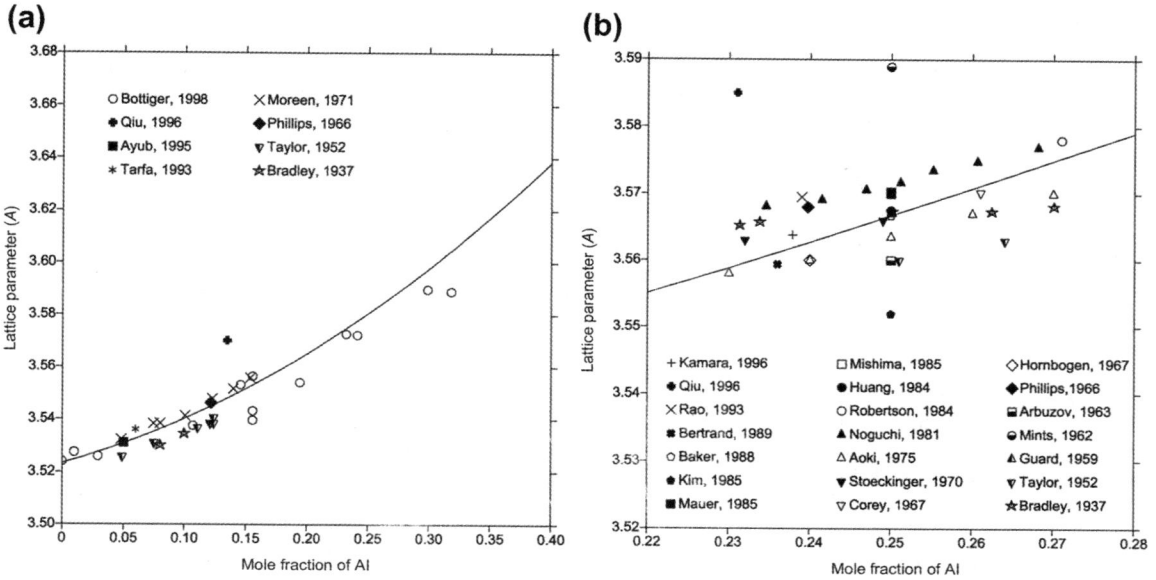

Figure 2 Room-temperature lattice parameter of (a) γ phase and (b) γ' phase in the Ni–Al system (Wang et al., 2004a).

27.2.6 First-Principles Calculations of Interfacial Energy

Interfacial energy and its anisotropy are important in determining not only the morphologies of a particle but also the driving force for microstructure coarsening. Experimentally, it is difficult to directly measure the interfacial energies along different orientations, and computationally, it is more expensive than calculations of bulk properties. Static lattice calculations at 0 K started in the 1960s by using pairwise interatomic potentials with the potential parameters fitted to the lattice parameters and cohesive energies of perfect elemental single crystals from experiments. There have been extensive studies on atomistic structures and energies of surfaces and interfaces, mainly grain boundaries, in metals using empirical or semiempirical potentials and more recently first-principles calculations.

Typically, the surface, grain boundary, and interfacial energies (σ) are obtained by comparing the total energy of a supercell containing two surfaces, grain boundaries or interfaces and that of a perfect crystal or perfect crystals with the same number of atoms, i.e.

$$\sigma = (E_{tot} - E_1 - E_2)/(2A) \tag{11}$$

where E_{tot} is the total energy of the configuration, E_1 and E_2 are the energies of the first and second perfect crystals, respectively, and A is the area of the interface. This calculation is relatively straightforward at 0 K although care has to be taken with regard to the possible interactions between the two interfaces in a computational cell as well as the separation between strain energy and interfacial energy due to the change in chemical environment at the interfaces depending on how the cell dimension relaxation is carried out. At finite temperatures, the change in grain boundary energy with temperature and the entropy contribution, mainly the vibrational entropy contribution, has to be included to obtain

Table 1   Interfacial energies from first-principles.

Material systems	Interfaces (planes)	Interfacial energy (mJ/m²)
Al/Al₃Sc (Asta et al., 1998)	{100}	192
	{111}	226
γ'-Ni₃Al/γ-Ni (Costa e Silva et al., 2007)	{100}	39.6
	{110}	63.8
β''-Mg₅Si₆/α-Al (Wang et al., 2007b)	$(\bar{3}20)_{Al}\|\|(001)_{\beta''}$	100
	$(130)_{Al}\|\|(100)_{\beta''}$	124
	$(001)_{Al}\|\|(010)_{\beta''}$	~300
bcc Ta (Mishin and Lozovoi, 2006)	{211} twin boundary	217
bcc Mo (Ogata et al., 2005)	{211} twin boundary	607
fcc Al (Ogata et al., 2005)	{111} twin boundary	60
fcc Cu (Ogata et al., 2005)	{111} twin boundary	21

the interfacial free energy as a function of temperature. For more details, please see a recent review (Mishin et al., 2010).

It should be pointed out that essentially all the existing calculations are limited to low-index surfaces and high-angle grain boundaries with short periodicity in simple metals due to the applications of periodic boundary conditions and the limit on the number of atoms in a computational cell. Energies of interfaces between different phases, for example between a precipitate phase and a matrix phase, or the interfacial energy typically require first-principles methods due to the difficulty of constructing reliable interatomic potentials involving several different atomic species and different phases. Similar to surfaces and grain boundaries, only special interphase boundaries involving low-index atomic planes for both phases are possible. Finally, all the existing atomistic/electronic structure calculations of surfaces and interfaces assume either pure metals or stoichiometric ordered compounds or including a single solute or point defect. It is technically difficult to include solution phases with concentrated solutes, that is the effect of compositions. Despite these limitations, atomistic/electronic calculations are useful to provide the thermodynamic data for interfaces since for most cases it is very difficult to measure the interfacial energies directly in experiments. **Table 1** lists a few examples of interfacial energies from first-principles calculations.

Another useful approach is to use the information about the thermodynamics of a system and the interfacial width to estimate the interphase boundary energies with the following expression,

$$\sigma = 1/2l\Delta f_{max} \tag{12}$$

where l is the interfacial width and Δf_{max} is the maximum difference between the nonequilibrium free energy as a function of composition and the equilibrium free energy of a two-phase mixture (**Figure 3**).

27.2.7 Atomic Diffusion Coefficients

Many of the processes in materials, including majority of microstructure evolution processes, are diffusion controlled, and thus the knowledge of diffusion coefficients in a given system is critical for the

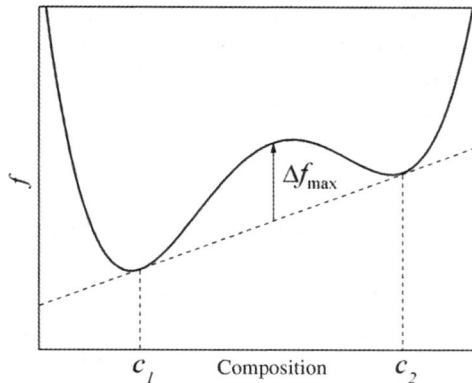

Figure 3 The schematic of free energy as a function of composition. Δf_{max} is the maximum difference between the nonequilibrium free energy and the equilibrium free energy (the dashed common tangent).

quantitative prediction of the rate of processes and kinetics of microstructure evolution. Methods of computing diffusion coefficients can be roughly separated into three categories: (1) direct MD simulations, (2) a combination of first-principles calculations and transition state theory (TST), and (3) a combination of first-principles calculations, rate theory, and MC simulations.

In an MD simulation, the movements of atoms are followed deterministically according to the Newton's equation of motion (Allen and Tildesley, 1989). From the positions of atoms as a function of time, one can calculate the displacements of all atoms, and thus the mean square displacements. The atomic diffusion coefficients can then be related to the mean square displacements through the relation

$$D = \overline{R^2}/2dt \tag{13}$$

where D is the diffusion coefficient, d is the dimensionality of a system, t is the time, and $\overline{R^2}$ is the mean square displacements. This method is applicable to determining the diffusion coefficients in materials with sufficiently high diffusion coefficients such as liquids so that atoms make many jumps within a given MD simulation on the order of picoseconds. But it is difficult for solids since atoms almost rarely make any jumps within an MD simulation timescale; the atoms are essentially moving around their equilibrium lattice positions, that is going through lattice vibrations rather than diffusional jumps from one lattice site to another. The exceptions are solid superionic conductors in which the diffusion coefficients are on the same order of magnitude as those in typical liquids. This approach suffers its usual limitations: the availability and reliability of interatomic potentials for classical MD simulations and the small number of atoms in ab initio MD simulations. Previous work on the calculation of temperature-dependent self-diffusion coefficients has either adopted (semi)empirical approaches (Varotsos and Alexopoulos, 1986; Neumann, 1987; Adams et al., 1989; Frank et al., 1996) or computationally-demanding classical and ab initio MD simulation techniques (Blochl et al., 1993; Sandberg et al., 2002; Milman et al., 1993).

The second approach, a more common one, to obtain tracer coefficients or diffusion coefficients of dilute solutes is to combine the first-principles calculations with transition rate theory (Katz et al., 1971; Sholl, 2007). This includes the evaluation of the activation energy barrier and preexponential factor using atomistic/first-principles calculations. According to the microscopic theory of diffusion, the macroscopic tracer diffusion coefficient, D, is simply related to the microscopic parameters including the jump distance and successful jump frequency. For example, for the most common vacancy diffusion mechanism in a cubic face-centered or body-centered crystal,

$$D = fa_0^2 C_v \Gamma \tag{14}$$

where f is the geometrical correlation factor, a_0 is the lattice parameter, C_v is the vacancy concentration,

$$C_v = \exp\left(\Delta S_f^{\mathrm{vib}}/k_B\right)\left(-\Delta H_f/k_B T\right) \tag{15}$$

where $\Delta S_f^{\mathrm{vib}}$ and ΔH_f are the vibrational entropy and enthalpy of vacancy formation.

The jump frequency Γ of a vacancy to a particular nearest-neighbor site in Eqn (14) can be related to the enthalpy of migration ΔH_m and an effective frequency v^* using the TST (Eyring, 1935; Vineyard, 1957)

$$\Gamma = \frac{\Pi_{i=1}^{3N-3} v_i}{\Pi_{j=1}^{3N-4} v_j'} \exp(-\Delta H/k_B T) = v^* \exp(-\Delta H_m/k_B T) \tag{16}$$

where v_i and v_j' are the normal vibrational frequencies at the equilibrium and transition states, respectively, for a system of N atoms and one vacancy. The product in the denominator specifically excludes the frequency corresponding to the unstable mode at the transition state. Therefore, the calculation of tracer diffusion coefficient D is reduced to the computation of the equilibrium lattice parameter a_0, the enthalpy of vacancy formation and atom migration, vibrational entropy of vacancy formation as well as the effective frequency v^*.

As an example, **Figure 4** shows the self-diffusion coefficients as a function of temperature for Al from the HA and QHA using the projector augmented wave potentials (Blochl, 1994; Kresse and Joubert, 1999) as implemented in the Vienna ab initio simulation package (Kresse and Furthmuller, 1996) with

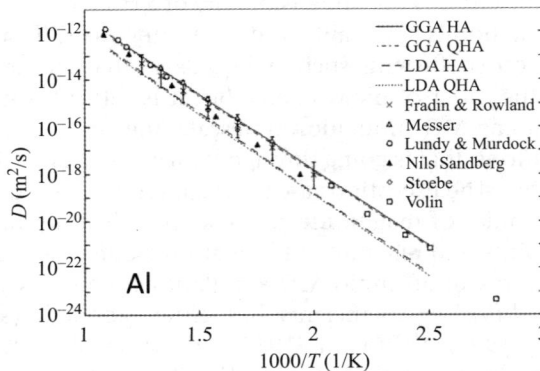

Figure 4 First-principles results of self-diffusion coefficients for Al. Results are shown for both the HA and the QHA using both LDA and GGA. Calculated results are compared with experimental results (Fradin and Rowland, 1967; Messer et al., 1974; Lundy and Murdock, 1962; Stoebe et al., 1965; Volin and Balluffi, 1968) and theoretical work (Sandberg et al., 2002), showing excellent agreement for the GGA (with "surface correction" included) calculated self-diffusion coefficient.

Table 2   Comparing frequency factor D_0 and activation energy Q of self-diffusion in copper within HA from LDA and GGA, with experimental data (Mantina, 2008).

Studies	D_0 (m^2/s)	Q (eV)	T (K)
GGA	$3.97\text{--}3.96 \times 10^{-6}$	1.67	
LDA	$1.19\text{--}1.22 \times 10^{-5}$	2.07	400–1000
EAM (Adams et al., 1989)	–	1.98 ± 0.03	
Experiment (Mehrer and Seeger, 1969)	1×10^{-5}	2.04	300–1000
Experiment (Rothman and Peterson, 1969)	7.8×10^{-5}	2.19 ± 0.01	600–1000
Experiment (Maier et al., 1973)	3.5×10^{-5}	8.84	574–905
Experiment (Weithase and Noack, 1974)	1.05×10^{-4}	9.13	845–1111
Experiment (Beyeler and Adda, 1965)	1.9×10^{-5}	8.52	973–1263
Experiment (Bowden and Balluffi, 1969)	3.1×10^{-5}	8.7	663–833

LDA (Ceperley and Alder, 1980),GGA (Perdew et al., 1996b) and the nudged elastic band method (Henkelman and Jonsson, 2000). The normal phonon frequencies are calculated using the direct force constant approach (Wei and Chou, 1992), as implemented in the ATAT (Van De Walle et al., 2002) package. **Table 2** lists the diffusion prefactor $D_0 = fa_0^2 v^* \exp(\Delta S_f^{vib}/k_B)$ and activation energy $Q = \Delta H_f + \Delta H_m$ along with experimental data in copper (Mantina, 2008).

This approach can be extended to impurity diffusion coefficients. The correlation factor for fcc systems in the presence of impurity or solutes is often approximated using the five-frequency model by Leclaire and Lidiard (Adams et al., 1989; LeClaire and Lidiard, 1956; Janotti et al., 2004; Krcmar et al., 2005). **Figure 5** shows the diffusion coefficients of Cu in fcc Al and the results agree reasonably well with existing experimental measurements (Mantina et al., 2009).

For nondilute and nonideal alloys, Van der Ven et al. proposed a Kubo-Green linear response formalism to calculate diffusion coefficients (Van der Ven and Ceder, 2000; Van der Ven et al., 2001; Van der Ven and Ceder, 2005b; Van der Ven and Ceder, 2005a). In principle, it is applicable to multicomponent alloys with any crystal structure and degree of atomic order. The dependence of the

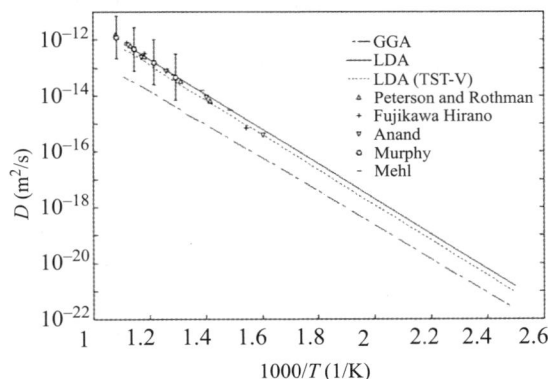

Figure 5   Diffusion coefficient with HA of different elements in fcc Al from simplified approach using LDA and GGA and from TST-V using LDA (without including surface corrections), in comparison to experimental data Cu in fcc Al (Peterson and Rothman, 1970, Fujikawa and Hirano, 1977, Mehl et al., 1941, Anand and Agarwala, 1972, Murphy, 1961).

activation barrier on its environment is determined by using the cluster expansion formalism. By combining environment-dependent transition rates and kinetic MC simulations with a cluster expansion thermodynamic description (Bortz et al., 1975; Bulnes et al., 1998), the diffusion coefficients can be obtained from a large number of trajectories of interacting diffusing particles.

A more empirical method to concentrated multicomponent alloys is to use the CALPHAD approach. Based on simple kinetic models it extrapolates the diffusivity information from experiments and first-principles calculations of relatively simpler systems to binary or multicomponent systems. The most basic diffusion coefficients are the tracer diffusion coefficients as the intrinsic and chemical diffusion coefficients can be calculated from tracer diffusivity when the thermodynamic model of the phase is available. The tracer diffusion coefficient (D_i^*) is related to the atomic mobility (M_i) by the Einstein equation

$$D_i^* = RTM_i \qquad (17)$$

with D_i^* expressed by $D_i^* = D_i^0 \exp(-Q_i/RT)$, where D_i^0 and Q_i are the prefactor and activation energy.

27.3   Stability and Evolution of Microstructures

To model the formation and evolution of microstructure, one has to digitize a microstructure. Depending on the description of a microstructure and its energetics or thermodynamics, the computational methods can be classified as (1) discrete atomistic methods such as MD and MC; (2) microscopic diffusion equations based on atomic probability distribution functions on a fixed lattice; (3) phase-field method; and (4) Monte Carlo Potts model.

27.3.1   Molecular Dynamics

In an MD simulation, a microstructure and its evolution is described by the actual atom positions and their changes with time (Allen and Tildesley, 1989). One starts with an initial set of atom positions corresponding to the initial microstructure at the atomic scale as well as the initial velocity for each atom determined by the temperature or thermal energy (k_BT) of the system. The movements of atoms, that is the atom positions and velocities of atoms as a function of time, are then determined by integrating the Newton's equation of motion,

$$\mathbf{m}_i \frac{\partial \mathbf{r}_i}{\partial t} = \mathbf{F}_i \qquad (19)$$

where $\mathbf{m}_i$, $\mathbf{r}_i$, and $\mathbf{F}_i$ are the mass, position vector and force on atom i, respectively, and t is time. Based on the problem of interest, one may perform MD simulations under different thermal and mechanical boundary conditions such as constant volume, constant pressure, constant temperature and pressure, or constant temperature and stress. Depending on how one determines the forces on each atom, for example classical or semiempirical interatomic potentials or from ab initio, the corresponding simulations are called classical, semiempirical or ab initio MD.

The main applications of MD include determining the energetics and mobility of a single defect such as a dislocation, crack, or grain boundary, investigating nanoscale grain formation and growth, and simulating deformation twinning.

The main advantage of MD simulations is its ability to capture the atomistic mechanisms involved in interface motion during microstructure evolution. The main disadvantage is its limit on the duration of simulation time and its spatial scale. The timescale limit is due to the fact that for solids the vibration periods of atoms are on the order of 10^{-13} s. To resolve atom vibrations, one has to use a time step which is a small fraction (typically 1/30 to 1/50) of the vibration periods for integrating the Newton's equations of motion. Therefore, typical simulation durations are on the order of picoseconds for 10^5–10^6 time steps. In addition, the reliability of interatomic potentials, and thus the accuracy of the forces on each atom, is always a concern for empirical or semiempirical interatomic potentials, while MD simulations directly using DFT to obtain the forces are computationally expensive and thus are limited to a small number of atoms (a few tens or hundreds) and thus atomic level structures.

27.3.2 Monte Carlo

There are two types of MC models, namely the Ising model in which a spin configuration evolves following the Glauber Dynamics (Glauber, 1963) of spin flipping and the lattice gas model in which atomic species exchange positions following the Kawasaki Dynamics (Kawasaki, 1974). Most of the simulations in materials sciences are carried out using the lattice gas model in which a microstructure is described by the atomic arrangement of different atomic species on a lattice (Binder, 1979). Although the lattice site displacements can be employed as an additional set of random variables for describing the elasticity effect or the effect of lattice vibrations on the total energy of an atomic configuration or microstructure, the majority of MC simulations were performed on fixed lattice positions.

A typical simulation starts with an initial distribution of atoms including vacancies on a lattice with known symmetry and periodic boundary conditions along all directions unless free surfaces are required, for example for thin films. For the vacancy mechanism of atom migration, the vacancy-atom exchange probability is given by

$$w = v \, \exp\left[-\frac{\Delta H_m}{k_B T}\right] = v \, \exp\left[-\frac{\Delta E_m}{k_B T}\right] \tag{20}$$

where v is the Debye frequency and ΔE_m is the migration energy, that is the energy increase on moving an atom, nearest-neighbor of a vacancy V, from its stable site to the saddle point position (**Figure 6**).

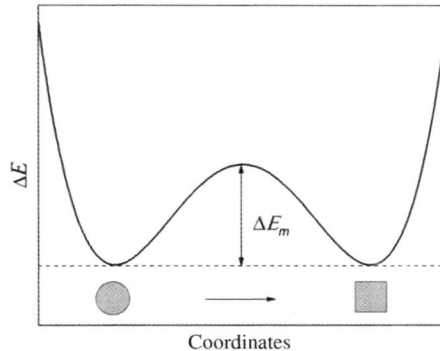

Figure 6 Schematic of the migration energy vs coordinates. The filled circle denotes an atom, and the filled square represents a vacancy. ΔE_m is the migration energy which is the difference between the saddle point and the stable site.

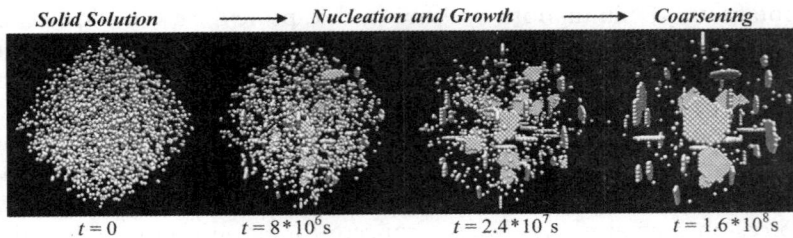

Figure 7 First-principles mixed-space expansion calculated evolution of GP zone with Al-1.0at%Cu at $T = 373$ K. Only Cu atoms are shown, and from the perspective view, it can be seen that precipitates from three variants of monolayer (100) plates consisting of pure Cu. Simulation times are shown in seconds (Wang et al., 2005a).

A significant amount of literature employed simplistic pair interaction models to determine the vacancy migration energy. However, to accurately describe the energetics of a particular alloy system, usually more accurate methods such as cluster expansion is required.

The evolution of the atomic configurations with time is performed according to either the simple Metropolis algorithm or the residence time method. The Metropolis importance sampling MC scheme involves (1) randomly choosing a vacancy jump among all possible jumps, (2) calculating ΔE_m for the vacancy jump, (3) generating a random number r between 0 and 1, (4) making a vacancy jump if $r < \exp(-\Delta E_m/k_B T)$, and (5) repeating the whole process. The Metropolis algorithm becomes inefficient at low temperatures as the acceptance rate for the vacancy jump dramatically decreases with T, and hence a vacancy may not make any jumps after many tries. Therefore, the residence time algorithm is typically employed in modeling atomic ordering and clustering processes. It requires the calculations of all the jump frequencies of the vacancy to its nearest-neighbor sites at a given moment. A vacancy always makes a jump at each step, and the probability of the vacancy making a jump to a specific site is determined by its jump frequency normalized by the total jump frequency of all possible jumps. Furthermore, it is possible to relate the MC time to real time in the residence time algorithm (see Fichthorn and Weinberg, 1991 for details). MC method is applicable to a wide range of problems including the formation of atomic clusters and nuclei during diffusional phase transformations. **Figure 7** shows an example of GP zone formation in Al–Cu obtained using a combination of MC simulations with a mixed-space expansion fitted to first-principles energetics (Wang et al., 2005a). However, it should be emphasized that the lattice MC method is only applicable to diffusional processes that involve atomic redistributions on a fixed lattice. The accuracy of prediction greatly depends on the input for the energetic or the rate constants for the atomic jumps.

27.3.3 Microscopic Diffusion Equations

Different from the MC algorithm which is designed to simulate the equilibrium ensemble and produces a series of instantaneous atomic configurations which appear along the simulated Markov chain, the method of microscopic diffusion equations describes a microstructure using a set of single, pair, and multisite probability distribution functions of species $\alpha_1, \alpha_2, ..., \alpha_n$ at lattice site $\mathbf{r}_1, \mathbf{r}_2, ...,$ and $\mathbf{r}_n$ (Vineyard, 1956)

$$P_{\alpha_1}(\mathbf{r}_1), \ P_{\alpha_1,\alpha_2}(\mathbf{r}_1, \mathbf{r}_2) \ ... \ P_{\alpha_1,\alpha_2,...,\alpha_n}(\mathbf{r}_1, \mathbf{r}_2, ..., \mathbf{r}_n)$$

which are the ensemble averages over time-dependent nonequilibrium atomic configurations. The evolution of the probability distribution functions is determined by a set of kinetic equations with transition probabilities determined by the interatomic interactions. The complexity of this approach increases rapidly as increasingly higher order probability distribution functions are included, and hence most of the existing applications of the microscopic diffusion equations included only single-site or pair probability distribution functions.

This approach has the following features (Chen and Simmons, 1994): (1) with the knowledge of atom–atom bond energies and the initial distributions, the temporal evolution of point and pair distribution functions obtained from the kinetic equations automatically describes the kinetics of long-range order, short-range order, phase separation, and coarsening; (2) the kinetics can be either nonlinear or linear with respect to thermodynamic driving forces, and hence microscopic diffusion equations can be applicable to systems with large driving forces; (3) the dependence of atomic mobility on local configuration is automatically taken into account; and (4) at equilibrium, it produces the same equilibrium distribution functions as derived from the cluster variation method.

The path probability method of Kikuchi (Kikuchi, 1966) is similar to the microscopic diffusion equations at the same order of approximation. In the single-site approximation in which the structure state of an alloy and its evolution kinetics are described by the temporal evolution of the point probability distribution function, the set of microscopic diffusion equations are essentially the mean-field kinetic equations proposed by Martin (1990) and Penrose (1991) for a binary alloy through the direct exchange mechanism (for a vacancy mechanism, it is the exchange between an atom and a vacancy) (**Figure 8**)

$$\frac{dP_A(\mathbf{r}, t)}{dt} = \sum_{\delta} P_B(\mathbf{r}, t) P_A(\mathbf{r} + \delta, t) v e^{-[\phi_s - \phi(\mathbf{r}+\delta,t)]/kT} - \sum_{\delta} P_A(\mathbf{r}, t) P_B(\mathbf{r} + \delta, t) v e^{-[\phi_s - \phi(\mathbf{r},t)]/kT} \qquad (21)$$

where $P_A(\mathbf{r}, t)$ is the single-site occupation probability of A at lattice site $\mathbf{r}$ for a given time t, $P_B(\mathbf{r} + \delta, t)$ is the single-site occupation probability of B at lattice site $\mathbf{r} + \delta$ which is the neighboring site of $\mathbf{r}$, k is the Boltzmann constant, T is temperature, v is the jump frequency, $\phi(\mathbf{r})$ and $\phi(\mathbf{r} + \delta)$ are the potential of lattice sites $\mathbf{r}$ and $\mathbf{r} + \delta$, respectively, and ϕ_S is the energy barrier between these two sites.

In the microscopic diffusion theory of Khachaturyan (1968), the evolution of a nonequilibrium distribution of single-site occupation probability function toward equilibrium is described by the Onsager-type diffusion equations in which the rate of change of the single-site occupation probability is

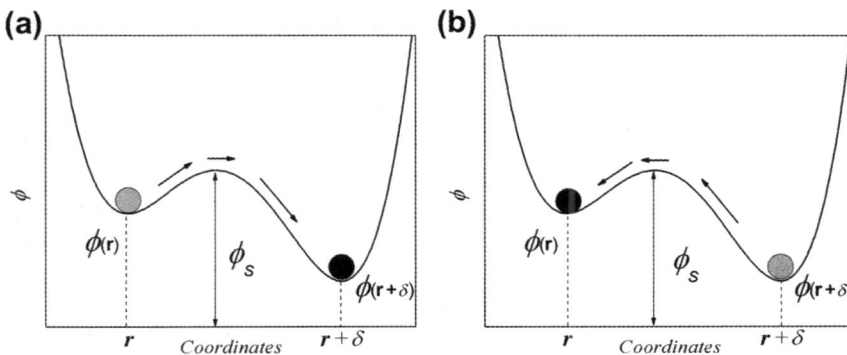

Figure 8 The schematic of the direct exchange mechanism. (a) Jump from a relative metastable site to a state site, (b) jump from a stable state to a metastable one.

linearly proportional to the thermodynamic driving force defined as the variational derivative of total free energy with respect to the single-site occupation probability function,

$$\frac{dP_A(\mathbf{r}, t)}{dt} = \sum_{\mathbf{r}'} L(\mathbf{r} - \mathbf{r}') \frac{\partial F}{\partial P_A(\mathbf{r}')} \tag{22}$$

where $P_A(\mathbf{r}, t)$ is the single-site occupation probability at lattice site $\mathbf{r}$ for a given time t. The temporal evolution of the probability distribution functions describes the kinetics of diffusional processes and microstructure evolution in an alloy. At the continuum limit, the set of microscopic diffusion equations are reduced to the Cahn–Hilliard diffusion equation (Cahn and Hilliard, 1958) for the temporal evolution of composition profiles and the Allen–Cahn relaxation equation (Allen and Cahn, 1979) for the evolution of long-range order distributions.

There have been numerous applications of microscopic diffusion equations to modeling the kinetics of concurrent atomic ordering and clustering processes during precipitation of ordered intermetallics in alloys. It automatically describes the faster ordering kinetics which only involves a diffusion range on the order of nearest-neighbor distances than phase separation which requires a diffusion distance on the order of composition domains. The effect of elastic stress during a phase transformation can be incorporated by introducing the dependence of lattice parameters or transformation strains on the local composition or order parameter which can be determined from the single-site occupation probability distribution. However, the kinetic equations are deterministic, and thus microscopic diffusion equations cannot describe activated processes such as nucleation although it is also possible to add random noises to the kinetic equations to mimic thermal fluctuations. Furthermore, similar to MC simulations, they are only applicable to diffusional processes on a fixed crystal lattice.

27.3.4   Phase-Field Method

In contrast to atomistic models or models based on a fixed lattice, a set of continuous field variables are used to describe a microstructure in a phase-field model. The field variables are uniform or nearly uniform inside a phase or domain away from the interfacial or wall regions. They vary continuously across the interfaces, and thus there is a thickness associated with an interface as opposed to the conventional sharp-interface description. They can be either conserved or nonconserved, depending on if they satisfy the local conservation law, $\partial \phi / \partial t = -\nabla \cdot \mathbf{J}$ where ϕ is a field variable and $\mathbf{J}$ is the corresponding flux. For example, composition and temperature fields are both conserved while long-range order parameter fields describing ordered domain structures are nonconserved. Field variables can be physical or artificial. Physical fields refer to well-defined order parameters which can be experimentally measured such as long-range order parameter or composition. On the other hand, artificial fields are introduced for the sole purpose of avoiding tracking the interfaces during a microstructure evolution. They can be a single component or multicomponent. An ordered phase such as the B2 phase based on a bcc lattice can be characterized by a single long-range order parameter while the physical characterization of an L1$_2$ ordered phase on an fcc lattice requires an order parameter with three components (Braun et al., 1997; Zhu et al., 2002). The description of many practical microstructures requires more than one type of order parameters. For example, precipitation of an ordered intermetallic phase in a disordered matrix involves both ordering and compositional clustering, and thus the description of ordered precipitate microstructures requires both composition and nonconserved order parameter fields.

In the phase-field model (Cahn and Hilliard, 1958), the total free energy of an inhomogeneous microstructure in Eqn (1) is replaced by

$$F = \int_V f_l^0 dV + \int_V \left[f_l - f_l^0 + f_g \right] dV + \int_V f_e dV \tag{23}$$

where f_l^0 and f_l is the equilibrium and nonequilibrium local bulk chemical free energy densities that are functions of field variables, f_g is the gradient energy density, and f_e represents the total energy due to nonlocal long-range interactions.

These long-range interactions and thus the interaction energy can be obtained by solving the corresponding mechanical, electrostatic, and magnetostatic equilibrium equations for a given microstructure. For example, the mechanical equilibrium equation is given by

$$\frac{\partial \sigma_{ij}}{\partial r_j} = 0 \quad \text{with} \quad \sigma_{ij}(\mathbf{r}) = \lambda_{ijkl}(\mathbf{r})\left[\varepsilon_{kl}(\mathbf{r}) - \varepsilon_{kl}^o(c, n, \ldots)\right] \tag{24}$$

where σ_{ij} is the local elastic stress, r_j is the jth component of the position vector, $\mathbf{r}$, $\lambda_{ijkl}(\mathbf{r})$ is the elastic stiffness tensor which varies with space, $\varepsilon_{jk}(\mathbf{r})$ is the total strain state at a given position in a microstructure, and ε_{kl}^0 is the local stress-free strain which is also a function of position through its dependence on field variables. The resulted elastic energy is a function of phase-field variables and thus the microstructure (Khachaturian, 1983). For the case of homogeneous approximation and periodic boundary conditions, it was shown by Khachaturyan (Khachaturyan, 1969) that an analytical solution for the displacements, strains, and thus the strain energy could be obtained in the Fourier space. For systems with small elastic homogeneity, first-order approximations may be employed (Onuki, 1989; Sagui et al., 1998). For large elastic inhomogeneities, a number of approaches have been proposed for obtaining elastic solutions (Hu and Chen, 2001a; Leo et al., 1998; Wang et al., 2002; Zhu et al., 2001a). The interactions between precipitates and structural defects such as dislocations can be described using essentially the same approach (Hu and Chen, 2001b; Hu and Chen, 2002; Wang et al., 2001b; Wang et al., 2001a).

The temporal and spatial evolution of the field variables follows a set of kinetic equations. All conserved fields, c_i, evolve with time according to the Cahn–Hilliard equation (Cahn and Hilliard, 1958) whereas the nonconserved fields, η_p, are governed by the Allen–Cahn equation (Allen and Cahn, 1979), that is

$$\frac{\partial c_i(\mathbf{r}, t)}{\partial t} = \nabla \left(M_{ij} \nabla \frac{\delta F}{\delta c_j(\mathbf{r}, t)} \right) \tag{25}$$

$$\frac{\partial \eta_p(\mathbf{r}, t)}{\partial t} = -L_{pq} \frac{\delta F}{\delta \eta_q(\mathbf{r}, t)} \tag{26}$$

where M_{ij} and L_{pq} are related to atom or interface mobility. F is the total free energy of a system which is a functional of all the relevant conserved and nonconserved fields given by Eqn (23).

The evolution profiles of the field variables, and thus the microstructure evolution, are obtained by numerically solving the systems of evolution equations subject to appropriate initial and boundary conditions. Most of the phase-field simulations employ the second-order finite difference discretization in space using uniform grids and the forward Euler method for time stepping to solve the phase-field equations for simplicity. Significant savings in computation time and improvement in numerical accuracy can be achieved by using more advanced numerical approaches such as the semi-implicit Fourier Spectral method (Chen and Shen, 1998; Zhu et al., 1999). In addition, spatially adaptive schemes can be implemented by working with both a computational space with uniform grids and a real space with adaptive grids that allow one to maintain the applicability of the spectral codes (Feng et al., 2006). For systems with complicated geometry of surfaces, finite-element method is usually the method of choice although there have been recent attempts to treat complicated boundary geometries using the smooth boundary method (Bueno-Orovio and Perez-Garcia, 2006).

The phase-field method has been applied to a wide variety of different processes including solidifications, grain growth, precipitation reactions, ferroelastic transitions, and many others (Chen, 2002; Boettinger et al., 2002; Thornton et al., 2003; Moelans et al., 2008; Chen, 2008; Wang and Li, 2010; Steinbach, 2009; Provatas and Elder, 2010). **Figure 9** shows an example of the microstructure evolution during precipitation of gamma prime particles in an Ni–Al–Mo ternary alloy generated by three dimensional phase-field simulation (Wang et al., 2008). Phase-field method offers a number of advantages among other continuum-level computational approaches. It can describe different processes such as phase transformations (driven by bulk free energy reduction) and particle coarsening (driven by interfacial energy reduction) within the same formulation. In addition, it is rather straightforward to incorporate the effect of coherency and applied stresses as well as electrical and magnetic fields in the phase-field model. However, it relies on more fundamental calculations such as first-principles calculations or experimental data for the input parameters. Furthermore, the physical size that a phase-field simulation can handle is, in many cases, limited by the usually small physical width of real interfaces in microstructures as compared with phase and domain sizes. In addition, although there are ways to model nucleation within the phase-field method (Poduri and Chen, 1996; Roy et al., 1998; Simmons et al., 2000; Granasy et al., 2002; Castro, 2003; Heo et al., 2010), significant effort is still required to establish a robust, physical, and quantitative approach to induce both homogeneous nucleation in the bulk and heterogeneous nucleation around defects. Finally, phase-field method does not provide information about the atomistic mechanisms of interface motion. However, this shortcoming can be overcome by using the phase-field crystal method. The phase-field crystal model describes the crystal structures and microstructures on atomic or subatomic length and diffusive timescales using an atomic density function (Elder and Grant, 2004; Elder et al., 2002). The main advantage of this approach is its ability to model both plastic and elastic deformations at the subatomic scale and with a timescale larger than atomistic MD simulations. The main disadvantage, however, is the small length scale and the difficulty to construct the local free energy density for a real material.

27.3.5 Mesoscopic Potts Model

In a Monte Carlo Potts model, a microstructure is discretized using a regular grid (Potts, 1952; Wu, 1982). Essentially all the simulations based on the Monte Carlo Potts model were performed to study grain growth (Anderson et al., 1984; Wejchert et al., 1986). Each grid point is labeled by an integer number representing the orientation of a grain. The total energy is determined by essentially counting how many nearest-neighbors having different grain numbers, that is

$$E = \sum_{i=1}^{q} S_i \gamma_i = \sum_{i=1}^{N} \sum_{j=1}^{n} J \delta_{Q_i Q_j} \tag{27}$$

where J represents the grain boundary energy per unit grid area, δ is the Kronecker delta function. The evolution of a grain structure is driven by the total grain boundary energy reduction and can be carried

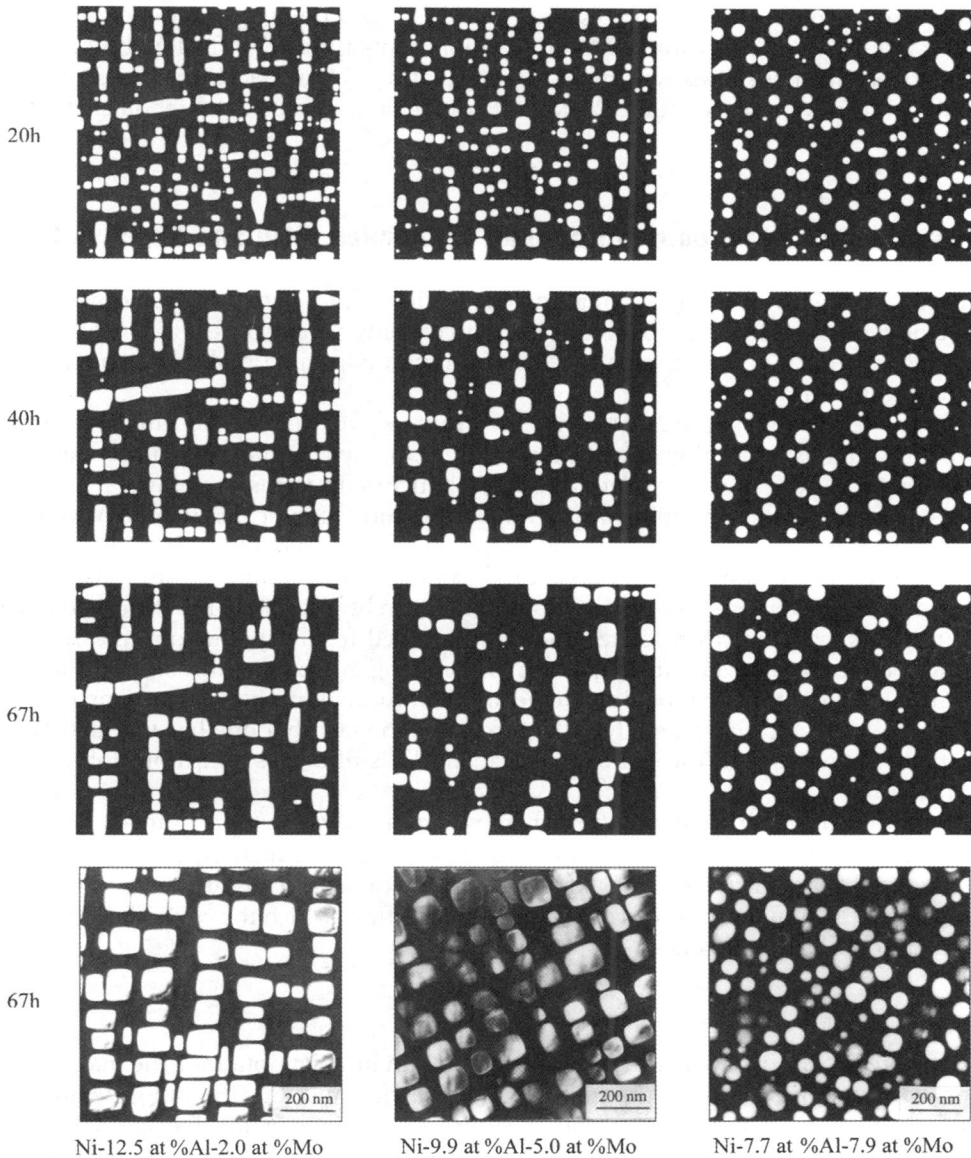

Figure 9 Microstructure evolution of the γ' precipitates in Ni–Al–Mo alloys at 1048 K. Figures in the bottom row are from experiments and others from 2D phase-field simulations (Wang et al., 2008).

out through the conventional Metropolis algorithm or the resident time algorithm as aforementioned in the atomistic MC simulations. The main advantage of this approach lies in its simplicity as well as computational efficiency. One is able to carry out very large-scale simulations of grain growth in three dimensions. It is also not difficult to introduce the contribution of strain energy to the total energy. It is possible to introduce diffusional exchanges among different grid points in a Monte Carlo Potts model although it is less convenient to carry out computational simulations of diffusional processes than the phase-field model which directly solves the diffusion equations. However, comparisons of grain growth and coarsening kinetics from the Monte Carlo Potts model and phase-field method show very similar results (Tikare et al., 1998).

27.4   Responses of a Microstructure under an Applied Field and Effective Properties

One of the eventual goals in computational materials science and engineering is to predict the properties of microstructures. While it is possible to directly compute the structures and properties of a single crystal using electronic/atomistic approaches, determining the properties of a microstructure usually requires continuum approaches.

Generally, there are two approaches that one can use to obtain the effective properties of a microstructure. One is to employ an effective medium theory or a mechanistically based theory to obtain expressions for effective properties of a microstructure in terms of volume fractions, average measures of morphologies, and the properties of the individual constituents. Various effective medium theories as well as computational approaches are discussed in much detail in the monograph (Torquato, 2002). The advantage of an analytical approach is its simplicity and fast evaluation of properties, and thus the temporal evolution of properties can be obtained once the microstructures are generated using a specific microstructure model or obtained from experimental measurements and reconstruction. For example, one can simply use the evolving microstructures obtained from a phase-field simulation in a constitutive model to predict the evolution of mechanical or transport properties. The disadvantage of using such an analytical approach is the fact that one has to make rather severe approximations with regard to the spatial and statistical distributions and morphologies of phases and domains.

The other approach, which is computationally most common, is to use the finite-element method to solve the three-dimensional full-field equations, for example the elasticity equation for describing the mechanical response of a microstructure. For example, one can subject a microstructure to an applied stress and evaluate the strain response or stress distributions from which the effective elastic moduli can be determined from the Hooke's law,

$$\sigma_i = \overline{C}_{ij} \varepsilon_j \tag{28}$$

where σ_i, $\overline{C}_{ij}$, ε_j are stress, effective elastic modulus, and strain in Voigt notation. The main disadvantage of the finite-element approach is its computational inefficiency, and thus it is not generally possible to perform the evaluation of properties of all the microstructures generated using a computational microstructure model as a function of time, that is to predict the temporal evolution of properties.

Although phase-field modeling is generally employed to model the microstructure evolution, it is possible to obtain the effective properties of a microstructure within the phase-field description (Zhu et al., 2001a). To obtain the effective property of a microstructure, one applies a driving force, for example apply stress or strain, or a temperature gradient, or a chemical potential gradient, to a microstructure, and then measure the response at mechanical equilibrium or at steady-state temperature or chemical potential distributions. For example, the effective diffusion coefficient of a grain structure was computed using a combination of phase-field description of a grain structure and a numerical solution of a diffusion equation with its diffusion coefficient dependent on the phase-field

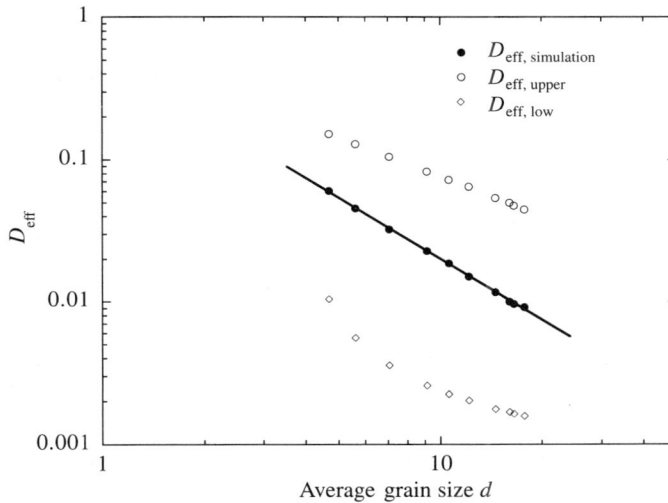

Figure 10 Effective diffusivity D_{eff} evolution during a single-phase grain growth: (a) D_{eff} as a function of evolution time; (b) D_{eff} as a function of average grain size d (Zhu et al., 2001b).

parameters. **Figure 10** shows an example of diffusion coefficient as a function of grain size (Zhu et al., 2001b). The same approach has been used to obtain the effective elastic moduli of a microstructure in Mg polycrystalline (Sheng et al., 2012). Since the properties can, in general, be much more efficiently evaluated within the phase-field approach than a separate finite-element calculation, one can model the evolution of microstructure and properties within the same phase-field simulation. Recent advances show that it is also possible to directly connect phase-field description of a microstructure and crystal plasticity to study the coupling of plastic deformation and microstructure evolution, that is the plasticity of an evolving microstructure and the effect of deformation on phase transformations and microstructure evolution (Guo et al., 2005; Uehara et al., 2007; Yamanaka et al., 2008; Gaubert et al., 2010; Kundin et al., 2010; Ammar et al., 2011).

27.5 Summary

This chapter briefly discusses some of the common computational approaches that have been applied to predicting the processing–structure–property relationships in metallic alloys, that is computational metallurgy. These approaches range from prediction and modeling of structures and thermodynamic/kinetic properties of perfect crystals and isolated defects in single-component, binary and multicomponent systems using a combination of first-principles calculations and CALPHAD method, simulation of microstructure evolution using either atomistic or continuum approaches such as MD, MC, and phase-field methods, and direct computation of mechanical and physical properties of three-dimensional microstructures using mean-field theories, finite-element method or phase-field method. Predictive modeling and simulation of materials structures and properties usually requires an integrated approach involves a combination of two or more computational approaches. It is hoped that the brief discussions of the main computational materials approaches and their advantages and disadvantages will make it easier for a novice to enter the field of computational materials science and engineering or computational metallurgy.

References

Adams, J.B., Foiles, S.M., Wolfer, W.G., 1989. Self-diffusion and impurity diffusion of fcc metals using the five-frequency model and the embedded atom method. J. Mater. Res. 4, 102–112.

Alder, B.J., Wainwright, T.E., 1959. Studies in molecular dynamics. I. General method. J. Chem. Phys. 31, 459.

Allen, M.P., Tildesley, D.J., 1989. Computer Simulation of Liquids. Oxford University Press.

Allen, S.M., Cahn, J.W., 1979. A microscopic theory for antiphase boundary motion and its application to antiphase domain coarsening. Acta Metall. 27, 1085–1095.

Ammar, K., Appolaire, B., Cailletaud, G., Forest, S., 2011. Phase field modeling of elasto-plastic deformation induced by diffusion controlled growth of a misfitting spherical precipitate. Philos. Mag. Lett. 91, 164–172.

Anand, M.S., Agarwala, R.P., 1972. Diffusion of cobalt in aluminium. Philos. Mag. 26, 297–309.

Anderson, M.P., Srolovitz, D.J., Grest, G.S., Sahni, P.S., 1984. Computer simulation of grain growth–I. Kinetics. Acta Metall. 32, 783–791.

Andersson, J.O., Guillermet, A.F., Hillert, M., Jansson, B., Sundman, B., 1986. A compound-energy model of ordering in a phase with sites of different coordination numbers. Acta Metall. 34, 437–445.

Asta, M., Foiles, S.M., Quong, A.A., 1998. First-principles calculations of bulk and interfacial thermodynamic properties for fcc-based Al–Sc alloys. Phys. Rev. B 57, 11265.

Balluffi, R.W., Allen, S.M., Carter, W.C., Kemper, R.A., 2005. Kinetics of Materials. Wiley-Interscience.

Baroni, S., De Gironcoli, S., Dal Corso, A., Giannozzi, P., 2001. Phonons and related crystal properties from density-functional perturbation theory. Rev. Mod. Phys. 73, 515.

Baroni, S., Moroni, S., 2005. Computer simulation of quantum melting in hydrogen clusters. Chem. Phys. Chem. 6, 1884–1888.

Beyeler, M., Adda, Y., 1965. Physics of solids at high pressures: proceedings. In: Tomizuka, C.T., Emrick, R.M. (Eds.), Academic Press.

Binder, K., 1979. Monte Carlo Methods. Wiley. Online Library.

Blaha, P., Schwarz, K., Sorantin, P., Trickey, S.B., 1990. Full-potential, linearized augmented plane wave programs for crystalline systems. Comput. Phys. Commun. 59, 399–415.

Blochl, P.E., 1994. Projector augmented-wave method. Phys. Rev. B 50, 17953.

Blochl, P.E., Smargiassi, E., Car, R., et al., 1993. First-principles calculations of self-diffusion constants in silicon. Phys. Rev. Lett. 70, 2435.

Boettinger, W.J., Warren, J.A., Beckermann, C., Karma, A., 2002. Phase-field simulation of solidification 1. Annu. Rev. Mater. Res. 32, 163–194.

Bortz, A.B., Kalos, M.H., Lebowitz, J.L., 1975. A new algorithm for Monte Carlo simulation of Ising spin systems. J. Comput. Phys. 17, 10–18.

Bowden, H.G., Balluffi, R.W., 1969. Measurements of self-diffusion coefficients in copper from the annealing of voids. Philos. Mag. 19, 1001–1014.

Braun, R.J., Cahn, J.W., Mcfadden, G.B., Wheeler, A.A., 1997. Anisotropy of interfaces in an ordered alloy: a multiple-order-parameter model. Philos. Trans. R. Soc. Lond. A: Math. Phys. Eng. Sci. 355, 1787–1833.

Broughton, J.Q., Abraham, F.F., Bernstein, N., Kaxiras, E., 1999. Concurrent coupling of length scales: methodology and application. Phys. Rev. B 60, 2391.

Bueno-Orovio, A., Perez-Garcia, V.M., 2006. Spectral smoothed boundary methods: the role of external boundary conditions. Numer. Methods Partial Differ. Equ. 22, 435–448.

Bulnes, F.M., Pereyra, V.D., Riccardo, J.L., 1998. Collective surface diffusion: n-fold way kinetic Monte Carlo simulation. Phys. Rev. E 58, 86.

Cahn, J.W., Hilliard, J.E., 1958. Free energy of a nonuniform system. I. Interfacial free energy. J. Chem. Phys. 28, 258.

Castro, M., 2003. Phase-field approach to heterogeneous nucleation. Phys. Rev. B 67, 35412.

Ceperley, D.M., Alder, B.J., 1980. Ground state of the electron gas by a stochastic method. Phys. Rev. Lett. 45, 566–569.

Chen, L.Q., 2002. Phase-field models for microstructure evolution. Annu. Rev. Mater. Res. 32, 113–140.

Chen, L.Q., 2008. Phase-field method of phase transitions/domain structures in ferroelectric thin films: a review. J. Am. Ceram. Soc. 91, 1835–1844.

Chen, L.Q., Shen, J., 1998. Applications of semi-implicit Fourier-spectral method to phase field equations. Comput. Phys. Commun. 108, 147–158.

Chen, L.Q., Simmons, J.A., 1994. Microscopic master equation approach to diffusional transformations in inhomogeneous systems–single-site approximation and direct exchange mechanism. Acta Metall. Mater. 42, 2943–2954.

Chen, L.Q., Wolverton, C., Vaithyanathan, V., Liu, Z.K., 2001. Modeling solid-state phase transformations and microstructure evolution. MRS Bull. 26, 197–202.

Christian, J.W., 2002. The Theory of Transformations in Metals and Alloys: Part I + II. Elsevier.

Costa E Silva, A., Ågren, J., Clavaguera-Mora, M.T., et al., 2007. Applications of computational thermodynamics-the extension from phase equilibrium to phase transformations and other properties. Calphad 31, 53–74.

Curtarolo, S., Morgan, D., Ceder, G., 2005. Accuracy of ab initio methods in predicting the crystal structures of metals: a review of 80 binary alloys. Calphad 29, 163–211.

Curtarolo, S., Morgan, D., Persson, K., Rodgers, J., Ceder, G., 2003. Predicting crystal structures with data mining of quantum calculations. Phys. Rev. Lett. 91, 135503.

Curtin, W.A., Miller, R.E., 2003. Atomistic/continuum coupling in computational materials science. Model. Simul. Mater. Sci. Eng. 11, R33.

Eberhart, M., 1994. Computational metallurgy. Science (Washington, DC); (United States) 265.

Elder, K.R., Grant, M., 2004. Modeling elastic and plastic deformations in nonequilibrium processing using phase field crystals. Phys. Rev. E, Stat. Nonlin. Soft Matter Phys. 70, 051605.1–051605.18.

Elder, K.R., Katakowski, M., Haataja, M., Grant, M., 2002. Modeling elasticity in crystal growth. Phys. Rev. Lett. 88, 245701.1–245701.4.

Eyring, H., 1935. The activated complex in chemical reactions. J. Chem. Phys. 3, 107.

Feng, W.M., Yu, P., Hu, S.Y., et al., 2006. Spectral implementation of an adaptive moving mesh method for phase-field equations. J. Comput. Phys. 220, 498–510.

Fichthorn, K.A., Weinberg, W.H., 1991. Theoretical foundations of dynamical Monte Carlo simulations. J. Chem. Phys. 95, 1090.

Fontaine, D.D., 1994. Cluster approach to order-disorder transformations in alloys. Solid State Phys. 47, 33–176.

Fradin, F.Y., Rowland, T.J., 1967. NMR measurement of the diffusion coefficient of pure aluminum. Appl. Phys. Lett. 11, 207–209.

Frank, W., Breier, U., Elsasser, C., Fahnle, M., 1996. First-principles calculations of absolute concentrations and self-diffusion constants of vacancies in lithium. Phys. Rev. Lett. 77, 518–521.

Fujikawa, S., Hirano, K., 1977. Diffusion of 28Mg in aluminum. Mater. Sci. Eng. 27, 25–33.

Gaubert, A., Le Bouar, Y., Finel, A., 2010. Coupling phase field and viscoplasticity to study rafting in Ni-based superalloys. Philos. Mag. 90, 375–404.

Glauber, R.J., 1963. Time dependent statistics of the Ising model. J. Math. Phys. 4, 294.

Granasy, L., Borzsonyi, T., Pusztai, T., 2002. Crystal nucleation and growth in binary phase-field theory. J. Cryst. Growth 237, 1813–1817.

Guo, X.H., Shi, S.Q., Ma, X.Q., 2005. Elastoplastic phase field model for microstructure evolution. Appl. Phys. Lett. 87, 221910.

Henkelman, G., Jonsson, H., 2000. Improved tangent estimate in the nudged elastic band method for finding minimum energy paths and saddle points. J. Chem. Phys. 113, 9978.

Heo, T.W., Wang, Y., Bhattacharya, S., et al., 2011. A phase-field model for deformation twinning. Philos. Mag. Lett. 91, 110–121.

Heo, T.W., Zhang, L., Du, Q., Chen, L.Q., 2010. Incorporating diffuse-interface nuclei in phase-field simulations. Scr. Mater. 63, 8–11.

Hohenberg, P., Kohn, W., 1964. Inhomogeneous electron gas. Phys. Rev. 136, B864.

Hoyt, J.J., Sadigh, B., Asta, M., Foiles, S.M., 1999. Kinetic phase field parameters for the Cu–Ni system derived from atomistic computations. Acta Mater. 47, 3181–3187.

Hu, S.Y., Chen, L.Q., 2001a. A phase-field model for evolving microstructures with strong elastic inhomogeneity. Acta Mater. 49, 1879–1890.

Hu, S.Y., Chen, L.Q., 2001b. Solute segregation and coherent nucleation and growth near a dislocation—a phase-field model integrating defect and phase microstructures. Acta Mater. 49, 463–472.

Hu, S.Y., Chen, L.Q., 2002. Diffuse-interface modeling of composition evolution in the presence of structural defects. Comput. Mater. Sci. 23, 270–282.

Janotti, A., Krcmar, M., Fu, C.L., Reed, R.C., 2004. Solute diffusion in metals: larger atoms can move faster. Phys. Rev. Lett. 92, 85901.

Jansen, H.J.F., Freeman, A.J., 1984. Total-energy full-potential linearized augmented-plane-wave method for bulk solids: electronic and structural properties of tungsten. Phys. Rev. B 30, 561.

Jiang, C., Wolverton, C., Sofo, J.O., Chen, L.Q., Liu, Z.K., 2004. First-principles study of binary bcc alloys using special quasirandom structures. Phys. Rev. B 69, 214202.

Katz, L., Guinan, M., Borg, R.J., 1971. Diffusion of H_2, D_2, and T_2 in single-crystal Ni and Cu. Phys. Rev. B 4, 330–341.

Kaufman, L., Bernstein, H., 1998. Computer Calculation of Phase Diagram Academic, New York, 1970. 2 N. Saunders and AP Miodownik. Calphad (Calculation of Phase Diagrams): A Comprehensive Guide.

Kawasaki, K. (Ed.), 1974. Phase Transitions and Critical Phenomena. Academic, London.

Khachaturian, A.G., 1983. Theory of Structural Transformations in Solids. John Wiley and Sons, New York, NY.

Khachaturyan, A.G., 1968. Microscopic theory of diffusion in crystalline solid solutions and the time evolution of the diffuse scattering of X rays and thermal neutrons. Soviet Phys. Solid State 9, 2040–2046.

Khachaturyan, A.G., 1969. Elastic-interaction potential of defects in a crystal. Soviet Phys. Solid State 11, 118–123.

Kikuchi, R., 1966. The path probability method. Suppl. Prog. Theor. Phys., 1–64.

Kissavos, A.E., Shallcross, S., Meded, V., Kaufman, L., Abrikosov, I.A., 2005. A critical test of ab initio and CALPHAD methods: the structural energy difference between bcc and hcp molybdenum. Calphad 29, 17–23.

Kohn, W., Sham, L.J., 1965. Self-consistent equations including exchange and correlation effects. Phys. Rev. 140, A1133–A1138.

Krcmar, M., Fu, C.L., Janotti, A., Reed, R.C., 2005. Diffusion rates of 3d transition metal solutes in nickel by first-principles calculations. Acta Mater. 53, 2369–2376.

Kresse, G., Furthmuller, J., 1996. Efficient iterative schemes for ab initio total-energy calculations using a plane-wave basis set. Phys. Rev. B 54, 11169.

Kresse, G., Joubert, D., 1999. From ultrasoft pseudopotentials to the projector augmented-wave method. Phys. Rev. B 59, 1758.

Kundin, J., Emmerich, H., Zimmer, J., 2010. Three-dimensional model of martensitic transformations with elasto-plastic effects. Philos. Mag. 90, 1495–1510.

Leclaire, A.D., Lidiard, A.B., 1956. LIII. Correlation effects in diffusion in crystals. Philos. Mag. 1, 518–527.

Leo, P.H., Lowengrub, J.S., Jou, H.J., 1998. A diffuse interface model for microstructural evolution in elastically stressed solids. Acta Mater. 46, 2113–2130.

Liu, Z.K., Chen, L.Q., Raghavan, P., et al., 2004. An integrated framework for multi-scale materials simulation and design. J Comput. Aided Mater. Des. 11, 183–199.

Lundy, T.S., Murdock, J.F., 1962. Diffusion of Al26 and Mn54 in Aluminum. J. Appl. Phys. 33, 1671–1673.

Maier, K., Bassani, C., Schule, W., 1973. Self-diffusion in copper between 359 and 632 °C. Phys. Lett. A 44, 539–540.

Mantina, M., 2008. A first-principles methodology for diffusion coefficients in metals and dilute alloys. Mater. Sci. Eng.. State College, The Pennsylvania State University.

Mantina, M., Wang, Y., Chen, L.Q., Liu, Z.K., Wolverton, C., 2009. First principles impurity diffusion coefficients. Acta Mater. 57, 4102–4108.

Martin, G., 1990. Atomic mobility in Cahn's diffusion model. Phys. Rev. B 41, 2279.

Mehl, R.F., Rhines, F.N., Von Den Steinen, K.A., 1941. Diffusion in alpha solid solution of Al. Metals Alloys 13, 41–44.

Mehrer, H., Seeger, A., 1969. Interpretation of self-diffusion and vacancy properties in copper. Phys. Status Solidi (b) 35, 313–328.

Messer, R., Dais, S., Wolf, D., 1974. In: Allen, P.S., Andrew, E.R., Bates, C.A. (Eds.), The Proceedings of the 18th Ampere Congress. Nottingham, England, North-Holland, Amsterdam.

Milman, V., Payne, M.C., Heine, V., et al., 1993. Free energy and entropy of diffusion by ab initio molecular dynamics: alkali ions in silicon. Phys. Rev. Lett. 70, 2928–2931.

Mishin, Y., Asta, M., Li, J., 2010. Atomistic modeling of interfaces and their impact on microstructure and properties. Acta Mater. 58, 1117–1151.

Mishin, Y., Lozovoi, A.Y., 2006. Angular-dependent interatomic potential for tantalum. Acta Mater. 54, 5013–5026.

Moelans, N., Blanpain, B., Wollants, P., 2008. An introduction to phase-field modeling of microstructure evolution. Calphad 32, 268–294.

Murphy, J.B., 1961. Interdiffusion in dilute aluminium-copper solid solutions. Acta Metall. 9, 563–569.

Neumann, G., 1987. A model for the calculation of monovacancy and divacancy contributions to the impurity diffusion in noble metals. Phys. Status Solidi (b) 144, 329–341.

Ogata, S., Li, J., Yip, S., 2005. Energy landscape of deformation twinning in bcc and fcc metals. Phys. Rev. B 71, 224102.

Onuki, A., 1989. Long-range interactions through elastic fields in phase-separating solids. J. Physical Soc. Japan 58, 3069–3072.

Ozolins, V., 2009. First-principles calculations of free energies of unstable phases: the case of fcc W. Phys. Rev. Lett. 102, 065702.

Ozolins, V., Wolverton, C., Zunger, A., 1998. Cu–Au, Ag–Au, Cu–Ag, and Ni–Au intermetallics: first-principles study of temperature-composition phase diagrams and structures. Phys. Rev. B 57, 6427–6443.

Penrose, O., 1991. A mean-field equation of motion for the dynamic Ising model. J. Stat. Phys. 63, 975–986.

Perdew, J.P., Burke, K., Ernzerhof, M., 1996a. Generalized gradient approximation made simple. Phys. Rev. Lett. 77, 3865–3868.

Perdew, J.P., Burke, K., Wang, Y., 1996b. Generalized gradient approximation for the exchange-correlation hole of a many-electron system. Phys. Rev. B Condens. Matter 54, 16533–16539.

Perdew, J.P., Yue, W., 1986. Accurate and simple density functional for the electronic exchange energy: generalized gradient approximation. Phys. Rev. B 33, 8800–8802.

Peterson, N.L., Rothman, S.J., 1970. Impurity diffusion in aluminum. Phys. Rev. B 1, 3264.

Pickett, W.E., 1989. Pseudopotential methods in condensed matter applications. Comput. Phys. Rep. 9, 115–197.

Poduri, R., Chen, L.Q., 1996. Non-classical nucleation theory of ordered intermetallic precipitates-application to the Al–Li alloy. Acta Mater. 44, 4253–4259.

Potts, R.B., 1952. Some Generalized Order-Disorder Transformations. Cambridge Univ Press.

Provatas, N., Elder, K., 2010. Phase-Field Methods in Materials Science and Engineering. Wiley-VCH, Berlin.

Rahman, A., 1964. Correlations in the motion of atoms in liquid argon. Phys. Rev. 136, 405–411.

Rothman, S.J., Peterson, N.L., 1969. Isotope effect and divacancies for self-diffusion in copper. Phys. Status Solidi (b) 35, 305–312.

Roy, A., Rickman, J.M., Gunton, J.D., Elder, K.R., 1998. Simulation study of nucleation in a phase-field model with nonlocal interactions. Phys. Rev. E 57, 2610–2617.

Ryckaert, J.P., Ciccotti, G., Berendsen, H.J.C., 1977. Numerical integration of the cartesian equations of motion of a system with constraints: molecular dynamics of n-alkanes. J. Comput. Phys. 23, 327–341.

Sagui, C., Orlikowski, D., Somoza, A.M., Roland, C., 1998. Three-dimensional simulations of Ostwald ripening with elastic effects. Phys. Rev. E 58, 4092–4095.

Sanchez, J.M., Ducastelle, F., Gratias, D., 1984. Generalized cluster description of multicomponent systems. Physica A: Stat. Theor. Phys. 128, 334–350.

Sandberg, N., Magyari-Kope, B., Mattsson, T.R., 2002. Self-diffusion rates in Al from combined first-principles and model-potential calculations. Phys. Rev. Lett. 89, 65901.

Saunders, N., Miodownik, A.P., 1998. CALPHAD (Calculation of Phase Diagrams): A Comprehensive Guide. A Pergamon Title.

Schwerdtfeger, P., 2011. The pseudopotential approximation in electronic structure theory. Chem. Phys. Chem. 12, 3143–3155.

Sheng, G., Bhattacharyya, S., Zhang, H., et al., 2012. Effective elastic properties of polycrystals based on phase-field desription. Mater. Sci. Eng. A 554, 67–71.

Shin, D., Arroyave, R., Liu, Z.-K., Van De Walle, A., 2006. Thermodynamic properties of binary hcp solution phases from special quasirandom structures. Phys. Rev. B 74, 024204.

Sholl, D.S., 2007. Using density functional theory to study hydrogen diffusion in metals: a brief overview. J. Alloys Compd. 446, 462–468.

Simmons, J.P., Shen, C., Wang, Y., 2000. Phase field modeling of simultaneous nucleation and growth by explicitly incorporating nucleation events. Scr. Mater. 43, 935–942.

Steinbach, I., 2009. Phase-field models in materials science. Model. Simul. Mater. Sci. Eng. 17, 073001.

Stocks, G.M., Temmerman, W.M., Gyorffy, B.L., 1978. Complete solution of the Korringa–Kohn–Rostoker coherent–potential–approximation equations: Cu–Ni alloys. Phys. Rev. Lett. 41, 339–343.

Stoebe, T.G., Gulliver Ii, R.D., Ogurtani, T.O., Huggins, R.A., 1965. Nuclear magnetic resonance studies of diffusion of Al27 in aluminum and aluminum alloys. Acta Metall. 13, 701–708.

Sundman, B., Ågren, J., 1981. A regular solution model for phases with several components and sublattices, suitable for computer applications. J. Phys. Chem. Solids 42, 297–301.

Tadmor, E.B., Ortiz, M., Phillips, R., 1996. Quasicontinuum analysis of defects in solids. Philos. Mag. A 73, 1529–1563.

Thornton, K., Ågren, J., Voorhees, P.W., 2003. Modelling the evolution of phase boundaries in solids at the meso- and nano-scales. Acta Mater. 51, 5675–5710.

Tikare, V., Holm, E.A., Fan, D., Chen, L.Q., 1998. Comparison of phase-field and Potts models for coarsening processes. Acta Mater. (USA) 47, 363–371.

Torquato, S., 2002. Random Heterogeneous Materials: Microstructure and Macroscopic Properties. Springer Verlag.

Uehara, T., Tsujino, T., Ohno, N., 2007. Elasto-plastic simulation of stress evolution during grain growth using a phase field model. J. Cryst. Growth 300, 530–537.

Vaithyanathan, V., Wolverton, C., Chen, L.Q., 2002. Multiscale modeling of precipitate microstructure evolution. Phys. Rev. Lett. 88, 125503.

Van De Walle, A., Asta, M., Ceder, G., 2002. The alloy theoretic automated toolkit: a user guide. Calphad 26, 539–553.

Van De Walle, A., Ceder, G., 2002. The effect of lattice vibrations on substitutional alloy thermodynamics. Rev. Mod. Phys. 74, 11.

Van De Walle, A., Ceder, G., Waghmare, U.V., 1998. First-principles computation of the vibrational entropy of ordered and disordered Ni 3 Al. Phys. Rev. Lett. 80, 4911–4914.

Van Der Ven, A., Ceder, G., 2000. Lithium diffusion in layered Li_xCoO_2. Electrochem. Solid State Lett. 3, 302–304.

Van Der Ven, A., Ceder, G., 2005a. Diffusion and configuration order in multicomponent solids. In: Yip, S. (Ed.), Handbook of Materials Modelling. Springer, Netherlands, pp. 367–394, 978-1-4020-3286-8.

Van Der Ven, A., Ceder, G., 2005b. First principles calculation of the interdiffusion coefficient in binary alloys. Phys. Rev. Lett. 94, 045901.

Van Der Ven, A., Ceder, G., Asta, M., Tepesch, P.D., 2001. First-principles theory of ionic diffusion with nondilute carriers. Phys. Rev. B 64, 184307.

Varotsos, P.A., Alexopoulos, K.D., 1986. Thermodynamics of Point Defects and Their Relation with Bulk Properties. North-Holland Amsterdam.

Vegard, L., 1921. Die Konstitution der Mischkristalle und die Raumfüllung der Atome. Z. Phys. A, Hadrons and Nuclei 5, 17–26.

Vineyard, G.H., 1956. Theory of order-disorder kinetics. Phys. Rev. 102, 981.

Vineyard, G.H., 1957. Frequency factors and isotope effects in solid state rate processes. J. Phys. Chem. Solids 3, 121–127.

Volin, T.E., Balluffi, R.W., 1968. Annealing kinetics of voids and the self-diffusion coefficient in aluminum. Phys. Status Solidi (b) 25, 163–173.

Wang, J., Wolverton, C., Muller, S., Liu, Z.K., Chen, L.Q., 2005a. First-principles growth kinetics and morphological evolution of Cu nanoscale particles in Al. Acta Mater. 53, 2759–2764.

Wang, T., Chen, L.Q., Liu, Z.K., 2005b. SQS for L12.

Wang, T., 2006. An integrated approach for microstructure simulation: application to Ni–Al–Mo alloys. Mater. Sci. Eng.. State College, The Pennsylvania State University.

Wang, T., Sheng, G., Liu, Z.K., Chen, L.Q., 2008. Coarsening kinetics of gamma' precipitates in the Ni–Al–Mo system. Acta Mater. 56, 5544–5551.

Wang, T., Zhu, J., Chen, L., Liu, Z.K., Mackay, R.A., 2004a. Modeling of lattice parameter in the Ni–Al system. Metall. Mater. Trans. A 35, 2313–2321.

Wang, Y., Curtarolo, S., Jiang, C., et al., 2004b. Ab initio lattice stability in comparison with CALPHAD lattice stability. Calphad 28, 79–90.

Wang, Y., Liu, Z.K., Chen, L.Q., 2004c. Thermodynamic properties of Al, Ni, NiAl, and Ni3Al from first-principles calculations. Acta Mater. 52, 2665–2671.

Wang, Y., Li, J., 2010. Phase field modeling of defects and deformation. Acta Mater. 58, 1212–1235.

Wang, T., Chen, L.Q., Liu, Z.K., 2007a. Lattice parameters and local lattice distortions in fcc-Ni solutions. Metall. Mater. Trans. A 38, 562–569.

Wang, Y., Liu, Z.K., Chen, L.Q., Wolverton, C., 2007b. First-principles calculations of β''-Mg_5Si_6/α-Al interfaces. Acta Mater. 55, 5934–5947.

Wang, Y., Wang, J.J., Zhang, H., et al., 2010. A first-principles approach to finite temperature elastic constants. J. Phys. Condens. Matter 22, 225404.

Wang, Y.U., Jin, Y.M., Cuitino, A.M., Khachaturyan, A.G., 2001a. Application of phase field microelasticity theory of phase transformations to dislocation dynamics: model and three-dimensional simulations in a single crystal. Philos. Mag. Lett. 81, 385–393.

Wang, Y.U., Jin, Y.M., Cuitino, A.M., Khachaturyan, A.G., 2001b. Phase field microelasticity theory and modeling of multiple dislocation dynamics. Appl. Phys. Lett. 78, 2324–2326.

Wang, Y.U., Jin, Y.M., Khachaturyan, A.G., 2002. Phase field microelasticity theory and modeling of elastically and structurally inhomogeneous solid. J. Appl. Phys. 92, 1351–1360.

Wei, S., Chou, M.Y., 1992. Ab initio calculation of force constants and full phonon dispersions. Phys. Rev. Lett. 69, 2799–2802.

Wei, S.H., Ferreira, L.G., Bernard, J.E., Zunger, A., 1990. Electronic properties of random alloys: special quasirandom structures. Phys. Rev. B 42, 9622.

Weithase, M., Noack, F., 1974. Kernrelaxations-Untersuchung der Selbstdiffusion in festem Kupfer. Z. Phys. A, Hadrons and Nuclei 270, 319–327.

Wejchert, J., Weaire, D., Kermode, J.P., 1986. Monte Carlo simulation of the evolution of a two-dimensional soap froth. Philos. Mag. B 53, 15–24.

Wimmer, E., Krakauer, H., Weinert, M., Freeman, A.J., 1981. Full-potential self-consistent linearized-augmented-plane-wave method for calculating the electronic structure of molecules and surfaces: O_2 molecule. Phys. Rev. B 24, 864–875.

Wolverton, C., Yan, X.Y., Vijayaraghavan, R., Ozolins, V., 2002. Incorporating first-principles energetics in computational thermodynamics approaches. Acta Mater. 50, 2187–2197.

Wolverton, C., Zunger, A., 1995. Ising-like description of structurally relaxed ordered and disordered alloys. Phys. Rev. Lett. 75, 3162–3165.

Woodward, C., 2005. First-principles simulations of dislocation cores. Mater. Sci. Eng. A 400-401, 59–67.

Wu, F.Y., 1982. The Potts model. Rev. Mod. Phys. 54, 235–268.

Yamanaka, A., Takaki, T., Tomita, Y., 2008. Elastoplastic phase-field simulation of self-and plastic accommodations in Cubic-tetragonal martensitic transformation. Mater. Sci. Eng. A 491, 378–384.

Zhu, J., Chen, L.Q., Shen, J., 2001a. Morphological evolution during phase separation and coarsening with strong inhomogeneous elasticity. Model. Simul. Mater. Sci. Eng. 9, 499.

Zhu, J., Chen, L.Q., Shen, J., Tikare, V., 2001b. Microstructure dependence of diffusional transport. Comput. Mater. Sci. 20, 37–47.

Zhu, J., Chen, L.Q., Shen, J., Tikare, V., 1999. Coarsening kinetics from a variable-mobility Cahn-Hilliard equation: application of a semi-implicit Fourier spectral method. Phys. Rev. E 60, 3564.

Zhu, J.Z., Liu, Z.K., Vaithyanathan, V., Chen, L.Q., 2002. Linking phase-field model to CALPHAD: application to precipitate shape evolution in Ni-base alloys. Scr. Mater. 46, 401–406.

Zunger, A., Wei, S.H., Ferreira, L.G., Bernard, J.E., 1990. Special quasirandom structures. Phys. Rev. Lett. 65, 353–356.

Biography

Long-Qing Chen is a distinguished professor of materials science and engineering, Engineering Science and Mechanics, and Mathematics at Penn State University. He received his B.S., M.S., and Ph.D. degrees from Zhejiang University, Stony Brook University, and MIT, respectively, all in materials science and engineering. Chen has interests in computational materials science of phase transformations and microstructure evolution using the phase-field method. He has received the TMS EMPMD Distinguished Scientist/Engineer Award (2011), and he is a Fellow of Guggenheim, APS, ASM, and MRS.

Yijia Gu is currently pursuing his Ph.D. in Materials Science and Engineering at the Pennsylvania State University. He received a B.S. in Materials Science and Engineering from Shandong University (2006), a M.S. in Materials Physics and Chemistry from Shanghai Jiao Tong University (2009). His doctoral research under the supervision of Prof. Long-Qing Chen is dedicated to modeling the flexoelectric effect in ferroelectric single crystals. He is a recipient of research awards from the American Ceramics Society (2013) and the Penn State Department of Materials Science and Engineering (2012). He has co-authored $\sim$20 publications in peer-reviewed journals.

INDEX

Ist Volume: Pages 1–1072
IInd Volume: Pages 1073–2008
IIIrd Volume: Pages 2009–2836

B

O

图书在版编目（ＣＩＰ）数据

物理冶金学：第 5 版：全 3 册／（美）戴维·劳克林
（David E. Laughlin），（日）宝野和博（Kazuhiro Hono）主编. --长沙：
中南大学出版社，2017.9
  ISBN 978 - 7 - 5487 - 3020 - 0

  Ⅰ.①物… Ⅱ.①戴… ②宝… Ⅲ.①物理冶金学
Ⅳ.①TG11

  中国版本图书馆 CIP 数据核字（2017）第 238825 号

物理冶金学(第 5 版)(全 3 册)
WULIYEJIN XUE (DI 5 BAN)(QUAN 3 CE)

David E. Laughlin　　Kazuhiro Hono　主编

□责任编辑　胡　炜　何运斌
□责任印制　易红卫
□出版发行　中南大学出版社
　　　　　　社址：长沙市麓山南路　　　　邮编：410083
　　　　　　发行科电话：0731 - 88876770　传真：0731 - 88710482
□印　　装　湖南众鑫印务有限公司

□开　　本　787 × 1092　1/16　□印张 186.5　□字数 6204 千字
□版　　次　2017 年 9 月第 1 版　□2017 年 9 月第 1 次印刷
□书　　号　ISBN 978 - 7 - 5487 - 3020 - 0
□全 3 册定价　800.00 元

国际材料前沿丛书
International Materials Frontier Series

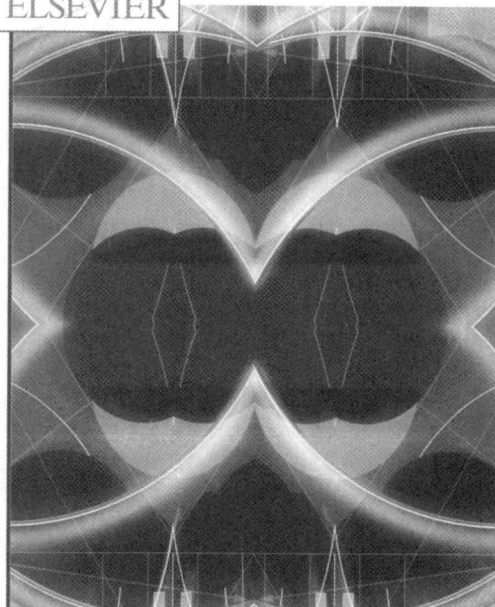

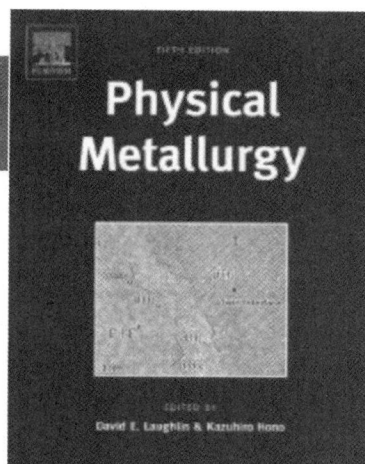

Physical
Metallurgy

FIFTH EDITION

EDITED BY
David E. Laughlin & Kazuhiro Hono

David E. Laughlin
Kazuhiro Hono

物理冶金学（第5版）
Physical Metallurgy (Fifth Edition)

（中）
影印版

中南大学出版社
www.csupress.com.cn
·长沙·

图字:18 - 2017 - 166 号

内容简介

　　该书全面系统地涵盖了材料科学领域中的相关知识，深刻地描述和解释了物理冶金学中的大多数方法，其中轻合金物理冶金、钛合金物理冶金、原子探针场离子显微镜、计算冶金和取向成像显微镜等都是当前科技发展的前沿。全书分为 3 册(上、中、下)：上册包含第 1 章 ~ 第 9 章的内容，中册包含第 10 章 ~ 第 19 章的内容，下册包含第 20 章 ~ 第 27 章的内容。

　　该书为材料类的经典书籍，第 5 版在第 4 版(1996 年)的基础上做了全面修改和扩充，且新增了几个主题以反映过去 18 年来物理冶金学的新进展。

　　本书可供相关学科领域(冶金、材料、物理、化学、生物医学等)的科研人员、工程技术人员使用，也可作为本科生、研究生等的参考用书。

作者简介

David E. Laughlin 美国匹兹堡卡内基梅隆大学教授，TMS 和 ASM International 的会士。他的主要研究方向为材料结构的电子显微镜观察、相变和磁性材料。1969 年获得美国德雷塞尔大学冶金工程专业学士学位，1973 年获得美国麻省理工学院冶金和材料科学专业博士学位。他从 1974 年起在美国卡内基梅隆大学任教。他发表了超过 450 篇同行评审的研究论文，并且共有 11 项美国专利。

Kazuhiro Hono 日本国立材料研究所研究员，磁性材料部主任，筑波大学教授。主要研究方向是金属材料，特别是磁性材料微观结构与性能的关系。1982 年获得日本东北大学学士学位，1988 年获得美国宾夕法尼亚州立大学材料科学与工程博士学位。

序

  这套 3 卷本的《物理冶金学》，是剑桥大学 Robert W. Cahn 教授和哥廷根大学 Peter Haasen 教授的享有崇高声望的同类著作的第 5 版。《物理冶金学》的第 1 版于 1965 年以单卷本的形式出版。关于本系列出版物的历史请参见第 4 版《序》。本系列权威参考工具书提供了物理冶金学——材料科学与工程领域的最大学科的全面知识。本系列著作广泛而深刻地描述和解释了物理冶金学的大多数方法。书中的每篇文章或由新作者重写，或由第 4 版作者单独或联合新的合作者全面修改和扩充。

  在《物理冶金学》第 1 版的《序》中，主编之一 R. W. Cahn 教授说："物理冶金学是现代材料科学赖以蓬勃发展的根源。"（R. W. Cahn，1965，《物理冶金学》，北荷兰出版公司，阿姆斯特丹 – 伦敦）。50 年过去了，这一说法仍然正确。本版的两位主编均是作为物理冶金学家培养的，但我们通常称自己为材料科学家。事实上，我们各自所在的部门并不使用"冶金学"一词。但材料科学的核心概念（有时称为"理论框架"），即材料的性能取决于其加工工艺和由此产生的微观结构，直接来源于物理冶金学和加工冶金学。若想了解材料科学的详尽历史，请查看 R. W. Cahn 主编的《材料科学的未来》，并与 R. F. Mehl 主编的《金属科学简史》进行对比。

  在本系列著作的第 1 版和第 2 版中，R. F. Mehl 在《物理冶金学的发展历史》一文中写道："物理冶金学和提取冶金学交织在一起。从历史的角度看，在很长一段时间内，冶金学的两个分枝构成的一门统一的艺术，并由同样的艺术家实现。"早在 2014 年面世的 3 卷本《冶金学论丛》，由 Seshadri Seetharaman 主编，同样由 ELSEVIER 出版，是本系列著作的同类出版物。这两套共 6 卷的著作无疑地全面覆盖了冶金学。

  《物理冶金学》第 5 版是继第 4 版（1996 年）大约 18 年之后出版的。2007 年，创始主编不幸辞世，延缓了第 5 版的面世。最后，由我们负责第 5 版的出版工作。

  本版作者具有更多的国际元素。21 世纪通信的便捷确实使编辑工作变得容易。然而，这似乎并没有加快所有作者写作的进程！

  本版增加了几个主题，以反映过去 18 年来物理冶金学的最新进展。部分章节由第 4 版的相同作者撰写，但均进行了更新，涵盖了新的论题和方法。我们感谢 45 名作者，他们各自勤奋地完成了有关章节的撰写和校对工作。这套著作获得的声誉当然应主要归功于作者而不是现任主编。当今，在我们的研究机构中，"精打细算的人"并不总是认可撰写类

似套书中的某一章节这样的工作。因此我们对每一位作者心存感激，他们将个人利益置之度外，详细报道他们精通的研究工作，而不在意其是否被 Web of Science 收录。

1970 年，本套著作主编之一（DEL）参加博士生入学资格考试，使用的教材就是《物理冶金学》的第 1 版。他从来不曾想过，近 50 年后他会亲自主编这套深受好评的《物理冶金学》著作。

我们愿将这 3 卷的著作献给我们的前任主编 Robert W. Cahn 教授和 Peter Haasen 教授。我们相信，我们在续写这套专著中付出的努力，可以达到他们曾经制定的高标准。

David E. Laughlin

Kazuhiro Hono

目　录

PHYSICAL METALLURGY

VOLUME II

FIFTH EDITION

PHYSICAL METALLURGY

VOLUME II

FIFTH EDITION

EDITED BY

DAVID E. LAUGHLIN

ALCOA Professor of Physical Metallurgy
Materials Science and Engineering
Carnegie Mellon University
Pittsburgh, PA, USA

KAZUHIRO HONO

Magnetic Materials Unit
National Institute for Materials Science
Tsukuba-city Ibaraki, Japan

ELSEVIER

AMSTERDAM • WALTHAM • HEIDELBERG • LONDON • NEW YORK
OXFORD • PARIS • SAN DIEGO • SAN FRANCISCO • SYDNEY • TOKYO

Elsevier
Radarweg 29, PO Box 211, 1000 AE Amsterdam, Netherlands
The Boulevard, Langford Lane, Kidlington, Oxford OX5 1GB, UK
225 Wyman Street, Waltham, MA 02451, USA

Fifth edition

Notice
No responsibility is assumed by the publisher for any injury and/or damage to persons or property as a matter of products liability, negligence or otherwise, or from any use or operation of any methods, products, instructions or ideas contained in the material herein. Because of rapid advances in the medical sciences, in particular, independent verification of diagnoses and drug dosages should be made.

British Library Cataloguing in Publication Data
A catalogue record for this book is available from the British Library

Library of Congress Cataloging in Publication Data
A catalog record for this book is available from the Library of Congress

Volume II ISBN: 978-0-444-59597-3
SET ISBN: 978-0-444-53770-6

For information on all Elsevier publications
visit our website at www.store.elsevier.com

Printed and bound in the UK

Working together
to grow libraries in
developing countries

www.elsevier.com • www.bookaid.org

CONTENTS

LIST OF CONTRIBUTORS TO VOLUME II

S.S. Babu
The University of Tennessee, Knoxville, Tennessee, USA

C.Y. Barlow
Institute for Manufacturing, University of Cambridge, Cambridge, UK

Katayun Barmak
Columbia University, New York, NY, USA

Joël Bonneville
Département de Physique et Mécanique des Matériaux, Institut P', Université de Poitiers, CNRS UPR 3346, France

Hamish L. Fraser
Center for Electron Microscopy and Analysis (CEMAS), Department of Materials Science and Engineering, The Ohio State University, Columbus, OH, USA

N. Hansen
Danish–Chinese Center for Nanometals, Section for Materials Science and Advanced Characterization, Technical University of Denmark, Roskilde, Denmark

K. Hono
National Institute of Materials Science, Tsukuba, Japan

James M. Howe
Department of Materials Science and Engineering, University of Virginia, Charlottesville, VA, USA

Gernot Kostorz
ETH Zurich, Department of Physics, Auguste-Piccard-Hof 1, Zurich, Switzerland

Campbell Laird
Department of Materials Science and Engineering, University of Pennsylvania, Philadelphia, PA, USA

David E. Laughlin
Department of Materials Science and Engineering, Carnegie Mellon University, Pittsburgh, PA, USA

David W. McComb
Center for Electron Microscopy and Analysis (CEMAS), Department of Materials Science and Engineering, The Ohio State University, Columbus, OH, USA

Michael E. McHenry
Department of Materials Science and Engineering, Carnegie Mellon University, Pittsburgh, PA, USA

Pedro Peralta
School for Engineering of Matter, Transport and Energy, Arizona State University, Tempe, AZ, USA

W.T. Reynolds
Virginia Tech, Blacksburg, VA, USA

David Rodney
Institut Lumière Matière, Université Lyon 1, CNRS UMR 5306, France

Anthony D. Rollett
Carnegie Mellon University, Pittsburgh, PA, USA

G. Spanos
The Minerals, Metals, and Materials Society,
Warrendale, PA, USA

Robert E.A. Williams
Center for Electron Microscopy and Analysis
(CEMAS), Department of Materials Science and
Engineering, The Ohio State University, Columbus,
OH, USA

PREFACE TO THE FIFTH EDITION

These three volumes represent the fifth edition of *Physical Metallurgy*, a prestigious and famous family formerly edited by Robert Cahn (University of Cambridge) and Peter Haasen (Universität Göttingen). Physical Metallurgy was first published as a single volume in 1965. See the preface to the fourth edition for a history of this series. It is an authoritative reference tool, providing a complete knowledge set in Physical Metallurgy, the largest discipline in the fields of Materials Science and Engineering. This series describes and explains most aspects of physical metallurgy across the full breadth and in considerable depth. Each article has been either rewritten by new authors, or thoroughly revised and expanded, either by the 4th edition authors alone or jointly with new co-authors.

In the preface to the first edition of Physical Metallurgy, the founding editor of this series stated that "Physical metallurgy is the root from which the modern science of materials has principally sprung." (R. W. Cahn (1965), Physical Metallurgy North Holland Publishing Company, Amsterdam-London). Over the next five decades this has continued to ring true. While both of the editors of this edition were educated as physical metallurgists, nowadays it is more common to call ourselves Materials Scientists, and indeed our respective departments do not utilize the word "metallurgy." But the core concept (or sometimes called its paradigm) of Materials Science, that the properties and performance of materials have their origin in their processing and resulting microstructure, is derived directly from Physical and Process Metallurgy. For an exhaustive history of Materials Science see "The Coming of Materials Science" by R. W. Cahn and compare this to "A Brief History of the Science of Metals," by R. F. Mehl.

In the article "The Historical Development of Physical Metallurgy" by R. F. Mehl, which appeared in the first two editions of this series, Mehl wrote: "*Physical Metallurgy* has been ... interwoven with *Extractive Metallurgy*, and for a long time, historically, these two branches constituted a common art, practiced by the same artisans..." Early in 2014 there appeared the three volume *Treatise on Process Metallurgy*, edited by Seshadri Seetharaman and also published by ELSEVIER, which may be said to be the companion to this set. These six volumes certainly cover Metallurgy comprehensively.

This fifth edition of Physical Metallurgy is published some eighteen years after the 4th edition of this series, which was published in 1996. The lamented death of the founding editor in 2007 slowed down the appearance of this 5th edition. Finally we are ready to present the 5th edition.

This edition has a more international flavor to the listing of authors. Indeed in this 21st Century the ease with which correspondence can be sent makes the task of editing easier. It does not seem to speed up the writing and response of all authors however!

Several new subjects were added in this edition to update the progress in *physical metallurgy* in the last 18 years. Several of the chapters are written by the same authors as those in the fourth edition; but they have all been updated to include new topics and approaches.

We do thank the 45 authors for their hard work and diligence to get their chapters and proofs in. It is of course to authors, more than the current editors, that a series such as this gets its reputation. In a day when "bean counters" in our institutions do not always appreciate the work that goes into writing a chapter in a series such as this, we are grateful that each of the authors put that aside and wrote up work about which they are experts, whether or not it gets indexed in the Web of Science!

In 1970 one of the editors of these volumes (DEL) studied for his qualifying examinations for entrance into the Ph.D. program from the first edition of the series of *Physical Metallurgy*. Little did he suspect that nearly five decades later he would be editing the 5th edition to this well received series on *Physical Metallurgy*.

We wish to dedicate these volumes to our predecessor editors: Prof. Robert W. Cahn and Prof. Peter Haasen. We trust that our efforts in the continuation of this series will be up to the high standards which they have set.

David E. Laughlin
ALCOA Professor of Physical, Metallurgy,
Department of Materials Science and Engineering,
Carnegie Mellon University, Pittsburgh, PA USA

Kazuhiro Hono
ZNIMS Fellow, Naitonal Institute for
Materials Science, Tsukuba, 305-0047, Japan

PREFACE TO THE FOURTH EDITION

The first, single-volume edition of this Work was published in 1965 and the second in 1970; continued demand prompted a third edition in two volumes which appeared in 1983. The first two editions were edited by myself alone, but in preparing the third, which was much longer and more complex, I had the crucial help of Peter Haasen as co-editor. The third edition came out in 1983, and sold steadily, so that the publishers were motivated to propose the preparation of yet another version of the Work; we began the joint planning for this in early 1992. We agreed on the changes and additions we wished to make: the responsibility for commissioning chapters was divided equally between us, but the many policy decisions, made during a series of face-to-face discussions, were very much a joint enterprise. Peter Haasen was able to commission all the chapters which he had agreed to handle, and this task (which involved detailed discussions with a number of authors) was completed in early 1993. Thereupon, in May 1993, my friend of many years was suddenly taken ill; the illness worsened rapidly, and in October of the same year he died, at the early age of 66. When he was already suffering the ravages of his fatal illness, he yet found the resolve and energy to revise his own chapter and to send it to me for comments, and to modify it further in the light of those comments. He was also able to examine, edit and approve the revised chapter on dislocations, which came in early. These were the very last professional tasks he performed. Peter Haasen was in every sense co-editor of this new edition, even though fate decreed that I had to complete the editing and approval of most of the Chapter I am proud to share the title-page with such an eminent physicist.

The first edition had 22 chapters and the second, 23. There were 31 chapters in the third edition and the present edition has 32. The first two editions were single volumes, the third had to be divided into two volumes, and now the further expansion of the text has made it necessary to go to three volumes. This fourth edition is nearly three times the size of the first edition 30 years ago; this is due not only to the addition of new topics, but also to the fact that the treatment of existing topics has become much more substantial than it was in 1965. There are those who express the conviction that physical metallurgy has passed its apogee and is in steady decline; the experience of editing this edition, and the problems I have encountered in holding enthusiastic authors back from even more lengthy treatments (to avoid exceeding the agreed page limits by a wholly unacceptable margin), have shown me how mistaken this pessimistic assessment is! Physical metallurgy, the parent discipline of materials science, has maintained its central status undiminished.

The first three editions each opened with a historical overview. We decided to omit this in the fourth edition, for two main reasons: the original author had died and it would have fallen to others to revise his work, never an entirely satisfactory proceeding; it had also become plain (especially from the reaction of the translators of the earlier editions into Russian) that the overview was not well balanced

between different parts of the world. I am engaged in writing a history of materials science, as a separate venture, and this will incorporate proper attention to the history of physical metallurgy as a principal constituent. — It also proved necessary to leave out the chapter on superconducting alloys: the ceramic superconductor revolution has virtually removed this whole field from the purview of physical metallurgy. — Three entirely new topics are treated in this edition: one is oxidation, hot (dry) corrosion and protection of metallic materials, another is the dislocation theory of the mechanical behavior of intermetallic compounds. The third new topic is a leap into very unfamiliar territory: it is entitled "A Metallurgist's Guide to Polymers". Many metallurgists — including Alan Windle, the author of this chapter — have converted in the course of their careers to the study of the more physical aspects of polymers (regarded by many materials scientists as *the* "materials of the future"), and have had to come to terms with novel concepts (such as "semicrystallinity") which they had not encountered in metals: Windle's chapter is devoted to analysing in some depth the conceptual differences between metallurgy and polymer science, for instance, the quite different principles which govern alloy formation in the two classes of materials. I believe that this is the first treatment of this kind.

Six of the existing chapters (now numbered 1, 4, 21, 22, 27, 30) have been entrusted to new authors, while another five chapters have been revised by the previous authors with the collaboration of additional authors (8,13,16, 17, 19). Chapter 19, originally entitled "Alloys rapidly quenched from the melt" has been broadened and retitled "Metastable states of alloys". A treatment of quasicrystals has been introduced in the form of an appendix to Chapter 4, which is devoted to the solid-state chemistry of intermetallic compounds; this seemed appropriate since quasicrystallinity is generally found in such compounds. — Only three chapters still have the same authors they had in the first edition, written some 32 years ago.

27 of the 29 new versions of existing chapters have been substantially revised, and many have been entirely recast. Two Chapters (11 and 25) have been reprinted as they were in the third edition, except for corrected cross-references to other chapters, but revision has been incorporated in the form of an Addendum to each of these chapters; this procedure was necessary on grounds of timing.

This edition has been written by a total of 44 authors, working in nine countries. It is a truly international effort.

I have prepared the subject index and am thus responsible for any inadequacies that may be found in it. I have also inserted some cross-references between chapters (internal cross- references within chapters are the responsibility of the various authors), but the function of such cross-references is better achieved by liberal use of the subject index.

As always, the editors have been well served by the exceedingly competent staff of North—Holland Physics Publishing (which is now an imprint of Elsevier Science B.V. in Amsterdam; at the time of the first two editions, North—Holland was still an independent company). My particular thanks go to Nanning van der Hoop and Michiel Bom on the administrative side, to Ruud de Boer who is responsible for production and to Chris Ryan and Maurine Alma who are charged with marketing. Mr. de Boer's care and devotion in getting the proofs just right have been extremely impressive. My special thanks also go to Professor Colin Humphreys, head of the department of materials science and metallurgy in Cambridge University, whose warm welcome and support for me in my retirement made the creation of this edition feasible. Finally, my thanks go to all the authors, who put up with good grace with the numerous forceful, sometimes impatient, messages which I was obliged to send in order to "get the show on the road", and produced such outstanding chapters under pressure of time.

I am grateful to Dr. W.J. Boettinger, one of the authors, and his colleague Dr. James A. Warren, for kindly providing the computer-generated dendrite microstructure that features on the dust-cover.

The third edition was dedicated to the memory of Robert Franklin Mehl, the author of the historical chapter and a famed innovator in the early days of physical metallurgy in America. I would like to dedicate this fourth edition to the memory of two people: my late father-in- law, *Daniel Hanson* (1892–1953), professor of metallurgy at Birmingham University for many years, who did more than any other academic in Britain to foster the development and teaching of modern physical metallurgy; and the physical metallurgist and scientific publisher — and effective founder of Pergamon Press — *Paul Rosbaud* (1896–1963), who was retained by the then proprietor of the North–Holland Publishing Company as an adviser and in 1960, in the presence of the proprietor, eloquently urged upon me the need for a new, advanced, multiauthor text on physical metallurgy.

Robert W. Cahn
Cambridge
November 1995.

PREFACE TO THE THIRD EDITION

The first edition of this book was published in 1965 and the second in 1970. The book continued to sell well during the 1970s and, once it was out of print, pressure developed for a new edition to be prepared. The subject had grown greatly during the 1970s and R. W. C. hesitated to undertake the task alone. He is immensely grateful to P. H. for converting into a pleasure what would otherwise have been an intolerable burden!

The second edition contained 22 chapters. In the present edition, 8 of these 22 have been thoroughly revised by the same authors as before, while the others have been entrusted to new contributors, some being divided into pairs of chapters. In addition, seven chapters have been commissioned on new themes. The difficult decision was taken to leave out the chapter on superpure metals and to replace it by one focused on solute segregation to interfaces and surfaces—a topic that has made major strides during the past decade and which is of great practical significance. A name index has also been added.

Research in physical metallurgy has become worldwide and this is reflected in the fact that the contributors to this edition live in no fewer than seven countries. We are proud to have been able to edit a truly international text, both of us having worked in several countries ourselves. We would like here to express our thanks to all our contributors for their hard and effective work, their promptness and their angelic patience with editorial pressures!

The length of the book has inevitably increased, by 50% over the second edition, which was itself 20% longer than the first edition. Even to contain the increase within these numbers has entailed draconian limitations and difficult choices; these were unavoidable if the book was not to be priced out of its market. Everything possible has been done by the editors and the publisher to keep the price to a minimum (to enable readers to take the advice of G. Chr. Lichtenberg (1775): "He who has two pairs of trousers should pawn one and buy this book".).

Two kinds of chapters have been allowed priority in allocating space: those covering very active fields and those concerned with the most basic topics such as phase transformations, including solidification (a central theme of physical metallurgy), defects, and diffusion. Also, this time we have devoted more space to experimental methods and their underlying principles, microscopy in particular. Since there is a plethora of texts available on the standard aspects of X-ray diffraction, the chapter on X-ray and neutron scattering has been designed to emphasize less familiar aspects. Because of space limitations, we regretfully decided that we could not include a chapter on corrosion.

This revised and enlarged edition can properly be regarded as to all intents and purposes a new book.

Sometimes it was difficult to draw a sharp dividing line between physical metallurgy and process metallurgy, but we have done our best to observe the distinction and to restrict the book to its intended

theme. Again, reference is inevitably made occasionally to nonmetallics, especially when they serve as model materials for metallic systems.

As before, the book is designed primarily for graduate students beginning research or undertaking advanced courses, and as a basis for more experienced research workers who require an overview of fields comparatively new to them, or with which they wish to renew contact after a gap of some years.

We should like to thank Ir J. Soutberg and Dr A. P. de Ruiter of the North-Holland Publishing Company for their major editorial and administrative contributions to the production of this edition, and in particular we acknowledge the good-humored resolve of Dr W. H. Wimmers, former managing director of the Company, to bring this third edition to fruition. We are grateful to Dr Bormann for preparing the subject index. We thank the hundreds of research workers who kindly gave permission for reproduction of their published illustrations: all are acknowledged in the figure captions.

Of the authors who contributed to the first edition, one is no longer alive: Robert Franklin Mehl, who wrote the introductory historical chapter. What he wrote has been left untouched in the present edition, but one of us has written a short supplement to bring the treatment up to date, and has updated the bibliography. Robert Mehl was one of the founders of the modern science of physical metallurgy, both through his direct scientific contributions and through his leadership and encouragement of many eminent metallurgists who at one time worked with him. We dedicate this third edition to his memory.

Robert W. Cahn, Paris
Peter Haasen, Göttingen.
April 1983.

PREFACE TO THE FIRST AND SECOND EDITIONS

This book sets forth in detail the present state of physical metallurgy, which is the root from which the modern science of materials has principally sprung. That science has burgeoned to such a degree that no author can do justice to it at an advanced level; accordingly, a number of well-known specialists have consented to write on the various principal branches, and the editor has been responsible for preserving a basic unity among the expert contributions. This book is the first general text, as distinct from research symposium, which has been conceived in this manner. While principally directed at senior under-graduates at universities and colleges of technology, the book is therefore also appropriate for post-graduates and particularly as a base for experienced research workers entering fields of physical metallurgy new to them.

Certain topics have been left to one side or treated at modest length, so as to limit the size of the book, but special stress has been placed on others, which have rarely been accorded much space. For instance, a good deal of space is devoted to the history of physical metallurgy, and to point defects, structure and mechanical properties of solid solutions, theory of phase transformations, recrystalliza-tion, superpure metals, ferromagnetic properties, and mechanical properties of two-phase alloys. These are all active fields of research. Experimental techniques, in particular diffraction methods, have been omitted for lack of space; these have been ably surveyed in a number of recent texts. An exception has however been made in favor of metallographic techniques since, electron microscopy apart, recent innovations have not been sufficiently treated in texts.

Each chapter is provided with a select list of books and reviews, which will enable readers to delve further into a particular subject. Internal cross-references and the general index will help to tie the various contributions together.

I should like here to acknowledge the sustained helpfulness and courtesy of the publisher's staff, and in particular of Mr A. T. G. van der Leij, and also the help provided by Prof P. Haasen and Dr T. B. Massalski in harmonizing several contributions.

Brighton, June 1965 (and again 1970) R. W. Cahn

ABOUT THE EDITORS

David E. Laughlin is the ALCOA Professor of Physical Metallurgy in the Department of Materials Science and Engineering at Carnegie Mellon University, Pittsburgh, PA. He obtained his B.S. in Metallurgical Engineering from Drexel University in 1969 and his Ph.D. in Metallurgy and Materials Science from MIT in 1973. He has taught at CMU since 1974. He is Principal Editor of *Metallurgical and Materials Transactions* and has co-edited eight books. His research has centered on the structure of materials as observed by electron microscopy, phase transformations and magnetic materials. He has published more than 450 peer reviewed research papers and is co-inventor on eleven US patents. Laughlin is a Fellow of TMS and ASM International.

Kazuhiro Hono is NIMS Fellow and Director of Magnetic Materials Unit at National Institute for Materials Science and Professor of Materials Science at the University of Tsukuba. He obtained his B.S. from Tohoku University in 1982, and Ph.D. in Materials Science and Engineering from the Pennsylvania State University in 1988. His research interests are mirostructure-property relationships of metallic materials, in particular of magnetic materials.

ABOUT THE EDITORS

10 Microstructure of Metals and Alloys

G. Spanos, The Minerals, Metals, and Materials Society, Warrendale, PA, USA
W.T. Reynolds, Virginia Tech, Blacksburg, VA, USA

10.1 Introduction

This chapter builds upon the excellent overview on "Microstructure" in Chapter 9 of the 4th edition of *Physical Metallurgy*, by H. Glieter (Gleiter, 1996). That chapter provided an in-depth assessment of the elements, characterization, and development of microstructure, and ended with a discussion on nanocrystalline materials. That comprehensive work in many ways provided a strong foundation for the present chapter.

Since publication of that 4th edition of *Physical Metallurgy* nearly 14 years ago (Cahn and Haasen, 1996), there have been many breakthroughs in the characterization, analysis, and modeling of microstructure in metallic materials, and probably none more important than the rapid evolution of three-dimensional (3D) materials analyses (Spanos, 2006; Uchic, 2006; Thornton and Poulsen, 2008; Kral et al., 2004; Spanos et al., 2010). This chapter provides an overview of microstructure in metallic materials that captures many of these new developments, with particular emphasis on 3D microstructural analyses of microstructures developed during *solid-state* evolution in metallic systems.

Including all microstructural forms and elements of microstructure in metallic materials is well beyond the scope of a single chapter; the focus here is thus on providing an overview and some representative examples centered about the characterization, description, and classification of 3D solid-state metallic microstructures. The reader is referred to any number of excellent texts that describe in detail lower dimensionality elements of microstructure in metals, such as point defects ("zero-dimensional") (Shewmon, 1989), line defects (one-dimensional) (Hirth and Lothe, 1982), and interfaces (two-dimensional) (Howe, 1997; Bollmann, 1982; Sutton and Balluffi, 1995).

10.1.1 Definition of "Microstructure" in Metallic Materials

Gleiter suggested that "the microstructure of crystalline materials has been defined by the type, structure, number, shape and topological arrangement of phases and/or lattice defects which are in most cases not part of the thermodynamic equilibrium structure" (Gleiter, 1996). A similar definition and further description of "microstructure" will be employed here, as follows. Microstructure in metallic materials consists of the distribution and topological arrangement of grains, phases, interfaces, and other defects in three dimensions. A metal microstructure is a direct result of the alloy chemistry and processing history, and it dictates the final properties and performance of the alloy, and/or any component made from it. Microstructure in the current context is not limited to features that are in the micron size range, but is used broadly to represent feature sizes spanning from angstroms to hundreds of microns.

10.1.2   Overview of the Sections in this Chapter

Because of the importance and rapid growth of accurate 3D characterization and quantification techniques to understand real (not just idealized) microstructures in three dimensions (Spanos, 2006; Uchic, 2006; Thornton and Poulsen, 2008; Kral et al., 2004; Spanos et al., 2010), Section 10.2 below provides an overview of the methodologies for reconstructing, visualizing, quantifying, and analyzing 3D metallic microstructures. A number of 3D microstructures characterized by these techniques are also highlighted and discussed. That section thus provides a foundation for understanding the various elements and classifications of metallic microstructures in three dimensions, and acquaints readers with the tools and methodologies available for studying microstructures in three dimensions. Section 10.3 begins with a brief discussion of how various microstructures are described, and then considers specific solid-state microstructures with an emphasis on the morphology and crystallography of the micro-constituents. Finally, a summary and brief discussion of the importance of microstructure in dictating materials properties and design is provided in Section 10.4.

10.2   3D Microstructure and Microstructural Analyses

As alluded to above, the past 15 years have seen rapid growth in the development of 3D techniques for the characterization, visualization, quantification, and modeling of 3D microstructures in materials (Mangan and Shiflet, 1994; Mangan et al., 1997; Wolfsdorf et al., 1997; Yokomizo et al., 2003; Kral and Spanos, 1999; Chawla and Sidhu, 2007; Spanos et al., 2008; Rowenhorst et al., 2010; Herbig et al., 2011; Kelly and Miller, 2007; Hono, 2002). In fact, within the last decade there have been four comprehensive review articles that have provided snapshots of the state of the art in the characterization and analysis of 3D microstructures (Spanos, 2006; Uchic, 2006; Thornton and Poulsen, 2008; Kral et al., 2004; Spanos et al., 2010).

Because of the importance of accurate 3D characterization and quantification in understanding and controlling (through processing) microstructures in three dimensions, an overview of the methodologies for characterizing and quantifying microstructures in 3D is provided here. This will specifically provide a foundation with examples for understanding the elements and classification of metallic microstructures discussed in Section 10.3 of this chapter. These 3D methodologies include experimental 3D microstructure characterization techniques; related computationally based 3D reconstruction, visualization, and quantitative analysis methods; and 3D microstructural modeling techniques that use experimentally determined 3D microstructures for model input and validation. Traditional two-dimensional (2D) methods of microstructure characterization and analysis will be briefly considered first to provide a partial explanation for the recent surge in 3D microstructure studies.

10.2.1   2D Microstructural Characterization and Stereology for Determining 3D Elements of Microstructure

Traditional 2D metallographic and microscopy techniques analyze single planes of observation (or very thin sections) in metallic materials, and often use quantitative stereological procedures to estimate parameters that describe 3D features of the microstructure. These techniques are especially valuable for measuring *average* parameters in systems with relatively simple microstructures.

Typical 2D characterization methods include (but are not limited to) optical microscopy and scanning electron microscopy (SEM) of a plane of polish, the more conventional transmission electron microscopy (TEM) techniques that provide a projected image through a foil of material approximately 100 nm or less in thickness, and X-ray mapping techniques that are mostly limited to the surface of a material (e.g. Laue diffraction, low-energy electron diffraction, energy dispersive-based X-ray mapping). There are a number of excellent reference books on stereological methods for extracting 3D microstructural information from 2D data (Russ and Dehoff, 2000; Underwood, 1981; Rhines, 1967); so these procedures will not be discussed here. Key stereological parameters that can be measured from such techniques can often be categorized, though, based on the "dimensionality" of the correlation level of the relevant features within the material microstructure (i.e. "n-point" correlations). Some of the key parameters that can be measured by such stereological procedures include volume fraction, particle/grain size (1-point correlations), and shape (e.g. aspect ratio) or texture (2-point correlations).

The degree of accuracy of any 3D microstructure analysis from 2D data is always limited by the size of the 2D data set and the complexity of the microstructural features being analyzed. For example, measurement of average particle size or volume fraction from 2D observations of a uniformly distributed array of spherical particles in a material can provide a reasonable estimate of the "averaged" 3D microstructural parameters. On the other hand, determining the size and shape *distributions* of particles or grains in the more complex microstructures typically found in most real materials is only possible by using 3D techniques. Complex microstructures might include, for instance, arrays of interconnected or heterogeneously distributed particles, or perhaps grains that have an unknown shape, bimodal distribution of sizes, or morphological texture. In cases like these, using stereology to infer 3D information from 2D data leads to oversimplification. Direct measurement of the desired 3D information using 3D techniques is essential to accurately characterize these complex microstructures.

10.2.2 3D Microstructural Characterization and Analysis

A number of characterization techniques are now available to determine the morphology, topology, crystallography, and/or interconnectivity of metallic microstructures in three dimensions. These techniques apply over the wide range of length scales represented schematically in **Figure 1**. Most of these methods are destructive, and thus provide a snapshot of a microstructure in time. Only X-ray techniques (both X-ray tomography and X-ray diffraction microscopy) and in some cases TEM tomography can be considered nondestructive, and thus allow for the possibility of in situ (or ex situ) observation of the evolution of the same 3D microstructure with time. Such observations can occur during experiments that subject the material to heating/cooling, and/or deformation.

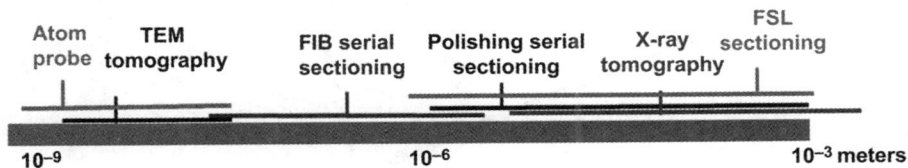

Figure 1 Schematic representation of 3D characterization techniques as function of length scale. Courtesy of D. J. Rowenhorst, Naval Research Laboratory. (For color version of this figure, the reader is referred to the online version of this book.)

The remainder of this section provides an overview of the capabilities and utility of techniques for reconstructing microstructures in 3D and shows some representative examples of 3D microstructures that have been imaged, characterized, and quantified. Somewhat more depth will be provided on one methodology—serial sectioning by mechanical polishing—with which the authors have had significant experience.

10.2.2.1   3D Analyses of Metallic Microstructures by X-ray Tomographic Techniques

X-ray tomography is a well-known methodology for characterizing materials in three dimensions. Historically, it has been well suited to studying materials with large differences in absorption (which often correlates with density) between the microstructural features and the matrix within which they lie. For example, there have been many X-ray tomographic studies of voids within a metal matrix (Everett et al., 2002; Fonda et al., 2007), inclusions and other second phase particles within metals (Fonda et al., 2007) (see **Figure 2**), and fiber-reinforced composites (Withers et al., 2006; Mcdonald et al., 2003). In addition to the continual improvements in image analysis and "cleaning" techniques to remove background and anomalies, newer tomography-based techniques (i.e. beyond conventional phase contrast X-ray tomography), such as X-ray holo-tomography (Ludwig et al., 2009a), and X-ray diffraction contrast tomography (Ludwig et al., 2009b) have been developed. These advances can result in greater spatial and/or mass/absorption resolution, and often provide for more accurate quantification of microstructural features. In this regard, the spatial resolution is now on the order of a couple microns, and with techniques such as X-ray diffraction contrast tomography, reconstruction of the grains of single-phase metallic materials is now possible. For example, the full reconstruction by diffraction contrast tomography of hundreds of beta-titanium grains within a beta-stabilized titanium alloy is shown in **Figure 3**.

A second type of X-ray technique for determining 3D microstructures that is differentiated from X-ray tomography is X-ray diffraction microscopy (Poulsen, 2004; Lienert et al., 2007). With this technique, in addition to providing quantitative 3D information on the morphology, size, and interconnectivity of grains in metallic materials, the full crystallographic orientation for each grain is

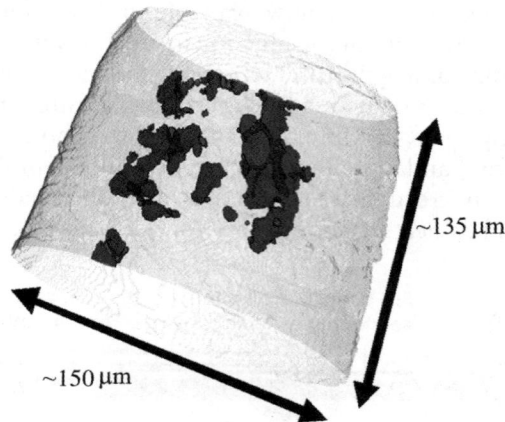

Figure 2   3D X-ray tomographic reconstruction of sigma precipitates (blue) and microvoids (red) in a super austenitic stainless steel (AL6XN). From Fonda et al., 2007. (For interpretation of the references to color in this figure legend, the reader is referred to the online version of this book.)

100 μm

Figure 3 Diffraction contrast tomography (DCT) reconstruction of beta-titanium grains in a beta-stabilized titanium alloy (Ti–21S). Reprinted with permission from Ludwig et al., 2009b, Copyright 2009, American Institute of Physics. (For color version of this figure, the reader is referred to the online version of this book.)

obtained. This methodology is especially powerful when it is desirable to run deformation or failure models on experimentally determined reconstructions of 3D grains (Lewis and Geltmacher, 2006; Quey and Dawson, 2011; Venkatramani et al., 2007).

Since these X-ray techniques are nondestructive, they also can be employed in combination with in situ furnaces or strain stages, or used in conjunction with ex situ techniques in which the specimen is taken back and forth between the furnace or strain device and the X-ray tomography (or X-ray diffraction microscopy) system (Schmidt et al., 2004; Lienert et al., 2009; Poulsen et al., 2003; Margulies et al., 2001). This flexibility provides for direct tracking of the temporal evolution of 3D grains as a result of mechanical strain or fatigue (Lienert et al., 2009; Withers et al., 2006; Moffat et al., 2007) and/or heating (Schmidt et al., 2004; Mckenna, 2010). This type of temporal 3D data is invaluable input for testing, verifying, and enhancing simulation models of 3D grain evolution and material failure (Lewis and Geltmacher, 2006; Dawson et al., 2002; Manchiraju et al., 2008; Herbig et al., 2011; McKenna et al., 2009; Chen, 2002; Ni et al., 2008) as well as phase-field models of grain evolution because of applied thermal treatments (Mckenna, 2010).

10.2.2.2 Femtosecond Laser Serial Sectioning Analysis of Metallic Microstructures
3D analyses of metallic materials by femtosecond laser sectioning is in its nascent stages, and this technique is especially suited for reconstructing large volumes of material ($\sim mm^3$) within tractable times (hours) (Echlin et al., 2011). The general principle involves using a femtosecond laser to ablate large lateral areas of material rapidly in conjunction with some imaging and/or analysis technique (such as optical microscopy) for each 2D slice. The process is repeated to generate data that are subsequently reconstructed in three dimensions. **Figure 4** shows a schematic representation of the arrangement with optical microscopy, and a built-in laser-induced breakdown spectroscopy detector to provide quantitative chemical information (Echlin et al., 2011). This technique has been used to

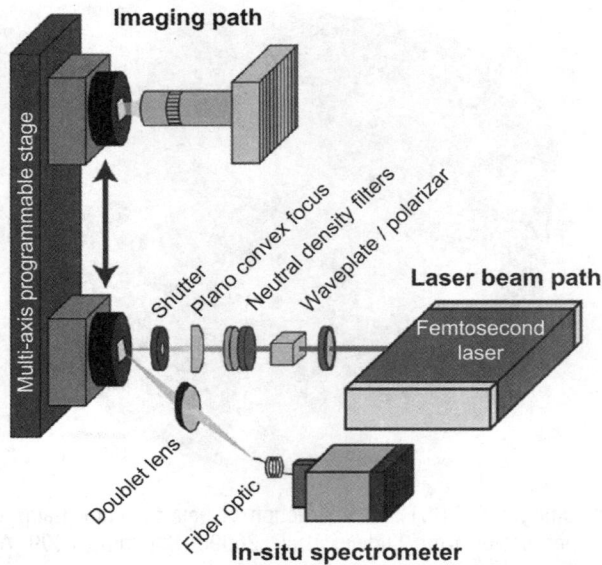

Figure 4 Schematic depiction of the femtosecond laser-aided serial sectioning technique showing a sample affixed to a four-axis programmable stage that translates between the optical imaging path and the laser machining path, with laser-induced breakdown spectroscopy diagnostic equipment. From Echlin et al., 2011. (For color version of this figure, the reader is referred to the online version of this book.)

study the 3D size and spatial distribution of TiN particles in a titanium-modified 4330 steel (**Figure 5**; Echlin et al., 2011).

10.2.2.3   Serial Sectioning of Metallic Microstructures by Mechanical Polishing (and/or milling)

A thorough discussion on this topic can be found in a recent review of "3D Microstructure Representation" (Spanos et al., 2010). Serial sectioning of metallic microstructures by mechanical polishing dates back at least to 1918, when the 3D microstructure of pearlite in steel was studied by serial sectioning (Forsman, 1918). In that work, light micrographs were projected onto cardboard layers, from which solid models of the pearlite were reconstructed. Subsequently, a number of other researchers used various ingenious methods to reconstruct metallic microstructures from serial sections. These included (1) producing a motion picture to show slices through a pearlite colony (Hillert, 1962), (2) building a 3D Plexiglas model to show the eutectic fault structure in a Cu–Al alloy (Hopkins and Kraft, 1965), and (3) using mostly hand-drawn sketches to manually reconstruct the shapes of grain boundary precipitates in an Ag–Al alloy (Hawbolt and Brown, 1967), voids in a sintered copper powder (Barrett and Yust, 1967), and grain boundary precipitates in brass (Ziolkowski, 1985).

Beginning in the early 1990s, advances in computerized 3D reconstruction, image processing, and visualization algorithms and software enabled high-fidelity computer-generated reconstructions and advanced quantitative 3D analyses from serial sections of metallic microstructures. In particular, in 1994 a significant step forward was made by two sets of independent researchers (Mangan and Shiflet, 1994; Weiland et al., 1994)—computerized, smoothed, digital 3D reconstructions were made of pearlite colonies in a high manganese steel (Mangan and Shiflet, 1994), and of the recrystallized grain

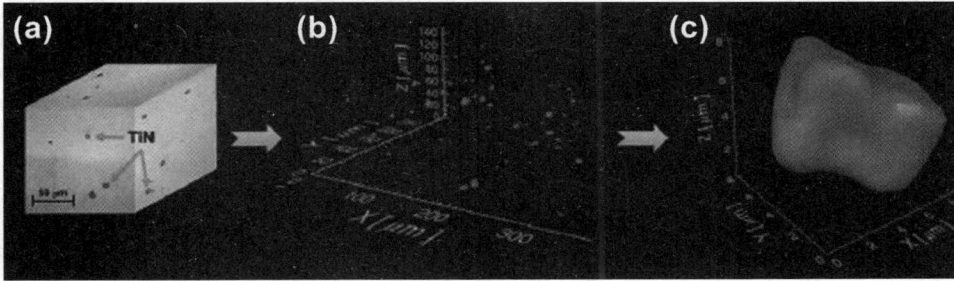

Figure 5 3D reconstructions obtained by femtosecond laser serial sectioning of a titanium-modified 4330 steel. (a) A representative orientation view of three orthogonal optical 2D images. Red TiN particles are shown in both this optical micrograph image and in the reconstructions. (b) 3D reconstructions of TiN particles in the 4330 steel matrix. (c) Single TiN particle reconstruction with near cuboidal morphology and a size of ~ 6 μm. From Echlin et al., 2011. (For color version of this figure, the reader is referred to the online version of this book.)

structure in an Al–Mn alloy (Weiland et al., 1994). Since then, continued advances in computer software and hardware, and in both manual and automated experimental serial sectioning techniques have led to a rapid and continuous growth in the study of metallic microstructures in three dimensions.

The key steps in serial sectioning by mechanical techniques include material removal, etching (if required), imaging, placing fiducial marks, alignment of the images, optimizing the number of serial sections required, and "segmentation" of the images (Spanos et al., 2010). These steps are depicted schematically in the flowchart in **Figure 6** (Kral et al., 2004). Reconstruction, visualization, and quantitative analysis of the 3D microstructures generated from such serial sections are considered in more general terms (common to a number of other 3D reconstruction methodologies as well) in Sections 10.2.3 and 10.2.4 below.

The most common method for mechanically removing material in metals and alloys is mechanical polishing, either by manual/semiautomated methods (Mangan and Shiflet, 1994; Kral and Spanos, 1999) or by automated robot-assisted techniques (Spowart, 2006; Spowart et al., 2003). A second

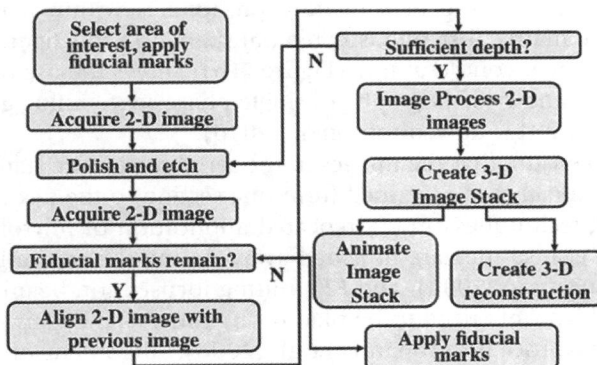

Figure 6 A schematic description of the process of serial sectioning by mechanical polishing (from Kral et al., 2004). Reprinted with permission of ASM-International (www.asminternational.org); all rights reserved.

mechanical technique that has been employed with great success is micro milling (Wolfsdorf et al., 1997). In either case, the choice of the polishing media (polishing solution, polishing surface, micro milling head) and polishing parameters (e.g. wheel speed, pressure) depends on a variety of factors. These factors include (1) the material being sectioned, (2) the depth, number, and removal rate of the sections, and (3) the final surface quality required—which is often dependent on the imaging technique. After completing the material removal for a given section, some applications require etching before imaging (Kral and Spanos, 1999; Spanos et al., 2008), while others require no etching (Alkemper and Voorhees, 2001; Wolfsdorf et al., 1997). This is dependent on both the material being studied and the microstructural feature(s) of interest.

Various techniques have been employed for imaging the individual sections (sometimes in combination), depending on the microstructural feature(s) to be reconstructed in 3D. These include light/optical microscopy (Wolfsdorf et al., 1997; Kral and Spanos, 1999; Spowart, 2006), secondary electron imaging in the SEM (Mangan et al., 1997; Mangan and Shiflet, 1994), and electron backscatter diffraction (EBSD) mapping (Uchic et al., 2006; Spanos et al., 2008). The choice of imaging technique is dictated not only by the resolution required to see the features of interest but also by the lateral area (and thus final volume in 3D) of material to be sampled. The amount of material sampled is critical to obtaining the required number of microstructural features that need to be reconstructed to provide for quantification and statistically relevant data sets for the problem being studied. For instance, to determine a grain size distribution with reasonable statistical significance, thousands of grains may be required (Rowenhorst et al., 2010), while to obtain qualitative information on precipitate morphology in three dimensions, many fewer grains can be sampled (Wu and Enomoto, 2002; Hackenberg et al., 2002; Kral and Spanos, 2005). Finally, the specific attributes of the features being reconstructed also dictate the imaging method. For example, in problems in which crystallography or texture is critical, EBSD mapping has been employed for at least some of the individual sections (Rowenhorst et al., 2006a; Spanos et al., 2008; Uchic et al., 2006).

Some examples of 3D metallic microstructures reconstructed by serial sectioning in conjunction with mechanical polishing material removal techniques are provided in **Figure 7**. These include a reconstruction in which the imaging technique was solely light microscopy and thus does not contain any crystallographic information (**Figure 7(a)**), as well as one that employed EBSD mapping of a number of the sections, to provide for crystallographic information in 3D as well (**Figure 7(b)**). The first case (**Figure 7(a)**) is taken from a high carbon manganese steel, and demonstrates the morphology, interconnectivity, and nucleation site of cementite precipitates within a face-centered cubic (fcc) austenite matrix grain (the matrix austenite is made transparent in this figure to show the cementite) (Kral and Spanos, 1999). The second example (**Figure 7(b)**) shows the use of this technique to study the morphology, topology, and crystallography of single-phase space-filling grains in a beta-stabilized titanium alloy (Lewis et al., 2011; Rowenhorst et al., 2010).

The most direct way to ensure that the images in the serial sectioning stack are properly aligned is by embedding fiducial marks that are retained from one section to the next. The two most common types of fiducial marking techniques are (1) repeated application of microhardness indents, which must be replaced often because they are polished away during the sectioning procedure (Kral and Spanos, 1999; Wu and Enomoto, 2002), and (2) milling focused ion beam (FIB) channels into the sides of the specimen, which only need to be placed one time, as they remain visible in the images throughout the sectioning procedure (Spanos et al., 2008). Alignment of the images is then performed by aligning the fiducial marks from section to section, using any of a variety of available software packages (e.g. IDL^{TR}, Adobe Photoshop^{TR}, ImageJ^{TR}, IMOD^{TR}). Alternatively, some experimental techniques provide relatively high resolution of specimen positioning between

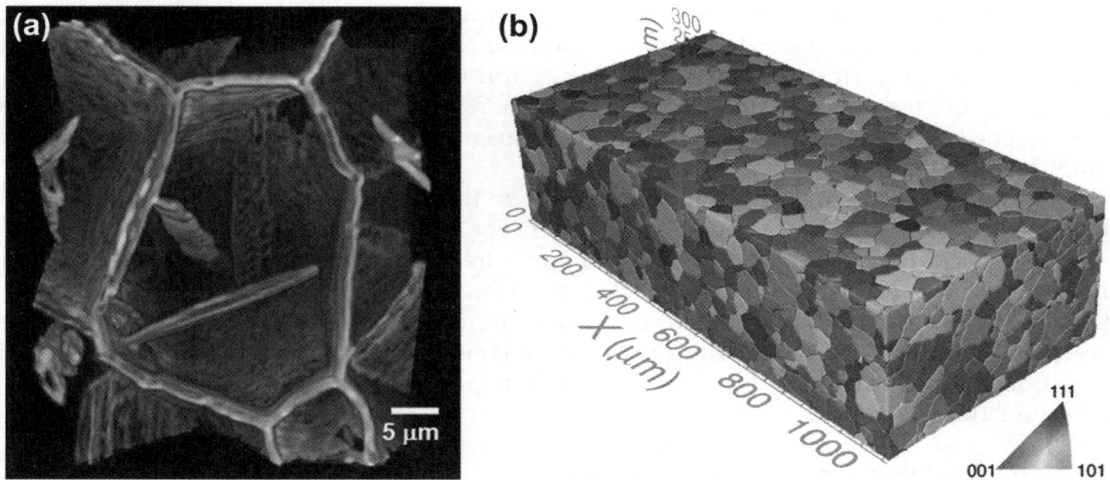

Figure 7 Examples of 3D microstructures reconstructed in metallic alloys by serial sectioning in conjunction with mechanical polishing material removal techniques. (a) 3D reconstruction of cementite precipitates coating austenite grain boundaries and extending into the austenite grains in a high-carbon, high-manganese austenitic steel. In this figure cementite is opaque and austenite is transparent (from Kral and Spanos, 1999). (b) 3D reconstruction of 4300 beta-titanium grains in a beta-stabilized titanium alloy (Ti–21S). Serial sectioning by mechanical polishing was performed in conjunction with EBSD analysis, allowing for the crystallographic orientation of each grain as well (from Lewis et al., 2011). Each grain is colored according to the crystallographic axis aligned with the z-direction. (For color version of this figure, the reader is referred to the online version of this book.)

sections, and either reduce or eliminate the need for fiducial markers (Wolfsdorf et al., 1997; Spowart, 2006).

Determining the number of sections required, and the depth between sections, is important for optimizing the time-intensive sectioning process, while at the same time guaranteeing the required amount of fidelity to properly reconstruct the features of interest (Spanos et al., 2010). That is, the length scale of the critical features of interest must be taken into account, as well as the volume of material required (i.e. the number of features reconstructed for statistical relevance). These two considerations taken together can often be considered a "trade-off" that must be balanced. As a rough rule of thumb, the choice of section depth is on the order of a minimum of 10 sections per critical feature being reconstructed, while the volume of material needed then determines the number of sections, within limitations set by the sectioning methodology (particularly the time to remove each section).

Once the series of raw 2D images in the stack are obtained, in almost all cases image-processing techniques must be employed to clearly delineate the features to be reconstructed in 3D. This is typically referred to as "image segmentation" and often involves segmenting each 2D image into specific regions of interest (e.g. for the identification of individual grains separated by grain boundaries), but can also be applied to the entire 3D volume in certain cases. There are many algorithms and techniques that are being developed and applied to image segmentation, and this is a very active area of 3D materials analysis (Spanos et al., 2008, 2010; Russ, 1998). Image segmentation is one of the most challenging, time-consuming, and critical steps in the 3D reconstruction of metallic microstructures. Segmentation applies to any serial sectioning technique (e.g. femtosecond laser material removal and FIB milling) used in conjunction with an imaging technique that provides grayscale or more continuous image contrast. Grayscale images are produced by optical microscopy and secondary electron imaging in an SEM, and therefore require segmentation. On the other hand, EBSD orientation maps are images of pixels generated as discrete values assigned to orientations in the original 2D image plane, and these maps do not need an additional segmentation step.

Another successful mechanical sectioning technique that has been employed to study metallic microstructures is micro milling. Alkemper and Voorhees (2000) employed a micro milling apparatus to study a directionally solidified Al-15wt%Cu alloy. In addition to the micro milling stage and milling tool, the experimental apparatus included a linear variable differential transformer to align the specimen (and thus the images) from section to section, eliminating the need for fiducial markers to register the sections. This methodology allowed a comparison of 3D morphologies and spatial distributions of microstructures between samples that had been coarsened for different times (see also Section 10.3.1.2 below).

10.2.2.4 FIB Serial Sectioning of Metallic Microstructures

Serial sectioning also can be accomplished by employing the ion beam in FIB instruments to remove material between sections (Kral et al., 2004; Spanos et al., 2010; Uchic et al., 2007). FIB serial sectioning, or tomography, enables 3D analysis of features at an intermediate length scale—smaller than those accessible to sectioning by mechanical methods, but larger than that associated with techniques such as TEM tomography or atom probe tomography (APT) (see below). Although TEM and APT provide much higher resolution (see **Figure 1**), they typically involve much smaller volumes of material sampled. By employing "dual beam FIB" instruments, which contain both an ion beam and an electron beam, various imaging modes can be used in conjunction with the ion beam sectioning, depending on the microstructural features to be reconstructed in 3D. These include secondary ion imaging, secondary electron imaging, and EBSD mapping. A number of reviews and articles have considered the details involved and type of results that can be obtained with FIB serial sectioning (Kral et al., 2004; Spanos et al., 2010; Uchic et al., 2006, 2007), and many investigations have used this methodology to study 3D microstructures in a variety of (mostly metallic) materials systems (Inkson et al., 2001a; Zaefferer, 2005; Groeber et al., 2006; Konrad et al., 2006; Mulders and Day, 2005; Petrov et al., 2007; Patkin and Morrison, 1982; Dunn and Hull, 1999; Phaneuf and Li, 2000; Inkson et al., 2001b; Sakamoto et al., 1998). Two examples of 3D reconstructions of metallic microstructures by FIB techniques are provided in **Figure 8**. These demonstrate two different types of 3D microstructural data sets that can be reconstructed using this method: (1) second phase precipitates within a parent matrix (**Figure 8(a)**) and (2) space-filling grains within a single-phase metallic microstructure (**Figure 8(b)**). The ability to include crystal-lographic information in the 3D reconstruction through EBSD mapping of the sections in the dual beam FIB is demonstrated in the reconstruction in **Figure 8(b)**. Both of these cases are from Ni-based superalloys, and the first shows the morphology and complex 3D interconnectivity of gamma' precipitates within an fcc Ni-based matrix—see **Figure 8(a)**. The second reconstruction (**Figure 8(b)**) shows the morphology, topology, and crystallography of single-phase space-filling fcc grains in the alloy IN-100.

10.2.2.5 Electron Tomography

Electron tomography employing a TEM enables observation of 3D microstructures at a high spatial resolution (in the nanometer range), but it restricts the 3D observations to small volumes of material

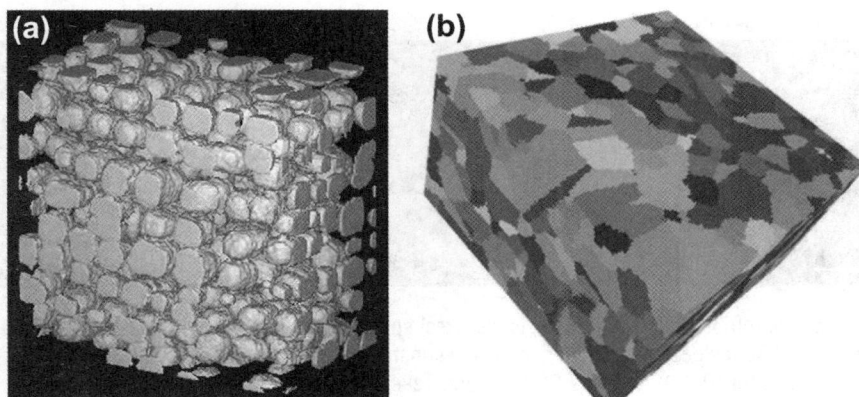

Figure 8 Examples of 3D metallic microstructures reconstructed by serial sectioning in conjunction with FIB material removal techniques. (a) 3D volume reconstruction of gamma' precipitates in a 70Ni–20Cr–10Al (wt%) alloy. This reconstruction was generated from a stack of images collected from a dual beam FIB–SEM, and the size scale of the cuboidal particles represented here are on the order of 0.5 microns on a side (from Kral et al., 2004; reprinted with permission of ASM-International (www.asminternational.org), all rights reserved). (b) 3D reconstruction of Ni-based superalloy IN-100, created from FIB serial sections in conjunction with EBSD data. The dimensions of the parallelepiped volume are 41 × 41 × 29 μm, and the cubic voxel dimension is 0.25 μm. From Uchic et al., 2007. (For color version of this figure, the reader is referred to the online version of this book.)

because the sample thickness must be thin enough to be electron transparent (typically several tens of nanometers). The steps for TEM tomography involve taking a series of 2D images as a sample is rotated through a large angle. Each image in the series is a 2D projection through the sample in a different orientation. A 3D representation of the sample is reconstructed from the series of 2D images using one of several mathematical inverse techniques.

Although electron tomography has been used on biological specimens for more than 50 years (Marco et al., 2004), its application to crystalline solids has been limited by the orientation-dependent contrast effects typically experienced in these materials. Unlike absorption contrast used for biological samples, the diffraction contrast common in crystalline solids changes dramatically with sample orientation, and these changes defeat the reconstruction algorithms used to generate a 3D represen-tation. The development of microscopes capable of imaging with alternate contrast mechanisms has made it possible to overcome this problem in recent years.

Two common imaging strategies used for electron tomography of metallic microstructures are high angle annular dark field (HAADF) imaging in the scanning transmission electron microscopy (STEM) mode, also known as Z-contrast imaging, and energy-filtered imaging in the TEM mode using electron energy loss spectrometry. Both approaches eliminate or greatly reduce contrast changes arising from diffraction as the sample is tilted. **Figure 9** provides an example of a 3D reconstruction generated with conventional STEM and careful tilting to avoid contrast changes; it shows the 3D spatial arrangement of a network of dislocations in an austenitic steel (Hata et al., 2011). From such STEM tomography reconstructions, the line direction, dislocation intersections and interactions, and dislocation density can be directly determined; in this case, the dislocation density was evaluated to be 4×10^{13} m^{-2}. When evaluated from 2D TEM/STEM images, influences of specimen thickness and nonisotropic dislocation arrangements must be taken into account,

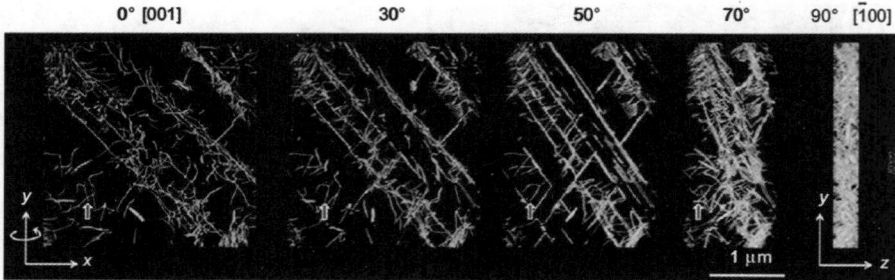

Figure 9   Dual-axis STEM tomography from an austenitic steel specimen. Specimen-tilt angles are denoted in each of the images. Projected views of 3D dislocation arrangements are reconstructed from two bright field (BF) STEM tilt series acquired under two-beam conditions for (0 2 0) and (2 0 0) g-vectors. Taken from Hata et al., 2011.

whereas these uncertainties are removed by using STEM tomography. Analysis of the 3D reconstruction in **Figure 9** also showed that within 20–50 nm of the specimen surfaces in this austenitic steel the dislocation density was reduced about 20–30%, which was at least partially attributed to stress relaxation during the TEM specimen preparation process (Hata et al., 2011). Observations of dislocations by STEM tomography thus allow for more direct 3D evaluation of dislocations than is possible by 2D TEM/STEM techniques alone.

10.2.2.6   3D Atom Probe Tomography (APT) for Studying 3D Metallic Microstructures at the Atomic Scale

3D APT (Seidman, 2007; Hono, 1999; Kelly and Miller, 2007) provides for the visualization and quantitative analysis of 3D microstructures at the atomic level (see **Figure 1**). A very brief description of the utility of this technique is provided here to put it in the context of the other techniques available for analyzing 3D microstructures in metals. With continuously improving techniques such as the local electrode atom probe (LEAP) (Kelly et al., 2004; Kelly and Miller, 2007) and pulsed laser atom probe (Kelly and Miller, 2007), the size of today's Atom Probe Tomography (APT) data sets is typically 100 nm in diameter, and can be an order of magnitude larger in depth. These dimensions enable sampling volumes of about 10^6 to 10^7 nm^3 containing 10^7 to 10^8 atoms. 3D APT is especially useful for analyzing composition profiles and segmenting microstructural features, which possess differences in composition from the phases within which they are embedded (Marquis and Hyde, 2010). An example of a 3D atom probe tomographic reconstruction and related analysis is shown in **Figure 10** (Kelly et al., 2007), which demonstrates both the capability for visualizing 3D reconstructions of the atoms, **Figure 10(a)**, and the corresponding quantitative compositional analysis, **Figure 10(b)**.

10.2.3   3D Reconstruction Methodologies for Analyzing Microstructures

None of the 3D characterization methods described in Section 10.2.2 can be implemented without the aid of computer-assisted 3D reconstruction and image analysis algorithms and codes that produce 3D images and 3D data sets of the microstructure for subsequent interpretation. With the exception of 3D APT, which uses atomic mass and position information obtained from a time-of-flight mass spectrometer to reconstruct a 3D image, all of the 3D characterization techniques reconstruct 3D microstructures from a number of individual planes or angles of observation. These could be, for instance, a series of optical micrographs, SEM micrographs, EBSD maps, projections from various angles of tilt in a TEM, or intensities on an X-ray detector. The final 3D reconstructions of the microstructure in these

(a)

Analysis direction →

30 nm

Interface 1 → ← Interface 2

SiGe   Si   SiGe

Si   Ge

(b)

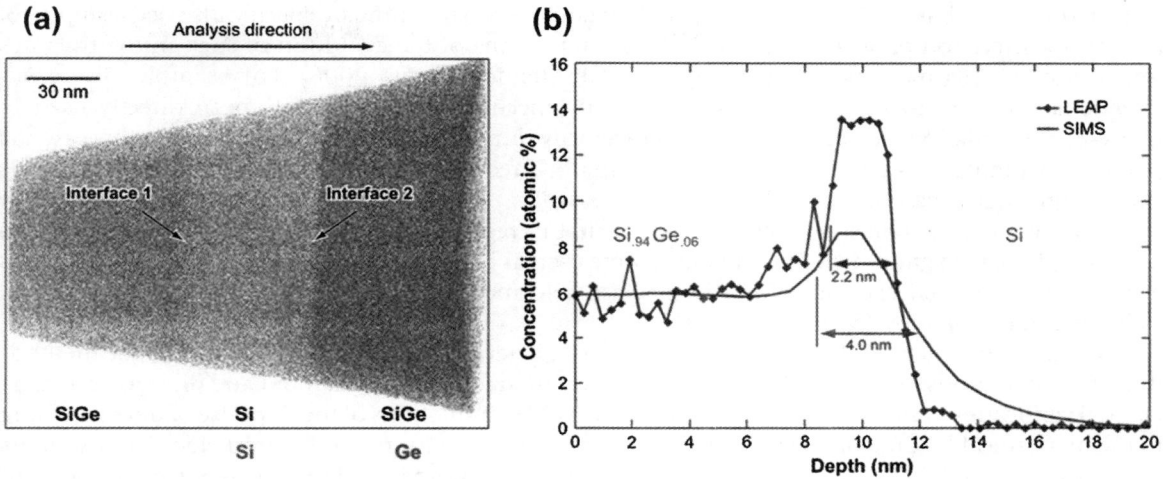

Figure 10 APT analysis of a SiGe–Si multilayer stack, demonstrating the imaging and analytical capabilities of 3D APT. (a) Side view of an APT image of CVD Si/SiGe multilayer films (courtesy of Sean Corcoran, Intel Corporation). (b) Concentration profile at Interface 1, obtained by APT and SIMS. The APT profile is taken from 10 nm on either side of Interface 1. Taken from Kelly et al., 2007. (For color version of this figure, the reader is referred to the online version of this book.)

cases typically take one of two forms: (1) a 3D array of voxels, or volume elements, in which each individual volume element is assigned a value of a specific phase, grain, domain, or other microstructural entity (Spanos et al., 2010), or (2) a surface mesh for each of the features of interest (e.g. grains or precipitates). In the latter case, the interface of each microstructural entity is typically constructed by a set of vertex points whose connections are described by a network of polygons (often triangles) to make a continuous surface (Spanos et al., 2010).

10.2.4 "Microstructure-Based" or "Image-Based" Modeling of 3D Microstructures

In addition to the great advances in the reconstruction, visualization, and quantitative analysis of experimental microstructures in three dimensions over the past 15 years, there has been a surge in 3D predictive modeling of microstructural evolution in which real, experimentally determined 3D microstructures are used as input for the initial conditions of the models, and/or for direct comparisons between the predicted and the experimentally determined 3D microstructures at various stages of evolution. This type of experimentally based modeling has been referred to as both "image-based" modeling (Everett et al., 2002; Lewis and Geltmacher, 2006; Qidwai et al., 2009) and "microstructure-based" modeling (Chawla and Sidhu, 2007) (for consistency, the term "image-based" modeling will be used in this context here). This represents a very powerful nexus of 3D characterization and modeling techniques that allows for much greater accuracy, utility, and improvements in the predictive models of microstructure and property evolution in 3D.

The most common types of image-based modeling of microstructural evolution are centered about response to externally applied thermal (Mckenna, 2010) or mechanical (Lewis and Geltmacher, 2006; Efstathiou et al., 2010; Qidwai et al., 2009) loads. In this regard, two specific types of image-based 3D microstructural evolution models that have been employed to study metallic microstructures include phase-field modeling and finite element method (FEM) modeling, respectively.

Experimentally based 3D reconstructions of metallic microstructures are specifically used as input to, and/or for direct comparisons with, phase-field simulations of the evolution of such microstructures under applied thermal exposure (Mckenna, 2010; Mendoza et al., 2006). For example, 3D reconstructions taken from serial sections developed by mechanical milling have been directly used in phase-field calculations to study the interfacial velocity during solid–liquid coarsening; in this case the phase-field simulations employed the experimental 3D reconstructions as the initial condition in the simulation (Mendoza et al., 2006). Such image-based 3D phase-field modeling is also used to study solid-state grain growth in single-phase materials. In this regard, 3D X-ray tomographic reconstructions of single-phase beta grains in a titanium alloy were used as both initial input and direct comparison (as a function of thermal exposure time) in a phase-field model of solid-state grain growth in a beta-stabilized titanium alloy (Mckenna, 2010).

Real 3D microstructures that have been imaged, characterized, and analyzed by some of the methods described in Sections 10.2.2 and 10.2.3 above are also used for both initial input and direct comparison (model verification, validation, and enhancement) in FEM simulations of the response of microstructure to the mechanical loading of metals and alloys. Such FEM simulations are a critical element to accurate and efficient material and process design, particularly in an approach referred to as integrated computational materials engineering (ICME) (Pollock and Allison, 2008; Allison et al., 2006; Collins et al., 2011). The experimental methodologies that are used to develop the 3D data for such simulations include serial sectioning by mechanical polishing, mechanical milling, and/or FIB milling, and X-ray tomographic reconstruction methods. These FEM simulations often involve anisotropic elasticity in the small strain regime to study the microstructural "hot spots" that are expected to be the initiation points for plastic deformation (Geltmacher et al., 2007; Lewis and Geltmacher, 2006; Spanos et al., 2008). Image-based 3D FEM models involving crystal plasticity are typically used to examine the evolution of microstructures at strains beyond the elastic regime to determine among other factors the microstructural features leading to highest local regions of plastic strain, and those that also might initiate failure in a material component (Zaafarani et al., 2006; Qidwai et al., 2009; Lewis and Geltmacher, 2006; Proudhon, 2011). Such image-based 3D FEM simulations are performed under various types of loading (e.g. torsion, tension, compression) to simulate the microstructural response of metals to different conditions either during processing (e.g. cold rolling, drawing) or during in-service use of metallic components. In this regard, these simulations of microstructural and property evolution are also a critical component in ICME approaches to material and manufacturing design (Proudhon, 2011; Helm et al., 2011; Allison et al., 2006), and are used in conjunction with many other predictive models over a wide range of length scales and engineering subdisciplines (Allison et al., 2006; Gottstein and Mohles, 2011).

10.3   Specific Classes of Microstructures and Microstructural Evolution

In Section 10.2 some examples of real 3D microstructures that have been directly determined in a number of metallic systems were shown, and the techniques that are employed to reconstruct, visualize, and quantitatively analyze them in three dimensions were discussed to provide a foundation upon which to examine microstructures in metallic materials. After briefly discussing how various microstructures are typically described and characterized (e.g. by their morphology, crystallography, chemistry), this section will analyze in more depth the specific types of metallic microstructures within different microstructural categories.

How is microstructure described? A first step in microstructural analysis involves selecting a method for describing the nonuniform features in the material. These features may be associated with a physical

Table 1 Elements of microstructure

Microstructure descriptor	Examples of relevant parameters
Composition	Concentration: $C(x,y,z)$
Crystal structure	Bravais lattice, point group, space group, prototype, Pearson symbol, order parameter: $\eta(x,y,z)$
Orientation/misorientation	Rodriques vector, Euler angles, orientation relationship, misorientation angle
Domain property	Elastic modulus, magnetization, electric polarization, lattice parameter

characteristic like chemical concentration or crystal structure, or they might be related to a property like modulus or magnetization. In metallic microstructures the features have dimensionality (one-, two-, or three-dimensional features like dislocations, boundaries, and inclusions) and they usually have a length scale associated with a structural feature of the material, such as the grains, for example. The grains themselves are then described by parameters such as size, shape or aspect ratio, and orientation. The dimensionality of microstructural features is typically described in terms of point defects (zero-dimensional), line defects (one-dimensional), interfaces or surfaces (2D), and 3D microstructural entities. The focus of this chapter is on the description, classification, observation, and characterization of 3D solid-state metallic microstructures. The reader is referred to any number of excellent texts that describe in-depth point defects (Shewmon, 1989), line defects (Hirth and Lothe, 1982), and interfaces (Howe, 1997; Bollmann, 1982; Sutton and Balluffi, 1995) in metals, as well as to Chapter 6, Volume I and Chapters 14, 16 of this volume.

The types of microstructural descriptors that are used to describe an individual microstructural feature include composition, crystal structure, orientation/misorientation, and local properties of a given domain. These descriptors, along with some relevant parameters used to evaluate them, are summarized in **Table 1**.

Specific types of metallic microstructures within different microstructural categories associated with the formation and evolution of solid-state microstructures in metals and alloys are now considered, with an emphasis on microstructures that have been characterized and analyzed in three dimensions. It is useful to keep in mind that microstructures are often associated with specific transformations, but transformations do not produce unique microstructures; multiple paths can produce a particular type of microstructure.

10.3.1 Metallic Microstructures whose Evolution is Driven Predominantly by Interfacial Energy Minimization

10.3.1.1 Single-Phase Materials

The internal microstructure in single-phase solid-state metals and alloys consists of grain networks composed of space-filling polyhedra, and such grain structures have been studied in considerable detail (Smith, 1952; DeHoff and Liu, 1985; Glicksman, 2005). The size, shape, texture, and topological arrangement of the grains play a critical role in dictating the mechanical and other physical properties of metallic materials. There is an especially strong influence of grain size on mechanical properties, often expressed through some type of Hall–Petch relationship (Hall, 1951; Petch, 1953) of the general form:

$$\sigma = \sigma^\circ + Kd^{-1/2} \tag{1}$$

where σ represents yield stress, d is an average grain size, and K and σ° are materials (and temperature) dependent constants. While experimental observations of the value of the exponent can lie in a range somewhat below or above $-1/2$, this is a reasonable approximation for most metals. On the other hand, for nanomaterials (e.g. grain sizes on the order of 20 nm or less), this relationship can break down (Schiotz et al., 1998;Carlton and Ferreira, 2007).

Coarsening of the grains in single-phase materials, more commonly called grain growth, has been a topic of considerable interest because of the strong dependence of mechanical properties on grain size. There have been many in-depth investigations of grain growth centered about both theoretical treatments and experimental studies (Smith, 1948; von Neumann, 1952; Mullins, 1956; Cahn, 1967; Krill and Chen, 2002; Wakai et al., 2000; Glicksman, 2005; MacPherson and Srolovitz, 2007; Rios and Glicksman, 2008; Rowenhorst et al., 2010). Although a thorough discussion of the full details of these investigations is beyond the scope of this chapter, a brief description of the salient features of grain growth and topology will be provided here as they relate to 3D microstructures in single-phase metallic materials.

The overall driving force for grain growth is the reduction in interfacial energy, and the local grain boundary curvatures and grain topologies (i.e. number of grain faces) dictate the local velocity and direction of movement of a given grain boundary. The general form for the relationship between the average perpendicular velocity of grain boundaries (v) is related to both a grain boundary mobility (M) and a driving force (F) term, and can be expressed as follows:

$$v = MF, \tag{2}$$

where the driving force term is typically represented by an expression of the form:

$$F = \gamma(1/r_1 + 1/r_2) \tag{3}$$

where r_1 and r_2 are the two principal radii of curvature of the grain boundary, and γ is the grain boundary energy.

The topology of the grains is also directly related to the migration of grain boundaries, and in this regard von Neumann (1952) and Mullins (1956) established a relationship between the growth (or shrinkage) of a given grain and the number of sides of that grain. This relationship is applicable in two dimensions, and indicates that grains with more than six sides will have a positive growth rate (growth), grains with less than six sides will have a negative growth rate (shrinkage), and those grains with exactly six sides will have a zero growth rate (will not grow or shrink until they undergo a topological change).

This relationship is known as the von Neumann–Mullins law (von Neumann, 1952; Mullins, 1956), and can be expressed as follows:

$$dA/dt = \pi M\gamma/(n - 6) \tag{4}$$

where A is the interfacial area of the grain, and n is the number of sides.

MacPherson and Srolovitz subsequently reported a more generalized 3D relationship between the rate of growth (or shrinkage) of the grains and their topology (MacPherson and Srolovitz, 2007).

Most of the theories of grain growth (whether based on 2D or 3D grain ensembles) have been tested by comparisons with 2D data because in the past it has been difficult to obtain 3D experimental data from statistically significant samplings of grains. Advances in 3D characterization techniques described in Section 10.2 above, however, have made possible direct comparisons between theories and experimentally determined 3D grain structures in metals (Rowenhorst et al., 2010; Ludwig et al., 2009a). For instance, Rowenhorst et al. (2010) used serial sectioning to reconstruct and analyze quantitatively more than 2000 beta-titanium grains in a beta-stabilized titanium alloy (Ti–21S). This data set was developed with enough fidelity that the shape, topology (number of faces), and curvature of the individual faces of all the grains in the ensemble were able to be determined in 3D. This type of data not only provides for the direct observation and qualitative study in three dimensions of such space-filling grain structures (**Figure 11(a) and 11(b)**) but also allows for quantitative determination of the distributions of grain size and topology, as shown in **Figure 11(c) and 11(d)**. In this regard, these experimental results were analyzed and directly compared with a number of grain growth models, and the experimental distributions in both grain size and number of grain faces were shown to match relatively closely with the predictions of steady-state size distributions from various simulations and analytical theories (Rowenhorst et al., 2010) (see **Figure 11(c) and 11(d)**). It was found that on average, the grains with an average of 15.5 faces have a zero integral mean curvature (the zero growth rate condition; see **Figure 11(e)**), which is higher than the predicted value of 13.4 faces. This difference was suggested to be because of nearest neighbor effects within the grain network (Rowenhorst et al., 2010).

10.3.1.2 *Multiphase Materials: Two-Phase Microstructures Resulting from Coarsening*

Precipitate coarsening in multiphase materials, also known as Ostwald ripening, is similar to grain growth; in that the overall driving force is the reduction in interfacial energy. But in this case, in addition to the local curvature, the composition gradient and relevant diffusivities affect the local velocity of a given particle boundary during grain coarsening. The reader is referred to several review articles (Voorhees, 1992; Wang and Glicksman, 2007) and key investigations (Lifshitz and Slyozov, 1961; Wagner, 1961; Ardell, 1972; Hardy and Voorhees, 1988; Kammer and Voorhees, 2006; Marsh and Glicksman, 1996; Mendoza et al., 2004, 2006; Rowenhorst et al., 2006b; Wang et al., 2010, 2006) for details of theoretical and experimental treatments of two-phase coarsening. The following brief description of some observations of two-phase coarsening highlights the direct 3D characterization and analysis of such metallic microstructures.

The most detailed and direct 3D observations have been in investigations by Voorhees et al. who used a serial sectioning and 3D reconstruction technique involving mechanical micro milling (Alkemper and Voorhees, 2000) (see Section 10.2.2.3). These studies showed that the 3D morphologies and interconnectivities of coarsened dendritic structures can be quite complex (Alkemper and Voorhees, 2000, 2001; Mendoza et al., 2003; Mendoza et al., 2004; Rowenhorst et al., 2006b). A quantitative topological analysis on directionally solidified microstructures of Al-15 wt pct Cu coarsened at 553 C with a solid volume fraction of 74%, for example, showed that the "genus" of the dendrites (a measure of shape) decreases with coarsening time because of simplification of the topological structure, while the number of liquid droplets concurrently increases with coarsening time. These 3D experimental results were also used as initial input for phase-field calculations of interfacial velocity (Mendoza et al., 2006).

Although there have been a limited number of direct 3D observations of coarsened microstructures, further development of such experimental data is expected because it provides statistically relevant data sets of particle shape and size distributions as well as interconnectivity information critical for direct comparison with modern theories of Ostwald ripening.

10.3.2 Microstructures whose Evolution is Driven Predominantly by Chemical Driving Force (Solid-State Phase Transformations)

10.3.2.1 *Martensite—Composition Invariant Transformation*

10.3.2.1.1 *Ferrous Martensite*

Martensitic transformations are considered in much more detail in many articles on the subject (Marder and Krauss, 1967; Sandvik and Wayman, 1983; Wayman, 1984). A very brief consideration of martensitic microstructures will be provided here, with emphasis on differentiation from some of the other microconstituents observed in metals, and a focus on direct 3D observations of martensite.

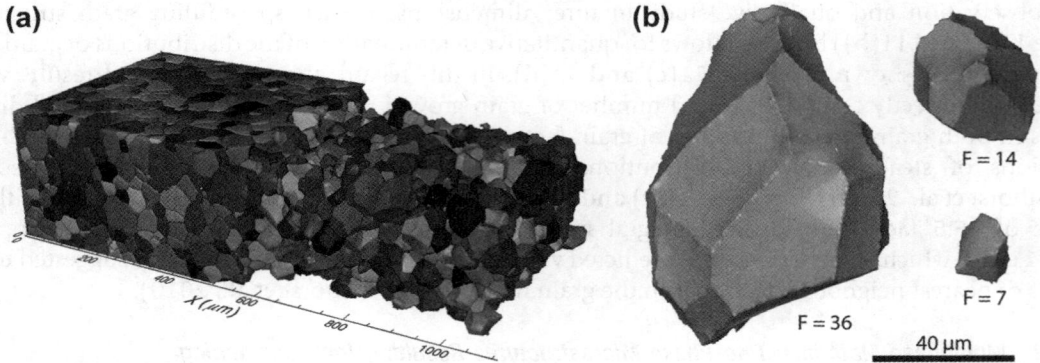

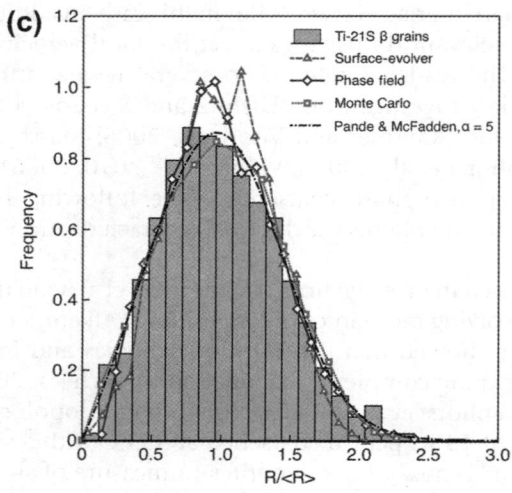

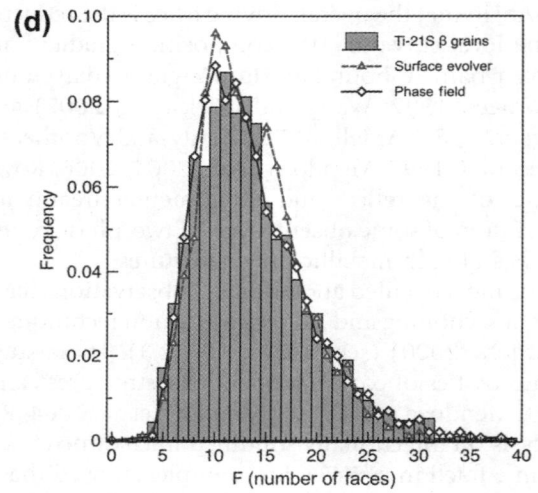

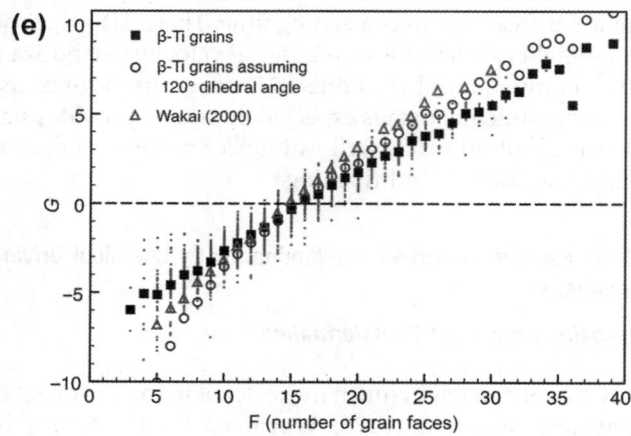

Martensitic structures in steels (i.e. ferrous martensites) have been the most widely studied, and there is a plethora of information available on the crystallography, morphology, and formation mechanism of such martensites. In ferrous martensites, the classic picture is one in which low-carbon martensite is highly dislocated and has a "lath" structure, while higher carbon martensite is typically termed "plate" martensite and is often highly twinned (Marder and Krauss, 1967). Yet almost all studies of the morphology and habit plane of martensite have been made with essentially 2D techniques. An exception is a serial sectioning and 3D reconstruction study that was performed on coarse martensite formed in a high-strength low-alloy steel, revealing the true 3D morphology and crystallography of the coarse martensite crystals (Rowenhorst et al., 2006a). In that study EBSD was employed in conjunction with serial sectioning and 3D reconstruction to provide for the crystallographic analysis. The overall shape of the coarse martensite crystals was often found to correspond to a long lath-shaped wedge, with a small facet opposite the sharp edge of the wedge, two parallel facets on opposite broad faces, and an angled facet on one of the broad faces—see **Figure 12**. The crystallographic facet/habit planes were also reported (Rowenhorst et al., 2006a). Similar 3D studies are now required in other alloys to determine the true 3D morphology and facet/habit planes enclosing martensite crystals in a variety of metallic alloy systems and to compare these results with theories of martensite formation and crystallography.

10.3.2.1.2 Shape Memory Martensite

The morphology of shape memory martensite is related to the shape change associated with the structural transformation, how this strain is accommodated, the presence of plastic strain in the parent phase, and any applied stresses or elastic constraints during transformation. Martensite in shape memory alloys is reversible, and the interface separating the martensite phase from its parent phase is highly mobile. These factors lead to a variety of morphologies ranging from bulk single crystals (when a parent single crystal is transformed through the migration of a single interface) to multiple variants of laths and plates arranged in irregular patterns or highly organized periodic arrays.

To date, characteristics of these martensite microstructures have been inferred from 2D data or projected information from TEM thin foils. In principle, the 3D morphology of shape memory martensite can be determined using the techniques described in Section 10.2, but it is important to keep in mind that the high mobility of the interfaces of reversible martensite introduces some significant challenges. The microstructure of reversible martensite can change in response to changes in elastic stress, and consequently, free surfaces introduced by serial sectioning or thin sample techniques can be expected to alter the morphology of the martensite under observation.

◀───

Figure 11 3D experimental results, analyses, and comparison to theories and simulations of the 3D topology and morphology of 2098 beta-titanium grains in a beta-stabilized titanium alloy (Ti–21S) reconstructed by serial sectioning techniques (taken from Rowenhorst et al., 2010). (a) Reconstruction of 4380 grains from 200 serial sections of which 2098 grains were analyzed. The entire reconstructed volume is $1115 \times 516 \times 300$ microns^3. (b) Sampling of a few of the individual grains, showing varying topology and size. (c) The spherical equivalent radius grain size distribution normalized by the average grain size. The results from a Surface-Evolver simulation (Wakai et al., 2000), a Phase-Field simulation (Krill and Chen, 2002), a Monte-Carlo simulation (Zollner and Streitenberger, 2006), and an analytical theory (Pande and McFadden, 2010) are plotted for comparison. (d) Distribution in the number of faces per grain. The results from a Surface-Evolver simulation (Wakai et al., 2000), and a Phase-Field simulation (Krill and Chen, 2002), are plotted for comparison. (e) The normalized integral mean curvature of the grain faces, G, as a function of the number of faces, F, per grain. The small symbols represent each individual grain while large symbols represent the average value of G for each face class. (For color version of this figure, the reader is referred to the online version of this book.)

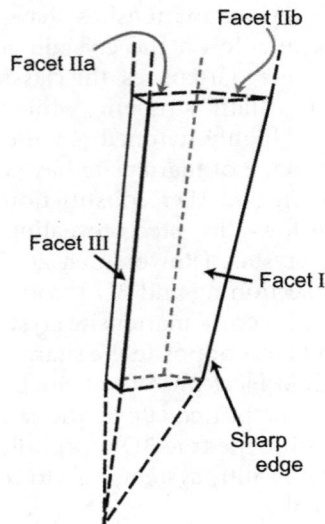

Figure 12 Schematic representation of a coarse martensite morphology formed in a high-strength low-alloy steel. The shape consists of a long lath-shaped wedge, with a small facet opposite the sharp edge of the wedge (facet III), two parallel facets on the broad faces (facets I and IIa), and an angled facet on the one broad face (facet IIb). Taken from Rowenhorst et al., 2006a.

10.3.2.2 Microstructures Characterized by a Composition Change
10.3.2.2.1 Solid-State Precipitates Formed during Aging

The 3D morphology of precipitates has been assessed using multiple techniques. **Figure 13** shows examples of two techniques applied to different types of precipitates formed during aging. In each of these particular cases, the crystal structure and orientation of the precipitates and parent phase are sufficiently similar that these traits are not particularly useful for revealing the microstructure. Instead, the relevant microstructural descriptor is composition. **Figure 13(a)** is a perspective view of gamma′ precipitates in a Ni–5.2 Al–14.2 Cr at% alloy aged at 873 K for 256 h, and observed using LEAP APT (Sudbrack et al., 2006). These precipitates were revealed by the concentration difference between the precipitates and the matrix. A relatively large number of small precipitates can be sampled with the LEAP technique making it practical to measure quantities like mean precipitate size, number density, volume fraction, interprecipitate spacing, and particle interconnectivity.

For comparison, **Figure 13(b)** shows gamma′ precipitates in UMF-20, a Ni-based superalloy reconstructed from serial sections made with FIB and imaged with SEM (MacSleyne et al., 2009). These gamma′ precipitates, each of which have a different color in **Figure 13(b)**, are roughly 50 times larger than those in 13(a). Nevertheless, the larger volume of material sampled with FIB serial sectioning made it possible to analyze enough precipitates to obtain useful statistical data. MacSleyne et al. used the precipitate aspect ratio distribution quantified by normalized moment invariants and a shape quotient parameter to describe objectively the complex precipitate shapes.

In another example of 3D analyses of precipitate microstructures formed during aging, precipitate morphologies in peak-aged Al 2024 were determined with HAADF STEM tomography (Feng et al., 2011). Concentration differences between the phases produced the HAADF STEM contrast used to generate tomographic reconstructions. Several different types of precipitates were elucidated: rod-like

(a)

(b)

LEAP™ tomograph analysis

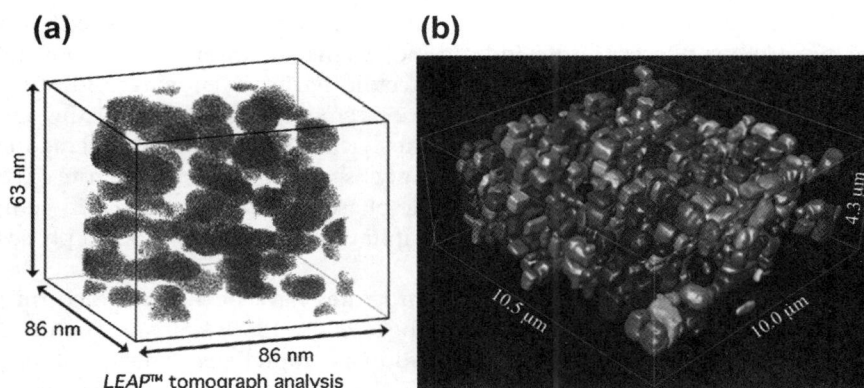

Figure 13 Examples of 3D microstructural characterization of precipitates formed during aging. (a) A perspective view of gamma' precipitates in a Ni–5.2 Al–14.2 Cr at% alloy aged at 873 K for 256 h, and observed using LEAP APT (Sudbrack et al., 2006). (b) Gamma' precipitates in UMF-20, a Ni-based superalloy; reconstructed from serial sections made with FIB and imaged with SEM (Macsleyne et al., 2009). Note: the colors in Figure 13(b) do not have direct crystallographic significance here, and instead were used to highlight the individual particles in this study of precipitate morphology. (For color version of this figure, the reader is referred to the online version of this book.)

$Al_{20}Cu_2Mn_3$ dispersoids, needle-like S precipitates within aluminum grains, and spherical/lenticular S precipitates along a grain boundary. The large number of precipitates in the reconstructed volume enabled statistical analysis of the precipitate characteristics. The 3D information also made it possible to determine the nonuniform arrangement of precipitates in the vicinity of the grain boundary and to measure the areal number density of precipitates on the grain boundary, as well as the precipitate spacing, without the projection artifacts found in conventional TEM images.

10.3.2.2.2   Proeutectoid Precipitation

Steels (Fe–C-based alloys) represent the classic eutectoid decomposition alloy system, are ubiquitous in their engineering application throughout society, and have been studied in detail by metallurgists for more than 100 years (Sorby, 1887; Forsman, 1918; Mehl et al., 1933; Greninger and Troiano, 1940; Hultgren, 1947; Aaronson, 1962; Bhadeshia, 2001). The major categories of microstructure corresponding to eutectoid decomposition products will be briefly considered here (in Sections 10.3.2.2.2–10.3.2.2.4), as they provide for a representative cross-section of common solid-state microstructures formed in metallic systems, and include a wide variety of morphological forms. Particular emphasis is placed here on the 3D characteristics of these microstructures. Although such transformation products will be examined primarily in steels, these microconstituents and some of the discussions thereof also generally apply to a number of other alloy systems as well, such as Ti-based (Lee and Aaronson, 1988; Furuhara, 1990) or Cu-based (Li et al., 1995) systems.

The high temperature parent phase in steels is austenite and has a fcc crystal structure. In Fe–C alloys and/or plain carbon steels, upon relatively slow to intermediate cooling rates (or appropriate isothermal heat treatment conditions) the austenite decomposes to product phases of ferrite and the carbide phase, cementite. Ferrite possesses a body-centered cubic crystal structure, while cementite has a complex orhthorhombic crystal structure. In more highly alloyed steels containing elements such as

Cr, W, Mo, or V, alloy carbides of various crystal structures can form. Upon the decomposition of austenite, these two product phases (ferrite and carbide) arrange themselves in various microstructural forms, including pearlite, bainite (see below), proeutectoid ferrite, or proeutectoid carbide. The term "proeutectoid" was originally used to refer to ferrite or cementite formed from the austenite phase above the eutectoid temperature, in the austenite-plus-ferrite or austenite-plus-cementite two-phase field of the Fe–C phase diagram. It is now also commonly used to refer to either ferrite or cementite that forms directly from the high-temperature austenite phase, below the eutectoid temperature (in a metastable fashion), without the accompanying precipitation of the other product phase (Spanos and Kral, 2001).

Some of the most in-depth studies of 3D microstructures have been performed on proeutectoid cementite in high-carbon steels (Kral et al., 2000; Mangan et al., 1999; Kral and Spanos, 1999, 1997); a detailed discussion of these studies has been included in a comprehensive review of the proeutectoid cementite transformation (Spanos and Kral, 2009). These proeutectoid cementite investigations are now considered here in the context of providing important information on the true 3D characteristics of solid-state microstructures, which may indeed have broader applicability to other metallic alloy systems as well.

Full 3D reconstruction and analysis of more than 200 Widmanstatten (or elongated) proeutectoid cementite precipitates in a high-carbon Fe–C–Mn alloy have shown that although from 2D observations of single planes of polish the majority of these crystals appeared to be *intra*-granular, in three dimensions all of the cementite precipitates made contact somewhere with the original austenite grain boundaries (see **Figure 7(a)**) (Kral and Spanos, 1999). Additionally, two predominant 3D morphologies of Widmanstatten proeutectoid cementite have been observed: (1) monolithic plates often stacked in a face-to-face arrangement (Spanos and Kral, 2009; Kral et al., 2000; Kral and Spanos, 1999)—as shown in **Figure 14**, and agglomerations of fine cementite laths (~25 microns in width) stacked in an edge-to-edge fashion (Spanos and Kral, 2009; Kral and Spanos, 1999)—see **Figure 15**. When viewed at relatively low resolution by either 2D optical microscopy or 3D characterization based on reconstructed stacks of 2D optical micrographs, these agglomerates of cementite crystals can take on an overall external shape of either plates or laths. Such observations of external morphology can in many cases though mask the actual underlying 3D morphology of the individual crystals (Spanos and Kral, 2009; Kral and Spanos, 1999), and in that regard should be correlated with higher resolution SEM and/or TEM studies. This type of correlation has been performed on deeply etched specimens, which provide for observation of 3D character as well (Kral et al., 2000; Mangan et al., 1999). By combining EBSD analysis with subsequent deep etching of the austenite matrix away from the cementite, the orientation relationships (ORs) of these proeutectoid precipitates were directly correlated to SEM observations of the 3D cementite (C) morphology (Mangan et al., 1999). It was shown that proeutectoid cementite *plates* typically possessed the Pitsch orientation relationship (OR) (Pitsch, 1963, 1962) with the austenite (A) lattice,

$$[100]_C // [5\bar{5}4]_A$$

$$[010]_C // [110]_A$$

$$[001]_C // [\bar{2}25]_A$$

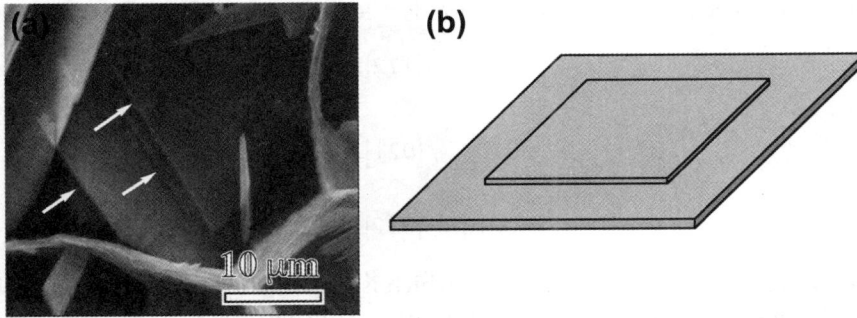

Figure 14 (a) SEM micrograph of a deeply etched specimen in which the austenite matrix was completely etched away, showing three monolithic cementite plates (arrowed) sitting atop one another in a face-to-face arrangement (taken from Mangan et al., 1999). (b) Schematic representation of two cementite plates stacked in a face-to-face arrangement. Taken from Spanos and Kral, 2009.

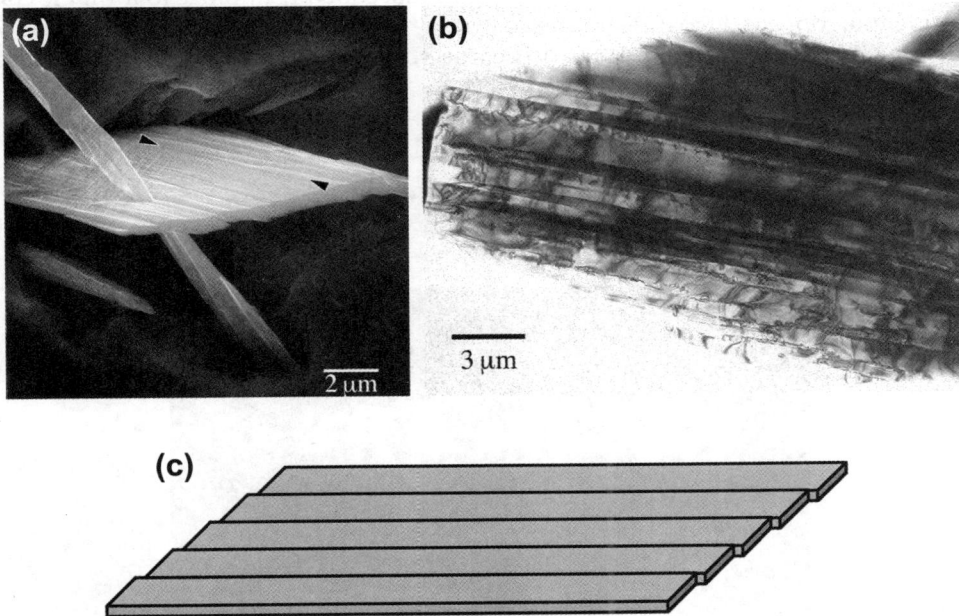

Figure 15 (a) SEM micrograph of a deeply etched specimen in which austenite was completely etched away, showing striations (e.g. see arrows) exhibited by conglomerate of cementite laths (taken from Mangan et al., 1999). (b) TEM micrograph of a deeply etched specimen showing Widmanstatten cementite composed of fine lath-shaped subunits, shown by convergent beam electron diffraction (CBED) to have small misorientations between the fine laths (taken from Kral and Spanos, 1999). (c) Schematic representation of multiple fine lath-shaped subunits stacked in an edge-to-edge arrangement, forming an overall conglomerate of laths. Taken from Spanos and Kral, 2009.

while cementite *laths* obeyed the Farooque–Edmonds OR (Farooque and Edmonds, 1990).

$$[100]_C // [112]_A$$

$$[010]_C // [02\overline{1}]_A$$

$$[001]_C // [\overline{5}12]_A$$

This indicates that the OR of such precipitates, which is presumably dictated at the nucleation stage, sets the conditions for the evolution of the 3D morphology during growth, into either the lath shape or the plate shape.

Intergranular proeutectoid cementite precipitates that are predominantly confined to the austenite grain boundaries (i.e. they do not grow appreciably into the austenite matrix grains) have been shown to possess a unique 3D shape, which is not readily observable by 2D observations (Kral and Spanos, 1999, 1997). That is, 2D characterization techniques typically result in description of the morphology of grain boundary proeutectoid precipitates as "allotriomorphs" at early stages (Aaronson, 1962), often idealized as having a shape approximating that of a "bulged pancake" (Aaronson, 1962) or double spherical cap (Bradley et al., 1977; Hawbolt and Brown, 1967) (see **Figure 16(a)**). (In the case of cementite, it has been inferred that these precipitates subsequently impinge to form the continuous films of cementite observed coating austenite grain boundaries at later transformation times (Aaronson, 1962; Heckel and Paxton, 1960).) The true 3D morphology of such cementite though can

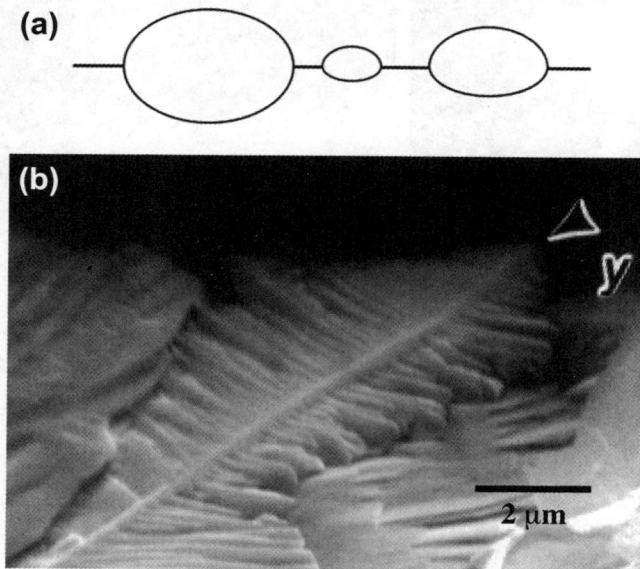

Figure 16   (a) Schematic representation of grain boundary allotriomorph morphology approximated as a double spherical cap. (b) SEM micrograph of grain boundary cementite with austenite deep-etched away, showing solid-state grain boundary cementite dendrites with a central spine (see arrow) and secondary arms. Taken from Kral and Spanos, 1997.

be quite different. As observed by deep etching experiments, grain boundary cementite formed in a hypereutectoid Fe–C–Mn alloy has been shown to posses a dendritic or fernlike shape, in which the growth of the solid-state cementite dendrite arms is confined predominantly within the austenite grain boundary planes (Kral and Spanos, 1997, 1999) (see **Figure 16(b)**). The probability that an austenite grain boundary plane would lie parallel to (or very nearly parallel to) the sectioning plane during 2D observation is minimal; thus, even at very early stages of transformation (before the cementite dendrites impinge to form a film), almost all 2D observations/sections would be at some glancing angle through the dendrite arms, and the actual 3D dendritic morphology would not be deduced.

Although the most complete set of 3D observations of proeutectoid precipitates has been made for cementite in high-carbon steels (Mangan et al., 1997; Spanos and Kral, 2009; Kral et al., 2000; Mangan et al., 1999; Kral and Spanos, 1999, 1997), the degree to which these observations can be generalized to other alloy systems has yet to be determined, in large part due to the lack of such 3D data in other materials. Nevertheless, one other type of proeutectoid precipitate that has been studied in 3D by a number of researchers is ferrite formed in lower carbon steels, which will now be considered particularly in the context of comparison to the 3D observations of cementite.

Dube's seminal system of morphological classification of proeutectoid ferrite (Dubé et al., 1958; Dubé, 1948) (**Figure 17**), which was later extended to describe proeutectoid cementite as well (Aaronson, 1962), was based on 2D observations. Although many more advanced and higher resolution studies of proeutectoid ferrite (employing SEM and TEM as opposed to optical microscopy) followed after that classification, they were still mostly limited to 2D analyses. While 3D studies of proeutectoid cementite (see above) have shown it to differ significantly from the descriptions in Dube's morphological classification system (Spanos and Kral, 2009), the true 3D nature of ferrite has been shown to differ not only from Dube's 2D descriptions but also from the 3D observations of cementite. In particular, direct 3D observations have been made on three different types of ferrite precipitates: (1) intragranular "degenerate ferrite" (Hackenberg et al., 2002; Wu and Enomoto, 2002), (2) intragranular Widmanstatten ferrite formed on inclusions (Yokomizo et al., 2003), and (3) grain boundary ferrite (i.e. ferrite formed on austenite grain boundaries) (Kral and Spanos, 2005).

"Degenerate ferrite" is a term coined to describe ferrite that has a mottled or degenerate appearance when observed by light microscopy, and this form of ferrite is observed in only certain alloys and temperature regimes (Boswell et al., 1986; Reynolds-Jr. et al., 1990; Reynolds et al., 1990). TEM observations have suggested that this morphological variant is produced by the sympathetic nucleation of multiple ferrite crystals atop one another (Reynolds-Jr. et al., 1990), and subsequent 3D reconstructions by serial sectioning indicate that degenerate ferrite is indeed composed of agglomerates of many fine rod-shaped ferrite crystals (Hackenberg et al., 2002; Wu and Enomoto, 2002). When nucleated on inclusions, intragranular ferrite has also been shown to posses a Widmanstatten morphology of plate-like protuberances in 3D, emanating in an almost "starburst" pattern from the central inclusions (Yokomizo et al., 2003).

Grain boundary nucleated ferrite has been studied in 3D by serial sectioning in an Fe-0.12 wt pct C-3.28 wt pct Ni alloy, revealing actual 3D morphologies that differ in various ways from the simplified versions depicted in Dube's 2D classification system (Kral and Spanos, 2005)—see **Figure 18**. The 3D morphological variants of grain boundary proeutectoid ferrite observed in that study include (1) primary and secondary ferrite "spikes" (Figure 18-1), (2, 3) secondary ferrite plates and laths (Figures 18-2 and 18-3, respectively), (4) secondary ferrite sawteeth (Figure 18-4), and (5) ellipsoidal shaped ferrite allotriomorphs (Figure 18-5). In this context, "primary" refers to elongated or "Widmanstatten" ferrite crystals emanating directly from the austenite grain

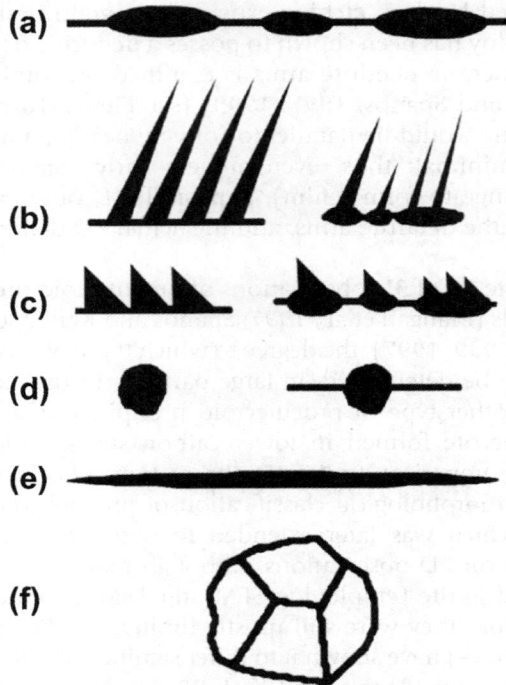

Figure 17 Dube's morphological classification system (Dubé et al., 1958; Dubé, 1948), as modified by (and reproduced from) Aaronson (Aaronson, 1962). (a) Grain boundary allotriomorphs, (b) primary and secondary Widmanstatten sideplates, (c) primary and secondary Widmanstatten sawteeth, (d) idiomorphs, (e) intragranular Widmanstatten plates (or needles), and (f) massive structures.

boundaries, while "secondary" refers to crystals that form atop previously formed ferrite grain boundary allotriomorphs. These 3D proeutectoid ferrite shapes also differ in many ways from those of cementite described above, with probably the most prominent difference being that the morphology of grain boundary proeutectoid ferrite is that of individual allotriomorphs possessing a 3D shape best approximated as a prolate ellipsoid (Figure 18-5) (Kral and Spanos, 2005). This is in contrast to the grain boundary proeutectoid cementite shown to possess a branched solid-state dendritic structure (see **Figure 16(b)** and related discussion above) (Kral and Spanos, 1997; Spanos and Kral, 2009).

It is emphasized that the features discussed here are based on 3D investigations in a limited number of alloy systems. There is thus a great need for further 3D study in many more alloys and temperature ranges, in both steels and nonferrous metallic alloy systems.

10.3.2.2.3 Lamellar Microstructures

Lamellar microstructures develop from the formation of (usually) two phases arranged as a collection of alternating parallel plates. The morphology typically arises when one or more phases transform to two compositionally distinct product phases with the same average composition. The composition of the individual product phases complement each other in the sense that one is solute rich while the

	2D example	3D reconstruction	3D schematic
(1) Primary spikes			
(2) Secondary plate			
(3) Secondary plate/lath			
(4) Secondary sawtooth			
(5) Allotrimorph			

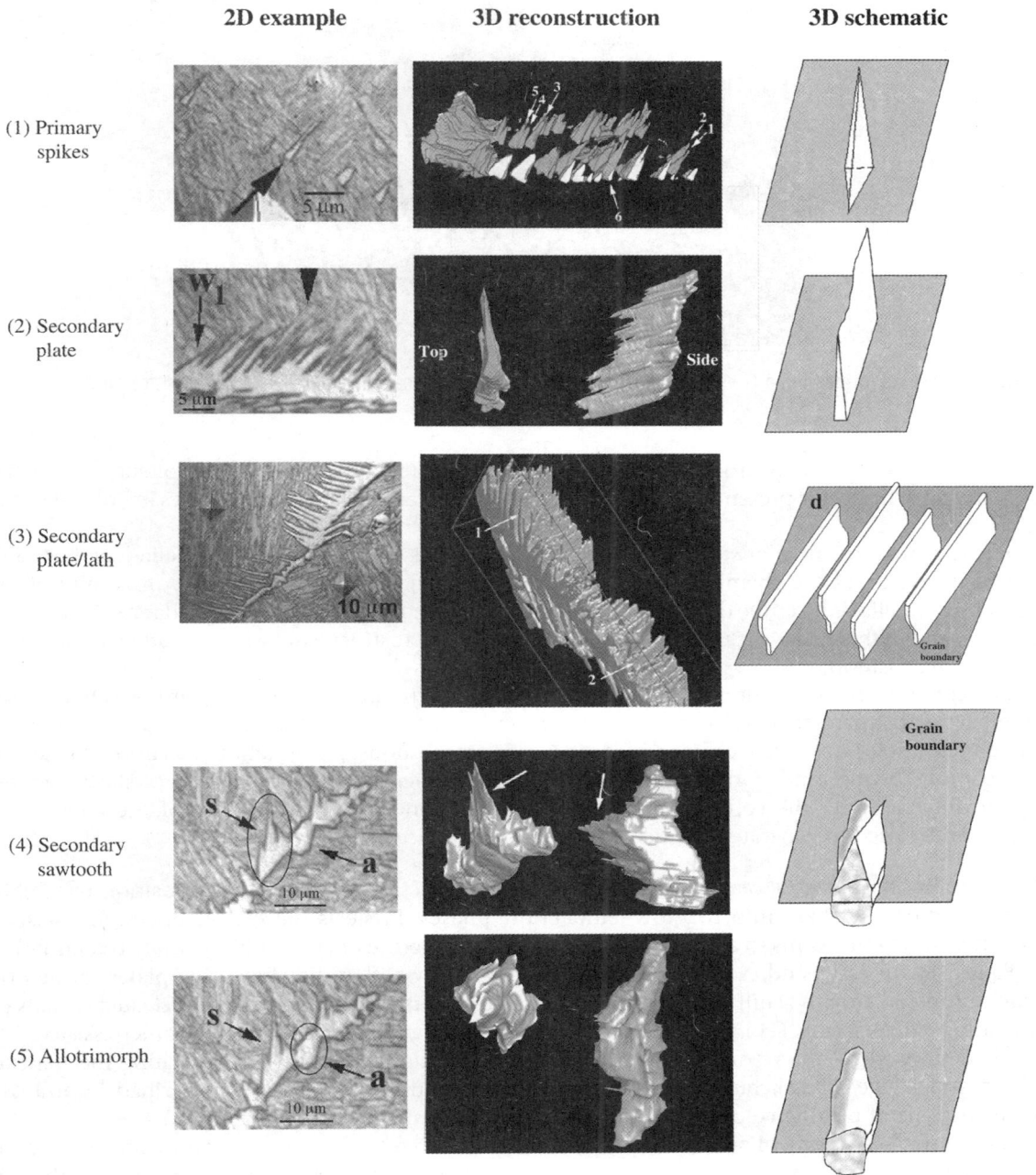

Figure 18 A partially revised system of morphological classification of proeutectoid ferrite in three dimensions. Taken from Kral and Spanos, 2005.

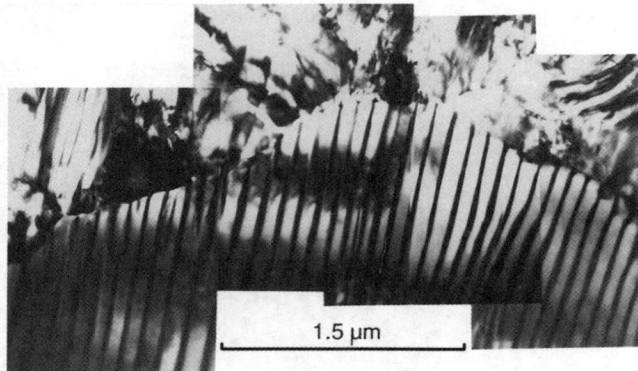

Figure 19   Pearlite formed in an Fe-0.95wt%C-1.93wt%Mn steel isothermally reacted at 600 C. Taken from Spanos et al., 1990b.

other is solute poor, relative to the average composition of the alloy. An example of such a lamellar microstructure in steel is presented in **Figure 19**, and in this case the alternating plates are of cementite and ferrite.

During growth of lamellar structures, the individual plates in a collection of lamella lengthen, and each plate consumes the chemical specie rejected by the adjacent phase. The close proximity of the alternating lamella reduces the distance over which solute is redistributed as the lamellar microstructure develops, and in this sense, the product phases of lamellar microstructures assist each other and grow in a cooperative fashion.

The lamellar microstructure produced by eutectoid reactions is known as pearlite (**Figure 19**). Lamellar microstructures are also associated with the redistribution of solute behind moving grain boundaries; these are called cellular reactions or discontinuous precipitation reactions. Since the compositions of the phases in lamellar microstructures differ and their respective crystal structures usually differ as well, the composition, crystal structure, and/or misorientation of the constituent phases can be used to generate 3D microstructural information.

10.3.2.2.3.1 Pearlite (Lamellar Eutectoid) Pearlite forms as the product of eutectoid decomposition of a high-temperature parent phase into two low-temperature phases. In steels the high-temperature phase is austenite, and it transforms to alternating lamellae/plates of ferrite plus carbide (typically cementite)—see **Figure 19**. In early studies there was some debate as to whether the alternating plates formed by renucleation and growth (Hull and Mehl, 1942) or by branching from the initially nucleated crystals of cementite and/or ferrite (Hillert, 1962). A serial sectioning study in which computer-assisted 3D reconstructions were not yet available suggested that pearlite formed by the branching mechanism (Hillert and Lange, 1962), but it was not until later that this mechanism was verified by full 3D reconstructions of pearlite colonies (DeGraef et al., 2006; Mangan and Shiflet, 1994).

Thus, in steels, a colony of pearlite originates from a single crystal of cementite and a single crystal of ferrite. The lamellar morphology arises from branching of at least one of the phases during growth such that an entire pearlite colony consists of essentially two crystals (one for each phase). The spacing of the lamella is dependent on the temperature and alloy system, and finer spacings often lead to increased strength of this microconstituent. The growth kinetics depend upon the path for solute redistribution (i.e. volume diffusion and/or interphase boundary diffusion) and the lamellar spacing. It has also been shown that there is a kinetic/structural constraint at a migrating pearlite boundary that is related to the lattice matching between the product phases and the parent phase—that is, growth ledges migrate at the leading pearlite:austenite interface (Hackney and Shiflet, 1987a,b).

10.3.2.2.3.2 Cellular Reaction (Discontinuous Precipitation) Although these lamellar microstructures can closely resemble pearlite, the transformation mechanisms and number of phases involved differ. Rather than formation of two product phases by eutectoid decomposition (as in the case of pearlite), the cellular reaction involves the cooperative growth of crystals of one new precipitate phase and the solute-depleted matrix phase into a more supersaturated matrix. The cellular product often consumes a metastable precipitate phase that has high strain energy because of coherency. The reaction generally initiates at relatively mobile high-angle grain boundaries and produces discrete "cells" or "colonies" of the lamellar microstructure. Solute redistribution takes place along the advancing cell boundary, which in this case is also the original grain boundary at which the cellular reaction initiated (Aaronson et al., 2010). There have been numerous studies of cellular microstructures, but most have employed 2D observation methods. At least two investigations though have used serial sectioning and 3D reconstruction to visualize and analyze the 3D characteristics of the cellular transformation, specifically in Cr–Ni and Cu–Ti alloys (Matsuoka et al., 1994; Mangan and Shiflet, 1997).

10.3.2.2.4 Nonlamellar Eutectoid Microstructures
Bainite is a term that was coined more than 50 years ago to refer to non-lamellar products of eutectoid decomposition observed in steels (Hultgren, 1920; Davenport and Bain, 1930; Greninger and Troiano, 1940). It has since been applied to other alloy systems (such as Ti-based alloys) as well (Lee and Aaronson, 1988; Furuhara, 1990). In the past 50 years there have been a great number of studies and reviews on bainite (Aaronson, 1969; Aaronson et al., 2002; Reynolds-Jr. et al., 1991; Bhadeshia, 2001, 1984; Sandvik, 1982), and the reader is referred to those investigations for details of the formation mechanisms, crystallography, properties, and practical applications of this microconstituent. It should be mentioned at the outset that for the past 30 years or so there has been a somewhat spirited debate as to the relative roles of diffusion and shear in the formation of bainite (Aaronson, 1969; Aaronson et al., 2002; Reynolds-JR. et al., 1991; Bhadeshia, 2001, 1984), and although the present authors subscribe to the diffusionist viewpoint (Reynolds-Jr. et al., 1991), this issue will not be debated here. Instead, a brief overview of bainite will be provided in the context of describing the overall appearance and formation of this microstructure, its uniqueness from other microstructures, and some different morphological forms that it may take.

In steels, the two eutectoid phases comprising bainite are ferrite as the majority phase and some type of carbide (cementite, or some alloy carbide) as the minority phase. During early observations in steels, bainite was recognized to take on two different forms—either upper bainite or lower bainite, formed at higher and lower temperatures (both below the eutectoid temperature), respectively. Examples of these two types of bainite are presented in **Figures 20** and **21**. Upper bainite generally has an apparent (2D) morphology characterized by sheaves of laths, or needles, with the carbides lying parallel to the lath axis and at the lath boundaries (Pickering, 1967; Oblak and Hehemann, 1967; Spanos et al., 1990b) (**Figure 20**). Lower bainite typically has a plate (or lath) morphology in which the carbides lie at angle of approximately 55° to the plate/lath axis (Hehemann, 1970; Modin and Modin, 1955; Spanos et al., 1990a)—see **Figure 21**.

On the generalized definition of bainite (particularly espoused from the diffusionist viewpoint (Aaronson, 1969; Aaronson et al., 2002; Reynolds-Jr. et al., 1991)), "nodular bainite" is described as a nonlamellar product of eutectoid decomposition with a more equiaxed external morphology (see **Figure 22(a)**), and when such an equiaxed nonlamellar product is formed at austenite grain

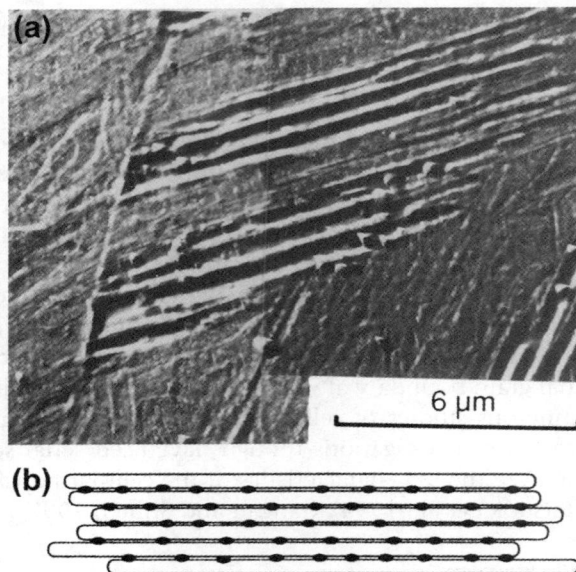

Figure 20   Upper bainite. (a) A carbon replica of upper bainite formed in an Fe-0.1wt%C2.9wt% Mn alloy isothermally reacted at 450 C (taken from Spanos et al., 1990b). (b) Schematic representation of upper bainite. Taken from Reynolds-Jr. et al., 1991.

boundaries, it has been referred to as "columnar bainite" (**Figure 22(b)**) (Reynolds-Jr. et al., 1991). Finally, the term inverse bainite was coined to describe bainite that is formed in hypereutectoid steels and is originally initiated by a proeutectoid cementite spine upon which the eutectoid decomposition of austenite (to ferrite plus cementite) evolves (Hillert, 1957; Spanos et al., 1990b; Reynolds-Jr. et al., 1991)—see **Figure 23**.

Although the majority of observations of these various forms of bainite have been in steels, these microstructures have been reported in other metallic systems as well, including Ti-based (Lee and Aaronson, 1988; Furuhara, 1990) and Cu-based (Li et al., 1995) alloys. Almost all observations of bainite morphology and crystallography to date though have been based on essentially 2D observations by optical microscopy, SEM, or TEM. (There have been some 3D atom probe analyses of small volumes within bainitic microstructures, but these have usually been centered about determining the carbon concentration of the bainitic ferrite (Peet et al., 2004)). Future 3D investigations are thus needed to help elucidate the true 3D nature of bainitic microconstituents.

10.3.2.2.5   *Microstructures Formed by Spinodal Decomposition*

The understanding of spinodal decomposition microstructures has developed in a way somewhat different from those of other transformation products. Theoretical treatments of spinodal decomposition consider the time evolution of concentration fluctuations that have a small spatial wavelength and small concentration amplitude (Hillert, 1955; Cahn and Hilliard, 1959; Cahn, 1961, 1962; Hillert, 1961). These characteristics render the early stages of spinodal decomposition difficult to study by image-based microscopy techniques because these methods typically cannot resolve small concentration differences a few nanometers in size, within a structurally uniform material. For this reason, scattering techniques and

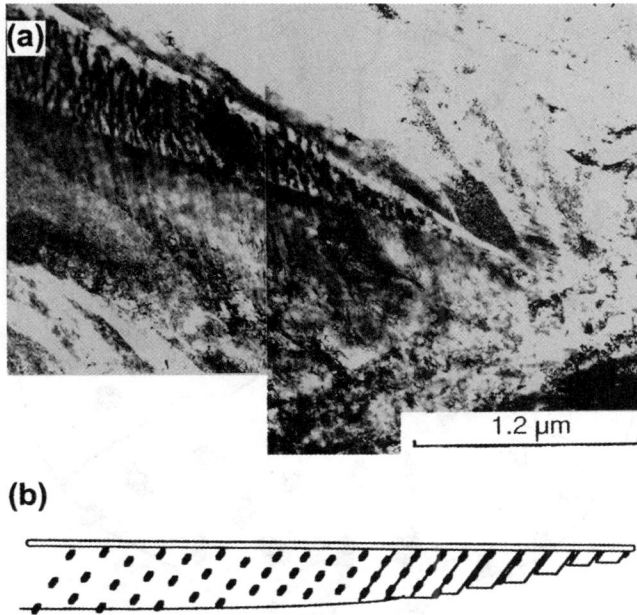

Figure 21 Lower bainite. (a) Transmission electron micrograph of lower bainite formed in an Fe-0.95wt%C1.93wt% Mn alloy isothermally reacted at 250 C (taken from Spanos et al., 1990b). (b) Schematic representation of lower bainite. Taken from Reynolds-Jr. et al., 1991.

atom probe field ion microscopy spectroscopy (AP/FIM) were the primary tools for analyzing spinodal microstructures (Haasen and Wagner, 1985; Miller et al., 1986). These techniques have aspects of direct 3D microstructure characterizations in that they measure concentration wavelengths in 3D space, even though TEM and FIM *images* of spinodal microstructures were initially in two dimensions. With the inclusion of position-sensitive detectors on atom probes, full 3D imaging of spinodal microstructures became possible (Miller et al., 1995; Danoix et al., 2004), and measuring the concentration amplitude and wavelength as well as visualizing the connectivity of the phases in spinodal microstructures in three dimensions with APT is now relatively straightforward (Eiedenberger et al., 2010).

10.4 Concluding Remarks

Internal microstructure controls the properties of metallic alloys, and these properties can be optimized by manipulation of the microstructure through thermomechanical processing (Dieter, 1961; Krauss, 1990; Bhadeshia, 2001; Davis et al., 1990). The properties in turn dictate the final performance of metallic components and platforms used in a variety of applications, from automobiles to airplanes, to ships, to construction for infrastructure, to medical devices, to a wide array of consumer products. A detailed knowledge of microstructure in metals and alloys is thus vital for developing these materials for a wide variety of product and manufacturing sectors.

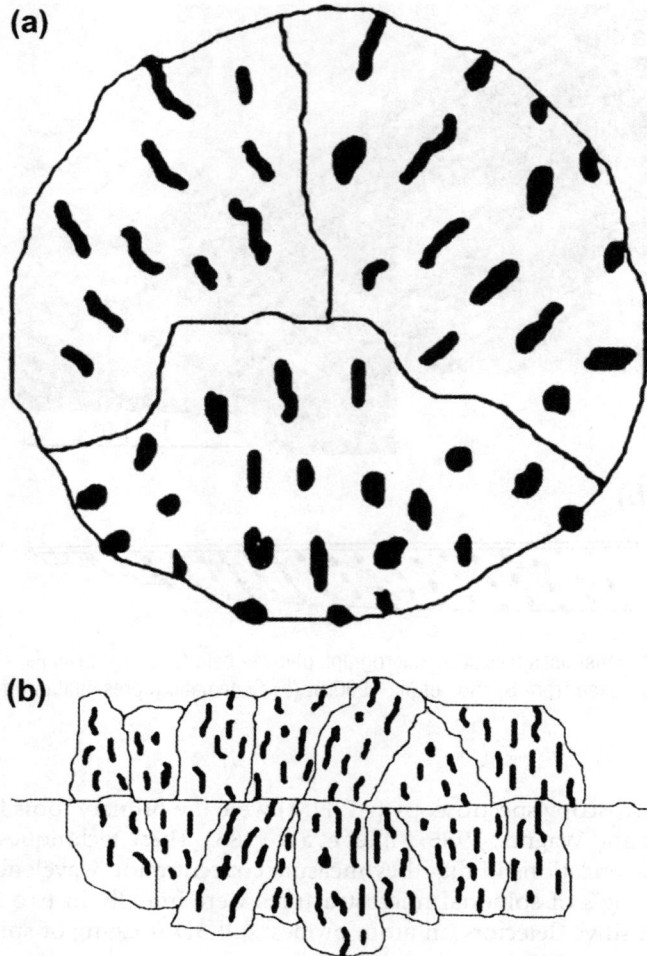

Figure 22 Nonlamellar eutectoid decomposition product with a more equiaxed external morphology—schematic representations. (a) Nodular bainite. (b) Columnar bainite. Taken from Reynolds-Jr. et al., 1991.

Figure 23 Schematic representation of inverse bainite. Taken from Reynolds-Jr. et al., 1991.

Quantitative characterization of microstructure is particularly important in a relatively new thrust area in materials development and engineering referred to as Integrated Computational Materials Engineering (ICME) (Allison et al., 2006; Collins et al., 2011; Pollock and Allison, 2008), and/or through process modeling (Gottstein and Mohles, 2011). In this methodology, predictive simulations of materials processing and properties are used in conjunction with a variety of other predictive models over a wide range of length scales and engineering disciplines to design and deploy materials components and manufacturing processes at greatly reduced cost and time investment (Allison et al., 2006; Pollock and Allison, 2008). This approach has especially been employed in the development of metallic-based components, and accurate quantitative experimental characterization of the microstructure has been shown to be a critical component for model input, validation, and verification (Collins et al., 2011; Allison et al., 2006; Gottstein and Mohles, 2011; Pollock and Allison, 2008). Concurrently, the vital role of characterization and analyses of materials microstructures in *three dimensions* (as opposed to the conventional 2D techniques typically employed) has been elucidated and highlighted within the materials community (Uchic, 2006; Spanos, 2006; Thornton and Poulsen, 2008; Spanos et al., 2010).

Considering the aforementioned importance of microstructure and accurate microstructural characterization in materials design, this review has provided an overview of the methodology and utility of a number of techniques for quantitatively analyzing metallic microstructures in three dimensions, as well as some specific examples of 3D microstructural data (such as morphology, spatial distribution of phases, and crystallography) in some specific classes of microstructures in metals and alloys.

Acknowledgments

GS gratefully acknowledges IT and other support from The Minerals, Metals, and Materials Society (TMS) for this effort; WTR acknowledges support from the U.S. Department of Energy.

References

Aaronson, H.I., Spanos, G., Reynolds-JR., W.T., 2002. A progress report on the definitions of bainite. Scripta Materialia 47, 139–144.

Aaronson, H.I., Enomoto, M., Lee, J.K., 2010. Cellular Reaction. Mechanisms of Diffusional Transformations in Metals and Alloys. CRC Press, Boca Raton, FL.

Aaronson, H.I., 1962. Proeutectoid ferrite and cementite reactions. In: Zackay, V.F., Aaronson, H.I. (Eds.), The Decomposition of Austenite by Diffusional Processes. Interscience, NY.

Aaronson, H.I., 1969. The Mechanism of Phase Transformations in Crystalline Solids. Inst. of Metals, London.

Alkemper, J., Voorhees, P.W., 2000. Quantitative serial sectioning analysis. Journal of Microscopy 201, 1–8.

Alkemper, J., Voorhees, P.W., 2001. Three-dimensional characterization of dendritic microstructures. Acta Materialia 49, 987–1902.

Allison, J., Backman, D., Christodoulou, L., 2006. Integrated computational materials engineering: a new paradigm for the global materials profession. JOM 11, 25–27.

Ardell, A.J., 1972. Acta Metallurgica 20, 61–71.

Barrett, L.K., Yust, C.S., 1967. Progressive shape changes of the void during sintering. Transactions of AIME 239, 1172–1180.

Bhadeshia, H.K.D.H., 1984. Phase Transformations in Ferrous Alloys. TMS-AIME, Warrendale, PA.

Bhadeshia, H.K.D.H., 2001. Bainite in Steels—Transformation, Microstructure and Properties. Institute of Materials Communications Ltd.

Bollmann, W., 1982. Crystal Lattices, Interfaces, Matrices. Bollmann, Geneva.

Boswell, P.G., Kinsman, K.R., Shiflet, G.J., Aaronson, H.I., 1986. In: Antolovich, S.D., Richie, R.O., Gerberich, W.W. (Eds.), Mechanical Properties and Phase Transformations in Engineering Materials. TMS-AIME, Warrendale, PA.

Bradley, J.R., Rigsbee, J.M., Aaronson, H.I., 1977. Growth kinetics of grain boundary ferrite allotriomorphs in Fe–C alloys. Metallurgical Transactions A: Physical Metallurgy and Materials Science 8, 323–333.

Cahn, R.W., Haasen, P. (Eds.), 1996. Physical Metallurgy, fourth ed. Elsevier Science.

Cahn, J.W., Hilliard, J.E., 1959. Free energy of a nonuniform system. III. Nucleation in a two-component incompressible fluid. Journal of Chemical Physics 31, 688–699.

Cahn, J.W., 1961. On spinodal decomposition. Acta Metallurgica 9, 795–801.

Cahn, J.W., 1962. Coherent fluctuations and nucleation in isotropic solids. Acta Metallurgica 10, 907–913.

Cahn, J.W., 1967. The significance of average mean curvature and its determination by quantitative metallography. Transactions of The Metallurgical Society of AIME 239, 610.

Carlton, C., Ferreira, P.J., 2007. What is behind the inverse Hall–Petch behavior in nanocrystalline materials? (Mater. Res. Soc. Symp. Proc. 976E) In: Lilleodden, E., Besser, P., Levine, L., Needleman, A. (Eds.), Size Effects in the Deformation of Materials—Experiments and Modeling. Materials Research Society, Warrendale, PA.

Chawla, N., Sidhu, R.S., 2007. Microstructure based modeling of deformation in Sn-rich Pb-free solder alloys. Journal of Materials Science: Materials and Electronics 18, 175–189.

Chen, L.-Q., 2002. Phase-field models for microstructural evolution. Annual Review of Materials Research 32, 113–140.

Collins, P., Allison, J.A., Spanos, G., 2011. Proceedings of the 1st World Congress on Integrated Computational Materials Engineering (ICME). The Minerals, Metals & Materials Society (TMS); Wiley.

Danoix, F., Auger, P., Blavette, D., 2004. Hardening of aged duplex stainless steels by spinodal decomposition. Microscopy and Microanalysis 10, 349–354.

Davenport, E.S., Bain, E.C., 1930. Transformation of austenite at constant subcritical temperatures. Transactions of AIME 90, 117–154.

Davis, J.R., Allen, P., Lampman, S.R., Zorc, T.B., Henry, S.D., Daquila, J.L., Ronke, A.W., Jakel, J., O'keefe, K.L. (Eds.), 1990. ASM Handbook Properties and Selection of Nonferrous Alloys and Special Purpose Materials. ASM International, Materials Park Ohio.

Dawson, P.R., Mika, D.P., Barton, N.R., 2002. Finite element modeling of lattice misorientations in aluminum polycrystals. Scripta Materialia 47, 713–717.

Degraef, M., Kral, M.V., Hillert, M., 2006. A modern 3-D view of an "Old" pearlite colony. JOM 58, 25–28.

Dehoff, R.T., Liu, G.Q., 1985. On the relation between grain size and grain topology. Metallurgical Transactions A: Physical Metallurgy and Materials Science 16A, 2007–2011.

Dieter, G.E., 1961. Mechanical Metallurgy. McGraw-Hill.

Dubé, C.A., Aaronson, H.I., Mehl, R.F., 1958. La formation de la ferrite proeutectoide dans les aciers au carbon. Revue de Metallurgie. 55, 201–210.

Dubé, C. A., 1948. Ph.D. Dissertation, Carnegie-Mellon University.

Dunn, D.N., Hull, R., 1999. Reconstruction of three-dimensional chemistry and geometry using focused ion beam microscopy. Applied Physics Letters 75, 3414–3416.

Echlin, M.P., Husseini, N.S., Nees, J.A., Pollock, T.M., 2011. A new femtosecond laser-based tomography technique for multiphase materials. Advanced Materials 23, 2339–2342.

Efstathiou, C., Boyce, D., Park, J.-S., Lienert, U., Dawson, P.R., Miller, M.P., 2010. A method for measuring single-crystal elastic moduli using high-energy X-ray diffraction and a crystal-based finite element model. Acta Materialia 58, 5806–5819.

Eiedenberger, E., Schober, M., Staron, P., Caliskanoglu, D., Leitner, H., Clemens, H., 2010. Spinodal decomposition in Fe-25 at%Co-9 at%Mo. Intermetallics 18, 2128–2135.

Everett, R.K., Geltmacher, A.B., Simmonds, K.E., 2002. 3D image-based modeling of void interactions in HY100 steel. In: Kahn, A.S., Lopez-Pamies, O. (Eds.), Plasticity, Damage, and Fracture at Macro, Micro, and Nano Scales. Neat Press, pp. 699–701.

Farooque, M., Edmonds, D.V., 1990. The orientation relationships between Widmanstatten cementite and austenite. In: Hobbs, L.W., Sinclair, R. (Eds.), Proc. XIIth Intl. Congress for Electron Microscopy. San Francisco Press, San Francisco, CA, pp. 910–911.

Feng, Z.Q., Yang, Y.Q., Huang, B., Luo, X., Li, M.H., Chen, Y.X., Han, M., Fu, M.S., Rub, J.G., 2011. STEM-HAADF tomography investigation of grain boundary precipitates in Al–Cu–Mg alloy. Materials Letters 65, 2808–2811.

Fonda, R.W., Lauridsen, E., Ludwig, W., Tafforeau, P., Spanos, G., 2007. 2D and 3D analyses of sigma precipitates and porosity in a super-austenitic stainless steel. Metallurgical and Materials Transactions A: Physical Metallurgy and Materials Science 38A, 2721–2726.

Forsman, O., 1918. Undersökning av rymdstrukturen hos ett kolstål av hypereutektoid sammansättning. Jernkontorets Annaler 102, 1–30.

Furuhara, T., Lee, H.J., Menon, E.S.K., Aaronson, H.I., 1990. Interphase boundary structures associated with diffusional phase transformations in Ti-base alloys. Metallurgical and Materials Transactions A: Physical Metallurgy and Materials Science 21A, 1627–1643.

Geltmacher, A.B., Lewis, A.C., Rowenhorst, D.J., Qidwai, M.A., Spanos, G., 2007. Three-dimensional characterization and mesoscale mechanical modeling of a beta titanium alloy. In: Niinomi, M., Akiyama, S., Hagiwara, M., Ikeda, M., Maruyama, K. (Eds.), Poceedings of the 11th World Conference on Titanium: Ti-2007 Science and Technology, 3 June 2007. Japan Institute of Metals, Kyoto, Japan, pp. 475–478.

Gleiter, H., 1996. Microstructure. In: Cahn, R.W., Haasen, P. (Eds.), Physical Metallurgy, Fourth ed. Elsevier Science. (Chapter 9).

Glicksman, M.E., 2005. Analysis of 3-D network structures. Philosophical Magazine 85, 3–31.

Gottstein, G., Mohles, V., 2011. From processing to properties: through-process modeling of aluminum sheet fabrication. In: Collins, P., Allison, J.A., Spanos, G. (Eds.), Proceedings of the 1st World Congress on Integrated Computational Materials Engineering (ICME). The Minerals, Metals & Materials Society (TMS); Wiley.

Greninger, A.B., Troiano, R., 1940. Transactions of AIME 140, 307–331.

Groeber, M., Haley, B., Uchic, M., Dimiduk, D.M., Ghosh, S., 2006. 3D reconstruction and characterization of polycrystalline microstructures using a FIB–SEM system. Materials Characterization 57, 259–273.

Haasen, P., Wagner, R., 1985. Application of analytical field-ion microscopy to the decomposition of alloys. Annual Review of Materials Science 15, 43–78.

Hackenberg, R.E., Nordstrom, D.P., Shiflet, G.J., 2002. Morphology and the three-dimensional structure of ferrite formed below the bay in an Fe–C–W alloy. Scripta Materialia 47, 357–361.

Hackney, S.A., Shiflet, G.J., 1987a. Acta Metallurgica 35, 1007–1017.

Hackney, S.A., Shiflet, G.J., 1987b. Acta Metallurgica 35, 1019.

Hall, E.O., 1951. Proceedings of the Physical Society, London, Section B 64, 747–753.

Hardy, C., Voorhees, P.W., 1988. Ostwald ripening in a system with a high volume fraction of coarsening phase. Metallurgical Transactions A: Physical Metallurgy and Materials Science 19A, 2713–2721.

Hata, S., Miyazaki, H., Miyazaki, S., Mitsuhara, M., Tanaka, M., Kaneko, K., Higashida, K., Ikeda, K., Nakashima, H., Matsumura, S., Barnard, J.S., Sharp, J.H., Midgley, P.A., 2011. Ultramicroscopy 111, 1168–1175.

Hawbolt, E.B., Brown, L.C., 1967. Grain boundary precipitation in Ag-5.64 wt pct Al alloys. Transactions of AIME 239, 1916–1924.

Heckel, R.W., Paxton, H.W., 1960. Rates of growth of cementite in hypereutectoid steels. Transactions of AIME 218, 799–806.

Hehemann, R.F., 1970. Phase Transformations. ASM, Metals Park, OH.

Helm, D., Butz, A., Raabe, D., Gumbsch, P., 2011. Microstructure-based description of the deformation of metals: theory and application. In: Collins, P., Allison, J.A., Spanos, G. (Eds.), Proceedings of the 1st World Congress on Integrated Computational Materials Engineering (ICME). The Minerals, Metals & Materials Society (TMS); Wiley.

Herbig, M., King, A., Reischig, P., Proudhon, H., Lauridsen, E.M., Marrow, J., Buffiere, J., Ludwig, W., 2011. 3D growth of a short fatigue crack within a polycrystalline microstructure studied using combined diffraction and phase-contrast X-ray tomography. Acta Materialia 59, 590–601.

Hillert, M., Lange, N., 1962. In: Zackay, V.F., Aaronson, H.I. (Eds.), Unpublished Research, Cited by M. Hillert in Decomposition of Austenite by Diffusional Processes. Interscience Publ., New York, p. 197.

Hillert, M., 1955. A Theory of Nucleation for Solid Metallic Solutions. Sc. D. MIT.

Hillert, M., 1957. Jernkontorets Annaler 141, 757–789.

Hillert, M., 1961. A solid solution model for inhomogeneous systems. Acta Metallurgica 9, 525–535.

Hillert, M., 1962. The formation of pearlite. In: Zackay, V.F., Aaronson, H.I. (Eds.), The Decomposition of Austenite by Diffusional Processes. Interscience, New York.

Hirth, J.P., Lothe, J., 1982. Theory of Dislocations. Krieger Publishing Company, Malabar, FL.

Hono, K., 1999. Atom probe microanalysis and nanoscale microstructures in metallic materials. Acta Materialia 47, 3127–3145.

Hono, K., 2002. Nanoscale microstructural analysis of metallic materials by atom probe field ion microscopy. Progress in Materials Science 47, 621–729.

Hopkins, R.H., Kraft, R.W., 1965. A rapid technique for observation of three-dimensional microstructures. Transactions of AIME 233, 1526–1532.

Howe, J.M., 1997. Interfaces in Materials: Atomic Structure, Kinetics and Thermodynamics of Solid-Vapor, Solid-Liquid and Solid-Solid Interfaces. John Wiley & Sons, NY.

Hull, F.C., Mehl, R.F., 1942. Transactions of ASM 30, 381.

Hultgren, A., 1920. A Metallographic Study on Tungsten Steels. John Wiley, New York, NY.

Hultgren, A., 1947. Transactions of ASM 39, 915–989.

Inkson, B.J., Mulvihill, M., Möbus, G., 2001a. 3D determination of grain shape in a FeAl-based nanocomposite by 3D FIB tomography. Scripta Materialia 45, 753–758.

Inkson, B.J., Steer, T., Möbus, G., Wagner, T., 2001b. Subsurface nanoindentation deformation of Cu–Al multilayers mapped in 3D by focused ion beam microscopy. Journal of Microscopy 201, 256–269.

Kammer, D., Voorhees, P.W., 2006. The morphological evolution of dendritic microstructures during coarsening. Acta Materialia 54, 1549–1558.

Kelly, T.F., Miller, M.K., 2007. Atom probe tomography. Review of Scientific Instruments 78, 031101.1–031101.20.

Kelly, T.F., Gribb, T.T., Olson, J.D., Martens, R.L., Shepard, J.D., Wiener, S.A., Kunicki, T.C., Ulfig, R.M., Lenz, D.R., Strennen, E.M., Oltman, E., Bunton, J.H., Strait, D.R., 2004. First data from a commercial local electrode atom probe (LEAP). Microscopy and Microanalysis 10, 373–383.

Kelly, T.F., Larson, D.J., Thompson, K., Olson, J.D., Alvis, R.L., Bunton, J.H., Gorman, B.P., 2007. Atom probe tomography of electronic materials. Annual Review of Materials Research 37, 681–727.

Konrad, J., Zaefferer, S., Raabe, D., 2006. Investigation of orientation gradients around a hard Laves particle in a warm-rolled Fe$_3$Al-based alloy using a 3D EBSD–FIB technique. Acta Materialia 54, 1369–1380.

Kral, M.V., Spanos, G., 1997. Three dimensional morphology of cementite precipitates. Scripta Materialia 36, 875–882.

Kral, M.V., Spanos, G., 1999. Three dimensional analysis of proeutectoid cementite precipitates. Acta Metallurgica 47, 711–724.

Kral, M.V., Spanos, G., 2005. Three dimensional analysis and classification of grain boundary nucleated proeutectoid ferrite precipitates. Metallurgical and Materials Transactions A: Physical Metallurgy and Materials Science 36A, 1199–1207.

Kral, M.V., Mangan, M.A., Spanos, G., Rosenberg, R.O., 2000. Three-dimensional analysis of microstructures. Materials Characterization 45, 17–23.

Kral, M.V., Ice, G.E., Miller, M.K., Uchic, M.D., Rosenberg, R.O., 2004. Three-dimensional microscopy. In: Vandervoort, G.F. (Ed.), ASM International Metallography and Microstructures Handbook. ASM International, Materials Park, OH.

Krauss, G., 1990. Steels: Heat Treatment and Processing Principles. ASM International, Metals Park, OH.

Krill, C., Chen, L., 2002. Acta Materialia 50, 3057–3073.

Lee, H.J., Aaronson, H.I., 1988. Acta Metallurgica 36, 1141.

Lewis, A.C., Geltmacher, A.B., 2006. Image-based modeling of the response of experimental 3D microstructures to mechanical loading. Scripta Materialia 55, 81–86.

Lewis, A.C., Qidwai, S.M., Rowenhorst, D.J., Geltmacher, A.B., 2011. Correlation between crystallographic orientation and mechanical response in a three-dimensional beta-Ti microstructure. Journal of Materials Research 26, 957–964.

Li, C.M., Fang, H.S., Wang, J.J., Zheng, Y.K., 1995. Study of ledge structures of bainites in the Cu-base 27.1Zn–3.6Al alloy. Materials Letters 22, 233–236.

Lienert, U., Almer, J., Jakobsen, B., Pantleon, W., Poulsen, H.F., Hennessy, D., Xiao, C., Suter, R.M., 2007. 3-Dimensional characterization of polycrystalline bulk materials using high-energy synchrotron radiation. Materials Science Forum 539-543, 2353–2358.

Lienert, U., Brandes, M., Bernier, J.V., Weiss, J., Shastri, S., Mills, M.J., Miller, M.P., 2009. In-situ single grain peak profile measurements on Ti–7Al during tensile deformation. Materials Science & Engineering, A: Structural Materials: Properties, Microstructure and Processing 524, 46–54.

Lifshitz, I.M., Slyozov, V.V., 1961. Journal of Physics and Chemistry of Solids 19, 35–50.

Ludwig, W., King, A., Reischig, P., Herbig, M., Lauridsen, E.M., Schmidt, S., Proudhon, H., Forest, S., Cloetens, P., Rolland du Roscoat, S., Buffiere, J.-Y., Marrow, T.J., Poulsen, H.F., 2009a. New opportunities for 3D materials science of polycrystalline materials at the micrometre lengthscale by combined use of X-ray diffraction and X-ray imaging. Materials Science & Engineering, A: Structural Materials: Properties, Microstructure and Processing 524, 69–76.

Ludwig, W., Reischig, P., King, A., Herbig, M., Lauridsen, E.M., Johnson, G., Marrow, T.J., Buffière, J.-Y., 2009b. Three-dimensional grain mapping by X-ray diffraction contrast tomography and the use of Friedel pairs in diffraction data analysis. Review of Scientific Instruments 80, 033905.

Macpherson, R., Srolovitz, D., 2007. The von Neumann relation generalized to coarsening of three-dimensional microstructures. Nature Materials 446, 1053–1055.

Macsleyne, J., Uchic, M.D., Simmons, J.P., De Graef, M., 2009. Three-dimensional analysis of secondary gamma' precipitates in René-88 DT and UMF-20 superalloys. Acta Materialia 57, 6251–6267.

Manchiraju, S., Kirane, K., Ghosh, S., 2008. Dual-time scale crystal plasticity FE model for cyclic deformation of Ti alloys. Journal of Computer-Aided Material Design 14, 47–61.

Mangan, M.A., Shiflet, G.J., 1994. Three-dimensional reconstruction of pearlite colonies. In: Johnson, W.C., Howe, J.M., Laughlin, D.E., Soffa, W.A. (Eds.), Solid–Solid Phase Transformations Proceedings. TMS, Warrendale, PA, pp. 547–552.

Mangan, M.A., Shiflet, G.J., 1997. Three dimensional investigation of Cu–Ti discontinuous precipitation. Scipta Materialia 37, 517–522.

Mangan, M.A., Lauren, P.D., Shiflet, G.J., 1997. Three-dimensional reconstruction of Widmanstatten plates in Fe-12.3Mn-0.8C. Journal of Microscopy 188, 36–41.

Mangan, M.A., Kral, M.V., Spanos, G., 1999. Correlation between the crystallography and morphology of proeutectoid Widmanstatten cementite precipitates. Acta Materialia 47, 4263–4274.

Marco, S., Boudie, T., Messaoudi, C., Rigaud, J.-L., 2004. Electron tomography of biological samples. Biochemistry (Moscow) 69, 1219–1225.

Marder, A.R., Krauss, G., 1967. Transactions of ASM 60, 651–660.

Margulies, L., Winther, G., Poulsen, H.F., 2001. In Situ measurement of grain rotation during deformation of polycrystals. Science 291.

Marquis, E., Hyde, J., 2010. Applications of atom-probe tomography to the characterization of solute behavior. Materials Science and Engineering R: Reports 69, 37–62.

Marsh, S.P., Glicksman, M.E., 1996. Kinetics of phase coarsening in dense systems. Acta Materialia 44, 3761–3771.

Matsuoka, S., Mangan, M.A., Shiflet, G.J., 1994. Morphological development of cellular colonies in a 19Cr–5Ni austenite steel. In: Johnson, W.C., Howe, J.M., Laughlin, D.E., Soffa, W.A. (Eds.), Solid–Solid Phase Transformations. TMS, Farmington, Warrendale, PA, pp. 521–526.

Mcdonald, S., Preuss, M., Maire, E., Buffiere, J., Mummery, P., Withers, P., 2003. X-ray tomographic imaging of Ti/SiC composites. Journal of Microscopy-Oxford 209, 102–112.

Mckenna, I.M., Gururajan, M.P., Voorhees, P.W., 2009. Phase field modeling of grain growth: effect of boundary thickness, triple junctions, misorientation, and anisotropy. Journal of Materials Science 44, 2206–2217.

Mckenna, I. M., 2010. Three-Dimensional Anisotropic and Isotropic Grain Growth Simulations with Comparisons to Experiments – Ph.D. Thesis. Northwestern University.

Mehl, R.F., Barrett, C.S., Smith, D.W., 1933. Studies upon the Widmanstatten structure, IV—the iron–carbon alloys. Transactions of AIME 105, 215–258.

Mendoza, R., Alkemper, J., Voorhees, P.W., 2003. The morphological evolution of dendritic microstructures during coarsening. Metallurgical and Materials Transactions A: Physical Metallurgy and Materials Science 34, 481–489.

Mendoza, R., Savin, I., Thornton, K., Voorhees, P.W., 2004. Topological complexity and the dynamics of coarsening. Nature Materials 3, 385–388.

Mendoza, R., Thornton, K., Savin, I., Voorhees, P.W., 2006. The evolution of interfacial topology during coarsening. Acta Materialia 54, 743–750.

Miller, M.K., Horton, L.L., Spooner, S., 1986. A comparison of characteristic distance measurements by AP, TEM and SANS. Journal de Physique C2, 409–414.

Miller, M.K., Hyde, J.M., Hetherington, M.G., Cerezo, A., Smith, G.D.W., Elliott, C.M., 1995. Spinodal decomposition in Fe–Cr alloys: experimental study at the atomic level and comparison with computer models—I. introduction and methodology. Acta Metallurgica et Materialia 43, 3385–3401.

Modin, H., Modin, S., 1955. Jernkontorets Annaler 139, 480–515.

Moffat, A.J., Mellor, B.G., Sinclair, I., Reed, P.A.S., 2007. The mechanisms of long fatigue crack growth behaviour in Al–Si casting alloys at room and elevated temperature. Materials Science and Technology 23, 1396–1401.

Mulders, J.J.L., Day, A.P., 2005. Three-dimensional texture analysis. Materials Science Forum 495-497, 237–242.

Mullins, W.W., 1956. Two-dimensional motion of idealized grain boundaries. Journal of Applied Physics 27, 900–904.

Ni, Y., Jin, Y.M., Khachaturyan, A.G., 2008. Domain structure produced by confined displacive transformation and its response to the applied field. Metallurgical and Materials Transactions A: Physical Metallurgy and Materials Science 39A, 1658–1664.

Oblak, J.M., Hehemann, R.F., 1967. Transformations and Hardenability in Steels. Climax Molybdenum Co, Ann Arbor, MI.

Pande, C., Mcfadden, G., 2010. Acta Materialia 58, 1037–1044.

Patkin, A.J., Morrison, G.H., 1982. Secondary ion mass spectrometric image depth profiling for three-dimensional elemental analysis. Analytical Chemistry 54, 2–5.

Peet, M., Babu, S.S., Miller, M.K., Bhadeshia, H.K.D.H., 2004. Three-dimensional atom probe analysis of carbon distribution in low-temperature bainite. Scripta Materialia 50, 1277–1281.

Petch, N.J., 1953. Journal of the Iron and Steel Institute, London, 25–28.

Petrov, R., Garcia, O.L., Mulders, J.J.L., Reis, A.C.C., Bae, J.-H., Kestens, L., Houbaert, Y., 2007. Three dimensional microstructure–microtexture characterization of pipeline steel. Materials Science Forum 550, 625–630.

Phaneuf, M.W., Li, J., 2000. FIB techniques for analysis of metallurgical specimens. 13-Aug. 17, 2000, Philadelphia, Pa. Cambridge Univerisity Press (Cambridge, England). Microscopy and Microanalysis, 524–525.

Pickering, F.B., 1967. Transformations and Hardenability in Steels. Climax Molybdenum Co, Ann Arbor, MI.

Pitsch, W., 1962. Der orientierungs zusammenhang zwischen zementit und autenit. Acta Metallurgica 10, 897–900.

Pitsch, W., 1963. Der kristallographischen eigenschaftern der zementit ausscheidung in autenit. Archiv fur das Eisenhuttenwsen 34, 381–390.

Pollock, T. M., Allison, J. A., 2008. Integrated Computational Materials Engineering: A Transformational Discipline for Improved Competitiveness and National Security. Washington, D.C.

Poulsen, H.F., Margulies, L., Schmidt, S., Winther, G., 2003. Lattice rotations of individual bulk grains—part I: 3D X-ray characterization. Acta Materialia 51, 3821–3830.

Poulsen, H.F., 2004. Three-Dimensional X-ray Diffraction Microscopy: Mapping Polycrystals and Their Dynamics. Springer-Verlag, Berlin.

Proudhon, H., 2011. Large scale finite element computations using real grain microstructures. In: Collins, P., Allison, J.A., Spanos, G. (Eds.), Proceedings of the 1st World Congress on Integrated Computational Materials Engineering (ICME). The Minerals, Metals & Materials Society (TMS); Wiley.

Qidwai, M.A.S., Lewis, A.C., Geltmacher, A.B., 2009. Using image-based computational modeling to study microstructure–yield correlations in metals. Acta Materialia 57, 4233–4247.

Quey, R., Dawson, P.R., 2011. Large-scale 3D random polycrystals for the finite element method: generation, meshing and remeshing. Computer Methods in Applied Mechanics and Engineering 200, 1729–1745.

Reynolds, W.T., Liu, S.K., Li, F.Z., Hartfield, S., Aaronson, H.I., 1990. Investigation of the generality of incomplete transformation to bainite in Fe–C–X alloys. Metallurgical Transactions A: Physical Metallurgy and Materials Science 21, 1479–1491.

Reynolds-JR., W.T., Li, F.Z., Shui, C.K., Aaronson, H.I., 1990. Incomplete transformation phenomenon in Fe–C–Mo alloys. Metallurgical Transactions A: Physical Metallurgy and Materials Science 21, 1433–1463.

Reynolds-JR., W.T., Aaronson, H.I., Spanos, G., 1991. A summary of the present diffusionist views on bainite. Materials Transactions of the Japan Institute of Metals 32, 737–746.

Rhines, F.N., 1967. Stereology. Springer-Verlag, New York.

Rios, P., Glicksman, M., 2008. Acta Materialia 56, 1165–1171.

Rowenhorst, D.J., Gupta, A., Feng, C.R., Spanos, G., 2006a. 3D crystallographic and morphological analysis of coarse martensite combining EBSD and serial sectioning. Scripta Materialia 55, 11–16.

Rowenhorst, D.J., Kuang, J., Thornton, K., Voorhees, P.W., 2006b. Three-dimensional analysis of particle coarsening in high volume fraction solid–liquid mixtures. Acta Materialia, 2027–2039.

Rowenhorst, D.J., Lewis, A.C., Spanos, G., 2010. Three-dimensional analysis of grain topology and interface curvature in a beta titanium alloy. Acta Materialia 58, 5511–5519.

Russ, J.C., Dehoff, R.T., 2000. Practical Stereology. Kluwer Academic/Plenum Publishers, New York.

Russ, J.C., 1998. The Image Processing Handbook. CRC Press.

Sakamoto, T., Cheng, Z., Takahashi, M., Owari, M., Nihei, Y., 1998. Development of an ion and electron dual beam focused beam apparatus for three-dimensional microanalysis. Japan Journal of Applied Physics 37 (1), 2051–2056.

Sandvik, B.P.J., Wayman, C.M., 1983. Characteristics of lath martensite: part II. The martensite–austenite interface. Metallurgical Transactions A: Physical Metallurgy and Materials Science 14A, 823–834.

Sandvik, B.P.J., 1982. The bainite reaction in Fe–Si–C alloys: the secondary stage. Metallurgical Transactions A: Physical Metallurgy and Materials Science 13A, 789–800.

Schiotz, J., Tolla, F.D.D., Jacobsen, K.W., 1998. Softening of nanocrystalline metals at very small grains. Nature Materials 391, 561.

Schmidt, S., Nielsen, S.F., Gundlach, C., Margulies, L., Huang, X., Jensen, D.J., 2004. Watching the growth of bulk grains during recrystallization of deformed metals. Science 305, 229–232.

Seidman, D., 2007. Three-dimensional atom-probe tomography: advances and applications. Annual Review of Materials Research 37, 127–158.

Shewmon, P.G., 1989. Diffusion in Solids. The Minerals, Metals & Materials Society, Warrendale, PA.

Smith, C.S., 1948. Grains, phases, and interfaces—an interpretation of microstructure. Transactions of the American Institute of Mining and Metallurgical Engineers 175, 15–51.

Smith, C.S., 1952. Grain Shapes and Other Metallurgical Applications of Topology. Metal Interfaces. American Society for Metals, Cleveland, OH.

Sorby, H.C., 1887. Journal of the Iron and Steel Institute, London 30, 255.

Spanos, G., Kral, M.V., 2001. Proeutectoid cementite. In: Buschow, K.H.J., Cahn, R.W., Flemings, M.C., Ilschner, B., Kramer, E.J., Mahajan, S. (Eds.), Encyclopedia of Materials: Science and Technology, second ed. Elsevier Science Ltd, Oxford, U.K.

Spanos, G., Kral, M.V., 2009. The proeutectoid cementite transformation in steels. International Materials Reviews 54, 19–47.

Spanos, G., Fang, H.S., Aaronson, H.I., 1990a. A mechanism for the formation of lower bainite. Metallurgical Transactions A: Physical Metallurgy and Materials Science 21A, 1381–1390.

Spanos, G., Fang, H.S., Sarma, D.S., Aaronson, I., 1990b. Influence of carbon concentration and reaction temperature upon bainite morphology in Fe–C–2 pct Mn alloys. Metallurgical Transactions A: Physical Metallurgy and Materials Science 21A, 1391–1412.

Spanos, G., Rowenhorst, D.J., Lewis, A.C., Geltmacher, A.B., 2008. Combining serial sectioning, EBSD analysis, and image-based finite element modeling. MRS Bulletin 33, 597–602.

Spanos, G., Rowenhorst, D.J., Kral, M.V., Voorhees, P., Kammer, D., 2010. 3D microstructure representation. In: Furrer, D., Semiatin, S.L. (Eds.), ASM Handbook—Modeling and Simulation: Processing of Metallic Materials. ASM.

Spanos, G., 2006. Viewpoint set on 3D characterization and analysis of materials. Scripta Materialia 55, 1–86.

Spowart, J.E., Mullens, H.M., Puchala, B.T., 2003. Collecting and analyzing microstructures in three dimensions: a fully automated approach. JOM 55, 35–37.

Spowart, J.E., 2006. Automated serial sectioning for 3-D analysis of microstructures. Scripta Materialia 55, 5–10.

Sudbrack, C.K., Yoon, K.E., Noebe, R.D., Seidman, D.N., 2006. Temporal evolution of the nanostructure and phase compositions in a model Ni–Al–Cr alloy. Acta Materialia 54, 3199–3210.

Sutton, A.P., Balluffi, R.W., 1995. Interfaces in Crystalline Materials. Oxford University Press, New York.

Thornton, K., Poulsen, H.F., 2008. Special issue on: three-dimensional materials science. MRS Bulletin 33, 587–629.

Uchic, M.D., Groeber, M.A., Dimiduk, D.M., Simmonds, J.P., 2006. 3D microstructural characterization of nickel superalloys via serial-sectioning using a dual beam FIB–SEM. Scripta Materialia 55, 23–28.

Uchic, M.D., Holzer, L., Inkson, B.J., Principe, E.L., Munroe, P., 2007. Three-dimensional microstructural characterization using focused ion beam tomography. MRS Bulletin 32, 408–416.

Uchic, M.D., 2006. Special issue on 3-D characterization: methods and applications. JOM 58, 24–52.

Underwood, E.E., 1981. Quantitative Stereology. Addison-Wesley Pub. Co.

Venkatramani, G., Ghosh, S., Mills, M., 2007. A size-dependent crystal plasticity finite-element model for creep and load shedding in polycrystalline titanium alloys. Acta Materialia 55, 3971–3986.

von Neumann, J., 1952. Discussion. Metal Interfaces. Cleveland OH: American Society for Metals.

Voorhees, W., 1992. Ostwald ripening of two-phase mixtures. Annual Review of Materials Science 22, 197–215.

Wagner, C., 1961. Zeitschrift fuer Elektrochemie 65, 581–591.

Wakai, F., Enomoto, N., Ogawa, H., 2000. Acta Materialia 48, 1297–1311.

Wang, K.G., Glicksman, M.E., 2007. In: Groza, J.R., Shackelford, J.E., Lavernia, E.J., Powers, M.T. (Eds.), Osawal Ripening in Materials Processing. Processing Handbook. CRC, Boca Raton.

Wang, K.G., Glicksman, M.E., Lou, C., 2006. Correlations and fluctuations in phase coarsening. Physical Review E: Statistical, Nonlinear, and Soft Matter Physics 73.

Wang, K.G., Ding, X., Chang, K., Chen, L.Q., 2010. Phase-field simulation of phase coarsening at ultrahigh volume fractions. Journal of Applied Physics 107 (061801), 1–8.

Wayman, C.M., 1984. In: Marder, A.R., Goldstein, J.I. (Eds.), Phase Transformations in Ferrous Alloys. TMS-AIME, Warrendale, PA.

Weiland, H., Rouns, T.N., Liu, J., 1994. The role of particle stimulated nucleation during recrystallization of an aluminum–manganese alloy. Zeitschrift fuer Metallkunde 85, 592–597.

Withers, P., Bennett, J., Hung, Y., Preuss, M., 2006. Crack opening displacements during fatigue crack growth in Ti–SiC fibre metal matrix composites by X-ray tomography. Materials Science and Technology 22, 1052–1058.

Wolfsdorf, T.L., Bender, W.H., Voorhees, P.W., 1997. The morphology of high volume fraction solid–liquid mixtures: an application of micro-structural tomography. Acta Materialia 45, 2279–2295.

Wu, K.M., Enomoto, M., 2002. Three-dimensional morphology of degenerate ferrite in an Fe–C–Mo alloy. Scripta Materialia 46, 569–574.

Yokomizo, T., Enomoto, M., Umezawa, O., Spanos, G., Rosenberg, R.O., 2003. Three-dimensional distribution, morphology and nucleation site of intragranular ferrite formed in association with inclusions. Materials Science and Engineering A 344, 261–267.

Zaafarani, N., Raabe, D., Singh, R.N., Roters, F., Zaefferer, S., 2006. Three-dimensional investigation of the texture and microstructure below a nanoindent in a Cu single crystal using 3D EBSD and crystal plasticity finite element simulations. Acta Materialia 54, 1863–1876.

Zaefferer, S., 2005. Application of orientation microscopy in SEM and TEM for the study of texture formation during recrystallisation processes. Materials Science Forum 495-497, 3–12.

Ziolkowski, P. M., 1985. The three dimensional shapes of grain boundary precipitates in a/b brass. Masters Thesis, Mich. Tech.

Zollner, D., Streitenberger, P., 2006. Scripta materialia 54, 1697–1702.

Biography

George Spanos received his B.S., M.E., and Ph.D. degrees in Metallurgical Engineering and Materials Science from Carnegie Mellon University. In 1989 he joined the Naval Research Laboratory as a staff scientist, in 1994 was promoted to Section Head at NRL, and in June of 2010 joined The Minerals, Metals and Materials Society (TMS) as their Technical Director. Dr. Spanos is author/co-author of 95 publications which have been cited more than 2,500 times, in the fields of phase transformations, processing-structure-property relationships, 3D materials analyses, and Integrated Computational Materials Engineering (ICME).

William T. Reynolds Jr., is a Professor of Materials Science and Engineering at Virginia Polytechnic Institute and State University (Virginia Tech) in Blacksburg, VA, USA. He received B.S., M.S. and Ph.D. degrees in Metallurgical Engineering and Materials Science from Carnegie Mellon University. He joined Virginia Tech in December 1988, and was Visiting Professor from 1996 to 1997 at the National Institute for Materials Science, Tsukuba, Japan. He is the Director of the Nanoscale Characterization and Fabrication Laboratory, Virginia Tech's centralized instrumentation facility. His research activities are in the area of physical metallurgy and include investigations of the thermodynamics and kinetics of phase transformations.

11 Orientation Mapping

Anthony D. Rollett, Carnegie Mellon University, Pittsburgh, PA, USA
Katayun Barmak, Columbia University, New York, NY, USA

11.1 Orientation Mapping in the Scanning Electron Microscope

In the scanning electron microscope (SEM), electrons interact with crystals in various ways. We will describe the most popular method, EBSD, along with the lesser known electron channeling and X-ray pattern methods. The main components of a typical system are shown as additional external components, **Figure 1**, connected to an SEM. The camera assembly contains the scintillation (phosphor) screen, optics and charge-coupled device (CCD) camera. The control unit with computer acquires the diffraction patterns via communication with the camera and control of the electron beam, as well as analyzing them.

11.1.1 EBSD

Electron back scatter diffraction (EBSD) as an add-on to an SEM is now widespread in the materials community. In fact there are two books that have been produced with extended articles about various aspects of EBSD (Schwartz et al., 2000, 2009). In general terms it is capable of mapping the crystallographic orientation across an area as large as can be prepared in a sufficiently damage-free state (Wright, 1993). A large tilt angle (away from normal incidence of the electron beam on the specimen surface) of order 70° is required to maximize the intensity of backscattered electrons, **Figure 2**. This combined with the limited penetrating power of electrons (compared with, say, X-rays) means that the diffraction patterns originate from a layer on the order of 20 nm from the surface. This helps to illustrate the need for a damage-free surface to obtain diffraction patterns of sufficiently high quality that they can be indexed using automated methods. Consequently, surface preparation must culminate in a state that is completely free of mechanical damage and chemical contamination (especially oxide films) and flat (see below). Although well-annealed materials are most easily imaged in this fashion because of their low dislocation densities, deformed materials can be also be examined. In general softer materials deformed at elevated temperatures are easier than hard materials deformed at low temperatures. Electrically conductive materials are easier to work with than electrical insulators although thin surface coatings can mitigate charging problems without eliminating diffraction, as can low vacuum conditions as in environmental SEMs.

Physical Metallurgy. http://dx.doi.org/10.1016/B978-0-444-53770-6.00011-3

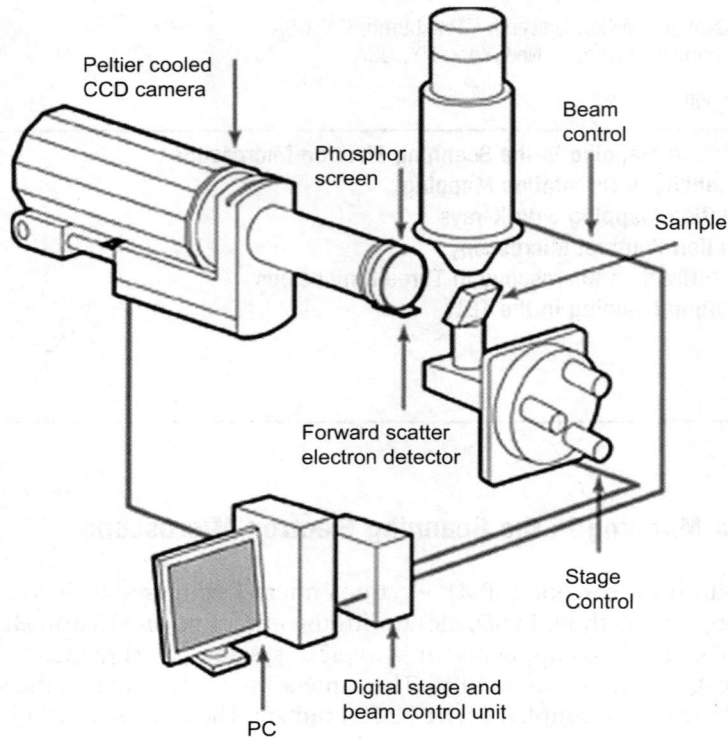

Figure 1   Schematic of the connections between the components of an EBSD system. After: www.ebsd.com.

Under optimal circumstances, the spatial resolution is of the order of 0.1 microns (Zaefferer, 2007) and the orientation accuracy is of the order of 0.5° (Brough et al., 2006). Recalling that the typical angle of incidence is 70°, it is obvious that the scattering volume is anisotropic with respect to directions in the sample and that it is elongated pointing down the sample surface, that is perpendicular to the tilt axis. This has a natural consequence that the spatial resolution is also anisotropic and is significantly better parallel to the tilt axis and worse perpendicular. Zaefferer studied the scattering in some detail (Zaefferer, 2007) and estimated that the resolution is three times better parallel to the tilt axis than perpendicular. Another consequence of the large incidence angle is that the specimen surface must be flat because even slight variations in the surface normal change the effective angle of incidence. With too much variation in surface height certain areas may not produce diffraction patterns and other areas will exhibit distortions that degrade the quality of the resulting orientation map. Notwithstanding the various limitations outlined above, the EBSD technique has been applied to an enormous range of materials problems and has served to expand the range of questions that can be examined about materials behavior.

11.1.2   EBSD: History

In brief, the history of the technique is that a combination of computer automation and digital camera technology enabled automation and thus mapping with enough points to make imaging practicable.

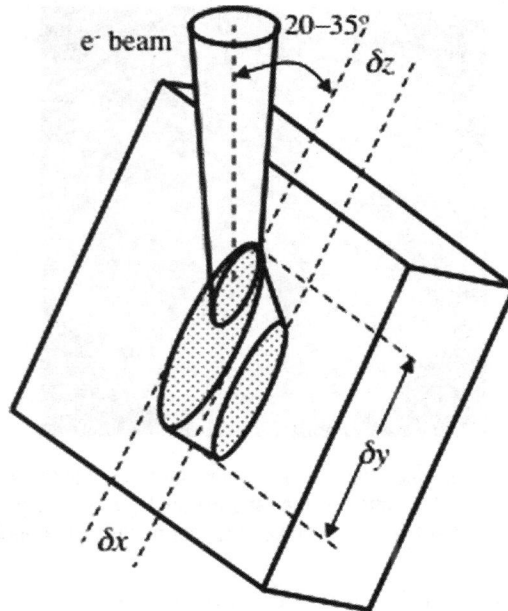

Figure 2 Diagram of interaction of electron beam with tilted specimen surface. Note how the oblique interaction spreads out the beam on the surface.

Before the 1990s, the situation was one in which diffraction patterns and thus crystallographic orientations were available, almost from the beginning of electron microscopy, but acquiring maps of more than a few dozen points was impracticable. Once diffraction patterns could be acquired automatically then it was possible to also analyze or "index" the patterns in the computer. The introduction and development of CCD cameras has been important mainly because of increase in the speed of the technique that it has enabled. At the time of writing, it is commonplace to acquire images with many millions of pixels (points). There are several names and acronyms for the method, including backscatter Kikuchi diffraction, electron backscatter patterns (EBSP) and orientation imaging microscopy™. Since the latter is a registered trademark (of one of the vendors of such systems), its use is deprecated in scientific publications. Of these names, EBSD has become the most widely accepted in the literature and is the recommended acronym.

11.1.3 EBSD: Indexation of Diffraction Patterns

Each individual point is associated with its own diffraction pattern. As the example in **Figure 3** shows, the information is sparse in the sense that the intensity of interest is concentrated in bands that are delimited by nearly parallel lines. By "nearly parallel lines" we refer to the fact that each diffracting plane gives rise to a pair of cones that intersect the (flat) scintillation screen as hyperbolae but the geometry is such that the lines of intersection are very nearly straight. Several different ways of reducing the data in each image were tried when the method was first being automated such as the Burns algorithm (Burns et al., 1986) but only one, the Hough transform (Hough, 1962; Duda and Hart, 1972), was found to be effective because it happens to be well suited to finding pairs of lines that span

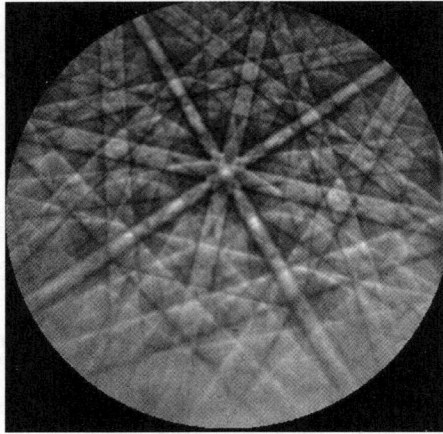

Figure 3 Example of an EBSD pattern showing many bands and low-index zone axes where bands intersect. For the Hough transform, the radius, *r*, is measured from the pattern center and the angle, θ, is measured as the azimuthal angle from a reference direction whose tail is located at the pattern center and that points to the right.

an entire image. The basic concept of the Hough transform is to convert each line to a point in a polar coordinate space or "Hough Space," where θ represents the inclination angle of the line from the horizontal and *r* the distance from the origin. The numerical transformation of an image uses an accumulator diagram where the intensity at a point in the diffraction pattern with coordinates (x,y) is accumulated at points along the line $r(\theta) = x \cos(\theta) + y \sin(\theta)$. In general, a threshold is applied such that only points are only transferred to (or accumulated in) the Hough space if their intensity is greater than a certain intensity, with the aim of reducing noise (2011a).

The net effect of the Hough transform, as applied to EBSD patterns, is to convert diffraction pattern lines into Hough space peaks whose position corresponds to the inclination angle and distance from the origin of the line. Each peak can then be identified using a mask function according to standard image segmentation techniques; the fact that each set of diffracting planes gives rise to a pair of lines simply means that the peak has a characteristic "butterfly" shape. The advantage of being able to identify diffracting planes as points associated with angles is that any given crystal lattice gives rise to a fixed set of interplanar angles. By "angles" we refer to the azimuthal angles associated with the bands in the diffraction pattern. This in turn means that identifying Miller indices for each peak can be reduced to recognizing certain combinations of these interplanar angles. Given that all these procedures are performed on images with a finite resolution and that there is a strong emphasis on speed to be able to work with very large images ($>10^6$ pixels), the main limitation on accuracy is the resolution used to define the Hough transform and the signal-to-noise ratio of the data itself that determines how many peaks can be reliably identified. Having said this, the resolution of the CCD camera in relation to the specimen to scintillation screen distance as well as the binning used for pattern acquisition can also affect accuracy. It is also noteworthy that the basic description given here emphasizes the position of each diffracting plane but there is also useful information in each band, for example width of the band, which could be used in the indexing process. Schmidt and Olesen point out (Schmidt and Olesen, 1989) that "any electron channeling pattern (ECP) exhibits a unique combination of angles between bands and their widths." They developed a software called CHANNEL that used the width of each pair of bands to compute its lattice spacing and determined the lattice orientation associated with a given diffraction pattern by analyzing a set of three bands. Although developed for indexing ECPs, the method is applicable to EBSD patterns. Later in the article we describe a technique for automated orientation determination in the transmission electron microscope (TEM), which relies on pattern matching in place of determination of the geometry of the pattern. This relies on the recent increase in computer power in comparison to what was available when the EBSD indexing methods were developed in the early 1990s. There is no reason that pattern matching could not be applied to indexing EBSD patterns.

The quality of indexing is sensitive to many factors. One in particular that should be mentioned is that of the determination of the pattern center. The pattern center is simple to understand as the point on any diffraction pattern that corresponds to the perpendicular intersection of a ray from the origin of the diffracting volume in the specimen with the scintillation screen surface, **Figure 4**. It can be determined by (1) shadow casting methods (Venables and Binjaya, 1977), (2) using a known crystal orientation (Dingley et al., 1987), (3) iterative pattern fitting (Lassen and Bilde-Sørensen, 1993), and (4) varying the position of the screen with respect to the sample. In practice moving the detector is most commonly used but it is difficult to determine it with high accuracy. The latter is particularly essential if one wishes to determine elastic strains, which is feasible under some circumstances (Wilkinson, 1996; Wilkinson et al., 2006; Kacher et al., 2009; Britton et al., 2010). Several authors have discussed pattern center determination in detail and a recent analysis of the problem (Maurice et al., 2011) suggests that high accuracy depends on taking into account all possible distortion factors, namely translation of the detector as it is moved, rotation of the detector as the distance is varied, and distortions in the optical system used to acquire the diffraction pattern.

No orientation map is ever perfect. Any given pixel may have had a low signal-to-noise ratio, which is apparent as low image quality or band contrast. The indexing may have low confidence associated with it perhaps because the beam was straddling a grain boundary and two diffraction patterns were present.

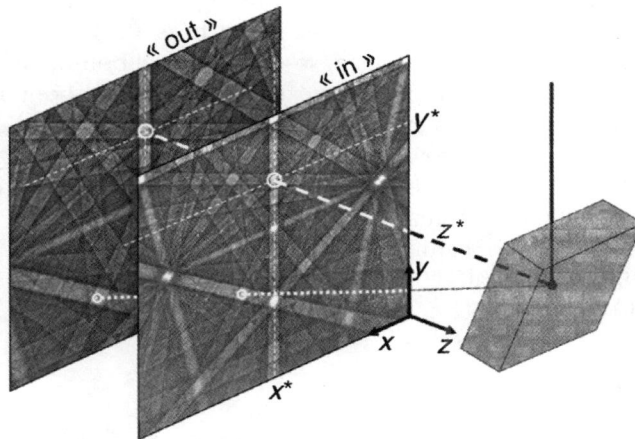

Figure 4 Diagram from Maurice et al. (2011) showing that there is a direction emanating from the point of intersection of the electron beam with the sample surface that is perpendicular to the surface. Detecting the point of intersection with the detector surface, which is the pattern center, can be accomplished by acquiring patterns at two or more values of z^*, that is the sample to detector distance. High accuracy requires allowance for camera lens distortions, rotation of the detector with respect to distance, and nonorthogonality of the detector with respect to the direction that defines the pattern center (dashed line in the diagram).

Even more subtle than indexing problems are pseudo-symmetry ambiguities. If the orientation is such that a low-index zone lies in the center of the pattern and the solid angle subtended by the pattern is not large enough, there may be two possible solutions that are twin related. They may appear in the map as a "pepper-and-salt" subdivision of certain grains or a high apparent density of twins. All this means that the operator or scientist has to make decisions about how to deal with "bad points". Standard "cleanup" procedures include erosion–dilation steps that use the assumption that neighboring orientations are highly likely to be reasonable substitutes for a poorly indexed point. The main responsibility of the operator is to document and report all such cleanup steps applied to the map.

11.2 Applicability of Orientation Mapping

Before describing some of the other orientation mapping techniques, it is appropriate to pause and provide some examples of why it is such a useful aspect of characterization. These examples are dominated by EBSD applications but examples will be given for the other techniques as well. One basic point is that our ability to quantify grain size has been transformed by orientation mapping. In fact, an ASTM standard, E2627, is available for measuring grain size via EBSD in fully recrystallized materials. Instead of relying on chemical or physical means of etching grain boundaries to produce image contrast and then having to use expert judgment to fill in missing boundaries where grain shape suggests them, orientation mapping, at least in annealed polycrystals, provides unambiguous quantitative measures of difference in orientation. Provided that the fraction of nonmeasured points is small, the uncertainty in grain size determination associated with image cleanup is small.

11.2.1 Grain Boundary Character

Another application of orientation mapping is in determining the crystallographic nature of the grain boundaries in a material. Certain types of processing, such as those generically known as grain boundary engineering, can induce high fractions of so-called special boundaries in face-centered cubic (fcc) metals such as nickel and copper. In this context the term special boundary means one that is crystallographically close to a coincident site lattice (CSL) type (Sutton and Balluffi, 1995). Suffice it to say that orientation mapping readily provides data on the fraction of such boundaries and this is important because high fractions of special boundaries have been shown to provide enhanced corrosion resistance (Gao et al., 2005; Gertsman and Bruemmer, 2001; Lehockey and Palumbo, 1997; Li et al., 2009b; Lin et al., 1997, 2010; Palumbo et al., 1991; Randle, 1999, Randle, 2004; Randle et al., 2005, Randle et al., 2008; Rohrer et al., 2006; Souaï et al., 2010; Souai et al., 2010; Tan et al., 2006; Thaveeprungsriporn and Was, 1997).

11.2.2 Phase Identification

Phase identification is feasible with EBSD although the need to index patterns from two different possible phases at each point means that more care (and accuracy) is needed when setting the data processing parameters. One example of the application is to measuring the volume fractions of martensite and austenite in stainless steels. Comparisons have been made with measurements from both optical micrographs and X-ray analysis. As is so often the case, provided that the surface preparation is good enough to give a high fraction of high-quality points then the orientation mapping not only provides the standard phase fractions but also gives information about orientation relationships

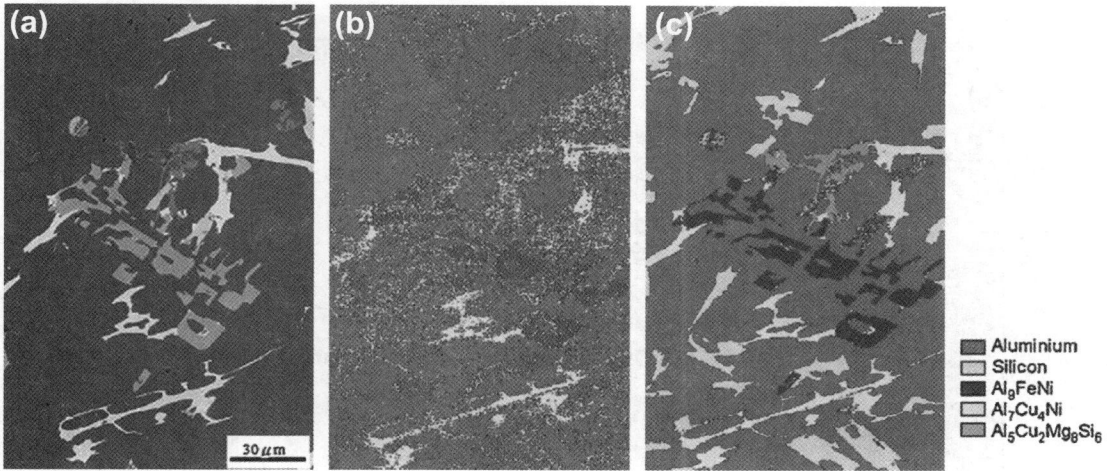

Legend:
- Aluminium
- Silicon
- Al_9FeNi
- Al_7Cu_4Ni
- $Al_5Cu_2Mg_8Si_6$

Figure 5 (a) SEM image showing the area of the scan in the alloy; (b) original EBSD phase mapping; and (c) EBSD phase mapping by combining EBSD and Energy Dispersive Mapping (EDX). (For color version of this figure, the reader is referred to the online version of this book.)

between phases (El-Dasher and Torres, 2006; Dingley, 2004). Note that local measurement of chemical composition through energy dispersive spectroscopy, for example, can provide very complementary information about the phases present in a material. In a recent example, Chen and Thompson analyzed an aluminum-based casting alloy, which exhibits a variety of second phases. **Figure 5** shows a sequence from Chen and Thomson in which the SEM image, **Figure 5(a)**, indicates more than two phases based on grayscale contrast and the EBSD map, **Figure 5(b)**, shows at least three phases but with significant uncertainty as the pixilation reveals. When the two techniques are combined, however, the resulting map is clearly delineated with respect to phases (Chen and Thomson, 2010).

11.2.3 Orientation Relationships

Orientation relationships between phases are usually thought of in terms of specific alignments of both interface surfaces and lattice directions. Orientation mapping on a surface provides only lattice relationships directly, although the final example illustrates how stereology can be used to obtain more complete information. Where orientation relationships are the result of a phase transformation with multiple variants based on crystal symmetry, however, the possibility exists to infer orientations of the parent phase (before transformation) provided that spatially coherent sets of the transformed phase can be identified. This has been investigated in titanium alloys (Humbert et al., 1995; Gey and Humbert, 2002; Gey et al., 2005), zirconia (Humbert et al., 2010), and shape-memory alloys (Li et al., 2011) and is also the subject of active research in steels (Lischewski and Gottstein, 2011; Zhang et al., 2007; Vari et al., 2012). Fukuno and Tsurekawa reported on orientation relationships in an Fe–Ni alloy between the austenite and ferrite phases (Fukino and Tsurekawa, 2008) observed during heating with in situ microscopy. **Figure 6** shows that both the Kurdjumov-Sachs (KS) and the Nishiyama-Wasserman orientation relationships were evident in this case. They also reported that nucleation of the austenite phase occurred preferentially at triple junctions at which all the adjoining boundaries were not special

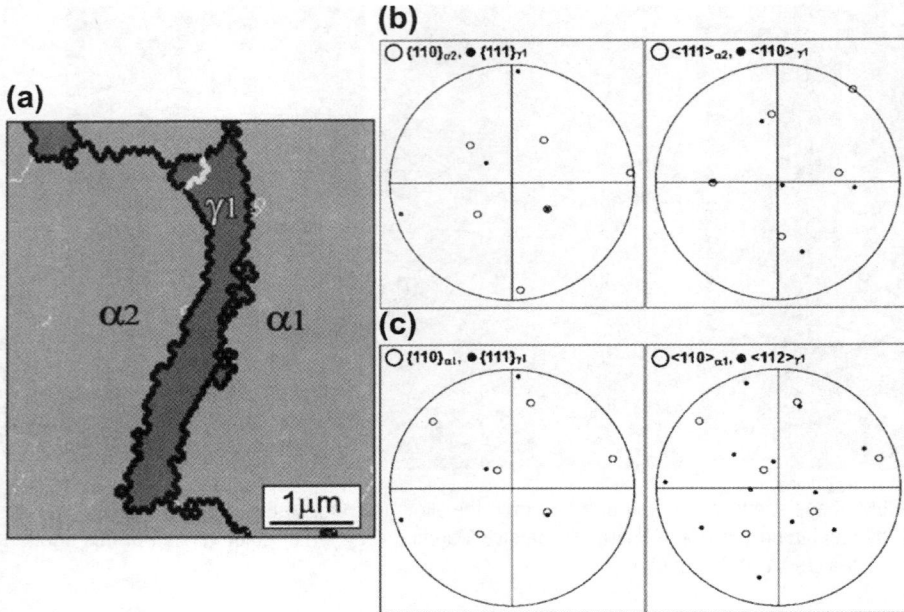

Figure 6 The SEM image (a) is labeled with the grain boundary austenite observed in situ at 883 K. The corresponding overlapped pole figures (b) show a KS relationship between α2 (left) and γ1 and (c) an NW relationship between α1 (right) and γ1.

boundaries, that is CSL boundaries with $\Sigma < 31$. They also observed the formation of coherent twin boundaries in the austenite phase during its formation.

11.2.4 Stereology for Five-Parameter Grain Boundary Character

The last example is slightly more complex in that it illustrates how stereology can be exploited to obtain more complete information about interfaces than is available directly from orientation mapping on surfaces. Grain boundaries have five degrees of freedom, in that a complete description needs to include both the lattice misorientation and the boundary normal. EBSD provides the lattice misorientation directly but the trace of the boundary provides information on only one of the two degrees of freedom associated with the normal. In randomly oriented polycrystals, however, the frequency of occurrence of the trace of any given boundary plane is known from stereology (or, more correctly, geometric probability) and measurements of boundary traces can be integrated via a suitable accumulator diagram to provide data on all five grain boundary parameters (Saylor et al., 2003). Such distributions provide relative populations of different grain boundary types that are very closely related to relative grain boundary energies (Li et al., 2009a).

11.2.5 Orientation Mapping in the Scanning Ion Microscope

Microscopes are now available that use ions, such as helium ions, to bombard surfaces and generate images (2011c). The substantially larger mass of helium ions compared with electrons means that the beam is less affected by diffraction than in the SEM. Also the beam current is lower, typically by a few orders of magnitude than in a typical SEM. Just as the depth of penetration of electrons into a crystal is a function of the beam direction with respect to the crystal axes, so too is channeling of ions a function of crystallographic direction. The contrast in the images of the diffraction patterns is controlled by the critical channeling angle, as opposed to the Bragg angle that applies to electron diffraction, so the mechanism is different from EBSD (2011b). The appearance of the patterns is, however, rather similar to EBSPs, with lines crossing the image, each of which corresponds to a particular crystallographic plane. Such patterns are also known as *ion-blocking* patterns and have been known for a long time thanks to the work of Tulinov (1965), Tulinov et al. (1965), Barrett et al. (1969), and Barrett (1969, 1970).

Regardless of the mechanism, ion channeling can be used to generate channeling patterns and analysis of these patterns can be automated in a similar manner to EBSD to provide orientation maps. One advantage of using He ions is that their small mass means that damage is negligible, in contrast to other ions such as the Ga ions commonly used for cutting and erosion in dual-beam instruments. Veligura et al. show how indexing channeling patterns from a gold thin film can be used to generate orientation maps (Veligura et al., 2012). Although the film is strongly textured with ⟨111⟩ parallel to the film normal, the orientation map, **Figure 7**, is colored in a way that reveals that the channeling patterns can be indexed to measure the full orientation of the grains. The map covers about 10 microns on a side, which means that the grain size is slightly less than 1 micron. This also suggests that the spatial resolution of the technique is as good as EBSD itself. The speckle apparent in the orientation map, however, indicates that the method may suffer from similar pseudo-symmetry problems as can happen in EBSD indexing.

Figure 7 Orientation map of a gold thin film with < 111 > texture; the color scale shows the azimuthal angle around the [111] surface normal for a 35° sample tilt at which a < 110 > direction is aligned with the incoming helium ion beam. The micrograph covers approximately 10 μm on each side. Note that annealing twins appear to be present in many grains. (For color version of this figure, the reader is referred to the online version of this book.)

11.2.6　Geometrically Necessary Dislocations and Disclinations

The orientation maps can be treated as fields. From a continuum perspective, geometrically necessary dislocations (GNDs) are those that are required to support a particular curvature in the crystallographic lattice at any given point in a deformed structure (Ashby, 1970). The fundamental equation of continuum dislocation theory establishes a link between the elastic distortion tensor β^e and the dislocation tensor α (Kröner, 1958):

$$\alpha = \operatorname{curl} \beta^e. \tag{1}$$

This can be simplified to a form that relates the dislocation tensor to the curvature of the elastic strain and lattice orientation (Sun et al., 2000):

$$\alpha_{ij} = e_{ikl}\left(\varepsilon^e_{jl,k} + g_{jl,k}\right) \tag{2}$$

where e_{ikl} are components of the permutation tensor, $\varepsilon^e_{jl,k}$ is the gradient of the infinitesimal elastic strain, and $g_{jl,k}$ is the gradient in lattice orientation. If the gradients in the elastic strain are negligible, Nye's original formulation of the dislocation tensor is recovered (Nye, 1953):

$$\alpha_{ij} = e_{ikl}g_{jl,k}, \tag{3}$$

thus providing a direct relationship between the measured crystallographic orientation gradient and the dislocation tensor. This can also be written as $\alpha = \operatorname{curl}(g)$. Nye's paper also shows how the dislocation tensor and the local dislocation network are connected, which is formally expressed as

$$\alpha_{ij} = \sum_{k=1}^{K} \rho^k b_i^k \hat{z}_j^k \tag{4}$$

where the sum is over all the K dislocation types present in the material, ρ^k denotes the density of the dislocation of type k, and b^k and $\hat{z}^k$ denote the Burgers vector and unit line direction of the specific dislocation type. Nye originally solved this relation for the trivial case of the dislocation systems of a simple cubic lattice. Application of Eqn (4) to materials with other crystal structures is nontrivial in that no unique solution is possible; lower bound estimates on dislocation densities have been obtained (Sun et al., 2000; El-Dasher et al., 2003). In more detail, it is typical to make the line direction discrete, that is either edge (perpendicular to the Burgers vector) or screw type (parallel to b). In fcc metals with six distinct <110> Burgers vectors, this gives 18 dislocation types because each <110> edge can exist on two different slip planes. The analytical challenge is, therefore, the undetermined nature of the equation because there are more free variables on the RHS than the nine (derived) quantities on the LHS. Generally speaking, numerical approaches such as simplex method must be resorted to, combined with a constraint such as minimization of the total dislocation line length (Pantleon, 2008). An alternative approach that avoids recasting the Nye tensor in terms of lattice dislocations is the "weighted Burgers vector" (WBV) as devised by Wheeler et al. (2009). The WBV is a vector that corresponds to the third column of the Nye tensor and represents the dislocation density threading through the plane of the EBSD map from which it is calculated, **Figure 8**. The EBSD map was measured on a sample of Magnox alloy containing 0.9% Al and 0.005% Be, balance Mg, that was deformed to 30% strain at $200\,^\circ\text{C}$ (0.51 of melting point) and $1.9 \times 10^{-4}\,\text{s}^{-1}$. The authors make the point that the WBV has the

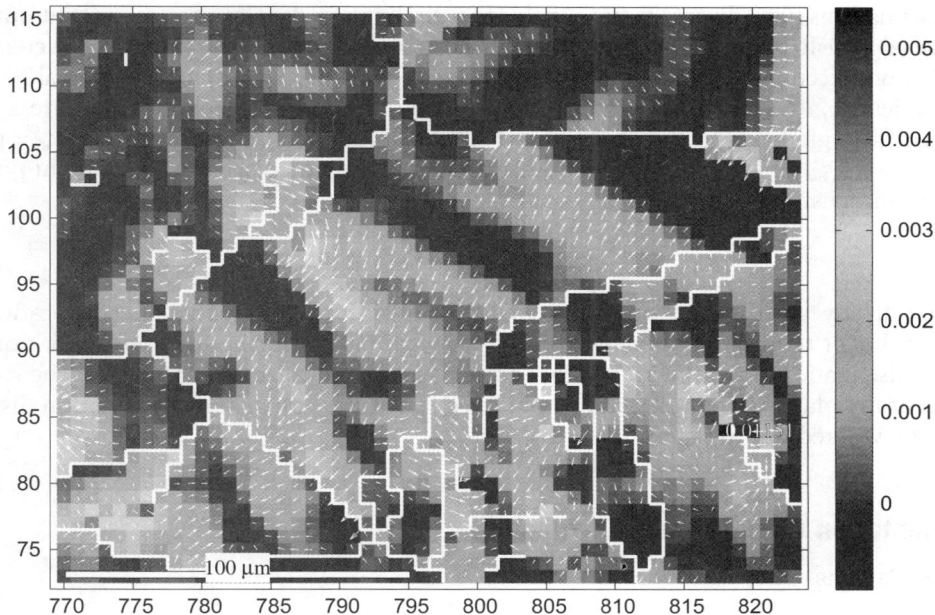

Figure 8 This figure shows the WBV as devised by Wheeler et al. (2009) with units of μm^{-1}. The WBV is a vector that corresponds to the third column of the Nye tensor and represents the dislocation density threading through the plane of the EBSD map from which it is calculated. The color corresponds to the WBV magnitude and the arrows are the WBV projected onto the section for values >0.002 μm^{-1}. White lines indicate boundaries $>5°$. (For color version of this figure, the reader is referred to the online version of this book.)

advantage of being frame invariant and it conveys more information than the local orientation gradient that is typically available in EBSD software packages, such as the kernel average misorientation (KAM).

Note that Frank's analysis for low-angle boundaries can also be used to calculate the required dislocation densities between neighboring points (Frank, 1950; Bilby, 1955). It assumes that the orientation gradient is concentrated on a plane between the points, which means that it is fundamentally the same in that one recovers a surface dislocation at a particular limit. The surface dislocation must still be decomposed into lattice dislocations in the same manner as described above. An even simpler approach, which has been used frequently for estimating stored energy in dislocation cell (or subgrain) networks, is to extract the misorientation angle between adjacent points and apply the Read–Shockley relationship to estimate the energy of the associated low-angle grain boundary (Theyssier and Driver, 1999).

A large literature has grown up around the characterization of GND content, partly in response to the interest in strain gradient theories that relate the higher flow stresses observed in small specimens and in submicron grain-sized materials (Acharya and Bassani, 2000; Shu and Fleck, 1999). An interesting alternative to analyzing orientation gradients that originated in theory for grain boundary structure is that of disclinations. Disclinations are an alternative approach to accommodating orientation gradients. They are more controversial because there are very few direct observations of disclinations, by contrast to lattice dislocations. Disclinations can be thought of as wedges on a continuum basis.

Isolated disclinations have divergent strain fields but disclination dipoles are similar (but not the same as) arrays of edge dislocations. Disclination analysis has been developed by Romanov (Gertsman et al., 1989) and more recently by Fressengeas et al. (Fressengeas et al., 2011; Upadhyay et al., 2011). The disclination density, θ, at any point can be estimated from the curl of the elastic curvature, κ_e, which is itself derived from gradients in the orientation field. Defining the elastic curvature as $\kappa_e = \mathrm{grad}(\vec{\omega}_e)$, where $\vec{\omega}_i^e = -e_{ijk}\Delta g_{jk}$ and the misorientation Δg is taken between adjacent measurement points, then the disclination density is given by

$$\theta_{ij} \approx e_{jkl}\Delta\omega_i/\Delta x_l/\Delta x_k. \tag{5}$$

Note that at any given point, one must make a decision as to whether the dislocation density description is better than the disclination density or vice versa. A further complication, not explored here, is that disclination dipoles with not too large spacings minimize elastic stresses. The strategies to achieve physically plausible representations of lattice curvature with general disclination distributions have yet to be worked out in detail.

11.3   Orientation Mapping with X-rays

11.3.1   Laue Patterns

The simplest method of orientation mapping with X-rays is very similar to the electron-based EBSD method. A white X-ray beam (i.e. with a wide range of wavelengths present in the incident beam) is focused on the surface of a specimen. This generates a Laue pattern that can be readily indexed thus providing an orientation at the point illuminated. Whether transmission or reflection is used depends on the specimen thickness since penetration depths vary strongly with atomic number. If a conventional X-ray source is used the main limitations are that the smallest spot size that is readily achievable with, say, capillary optics is of order 10 microns. This means that the spatial resolution is coarse compared with SEM-based methods. Also, the intensity is such that counting times on the order of several seconds to a minute are required to obtain a sufficiently high signal-to-noise ratio in the Laue pattern. One advantage under some circumstances is that X-rays are significantly more penetrating than electrons, which means that the information obtained from a greater depth than the essentially near-surface nature of the SEM-based methods. It also means that surface preparation is far less critical than for electron diffraction. Also there is no need to enclose the specimen in a vacuum chamber, which means that specimens can be as large as the positioning system can handle.

11.3.2   Automated Mapping with Laue Patterns

The group at the University of Metz led by A. Tidu has implemented such a system using an unfiltered conventional X-ray source with a Mo target (to obtain a white beam), a collimator that delivers a spot with diameter about 200 microns and a Bruker General Area Detector Diffraction System. A Laue pattern is recorded at each point on a grid on the surface. The XMAS software program (developed by N. Tamura at the Advanced Light Source, Berkeley, California) was used to index the patterns (Kunz et al., 2009). The X-ray intensity and collimation are such that each point requires of the order of 30 s which means that many hours are required to accumulate a few thousand points. The system has been successfully used to index orientations in a heavily deformed sample of aluminum for which EBSD was infeasible (Tidu, 2012).

11.3.3 Differential Aperture X-ray Microscopy

The restriction to surface characterization with conventional laboratory-scale X-ray sources is removed by using a synchrotron source of X-rays, which provides higher energies and substantially higher intensities. The best known of these is the differential aperture X-ray microscopy (DAXM) method pioneered by the group at Oak Ridge (Ice and Larson, 2000). In basic terms, polychromatic synchrotron radiation is extracted from the accelerator in the energy range 8–22 keV. The experiments are conducted at the Advanced Photon Source at Argonne National Laboratory. Note that the intensity is finite over a range of wavelength so that the Laue technique can be used, that is one diffraction pattern per point (in contrast to the some of the other methods to be described). This gives an effective penetration up to 100 microns, depending of course on the absorption coefficient of the material. Combinations of mirrors and lenses are used to collimate the beam onto the specimen surface. The high intensity shortens the time required to accumulate a diffraction pattern at each point and the small spot size permits high spatial resolution. Crucially, however, the provision of a wire, made of platinum or tungsten to have high absorption, that passes between the specimen and the detector provides depth information. **Figure** 9 shows the geometry of the method, which explains that correlating which diffraction spots weaken depending on the position of the wire allows the depth position of the diffracting material to be resolved. The method is a differential aperture method because it is equivalent in some sense to a pinhole camera but the moving wire differentiates between one location and another as it blocks diffracted beams from reaching the detector. Notwithstanding the high resolution and high orientation accuracy achieved with this technique, it remains a near-surface method because of the limited penetrating power of the X-rays in the energy range that is feasible for the method, that is 22 keV. The DAXM method has the ability to map out orientations in 3D with high enough spatial resolution that dislocation structures can be imaged and quantified. This in turn allows the dislocation network to be analyzed in terms of GND content (Yang et al., 2004a,b).

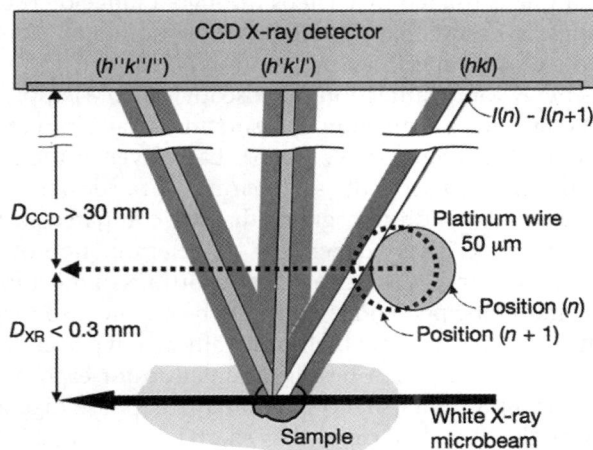

Figure 9 Diagram of the technique used to scan a highly absorbing wire over a sample surface and block diffracted beams in a known sequence. This permits spatial information to be obtained from the illuminated spot (Larson et al., 2002).

11.3.4   True 3D Mapping Methods

To map orientation in bulk specimens of standard materials such as aluminum, copper, iron, nickel and so on, it is necessary to use high-energy X-rays above about 50 keV. Here the very limited set of third-generation synchrotrons is essential as sources of such radiation. The group at the Risø Laboratory in Denmark pioneered the application to orientation mapping by developing a method in which a flat monochromatic beam that illuminates a layer in a specimen up to 1 mm in thickness. The specimen is rotated in front of the beam to develop all the available diffracted beams from all orientations present in the illuminated layer. At each rotation position, at least two detector positions in terms of specimen to detector separation are used to acquire diffraction patterns (Poulsen, 2004).

11.3.5   High-Energy X-ray Diffraction Microscopy

This section describes a particular realization of the general approach to reconstructing the orientation map of a material from diffraction data acquired with high-energy X-rays that can penetrate the full cross-section of a sample up to about 1 mm thickness. This means working with beams with energies between 10 and 100 keV that may be focused or parallel, monochromated or broad spectrum (white) and so on. One essential difference between such a technique and most of the methods discussed in this chapter is that illumination of a volume with many grains means it is infeasible to obtain an individual diffraction pattern for each point. Instead, one must infer the orientation of each point by fitting to the entire set of points in real space that have contributed to a (large) set of diffraction patterns. To set the scene for this method, it is convenient to contrast "far-field" from "near-field" approaches. In far-field synchrotron microscopy, the detector is placed of order 1 m away from the sample and only a single Bragg peak is imaged. This gives excellent resolution of the peak and its internal structure but no spatial information. Orientation mapping in 3D, however, requires the near-field approach in which the detector (or effective imaging surface) is placed at distances of a few millimeters from the specimen, such that many diffraction peaks are acquired up to a high order of reflection. A flat monochromatic beam illuminates the entire cross-section of a sample, which is generally a wire no more than 1 mm in diameter, **Figure 10**. The sample is rotated in 1° steps through a range of 180°, and a diffractogram is acquired at each rotation angle to ensure that all points in the illuminated volume contribute equally to the data set as a whole. Note that no specimen preparation is required.

The key feature of high-energy X-ray diffraction microscopy (HEDM) approach is forward modeling whereby each spatial point contributes diffraction peaks to an overall simulated diffractogram for each rotation angle (Hefferan et al., 2009; Suter et al., 2006). Each layer in the sample is divided up into a regular grid, **Figure 11**, with a point spacing of 1–5 microns. Diffraction from each point in the spatial grid is simulated to produce simulated diffractograms that are compared with the experimental data. Successful application of the method is very dependent on determination of experimental parameters such as the specimen – detector distance. Corrections for distortions of the diffractograms caused by, for example, the detector plane not being perpendicular to the beam, are also very important. One reason for the small standoff distances is to ensure that high-order diffraction peaks are included in the analysis. Simulated annealing[1] is used to determine a best-fit orientation for each point. In contrast to some minimization problems that exhibit broad shallow minima (e.g. phase diagram prediction from thermodynamic functions), this problem is characterized by sharp local minima in terms of the shape of the

---

[1] Simulated annealing refers to the numerical technique in which small randomly chosen changes are made and a Monte Carlo algorithm is used to decide which changes are kept, based on the change in the objective function.

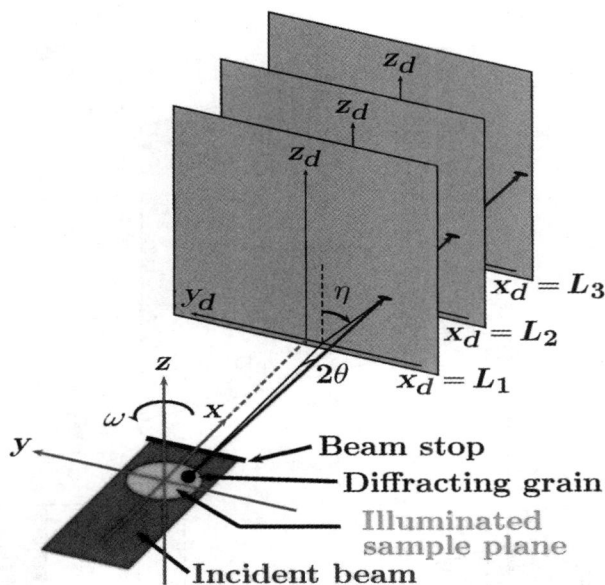

Figure 10   Diagram of the geometry of the high-energy X-ray diffraction microscopy method. The specimen is located on the z-axis, which is also a rotation axis in omega. The X-rays are collimated to a flat beam approximately 2 μm thick, shown in red, that illuminates a complete cross-section of the sample. For each rotation position, at least two diffractograms are acquired at different standoff distances of the order of a few millimeters, with an optional third standoff (Hefferan et al., 2009). (For interpretation of the references to color in this figure legend, the reader is referred to the online version of this book.)

cost function in orientation space. Accordingly the current algorithm (Li and Suter, 2013) uses adaptive gridding in orientation space to compute the cost function only at the locations of the local minima.

For well-annealed samples, the method produces high-quality results, as illustrated in **Figure 12** for a sample of high-purity nickel. The fraction of diffracted peaks that are correctly reproduced by the forward modeling provides a measure of the confidence in the result. For the example shown, the confidence is high in the bulk of each grain and decreases near boundaries, as one might expect.

Recent advances in the HEDM technique have included analysis of a lightly deformed aluminum sample and its microstructural evolution under annealing (Hefferan, 2012).

11.4   Diffraction Contrast Microscopy

Another approach to the challenge of producing images of polycrystalline samples that have a high degree of internal perfection (low mosaic spread) is to exploit the ability to capture detailed images of multiple diffracted spots (Ludwig et al., 2010). A parallel monochromatic beam from a synchrotron source is used to illuminate a cylindrical volume of interest, **Figure 13(a)**. The sample is rotated through 360° in front of the beam so as to generate a large number of diffractograms. As with the other methods described here, a typical data set contains thousands of images and hundreds of thousands of diffraction spots. Each diffracted beam from an individual grain produces a spot that is a 2D projection of its shape. A simultaneous solution procedure is then performed on the data that finds sets of diffracted spots that

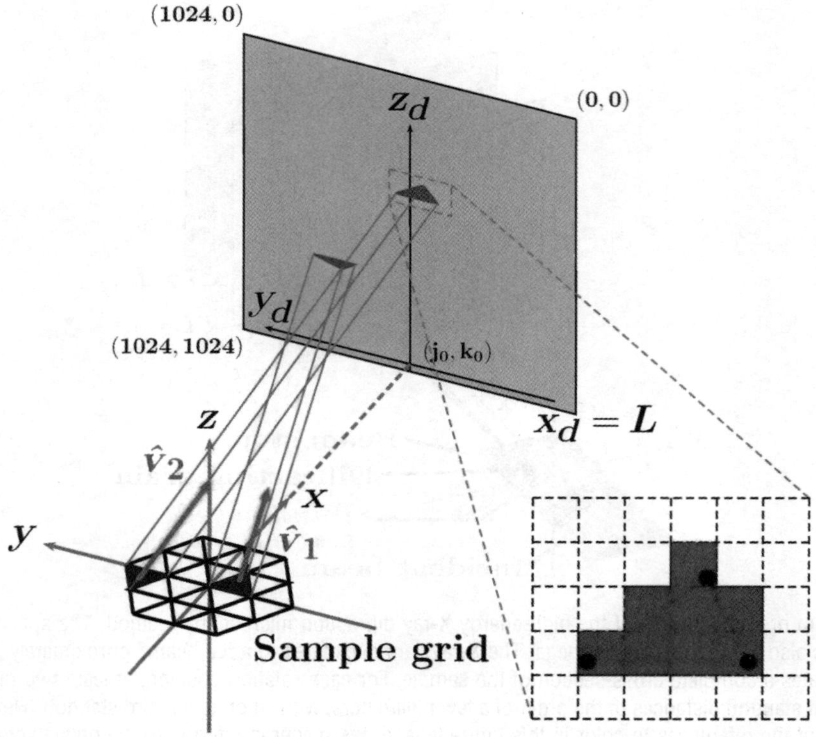

Figure 11   Illustration of the relationships between the grid of points used to index orientations in each illuminated layer of the sample and the pixels in the detector. Illuminated pixels are colored red. The forward modeling algorithm seeks to match all illuminated spots in all diffractograms, thereby indexing all the orientations in the sample grid (Hefferan et al., 2009). (For interpretation of the references to color in this figure legend, the reader is referred to the online version of this book.)

are consistent with individual grains. A key step in this process is identifying Friedel pairs of diffraction spots (i.e. hkl and $-h-k-l$) that match up (Ludwig et al., 2009), **Figure 13(b)**; it is this step that sets the requirement for a 360° rotation of the specimen in front of the beam. Detecting such pairs aids in identifying the Bragg angles associated with spots and the common axis between them helps with locating the grain position spatially within the sample. Finally the grain shapes are reconstructed from the sets of multiple projected shapes obtained for each grain using algebraic reconstruction techniques (Ludwig et al., 2008). One distinctive advantage of the technique is its ability to produce a 3D image of a polycrystalline sample in a short time compared with the other 3D techniques described here.

11.5 X-ray Diffraction Microscopy in Three Dimensions

The group at the Risø National Laboratory in Denmark has developed, in collaboration with scientists at ESRF, a suite of synchrotron-based methods for orientation mapping of polycrystalline materials in three dimensions. The method has undergone several significant changes in recent years as the sophistication of the analysis has increased over time. One early method used conical slits to isolate the diffracted beams

(a) **(b)**

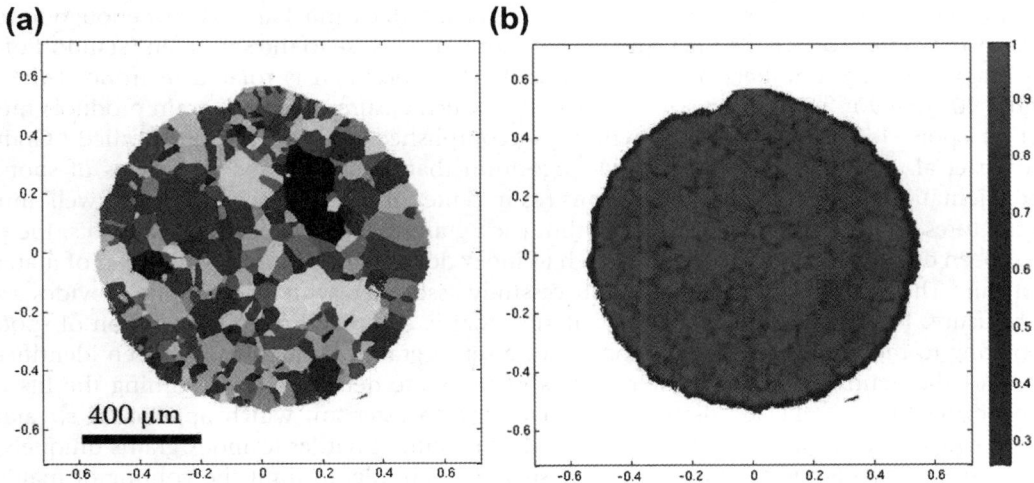

Figure 12 The image on the left is one section through a sample of high-purity nickel, colored by orientation. The corresponding confidence map is shown on the right where red indicates that a high fraction of the diffraction peaks in each diffractogram are matched, whereas blue indicates a low fraction (Hefferan et al., 2009). (For interpretation of the references to color in this figure legend, the reader is referred to the online version of this book.)

(a) **(b)**

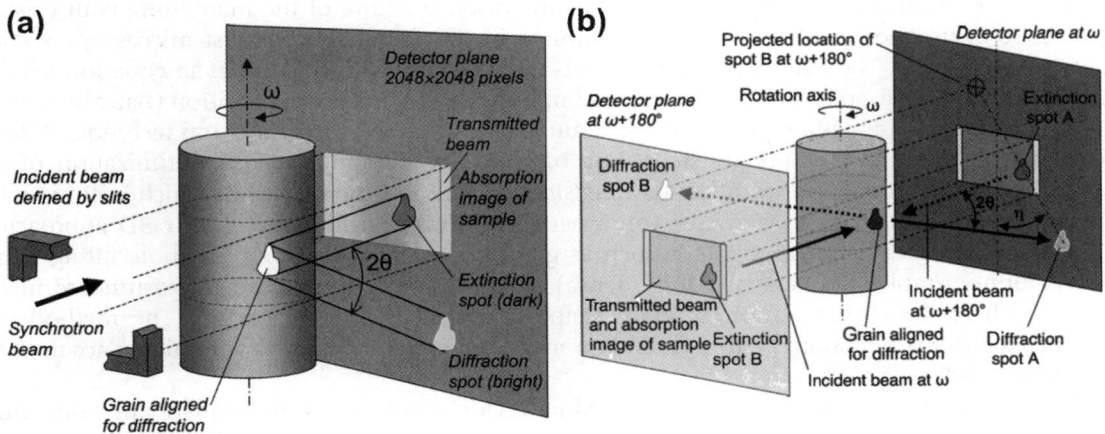

Figure 13 (a) Diagram of the illumination of a cylindrical sample with a monochromatic synchrotron beam (Ludwig et al., 2009). (b) Diagram showing the value of detecting Friedel pairs, which define both the Bragg angle and the spatial location of the diffracting grain (Ludwig et al., 2009).

from a small volume of a sample (Poulsen et al., 2003). The geometry of the slits had to be matched to the available reflections of a particular material, obviously. In the cited work, the lattice rotations of a number of grains in aluminum were tracked during plastic deformation under tensile loading. The results were compared against the predictions of several different models of crystal plasticity.

In recent times, X-ray diffraction microscopy in three dimensions (3DXDM) has used wide-beam illumination in a very similar manner to that described above for HEDM. With a beam wide enough to illuminate the complete cross-section of a specimen (of order 1 mm) and narrow enough to resolve features within grains (of order 1 μm), an area detector placed close to the specimen (standoff of order 10 mm) captures large numbers of diffraction spots. The specimen is rotated in front of the beam through $-90°$ to $+90°$, with a step size of order 0.5°, which ensures that each grain produces multiple diffraction spots. Identifying grains is typically accomplished with a procedure called "Graindex" (Lauridsen et al., 2001). This is an iterative algorithm that seeks to associate groups of spots with specific orientations and with specific positions (as in center of mass of each grain). For well-annealed microstructures with small mosaic spread within each grain, hence nonoverlapping spots, the procedure has been demonstrated to be rapid enough to index dozens of grains in a given layer of material in about 1 min. The fraction of spots that are successfully associated with orientations provides a confidence measure. There is also a uniqueness criterion that is computed from the fraction of spots that could belong to more than one orientation. Once a set of grain orientations has been identified, the locations of the grains in terms of their centers of mass are determined by refining the fits to the observed diffractograms. Plastic deformation gives rise to asterism, which appears as streaking of the spots, which in turn causes spots to overlap, thus making it harder to index grains uniquely. This problem can be compensated by using conical slits, for example, to limit the volume of material in view. This indexing procedure (Graindex) has been replaced in the past few years by ImageD11 (2012b) and Grainspotter (2012a). These latter are freely available programs that are still under development and lack comprehensive descriptions as yet.

Identifying orientations is a first step toward building up a 3D orientation map. There are various ways in which this can be done. One recent proposed method is known as the *discrete algebraic reconstruction technique* (DART), which exploits the discrete nature of the map along with existing algebraic reconstruction methods already mentioned in the diffraction contrast microscopy section (Ludwig et al., 2008). The DART approach embeds the geometry of diffraction in an equation relating intensity on the detector plane to density in the sample, both as a function of position (Batenburg et al., 2010). This part of the method is known as a simultaneous iterative reconstruction technique (SIRT). The algorithm employs two update steps, one of which is the simultaneous optimization of the positions of grains in relation to diffracted intensity or SIRT, and the second of which optimizes the microstructure in the reconstructed area (on a per-slice basis) or volume (if a fully 3D approach is used). The latter uses heuristic methods such as grain boundary smoothing and hole filling (since polycrystalline samples are generally fully dense) to refine the reconstructed microstructure in the locations where the determination of grain shape (from SIRT) is least certain. The method was demonstrated to work on both phantoms and on measured orientation maps with higher accuracy for equivalent computation times.

3DXDM in its various forms has been applied to a wide variety of problems involving orientation mapping. One has been the measurement of the nucleation and growth of individual grains during recrystallization (Larsen et al., 2005), which has illuminated the highly irregular nature of the growth process, in contrast to nearly all computer simulations. Another application has been the measurement of lattice rotations (Sorensen et al., 2006), as mentioned above, where the high accuracy of orientation determination is helpful when strains are limited to small values (of order 10%). More recently, the method(s) have been extended to measuring elastic strains on a per-grain basis (Oddershede et al., 2011, 2010), which has demonstrated the capability of 3DXDM to measure elastic strains as the full tensor. In the IF steel study, simultaneous fitting of all 96 illuminated grains with respect to position, orientation and stress state (Oddershede et al., 2010) was demonstrated. This is an impressive achievement in that elastic strain measurement to a resolution of order 20×10^{-6} was demonstrated, along with the variations in strain under tensile loading that were shown to be close to linear under the purely elastic loading conditions used. The 96 grains were all located within a particular layer in the cylindrical specimen. They extended this study with a combination of near-field and far-field detectors and measured the orientations and positions of about 2000 grains in the sample, thereby demonstrating that the 3D approach has the power to resolve all these quantities at the grain scale. In the study of a nominally well-annealed polycrystalline copper sample (Oddershede et al., 2011) some unexpected nonzero strains were measured in a set of about 1000 grains. This material raised some new challenges for the indexing because of the presence of annealing twins; twin-related adjacent orientations result in significant overlap of diffraction peaks, which necessitated iterative use of the GrainSpotter algorithm to correctly identify all grains in the illuminated volume. Again, the sample was subjected to tensile loading to a plastic strain of 1.5% and the full elastic strain tensor was measured for all 1000 grains. The error bar reported for the axial (elastic) strains was of order 10^{-4}. Significant variations in elastic strain were reported as a function of grain orientation as might be expected for the anisotropy in the elastic moduli in Cu and the anisotropy of the plastic response.

Notwithstanding the importance of the technique, it is important to keep in mind that 3DXDM is limited to mapping orientation at the scale of grains, in contrast to the true grid-based orientation mapping implemented in the HEDM method.

11.6   Orientation Mapping in the TEM

The main reason to perform orientation mapping in the TEM is that the spatial resolution of the SEM-based methods is limited to 20–80 nm by the interaction volume of electrons in bulk materials (Zaefferer, 2000). The use of thin electron-transparent foils in the TEM means that far better spatial resolution can be achieved (Wu and Zaefferer, 2009).

Methods for orientation mapping in the TEM fall into two broad categories, depending on whether beam scanning is done in real space, as in SEM-EBSD or in reciprocal space by changing electron beam tilt and rotation. Thus we have (A) scan in real space, record reciprocal space (i.e. the diffraction pattern) and (B) scan in reciprocal space, record real space (i.e. the image). A hybrid of (A) and (B) is also used (Fundenberger et al., 2003; Rauch et al., 2010; Dingley, 2006; Rauch et al., 2008).

(A) Scan in Real Space, Record Reciprocal Space Methods

(I) Convergent Beam Electron Diffraction Method

Convergent beam electron diffraction (CBED) makes use of a convergent, rather than a parallel, beam to form diffraction patterns. The beam is scanned is across the field of view and at each location a CBED pattern is recorded. The disks of diffraction in CBED patterns are nonoverlapping disks (Kossel–Möllenstedt patterns) for small convergence angles and overlapping disks (Kossel patterns) for large convergence angles. In addition, the patterns can include the zero-order Laue zone (ZOLZ) intensity bands and the higher order Laue zone HOLZ-Kikuchi line pairs (Morawiec, 1999; Fundenberger et al., 2003). The fine structure of the CBED patterns is not made use of in this technique. Nevertheless, the authors retain the designation of CBED to denote the distinction associated with the diffraction technique and the convergent nature of the electron beam. By stripping out the spots and concentrating on the HOLZ-Kikuchi lines, an analysis similar to that of EBSD indexing can be applied to determine the orientation. The orientation accuracy of 0.2° is good compared with EBSD and the spatial resolution achievable is about 10 nm, which is about 10 times finer than that of EBSD. An example of CBED

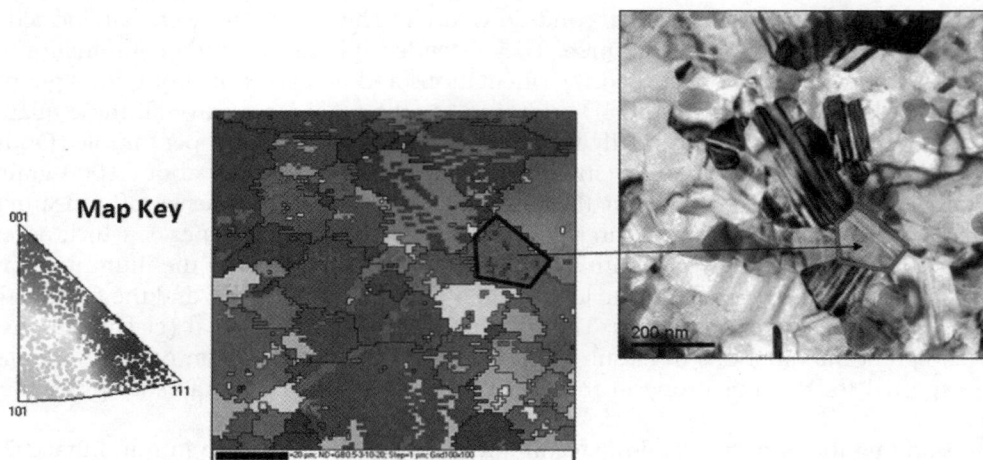

Figure 14   CBED method: the figure shows the inverse pole figure map in sample normal direction (left) and the corresponding bright-field image (right) for a 40-nm-thick Cu film imaged in plan view. (For color version of this figure, the reader is referred to the online version of this book.)

orientation map and the corresponding bright-field image of a 40-nm-thick fine-grained Cu film are shown in **Figure 14**.

(II) PED Method

In the precession electron diffraction or precession-enhanced electron diffraction (PED) method, a nanosized beam, ~ 1–30 nm, is used. The size of the beam depends on whether a field emission source or a LaB_6 filament is used (Moeck et al., 2011). The beam is nearly parallel or has a small angle of convergence and it scans the field of view from left to right and from top to bottom. The scanning step size can be <1 nm or as high as tens of nanometer. The combination of beam size and step size determines the spatial resolution. A spatial resolution of <5 nm is possible using this method. At each beam location, the spot diffraction pattern is recorded. The advantage of spot diffraction patterns over Kikuchi patterns or CBED patterns is that they are obtainable even in heavily deformed materials. Rather than analyzing the geometry of the pattern directly, pattern matching (via cross-correlation) is used to identify each pattern as being associated with a particular orientation in a catalogue of precalculated templates. The orientation that gives rise to the highest correlation index is selected as the most probable orientation for the given pixel, **Figure 15**. The number of precalculated patterns, and thus the angular resolution of this mapping method, depends on the angular step size used in calculating the templates. So, for example, for a cubic system and for an orientation angular resolution of approximately of $1°$, fewer than 1500 templates need to be calculated for cross-correlation with the collected spot diffraction patterns. The template bank must be generated for each specific microscope and material. The cross-correlation of the acquired patterns with the precalculated templates requires significant corrections for the oblique view that the (often used) externally mounted CCD camera has of the scintillation screen. The use of an externally mounted camera allows for significantly faster data collection (at 180 frames per second) compared with in-line cameras currently used in many TEMs. There is also a reliability index (roughly equivalent to the confidence index in TSL™ EBSD software)

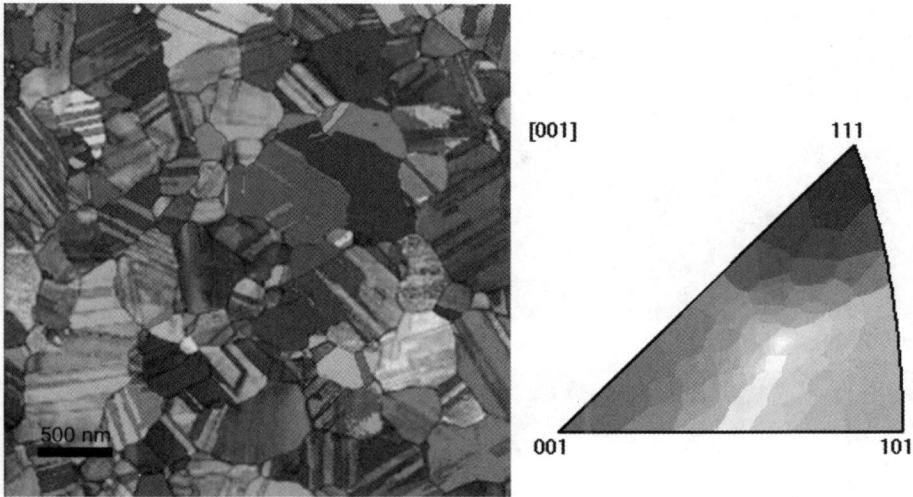

Figure 15 PED method: Inverse pole figure (IPF) map in the sample normal (z) or $[001]_S$ direction (left) and key (right) for a 111-nm-thick Cu film imaged in plan view. Superimposed on the IPF is the grayscale correlation index map, with black representing the lowest index and white the highest index. Beam precession was not used for this scan. (For color version of this figure, the reader is referred to the online version of this book.)

that is based on the ratio of the correlation indices between the best-fit pattern and the second best candidate pattern. The reliability index provides a measure of the goodness of the assigned orientation.

When precession is employed, the focused beam is scanned at a constant angle around the optic axis, making this a hybrid (A) + (B) method. Below the specimen, the beam is descanned. The combination scan above-descan below, or double conical scanning with the pivot point in the sample plane, is equivalent to precessing the specimen around a stationary focused beam and results in a stationary spot diffraction pattern. Unlike CBED patterns aligned on a zone axis, the precessed patterns include many reflections intercepted by the Ewald sphere, and the diffracted intensities are determined by integration through the Bragg condition (Vincent and Midgley, 1994). In other words, precession reduces (nonsystematic) dynamical diffraction effects and brings the intensity of the spots closer to the kinematical values. Improved quality and signal-to-noise ratio resulting from the use of precession (typically 0.1–0.9°) significantly enhances the effective orientation accuracy and so it is generally worth the additional effort in setting up the scanning procedure (Rauch et al., 2010; Vincent and Midgley, 1994; Oleynikov et al., 2007), **Figure 16**. However, the use of precession also increases data acquisition times, so it must be balanced against the disadvantage of sample drift, which inevitably limits the accuracy and resolution of any mapping method.

Finally it is worth noting that essentially all TEMs exhibit some rotation between a direct image and the corresponding diffraction pattern. This means that some calibration is required to determine this offset (Darbal et al., 2009). However, once this calibration is performed, all the analysis software and methods used for determining grain boundary character and interface boundary character distributions from SEM-EBSD maps can be readily applied to the orientation maps obtained by the PED method, **Figure 17**. The commercially available combined PED hardware and software system is known as ASTAR™ by NanoMEGAS, which, over the few years since its introduction, has proven to be the most robust and successful orientation mapping method in the TEM.

Mayenite crystal ($Ca_{12}Al_{14}O_{33}$) : space group I-43d

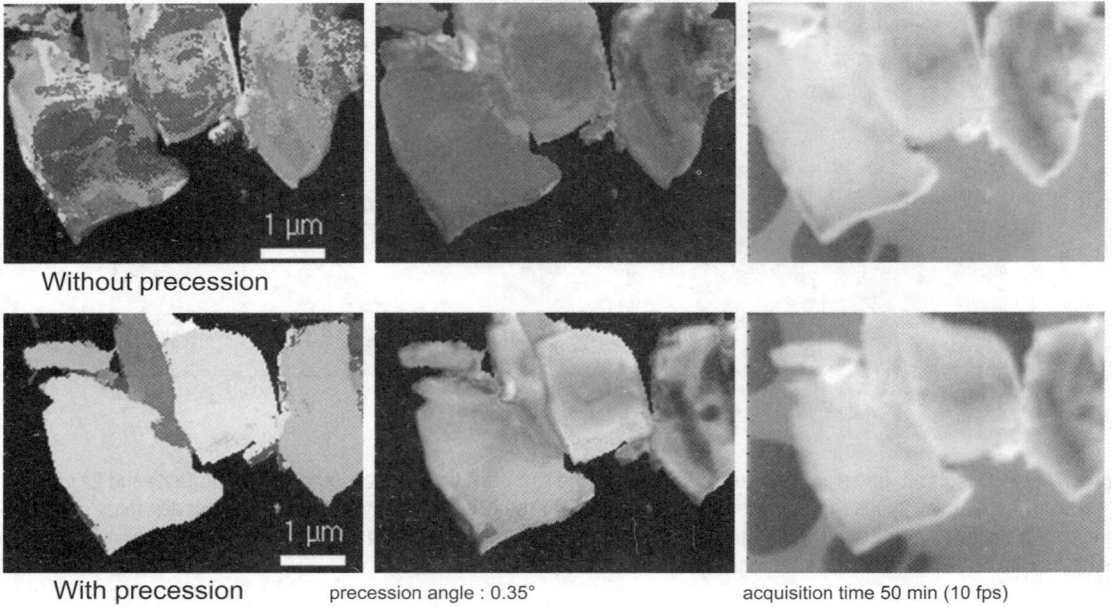

Without precession

With precession   precession angle : 0.35°   acquisition time 50 min (10 fps)

Figure 16   Example of orientation mapping at the nanoscale, taken from Rauch et al. (2008). (For color version of this figure, the reader is referred to the online version of this book.)

Figure 17   PED method: inverse pole figure map in the sample normal direction of a 46-nm-thick Cu film imaged in plan view. The coherent twin boundaries (60° misorientation about [111] with {111} as the boundary plane), as determined by trace analysis with a 10° tolerance, are marked with black lines. A precession of 0.4° was used in collecting the data. The color key is given in **Figure 15**. (For color version of this figure, the reader is referred to the online version of this book.)

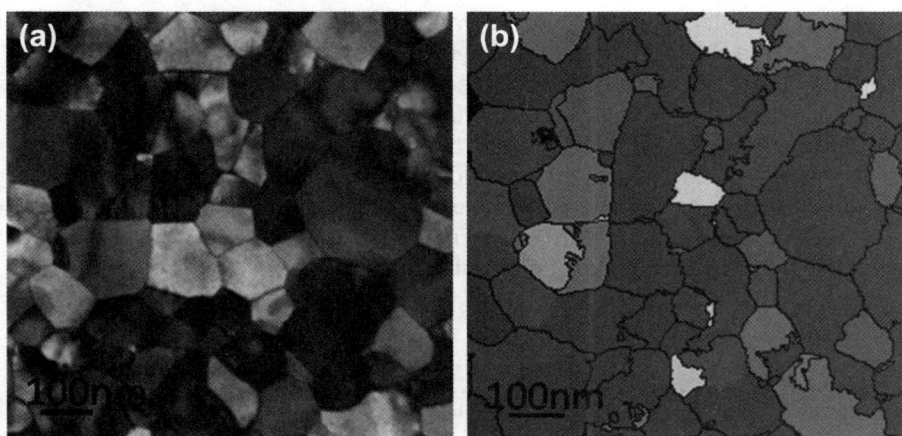

Figure 18 ACT method: (a) bright-field transmission electron micrograph; (b) inverse pole figure map in the sample normal direction for a 50-nm-thick Pt film imaged in plan view. (For color version of this figure, the reader is referred to the online version of this book.)

(B) Scan in Reciprocal Space, Record Real Space Methods

(I) The Automated Crystallography in the TEM Method

The automated crystallography in the TEM (ACT) method is an approach to orientation mapping that relies on accumulating dark-field images for a given area of the specimen and reconstructing diffraction patterns at each location in the imaged area (Dingley, 2006). Unlike the methods under category (A), the ACT method uses a broad parallel beam that illuminates the entire field of view simultaneously. The electron beam is controlled (generally via external circuitry) so as to precess around a cone at preset angular intervals up to a few degrees to normal incidence (hollow cone illumination). The use of an objective aperture limits the angular range of diffracted beams that give rise to the dark-field image at each beam tilt and rotation. The resulting data set is a few thousand dark-field images for each field of view. For example, by using 30 tilt angles and $4°$ rotation steps, 2700 images are collected. A polar plot (i.e. as function of tilt and rotation of the intensity for a given pixel, i.e. location in the field of view) as a function of beam tilt or radial coordinate and rotation or azimuthal angle is a reconstructed spot diffraction pattern (RDP). In the earlier versions of the indexing software, the RDPs were indexed using triplets of spots and the pattern center plus a voting scheme to determine crystal orientations. However, this approach to indexing proved to have limited success. Later versions made use of precalculated reciprocal lattice nets and proved somewhat more successful, but still only accounted for the geometry of the net and not the intensity of the diffraction spots. The spatial resolution of the method is better than 10 nm for strain-free samples and the angular resolution is $\sim 1°$ (Dingley, 2006). An example of the bright-field image and the orientation map of a nanocrystalline Pt film are shown in **Figure 18**.

(II) 3D Mapping in the TEM

Even though the technique is in its infancy at the time of writing, we mention three-dimensional mapping in the TEM because it builds on existing techniques and offers the promise of connecting the mesoscopic characterization length scales emphasized in this article with the near-atomic scale of the TEM. The technique is an extension of the ACT method of hollow conical dark-field imaging with the addition of sample tilting in a tomographic holder and has been demonstrated by the Risø group led by Liu et al. (2011). To generate a 3D orientation map, more than 100,000 dark-field images are required. The angular resolution of the orientation maps is determined by the angular step size in the conical scanning and in sample tilting. The spatial resolution depends on the magnification used for imaging. The orientation of each voxel in the sample is obtained by using many of the computational methods directly from previous work on 3D orientation mapping with synchrotron X-rays, as described above for 3DXDM, while also accounting for the dynamical nature of electron diffraction in the TEM. This method differs from ACT in that it does not attempt to reconstruct diffraction patterns for indexing, but rather simulates the images directly.

Summary

Orientation mapping is widely used in the materials science and engineering community. It provides information on polycrystalline materials that is rich in orientation, orientation relationships, phase composition, orientation gradients, and so on. It has been used to gain new insight into problems involving deformation, precipitation, coarsening, fatigue, and so on. As a tool associated with scanning electron microscopy it is regarded as a commonplace, if not quite essential to any materials research laboratory. It has caused the community to reevaluate how it thinks of the meaning of terms such as "grain" and "grain size" because segmentation of images into grains can be based on quantitative criteria on orientation difference in place of image contrast. Orientation mapping is also regularly practiced in three dimensions with the aid of synchrotrons, although general use will have to await the advent of laboratory-scale sources of high-energy X-rays. Users of the method should always be cautious about the quality of their data because the software packages that acquire and process the images make it easy to produce colorful maps. Another caution is that associated with the cliché "typical microstructure," by which we mean the temptation to publish an orientation map of a material that contains a number of grains that is statistically insufficient to support the conclusions drawn.

Acknowledgments

ADR is indebted to many colleagues for their input to and criticism of this article. These include N. Bozzolo, L. Germain, N. Gey, S.I. Wright, G.S. Rohrer, D. Haeffner, U. Lienert, S.-B. Lee, M. Nowell, D. Juul Jensen, H. Poulsen, R.M. Suter, S.F. Li, C. Hefferan, J. Lind, B. El Dasher, and R.D. Doherty. KB thanks colleagues and students E. Bouzy, K.R. Coffey, D. Dingley, P. Ferreira, J.-J. Fundenberger, J.S. Lecomte, A. Morawiec, N.T. Nuhfer, D.C. Berry, A. Darbal, K. J. Ganesh, X. Liu, T. Sun, and B. Yao for the preparation of samples, data collection, and data analysis.

References

2011a. Hough Transform [Online]. Available: http://en.wikipedia.org/wiki/Hough_transform (accessed 04.07.11.).
2011b. Ion Beam Microscopy [Online]. Available: http://www.crystaltexture.com/ibp_1.htm.
2011c. Scanning Helium Ion Microscope [Online]. Available: http://en.wikipedia.org/wiki/Scanning_Helium_Ion_Microscope.
2012a. Grain Spotter [Online]. Available: http://sourceforge.net/apps/trac/fable/wiki/grainspotter (Accessed January 2012).
2012b. ImageD11 [Online]. Available: http://sourceforge.net/apps/trac/fable/wiki/imaged11 (Accessed January 2012).
Acharya, A., Bassani, J.L., 2000. Lattice incompatibility and a gradient theory of crystal plasticity. J. Mech. Phys. Solids 48, 1565–1595.
Ashby, M., 1970. The deformation of plastically non-homogeneous materials. Philos. Mag. 21, 399–424.
Barrett, C.S., 1969. Line intensities in proton scattering patterns. Trans. Metall. Soc. AIME 245, 429.
Barrett, C.S., 1970. Ion beam scattering applied to crystallography. Naturwissenschaften 57, 287.

Barrett, C.S., Mueller, R.M., White, W., 1969. Proton blocking patterns for Hcp and Wurtzite structures. Trans. Metall. Soc. AIME 245, 427.

Batenburg, K.J., Sijbers, J., Poulsen, H.F., Knudsen, E., 2010. DART: a robust algorithm for fast reconstruction of three-dimensional grain maps. J. Appl. Cryst. 43, 1464–1473.

Bilby, B.A., 1955. Types of Dislocation Sources. Conference on Defects in Crystalline Solids. Univ. Bristol, Bristol, England. Physical Society, London, pp. 124–133.

Britton, T.B., Maurice, C., Fortunier, R., Driver, J.H., Day, A.P., Meaden, G., Dingley, D.J., Mingard, K., Wilkinson, A.J., 2010. Factors affecting the accuracy of high resolution electron backscatter diffraction when using simulated patterns. Ultramicroscopy 110, 1443–1453.

Brough, I., Bate, P.S., Humphreys, F.J., 2006. Optimising the angular resolution of EBSD. Mater. Sci. Technol. 22, 1279–1286.

Burns, J.B., Hanson, A.R., Riseman, E.M., 1986. Extracting straight lines. IEEE Trans. Pattern Anal. Mach. Intell. 8, 425–455.

Chen, C.L., Thomson, R.C., 2010. The combined use of EBSD and EDX analyses for the identification of complex intermetallic phases in multicomponent Al-Si piston alloys. J. Alloys Compd. 490, 293–300.

Darbal, A., Barmak, K., Nuhfer, T., Dingley, D.J., Meaden, G., Michael, J., Sun, T., Yao, B., Coffey, K.R., 2009. Orientation imaging of nano-crystalline platinum films in the TEM. Microsc. Microanal. 15, 1232–1233.

Dingley, D., 2004. Progressive steps in the development of electron backscatter diffraction and orientation imaging microscopy. J. Microsc. (Oxford) 213, 214–224.

Dingley, D.J., 2006. Orientation imaging microscopy for the transmission electron microscope. Microchim. Acta 155, 19–29.

Dingley, D.J., Longden, M., Weinbren, J., Alderman, J., 1987. Online analysis of electron back scatter diffraction patterns. 1. Texture analysis of zone refined polysilicon. Scanning Microscopy 1, 451–456.

Duda, R.O., Hart, P.E., 1972. Use of Hough transformation to detect lines and curves in pictures. Commun. Acm 15, 11.

El-Dasher, B.S., Torres, S.G., 2006. Second phase precipitation in as-welded and solution annealed Alloy 22 welds. J. Press. Vess. Technol. Trans. ASME 128, 644–647.

El-Dasher, B.S., Adams, B.L., Rollett, A.D., 2003. Viewpoint: experimental recovery of geometrically necessary dislocation density in polycrystals. Scr. Mater. 48, 141–145.

Frank, F.C., 1950. The Resultant Content of Dislocations in an Arbitrary Intercrystalline Boundary. Symposium on the Plastic Deformation of Crystalline Solids. Office of Naval Research, Pittsburgh, p. 150.

Fressengeas, C., Taupin, V., Capolungo, L., 2011. An elasto-plastic theory of dislocation and disclination fields. Int. J. Solids Struct. 48, 3499–3509.

Fukino, T., Tsurekawa, S., 2008. In-situ SEM/EBSD observation of alpha/gamma phase transformation in Fe–Ni alloy. Mater. Trans. 49, 2770–2775.

Fundenberger, J.J., Morawiec, A., Bouzy, E., Lecomte, J.S., 2003. Polycrystal orientation maps from TEM. Ultramicroscopy 96, 127–137.

Gao, Y., Kumar, M., Nalla, R.K., Ritchie, R.O., 2005. High-cycle fatigue of nickel-based superalloy ME3 at ambient and elevated temperatures: role of grain-boundary engineering. Metall. Mater. Trans. A-Phys. Metall. Mater. Sci. 36A, 3325–3333.

Gertsman, V.Y., Bruemmer, S.M., 2001. Study of grain boundary character along intergranular stress corrosion crack paths in austenitic alloys. Acta Mater. 49, 1589–1598.

Gertsman, V.Y., Nazarov, A.A., Romanov, A.E., Valiev, R.Z., Vladimirov, V.I., 1989. Disclination structural unit model of grain-boundaries. Philos. Mag. A-Phys. Condensed Matter Struct. Defects Mech. Properties 59, 1113–1118.

Gey, N., Humbert, M., 2002. Characterization of the variant selection occurring during the alpha ->beta ->alpha phase transformations of a cold rolled titanium sheet. Acta Mater. 50, 277–287.

Gey, N., Petit, B., Humbert, M., 2005. Electron backscattered diffraction study of epsilon/alpha martensitic variants induced by plastic deformation in 304 stainless steel. Metall. Mater. Trans. A-Phys. Metall. Mater. Sci. 36A, 3291–3299.

Hefferan, C.M., 2012. Measurement of Annealing Phenomena in High Purity Metals with Near-Field High Energy X-ray Diffraction Microscopy. PhD. Carnegie Mellon University.

Hefferan, C.M., LI, S.F., Lind, J., Lienert, U., Rollett, A.D., Wynblatt, P., Suter, R.M., 2009. Statistics of high purity nickel microstructure from high energy X-ray diffraction microscopy. CMC—Comput. Mater. Con. 14, 209–219.

Hough, P.V.C., 1962. Method and Means for Recognizing Complex Patterns. USA Patent Application.

Humbert, M., Wagner, F., Moustahfid, H., Esling, C., 1995. Determination of the orientation of a parent beta-grain from the orientations of the inherited alpha-plates in the phase-transformation from body-centered-cubic to Hexagonal close-packed. J. Appl. Cryst. 28, 571–576.

Humbert, M., Gey, N., Patapy, C., Joussein, E., Huger, M., Guinebretiere, R., Chotard, T., Hazotte, A., 2010. Identification and orientation determination of parent cubic domains from electron backscattered diffraction maps of monoclinic pure zirconia. Scr. Mater. 63, 411–414.

Ice, G.E., Larson, B.C., 2000. 3D X-ray crystal microscope. Adv. Eng. Mater. 2, 643–646.

Kacher, J., Landon, C., Adams, B.L., Fullwood, D., 2009. Bragg's Law diffraction simulations for electron backscatter diffraction analysis. Ultramicroscopy 109, 1148–1156.

Kröner, E., 1958. Continuum Theory of Dislocations and Self-stresses. College of Engineering, University of Florida, Gainesville, FL.

Kunz, M., Tamura, N., Chen, K., Macdowell, A.A., Celestre, R.S., Church, M.M., Fakra, S., Domning, E.E., Glossinger, J.M., Kirschman, J.L., Morrison, G.Y., Plate, D.W., Smith, B.V., Warwick, T., Yashchuk, V.V., Padmore, H.A., Ustundag, E., 2009. A dedicated superbend X-ray microdiffraction beamline for materials, geo-, and environmental sciences at the advanced light source. Rev. Sci. Instrum. 80.

Larsen, A.W., Poulsen, H.F., Margulies, L., Gundlach, C., Xing, Q.F., Huang, X.X., Jensen, D.J., 2005. Nucleation of recrystallization observed in situ in the bulk of a deformed metal. Scr. Mater. 53, 553–557.

Larson, B.C., Yang, W., Ice, G.E., Budai, J.D., Tischler, J.Z., 2002. Three-dimensional X-ray structural microscopy with submicrometre resolution. Nature 415, 887–890.

Lassen, N.C.K., Bilde-Sørensen, J.B., 1993. Calibration of an electron backscattering pattern set-up. J. Microsc. (Oxford) 170, 125–129.

Lauridsen, E.M., Schmidt, S., Suter, R.M., Poulsen, H.F., 2001. Tracking: a method for structural characterization of grains in powders or polycrystals. J. Appl. Cryst. 34, 744–750.

Lehockey, E.M., Palumbo, G., 1997. On the creep behaviour of grain boundary engineered nickel. Mater. Sci. Eng. A Struct. Mater. Properties Microstruct. Process. 237, 168–172.

Li, J., Dillon, S.J., Rohrer, G.S., 2009a. Relative grain boundary area and energy distributions in nickel. Acta Mater. 57, 4304–4311.

Li, Q.Y., Cahoon, J.R., Richards, N.L., 2009b. Effects of thermo-mechanical processing parameters on the special boundary configurations of commercially pure nickel. Mater. Sci. Eng. A Struct. Mater. Properties Microstruct. Process. 527, 263–271.

Li, S.F., Suter, R.M., 2013. Adaptive reconstruction method for three-dimensional orientation imaging. Journal of Applied Crystallography 46, 512–524.

Li, Z.B., Zhang, Y.D., Esling, C., Zhao, X., Zuo, L., 2011. Determination of the orientation relationship between austenite and incommensurate 7M modulated martensite in Ni–Mn–Ga alloys. Acta Mater. 59, 2762–2772.

Lin, P., Palumbo, G., Aust, K.T., 1997. Experimental assessment of the contribution of annealing twins to CSL distributions in FCC materials. Scr. Mater. 36, 1145–1149.

Lin, P., Provenzano, V., Palumbo, G., Gabb, T., Telesman, J., 2010. Grain boundary engineering the mechanical properties of Allvac 718Plus(tm) for aerospace engine applications. In: Ott, E.A., Groh, J.R., Banik, A., Dempster, I., Gabb, T.G., Helmink, R., Liu, X., Mitchell, A., Sjoberg, G.P., Wusatowska-Sarnek, A. (Eds.), 7th International Symposium on Superalloy 718 and Derivatives. Seven Springs. The Materials, Metals, & Materials Society, p. 243.

Lischewski, I., Gottstein, G., 2011. Nucleation and variant selection during the alpha-gamma-alpha phase transformation in microalloyed steel. Acta Mater. 59, 1530–1541.

Liu, H.H., Schmidt, S., Poulsen, H.F., Godfrey, A., Liu, Z.Q., Sharon, J.A., Huang, X., 2011. Three-dimensional orientation mapping in the transmission electron microscope. Science 332, 833–834.

Ludwig, W., Schmidt, S., Lauridsen, E.M., Poulsen, H.F., 2008. X-ray diffraction contrast tomography: a novel technique for three-dimensional grain mapping of polycrystals. I. Direct beam case. J. Appl. Cryst. 41, 302–309.

Ludwig, W., Reischig, P., King, A., Herbig, M., Lauridsen, E.M., Johnson, G., Marrow, T.J., Buffiere, J.Y., 2009. Three-dimensional grain mapping by x-ray diffraction contrast tomography and the use of Friedel pairs in diffraction data analysis. Rev. Sci. Instrum. 80.

Ludwig, W., King, A., Herbig, M., Reischig, P., Marrow, J., Babout, L., Lauridsen, E.M., Proudhon, H., Buffiere, J.Y., 2010. Characterization of polycrystalline materials using synchrotron X-ray imaging and diffraction techniques. JOM 62, 22–28.

Maurice, C., Dzieciol, K., Fortunier, R., 2011. A method for accurate localisation of EBSD pattern centres. Ultramicroscopy 111, 140–148.

Moeck, P., Rouvimov, S., Rauch, E.F., Veron, M., Kirmse, H., Hausler, I., Neumann, W., Bultreys, D., Maniette, Y., Nicolopoulos, S., 2011. High spatial resolution semi-automatic crystallite orientation and phase mapping of nanocrystals in transmission electron microscopes. Cryst. Res. Technol. 46, 589–606.

Morawiec, A., 1999. Automatic orientation determination from Kikuchi patterns. J. Appl. Cryst. 32, 788–798.

Nye, J., 1953. Some geometrical relations in dislocated crystals. Acta Metall. 1, 153.

Oddershede, J., Schmidt, S., Poulsen, H.F., Sorensen, H.O., Wright, J., Reimers, W., 2010. Determining grain resolved stresses in polycrystalline materials using three-dimensional X-ray diffraction. J. Appl. Cryst. 43, 539–549.

Oddershede, J., Schmidt, S., Poulsen, H.F., Margulies, L., Wright, J., Moscicki, M., Reimers, W., Winther, G., 2011. Grain-resolved elastic strains in deformed copper measured by three-dimensional X-ray diffraction. Mater. Charact. 62, 651–660.

Oleynikov, P., Hovmoller, S., Zou, X.D., 2007. Precession electron diffraction: observed and calculated intensities. Ultramicroscopy 107, 523–533.

Palumbo, G., King, P.J., Aust, K.T., Erb, U., Lichtenberger, P.C., 1991. Grain-boundary design and control for intergranular stress-corrosion resistance. Scr. metall. mater. 25, 1775–1780.

Pantleon, W., 2008. Resolving the geometrically necessary dislocation content by conventional electron backscattering diffraction. Scr. Mater. 58, 994–997.

Poulsen, H.F., Margulies, L., Schmidt, S., Winther, G., 2003. Lattice rotations of individual bulk grains - Part I: 3D X-ray characterization. Acta Mater. 51, 3821–3830.

Poulsen. Three-Dimensional X-Ray Diffraction Microscopy: Mapping Polycrystals and their Dynamics (Springer Tracts in Modern Physics); ISBN-13: 978-3540223306, 2004, 156 pp.

Randle, V., 1999. Mechanism of twinning-induced grain boundary engineering in low stacking-fault energy materials. Acta Mater. 47, 4187–4196.

Randle, V., 2004. Twinning-related grain boundary engineering. Acta Mater. 52, 4067–4081.

Randle, V., Hu, Y., Rohrer, G.S., Kim, C.S., 2005. Distribution of misorientations and grain boundary planes in grain boundary engineered brass. Mater. Sci. Technol. 21, 1287–1292.

Randle, V., Rohrer, G.S., Miller, H.M., Coleman, M., Owen, G.T., 2008. Five-parameter grain boundary distribution of commercially grain boundary engineered nickel and copper. Acta Mater. 56, 2363–2373.

Rauch, E.F., Portillo, J., Nicolopoulos, S., Bultreys, D., Rouvimov, S., Moeck, P., 2010. Automated nanocrystal orientation and phase mapping in the transmission electron microscope on the basis of precession electron diffraction. Z. Kristallogr. 225, 103–109.

Rauch, E.F., Veron, M., Portillo, J., Bultreys, D., Maniette, Y., Nicolopoulos, S., 2008. Automatic crystal orientation and phase mapping in TEM by precession diffraction. Microsc. Anal. 93, S5–S8.

Rohrer, G.S., Randle, V., Kim, C.S., Hu, Y., 2006. Changes in the five-parameter grain boundary character distribution in alpha-brass brought about by iterative thermomechanical processing. Acta Mater. 54, 4489–4502.

Saylor, D.M., Morawiec, A., Rohrer, G.S., 2003. Distribution of grain boundaries in magnesia as a function of five macroscopic parameters. Acta Mater. 51, 3663–3674.

Schmidt, N.H., Olesen, N.O., 1989. Computer-aided determination of crystal-lattice orientation from electron-channeling patterns in the sem. Can. Mineral. 27, 15–22.

Schwartz, A.J., Kumar, M., Adams, B.L. (Eds.), 2000. Electron Backscatter Diffraction in Materials Science. Kluwer, New York, NY.

Schwartz, A.J., Kumar, M., Adams, B.L., Field, D.P., 2009. Electron Backscatter Diffraction in Materials Science. Springer, New York.

Shu, J.Y., Fleck, N.A., 1999. Strain gradient crystal plasticity: size-dependent deformation of bicrystals. J. Mech. Phys. Solids 47, 297–324.

Sorensen, H.O., Jakobsen, B., Knudsen, E., Lauridsen, E.M., Nielsen, S.F., Poulsen, H.F., Schmidt, S., Winther, G., Margulies, L., 2006. Mapping grains and their dynamics in three dimensions. Nucl. Instrum. Methods Phys. Res. Section B Beam Interact. Mater. Atoms 246, 232–237.

Souai, N., Bozzolo, N., Naze, L., Chastel, Y., Loge, R., 2010. About the possibility of grain boundary engineering via hot-working in a nickel-base superalloy. Scr. Mater. 62, 851–854.

Souaï, N., Logé, R., Chastel, Y., Bozzolo, N., Maurel, V., Nazé, L., 2010. Effect of thermomechanical processes on $\Sigma 3$ grain boundary distribution in a nickel base superalloy. In: Chandra, T. (Ed.), Materials Science Forum. Trans Tech Publications, Berlin, Germany, pp. 2333–2338.

Sun, S., Adams, B.L., King, W.E., 2000. Observations of lattice curvature near the interface of a deformed aluminium bicrystal. Philos. Mag. A-Phys. Condensed Matter Struct. Defects Mech. Properties 80, 9–25.

Suter, R.M., Hennessy, D., Xiao, C., Lienert, U., 2006. Forward modeling method for microstructure reconstruction using x-ray diffraction microscopy: single-crystal verification. Rev. Sci. Instrum. 77.

Sutton, A.P., Balluffi, R.W., 1995. Interfaces in Crystalline Materials. Clarendon Press, Oxford, UK.

Tan, L., Sridharan, K., Allen, T.R., 2006. The effect of grain boundary engineering on the oxidation behavior of INCOLOY alloy 800H in supercritical water. J. Nuclear Mater. 348, 263–271.

Tari, V., Rollett, A.D., Beladi, H., 2013. Back calculation of parent austenite orientation using a clustering approach. J. Appl. Cryst. 46, 210–215.

Thaveeprungsriporn, V., Was, G.S., 1997. The role of coincidence-site-lattice boundaries in creep of Ni-16Cr-9Fe at 360 degrees C. Metall. Mater. Trans. A-Phys. Metall. Mater. Sci. 28, 2101–2112.

Theyssier, M.C., Driver, J.H., 1999. Recrystallization nucleation mechanism along boundaries in hot deformed Al bicrystals. Mater. Sci. Eng. A272, 73–82.

Tidu, A., 2012. Orientation Mapping with Laboratory X-rays. Metz, France.

Tulinov, A.F., 1965. A certain effect attending nuclear reactions on single crystals and its use in different physical investigations. Dokl. Akad. Nauk SSSR 162, 546.

Tulinov, A.F., Kulikaus, V.S., Malov, M.M., 1965. Proton scattering from a tungsten single crystal. Phys. Lett. 18, 304.

Upadhyay, M., Capolungo, L., Taupin, V., Fressengeas, C., 2011. Grain boundary and triple junction energies in crystalline media: a disclination based approach. Int. J. Solids Struct. 48, 3176–3193.

Veligura, V., Hlawacek, G., Van Gastel, R., Zandvliet, H.J.W., Poelsema, B., 2012. Channeling in helium ion microscopy: mapping of crystal orientation. Beilstein J. Nanotechnol. 3, 501–506.

Venables, J.A., Binjaya, R., 1977. Accurate micro-crystallography using electron backscattering patterns. Philos. Mag. 35, 1317–1332.

Vincent, R., Midgley, P.A., 1994. Double conical beam-rocking system for measurement of integrated electron-diffraction intensities. Ultramicroscopy 53, 271–282.

Wheeler, J., Mariani, E., Piazolo, S., Prior, D.J., Trimby, P., Drury, M.R., 2009. The weighted Burgers vector: a new quantity for constraining dislocation densities and types using electron backscatter diffraction on 2D sections through crystalline materials. J. Microsc. 233, 482–494.

Wilkinson, A.J., 1996. Measurement of elastic strains and small lattice rotations using electron back scatter diffraction. Ultramicroscopy 62, 237–247.

Wilkinson, A.J., Meaden, G., Dingley, D.J., 2006. High-resolution elastic strain measurement from electron backscatter diffraction patterns: new levels of sensitivity. Ultramicroscopy 106, 307–313.

Wright, S.I., 1993. A review of automated orientation imaging microscopy (OIM). J. Comput. Assist. Microsc. 5, 207–221.

Wu, G.L., Zaefferer, S., 2009. Advances in TEM orientation microscopy by combination of dark-field conical scanning and improved image matching. Ultramicroscopy 109, 1317–1325.

Yang, W., Larson, B.C., Pharr, G.M., Ice, G.E., Budai, J.D., Tischler, J.Z., Liu, W.J., 2004a. Deformation microstructure under microindents in single-crystal Cu using three-dimensional X-ray structural microscopy. J. Mater. Res. 19, 66–72.

Yang, W., Larson, B.C., Tischler, J.Z., Ice, G.E., Budai, J.D., Liu, W., 2004b. Differential-aperture X-ray structural microscopy: a submicron-resolution three-dimensional probe of local microstructure and strain. Micron 35, 431–439.

Zaefferer, S., 2000. New developments of computer-aided crystallographic analysis in transmission electron microscopy. J. Appl. Cryst. 33, 10–25.

Zaefferer, S., 2007. On the formation mechanisms, spatial resolution and intensity of backscatter Kikuchi patterns. Ultramicroscopy 107, 254–266.

Zhang, Y.D., Esling, C., Calcagnotto, M., Zhao, X., Zuo, L., 2007. New insights into crystallographic correlations between ferrite and cementite in lamellar eutectoid structures, obtained by SEM-FEG/EBSD and an indirect two-trace method. J. Appl. Cryst. 40, 849–856.

Biography

Dr. Rollett has been a Professor of Materials Science & Engineering at Carnegie Mellon University (CMU) since 1995 and was the Department Head 1995-2000. Prior to CMU he worked for the University of California at the Los Alamos National Laboratory (1979-2005). He spent ten years in management with five years as a Group Leader (and then Deputy Division Director) at Los Alamos, followed by five years as Department Head at CMU (1995-2000). He has a research group of about ten students. The main focus of his research is on the measurement and computational prediction of microstructural evolution especially in three dimensions. His interests include strength of materials, constitutive relations, microstructure, texture, anisotropy, grain growth, recrystallization, formability and stereology.

Prof. Rollett's honors include an Award for Technology Transfer from the Federal Laboratories Consortium in 1989, Fellow of ASM-International in 1996, Fellow of the Institute of Physics (UK) in 2004, the Howe Medal (Best Paper in Metallurgical & Materials Transactions) in 2004, Fellow of TMS in 2011, Brahm Prakash Professor at the Indian Institute of Science (Bangalore) in 2011, Chercheur d'Excellence (Outstanding Researcher) at the University of Lorraine, Metz, France in 2012, and the Cyril Stanley Smith award from TMS in 2014. He was the Chair of the International Conference on Texture (ICOTOM-15), which was held on campus at CMU June 2-6, 2008 and is a member of its International Scientific Committee. From 2001-2013 he was the Chair of the International Committee of the conference on Grain Growth and Recrystallization that is held every three years; the next meeting will be in Pittsburgh in 2016. He was a co-Chair of the 13th International Conference on Aluminum and its Applications, which was held on campus at CMU in June 2012. He is a co-author of the texture analysis package popLA, and the polycrystal plasticity code, LApp; he is also a contributor to the well-known textbook Texture & Anisotropy edited by Kocks, Tomé and Wenk. He is an author on over 160 scholarly articles, many of which are cited frequently in the literature.

Katayun Barmak obtained her B.A and M.A. degrees in Natural Sciences from the University of Cambridge, England in 1983 and 1985, respectively. She was the recipient of an AT&T Foundation Fellowship and completed her M.S. in Metallurgy and Ph.D. in Materials Science at the Massachusetts Institute of Technology in 1985 and 1989, respectively. Prior to joining Lehigh University in 1992, she spent three years at IBM T. J. Watson Research Center and IBM East Fishkill. She joined Carnegie Mellon University in 1999 and was promoted to the rank of Full Professor in 2002. She received the National Young Investigator award in 1994 and the Creativity Award in 2001, both from the National Science Foundation. She received a Deutsche Forschungsgemeinschaft Fellowship in 1994. She was a Chair of the Spring 1999 meeting of the Materials Research Society (MRS) and served on the MRS Council 1998-2000. She was the associate editor of the Journal of Electronic Materials 2007-2013. She co-edited and authored three chapters in a book on Metallic Films that appeared 2014. Barmak joined Columbia University in 2011 as the Philips Electronics Chair of Applied Physics and Applied Mathematics, and Materials Science and Engineering. She became the director of the Materials Science and Engineering Program in 2013.

12 Transmission Electron Microscopy for Physical Metallurgists

Hamish L. Fraser, David W. McComb, and Robert E.A. Williams, Center for Electron Microscopy and Analysis (CEMAS), Department of Materials Science and Engineering, The Ohio State University, Columbus, OH, USA

12.1 Introduction

The aim of this chapter is to provide the reader with an outline of the techniques most commonly used to characterize samples taken from studies involving physical metallurgy. In the main, because this is a chapter, rather than a book, full and comprehensive details on all techniques cannot be provided, references are made to existing texts and papers that will fill in the details that are not provided here. There are three books, which are heavily referenced (Hirsch et al., 1977; Loretto, 1984; Williams and Carter, 2009), and the reader is highly recommended to read these volumes. It is the attempt of the coauthors to outline techniques, and then provide ways in which difficulties in interpretation may be avoided.

Transmission electron microscopy (TEM) and the allied technique, scanning transmission electron microscopy (STEM), have provided physical metallurgists with a revolutionary increase in the ability to characterize samples at the micro- and nanoscales, scales where much of the "action" takes place in terms of materials' behavior and performance. Apart from scale, the techniques provide information gained from various imaging and diffraction methods, and analytical spectroscopies. This wide range of characterization possibilities makes the TEM and STEM instruments of great value to the metallurgist. A number of these possibilities are discussed below.

12.2 Electron Diffraction in the (S)TEM

One of the powerful aspects of TEM, particularly for studies of metallic materials, is the ability to not only image microstructures and defects, but also to use diffraction to gain information about crystal structure, crystallography, lattice parameters, etc. of the constituent phases in the microstructures. Diffraction is a technique well known to physical scientists, most usually involving X-rays and neutrons as the incident probes, which are subsequently diffracted by the sample. Diffraction has been well described in a number of books, of which perhaps the most well known and used is that of Cullity and Stock (2001). Briefly, for crystalline materials, diffracted beams exit the material in particular directions described by Bragg's Law, given by

$$n\lambda = 2d_{hkl}.\sin(\theta_{hkl}) \tag{1}$$

where n is an integer, λ is the wavelength of the incident radiation, d_{hkl} is the interplanar spacing of the $\{hkl\}$ planes and θ_{hkl} is the angle between the incident probe and the $\{hkl\}$ planes. As is well known, this law is based on the fact that constructive interference, where radiation scattered by adjacent planes is in phase, occurs when the path differences between scatterings by these adjacent planes are integer numbers of the wavelength of the radiation. When such constructive interference occurs, it is said that the Bragg condition for the given diffracting planes is met. Of course, for complete destructive interference to occur for all other angles of incidence, it assumed that a very large number (formally infinite) of parallel planes are involved in the event. While this assumption is reasonably met for X-ray and neutron diffraction from bulk samples, it is *not* typically met for electron diffraction in thin samples, as in TEM studies. The consequences of this are discussed below. As an aside, in the literature, as in the text below, the terms diffracted beam and reflection are used frequently to mean the same, i.e. a scattering event where Bragg's Law is satisfied.

It is noted that Bragg's Law involves the term d_{hkl}, the interplanar spacing of the $\{hkl\}$ planes. If the cell used to describe, or define, a given crystal is primitive (i.e. has only one lattice point per cell), then diffracted intensities would be observed from every plane $\{hkl\}$ in the crystal, at angles prescribed by Bragg's Law. However, for convenience, physical metallurgists prefer to reference crystal structures to simple unit cells (e.g. cubic cells rather than rhombohedral ones) that often contain more than one lattice point per cell, e.g. *fcc* (four lattice points per cell) and *bcc* (two lattice points per cell). The presence of these lattice points in the cell causes interference effects for scattering by certain planes in the structure when described by these unit cells, and so although Bragg's Law may predict a scattering event for a given $\{hkl\}$ plane at a given Bragg angle, the intensity exiting the sample in that direction may be diminished or zero. To predict the intensity of given diffracted beams from unit cells, reference is made to the Structure Factor, F_{hkl} defined as:

$$F_{hkl} = \sum_i f_i \exp(2\pi i(hx_i + ky_i + lz_i)), \tag{2}$$

where the sum is over the number (i) of lattice points in the unit cell, f_i are the atomic scattering factors of the atom on the ith lattice site in the cell, and x_i, y_i, and z_i are the positions of the ith lattice point in the given cell. Substituting the coordinates of the lattice points in a unit cell for a given structure yields a set of rules for "allowed" reflections. For example, for diffracted beams (i.e. those that are at the appropriate Bragg angle for their interplanar spacing) in *fcc* crystals, intensity will only be present in those diffracted beams for which the indices of the diffracting planes, h, k and l, are either all even or all odd. The various rules can be found in Loretto (1984).

12.2.1 Selected Area Diffraction

In the early days of TEM, it was most common to select an area of the sample from which to record the diffraction pattern by use of an aperture positioned below the sample in a typical TEM, at an intermediate image plane. This is known as selected area diffraction (SAD). In principle, the aperture allows only those electrons illuminating the area it defines to contribute to the diffraction pattern observed on the viewing screen. However, because of spherical aberration, there is a slight error between the area selected by the aperture and the area actually illuminated. In addition, if the sample is not at the eucentric height, the first image plane will not coincide with the plane of the selected area aperture, leading to a further error. These errors limit the smallest possible area from which SAD patterns can be obtained to be about 1–5 μm in diameter. One characteristic of SAD is that, in principle, the incident beam involves parallel illumination. This means that, from the point of view of the incident electron beam, the orientation of diffracted intensities will be exactly described by Bragg's Law, i.e. delta functions of intensity at specific angles of diffraction. It is shown below that this is not necessarily realized when effects of the sample itself are taken into account.

The first step in using electron diffraction in metallurgical studies involves predicting what diffraction patterns look like, so that recorded patterns may be interpreted, or the form of patterns expected from a crystal when the incident beam is parallel to a given crystallographic direction may be predicted. As most readers will have experienced during introductory courses on X-ray diffraction, Bragg's Law may be represented graphically by the Ewald Sphere construction, a construction drawn in reciprocal space, or superimposed on the reciprocal lattice. The reciprocal lattice is described and defined in Williams and Carter (2009); it is a lattice of points that each refer to particular sets of planes rather than to atomic positions, as in the real space lattice. The vector that connects the origin of the reciprocal lattice to a given reciprocal lattice point representing the plane (hkl) in the real lattice is g_{hkl}. Two properties of the reciprocal lattice are particularly important, the first being that the vector g_{hkl} in the reciprocal lattice is parallel to the normal to the (hkl) plane in the real lattice, and the second being that the magnitude of g_{hkl} is the reciprocal of the interplanar spacing of the (hkl) plane. The Ewald Sphere, of radius $1/\lambda$, passes through the origin of the reciprocal lattice; the radius connecting the origin of the sphere to the origin of the reciprocal lattice is parallel to the direction of the incident radiation. Whenever the surface of the sphere intersects a reciprocal lattice point, Bragg's Law will be satisfied, and the beam will be diffracted by those planes that are depicted by the intersected reciprocal lattice point. The intensity of this diffracted beam will be determined by the value of the Structure Factor, and so if a reciprocal lattice is drawn containing only points for which the Structure Factor is nonzero, then any intersected point will give rise to a diffracted beam. The diffracted beam is referred to as g_{hkl}. Reciprocal lattice points, and therefore potential diffracted beams, that are contained in a plane normal to the incident beam direction including the origin of the reciprocal lattice are said to be in the zeroth order Laue zone (ZOLZ), and the successive parallel layers of points are said to be in the higher (i.e. first, second, third, etc.) order Laue zones (HOLZ), as shown in **Figure 1**.

12.2.1.1 *Effect of Sample Thickness*

In TEM, the electron wavelength is very small, which means that the radius of the Ewald Sphere will be very large. Therefore, in the vicinity of the origin of the reciprocal lattice, the Ewald Sphere will be fairly flat (very large curvature), and so it has often been assumed that the form of a diffraction pattern may be predicted rather simply by taking a plane section through the reciprocal lattice, normal to the direction of the incident beam. This approximation appeared to be valid in the earlier days of microscopes, where the angular range exhibited in diffraction patterns was rather limited, but the curvature is very much evident in diffraction patterns observed in modern microscopes where the angular range can be relatively large. Thus, depending on the camera length selected, reflections caused by the intersection of the Ewald Sphere with points in the higher order Laue zones may be observed clearly, as shown in **Figure 2**. Interestingly, a significant number of reflections from the ZOLZ are observed in any given diffraction pattern, see example in **Figure 3**. This is of interest because the basis of Bragg's Law is that diffracted beams will arise from constructive interference in one direction only (defined by the Bragg angle), and for all other angles, destructive interference will reduce the scattered intensity to zero. The reason why many reflections in the ZOLZ are observed is because of the fact that in the thin dimension of the thin

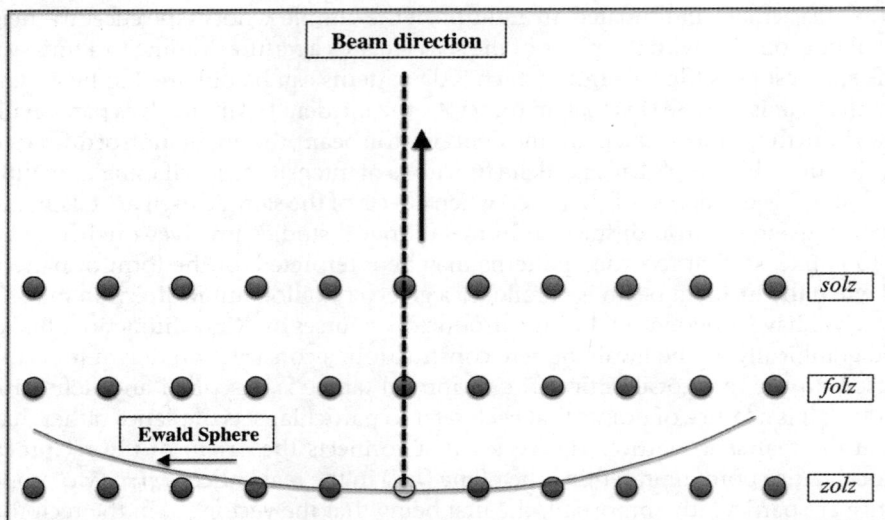

Figure 1   Schematic diagram showing the Ewald Sphere construction, showing reciprocal lattice points in the higher order Laue zones (i.e. zeroth-order (*zolz*), first-order (*folz*) and second-order (*solz*) Laue zones). (For color version of this figure, the reader is referred to the online version of this book.)

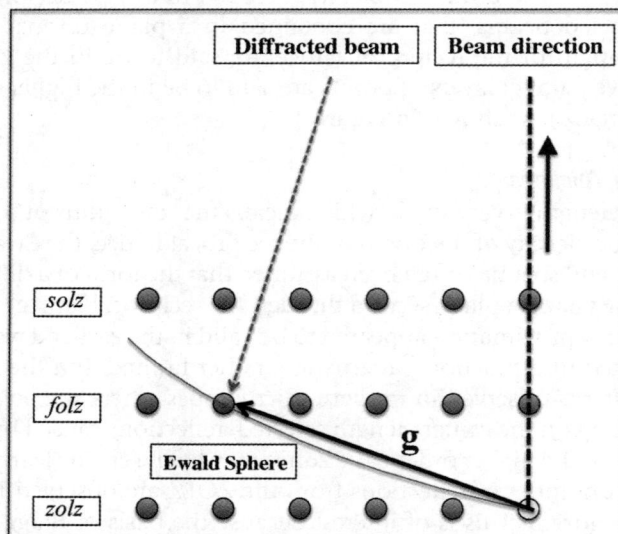

Figure 2   Schematic diagram showing the Ewald Sphere intersecting a reciprocal lattice point in the *folz*. For this reflection, Bragg's Law would be satisfied. (For color version of this figure, the reader is referred to the online version of this book.)

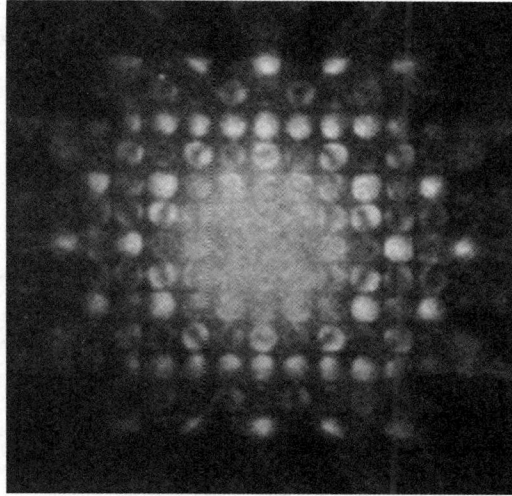

Figure 3 An electron diffraction pattern showing a significant number of excited reflections from the *zolz* despite the curvature of the Ewald Sphere. The diffraction maxima appear as disks because the incident beam is somewhat convergent.

foil typically employed in TEM there will be a rather small number of scattering planes in that direction, leading to incomplete destructive interference, and hence scattered intensity in directions that deviate from the exact Bragg angle, i.e. in the direction of, and near to, the thin dimension of the sample. These violations of Bragg's Law are represented by extending the reciprocal lattice points in the thin dimension of the crystal and are known as reciprocal lattice rods, or relrods. If the Ewald Sphere intersects a relrod (as depicted in **Figure 4**), then a reflection will result, the intensity of which will decrease as the angle from the point of intersection of the Sphere and the center of the reciprocal lattice point increases. For this reason, the simple assumption that SAD pattern can be predicted by noting the positions of reciprocal lattice points in a plane section normal to the incident beam direction appears to be valid, despite the curvature of the Ewald Sphere.

12.2.1.2 *Effect of Small Volumes*

Following on from the discussion about the effect of sample thickness, consider the situation when a thin foil contains small precipitates. These precipitates, or similar features consisting of small volumes (e.g. faults, twins, small precipitates, etc.), will have one or more thin dimensions, and for each, there will be insufficient numbers of planes to effect complete destructive interference, and so extensions of the reciprocal lattice points will occur, such that diffraction may result even when the exact Bragg Angle is not met. The actual extensions of the reciprocal lattice points will depend on the shape of the small objects, with extensions being parallel to the thin dimension(s) of the given particle; for example, a small thin disk (e.g. a GPB zone in a precipitation hardened Al alloy) would give rise to a rod-shaped extension perpendicular to the surface of the disk (as well as in the plane of the disk if the diameter is small). The result of the extension in reciprocal space on the resulting diffraction pattern will depend on the orientation of the small volume. Thus, if the extension is parallel to the incident beam (similar to the effect for sample thickness discussed above), when the Ewald Sphere intersects the extension, it will give rise to a diffracted beam (of low intensity because of the small volume of the scattering material).

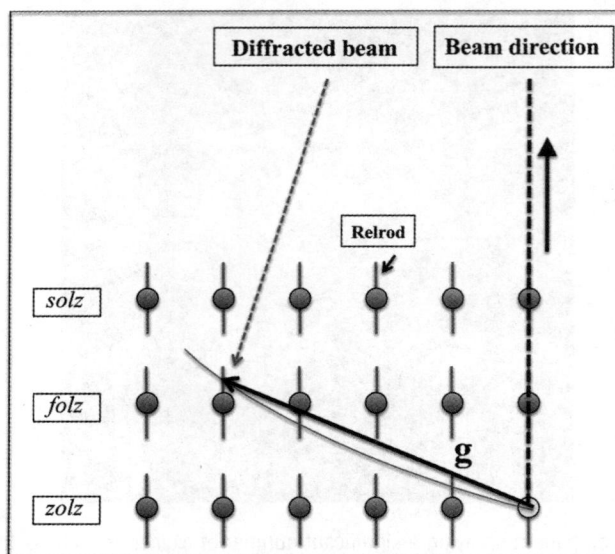

Figure 4 Schematic diagram of the Ewald Sphere construction showing relrods on the various reciprocal lattice points. Where the Sphere intersects the relrods (here at a reciprocal lattice point in the folz), there will be a resulting diffraction maximum. (For color version of this figure, the reader is referred to the online version of this book.)

If the object is oriented such that the thin dimension is highly inclined to the incident beam direction, e.g. a disk-shaped precipitate is inclined with the surface of the disk parallel to the incident beam, then diffraction can occur over a relatively significant angle defined by the intersection of the Ewald Sphere with the reciprocal lattice extension, an example being given in **Figure 5**; the thin plates are GPB zones in an Fe–Mo alloy (Williams and Carter, 2009). This would give rise to a streak of intensity in the diffraction pattern. The thinnest such object would be a thin twin or a stacking fault. Diffraction from such a single fault or thin twin would be of very low intensity; however, if the sample contains a large density of faults or thin twins, fairly intense streaks of intensity in the diffraction pattern may result.

12.2.1.3 *Additional Sources of Diffraction Maxima*
12.2.1.3.1 *Double Diffraction*
It is been tacitly assumed that the incident beam may be scattered constructively by a given set of planes producing a diffracted beam, g_1. These scattered electrons may then be incident upon another set of planes, represented by g_2, at the appropriate Bragg angle, and so be Bragg scattered once more. This second diffracting event results in a reflection given by $g_1 + g_2$. This is known as double diffraction, but is more generally a form of multiple scattering. The question is then: do these double diffracting events give rise to extra reflections in SAD patterns? There are two ways in which the expected reflections from double diffraction may be predicted. Firstly, all g_{hkl} vectors in a given SAD may be added together resulting in the end point of the combined vectors being located either where there are existing reciprocal lattice points, or where prior to double diffraction, no reciprocal lattice points exist. Obviously, in the latter case, an extra reflection would result, in a location of a kinematically forbidden reflection. Secondly, the origin of the reciprocal lattice may be translated to each of the other reciprocal

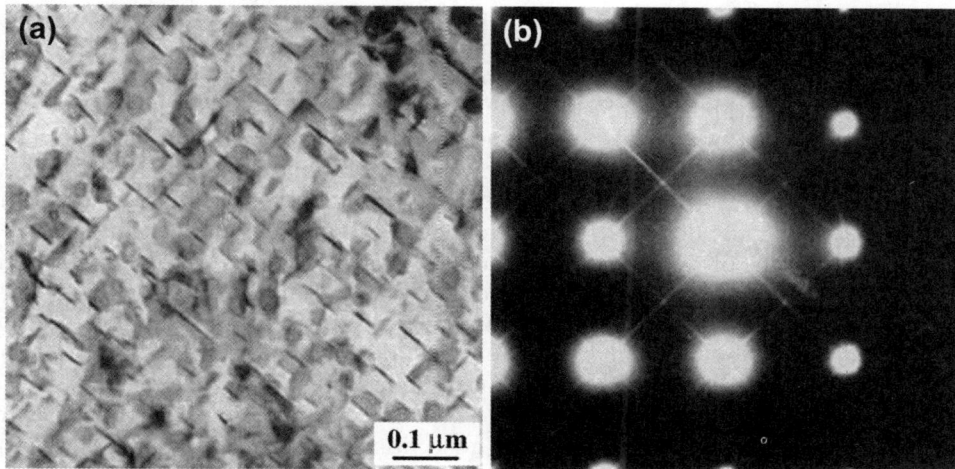

Figure 5 Example of diffraction effects from thin plate-like GPB zones in an Fe-Mo alloy. (a) electron micrograph showing the array of thin particles. (b) diffraction pattern exhibiting streaks of intensity in directions normal to the thin dimension of the precipitates. Taken from Williams and Carter (2009).

lattice points, and again any reciprocal lattice points that do not line up with the original pattern of reflections would result in an intensity contribution from double diffraction. These two methods actually are ways of doing the same function.

Most foils of metallurgical samples are relatively thick, and so this form of scattering occurs frequently. However, in the vast majority of cases for single-phase materials, the end points of the resultant vectors, e.g. $g_1 + g_2$, normally lie on the end points of existing vectors g_{hkl}, and hence there is no manifestation of the multiple scattering event. There are some well-known exceptions to this, for example, the case of *hcp* metals. When the beam is oriented parallel to the direction $\langle 11\bar{2}0 \rangle$, the pattern of reflections marked • in **Figure 6** would be predicted from a plane section through the reciprocal lattice. However, those marked **x** result from double diffraction (for example, $g_{10\bar{1}1} + g_{\bar{1}010} = g_{0001}$).

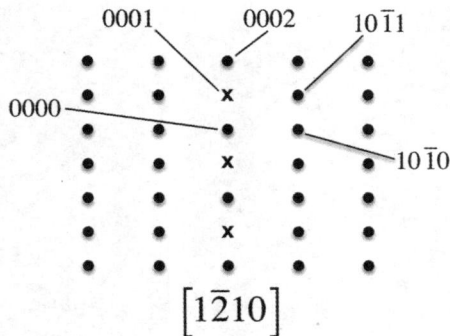

Figure 6 Schematic diffraction pattern in a close-packed hexagonal material with a beam direction parallel to [$1\bar{2}10$]. The reflections ±0001, ±0003, etc. are kinematically forbidden but appear through double diffraction (they are marked with the symbol "x").

The reflection corresponding to g_{0001} is a forbidden reflection for this crystal structure (i.e. a value of zero for the Structure Factor, F_{0001}), but in this beam direction, is present with relatively weak intensity from double diffraction from two allowed reflections (in the example, both $10\bar{1}1$ and $\bar{1}010$ are allowed reflections). Extra (forbidden) reflections also occur in diffraction patterns observed when the incident beam is parallel to either $\langle 11\bar{2}3 \rangle$ and $\langle 01\bar{1}1 \rangle$; the reader should verify this. It is important to note that these extra reflections occur by double diffraction involving scattering from two sets of planes that are oriented to the Bragg condition. A check may be made to establish that the observed "forbidden" reflections (in the example, the 0001 reflection) are indeed produced by double diffraction (and not, for example, from some type of ordering of the structure). Thus, the sample may be tilted about an axis parallel to the extra reflection, such that the reflections in that systematic row (here 0002, 0004, 0006, etc.) remain at the Bragg condition, but the other reflections responsible for the double diffraction effect (e.g. here, $10\bar{1}1$ and $\bar{1}010$) would be tilted away from the Bragg condition. Then, they would no longer be involved in a double diffraction event, and the extra reflections would be extinguished.

The discussion above has involved the case of double diffraction in single-phase materials. In contrast to the fairly rare occurrences of extra reflections arising in these types of samples, the opposite is true for multiphase materials. Thus, the situation when one or more second phases are present usually results in extra reflections being observed, when electrons diffracted by a set of planes in one phase are diffracted by a set of planes from a different phase. In fact, many extra reflections may result as a result of multiple scattering between two or more phases. An interesting example involves the case of precipitation of ζ-Ti_5Si_3 in Ti–Si alloys (Chumbley et al., 1988). It was found that some of the silicides formed as hexagonal plates in a matrix of α-Ti, with the habit of the plates being parallel to $(0001)_\alpha$. An SAD pattern with the electron beam parallel to $[0001]_\alpha$ is shown in **Figure 7(a)**. As can be seen, there are a very large number of reflections, and this represents a fairly arduous task in terms of interpretation (i.e. indexing the pattern). The influence of double diffraction was assessed by following the procedures outlined above (specifically by adding all possible combinations of reflections) and was performed on a computer. Of course, some assumptions must be made concerning the possible orientation relationships (OR) that were deduced from the diffraction patterns, which

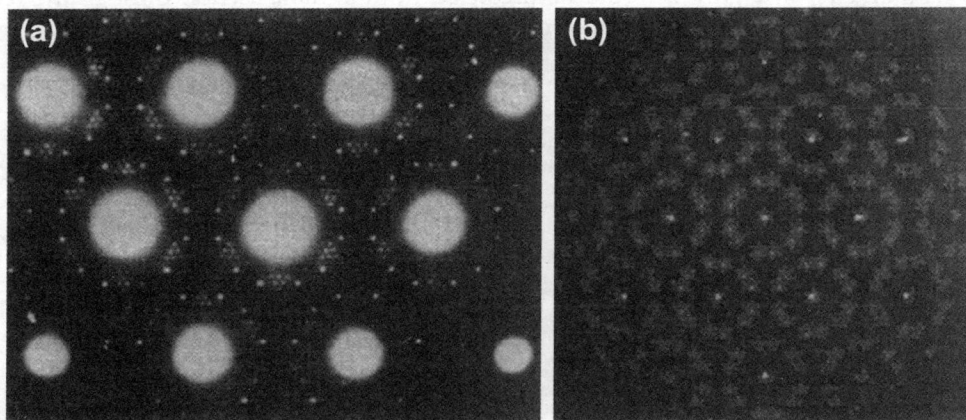

Figure 7 (a) A SAD diffraction pattern recorded from a Si sample containing silicides. There are a large number of reflections apart from those from the two phases. (b) a computer-generated pattern taking into account double diffraction between the two phases. Taken from Chumbley et al. (1988).

revealed that the hexagonal facets were parallel to $\{\bar{1}010\}_\xi$ and $\{21\bar{3}0\}_\alpha$, leading to an assumed OR given by:

$$(0001)_\alpha //(0001)_\xi \quad \text{and} \quad \langle 21\bar{3}0 \rangle_\alpha //\langle 10\bar{1}0 \rangle_\xi.$$

The results when predicting double diffraction from the presence of two variants of this OR and the α-Ti matrix are shown in **Figure 7(b)**. As can be seen, a reasonable facsimile of the experimental pattern is obtained. There are, of course, small differences that arise from the fact that exact orientations are not known, and also the structure factors of the reflections from the silicide phase and intensities from volume fraction differences are not known. It can be seen that double diffraction effects can produce quite complex patterns, but that these can often be indexed correctly when using simulation of double diffraction effects.

12.2.1.3.2   Long- and Short-Range Ordering

Long- and short-range ordering leads to the presence of either diffuse scattered intensities and/or extra reflections in locations that would normally be kinematically forbidden in the disordered structure. Consider first, the case of long-range ordering. The Structure Factor is used to predict intensities from unit cells that are ordered (i.e. superlattice structures). For example, consider an ordered crystal AB with the B2 (CsCl) structure, such that all corner atoms are occupied by A atoms and the body-centered sites are occupied by B atoms, then the Structure Factor would inform that fundamental reflections would occur for diffracting planes with indices, hkl, such that the sum $h + k + l$ is even (in this case, the intensity (i.e. the square of the amplitude scattered $((F_{hkl})^2)$ would be proportional to $((f_A + f_B)^2)$, and superlattice reflections (of diminished intensity) would occur when $h + k + l$ is odd (in this case, the diminished intensity would be proportional to $((f_A - f_B)^2)$. The intensity of the superlattice relection can often be very weak, depending on the differences of the atomic scattering amplitudes between the sublattices, and so missed when making observations on a fluorescent screen. It is important that long-term exposures, or long integration times, be used to record the diffraction pattern.

Short-range ordering (SRO) leads to diffuse scattered intensities in the diffraction pattern, occurring usually about locations where a reflection from a fully ordered version of the SRO would occur. The intensity of the diffuse scattering will depend on the degree of ordering, and is often difficult to observe because of the relatively significant, and changing, background intensity in the diffraction pattern. It would be advantageous to simulate the background intensity, and then subtract this from the recorded pattern to reveal the diffuse-scattered intensity. However, such simulation is difficult to do, and a fairly crude experimental approach often will suffice. An example is given from a study of the alloy Ti–6Al–4V, where samples had been slowly cooled from 930 °C to room temperature (Wu et al., 2013). The sample was tilted to maximize the intensity of the superlattice reflection, with the electron beam essentially parallel to $[1\bar{2}10]$. The result is shown in **Figure 8(a)**; as can be seen, there is a diffuse scattering in the location of the $10\bar{1}0$ superlattice reflection (the diffraction pattern has been doubly exposed, once with the objective aperture inserted for dark-field (DF) imaging); however, the resulting DF image was not convincing. To ensure that the diffuse scattering is real, the background was subtracted by using ImageJ to provide a line profile of the background intensity adjacent to the diffuse scattering which was then subtracted from a profile taken through the diffuse scattering, and the result is shown in **Figure 8(b)**. Although this is a somewhat approximate way of effecting background subtraction, the result is qualitatively useful.

12.2.1.3.3   Twinning

Another way that extra reflections, over and above those predicted by the Structure Factor (and of course, Bragg's Law), are observed is from the presence of a twin. These extra reflections are observed

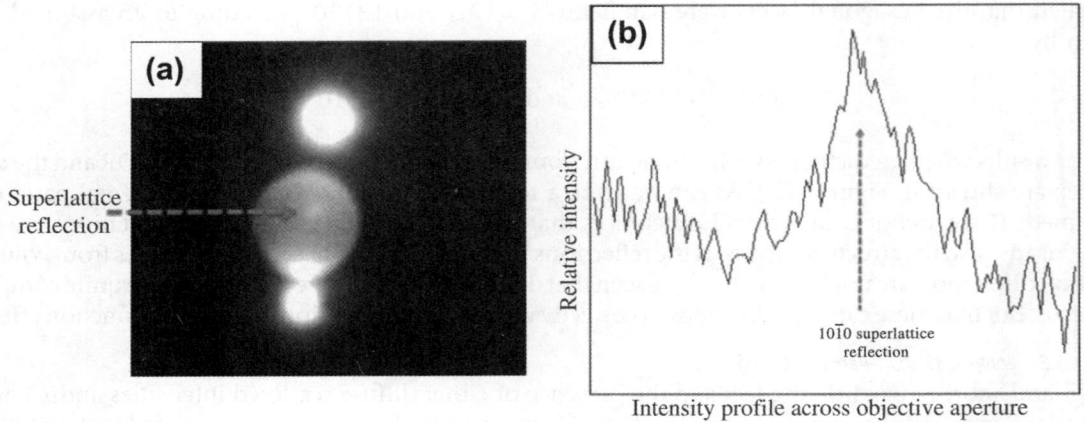

Figure 8 (a) Diffraction pattern recorded from an aged sample of Ti-6-4. Diffuse scattering is apparent in the location of the superlattice reflection (the diffraction pattern has been doubly exposed, once with the objective aperture inserted for dark-field imaging). (b) The background intensity in the diffraction pattern has been subtracted by using ImageJ to provide a line profile of the background intensity adjacent to the diffuse scattering. The diffuse maximum is clearly observed. Taken from Wu et al. (2013).

when the composition plane, i.e. the interface plane, contains the electron beam direction. In this geometry, the extra reflections from twinning are produced by rotating the pattern, formed with the beam parallel to a given zone axis, by 180° about the normal to the composition plane. An example is shown in **Figure 9**; here, the sample is γ', the ordered phase based on Ni_3Al, and the rotation is about

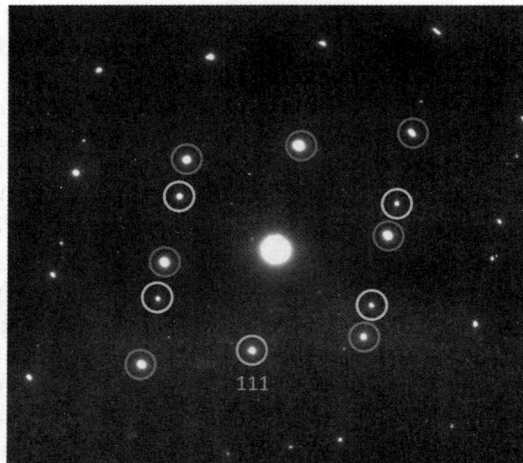

Figure 9 A selected area diffraction pattern recorded from γ', the ordered phase based on Ni_3Al. Some of the fundamental reflections are marked by red circles, and some twin reflections are marked by green circles. The twin reflections are a result of a 180° rotation about the ⟨111⟩ systematic row (marked). (For interpretation of the references to color in this figure legend, the reader is referred to the online version of this book.)

the $\langle 111 \rangle$ systematic row (marked). Some of the fundamental reflections are marked by red circles, and some twin reflections are marked by green circles. The diffraction event described above requires that the interface plane of the twin is oriented such that it contains the beam direction. In microscopes with limited tilt, it may not be possible to effect this crystallographic arrangement. In this case, diffraction patterns from either side of a twin must be recorded in two different beam directions, and the orientation data plotted on a stereographic projection, from which the twin may be identified. This latter procedure is somewhat cumbersome, and these days can be avoided by the use of orientation imaging (i.e. electron backscattering diffraction) in a dual-beam FIB in which a thin membrane may be excised in an orientation such that the normal to the twin plane is itself normal to the thin dimension of the membrane.

12.2.1.3.4 *Other Sources of Scattered Intensity*
Diffuse scattering may be produced by a number of events occurring in samples. Examples of these involve clustering of solute, pretransformation phenomena, and shear transformations (e.g. martensite and the ω-phase transformation). Satellite reflections also can occur when compositional modulations are present, for example, when phase separation occurs via spinodal decomposition. These various diffraction events are well described in Loretto (1984) and Williams and Carter (2009).

12.2.1.4 *Kikuchi Diffraction*
Very often, electron diffraction patterns exhibit intensity maxima produced in the ways described above, and also sets of parallel pairs of lines, one being above background (excess intensity) and the other below background intensity (deficient intensity). These pairs of lines are called Kikuchi lines. They are formed by the elastic scattering of inelastically scattered electrons, as described in a simple way in the following. Consider **Figure 10(a)**. The electron beam enters the crystal, and at some arbitrary point P it is inelastically scattered. The electrons will be scattered through angles determined by the particular inelastic scattering event, but in general, the intensity will be peaked in the forward scattered direction. Some of these electrons, along PE and PD, will be oriented correctly for diffraction by the planes, $\{hkl\}$, and undergo elastic scattering as defined by Bragg's Law, as depicted in the figure. Because the ray path-marked PE is closer to the forward-scattering direction than the path PD, the intensity scattered into EE′ will be greater than that scattered into DD′. Importantly, the background intensity about the ray path EE′ will be similar to that in PD (i.e. ray paths in similar directions), and for DD′, the background will be similar to that in PE. Hence, the electrons diffracted along EE′ will be above background intensity, and those along DD′ will be below background intensity.

Because the electrons that suffer an inelastic event may be scattered over very wide angles, and because in a real crystal there will be a very large number of scattering points, P, throughout the sample, such scattered electrons will be incident on $\{hkl\}$ over essentially all the solid angle defined by Bragg's Law for the given planes. Consequently, excess and deficient rays will be generated over the same solid angle, and this will give rise to cones of excess and deficient intensities, as shown schematically in **Figure 10(b)**. The intersection of these cones with the viewing screen, or the detection device (photographic film, image plates, or CCD camera), results in pairs of lines, known as Kikuchi lines. These lines are, in principle, curved conic sections, but because of the limited angular range represented in the diffraction patterns that are recorded, they appear usually to be relatively straight lines.

It is clear from the diagram in **Figure 10(b)** that the angular separation of the excess and deficient lines is $2\theta_B$, i.e. the same angle that electrons are scattered elastically by $\{hkl\}$. Actually, the Bragg angle

(a)

(b)

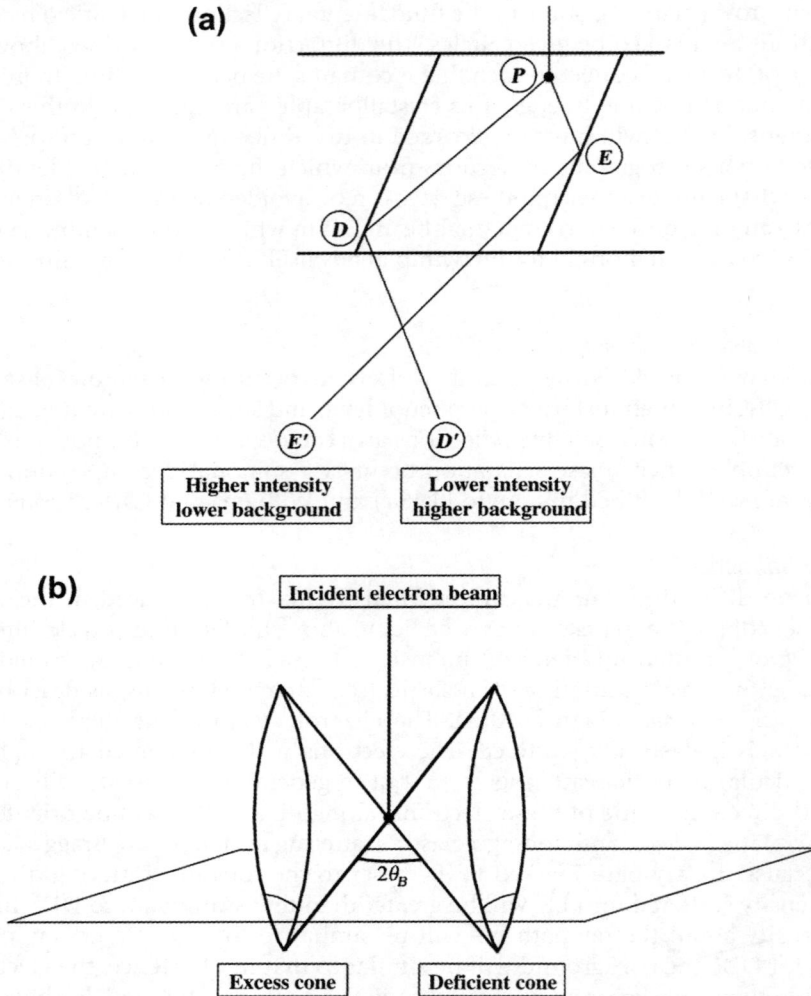

Figure 10   (a) and (b) Schematic diagrams depicting the formation of Kikuchi lines. See text for explanation.

for the electrons that are first inelastically scattered will be very slightly different from that for electrons that are first elastically scattered, because some energy will be lost in the inelastic event, modifying the wavelength of the electrons, but this very small change may be ignored. Hence, for every reflection in a diffraction pattern, there will be a corresponding pair of Kikuchi lines, one of excess, and the other of deficient, intensity. The presence of these lines is of great importance to the electron microscopist, as is explained in the following. Suppose that a given set of planes is oriented for Bragg diffraction. As the crystal is tilted, the intensity of the diffracted beam will decrease to zero, moving very slightly as defined by the shape of the relrod associated with sample and phase morphology and dimensions. Because the intensity resulting from Bragg diffraction in thin samples is not described by a delta function, but exists over a small range of angles, it is not possible to know the exact crystal orientation of a sample (for example, whether or not the foil is oriented exactly for diffraction by the {*hkl*} planes). Kikuchi lines act differently as a sample is tilted. Thus, as noted above, the initial inelastic scattering occurs over a wide range of angles. Therefore, as the sample is tilted, there are always (different) inelastically scattered electrons that will be at the Bragg angle for diffraction by the given {*hkl*} planes, and so the Kikuchi lines will move but remain present in the diffraction pattern. It is as if the Kikuchi lines were fixed to the crystal. Because the Kikuchi lines move, and they maintain their intensity (excess and deficient), as the sample is tilted, they provide a means of determining the crystal orientation with great accuracy. The applications of this phenomenon are described below.

12.2.1.4.1 *Controlled Tilting Experiments*

The Kikuchi lines form patterns throughout reciprocal space that appear as maps that connect various zone axes. An example of such a "Kikuchi map" for two crystallographic triangles for an *fcc* crystal is shown in **Figure 11(a)**. Note that although this map resembles a real space stereographic projection, there is a major difference. Thus, inspection of the crystallographic indices reveals that they appear to conform to a left-handed, rather than right-handed, set. For example, the pole [011] is positioned on the left of [001], rather than on the right, as with a conventional stereogram. The reason for this is as follows. Suppose a crystal is oriented with [001] parallel to the incident beam direction, and a tilt of 45° about [200] is applied such that the [011] becomes parallel to the beam direction. On a real space stereogram, the operation would cause the [011] pole to move from the right to left, as shown in **Figure 11(b)**; however, when observing the diffraction pattern during this titling operation, the pattern would appear to move from left to right, consistent with the notion that the Kikuchi lines are fixed to the crystal. This is an important concept, particularly when performing defect analysis where

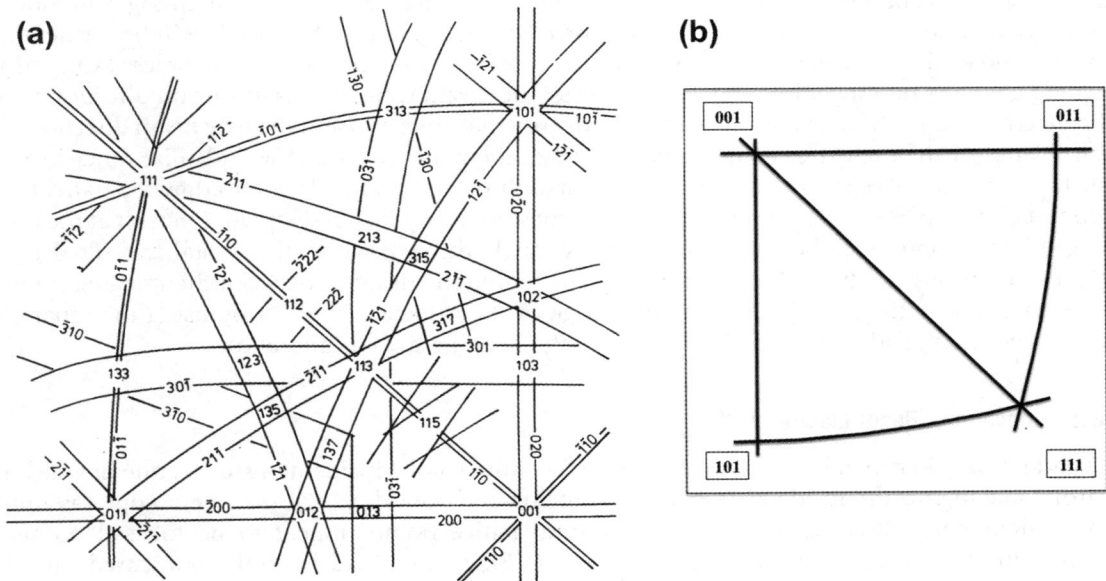

Figure 11 (a) A Kikuchi map for fcc crystal (Taken from Loretto (1984)). (b) The corresponding real-space stereogram. See text for explanation.

the nature and sign of diffraction vectors must be precise. Various Kikuchi maps may be found in Loretto (1984).

12.2.1.4.2 Determination of the Deviation Parameter

As noted in the discussion of defect contrast, it is often necessary to tilt a sample away from a given pole, about a given **g**-vector, and then to orient the sample such that there is a small positive deviation from the exact Bragg condition. In the absence of Kikuchi lines, this would be a difficult procedure, since the intensity of the Bragg reflections varies with both specimen thickness and orientation. However, as Kikuchi lines act as though they are fixed to the crystal, the deviation from the Bragg condition is revealed by the displacement of the excess Kikuchi line from the given Bragg reflection. The relationship between the deviation parameter, s, and the position of the Kikuchi lines is shown graphically on the Ewald Sphere construction in **Figure 12**. Simple geometrical considerations (Williams and Carter, 2009) result in an expression that permits the determination of the magnitude of the deviation parameter given by:

$$|S| = \frac{x}{R}\lambda g^2, \tag{3}$$

where x and R are measured from the diffraction pattern, and λ is the wavelength of the electron. By convention, the sign of the vector s is positive when the endpoint of the vector **g** lies within the Ewald Sphere (as in **Figure 12**).

12.2.1.4.3 Exact Crystal Orientation Determination

It is also often important to determine the exact beam direction when recording images of defects. Here, for example, when simulating images Kikuchi lines play a significant role. It is noted that usually in a defect analysis experiment, the sample is first tilted to a given zone axis, containing a number of reflections, and hence potential imaging conditions. For each **g** vector, the sample will be further tilted by a few degrees about the given **g**, such that the other diffraction maxima are more or less extinguished (i.e. their **g** vectors no longer intersect, or lie near to, the Ewald sphere). It is this beam direction that is to be determined, i.e. what is the angle between the original zone axis and the new beam direction. The Kikuchi lines of the excited **g** of course pass through the zone axis; there will also be other inclined Kikuchi lines that correspond to other weakly excited diffracted beams. The lines drawn parallel to the midpoint of each pair (i.e. excess and deficient) of these various Kikuchi lines will intersect at the center of the original zone axis. The beam direction (i.e. angle through which the crystal was tilted to the imaging condition) can then be determined by converting the distance between the imaging point to the center of the zone axis, either by a knowledge of the camera length, or by using the separation between the excess and deficient Kikuchi lines of the imaging **g**, as a calibration.

12.2.2 Convergent Beam Electron Diffraction

SAD patterns are, in general, obtained using an essentially parallel incident beam, i.e. one in which the electrons making up the incident probe are parallel to the optical axis of the microscope (assuming correct alignment!). Because of this, the reciprocal lattice points appear to be focused points of intensity. In convergent beam electron diffraction (CBED), the incident probe is focused onto the sample, and so is a convergent beam with convergence angle, 2α. To account for this convergence, i.e. variation in angle of incidence, the reciprocal lattice point at the origin of the lattice, corresponding to the incident beam in diffraction patterns, becomes a disk of diameter 2α, and of course, all other

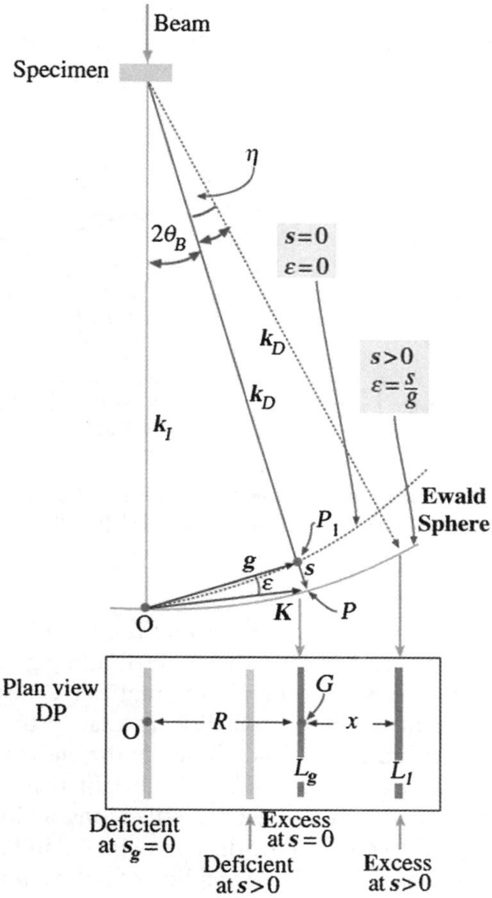

Figure 12 The relationship between the deviation parameter, *s*, and the position of the Kikuchi lines shown on the Ewald Sphere construction. Taken from Williams and Carter (2009). (For color version of this figure, the reader is referred to the online version of this book.)

reciprocal lattice points also become disks of diameter 2α. This provides an advantage because it reveals (useful) intensity information within the disks that in a parallel beam situation is not available. The advent of the use of CBED coincided (almost) with technological advances in the electron optics of TEMs that permitted much wider angles of diffraction space to be observed. This allows observation of the intersection of the Ewald sphere with reciprocal lattice points, or disks in CBED, in *holz* layers, which provides three-dimensional information about the structure of the sample, rather than the two-dimensional information gained from SAD, as shown in **Figure** 4.

Consider the intersection of the Ewald sphere with reciprocal lattice disks in the *holz* layers; as an example, an intersection with the first order Laue zone (*folz*) is depicted schematically in **Figure 13**. In **Figure 13(a)**, the Ewald sphere intersects the given reciprocal lattice disk, say the *uvw* reflection, in the *folz* layer through the center of the disk. This gives rise to diffraction as, for all points along the line of intersection, Bragg's Law will be satisfied (i.e. as per the Ewald Sphere construction). The line of

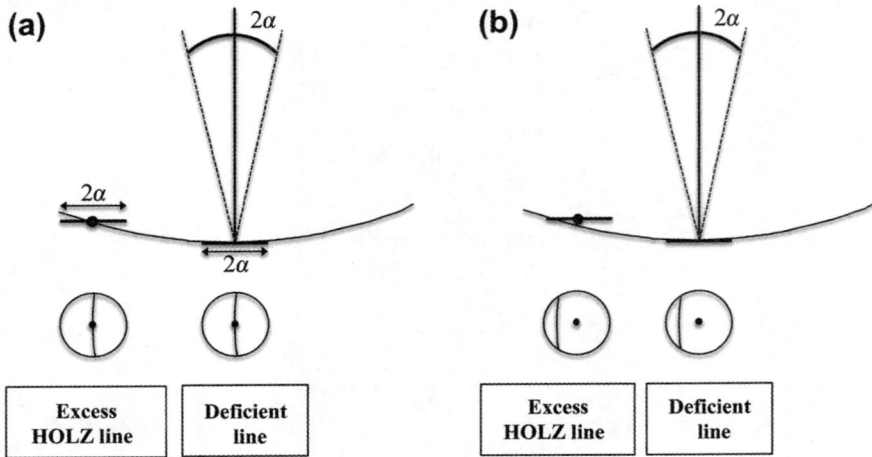

Figure 13 Schematic diagram depicting the formation of *holz* lines in the central disk of a CBED pattern. (a) Showing the intersection of the Ewald Sphere through the center of a reciprocal lattice disk (i.e., the *uvw* disk) and (b) at a location within the disk but displaced by a small angle.

intersection will appear to be straight, despite the fact that the Ewald sphere is curved because of the scale of the diagram. In the case of holz reflections, the corresponding **g**-vectors, e.g. *uvw* in the example, are rather long and so the scattering is, in general, kinematical. Consequently, a dark line (i.e. an absence of intensity) will be present in the central disk, in exactly the same location as the line of intersection of the Ewald sphere with the *uvw* disk. These are the electrons which have been diffracted (kinematically); the dark line produced in the *central*, or bright field, disk is known as a *holz* line. Interestingly, the Ewald sphere can intersect the holz reflections at locations such that the line of intersection does not pass through the center of the disk, as depicted in **Figure 13(b)**. Of course, Bragg's Law is satisfied and diffraction will occur, resulting in a line of deficient intensity in the central disk, as noted in the figure. Because of this, normally a central disk will contain a series of dark lines that correspond to the diffraction events in the *holz* layers, an experimental example being given in **Figure 14**. The pattern of lines in the experimental pattern may be compared with those simulated on the basis of the kinematical theory and using the dynamical theory, as indicated in the figure (Swaminathan et al., 1997). The latter simulation is a very good facsimile of the experimental pattern. When observing *holz* lines in a bright-field (BF) disk, they often appear to continue outside the disk itself. These lines outside the reciprocal lattice disks are simply the deficient Kikuchi lines for the particular reflection that has produced the *holz* line.

12.2.2.1 *Information Contained in CBED Patterns*

For physical metallurgists, there are three sources of information from CBED that are particularly useful. These are:

12.2.2.1.1 *Intensity Variations in the Bright Field Disk*

There are two types of intensity in the bright field disk, one being the *holz* lines, and the other a varying background intensity. In general, only *holz* lines yield useful information, arising from firstly their position (leading to local lattice parameter determinations, see below), and secondly, the symmetry

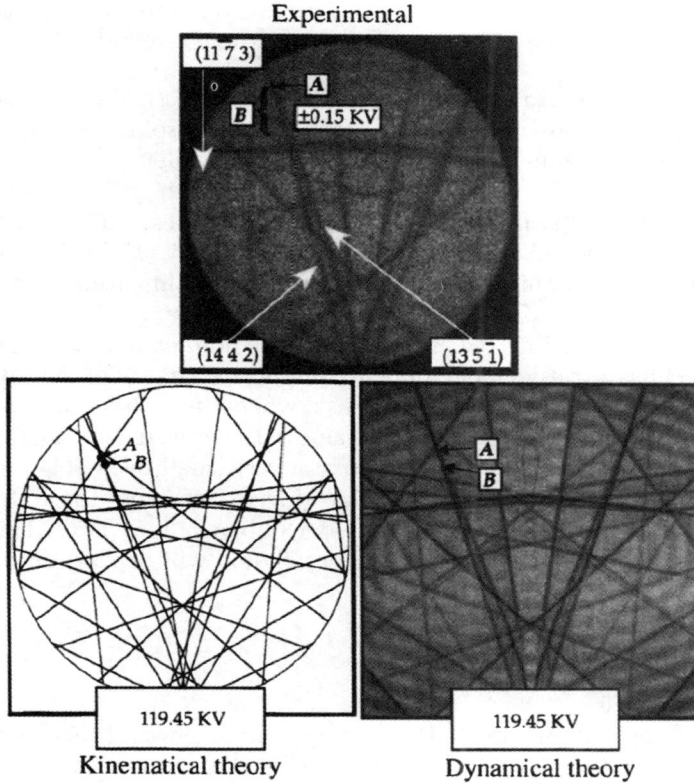

Figure 14 An experimentally CBED pattern recorded from a Si sample showing the BF disk. Note the various *holz* lines, and the "bent" lines that are dynamically split. Also shown are simulations of the pattern of *holz* lines using both kinematical and dynamical formulations. Note that the dynamical simulation predicts the dynamic splitting of the *holz* lines. Taken from Swaminathan et al. (1997).

exhibited by the pattern of the lines (useful in crystal structure determinations, see below). Because the *holz* lines are formed from diffraction by reciprocal lattice points in the higher order Laue zones, the information they yield is three dimensional in nature. The symmetry exhibited by the pattern of *holz* lines is known as the "bright-field symmetry". Sometimes, the *holz* lines observed in BF disks appear not to be straight, but seem to be somewhat curved (see **Figures** 14 and 16). In fact, the lines are not, in fact, curved, but are the result of dynamical scattering of two slightly inclined lines that are straight at their extremities, i.e. near the edge of the disks. This is a rather subtle point; the *holz* lines are said to be, "dynamically split." Note that in **Figure** 14, the experimental pattern exhibits some *holz* lines that are dynamically split, and these are not predicted by the kinematical calculation but are shown in the dynamical simulation (**Figure** 14).

12.2.2.1.2 Intensity Variations in the Pattern Obtained from then Zero-Order Laue Zone
The various reciprocal lattice disks in the zero-order Laue zone may contain intensity information (depending on the thickness of the foil). These intensities represent a two-dimensional projection of the

crystal potential in the given zone axis. The symmetry exhibited is known as the "projection diffraction symmetry", and can be useful in crystal structure determinations, see below.

12.2.2.1.3 Intensity Variations from Intersections of the Ewald Sphere and the Reciprocal Lattice Disks in the Higher Order Laue Zones
As noted above, Bragg diffraction will occur whenever the Ewald sphere intersects a reciprocal lattice disk in the higher order Laue zones. Usually, a bright line of intersection results, i.e. the electrons diffracted (often kinematically) from the bright field disk. As with SAD, the intensities in these diffracted beams will vary with both orientation and specimen thickness, and so are unsuitable for use in determining values of the Structure Factor. However, the symmetry exhibited by the intensities in the various beams is a very useful parameter; this information is three dimensional in nature and is known as the "whole pattern symmetry."

Before considering the applications of CBED, it should be pointed out that the value of the incident beam convergence (2α) has a considerable influence on the appearance of the CBED pattern. Thus, as noted above, the reciprocal lattice points appear as disks when a beam has a finite value of convergence, and the diameter of these disks will be given by the value of the convergence, 2α. Consequently, for very small values of convergence, the pattern will consist of very small disks (**Figure 15(a)**), resembling a "normal" SAD pattern (because the small value of convergence is not too different from a parallel beam). For larger values of the convergence, the disks will be relatively large (**Figure 15(b)**), and the disks may even overlap. The value of the convergence chosen will depend on the given application of CBED.

12.2.2.2 Applications of CBED
There are a number of applications of CBED that are useful to physical metallurgists, which include local lattice parameter measurements, crystal structure determination and local specimen thickness determination. In this chapter, the first two are described; a detailed summary of foil thickness determination is given in Williams and Carter (2009) with original references.

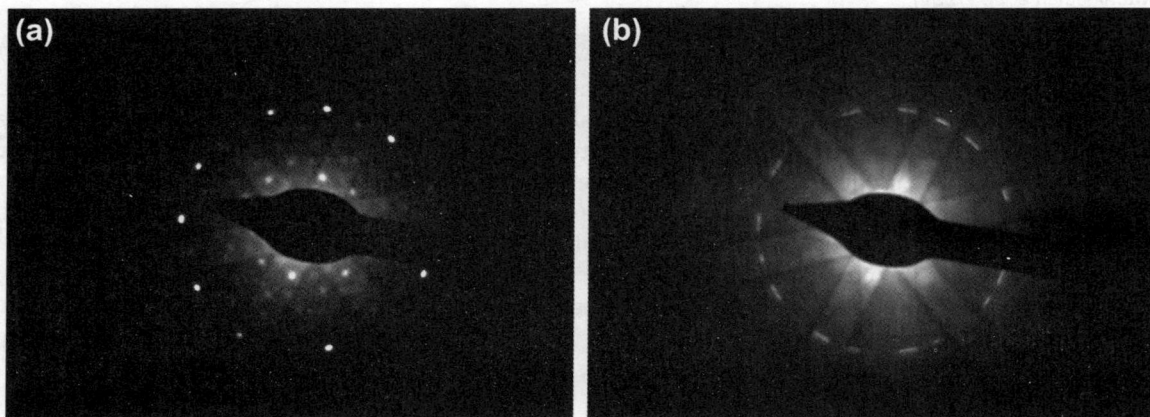

Figure 15   The effect of the beam convergence on the resulting CBED patterns. CBED diffraction patterns recorded with (a) a relatively small convergence angle, and (b) a larger convergence angle.

12.2.2.2.1   Local Lattice-Parameter Measurements

It was noted that (deficient) *holz* lines appear in the BF disk, and that these lines are the result of diffraction by the *holz* reflections. The individual *holz* lines may be indexed from simple geometry (Loretto, 1984), assuming the crystal structure and lattice parameters of the sample are known. Because the lines correspond to diffraction by reflections with large **g**-vectors, their positions in the bright field disk may vary quite significantly with small changes in lattice parameters, i.e. the lines may be shifted from their position expected on the basis of the published lattice parameters. These shifts in the position of the lines may be used to make local (i.e. in the region of the sample illuminated by the convergent beam) lattice parameter determinations. Thus, a simple computer program may be written that simulates where the *holz* lines lie in a given bright field disk, and these simulations may be compared with the experimental pattern. By varying the lattice parameters in the program, a best fit between simulation and experiment may be achieved, yielding values for the local lattice parameters.

As an example of the determination of lattice parameters, consider the patterns shown in **Figure 16**, recorded from the phase γ', based on Ni$_3$Al with the L1$_2$ crystal structure in a directionally solidified ternary eutectic Ni-base alloy, Ni–12.8Al–22.2Mo, compositions in at.% (Kaufman et al., 1986). In the example, the electron beam direction is parallel to $\langle 114 \rangle_{\gamma'}$. The experimental bright field disk, and the simulated (kinematical) position of the *holz* lines is shown in **Figure 16 (a and b)**, respectively. The lattice parameters used for the simulation were $a = b = c = 0.357$ nm, and $\alpha = \beta = \gamma = 90°$; these values represent "stress-free" γ' (compared with the locally distorted γ' discussed below). It should be noted that both the experimental and simulated patterns exhibit a single mirror plane expected on the basis of the point group $m3m$.

The CBED technique appears to be very powerful. However, as is always the case in TEM, it is important to remember that the sample is a very thin membrane, and so its exact configuration may have changed compared with the original bulk piece from which it was excised. This is particularly the case when local distortions are involved, since the proximity of the foil surfaces may lead to relaxations of those distortions. Consider the case of the Ni-base alloy discussed above. In this alloy, the phase γ' is found to form about Mo rods; upon cooling from elevated temperatures, stresses are developed

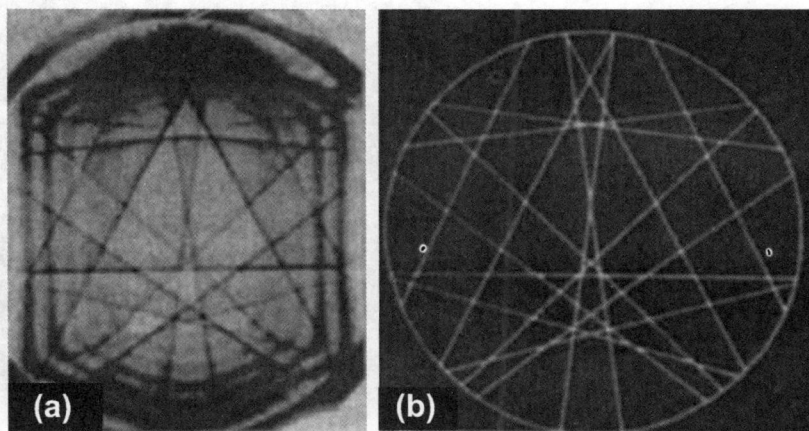

Figure 16   (a) A bright-field disk from a CBED pattern recorded from a sample of the γ' phase in a ternary eutectic γ, γ', α-Mo alloy (Ni-12.8Al-22.2Mo, at. %) with the electron beam parallel to $\langle 114 \rangle$. (b) A simulated (kinematical) pattern from which the lattice parameters of the γ' phase were deduced. Taken from Kaufman et al. (1986).

between the phases because of the very large difference in their thermal expansion coefficients. Because of this, the γ' is expected to be locally distorted, and this should be manifested by changes in the patterns of *holz* lines in bright field disks (the orientation relationship between the two phases is given by $[001]_{\gamma'}//[001]_{\alpha-Mo}, (100)_{\gamma'}//(110)_{\alpha-Mo}$). Sets of experimental and simulated patterns are shown in **Figure 17**, and it can be seen that the patterns of lines vary considerably, as expected. For example, consider the patterns in **Figure 17(b and g)**. As can be seen, the single mirror plane is broken, by the local lattice distortions from the stresses between the two phases developed on cooling. The local lattice parameters from the simulated pattern are $a = 0.3585$ nm, $b = c = 0.3563$ nm, i.e. a tetragonal distortion, with **a** being the unique axis. The patterns in **Figure 17(c and h)** are less distorted than those in **Figure 17(b and g)** as expected, since they correspond to a point in the γ' farther from the α-Mo/γ' interface. The patterns in **Figure 17(d and e)** are distorted in the opposite sense (i.e. **b** would be the unique axis of the tetragonal distortion), as expected from the form of the stresses generated between

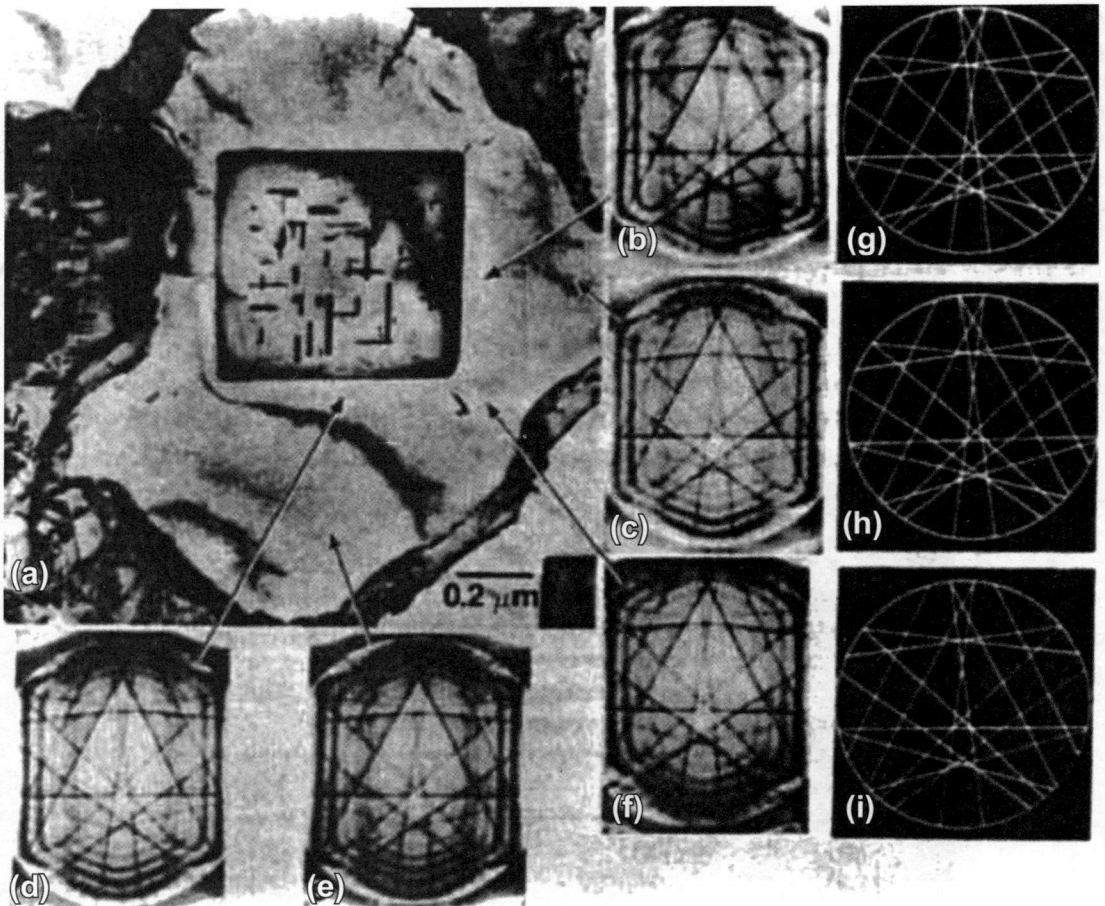

Figure 17   (a) Image, (b–f) experimentally-recorded bright-field disks of CBED patterns, and (g–i) simulated patterns, recorded from the sample as in **Figure 16**.

the two phases. Finally, the pattern obtained, **Figure 17(f)** with the beam positioned parallel to a diagonal of the α-Mo rod, the mirror plane is present, since the γ' in these locations would be equally distorted by the stresses between the two phases.

It appears then that this CBED technique is able to determine the expected lattice distortions in the present example. However, it is of interest to consider the actual values that are obtained for the lattice parameters from simulation. Thus, the **c** axis of the γ' is parallel to the long direction of the α-Mo rod; because of the very small thermal contraction of the rod, it is expected that the γ' will be distorted such that its local **c** lattice parameter would be *larger* than in the unstressed condition. This turns out not to be the case; thus, in the simulations, the **c** lattice parameter is always smaller than the unstressed value. Indeed, if the various sets of lattice parameters are used to calculate the local values of the volumes of unit cells, it turns out that the resultant values are essentially equal. It appears that significant relaxations are occurring in the thin direction of the sample (i.e. in the direction of the c-axis), because of the proximity of the foil surfaces, such that the results obtained of the local lattice distortions represent those in the thin foil rather than the bulk sample. Such surface relaxations can provide data that is not representative in various types of material. For example, there has been considerable interest in strained layer superlattices (SLSs) in device materials. For these materials, it is important to assess the extent of the strain at the interfaces, and one way to provide that assessment would be to measure the distortions (via local lattice parameters) using CBED. However, surface relaxations in thinned SLS samples cause rather improbable results to be obtained, e.g. Si with monoclinic symmetries has been observed, even in foils as thick as 1 μm (Maher et al., 1987). Because of this phenomenon, the results obtained by application of CBED for local lattice parameter determinations must be interpreted with considerable care.

12.2.2.2.2 Crystal Structure Determination

Crystal structure determinations using diffraction (e.g. X-ray) involves noting the angles of the diffracted beams combined with measurements of the intensities of the reflections. In this way, Bragg's Law is used to deduce interplanar spacings and so lattice parameters, and the intensities lead to Structure Factor information. When using TEM, qualitative structure determinations using SAD, especially when including compositional information from either or both X-ray energy dispersive spectroscopy (XEDS) and electron energy loss spectroscopy (EELS), are sometimes possible. However, the problem with using this approach in TEM is that the necessary use of thin foils results in variations in the intensities of reflections because of thickness and orientation. Hence, it is not usually possible to perform *quantitative* and *rigorous* crystal structure determinations using SAD. Fortunately, CBED provides both a rigorous and often fairly straightforward way to gain structural information.

The approach in CBED is based on observations of the various symmetries exhibited in the diffraction patterns leading to the point group of the crystal. Then other information may be available and used to refine the analysis to deduce the space group. The background theoretical description of this CBED technique is rather involved and not described in this chapter. The reader is referred to an original publication from the Steeds group in Bristol (Buxton et al., 1976), and a paper regarding the use of CBED for structure determination (Steeds, 1979). The methodology usually involves five steps:

(1) Recording of one or more zone axis CBED patterns at various camera lengths to permit observation of the projection diffraction, bright field and whole pattern symmetries.
(2) Determination of the point group from the various symmetries exhibited.
(3) Identification of any kinematically forbidden reflections.
(4) Determination of the type of unit cell.
(5) Determination of the space group.

As noted, the theoretical underpinning of this technique is not described here. Rather, the practical approach regarding how the analysis is done is described. There are three tables and one figure that are the essential toolkit required to perform the necessary analyses. The tables, given in Buxton et al. (1976), are:

Table 1, relating the observed symmetries in CBED patterns to possible diffraction groups;

Table 2, relating diffraction groups to possible crystal point groups; and,

Table 3, displaying the possible diffraction groups that may be exhibited in given electron-beam directions for the various point groups.

The figure is given in **Figure 18**, taken from Henry and Lonsdale (1952), and displays the possible screw axes and glide planes for the various point groups.

   The use of these tables is described here by reference to an example involving the identification of the phase Ni_3Mo in a binary Ni–25Mo (at%) alloy (Kaufman et al., 1983).

   Step 1: *Recording the CBED patterns*: CBED patterns are collected with the electron beam parallel to low-index zone axes; the question is, how many zone axes, especially as it was originally claimed that perhaps one zone axis would be sufficient. While in some very few cases, one zone axis will suffice, the answer is: as many zone axes as is required! This will become evident in the example, where zone axis diffraction patterns were recorded parallel to [100], [010], [001], [101], [110], and [011] (the zone axis identification being confirmed after the complete analysis). Here, it can be noted that the angle between the first three zone axes was mutually 90°, and they are shown in **Figure 19(a–f)**. In the case of the [010] and [001] zone axes, no useful information (i.e., a pattern of *holz* lines) was displayed in the bright field symmetry. This raises an interesting point about the visibility of *holz* lines in bright field disks. In actual fact, these lines are present in essentially all bright field disks (from crystalline materials) but their visibility depends on the overall signal-to-noise ratio (SNR), and hence the background intensity. For example, viewing Si with the electron beam parallel to ⟨100⟩ at room temperature and at an accelerating voltage of ≈ 100 kV, no *holz* lines are visible on the screen or on photographic film. However, if the temperature is reduced in a cold stage (to reduce inelastic scattering contributions to the background), or if a *ccd*-camera is used to record the image (where the background can be arbitrarily reduced), then the lines become visible. In the present example, neither a cold stage nor a *ccd*-camera was available.

   Step 2: *Deducing the point group*: Consider the symmetries displayed in the following figures:

(1) From **Figure 19(a)**, the projection diffraction symmetry (i.e. the zero-order Laue zone symmetry, excluding the *holz* lines in the bright field disk) is *2mm* (i.e. two mirror planes in the pattern and the zone axis is a two-fold axis of symmetry). Using **Table 1**, the projection diffraction group is $2mm1_R$, with possible diffraction groups being $2m_Rm_R$, $2mm$, 2_Rmm_R and $2mm1_R$. The symmetries exhibited by the bright disk and the whole pattern are both $2mm$, so that from **Table 1**, the possible diffraction groups are reduced to $2mm$ and $2mm1_R$. The same reasoning applies to the results from **Figure 19(b and c)**.

(2) For the zone axes in **Figure 19(d–f)**, the projection diffraction symmetry is $2mm$, the whole pattern m, and the bright field symmetry is also m. For each of these zone axes, a–d, again referring to **Table 1**, the symmetry information is consistent with a deduced diffraction group of 2_Rmm_R. **Table 2**, which relates point groups with their diffraction groups, is now used.

For the zone axes ⟨100⟩ and ⟨110⟩, the table yields the possible point groups, listed here in **Table 4**. For the point groups $4/m$, $4/mmm$, $\bar{3}m$, $6/m$, $\bar{6}m2$, $6/mmm$, $m3$ and $m3m$, all would require higher symmetries to be exhibited in the zone-axis patterns (the expected symmetries can be noted from **Table 3**). For example, for the point group $4/m$, while a deduced diffraction group of 2_Rmm_R would be

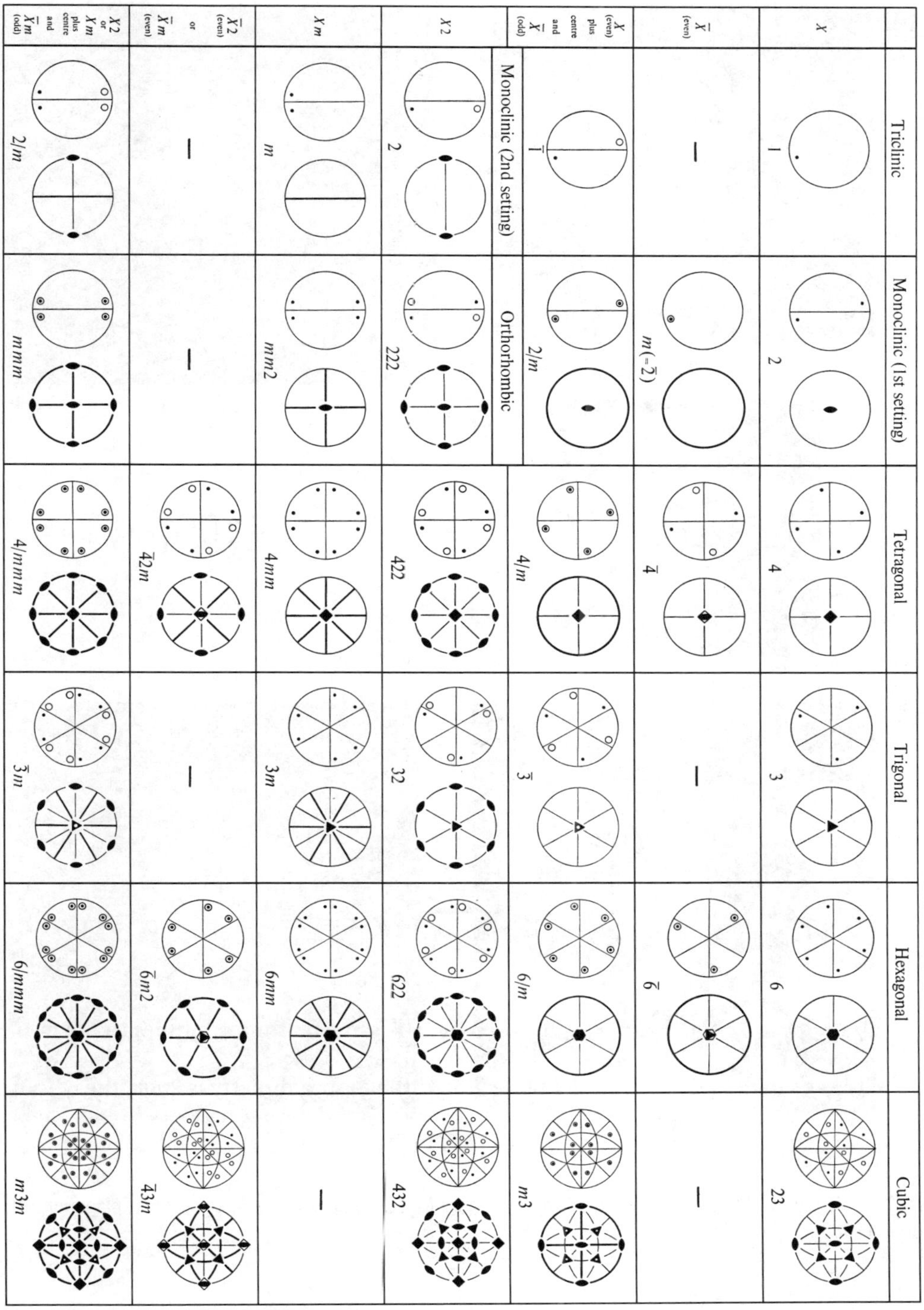

Figure 18 Diagram depicting the possible screw axes and glide planes for the various point groups. Taken from Henry and Lonsdale (1952).

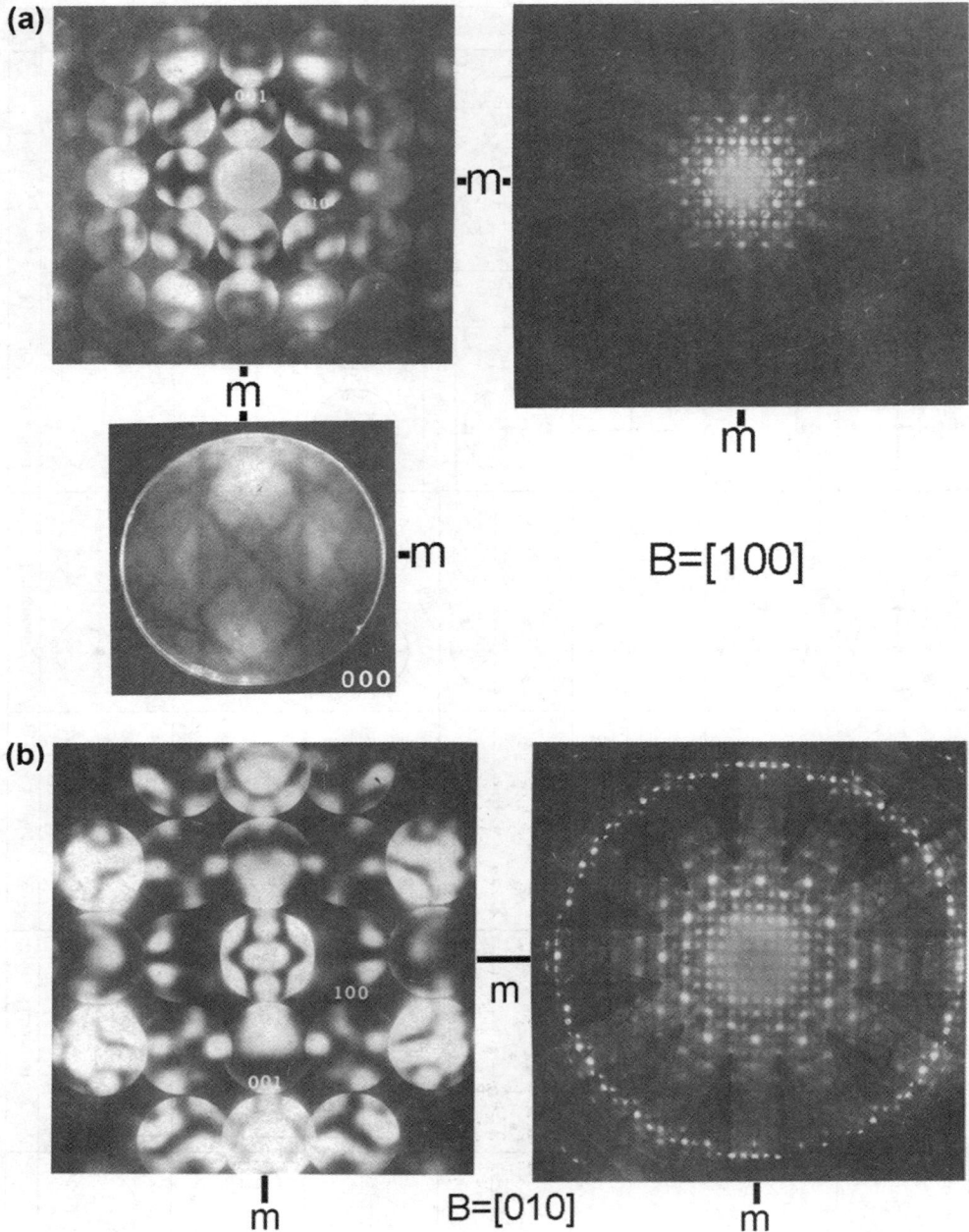

Figure 19 (a–f) Various CBED patterns recorded from a binary Ni–25Mo at.% alloy. See text for explanation. Taken from Kaufman et al. (1983).

(c)

B=[001]

(d)

B=[010]

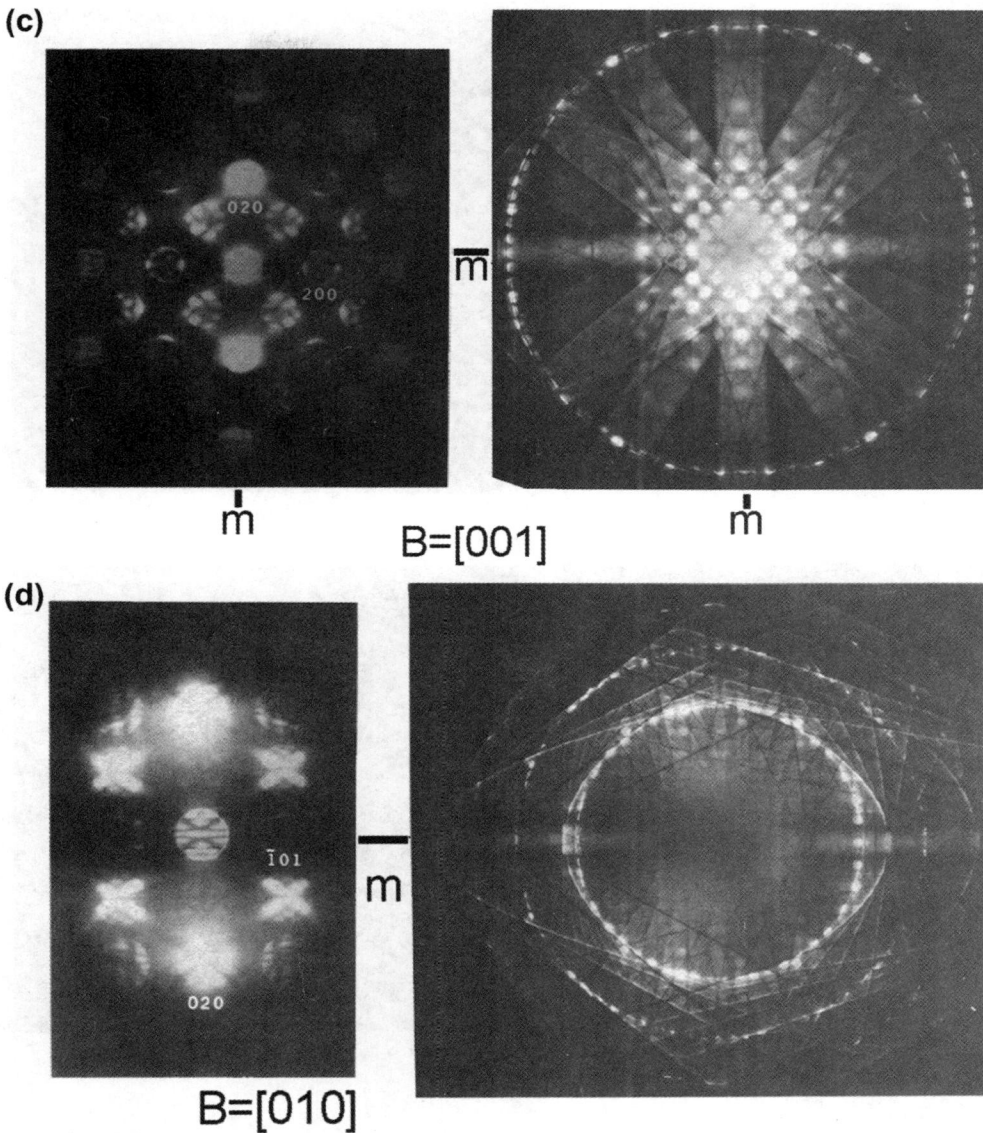

Figure 19   (*continued*).

expected with the electron beam parallel to ⟨110⟩, a diffraction group of 41_R would be expected when the electron beam is parallel to [001]; for this latter diffraction group, from **Table 1**, a fourfold axis of symmetry would be expected to be observed in both the whole pattern and bright field symmetries. Because these higher order symmetries were not observed, these point groups can be discounted. It follows that the only point group consistent with both deduced diffraction groups is *mmm*.

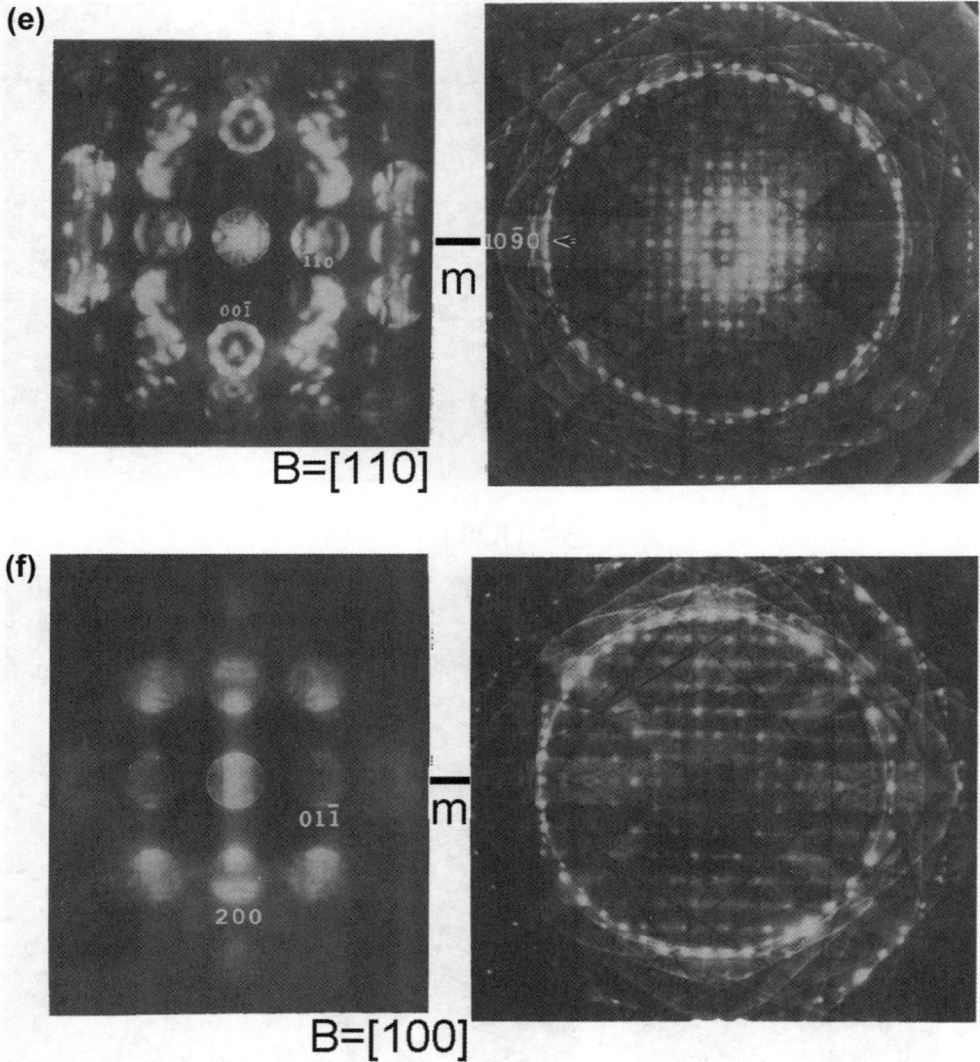

(e) B=[110]

(f) B=[100]

Figure 19 (*continued*).

Step 3. *Identification and interpretation of kinematically forbidden reflections*: There are two ways in which a kinematically forbidden reflection may be identified. Firstly, if such a reflection were to be present, the intensity observed would occur by double diffraction. Titling the thin foil about the systematic row containing the suspected forbidden reflection would result in its disappearance (as the necessary reflections which are doubly diffracted into this **g** are no longer excited). Secondly, very often, kinematically forbidden reflections contain lines of deficient intensity running parallel to the systematic row containing the reflection, or a dark cross. These are known as dynamic absences, and examples are immediately apparent in the 010 and 100 reflections, **Figure 19(a and b)**, respectively. These dynamic

Table 1 Pattern symmetries, where a dash appears in column 7, the special symmetries can be deduced from columns 5 and 6 of this table

diffraction group	bright field	whole pattern	dark field general	dark field special	±G general	±G special	projection diffraction group
1	1	1	1	none	1	none	1_R
1_R	2	1	2	none	1	none	
2	2	2	1	none	2	none	21_R
2_R	1	1	1	none	2_R	none	
21_R	2	2	2	none	21_R	none	
m_R	m	1	1	m	1	m_R	$m1_R$
m	m	m	1	m	1	m	
$m1_R$	2mm	m	2	2mm	1	$m1_R$	
$2m_R m_R$	2mm	2	1	m	2	—	$2mm1_R$
$2mm$	2mm	2mm	1	m	2	—	
$2_R mm_R$	m	m	1	m	2_R	—	
$2mm1_R$	2mm	2mm	2	2mm	21_R	—	
4	4	4	1	none	2	none	41_R
4_R	4	2	1	none	2	none	
41_R	4	4	2	none	21_R	none	
$4m_R m_R$	4mm	4	1	m	2	—	$4mm1_R$
$4mm$	4mm	4mm	1	m	2	—	
$4_R mm_R$	4mm	2mm	1	m	2	—	
$4mm1_R$	4mm	4mm	2	2mm	21_R	—	
3	3	3	1	none	1	none	31_R
31_R	6	3	2	none	1	none	
$3m_R$	3m	3	1	m	1	m_R	$3m1_R$
$3m$	3m	3m	1	m	1	m	
$3m1_R$	6mm	3m	2	2mm	1	$m1_R$	
6	6	6	1	none	2	none	61_R
6_R	3	3	1	none	2_R	none	
61_R	6	6	2	none	21_R	none	
$6m_R m_R$	6mm	6	1	m	2	—	$6mm1_R$
$6mm$	6mm	6mm	1	m	2	—	
$6_R mm_R$	3m	3m	1	m	2_R	—	
$6mm1_R$	6mm	6mm	2	2mm	21_R	—	

absences are due to either a twofold screw axis parallel to the systematic row containing the dynamic absence, or a glide plane containing both the direction parallel to the systematic row and the beam direction.

As noted above, a dynamic absence in a given reflection may be interpreted on the basis of either a twofold screw axis parallel to the systematic row containing that reflection, or as a glide plane containing both the direction of the systematic row and the electron beam direction. **Figure 18** (displaying the various screw axes/glide planes for the point groups) gives details of the possible combinations of screw axes and glide planes, and for the point group *mmm* (i.e. for the example being discussed), the

Table 2   Relation between the diffraction groups and the crystal point groups

Diffraction group	1	1̄	2	m	2/m	222	mm2	mmm	4	4̄	4/m	422	4mm	4̄2m	4/mmm	3	3̄	32	3m	3̄m	6	6̄	6/m	622	6mm	6̄m2	6/mmm	23	m3	432	4̄3m	m3m
6mm1$_R$																											X					
3m1$_R$																										X						
6mm																									X							
6m$_R$m$_R$																								X								
61$_R$																							X									
31$_R$																						X										
6																					X											
6$_R$mm$_R$																				X												X
3m																			X												X	
3m$_R$																		X												X		
6$_R$																	X												X			
3																X												X				
4mm1$_R$															X																	X
4$_R$mm$_R$														X																	X	
4mm													X																			
4m$_R$m$_R$												X																		X		
41$_R$											X																					
4$_R$										X																						
4									X																							
2mm1$_R$								X							X												X		X			X
2$_R$mm$_R$					X															X												
2mm							X							X																		
2m$_R$m$_R$						X						X												X				X		X		
m1$_R$							X						X												X						X	
m				X															X			X										
m$_R$			X						X	X											X					X						
21$_R$					X																											
2$_R$		X			X			X			X				X		X			X			X				X		X			X
2			X															X														
1$_R$				X																												
1	X		X	X		X	X		X	X			X	X		X		X	X		X	X		X	X	X		X		X	X	

possible symmetry elements are shown in **Figure 20**. As can be seen, twofold screw axes parallel to, and glide planes normal to, ⟨100⟩ are permitted.

The following dynamic absences are present in the zone-axis diffraction patterns (**Figure 19(a–f)**):

(1) A dynamic absence may be observed in the *010* reflection in both the [100] and [101] patterns.
(2) Similarly, a dynamic absence may be observed in the *100* reflection in the [010] pattern.
(3) In the [110] pattern, a dynamic absence is observed in the 101̄0 reflection.

Table 3   Zone axis symmetries

point group	$\langle 111 \rangle$	$\langle 100 \rangle$	$\langle 110 \rangle$	$\langle uvo \rangle$	$\langle uuw \rangle$	$[uvw]$
m3m	$6_R mm_R$	$4mm1_R$	$2mm1_R$	$2_R mm_R$	$2_R mm_R$	2_R
$\bar{4}3m$	$3m$	$4_R mm_R$	$m1_R$	m_R	m	1
432	$3m_R$	$4m_R m_R$	$2m_R m_R$	m_R	m_R	1

point group	$\langle 111 \rangle$	$\langle 100 \rangle$	$\langle uvo \rangle$	$[uvw]$
m3	6_R	$2mm1_R$	$2_R mm_R$	2_R
23	3	$2m_R m_R$	m_R	1

point group	$[0001]$	$\langle 11\bar{2}0 \rangle$	$\langle 1\bar{1}00 \rangle$	$[uv.o]$	$[uu.w]$	$[u\bar{u}.w]$	$[uv.w]$
6/mmm	$6mm1_R$	$2mm1_R$	$2mm1_R$	$2_R mm_R$	$2_R mm_R$	$2_R mm_R$	2_R
$\bar{6}m2$	$3m1_R$	$m1_R$	$2mm$	m	m_R	m	1
6mm	$6mm$	$m1_R$	$m1_R$	m_R	m	m	1
622	$6m_R m_R$	$2m_R m_R$	$2m_R m_R$	m_R	m_R	m_R	1

point group	$[0001]$	$[uv.o]$	$[uv.w]$
6/m	61_R	$2_R mm_R$	2_R
$\bar{6}$	31_R	m	1
6	6	m_R	1

point group	$[0001]$	$\langle 11\bar{2}0 \rangle$	$[u\bar{u}.w]$	$[uv.w]$
$\bar{3}m$	$6_R mm_R$	21_R	$2_R mm_R$	2_R
3m	$3m$	1_R	m	1
32	$3m_R$	2	m_R	1

point group	$[0001]$	$[uv.w]$
$\bar{3}$	6_R	2_R
3	3	1

point group	$[001]$	$\langle 100 \rangle$	$\langle 110 \rangle$	$[uow]$	$[uvo]$	$[uuw]$	$[uvw]$
4/mmm	$4mm1_R$	$2mm1_R$	$2mm1_R$	$2_R mm_R$	$2_R mm_R$	$2_R mm_R$	2_R
$\bar{4}2m$	$4_R mm_R$	$2m_R m_R$	$m1_R$	m_R	m_R	m	1
4mm	$4mm$	$m1_R$	$m1_R$	m	m_R	m	1
422	$4m_R m_R$	$2m_R m_R$	$2m_R m_R$	m_R	m_R	m_R	1

point group	$[001]$	$[uvo]$	$[uvw]$
4/m	41_R	$2_R mm_R$	2_R
$\bar{4}$	4_R	m_R	1
4	4	m_R	1

point group	$[001]$	$\langle 100 \rangle$	$[uow]$	$[uvo]$	$[uvw]$
mmm	$2mm1_R$	$2mm1_R$	$2_R mm_R$	$2_R mm_R$	2_R
mm2	$2mm$	$m1_R$	m	m_R	1
222	$2m_R m_R$	$2m_R m_R$	m_R	m_R	1

point group	$[010]$	$[uow]$	$[uvw]$
2/m	21_R	$2_R mm_R$	2_R
m	1_R	m	1
2	2	m_R	1

point group	$[uvw]$
$\bar{1}$	2_R
1	1

Table 4   All possible crystal point groups tabulated for the given diffraction groups. The crystal is orthorhombic and therefore the only consistent point group is mmm

		Possible crystal-point groups										
Zone axis	Deduced diffraction group	2/m	mm2	mmm	4/m	4/mmm	$\bar{3}m$	6/m	6m2	6/mmm	m3	m3m
⟨100⟩	2mm1$_R$			×		×				×	×	×
	2mm		×						×			
⟨110⟩	2$_R$mm$_R$	×		×	×	×	×	×		×	×	×

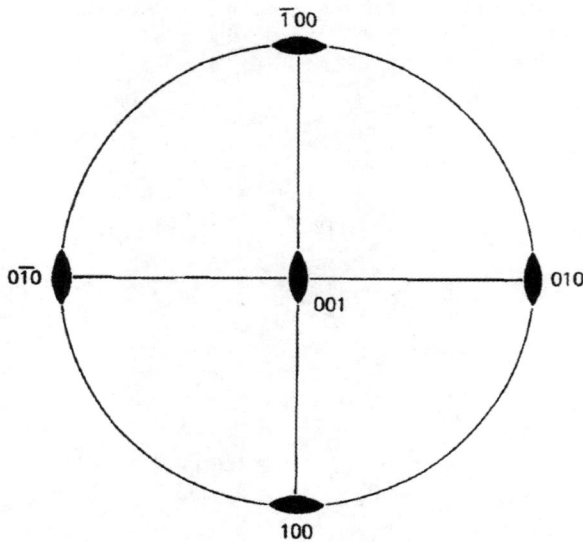

Figure 20   A diagram depicting the possible screw axes and glide planes for the *mmm* point group. Taken from Kaufman et al. (1983).

The first of these can be interpreted as being due to a twofold screw axis parallel to [010] because it is observed in both patterns, [010] being the axis of tilting between the two zone axes. It is also permitted in the *mmm* point group (see **Figure 20**). The dynamic absence in the *100* reflection in the [010] zone-axis pattern may be interpreted as either a twofold screw axis parallel to [100], or a glide plane with normal [001]. Although both are permitted by the point group (**Figure 20**), it is possible to distinguish between these possibilities. Thus, a convergent beam pattern was recorded near to a [011] zone axis, where the foil had been tilted slightly about an axis parallel to [0$\bar{1}$1] in order to orient the $\bar{5}$00 reflection to the Bragg condition. As observed in **Figure 21**, a dynamic absence is observed. This can only be interpreted as being due to a twofold screw axis parallel to [100], because the alternative, a glide plane with normal [0$\bar{1}$1] is not permitted in the *mmm* point group (**Figure 20**). Finally, the dynamic absence in the 10$\bar{9}$0 reflection can only be due to a glide plane with normal [001], as the alternative is not permitted for this point group. In summary, there is evidence for two

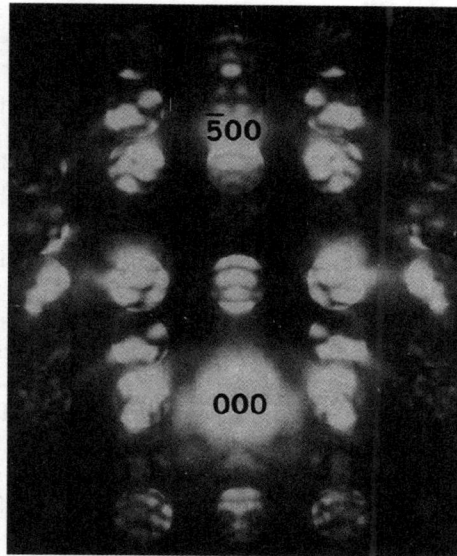

Figure 21   A convergent beam pattern recorded near to a [011] zone axis, where the foil had been tilted slightly about an axis parallel to [0$\bar{1}$1] in order to orient the reflection to the Bragg condition. Note that a dynamic absence is observed in that disk. Taken from Kaufman et al. (1983).

twofold screw axes, parallel to [010] and [100], and also a (001) glide plane. These various dynamic absences lead to a rule for kinematically forbidden reflections, namely $h00$, $0k0$, h or $k = 2n + 1$; $hk0$, $h + k = 2n + 1$.

Step 4. *Determining the type of unit cell*: It is sometimes possible to identify the nature of the unit cell, i.e. whether it is primitive, face-centered, or body-centered from the convergent beam patterns recorded parallel to $\langle 001 \rangle$. This is done by projecting the reciprocal lattice disks from the *folz*, or *solz*, onto the *zolz* reflections. If these reciprocal lattice disks lie directly on the reciprocal lattice points in the *zolz*, then the unit cell is most probably primitive. If these *holz* reciprocal lattice points lie at the face-centered positions (in projection), then this would be consistent with a face-centered reciprocal lattice and so a body-centered cell in real space. Similarly, if the *holz* reflections project to body-centered positions in the *zolz*, then the real space cell will be face-centered. In the case of the present example, the cell appears to be primitive.

Step 5. *Determination of the space group*: The space group is now determined by making use of the various pieces of the puzzle. In the case of the example, the space group is *Pmmm*, which follows from the point group, *mmm*, the cell type, primitive, and the rules derived for the kinematically forbidden reflections from the dynamic absences, i.e. $h00$, $0k0$, h or $k = 2n + 1$; $hk0$, $h + k = 2n + 1$. Recall that all of this information has been derived *without making detailed intensity measurements* as in the case of X-ray diffraction.

The example that was chosen was rather detailed, but does show how the various steps in the approach lead to a number of clues about the structure. The approach does avoid the difficulty associated with the variation of diffracted intensity as a function of both sample thickness and orientation when using a thin foil in the TEM.

12.3   Imaging in Transmission Electron Microscopy

There are a number of imaging modes that may be exercised in the (S)TEM instrument and these have been described in detail in a number of books, for example Williams and Careter (2009). Many of these rely on the introduction of an objective aperture to exclude one or more beams from reaching the image, so that contrast from scattering events may be observed. For materials sciences (i.e. physical metallurgy), the imaging technique most employed over the years has been that involving diffraction contrast, where the contrast observed has as its origin the influence of crystal defects in the foil on the diffraction process. For example, the bending of planes about a dislocation core can cause these planes to be tilted such that Bragg's Law is satisfied in this local region, causing electrons to be diffracted. More recently, with the advent of microscopes with high spatial resolution and useful STEM systems, high-resolution transmission electron microscopy (HRTEM) and high-resolution scanning transmission electron microscopy (HRSTEM) have been increasingly applied to the characterization of samples in materials sciences. For metallic samples, as will be seen, there are several difficulties associated with the application of these techniques compared with oxide and semiconductor/device materials, including the fact that metallic thin foils are usually not very flat and that the atomic spacings are often somewhat smaller than those of oxide and semiconducting materials. In the following, because diffraction contrast imaging has been so well described in the literature, emphasis here is placed on a general understanding of the basis of the technique, and where common mistakes that may be made are identified. This is followed by a short description of high-resolution imaging.

12.3.1   The Origin of Fringe-Like Contrast

One of the most immediate features that can be noted of an image of a dislocation inclined in a thin foil is that it can exhibit fringe-like contrast, as shown in **Figure 22**. As such observations are made in

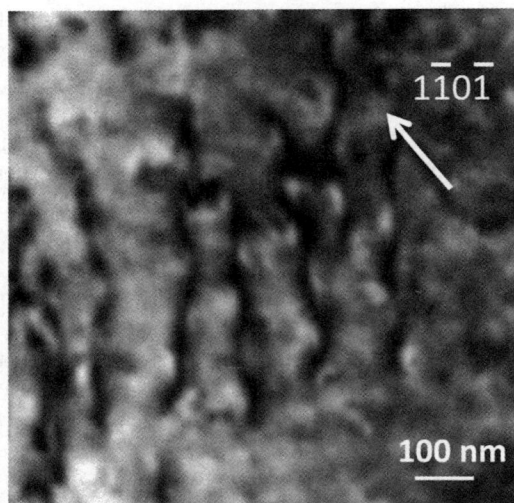

Figure 22   Bright-field TEM micrograph of dislocations from a sample of Ti imaged with **g** = $1\bar{1}0\bar{1}$ with a small deviation from Bragg. Note the "zigzag" contrast.

increasingly thick foils, this fringe-like contrast decreases in intensity, so that, for example, a dislocation image tends to resemble a dark line, actually a more intuitive image of a dislocation. It is important to understand the origin of the contrast in order to be able to interpret images correctly.

The most simple account of electron diffraction, and so imaging, involves the kinematical theory, which is based on contrast being derived from a single scattering event, i.e. there is no multiple scattering and no interaction between the direct and diffracted beams. A useful account of the kinematical theory of image contrast is given in Hirsch et al. (1977), where it is predicted that the intensity of the scattered electrons varies as $\frac{\sin^2(\pi ts)}{(\pi s)^2}$, where t is the foil thickness and s is the magnitude of the vector describing the deviation of the foil from the exact Bragg condition for the given set of diffracting planes. It follows that the diffracted intensity, and so the intensity of the direct, forward scattered beam (i.e. the BF beam), varies in a sinusoidal manner with both foil thickness and orientation. It would be expected then that contrast would be observed in an image of a thin foil when either the thickness and/or the orientation of the foil would vary. In a perfect crystal (i.e. one without crystal defects present), it would not be surprising to observe thickness fringes and/or bend contours, as shown in **Figure 23(a and b)**. The problem is that the details of these images cannot be explained in detail based on the kinematical theory. For example, it may be noted in **Figure 23(a)** that the contrast associated with the thickness fringes tends to disappear, despite the fact that the foil is still increasing in thickness and that there is still considerable background intensity in the image (and crystal defects still may be observed clearly). The inability of the kinematical theory to predict image contrast requires a more detailed theory to be considered, which accounts for multiple scattering events and dynamical interactions between the BF and various diffracted beams.

The dynamical theory of electron diffraction provides a fairly accurate description of electron scattering, and so accounts for image contrast quantitatively. For a detailed description of the theory, the reader is referred to De Graef (2003); here, a "physical" picture is described in the absence of a derivation of the theory (essentially, no equations!). In the dynamical theory, the incident electrons, represented by a wave vector **K** (of magnitude $1/\lambda$, the electron wavelength, after adjustment for the refractive index effect, which occurs at the entrance surface of the foil), interact with the periodic crystal potential. Applying quantum mechanics (i.e. Schrödinger's equation) to this problem of fast electrons interacting with crystal electrons results in a solution that reveals that inside the crystal, the incident electrons consist of an infinite number of Bloch waves, each characterized by an amplitude C_g and wave vector k; the number of Bloch waves is taken to be equal to the number of diffracted beams (including in that number the BF beam). Frequently, defect analysis is performed with only two beams with significant intensity (the so-called two-beam condition, i.e. with bright- and DF beams), and so there would be two significant Bloch waves excited. While the concept of forward scattered and diffracted beams may be simple to grasp, where are the Bloch waves, and what relationship do they have with the various diffracted beams? It turns out from the theory that each Bloch wave contributes a certain amplitude to each of the scattered beams (e.g. in the BF and diffracted beams), that amplitude being given by the C_g term, where "g" refers to the given scattered beam (i.e., g=0 being the BF and g=g being the DF beam). So, each diffracted beam will have an intensity equal to the sum of the contributions from each of the Bloch waves. It should be noted that each Bloch wave is characterized by its wave vector, each of magnitude $1/\lambda$, and so it might be considered that the energy of each of the Bloch waves would differ, involving energy loss. This is not the case, as the potential energy of each of the Bloch waves also differs, such the total energy of each is unchanged. Since a given Bloch wave provides amplitude to each of the scattered beams, it may be recognized that dynamical interactions between the scattered beams occurs through variations in the term C_g.

Consider a two-beam condition, i.e. one with the BF beam and one strongly diffracted beam. It follows that there will be contributions from two Bloch waves (denoted 1 and 2, respectively). In each

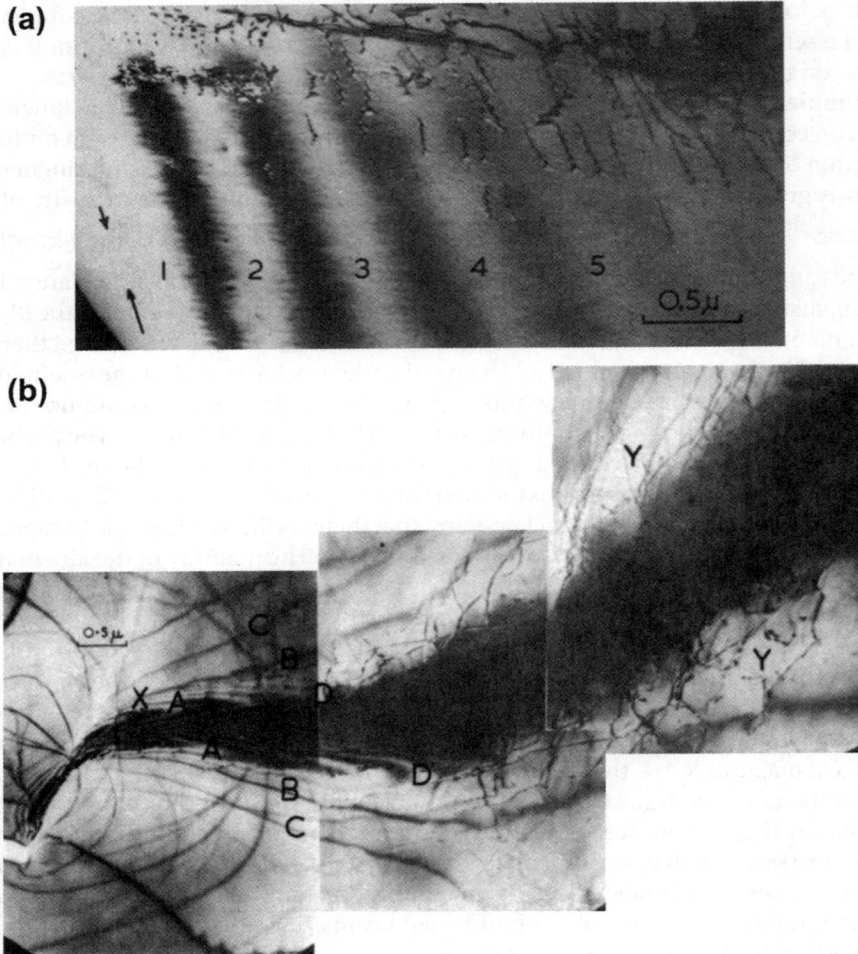

Figure 23 Images of "perfect" crystals showing the intensity variation in (a) thickness and (b) orientation (i.e. a bend contour). Taken from Hirsch et al. (1977).

beam, there will be two waves, each with a given wavelength, and so it is expected that in the resulting intensities of the *beams*, there will be a beating effect between the Bloch waves of different wavelength. It is this beating effect that gives rise to fringes in images, for example, to thickness fringes, as shown in **Figure 23(a)**. It may be shown that the two-beam extinction distance (i.e. the increase in foil thickness required for the intensity in thickness fringes to vary from one maximum to the next) for a given reflection (referenced by its reciprocal lattice vector, **g**), ξ_g, is at a maximum at the Bragg condition, and decreases with deviation from the Bragg orientation. From the dynamical theory, both the wave vectors and wave amplitudes of the Bloch waves in the various diffracted beams vary with orientation, and so bend contours occur in foils that are not flat (i.e. with variations of orientation). It turns out (from theory) that the *excitation* of each Bloch wave varies with orientation, and is given by the amplitude of the given Bloch wave in the BF beam (i.e., C_0). Again, from the theory, the excitation of Bloch wave 1 will be at a maximum if the foil is oriented to a negative deviation from a given Bragg condition, whereas Bloch wave 2 has its maximum excitation at a positive deviation from that Bragg condition, as depicted in **Figure 24**. The spatial distribution of the Bloch waves may be plotted over the scattering planes, with the result shown in **Figure 25**, from which it may be seen that Bloch wave 1 has a maximum intensity over the scattering planes. It is in the vicinity of the atomic centers that much of the inelastic scattering occurs, so that with increasing thickness of foil, the intensity remaining in Bloch wave 1 is expected to be significantly reduced by such scattering of electrons outside the objective aperture. In contrast, Bloch wave 2 is located mainly between the scattering planes, and away from inelastic scattering centers. Consequently, the intensity of this Bloch wave, as the foil thickness is increased, is not expected to decrease in the same way as for Bloch wave 1. This apparent difference in the reduction in intensity of given Bloch waves is known as "anomalous absorption" (note that absorption of fast electrons in thin foils is not considered to occur, but the term is used to describe scattered electrons that are stopped by the objective aperture).

The presence of Bloch waves also explains the anomalies observed in the images in **Figure 23**. Consider the thickness fringes in **Figure 23(a)**. The fringes are formed by the beating effect from the difference in wavelengths of the two prominent Bloch waves. The fringes are of strong contrast in the thin parts of the sample, since the attenuation of Bloch wave 1, for the given crystal orientation, will not be significant. However, as the thickness increases, the intensity in Bloch wave 1 will be increasingly reduced, and the beating effect will be far less significant. However, although the beating effect ceases to be observed, there is still significant intensity in Bloch wave 2, and so there is considerable intensity available for observation of crystal defects. Regarding bend contours, their behavior may be understood by reference to the rocking curve shown in **Figure 26**. Thus, for negative deviations from the Bragg condition, **Figure 24** is consistent with Bloch wave 1 being highly excited, but, of course, strongly "absorbed" as expected from anomalous absorption. Consequently, the overall intensity for these orientations is expected to be very low. In contrast, for orientations corresponding to positive deviations from Bragg, Bloch wave 2 is highly excited and weakly absorbed, resulting in a maximum in the

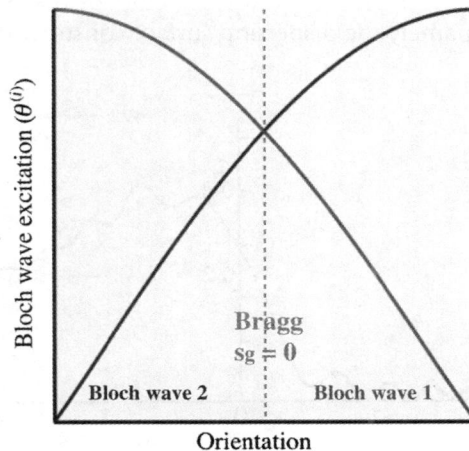

Figure 24 Schematic diagram showing the dependence of the excitation of the two Bloch waves as a function of orientation (about the Bragg position).

(a)   **(b)**

$k^{(1)}$ $k^{(1)} + g$ $k^{(2)}$ $k^{(2)} + g$

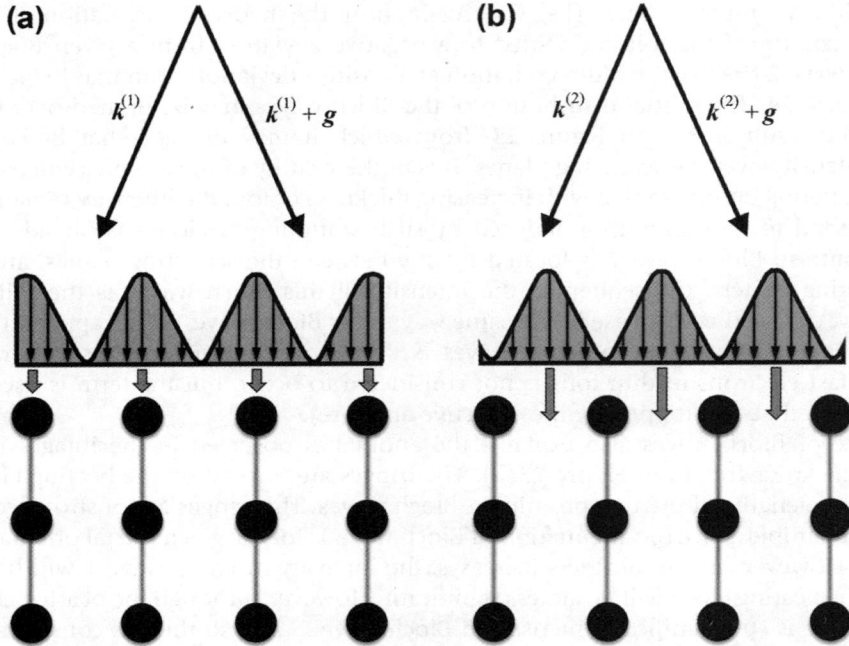

Figure 25 Schematic diagram showing the spatial distribution of the Bloch waves plotted over the scattering planes (a) the spatial distribution of intensity in Bloch wave 1; (b) the distribution in Bloch wave 2. Taken from Williams and Carter (2009).

intensity. Hence, an asymmetric appearance of the rocking curve is observed. It should be emphasized that the behaviors of thickness fringes and bend contours (rocking curves) are very difficult to explain without reference to the dynamical theory and the role of Bloch waves.

There is one last consideration involving the wave vectors of the Bloch waves of which physical metallurgists should be aware, namely the dispersion surface construction, required to understand the

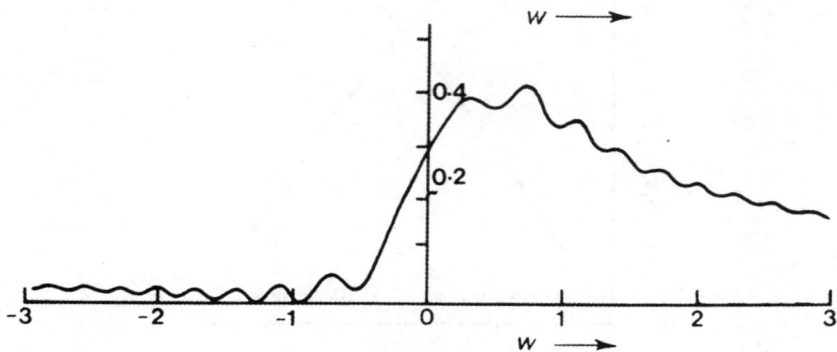

Figure 26 A rocking curve showing the scattered intensity of the bright-field beam as a function of orientation about the Bragg condition. Taken from Hirsch et al. (1977).

contrast exhibited by stacking faults. In solid state, or metal physics, courses, one is very familiar with the notion of forbidden energy gaps and Brillouin zones and Brillouin zone boundaries when plotting energy in "k" space (k being the wave number or wave vector). A similar consideration of electron energy in wave vector space for the high energy electrons in a microscope results in a diagram of a dispersion surface such as that shown in **Figure 27**, corresponding to a two-beam condition and consisting of two branches that depict the forbidden energy gap. The branches are simply the loci of the ends of the two wave vectors of the two prominent Bloch waves (two-beam condition). At the Bragg condition, the separation of the branches is at a minimum, and the way in which the Bloch wave vectors vary with orientation can clearly be noted, i.e. the separation of the branches increases with deviation from the Bragg condition, and so the wave vectors change in value accordingly. Actually, the extinction distance discussed above varies as the inverse of the separation of the branches of the dispersion surface, such that as noted above, it is at a maximum value when the Bragg condition is met (i.e. minimum spacing between the branches), and decreases with deviations from the Bragg condition. Defects can cause intraband and interband scattering, i.e. scattering within a given branch or between two branches. This point is discussed below when contrast in images of stacking faults is considered.

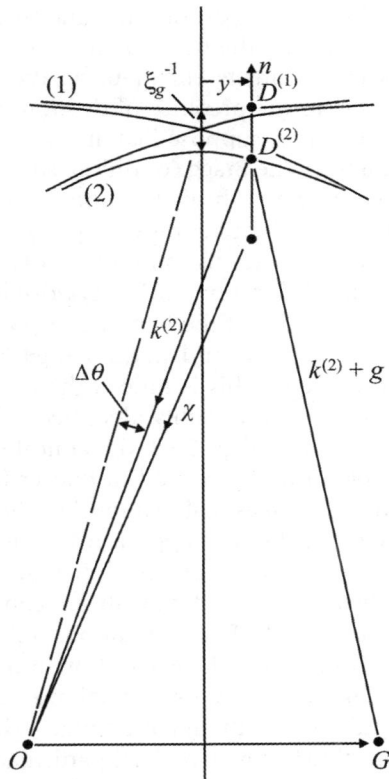

Figure 27 A schematic diagram of the dispersion surface for two Bloch waves (i.e. two-beam condition) near to the Bragg condition. Taken from Hirsch et al. (1977).

12.3.2   Diffraction Contrast

As noted above, imaging using diffraction contrast is based on the influence of crystal defects in the foil on the diffraction process, and, of course, the use of an objective aperture. Generally, there are two types of defects that may be considered, one set involving planar defects, e.g. stacking faults, in which the displacement of the lattice is discontinuous, occurring at the plane of the defect, and the other set, where the displacement field associated with the defect is continuous, e.g. dislocations. In most of the imaging experiments, two-beam conditions (i.e. the BF and one diffracted beam) are employed, so that the influence of the given displacements of the defect on a given diffracting plane may be determined. The dynamical theory of electron diffraction may be used to predict the way in which these various displacement fields influence the diffraction process, and so imaging, and this has resulted in a series of rules that may be used to interpret images recorded in a given experiment. These have been clearly described in Loretto (1984), which gives an excellent and straightforward account of diffraction contrast. In the following, defect analysis using this contrast mechanism is described, and solutions to possible ambiguities outlined.

12.3.2.1   *Characterization of Dislocations*

Dislocations are crystal defects with continuously varying displacement fields; the most significant displacements are in the vicinity of the core of a dislocation, and these decrease with distance from the defect. The major displacement is defined by the Burgers vector, **b**, of the dislocation. In this text, the Burgers vector is defined by drawing a Burgers circuit using the FS/RH convention. The simplest explanation of diffraction contrast due to the presence of dislocations is to consider the effect of the displacement on the diffracting planes. Thus, suppose that, in a given crystal, a set of diffracting planes is oriented with a sufficient deviation from the Bragg condition, such that there is little or no diffraction into the diffracted beam. If the displacement of the dislocation is such as to bend the diffracting planes into the Bragg condition, then diffraction will occur, providing a reduced intensity in the BF beam (i.e., a dark image of the dislocation). Of course, bearing in mind the discussion above regarding fringe-like contrast, one would expect that as the dislocation varies its position with depth in the foil, e.g. an inclined defect, and local orientation because of the bending of planes, that variations in the beating effect between the two Bloch waves would occur, and, in thinner parts of the foil, there would be fringe-like contrast exhibited in the defect's image. This is indeed the case, as exhibited by the dislocation images in **Figure 22**; this type of contrast is very readily identified, being characterized by the so-called zigzag alternating black and white intensity lobes. Such dynamical contrast effects (i.e. the fringe-like contrast) would be expected to be significantly reduced in thicker foils when anomalous absorption would result in only one Bloch wave having reasonable intensity. The actual form of the image contrast in thinner foils is quite complicated and the best approach to interpreting the exact nature of such images is to use image simulation based on the dynamical theory.

Images of dislocations are normally recorded under two-beam conditions, with the diffracting planes oriented with a small positive deviation from the Bragg condition. This condition is depicted on the Ewald Sphere diagram shown in **Figure 28**. The small positive deviation from the Bragg condition is denoted by the vector s, the deviation parameter introduced in the diffraction section of this chapter. To understand why a two-beam condition is employed, with the BF and one diffracted beam excited, it is important to note the influence of the defect on the diffraction behavior of those particular planes (see defect characterization below). The diffracting planes are oriented with a small positive deviation from Bragg in order to maximize the contrast of the scattering by the defect. Thus, as seen in the rocking curve depicted schematically in **Figure 26**, because of anomalous absorption and the excitation of Bloch wave two for positive deviations

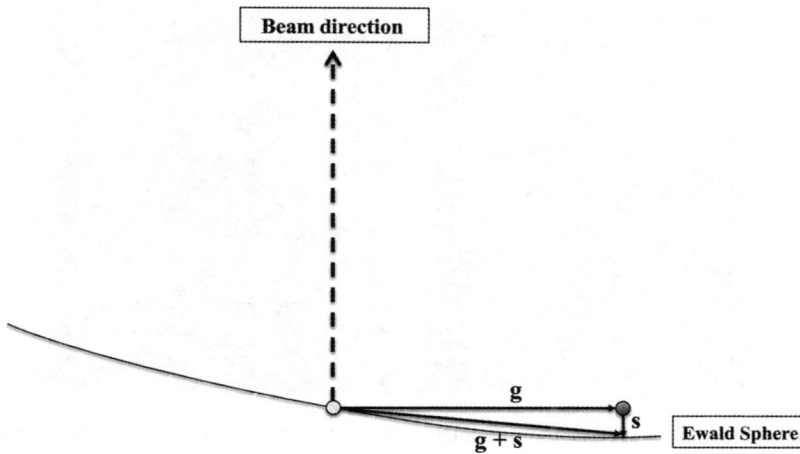

Figure 28 The condition for imaging involving two-beam conditions, with the diffracting planes oriented with a small positive deviation from the Bragg condition, is depicted on an Ewald Sphere diagram.

from Bragg, the intensity is maximized in this orientation. This would be the background intensity in the image. A dislocation would, in general, cause the diffracting planes to be locally tilted into the Bragg condition, where the intensity is lower than that for positive deviations (see **Figure 26**). Hence, there will be a maximum in the contrast between the dark image of the defect and the bright background intensity.

Dislocation characterization generally involves a determination of the Burgers vector and the dislocation line direction. The latter exercise involves crystallographic analysis, involving the imaging of the defect line in two different incident beam directions. The real direction lies in both of the two planes that each contains the projected direction and the beam direction for each of the imaging cases, i.e. it will be the line of intersection of those two planes. This exercise is described in Loretto (1984), and is most easily done using a stereographic projection, or a computer version of the stereographic analysis as in *Desktop Microscopist.*

The determination of the Burgers vector, **b**, involves the use of the so-called invisibility criterion. This criterion is based on the rendering of a dislocation invisible by imaging using a set of diffracting planes (represented by their reciprocal lattice vector, **g**, which is also their normal) which *contain* the displacement, i.e. in principle, essentially no bending of the diffracting planes takes place and the planes remain undistorted. In this case, the product $\mathbf{g} \cdot \mathbf{b} = 0$. If such invisibilities can be obtained for two different **g** vectors, then the direction of the Burgers vector will be given by the cross-product of the two diffracting plane normal, $\mathbf{g}_1$ and $\mathbf{g}_2$, i.e. $\mathbf{b} = \mathbf{g}_1 \times \mathbf{g}_2$. For a screw dislocation (in an elastically isotropic medium), the Burgers vector is the only displacement associated with the defect.

This is not the case for edge dislocations, for which, in an elastically isotropic medium, there are two stress fields about the dislocation, one described by the vector **b**, i.e. the shear component of the stress-field about the dislocation, and one parallel to the cross product of the Burgers vector and the dislocation line direction, **u**, i.e. a displacement parallel to $\pm(\mathbf{b} \times \mathbf{u})$, i.e. the hydrostatic states of stress about the core, normal to the slip plane. For an edge dislocation to be invisible in an image requires that simultaneously both $\mathbf{g} \cdot \mathbf{b} = 0$ and $\mathbf{g} \cdot (\mathbf{b} \times \mathbf{u}) = 0$, in other words, both types of displacement, shear and hydrostatic, must be contained in the diffracting planes. For the cases when $\mathbf{g} \cdot \mathbf{b} = 0$ and $\mathbf{g} \cdot (\mathbf{b} \times \mathbf{u}) \neq 0$, the dislocation is characterized by a double image, often referred to as residual contrast, as shown in **Figure 29**. In this image, the defects are vacancy loops in the

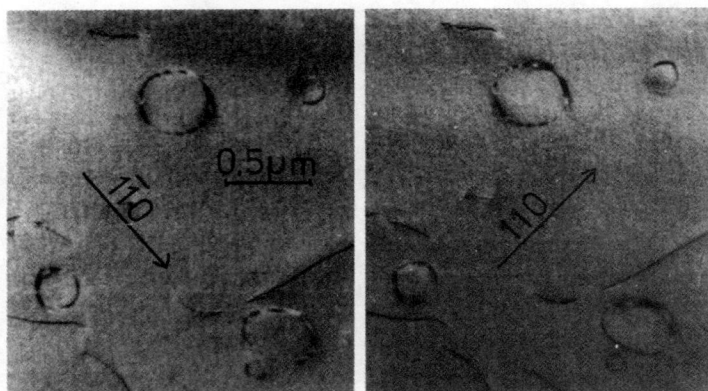

Figure 29   Bright-field TEM images of edge dislocation loops in NiAl with **b** = [001]. Note that for both images, the product **g**·**b** = 0. See text for explanation.

intermetallic compound NiAl, with Burgers vectors given by **b** = [001]. As the loops clearly lie in (001), i.e. with normals parallel to the incident beam direction ([001]), the dislocations are edge loops (i.e. for all segments of the loop). For these loops imaged as shown in **Figure 29**, **g**·**b** = 0. As can be seen, for almost all parts of the loops, double images are exhibited (contrast arising from the hydrostatic stress components, where **g**·(**b** × **u**) ≠ 0). However, the segments that are perpendicular to the **g** vector (**g** = 220) show essentially complete invisibility, i.e. for which **g**·(**b** × **u**) = 0 (at the points of invisibility, the line directions are given by **u** = [110], so that **b** × **u** = [1$\bar{1}$0], and **g**·(**b** × **u**) = [220].[1$\bar{1}$0] = 0). The term "residual" is used because, in general, the contrast from this component of the stress-field about the defect is somewhat less than that from the shear component. What role does fringe-like contrast play regarding these residual contrast images? In fact, they produce unique images. Thus, for the case of a dislocation inclined in a relatively thin foil (typically six or fewer extinction distances), and for which **g**·**b** = 0 and **g**·(**b** × **u**) ≠ 0, the contrast is as shown in **Figure 30**, i.e. the dislocation marked "A". As can be seen, the contrast appears as a series of lobes of intensity positioned discontinuously along the projected line of the defect, i.e. double dotted lobes, in contrast to the zigzag images exhibited by dislocations with **g**·**b** ≠ 0, e.g., the image of the dislocation at B in **Figure 30**. The "double-dotted" images are unique and readily identified, and immediately tell the microscopist that for those dislocations exhibiting this contrast feature, **g**·**b** = 0. This discussion concerning the images obtained for edge dislocations also applies with varying degrees to those that have mixed character.

An example of a Burgers vector analysis is given in **Figure 31**. The visibilities of the dislocations marked A and B in the figure are listed in **Table 5**, from which it may be deduced that for dislocation A: *b* = [100] and for dislocation B: *b* = [010]. While these invisibilities are readily observed, such unambiguity is not typical! There are a number of reasons why dislocation images are not readily interpretable and some of these are discussed below, after a discussion of image position relative to object position.

It is of interest to consider computed intensity profiles of a dislocation image. Two such profiles, taken from Howie and Whelan (1962), and computed for screw dislocations lying in the middle of a foil of eight extinction distances in thickness, with **g**·**b** = 1, are shown in **Figure 32(a)**, the foil is

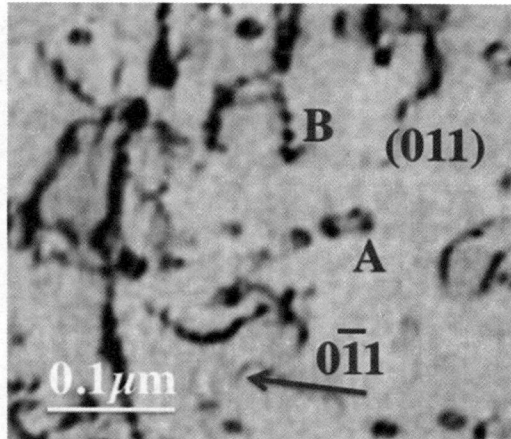

Figure 30   An inclined dislocation in NiAl marked "A", imaged with $\mathbf{g}\cdot\mathbf{b} = 0$, but $\mathbf{g}\cdot(\mathbf{b} \times \mathbf{u}) \neq 0$, exhibiting contrast that appears as a series of lobes of intensity positioned discontinuously along the projected line of the defect, i.e. double-dotted lobes, in contrast to the zigzag images exhibited by dislocations with $\mathbf{g}\cdot\mathbf{b} \neq 0$.

oriented to the exact Bragg condition, whereas in **Figure 32(b)**, the foil is oriented to a positive deviation from Bragg given by $w = 0.3$ (w, is a dimensionless deviation parameter given by $w = s\cdot\xi_g$). There are two features of import, firstly, it may be noted that the minimum in the intensity, i.e. the apparent image of the core of the dislocation, is positioned to one side of the actual core (in both profiles in the figure, to the left of the core). If the sign of the g-vector were to be reversed, the position of the image would switch to the other side of the core. It is immediately clear that the image of a dislocation does not correspond exactly with the position of the defect, and that this separation of the image from the actual position increases with the deviation from Bragg (i.e. the magnitude of s), as can be noted by consideration of the two profiles, one with $w = 0$ and the other with $w = 0.3$. In fact, the amount of the shift in position of the image relative to the actual location is proportional to the magnitude of the product of $(\mathbf{g}\cdot\mathbf{b})s$; recall that if the sign of $(\mathbf{g}\cdot\mathbf{b})s$ is reversed, for example by reversing the sign of the $\mathbf{g}$-vector, the position of the image would switch to the other side of the actual defect. This phenomenon can be used to advantage, as explained in the following.

Consider the interstitial loop depicted schematically in **Figure 33**. The Burgers vector of the loops is drawn assuming the defects are viewed from above the sample; hence also, the upward drawn normal for the electron beam direction. The sense of the bending of the diffracting planes at the core of the dislocation loop is shown. For the loop in **Figure 33(a)**, with $\mathbf{g}$ pointing to the left, it is the bending of the planes on the lower left and upper right of the loop that result in the end point of the $\mathbf{g}$-vector being rotated toward the reciprocal lattice point such that Bragg's Law is satisfied. In this case, the image of the loop will correspond to "outside" contrast, i.e. the image of the dislocation will fall outside the actual position, and therefore, will appear larger than the actual defect. Note for this configuration, assuming that an image will be recorded conventionally with $|s| > 0$ (positive deviation from Bragg), the value of $(\mathbf{g}\cdot\mathbf{b})s > 0$. In contrast, for **Figure 33(b)** with the $\mathbf{g}$-vector pointing to the right, it will be the Bragg planes at the bottom right and top left of the loop that will rotate the end point of the reciprocal lattice vector into the Bragg condition, and so the image will be of "inside" contrast, with the image being smaller than the actual size. It is

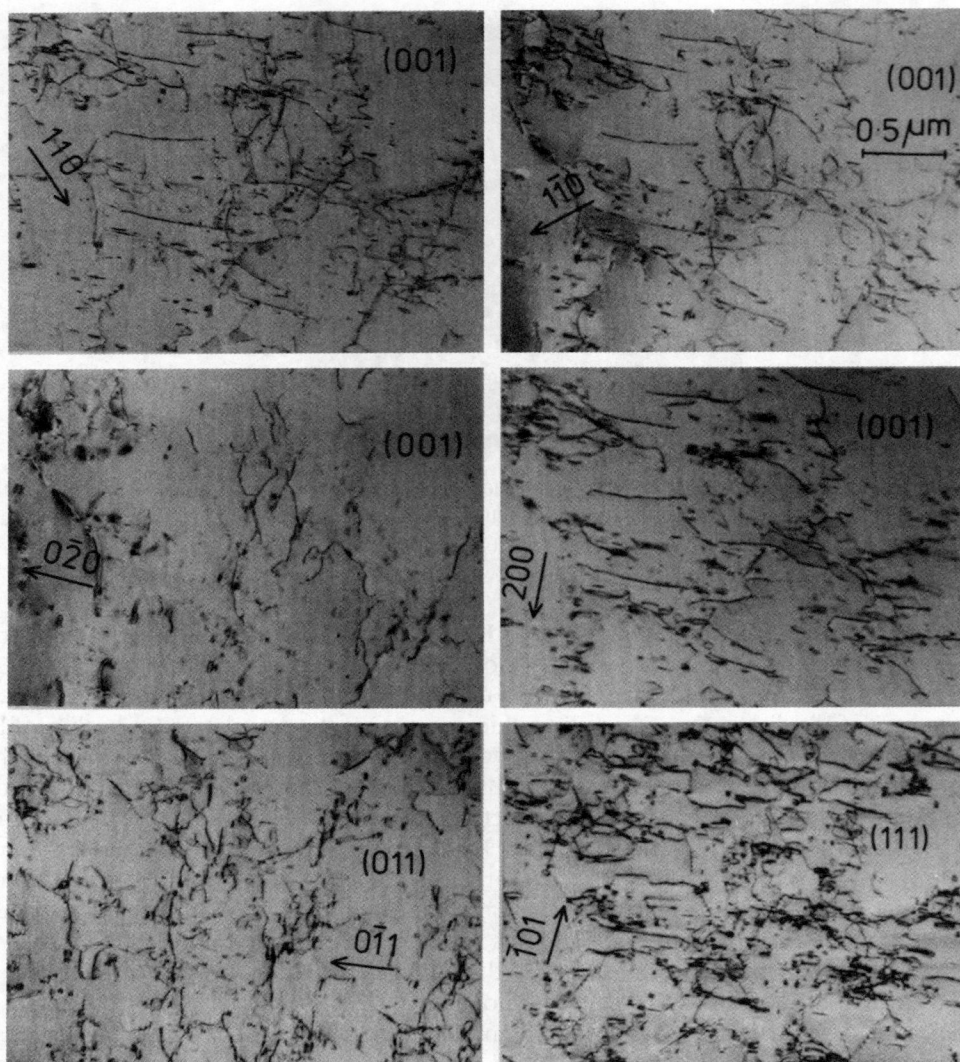

Figure 31   An example of a Burgers vector analysis. See text for explanation.

this "inside" and "outside" contrasts that may be used to determine the nature of the loop. Two points are made regarding this analysis:

(1) Firstly, simulation of images using the dynamical theory teaches that for $(\mathbf{g} \cdot \mathbf{b})s > 0$, *for viewing from above the specimen* (i.e. as if viewing the image on the microscope screen; viewing from below would reverse this analysis), loops will exhibit *outside* contrast, and for $(\mathbf{g} \cdot \mathbf{b})s < 0$, inside contrast will result.

(2) Secondly, it is important to know the sense of inclination of a loop (by simple tilting experiments in the microscope), as is apparent in **Figure 33**. Thus, consider **Figure 33(b and c)**. The two schematic loops, one an interstitial and the other a vacancy loop, lying on inclined planes, will both

Table 5   Visibility/invisibility condition for dislocations *A* and *B* under given **g** vectors

	A	B
110	In	In
1$\bar{1}$0	In	In
0$\bar{2}$0	Out	In
200	In	Out
0$\bar{1}$1	Out	In
$\bar{1}$01	In	Out

(a)

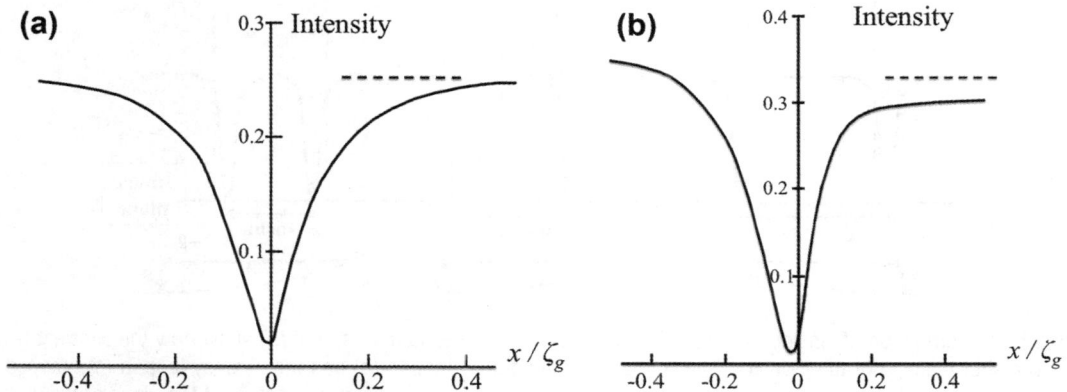

(b)

Figure 32   Image profiles of a screw dislocation positioned in the middle of a foil of eight extinction distances in thickness, with **g.b**=1. (a) foil oriented such that w=0, and (b) with w=0.3. See text for explanation. Taken from Howie and Whelan (1962).

give rise to inside contrast for the diffracting vector depicted. Hence, it would not be possible to distinguish between these two without a knowledge of the actual inclination of the planes of the loops.

An example of the determination of the nature of a dislocation loop using ±**g** imaging is shown in **Figure 34**. Thus, **Figure 34(a)** shows a number of dislocation loops, imaged with an incident beam direction near to [103] and with **g** = 0$\bar{3}$1 in a sample of NiAl after being lightly deformed at 550 °C. A diffraction contrast experiment has shown (not here) that for these loops, **b** = ±[001]; the question then is, are the loops interstitial in nature, with **b** = [001], or vacancy in nature, with **b** = [00$\bar{1}$]? **Figure 34(b)** shows the same area of the foil but imaged with **g** = 03$\bar{1}$ (i.e. with the sign of the **g** in **Figure 34(a)** reversed). Careful observation of the same loops in each of the micrographs reveals that the loops in **Figure 34(a)** exhibit "inside" contrast, whereas those in **Figure 34(b)** exhibit "outside" contrast. Suppose that it is assumed that **b** = [001]; for the case of **Figure 34(a)**, an image recorded with **g** = 0$\bar{3}$1 and a positive value of the deviation parameter, *s*, the product (**g·b**)*s* > 0, and so outside contrast would be expected. This is not in agreement with experiment (the contrast exhibited is "inside"), and so the assumption that **b** = [001] is incorrect. If it is assumed that **b** = [00$\bar{1}$], then with **g** = 0$\bar{3}$1, the product (**g·b**)*s* < 0, and the contrast is as expected, namely "inside" contrast. The identity of the loops is confirmed as being vacancy in nature.

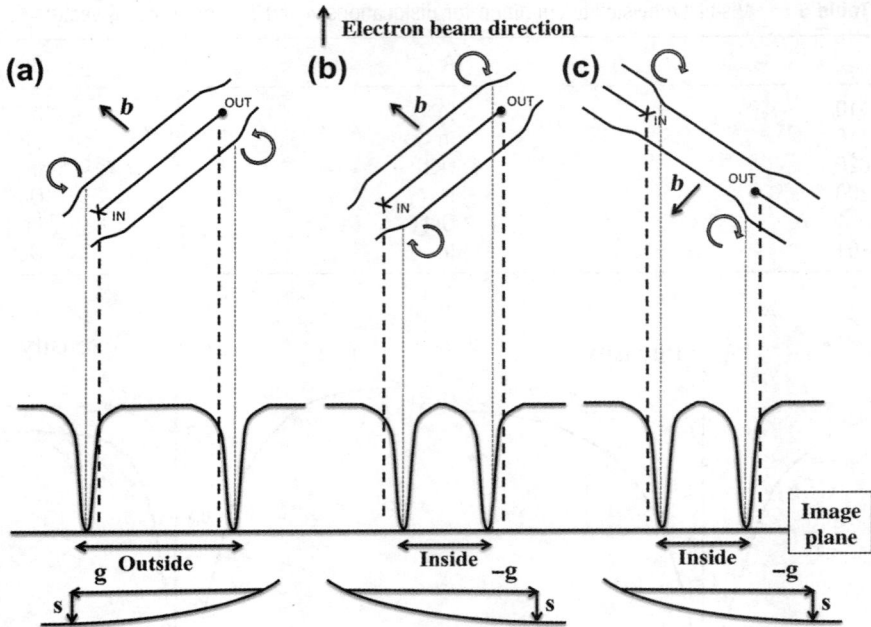

Figure 33 The formation of "inside" and "outside" contrast for inclined loops. (a) and (b) show how the contrast reverses upon reversal of the sign of the **g**-vector. The contrast of the intrinsic loop in (b) and the vacancy loop in (c), imaged with the same g-vector, is the same, illustrating the importance of knowing the sense of inclination of the plane of the loop. "In" refers to the positive direction of the dislocation, i.e., the positive sense of the dislocation line is into the page.

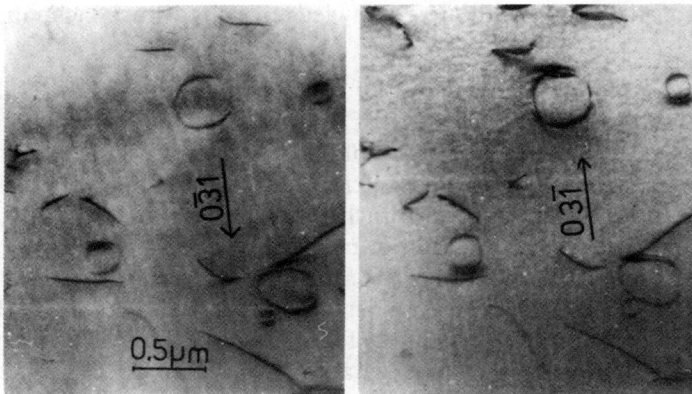

Figure 34 An example of an experiment aimed at determining the nature of the dislocation loops shown. The change in image position (i.e. inside or outside contrast) with $+\mathbf{g} = 0\bar{3}1$ is consistent with these defects being vacancy in nature. See text for further details.

12.3.2.1.1   Distinguishing between Pairs of Like-Signed Dislocations or Dipoles

As has been noted, the image of a dislocation will tend to lie on one side of the actual defect line or the other, the position depending on the sign of $(\mathbf{g} \cdot \mathbf{b})s$. This behavior has a second application in defect analysis, in addition to the determination of the nature of loops, as discussed above. Thus, often is the case when dislocations appear in micrographs in the form of pairs. It is important to know whether these dislocations have parallel Burgers vectors, or antiparallel, i.e. in this latter case, if they are dipoles. This determination is simply effected by recording two images of the defects, one with a given set of planes, $\mathbf{g}$, oriented for diffraction, and then one with the sign of $\mathbf{g}$ reversed. For a pair of like-signed dislocations, the images will both shift in the same direction upon reversal of the sign of $\mathbf{g}$, i.e. the spacing between the dislocations will not change, whereas for dipoles, the shift in image position will be opposite such that the spacing will change. Remember, the degree of image shift depends on magnitude of the product of $(\mathbf{g} \cdot \mathbf{b})s$, so that the use of $\mathbf{g}$-vectors with large magnitudes will result in larger, perhaps more discernable, image shifts, making the analysis less ambiguous. An example of this analysis is shown in **Figure 35** for dislocations in a deformed sample of the intermetallic compound Ti_3Al (Court et al., 1990); note, the images have been recorded using weak-beam DF conditions, as described below. In **Figure 35(a)**, two pairs of superdislocations (Ti_3Al is an ordered compound with the crystal structure DO_{19}) may be observed, at A and B, respectively. Each of these superdislocations is dissociated into superpartial pairs (the bright and weak lines, but very closely spaced). Upon reversing the sign of $\mathbf{g}$, see **Figure 35(b)**, the spacing between the superdislocations at A increases, while that at B decreases. Clearly, these superdislocations are dipoles, of opposite sign. Note, however, that the spacing between the superpartial dislocations *does not* change, as would be expected as these superpartial pairs have the same sign. This is a very simple experiment to perform in the TEM, but yet yields a very important result.

12.3.2.1.2   Interpretation of Double Images

It has been noted above that for edge (or mixed) dislocations, double images may result when $\mathbf{g} \cdot \mathbf{b} = 0$ with $\mathbf{g} \cdot (\mathbf{b} \times \mathbf{u}) \neq 0$. Such double images also may occur when, for a given dislocation, the condition $\mathbf{g} \cdot \mathbf{b} = 2$ occurs. An image profile is shown in Figure 36(a), from Howie and Whelan (1962), for the case

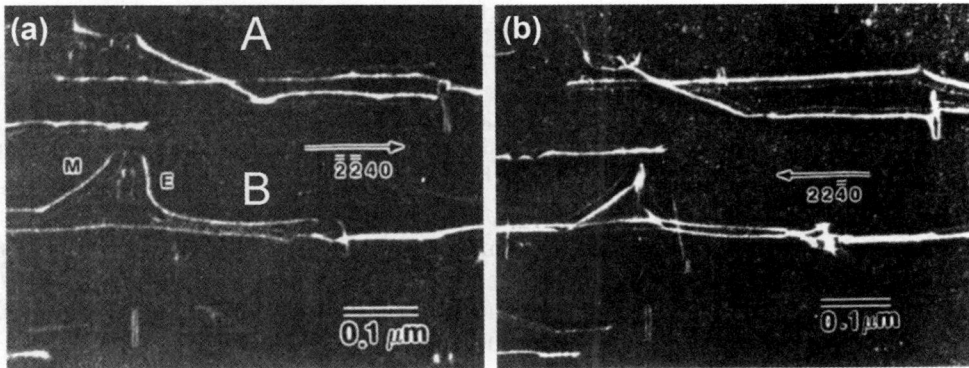

Figure 35   WBDF images of dislocation configurations in the intermetallic compound Ti_3Al, where the sign of the **g**-vector has been reversed between the images, in (a) and (b) respectively. The image shifts upon reversal of the sign of the **g**-vector is consistent with the superdislocations being present as dipoles, and the (assumed) superpartials having parallel Burgers vectors. Taken from Court et al. (1990).

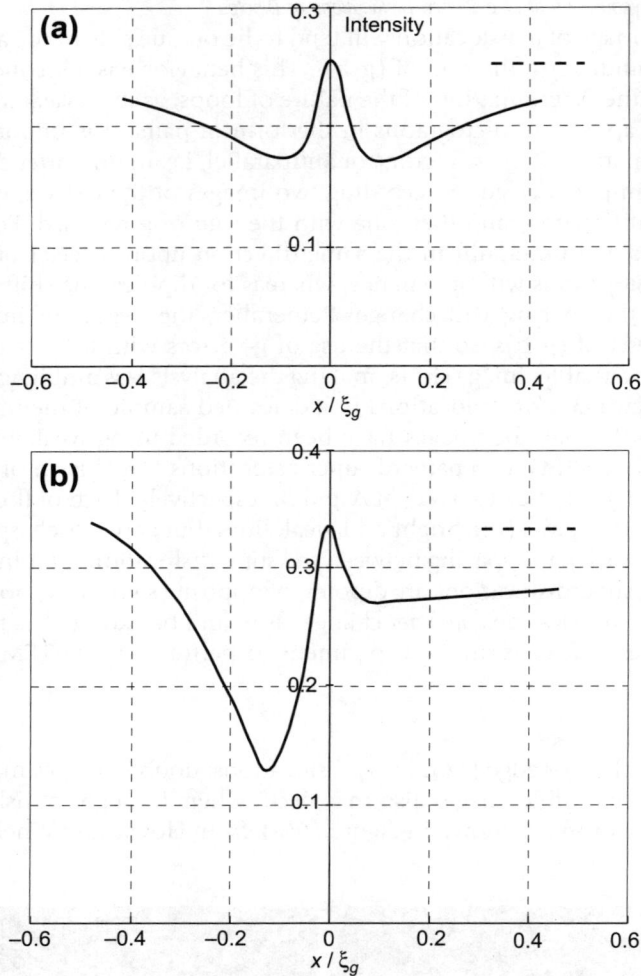

Figure 36 Profiles of dislocation images for the case for a dislocation with $\mathbf{g} \cdot \mathbf{b} = 2$. (a) With $w = 0$, where $w = s.\xi_g$ (i.e. with $s \approx 0$), and (b) with $w = 0.3$. As can be seen, the double image exhibited when oriented close to Bragg tends to a single image when deviated far from Bragg. Redrawn from Howie and Whelan (1962).

of a dislocation with $\mathbf{g} \cdot \mathbf{b} = 2$ with deviation parameter, w, given by $w = 0$, and it is clear that there is a double image. It is obviously important to be able to discern between these two possibilities ($\mathbf{g} \cdot \mathbf{b} = 0$ and $\mathbf{g} \cdot \mathbf{b} = 2$). This may be achieved by increasing the value of the deviation parameter, i.e. by tilting the foil away from the Bragg condition. In the case of the defect imaged under $\mathbf{g} \cdot \mathbf{b} = 0$ with $\mathbf{g} \cdot (\mathbf{b} \times \mathbf{u}) \neq 0$, both images will reduced in contrast. Differently, for the defect imaged such that $\mathbf{g} \cdot \mathbf{b} = 2$, increasing the deviation from Bragg will result in one of the double images markedly decreasing in contrast, with eventually only one of the double images being visible. The image profile for such an imaging condition, with the deviation parameter given by $w = 0.3$, is shown in Figure 36(b); note that the contrast of the image on the right of the dislocation has markedly reduced. So, the distinction between double images, from dislocations imaged with $\mathbf{g} \cdot \mathbf{b} = 0$ with $\mathbf{g} \cdot (\mathbf{b} \times \mathbf{u}) \neq 0$, and $\mathbf{g} \cdot \mathbf{b} = 2$, may be made by titling the thin foil such that the deviation from the Bragg condition is increased.

12.3.2.1.3 Contrast from Partial Dislocations

In the discussion above, dislocations have been considered to be invisible if the product $\mathbf{g} \cdot \mathbf{b} = 0$ (and for edge dislocations, also with $\mathbf{g} \cdot (\mathbf{b} \times \mathbf{u}) = 0$); those dislocations with $\mathbf{g} \cdot \mathbf{b} = 1$ and 2 are said to be visible. The magnitude of the product of $\mathbf{g} \cdot \mathbf{b}$ for dislocations with small Burgers vectors, such as partial dislocations in *fcc* materials, can be greater than zero, but have a fractional value less than one. Computer simulations using the dynamical theory of electron diffraction have shown that dislocations with Burgers vectors such that $\mathbf{g} \cdot \mathbf{b} \leq 1/3$ are essentially invisible (Silcock and Tunstall, 1964). This result is illustrated by noting the images of dissociated dislocations in an *fcc* alloy, such as Cu–7%Al, as shown in **Figure 37** (taken from Howie and Whelan, 1962). It would be expected that the defect configuration would involve two partial dislocations separated by a stacking fault. As can be seen, fairly strong contrast is observed from a dislocation on one edge of a fault, whereas the dislocation on the other end of the same fault is essentially invisible. Suppose that the dislocation has a Burgers vector given by $\mathbf{b} = 1/2[0\bar{1}1]$, and is dissociated into Shockley partials on the (111) plane. The dissociation reaction would involve:

$$\mathbf{b} = 1/2[0\bar{1}1] = 1/6[1\bar{2}1] + 1/6[\bar{1}\bar{1}2].$$

Suppose that an image is obtained for $\mathbf{g} = 1\bar{1}1$, then it follows that for the partial dislocation with $\mathbf{b} = 1/6[1\bar{2}1]$, the value of $\mathbf{g} \cdot \mathbf{b} = 2/3$, whereas for the partial with $\mathbf{b} = 1/6[\bar{1}\bar{1}2]$, the value of $\mathbf{g} \cdot \mathbf{b} = 1/3$. In this case, the first partial would be visible and the second partial would be essentially invisible. This analysis shows that it is important to interpret "invisibilities" with care! Another example, involving imaging with diffraction vectors parallel to <200>, is given by Silcock and Tunstall (1964).

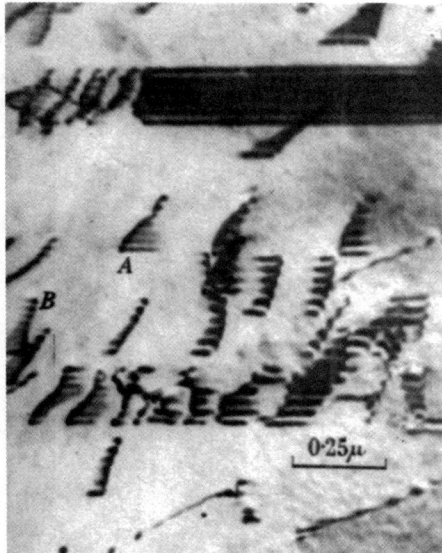

Figure 37   Images of dissociated dislocations in an *fcc* alloy, Cu–7%Al. Note that the partial dislocations on one end of the faults exhibit strong contrast whereas the image of the partial dislocations on the other end of the faults is essentially invisible. The g-vector was not marked on the original figure. See text for explanation. Taken from Howie and Whelan (1962).

12.3.2.1.4   Contrast of Defects in Elastically Anisotropic Materials

The discussion of defect images above has involved the influence of stress-fields about dislocations on the diffraction process. These stress-fields have been considered to be those expected in an elastically isotropic material. Most materials are, in fact, elastically anisotropic, with atom positions around a dislocation being somewhat displaced from those expected for isotropic elasticity, and modern versions of computer simulation of images take this into account (see discussion in Head, Loretto and Humble (1967)). However, it is important to have some notion of the degree to which the material being studied is anisotropic, so it would be expected that anomalies in the contrast will probably arise from this property and should be taken into account.

12.3.2.2   Characterization of Planar Defects

Planar defects such as intrinsic stacking faults in *fcc* crystals are characterized by displacement vectors that impose a shift in the position of lattice planes at the defect location. This is rather different from the continuous strain fields, characteristic of dislocations. For example, for the case of intrinsic faults in *fcc* crystals, the displacement vector, $\mathbf{R}$, is given by $\mathbf{R} = 1/6\langle 112 \rangle$. The presence of the fault vector is to displace, or distort, lattice planes as they traverse the fault. Regarding imaging of the faults, it is the displacement/distortion of the diffracting planes ($\mathbf{g}$) that leads to contrast. In an analogous way to considering the distortion of Bragg planes by dislocations (i.e. by considering the value of the vector product, $\mathbf{g} \cdot \mathbf{b}$), for planar defects it is the value of the vector product $\mathbf{g} \cdot \mathbf{R}$ that is examined.

There are four important values of this product. The first is $\mathbf{g} \cdot \mathbf{R} = 1/2$; for this value of the product, so-called π-fringes may be observed. These fringes may be observed in ordered crystals, when antiphase boundary (*apb*) ribbons are imaged with a superlattice reflection, with the crystal oriented to the Bragg condition ($s = 0$). Consider the example of a B2 compound; the displacement vector for an *apb* in this material would be $\mathbf{R} = 1/2\langle 111 \rangle$. Suppose the *apb* with $\mathbf{R} = 1/2[111]$ is imaged with a superlattice reflection, e.g. $\mathbf{g} = [100]$, then in this case, $\mathbf{g} \cdot \mathbf{R} = 1/2$, and the image would consist of a set of π-fringes. Again, imaging with ($s = 0$), BF and DF images are complementary, where in the BF image, the central fringe has intensity above background, and conversely for DF imaging, the central fringe has intensity below background. The fringe spacing is $\xi_g/2$, and the intensity of the outermost fringe is determined by the foil thickness. It is very uncommon for physical metallurgists to make use of π-fringe imaging, as for most metallic samples, the structure factors of superlattice reflections are rather small, and the extinction distances correspondingly large.

The three other values of $\mathbf{g} \cdot \mathbf{R}$ are rather more important in terms of imaging faults. The first of these is $\mathbf{g} \cdot \mathbf{R} = 0$, where the fault vector is perpendicular to the normal to the diffracting planes (**Figure 38(a)**).

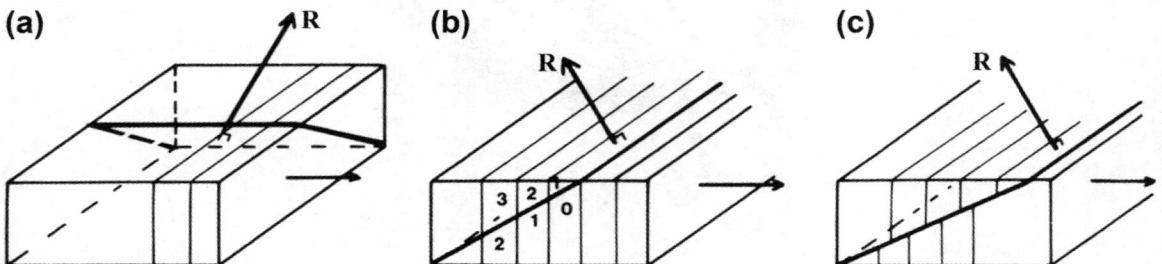

Figure 38   Schematic diagram showing the physical situation corresponding to the important values of $\mathbf{g} \cdot \mathbf{R}$ for imaging of stacking faults. For (a) $\mathbf{g} \cdot \mathbf{R} = 0$; for (b) $\mathbf{g} \cdot \mathbf{R} = n$; and for (c) $\mathbf{g} \cdot \mathbf{R} \neq n$. Taken from Loretto (1984).

In this case, the displacement vector does not distort the diffracting planes, and the fault will be invisible. The second is $\mathbf{g}\cdot\mathbf{R} = n$, where n is an integer. In this case, although the diffraction vector and fault vector are not mutually perpendicular, the diffracting planes remain continuous and undistorted, as shown in **Figure 38(b)**. The fault is invisible for this imaging condition. The third, and important vector product, is for $\mathbf{g}\cdot\mathbf{R} = a$ fractional value. For example, for a fault vector $\mathbf{R} = 1/6\langle 112 \rangle$ and $\mathbf{g} = 1\bar{1}1$, then $\mathbf{g}\cdot\mathbf{R} = 1/3$. For these fractional values of the vector product, the fault is visible in TEM images. The reason for the visibility of the faults is because in this imaging condition, the fault vector distorts the diffracting planes, such that there is a discontinuity of the Bragg planes at the position of the fault, as shown in **Figure 38(c)**.

The images of faults when $\mathbf{g}\cdot\mathbf{R}$ is a fractional value consist of a set of fringes running parallel to the surface of the thin foil, along the length of the fault. From computer simulations of images using the dynamical theory, there are some "rules" that one must follow in order to be able to interpret the fault images; these "rules" have been discussed by Loretto (1962), and are summarized briefly here. They are as follows:

(1) Record the images of the fault in BF and DF, with $s = 0$.
(2) Images in BF and DF consist of a series of fringes; for BF, the intensities of these fringes are symmetrical about the central fringe. For DF, the fringes are asymmetrical, i.e. if the outermost fringe at the top surface (i.e. that surface upon which the electrons are incident) is above background intensity (bright fringe), then the outermost fringe at the exit (bottom) surface of the foil will be below background intensity (dark fringe).
(3) At the intersection of the fault with the top of the foil, BF and DF images will appear essentially the same (i.e. both have either a bright or dark fringe). At the intersection of the fault with the bottom surface, these images will be complementary. From this, it will be immediately obvious which side of the fault corresponds to the top or bottom of the foil, and this is an important information that permits the orientation of the fault to be determined.
(4) When considering the fractional value of the vector product, regarding the form of the image, 1/3 and 4/3 yield identical images, and for $-1/3$ and 2/3, the images are also identical (but different from those with $\mathbf{g}\cdot\mathbf{R} = 1/3$, see next "rule"). Therefore, it is useful to reduce the value of the vector product to either 1/3, or $-1/3$. The reason for this is that the effect of the fault is to introduce a phase shift given by $2\pi\mathbf{g}\cdot\mathbf{R}$, and the subtraction of π makes no difference on this phase shift.
(5) From the computer simulations, a result is that for $\mathbf{g}\cdot\mathbf{R} = 1/3$, the outermost fringe in a BF image will be bright, whereas for $\mathbf{g}\cdot\mathbf{R} = -1/3$, the outermost fringe will be dark.
(6) It is very important that the actual sign of the diffracting vector, $\mathbf{g}$, be known. This is not normally the case for dislocation analyses, but is very important here.
(7) The fringes in the central part of the fault image tend to fade, and eventually disappear, as the foil thickness increases.

Determination of the nature of a fault: As an example, a determination of the nature of a fault in *fcc* will be outlined (images taken from Williams and Carter (2009)). The following convention should be followed: the sign of the fault vector is given by the relative motion of the bottom part of the foil below the fault when forming the fault. For example, the formation of an intrinsic fault (vacancies) occurs by an *upward* relative motion of the bottom of the foil (because essentially the agglomeration of the vacancies corresponds to the removal of part of the plane of atoms), so that the fault vector is upwardly drawn. The opposite is true for an extrinsic fault (interstitials). From this, the nature of a fault is assessed by determining whether the fault vector is upwardly or downwardly drawn. As an example, a series of images of a given stacking fault recorded in BF and DF, for $\pm\mathbf{g} = 0\bar{2}2$, $s = 0$, and with a beam direction close to [111], are given in **Figure 39**. Assume that diffraction contrast experiments (imaging with various diffraction vectors to record invisibilities) have been done such that it is known that the fault vector is $\mathbf{R} \pm 1/3\langle \overline{1}\overline{1}1 \rangle$, i.e. the sign of the vector is not known. It is noted that for imaging with $\mathbf{g} = 02\overline{2}$, the BF image (**Figure 39(a)**) exhibits a dark outermost fringe; the DF image (**Figure 39(d)**) for this diffracting vector exhibits a dark outermost fringe at the lower edge of the fault image, from which it may be concluded that this is the intersection of the fault with the top surface of the foil. As a check, it is noted that when imaging with $\mathbf{g} = 0\overline{2}2$, in BF (**Figure 39(b)**), the outermost fringes are bright, and in DF (**Figure 39(c)**), the bright fringe occurs at the lower edge of the image, consistent with the images with $\mathbf{g} = 02\overline{2}$. The fault is then inclined such that the projection of the upward-drawn normal lies perpendicular to the fault images towards the top of the page. Given that the normal to the plane of projection is [111] (i.e. the beam direction), the upward-drawn normal is [11$\overline{1}$]. Consider **Figure 39(a)** for which $\mathbf{g} = 02\overline{2}$; if it is assumed that $\mathbf{R} = 1/3[11\overline{1}]$, then $\mathbf{g} \cdot \mathbf{R} = 4/3$, equivalent to 1/3. In this case, the outermost fringe should be bright (see "rules" above), but is actually dark. Hence, the assumption

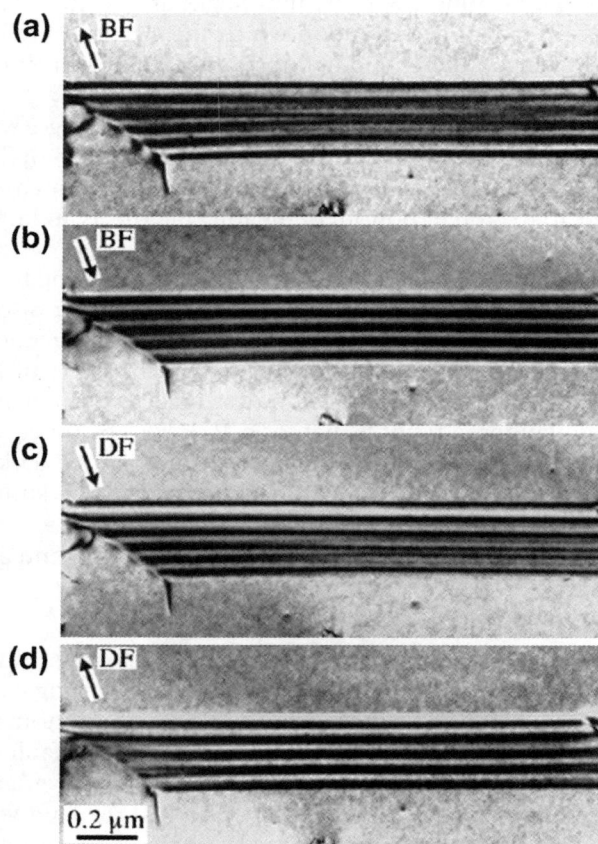

Figure 39   An example of the use of diffraction contrast to identify the nature of a stacking fault. The images labeled (a)–(d) are of the same fault, with the diffraction vector depicted by the arrows and with BF and DF referring to bright and dark field, respectively. See text for detailed procedure of the analysis. Images taken from Williams and Carter (2009).

concerning the fault vector is incorrect, and in fact, the correct fault vector is $\mathbf{R} = 1/3[\bar{1}1\bar{1}]$. Since this is a downward-drawn normal for the imaging conditions in **Figure 39**, the fault is extrinsic in nature.

In thicker samples, the fringes from the center of faults tend to decrease in intensity until there is essentially only background intensity from this region of the image. The cause of this trend involves anomalous absorption. Consider the diagram shown in **Figure 40(a)**, which shows schematically the scattering events that occur when the electrons are scattered at the fault. Thus, the electrons in Bloch wave 1, with wave vector $\mathbf{D}_1$, will undergo interband scattering and produce a Bloch wave with wave vector $\mathbf{D}'_2$, rather than $\mathbf{D}_2$ because of the inclination of the stacking fault. Similarly, the electrons in Bloch wave 2, with wave vector $\mathbf{D}_2$, will undergo interband scattering and produce a Bloch wave with wave vector $\mathbf{D}'_1$. To explain the various contrast features observed, consider **Figure 40(b)**. At the foil surface, assuming two-beam conditions, two Bloch waves will propagate through the foil. Bloch wave 1 (denoted by its wave vector $\mathbf{D}_1$ in the foil) will be increasingly absorbed because of anomalous absorption. When the electrons in Bloch wave 2 intersect the stacking fault near the bottom of the foil, interband scattering [see schematic diagram in **Figure 40(b)**] occurs, such that two Bloch waves, Bloch wave 2, with wave vector $\mathbf{D}_2$, and Bloch wave 1 with wave vector given by $\mathbf{D}'_1$, then propagate toward the exit surface of the foil. These two Bloch waves will beat against each other and so yield fringes in the image. When the interaction of the electron beam with the stacking fault occurs in the vicinity of the top surface of the foil, both original Bloch waves, with vectors $\mathbf{D}_1$ and $\mathbf{D}_2$, will scatter at the fault, leading to two additional Bloch waves, with vectors $\mathbf{D}'_1$ and $\mathbf{D}'_2$, i.e. there will be four Bloch waves present. Both wave vectors $\mathbf{D}_1$ and $\mathbf{D}'_1$ will be absorbed by anomalous absorption, such that only the two Bloch waves with vectors $\mathbf{D}_2$ and $\mathbf{D}'_2$ will exit the foil. These two waves will beat against one another, again yielding

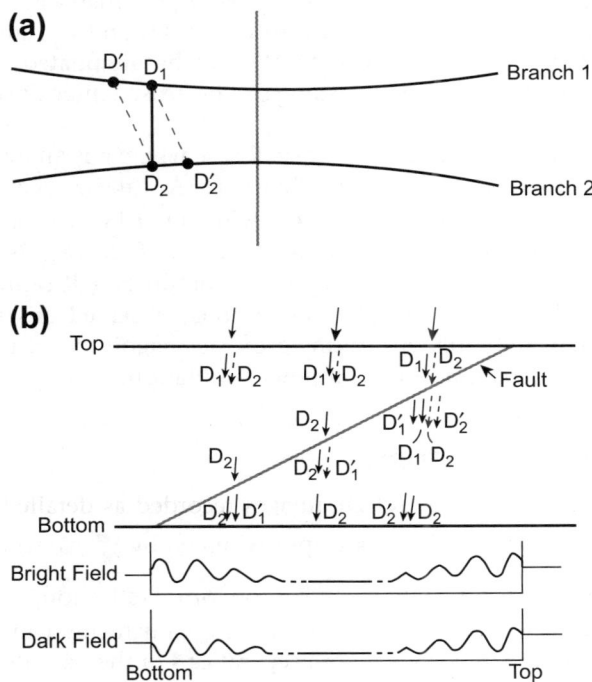

Figure 40 Schematic diagrams illustrating scattering events when electrons interact with stacking faults (a), and the formation of images of inclined faults in foils of reasonable thickness (b). See text for full details. Redrawn from Hirsch et al. (1977).

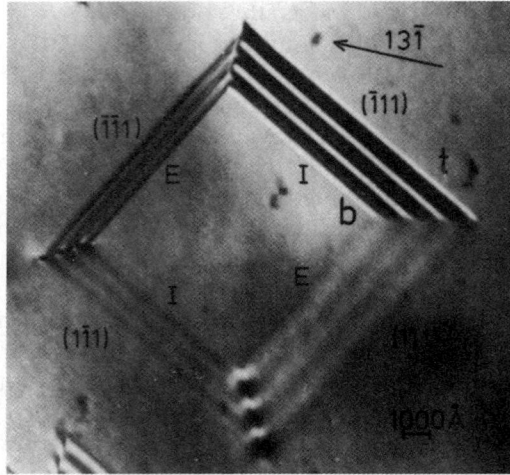

Figure 41 A set of four inclined faults in Si, showing residual contrast on two of the faults, revealing the presence of supplemental displacements at the faults. See text for details. Taken from Loretto (1984).

fringe contrast in the image. For electrons propagating through the foil such that they interact with the fault in the central regions of the foil, only Bloch wave 2, with vector D_2 will scatter at the fault, since Bloch wave 1 will have been absorbed by anomalous absorption. Bloch wave 2 will undergo interband scattering at the fault, such that two Bloch waves, with vectors D_1' and D_2, are present. However, again because of anomalous absorption, the wave with D_1' will be attenuated, and hence there will be a reduced beating effect between Bloch waves, leading to a decreased intensity of the fringes for this part of the image.

It is noted above that for imaging faults such that $g \cdot R = n$, where n is an integer, the fault will not be visible. Consider the set of faults in Si shown in **Figure 41**. As can be seen, two of the faults exhibit strong contrast; the fault on $(\bar{1}11)$ is intrinsic and that lying on $(\bar{1}\,\bar{1}1)$ is extrinsic, in nature. The faults lying on $(1\bar{1}1)$ and (111), and exhibit weak contrast; they have fault vectors given by $R = \pm 1/3[1\bar{1}1]$ and $R = \pm 1/3[111]$. For both of these faults, the vector products $g \cdot R$ both yield a value of ± 1. In principle, then, the faults should be invisible. The contrast observed arises from a supplementary displacement associated with the fault. An estimate of the magnitude of the displacement may be determined by use of image matching using computer simulation.

12.3.3   Weak-Beam Dark-Field (WBDF) Imaging

The widths of dislocation images in BF and DF images recorded as detailed above, i.e. with a small positive deviation from the Bragg condition, are approximated by $\xi_g^{\text{eff}}/3$, where $\xi_g^{\text{eff}} = \dfrac{\xi_g}{\sqrt{(1+s^2\xi_g^2)}}$ (recall that ξ_g is the extinction distance for the planes g). For low-order reflections, i.e. those typically used to image dislocations, the widths will be typically ≈ 10 nm. In many metallic materials with the *fcc* or *hcp* crystal structures, dislocations are dissociated, with separations of the partials being typically <10 nm. Hence, these partial dislocations are not revealed in two-beam BF and DF imaging, and it is necessary to employ a technique for which the image widths are markedly reduced. Such narrow dislocation widths would result if ξ_g^{eff} were to have a small value, and this can be achieved by increasing the value of s, i.e. with a very large deviation from Bragg. For example, if $\xi_g = 40$ nm and $s = 0.2$ nm^{-1}, then $\xi_g^{\text{eff}} = 5.0$ nm, such that the image width would be ≈ 1.5 nm. A simple physical picture of the effectiveness of large deviations from Bragg on the reduction of image width is as follows. It was described in the introduction to this section (Diffraction Contrast) that distortions associated with a dislocation may result in the bending of potential diffracting planes into the Bragg condition. Typically, two-beam conditions are set with the foil oriented to a small positive deviation from Bragg. It is reasonable under these conditions to consider that, with only a small deviation from the Bragg condition, a reasonable number of planes will be tilted into the Bragg orientation such that a fairly wide image results. When the foil is set to a very large deviation from Bragg, it will be only a few planes very near to the dislocation core that will be sufficiently tilted, by the distortions associated with the defect, that they give rise to diffraction, yielding a very narrow image width.

Consider the rocking curve in **Figure 26**. If BF imaging is employed, a very large deviation from Bragg would result in a low background intensity (i.e. to the right-hand side of the curve), such that a diffraction event (caused by the dislocation) would yield a dark image on a low-intensity background, i.e. a weak but thin image. However, if the foil is oriented to the same large deviation from Bragg, and then imaged in DF, the background will be of very low intensity, but the image will be above background (a thin white image against an essentially black background). In this way, the contrast exhibited by the defect will be optimized, and the image will be very thin. Hence the name: WBDF imaging. Another advantage of WBDF imaging involves the large deviation from Bragg. Thus, under these conditions, there is minimal dynamical interaction between the BF and DF beams so that once the diffraction event occurs, the intensity in the DF beam is retained.

The technique, including details of measurement of the deviation parameter, is well documented in several texts, (e.g. Loretto, 1984; Williams and Carter, 2009). The setting up of conditions for WBDF imaging is fairly straightforward. Most usually, the foil will be oriented for diffraction by **g** (two-beam conditions with a slight positive deviation from Bragg). The DF beam is then tilted onto the optical axis; as shown in **Figure 42**, when the DF beam is tilted in this way, the beam 3**g** is excited. In this way, images in the weakly excited DF beam **g** may be recorded, obviously with a very large deviation from Bragg. This imaging condition is termed **g**/3**g**, i.e. images recorded with **g** parallel to the optical axis, with 3**g** at, or near, the Bragg condition. If the image width is till not sufficiently thin, then the foil can be tilted further to increase the deviation from Bragg, but with a decrease in intensity. An example of WBDF imaging is shown in **Figure 35**.

12.3.4 High Resolution (S)TEM Imaging

It may be noted that this chapter does not contain a description of high-resolution imaging in the TEM. This is because in most recent research involving high-resolution imaging of metallic samples, STEM imaging has been the preferred mode of characterization. This is mainly because of the improvements of spatial resolution afforded by STEM in modern electron microscopes, and also because the images in STEM are directly interpretable (generally), whereas high resolution in the TEM, accessed by phase contrast imaging, requires the recording of through-focal series of images, followed by object reconstruction. Readers interested in high-resolution imaging in TEM are directed to the provided references: De Graef (2003), Spence (2003), and Busek et al. (1988).

STEM has become a well-established characterization technique over the past 50 years, with a number of publications and books written to cover the subject in much detail. The purpose of this

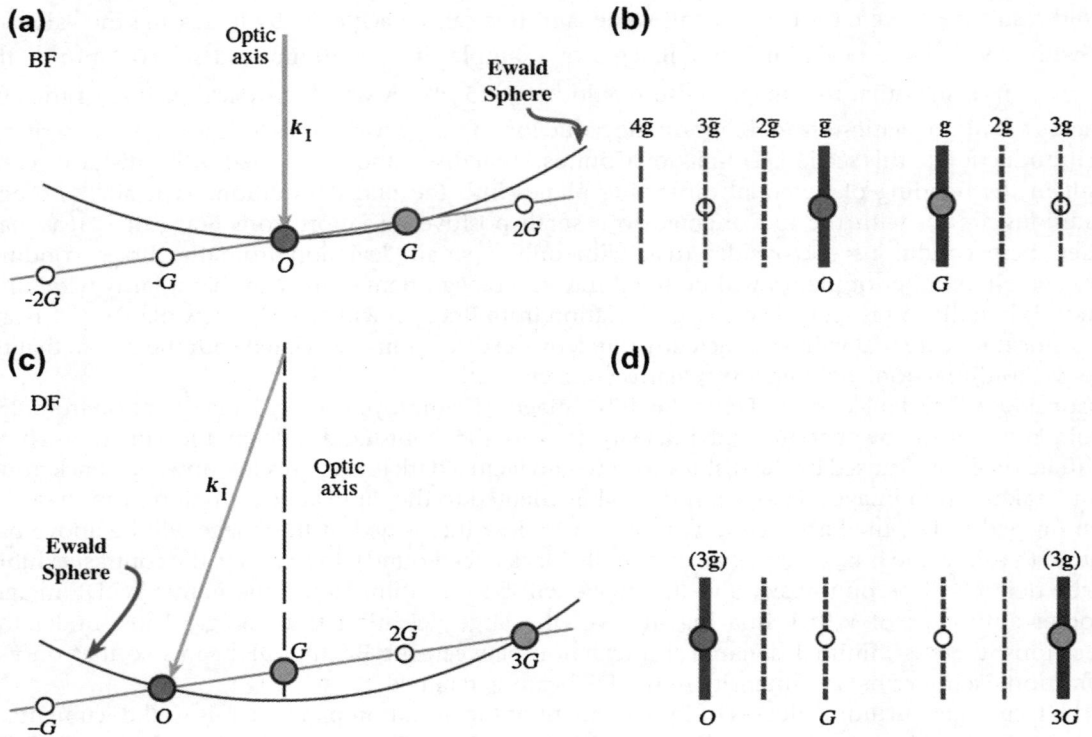

Figure 42   Schematic diagram showing the result of setting up conventional two-beam diffracting conditions (a) and (b), and the resulting imaging conditions when the beam **g** is tilted onto the optical axis (c) and (d). The result is a condition where images are recorded in dark-field but at a very large deviation form Bragg. Taken from Williams and Carter (2009). (For color version of this figure, the reader is referred to the online version of this book.)

chapter is to review *some* of the principles underlying imaging and characterization in the STEM as related to metallic specimens. Similar to a scanning electron microscope, STEM operates by scanning a condensed probe across the specimen in a raster pattern, while simultaneously collecting various signals for image magnification and/or chemical analysis, synchronized with beam position. For comparison, STEM also shares the *principle of reciprocity* with conventional transmission electron microscopy (CTEM) (Cowley, 1969; Zeitler and Thomson, 1970). This implies that the imaging processes, and so resultant images, will be essentially the same in CTEM and STEM.

   A relevant question is why would physical metallurgists gain an advantage from using STEM for imaging purposes, e.g. for defect characterization. To answer this question, it is necessary to consider the expected impact of STEM when the technique was first introduced. Thus, in the late 1960s and early 1970s, there was a very significant effort to promote high-voltage electron microscopy (HVEM). One of the reasons for this promotion was the expected increase in useable foil thickness when using electrons accelerated to 1 MV—a significant advantage for metallic samples. However, the instruments were expensive and somewhat difficult to operate and maintain. Working with soft materials, Albert Crewe, in Chicago, led an effort to illustrate the advantages of STEM regarding imaging. He made the observation that, in the absence of post-specimen lenses, chromatic aberration played no role in image degradation, and so the useable foil thickness could be increased. On this basis, the hard materials' community was quick to promote STEM as an alternative to HVEM regarding useable foil thickness. This enthusiasm was quenched by a work that showed that for materials with high atomic numbers, loss of intensity is the limiting factor (Fraser and Jones, 1975), and for defect analysis using diffraction contrast, the useable foil thickness is limited by beam spreading in thick foils with a resulting loss of information for scattering processes occurring in the thick regions (Fraser et al., 1977). Indeed, as imaging in crystalline materials is generally not limited by chromatic aberration, this result is not surprising. More recently, there has been resurgence in the application of STEM imaging in diffraction contrast experiments (Phillips et al., 2011a). This research has shown that working within the "thick-foil" limit, there are some advantages that may be accrued by the use of STEM regarding defect visibilities, and these are discussed below. In addition to this application of STEM, the advent of high-angle annular dark-field (HAADF) imaging, which permits contrast arising from differences in atomic number (so-called *Z-contrast*) to be exploited, presents several advantages to the characterization of metallic samples. These two modes of STEM imaging will be discussed below.

12.3.4.1 Instrumental Considerations

Figure 43 shows a schematic diagram of the electro-optical configuration used when operating the microscope in an STEM condition. Most STEM microscopes have a two-condenser lens system; however, some newer systems are available with a three-condenser lens system. The condenser lens

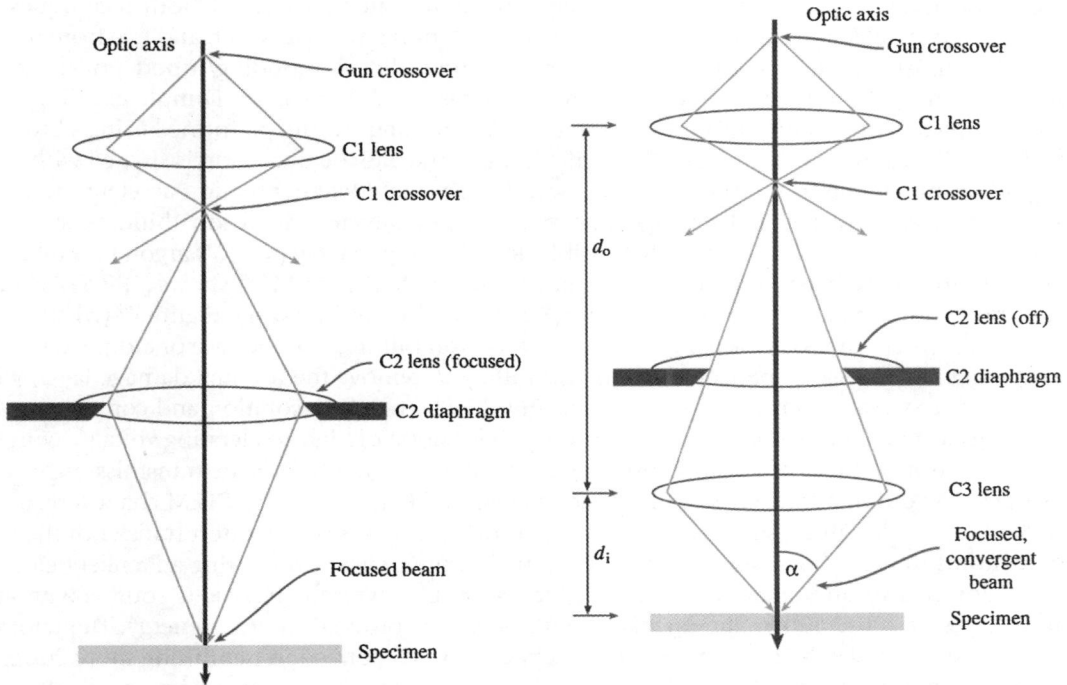

Figure 43 A schematic diagram of the electro-optical configuration used when operating the microscope in an STEM condition. Taken from Williams and Carter (2009). (For color version of this figure, the reader is referred to the online version of this book.)

system is optimized to focus the electron beam to form as small a probe as possible upon an electron-transparent sample as discussed in earlier CBED sections. Indeed, following the optimization of the condenser lens system, demagnification of the finite-sized, electron source to form a sub-angstrom probe incident on the sample, is achievable in certain STEM microscopes. Generally, the objective lens will have minimal effect on practical STEM imaging conditions; however, proper objective lens astigmation is required for viewing the smallest probe diameter for alignment purposes. For STEM, the C2 aperture is used as a physical cutoff to reduce aberrations that will lead to significant blurring of the probe, thus increasing the ultimate achievable STEM resolution. Moreover, the C2 aperture will reduce the number of electrons reaching the sample and therefore, defines the available beam current. Ideally, the aperture should be chosen to give the optimal probe size with maximum useable current. It should be noted that modern STEMs may come equipped with a monochromator, located just below the electron source, that, apart from allowing the energy spread in the incident probe to be reduced, also may provide a range of currents for experimental optimization.

Lastly, scan coils are arranged to provide precise beam control allowing for a variety of scattered signals to be detected and plotted as a function of probe position. Modern STEM systems combine the ability to precisely scan the beam in-line profiles and reduced areas while incorporating drift correction for data fidelity. Commonly collected STEM signals include, but are not limited to, annular dark field imaging (ADF), HAADF, XEDS, EELS and on-axis BF imaging (BF).

12.3.4.2 Sample Preparation and Beam Damage to Sample

Sample preparation for CTEM and STEM is very similar in nature. Thus, for both techniques it is desirable to have an electron transparent sample that is <50 nm in thickness, flat, and free from residual damage. The most common methods are electropolishing, dimple grinding/tripod polishing, and focused ion beam (FIB) milled samples. Electropolishing is a well known art. Dimple grinding/tripod polished samples are mechanically prepared, but the final thinning is again performed using a broad ion beam of 2 kV. FIB samples provide for site-specific extraction while the final form is shaped with a high fidelity ion beam ranging from 30 to 2 kV. It is well known that lowering the ion-accelerating voltage will reduce the resultant residual damage/amorphous layer, therefore final ion thinning should be performed at as low a current as practically possible. Recently, low-energy (<1 kV) argon ion mills, with imaging detectors, have been developed specifically for this task. **Figure 44(a)** shows a BF-STEM image of dislocations and ion damage (black dots) after 30 kV Ga^+ ion milling while **Figure 44(b)** shows the marked improvement that is achieved by low-energy Ar^+ ion milling. For the case of atomic resolution STEM imaging, it is critical to perform low-energy milling to remove the residual damage layer, which will act to diffusely scatter the small STEM probe and thereby reduce resolution and contrast.

Regarding electron-induced damage, the modern combination of high accelerating voltages with high brightness electron sources tends to cause significant damage to samples in transmission electron microscopes. The dominant mechanism by which metals are damaged during STEM characterization is by knock-on or displacement/sputtering damage. This process occurs by the direct transfer of the beam energy to atoms in the solid, displacing them from their atomic sites and creating a Frenkel defect (i.e. a combination of a vacancy and an interstitial). Modern STEMs typically have FEG sources with accelerating voltages of 200–300 kV, and knock-on damage is proportional to beam energy. The knock-on damage problem can also be exacerbated due to advances in Cs-correction permitting more current in highly focused probes. Reduced dwell times are often possible with modern detectors, which reduce damage for a given accelerating voltage. Quite often in metallic samples, damage is manifested as secondary defects, e.g., dislocation loops formed by the clustering of vacancies, so that damage can represent a real physical limit for STEM analysis. After a metallic specimen is modified, due to electron

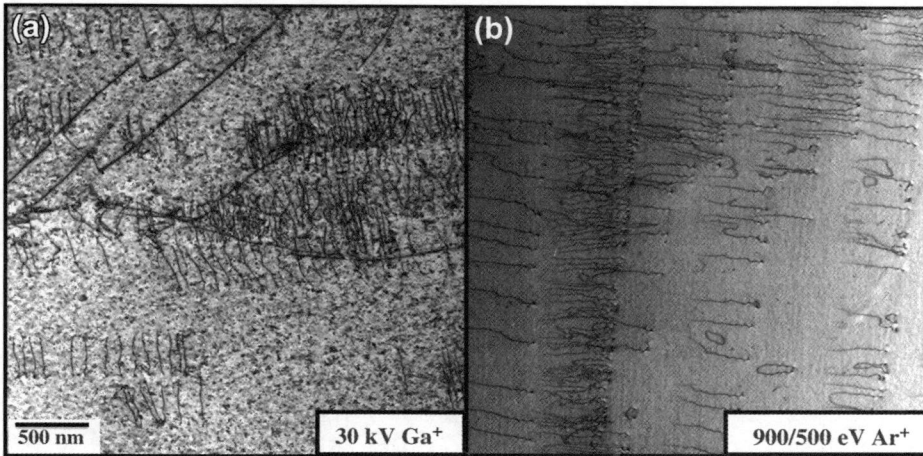

Figure 44 A comparison of the damage incurred during specimen preparation using a dual-beam FIB. (a) Image recorded in a sample prepared using a Ga^+ ion beam with an accelerating voltage of 30 kV, and (b) the image after a further exposure to Ar^+ ions accelerated to 900 and 500 V.

beam interaction, the sample is in some way no longer representative of its parent material and interpretation of data is more difficult, if not impossible. Caution must be taken to ensure that crystal defects are intrinsic to the material rather than introduced by the act of observing the materials in the STEM.

The inelastic scattering events that give rise to useful signals may produce deleterious side effects due to electron-beam damage. Prolonged imaging times may result in defect formation such that the foil is sufficiently modified and so not representative of the bulk material from which is was excised. Therefore, damage can represent a real physical limit for STEM analysis.

12.3.4.3   Imaging Using STEM

12.3.4.3.1   Detector Configurations

Figure 45 is a schematic diagram depicting a common detector setup that allows for collection of a large angular range of scattered electrons. Dependent on the individual microscope configuration, all of these detectors may or may not be installed for data collection. An advantage to having this type of detector arrangement is the ability to collect the signals, scattered at various angular ranges, simultaneously for direct comparison. As discussed below, two other characterization techniques that are readily available with STEM are EELS and XEDS. On this schematic diagram, the XEDS detector is located above the sample and the EELS spectrometer entrance is located on-axis, below the BF detector. Obviously, the BF detector must be extracted from the beam path for the signal to reach the spectrometer, however a configuration can be realized such that ADF, HAADF, EELS and XEDS signals may be simultaneously collected.

12.3.4.3.2   Diffraction Contrast Using STEM

One disadvantage associated with the diffraction contrast technique for defect analysis involves the very basis of the technique, namely dynamical scattering. For example, when characterizing dislocation substructures, the presence of bend contours can obscure large parts of an image. Mills and colleagues (e.g. Phillips et al., 2011b) have shown that significant advantages may be accrued by using STEM to reduce the extent of the dynamical contrast, while retaining the ability to characterize defects following the "rules" of diffraction contrast. An example from their work is shown in **Figure 46**, which compares

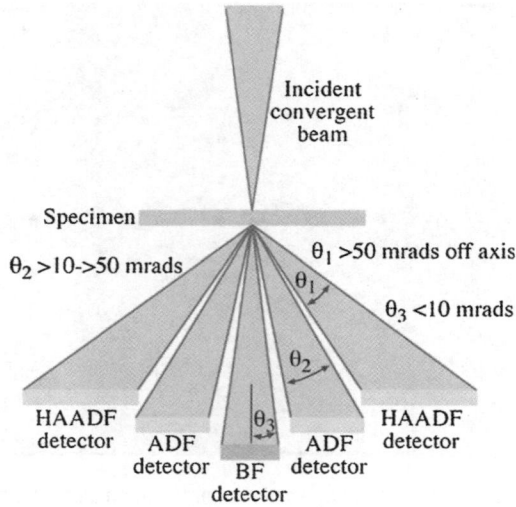

Figure 45   A schematic diagram depicting a common detector setup that allows for collection of a large angular range of scattered electrons after passing through the specimen. Taken from Williams and Carter (2009). (For color version of this figure, the reader is referred to the online version of this book.)

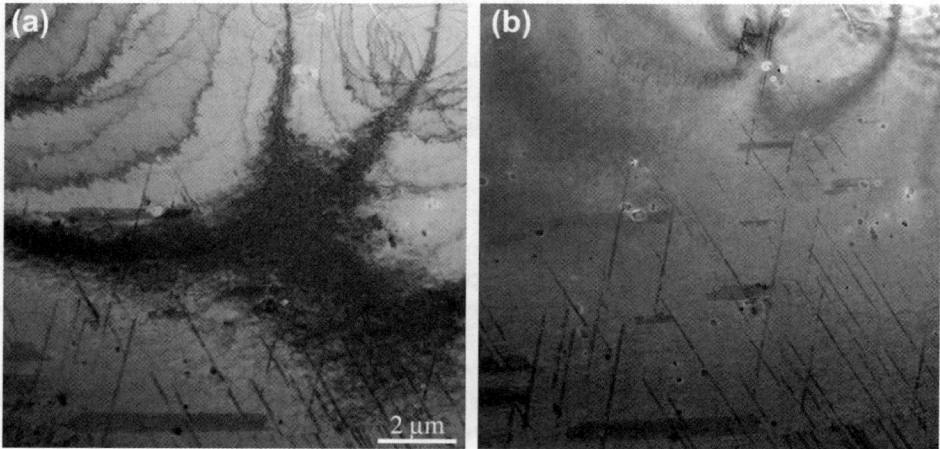

Figure 46   A comparison of the diffraction contrast exhibited when imaging in (a) TEM and (b) STEM. The reduced influence of the bend contours is immediately obvious. Taken from Phillips et al. (2011b).

the same area imaged conventionally in TEM with that in STEM. The advantage regarding image clarity is clear. Of course, this type of imaging relies on diffraction contrast, and will make use of both bright-field and DF images. Hence, most usually, the signals will be collected on the BF and ADF detectors (see **Figure 47**). A significant advantage of such STEM imaging involves the ability to vary the camera length. This permits a deliberate choice of the angular variation of the scattered electrons that are incident on the various detectors, shown schematically in **Figure 47**, for subsequent image formation.

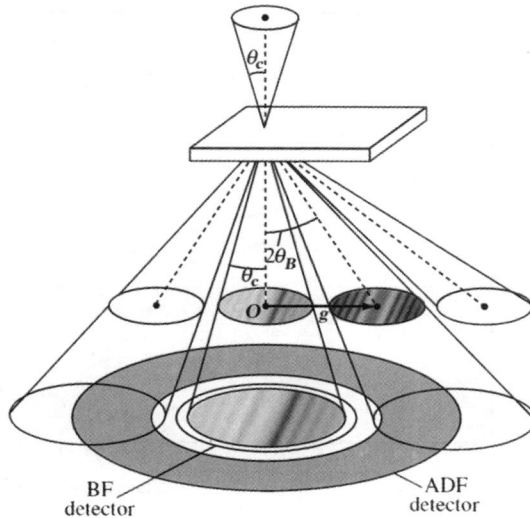

Figure 47   Schematic diagram depicting the arrangement of detectors for STEM imaging. Taken from Phillips et al. (2011b).

For example, in this way, it is possible to ensure that DF images are formed by electrons that have been diffracted rather than having significant proportions of the signal being from those electrons that are incoherently scattered.

12.3.4.3.3 Z–Contrast Imaging

In the discussion above, clearly diffraction effects are extremely important, and are the origin of the contrast observed. In order to detect electrons with the appropriate "signal", i.e. from diffraction, only those electrons scattered over relatively small angles are collected. It is, however, well known that the coherency of the scattered electrons is related to their angle of scattering (θ). As this angle increases, the degree of coherency decreases, and so electrons that are Rutherford scattered to high angles ($>5°$) are incoherent. In this high angular regime, diffraction effects, such as bend contour and thickness fringes, should provide a negligible contribution to the image whereas it is expected that the contrast will be effected by variations in atomic number (Z) in the specimen. A rather simple way to assess the characteristics of cross-sectional dependence on scattering is shown in **Figure 48**, which shows the variation of the screened Rutherford cross-section with scattering angle for (a) three different elements and (b) two different beam energies. The behavior of Cu shows the cross-section decreases by several orders of magnitude as the scattering angle increases; therefore, scattering is most likely to occur in the forward direction and decrease fairly rapidly above a few degrees. This can be observed by collecting a BF, ADF and HAADF image for equal dwell times. Indeed it is possible to saturate the BF image while collecting no useable signal with the HAADF. Furthermore, increasing Z from copper to gold can increase the cross-section by a factor of approximately 10.

Z-contrast imaging corresponds to those observations when image intensity is related to alloy composition (Z). Generally, the intensity of these electrons scattered through large angles varies approximately as Z^2. Unlike HRTEM-phase contrast imaging, HAADF imaging can be used to form directly interpretable images from a crystalline specimen, e.g. spatially resolved images of atomic columns in a sample, in which the image contrast is due solely to the value of Z, using the HAADF

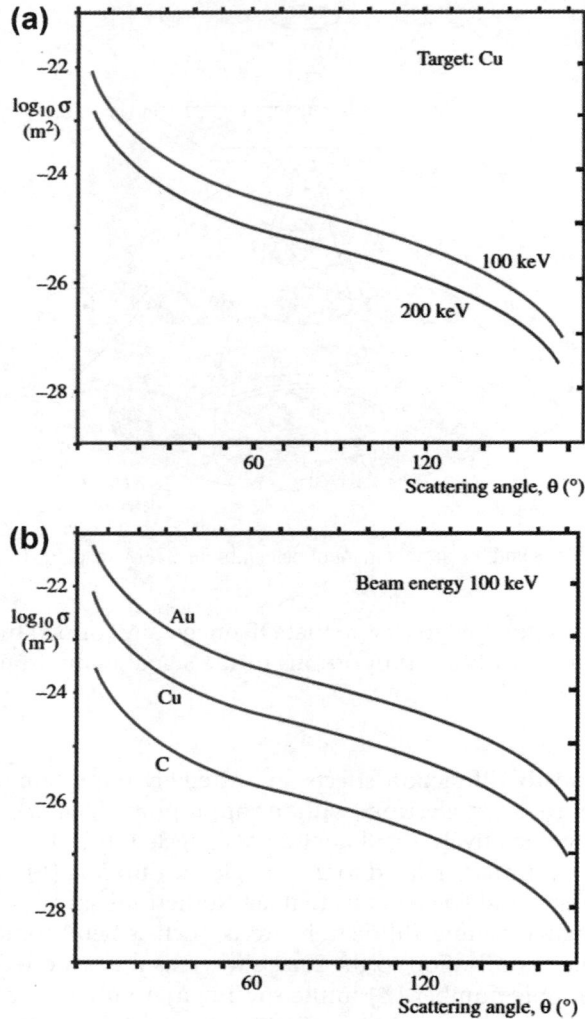

Figure 48 Diagrams to illustrate the characteristics of cross-sectional dependence on scattering through the variation of the screened Rutherford cross-section with scattering angle for (a) two different beam energies (for Cu: 100 keV, 200 keV) and (b) three different elements (Au, Cu, C). Taken from Williams and Carter (2009).

detector for signal collection. Hence, the need to record through-focal series of micrographs, and then reconstruct the object, is obviated. An example of the use of such spatially resolved images involves the high-resolution STEM (HRSTEM) image in **Figure 49**, which shows a particle of the ω-phase, which has transformed from the surrounding β-phase in a Ti–Mo alloy (Nag et al., 2011). It has traditionally been thought that the ω-phase forms by a shear transformation; the blue box outlines the ω motif for this beam direction, the green line shows the expected orientation of the two atomic columns within the box, and the yellow line approximates the actual orientation of the columns. The locations of the atomic columns in the athermal ω-phase are consistent with a partial collapse, and subsequent research, again involving HRSTEM imaging, reveals that the transformation is actually a mixed-mode one, involving both shear and diffusion. In this case, HRSTEM (HAADF) imaging proved invaluable.

In addition to showing atomic-resolution detail, Z-contrast images also provide qualitative, spatially resolved compositional information at interfaces and ordered alloys as shown in **Figure 50**. In this example, the study was of interfaces between the γ (based on a Ni solid solution) and γ' (based on Ni$_3$Al) phases in a Ni-base superalloy (Srinivasan et al., 2009). Thus, the low magnification HAADF image in **Figure 50(a)** shows the interfacial region between the γ'-phase, on the left, and the γ-phase on the right; as expected, the image of γ'-phase is of lower intensity than that of the γ-phase because of the partitioning of elements with higher atomic numbers, elements (e.g. Cr, Co, W, Nb) to the latter phase. To probe the compositional and structural transition from disordered γ to ordered γ', an intensity profile was plotted across the interface by averaging the column intensities in the direction of the arrow over the box ABCD in **Figure 50(b)**. The profile of average intensity across atomic columns lying parallel to AB (along the {002} planes) versus the perpendicular distance has been plotted in **Figure 50(c)** Again, the higher background intensity in the γ matrix is due to the partitioning of heavy alloying elements. In addition to the expected superlattice contrast in the γ-phase and lack of it in the γ-phase, a transition zone is observed at the γ/γ interface in **Figure 50(c)**. The ratio of the intensity of each atomic column to its adjacent column on the right can be used as a qualitative measure of the presence or absence of long-range chemical ordering, as shown in **Figure 50(d)**. Thus, for the γ-phase, this ratio remains essentially constant at a value near 1, since the compositions of atomic columns are assumed to be equal in a random solid solution, while for the ordered γ phase the ratio alternates between ≈ 1.1 and 0.9 due to the differences in the site occupancies. In **Figure 50(d)** a clear order–disorder transition zone can be defined at the γ/γ interface (d_1) where the long-range order decreases roughly over 6–8 atomic {002} planes (approximately 1–1.3 nm in width). This region of structural change between the ordered and disordered condition has been revealed by HAADF imaging.

It is of interest to consider to what degree these various techniques may be quantitative. It turns out that generally, the detectors are not uniform regarding electron-collecting efficiency. Recently, an

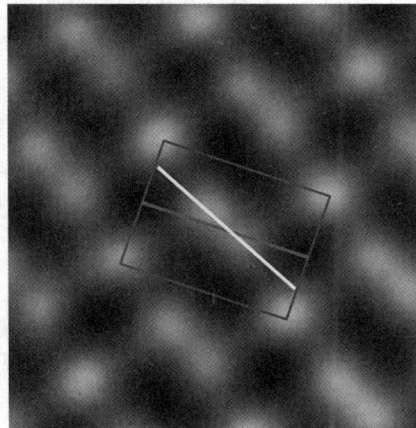

Figure 49   HAADF image of the athermal ω-phase in a Ti–Mo alloy. The images reveal that the transformation to the "stable" structure of ω is not complete and the shear process has only partially occurred. Taken from Nag et al. (2011).

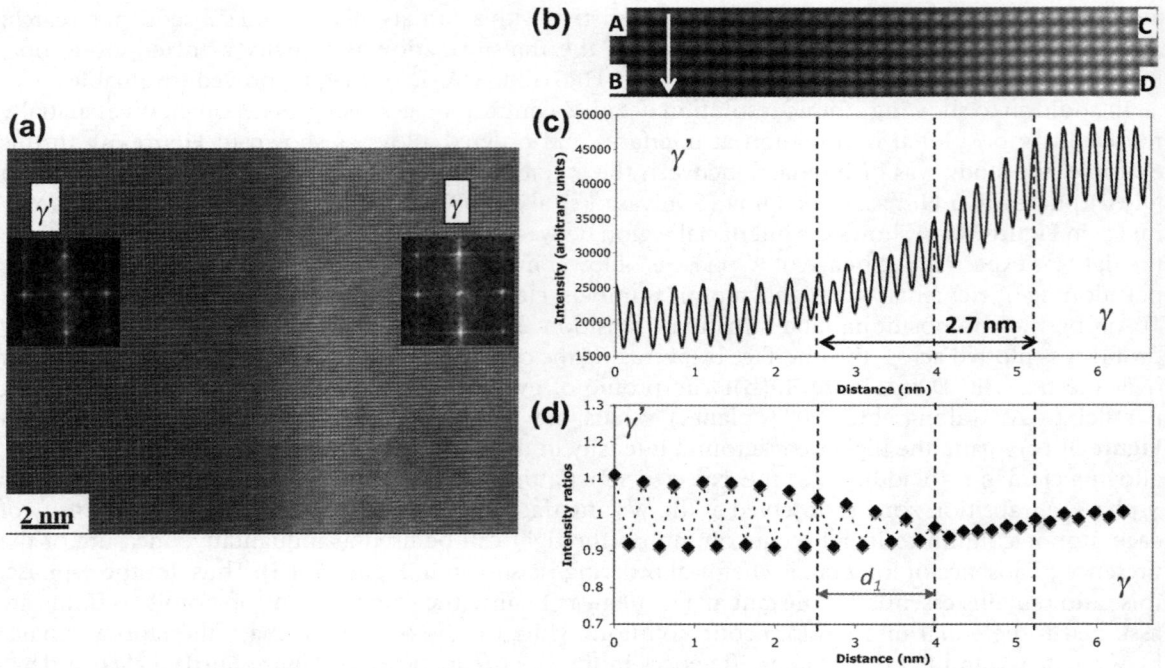

Figure 50 (a) and (b) HAADF images of the interface between the γ and γ' phases in a Ni-base superalloy. The intensities in the HAADF image have been used to generate the profiles shown in (c) and (d). See text for explanation. Taken from Srinivasan et al. (2009).

HAADF detector was calibrated such that it would provide a linear signal as an output (LeBeau and Stemmer, 2008). When combined with accurate thickness measurements, this allowed for quantitative HAADF Z-contrast imaging. In the future, this may become a more readily available technique that is provided on a number of systems. Currently, this is not case. The advancement of STEM imaging and optics, along with the inception of Cs (probe aberration) correctors, has produced the highest resolution, most directly interpretable HR-STEM images to date.

12.4 Electron Energy-Loss Spectroscopy (EELS)

As described earlier in this review, fast electrons interacting with a thin sample in the TEM can be scattered elastically or inelastically. In electron energy-loss spectroscopy (EELS), elastically scattered electrons (which have lost a negligible amount of energy) are contained in the "zero-loss peak (ZLP)". The inelastically scattered electrons, which have lost energy due to interactions with the specimen, can be analyzed by the electron energy-loss spectrometer, giving detailed information on the local chemistry, electronic structure and bonding of the sample (Garvie et al., 1994; Brydson, 2001; Keast et al., 2001).

12.4.1 The Electron Energy-Loss Spectrum

Figure 51 shows typical electron energy-loss spectra to illustrate the principal spectral features. The energy-loss spectrum can be divided into three main regions: the ZLP (at 0 eV), the low-loss region

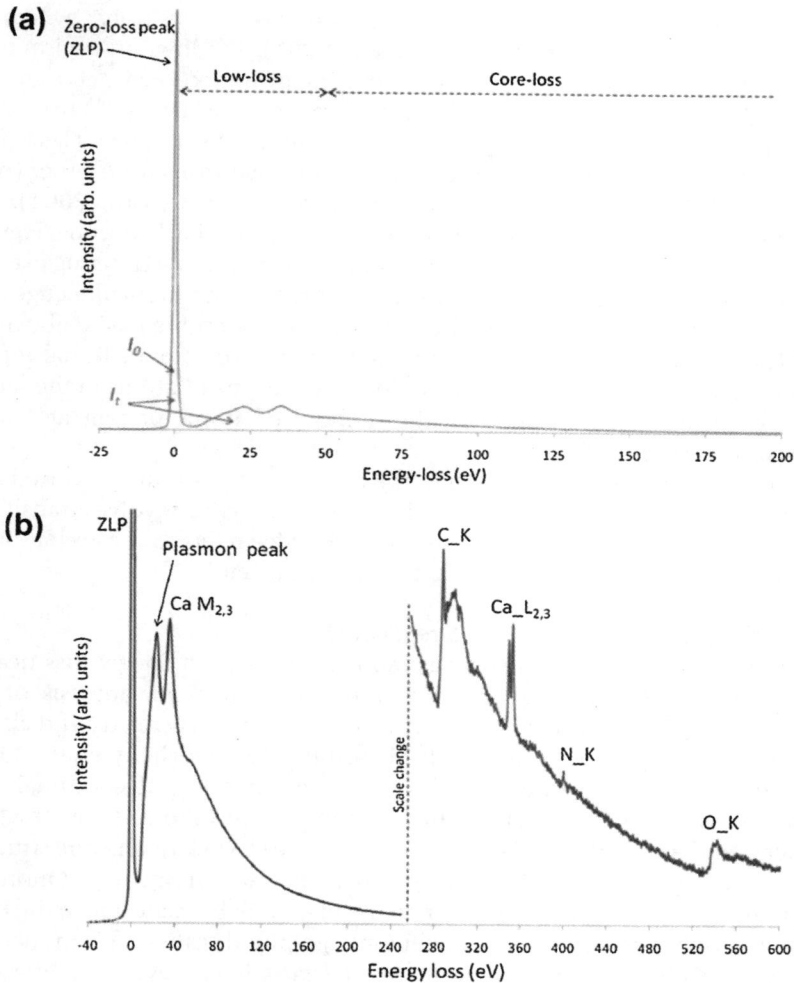

Figure 51   (a) Typical EELS spectrum of ivory dentine showing the intense zero-loss peak (ZLP) and the low-loss region (containing the plasmon peak), up to 50 eV, followed by the core-loss region at higher energies. I_0 denotes the area under the ZLP and I_t represents the total area under the spectrum; (b) magnified view of the low-loss (blue line) and core-loss (green line) spectra. From Jantou (2009). (For interpretation of the references to color in this figure legend, the reader is referred to the online version of this book.)

(typically up to 50 eV) and the core-loss region (from 50 eV onward). The ZLP is usually the most intense feature on the spectrum. As the name implies, it contains electrons that have lost no measurable amount of energy. The full-width at half maximum (FWHM) of the ZLP is a measure of the energy resolution of the spectrum. The energy resolution mainly depends on the energy spread of the electron beam, which in turn depends on the electron source. A Schottky FEG source gives a typical energy spread of 0.6–0.8 eV while a cold FEG gives 0.3–0.5 eV. Use of a monochromator can improve the energy resolution to 0.2 eV or less.

The low-loss region is dominated by peaks associated with plasmons, which correspond to a collective oscillation of the valence electrons (Brydson, 2001). The low-loss region may also contain single-electron excitations from the valence band to low-energy unoccupied electronic states above the Fermi level. The low-loss region can be used to obtain the relative thickness (t/λ) of a sample and can be used to deconvolute the effect of plural inelastic scattering in thick specimens (Egerton, 2011).

At higher energy losses, the incident electrons excite atomic electrons from inner (or core) shells to unoccupied energy levels (Egerton, 2011; Williams and Carter, 2009; Brydson, 2001). This leads to the appearance of ionization "edges" superimposed on a decreasing background (**Figure 51(b)**). The background on an EELS spectrum arises from contributions from plural-scattering events and from the tail of core-loss edges lying at lower energies. The threshold energy of a particular edge is determined by the energy required to promote an inner-shell electron to an unoccupied state above the Fermi level within an atom. The EELS technique can therefore be used to give compositional information as the presence of an edge on the spectrum can be used to identify the type of atoms in the sample. Moreover, the intensity of a particular edge depends on the number of atoms present and may be used for quantitative analysis (Egerton, 2011).

EELS ionization edges are labeled according to the type of transition that occurred and the electron shells (i.e. K, L, M) and subshells (i.e. s, p, d) involved. Thus, a K-edge involves transitions of electrons from 1s energy levels in the unexcited atom, whereas an L_1-edge refers to the excitation from 2s levels, an L_2-edge involves excitation from $2p_{1/2}$ levels, L_3- from $2p_{3/2}$ etc.

12.4.1.1 *Electron Energy-Loss Near-Edge Structure (ELNES)*

The fine structure observed on core-loss edges is termed the electron energy-loss near-edge structure (ELNES) (Egerton, 2011). Provided the energy resolution is sufficient, analysis of the ELNES can provide information on the bonding environments, oxidation states, chemistry and electronic structure with high spatial resolution (Garvie et al., 1994; McComb, 1996; Docherty et al., 2001). The ELNES region can extend up to 30 eV above the edge threshold, although the effect of final state broadening decreases the usefulness of the information further from the threshold. Fine structure beyond the ELNES region is termed the extended energy-loss fine structure (EXELFS). The fine structure on an EELS edge arises because, during the excitation process, the incident electron can impart more energy than the critical energy required to promote a core electron to the first available state above the Fermi level. If the core electron energy is only a few eV greater, it will undergo plural scattering from neighboring atoms. This scattering is responsible for the ELNES. If the core electron has an even greater excess of energy, it will be scattered only once. This single-scattering event is the cause of the EXELFS. This region is usually weak and may be obscured by other edges lying at higher energies (Williams and Carter, 2009).

The energy imparted to the core electron beyond the ionization threshold energy opens up a range of possible energy levels above the Fermi level where the electron may reside. However, although the excited electron can reside in a range of unfilled final states, it does not do so with equal probability: some available states are more likely to be filled than others as there are more states within certain energy ranges than others (Egerton, 2011). This nonuniform distribution of available electron energy levels is termed the unoccupied density of states (DOS), $\rho(E)$. Due to the higher probability of excited electrons residing in certain empty states above the Fermi level, the intensity in the core-loss edge (above the edge threshold) is greater at the corresponding energy losses: the excited electron preferentially fills an energy region where the DOS is higher. This is the reason why the ELNES is often referred to as a tool to probe the electronic states above the Fermi level of the excited atom (Brydson, 2001). Hence the importance of studying the ELNES, as the DOS is directly related to how the atoms present in a specimen interact to form bonds with each other.

The intensity of a peak in the ELNES is related to the probability of transition occurring between an initial state wavefunction, i, and a final state wavefunction, f. Derived from Fermi's golden rule this can be expressed in terms of a double differential cross-section for inelastic scattering, which depends on the transition matrix element and the unoccupied DOS as shown in Eqn (4).

$$\frac{d^2I}{d\Omega dE} = \frac{4\gamma^2}{a_0^2 q^4} |\langle f|\exp(iq \cdot r)|i\rangle|^2 \rho(E).$$ (4)

The transition matrix element is generally a slowly varying function of energy over the region up to about 30 eV above the edge threshold (Egerton, 2011). Therefore, the ELNES can be attributed to the structure of $\rho(E)$. In Eqn (4), γ is the relativistic factor, a_0 the Bohr radius, q the momentum transfer. The exponential term can be expanded as $[1 + i(q \cdot r) + higher\ order\ terms]$. The unity term gives a zero value for the matrix element since the initial- and final-state wavefunctions are orthogonal. The second term results in a nonzero matrix element only if the initial- and final-state wavefunctions have different symmetries. This manifests itself as the optical dipole selection rule, which requires a change in the orbital angular momentum quantum number, ℓ, in order for a transition to occur (Egerton, 2011):

$$\Delta\ell = \pm 1.$$ (5)

The higher order terms are very small and have no significant effect of the structure of the ELNES (Egerton, 2011). It is now apparent that $\rho(E)$ is the site- and symmetry-projected unoccupied DOS: Symmetry-projected due to the dipole selection rule and site-projected since the initial state wavefunction is highly localized on the atomic site of excitation and will be dependent on the local atomic site coordination, valence and bonding. Any changes in the local chemical or coordination environment will be reflected in changes in the ELNES. The electron energy-loss spectrum is typically collected within a small angular range so that the dipole selection rule applies (Egerton, 1981). The ELNES shape of a K edge is determined by the density of p-like states, whereas the ELNES associated with an L edge depends on the density of d-like (and to a lesser extent s-like) states (Brydson, 2001).

12.4.1.2 Low-Loss Spectroscopy

The low-loss part of a spectrum of a thin specimen contains the majority of inelastically scattered electrons and has the advantage of possessing a high signal. The energy scale of low-loss spectra is easily calibrated by using the ZLP. Since both single electron and collective excitations contribute to the low-loss signal it is not appropriate to describe the spectrum using Eqn (4). In some cases, it is informative to use a dielectric formulation from which it can be shown that the low-loss spectrum is proportional to $Im[-1/\varepsilon(E)]$, where $\varepsilon(E)$ is the complex dielectric response function. It is possible to utilize a Kramers–Kronig transformation to extract both the real and imaginary parts of $\varepsilon(E)$ from the low-loss spectrum and thus differentiate between peaks in the spectrum that are associated with single and collective excitations.

The low-loss region can give valuable information about the relative thickness (t/λ, where t is the thickness and λ the mean free path) of the region being probed by the electron beam (Egerton, 2011). If the exact composition is known, the absolute thickness may be calculated. The mean free path depends on the incident beam energy and the chemical composition of the sample. The easiest method to determine the relative thickness of a region of the sample defined by the incident beam is to record an

EEL spectrum and use an integration procedure, termed the "log-ratio method", to compare the area under the ZLP, I_0, with the total area, I_t, under the spectrum (Egerton, 2011; Malis et al., 1988):

$$\frac{t}{\lambda} = \ln\left(\frac{I_t}{I_0}\right). \tag{6}$$

In practice, the intensity of the spectrum decreases rapidly with increasing energy loss and therefore it is often approximated to the intensity of the low-loss region. For EELS analysis in the TEM, t/λ should typically be 0.2–0.4. If t/λ is less than 0.2, the SNR may be too low, whilst for t/λ much greater than 0.4, plural scattering may contribute to the spectrum and obscure features in the fine structure of the ionization edges.

12.4.1.3 Core-Loss Spectroscopy

There are several key processing steps that need to be undertaken before fine structure on a core-loss edge can be reliably analyzed (Egerton, 2011). **Figure 52** shows an example of two core-loss EEL spectra acquired from hydroxyapatite crystals in a section of ivory dentine prepared by ultramicrotomy. Since the ZLP is not in the acquired spectrum energy calibration is less straightforward than with low-loss spectra. In principle, the zero-loss detector channel should calibrated to the voltage offset applied to the spectrometer drift tube; however, if the energy stability is poor or if there are spectral shifts due to magnetic fields this may not be reliable. In **Figure 51** the energy scale for core-loss spectroscopy was calibrated using the known threshold energy of the carbon K-edge of the amorphous carbon support film of the TEM grid. After calibration the background intensity is subtracted in order to remove contributions from the zero-loss and plasmon peaks, as well as from tails of edges lying at lower energies (Egerton, 2011). An energy window of several tens of eV wide (i.e. 50–70 eV) is placed just before the edge onset. This corresponds to the red box just before the phosphorus and carbon edges in **Figure 52**. The background signal is fitted using an inverse power-law of the form, AE^{-r}, where E is energy-loss, and A and r are fitting constants. For the calcium $L_{2,3}$ signal, a smaller background window (i.e. 20 eV) is fitted in order to remove contributions from the carbon edge, so that only the signal that is produced by calcium in the specimen remains. A good indication that the background has been subtracted correctly is that the intensity before the edge onset (on the background subtracted edge) fluctuates only slightly around zero intensity (**Figure 52**).

If the sample is too thick ($t/\lambda > 0.4$) deconvolution procedures must be carried out to remove the contribution of plural scattering. Plural scattering arises from the contribution of combined energy losses from core and valence electron excitations. This increases the intensity in the higher energy region of an ionization edge, can mask the fine structure and can make the background signal on subsequent edges deviate significantly from the power law model. In order to deconvolve a core-loss spectrum, the Fourier-ratio method is used. This requires collection of both the low- and core-loss spectra from the same region of the sample, under identical conditions (i.e. eV/channel, convergence and collection semiangles). Once the core-loss signal is collected, the first step is to isolate the edge of interest and remove the background intensity. The low-loss spectrum and background-subtracted edge are then Fourier-transformed. The core-loss spectrum Fourier transform is divided by the low-loss Fourier transform and the result is inverse Fourier transformed to yield the desired deconvolved spectrum. However, recording low-loss and core-loss spectra under identical conditions is in many cases extremely challenging. It is often the case that under experimental conditions required for good SNR in the core-loss spectrum it is impossible to use a short enough acquisition to avoid saturation of the signal from

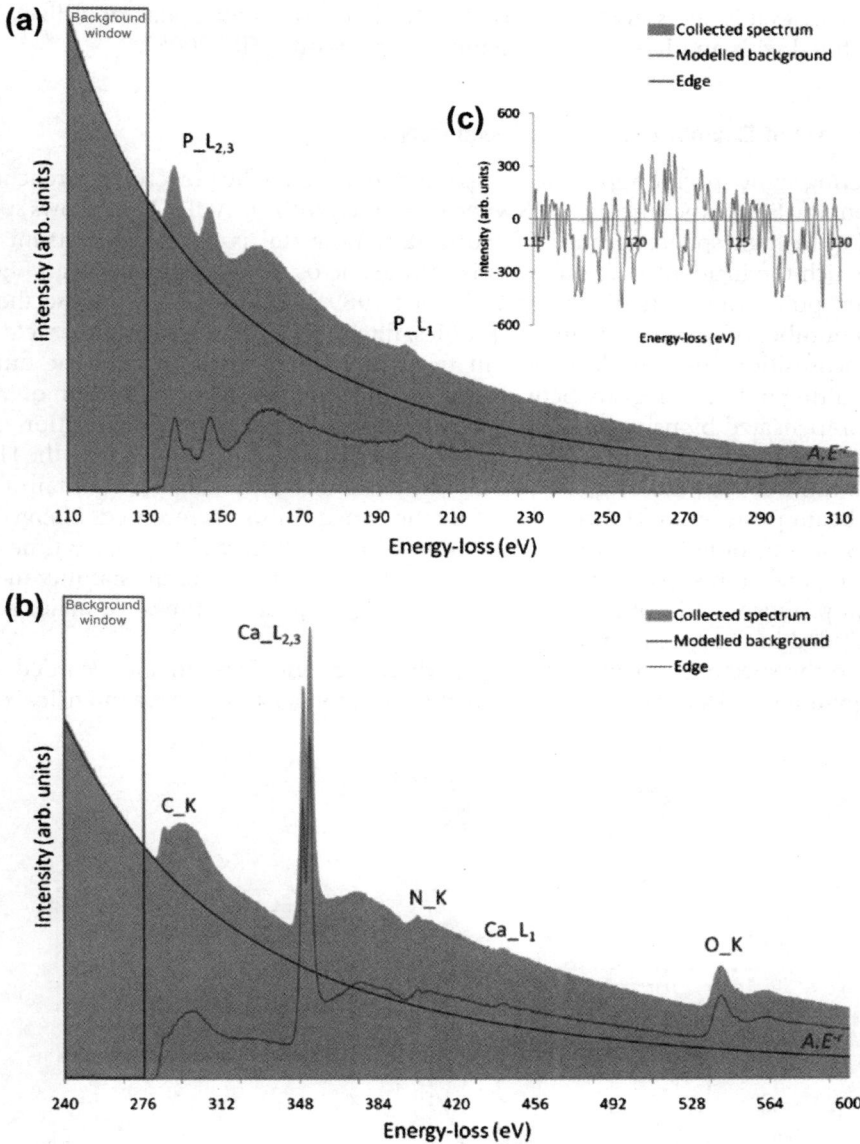

Figure 52 Example of core-loss spectra, showing the signal from (a) phosphorus and (b) carbon, calcium and oxygen. The edges (green line) are extracted after removal of the background signal (red line); (c) is a magnified view of the background signal after background subtraction from (a). Taken from Jantou (2009). (For interpretation of the references to color in this figure legend, the reader is referred to the online version of this book.)

the ZLP. In such cases it is necessary to sacrifice the SNR in the core-loss signal, or utilize a spectrometer system that is fitted with an ultrafast electrostatic shutter (Scott et al., 2008).

12.4.2   Applications of Electron Energy-Loss Spectroscopy

Before considering some specific applications it is worth discussing the powerful approaches that can be used to carry out the EELS experiments. EELS when combined with HAADF-STEM allows spectroscopy to be performed with high spatial resolution. Particularly powerful is to use a spectrum imaging (SI) approach in which the incident electron beam is scanned across the specimen and an EEL spectrum is collected at every point (x,y) (Jeanguillaume and Colliex, 1989). The result is a 3D dataset that can be post-processed in a number of ways to obtain line profiles, filtered images, thickness maps, etc. (**Figure 53**).

During SI acquisition the sample may drift from its original position over the duration of the acquisition resulting in loss of registry between the spectra and their relative position on the specimen. This can be compensated by using spatial drift correction software. During acquisition of a spectrum image, data collection is paused so that the spatial drift region can be scanned using the HAADF signal. A cross-correlation between the original and the drifted image is calculated to obtain the correction vector for the beam position on the sample before the SI acquisition is resumed. Energy drift can also occur during SI acquisition—effectively the position of the ZLP may change over time due to insta-bilities in the high-tension supply or thermal effects which affect the current stability in the magnetic prism. This can produce a drift of a several eV during long acquisition times—a major problem if the ZLP is not included in the spectrum image.

In contrast to the spectrum imaging technique where individual spectra are recorded at each probe position, energy-filtering TEM data consist of a two-dimensional image whose intensity is a function of

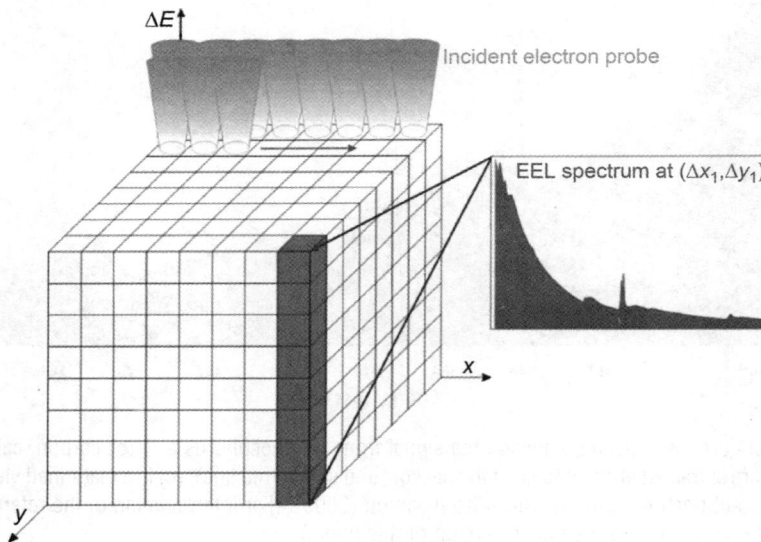

Figure 53   Three-dimensional data cube illustrating the principle of STEM "spectrum imaging". The electron beam (shown in green) is scanned across the sample and an EEL spectrum is collected at each data point (x,y). The green arrow represents the scan direction. Taken from Jantou (2009). (For interpretation of the references to color in this figure legend, the reader is referred to the online version of this book.)

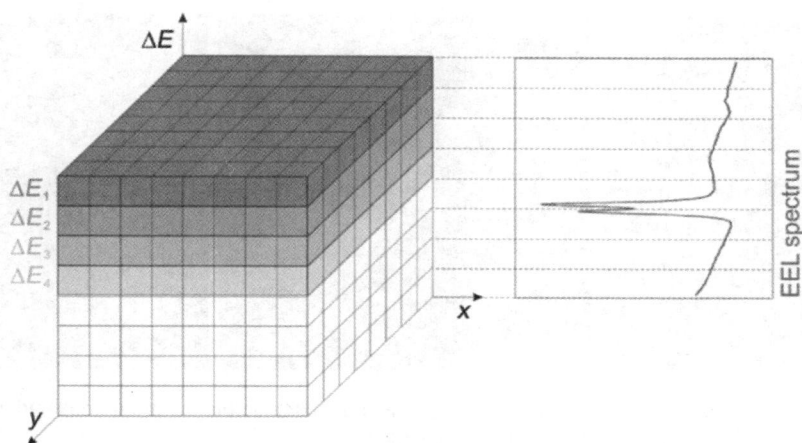

Figure 54 Illustration of the principles of EFTEM, or "image-spectroscopy". Energy-filtered images are acquired containing a narrow energy range ΔE_x. The 3D data cube is filled one energy plane at a time. Taken from Jantou (2009). (For color version of this figure, the reader is referred to the online version of this book.)

energy loss (Reimer et al., 1995; Reimer, 1998). This technique is illustrated schematically in **Figure 54**. Energy-filtered images are recorded sequentially at different energy losses ΔE_x. The width of the energy-selecting slit determines the energy resolution of the energy-filtered images. Using similar post-acquisition procedures to spectrum imaging, the image intensity at each pixel in an energy-filtered image can be converted to mass or concentration, in order to obtain information about the spatial distribution of a particular element. Both thickness and elemental mapping can be achieved by adjusting the energy shift so that only the low-loss signal or a chosen ionization edge, respectively, pass through the energy-selecting slit.

12.4.2.1 *Zero-Loss Filtering and Thickness Mapping*

Traditionally EFTEM mapping using the low-loss part of the EEL spectrum has been used for zero-loss filtering (**Figure 54**) and thickness mapping (**Figure 55**). Conventional BF TEM images contain both elastic and inelastic contributions, and the latter increases with sample thickness. By selecting a narrow energy slit (e.g. 5 eV) and placing the energy window over the ZLP, the majority of the inelastic component is filtered out. This significantly improves the image contrast (although the overall image intensity is lower) and resolution due to the reduction in chromatic aberrations. Such images are called zero-loss filtered or elastic images (Brydson, 2001). A thickness map (**Figure 55**) can be obtained by acquiring both an unfiltered and zero-loss filtered image from the same region of interest. Unfiltered images are acquired with the energy-selecting slit removed so that both elastic and inelastic electrons fall onto the spectrometer detector. By processing the intensity of both images at each pixel using Eqn (6) a map of the relative thickness (t/λ) can be obtained.

Zero-loss filtered images and thickness maps can also be directly obtained by post-processing of an STEM-EELS spectrum image. This has significant advantages over the traditional EFTEM approach as it allows significantly more user intervention in the calculation procedures to ensure that any energy drift in the SI is compensated for by alignment of the zero loss peak, to monitor the quality of the fitting and

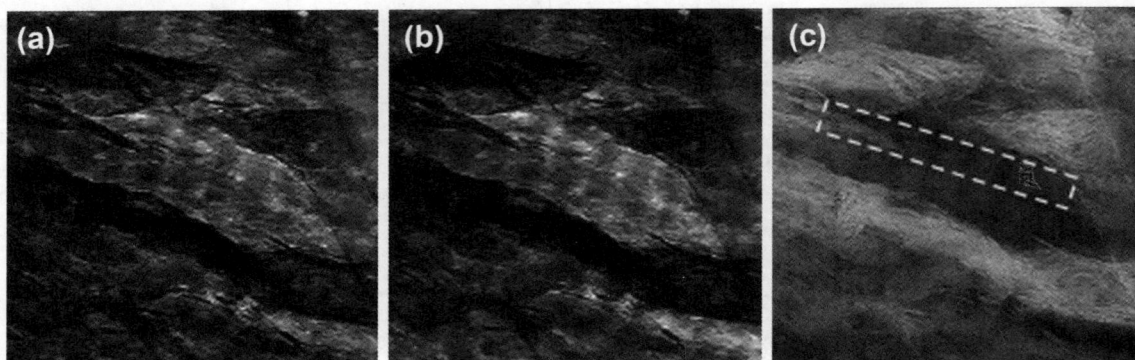

Figure 55 Mineralized collagen fibril from ivory dentine, prepared by ultramicrotomy, in (a) unfiltered and (b) zero-loss filtered (or elastic) mode; (c) shows a t/λ map of the same region.

subtraction of the ZLP and to correct for other instrumentation artifacts such as gain nonuniformities and variations in beam current.

12.4.2.2   Plasmon Mapping

The most intense region of the low-loss spectrum involves plasmon excitations that arise from the interaction of the electron beam with the valence electron density of the specimen. In a simple approach, the valence electrons in a solid can be treated as free electrons constituting a free-electron gas (Egerton, 2011). The response of the electron gas to an electromagnetic field, such that generated by a charged particle, can be modeled using the classical Drude–Lorentz model (Egerton, 2011). This approach combines the free-electron Drude model of metals that considers the valence electrons of a metal as free particles, with the Lorentz's dipole oscillator model for which the electrons oscillate collectively in response to the perturbing electromagnetic field. Plasmon excitation involves a collective response of the valence electrons that appear to oscillate as a result of the interaction with the high-energy incident electrons. The electrons in the sample oscillate at a characteristic angular frequency (ω_p), which is obtained by solving the equation of motion of the displacement of the electrons due to an oscillating electric field E. This resonance frequency can be written as:

$$\omega_p = \left(\frac{ne^2}{\varepsilon_0 m}\right)^{1/2}, \tag{7}$$

where n is the valence electron density, e and m are the charge and mass of the electron, respectively, and e_0 is the permittivity of free space. The energy loss associated with the excitation of the plasmon is given by:

$$E_p = h\omega_p = h\left(\frac{ne^2}{\varepsilon_0 m}\right)^{1/2}. \tag{8}$$

A number of low atomic number metals and their alloys can be considered as free-electron type materials. In other metals the band structure and properties deviate significantly from free electron-like behavior. This can be accounted for by modification of the theory by introduction of damping terms, etc; however, the fundamental proportionality between the E_p and $n^{1/2}$ is not changed (Raether, 1965).

Table 6   Alloys in which the variation in plasmon energy loss E_p has been measured as a function of composition. Taken from Williams and Egerton (1976).

Alloy	Range (at.%)	E_p (eV) variation with fractional concentration C	References
A1–Mg	0–100	$E_p = 15.3 - 5.0C_{Mg}$	(1) Spalding and Metherell (1968)
A1–Mg	0–8	$E_p = 15.3 - 4.4C_{Mg}$	(2) Hobbert et al. (1972
Mg–A1	0–9	$E_p = 10.61 + 5.9C_{Al}$	(3) Porter and Edington (1976b)
A1–Cu	0–2	Non linear	(4) Spalding et al. (1969)
A1–Cu	0–2	$E_p = 15.3^a - 10C_{cu}$	(5) Doig and Edington (1973)
A1–Cu	0–17.3	$E_p = 15.3 + 4.0C_{cu}$	(6) Williams and Edington (1976b)
A1–Zn	0–30	$E_p = 15.3^a - 0.2C_{zu}$	(7) Cool and Cundy (1969)
A1–Ag	0–6	$E_p = 15.3^a + 1.6C_{Ag}$	(8) Porter et al. (1974)
A1–Li	0–25	$E_p = 15.3^a - 4.0C_{Li}$	(9) Williams and Edington (1974)
A1–Ge	0–10	$E_p = 15.3^a + 0.1C_{Ge}$	(10) Williams (1975)
A1–Zn–Mg	0–4	$E_p = 15.3^a - 4.7C_{Mg}$	(11) Doig et al. (1973)

[a]Normalized to 15.3 eV energy loss for pure aluminum.

Williams and Edington (1976) reported a wide range of Al and Mg alloys where the shift in E_p as a result of alloy composition varies linearly with composition, x:

$$E_p(x) = E_p(0) + x(dE_p/dx) \tag{9}$$

These data are reproduced in **Table 6** with selected results displayed graphically in **Figure 56**. The sharp plasmon peaks observed in these alloys facilitates this analysis but since E_p is essentially measured as the difference between the center of mass of two peaks in a spectrum i.e. the ZLP and the plasmon peak, it is possible to get accuracies of <100 meV even if the energy resolution achievable on the microscope is poor. This method has been used to study the volume fraction of precipitates in aluminum alloys (Tremblay and L'Esperance, 1994), to evaluate composition in aluminosilicate catalysts (McComb and Howie, 1990) and to investigate hydrogen content in zirconium–niobium alloys (Woo and Carpenter, 1992).

12.4.2.3   Elemental Mapping

The ability to obtain two-dimensional distributions of elements in a specimen can be enormously useful in material's characterization and EELS performed either in TEM or STEM mode allows this to be achieved. In EFTEM elemental maps are generated by imaging with electrons that have lost energies corresponding to the characteristic inner-shell ionization edges of interest. In order to generate such a map, the "three-window method" is used (Williams and Carter, 2009; Brydson, 2001; Egerton, 2011). For quantitative mapping this method normally requires two "pre-edge" images and one "post-edge" image (**Figure 57**). The post-edge image contains information about the spatial distribution and concentration of the chosen element, although with the additional back-ground intensity, which must be removed. To evaluate the contribution of the background, the pre-edge images are acquired at slightly different energies. They are used to produce an extrapolated background image, assuming that the corresponding background intensities follow an inverse power law. An algorithm is used to subtract the background image from the post-edge image to generate an elemental map where the intensity at each pixel is directly related to the areal density. Using large

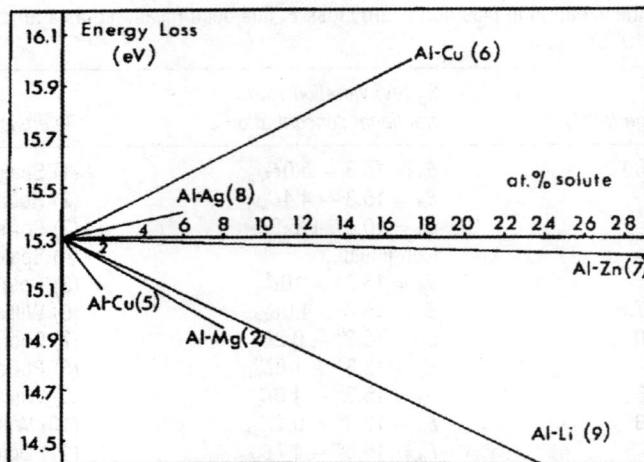

Figure 56 A selection of experimental data showing the variation of plasmon energy of pure Al with various additions of solute. All values of pure Al are normalized 15.3 eV. Taken from Williams and Egerton (1976).

energy windows improves the reliability of the background fit. However, care must be taken so that they do not include contributions from neighboring edges in the pre- and post-edge regions. Because the pre-edge and post-edge images are recorded sequentially during the three-window procedure, it is important to ensure that spatial drift does not occur between successive acquisitions—a cross-correlation algorithm is used to ensure that all energy-filtered images are spatially registered with respect to each other.

As discussed previously maps of this nature can also be directly obtained by utilizing the spectrum-imaging approach. Since data are acquired over the full desired spectral range at every pixel this allows post-processing to obtain elemental maps. Again there are significant advantages to this approach compared with EFTEM imaging since the processing steps can be easily modified to deal with complexities such as overlapping edges or instrument artifacts. Full quantification requires the low-loss spectrum to be available for every pixel to allow deconvolution of plural scattering and normalization of thickness to obtain atomic densities (Scott et al., 2008). This method is termed "dual EELS" and requires ultrafast shuttering to allow the low-loss spectrum image to be collected under the same angular conditions as the core-loss spectrum image.

12.4.2.4 Chemical Bonding
The use of ELNES as a probe of local chemical environment has proved extremely powerful in the study of inorganic compounds, organic materials, polymers, composites and many other systems. In metals and alloys the unoccupied DOS is generally a relatively slowly varying function of energy that makes the use of ELNES less appealing than in other materials. However, ELNES can be a very powerful approach to investigate segregation to grain boundaries and interfaces, as well as the chemistry of precipitates in metals and alloys. Finally, extended energy-loss fine structure (EXELFS) has been used to investigate interatomic spacings in a number of intermetallics. Fourier analysis of the EXELFS data yields a radial distribution function for a particular element and it has been shown that this is a sensitive probe of changes in the interatomic spacings.

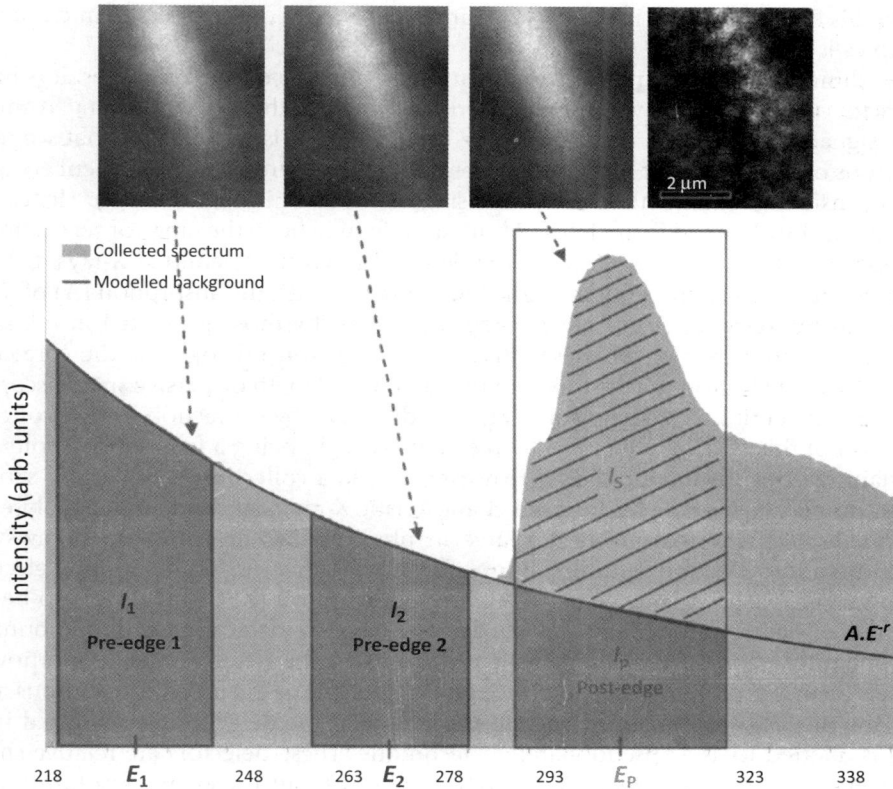

Figure 57 Illustration of the "three-window method" on the carbon K edge for EFTEM elemental mapping. An extrapolated background image is calculated using the power-law model from the two pre-edge images and subtracted from the post-edge image. This results in an elemental map with pixel intensity values that are proportional to the actual concentration of the chosen element. (For color version of this figure, the reader is referred to the online version of this book.)

12.5   X-ray Energy Dispersive Spectroscopy (XEDS) in a (S)TEM

There has been much written about the use of detection of X-rays with energies (or wavelengths) characteristic of given atoms, both for bulk samples (electron microprobe analysis) and surface analysis (Auger electron spectroscopy). These techniques make use of the fact that fast electrons incident on a given solid sample will interact with the electrons in the solid sample in a number of ways; one such interaction involves the ionization of an atom by ejecting an electron from a given energy level, such that subsequently an electron from a higher energy level will drop into the hole in the original energy level, releasing the difference in energy between the two levels as either an X-ray photon or an Auger electron. In general, the probability for the release of a photon increases with atomic number, and oppositely for an Auger electron; these are competitive events. Because electron microprobe analysis is a very well-established technique, its basics are not covered here; the reader is referred to a number of treatments in the literature (Reed, 1997; Goldstein et al., 1981). Here, the focus is on aspects pertaining to performing this kind of analysis in the (S)TEM, and a knowledge of the basics of microprobe analysis for bulk samples is assumed.

In a way of summary, when using electron microprobe analysis on bulk samples, it is important to be aware of factors that impact the volume of material from which the signal is generated and those that influence the signal collected by a detector. In the former case, it is well known that scattering of the incident electrons occurs, such that the volume over which the electrons have sufficient energy to ionize host atoms is considerably larger than the probe size, as depicted in **Figure 58(a)**. Hence, the spatial resolution that may be obtained in analyses of bulk samples will be in the range of several micrometers. There are a number of factors that influence the detected intensity of emitted X-rays, the three most important being the atomic number of the atoms in the sample (Z), the absorption (A) of X-rays by the sample itself, and fluorescence (F) of given X-rays of interest by those generated in the sample. It is possible to correct, more or less, for these effects in the electron microprobe, the corrections being known as ZAF. In most electron microprobes, in the main wavelength dispersive spectroscopy (WDS) is employed, where the emitted characteristic X-rays are dispersed by diffraction by a set of crystals. The energy resolution of this type of detector is then extremely high, being a few electron volts, and this is one of the main reasons for its popularity. However, the data collection is done in a serial manner, having to set a small range of angles for each characteristic X-ray peak, and so the analyses are rather slow. Additionally, the detectors are rather large in physical size, and this can be awkward when incorporating them into electron optical columns for TEM. Rather, for both scanning electron microscopes (SEM) and (S)TEMs, it is more usual for energy dispersive (XEDS) detectors to be employed. These make use of semiconductor devices to detect X-rays by determining the amount of charge developed from interaction of the X-rays with the atoms in the detector, i.e. proportional to their energies. This is a fairly rapid process, and although the energies of the incoming photons are detected individually, the process has the appearance to the user as if the detection was effected in a parallel manner, and is referred to as a "pseudoparallel" technique. These detectors are relative small in size, which is convenient for coupling to electron optical columns, but the energy resolution is not high, being about 140 V, or so. Since the emission of X-rays from the sample occurs essentially in all directions from the sample surfaces, the solid angle subtended by the XEDS detector is rather small, being typically $\approx 0.13-0.3\mathrm{sr}$. This does represent a limitation in terms of count rate, especially for thin samples as used in TEM, and this issue is discussed below. Some new detectors have recently been introduced, with significantly larger collection angles, and these are also discussed below.

12.5.1   Coupling XEDS with (S)TEM

There are, on the one hand, some significant advantages when coupling XEDS with (S)TEM, and, on the other hand, there are some spectral details that must be taken into account. These are discussed in turn.

12.5.1.1   *Advantages*

An immediate advantage in the (S)TEM is accrued by the use of thin foils as samples. Thus, because these foils are usually in the range of 3–300 nm in thickness, the electron beam is broadened by multiple scattering to a far smaller degree than that which occurs in bulk samples, as depicted in **Figure 58(b and c)**. Hence, the spatial resolution of the analyses will be considerably higher than that achievable when using bulk samples. It will be shown below that beam broadening is still an important issue when considering spatial resolution, but nevertheless, the use of thin foils represents a significant advantage regarding this parameter.

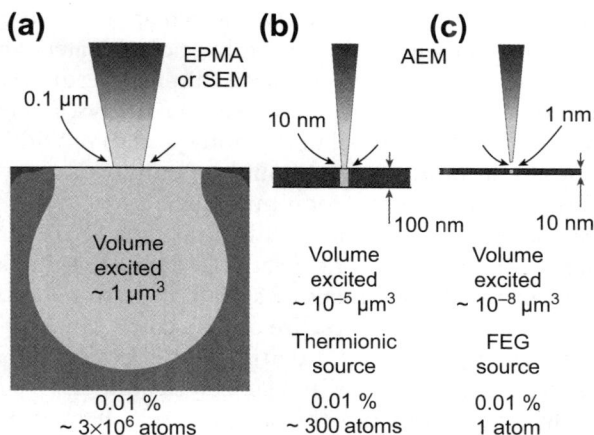

Figure 58 Schematic diagram illustrating the difference in beam spreading between that in bulk samples with that in thin foils. (a) schematic of a bulk sample; (b) schematic of a relatively thick foil; (c) schematic of a very thin foil. Redrawn from Williams and Carter (2009). (For color version of this figure, the reader is referred to the online version of this book.)

The second advantage of the use of thin foils involves the very much reduced effects of fluorescence and absorption which, as noted above, are two of the three most important correction factors when using bulk samples. It is still important to account for the influence of these two phenomena, but their correction is much more simple in thin foil analyses compared with the situation with bulk samples.

The third advantage involves the low intensity of the background spectrum (i.e. the continuous X-ray spectrum, rather than the characteristic spectrum together with multiply scattered photons), and the relative ease with which it can be modeled and subsequently subtracted. This permits the counts in the characteristic peaks to be assessed relatively simply.

12.5.1.2 *Spectral Details*

While the advantages described above are significant and attractive regarding *local* compositional determinations in samples, there are, of course, some disadvantages that must be taken into account when coupling XEDS with (S)TEM. They involve contributions to the collected spectra from artifacts that do not arise from the region of the foil under study. The first of these has been largely made irrelevant by the manufacturers, and involves contributions to a measured spectrum by radiation that is not included in the focused incident probe. It is manifested by placing the probe in a hole in the sample, and determining the presence of any spectral contributions. This is the so-called "*in-h*ole" count and is caused by radiation not focused in the incident probe, and so is generated over the sample as a whole. As noted, this effect has been mitigated in modern microscopes. However, a quick check is probably in order!

The second spectral issue involves the presence of "system peaks." These are contributions to measured spectra from interactions of scattered electrons and X-rays fluoresced by scattered electrons with the environment around the sample. The most usual is a Cu peak, arising from the use of specimen cups fabricated from Cu and Cu grids. It is most usual to make use of sample cups fabricated from Be to reduce the presence of such system peaks. Modern microscopes can also have liners on the bottom pole-piece, fabricated from low atomic-number materials to reduce the probability of back-scattered electrons or fluoresced photons from interacting with the sample and the environment about the sample, and so providing unwanted contributions to the measured spectra.

The third problem with coupling XEDS with (S)TEM, which applies also to EELS analyses, involves the nature of the thin foils used, and their relationship to the actual material being studied. Thus, most specimen preparation techniques that employed to make thin foils result in the production of surface layers, and various forms of damage. Regarding compositional analyses, the surface layers may be of a composition different from that of the material under study, and so provide aberrant contributions to the measured spectra. There are techniques that should be employed to reduce the extent of these surface-modified regions, such as low-energy Ar-ion milling.

The fourth problem associated with the use of thin foils in a (S)TEM regarding compositional analyses involves specimen contamination. The relatively high electron flux associated with high brightness sources can result in the development of surface contamination in the region where the focused probe is incident on the sample. The presence of such contamination causes the electron beam to be scattered, and so broadened, which has a detrimental effect on the spatial resolution of the technique. In modern microscopes, the vacuum, and general cleanliness, is such that most commonly, such contamination generally originates from the sample itself. It may be avoided by employing meticulously clean methods of sample preparation.

12.5.2 Spatial Resolution

As has been noted above, spatial resolution is expected to be degraded by beam spreading. This has been documented by a series of experiments using XEDS, EELS and atom probe tomography, to determine the compositional intermixing in nanoscale metallic multilayers (Genc et al., 2009). The sample consisted of alternating layers of Nb and Ti, with a thickness of approximately 2 nm/layer, as shown in **Figure 59(a)**. The samples were prepared in cross-section so that the interfaces between the layers were essentially parallel to the incident probe in the (S)TEM foils. The compositional profiles obtained in each of the three techniques are shown in **Figure 59(b–d)**. It can be seen that the profiles obtained from both EELS and the atom probe are very similar, indicating that the Nb intermixes in the Ti layers, but very little Ti intermixes in the Nb layers. In marked contrast, the XEDS profile implies significant intermixing in both layers. The explanation for this difference involves the effect of beam spreading in the case of XEDS, where the electron probe spreads into the adjacent layers providing a false contribution to the measured spectrum. EELS does not suffer from this particular effect of beam spreading since the volume from which the signal is generated in this latter technique is determined by the aperture at the entrance to the EELS spectrometer rather than by beam spreading in the foil. From these experiments, it is clear that beam spreading plays a very significant role in determining spatial resolution when coupling XEDS with (S)TEM.

Although it is clear that multiple scattering leading to beam spreading influences *spatial resolution* in XEDS, compared with the case of EELS, in a (S)TEM, there are other spectral measurements that have been made where it is shown that XEDS has a significant advantage over EELS regarding the effects of multiple scattering. Consider the spectra shown in **Figure 60(a and b)**, recorded from a thin foil of boron nitride using both EELS and XEDS, as a function of foil thickness (Fraser et al., 2011). It is very clear that the effect of multiple scattering as the foil thickness is increased appears to severely degrade the EELS spectra whereas having considerably less influence on the spectra recorded using XEDS. This result can be understood in terms of the difference in scattering cross-sections for electrons and X-rays. Thus, it is increasingly likely that electrons that have lost energy in an initial inelastic event may scatter again, losing more energy, as they traverse through the foil. In contrast, when a characteristic photon is generated, it is much less likely to be scattered by the sample (much smaller scattering cross-section), and so the "signal" will not be influenced by multiple scattering to the same degree. Of course, in this

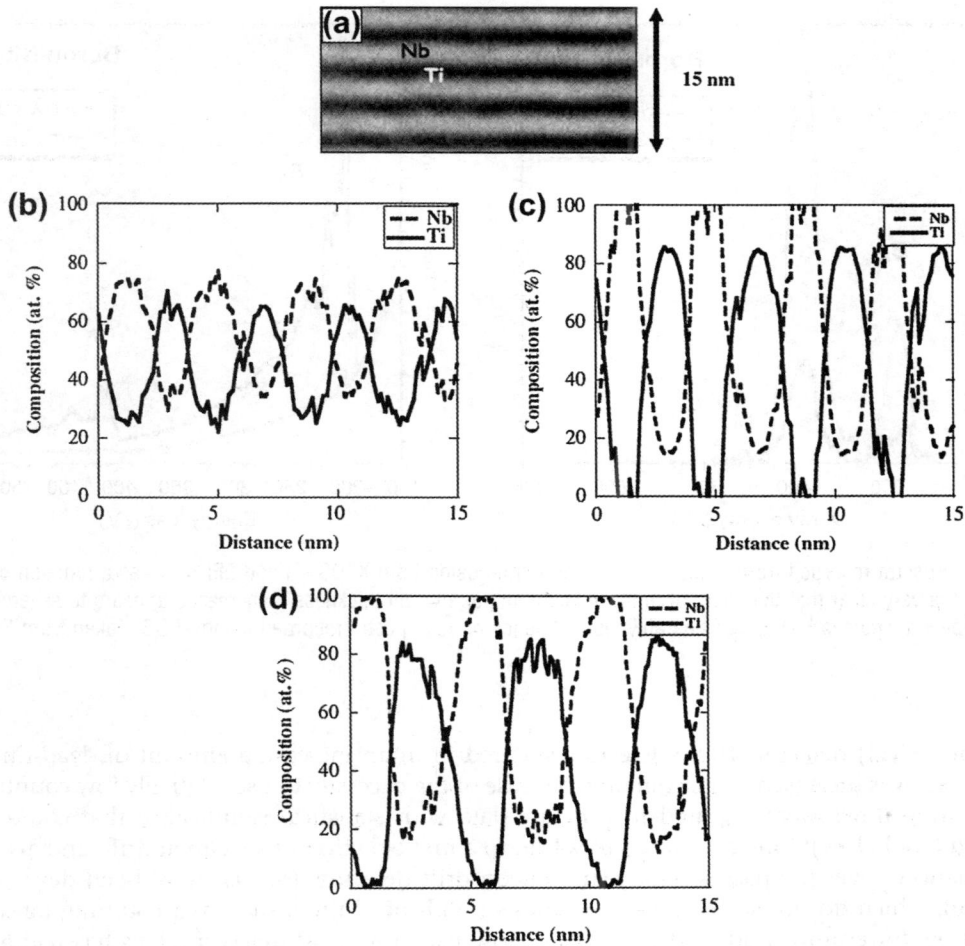

Figure 59 Comparison of the spatial resolution for compositional analysis of XEDS, EELS and atom probe tomography. The sample used for the test consisted of alternating layers of Nb and Ti, with a thickness of approximately 2nm/layer, as shown in (a). The compositional profiles obtained across the layer interfaces in each of the three techniques are shown in (b–d). Taken from Genc et al. (2009). (For color version of this figure, the reader is referred to the online version of this book.)

example, the foil is single phase, and spatial resolution is not an issue. Also, as the energies of the B and N peaks in the XEDS spectra are fairly similar, so that absorption as the foil thickness is increased is also not a significant issue.

12.5.3 New Detector Schemes

Recently, there have been two advances in systems coupling XEDS with (S)TEM, which are making a revolutionary impact on the field of analytical TEM. These advances involve the detector itself and the solid angle of detection. Regarding the detectors, in the past, XEDS analysis was usually involved in

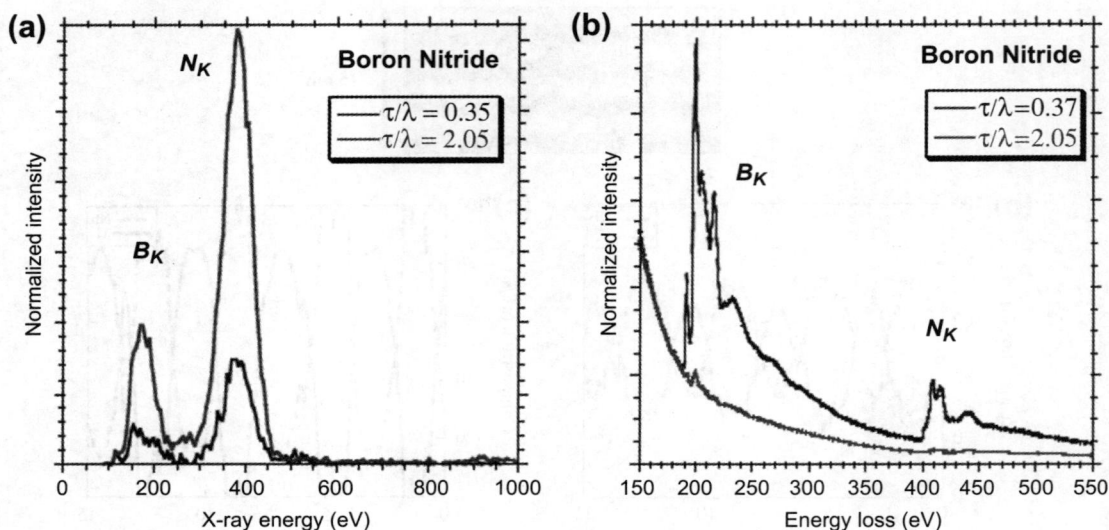

Figure 60 Spectra recorded from a thin foil of boron nitride using both XEDS (a) and EELS (b), as a function of foil thickness. It is very clear that the effect of multiple scattering as the foil thickness is increased appears to severely degrade the EELS spectra whereas having considerably less influence on the spectra recorded using XEDS. Taken from Fraser et al. (2011).

the use of a Si(Li) detector. These detectors suffered from an increasing amount of dead-time as the counting rate was increased. Consequently, because of the necessity to use relatively low counting rates, long counting times were required to generate data with statistical significance that could then be quantified (see below); long counting times lead to a marked effect of specimen drift, and also sample contamination. Over the past several years, silicon drift detectors (SDDs) have been developed and introduced, which do not suffer the same problem with dead-time, such that counting rates may be increased by, for example, 50–100 times over those used for Si(Li) detectors. This has had a marked influence on the efficient generation of meaningful XEDS data.

As has been discussed above, most usually the solid angle of collection for an XEDS detector in a (S)TEM is typically ≈0.13–0.3 sr. It has been speculated for some time that major advantages would be accrued by developing a detection scheme where the solid angle of collection would be approaching 1 sr. This has recently been done and is now offered on commercial machines, e.g. FEI's SuperX detector system. In this detection scheme, four SDD detectors are used to provide a solid angle of collection of approximately 0.9 sr, which when combined with the ability of these SDD detectors to accept very high count rates during data acquisition, provides a dramatic improvement in XEDS analysis. An example of the use of this detection scheme to generate elemental maps in a high entropy alloy is shown in **Figure 61** (Welk, et al., 2013). The acquisition time to capture the data required to produce these maps was less than 5 min. This reduction in acquisition time, from the increased counting rates possible and the very large solid angle of collection afforded by the detection scheme, has profound implications for analytical TEM. Thus, because statistically meaningful data can be recorded in short periods of time, specimen drift and sample contamination become factors of considerably reduced importance.

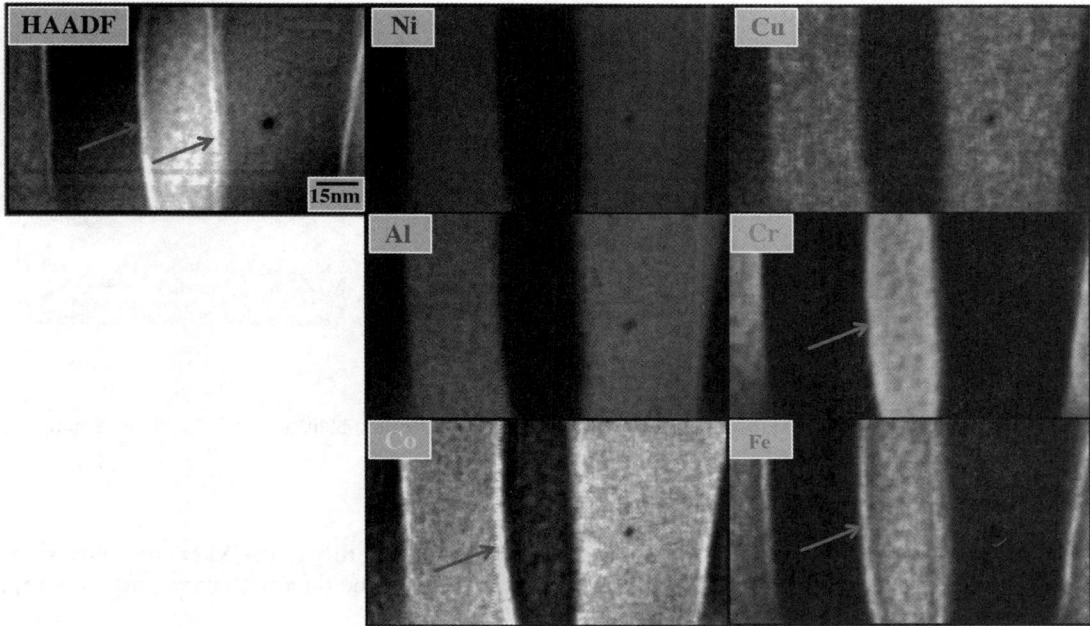

Figure 61   An example of the use of the new XEDS detection scheme, with very large solid angle of collection, to generate elemental maps in a high entropy alloy. All the maps shown were recorded in less than 5 min. Taken from Welk, et al., 2013.

This new detection scheme has been used to advantage in terms of spatially resolved elemental images. **Figure 62** shows an atomically resolved elemental map of a sample of strontium titanate ($SrTiO_3$) (d'Alfonso et al., 2010), with the electron beam being parallel to $\langle 100 \rangle$; the oxygen signal in this case is rather weak, as expected form the very low probability of producing an X-ray photon over an Auger electron for elements with very low atomic number, and perhaps the increased absorption of relatively low-energy photons in the sample. This remarkable image obtained using XEDS is at first take rather surprising, in view of the less than optimum spatial resolution of the technique compared, for example, with EELS (see discussion above). However, the details of the experiment must be considered. Thus, the intensities plotted are from windows drawn about the characteristic peaks following subtraction of background. The electrons in the focused probe when incident on a Sr atomic column, for example, will be drawn in to the atomic column by the deep potentials of the Sr atoms, and ionizations will occur. However, with increasing depth in the foil, the electrons will tend to be scattered away from the given atomic column, for example by phonon scattering, and the rate of generation of photons from Sr atoms, in the given column, will decrease. Hence, it is expected that most of the signal used to form the elemental maps will come from interactions with atoms in the crystal near to the electron entrance surface. Moreover, it is also possible that a fraction of the signal ascribed to a given column may well be coming from ionization events in neighboring atomic columns, the ionizations being caused by electrons that have been phonon-scattered through relatively large angles (Forbes et al., 2012). The photons generated will have a low probability of being absorbed by the sample, and a relatively large fraction of those generated and exiting the foil via the entrance surface will be detected

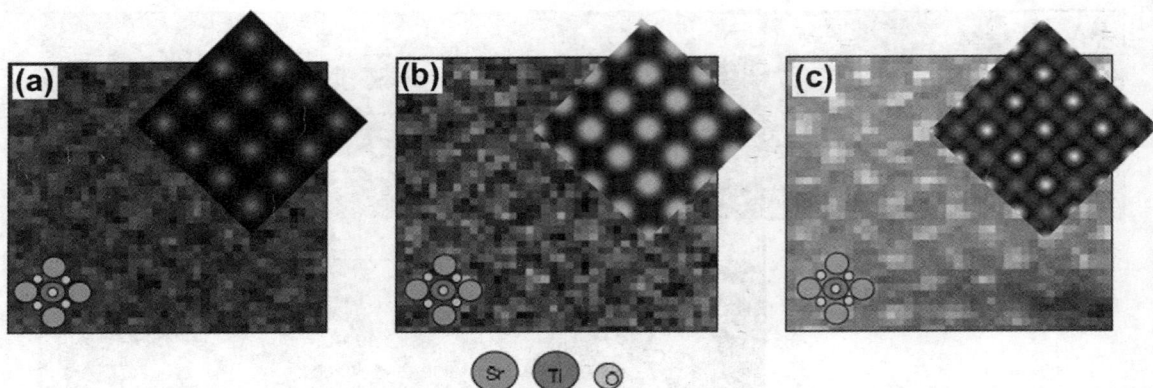

Figure 62 An atomically resolved XEDS elemental map of a sample of strontium titanate ($SrTiO_3$) with the electron beam being parallel to $\langle 100 \rangle$. Taken from d'Alfonso et al. (2010).

by the new spectrometers with large solid angle of collection. This ability to use XEDS to image samples at the atomic scale is very exciting, but it is important to interpret the data with care, and contemplate from where in the sample the signal is generated.

12.5.4   Quantification

The quantification of XEDS data obtained using (S)TEM was considered, at first thought, to be a rather more simple task than the procedures developed for bulk samples (i.e. processing data from the electron microprobe involving the ZAF corrections) since, as noted above, in thin samples, the importance of fluorescence and absorption was expected to be somewhat reduced. The first task in any quantification procedure involves the subtraction of the background spectrum, and this is indeed a fairly simple procedure, in part because the background intensity is relatively low and mostly a very small fraction of the characteristic spectrum. Usually, it involves either fitting the background using Kramers' method (Kramer, 1923), or filtering the background by passing a top-hat filter through the data. These methods are described in some detail by Williams and Carter (2009).

While background subtraction is apparently fairly straight forward, this cannot be said about the next important steps of quantification, i.e. the interpretation of the characteristic spectrum. This has proven to be difficult, and there is even today no consensus concerning the procedures that should be used to quantify data—it is still a work in progress. However, an excellent review has been provided by Watanabe and Williams (2006), to which the reader is referred. In summary, the most simple, and most used, approach to quantification of XEDS data obtained from thin foils involves the approach of Cliff and Lorimer (1975), involving the use of k-factors. For a sample containing the elements A and B, the compositions, C_A and C_B, are related to the measured intensities, I_A and I_B by the following simple relationship:

$$\frac{C_A}{C_B} = k_{AB}\frac{I_A}{I_B},$$

$$(10)$$

where k_{AB} is the so-called "k-factor." There are two approaches to developing values for the appropriate k-factors in a given experiment, one involving the use of standards and the other standardless, i.e. values are derived by calculation. Of course, in principle, it would be expected that the most simple and accurate approach would be the former, involving standards. However, in order to be useful standards, they must resemble as close as possible the same (scattering) environment as the sample, and their composition must be known exactly. These requirements are rather restrictive and a comprehensive "set" of standards for k-factors has not been produced. This is the subject of on-going effort and deliberation, so the reader is urged to follow the latest literature!

In an attempt to develop a more rigorous approach to quantification using standards, Watanabe and Williams (2006) have developed a technique using pure elemental standards, the so-called ζ-factor method. Obviously, this method has the tremendous advantage of the availability of most pure elemental standards, but does require a measure of the absolute value of the beam current incident on a sample, which necessitates the use of a Faraday cup. These are not available in essentially any commercial instrument, and because of this, this technique has not received the attention it deserves.

Regarding calculation of k-factors, this is the most applied method used today, and a procedure for performing such calculations is provided by commercial vendors (the beloved "quant" button!). Calculations of the k-factors mimic to a certain degree the same procedures used in microprobe analysis, i.e. the ZAF corrections, but taking into account approximations that are afforded by the use of thin foils. The problem with this approach is that the calculation of k-factors should take into account the scattering environment of the sample under study (similar to the requirement for the use of standards), and the provision of a universal value of a k-factor is not always appropriate. For example, measurement of Al compositions in Al–Cu alloys will be rather different from their measurement in Al–Co–W alloys. To take these types of consideration into account, a hybrid system is being developed (Zaluzec, 2013), where initially standards are used to guide subsequent computation of k-factors. This is not published work at present.

When approaching quantification, it is important to consider the measured data, i.e. the spectrum, with regards to the spatial origin of the various contributing counts. Most usually, the sample will consist of more than one phase, and so it is important to consider the effects of beam spreading, which may result in spectral contributions from areas, or phases, other than that of interest. It has been claimed many times that for XEDS analysis, one should use thick samples, and so increase the signal so that statistically meaningful counts (e.g. more than 10k–100k counts per peak) may be collected. The problem is that one must account for beam spreading when using such thick foils, and in many cases, this suggestion actually will lead to increased errors in analysis. With the adoption of SDD detectors and possible large solid angles of collection, it is now possible to use thin foils, permitting an optimization of spatial resolution, and still collect large and relevant amounts of data that may lead to useful quantitative results. So, the claim here is to use "thin" foils for XEDS analysis!

In summary, it is important for users to be very much aware that the procedures provided commercially are, on the one hand, useful in order to gain an approximation of the true composition of a sample, but on the other hand, there are a number of uncertainties that must be taken into account to ensure that significant errors in terms of quantification are avoided.

Acknowledgments

The authors express their thanks for help with the preparation of this chapter to Dr. G. Babu Viswanathan, Sam Kuhr, Michael Presley and Michael Loretto.

References

Brydson, R., 2001. Electron energy loss spectroscopy. BIOS Scientific.

Buseck, P., Cowley, J., Eyring, L., 1989. High-Resolution Transmission Electron Microscopy: and Associated Techniques. Oxford University Press.

Buxton, B.F., Eades, J.A., Steeds, J.W., Rackham, G.M., 1976. The symmetry of electron diffraction zone axis patterns. Philos. Trans. R. Soc. London A281, 171.

Chumbley, L.S., Muddle, B.C., Fraser, H.L., 1988. The crystallography of the precipitation of Ti_5Si_3 in Ti–Si alloys. Acta Metall. 36 (6), 299.

Cliff, G., Lorimer, G.W., 1975. The quantitative analysis of thin specimens. J. Microsc. 103, 203.

Cook, R.F., Cundy, S.L., 1969. Plasmon energy losses in Al-Zn alloys. Philos. Mag. 20 (166), 665–673.

Court, S.A., Löfvander, J.P.A., Loretto, M.H., Fraser, H.L., 1990. The influence of temperature and alloying additions on the mechanisms of plastic deformation of Ti_3Al. Phil. Mag.A 61, 109.

Cowley, J.M., 1969. Image contrast in a transmission scanning electron microscope. Appl. Phys. Lett. 15, 58.

Cullity, B.D., Stock, S.R., 2001. Elements of X-ray Diffraction, third ed. Prentice Hall.

d'Alfonso, A.J., Freitag, B., Klenov, D., Allen, L.J., 2010. Atomic resolution chemical mapping using energy dispersive X-ray spectroscopy. Phys. Rev. B 81, 100101(R).

De Graef, M., 2003. Introduction to Conventional Transmission Electron Microscopy. Cambridge University Press.

Docherty, F.T., Craven, A.J., McComb, D.W., Skakle, J., 2001. ELNES investigations of the oxygen K-edge in spinels. Ultramicroscopy 86 (3–4), 273–288.

Doig, P., Edington, J.W., 1973. Low-temperature diffusion in A1–7 wt.% Mg and A1–4 wt.% Cu alloys. Philos. Mag. 28 (5), 961–970.

Doig, P., Edington, J.W., Hibbert, G., 1973. Measurement of Mg supersaturations within precipitate-free zones in Al-Zn-Mg alloys. Philos. Mag. 28 (5), 971–981.

Egerton, R.F., 1981. Energy-loss spectrometry with a large collection angle. Ultramicroscopy 7 (2), 207–210.

Egerton, R.F., 2011. Electron Energy-loss Spectroscopy in the Electron Microscope, third ed. Plenum, New York.

Forbes, B.D., d'Alfonso, A.J., Williams, R.E.A., Srinivasan, R., Fraser, H.L., McComb, D.W., Freitag, B., Klenov, D.O., Allen, L.J., 2012. Contribution of thermally scattered electrons to atomic resolution elemental maps. Phys. Rev. B 86, 024108.

Fraser, H.L., Jones, I.P., 1975. A note on the increase in usable foil thickness in scanning transmission electron microscopy. Philos. Mag. 31, 255.

Fraser, H.L., Loretto, M.H., Jones, I.P., 1977. Limiting factors in specimen thickness in conventional and scanning electron microscopy. Philos. Mag. 35, 159–176.

Fraser, H.L., Klenov, D.O., Wang, Y.C., Cheng, H., Zaluzec, N.J., 2011. On the performance of XEDS and EELS in the AEM: 25 years later. Microsc. Microanal. 17 (Suppl. 2).

Garvie, L.A.J., Craven, A.J., Brydson, R., 1994. Use of electron-energy-loss near-edge fine-structure in the study of minerals. Am. Mineral. 79 (5–6), 411–425.

Genc, A., Banerjee, R., Thompson, G.B., Maher, D.M., Johnson, A.W., Fraser, H.L., 2009. Complementary techniques for the characterization of thin film Ti/Nb multilayers. Ultramicroscopy 109 (10), 1276.

Goldstein, J.I., Newbury, D.E., Echlin, P., Joy, D.C., Fiori, C., Lifshin, E., 1981. Scanning Electron Microscopy and X-ray Microanalysis. Plenum Press.

Head, A.K., Loretto, M.H., Humble, P., 1967. Phys. Stat. Sol., 20, 505.

Henry, Norman FM., Lonsdale, Kathleen, 1952. International tables for X-ray crystallography. Kynoch Press.

Hibbert, G., Edington, J.W., Williams, D.B., Doig, P., 1972. The variation of plasma energy loss with composition in dilute aluminium-magnesium solid solutions. Philos. Mag. 26 (6), 1491–1494.

Hirsch, P., Howie, A., Nicholson, R., Pashley, D., Whelan, M., 1977. Electron Microscopy of Thin Crystals. Robert Krieger Publishing Co. Inc, Malabar, Florida.

Howie, A., Whelan, M.J., 1962. Proc. Roy. Soc. A267, 206.

Jantou, V., Turmaine, M., West, G.D., Horton, M.A., McComb, D.W., 2009. Focused ion beam milling and ultramicrotomy of mineralised ivory dentine for analytical transmission electron microscopy. Micron 40 (4), 495–501.

Jeanguillaume, C., Colliex, C., 1989. Spectrum-image—the next step in EELS digital acquisition and processing. Ultramicroscopy 28 (1–4), 252–257.

Kaufman, M.J., Loretto, M.H., Eades, J.A., Fraser, H.L., 1983. A study of a cellular phase transformation in the ternary Ni–Al–Mo alloy system. Metall. Trans. 14A, 1561.

Kaufman, M.J., Pearson, D.D., Fraser, H.L., 1986. The use of convergent beam electron diffraction to determine local lattice distortions in Ni-base superalloys. Philos. Mag. 54 (1), 79.

Keast, V.J., Scott, A.J., Brydson, R., Williams, D.B., Bruley, J., 2001. Electron energy-loss near-edge structure—a tool for the investigation of electronic structure on the nanometre scale. J. Microsc. Oxford 203, 135–175.

Kramers, H.A., 1923. On the theory of X-ray absorption and of the continuous x-ray spectrum. Philos. Mag. 46, 836.

LeBeau, J.M., Stemmer, S., 2008. Experimental quantification of annular dark-field images in scanning transmission electron microscopy. Ultramicroscopy 108, 1653–1658.

Loretto, M.H., 1984. Electron Beam Analysis of Materials. Chapman and Hall, London.

Maher, D.M., Fraser, H.L., Humphreys, C.J., Knoell, R.V., Bean, J.C., 1987. Detection and measurement of local distortions in a semiconductor layered structure by convergent-beam electron diffraction. Appl. Phys. Lett. 50, 574.

Malis, T., Cheng, S.C., Egerton, R.F., 1988. EELS log-ratio technique for specimen-thickness measurement in the TEM. J. Electron Microsc. Tech. 8 (2), 193–200.

McComb, D.W., Howie, A., 1990. Characterization of zeolite catalysts using electron-energy loss spectroscopy. Ultramicroscopy 34 (1–2), 84–92.

McComb, D.W., 1996. Bonding and electronic structure in zirconia pseudopolymorphs investigated by electron energy-loss spectroscopy. Phys. Rev. B 54 (10), 7094–7102.

Nag, S., Devaraj, A., Srinivasan, R., Williams, R.E.A., Gupta, N., Viswanathan, G.B., Tiley, J.S., Banerjee, S., Srinivasan, S.G., Banerjee, R., Fraser, H.L., 2011. Novel mixed-mode phase transition involving a composition-dependent displacive component. Phys. Rev. Lett. 106 (24), 245701.

Perkins, J.M., Fearn, S., Cook, S.N., Srinivasan, R., Rouleau, C.M., Christen, H.M., West, G.D., Morris, R.J.H., Fraser, H.L., Skinner, S.J., Kilner, J.A., McComb, D.W., 2010. Anomalous oxidation states in multilayers for fuel cell applications. Adv. Funct. Mater. 20 (16), 2664–2674.

Phillips, P.J., Brandes, M.C., Mills, M.J., DeGraef, M., 2011a. Diffraction contrast STEM of dislocations: imaging and simulations. Ultramicroscopy 111, 1483.

Phillips, P.J., Mills, M.J., De Graef, M., 2011b. Systematic row and zone axis STEM defect image simulations. Philos. Mag. 91, 2081.

Porter, D.A., Doig, P., Edington, J.W., 1974. The variation of plasma energy loss with composition in dilute Al-Ag alloys. Philos. Mag. 29 (2), 437–440.

Raether, H., 1965. Solid state excitations by electrons. Springer Tracts Mod. Phys. 38, 84.

Reed, S.J.B., 1997. Electron Microprobe Analysis. Cambridge University Press.

Reimer, L., Egerton, R.F., Deininger, C., Hofer, F., Jouffrey, B., Krahl, D., Leapman, R.D., Mayer, J., Rose, H., Schattschneider, P., Spence, J.C.H., 1995. Energy-filtering Transmission Electron Microscopy. Springer.

Reimer, L., 1998. Energy-filtering imaging and diffraction. Mater. Trans. Jim 39 (9), 873–882.

Scott, J., Thomas, P.J., Mackenzie, M., McFadzean, S., Wilbrink, J., Craven, A.J., Nicholson, W.A.P., 2008. Near-simultaneous dual energy range EELS spectrum imaging. Ultramicroscopy 108, 1586–1594.

Self, P.G., O'Keefe, M.A., 1988. In: Buseck, P.R., Cowley, J.M., Eyring, L. (Eds.), High Resolution Transmission Electron Microscopy and Associated Techniques.

Silcock, J.M., Tunstall, W.J., 1964. Partial Dislocations associated with NbC Precipitation in Austenitic Stainless Steels. Philos. Mag. 10, 361.

Spalding, D.R., Metherell, A.J.F., 1968. Plasmon losses in Al-Mg alloys. Philos. Mag. 18 (151), 41–48.

Spalding, D.R., Villacrana, R.E., Chadwick, G.A., 1969. A study of copper distribution in lamellar Al–CuAl2 eutectics using an energy analysing electron microscope. Philos. Mag. 20 (165), 471–488.

Spence, J.C.H., 2003. High-resolution Electron Microscopy, vol. 60. Clarendon Press.

Srinivasan, R., Banerjee, R., Viswanathan, G.B., Tiley, J., Dimiduk, D.M., Fraser, H.L., 2009. Atomic scale structure and chemical composition across order-disorder interfaces. Phys. Rev. Lett. 102 (8), 086101.

Steeds, J.W., 1979. Convergent beam electron diffraction. In: Hren, J.J., Goldstein, J.I., Joy, D.C. (Eds.), Introduction to Analytical Electron Microscopy. Plenum Press, New York.

Swaminathan, S., Altynov, S., Jones, I.P., Zaluzec, N.J., Maher, D.M., Fraser, H.L., 1997. Precise and accurate refinements of the 220 structure factor for silicon by the systematic-row CBED method. Ultramicroscopy 69, 169.

Tremblay, S., L'Esperance, G., 1994. Volume fraction determination of secondary phase particles in aluminum thin foils with plasmon energy shift imaging. In: Proceeding of ICEM-13, Paris, vol. I, pp. 627–628.

Watanabe, M., Williams, D.B., 2006. The quantitative analysis of thin specimens: a review of progress from the Cliff–Lorimer to the new ζ-factor method. J. Microsc. 221, 9.

Welk, B., Williams, R.E.A., Viswanathan, G., Gibson, M., Liaw, P., Fraser, H.L.F., 2013. Nature of the interfaces between the constituent phases in the high entropy alloy CoCrCuFeNiAl. Ultramicroscopy 134, 193–199.

Williams, D.B., 1975. Unpublished research.

Williams, D.B., Carter, C.B., 2009. Transmission Electron Microscopy, second ed. Springer.

Williams, D.B., Edington, J.W., 1976. High resolution microanalysis in materials science using electron energy-loss measurements. J. Microsc. 108, 113–145.

Williams, D.B., Edington, J.W., 1976. Microanalysis of splat quenched Al-Cu alloys. Philos. Mag. 34 (2), 235–242.

Williams, D.B., Edington, J.W., 1974. Microanalysis of Al-Li alloys containing fine δ'(Al3Li) precipitates. Philos. mag. 30 (5), 1147–1153.

Woo, O.T., Carpenter, G.J.C., 1992. EELS characterization of zirconium hydrides. Microsc. Microanal. Microstruct. 3, 35.

Wu, Z., Qiu, C., Venkatesh, V., Williams, R.E.A., Viswanathan, G.B., Thomas, M., Nag, S., Banerjee, R., Fraser, H.L., Loretto, M.H., 1988. The influence of precipitation of Alpha2 on properties and microstructure in TIMETAL 6-4. Met. Trans. A 44, 1706.

Zaluzec, N. J., Private communication, 2013.

Zeitler, E., Thomson, M.G.R., 1970. Scanning transmission electron microscopy. Optik 31 (258–280), 359.

Biography

Dr. Fraser graduated from the University of Birmingham (UK) with the degrees of B.Sc. (1970) and Ph.D. (1972). He was appointed to the faculty of the University of Illinois in 1973 (Assistant, Associate and Full Professor), before moving in 1989 to the Ohio State University (OSU) as Ohio Regents Eminent Scholar and Professor. He was appointed as a Senior Research Scientist at the United Technologies Research Center from 1979–1980. He has also been a Senior von Humboldt Researcher at the University of Göttingen, a Senior Visitor at the University of Cambridge, a visiting professor at the University of Liverpool, and spent a sabbatical leave at the Max-Planck Institut für Werkstoffwissenscahften in Stuttgart. He has been an Honorary Professor of Materials and Technology at the University of Birmingham since 1988. In 2014, he was recognized as an Honorary Professor at the Nelson Mandela Metropolitan University in Port Elizabeth, South Africa. At present, he serves as Director of the Center for the Accelerated Maturation of Materials (CAMM) at OSU. He has been a member of the National Materials Advisory Board and the US Air Force Scientific Advisory Board. He has consulted for a number of national laboratories and several industrial companies. He is a Fellow of TMS, ASM, IOM^3 (UK), and MSA. He has published over 380 papers in scholarly journals, and given over 280 invited presentations. He has graduated 48 doctoral students and 36 students graduating with the degree of M.S.

David William McComb is an Ohio Research Scholar and Professor of Materials Science and Engineering at The Ohio State University. A chemistry graduate from the University of Glasgow, David did his Ph.D. in Physics at the University of Cambridge. David is an expert in the development and application of electron energy-loss spectroscopy (EELS) as a sub-nanometre scale probe of chemistry, structure and bonding. He has extensive experience in the application of EELS to the study of problems in solid-state chemistry and materials science including structural and compositional variations in high-k oxides, short range magnetic order in transition metal oxides, interfaces in fuel cells, photovoltaics, multiferroics and biomaterials. He is a fellow of the Royal Society of Chemistry and the Institute of Materials, and former co-director of the London Centre for Nanotechnology as well as Director of Research and Deputy Head of the Department of Materials at Imperial College London. In October 2011 he joined The Ohio State University as the founding director of the Center for Electron Microscopy and AnalysiS (CEMAS). This multidisciplinary facility was conceived to drive the application of state of the art electron microscopy techniques to strategic research challenges in the physical, engineering, life and medical sciences.

Robert E. A. Williams is currently the inorganic Senior Research Officer at the Center for Electron Microscopy and Microanalysis at The Ohio State University in Columbus, Ohio. Robert received a Ph.D. from The Ohio State University in materials science and engineering by applying high-resolution analytical transmission electron microscopy to novel transformation pathways in titanium alloys. Robert has more than 13 years experience working with and operating various electron microscopes. Previously, Robert was a post-doctoral fellow in the Campus Electron Optics Facility followed by a short stint as a research assistant professor at the University of North Texas. Following these experiences, Robert joined the newly formed Center for Microscopy and Analysis in 2013. He was recipient of the 2011 ASM Henry Marion Howe Medal and the 2012 TMS Champion H. Mathewson Award.

13 X-ray and Neutron Scattering

Gernot Kostorz, ETH Zurich, Department of Physics, Auguste-Piccard-Hof 1, Zurich, Switzerland

13.1 Introduction

There are many textbooks and monographs on X-ray and neutron scattering from condensed matter (some are included in the references), and the number of publications is immense. This short chapter cannot give a complete account of the history (which now exceeds 100 years for X-rays) or the current state of the field. The basic theory, standard methods and many of the more classical applications are well described in various textbooks and only a brief survey of the fundamentals of scattering will be given. The main purpose of the present chapter will be to demonstrate that there is much more we can "learn [about metallic materials] from scattering experiments besides the average structure" (Schwartz and Cohen, 1987; Ch. 7).

One immediate question concerns the range of stability of a given structure, and scattering experiments are helpful in revealing and analyzing the formation and transformation of phases. There are classical methods (e.g. powder diffraction) and very sophisticated, more recent techniques (e.g. quasielastic neutron scattering, diffraction of coherent X-rays) to study transformations and phase separation in metallic systems. Some examples will be discussed. Much space will be devoted to studies of inhomogeneities, i.e. deviations from the average structure (point defects, clusters, short-range order (SRO), precipitates, etc.).

In Section 13.2, some general remarks on the scattering response from a crystalline material containing defects are followed by a discussion of X-rays and neutrons as the two types of radiation now commonly used for scattering studies of essentially bulk materials and their surfaces and interfaces. Scattering of low-energy electrons will not be covered, while high-energy electron diffraction is discussed in the chapter on transmission electron microscopy. Throughout the present chapter, scattering intensities are expressed in terms of scattering cross-sections (differential cross-sections without energy analysis, double-differential cross-sections for scattering experiments with energy analysis), mostly applicable to both X-rays and neutrons. However, the reader should not underestimate the difficulties in converting measured intensities to absolute cross-sections. Calibration procedures and corrections concerning absorption, background, polarization, etc. may introduce considerable errors. These problems and questions of instrument design and optimization can only be mentioned occasionally. The subsequent sections (13.3–13.6) cover recent applications, and the subject matter has been divided according to simple criteria based on the various scattering phenomena. Diffraction and elastic scattering at and near Bragg peaks, between Bragg peaks and near the incident beam are treated in

Sections 13.3–13.5, and Section 13.6 gives a few examples of inelastic and quasielastic scattering. Theoretical expressions are kept to a minimum, but some are required to outline the main effects.

There are several systematically elegant and quite general schemes to describe the same scattering phenomena (starting, e.g. from lattice sums, correlation functions or convolutions). No formal elaboration of this type will be attempted. Results of the kinematical theory and of the equivalent first-order Born approximation adapted to simple but manageable cases will mostly be relied upon. Elements of dynamic diffraction related to transmission electron microscopy are given in Chapter 12 of this volume.

13.2   Scattering from Real Crystals

13.2.1   General Predictions of the Kinematical Theory

We take a sample exposed to a coherent plane wave train of wavelength λ (along its wave vector $\mathbf{k}_0$, $|\mathbf{k}_0| = k_0 = 1/\lambda$) and of amplitude A_0 (for simplicity taken as a scalar quantity—for X-rays, the vector nature of the amplitude leads to some formal changes). Then $|A_0|^2$ is the incident intensity per unit area. Kinematical diffraction is based on weak elastic and coherent scattering from individual scattering centers (electrons for X-rays, nuclei and local magnetic moments for neutrons). The scattering intensity on a detector analyzing a solid angle $\Delta\Omega$ in the direction of the scattering vector $\mathbf{k}_s$ at a distance far away from the sample is then controlled by the scattering (diffraction) amplitude resulting from the coherent superposition of the individual scattered wavelets at all N scattering centers exposed to the beam at positions $\mathbf{r}_n$

$$F(\boldsymbol{\kappa}) = \sum_n b_n \exp(-2\pi i\,\boldsymbol{\kappa}\cdot\mathbf{r}_n), \tag{1}$$

where b_n denotes the individual scattering amplitudes (often taken as real, but generally complex). For X-rays, atomic amplitudes are the phase-weighted sum over the amplitudes of all associated electrons, and b will usually depend on $\boldsymbol{\kappa}$ defined by

$$\boldsymbol{\kappa} = \mathbf{k}_s - \mathbf{k}_0 \tag{2}$$

For thermal neutrons, scattering by nuclei (nuclear scattering) is practically independent of $\boldsymbol{\kappa}$.

For elastic scattering, $|\mathbf{k}_s| = |\mathbf{k}_0|$ and $|\boldsymbol{\kappa}| = \kappa = 2k_0 \sin\theta$ where θ is half the scattering angle. Thus, the scattering intensity detected in $\Delta\Omega$ [= area/(distance)^2] is

$$I_s(\boldsymbol{\kappa}) = |A_0|^2 |F(\boldsymbol{\kappa})|^2 \Delta\Omega \tag{3}$$

The properties of the sample are entirely contained in $|F(\boldsymbol{\kappa})|^2$, which has the dimension of an area and is called the differential (elastic) cross-section $d\sigma/d\Omega$. The scattering amplitudes b_n must therefore have the dimension of a length [for X-rays, b is often called f and given in units of the classical electron radius, $e^2(4\pi\varepsilon_0 m_e c^2)^{-1}$]. The differential cross-section can be worked out for any static (and, if inelastic scattering is included, also dynamic) system of scattering objects.

For a perfect crystal, all atoms are arranged on the nodes of a lattice, i.e. (we consider a monatomic crystal for simplicity),

$$\mathbf{r}_n = n_1\mathbf{a}_1 + n_2\mathbf{a}_2 + n_3\mathbf{a}_3, \tag{4}$$

where $1 \leq n_j \leq N_j$ and N_j is the number of atoms in the j-direction, while the $\mathbf{a}_j$ are the lattice constants (in three dimensions). As explained in more detail in standard texts on crystallography, in reciprocal space (of which κ is a vector), the reciprocal lattice points $\mathbf{g}$ corresponding to the real-space lattice (eqn (4)) are given by multiples of the reciprocal basic vectors $\mathbf{b}_j$,

$$\mathbf{g}_{hkl} = h\mathbf{b}_1 + k\mathbf{b}_2 + l\mathbf{b}_3 \tag{5}$$

with the Miller indices h, k, l.

Bragg scattering (a diffraction peak) appears if $\kappa = \mathbf{g}_{hkl}$. The Bragg peaks are characteristic of the specific structure. For a monatomic crystal, $|F(\kappa)|^2$ [eqn (1)] is solely controlled by the sum of all the exponentials. If κ is exactly equal to a Bragg vector $\mathbf{g}_{hkl}$, all N amplitudes add up precisely. If we admit a slight deviation $\mathbf{s}$ from $\mathbf{g}_{hkl}$, i.e.

$$\kappa = \mathbf{g}_{hkl} + \mathbf{s}, \quad |\mathbf{s}| \ll |\mathbf{g}_{hkl}|, \tag{6}$$

the scattering intensity decreases rapidly with distance from the Bragg peak, with a half-width essentially proportional to $(N_\kappa a_\kappa)^{-1}$ along any direction of κ, i.e. the smaller the crystal (or crystallite for polycrystals) size, the broader the Bragg peak. Since this "finite-size effect" broadening is independent of the magnitude of $\mathbf{g}_{hkl}$, its contribution to a potential broadening caused by many other effects can be extracted by comparing the width of various Bragg peaks at their Bragg angles θ_{hkl}. For a statistically isotropic (with regard to Bragg scattering) powder, we thus have a method of determining crystallite sizes, since a constant $\Delta\kappa_g$ means:

$$\Delta(2\theta_{hkl}) = \frac{\lambda}{L \cos \theta_{hkl}} \tag{7}$$

with L as the average crystallite diameter. However, experimental conditions will severely limit the range over which peak widths are a direct measure of the crystallite size according to eqn (7). If L is too large, we may reach the resolution limit of the experiment, and if L is too small (≤ 100 Å $= 10$ nm), it becomes difficult to separate the diffraction lines [the detailed discussion of these aspects in the book by Barrett and Massalski (1980) is still worth reading]. The increased interest in nanosized and nano-structured materials has refueled discussions especially at the lower end of the size range, while the improved instrumental resolution of X-ray diffractometers at synchrotron radiation facilities has opened the way to "measure" sizes to almost 100 μm by diffraction techniques.

For large (say > 100 μm) crystals, the size broadening may be ignored, and the differential cross-section for Bragg scattering may be written as

$$\left(\frac{d\sigma}{d\Omega}\right)_B = N|F|^2 \frac{\delta(\kappa - \mathbf{g})}{V_c} \tag{8}$$

where $\delta(\kappa - \mathbf{g})$ is the (three-dimensional) delta function and V_c is the unit-cell volume. However, a line or peak broadening is very frequently found in real crystals—caused by imperfections. Before addressing some of these cases, the basic expression for $F(\kappa)$ must be generalized to include cases with unit cells containing several (different) atoms. For this purpose, the vectors $\mathbf{r}_n$ now take the role of denoting the positions of the elementary cells, and the positions of all atoms are given by

$$\mathbf{r}_{nm} = \mathbf{r}_n + \mathbf{r}_m \tag{9}$$

where the $\mathbf{r}_m$ are position vectors for the M atoms in the unit cell ($1 \leq m \leq M$) with scattering amplitudes b_m.

The scattering amplitude $F(\kappa)$ may now be written as

$$F(\kappa) = F_L(\kappa) \cdot F_s(\kappa) \tag{10}$$

where

$$F_L(\kappa) = \sum_n \exp(-2\pi i\kappa \cdot r_n) \tag{11}$$

is the (dimensionless) "lattice amplitude", which is independent of the detailed arrangement of atoms in the unit cell. The quantity $F_s(\kappa)$ is the "structure amplitude",

$$F_s(\kappa) = \sum_m b_m \exp(-2\pi i\kappa \cdot r_m). \tag{12}$$

Its square, the structure factor, is the essential quantity in the determination of crystal structures.

As is well known, atoms in a crystal are never at rest, and thermal vibrations lead to a reduction of Bragg peak intensities with increasing temperature and increasing magnitude of κ, described by the thermal Debye–Waller factor. This omnipresent effect has to be distinguished from static displacements that may be different in different unit cells. Also, the population of lattice sites may vary from cell to cell, and compositional variations over larger distances may also occur. The structure amplitude of the n-th cell thus reads

$$F_{sn}(\kappa) = \sum_m b_{nm} \exp(-2\pi i\kappa \cdot r_{nm}) \exp[-2\pi i\kappa \cdot (u_{nm} - u_{n1})], \tag{13}$$

with displacements u_{nm} (u_{n1} is the displacement of the first atom in the unit cell) and

$$F(\kappa) = \sum_n F_{sn}(\kappa) \exp(-2\pi i\kappa \cdot r_n) \exp(-2\pi i\kappa \cdot u_{n1}). \tag{14}$$

More generally, an additional term may be added if interstitial sites are variably occupied and induce displacements.

For specific defects, it is useful to express displacements and structure amplitudes in terms of the properties of individual defects and their mutual arrangement. If t enumerates all possible positions for a defect, and u_{tn} is the displacement vector at position n related to the defect at t, the total displacement u_n may be expressed as

$$u_n = \sum_t c_t u_{tn}, \tag{15}$$

where

$$c_t = \begin{cases} 1 & \text{if a defect is present at site } t \\ 0 & \text{if no defect is present at site } t. \end{cases} \tag{16}$$

This simple superposition is certainly justified for low defect concentrations but may be questionable in the case of concentrated alloys.

The corresponding expression for the structure amplitude of the nth cell reads

$$F_{sn} = F_{s0} + \sum_t c_t \Delta F_{stn}, \tag{17}$$

where F_{s0} is the structure amplitude of the defect-free crystal. Replacing u_n and F_{sn} in the general expression (14) by eqns (15) and (17), the mean value for $|F(\kappa)|^2$, i.e. the scattering cross-section $d\sigma/d\Omega$

of the scattering ensemble, calculated for a random distribution of defects of concentration c, is (we follow Krivoglaz, 1969, 1996, whose books still deal with these matters in a most coherent and elucidating way):

$$\frac{d\sigma}{d\Omega} = \sum_{n,n'} \{\exp[-2\pi i\kappa \cdot (\mathbf{r}_n - \mathbf{r}_{n'})]\} e^{-\Im} \left\{ |\overline{F}_s|^2 + c \sum_t \Delta F_{stn} \Delta F_{stn'}^* \exp[-2\pi i\kappa \cdot (\mathbf{u}_n - \mathbf{u}_{tn'})] \right\} \quad (18)$$

where

$$\Im = c \sum_t \{1 - \exp[-2\pi i\kappa \cdot (\mathbf{u}_{tn} - \mathbf{u}_{tn'})]\} \left[1 + \frac{1}{F_{s0}} (\Delta F_{stn} + \Delta F_{stn'}) \right], \quad (19)$$

and $\overline{F}_s$ is the average structure amplitude of the crystal with defects:

$$\overline{F}_s = F_{s0} + c \sum_t \Delta F_{stn}. \quad (20)$$

It is now possible to split $d\sigma/d\Omega$ into two terms, a Bragg-like term with sharp maxima at the Bragg positions, and a second term describing scattering for all other scattering vectors. Taking into account that for the Bragg peaks, $d\sigma/d\Omega$ from the double sum of eqn (18) is controlled by terms corresponding to large distances $\rho = |\mathbf{r}_n - \mathbf{r}_{n'}|$, we may write:

$$\left(\frac{d\sigma}{d\Omega} \right)_B = N |\overline{\overline{F}}_s|^2 e^{-2M} \frac{\delta(\kappa - \mathbf{g}_{hkl})}{V_c}, \quad (21)$$

where $2M$ is the real part of $\Im$ according to eqn (19) for $\rho \to \infty$:

$$2M = c \lim_{\rho \to \infty} \sum_t \{1 - \cos[2\pi\kappa \cdot (\mathbf{u}_{tn} - \mathbf{u}_{tn'})]\} \left[1 + \frac{\Delta F_{stn} + \Delta F_{stn'}}{F_{s0}} \right]. \quad (22)$$

Equation (21) defines a generalized Debye–Waller factor e^{-2M}. The reciprocal lattice vectors $\mathbf{g}_{hkl}$ occurring in eqn (21) are defined in the average lattice containing all its defects. Defects may thus cause a shift of Bragg peaks because the average lattice constant changes, but also a reduction in intensity due to the factor e^{-2M} [eqns (21) and (22)]. As long as M remains finite, Bragg peaks will still be present, but cases where M tends to infinity may also exist.

If we subtract the Bragg intensity according to eqn (21) from the total scattering cross-section, eqn (18), we obtain the diffuse scattering cross-section

$$\left(\frac{d\sigma}{d\Omega} \right)_d = \frac{d\sigma}{d\Omega} - \left(\frac{d\sigma}{d\Omega} \right)_B. \quad (23)$$

The cross-section $(d\sigma/d\Omega)_d$ contains no δ-function but varies smoothly with κ, even for an infinitely large crystal. Krivoglaz (1969, 1996) distinguishes two types of defects depending on whether the Bragg intensities are reduced to zero or not (eqn (21) with $M \to \infty$).

The limiting behavior of $2M$ [eqn (22)] may be discussed by considering a displacement $\mathbf{u}_{tn}$ that decreases rapidly toward zero with increasing distance between t and n. Then, only one of the displacements $\mathbf{u}$ and one of the ΔF_s in eqn (22) are markedly different from zero, and:

$$M = c \sum_t [1 - \cos(2\pi\,\boldsymbol{\kappa}\cdot\mathbf{u}_{tn})]\left(1 + \frac{\Delta F_{stn}}{F_{s0}}\right). \tag{24}$$

The convergence of the sum depends on the contributions from large distances between n and t. There, the size of $\mathbf{u}_{tn}$ is small and the cosine can be expanded, i.e. $1 - \cos(2\pi\kappa\cdot\mathbf{u}_{tn}) = (2\pi\kappa\cdot\mathbf{u}_{tn})^2/2$, and if one writes $\mathbf{u}_{tn}$ as a continuous function of $\mathbf{r}' = \mathbf{r}_t - \mathbf{r}_n$, the convergence of the integral $\int(\kappa\cdot\mathbf{u}_{tn})^2 d^3\mathbf{r}'$, $\infty \geq r' > r_0'$ ($\gg$ lattice constant), will assure a finite value of M. We see that for a large class of defects where u decreases as $(r')^{-2}$ or faster, M remains finite. Other defects, e.g. straight dislocations, small-angle grain boundaries and stacking faults, cause the Debye–Waller factor to decrease to very small values for large crystals, and the total scattering must be termed diffuse—although it will be concentrated near the original Bragg positions ("line broadening"), but not in a δ-like fashion.

An experimental distinction between Bragg intensity (eqn (21)) and diffuse intensity (line broadening and scattering far away from Bragg peaks, at any position of reciprocal space), discussed here for systems with a random distribution of equivalent defects, will not always be possible, since the width of the Bragg peak is also affected by the resolution function of the instrument.

Correlations in the arrangement of defects may reduce long-range displacements and modify the expected effects, as will the presence of different types of defects. **Figure 1** shows schematically how the scattering of an ideal crystal is modified by defects.

13.2.2   X-rays and Neutrons

Apart from electrons (as used in the electron microscope), X-rays (and γ-rays) and thermal neutrons are frequently used for structure determination and the study of defects as they provide the appropriate range of wavelengths for such investigations. Table 1 gives some of the important properties of both types of radiation. For the wavelengths of interest in normal diffraction work (0.5–20 Å), corresponding photon energies are in the range of about 1–40 keV (higher energies are now available at synchrotron radiation facilities for special applications, see below) whereas neutron energies are between 0.85 and 400 meV. Excitations in condensed matter (phonons, magnons, etc.) are in the range of a few meV and above. The relative energy change of X-rays scattered inelastically (with energy loss or gain) by any sample is then very small (say $<10^{-6}$) though it can be resolved in Mössbauer or high-resolution experiments at synchrotron radiation sources. In contrast, neutrons can experience an appreciable

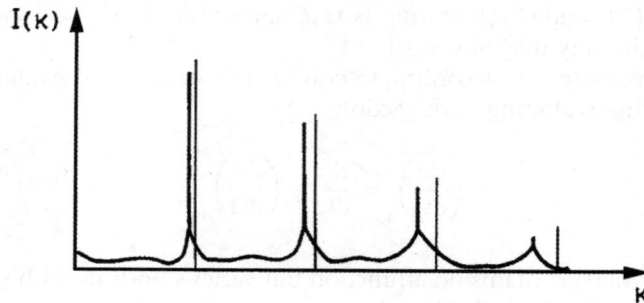

Figure 1   Scattering from a crystal containing defects. Bragg positions of a perfect crystal are indicated by thin vertical lines.

relative change in energy, so that elastic (no energy change) and inelastic scattering can be easily distinguished (see below, Section 13.2.4). Another important difference arises from the magnetic moment of the neutron, which interacts with the local magnetization density. This leads to often appreciable magnetic scattering, which has very important applications in the study of magnetic substances (see Section 13.2.3). Finally, absorption differs considerably for the two types of radiation. (Measurements of absorption can also yield unique insight in the structure of matter, e.g. in EXAFS = extended X-ray absorption fine structure, see Section 13.5.4).

The linear absorption coefficient Σ_t is defined by

$$\ln(I/I_0) = -\Sigma_t D_s, \tag{25}$$

where I_0 and I are the intensities of incident and transmitted beam, respectively, and D_s is the sample thickness. The absorption coefficient has the dimension of 1/length or area/volume, and can be understood as a total macroscopic removal cross-section, as is common in neutron scattering. In the X-ray literature, the so-called mass absorption coefficient is defined by $\rho_m^{-1}\Sigma_t$ with ρ_m = (mass) density. Its dimensions are area/mass, and values are independent of the thermodynamic state of the sample material.

All beam-attenuating processes (including coherent and incoherent scattering) are included in Σ_t, but for X-rays, the excitation of fluorescence radiation can be singled out as the most important true absorption mechanism. When the energy of the incident X-rays approaches a resonance energy in the electronic states, an absorption edge is observed, corresponding to the excitation of electrons in the K, L, etc., levels (see **Figure 2**). The electrons associated with the absorption move out of phase from the others, and destructive interference reduces the atomic scattering factor by the equivalent of twice the number of resonance electrons (e.g. at least a total of four at the K edge).

The atomic scattering factor has to be corrected according to

$$f_a = f_{ao} + f' + if'', \tag{26}$$

where f' and f'' are the real and the imaginary part of the Hönl corrections. The wavelength dependence of f' and f'' is shown schematically in **Figure 3** [see *International Tables for Crystallography* (2006). Vol. C,

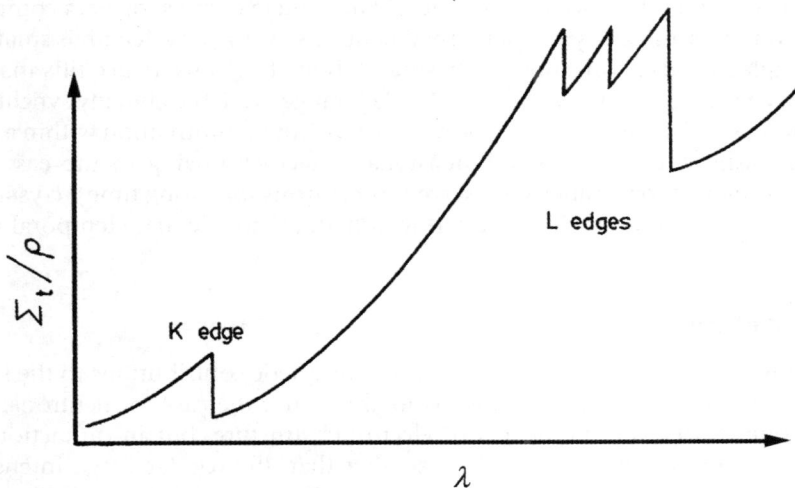

Figure 2 Mass absorption coefficient Σ_t/ρ for X-rays as a function of wavelength near the K and L edges of an element (schematic).

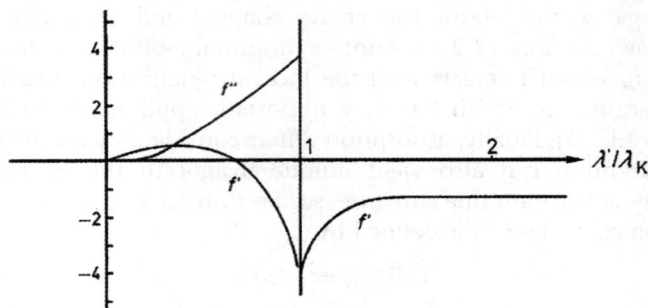

Figure 3 Variation of real and imaginary components f' and f'' of the Hönl corrections as a function of X-ray wavelength near the K absorption edge (at λ_K) of an element. After Cowley (1981).

Section 4.2.6, pp. 241–258, for details and tabulated values]. As synchrotron radiation sources now provide strong X-ray beams with continuously tunable wavelength, diffraction work in the neighborhood of absorption edges can take advantage of the variation of specific scattering factors.

According to eqn (25), $\Sigma_t D_s = 1$ corresponds to a reduction in intensity by a factor $1/e$, which gives an estimate for the typical thickness of a sample (it is the optimal thickness for measurements in transmission). Generally, neutrons penetrate matter more easily than X-rays. Even Cd, a material frequently used for shielding purposes in thermal neutron work, has a smaller absorption cross-section for neutrons than most common metals for X-rays. On the other hand, lead, a good shielding material for X-rays, is almost transparent to neutrons. [See Bacon (1975); Schwartz and Cohen (1987) for more details on the different properties of X-rays and neutrons.]

High-energy X-rays (about 100 keV or more) are progressively more weakly absorbed by matter, and bulk samples can be investigated. Diffraction experiments with the 412 keV γ-radiation from radioactive gold were initiated in Grenoble in the nineteen sixties [see Schneider (1981) for a detailed account] as a means to study the mosaic structure of large single crystals, of sizes comparable to those used for monochromators and analyzers of thermal neutrons. As the wavelength is small (0.03 Å in this case), the Bragg angles are also correspondingly small (about 1°). Thus, lattice tilts may be monitored directly for large crystals. Nowadays, in the 100–150 keV range available at many synchrotron radiation facilities, diffraction experiments in transmission yield a wealth of information within a scattering range of a few degrees, easily covered by two-dimensional detectors. Owing to the easy penetration of windows and shielding materials (an asset reserved to neutrons for a long time), crystal growth, phase transformations etc. can be studied in situ, often with better than adequate temporal resolution.

13.2.3 Magnetic Scattering

For X-rays (photons), the polarization dependencies of magnetic contributions to the scattering length are different for spin and orbital magnetisms, while this is not the case for neutrons. X-rays are thus a more subtle probe concerning the details of electronic structure, but in diffraction, the magnetic intensities are six to eight orders of magnitude smaller than the regular Bragg intensities. Since the pioneering experiment (using a laboratory source) by De Bergevin and Brunel (1972) who identified magnetic Bragg peaks in antiferromagnetic NiO, synchrotron radiation sources have provided beams of increasingly higher power, and magnetic Bragg peaks are now well accessible, and in combination with spectroscopic methods, very complex situations of magnetic order can be resolved [see, e.g. the work on $Ca_3Co_2O_6$ by Agrestini et al. (2008)].

For neutrons, nuclear scattering and magnetic scattering are of similar intensity in many cases, and thus neutrons are most convenient for the study of magnetic structures and especially magnetic defects. The magnetic moment of the neutron has a value of $\gamma\mu_n$ (see **Table 1**) with $\gamma = -1.913$. Owing to the dipole interaction between neutron and local magnetic induction (see Gurevich and Tarasov, 1968; Lovesey, 1984)] only the component $\mathbf{M}^\perp(\kappa)$ perpendicular to the scattering vector κ, of the Fourier transform (FT) of the local magnetization density, $\mathbf{M}(\kappa)$, contributes to the scattering. Therefore, a magnetic interaction vector $\mathbf{q}_M$ may be defined by

$$\mathbf{q}_M = \mathbf{M}^\perp(\kappa)/|\mathbf{M}(\kappa)|. \tag{27}$$

The magnetic scattering length, commonly denoted by $\mathbf{p}$, is a vector:

$$\mathbf{p} = p\mathbf{q}_M \tag{28}$$

with p proportional to $|\mathbf{M}(\kappa)|$. For magnetic scattering due to unpaired spins (without orbital moments), p is given by

$$p(\kappa) = \frac{e^2}{m_0c^2}|\gamma|Sf_{\text{mag}}(\kappa). \tag{29}$$

Here, e^2/m_0c^2 is the classical electron radius f_0 known from X-ray scattering ($f_0 = 2.8 \times 10^{-15}$ m, see Section 13.2.1), S is the total number of unpaired spins, and $f_{\text{mag}}(\kappa)$ is the FT of the spin density, normalized to $f_{\text{mag}}(0) = 1$. As the number μ of Bohr magnetons ($\mu_B = eh/m_0c$) equals $2S$, we can write

$$p(\kappa)[10^{-14}\text{m}] = 0.27\mu f_{\text{mag}}(\kappa). \tag{30}$$

For not too large values of κ, p is comparable to the values of the nuclear scattering amplitude b (for the pure elements Fe, Co, Ni one has $b = 0.96, 0.28, 1.03$ and $p(0) = 0.6, 0.47$ and 0.16, respectively, all in units of 10^{-14} m). The magnetic scattering factor $f_{\text{mag}}(\kappa)$ falls off more rapidly with $(\sin\theta)/\lambda$ than the atomic scattering factor for X-rays (see Section 13.2.1), as only a few electrons in an outer shell contribute to f_{mag}.

Table 1   Some Properties of X-rays and Neutrons

Property	Value[a]	
	For X-rays (photons)	For neutrons
mass m [kg]	0	1.675×10^{-27}
momentum	h/λ	$h/\lambda = mv$
energy	hc/λ	$h^2/2m\lambda^2$
scattering length	Zf_0(for $\kappa\rightarrow0$)	b(nuclear)
absorption	$\propto Z^4\lambda^3$ (strong)	$\propto \lambda$ (mostly weak)
speed v [m/s][b]	c	$437\ E^{1/2}$
wavelength λ [Å][b]	$12.4/E$	$3956/v$
magnetic moment	0	$-1.913\ \mu_n$
frequency ν [s^{-1}][b]	$2.42 \times 10^{17}\ E$	$2.42 \times 10^{11}\ E$

[a] h = Planck's constant, c = velocity of light, f_0 = classical electron radius, μ_n = nuclear magneton.
[b] For photons, E in keV; for neutrons, E in meV.

The total scattering of a magnetic material will show a combination of nuclear and magnetic scattering and will also depend on the polarization of the incident beam. **Figure 4** shows the scattering geometry, assuming a fully polarized neutron beam with the polarization vector parallel to the z-axis (up: $+$, down: $-$). If the scattered beam is analyzed along the same direction, four different scattering cross-sections can be defined ($++$, $+-$, $-+$ and $--$). In an otherwise perfect crystal, the nuclear unit-cell structure factor, $F_s(\kappa)$ [see eqn (12)], can be combined with the magnetic structure amplitude,

$$\mathbf{F}_M^\perp(\kappa) = \sum_m \mathbf{q}_{Mm} p_m \exp(-2\pi\, i\kappa \cdot \mathbf{r}_m), \tag{31}$$

where $\mathbf{q}_{Mm}$ (cf. eqn (27)) is defined according to the direction of the moment at site m, and p_m at each site is given by eqn (30). This yields $|F_s + F_{Mz}^\perp|^2$ for ($++$), $|F_{Mx}^\perp|^2 + |F_{My}^\perp|^2$ for ($+-$) and ($-+$), and $|F_s - F_{Mz}^\perp|^2$ for ($--$), where the $F_{Mi}^\perp$ are the Cartesian components of $\mathbf{F}_M^\perp$.

Unpolarized beams may be assumed to be composed of 50% positive and 50% negative polarizations, and the scattering cross-section (per atom) without polarization analysis is $d\sigma/d\Omega = |F_s|^2 + |F_M|^2$. For a simple ferromagnet, $\mathbf{q}_{Mm}$ is the same for all sites, and with $\mathbf{F}_M^\perp = \mathbf{q}_M F_{mag}$ (see eqn (31)),

$$\frac{d\sigma}{d\Omega} = |F_s|^2 + |F_{mag}|^2 \sin^2\alpha, \tag{32}$$

where α is the angle between κ and $\mathbf{M}$ (see **Figure 4**). By measuring the scattering at different values of α (varying the external magnetic field or the scattering geometry), nuclear and magnetic contributions can be separated. Equation (32) also applies to antiferromagnetic materials if we define α relative to the direction of the moment at site one and write $F_{mag} = \sum_m \hat{\sigma}_m p_m \exp(-2\pi\, i\kappa \cdot \mathbf{r}_m)$ where $\hat{\sigma}_m$ is either $+1$ or -1, depending on whether or not the moment at site m is parallel or antiparallel to $\mathbf{q}_1$. **Figure 5** shows the consequences of ferro- and antiferromagnetic orders in the bcc lattice ($\alpha \neq 0$). Additional peaks ("magnetic peaks") occur for antiferromagnetic order, whereas in ferromagnets, the nuclear peaks are enhanced.

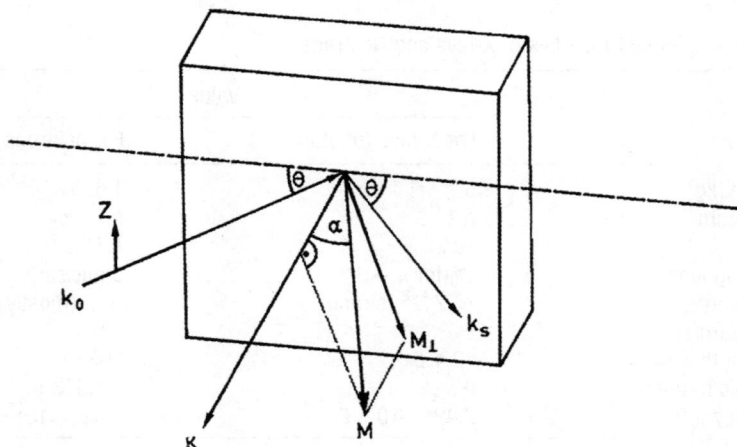

Figure 4 Scattering geometry for magnetic scattering. The incident neutrons are polarized along the z-axis.

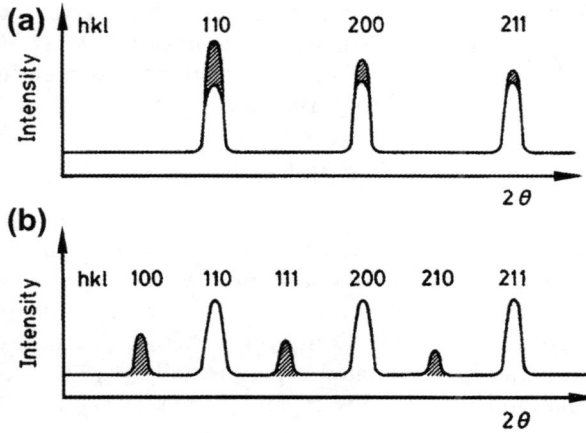

Figure 5 Schematic neutron diffraction patterns from a b.c.c. poylcrystalline (a) ferromagnet and (b) antiferromagnet. The shaded areas represent the magnetic scattering contributions which decrease with 2θ because f_{mag} decreases. After Schwartz and Cohen (1987).

If a polarized beam is used, the cross-section for scattering without polarization analysis is the sum of, e.g. $(++)$ and $(+-)$ scattering and contains an interference term between nuclear and magnetic scattering with F_s and $\mathbf{F}_M^{\perp}$ appearing unsquared:

$$\left(\frac{d\sigma}{d\Omega}\right)_{P_0} = |F_s|^2 + 2F_s\mathbf{F}_M^{\perp} \cdot \mathbf{P}_0 + |\mathbf{F}_M^{\perp}|^2, \tag{33}$$

where $\mathbf{P}_0$ is a unit vector indicating the direction of polarization. "Flipping" $\mathbf{P}_0$ from $+1$ to -1, we have a very sensitive method to measure $\mathbf{F}_M^{\perp}$ and consequently $\mathbf{F}_{mag}(\kappa)$, $p_m(\kappa)$ or $f_{mag}(\kappa)$. Equation (33), if generalized as indicated by eqn (20) for structural disorder, is also the basis for a separation of structural and magnetic disorder in alloys.

Additional neutron scattering also occurs for a randomly oriented ensemble of magnetic moments. If they are all of identical magnitude, e.g. related to equal spins S, Halpern and Johnson (1939) derived the original result for such paramagnetic scattering:

$$\left(\frac{d\sigma}{d\Omega}\right)_p = \frac{2}{3}S(S+1)f_0^2\gamma^2 f_{mag}^2. \tag{34}$$

This scattering is similar to the Laue scattering term in diffuse scattering. Deviations from the simple monotonic κ-dependence are of interest in both cases since they relate to correlations in the atomic or the spin arrangement.

There are many more special features of magnetic neutron scattering than this brief discussion can indicate (see, e.g. Brown, 1979; Hicks, 1979, for more details). Many complex magnetic structures have been studied in recent years.

13.2.4 Inelastic and Quasielastic Scattering

As mentioned in Section 13.2.2, the relative energy gain or loss of X-rays scattered from a sample with lattice vibrations (phonons) is very small. Near an absorption edge, resonant Raman scattering (see Sparks, 1974) occurs just below the edge, and incoherent fluorescence radiation is emitted above the edge. Compton scattering is another inelastic scattering process of X-rays. From the conservation of energy and momentum for the scattering of a photon from an individual electron, the wavelength shift of Compton-modified radiation is (for electrons assumed at rest):

$$\Delta\lambda\left[\overset{\circ}{A}\right] = 0.0243(1 - \cos 2\theta) \tag{35}$$

independent of the wavelength of the incident radiation. The scattering is incoherent as there is no fixed phase-relationship between the different inelastic scattering events. The relative contribution of Compton scattering to the total scattering is given by $\left(1 - \sum_j f_{aj}^2/Z\right)$, where f_{aj} is the form factor (normalized to one) of each of the Z electrons of an atom. Compton scattering increases the background in X-ray diffraction experiments, but it can be eliminated experimentally (e.g. with a monochromator in the diffracted beam) or by calculation [see, e.g. Schwartz and Cohen, 1987]. Detailed analysis of the energy distribution of Compton scattering radiation provides information on the momentum distribution of electrons in condensed matter, one of the few techniques also applicable to alloys.

Much smaller (absolute) energy transfers can be detected by neutron scattering, either by analyzing the change of wavelength with a single crystal or the change of neutron momentum by time-of-flight methods. Here, we consider one-phonon scattering only [for a complete account see, e.g. Lovesey (1984); Bacon (1975)]. **Figure 6** shows two possible scattering configurations in reciprocal space. The scattering vector κ can now be written as

$$\kappa = \mathbf{k}_s - \mathbf{k}_0 = \mathbf{g} + \mathbf{q}, \tag{36}$$

where $\mathbf{q}$ is the phonon wave vector, counted from the nearest reciprocal lattice point. If the wave vector after scattering is $\mathbf{k}_{s1}$, the neutron has lost energy ($k_{sl} < k_0$), and a phonon has been created. For $\mathbf{k}_{s2}$, a phonon has been annihilated. As the neutron momentum is h/λ, eqn (36) states the conservation of momentum in the scattering process ($\lambda_q = 1/q$ is the phonon wavelength). Simultaneously, the conservation of energy,

$$h^2/2m_n\lambda_0^2 - h^2/2m_n\lambda_s^2 = h\nu_p, \tag{37}$$

defines the energy of the phonon participating in the scattering (λ_0 and λ_s denote the wavelengths of the incident and the scattered neutrons). There are only discrete values of phonon frequency ν_p for

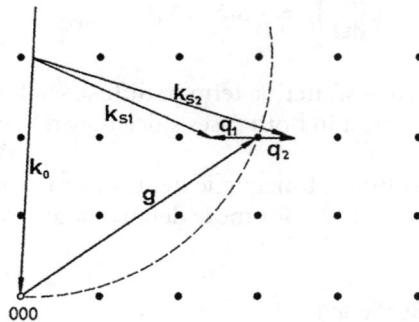

Figure 6   Two possible inelastic scattering events in a plane of reciprocal space involving the creation ($\kappa = \kappa_{s1} - \kappa_0$) and the annihilation ($\kappa = \kappa_{s2} - \kappa_0$) of a phonon (here, $\mathbf{k} = \kappa$, and $\mathbf{g}$ is a Bragg vector).

a given **q**, and appropriate scans can be designed to obtain directly the phonon dispersion curves of a crystal.

Apart from the coherent inelastic scattering processes, analysis of the incoherent inelastic scattering of neutrons may often be of interest. True incoherent scattering processes are due to the interaction of the neutron spin with nuclear spins $I \neq 0$. The scattering length of the compound nucleus depends on its total spin, which is $I + 1/2$ or $I - 1/2$. The tabulated values of coherent scattering lengths and incoherent cross-sections for individual isotopes represent properly weighted averages (for unpolarized nuclei). Natural elements are frequently a mixture of different isotopes, each with its own nuclear spin. Coherent scattering lengths and spin-incoherent cross-sections are simply arithmetic averages, but owing to the random distribution of nuclei with different coherent scattering lengths over the sites of the sample, another κ-independent term, $\sigma_i = 4\pi\{\overline{b^2} - (\overline{b})^2\}$, occurs in complete analogy to the monotonic Laue-scattering term in diffuse X-ray scattering (see Section 13.4). For neutrons, this part is included in the total value of incoherent scattering of an element [see the discussion of incoherent neutron scattering from multielement materials by Glinka (2011)]. Incoherent neutron scattering cross-sections are quite large in several cases, and for coherent scattering experiments (see Section 13.2.5), they may impose severe limitations. However, similar to the case of Compton scattering and electrons, energy analysis of the incoherently scattered neutrons will reveal details of the motion of nuclei. The incoherent scattering function can be calculated for different processes (see, e.g. Bacon, 1975; Lovesey, 1984). If the energy transfer is small, the scattering is called quasielastic. In condensed matter, the motion of atoms is restricted by the environment, and quasielastic neutron scattering has become a widely used technique to study atomic and molecular motion, especially if hydrogen with its high incoherent cross-section is involved. Especially in crystalline solids, a study of the quasielastic linewidth (coherent or incoherent) may help to identify diffusion mechanisms (see Section 13.6.2).

13.2.5 Some Experimental Considerations

Incoherent scattering as a source of background and absorption was already mentioned in Section 13.2.4. Extensive discussions on general experimental problems in X-ray and neutron scattering can be found in the book by Schwartz and Cohen (1987). Some details relating to particular applications can be found in subsequent sections. Here, we state only a few quite general points.

Laboratory X-ray sources with rotating anodes are frequently used, mostly with Cu or Mo targets at powers of the order of 10 kW. At a very high power of 100 kW (reached by Haubold, 1975) one obtains 4×10^{16} quanta/s cm^2 sr (CuK$_\alpha$). Higher luminosities are obtained (originally as a by-product, now from dedicated sources) from electron or positron synchrotrons and storage rings. Depending on the energy of the electrons or positrons, synchrotron radiation emerges in a wavelength range from infrared to about 0.1 Å. Synchrotron radiation is continuous and very intense (up to some 10^{20} quanta/s sr, integrated over the vertical direction). The radiation is highly collimated perpendicular to the orbital plane. In the GeV region, the divergence is about 10^{-4} rad. In the plane of the ideal orbit, synchrotron radiation is 100% polarized with the electrical vector parallel to this plane. There is a well-defined time structure of the beam with pulse durations as short as 100 ps with repetition rates of 1 MHz or more, depending on the number of bunches in the ring. The use of synchrotron radiation for materials studies is currently experiencing a very rapid growth (see, e.g. Baruchel et al., 1993; Reimers et al., 2008).

Compared even with a classical sealed X-ray tube, neutron sources are not very powerful. **Figure 7** shows a comparison of several X-ray and neutron sources. As originally suggested by Maier-Leibnitz (1966), the momentum space density $p(\mathbf{k}_0)$ (hk_0, where h is Planck's constant, is the linear momentum of a particle) is

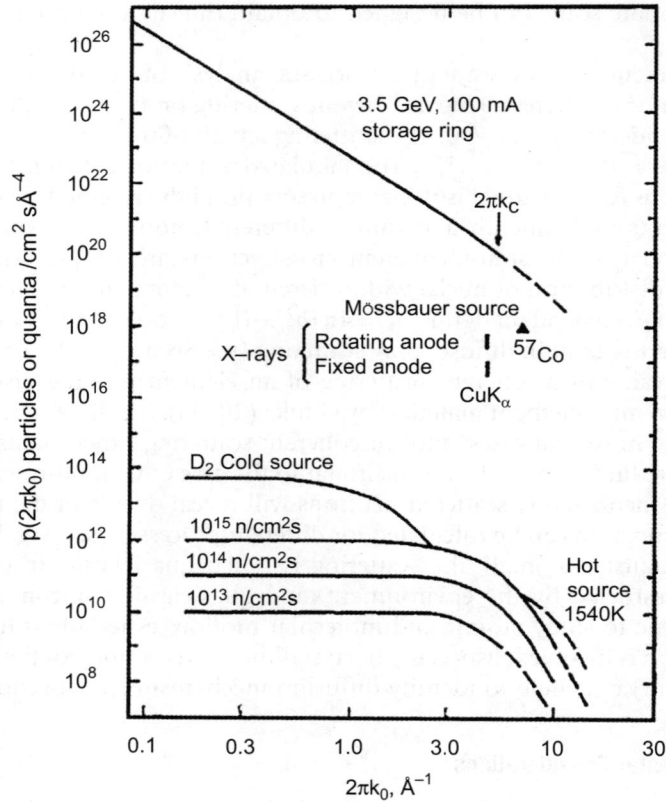

Figure 7 Comparison of phase space densities of different sources of X-rays and thermalized neutrons.

an adequate quantity for comparison, as the count rate $\dot{Z}$ at a detector is proportional to $p(k_0)$ multiplied by the momentum space elements d^3k_0 and d^3k_s that can be optimized instrument parameters (within the bounds of $\mathbf{k}$ imposed by the properties of the source), i.e. $\dot{Z} \sim p(\mathbf{k}_0)S(\kappa, \nu)\, d^3k_0\, d^3k_s$, where $S(\kappa, \nu)$ is the scattering law to be studied. We see that $p(\mathbf{k}_0)$ for neutrons is several orders of magnitude lower than the values of all X-ray sources. On the other hand, restrictions on the choice of parameters of the incident beam (wavelength/energy range or vertical collimation), which are common for X-rays, are less severe for neutrons where d^3k_0 can be chosen more flexibly to match the resolution requirements for a given scattering law. This may compensate in part for the lower flux at the sample position.

Traditionally, thermal neutrons are produced by fission and subsequent moderation, i.e. in a nuclear reactor. A steady-state thermal flux of about 10^{15} n/cm^2 is virtually impossible to surpass as heat removal is a limiting factor. With hot and cold sources, the Maxwellian spectrum of neutrons may be shifted to smaller or larger wavelengths (see **Figure 7**). Pulsed neutron sources may provide a higher peak flux. If the pulsed structure of the neutron beam is maintained after moderation, time-of-flight experiments are advantageous at pulsed sources because the relevant flux at the sample will be higher than the average flux. The production of pulsed neutrons is usually based on the use of charged particles from accelerators (spallation). A variety of spallation sources are now operational, and more powerful ones are planned.

Apart from film techniques (with a converter foil for neutrons), the detection of the scattered radiation is based on electronic counting circuits attached to gas or solid-state detectors. For X-rays, gas-filled proportional "counters" and solid-state scintillation "counters" detect the incident quanta via the ionization of a gas or the production of photoelectrons by scintillation photons. The energy resolution is poor in both detector types, typically about 20% in the 10 keV range for gas detectors and 50% for scintillation detectors. Solid-state detectors based on electron–hole pair production in doped (Ge or Si with Li) or intrinsic (Ge) semiconductors have a theoretical resolution of about 1% at 10 keV, about 200 eV in practice.

Another important aspect for many experiments is the spatial resolution of large detectors. Linear position-sensitive detectors with a resolution of about 30 μm employ a resistive wire, and peak heights or pulse shapes are analyzed as a function of position of the detected event. Two-dimensional gas-filled detectors for X-rays employ (sets of) mutually perpendicular wires and different electronic techniques to locate the detected events. Charge-coupled devices (CCD detectors) use solid-state electronics to read out incident photons directly or from a scintillation screen, thus providing two-dimensional detection. So-called pixel detectors consisting of arrays of single-photon semiconductor detectors bonded to Si sensors [e.g. pixel size 217 μm × 217 μm, module of 366 × 157 pixels, active area of about 80 mm × 34 mm, see Broennimann et al. (2006); Kraft et al. (2009); microstrip detectors see Bergamaschi et al. (2010)] are now more frequently used. The simultaneous measurement of scattering over a large solid angle is (apart from collecting many Bragg reflections in structure determinations) of particular importance for weak scattering signals but also for in situ kinetic studies.

Neutrons can be detected only after they have participated in a nuclear reaction with the emission of charged particles or γ-rays. As beam dimensions are usually much larger in neutron scattering than in X-ray scattering, the use of large arrays of individual detectors is possible without a loss of resolution (and with time-of-flight resolution, i.e. energy resolution for each desired detector). Position-sensitive detectors have been developed for diffraction work; planar detectors, e.g. for work on single crystals and small-angle scattering (SAS), and curved ones for recording powder patterns.

13.3 Bragg Peaks and Vicinity

Scattering from real crystals may occur anywhere in reciprocal space but Bragg peaks, though modified, will remain a predominant feature as long as an average lattice can be defined. A diffraction pattern, after all instrumental corrections, reflects the distribution of scattering matter within the (average) unit cell of a substance, and the atomic coordinates (location of the centers of scattering objects) as well as the scattering length density distribution (electrons with X-rays, nuclei and magnetic moments with neutrons) can be determined from an analysis of Bragg peaks (see, e.g. Lipson and Cochran, 1953; Warren, 1969, 1999; Schwartz and Cohen, 1987). Particular Bragg peaks occur for different phases in a sample, and diffraction methods are thus essential in the study of phase diagrams and phase transitions. The orientation distribution of Bragg peaks for a polycrystalline sample reveals its texture (see, e.g. Barrett and Massalski, 1980). Making use of highly collimated high-energy beams at a synchrotron radiation source, it becomes possible to analyze individual grains in a three-dimensional environment (see Poulsen, 2004). For example, Schmidt et al. (2008) have studied grain arrangements in an Al-0.1% Mn alloy after deformation and annealing. An X-ray beam of an energy of 50 keV was focused vertically to a spot of 8 μm, and a disk of a height of 350 μm in the middle of the cylindrical sample was mapped by consecutively illuminating layers in steps of 10 μm (see **Figure 8**). From 62 diffraction images per layer (the acquisition time for the full 3-D map was 6 h) and several evaluation steps, the three-dimensional grain morphology as well as the crystallographic orientation were determined for 483

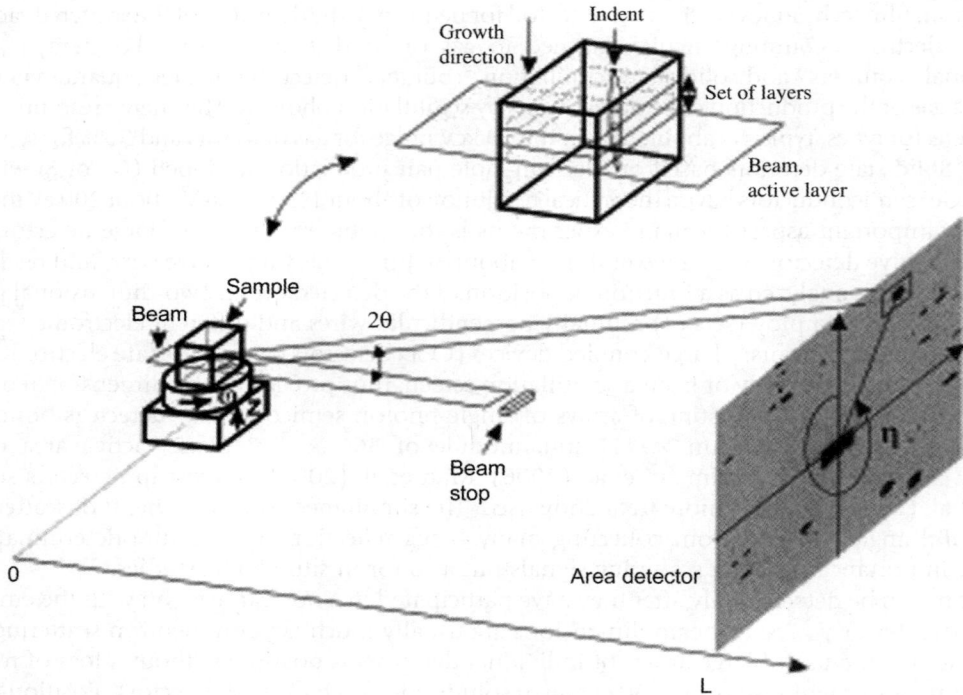

Figure 8 Schematic view of the experimental setup for a three-dimensional analysis of grain growth in Al-0.1% Mn by a highly focused X-ray beam. From Schmidt et al. (2008).

grains prior to annealing. After annealing, significant grain growth had taken place, with only 27 remaining grains in the same volume. **Figures 9** and **10** show the results.

13.3.1   Peak Shifts

Changes in peak position may be caused by residual stresses, by faulting on certain crystallographic planes (Warren, 1969; Wagner, 1966; Cowley, 1981), or by lattice parameter changes as a function of alloying or of defect concentration. Whereas a lattice parameter change affects all crystallographically equivalent Bragg peaks in the same way, the other two causes may lead to different shifts depending on the orientation of the reflecting planes relative to a preferred axis of the disturbance. **Figure 11** illustrates this for a polycrystal with a surface under compression. As residual stresses and applied stresses combined determine load-carrying capabilities, X-ray methods in this field have received considerable attention (e.g. Barrett and Massalski, 1980; Cullity, 1977; Noyan and Cohen, 1987; Reimers et al., 2008).

As the penetration depth of standard X-rays is limited and strongly depends on wavelength, only near-surface regions are accessible. In the past, destructive layer-by-layer studies were thus common to arrive at depth-dependent strain information. This as well as the occasionally reported use of different wavelengths without removal of material requires some caution in the interpretation. Manns et al. (2009) have implemented a method to determine residual stresses in real space directly by means of small gauge volumes, defined by beam-limiting masks. With long exposure times, suitable data were

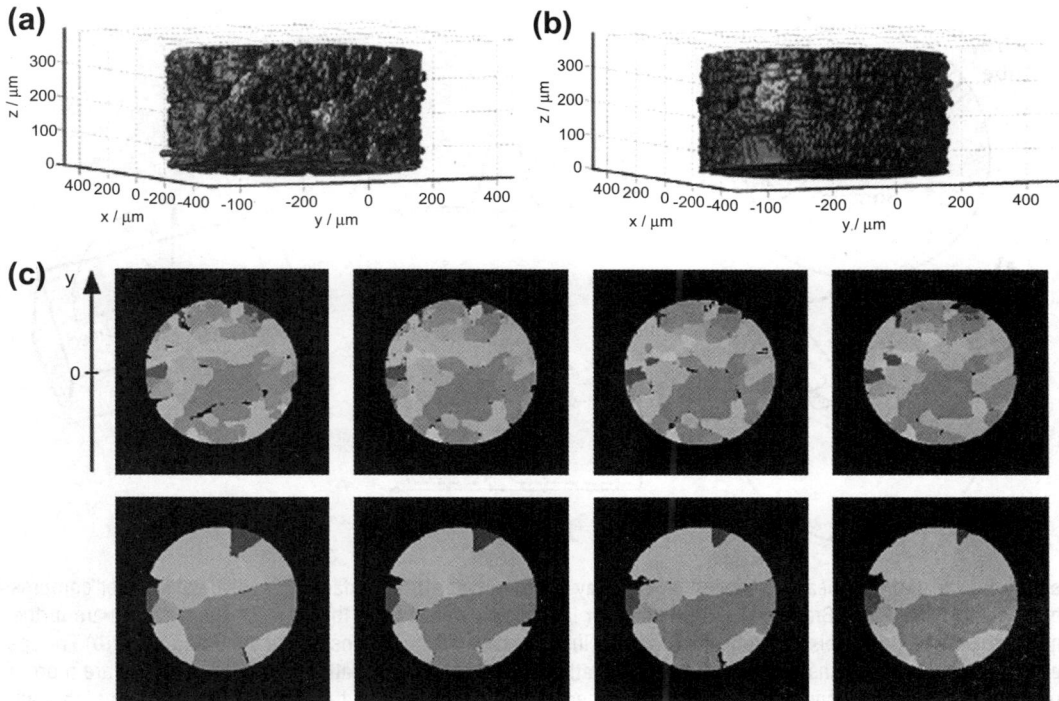

Figure 9 Results of the three-dimensional grain mapping of Al-0.1% Mn after annealing, (a) after a first annealing step, (b) after final annealing. (c) Four layers, 10 μm apart, from the middle of the probed volume after the first (top row) and final (bottom row) annealing step (pixel size 5 μm × 5 μm, different colors indicate different orientations, black areas inside the sample areas are artifacts). From Schmidt et al. (2008). (For color version of this figure, the reader is referred to the online version of this book.)

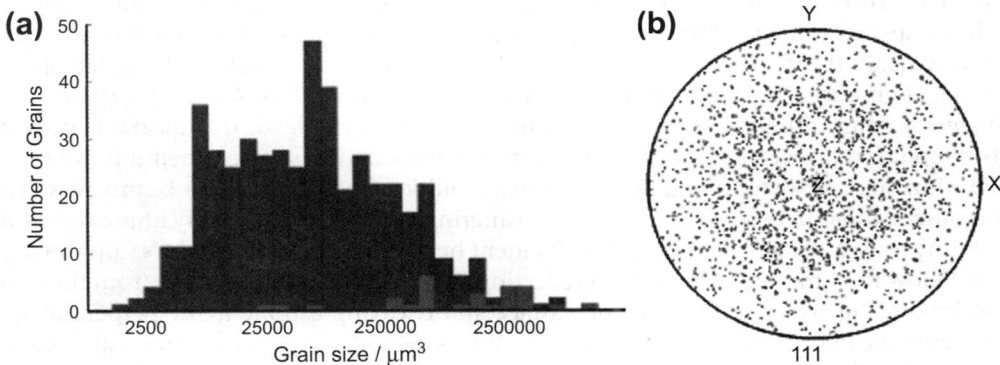

Figure 10 Grain characteristics of a sample of Al-0.1% Mn after two annealing steps (see **Figure 9**). (a) Grain volume distribution after the first annealing step (blue and red) and remaining subset (red) after the final annealing. (b) (111) pole figure with respect to the sample axes shown in **Figure 9** (a, b), colors as in (a). From Schmidt et al. (2008). (For interpretation of the references to color in this figure legend, the reader is referred to the online version of this book.)

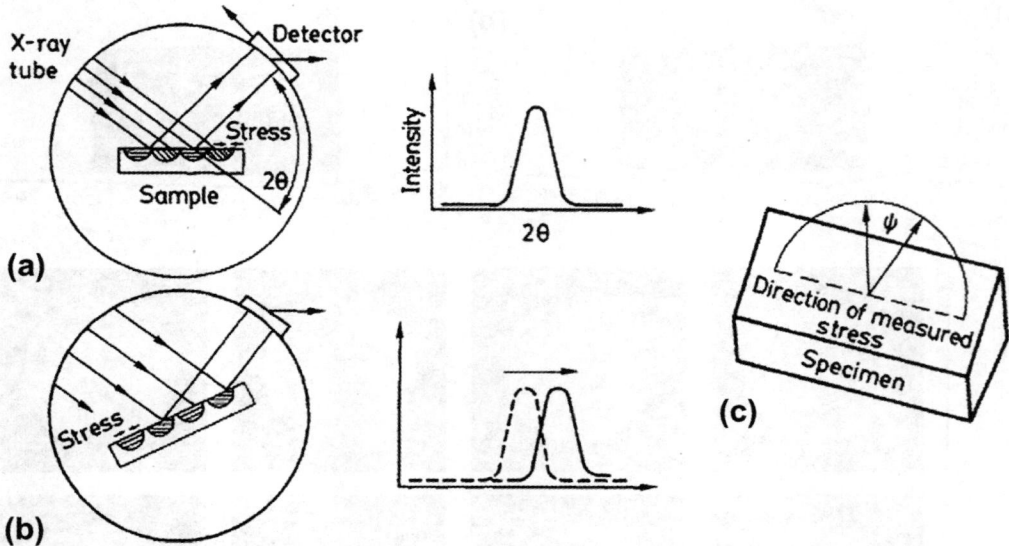

Figure 11   Residual stress measurements with an X-ray diffractometer for a surface of a polycrystal under compression (schematic, not to scale). (a) Bragg's law is satisfied for lattice planes parallel to the surface. These planes are further apart than in the stress-free state (Poisson's effect). Their spacing is obtained from the position of the Bragg peak. (b) The specimen has been tilted, and other grains now present suitable lattice planes for Bragg scattering. As these planes are more nearly perpendicular to the compressive stress, they are less separated than in (a), and the Bragg peak moves to higher angles. (c) The direction of the measured stress is given by the intersection of the circle of tilt and the surface. After M.R. James and Cohen (1980).

recorded even at a laboratory source. With focused white-beam synchrotron radiation, Kassner et al. (2009) have analyzed internal strains in deformed Cu single crystals at submicrometer resolution.

Neutron diffraction averages over larger volumes, and the nondestructive measurement of internal stresses in "real" components has seen many application at specially equipped "strain scanners". The method has a special value for large components and "hidden" stress concentrations, e.g. near welds [see Woo et al. (2008) for a friction-stir-processed AZ31B magnesium alloy, and Holden et al. (2006) for a survey]. High-energy synchrotron X-rays offer new options, see Withers et al. (2008).

The preferred Bragg diffraction of X-rays from near-surface regions leads to important applications in the study of epitaxial layers and heterostructures, as accommodation stresses, their inhomogeneity and their relaxation may be determined. With the availability of powerful X-ray beams at synchrotron radiation sources, the surface sensitivity for Bragg scattering has been considerably improved (several nm instead of μm) near grazing incidence. For an incident beam that impinges on the sample at a glancing angle at or below the limiting angle of total reflection, diffraction peaks may arise from the evanescent wave traveling in the crystal parallel to the surface and decaying exponentially perpendicular to the surface. Grazing incidence scattering of X-rays (and neutrons) offers many experimental possibilities in surface science [see Feidenhans'l (1989); Zabel and Robinson (1992); and Dosch (1992, 1993)]. As an example of evanescent X-ray Bragg scattering we quote the studies on depth-resolved near-surface ordering of Cu_3Au (Zhu et al., 1988; Dosch et al., 1991). An X-ray study of the crystal truncation rods of Cu layers on Ni(100) by Rasmussen et al. (1997) showed that upon increasing thickness of the Cu layer, embedded {111} wedges appear and provide relaxation of the mismatch strain.

13.3.2 Peak Broadening and Intensity Changes

In Section 13.2.1, we have already mentioned a broadening due to sample size that could be relevant for powders [eqn (1)]. However, for bulk samples, we have tacitly ignored size broadening and written the Bragg cross-section as a delta function. Thus, for a well-collimated X-ray beam of characteristic radiation, the width of a Bragg peak should be controlled by the natural linewidth, since Bragg's law yields (θ being a Bragg angle):

$$\Delta(2\theta) = 2\frac{\Delta\lambda}{\lambda}\tan\theta. \tag{38}$$

With $\Delta\lambda/\lambda = 5 \times 10^{-4}$, $\Delta(2\theta)$ is about 1 min of arc at $2\theta = 30°$ but increases dramatically for $\theta \to 90°$, and eqn (38) can be experimentally confirmed for large perfect crystals. However, quite frequently, the Bragg peaks are much broader (many minutes at moderate Bragg angles), and only crystals with such "mosaicity" will actually approach the predictions of the kinematical theory as may be judged from the integrated reflecting power (which should be proportional to $|F_s|^2$). According to Darwin (1922) (see Zachariasen, 1945; James, 1963), the kinematical theory is valid for an ideal mosaic crystal consisting of small perfect domains that are tilted against each other by small angles (a few minutes). According to this concept, the effect of coherent multiple reflection of the same beam can be reduced to just the planes within one "mosaic block" and finally neglected (i.e. primary extinction plays no role). The dynamical theory of diffraction shows that the tolerable size of the blocks depends on the wavelength of the radiation used. As the extinction length (for Bragg reflections) is proportional to $(\lambda|F_s|)^{-1}$, a sample of given thickness or mosaicity will approach the case of the ideal mosaic crystal with decreasing wavelength.

Mosaicity in a real crystal is a merely formal concept for peak broadening that may be caused by various defects, e.g. dislocations. Referring to crystals containing dislocations in connection with Bragg peaks appears contradictory since eqn (24) and the subsequent discussion imply that dislocations cause Bragg peaks to disappear as their strain fields decrease proportionally to $(r')^{-1}$. However, this would hold only for a random arrangement of dislocations in a large crystal. As Wilkens (1969) has shown, a random distribution is not a very suitable model for the calculation of Bragg peak broadening (or of the elastic energy that would monotonically increase with crystal size). A so-called restrictedly random distribution of dislocations was used to calculate the line broadening and the peak profile more realistically (Wilkens, 1970, 1975) from the kinematical theory. The peak width $\Delta\kappa$ is proportional to $g_{hkl}\rho_d \ln M_e$, where ρ_d is the dislocation density and M_e is a parameter influencing the peak shape. It is given by $M_e = \rho_d^{1/2}R_e$ where R_e is an effective outer cutoff radius, indicating the range over which a random dislocation arrangement can be admitted. Beyond this distance, long-range stresses should compensate each other. These ideas have been adopted by a number of researchers and are extensively used, with some modifications, to separate dislocation-related peak shifts and shapes from size broadening and other effects, including recent procedures of "whole profile fitting" or "... modeling" (see Ribárik et al., 2001; Ribárik and Ungár, 2010; Scardi and Leoni, 2002; Scardi, 2008). Nano-structured materials have received much attention in recent years, and the analysis of the resulting very broad diffraction peaks requires special attention. Generally, dislocations contribute less and planar faults occur more frequently [see, e.g. Page et al. (2011) for a pair distribution function (PDF) approach]. Rocking-curve broadening, in contrast to the common peak broadening along the scattering vector, is only very indirectly related to dislocation densities. It rather reflects the spread of lattice rotations around the rocking axis. Active slip systems may thus be identified from rocking curves taken around different axes.

An analysis of the "static Debye–Waller factor" [see eqn (24)] may be useful to determine displacements around defects (see Krivoglaz, 1969; Dederichs, 1973; Trinkaus, 1973; Dietrich and Fenzl, 1989a). In a dilute alloy, ($c \ll 1$), eqn (24) simplifies to

$$M = c \sum_n [1 - \cos(2\pi\, \boldsymbol{\kappa} \cdot \mathbf{u}_n)], \tag{39}$$

or even to $2M = c\sum_n (2\pi\boldsymbol{\kappa}\cdot\mathbf{u}_n)^2$ if $|2\pi\boldsymbol{\kappa}\cdot\mathbf{u}_n| \ll 1$. The $\mathbf{u}_n$ are displacements caused by a defect taken at the origin, and the summation extends over all cells.

Although the method seems simple, it has not been used too often. Chen et al. (1980) studied martensite, where interstitially dissolved carbon causes a tetragonal distortion, and M_{002} is expected to be different from M_{200} or M_{020}. The authors have evaluated the corresponding (X-ray) intensity ratios for martensite formed in quenched fine-grained Fe-18 wt.% Ni-(0.9/1.15)wt.% C. A similar study of cubic NbH_x was performed by Metzger et al. (1982) who obtained $u_1 = (0.100\pm0.007)$ Å for H occupying tetrahedral sites, a value very close to theoretical results based on lattice theory. In binary ordered crystals containing dilute solutes, e.g. Ni_3Al studied by Morinaga et al. (1988), the static atomic displacements in the two sublattices may be determined individually from the analysis of a large number of fundamental and superstructure peaks. For alloying elements substituting mainly for Al in Ni_3Al, it was found that there is a reverse correlation between Al and Ni displacements and that the average lattice constant is primarily controlled by the Al displacements. Rao et al. (1993) found that the addition of various amounts of hafnium, zirconium and titanium caused the Debye–Waller factors of Ni_3Al to increase linearly, while boron addition resulted in a nonlinear increase.

In these applications, a separation of static and dynamic displacements relies on the thermal displacements known from the defect-free reference state. Neutron and X-ray results for pure Cu_3Au showed a very strong increase of the Debye–Waller exponent when T_c, the critical temperature for appearance of long-range order, was approached from either side (see Bardhan et al., 1977). From the difference in the intensity ratios of fundamental and superstructure/diffuse peaks, individual mean-square displacements for Au and Cu [based on an expansion of eqn (39) for M] were calculated, and large displacements of Au near T_c were found. From a wealth of other information, including phonon dispersion curves (Chen et al., 1977) near T_c, the authors concluded that these displacements are static and related to a high density of antiphase boundaries.

13.3.3 Diffuse Scattering near Bragg Peaks

We now turn to the scattering cross-section given formally by eqn (23), i.e. continuous scattering for $\boldsymbol{\kappa} \neq \mathbf{g}$. There is no general analytical solution, and approximations have to be introduced to obtain manageable expressions.

As pointed out in Section 13.2.4, thermal vibrations (phonons) are always present in a crystal, and their effect on the Bragg intensity is usually contained in the Debye–Waller factor. The resulting diffuse intensity is called thermal-diffuse scattering (TDS) and has been treated extensively (e.g. James, 1963; Warren, 1969, 1999; Bacon, 1975; Willis and Pryor, 1975; see also Xu et al., 2009). TDS occurs for all values of $\boldsymbol{\kappa}$ but it increases with κ and is particularly pronounced near reciprocal lattice points. This is also true for static displacements but there are some important differences regarding the symmetry of diffuse scattering in reciprocal space.

The theory of diffuse scattering from point defects and clusters has been considered by Krivoglaz (1969, 1996); Dederichs (1971, 1973), Trinkaus (1972) and Dietrich and Fenzl (1989a,b). Larson

(2009) has recently reviewed the field. For small defect concentrations c one obtains from eqns (18 and 19) the diffuse-scattering cross-section:

$$\left(\frac{d\sigma}{d\Omega}\right) = cN|F(\kappa)|^2, \tag{40}$$

with $F(\kappa) = \Delta f_B + f_A \sum_n \exp(-2\pi i\kappa \cdot \mathbf{r}_n^a)[\exp(-2\pi i\kappa \cdot \mathbf{u}_{0n}) - 1], \tag{41}$

where $\mathbf{u}_{0n}$ is the displacement at site n due to one single object at the origin, $\mathbf{r}_n^a$ denotes the position vector in the average lattice, and $\Delta f_B, f_A$ are the scattering lengths of defect and host atoms (one atom per unit cell assumed for simplicity), including thermal and static Debye–Waller factors. The value of Δf_B depends on the type of defect (e.g. Dederichs, 1973):

$$\Delta f_B = \begin{cases} f_I \exp(-2\pi i\kappa \cdot \mathbf{r}_1) & \text{for interstitial impurities,} \\ f_B - f_A & \text{for substitutional impurities,} \\ -f_A & \text{for vacancies,} \\ f_A[2\cos(2\pi\kappa \cdot \mathbf{r}_1) - 1] & \text{for split (self-)interstitials,} \end{cases} \tag{42}$$

where $\mathbf{r}_1$ is the position of the interstitial (or of one of the split interstitials). For κ close to a Bragg peak ($\kappa = \mathbf{g} + \mathbf{s}$, and $s \ll g$), but also, with a different reasoning, for any Δf_B as long as $|2\pi\kappa \cdot \mathbf{u}_{0n}| \ll 1$, eqn (41) yields:

$$F(\kappa) = \Delta f_B - 2\pi i f_A \kappa \cdot \sum_n \mathbf{u}_{0n} \exp(-2\pi i\kappa \cdot \mathbf{r}_n^a), \tag{43}$$

and finally, for the Huang-scattering region ($s \ll g$, Δf_B neglected, N = number of exposed atoms):

$$\left(\frac{d\sigma}{d\Omega}\right)_{Hd} = cN f_A^2 |2\pi\mathbf{g} \cdot \mathbf{u}(\kappa)|^2, \tag{44}$$

where $\mathbf{u}(\kappa)$ is the FT of the displacement field (which is always centrosymmetric at large distances). This can be written as [only $\mathbf{s} = \kappa - \mathbf{g}$ is relevant, since $\mathbf{u}(\kappa)$ is periodic in reciprocal space]:

$$\mathbf{u}(\kappa) = \mathbf{u}(\mathbf{s}) = \frac{i}{sV_c} S(\hat{\mathbf{s}}) \cdot [P \cdot \hat{\mathbf{s}}]. \tag{45}$$

Here, S is a tensor containing combinations of elastic compliances depending on the orientation of $\hat{\mathbf{s}}$ where $\hat{\mathbf{s}}$ is a unit vector in the direction of $\mathbf{s}$. The tensor P is the dipole force tensor of the defect with components

$$P_{jk} = P_{kj} = \sum_n p_{jn} x_{kn}, \tag{46}$$

where p_{jn} are the components of the forces $\mathbf{p}_n$ and x_{kn} the components of the position vectors $\mathbf{r}_n$. We note that the volume change caused by defects is proportional to the trace of P, and the lattice parameter change $\Delta a/a$ can be obtained from

$$\frac{\Delta V}{V} = 3\frac{\Delta a}{a} = \frac{c}{V_c} \frac{\text{Tr } P}{C_{11} + 2C_{12}} \tag{47}$$

for a cubic crystal with the elastic constants C_{11}, C_{12}. Equation (47) is valid for isotropic and anisotropic defects, provided the latter show no preferred orientational arrangement relative to one of the cubic axes (see Peisl et al., 1974).

Substituting $\mathbf{u}(\boldsymbol{\kappa})$ in eqn (44) by expression (45), the Huang intensity is:

$$\left(\frac{d\sigma}{d\Omega}\right)_{Hd} = \frac{cN}{V_c^2} f_A^2 \left(\frac{g}{s}\right)^2 |2\pi\widehat{\mathbf{g}}\cdot\{S(\widehat{\mathbf{s}})\cdot[\mathbf{P}\cdot\widehat{\mathbf{s}}]\}|^2. \tag{48}$$

We see that Huang scattering has the same general $\boldsymbol{\kappa}$ dependence as first-order TDS for $\boldsymbol{\kappa} \to \mathbf{g}$, especially inversion symmetry around $\mathbf{g}$, but the intensity distribution around a reciprocal lattice point varies according to $\mathbf{g}\cdot\mathbf{u}(\mathbf{s})$. For example, an isotropic defect with three equal, mutually perpendicular double forces $\mathbf{P}_0$ in an isotropic cubic crystal has

$$\mathbf{u}(\mathbf{s}) = \frac{iP_0\mathbf{s}}{s^2 V_c C_{11}}, \tag{49}$$

and it follows from eqn (44) that there is a zero-intensity plane for $\mathbf{s}$ perpendicular to $\boldsymbol{\kappa} \approx \mathbf{g}$. **Figure 12**(a) illustrates this case. Double-force tensor, defect configuration and lines of equal intensity in reciprocal space are indicated. In **Figure 12(b, c, d)** some other cases are shown where the defect has a preferred axis ("anisotropic" defect, in an isotropic matrix). Because the defect axis can be oriented along several crystallographically equivalent directions, the scattering cross-section [eqn (44) or (48)] has to be averaged over all these orientations. As the orientation of a zero-intensity plane (nodal surface) is determined by the axis of the defect, nodal planes will then be reduced to a nodal line or disappear entirely. Isointensity surfaces in anisotropic crystals look deformed and rotated but additional nodal planes can also appear (Dederichs, 1973).

As the Huang-scattering cross-section is proportional to the square of the dipole tensor [eqn (48)], interstitials will scatter considerably more than vacancies (e.g. for Cu, the ratio exceeds 20). A Huang-scattering study of a sample containing Frenkel pairs will therefore yield information predominantly related to interstitials. In a classical series of experiments on electron-irradiated Al and Cu single crystals, Ehrhart et al. (1974) have analyzed the Huang scattering from as-irradiated and step-annealed samples

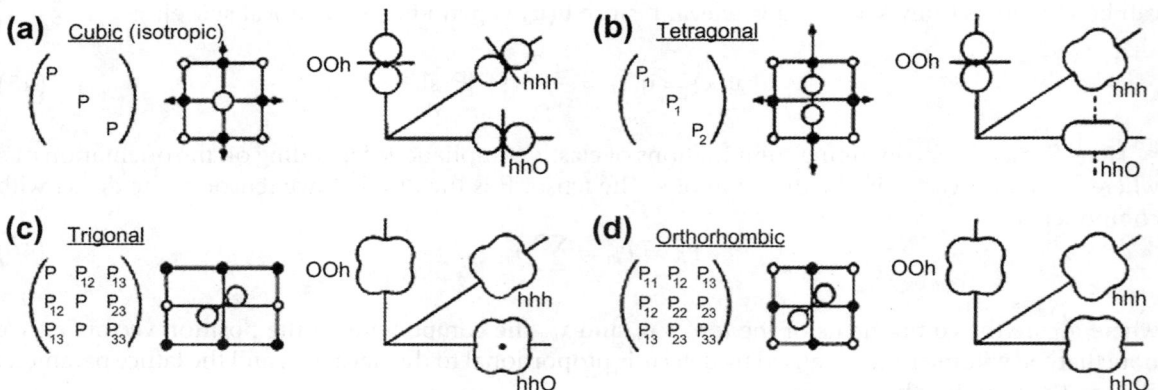

Figure 12 (a–d) Schematic contours of equal scattering intensity (right column) for various defect symmetries in a cubic lattice (central column), for anisotropic defects averaged over all crystallographically equivalent defect orientations. The column on the left indicates the form of the dipole force tensor. After Peisl (1976).

in terms of single defects and clusters. Clustering of defects leads to an increase of Huang scattering [proportional to the number of clustering defects, z, if the dipole strengths superpose linearly, see eqn (48)]. Interstitial clusters ($z \approx 3$) in Al irradiated with fast neutrons at low temperatures (4.6 K) have been identified by Von Guérard et al. (1980) who compared Huang scattering from these samples with single-defect scattering from electron-irradiated Al. The analysis yields further details on the size (radius about 50 Å) of displacement cascades, the number of defects per cascade (about 200), overlap at larger irradiation doses and annealing behavior.

The analysis of this type of defect scattering (near Bragg peaks) requires otherwise perfect crystals (low mosaicity) and sufficiently large samples for classical X-ray sources and even more so for neutrons. Low temperatures are advisable to reduce TDS (for X-rays; in neutron scattering, an experimental separation is possible). With all these prerequisites, good examples are scarce. More recent X-ray work by Stoller et al. (2007) revealed <100>-oriented dumbbell-type interstitial defect clusters in a neutron-irradiated Fe single crystal. With the narrow beams now available at synchrotron radiation sources, Huang scattering from individual grains in the micrometer size range (that may be quite perfect, especially after recrystallization) may be studied, as has been demonstrated by Specht et al. (2010) for a proton-irradiated iron foil. As an example of displacement scattering in an oxide, we mention the combined neutron and (synchrotron radiation) X-ray scattering study of single crystals of the relaxor ferroelectric $PbMg_{1/3}Ta_{1/3}O_3$ by Cervellino et al. (2011) where further references may be found to previous work. In this material, slowly fluctuating polar nanoregions freeze below about 100 K, and orientation-dependent static diffuse Huang-type scattering has been analyzed and modeled. The authors relate the scattering to displacements of the Pb ions due to the development of a double-well potential with decreasing temperature.

13.4 Between Bragg Peaks

We now extend the range of κ and look for intensity at some distance from the Bragg peaks of a given structure, and find diffuse intensity due to TDS and diffuse scattering from static displacements if defects are present. This latter scattering (occasionally called structural diffuse scattering) is particularly sensitive to the symmetry of displacements in the immediate vicinity of defects and can help to confirm (or eliminate) certain defect models. In addition to this "positional disorder", alloys will also exhibit "compositional disorder". Whereas the Bragg peaks are controlled simply by the average structure factor of the unit cell, diffuse scattering will reflect deviations from a random arrangement (short-range order or clustering). Basic interactions in alloys can thus be studied. Finally, new Bragg peaks may appear when (chemical or magnetic) order or (reconstructive or displacive) phase transformations occur.

The experimental requirements for diffuse scattering between Bragg peaks are somewhat different from those for scattering near Bragg peaks. There, very good κ resolution was desirable. Now, as small intensities are to be monitored over a wide κ range, resolution demands may often be relaxed in favor of higher intensity, and simultaneous observations at various values of κ are advantageous. Multi-detector arrangements or position-sensitive detectors are therefore used wherever possible.

13.4.1 Displacement Scattering

The diffuse scattering for κ not too close to a Bragg peak depends on the displacements in the immediate vicinity of the defect. In the dilute limit, $c \ll 1$, eqn (40) is still valid with $F(\kappa)$ according to eqn (41). However, eqn (43) may not be valid for all displacements, and we write

$$F(\kappa) = \Delta f_B + f_A \sum_n \exp(-2\pi i \kappa \cdot r_n^a)[\exp(-2\pi i \kappa \cdot u_{0n}) - 1 + 2\pi i \kappa \cdot u_{0n}] - 2\pi i f_A \kappa \cdot u(\kappa) \qquad (50)$$

Thus, if there are any $\mathbf{u}_{0n}$ too large to permit an expansion of the exponential, they are taken into account by the second term of eqn (50). If $|2\pi\boldsymbol{\kappa}\cdot\mathbf{u}_{0n}| \ll 1$, eqn (50) reduces to eqn (43). The $\mathbf{u}_{0n}$ may be obtained from lattice statics (see, e.g. Dederichs et al., 1978). A set of (virtual) forces (called Matsubara–Kanzaki forces) is assumed to be applied to the neighboring atoms of a defect in order to produce the displacements $\mathbf{u}_{0n}$ within the harmonic approximation, i.e. with the dynamical matrix of the ideal, defect-free crystal.

In reciprocal space, we have:

$$\mathbf{f}(\boldsymbol{\kappa}) = \boldsymbol{\phi}(\boldsymbol{\kappa})\cdot\mathbf{u}(\boldsymbol{\kappa}), \tag{51}$$

where $\mathbf{f}(\boldsymbol{\kappa})$ is the FT of the virtual forces $\mathbf{f}_{0n}$ and $\boldsymbol{\phi}(\boldsymbol{\kappa})$, the dynamical matrix (inverse of the FT of the lattice Green's function), contains combinations of squares of phonon frequencies. In symmetry directions, $\boldsymbol{\phi}(\boldsymbol{\kappa})$ can be expressed in terms of the three acoustic phonon branch frequencies and v_L^2 and v_{T1}^2, v_{T2}^2. Although the actual displacements may sometimes be more interesting, the concept of virtual forces has the advantage of giving a more rapidly converging method to calculate the scattering of (model) defects as the relevant forces may be restricted to the immediate surroundings of the defect [see Bauer (1979) for further discussions]. **Figure 13** illustrates the sensitivity of diffuse scattering between Bragg peaks to the details of the displacement field near the defect. Experiments with X-rays have first been reported for electron-irradiated Al by Haubold (1975) and subsequently for Cu by Haubold and Martinsen (1978). These experiments clearly confirmed the dumbbell configuration in <100> direction for the self-interstitial.

Early displacement scattering studies with neutrons have been reviewed by Bauer (1979). Although the incident beam intensities are much smaller than with X-rays, experiments are feasible, and the

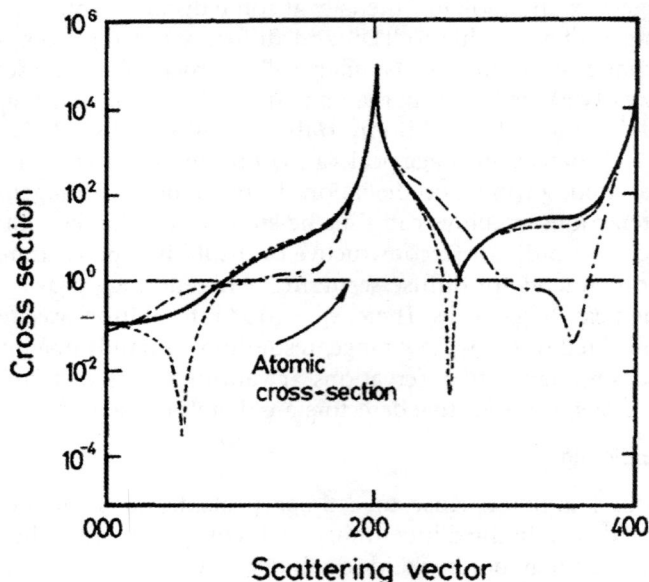

Figure 13 Calculated cross-sections for diffuse scattering by self-interstitials in Al. Solid line: ⟨100⟩ dumbbell, dashed-dotted line: tetrahedral, dotted line: octahedral configuration. After Haubold (1975).

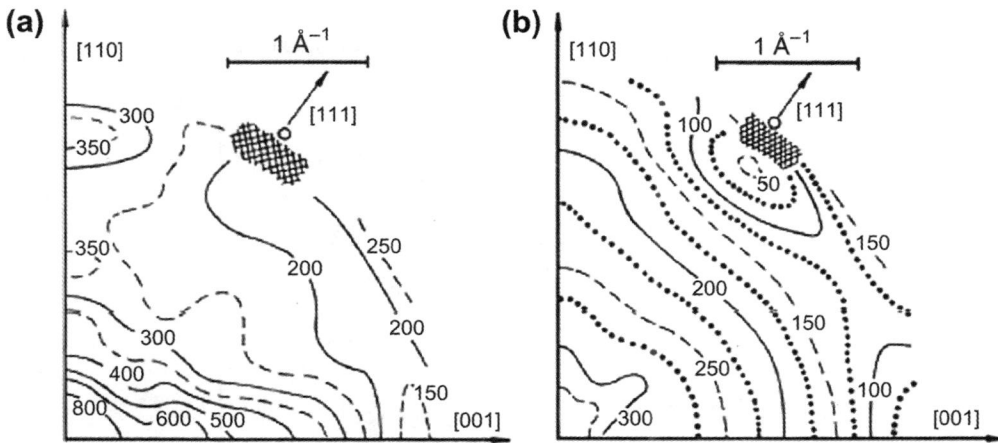

Figure 14   Lines of equal elastic diffuse scattering cross-section (in mbarn/sr per Cu atom) for Al-0.8 at.% Cu in the $(1\bar{1}0)$ plane of the reciprocal lattice for (a) $T = 300$ K and (b) $T = 800$ K. Data in the cross-hatched regions have been omitted. Scale: 1 Å$^{-1} = 2\pi\kappa$. After Bauer (1979).

energy analysis usually achieved with a chopper allows one to extend diffuse scattering measurements to higher temperatures. Several dilute alloys with low incoherent scattering cross-sections have been studied (Al, Nb, Pb are favorable solvents). **Figure 14** shows diffuse elastic scattering contours for an Al-0.8 at.% Cu single crystal. The scattering at 300 K shows signs of precipitation, the results at 800 K are indicative of a random distribution of defects. **Figure 15** shows the scattering cross-sections along the three main symmetry directions, compared with calculated values from a model with radial virtual forces acting on the nearest and next-nearest neighbors of the substitutional Cu solute. As the strength of the dipole force tensor is given by the lattice parameter change and c is known, no adjustable parameters remain in a nearest-neighbor model, and the forces f_0 must be -4.22×10^{-10} N, directed toward the Cu atom (lattice contraction). For this and any other model with centrosymmetric forces (inversion symmetry), the distortion scattering must be zero halfway between the Bragg peaks. For a centrosymmetric defect, one thus obtains just a diffuse scattering contribution from Δf_B (see eqn (50)) at these particular positions. **Figure 15** shows that for Cu in Al, inversion symmetry is fully compatible with the results. The nearest-neighbor forces f_0 alone yield quite a satisfactory fit. Detailed analysis of the slope near $\kappa = g/2$ led to the best fit indicated in the figure.

Diffuse neutron scattering results from interstitially dissolved deuterium (replacing hydrogen to reduce the incoherent scattering, which is much higher for protons) in niobium studied by Dosch and Peisl (1986) and Dosch et al. (1987, 1992) showed that a static description of the solute was not sufficient. A finite probability for D in a mobile state, appearing delocalized over three tetrahedral sites, had to be admitted. The temperature-dependent jump rate of the diffusing defect (D) was found to be much larger than previously assumed, and this was confirmed by quasielastic neutron scattering (Dosch et al., 1992).

13.4.2   Short-range Order

So far, we have considered dilute solutions of defects and assumed a random distribution. This approach yields simple scattering laws and is sufficient for many cases. However, in concentrated solid solutions, the notion of single defects is difficult to maintain, and additional scattering effects will occur

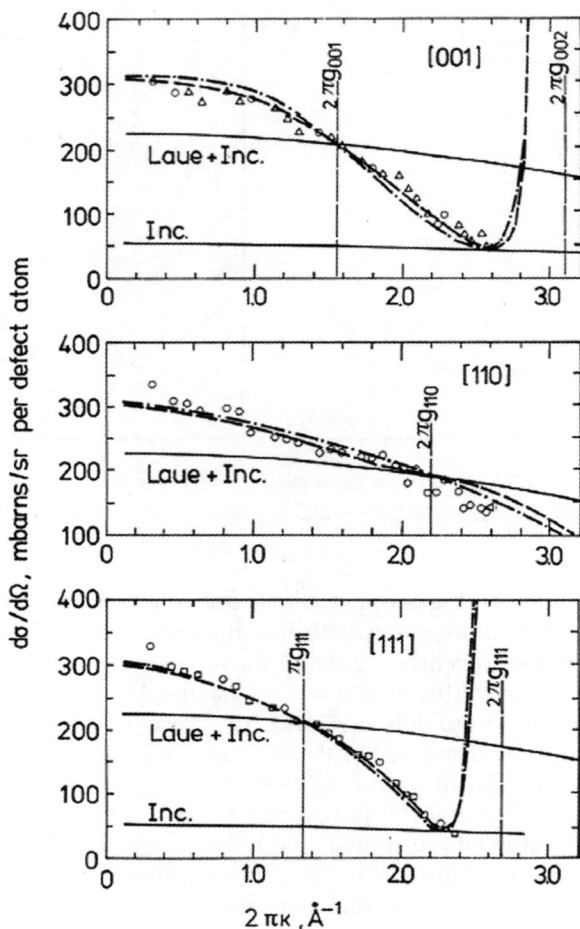

Figure 15 Comparison of measured cross-sections for elastic diffuse scattering from Al-0.8 at.% Cu at 800 K with the results of model calculations. The solid lines marked "Laue + Inc." indicate the expected behavior without static displacements (but including the Debye–Waller factor). Dashed lines: model with forces on nearest neighbors only; dashed dotted lines: "best fit" with forces f_1 on nearest and f_2 on next-nearest neighbors, $f_2/f_1 = -0.2^{+0.2}_{-0.1}$. The symbols represent experimental data. After Bauer (1979).

if there is any deviation from a random occupation of lattice sites by the different scattering centers, i.e. short-range order (SRO) occurs.

Much work has been devoted to binary substitutional alloys. If we neglect positional disorder, eqn (40) for diffuse scattering (i.e. $\kappa \neq g$) can be replaced by (see Krivoglaz, 1969)

$$\left(\frac{d\sigma}{d\Omega}\right)_{SRO} = N|c(\kappa)|^2 (f_B - f_A)^2 \qquad (52)$$

for an A–B alloy of concentration $c = c_B$. (We consider alloys with one sublattice only. More general expressions for several sublattices, important for SRO in intermetallic compounds, hydrogen in metals, nonstoichiometric compounds etc., are given, e.g. by Hayakawa and Cohen (1975) and Bauer (1979). For some further discussions relating to ternary alloys, see Cohen (1970) and De Fontaine (1971, 1979). In eqn (52), Debye–Waller factors are again included in the scattering lengths f_A, f_B. The term $|c(\kappa)|^2$ contains the FT of the compositional fluctuations (of component B),

$$c(\kappa) = \frac{1}{\sqrt{N}} \sum_n (c_{Bn} - c_B)\exp(-2\pi \, i\kappa \cdot \mathbf{r}_n^a), \tag{53}$$

and c_{Bn} is defined, as in eqn (16), to be one if a B atom is present at site n and zero otherwise. In the limit of small concentrations, $N|c(\kappa)|^2 \to Nc$ as in eqn (40). For a random distribution of B atoms of arbitrary concentration, $N|c(\kappa)|^2 \to Nc(1-c)$, and eqn (52) then yields the well-known monotonic Laue scattering $\sim c(1-c)(f_B - f_A)^2$. Deviations from a random distribution will manifest themselves in a modulation of this term. If displacements were negligible, $|c(\kappa)|^2$ could be obtained directly from diffuse scattering measurements, but in most cases, displacements have to be taken into account.

To illustrate the procedures involved, **Figure 16** shows a classical experimental result for a polycrystalline Cu-47.5 at.% Ni sample studied with neutrons. A unique possibility offered by neutrons has been exploited here. Some isotopes have a negative scattering length, and for one (or more, if one changes the isotopic composition of the alloying partners) chemical composition of a binary alloy, it is thus possible to obtain a "null-matrix alloy" with an average scattering length $\bar{f} = 0$. This eliminates Bragg peaks and Huang-type scattering (see eqn (44) with $\bar{f}$ instead of f_A) and enhances the "contrast" $(f_B - f_A)^2$ for the SRO term [eqn (52)]. The alloy used by Mozer et al. (1968) contained natural Cu and

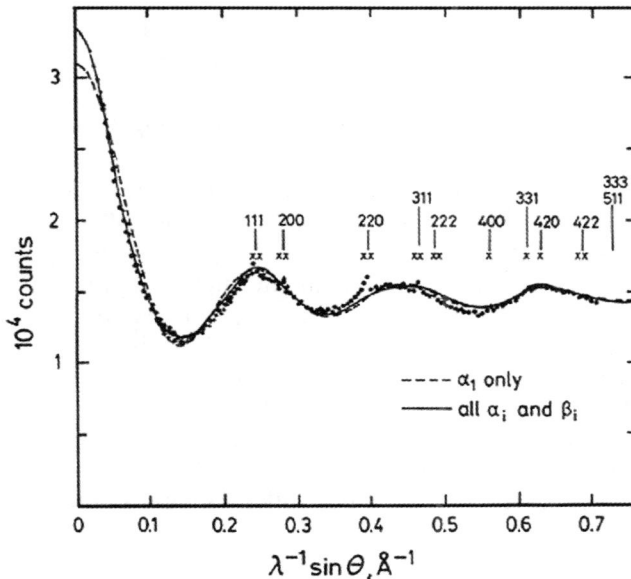

Figure 16 Diffuse neutron scattering from polycrystalline Cu-47.5 at.% ^{62}Ni, furnace-cooled from 1021°C. The data indicated by crosses were taken with a neutron wavelength $\lambda = 1.951$ Å, those indicated by solid circles with $\lambda = 1.024$ Å. The Bragg positions are marked. The dashed curve is a fit with one SRO parameter only, the solid curve is the "best fit" described in the text. After Mozer et al. (1968).

Ni enriched to 99% in ^{62}Ni. In **Figure 16**, only very small Bragg intensities are visible, but a very pronounced modulation of the diffuse scattering indicates a nonrandom distribution of the alloying elements on the f.c.c. lattice sites (this alloy shows "negative" SRO, i.e. clustering).

It is common to describe SRO in terms of the Warren–Cowley parameters α_{0n} (Cowley, 1950; see, e.g. Schwartz and Cohen, 1987; Welberry and Butler, 1994 for a survey of other descriptions). They are related to $|c(\kappa)|^2$ by:

$$|c(\kappa)|^2 = c(1 - c)\alpha(\kappa), \tag{54}$$

where

$$\alpha(\kappa) = \sum_n \alpha_{0n}\exp(-2\pi i\kappa \cdot \mathbf{r}_{0n}). \tag{55}$$

The short-range-order coefficients α_{0n} are defined using "conditional probabilities" P_{0n}^{AB} (indicating the probability that a B atom is at position n if an A atom is at the origin), P_{0n}^{BB} etc.

$$\alpha_{0n} = \frac{P_{0n}^{BB} - c}{1 - c} = \frac{c - P_{0n}^{AB}}{c}. \tag{56}$$

Obviously, $\alpha_{00} = 1$, independent of the state of SRO, and for a random distribution $\alpha_{0n} = 0$ for all $n \neq 0$. In cubic structures, the α_{0n} are usually labeled per coordination shell i or with indices lmn referring to all combinations of coordinates for positions within one shell.

For a polycrystalline sample, the diffuse SRO scattering [eqns (52–55)], averaged over all orientations (e.g. Warren, 1969, 1999) is

$$\left(\frac{d\sigma}{d\Omega}\right)_{SRO} = Nc(1 - c)(f_B - f_A)^2 \sum_i Z_i\alpha_i\frac{\sin(2\pi\kappa r_i)}{2\pi\kappa r_i}, \tag{57}$$

where Z_i is the number of neighbors in the ith shell with radius r_i (the sum includes $i = 0$). The dashed curve in **Figure 16** is a fit according to eqn (57) with only one adjusted SRO parameter ($\alpha_1 = 0.131$). This fit already looks quite reasonable and indicates that short-range clustering is present in this alloy. A positive SRO parameter indicates more BB pairs than expected from the average concentration. This leads to a reduction of diffuse scattering (the Laue scattering intensity, see after eqn (53)) between Bragg positions and an enhancement near Bragg positions (including 000, i.e. $\kappa = 0$). The deviations from the simple α_1 fit visible in **Figure 16** cannot be explained by contributions of more α_i terms alone (except for κ near zero). As discussed in Sections 13.3 and 13.4, displacements from ideal lattice sites introduce diffuse scattering (Huang scattering symmetric around Bragg peaks, but also "size effect" scattering between Bragg peaks [the leading intensity term resulting from $F(\kappa)$ according to eqn (50) is antisymmetric around reciprocal lattice points]. Both contributions usually require attention for concentrated alloys, and the treatment, even for the approximation of small displacements, becomes rather involved [see, e.g. Warren, 1969, 1999; Schwartz and Cohen, 1987).

In the approximation $\exp[-2\pi i\kappa \cdot (\mathbf{u}_n - \mathbf{u}_{n'})] \approx 1 - 2\pi i\kappa \cdot (\mathbf{u}_n - \mathbf{u}_{n'})$, and neglecting the Huang term, the diffuse scattering can be written as (see Warren, 1969, 1999)

$$\left(\frac{d\sigma}{d\Omega}\right)_d = Nc(1 - c)(f_B - f_A)^2 \sum_{ij}(\alpha_i - 2\pi i\beta_i\kappa \cdot \mathbf{r}_{ij})\exp(-2\pi i\kappa \cdot \mathbf{r}_{ij}), \tag{58}$$

where i indicates the shell and j the position in the ith shell, and the "size effect coefficients" are given by

$$\beta_i = \frac{f_A}{f_A - f_B}\left(\frac{1-c}{c} + \alpha_i\right)\varepsilon_{AA}^i - \frac{f_B}{f_A - f_B}\left(\frac{c}{1-c} + \alpha_i\right)\varepsilon_{BB}^i, \tag{59}$$

where the $\varepsilon_{\mu\nu}^i$ are defined as the average relative deviations of the distance between a μ atom at the origin and a ν atom in the ith shell,

$$r_{\mu\nu,i} = r_i^a\left(1 + \varepsilon_{\mu\nu}^i\right). \tag{60}$$

Equation (58), averaged for polycrystals, was used by Mozer et al. (1968) to fit the data in **Figure 16**. The "best fit", shown by the continuous curve, was obtained with nine adjustable values α_i and size effect coefficients β_i for five shells. The parameter $\alpha_2 = -0.008$ indicates a slight preference for unlike second-nearest neighbors (α_1 settles at 0.121). Several other studies followed, including also some ternary alloys [see Vrijen et al. (1980) for references and the kinetics of clustering in Cu–Ni].

Early work on single crystals (which are required for quantitative SRO studies if Bragg peaks are present) was also evaluated based on linear size-effect corrections. The different symmetries of the α_i and the β_i terms in eqn (58) make such a procedure attractive [see Sparks and Borie (1966), Moss and Clapp (1968) and Schmatz (1973) for surveys of early experimental results]. As seen above, for neutron scattering from a null-matrix alloy, this treatment is sufficient. A null-matrix single crystal of ^{62}Ni-48 at.% Pt has been studied by Rodriguez et al. (2006), and **Figure 17** shows experimental maps of the diffuse scattering intensity in the scattering planes perpendicular to [001] and [1$\bar{1}$0]. For comparison,

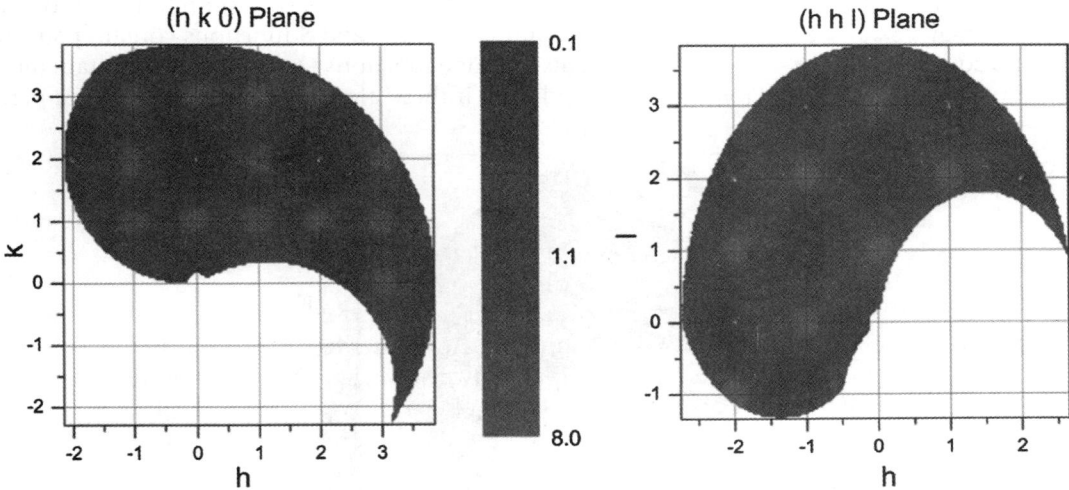

Figure 17 Experimental maps of the diffuse scattering intensity from a ^{62}Ni-48 at.% Pt single crystal (logarithmic intensity scale). The tiny dots are the very weak Bragg reflections due to a slight deviation from the null-matrix–alloy composition. Displacement scattering is weak in this alloy, but the plot on the right shows the shifts of the diffuse SRO peaks due to the antisymmetric displacement scattering term (thermal diffuse scattering was eliminated by time-of-flight analysis of the scattered neutrons). From Rodriguez et al. (2006). (For color version of this figure, the reader is referred to the online version of this book.)

Figure 18 shows the data for the plane perpendicular to $[1\bar{1}0]$, obtained from a crystal with natural Ni and Pt. The diffuse scattering between the Bragg peaks is hardly visible at this scale which is controlled by the intense Bragg peaks. Huang scattering near the peaks is now also visible (TDS removed). Rodriguez et al. (2006) analyze their results in terms of effective pair interactions (see below) and local displacements, fitting 35 SRO parameters and 15 (linear, radial) displacement parameters.

The extension of the evaluation scheme to include size-effect contributions up to second order, i.e. starting from

$$\exp[-2\pi i\kappa\cdot(\mathbf{u}_n - \mathbf{u}_{n'})] \approx 1 - 2\pi i\kappa\cdot(\mathbf{u}_n - \mathbf{u}_{n'}) - 2\pi^2[\kappa\cdot(\mathbf{u}_n - \mathbf{u}_{n'})]^2 \tag{61}$$

is due to Borie and Sparks (1971) and Gragg and Cohen (1971).

The "Borie–Sparks approach" is fully described in textbooks (e.g. Schwartz and Cohen, 1987; Cowley, 1981). For cubic crystals, the total diffuse intensity, divided by $Nc(1 - c)(f_B - f_A)^2$ (the result is then given in "Laue units"), can be written in this approximation as [see Schwartz and Cohen (1987); a corresponding treatment for hexagonal crystals can be found in Yu et al. (2004)]

$$I_d = I_\alpha + h_1 Q_x + h_2 Q_y + h_3 Q_z + h_1^2 R_x + h_2^2 R_y + h_3^2 R_z + h_1 h_2 S_{xy} + h_2 h_3 S_{yz} + h_3 h_1 S_{zx}, \tag{62}$$

where I_α is the SRO term,

$$I_\alpha = \sum_{l,m,n} \alpha_{lmn} \exp[-2\pi i(h_1 l + h_2 m + h_3 n)] \tag{63}$$

(with l, m, n from $-\infty$ to $+\infty$), h_1, h_2, h_3 are the components of the scattering vector in units of the reciprocal lattice vectors $\mathbf{b}_i$ [see Section 13.2.1] and the l, m, n coordinates are measured in units of the lattice vectors $\mathbf{a}_i$. The functions Q are the size-effect terms containing corresponding components of the displacements β_{lmn} [see eqns (59) and (60)]. Similarly, the R and S functions contain pair probabilities and squares and products of displacements. All nine functions Q, R and S also contain ratios of atomic form factors, $f_A/(f_B - f_A)$, $f_B/(f_B - f_A)$, and only if these are constant, Q, R and S are periodic

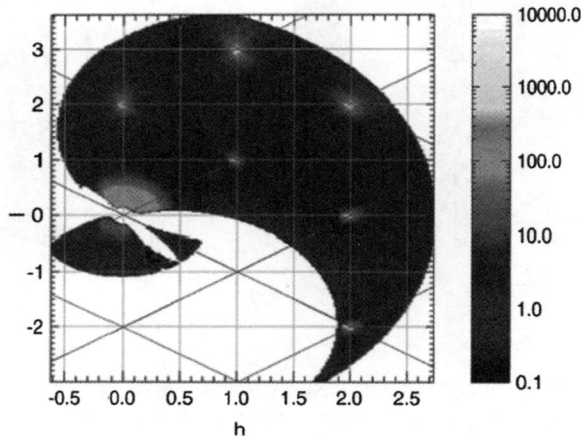

Figure 18 Elastic neutron scattering in the plane perpendicular to $[1\bar{1}0]$ of a natural-abundance Ni-48 at.% Pt single crystal (in arbitrary units). Only the Bragg scattering and Huang scattering are clearly visible. From Rodriguez et al. (2006). (For color version of this figure, the reader is referred to the online version of this book.)

functions in reciprocal space. **Figure 19** shows schematically the three contributions along $[h_1 00]$. Using the symmetry properties of Q, R and S, it is now possible to construct a minimum volume in reciprocal space where the diffuse intensity must be measured in order to form appropriate sums and differences for certain sets of reciprocal lattice points that will yield the various terms of eqn (62) separately (see Schwartz and Cohen, 1987). **Figure 20** shows the minimum volume in reciprocal space for f.c.c. crystals. It corresponds to one eighth of a unit cell in reciprocal space and has been placed in a region accessible in reflection geometry (lower values of κ may be accessible and useful, since the displacement terms become smaller, for transmission geometry, i.e. for neutrons and high-energy X-rays). The method was first applied to diffuse X-ray scattering from a clustering system (Al-5 at.% Ag) by Gragg and Cohen (1971).

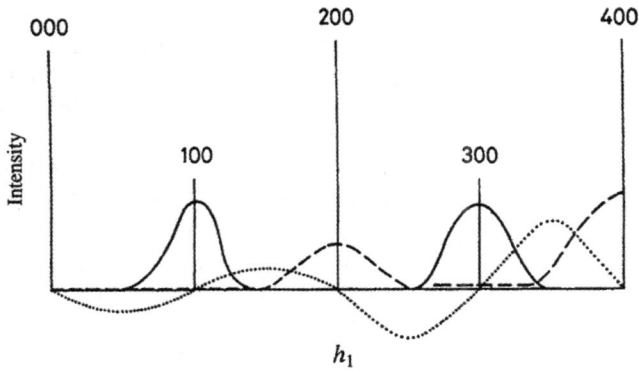

Figure 19 Schematic of the diffuse intensity components according to eqn (74) after Schwartz and Cohen (1987). The intensity scale is not the same for the three components. Continuous line:, I_α, dashed line: $h_1 Q_x$, dotted line: $h_1^2 R_x$.

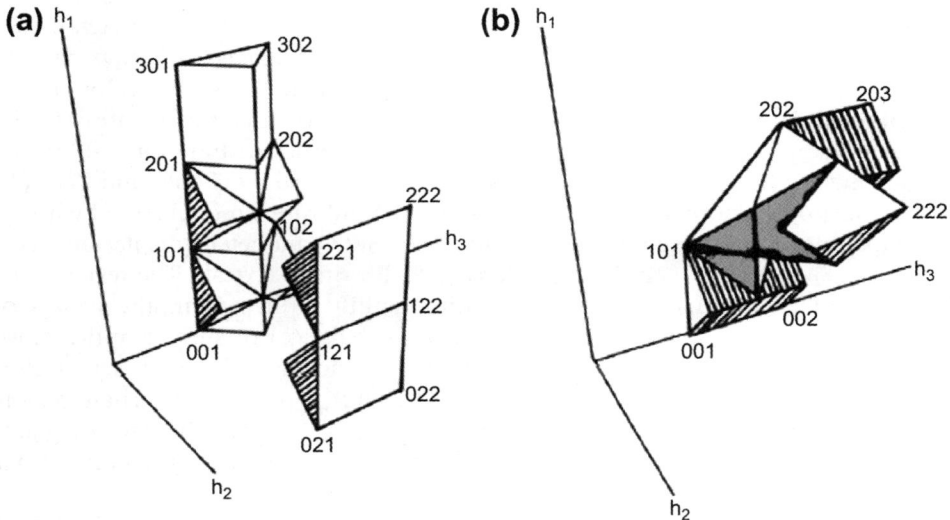

Figure 20 (a) Typical volumes in reciprocal space required to separate the diffuse intensity from an f.c.c. crystal into components due to local order and due to atomic displacements (to second order in the displacements). (b) As (a), after symmetry of intensity in reciprocal space across the planes $h_i = h_j$ has been employed. After Hayakawa and Cohen (1975).

Apart from the fact that the linear combination of the different $\varepsilon^i_{\mu\nu}$ cannot be separated, the assumption of constant ratios of the atomic form factors is the main limitation of the Borie–Sparks approach if X-rays are used (see Tiballs, 1975; Georgopoulos and Cohen, 1977, 1979), whereas in this respect, it is ideally suited for neutron scattering. To exploit the substantial variation of the atomic form factors across the experimental κ range for the X-ray case, Tiballs (1975) proposed to write the functions Q, R and S of the Borie–Sparks approach as products of the κ-dependent ratios of form factors and new functions Q', R', and S' that are strictly periodic in reciprocal space. This requires 25 Fourier series instead of 10 in the Borie–Sparks method. Georgopoulos and Cohen (1977, 1979) developed a separation method based on this approach. The procedure is considerably more laborious than the Borie–Sparks method. A detailed error analysis is required in each particular case. The extra effort may be worthwhile especially for clustering systems. Apart from more reliable values for the SRO parameters α_i, the method also yields individual (AA, BB, AB) values for the displacement terms. Of course, using least-squares fitting procedures to adapt a real-space object to the scattering data is also possible and has been used in various forms since the early work by Williams (1972, 1976). The analysis in reciprocal space suggested by Krivoglaz (1969, 1996) may also be instructive.

With the availability of continuously tunable X-ray wavelengths at synchrotron radiation sources, the anomalous scattering variation of the atomic form factors offers additional advantages since with a proper choice of f_A, f_B, the thermal diffuse background may be removed, and a reliable separation of SRO and individual displacement terms is achieved by working with three wavelength (three-lambda method) (Ramesh and Rameseshan, 1971; see, e.g. Ice et al., 1992; Reinhard et al., 1992 for b.c.c. Fe–Cr, Schönfeld, 1994, 2008; Jiang et al., 1996; Schönfeld, 1999).

From a given set of SRO parameters, the local atomic arrangement in SRO structures can be simulated on a computer by rearranging many thousands of atoms on a given lattice until sufficient agreement is obtained with the SRO parameters (Gehlen and Cohen, 1965; Gragg et al., 1971). The simulated structure can then be analyzed in terms of specific atomic arrangements. It is not yet possible to completely simulate local atomic arrangements and displacements, though some progress has been reported (Kyobu et al., 1994; Schönfeld et al., 2001). If the set of SRO parameters belongs to an equilibrium state, they may be used to calculate basic alloy parameters (i.e. effective pair interactions V_{lmn}). Apart from approximate analytical methods (see Reinhard and Moss, 1994; for an assessment), exact numerical calculations are now feasible, e.g. based on the inverse cluster variation method due to Gratias and Cénédèse (1985) (see, e.g. Mehaddene et al., 2004; for a study on Fe–Pd) or the more commonly used inverse (reverse) Monte Carlo modeling introduced by Gerold and Kern (1987). The effective pair interactions are obtained from a real-space (equilibrium) model crystal whose scattering function is compatible with the experimentally determined SRO parameters [displacements are treated separately, see, e.g. Steiner et al. (2005), for a Pt-47 at.% Rh single crystal]. The reverse Monte Carlo method makes use of the fact that for any equilibrium quantity of the system, the average over many fluctuations (weighted according to conditions of detailed balance) of such quantity must be zero. From many (virtual) pair exchanges in the model crystal, one thus obtains a set of effective pair interactions V_{lmn} (defined by the difference between $V^{AA}_{lmn} + V^{BB}_{lmn}$ and $2V^{AB}_{lmn}$). These may in turn be used in Monte Carlo simulations for other situations (see **Figure 21** for a schematic diagram), and, of course, be tested against first-principle calculations of alloys properties [see, e.g. Schönfeld et al. (2012) for a comparative study of equiatomic Au–Pd].

Early experimental work on SRO (and long-range order) had concentrated on Cu–Au, probably the most extensively studied binary system (e.g. Cowley, 1950; Moss and Clapp, 1968; Bardhan and Cohen, 1976; Bessière et al., 1986; Ohshima et al., 1986; Butler and Cohen, 1989; Schönfeld et al., 1999). SRO in many more f.c.c. binary alloys has been analyzed during the last five decades, and

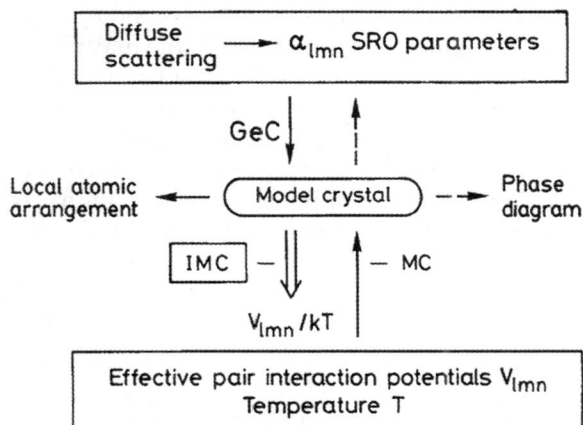

Figure 21 Schematic of the steps involved in the inverse Monte Carlo method (IMC) and the Monte Carlo method (MC). GeC = simulation of a model crystal according to Gehlen and Cohen (1965). After Gerold and Kern (1987).

extensive reviews are available (e.g. Schönfeld, 1994, 1999; Schweika, 1998; Schönfeld and Kostorz, 2009 especially for Ni–Au, Ni–Re and Cu–Mn). We concentrate here on the diffuse X-ray scattering study by Steiner et al. (2005) of a Pt–Rh single crystal using synchrotron radiation ($\lambda = 0.534$ Å) and a laboratory rotating anode source emitting Mo–K_α ($\lambda = 0.71$ Å). The contrast $|f_{Pt} - f_{Rh}|^2$ is enhanced by 39% if synchrotron radiation at 0.534 Å (18 eV below the Rh–K-absorption edge) is used. The melting temperatures of both components exceed 2000 K, and the known phase diagram shows continuous solid solubility (f.c.c. structure). However, Raub (1959) had argued that a solid-solubility gap should appear around Pt-50 at.% Rh below about 1030 K. This suggestion was based on a comparison of several homologous systems (Ir–Pt, Ir–Pd, Pd–Rh), but no experimental evidence was available. In contrast, ab initio calculations by Lu et al. (1995) and a bond-order simulation mixing model by Pohl and Albe (2009) indicated a tendency toward local order. This prompted an experimental investigation. A single crystal of Pt-47.0(5) at.% Rh was grown and heat treated to obtain a state corresponding to thermal equilibrium at 923 K (verified by electrical conductivity measurements). Diffuse scattering (at room temperature) was measured at the ESRF (Grenoble) ID1 beamline, h varying from 0.8 to 5.0, at 4600 positions, with 4000–10 000 counts in 20 s per position. Further data at 15 800 positions with h between 1.4 and 7.5 ($\Delta h = 0.1$) were taken at the laboratory source, with 10 000–30 000 counts in 220 s. These additional measurements including larger h values (in a more favorable signal-to-noise situation) turned out to be decisive. For a reliable evaluation, a calibration to obtain absolute scattering cross-sections is required and was performed using a polystyrene sample. Static and dynamical Debye–Waller factors were calculated (see, e.g. Warren, 1969, 1999; Krivoglaz, 1996) from the experimentally determined elastic constants, and thermal diffuse scattering (to third order) and Compton scattering were calculated and subtracted from the experimental data. Small-angle neutron scattering was also performed (at the ILL Grenoble and the Paul Scherrer Institute, Villigen, Switzerland) to check for decomposition tendencies. The scattering intensity was constant and below one Laue unit, i.e. no clustering was found. All the data were used to complete an analysis following the procedures described above (both the Borie–Sparks and the Georgopoulos–Cohen approaches were used, with subtle differences in the results). The isolated SRO scattering term I_{SRO} shows slight modulations with maxima around $1\frac{1}{2}$ 0, but also along <100>. The calibration and separation treatment yields a value for α_{000} close to 1, as required. The first SRO parameter is negative (-0.046 to -0.063), indicating very weak local order (only about 6% of the maximum possible value of $|\alpha_{110}|$ for this composition, see eqn (56)]. With the complete sets of SRO and displacement parameters [and the other scattering contributions) a reconstruction of the measured intensities becomes possible, e.g. for a (100) plane as shown in **Figure 22**. The SRO parameters α_{hkl} were used to generate corresponding model crystals ($32 \times 32 \times 32$ f.c.c. unit cells), and the atomic arrangements were then compared with those in a completely random model crystal of the same composition. **Figure 23** shows the coordination polyhedron of the nearest neighbors around a Rh atom. The numbers in the figure serve to

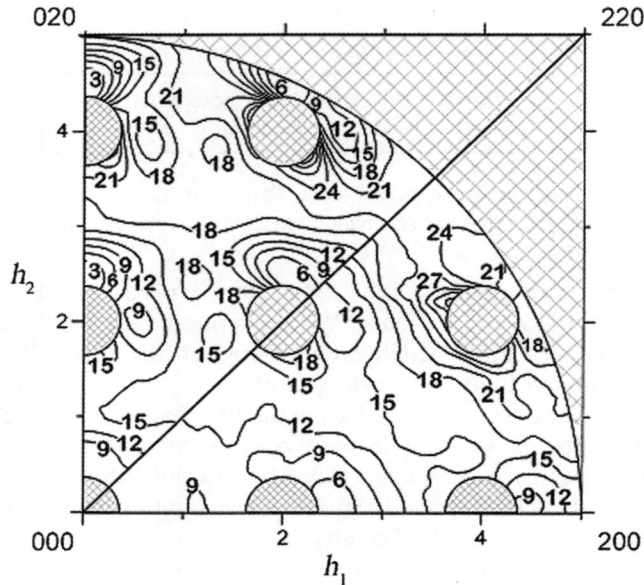

Figure 22 Elastic and inelastic diffuse X-ray scattering in a (001) plane of a Pt-47 at.% Rh single crystal. Data were obtained with X-rays of an energy of 23.201 keV at ESRF ($h_1 > h_2$). The hatched areas around the Bragg positions indicate the regions of largest thermal diffuse scattering. Lines of equal intensity are given in 0.1 Laue units. For $h_1 < h_2$, the recalculated diffuse scattering is shown using SRO and displacement parameters determined from the experiment. From Steiner et al. (2005).

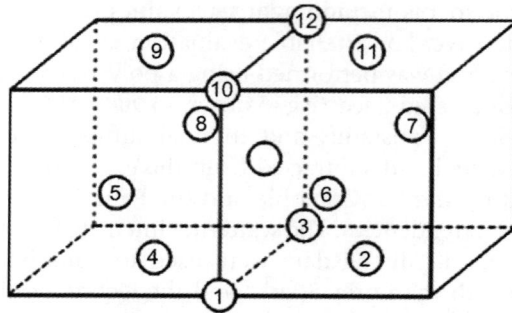

Figure 23 Coordination polyhedron of the nearest neighbors in the f.c.c. structure with site numbering.

distinguish the 144 different configurations enumerated by Clapp (1971), for example $C8$: Pt occupies sites 5, 6, 12 (and Rh the other sites), $C16$: Pt at 5, 6, 7, 8; $C128$: Pt at 1, 2, 3, 5, 8, 10, 11, 12. An enhanced abundance of $C8$ is typical for order of the type (i.e. compatible with a long-range ordered structure) $PtRh_4(D1_a)$, while $C16$ resembles $PtRh_3(D0_{22})$ and $PtRh_5(X_2)$, and $C128$ indicates the structure "40" (PtRh) (Lu et al., 1995). For $C128$, an abundance of 10% was found in the short-range ordered model crystals, about 220% of the value of an ideal solid solution.

The effective interaction parameters determined by use of the reverse Monte Carlo method [about 820 000 virtual exchanges, Steiner et al. (2005)] are small. For example, V_{110} amounts to 7.6–11.9 meV. Nevertheless, the calculated ground-state ($T = 0$) energies of various ordered structures also show a preference of structure "40". Monte Carlo simulations based on this structure yield a critical temperature for the order–disorder transition at 170–200 K. Thus, these experiments indicate that decomposition is very unlikely in this system, and although diffuse scattering indicates a tendency toward order, the critical temperature is too low to observe the establishment of long-range order at normal kinetic conditions. Similarly low order–disorder temperatures were derived from diffuse scattering studies of several other binary alloys, e.g. $T_c \cong 330$ K for Cu_3Zn ($L1_2$ structure, Reinhard et al., 1990), $T_c \cong 115$–165 K for Ag_3Au ($L1_2$, Schönfeld et al., 1992), $T_c \cong 135$ K for Ag_5Al (structure A_5B, Yu et al., 1997), $T_c \cong 230$–280 K for AuPd (long-period superstructure, Schönfeld et al., 2012).

Diffuse scattering of X-rays from synchrotron radiation sources has without doubt led to considerable improvements in the understanding of basic alloy properties in recent years. The variation of scattering contrast from the same sample by simply changing the wavelength remains unique and may also be useful for ternary and multicomponent alloys (see also Section 13.5). Also, quantitative diffuse near-surface scattering experiments (under grazing incidence) from alloys have now become possible (see Kimura et al., 1997; for Cu-25 at.% Au, Steiner et al., 2006; for Pt-48 at.% Rh, and Engelke et al., 2010; for Ni-23 at.% Pt).

With neutrons, TDS can be separated experimentally, and thus, elastic diffuse scattering can be obtained in situ at elevated temperatures (i.e. at equilibrium). As an example, **Figure 24**(a) shows the elastic diffuse scattering intensity in the (100) plane, as measured by use of a triple-axis neutron spectrometer for two alloy compositions [one at elevated temperature, the other at room temperature, Bucher et al. (2002)]. The incoherent scattering was reduced and the contrast for SRO scattering further enhanced by the use of the [58]Ni isotope (natural Ti has a negative scattering length for thermal neutron which is already favorable per se). These very reliable results lead to the I_{SRO} reconstructions shown in **Figure 24**(b). Very strong diffuse peaks of 100-type indicate $D0_{22}$ and $L1_2$ as preferred ordered structures. In the modeled crystals, a strong enhancement of the $C16$ configuration is found (and also $C17$, positions 4,6,7,9 occupied by Ti around a Ni atom). The effective pair interactions from reverse Monte Carlo calculations are also correspondingly large; for example ($\pm 10\%$) $V_{110} = 45$ meV, $V_{200} = -24$ meV.

Apart from their fundamental importance for comparison with first-principle calculations, SRO parameters and effective pair interactions may also control the mechanical properties of locally ordered alloys. It is well known that single-phase f.c.c. alloys, starting at a certain solute concentration (a few percent), show localized glide rather than distributed, disperse glide typical of pure f.c.c. metals. The so-called planar glide is related to the release of a large number of dislocations either on a single plane or on closely spaced neighboring planes, accompanied by a transitory instability. Although a decrease of the stacking-fault energy may also be responsible for an enhanced planarity of glide (but Cu–Mn despite coarse glide shows no change of stacking-fault energy), a more convincing explanation may be based on the extra energy required for a first dislocation to move across a plane of a short-range ordered crystal (Gerold and Karnthaler, 1989; Kostorz, 1995). Following a proposal of Büchner and

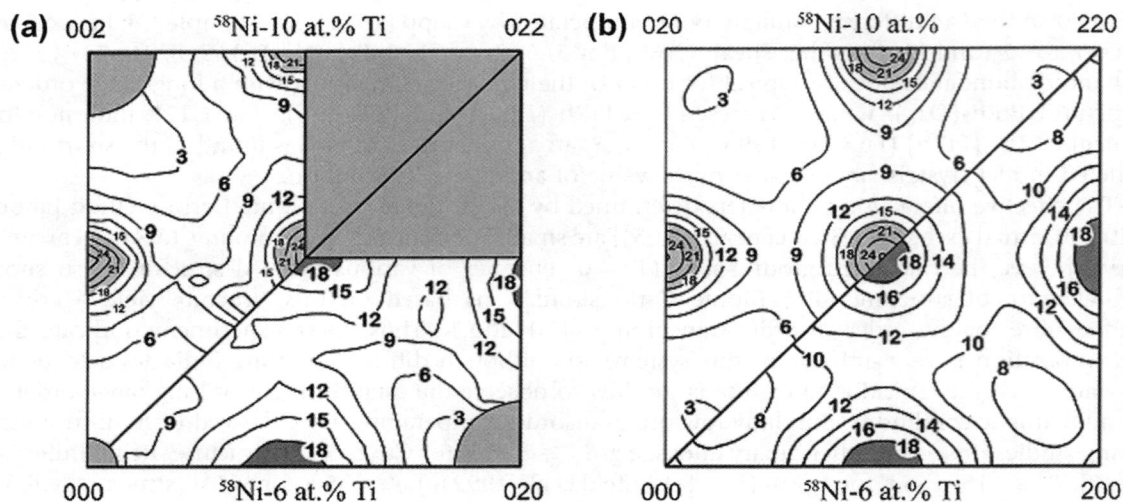

(a) 002 ^{58}Ni-10 at.% Ti   022   **(b)** 020 ^{58}Ni-10 at.%   220

000   ^{58}Ni-6 at.% Ti   020   000 ^{58}Ni-6 at.% Ti   200

Figure 24 (a) Diffuse elastic neutron scattering in the (100) plane of single crystals of Ni-10 at.% Ti (measured at 1100 K) and Ni-6 at.% Ti (measured at room temperature after quenching from 780 K). (b) Short-range-order scattering intensity recalculated with the SRO parameters determined from the experimental data. After Bucher (1999) and Bucher et al. (2002).

Pitsch (1985) one calculates (Schwander et al., 1992) the energy stored upon each elementary shear by the motion of a dislocation ({111} planes for a Burgers vector of type <110>/2 for the f.c.c. structure), keeping track of the changes in neighborhood and the associated effective pair interactions. This energy is largest for the first dislocation and converges toward a final value γ_d (the diffuse antiphase boundary energy) once the range of significant SRO parameters is exceeded by the passage of further dislocations. A few values of γ_d (in energy per unit area) are plotted in **Figure 25** for some Cu-rich alloys [Cu–Al: diffuse X-ray scattering, Schönfeld et al. (1996), Cu–Mn, Cu–Zn: diffuse neutron scattering, Roelofs et al. (1995) and Reinhard et al. (1990)]. One sees that the transition from wavy to planar slip occurs between 2 and 3 mJ m^{-2}. This interpretation relies on a static concept (i.e. no diffusion occurs).

There is not much comparable quantitative diffuse scattering work on any other structures but f.c.c., apart from the work of Reinhard et al. (1992) on b.c.c. Fe-47 at.% Cr (X-ray three-lambda method) and by Yu et al. (2004) on h.c.p. Ag-36.7 at.% Al. Mirebeau and Parette (2010) have recently reinvestigated diffuse neutron scattering from Fe-(1–15) at.% Cr with improved precision, and the SRO parameters are found to change sign [ordering tendency changing to clustering at $c_{Cr} = 0.110(5)$], in agreement with first-principles calculations taking into account magnetic contributions. Diffuse neutron scattering from a Cu-40 at.% Zn single crystal quenched from the b.c.c. β-phase was measured and evaluated by Zolliker et al. (1992). The Borie–Sparks method (displacements to second order included, Bragg peaks of the rapidly developed ordered β'-phase excluded) showed that the excess Cu (with respect to the stoichiometric CuZn composition) in this (nonequilibrium) situation is nearly randomly distributed on the Zn sublattice.

In ternary substitutional alloys, three independent pair correlations must be considered for SRO, and even without displacements, they cannot be separated from a single scattering experiment. Only very few attempts have been reported following the procedures outlined above. The local atomic arrangement in a stainless-steel Fe-23 at.% Ni-21 at.% Cr was investigated by Cénédèse et al. (1984) who used

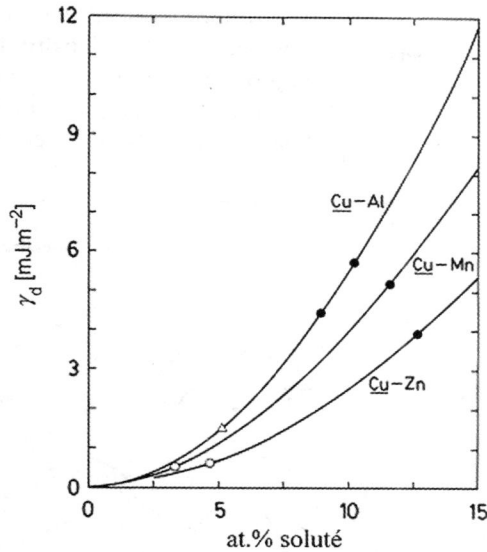

Figure 25 Diffuse antiphase boundary energy γ_d as a function of concentration of several solutes in Cu. Continuous lines are calculated from results of diffuse scattering. The open circles indicate distributed glide, the full dots stand for planar glide, and the triangle indicates a mixed situation. The transition from distributed to planar glide—which could also depend on other parameters—takes place between γ_d values from 1 to 3 mJ m^{-2}.

thermal neutron diffuse scattering from three single crystals of different isotopic compositions (at constant chemical composition), thus providing the required contrast variation. Local order between Ni and Cr was revealed, and the displacements were found to be small. A similar study, with one single crystal of Fe-23 at.% Ni-7 at.% Cr exposed to synchrotron radiation of different wavelengths, was performed by Marty et al. (1994), with useful results for effective pair interactions. For (ternary) quasicrystals, the analysis of diffuse scattering becomes more complex since, e.g. phasons, phason–phonon coupling, the presence of vacancies have to be considered. An extensive study by Abe et al. (2007) of a decagonal $Al_{72}Ni_{18}Fe_{10}$ quasicrystal concentrates on chemical SRO using similar procedures as described above. A different approach based on three-dimensional difference (between full diffraction pattern and isolated diffuse scattering) pair distribution functions (PDFs) has been used by Schaub et al. (2010) for decagonal Al–Cu–Co where structural disorder was predominant. Diffuse scattering is much more complex for quasicrystals than for crystals and there remain several open questions [see De Boissieu and Francoual (2009); Weber and Steurer (2009); and Abe (2009) for further information].

Three independent pair correlations are also required to describe an amorphous binary alloy. Metallic Cu–Zr glasses of various compositions have frequently been studied. The work by Ma et al. (2009) quotes the related literature. Ma et al. (2009) used amorphous Cu–Zr ribbons about 50 μm thick and 5 mm wide of six different compositions ($0.355 \leq c_{Zr} \leq 0.60$) in their combined synchrotron X-ray and (time-of-flight) neutron scattering study. For $c_{Zr} = 0.355$, samples intended for neutron scattering were made from four types of Cu; natural Cu, the isotopes ^{63}Cu and ^{65}Cu, and a mixture of natural Cu and ^{63}Cu to match the scattering length of Zr. Thus, the contributions of the three pair correlations had different weights in five different measurements. The experimental scattering functions were Fourier transformed to obtain reduced PDFs $G(r) = 4\pi r \left[\rho_n(r) - \rho_0\right]$ where $\rho_n(r)$ is the local number density and ρ_0 the average number density. **Figure 26** shows $G(r)$ for the various Cu-35.5 at.% Zr samples. Nearest-neighbor correlations were extracted from the first PDF peak by assuming weighted partial PDFs with $T_{ij}(r) = 4\pi r \rho_{ij}(r)$ (i, j are the atomic species) resulting from a convolution of a Gaussian distribution and an exponential decay. **Figure 27** shows the best fit

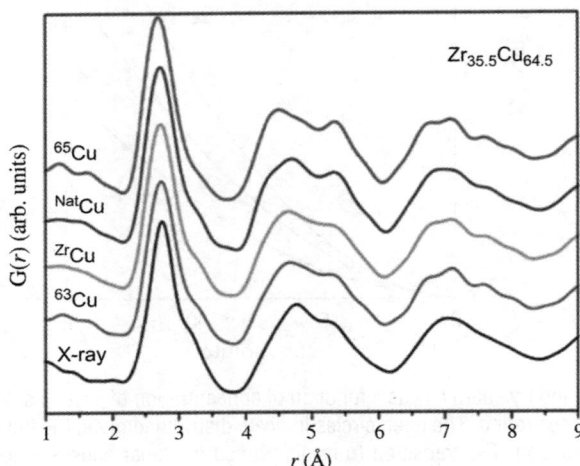

Figure 26   The reduced pair distribution functions $G(r)$ obtained from neutron diffraction of the Cu-35.5 at.% Zr metallic glass containing ^{65}Cu, ^{Nat}Cu, ^{Zr}Cu, and ^{63}Cu, respectively, and from X-ray diffraction of the sample containing natural Cu. From Ma et al. (2009). (For color version of this figure, the reader is referred to the online version of this book.)

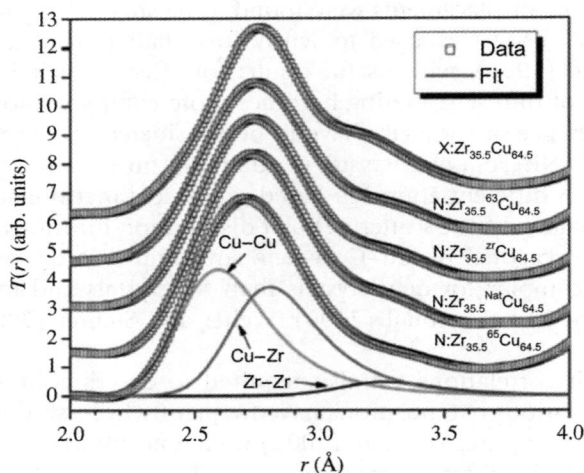

Figure 27   The best fit of a linear combination of convoluted Gaussian functions to the neutron (N, with isotope substitution as indicated) and X-ray (X) results for the functions $T(r)$ of the Cu-35.5 at.% Zr alloy. The extracted Zr–Zr, Zr–Cu, and Cu–Cu partials are also shown at the bottom. From Ma et al. (2009). (For color version of this figure, the reader is referred to the online version of this book.)

involving all five peaks and the corresponding partials for the sample containing ^{65}Cu. From the functions $T_{ij}(r)$ the partial coordination numbers were obtained by integration, and appropriately defined Warren–Cowley SRO parameters were calculated. The authors found that there is some preference for Zr around a Cu atom. This feature has an impact on glass-forming ability [see Ma et al. (2009) for more details].

Returning to crystalline alloys, diffuse scattering from those with a tendency toward clustering ($\alpha_1 > 0$) is much more difficult to evaluate than for those with ordering tendency, since the relevant intensity is concentrated around the Bragg peaks where all the other contributions are also present [for work on Ni–Au, see Portmann et al. (2003); Schönfeld and Kostorz (2009) for Al–Cu, Al–Ag etc., e.g. Kostorz (1998); Schönfeld (1999)]. However, there is also diffuse scattering around the 000 peak, and displacement scattering terms diminish. For the study of decomposition in alloys, small-angle scattering is therefore an important tool.

13.5 Near the Incident Beam

This section will be mostly concerned with the region around the origin of the reciprocal lattice, i.e. $\kappa \ll g_m$ where g_m is the modulus of the smallest reciprocal lattice vector (for crystals; for noncrystalline matter, the position of the first maximum of the structure function serves as reference). This field is commonly called low- or small-angle scattering, which means diffuse scattering for small κ, not just Bragg diffraction at small angles as in the transmission electron microscope or with hard X-rays. After a summary of the principles of the method, some applications to the study of inhomogeneities in alloys and of defect clusters will be presented. Finally, some other effects and methods for κ near 0 will be discussed.

13.5.1 Small-angle Scattering

Small-angle scattering (SAS) is caused by the variation of scattering length density over distances exceeding the normal interatomic distances in condensed matter. If κ_{max} is the maximum value of the scattering vector accessible in an SAS experiment, structural details on a scale smaller than about $1/\kappa_{max}$ will not be resolved, and the discrete arrangement of scattering centers (atoms or nuclei) can be replaced by a continuous distribution of scattering length over volumes of about $(\kappa_{max})^{-3}$. Sums as found in normal crystallographic expressions may then be replaced by integrals, and the theory of SAS specializes on the evaluation of these integrals for a variety of cases. For a full account of the theoretical principles, developed for X-rays but equally valid for neutrons, see Guinier and Fournet (1955); Beeman et al. (1957); Guinier (1963); Gerold (1967); and Porod (1982), for neutrons see Schmatz (1978) and Kostorz (1979). The basic expression for the SAS cross-section is

$$\left(\frac{d\sigma}{d\Omega}\right) = \left| \int_{V_s} \rho_f(\mathbf{r}) \exp(-2\pi i \mathbf{\kappa} \cdot \mathbf{r}) d^3\mathbf{r} \right|^2 , \tag{64}$$

where the integration extends over the sample volume, V_s and $\rho_f(\mathbf{r})$ is the scattering length density, locally averaged as indicated above. It is also useful to write

$$\rho_f(\mathbf{r}) = \Delta\rho_f(\mathbf{r}) + \bar{\rho}_f, \tag{65}$$

where $\overline{\rho}_f$ is averaged over distances much larger than $1/\kappa_{min}$, and κ_{min} is the smallest κ value accessible in the experiment. For the κ range considered, only $\Delta\rho_f(\mathbf{r})$ will contribute to the scattering, i.e.:

$$\left(\frac{d\sigma}{d\Omega}\right) = \left| \int_{V_s} \left[\Delta\rho_f(\mathbf{r})\right] \exp(-2\pi i\boldsymbol{\kappa}\cdot\mathbf{r}) d^3\mathbf{r} \right|^2 , \tag{66}$$

Equation (64) or (66) may be used as a starting point for analytical or numerical calculations of scattering for any imaginable model distribution. With the help of the equations of Section 13.2.3, magnetic SAS may also be included.

The simplest and very widely used approach to evaluate SAS is based on the two-phase model, assuming small particles with a homogeneous scattering-length density ρ_{fp} embedded in a homogeneous matrix of scattering-length density ρ_{fm} ($\rho_{fm} = 0$ covers the case of small particles in vacuo). One can write

$$\frac{d\sigma}{d\Omega} = \left\langle \sum_{\mu=1}^{N_p} |F_{p\mu}(\boldsymbol{\kappa})|^2 \right\rangle + \left\langle \sum_{\mu=1}^{N_p} \sum_{\substack{\nu=1 \\ \nu\neq\mu}}^{N_p} F_{p\mu}(\boldsymbol{\kappa}) F_{p\nu}^*(\boldsymbol{\kappa}) \exp\left[-2\pi i\boldsymbol{\kappa}\cdot(\mathbf{R}_\mu - \mathbf{R}_\nu)\right] \right\rangle, \tag{67}$$

where $F_{p\mu}$ is the single-particle scattering amplitude of the μth particle whose volume is $V_{p\mu}$,

$$F_{p\mu}(\boldsymbol{\kappa}) = (\rho_{fp} - \rho_{fm}) \int_{V_{p\mu}} \exp(-2\pi i\boldsymbol{\kappa}\cdot\mathbf{r}) d^3\mathbf{r} \tag{68}$$

Equation (67) is still too complicated to be of general use. One often assumes a size (and/or orientation) distribution of equally shaped particles and some ad hoc particle distribution function to treat the interparticle interference term. We simplify the presentation assuming N_p identical particles without spatial correlations (dilute limit), but—if they are anisometric—all aligned in the same direction. Then

$$\left(\frac{d\sigma}{d\Omega}\right) = N_p(\rho_{fp} - \rho_{fm})^2 V_p^2 S(\boldsymbol{\kappa}), \tag{69}$$

where $S(\boldsymbol{\kappa})$ is the single-particle scattering function, and

$$S(\boldsymbol{\kappa}) = \left| \frac{1}{V_p} \int_{V_p} \exp(-2\pi i\boldsymbol{\kappa}\cdot\mathbf{r}) d^3\mathbf{r} \right|^2 . \tag{70}$$

The function $S(\boldsymbol{\kappa})$ can be calculated for a variety of particle shapes (see, e.g. Porod, 1982; Volkov and Svergun, 2003) and depends on the direction of $\boldsymbol{\kappa}$ relative to any given axis of the particle, except for spheres. For all other particle shapes, $S(\boldsymbol{\kappa})$ has to be averaged for a given distribution of orientations. Only if the particle orientation distribution is isotropic, $\boldsymbol{\kappa}$ can be replaced by κ for anisometric particles.

In many real scattering systems, the single-particle approach is not valid since particle sizes and possibly particle shape as a function of size may vary and spatial as well as orientational correlations must be considered. Generalized cases can be treated only by finite-element methods, but several approximations have been suggested, e.g. evaluation of size distributions of spheres or other uncorrelated particles of known shape, or linear distance distributions of polydisperse spheres (Guinier, 1963). Methods and problems are extensively discussed, e.g. by Porod (1982); Mittelbach and Glatter (1998); Pedersen (1999); Volkov and Svergun (2003). If eqn (69) can be used (if required and possible, after correction for interparticle interference), the evaluation of SAS yields important parameters of the scattering systems. We summarize some general properties.

13.5.1.1 Some Special Scattering Functions

The simplest case, scattering from a spherical particle, is known since Lord Rayleigh (1842–1919). The single-particle scattering function reads

$$S_p(\kappa R_s) = 9 \left[\frac{\sin(2\pi\kappa R_s) - 2\pi\kappa R_s \sin(2\pi\kappa R_s)}{(2\pi\kappa R_s)^3} \right], \tag{71}$$

where R_s = radius of the sphere. This function decreases to zero at $2\pi\kappa R_s = 4.493$ and reaches a second maximum (the first one obviously being at $\kappa = 0$) at $2\pi\kappa R_s = 5.765$, but the intensity is less than 1% of that at $\kappa = 0$. Further zeros and maxima may be seen for almost monodisperse systems, but rarely in alloys.

For several anisometric particles, approximate expressions may be obtained [the transition from spheres to cuboids has been described analytically by Schneider et al. (2000)]. For example, for cylinders of diameter $2a$ and length $2h$, with axes isotropically averaged,

$$S_p \propto (\kappa h)^{-1} \exp(-\pi^2 \kappa^2 a^2) \tag{72}$$

if $2\pi\kappa h \gg 1 \gg 2\pi\kappa a$ (rods), and

$$S_p \propto (\kappa^2 a^2)^{-1} \exp\left(-\frac{4\pi^2\kappa^2 h^2}{3}\right) \tag{73}$$

if $h \ll a$ (disks) and $\kappa h < 1 \ll \kappa a$.

Thus, special features of the scattering objects may be identified if the κ range is properly chosen.

13.5.1.2 Extrapolation to κ = 0

If the extrapolation can be performed reliably (experimental problems at very small angles may interfere), eqn (69) yields [because $S(0) = 1$]:

$$\left(\frac{d\sigma}{d\Omega}\right)(\kappa \to 0) = N_p(\rho_{fp} - \rho_{fm})^2 V_p^2. \tag{74}$$

Three parameters determine $d\sigma/d\Omega(\kappa \to 0)$: the number of particles (i.e. the particle density $n_p = N_p/V_s$), the scattering contrast and the particle size. Combined with information from other parts of the scattering curve, their values may be obtained separately.

13.5.1.3   *Guinier Approximation*

For any particle shape, the scattering function at small values of κa (where a is the relevant size of the particle) can be approximated by an exponential function (Guinier, 1939, 1963),

$$S(\kappa) = \exp\left(-4\pi^2\kappa^2 R_\kappa^2\right), \tag{75}$$

where R_κ is the average inertial radius (Guinier and Fournet, 1955) of the particle defined by

$$R_\kappa^2 = \frac{1}{V_p} \int\limits_{V_p} r_\kappa^2 q(r_\kappa)\mathrm{d}r_\kappa, \tag{76}$$

where $q(r_\kappa)$ is the geometrical cross-section of the particle perpendicular to a distance vector $\mathbf{r}_\kappa$ parallel to the direction of κ, with its origin inside the particle (defined by $\int r_\kappa q(r_\kappa)\mathrm{d}r_\kappa = 0$). The scattering function reflects the anisometry of a particle, since its decrease with increasing κ is the steeper the larger R_κ. For randomly oriented particles:

$$S(\kappa) = \overline{S(\kappa)} = \exp\left(-4\pi^2\kappa^2 R_G^2/3\right), \tag{77}$$

with the radius of gyration R_G defined by

$$R_G^2 = \frac{1}{V_p} \int\limits_{V_p} r^2\mathrm{d}^3\mathbf{r}. \tag{78}$$

Confusion often arises if the term "Guinier radius" is used. It may signify the radius of a sphere, R_s, calculated from R_G, but sometimes R_G itself. For a sphere, $R_G^2 = 3R_s^2/5$, for an ellipsoid with axes $2a, 2b, 2c, R_G^2 = (a^2 + b^2 + c^2)/5$. For other particle shapes, see e.g. Glatter (1982).

The Guinier approximation results from an expansion of the exponential function in eqn (70) and a reinterpretation of the resulting series for $S(\kappa)$ as an exponential [eqn (75)]. The term proportional to κ^2 in the Guinier approximation is exact for any particle shape. For spheres, the term with κ^4 from eqn (77) is correct to better than 10%, and higher-order terms are in fair agreement. The Guinier approximation is therefore acceptable for $2\pi R_G\kappa \leq 1.2$. For not too strongly anisometric particles, a wider range may even be possible, whereas for extremely anisometric particles (e.g. rods or platelets) the higher-order terms will contribute at smaller values of κ (see, e.g. Porod, 1982).

13.5.1.4   *Integrated Intensity*

From eqn (66), one obtains a quantity Q by integration,

$$Q = \frac{1}{V_s} \int \frac{\mathrm{d}\sigma}{\mathrm{d}\Omega}(\kappa)\mathrm{d}^3\kappa = \overline{\left\{\Delta\rho_f(\mathbf{r})\right\}^2}, \tag{79}$$

where the integration extends over the entire reciprocal space ($0 < \kappa < \infty$) and the bar denotes averaging over the sample. As Q represents the mean-square fluctuation of the scattering length density of the system, which is insensitive to detailed structural features, it is sometimes (see Porod, 1982) called "invariant". In the two-phase model:

$$Q = \left(\rho_{fp} - \overline{\rho}_f\right)\left(\overline{\rho}_f - \rho_{fm}\right) \tag{80}$$

or

$$Q = C_p(1 - C_p)\left(\rho_{fp} - \rho_{fm}\right)^2, \tag{81}$$

where $C_p = N_p V_p / V_s$ is the volume fraction of the particles. The invariant is, as eqn (79) indicates, the FT of the scattering cross-section for $r = 0$, i.e. a special value $\gamma(0)$ of a correlation function $\gamma(r)$ (see Porod, 1982) which, however, has a simple meaning only for dilute isotropic systems.

13.5.1.5 Characteristic Length
An average correlation length L_c can be defined by (the bar denotes an average of all directions of κ if the system is anisotropic):

$$L_c = \frac{2\pi}{V_s Q} \int \kappa \overline{\frac{d\sigma}{d\Omega}}(\kappa) d\kappa, \tag{82}$$

with Q as defined in eqn (79), for any scattering system. In the two-phase model, L_c can be interpreted as the mean length of all lines passing through all points in all directions. The average of all the chords is L_p, the characteristic length, given by

$$L_p = 2\pi V_p \int \kappa \overline{S(\kappa)} d\kappa. \tag{83}$$

13.5.1.6 The Porod Approximation
For scattering systems with well-defined interfaces between the two phases (see Porod, 1951, 1982), the final slope of the scattering function is proportional to κ^{-4}. For particles of any shape (κ must be larger than the inverse of the shortest dimension of the particle):

$$\overline{S(\kappa)} \approx (2\pi)^{-3} A_p V_p^{-2} \kappa^{-4}, \tag{84}$$

where A_p is the surface area of the particle. Inserting eqn (84) into eqn (69), we get:

$$\frac{d\sigma}{d\Omega} = (2\pi)^{-3} N_p A_p \left(\rho_{fp} - \rho_{fm}\right)^2 \kappa^{-4}; \tag{85}$$

the cross-section becomes proportional to the total surface area of the scattering particles.

The Porod approximation was originally derived for particles with sharp interfaces, but without sharp edges or corners. Many further extensions have since been considered, e.g. for anisotropic systems (see Ciccariello et al., 2000, 2002; Ciccariello, 2010), for diffuse interfaces (Ciccariello and Sobry, 1997), for rough interfaces and surfaces (Sinha et al., 1988; Sinha, 2009; grazing incidence), for fractal systems (Schmidt, 1991; Teixeira, 1988), and if the scattering intensity is large enough in this range of weak scattering (note that the SAS intensity as discussed here is never the only intensity measured experimentally; corrections for background intensities, incoherent scattering and other scattering contributions are particularly crucial in this range), single-particle characteristics may be obtained even for less dilute systems, since interparticle interference effects are reduced at larger κ. If the scattering intensity follows a power of κ different from -4, this is usually taken as an indication of self-similar surface or mass fractals, but power law scattering deviating from a power -4 may also be caused by specific size distributions. For some critical discussions of the impact of size distributions and fractal dimensions, see Hammouda (2010) and Deschamps and De Geuser (2011).

The advantage of the Guinier approximation is that the radius of gyration, i.e. a size parameter, can be determined from uncalibrated SAS data [cf. eqns. (75) and (77), inserted into eqn (69)]. The combination of several of the above relationships [eqns (74) and (75) or (77), (80, 81), (83) and (85)] allows one to evaluate shape, size, number and composition of uniform particles from precise measurements over a sufficiently large range of κ. For example, eqns (81) and (85) yield the surface-to-volume ratio of particles, again without calibration of the measured intensities.

Although X-rays are more widely available than neutrons, the use of neutrons in SAS studies of metallic systems is basically more generally applicable for bulk studies in transmission geometry. The most severe restriction for X-rays stems from the absorption properties of materials. For CuK_α radiation, the optimum thickness (see eqn (25), $\Sigma_t D_s = 1$) of a pure Al sample is $D_s = 76$ µm, but less than 10 µm for many other metals (atomic numbers 23–27 and above 41, except 55). With MoK_α, a thickness of 713 µm is ideal for Al, but for most of the heavier metals, 20–100 µm should not be exceeded. Although with hard synchrotron X-ray beams, thicker samples may be used, the sample volume may still not be representative of the bulk material. Scattering from any surface irregularities and sample environment (windows, heat shields etc.) must be carefully considered for all SAS experiments, but especially for thin samples. Another difficulty arises from double Bragg scattering which can obscure the SAS effects and sometimes exceed them. The obvious remedy, increasing the incident wavelength above the Bragg cut-off, works well for neutrons but not for X-rays as absorption becomes prohibitively large. Double Bragg scattering may then only be avoided using properly oriented single crystals. Finally, inelastic scattering may be appreciable at high sample temperatures, and if an energy-resolved separation is required, neutrons remain more adequate.

The applications of SAS of X-rays and neutrons to metallurgical problems are numerous and have frequently been reviewed (Gerold and Kostorz, 1978; Kostorz, 1979, 1982, 1988a, c, 1991, 1992, 1993, 2001, 2003; Wagner et al., 2001; Williams et al., 1993; Fratzl, 2003; Simon, 2007). From a simple verification of sample homogeneity to the determination of sizes, size distributions and interparticle interference effects, the degree of complexity of evaluation procedures varies from simple analytical methods to very involved fitting routines and modeling. In the following, a few examples of SAS research related to alloys and defects will be given.

13.5.2 Alloys

Guinier (1938, 1939) discovered SAS from various Al-rich alloys, e.g. Al–Cu and Al–Ag, and found evidence for the existence of very small coherent precipitates, now known as Guinier–Preston zones (GP zones, Preston, 1938). An important problem that still has not been solved completely is the question of how phase separation is initiated and progresses during the early stages when large parts of the sample are still in a supersaturated state. The SAS technique is sensitive to small (in scale and in amplitude) compositional variations, bridging and overlapping with the domains accessible by field-ion microscopy and transmission electron microscopy. Many SAS experiments have therefore been performed to study phase separation, but also coarsening reactions, precipitate parameters in relation to other properties, and dissolution of precipitates (see the reviews quoted above). For X-rays, Al-rich alloys have been most suitable, and very few other systems have been investigated until very recently when synchrotron radiation sources became available. Initial experiments with neutrons concentrated on Al alloys, too, but many other systems have been studied during the last 25 years. Whereas for intensity reasons, slit collimation was initially used for the incident beam in transmission geometry (with single detectors), point collimation is nowadays common for X-rays and neutrons (with two-dimensional position-sensitive detectors), except for the double-crystal (Bonse–Hart) arrangement for ultrasmall-angle scattering. **Figure 28** shows the standard setup, and **Figure 29** shows the layout of an ultrasmall-angle-scattering instrument (the lowest values of κ accessible on such instruments are in the range of 10^{-5} Å^{-1}, see Iwase et al., 2007).

As an early example, **Figure 30** shows two SAS patterns obtained for an Al-6.8 at.% Zn single crystal (see Bubeck et al., 1985; Kostorz, 2003). Many details of the individual scattering objects are averaged out or even remain hidden when polycrystals or powders are used in SAS. In the figure, the maxima in the scattering intensity (along <001>) are due to interparticle interference of small Zn-rich pre-precipitates (GP zones) coherently embedded in the matrix. Elastic interactions induce a preferential alignment of these zones along the elastically soft directions. The scattering at larger scattering vectors is mainly controlled by the single-particle scattering function. Model calculations including a particle size distribution and an interparticle distance distribution yield an average radius of spherical zones of about 1.6 nm for the situation shown in **Figure 30**(a). In **Figure 30**(b), the peaks have moved to smaller values of κ_{001} (indicating larger zone separation), and the azimuthal dependence of the scattering intensity near the periphery of the detector plane relates to the well-known transition from spherical to oblate ellipsoidal shape for zones exceeding a certain size (about 2 nm, depending somewhat on the details of the thermal treatment).

A peak of the SAS intensity at a value $\kappa_m \neq 0$ is quite common for decomposing or decomposed alloys containing—in their fully decomposed metastable or stable state—a precipitate fraction of a few percent ($C_p \leq 7\%$ in the present example). An early appearance of a peak has sometimes been taken as evidence for the mechanism of spinodal decomposition but this feature of SAS alone is insufficient to distinguish between concentration fluctuations and well-defined homogeneous particles. A SAS peak

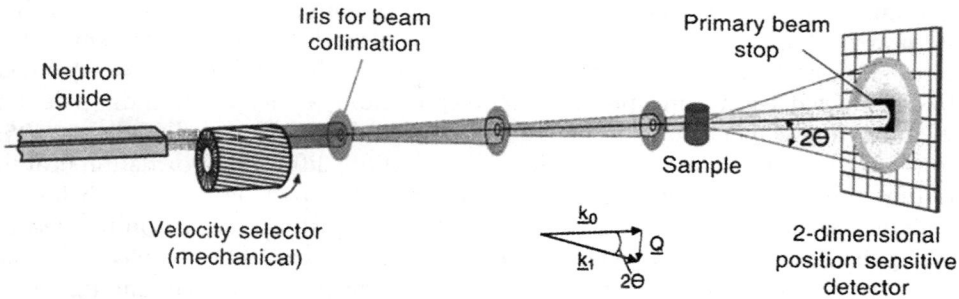

Figure 28 Schematic view of a small-angle (neutron) scattering instrument ($\mathbf{Q} \equiv \mathbf{\kappa}$). The beam collimation section will of course be different for X-rays. Courtesy of J. Kohlbrecher, PSI Villigen, Switzerland. (For color version of this figure, the reader is referred to the online version of this book.)

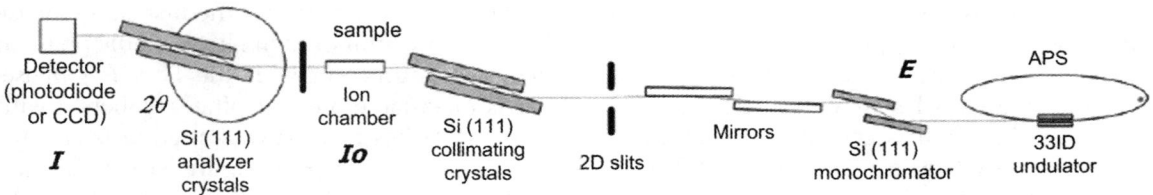

Figure 29 Schematic side view of the UNICAT ultrasmall-angle X-ray scattering instrument at sector 33ID (insertion device: undulator) of the Advanced Photon Source (APS) at Argonne National Laboratory, Illinois, USA. The standard collimation is one-dimensional, i.e. in slit geometry, which requires deconvolution of the measured data. From Ilavsky et al. (2004).

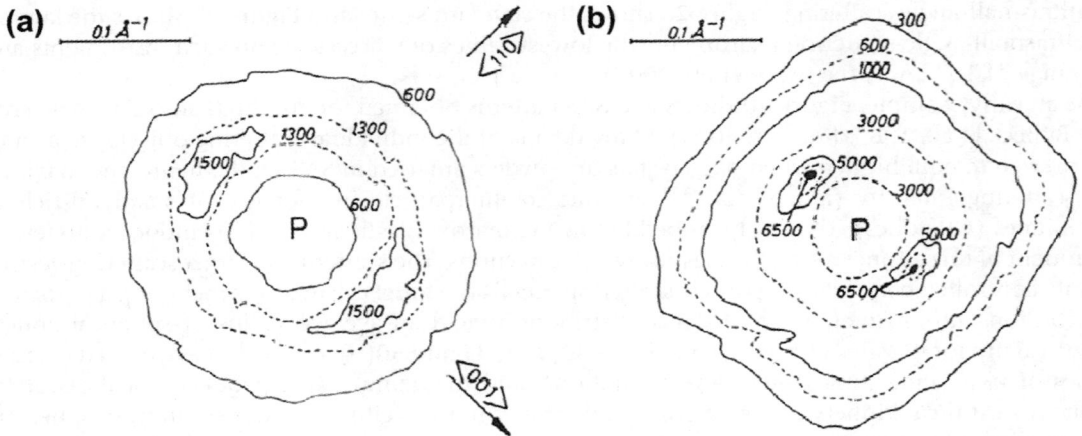

Figure 30 Small-angle scattering patterns for an Al-6.8 at.% Zn single crystal with {110} faces perpendicular to the incident beam. Some other crystallographic orientations are indicated. After quenching, the crystal was aged in situ at room temperature for (a) 2 h and (b) 7.5 h. The azimuthal dependence of the maximum intensity is visible. After Bubeck et al. (1985).

will of course also occur in a two-phase system as the first maximum of the interparticle interference function (see Guinier and Fournet, 1955; Glatter, 1982), just demonstrated. Quantitative kinetic measurements with a variation of important parameters (initial concentration, quenching conditions and aging conditions) are necessary and must be compared with specific predictions of the different theoretical models. Supporting evidence from other methods, especially transmission electron microscopy or atom-probe field-ion microscopy, should be sought. Many other cases producing a SAS peak can be constructed, e.g. a "three-phase" system representing particle, surroundings depleted of the alloying element and supersaturated matrix (introduced by Walker and Guinier, 1953). Shell compositions may, of course take on other values if segregation, diffusion or oxidation near the surface or interface of nanoparticles is involved. Examples for core–shell particles are given below.

As another example of the use of SAS, we refer to the long-lasting research on the decomposition kinetics of Ni-rich alloys (this part is essentially taken from Kostorz, 2003). Ni-base "superalloys" are widely used in structural applications owing to their high-temperature strength and their resistance to corrosion. Much of the high-temperature performance of these alloys is due to the presence of ordered ($L1_2$ structure) coherent precipitates in the face-centered cubic matrix. The technological development of multicomponent Ni-base superalloys is quite advanced, although even for such important binary systems as Ni–Al and Ni–Ti, the details of the microstructural evolution are still not entirely understood. In contrast to Ni–Al where the $L1_2$ precipitates (called the γ' phase) are the first stable phase appearing in the binary phase diagram beyond the solid-solubility limit of Al in Ni, the stable phase in Ni-rich Ni–Ti is the hexagonal $D0_{24}(\eta)$ phase, which is semicoherent with the f.c.c. (γ) phase. $L1_2$-structured particles are nevertheless found in Ni–Ti as metastable pre-precipitates, coherent with the matrix, but with larger misfit strains than in Ni–Al. As Ti has a negative scattering length for neutrons, a very large signal is expected from a large $|\rho_{fp} - \rho_{fm}|$ in eqn (68). **Figure 31** shows some results of an in situ aging experiment performed at the D11 instrument of the ILL, Grenoble, France. In **Figure 32**, the results of a polycrystalline sample are compared with 40° sector-averaged scattering taken from **Figure 31** (10 h) along [001] and [110]. In an early series of in situ SAS measurements, the kinetics of phase separation in a Ni-11.5 at.% Ti single crystal were followed at 580°C for up to 10 h,

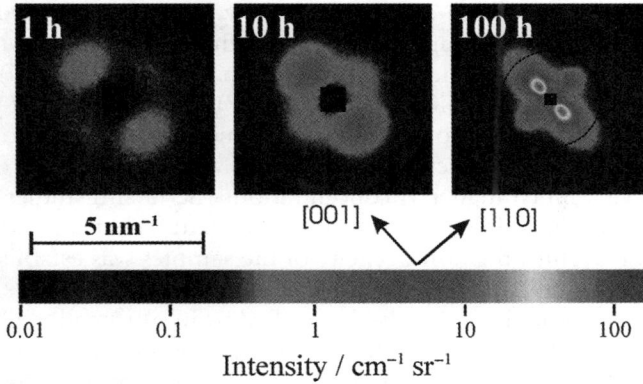

Figure 31 Small-angle neutron scattering patterns of a Ni-11.3 at.% Ti single crystal aged in the neutron beam at 600°C (intensities are given in macroscopic differential cross-sections). The inner part of the right pattern was taken at a sample-to-detector distance of 4 m while all the other measurements were taken at 2 m. The wavelength of the incident neutrons was 8. The average diameter of the precipitates increases from about 60 Å after 10 h to about 140 Å after 100 h of aging. From Kostorz (2003). (For color version of this figure, the reader is referred to the online version of this book.)

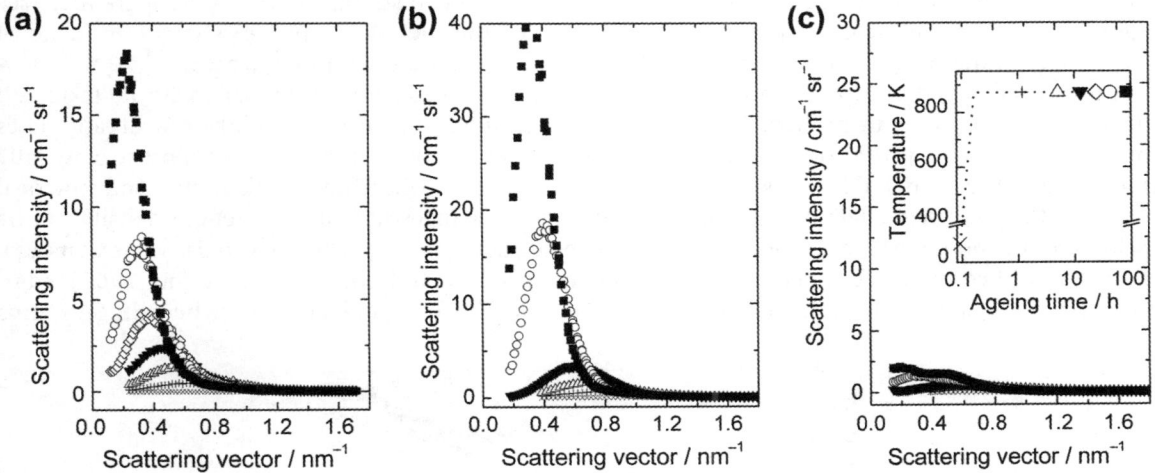

Figure 32 Comparison of the small-angle scattering profiles from the measurement on (a) polycrystalline Ni-11.3 at.% Ti with 40° sector-averaged scattering profiles along (b) <100> and (c) <110> directions of a Ni-11.3 at.% Ti single crystal. Both samples were aged at 600°C for up to 100 h (see inset in (c)). (The scale indicates the range of $2\pi\kappa$.) From Kompatscher et al. (2002).

and a plateau was found in the integrated SAS, see eqn (80), toward the end of the experiment (Cerri et al., 1990). Its evaluation resulted in a Ti concentration of less than 20 at.% in the Ti-rich zones. The aging times available for in situ studies were too short to check whether a further increase toward the normally assumed 25 at.% Ti content was possible. An increase was subsequently found using a series of samples aged for progressively longer periods, with the result that after an initial stage where γ'' forms, the second stage, γ', corresponds to (22 ± 2) at.% Ti in the zones (Vyskocil et al., 1997). Another in situ study (Kompatscher et al., 2000) confirmed the lower Ti plateau at 600°C and also at 630°C (some of the results are summarized in **Figure 33**). In the many other published investigations, reviewed to a large extent by Vyskocil et al. (1997) and Kompatscher et al. (2000), including a variety of techniques other than SAS as well, two different successive metastable states had never been identified in one experiment. In the X-ray diffraction study of Hashimoto and Tsujimoto (1978), Ti concentrations of less than 20% were found after very extensive aging, where the neutron SAS results shown here indicate that γ' must have a much higher Ti concentration. The in situ studies of Kompatscher et al. (2000) employed special furnaces designed for quick temperature changes and in situ solution treatments at high temperatures. Thus, the homogeneity of the samples was established directly by monitoring the SAS before the beginning of any phase-separation sequence.

The appearance of the stable η phase can also be studied by in situ small-angle neutron scattering at sufficiently high temperatures. Very dramatic effects are observed if single crystals are used (Kompatscher et al., 2000; 2003). Some examples of scattering patterns obtained for a Ni-12.3 at.% Ti single crystal during aging at 930°C are shown in **Figure 34**. The incident beam is along a <110> direction, and <100> lies in the horizontal direction. The scattering patterns show that the $L1_2$-related metastable phase forms first. This is visible from the strong intensity along <100>, which is also observed at lower temperatures where η never forms. The scattering anisotropy is related to elastic interactions of the coherent, metastable precipitates. Additional sharp streaks along the <111> directions appear after about 10 min. They are due to the formation of lamellae of η phase parallel to the cubic {111} planes. The onset time for η formation has been studied at several temperatures (see **Figure 35**). Further growth of η leads to a reversion of the metastable phase. The current state of knowledge on the metastable miscibility gap in Ni-rich Ni–Ti is shown in **Figure 36**.

Anisotropies of a decomposing alloy will, of course, not be visible for a polycrystal without texture, if a large number of grains are probed. Polycrystalline Al–Zn alloys (on the Al-rich side) are ideal candidates for an experimental verification of many theoretical predictions as the matrix is almost elastically isotropic and the Zn-rich GP zones are—at least initially—fully coherent with the matrix and spherical because of a small size effect. Under these circumstances, changes in scattering length density can be exclusively attributed to compositional changes (displacement effects can be neglected). Some years ago, an oversimplified interpretation of the original paper on spinodal decomposition (Cahn, 1961) led to a search for special features at the spinodal line and for the "linear spinodal regime" where the SAS curve

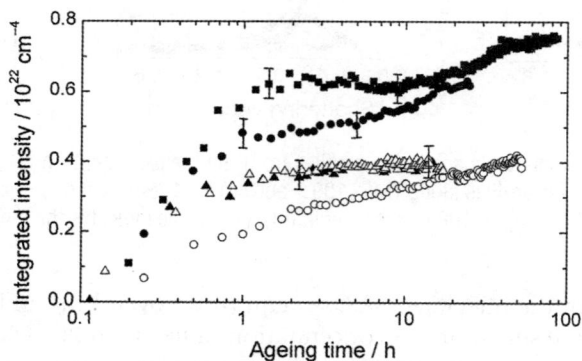

Figure 33 Integrated small-angle neutron scattering intensity as a function of aging time for Ni-11.3 at.% Ti aged at 600°C (filled squares), 630°C (filled circles) and 680°C (filled triangles), for Ni-11.1 at.% Ti at 680°C (open triangles) and for Ni-10.1 at.% Ti at 630°C (open circles).

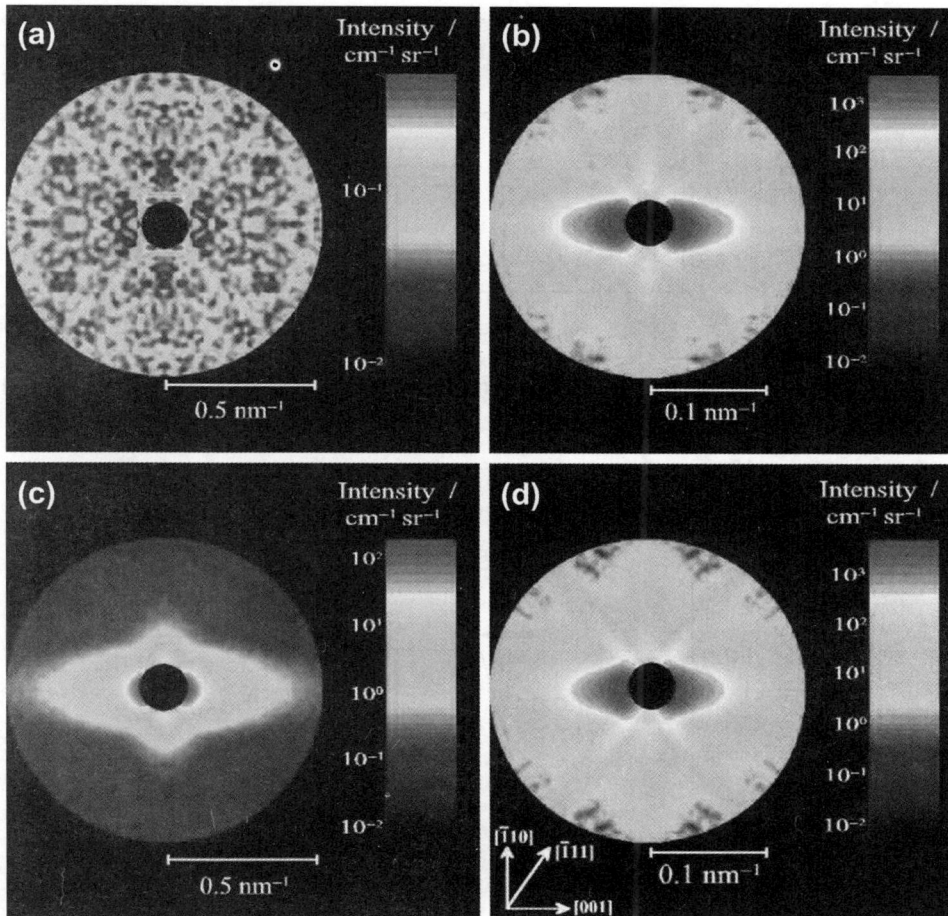

Figure 34 Symmetrized small-angle neutron scattering patterns of a Ni-12.3 at.% Ti single crystal, (a) solution treated at 1440 K, and subsequently aged in situ at 1200 K for (b) 6 min, (c) 26 min, (d) 47 min. The incident beam was parallel to a face diagonal of the f.c.c. unit cell. (The scale indicates the range of $2\pi\kappa$). From Kompatscher (2001). (For color version of this figure, the reader is referred to the online version of this book.)

should show a time-invariant peak at a specific value of κ, κ_m, an exponential growth of the SAS intensity for all $\kappa < \sqrt{2}\kappa_m$ and a decay for larger values of κ. The overwhelming evidence of SAS experiments (in accordance with other experimental and more elaborate, more realistic theoretical results, see Binder and Fratzl, 2001; Wagner et al., 2001) indicates that for metallic systems, a linear regime is found only very exceptionally (and may then be explained by partial phase separation during quenching), and that the search for singularities at the spinodal line may be pointless in a system with short-range interactions. In these cases, the spinodal singularity disappears and there is a gradual transition from the nucleation regime to spinodal decomposition in a region of the phase diagram where the free-energy barrier of a nucleus becomes comparable to the thermal energy, kT. A sequence of scattering curves as shown, e.g. in **Figure 32**(a) is thus more typical. Nonlinear extensions of the spinodal theories, but many other models as well, may yield similar sequences of scattering curves (see also Binder and Fratzl, 2001).

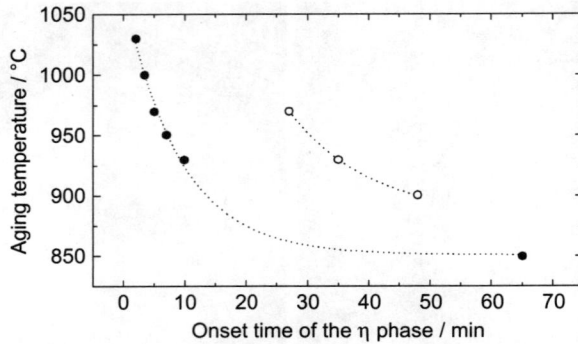

Figure 35 Onset time of streak formation versus aging temperature for Ni-11.7 at.% Ti (filled circles) and for Ni-11.3 at.% Ti (open circles). The lines are to guide the eye.

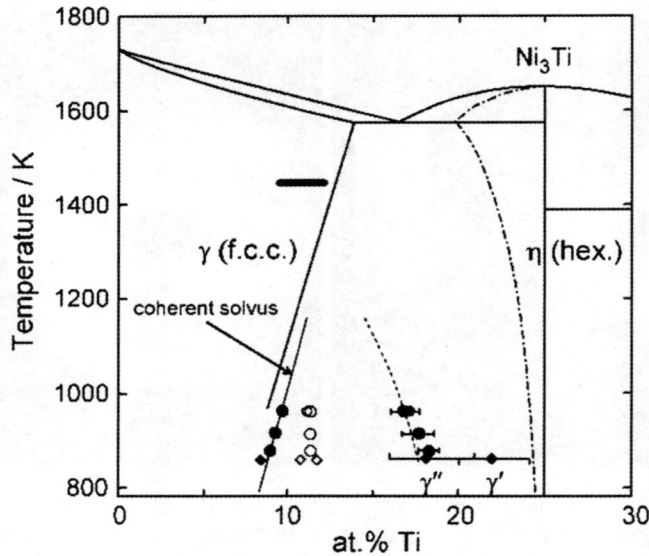

Figure 36 Ni-rich part of the Ni–Ti phase diagram (Massalski, 1990), including the approximate loci of the metastable (coherent) miscibility gap and the matrix and precipitate concentrations (filled symbols) evaluated from small-angle neutron scattering of alloys (with the average concentrations indicated by open symbols) aged at four different temperatures (after Vyskocil et al., 1997; Kompatscher et al., 2000).

A general analysis of the scattering curves, which may sometimes be capable of distinguishing different regimes of the decomposition process, is a test of self-similarity (see Fratzl and Lebowitz, 1989). If a system evolves under the influence of a single characteristic length $L(t)$, a scaling test of the SAS function $I(x, t)$ (we use I for simplicity for the cross-section) according to

$$I(\kappa, t)/Q(t) = L^3(t)s(x, t), \tag{86}$$

where t is time, Q is the integrated intensity as defined in eqn (79) and $x = 2\pi L(t)\kappa$, yields a scaled scattering (or "structure") function $s(x)$ if $s(x,t)$ does not change with time. The characteristic length $L(t)$

may be obtained from the radius of gyration of a Guinier plot [see eqn (77)], or from the peak position (κ_m), $L_m = (2\pi\kappa_m)^{-1}$, or as the inverse of the first moment of the scattering curve. If scaling holds, these different parameters should not give different results (but this test is rarely made). Power laws for $\kappa_m(t)$ and the peak intensity, $I_m(t)$, are also frequently tested, as they hold approximately in several model calculations, at least for some time interval.

$$\kappa_m(t) = At^{-a}, \quad I_m(t) = Bt^b \tag{87}$$

with constants A, B, and constant exponents a, b. If Q is constant, $b = 3a$, cf. Equation (81). Scaling is usually observed for advanced stages of decomposition when matrix and precipitate have almost reached their final concentrations. There is an agreement that in the dilute "droplet" region, i.e. for isolated precipitates, the exponents a, b coincide with those already predicted by Lifshitz and Slyosov (1961) and Wagner (1961), i.e. $a = 1/3$, $b = 1$ for coarsening under the sole influence of an isotropic interfacial energy. In real systems, however, coherent precipitates introduce strain fields, and the resulting elastic interactions may introduce different growth laws. Early stages are expected to follow different time laws. The two regimes (γ'', γ') in Ni–Ti described above were also distinguished by a scaling analysis of the scattering curves of polycrystals by Vyskocil et al. (1997).

For kinetic studies with neutrons, a few minutes are usually necessary to accumulate statistically relevant data (see He et al., 2010; for a recent in situ SAS study of precipitation in deformed Fe-rich Fe–Cu and Fe–Cu–B–N alloys). Synchrotron radiation X-rays are sufficiently intense to follow decomposition processes with a time resolution of a few seconds (time resolution may be much better, even 10 ms at third-generation sources, for other applications).

The work of Deschamps et al. (2007) may serve as an example of time-resolved synchrotron X-ray studies. The formation of metastable $L1_2$-type Al_3Zr precipitates in Al-0.09 at.% Zr-0.03 at.% Sc alloys quenched from 630°C to room temperature was followed by SAS at temperatures between 400 and 430°C, employing different heating rates. The precipitates have a core–shell structure with a higher Zr content in the shell than in the core. The SAS curves can be fitted to a three-phase model (see Section 13.5.1) with three distinct compositions for matrix, core and shell (see also Dubey et al., 1991; Kostorz, 2001; for the core–shell structure of Ag-rich GP zones in Al-3 at.% Ag from SAS, later confirmed by HAADF-STEM, Erni et al., 2003). **Figure 37** shows a summary of the results obtained for the heating rate 10 K min^{-1}.

In a recent study of the evolution of microstructure of an Al-2.5Cu-1.5 Mg (wt.%) alloy known for its rapid age hardening, Deschamps et al. (2011) combined nuclear magnetic resonance and in situ synchrotron X-ray SAS. **Figure 38** shows that new Cu–Mg-rich clusters appear within seconds upon heating to 200°C. The plateau of cluster size corresponds to the plateau found for the hardness of the alloy. With the beginning of the formation of the S-phase (Al_2CuMg) after about 20 min at 200°C, the SAS patterns show strong streaking due to the anisometric shape of these precipitates (the beam size of 200 µm is comparable to the grain size of the polycrystalline material, so the beam explores only a few crystallites). The subsequent increase of the radius of gyration corresponds to the increase of hardness toward peak hardening. For further details and an in-depth discussion, see the original paper.

With reduced diameters of the incident beam, mapping (scanning) of the SAS signal can be performed with good resolution. **Figure 39** gives an example of a study of the microstructure of a weld (Deschamps et al., 2009) employing, along with several other techniques, SAS mapping. The grid size was 200 µm, and the figure caption gives further details. A yet broader study of a friction stir weld of the Al–Li AA2199 alloy includes a similar SAS approach (Steuwer et al., 2011).

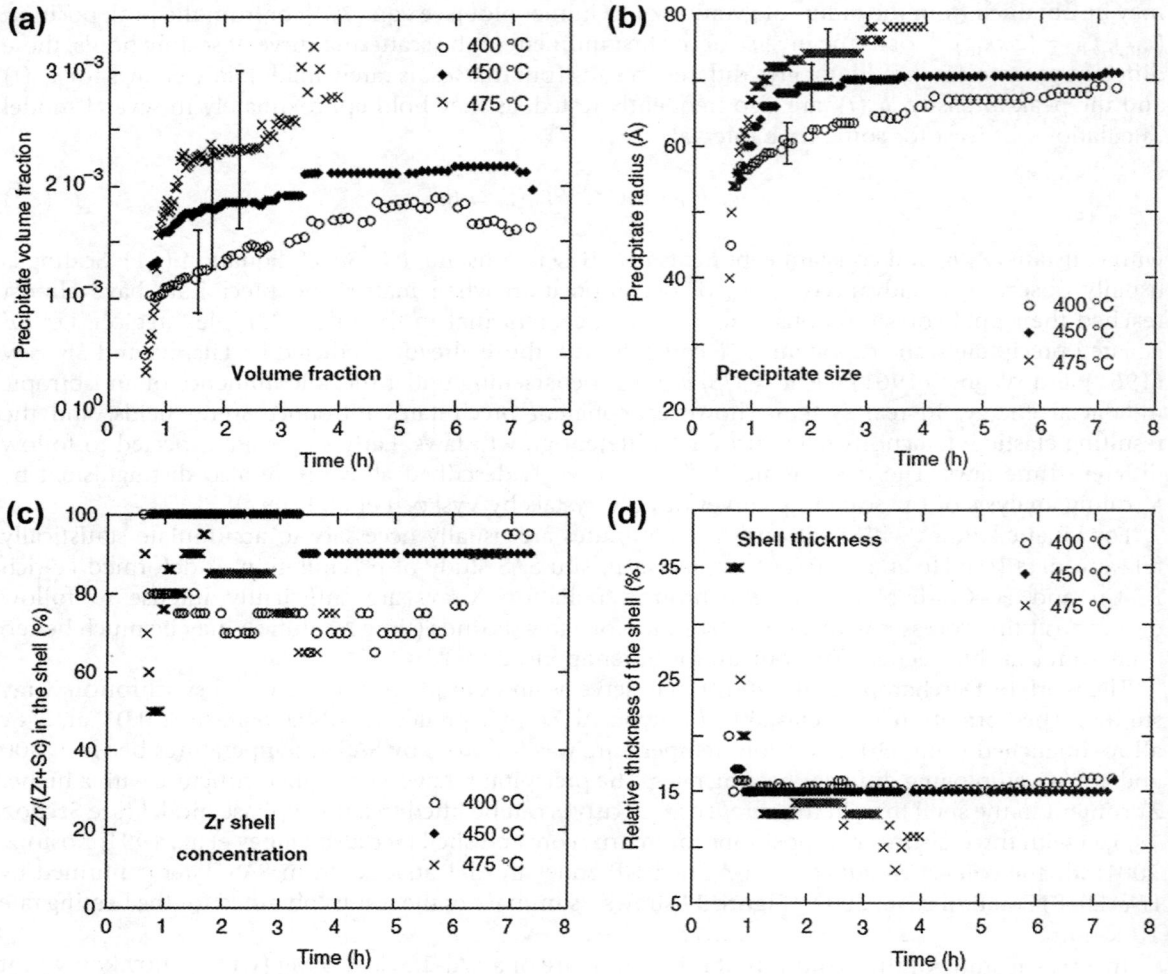

Figure 37 Evolution of the microstructural parameters of an Al-0.09 at.% Zr-0.03 at.% Sc alloy. (a) Volume fraction; (b) radius; (c) Zr concentration in the shell; and (d) shell thickness during heat treatments at 400, 450 and 475°C. In all three cases, the heat treatment temperature was reached with a heating rate of 10 K min^{-1}. Error bars correspond to 30 and 5% uncertainty on volume fraction and size. From Deschamps et al. (2007).

The high intensity of synchrotron radiation also opens up the possibility of exploiting the anomalous scattering terms [see eqn (26)] by tuning the incident wavelength to appropriate values. This may be useful in binary alloys to distinguish between surface and volume effects as the scattering varies proportional to the square of the average scattering length in the former case and proportional to the square of the difference in scattering lengths in the latter—e.g. see Simon et al. (1992), who showed that a (neutron) SAS signal from a thin-foil of an isotope-enriched Fe–Ni Invar alloy (Wiedenmann et al., 1989) was most likely not related to phase separation. Contrast enhancement may also be useful for some binary alloys, see Fratzl et al. (1992) for an example (a nucleation regime was identified in dilute Cu-rich Cu–Fe alloys). The anomalous dispersion also allows decomposition

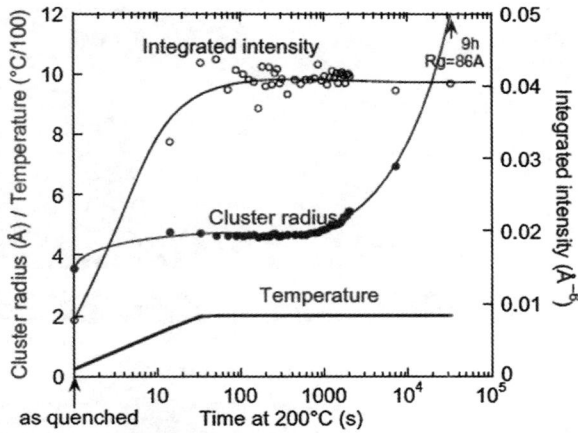

Figure 38 Evolution of cluster size [two-phase model, radius of gyration R_G, see eqn (77)] and integrated intensity determined from small-angle X-ray scattering patterns of an Al-2.5Cu-1.5 Mg (wt.%) alloy heated to and aged at 200°C (temperature ramp and plateau are also indicated). From Deschamps et al. (2011). (For color version of this figure, the reader is referred to the online version of this book.)

in ternary alloys to be analyzed. Simon and Lyon (1994) give a review of early work, see also Simon (2007). While in a structurally disordered alloy with three components, there are six independent partial scattering functions (see Section 13.4.2), for substitutional solid solutions, this number is reduced to three. The SAS cross-section may then be written proportional to

$$I(\kappa) = |f_A - f_C|^2 S_{AA} + \text{Re}\{(f_A - f_B)(f_B - f_C)^*\}S_{AB} + |f_B - f_C|^2 S_{BB} \qquad (88)$$

(A, B, C are the three components, density variations are neglected).

A classical result is shown in **Figure 40**. The three independent functions S_{NiNi}, S_{FeFe} and S_{FeNi} for Cu-42.5 at.% Ni-15 at.% Fe were obtained from SAS data taken near the K edges of Fe and Ni, yielding an overdetermined set of equations of the type given in eqn (88). The values for S_{CuCu} were calculated from the other $S_{\mu\nu}$. If a two-phase description is adopted, $S_{FeNi}^2 = S_{NiNi}S_{FeFe}$ should hold, which is obviously not fulfilled. The best interpretation (Simon and Lyon, 1994) is based on assuming an interfacial segregation layer of Fe between coherent Cu- and (Ni, Fe)-rich domains.

Based on a combination of several methods including in situ anomalous X-ray SAS, Marlaud et al. (2010a) were able to evaluate the composition of precipitates as a function of time of aging in a stepped aging process for three (essentially) quaternary aluminum alloys of technical relevance in the airplane industry (AA7150, AA7449 and an experimental alloy "PA" with some Sc). The main components in all three alloys are Al with 2.8, 3.6, 4.6 at.% Zn, 1.0, 0.8, 0.7 at.% Cu, 2.7, 2.6, 2.4 at.% Mg, respectively. They also contain small quantities of Fe, Si, Zr, Ti (and 0.1 at.% Sc for alloy PA). **Figure 41** shows $\kappa^2 I(\kappa)$ for the peak-aged state of alloy AA7150. The contrast actually decreases when the energy of the X-ray beam (given in eV) is increased from below the K edges of Cu and Zn. This contrast change, however, enters the integrated intensity which can be used, with additional assumptions on the Al and Mg content of the (η') precipitates [Mg(Zn, Mg, Al)$_2$ with $c_{Al} = 0.15$], to work out the Cu and Zn concentrations as a function of aging time. **Figure 42** shows that very little Cu and about 50 at.% Zn are present after the 120°C aging step. Marlaud et al. (2010a) conclude from an extensive analysis that the

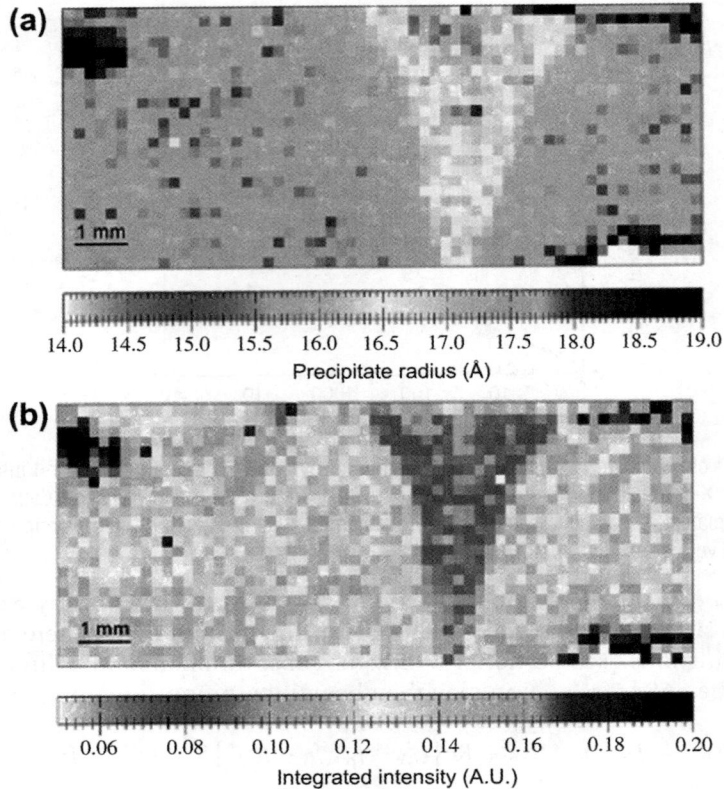

Figure 39 Maps of (a) precipitate size in Å (radius of gyration); (b) integrated intensity in the plane normal to the welding direction, as determined from a synchrotron X-ray small-angle scattering study of an electron-beam weld of an Al-4.5%Zn-1%Mg (wt.%) alloy after a T6 postwelding heat treatment. The area of the weld nugget can be clearly distinguished from the base material. The precipitate size varies by less than 10%, whereas the integrated intensity in the nugget area is little more than 50% of that in the base material. These differences can be matched very well with the changes in mechanical properties across the weld. From Deschamps et al. (2009). (For color version of this figure, the reader is referred to the online version of this book.)

apparent plateaus reached in the precipitate compositions are related to slow diffusion of Cu rather than to reaching phase equilibrium. For more details, see the original paper and further studies by Marlaud et al. (2010b).

By working with three different wavelengths in the vicinity of an absorption edge, Goerigk and Mattern (2009) showed how, apart from a standard distinction of compositional fluctuations and phase separation from the κ-dependence of the SAS intensity (κ^{-2} vs κ^{-4}), the separation of the form factor of the pure resonant scattering of the atomic species selected can reveal the compositional extent of the fluctuations. **Figure 43** shows the results for the YK edge of amorphous and crystallized Ni-16 at.% Nb-16 at.% Y. For Y, 4% of the total number of atoms (i.e. 0.6 of 16 at.%) and similarly for Nb, 12% (1.9 of 16 at.%) were found to be frozen in as fluctuations in the as-quenched sample.

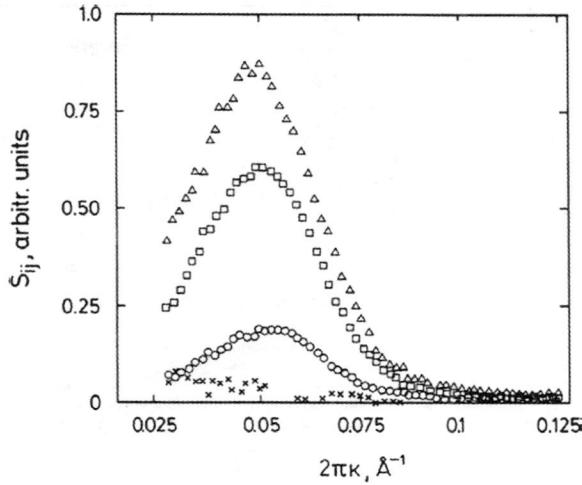

Figure 40 Partial scattering functions of polycrystalline Cu-42.5 at.% Fe aged at 500°C for 56 h $\triangle$: S_{CuCu}, $\square$: S_{NiNi}, o: S_{FeFe} x: S_{FeNi}. From Simon and Lyon (1994).

Magnetic clusters and precipitates can be studied with neutron SAS. The magnetic structure factor, eqn (31), can, in analogy to nuclear scattering, be reinterpreted by introducing an integration over fluctuations of the local magnetization density rather than discrete magnetic moments:

$$\mathbf{M}(\kappa) = \frac{1}{V_s} \int_{V_s} \mathbf{M}(\mathbf{r}) \exp(-2\pi i \kappa \cdot \mathbf{r}) d^3 \mathbf{r}. \tag{89}$$

A homogeneously magnetized sample will therefore not show any magnetic SAS, but fluctuations in the orientation and/or the magnitude of $\mathbf{M}(\mathbf{r})$ will be revealed. A number of useful applications of eqn (32) for unpolarized neutrons are possible. In the simple case of a magnetic phase embedded in a nonmagnetic matrix or vice versa, the ratio $I^{\perp}/I^{\parallel}$ is independent of the magnitude of κ for complete saturation. Any deviation is an indication of incomplete saturation, i.e. some local variations of magnetization still exist (then spin-flip scattering will occur and an analysis is worthwhile), or of another scattering object of different magnetic contrast and size. Using measurements of $I^{\perp}/I^{\parallel}$, Beaven et al. (1986) were able to distinguish scattering contributions from small voids in the presence of somewhat larger Cu precipitates in Fe (the ratio $I^{\perp}/I^{\parallel}$ is 1.4 for voids in ferromagnetic Fe, but 11 for Cu precipitates). The Fe-rich Fe–Cu alloys have frequently been studied under different conditions (see Staron et al., 2010; Schober et al., 2011 and references therein). The recent time-resolved neutron SAS study of Fe-1 at.% Cu by He et al. (2010) aimed at revealing influences of prestraining on aging. The samples were solution treated for 1 h at 850°C and quenched. **Figure 44** shows SAS patterns for the alloy after 24% tensile deformation, before and after the aging treatment. The magnetic field was in the horizontal direction. Equation (32) was fitted to these patterns to obtain the nuclear and magnetic scattering. **Figure 45** shows the corresponding scattering curves before and after aging, also including a sample without deformation. The scattering before aging results from a variety of objects (mainly grain boundaries and large inclusions) on different length scales, but the Cu-rich nanoscale (mean radius after 12 h about 14 Å) precipitates clearly contribute additional scattering (during and) after

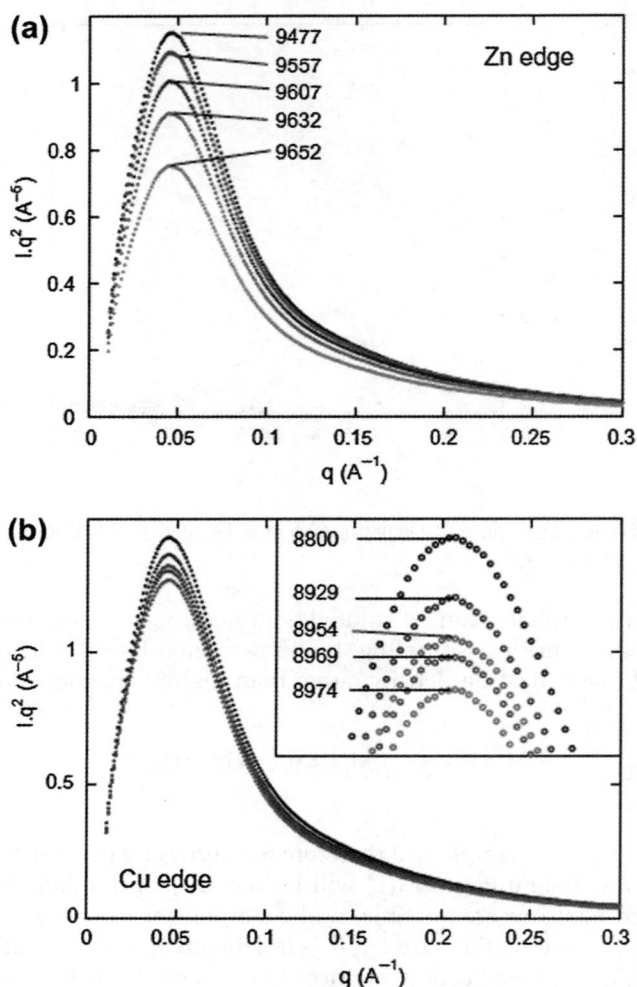

Figure 41 Anomalous X-ray small-angle scattering curves $I(\kappa)\,\kappa^2$ (q in the figure corresponds to $2\pi\kappa$) near the K absorption edges of (a) Zn and (b) Cu for a peak-aged AA7150 alloy. The gap occurring between 3.5 and 5 h is due to an accidental loss of beam-power. Data obtained at beamline BM02-D2AM at the ESRF, Grenoble, France (energies in eV, note that $q = 2\pi\kappa$). From Marlaud et al. (2010a). (For color version of this figure, the reader is referred to the online version of this book.)

aging. Whereas the undeformed sample shows roughly the expected proportionality of the nuclear and magnetic scattering increments, the deformed sample's nuclear scattering shows an additional enhancement at low values of κ, which is not seen at the same level in the magnetic scattering. The authors relate this extra scattering to the precipitation of Cu near dislocations and their correlated arrangements after deformation. In inhomogeneous alloys, $M(\kappa)$ may vary considerably over a wide range of temperatures, since the Curie temperature depends on the chemical composition. Without polarization analysis, magnetic scattering as a function of temperature will give a first indication of the extent and magnitude of magnetized regions.

(a)

(b)

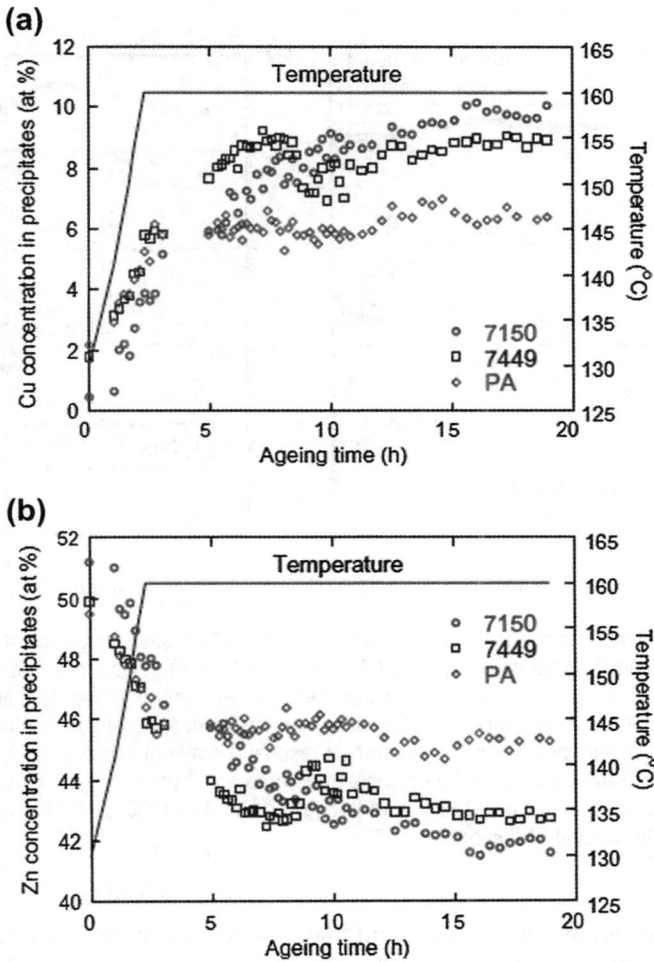

Figure 42 (a) Cu and (b) Zn concentration in the precipitates forming in three different quaternary AA-type Al alloys as a function of aging time at 160°C (the temperature ramp from a preceding aging step at 120°C for 22 h is indicated). From Marlaud et al. (2010a). (For color version of this figure, the reader is referred to the online version of this book.)

Measurements of magnetic neutron SAS in zero field for bulk ferromagnets are not advisable if domain walls are present as these will introduce multiple refraction. However, the scattering due to fluctuations of magnetic moments over length scales accessible to neutron SAS may be easily followed in (magnetically) inhomogeneous systems (giant clusters, superparamagnetism, spinglass behavior, magnetic nanoparticles, etc.) at various magnetic-field strengths. Löffler et al. (2005) investigated the magnetization of nanostructured Fe and Co by neutron SAS at variable magnetic field strength. **Figure 46** shows some results for nanostructured Fe. The enhancement of scattering along the magnetic field, at intermediate field strengths, is unexpected. Intergrain exchange coupling between neighboring (magnetically anisotropic) grains intervenes at small grain size (below the magnetic domain wall thickness of the bulk material). The random anisotropy model as applied to nanostructured materials

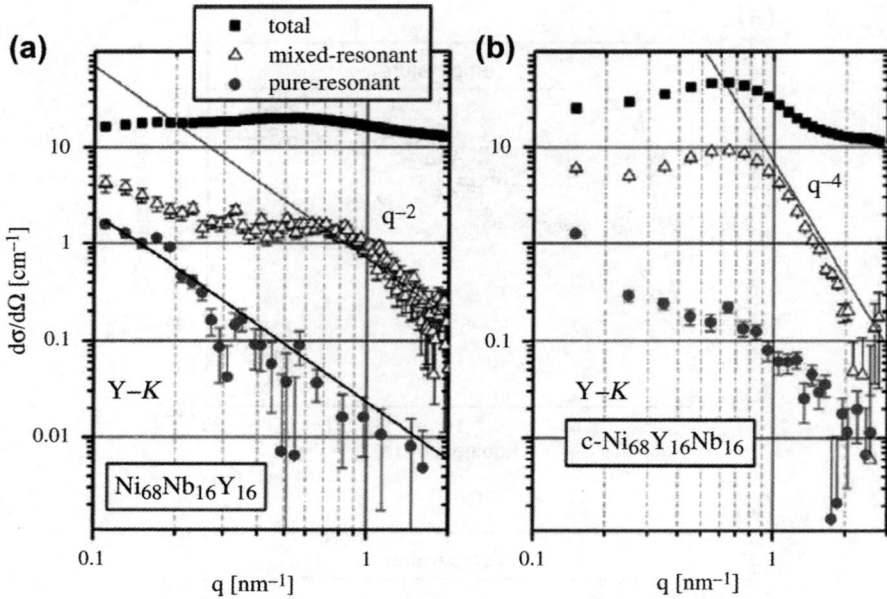

Figure 43 Anomalous small-angle scattering near the *K* edge of Y (17038 eV, Jusifa beamline at Hasylab, Hamburg) for (a) amorphous (melt spinning) and (b) partially crystallized Ni-16 at.% Nb-16 at.% Y. Squares: total scattering; triangles: separated scattering, obtained from the difference of measurements at two X-ray energies; this eliminates an isotropic background stemming mainly from the fluorescence of Ni; dots: pure resonant scattering of Y, obtained from two differences between measurements at three energies. The separated (mixed-resonant) scattering shows clearly how a κ^{-2} power at larger κ (correlation length ca. 122 Å) for the as-quenched sample turns onto a κ^{-4} power for the partially crystallized (mean crystallite size ca. 50 Å) sample. (Note that $q = 2\pi\kappa$.) From Goerigk and Mattern (2009). (For colour version of this figure, the reader is referred to the online version of this book.)

(see Herzer, 1992) serves as a basis to understand that magnetic domains extending over several grains may not align easily along the external field for intermediate field strength. More results on ferromagnetic nanostructured materials, SAS and micromagnetism have been summarized by Michels and Weissmüller (2008).

When cooling a sample from above the Curie temperature of a ferromagnet, critical magnetic SAS of neutrons yields immediate access to the magnetic correlation length (the neutron energy should be large enough to include all the allowed energy losses of this quasielastic scattering). Fischer et al. (2007) have used an aerodynamic (containerless) levitation technique combined with laser heating to record critical neutron SAS of a Co–Pd alloy. At high temperatures, Co and Pd show complete mutual solid solubility (f.c.c.); and the Curie temperature of Co-20 at.% Pd is close to 1000°C (the composition is of interest because signs of magnetic order have also been found in the undercooled liquid state, see Bührer et al., 2000). From Ornstein–Zernike plots (I^{-1} vs κ^2), Fischer et al. (2007) determined the magnetic correlation length ξ (of solid Co-20 at.% Pd) as a function of temperature up to about 1400°C and from a fit according to

$$\xi(t) = \xi_0 t^{-\nu},\tag{90}$$

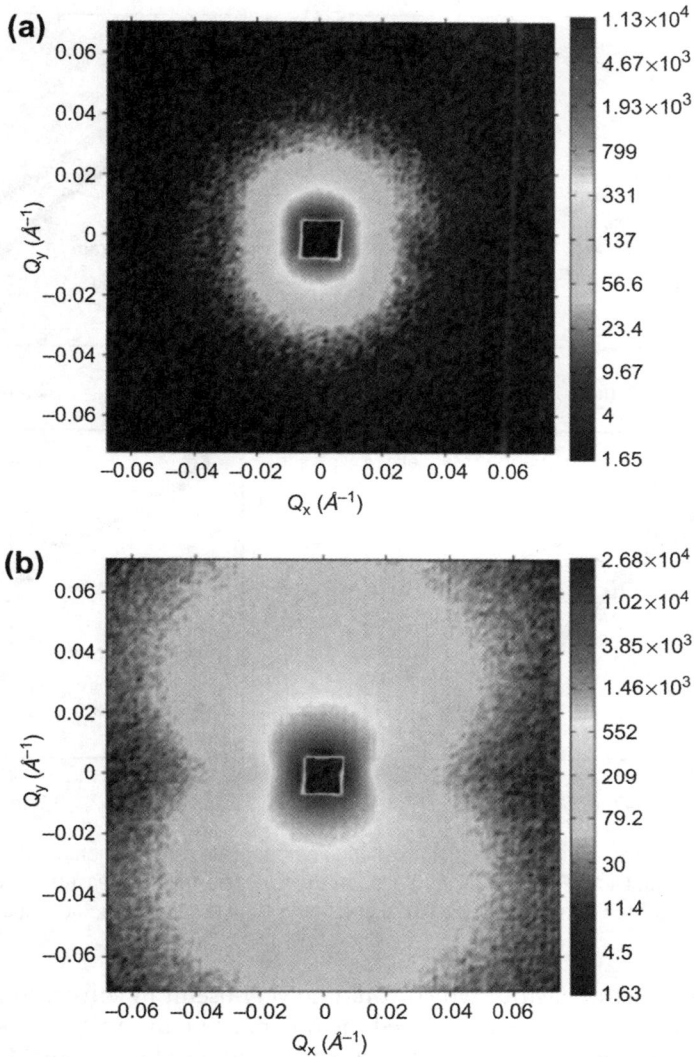

Figure 44 Small-angle neutron-scattering patterns of an Fe-1 at.% Cu alloy with 24% tensile deformation (a) before and (b) after 12 h of aging at 550°C. A magnetic field of 1.1 T was applied horizontally (along $Q_x = 2\pi\kappa_x$). From He et al. (2010). (For colour version of this figure, the reader is referred to the online version of this book.)

where $t = (T - T_c)/T_c$, a critical exponent $\nu = 0.76(5)$ resulted. This result is compatible with the value of 0.707 of the Heisenberg model. The critical temperature, with some uncertainty concerning the temperature measurement, was found to be about 1020°C.

Examples of work with polarized neutrons are still scarce, although there is much information to be gained [see the lines following eqn (31)]. Without polarization analysis, the difference between the scattering intensities of positive and negative polarization of the incident beam contains the product of

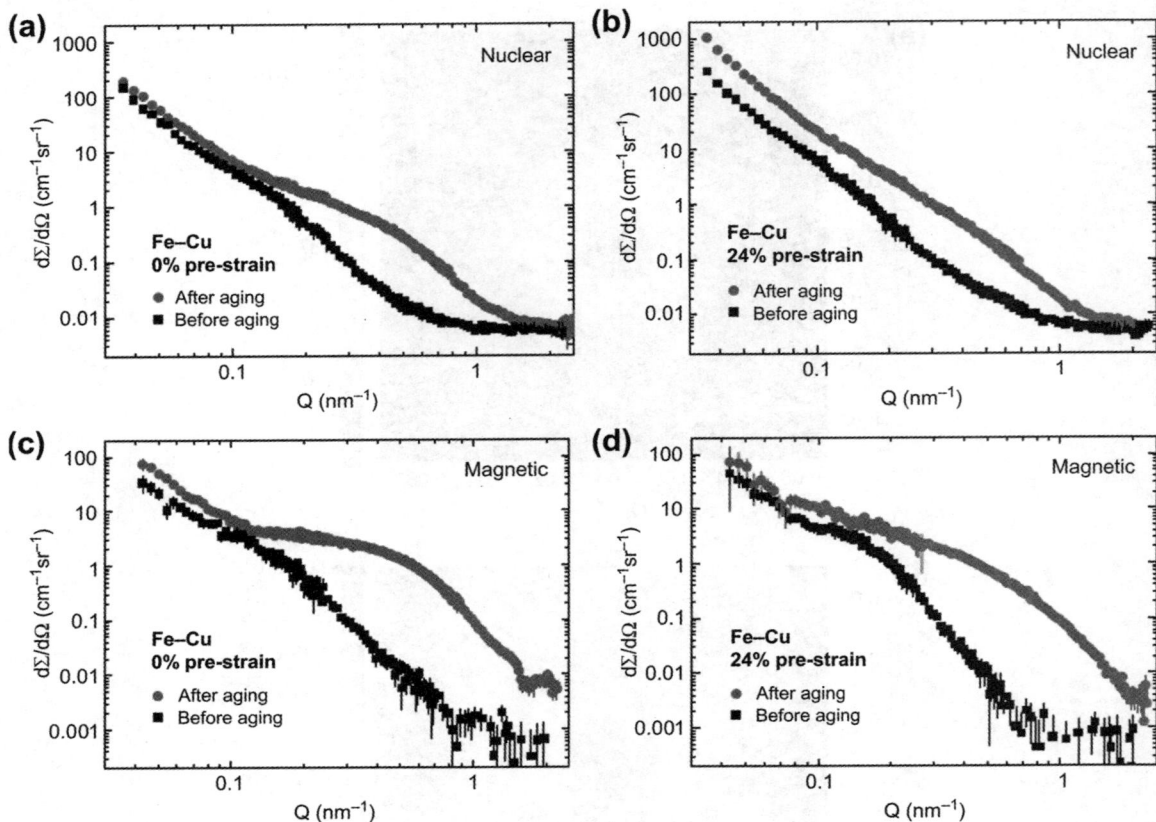

Figure 45 Nuclear and magnetic neutron small-angle scattering components as a function of the modulus of the scattering vector ($Q = 2\pi\kappa$) for the Fe–Cu alloy (**Figure 44**) with 0 and 24% tensile deformation measured at room temperature before and after aging at 550 °C for 12 h. From He et al. (2010). (For color version of this figure, the reader is referred to the online version of this book.)

nuclear and magnetic structure factors, which could be very useful in constructing the models of the scattering objects. Oku et al. (2009) have studied the polarized neutron SAS of spherical Fe-11 at.% N nanoparticles made from magnetite particles by reduction and nitration. **Figure 47** shows an example. The authors fit a model of a magnetic core (about 122 Å in diameter in this case) and a nonmagnetic shell (34 Å thick) consisting of Y and Al compounds used to protect the particles from oxidation and sintering. They work out the particle parameters for several cases including several degrees of (forced) oxidation.

Polarization analysis in SAS is more demanding, since the analyzer must cover a large angular range. Thus, while a supermirror cavity may be used for the incident beam, for polarization analysis, a ^{3}He spin-polarized analyzer was found to be more suitable and used at NIST beamline NG3 by Krycka et al. (2009). In addition to a clear separation of magnetic and nuclear scattering, a three-dimensional analysis of the magnetism of magnetite nanoparticles was possible.

Neutron depolarization can be monitored for the transmitted beam and is primarily due to magnetic inhomogeneities (domains, fluctuations) from the micrometer to millimeter range (depending on the κ-range covered by the spin analyzer/detector) and is thus complementary to SAS and, in an

(a)

(b)

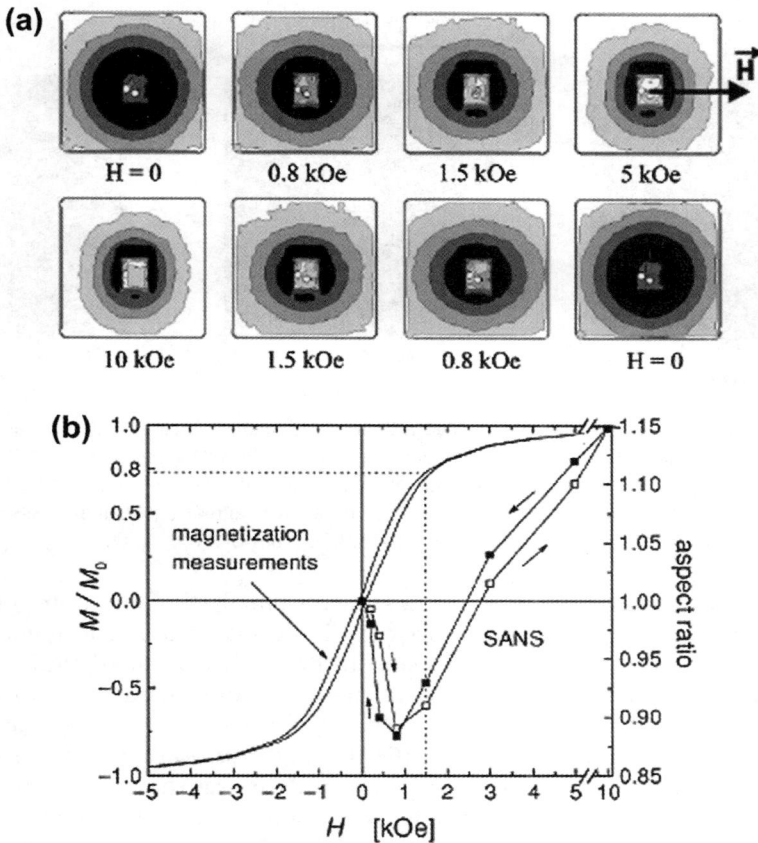

Figure 46 Neutron small-angle scattering (SANS) from a nanostructured Fe (120 Å grain size from SANS) sample. (a) Lines of equal intensity on a two-dimensional detector are shown for measurements with a magnetic field applied in the horizontal direction. Different field strengths were applied along the hysteresis loop shown in (b). The range of κ extends to about 2.5×10^{-3} Å^{-1}. The wavelength of the incident neutrons was 8.0(8) Å and the detector distance was 16 m. (b) Magnetic hysteresis loop at room temperature (M_0 is the saturation magnetization) and ratio of the vertical to the horizontal axis of the lines of equal intensity, taken at $\kappa = 8 \times 10^{-4}$ Å^{-1}. From Löffler et al. (2005).

overlapping κ-range, will yield similar results as (magnetic) ultraSAS [see Wagner and Bellmann (2007) for a comparison of both methods and further references].

13.5.3 Defects

Any local change in scattering length density will yield a SAS signal according to eqn (66). Therefore, even in chemically homogeneous materials, density changes of mesoscopic extent may be analyzed by SAS. In metallic materials, dislocations, agglomerates of point defects and, especially in nanostructured materials, grain boundaries may be candidates. Such objects may also be chemically "decorated" via segregation and precipitation. These effects are included in Section 13.5.2, with arrangements of scattering objects controlled by the underlying microstructure. Here, we are concerned with the "pure" defects.

Magnetic field vector

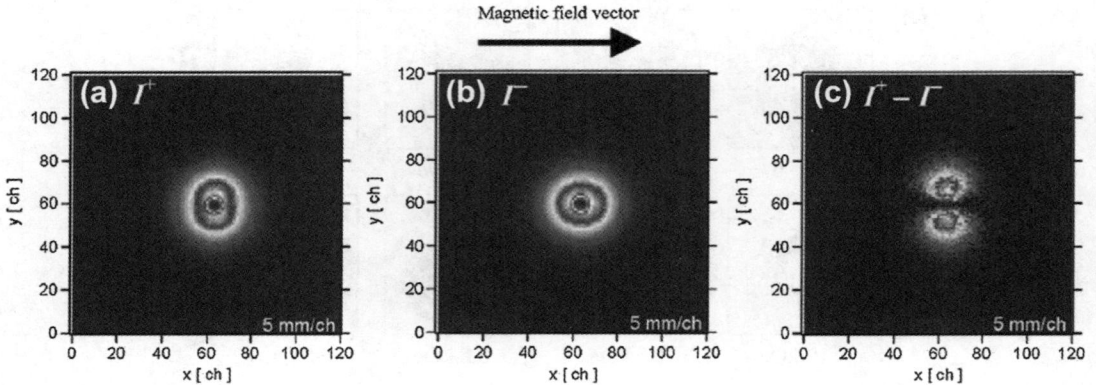

Figure 47   (a, b) Small-angle scattering patterns obtained with polarized neutrons of a sample with Fe–N nanoparticles of an average size of 190 Å. The wavelength of the neutrons was 6.5 Å ($\pm15\%$), the κ-range was explored to about 0.14 Å^{-1}, and a magnetic field of 1 T was applied in the horizontal direction perpendicular to the incident beam to obtain magnetic saturation. (a) Positive, (b) negative incident polarization. (c) The difference between I^+ and I^- shows the cross-term of magnetic and nuclear scattering which is proportional to $\sin^2\alpha$ (see Section 13.2.3). From Oku et al. (2009).

The local density changes around dislocations, known from elastic continuum theory, lead to very small scattering cross-sections. Since the presence of dislocations simultaneously increases the probability of double Bragg scattering if wavelengths below the Bragg cutoff are used, initial attempts to reveal X-ray SAS from dislocations failed. With long-wavelength neutrons, Atkinson and Lowde (1957) and Atkinson (1959) succeeded to find the correct intensities for deformed pure metals. This stimulated theoretical work, mostly on the scattering properties of single dislocations, dipoles and loops, while experimentally convincing experiments remained scarce. There was, however, experimental evidence that dislocation scattering is strongly anisotropic (within linear elasticity, screw dislocations do not scatter), since it vanishes if the scattering vector is parallel to the Burgers vector or the dislocation line. This early period has been reviewed by Schmatz (1975). The SAS cross-sections are so small that even for high dislocation densities, surface irregularities or a few large inclusions or pores may easily obscure the scattering pattern.

In magnetic crystals, magnetoelastic coupling leads to an additional SAS term (for unpolarized neutrons), and deformed single crystals of Ni (Anders et al., 1984) and Fe (Göltz et al., 1986) have been studied making use of the additional intensity and its symmetry properties near magnetic saturation. Despite the magnetic scattering, experiments of this type are very tedious and require a high incident neutron flux.

More recently, Thomson et al. (1999) and Long and Levine (2005) have extended theoretical considerations of SAS by dislocations to include dislocation walls. The authors developed and tested expressions for the ultrasmall-angle range and found that κ should be below 10^{-3} Å^{-1} for the relevant length scales. Several Al single crystals were deformed in situ in an X-ray ultrasmall-angle device at NIST, USA. With incident X-rays of 1.76 Å wavelength, several orientations were investigated for each crystal, avoiding the excitation of Bragg reflections for the SAS measurements. It was possible to extract the interface width of dislocation wall boundaries (decreasing from 1.1 μm to 400 Å with increasing strain; see Long and Levine, 2005 for further details).

The agglomeration of point defects should lead to more easily measurable SAS effects as the scattering contrast is much higher. The structure and annealing behavior of voids in Al single crystals, produced by irradiation with fast neutrons, present a classical example (Hendricks et al., 1977; Lindberg et al., 1977). X-rays, neutrons, transmission electron microscopy and positron annihilation were combined to obtain a rather complete picture of the microstructure of these samples. From the SAS point-of-view, the two-phase model could be employed, and no interparticle interference effects were visible at volume fractions of about 1% and void sizes of several 100 Å. The voids in irradiated Al showed faceting attributed to truncated octahedra with {111} faces. Voids formed in quenched and annealed β'-NiAl (B2 structure) single crystals of strictly stoichiometric composition showed faceting on {110} planes, forming a rhombic dodecahedron (Epperson et al., 1978).

The void concentrations in the studies mentioned above, about 0.1–1%, were easily measured in otherwise undamaged single crystals. Among the factors influencing mechanical properties of metals and alloys, void formation during fatigue or high-temperature deformation presents a major problem, and it is desirable to recognize void formation as early as possible. SAS is a useful method if other scattering contributions are not prohibitively large. This is, unfortunately, not always the case for technologically important alloys, since in their applications, high temperatures or radiation environment may induce various simultaneous changes of the microstructure. In ferritic/martensitic Cr (Mo) steels containing carbides, fracture is believed to be associated with the formation of voids at the matrix–particle interfaces (see Pan et al., 2010). In a scanning SAS experiment along the tensile axis during plastic deformation at room temperature, Pan et al. (2010) found an increase of SAS intensity with increasing strain and with decreasing distance to the final necking point for a model alloy Fe–9Cr–0.5C (mass-%). The scattering increment followed the Porod law [eqn (85); since the κ-range covered only rather large values, it can only be estimated that the scattering objects were larger than about 100 Å]. A plot of the scattering intensity vs. κ^{-4} yields a measure for the total surface area of the scattering objects (see **Figure** 48). Since other microstructural changes were minimal and the contrast

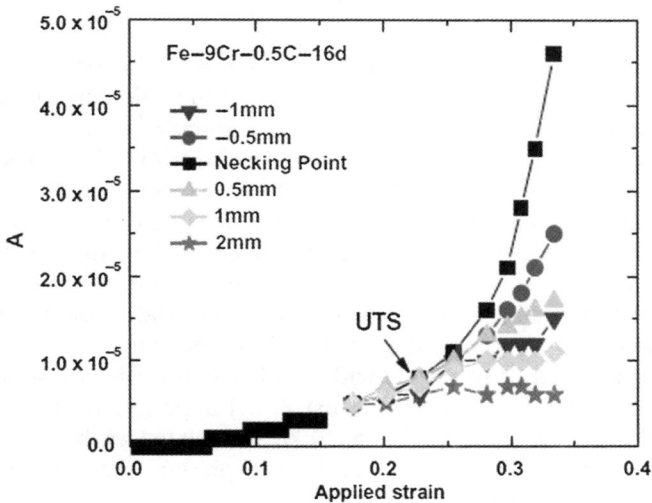

Figure 48   Porod constant 'A' in arbitrary units as a measure of the total surface area attributed to voids in an Fe–9Cr–0.5C alloy containing carbides (about 0.5 µm diameter), as a function of strain from X-ray SAS measurements with a 100 µm × 100 µm beam ($\lambda = 0.15$ Å, the tensile sample was 760 µm thick) at various positions along the axis of the sample (distances are given with respect to the ultimate necking point). From Pan et al. (2010). (For color version of this figure, the reader is referred to the online version of this book.)

for voids is large, the extra scattering was attributed with some confidence to voids. A similar case has been presented by Coppola et al. (2009) who investigated the microstructural effects of neutron irradiation in Eurofer97 steel (9Cr, 0.01C, 1W, 0.2V) by neutron SAS and found from the magnetic and nonmagnetic part of the scattering that at 300°C, upon an increase of the radiation dose from 2.5 dpa to 8.4 dpa, nonmagnetic defects (in the size range of 20–30 Å) grow, increasing in volume fraction by a factor of 2, while their average size remains nearly unchanged. Voids have also been seen in transmission electron microscopy, but the authors point out that the possibility of helium-filled bubbles remains to be checked. Grain boundary cavitation during creep is another source of failure. Bouchard et al. (2002) investigated various creep specimens of type-316H stainless pressure vessel steel by neutron SAS. The very large size distribution (up to radii of 2 μm) required the use of standard and ultrasmall angle scattering devices to obtain the essential scattering contributions. The evolution of creep damage could thus be more reliably described.

13.5.4   Special Topics

13.5.4.1   *Large Lattice Constants*
A special case of small-κ scattering is Bragg diffraction from structures with large lattice constants or large-scale periodicity, including magnetic structures. The flux line lattice of type II superconductors with a lattice constant typically around 1000 Å and first revealed by neutron scattering by Cribier et al. (1964) has subsequently been studied in great detail for many different materials and conditions. In niobium, not only morphological changes, but also the dynamics of flux-line lattices (by stroboscopic neutron SAS), have become accessible by sophisticated experimentation (see Mühlbauer et al., 2011 and references therein). Flux line lattices have also been found (since Forgan et al., 1990) and studied in several high-temperature superconductors (see recent work and references by Li et al., 2011).

13.5.4.2   *Grazing Incidence*
In grazing incidence, specular reflection (of X-rays and neutrons) has been known and used to study surface roughness, surface layers and multilayers for a long time. Such investigations are also essential in the development of X-ray and neutron mirrors. Owing to intense neutron beams at high-flux reactors and in more recent years, the improved collimation at synchrotron radiation sources, the off-specular SAS at and near grazing incidence (pioneered by Levine et al., 1989; see also Naudon et al., 1994) has seen a lively development. It serves to identify agglomerates or pores and any nanoscale objects of suitable contrast at and near surfaces, including time-resolved growth. The field has recently been reviewed by Sinha (2009). As an example, **Figures 49** and **50** show the experimental scheme and some results of a study by Renaud et al. (2003) of the growth of Pd islands on a MgO surface (supporting online material includes a movie showing the growth of the SAS signal with increasing coverage of the surface.) The accurate evaluation of scattering with incidence near the critical angle of total reflection must often be extended beyond the kinematic (first Born) approximation. The distorted wave Born approximation is useful in many cases (see Sinha, 2009, and references therein); with densely arranged three-dimensional objects on a flat surface—as in the present example—multiple interfaces are involved and require further extensions (see Lazzari et al., 2007).

13.5.4.3   *Refraction and Multiple Scattering*
At small angles, multiple refraction in two-phase systems may obscure scattering. An experimental separation of refraction and scattering is in principle possible by varying the incident wavelength. In dense two-phase systems (volume fractions of a few percent may suffice), multiple scattering must be

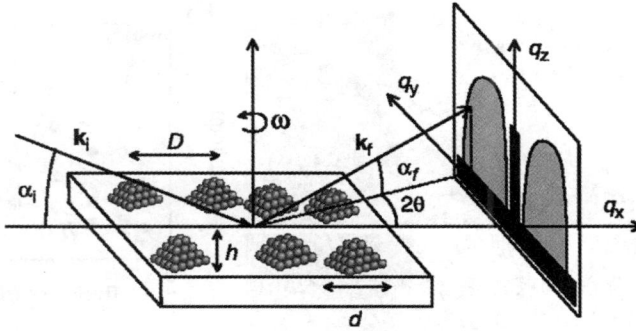

Figure 49 Schematic view of grazing-incidence small-angle scattering as described by Renaud et al. (2003). The wave vector of the incident beam is in the *xz* plane. From the angles α_i, α_f, and 2θ, the three components of the scattering vector κ (**q** in the figure) can be calculated. The sample (here MgO with Pd nanodeposits) can be rotated (angle ω) around its surface normal. A beam stop protects the two-dimensional detector from the direct and reflected beams. From Renaud et al. (2003).

considered, especially with long wavelengths (neutron SAS). Berk and Hardman-Rhyne (1985, 1988) have considered both, multiple scattering and refraction and developed a multiple-scattering treatment of neutron SAS data from dense powders and porous systems. Allen et al. (2001), based on these ideas, analyzed the neutron SAS of yttria-stabilized zirconia plasma-sprayed thermal barrier coatings. The two-dimensional multiple scattering and in particular the Porod regime (where anisometric shapes are clearly revealed) showed anisotropic intensity distributions depending on the angle between spray direction and incident beam. Contributions from three different types of pores, intrasplat cracks, intersplat lamellar pores and globular pores, were separated. For a study of thermally sprayed metallic ("NiCrAlY") deposits, see Keller et al. (2002).

13.5.4.4 Radiography
A very simple method of monitoring changes in the microstructure of a sample is to measure its transmission. As indicated by eqn (25), the transmission is a function of the total removal cross-section, i.e. scattering into any direction outside the range covered by the transmission monitor appears like additional absorption. With appropriate local resolution, this may be used for different types of mapping (radiography). With neutrons, large structures may be studied, especially for the distribution of hydrogen-containing matter.

13.5.4.5 Absorption Spectroscopies
X-ray absorption spectroscopies (not only energy-resolved transmission, but also the analysis of fluorescence and emitted electrons) can now be performed with great accuracy. The EXAFS covers a range of up to about 1000 keV above the absorption edge (e.g. the K edge, see **Figure 2**). The final state of the photoelectron of the excited atom is modified by backscattered wavelets from the surrounding atoms, which leads to modulations of the absorption coefficient (and connected emissions), see, e.g. Stern (1974). One defines "the EXAFS" as

$$\chi(E) = \frac{\mu(E) - \mu_0(E)}{\mu_0(E_0)}, \tag{91}$$

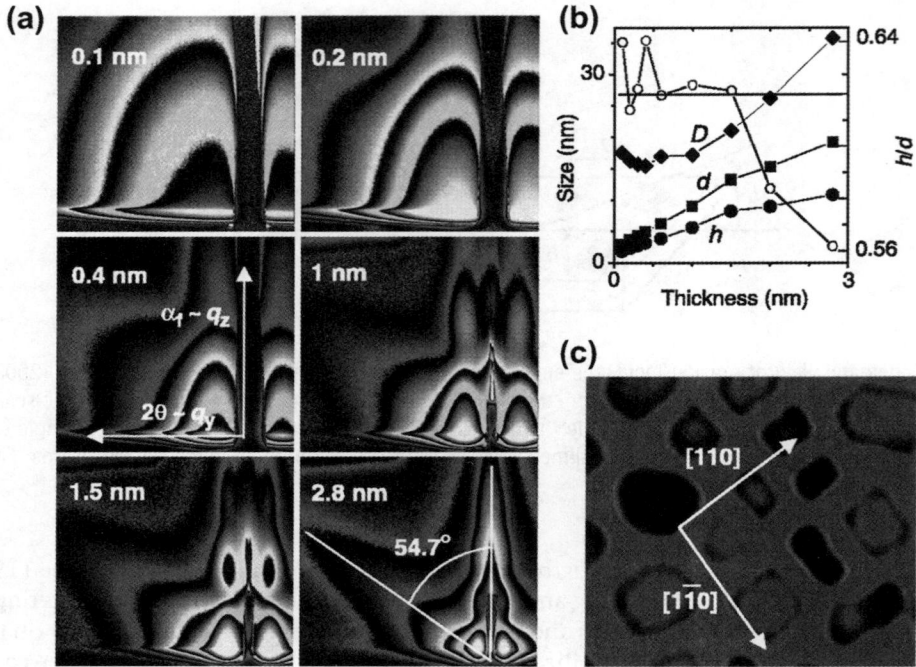

Figure 50 In situ grazing-incidence X-ray small-angle scattering measurements during Pd deposition on MgO(001) at 650 K (Renaud et al., 2003). (a) Scattering patterns patterns for different thicknesses (equivalent coverages), with the incident beam along the MgO[110] direction. The vertical direction is perpendicular to the sample surface, and the horizontal direction is parallel. The intensity scale is logarithmic. Colors separated by black shades correspond to orders of magnitude. The first images were recorded in a few seconds with a large beam stop, whereas the last were recorded with a narrower beam stop in some tens of milliseconds. (b) Left scale: the evolution of island spacing D (solid diamonds), lateral size d (solid squares), and height h (solid circles) with equivalent thickness, as deduced from the SAS data, with a truncated octahedron as the average island shape. Right scale: evolution of the aspect ratio h/d (open circles). (c) Ex situ transmission electron microscopy image (250 nm by 250 nm) of the last deposit of a thickness of 2.8 nm, after carbon coating and substrate dissolution. Polydispersity increases with thickness. From Renaud et al. (2003). (For color version of this figure, the reader is referred to the online version of this book.)

where $\mu(E)$ is commonly used for the linear absorption coefficient [$=\sum_t$ in eqn (25)], $\mu_0(E)$ is a smooth "bare-atom" background (to be fitted) and $\mu_0(E_0)$ is the absorption edge value (with E_0 adjustable, within limits).

As a function of k (assuming s-waves) we write

$$\chi(k) = \sum_j \frac{N_j f_j(k) e^{-2R_j/\lambda(k)} e^{-8\pi^2 k^2 \langle u_j^2 \rangle}}{2\pi k R_j^2} \sin\left[4\pi k R_j + \delta_j(k)\right], \qquad (92)$$

where the N_j are the neighbors at sites j, $f_j(k)$ are the backscattering amplitudes, R_j are the positions of neighboring atoms, $\lambda(k)$ is the electron mean free path, u_j denotes the thermal and static displacements, and δ_j are the phase shifts. An additional factor is needed to allow for the effect of relaxations around the absorbing atom. In the common EXAFS regime, $\lambda(k)$ is so small that only contributions from nearest and next-nearest neighbors matter. For very small $E - E_0 (<5 \text{ eV})$, the range is larger and not only more neighbors, but also multiple scattering events have to be included in a more appropriate modeling (this is the range of XANES = X-ray absorption near-edge structure or near-edge X-ray absorption fine structure). Many further details can be found in the book by Als-Nielsen and McMorrow (2001), and the web site http://xafs.org hosts and gives links to many tutorials and overviews.

Apart from being species dependent and sensitive to the oxidation state of the excited atom, EXAFS is often able to identify nearest-neighbor species and coordination. After Fourier transformation of $\chi(k)k^n$ ($n = 2$, 3—this multiplication serves to amplify the oscillation), filters may be set in the resulting radial function to obtain a back-transformation for the first or second shell (and more if possible). With calculated values for $f(k)$, $\delta(k)$ and further fitting, coordination numbers (important especially for noncrystalline materials), distances, distance distributions, atomic species, Debye–Waller factors etc. can be obtained. At low temperatures, nearest-neighbor distances may be determined with a precision of 1 pm. The potential and methodology to obtain coordination numbers and displacements in several crystalline Al–Cu alloys have been analyzed in the classical work of Lengeler and Eisenberger (1980). A recent example concerns the short-range ordering of Cr and N in nitrogen-expanded austenitic steel ASI 316 (Oddershede et al., 2010). Single-phase austenitic foils of 7.5 μm thickness were nitrided at 718 K to a nitrogen-per-metal-atom ratio $y = 0.51$ (without the formation of the CrN phase) and EXAFS measurements were taken at MAX-lab, Lund, Sweden for the unnitrided, the nitrided, and the denitrided ($y = 0.14$) states, at the Fe, Ni, and Cr K-edges.

Figure 51(a) shows the Fourier-transformed results (arbitrary scaling). They show that the local environment around Cr has considerably changed upon nitriding and that these changes are only partially reversed in the denitrided state. Model calculations involving a locally adjustable cubic lattice parameter, the coordination number of nitrogen, the edge energy E_0 and the Debye–Waller factor were fitted to all radial functions. The results for Cr are shown in **Figure 51**(b). These results show that there is a strong local interaction of Cr and N with a correspondingly large local expansion, but without the formation of a defect of lower symmetry. The nitrogen remaining in the denitrided state is trapped by Cr (with a coordination number of about 3 which implies that on average 3 Cr atoms share each N atom as nearest neighbor).

For nanocrystalline and noncrystalline materials, EXAFS as a sensitive local probe is an ideal complementary tool to X-ray and neutron diffraction. The local atomic arrangements and their evolution are important for the properties of metallic glasses where topological and chemical SRO may be separated by EXAFS. Solute-centered atomic clusters (see Miracle, 2004) of various types can now be modeled and fitted to diffraction, EXAFS, and XANES results in combination with other structural and physical properties. As a recent example, **Figure 52** from the work of Antonowicz et al. (2012) shows a comparison of experimental XANES and EXAFS results for amorphous Cu-35 at.% Zr and theoretical values obtained from ab initio molecular dynamics calculations for several interconnected superclusters based on icosahedral clusters. The Cu-centered icosahedral cluster Cu_8Zr_5 shown in the figure is a dominant, but not the only, short-range ordered motif in the population of superclusters. Topological ordering seems to be more decisive in this alloy than chemical ordering tendencies.

To explore the local environment in a liquid, Jacobs and Egry (1999) combined electromagnetic levitation with the EXAFS method and investigated the SRO of a metallic melt (Co-20 at.% Pd) in a wide temperature range, including deeply undercooled (more than 300 K) states. Solid and liquid states could be compared.

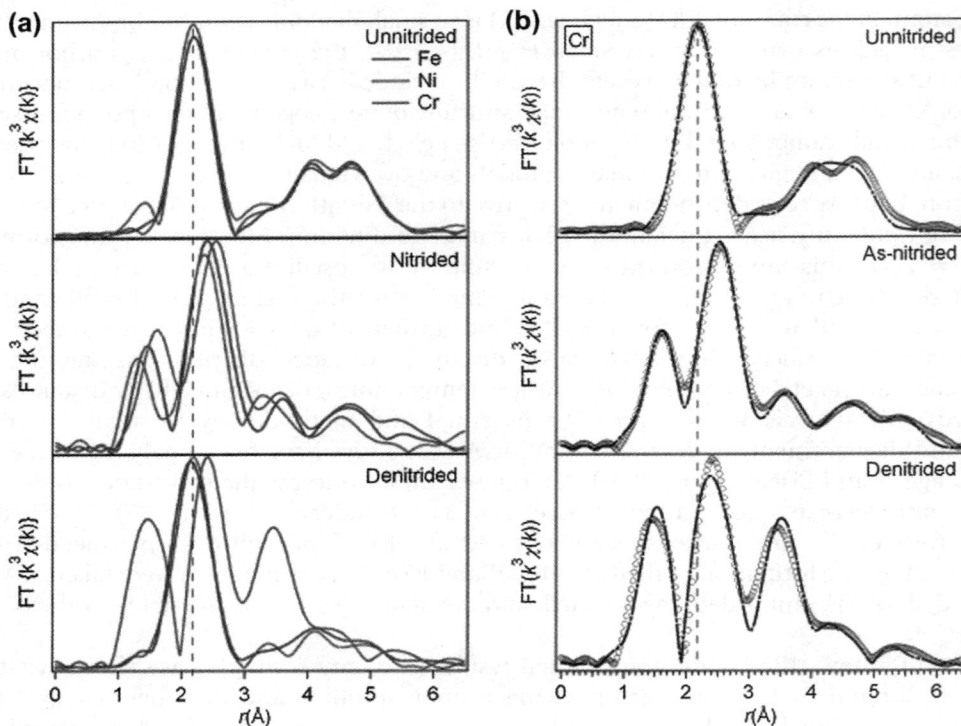

Figure 51   (a) EXAFS results shown as Fourier transforms (FT) of $k^3\chi(k)$ as a function of distance (r) from the reference atom (Cr, Fe, Ni, K-edges) in austenitic nitrogen-stabilized steel in the unnitrided, as-nitrided and denitrided states. (b) Experimental (circles) and modeled (lines, see text and original for details) radial functions for Cr in the unnitrided, as-nitrided and denitrided samples. From Oddershede et al. (2010). (For color version of this figure, the reader is referred to the online version of this book.)

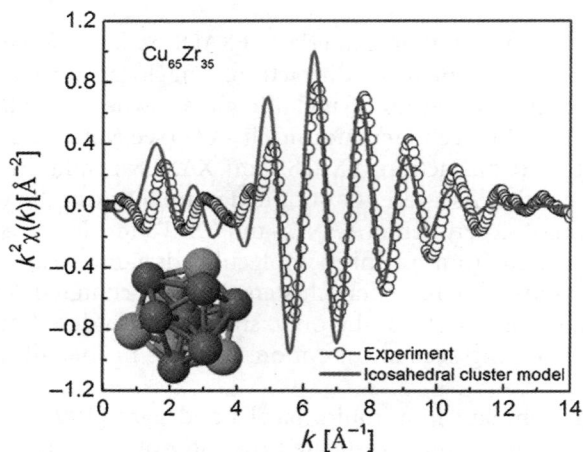

Figure 52   Experimental EXAFS $k^2\chi(k)$(symbols, a factor 2π is included in k) and theoretical function (continuous line) calculated based on molecular-dynamics results for the icosahedron-like Cu-centered cluster (Cu_8Zr_5) shown in the inset. From Antonowicz et al. (2012). (For color version of this figure, the reader is referred to the online version of this book.)

13.5.4.6 Coherent X-rays

At current third-generation synchrotron radiation sources and future X-ray free electron lasers, intense coherent X-ray beams are becoming available. Several methods known from optics can be transposed into the X-ray regime and will provide new tools also for physical metallurgy. X-ray diffraction imaging (Sayre, 1952; Miao et al., 1999) also works at small angles and allows nanoscale structures to be imaged. Zhang et al. (2011) describe the development of ultrasmall-angle X-ray scattering–X-ray photon correlation spectroscopy in Bonse–Hart geometry that will close the gap between optical and classical X-ray photon correlation spectroscopy (especially for metallic samples). Thus, static and dynamic properties at mesoscopic scales can be revealed.

13.6 Energy Transfers

As mentioned in Section 13.2.4, inelastic neutron scattering (coherent and incoherent), owing to the comparable energy ranges of thermal (including cold and hot) neutrons and many excitations in materials, is a versatile method to study phonons [eqns (36) and (37)] whereas incoherent (as well as coherent) energy-resolved quasielastic scattering of neutrons reveals diffusive motion. A few examples will now be given. The relative energy transfer between excitations in solids and X-rays is much smaller. Thermal diffuse scattering is thus usually not separable from elastic scattering, but the high energy resolution available at third-generation synchrotron radiation X-ray facilities and the different selection rules for inelastic scattering of neutrons and X-rays may call for some X-ray studies as well (with an energy resolution of 2–3 meV; see, e.g. Scopigno et al., 2006). The use of coherent X-rays at advanced sources is offering new possibilities to study diffusive motion.

13.6.1 Phonons in Real Crystals

The dynamic properties of crystals are very sensitive to lattice defects and change dramatically near phase transitions. Early work on phonons and defects has been discussed by Nicklow (1979), and on phonons and phase transformations, e.g. by Currat and Pynn (1979); Schober and Petry (1993) and Petry (1995). A recent survey by Fultz (2010) may serve as an introduction to the use of inelastic neutron scattering in the context of vibrational thermodynamics of materials.

If scattering events are to be tracked down as illustrated in **Figure 6**, single crystals are required (coherent inelastic scattering). For a determination of the phonon density of states, the complete spectrum of energy transfers is measured (incoherent and coherent inelastic scattering, usually by time-of-flight methods). Since changes in the phonon density of states of a material reveal anharmonic contributions to the entropy, such measurements help to unravel the sometimes complex interplay of different entropy contributions. For example, Kresch et al. (2007) have studied the temperature dependence of the phonon density of states of pure Ni. **Figure 53** shows results for temperatures up to 1275 K. Detailed analysis and modeling show that the longitudinal nearest-neighbor force constant decreases monotonically with increasing temperature. Decreases of the nearest-neighbor forces turn out to be mainly responsible for the shift of the density of states to lower energy (see **Figure 53**). The observed broadening was attributed to phonon–phonon interactions. It was possible from these measurements to estimate the entropy contribution from anharmonicity which in turn sets limits upon the value of the magnetic entropy part (for details, see Kresch et al., 2007). Similar work on the austenitic and martensitic phases of Fe-29 at.% Ni (Delaire et al., 2005) shows a stiffening of the martensite phonons and a shift to higher energies for transverse and longitudinal phonons (there is no elastic softening in these transitions). If components in an alloy have different phonon scattering cross-sections, they are weighted differently in the neutron scattering spectra. After corrections for this effect, it

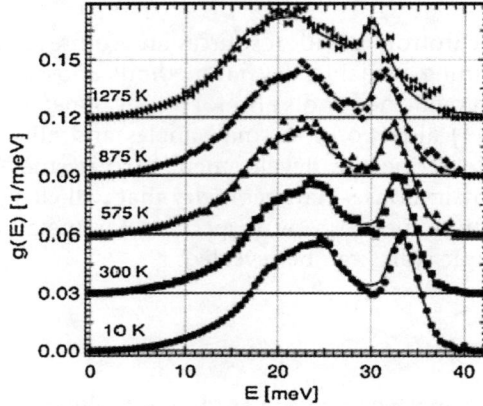

Figure 53 Phonon density of states for nickel at temperatures indicated. Markers are experimental data, and lines are fits. An increase in phonon lifetime broadening and the shifting of modes to lower energies with increasing temperature is evident. The curves are successively offset by 0.03 meV^{-1}. From Kresch et al. (2007).

still appears that the martensite phonon density of states has more low-energy modes than austenite (Delaire et al., 2005).

Crystallite size is another parameter influencing the phonon density of states, especially in the nanosize regime. Results obtained by Trampenau et al. (1995) for nanocrystalline Ni (average particle size 10 nm) are shown in **Figure 54**. Upon compaction, there is a strong change due to the grain boundaries (**Figure 54**(a)). The non-compacted powder shows only a moderate enhancement of the density of states at low energies (**Figure 54**(b vs. c)). This suggests a separation into two components (**Figure 54**(d)). While this Ni sample was prepared from single nano-particles, Frase et al. (1998)

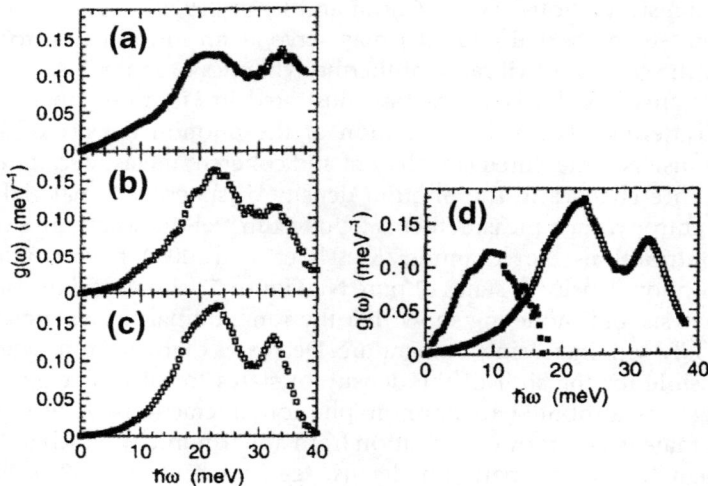

Figure 54 Phonon density of states of (a) nanocrystalline Ni, (b) crystalline Ni and (c) coarse-grained Ni, (d) of the boundary component (full squares, scaled by a factor of 8) and the crystalline part (open squares). From Trampenau et al. (1995).

investigated ball-milled nanostructured Ni_3Fe powder. **Figure 55** shows a comparison of the as-ball-milled material (submicrometer-size particles with nanograins of about 6 nm) with the same material after annealing at 600°C. The authors calculated the as milled density curve by convoluting each phonon frequency of the spectrum after annealing with a damped harmonic oscillator function (two examples are shown in the center of the plot) with the only adjustable parameter being the oscillator quality factor. This work also reports that the grain-boundary component of the phonon density of states is very sensitive to details of the grain-boundary structure. Already mild anneals and also an initial cool down to 10 K reduced the broadening of the spectra. The authors therefore attribute the broadening to the unrelaxed parts of the grain boundaries.

A study of the phonon density of states of hydrogen trapped at dislocations in deformed Pd–H by Heuser et al. (2008) shows different vibrational features at 4 K and 295 K. At low temperatures, the spectrum indicates the presence of the β-hydride phase whereas at room temperature, the density of states of the α-Pd-H solid solution is found. This result shows that an α → β phase transition takes place in the strained environment of the dislocations.

Measuring phonon dispersion requires scans of energy transfer ($h\nu$) at constant κ or κ scans at constant energy transfer. This is a traditional domain of neutron scattering. Although each available and experimentally practicable crystal has been characterized by its phonon dispersion laws, experiments with metallic crystals involving defects or phase transitions are relatively scarce. A popular approach is to look for phonon softening in the vicinity of displacive phase transformations. Low-energy phonons would soften where elastic constants also show a softening (static instability), e.g. in Al5 compounds.

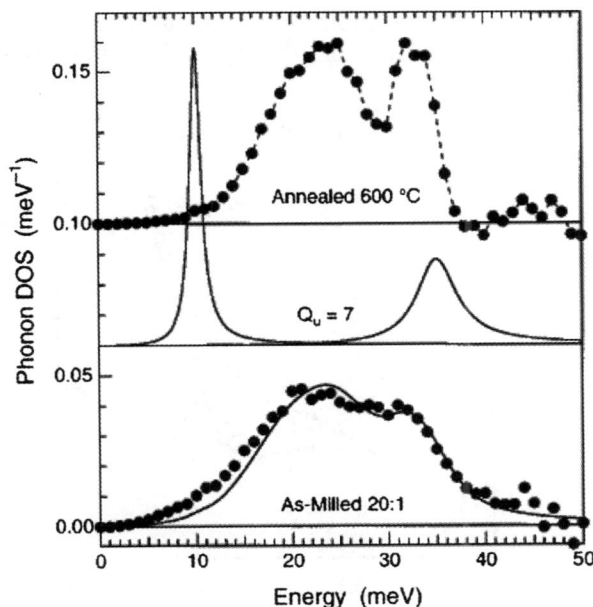

Figure 55 Experimental phonon density of states (full circles) for as-milled powder of Ni_3Fe (bottom) and the powder annealed at 600°C (top). The solid curve at the bottom is the result of the convolution of the experimental data from the annealed sample with a damped harmonic oscillator function (quality factor 7). The characteristic shape of the oscillator function is shown in the center of the figure for resonances at 10 and 35 meV. From Frase et al. (1998).

Such instabilities are expected for pure b.c.c. metals undergoing a structural transformation. For the group-IV metals, the transition temperatures lie far above room temperature, and if a single crystal is to be studied, it must always be kept above the transition. In a specially designed furnace, crystals of b.c.c Ti, Zr, and Hf have been grown and investigated at high temperatures by neutron inelastic scattering (see Petry, 1995; and references therein). The phonon branches show dynamical (but no static) precursor effects for the displacements necessary to reach the h.d.p. α-phase and the ω-phase (found under pressure or by alloying these metals).

In many cases, special phonons of a given wavelength soften as a consequence of Kohn anomalies (see Kohn, 1959). In two and three dimensions, Kohn anomalies, i.e. singularities in the phonon dispersion curves due to the abrupt change of the screening response of conduction electrons when wave vectors at the Fermi surface are involved, depend on the topology of the Fermi surface. Sections of low curvature (flat pieces) are particularly effective. A structural phase transformation may thus be induced by the condensation ("softening") of phonons of the appropriate wave vectors to create the new structure. An instructive case is β′-NiAl which shows a broad shallow dip in the transverse acoustic phonon branch with [110] propagation and [1$\overline{1}$0] polarization (see Mostoller et al., 1989) for the stoichiometric composition. Upon cooling, these phonon modes do not condense, and there is no structural transition. For the off-stoichiometric composition Ni-37.5 at.% Al, a martensitic phase transformation occurs at about 80 K [the martensitic start temperature M_s varies from 0 K to 600 K, approximately linearly with off-stoichiometry between about 60 at.% and 68 at.% Ni, see Ochiai and Ueno (1988)]. The transverse phonon branch shows a deeper dip when the transformation temperature is approached (Tanner et al., 1990), as shown in **Figure 56**. The transformation sets in before the condensation is complete (the corresponding elastic shear constant C' is also somewhat reduced but not to zero), indicating that the transformation must be triggered at strain defects, e.g. dislocations (Tanner et al., 1990). Application of an external stress leads to a shift

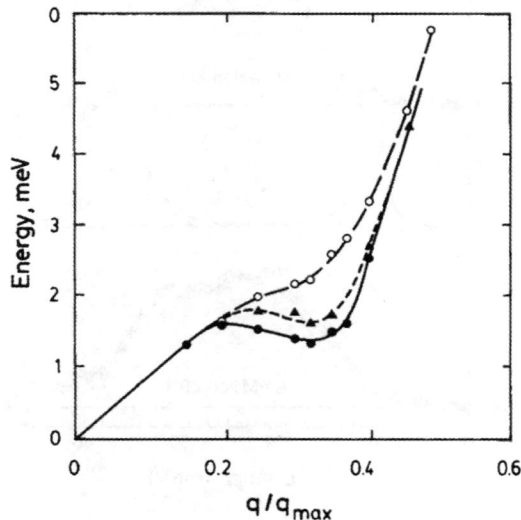

Figure 56   Neutron scattering results (energy transfer $E = h\nu$) for the transverse acoustic phonon branch in [110] direction ([1$\overline{1}$0] polarization) of Ni-37.5 at.% Al (B2 structure) at 290 K (open circles, 150 K triangles) and 85 K (full circles). After Tanner et al. (1990).

of the dip in the dispersion curve (Shapiro et al., 1993). Zhao and Harmon (1992) have calculated the band structure of NiAl around the stoichiometric composition and related the martensitic transformation to Fermi nesting and strong electron–phonon interactions; local variations in concentration and thus in order parameter could trigger the transformation. The subject has since been taken up by several groups (see, e.g. Isaeva et al., 2004; and references therein). Strong Fermi nesting corresponding to the wave vector of phonon softening has been confirmed experimentally by Dugdale et al. (2006) for Ni-38 at.% Al based on Compton scattering. In contrast to NiAl, the isostructural shape-memory alloy NiTi shows a static instability when M_s is approached, and the same phonon branch collapses entirely (see Tietze et al., 1984; Manley et al., 2008). Other examples of softening of the same transverse acoustic branch are the Fe-rich Invar alloys Fe–Ni, Fe–Pd, Fe–Pt (see, e.g. Kästner et al., 1999). Recently revived interest in "superelastic" Heusler alloys (for their magnetic shape memory properties) has led to a number of lattice dynamics studies including neutron scattering, e.g. on Ni_2MnGa (Shapiro et al., 2007), Ni_2MnAl (Mehaddene et al., 2008), and Ni-36 at.% Mn-16 at.% In (Moya et al., 2009).

As an example for defect-induced phonon changes, **Figure 57** shows the frequency distribution for several phonon wave vectors in a Cu single crystal irradiated with thermal and fast neutrons to produce Frenkel pairs (concentration about 1.3×10^{-4}, Nicklow, 1979). The measured phonon peaks are shifted relative to those measured for defect-free Cu, and an additional component on the high-frequency side occurs. Subsequent annealing results in an elimination of the peak shifts (72 K), but the additional high-frequency structure is only removed at 800 K. Part of the results can be explained by a resonant coupling of phonons to the librational modes of the split interstitial, but other defects, possibly small vacancy clusters, also contribute (see Dederichs and Zeller, 1979; for details).

Larger interstitial defect concentrations are present in metal-hydrogen systems (for coherent neutron scattering, H is replaced by D to avoid the high incoherent background of H), and local modes have frequently been reported in the literature.

We close this subsection by referring to a synchrotron X-ray study of high-frequency phonons in a bulk metallic glass (Serrano et al., 2010). These results may be significant for the understanding of brittleness in these materials. **Figure 58** displays selected inelastic X-ray scattering spectra taken with an energy resolution of 3 meV at different values of the scattering vectors κ. Excitations are observed at positive and negative energy as satellites of the central elastic line, corresponding to the creation and annihilation of longitudinal acoustic modes. The energies of the excitations increase linearly with increasing κ at low values of κ and start decreasing near q $(=2\pi\kappa)=14$ nm^{-1}. The dispersion relation thus resembles that of a crystal, if half the distance to the first maximum in the diffraction pattern (at $2\pi\kappa = 28.47$ nm^{-1}) is taken as defining the size of the Brillouin zone.

13.6.2 Diffusive Motion

Quasielastic neutron scattering as a means of studying atomic motion (and low-energy excitations) has been used since the 1960s (see Springer, 1977; Sköld et al., 1979 for an introduction and surveys on hydrogen diffusion in solids). A recent tutorial introduction by Embs et al. (2010) gives basic information. An extensive account of the field can be found in the book by Hempelmann (2000). Incoherent quasielastic scattering is somewhat simpler to interpret, but coherent quasielastic scattering may also be evaluated (see Springer, 1977; Lovesey, 1984; Petry et al., 1991). The availability of backscattering and spin-echo spectrometry (Mezei, 1980; Mezei et al., 2003) allows changes of neutron energies to be detected between 10^{-5} and 10^{-9} eV, corresponding to atomic motion on a time scale of 10^{-11} and 10^{-7} s, or diffusion coefficients in the range of 10^{-7} to 10^{-11} cm^2 s^{-1}.

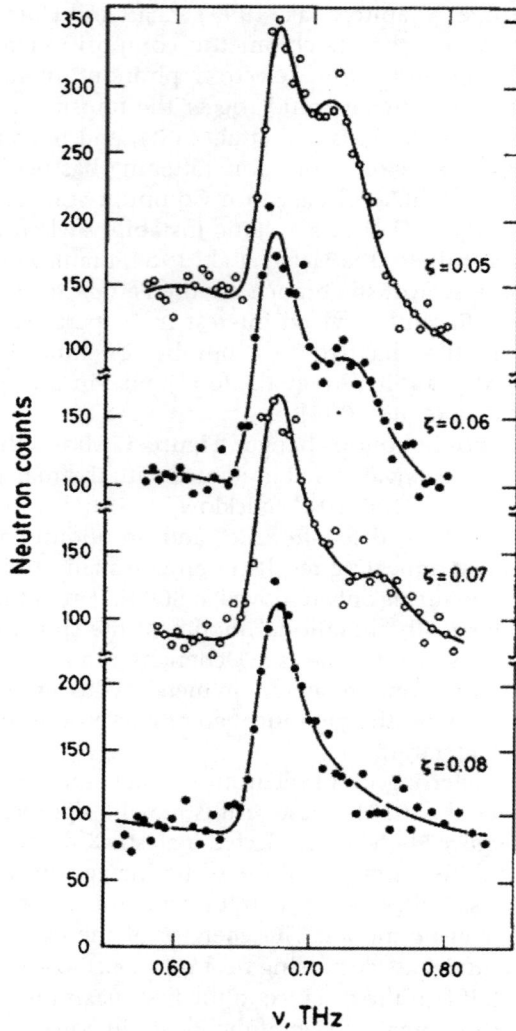

Figure 57 The frequency distributions of neutrons scattered from irradiated Cu at 4 K for $\mathbf{q} = (0, 0, \varsigma)/a$. Open and solid circles are experimental data, the lines are fits including resonant coupling of librational modes of the split interstitial. After Nicklow et al. (1979).

Hydrogen with its large incoherent scattering cross-section is an ideal candidate for such studies. A recent example is the work by Skripov et al. (2011) on coarse-grained and nanostructured $ZrCr_2H_3$, a potential hydrogen storage material, for which the grain-size dependence of absorption and desorption of hydrogen is of interest. **Figure 59**(a) shows a typical quasielastic spectrum and its reconstruction from three components, the elastic line given by the resolution function of the instrument, and two Lorentzians (although one was enough at lower temperatures, a second quasielastic component was required to fully fit the data at higher temperatures). In **Figure 59**(b), the half-width of the narrow component is shown as a function of κ (in the Figure, $Q = 2\pi\kappa$) for the

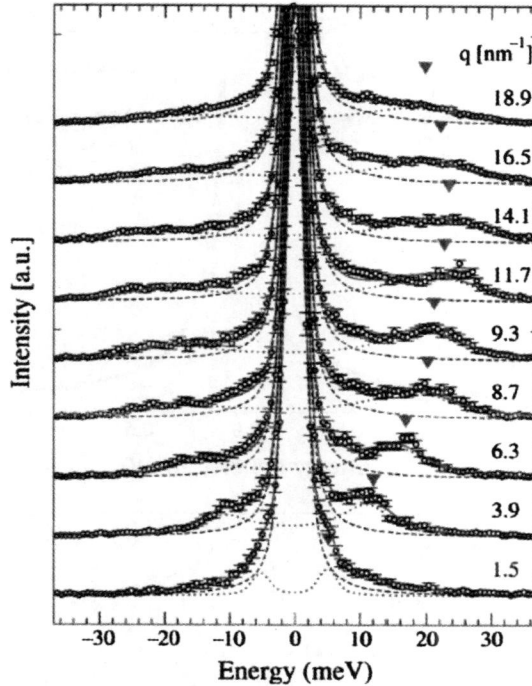

Figure 58 Inelastic X-ray scattering spectra for selected values of the scattering vector. The inverted triangles indicate the maximum of the longitudinal current correlation function in the damped harmonic oscillator model, the dashed and dotted curves show the elastic and inelastic contributions, and the solid (light gray) curves correspond to the best fit to the spectra. From Serrano et al. (2010).

coarse-grained material. The curves are fits according to the very frequently adequate jump-diffusion model of Chudley and Elliott (1961), that gives the half-width Γ as a function of κ (for statistically random orientations);

$$\Gamma(\kappa) = \frac{h}{2\pi\tau_d} \left(1 - \frac{\sin(2\pi\kappa L)}{2\pi\kappa L}\right) \quad (93)$$

Here, L is the jump distance and τ_d^{-1} the jump frequency of an elementary diffusion jump (all jumps are assumed to be equal, but extensions with distributions of jump rates and jump distances are possible). It may be noted that in all free diffusion models, Γ will follow a κ^2 dependence for small κ (which means long-range diffusion), and the prefactor is a measure of the diffusion coefficient, whereas the κ-dependence at larger κ reveals microscopic details of the diffusion mechanism. From the Chudley–Elliott fits, one may extract the jump rates at different temperatures. The results for the present case are shown in **Figure 60**, plotted vs. reciprocal temperature. It becomes clear that the slower process in all materials studied follows an Arrhenius law. The activation energies obtained from the fits (142 ± 4 meV for coarse grained, 153 ± 6 for ball-milled material) do not differ very much, but the jump rates are considerably smaller for the ball-milled samples. The jump distances are close to 2 Å in all cases. This slow process is attributed to

(a)

(b)

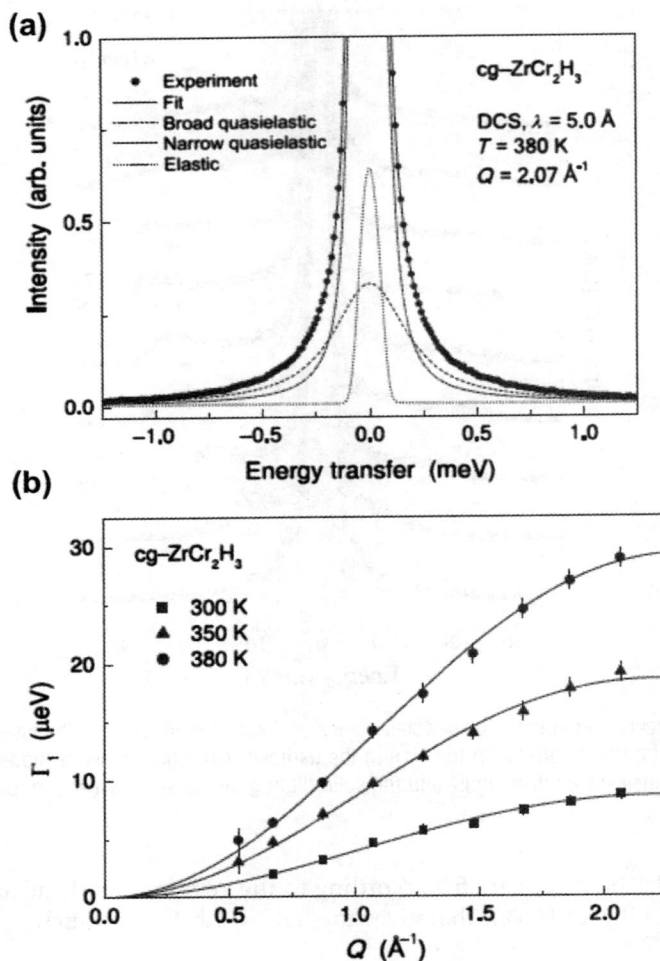

Figure 59 Quasielastic neutron scattering of hydrogen in $ZrCr_2H_3$. (a) The scattering spectrum for coarse-grained material measured at $T = 380$ K and Q $(=2\pi\kappa) = 2.07$ Å^{-1}. The circles are the experimental points interpolated to the uniform energy grid. The full curve shows the fit of the three-component reconstruction to the data. The three components are indicated in the figure. (b) The half-width at half-maximum Γ of the narrow Lorentzian component for coarse-grained material as a function of Q at $T = 300$, 350, and 380 K. The full curves show the fits of the Chudley–Elliott model (eqn (93)) to the data. From Skripov et al. (2011).

long-range diffusion whereas the higher jump rates are related to localized motion across neighboring interstitial sites within the cubic $C15$-type structure (for further details and references, see Skripov et al., 2011).

Self-diffusion in solid metals can also be studied at temperatures not too far below the melting point if scattering and absorption cross-sections are favorable. The fast self diffusion in b.c.c. β-Ti was studied by Petry et al. (1988) and Vogl et al. (1989) on a backscattering spectrometer directly after crystal growth in situ (to avoid the transition to the α-phase at 882°C). **Figure 61** shows the results

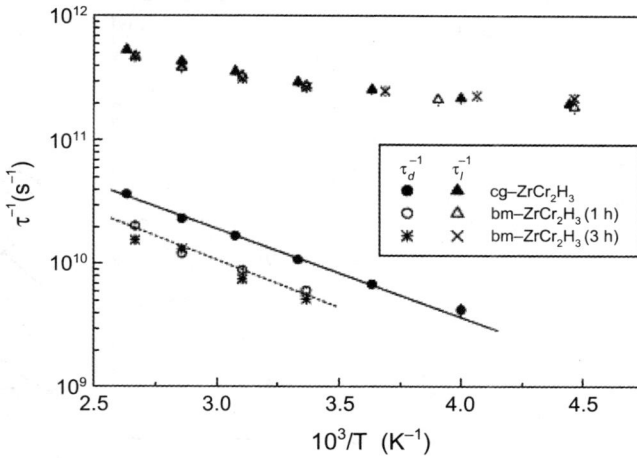

Figure 60 The hydrogen jump rates τ^{-1} (subscript d for the slow process, i for the fast process) as functions of the inverse temperature in $ZrCr_2H_3$. The inset shows the key to the symbols (with cg = coarse-grained and bm = ball-milled materials after two different milling times). The lines are Arrhenius fits to the data of cg and bm(1 h) materials. From Skripov et al. (2011).

obtained at 1460 and 1530°C along with a visualization of the two jump vectors used for the fitted curves. Such fits can be attempted for all reasonable combinations of jump mechanisms, and the best fits are shown; in **Figure 61**(a) the admixture of next-nearest neighbors tends to improve the fit for some orientations, but in **Figure 61**(b), the nearest-neighbor jumps are better. The authors conclude that, if any, there might be 10–15% of direct next-nearest-neighbor jumps. The diffusion coefficients are in the range of 10^{-7} to 10^{-8} cm^2 s^{-1}. Spin-echo measurements have also been performed down to 1100°C (polycrystals) where the diffusion coefficient is only about 10^{-9} cm^2 s^{-1}. The phonon softening observed in the inelastic scattering measurements (see Section 13.6.1) provides a means to explain the low migration enthalpy for nearest-neighbor jumps (for details see Petry et al., 1989). Diffusion of Fe, Co, Ni and other metallic solutes in group-IV metals is at least two orders of magnitude faster than self-diffusion. In a similar study of dilute Zr-rich Zr–Co, the vacancy mechanism was found to be predominant (Petry et al., 1987) but only an enhancement of diffusivity by a factor of 10 was found via this path, calling for another mechanism, e.g. very fast diffusion via interstitial sites. A hopping mechanism over interstitial sites has also been identified by quasielastic neutron scattering for Cu and Ni in an Al-35 at.% Cu-15 at.% Ni alloy with a long-range orderd structure (Dahlborg et al., 2000).

In the domain of X-rays, Mössbauer spectroscopy has long been the only method to study diffusive motion with atomic resolution. With third-generation synchrotron X-ray sources and upcoming free-electron lasers, nuclear resonant scattering and X-ray photon correlation spectroscopy offer new possibilities. In their report on the observation of speckle by diffraction with (partially) coherent X-rays off a superstructure peak of Cu_3Au, Sutton et al. (1991) already mentioned the possibilities of studying dynamical processes at the atomic scale in the time domain using methods known from optical correlation spectroscopy. Only recently, Leitner et al. (2009) reported the first results on a diffusion problem in an alloy. These results will now briefly be presented.

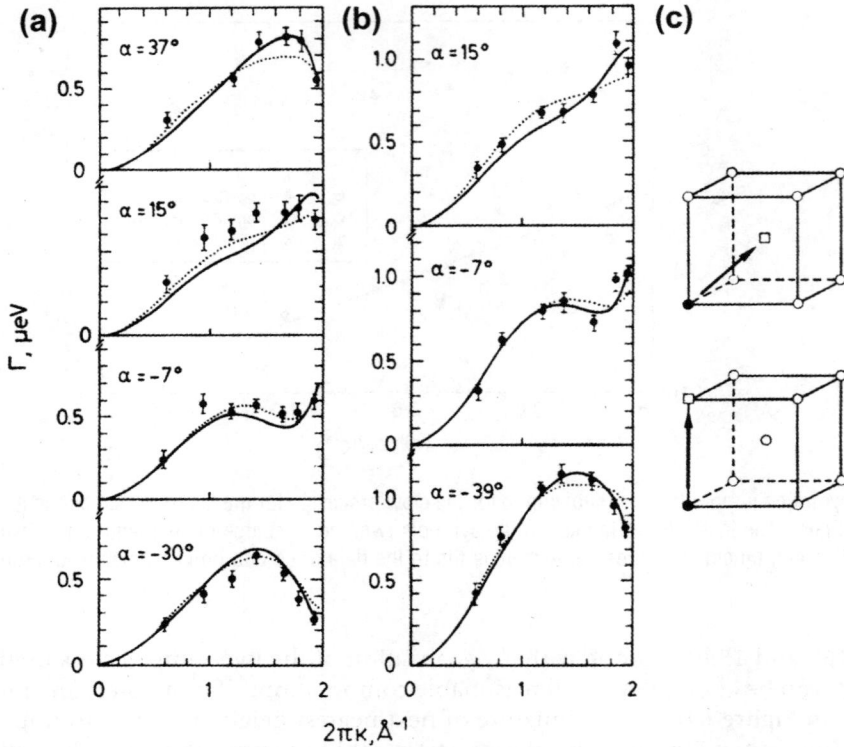

Figure 61 Quasielastic line broadening Γ (full width at half maximum) for two single crystals of Ti, at (a) 1460°C and (b) 1530°C, as a function of the modulus of the scattering vector for different angles α with respect to the lattice axes (two different scattering planes). The solid curve is calculated for nearest-neighbor jumps, the dotted curve for 75% nearest-neighbor jumps and 25% direct next-nearest-neighbor jumps. (c) Schematic of the two types of jump used in the fits of (a) and (b). From Vogl et al. (1989).

Since scattering intensity is related to real-space correlations, autocorrelations of intensity reflect the properties of $G(\mathbf{x}, \Delta t)$ where $\mathbf{x}$ is a distance vector and Δt the time segment considered (for correlation functions in general, see Van Hove, 1954; e.g. Lovesey, 1984; Sutton, 2006). The time-dependent autocorrelation function of the intensity is given by

$$G(\kappa, \Delta t) = \frac{\langle I(\kappa, t)I(\kappa, t + \Delta t)\rangle}{\langle I(\kappa)\rangle^2} \tag{94}$$

For a stationary ensemble, the beginning of a measurement is arbitrary, and the averages can be taken over absolute time for any κ. Within the Chudley–Elliott model for atomic jumps, a decay time $\tau(\kappa)$ controls the quasieleastic linewidth,

$$G(\kappa, \Delta t) = 1 + \beta \exp\left[-2(\Delta t)\tau^{-1}(\kappa)\right] \tag{95}$$

where β is the coherency factor and τ is the decay time.

With a 7 μm × 10 μm spatially partially coherent beam of a wavelength of 1.55 Å, and temporal coherence assured by an energy resolution of almost 1 meV, Leitner et al. (2009) took time series of typically 2 h with 10 s exposure per frame with a CCD detector (1300 × 1340 pixels of 20 μm², count rates of about 3 photons per hour and pixel). The sample was a Cu-10 at.% Au alloy at temperatures ranging from about 530 to 555 K. This alloy's short-range order had been well documented by Schönfeld et al. (1999), and effective pair interactions are known. Within linear-response theory, $\tau(\kappa)$ can be related to the SRO intensity $I_{SRO}(\kappa)$ (Leitner et al., 2009; see also Leitner and Vogl, 2011 for further derivations and discussions),

$$\tau(\kappa) = \tau_0 \frac{I_{SRO}(\kappa)}{1 - \sum_j p_j \cos(2\pi\kappa \cdot x_j)} \qquad (96)$$

where τ_0 is the mean time for site exchanges and p_j is the jump probability for jump vector x_j. **Figure 62**(a) shows the SRO intensity map (I_{SRO} from Schönfeld et al., 1999) and the various κ positions where measurements were taken. **Figure 62**(b) shows some results. **Figure 63** gives a summary of the experimental correlation times for the three scans indicated in **Figure 60**(a) along with fits leading to the conclusion that a nearest-neighbor model (with only one free parameter, τ_0) is most suitable. The maxima of τ are related to the SRO and show, for the first time in an alloy, the effect of local stability on decay times (see de Gennes, 1959). It is important to note the extension of time

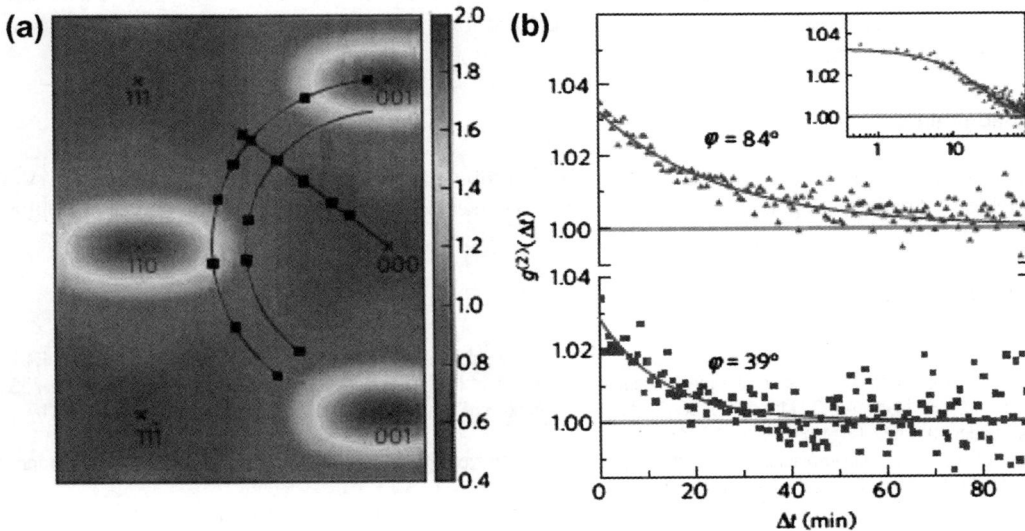

Figure 62 Diffusive motion in short-range ordered Cu-10 at.% Au studied with coherent X-rays (Leitner et al., 2009). (a) The static scattering intensity $I_{SRO}(\kappa)$ in the ($1\bar{1}0$) plane with data from Schönfeld et al. (1999) and projection (black squares) of the measurement positions for the quasielastic linewidth [see eqn (95)]. (b) Two examples of measurements at fixed temperature 543 K, at fixed scattering angle $2\theta = 25°$ and azimuthal angles $\varphi = 84°$ and $\varphi = 39°$ (see (a) for angles) with fitted exponential decays. Although the modulus of the scattering vector is the same, the fitted correlation times 52 ± 5 min for 84° and 28 ± 4 min for 39° differ by almost a factor of two. The inset shows the measurement at $\varphi = 84°$ plotted on a logarithmic time scale. From Leitner et al. (2009). (For color version of this figure, the reader is referred to the online version of this book.)

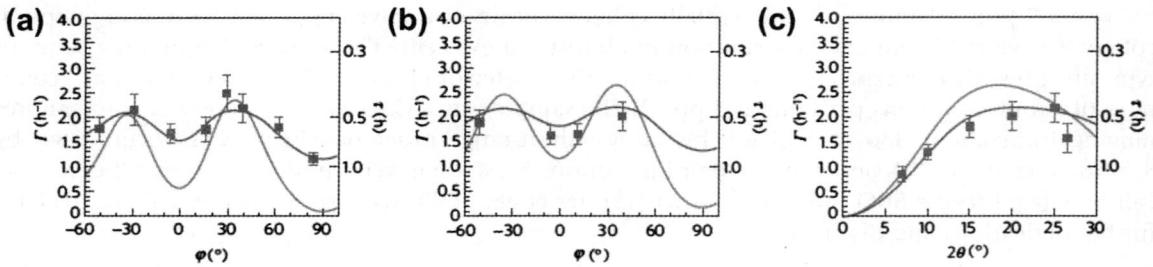

Figure 63 Diffusive motion in short-range ordered Cu-10 at.% Au studied with coherent X-rays (Leitner et al., 2009). Experimental correlation times together with fits according to eqn (96) for three one-dimensional scans through reciprocal space. The error bars are standard errors of the fitted correlation times. (a) Azimuthal scan with fixed scattering angle $2\theta = 25°$, corresponding to $|\kappa| = 0.279$ Å^{-1}, (b) Azimuthal scan with fixed scattering angle $2\theta = 20°$, corresponding to $|\kappa| = 0.224$ Å^{-1}, (c) 2θ scan with fixed azimuthal angle $\varphi = 39°$ (see **Figure 62** for the angles). The solid lines are fits according to eqn (96), with the factor τ_0 as the only adjustable parameter. A model with nearest-neighbor exchanges (blue line) reproduces the measurements well, whereas a model with second-nearest-neighbor jumps (green line) is not suitable. From Leitner et al. (2009). (For interpretation of the references to color in this figure legend, the reader is referred to the online version of this book.)

scales accessible by this type of experiment, which translates to diffusion constants of the order of 10^{-20} cm^2 s^{-1}. This opens a wide field, so far inaccessible to the exploration of basic diffusion mechanisms, and with the advent of more powerful coherent hard X-ray sources, much more work on related issues can be expected.

Acknowledgments

The author is grateful to many colleagues who have contributed to the field, especially to J.B. Cohen, V. Gerold, B. Schönfeld, C.J. Sparks, G.Vogl and many students and coworkers who have been a constant source of inspiration.

References

Abe, H., 2009. Diffuse scattering from quasicrystals—short-range order, atomic size effect, and phason fluctuations. In: Barabash, R.I., Ice, G.E., Turchi, P.E.A. (Eds.), Diffuse Scattering and the Fundamental Properties of Materials. Momentum Press, LLC, New Jersey, pp. 259–281.

Abe, H., Yamamoto, K., Matsuoka, S., Matsuo, Y., 2007. Atomic short-range order in an $Al_{72}Ni_{18}Fe_{10}$ decagonal quasicrystal studied by anomalous X-ray scattering. J. Phys. Condens. Matter 19 (466201), 15.

Agrestini, S., Chapon, L.C., Daoud-Aladine, A., Schefer, J., Gukasov, A., Mazzoli, C., Lees, M.R., Petrenko, O.A., 2008. Nature of the magnetic order in $Ca_3Co_2O_6$. Phys. Rev. Lett. 101 (097207), 4.

Allen, A.J., Ilavsky, J., Long, G.G., Wallace, J.S., Berndt, C.C., Herman, H., 2001. Microstructural characterization of yttria-stabilized zirconia plasma-sprayed deposits using multiple small-angle neutron scattering. Acta Mater. 49, 1661–1675.

Als-Nielsen, J., McMorrow, D., 2001. Elements of Modern X-ray Physics. Wiley, New York.

Anders, R., Giehrl, M., Röber, E., Stierstadt, K., Schwahn, D., 1984. Temperature dependence of magnetic neutron scattering by dislocations. II. Solid State Commun. 51, 111–113.

Antonowicz, J., Pietnoczka, A., Drobiazg, T., Almyras, G.A., Papageorgiou, D.G., Evangelakis, G.A., 2012. Icosahedral order in Cu–Zr amorphous alloys studied by means of X-ray absorption fine structure and molecular dynamics simulations. Philos. Mag. 92, 1865–1875.

Atkinson, H.H., 1959. Small-angle scattering of X-rays and neutrons from deformed metals. J. Appl. Phys. 30, 637–645.

Atkinson, H.H., Lowde, R.D., 1957. Small-angle scattering from deformed metals. Philos. Mag. 2, 589–590.

Bacon, G.E., 1975. Neutron Diffraction. Clarendon Press, Oxford.

Bardhan, P., Cohen, J.B., 1976. A structural study of the alloy Cu_3Au above its critical temperature. Acta Crystallogr. A 32, 597–614.

Bardhan, P., Chen, H., Cohen, J.B., 1977. Premonitory effects in Cu_3Au near the order–disorder transformation. Philos. Mag. 35, 1653–1666.

Barrett, C.S., Massalski, T.B., 1980. Structure of Metals. Pergamon Press, Oxford.

Baruchel, J., Hodeau, J.L., Lehmann, M.S., Regnard, J.R., Schlenker, C. (Eds.), 1993. Neutron and Synchrotron Radiation for Condensed Matter Studies, vol. III. Les Editions de Physique, Les Ulis, France.

Bauer, G.S., 1979. Diffuse elastic neutron scattering. In: Kostorz, G. (Ed.), Neutron Scattering. Academic Press, New York, pp. 291–336.

Beaven, P.A., Frisius, F., Kampmann, R., Wagner, R., 1986. Analysis of defect microstructures in irradiated ferritic alloys. In: Janot, C., Petry, W., Richter, D., Springer, T. (Eds.), Atomic Transport and Defects in Metals by Neutron Scattering. Springer, Berlin, pp. 228–234.

Beeman, W.W., Kaesberg, P., Anderegg, J.W., Webb, M.B., 1957. Size of particles and lattice defects. In: Flügge, S. (Ed.), Handbuch der Physik. Springer, Berlin, pp. 321–323.

Bergamaschi, A., Cervellino, A., Dinapoli, R., Gozzo, F., Henrich, B., Johnson, I., Kraft, P., Mozzanica, A., Schmitt, B., Shi, X., 2010. The MYTHEN detector for X-ray powder diffraction experiments at the Swiss Light Source. J. Synchrotron Radiat. 17, 653–668.

Berk, N.F., Hardman-Rhyne, K.A., 1985. Characterization of alumina powder using multiple small-angle neutron scattering. I. Theory. J. Appl. Crystallogr. 18, 467–472.

Berk, N.F., Hardman-Rhyne, K.A., 1988. Analysis of SAS data dominated by multiple scattering. J. Appl. Crystallogr. 21, 645–651.

Bessière, M., Calveyrac, Y., Lefèbvre, S., Gratias, D., Cénédèse, P., 1986. Study of local order in $Au_{75}Cu_{25}$ and $Au_{70}Cu_{30}$. Results of diffuse scattering measurements. J. Phys. 47, 1961–1976.

Binder, K., Fratzl, P., 2001. Spinodal decomposition. In: Kostorz, G. (Ed.), Phase Transformations in Materials. Wiley-VCH, Weinheim, pp. 409–480.

Borie, B., Sparks, C.J., 1971. The interpretation of intensity distributions from disordered binary alloys. Acta Crystallogr. A 27, 198–201.

Bouchard, P.J., Fiori, F., Treimer, W., 2002. Characterisation of creep cavitation damage in a stainless steel pressure vessel using small angle neutron scattering. Appl. Phys. A–Mater. Sci. Process. 74, S1689–S1691.

Broennimann, Ch., Eikenberry, E.F., Henrich, B., Horisberger, R., Huelsen, G., Pohl, E., Schmitt, B., Schulze-Briese, C., Suzuki, M., Tomizaki, T., Toyokawa, H., Wagner, A., 2006. The PILATUS 1M detector. J. Synchrotron Radiat. 13, 120–130.

Brown, P.J., 1979. Neutron crystallography. In: Kostorz, G. (Ed.), Neutron Scattering. Academic Press, New York, pp. 69–130.

Bubeck, E., Gerold, V., Kostorz, G., 1985. Small-angle neutron scattering from Guinier–Preston zones in Al-6.8 at.% Zn single crystals. Cryst. Res. Technol. 20, 97–103.

Bucher, R., 1999. Lokale Atomanordnungen in Ni–Ti und Fe–Al bei hohen Temperaturen. Ph.D. Thesis, ETH Zürich.

Bucher, R., Demé, B., Heinrich, H., Kohlbrecher, J., Kompatscher, M., Kostorz, G., Schneider, J.M., Schönfeld, B., Zolliker, M., 2002. In-situ neutron scattering studies of order and decomposition in Ni-rich Ni–Ti. Mater. Sci. Eng. A 324, 77–81.

Büchner, A.R., Pitsch, W., 1985. Solid solution strengthening by short range order. Z. Metallkd. 76, 651–656.

Bührer, C., Beckmann, M., Fähnle, M., Grünewald, U., Maier, K., 2000. The liquid ferromagnet $Co_{80}Pd_{20}$ and its critical exponent γ. J. Magn. Magn. Mater. 212, 211–224.

Butler, B.D., Cohen, J.B., 1989. The structure of Cu_3Au above the critical temperature. J. Appl. Phys. 65, 2214–2219.

Cahn, J.W., 1961. On spinodal decomposition. Acta Metall. 9, 795–801.

Cénédèse, P., Bley, F., Lefèbvre, S., 1984. Diffuse scattering in disordered ternary alloys: neutron measurement of local order in a stainless steel $Fe_{0.56}Cr_{0.21}Ni_{0.23}$. Acta Crystallogr. A 40, 228–240.

Cerri, A., Schönfeld, B., Kostorz, G., 1990. Decomposition kinetics in Ni–Ti alloys. Phys. Rev. B 42, 958–960.

Cervellino, A., Gvasaliya, S.N., Zaharko, O., Roessli, B., Rotaru, G.M., Cowley, R.A., Lushnikov, S.G., Shaplygina, T.A., Fernandez-Dias, M.T., 2011. Diffuse scattering from the lead-based relaxor ferroelectric $PbMg_{1/3}Ta_{2/3}O_3$. J. Appl. Crystallogr. 44, 603–609.

Chen, H., Cohen, J.B., Ghosh, R., 1977. Spinodal ordering in Cu_3Au. J. Phys. Chem. Solids 38, 855–857.

Chen, P.C., Hall, B.O., Winchell, P.G., 1980. Atomic displacements due to C in Fe–Ni–C martensite. Metall. Trans. A 11, 1323–1331.

Chudley, C.T., Elliott, R.J., 1961. Neutron scattering from a liquid on a jump diffusion model. Proc. Phys. Soc. 77, 353–361.

Ciccariello, S., 2010. Asymptotic analysis of small-angle scattering intensities of plane columnar layers. J. Appl. Crystallogr. 43, 1377–1388.

Ciccariello, S., Sobry, R., 1997. Diffuse interfaces and small-angle scattering intensity behaviour. J. Appl. Crystallogr. 30, 1026–1035.

Ciccariello, S., Schneider, J.M., Schönfeld, B., Kostorz, G., 2000. Generalization of Porod's law of small-angle scattering to anisotropic samples. Europhys. Lett. 50, 601–607.

Ciccariello, S., Schneider, J.M., Schönfeld, B., Kostorz, G., 2002. Illustration of the anisotropic Porod law. J. Appl. Crystallogr. 35, 304–313.

Clapp, P.C., 1971. Atomic configurations in binary alloys. Phys. Rev. B 4, 255–270.

Cohen, J.B., 1970. The order–disorder transformation. In: Phase Tranformations. ASM, Metals Park, Ohio, pp. 561–620.

Coppola, R., Lindau, R., May, R.P., Moslang, A., Valli, M., 2009. Investigation of microstructural evolution under neutron irradiation in Eurofer97 steel by means of small-angle neutron scattering. J. Nucl. Mater. 386, 195–198.

Cowley, J.M., 1950. X-ray measurement of order in single crystals of Cu_3Au. J. Appl. Phys. 21, 24–30.

Cowley, J.M., 1981. Diffraction Physics, second ed. North-Holland, Amsterdam.

Cribier, D., Jacrot, B., Madhav Rao, L., Farnoux, B., 1964. Mise en evidence par diffraction de neutrons d'une structure periodique du champ magnetique dans le niobium supraconducteur. Phys. Lett. 9, 106–107.

Cullity, B.D., 1977. Elements of X-ray Diffraction. Addison-Wesley, Reading, MA.

Currat, R., Pynn, R., 1979. Phonons and phase transitions. In: Kostorz, G. (Ed.), Neutron Scattering. Academic Press, New York, pp. 131–189.

Dahlborg, U., Howells, W.S., Calvo-Dahlborg, M., Dubois, J.M., 2000. Atomic motions in the crystalline $Al_{50}Cu_{35}Ni_{15}$ alloy. J. Phys. Condens. Matter 12, 4021–4041.

Darwin, C.G., 1922. The reflexion of X-rays from imperfect crystals. Philos. Mag. 43, 800–829.

De Bergevin, F., Brunel, M., 1972. Observation of magnetic superlattice peaks by X-ray diffraction on an antiferromagnetic NiO crystal. Phys. Lett. A 39, 141–142.

De Boissieu, M., Francoual, S., 2009. Diffuse scattering and phason modes in icosahedral quasicrystals. In: Barabash, R.I., Ice, G.E., Turchi, P.E.A. (Eds.), Diffuse Scattering and the Fundamental Properties of Materials. Momentum Press, LLC, New Jersey, pp. 211–237.

De Fontaine, D., 1971. The number of independent pair-correlation functions in multicomponent systems. J. Appl. Crystallogr. 4, 15–19.

De Fontaine, D., 1979. Configurational thermodynamics of solid solutions. Solid State Phys. 34, 74–274.

Dederichs, P.H., 1971. Diffuse scattering from defect clusters near Bragg reflexions. Phys. Rev. B 4, 1041–1050.

Dederichs, P.H., 1973. Theory of diffuse X-ray scattering and its application to study of point defects and their clusters. J. Phys. F: Met. Phys. 3, 471–496.

Dederichs, P.H., Zeller, R., 1979. Dynamical properties of point defects in metals. In: Hohler, G., Niekisch, E.A. (Eds.), Point Defects in Metals II. Springer, Berlin, pp. 1–170.

Dederichs, P.H., Lehmann, C., Schober, H.R., Scholz, A., Zeller, R., 1978. Lattice theory of point defects. J. Nucl. Mater. 69/70, 176–199.

Delaire, O., Kresch, M., Fultz, B., 2005. Vibrational entropy of the $\gamma-\alpha$ martensitic transformation in $Fe_{71}Ni_{29}$. Philos. Mag. 85, 3567–3583.

Deschamps, A., De Geuser, F., 2011. On the validity of simple precipitate size measurements by small-angle scattering in metallic systems. J. Appl. Crystallogr. 44, 343–352.

Deschamps, A., Lae, L., Guyot, P., 2007. In situ small-angle scattering study of the precipitation kinetics in an Al–Zr–Sc alloy. Acta Mater. 55, 2775–2783.

Deschamps, A., Ringeval, S., Texier, G., Delfaut-Durut, L., 2009. Quantitative characterization of the microstructure of an electron-beam welded medium strength Al–Zn–Mg alloy. Mater. Sci. Eng. A 517, 361–368.

Deschamps, A., Bastow, T.J., De Geuser, F., Hill, A.J., Hutchinson, C.R., 2011. In situ evaluation of the microstructure evolution during rapid hardening of an Al-2.5Cu-1.5Mg (wt.%) alloy. Acta Mater. 59, 2918–2927.

Dietrich, S., Fenzl, W., 1989a. Correlations in disordered crystals and diffuse scattering of X rays or neutrons. Phys. Rev. B 39, 8873–8899.

Dietrich, S., Fenzl, W., 1989b. Diffuse scattering of X rays or neutrons from binary alloys and null matrices. Phys. Rev. B 39, 8900–8906.

Dosch, H., 1992. Evanescent X-rays probing surface dominated phase transitions. Int. J. Mod. Phys. B 6, 2773–2808.

Dosch, H., 1993. Critical Phenomena at Surfaces and Interfaces—Evanescent X-ray and Neutron Scattering. In: Springer Tracts in Modern Physics, vol. 126. Springer, Berlin, Heidelberg.

Dosch, H., Peisl, J., 1986. Local defect structure of highly mobile deuterium in niobium. Phys. Rev. Lett. 56, 1385–1388.

Dosch, H., Peisl, J., Dorner, B., 1987. Coherent quasielastic neutron scattering from NbD_x. Phys. Rev. B 35, 3069–3081.

Dosch, H., Mailänder, L., Reichert, H., Peisl, J., Johnson, R.L., 1991. Long-range order near the $Cu_3Au(001)$ surface by evanescent X-ray scattering. Phys. Rev. B 43, 13172–13186.

Dosch, H., Schmid, F., Wiethoff, P., Peisl, J., 1992. Lattice-distortion-mediated local jumps of hydrogen in niobium from diffuse neutron scattering. Phys. Rev. B 46, 55–68.

Dubey, Ph.A., Schönfeld, B., Kostorz, G., 1991. Shape and internal structure of Guinier–Preston zones in Al–Ag. Acta Metall. Mater. 39, 1161–1170.

Dugdale, S.B., Watts, R.J., Laverock, J., Major, Zs., Alam, M.A., Samsel-Czekala, M., Kontrym-Sznajd, G., Sakurai, Y., Itou, M., Fort, D., 2006. Observation of a strongly nested Fermi surface in the shape-memory alloy $Ni_{0.62}Al_{0.38}$. Phys. Rev. Lett. 96, 6(046406).

Ehrhart, P., Schilling, W., Haubold, H.-G., 1974. Investigation of point defects and their agglomerates in irradiated metals by diffuse X-ray scattering. Adv. Solid State Phys. 14, 87–110.

Embs, J.P., Juranyi, F., Hempelmann, R., 2010. Introduction to quasielastic neutron scattering. Z. Phys. Chem. 224, 5–32.

Engelke, M., Schönfeld, B., Ruban, A.V., 2010. Near-surface microstructure of Pt-23 at.% Ni: grazing incidence diffraction and first-principles calculations. Phys. Rev. B 81, 13, 054205.

Epperson, J.E., Gerstenberg, K.W., Berner, D., Kostorz, G., Ortiz, C., 1978. Voids formed in quenched and annealed NiAl. Philos. Mag. A 38, 529–541.

Erni, R., Heinrich, H., Kostorz, G., 2003. On the internal structure of Guinier–Preston zones in Al-3 at.% Ag. Philos. Mag. Lett. 83, 599–609.

Feidenhans'l, R., 1989. Surface structure determination by X-ray diffraction. Surf. Sci. Rep. 10, 105–188.

Fischer, H.E., Hennet, L., Cristiglio, V., Zanghi, D., Pozdnyakova, I., May., R.P., Price, D.L., Wood, S., 2007. Magnetic critical scattering in solid $Co_{80}Pd_{20}$. J. Phys. Condens. Matter 19, 5(415106).

Forgan, E.M., Fault, D.McK., Mook, H.A., Timmins, P.A., Keller, H., Sutton, S., Abell, J.S., 1990. Observation by neutron diffraction of the magnetic flux lattice in single-crystal $YBa_2Cu_3O_{7-\delta}$. Nature 343, 735–737.

Frase, H., Fultz, B., Robertson, J.L., 1998. Phonons in nanocrystalline Ni_3Fe. Phys. Rev. B 57, 898–905.

Fratzl, P., 2003. Small-angle scattering in materials science–a short review of applications in alloys, ceramics and composite materials. J. Appl. Crystallogr. 36, 397–404.

Fratzl, P., Lebowitz, J.L., 1989. Universality of scaled structure functions in quenched systems undergoing phase separation. Acta Metall. 37, 3245–3248.

Fratzl, P., Yoshida, Y., Vogl, G., Haubold, H.G., 1992. Phase-separation kinetics of dilute Cu–Fe alloys studied by anomalous small-angle X-ray scattering and Mössbauer spectroscopy. Phys. Rev. B 46, 11323–11331.

Fultz, B., 2010. Vibrational thermodynamics of materials. Prog. Mater. Sci. 55, 247–352.

Gehlen, P.C., Cohen, J.B., 1965. Computer simulation of the structure associated with local order in alloys. Phys. Rev. 139, A844–A855.

De Gennes, P.G., 1959. Liquid dynamics and inelastic scattering of neutrons. Physica 25, 825–839.

Georgopoulos, P., Cohen, J.B., 1977. The determination of short range order and local atomic displacements in disordered binary solid-solutions. J. Phys. 12. C 7 191–196.

Georgopoulos, P., Cohen, J.B., 1979. Direct methods of analyzing diffuse scattering. In: Cowley, J.M., Cohen, J.B., Salamon, M.B., Wuensch, B.J. (Eds.), Proc. Conf. Kailua Kona, Hawai. AIP, New York, pp. 21–29.

Gerold, V., 1967. Application of small-angle X-ray scattering to problems in physical metallurgy and metal physics. In: Brumberger, H. (Ed.), Small-angle X-ray Scattering, Proc. Conf. Syracuse University. Gordon and Breach, New York, pp. 277–317.

Gerold, V., Karnthaler, H.P., 1989. On the origin of planar slip in f.c.c. alloys. Acta Metall. 37, 2177–2183.

Gerold, V., Kern, J., 1987. The determination of atomic interaction energies in solid solutions from short range order coefficients—an inverse Monte-Carlo method. Acta Metall. 35, 393–399.

Gerold, V., Kostorz, G., 1978. Small-angle scattering applications to materials science. J. Appl. Crystallogr. 11, 376–404.

Glatter, O., 1982. Data treatment. In: Glatter, O., Kratky, O. (Eds.), Small Angle X-ray Scattering. Academic, London, pp. 119–165.

Glinka, C.J., 2011. Incoherent neutron scattering from multi-element materials. J. Appl. Crystallogr. 44, 618–624.

Goerigk, G., Mattern, N., 2009. Critical scattering of Ni–Nb–Y metallic glasses probed by quantitative anomalous small-angle X-ray scattering. Acta Mater. 57, 3652–3661.

Göltz, G., Kronmüller, H., Seeger, A., Scheuer, H., Schmatz, W., 1986. Investigation of the dislocation arrangements in plastically deformed Fe single crystals using magnetic small-angle scattering of neutrons. Philos. Mag. A 54, 213–235.

Gragg, J.E., Bardhan, P., Cohen, J.B., 1971. The "Gestalt" of local order. In: Mills, R.E., Ascher, E., Jaffee, R.I. (Eds.), Critical Phenomena in Alloys, Magnets and Superconductors. McGraw-Hill, New York, pp. 309–341.

Gragg, J.E., Cohen, J.B., 1971. The structure of Guinier–Preston zones in aluminum-5 at.% silver. Acta Metall. 19, 507–519.

Gratias, D., Cénédèse, P., 1985. A cvm approach of multiplet correlation functions in substitutional solid solutions. J. Phys. Colloq. 46. C 9 149–153.

Guinier, A., 1938. Structure of age-hardened aluminium–copper alloys. Nature 142, 569–570.

Guinier, A., 1939. La diffraction des rayons X aux très petits angles; application à l'étude de phénomènes ultramicroscopiques. Ann. Phys. 12, 161–237.

Guinier, A., 1963. X-ray Diffraction. Freeman, San Francisco.

Guinier, A., Fournet, G., 1955. Small Angle Scattering of X-rays. Wiley, New York.

Gurevich, I.I., Tarasov, L.V., 1968. Low Energy Neutron Physics. North-Holland, Amsterdam.

Halpern, O., Johnson, M.H., 1939. On the magnetic scattering of neutrons. Phys. Rev. 55, 898–923.

Hammouda, B., 2010. Analysis of the Beaucage model. J. Appl. Crystallogr. 43, 1474–1478.

Hashimoto, K., Tsujimoto, T., 1978. X-ray diffraction patterns and microstructures of aged Ni–Ti alloys. Trans. Jpn. Inst. Met. 19, 77–84.

Haubold, H.-G., 1975. Measurement of diffuse X-ray scattering between reciprocal-lattice points as a new experimental method in determining interstitial structures. J. Appl. Crystallogr. 8, 175–183.

Haubold, H.-G., Martinsen, D., 1978. Structure determination of self-interstitials and investigation of vacancy clustering in copper by diffuse X-ray scattering. J. Nucl. Mater. 69/70, 644–649.

Hayakawa, M., Cohen, J.B., 1975. Equations for diffuse scattering from materials with multiple sublattices. Acta Crystallogr. A 31, 635–645.

He, S.M., van Dijk, N.H., Paladugu, M., Schut, H., Kohlbrecher, J., Tichelaar, van der Zwaag, S., 2010. In situ determination of aging precipitation in deformed Fe–Cu and Fe–Cu–B–N alloys by time-resolved small-angle neutron scattering. Phys. Rev. B 82, 14(174111).

Hempelmann, R., 2000. Quasielastic Neutron Scattering and Solid State Diffusion. University Press (Oxford Science Publications), Oxford.

Hendricks, R.W., Schelten, J., Lippmann, G., 1977. Studies of voids in neutron-irradiated aluminum single crystals. 3. Determination of void surface properties. Philos. Mag. 36, 907–921.

Herzer, G., 1992. Nanocrystalline soft magnetic materials. J. Magn. Magn. Mater. 112, 258–262.

Heuser, B.J., Udovic, T.J., Ju, H., 2008. Vibrational density of states measurement of hydrogen trapped at dislocations in deformed PdH0.0008. Phys. Rev. B 78, 5(214101).

Hicks, T.J., 1979. Magnetic inhomogeneities. In: Kostorz, G. (Ed.), Neutron Scattering. Academic, New York, pp. 337–380.

Holden, T.M., Suzuki, H., Carr, D.G., Ripley, M.I., Clausen, B., 2006. Stress measurements in welds: problem areas. Mater. Sci. Eng., A 437, 33–37.

Ice, G.E., Sparks, C.J., Habenschuss, A., Shaffer, L.B., 1992. Anomalous X-ray scattering measurement of near-neighbor individual pair displacements and chemical order in Fe22.5Ni77.5. Phys. Rev. Lett. 68, 863–866.

Ilavsky, I., Jemian, P., Allen, A.J., Long, G.G., 2004. Versatile USAXS (Bonse-Hart) facility for advanced materials research. In: Warwick, T., Arthur, J., Padmore, H.A.; Stohr, J. (Eds.), 2004. 8th International Conference on Synchrotron Radiation Instrumentation (SRI 2003), San Francisco, CA, August 25–29, 2003, AIP Conference Proceedings, vol. 705. AIP, College Park, MD, USA, pp. 510–513.

Isaeva, E.I., Lichtenstein, A.I., Vekilov, Y.K., Smirnova, E.A., Abrikosov, I.A., Simak, S.I., Ahuja, R., Johansson, B., 2004. Ab initio phonon calculations for L_{12} Ni$_3$Al and B2 NiAl. Solid State Commun. 129, 809–814.

Iwase, H., Koizumi, S., Suzuki, J., Oku, T., Sasao, H., Tanaka, H., Shimizu, H.M., Hashimoto, T., 2007. Wide-q observation from 10^{-4} to 2.0 Å^{-1} using a focusing and polarized neutron small-angle scattering spectrometer, SANS-J-II. J. Appl. Crystallogr. 40, s414–s417.

Jacobs, G., Egry, I., 1999. EXAFS studies on undercooled liquid Co$_{80}$Pd$_{20}$ alloy. Phys. Rev. B 59, 3961–3968.

James, R.W., 1963. The Optical Principles of the Diffraction of X-rays, third ed. Bell, London.

James, M.R., Cohen, J.B., 1980. The measurement of residual stresses by X-ray diffraction techniques. Part A. In: Herman, H. (Ed.), 1980. Treatise on Materials Science and Technology, vol. 19. Academic, New York, pp. 2–30.

Jiang, X., Ice, G.E., Sparks, C.J., Robertson, L., Zschack, P., 1996. Local atomic order and individual pair displacements of Fe$_{46.5}$Ni$_{53.5}$ and Fe$_{22.5}$Ni$_{77.5}$ from diffuse X-ray scattering studies. Phys. Rev. B 54, 3211–3226.

Kassner, M.E., Geantil, P., Levine, L.E., Larson, B.C., 2009. Long-range internal stresses in monotonically and cyclically deformed metallic single crystals. Int. J. Mater. Res. 100, 333–339.

Kästner, J., Neuhaus, J., Wassermann, E.F., Petry, W., Hennion, B., Bach, H., 1999. TA(1)[110] phonon dispersion and martensitic phase transition in ordered alloys Fe$_3$Pt. Eur. Phys. J. B 11, 75–81.

Keller, T., Wagner, W., Allen, A., Ilavsky, J., Margadant, N., Siegmann, S., Kostorz, G., 2002. Characterization of thermally sprayed metallic NiCrAlY deposits by multiple small-angle scattering. Appl. Phys. A 74 (Suppl.), S975–S977.

Kimura, M., Cohen, J.B., Chandavarkar, S., Liang, K., 1997. Short-range ordering of Cu$_3$Au above T$_c$ in the topmost 80 Å of a (001) face. J. Mater. Res. 12, 75–82.

Kohn, W., 1959. Image of the Fermi surface in the vibration spectrum of a metal. Phys. Rev. Lett. 2, 393–394.

Kompatscher, M., Schönfeld, B., Heinrich, H., Kostorz, G., 2000. Small-angle neutron scattering investigation of the early stages of decomposition in Ni-rich Ni–Ti. J. Appl. Crystallogr. 33, 488–491.

Kompatscher, M., 2001. Phase Separation in Ni-rich Ni-Ti: the metastable states. Ph.D. Thesis, ETH Zürich.

Kompatscher, M., Bär, M., Hecht, J., Muheim, C., Kohlbrecher, J., Kostorz, G., Wagner, W., 2002. A high temperature cell for in situ small-angle neutron scattering studies of phase separation in alloys. Nucl. Instrum. Methods Phys. Res. A 495, 40–47.

Kompatscher, M., Schönfeld, B., Heinrich, H., Kostorz, G., 2003. Phase separation in Ni-rich Ni–Ti: the metastable states. Acta Mater. 51, 165–175.

Kostorz, G., 1979. Small-angle scattering and its applications to materials science. In: Kostorz, G. (Ed.), Neutron Scattering. Academic, New York, pp. 227–289.

Kostorz, G., 1982. Inorganic substances. In: Glatter, O., Kratky, O. (Eds.), Small-angle X-ray Scattering. Academic, London, pp. 467–498.

Kostorz, G., 1988a. Small-angle scattering in materials science. Macromol. Chem. 15, 131–151.

Kostorz, G., 1988c. Experimental studies of ordering and decomposition processes in alloys. In: Komura, S., Furukawa, H. (Eds.), Dynamics of Ordering Processes in Condensed Matter. Plenum, New York, London, pp. 199–210.

Kostorz, G., 1991. Small-angle scattering studies of phase separation and defects in inorganic materials. J. Appl. Crystallogr. 24, 444–456.

Kostorz, G., 1992. Scattering from structural defects. In: Fontana, M., Rusticelli, F., Coppola, R. (Eds.), Industrial and Technological Applications of Neutrons. Soc. Italiana di Fisica, Bologna, pp. 85–111.

Kostorz, G., 1993. Phase separation and structural defects by scattering techniques. Phys. Scr. T 49, 636–643.

Kostorz, G., 1995. Short-range order, slip coarsening and slip instabilities in alloys. In: Ananthakrishna, G., Kubin, L.P., Martin, G. (Eds.), Non-linear Phenomena in Materials Science III, Solid State Phenomena. Trans Tech Publications, Switzerland, pp. 187–194.

Kostorz, G., 1998. Metastable precipitates in aluminium alloys. In: Ciach, R. (Ed.), Advanced Light Alloys and Composites. Kluwer, Dordrecht, The Netherlands, pp. 221–232.

Kostorz, G., 2001. Small-angle scattering and the decomposition of alloys. In: Morawiec, H., Stróż, D. (Eds.), Applied Crystallography XVII. World Scientific, Singapore, pp. 65–78.

Kostorz, G., 2003. Small-angle neutron scattering and phase separation in alloys. Z. Kristallogr. 218, 154–159.

Kraft, K.P., Bergamaschi, A., Broennimann, Ch., Dinapoli, R., Eikenberry, E.F., Henrich, B., Johnson, I., Mozzanica, I.A., Schlepütz, C.M., Willmott, P.R., Schmitt, B., 2009. Performance of single-photon-counting PILATUS detector modules. J. Synchrotron Radiat. 16, 368–375.

Kresch, M., Delaire, O., Stevens, R., Lin, J.Y.Y., Fultz, B., 2007. Neutron scattering measurements of phonons in nickel at elevated temperatures. Phys. Rev. B 75, 7(104301).

Krivoglaz, M.A., 1969. Theory of X-ray and Thermal Neutron Scattering by Real Crystals. Plenum, New York.

Krivoglaz, M.A., 1996. X-ray and Neutron Diffraction in Nonideal Crystals. Springer, Berlin.

Krycka, K.L., Booth, R., Borchers, J.A., Chen, W.C., Conlon, C., Gentile, T.R., Hogg, C., Ijiri, Y., Laver, M., Maranville, B.B., Majetich, S.A., Rhyne, J.J., Watson, S.M., 2009. Resolving 3D magnetism in nanoparticles using polarization analyzed SANS. Physica B 404, 2561–2564.

Kyobu, J., Murata, Y., Morinaga, M., 1994. Computer simulation of local atomic displacements in alloys. Application to Guinier–Preston zones in Al–Cu. J. Appl. Crystallogr. 27, 772–781.

Larson, B.C., 2009. X-ray diffuse scattering near Bragg reflections for the study of clustered defects in crystalline materials. In: Barabash, R.I., Ice, G.E., Turchi, P.E.A. (Eds.), Diffuse Scattering and the Fundamental Properties of Materials. Momentum Press, LLC, New Jersey, pp. 139–160.

Lazzari, R., Leroy, F., Renaud, G., 2007. Grazing-incidence small-angle X-ray scattering from dense packing of islands on surfaces: development of distorted wave Born approximation and correlation between particle sizes and spacing. Phys. Rev. B 76, 14, 125411.

Leitner, M., Vogl, G., 2011. Quasi-elastic scattering under short-range order: the linear regime and beyond. J. Phys. Condens. Matter 23, 8(254206).

Leitner, M., Sepiol, B., Stadler, L.M., Pfau, B., Vogl, G., 2009. Atomic diffusion studied with coherent X-rays. Nat. Mater. 8, 717–720.

Lengeler, B., Eisenberger, P., 1980. Extended X-ray absorption fine structure analysis of interatomic distances, coordination numbers, and mean relative displacements in disordered alloys. Phys. Rev. B 21, 4507–4520.

Levine, J.R., Cohen, J.B., Chung, Y.W., Georgopoulos, P., 1989. Grazing-incidence small-angle X-ray scattering: new tool for studying thin film growth. J. Appl. Crystallogr. 22, 528–532.

Li, Y., Egetenmeyer, N., Gavilano, J.L., Barišić, N., Greven, M., 2011. Magnetic vortex lattice in $HgBa_2CuO_{4+\delta}$ observed by small-angle neutron scattering. Phys. Rev. B 83, 5(054507).

Lifshitz, I.M., Slyosov, V.V., 1961. The kinetics of precipitation from supersaturated solid solutions. J. Phys. Chem. Solids 19, 35–50.

Lindberg, V.W., McGervey, J.D., Hendricks, R.W., Triftshäuser, W., 1977. Annealing studies of voids in neutron irradiated aluminum single crystals by positron annihilation. Philos. Mag. 36, 117–128.

Lipson, H., Cochran, W., 1953. The Determination of Crystal Structures. Bell, London.

Löffler, J.F., Braun, H.B., Wagner, W., Kostorz, G., Wiedenmann, A., 2005. Magnetization processes in nanostructured metals and small-angle neutron scattering. Phys. Rev. B 71, 15(134410).

Long, G.G., Levine, L.E., 2005. Ultra-small-angle X-ray scattering from dislocation structures. Acta Crystallogr. A 61, 557–567.

Lovesey, S.W., 1984. In: Theory of Neutron Scattering from Condensed Matter, vol. 2. Clarendon, Oxford.

Lu, Z.W., Klein, B.M., Zunger, A., 1995. Ordering tendencies in Pd–Pt, Rh–Pt, and Ag–Au alloys. J. Phase Equilib. 16, 36–45.

Ma, D., Stoica, A.D., Wang, X.-L., Lu, Z.P., Xu, M., Kramer, M., 2009. Efficient local atomic packing in metallic glasses and its correlation with glass-forming ability. Phys. Rev. B 80, 6(014202).

Maier-Leibnitz, H., 1966. Grundlagen für die Beurteilung von Intensitäts- und Genauigkeitsfragen bei Neutronenstreumessungen. Nukleonik 8, 61–67.

Manley, M.E., Asta, M., Lashley, J.C., Retford, C.M., Hults, W.L., Taylor, R.D., Thoma, D.J., Smith, J.L., Hackenberg, R.E., Littrell, K., 2008. Soft-phonon feature, site defects, and a frustrated phase transition in $Ni_{50}Ti_{47}Fe_3$: experiments and first-principles calculations. Phys. Rev. B 77, 7 (024201).

Manns, T., Rothkirch, A., Scholtes, B., 2009. Residual stress determination in surface treated alumina samples applying beam limiting masks. Powder Diffr. 24 (Suppl. 1), S77–S81.

Marlaud, T., Deschamps, A., Bley, F., Lefebvre, W., Baroux, B., 2010a. Influence of alloy composition and heat treatment on precipitate composition in Al–Zn–Mg–Cu alloys. Acta Mater. 58, 248–260.

Marlaud, T., Deschamps, A., Bley, F., Lefebvre, W., Baroux, B., 2010b. Evolution of precipitate microstructures during the retrogression and re-ageing heat treatment of an Al–Zn–Mg–Cu alloy. Acta Mater. 58, 4814–4826.

Marty, A., Cénédèse, P., Bessière, M., Lefèbvre, S., Calvayrac, Y., 1994. Local chemical order in a $(Ni_3Fe)_{0.93}Cr_{0.07}$ single crystal. Phys. Rev. B 49, 15626–15636.

Massalski, T.B. (Ed.), 1990. Binary Alloy Phase Diagrams. ASM International, Materials Park, OH.

Mehaddene, T., Neuhaus, J., Petry, W., Hradil, K., Bourges, P., Hiess, A., 2008. Interplay of structural instability and lattice dynamics in Ni_2MnAl. Phys. Rev. B 78, 9 (104110).

Metzger, H., Behr, H., Peisl, J., 1982. The static Debye–Waller factor of defect induced lattice displacements. Z. Phys. B 46, 295–299.

Mezei, F. (Ed.), 1980. Neutron Spin Echo. Springer, Berlin.

Mezei, F., Pappas, C., Gutberlet, T. (Eds.), 2003. Neutron Spin Echo Spectroscopy—Basics, Trends and Applications. Springer, Berlin.

Miao, J.W., Charalambous, P., Kirz, J., Sayre, D., 1999. Extending the methodology of X-ray crystallography to allow imaging of micrometre-sized non-crystalline specimens. Nature 400, 342–344.

Michels, A., Weissmüller, J., 2008. Magnetic-field-dependent small-angle neutron scattering on random anisotropy ferromagnets. Rep. Prog. Phys. 71, 37 (066501).

Miracle, D.B., 2004. A structural model for metallic glasses. Nat. Mater. 3, 697–702.

Mirebeau, I., Parette, G., 2010. Neutron study of the short range order inversion in $Fe_{1-x}Cr_x$. Phys. Rev. B 82, 5, 104203.

Mittelbach, R., Glatter, O., 1998. Direct structure analysis of small-angle scattering data from polydisperse colloidal particles. J. Appl. Crystallogr. 31, 600–608.

Morinaga, M., Soner, K., Kamimura, T., Othaka, K., Yukawa, N., 1988. X-ray determination of static displacements of atoms in alloyed Ni_3Al. J. Appl. Crystallogr. 21, 41–47.

Moss, S.C., Clapp, P.C., 1968. Correlation functions of disordered binary alloys III. Phys. Rev. 171, 764–777.

Mostoller, M., Nicklow, R.M., Zwehner, D.M., Lui, S.C., Mundenar, J.M., Plummer, E.W., 1989. Bulk and surface vibrational modes in NiAl. Phys. Rev. B 40, 2856–2872.

Moya, X., Gonzalez-Alonso, D., Manosa, L., Planes, A., Garlea, V.O., Logrosso, T.A., Schlagel, D.L., Zaretsky, J.L., Aksoy, S., Acet, M., 2009. Lattice dynamics in magnetic superelastic Ni–Mn–In alloys: neutron scattering and ultrasonic experiments. Phys. Rev. B 79, 7(214118).

Mozer, B., Keating, D.T., Moss, S.C., 1968. Neutron measurement of clustering in the alloy CuNi. Phys. Rev. 175, 868–867.

Mühlbauer, S., Pfleiderer, C., Böni, P., Forgan, E.M., Brandt, E.H., Wiedenmann, A., Keiderling, U., Behr, G., 2011. Time-resolved stroboscopic neutron scattering of vortex lattice dynamics in superconducting niobium. Phys. Rev. B 83, 12(184502).

Naudon, A., Goudeau, P., Halimaoui, A., Lambert, B., Bomchil, B., 1994. Small-angle X-ray scattering study of anodically oxidized porous silicon layers. J. Appl. Phys. 75, 780–784.

Nicklow, R.M., 1979. Phonons and defects. In: Kostorz, G. (Ed.), Neutron Scattering. Academic Press, New York, pp. 191–226.

Nicklow, R.M., Crummett, W.P., Williams, J.M., 1979. Neutron-scattering measurements of phonon perturbations in irradiated copper. Phys. Rev. B 20, 5034–5043.

Noyan, I.C., Cohen, J.B., 1987. Residual Stress—Measurement by Diffraction and Interpretation. Springer, New York.

Ochiai, S., Ueno, M., 1988. Composition dependence of the MS temperature in the beta′-NiAl compound. J. Jpn. Inst. Met. 52, 157–162.

Oddershede, J., Christiansen, T.L., Ståhl, K., Somers, M.A., 2010. Extended X-ray absorption fine structure investigation of nitrogen stabilized expanded austenite. Scr. Mater. 62, 290–293.

Ohshima, K., Harada, J., Moss, S.C., 1986. On the separation of split diffuse intensity maxima from a disordered Cu-Au alloy by an X-ray counter method. J. Appl. Crystallogr 19, 276–8278.

Oku, T., Kikuchi, T., Shinohara, T., Suzuki, J.-I., Ishii, Y., Takeda, M., Kakurai, K., Sasaki, Y., Kishimoto, M., Yokoyama, M., Nishihara, Y., 2009. Small-angle polarized neutron scattering study of spherical $Fe_{16}N_2$ nano-particles for magnetic recording tape. Physica B 404, 2575–2577.

Page, K., Hood, T.C., Proffen, Th., Nader, R.B., 2011. Building and refining complete nanoparticle structures with total scattering data. J. Appl. Crystallogr. 44, 327–336.

Pan, X., Wu, X., Chen, X., Mo, K., Almer, J., Haeffner, D.R., Stubbins, J.F., 2010. Temperature and particle size effects on flow localization of 9–12%Cr ferritic/martensitic steel by in situ X-ray diffraction and small angle scattering. J. Nucl. Mater. 398, 220–226.

Pedersen, J.S., 1999. Analysis of small-angle scattering data from micelles and microemulsions: free-form approaches and model fitting. Curr. Opin. Colloid Interface Sci. 4, 190–196.

Peisl, H., 1976. Defect properties from X-ray scattering experiments. J. Phys. 37. C7–C47–C7–53.

Peisl, H., Balser, R., Peters, H., 1974. Change of crystal dimensions due to alignment of anisotropic elastic dipoles. Phys. Lett. A 46, 263–264.

Petry, W., 1995. Dynamical precursors of martensitic transitions. J. Phys. IV Colloque C2, supplément au Journal de Physique III 5. C2–15-C2–28.

Petry, W., Vogl, G., Heidemann, A., Steinmetz, K.H., 1987. A quasi-elastic neutron scattering (QNS) study. Philos. Mag. A 55, 183–201.

Petry, W., Flottmann, T., Heiming, A., Trampenau, J., Alba, M., Vogl, G., 1988. Atomistic study of anomalous self-diffusion in bcc β-titanium. Phys. Rev. Lett. 61, 722–725.

Petry, W., Bartsch, E., Fujara, F., Kiebel, M., Sillescu, H., Farago, B., 1991. Dynamic anomaly in the glass transition region of orthoterphenyl. A neutron scattering study. Z. Phys. B (Condensed Matter) 83, 175–184.

Petry, W., Heiming, A., Trampenau, J., Vogl, G., 1999. On the diffusion mechanism in the bcc phase of the group 4 metals. Defect and Diffusion Forum 66-69, 157–174.

Pohl, J., Albe, K., 2009. Phase equilibria and ordering in solid Pt–Rh calculated by means of a refined bond-order simulation mixing model. Acta Mater. 57, 4140–4147.

Porod, G., 1951. Die Röntgenkleinwinkelstreuung von dichtgepackten kolloiden Systemen – I. Teil. Kolloid-Z. 124, 83–114.

Porod, G., 1982. General theory. In: Glatter, O., Kratky, O. (Eds.), Small-angle X-ray Scattering. Academic Press, London, pp. 17–51.

Portmann, M.J., Schönfeld, B., Kostorz, G., Altorfer, F., Kohlbrecher, J., 2003. Evaluation of diffuse neutron scattering at elevated temperatures and local decomposition in Ni–Au. Phys. Rev. B 68, 4(012103).

Poulsen, H.F., 2004. Three-dimensional X-ray Diffraction Microscopy. Springer, Berlin.

Preston, G.D., 1938. Structure of age-hardened aluminium–copper alloys. Nature 142, 570.

Ramesh, T.G., Rameseshan, S., 1971. Determination of the static displacement of atoms in a binary alloy system using anomalous scattering. Acta Crystallogr. A 27, 569–572.

Rao, P.V.M., Murthy, K.S., SurYanarayana, S.V., Naidu, S.V.N., 1993. X-ray determination of the Debye–Waller factors and order parameters of Ni_3Al alloys. J. Appl. Crystallogr. 26, 670–676.

Rasmussen, F.B., Baker, J., Nielsen, M., Feidenhansl, R., Johnson, R.L., 1997. Unusual strain relaxation in Cu thin films on Ni(001). Phys. Rev. Lett. 79, 4413–4416.

Raub, E.J., 1959. Metals and alloys of the platinum group. J. Less-Common Met. 1, 3–18.

Reimers, W., Kaysser-Pyzalla, A., Schreyer, A., Clemens, H. (Eds.), 2008. Neutrons and Synchrotron Radiation in Engineering Materials Science. Wiley-VCH, Weinheim.

Reinhard, L., Moss, S.C., 1994. Recent studies of short-range order in alloys: the Cowley theory revisited. Ultramicroscopy 52, 223–232.

Reinhard, L., Schönfeld, B., Kostorz, G., Bührer, W., 1990. Short-range order in α-brass. Phys. Rev. B 41, 1727–1734.

Reinhard, L., Robertson, J.L., Moss, S.C., Ice, G.E., Zschack, P., Sparks, C.J., 1992. Anomalous-X-ray-scattering study of local order in bcc $Fe_{0.53}Cr_{0.47}$. Phys. Rev. B 45, 2662–2676.

Renaud, G., Lazzari, R., Revenant, C., Barbier, A., Noblet, M., Ulrich, O., Leroy, F., Jupille, J., Borensztein, Y., Henry, C.R., Deville, J.-P., Scheurer, F., Mane-Mane, J., Fruchart, O., 2003. Real-time monitoring of growing nanoparticles. Science 300, 1416–1419.

Ribárik, G., Ungár, T., 2010. Characterization of the microstructure in random and textured polycrystals and single crystals by diffraction line profile analysis. Mater. Sci. Eng., A 528, 112–121.

Ribárik, G., Ungár, T., Gubicza, J., 2001. MWP-fit: a program for multiple whole-profile fitting of diffraction peak profiles by *abinitio* theoretical functions. J. Appl. Crystallogr. 34, 669–676.

Rodriguez, J.A., Moss, S.C., Robertson, J.L., Copley, J.R.D., Neumann, D.A., Major, J., 2006. Phys. Rev. B 74, 104–115.

Roelofs, H., Schönfeld, B., Kostorz, G., Bührer, W., 1995. Atomic short-range order in Cu-17 at.% Mn. Phys. Status Solidi B 187, 31–42.

Sayre, D., 1952. Some implications of a theorem due to Shannon. Acta Crystallogr. 5, 843–843.

Scardi, P., 2008. Recent advancements in whole powder pattern modelling. Z. Kristallogr. 27, 101–111.

Scardi, P., Leoni, M., 2002. Whole powder pattern modelling. Acta Crystallogr. A 58, 190–200.

Schaub, P., Weber, T., Steurer, W., 2010. Analysis and modelling of structural disorder by the use of the three-dimensional pair distribution function method exemplified by the disordered twofold superstructure of decagonal Al–Cu–Co. J. Appl. Crystallogr. 44, 134–149.

Schmatz, W., 1973. X-ray and neutron scattering studies on disordered crystals. In: Herman, H. (Ed.), Treatise on Materials Science and Technology. Academic, New York, pp. 105–229.

Schmatz, W., 1975. Neutron small-angle scattering from dislocations. La Rivista del Nuovo Cimento 5, 398–422.

Schmatz, W., 1978. Disordered structures. In: Dachs, H. (Ed.), Neutron Diffraction. Springer, Berlin chapter 5.

Schmidt, P.W., 1991. Small-angle scattering studies of disordered, porous and fractal systems. J. Appl. Crystallogr. 24, 414–435.

Schmidt, S., Olsen, U.L., Poulsen, H.F., Sørensen, H.O., Lauridsen, E.M., Margulies, L., Maurice, C., Juul Jensen, D., 2008. Direct observation of 3-D grain growth in Al-0.1% Mn. Scr. Mater. 59, 491–494.

Schneider, J.R., 1981. Applications of gamma-ray diffractometry. Nucl. Sci. Appl. Ser. A 1, 227–276.

Schneider, J.M., Schönfeld, B., Demé, B., Kostorz, G., 2000. Shape of precipitates in Ni–Al–Mo single crystals. J. Appl. Crystallogr. 33, 465–468.

Schober, M., Petry, W., 1993. Lattice vibrations. In: Gerold, V. (Ed.), Materials Science and Technology, vol. 1, Structure of Solids, pp. 289–355.

Schober, M., Eidenberger, E., Staron, P., Leitner, H., 2011. Critical consideration of precipitate analysis of Fe-1 at.% Cu using atom probe and small-angle neutron scattering. Microsc. Microanal. 17, 26–33.

Schönfeld, B., 1994. Short range order and pair interactions in binary nickel alloys. In: Turchi, P.E.A., Gonis, A. (Eds.), Statics and Dynamics of Alloy Phase Transformations. Plenum, New York, pp. 175–178.

Schönfeld, B., 1999. Local atomic arrangements in binary alloys. Prog. Mater. Sci. 44, 435–543.

Schönfeld, B., Kostorz, G., 2009. Elastic diffuse scattering of alloys: status and perspectives. In: Barabash, R.I., Ice, G.E., Turchi, P.E.A. (Eds.), Diffuse Scattering and the Fundamental Properties of Materials. Momentum Press, LLC, New Jersey, pp. 119–137.

Schönfeld, B., Traube, J., Kostorz, G., 1992. Short-range order and pair potentials in Au–Ag. Phys. Rev. B 45, 613–621.

Schönfeld, B., Roelofs, H., Malik, A., Kostorz, G., Plessing, J., Neuhäuser, H., 1996. The microstructure of Cu–Al. Acta Mater. 44, 335–342.

Schönfeld, B., Portmann, M.J., Yu, S.Y., Kostorz, G., 1999. The type of order in Cu-10 at.% Au—evidence from the diffuse scattering of X-rays. Acta Mater. 47, 1413–1416.

Schönfeld, B., Kostorz, G., Celino, M., Rosato, V., 2001. Static atomic displacement in Ni-rich Ni–Al. Europhys. Lett. 54, 482–487.

Schönfeld, B., Roelofs, H., Kostorz, G., Robertson, J.L., Zschack, P., Ice, G.E., 2008. Static atomic displacements in Cu–Mn measured with diffuse X-ray scattering. Phys. Rev. B 77, 144202-1–144202-8.

Schönfeld, B., Sax, C.R., Ruban, A.V., 2012. Atomic ordering in Au-(42 to 50) at.% Pd: a diffuse scattering and first-principles investigation. Phys. Rev. B 85, 11(014204).

Schwander, P., Schönfeld, B., Kostorz, G., 1992. Configurational energy change caused by slip in short-range ordered Ni–Mo. Phys. Status Solidi B 172, 73–85.

Schwartz, L.H., Cohen, J.B., 1987. Diffraction from Materials, second ed. Springer, Berlin.

Schweika, W., 1998. Disordered alloys. Diffuse scattering and Monte Carlo simulations (Lecture Notes in Physics). Springer, Berlin and Heidelberg.

Scopigno, T., Suck, J.-B., Angelini, R., Albergamo, F., Ruocco, G., 2006. High-frequency dynamics in metallic glasses. Phys. Rev. Lett. 96, 4(135501).

Serrano, J., Pineda, E., Bruna, P., Labrador, A., Le Tacon, M., Krisch, M., Monaco, G., Crespo, D., 2010. High frequency dynamics of BMG determined by synchrotron radiation: a microscopic picture. J. Alloys Compd. 495, 319–322.

Shapiro, S.M., Svensson, E.C., Vettier, C., Hennion, B., 1993. Uniaxial-stress dependence of the phonon behavior in the premartensitic phase of $Ni_{62.5}Al_{37.5}$. Phys. Rev. B 48, 13223–13229.

Shapiro, S.M., Vorderwisch, P., Habicht, K., Hradil, K., Schneider, H., 2007. Observation of phasons in the magnetic shape memory alloy Ni_2MnGa. Europhys. Lett. 77, 56004.

Simon, J.P., 2007. Contribution of synchrotron radiation to small-angle X-ray scattering studies in hard condensed matter. J. Appl. Crystallogr. 40, S1–S9.

Simon, J.P., Lyon, O., 1994. Anomalous small-angle scattering in materials science. In: Materlik, G., Sparks, C.J., Fischer, K. (Eds.), Resonant and Anomalous X-ray Scattering. North-Holland, Amsterdam, pp. 305–322.

Simon, J.P., Mainville, J., Ziegler, B., Lyon, O., 1992. A multilayer analyser suitable for linear detection of anomalous small-angle X-ray scattering. J. Appl. Crystallogr. 25, 785–788.

Sinha, S.K., 2009. Diffuse scattering from surfaces, interfaces and thin films. In: Barabash, R.I., Ice, G.E., Turchi, P.E.A. (Eds.), Diffuse Scattering and the Fundamental Properties of Materials. Momentum Press, LLC, New Jersey, pp. 3–34.

Sinha, S.K., Sirota, E.B., Garoff, S., Stanley, H.B., 1988. X-ray and neutron scattering from rough surfaces. Phys. Rev. B 38, 2297–2311.

Sköld, K., Mueller, M.H., Brun, T.O., 1979. Hydrogen in metals. In: Kostorz, G. (Ed.), Neutron Scattering. Academic Press, New York, pp. 423–460.

Skripov, A.V., Udovic, T.J., Rush, J.J., Uimin, M.A., 2011. A neutron scattering study of hydrogen dynamics in coarse-grained and nanostructured ZrCr2H3. J. Phys. Condens. Matter 23, 8(065402).

Sparks, C.J., 1974. Inelastic resonance emission of X-rays: anomalous scattering associated with anomalous dispersion. Phys. Rev. Lett. 33, 262–265.

Sparks, C.J., Borie, B., 1966. Methods of analysis for diffuse X-ray scattering modulated by local order and atomic displacement. In: Cohen, J.B., Hilliard, J.E. (Eds.), Local Atomic Arrangements Studied by X-ray Diffraction. Gordon and Breach, New York, pp. 5–50.

Specht, E.D., Walker, F.J., Liu, Wenjun, 2010. X-ray microdiffraction analysis of radiation-induced defects in single grains of polycrystalline Fe. J. Synchrotron Radiat. 17, 250–256.

Springer, T., 1977. Molecular rotations and diffusion in solids, in particular hydrogen in metals. In: Lovesey, S.W., Springer, T. (Eds.), Dynamics of Solids and Liquids by Neutron Scattering. Springer, Berlin, pp. 255–300.

Staron, P., Eidenberger, E., Schober, M., Sharp, M., Leitner, H., Schreyer, A., Clemens, H., 2010. In situ small-angle neutron scattering study of the early stages of precipitation in Fe-25 at.% Co-9 at.% Mo and Fe-1 at.% Cu at 500°C. J. Phys. Conf. Ser. 247, 8(012038).

Steiner, Ch., Schönfeld, B., Portmann, M.J., Kompatscher, M., Kostorz, G., Mazuelas, A., Metzger, T., Kohlbrecher, J., Demé, B., 2005. Local order in Pt-47 at.% Rh measured with x-ray and neutron scattering. Phys. Rev. B 71, 7(104204).

Steiner, Ch., Schönfeld, B., Van der Klis, M.M.I.P., Kostorz, G., Patterson, B.D., Willmott, P.R., 2006. Near-surface microstructure of Pt–Rh(110) and (111) surfaces. Phys. Rev. B 73, 7(174205).

Stern, E.A., 1974. Theory of the extended X-ray-absorption fine structure. Phys. Rev. B 10, 3027–3037.

Steuwer, A., Dumont, M., Altenkirch, J., Birosca, S., Deschamps, A., Prangnell, P.B., Withers, P.J., 2011. A combined approach to microstructure mapping of an Al–Li AA2199 friction stir weld. Acta Mater. 59, 3002–3011.

Stoller, R.E., Walker, F.J., Specht, E.D., Nicholson, D.M., Barabash, R.I., Zschack, P., Ice, G.E., 2007. Diffuse X-ray scattering measurements of point defects and clusters in iron. J. Nucl. Mater. 367–370, 269–275.

Sutton, M., 2006. X-ray intensity fluctuation spectroscopy. In: Hippert, F., Geissler, E., Hodeau, J., Lelièvre-Berna, E., Regnard, J. (Eds.), Neutron and X-ray Spectroscopy. Springer, Berlin, pp. 297–318.

Sutton, M., Mochrie, S.G.J., Greytag, T., Nagler, S.E., Berman, L.E., Held, G.A., Stephenson, G.B., 1991. Observation of speckle by diffraction with coherent X-rays. Nature 352, 608–610.

Tanner, L.E., Shriver, D., Shapiro, S.M., 1990. Electron microscopy and neutron scattering studies of premartensitic behavior in ordered Ni–Al β_2 phase. Mater. Sci. Eng. A 127, 205–213.

Teixeira, J., 1988. Small-angle scattering by fractal systems. J. Appl. Crystallogr. 21, 781–785.

Thomson, R., Levine, L.E., Long, G.G., 1999. Small-angle scattering by dislocations. Acta Crystallogr. A 55, 433–447.

Tiballs, J.E., 1975. The separation of displacement and substitutional disorder scattering: a correction for structure-factor ratio variation. J. Appl. Crystallogr. 8, 111–114.

Tietze, H., Müllner, M., Renker, B., 1984. Dynamical properties of premartensitic NiTi. J. Phys. C 17, L529–L532.

Trampenau, J., Bauszus, K., Petry, W., Herr, U., 1995. Vibrational behaviour of nanocrystalline Ni. Nanostruct. Mater. 6, 551–554.

Trinkaus, H., 1972. On determination of the double-force tensor of point defects in cubic crystals by diffuse X-ray scattering. Phys. Status Solidi B 51, 307–319.

Trinkaus, H., 1975. Scattering of X-rays from crystals containing a random distribution of defects. Z. Naturforsch. 28a, 980–994.

Van Hove, L., 1954. Correlations in space and time and Born approximation scattering in systems of interacting particles. Phys. Rev. 95, 249–262.

Vogl, G., Petry, W., Flottmann, T., Heiming, A., 1989. Direct determination of the self-diffusion mechanism in bcc β-titanium. Phys. Rev. B 39, 5025–5034.

Volkov, V.V., Svergun, D.I., 2003. Uniqueness of *ab initio* shape determination in small-angle scattering. J. Appl. Crystallogr. 36, 860–864.

Von Guérard, B., Grasse, D., Peisl, J., 1980. Structure of defect cascades in fast-neutron-irradiated aluminum by diffuse X-ray scattering. Phys. Rev. Lett. 44, 262–265.

Vrijen, J., Aalders, J., van Dijk, C., Radelaar, S., 1980. Neutron scattering study on the kinetics of clustering in CuNi alloys. Phys. Rev. B 22, 1503–1514.

Vyskocil, P., Pedersen, J.S., Kostorz, G., Schönfeld, B., 1997. Small-angle neutron scattering of precipitates in Ni-rich Ni–Ti alloys – I. Metastable states in poly- and single crystals. Acta Mater. 45, 3311–3318.

Wagner, C., 1961. Theorie der Alterung von Niederschlägen durch Umlösen (Ostwald-Reifung). Z. Elektrochem. 65, 581–591.

Wagner, C.N.J., 1966. Analysis of the broadening and changes in position of peaks in an X-ray powder pattern. In: Cohen, J.B., Hilliard, J.E. (Eds.), Local Atomic Arrangements Studied by X-ray Diffraction. Gordon and Breach, New York, pp. 219–269.

Wagner, V., Bellmann, D., 2007. Bulk domain sizes determined by complementary scattering methods in polycrystalline Fe. Physica B 397, 27–29.

Wagner, R., Kampmann, R., Voorhees, P.W., 2001. Homogeneous second-phase precipitation. In: Kostorz, G. (Ed.), Phase Transformations in Materials. Wiley-VCH, Weinheim, pp. 309–407.

Walker, C.B., Guinier, A., 1953. An X-ray investigation of age hardening in Al–Ag. Acta Metall. 1, 568–577.

Warren, B.E., 1969. X-ray Diffraction. Addison-Wesley, Reading, MA.

Warren, B.E., 1999. X-ray Diffraction. Dover, New York.

Weber, T., Steurer, W., 2009. Structural disorder in quasicrystals. In: Barabash, R.I., Ice, G.E., Turchi, P.E.A. (Eds.), Diffuse Scattering and the Fundamental Properties of Materials. Momentum Press, LLC, New Jersey, pp. 239–258.

Welberry, T.R., Butler, B.D., 1994. Interpretation of diffuse X-ray scattering via models of disorder. J. Appl. Crystallogr. 27, 205–231.

Wiedenmann, A., Wagner, W., Wollenberger, H., 1989. Thermal decomposition of Fe-34 at.% Ni between 625°C and 725°C. Scr. Metall. 23, 603–605.

Wilkens, M., 1969. The mean square stress $<\sigma^2>$ for restrictedly random distributions of dislocations in a cylindrical body. Acta Metall. 17, 1155–1159.

Wilkens, M., 1970. The determination of density and distribution of dislocations in deformed single crystals from broadened X-ray diffraction profiles. Phys. Status Solidi A 2, 359–370.

Wilkens, M., 1975. Quantitative interpretation of X-ray line broadening of plastically deformed crystals. J. Appl. Crystallogr. 8, 191–192.

Williams, R.O., 1972. ORNL Report No. 4828. National Laboratory, Oak Ridge, TN.

Williams, R.O., 1976. ORNL Report No. 5140. National Laboratory, Oak Ridge, TN.

Williams, C., May, R.P., Guinier, A., 1993. Small-angle scattering of X-rays and neutrons. In: Lifshin, E. (Ed.), Materials Science and Technology. Materials Characterization, vol. 2B. VCH Verlagsgesellschaft, Weinheim, pp. 611–656.

Willis, B.T.M., Pryor, A.W., 1975. Thermal Vibrations in Crystallography. Cambridge University Press, London.

Withers, P.J., Turski, M., Edwards, L., Bouchard, P.J., Buttle, D.J., 2008. Recent advances in residual stress measurement. Int. J. Pressure Vessels Piping 85, 118–127.

Woo, W., Choo, H., Prime, M.B., Feng, Z., Clausen, B., 2008. Microstructure, texture and residual stress in a friction-stir-processed AZ31B magnesium alloy. Acta Mater. 56, 1701–1711.

Xu, R., Hong, H., Chiang, T.-C., 2009. Probing phonons and phase transitions in solids with X-ray thermal diffuse scattering. In: Barabash, R.I., Ice, G.E., Turchi, P.E.A. (Eds.), Diffuse Scattering and the Fundamental Properties of Materials. Momentum Press, LLC, New Jersey, pp. 161–178.

Yu, S.Y., Schönfeld, B., Kostorz, G., 1997. Diffuse X-ray scattering of Ag-13.4 at.% Al. Phys. Rev. B 56, 8535–8541.

Yu, S.Y., Schönfeld, B., Heinrich, H., Kostorz, G., 2004. Short-range order in h.c.p. Ag–Al. Prog. Mater. Sci. 49, 561–579.

Zabel, H., Robinson, I.K. (Eds.), 1992. Surface X-ray and Neutron Scattering. Springer Proceedings in Physics, vol. 61. Springer, Berlin.

Zachariasen, W.H., 1945. Theory of X-ray Diffraction in Crystals. Wiley, New York.

Zhang, F., Allen, A.J., Levine, L.E., Ilavsky, J., Long, G.G., Sandy, A.R., 2011. Development of ultra-small-angle X-ray scattering–X-ray photon correlation spectroscopy. J. Appl. Crystallogr. 44, 200–212.

Zhao, G.L., Harmon, B.N., 1992. Phonon anomalies in β-phase Ni_xAl_{1-x} alloys. Phys. Rev. B 45, 2818–2824.

Zhu, X.-M., Feidenhans'l, R., Zabel, H., Als-Nielsen, J., Du, R., Flynn, C.P., Grey, F., 1988. Grazing-incidence X-ray scattering on the $Cu_3Au(111)$ phase transition. Phys. Rev. B 37, 7157–7160.

Zolliker, M., Bührer, W., Schönfeld, B., 1992. Inelastic scattering and elastic diffuse scattering from the shape-memory alloy β'-Cu–Zn. Physica B 180/181, 303–305.

Biography

Gernot Kostorz, Professor Emeritus of Physics at ETH Zurich, German citizen, studied physics at the University of Göttingen. He held research positions at Argonne National Laboratory, USA, Institut Laue-Langevin, Grenoble, France, and Max Planck Institute (Metal Research), Stuttgart. His research and teaching focused on the relationship between microstructure and properties of real solids, using various methods. In 1980, he took a chair in Physics at ETH Zurich. With his students and colleagues, he has published more than 250 scientific papers, and he (co-)edited numerous books and conference volumes. He is an honorary member of several learned societies, and in 2005, he received the "Heyn Denkmünze" of the DGM. He served as Editor-in-Chief of the journals of the International Union of Crystallography (2005–2012) and continues as Co-Editor of "Journal of Applied Crystallography".

14 Structure, Composition and Energy of Solid–Solid Interfaces

James M. Howe, Department of Materials Science and Engineering, University of Virginia, Charlottesville, VA, USA

14.1 Introduction

It is fairly safe to say that with the current status of (scanning) transmission electron microscopy [(S)TEM] and atom probe tomography, it is presently possible to determine the structure and composition of most material interfaces at atomic resolution (see e.g. Ene et al., 2005; Kelly and Miller, 2007; Varela et al., 2012). The purpose of this chapter is not to recount the many interfaces that have been analyzed in recent years, but to provide a thorough description of the science of interfaces, both homophase and heterophase, using analytical models, and to illustrate the validity of these descriptions as revealed by experimental and computational results. This description concentrates on the structure, composition and interfacial energy of the interfaces. There are many other important interfacial properties that could be discussed, such as growth kinetics, strength, roughening and melting, but they are discussed elsewhere (Wolf and Yip, 1992; Howe, 1993a, 1997; Gottstein and Shvindlerman, 2010). It is also the intent of this chapter to reveal areas that need further exploration in the future. The chapter starts by considering the structure and composition of homophase boundaries, the main class being grain boundaries, and this is followed by heterophase boundaries, which comprise a wide range of interface types varying from structurally coherent and chemically diffuse to structurally incoherent and chemically sharp. The chapter begins with an overview of different types of homophase and heterophase interfaces, to acquaint the reader with some important principles governing all types of interfaces, as well as to illustrate the variations in structure and composition that one can expect at interfaces in solids.

14.1.1 Types of Solid–Solid Interfaces

There are various ways of classifying solid–solid interfaces. A solid–solid interface that arises from an orientation difference or a translation between two crystals of the same phase across an interface is referred to as a homophase interface. Examples of homophase interfaces include interfaces such as grain boundaries, twin boundaries and stacking faults. These interfaces are discussed in Section 14.2. Strictly speaking, antiphase boundaries (APBs) in ordered alloys also fall within this definition, but because the displacement associated with APBs leads to wrong nearest-neighbor compositional bonds across the interface, this type of interface is treated as a heterophase interface after discussion of the Cahn–Hilliard (CH) (Cahn and Hilliard, 1958) theory of interfacial structure. Interfaces that separate two crystals that differ in composition, Bravais lattice or both composition and lattice are referred to as heterophase interfaces. These interfaces are discussed in Section 14.3. Heterophase interfaces are often divided into three classes based on the degree of atomic matching or coherency across the interface, including (Christian, 1975; Howe et al., 2000):

(i) fully coherent interfaces, where there is complete continuity of atoms and planes across the interface between the two phases,
(ii) partly coherent interfaces, in which the disregistry between two crystal structures across the interface is accommodated by periodic misfit dislocations in the interface, and
(iii) incoherent interfaces, where atomic matching is sufficiently poor that there is no correspondence of atoms or planes across the interface even locally.

Fully coherent (unrelaxed) and partly coherent (relaxed) planar interfaces are commonly found in lattice-matched heterostructures, such as in semiconductor devices, for example, while a good example of an incoherent interface is illustrated by interfaces associated with the massive transformation. The terms commensurate and incommensurate are also used to describe interfaces that possess long-range order, such as fully coherent interfaces, and those which possess no long-range order, such as incoherent interfaces. Partly coherent interfaces are referred to as discommensurate interfaces in this terminology. A thorough discussion of the terminology of different types of interfaces, particularly as they apply to bulk, semi-bulk and thin-film materials, is provided by Wolf (1992). This chapter uses the commonly accepted definitions of coherent, partly coherent and incoherent homophase and heterophase interfaces described above. These are sufficiently general to include all different types of crystal interfaces in physical metallurgy, whether they lie parallel to low-index or high-index crystal planes. Zhang and Yang (2011) have proposed a fourth class of interface, which is intermediate between partly coherent and incoherent interfaces, called a CS-coherent interface, but this distinction is not used herein. The following sections examine the atomic structures of each of these different types of interfaces in detail.

14.1.2   Atomic Matching at Solid–Solid Interfaces

If one considers the different possible solid–solid homophase and heterophase interfaces that can form in metals by combining both ordered and disordered alloys along low-index planes, a very long list results. If the possibility of combining different types of materials across the interface, for example metals, ceramics, semiconductors and polymers, the list becomes almost endless. This may cause one to wonder how it is possible to understand and quantify the enormous variety of potential interfaces. Fortunately, all of these different interfaces seek to minimize their interfacial free energy and this is usually accomplished by maximizing atomic matching and minimizing elastic strain at the interface. Thus, there is a strong tendency for the planes and directions with the highest atomic densities to align across solid–solid interfaces in alloys, as well as between dissimilar materials. This behavior has been demonstrated for dissimilar material interfaces such as metal–ceramic interfaces (Howe, 1993a) and even ceramic–polymer interfaces (Mauritz et al., 1978), illustrating its wide range of applicability. Unfortunately, while this fundamental principle is relatively easy to understand physically, it is not always easy to quantify or predict analytically when defects, elastic strains and different types of atomic bonds are present at an interface.

High-resolution TEM (HRTEM) has been an invaluable technique for determining the atomic structures of homophase and heterophase interfaces, that is internal interfaces, in materials (Sinclair et al., 1990; Howe and Mahon, 1989; Wolf and Merkle, 1992). To illustrate the fundamental principle of minimizing the interfacial energy by maximizing atomic bonding and matching of close-packed planes and directions across solid–solid interfaces, the following sequence of figures shows experimental HRTEM images of a number of different types of solid–solid interfaces. Metal alloys [mostly face-centered cubic (FCC)] are used in these examples but identical interfaces can be found in semiconductor and ceramic materials as well, and references to these are included in the discussion of each interface. All of the HRTEM images were obtained by orienting the interfaces edge-on along low-index <110>, <111> and < 11$\overline{2}$0 > directions in the FCC, body-centered cubic (BCC) and hexagonal close-packed (HCP) crystals, respectively. The white dots in the images can be imagined to represent the projections of the atomic columns in the structures viewed along these directions.

Two homophase interfaces are shown in **Figures 1 and 2**. **Figure 1** shows an HRTEM image of a {111} twin interface in FCC Cu (Luzzi, 1991). The {111} planes are edge-on and horizontal in the image so that one can see the ABCABC... stacking of the close-packed planes progressing vertically in the image. Note the mirror symmetry across the twin interface indicated by an arrow. Since the atom planes across the twin interface only change their stacking order and there are no missing bonds across the interface, one expects this to be a relatively low-energy interface. Twins are a common planar defect in many different materials (Murr, 1975; Niewczas, 2007).

Figure 2(a) shows a symmetrical tilt grain boundary in Au (Wolf and Merkle, 1992). Comparison with the adjacent hard-sphere model in **Figure 2(b)** makes it possible to imagine that this boundary could be formed by joining two vicinal (111) surfaces, each containing [1$\overline{1}$0] steps. When the surfaces are placed in contact and bonding occurs across the interface, the (111) steps become dislocations with associated elastic strain fields. Between the dislocations, the atoms adopt a configuration that is similar to the close-packed planes in both grains. The resulting interface thus has periodic regions of good matching (as evidenced by the continuity of the other set of (11$\overline{1}$) planes and the (002) planes across the interface) with the mismatch being accommodated by regions of localized strain at the dislocation cores in between. Because of the high proportion of

Figure 1   HRTEM image of a Σ = 3 {111} twin in Cu imaged along a <110> zone axis. The ABC... stacking of {111} planes is indicated in the figure. Adapted from Luzzi (1991) with permission.

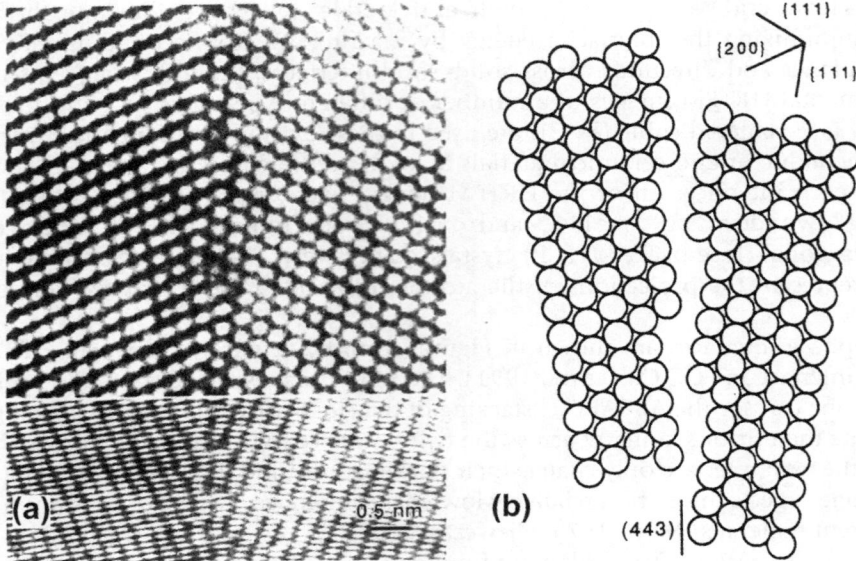

Figure 2 (a) HRTEM image of a (443) symmetrical tilt grain boundary in FCC Au. Note the misfit localization and the strong tendency to maintain coherency between the close-packed planes. The compressed image at the bottom illustrates these features more clearly. Adapted from Wolf and Merkle (1992) with permission. (b) Hard-sphere model of a (443) symmetrical tilt grain boundary composed of two vicinal {111} surfaces. The two grains are slightly separated for clarity.

atomic matching across the interface when the crystals are joined, one expects it to have an energy that is lower than the combined energy of the two vicinal (443) surfaces but not as low as the twin interface in **Figure 1**. Without any calculation, one can argue that the energy of a symmetrical tilt grain boundary should increase proportional to the rotation angle, since the density of steps (dislocations) increases with angle from the singular (111) surface orientation, as occurs for solid surfaces (Blakely, 1973).

Examples of several different types of heterophase interfaces are shown in the following figures. **Figure 3** shows an HRTEM image of an Ag-rich Guinier-Preston (GP) zone in an Al–Ag alloy (Howe and Gronsky, 1986). The Ag and Al atoms are almost the same size so that clustering of Ag atoms (darker region) leads to a precipitate that has the same crystal structure as the Al matrix but a different composition. This type of orientation relationship (OR) between phases is called a cube-on-cube OR. There is a tendency for these precipitates to facet along the {111} and {200} planes at low temperatures and one facet along {111} is indicated by arrows. This interface represents a perfectly coherent interface with a change in composition only across the interface. The behavior of such an interface can be quantified using the CH (Cahn and Hilliard, 1958) and discrete lattice plane models (Lee and Aaronson, 1980) of interfacial structure. Many other examples of coherent interfaces, where there is a change in composition without a change in lattice type across the interface, can be found in semi-conductor materials such as Si–SiGe or GaAs–AlGaAs multilayers, for example (Ourmazd et al., 1989), as well as for other precipitates (Howe et al., 2007; Yoon et al., 2007).

In contrast to **Figure 3**, the (111) Co–Ni martensite interface in **Figure 4** is an example of a perfectly coherent interface where there is a change in Bravais lattice but no change in composition across the

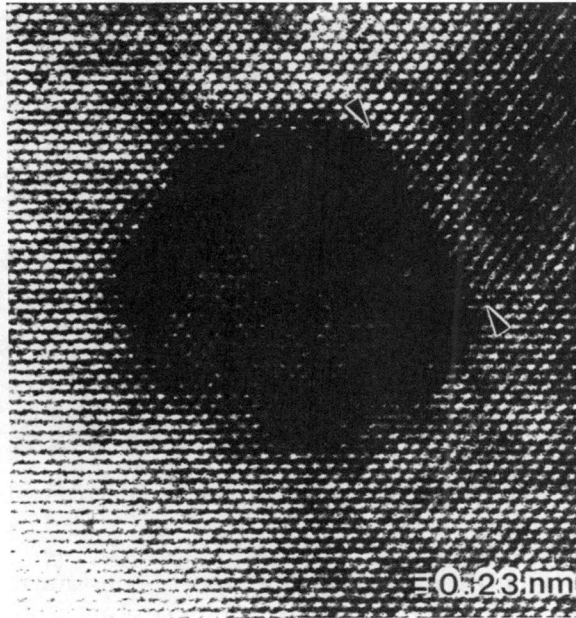

Figure 3 HRTEM image of an Ag-rich G.P. zone in an Al–Ag alloy. Arrows indicate a {111} facet. Adapted from Howe and Gronsky (1986) with permission.

interface (Howe, 1993b). Note that this interface is atomically flat because any perturbation in the interface produces a high elastic distortion that is not favorable energetically. Unlike the G.P. zone in the previous figure where the Ag atoms make the image dark, the white spots have the same intensity in both phases at the interface in **Figure 4** because there is no difference in composition, similar to the twin boundary in **Figure 1**.

An example of a partly coherent interface between two materials with the same Bravais lattice and cube-on-cube OR, but different compositions and lattice parameters, is illustrated by the (111) Cu–Ag interface in **Figure 5** (Rao, 1994). In this case, the lattice parameters of Cu and Ag differ by about 12% so that misfit dislocations are necessary about every nine planes along the interface and two such dislocations are indicated by arrows in the figure. Again notice that the interface is atomically flat and that there is excellent atomic matching across the interface between the misfit dislocations, which accommodate the mismatch locally at the interface. Similar examples can be found in semiconductor materials such as CdTe on GaAs, for example (Schwartzman, 1990).

Figure 6 shows an example of an incoherent interface between a faceted ZrN precipitate in Zr in a Zr–N alloy (Li et al., 2004). Here the ZrN phase is atomically flat along the {111} interface, while the adjacent Zr matrix presents a high-index plane to the interface. The OR between the two phases across the interface is high index and given approximately as $[450]_{Zr}//[\bar{1}0\bar{1}]_{ZrN}$; $(002)_{Zr}//(\bar{1}31)_{ZrN}$. Notice that all of the low-index planes, that is, the $(\bar{1}11)$, $(11\bar{1})$ and (020) planes, in the $[\bar{1}0\bar{1}]$ ZrN grain are visible in **Figure 6**, while only one set of planes, that is, the (002) planes, is visible in the [450] Zr matrix. There is no evidence of any planar matching or localized strain at the interface. This is a fairly

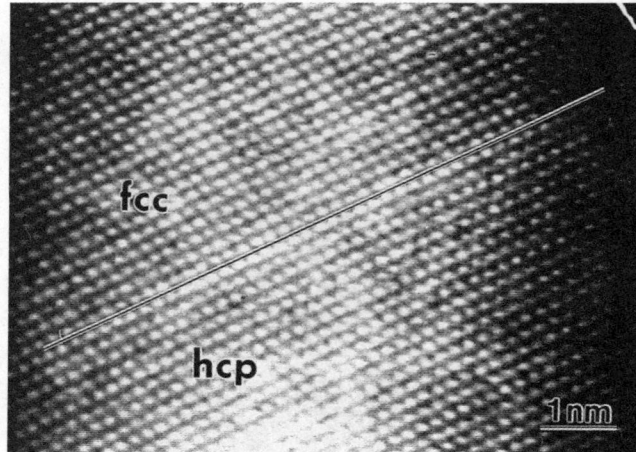

Figure 4   HRTEM image of an FCC–HCP martensite interface in Co–Ni alloy. Adapted from Howe (1993b) with permission.

common type of incoherent interface found in the massive transformation; some random grain boundaries (RGB's) can be quite similar. Another classic example of an incoherent interface is found for Ge particles in an Al–Ge alloy (Dahmen, 1994a). In this case, the Al and Ge lattice parameters differ by more than 30%. In addition to this misfit, the directional covalent bonding in Ge is much different than the metallic bonding in Al. In spite of this lack of atomic matching across the interface, the interface

Figure 5   HRTEM image of a partly coherent {111} Cu–Ag interface in a Cu–Ag alloy aged to grow Ag precipitates in the Cu matrix. The Ag precipitates facet along low-index planes such as the {111} plane indicated by a line. Arrows indicate the cores of periodic misfit dislocations in the interface. Adapted from Rao (1994) with permission.

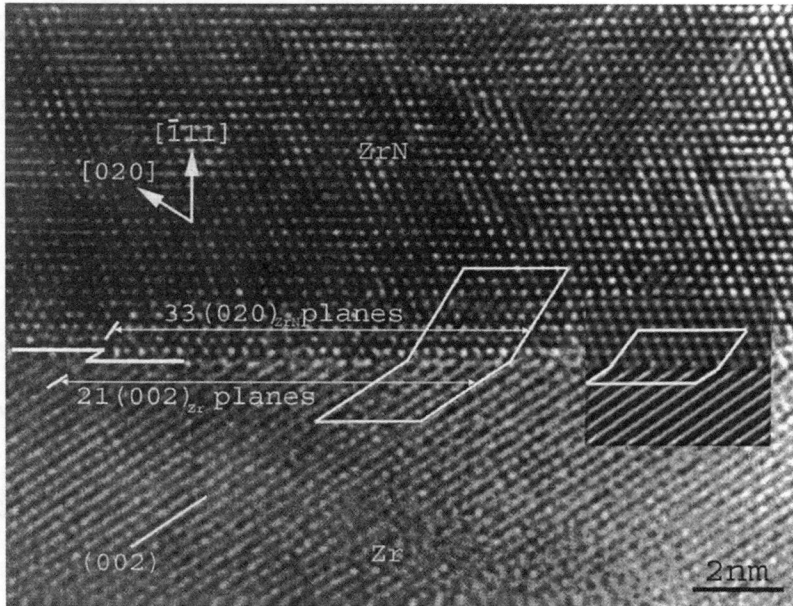

Figure 6   HRTEM image of an incoherent (incommensurate) {111} ZrN interface in a Zr–N alloy nitrided to grow ZrN grains from the surface. A simulated HRTEM image of the interface is superimposed and various other features indicated by white lines are discussed later in the text. Adapted from Li et al. (2004) with permission.

plane (IP) parallels the common close-packed {111} planes in the two phases, indicating a tendency for both phases to minimize the interfacial energy by faceting along common close-packed planes. The coherent and partly coherent interfaces shown above are generally more common than the incoherent interface shown in **Figure 6**.

In the previous heterophase interfaces, the IP was atomically flat and parallel to one or both of the common close-packed planes in the crystals (except perhaps for the G.P. zone, where the interface was not so well defined for reasons which will become apparent with the CH theory). In the final example, consider the possibility of having an interface that is not parallel to low-index crystal planes. **Figure 7** shows the $(474)_\gamma$ habit plane of an ordered BCC (B2–TiAl) precipitate in an ordered face-centered tetragonal (FCT) γ-TiAl matrix (Das et al., 1996). The OR is such that one set of close-packed planes and directions is common to both phases and the high-index $(474)_\gamma$ habit plane is formed by introducing atomic steps spaced only a few atoms apart along the common close-packed planes. It is relatively easy to imagine that the interface in **Figure 7** could lie parallel to any arbitrary crystal plane by varying the density (i.e. the spacing) of the atomic steps along the common close-packed planes in the interface, similar to a vicinal solid–vapor interface. There is considerable experimental evidence that this in fact occurs. However, unlike the solid–vapor case, these steps have an associated elastic strain field that accommodates the shape change between the two crystal structures and therefore, increasing the density of steps increases the interfacial energy, similar to the case of the symmetrical tilt grain boundary in **Figure 2**. High-index habit planes with atomic ledges such as those in **Figure 7** occur in both diffusional and martensitic transformations

Figure 7   HRTEM image of the $(474)_\gamma$ interface between ordered BCC and FCT structures of TiAl in a Ti–Al–Mo alloy. Adapted from Das et al. (1996) with permission.

(Mahon et al., 1989; Howe et al., 2009). Such interfaces have been observed in semiconductors (Batstone, 1992) and ceramics (Norton and Carter, 1992) and appear to be quite general. Interfacial defects that possess both step and dislocation character have been termed disconnections (Pond and Hirth, 1994; Hirth and Pond, 1996; Howe et al., 2009), and these defects can vary from pure step to pure dislocation character.

In summary, the previous HRTEM images and accompanying descriptions illustrate that there is a strong tendency for solid–solid interfaces in materials to lie along close-packed planes to minimize their interfacial energy. The interfaces tend to be planar for the same reason. Different phases also display a strong tendency to align their close-packed directions across the interface. This creates good atomic matching, that is low-energy interfaces, between the materials and leads to the familiar ORs that one finds in many materials. In cases where it is not energetically favorable to have parallel close-packed planes, there is still a substantial driving force to produce atomic row matching along relatively close-packed rows of atoms (Frank, 1953; Dahmen, 1982; Fecht and Gleiter, 1985; Howe, 1997; Zhang and Kelly, 2009). It is also possible to produce high-index solid–solid interfaces by forming vicinal interfaces, that is a terrace–ledge–kink structure on low-index planes (Howe et al., 1987), similar to solid–vapor interfaces (Burton et al., 1950).

14.2 Structure and Composition of Homophase Interfaces

14.2.1 Grain Boundary Structure

A number of different analytical models have been proposed to describe the structure and properties of grain boundaries. The models vary both in their approach and degree of sophistication. Several of the more important ones are discussed in the following sections. The discussion progresses in approximate order of historical development, beginning with the early interpretation of low-angle grain boundaries in terms of dislocation networks by Read and Shockley (1950); Read (1953), and following with comprehensive atomistic simulations by Wolf and Merkle (1992); Wolf (1992). Most of the analytical treatments that have been developed for grain (homophase) boundaries can be similarly applied to interphase (heterophase) interfaces. To emphasize the generality of the methods and to avoid duplication in subsequent sections on heterophase interfaces, extension of the treatments from homophase to heterophase interfaces is included with the discussion of various models in this section on homophase interfaces.

14.2.1.1 Dislocation Models

14.2.1.1.1 Symmetrical Tilt Grain Boundary

One of the simplest types of grain boundaries to visualize is a symmetrical tilt grain boundary, where two grains on either side of the boundary are related by symmetrical rotations about an axis lying in the boundary plane. **Figure 8** illustrates such a boundary in a simple cubic structure, where the boundary was formed by joining two crystals having surfaces that are rotated from a cube plane by $\pm\theta/2$ about a <100> axis. When the two surfaces in **Figure 8(a)** are joined, the steps on them become edge dislocations of Burgers vector **b**, where the magnitude of the Burgers vector $b = |\mathbf{b}|$ is equal to the step height h. A similar tilt boundary for an atomistic model with (111) surfaces and a $[1\bar{1}0]$ tilt axis was previously shown in **Figure 2(b)**. Joining the crystals in **Figure 8(b)** requires only elastic strain except where planes of atoms terminate at the boundary in an edge dislocation, indicated by the symbol $\perp$.

From the geometry in **Figure 8(b)**, the number of dislocations per unit length in the grain boundary is given by

$$1/D = (2 \sin \theta/2)/\mathrm{b} \tag{1a}$$

or when θ is small by

$$1/D \sim \theta/\mathrm{b}, \tag{1b}$$

where D is the distance between the grain boundary dislocations. This boundary constitutes a wall of dislocations and Read and Shockley, 1950) have calculated the energy of such an array, situated in an infinite medium of shear modulus μ and Poisson's ratio v. An important consideration in their analysis is that each dislocation has an elastic strain energy per unit length E_D, which varies according to the spacing between the dislocations D. Thus, the elastic strain energy must be integrated as a function of θ to find its angular dependence. The grain boundary energy γ_{GB} is then given by the product of the

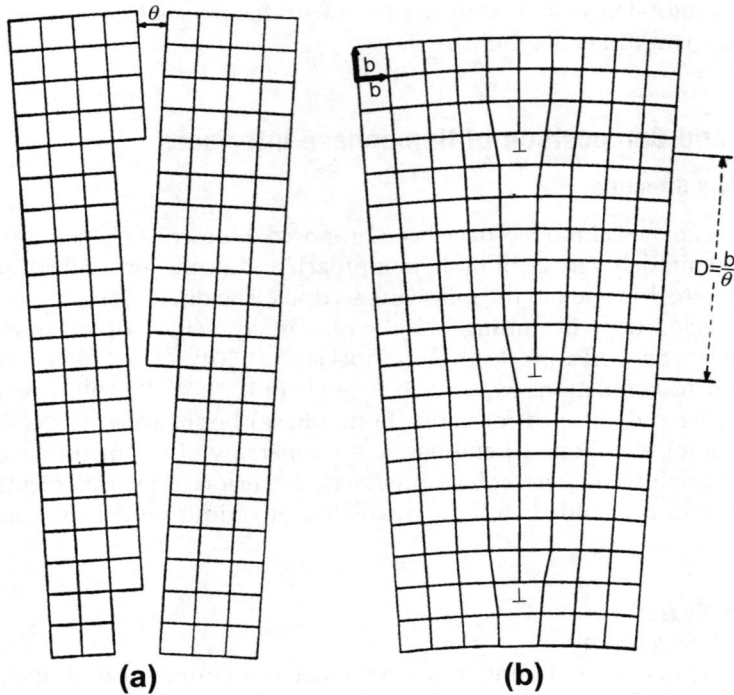

Figure 8 A symmetrical tilt grain boundary in a simple cubic crystal. The plane of the figure is parallel to a cube face and normal to the axis of rotation of the two grains. (a) The two grains are rotated by an angle θ and (b) joined to form a bicrystal (grain boundary). Adapted from Read (1953) with permission.

energy per unit length of the dislocations E_D times the density of the dislocations as $\gamma_{GB} = E_D(1/D) = E_D(\theta/b)$. The resulting expression for the energy per unit area of the grain boundary γ_{GB} is given by

$$\gamma_{GB} = E_o\theta(A_o - \ln\theta), \tag{2a}$$

where

$$E_o = \mu b/4\pi(1-v) \text{ and} \tag{2b}$$

$$A_o = 1 + \ln(b/2\pi r_o) \tag{2c}$$

and r_o represents the core energy of a single boundary dislocation. The grain boundary energy γ_{GB} has units of energy/area (mJ/m^2). The term A_o in parentheses in Eqn (2a) depends on the total core energy of the dislocation per unit area of the boundary, while the $\ln\theta$ term accounts for the elastic energy of the boundary. The first term represents a constant energy per dislocation and it leads to an increase in the energy, which is proportional to the density of the dislocations. The second term decreases as θ

increases, because the stress fields of the dislocations overlap and cancel one another as D decreases. Hence, this equation can only be applied to boundaries having a small angle of tilt ($\theta \sim 10\text{--}15°$) such that the cores of the dislocations do not overlap. Such boundaries are referred to as low-angle tilt boundaries. Also note that the Read–Shockley equation (Eqn (2a)) shows that it is energetically favorable for two low-angle tilt boundaries with misorientations θ_1 and θ_2 to combine to form a single tilt boundary with angle ($\theta_1 + \theta_2$). In addition, differentiation of Eqn (2a) predicts that γ_{GB} has a maximum at $\theta = \exp(A_0 - 1)$.

A plot of Eqn (2a) is shown in **Figure 9** (Kelly and Groves, 1970), where it is seen that the cusp at $\theta = 0$ is very sharp when the boundary disappears, that is $d\gamma_{GB}/d\theta$ becomes infinite at $\theta = 0$. This is because a long-range stress field is established when an isolated step is pressed into another crystal to make a grain boundary dislocation. Thus, the energy of a tilt boundary rises steeply as its angle increases from zero because the strain field of each dislocation spreads out to very large distances in the crystal when they are widely separated. In the tilt boundary, the slope becomes less steep as the stress fields approach one another and begin to cancel, and eventually the slope becomes almost constant at high tilt angles. Using this dislocation concept of a low-angle symmetrical tilt grain boundary, one sees that the boundary has characteristics that are similar to vicinal solid–vapor interfaces (Howe, 1997), such as a low-energy cusp at low tilt angles, but that the nature of the elastic strains is an additional important part of this solid–solid interface.

Although the dislocation model was developed for low-angle grain boundaries where the dislocation cores are well defined and surrounded by regions of relatively well-bonded crystal, it can be extended on a purely geometrical basis to high-angle tilt boundaries. If the angular tilt is continued beyond the low-angle regime, the energy is expected to increase only slightly as the density of dislocation cores increases and shallow cusps are expected to occur at particular angles of tilt where the

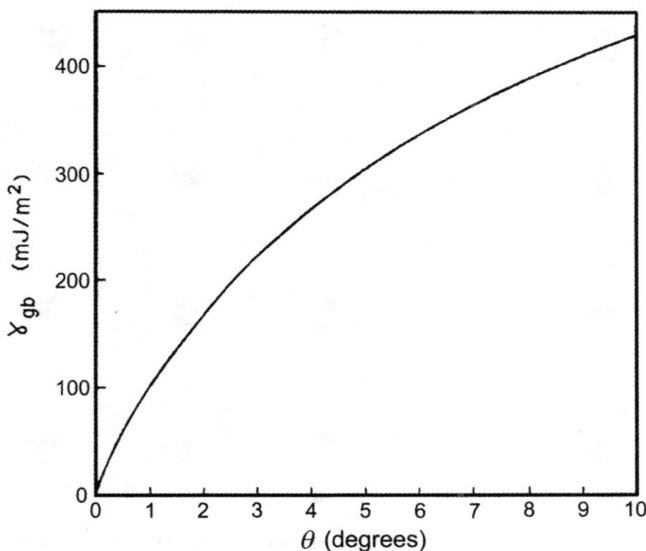

Figure 9 Energy of a tilt grain boundary as a function of tilt angle θ with $E_0 = 1450$ mJ/m^2 (appropriate for Cu) and $A_0 = 0$. Adapted from Read (1953) with permission.

dislocations are uniformly spaced. According to the formula for a simple cubic lattice, this occurs when $\cot \theta/2 = 2n$, where n is an integer. For example, when $\cot \theta/2 = 14$, there is one dislocation on every seventh cube plane, or when $\cot \theta/2 = 2$ (or $\theta = 53°$), the structure is particularly simple and forms a twin plane, as shown in **Figure 10** (Shewmon, 1965). In this case, the twin plane is (210) and the atoms in the twin boundary lie on the lattices of both grains. The density of atoms on the boundary plane is high and the resulting grain boundary energy is low. This type of boundary is sometimes referred to as a special high-angle grain boundary.

A uniform dislocation spacing in a high-angle grain boundary such as in **Figure 10** only results when the dislocation spacing is an integral number of lattice planes terminating at the boundary. When the misorientation changes in a symmetrical high-angle tilt grain boundary, the dislocation spacing ideally varies from uniform to nonuniform. A nonuniform grain boundary can be described as consisting of a uniform dislocation array and a superimposed nonuniform array. For example, a nonuniform 60° symmetrical tilt grain boundary can be thought of as a 53° tilt boundary with a superimposed 7° tilt boundary. With further tilt to 62°, the boundary once again becomes uniform, but with a higher dislocation density. The variation in energy of such symmetrical tilt boundaries has been calculated assuming that the energy consists of the sum of the two superimposed dislocation arrays and is shown in **Figure 11** (Gleiter, 1971). Note the cusp at 53° tilt in **Figure 11** corresponding to the (210) boundary in **Figure 10**.

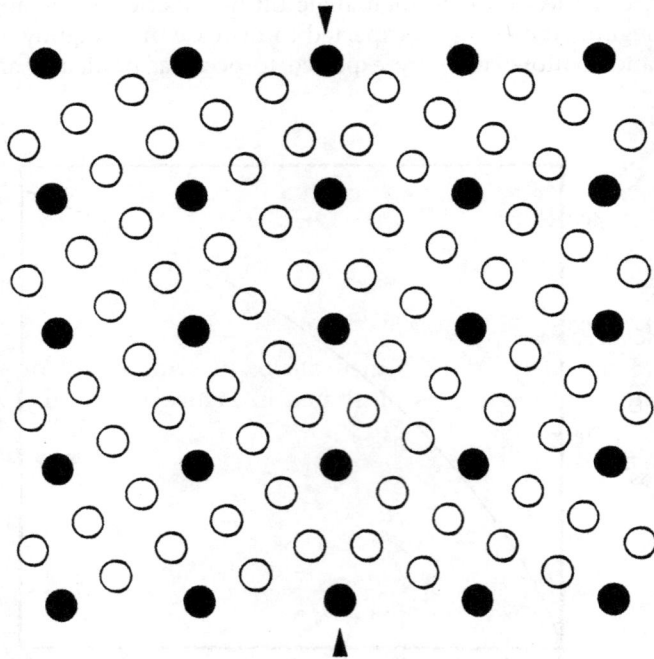

Figure 10   A good matching high-angle tilt boundary in a simple cubic lattice. The (210) twin boundary (arrows) is normal to the plane of the figure and the dark circles represent atoms that lie on points of lattices of both grains. Adapted from Shewmon (1965) with permission.

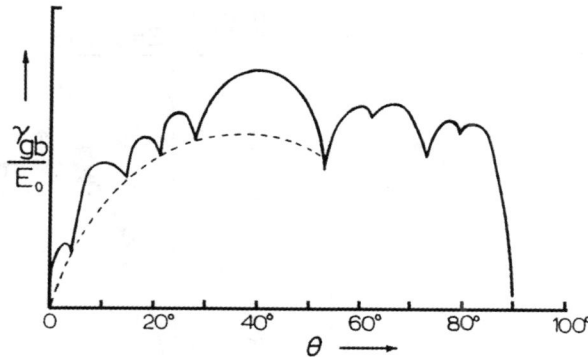

Figure 11 Relative grain boundary energy versus misorientation angle for symmetrical tilt boundaries, which vary in uniformity of dislocation arrays. Cusps correspond to uniform arrays. Adapted from Gleiter (1971) with permission.

Both atomistic calculation and experimental data suggest that the cusps shown in **Figure 11** exist. For example, **Figure 12(a) and (b)** compare calculated and experimental data for the grain boundary energies of symmetrical <100> and <110> grain boundaries in FCC Al, using a Morse potential for Al (Hasson et al., 1972; Balluffi, 1982). The atomistic calculations are similar to, but more realistic than, the pure dislocation model described above because they allow for atomic relaxations and translations along the boundary that are not included in the dislocation model. The experimental and calculated curves in **Figure 12** are in generally good agreement, although the small cusps associated with the <100> tilt boundary in **Figure 12(a)** are not evident experimentally in **Figure 12(b)** because of their relatively small magnitude compared with the scatter in the data. In contrast, the large cusps at about 70 and 129° in **Figure 12(c) and (d)** are clearly evident and these are associated with low-energy {111} and {113} twin boundaries, which are possible with rotation about the <110> axis. The {111} twin boundary was shown previously in **Figure 10**. (The notation Σ indicated along the top of the figures is explained in Section 14.2.1.3.1).

14.2.1.1.2 Asymmetrical Tilt Grain Boundary

The tilt boundary in **Figure 8(b)** can be turned out of its symmetrical orientation by rotating it about the tilt axis through an angle ϕ as illustrated in **Figure 13** (Read, 1953). In this case, the angle ϕ is the angle between the boundary plane and the mean [100] direction of the two grains. The boundary makes an angle of $(\phi + \theta/2)$ with the [100] direction in one grain and an angle of $(\phi - \theta/2)$ with the [100] direction in the other grain. **Figure 8** is a thus a special case where $\phi = 0$ or 90°. When the boundary becomes asymmetrical, edge dislocations with extra planes that are normal to those of the original set are introduced and it can be shown that their spacing is given by Read (1953):

$$D_2 = b_2/(\theta \sin \phi) \tag{3a}$$

while the spacing of the original set is reduced to

$$D_1 = b_1/(\theta \cos \phi). \tag{3b}$$

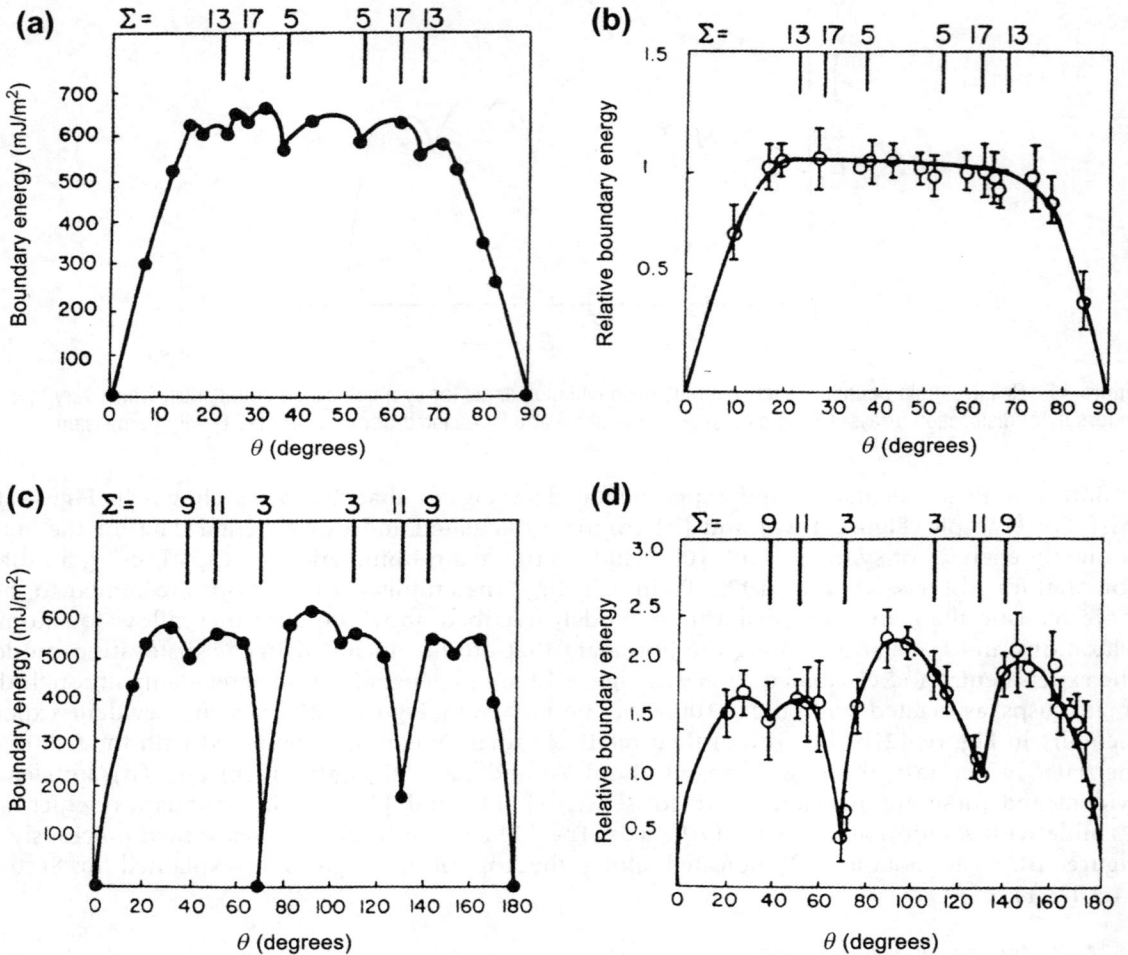

Figure 12   (a,c) Calculated and (b,d) experimental grain boundary energies as a function of misorientation angle for (a,b) <100> and (c,d) <110> symmetrical tilt boundaries in Al. Adapted from Hasson et al. (1972) and Balluffi (1982) with permission.

The new dislocations increase the energy of the boundary sharply as ϕ increases from zero because they are far apart. Read and Shockley (1950) showed that the energy of this boundary has the same form as the symmetrical tilt boundary with an additional dependence on ϕ, and that a sharp cusp exists when $\phi = 0$.

14.2.1.1.3 Twist Grain Boundary
A more drastic change in the grain boundary structure is produced by rotating the tilt boundary in **Figure 8** through 90° about an axis in the plane of the boundary and normal to the tilt axis. The boundary is then normal to the <100> axis about which the two grains are rotated and it is called

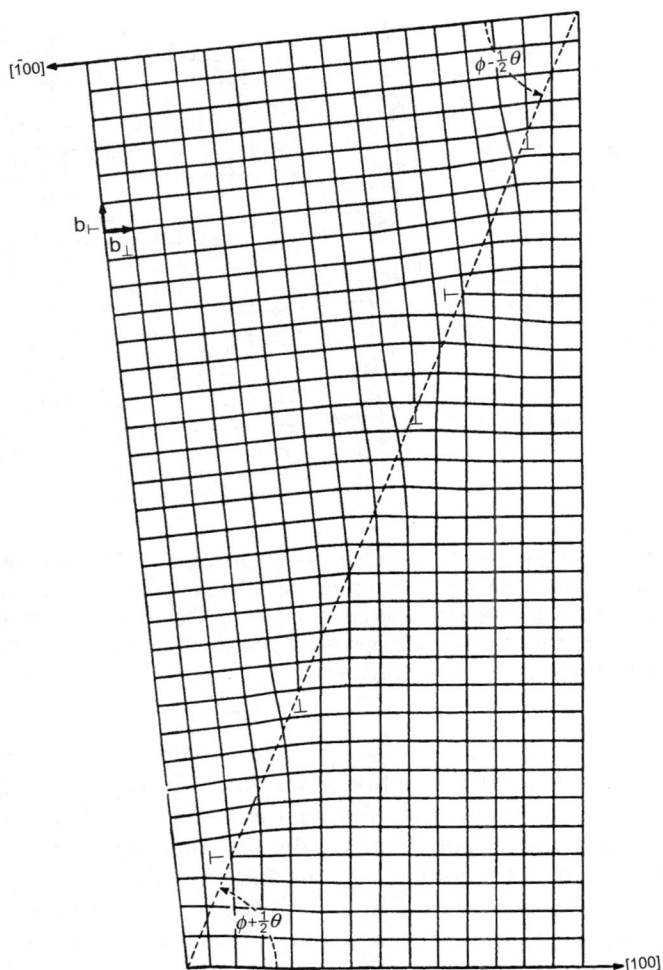

Figure 13 The same grain boundary as in **Figure 8** except that the plane of the boundary makes an arbitrary angle ϕ with the mean (010) planes in the two grains. Note the new set of perpendicular dislocations introduced into the boundary. Adapted from Read (1953) with permission.

a twist boundary. It consists of a grid of screw dislocations, as shown in **Figure 14** (Read, 1953). The screw dislocations are visible as regions of poor matching where there is a shear displacement parallel to the lines of distortion. It can be shown (Kelly and Groves, 1970) that if a vector is drawn parallel to a <100> direction in the mean lattice between the two grains in a twist boundary such as **Figure 14**, the number of <010> screw dislocations, which intersects per unit length, is given by the same formula as that for the number of edge dislocations in a tilt boundary in Eqn (1), or $1/D = (2 \sin \theta/2)/b$, where b is the magnitude of the Burgers vector of the screw dislocation. Thus, the energy of a twist boundary increases with the angle of twist in the same general way as the energy of a tilt boundary increases with the angle of tilt.

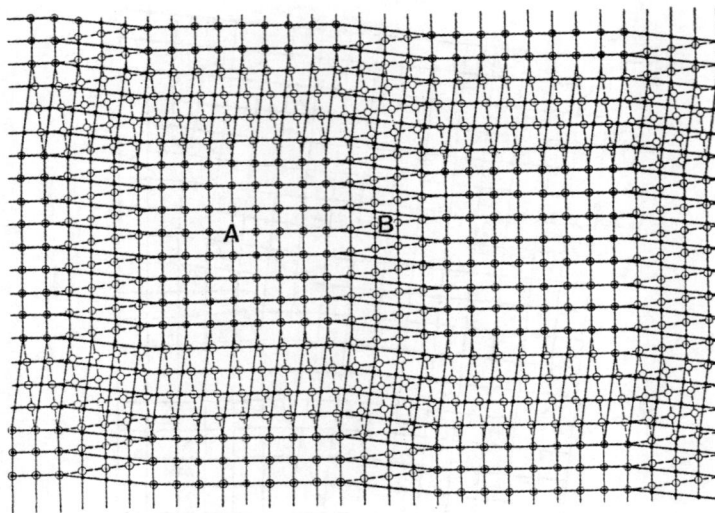

Figure 14   A pure (relaxed) twist grain boundary between two simple cubic crystals. The boundary is in the plane of the figure and the two grains have a small rotation about the cube axis normal to the boundary. The open circles represent atoms just above the plane of the boundary in one grain and the solid circles atoms just below in the opposite grain Adapted from Read (1953) with permission. Atoms at the interface have been relaxed to produce regions of good atomic matching (A) separated by regions of poor matching (B), which are the screw dislocations.

As for tilt boundaries, the energy of a twist boundary should be cusped at angles of twist at which the atoms fit well together in the boundary, although atomistic calculation has shown that these cusps are generally much shallower. For example, the 53° rotation about <100> that produced a cusp in the energy of the cubic tilt boundary also produces a twist boundary normal to <100> on which the atoms fit uniformly, as it should since the misorientation between the crystals is the same. This is shown in **Figure 15** (Shewmon, 1965), where the net of lattice points common to both grains and the similar net of coincident points lying on the corresponding tilt boundary are indicated in each figure. Thus, both boundaries lie on a plane of the coincident atom sites in the two crystals.

Unlike the edge dislocation networks shown in **Figures 2 and 8**, it is not possible to reveal the atomic structure of screw dislocation networks by HRTEM because their Burgers vector lies parallel to the line direction of the dislocation, that is there is no displacement of the atomic columns perpendicular to the viewing direction when the boundary is edge-on. However, these dislocations can be observed by diffraction contrast imaging techniques in the TEM (Hirsch et al., 1979; Forwood and Clarebrough, 1991) (as can edge or mixed dislocations). **Figure 16(a)** shows a bright-field TEM image of the dislocation structure in a (001) pure twist low-angle boundary in Au (Schober and Balluffi, 1970; Tan et al., 1975). The interface is oriented normal to the viewing direction and the dislocations are visible as dark lines in the interface. The boundary consists of an orthogonal array of screw dislocations in agreement with the model shown in **Figure 14**. In their investigation, Schober and Balluffi, (1970) examined a series of such twist boundaries with misorientations of 1–9° and used contrast analyses to establish that the two sets of dislocations in the boundary had Burgers vectors $\mathbf{b} = 1/2[110]$ and

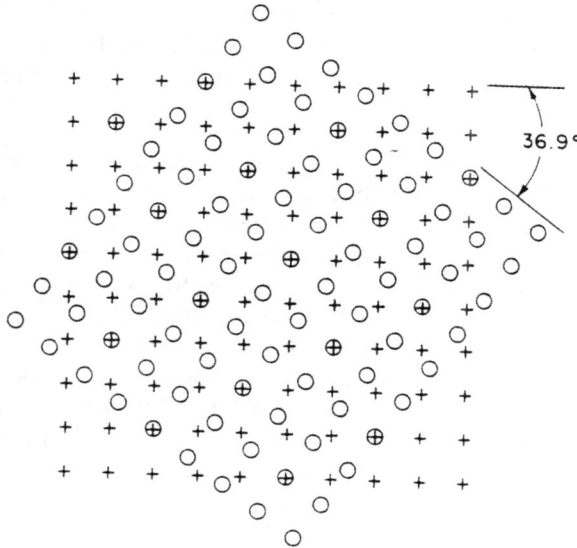

Figure 15 Twist boundary of good fit in a simple cubic lattice. The boundary is parallel to the plane of the figure. One lattice is indicated by circles and the other by crosses so that the coincident positions are easily visible. Adapted from Shewmon (1965) with permission.

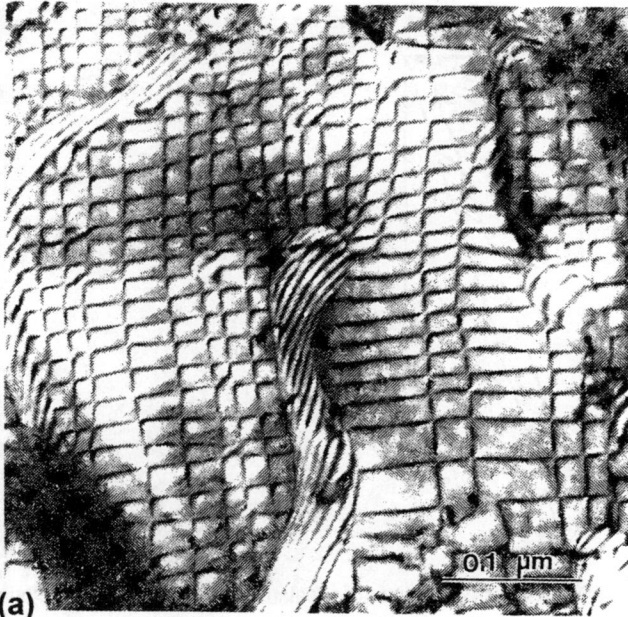

(a)

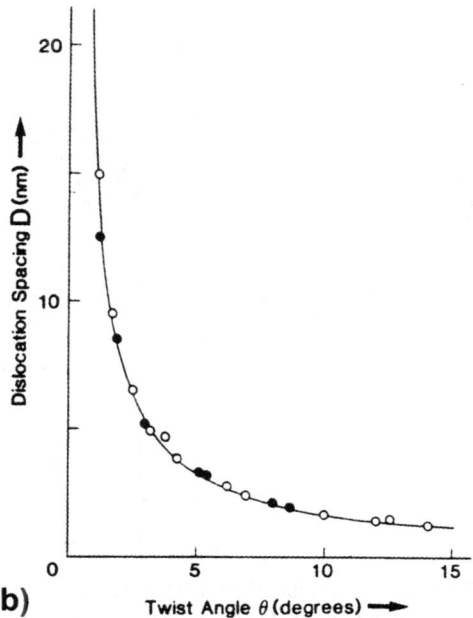

(b)

Figure 16 (a) Bright-field TEM image of a cross-grid of screw dislocations in a pure twist boundary ($\theta \sim 1°$) in Au (from [34]). (b) Comparison of experimental and calculated spacings of dislocations for (001) twist boundaries in Au. Adapted from Forwood and Clarebrough (1991) with permission. The filled circles are the results of Schober and Balluffi (1970) and the open circles are from the results of Tan et al. (1975).

$1/2[1\bar{1}0]$. Their results showed that the screw dislocation spacing in the grain boundary varied with θ as expected from the simple relation $D = b/\theta$, as illustrated in **Figure 16(b)**. The same type of data were also generated for low-angle tilt boundaries and found to agree with the predictions of Eqn (1).

14.2.1.1.4 *Degrees of Freedom of a General Grain Boundary*

The symmetrical tilt boundary in **Figure 8** is a special type of boundary that can be specified by the axis of tilt and the single angle θ. A general grain boundary has five degrees of freedom (DOF), or five quantities that are required to define it. These are often referred to as the five macroscopic DOF of a general grain boundary. Three values are required to specify the rotation that brings the lattice of one grain parallel to the other, and two more values are required to specify the orientation of the boundary plane with respect to one of the grains. These features are illustrated with respect to **Figure 17** (Read, 1953). In this figure, one can imagine making a wedge-shaped cut in a crystal along two arbitrary planes, taking out the wedge-shaped material between the two cuts, and bringing the two grains together with a twist. The resulting boundary has five DOF. If l is taken as a unit vector parallel to the axis of relative rotation of the two grains, the rotation between the two grains can be represented by the vector:

$$\mathbf{u}_{l\theta} = \mathbf{l}\theta \tag{4}$$

which has three independent components or DOF. The orientation of the boundary may then be specified by a unit vector **n** normal to the boundary plane, which requires two additional DOF. A pure

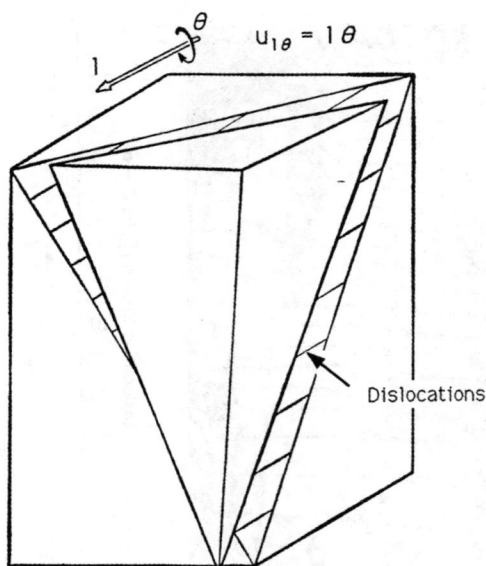

Figure 17 Formation of a general grain boundary with five DOF. In this figure, the misorientation is specified by a rotation θ about a common axis I in both grains. In a pure tilt boundary, I lies in the plane of the boundary and leads to a set of parallel edge dislocations, indicated by lines on the exposed crystal surface. The vector I is normal to the boundary plane in a pure twist boundary. Adapted from Read (1953) with permission.

twist boundary is thus defined as one where l is parallel to **n**, and for a pure tilt boundary l is perpendicular to **n**. The quantity lθ is often called the axis–angle pair for the grain boundary (Randle, 1996).

There are various ways of expressing the crystallography of a grain boundary using the five DOF (Wolf (1992)). One method that is similar to **Figure 17**, but emphasizes the importance of the interface plane relative to the two grains, is called the interface plane (IP) notation. In this description, one creates a general grain boundary by a twist rotation θ about the common grain boundary plane normal, which is parallel to the surface normals, n_1 and n_2, of the two grains that form the grain boundary. This interface plane notation developed for grain boundaries is similar to the notation that is often used to describe the IP, or OR, between two different phases at a heterophase interface. In the case of two different phases α and β, the interface is often specified by a set of parallel (hkl) planes and [uvw] directions in the two phases as

$$(hkl)_\alpha \parallel (hkl)_\beta; \quad [uvw]_\alpha \parallel [uvw]_\beta, \tag{5}$$

where the symbol $\parallel$ is used to indicate the parallel relationship and the [uvw] directions are contained in the parallel adjoining (hkl) planes (Edington, 1975). Specifying a set of parallel directions within the (hkl) planes is similar to specifying a rotation angle θ in the interface plane notation. It is often found that heterophase interfaces do not lie along the parallel set of planes in the two phases given by Eqn (5). In this case, it is common to specify the OR between the two phases according to Eqn (5) and then to denote the IP as a (hkl) plane in either (or both) of the phases. For the situation of precipitates in a matrix, the matrix plane is often chosen. An example of such notation is shown by the $(474)_\gamma$ interface in **Figure 7** and additional examples are discussed in the sections on heterophase interfaces.

In addition to the five macroscopic DOF, there are three independent translational (or so-called microscopic) DOF for a grain boundary involving translations $T = (T_x, T_y, T_z)$ parallel (x,y) and perpendicular (z) to the IP. Computer simulations and HRTEM have shown that such translations frequently occur at grain boundaries. From a thermodynamic viewpoint the z-component of T perpendicular to the interface is important because it accounts for any volume expansion in the interface. Such an excess free volume at the interface is expected to be closely related to its excess free energy and to give rise to stresses near the interface that are similar to the well-known surface stress (Somorjai, 1972; Broughton and Gilmer, 1983; Cammarata, 1994). It is later shown in a section on atomistic modeling of grain boundaries that this excess free volume per unit area of the interface ΔV_i is an important parameter related to the grain boundary energy.

14.2.1.1.5 *Frank's Formula for the Dislocation Content of a Boundary*

The dislocation content of a general grain boundary can be determined according to the theory of Frank (1951). The same procedure can be applied to heterophase interfaces where there is a change in crystal structure across the interface (Jesser, 1973; Christian, 1975). This theory is illustrated for a grain boundary geometrically in **Figure 18** (Forwood and Clarebrough (1991)). In **Figure 18(a)**, a reference lattice has been cut along AA' by a plane with a normal specified by the vector **n**, so as to divide the lattice into two crystals represented by + and −. In **Figure 18(b)**, crystal + is rotated by an angle $+\theta/2$ and crystal − by an angle of $-\theta/2$ in a right-handed sense about an axis defined by a unit vector l passing through the lattice point O and directed into the page. In **Figure 18(c)** these two misoriented lattices are extended until they join at the original cut, forming a grain boundary. A vector **x**, which can

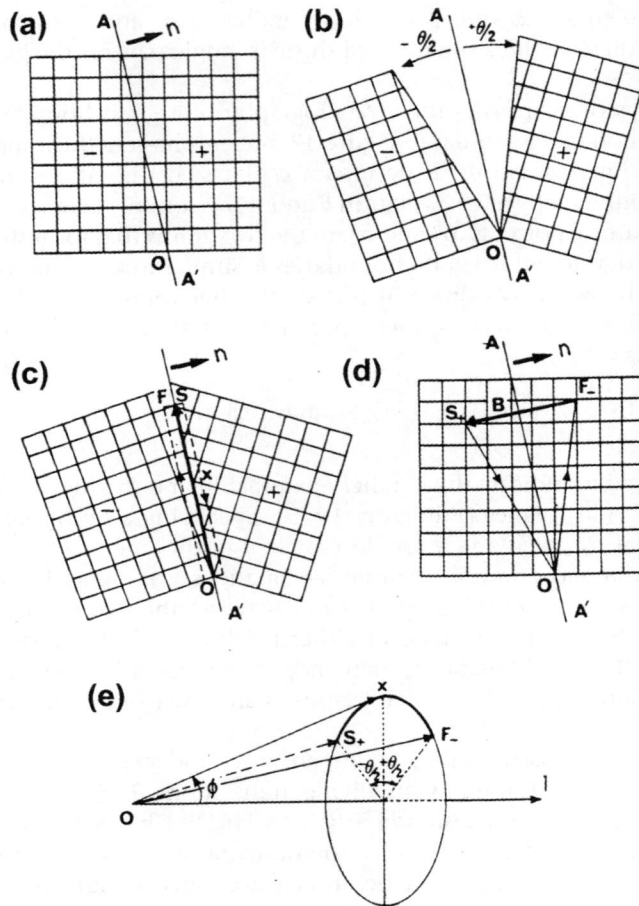

Figure 18 (a–e) Illustration of the derivation of the net Burgers vector **B** crossing a vector **x** in a planar grain boundary AA′ with unit normal **n**, where lattice + is rotated with respect to lattice − by an angle θ in a right-handed sense about an axis **l** directed into the plane of the page through the point O. Adapted from Forwood and Clarebrough (1991) with permission.

have any direction in this boundary plane, is then chosen to extend from the origin O over several unit cells, as illustrated in **Figure 18(c)**. The net Burgers vector of the dislocations in the grain boundary which intersects **x** can be determined by comparing a Burgers circuit containing the vector **x** in the bicrystal with an equivalent circuit in the reference lattice. This is done using the finish-to-start right-hand (FS/RH) convention (Read, 1953; Hirth and Lothe, 1982) with the closure failure being made in the good crystal. The circuit in the bicrystal (**Figure 18(c)**) is a closed circuit made in a right-handed sense around an axis parallel to $(\mathbf{x} \times \mathbf{n})$, where **n** is defined as pointing from crystal − into crystal +. This circuit is also made in a right-handed sense with respect to the rotation axis **l**. The circuit starts at S, the end point of the vector **x**, extends through crystal + to the origin O and then returns through crystal − to the point F, which is coincident with S. In the reference lattice redrawn in **Figure 18(d)**, the first part of the circuit SO in crystal + is represented by S_+O and the second part of the circuit, OF in crystal −, is represented by OF_. There is clearly a closure failure F_S$_+$ = B in the reference lattice and this defines the net Burgers vector of those dislocations contained in the boundary which are intersected by the vector x. In general the vector x makes an angle ϕ with l as shown in **Figure 18(e)**, so the resulting vector B has a magnitude given by

$$|\mathbf{B}| = |\mathbf{x}|2 \sin (\theta/2) \sin \phi \tag{6a}$$

with a direction along $(\mathbf{x} \times \mathbf{l})$. Furthermore, since

$$|\mathbf{x} \times \mathbf{l}| = |\mathbf{x}| \sin \phi, \tag{6b}$$

then

$$\mathbf{B} = (\mathbf{x} \times \mathbf{l})2 \sin (\theta/2) \tag{6c}$$

which for small θ yields

$$\mathbf{B} = (\mathbf{x} \times \mathbf{l})\theta. \tag{6d}$$

Equation (6d) is the general formula derived by Frank (1951) for the net Burgers vector of the dislocations required geometrically to accommodate the misorientation at a general grain boundary. The quantity B is the sum of the Burgers vectors of all the dislocations intersected by x or

$$\mathbf{B} = \sum_i n_i \mathbf{b}_i, \tag{7}$$

where n_i is the number of dislocations of Burgers vectors $\mathbf{b}_i$ cut by x. This summation can be extended to cases where the net Burgers vector B of Eqn (7) arises from two or more independent arrays of grain boundary dislocations with noncoplanar Burgers vectors b, such that the summation in Eqn (7) equals Eqn (6).

Several points should be made regarding Eqns (6) and (7). First, they apply only to boundaries that are essentially flat and have no long-range stress field, that is the elastic distortion is restricted to the region close to the dislocations. Second, the equations do not uniquely determine the dislocations present at the boundary or their pattern for a given crystal and boundary. Thus, a variety of possible dislocation structures could exist, with the most probable one being the one with the lowest energy. Third, the density of a given set of dislocations in a boundary is directly proportional to θ for small θ. Fourth, each set of dislocations is assumed to be straight, equally spaced and parallel even for a boundary containing several sets of dislocations with different Burgers vectors. Lastly, a general boundary requires three sets of dislocations with three noncoplanar Burgers vectors at the boundary and Frank's formula can be applied to analyze all possible cases, either to determine the possible dislocation arrangements if n, l and θ are known or, conversely, to find the orientation if the dislocation content B is specified. These cases are analyzed elsewhere (Read, 1953; Forwood and Clarebrough, 1991) and are not discussed further here.

14.2.1.2 O-Lattice Formulation

The Read and Shockley dislocation model for low-angle tilt and twist grain boundaries discussed in Section 14.2.1.1 is appealing, because it is relatively easy to understand both physically and analytically, and because it has been shown to be experimentally correct for low-angle grain boundaries.

Unfortunately, the dislocation model becomes unphysical for high tilt angles and an alternative description for the structure and energy of high-angle grain boundaries is required. Although several different approaches to this problem have been tried, none has the sort of predictive capability and accuracy that is usually desired. In fact, extensive atomistic simulations have been required to provide sufficient data to explain the properties of high-angle grain boundaries in a simple physical manner (Wolf and Merkle, 1992; Wolf, 1992).

This article defers discussion of the atomistic calculations of grain boundary structure and energies until two methods for quantifying interfacial structure that have some popularity and relate to the previous dislocation description of interfacial structure are considered. These are the so-called O-lattice and coincident site lattice/displacement shift complete (CSL/DSC) descriptions of grain and interphase boundary structure. Both treatments are geometrical models based on the matching of atoms across an interface. These theories build on the Read and Shockley dislocation model and Frank's equation for the dislocation content of a grain boundary and hence, they are examined in some detail before atomistic simulations. The O-lattice and CSL/DSC theories are developed for grain boundaries for simplicity, but examples are provided to demonstrate that they can also be applied to heterophase interfaces.

Bollmann developed a technique for analyzing the structure of grain and interphase interfaces that is quite general and has a number of useful properties (Bollmann, 1970, 1972; Smith and Pond, 1976; Mou and Aaronson, 1994; Gu and Zhang, 2010). His method is based on the concept of the O-lattice, which describes the matching and mismatching of oriented lattices at an interface. This theory is developed with reference to Frank's formula for the Burgers vector content of a general grain boundary in **Figure 18**, and shows that in fact, Bollmann's O-lattice equation is mathematically similar to Frank's formula in Eqn (6).

Referring to **Figure 18(d)**, the net Burgers vector **B** of the dislocations in the boundary that are intersected by the vector **x** is given by

$$\mathbf{B} = F_-S_+ = OS_+ - OF_-.$$

If $\mathbf{R}_+$ is defined as the rotation tensor $(l, +\theta/2)$ that transforms the reference lattice into +, and $\mathbf{R}_-$ as the rotation tensor $(l, -\theta/2)$ that transforms the reference lattice into lattice −, then

$$OS_+ = \mathbf{R}_+^{-1}\mathbf{x} \text{ and } OF_- = \mathbf{R}_-^{-1}\mathbf{x}$$

so that

$$\mathbf{B} = \left(\mathbf{R}_+^{-1} - \mathbf{R}_-^{-1}\right)\mathbf{x}. \tag{8a}$$

In Eqn (8c), **B** and **x** are expressed with respect to the reference lattices in **Figure 18(a) and (d)**. Alternatively, if one of the lattices is chosen as the reference lattice, say crystal +, and **B** and **x** are expressed in this lattice, then Eqn (8c) becomes

$$\mathbf{B} = \left(\mathbf{I} - \mathbf{R}^{-1}\right)\mathbf{x}, \tag{8b}$$

where **R** is the rotation tensor which transforms lattice + into lattice −, and **I** is the identity matrix.

The advantage of expressing the relationship between the two crystals by Eqn (8b) is that to heterophase interfaces by replacing the rotation tensor **R** with a general deformation tensor **A**, which transforms the lattice of crystal + into the lattice of crystal −, and involves a strain as well as a rotation

(Bilby, 1955a; Christian, 1975; Jesser, 1973). Thus, with respect to crystal + the net Burgers vector of those dislocations contained in the heterophase interface which are intersected by a vector **x** is given by

$$\mathbf{B} = \left(\mathbf{I} - \mathbf{A}^{-1}\right)\mathbf{x}. \tag{8c}$$

The properties of Eqn (8c) are similar to those for the grain boundary described in Eqns (6) and (7) in that for a general heterophase interface, **B** must be the resultant Burgers vector of at least three independent arrays of interfacial dislocations with noncoplanar Burgers vectors. These three independent arrays expressed by the summation in Eqn (7) must equal Eqn (8c). This similarity indicates that the arrays of dislocations in heterophase interfaces can be considered in the same way as for grain boundaries.

Given this background it is now possible to derive the basic equation of O-lattice theory after having expressed Eqn (8c) in a way that is applicable to any type of interface through a matrix **A**, which relates the two crystal lattices. If two misoriented crystal lattices, specified here as 1 and 2 to indicate that they may be different lattice types, are allowed to interpenetrate, there is a periodic set of points in space (not generally lattice points of either lattice) where, for each point, the internal coordinates in a cell of lattice 1 are identical with the internal coordinates in a cell 1 of lattice 2. This set of points defines Bollmann's O-lattice (Bollmann, 1970). In terms of a general deformation tensor **A** that transforms lattice 1 into lattice 2, a point defined by a vector $\mathbf{x}^{(2)}$ in lattice 2 is generated from a point defined by a vector $\mathbf{x}^{(1)}$ in lattice 1 according to

$$\mathbf{x}^{(2)} = \mathbf{A}\,\mathbf{x}^{(1)}. \tag{9a}$$

A point on the O-lattice is therefore defined by a vector $\mathbf{x}^{(o)}$ when $\mathbf{x}^{(2)}$ differs from $\mathbf{x}^{(1)}$ by a translation vector $\mathbf{b}^{(L)}$ of lattice 1, that is when

$$\mathbf{x}^{(2)} = \mathbf{x}^{(o)} = \mathbf{x}^{(1)} + \mathbf{b}^{(L)} = \mathbf{A}\,\mathbf{x}^{(1)}. \tag{9b}$$

From this set of equations, one obtains

$$\mathbf{b}^{(L)} = \left(\mathbf{I} - \mathbf{A}^{-1}\right)\mathbf{x}^{(o)} \tag{9c}$$

which is the basic equation of O-lattice theory (Bollmann, 1970), and also called the Frank–Bilby equation (Frank, 1950; Bilby, 1955b). This equation gives the O-lattice vectors in terms of the vectors of crystal lattice 1. In this equation, $\mathbf{b}^{(L)}$ corresponds to **B** in Eqn (6) or Eqn (8c) and $\mathbf{x}^{(o)}$ corresponds to **x**. Thus, the vectors $\mathbf{x}^{(o)}$ cross dislocations with Burgers vectors summing to $\mathbf{b}^{(L)}$, referred to lattice 1.

The following examples derive the O-lattice solution for two simple cubic crystal lattices that are rotated about the [001] axis referring to the coordinate system of lattice 1, as illustrated in **Figure 19**. This example follows the work of Smith and Pond (1976). The transformation tensor **A** is a rotation **R**, which can be written as

$$\mathbf{A} = \mathbf{R} = \begin{bmatrix} \sin\theta & -\sin\theta & 0 \\ \sin\theta & \cos\theta & 0 \\ 0 & 0 & 1 \end{bmatrix} \tag{10a}$$

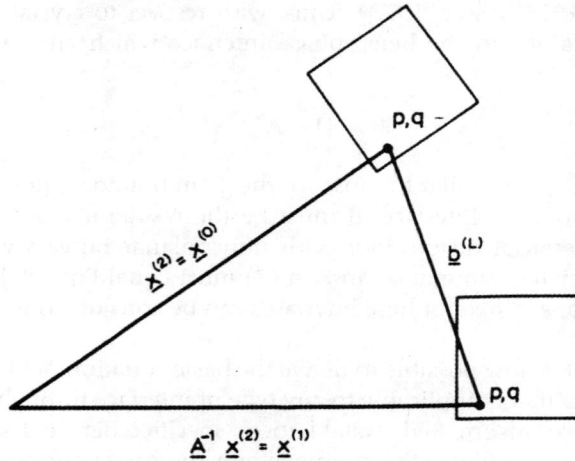

Figure 19   Diagram showing the difference vector $\mathbf{b}^{(L)}$ between $\mathbf{x}^{(1)}$ and $\mathbf{x}^{(2)}$ for a simple cubic unit cell having an O-point with internal coordinates (p,q). Adapted from Smith and Pond (1976) with permission.

so that

$$\mathbf{A}^{-1} = \begin{bmatrix} \cos\theta & \sin\theta & 0 \\ -\sin\theta & \cos\theta & 0 \\ 0 & 0 & 1 \end{bmatrix} \tag{10b}$$

and

$$(\mathbf{I} - \mathbf{A}^{-1}) = \begin{bmatrix} 1 - \cos\theta & -\sin\theta & 0 \\ \sin\theta & 1 - \cos\theta & 0 \\ 0 & 0 & 0 \end{bmatrix}. \tag{10c}$$

Substituting $(\mathbf{I} - \mathbf{A}^{-1})$ into Eqn (9c) gives

$$\begin{bmatrix} 1 - \cos\theta & -\sin\theta & 0 \\ \sin\theta & 1 - \cos\theta & 0 \\ 0 & 0 & 0 \end{bmatrix} \begin{bmatrix} x_1 \\ x_2 \\ x_3 \end{bmatrix} = \begin{bmatrix} b_1 \\ b_2 \\ b_3 \end{bmatrix}, \tag{11}$$

where x_1, x_2 and x_3 and b_1, b_2 and b_3 are the components of $\mathbf{x}^{(o)}$ and $\mathbf{b}^{(L)}$. This expression can be written explicitly as three linear equations:

$$(1 - \cos\theta)x_1 - (\sin\theta)x_2 + 0x_3 = b_1, \tag{12a}$$

$$(\sin\theta)x_1 + (1 - \cos\theta)x_2 + 0x_3 = b_2, \tag{12b}$$

$$0x_1 + 0x_2 + 0x_3 = b_3, \tag{12c}$$

where $b_3 = 0$ and x_3 can have any value. Equation (12c) means that there are no uncanceled components of the Burgers vector parallel to the rotation axis of the grain boundary x_3. Therefore, the O-lattice consists of lines going through points $x_1^{(o)}$ and $x_2^{(o)}$, which have discrete values. The primitive vectors of the O-lattice in the plane perpendicular to the rotation axis, that is for a pure twist boundary, can be found by assuming a Burgers vector of unity and solving Eqns (12a) and (12b) simultaneously as

$$\begin{bmatrix} 1 - \cos\theta & -\sin\theta \\ \sin\theta & 1 - \cos\theta \end{bmatrix} \begin{bmatrix} x_1 \\ x_2 \end{bmatrix} = \begin{bmatrix} b_1 \\ b_2 \end{bmatrix} \tag{13}$$

so

$$x_1^{(o)} = \begin{bmatrix} x_1 \\ x_2 \end{bmatrix} = \begin{bmatrix} 1/2 & 1/2\cot(\theta/2) \\ -1/2\cot(\theta/2) & 1/2 \end{bmatrix} \begin{bmatrix} 1 \\ 0 \end{bmatrix} \tag{14}$$

and similarly for $x_2^{(o)}$, and the primitive vectors are

$$x_1^{(o)} = 1/2, \; -1/2\cot\theta/2, \tag{15a}$$

$$x_2^{(o)} = 1/2\cot\theta/2, \; 1/2. \tag{15b}$$

Note that $x_1^{(o)}$ and $x_2^{(o)}$ are continuous functions of θ and that when θ is small, Eqn (15) reduces to the result given by Frank's formula in Eqns (1b) and (6d), that is, the dislocation spacing D is given by b (unity in this case) divided by θ. Also notice that there are particular values where the $\cot\theta/2$ in Eqn (15) is rational, for example when $\cot\theta/2$ equals an integer n. This situation was illustrated previously in **Figure 10**, where $\cot\theta/2 = 14$. Whenever the $\cot\theta/2 = 2n$ for both $x_1^{(o)}$ and $x_2^{(o)}$, that is, it is even for both, the O-point has integral coordinates and is thus is a crystal lattice site. When either or both of the O-points are odd, the O-point is either in the center of an edge or in the center of a crystal cell. This illustrates the general result that points of the O-lattice need not coincide with crystal lattice sites.

In the derivation of Eqn (9c) it was assumed that the points defined by the vectors $x^{(o)}$ have identical internal coordinates in both lattices 1 and 2 and the $b^{(L)}$ vectors are all lattice translation vectors. Physically, this means that the interface under consideration is composed of perfect crystal dislocations. In the section that follows, it is shown that other dislocations with Burgers vectors that are less than a full lattice translation are possible at an interface. This does not mean that the O-lattice theory is incorrect but rather that the choice of Burgers vectors at an interface is ambiguous and some additional criteria must be considered to determine the most favorable Burgers vectors at the interface. This is a weakness of the O-lattice theory in terms of its predictive capabilities. Another shortcoming is that the theory can be used to analyze an interface if A and n and/or the dislocation Burgers vectors $b^{(L)}$ are known, but it cannot be used to predict n or $b^{(L)}$ without knowing A and using some other criterion.

The physical interpretation of the O-lattice in terms of grain boundaries is that the O-points in Eqn (9c) are points of geometric registry (minimum strain) between crystal lattices 1 and 2. Between each of these O-points there is an accumulating disregristry, which reaches the value $b^{(L)}$ at a neighboring O-point. Thus, it is imagined that the misfit between any set of O-points is concentrated onto planes between the O-points just as the misfit between two identical but slightly rotated lattices relaxes into a low-angle boundary network of dislocations, as in **Figure 14**. In the example above, where the axis of rotation was parallel to [001] in the two lattices, the O-lattice construction yields O-lines parallel to the rotation axis with two orthogonal planes of dislocations bisecting the space between the O-lines

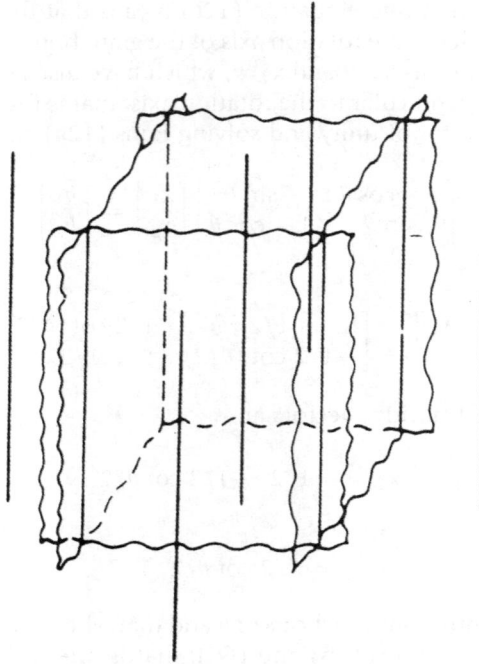

Figure 20   Schematic representation of the O-lines and dislocation cell walls associated with the O-lattice of two simple cubic lattices rotated about a common [001] axis. Adapted from Smith and Pond (1976) with permission.

as illustrated in **Figure 20** (Smith and Pond, 1976). Note that a plane taken perpendicular to the O-lines in this figure would look exactly like **Figure 14**. In the case of a general grain or interphase boundary, the set of planes on which misfit is condensed defines a three-dimensional cell structure with each cell enclosing an O-point. Consequently, wherever the IP cuts a cell wall there is a dislocation with Burgers vector $b^{(L)}$.

Although this section is primarily concerned with the structure of grain boundaries and therefore a rotation **R** between two crystals, it is convenient to illustrate the use of the O-lattice theory for a different kind of transformation tensor **A**, in which there is a simple dilatation Δa between two simple cubic crystal lattices. This also serves to emphasize the generality of the O-lattice formulation to other types of transformations. A second example of the O-lattice applied to a rotation between two crystals is presented in the next section.

Figure 21(a) (Smith and Pond, 1976) shows schematically a (001) projection of the O-lattice between two simple cubic lattices with the same orientation but different lattice parameters, where crystals 1 and 2 are represented by dots and crosses, respectively, and the O-points are circled. The two lattices are related by a simple dilatation Δa and inspection shows that the O-lattice itself is a simple cubic lattice. The $b^{(L)}$ vectors lie between the O-points, which are indicated by circles in the figures. The dislocation spacing in the O-lattice is given as $(1 + \Delta a)/\Delta a$ (which is equivalent to the inverse of the misfit). **Figure 21(b)** shows the planes that bisect the O-lattice vectors perpendicular to the plane of the figure. These form square cell walls around the O-lattice points in three dimensions. In **Figure 21(c)** an interface path is chosen so that it passes through as many O-points as possible to

(a)

(b)

(c)

(d)

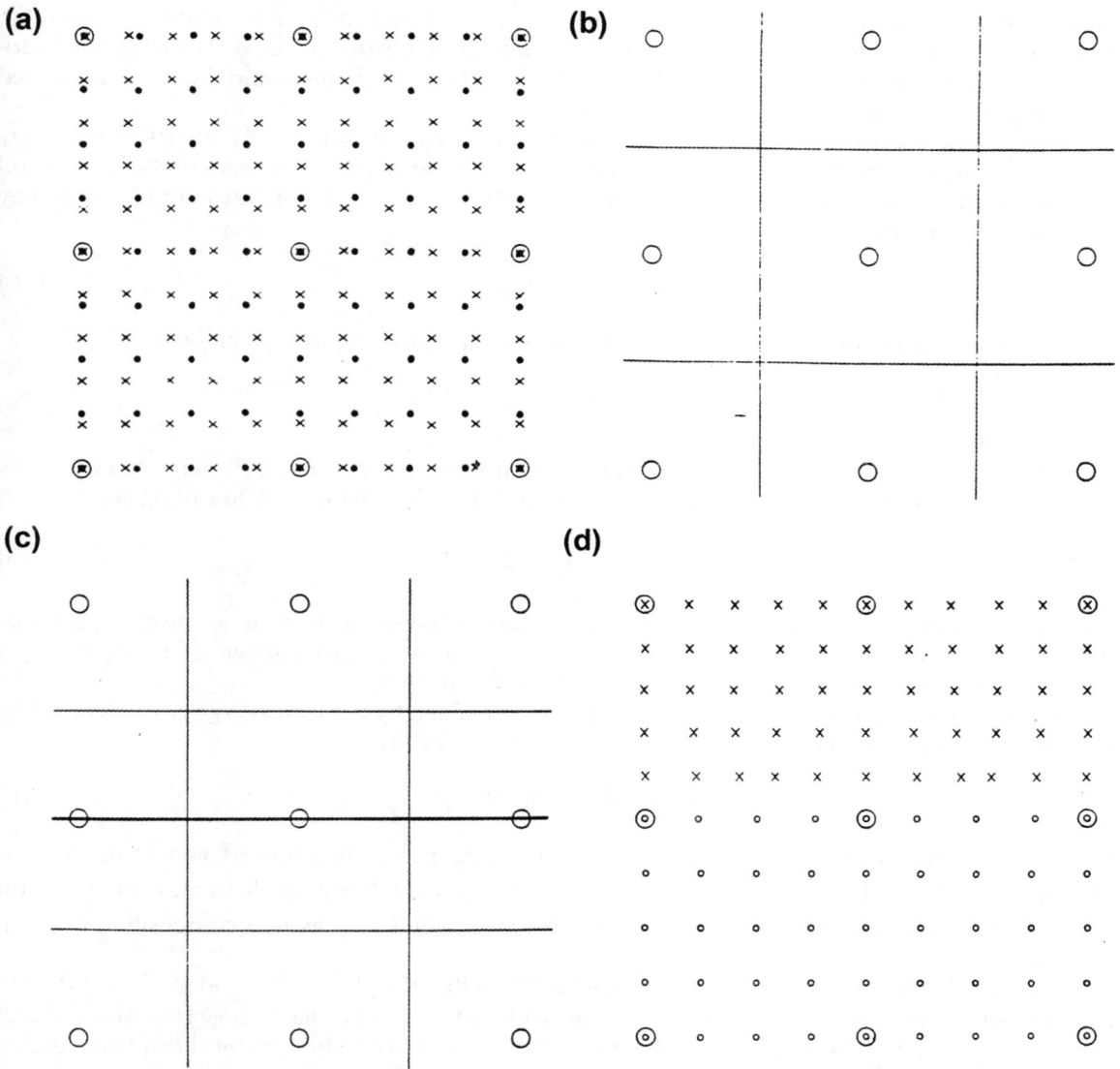

Figure 21 Illustration of the O-lattice construction for a dilatation between two simple cubic lattices: (a) (001) projection of the O-lattice, (b) cells walls drawn midway between the O-points, (c) interface path through the highest density of O-points, and (d) section showing dislocations at the interface. Adapted from Smith and Pond (1976) with permission.

maximize the amount of good matching in the interface. When crystal 1 is removed from above the IP in **Figure 21(c)** and crystal 2 is removed from the IP downward, the interface in **Figure 21(d)** results. In this interface, the O-points are regions of good matching and these are separated by regions where relaxation of the interface into dislocations has occurred (where the cell walls were crossed). This illustrates the O-lattice construction, which can be performed analytically using the basic O-1attice equation (Eqn (9c)). Gu and Zhang (2010) and Zhang and Yang (2011) have provided a thorough comparison between the use of the O-lattice and Frank–Bilby treatments for determining the dislocation content of interfaces, showing the advantages of the O-lattice method, and the reader is referred to these papers for further details.

Before closing this section, it is worth showing how the matrices in Eqns (8) and (9), which were developed in real space, can be expressed in reciprocal space, since this is the basis of the $\Delta\mathbf{g}$ method mentioned later. In reciprocal space, the vectors $\mathbf{g}^{(1)}$ and $\mathbf{g}^{(2)}$ in two different lattices can be related by the same transformation strain in Eqn (9a) as

$$\mathbf{g}^{(1)} = (\mathbf{A}^{-1})\mathbf{g}^{(2)}. \tag{16}$$

The displacement between a pair of correlated reciprocal lattice vectors is given by

$$\Delta\mathbf{g} = \mathbf{g}^{(1)}\left(\mathbf{I} - \mathbf{A}^{-1}\right) = \mathbf{g}^{(1)}\mathbf{T}, \tag{17a}$$

where $\mathbf{T} = (\mathbf{I} - \mathbf{A}^{-1})$, and it is assumed that $\Delta\mathbf{g}$ only connects vectors $\mathbf{g}^{(1)}$ and $\mathbf{g}^{(2)}$, which relate primary misfit strains between the two lattices and not any $\Delta\mathbf{g}$ that are otherwise related. Rewriting Eqn (17a) as

$$\left(\mathbf{T}^{-1}\right)\Delta\mathbf{g} = \mathbf{g}^{(1)}, \tag{17b}$$

it becomes apparent that Eqn (17b) defines an O-lattice transformation in reciprocal space that corresponds to Eqn (9c) in real space. Therefore, the displacement vector $\Delta\mathbf{g}$ defined by Eqn (17a) is a reciprocal vector of the primary O-lattice (Zhang and Purdy, 1993).

A set of periodic principal primary O-lattice planes is defined by a principal $\Delta\mathbf{g}$ vector, denoted by $\Delta\mathbf{g}^{\mathrm{P}}$, which is associated with $\mathbf{g}^{(1)}$ by Zhang and Weatherly (2005)

$$\Delta\mathbf{g}^{\mathrm{P}} = \mathbf{T}\mathbf{g}^{\mathrm{P}(1)} = \mathbf{g}^{\mathrm{P}(1)} - \mathbf{g}^{\mathrm{P}(2)} \tag{18}$$

with $\mathbf{g}^{\mathrm{P}(1)}$ representing a set of principal planes containing at least two Burgers vectors in lattice 1, while $\mathbf{g}^{\mathrm{P}(2)}$ is the vector in lattice 2 correlated to $\mathbf{g}^{\mathrm{P}(1)}$ by Eqn (16). It is possible to measure the vector $\Delta\mathbf{g}^{\mathrm{P}}$, since $\mathbf{g}^{\mathrm{P}(1)}$ and $\mathbf{g}^{\mathrm{P}(2)}$ can often be recognized from low-index spots in a composite diffraction pattern.

An unique character of the interface normal to $\Delta\mathbf{g}^{\mathrm{P}(1)}$ is that the planes $\mathbf{g}^{\mathrm{P}(1)}$ and $\mathbf{g}^{\mathrm{P}(2)}$ should be in perfect registry at the interface. This means that the planes should meet edge to edge (as discussed later in Section 14.3.4.1 and illustrated in **Figure 79**). This behavior can also be seen by allowing the planes defined by $\mathbf{g}^{\mathrm{P}(1)}$ and $\mathbf{g}^{\mathrm{P}(2)}$ to overlap and interfere to form Moiré fringes, which will lie normal to $\Delta\mathbf{g}^{\mathrm{P}(1)}$ and mark the locations where the planes intersect (Zhang and Weatherly, 2005; Nie, 2004). This property of the Moiré planes is completely general, and it is not necessary that corresponding $\mathbf{g}^{(1)}$ and $\mathbf{g}^{(2)}$ be related by any particular strain.

14.2.1.3 Coincident Site and Displacement Shift Complete Lattices

The previous O-lattice theory provides a convenient mathematical technique for analyzing possible dislocation structures between two arbitrary lattices in space when their OR is known. It is a useful technique, but it has limitations in that it cannot predict certain aspects of interfacial structure that are commonly found at interfaces, such as steps and partial dislocation structures. In addition, as with the

Read and Shockley analysis, its meaning becomes unclear in the case of high-angle grain boundaries. An alternate technique for quantifying the interfacial structure of grain and interphase boundaries for any angle of tilt or orientation is provided by the CSL and DSC lattice constructions discussed in this section. The CSL/DSC theory was actually developed before O-lattice theory, but the O-lattice was introduced first in this chapter because of its close relation to Frank's equation for the dislocation content of a grain boundary in Section 14.2.1.1.5. In this section, the CSL/DSC construction is developed for grain boundaries using a pure tilt boundary for purposes of illustration, and then it is shown that the CSL/DSC framework can be readily extended to include heterophase interfaces as well.

14.2.1.3.1 Coincident Site Lattice
Figures 10 and 15 showed that when two identical interpenetrating lattices are rotated from initial coincidence around a lattice point, there are certain discrete rotation angles where lattice points other than the origin coincide. An example that is often found in the literature is shown in **Figure 22(a)** (Fishmeister, 1985), where two simple cubic lattices, one indicated by solid dots and the other by open circles, are outlined by a dashed and solid square, respectively. The two lattices have been rotated 36.9° about an axis perpendicular to the plane of the figure as in **Figure 15** (and **Figure 10**), and the pattern that results is often referred to as the dichromatic pattern. In this case $\cot \theta/2 = 3$ and the [310] vector in one lattice is coincident with the $[3\bar{1}0]$ vector in the other lattice. This rotation causes one-fifth of the lattice points of the simple cubic crystal to coincide and this is true for both FCC and BCC lattices as well. The coincident points themselves form a lattice, called the CSL, and this is indicated by filled circles and outlined in **Figure 22(b)**. The lattice vectors of the CSL are given as

$$1/2[310]_1 \| 1/2[3\bar{1}0]_2, 1/2[\bar{1}30]_1 \| 1/2[130]_2 \quad \text{and} \quad 1/2[12\bar{1}]_1 \| 1/2[21\bar{1}]_2.$$

The CSL is characterized by Σ, the inverse density of coincident sites, alternatively expressed as the ratio of the area of the coincident lattice cell to that of the original lattice. For the example shown in **Figure 22**, $\Sigma = 5$, since 2 out of every 10 atoms, that is the corner atoms, in each unit cell of the CSL lattice are coincident. Any rational <uvw> lattice vector can be used to generate a CSL and it has been shown (Ranganathan, 1966; Mykura, 1980) that all possible CSL's in the cubic system can be described by the function

$$\Sigma = x^2 + Ny^2, \tag{19a}$$

where x and y are nonnegative integers representing the Cartesian coordinates of the lattice point joined to the origin, and $N = u^2 + v^2 + w^2$. A CSL is always generated for a rotation of 180° about a rational direction <uvw>. When the value of Σ determined from Eqn (19a) is even, it must be divided by 2 until an odd number results. For example, a <310> vector where $\Sigma = 10$ and a <210> vector where $\Sigma = 5$ generate the same $\Sigma = 5$ CSL shown in **Figure 22(b)**. Thus the $\Sigma = 10$ CSL must be divided by 2 to yield the $\Sigma = 5$ CSL in **Figure 22(b)**. The angle of misorientation corresponding to a particular CSL is given by

$$\theta = 2 \tan^{-1}\left(y/x\right)N^{1/2}, \tag{19b}$$

where $\theta = 180°$ corresponds to $x = 0, y = 1$. Thus, the rotation of 180° around <uvw> in a cubic system gives rise to a CSL of $\Sigma = u^2 + v^2 + w^2$ if N is odd, or $N/2$ if N is even. Since the same values of Σ may be generated by Eqn (19a) from different sets of x, y and N, usually, both Σ and the axis–angle pair (Eqn (4)) are specified in CSL nomenclature, for example $\Sigma = 5$, 36.9°/[001] for **Figure 22(b)**. The

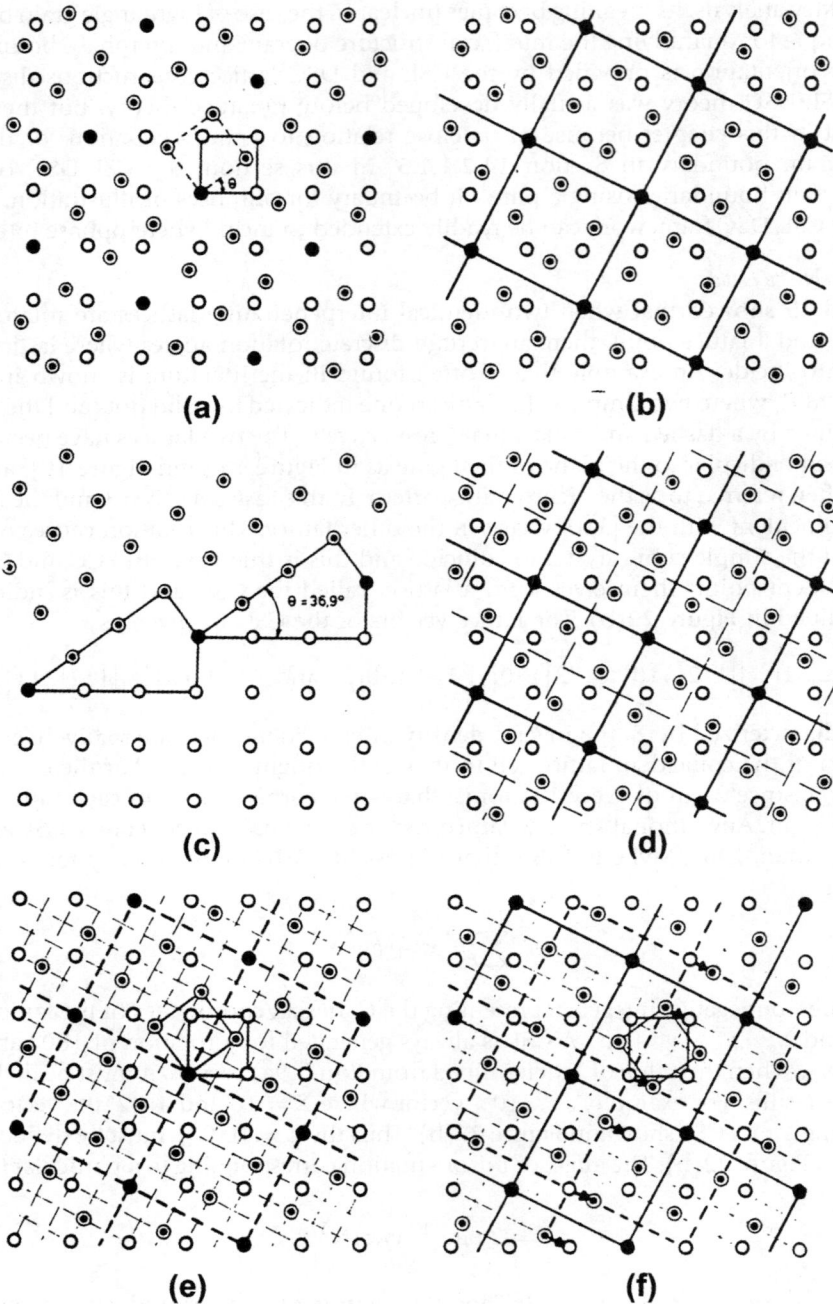

Figure 22 (a) Dichromatic pattern produced by rotating two cubic crystals 36.9° about [001]. (b) $\Sigma = 5$ CSL outlined in the dichromatic pattern. (c) A possible $(130)_1$, $(130)_2$ grain boundary generated from the CSL. (d) The O-lattice (=CSL plus the dashed lines). (e) The DSC lattice of the $\Sigma = 5$ dichromatic pattern. (f) The dichromatic pattern after translation by a DSC vector; dashed lines indicate old CSL and solid lines indicate new CSL. Adapted from Fishmeister (1985) with permission.

inclination of the boundary is then specified by the Miller indices of the boundary plane in both lattices, for example $(hkl)_1 \| (hkl)_2$, similar to the notation used in Eqn (5). For the symmetric tilt boundary in **Figure 22(c)** the complete CSL notation is $\Sigma = 5$ (310) 36.9°/[001], where only one (hkl) is necessary because the boundary is symmetric.

In **Figure 22(c)**, a (310) boundary plane was chosen and opposite lattice types were discarded from either side of the boundary, leaving a grain boundary composed of typical structural units, as indicated in the figure. While the density of coincident sites depends only on the OR between two crystals, the density of coincident sites at a boundary depends on the choice of the boundary plane intersecting the CSL. Special grain boundaries have a high density of coincident sites and therefore a low value of Σ. It seems reasonable to expect that the energy of a grain boundary should be proportional to Σ, but except in the case of some special grain boundaries, it has been shown that there is no simple correlation between the two. This is a disappointment of the CSL theory, but it is now known that this lack of correlation is because of volume expansions and translations that frequently occur at grain boundaries to minimize the energy of the boundary. Such relaxations cannot be predicted from purely geometric models.

Figure 23 shows an HRTEM image of a $\Sigma = 5$ (310) 36.9°/[001], symmetrical tilt grain boundary in cubic NiO (Merkle, 1995). The atomic columns appear as dark spots in this image and a regular pattern of structural units at the boundary is evident, as predicted by the CSL construction in **Figure 22(c)**. However, note that there is a lack of mirror symmetry across the boundary in the HRTEM image that is not present in the unrelaxed CSL model in **Figure 22(c)**. This is because of slight translation of the boundary involving the three microscopic DOF discussed in Section 14.2.1.4, and it is such translations that limit the geometric CSL model from accurately describing the energies of grain boundaries. Thus, to fully understand the correlation between the structure and energy of high-angle grain boundaries, it has been necessary to employ extensive atomistic calculation, as described in Section 14.2.2.5. Nevertheless, the CSL theory does provide a simple and useful geometric model for understanding interfacial structure in terms of atom matching and mismatching between two crystals across the interface for any degree of misorientation, and it has been used to describe the dislocation structure of solid–solid interfaces with considerable success.

Any lattice vectors of the CSL are solutions $\mathbf{x}^{(o)}$ of the O-lattice equation (Eqn (9c)) because at each independent site the rotational displacements between the two lattices amount to a lattice translation $\mathbf{b}^{(L)}$. However, not all O-lattice vectors $\mathbf{x}^{(o)}$ are CSL vectors. This point is illustrated by the

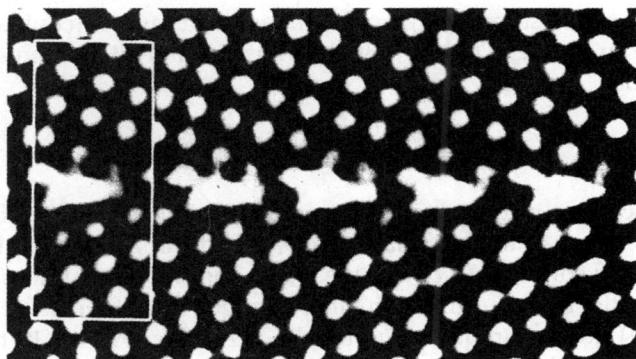

Figure 23 HRTEM image of a $\Sigma = 5$ grain boundary in NiO viewed parallel to the <001> tilt axis. A digitally averaged image in the inset shows a representative structural unit of the boundary, which is slightly asymmetric. Adapted from Merkle (1995) with permission.

complete O-lattice for the $\Sigma = 5$ boundary shown in **Figure 22(d)**. When the orientation between the two lattices is not exactly equal to one necessary for a CSL, all the coincident sites (except the center of rotation) are lost. In contrast, the O-lattice is still maintained with lattice vectors $\mathbf{x}^{(o)}$, which are now irrational although close to the rational CSL vectors. Thus, the O-lattice changes continuously between the discrete CSLs.

14.2.1.3.2   DSC Lattice

Geometrically, high-angle grain boundaries can be treated as small deviations from the nearest CSL. They are then similar to low-angle grain boundaries, where small deviations from the perfect single crystal (i.e. $\Sigma = 1$ CSL) are accommodated by lattice dislocations. Any deviation from a CSL is accommodated by lines of high local distortion, or dislocations, separating regions of undistorted CSL. In the case of low-angle grain boundaries, the dislocations are usually called primary dislocations and the boundary between them is perfect crystal ($\Sigma = 1$). In the case of high-angle grain boundaries, the dislocations are called secondary dislocations and the boundary between them is perfect CSL.

The DSC lattice defines the possible Burgers vectors of secondary dislocations (Grimmer et al., 1974; Grimmer, 1974). These are perfect grain boundary dislocations that conserve the structure of the optimal coincidence orientation boundary. The DSC lattice acquired its name because it is a lattice of pattern conserving displacements, that is, a displacement of one crystal with respect to the other by a DSC vector restores the dichromatic pattern but with shifted coincidence sites. Thus, it causes a pattern shift that is complete. This feature is illustrated with reference to the $\Sigma = 5$ CSL in **Figure 22(e) and (f)**. In **Figure 22(e)**, the DSC lattice was constructed by defining the coarsest lattice that contained all of the sites of crystals 1 and 2 in the coincidence orientation as sublattices. The DSC lattice is indicated by the fine square mesh in **Figure 22(e)**. Note that the DSC lattice points are not all translationally equivalent in the surroundings of crystals 1 and 2. When lattice 1 is shifted by the DSC vector indicated by an arrow in **Figure 22(f)**, the dichromatic pattern of the lattices is restored, but the resulting CSL (solid line) is shifted from its original position (dashed line).

The utility of the DSC construction is that any misorientation from the perfect $\Sigma = 5$ CSL orientation can be accommodated by these secondary dislocations with areas of perfect $\Sigma = 5$ boundary in between. The secondary dislocations thus occur when the primary dislocations are so close that they are not physically separate. The DSC dislocations then accommodate the irregularity in the periodicity of the primary dislocations, similar to the situation previously discussed with reference to **Figure 11** using only primary dislocations. The secondary grain boundary dislocations required to accommodate the small misorientation from the exact CSL can be described by an equation equivalent to Eqn (6c) for low-angle boundaries, or

$$\mathbf{B}_s = (\mathbf{x} \times \mathbf{q}) \, 2 \sin{(\theta_{CSL}/2)}, \tag{20}$$

where $\mathbf{B}_s$ is the net Burgers vector content of the secondary grain boundary dislocations intercepted by any vector $\mathbf{x}$ lying in the plane of the boundary, and θ_{CSL} is the angular departure from the exact CSL orientation about an axis $\mathbf{q}$ common to one of the grains and the CSL. An important property of Eqn (20) is that it does not suffer from the ambiguity in the uniqueness of the Burgers vectors discussed for Eqn (6c). Equation (20) gives a unique value for the net Burgers vector $\mathbf{B}_s$ of the secondary grain boundary dislocation network because it is a property of the intersection of the translational symmetry elements of the two crystals and this is unique.

Formally, the DSC lattice can be defined as the lattice of difference vectors (i.e. vectors linking the lattice sites of crystal 1 to crystal 2) between lattices 1 and 2 in the exact coincidence orientation.

Analytically, this can be expressed as

$$\mathbf{d}^{(\mathrm{DSC})} = \mathbf{x}^{(2)} - \mathbf{x}^{(1)} = \left(\mathbf{I} - \mathbf{A}^{-1}\right)\mathbf{x}^{(2)}, \tag{21}$$

where $\mathbf{d}^{(\mathrm{DSC})}$ is a DSC lattice vector and the other symbols have the same meaning as in Eqn (10). (A reciprocal relation that exists between the CSL and DSC lattices is such that as Σ increases, the magnitude of the primitive DSC vectors decreases.) Equation (21) can be solved graphically if Σ is not too high by drawing the two lattices in the coincidence orientation and position and finding the DSC vectors by inspection, as in **Figure 22(d)**. In this case, the lattice vectors of the DSC lattice are given as

$$1/10[310]_1, \ 1/10[\overline{1}30]_1 \quad \text{and} \quad 1/10[21\overline{5}]_1.$$

Otherwise, the DSC vectors must be found analytically.

Schober and Balluffi (1970) were the first investigators to provide strong experimental evidence for the CSL model of grain boundary structure by demonstrating that secondary grain boundary dislocations accommodated departures from CSL orientations. Their original results on twist boundaries in Au as well as those obtained in subsequent investigations are summarized in **Figure 24(a)** (Schober and Balluffi, 1970; Babcock and Balluffi, 1987). This figure shows the variation in dislocation spacing with twist angle θ for both primary and secondary grain boundary dislocations. The spacing D_{Si} of the secondary ith grain boundary dislocations in the crossed grids varied with angular departure θ_{CSL} from each of the exact CSL orientations according to the same relationship given for primary dislocations in Eqn (1b), or

$$D_{Si} = |\mathbf{b}_{Si}|/\theta_{\mathrm{CSL}}, \tag{22}$$

where $\mathbf{b}_{Si}$ is the Burgers vector of the ith secondary grain boundary dislocation. The Burgers vectors that were found to satisfy Eqn (22) were the basis DSC vectors that were compatible with line directions of the screw dislocations. Thus, for the near $\Sigma = 5$, $\Sigma = 13$ and $\Sigma = 17$ boundaries, the Burgers vectors found for the secondary grain boundary dislocations were of the type $1/10<310>$, $1/26<510>$ and $1/34<530>$, respectively. In addition, for the case of the near $\Sigma = 5$ boundaries, the Burgers vectors were found to agree with the contrast observed in TEM images of the interfaces, as illustrated in **Figure 24(b)**. Note the weak contrast of the dislocations because of their small Burgers vectors compared with **Figure 16(a)**.

It is important to note that the DSC lattice contains both lattices 1 and 2 in the coincidence orientation and position as sublattices. Therefore, it is continuous across any grain (or interphase) boundary and provides a suitable reference lattice on which Burgers circuits may be drawn (Hirth and Balluffi, 1973). This is important because various defects such as dislocations and steps can occur at grain and interphase boundaries, and one needs a method to quantify the nature of these defects. The next section examines the character of steps in grain boundaries using the DSC lattice as our reference lattice. This is followed by extension of the CSL/DSC construction to interphase boundaries to emphasize some important similarities, and then a related but more comprehensive development incorporating disconnections in interfaces is presented.

14.2.1.3.3 Structure and Properties of Grain and Interphase Boundary Line Defects
The topology of line defects that exist in grain boundaries in cubic materials is readily established within the framework of the DSC model (Balluffi et al., 1982; Hirth and Balluffi, 1973). The most

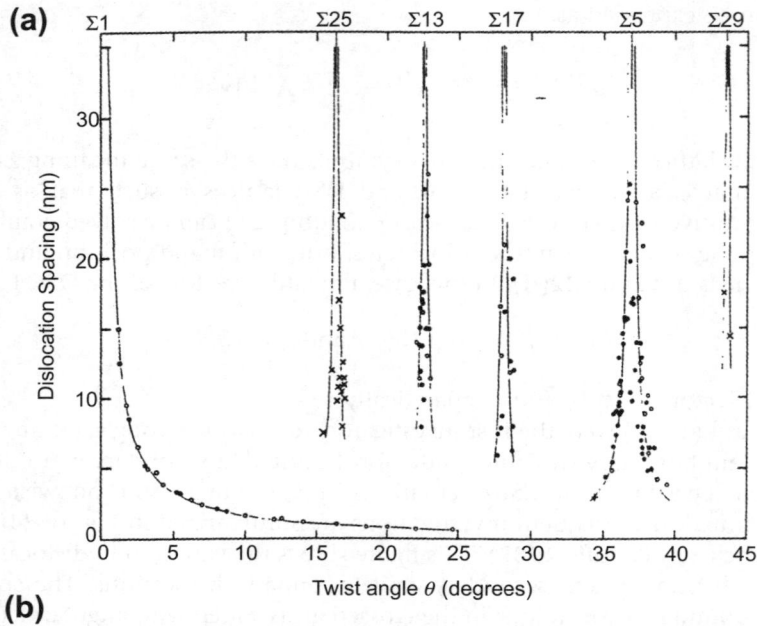

Figure 24 (a) Variation of dislocation spacing with twist angle for primary and secondary grain boundary dislocations in Au Adapted from Schober and Balluffi (1970), Babcock and Balluffi (1987) and Forwood and Clarebrough (1991) with permission. (b) TEM image showing a cross-grid of secondary grain boundary dislocations in a near $\Sigma = 5$ (001) twist boundary in Au. Adapted from Schober and Balluffi (1970) with permission.

general type of line defects possess both dislocation and step character, which can be defined by the Burgers vector **b** and the step vector **s**, respectively. (The term ledge is often used in the literature instead of step in keeping with the terrace–ledge–kink model of crystal growth that originated for surfaces, but here step is used in keeping with the topological treatments developed for grain and interphase boundaries.) Possible line defects therefore include those with both dislocation and step character and special ones that are either pure dislocations or pure steps. An alternative description of defect topology that is particularly useful for more complicated interphase boundaries has been developed by Pond (1989) and Pond and Hirth (1994), and this is discussed after the DSC model, which is convenient because it builds on the preceding CSL/DSC treatments and establishes a framework for understanding the more general description.

If the dislocation character of a line defect is that of a perfect grain boundary dislocation, then the Burgers vector must be that of the DSC lattice, since this is composed of all lattice translation vectors of lattice 2 with respect to lattice 1 that conserve the atomic (dichromatic) pattern. The step vector then corresponds to a shift in space of the above pattern that occurs when lattice 2 is translated with respect to lattice 1 by the Burgers vector. These step vectors are always vectors of the crystal lattices and therefore also of the DSC lattice. The topological properties of these line defects in grain boundaries can be revealed with the DSC framework using the examples shown in **Figure 25** (Balluffi et al., 1982).

In **Figure 25**, the Burgers vector of the dislocation is seen to be a vector of the DSC lattice as are step vectors s_1 and s_2 in lattices 1 and 2, respectively. The Burgers vector of the defect on the left can be readily obtained by constructing a Burgers circuit using the DSC lattice as the reference lattice as described by Hirth and Balluffi (1973). The Burgers vector and step vectors are related by

$$\mathbf{b} = \mathbf{s}_1 - \mathbf{s}_2. \tag{23}$$

The heights of the steps in the boundary plane (with their appropriate signs) are given by $s_1 \cdot n$ and $s_2 \cdot n$ for lattices 1 and 2, respectively, where n is a unit vector normal to the boundary plane and pointing from crystal 1 into crystal 2. In **Figure 25**, $s_1 \cdot n = s_2 \cdot n$ and the effective step height produced by the line defect in the relaxed boundary structure is given by $s_1 \cdot n = s_2 \cdot n$. When $s_1 \cdot n \neq s_2 \cdot n$, the effective step height may be taken as $1/2(s_1 + s_2) \cdot n$.

A line defect that is a pure step without dislocation character is obtained in the special case when $s_1 - s_2 = \mathbf{b} = 0$. An example is the pure step at AB in **Figure 25**, where $s_1 = s_2$ is a vector of the CSL.

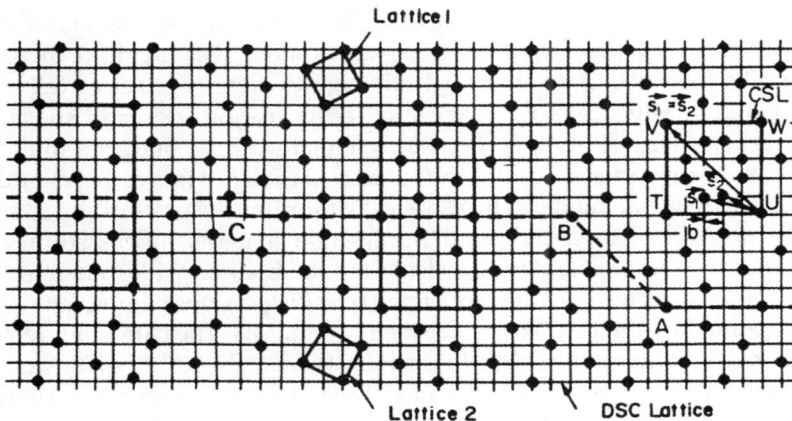

Figure 25 Line defects in a $\Sigma = 5$ symmetrical tilt grain boundary produced by 36.9° rotation around [001]. A line defect possessing both dislocation and step character is shown on the left at C and a pure step is shown at AB. The Burgers vector and step vector of the two line defects are shown in the diagram in the upper right. Adapted from Balluffi et al. (1982) with permission.

Conversely, a line defect that is a pure grain boundary dislocation without step character is obtained in the special case when $s_1 \cdot n = s_2 \cdot n = 0$. An example would be a grain boundary dislocation lying in a boundary plane parallel to the VWUT plane of the CSL shown in **Figure 25**, possessing the same Burgers vector **b** as the dislocation shown at C. These line defects possess a number of important topological properties that are summarized below.

(i) Movement in the plane of the boundary of a line defect that possesses dislocation character (which may require glide and climb) causes lattice 2 to translate with respect to lattice 1.

(ii) Movement of a line defect that possesses step character causes the interface to translate with respect to lattices 1 and 2.

(iii) Climb of line defects possessing dislocation character allows the boundary to act as a source or sink for point defects.

(iv) Lattice dislocations $b^{(L)}$ that impinge on the boundary may dissociate into an integral number of grain boundary line defects b_i because the Burgers vectors of the lattice dislocations and the interfacial dislocations are all vectors of the DSC lattice and must obey the conservation relation $b^{(L)} = n_i b_i$, where n_i is the number of grain boundary dislocations of type i.

(v) Lattice dislocations that cut through a boundary leave behind a line defect with dislocation character possessing a Burgers vector given by the DSC lattice.

The description of line defects in a grain boundary (homophase interface) by the DSC lattice can be readily extended to general heterophase interfaces. The details of this construction have been thoroughly discussed (Balluffi et al., 1982; Bonnet and Durand, 1975), and only the end results are presented here to show that dislocations and steps in general heterophase boundaries can be treated within the same DSC framework previously developed for grain boundaries. This construction provides a method to understand and quantify the character of line defects at heterophase interfaces in much the same way that one quantifies lattice dislocations and defects at grain boundaries. It has limitations that can be overcome by a more comprehensive treatment based on bicrystal symmetries (Pond, 1989) and this is discussed in the next subsection.

Figures 26 and 27 (Balluffi et al., 1982) show partly coherent planar and inclined interphase interfaces between two lattices (phases), respectively. The lattices can be assumed to be related by the transformation matrix **A** in Eqn (9a). The vectors **b**, s_1 and s_2 in these figures have the same meanings as in **Figure 25**. The cells indicated as M_1 and M_2 in the figures identify a pair of cells in the strain-free lattices (phases) 1 and 2, which almost match each other in both size and shape. These cells are usually but not always sublattices of the respective crystal lattices. The fine lines in the figures indicate the DSC lattice formed from lattices 1 and 2 in the same way as in **Figures 22(e) and 25**.

The line defects in the general interfaces have basically the same dislocation and step features as their counterparts in the special case of grain boundaries in cubic materials discussed previously. The step vector **s** is defined in the same way and for the defects in the boundary illustrated in **Figure 26**, when $s_1 \cdot n = s_2 \cdot n = 0$. These line defects are therefore interfacial dislocations without any step character and a Burgers vector **b** defined by the DSC lattice between M_1 and M_2. They are often referred to as misfit dislocations and this is a typical partly coherent interface (Christian, 1975), as illustrated previously in **Figure 5**. In the general case, line defects can have both dislocation and step character as seen in **Figure 27**, where a boundary is shown that differs from the boundary in **Figure 26** only by a rotation of the IP. In this case $s_1 \cdot n \neq s_2 \cdot n \neq 0$ for one type of line defect (indicated by a vertical T) and $s_1 = 0$ and $s_1 \cdot n \neq 0$ for the other (indicated by a horizontal T). A step is therefore evident at each defect. Note that unlike the atom positions around the two dislocations in **Figure 26**, where a terminating atom plane is evident in lattice 2, the atom planes are completely coherent around the steps (indicated by a vertical T)

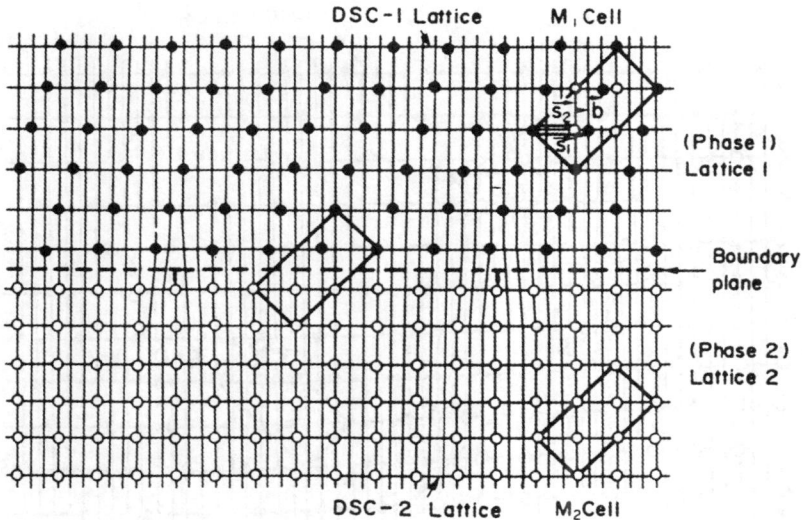

Figure 26 The interphase boundary between two different lattices (phases) M_1 and M_2, which form a planar interface containing interfacial dislocations with a spacing determined by O-lattice theory. The Burgers vector and step vectors of the interfacial dislocations are shown in the diagram in the upper right. The DSC lattice is indicated by fine lines in the figure. Adapted from Balluffi et al. (1982) with permission.

for which $s_1 \cdot n \neq 0$ in **Figure 27**, although a distortion, or DSC dislocation, is associated with each step. Such defects are commonly found at high-index interfaces in both martensitic (Howe and Mahon, 1989; Hirth and Pond, 1996; Pond et al., 2003; Ma and Pond, 2007) and diffusional (Das et al., 1996; Howe et al., 2009) solid–solid transformations and are often referred to as transformation dislocations or structural ledges (Christian, 1975; Hall et al., 1972; van der Merwe et al., 1991) depending on their role in the interface. The steps shown previously in **Figure 7** possess this type of defect character. The converse is true for the defect in the center of **Figure 27** (Balluffi et al., 1982), where $s_1 \cdot n = 0$ and a terminating atom plane is evident in lattice 2.

Some important concepts to appreciate regarding **Figures 26 and 27** are that when the lattice deformation **A** between two phases involves a simple shear and an expansion in the shear plane, as is commonly found in many martensitic and diffusional transformations, the partly coherent interface and associated dislocations that result from the expansion evident in **Figure 26** can be relieved by reorienting the IP so that the lattices match across the interface, as done in **Figure 27**. Notice that there has been no reorientation of the lattices in **Figures 26 and 27**, but only a change in the orientation of the IP, now angled from top to bottom across the figure. These steps also have the ability to transform lattice 2 to lattice 1 (or vice versa) as they move along the IP perpendicular to **n**. If there is also an expansion or contraction of one lattice relative to the other along the vertical direction in **Figure 26**, that is, normal to the shear plane, this can be accommodated by a set of dislocations such as the one evident in the center of **Figure 27**. Since this step contains a lattice dislocation, it cannot move without diffusion of atoms or vacancies to the step. Hence the former defect can glide while the latter must climb. Of course, the distribution of either type of defect and their ability to glide or climb can also be altered by a slight rotation of one lattice relative to the other, and phases usually have the ability to manipulate these variables during formation. It is also important to keep in mind that that **Figures 26 and 27** are

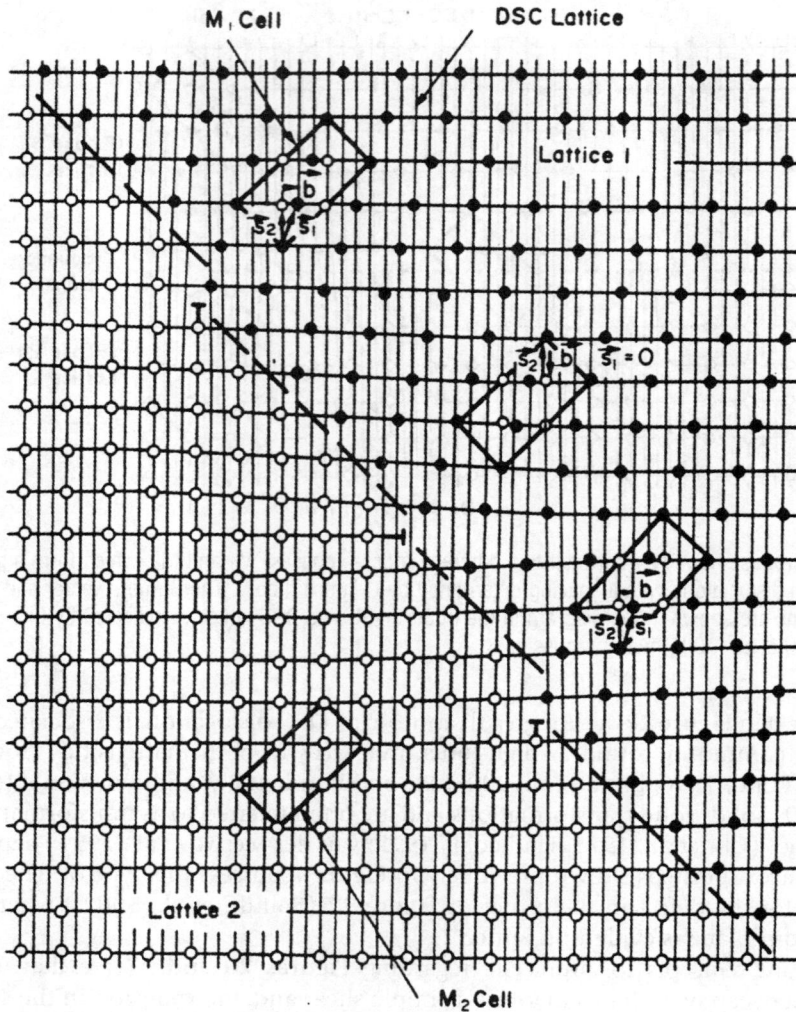

Figure 27 An interphase boundary between lattices 1 and 2 in **Figure 26**, but with the boundary plane at a different inclination. Two types of line defects are present possessing both dislocation and step character. The Burgers vectors and step vectors are illustrated in each case. Adapted from Balluffi et al. (1982) with permission.

two-dimensional representations where it is assumed that the lattices have the same dimension in the direction normal to the figures. It is possible that this dimension is also different between lattices 1 and 2, requiring more complicated interfacial structures, although the same considerations apply. Because these structural features are difficult to quantify, particularly regarding the diffusional fluxes that may be required for interfacial motion, the concept of disconnections discussed in the next section presents a more general and quantitative way to characterize heterophase interfaces.

Just as for a grain boundary, it is also possible to have pure interphase boundary steps without DSC lattice dislocation character. For example, a step in the interface in **Figure 26** that joins the lattice sites delineated by the bottom edges of cells M_1 and M_2 would produce a pure interphase boundary step. In this case, $s_1 = s_2$ and $b = s_1 - s_2 = 0$, although in the DSC lattice framework the step may have an associated weak, long-range stress field if isolated in the boundary without complete relaxation by other DSC lattice dislocations in the interface. Since the line defects in general interphase boundaries have basically the same dislocation and step features as their counterparts in grain boundaries in cubic crystals, one expects them to possess the same basic topological properties summarized previously for grain boundaries.

In summary, the DSC lattice framework provides a means of quantifying the various types of line defects found in grain and interphase boundaries in materials. In many cases the DSC lattice can be obtained by simple graphical construction based on the atomic positions in the structures, as in the examples shown here. The Burgers vectors and step vectors of interfacial line defects can then be obtained directly from the construction. In cases where graphical construction is difficult, the problem can be solved by calculation or matrix methods. The main limitations of the DSC method are that it is a geometric construction and thus unable to account for local atomic relaxations that can occur at interfaces, and it represents an approximation in near-coincidence situations when there are two different types of lattices, as in **Figure 26**, for example. Pond (1989), Pond and Hirth (1994) and Hirth and Pond (1996) have developed a method for quantifying the nature of line defects in interfaces based on the space-group symmetries of the adjacent crystals. This method is completely general and does not suffer from the limitations of the DSC model above in its capability to quantify the character of interfacial defects. It also provides a physical picture of the interphase boundary, so it is discussed further below.

14.2.1.4 *Disconnections in Heterophase Interfaces*

Pond (1989) has shown that the topological properties of interfacial defects can be quantified based on the consequences of symmetry breaking between adjacent crystals at an interface. The admissible defects at the interface, that is those that separate crystallographically equivalent interface structures, are characterized by operations that are given by the product of two symmetry operations, one from each of the two crystals. The range of distinct defects that can arise in a given interface depends on the extent of symmetry breaking when the bicrystal is formed.

Consider a bicrystal where a planar surface of one crystal, designated "white" (λ), abuts the surface of a "black" (μ) one. The topological character of an admissible defect that can be superimposed on this reference structure is given by a combination of symmetry operators, one from each of the crystals (Howe et al., 2009). Using the matrix formalism for symmetry operators set out in the International Tables for Crystallography (Hahn, 1983), one finds that the operator characterizing a defect is given by

$$\mathscr{Q}_{ij} = \mathscr{W}(\lambda)_i \, \mathscr{W}*(\mu)_j^{-1}, \tag{24}$$

where $\mathscr{W}(\lambda)_i$ and $\mathscr{W}(\mu)_j$ are the relevant operators and the asterisk implies that $\mathscr{W}(\mu)_j$ has been expressed in the λ coordinate frame. The set of defects defined by Eqn (24) includes dislocations that have reached the interface from either of the crystals and also a range of other defects peculiar to the interface. In the former case, a perfect white crystal dislocation is represented by $\mathscr{Q}_{ij} = \mathscr{W}(\lambda)_i = (\mathbf{I}, \mathbf{t}(\lambda)_i)$, where $\mathbf{I}$ represents the identity operation and $\mathbf{t}(\lambda)_i$ is the translation vector equal to the dislocation's Burgers vector $\mathbf{b}$. An interfacial defect arises, for example, when dislocations from both crystals coincide at the interface; then $\mathscr{Q}_{ij} = (\mathbf{I}, \mathbf{t}(\lambda)_i - \mathbf{t}*(\mu)_j)$, or, in other words, the defect exhibits dislocation character with $\mathbf{b} = \mathbf{t}(\lambda)_i - \mathbf{t}*(\mu)_j$. Hirth and Pond (1996) have defined the overlap step height h associated with such a defect to be the smaller of $\mathbf{n} \cdot \mathbf{t}(\lambda)_i$ and $\mathbf{n} \cdot \mathbf{t}*(\mu)_j$, where $\mathbf{n}$ is the unit

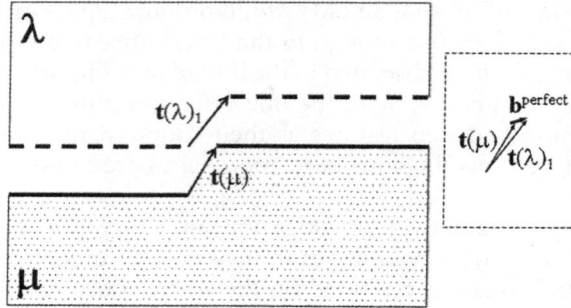

Figure 28 Schematic illustration of a perfect disconnection formed by bonding incompatible surface steps on the λ and μ crystals (Pond, 1989). If the λ and μ surfaces to the left of the defect are bonded first, a Volterra operation Q_{ij} is needed to correctly juxtapose the surfaces on the right-hand side before joining to form a degenerate interface. Adapted from Howe et al. (2009) with permission.

normal to the interface pointing into the λ crystal. When h is finite, the defect is a disconnection characterized by ($\mathbf{b}$, h), as depicted schematically in **Figure 28** (Howe et al., 2009). Disconnections are perfect when they separate energetically degenerate regions of interface, and partial otherwise (Pond, 1989). Since there is no overlap step when only one crystal dislocation is present, such a defect is characterized by ($\mathbf{b}$, 0). The long-range elastic field is that of the dislocation. The overlap step produces no long-range elastic field, but it does produce an offset of the interface. Hence, for arrays of disconnections, the interface can be described by the superposition of the fields of the individual disconnections and the steps can be considered to lie normal to the adjoining terraces.

Hirth and Pond (1996), Pond et al. (2003), Ma and Pond (2007) and Howe et al. (2009) have extensively applied this approach to heterophase interfacial structures developed during martensitic and diffusional phase transformations. This approach differs from the former O-lattice and CSL/DSC approaches in that it is based on minimizing the elastic fields associated with potential defect arrays at the interface as well as balancing any diffusional fluxes that are required (or not) to move the interface. Thus, it is a more mechanistic-based approach to heterophase interfacial structure. The following description is provided to illustrate its main concepts and utility. It should be mentioned that there other competing treatments for describing the same, or similar, interfacial structures developed during phase transformations, and these are discussed later in the section on heterophase interfaces.

Coherent interfaces between distinct phases generally require the adjacent crystals to be strained, and such strains can be removed by the superposition of an appropriate network of line defects (Frank and van der Merwe, 1949). The following treatment shows how to determine such network configurations when these comprise both dislocations ($\mathbf{b}$, 0) and disconnections ($\mathbf{b}$, h).

For quantitative modeling, one must define the bicrystal reference structure employed, since this in turn determines the parameters ($\mathbf{b}$, 0) and ($\mathbf{b}$, h) of the defects in the strain accommodating network. In the present context the use of two references, the natural and the coherent, is convenient. In the natural reference state, the two crystals exhibit their unstrained unit-cell dimensions (at the chosen temperature), and they are relatively disposed with selected misorientation and IP—the latter as the terrace plane. Later, for actual interfaces containing steps, the average IP is defined as the habit plane, but the planar regions between defects still corresponds to the terrace plane. In the coherent reference state, both crystals are homogeneously strained to produce a coherent terrace. An illustration of natural and coherent reference states is depicted in **Figure 29** (Howe et al., 2009). (The term reference state is used

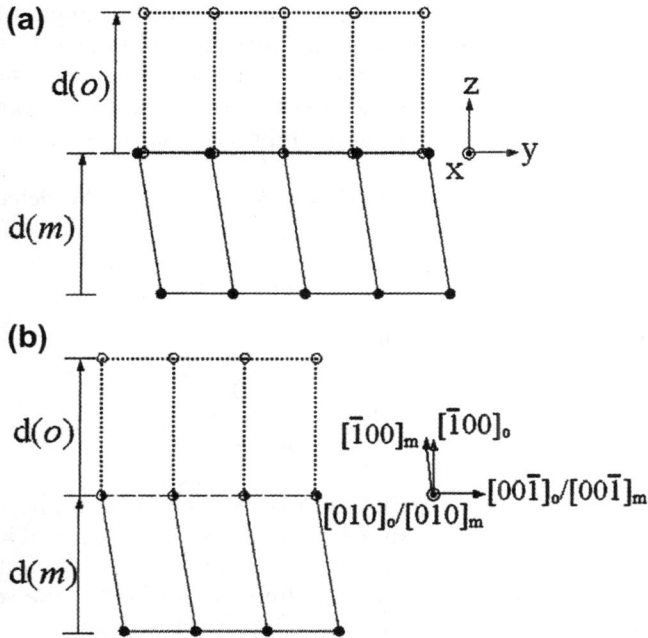

Figure 29 Schematic view of the $(100)_o \| (100)_m$ terrace plane between orthorhombic and monoclinic ZrO_2 (Pond et al., 2007). (a) The interface is incoherent and the crystals exhibit their natural lattice parameters (the monoclinic cell shown comprises two crystal unit cells). (b) The interface is coherent following uniaxial strains of the two crystals. Adapted from Howe et al. (2009) with permission.

here to mean either a reference bicrystal structure, as above, or a reference dichromatic pattern, that is, the pattern where the λ and μ lattices interpenetrate (Pond and Vlachavas, 1983).

The defect content of the coherent with respect to the natural state, $\mathbf{B}^c$, can be formally quantified using the Frank–Bilby equation (Eqn (10c)). Let $_c\mathbf{\Lambda}_n$ and $_c\mathbf{M}_n$ be matrices representing the homogeneous deformations necessary to transform the natural λ and μ crystals, respectively, into their coherent forms. Then

$$\mathbf{B}^c = (_c\mathbf{\Lambda}_n^{-1} - {_c}\mathbf{M}_n^{-1})\mathbf{x}, \tag{25}$$

where $\mathbf{x}$ is a probe vector in the coherent structure. (The Frank–Bilby equation is equivalent to Eqn (24), where $\mathscr{W}(\lambda)_i$ and $\mathscr{W}*(\mu)_j$ represent the probe vectors after mapping into the λ and μ crystals, respectively, of the natural reference (Pond and Hirth, 1994) The defects in the coherent interface, **Figure 29(b)**, can be regarded as smeared into a distribution of continuous infinitesimal defects (Bilby et al., 1955) referred to as the coherency dislocation content, signified by the superscript $\mathbf{B}^c$.

In real bicrystals of significant size, coherency strains are accommodated by networks of discrete defects (Hirth and Lothe, 1982). Thus, superimposing a network of localized defects with total content $-\mathbf{B}^c$ onto the coherent reference would remove long-range coherency strains from the interface. The total dislocation content with respect to the natural reference state would be zero and only short-range strains would remain. In practice, it may not be possible to construct networks comprising admissible defects in the coherent reference that sum exactly to $-\mathbf{B}^c$, but networks can be identified where any excess Burgers vector content corresponds to the introduction of low-angle tilt or twist boundaries. Thus, small angular deviations from the reference state may arise, but the λ and μ crystals would retain their natural cell dimensions away from the interface. The following analyses consider dislocation networks where no ancillary tilts or supplementary twists arise, that is any Burgers vectors normal to the habit plane are suppressed, and these can be superimposed later, together with associated tilts and twists.

The terrace coherency strain is expressed as $_n\mathbf{E}_c = (_c\mathbf{\Lambda}_n^{-1} - _c\mathbf{M}_n^{-1})$. The defect content necessary to remove the coherency strain, $\mathbf{B} = -\mathbf{B}^c$, is given by

$$\mathbf{B} = -_n\mathbf{E}_c\mathbf{x}, \tag{26}$$

where $\mathbf{x} = [x, y, 0]$. Furthermore, one can write

$$_n\mathbf{E}_c = \begin{pmatrix} e_{xx} & 0 & 0 \\ 0 & e_{yy} & 0 \\ 0 & 0 & 0 \end{pmatrix}, \tag{27}$$

where e_{xx} and e_{yy} are the principal strains in the (x,y) terrace plane. The components e_{xz}, e_{yz}, and e_{zz} are free because of the suppression of defect components b_z, so two or three sets of independent defects are required to fix the other three components as described below. The total defect content of the strain-relieving network $\mathbf{B}$ is a sum of Burgers vectors of admissible defects in the coherent reference; one can obtain these Burgers vectors either directly from Eqn (24) with the coherent λ and μ unit-cell dimensions or graphically from the coherent dichromatic pattern. The following considers some simple examples of networks consistent with Eqn (26) for dislocations having no component of $\mathbf{b}$ normal to the terrace plane. Depending on the multiplicity of feasible Burgers vectors in a dichromatic pattern and their orientation with respect to the principal axes, there may be more than one network capable of relieving the strain $_n\mathbf{E}_c$. Equation (26) shows that coherency strains where both e_{xx} and e_{yy} are finite can be accommodated by two-array networks if both sets of dislocations have Burgers vectors of the form $\mathbf{b} = [b_x, b_y, 0]$ and their line directions and spacings are unconstrained. The condition $e_{xy} = 0$ is satisfied if the two-dislocation network has no twist component. Two-array networks are important in phase transformations and are treated further in terms of Eqn (26) below. However, there are circumstances where a single array may suffice, and other situations where three arrays are needed. Some simple examples are outlined below.

In the case where $e_{xx} \neq 0$ and $e_{yy} = 0$ and an admissible dislocation exists with $\mathbf{b} = [b_x,0,0]$, a single array of defects in edge orientation with spacing b_x/e_{xx} would accommodate the coherency strain. In the case of balanced biaxial strain, $e_{xx} = e_{yy}$, as occurs, for example at epitaxial $\{111\}$ interfaces between FCC crystals, a hexagonal network of three arrays of edge dislocations with $\mathbf{b} = 1/2\langle 110\rangle$ is needed (Hirth and Lothe, 1982). The principal axes are indeterminate in this case, so x may be arbitrarily chosen parallel to one of the three $\langle 110\rangle$ directions. Thus, one edge array with $\mathbf{b} = [b_x,0,0]$ accommodates e_{xx} while the other two, in combination, accommodate e_{yy}. Of these three sets, only two are independent (Groves and Kelly, 1963), in accord with the statement following Eqn (26). If the Peierls structure of the interface constrains the dislocation spacing d to not quite equal b_x/e_{xx}, for example, sets of defects with increasing spacings can be arranged to remove coherency strains on average, analogous to the technique for describing an irrational number as a series of rational fractions.

Returning to the geometry of two-array networks, let the Burgers vectors of the dislocations be $\mathbf{b}^p$ and $\mathbf{b}^q$. One selects a unit probe vector $\mathbf{x}^p$ parallel to the p-disconnection line, so that only the q-dislocation array is cut so that Eqn (26) leads to the dislocation density $\mathbf{B}^q = -_n\mathbf{E}_c\,\mathbf{x}^p$. Similarly, the dislocation density $\mathbf{B}^p = -_n\mathbf{E}_c\,\mathbf{x}^q$ can be established using a unit vector parallel to the q-dislocation line direction $\mathbf{x}^q$. One can either solve these equations to find $\mathbf{x}^p$ and $\mathbf{x}^q$ assuming $\mathbf{b}^p$ and $\mathbf{b}^q$, or vice versa. In the former case one has

$$\mathbf{x}^q = \frac{-(_n\mathbf{E}_c)^{-1}\mathbf{b}^p}{\left|(_n\mathbf{E}_c)^{-1}\mathbf{b}^p\right|} \tag{28a}$$

and

$$\mathbf{x}^p = \frac{-(_n\mathbf{E}_c)^{-1}\mathbf{b}^q}{\left|(_n\mathbf{E}_c)^{-1}\mathbf{b}^q\right|}. \tag{28b}$$

The line senses of the dislocations, $\mathbf{l}^p$ and $\mathbf{l}^q$, are either parallel or antiparallel to $\mathbf{x}^p$ and $\mathbf{x}^q$, respectively; to be consistent with the FS/RH convention $\mathbf{l}^p \times \mathbf{x}^q$ and $\mathbf{l}^q \times \mathbf{x}^p$ must be parallel to $\mathbf{n}$, that is, point toward the (upper) λ crystal. Furthermore, since $\mathbf{B}^p = \dfrac{\mathbf{b}^p}{d^p}\sin(\theta^p - \theta^q)$ and $\mathbf{B}^q = \dfrac{\mathbf{b}^q}{d^q}\sin(\theta^p - \theta^q)$, where d^p and d^q are the array spacings and θ^p and θ^q are the angles subtended by $\mathbf{l}^p$ and $\mathbf{l}^q$ from a common direction, respectively, one has

$$d^p = \frac{|\mathbf{b}^p|}{|(_n\mathbf{E}_c)\mathbf{v}^q|}\sin(\theta^p - \theta^q) \tag{29a}$$

and

$$d^q = \frac{|\mathbf{b}^q|}{|(_n\mathbf{E}_c)\mathbf{v}^p|}\sin(\theta^p - \theta^q). \tag{29b}$$

To illustrate the use of Eqns (28) and (29), consider the case where $e_{xx} = -e_{yy}$. Several authors have discussed this case (Matthews, 1974; Hirth, 1993). Matthews (1974) considered the case of two orthorhombic crystals, where the natural and coherent reference states are shown projected along [001] in **Figure 30** (Howe et al., 2009). The matrix $_n\mathbf{E}_c = (_c\Lambda_n^{-1} - _c\mathbf{M}_n^{-1})$ takes the form

$$_n\mathbf{E}_c = \begin{pmatrix} \varepsilon & 0 & 0 \\ 0 & -\varepsilon & 0 \\ 0 & 0 & 0 \end{pmatrix} \tag{30}$$

with $\varepsilon = (1 + e_{xx}(\lambda)) - (1 + e_{xx}(\mu)) = -(1 + e_{yy}(\lambda)) - (1 + e_{yy}(\mu))$, where the strains $e_{xx}(\lambda)$, are defined in **Figure 30**. Taking the pair $\mathbf{x}^p = [100]_{\text{coh}}$ and $\mathbf{x}^q = [010]_{\text{coh}}$, substitution into Eqns (28a) and (28b) shows that $\mathbf{b}^p = [010]_{\text{coh}}$ with $\mathbf{l}^p = [100]_{\text{coh}}$, and $\mathbf{b}^q = [\bar{1}00]_{\text{coh}}$ with $\mathbf{l}^q = [0\bar{1}0]_{\text{coh}}$. Using Eqns (29a) and (29b) one obtains the spacings of these orthogonal edge dislocation arrays as $d^p = d^q = b_e/\varepsilon$, where $b_e = |\mathbf{b}^p| = |\mathbf{b}^q|$. The resulting network is depicted in **Figure 31(a)** (Howe et al., 2009), where it is seen that the extra half-planes of the q-dislocations are located in the μ crystal and the p-dislocations in the λ crystal. Alternatively, taking $\mathbf{x}^p = 1/\sqrt{2}[110]_{\text{coh}}$ and $\mathbf{x}^q = 1/\sqrt{2}[\bar{1}10]_{\text{coh}}$, one obtains $\mathbf{b}^p = [110]_{\text{coh}}$ with $\mathbf{l}^p = [110]_{\text{coh}}$, and $\mathbf{b}^q = [\bar{1}10]_{\text{coh}}$ with $\mathbf{l}^q = [1\bar{1}0]_{\text{coh}}$. In this case, $d^p = d^q = b_s/\varepsilon$, where $b_s = \sqrt{2}b_e$. The p-dislocations in the network are right-handed screws and the orthogonal q-dislocations

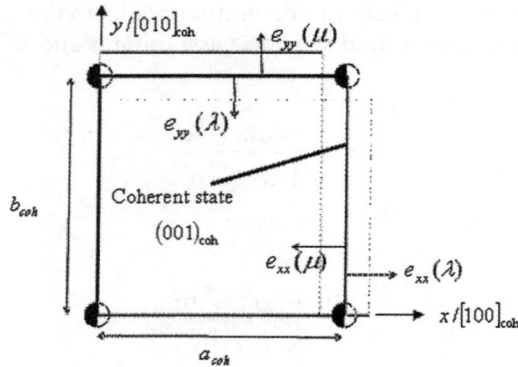

Figure 30 Reference states for two orthorhombic crystals. The [001] projection direction is normal to the IP. The natural λ and μ crystal unit cells are depicted by dashed and full lines respectively; the coherent unit cell is depicted by bold lines. Adapted from Howe et al. (2009) with permission.

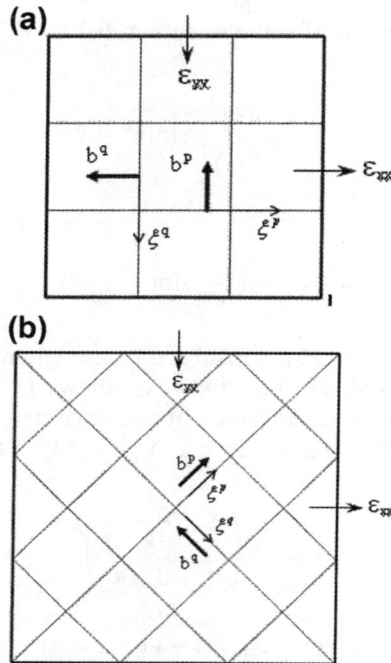

Figure 31 Dislocation networks accommodating equal and opposite principal strains: (a) edge dislocations, and (b) screw dislocations (not to scale). Adapted from Howe et al. (2009) with permission.

are left-handed screws, as depicted in **Figure 31(b)**. Either configuration in **Figure 31** fully accommodates the strains e_{xx} and e_{yy}.

The Burgers vectors of the dislocations discussed above are parallel to the terrace plane. In phase transformations, network defects sometimes include edge components normal to the interface b_z that do not contribute to strain accommodation in the terrace plane. An array of such defects produces tilt, introducing a rigid body rotation between the crystals about an axis parallel to the defect line direction, or the tilt axis. This changes the misorientation relative to the reference coherent state (Pond et al., 2003). The field of this array can be superposed on the dislocation array in the terrace plane (**Figure 30**), thereby compensating coherency without introducing cross-terms in the total elastic distortion field. In the isotropic elastic case, the tilt rotation φ is partitioned equally between the two crystals. Otherwise, the rotation is partitioned between the two crystals depending on their relative elastic compliances (Hirth et al., 1979).

The defect content producing the ancillary tilt rotation can be incorporated into Eqn (26), as follows:

$$\mathbf{B} = (-{_n}\mathbf{E_c} + \mathbf{R})\mathbf{x} \tag{31a}$$

and

$$\mathbf{R} = \left(\mathbf{R}^\lambda\right)^{-1} - \left(\mathbf{R}^\mu\right)^{-1}, \tag{31b}$$

where $\mathbf{R}^\lambda$ and $\mathbf{R}^\mu$, each equal in magnitude to $\varphi/2$, are the partitioned rotations of the λ and μ crystals. The matrix $(-{_n}\mathbf{E_c} + \mathbf{R})$ defines the closure failure of a Burgers circuit associated with the vector $\mathbf{x}$. If two or more arrays of defects are present in a network, tilts may be associated with each one. There are several observations of such tilts in the literature, especially in epitaxial systems (Beanland et al., 1994).

In the case of phase transformations, twist arrays can be added if they lower the energy of the interface. The Burgers vectors and rotation would still be given by Eqn (31) and φ would be the total rotation about the normal to the IP partitioned between the two crystals. The interface remains coherent in the regions between the dislocations in the final network. Matthews (1974) considered the case depicted in **Figure 31(b)** where superposition of a twist array of right-handed screw dislocations results in a decrease of the p-dislocation spacing and an increase in that of the q-dislocation spacing. A purely p-dislocation network is possible in the limit. Thus, tilt and twist arrays may be present separately or in combination in an interface.

Further considerations are necessary when one of the defect arrays in a network comprises disconnections because of their step character h. An array of disconnections causes the resultant IP to deviate from the terrace plane orientation. This IP is often referred to as the habit plane in phase transformations. The deviation of the habit plane from the terrace plane depends upon the line direction and spacing of the disconnection array. In other words, a probe vector $\mathbf{x}$ to be used in the Frank–Bilby equation (Eqn (25)) is not independent of $\mathbf{B}$ because of the coupled topological properties of $\mathbf{b}$ and h associated with disconnections. Also, one must consider whether the line directions of the dislocations $\mathbf{l}^L$ (the superscript L signifies a lattice invariant deformation (LID)), or disconnections $\mathbf{l}^D$, are free parameters. For example, if the dislocations in an array glide to the habit plane from one of the crystals and are sessile therein, $\mathbf{l}^L$ is fixed along the direction of the intersection of the habit and glide planes $\mathbf{l}^G$. In contrast, if dislocation climb is permitted $\mathbf{l}^L$ may not be constrained in this way. Also, if the LID defects are twinning dislocations (disconnections), they will be sessile whether or not climb can occur, since any motion in the IP would create a high-energy fault. A practical procedure to take these factors into account (Pond et al., 2007) is to temporarily suppress the z components of $\mathbf{b}^D$ and $\mathbf{b}^L$ and find a provisional solution to Eqn (26), treating $\mathbf{l}^D$ and $\mathbf{l}^L$ as unconstrained parameters. The step height h of the disconnections is also temporarily set to zero, and a provisional terrace network determined. Using the current network parameters, one can find an updated habit plane by introducing the disconnection steps. The updated components b'_z of the defects (the coordinate frame based on the habit plane is $x'y'z'$)

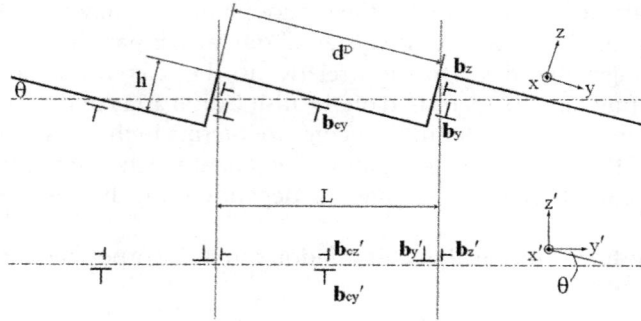

Figure 32 Schematic illustration showing the disconnection content of an interface, with Burgers vector components resolved in the terrace (upper) and habit plane (lower) frames (Pond et al. (2003)). The terrace plane is inclined at an angle θ to the horizontal habit plane. Coherency strain is represented by the equivalent "coherency" defect content, $\mathbf{b}_c$. The x-axis points out of the page. Adapted from Howe et al. (2009) with permission.

and the current line directions can be calculated. The procedure is repeated iteratively and utilizing crystal plasticity theory, one can determine the residual coherency strain at any stage until an acceptable value is reached.

To illustrate this procedure, consider a simple case where $e_{xx} = 0$ and the finite strain e_{yy} is accommodated by a single array of disconnections with $\mathbf{l}^D$ parallel to x for which $\mathbf{b}^D = [0, b_y, b_z]_{\text{coh}}$, as shown in **Figure 32** (Howe et al., 2009). In this case (Pond et al., 2003, 2007) an exact answer can be obtained without iteration. The equilibrium habit plane is inclined to the terrace plane by the angle θ obtained from

$$-e_{yy} = (b_y \tan \theta + b_z \tan^2 \theta)h^{-1}. \tag{32}$$

An ancillary rotation about x, ϕ_D is equal to

$$\phi_D = 2\sin^{-1}\left[(b_z \cos \theta - b_y \sin \theta - e_{yy}h \cos \theta) \sin \theta/2h\right]. \tag{33}$$

(Note that Kelly and Qui (2011) have shown that a simpler version of Eqns (32) and (33) results if the step height h in the topological model is defined at the average of the μ and λ crystal step heights, rather than the overlap step height, in which case the habit planes also provide better agreement with the phenomenological theory of martensite crystallography (PTMC).) Because of the equal partitioning of ϕ_D between the crystals, the apparent habit plane inclination for an observer in the λ crystal far from the interface and unable to resolve the near-field elastic strains of the discrete defects (or an experimentalist using optical microscopy and X-ray diffraction) is $(\theta + \phi_D/2)$. Conversely, it is $(\theta - \phi_D/2)$ for an observer in the μ crystal. The inclined interface in **Figure 32** is analogous to the one shown previously in **Figure 27**, but determined from a much different approach.

If e_{xx} is not zero, it can be accommodated by an array of dislocations from, say, the μ crystal with components of $\mathbf{b}^L = [b_x, 0, b_z]_{\text{coh}}$. Formally, the iterative refinement procedure outlined above is needed to find the final configuration on the habit plane, and this is shown schematically in **Figure 33**. The terraces remain coherent between the defects and the distortion diminishes with distance from the interface. An ancillary tilt of the phases about $\mathbf{l}^L$ arises as outlined above since $b_z \neq 0$.

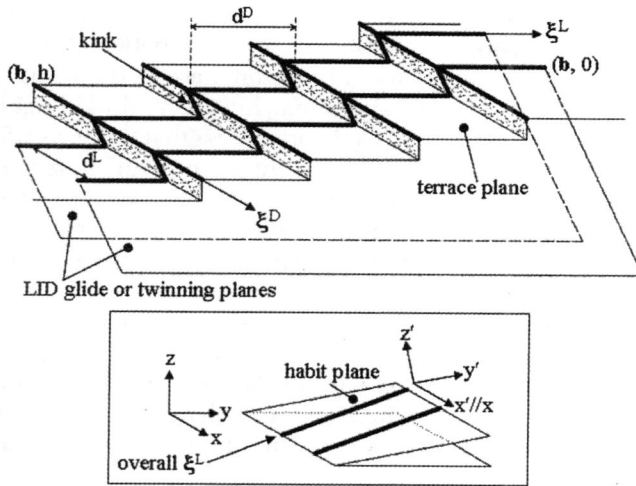

Figure 33 Schematic illustration of an interphase interface showing the terrace segments and defect arrays (Pond et al., 2007). Coherently strained terraces are reticulated by arrays of disconnections (**b**, h) with spacing d^D and crystal slip dislocations (**b**, 0) in the (lower) μ crystal. The terrace and habit (primed) coordinate frames are shown and the line directions of the disconnections, ξ^D, and dislocations, ξ^L, are parallel to x and close to y' respectively. Adapted from Howe et al. (2009) with permission.

The preceding analyses have emphasized some solutions to Eqn (26) that correspond to important examples of phase transformations in the literature. In general, with b_z still suppressed, Eqn (26) can be satisfied by two independent arrays of disconnections and LID dislocations, plus a twist rotation, as demonstrated for martensite in Fe (Ma and Pond, 2007), or by three independent arrays plus the condition of no twist rotation. Sargent and Purdy (1975) showed that three in-plane Burgers vectors are required for a general misfit case. This is also true for three in-plane Burgers vector components, since the normal components produce pure rotations. Reduction of the three-array requirement to two is analogous to the reduction of a two-array requirement to one when a twist rotation is added in the Matthews example previously discussed. For example, if one array is sessile (e.g. a single twin or glide LID on a plane inclined to the terrace plane) then three independent arrays are required except for a special line direction of the LID in the terrace plane. Also the iteration previously described when steps are introduced into the solution can lead to components of **b** that vary with step spacing, which again can lead to the need for more independent systems.

The following summarizes implementation of the topological model to find the defect structure of a two-dimensional transformation interface. One first seeks a matched terrace plane with small misfit as these display energy minima (Dahmen, 1982; van der Merwe, 1982). The two-dimensional coherency of the terrace plane is essential for diffusionless transformations. The crystals adjoining the terrace plane are then uniformly strained to achieve a coherent interface. The crystal lattices are subsequently interpenetrated to form a dichromatic pattern, which is the coherent reference state. Inspection of the reference state allows one to select candidate disconnections (**b**, h) and dislocations (**b**, 0) much like one selects DSC dislocations in **Figures 26 and 27**, for example. For simple systems, one can then deduce the correct disconnection and LID spacings and line directions to relieve all long-range coherency strains, and thus determine the orientation of the habit plane and the ancillary rotation φ. For less simple systems one can proceed to solve Eqn (26) to determine the spacings and line directions. In any case, tilt or twist rotations can be superposed if needed, partitioning the rotations between the two crystals.

The concept of the habit plane as an invariant plane interface in the PTMC (Wechsler et al., 1953; Bowles and Mackenzie, 1954; Christian, 1975) is a matching condition in rigid crystal modeling. The topological model, which is based on dislocation theory, shows that this is a sufficient but not necessary condition for the existence of glissile partly coherent interfaces with short-range strain fields. A very important aspect of the topological model is its ability to take into account any diffusive fluxes that are involved in interface movement and thereby, the glide and/or climb components involved in a particular phase transformation (Pond and Sarrazit, 1996). These are not accounted for in the O-lattice and CSL/DSC analyses, for example.

To calculate the diffusive flux of material associated with the motion of a dislocation or discon-nection along an interface it is necessary to take into account any differences of composition and density between the two phases. Let the interface be the plane (x,y) in a Cartesian coordinate frame, and consider a defect parallel to x moving in the direction y. For a length of defect l moving a distance δy the number of (substitutional) atoms of species A, δN_A, that must diffuse to/away (+ve/−ve) from the defect core is given by

$$\delta N_A = l(h\Delta X_A + b_z X_A)\delta y, \tag{34}$$

where $\Delta X_A = X(\lambda)_A - X(\mu)_A$ is the difference between the number of A atoms per unit volume in the λ and μ crystals, X_A is either $X(\lambda)_A$ or $X(\mu)_A$ (that with the larger of $\mathbf{n} \cdot \mathbf{t}(\lambda)_i$ and $\mathbf{n} \cdot \mathbf{t}^*(\mu)_j$), and b_z is the component of $\mathbf{b}$ normal to the IP (Hirth and Pond, 1996). Conservative motion of disconnections occurs only when δN_i is zero for each species present. When δN_i is nonzero, long-range diffusion must occur to remove or add this amount as the defect moves. For the case of disconnections in interphase interfaces, this only arises for motion along coherent interfaces, that is where the interface exhibits two-dimensional periodicity (Pond and Celotto, 2003). In these circumstances, atoms can be transferred from λ to μ in a relatively orderly manner by a combination of shear and/or shuffling. While the number of species present is thereby conserved, the volume occupied by them may not be since the crystal densities are usually not identical. Thus, interfacial coherency is seen to be important not only from an energetic point of view (Dahmen, 1982; van der Merwe, 1982) but also from a mechanistic perspective in site-conserving transformations (Howe, 1994a; Muddle et al., 1994). Equation (34) can be quantized in terms of elementary kink jumps as described by Howe et al. (2009), but that is not described here.

If one takes the time derivative of Eqn (34), one finds that the current of atoms produced at a moving disconnection is

$$I_A = l(h\Delta X_A + b_z X_A)v^D, \tag{35}$$

where v^D is the steady-state velocity of a disconnection. This current equals the diffusion flux J_A of A multiplied by the area enclosing the disconnection through which the flux is passing, as evident from Eqn (35) when treated in the moving frame of reference. The rate of transformation expressed as a velocity v normal to the habit plane is given by $v = v^D \sin\theta$, and $\tan\theta = h/d^D$ (refer to **Figures 32 and 33**). Here d^D need not equal the ideal equilibrium spacing since it can be kinetically controlled. Thus, the rate of transformation is given by a solution of the diffusion equations for J_A, together with a kinetic analysis for d^D. For the special case of a martensite transformation, which may involve only one chemical compo-nent, Eqn (34) has $\delta N_i = 0$ for all atomic species. That is, $b_z = -h\Delta X_i/X_i$ for each species. In this circumstance the components b_z of the dislocations can move conservatively along with the discon-nection. In other words, the number of atoms of each species present is the same in the elemental cells $v(\lambda)$ and $v(\mu)$, even though these cells may have different volumes, and climb is not required.

14.2.1.5 Atomistic Modeling of Grain Boundary Structure and Energy

Having considered the properties of grain and interphase boundary line defects, the discussion now returns to determining the atomic structure and energy of grain boundaries. The main purpose of atomistic investigations of grain boundary structure is to relate the structure and grain boundary energy γ_{GB} to the grain misorientation and boundary plane. The calculations are usually performed by establishing an initial relationship between two grains using the five macroscopic DOF discussed in Section 14.2.1.1.4. An atomic potential is chosen and the bicrystal is then allowed to relax into a minimum-energy configuration through a suitable algorithm (Wolf and Merkle, 1992; Foiles, 1992). Relative translations of the two lattices through the three microscopic DOF and/or local atomic rear-rangements are all allowed as part of the minimization process. Initially, there was some question as to the validity of the interatomic potentials used in the calculations, the approximations used to reach equilibrium and whether calculated energy minima corresponded to real energy minima, but these uncertainties have been largely relieved by extensive comparison between calculated and observed grain boundary structures (Wolf, 1992; Wolf and Merkle, 1992).

Two important fundamental insights have arisen from calculations of the structure and energies of grain boundaries in metals. First, it has been shown that the structure of grain boundaries can be described as composed of several structural (or polyhedral) units, which are repeated in a particular sequence along the boundary, depending on the type of boundary and angle of misorientation. This result is appealing because of its conceptual simplicity and because it relates to certain properties of boundaries such segregation and diffusion. Secondly, it has been shown that the grain boundary energy is directly correlated with the volume expansion per unit area of the boundary, and that this feature is also related to the number of nearest-neighbor broken bonds at the interface. These studies have also elucidated the reason why the CSL model does not accurately predict the energies of grain boundaries except in special orientations. Each of these subjects is examined in greater detail below since they provide an atomistic picture that relates the structure of grain boundaries to their energy and also to their segregation tendency, which is discussed in a following section.

Finally, an important aspect of grain boundaries is not only their structure and energy but also their population in real materials, since this often governs the important macroscopic properties of interest. Significant progress has been made recently which shows that the population of grain boundaries is inversely correlated to their energy, and that the most common grain boundary planes are typically low-index low surface-energy planes, such as $\{111\}$ in FCC Al and Ni, or $\{100\}$ in other cubic ceramics (see, e.g. Holm et al., 2011; Rohrer, 2011). These studies emphasize the key role of the grain boundary plane in determining the resulting energy, mobility and population of the boundaries, and they deemphasize the role of the CSL as an important factor in determining resulting grain boundary structures except in certain cases. These studies also reveal important insight into the nature of incoherent interphase boundaries, such as that shown in **Figure 6**, where one of the boundary planes is a low-energy $\{111\}$ plane, which dictates the orientation and energy of the interface. Hence, this section concludes with a brief discussion of these recent findings.

14.2.1.5.1 Structural Unit Model

An example of the structural unit (polyhedral) model is shown in **Figure 34** for a series of symmetrical [001] tilt grain boundary structures calculated using a pair potential appropriate for FCC Cu (Sutton

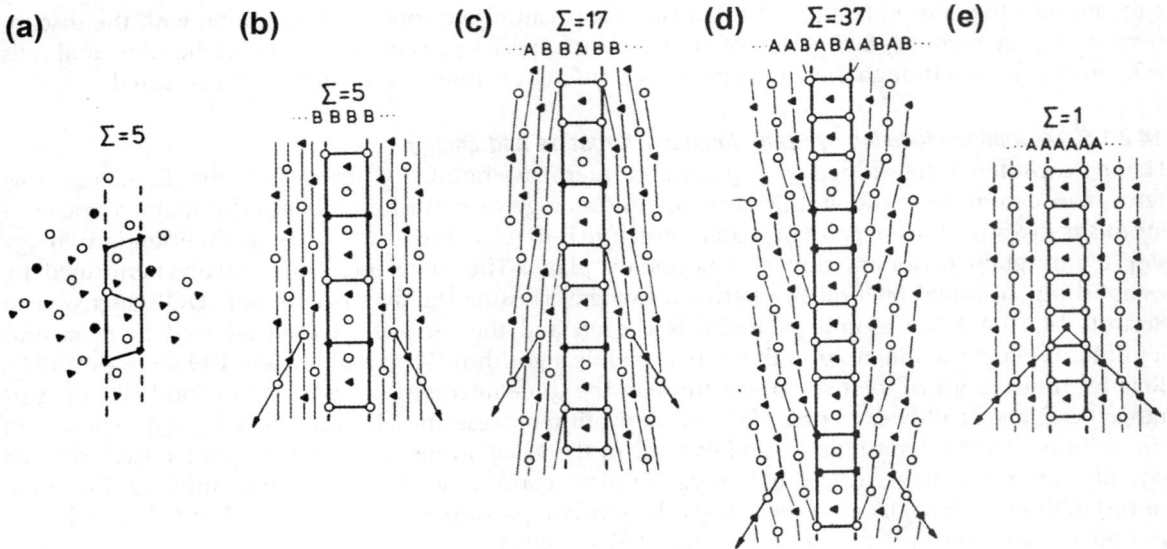

Figure 34   (a) Unrelaxed, and (b–e) relaxed core structures of symmetrical [001] tilt boundaries in Cu in terms of A and B structural units. The open circles indicate atoms at $z = 0$ and the filled triangles at $z = 1/2$. Adapted from Balluffi (1982) and Fishmeister (1985) with permission.

and Vitek, 1983). These tilt boundaries are seen to be composed of various mixtures of the basic structural units labeled B and A in **Figure 34(b)** and **(e)**, respectively. Boundaries composed entirely of a single structural unit (such as **Figure 34(b)** and **(e)**) are special boundaries. Examination of **Figure 34(c)** shows that each A-unit in the $\Sigma = 17$ boundary corresponds to the termination of two symmetrical (120) planes. Hence, each A-unit in the boundary represents a dislocation with $\mathbf{b} = 1/5$ [120]. This is a DSC lattice vector of the $\Sigma = 5$ coincidence lattice shown in **Figure 34(b)**. The $\Sigma = 17$ grain boundary can thus be considered as an $\Sigma = 5$ boundary with A-units interspersed between every two B-units. This agrees with the picture of DSC dislocations accommodating additional misfit between CSL orientations developed in Section 14.2.1.3.2, but now displayed in terms of the A and B structural units derived from atomistic calculations of grain boundary structure. The A-units are elements of a 90° boundary in the undisturbed lattice in **Figure 34(e)**, for which $\Sigma = 1$ and the unit cells of the DSC and the crystal lattice are identical. When the B-units occur as a minority component among the A-units in the $\Sigma = 37$ boundary in **Figure 34(d)**, they mark the termination of two (110) planes, or an ordinary lattice dislocation with $\mathbf{b} = 1/2[110]$ in each crystal.

The picture that emerges from the structural unit model (Sutton and Vitek, 1983; Balluffi and Bristowe, 1984; Bishop and Chalmers, 1986) is that the range of all possible misorientations between two grains can be viewed as divided into sectors, each delimited by two special boundaries, such as $\Sigma = 5$ (B-units) and $\Sigma = 1$ (A-units) above. Within each sector, the intermediate boundaries consist of structural units taken from the delimiting special ones, arranged in an ordered sequence and in proportions given by a linear rule of mixtures, as in **Figure 34(c)** and **(d)**. The stress fields surrounding the relaxed boundaries in **Figure 34(c)** and **(d)** agree with the presence of primary and secondary dislocations and therefore are equivalent to the previous dislocation description of grain boundary structure (Pond et al., 1979). It is important to note that the structural units formed in the relaxed grain boundaries in **Figure 34(b)** can be identified in a distorted shape in the unrelaxed boundary structure shown in **Figure 34(a)**.

Another interesting feature of the structural unit model is that the various structural units formed in relaxed grain boundaries correspond to the same type of polyhedra that are the basic units of the liquid structure (Bernal, 1964; Spaepen, 1975). An example of this has been discussed for the $\Sigma = 13$ (001) twist grain boundary in FCC Ag, where octahedral and archimedean antiprism configurations are found in the grain boundary (Sass and Bristowe, 1980). Such data indicate that the nearest- and next-nearest neighbor environments at grain boundaries can be similar to those found in liquids and therefore that one might think of segregation to such boundaries in terms of the equilibrium partition ratio k_e used in solidification (Flemings, 1974). In addition, the polyhedra formed in liquids tend to maximize nearest-neighbor coordination and as shown in the following section, this has an important effect on the grain boundary energy.

14.2.1.5.2 *Structure–Energy Correlation*

The results presented in this section are based on a review by Wolf and Merkle (1992). These investigators performed an extensive comparison between atomistic modeling and HRTEM observations of grain boundary structure. This section emphasizes aspects of their results that are relevant to the dislocation and structural unit models of grain boundary structure discussed previously. Many of the data compare results obtained from atomistic calculations using both many-body embedded-atom method (EAM) potentials for Au and the Lennard–Jones (LJ) pair potential for Cu. Only the results for these FCC metals are shown here, but Wolf and Merkle (1992) show a similar series of data for BCC metals.

Figure 35 (Wolf and Merkle, 1992) shows the grain boundary energy γ_{GB} as a function of angle of misorientation ψ for <110> symmetric tilt grain boundaries (STGBs) in FCC Au and Cu. Note that both metals give qualitatively similar results and that the energies of Cu are generally higher than those

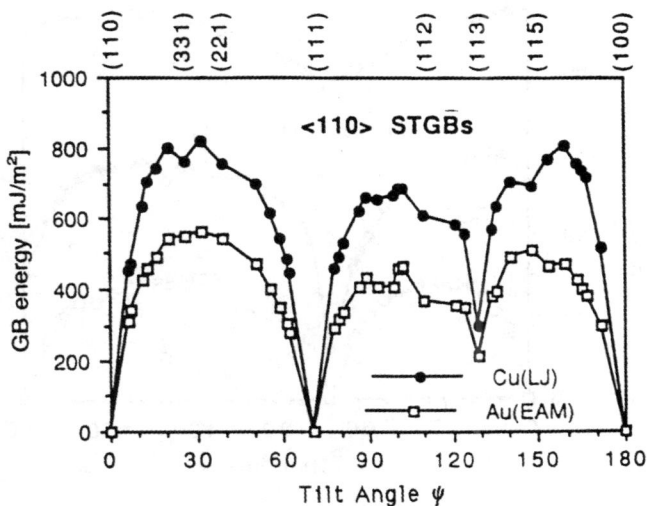

Figure 35 The energies of symmetric <110> tilt grain boundaries in Cu and Au as a function of angle of misorientation. Adapted from Wolf and Merkle (1992) with permission.

of Au, as expected based on the relative strengths of the atomic bonds, that is, melting temperatures of the two metals. Also note that these data are similar to those shown previously for Al in **Figure 12(c)** and 12(d). All of these metals display deep cusps when the low-index {111}, {100} and {110} planes are parallel across the boundary, as indicated by the symmetrical tilt grain boundary planes {hkl} at the top of the figure. These data indicate that there is a correlation between the atomic density of the planes comprising the grain boundary and the energy of the boundary. This correlation has a similar trend to solid–vapor interfaces, although there are clearly additional cusps present in the grain boundary data (e.g. at (113)) that are absent in the case of solid–vapor interfaces (Heyraud and Metois, 1980; Howe, 1997). An alternate way of expressing this correlation is to note that there is an inverse relationship between the interplanar spacing of the planes parallel to the grain boundary and the grain boundary energy, with widely spaced low-index planes such as {111} and {100} generally displaying lower grain boundary energies. This has an important consequence on the grain boundary population discussed in the next section.

The relationship between the energy of symmetrical tilt grain boundaries and surfaces is further illustrated in **Figure 36** (Wolf and Merkle, 1992), where the results for symmetrical tilt grain boundaries and random grain boundaries (RGBs) are compared with the surface energy (expressed in terms of $2\gamma_{sv}$, the energy necessary to separate a grain boundary into two surfaces of Au, as in an ideal cleavage experiment). Again one sees a correlation between the deep cusps associated with low-index planes, but the absence of additional cusps in the case of surfaces, suggesting that the behavior of grain boundaries is more complicated than that of surfaces. In these experiments, an RGB is defined such that all interactions across the grain boundary are assumed to be random with the constraint that as in actual boundaries, the atoms are assumed to lie in well-defined planes parallel to the boundary. The similarity

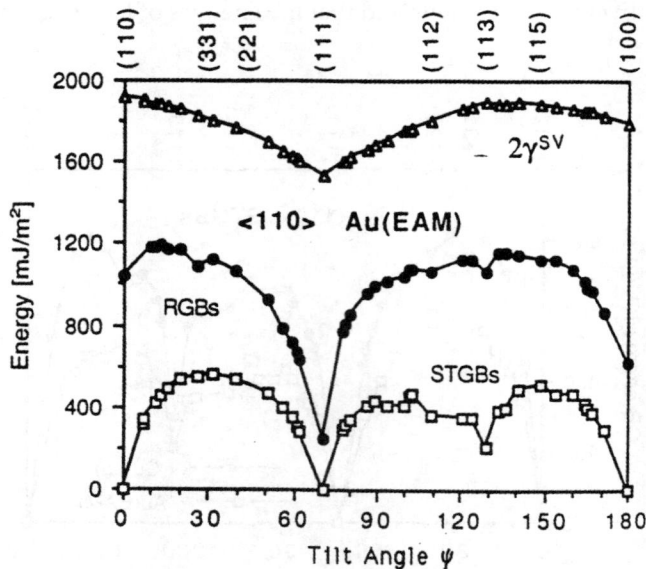

Figure 36 Comparison of the energies of RGBs and STGBs with those of the free surfaces for <110> Au. Note that the energy of two free surfaces $2\gamma^{SV}$ is plotted and that this corresponds to the RGB limit for a grain boundary expansion $\Delta V_i \rightarrow \infty$. Adapted from Wolf and Merkle (1992) with permission.

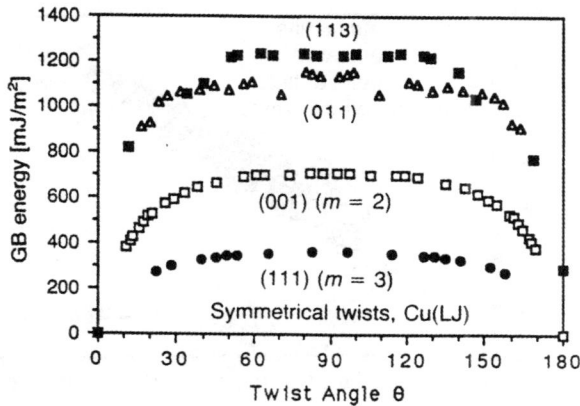

Figure 37 Energies of symmetrical twist boundaries versus twist angle for the four densest planes of FCC Cu obtained with an LJ potential. The only significant cusps occur at the beginning and endpoints of the misorientation range where there are small planar unit cells in the boundary. Adapted from Wolf and Merkle (1992) with permission.

of the RGB model with the symmetrical tilt model again indicates a correlation between the interplanar spacing parallel to the boundary and the grain boundary energy, particularly for the three most densely packed planes. Note the large reduction in energy that occurs when the two surfaces are joined to form an RGB, because of the elimination of broken bonds at the interface. Average STGB and RGB grain boundary energies for Au are on the order of 0.5 and 1 J/m^2, respectively, and maybe 50% higher for Cu, providing an estimate for typical values for metals.

Figure 37 shows the energies γ_{GB} of symmetrical twist boundaries versus twist angle ψ for LJ Cu for the (111), (001), (011) and (113) grain boundary planes (Wolf and Merkle, 1992). In these figures, $\psi = 180°/n$, where n indicates a possible rotation symmetry for the planar unit cell of the grain boundary. These data clearly show that the lowest grain boundary energies are associated with the most densely packed planes of the crystal and that the curves are rather featureless over a large range of twist angle, in comparison to the tilt boundaries in **Figure 35**. The only significant cusps at the beginning and endpoints of the twist misorientation range indicate that in the case of twist boundaries, small planar unit-cell areas in the boundary correspond to low energies. The lack of variation in energy with twist angle except at the endpoints and the strong dependence on the interplanar spacing indicates that it is the short-range repulsion of the atoms and the volume expansion at the interface that dominate the grain boundary energy. Thus, the formation of twist grain boundaries on planes that are widely spaced and contain a high density of atoms with small planar unit cells leads to lower energy grain boundary structures.

Wolf (1990) and Wolf and Merkle (1992) also examined the correlation between the grain boundary energy and the volume expansion per unit area at the grain boundary $\Delta V_{GB}/a$ (normalized by the lattice parameter a) for symmetrical and asymmetrical tilt and twist grain boundaries in Au and Cu, and their results are compiled in **Figure 38** (Wolf and Merkle, 1992). These data indicate a nearly linear relationship between $\Delta V_{GB}/a$ and the grain boundary energy. This appears to be a general relationship that exists for all types of grain boundaries. Another important correlation obtained from the same study was that there is an approximately linear relationship between the grain boundary energy γ_{GB} and the number of nearest-neighbor broken bonds per unit grain boundary area. This suggests that there is a correlation between the volume expansion at the grain boundary and the number of nearest-neighbor

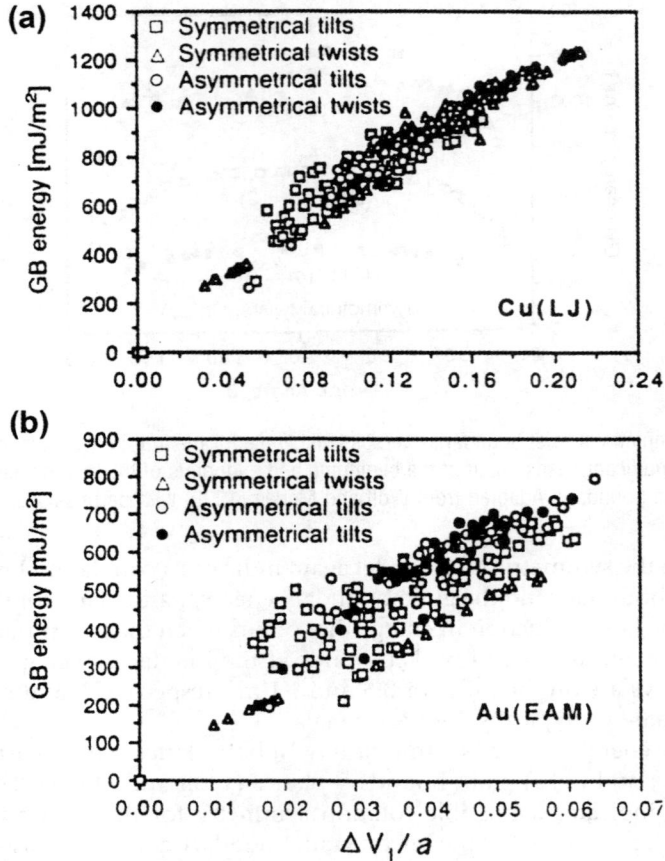

Figure 38 Correlation between the grain boundary energy and volume expansion per unit area (normalized by the lattice parameter) for symmetrical and asymmetrical tilt and twist grain boundaries in FCC metals using an (a) LJ Cu potential and (b) EAM Au potential. Adapted from Wolf and Merkle (1992) with permission.

broken bonds for the two FCC potentials. The physical basis for this correlation is that as the number of broken bonds (or roughness of the interface) increases, the ability of the two surfaces comprising the grain boundary to obtain good matching decreases, leading to a volume expansion at the grain boundary, with a corresponding increase in the grain boundary energy.

14.2.1.5.3 Grain Boundary Character Distribution

The grain boundary data in the previous section clearly show that low-energy boundaries are associated with low-index planes at the boundaries. One feature that was not studied by Wolf and Merkle (1992), but is an important factor in both the grain boundary character distribution and the behavior of interphase boundaries found in the massive transformation, for example, is the energy of grain boundaries when one plane at the boundary is low index, for example $\{111\}$, while the opposing plane is high index, for example $\{11\ 3\ 20\}$. An interphase boundary of this type was previously shown in **Figure 6**. One presumes that such a grain boundary displays an asymmetric energy cusp, similar to interphase boundaries (Reynolds and Farkas, 2006) with values less than RGBs and somewhat comparable to the (113) STGBs in **Figures 35 and 36**, and low-to-intermediate volume expansions in **Figure 37**.

During the past dozen years or so, techniques have been developed to measure the distribution of grain boundaries in polycrystals as a function of both lattice misorientation and grain boundary plane (El Dasher et al., 2004; Saylor et al., 2004; Rohrer, 2007; Dillon and Rohrer, 2009; Holm et al., 2011; Rohrer, 2011). The most significant finding from these studies are that grain boundary plane distributions are anisotropic, that the most common grain boundary planes are those with low surface energies, that the grain boundary population is inversely correlated with the grain boundary energy, and lastly, that the CSL number Σ is a poor predictor of the grain boundary energy and population.

An example that demonstrates this behavior is shown in **Figure 39**, where the grain boundary plane distributions for the $\Sigma 3$ grain boundaries (i.e. a misorientation of 60° around [111]) in MgO, Al and Ni are plotted as stereographic projections measured in multiples of a random distribution (MRD) such that values greater than unity signify grain boundary types that are observed more frequently than expected (Rohrer, 2007). All three distributions peak at the position of the pure twist configuration, where the grain boundary plane is perpendicular to the misorientation axis, that is the position of the coherent (111) twin boundary. The three materials exhibit different degrees of anisotropy and the intensity of the peak at the coherent twin orientation is inversely related to the stacking fault energy (see Section 14.2.2). Even in Al, which is relatively isotropic, many peaks exceed 3.0 MRD. These studies and others have found that the favored grain boundary planes typically correspond to low-index orientations and are present in greater number densities and larger average areas than other types of grain boundaries. In most situations, the most frequently adopted grain boundary orientation is a low-index plane with a low surface energy, as shown in **Figure 40** (Rohrer, 2007). These observations indicate that the grain boundary character distribution develops deterministically based on the relative energies of the boundaries. This is similar to crystal growth, where slow-growing crystal facets dominate the crystal growth morphologies (Frank, 1958; Du, 2006). They also indicate that while the CSL construction previously described is useful for visualizing grain boundary structures, the CSL density is a poor indicator of the actual grain boundary population.

The previous sections examined the structure and properties of grain boundaries from a number of different viewpoints and it is worth summarizing some of the more important conclusions and relationships that arise from these treatments. First, it is important to remember that regardless of whether

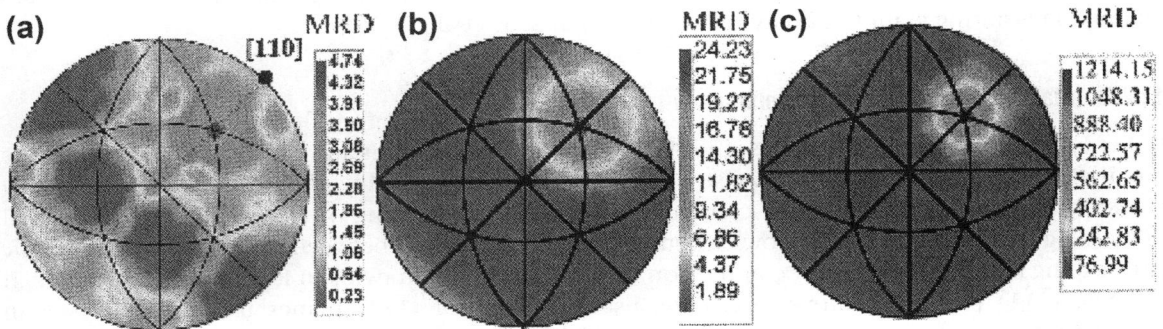

Figure 39   Grain boundary plane distributions for the $\Sigma = 3$ misorientation (this is a misorientation of 60° around [111]) in: (a) MgO, (b) Al and (c) Ni. In each case, the distribution peaks at the orientation of the {111} plane, indicating that most of the grain boundaries are perpendicular to this misorientation axis. Adapted from Rohrer (2007) with permission.

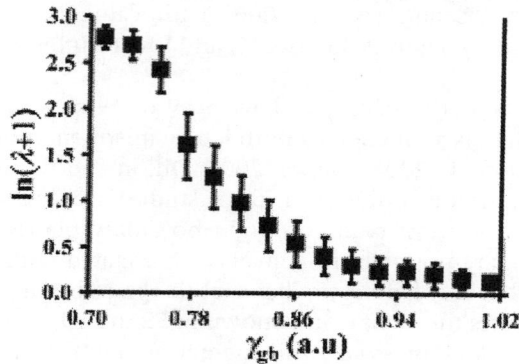

Figure 40 The logarithm of the grain boundary population λ in MgO as a function of the grain boundary energy γ_{GB}. The mean values are represented by the points; the bars indicate one standard deviation. Adapted from Rohrer (2007) with permission.

the Read–Shockley, CSL/DSC or structural unit model is used to interpret the structure of grain boundaries, they all model the boundary as being composed of areas of good and poor matching, with the poor matching comprised of dislocations. Areas of good matching are generally associated with low energy while those with poorer matching generally have higher energy. This relationship is not always simple, but it is quantifiable in terms of the dislocation models and atomistic calculations presented. Second, while the geometric models are useful for understanding the possible structures of grain boundaries, they are not usually accurate at predicting the energies associated with grain boundaries except in certain special cases. Here, atomistic calculations are far more capable in predicting favored grain boundary structures, and these have been found to be associated with low-index low-energy planes across the boundary. This prediction has been experimentally found in that the most common grain boundary planes in real FCC base materials as determined by quantitative stereological methods are those that contain at least one low-index low-energy plane such as {111} or {100}, and that this feature dominates the resulting grain boundary distribution. This is not a feature that is readily apparent from O-lattice or CSL treatments, but can be studied through atomistic calculation. Although it is not discussed here, the low mobilities often associated with such boundaries are also important in establishing the resulting grain boundary distribution (Janssens et al., 2006).

14.2.2 Stacking Faults and Twin Boundaries

Other important types of homophase interfaces include stacking faults and twin boundaries and these are briefly described here. The atomic structure of stacking faults and twin boundaries can be readily envisioned in FCC crystals, which consist of an ABCABC... stacking of close-packed ($\bar{1}11$) crystal planes. **Figure 41** (Murr, 1975) shows schematically the distinction between intrinsic (i) and extrinsic (e) stacking faults and shows for comparison the relationship with twins in FCC metals and alloys. If a single ($\bar{1}11$) plane is removed from the usual stacking of ($\bar{1}11$) planes in an FCC crystal in **Figure 41(a)**, an intrinsic stacking fault with an ABCBC... stacking sequence occurs, as illustrated in **Figure 41(b)**. Conversely, if an extra ($\bar{1}11$) plane of atoms is inserted into the crystal, an extrinsic stacking fault with an ABCBABC... stacking sequence results and this is also shown in **Figure 41(b)**. An alternate description of the extrinsic fault is that it arises from the superposition of two intrinsic

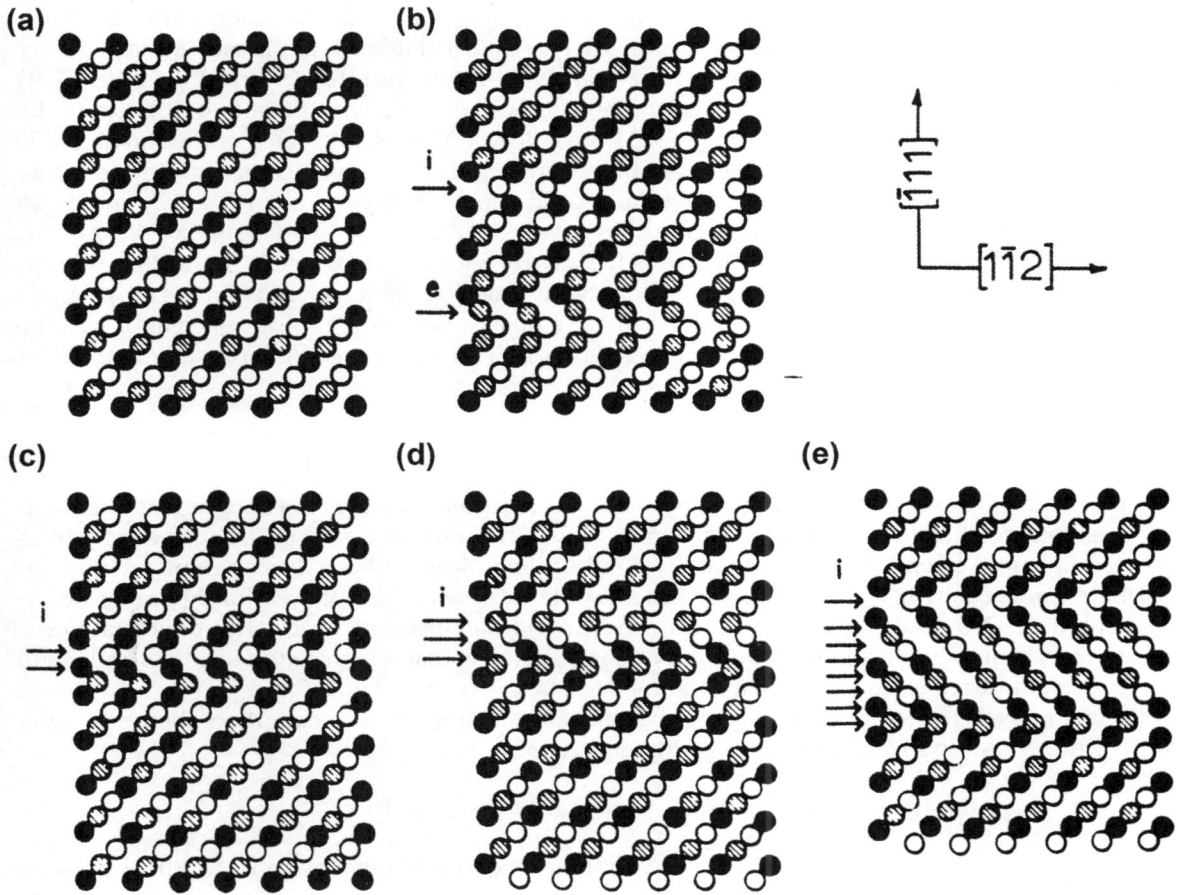

Figure 41 Idealized view of stacking faults and their development from perfect crystal in (a) to an *n*-layer twin in (e) in the close packed FCC structure. The viewing direction is along [110]. Adapted from Murr (1975) with permission.

faults, as illustrated in **Figure 41(c)**. If intrinsic faults occur on every $(\bar{1}11)$ plane then thin twins form within the crystal, as illustrated in **Figure 41(d)** and **(e)**. Based on the number of nearest-neighbor bonds that are in the wrong stacking sequence, the energies of intrinsic and extrinsic stacking faults are expected to be similar and the twin boundary energy is expected to be only about half of the intrinsic stacking fault energy. In the case of two twinned crystals, if a $(\bar{1}11)$ IP is defined and the top half of the crystal is rotated $180°$ about the interface normal, a coherent twin boundary with an ABCACBA... stacking like that in **Figure 41(e)** results. This is a special type of $\Sigma = 3$ grain boundary and an example of this was shown in **Figure 1**. It is found that the intrinsic stacking fault energy γ_{ISF}, extrinsic stacking fault energy γ_{ESF}, and the coherent twin boundary energy γ_T, generally obey the relationship

$$\gamma_{ISF} = \gamma_{ESF} = 2\gamma_T, \tag{36}$$

where the fault energies are expressed as the energy per atomic area in the fault. Typical values for γ_T range from 8 to 161 mJ/m^2 for Ag and Pt, respectively, with metals such as Cu and Al having intermediate values of 24 and 75 mJ/m^2, respectively (Gallagher, 1970; Murr, 1975; Hirth and Lothe, 1982).

A coherent FCC–HCP interface is similar to the twin boundary shown in **Figure 41(e)**, except that the stacking of close-packed ($\bar{1}11$) planes goes from ABCABC... to ABAB... across the interface. The IP can be uniquely identified by considering first nearest-neighbor bonds in the stacking sequence. Since the atomic configuration is the same until second nearest-neighbors, the FCC–HCP interphase boundary energy is found to be the nearly the same as the coherent twin boundary energy. For example, the FCC–HCP interface in **Figure 3** has an energy of approximately 15 mJ/m^2 (Hitzenberger et al., 1985; Ramanujan, 1990). An intrinsic stacking fault as in **Figure 41(b)** can be considered as the first layer of an HCP phase in FCC, but if one uses first nearest-neighbor bonding as a criterion to identify the interface, then there is no HCP phase between the two interfaces for an intrinsic stacking fault.

14.2.3   Segregation to Grain Boundaries

The thermodynamics of equilibrium grain boundary segregation in binary alloys are similar to those of surfaces, and prediction of segregation to grain boundaries can be described using the Langmuir–McLean relation often used for surfaces (Hondros and Seah, 1988). For example, **Figure 42** (Hondros and Seah, 1988) shows that the Langmuir–McLean relation is obeyed for the segregation of 8–12 ppm of O to grain boundaries in Mo over a wide temperature range. The basic features of equilibrium grain boundary segregation are evident in this plot, that is segregation increases as the bulk solute content increases and as the temperature decreases.

Seah and Hondros (1973) have used a solid-state analog of gas adsorption theory to write a predictive equation for segregation to grain boundaries as

$$X_{GB}/(1 - X_{GB}) = (X_s/X_B^{sat})\exp(-\Delta G'/RT), \tag{37a}$$

where X_{GB} is the fraction of the grain boundary monolayer covered with segregant, X_s is the bulk mole fraction of solute B, X_B^{sat} is the bulk solid solubility, and

$$\Delta G' = \Delta G_{GB} - \Delta G_B, \tag{37b}$$

where ΔG_{GB} is the free energy for grain boundary segregation and ΔG_B is the free energy of solution of the alloying element in the bulk. Thus, $\Delta G'$ represents the free energy difference between solute that is in solution in the alloy versus solute that is segregated to the grain boundary. There is a tendency for segregation when $\Delta G_{GB} > \Delta G_B$ so that $\Delta G'$ is negative. As with surfaces, it is useful to define a parameter called the grain boundary enrichment ratio β_{GB}, which represents the ratio X_{GB}/X_s in the dilute limit. For grain boundaries, this quantity is given as

$$\beta_{GB} = (1/X_B^{sat})\exp(-\Delta G'/RT). \tag{38a}$$

A plot of β_{GB} versus the atomic solid solubility X_B^{sat} is shown in **Figure 43** (Hondros and Seah, 1988) for a number of experimental data at various temperatures. As for the case of surfaces, there is good overall agreement between the grain boundary enrichment ratio and the solid solubility, so that the grain boundary enrichment for systems in which experimental data are not available can be estimated from the solvus line in the corresponding binary phase diagram and **Figure 43**.

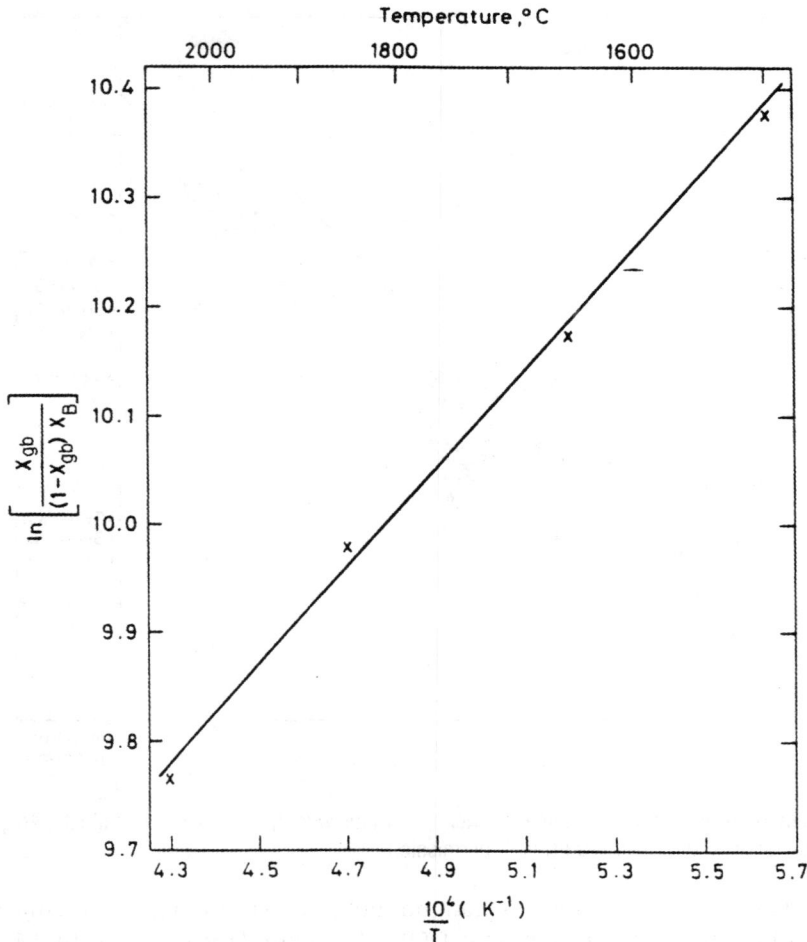

Figure 42 Langmuir–Mclean plot for O grain-boundary segregation in Mo measured by AES. Adapted from Hondros and Seah (1988) with permission.

Based on the data in **Figure 43**, it was determined that, although ΔG_{GB} values typically range from 0 to -80 kJ/mol, $\Delta G'$ has a mean value of -10 ± 6 kJ/mol, so that

$$\Delta G_{GB} = \Delta G_B - (10 \pm 6) \text{ kJ/mol.} \tag{38b}$$

This treatment is valid only when the bulk molar solute fraction is less than the solid solubility. If excess solute content is also present so that a second phase appears, the solute content is limited to the solid solubility. The situation becomes more complicated in ternary and higher order alloys and these are discussed elsewhere (Hondros and Seah, 1988; Briant, 1992).

Although the same physical phenomena used in analytical models for surface segregation can be used to develop expressions for the quantity ΔG_{GB} in grain boundary segregation, the problem is much

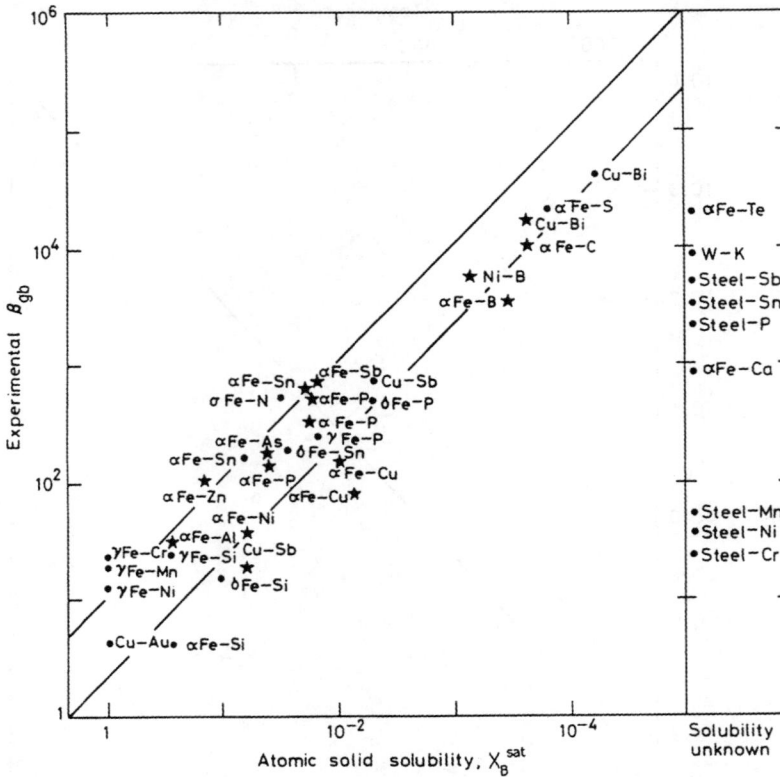

Figure 43 Correlation of measured grain boundary enrichment ratios with the atomic solid solubility. Adapted from Seah and Hondros (1973) and Hondros and Seah (1988) with permission.

more complicated because of the variety of grain boundary structures that are possible as a function of the five macroscopic and three microscopic DOF of a boundary (Section 14.2.1.1.5). There are a number of experimental data which demonstrate that segregation to grain boundaries is orientation dependent (e.g. **Figure 44**) and indicate that high-angle boundaries may exhibit more segregation than low-angle or special grain boundaries, but these relationships are only qualitative and not fully understood (Fraczkiewicz and Biscondi, 1985; Hofmann, 1990). As a result, there is no quantitative model for grain boundary segregation that can account for the role of boundary structure and solute properties on segregation. Atomistic simulations have been beneficial in elucidating some of the factors involved in grain boundary segregation and these are discussed further below.

In general, alloys that display surface segregation also display grain boundary segregation, but often to a lesser extent. This is illustrated for the case of Sn in Fe in **Figure 45** (Seah and Lea, 1975; Hondros and Seah, 1988). This behavior can be rationalized by remembering that a grain boundary generally contains more free volume and some number of broken bonds (refer to **Figures 34 and 38**) like a free surface, although to a lesser extent than a surface. By analogy with surfaces, it seems reasonable to expect that the equilibrium partition ratio k_e from the phase diagram, that is, the ratio of the solid and liquid compositions in equilibrium at the solid–liquid interface at temperature C_s^e / C_l^e (Burton and Machlin, 1976), might be a reasonable indicator of grain boundary segregation, particularly given the similarity of the bonding

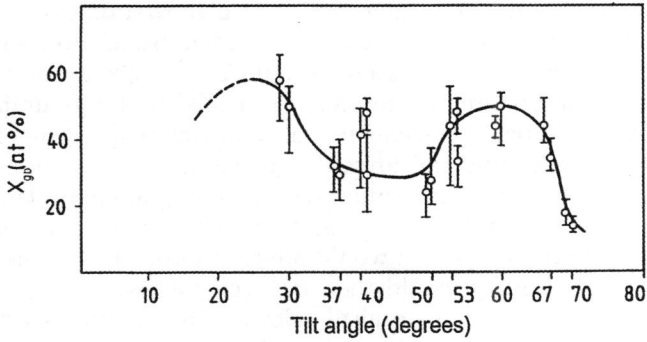

Figure 44   Grain boundary segregation of Bi in Cu as a function of tilt angle for Cu <100> tilt bicrystals determined by AES. Adapted from Fraczkiewicz and Biscondi (1985) and Hofmann (1990) with permission.

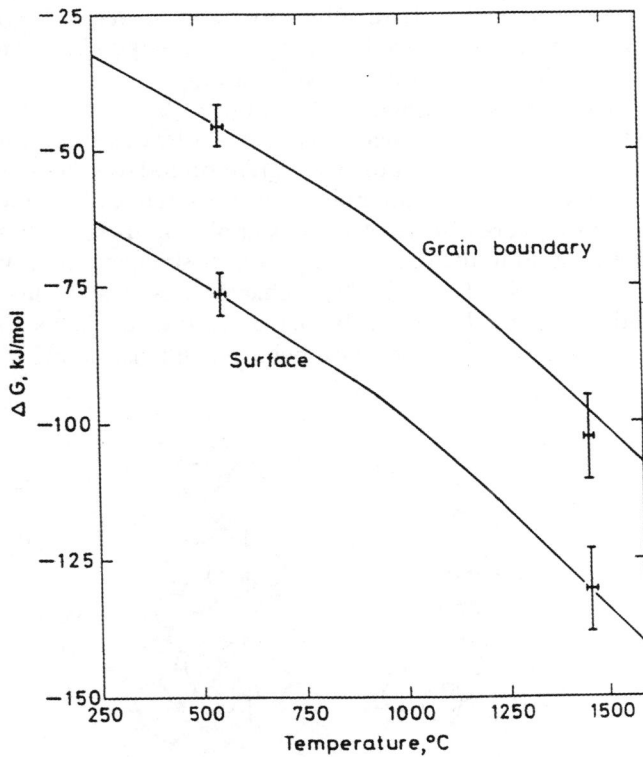

Figure 45   Predicted temperature dependence of the free energies of surface and grain boundary segregation of Sn in Fe as a function of temperature, with experimental data at 550° and 1420 °C. Adapted from Seah and Lea (1975) and Hondros and Seah (1988) with permission.

environments in grain boundary structural units and the polyhedral units in a dense random packed (DRP) liquid (Ashby and Spaepen, 1978; Pond et al., 1978), but this relationship has not been tested.

Qualitatively, it has been argued that once a segregating atom reaches a grain boundary, it remains there because of the favorable bonding environment provided by the boundary (Sutton and Vitek, 1982). In many systems, the segregating element is a nonmetal, which tends to form stoichiometric compounds with the metal. The structural units that make up such compounds often bear a close resemblance to the structural units that are commonly found in grain boundaries (**Figure 34**). Hence, a segregating species may find an atomic geometry at the grain boundary that is much more compatible with the type of bonding environment that it would prefer to form with the metal and this causes it to stay in the boundary. The boundary would then be decorated with such segregated units. Some evidence for this explanation comes from atomistic calculations of grain boundary segregation, which clearly indicate site selectivity of the segregating species. For example, Sutton and Vitek (1982) studied the segregation behavior of Bi (a strong segregant and embrittler) and of Ag (a weak segregant) in Cu and Au to symmetric $\Sigma = 5$ (210) and $\Sigma = 17$ (530)/[001] tilt boundaries (refer to **Figure 46**) by atomistic calculation. The energies of segregation of the elements to various sites in the boundary were found to vary widely, as illustrated in **Figure 46** (Sutton and Vitek, 1982). For a given structural unit, the segregation energy also varied considerably with its surroundings in the boundary, that is the boundary type. The large Bi atoms generally went into those sites that were surrounded by a tensile stress field in the unsegregated boundary. In terms of a hard-sphere model, it segregated to the most spacious site. Segregation of Bi to Cu was also accompanied by an expansion perpendicular to the boundary, which can be interpreted as a weakening of the boundary cohesion.

Several Monte Carlo calculations of segregation to pure <100> twist boundaries in Ni–Cu and Pt–Au alloys using EAM potentials have revealed a detailed picture of segregation in these alloys (Wang et al., 1993; Seki et al., 1991; Foiles, 1989). Pure twist grain boundaries are particularly convenient to study because the grain boundary contains an array of identical screw dislocations as a function of tilt angle up to about 35° (refer to **Figures 14 and 16**, for example). In these studies, the concentrations of the segregating species were found to increase linearly with twist angle until saturation occurred at the boundaries at relatively high angles ($\sim 35°$). This behavior was related directly to the dislocation density in the twist boundaries rather than to the particular value of Σ. It is a significant finding because it provides a simple explanation for the segregation behavior of twist boundaries.

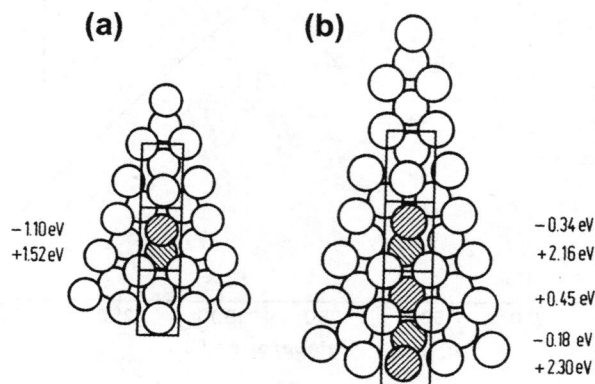

Figure 46 Segregation enthalpies (in eV/atom) for Bi atoms in Cu tilt boundaries when (a) $\Sigma \sim 5$ and (b) $\Sigma \sim 17$. Adapted from Sutton and Vitek (1982) and Fishmeister (1985) with permission.

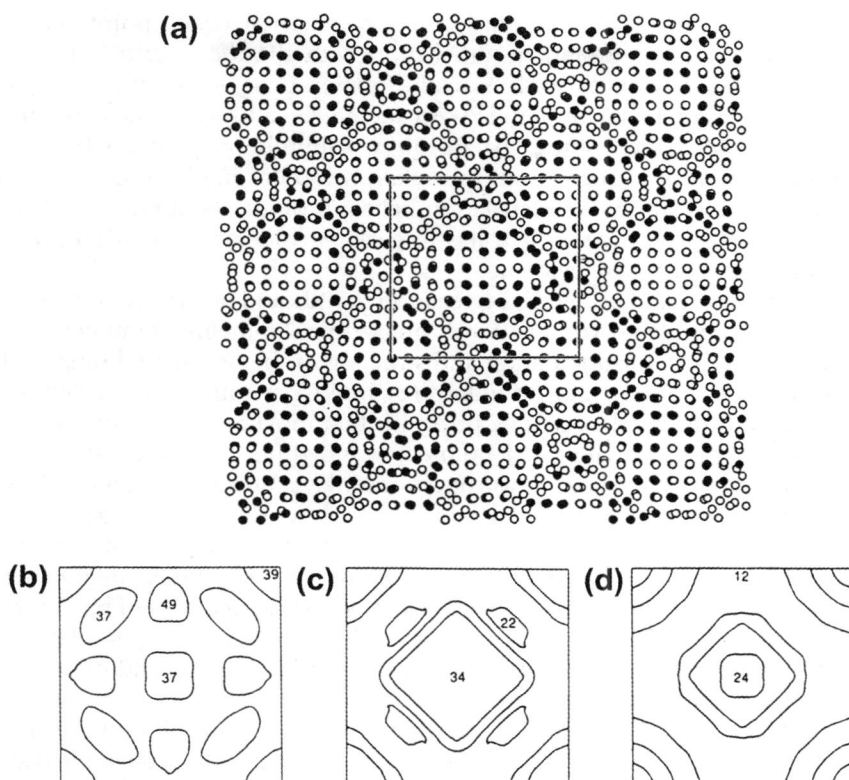

Figure 47 (a) Atomic positions projected on the boundary plane for a Σ ~ 61 twist boundary in a Ni-10 at.% Cu alloy. The first two atomic layers on either side of the boundary are shown. The open circles are Ni atoms and the filled circles are Cu. The screw dislocation network is located in the regions of poor matching between the crystals (compare with **Figure 14**). (b–d) Contour plots of the Cu concentration as a function of position in the plane of the first (b), second (c) and third (d) atomic layers of the boundary. The plots correspond to the central unit cell outlined in (a). The numbers indicate the compositions at extremums in at.% Cu and the contour spacing is 4.4 at.%. Adapted from Foiles (1989) with permission.

In addition, the segregating species were found to lie within two or three planes on either side of the interfaces, and to depend on the strain fields of the screw dislocations, as illustrated for the Ni–Cu system in **Figure 47** (Foiles, 1989). The amount of segregant also increased in proportion to $\Delta G'$ in Eqn (37) and decreased with increasing temperature in the studies, as expected from thermodynamics. An additional interesting feature in these studies was that in the Pt–Au alloy, the solute was highest in the dislocation cores, while in the Ni–Cu alloy, the reverse trend was found. In both alloy systems, the positive heats of solution indicate a tendency for the elements to separate and thus, to segregate to the boundary. In the Pt–Au alloy, it was found that Au segregated to the dislocation cores, where the atoms could relieve the high hydrostatic tensile and shear stresses associated with the dislocations. In contrast, in the Ni–Cu alloy, a volume expansion at the boundary (ΔV_{GB}) left the regions of good matching in tension, which could be relieved by segregation of the elastically softer Cu atoms to these regions in between the dislocation cores. This effect is illustrated in **Figure 47**, where the Cu concentration in the second and third planes away from the boundary (**Figure 47(c)** and **(d)**) is highest in the center and at the corners, between the screw dislocation cores. These results point to the importance of the elastic properties of the solute and its interaction with the local grain boundary structure in determining segregation behavior, as well as to the mixing behavior of the elements. Comparing **Figure 38** with the segregation behavior of the Pt–Au and Ni–Cu alloys also indicates that there may be a direct correlation between the volume expansion and number of broken bonds at twist grain boundaries and the amount of segregant, but this relationship has not been examined. Clearly, more work needs to be done to quantify the segregation behavior of grain boundaries, although the same basic trends as in surface segregation, which has received considerable attention (Wynblatt and Ku, 1977; Hondros and Seah, 1988), are evident.

One additional feature of grain boundary segregation that should be mentioned is the recent finding that grain boundaries can undergo various transitions below the melting temperature, related to the type and amount of segregant, misorientation and structure of the boundary (Tang et al., 2006). This type of behavior has been called grain boundary (or interface) complexions (Dillon et al., 2007). The complexion can be considered a separate phase in thermodynamic equilibrium at the boundary. One complexion is the clean unsegregated boundary, and different phases include submonolayer absorption in the boundary core, bilayer adsorption, multilayer absorption, an equilibrium thickness intergranular film and a wetting intergranular film. Transitions between grain boundary complexions can be mapped onto a phase diagram as a function of temperature and composition (Dillon et al., 2007). The general behaviors predicted from thermodynamic analysis of this phenomenon (Tang et al., 2006) are that large misorientation boundaries are expected to undergo either first-order or continuous complexion transitions below the eutectic point in a single or two-phase region of a binary alloy, and that boundaries with relatively small misorientations will remain ordered and unsegregated up to the solidus line or eutectic temperature. The presence of interface complexions can be useful for explaining a number of important materials issues such as abnormal grain growth, liquid metal embrittlement, and sintering behavior. Several different metal and ceramic systems have been studied and analyzed in terms of this phenomenon with considerable success, and the reader is referred to the following references for more details (Luo, 2009; Shi and Luo, 2009; Dillon et al., 2010; Luo et al., 2011; Asl and Luo, 2012).

14.3   Structure and Composition of Heterophase Interfaces

Section 14.2 focused on homophase interfaces, where there was a structural discontinuity between two crystals of the same phase, as illustrated in **Figures 1 and 2**, for example. **Figure 4** shows an FCC–HCP heterophase interface that is similar to the stacking fault and twin boundaries discussed in Section 14.2.2, since the two phases on opposite sides of the interface have the same composition, but different crystal structures. This represents one limiting type of heterophase boundary, where there is a change in Bravais lattice, but no change in composition across the interface. In this case, the interface was found to have an energy similar to twin boundaries in metals (see Section 14.2.2).

The present section concentrates on the structure of the other types of heterophase interfaces, where there is a change in composition, or both composition and lattice, across the interface between two different phases. It begins by considering a planar coherent interface between two different phases that have the same Bravais lattice and orientation, but differ in composition, as illustrated by the G.P. zone in **Figure 3**. This demonstrates the principles that determine the compositional profile across such an interface, another limiting example of a heterophase interface. Then the structural and compositional changes across a partly coherent interface such as **Figure 5** are analyzed. This is followed by a discussion of the OR between two crystals that have different Bravais lattices, since this is a prerequisite for discussing the structure of any heterophase interface. An inclined heterophase interface such as **Figure 7** is then analyzed. Such interfaces are more complicated because one must account for structural defects, such as dislocations and steps in the interface, and partitioning of elastic energy along the interface. The structure of one-dimensionally commensurate and incoherent interfaces (like that shown in **Figure 6**) are then described. Section 14.3 concludes by discussing segregation to heterophase interfaces in the context of a regular solution model.

14.3.1 Coherent Interphase Boundary Structure and Energy

The treatments discussed in this section apply to the special case of two phases that have the same crystal structure and OR, but differ in composition. Three treatments of this type of interface are described, two of which are based on a broken bond model and a third that uses a continuum approach to the problem. The theories are described in chronological order, which is convenient in the development of this section.

14.3.1.1 Becker Model

The simplest compositional model of a fully coherent interface is shown in **Figure 48** (Swalin, 1972). Here the phases α and β are assumed to be compositionally invariant up to the interface, which is a step function in composition. The earliest calculation of the interfacial energy of this type of interface between two phases is due to Becker (1938). He employed a nearest-neighbor broken bond model to derive the coherent solid–solid interphase boundary energy γ_c^{SS} assuming that both phases are homogeneous up to the interface. The energy is given as

$$\gamma_c^{SS} = N_{s\{hkl\}} z_j / \left(X_\alpha - X_\beta\right)^2 \varepsilon, \tag{39}$$

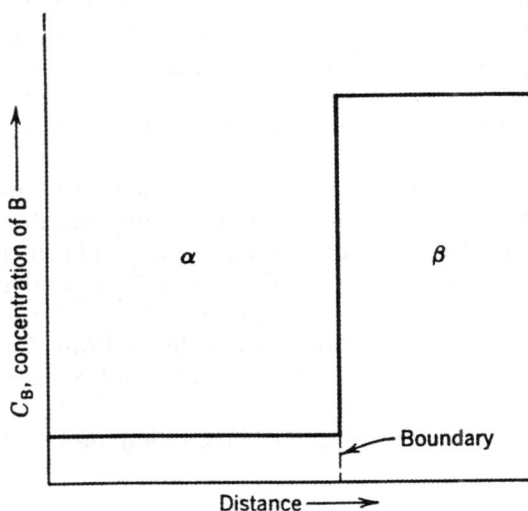

Figure 48 Illustration of composition versus distance across an ideal planar boundary between the α and β phases. Adapted from Swalin (1972) with permission.

where $N_{s\{hkl\}}$ is the number of atoms per unit area on the $\{hkl\}$ IP, z_j is the coordination number across the interface, X_α and X_β are the equilibrium atom fractions of solute B in the α and β phases, respectively, and ε is defined according to the regular solution model as

$$\varepsilon = \varepsilon_{AB} - 1/2(\varepsilon_{AA} - \varepsilon_{BB}), \tag{40}$$

where ε_{AA}, ε_{BB} and ε_{AB}, are the AA, BB and AB bond energies, respectively, given as negative quantities by considering zero energy as the state where the atoms are separated at infinity. These quantities can be found from the heat of mixing ΔH_{mix} for a regular solution, which is given by

$$\Delta H_{mix} = \Omega X_\alpha X_\beta, \tag{41a}$$

where the interaction parameter

$$\Omega = z N_A \varepsilon \tag{41b}$$

and z is the nearest-neighbor coordination (12 for FCC and HCP crystals) and N_A is Avagadro's number. If $\varepsilon = 0$ in Eqn (40), then $\Delta H_{mix} = 0$ in Eqn (41a) and the solution is ideal. If $\varepsilon < 0$ in Eqn (40), the atoms in the solution preferred to be surrounded by atoms of opposite type, and the alloy tends to form an ordered solution. Conversely, if $\varepsilon > 0$, the atoms preferred to be surrounded by atoms of the same kind and this is referred to as a phase-separating solution. Real solutions that closely obey Eqn (41a) are known as regular solutions (Lupis, 1983). Since Ω is independent of composition, ΔH_{mix} is a parabolic function of composition with a maximum at $X_\alpha = X_\beta = 0.5$, and a critical temperature T_c that is related to the regulation solution (interaction) parameter Ω as

$$T_c = 2X_c^B(1 - X_c^B)\Omega/R, \tag{42}$$

where X_c^B is the mole fraction B at the critical composition and R is the Universal gas constant (or Boltzmann's constant k_B can be used on an atomic basis). Equation (42) shows that Ω must be positive for T_c to be positive, and that T_c increases as ε_{AB} becomes less negative in Eqn (40), that is, as the energy of the AB bond pair decreases relative to the average of the AA and BB pairs. These quantities are introduced here because they will be used in subsequent models of interfacial structure.

To illustrate the use of Eqn (39) consider the Au–Ni system (Massalski et al., 1986). This alloy has a miscibility gap with a critical temperature $T_c = 810.3\,°C$ and critical composition $X_c^B = 0.706$. The following steps are taken to calculate the energy of a coherent $\{111\}$ interphase boundary at 400 °C. From the phase diagram, $X_\alpha = 0.075$ and $X_\beta = 0.985$ at 400 °C so that $(X_\alpha - X_\beta)^2 = 0.828$. The lattice parameter of Au $= 0.40$ nm so $N_{s\{hkl\}} = 4/\sqrt{3}(0.4 \times 10^{-9}\,\text{nm})^2 = 1.44 \times 10^{19}$ atoms/m^2 and $z_j = 3$ for the $\{111\}$ plane. One can find Ω (and thus ε from T_c and X_c^B) using Eqn (42) as $\Omega = (8.315\,\text{J/mol K})(1083.3\,\text{K})/2(0.706)(0.295) = 21,695$ J/mol. Since $\varepsilon = \Omega/12N_A$ (Eqn (41b):

$$\gamma_c^{SS}(\text{Au}-\text{Ni}) = (1.44 \times 10^{19}\,\text{atoms/m}^2)3(0.828)(21,695\,\text{J/mol})/12(6.022 \times 10^{23}\,\text{J/mol})$$
$$= 0.107\,\text{J/m}^2.$$

This is a fairly high value for a coherent interphase boundary energy, mainly because of the high positive enthalpy of mixing ΔH_{mix} in the Au–Ni system.

The main limitation of Eqn (39) is that it assumes that the interface is infinitely sharp as illustrated in **Figure 48**, and it does not include the interphase boundary width as a thermodynamic variable. The gradient energy theory and discrete lattice plane (DLP) models which follow show that this assumption is only valid for alloys with a large positive enthalpy of mixing ($\Omega \gg 0$) at 0 K. Thus, while this equation is appealing because of its simplicity and it can be used to obtain an estimate of the interphase boundary energy, it does not properly describe the underlying physics associated with most interphase boundaries.

14.3.1.2 Cahn-Hilliard (CH) Gradient Energy Model

Cahn and Hilliard (1958) developed a continuum description of the coherent interphase boundary energy by considering a flat interface of area A between two isotropic phases α and β of composition C_α and C_β, and assuming that the free energy of nonequilibrium material of composition intermediate between C_α and C_β can be represented by a continuous free energy function $G_0(C)$, as in the regular solution model, as shown in **Figure 49** (Cahn and Hilliard, 1958; Swalin, 1972). They express the total free energy G of the solution as a multivariable Taylor's series expansion about G_0, the free energy per molecule of a solution of uniform composition C, and considering only a one-dimensional composition change across the interface and neglecting derivative terms higher than the second, they obtain

$$G = AN_v \int_{-\infty}^{+\infty} [G_0(C) + K(dC/dx)^2]\,dx, \tag{43}$$

where N_v is the number of molecules per unit volume and

$$K = [\partial^2 G/\partial C \nabla^2 C]_0 + [\partial^2 G/\partial |\nabla C|^2]_0. \tag{44}$$

Equation (43) indicates that to first approximation, the free energy of a small volume of nonuniform solution can be expressed as the sum of two contributions, one being the free energy that this volume would have in a homogeneous solution ($G_0(C)$) and the other being a gradient energy contribution ($K\,(dC/dx)^2$), which is a function of the local composition.

The specific interfacial free energy γ_c^{SS} is by definition the difference per unit area between the actual free energy of the system and that which it would have if the properties of the phases were continuous throughout. Thus, it can be written as

$$\gamma_c^{SS} = N_v \int_{-\infty}^{+\infty} [\Delta G(C) + K(dC/dx)^2]\,dx, \tag{45}$$

where

$$\Delta G(C) = G_0(C) - [C\mu_B^e - (1 - C)\mu_A^e] \tag{46}$$

and μ_B^e and μ_A^e are the chemical potentials per atom of the species A and B in the α or β phases at equilibrium. The quantity $\Delta G(C)$ may therefore be referred to as the free energy per atom of transferring material from an infinite reservoir of composition C_α or C_β to material of composition C. According to Eqn (45), the more diffuse the interface is, the smaller will be the contribution of the gradient energy term $K(dC/dx)^2$ to γ^{SS}_c. However, this decrease in energy can only be achieved by introducing more material at the interface of nonequilibrium composition and thus at the expense of increasing the integrated value of $\Delta G(C)$. At equilibrium, the composition variable will be such as to minimize the integral of this equation. (This is equivalent to the requirement that the chemical potentials be constant throughout the system.)

To obtain a composition profile with a stationary value and a minimum value for γ^{SS}_c, the following condition must hold

$$\Delta G\{C\} = K(dC/dx)^2. \tag{47}$$

Changing the variable of integration from x to C and substituting Eqn (47) into Eqn (45) then yields

$$\gamma^{SS}_c = 2N_v \int\limits_{c\alpha}^{c\beta} [K\Delta G(C)]^{1/2}dC. \tag{48}$$

Cahn and Hilliard (1958) examine Eqn (48) in two forms: (i) to determine the functional dependence of γ^{SS}_c on temperature in the immediate vicinity of the critical temperature T_c at which the two phases attain the same critical composition C_c, and (ii) to determine the absolute value of γ^{SS}_c and its temperature dependence outside the range $T \sim T_c$ using a regular solution model to evaluate the gradient term K and the free energy function $\Delta G(C)$.

In the first case, if G_o is expanded in a Taylor's series about C_o and K is assumed to be continuous and nonvanishing in the vicinity of the critical point, the interfacial energy is evaluated as

$$\gamma^{SS}_{c(T \sim T_c)} = \left(2\sqrt{2}N_v/3\sigma\right)K^{1/2}\xi^{3/2}(T_c - T)^{3/2}, \tag{49}$$

where σ and ξ are inherently positive constants defined by the derivatives of G_o evaluated at $C = C_c$ and $T = T_c$ (Cahn and Hilliard, 1958). Therefore, at the critical temperature, the interfacial energy is proportional to $(T_c - T)^{3/2}$. Furthermore, if the composition across the interface is such that $dC/dx = (\Delta G/K)^{1/2}$, then inspection of the ΔG function in **Figure 49** indicates that to satisfy this equation the composition profile must be sigmoidal in shape as shown in **Figure 50**. Near the critical temperature, the thickness L of the interface is obtained as

$$L_{(T \sim T_c)} = 2[2K/\xi(T_c - T)]^{1/2}, \tag{50}$$

where it becomes evident that the thickness of the interface increases with increasing temperature and becomes infinite at the critical temperature. Note that this analysis assumes that K in the gradient energy term is a positive quantity.

In the second case, if the properties of a regular (R) solution are used, then an expression for the free energy referred to a standard state of an equilibrium mixture of α and β becomes

$$\begin{aligned}\Delta G_R &= G_R(C) - G_R(C_e)\\ &= \Omega(C - C_e)^2 + k_B T\{C\ln(C/C_e) + (1 - C)\ln[(1 - C)/(1 - C_e)]\},\end{aligned} \tag{51}$$

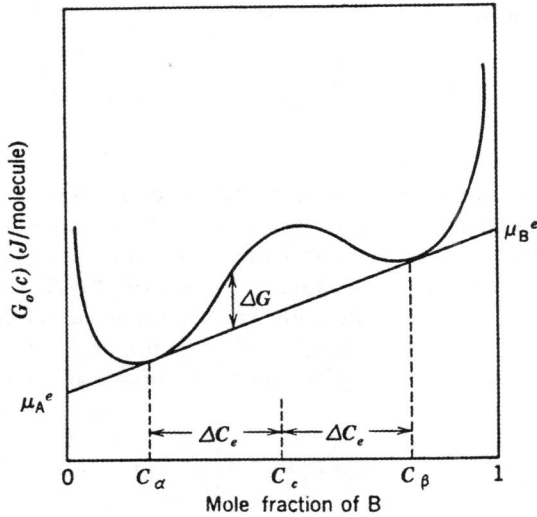

Figure 49 The function $G_0(C)$ for T less than the critical temperature T_c. Adapted from Cahn and Hilliard (1958) and Swalin (1972) with permission.

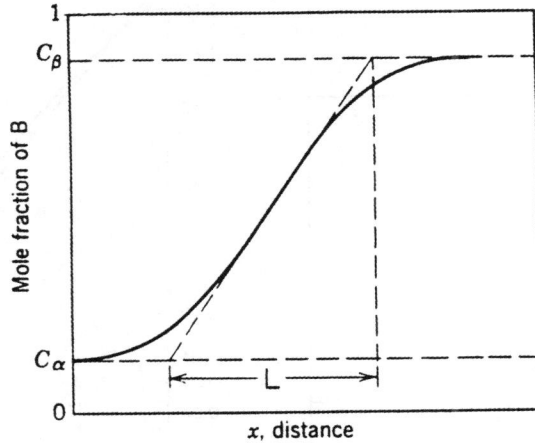

Figure 50 Composition of α and β in the vicinity of the interphase boundary. Note the difference between this profile and that shown in **Figure 48**. Adapted from Cahn and Hilliard (1958) and Swalin (1972) with permission.

where C_e can be set equal to either of the compositions C_α or C_β as indicated in **Figure 49**, and Ω is the regular solution parameter (Eqn (41)(b)). If Eqn (51) is differentiated and the values for σ and ξ mentioned above are substituted, it is found that a general expression for the interfacial free energy is given as

$$\gamma_{cR}^{SS} = 2N_v x_R k_B T_c \gamma_{cr}^{SS}, \tag{52a}$$

where γ_{cr}^{SS} is a reduced interfacial energy defined by

$$\gamma_{cr}^{SS} = \int\limits_{c\alpha}^{c\beta} (\Delta G_R / k_B T_c)^{1/2} dC \tag{52b}$$

and the parameter x_R has the dimensions of length and represents a root-mean-square effective interaction distance for the energy in a concentration gradient. The parameter x_R is sensitive to the exact nature of long-range molecular interactions and can vary from $r_e/\sqrt{3}$, where r_e is the interatomic distance for a simple nearest-neighbor interaction, to a value of $(11/7)^{1/2} r_e$ for a 6–12 potential.

The integral in Eqn (52b) has been evaluated numerically and is plotted in **Figure 51(a)** as γ_{cr}^{SS} versus T/T_c and in **Figure 51(b)** as log (γ_{cr}^{SS}) versus $\log(l - T/T_c)$ (Cahn and Hilliard, 1958). An approximate expression for γ_{cr}^{SS} that is valid over the whole temperature range was obtained from these data and is given by

$$\gamma_{cR}^{SS} \sim 2 N_v x_R (k_B T_c)^{1/2} \left\{ \left[\pi \Delta C_e (\Delta G_R^{max})^{1/2} \right] \big/ 2 \right\} \{ 1 - (\pi/2 - 4/3)(T/T_c) \}, \tag{53a}$$

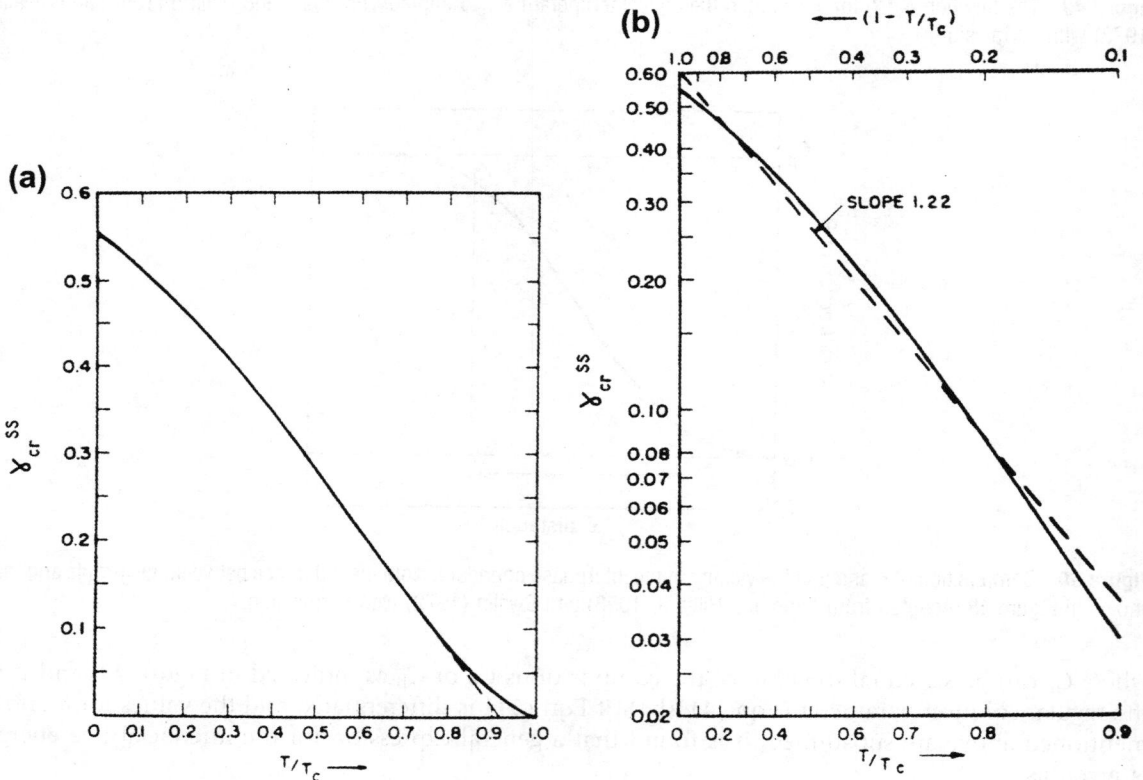

Figure 51 Reduced interfacial free energy γ_{cr}^{SS} versus T/T_c for a regular solution plotted on (a) linear and (b) logarithmic scales. Adapted from Cahn and Hilliard (1958) with permission.

where

$$\Delta G_R^{max} = k_B T_c \Delta C_e^2 + k_B T \{C_c \ln C_c / C_\alpha + (1 - C_c) \ln(1 - C_c)/(1 - C_\alpha)\}, \tag{53b}$$

$$\Delta C_e = (C_c - C_\alpha) \tag{53c}$$

and C_c is the critical composition. In the case of a symmetric miscibility gap $C_c = 0.5$, as shown in **Figure 49**. This equation can be used with Eqn (41b) to calculate γ_{cr}^{SS} with an error of less than 1%. As in the previous case, the interface thickness can be defined in the case of a regular solution model and is given by

$$L_R/x_R = \sqrt{2}\{-1 - [T\ln 4 C_e(1 - C_e)]/[(T_c/(1 - 2C_e)]\}^{1/2} \tag{54}$$

For a regular solution the interface profile is symmetric about $C = 0.5$ (**Figure 50**). This quantity has been calculated numerically and is plotted versus T/T_c in **Figure 52** (Cahn and Hilliard, 1958). When $T \sim T_c$ an equation similar to Eqn (50) results for the regular solution model as

$$L_R/x_{R(T \sim T_c)} = 2[T_c/(T_c - T)]^{1/2}. \tag{55}$$

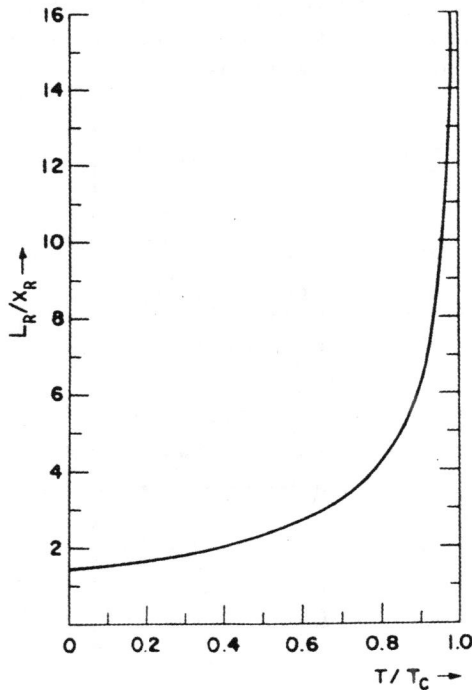

Figure 52 Plot of interfacial width L_R/X_R versus T/T_c for a regular solution. Adapted from Cahn and Hilliard (1958) with permission.

14.3.1.3 Discrete Lattice Plane (DLP) Model

More recently, Lee and Aaronson (1980) developed a DLP regular solution model to calculate the interphase boundary energy and interfacial concentration profiles across coherent interphase boundaries. Although the DLP model was originally developed by Wynblatt and Ku (1977) to analyze surface segregation, these two processes are physically analogous, since the movement of atoms to or away from a coherent interface in an attempt to reach an equilibrium state is similar to segregation to a surface that is initially not in an equilibrium configuration. The problem is thus reduced to calculating ΔH and ΔS, the enthalpy and entropy changes accompanying the net atomic movements required to reach the equilibrium configuration of the α and β phases from an initial state. The homogeneous α phase is taken as the initial or reference state in this treatment for convenience.

In this model, the system is assumed to consist of n solute (B) atoms in N atomic sites. When equilibrium is achieved throughout the system, the ith layer of the boundary region parallel to the IP contains n_i of B atoms per unit area at the interface $N_{s\{hkl\}}$, such that x_i is the atom fraction of B in the ith boundary layer, as illustrated in **Figure 53** (Lee and Aaronson, 1980). This figure also shows how the nearest-neighbor atoms are distributed around an atom P on the ith boundary plane, where z_l is the lateral coordination number and z_j is the coordination number to the nearest-neighboring atoms in the j plane. The total coordination number z is thus

$$z = z_l + 2\Sigma z_j. \tag{56}$$

The average bond enthalpy change attending the breaking of bonds between a B atom in the bulk α phase with composition X_α and its nearest-neighbors and the forming of bonds between this atom and its nearest-neighbors in the ith boundary layer is

$$1/2\left\{ \sum_j \left(X_{i+j} + X_{i-j}\right)z_j + X_i z_l - X_\alpha z \right\}\left(\varepsilon_{BB} - \varepsilon_{AB}\right) \tag{57a}$$

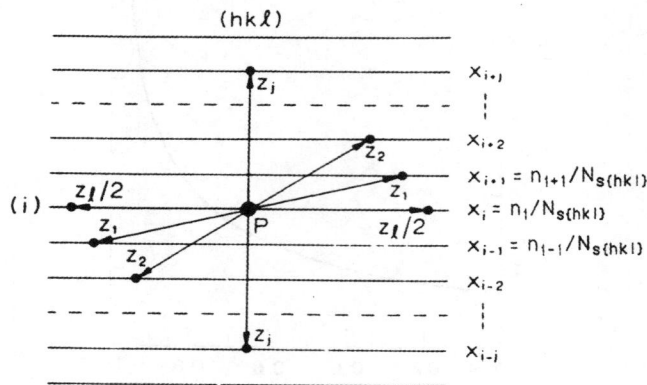

Figure 53 Diagram showing the distribution of z nearest-neighbors about atom P in adjacent planes parallel to (hkl). Z_j is the unidirectional vertical coordination-number to the jth plane as determined from the plane i. Adapted from Lee and Aaronson (1980) with permission.

and a similar expression can be written for the bond enthalpy change undergone by an A atom displaced from the ith α boundary plane to the bulk as

$$1/2\left\{\sum_j (X_{i+j} + X_{i-j})z_j + X_i z_1 - X_\alpha z\right\}(\varepsilon_{AB} - \varepsilon_{AA}). \tag{57b}$$

The sum of Eqn (57) represents the enthalpy change associated with the segregation of a B atom to the ith boundary plane:

$$\Delta h_i = \varepsilon\left\{X_\alpha z - X_i z_1 - \sum_j (X_{i+j} + X_{i-j})z_j\right\}, \tag{58}$$

where $\varepsilon = \varepsilon_{AB} - 1/2\,(\varepsilon_{AA} - \varepsilon_{BB})$, that is Eqn (40), and the total bond enthalpy required to reach the equilibrium state from the reference state ΔH is thus

$$\Delta H = N_{s\{hkl\}}\sum_i (X_i - X_\alpha)\Delta h_i, \tag{59}$$

where the sum is used to denote summation over the entire boundary region. The binding enthalpies are dependent on orientation through z_j and z_1 in Eqn (58).

The entropy change attending segregation to the boundary is because of the configurational entropy difference between the equilibrium state and the reference state ΔS and is given by

$$\Delta S = -N_{s\{hkl\}}k_B \sum_i \{X_i \ln(X_i/X_\alpha) + (1 - X_i)\ln[(1 - X_i)/(1 - X_\alpha)]\}. \tag{60}$$

The equilibrium solute distribution in the boundary region is achieved when

$$\partial(\Delta G)/\partial X_i = 0, \tag{61}$$

where the Gibbs free energy ΔG is given by

$$\Delta G = \Delta H - T\Delta S \tag{62}$$

and Eqn (59) is substituted for ΔH and Eqn (60) for ΔS. Equation (61) then yields a series of difference equations which must be solved simultaneously for X_i according to

$$2\varepsilon\left\{X_\alpha z - X_i z_1 - \sum_j (X_{i+j} + X_{i-j})z_j\right\} - k_B T\{\ln[1/(X_i - 1)] - \ln[(1/(X_\alpha - 1)] = 0. \tag{63}$$

The summation is performed over all boundary planes whose composition differs significantly from either X_α or $(1 - X_\alpha)$ and a computer is used to evaluate the X_i from this equation.

As in the previous CH analysis (Cahn and Hilliard, 1958) the interfacial energy is defined as the free energy difference between an equilibrated mixture of α and β, and homogeneous α and β of equilibrium composition, which are continuous up to the interface. When ΔH and ΔS are substituted into Eqn

(62) for the total free energy ΔG above, and appropriate mathematical rearrangements are performed (Lee and Aaronson, 1980), the coherent boundary energy becomes

$$\gamma_c^{SS} = N_{s\{hkl\}} \sum_i \left\{ -\varepsilon(X_i - X_\alpha)^2 z + \varepsilon \sum_j (X_i - X_{i+j})^2 z_j + k_B T \{X_i \ln(X_i/X_\alpha) \right.$$

$$\left. +(1 - X_i) \ln[(1 - X_i)/1 - X_\alpha]\} \right\}. \tag{64}$$

When $T = 0$ K, Eqn (64) is simplified by noting that the interface is no longer diffuse since both X_α and the $k_B T$ terms become equal to zero and thus so must all the values for X_i in the α phase. Similarly, when the β phase is considered, X_α and all the values of X_i are replaced by $-X_\beta$ and $-X_j$ and these terms must be unity at 0 K. Hence, the concentration difference across the interface jumps directly from zero to unity and the only remaining term in Eqn (64) at $T = 0$ K is

$$\gamma_c^{SS} = N_{s\{hkl\}} \varepsilon \sum_j (X_i - X_{i+j})^2 z_j. \tag{65a}$$

Note that this is the same as Eqn (41). Under this condition, when both planes i and $(i + j)$ are in the same phase $(X_i - X_{i+j})^2 = 0$ and when i is in the α phase and plane $(i + j)$ is in the β phase, $(X_i - X_{i+j})^2 = 1$. Thus Eqn (65a) reduces to

$$\gamma_c^{SS} = N_{s\{hkl\}} \varepsilon \sum_j z_j. \tag{65b}$$

and γ_c^{SS} at 0 K is simply proportional to the number of broken bonds when the interphase boundary is ruptured exactly along the interface. This is similar to the Becker model in the limit of no solubility.

The summation over z_j in Eqns (64) and (65) indicates that the energy varies as a function of orientation. To investigate the orientation dependence of γ_c^{SS} and the resulting concentration profile, it is necessary to evaluate $N_{s\{hkl\}}$, z_1 and z_j as a function of orientation for a particular value of ε. This was performed for an FCC crystal system using a regular solution model with $\varepsilon = 0.167\ k_B T_c$, where the calculations can be confined to within the unit stereographic triangle because of symmetry. The reader is referred to the original references for further information on evaluating these quantities (Mackenzie et al., 1962; Lee and Aaronson, 1980).

Figure 54 (Lee and Aaronson (1980)) shows the concentration profiles in one phase at three temperatures for the (100), (110) and (111) planes only. The profile for the other half of these interfaces is an inverted mirror image of those shown in this figure because of the symmetry relation across the interface (refer to **Figure 50**). The profiles are given in units of the lattice parameter a of the FCC phase. Note that the concentration profile is controlled solely by T/T_c and is independent of boundary orientation above about 0.6 T/T_c, in agreement with the previous results of Cahn and Hilliard (1958). This situation applies at relatively high value of T/T_c when the concentration gradient is small across the boundary zone. Comparing the curves for 0.5 T/T_c in **Figure 54** shows that there is some difference between the two treatments at 0.5 T/T_c when the concentration gradient is relatively steep. The concentration profiles in **Figure 54** would continue to flatten as the temperature is raised until T_c, where the plot would appear as a horizontal line at $X_i \sim 0.5$.

At $T \sim 0$ K, Eqn (65) yields γ_c^{SS} directly by summing over the planes j. At higher temperatures, Eqn (63) must be solved for X_i for various $\{hkl\}$ planes and the results incorporated into Eqn (64). The

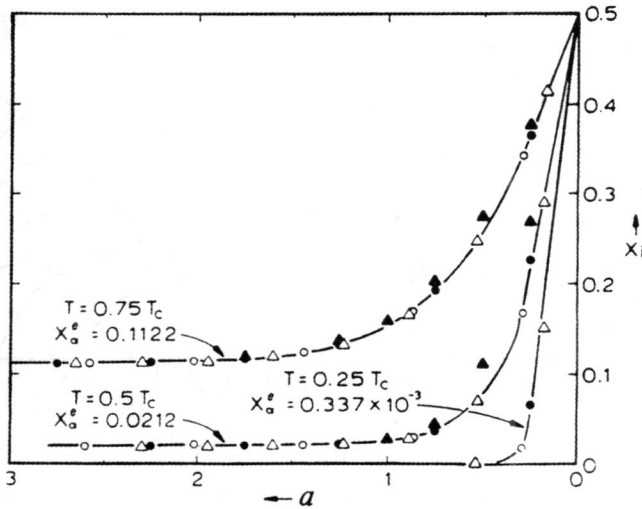

Figure 54 Half concentration profile normal to the α–β boundary. Filled circles represent (100), open circles (111), open triangles (110) and filled triangles are from the CH (continuum) analysis. Adapted from Lee and Aaronson (1980) with permission.

resulting calculations for γ_c^{SS} are shown in the form of contour plots in the stereographic triangle in **Figure 55** for $T/T_c \sim 0$, 0.25 and 0.5, respectively (Lee and Aaronson, 1980). These results are normalized to unity at (100) and the result for 0 K in **Figure 55(a)** is similar to that for the surface energy of a pure FCC crystal (Nicholas, 1968). As T increases in **Figure 55(b) and (c)**, γ_c^{SS} is not a simple function of the number of bonds across the interface and the energy contours do not consist of spheres centered about (210) as in **Figure 55(a)**. In addition, the degree of anisotropy decreases with temperature from about 1.30 at 0 K, to 1.14 at 0.25 T_c, to 1.03 at 0.5 T_c and to less than 1.006 at 0.75 T_c (data not shown). These results are similar to those for the surface energies of metals versus temperature (Heyraud and Metois, 1980, 1983). Such a small anisotropy at 0.75 T_c indicates that at higher temperatures, γ_c^{SS} is effectively orientation independent. Note that a cusp exists at (111) for all of the temperatures investigated and that the maximum energy moves from (210) toward (100) such that the maximum occurs at (100) at and above 0.5 T_c. This is an unexpected result since most experimental studies of the surface energy show that the maximum surface energy is located at approximately the center of the unit triangle.

Lee and Aaronson (1980) have shown that the DLP formalism can be used to derive the regular solution version of the CH continuum equation for γ_c^{SS} and they compare the CH theory (curve c), DLP model (curves a and b) and Becker theory (curve d) in **Figure 56** for the (111) and (100) interfaces (Lee and Aaronson, 1980). Above about 0.7 T_c, the CH and DLP theories converge, which supports the equivalency of these models as well as the isotropic nature of γ_c^{SS} at high temperatures. Comparison of the CH and DLP diffuse interface theories with Becker's result for an abrupt (100) interface shows that a significant reduction in energy occurs as an interface becomes diffuse and therefore, that Becker's result is only valid for $T < 0.2 \, T_c$.

Lee and Aaronson (1980) also compared the contributions of the two components with the interfacial free energy in Eqn (64) as a function of temperature. The two components in Eqn (64) consist of: (i) the first and last terms, which represent the free energy difference between the homogeneous metastable solution with solute concentration X_i and the equilibrium homogeneous solution of

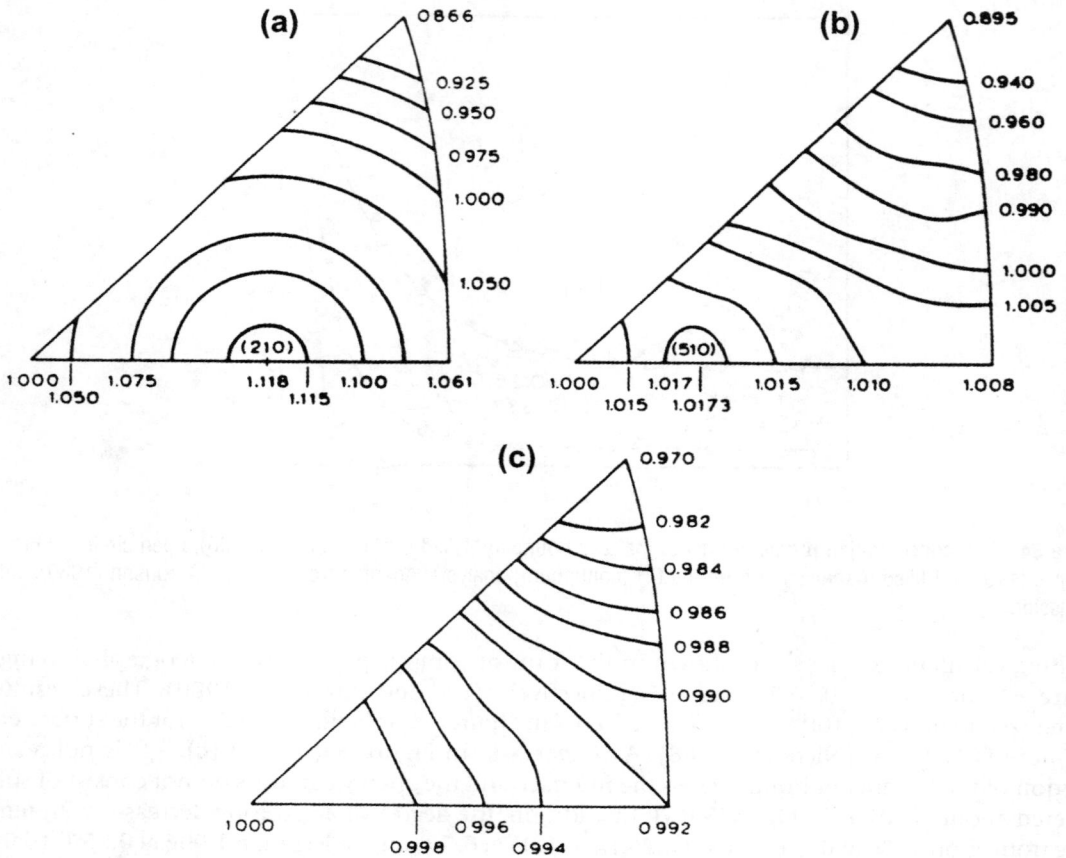

Figure 55 γ_c^{SS} Plots for: (a) 0 K, (b) 0.25 T_c and (c) 0.5 T_c. The results are normalized to unity at (100). Adapted from Lee and Aaronson (1980) with permission.

concentration X_{α}, and (ii) the middle term, which is the free energy change associated with the concentration difference between adjacent parallel planes in the interfacial region. For a solid solution with a positive heat of mixing, as required to form a miscibility gap, both terms are always positive. **Figure 57** (Lee and Aaronson, 1980) shows the contributions of these two terms to the interfacial energy as a function of temperature, where the first and last terms in Eqn (64) are labeled $\Sigma\Delta G_i$ and the middle term is identified as $\Sigma(\Delta X_{i-1})^2$. The free energy increase because of segregation to the interface is maximum at 0.5 T/T_c, while that associated with the concentration difference decreases steadily with temperature. The two contributions are nearly equal above about 0.6 T/T_c and the sum of the two curves in **Figure 57** gives curve (a) in **Figure 56**.

It is possible to use a Wulff construction (Wulff, 1901; Herring, 1951; Hirth, 1965) to determine the equilibrium shapes of coherent precipitates in solids, such as the G.P. zone in **Figure 3**. **Figure 58** shows the equilibrium shapes of coherent FCC precipitate as a function of temperature obtained from the γ_c^{SS} plots in **Figure 55** (LeGoues et al., 1983). At 0 K the shape determined from the Wulff construction is seen to be entirely faceted, as shown by the two-dimensional plot in **Figure 58(a)** and the faceted particle in **Figure 58(b)**. The larger facets are {111} and the smaller ones are {100}. At 0.25 T_c in **Figure 58(c)**, the Wulff shape becomes a faceted sphere, with the {111} facets still larger than {100}. Increasing the

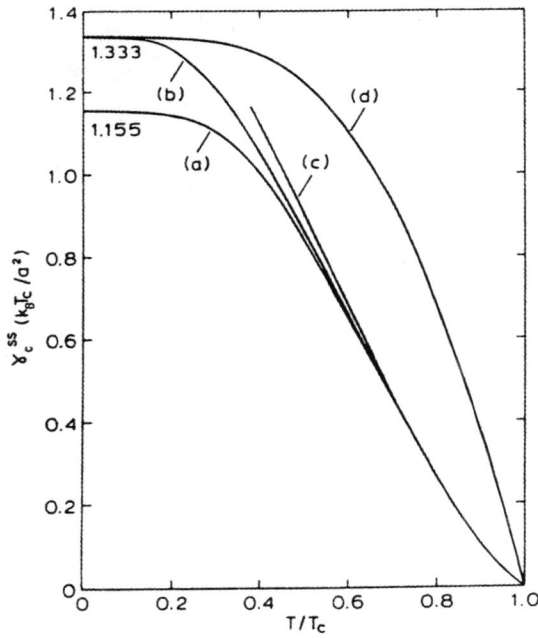

Figure 56   Comparison of the values of γ_c^{SS} in an FCC crystal. Curves (a) and (b) are for (111) and (100) from the DLP model, curve (c) is from Eqn (53) and curve (d) is for Eqn (41). Adapted from Lee and Aaronson (1980) with permission.

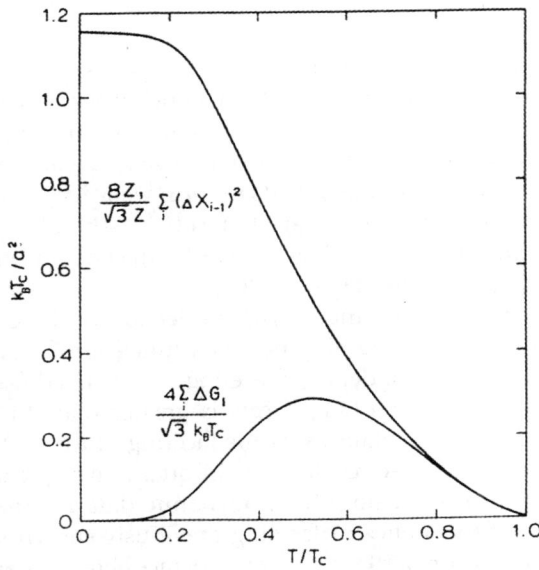

Figure 57   Comparison of the two contributions to γ_c^{SS} for the (111) FCC interface. Adapted from Lee and Aaronson (1980) with permission.

(a)

(b)

(c)

(d)

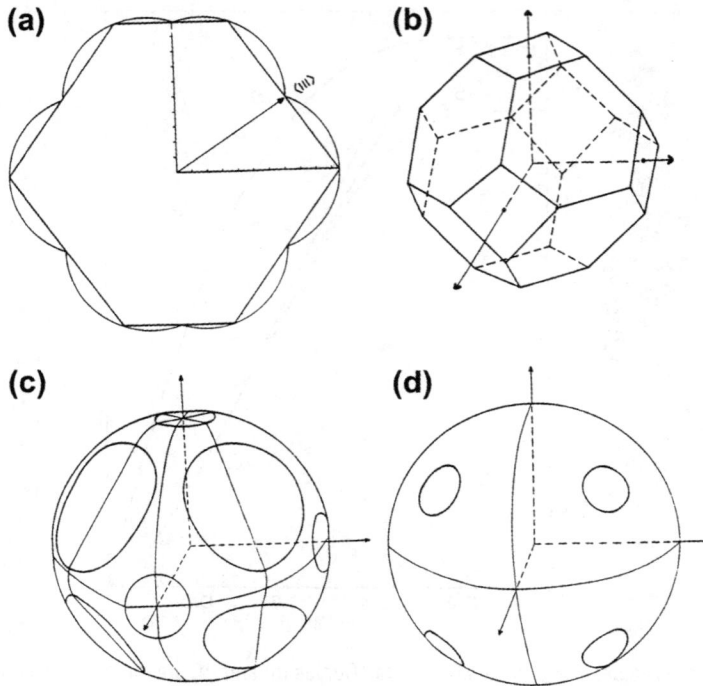

Figure 58   (a) (110) Section of the polar γ plot for an FCC crystal at 0 K and the corresponding Wulff construction. The corresponding three-dimensional precipitate shapes at 0 K, 0.25 T_c and 0.5 T_c are shown in (b), (c) and (d), respectively. Adapted from LeGoues et al. (1983) with permission.

temperature to 0.5 T_c is seen to completely eliminate the {100} facets and considerably reduce the {111} facets as shown in **Figure 58(d)**. The shape at 0.75 T_c (data not shown) is essentially spherical.

An example of the roughening of Ag-rich G.P. zones as a function of temperature in the Al–Ag system (**Figure 3**) is shown in **Figures 59 and 60** (Alexander et al., 1984). **Figure 59** shows <110> bright-field TEM images of two samples aged at 160° and 350 °C to form well-developed G.P. zones. Note that the G.P. zones appear more angular in the sample aged at 160 °C. The diffraction pattern shown in the figure exhibits diffuse scattering in the {111} and {100} directions around the fundamental spots because of the {111} and {100} facets. **Figure 60(a)** shows the measurement technique used to determine the percentage of faceting from the <110> projection of the G.P. zones and **Figure 60(b)** shows the experimental results on faceting obtained as a function of aging temperature. These data clearly show the strong temperature dependence of faceting as calculated from the regular solution DLP model in **Figure 58**. The percentage of {111} facets is greater than {100} and persists to higher temperatures, as predicted from the calculations. Some faceting of the G.P. zones remained up to the solvus, which is just above 350 °C. These results were in qualitative agreement with regular solution DLP calculations that were performed using thermodynamic data for the Al–Ag system (Alexander et al., 1984). The temperature dependence of faceting previously shown for the coherent interphase boundary, previously also occurs for partly coherent and incoherent interfaces although only a few examples, such as Al–Ge (Lours et al., 1992), have been documented so far.

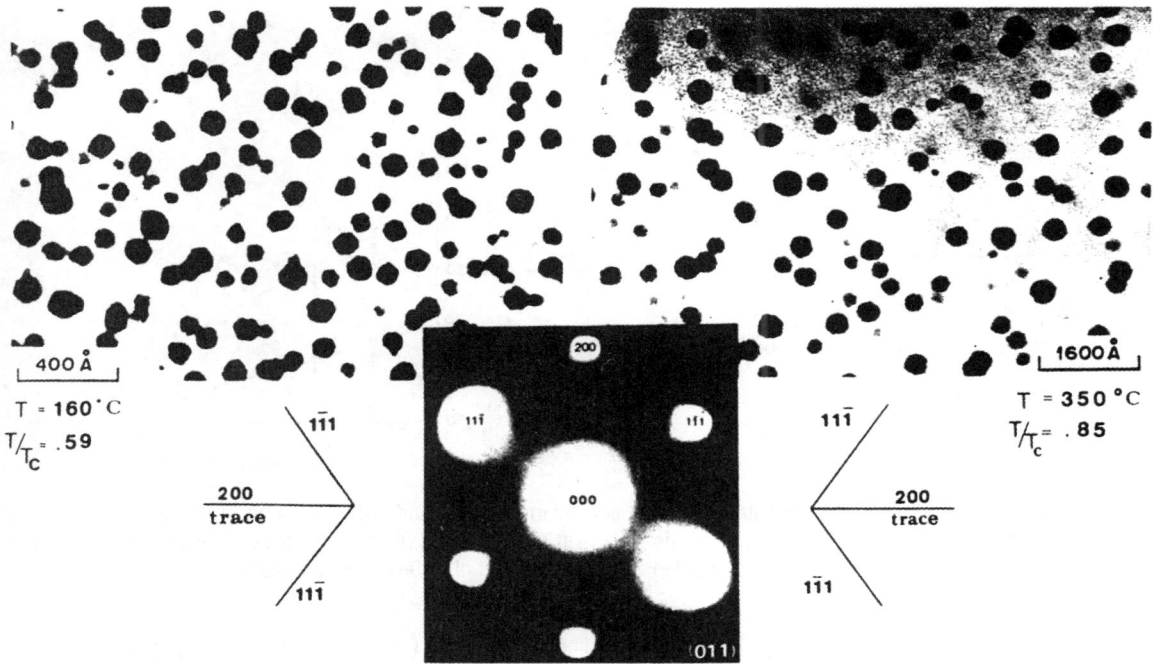

Figure 59 Bright-field <110> TEM images of G.P. zones aged at 160 °C (0.59 T_c) and 350 °C (0.85 T_c) in Al–Ag alloy. Adapted from Alexander et al. (1984) with permission.

14.3.2 Antiphase Boundaries

An antiphase boundary (APB) separates two domains of the same ordered phase (Marcinkowski, 1963; Kikuchi and Cahn, 1979). It results from symmetry breaking that occurs during ordering processes, which can start at different locations in a disordered lattice. An APB forms when two such regions contact so that they display wrong compositional bonds across the interface, as illustrated in **Figure 61** (Haasen, 1986). Dislocations with Burgers vectors that are not translation vectors of the ordered superlattice can also create APBs as they move through an ordered phase (Marcinkowski, 1963).

APBs are quite similar to the coherent interphase boundaries that were discussed in Section 14.3.1.3. One can readily calculate the interfacial energy associated with an APB using a nearest-neighbor bonding model similar to the DLP model, and it is possible to envision the temperature dependence of the APB width by referring to the CH/DLP treatments for coherent interphase boundaries. This section examines APBs in an ordered FCC structure in some detail to illustrate these features. This is followed by some examples of the behavior of APBs in FCC alloy systems.

The $L1_2$ structure, or Cu_3Au-type superlattice, is shown in **Figure 62** (Haasen, 1986). The disordered FCC structure transforms to a lower symmetry on ordering, that is to a simple cubic structure with four atoms or sublattices per primitive lattice point. A B-atom occupies one sublattice site in the ordered structure and A atoms occupy the other three sites. Thus, four types of ordered domains can occur on the four sublattices in a crystal: type I with B at 0,0,0; type II with B at 0,1/2,1/2; type III with B at 1/2,0,1/2; and type IV with B at 1/2,1/2,0. An APB is produced by an antiphase vector $\mathbf{u}_i$, which brings about the displacement of one atomic species from one sublattice to another. The antiphase vectors

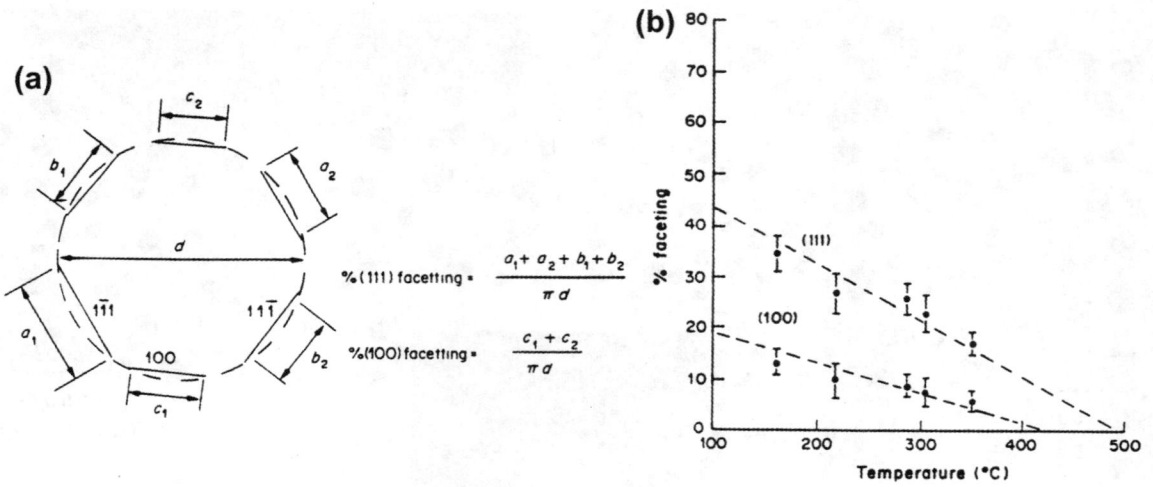

Figure 60 Measurement technique for determining the percentage of {111} and {100} facets on the G.P. zones. (b) Percent faceting versus temperature using the measurement technique in (a). The error bars represent the standard deviation of the experimental measurements. Adapted from Alexander et al. (1984) with permission.

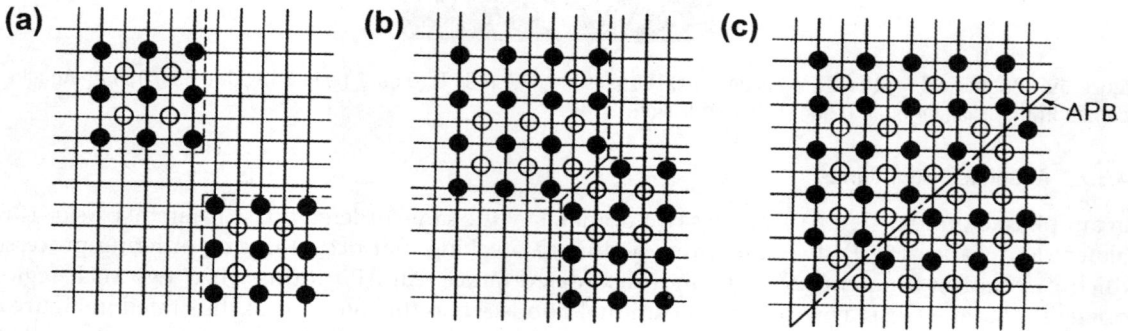

Figure 61 Formation of an APB when ordered regions in which A (open circles) and B (filled circles) atoms occupying different sublattices grow together: (a) nucleation of ordered domains on different sublattices, (b) contact of domains, and (c) the resulting APB (dashed line). Adapted from Haasen (1986) with permission.

correspond to perfect dislocation Burgers vectors in the disordered crystal and are of the type $a/2 <110>$ as illustrated in **Figure 62**.

An APB can be sufficiently described by u_i and the normal $\mathbf{n}$ to the boundary plane. If u_i lies in the boundary plane, then

$$\mathbf{u_i} \cdot \mathbf{n} = hu + kv + lw = 0 \tag{66}$$

where [uvw] are the components of the antiphase vector $\mathbf{u_i}$ and (hkl) are the Miller indices denoting the plane of the APB (Flinn, 1960). In this case, the APB is produced by a lattice translation in the boundary plane and there is no net increase or decrease in the number of A and B atoms at the interface. This type

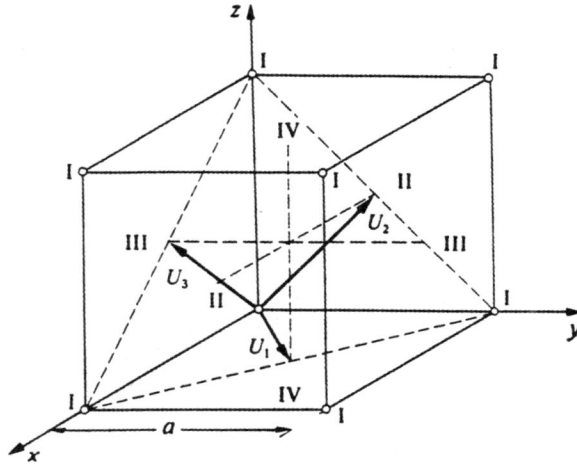

Figure 62 Unit cell of the Cu_3Au superlattice. Open and filled circles represent A and B atoms, respectively. The vectors show equivalent ways of moving a B atom from a type I site to type II, III and IV sites. Adapted from Haasen (1986) with permission.

of APB is called a conservative or type-1 APB and a relatively small APB energy γ_{APB} is expected for this type of interface. In the second case

$$\mathbf{u_i} \cdot \mathbf{n} \neq 0 \tag{67}$$

and a nonconservative, or type-2, APB boundary results. In this case, it is necessary to remove either a plane of A or B atoms to form the APB. This leads to an excess or deficit of A–A or B–B bonds at the interface and a higher energy is expected for this type of APB. **Figure 62** shows the four sublattices (I–IV) for the $L1_2$ structure with the three associated vectors $\mathbf{u_i} = a/2[110]$, $a/2[101]$ and $a/2[011]$. For an APB with $\mathbf{n} = [001]$, a conservative (type-I) APB is produced by $\mathbf{u_i}$ and the others are nonconservative (type 2).

The APB energy γ_{APB} can be estimated as a function of the orientation of the APB, that is of the boundary normal $\mathbf{n}$, using a nearest-neighbor bonding model, similar to the previous coherent interphase boundary interfaces. This provides a reasonable estimate of the APB energy at 0 K, since the contribution of higher order neighbors to the energy is less than 10%. A complete derivation of the equations for APBs in several different FCC and BCC ordered structures is provided by Flinn (1960) and Marcinkowski (1963). The derivations are similar to those used in Eqn (59) and only the end result for the case of the $L1_2$ structure in **Figure 62** is summarized here.

In the case of an $L1_2$ structure with the pair exchange energy ε, the APB energy for $\mathbf{u_i} = a/2[110]$ is given as

$$\gamma_{APB} = 2\varepsilon h \pounds^2 / a^2 \left(h^2 + k^2 + l^2 \right)^{1/2} \tag{68}$$

where $\pounds$ is the long-range order parameter, which varies from 1 for a fully ordered structure at 0 K to 0 for a completely random structure above T_c, a is the lattice parameter, and assuming that $h \geq k$. For a conservative $\mathbf{n} = [001]$ APB with $\pounds = 1$, $\gamma_{APB} = 0$, although the APB energy remains finite when second nearest-neighbor interactions are taken into account. The orientation dependence of γ_{APB} for the $L1_2$ structure is shown in **Figure 63** (Haasen, 1986) by lines of constant energy in a stereographic projection

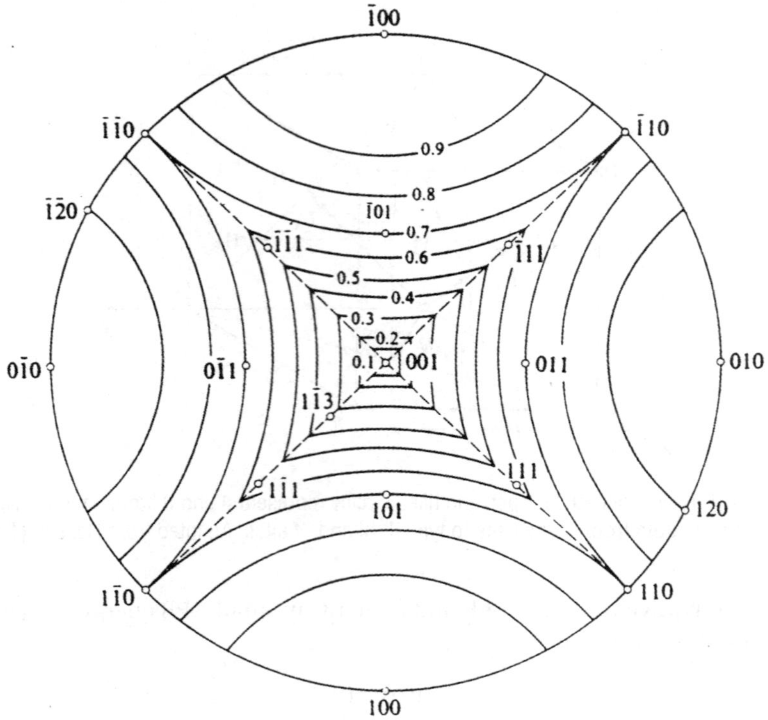

Figure 63 Lines of constant APB energy created by $\mathbf{u}_1 = a/2[110]$ in Cu_3Au as a function of the plane normal [hkl]. Adapted from Flinn (1960) with permission.

of the possible **n**. There is a minimum at $\mathbf{n} = [001]$, maxima at [100] and [010], with saddle points at [110] and [1$\bar{1}$0]. Antiphase boundary energies are expected to be relatively low and can range from just a few (i.e. 3–5 mJ/m^2) in FeAl (Krzanowski and Allen, 1984) to approximately 25–40 mJ/m^2 for the (111) APB energy in Cu_3Au, depending on temperature (Flinn, 1960; Kikuchi and Cahn, 1979).

According to **Figure 63**, APBs in $L1_2$ structures should lie parallel to the cube planes and this has been observed experimentally in alloys where nearest-neighbor bonding interactions are the dominant component to the APB energy. For example, **Figure 64** shows conventional and HRTEM images of APBs in Cu_3Au (Loiseau et al., 1995). The viewing direction is [001] and the conservative APBs labeled 2 and 3 in **Figure 64(a)** clearly lie along the cube planes. The white spots in the HRTEM image in **Figure 64(b)** are the projections of the atomic columns in the structure and the $a/2[110]$ displacement of the white spots across the APBs is directly visible in the image. More accurate expressions for the APB energy that account for the entropy contribution and the presence of extra A or B atoms at nonconservative APBs have been obtained by Kikuchi and Cahn (1979) and Loiseau et al. (1995). At low temperatures the entropic contribution is small, but it can have a substantial effect on the APB profile at temperatures approaching T_c.

The degree of order across an APB as a function of temperature can be represented schematically by an order parameter profile, as shown in **Figure 65** (Loiseau et al., 1995). At 0 K in **Figure 65(a)**, the order parameter £ jumps discontinuously from -1 to $+1$ across the APB, similar to the composition profile across a coherent α–β interphase boundary at 0 K in **Figure 48**. As the temperature is raised, the

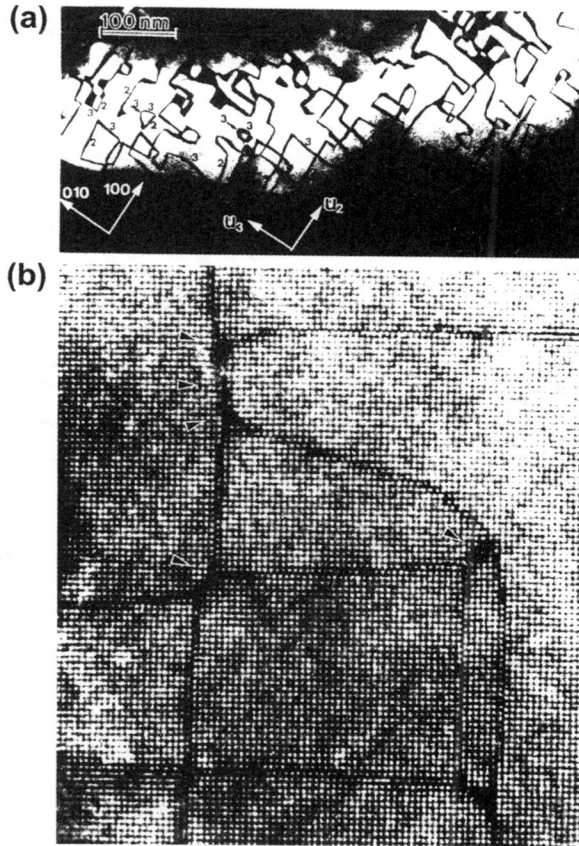

Figure 64 (a) (110) Dark-field TEM image of APBs in Cu_3Au projected along [001]. The numbers 2 and 3 indicate APBs with antiphase vectors $\mathbf{u}_2$ and $\mathbf{u}_3$. Examples of nonconservative APBs are indicated by arrows. (b) [001] HRTEM image showing conservative APBs along (100) and (010) joining at four-point and triple-point junctions (arrows). Adapted from Loiseau et al. (1995) with permission.

long-range order parameter decreases and the APB obtains a finite width $\pounds_{APB}$. The order parameter profile across the APB then looks like T_1 in **Figure 65(b)**. This profile has the same shape as the composition profile for the coherent α–β interface in **Figures 50 and 54**. The long-range order parameter continues to decrease and the APB width continues to increase with temperature, that is T_2 in **Figure 65(b)**, until at T_c, $\pounds$ approaches 0, $\pounds_{APB}$ approaches ∞, and the APB vanishes. This process has been verified experimentally for the DO_3 – B2 ordering transformation in an Fe-27%Al alloy (Loiseau et al., 1995; Stinchcombe, 1983). The behavior of an APB in the case of a first-order order–disorder transformation is more complicated than for a second-order transformation, but it can be qualitatively described for $L1_2$ phase as follows (Kikuchi and Cahn, 1979; Loiseau et al., 1995). At 0 K, the order parameter changes abruptly from its value in one domain to that in the adjacent domain across the APB, as shown in **Figure 65(c)**. As the temperature is increased to slightly below T_c, the APB progressively splits into two new interphase boundaries (IPBs) of thickness $\pounds_{IPB}$ bounding a region of width L_0, where the order parameter is zero **Figure 65(d)**. In essence, the APB has become perfectly wet by the

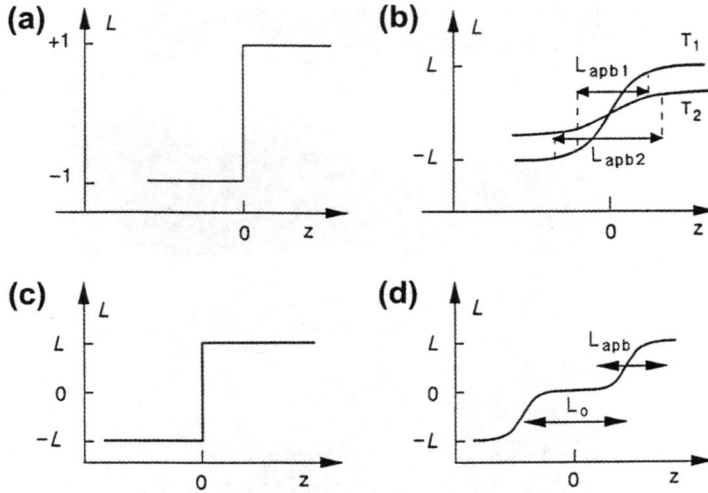

Figure 65 (a,b) Evolution of the order parameter profile of an APB for a second-order transformation with $T_1 < T_2$ (from [116]). (c,d) Evolution of the order parameter profile of an APB for a first-order transformation. Adapted from Loiseau et al. (1995) with permission.

disordered FCC phase and the ordered and disordered phases coexist. As T_c is approached, $\pounds_{APB}$ remains finite but $\pounds_0$ diverges so that the disordered area becomes large compared with the domain size. Evidence for this phenomenon has been shown by TEM in several L1$_2$ alloys, for example Cu-17%Pd alloy (Loiseau et al., 1995).

14.3.3 Partly Coherent Interphase Boundaries

Coherent interphase boundaries form in solids when the lattice parameters of the α and β phases are similar and there is only a small amount of misfit across the interface. This situation frequently occurs with small or plate-shaped precipitates and lattice-matched semiconductors, for example, and two examples of coherent interphase boundaries were shown previously in **Figures 3 and 4**. In many other cases, there is sufficient lattice mismatch that coherent interfaces relax to an unstrained condition with an array of misfit dislocations at the interface. This type of interface was demonstrated by the {111} Cu–Ag interface in **Figure 5**.

When an interface relaxes and acquires a series of misfit dislocations, it is referred to as a partly coherent (or discommensurate) interface. The energy of a partly coherent interface between two phases α and β that differ in composition can be considered to contain two parts, one associated with the structural defects at the interface and the other because of the unfavorable compositional bonds across the interface. In the previous section on coherent interphase boundaries (Section 14.3.1), several procedures were developed to calculate the interfacial energy because of compositional differences across an interface. The present section describes a treatment that was developed to quantify the elastic strain energy associated with a partly coherent interface.

Turnbull (1955) suggested that when the misfit is small, it is reasonable to assume that the interfacial energy of a partly coherent interface γ_{sc}^{SS} can be determined by adding the compositional γ_c^{SS} and structural γ_s^{SS} (elastic strain) components together to obtain the total interfacial energy. In equation form this can be expressed as

$$\gamma_{sc}^{SS} = \gamma_c^{SS} + \gamma_s^{SS}. \tag{69}$$

This section begins by determining the structural component of the interphase boundary energy. The structural component is then combined with the compositional component to obtain the total interphase boundary energy γ_{sc}^{SS} of a partly coherent interface according to Eqn (69). The relative magnitudes of the compositional and structural components to the interphase boundary energy are also compared in an FCC–FCC system as a function of misfit and temperature.

14.3.3.1 Geometry of Partly Coherent Interfaces

In this section, the subscripts 1 and 2 are used to indicate the lattice parameters of the α and β phases, respectively, which are both assumed to be semi-infinite in extent except when noted otherwise. For simplicity, the α and β phases are assumed to be simple cubic crystals that are aligned across the interface and differ only in lattice parameter, as illustrated in **Figure 66** (Smith and Shiflet, 1987). The approach of Frank and van der Merwe (1949) and van der Merwe (1950, 1963) is then followed in analyzing the resulting interfacial structure and energy, as summarized in a review by Aaronson et al. (1970).

In most cases, one is interested in the situation where there is relatively strong bonding across the α–β interface, so that relaxation of the interface occurs and the misfit is localized at dislocations in the interface, as illustrated in **Figure 66**. Between the dislocations, the planes in the two lattices match perfectly. Designating the lattice parameter of the α phase as a_1 and that of the β phase as a_2, the repeat distance D_δ between the dislocations (or alternatively, between the regions of perfect matching) is given by

$$D_\delta = n\,a_1 = (n+1)a_2, \tag{70}$$

where n represents an integral number of spacings in either lattice. Since the interfacial dislocations are derived from the difference in lattice parameters across the interface, they are commonly called misfit dislocations. Unlike isolated dislocations, which increase the energy of a single-phase material, these dislocations are an energy-reducing feature of the structure. It is sometimes convenient to quantify the misfit between two lattices in terms of a reference lattice that is an average of the two lattices rather than

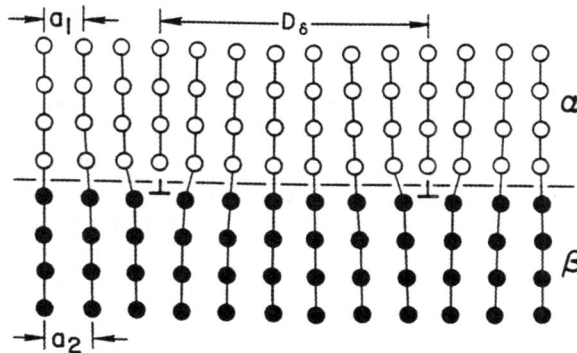

Figure 66 Atomic model of a partly coherent interface in a simple cubic crystal where $a_\beta > a_\alpha$. Adapted from Aaronson et al. (1970) with permission.

in terms of one or the other of the two lattices. If one uses an average reference lattice, its lattice parameter is given by

$$a' = (a_1 + a_2)/2 \qquad (71)$$

and with this, Eqn (70) becomes

$$D_\delta = (n + 1/2)a'. \qquad (72)$$

Noting that

$$n = a_2/(a_1 - a_2) \qquad (73)$$

and substituting into Eqn (71) leads to the following expression for the repeat distance

$$D_\delta = (a_1 + a_2)^2/4(a_1 - a_2). \qquad (74)$$

The misfit δ between the α and β lattices can then be defined by the ratio of the reference lattice parameter to the repeat distance, or $\delta = a'/D_\delta$. Thus

$$\delta = 2(a_1 - a_2)/(a_1 + a_2) = (a_1 - a_2)/a' \qquad (75)$$

and the misfit is given as the ratio of the difference in the α and β lattice parameters to the parameter of the reference lattice. The lattice parameters are given in units of length (nanometers) so that δ is dimensionless. The misfit δ can also be referred to one of the crystal lattices instead of an average reference lattice, so that $\delta = (a_1 - a_2)/a_1$, for example. Rewriting Eqn (74) on the basis of Eqn (75) gives the misfit dislocation spacing at the interphase boundary as

$$D_\delta = a'/\delta = b/\delta, \qquad (76)$$

where b is the magnitude of the edge component of the Burgers vector **b** of the misfit dislocation. Physically, this equation is analogous to Eqn (1b) developed for a grain boundary, with the rotation θ replaced by the misfit δ.

14.3.3.2   *Energy of Partly Coherent Interfaces*

The energy of a partly coherent interface was first analyzed by Frank and van der Merwe (1949) and van der Merwe (1950). In their treatment, misfit is considered in only one direction, as in **Figure 66**, and atomic relaxation is permitted in the vicinity of the misfit dislocations. The interaction of the atoms across the interphase boundary is assumed to occur according to a sinusoidal force law and the interaction within a given phase is treated on the basis of an elastic continuum. Their approach yields the following expression for γ_s^{SS}, the misfit dislocation energy at the α–β boundary (van der Merwe, 1963):

$$\gamma_s^{SS} = \left(\mu a'/4\pi^2\right)\left\{1 + \Lambda - \left(1 + \Lambda^2\right)^{1/2} - \Lambda \ln\left[2\Lambda\left(1 + \Lambda^2\right)^{1/2} - 2\Lambda^2\right]\right\}, \qquad (77a)$$

where

$$\Lambda = 2\pi\delta \left(C_8/\mu\right), \qquad (77b)$$

$$1/C_8 = [(1 - \nu_\alpha)/\mu_\alpha] + [(1 - \nu_\beta)/\mu_\beta] \qquad (77c)$$

and μ is the shear modulus at the interface, μ_α and μ_β are the shear moduli and ν_α and ν_β are Poisson's ratios in the α and β phases, respectively. The C_8 term accounts for elastic interactions within each crystal while μ accounts for such interactions across the α–β interface. When these individual characteristics are suppressed by letting $\nu_\alpha = \nu_\beta = 1/3$ and $\mu_\alpha = \mu_\beta = \mu$, then $\Lambda = 3\pi\delta/2$ and γ_s^{SS} is only a function of μ and δ.

The solid curve in **Figure 67(a)** (van der Merwe, 1963) shows the variation of γ_s^{SS} with δ under these conditions. It is particularly important to note the rapid increase in D_δ with δ when $\delta < 0.01$ and the small dependence of γ_s^{SS} on δ at larger misfits. Also note the energy reaches a maximum of $\mu a'/4\pi^2$, or the first term in Eqn (77a). van der Merwe (1963) has shown that at least 98% of the total elastic strain energy is stored within the region $x \leq D_\delta/2$, where x is the distance normal to the interface.

It is common to find misfit in two mutually perpendicular directions as in the {100} interphase boundary in semiconductor crystals, and also in three directions 120° apart as between the close-packed {111} and (0001) planes of cubic and hexagonal crystals. Thus it is possible to have interfacial

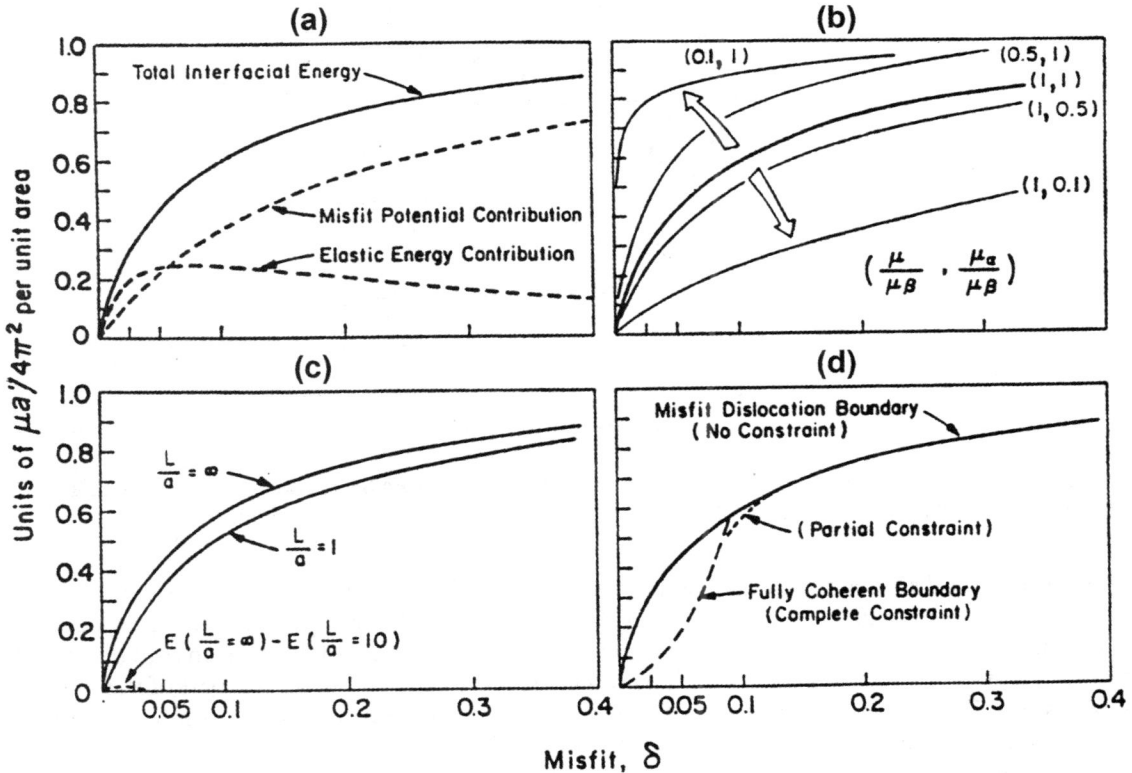

Figure 67 Misfit dependence of the dislocation interphase boundary energy γ_s^{SS} as influenced by: (a) contributions of the misfit potential and elastic energy, (b) the relative hardness (μ_α/μ_β) and interfacial bonding strength (μ/μ_β), (c) the relative thickness of one component, and (d) the relative elastic strain accommodation. Adapted from van der Merwe (1963) and Aaronson et al. (1970) with permission.

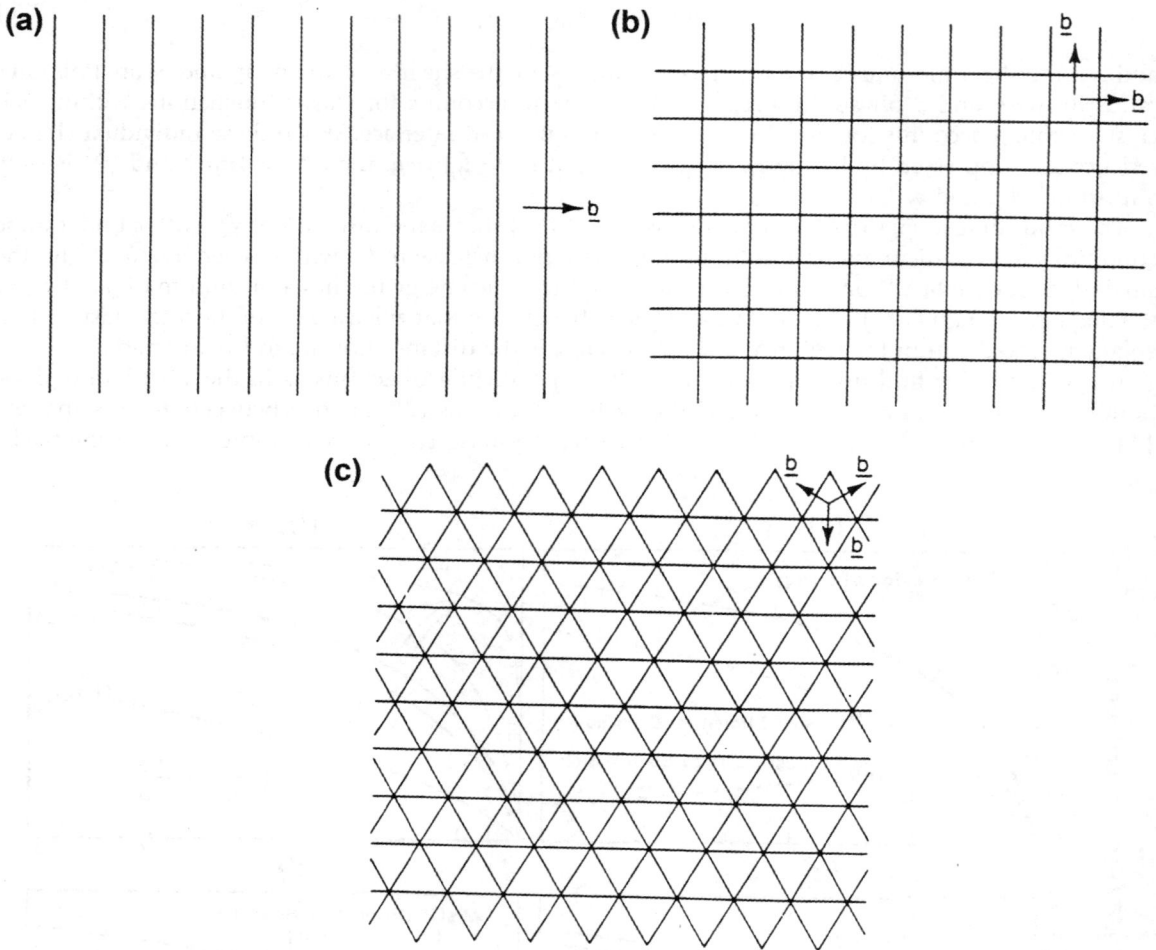

Figure 68 Misfit dislocation network for (a) one set of dislocations as in **Figure 66**, (b) two sets of perpendicular dislocations, and (c) three sets of dislocations at 120°. Adapted from Cullen et al. (1973) with permission.

dislocation networks such as those illustrated in **Figure 68** (Cullen et al. 1973). When there are two perpendicular sets of parallel edge dislocations (**Figure 68(b)** each having an energy expressed by Eqn (77), to first approximation the interaction energy between the two sets of dislocations can be neglected and the energy of the two sets of dislocations can be added to give the structural part of the interfacial energy as γ_s^{SS}. The example of three sets of dislocations in **Figure 68(c)** is more complicated since the Burgers vectors have mutual components and the interaction energy is not negligible.

It is also important to consider the effects of the other factors such as the elastic moduli on the interfacial energy γ_s^{SS}. The ratio μ_α/μ_β is referred to as the relative hardness (or strength) of the two crystals and the ratio μ/μ_β is taken to represent the strength of the interfacial bond. **Figure 67(b)** shows that for a constant value of the strengths of the phases (μ_α/μ_β), increasing the strength of the interfacial bond (μ/μ_β) by a factor of 10 sharply decreases the initial slope of the γ_s^{SS} versus δ curve and greatly extends the range over which γ_s^{SS}

increases appreciably with δ. In comparison, for a constant value of μ/μ_β, increasing μ_α/μ_β produces an increase in the curve and a decrease in the range where γ_s^{SS} depends on δ.

Another situation that is important to consider, is when the thickness L of one of the phases, say the α phase, is less than the interdislocation spacing D_δ. This situation is encountered in thin films and also applies to small misfitting precipitates in a matrix. **Figure 67(c)** shows that when the α and β phases differ only in lattice parameter, γ_s^{SS} is less in a monatomic film than in one of semi-infinite thickness only at small values of δ. When $L/a \cong 20$ the difference becomes negligible throughout the range of misfit.

When both L and δ are sufficiently small, a fully coherent interface has a lower energy than one with an equilibrium arrangement of misfit dislocations. A qualitative argument upon which this conclusion is based is illustrated in **Figure 67(d)**. The energy of a fully coherent interface increases proportional to δ^2 (Christian, 1975) and therefore increases slowly at small misfits, while that of a misfit dislocation boundary exhibits a rapid initial rise (**Figure 67(a)**). Hence, a misfit dislocation interface does not have a lower energy until a critical value of the misfit is exceeded. Quantitatively, the elastic strain energy E_{el} required to deform a thin film homogeneously in one direction is given by

$$E_{el} = \left[\mu_\alpha(1 - \nu_\alpha)/L\delta'^2\right]\Big/(1 - 2\nu_\alpha), \tag{78}$$

where $\delta' = (a_1 - a_2)/a_1 \cong \delta$ for small values of misfit, a_1 is the lattice parameter of the thin film and a_2 is the lattice parameter of the substrate. The critical value of the misfit δ'_c occurs when E_{el} in Eqn (78) is equal to γ_s^{SS} in Eqn (77a). Letting $\nu_\alpha = \nu_\beta = \nu = 1/3$, $\mu_\alpha = \mu_\beta = \mu$ and considering the misfit to be sufficiently small that $\Lambda \cong 0$ yields the following relationship for the thickness of α phase corresponding to δ'_c (van der Merwe, 1963):

$$L_c = 3a_1\left[1 - \ln(\pi\delta'_c)\right]/(16\pi\delta'_c). \tag{79}$$

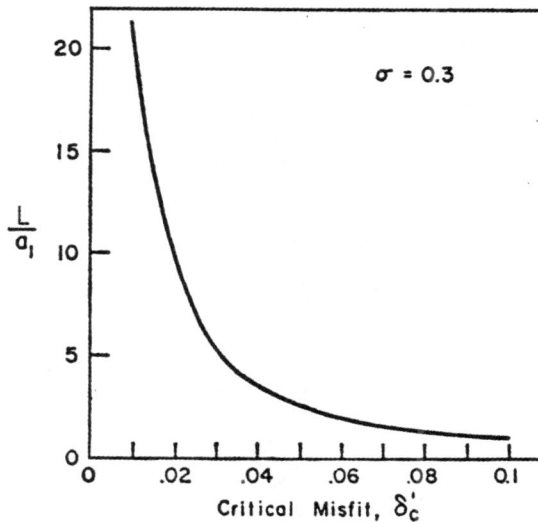

Figure 69 Dependence of the critical misfit δ_c' on the ratio of the thickness of one phase L to the interatomic spacing of the phase a_1. Adapted from van der Merwe (1963) and Aaronson et al. (1970) with permission.

Figure 69 (van der Merwe, 1963; Aaronson et al., 1970) shows the dependence of δ'_c on the ratio L/a_1 when $\mu_\alpha = \mu_\beta = \mu$ and $\nu = 1/3$. This plot shows that it is possible to maintain full coherency of interfaces for significant thicknesses only for small values of the misfit, that is when δ'_c is on the order of 1% or less. Since the generation of misfit dislocations requires an activation energy, the fully coherent state is found to be retained in metastable equilibrium until $(\pi/2)\delta'_c$. The value of L_c depends on the ratios of the elastic moduli so for a given value of L/a_1, the value of the critical misfit δ'_c increases with the bonding strength (μ/μ_β) for a given level of relative hardness (μ_α/μ_β) and with decreasing relative hardness for a given level of bond strength, both in accord with qualitative expectation.

Examples of misfit dislocations at heterophase interfaces abound and have been compiled in several reviews (Aaronson et al., 1970; Shiflet, 1986; Hull and Bean, 1992). Just one example of a typical interphase boundary dislocation structure is shown here to illustrate its appearance in the TEM. **Figure 70** (Cline et al., 1971) shows the interfacial dislocation structure formed at the interface between a Cr rod (dark) and the NiAl matrix (light) in a directionally solidified Cr/NiAl eutectic alloy. Both phases in this system are cubic and they are in a parallel OR. The misfit is small in this system and the spacing of the dislocations with $\mathbf{b} = a\langle 100 \rangle$ was found to be quite regular ($D_\delta \sim 120$ nm), as evident in the figure. It is often found that experimentally measured dislocation spacings are greater than those expected from equilibrium elasticity calculations and this has been attributed to the difficulty of nucleating misfit dislocations at interfaces (van der Merwe, 1950; Aaronson et al., 1970).

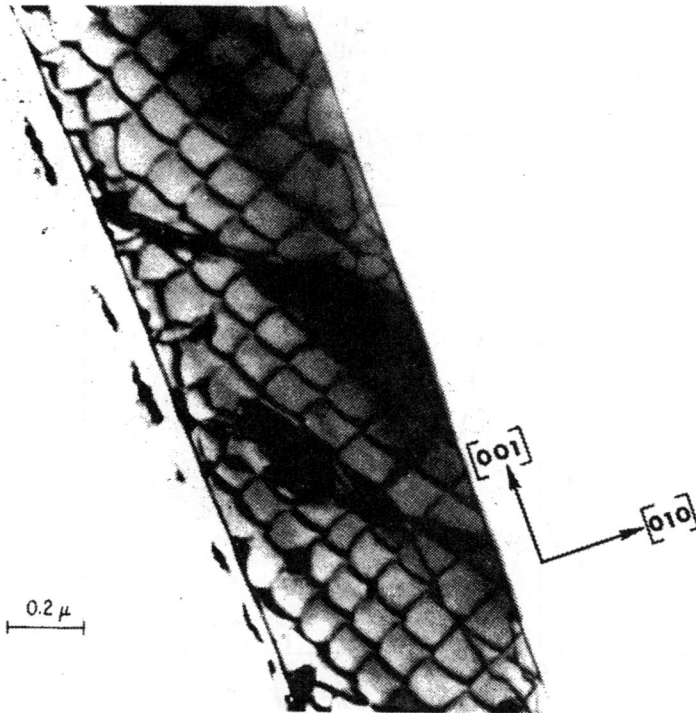

Figure 70 Bright-field TEM image illustrating a dislocation network in a directionally solidified Cr–NiAl eutectic quenched from 1200 °C. The network is rotated 45° with respect to the rod axis. Adapted from Cline et al. (1971) with permission.

It is worth noting the similarity of Eqn (77a) with the Read–Shockley formula for grain boundaries in Eqn (2a). For well-separated dislocations and phases with similar elastic moduli where $\mu_\alpha = \mu_\beta = \mu$, Λ reduces to $\Lambda \sim \pi a'/D_\delta(1-\nu)$. Further, for $D_\delta \gg a'$, higher powers of Λ may be neglected and if the misfit δ is substituted for a'/D_δ, then Eqn (77a) simplifies to (Smith and Shiflet, 1987):

$$\gamma_s^{SS} = [\mu a'\delta/4\pi(1-\nu)](A_o - \ln \delta), \qquad (80)$$

where A_o is a dislocation core energy. If δ is replaced by θ, the factor $(1-\nu)$ is replaced by unity, and a' is replaced by b for a homophase system, Eqn (80) becomes the formula for a low-angle twist boundary with a rotation angle θ (Eqn (2)). Hence, the origin of the interfacial dislocation energies is similar in both homophase and heterophase interfaces.

14.3.3.3 Comparison of the Compositional and Structural Components of the Interphase Boundary Energy

Sections 14.3.1 and 14.3.3.2 developed procedures to calculate the compositional and structural components of the interphase boundary energy for coherent and partly coherent interfaces, respectively. As pointed out at the start of Section 14.3.3 (Eqn (69)), Turnbull (1955) suggested that these two components could be added to obtain the total interphase boundary energy of a partly coherent interface. This procedure has been applied to a few interfaces (Stephens and Purdy, 1975; Aaronson et al., 1968; Spanos, 1989). This section shows the results from the most complete analysis of this type, which was performed for a party coherent FCC–FCC interface using lattice parameters and elastic constants appropriate for Cu as a function of temperature and assuming a symmetric regular solution with a critical temperature of 1200 K (corresponding to a regular-solution parameter of 19.9 kJ/mol). Further details of the calculation are described by Spanos (1989).

The basic procedure employed in the study was the following:

(i) The O-lattice method (Section 14.3.2.2) was used to determine the dislocation structure of the interphase boundaries between two FCC crystals of identical orientation but with different lattice parameters as a function of boundary orientation.

(ii) The structural energy of these interfaces was determined by employing elastic energy calculations for dislocation arrays due to Hirth and Lothe (1982) (HL) and also according to the van der Merwe analysis (Eqn (77)) for the (100) interface.

(iii) The compositional component to the interphase boundary energy was calculated using the regular solution DLP model described in Section 14.3.3 (Eqn (64)) and this was compared with the structural energy obtained for the same boundary as a function of temperature.

(iv) Polar γ-plots were developed for each of these energies to compare their anisotropies as a function of temperature.

(v) A Wulff construction (Herring, 1951) was performed on each polar γ-plot for the purpose of comparing the equilibrium shapes yielded by the structural and compositional energies as a function of temperature.

(vi) The polar γ-plots and the Wulff constructions were evaluated for the sum of the structural and compositional interfacial energies.

Table 1 compares the magnitudes of the compositional interfacial energy γ_c^{SS} with the structural energy γ_s^{SS} for the DLP and HL methods of calculation for the $\{100\}_{FCC}\|\{100\}_{FCC}$ and $\{111\}_{FCC}\|\{111\}_{FCC}$ interfaces as a function of reduced temperature T/T_c. These data show that at relatively low misfits (0.2%) the compositional contribution to the interfacial energy dominates, but that for higher values of misfit (2.0%), the structural contribution γ_s^{SS} can be more than twice γ_c^{SS}. With an increase in

Table 1 Comparison of the DLP model compositional energy and HL structural energy for $\{100\}_{FCC}\|\{100\}_{FCC}$ and $\{111\}_{FCC}\|\{111\}_{FCC}$ interfaces as a function of reduced temperature T/T_c; RT = room temperature. From Spanos (1989)

T/T_c	γ_C^{SS}		$\gamma_S^{SS}(\delta^{RT} = 0.2\%)$		$\gamma_S^{SS}(\delta^{RT} = 2.0\%)$	
	$\{100\}$	$\{111\}$	$\{100\}$	$\{111\}$	$\{100\}$	$\{111\}$
0.25	161.9	144.9	45.5	47.3	298.2	318.1
0.50	111.3	107.5	40.1	41.7	265.9	283.2
0.75	45.2	45.3	29.7	30.8	207.2	219.7
0.90	12.4	12.4	21.1	21.9	147.6	155.6

temperature, there is a large decrease in the compositional part of the interfacial energy as in **Figure 56**, while the structural energy drops off much more slowly. Thus, at higher temperatures, the compositional component should provide a much smaller contribution to the total interfacial energy of a partly coherent interface γ_{sc}^{SS} than the structural component. The structural energy is lowest for the $\{100\}$ interface while the $\{111\}$ interface has the lowest compositional interfacial energy.

The temperature dependence of the anisotropy of the compositional component of the energy for these interfaces was also found to be much greater than for the structural part. This is illustrated by the $\{011\}$ polar γ-plots for the compositional and structural parts of the interfacial energy shown in **Figure 71** (Spanos, 1989). These plots were constructed from 180 different values for γ_s^{SS}. The absence of facets on the two-dimensional Wulff construction of the compositional polar γ-plot at $T/T_c = 0.9$ follows directly from the rapid decrease in the anisotropy of γ_c^{SS} as T_c is approached (LeGoues et al., 1983). In contrast, at the same temperature, facets are still present on the two-dimensional Wulff construction for γ_s^{SS}. Also, the bumpiness of the γ_c^{SS} plot is because of the fact that rather small changes in orientation can result in quite different dislocation configurations at the interface (Spanos, 1989). In these studies, the HL analysis (Hirth and Lothe, 1982) for γ_s^{SS} was found to be more reliable at high values of misfit because it accounted for interaction of the misfit dislocations more rigorously than in the van der Merwe (1963) analysis.

Results from analyses of the $\{100\}_{FCC}\|\{100\}_{FCC}$ interface are shown graphically in **Figure 72** (Spanos 1989) to illustrate the relative magnitudes of γ_c^{SS}, γ_s^{SS} and γ_{sc}^{SS} (Eqn (69)) as a function of T/T_c for two levels of room temperature misfit δ_{RT}. The structural component is greater than the compositional component at all levels of T/T_c for $\delta_{RT} \geq 2\%$. On the other hand, at $\delta_{RT} = 0.2\%$, $\gamma_c^{SS} > \gamma_s^{SS}$ for all values of T/T_c less than about 0.82. It is important to note that γ_c^{SS} depends on the temperature and shape of the miscibility gap in the regular-solution model and this can change appreciably from one system to another, thus influencing the ratio $\gamma_c^{SS}/\gamma_s^{SS}$. These results suggest that γ_c^{SS} is most likely to influence the facets which form at interfaces at temperatures below 0.25 T_c, while at temperatures greater than 0.75 T_c where there is no compositional anisotropy, the structural part of the interfacial energy may lead to interfaces that are still heavily faceted. At intermediate temperatures, both components contribute significantly to the equilibrium shape of the facets at heterophase interfaces.

14.3.3.4 *Atomistic Modeling of Partly Coherent Interphase Boundaries*

Several detailed calculational and experimental studies have been performed on the structure and energy of partly coherent interphase boundaries in metals (Gao et al., 1989; Rogers et al., 1990). The atomistic studies were performed on a number of low-index orientations between two metals using EAM potentials and the Monte Carlo method was employed to include temperature in some

(a)

(b)

(c)

(d)

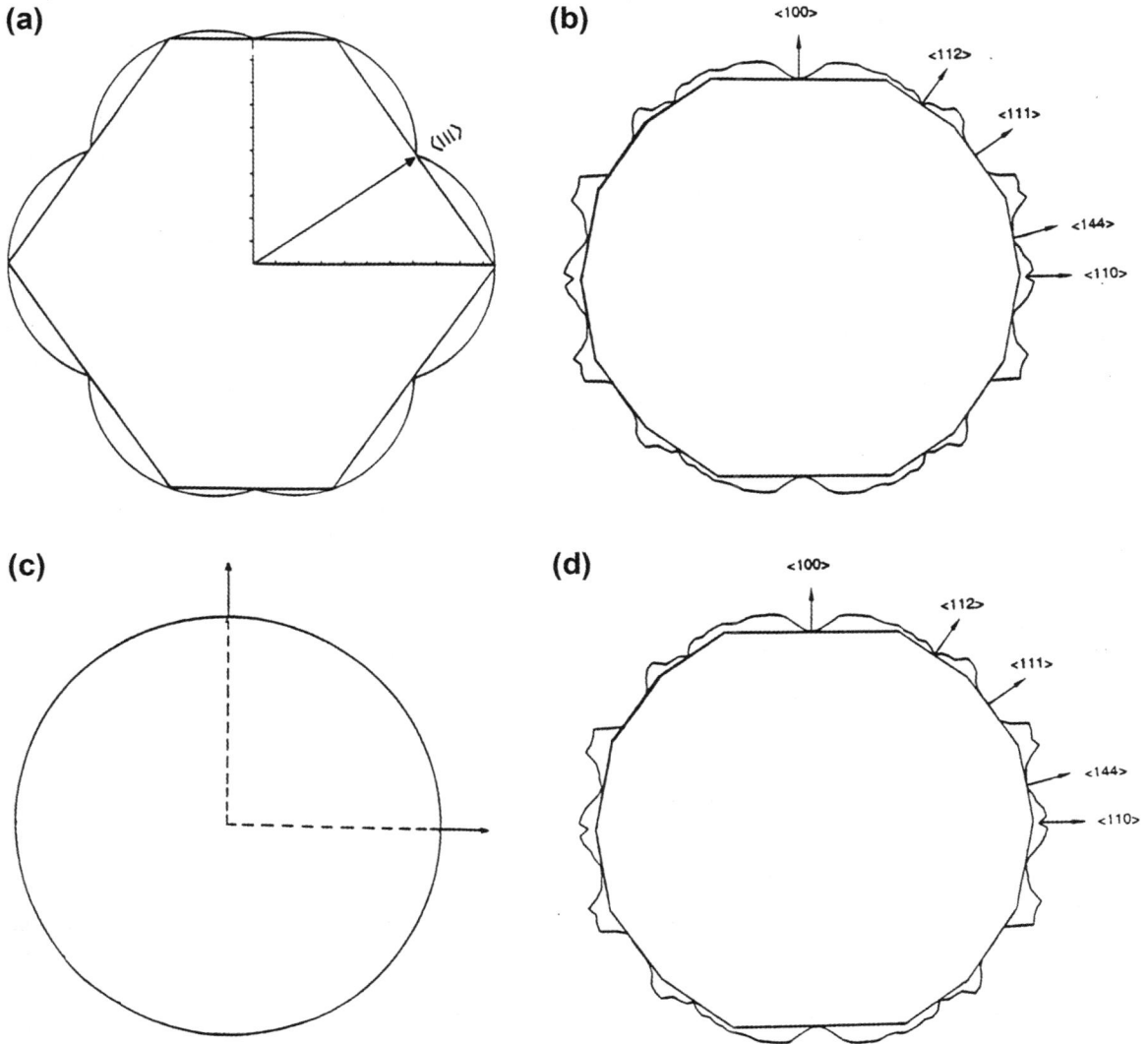

Figure 71   Comparison of (011) sections through polar γ plots for γ_c^{SS} and γ_s^{SS} estimated from the DLP model and HL analysis at (a,b) $T/T_c = 0$ and (c,d) $T/T_c = 0.9$. The plots were normalized by the minimum value of the plot. Adapted from Spanos (1989) with permission.

calculations (Rogers et al., 1990). The predictions of the EAM calculations at 0 K were confirmed experimentally using the rotating crystallite method (Fecht and Gleiter, 1985; Herrmann et al., 1976). In this technique, small metal crystals ($\sim$0.1 μm) deposited on a single-crystal metal substrate are annealed until they rotate into their equilibrium OR. Some of the more important findings from these studies are summarized below.

Figure 73 (Gao et al., 1989) shows a plot of the interfacial energy γ_{sc}^{SS} versus rotation angle θ for a (001) twist interphase boundary between Ag and Ni with a cube-on-cube OR. This twist interphase

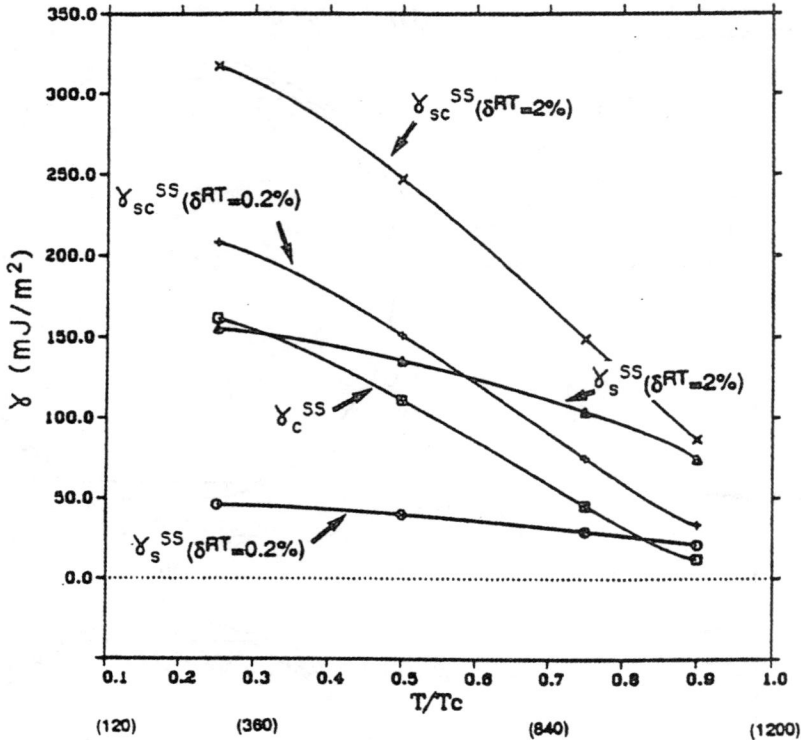

Figure 72   The compositional interfacial energy γ_c^{SS} (Eqn (64)) the van der Merwe structural energy γ_s^{SS} (Eqn (77)) and the total interfacial energy $\gamma_{sc}^{SS}(= \gamma_c^{SS} + \gamma_s^{SS})$ of the $\{100\}_{FCC}\|\{100\}_{FCC}$ interface as a function of temperature T/T_c. The absolute temperatures in Kelvin are shown in parentheses. Adapted from Spanos (1989) with permission.

boundary is similar to the twist grain boundary shown in **Figure 14**, except that the crystals on either side of the interface are Ag and Ni, and because Ag and Ni have significantly different lattice parameters ($a_{Ag} = 0.4086$ nm and $a_{Ni} = 0.3524$ nm), there is an orthogonal set of edge-type misfit dislocations present in the interface when $\theta = 0$. This initial dislocation structure then transforms into a set of mixed dislocations with an increasing screw component as θ increases. The elements Ag and Ni have a large positive enthalpy of mixing and should display very limited mixing across the interface.

Several features are apparent from **Figure 73**. One is that the interfacial energy varies with twist angle, similar to the grain boundaries shown in **Figures 12 and 37**. In addition, there is a deep energy minimum at $\theta = 26.56°$ and several shallower minima at other orientations. The deep minimum at $\theta = 26.56°$ corresponds to a $\Sigma = 5/4$ coincidence boundary, as illustrated in **Figure 74** (Gao et al., 1989). (The notation 5/4 is the ratio of the inverse density of coincident sites in crystals 1 and 2.) Thus, coincidence boundaries can occur at heterophase interfaces, similar to grain boundaries, and these may be low-energy orientations. This feature is further illustrated by the similarity of **Figure 74** with **Figure 22**. It is also important to notice that most of the interfacial energies of the Ag/Ni (001) twist interface lie between about 800 and 900 mJ/m^2 and are on the same order of magnitude as the twist grain boundary energies in **Figure 37**. Thus, these are fairly high-energy interfaces. Unlike the homophase twist grain boundaries, there is no energy minimum at $\theta = 0°$ because the interphase boundary is still partly coherent.

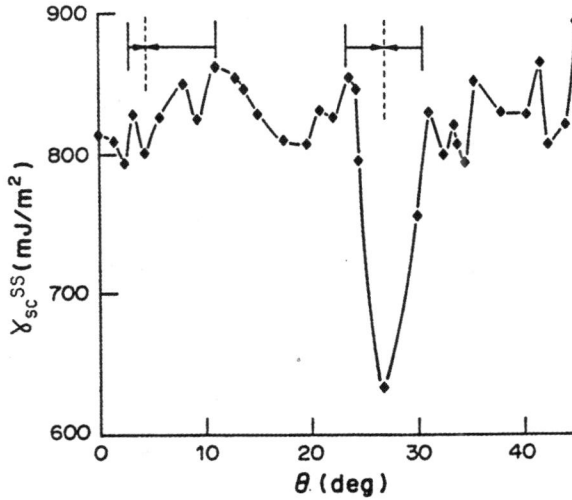

Figure 73 Relaxed interfacial energies as a function of twist angle for a (001) partly coherent interface between Ag and Ni with a cube-on-cube OR. Adapted from Gao et al. (1989) with permission.

Gao et al., (1989) also examined the detailed atomic relaxations that occurred at the Ag–Ni interface as a function of twist angle and found that relaxation occurs in both the Ag and Ni crystals. In particular, atoms located at coincidence sites relaxed perpendicular to the boundary but not parallel to it. All other atoms had components of relaxation both parallel and perpendicular to the boundary. These relaxations gave rise to a corrugation of the atomic layers parallel to the interface that extended four to five atomic planes into the crystals. Different degrees of strain localization were observed for the Ag and Ni layers because of the different elastic properties of the metals. The misfit dislocation networks always passed midway between the coincident points. The rotating crystal experiments also performed by these investigators confirmed the presence of a deep energy minimum at $\theta = 26.56°$ in agreement with the interfacial energy calculations.

In a second set of calculations and experiments, Gao et al. (1989) determined the interfacial energies of other possible partly coherent interfaces and ORs between Ag and Ni crystals. A summary of their results is given in **Table 2**. The most important features for these low-index interfaces are listed below.

(i) The lowest interfacial energy is found for the $(111)_{Ag}\|(111)_{Ni}$ interface with $\theta = 0°$, that is, with parallel close-packed atomic rows, and the second lowest occurs for the $(111)_{Ag}\|(001)_{Ni}$ interface with parallel close-packed atomic rows.

(ii) The interfaces with parallel close-packed atomic rows, that is with $\theta = 0°$, are always energetically favored except for the $(111)_{Ag}\|(110)_{Ni}$ interface.

(iii) The interfaces formed by joining low-energy solid–vapor surfaces, for example the (111) planes, are associated with low interfacial energies.

(iv) Inverting the crystals at the interface, for example changing the $(111)_{Ag}\|(001)_{Ni}$ interface to the $(111)_{Ni}\|(001)_{Ag}$ interface, changes the energy.

These results are entirely consistent with the interfacial behavior discussed for grain boundaries in Section 14.2.1.5.2, emphasizing the importance of nearest-neighbor bonding in determining the ORs and interfacial energies at interfaces. Notice that the interfacial energy of the $(111)_{Ag}\|(111)_{Ni}$ interface

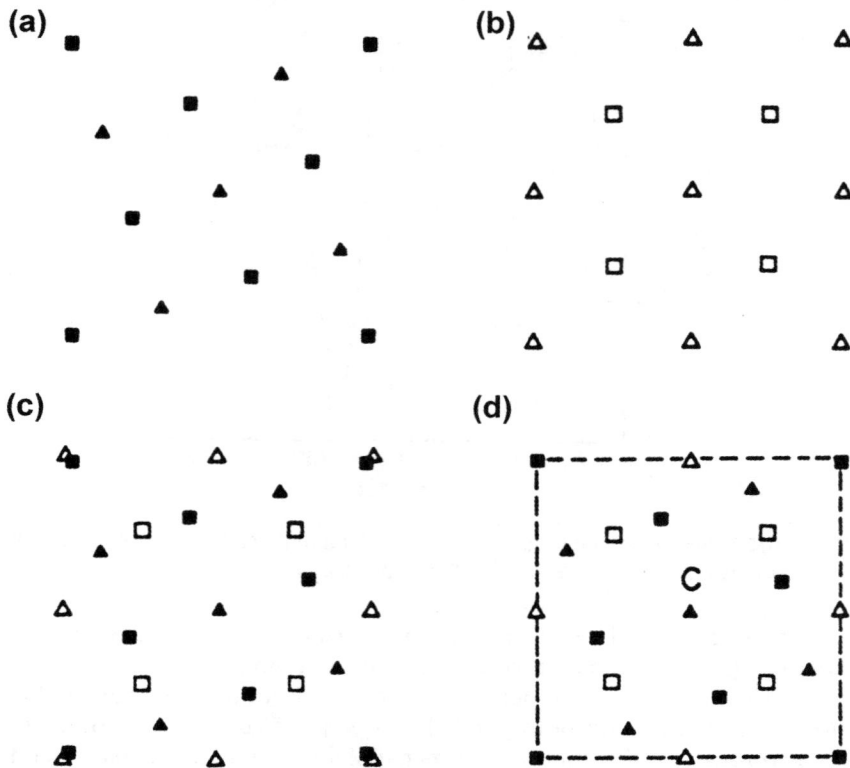

Figure 74   Unrelaxed structure for a $\Sigma = 5/4$ (001) twist interphase boundary between Ag and Ni. (a) Unit cell of the Ni layer, (b) unit cell of the Ag layer, (c) two-dimensional near coincidence structure with a 3.3% misfit between the crystals, and (d) exact coincidence at the interface after uniform deformation to accommodate the misfit. Adapted from Gao et al. (1989) with permission.

Table 2   Calculated interfacial energies of various low-index Ag–Ni interfaces. From Gao et al. (1989)

Ag	(001)		(110)		(111)	
Ni	0°	90°	0°	90°	0°	90°
(001)	814	814	1124	1124	437	437
(110)	995	995	828	1214	988	718
(111)	670	670	670	960	416	468

($416 \, mJ/m^2$) is about half that of the other interfaces. Crystallite rotation experiments performed for Ag droplets on Ni substrates confirmed the predictions of the EAM calculations in these orientations shown in **Table 2**. The atomic relaxations at these interfaces are examined in detail by Gao et al. (1989) and not discussed here.

One last result of EAM Monte Carlo calculations that were used to examine the structure and properties of a partly coherent Cu–Ag (001) interphase boundary with the crystals in a cube-on-cube

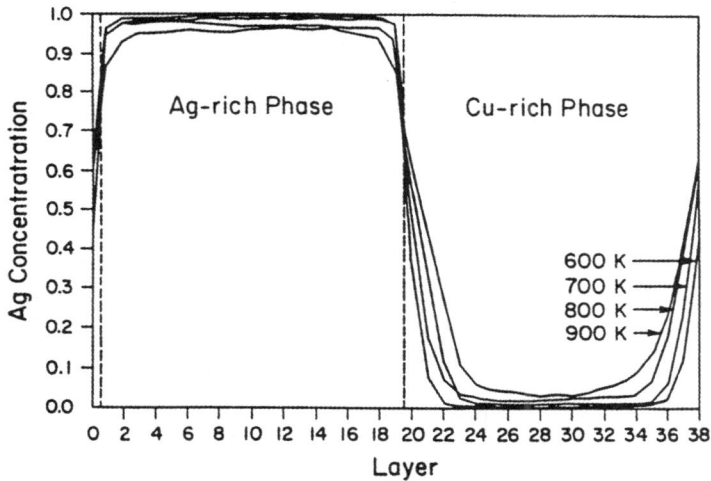

Figure 75 Ag concentration in each atomic layer as a function of layer number, for four different temperatures. The vertical dashed lines represent the locations of the physical interfaces, where the number of atoms per layer changes from that appropriate to a Cu-rich slab to that of an Ag-rich slab. Adapted from Rogers et al. (1990) with permission.

OR (Rogers et al., 1990) is discussed. This interface is interesting because Cu and Ag display mutual solid solubility with temperature and therefore, there should be some compositional mixing at the interface superimposed on the dislocation network necessary to accommodate misfit between the crystals. This is in fact observed in the calculations, as illustrated in **Figure 75** (Rogers et al., 1990). In this figure, it can be seen that the width of the composition profile associated with the interface increases with increasing temperature in the range of 600–900 K, in accordance with the predictions of both the continuum and DLP models of coherent interfaces (Section 14.4.2). In contrast to the profile of a perfectly coherent interface (**Figure 4.3**), the composition profile of the partly coherent interface is asymmetric in **Figure 75**. This interface is associated with a Gibbsian excess of Ag if the position of the dividing surface is chosen at the physical interface. Detailed examination of the atomic positions revealed that the distribution of Ag and Cu in the layers was not uniform, with the larger Ag atoms tending to cluster in regions of energy at the interface and the smaller Cu atoms in regions of compression. It was also observed that the interfacial dislocations were localized and symmetrical at 600 K, but that they tended to wander with increasing temperature and appeared to loose their identity between 800 and 900 K, possibly indicating the transition to an incoherent interface in that temperature range. The overall features of this interface are similar to those revealed by the regular-solution calculation discussed in Section 14.3.1.

14.3.4 Interfaces Between Phases With Different Bravais Lattices

The previous Sections 14.3.1–14.3.3 considered coherent and partly coherent interfaces between two phases that had the same crystal structure but different compositions. The phases also had the same orientation, often referred to as a parallel, or cube-on-cube OR. Many heterophase interfaces form between two phases that have different Bravais lattices and one needs to understand the behavior of these types of interfaces. They are the most complicated interfaces, but it is shown that they display similar behavior in an attempt to minimize the interphase boundary energy. As before, our discussion is limited to metal–metal interfaces, but it is important to realize that the same principles and behavior apply to other heterophase interfaces involving various combinations of metal, ceramic, semiconductor and polymer materials as well (Mauritz et al., 1978; Fecht and Gleiter, 1985; Sinclair et al., 1990; Howe, 1993a).

Experimental studies of crystal growth show quite clearly that preferred ORs exist between different crystal structures when they are forced into intimate contact. This phenomenon has been extensively studied (Bruce and Jaeger, 1978; Dahmen, 1982; van der Merwe, 1982; Kato, 1991; Fecht and Gleiter, 1985), and it turns out that the experimental observations can be explained by purely geometrical atomic row matching between the two structures. The dominant principle that applies is that two structures usually try to match their most densely packed planes and directions in that plane to minimize the interfacial free energy, as first proposed by Frank (1953).

Consider the case of a bicrystal formed by the placement of a close-packed FCC (111) monolayer with a nearest-neighbor distance a_{FCC} on top of a close-packed (110) substrate of a BCC structure with nearest-neighbor distance a_{BCC}. For the orientation shown in **Figure 76(a)** (Dahmen, 1982), one sees that consecutive atomic rows parallel to the y axis of the two lattices can match if the monolayer is slightly expanded along the x axis, or conversely, matching along x can occur with a contraction along y. Alternatively, for one particular value of $r = a_{FCC}/a_{BCC} = 1.0887$, it is possible to achieve matching along the most close-packed atomic rows of the structures by rotating the monolayer through 5.26°, as shown in **Figure 76(b)**. This is in fact, what is often found experimentally (Dahmen, 1982; Kato, 1991).

It seems reasonable that the atomic row matching condition must be related to the requirement that the maximum number of monolayer atoms sit in the minima of the substrate corrugation potential.

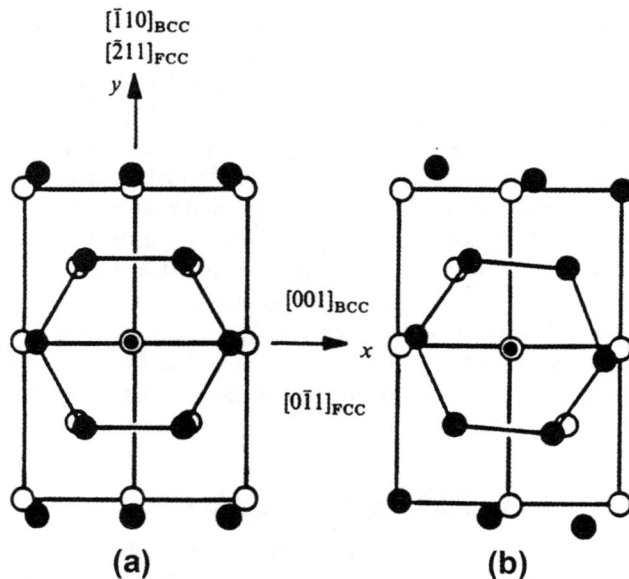

Figure 76 Schematic of an FCC (111) monolayer (filled circles) on a BCC (110) substrate (open circles). (a) The $[01\bar{1}]_{FCC}$ direction is parallel to $[001]_{BCC}$ and (b) a 5.26° rotation relative to (a). The lattice constants were chosen to produce atomic row matching in (b). Adapted from Dahmen (1982) with permission.

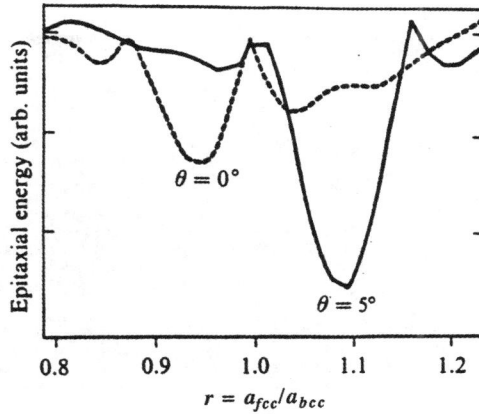

Figure 77   Calculation of the total interaction energy for rigid FCC (111)–BCC (110) epitaxy as a function of the ratio a_{FCC}/a_{BCC} for two angles of orientation θ relative to **Figure 76**. Adapted from Kato (1991) with permission.

This turns out to be true. To describe this process further, imagine that each pair of atoms in the monolayer and substrate interacts with one another by a conventional 6–12 LJ potential. If one evaluates the total energy of the $(111)_{FCC}\|(110)_{BCC}$ system as a function of the parameter $r = a_{FCC}/a_{BCC}$ and the orientation angle θ at the equilibrium separation of the lattices, the energy turns out to be relatively independent of r except at the angles $\theta = 0$ and $5.26°$, as illustrated in **Figure 77** (Ramirez et al., 1984). The deepest minimum corresponds precisely to the close-packed atom row matching condition. The OR between the FCC and BCC lattices is called the Kurdjumov–Sachs (KS) OR, named after the founders (Kurdjumov and Sachs, 1930), and it is commonly found between FCC and BCC phases in steels and other metal alloys. The OR is expressed fully as

$$(111)_{FCC} \| (110)_{BCC}; \quad [0\bar{1}1]_{FCC} \| [1\bar{1}1]_{BCC}. \tag{81a}$$

The principal minimum for $\theta = 0°$, which corresponds to the so-called Nishiyama–Wasserman (NW) OR (Wassermann, 1933; Nishiyama, 1934), is simply the atomic row matching condition parallel to the x axis in **Figure 76(a)**. The LJ potential predicts no distinct minimum for row matching along the y axis.

The KS and NW ORs are observed for many metal thin-film systems (Bauer, 1983; Kato et al., 1989). This does not mean, however, that the natural lattice constants are in the precise ratios predicted by **Figure 77**. Generally, there is some nonzero misfit δ and in this case, epitaxy occurs when the film distorts to achieve atomic row matching. As the strain energy become significant, a variety of atomic relaxations can occur, such as the formation of domain walls and interfacial dislocations in the case of thicker films, as described in the previous section (Stoop and van der Merwe, 1982).

The KS OR mentioned above matches parallel close-packed planes and directions between the FCC and BCC phases. A similar situation occurs for the case of HCP and BCC crystals, where the close-packed planes have the same symmetries. When the parallel close-packed planes and directions align in the HCP–BCC case, the resulting OR is called the Burgers OR (Burgers, 1959) and it is given as

$$(0001)_{HCP} \| (110)_{BCC}; \quad [11\bar{2}0]_{HCP} \| [1\bar{1}1]_{BCC}. \tag{81b}$$

Figure 78 Experimental HRTEM image of a γ' (Ag$_2$Al) precipitate interface in an Al–Ag alloy with a superimposed simulated image of the interface. The position of the interface between the FCC and HCP phases is indicated by a line. Adapted from Howe et al. (1987) with permission.

An even more perfect example of matching close-packed planes and directions between two crystals can occur for the FCC–HCP case, where the (111) and (0001) planes have the same three-fold symmetry. In the case of a martensitic transformation the two phases may have the same composition, but different crystal structures, such as for the FCC–HCP interface previously shown for the Co–Ni system in **Figure 4**. In a diffusional transformation, the phases may vary in both composition and lattice, as illustrated by a similar example for the FCC–HCP interface in an Al–Ag alloy shown in **Figure 78** (Howe et al., 1987). In both **Figures 4 and 78**, the viewing direction is $[10\bar{1}]_{FCC}\|[11\bar{2}0]_{HCP}$ and the OR between the phases is

$$(111)_{FCC} \parallel (0001)_{HCP}; \quad [11\bar{1}]_{FCC} \parallel [11\bar{2}0]_{HCP}. \tag{81c}$$

This is called the Shoji–Nishiyama (SN) OR (Nishiyama, 1978) and the parallel close-packed planes are edge-on and horizontal in **Figure 78**. The IP (often called the habit plane when referring to plate or lath-shaped precipitates) is the parallel close-packed planes in the two structures. Since there are no broken bonds in the FCC–HCP interfaces in **Figures 4 and 78**, and only wrong compositional bonds in **Figure 78**, these interphase boundaries are expected to have low energies, and values of approximately 15 and 30 mJ/m^2 at room temperature, respectively, have been reported (Ramanujan, 1990; Hitzenberger et al., 1985).

In all of the FCC, BCC, HCP relationships described above, the basic arrangement of atoms in the close-packed planes is a rhombus, and aligning the close-packed planes in the crystals across the interface matches the rhombohedral symmetries in the structures (van der Merwe, 1982; Fecht and Gleiter, 1985; Shiflet and van der Merwe, 1994). Aligning the close-packed rows of atoms in these planes, that is atomic row matching, further reduces the energy of the interfaces, as previously shown (**Figure 77**). This feature has been thoroughly demonstrated in energetic calculations by van der Merwe, 1982 and others (Kato et al., 1989; Ramirez et al., 1984; Gotoh et al., 1989), who show that the ORs above minimize the elastic strain energy of the interfaces, and the relationships among these various ORs have been analyzed by Dahmen (1982). Thus, the interfacial energy of most heterophase interfaces is minimized through atomic row matching generally, and this criterion usually determines the equilibrium OR between two phases. What is not so obvious from these treatments, which focus on matching of the close-packed planes in the structures, is the consequences of atomic row matching when the interphase boundary (habit plane) is not parallel to the close-packed planes. An interface of this type was shown in **Figure 7**, where it is seen that the average $(474)_\gamma$ IP, which is approximately horizontal in the figure, is rotated about 30° counterclockwise from the (very nearly) parallel close-packed planes in the FCC and BCC phases in the figure, which run continuously across the atomic steps in the interface. This type of inclined interface has received considerable attention because it is commonly found in both martensitic and diffusional transformations (Howe, 2006; Howe et al., 2009; Zhang and Kelly, 2009) and it is discussed in the following sections. Frank (1953) recognized that such interfaces match close-packed atom rows across the interface even though the IPs are inclined to the close-packed planes in the crystals. A more recent realization, is that in the absence of being able to match close-packed rows of atoms across a high-index interface, two phases still try to achieve atom row matching between planes of atoms that are not close packed (Howe et al., 2002; Nie and Muddle 2002; Reynolds et al., 2003; Reynolds and Farkas, 2006). This is discussed in a subsequent Section 14.3.6.4.

14.3.4.1 Orientation Relationship Determination
Determining the OR between two phases is a fundamental problem in interface science. Since knowledge of the OR between two phases is essential for determining the IP (or habit plane), various methods that have been proposed to determine the OR are described first, with particular emphasis being placed on a recent methodology developed by (Gautam and Howe 2011, 2013). This is followed by a discussion of high-index IPs.

Experimental studies demonstrate that most systems, such as thin films on a substrate or precipitates in a matrix, favor certain ORs (Zhang and Weatherly, 2005; Rong, 2012). This preference minimizes the energy during nucleation as well as the total energy of the system, by forming low-energy interfaces (Aaronson et al., 1970; Plichta et al., 1976, 1980). A number of different theoretical models based on interfacial energy minimization have been proposed to identify the preferred OR and/or IP (or habit plane); see, for example (Howe, 1997; Sutton and Balluffi, 1995; Zhang and Weatherly, 2005). However, the complex interdependence of the OR and IP on the crystallography and composition of a system makes it difficult to include all the relevant factors in a single model. The advantage of a geometric approach to this problem lies in its ability to identify possible candidates for preferred ORs. Once these orientations are identified, the information can be used with other modeling techniques to obtain quantitative information, such as the interfacial structure and energy. In the case of most geometric models, where it is not possible to calculate the interfacial energy explicitly (discrete lattice models (Wynblatt and Ku, 1977; Lee and Aaronson, 1980) being an exception), a criterion for the minimum interfacial energy is implemented by minimizing the atomic mismatch between the two crystals at the interface. This choice of minimizing the lattice misfit at the interface can be explained by considering various factors that contribute toward the energy of an interface. As seen previously, the interfacial energy can be subdivided into chemical and structural components as in Turnbull (1955). The chemical component results from contributions, such as the nature and density of atomic bonds, the diffuseness of the interface, while the structural component arises mainly because of geometrical considerations between two crystals, for example the elastic misfit strain and/or interface dislocations.

Optimizing atomic mismatch at the interface usually achieves the combined objectives of simultaneously minimizing the number of broken or wrong bonds (chemical) as well as the misfit strain (structural) at the interface, and therefore, such configurations can be considered as strong candidates for preferred ORs and IPs. Based on this argument, a number of different geometric approaches have been proposed to calculate preferred ORs and/or IPs between two crystals. Some of these approaches, such as the O-lattice, CSL and topological model, were previously described. All of these models begin by assuming that the close-packed planes in the phases are parallel, or nearly so, so that the OR is not a completely free variable. There are several other models that make the same assumption, and these are briefly summarized below. This summary is followed by a more thorough description of a method that was recently developed, which is able to predict the OR (or the most likely few candidates) in almost all cases, given only the two crystal structures of the phases as input. In other words, it provides the OR without making any assumptions about the crystal structures involved.

The CSL lattice described in Section 14.2.1.3.1, was one of the earliest attempts to study ORs and IPs using a geometric approach. As mentioned, the model was initially used to study grain boundaries (Bollmann, 1970; Randle, 1996), and with some modifications, was applied to heterophase interfaces (Liang and Reynolds, 1998; Chen and Reynolds, 1997). The CSL model is based upon the observation that certain rotation operations lead to partial self-coincidence of a crystal (Kronberg and Wilson, 1949). The network of overlapping lattice sites defines the CSL. The orientation with highest density of coincident lattice sites is identified as the preferred OR and the plane with highest density of such sites represents a low-energy interface (Bollmann, 1970). The shortcoming of this method is that meaningful CSLs can be formed only for limited orientations of the two crystals and therefore the model samples only discrete points in orientation space. The discrete nature of CSL can be avoided by considering coincidence of equivalent points (points with same fractional coordinates in two crystals) instead of lattice points to construct the space lattice. This generalization, known as the O-lattice theory, which was described in Section 14.2.1.2, has been quite successful in predicting the structure of interphase boundaries containing misfit dislocations between two crystals with a known OR, or when there is near alignment of parallel close-packed planes (Zhang and Weatherly, 2005; Gu and Zhang, 2010; Zhang and Yang, 2011). It is difficult to apply to study the structure of a general interphase boundary though, when there is substantial rotation between the planes across the interface, the planes are not close packed, and/or which may contain arrays of disconnection defects (Hall et al., 1972; Furuhara and Aaronson, 1991; Furuhara et al., 1991; Howe et al., 2009), which are of substantial importance in determining the IP and energy.

Hall et al. (1972) showed that introducing steps at a partially coherent and flat interface between FCC–BCC phases significantly increases the density of coherent atoms at the interface. This observation motivated the structural ledge approach to identify the orientation and structure of low-energy interfaces between phases, by assuming a known OR that aligns the close-packed planes between the phases. The approach is based on maximizing the atomic matching across the interface by introducing structural ledges at the interface (Rigsbee and Aaronson, 1979). Similarly, van der Merwe et al. (1991) and van der Merwe and Shiflet (1994); Shiflet and van der Merwe (1994) calculated the interfacial energy while accounting for both structural ledges and misfit dislocations at the interface, and found a ledged interface to be more stable than a planar interface for all meaningful values of misfit.

One of the more successful methods to determine the IP between two phases with a known OR is based on the relative orientation of reciprocal lattice vectors (electron diffraction spots). According to this method, the IP is defined by the plane normal Δg, which is the difference between two near-coincident, low-index reciprocal lattice vectors (Zhang and Purdy, 1993; Zhang and Weatherly, 1998). This criterion for interface selection eliminates lattice mismatch in one-dimension and therefore offers a very effective way of further reducing the lattice misfit, by optimizing the interface orientation, once the OR between the two phases is fixed. The Δg criterion has been found to be quite accurate in identifying the IP in various alloy systems, and therefore, it is not surprising that this concept, with some variations, has been used in other geometric models such as edge-to-edge (ETE) matching. The ETE approach calculates the preferred OR and IP by matching a pair of directions with similar atom spacings in two close-packed, or nearly close-packed, planes between two crystals. The interplanar spacings of the two planes are also similar, and this minimizes the lattice mismatch at the interface (Frank, 1953; Fecht and Gleiter, 1985; Zhang and Kelly, 1998, 2009; Kelly and Zhang, 1999). The IP calculated using this approach is considered to be approximate and requires application of Δg method to determine the exact interface orientation. Rong (2012) provides a complete summary of the ETE method including numerous examples. Nie (2004, 2005) proposed a similar approach of plane edge matching for OR determination. The procedure involves identification of closely packed planes of the two crystals that can be related by a shear and/or dilatation. A planar interphase boundary between two phases is then defined by the Moiré plane resulting from intersection of at least two sets of the identified closely packed planes.

An approach based on strain energy minimization is widely used to study the crystallography of martensitic transformations. The invariant line strain is one such approach, which uses elastic strain minimization in one direction as the criterion to determine the OR between two phases (Dahmen, 1982; Luo and Weatherly, 1987). Implementation of the model usually assumes the close-packed planes or directions of the two phases to be parallel and determines the orientation resulting in an invariant line as the preferred OR. The crystallographic direction parallel to the invariant line is considered as the direction of easy growth. This model leaves the IP indeterminate, and requires invoking another constraint such as the existence of a second direction with a small misfit, to determine the IP. Similarly, the PTMC considers the invariant plane of the total shape transformation as the IP between the parent and product phases in a martensitic transformation (Bowles and Mackenzie, 1954; Wechsler et al., 1953). Disagreement between the PTMC results and experimental observations (Wayman, 1964) led to introduction of a dilatational parameter (Wayman, 1964; Mackenzie and Bowles, 1957) to improve agreement. Another source of disagreement between experiment and PTMC results is attributed to relaxation of the interface into a glissile defect structure resulting in strains and rotations, which in turn may require a change in OR and IP (Pond et al., 2003).

Pond and Hirth (1994); Pond et al., (2003, 2006) have proposed a topological model to calculate the OR and IP between a parent and product phase in martensitic transformations. This model was discussed in Section 14.2.1.4 and illustrated in **Figure 33**, for example. The model was motivated by the experimental observation of the atomic scale structure of martensitic interfaces, which consists of coherent terraces reticulated by an array of line defects (Wayman, 1964; Chen and Chiao, 1985; Mahon et al., 1989; Howe et al., 2009). The approach involves identifying planes with an equal area-density of atoms projected on to the plane. These planes form the coherent terrace of the interface and the coherency strain in one direction is accommodated by an array of parallel disconnection while twins or dislocations may accommodate the strain in direction parallel to disconnection lines. Because of the step character of the disconnections the overall habit plane is inclined to the terrace planes (Pond et al., 2003).

The models above follow procedures for lattice or atom matching in direct space. A similar approach has been implemented in reciprocal space that attempts to match the lattice periodicity while using the reciprocal lattice of the two phases. van der Merwe (1982) proposed a two-dimensional model based on the reciprocal lattices of the two crystals to determine ORs in epitaxial systems. This concept was later extended and applied to explain some of the commonly observed ORs between precipitates and matrices in solids (Braun, 1987; Braun and van der Merwe, 2002). In this approach, the substrate surface is replaced by a two-dimensional potential surface and the overgrowth (thin film) is modeled as a monolayer of atoms with the same structure as the bulk phase. Numerical implementation of the model involves rotating a rigid reciprocal lattice of the overgrowth with respect to the substrate surface to identify a low-energy orientation. Generalization by Braun (1987) relaxed the rigid lattice constraint by allowing for homogeneous strain in the overgrowth. The homogeneous nature of the strain was further relaxed using the finite element method to account for local strains and thereby, misfit dislocations in the model. Results obtained from this model show that energy minima correspond to orientations where the wave vector of the Fourier expansion coincides with a reciprocal lattice translation vector of the substrate, which is similar to atom row matching in direct space.

The static distortion wave model calculates the interaction energy between a substrate and overgrowth using a single atom interaction potential. According to this model, the average adatom energy (U) as a function of orientation can be approximated as (Novaco and McTague, 1977; McTague and Novaco, 1979; Markov and Stoyanov, 1987)

$$U = V_0 + \sum_G \sum_g V_G \delta_{G,g} - \frac{1}{2\Omega} \sum_G \sum_g \sum_{\Delta g \neq 0} V_G^2 \delta_{G,g+\Delta g} \left[\frac{(G \times \Delta g)^2}{\mu} + \frac{(G \cdot \Delta g)^2}{K + \mu} \right] \frac{1}{\Delta g^4}, \tag{82}$$

where V_0 is a constant background potential, V_G (negative in magnitude) is the Fourier coefficient and is related to the reciprocal vector G of the substrate, $1/K$ is the two-dimensional compressibility of a monolayer, μ is the Lamé constant, and Δg is the deviation of G from g (a reciprocal lattice vector of the overgrowth), that is $\Delta g = 0$ for $G = g$. The first term in Eqn (82) is constant, while the second and third terms are orientation dependent. The second term in the equation calculates the overlapping potential for reciprocal lattice pairs that are coincident, while the third term calculates the overlapping potential for near-coincident reciprocal lattice points. The energy minimization for a given substrate overgrowth combination requires minimizing the second and third terms of the equation. The absolute value of the second term is much larger than that of the third term; however, it goes to zero if the reciprocal lattice vectors (G and g) are not coincident. Hence, the energy is minimized for an orientation that results in the maximum number of coincident reciprocal lattice points. The condition of coincident reciprocal lattice points, easily satisfied in homophase systems, is not assured for heterophase systems. In such situations, the third term of the equation dictates the energy minimization. The third term requires minimization of the deviation, Δg, between a pair of reciprocal lattice spots. It is important to mention that the energy will be lower when multiple pairs of reciprocal lattice points satisfy this criterion rather than for near-coincidence of only a few spots.

Ikuhara and Pirouz (1996) and Pirouz et al. (1998) proposed a simplified model that is similar to the reciprocal lattice model of Braun and the static distortion wave approach. In this model, the OR is based on the near-coincidence of reciprocal lattice points in three dimensions. This criterion finds support from the observation made by van der Merwe and Braun that the energy minima always correspond to orientations where wave vectors of the Fourier expansion (substrate reciprocal vector) coincide with reciprocal lattice translation vectors of the overgrowth. Similarly, the static distortion wave model indicates that low-energy interfaces correspond to orientations resulting in near coincidence of multiple reciprocal lattice points. Numerical implementation of this model involves generating the three-dimensional reciprocal lattices of the two crystals. The dimensionless reciprocal lattice points are then replaced with reciprocal lattice spheres of fixed radius with their centers at the reciprocal lattice points. One of the reciprocal lattices is then rotated with respect to the other and the common volume between all overlapping spheres is added. The overlapping volume between two reciprocal lattice spheres represents the extent of overlap of the potential fields associated with two reciprocal lattice points. The orientation that maximizes the total overlap volume is selected as the preferred OR. The advantage of such as approach over the previous two-dimensional reciprocal lattice model lies in the relative ease with which one can explore the entire orientation space for the preferred OR. However, replacing the interaction potential associated with each of the reciprocal lattice points (coefficients of the wave vectors of the Fourier series in the Braun model) with spheres of the same size effectively assumes that the potential distribution for all of the planes in both crystals is the same.

While these geometric models for determining the OR and IP in crystalline solids have been successful in explaining experimentally observed ORs and IPs in many material systems (see, e.g. Zhang and Weatherly, 2005; Zhang and Kelly, 2009), none implicitly accounts for the atomic species matching across the interface and therefore, the models tend to minimize the interfacial energy by maximizing the number of atomic bonds across the interface, regardless of the type of atoms or nature of their bonds. These limitations were recently overcome by Gautam and Howe (2011, 2013), who developed a model to determine the OR and IP between two crystalline solids given only their crystal structures as input. The approach is based on the near coincidence of diffraction spots in three dimensions and is an extension of Ikuhara's reciprocal lattice method (Ikuhara and Pirouz, 1996). The model includes successful concepts of existing geometric models, but also incorporates atom types to develop a comprehensive approach to predicting preferred ORs and IPs. The procedure to find the OR involves

(i) constructing the three-dimensional reciprocal lattices of the two crystals,
(ii) calculating the X-ray diffraction intensity, that is the structure factor, associated with each reciprocal lattice point,
(iii) distributing the calculated intensities about each reciprocal lattice point,
(iv) rotating one reciprocal lattice with respect to the other and calculating the total overlapping intensity of the diffraction spots, and
(v) finding the OR that gives the maximum total overlapping intensity.

The criterion for OR selection is based on the assumption that the intensity of the diffraction spot from a plane is also a measure of the electron density associated with the atomic plane. This is consistent with the static distortion wave and Braun models above, which show that the energy of an interface is proportional to the sum of the overlapping atomic potentials as (Novaco and McTague, 1977; McTague and Novaco, 1979; Markov and Stoyanov, 1987) so that interfaces with high overlapping potentials have a low interfacial energy. Based on this observation and the assumption of diffraction intensity as a measure of the atomic potential, the preferred OR is identified as the orientation that maximizes the total overlapping intensity. Since the bonding information, for example atom types, bond lengths and angles, is inherently included in the structure factor calculations through the crystal structure data, the nature and strength of the atomic interactions in the phases is taken into account in the diffraction intensity calculation. For example, the tetrahedral bonding in Si is reflected in the structure factor of a diamond–cubic crystal, which has different reflections compared with a close-packed FCC crystal.

Maximizing the total overlap intensity to determine the OR between two phases effectively attempts to minimize the three-dimensional misfit between the phases. This is achieved whenever the orientation between the two crystals includes three or more noncoplanar overlapping spot pairs. This criterion for OR selection is effectively similar to the static distortion wave model, where the energy of an interface is lower for orientations with many near-coincident reciprocal points as opposed to a few coincident points. Therefore, if a small deviation from an orientation allowing the exact coincidence of a pair of high-intensity spots (usually close-packed planes) significantly increases the number of overlapping spot pairs, then the OR selection criterion favors such a deviation. This effectively achieves the atom row matching condition previously described. This situation differs from models based on one or two-dimensional atomic mismatch optimization, that is almost all other models, where the close-packed planes or directions are usually assumed to be parallel.

Once the OR between two phases has been determined, the corresponding IPs can be calculated from the relative positions of the overlapping spots. Mismatch between two overlapping reciprocal lattice spots can be given by the difference in their position vectors g_1 and g_2. In direct space this condition can be visualized as an interface formed by the edge-on matching of the atomic planes corresponding to the two vectors. There are numerous possible ways to cut an interface between these two matching planes whose spatial relationship is fixed by the OR. One method is to draw an interface such that the mismatch vector, $\Delta g (= g_1 - g_2)$ is normal to the interface formed by edge-on matching of the two atomic planes. As shown in **Figure 79** (Gautam and Howe, 2011), such an interface will result in exact matching between the two atomic planes at the interface, that is, ETE matching. This procedure for IP selection is similar to Δg approach proposed by Zhang and Purdy (1993; Zhang and Weatherly (2005), and can be directly combined with their approach without modification. Since optimal matching in three dimensions is taken into account in the OR determination, this ensures that the atomic mismatch along other directions, if present, is small. Hence, choosing a Δg plane as an IP eliminates mismatch along one direction while the OR determination ensures small misfits along other directions (in the plane).

Three-dimensional matching of the diffraction spots generally results in more than one overlapping spot pair, depending upon the lattice parameters of the two crystals. The energy of an interface (both chemical and structural) can be qualitatively assessed from the magnitudes of the average reciprocal lattice vectors g_1 and g_2 of the overlapping diffraction spot pairs, since these relate to the corresponding atomic potentials. For a given OR, one can draw the Δg vectors associated with each overlapping spot pair and scale their magnitude according to the corresponding average reciprocal lattice vector $((g_1 + g_2)/2)$ of the overlapping spot pair. The resulting vectors effectively represent the orientations of all possible interfaces between the precipitate and matrix, with the magnitudes of the vectors being proportional to the energies of the interfaces. An interface corresponding to each spot pair is thus a possible interface and can be included in calculation of a precipitate shape. **Figure 80** illustrates all possible Δg vectors in two dimensions for a cube-on-cube OR between two crystals with different lattice parameters a.

The plot in **Figure 80** (Gautam and Howe, 2011) in some respects is similar to the Wulff plot (Herring, 1951, 1953), and therefore, using a similar approach, the smallest volume enclosed by the perpendicular bisector planes of all of the g vectors should correspond to the equilibrium shape of a precipitate (Wulff, 1901). It has been shown that for faceted particles, the area (or length) of a crystal facet varies inversely with the surface or interfacial energy, and the shapes of faceted crystals are therefore dominated by facets having low energy. The model of Gautam and Howe (2011) incorporates

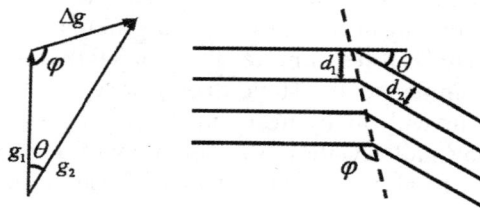

Figure 79 Schematic showing the orientations of two reciprocal lattice vectors g_1 and g_2, and their corresponding planes with interplanar spacings of d_1 and d_2, respectively. The interface depicted by the dashed line is normal to the vector Δg and matches the interplanar spacings d_1 and d_2 at the interface. Adapted from Gautam and Howe (2011) with permission.

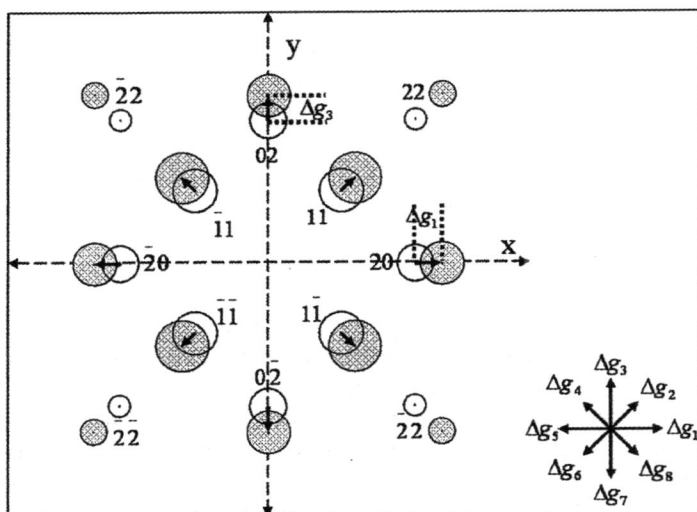

Figure 80 Overlapping reciprocal lattice spheres projected in two dimensions for a cube-on-cube OR between two FCC crystals with 10% lattice mismatch. The mismatch vectors $\Delta\mathbf{g}$ also represent the IPs, that is the IP-normals, formed from the overlapping diffraction spots. Adapted from Gautam and Howe (2011) with permission.

this phenomenon by employing a Voronoi or Weigner–Seitz construction (Wigner and Seitz, 1933; Slater, 1934) and using the space lattice defined by the set of $\Delta\mathbf{g}$ vectors from overlapping diffraction spot pairs. An example of this construction based on the OR defined in **Figure 80** is shown in **Figure 81** (Gautam and Howe, 2011).

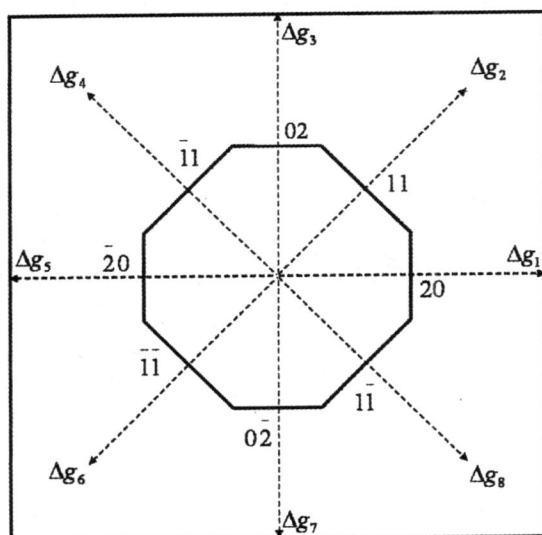

Figure 81 Voronoi construction on a space lattice defined by the $\Delta\mathbf{g}$ vectors in **Figure 80**. The magnitude of these vectors is proportional to the average $\mathbf{g}$ vector of the corresponding overlapping spots. The area enclosed by solid lines is the approximate precipitate shape. Adapted from Gautam and Howe (2011) with permission.

It is important to note that while the method above was described for determining the interfaces between a precipitate and matrix, the same procedures can be applied to grain boundaries in homo-phase systems (Randle, 1996; Gottstein and Shvindlerman, 2010), high-index interphase boundaries such as in the massive transformation (Li et al., 2004; Yanar et al., 2002), or thin films (Braun and van der Merwe, 2002).

Having determined the OR using a method such as the one above, it is now possible to further describe aspects of high-index interfaces between the two crystals, again using some of the previous methods described, or using other analyses that are described below.

14.3.4.2   High-Index Interface (Habit) Planes

A number of different models have been proposed to explain the origin of high-index interphase boundaries that occur between phases in martensitic and diffusional transformations. The topological model, which is particularly well suited to martensitic transformations because of its condition of zero flux in Eqn (35) as well because of its mechanistic approach to interfacial structure, was previously described in some detail in Section 14.2.1.4. This section describes several different approaches to describe the formation of high-index IPs that provide additional insight into the physics governing this process. Other treatments that have been used with considerable success, but which do not provide additional physical insight, are summarized below, but not described in detail.

Dahmen (1982) has shown that the rows of atoms that superimpose between the two crystals in **Figure 76(b)** can be viewed as an invariant line between the two structures. In a continuum sense, an invariant line is a direction that is both undistorted and unrotated when one crystal structure transforms into another. This is illustrated schematically in **Figure 82** (Dahmen, 1982). In **Figure 76(a)** and **Figure 82(a)** one can imagine that a unit cell of the BCC lattice (open circles) is transformed into a unit cell of the FCC lattice (filled circles) by an expansion e_{22} along the y direction and a contraction e_{11} along the x direction. This defines both the lattice correspondence and the transformation strain. The two axes x and y are the principal axes of the transformation and this operation may be written as a diagonal matrix:

$$A = \begin{bmatrix} e_{11} & 0 \\ 0 & e_{22} \end{bmatrix}$$

(83)

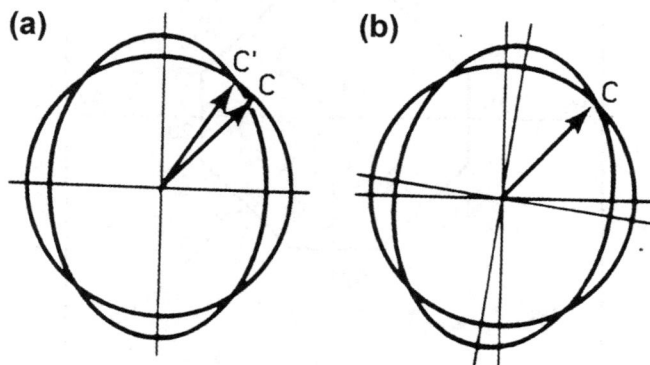

(a)   **(b)**

Figure 82   The lattice strain of **Figure 76** in continuum representation. (a) The circle (BCC plane) is transformed into an ellipse (FCC or HCP plane). (b) A small rotation bringing **C** and **C′** into coincidence produces an invariant line. Adapted from Dahmen (1982) with permission.

when related to the principal coordinate system. The length of a vector $\mathbf{u}$ changes to $|\mathbf{v}| = |\mathbf{Au}|$ during the transformation. This is the well-known description of the deformation of a circle ($u_x^2 + u_y^2 = 1$) into an ellipse ($v_x^2/a + v_y^2/b = 1$) with major axes a and b. An illustration of the continuum deformation corresponding to **Figure 76(a)** is shown in **Figure 82**, where the circle represents the BCC plane and the ellipse the FCC plane. Any radius vector on the circle changes both length and direction during the transformation. A special vector is the one ending at $\mathbf{C}$ because it changes direction from $\mathbf{C}$ to $\mathbf{C'}$ but its length is preserved. If the directional change from $\mathbf{C}$ to $\mathbf{C'}$ is compensated by a rigid body rotation of the transformed structure (ellipse) it becomes an invariant line (i.e. an undistorted and unrotated direction during the transformation). In terms of matrix notation, the total deformation in **Figure 82(b)** can be decomposed into the shape deformation $\mathbf{A}$ and a rigid body rotation $\mathbf{R}$, or $\mathbf{RA}$. Because the rows of atoms along the $[011]_{FCC}\|[111]_{BCC}$ direction in **Figure 76(b)** exactly superimpose they satisfy this criterion, that is they form an invariant line on an atomic level.

Because of the special choice of lattice parameters in **Figure 76(b)**, the invariant line coincides with the low-index close-packed direction. This is not usually the case and in general the invariant line is a nonrational direction. Dahmen (1982) used matrix algebra to find the angle of rotation that is necessary to form an invariant line as a function of the ratio of the principal distortions e_{11} (a) and e_{22} (b) in the two-dimensional case. Since e_{11} and e_{22} have a constant ratio for a given FCC–BCC or HCP–BCC transformation, the angle θ simply becomes a function of the lattice parameter ratios $r = a_{FCC}/a_{BCC}$ or $\sqrt{2}a_{HCP}/a_{BCC}$. This function is shown in **Figure 83** (Dahmen, 1982; Kato, 1991), where it can be seen that θ varies rapidly from 0° at the exact NW OR for $r < 1/\sqrt{2}$ and $r > \sqrt{3}/2$, to a wide maximum at the exact KS OR at 5.26° for $3/4 < r < \sqrt{6}/3$. Thus, the NW and KS ORs cover most of the possible rotations between the close-packed planes and that is why they are the dominant ORs found experimentally (Dahmen, 1982; Kato, 1991). The energetic calculations by van der Merwe (1982) show that the NW and KS orientations are also the lowest energy orientations. It has been shown that the formation of an invariant line as described above is exactly analogous to forming O-lines in the interface (**Figure 20**) and that the same result can be obtained by either theory (Sasjima et al., 1989; Mou, 1994; Weatherly and Zhang, 1994).

It is possible to have a fully coherent FCC–HCP interface if the atomic sizes are similar in the two phases, as in the Al–Ag system in **Figure 78**. In contrast, even if the atom sizes are the same, it is not possible to have fully coherent FCC–BCC and HCP–BCC heterophase interfaces because the six-fold

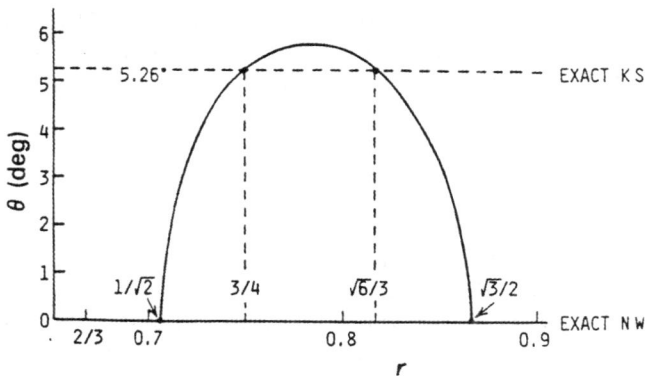

Figure 83   Rotation angle θ necessary to produce an invariant line by rotation around the normal to the close-packed planes as a function of the lattice parameter ratio r. Adapted from Dahmen (1982) and Kato (1991) with permission.

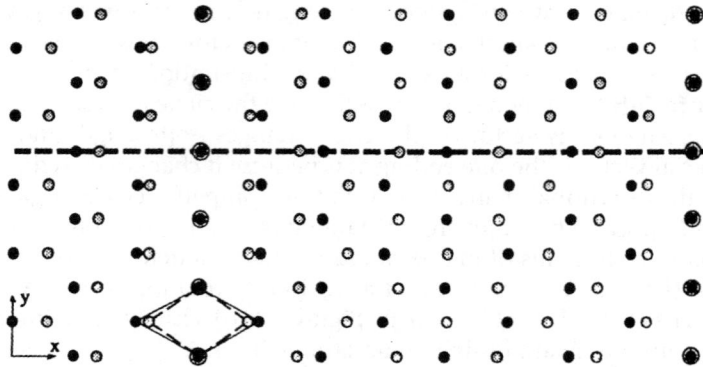

Figure 84   A schematic diagram representing the rhombic surface unit cells of a {111} FCC plane (filled circles) and a {110} BCC plane (open circles) when $r = a_{FCC}/a_{BCC} = 1.15$. Adapted from Shiflet and van der Merwe (1994) with permission.

symmetries of the close-packed {111} FCC and (0001) HCP planes do not match the two-fold symmetry of the {110} BCC plane. Hence, these interfaces must be partly coherent in some directions. This feature is illustrated schematically in **Figure 84** (Shiflet and van der Merwe, 1994), where the lattice parameter ratio has been adjusted so that the rhombohedral patterns of atoms in the FCC (filled circles) and BCC (open circles) have exactly the same spacing along the y-direction (i.e. $r = a_{FCC}/ a_{BCC} = 1.15$). This produces atomic row matching parallel to the x-direction as indicated by the heavy dashed line. Normal to this direction are rows of perfectly matched atoms (invariant lines or O-lines) with complete disregistry in between. The regions of disregistry correspond to edge misfit dislocations in the interface, parallel to the invariant line. This situation is analogous to the partly coherent interfaces shown in **Figure 21(d)** and **Figure 66**, except that the phases on either side of the interface have different lattices.

It is possible to transform the partly coherent interface containing perfect lattice misfit dislocations in **Figure 84** into a partly coherent interface containing partial dislocations (with Burgers vectors defined by the DCS lattice as in **Figure 27**, for example), by rotating the IP about the invariant line axis, so that it is no longer parallel to the common close-packed planes. **Figure 85** (Dahmen, 1987) shows four different interfacial structures that could form between two crystals that are related by a transformation strain that includes a simple shear e_{12} and an expansion e_{11} in the shear plane, so that the two-dimensional transformation matrix can be written as (Howe and Smith, 1992; Dahmen, 1994b); Xiao and Howe, 2000):

$$A = \begin{bmatrix} e_{11} & e_{12} \\ 0 & 1 \end{bmatrix}. \tag{84}$$

This deformation changes a square (top crystal in **Figure 85**) into a rhombus (bottom crystal in **Figure 85**) and leaves the spacing of the planes normal to the direction of the shear equal. This type of deformation is appropriate to the FCC–BCC transformation and in fact, the lines that form the square lattice in **Figure 85** can be imagined to be the {110} planes of the BCC phase and the set of lines that form the rhombus in the opposite crystal can be imagined as the {111} planes of the FCC phase (Hall et al., 1972; Howe and Smith, 1992). Thus, this schematic can be imagined to represent a general FCC–BCC interface.

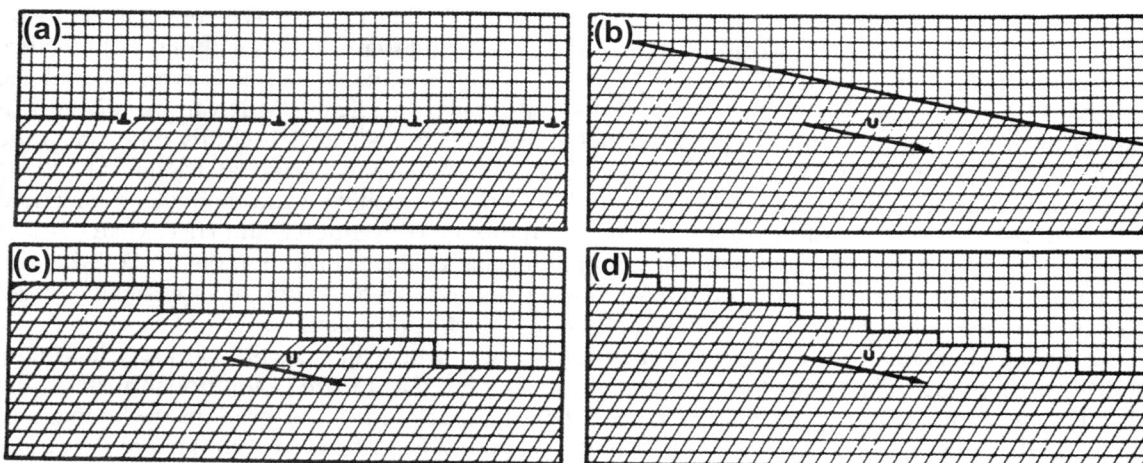

Figure 85 Possible interfaces for a transformation described by a shear e_{12} coupled with an expansion e_{11} in the shear plane. (a) The expansion is accommodated by a set of perfect lattice (misfit) dislocations. (b) The interface lies along an invariant line **u** and no dislocations are necessary. (c) The invariant line is resolved onto every other close-packed plane as in the FCC–HCP transformation where each step (disconnection) is an $a/6<112>$ partial dislocation. (d) The invariant line is resolved onto every close-packed plane as in an FCC–BCC transformation where each step contains an $a/12<112>$ partial dislocation. Adapted from Dahmen (1987) with permission.

Figure 85(a) shows the interface parallel to the shear plane (the parallel {111} FCC and {110} BCC planes), with the misfit in this plane accommodated by an array of $a/2<110>$ lattice dislocations in the square (BCC) lattice. This interface is analogous to that in **Figure 84** (and also **Figure 26**). In **Figure 85(b)**, the IP has been rotated through an angle θ that is related to the shear strain e_{12} and the dilatational strain e_{11} by Dahmen (1982), Howe and Smith (1992), Dahmen (1994b) and Xiao and Howe (2000)

$$\tan \theta = e_{11}/e_{12}. \tag{85}$$

This produces an interface along a high-index direction **u**, which is an undistorted and unrotated direction in the two crystals, that is it is an invariant line. If this invariant line is resolved onto every other close-packed plane, the staggered array of steps shown in **Figure 85(c)** is formed. This type of interface might correspond to the FCC–HCP transformation (e.g. **Figures 4 and 78**), where a Shockley partial dislocation (disconnection) glides along every other {111} plane to accomplish the transformation (Howe et al., 1987). In the FCC–HCP case, each disconnection contains an $a/6<112>$ displacement, which accomplishes the structural part of the transformation. If instead, the invariant line is resolved onto every close-packed plane, the staggered array in **Figure 85(d)** results. This is the case of an FCC–BCC transformation interface with every disconnection containing an $a/12<112>$ displacement (Mahon et al., 1989). An example of such an interface is shown in **Figure 7**.

In the area of diffusional transformations, the steps in **Figure 82(d)** are often called structural ledges because they are considered an intrinsic part of the interfacial structure (Hall et al., 1972; van der Merwe et al., 1991). Experimental evidence suggests that they are relatively immobile and that growth occurs locally at larger ledges or perturbations that move a long the interface (Furuhara et al., 1995). In martensitic transformations, such disconnections are highly mobile and accomplish the required structural transformation (Christian, 1975; Mahon et al., 1989). Thus, they are often called transformation dislocation ledges. It is important to note that all of the linear defects shown in **Figure 85** can be characterized structurally according to their dislocation and step contents within the DSC lattice (Section 14.2.1.3.2) or topological (Section 14.2.1.4) models and illustrated in **Figures 27 and 32**, respectively. Crystallographically, the disconnections illustrated in **Figure 85(d)** can be identical in martensitic and diffusional transformations, although their kinetic behavior is clearly different in the two types of transformation (Howe, 2006). Kajiwara et al., 1996 and Ogawa and Kajiwara (2004) suggested that the steps across disconnections are diffuse in martensitic transformations as compared with sharp in diffusional transformations. In contrast, thorough conventional and HRTEM analyses of the interfacial structures associated with diffusional and martensitic transformations in Fe–Ni alloys (Shibata et al., 2010; Moritani et al., 2002) have shown that this is probably not the case, and the reader is referred to these latter papers for good examples of the interfacial structures.

The interface in **Figure 85** is only a two-dimensional sketch, where the behavior of the crystals in the third dimension has been omitted for simplicity. In many interfaces there is additional misfit in the third dimension and this can be accommodated by another set of misfit dislocations in the interface parallel to the invariant line direction **u** illustrated in **Figure 86(a)** (Dahmen, 1987). Note the equivalence of this interface to the one shown in **Figure 33**, and that both of these interfaces are the same as the original stepped model proposed for the FCC–BCC (austenite–ferrite) transformation by Rigsbee and Aaronson (1979). Alternatively, these dislocations can be transformed into a second set of steps on the close-packed planes perpendicular to the first set, producing a second high-index direction **v** in the interface, as illustrated in **Figure 86(b)**. Crystallographic analyses that take into account all possible deformations in both two and three dimensions have been performed by Xiao and Howe (2000) and Gu and Zhang (2010), respectively, and the reader is referred to these papers for more thorough discussions. Rong (2012) also provides instructive examples of this approach in two and three dimensions based on the work of Luo and Weatherly (1987) and others.

Still more complicated situations can be envisioned and are found at heterophase interfaces in materials. For example, suppose that in addition to the shear and dilatation in **Figure 86**, there is a dilatation (misfit) normal to the shear plane so that the close-packed planes that are parallel to the interface in **Figure 86(a)** no longer have the same interplanar spacing. This misfit can be accommodated either by inserting misfit dislocations periodically along the interface, or by partitioning the strain evenly along the interface (Hall et al., 1972; Howe and Smith, 1992; Shiflet and van der Merwe, 1994; Howe et al., 2009). Either case is equivalent to performing a small rigid body rotation between the two crystals about an axis that is normal to the plane of the figure as mentioned previously in Section 14.2.1.4. Both cases lead to rotation between the two phases and have been observed experimentally.

Figure 86 showed schematically how a partly coherent planar interface between two crystals could transform into a disconnection interface containing an invariant line by rotation of the IP. The energetics of this process have been calculated by van der Merwe et al. (1991) and shown to be favorable for small misfits and large pattern advances when the interface is stepped. The idea of pattern advance created by a stepped interface is illustrated in **Figure 87** (Howe and Smith, 1992). **Figure 87(a)** shows an invariant line formed between two crystals and is similar to **Figure 85(b)**. In **Figure 87(b)**, the same interface is shown on an atomic level, where steps spaced periodically along the interface are evident. Notice that when crystal B steps down one plane parallel to the interface at a step, a pattern advance of the B lattice a distance of $\delta b_{[100]}$ to the left occurs, as indicated in the figure. According to theory van der Merwe et al. (1991), the center-to-center distance of consecutive terraces $D_{a[100]}$ along the [100] direction in the close-packed plane is given as

$$D_{a[100]} = -\delta b_{[100]}/\left(b_{[100]} - 1\right), \tag{86}$$

(a)

(b)

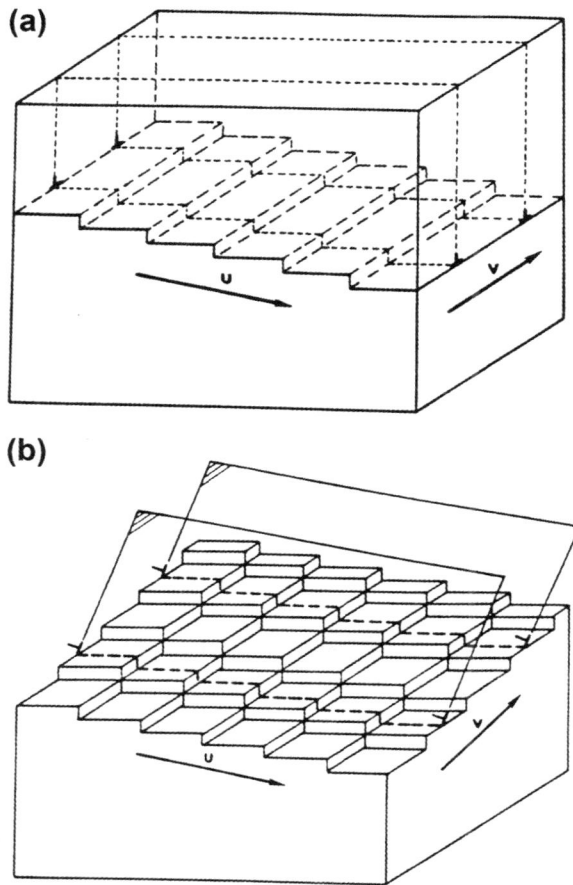

Figure 86   (a) The interface in **Figure 85(d)** in perspective with the crystal lattices omitted for clarity. Additional misfit in the direction **v** is accommodated by a set of lattice misfit dislocations similar to those in **Figure 85(a)**. (b) A geometrically glissile invariant plane–strain interface as in a martensite transformation. The direction **v** is now high-index and an additional simple shear is accommodated by a set of dislocations, which are glissile in the interface. Adapted from Dahmen (1987) with permission.

where $\delta b_{[100]}$ is the forward displacement of the B atomic step and $b_{[100]}$ is the lattice spacing of the B crystal along the [100] direction, as indicated in **Figure 87(b)**. The angle between [100] and the habit plane is then given by

$$\tan \theta = a_{\perp}/D_{a[100]}, \tag{87}$$

where $a_{\perp}$ is the lattice spacing normal to the parallel close-packed planes. Equation (87) produces a result identical to Eqn (85) for the transformation strain in Eqn (84), but now based on atomic pattern matching across the interface. This approach to interfacial structure has been compared with the invariant line theory (Howe and Smith, 1992; Mou, 1994) and can be useful for analyzing high-index stepped interfacial structures (Mou and Aaronson, 1994; Rao et al., 1994).

(a)

(b)

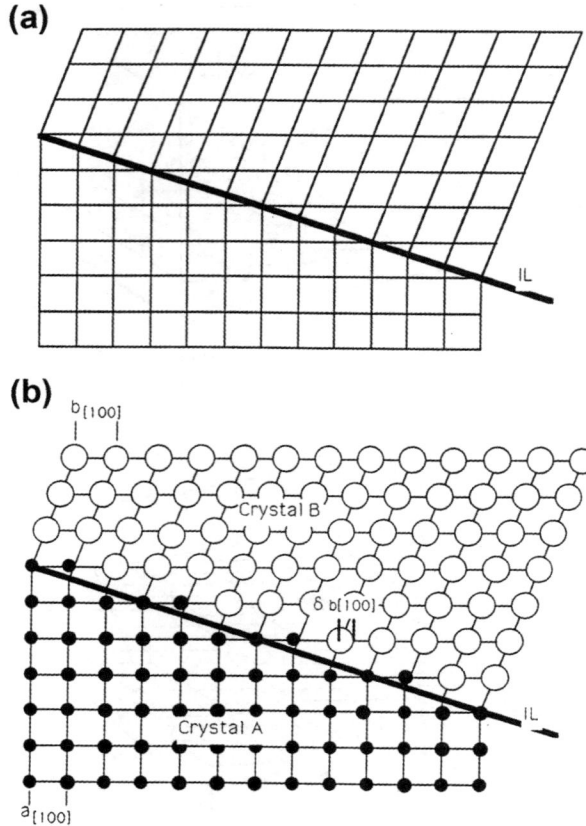

Figure 87 (a) Illustration of the invariant line formed because of a transformation involving a simple shear plus uniaxial expansion in the shear plane, and (b) atomic model of the interface constructed according to the pattern advance (structural ledge) treatment. Adapted from Howe and Smith (1992) with permission.

To illustrate the use of the invariant line and structural ledge treatments, consider the interface shown in **Figure 87**. Crystal B is produced from crystal A by a simple shear $e_{12} = 0.34$ accompanied by a 10% expansion in the (001) shear plane such that $e_{11} = 1.1$. The corresponding transformation matrix, Eqn (84) is

$$\mathbf{A} = \begin{bmatrix} 1.1 & 0.34 \\ 0 & 1 \end{bmatrix}.$$

When these values are substituted into Eqn (85), the angle between the [100] direction and the invariant line is given by $\theta = 16.7°$. The resulting interface between the two lattices is shown in **Figure 87(a)**. The structural ledge treatment then gives the center-to-center distance between consecutive (001) terraces along the [100] direction from Eqn (86) as

$$D_{a[100]} = 0.34/(1.1 - 1) = 3.34,$$

or 3.34 $[100]_A$ lattice planes. The angle between the [100] direction and the habit plane is then given by $\theta = \tan^{-1}(1/3.34) = 16.7°$, which is identical to the answer above for the invariant line theory. This interface is shown in **Figure 87(b)**.

There are three other common procedures in the literature, the ETE matching analysis (Kelly and Zhang, 1999; Zhang and Kelly, 2009), the Moiré (Nie and Muddle, 2002; Nie, 2004) and **Δg** methods (Zhang and Purdy, 1993), which have been used to analyze and predict ORs and IPs with considerable success. These were briefly described in Section 14.3.4.1, with the concept of **Δg** having been introduced in Section 14.2.1.2. All are successful because they rely on matching of close-packed, or nearly close-packed, planes of atoms meeting edge to edge across an interface, as illustrated previously in **Figure 79**. ETE planar matching can be performed by considering the spacings of the planes and the angle between them (as well as the distances between the atoms in the planes in the ETE model), in real or reciprocal space as in the Moiré and **Δg** methods, respectively. Rong (2012) provides an excellent summary of the ETE method including many examples for FCC/BCC/HCP systems, as well as the O-lattice and **Δg** methods, and the reader is referred to this for additional examples. In essence, each of these models are different ways of describing planar matching across, or the formation of an invariant line along, an interface as illustrated in **Figure 79**, as well as **Figures 85(b), (d) and 87**. Because these methods rely on the same concept illustrated in **Figure 79**, they are not described further here.

Before closing this section, it is important to mention that the O-lattice, CSL and DSC theories discussed in Sections 14.2.1.2 and 14.2.1.3, the invariant line and pattern-matching theories described in Section 14.3.4.2, as well as the ETE, Moiré, **Δg** methods just mentioned, and the PTMC (Christian, 1975; Wechsler et al., 1953; Bowles and Mackenzie, 1954) have all been utilized with various degrees of success. Each approach has its strengths and weaknesses, requires certain input, and so on, and most yield similar, if not identical, results. The reader can consult various references (Smith and Pond, 1976; Hirth and Balluffi, 1973; Pond and Hirth, 1994; Smith and Shiflet, 1987; Mou, 1994; Howe and Smith, 1992; Pond et al., 2003; Nie, 2004; Zhang and Weatherly, 2005; Zhang and Kelly, 2009; Zhang and Yang, 2011) for more detailed descriptions of the various theories and their strengths and limitations. It is also important to note that all of the energetic calculations performed for the stepped interfaces described above only included the structural (elastic) energy contribution (γ_s^{SS}) to the interphase boundary energy. Only a few calculations have been performed that include the contribution (γ_c^{SS}) because of wrong compositional bonds across stepped interphase boundaries, such as those in **Figures 7, 85(d) and 87 (b)**, and some of these are discussed in the following section. Lastly, the ETE model just mentioned and illustrated in **Figure 79**, has also been found to be useful in describing faceted interfaces where high-index planes that do not contain close-packed rows of atoms meet across the interface. These interfaces are considered to be commensurate in one dimension and constitute an important type of interface (Howe et al., 2000), so they are discussed further after the next section on atomistic calculation of high-index interfaces.

14.3.4.2.1 *Atomistic Calculations*

Only a limited number of atomistic calculations of the interfacial properties of high-index interfaces between crystals that have a well-defined OR but different Bravais lattices and compositions have been performed. Most of the calculations have been performed for FCC–BCC interfaces similar to those shown in **Figures 7 and 87**, since this represents an important and fairly common type of interface. Some of the results from atomistic calculations of the high-index interfaces are summarized below.

Figure 88(a) shows an atomistic model of a high-index FCC–BCC interface (Yang and Johnson, 1993). The interface was constructed using an NW OR between the phases and an EAM Fe potential (appropriate for an FCC–BCC martensite interface, for example). Approximate interfacial energies were

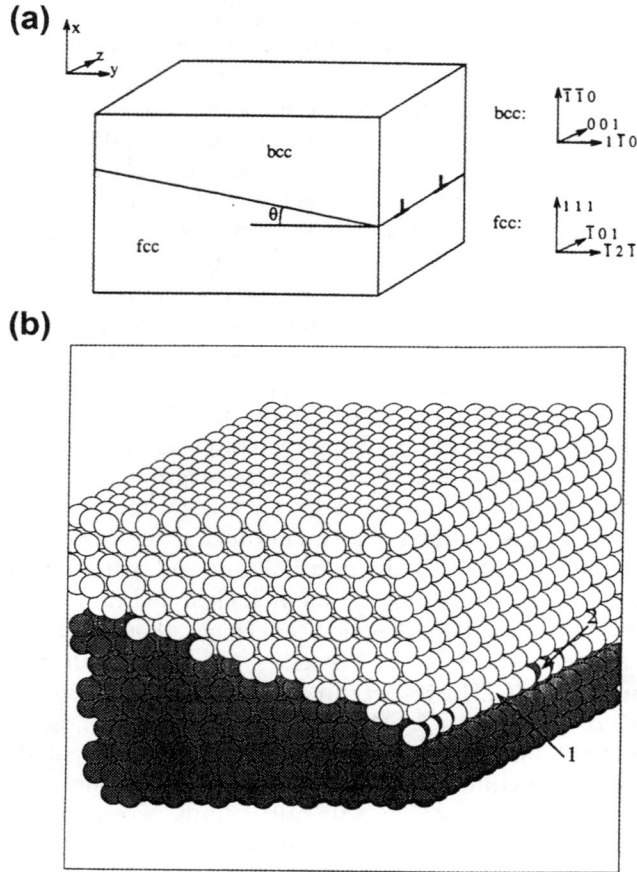

Figure 88 (a) Schematic drawing of an FCC–BCC interface with the NW OR, and (b) atomic model of the interface with $\theta = 13°$ with respect to the close-packed planes. The numbers 1 and 2 in (b) indicate coherent regions and misfit dislocations in the interface, respectively. Adapted from Yang and Johnson (1993) with permission.

obtained with fully relaxed simulation conditions for interfaces with orientations that varied from 0 to 26° with respect to the parallel close-packed $(111)_{FCC}\|(110)_{BCC}$ planes. The calculations showed that an interface with monatomic ledges, inclined about 13° with respect to the parallel close-packed planes, yielded the lowest interfacial energy with a value of about 240 mJ/m^2. This interface is shown in **Figure 88(b)** and it is the maximum coherency interface observed experimentally (Rigsbee and Aaronson, 1979). A similar calculation for a $(121)_{FCC}$ interface between FCC and BCC Fe phases with a KS OR yielded an interphase boundary energy of 179 mJ/m^2 (Yang and Johnson, 1993). Both of these interfacial energies are considerably lower than the energies of the partly coherent interfaces modeled using EAM potentials in Section 14.3.3.4, indicating that it is energetically favorable for partly coherent interfaces to form coherent stepped interfaces, as illustrated schematically in **Figure 85**.

Another study of this type was performed for a $(121)_{FCC}$ interface between Ni and Cr crystals (and alloys) in a KS OR using EAM potentials (Chen et al., 1997). A projection of the atomic structure of this interface for pure Ni and Cr is shown in **Figure 89** (Chen et al., 1997). This study is particularly

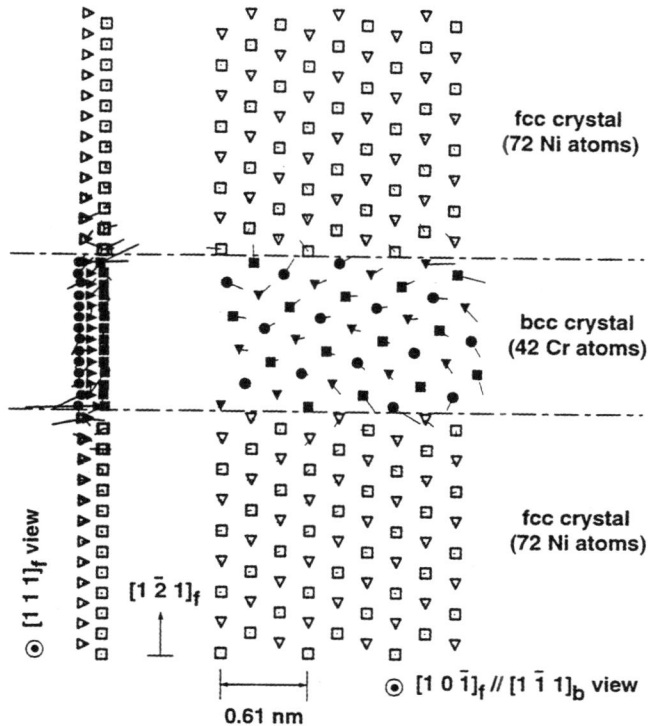

Figure 89 Two views of the relaxed structure of a simulation block of the (121)$_{FCC}$ interface for pure FCC Ni and BCC Cr crystals. Different shapes (square, triangle and circles) are used to represent atoms on different (202)$_{FCC}$ and (222)$_{BCC}$ planes parallel to the paper. The final positions of Ni and Cr atoms are represented by empty and filled shapes, respectively. Relative relaxations indicated by lines attached to the atoms are enlarged 10 times to show their directions. Adapted from Chen et al. (1997) with permission.

interesting because FCC Ni can accommodate up to almost 50 at.% Cr (the heat of mixing is about 0.06 eV/atom) and therefore it is possible to examine the interfacial structure and energy as a function of Cr concentration in the FCC phase. It was found that the interfacial energy varies only slightly with Cr concentration, decreasing from 216 to 200 mJ/m^2 as the Cr concentration increases from 0 to 50 at.%. Since this value is close to the value of 179 mJ/m^2 found for the same (121)$_{FCC}$ interface with an Fe potential above, and it does not vary more than about 10% with composition, these results indicate that the major portion of the interfacial energy (80–90%) is because of coherency strains at the steps, with the remaining part because of wrong compositional bonds across the interface. Another more recent calculation of these types of interfaces (Gu and Zhang, 2011) compares the interfacial energies of pure Fe with Fe–Cu and Ni–Cr near the KS and NW ORs, finding that the compositional contribution is higher in the case of Fe–Cu and that the near-KS ORs have lower interfacial energies than the NW ORs, even though the dislocations in the interface are closer together, the dislocation densities are higher.

14.3.5 One-Dimensionally Commensurate Interfaces

A type of interface that is intermediate between the partly coherent interfaces described in the previous sections, where coherent patches of atoms and interfacial defects are readily identified in the interface, and the incoherent interface that is described in the next section, is a one-dimensionally commensurate interface (Howe et al., 2000; Howe et al., 2002; Nie and Muddle, 2002), often called an ETE interphase boundary (Reynolds and Farkas, 2006). The difference between these ETE boundaries and the ones discussed in Section 14.3.4.2, is that the matching of atoms along the plane edges that meet ETE in the boundary is generally poor. This means that there are no good matching patterns of atoms in the boundary. Some atom matching that occurs can that described by near-coincident site (NCS) matching of atoms in the interface, but these atoms are not regular and the edges of the planes appear more comingled than atom matched. Near-coincidence atoms are usually defined as being within 15% of the nearest-neighbor distance of the atoms on the other phase (Chen and Reynolds, 1997; Liang and Reynolds, 1998). The habit planes of these interfaces are often high index, or irrational, but they can occur along low-index planes as well. The planes that meet edge to edge might also involve fairly close-packed planes of atoms. One can imagine that the energies of one-dimensionally commensurate interfaces, or incoherent interfaces in the next section, might increase proportional to the density of NCS atoms in the plane, but this possibility has not been thoroughly explored. A description of the structure and energies of a one-dimensional interface that was examined in some detail for the Ti–Al system is described below.

A dichromatic complex (Pond and Hirth, 1994) was first computer-modeled using the atoms from interpenetrating α_2 (ordered HCP) and γ_m (ordered FCT) crystals in a Ti–Al alloy (lattice parameters $a = b = 0.575$, $c = 0.473$ nm for α_2 and $a = b = 0.398$, $c = 0.408$ nm for γ_m) arranged in one of the irrational ORs observed at a planar α_2–γ_m boundary by Nie and Muddle (2002) (their **Figure 5**):

$$(200)_{\gamma_m} 1.8 \text{ deg} \quad \text{from} \quad (2\bar{2}00)\alpha_2 \quad \text{toward} \quad (0002)\alpha_2,$$

$$[0\bar{1}2]_{\gamma_m} 0.5 \text{ deg} \quad \text{from} \quad [\bar{1}\bar{1}20]\alpha_2 \quad \text{toward} \quad [0001]\alpha_2.$$

A $4 \times 6 \times 3$ nm block of the dichromatic complex is viewed along the $[1.00 \ \bar{3}.65 \ 2.65]_{\gamma m} = [\bar{2}.39 \ 3.24 \ 5.63 \ \bar{1}.70]_{\alpha2}$ direction in **Figure 90(a)** (Reynolds et al., 2003). This reveals the orientation of the $(111)_{\gamma m}$ and the $(2\bar{2}01)_{\alpha2}$ planes that meet ETE at the boundary plane. The schematic reciprocal lattice vectors in **Figure 90(a)** indicate the orientation of these planes. The orientation of the interphase boundary plane is horizontal and perpendicular to Δg, which in this case is given by the difference between $g = (111)_{\gamma m}$ and $g = (2\bar{2}01)_{\alpha2}$. The γ_m atoms are displayed as small black dots, the α_2 atoms are slightly larger, gray dots, and the NCSs are displayed as large circles. The NCS lie in horizontal arrays formed at the intersections of the $(111)_{\gamma m}$ and the $(2\bar{2}01)_{\alpha2}$ planes. These intersections occur at regular intervals in the vertical direction of the **Figure 90(a)** the Moiré planes (Nie and Muddle, 2002) and the NCS arrays are thus periodic in the direction perpendicular to the boundary plane.

A thin dichromatic slab $11 \times 11 \times 0.2$ nm in size, oriented parallel to the boundary plane is shown in perspective in **Figure 90(b)**. This slab reveals the geometric matching in the boundary. The boundary plane is irrational and is approximately the $(\bar{2}.13 \ 1.00 \ 2.18)_{\gamma m}$, or $(\bar{2}.25 \ 1.00 \ 1.25 \ 1.25)_{\alpha2}$. The boundary plane is viewed down the edges of the $(111)_{\gamma m}$ and the $(2\bar{2}01)_{\alpha2}$ planes that meet edge to edge at the boundary. The traces of these planes are the parallel rows of atoms commingled from the two phases, running from the front to the back of the boundary plane in **Figure 90(b)**. It is evident that the NCS are not continuous along the edges of the $(111)_{\gamma m}$ and the $(2\bar{2}01)_{\alpha2}$ planes despite the relatively high atomic density of these planes. On the other hand, the dense parallel rows of atoms commingled from both phases, shown in perspective in **Figure 86(b)**, should provide a distinctly low-energy interface relative to a boundary plane located between Moiré planes. That this interface is part of

(a)

(b)

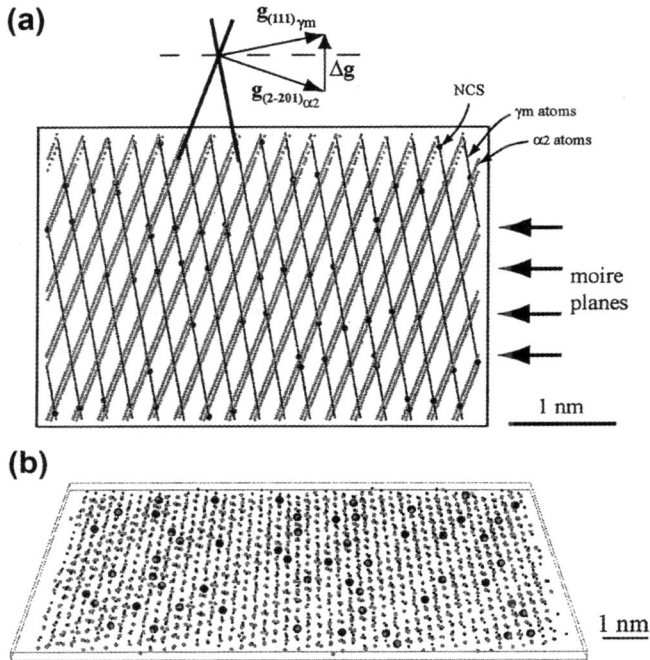

Figure 90 The arrangement of NCS (large circles), γ_m atoms (black), and α_2 atoms (gray) in a dichromatic complex corresponding to one of Nie and Muddle's (2002) observed ORs between the parent phase (α_2) and the product massive phase (γ_m) in Ti–Al. (a) The dichromatic complex viewed parallel to the boundary plane (horizontal) and both the $(111)\gamma_m$ and the $(\bar{2}\bar{2}01)\alpha_2$ planes that meet edge to edge at the boundary plane. (b) Thin slice of the dichromatic complex 0.2 nm thick oriented parallel to the boundary plane, centered about a Moiré plane (arrows in (a)). (a) And (b) differ in scale. If the boundary plane is between Moire planes, there are no NCS and atoms in (b) appear random. Adapted from Reynolds et al. (2003) with permission.

a micron-size planar boundary (**Figure 5**, Nie and Muddle, 2002) indicates that its structure provides a significant barrier to boundary migration.

The interphase boundary energy of the interface in **Figure 90** was calculated using molecular statics and dynamics simulations based upon the EAM and the dependence of the energy on boundary orientation was also explored (Reynolds and Farkas, 2006). It was found that the energy of the annealed and relaxed boundary was 1.34 J/m^2, which was about 2.6 times higher than the energy of a coherent boundary between the two phases in a SN OR (Section 14.3.4) and is similar to the energy of a RGB (see, e.g. **Figures 36 and 37**). Rotation of the interphase boundary about the ETE direction and the commensurate directions perpendicular to this showed that the energy of the boundary was lowest when it was oriented in an ETE configuration parallel to the Moiré plane and hence, the boundary lies in an energy cusp, that is tilting the boundary from a Moiré plane increases its energy. It was also found that there is no strain localization within the boundary and although it is not structurally singular, it does not appear to dissociate into other low-energy configurations.

14.3.6 Incoherent Interfaces

Figure 91 (Smith and Shiflet, 1987) shows a classic example of an incoherent (incommensurate) interphase boundary, which would have an interphase boundary energy γ_i^{SS}. It is different from the

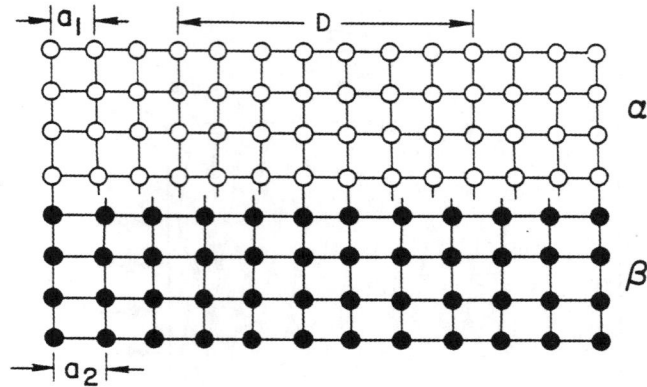

Figure 91 Atomic model of an incoherent interface in a simple cubic crystal where $a_\beta > a_\alpha$ and there is weak interaction across the interface. Adapted from Smith and Shiflet (1987) with permission.

partly coherent interface in **Figure 66** in that the bonding across the interface is sufficiently poor and/or the atomic planes are so unevenly matched that there is no local relaxation of the misfit into dislocations at the interface. Instead, the misfit is spread evenly across the entire interface and the two phases retain their bulk structures up to the interfacial plane, where they terminate abruptly. The $\{111\}$ interface between Al and Ge in a cube-on-cube OR is an example of this type of incoherent interface (Dahmen, 1994a). Such interfaces are expected to have a high interfacial energy. An upper limit on γ_i^{ss} can be obtained by adding the surface energies of the two crystallographic faces (hkl) of the α and β phases. Of course there is some overlap of electron density among the atoms at the interface so the energy is likely to be less than the sum of the free surface energies, but still high, similar or greater to the one-dimensional commensurate interface just discussed, or the RGB also mentioned in Section 14.2.1.5.2.

Howe et al., 2000) proposed a classification scheme for three different types of incoherent interphase boundaries based upon the OR and IP between the two phases, that is

 (i) high-index OR with no low-index conjugate IPs,
 (ii) low-index OR in at least one direction with ill-matched rational IPs, and
(iii) high-index OR with a low-index IP in only one phase.

These three types of incoherent interfaces are illustrated in **Figure 92** (Howe et al., 2000). The incoherent interface in **Figure 91** is the second type of interface when a pair of low-index directions is parallel between the phases across the interface, as illustrated in **Figure 92(d)**. The third type of incoherent interface was previously shown in **Figure 6** and this may be the most common type of incoherent interface. Here the interphase boundary follows a close-packed, or nearly close-packed, plane in one phase, while the other phase matches a high-index or irrational plane across the interface. This type of idomorphic interface, called a type-3 incoherent interface in the literature (Li et al., 2004, 2006), has received considerable attention in the phase transformations community, particularly with regard to the massive transformation (Aaronson, 2002, 2006; Massalski et al., 2006; Aaronson and Reynolds, 2006). In essence, this type of incoherent interface can be considered analogous to the grain boundaries discussed in Section 14.2.1.5.3. If the interphase boundary is thought of as being composed of two crystal surfaces, the boundary can achieve a lower energy when

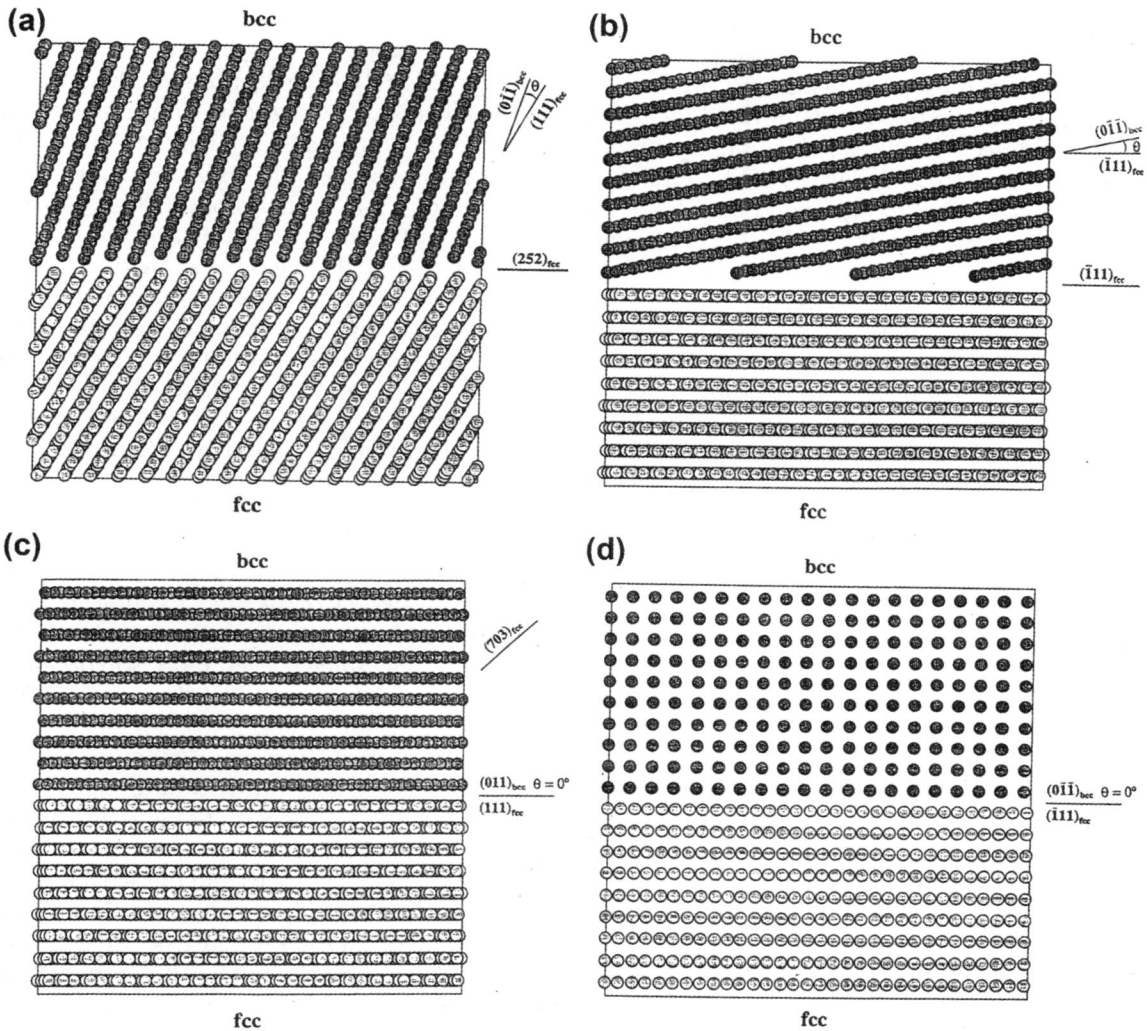

Figure 92 (a) Geometric atomic model of an incoherent interphase boundary between BCC (top, $a = 0.293$ nm as for Fe) and FCC (bottom, $a = 0.362$ nm as for Cu) crystals. The OR between the two crystals is $(\bar{1}\bar{7}\bar{2})_{BCC}\|(\bar{7}03)_{FCC}$ and $[\bar{5}1\bar{5}]_{BCC}\|[3\bar{4}7]_{FCC}$, the IP is $(252)_{FCC}$ and the viewing direction is $[3\bar{4}7]_{FCC}$. The $(0\bar{1}\bar{1})_{BCC}$ and $(\bar{1}11)_{FCC}$ planes are inclined at an angle θ, as indicated in the figure. (b) The same BCC and FCC crystals and OR as in (a), but with the interface rotated so that it is a rational $(\bar{1}11)_{FCC}$ habit plane in the FCC crystal. (c) The same BCC and FCC crystals as in (a), but with a rational OR in the direction perpendicular to the low-index $(0\bar{1}\bar{1})_{BCC}$ and $(\bar{1}11)_{FCC}$ conjugate habit planes. The OR between the two crystals is: $(0\bar{1}\bar{1})_{BCC}$ and $(\bar{1}11)_{FCC}$ and $[\bar{5}1\bar{5}]_{BCC}\|[3\bar{4}7]_{FCC}$, the IP is $(\bar{1}11)_{FCC}$ and the viewing direction is $[3\bar{4}7]_{FCC}$. The $(0\bar{1}\bar{1})_{BCC}$ and $(\bar{1}11)_{FCC}$ planes are parallel so $\theta = 0°$, as indicated in the figure. Note that θ does not represent the angle of twist in the plane of the interface, but the angle of inclination between the close-packed planes in the two crystals. (d) The same as (c), but with a set of low-index $[\bar{1}00]_{BCC}\|[1\bar{1}2]_{FCC}$ directions parallel across the interface and the viewing direction as $[1\bar{1}2]_{FCC}$. Adapted from Howe et al. (2000) with permission.

one of the phases presents a low-index low-energy surface, as compared with both phases presenting high-index high-energy surfaces.

Perhaps the most comprehensive structural (and compositional) investigations of this type of incoherent interface has been performed for ZrN grains growing in Zr, as in **Figure 6**. Conventional and HRTEM, diffraction and crystallographic analyses, and NCS atom modeling were all used to investigate the atomic structure of a planar $\{111\}_{ZrN}$ interface between the Zr and ZrN phases in a Zr-33.9 at.% N alloy (Li et al., 2004, 2006). The Zr and ZrN phases were found to have a high-index OR and display poor atomic matching across the $\{111\}_{ZrN}$ interface in the HRTEM images and using NCS atom modeling. The NCS contained within a 0.4 nm slice centered about the IP, that is the approximate near-coincidence sites in the interface, are shown in **Figure 93(a)**, viewed over an expanded area of interface 24 nm × 24 nm wide (Li et al., 2004, 2006). It is apparent from **Figure 93(a)** that the NCS in the interface are isolated and irregularly spaced over the entire interface area. This result contrasts with the regular arrays of NCS that are typically found with crystals that have nearly parallel closed-packed planes and directions, even with high-index interphase boundaries (Liang and Reynolds, 1998; Miyano et al., 2000). In addition, no sets of planes in the two phases were found to meet edge to edge at the interface, as in the previous one-dimensional commensurate interfaces and there was no evidence of strain localization in conventional or HRTEM images.

These results show that the $\{111\}_{ZrN}$ interface is atomically incoherent and that atomic faceting in the ZrN phase, rather than atomic matching at the interface, is the dominant factor in determining the $\{111\}_{ZrN}$ IP. Both the experimental HRTEM images and an atomic model of the interface indicated that six $(002)_{Zr}$ planes corresponded to nine $(020)_{ZrN}$ planes across the interface, so that it was mathematically rational. This particular interface was faceted at the micrometer level and also contained atomic steps, similar to partly coherent interfaces, indicating that such features cannot be used to define partial coherency at interfaces.

In other studies on the Zr–ZrN system, the same investigators found that certain other incoherent $\{111\}_{ZrN}$ and $\{110\}_{ZrN}$ interfaces can display patches of NCS matching, but there are many non-matching atoms within a given NCS patch and the NCS patches that are present do not repeat in the IP, as in partly coherent interphase boundaries, as shown in **Figure 93(b)**. Furthermore, the area of NCS matching in the interface was far less than the area of NCS nonmatching regions and the separation of adjacent NCS patches was approximately 8 nm. In addition, the $\{110\}_{ZrN}$ interface was atomically rough, indicating that there may be both atomically faceted and atomically rough type-3 incoherent interfaces. These features indicate that there may be various other types of partial NCS matching in incoherent interfaces, but the possible behaviors and energetics of such boundaries remain to be determined.

14.3.7 Kinks in Disconnections at Solid–Solid Interfaces

Before concluding this section, it is worth mentioning that most of the interphase boundaries previously discussed have an additional level of detail associated with them that has not been described, and that is the presence and/or the necessity of kinks in steps or disconnections on the interfaces (Laird and Aaronson, 1969; Howe et al., 1987; Howe and Prabhu, 1990; Howe, 1994a,b; Howe, et al., 2009). Such kinks are often necessary for movement of solid–solid interfaces in the same way as for solid–vapor interfaces (Howe et al., 1985; Howe, 1997). In the solid–vapor case, this preferred attachment can be explained by a simple broken bond argument. The same rational applies to solid–solid interfaces except that in the solid–solid case, wrong compositional bonds occur across the interface instead of broken bonds, and it is also necessary to include the strain energy contribution or

(a)

(b)

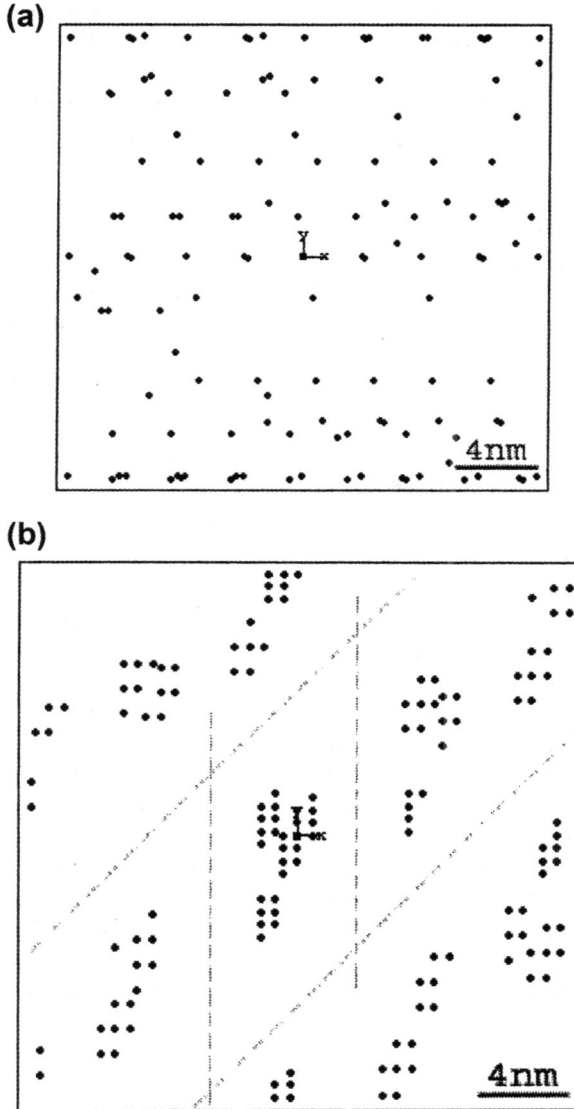

Figure 93   (a) The Zr–ZrN NCS projected onto the $(\bar{1}11)_{ZrN}$ IP, using the OR specified in the text and (b) another incoherent $(111)_{ZrN}$ interface associated with a different OR and showing patches of NCS in the interface. Adapted from Li et al. (2004), (2006) with permission.

dislocation character along a disconnection, which requires bond stretching and is very unfavorable energetically. Thus, motion of coherent and partly coherent interfaces between crystals with different crystal structures rarely occurs by a continuous growth mechanism. Migration of these coherent and partly coherent interfaces requires nucleation of steps or disconnections that propagate by kinks, just like the vicinal solid–vapor.

In FCC, BCC and HCP systems, disconnections can often minimize both their elastic and compositional energies by aligning along the close-packed directions in the crystals and this is observed experimentally. Thus, the ORs discussed in Section 14.3.4 tend to align the close-packed planes and directions between the crystals and the additional considerations just mentioned tend to further insure that disconnections are aligned along the relatively low-index close-packed directions in the interfaces. Extra energy required to perturb the disconnections provides an activation barrier to kink nucleation in solids, and steps and disconnections can become immobile in the absence of kinks (Aaronson et al., 1970; Howe and Prabhu, 1990). However, once a kink forms, it is favorable for atoms to attach to it rather than to attach randomly along the step or disconnection because it does not further increase the length of the defect or the number of wrong bonds, which is analogous to the solid–vapor situation. It has also been suggested that extra strain associated with kinks helps facilitate substitutional diffusion across the interface at these sites, (Howe et al., 1987). These concepts apply whether the disconnection moves with or without a corresponding compositional change (Howe, 1994a, b; Howe et al., 2009).

14.3.8 Segregation to Heterophase Interfaces

Equilibrium segregation to interphase boundaries has been treated theoretically for the case of a coherent interface in ternary alloys using the continuum and DLP models described in Section 14.3.1.3 (Dregia and Wynblatt, 1991). In addition, analysis of segregation to partly coherent interfaces in a ternary system containing two phases with a parallel OR but different compositions has been performed by atomistic modeling (Dregia et al., 1987; Bacher et al., 1991; Rao, 1994). Unfortunately, there are only a few experimental data available for direct comparison with these models (Rao et al., 1993; Rao, 1994). This section summarizes the main features of the DLP model for segregation to a coherent interphase boundary in a ternary alloy because it follows directly from the DLP model of coherent binary interfaces discussed in Section 14.3.1.3. It then describes some results obtained from regular solution DLP calculation, atomistic modeling and compositional profiling, on segregation of Au to the Ag–Cu interface in an Ag–Cu–Au alloy, since this represents one of the most complete and systematic investigations performed on a ternary alloy to date. The continuum approach to segregation is not described here, but it equivalent to the DLP treatment, just as in Section 14.3.1.3.

The DLP model developed in Section 14.3.1.3 can be readily extended to treat interfaces in a ternary FCC alloy. One begins by defining the coordination numbers z, z_l and z_j exactly as in Eqn (56). In a ternary regular solution system, the regular solution parameters can be expressed in terms of the nearest-neighbor bond energies as

$$\Omega_{ij}/zN_A = \varepsilon_{ij} - 1/2(\varepsilon_{ii} + \varepsilon_{jj}) \quad (i,j = 1,2,3), \tag{88}$$

where ε_{ij} (ε_{ii} or ε_{jj}) are the energies of the bonds connecting nearest-neighbor pairs of atoms of types i and j, as in Eqn (40). The composition of the jth atomic layer in the system parallel to the IP is then given by the equations:

$$C_3^j/C_1^j = C_3^\alpha/C_1^\alpha \exp\left(\Delta H_{31}^j\right)/k_B T, \tag{89a}$$

$$C_3^j/C_2^j = C_3^\alpha/C_2^\alpha \exp\left(\Delta H_{32}^j\right)/k_B T, \tag{89b}$$

where C_3^α and C_i^α are the atom fractions of component i in the jth layer and in the bulk of the α phase, respectively, and the ΔH_{ij}^j are the enthalpies of segregation, which may be expressed as

$$\Delta H_{31}^j = 2\Omega_{13}\left[z_j\left(C_1^{j+1} + C_1^{j-1}\right) + z_1 C_1^j - z\, C_1^\alpha\right] - \Omega\left[z_j\left(C_2^{j+1} + C_2^{j-1}\right) + z_1 C_2^j - z\, C_2^\alpha\right], \tag{90a}$$

$$\Delta H_{32}^j = 2\Omega_{23}\left[z_j\left(C_2^{j+1} + C_2^{j-1}\right) + z_1 C_2^j - z\, C_2^\alpha\right] - \Omega\left[z_j\left(C_1^{j+1} + C_1^{j-1}\right) + z_1 C_1^j - z\, C_1^\alpha\right], \tag{90b}$$

where

$$\Omega = \Omega_{12} - \Omega_{13} - \Omega_{23}. \tag{91}$$

Equation (90) for the enthalpies of segregation are found in a manner analogous to Eqn (59) in the binary alloy. The bulk composition of the α phase was used as a reference in Eqns (89) and (90) but any plane in the two-phase system could have been used as a reference. The terms containing Ω in Eqn (91) are the ternary contributions to the enthalpies of segregations and these vanish in the limit of a binary system. Since a pair-wise nearest-neighbor bond model was used, the ternary term Ω is composed of three binary regular solution parameters. Thus, to apply this model, one must extract values of these parameters from measured or calculated values of the heats of mixing (Eqn (41)) of the three binary systems.

This regular solution model is useful because it provides a simple framework for predicting interfacial compositions from the thermodynamic properties of the bulk phases. As an example, we will use the model to calculate the equilibrium composition profiles normal to planar {100} and {111} interfaces in the Ag–Cu–Au system dilute in Au, following the work of Dregia and Wynblatt, (1991). These interfaces differ only in their coordination numbers, which are z_1, $z_j = 4$ for {100} and $z_1 = 6$, $z_j = 3$ for {111}. The regular solution parameters Ω_{ij} were extracted from the measured enthalpies of mixing for Ag–Cu, Cu–Au and Ag–Au binary alloys (Hultgren et al., 1973). Since the experimental values of Ω_{CuAu} and Ω_{AgAu} are nearly equal, they were set equal to their average value, which reduces the ternary problem to a modified binary one. The values of the parameters used were $\Omega_{AgCu} = 0.32$ eV and $\Omega_{CuAu} = \Omega_{AgAu} = -0.175$ eV.

The first step in the determination of the segregation profile is the solution of the bulk equilibrium equations. In a three-component system that consists of two phases in equilibrium at constant temperature and pressure, if the interface is neglected and only the equilibrium between the two bulk phases is considered, then equilibrium requires equality of the chemical potentials of each component in the two phases. The conditions of equilibrium are thus

$$\mu_i^\alpha = \mu_i^\beta = \text{constant} \quad (i = 1, 2, 3),$$

where μ_i are the chemical potentials of the ith component in the α and β phases. Since four composition variables are required to specify the system (two atom fractions in each phase) and only three relations are available from Eqn (92), one composition variable can be selected arbitrarily, that is there is one compositional DOF in the system. For a ternary regular solution, the chemical potentials may be written as (Meijering, 1950)

$$\mu_1 = k_B T(\ln C_1) + z[\Omega_{12}C_2 + \Omega_{13}C_3 - (\Omega_{12}C_1 C_2 + \Omega_{13}C_1 C_3 + \Omega_{23}C_2 C_3)], \tag{93}$$

where the C_i are the bulk atom fractions of the three components in the solution and Ω_{ij} are given by Eqn (88). Thus, it is possible to determine the compositions of the coexisting phases by selecting a value for C_i in one of the phases and substituting relations such as Eqn (93) for each of the components into Eqn (92).

In the Ag–Cu–Au alloy, bulk equilibrium was first established for the binary Ag–Cu system. Next, the atom fraction of Au was fixed in one of the phases and the remaining atom fractions were determined by solution of the equations for the ternary system. The available DOF is thus consumed by fixing the atom fraction of Au in one of the phases. This procedure consists of finding three atom fractions from three equations that express uniformity of the chemical potentials of the three components (Eqn (92)). This step is simplified by the equality of Ω_{CuAu} and Ω_{AgAu}, which necessitates that the atom fractions of Au be equal in the two phases. The equilibrium bulk phases are used as a starting point in the solution for the interfacial composition profiles. The two phases are represented by two contiguous slabs of {100} or {111} layers with one slab assigned the composition of the Cu-rich phase and the other the composition of the Ag-rich phase. The individual layer compositions are adjusted iteratively by continued substitution into the equations of equilibrium (Eqn (89)) with the free energies of segregation calculated by Eqn (90). This iterative process is continued until the composition profiles converge to a stationary value.

The equilibrium composition profiles calculated for the Ag–Cu–Au system at 500 K are illustrated in **Figure 94** (Dregia and Wynblatt, 1991). These profiles display the following important features. The Au profile is symmetric while the Cu and Ag profiles are anti-symmetric. This is a direct consequence of the equality of the alloy interaction parameters Ω_{CuAu} and Ω_{AgAu}. The profiles show negligible solid solubility of Cu in Ag and vice versa, which is because of the large positive value for the enthalpy of mixing in the Ag–Cu system. The profiles also show composition oscillations extending into both phases, which result from the tendency toward ordering because of the negative enthalpies of mixing in the Cu–Au and Ag–Au systems. As evident from **Figure 94**, Au segregated to the interface, and about four atomic layers are substantially different in composition than the bulk phases at this temperature. The overall driving force for segregation in a ternary system depends on the relationships among the

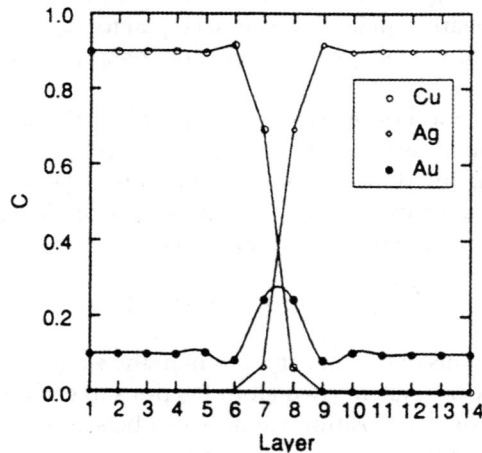

Figure 94 Calculated equilibrium profile showing segregation of Au at a coherent {100} interphase boundary in Ag–Cu–Au at 500 K. The bulk compositions of the two phases are ($C_{Ag} = 0.003$, $C_{Au} = 0.1$, $C_{Cu} = 0.897$) and ($C_{Ag} = 0.897$, $C_{Au} = 0.1$, $C_{Cu} = 0.003$). Adapted from Dregia and Wynblatt (1991) with permission.

various binary solution parameters in Eqn (90) and may not be immediately apparent. However, the overall segregation of Au at the Ag–Cu interface can be rationalized as because of the replacement of unfavorable Ag–Cu bonds (with positive Ω_{AgCu}) by more favorable Cu–Ag and Ag–Au bonds (with negative Ω_{CuAu} and Ω_{AgAu}).

The dependence of the segregation on bulk composition, temperature and interface orientation is shown in **Figure 95** (Dregia and Wynblatt, 1991). At a given temperature, the Au excess initially increases with increasing bulk composition, reaches a maximum and ultimately vanishes when the bulk composition falls outside of the miscibility gap in a single-phase region and the interface disappears. The strength of segregation and the miscibility gap limit increase with decreasing temperature as expected. The interface orientation affects the segregation through the parameter z_j, which is larger for less-densely packed layers. Since the free energy of segregation in Eqn (90a) is proportional to z_j, there is more segregation at the less densely packed {100} interface. The interfacial energy as a function of bulk composition and temperature was also calculated from the data in a manner similar to that discussed in Section 14.2.3 and compared with the Gibbs adsorption equation (Lupis, 1983, Howe, 1997). Generally good agreement was obtained except at low temperatures and compositions where the interfacial composition gradients were largest. These data demonstrate the usefulness of the nearest-neighbor regular solution model for predicting interfacial segregation at interphase boundaries. The authors also included the effects of coherency strain energy into the regular solution calculations and found that it introduces asymmetry into the composition profiles (Dregia and Wynblatt, 1991).

Bacher et al. (1991) subsequently performed EAM Monte Carlo calculations of the same Cu–Ag–Au interfaces and found additional interesting effects. The atomistic calculations showed that a square interfacial dislocation network formed at the Ag–Cu {001} interface because of the approximately 10% misfit between the two phases, and that the Ag-rich phase tended to penetrate into the Cu-rich phase in between the square dislocation array to form pyramidal protrusions which were faceted along the {111} planes. In essence, the smaller Cu atoms tended to cluster along the interfacial dislocations in regions of compression while the larger Ag atoms tended to occupy regions of tension. As in the

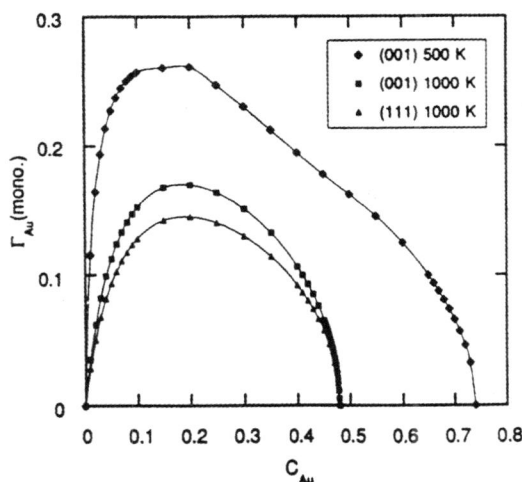

Figure 95 Variation of interfacial excess of Au (in units of monolayers) with bulk composition. Segregation is enhanced with decreasing temperature and planar density of atoms at the interface. Adapted from Dregia and Wynblatt (1991) with permission.

previous regular solution calculations, there was a clear preference for Au atoms to occupy sites between Cu-rich and Ag-rich regions at the interface. However, because of the pyramidal shape of the protrusions at the interface in the Monte Carlo calculations, the Au segregation appeared smeared out into the Cu-rich side of the interface creating an asymmetric profile. The values for Ω_{CuAu} and Ω_{AgAu} are also not exactly equal in the Monte Carlo calculations, and this introduces additional asymmetry into the profiles that is not present in the previous regular solution calculations.

Lastly, an example is shown of actual experimental data obtained on an Ag–Cu–Au {100} interface (Dregia et al., 1987). In this investigation, a single-crystal {100} partly coherent Ag–Cu interface was produced by vapor deposition onto a NaCl {001} substrate using alternating layers of Cu, Ag and Au. After deposition, the films were annealed to achieve equilibrium and the distribution of the elements across the interface was determined by scanning Auger electron spectroscopy (AES), using a crater-edge profiling technique. In this technique, a shallow crater is sputtered through the interface. The low angle of intersection where the crater cuts the interface effectively magnifies it by several orders of magnitude, allowing composition profiles to be determined across the interface by AES (Johannessen et al., 1979). The results of an Auger line scan along the surface of the sputtered crater for the Ag–Cu–Au interface are shown in **Figure 96** (Dregia et al., 1987). The results show that Au is segregated to the interphase boundary, in qualitative agreement with the calculated profile in **Figure 94**. Some elemental mixing associated with the sputtering process and the limited spatial resolution of the technique prevent more quantitative comparison between experiment and theory. In the future, HRTEM may provide a more quantitative comparison (Howe, 1994a, b; Luo et al., 2011). This example of the Ag–Cu–Au system clearly illustrates the utility of the regular solution model and atomistic modeling in predicting segregation to interphase boundaries.

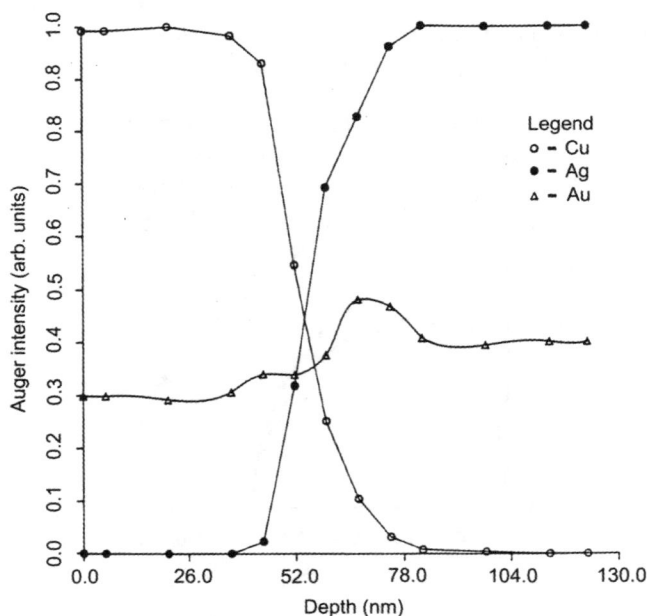

Figure 96   Auger intensity profiles obtained across an Ag–Cu interphase boundary in annealed Ag–Cu thin films containing Au. Adapted from Dregia et al. (1987) with permission.

It is worth commenting on one last feature of interphase boundary segregation, and that is the close relationship with grain boundary complexions mentioned previously, at the end of the section on grain boundary segregation (Baram et al., 2011).

Acknowledgments

I am grateful to all of the students and colleagues who have contributed to the research cited in this article, and to Matt Schneider for his help in preparing the manuscript. I also gratefully acknowledge the National Science Foundation for support of this research over many years and currently under Grant DMR-1106230.

References

Aaronson, H.I., Reynolds, Jr., W.T. (Eds.), 2006. Metall. Mater. Trans. A 37, 961.
Aaronson, H.I., Clark, J.B., Laird, C., 1968. Met. Sci. J. 2, 155.
Aaronson, H.I., Laird, C., Kinsman, K.R., 1970. In: Aaronson, H.I. (Ed.), Phase Transformations. American Society for Metals, Metals Park, OH, p. 313.
Aaronson, H.I., 2002. Metall. Mater. Trans. A 33, 2445.
Aaronson, H.I., 2006. Metall. Mater. Trans. A 37, 803.
Alexander, K.B., LeGoues, F.K., Aaronson, H.I., Laughlin, D.E., 1984. Acta Metall. 32, 2241.
Ashby, M.F., Spaepen, F., 1978. Scr. Metall. 12, 193.
Asl, K.M., Luo, J., 2012. Acta Mater. 60, 149.
Babcock, S.E., Balluffi, R.W., 1987. Philos. Mag. 55, 643.
Bacher, P., Wynblatt, P., Foiles, S.M., 1991. Acta Metall. Mater. 39, 2681.
Balluffi, R.W., Bristowe, P.D., 1984. Surf. Sci. 144, 28.
Balluffi, R.W., Brockman, A., King, A.H., 1982. Acta Metall. 30, 1453.
Balluffi, R.W., 1982. Metall. Trans. A 13, 2069.
Baram, M., Chatain, D., Kaplan, W.D., 2011. Science 8, 206.
Batstone, J.L., 1992. In: Wolf, D., Yip, S. (Eds.), Materials Interfaces: Atomic-level Structure and Properties. Chapman Hall, London, p. 316.
Bauer, E., 1983. Appl. Surf. Sci. 11/12, 479.
Beanland, R., Kiely, C.J., Pond, R.C., 1994. In: Handbook on Semiconductors. North-Holland, Amsterdam, p. 1149.
Becker, R., 1938. Ann. Phys. 32, 128.
Bernal, J.D., 1964. Proc. R. Soc. London, Ser. A 280, 299.
Bilby, B.A., Bullough, R., Smith, E., 1955. Proc. R. Soc. London, Ser. A 231, 263.
Bilby, B.A., 1955a. Defects in Crystalline Solids. The Physical Society, London, UK. p. 33.
Bilby, B.A., 1955b. In: Report of the Conference on Defects in Crystalline Solids. Physical Soc., London, p. 124.
Bishop, G.H., Chalmers, B., 1986. Scr. Metall. 2, 133.
Blakely, J.M., 1973. Introduction to the Properties of Crystal Surfaces. Pergamon Press, Oxford.
Bollmann, W., 1970. Crystal Defects and Crystalline Interfaces. Springer-Verlag, Berlin, DE.
Bollmann, W., 1972. Surf. Sci. 31, 1.
Bonnet, R., Durand, F., 1975. Philos. Mag. 32, 997.
Bowles, J.S., Mackenzie, J.K., 1954. Acta Metall. 2 (129), 138.
Braun, M.W.H., van der Merwe, J.H., 2002. Metall. Mater. Trans. A 33, 2485.
Braun, M.W.H., 1987. Ph. D. Thesis. University of Pretoria.
Briant, C.L., 1992. In: Wolf, D., Yip, S. (Eds.), Material Interfaces: Atomic-level Structure and Properties. Chapman Hall, London, UK, pp. 436–469.
Broughton, J.Q., Gilmer, G.H., 1983. Acta Metall. 31, 845.
Bruce, L.A., Jaeger, H., 1978. Philos. Mag. A 38, 223.
Burgers, W.G., 1959. Physica 1, 561.
Burton, J.J., Machlin, E.S., 1976. Phys. Rev. Lett. 37, 1433.
Burton, W.K., Cabrera, N., Frank, F.C., 1950. Philos. Trans. R. Soc. London, Seri. A 243, 299.
Cahn, J.W., Hilliard, J.E., 1958. J. Chem. Phys. 258, 258.
Cammarata, R.C., 1994. Prog. Surf. Sci. 46, 1.

Chen, I.-W., Chiao, Y.-H., 1985. Acta Metall. 33, 1827.

Chen, J.K., Reynolds Jr., W.T., 1997. Acta Mater. 45, 4423.

Chen, J.K., Farkas, D., Reynolds Jr., W.T., 1997. Acta Metall. Mater. 47, 4415.

Christian, J.W., 1975. The Theory of Transformations in Metals and Alloys, Part I – Equilibrium and General Kinetic Theory. Pergamon Press, Oxford. pp. 362, 460.

Cline, H.E., Walter, J.L., Koch, E.F., Osika, L.M., 1971. Acta Metall. 19, 405.

Cullen, W.H., Marcinkowski, M.J., Das, E.S.P., 1973. Surf. Sci. 36, 395.

Dahmen, U., 1982. Acta Metall. 30, 63.

Dahmen, U., 1987. Scr. Metall. 21, 1029.

Dahmen, U., 1994a. Micros. Soc. Amer. Bull. 24, 341.

Dahmen, U., 1994b. Encyclopedia of Physical Sciences and Technology, vol. 10. Academic Press, San Diego, CA, 319.

Das, S., Howe, J.M., Perepezko, J.H., 1996. Metall. Mater. Trans. A 27, 1623.

Dillon, S.J., Rohrer, G.S., 2009. Acta Mater. 57, 1.

Dillon, S.J., Tang, M., Carter, W.C., Harmer, M.P., 2007. Acta Mater. 55, 6208.

Dillon, S.J., Harmer, M.P., Rohrer, G.S., 2010. J. Am. Ceram. Soc. 93, 1796.

Dregia, S.A., Wynblatt, P., 1991. Acta Metall. Mater. 39, 771.

Dregia, S.A., Wynblatt, P., Bauer, C.L., 1987. J. Vac. Sci. Technol., A 5, 1746.

Du, D., 2006. Ph.D. Thesis. University of Michigan, Ann Arbor, MI.

Edington, J.W., 1975. Practical Electron Microscopy in Materials Science. In: Electron Diffraction in the Electron Microscope, vol. 2. Philips Technical Library, Eindhoven, Amsterdam, NL, 38.

El Dasher, B.S., Pang, Y., Miller, H.M., Wynblatt, P., Rollett, A.D., Rohrer, G.S., 2004. J. Am. Ceram. Soc. 87, 724.

Ene, C.B., Schmitz, G., Kirchheim, R., Huetten, A., 2005. Acta Mater. 53, 3383.

Fecht, H.J., Gleiter, H., 1985. Acta Metall. 33, 557.

Fishmeister, H.F., 1985. J. Phys., Colloq. C4 (45), 3.

Flemings, M.C., 1974. Solidification Processing. McGraw-Hill, New York.

Flinn, P.A., 1960. Trans. Metall. Soc. AIME 218, 145.

Foiles, S.M., 1989. Phys. Rev., B 40, 11502.

Foiles, S.M., 1992. In: Gonis, A., Stocks, G.M. (Eds.), Equilibrium Structure and Properties of Surfaces and Interfaces. Plenum Press, New York, p. 89.

Forwood, C.T., Clarebrough, L.M., 1991. Electron Microscopy of Interfaces in Metals and Alloys. Adam Hilger Publishing Ltd, Bristol, UK. p. 128.

Fraczkiewicz, A., Biscondi, M., 1985. J. Phys. 46, C4–C497.

Frank, F.C., van der Merwe, J.H., 1949. Proc. R. Soc. London, Ser. A 198 (205), 216.

Frank, F.C., 1950. In: Report of the Symposium on the Plastic Deformation of Crystalline Solids. Carnegie Institute of Technology, Pittsburgh, PA, p. 150.

Frank, F.C., 1951. Philos. Mag. 42, 801.

Frank, F.C., 1953. Acta Metall. 1, 15.

Frank, F.C., 1958. Growth and Perfection of Crystals. JohnWiley & Sons, New York.

Furuhara, T., Aaronson, H.I., 1991. Acta Metall. 39, 2857.

Furuhara, T., Howe, J.M., Aaronson, H.I., 1991. Acta Metall. Mater. 39, 2873.

Furuhara, T., Wada, K., Maki, T., 1995. Metall. Mater. Trans. A 26, 1971.

Gallagher, P.C.J., 1970. Metall. Mater. Trans. A 1, 2429.

Gao, Y., Dregia, S.A., Shewmon, P.G., 1989. Acta Metall. 37 (1627), 3165.

Gautam, A.R.S., Howe, J.M., 2011. Philos. Mag. A 91, 3203.

Gautam, A.R.S., and Howe, J.M., 2013 Philos. Mag. A 93, 3472.

Gleiter, H., 1971. Phys. Status Solidi B 45, 9.

Gotoh, Y., Uwaha, M., Yahokura, E., Arai, I., 1989. In: Mrs International Meeting on Advanced Materials, vol. 10. Materials Research Society, Pittsburgh, PA, p. 479.

Gottstein, G., Shvindlerman, L.S., 2010. Grain Boundary Migration in Metals thermodynamics, kinetics, applications, second ed. CRC Press, Taylor & Francis Group, Boca Raton, FL.

Grimmer, H., Bollmann, W., Warrington, D.H., 1974. Acta Crystallogr. 30, 197.

Grimmer, H., 1974. Scr. Metall. 8, 1221.

Groves, G.W., Kelly, A., 1963. Philos. Mag. 8, 877.

Gu, X.-F., Zhang, W.-Z., 2010. Philos. Mag. 90 (3281), 4503.

Gu, X.-F., Zhang, W.Z., 2011. Solid State Phenom. 172–174, 25.

Haasen, P., 1986. Physical Metallurgy, second ed. Cambridge University Press, Cambridge, UK. p. 157.

Hahn, T. (Ed.), 1983. International Tables for Crystallography. Reidel, Dordrecht, Germany.

Hall, M.G., Aaronson, H.I., Kinsman, K.R., 1972. Surf. Sci. 31, 257.

Hasson, G., Boos, J.-Y., Herbeuval, I., Miscondi, M., Goux, C., 1972. Surf. Sci. 31, 115.

Herring, C., 1951. Phys. Rev. 82, 87.

Herring, C., 1953. In: Gomer, R., Smith, C.S. (Eds.), Structures and Properties of Solid Surfaces. University of Chicago Press, Chicago, p. 5.

Herrmann, G., Gleiter, H., Baro, G., 1976. Acta Metall. 24, 353.

Heyraud, J.C., Metois, J.J., 1980. Acta Metall. 28, 1789.

Heyraud, J.C., Metois, J.J., 1983. Surf. Sci. 128, 334.

Hirsch, P.B., Howie, A., Nicholson, R.B., Pashley, D.W., 1979. Electron Microscopy of Thin Crystals. R. E. Krieger Publishing Company, Malabar, FL.

Hirth, J.P., Balluffi, R.W., 1973. Acta Metall. 21, 929.

Hirth, J.P., Lothe, J., 1982. Theory of Dislocations, second ed. John Wiley & Sons, New York, NY. pp. 310, 341, 393, 703.

Hirth, J.P., Pond, R.C., 1996. Acta Mater. 44, 4749.

Hirth, J.P., Barnett, D.M., Lothe, J., 1979. Philos. Mag. A 40, 39.

Hirth, J.P., 1965. In: Mueller, W.M. (Ed.), 1965. Energetics in Metallurgical Phenomena, vol. 2. Gordon and Breach, New York, p. 1.

Hirth, J.P., 1993. J. Mater. Res. 8, 1572.

Hitzenberger, C., Karnthaler, H.P., Korner, A., 1985. Acta Metall. 33, 1293.

Hofmann, S., 1990. In: Dowben, P.A., Miller, A. (Eds.), Surface Segregation Phenomena. CRC Press, Inc, Boca Raton, FL, p. 107.

Holm, E.A., Rohrer, G.S., Foiles, S.M., et al., 2011. Acta Mater. 59, 5250.

Hondros, E.D., Seah, M.P., 1988. In: Cahn, R.W., Haasen, P. (Eds.), Physical Metallurgy. Elsevier Science Publishing Company, Amsterdam, NL, p. 888.

Howe, J.M., Gronsky, R., 1986. In: Bailey, G.W. (Ed.), Proc. 44th Annual Meeting of the Electron Microscopy Society of America. San Francisco Press, San Francisco, CA, p. 552.

Howe, J.M., Mahon, G.J., 1989. Ultramicroscopy 30, 132.

Howe, J.M., Prabhu, N., 1990. Acta Metall. 38 (881), 889.

Howe, J.M., Smith, D.A., 1992. Acta Metall. Mater. 40, 2343.

Howe, J.M., Aaronson, H.I., Gronsky, R., 1985. Acta Metall. 33 (639), 649.

Howe, J.M., Dahmen, U., Gronsky, R., 1987. Philos. Mag. A 56, 31.

Howe, J.M., Aaronson, H.I., Hirth, J.P., 2000. Acta Metall. 48, 3977.

Howe, J.M., Reynolds Jr., W.T., Vasudevan, V.K., 2002. Metall. Mater. Trans. A 33, 2391.

Howe, J.M., Gautam, A.S.R., Chatterjee, K., Phillipp, F., 2007. Acta Mater. 55, 2159.

Howe, J.M., Pond, R.C., Hirth, J.P., 2009. Prog. Mater. Sci. 54, 792.

Howe, J.M., 1993a. Int. Mater. Rev. 38, 257.

Howe, J.M., 1993b. In: Wayman, C.M., Perkins, J. (Eds.), Proc. of the International Conference on Martensitic Transformations (ICOMAT-92). Monterey Institute of Advanced Studies, Carmel, CA, p. 185.

Howe, J.M., 1994a. Metall. Mater. Trans. A 25, 1917.

Howe, J.M., 1994b. Philos. Mag. Lett. 70, 111.

Howe, J.M., 1997. Interfaces in Materials: Atomic Structure, Kinetics and Thermodynamics of Solid–vapor, Solid–liquid and Solid–solid Interfaces. John Wiley & Sons, New York.

Howe, J.M., 2006. Mater. Sci. Eng., A 438, 35.

Hull, R., Bean, J.C., 1992. Crit. Rev. Solid State Mater. Sci. 17, 507.

Hultgren, R., Desai, P.D., Hawkins, D.T., Gleiser, M., Kelley, K.K., 1973. Selected Values of Thermodynamic Properties of Binary Alloys. American Society for Metals, Metals Park, OH.

Ikuhara, Y., Pirouz, P., 1996. Mater. Sci. Forum 207–209, 121.

Janssens, K.G.F., Olmsted, D., Holm, E.A., Foiles, S.M., Plimpton, S.J., Derlet, P.M., 2006. Nat. Mater. 5, 124.

Jesser, W.A., 1973. Phys. Status Solidi A 20, 63.

Johannessen, J.W., Spicer, W.E., Gibbons, J.F., Plummer, J.D., Taylor, N.J., 1979. J. Appl. Phys. 49, 4453.

Kajiwara, S., Ogawa, K., Kikuchi, T., 1996. Philos. Mag. Lett. 4, 405.

Kato, M., Wada, M., Sato, A., Mori, T., 1989. Acta Metall. 37, 749.

Kato, M., 1991. Mater. Sci. Eng., A 146, 205.

Kelly, A., Groves, G.W., 1970. Crystallography and Crystal Defects. Addison-Wesley Publishing Company, Reading, MA p 350.

Kelly, T.F., Miller, M.K., 2007. Rev. Sci. Instrum. 78, 031101.

Kelly, P.M., Qui, D., 2011. Solid State Phenom. 172–174, 25.

Kelly, P.M., Zhang, M.-X., 1999. Mater. Forum 23, 41.

Kikuchi, R., Cahn, J.W., 1979. Acta Metall. 27, 1337.

Kronberg, M.L., Wilson, F.H., 1949. Trans. Metall. Soc. AIME 185, 50.

Krzanowski, E., Allen, S.M., 1984. Surf. Sci. 144, 153.

Kurdjumov, G., Sachs, G., 1930. Z. Phys. 64, 325.

Laird, C., Aaronson, H.I., 1969. Acta Metall. 17, 505.

Lee, Y.W., Aaronson, H.I., 1980. Acta Metall. 28, 539.

LeGoues, F.K., Aaronson, H.I., Lee, Y.W., Fix, G.J., 1983. In: Aaronson, H.I., Laughlin, D.E., Sekerka, R.F., Wayman, C.M. (Eds.), Proceedings of an International Conference on Solid–solid Phase Transformations. The Metallurgical Society of A.I.M.E, Warrendale, PA, p. 427.

Li, P., Reynolds Jr., W.T., Howe, J.M., 2004. Acta Mater. 52, 239.

Li, P., Howe, J.M., Reynolds Jr., W.T., 2006. Metall. Mater. Trans. A 37 (879), 895.

Liang, Q., Reynolds Jr., W.T., 1998. Metall. Mater. Trans. A 29, 2059.

Loiseau, A., Ricolleau, C., Potez, L., Ducastelle, F., 1995. In: Johnson, W.C., Howe, J.M., Laughlin, D.E., Soffa, W.A. (Eds.), Proceedings of the International Conference on Solid-to-solid Phase Transformations in Inorganic Materials (PTM'94). The Metals, Materials and Minerals Society, Warrendale, PA, p. 385.

Lours, P., Westmacott, K.H., Dahmen, U., 1992. In: Clark, W.A.T., Dahmen, U., Briant, C.L. (Eds.), Structure and Properties of Material Interfaces. Mrs Symp. Proc., 238. Materials Research Society, Pittsburgh, PA, p. 207.

Luo, C.P., Weatherly, G.C., 1987. Acta Metall. 35, 1963.

Luo, J., Cheng, H., Asl, K.M., Kiely, C.J., Harmer, M.P., 2011. Science 333, 1730.

Luo, J., 2009. Appl. Phys. Lett. 95, 071911.

Lupis, C.H.P., 1983. Chemical Thermodynamics of Materials. Elsevier, New York.

Luzzi, D.E., 1991. Ultramicroscopy 37, 180.

Ma, X., Pond, R.C., 2007. J. Nucl. Mater. 361, 313.

Mackenzie, J.K., Bowles, J.S., 1957. Acta Metall. 5, 137.

Mackenzie, J.K., Moore, A.J.W., Nicholas, J.F., 1962. J. Phys. Chem. Solids 23, 185.

Mahon, G.J., Howe, J.M., Mahajan, S., 1989. Philos. Mag. Lett. 59, 273.

Marcinkowski, M.J., 1963. In: Thomas, G., Washburn, J. (Eds.), Electron Microscopy and Strength of Crystals. John Wiley, Interscience Publishers, New York, NY, p. 333.

Markov, I., Stoyanov, S., 1987. Contemp. Phys. 28, 267.

Massalski, T.B., Okamoto, H., Subramanian, P.R., Kacprzak, L. (Eds.), 1986. Binary Alloy Phase Diagrams, vols. 1–3. ASM International, Metals Park, OH.

Massalski, T.B., Soffa, W.A., Laughlin, D.E., 2006. Metall. Mater. Trans. A 37, 825.

Matthews, J.W., 1974. Philos. Mag. 29, 797.

Mauritz, K.A., Baer, E., Hopfinger, J.A., 1978. J. Polym. Sci., Part D: Macromol. Rev. 13, 1.

McTague, J.P., Novaco, A.D., 1979. Phys. Rev., B 19, 5299.

Meijering, J.L., 1950. Philips Res. Rep. 5, 333.

Merkle, K.L., 1995. Interface Sci. 2, 311.

Miyano, N., Ameyama, K., Weatherly, G.C., 2000. ISIJ Int. 40, S199.

Moritani, T., Miyajima, N., Furuhara, T., Maki, T., 2002. Scripta Mater. 47, 93.

Mou, Y., Aaronson, H.I., 1994. Acta Metall. Mater. 42, 2133.

Mou, Y., 1994. Metall. Mater. Trans. A 25, 1905.

Muddle, B.C., Nie, J.F., Hugo, G.R., 1994. Metall. Mater. Trans. A 25, 1841.

Murr, L.E., 1975. Interfacial Phenomena in Metals and Alloys. Addison-Wesley Publishing Co, Reading, MA.

Mykura, H., 1980. Grain Boundary Structure and Kinetics. American Society for Metals, Metals Park, OH. p. 445.

Nicholas, J.F., 1968. Aust. J. Phys. 21, 21.

Nie, J.F., Muddle, B.C., 2002. Metall. Mater. Trans. A 33, 2381.

Nie, J.F., 2004. Acta Mater. 52, 795.

Nie, J.F., 2005. Scr. Mater. 52, 687.

Niewczas, M., 2007. In: Nabarro, F.R.N., Hirth, J.P. (Eds.), Dislocations in Solids. Elsevier B. V, Amsterdam, NL. (Chapter 75).

Nishiyama, Z., 1934. Sci. Rep. Res. Inst. Tohoku Univ. 23, 638.

Nishiyama, Z., 1978. Martensitic Transformations. Academic Press, New York, NY. p. 49.

Norton, M.G., Carter, C.B., 1992. In: Wolf, D., Yip, S. (Eds.), Materials Interfaces: Atomic-level Structure and Properties. Chapman Hall, London, p. 151.

Novaco, A.D., McTague, J.P., 1977. Phys. Rev. Lett. 38, 1286.

Ogawa, K., Kajiwara, S., 2004. Philos. Mag. 84, 2919.

Ourmazd, A., Taylor, D.W., Bode, M., Kim, Y., 1989. Science 246, 1571.

Pirouz, P., Ernst, F., Ikuhara, Y., 1998. Solid State Phenom. 55–60, 51.

Plichta, M.R., Rigsbee, J.M., Hall, M.G., Russell, K.C., Aaronson, H.I., 1976. Scr. Metall. 10, 1065.

Plichta, M.R., Perepezko, J.H., Aaronson, H.I., Lange, W.F., 1980. Acta Metall. 28, 1031.
Pond, R.C., Celotto, S., 2003. Int. Mater. Rev. 48, 225.
Pond, R.C., Hirth, J.P., 1994. In: Ehrenrich, H., Turnbull, D. (Eds.), 1994. Solid State Physics, 37. Academic Press, San Diego, CA, p. 288.
Pond, R.C., Sarrazit, F., 1996. Interface Sci. 4, 99.
Pond, R.C., Vlachavas, D., 1983. Proc. R. Soc. London, Ser. A 386, 95.
Pond, R.C., Smith, D.A., Vitek, V., 1978. Scr. Metall. 12, 699.
Pond, R.C., Smith, D.A., Vitek, V., 1979. Acta Crystallogr. A 35, 689.
Pond, R.C., Celotto, S., Hirth, J.P., 2003. Acta Mater. 51, 5358.
Pond, R.C., Ma, X., Hirth, J.P., 2006. Mater. Sci. Eng., A 438–440, 109.
Pond, R.C., Ma, X., Chai, Y.W., Hirth, J.P., 2007. In: Nabarro, F.R.N., Hirth, J.P. (Eds.). Dislocations in Solids, 13. Elsevier, Amsterdam, p. 225.
Pond, R.C., 1989. In: Nabarro, F.R.N. (Ed.), Dislocations in Solids. Elsevier Science Publishers, Amsterdam, NL. (Chapter 38).
Ramanujan, R., 1990. Ph.D. Thesis. Carnegie Mellon University, Pittsburgh, PA.
Ramirez, R., Rahman, A., Schuller, I.K., 1984. Phys. Rev. 830, 6208.
Randle, V., 1996. The Role of the Coincidence Site Lattice in Grain Boundary Engineering. Institute of Materials, London.
Ranganathan, S., 1966. Acta Crystallogr. 21, 197.
Rao, G., Zhang, D.-B., Wynblatt, P., 1993. Scr. Metall. Mater. 28, 459.
Rao, G., Howe, J.M., Wynblatt, P., 1994. Scr. Metall. Mater. 30, 731.
Rao, G., 1994. Ph.D. Thesis. Carnegie Mellon University, Pittsburgh, PA.
Read, W.T., Shockley, W., 1950. Phys. Rev. 78, 275.
Read, W.T., 1953. Dislocations in Crystals. McGraw-Hill Book Company, Inc, New York, NY. p. 155.
Reynolds Jr., W.T., Farkas, D., 2006. Metall. Mater. Trans. A 37, 865.
Reynolds Jr., W.T., Nie, J.F., Zhang, W.-Z., Howe, J.M., Aaronson, H.I., Purdy, G.R., 2003. Scr. Mater. 49, 405.
Rigsbee, J.M., Aaronson, H.I., 1979. Acta Metall. 27 (351), 365.
Rogers III, J.P., Wynblatt, P., Foiles, S.M., Baskes, M.I., 1990. Acta Metall. 38, 177.
Rohrer, G.S., 2007. JOM 59, 38.
Rohrer, G.S., 2011. J. Mater. Sci. 46, 5881.
Rong, Y., 2012. Characterization of Microstructures by Analytical Electron Microscopy (AEM). Higher education Press and (Heidelberg/Berlin: Springer-Verlag), Beijing. pp. 171–253.
Sargent, C.M., Purdy, G.R., 1975. Philos. Mag. 32, 27.
Sasjima, Y., Makagawa, S., Murai, Y., Ichimura, M., Imbayashi, M., 1989. In: Mrs International Meeting on Advanced Materials, vol. 10. Materials Research Society, Pittsburgh, PA, p. 485.
Sass, S.L., Bristowe, P.D., 1980. In: Balliffi, R.W. (Ed.), Grain Boundary Structure and Kinetics. American Society for Metals, Metals Park, OH, p. 531.
Saylor, D.M., El Dasher, B.S., Rollett, A.D., Rohrer, G.S., 2004. Acta Mater. 52, 3649.
Schober, T., Balluffi, R.W., 1970. Philos. Mag. 21, 109.
Schwartzman, A.F., 1990. In: Sinclair, R., Smith, D.J., Dahmen, U. (Eds.), 1990. Mrs Symp. Proc., 183. Materials Research Society, Pittsburgh, PA, p. 161.
Seah, M.P., Hondros, E.D., 1973. Proc. R. Soc. London, Ser. A 335, 191.
Seah, M.P., Lea, C., 1975. Philos. Mag. 31, 627.
Seki, A., Seidman, D.N., Oh, Y., Foiles, S.M., 1991. Acta Metall. 39 (3167), 3179.
Shewmon, P.G., 1965. Recrystallization, Grain Growth and Textures. American Society for Metals, Metals Park, OH. p. 165.
Shi, X., Luo, J., 2009. Appl. Phys. Lett. 94, 251908.
Shibata, T., Furuhara, T., Maki, T., 2010. Acta Mater. 58, 3477.
Shiflet, G.J., van der Merwe, J.H., 1994. Metall. Trans. A 24, 1895.
Shiflet, G.J., 1986. Mater. Sci. Eng. 81, 61.
Sinclair, R., Smith, D.J., Dahmen, U., 1990. High Resolution Electron Microscopy of Defects in Materials. Materials Research Society, Pittsburgh, PA.
Slater, J.C., 1934. Phys. Rev. 45, 794.
Smith, D.A., Pond, R.C., 1976. Int. Met. Rev. 205, 61.
Smith, D.A., Shiflet, G.J., 1987. Mater. Sci. Eng. 86, 67.
Somorjai, A., 1972. Principles of Surface Chemistry. Prentice-Hall, Englewood Cliffs, New Jersey.
Spaepen, F., 1975. Acta Metall. 23, 729.
Spanos, G., 1989, Ph.D. Thesis. Carnegie Mellon University, Pittsburgh, PA.
Stephens, D.E., Purdy, G.R., 1975. Acta Metall. 23, 1343.
Stinchcombe, R.B., 1983. Phase Transitions and Critical Phenomena. Academic Press, New York, NY. (Chapter 3).

Stoop, L.C.A., van der Merwe, J.H., 1982. Thin Solid Films 94, 341.
Sutton, A.D., Balluffi, R.W., 1995. Interfaces in Crystalline Materials. Clarendon Press, Oxford, UK.
Sutton, A.P., Vitek, V., 1982. Acta Metall. 30, 2011.
Sutton, A.P., Vitek, V., 1983. Philos. Trans. R. Soc., A 309 (1), 37–55.
Swalin, R.A., 1972. Thermodynamics of Solids, second ed. John Wiley & Sons, New York, NY.
Tan, T.Y., Sass, S.L., Balluffi, R.W., 1975. Philos. Mag. 31 (456), 575.
Tang, M., Carter, W.C., Cannon, R.M., 2006. Phys. Rev. Lett. 97, 075502/1.
Turnbull, D., 1955. Impurities and Imperfections. American Society for Metals, Metals Park, OH. p. 121.
van der Merwe, J.H., Shiflet, G.J., 1994. Acta Metall. Mater. 42, 1173, 1189, 1199.
van der Merwe, J.H., Shiflet, G.J., Stoop, P.M., 1991. Metall. Trans. A 22, 1165.
van der Merwe, J.H., 1950. Proc. R. Soc. London, Ser. A 63, 616.
van der Merwe, J.H., 1963. J. Appl. Phys. 34, 117,123.
van der Merwe, J.H., 1982. Philos. Mag. A 45, 127,145,159.
Varela, M., Gazquez, J., Pennycook, S.J., 2012. MRS Bull. 37, 29.
Wang, H.Y., Najafabadi, D.J., Lesar, R., 1993. Acta Metall. 41, 2533.
Wassermann, G., 1933. Arch. Eisenhutt. Wes. 16, 647.
Wayman, C.M., 1964. Introduction to the Crystallography of Martensitic Transformation. The Macmillan Company, New York.
Weatherly, G.C., Zhang, W.Z., 1994. Metall. Mater. Trans. A 25, 1865.
Wechsler, M.S., Lieberman, D.S., Read, T.A., 1953. Trans. Metall. Soc. AIME 197, 1503.
Wigner, E., Seitz, F., 1933. Phys. Rev. 43, 804.
Wolf, D., Merkle, D., 1992. In: Wolf, D., Yip, S. (Eds.), Materials Interfaces: Atomic-level Structure and Properties. Chapman Hall, London, p. 87.
Wolf, D., Yip, S. (Eds.), 1992. Material Interfaces: Atomic-level Structure and Properties. Chapman Hall, London, UK.
Wolf, D., 1990. J. Mater. Res. 5, 1708.
Wolf, D., 1992. In: Wolf, D., Yip, S. (Eds.), Materials Interfaces: Atomic-level Structure and Properties. Chapman Hall, London, pp. 1–17.
Wulff, G., 1901. Z. Krystall. Mineral 34, 449.
Wynblatt, P., Ku, R.C., 1977. Surf. Sci. 65, 511.
Xiao, S.Q., Howe, J.M., 2000. Acta Mater. 48, 3253.
Yanar, C., Wiezorek, J.M.K., Radmilovic, V., Soffa, W.A., 2002. Metall. Mater. Trans. A 33, 2413.
Yang, Z., Johnson, R.A., 1993. Modell. Simul. Mater. Sci. Eng. 1, 707.
Yoon, K.E., Noebe, R.D., Seidman, D.N., 2007. Acta Mater. 55, 1145.
Zhang, M.X., Kelly, P.M., 1998. Acta Mater. 46, 4617.
Zhang, M.-X., Kelly, P.M., 2009. Prog. Mater. Sci. 54, 1101.
Zhang, W.-Z., Purdy, G.R., 1993. Philos. Mag. A 68 (279), 291.
Zhang, W.-Z., Weatherly, G.C., 1998. Acta Mater. 46, 1837.
Zhang, W.-Z., Weatherly, G.C., 2005. Prog. Mater. Sci. 50, 181.
Zhang, W.Z., Yang, X.-P., 2011. J. Mater. Sci. 46, 4135.

Biography

James M. Howe received his Ph.D. in Materials Science from the University of California-Berkeley (1985) and joined Carnegie Mellon University as the Alcoa Assistant Professor of Physical Metallurgy. He moved to the University of Virginia (1991), where he was promoted to Full Professor (1999) and recognized with the Thomas Goodwin Digges Chair (2010). He is Director of the Nanoscale Materials Characterization Facility, and his research utilizes in-situ TEM techniques to study plasmons and the dynamic behavior of interfaces in materials. Dr. Howe received a von Humboldt Senior Research Award (1999), the ASM Materials Science Research Silver Medal (2000), and the TMS Champion H. Mathewson Research Medal (2005, 2009). He was a visiting professor at the University of Vienna and Osaka University, and is author or co-author of the books "Interfaces in Materials" (1997), "In-Situ Electron Microscopy" (2012), and "Transmission Electron Microscopy and Diffractometry of Materials" (4th ed. 2013).

15 Atom-Probe Field Ion Microscopy

K. Hono, National Institute of Materials Science, Tsukuba, Japan
S.S. Babu, The University of Tennessee, Knoxville, Tennessee, USA

Nomenclature

D Distance between an FIM tip and a phosphor screen

e Charge of electron

F Electric field on an FIM tip surface

G Gibbs free energy

H_c Coercivity

k Geometrical factor

$\langle K \rangle$ Average magnetocrystalline anisotropy

l Flight length

L_{ex} Exchange length

m Atomic mass

m/n Mass to charge ratio

M Magnification

M_s Saturation magnetization

Q_i Charge

r Radius of an FIM tip apex

T_g Glass transition temperature

T_x Crystallization temperature

v Velocity of ion

V_{dc} DC standing voltage applied to an FIM specimen

V_e Voltage of ion

V_p Pulse voltage

x Atomic fraction of a solute

α Pulse factor

δ Delay time

ΔT_x Supercooled liquid region

λ_s Saturation magnetostriction constant

σ Standard deviation

ADC Analogue to digital converter

AEM Analytical electron microscope

AC Alternative current

AP Atom probe

APT Atom probe tomography; APT is more commonly used for the 3DAP technique recently, but tomography is not used for atom-probe data presentation. Hence, in this chapter, 3DAP is adopted as general terminology for the atom probe technique with position information. Since all atom probes currently used are now equipped with position sensitive detector, "atom probe" (AP) can be used for 3DAP and APT

APFIM Atom-probe field ion microscope

DC Direct current

DSC Differential scanning calorimetry

ECOPoSAP Energy-compensated optical position sensitive atom probe

EDS Energy dispersive X-ray spectroscopy

EELS Electron energy loss spectroscopy

fcc Face centered cubic

FIB Focused ion beam

FIM Field ion microscope	**MCP** Microchannel plate
FINEMET Fe–Si–B–Nb–Cu nanocrystalline soft magnetic material	**OPoSAP** Optical position sensitive atom probe
GP zones Guinier–Preston zones	**PoSAP** Position sensitive atom probe
GPB zones Guinier–Preston–Bagaryatsky zones	**PSD** Position sensitive detector
HAADF High-angle annular dark field	**RE** Rare earth element
HREM High-resolution electron microscope	**TAP** Tomographic atom probe
	ToFAP Time-of-flight atom probe
LEAP Local electrode atom probe	**TDC** Time-to-digital converter
Ln Lanthanide metals	**3DAP** Three-dimensional atom probe

15.1   Introduction

More than fifty years have passed since Müller et al. invented the atom-probe field ion microscope (APFIM) in 1968 (Müller et al., 1968). The APFIM was originally developed as a tool for surface science; however, from the emerging stage of the technique, physical metallurgists realized that its use could solve many critical problems on the microstructures of metallic materials, in particular steels. By the 1980s, several groups started applying the atom probe (AP) technique for microstructural characterizations of various metallic materials. Besides metals, it was also applied to a wide variety of materials such as semiconductors and oxide superconductors. However, due to the limitation of the AP using voltage pulses for field evaporation, most of the early attempts of AP analyses of nonmetallic materials were limited only to feasibility demonstrations. Until 1990s, the most widely used AP was a time-of-flight atom probe (ToFAP). By applying high voltage pulses to a needle-like specimen (tip) for FIM, atoms are evaporated from the surface due to the presence of high electric field of the order of 10 MV/m as shown in **Figure 1a**.

The sharp needle-like specimen (tip) is required to achieve the electric field high enough to field-evaporated atoms from the apex of the tip, as well as, to obtain a projection field toward the detector. The field-evaporated atoms, i.e. positively charged ions, are accelerated toward the grounded FIM screen, which is about several centimeters away from the tip. Because of the presence of the projection field, the field-evaporated ions are radially emitted, so the projection magnification at the aperture is in the order of a few million. The ions that go through a small aperture of diameter 2 mm, referred as *"probe hole"*, are detected one-by-one using a microchannel plate (MCP) detector. Using the time of flight (ToF) from the tip to the detector, the identities of the ions (i.e. mass to charge ratio) are determined. Because of the high magnification, the effective size of the aperture on the tip apex is only ~2 nm. As a result, the AP collects field-evaporated atoms only within a selected area, and thereby sets the diameter of the *cylinder of analysis* to 2 nm. The height of the cylinder defines the depth of analysis from the initial tip surface. The local concentrations within this nanoscale cylinder are measured by counting the number of detected atoms with respect to total number of collected atoms. Since the depth scale is related to the number of collected atoms, the dimensionality of this concentration depth profile is one. Therefore, the ToF AP is now called *"one-dimensional atom probe"* or 1DAP in contrast to modern *"three-dimensional atom probe"* or 3DAP (see **Figure 1b**).

(a)

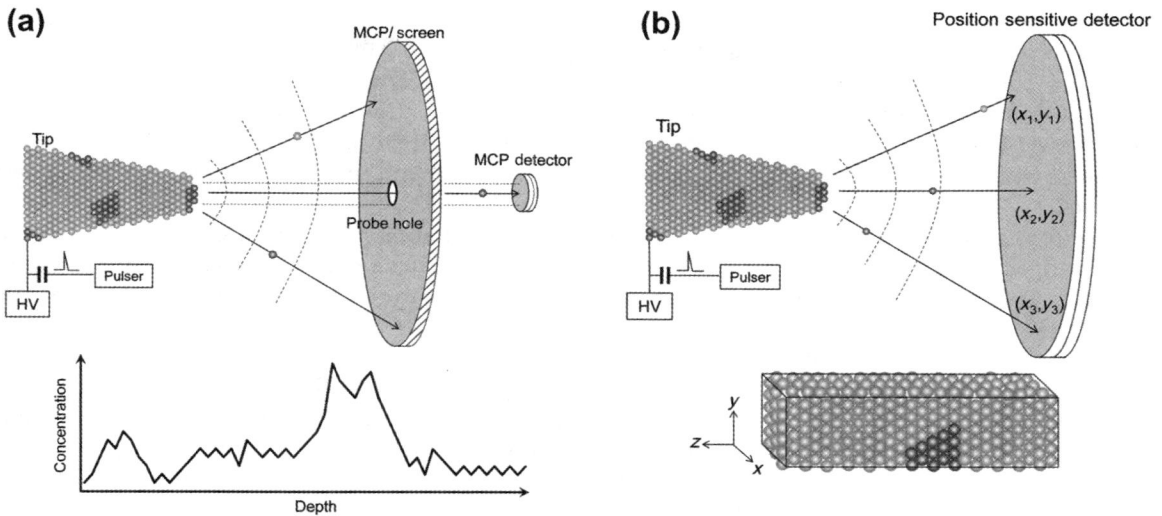

(b)

Figure 1 Schematic illustration of (a) a time-of-flight atom probe (ToFAP) and (b) three-dimensional atom probe (3DAP). Note that the actual radius of the tips are ~50 nm and in the figure it is exaggerated to emphasize the atomic feature of the field evaporation. (For color version of this figure, the reader is referred to the online version of this book.)

In 1978, Panitz et al. invented an *imaging atom probe* in which the positions of specific atoms were imaged by gating an MCP detector with timing signals (Panitz, 1978). The MCP detector is activated only at specific time intervals corresponding to ToF of specific ion. This imaging AP showed two-dimensional elemental mapping of only one type of ion in an alloy. Thus, the imaging analysis leads only to qualitative two-dimensional mapping of single alloying elements. Due to its poor quantitativeness and limited mass resolution, the imaging AP was not widely used for materials' characterization. On the other hand, the *position sensitive atom probe* (PoSAP), first developed by Cerezo et al. (1988) in 1988, adopted a *position sensitive detector* (PSD) in a ToFAP. This innovation succeeded in obtaining two-dimensional elemental maps (x- and y- coordinates) with a near-atomic resolution for all alloying elements as shown in **Figure 1 (b)**. These authors also demonstrated that locations of sequentially collected elemental ions could be used for three-dimensional tomography by assigning z-coordinates to the same and by using tomography software (Cerezo et al., 1989). This was the first implementation of 3DAP. However, the first generation PoSAP was not able to detect more than two atoms those triggered by one pulse event, because of the serial nature of the wedge-and-stripe-type PSD (Cerezo et al., 1988). This limitation forced the data acquisition speed to be slow, because the ionization rate had to be adjusted so that only one field ionization occurs for each pulse. Furthermore, the detection efficiency was somewhat lower in PoSAP compared to the modern 3DAP. As a result, multiple signals triggered by a single pulse had to be discarded. On the other hand, the *tomographic atom probe* (TAP) developed by Blavette et al. (1993) adopted a 96-channel multianode parallel detector, which made it possible to detect more than two atoms triggered by single pulse. The time resolution of each sequential event was also improved, and up to eight sequential atom arrivals can be detected using an eight 96-channel charge-sensitive analogue to digital converters (ADCs). In this detector, each of the channels is triggered by individual time event. The tomographic atom probe (TAP) improved the data acquisition speed significantly, and with ability to detect multihit ions, the detection efficiency was also improved compared to its PoSAP. The data obtained by the TAP clearly demonstrated that the reconstructed 3D tomography had atomic layer resolution for the first time (Deconihout et al., 1994; Blavette et al., 1998). As both PoSAP and TAP provide 3D tomography of atoms, the APFIM equipped with a PSD is now generally called a 3DAP. This technique is also referred as *atom probe tomography* (APT) in recent literature.

Since the conduction of voltage pulse is not possible in insulating materials, the applications of the AP technique had been limited to only electrical conductive materials, mainly metals, until recently. However, the recent successful implementation of pulsed laser to assist field evaporation for 3DAP analyses (Deconihout et al., 2007a; Gault et al., 2005; Cerezo et al., 2006; Schlesiger et al., 2010) expanded the application areas to a wide variety of materials including semiconductors (Kelly et al., 2007), insulator thin films (Pinitsoontorn et al., 2008) and even to bulk insulating ceramics (Chen et al., 2009a). In addition, the development of site specific specimen preparation method using the focused ion beam (FIB) technique (Thompson et al., 2007) made it possible to prepare tips from all types of materials, including powder (Srinivasarao et al., 2009), thin films (Larson et al., 2004a) and devices (Chiaramonti et al., 2008). The substantially reduced frequency of specimen ruptures in the ultraviolet laser-assisted mode was also critical to make the 3DAP technique practical for a wide range of materials (Hono et al., 2011a).

The aim of this chapter is to provide the basic principles of the 3DAP technique for those who are about to start using this technique for microstructure characterization of materials. In recent APT using commercial *local electrode atom probe* (LEAP), field ion microscopy (FIM) is rarely used. Nevertheless, in Section 15.2, the fundamentals of FIM is introduced, because fundamental understanding of physical phenomena relevant to FIM such as *field ionization* and *field evaporation* are essential for accurate interpretation of APT data. In Section 15.2, common data analyses methods are also described. In Section 15.3, scope of APFIM analyses with reference to physical phenomena is introduced to demonstrate its relevance with reference to other techniques. In Section 15.4, specific applications of APFIM techniques, both 1D and 3D, with reference to metals and alloys, ceramics and semiconductors are reviewed. In order to provide comprehensive review of the original research on materials problems, we intentionally did not limit the examples to recent APT work. Many early examples based on analyses are also presented. Although this chapter attempts to review the contribution of the APFIM technique to physical metallurgy, it is not comprehensive enough to cover the current technical advances in the atom-probe FIM itself. Therefore, readers are encouraged to read more comprehensive textbooks (Wagner, 1982; Miller and Smith, 1989; Tsong, 1990; Miller et al., 1996; Miller, 2000) on this topic. Recently, Gault et al. published a comprehensive textbook on AP tomography including useful data relevant for practical utilization of APT (Galt et al., 2012).

15.2   Overview of the Atom Probe Technique

15.2.1   High-Field Phenomena—Field Emission, Ionization, and Evaporation

The AP technique has developed based on the field ion microscope (FIM) that utilizes two high-field phenomena called "field ionization" and "field evaporation". By applying high voltage to sharp tips, electric field higher than $>10^9$ V/m is generated around the top. As a result, a radial isofield surface is formed from the tip apex to the grounded phosphor screen. This field causes either emission, or ionization or evaporation depending on the polarity and strength of the field (Miller, 2000), as shown in **Figure 2**.

The field emission is the emission of electrons from a surface of an anode by tunneling phenomena. To emit electron from a tip, a negative voltage must be applied. In one-dimensional model, an electron at a distance x from the anode surface feels the potential $V(x)$ expressed as

$$V(x) = -\frac{e^2}{16\pi\varepsilon_0}x^{-1} - eFx, \tag{1}$$

where e is the elemental charge, ε_0 is the permittivity of free space and F is the electric field strength. In this equation, the first term is the image potential and the second term means the external field. The potential has a hump in the vicinity of the surface and induces tunneling of electrons from the surface of the specimen to vacuum when the external field reaches ~ 1 V/nm as shown in **Figure 2(a)**. This phenomenon is used for field emission microscope (FEM), a predecessor to FIM as shown in **Figure 3**. Filed emission from metal tip is also used as electron source for modern field emission-type transmission electron microscope (FE-TEM) and field emission scanning electron microscope (FE-SEM).

When positive voltage is applied to a tip under the presence of an imaging gas such as H_2, He and Ne, the field ionization of the gas atoms occur at protruding sites on the tip surface as shown in **Figure 2 (b)** and **Figure 4**. The neutral gas atoms are polarized under the presence of electric field and they approach the metal surface due to the positive external field (**Figure 4**). In one-dimensional model (Galt et al.,

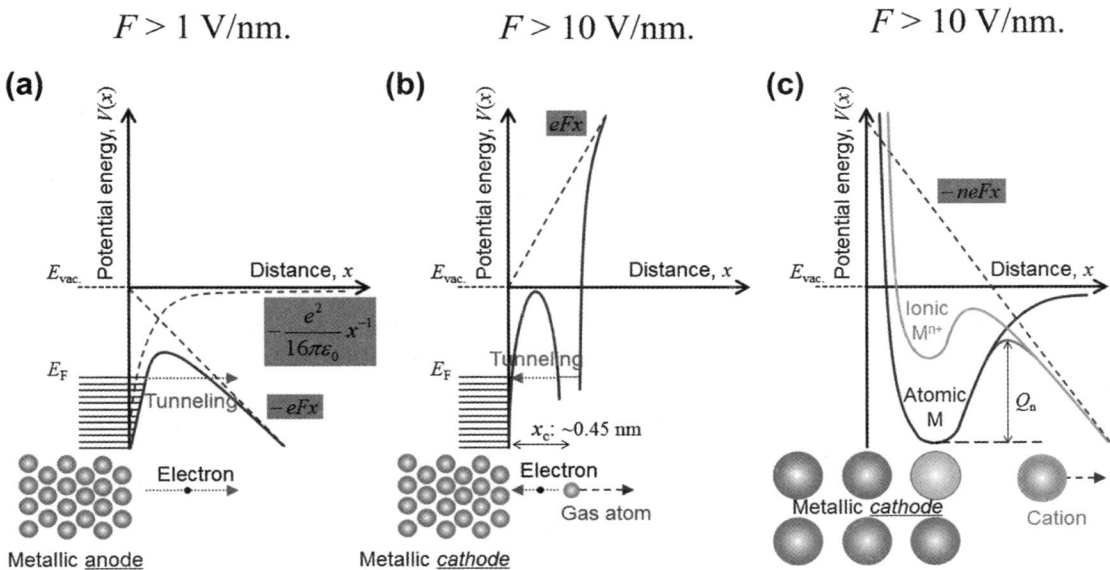

Figure 2 Schematic illustration of three phenomena that are observed under a high electrical field: (a) field emission, (b) field ionization and (c) field evaporation. (For color version of this figure, the reader is referred to the online version of this book.)

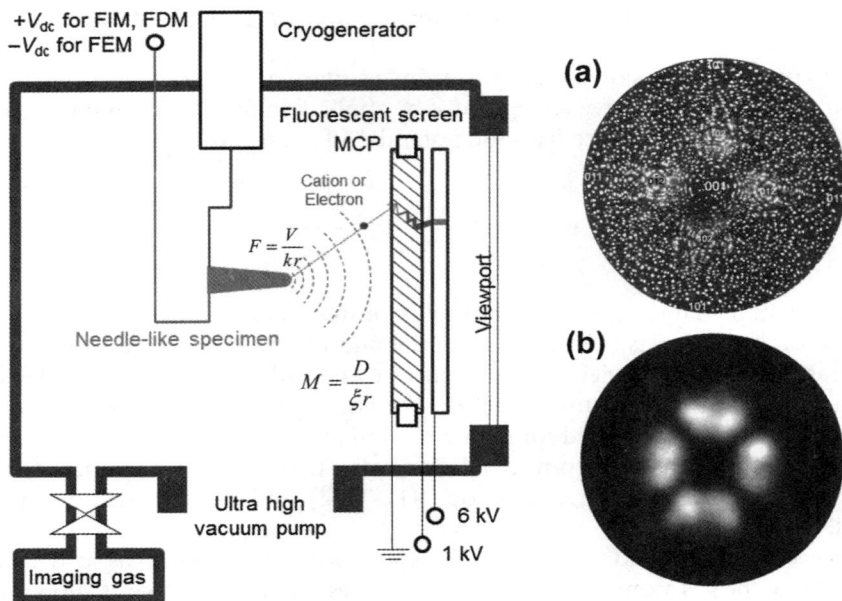

Figure 3 Schematic illustration of field ion microscopy (FIM), field desorption microscopy (FDM), and field emission microscopy (FEM). (a) Field Ion image of Ni using Ne as an imaging gas and (b) field emission image taken from exactly the same tip. (For color version of this figure, the reader is referred to the online version of this book.)

2012), the potential $V(x)$, which is felt by the electrons of the neutral species at a distance x from the cathode surface, is expressed as

$$V(x) = -\frac{e^2}{16\pi\varepsilon_0}x^{-1} + eFx + \frac{e^2}{4\pi\varepsilon_0|x_i - x|} + \frac{e^2}{4\pi\varepsilon_0(x_i - x)}, \qquad (2)$$

where x_i is the distance between the center of the positive ion and the image plane of the cathode. In this equation, the third and fourth terms correspond to the potential well of the electrode and the repulsion of the electron by the image force of the ion in the electrode, respectively. Although the rate of tunneling increases with decrease in the distance to the metallic electrode, the tunneling is forbidden at distances closer than a critical distance $x_c \sim 0.45$ nm. Because, at this distance, the energy level of electrons in the neutral species is lower than the Fermi level in the metallic electrode and all the energy levels are occupied by electrons. This region is known as forbidden zone (**Figure 2(b)**). The observation of field ionization is usually conducted in the presence of imaging gas under an electric field strength of $\sim 10^{10}$ V/m, which is ten times higher than that for field emission.

The field evaporation is defined as desorption of a surface atom on cathode as a cation. The ionic states of surface atoms are generally metastable with respect to the neutral atomic state in the absence of external electric field, but the application of positive electric field makes the ionic states more stable and prompts desorption of a surface atom as a cation caused by thermal activation over a reduced activation energy barrier. In one-dimensional charge exchange model (Gormer model)

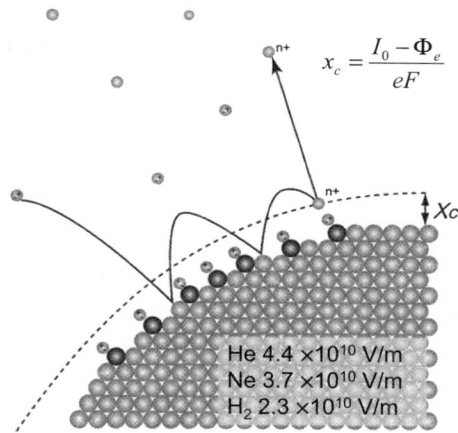

$$x_c = \frac{I_0 - \Phi_e}{eF}$$

He 4.4 ×10^{10} V/m
Ne 3.7 ×10^{10} V/m
H$_2$ 2.3 ×10^{10} V/m

Figure 4 Schematic illustration of field ionization process and ionization field for various imaging gases.

(Müller and Krishnaswamy, 1968), the reduced activation energy barrier is expressed as (Drachsel et al., 1989),

$$Q_n = \left(\Lambda + \sum_n I_n - n\phi \right) - \frac{(ne)^2}{16\pi\varepsilon_0} x^{-1} - neFx - \frac{B_a}{2} - \Delta E_a + \mu_a F + \frac{1/2(\alpha_a - \alpha_i)}{4\pi\varepsilon_0} F^2, \qquad (3)$$

where Λ is the sublimation heat of a neutral atom, I_n is the nth ionization energy of the atoms, ϕ is the work function of the emitting surface, x is the distance from the electrical surface of the metal at which the potential curves intersect, B_a and ΔE_a represent the broadening and shift of the adsorbate energy level due to the interaction with the anode, μ_a is the zero field dipole moment of the adsorbate and α_a and α_i are the polarizabilities in the neutrally bound and ionic states, respectively. In this model, Schottky hump lies inside the potential energy curve of the neutral state (**Figure 2(c)**). The generation of multicharged ions is explained by post-ionization, which is an additional field ionization of field-evaporated cation (Da Costa et al., 2005). The observation of field evaporation is usually conducted under ultrahigh vacuum (UHV) in the absence of imaging gas by a single-atom detector under an electric field strength of ∼ 10 V/nm. When field evaporation rate is fast, field desorption image can be observed.

15.2.2 Field Ion Microscope

APFIM is a combination of a FIM and a time-of-flight mass spectrometer. Thus, understanding the basic principle of FIM is essential for understanding the AP technique. In this section, only a brief description of FIM is given, and for more details, the reader is directed to other references (Miller and Smith, 1989; Tsong, 1990; Miller et al., 1996; Miller, 2000). FIM is a projection-type microscope with a magnification of a million. Using FIM, individual atoms on tip apex can be imaged by field ionization of imaging gas atoms at the protruding sites on the surface of a metal tip. A typical diameter of an FIM tip is approximately 100 nm, and a high voltage ranging from 5 to 20 kV is applied to the tip. The electric field F on the tip surface is given by

$$F = V/kr, \tag{4}$$

where k is an appropriate constant and r is the radius of a tip, so the field on a tip apex is approximately 20–30 V/nm. Under such a high electric filed, imaging gas atoms (e.g. Ne or He) introduced in the FIM chamber are ionized on the tip surface by the *field ionization* process (see **Figure 4**). The ionization field to obtain the best field ion image varies depending on imaging gas, i.e. 40, 35, and 19 V/nm for He, Ne, and H_2; thus, in order to observe Ne field ion image at a specimen voltage of 10 kV, the tip radius must be approximately 50 nm. The field ionization of the imaging gas atoms occurs preferentially at the site where the electric field is locally high, i.e. protruding site on hemispherical tip surface as shown in **Figure 4**, and these ions are accelerated toward an MCP detector normal to the isofield surface between the tip and the detector surface. When ions arrive at the surface of MCP, secondary electron shower is created, which forms bright spots on a phosphor screen. Thus, the FIM image is a projection of protruding atoms on a tip surface to an MCP, and its magnification M is given by

$$M = \beta D/r, \tag{5}$$

where β is a parameter to adjust the projection angle and D is the distance between the tip apex and the MCP. For Gnomic projection, $\beta = 1$ and for stereographic projection, $\beta = 0.5$ and the actual value varies depending on the shape of the isofield surface near the tip apex. A typical tip radius is 50 nm, so if D is 10 cm, the magnification will be in the range of a million.

Sharp tips of metals and alloys are prepared by electropolishing wires or square rods extracted from bulk. Oxide films often cover freshly prepared tips. As a result, their surface is atomically rough. However, above a certain magnitude of electric field, atoms from the specimens are evaporated to form singly or multiply charged ions. This *field evaporation* occurs from the protruding region where local electric field is higher. Thus, by continuing the field evaporation of surface atoms, hemispherical clean surface is automatically developed. This field evaporation phenomenon makes the atomic resolution of FIM possible, because FIM image, based on the radial projection of surface atoms, cannot be obtained unless the tip surface is atomically smooth hemisphere. The field evaporation also makes AP mass spectroscopy possible. The evaporation field of n-fold charged ions for metals is given by the following equation

$$F_{en} \approx \left(\Lambda + \sum_{i}^{n} I_i - n\varphi \right)^2 / n^3 e^3, \tag{6}$$

which should become minimum for a certain number of n. In most of the cases, the ions with the lowest calculated field correspond to the most frequently observed ions experimentally. The charge states also change depending on the specimen temperature, as well as photon energy of laser-assisted field evaporation. At lower temperature, the ratio of multiple ions tends to be more frequently observed. Note that the charge state in the field-evaporated ions do not have any correlation with the charge states in ionic compounds.

A typical FIM image obtained from a metallic surface shows highly symmetrical concentric ring patterns as shown in **Figure 3(a)** and **Figure 5(a)**. **Figure 5(a)** shows a He field ion image of tungsten. The simulated image (**Figure 5(b)**) and corresponding indices of the lattice plane, using stereographic projection, are shown in **Figure 5(c)**. Concentric rings show atomic steps on the hemispherical surface.

(a)

(b)

(c)

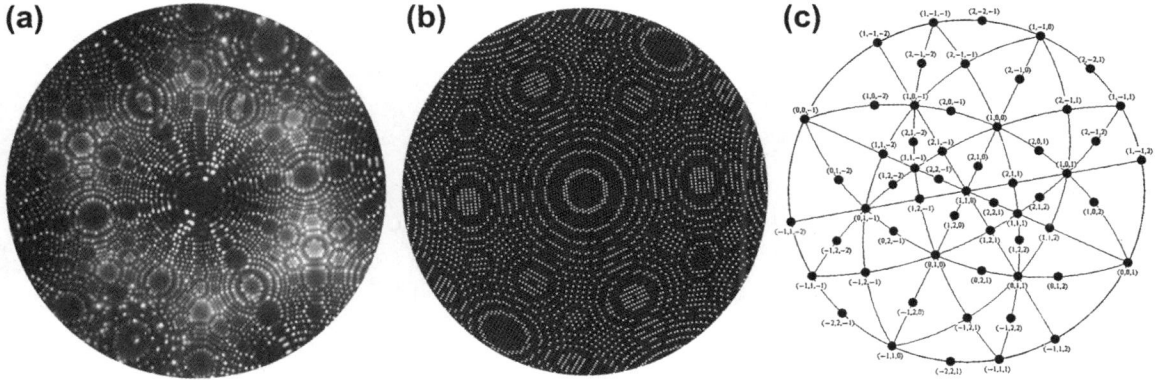

Figure 5 (a) Field ion microscope image of W using He as an imaging gas with the (001) pole in the center. (b) Computer-simulated projection image of surface atoms using a shell model and (c) (011) stereographic projection. This is B/W.

Since the centers of the low index planes have large terraces as shown in **Figure 4**, field ionization of imaging gas atoms does not occur from the interiors of the low index planes. In high index planes, all atoms on the surface can be imaged with an atomic resolution. Indexing planes can be readily done from the symmetry of the poles, i.e. 4-fold symmetry for cubic {001} planes, three for {111} and two for {011} planes. Since only protruding atoms contribute to image formation, FIM cannot image all the atoms on the surface; thus, FIM is not suitable for structural studies although it resolves individual atoms.

In order to estimate the magnification of tips, it is sometimes useful to calculate the tip radius. The methodology is outlined in **Figure 6 (a)**. If the number of concentric rings between two poles is n, the distance in z direction between the two poles is nd_{hkl}. Since the angle q between two poles is known, the tip radius r can be calculated from Eqn (7)

$$r = \frac{nd_{hkl}}{1 - \cos\theta}. \tag{7}$$

Since the evaporation field is unique to atomic species, the radius of tip is determined so that the surface field is equal to the evaporation field of atoms. Usually, tips have a taper as shown in **Figure 6(b)**.

As atomic layers are evaporated, the tip radius increases depending on the taper angle, α. If the tip radius increased from R_1 to R_2 after n atomic planes are field evaporated, the taper angle α can be determined using Eqn (8)

$$\sin\alpha = \frac{R_2 - R_1}{R_2 - R_1 + nd_{hkl}}. \tag{8}$$

Image contrast arises from various reasons in FIM images as shown in **Figure 7**. In **Figure 7(a)**, discontinuity of the regular pattern of FIM image is observed as indicated by arrowheads. This is due to the presence of a grain boundary (GB), where the orientations of two grains are different resulting in the discontinuity in the symmetric image. **Figure 7(b)** shows He ion image of pure aluminum. The

(a)

$$r = \frac{nd_{hkl}}{1 - \cos\theta}$$

(b)

$$\sin\alpha = \frac{R_2 - R_1}{R_2 - R_1 + nd_{hkl}}$$

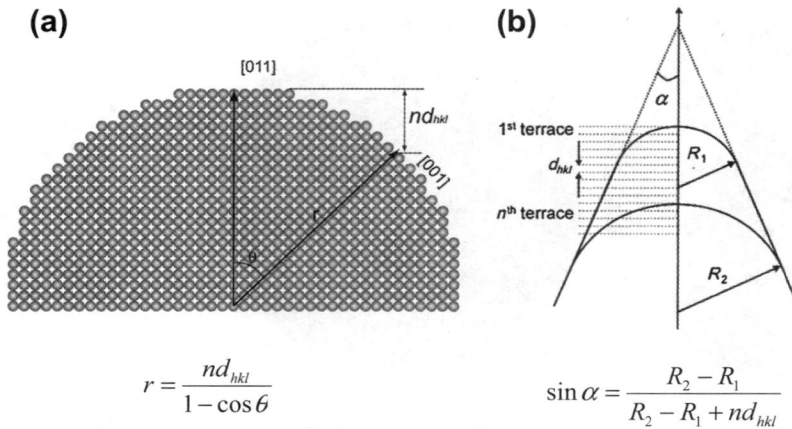

Figure 6   (a) Schematic illustration on how to calculate tip radius from the number of atom steps between (011) and (001) poles. (b) Schematic illustration of how tip radius increases as field evaporation of atomic planes occur. Reproduced from Miller MK. Atom probe tomography: analysis at the atomic level, Kluwer Academic, 2000.

{011} pole has brighter contrast, while {111} and {001} poles image darkly. This is because of the change in the local radius of the FIM tip brought about by the difference in surface energy of crystallographic planes. In fcc metals, the (111) and {001} planes have lower surface energy than {011}, and thus these surface tend to facet in the field-evaporated end form. FIM images obtained from alloys show contrast that is more complicated. This is because of the difference in both field ionization and field evaporation behavior of different chemical species. In a solid solution, contrast originating from the difference in the evaporation field appears. Solute atoms with higher evaporation field tend to image more brightly than the matrix atoms, because these solute atoms tend to retain on the surface.

A typical example of brightly imaging atoms from retained Fe atoms in Cu is shown in **Figure 7(c)**. Since the evaporation field of Fe is higher than that of Cu, Fe tends to preferentially retain on the surface, giving rise to a brighter contrast. Depending on the composition of the precipitate, two types of contrast are expected. **Figure 7(d)** shows δ' particles precipitated in an Al–Li alloy. Since Li atoms evaporate preferentially, Li-enriched δ' particles tend to be concave with respect to the matrix phase. These particles give darker contrast, because the electric field from the concave area is lower than the surrounding matrix. On the other hand, from Fe precipitates in Cr–Fe alloy are imaged with bright contrast as shown in **Figure 7(e)**. Since Fe has a higher evaporation field, Fe-enriched phase retains after the matrix atoms are evaporated, thus the precipitate particles protrude on the surface. Since protruding region causes a high electric field, such regions will image brightly. On the other hand, if precipitates are enriched with a solute with a lower evaporation field, they image darkly. The reason for the two types of contrast is schematically illustrated in **Figure 7(f)**. As seen from the above examples, various artifacts are expected from FIM images because of the difference in local radii. As mentioned before, it is possible to form image using field-evaporated atoms if the evaporation rate is fast enough to obtain sufficient brightness in microchannel place. Such images are called "field desorption image" as shown in **Figure 8(b)**.

On comparing FDM and FIM images from the same sample (see **Figure 8(a)**), large differences are observed due to trajectory differences. Dark and bright linear contrast appears connecting poles, which is called "zone decoration". This image contrast suggests that desorbed ion does not arrive to the lines

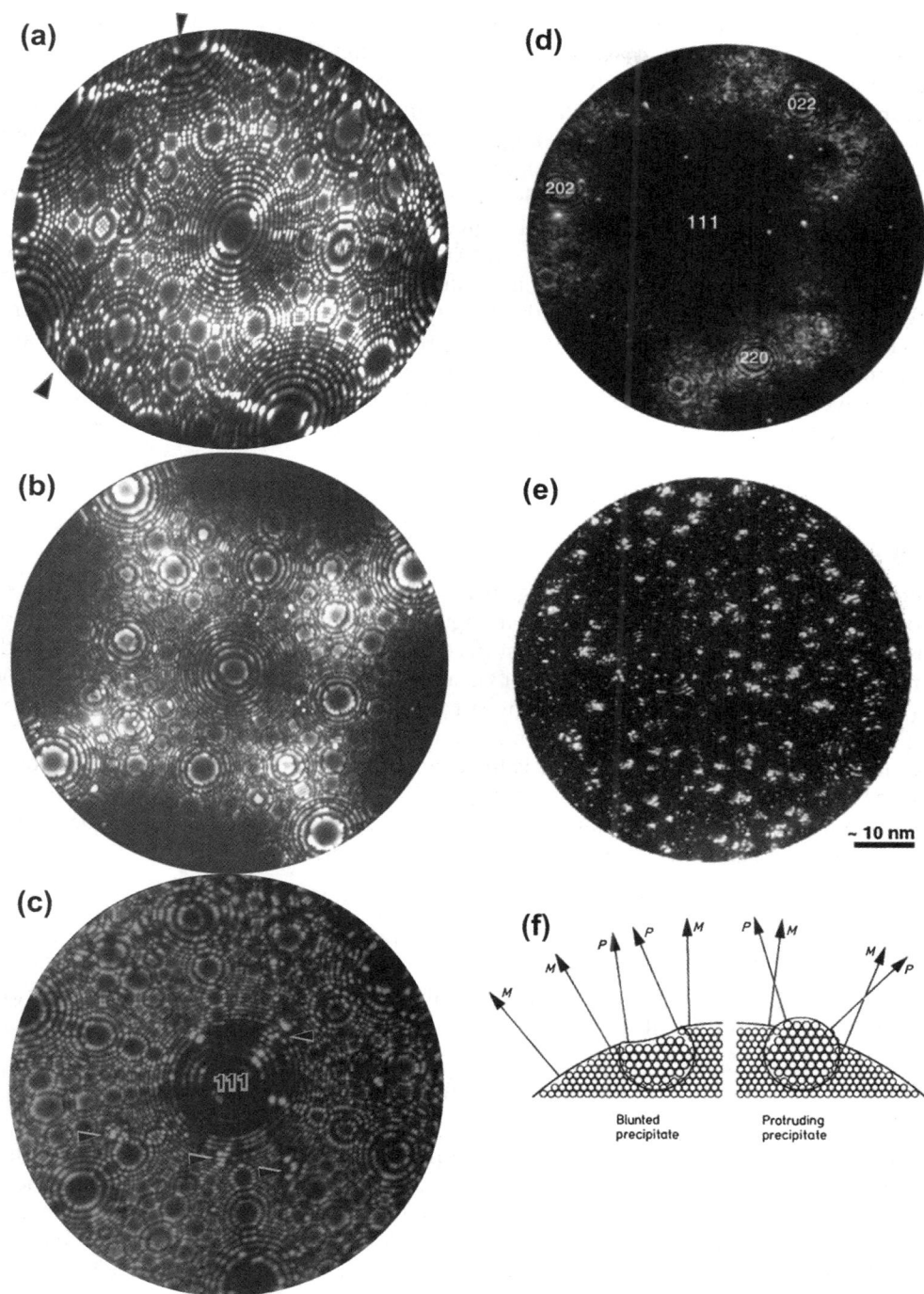

Figure 7 (a) He-field ion image of tungsten with a GB, (b) He-field ion image of aluminum, (c) Ne-field ion image of aged Cu–2Fe alloy, (d) He-field ion image of Al–6Li alloy, (e) Ne-field ion image of aged Cr–20Fe alloy, and (f) schematic illustration of blunted and protruding precipitates. Reprinted with permission from Hono (2002).

connecting zones. This suggests that there is a trajectory aberration when atoms are field evaporated, and that these trajectories are totally different from that of field ionized imaging gas. Field ionized imaging gas ions are projected to the screen from the field ionization zone near protruding atoms and their trajectory is normal to the isofield surface. On the other hand, surface atoms roll along the edge of terrace before field evaporation. This causes large trajectory aberration for field-evaporated ions. This phenomenon causes inaccurate positioning of atoms in reconstruction of 3DAP data.

Although the FIM images are to be used to determine precipitate size, it is better to use TEM or small angle scattering (SAS) because these techniques give statistically more reliable data. However, FIM observation is still useful for selecting the area for 3DAP analysis. By looking at FIM image, specimen can be eucentrically tilted so that desired pole is located in the center of the FIM screen or the PSD for 3DAP analysis; then AP data can be taken nearly along the selected zone axis. Considering the trajectory aberrations in field desorption image, i.e. AP tomography, selecting the region where field evaporation does not get much influence form the trajectory aberration is important. Also, it is essential to obtain a good FIM image before starting 3DAP analysis, because this assures that the specimen has nicely developed a clean hemispherical surface, which is essential for obtaining good AP results.

15.2.3 Early Atom Probe

By selecting a small region from the FIM image using an aperture called the *probe hole*, atoms can be collected only from the selected area as shown in **Figure 9**. On applying a high voltage pulse of a few tens nanoseconds, atoms are ionized from the surface by the *field evaporation phenomena*, and the ions that go through the probe hole fly to the detector via an electrostatic reflector called *reflectron*. The time-of-flights (ToF) of the ions evaporated from the specimen surface are measured by a time-to-digital converter (TDC) with a nanosecond resolution. The electrostatic energy of an ion is given by neV_e, where n is the number of charge of the ion, e is the charge of electron and V_e is the voltage of the ion.

Since atoms are ionized at the total voltage of $V_{dc} + \alpha V_p$, where α is a pulse factor close to unity, this factor varies depending on the transmittance of the pulse to the specimen. The kinetic energy of the ion is given by $(1/2)mv^2$, which leads to an equivalence condition shown in Eqn (9)

$$ne\left(V_{dc} + \alpha V_p\right) = \frac{1}{2}m\left(\frac{l}{t-\delta}\right)^2, \tag{9}$$

where t is time-of-flight, δ is a delay constant and l is a flight length. Thus, a mass to charge ratio of ion, m/n, can be measured with the following equation.

$$\frac{m}{n} = \frac{2e}{l^2}\left(V_{dc} + \alpha V_p\right)(t-\delta)^2, \tag{10}$$

According to this relationship, m/n may be determined accurately as long as time-of-flights can be measured accurately. In reality, however, αV_p has some distribution, because atoms are not necessarily ionized at the peak of the pulse height. Due to this effect, each ion contains some energy deficit, which causes errors in experimentally determined m/n values. Hence, an energy compensator, either a Poschenreider lens (Müller and Krishnaswamy, 1968) or a reflectron (Drachsel et al., 1989), is installed in the ToF-AP to compensate the energy deficit of ions. Without an energy compensator, the mass resolution ($m/\Delta m$ full width at half maximum) obtained by the time-of-flight mass analysis is

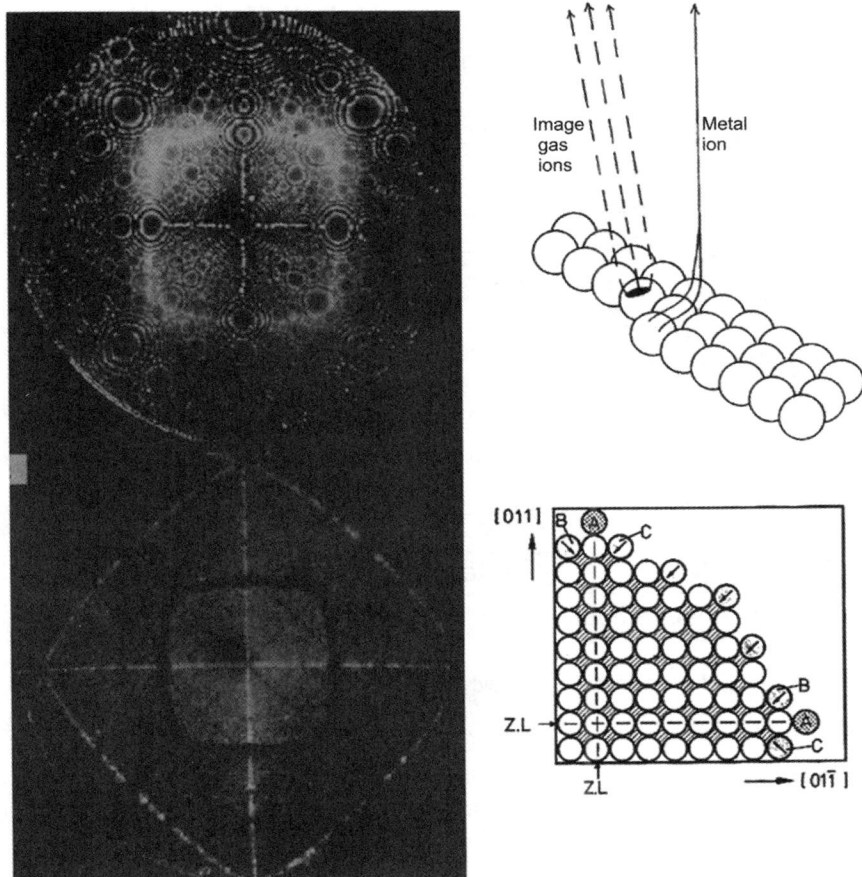

Figure 8 Field ion and desorption image and their trajectory difference. From Waugh et al. (1976).

limited to less than 100. By adopting a reflectron energy compensator, it is possible to achieve a mass resolution higher than 500.

The appearance of a typical reflectron energy-compensated ToF-AP is shown in **Figure 10**. The instrument is composed of four UHV chambers, air lock, specimen storage, FIM and reflectron chambers. Specimens are introduced into the specimen storage chamber through the air lock chamber. The background pressure of the FIM chamber must be maintained in the 10^{-9} Pa range, as the evaporation field is rather sensitive to the residual gas like H_2. To better control to the field evaporation, the specimen is cooled to a cryogenic temperature of about 25 K. In the FIM chamber, imaging gasses of He, Ne and H_2 can be introduced independently. Use of H_2 is very effective in removing surface oxide and developing the tip to an ideal end form for AP analysis, because H_2 reduces the evaporation field of most of metals. A mixture of imaging gases within FIM chamber can be achieved by introducing different types gases from independent gas lines. After obtaining good FIM images, imaging gases are usually pumped out of the chamber completely, and AP analysis is initiated by applying high-voltage pulses to a DC-standing voltage. In the reflectron-type APs, this voltage is typically 10 percent higher

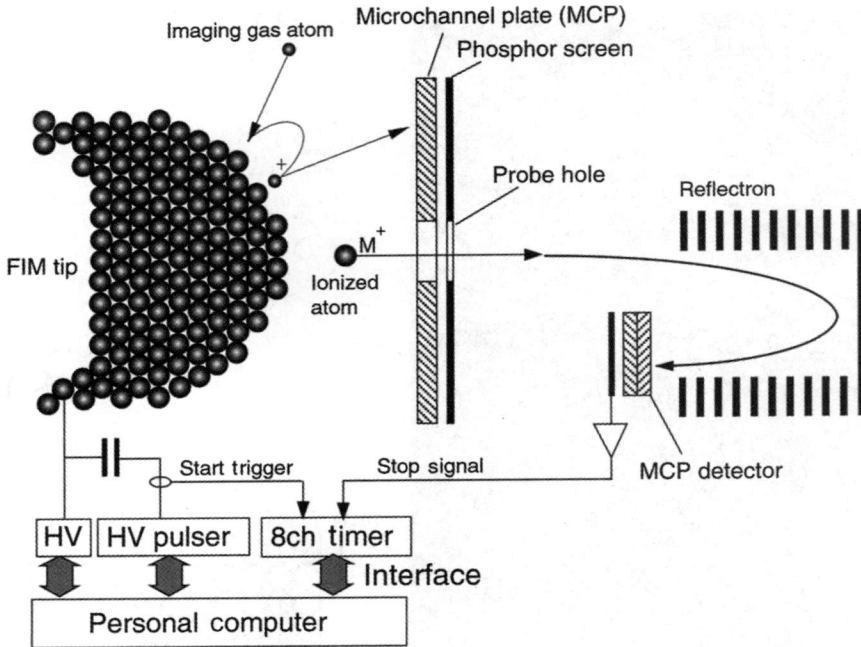

Figure 9 Schematic illustration for the principle of a field ion microscope (FIM) and a conventional atom probe (AP). From Hono (2002).

than the ionization voltage. The analysis was performed at a pulse repetition rate of 200–1000 Hz with a detection rate of 0.005 atom/pulse. These means that approximately 60 atoms are collected per min, 3600 atoms per hour and 36000 atoms per night. This is the typical rate of data acquisition in the one-dimensional AP. Faster evaporation rate often causes rupture of tips due to the stress originating from high electric field (Eaton and Bayuzick, 1978). Such rupturing of specimens due to high field stress is the most difficult part of AP experiment. Many commercial alloys such as martensitic steels and nanocrystalline alloys are extremely susceptible to rupture under the presence of high field stress, and many attempts had to be repeated until meaningful number of atoms can be sampled.

The data chain of the collected atoms can be converted to a one-dimensional depth profile by scaling the depth in proportion to the number of collected atoms. As mentioned earlier, this type of AP is now called the one-dimensional atom probe (1DAP) to differentiate it from the modern 3DAP. Since atoms are field evaporated from the surface regularly, it is possible to resolve atomic layers in the concentration–depth profile. **Figure 11** shows an integral concentration depth profile obtained from the (001) pole of $L1_2$ Ni_3Al intermetallic phase. This phase has alternating planes of pure Ni and $Ni_{50}Al_{50}$ as shown in the inset figure. In this integral depth profile, the number of detected Al atoms is plotted as a function of total number of detected atoms. Thus, the slope of the plot represents the local concentration of Al. Horizontal line corresponds to 100%Ni plane, and the plot having 50% Al corresponds to NiAl plane. As seen from this example, atom-probe concentration depth profile can have an atomic layer resolution when data are properly corrected.

On the other hand, the lateral resolution is determined by the size of the probe hole and the trajectory aberration. The trajectory aberration can be caused by a few atomic distance migration of atom prior to

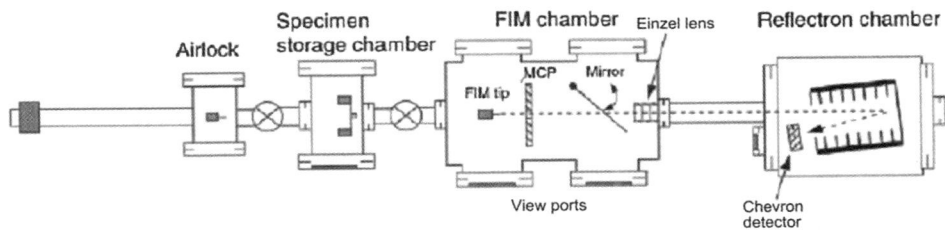

Figure 10 Appearance of an energy-compensated time-of-flight atom-probe with a reflectron energy compensator. From Hono (2002).

field evaporation. This occurs because the kink-site atoms on a low-index plane migrate to the edge of its lattice plane where they are exposed to a higher electric field when field evaporation occurs (Waugh et al., 1976). Because of this effect, the trajectories of the ions field evaporated from certain position of tip-apex diverge from that of field ionized ions. Thus, even if the object to be analyzed is completely covered by a probe hole, atoms imaged within the probe hole do not necessarily go through the aperture by field evaporation. The physical dimension of the probe hole is typically 2–3 mm in diameter, which corresponds to 2–5 nm with a magnification of a million. The selected area can be made smaller by increasing the distance between the specimen and the screen, but the statistical error of the depth profile will then be increased because the number of collected atoms within the selected area decreases.

Figure 12 shows schematic illustration of an FIM tip that contains fine particles and interphase interface. Atoms are collected from the volume covered by the probe hole called *cylinder of analysis*. As seen from this illustration, 1DAP data give correct composition of the fine particles only when they are larger than the probe hole. Therefore, accurate measurement of chemical compositions of small precipitates, which are less than 5 nm, is still challenging with the 1DAP technique. When it comes to the analysis of an interface, AP data give sharp concentration change only when the interface is cut in

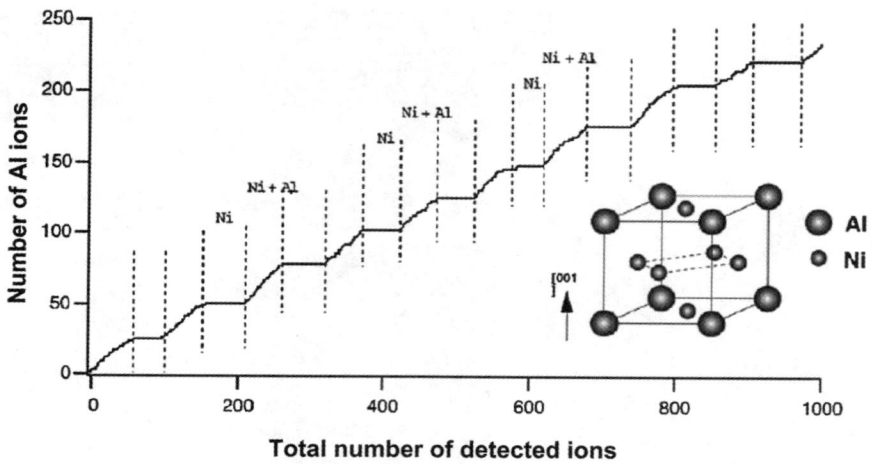

Figure 11 Integrated concentration depth profile obtained from the (001) pole of $L1_2$ Ni_3Al intermetallic phase. From Hono (2002).

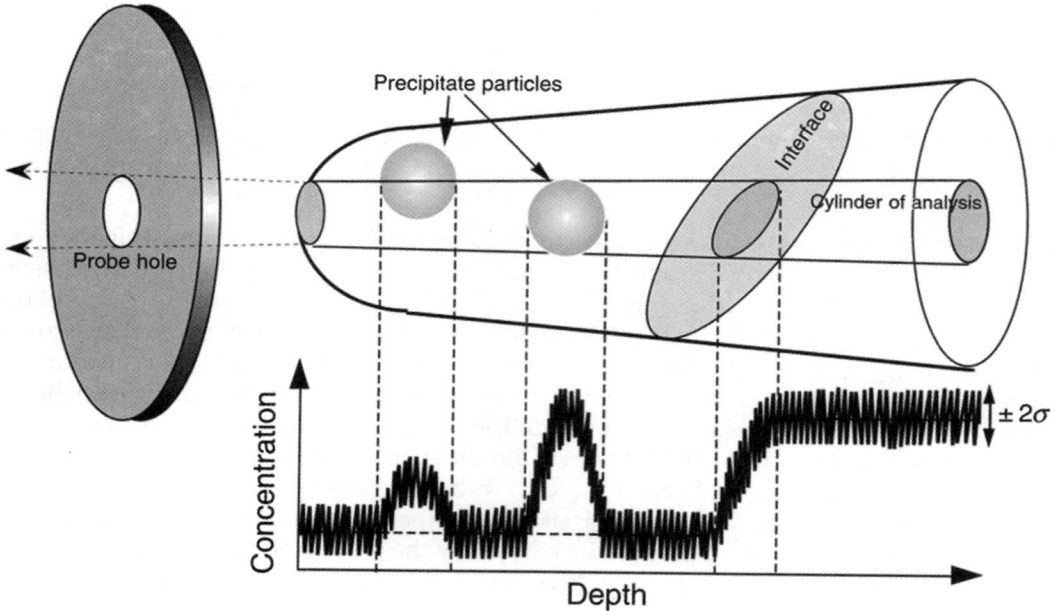

Figure 12 Schematic illustration of an FIM tip that contains fine particles and interphase interface and the cylinder of analysis using a conventional AP. From Hono (2002).

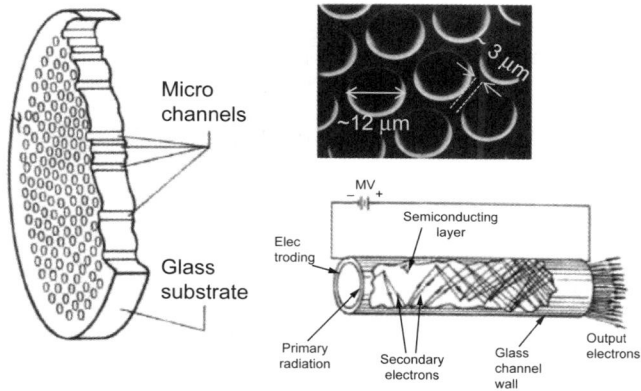

Figure 13 Open area ratio of microchannel plate detector, which determines the upper limit of the detection efficiency.

the normal direction. If the interface is inclined with respect to the cylinder of analysis as shown in the illustration, the apparent change in composition becomes diffuse even if it is sharp in reality. Therefore, it is obvious that 1DAP has limitations in order to get quantitative analyses.

It should be noted that it is not possible to obtain the information on short-range ordering in alloys by the 1DAP technique, because the data collected by an AP do not retain the information on the nearest neighbor atoms. In addition, the detection efficiency of an AP is not higher than 60% because the maximum detection efficiency of an MCP detector is determined by the open area ratio of the channels (see **Figure 13**). Because of this, the data obtained by the conventional AP do not contain any information on short-range ordering. On the other hand, the AP is rather sensitive to detect the formation of solute clusters in dilute alloys whose size is below the detection limit of high-resolution electron microscope (HREM). The detection sensitivity of clusters also depends on the solute concentration in the alloy.

15.2.4 Three-Dimensional Atom Probe

The 3DAP measure (x, y) coordinates of field-evaporated ions using a PSD as shown in **Figure 1(b)**. By measuring the time-of-flight and the coordinates of ions using a PSD, it is possible to map out two-dimensional elemental distribution with a near-atomic resolution. The lateral spatial resolution is limited by the trajectory aberration as in the case of the 1DAP. However, the error originating from the evaporation aberration does not exceed 0.2 nm. The elemental maps can be extended to the depth direction as atoms are ionized from the topmost surface, by which 3D tomography of atoms can be constructed. Since field evaporation occurs layer-by-layer in low index planes, the 3D tomography of atoms resolves atomic layers in the depth direction by assigning z-coordinate proportional to the number of detected atoms. The first 3DAP adopted a wedge-and-strip-type anode as PSD (Cerezo et al., 1988) as shown in **Figure 14(a)**. In front of this anode, doubly stacked MCP is placed. When an ion hits the MCP, secondary electron shower is generated and the charge originating from these electrons are divided into the three anodes. The position of the ion can be calculated from the amount of the charge on each anode as given by the Eqn (9).

$$x = \frac{2Q_1}{Q_1 + Q_2 + Q_3}; \quad y = \frac{2Q_2}{Q_1 + Q_2 + Q_3}, \tag{11}$$

(a)

$$x = \frac{2Q_1}{Q_1 + Q_2 + Q_3}$$
$$y = \frac{2Q_2}{Q_1 + Q_2 + Q_3}$$

3ch charge sensitive ADC

(b)

96ch charge ADC

$$x = f(Q_d/Q_{total})$$
$$y = f(Q_h/Q_{total})$$

(c)

$$y = t_1 - t_2$$

$$x = t_3 - t_4$$

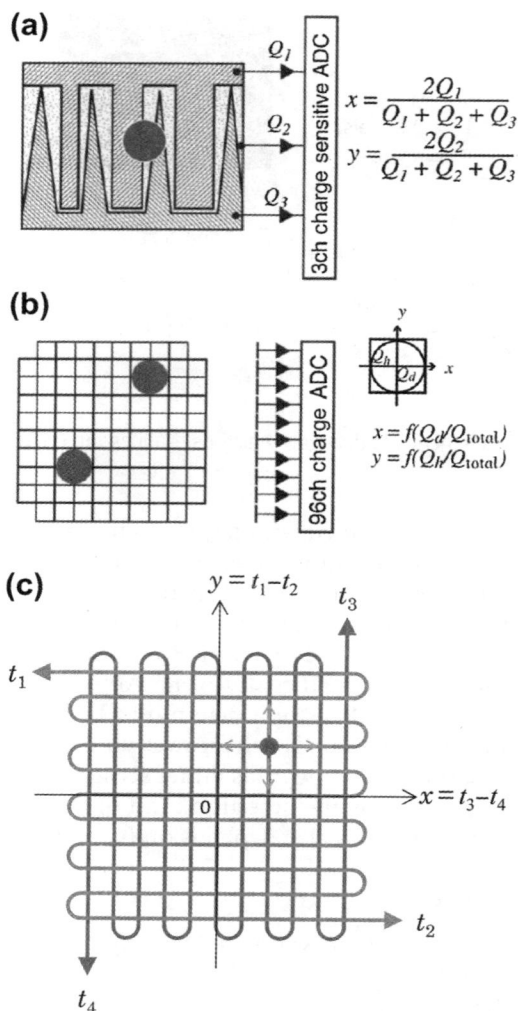

Figure 14 Position sensitive detectors used in (a) first generation position sensitive atom probe (Cerezo et al., 1988) and (b) tomographic atom probe and (c) delay-line detector (DLD). Reproduced from Miller (2000). (For color version of this figure, the reader is referred to the online version of this book.)

However, if more than two ions hit the detector at the same time, the position of these ions cannot be determined using this method. Because of this, the evaporation rate had to be maintained very low, so that the chance of multihit detection can be kept low. If more than two ions hit the detector, these ions had to be ignored. Thus, the detection efficiency obtained with this type of serial 3DAP is very low. In order to overcome this shortcoming, Blavette et al. (1993) developed a 3DAP with a multihit detection capability. In their detector, 96 square anodes of 1×1 cm^2 are placed with square arrays of 10×10 omitting 1 from each corner.

In front of this anode array, triply stacked MCP is placed. One ion hit causes large secondary electron shower, giving a charge cloud of 2–3 cm in diameter on the anodes. Such charge clouds are divided into multiple anodes, and the charge on each anode is measured using a 96-channel charge sensitive ADC as shown in **Figure 14(b)**. The position of the input ion can be determined by calculating the center of the charge cloud that spreads over several anodes on the detector. In this multianode system, multiple hits of ions on the detector can be individually detected as long as the charges do not overlap. Thus, even if more than two atoms with the same m/n arrive on the detector simultaneously, they can be detected separately. Eight 96-channel charge-sensitive ADC is installed in parallel, each of which is triggered by different ToF signals. Thus, ions with up to eight different m/n's can be detected even if they are evaporated with a single pulse event. Because of this parallel processing system, data can be acquired at a much faster evaporation rate than that was limited in the serial type 3DAP. The parallel 3DAP can be run at a much faster evaporation rate, so the data acquisition time has been substantially shortened. The position-sensitive detectors commonly used nowadays are delay-line detectors (DLDs) as shown in **Figure 14(c)** (Da Costa et al., 2005).

Detailed view of the DLD detector is also shown in **Figure 15**. The DLD consist of two layered delay lines crossing perpendicular each other. The position of one direction is determined by measuring the difference in arrival times from the impact of electron on a conductive wire. Since two delay lines cross orthogonally, (x, y) coordinates can be measured as

$$x = t_3 - t_4, \quad y = t_1 - t_2, \tag{12}$$

where $t_1 \sim t_4$ are time from the application of HV pulse to incoming of the signal to each anode of DLD as shown in **Figure 14(c)**. Since the time resolution of the DLD detector is very high, the modern 3DAP can be operated at frequencies higher than 100 kHz. Unlike 1DAP, the acceptance angle of PSD is large

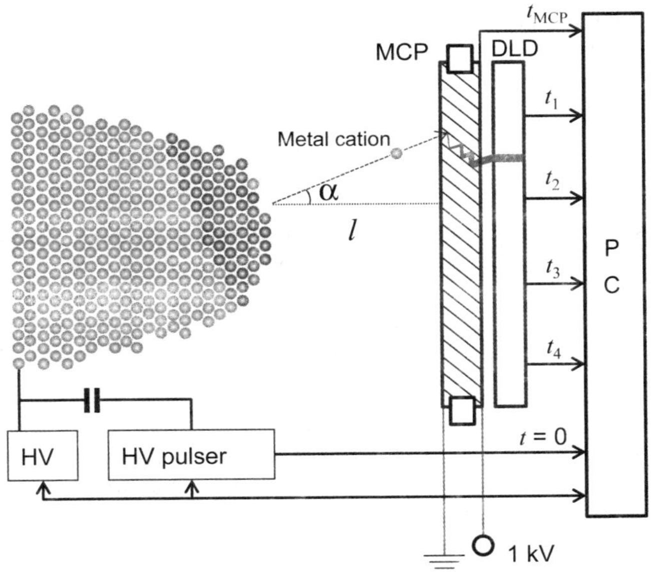

Figure 15 Schematic illustration of DLD detector design. (For color version of this figure, the reader is referred to the online version of this book.)

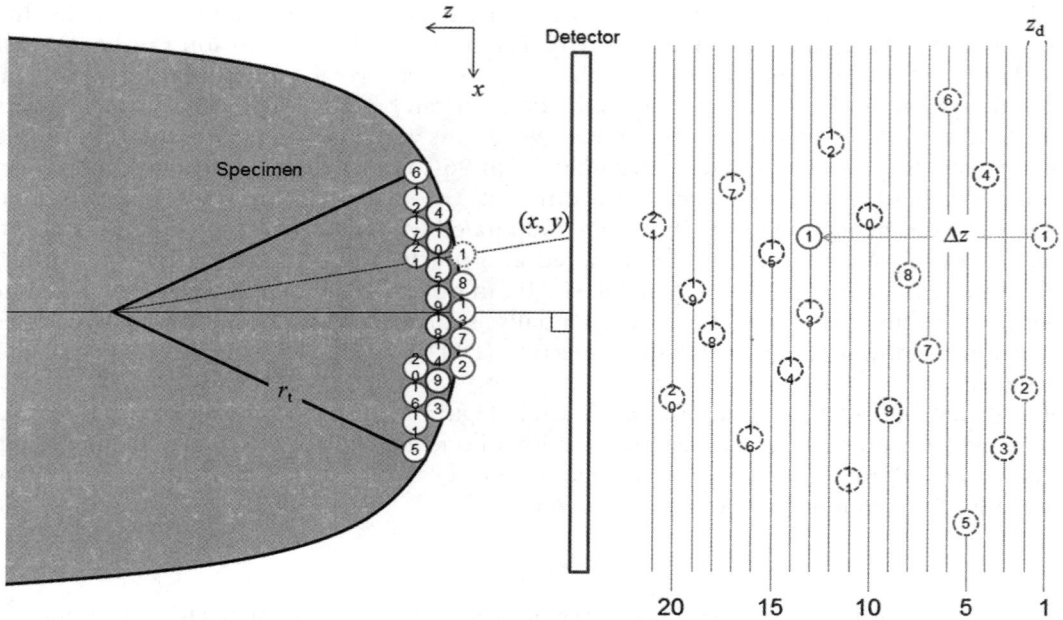

Figure 16 Schematic illustration for the acceptance angle and process of determination of the z-coordinate of each atom in 3DAP. (For color version of this figure, the reader is referred to the online version of this book.)

as shown in **Figure 16**. This causes difference in flight length depending on the position. However, the flight length or time-of-flight can be corrected based on the impact position. The incident angle results in dispersion of flight length. Here, the measured ToF t_{MCP} contains dispersion, which is related to the impact position and is expressed as

$$t_{MCP} = \sqrt{\frac{(m/n)}{2eV}} \{l^2 + x^2 + y^2\}. \tag{13}$$

So, the analyses software for three-dimensional APs usually has functions to calibrate this dispersion. The z-coordinate z_i of the nth atom detected is expressed as (Al-Kassab et al., 1996)

$$z_n = (n - 1)z_d + \Delta z = \sum_{i=1}^{n-1} \frac{\Omega_i}{\zeta A_D} + \left(r_t - \sqrt{r_t^2 - (x_i^2 + y_i^2)}\right), \tag{14}$$

where, Ω_i is the atomic volume of the ith ion in the phase under analysis, ζ is the detection efficiency, A_D is the effective area of the DLD and r_t is the curvature radius of the specimen. In this equation, the first term is the sum of the amount, which is added once an ion impacts the detector, and the second term is a correction, which is for the assumption that the specimen has a hemispherical end shape and the field evaporation occurs from the outside to inside in each atomic plane. **Figure 16** schematically shows the process of determination of the z-coordinate. In the practical analysis, the three-dimensional reconstruction of the detected ions is conducted by several methods based on applied voltage, cone shape, the specimen geometry and so on.

In a 3DAP, position information has to be preserved during the flights of field-evaporated ions. Thus, energy compensator was not installed in the early 3DAP design (Cerezo et al., 1988; Blavette et al., 1993). The mass resolution ($m/\Delta m$) typically obtained from a straight ToF AP was less than 100 full width at half maximum (FWHM), even after the correction of the time-of-flight based on the position. This mass resolution is significantly lower than that of an energy-compensated AP ($m/\Delta m \sim 1000$ FWHM). With a low mass resolution, ions with close mass-to-charge ratios cannot be separated. For example, a mass resolution higher than 200 is required for separating $^{93}Nb^{+++}$ ($m/n = 31$), $^{94}Zr^{+++}$ ($m/n = 31.33$) and $^{63}Cu^{++}$ ($m/n = 31.5$). In order to improve the mass resolution of 3DAP, Cerezo et al. (1994) adopted a wide acceptance angle reflectron energy compensator in the flight path of the PoSAP. With this innovative approach, he demonstrated that the position information could be preserved accurately even if ions are detected after a reflectron energy compensator. Based on this experiment, a prototype energy-compensated optical position sensitive atom probe (ECOPoSAP) was developed, which adopted an optical parallel detection system for PSD. One drawback of the energy-compensated 3DAP is its lower detection rate, as two meshes must be put at the entrance and exits of the reflection to attain a lens effect in the reflectron.

15.2.5 Laser-Assisted Atom Probe

The applications of conventional ToF atom probes have been limited to conductive materials because high-voltage (HV) nanosecond pulses do not transmit to tip apex through insulator specimens. Before the development of energy-compensated AP (Müller and Krishnaswamy, 1974), the poor mass resolution due to the energy deficit by HV-pulsed field evaporation was also a serious problem for ToF AP. In order to overcome this problem, Tsong et al. (Kellogg and Tsong, 1980; Tsong et al., 1982) developed pulsed-laser atom-probe (PLAP). They demonstrated that the mass resolution is substantially improved and Si can be successfully analyzed. However, due to the slow frequency of the pulsed lasers at that time and little merit in the analyses of metallic samples, PLAP has become dormant for some time.

In 2005, Deconihout et al. presented the results of 3DAP analyses using ultrashort laser pulse and demonstrated that 3D AP tomography can be obtained with superior mass resolution compared to HV pulse-operated 3DAP (Deconihout et al., 2007b). This work has reactivated the research on laser-assisted AP and successive investigations convincingly showed the merits of employing ultrashort lasers for 3DAP analyses (Gault et al., 2005; Cerezo et al., 2006; Gault et al., 2006). The experimental setup of the femtosecond laser-assisted 3DAP is illustrated in **Figure 17**. Ultrafast laser with variable wavelength is focused to tip apex. In this setup, the focus diameter of the laser beam was not accurately measured, but it was orders of magnitude larger than the tip radius (Gault et al., 2006). These investigations clearly demonstrated that fully controlled field evaporation with an atomic layer resolution is possible with ultrashort laser as long as the laser power is low enough to suppress the increase of the specimen temperature.

In addition, the mass resolution can be substantially improved compared to the voltage pulse AP. As shown in **Figure 18**, the mass ($m/\Delta m$) resolution of tungsten increased from 90 obtained with a voltage pulse to a mass resolution of 250 by the incorporation of infrared laser (1030 nm) pulsing. With a gradual decrease in wavelength from 515 to 257 nm, incorporation of new detectors (delay line detectors), and increasing the flight length, the mass resolution was improved closer to 3000. The dependence of laser wavelength on the analyses of semiconductors and insulators (see **Figure 19**) has also been demonstrated.

For example, the laser-assisted AP analyses of ZnO semiconductor show dramatic improvement in mass resolution with the use of shorter wavelength lasers (Ohkubo and Hono, 2010). ZnO is

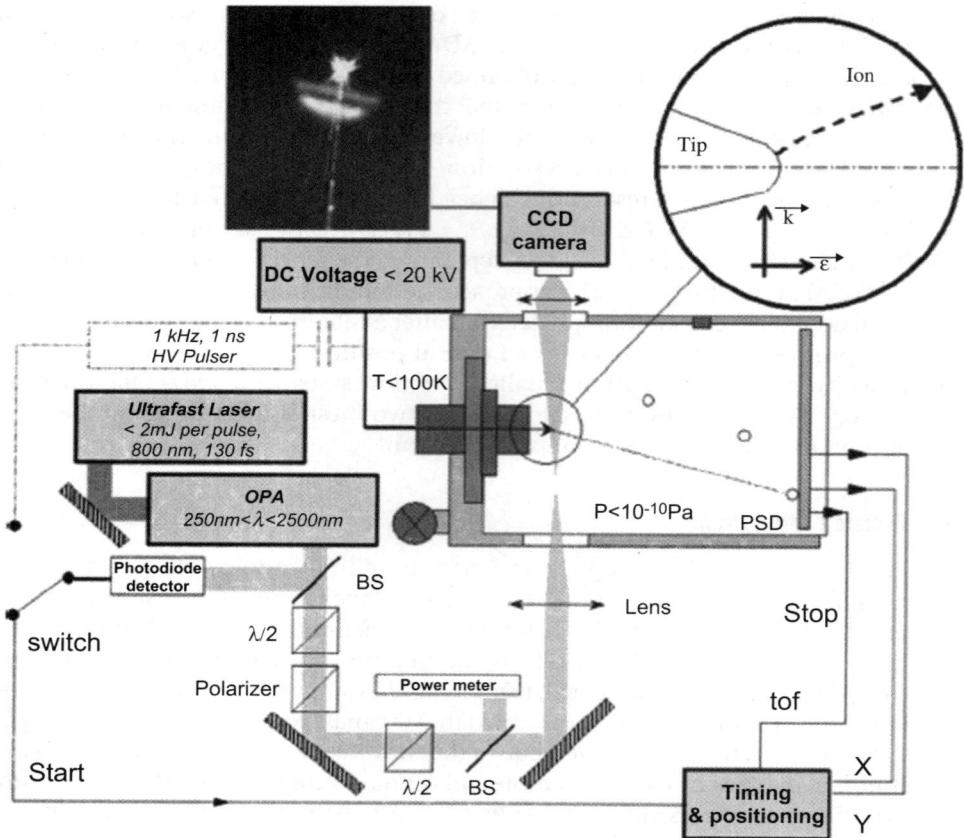

Figure 17 Experimental setup of laser-assisted 3DAP. Ref: B. Gault et al. Rev. Sci. Instrum. 77, 043705 (2006).

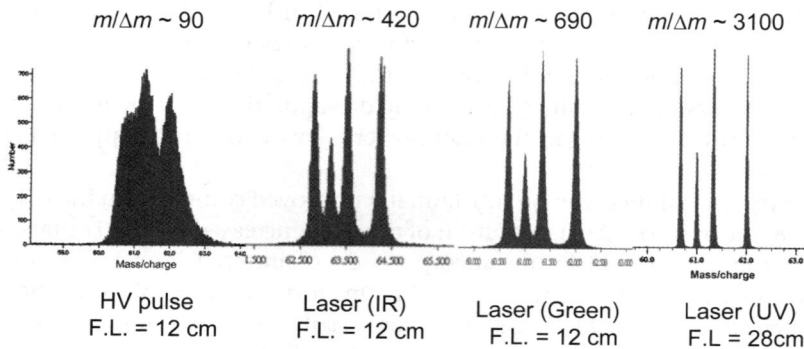

$m/\Delta m \sim 90$ $m/\Delta m \sim 420$ $m/\Delta m \sim 690$ $m/\Delta m \sim 3100$

HV pulse   Laser (IR)   Laser (Green)   Laser (UV)
F.L. = 12 cm   F.L. = 12 cm   F.L. = 12 cm   F.L = 28cm

Figure 18 Mass resolution of straight-type AP with flight lengths of 12 and 28 cm with voltage–pulse mode, and laser-assisted field evaporation with IR (1030 nm), Green (515 nm) and UV (375 nm) femtosecond laser. From Ohkubo and Hono (2010). (For color version of this figure, the reader is referred to the online version of this book.)

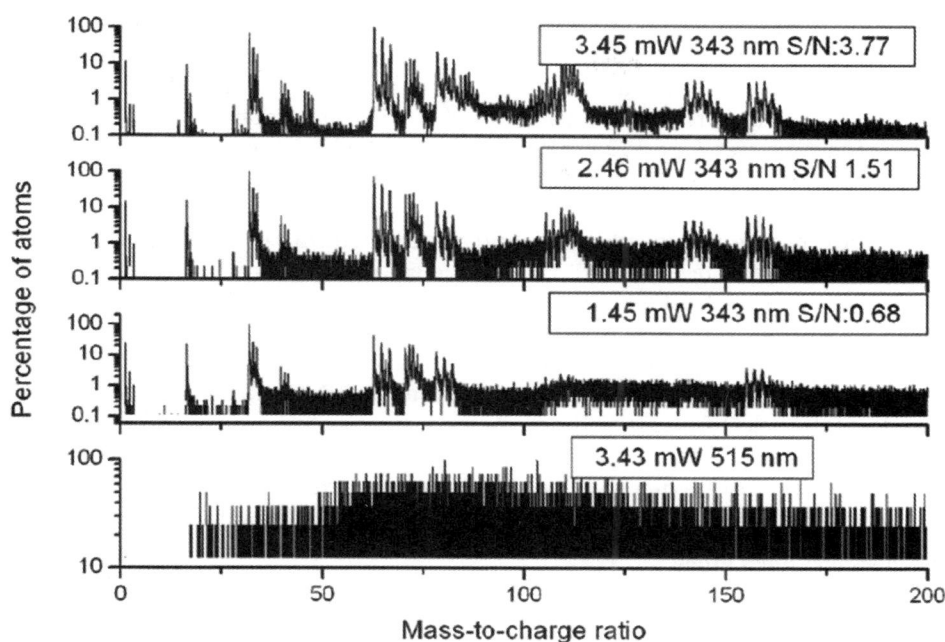

Figure 19 Mass spectrum improvement while analyzing ZnO semiconductor with a decrease in laser wavelength from 515 to 343 nm and an increase in laser power. From Ohkubo and Hono (2010).

a wide-band gap oxide semiconductor with the band gap of 3.3 eV ($\lambda = 375$ nm), which is located between the photon energy of 515 nm (2.4 eV) and 343 nm (3.6 eV) laser and the electric resistivity is $\sim 10^4$ Ωcm. **Figure 40** shows a series of mass spectra of ZnO obtained using 515 and 343 nm lasers. Laser-assisted field evaporation was confirmed for the 343 nm laser with ion mass peaks for single and molecular ions. Regardless of the power of the 343 nm laser, similar grade of mass resolution was obtained, i.e. the mass resolution (m/Δm) at full-width half-maximum (FWHW) was 456 for $^{64}Zn^{2+}$ and 524 for $^{64}Zn^{+}$. As 343 nm laser energy was increased from 0.7 to 1.7 µJ/pulse, the background noise at a high mass range decreased resulting in an overall improvement of the signal-to-noise (SN) ratio. However, in the case of 515-nm laser, the laser-assisted field evaporation was not confirmed even with pulse energy of 1.7 µJ/pulse. The explanation for this observation could be is as follows: the light absorption coefficient of ZnO is close to zero when the wave length is longer than the band gap; on the other hand, the absorption coefficient is as high as 2×10^5 cm^{-1} when the wavelength of the light is shorter than the band gap. This result indicates that sufficient laser absorption is essential to stimulate field evaporation by laser. This means that the usage of wavelength shorter than a bad gap is better to field evaporate oxides by the assistance of pulsed laser by phonon excitation.

Nowadays, 3DAP with UV lasers is applied to a variety of materials including semiconductors and insulators. Possibility of extending the AP technique to biological materials has been mentioned by Rajan (2011), based on successful imaging of DNA molecules by FIM technique. However, benefits of this approach have not been demonstrated yet.

There have been controversy on the mechanism of laser-assisted field evaporation, but experimental results on the mass resolution depending on the taper angle and thermal conductivities of FIM tips suggest the laser-assisted field evaporation is a thermal effect. This appears to be true for metals, but recent AP investigations using ultraviolet (UV) laser demonstrated precise AP tomography is possible even from bulk insulators. For example, Chen et al. demonstrated controlled field evaporation of insulator oxide, as well as, precise reconstruction for AP tomography of a zirconia-spinel nano-composite by using UV laser ($\lambda = 350$ nm) (Chen et al., 2009b). They also showed that the field evaporation rate is faster using UV laser compared to that obtained using green laser ($\lambda = 525$ nm) with the same power (Chen et al., 2011a). The mass resolution of oxide materials, whose thermal conductance is poor, can be improved using short wave length. Based on these experimental results (Tsukada et al., 2011a; Tamura et al., 2012), direct electron excitation was proposed as a possible mechanism for laser-assisted field evaporation using UV laser. Houard et al. (2010) attributed the improved mass resolution to the localized absorption of laser pulse at the tip apex. Because of the improved mass resolution using UV laser, 3DAP do not require energy compensator unless ultimate mass resolution higher than 1000 is required (Stender et al., 2007; Hono et al., 2011a). For higher mass resolution, wide-angle reflectron can be implemented; however, the detection efficiency of energy-compensated AP becomes substantially lower than that of straight-type 3DAP because meshes to flatten electrostatic field must be placed in both entrance and exit of the wide-angle reflectron.

15.2.6 Local Electrode Atom Probe

In conventional FIM and AP, a single tip faces grounded surface of MCP about 10 cm away. In order to obtain the electric field for field ionization of imaging gas atoms and for field evaporation for AP analysis, a high voltage ranging from 4 to 15 kV must be applied to the tip. This often causes insulator problems in the specimen holder. In addition, since the large electric field causes a large field stress at the tip apex, FIM and AP specimen often rupture during FIM observation or HV-pulsed AP analyses. In the AP community, the success rate of the analysis has been called "yield of successful analyses" and this yield is low for the samples containing defects such as dislocations, grain boundaries and interfaces formed by the two materials with different evaporation fields. One way to overcome this problem is to use a local electrode so that multiple specimens can be analyzed after another without new specimen loading. To achieve such analyses of multitip array, Nishikawa and Kimoto proposed a *scanning atom probe* (SAP) (Nishikawa and Kimoto, 1994), in which funnel shaped microextraction electrode or local electrode to be scanned near tip apex of multitip arrays. One advantage of SAP is that the electric field required for field evaporation can be achieved with relatively low voltage applied to the specimen, so the insulator can be downsized for high frequency electronics. Also negative voltage pulses can be applied to the local electrode instead of superimposing high-voltage pulse to the positive standing high voltage, so the termination for pulse transmittance is easier.

Using this SAP concept, Imago in US (now CAMECA) developed a commercial AP with a trade name of LEAP. A local electrode that can be manipulated with piezoelectric actuator and the detector position are shown in **Figure 20**. The adoption of the local electrode whose position can be adjusted using piezoelectric actuators, multiple array specimens can be installed. If one specimen ruptures, next specimen can be analyzed quickly by changing the location of the local electrode. There are two types of operation modes: one is with voltage pulse and another is for laser-assisted mode. The laser installed in the commercial system is picosecond laser. With faster electronics and fully automated operation system, LEAP got an overwhelming success as a commercial AP instruments, and now LEAP has become synonymous with a 3DAP instrument. The most up-to-data LEAP incorporates green and optional UV picoseconds-pulsed laser. The laser beam is focused to a few microns, which is one order magnitude smaller than the laboratory-made instruments, and the laser beam is automatically positioned so that

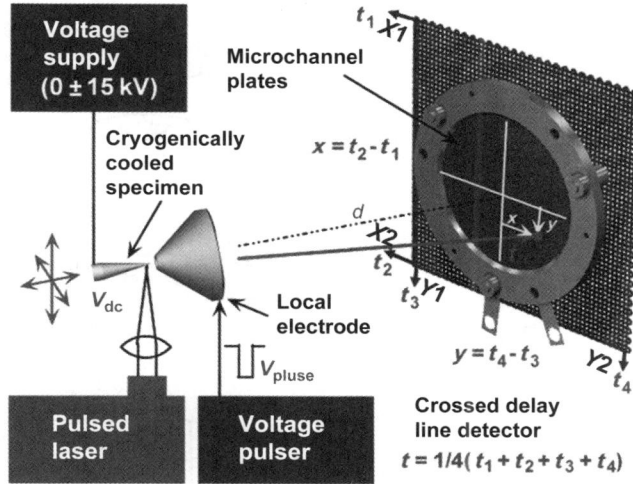

Figure 20 Illustration of the detector geometry and local electrode in a commercial local electrode AP. Reproduced from Miller (2000). (For color version of this figure, the reader is referred to the online version of this book.)

the fastest evaporation rate can be attained. There are both energy-compensated LEAP for higher mass resolution with a somewhat lower detection efficiency and narrower acceptance angle and straight type. Since LEAP can scan multiple-tip arrays, it has merit in the analysis of devices from which multiple tip arrays can be prepared with the recent FIB technique. With its easy operation and standardized software, LEAP is rapidly gaining customers throughout the world. As a result, the 3DAP, once a special technique for only specialist, is becoming a ubiquitous tool for nanoscale analysis tool for material scientists.

15.2.7 Specimen Preparation Techniques

In order to obtain an electric field high enough to field-evaporate atoms, the specimen for FIM observation and AP analysis must be a very sharp needle. These sharpened specimens are called FIM tips. Typical diameter of FIM tips at a typical operating voltage of 10 kV is approximately 100 nm. Most metallic materials can be fabricated to this shape by the standard electropolishing technique as shown in **Figure 21**. The starting specimen before electropolishing should be either wire or a rod with a square cross section. By electropolishing these either with DC or AC in a simple setting as shown in **Figure 21(a)**, most of metallic specimens can be made into an FIM tip. These electropolished tips, especially in aluminum and magnesium alloys, are usually covered with oxides. As mentioned earlier, these surface oxides are removed by field evaporation in FIM. Some specimens cannot be prepared with this simple polishing method, or in some cases, ruptured specimens need to be repolished for repeated analyses. For this purpose, a micro-electropolishing technique as shown in **Figure 21(b)** is used (Melmed and Carroll, 1984). An electrolyte is put in a small loop of Pt wire, and pre-prepared thin specimens are polished further using a drop of the electrolyte. By this method, local polishing is possible while monitoring the tip shape with an optical microscope. In order to polish away a few microns, few millisecond pulses can be applied to the Pt electrode (Fasth et al., 1967).

Amorphous alloys and nanocrystalline alloys, those used as nanocrystalline soft and hard magnetic materials, are usually prepared by melt spinning, as a result, these samples are in the form of thin

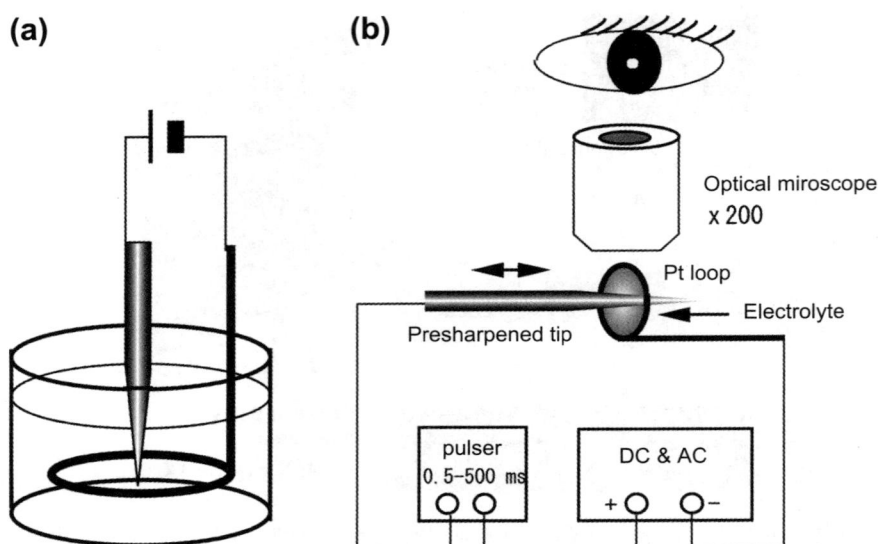

Figure 21   Schematic illustration of (a) a standard electropolishing setup to polish metal blank to a sharp-needle and (b) microelectropolishing setup by which local polishing can be performed while monitoring the specimen using an optical microscope. From Hono (2002).

ribbons with thickness ranging from 10 to 100 μm. For making AP specimens, these ribbon samples are prepared as blank rods by mechanical grinding first as shown in **Figure 22**. These ribbon samples are sandwiched with acrylic plates, then grinded from one side. Subsequently, these samples can be prepared to FIM needle by the standard micropolishing technique (see **Figure 21(b)**).

Thin film analysis had been a long challenge for AP, since the fabrication of FIM tips was difficult. In early days, these films were deposited on tungsten tips by electroplating, vapor deposition or sputtering (Hono et al., 1992c; Al-Kassab et al., 1996; Schleiwies and Schmitz, 2002; Gemma et al., 2009). Although interdiffusion of multilayer thin films can be studied using the films deposited on tips, the structure and properties of such films are different from those deposited on flat substrates. To prepare FIM tips from the films sputter deposited on flat substrates, Hasegawa et al. combined photolithography and microelectropolishing (Hasegawa et al., 1993). However, in this method, the films had to be physically removed from the substrate by etching Cu buffer layer. The controlled method of tip preparation from thin films on flat substrate, by using the FIB technique, was developed by Larson et al. (1998, 1999a, 1999b). Using annular ion beam from the top of film surface, they succeeded in preparing FIM tips from multilayer thin films in the perpendicular direction to the film. The AP tomography obtained from the specimen showed atomic layer resolution from multilayer thin films (Larson et al., 1999c, 2000a, 2000b). However, in this method, films had to be deposited on Si-island post array that were prefabricated by photolithography and reactive ion etching (Thompson et al., 2004). The films deposited on such Si posts were sharpened by the annular gallium (Ga) ion beams. By this way, multiple specimen arrays were fabricated using the Si post, so even if one tip is ruptured during the LEAP analysis, next analysis can be started by positioning the local electrode on top of the next tip without reloading a new tip specimen. One drawback of this method is that films have to be deposited on the Si post array specially prepared for a LEAP analysis.

Melt-spun ribbon $t \sim 50\ \mu m$

Sandwitch with acrylic plates

Mechanical grinding

Bar

$\sim 50\ \mu m \times \sim 50\ \mu m \times \sim 5\ mm$

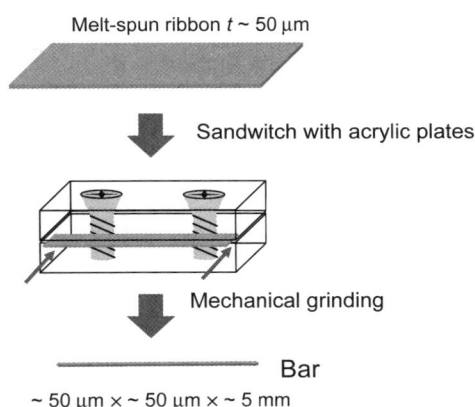

Figure 22 Schematic illustration of how to prepare blank specimen from melt-spun ribbon by mechanical grinding. The blank bar can be prepared to a tip using the standard electropolishing technique. (For color version of this figure, the reader is referred to the online version of this book.)

Since the device structures and properties change depending on substrates, analysis of the thin films that are specially prepared for AP analysis is not so useful to correlate the analysis results with the physical properties such as optical, magnetic and transport properties. Thus, there had been a strong need to prepare FIM tips from any specific area of thin films and their devices. By employing the FIB-based microsampling technique (US Patent Number 5,270,552, 1993) or the lift-out method that was developed for TEM specimen preparation, Thompson et al. demonstrated a site-specific FIM specimen preparation method by combining with the annular Ga-ion milling while monitoring a specimen with high-resolution scanning electron microscopy (Miller et al., 2005; Thompson et al., 2007). **Figure 23** shows how the specimen can be prepared from a specific site of thin films (Kodzuka et al., 2009).

In addition to the development of the above mentioned site-specific specimen preparation method, the implementation of ultrafast laser to 3DAP made the yield of AP analyses better. Because of this technical breakthrough, recent 3DAP applications have been extended to the characterization of device structures (Kelly et al., 2007; Hono et al., 2011a). In the previous studies, thin films for 3DAP analyses were specially grown on electrical conductive substrate such as doped Si; however, oxide substrates are widely used in the thin film community, i.e. single-crystalline oxide substrates such as MgO, rare-earth oxides such as LaAlO, and thermally oxidized Si substrates. Kodzuka et al. have recently shown that metal thin films deposited on insulator substrates such as thermally oxidized Si and MgO can be analyzed with an atomic layer resolution using a UV laser (Kodzuka et al., 2011).

Figure 24 shows the AP specimen prepared from a flat CuNi/Au thin film deposited on thermally oxidized Si. The microsampled thin film specimen with a-SiO$_2$ substrate was placed on top of a tungsten needle using the lift-out technique. There is a 3 mm thick a-SiO$_2$ between the CuNi/Au thin films, which is of interest for the analysis. Nevertheless, AP tomography was able to characterize nanoscale phase separation with Cu-rich and Ni-rich lamellae and corrugations at the CuNi/Au interface caused by the columnar growth grains of CuNi. The magnified Cu and Ni map obtained along the [001] direction perpendicular to the film resolves (002) atomic layers. This data convincingly show that atomic layer analysis of thin film is possible even if metallic or semiconducting films are grown on insulator substrates. This may open up much broader applications of 3DAP technique for thin films

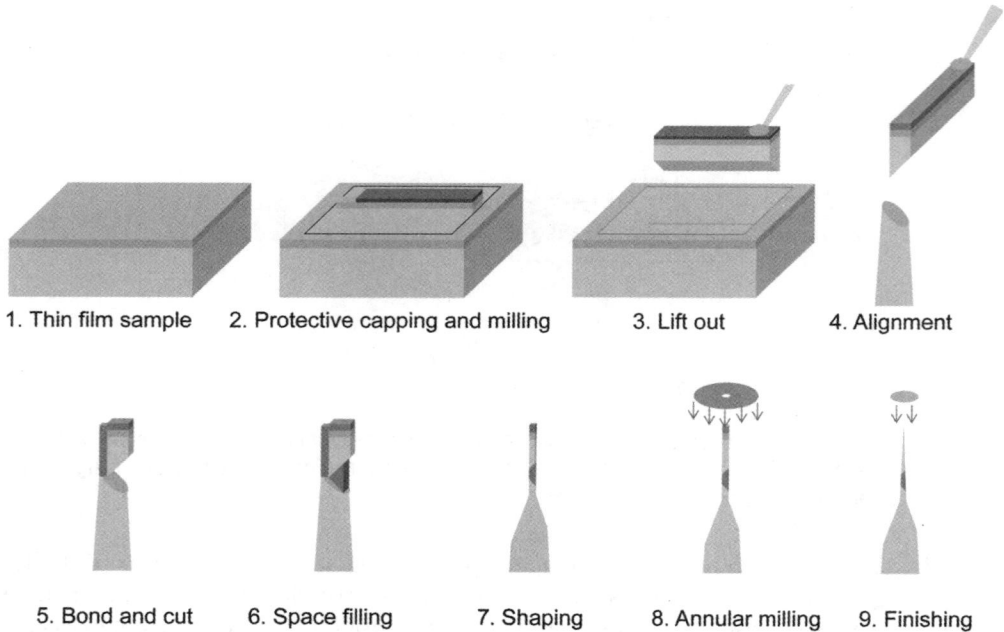

1. Thin film sample   2. Protective capping and milling   3. Lift out   4. Alignment

5. Bond and cut   6. Space filling   7. Shaping   8. Annular milling   9. Finishing

Figure 23 Site-specific FIM specimen preparation method using a Ga-ion-FIB machine with the lift-out technique. Microsampled specimens in step 4 are put onto a tungsten needle in step 4, thereafter the specimen is bonded with tungsten deposition. In the final stage, microsampled specimen is sharpened to a tip containing the interface near the tip apex. From Kodzuka et al. (2009). (For color version of this figure, the reader is referred to the online version of this book.)

studies. The site-specific specimen preparation method using the FIB lift-out technique can be applied not only to thin films but also to the standard metallic specimens including powders, grain boundaries and interfaces.

15.2.8   Data Reconstruction and Analyses Techniques

15.2.8.1   General Feature of 3DAP Data

Typical 3DAP results contain 10^6 to 10^8 data points [x, y, z and m/n]. **Figure 25** shows typical 3DAP data containing thin interphase boundaries. The thin phase boundaries are not perpendicular to the *probing direction* that is typically along the longitudinal direction of a tip. In such a specimen configuration, the accurate concentration depth profile across the interfaces cannot be determined using 1DAP as shown in **Figure 12**. On the other hand, concentration depth profiles can be obtained from any arbitrary region in the 3DAP data after acquiring data as shown in the inset box in **Figure 25(a)**. The selected box corresponds to the cylinder of analysis in a 1DAP analysis. While that the number of atoms contained in the cylinder of analysis in 1DAP data is small as the effective probe size is only a few nm, the number of atoms in the analysis box in 3DAP can be selected arbitrarily as long as the interface can be kept normal to the analysis direction as shown in **Figure 25(b)**. Then the standard errors due to counting the statistics in determining concentration depth profiles can be significantly reduced compared to that of 1DAP. When analyzing a small particle, one can determine the composition by selecting atoms only

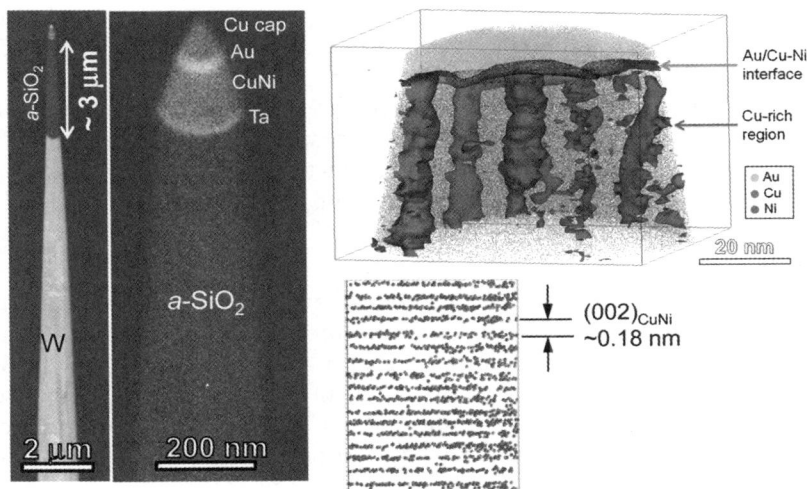

Figure 24 FIM specimen prepared from Ox-Si/Ta/CuNi/Au/Ni thin films deposited on thermally oxidized Si. 3 mm a-SiO$_2$ is left during the lift-out process to attach the film with the oxide substrate on the tungsten needle. Although the thickness of the oxide substrate is 3 mm, atomic resolution is obtained from the CuNi thin film. Courtesy of M. Kodzuka (2011). (For color version of this figure, the reader is referred to the online version of this book.)

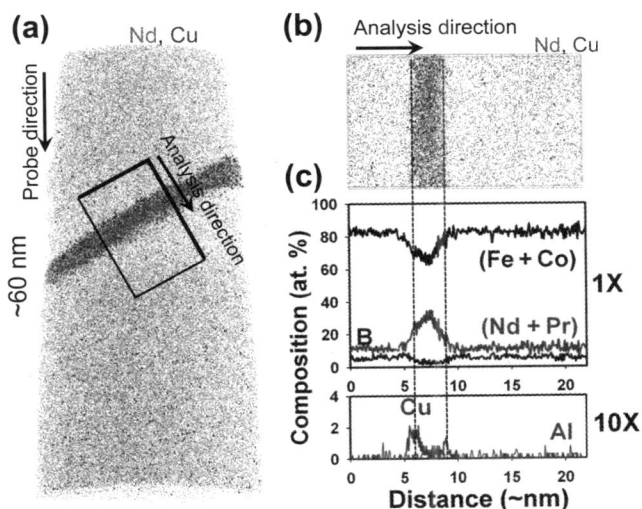

Figure 25 Schematic illustration of probing direction and selected volume for analysis in 3DAP data. (For color version of this figure, the reader is referred to the online version of this book.)

from the particle visualized in the 3DAP elemental mapping. Thus, the convolution effect can be eliminated in the 3DAP analysis, although a possibility of the artifact due to evaporation aberration remains.

Before we embark on discussions on data analyses methods, we need to make sure that these data are indeed accurate and true representation of the probed sample. Ongoing research focuses on the sensitivity of locating (x, y, z) field-evaporated ions as a function of detector efficiency (η), tip geometry, probing parameters (voltage, laser energy, laser wavelength) and type of AP (e.g. local electrode). The advances in these areas are briefly discussed here.

15.2.8.2   *Uncertainty of the Compositional Measurements*

Danoix et al. (2007a, 2007b) performed theoretical analyses of variance in the composition measurements by 1DAP and 3DAP. If we assume that the AP sample is indeed true representation of the overall material, then the standard deviation for a given concentration (C_o) of an element is given by $(C_o (1 - C_o)/N)^{0.5}$. In this equation, N is the total number of ions detected. Since the wide-angle 3D APs are capable of detecting more ions, these errors are expected to be low. However, if we are interested in the accuracy of the composition in a subset of the volume probed by 3DAP, the detector efficiency (η) and the number of atoms (m) within a representative volume must be considered. The efficiencies of currently available detectors are less than 0.6 and it will be somewhat lower in the energy-compensated AP because of the meshes placed on both entrance and exit of the reflectron energy compensator. For a given efficiency (η), the standard deviation for a given concentration (C_o) of an element within a probed volume (assuming $m > 100$) is given by $(C_o(1 - C_o)(1 - \eta)/N^{0.5})$. Danoix et al. (2007b) argue that if the detection efficiency can be made 100%, e.g. by using channeltron (Sakurai et al., 1984) detectors, there will be no errors in the compositional analyses! This is indeed the goal of new generation of AP s to improve the detector efficiency toward 100% (Miller et al., 2012). However, the challenge of locating (x, y, z) all 100% atoms on the surface of the tip remains as a big challenge and is the focus of ongoing work.

15.2.8.3   *Uncertainty in Atom Locations*

The point-projection methodology (Gault et al., 2008) for reconstruction of the 3DAP data from the ion positions in the detector is briefly reviewed below. The methodology is illustrated schematically in **Figure 26**.

The tip is assumed hemispherical in shape with a given radius of curvature (R). The location of an atom at the specimen tip is given by x-, y- and z-coordinates. On a field-evaporation, the ionized atom trajectory leads to scintillation at x_D, y_D-detector coordinates. The evaporation field lines can be traced back to the point "P" on the sample axis, which is at a distance ξR from the sample tip. The parameter ξ is known as compression factor (equal to 1 for radial projection and 2 for stereographic projection), which modifies the field lines depending on the tip geometry. The detector is placed at a distance "L" from the sample tip location.

Using the above geometry, the magnification (M) of the distance between the atoms can be calculated using the following equation:

$$M = \frac{L}{\xi R}. \tag{15}$$

In the above equation, ξ is the image compression factor (ICF), which in turn changes with radius of curvature during the progression of the AP analyses. Since one cannot measure R experimentally, R is

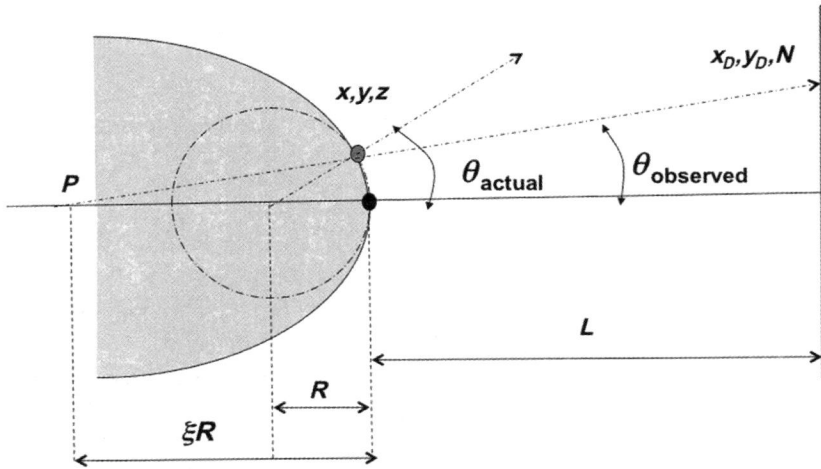

Figure 26 Schematic illustration of APFIM sample tip and detector, as well as, the point-project methodologies used for back calculation of ion locations based on the ion-hit data. Reproduced from Miller (2000). (For color version of this figure, the reader is referred to the online version of this book.)

related to the specimen voltage (V), electric field (F_e) and a geometric factor (k_f), using the following equation.

$$R = \frac{V}{k_f F_e}. \tag{16}$$

Using the (x_D, y_D)-detector coordinates and the magnification factor, the actual x, y location of the atom on the specimen surface can be calculated using the following equations:

$$x = \frac{X_D}{M}; \quad y = \frac{Y_D}{M}. \tag{17}$$

In order to calculate the "z"-coordinate of the atom on the specimen surface, an incremental in depth (dz), for each atom detected, is calculated using the following equation:

$$dz = \frac{\Omega}{\eta S_a}. \tag{18}$$

In the above equation, Ω is the atomic volume and η is the detection efficiency. The parameter S_a is the total surface area analyzed.

$$S_a = \frac{S_d}{M^2}. \tag{19}$$

In Eqn (19), S_d is the surface area of the PSD and M is the magnification. With the above relations and by assuming the electric field to be equal to evaporation field, the new expression for dz is given by the following equation.

$$dz = \frac{\Omega}{\eta S_a} = \frac{\Omega L^3 k_f^2 F_e^3}{\eta S_d \xi^2 V^2} \tag{20}$$

However, to take care of a changing curvature with evaporation sequences, a correction factor is applied as given by the Eqn (21).

$$dz = \frac{\Omega L^3 k_f^2 F_e^3}{\eta S_d \xi^2 V^2} + R\left(1 - \sqrt{1 - \frac{x^2 + y^2}{R^2}}\right). \tag{21}$$

The above equations demonstrate that the accuracy of the 3DAP data is related to image compression factor (ξ), geometric field factor (k_f) and the efficiency (η) of the detector.

One of the key assumptions of the 3DAP data reconstruction is the spherical evaporation surface of the sample tip. This assumption allows for a single point-project methodology and constant ICF. However, interrupted electron microscopy of AP samples (Haley et al., 2011) have shown that the evaporation surface does change with the progress of analyses. Therefore, arguments have been made, based on theoretical models, for using variable projection points, in conjunction with electric field effects (Larson et al., 2011a). Based on these analyses, the authors demonstrated that it is possible to retain planar features in the samples during the data analyses by deriving the ICF as a function of detector position, solid angle made by ion trajectory based on crystallographic and field conditions, and the z-coordinate of the sample. It is important to note that the goal to obtain automatic and accurate reconstruction methodologies (Gault et al., 2009; Moody et al., 2009; Niewieczerzai et al., 2012) are part of the on-going research and will become relevant in the goal to achieve 100% detection of atom identity and location (Miller et al., 2012). However, the challenge of making the detector capable of achieving this goal is indeed enormous. Additional complexities due to surface diffusion (field or thermally activated) of atoms also have to be considered (Gault et al., 2012). This is indeed a challenge for laser-assisted AP, where localized temperature increase can be significant for certain laser wavelengths. Recent work by Marquis and Gault (2008), based on earlier work by Kellogg (1981), suggests that it is also possible to estimate the tip temperature by analyzing the charge states with and without laser irradiation.

15.2.8.4 *Common Data Analyses Methods*

Once locations (x, y, z) and identifications of (m/n) of the ions are established, we have to interpret the data and extract useful information related to physical phenomena in materials. These methodologies are discussed in detail by Miller (Miller and Smith, 1989; Miller, 1999a) and available within commercial software (e.g. IVAS (http://www.cameca.com/support/ivas.aspx; CAMECA, May 2011)). We will discuss some of these methodologies briefly in this section, presenting typical examples from different materials. The transitions of some of these methodologies from 1D to 3D analyses are also highlighted.

15.2.8.4.1 *Mass Spectrum*

The first step in APFIM analyses is associating the measured m/n with respective elements. This involves plotting the mass-to-charge ratio (m/n) data in the form of frequency distribution. This graph is usually plotted with y-axis in log-scale to illustrate the presence of ions with low (ppm) concentrations. The peaks for each m/n bins in this spectrum are identified based on their natural ionization tendency. During a field evaporation, the ionization of individual atoms or molecules may occur. For example,

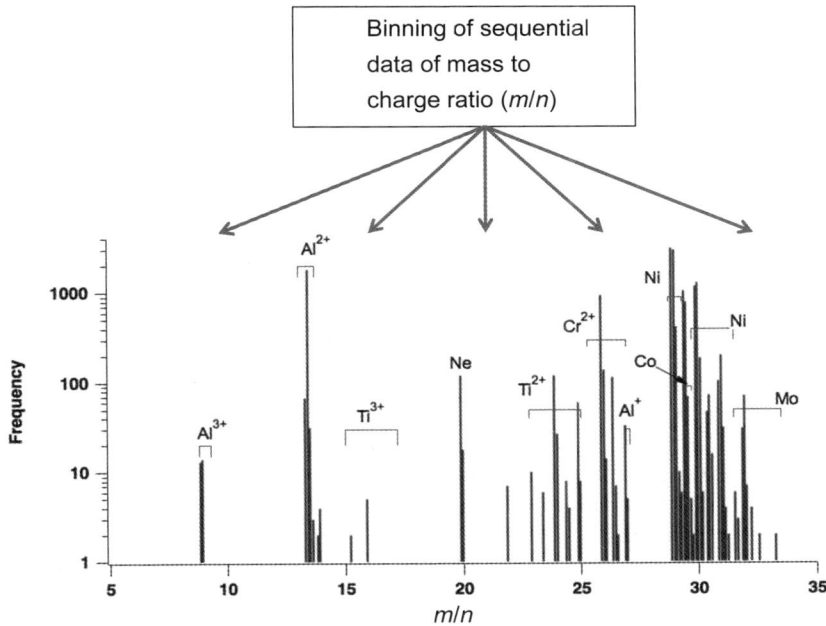

Figure 27 A typical mass-spectrum from a nickel-base superalloy, which is constructed from sequential m/n data, collected by a 3DAP. From Babu et al. (2001). (For color version of this figure, the reader is referred to the online version of this book.)

a typical mass spectrum from nickel-base alloy is shown in **Figure 27**. In this spectrum, the aluminum atoms ionize with one (Al^+) or two (Al^{2+}) positive charges. On the other hand, in steels, the ionization carbon may occur in individual (C) and molecular (C_2 and C_3) ions, i.e. in the form of C^{++}, C^+, C_3^{++}, C_2^+ and C_3^+. The ionization characteristics of all elements from periodic table are tabulated in many references (Miller, 2000). Most of the current data analyses software can identify the peaks automatically. However, it is always prudent to check these auto peak identification results by checking the overall composition of the sample. After identifying each peak [$(m/n)i$] and their respective number ($N_i^{(m/n)i}$) of ions, the mass spectrum can be used to calculate the average concentration of each elements ($C_i^{(m/n)i}$) in using simple summation Eqn (22).

$$C_i^{(m/n)i} = \frac{N_i^{(m/n)i}}{\sum_{i=1}^{n} N_i^{(m/n)i}}.$$ (22)

If we assume the sample is a homogeneous alloy and all the ions field evaporate equally, the above concentration should be close to the nominal concentration of each element in the alloy. If the overall composition determined form the mass spectrum shows a large deviation from the nominal composition of alloy, we can conclude that preferential evaporation of some of the constituent elements have occurred. Since the evaporation fields of elements are largely different, the atoms with high evaporation field tend to retain on the surface of specimen, which in turn leads to errors in overall composition. In the conventional AP and earlier 3DAP, high-voltage pulse is used to trigger the field evaporation for time-of-flight measurements. The ratio of the pulse with respect to the standing voltage, V_p/V_{dc}, is defined as *pulse fraction*, and if the pulse fraction is small, e.g. 0.1, the elements with lower evaporation field tend to be field evaporated by the standing voltage. These atoms are not synchronized for ToF measurements, so are miscounted. This causes systematic deviation of the composition determined by AP analysis from the nominal composition. If this occurs, the pulse fraction and temperature for analysis have to be optimized so that the overall composition matches with the nominal composition by minimizing the preferential evaporation of atoms. However, caution must be taken during this procedure, if the sample contains large precipitates or solute depleted or enriched region, because the composition of the sample determined by AP may not be consistent with the nominal composition. Additional complexity may arise due to ambiguity of identifying the m/n values. For example, the m/n values Fe^{++} and Si^+, Si^{++} and N^+ overlap. Under these circumstances, for average compositional measurements, one could assign certain number of atoms to one kind of atom based on their ionization characteristics.

The mass-spectrum can also be used to evaluate the performance of the AP instruments. The mass resolution is defined by $m/\Delta m$, where Δm is the width of a mass spectra at various peak heights, i.e. $m/\Delta m$ at full-width at half-maximum (FWHM) and full-width at the maximum (FWTM). The typical mass resolution for voltage-pulse straight-type AP is less than 500 that depends on flight length, and that for energy-compensated AP is larger than 1000. In the UV laser-assisted AP, the mass resolution close to 1000 was reported even from insulating ceramics material (Li et al., 2011b). The presence of large background data (i.e. m/n data that cannot be identified) could indicate uncontrolled field evaporation of atoms from the surface of the sample. In the case of voltage-based instruments, this may occur due to low pulse fraction or high specimen temperature. If imaging gas atoms react with specimen, similar catastrophic evaporation occurs (field corrosion, see Müller, Panitz, McLane, 1968). In laser-based ionization instruments, the background noise may increase due to improper selection of laser power and the wavelength of the laser in case of the oxide specimen (see **Figure 19**). The above mass resolution and S/N ratio are usually quoted as the instrument capability to probe different materials.

15.2.8.4.2 *Integrated Concentration Profile*

This method is the most direct approach to analyze the 1DAP data and is still effective to see slight deviation of local concentration from average value even in 3DAP data. This is done by plotting the number of specific ion detected as a function of the total number of ions as shown in **Figure 11**. The slope of the plot becomes local average concentration of atoms. For example, the number of detected C, Si and Mn ions as a function of total number of ions is plotted in **Figure 28**. These data were obtained from a cementite (Fe_3C) plate in a tempered martensite. The slopes of C and Mn outside the dotted line are nearly zero, indicating C and Mn are enriched in the area between the two dotted vertical lines. The data clearly show the depletion of silicon (as indicated by the slope of the Si line in the plot) and enrichment of Mn within the cementite phases. This is indeed in agreement with thermodynamic calculations (Babu et al., 1994). Such plots are usual in evaluating the partitioning of alloying elements and long-range diffusion of the same across the precipitating phases. However, this methodology is now superseded by Proxigram methods, a standard methodology within commercial software for analyzing 3DAP data. This methodology will be discussed later.

15.2.8.4.3 *Site Occupancy*

Another variant of integrated concentration profile is ladder plot. In this, the numbers of specific ions are plotted as a function of cumulative numbers two or more ions. This is very useful in evaluating the site occupancy of specific elements in an ordered structure. A typical application of this methodology with reference to preferential occupation of Cr atoms in Al lattice within a B-2 ordered Ni–Al is shown

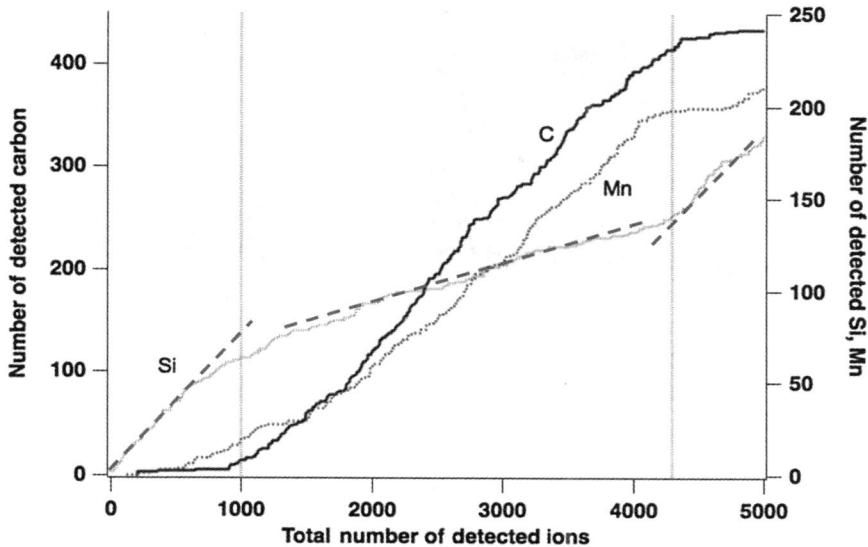

Figure 28 Ladder plots constructed from a 1D data from a Fe–C–Si–Mn steel show the presence of carbon- and manganese-enriched (cementite) region and Si-depleted cementite. The cementite phase is sandwiched by α-bcc matrix. From Babu, Hono, and Sakurai (1994). (For color version of this figure, the reader is referred to the online version of this book.)

in **Figure 29** (Frommeyer et al., 2004). The data were obtained by preferential analyses of the sample along the [001] direction of the B-2 ordered structure. In these plots, the nickel atoms are observed, with minimal detection of Al atoms and vice versa, during the AP analyses. This clearly shows ordered nature of NiAl compound. Interestingly, during these analyses, the Cr atoms are detected only in conjunction with aluminum atoms. This is clearly visible in three Al-rich planes. Based on the result, the Cr atoms are concluded to occupy the aluminum sites preferentially.

One of the problems with site occupation determination using ladder plot analyses is the preferential retention of atom and the double layer evaporation. The elements with higher evaporation rate tend to preferentially retain on terrace even after the evaporation of other atoms. In such a case, the solute atoms may be detected from another plane. In the case of $L1_0$-TiAl, Ti and Al layers tend to evaporate in double layer. This is because the evaporation field of Ti is higher than that of Al. While Al evaporates at a lower field, the evaporation rate of Ti is slower. Thus, having the Ti at the topmost surface, the under layer Al-plane shrinks with the topmost Ti plane. In such a case, AP analysis is not that effective in a site occupation. The site occupation study is possible only in limited cases where atomic planes evaporate layer by layer.

15.2.8.4.4 Concentration Profile

The concentration profiles are the most quoted technique in 1DAP and 3DAP. In this analysis, the detected ions are subdivided into small number of blocks with equal number of ions. This number of ions per block is then correlated to the real distance for each block based on calibration and molar volume. Then, the concentration of each element within each block is calculated sequentially. The calculated concentration is then plotted as a function of depth of analyses. Typical examples of concentration profiles from a solid solution and a phase-separated microstructure, within a Fe–Ni–Cr–Mo–Ti maraging

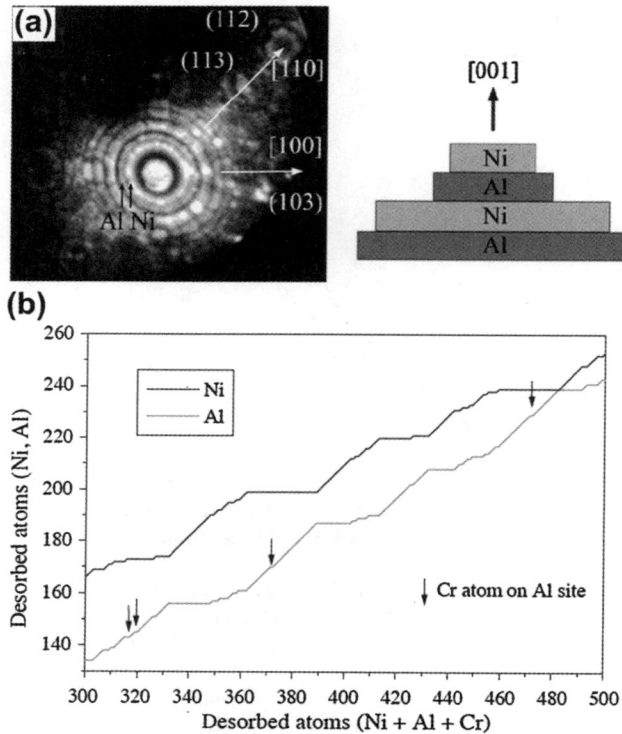

Figure 29 Demonstration of ladder plots to evaluate the site occupancy of tertiary elements in an ordered structured. In this case, the Cr atoms are found to be in the Al site within a B-2 ordered NiAl intermetallic compound. From Frommeyer et al. (2004).

stainless steel, are shown in **Figure 30**. As shown in the 3DAP map of Ti in **Figure 30(a)**, Ti distribution in the solution-treated sample is uniform, while Ti concentration fluctuation can be seen from the aged sample. To convert this to one-dimensional concentration profile, analysis volume is selected as shown in the inset. The concentration profiles determined form these analysis boxes are shown in **Figure 30(b)**. The scaling of the depth is rather delicate. If atomic planes are resolved, accurate scaling in the depth direction is possible. Otherwise, the depth scaling is automatically done by the data analysis software based on the expected radius of specimen depending on voltage, and the projection constant.

This depth scaling is rather arbitrary, so it is strongly encouraged to make correlative microscopy using TEM. Then, the scales of the microstructures such as average size of precipitate or wavelength of phase separation can be estimated. The parameters for 3D construction can be adjusted so the scaling of the AP data have good matching with the microscopy observation. Note that the all concentration profiles contain errors originating from the counting statistics. The standard deviation σ for concentration measured by AP data is given by

$$\sigma = \sqrt{\frac{x(1-x)}{n}}, \qquad (23)$$

where n is the number of atoms and x is the atomic fraction of the solute.

(a)

53 x 53 x 52 (nm)   71 x 71 x 59 (nm)

(b)

(c)

$\chi^2 = 16.99$
df = 19

$\chi^2 = 183017$
df = 19

(d)

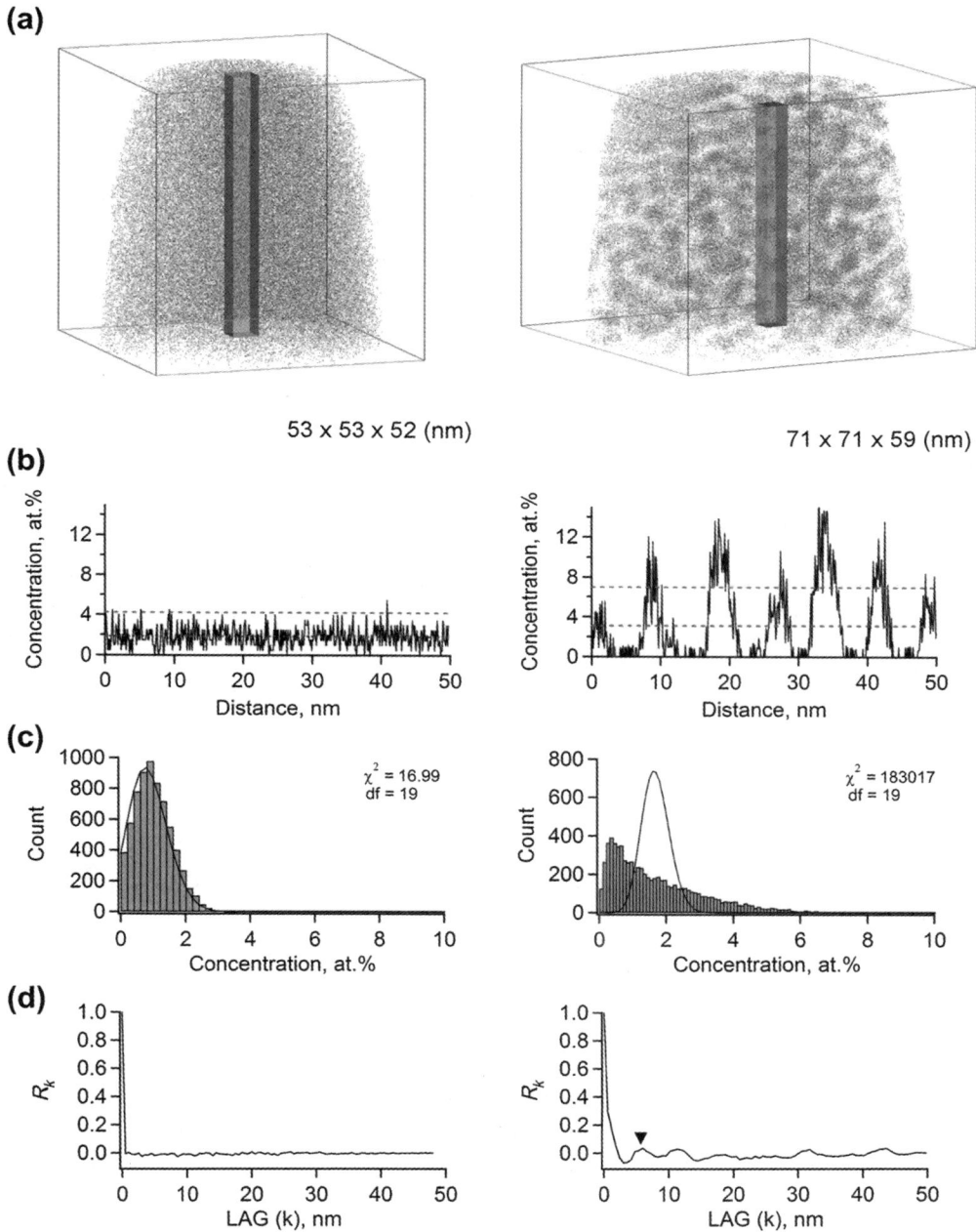

Figure 30   (a) Ti elemental maps obtained from Fe–Ni–Cr–Mo–Ti maraging stainless steel with and without short-term aging. (b) One-dimensional compositional profile analyzed from selected volume from sample without aging shows insignificant composition modulations since the variations are within statistical error. However, with aging, statistically significant composition modulations develop. This is inferred by the presence of variations that are beyond the statistical error. These modulations are then confirmed by (c) frequency distribution and (d) auto-correlation analyses. Courtesy: T. T. Sasaki, NIMS, Japan. (For color version of this figure, the reader is referred to the online version of this book.)

This means that the statistical errors expected from 2σ will become large when the block size, the number of atoms for calculating one concentration data point, is decreased. To make the spatial resolution for concentration profile to subnanometer dimension, we need to decrease n to around 100. If the solute concentration is 50 at.%, the standard deviation is 5% for the block size of 100 atoms, thus it is not possible to differentiate concentration fluctuations of less than 10 at.%. In order to decrease the statistical errors, we need to increase the block size. Then, naturally, we lose the spatial resolution in the concentration profile. On the other hand, if the solute concentration is 0.1 at.%, the standard deviation is only 0.3%, and the detection sensitivity of solute clusters is significantly improved. Thus, the AP is quite suitable for detecting clusters of the solute whose concentration is less than a few atomic percent. The block size is selected for optimizing both spatial resolution and statistical errors.

15.2.8.4.5 *Frequency Distribution*
Many times, before the onset of precipitation or extensive phase separation that can be identified by atom maps or iso-surfaces, statistical analyses are done using frequency distribution method to evaluate the tendency for phase separation or ordering. The frequency distribution is done by constructing histogram of all the calculated concentrations from every block or voxel within 1DAP or 3DAP data. The introduction to these analyses was shown in **Figure 30(c)**. If the frequency distribution is similar to binomial distribution, one can conclude that the data are representative of a random solid solution. In case of the data being sharper than binomial distribution, the data will be representative of ordering. If the frequency distribution is wider than binomial distribution, it will be indicative of phase separation. The above methodology was used to evaluate the onset of phase separation within ferrite matrix in an oxide dispersion strengthened alloy (Capdevila et al., 2008). The frequency distribution shows that it is broader than binomial distribution. This was later confirmed by the detailed analyses using Langer, Bar-on and Miller (LBM) method, as well as construction of 17 at.% iso-surface. Another example of 2D frequency distribution from a 3DAP data involving Cu–Ni alloy is shown in **Figure 31**. In this thin-film sample, as expected, the phase separation of Cu and Ni were observed.

15.2.8.4.6 *Autocorrelation*
Advanced and quantitative evaluations of tendency for phase separation in 1DAP or 3DAP data, as well as their sizes and periodicity, are usually done by autocorrelation. The autocorrelation parameter (R_k) is defined as the following (Miller et al., 1996):

$$R_k = \frac{N_b}{N_b - k} \frac{\sum_{i=1}^{N_b+k}(c_i - c_0)(c_{i+k} - c_0)}{\sum_{i=1}^{N_b}(c_i - c_0)^2}. \tag{24}$$

In the above equation, c_0 is the nominal concentration of the element in consideration, c_i and $c_i + k$ are the concentrations of ith and $(i+k)^{\text{th}}$ blocks or voxels and N_b is the total number of atoms in a block or voxel used for the calculation of ith or $(i+k)^{\text{th}}$ concentration, respectively. A typical plot calculated from 3DAP data from ODS data (Miller et al., 2011) is shown in **Figure 30(a)**. The first peak below the zero value of R_k, the size of these phase-separated regions grows from less than 10 nm (at 10 h) to a value greater than 10 nm. The reader is referred to the book by Miller et al. (1996) for more details.

15.2.8.4.7 *Image and Contour Map*
Similar to concentration profile in linear axes, it is possible to evaluate the concentration gradients in a 2D section from 3DAP data. In this analysis, 3D volume is divided into 2D slices of equal thickness

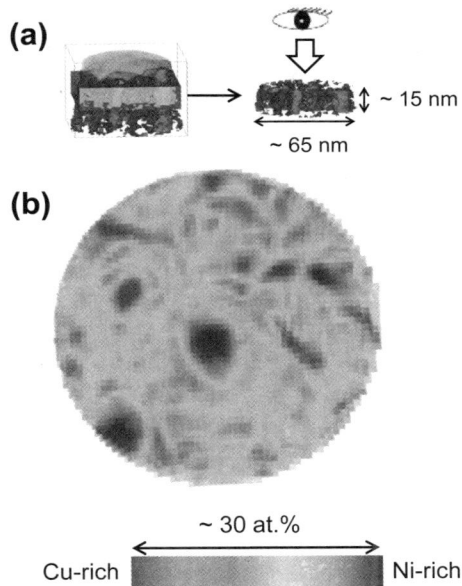

Figure 31 Typical use of image and contour map to analyze Cu–Ni multilayers. (For color version of this figure, the reader is referred to the online version of this book.)

along the z-direction of the data as shown in **Figure 31(a)**. Then, these (x, y) 2D slices are subdivided into small boxes. In the next step, the concentration of each element is calculated by using the same approach as linear concentration profile. Using the calculated concentration values for this 2D matrix, we can now construct an image map using a color scale reflecting the concentration of solute element as shown in **Figure 31(b)**. This visualization method provides conceptual understanding of phase transformations in complex alloys.

15.2.8.4.8 Iso-Surfaces

With the 3DAP data containing the location of the atom and its identification, it is now possible to visualize the shape of nanoscale precipitates within the data. In this method, coordinates of locations within the total volume of the analyzed region with equal concentrations are identified. Then these locations are connected by a smooth surface and visualized with computer software. A typical iso-surface constructed from 3DAP data obtained from a steel sample is shown in **Figure 32**. In this steel, nanoscale Cu and M_2C precipitates are distributed within martensitic matrix. This is visualized by construction of iso-surface of 8 at.% Cu and 1 at.% C. Imaging coherent Cu precipitates in steels are very difficult in TEM, but the iso-concentration contour show the uniform distribution of the Cu-rich precipitate within Fe. Using the visualization technique, the radius and the number density of these precipitates were also calculated. In certain cases, with small variations in compositions of precipitate (C_p) and matrix (C_m), the iso-surface concentrations are constructed with a concentration that is in between C_p and C_m.

15.2.8.4.9 Radial Proxigram

This is similar to the concentration profile but with a better statistical reliability. The concentration profiles are determined by counting the number of atoms normal to the analysis volume. This will give

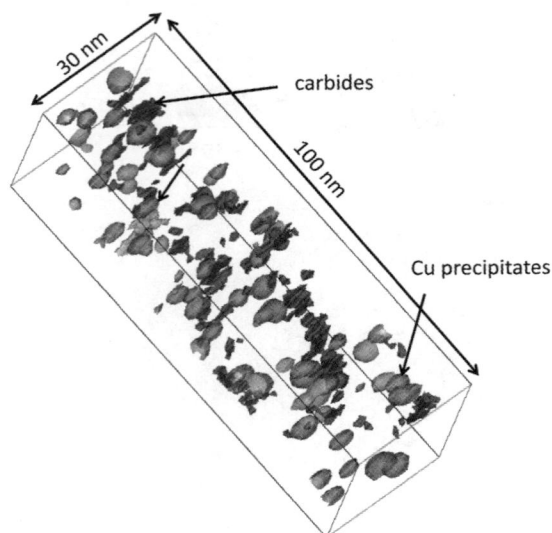

Figure 32 Analyses of 3-D distribution of Cu-precipitates and carbides within a martensitic matrix in alloyed steel revealed by iso-concentration surfaces. From Yu et al. (2010). (For color version of this figure, the reader is referred to the online version of this book.)

accurate profile when the interface is normal to the analysis volume. However, we often need to draw concentration profiles across curved interface of small precipitates as shown in **Figure 33**. In this case, the curvature near the interface causes convolution with the matrix. In addition, if the precipitates are in nanoscale dimension, the counting statistics become very poor. On the other hand, radial proxigram determine the concentration profile from the core of the precipitate to the matrix in a radial fashion. Once the precipitates are identified by iso-surfaces, the location of their center is calculated. Then, the concentration in spherical cells from this center is calculated and plotted. This methodology is described in **Figure 33** using the same data from **Figure 32** (Yu et al., 2011). One of the copper precipitates is identified and the concentration profiles of Fe, Cu and Ni are calculated. In this plot, the origin of the proxigram is at the right side of the plot. The Proxigram clearly shows that nanoscale copper precipitates are not 100%Cu and contain iron also. Such information can be used in modeling the nonequilibrium phase transformation kinetics.

15.2.8.4.10 Gibbsian Interfacial Excess
Equilibrium and nonequilibrium segregation of elements to interfaces or interphase boundaries have become integral part of rationalizing phase transformations and properties. AP is probably the best method to analyze the interfacial segregation quantitatively. In earlier work, specimen preparation was difficult, as the interface of interest must be located less than 100 nm away from a tip apex. Even if such a specimen can be prepared, interface composed of two phases with different evaporation field often led to specimen rupture. However, the site-specific specimen preparation method using the lift-out and annular ion milling has alleviated such difficulty dramatically. In addition, laser-assisted field evaporation improved the yield of successful analysis, so the analyses of interfaces have become a routine-type work using the modern AP technique. One of the common methods to evaluate the extent of

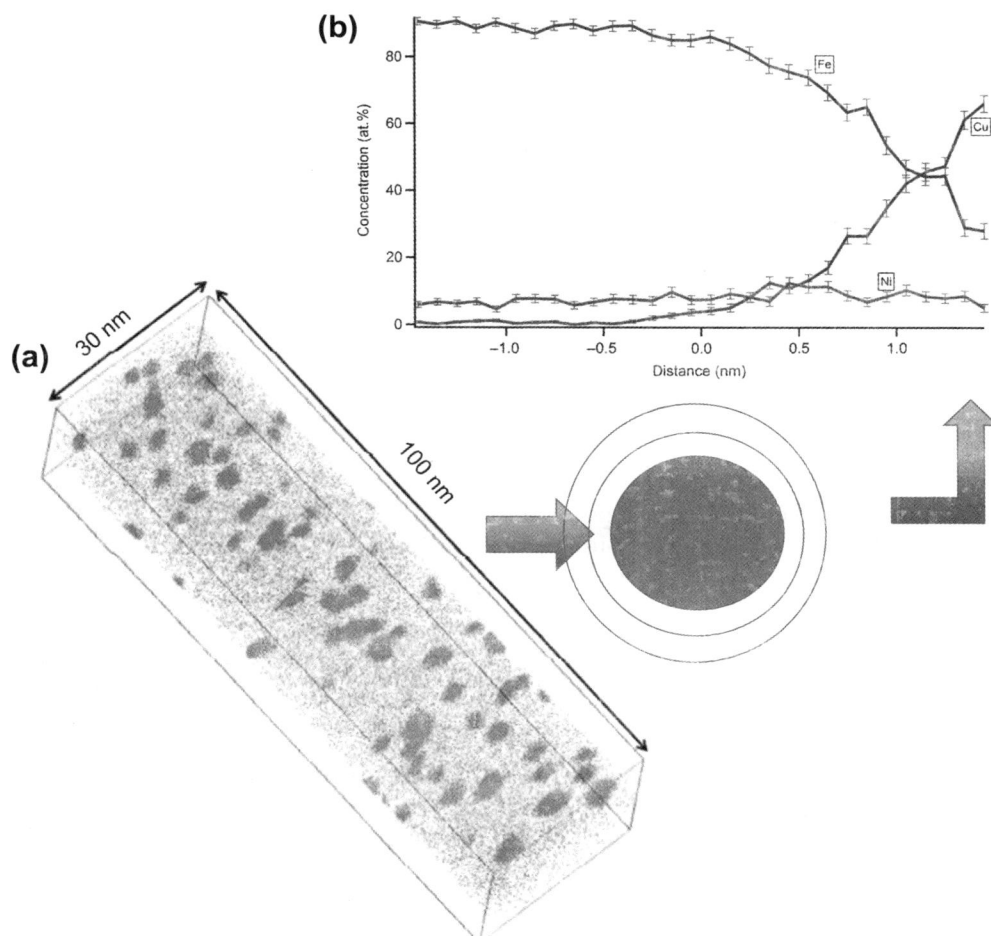

Figure 33 (a) Atom map of alloyed steel with nanoscale copper precipitates. (b) Proxigram constructed by analyzing the Fe, Cu and Ni data from the atom map across the precipitate/matrix interface. From Yu et al. (2010). (For color version of this figure, the reader is referred to the online version of this book.)

segregation in 1DAP or 3DAP data is to calculate the Gibbsian interfacial excess, which is defined by the following equation:

$$\Gamma_i = \frac{n_{ix}}{A}. \tag{25}$$

In the above equation, n_{ix} is the excess of solute atoms (determined by the difference between the total number of atoms and the number of atoms on either side of the interface) with the interface and A is the area of the interface. Using this estimate, one can describe whether or not this segregation is monolayer.

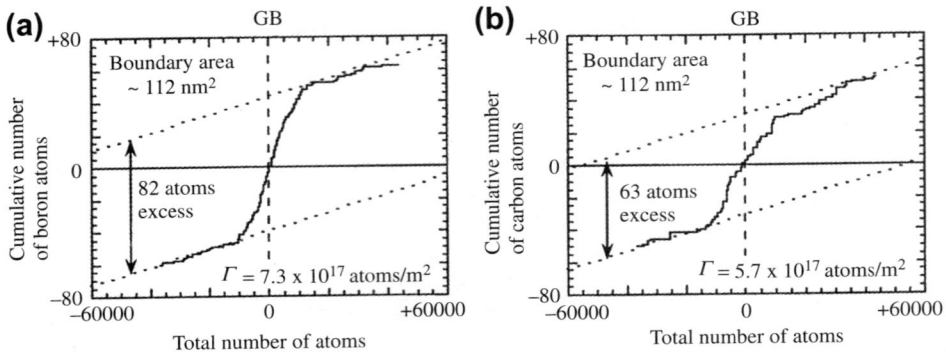

Figure 34   Measured Gibbsian interfacial excess of (a) boron and (b) carbon in an interfacial-free steel. From Seto et al. (1999).

A typical analysis (Seto et al., 1999) of boron and carbon segregations using 3DAP data is shown in **Figure 34**. In this data, by comparing the number of excess carbon and boron atoms, the relative east of boron segregation to grain boundaries were confirmed.

15.2.8.4.11   Radius of Gyration
The estimation of precipitates morphology from 3DAP data can be obtained by calculating the radius of gyration using the following formula. In this equation, the center of the precipitates is described by the coordinates $\bar{x}, \bar{y}$, and $\bar{z}$. The x_i, y_i and z_i are the locations of all the atoms within the precipitate and n is the number of total atoms.

$$I_g = \sqrt{\frac{\sum_{i=1}^{n} (x - \bar{x})^2 + (y - \bar{y})^2 + (z - \bar{z})^2}{n}}. \qquad (26)$$

Often this radius of gyration is multiplied by $\sqrt{5/3}$ into Guinier radius a true representation of the precipitate size. The above methodology was used to extract the precipitate distributions (see **Figure 35**) in a nickel-base superalloy (Karnesky et al., 2007). In this figure, the atom maps and the fitted ellipsoid shape of the precipitates are also shown. This analysis is noteworthy, because it allowed for the evaluation of the shape, as well as, crystallographic orientation of the precipitates based on identification of atomic locations.

15.2.8.4.12   Maximum Separation Method
In all the above analyses, it is assumed that the researcher can identify the precipitate or cluster features easily using the atom map or iso-surface method. However, in 3DAP data, this may prove to be subjective and therefore a robust and automatic analyses method is necessary. This problem is solved by maximum separation method (Miller, 2000). This is schematically explained in the diagram shown in **Figure 36**. Let us assume, we are looking for clusters of red and blue elements with a volume of given size d_{max}. At every red atom location (x, y, and z) within 3DAP data, a circular volume of diameter equivalent to that of d_{max} is defined. If we do not find another red or blue atom within that area, a decision is made to assign this red atom to be independent from the red atom. If we find another red or blue atom within this small volume, often calculated recursively, we assign them to be part of the cluster. By performing this analyses for all

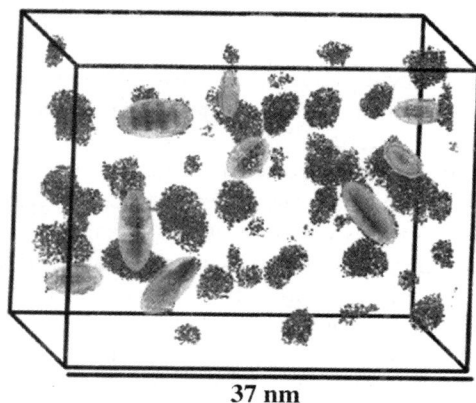

Figure 35 Quantification of gamma prime precipitate shape, size and also crystallographic orientation during coarsening in a nickel-base superalloy through data analyses and fitting 3D AP data. Reproduced form Miller (2000). (For color version of this figure, the reader is referred to the online version of this book.)

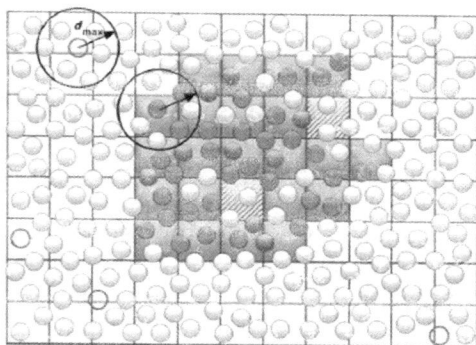

Figure 36 Schematic illustration of the method to perform maximum separation analyses to detect the small clusters or precipitates. Taken from Miller. Reproduced form Miller (2000). (For color version of this figure, the reader is referred to the online version of this book.)

detected elements, it is possible to define the cluster size, as well as, number density. Care must be taken during such analyses, for example by reducing the size of the cluster size one may create clusters, even if they are statistically uniform. If the concentration of solute is x, the chance of having the solute atom in the nearest position is zx, where z is the coordination number. If the concentration of solute is 2 at.%, the chance of having it in the nearest position in the fcc lattice is $0.02 \times 12 = 0.24$. With 24% of the chance, we will see dimers even in the case of homogeneous solid solution if we use this algorithm! If slightly large d_{max} is selected, this algorithm will create solute clusters in any solid solutions. In this sense, the application of this method should be limited for dilute alloys. To take care of this problem, Miller related the number of atoms associated with a feature, molar volume, and detection efficiency of the detector. In addition, care also must be taken to use this analysis for 3DAP data obtained with laser excitation since localized temperature increase may promote surface diffusion of element before the field evaporation. In this regard, the detected position of the ion may not be the real position within the lattice.

15.3 Scope of Atom-Probe Field-Ion Microscopy Technique

Physical metallurgy focuses on achieving targeted physical properties by manipulating phase consti-tutions (solute concentrations and crystal structures) and their spatial arrangements (micro/nano-structures) within a larger body of material. For example, published literature on metals and alloys demonstrate the role of clustering, segregation and precipitation processes (see **Figure 37**) in achieving desired microstructures that control mechanical and magnetic properties. Therefore, the characteriza-tion results have to be discussed in terms of its ability to reveal the mechanism of phase formation, microstructure development and microstructure-to-property relationships. In this section, a quick scope of applications of the AP technique is presented. It is important to note that AP is not an all-in-all inclusive characterization technique. It provides elemental distribution in materials with high spatial resolution, but it does not give any structural information. Therefore, the 3DAP techniques must be complemented with complementary techniques such as electron microscopy or X-ray diffraction for phase identification. As APT data often contain artifacts due to the local difference in tip radius stemming from the difference in evaporation field of constituent elements, the reconstructed data should be evaluated critically by comparing the results with structural information obtained by TEM.

Figure 38 **(a)** schematically illustrates specimen and electron beam interaction in TEM. The TEM image is the projection image of 3D nanostructure; hence even if two particles are spatially separated in 3D space, the projected image may show that they are in contact with each other. For chemical analysis, electron beam is converged to less than 1 nm diameter. Even if the beam size is smaller than the particle diameter, electron beam interacts with the matrix, so the EXD and EELS spectra are the convoluted information both from the precipitate and the matrix. On the other hand, APT gives quantitative chemical composition from nanometric volume if the data are as shown in **Figure 38(b)**. After getting 3D map from almost the same volume as the TEM foil, the positions of nearly 50% of atoms in the volume can be plotted. Thereafter, analysis volume can be selected so that the concentration can be calculated from only the selected precipitate. Then, the quantitative concentration of the nanosized precipitate can be determined without the convolution effect from the matrix phase. In particular, it is

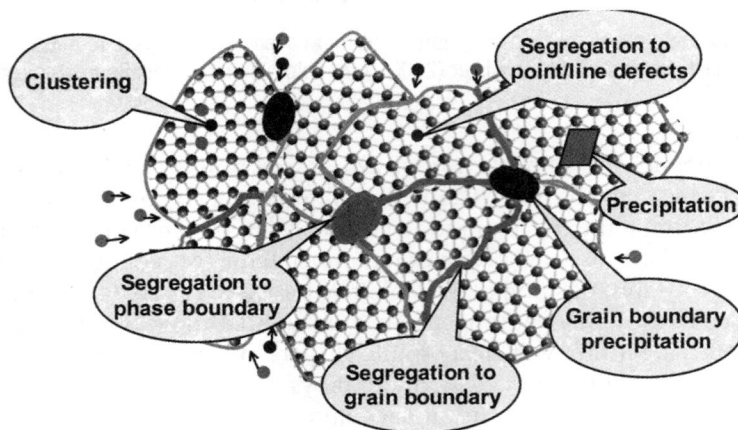

Figure 37 Some of the physical processes that can be characterized by AP field ion microscopy. (For color version of this figure, the reader is referred to the online version of this book.)

(a)

(b)

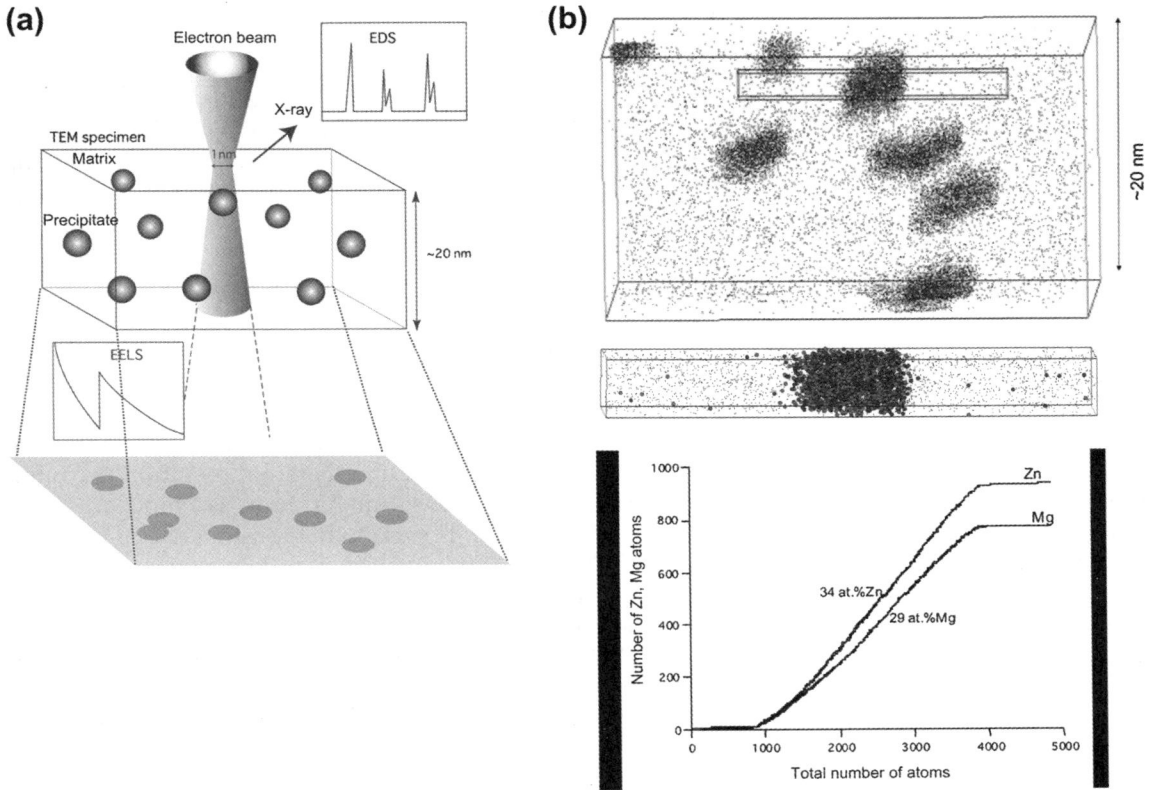

Figure 38 Illustration of (a) TEM thin foil specimen and projected image by TEM. Also, the interaction between focused electron beam and nanosized precipitates embedded in the matrix is shown. (b) 3DAP map of the same scale of specimen and nanosized precipitates embedded in the matrix. (For color version of this figure, the reader is referred to the online version of this book.)

suitable to detect atomic clusters and nanometric scale phase separations that are below the detection limit with TEM (see **Figure 38**).

15.3.1 Defects and Atomic Phenomena

15.3.1.1 Point Defects

Point defects in material include vacancy (e.g. quenched in vacancies in aluminum alloys), substitutional atoms (e.g. in alloys), self-interstitials (e.g. due to ion beam radiation), and impurity interstitials (e.g. carbon and nitrogen in steels). In earlier days, point defects in high melting-temperature refractory metals were directly observed using FIM. However, direct observations of vacancies are possible only from pure metals and this method is not applicable to any alloys with engineering interest such as steels, aluminum alloys and other structural metals. In an ordered crystal structure, the point defects may include antisites, e.g. the presence of antisite lattice defects within intermetallic compounds such as Ni_3Al.

Although Miller et al. (2011) have shown the feasibility of characterizing nanovoids through measurement of density variations, detection of point defects is beyond the capability of the AP technique. It should be also noted that FIM tips tend to be ruptured once void appears on tip apex. So the observation of voids and point defects will be limited for vary rare cases. One more reason for the unsuitability of the 3DAP technique for the characterization of voids and vacancies is the poor (0.6) detection efficiency of MCP detectors. Since the concentrations of point defects are in the order of 10^{-5} or less, one cannot make any distinction based on variations in density. Nevertheless, we can evaluate interaction of solute atoms with vacancies, in conjunction with positron annihilation technique.

For example, Honma et al. (2004a) characterized Al–5.1Li–1.9Cu–0.28Mg–0.2Ag (at.%) and Al–1.9Cu–0.30Mg–0.2Ag (at.%) alloys, after early stages of aging (solutionized, quenched and aged 15 s at 180 °C). In addition to the elemental mapping, ladder diagram, contingency table analysis of the 3DAP data (see **Figure 39**), they used coincidence Doppler broadening (CDB) and positron lifetime measurements. The 3D atom maps (see **Figure 39(a)**) showed preferential clustering of Cu, Mg and Ag atoms. In this case, since Mg and Ag concentration is only 0.1 at.% level, clustering of solute atoms can be distinguished in the raw data without filtering with the maximum separation method. To confirm this visual observation, quantitative ladder plots, i.e. plot of Cu, Mg and Ag atoms as a function of total number of detected ions, were drawn across one of the cluster. The results demonstrate the co-clustering of Mg and Ag.

It is interesting to note that for similar concentration of Mg and Ag, the tendency for clustering (see **Figure 39(b)**) was not observed in Al–Li–Cu–Mg–Ag alloys. These differences can be rationalized based on strong interaction between Mg and Ag atoms and vacancies to form clusters in Al–Cu–Mg–Ag alloy. Confirmation of this hypothesis was obtained by positron lifetime measurements. Higher positron lifetime in Al–Cu–Mg–Ag alloy (>200 ps) compared to that of Al–Li–Cu–Mg–Ag (<190 ps) alloy validated the presence of Mg–Ag–vacancy complexes. Since AP is not able to provide useful information on the vacancies and their interaction with solute, combined use of positron annihilation technique for getting information on vacancies and AP for getting information on solute clusters is very useful.

15.3.1.2 *Line Defects—Dislocations*

The line defects in crystals are dislocations. Both edge and screw dislocations have been imaged in an FIM, by preferential alignment of the crystallographic axis to the tip axis. These dislocations generate lattice strain fields around them. As a result, segregations of elements to dislocations are energetically favorable. These effects are well researched starting from classic work of Cottrell and Bilby (1949). Although, Cottrell atmosphere around dislocation is proven by other techniques, the direct measurement of elemental concentrations around dislocations eluded the researchers for a long time.

In 1999, Blavette et al. (1999) successfully characterized an edge dislocation in a boron-doped ordered Fe–Al alloy using 3DAP. In addition, they imaged and quantified the boron concentration around the dislocations using 3D atom map as shown in **Figure 40** and radial concentration distribution analyses, respectively. The (002) planes of B2-FeAl are atomically resolved so it is possible to count the number of atoms. The white line in the figure shows Burgers circuit. In the upper side, the number of (002) planes is 29, while that in the lower side is 28. This means there is an extra half plane in the upper part within the circuit. Near the dislocation core distinguished from the termination of an extra half plane, B atoms are segregated. This work was indeed a major milestone in 3DAP applications and paved a way for similar work on related to dislocations in other materials.

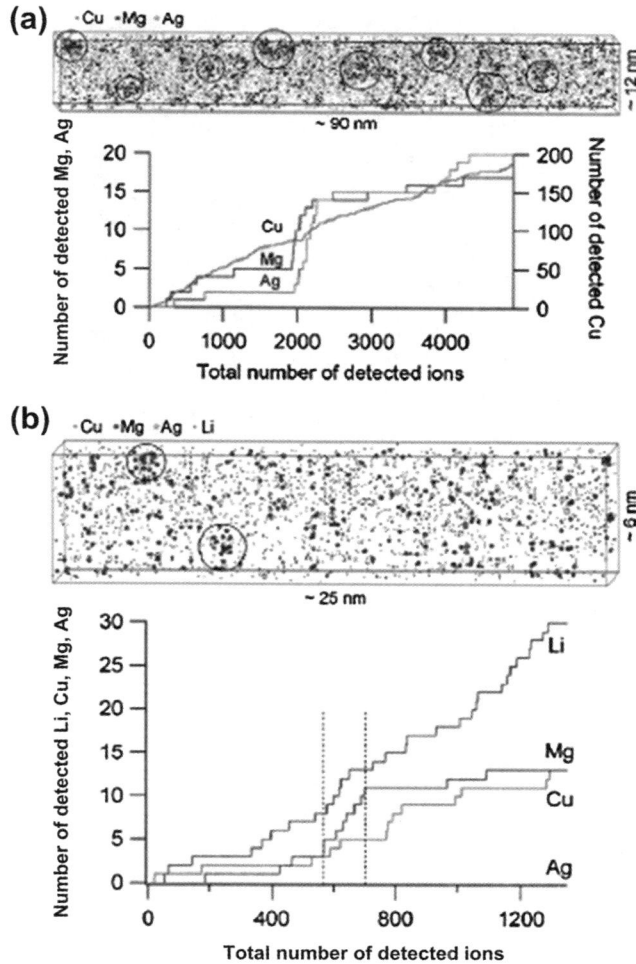

Figure 39 Evaluation of clustering phenomena in Al–Cu–Mg–Al alloys (a) without and (b) with Li shows marked difference at early stages of aging at 180°C. From Honma et al. (2004a). (For color version of this figure, the reader is referred to the online version of this book.)

The segregation of interstitial solute atoms along dislocations have been reported in several observations of 3DAP data; however, most of the case, direct confirmation of the presence of dislocations has not been confirmed because of the tediousness of observing a dislocation with TEM before 3DAP analysis. When solute atoms are observed with line feature in 3DAP data, they were interpreted as evidence for the solute segregation along dislocation cores. One such an example is shown in **Figure 41**, where c map gives line feature in a low-alloy steel. Even without the concurrent observation of crystallographic planes, these are regularly interpreted as Cottrell atmospheres (Duguay et al., 2010; Caballero et al., 2008). For example, the observation of linear features of elemental (Cr, W, B and C) enrichment (Miller et al., 2003) in $Fe–Cr–W–Y_2O_3$ ODS alloy was interpreted as supersaturation of dislocations with these elements.

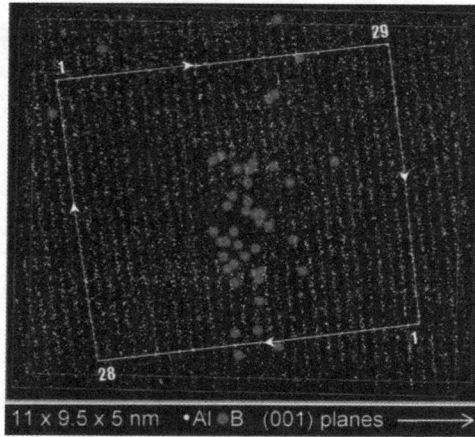

Figure 40   Direct observation of an edge dislocation in reconstructed atom map and individual identification of boron atoms in an ordered Fe–Al intermetallic is shown. From Blavette et al. (1999). (For color version of this figure, the reader is referred to the online version of this book.)

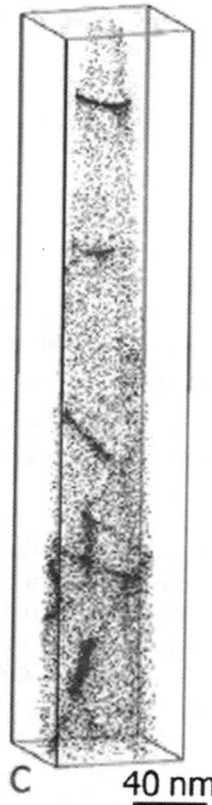

Figure 41   Atom map of C distribution in the LC(Cr,P) steel. From Pereloma et al. (2006). (For color version of this figure, the reader is referred to the online version of this book.)

15.3.1.3 Planar Defects

Planar defects include stacking faults (e.g. HCP staking within an fcc lattice), twins and antiphase boundaries (APB), e.g. interface within an ordered structure. The formation and stability of stacking faults and twins are directly relevant to the plastic deformation of alloys, and in the case of ordered alloys, antiphase boundaries also make critical role in deformation. In many cases, interstitial solute tend to segregate at stacking faults to reduce the stacking fault energy by Suzuki effect, and this has been a subjects for analytical studies in physical metallurgy. Experimental technique to prove the Suzuki effect requires very high spatial resolution and detection sensitivity. Kamio et al. reported evidence for Si segregation at a stacking fault in Cu–Si alloy using an energy dispersive X-ray spectroscopy (EDS) in a cold field emission-type transmission electron microscopy. Similar work has been performed for Co–Ni–Cr-based superalloy by Han et al. using FE-TEM. However, the poor statistics due to limited counting of EDS signals using nanosized electron beam was marginally enough to separate segregation from the noise level.

Cadel et al. (2004) studied these phenomena using TEM, FIM and 3DAP for B2-ordered FeAl intermetallic compound. In FeAl intermetallics, the addition of boron appear to stabilize the high-energy {001} stacking faults. A hypothesis is made that these stacking faults are stabilized due to a change in local composition or segregation, i.e. Suzuki effect. Using the sample in which planar defects are present with a sufficient density as shown in the TEM images in **Figure 42(a)**, they prepared an FIM tip in which stacking fault is located along the tip axis. The FIM images oriented along [001] direction also showed stacking faults in the center of the image. These planar defects were associated with brightly imaging atoms in **Figure 42(b)**, which were later identified as B. The width of the stacking fault associated boron segregation in the 3DAP reconstruction in **Figure 42(c)** is estimated to be 3 nm. This is indeed larger than the size of the defect observed in TEM and FIM. Therefore, the authors argued that this is due to local magnification effects in 3DAP as discussed in Section 15.2 as well as preferential evaporation of boron atoms from the surface during analyses. Using these arguments, the authors stated that the defect width is indeed a monolayer. The boron concentration at these planar defects in **Figure 42(d)** was measured to be 15 at.%.

The degradation of the spatial resolution due to the trajectory aberration is most pronounced when planer object for analysis is located perpendicular to the tip surface. This kind of analysis should be done using the tip in which planer objects are present perpendicular to the analysis direction; however, due to the difficulty of site specific specimen preparation containing stacking faults and dislocation, such analysis is still challenging. They are beyond the resolution of SEM image in FIB, so the selection of dislocation and stacking faults must be done using an STEM image.

The nature of APBs was studied in a B2-ordered NiAl alloy with 2at.% Cr additions (Fischer et al., 2001). The presence of APBs in this alloy was confirmed by FIM. Later, the elemental analyses along the [001] direction were performed using 1DAP. The data showed preferential segregation of Cr atoms to the APBs. The Cr atoms replace the Al atoms in the lattice. In addition, the Cr enrichment was diffuse around APB, i.e. with slow decay from a peak value as a function of distance from the APBs. These Cr-modified NiAl alloys also exhibited higher number density of superdislocations in comparison to NiAl alloys without Cr addition. The above work demonstrated the usefulness of identifying the crystallographic orientations in the TEM and FIM images and performing selective 3DAP analyses.

15.3.1.4 Surface Segregation and Reaction

One of the common uses of AP is to study surfaces, grain boundaries and interphase boundaries in wide range of materials (Thuvander and Andren, 2000). In fact, surface segregation and reaction had been very popular subjects for early days of 1DAP, because the AP was originally developed by surface

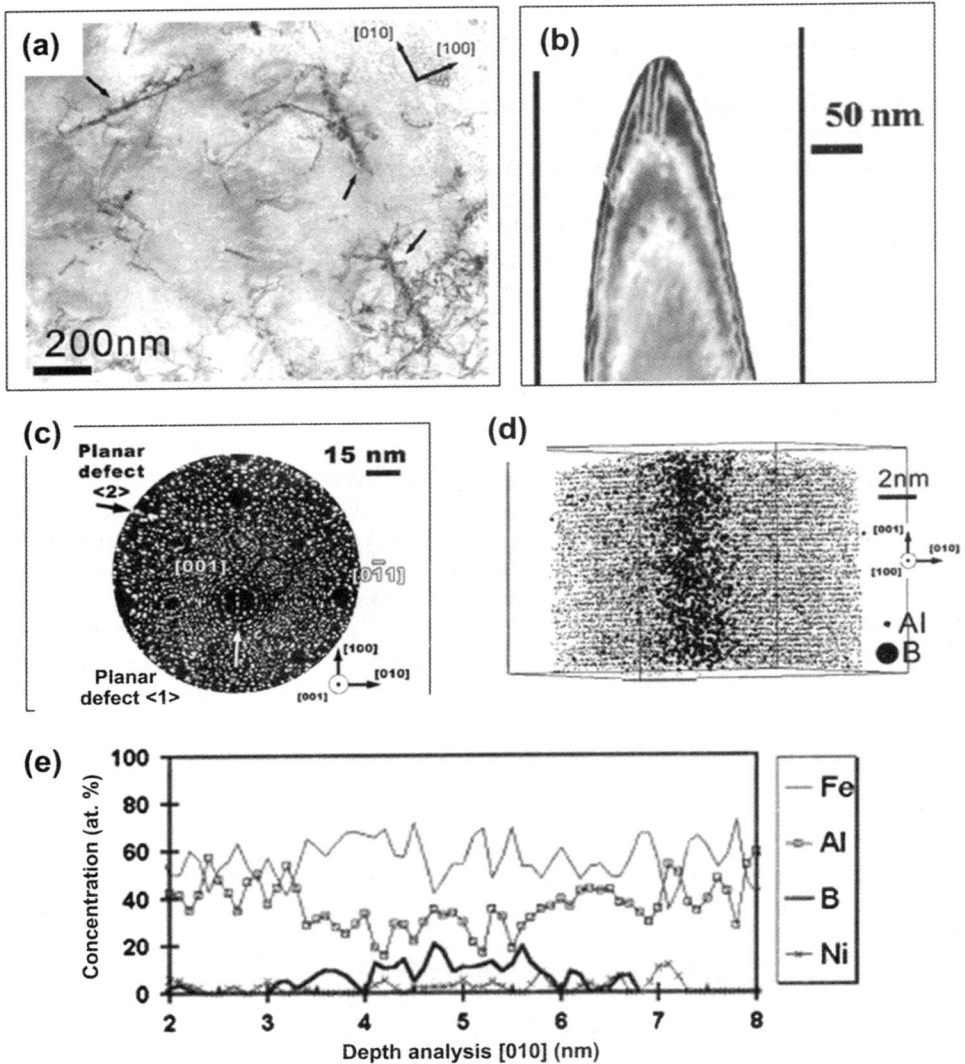

Figure 42 Comprehensive analyses of stacking faults and Suzuki effect in Fe–Al intermetallic alloy (a) TEM image; (b) TEM image of the atom probe specimen; (c) FIM image; (d) 3DAP data and (e) composition profile. From Cadel et al. (2004).

scientists. Many investigations have been done using a conventional AP. These early investigations were reviewed in the textbook by Tsong (1990). In this section, two examples of surface and GB segregation analysis using modern 3DAP are selected to illustrate this capability.

Since AP is surface-sensitive technique as has been demonstrated in the studies of surface segregation, it has also been employed for the studies of surface reaction such as oxidation. Oxidation and reduction reactions on a Pt–Rh alloy surface (Li et al., 2011a) were studied to understand the catalysis reaction on surfaces. The oxidation reaction at these surfaces may occur by either one of the following mechanisms: (a) The Rh atoms diffuse through oxide layer and reacts with oxygen at the free surface; (b) The diffusion of oxygen atoms to the metal-oxide interface occurs and oxidation of Rh atoms occurs at this interface. To validate the above mechanisms, the Pt–Rh alloys were oxidized at different temperatures. The oxidized surfaces were characterized using laser-assisted 3DAP (see **Figure 43**).

Typical 3D atom map shown in **Figure 43(a)**, measured after oxidation at 873 K, shows the presence oxides. This is indicated by the detection of Pt- and Rh-oxygen complexes on the surface. The concentration–depth profiles showed the depletion of Rh close to the interface and a diffusion profile as

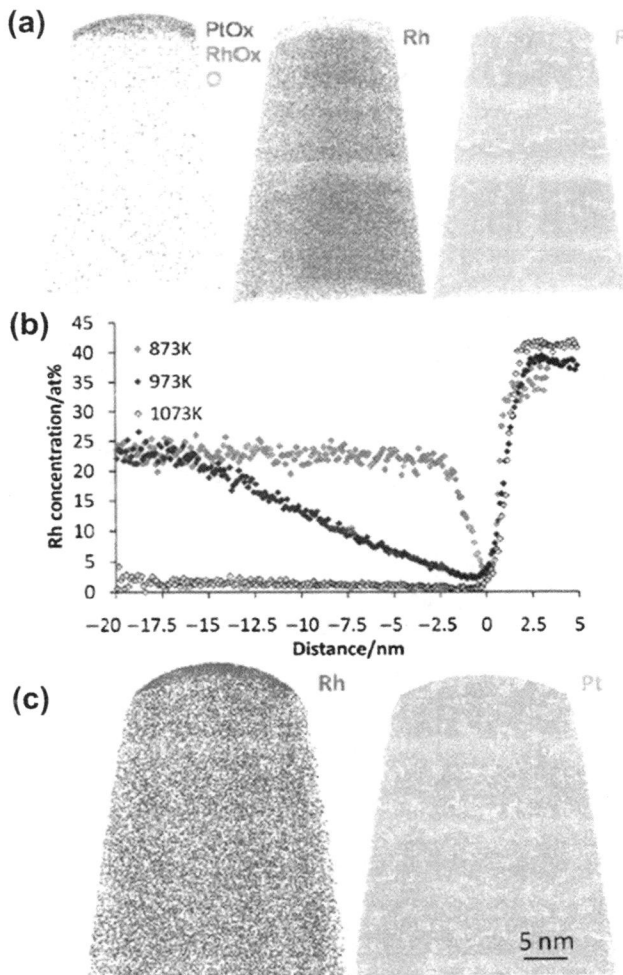

Figure 43   (a) 3D atom map of PT, Rh, and Pt/Rh oxide complex after oxidation at 873K; (b) measured concentration profile from the bulk metal to the oxide after oxidizing for 5 h at given temperatures; (c) 3D atom map after reduction reaction showing the presence of pure Rh on the surface. From Ohkubo and Hono (2010). (For color version of this figure, the reader is referred to the online version of this book.)

a function of depth from the interface into the bulk of the alloy. Based on this result, the authors concluded that the Rh oxidation occurs at the metal-oxide interfaces (see **Figure 43(b)**). In addition, they also found that the same oxide layer changes to pure Rh layer during reduction reaction (see **Figure 43(c)**). Such detailed analyses will be useful for developing computational models for solid–gas reactions at surfaces.

15.3.1.5   *Defects by Irradiation Damages*

Metals and alloys that are used as structural components for nuclear reactors received irradiation damages. During these conditions, excessive amount of defects including vacancies, dislocation loops, self-interstitials are introduced and phase transformation is accelerated due to the introduction of excess vacancies and defects. As a result, degradations in mechanical properties are commonly observed. Researchers at ORNL, USA and University of Rouen, France have rationalized these phenomena using AP. One of the difficulties associated with 3DAP analyses of irradiated samples is related to the requirement of stringent decontamination procedures, safety protocols and dedicated instrument.

An example 3DAP data from a steel sample exposed to neutron irradiation (see **Figure 44**) is presented in this section. Embrittlement occurs in Cr–Mo steels after exposure to neutron flux greater (Meslin et al., 2010; Pareige et al., 1993) than 10^{11} m^{-2} s^{-1} with total does ranging up to 10^{20} m^{-2}. This embrittlement is related to the formation Cu clusters and cosegregation of Mn and Si to grain boundaries. In certain Cr–Mo steel welds, after irradiation, the strength of weld and base metal increased with concurrent reduction in elongation and toughness. For example, the ductile to brittle transition temperature (DBTT) of base metal, before irradiation, was -49 °C. However, after irradiation

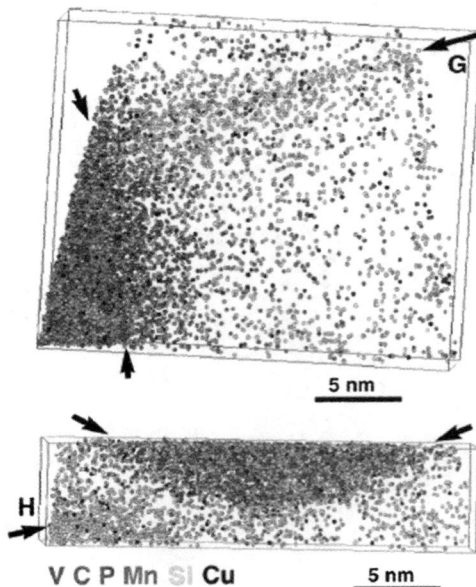

Figure 44 Atom map of only V, CP, P, Mn, SI and Cu elements measured in an irradiated Cr–Mo steel using 3DAP analyses of neutron. Reproduced form Miller (2000). (For color version of this figure, the reader is referred to the online version of this book.)

at 270 °C for 10 years with a flux of 3.8×10^{16} m^{-2} s^{-1}, the DBTT increased to 141 °C. In case of weld metal, the DBTT before irradiation was 7 °C. However, after neutron irradiation at the same flux and at 275 °C for 5 years, the DBTT increased to 123 °C. The mechanism for drastic degradation of properties was understood by using 3DAP by Miller et al. (2001). The 3DAP data showed (see Figure 44) cylindrical and spherical regions enriched with Mn and Si.

These are interpreted as segregation to dislocations and dislocation loops within a ferrite matrix created by radiation damage. Other analyses also showed cosegregation of phosphorous and carbon to the grain boundaries. The data were converted to the coverage based on Gibbsian excess methods. The monolayer coverage for phosphorus (30%) was higher than that of carbon (14%). Such results have provided greater insight for designing steel welded structures for new generation of nuclear power plants (Miller et al., 2002).

15.3.2 Phase Separation

In alloy system with strong miscibility gap, a single-phase alloy may separate into regions with two different concentrations. For example in Fe–Cr alloy system, below 900 K, the ferritic matrix will phase separate into Cr-lean (α) and Cr-rich (α') regions through spinodal decomposition. Because imaging compositional fluctuations of the atoms with similar atomic scattering factors within the same crystal lattice is difficult with TEM, the phase decomposition of Fe–Cr ferritic alloys and its derivatives became popular subject for AP analysis. These microstructural changes are often correlated to embrittlement. For example, Vitek et al. (1991) correlated the spinodal decomposition within the residual δ-ferrite to 475 °C (748 K) embrittlement in a 308 stainless steel weld.

Recently, similar microstructure (see Figure 45) evolution was found in a commercial (PM-2000) oxide dispersion-strengthened alloy (Capdevilla et al., 2008). This ferritic alloy was aged at 748 K up to 3600 h. The hardness was found to increase as a function of aging (see Figure 45(a)) time. The samples were characterized with 3DAP (see Figure 45(b)). The 3DAP data were processed to construct surfaces that have the 25 at.% concentrations. The 25 at.% Cr iso-surfaces show the gradual evolution of concentration fluctuation and coarsening of Cr-enriched region (see Figure 45(b)). The tendency for spinodal decomposition was validated through autocorrelation function analyses. The results confirmed the gradual development of concentration fluctuation (wavelength and amplitude) as a function of time, which is in agreement with theories of spinodal decomposition.

Care must be taken during interpretation of concentration fluctuations in 3DAP data as evidence for spinodal decomposition. This was pointed out by Clarke et al. (2009) while discussing their 3DAP data from U–Nb system. These authors found an increase in hardness while aging a quenched U–13 at.% Nb alloy at 300 °C, as a function of aging time. The 3DAP data from the as-quenched condition showed only random distribution of Nb. However, on aging, nonrandom distribution was observed associated with distinct Nb-rich and Nb-poor regions. Since they did not observe a gradual change in the amplitude and wavelength of concentration fluctuations as a function of time, the authors were *not* able conclude the validity of the spinodal decomposition in these alloys. It is noteworthy that 3DAP measurements relevant to spinodal decomposition can also be confirmed by other complementary techniques including small-angle neutron scattering (SANS) and Mössbauer spectroscopy.

15.3.3 Solute Clustering

In multicomponent alloys, solute clusters are often reported before noticeable formation of precipitates. Such examples were often seen in the pre-precipitation stage of aluminum alloys, but

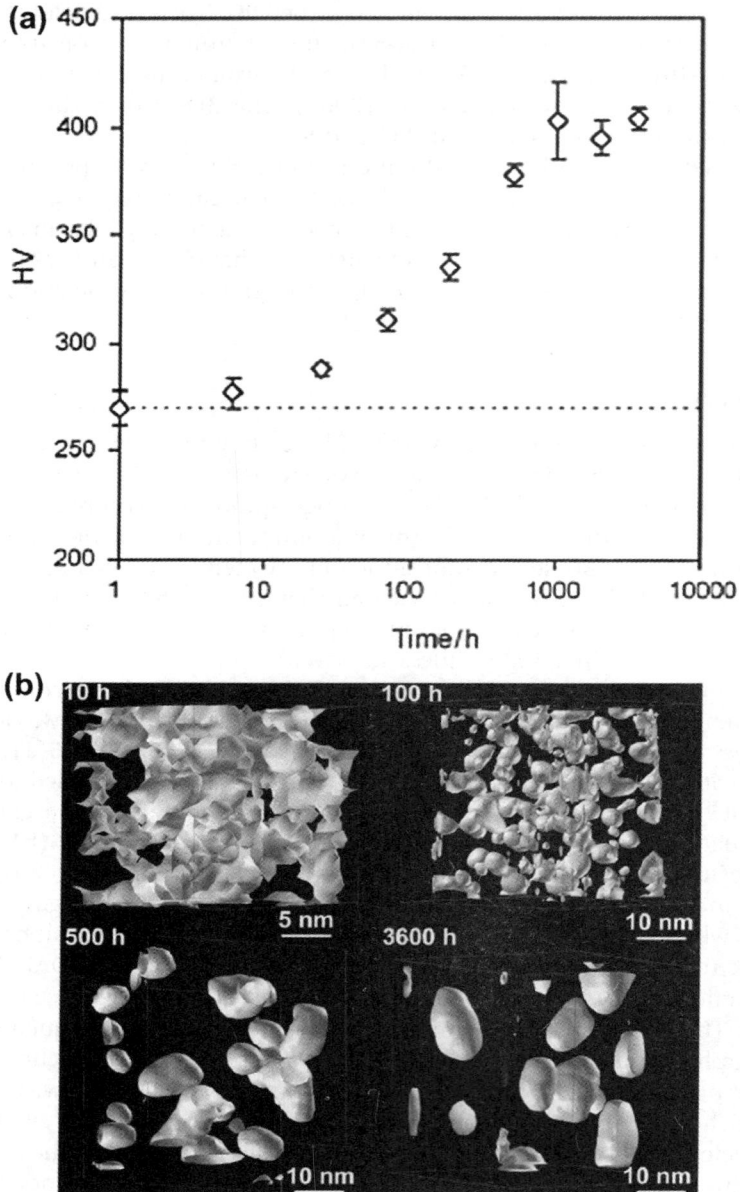

Figure 45 Spinodal Decomposition in a Fe-base powder metallurgy alloy on aging at 748 K: (a) Hardness change as a function of time; (b) iso-surface concentration based on 25 at.% Cr constructed using the 3DAP data from these samples aged for 10 to 3600 h. From Capdevilla et al. (2008).

trace Cu has also been known to form clusters in both thermal aging and under irradiation. Since the size of these clusters is often in the order of few nm or less, they are undetectable with transmission electron microscopy and. Since AP has single atom sensitivity, the clusters formed in diluted alloys can be distinctly detected by AP. Typical data for clustering effect were already presented with reference to the vacancy cluster formation (see previous section). Clustering in steels can occur either during embryonic stage before the nucleation of second phase or during nonequilibrium processes such as mechanical alloying and neutron irradiation. Cerezo et al. (2003) studied nanoscale copper precipitation in ferrite matrix in a Fe–Cu–Mn steel with a different nickel concentration, after an aging treatment at 638 K for 3000 h. In their 3-D AP analysis, the authors observed segregation of Ni and Mn at the interfaces between copper precipitate and iron matrix. It is hypothesized that these segregation reduces interfacial energy of the precipitate and iron matrix. In addition, during AP analysis of the samples before the precipitation event, nonrandom distribution of Mn (low-Ni alloy) and nonrandom distribution of both Ni and Mn (high-nickel alloy) were observed. Based on these observations and theoretical calculations, the authors concluded that Mn clusters or Ni–Mn clusters in Fe–Cu–Ni–Mn steels may act as heterogeneous nucleation site for copper. The clustering of Mn and Ni, as well as copper precipitation, is observed in pressure vessel steels exposed to neutron irradiation (Pareige et al., 1996; Auger et al., 1995). Recently, evidence for clustering of copper atoms has also been observed by Yu et al. (2010) in a high-strength steel during weld thermal cycling. However, it is not clear whether these clustering are indeed real or manifestation of preferential evaporation of copper atoms during laser-assisted field evaporation. In addition, the cluster analysis in commercial software (e.g. IVAS) has to be done very carefully. Selection of wrong parameters in the cluster search algorithm may lead to an unusually large number of detections.

15.3.4 Precipitation

The next popular 3DAP analysis is indeed evaluation of nanoscale precipitates in a wide range of materials. With the advent of 3DAP analyses, it is now possible to characterize the composition, number density and size distribution of precipitates. The above data can be used to evaluate the magnitude of strengthening. An example of data is presented below.

Recently, low-carbon martensitic steel (BA160) was developed for blast-resistance application. The steel achieves its strength through precipitation of nanoscale bcc-copper as well as Mo_2C precipitates. However, on welding these materials, the softening in heat-affected zone was observed. Using 3DAP, the softening was correlated to dissolution of nanoscale copper precipitates (Yu et al., 2010). The dissolution occurs during a typical weld thermal cycle with a peak temperature above the Ac_3 temperature. Using this data and thermodynamic models, a remediation method was proposed to solve this problem. A post-weld thermal cycle below the Ac_1 temperature was designed (Yu et al., 2011) to promote the copper precipitation within the martensite matrix. After this heat treatment, copious precipitation of nanoscale copper was confirmed by 3DAP (see **Figure 46**) and the strength in these regions was reinstated to its original levels. The above data showed that ferrite (bcc) matrix supersaturated with copper is highly unstable. It will be of interest to evaluate the feasibility of copper precipitation within the austenite (fcc) matrix. Preliminary evidence for the copper precipitation in austenitic phase was indeed proven by Isheim et al. (2008). The above results demonstrate the power of 3DAP analyses in guiding the materials processing to solve real-life problems. Specific examples of precipitation reactions in other alloy systems will be discussed in the next section.

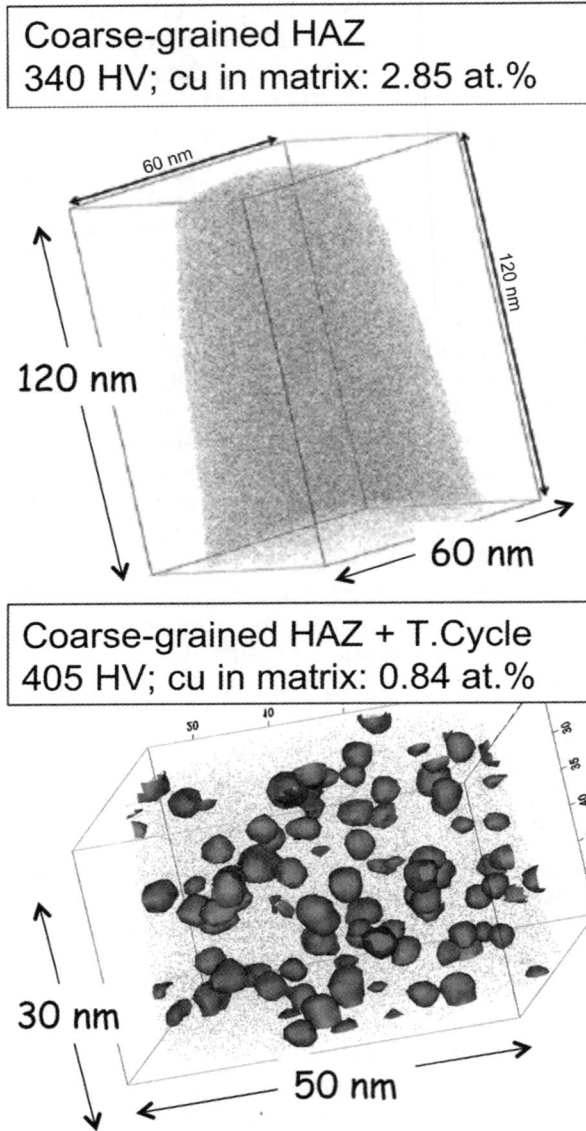

Figure 46 Comparison of atom probe analyses form a blast resistance steel samples: (top) after subjecting to an austenitizing weld thermal cycle, which is followed by (bottom) a secondary thermal cycle below the Ac_1 temperature. From Yu et al. (2011). (For color version of this figure, the reader is referred to the online version of this book.)

15.3.5   Interdiffusion

Interdiffusion is a process of diffusional exchange of atoms across two materials that are in contact. This is driven by the chemical potential gradient across the boundaries. Such phenomena are routinely observed during high-temperature service of dissimilar material welds (Babu, 2009) (e.g. low-alloy steels and stainless steels) and thermal barrier coatings (e.g. in turbine materials). Interdiffusion is also observed in multilayers and nanocomposites materials due to the processing conditions. Two examples from recent publications are discussed below.

Multilayers made up of ferromagnetic (FM) and antiferromagnetic (AFM) materials are being considered for perpendicular exchange bias at room temperature. One such multilayer is made up of (Pt/Co) FM material in contact with IrMn AFM material. Since other techniques (X-ray reflectometry, nano-SIMS and TEM) did not provide enough chemical resolution in depth direction, these multi-layers were characterized using laser-assisted 3DAP. A multilayer of a given sequence [Ta3 nm/ (Pt2 nm − Co0.4 nm − IrMn)$_7$/Pt10 nm] was made by DC magnetron sputtering (Larde et al., 2009). This multilayer was annealed at 200 °C for 30 min and cooled under a magnetic field of 2.4 kOe. Then the AP needle samples were prepared using FIB. The AP sample and the 3D atom maps are shown in **Figure 47**. For clarity, Ir and Mn atoms are shown with same color. The data suggest that the Ir and Mn atoms diffuse quickly into Co layer. However, their interdiffusion did not extend to Pt layers. The authors of this paper also validated these observations by analyzing possible artifacts that can be introduced during AP analyses.

When grain size decreases to the nanoscale dimension, the atomic transport through grain boundaries becomes dominant over the volume diffusion. Such diffusion is becoming the subject of recent studies since nanocrystalline materials becomes a popular subject in physical metallurgy (see Chapter 27, Volume III). The topologically necessary line-shaped defect formed by merging the three-grain boundaries is called "triple junctions" and these play important role in interdiffusion in multi-layers. Since AP tomography can show the location of triple junctions as well as the chemical composition in the specific sites in the tomography, it has a merit of analyzing diffusions governed by grain boundaries and triple junctions' diffusions.

Such an example is shown in **Figure 48** (Chellali, et al., 2012). Nanocrystalline thin-film speci-mens were prepared by sputter-deposition onto needle-shaped tungsten tips. These film layers also achieve curved shape by adopting the shape of the tip apex. The grain size of the films grown on the W tip was 20–30 nm. The 3D atomic reconstruction of the layer structure after annealing in **Figure 48(a)** shows Ni was enriched along grain boundaries of the Cu layer. In practically every sample annealed between 563 and 643 K, such clear indications of GB diffusion were observed. In this case, Ni markers nicely decorate the complex 3D GB morphology. As a result, triple junctions are reliably identified even on the length scale of a few nanometers. By cutting the local cylinders of analysis perpendicular to the GBs in the vicinity to a TJ as shown in the inset of **Figure 48(b)**, composition profiles across the GBs are determined. The GB segregation zone is reasonably approximated by a Gaussian compositional field, which has a considerable width of about 2.5 nm, much larger than the 0.5 nm expected from structural considerations. Based on these data, the authors deduced the GB diffusion constant.

15.3.6   Alloying by Plastic Deformation

In an effort to reduce the grain size of metals and alloys to nanoscale, many processes that rely on severe plastic deformation have been developed. These processes include accumulated roll bonding (ARB)

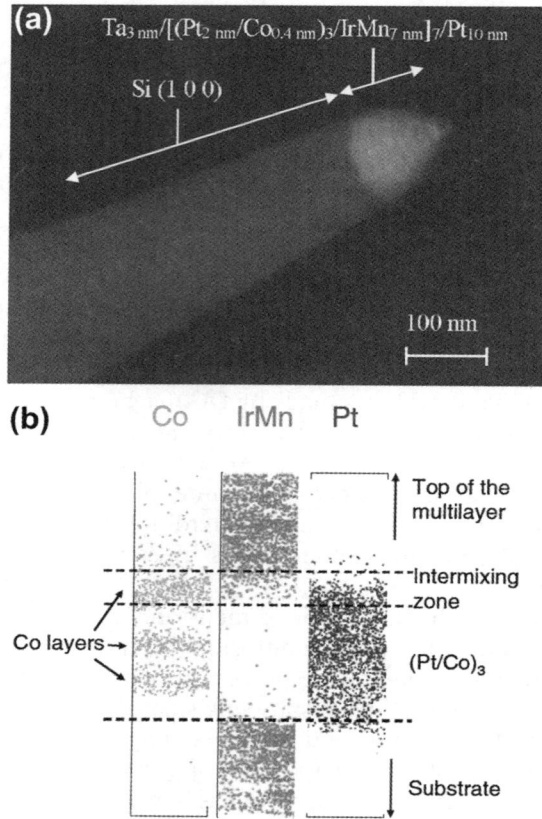

Figure 47 3DAP analyses of a multilayers made by DC magnetron sputtering: (a) 3DAP tip and (b) atom maps of Co, IrMn (denoted with same color) and Pt. From Larde et al. (2009). (For color version of this figure, the reader is referred to the online version of this book.)

(Saito et al., 1999), equal channel angular pressing (ECAP) (Valiev, 2001), and high-pressure torsional (HPT) straining (Islamgaliev et al., 1997), ball milling and wire drawing. Often these processes lead to dissolution of precipitates, enhanced solid solubility and formation of amorphous phase in certain alloy systems.

The extent of mutual solubility of Cu and Ag that can be achieved by ARB was investigated by 3DAP. The equilibrium concentration of Ag in the α-phase (Cu) is less than 5.0 at.%, while that of Cu in the β-phase (Ag) is less than 13.0 at.%. These eutectic lamellae were subjected to 5-cycles of accumulated roll bonding, equivalent to a total strain of 6.84. The FIM image (see **Figure 49(a)**) shows darkly (Cu-rich) and brightly (Ag-rich) imaging regions. The atom maps made from a large dataset is shown in **Figure 49(b)**. Region (c) has large lamellae of $\alpha + \beta$ microstructure. The data showed no large changes in mutual solubility in both Cu and Ag phases. In contrast, the region (d), with smaller lamellae (suggesting larger deformation strain), showed enhanced solubility of (40 at.%) Ag in Cu. Interestingly, there is no increased solubility of Cu in Ag. This clearly showed the ARB process could produce

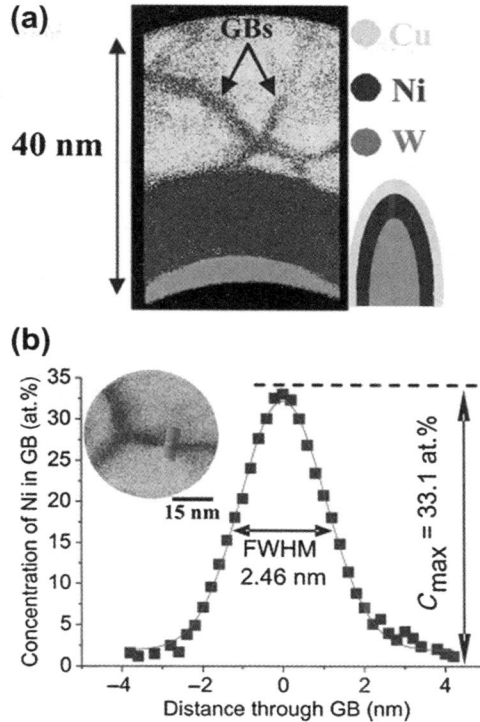

Figure 48 Atomic reconstruction of a Cu/Ni bilayer (7 million atoms) after annealing at 643 K for 45 min. Transport of Ni along GBs of Cu becomes obvious. (b) Local analysis of a GB close to a triple junction. Composition profile perpendicular to the GB matches a Gaussian distribution. (Inset clarifies position of the analysis cylinder used.) From Chellali (2012). (For color version of this figure, the reader is referred to the online version of this book.)

nonequilibrium solid solutions. A similar extended solid solubility was also reported in heavily wire drawn Ag–Cu wire, which is developed as a high-strength conductive wire (Ohsaki et al., 2005a; Raabe et al., 2009).

The decomposition of a secondary phase and dissolution of the decomposed solute to the matrix was also reported in heavily drawn pearlite wire (Hono et al., 2001) as well as in mechanically milled pearlite (Ohsaki et al., 2005b; Sauvage et al., 2009). **Figure 50** shows bright field TEM image of mechanically milled pearlite steel. The cementite phase (Fe_3C), which is originally present in the pearlite lamellae, is completely decomposed. The selected area diffraction pattern shows bcc single phase (the inner diffraction ring corresponds to Fe_3O_4 that form during mechanical milling). The FIM image shows nanocrystalline ferrite with bright contrast and grain boundaries with dim contrast. The 3DAP carbon map shows that only a small amount of carbon was dissolved in the ferrite and the majority of carbon atoms are segregated at grain boundaries. This kind of feature is commonly observed in heavily drawn pearlite steel. The ultimately deformed pearlite microstructure can be seen as "white etching layer" on rail surface, which is nanocrystalline ferrite analogous to mechanically milled pearlite (Zhang et al., 2006).

(a)

(b)

$13 \times 13 \times 113 \, nm^3$

ND

TD

(c)

$2 \times 2 \times 11 \, nm^3$

(d)

$3 \times 3 \times 10 \, nm^3$

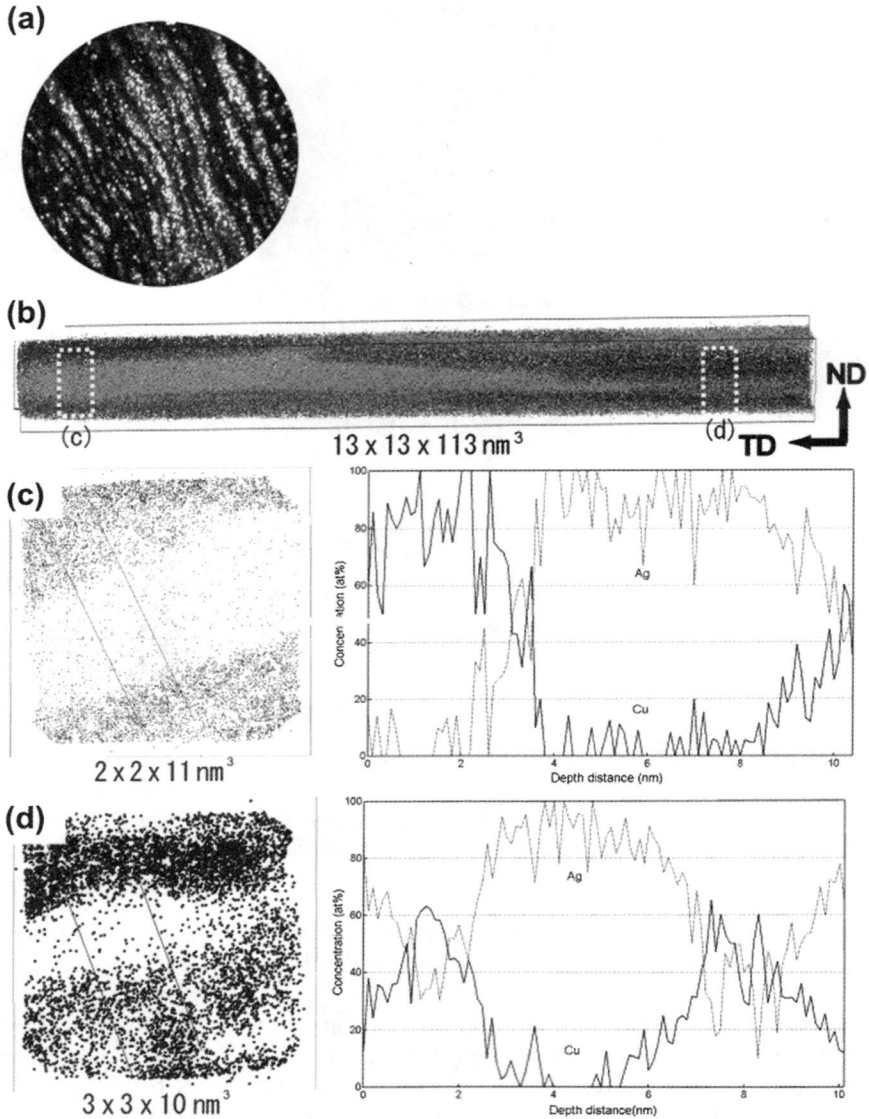

Figure 49 (a) FIM Image of a Cu–Ag sample after accumulated roll bonding; (b) 3DAP data showing (c) minimal intermixing in thick lamella regions (d) intense mixing in thin lamellae region. From Saito et al. (1999).

15.4 Specific Applications to Materials

In this section, specific applications of atom-probe FIM to wide range of materials are discussed. In the first part of the section, structural metals and alloys including steels, aluminum, magnesium, titanium and nickel alloys are discussed. In the second part, of the section nanostructured materials, metallic

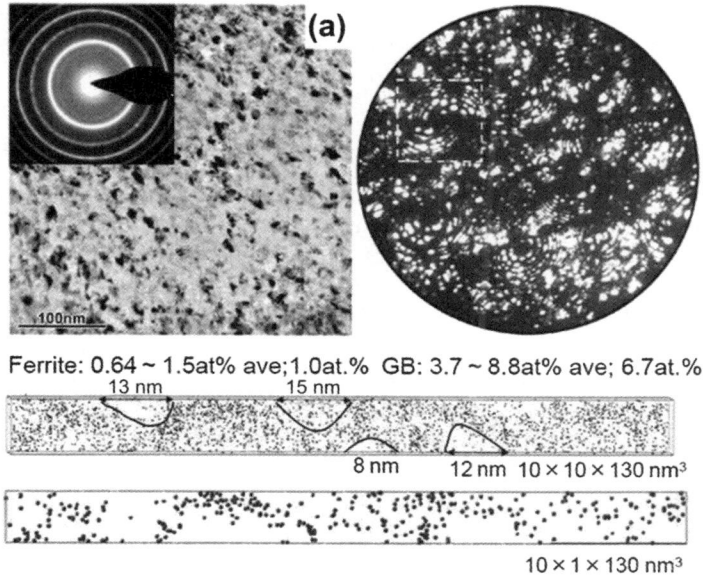

Figure 50 (a) TEM bright field image and (b) FIM image of mechanically milled pearlite. It is composed of randomly oriented nanocrystalline ferrite (BCC). The inner ring in the SAD pattern is due to oxide formation during mechanical milling. FIM shows randomly oriented nanocrystalline feature with grain boundaries imaging with dark contrast. The 3DAP maps of carbon show C is depleted from nanocrystals and are segregated at grain boundaries. Detailed analysis showed high carbon was dissolved in nanocrystalline ferrite. From Ohsaki et al. (2005b). (For color version of this figure, the reader is referred to the online version of this book.)

glass, magnetic materials, semiconductors and ceramics are presented. The section is not aimed at providing comprehensive literature review of all the published data with reference to these materials, rather a selection of few representative examples and highlights the importance of using AP as a tool to characterizing microstructure.

15.4.1 Steels

The application of AP FIM to materials science started from the characterization of steels at US steels by Brenner. AP was considered to be particular useful tool for steel research, as it can analyze local compositions of carbon and nitrogen, which are essential alloying elements for steels. The ability to control mechanical properties of steels for a specific application requires knowledge of microstructure evolution as a function of steel composition, material processing and service conditions. In addition to the interstitial elements such as C and N, major alloying elements in steels are Si, Mn, Cr, Mo and Ni (Honeycombe and Bhadeshia, 1995; Pickering, 1975; Gladman, 2002). The alloying elements are added in micro (<0.1 wt.%) or macro levels (>0.1 wt.%). In many occasions, the alloying elements cluster, precipitate homogeneously or heterogeneously, partition into different phases such as ferrite, austenite, and carbides, segregate to point and/or line defects or grain boundaries or interphase-boundaries (see Section 15.3). There is a need to understand these phenomena in wide range of steels including high-alloyed steels (Hattestrand and Andren, 1999; Hofer et al., 2000), stainless steels (Karlsson et al., 1982, 1988; Karlsson and Norden, 1988; Babu et al., 1995, 1996c), maraging steels for various reasons including the enhancement of creep-resistance and hot-cracking resistance. To be consistent, the published AP works related to steels are discussed in terms of physical processes that occur in steels.

15.4.1.1 Clustering in Steels

Similar to other alloys, clustering is a precursor to the nucleation of precipitates in steels. Since atomic level clusters are difficult to detect with TEM, AP has been most effectively employed in detecting solute cluster and substitutional–interstitial co-clusters. Solute clusters are often formed as precursor to precipitates. For example, atomic co-clusters of Ni and Ti were observed in a model Fe–Ni–Ti maraging steel (Shekhter et al., 2004) after aging the martensite at 550 °C for 5 s. The above conclusion was made by analyzing the 3D data set in high-volume and imaging the Ni and Ti atoms (see **Figure 51**), as well as, by performing contingency table. Interestingly, these clusters dissolved with the extended aging possibly due to the formation of Ni_3Ti intermetallic phases.

Lundin and Andren (Lundin and Andren, 1996) observed Mo–N clusters in the as-received martensitic microstructure of a Fe–Cr–Mo–W–V–C–B–N steel with a C/N ratio of 2:1. These clusters were noticed even with slow cooling rate from austenitization temperature. Based on this observation, the authors concluded that these clusters formed in the austenite before the formation of martensite. They also observed that these clusters dissolved to form the nitride precipitates while tempering the martensite microstructure. Similarly, Maruyama et al. also suggested the formation of niobium-nitrogen clusters in hot-deformed austenite based on AP analysis (Maruyama et al., 1996). Maruyama and Smith (2002) concluded that clustering of niobium and carbon in the austenite phase might lead to retardation of austenite recrystallization during hot processing. They observed rod-shaped nitrogen clusters in the martensite matrix that forms with the austenite phase during cooling. The formation of Nb–N and Mo–B clusters were also observed in a stabilized austenitic steel after aging at 500 °C by Källqvist and Andren (1999). They also noted that the tendency to observe these clusters depends on the local microstructure. For example, the Nb–C clusters were absent close to an Nb(CN) precipitate. They also hypothesized that the tendency for Mo–B clusters will be enhanced near GB, due to tendency of boron to segregate to grain boundaries. The above works show the importance of understanding overall processing effects to elucidate the observed clustering phenomena, rather than studying one single process step. With the advent of new generation of advanced high-strength steels

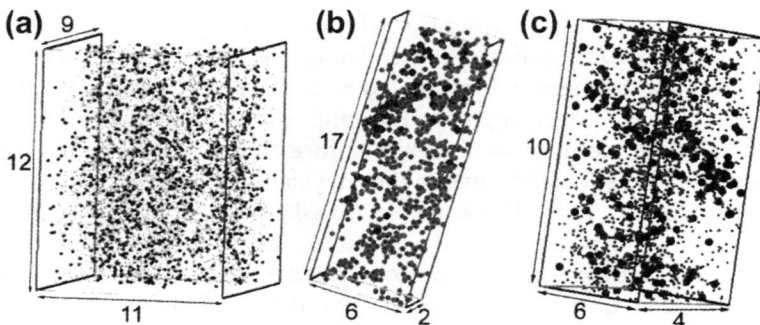

Figure 51 3D Atom map of Ni (gray spots) and Ti (dark spots) from Fe–Ni–Ti alloy after (a) quenching, (b) 5 s of aging at 550 °C and (c) 15 s at 550 °C; the dimensions on the side are in nm scale. From Shekhter et al. (2004).

based on austenitic microstructure, the above clustering phenomena provide another way to strengthen the steel.

Clustering of carbon alone has been observed during early stages of tempering martensite in steels (Miller et al., 1981, 1983). It has been speculated that these carbon clustering form due to spinodal-like decomposition in martensite (Taylor et al., 1989; Han et al., 2001). A carbon-enriched region in an untempered martensite in Fe–C–Mn–Si steel has also been reported using 1D AP (Babu et al., 1994). Other researchers have attributed this carbon clustering as segregation to dislocations (Ohmori and Tamura, 1992). There have been attempts to evaluate tendency for clustering of Fe–N steels. Experimental results indicate that even in Fe–N martensite, the nitrogen-enriched and depleted regions develop (vanGenderen et al., 1997) during tempering at 373 K.

Most of the clustering phenomena described above are driven by thermomechanical treatments. However, it is possible to obtain clusters during mechanical alloying operations, which drive the alloy to conditions far from equilibrium conditions. Larson et al. (2001), revealed the formation of 3–5 nm Ti–Y–O clusters in mechanically alloyed ($Fe - 12Cr - 3W - 0.4Ti + Y_2O_3$) steel. These clusters were found to be present in conjunction with high supersaturation of Y and O in the ferrite matrix too. Moreover, these clusters also exhibited remarkable resistance for coarsening at high temperature (Miller et al., 2003). It is possible that nonequilibrium materials processing such as mechanical milling will lead to increased solubility of elements like O, N, C and B in steels and eventually lead to the formation of metastable clusters or precipitates. Recent work (Miller and Parish, 2011), in similar ODS alloys, has confirmed the clustering of Ti, Y and O as well as segregation of C, P and Si to grain boundaries. After extended heat treatments, the formations of Ti(O, N, C) and Y_2TiO_7 precipitates were also confirmed. Recently, using EELS and LEAP analyses Cr-rich shell was found around Y–O clusters in other Fe–12Cr ODS alloys (de Castro et al., 2011; Marquis, 2008). Similar core–shell-type structure was also observed by Williams et al. in Eurofer-97 ODS Steel (Williams et al., 2010). All of the above processes show many microstructural pathways by which a supersaturated driven solid solution made by ball milling may decompose to reduce its free energy. Some of other effects due to severe plastic deformation have already been discussed in Section 15.3.

15.4.1.2 Segregation to Dislocations

As mentioned in Section 15.3, alloying elements often segregate to dislocations to reduce the overall strain fields around the dislocations. Carbon segregation to dislocations is known as "Cottrell atmosphere" and seeing it directly has been a long dream for physical metallurgists. Wilder et al. (2000) reported the direct observation of the Cottrell effect using 3DAP. They observed FIM tip using TEM to confirm that a dislocation is within tip, and then identify the position of a dislocation using FIM. Thereafter, they performed 3DAP analysis from the region where a dislocation was observed by FIM and corrected carbon mapping, thereby constructing 3DAP map of interstitial carbon Cottrell atmospheres around dislocations in iron. They reported that the atmospheres appear dispersed, having an asymmetric distribution and extend outward from the dislocation core to a distance of $\approx 7 \pm 1$ nm. The maximum carbon concentration observed within atmospheres was estimated to be $\sim 8 \pm 2$ at%C, approximately 21 ± 1 carbon atoms per atom plane along the dislocation line. Recently, Miller et al. (2003) and Miller and Parish (2011) observed the cosegregation of interstitial elements carbon, boron with substitutional elements including chromium and tungsten.

Pearlite consists of lamellae of ferrite (α-Fe) and cementite (Fe_3C), and there is little carbon in ferrite due to the low solubility of carbon in ferrite. However, by wire-drawing, strain is induced, and cementite is fragmented because of the lac of the plasticity, and some cementite is decomposed. Hinchliffe and Smith (2001) found carbon (2.5 at.%)-enriched regions in a pearlitic steel wire and attributed to the formation of Cottrell atmosphere. In an earlier research, Bhadeshia and Harada (Bhadeshia, 1993) measured large supersaturation of carbon in ferrite with fine dislocation cell structure in a high-strength (5 GPa) steel wire. Retardation of precipitation reactions in these wire were attributed to a binding force between carbon and dislocations (Read, 1997). Similar clustering of carbon with dislocations and carbon-enriched ferrite due to cementite fragmentation in a pearlitic steel wire during cold drawing operation have been suggested based on AP studies (Read et al., 1997). In a recent work, Hono et al. (2001) observed the decomposition of cementite in a pearlitic wire decomposed with an increase in drawing strain and the formation fibrous nano-sized ferrite with carbon enrichment.

Caballero et al. (2007) observed the segregation of carbon to dislocations within the bainitic ferrite close to its interface with austenite (see **Figure 52**). The atom map is full of linear features containing high carbon concentrations. One such feature was analyzed further (see arrow in **Figure 52**). It is interesting to note that the carbon-enriched region is indeed a linear feature and it does not correlate with any enrichment of Si, Mn and Mo. This result proves that the above phenomenon is driven by interstitial element diffusion to dislocation generated by the transformation strains.

Most of these elements segregate to the dislocation to allow for the minimization of lattice strain energy in the host matrix. Observation of cosegregation of two or more elements does not always indicate that these elements would prefer each other atoms as neighbors. There is also a possibility of competition between these elements to segregate to dislocations. Thermodynamics and kinetics of these segregation characteristics to dislocations are yet to be understood by theoretical models by coupling with thermomechanical processing parameters (Gavriljuk, 2003).

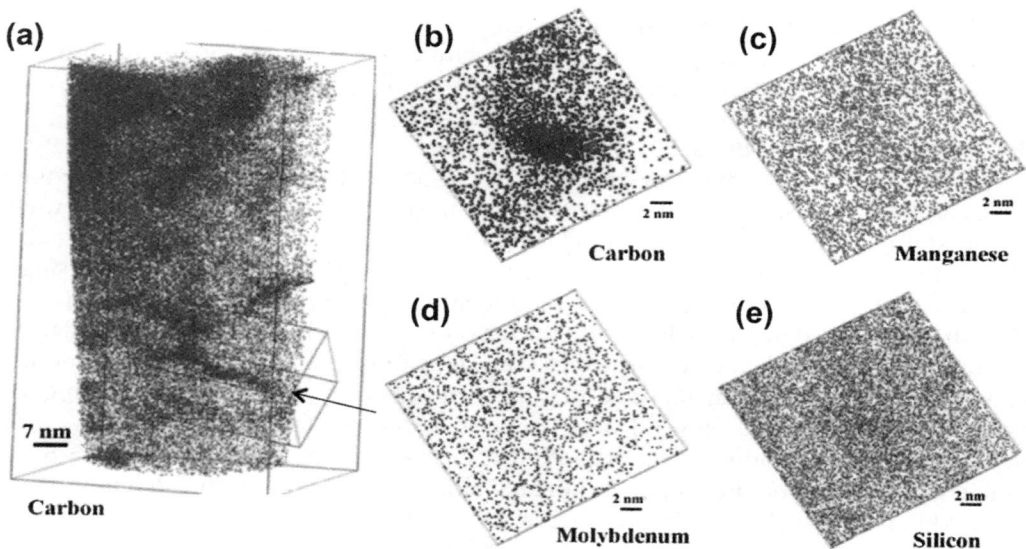

Figure 52 (a) 3D Atom map of carbon in a bainitic steel shows linear features. A selected area (marked by arrow) was taken and 2D projection atom map was created for various elements including (b) carbon, (c) Manganese, (d) Molybdenum, and (e) Silicon. From Caballero et al. (2007). (For color version of this figure, the reader is referred to the online version of this book.)

15.4.1.3 Segregation to Grain Boundaries and Interphase Boundaries

Application of FIM for investigation of segregation to grain boundaries and interphase boundaries are very common in literature. The segregation to grain/phase boundaries can be divided into two categories, i.e. equilibrium and nonequilibrium segregations (Hondros and Seah, 1983). Equilibrium segregation is completely reversible and is driven by thermodynamic considerations and is limited to one or two atomic spacing near the interface. In contrast, nonequilibrium solute-segregation is driven by the movement of vacancies to the boundaries. The mechanisms of dragging of solute-vacancy pairs (refer to Section 15.3) are system and process-specific. Both types of segregations have been studied by FIM (Ng et al., 1979) and AP field ion microcopy (Tsong et al., 1978). The GB and interface segregation in steels and other alloys are reviewed in reference (Faulkner, 1996).

The segregation of boron and carbon to interphase boundaries during weld solidification and subsequent cooling was investigated by Babu et al. (1996c). These researchers measured cosegregation of boron to the austenite–ferrite interface in the as-welded condition. It is important to note that the interphase forms during solidification conditions and may migrate during subsequent cooling after solidification (Kou, 2003). Since the solubility of boron is very low in both ferrite and austenite, the boron segregates to the interface. Interestingly the results also showed some enrichment of carbon at the interface. However, this carbon enrichment could also be due to the equilibrium interface concentration that is maintained during diffusion-controlled growth of ferrite into austenite. The boron atoms that segregated to this interphase boundary were incorporated into $M_{23}C_6$ carbide phase on high-temperature heat treatment. It is important to note that the analyses also show the presence of phosphorous close to the boundaries, which also has tendency to segregate to GB; however, kinetically C and B are favored. In other conditions, the phosphorus also may segregate to the grain or interphase boundaries. Lundin and Richarz (1995) and Lundin (1995) found phosphorus segregation to $M_{23}C_6$ and ferrite interphase boundaries. Kelly et al. (1999) observed segregation of boron and oxygen within a distance of 10 nm near to GB in a rapidly solidified 316 stainless steel. On aging, the authors also found that the boron and oxygen tend to be incorporated within the vanadium-rich nitride. Isheim et al. (2001) measured large tendency for segregation of vanadium at an interface between the Mo-rich precipitate and ferrite matrix in a Fe–Mo–V alloy. It is also important to differentiate between interfacial segregation and interfacial concentration stipulated by local equilibrium conditions (Bhadeshia, 1985).

Krakauer and Seidman (1998) characterized equilibrium segregation of silicon at ferrite–ferrite grain boundaries and related the same GB characteristics. This work related the Gibbsian interfacial excess to five geometric degrees of freedom to describe the GB. Cosegregation of boron and carbon (see **Figure 53**) to ferrite grain boundaries in interstitial-free steel, after different degrees of recrystallization, was also characterized by 3D AP (Seto et al., 1999). The data were attained by painstakingly careful tip preparation to get the GB on the tip of the sample, which includes trial and error electrolytic polishing, ion beam thinning and TEM observations. The results showed that even with low-bulk concentration of boron, there is a strong tendency for boron to segregate to ferrite grain boundaries. The data also suggest that even with intentional addition of Nb and Ti to the steel, there exists free carbon that cosegregates with boron to these boundaries. Comprehensive studies of nonequilibrium boron segregation to the grain boundaries were studied by Karlsson et al. using AP FIM and other characterization tools (Karlsson et al., 1982, 1988; Karlsson and Norden, 1988a, 1988b; Karlsson, 1988). These authors also used the results to develop computer simulations of the segregation behavior.

The above results show that segregations to interphase boundaries or grain boundaries are related to relative mobility of segregating species, i.e. faster the mobility of an atom in a host matrix, the grain

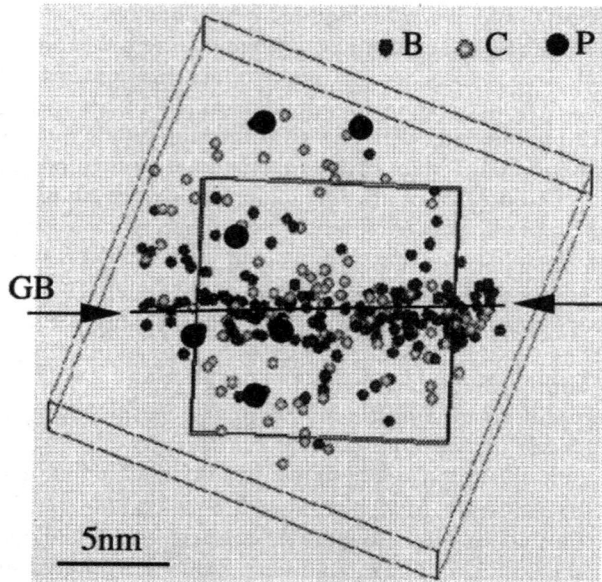

Figure 53 3D atom map from an interstitial free steel (Fe–0.002C–0.01Si–0.16Mn–0.007P–0.01S–0.0027N–0.04Ti–0.006Nb–0.007B) (wt.%) showing the cosegregation of B, C and P. Reproduced from Miller (2000).

boundaries of these matrix will be saturated with these atoms before competing atoms can arrive at the grain boundaries. In addition, the results show that the segregation at the interphase boundaries is often complicated by the concentration profile ahead of these interfaces during growth. Nevertheless, atom-probe FIM allows evaluating the competing segregation characteristics starting from a simple to complex alloy systems as well as correlation to the properties.

15.4.1.4 *Precipitation of Second Phases*

The most common approach of steels has been to allow for the precipitation of second phases during thermomechanical processing. These precipitation reactions remove either the carbon or nitrogen from solid solution by forming carbides or nitrides. These precipitation reactions lead to modification of mechanical properties as well as austenite GB pinning during processing. In other cases, the precipitation reaction leads to fine distribution of second phases that provide for strength in service, e.g. copper precipitation in steels (refer to Section 15.3). In other cases, microalloying elements are incorporated in carbides or nitrides and by which modify their stability. Most overwhelming application of atom-probe FIM has been to characterize carbides in steels (Thomson, 2000).

Murayama et al. (1999) characterized 17-4 PH stainless steels with 3D AP and found that fcc-copper precipitates formed within the δ-ferrite. In addition, on aging at 400 °C, coherent Cu-rich bcc phases were observed. The authors also found that with prolonged aging the iron concentration in these Cu-precipitates reduced and slowly transformed to fcc phase while spinodal decomposition of Cr supersaturate solid solution progresses in the bcc phases. The 3D tomography also showed that the G-phase was in direct contact with these copper precipitates. Other types of copper-containing maraging steels have also been studied by 3DAP to evaluate the microstructure evolution (Venker et al., 1998; Smith et al., 1998). Similar to the result by Murayama et al. (1999), the formation of Cu precipitate was found to be the main event for subsequent microstructure evolution. For example, the nickel-rich regions, near the Cu precipitates, evolve into γ' [Ni$_3$(Al,Ti)] phase. Since Cu has fcc structure and γ' phase has the L1$_2$ ordered structure, there is a good lattice match between these phases. Further aging leads to the precipitation of Mo-rich τ_1 phase. Similar results were reported by Stiller et al. (1998, 1996).

In many microalloyed steels, the segregation and precipitation phenomena may be related to each other. For example, in the boron-containing stainless steel weld, the boron segregates to the ferrite-austenite boundaries and on aging at high temperature, the boron gets incorporated into M$_{23}$C$_6$ precipitates. These M$_{23}$(C$_x$B$_{1-x}$)$_{26}$ precipitates are expected to be more stable than M$_{23}$C$_6$ precipitates. Indirect evidence for stability is inferred by the increase in creep strength of these alloys. A similar observation of M23(C$_x$B$_{1-x}$)26 precipitates in a martensite matrix of a boron-modified 9Cr-modified steel was reported by Hofer et al. (2000, 2002). This is in agreement with AP measurements from similar steels by other researchers (Hättestrand and Andren, 1999). Incorporation of boron in M$_{23}$C$_6$ in an austenitic stainless steel weld has also been reported (Babu et al., 1996c). Jayaram and Klue (Jayaram and Miller, 1998) found vanadium-rich carbonitride formation in Fe–9%Cr steel containing 2% W–V–Ta martensitic due to reaction with trace amounts of nitrogen present in the steel.

In some of the occasions, where the precipitates are coarsely distributed, AP analyses of these precipitates becomes difficult. In these conditions, however, one can estimate the volume fraction of precipitates by measuring the composition of the matrix. This technique was effectively used by Hättestrand and Andren to estimate the volume fraction of copper precipitates (Hättestrand and Andren, 2001a) in an 11% Cr steel and to estimate the volume fraction of VN and Laves phases in a 9% Chromium steel (Hättestrand and Andren, 2001b). However, this assumes that the precipitation events are indeed homogeneous within a given matrix.

15.4.1.5 *Alloying element partitioning during γ–α phase transformations*

In steels, the primary transformations that control the final microstructure are related to the transformation of austenite to various morphologies, including allotriomorphic, Widmanstätten, pearlite, bainite and martensite. Atom-probe FIM has been used to track the partition of substitutional and interstitial alloying elements between ferrite and austenite subjected to different heat treatments. The following section briefly highlights some of these researches.

15.4.1.5.1 *Allotriomorphic Ferrite*

One of the elusive goals of capturing the transition (Bhadeshia, 1985; Aaronson et al., 2004; Hutchinson et al., 2004; Zurob et al., 2009) from local equilibrium, where all elements partition, to paraequilibrium, where only carbon partitions, during reconstructive growth of allotriomorphic ferrite (γ to α) into austenite, has been attempted extensively. However, selection of the mobile or sessile interface from relatively coarse ferrite microstructure had been indeed very challenging. With the advent of site-specific specimen preparation method using FIB, these challenges have been overcome. The partitioning of C and Mn between ferrite and austenite in a Fe–0.18C–1.72Mn (wt.%) steel was studied after 50 s isothermal hold at 700 °C using 3DAP (Thuilier et al., 2006). The 3D atom maps of C and Mn across are shown in **Figure 54**. The maps clearly show the partitioning of carbon, while manganese is configurationally frozen on either side of austenite and ferrite. In a similar steel system, Fe–0.37C–3.0Mn–1.90Si wt.%, the austenite to ferrite transformation was studied after an isothermal hold at 656 °C for 1 h. The Mn enrichment was measured at α/γ interfaces using energy

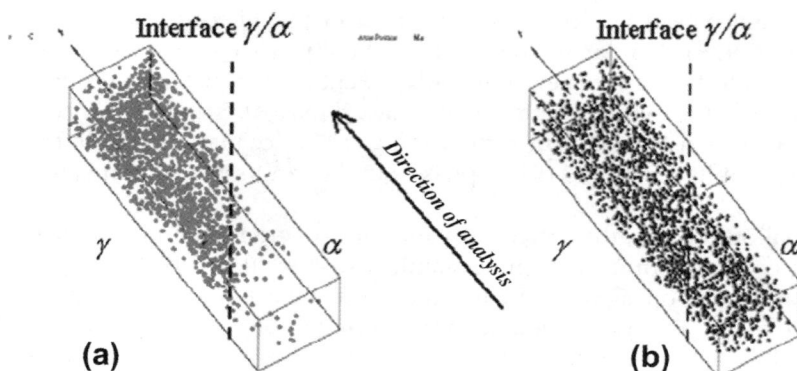

Figure 54 3D atom map of (a) carbon and (b) manganese across the γ/α interface from a steel that was transformed at 700 °C. The probed volume size is $10 \times 10 \times 36$ nm. The data show no discernible Mn peak at the interface. From Thuilier et al. (2006). (For color version of this figure, the reader is referred to the online version of this book.)

dispersive X-ray analyses in a TEM (Guo and Purdy, 2008). Interestingly, some of the interfaces showed Mn segregation at the interfaces and some interfaces did not show any segregation. This mixed result is attributed to different type of boundaries probed by the 3DAP, i.e. crystallographic orientation and mobility.

In Fe–C–Mo steels, the AP analyses and electron microscopy showed the accumulation of Mo at the α/γ interface (Humphreys et al., 2004a). The above data show that there is a need to perform comprehensive characterization of α/γ interfaces to rationalize the transition from local to para-equilibrium and vice versa. The characterization should include consistent driving force using the same alloy composition, the same interface mobility (e.g. using laser scanning confocal microscopy Terasaki and Komizo, 2011), crystallographic relationships (e.g. electron backscattered diffraction imaging), selected extraction of samples from these interfaces (e.g. using FIB) and coupled analyses of the interfaces using electron- and AP tomography. Now, during analyses of the carbon distribution in ferrite, martensite and cementite, care must be taken to avoid artifacts due to preferential trajectory differences in the evaporated ions during AP tomography (Kobayashi et al., 2011).

During austenite to ferrite transformation, in certain steels, the interphase precipitation of carbide or nitrides [e.g. V(CN)] may occur. Khalid and Edmonds (1992) have proposed pre-precipitation clustering of vanadium, carbon and nitrogen, before the sweeping of the α/γ interfaces, in model alloys. However, further analyses of the same phenomenon in engineering alloys (Khalod and Edmonds, 1993) suggest that the interphase precipitation can be triggered by precipitation at the interface as well as independent precipitation ahead of the interface. In addition to the above fundamental phase transformation issues, during thermomechanical processing (Pereloma et al., 2006) of these steels, it is possible to induce clustering of carbide forming elements such as Nb and C (Tomokhina et al., 2011). However, the onset of these clustering during austenitization or during austenite to ferrite transformation or within the transformed ferrite remains to be understood.

15.4.1.5.2 Pearlite

Element partitioning during pearlite ($\gamma \rightarrow \alpha +$ cementite) transformation has been extensively studied using AP, starting with the classic works by Miller et al. (1978, 1979). The data showed preferential partitioning of silicon into the ferrite matrix from cementite. One of the interesting methods to refine

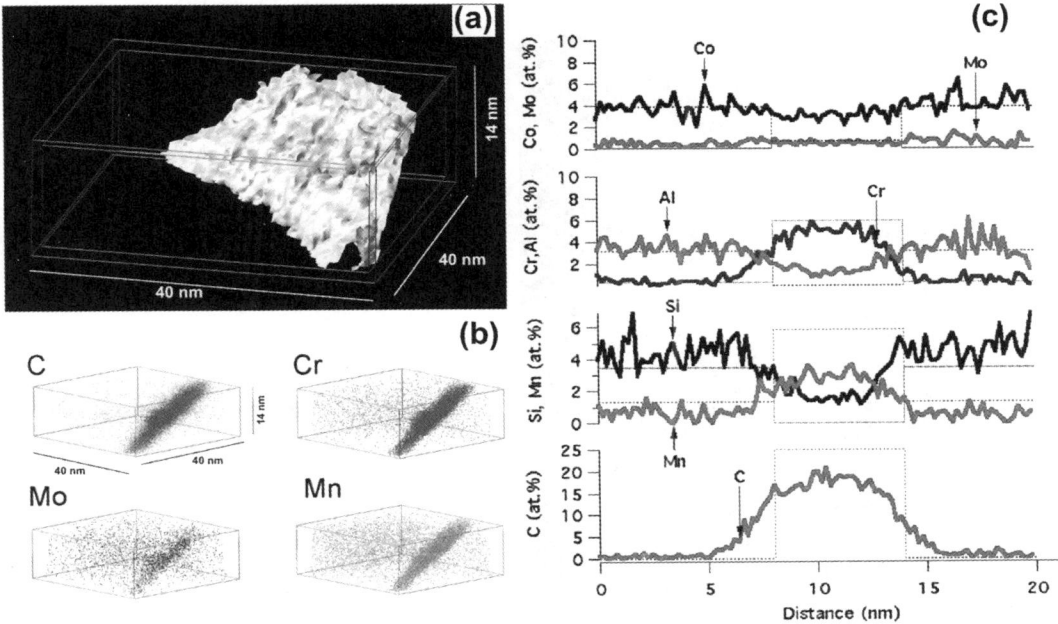

Figure 55 Iso-surface corresponding to 15 at.%C delineates the cementite lamella; (b) atom maps of C, Cr, Mo, Mn steel showing preferred partitioning of these elements to carbide; (c) concentration profile across the carbide and ferrite on either side demonstrating the preferential partitioning of Si to the ferrite matrix. From Jaramillo et al. (2005b). (For color version of this figure, the reader is referred to the online version of this book.)

the pearlitic microstructure is called patenting. Recently, an interesting idea of magnetic patenting was proposed by Jaramillo et al. (2005a). They studied the effect of high (30 T) magnetic field, on the austenite to pearlite transformation. The application of magnetic field accelerated the pearlite transformation kinetics as well as refined the lamella spacing (<100 nm). The partitioning of alloying elements between ferrite and cementite in this nanoscale pearlite was evaluated with 3DAP as shown in **Figure 55**. The iso-concentration surface corresponding to 15 at. % carbon outlines a small cementite plate in **Figure 55a**. The iso-concentration surface shows that the cementite plate thickness is less than 10 nm. The atom images in **Figure 55b** show preferential partitioning of Cr, Mo and Mn to the cementite. In addition, a concentration profile of all elements perpendicular to the cementite plate is shown in **Figure 55c**. The concentration profile clearly shows the partitioning of elements between ferrite and cementite has taken place, irrespective of rapid phase transformations (Jaramillo et al., 2005b) under magnetic field. The measured partitioning agrees with the trends predicted by thermodynamic calculations as well as published literature (Chance and Ridley, 1981; Bhadeshia, 1989; Ghosh and Olson, 2002).

15.4.1.5.3 Bainite

The AP has been extensively used to characterize the austenite/bainitic ferrite interfaces. The data have been used to validate different phase transformation mechanisms, i.e. reconstructive or displacive (Bhadeshia, 2001; Caballero et al., 2009). One of the key contributions of AP FIM is related to generation of data that support the incomplete reaction mechanism. As per this mechanism, the

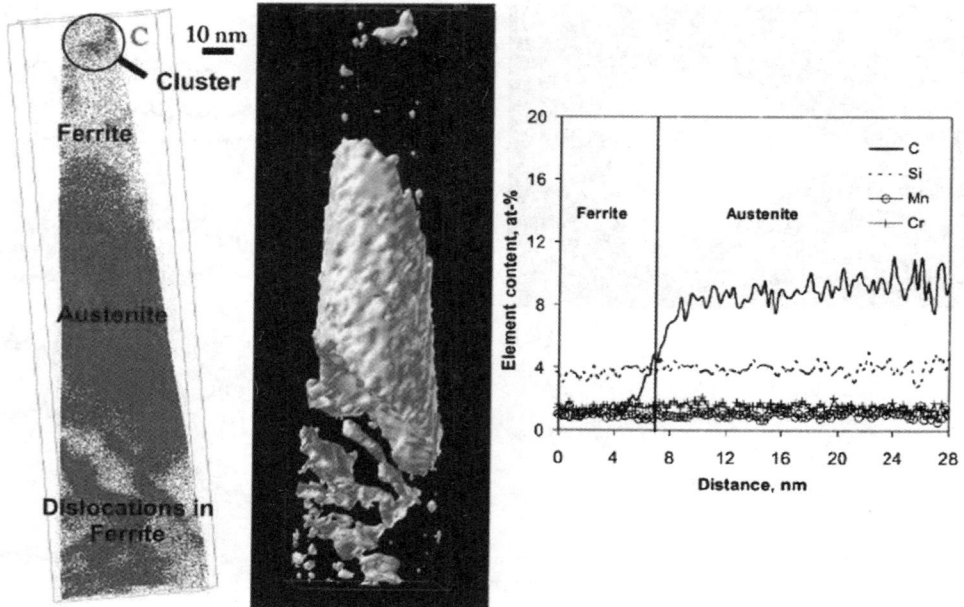

Figure 56 3D carbon atom map showing the presence of ferrite and austenite phase; (b) iso-surface corresponding to 10 at.% Carbon delineates the austenite and ferrite regions. (c) Concentration profiles of C, Si, and Mn, across the ferrite austenite boundary, show only carbon partitioning. From Caballero et al. (2010a). (For color version of this figure, the reader is referred to the online version of this book.)

displacive transformation of austenite to bainitic ferrite subunits stops as soon as the austenite carbon concentration reaches the T_0 limit, i.e. molar free energy of austenite is equal to that of ferrite for the same composition. Another key piece of evidence provided by the AP toward the bainitic transformation mechanism is the supersaturation of carbon in the ferrite. For example, Peet et al. (2004) observed ferrite carbon concentrations higher than that predicted by paraequilibrium thermodynamic calculations. Both phenomena have been recently reviewed by Caballero et al. (2010a). The paper provides a good snapshot of various bainitic transformation features as sown in **Figure 56**. The carbon atom maps show partitioning of carbon to austenite as well as segregation to dislocations in ferrite. The data also show negligible partitioning of substitutional alloying elements such as Cr, Si and Mn between ferrite and austenite. However, this mechanism cannot be rationalized by paraequilibrium transformation, since the carbon concentration in ferrite (Cabellero et al., 2010b) is more than that predicted by thermodynamic calculations. Accompanying diffraction data also demonstrated that the cessation of the austenite to bainite transformation, much before the carbon concentration reaches the equilibrium concentration dictated by α–γ phase boundaries.

15.4.1.5.4 Martensite
It is well known that athermal transformation of austenite to martensite by displacive mode does not lead to either substitution or interstitial solute partitioning. However, in medium-carbon steels, the formation of martensite from austenite becomes complicated by carbon partitioning to the austenite as well as auto tempering. Sherman et al. (2007) using transmission electron microscopy, Mössbauer spectroscopy, and 3DAP, evaluated the carbon redistribution within the martensite as well as the

stability of retained austenite. The retained austenite content decreased from 7 to 2% as the cooling rate increased from 25 °C/s to 560 °C/s. In all cases, the AP tomography showed clustering of carbon atoms within martensite laths. However, in slow cooling rates, stabilization of austenite due to partitioning of carbon from martensite to austenite is proposed. The above microstructure appears to be similar to the quench and partitioning steels as discussed in the next section.

15.4.1.5.5 Tempering of Martensite and Bainite
Tempering of martensite is the most important heat treatment given to industrial steels and has been extensively studied in the past. Since we need to characterize local carbon concentration in supersaturated solid solution during tempering, the AP made substantial contributions on understanding the microstructure evolution during tempering. The supersaturated martensite undergoes tempering through precipitation of carbides, nitrides and intermetallics. This has been described as a part of the precipitation in steels section. In many cases, austenite phase is not transformed to martensite fully and retained austenite and bainite are always present. Sherman et al. (2007) reported that carbon redistributes to the retained austenite films during continuous cooling. High-carbon steels can be quenched to martensite with 0–20% of retained austenite. On heating this mixture to an elevated temperature, the carbon from the martensite partitions into the retained austenite and stabilize it. This heat treatment, referred as quench and partitioning (Q&P), is utilized to obtain high strength and good toughness. Clark et al. studied the tendency for such carbon partitioning by AP tomography (Clarke et al., 2008), and ruled out the formation of bainite during Q&P.

Such thermodynamic equilibration between austenite and ferrite is also leveraged to produce transformation induced plasticity (TRIP) steels. In this case, the heat treatment is performed in high-Mn steel at high temperature between Ac_1 and Ac_3 to allow substitutional and interstitial atoms to diffuse. Dmitrieva et al. (2011) studied the partitioning characteristics in a Fe–Mn–No–Mo–Ti–Al steel by AP and compared the concentration profiles with those calculated by kinetic simulations using DICTRA. This classic paper shows the onset of local equilibrium at the martensite/austenite interfaces.

A new class of high-strength bainitic steel was (Caballero et al., 2009) developed by ausforming (isothermal heat treatment above the M_s) high-carbon Fe–Si–Mn steel. This steel produces nanoscale bainitic ferrite dispersed in high-carbon austenite matrix. The stability of this microstructure is very important for practical applications. Caballero et al. (2010c) studied the tempering resistance of the microstructure. They found that the tempering of the bainitic microstructure did not yield further enrichment of C in the austenite. This is in contrast to the Q&P process where the austenite enriches in carbon through partitioning from martensite. This suggests that the mechanisms for microstructure evolution during ausforming are different from that of the Q&P process. Extended tempering of this nanoscale bainitic microstructure leads to wide range of microstructures. Using 3DAP data, Caballero et al. (2008) rationalized this microstructural path due to local variations of carbon as the metastable bainitic microstructure transforms to equilibrium ferrite and cementite. This is schematically shown in **Figure 57**. Note that the microstructure evolution shown in **Figure 57** was revealed with the combined use of optical microscopy, hardness measurements, X-ray diffraction, transmission electron microscopy and 3DAP (Stone et al., 2008; Babu et al., 2005).

15.4.1.6 Microstructural Degradation during High-Temperature Service
3DAP has been extensively used for evaluating the stability of microstructure during service. One of the classic applications is related to spinodal decomposition of high-Cr ferrite in stainless steels (see Section 15.3). The reader is referred to classic papers by Danoix and Auger (2000) as well as Miller and Hyde

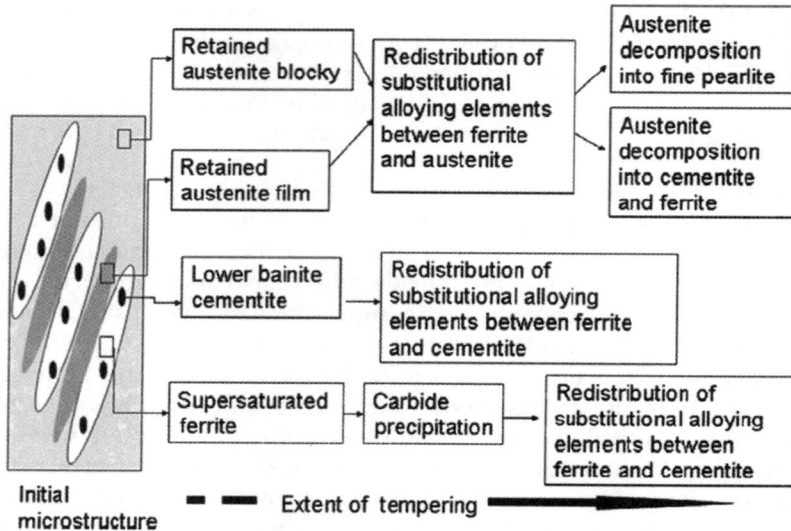

Figure 57 The above illustration provides overview of phase transformation events that can occur during tempering of steel that is related to local structure and composition. Therefore, care must be taken in interpreting the 3DAP data from single samples. From Caballero et al. (2008).

(Miller et al., 1995; Hyde et al., 1995a, 1995b). In addition, the segregation and precipitation of nanoscale precipitates were observed in steels while subjected to irradiation conditions (Burke et al., 2006; Miller and Burke, 1992; Miller et al., 2009).

15.4.1.7 Impact of 3DAP Analyses on Design of Steel for Targeted Properties

The above discussions showed that the alloying elements and heat treatment influence the microstructure evolution, which in turn influence the mechanical properties including service performance such as toughness and creep. The modification of microstructure evolution is brought about by clustering, interaction with dislocations, segregation to grain boundaries and modification of precipitation reactions. Although, the information provided by the AP analysis in understanding the above process, there is a need to translate the above data to real design of steels. This section attempts to show possible approaches to use these data in the design of steels.

The steel microstructure evolutions during primary processing, (melting and ingot processing), casting (sand-, die casting), thermomechanical processing (rolling, forging, etc.), fabrication (welding and machining) of structures, service (low- and high-temperature service) are related to composition of the steel, temperature, time and environmental conditions (gas composition). Therefore, additions of specific alloying elements to improve one specific property in all these steps must consider the sequence of steps that is involved in the processing. Atom-probe FIM observations of 9-Cr steels have shown that the incorporation of boron in the $M_{23}C_6$ carbides along the martensite lath boundaries lead to improved creep properties (Hofer et al., 2000). It is noteworthy that the bulk steel composition has only 0.051 at.% boron and it is indeed remarkable that boron the incorporation of boron into the $M_{23}C_6$ was detected.

The segregation of the trace boron to grain boundaries was also studied by APFIM (Karlsson et al., 1982, 1988; Karlsson and Norden, 1988a, 1988b; Karlsson, 1988). The extent of nonequilibrium segregation to grain boundaries depends on cooling rate (Hondros and Seah, 1983). On cooling below the martensite-start temperature, the austenite transforms to martensite by a displacive transformation mechanism. Because of the displacive transformation that does not allow for redistribution of solutes, the boron that is present in the austenite matrix may be trapped in the martensite matrix. In addition, some boron may be present along the prior austenite grain boundaries. The boron may also to segregate to lath boundaries since there is sufficient mobility of boron atoms in the martensite matrix. Seto et al. (1999) observed the segregation of boron and carbon to ferrite/ferrite grain boundaries in interstitial-free steel.

In fully tempered 9-Cr steel (Hofer et al., 2000), the presence of boron and carbon together at the austenite lead to copious precipitation of $M_{23}(C_xB_{1-x})_6$ carbides was observed at these boundaries. It is important to note that the APFIM analysis also indicated that there is still some boron enrichment in the martensite matrix (0.026 at.%) and is expected to redistribute during creep service. The transmission electron microscopy analyses of creep-tested sample indicated that different phases including MX (metal carbonitride) phase, Laves phase, and M_6C phase formed in the matrix during creep service. Interestingly, all of these phases had some dissolved boron in their constitution. Because of these precipitations, the dissolved boron in the martensite matrix reduced to very low levels (0.009 at.%). This example shows that it is possible to use the knowledge gained through AP microanalysis to optimize 9-Cr steels for high-temperature service. Additionally, the alloying must include all the elements that are intentionally added to improve the properties as well as the elements that lead to degradation of steel properties. For example, in some cases, the degrading behavior can be used to achieve particular end use, as demonstrated by Garcia et al. (2003), where they used strongly segregating and embrittling behavior of tin in steel to improve the machinability of steels. The above discussion clearly demonstrates that the 3DAP technique has provided critical and useful insight into the microstructure evolution in steels and will continue to be relevant to steel research.

15.4.2 Titanium Alloys

Titanium alloys are strengthened by changing morphology of α (HCP) phase within β (bcc) matrix by modifying composition and/or heat treatment parameters (Boyer, 1998). Compositional modifications include β-stabilizers such as V, Ta, Mo and Nb or α-stabilizers such as Al and O. Different morphologies including GB, colony, basket weave, martensitic and massive and grain size of α phase can be engineered by changing the cooling rate below the β-transus temperature. Depending on the undercooling below this temperature, diffusional and/or displacive (martensitic) transformations (Lütjering and Williams, 2010) may be triggered. In conjunction with the above microstructural modifications, mechanical working has been used for strengthening. Another important class of titanium alloys is titanium aluminides, Ti_3Al referred as α_2 and TiAl referred as γ, that provide high strength at high temperature. The role of atom-probe FIM to rationalize phase transformations and microstructure evolutions in titanium alloys are briefly reviewed in this section.

15.4.2.1 Partitioning of Alloying Elements and Phase Stability

The dissolution of oxygen in the α phase is a well-known phenomenon. Although one could use bulk analyses methods to measure oxygen concentration at high accuracy, 3D AP has been suggested as a way to evaluate the preferential partitioning of oxygen to α phases (Rissig et al., 2005). However, due to the problems related with specimen preparation and lack of calibration with respect to field-evaporation rates, the accuracy of AP measurements are limited to ±0.2 wt.%.

The role of heat treatment on the alloying element partitioning between α and β phases (Nag et al., 2009a, 2009b) in Ti–5Al–5Mo–5V–3Cr–0.5Fe (wt.%) alloy (Ti-5553) was investigated by 3DAP. For similar isothermal transformation temperatures, sluggish transformation was observed in the samples subjected to quenching treatment before the isothermal heat treatment. The AP samples after normal step-heat treatment were prepared by selective extraction using FIB machining as shown in **Figure 58a** and b. The LEAP analyses also showed asymmetric Mo segregation as shown in **Figure 58c** and d along the lath and GB α phases. Using all the data, the spatial morphology of the α phase is schematically shown in **Figure 58e**. The above data have been used to rationalize the microstructure evolution based on path-dependent phase stability and kinetics.

The characterization of ω phase in titanium alloys has been elusive in the past. Nag et al. (2009b) using electron microscopy and atom-probe FIM have explored the mechanism of ω phase precipitation. The heterogeneous nucleation of α phase at the β/ω interfaces was observed in a Ti–Nb–Zr–Ta alloy, confirming the earlier microscopy work by Ohmori et al. (2001). This heterogeneous nucleation of α phase is correlated with preferential partitioning of $Zr^{\dagger}$ into the β-phase during the formation of ω-phase as well as crystallographic coherency strains. Subsequent to this work, Devaraj et al. (2009) confirmed the formation of ω precipitates in a binary Ti–Mo alloy. This work reported that the core of the precipitates was slightly enriched in Mo, while the boundary of the precipitates was depleted in Mo. This result was used to postulate two-step β to ω transformation mechanism. In the first step, athermal ω nuclei may form displacively from β-phase, which is then followed by mixed mode of diffusional and displacive

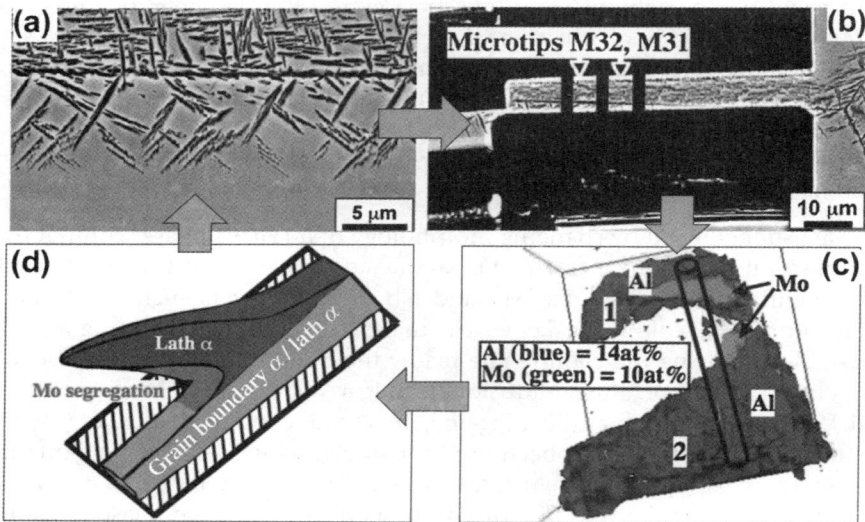

Figure 58   Analyses from Ti5553 alloy after heat treatment at 700 °C (a) Optical microscopy showing the b/b GB, as well as, GB a and lath a; (b) FIB extraction near the b/b GB; (c) iso-concentration surface from 3DAP data showing the Mo segregation near one side of alpha; (d) schematic representation of the microstructure and alloying element partitioning. From Nag et al. (2009a). (For color version of this figure, the reader is referred to the online version of this book.)

† Research on Ti–Al alloys does not support the preferential partition of Zr into α phase (Larson, 1999).

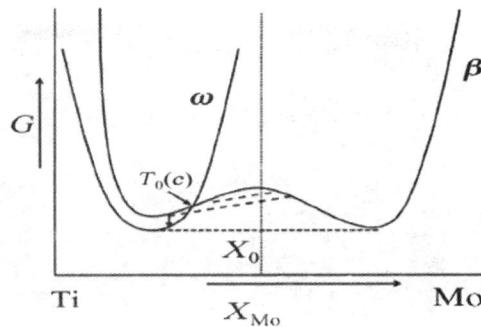

Figure 59 Schematic illustration of molar Gibbs free energy of β-phase in comparison to the ω phase. From Nag et al. (2009c).

growth. It is interesting to note that the above mechanism is similar to the one suggested by Olson et al. (1990) for rationalizing the austenite to bainite and martensite transformation. The above hypothesis was evaluated in Ti-5553 alloys also using TEM and 3DAP by Nag et al. Two different mechanisms for the ω-assisted formation of α-phase in titanium alloys was proposed (Nag et al., 2009c); one relies on the formation of α-phase close to the interface of ω/β interface and another relies on the displacive formation of α-phase within ω phase. Although transmission electron microscopy provided structural information, 3DAP analyses were crucial for obtaining spatial information of both ω and α phases.

Using the 3DAP results, the following hypothesis has been proposed for the transformation sequence involving β to ω that is followed by ω to α phases (Nag et al., 2011). An undercooled β phase might exhibit a double-well-type free energy versus composition as illustrated in **Figure 59**. Under such thermodynamic conditions, compositional modulation may be triggered to induce Mo-depleted and Mo-enriched regions. As the amplitude fluctuation increases, the Mo concentration in the β phase decrease beyond the C_0 point, at which molar free energy of ω and β are equal. To the left of this point, the β-phase becomes unstable with respect to the formation of hexagonal ω embryo by displacive transformation. With increasing time, the ω embryo may reject the Mo and evolve into a full precipitate. It is interesting to note the above mechanism resembles the congruent ordering and spinodal decomposition (Soffa and Laughlin, 1989; Khachaturyan et al., 1988) suggested for Al-base (Hono et al., 1992a; Okuda et al., 1998; Yu and Chen, 1992) and Ni-base (Babu et al., 2001) alloys. The extension of these instability principles to a wide range of alloys must consider the topology of the free energy versus concentration diagrams. Note that the use of schematic representation of molar Gibbs free energy to explain the phase transformation mechanism may not be realistic as it merely describes thermodynamic free energy without considering the transformation kinetics (Hono, 2002). However, the experimental results could be complemented with computational thermodynamic and kinetic simulations (Andersson et al., 2002).

15.4.2.2 *Partitioning and Site Occupancy in Titanium Intermetallics*

Titanium-based intermetallics such as TiAl (γ), Ti$_3$Al (α_2), and Ti$_2$AlNb provide low density and high strength at high temperature. Therefore, this alloy is ideal for aerospace applications. To stabilize the γ/α_2 microstructure, various alloying elements have been added. In order to understand the role of these alloying elements, extensive AP research has been done to analyze the site occupation of elements, relative partitioning of elements between γ and α_2 phases, segregation to interfaces and phase stability

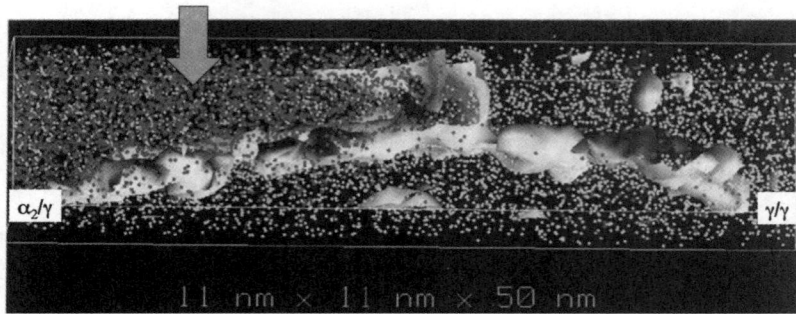

Figure 60 3D Atom map with Cr (yellow) and O (red) atoms obtained from a $Ti_{48}Al_{48}Cr_2Nb_2$ alloy shows the presence of O-rich a_2 phase (marked by arrow). The iso-surface with 3 at.% Cr shows the preferential segregation at a_2/g and g/g interfaces; Note: Oxygen was not added intentionally to the alloy. From Menand et al. (1998). (For color version of this figure, the reader is referred to the online version of this book.)

(Larson et al., 1997a, 1997b, 1997c, 1999d, 1999e; Menand, 1999; Larson and Miller, 1998). Ordering in γ-TiAl alloys has been analyzed by AP and the reduction in antiphase boundary energy (see Section 15.3) has been linked to the addition of Cr (Liu et al., 1991a). Menand et al. (1998) reported a beautiful 3DAP image of interface between α_2–γ in a $Ti_{48}Al_{48}Cr_2Nb_2$ alloy, in which the Cr segregation to both α_2–γ and γ–γ interface can be clearly seen as shown in **Figure 60**. The analyses also reveal the preferential partitioning of oxygen to α_2 phase.

As mentioned earlier, one of the challenges, with reference to site occupancy studies using 3DAP, is related to variations of evaporation fields for different elements and the data from site occupation studies may be less reliable. For example, the field evaporation behavior of the titanium was found to be much larger than that of aluminum and this tendency changes with crystallographic directions (Lefebvre et al., 2002). Boll et al. (2007) have developed a new algorithm (Atomic Vicinity) that allows for determination of site occupancy irrespective of evaporation fields. This methodology has been demonstrated by performing site occupancy of Nb in TiAl alloys.

15.4.3 Nickel Alloys

Nickel- base alloys are critical to high temperature and oxidation resistance applications within energy, aerospace and chemical industries. In nickel alloys, the fcc (denoted as γ) phase is strengthened either by solid solution or by precipitation. The precipitates may include carbides, ordered intermetallic precipitates (e.g. γ', γ'' and β–NiAl) (Reed, 2006). Similar to titanium aluminides, certain alloys are developed based on intermetallic compositions, known as nickel aluminides (Liu and Sikka, 1986). Majority of the AP research in this alloy system has been devoted to the investigation of site occupancy of other elements in the ordered compounds, partitioning between γ and precipitates as a function of their morphology that evolves during casting, heat treatment and welding. AP FIM work to evaluate the commercial nickel alloys was pioneered by Blavette et al. (Blavette and Auger, 1990; Blavette, 1992; Blavette et al., 1993), Haasen et al. (Liu et al., 1991b), Burke (Sieloff et al., 1986; Miller and Burke, 1993), Miller et al. (Miller and Jayaram, 1992; Jayaram et al., 1993) and Seidman et al. (Vanbakel et al., 1995). The onset of 3DAP allowed widespread application to commercial nickel base superalloys to interrogate the shape of the precipitates and associated concentration gradients. Few of the examples and their impact are discussed briefly below.

15.4.3.1 Site Occupation in Ordered Phases

In early stages of nickel aluminide development, similar to titanium aluminides, the room temperature ductility of these materials was too low. This hindered the processing of these materials into a required shape. Many dopants including B, Cu and Pd were added to improve the ductility (Hono et al., 1992b). However, the occupancy of these dopants was not known. Since the ALCHEMI method (Horita et al., 1995) was still under development, AP provided a viable option. In this technique, the field evaporation of atoms along certain crystallographic direction [e.g. <001>] is performed. The measured ions are analyzed and the location of dopant atoms are evaluated with respect to the site of major elements, e.g. Ni in face-centered site and Al in corner sites in L12-ordered Ni_3Al. The site occupancy analyses show that the Cu has a strong preference to Ni sites. Based on the above analyses, the ductilization of these intermetallics is related to the reduction of ordering energy by the addition of transition elements. The site occupation studies can also be used to understand diffusion characteristics in ordered compounds. For example, Ge is expected to occupy the Al sites, since Ni_3Ge are isomorphous with Ni_3Al. However, the site occupancy studies show that Ge atoms preferred the Ni–Al mixed planes (Hono et al., 1992d). This result was used to explain the anomalous rapid diffusion of Ge that is similar to Ni atoms. The site occupation data are also often used to evaluate the accuracy of computational models (Ruban and Skriver, 1997). For example, iron strongly prefers the Al site at low concentration and this tendency reduces with an increase in its concentration. These effects could be attributed to magnetic effects of iron atom (Almazouzi et al., 1997).

15.4.3.2 Alloying Element Partitioning and Morphology of the Precipitates

In commercial nickel-base superalloys, the high-temperature strength is related to relative fractions of γ and γ' phase, morphology as well as their composition. The composition also controls the lattice misfit between these phases. In early 1990s, extensive computational models have been developed to predict the phase fraction by either semi-empirical models or first principle calculations. However, these predictions had to be validated in nanoscale and APFIM has been used extensively for this purpose (Harada and Murakami, 1999; Murakami et al., 1994).

To understand the interaction between creep service conditions and microstructure evolution in nickel-base alloys, Jayaram and Miller (Jayaram et al., 1993) evaluated a model Ni–Mo–Ta–Al alloy using TEM and APFIM. The data show that the microstructure evolution during creep service conditions is more sensitive to the initial morphology of the γ' rather than the overall composition. In turn, these morphological differences could indeed be related to the local compositional variations between γ and γ' phases. To evaluate the morphology and composition of precipitates combined use of TEM and APFIM has always been stressed. For example, Miller and Burke (1993) studied the effect of cooling rate and subsequent heat treatment on the precipitation of $M_{23}C_6$ and γ' precipitates in X-750 alloys. They concluded that the composition of γ' is invariant and remained the same irrespective of the heat treatments. The same conclusion could not be made for very high cooling rates experienced during water quenching. In this alloy, due to lower concentration of Al (1.7 at.%), on water quenching the γ' formation was suppressed.

In contrast, in alloys (Babu et al., 2001) containing high aluminum concentration (12.1 at.%), the γ' formation cannot be suppressed even while water quenching. Under these conditions, the composition of the γ' was different from that obtained during slower cooling. A detailed 3DAP analysis of a sample from the water-quenched condition is presented in **Figure 61**. In this data, localized enrichment of Cr was observed within the γ' phase with wide variations of γ compositions. In addition, some of the γ' precipitates appear to be interconnected in space as shown in **Figure 61c**. Similar nonequilibrium precipitates and matrix composition was measured from Rene 95 (Ni–13wt.%Cr–8.0wt.%Co–3.5wt.%

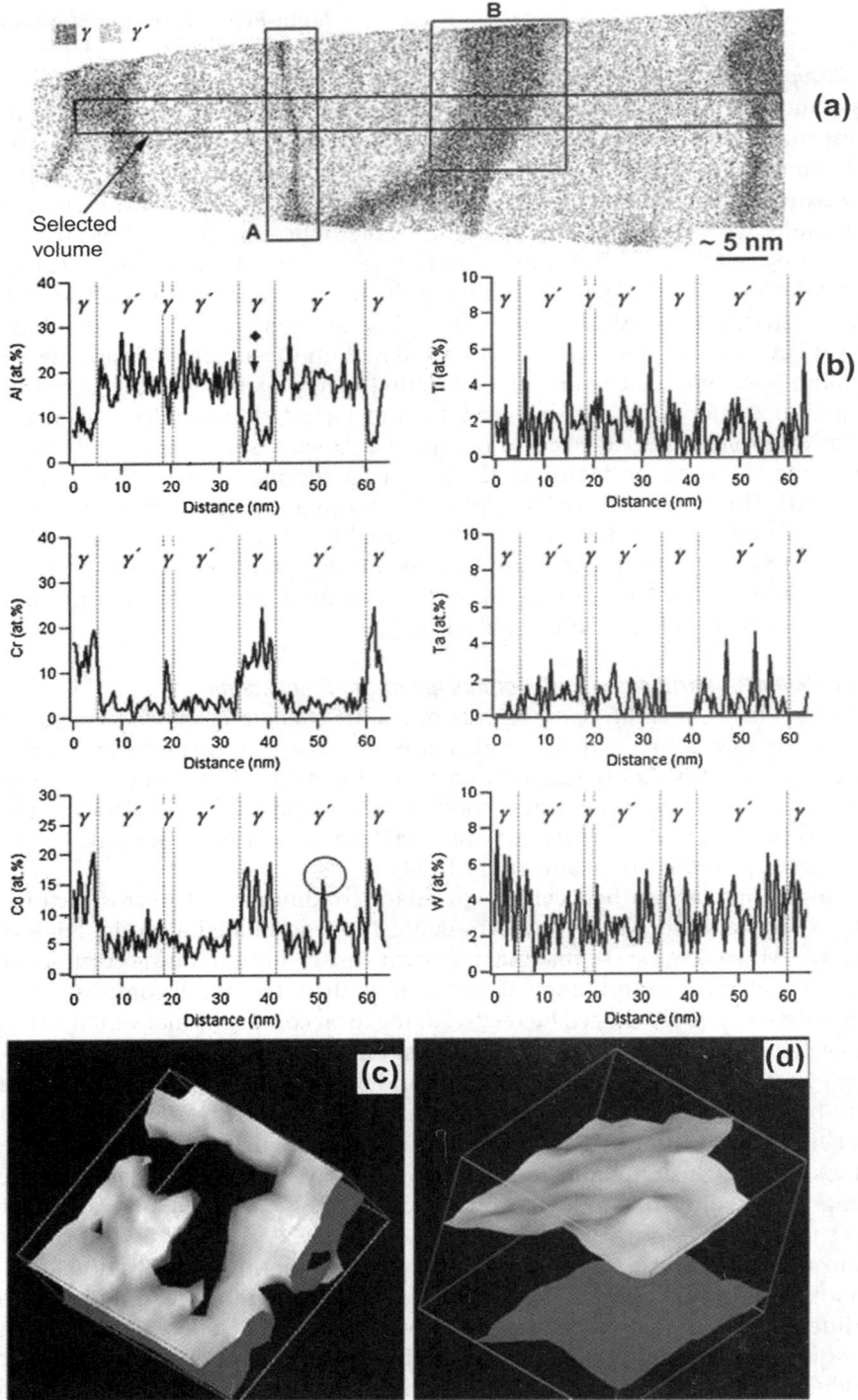

Figure 61 (a) 3D projection Cr atom from 3DAP data show the γ and γ' morphology (Cr-depleted regions) and highlights region A and B. (b) Concentration profile across the whole sample showing relative partitioning of all elements; iso-surface corresponding to 15 at.%Cr at (c) location A suggests that the γ' on either side are connected in space; while (d) at location B they are distinctly separated by a γ channel. From Babu et al. (2001).

Al–3.5wt%Mo–3.5wt.%Nb–3.5wt.%W) alloys subjected to continuous cooling in Rene 95 alloys have been performed by Hwang et al. (2009). The evolution of such features may be related to phase transformation mechanisms to be discussed in the next section. This cooling rate sensitivity of the microstructure is relevant to the development welding and joining processes (Babu et al., 1996a, 1996b; Babu et al., 1997) for repair and rejuvenation of gas-turbine parts.

Similarly, in PWA-1480 (Ni–5.0wt%Al–10.0Wt.%Cr–1.5wt.%Ti–5.0wt.%Co–12.0wt.%Ta–4.0wt.% W) single-crystal alloys under certain heat treatment conditions, small ultrafine (5–10 nm) γ precipitates precipitate within the γ' phase (Miller et al., 1994). These γ phases were enriched in Cr, Co and W. This unusual microstructure evolution can be attributed to large differences in elemental partitioning characteristics at different heat treatment temperature and cooling. Peculiar morphology of precipitates is also observed in Alloy 718-type (Ni–19wt.5Fe–18wt.%CR–5.0wt.%Nb–3.0wt.%Mo–1.0wt.% Ti–0.5wt.%Al–0.04wt.%C) alloys after certain heat treatment conditions. Under normal heat-treatment conditions, the alloy contains γ, γ' and γ'' precipitates. The compositions of these precipitates determined by APFIM were in good agreement with thermodynamic predictions (Miller et al., 1999). However, under certain heat treatments (Miller et al., 2002) or after service conditions, sandwich-like microstructures were found in these alloys. For example, these sandwich structures constitute $\gamma'/\gamma''/\gamma'$ or $\gamma''/\gamma'/\gamma''$ or γ'/γ'' morphologies. The formations of such precipitates have been studied by Cozar and Pineau (1973). In their works, the coating of γ' faces by γ'' was postulated. The above microstructure is expected to be very stable at high temperature. They also stressed the need for a critical (Ti + Al)/Nb ratio to be 0.9 to 1 to attain this "sandwich" or compact morphology. A classic image of the sandwich structure (Geng et al., 2007) in a 718 alloy subjected to extended service in an airplane engine (>10 000 h) is shown in **Figure 62**. The analyses was performed along the < 001 > direction where one can clearly observe the $\gamma''/\gamma'/\gamma''$ and γ''/γ' morphologies. The 3DAP data also show that there is a distinct aluminum enrichment at these interfaces and computational models suggest that this may be needed criteria for the formation and stability of these co-precipitates.

15.4.3.3 Phase Transformation Mechanisms

Extensive APFIM research has been performed in understanding the phase transformation mechanisms for the formation of γ'' and γ' precipitates from γ phase. For example, Miller et al. evaluated the formation mechanisms of primary and secondary precipitates in alloy 718 (Miller, 1999b). The role of GB segregation and subsequent precipitation of carbides was studied in a model Ni–Cr–Fe alloy. In this alloy, GB segregation of elements like C, B and N were observed. In addition, the GB precipitation of M_7C_3, $M_2(C,N)$, $M_{23}C_6$ were observed (Thuvander et al.). The authors also used SIMS analyses to understand the GB segregation over a large area (Thuvander and Stiller, 2000). The above study clearly demonstrates the need for prudent selection of characterization technique that leverages the APFIM technique with other techniques, rather than relying only on one technique.

The formation of γ' precipitate in γ phase requires compositional change and the ordering of fcc to $L1_2$ structure. Similar to the discussions related to ω and α phase formation from β phase in Ti alloys (Nag et al., 2011), various models have been postulated for the kinetic pathway for microstructure evolution. It is interesting to note that the γ' precipitate formation in γ phase is indeed similar to the δ' precipitation from supersaturated fcc solid solution in Al–Li alloys (Soffa and Laughlin, 1989; Khachaturyan et al., 1988). Based on the work by Soffa and Laughlin, the above transformation may occur by different pathways depending upon the supersaturation. Babu et al. (2001) proposed the transition form nucleation and growth mode to congruent ordering and spinodal decomposition to explain the observed γ' morphologies and composition. Sudbrack et al. (Sudbrack, 2004; Sudbrack et al., 2006, 2007, 2008; Booth-Morrison et al., 2008) have tried to address the above kinetic pathway issues by

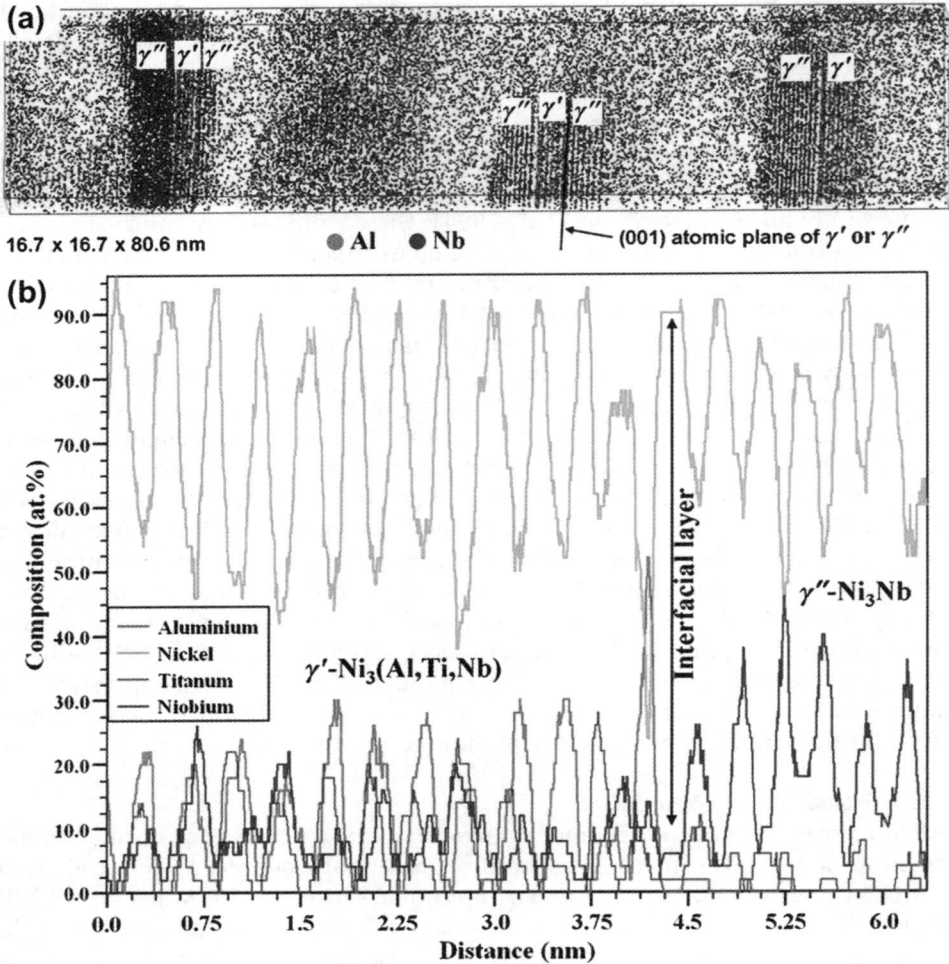

Figure 62 2D projection of the 3D atom map of Al and Nb from alloy 718 extracted from an engine that has been in serving greater than 10 000 h. (b) Composition profile across one of the γ' and γ'' interface shows enrichment of aluminum at the interface. From Geng (2007). (For color version of this figure, the reader is referred to the online version of this book.)

systematic 3DAP analyses and given a comprehensive treatment the transition from nucleation, growth and coarsening stages, including the morphological changes (sphere to cube and coalescence). It is important to note that the above phase transformation pathways are very sensitive to the supersaturation of elements that promote the formation of γ' phase. Recently, Viswanathan et al. (2011) tried to rationalize the kinetic pathways in Rene 88 alloy, which contains higher concentration of aluminum. 3DAP analyses shows interconnected Cr-rich and Cr-lean regions suggesting a phase separation. The electron diffraction information showed the ordering has occurred only within the small regions of these Cr-rich regions. Based on the above, the authors concluded that the phase transformation initiates by phase separation (compositional modulations) and ordering (formation of $L1_2$) structure. In contrast,

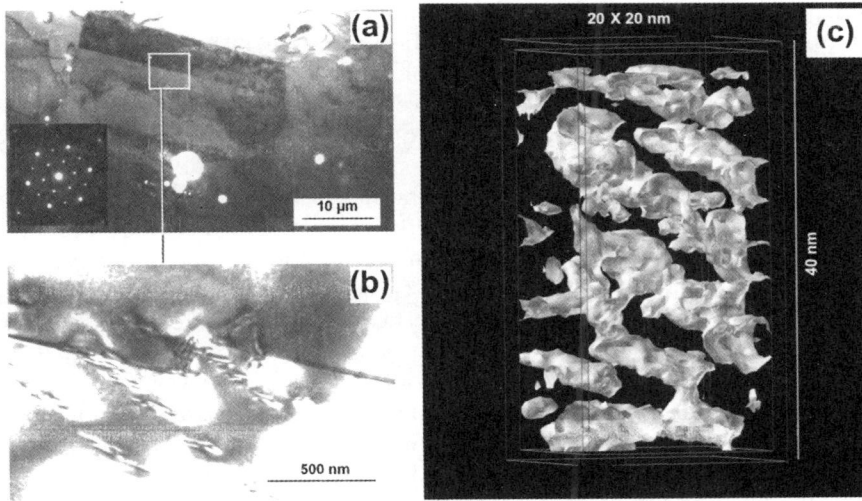

Figure 63 TEM and 3DAP results from heat treated samples. (*a*) Low-magnification picture shows the presence of γ grains and selected area diffraction pattern showing the presence of superlattice reflections. (*b*) High-magnification micrograph shows the presence of g GB as well as fine-speckle of γ' precipitates. (c) Iso-concentration map corresponding to 15 at.% aluminum shows the presence of interconnected γ'precipitates in the heat treated condition. From Babu et al. (2005).

preliminary evidence for uniform ordering and interconnection of γ' phase was seen in Haynes 214 alloy as shown in **Figure 63** (Baabu & et al., 2005). Rationalization of such subtle changes in kinetic pathways has to be related to the topology of free energy versus concentration curves as suggested by Soffa and Laughlin (1989). The above studies rely on ex situ characterization after the processing. Ideally, the above experiments have to be augmented with *in situ* scattering experiments using Synchrotron radiation to evaluate the time and temperature differential between these phase separation and ordering processes. This approach was used by Yu and Chen (1992) to evaluate the formation of congruent ordering and spinodal decomposition.

In addition to the above primary precipitations, there has been considerable interest in evaluating the clustering of Re atoms as well as enrichment at the γ/γ' interfaces (Mottura et al., 2010) to rationalize the observed creep strength improvement through Re additions. 3DAP analyses indicated that the Re clustering does not occur and the observed fluctuations are indeed random. At the same time, the authors also noticed an enrichment of Re near the interface. However, the authors postulated that this enrichment due to the movement of γ/γ' interfaces during cooling from operating temperature and not during service.

15.4.4 Aluminum Alloys

Aluminum alloys can be categorized into cast and wrought alloys. There are relatively small numbers of AP investigations on cast alloys, as the microstructure features are too coarse for AP analysis. On the other hand, most of wrought aluminum alloys are heat treated after wrought process and they contain nanosized precipitates in peak-aged conditions. As the size and the number density of the precipitates are suitable for AP analysis, there are a number of AP investigations on wrought Al alloys. Medium- and high-strength Al alloys are strengthened by precipitation hardening, and trace additions of solutes in binary and ternary base alloys influence the precipitation kinetics substantially. Precipitation in commercial aluminum alloys usually starts from the formation of solute clusters such as Guiner-Preston (GP) zones[§], and these evolve to more stable phases as they are aged for longer times. The nucleation barrier of the GP zones is significantly lower than those of the equilibrium phases, thus the formation of GP zones precedes the precipitation of equilibrium phase. GP zones and other metastable precipitates often work as heterogeneous nucleation sites for subsequent precipitation processes. The typical sizes of GP zones are in the order of tens of nanometers and their number density is in the order of 10^{23} m^{-3}. The quantitative chemical analyses of these nanosized precipitates are not possible by spectroscopic techniques on a transmission electron microscope as the spectroscopic information contains the signals-form matrix. While well-defined GP zones are observable using modern TEM, the observation of early clustering stage is still difficult. The AP has single-atom detection efficiency, and hence the solute clusters formed in dilute alloys can be clearly detected. In fact, many new findings regarding the pre-precipitation solute clustering in Al alloys have been made using 1DAP (Hono et al., 1986; Osamura et al., 1986; Hono et al., 1989; Brenner et al., 1991a; Hono et al., 1993; Hono et al., 1992e) and more recently with 3DAP. Although 1DAP can detect solute clusters in one-dimensional concentration depth profile or in ladder diagrams, 3DAP can show the morphology of clusters and precipitates and this is very useful for identifying the phases among various coexisting phases. In the following sections, AP studies of aluminum alloys are reviewed to supplement the previous review articles on AP studies of aluminum alloys (Ringer and Hono, 2000; Warren et al., 2000).

15.4.4.1 Al–Cu Alloy

Al–Cu alloy is the most fundamental system for the precipitation-hardening aluminum alloys. The precipitation sequence of this system is generally accepted as (Lorimer, 1978)

$$\text{supersaturated } \alpha \rightarrow \text{GP zones} \rightarrow \theta'' \rightarrow \theta' \rightarrow \theta.$$

Among these, all metastable phases precipitate lying on {001} planes, but the equilibrium phase θ precipitate incoherently with the matrix with many orientation relationships. GP zones are fully coherent plate substituting 100% Al with Cu on the (002) planes (Gerold, 1954). **Figure 64** (a) shows high-resolution TEM image of an Al–Cu alloy in which both GP zones and θ'' coexist. θ'' are composed of two Cu layers separated by three (002) Al layers. The ab inito calculations by Wolverton, (1999) suggest that these structures can be explained from the coherent phase boundary for the two-phase field between the Al-rich solid solution and a metastable Al_3Cu phase, which is precisely the ordered Al3Cu ground state. This clearly explains the transition from single layer 100%Cu (002) plane to the addition of another 100% Cu(002) layer with three-layer Al spacing. Although GP zones with multilayer Cu (002) were also observed, the above structures are the representative ones.

Chemical analysis of such a fine object is very challenging, as a monoatomic layer resolution is required. Direct AP analysis result showed that field evaporation behavior changes near the GP zones because Cu has a higher evaporation field than Al (Hono et al., 1986, 1989). Thus, Cu is strongly retained on the surface and the layer-by-layer evaporation behavior is deteriorated near GP zones. Bigot et al. (1996a) investigated the GP zones in Al-1.7 at.%Cu alloy using 3DAP, and succeeded in

[§] Guinier–Preston (GP) zones are defined as fully coherent precipitates with the same structure as the matrix phase. So they do not give extra diffraction unless there is internal order. Based on this definition, there is no clear distinction between solute clusters and G P zones. Solute clusters are the aggregates of solute atoms without distinct interfaces with the matrix. While GP zones give rise to contrast in TEM images, solute culsters do not give clear contrast in TEM image because they do not give sufficient strain field or atomic factor contrast. Based on this, solute clusters may be defined as the deviation of solute concentration larger than statistical fluctuations.

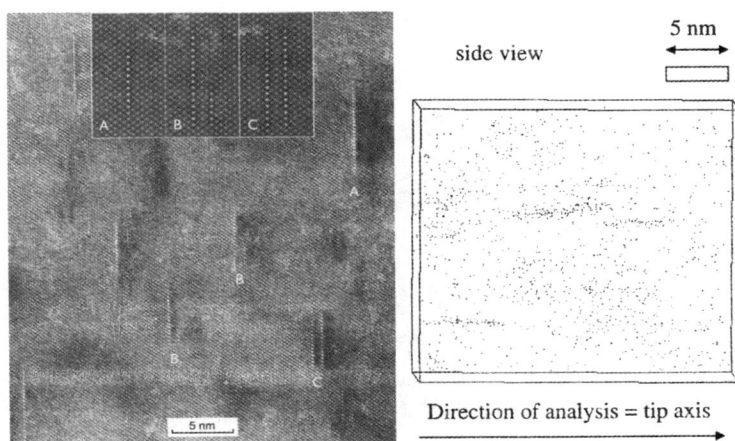

Figure 64 HRTEM image and 3DAP Cu map of GP zones in Al–Cu alloy. From Bigot et al. (1996a).

reconstructing thin Cu-rich plate on (002) planes as shown in **Figure 64(b)**. These data were obtained by analyzing the specimen in the (001) direction and the GP zones on $(200)_{Al}$ and $(020)_{Al}$ are successfully imaged, but somehow (002) plates were not observed. The concentration depth profile calculated from this data shows the maximum concentration of Cu is 35 at.%. They attributed this to the evaporation aberration. Assuming that the GP zone is single layer, they estimated the maximum Cu concentration to be 50 at.% or more. In principle, AP data give the highest spatial resolution in the depth direction of the analysis, because low index planes evaporate layer-by-layer. However, they did not succeed to reconstruct Cu layers from the GP zones analyzed in the vertical direction because of the trajectory aberration. The Cu concentration estimated from the analysis in the perpendicular to the plate was only 35 to 20 at.%, similar to the earlier work by 1DAP (Hono et al., 1989). They attributed this to the preferential retention of Cu atoms during the field evaporation process of the atoms with much higher evaporation field than the matrix. More recent AP tomography work by Biswas et al. (2011) reported similar problems in quantitative analysis of θ' precipitates; the composition of θ'' scatter in the range of 10–40% and they found segregation of Si at the coherent interface of θ' with the matrix. While FIM observation made a certain contribution in understanding the morphology of the GP zones, neither 1DAP nor 3DAP gave definite answer to the structure and chemical composition of the GP zones in Al–Cu alloys. If more rigorous control of field evaporation can be made by cooling down a specimen below 10 K under an UHV condition, the difference in evaporation field of Al and Cu may become closer. In such a case, real atomic layer analysis of plates embedded in a matrix may become possible. GP zones in the Al–Cu system are a good sample for benchmarking the performance of 3DAP.

15.4.4.2 Al–Cu Alloy with Trace addition of Sn

The influence of the trace additions of Cd, In and Sn on the precipitation kinetics of Al–Cu alloy is a classical subject of precipitation study in aluminum alloys (Hardy, 1950-51). The additions of these elements suppress the kinetics of GP zone formation, while it enhances the θ' precipitation. Strong solute-vacancy interaction suppresses the diffusion of Cu because vacancies that are necessary for diffusion of Cu are trapped by Sn (Kimura and Hashiguti, 1961). As to the enhanced kinetics of the θ' precipitation during high-temperature aging, several mechanisms have been proposed (Silcock et al., 1955;

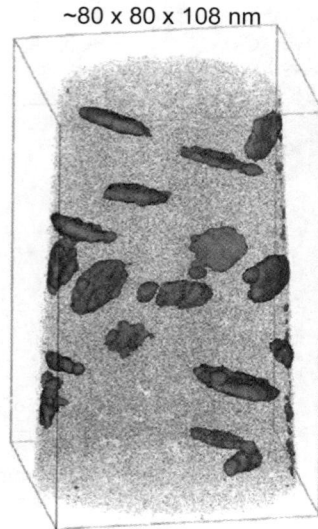

Figure 65   APT isodensity surfaces highlighting β-Sn precipitates and θ' phase in the Al–1.7Cu–0.01Sn alloy aged at 473 K (200 °C) for 180 s. The orange and green dots represent Cu atoms, respectively. From Honma et al. (2012). (For color version of this figure, the reader is referred to the online version of this book.)

Boyd and Nicholson, 1971; Sankaran and Laird, 1974), but TEM observation results suggests that the heterogeneous nucleation of θ' platelets on the Sn particles is the most probable mechanism (Kanno et al., 1980). Using 1DAP and TEM, Ringer et al. (1995) confirmed this result; in addition, they found evidence for Sn cluster formation in the early stage of aging. Since Sn atoms trap vacancies, they tend to form clusters in the early stage, leading to homogeneous distribution of Sn particles. **Figure 65** shows the AP tomography of Al–Cu–Sn alloy aged at 470 K for 180 s, which convincingly show most of the θ'' precipitates are associated with Sn precipitates (Honma et al., 2012).

15.4.4.3   Al–Cu(–Li) Alloy with Trace Addition of Mg and Ag
Trace additions of Mg and Ag in Al–Cu alloy leads to fine and uniform precipitation of plate-like Ω phase on the $\{111\}_{Al}$ planes in addition to the θ'' or θ' precipitates on the $\{001\}_{Al}$ planes (Polmear, 1964; Taylor et al., 1978). The Ω phase has a very similar structure to the equilibrium phase in Al–Cu binary alloy, θ (Garg and Howe, 1991). Thus, they are considered to be θ with a different habit plane. In Al–Cu–Li alloy, plate-like precipitate called T_1 phase (Al$_2$CuLi) that has a very similar structure to Ω precipitate with the presence of trace addition of Ag and Mg. The mechanism by which Ω and T_1 precipitate has been a subject of many investigations. Using a conventional AP, Hono et al. (1994) reported that Ag and Mg clusters form after very short-time aging (15 s at 180 °C). However, the role of the Mg–Ag co-clusters in the nucleation stage of the Ω phase was not clearly understood with the work using only 1DAP, because it did not reveal any information on the morphology of the clusters and their link with the subsequent precipitates.

In order to observe the microstructural evolution process from clusters to precipitates in an Al–Cu–Mg–Ag alloy precisely, Reich et al. (1998) revisited this alloy system using a 3DAP. **Figure 66(a)** shows 3DAP elemental map of solution treated Al–1.9Cu–0.3Mg–0.2Ag (at.%) alloy. All solute atoms, Cu, Mg and Ag, are homogeneously dissolved in the matrix forming a supersaturated solid solution.

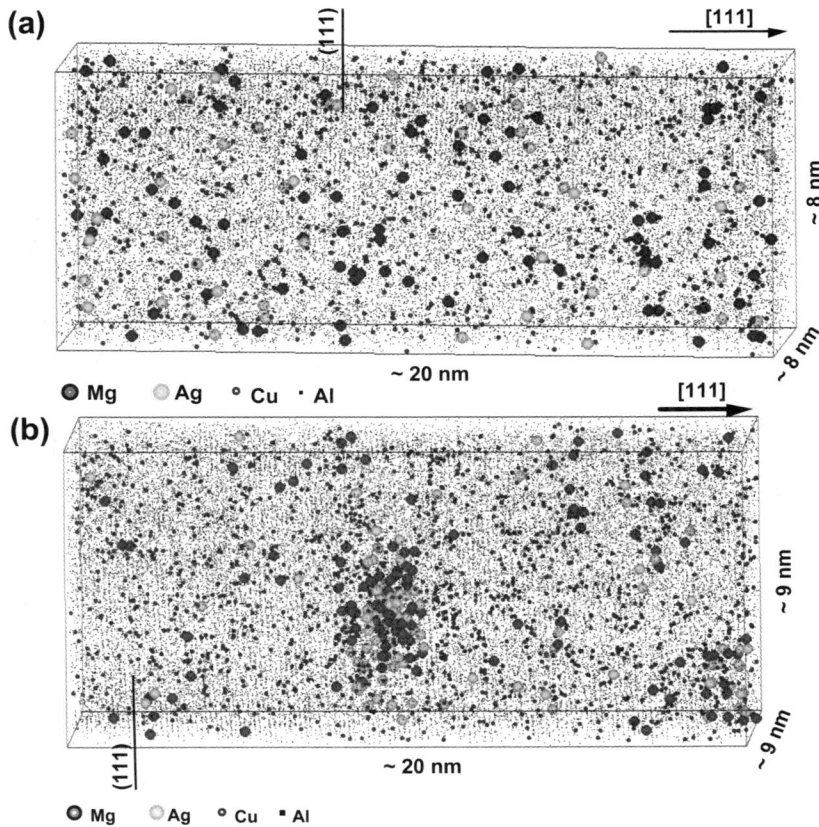

Figure 66 Atom probe tomography data from Al–1.7Cu–0.3Mg–0.2Ag alloy at (a) solution treated and (b) aged at 180 °C for 15 s. From Reich et al. (1998). (For color version of this figure, the reader is referred to the online version of this book.)

The layer pattern observed in the atom map are (111) planes. **Figure 66(b)** shows an elemental map obtained from the same alloy aged for 15 s at 453 K. An aggregate of Ag and Mg atoms is observed in the elemental map. The number of Ag and Mg atoms involved in the co-cluster ranges from 40 to 80 considering the detection efficiency of the MCP detector of $\sim$60%. The ratio of Mg to Ag atoms was found to be close to 1:1. The shape of the clusters is not well defined in this stage, and Cu atoms are not incorporated in this cluster. Unlike the previous conventional AP results (Hono et al., 1994), this 3DAP data clarifies the morphology of the co-clusters in the earliest stage of formation. Such clusters do not give any contrast in TEM images, and only the 3DAP result gave direct evidence for the presence of Ag–Mg co-clusters.

When aging continues, Cu atoms (see **Figure 67**) are also incorporated into the Mg–Ag co-clusters, then they evolve to plate-like shape on the {111} planes (Reich et al., 1998). In this stage, the plate-like precipitates on {111} do not have a distinct structure as Ω. The association of Cu atoms to Ag and Mg co-clusters is thought to cause a strain, and the formation of the {111} plate would be a result of minimizing the strain energy for the solute cluster containing $\sim$15at.% of Cu, $\sim$18at.% of Mg and Ag as predicted by Suh and Park (1995). Longer aging makes the plate-like feature more distinct, and the

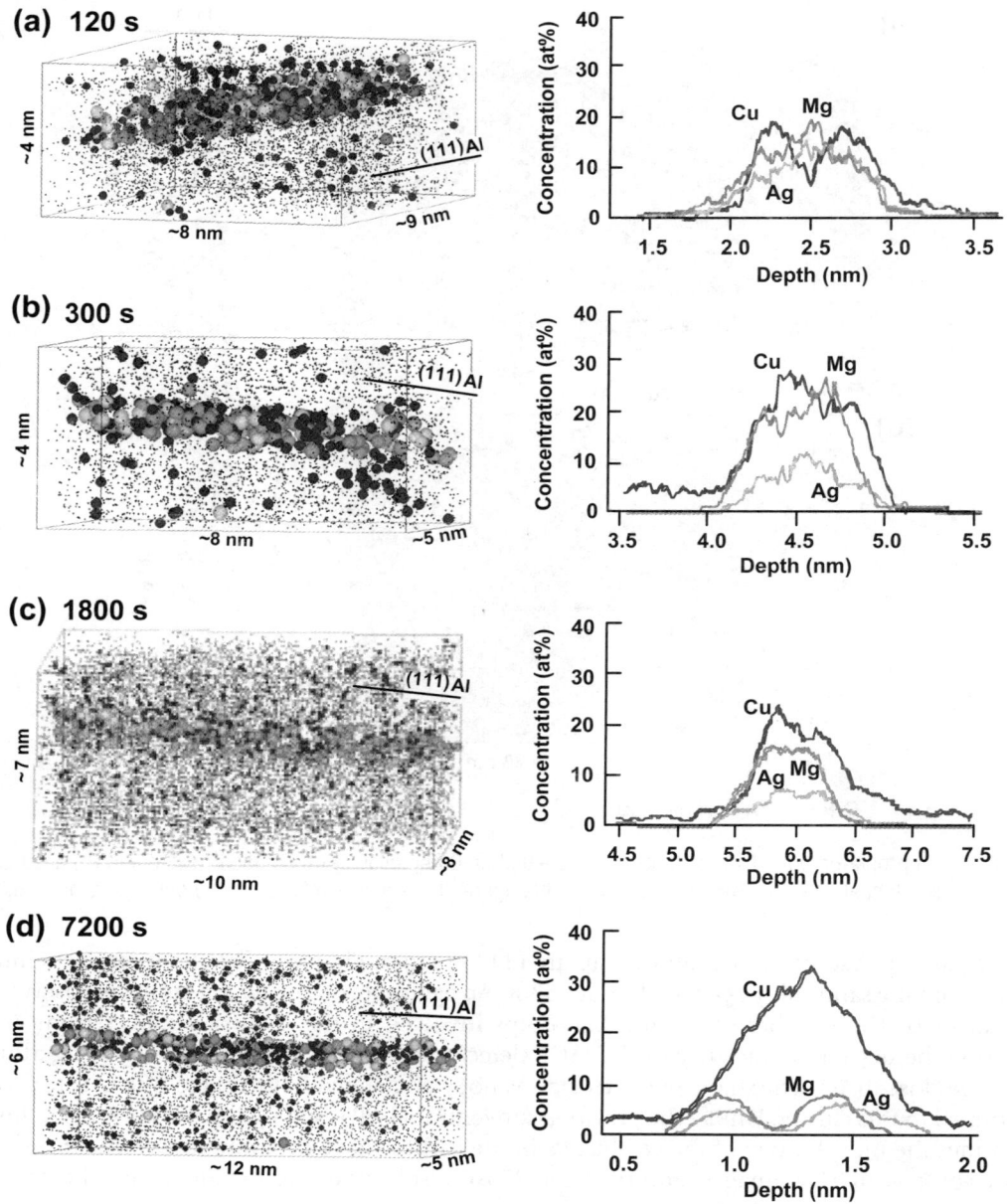

Figure 67 Atom probe tomography data from Al–1.7Cu–0.3Mg–0.2Ag alloy with further aging clearly shows the segregation of Mg and Ag atoms to the interface. From Reich et al. (1998). (For color version of this figure, the reader is referred to the online version of this book.)

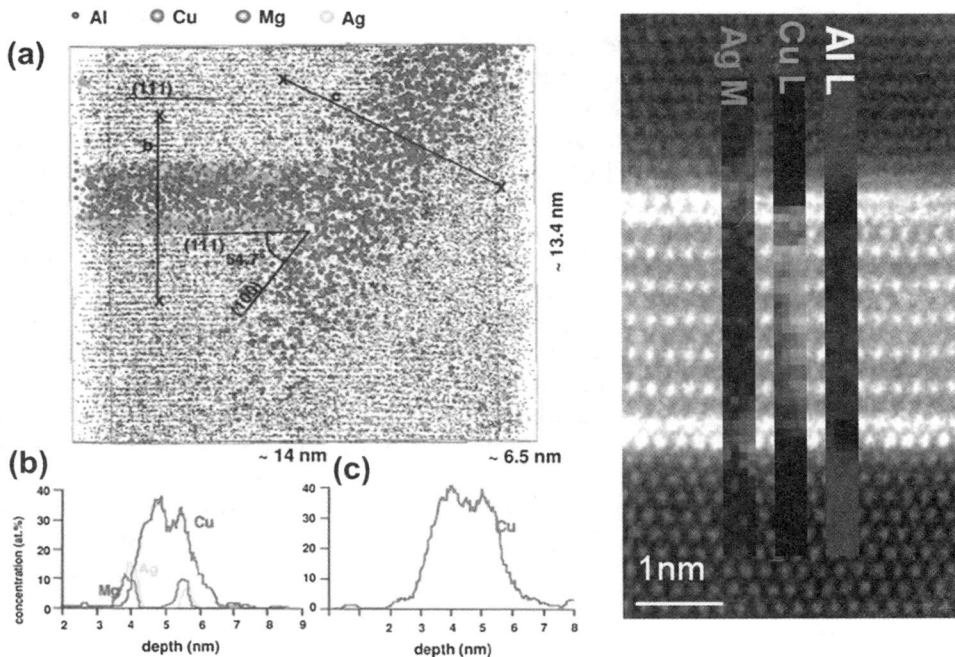

Figure 68 A typical 3DAP data obtained from an aged Al–1.7Cu–0.3Mg–0.2Ag alloy. From Hono (2002). (For color version of this figure, the reader is referred to the online version of this book.)

platelet starts to have a distinct structure as Ω. Initially formed Ω contains all Cu, Mg and Ag within the precipitate, and Mg and Ag atoms migrate from the inside to the broad interface of Ω/α as they grow.

Figure 68 shows a 3DAP elemental map obtained near the $\{111\}$ pole of the specimen aged for 10 h. The (111) atomic planes are resolved in this analysis. The plate-like precipitate lying on the (111) planes is a Ω precipitate, and the other platelet inclined by 54.5° from the Ω is a θ' precipitate on the (100). As seen from this example, 3DAP data reconstruct the morphology of the two plate-like precipitates on different habit planes quite accurately. The concentration depth profiles measured in the normal direction to the platelets are shown in **Figure 68(b) and (c)**. The concentration profiles across the Ω show that Ag and Mg atoms are no longer incorporated within the precipitate, but they are strongly segregated to the broad $\Omega/\alpha\{111\}$ interface. The segregation appears to be restricted to one or two atomic layers, and no Mg atoms are detected within the precipitate. The concentration of the Ω is close to 33 at.%Cu, which is consistent with the θ phase (Al$_2$Cu). This suggests that Ω is chemically equivalent to θ. The chemical composition of the θ' precipitate observed in **Figure 68(c)** is also close to 33 at.%Cu, but this precipitate does not contain any Ag and Mg atoms.

T_1 plate in an Al–5.0Li–2.25Cu–0.4Mg–0.1Ag–0.04Zr alloy was also analyzed by 3DAP together with an HREM (Murayama and Hono, 2001). As in the case of the Ω phase in the Al–1.9Cu–0.3Mg–0.2Ag alloy, Ag and Mg atoms were also found to be segregated at the T_1/α interfaces (see **Figure 68d**). Since T_1 precipitates are much thinner than W with a large aspect ratio, the segregations of Ag and Mg atoms visualized by 3DAP were not as clear as that observed in Ω. However, the concentration depth profile obtained in the normal direction to the T_1 plate suggests that the Ag and Mg atoms are segregated at the T_1/α interface rather than being partitioned in the precipitate. However, more recent APT work by Gault et al. reported that Ag and Mg are partitioned in the T_1 phase (Bault et al., 2011), which is inconsistent with the previous result.

Although segregation of Ag and Mg atoms was observed for both T_1/α and Ω/α interfaces, the clustering behaviors of Mg and Ag atoms in the Al–Cu–Li–Mg–Ag and Al–Cu–Mg–Ag alloys are quite different. Unlike the Al–Cu–Mg–Ag alloys, no evidence for Ag–Mg co-clusters has been obtained (Murayama and Hono, 2001). After several minutes of aging, only clustering of Mg atoms were observed. Thus, the mechanism of the nucleation of the T_1 appears to be different from that of Ω. The fact that Mg and Ag atoms do not form co-clusters suggests that Li additions affect the interactions between Mg and Ag atoms significantly. This observation is interesting when it is compared to the case of Al–Cu–Mg–Ag alloy, in which co-clusters of Mg–Ag atoms were observed after only several seconds at 180 °C (Hono et al., 1994; Reich et al., 1998). The absence of such Ag–Mg co-clusters in Al–Cu–Li–Mg–Ag alloys indicates that the interaction between Mg and Ag atoms is strongly affected by the presence of Li. Ag is known to have strong affinity with vacancies, because a high number density of dislocation loops commonly observed in Al–Cu–Mg alloys are not observed in the alloy containing a small amount of Ag. Although AP is capable of observing clusters of dilute solute atoms, it provides no information on vacancies and their interactions with solute. To obtain the information of vacancies, Nagai et al. pioneered the complementary use of positron annihilation measurements, Doppler broadening position spectroscopy and 3DAP (Nagai et al., 2001, 2002; Honma et al., 2004b).

15.4.4.4 *Al–Cu–Mg Alloys*

The precipitation process of Al–Mg–Cu alloys within the $\alpha + S$ (Al$_2$CuMg) two-phase region has been a subject of numerous studies, because this is the base system for one of the major commercial aluminum alloys, 2xxx series. The precipitation sequence of this alloy was described as:

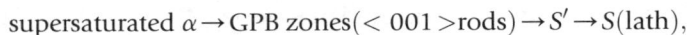

$$\text{supersaturated } \alpha \rightarrow \text{GPB zones}(< 001 > \text{rods}) \rightarrow S' \rightarrow S(\text{lath}),$$

by Silcock (1960-1961), but according to more recent work by Gupta et al. (1987), Radmilovic et al. (1989) and Ringer et al. (1997), the structural differences between S' and S are trivial, thus S' is simply denoted as S in recent papers. The age-hardening curve of this alloy has two distinct stages: the first one is a rapid hardness increase that occurs within 1 min aging at temperatures ranging from 100 to 240 °C as shown in **Figure 69** (Hardy, 1954-1955; Vietz and Polmear, 1966; Marceau et al., 2010). The hardness increase during this period accounts for approximately 60% of the total hardness increase observed at the peak hardness condition, and it was recently proposed that this quick age-hardening response may be used for bake hardening of body-sheet aluminum alloy (Ratchev et al., 1998). After this rapid hardening, the hardness curve has a long plateau ($\sim$ 100 h at 150 °C), then the specimen starts to harden again and reach the peak hardness (the second stage). The origin of this characteristic age-hardening behavior was widely studied in the past, and following Silcock (1960-1961), it has been generally accepted that the first stage is associated with the formation of GPB zones, and the second stage is attributed to the precipitation of S phase. Recently, using APFIM, Ringer et al. (1997) proposed that the initial rapid hardness increase is caused by clustering of a few Cu and Mg atoms, and the second stage hardening is due to the GPB zones, which start to form near the end of the hardness plateau. However, it is difficult to tell the solute clusters of few atoms from the concentration profile of the alloy containing solute atoms of more than a few atomic percent as statistical errors become large. Ratchev et al. (1998) attributed the rapid hardening to the formation of S'' phase (or S) on dislocations, by which dislocations are locked. Although the precipitation process of Al–Cu–Mg-based alloy has been

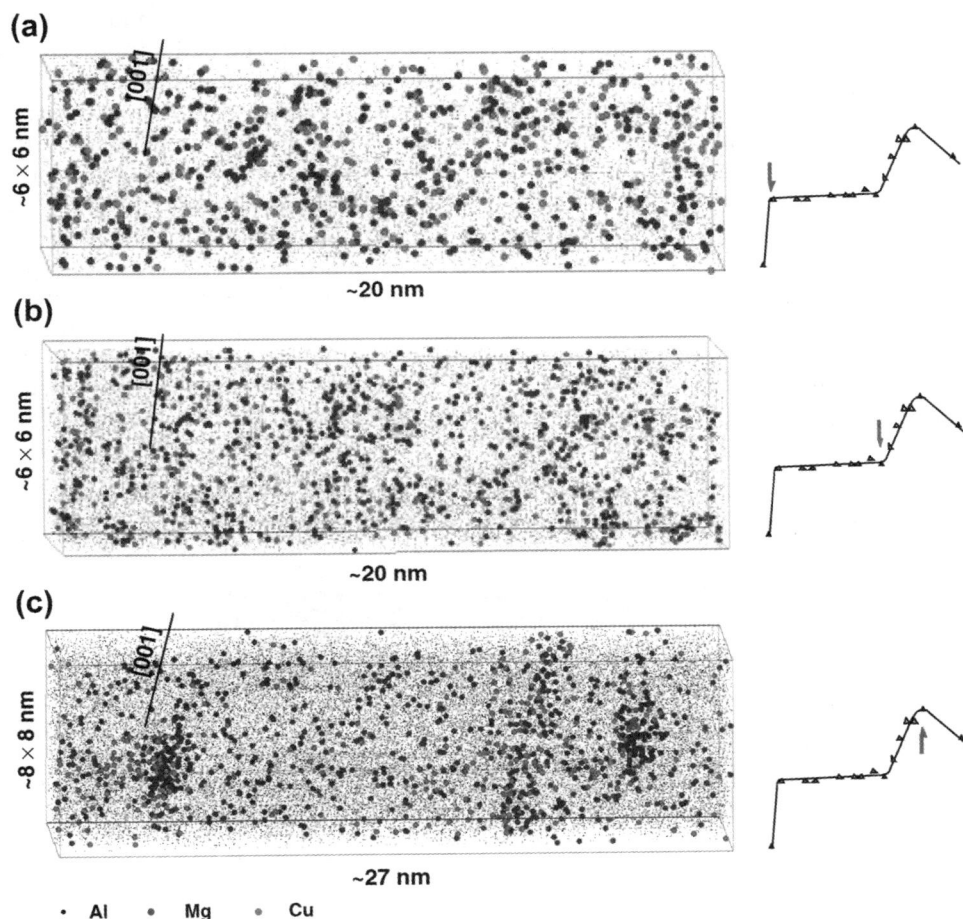

Figure 69 3DAP elemental map obtained from Al–1.1Cu–1.7 Mg alloy aged at 473 K for (a) 1 min, (b) 60 min and (c) 480 min. The corresponding aging stages are indicated by arrows in the hardness curve. From Marceau et al. (2010). (For color version of this figure, the reader is referred to the online version of this book.)

studied for a long time, there had been such controversies regarding the origin of the initial hardening response, and this is mainly because it has not been possible to observe solute distribution in the earliest stage of aging where hardness drastically changes.

Using 3DAP, Reich et al. (1999) observed the change in distribution of solute atoms in Al–1.1Cu–1.7 Mg alloy. **Figure 69** shows 3DAP elemental maps of Al, Mg and Cu atoms during aging at 473 K. The corresponding hardening stages are also indicated in the hardness–time curve. After 1 min aging, no clear evidence for Cu–Mg cluster formation is found as shown in **Figure 69(a)**. However, after aging for 60 min, which is the end of plateau and beginning of the second-stage hardening, 3DAP map show some tendency of Cu–Mg co-clustering. Statistical analysis of this data suggested there is a positive correlation between Cu and Mg atoms. At the peak hardness condition, Cu and Mg-enriched needle-like precipitates are observed clearly. These precipitates are attributed to GPB zones, which uniformly precipitate in the matrix. This series of elemental maps suggest that the initial rapid hardening that occurs within 1 min is neither due to the GPB zone formation nor to the formation of Cu–Mg co-clusters. Separate TEM observation indicated that only heterogeneous precipitation of S-phase along dislocations occurs during this short period of aging. However, at 423 K, no evidence for heterogeneous precipitation of S was found even after rapid hardening. Thus, they concluded that the initial hardening is most likely to originate from solute–dislocation interaction as a result of enrichment of Mg and Cu atoms. This proposal was later supported by CDB (Coincidence Doppler Broadening) of positron annihilation radiation study by Nagai et al. (2000), which strongly suggested that vacancy trapped by Mg atoms migrate to vacancy sinks, where vacancy-Mg–Cu complexes are formed after 1 min aging. The work has been further studied by 3DAP. Starink et al. found no indication of the presence of any particle with a distinct shape, such as GP/GPB zones, but they reported small Cu–Mg clusters after the room temperature aging. To study the density and composition of the clusters formed, the maximum separation method has been used. However, the measured value of cluster density depends on the parameters selected and is subject to significant errors due to the limited number of clusters (1–29) detected in any one analysis. To address these problems, Marceau et al. revisited this problem by APT. **Figure 70** shows the APT map of Cu and Mg obtained from the Al–1.1Cu–0.5 Mg (at.%) alloy that showed rapid hardness increase after aging for 60 s at 150 °C. The raw data shown in (a) does not show

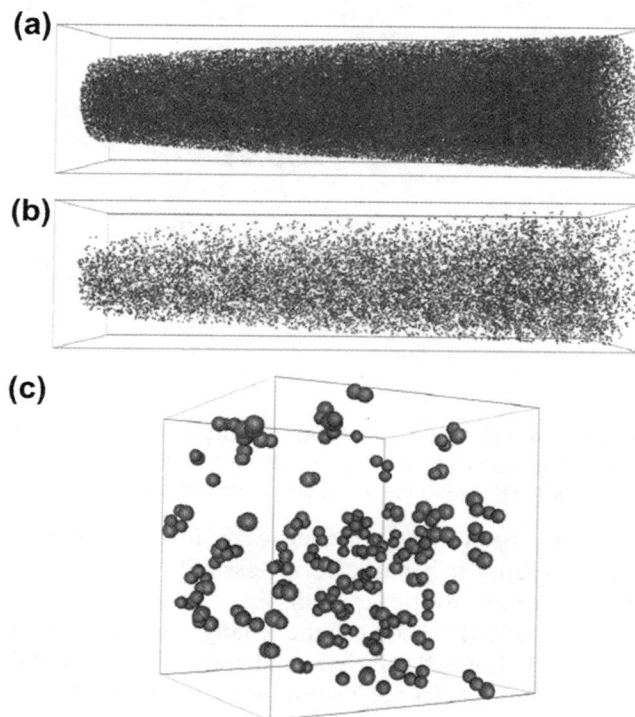

Figure 70   3D APT maps of Al–1.1Cu–0.5 Mg (at.%) after 60 s aging at 150 °C (orange = Cu atoms, blue = Mg atoms). (a) Raw data and (b) clustered data after application of the CL algorithm (these maps have dimensions 60 63 228 nm). (c) Enlarged 3D map (dimensions 10 10 10 nm) of individual clusters in (b). Ref: R.K.W. Marceau et al./Acta Materialia 58 (2010) 4923–4939). (For interpretation of the references to color in this figure legend, the reader is referred to the online version of this book.)

any indication of the presence of clusters and they attributed this to the small size of the clusters. By employing a refined maximum separation method called "Core-Linkage algorithm" (Stephenson et al., 2007) they selected the atoms that were classified as clustered atoms as shown in (b). One thing to note is that maximum separation method may create clusters artificially if the solute content is high. The alloy contains 1.1 and 0.5 at.% of Cu and Mg, so the distinction between clusters and solute atoms becomes a rather statistical concept. In fact, clusters were detected even in the as-quenched state and the cluster size is very small. The difference in the early work and this work would be only the difference of the definition of clusters. Assuming that the cluster analyses were valid, there is still no convincing explanation for the noticeable change in hardness in a short time. Marceau et al. (2010) also worked on the effect of secondary aging that at lower temperature after an artificial aging at elevated temperature and concluded that the secondary aging is due to the clusters of solute atom that supersaturated at lower temperature.

15.4.4.5 Al–Mg–Si Alloys

Al–Mg–Si alloys are one of the most widely used medium strength age-hardenable aluminum alloys, 6xxx series, and their precipitation processes and the kinetics of the age-hardening effect have been a subject of numerous studies since the 1960s (Thomas, 1961; Pashley et al., 1966, 1967; Ceresara et al., 1969/70). Due to the recent demand for the weight reduction of automobiles, the research interest in the kinetics of age hardening of Al–Mg–Si alloys has been revived for the last few decades (Moons et al., 1996). The precipitation sequence of this system is (Dutta and Allen, 1991):

$$\text{supersaturated } \alpha \rightarrow \text{solute cluster} \rightarrow \text{G.P. zones(spherical)} \rightarrow \beta''(\text{needle}) \rightarrow \beta'(\text{rod}) \rightarrow \beta(\text{Mg}_2\text{Si}).$$

However, there is no clear distinction between solute clusters and GP zones. The aluminum alloys for automobile body sheet applications have to be age-hardened during the paint-baking cycle, i.e. 30 min at around 175 °C. However, the age-hardening response after storage at room temperature (natural aging) becomes sluggish compared to that occurs right after a solution heat treatment. Pashley et al. (1967) studied the kinetics of the two-step aging process, and they proposed that the clusters that form during natural aging affect the age-hardening kinetics in the subsequent artificial aging. The presence of co-clusters of Mg and Cu was examined using the APFIM by Edwards et al. (1994, 1998), but naturally aged alloy was not examined in their work. Direct evidence for the presence of such clusters during natural aging was reported by Murayama et al. by 1DAP (Murayama et al., 1998). Subsequent 3DAP studies of Al–Mg–Si (Murayama and Hono, 1999) alloys revealed that the solute ratio in clusters, GP zones, and β'' precipitates vary depending on the alloy compositions, i.e. the atomic ratio of Mg:Si in these metastable precipitates is the same as that of the alloy composition. Because of this, larger number of clusters or GP zones are formed in the Si-excess alloy, resulting in larger age-hardening response. Although the atomic ratios in the clusters and the GP zones do not change during aging, the content of Si and Mg atoms increases as they grow. Based on these observations, they concluded that there is no discrete distinction between the clusters and the GP zones; the latter can be imaged with TEM as they are more concentrated with solute atoms. They also found that the atomic ratios in β'' precipitates are also the same as those in GP zones.

In commercial Al–Mg–Si alloys used for automobile body sheet, Cu is added as quaternary element to improve formability (Gupta et al., 1996), in which Q-phase coexists with β (Mg$_2$Si) in the equilibrium state (Chakrabarti et al., 1998). Cu addition in Al–Mg–Si alloy generally increases kinetics of precipitation and bake-hardening properties (Pashley et al., 1966). One explanation for this effect is the reduction of the solubility of Mg$_2$Si by addition of Cu (Collins, 1957/58). Other explanation is that Cu reduces the rate of migration of Mg and Si atoms, retarding the formation of Mg–Si clusters (Chatterjee and Entwistle, 1973). Miao and Laughlin (2000) pointed out that Q' precipitate following precipitation of β'', but this occurs only in the late stage of the precipitation sequence; thus the precipitation of Q' does not appear to explain the early aging stage relevant to bake hardening. In order to explain the effect of Cu, Murayama et al. (1999) studied Cu distribution in the early aging stage of Al–0.6Mg–1.2Si–0.4Cu alloy by 3DAP in conjunction with the bake-hardening conditions and reported that Cu atoms are not associated with the GP zones, but are partitioned in the β''. Based on this observation, they proposed that Cu primarily enhances the precipitation of the β'' phase, which is most effective precipitate for hardening, instead of the GP zones. Since then, the age-hardening processes in Al–Mg–Si alloys have been extensively studied by APT (Philos., 2007, 2006; Marceau et al., 2010; Appl. and Phys., 2009; Pogatscher et al., 2011; Torster et al., 2011; Mater., 2012a; Mater., 2013) but the conclusions are more or less the same as those reported in earlier work. **Figure 71** shows recent AP tomography results

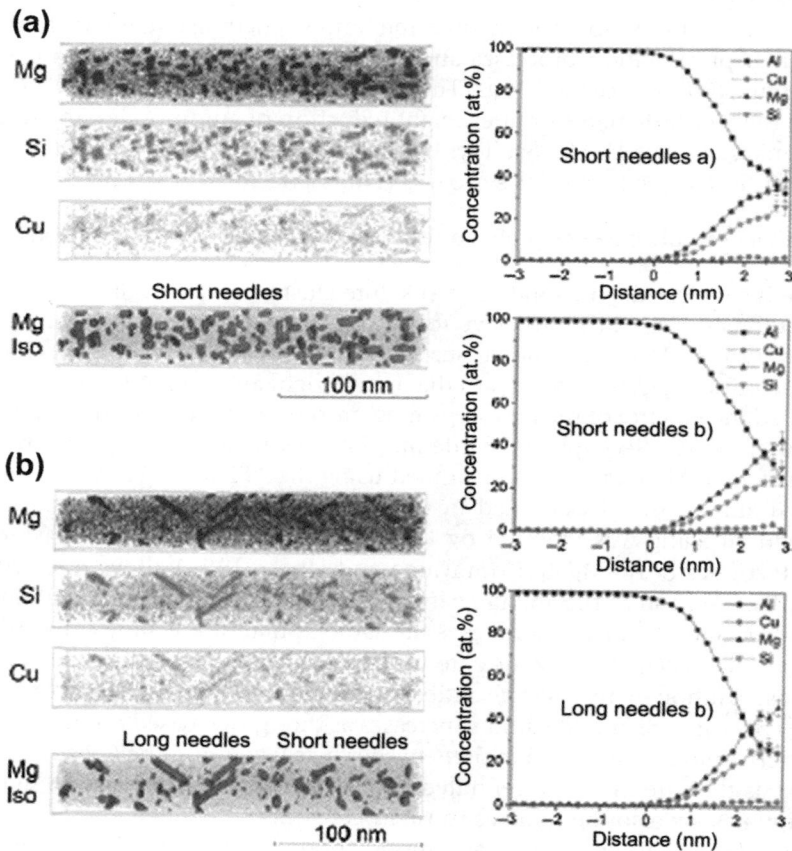

Figure 71 Three-dimensional (3D) reconstructions of the atom positions for Mg, Si and Cu in an Al–0.9Mg–0.6Si–0.1Cu (at.%) alloy and isoconcentration surfaces of Mg with corresponding proximity concentration profiles for GP zones and β'' precipitates. (a) Direct artificial aging and (b) artificial aging after long-term natural pre-aging. Ref: S. Pogatscher et al. Scr. Mater. 68 (2013) 158. (For color version of this figure, the reader is referred to the online version of this book.)

obtained from Al–0.9Mg–0.6Si–0.1Cu (at.%) alloy (a) peak-aged by direct artificial aging and (b) peak-aged after pre-aging. GP zones (spherical) and b'' precipitates can be clearly recognized based on the difference in morphology. These data have clearly shown that pre-aging enhances the precipitation of b'' with respect to the GP zones. This explains the enhanced peak-hardness in the two-step aged Al–Mg–Si-based alloy. The proximity concentration profiles show there's not much difference in the compositions of GP zones and β'' precipitates.

15.4.4.5 Al–Zn–Mg alloys

Al–Zn–Mg-based alloys are widely used as 7000 series high-strength aluminum alloys, and their precipitation sequence is accepted as (Lorimer and Nicholson, 1966)

$$\text{supersaturated } \alpha \rightarrow \text{GP zones(spherical)} \rightarrow \eta'(\text{plate}) \rightarrow \eta(\text{Zn}_2\text{Mg}).$$

For alloys with Zn:Mg $\sim$ 1:1, the equilibrium phase is $T((\text{Al,Zn})_{49}\text{Mg}_{32})$, and its metastable phase, T', precipitates prior to the precipitation of the T phase (Lorimer, 1978). Since the major contributor of strengthening is η' and their density is strongly influenced by their precursors, the compositions of GP zones and η' have been a subject of many investigations using conventional AP (Ortner et al., 1988; Ortner and Grovenor, 1988; Brenner et al., 1991b; Hono et al., 1992f; Warren et al., 1992). In these 1DAP investigations, substantial amount of Al was commonly detected from the precipitate particles and the Zn:Mg ratio of the precipitates determined by 1DAP significantly deviates from the stoichiometry of η, MgZn_2. Large amount of Al in the GP zones and η' particles was initially attributed to the matrix contribution, i.e. it was due to the Al atoms collected from the matrix phase when the probe hole cover both the matrix and precipitates due to the smallness of the precipitates. In the random area analysis mode using 1DAP, it is difficult to judge whether or not atoms were corrected from only the precipitates. Unlike 1DAP, it is possible to select the analysis area exclusively from inside the precipitates. However, substantial amount of Al atoms is detected even from relatively large particles in which matrix contribution due to the evaporation aberration effect is not expected. A typical example of 3DAP analysis result of η' precipitates is shown in **Figure 72**. Two precipitates are observed in this 3DAP data—the Zn:Mg ratios in these two precipitates are different. The one having Zn:Mg ratio 2:1 is interpreted as η, and the other is η'. Although the compositions were determined from the atoms only within the precipitate, almost 40 at.%Al is detected from these precipitates. Bigot et al. (1996b, 1997) also reported that a substantial amount of Al is incorporated in these phases and that the Zn:Mg ratio of η is not consistent with that is expected from the stoichiometry of Zn_2Mg. More recent 3DAP work by Maloney et al. (1999) reported that the Zn:Mg ratio in equilibrium phase η is 2:1. On the other hand, the Zn:Mg ratio of the GP zones and η' lie between 1:1 and 1.5:1, and they found that it has a good correlation with the solute ratio of the parent alloy as shown in **Figure 72**. More recent 3DAP work by Stiller et al. (1999) reported similar results.

Gradual enrichment of solute in metastable phases while maintaining the solute ratio constant is consistent with the results obtained in Al–Mg–Si alloy (Murayama and Hono, 1999), where the solute ratio of Mg:Si has good correlation with the alloy composition. Since GP zones are highly metastable aggregates of solute atoms, it may be reasonable that these clusters have the same atomic ratio with the alloy composition. The clusters are formed by the aggregation of available atoms nearby, keeping the structure isomorphous with the matrix. Thus, the number of available atoms within the diffusion field may determine their composition. As aging goes on, the clusters increase the solute content, and when solute content become high enough to form a distinct phase, they transform structurally to more stable phase having more stoichiometric composition. Recent 3DAP analysis results of solute clusters, GP

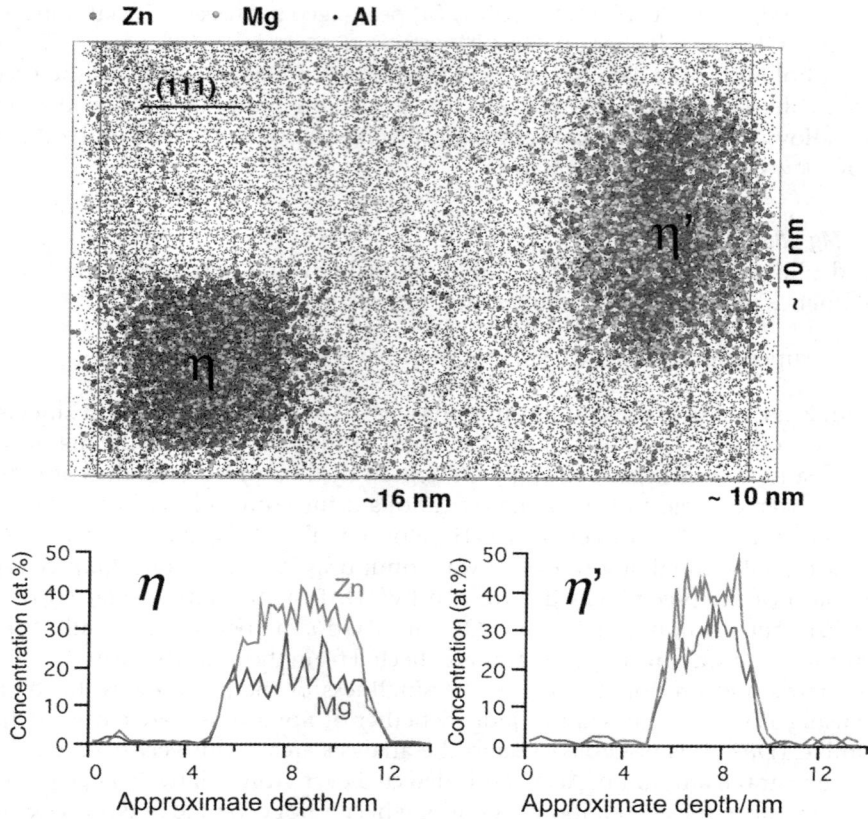

Figure 72 3DAP elemental map of Al–2.1Zn–1.7 Mg alloy aged at 150 °C for 45 h. The concentration depth profiles calculated from the two precipitates are shown. Ref: SK Maloney, K Hono, IJ Polmear, SP Ringer. Scr. Mater. 41, 1031 (1999). (For color version of this figure, the reader is referred to the online version of this book.)

zones and metastable phases as well as stable phases strongly suggest this view. Precipitation sequence is a continuous evolution from solute clusters to equilibrium phases, initially accompanied by solute enrichment followed by structural transformation.

15.4.5   Magnesium Alloys

More than 95% of magnesium alloys are used as cast alloy. Cast alloys usually contain various types of intermetallic compounds at grain boundaries, which are useful in suppressing the GB slip at elevated temperature for improving creep properties. Since the microstructures of cast alloys are rather coarse, they had not become subjects for AP studies. Recently, the research interest in age-hardenable magnesium alloys has been revived because of increasing demand of high-strength wrought magnesium alloys for weight reduction of transportation vehicles (Hono et al., 2010). The age-hardening response of Mg alloys can be promoted by the trace additions of alloying elements (microalloying) and by pre-aging (Hall, 1968; Park et al., 2003; Bettles et al., 2004; Mendis et al., 2008; Mendis et al.,

2009; Sasaki et al., 2006a; Nie and Muddle, 1997; Gao et al., 2005; Oh et al., 2005a; Oh-ishi et al., 2009a; Nie et al., 2005; Oh-ishi et al., 2008); therefore the demand of the quantitative analyses of alloying elements in the near atomic resolution to understand the solute clustering and precipitation processes is increasing for the development of wrought magnesium alloys. In Mg–RE–Zn (RE: rare earth) and Mg–Ca–Zn alloys, the precipitations of internally ordered GP zones were recently reported (Gao et al., 2005; Oh et al., 2005a; Oh-ishi et al., 2009a; Nie et al., 2005; Ping et al., 2003; Nie et al., 2008), but the quantitative analysis of the composition of these solute clusters is not possible using the spectroscopy techniques on transmission electron microscopes (TEMs). AP tomography is an ideal experimental technique to detect solute clusters in alloys, but a limited number of AP investigations had been reported on magnesium alloys until recently. This is because the FIM of magnesium alloys had been thought to be difficult because of the low evaporation field. Another reason was the difficulty in specimen preparations using electropolishing. The first 3DAP analysis of magnesium alloy was applied to the analysis of small monoatomic layer precipitates in a Mg–RE–Zn alloy using a voltage–pulse AP in 2002 (Ping et al., 2002). This work made the best use of the atomic layer resolution of the AP technique, by which periodic stacking of Y-rich and Zn-rich layers on a long period-ordered structure was quantitatively analyzed. Thereafter, 3DAP has been employed to the analyses of fine precipitates in Mg–RE–Zn–Zr (Ping et al., 2003), Mg–Ca–Zn (Oh et al., 2005a), Mg–Gd–Zn (Nie et al., 2008), Mg–Zn–Al (Oh-ishi et al., 2008), Mg–Zn–Ag–Ca(–Zr) (Mendis et al., 2009) with voltage pulse or laser-assisted 3DAP. However, the yield of successful analysis using the voltage-pulse AP and even with pulsed laser mode with infrared (IR) laser was very low. This yield of success has recently been improved by employing ultraviolet (UV) femtosecond laser to assist field evaporation to carry out the AP analyses of magnesium alloys (Oh-ishi et al., 2011).

15.4.5.1 Microalloying Effect in Mg Alloys

The Mg–Zn-based ZK60 [Mg–2.4Zn–0.16Zr (at.%)] alloy is a commercially used wrought Mg alloy with a relatively high tensile yield strength of ~ 240 MPa. The age-hardening response of this alloy is minor because of the coarse dispersion of rod-like precipitates along c-axis of hcp Mg. Recently, Mendis et al. reported that the age-hardening response of an Mg–2.4Zn alloy was substantially enhanced by a combined addition of Ag and Ca (Mendis et al., 2009). Three-dimensional AP analysis revealed that Ca atoms co-segregate with Zn atoms in the early stage, and they partition into the β' phase with Ag in the peak-aged condition. These microalloyed elements were found to refine the dispersion and the size of the precipitates. Later, more detailed analysis of the effect of Ag and Ca microalllying to the Mg–Zn alloy has been performed by Bhattacharjee et al. using 3DAP (Bhattacharjee et al., 2013). They compared the precipitation sequences in Mg–2.4Zn alloys aged at two temperatures. While Zn-rich GP zones were observed in the sample aged at low temperature, i.e. 70 °C, β'' particles were found to precipitate directly from a supersaturated solid solution at high temperature, i.e. 160 °C. The formation of the Zn-rich GP zones by low-temperature aging was also reported by Oh-ishi et al. in Mg–Zn–Al alloy (Oh-ishi et al., 2008). These GP zones that formed at low temperature (pre-aging) can act as the nucleation sites for β'_1 precipitates, thereby increasing the number density of the precipitates on artificial aging at high temperature (two-step aging).

Figure 73 shows 3DAP map of Zn, Ag, Ca in a Mg–2.4Zn–0.1Ag–0.1Ca (at%) alloy (ZQX600) aged at 160° together with the age-hardening curve and a TEM image of the sample aged for 8 h. After 10 min aging, there is no evidence for solute clusters. However, after 30 min aging, just before the age hardening becomes evident, a large number density of Zn-rich GP zones associated with Ag atoms are observed. This observation is interesting, as the GP zones are not observed at this temperature in the binary Mg–Zn alloy. After 8 h aging, rod-shaped β'_1 precipitates are observed with a much higher

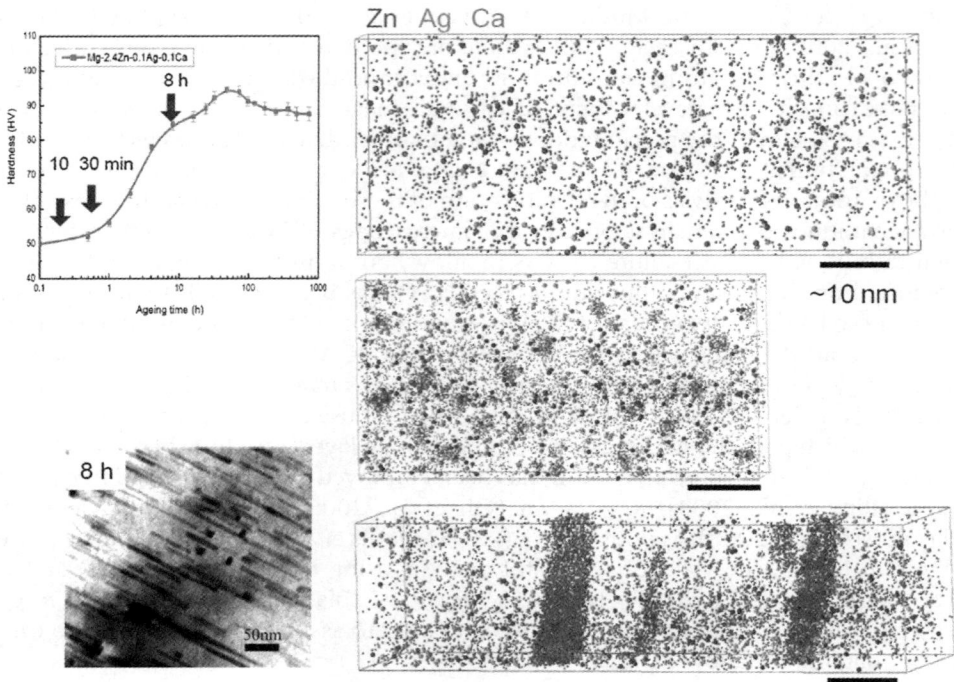

Figure 73 Age-hardening curve and 3DAP map of Zn, Ag and Ca for the Mg–Zn–Ca–Ag alloy (ZQX600) aged at 160 °C for (a) 10 min, (b) 30 min, and (c) 8 h. Zn-rich solute clusters or GP zones associated with Ag form after 30 min aging (b) from the initially homogenous solid solution (a) before distinct age-hardening occurs. Rod-like β'' precipitates are observed when age-hardening appear. From Bhattacharjee et al. (2013). (For color version of this figure, the reader is referred to the online version of this book.)

density compared with that in the directly aged samples. Base on these results, they concluding that the addition of Ag and Ca shift the solvus for GP zones to the lower Zn side in the Mg–Zn system.

In commercial ZK60 alloys, a trace amount of Zr has been added to refine the grain size of cast alloy. So it was generally believed that Zr does not give much influence on age hardening. However, Mendis et al. reported a small enhancement of the peak hardness in the Zr-containing Mg–2.4Zn–0.1Ag–0.1Ca–0.16Zr (at.%) alloy (ZKQX6000) compared to the Zr-free ZQX600. Oh-ishi et al. (2009b) investigated the mechanical properties of extruded ZQX600 and ZKQX6000 alloys and reported the partitioning of Zr atoms in the β'_1 precipitates in the alloy extruded at 350 °C. More importantly, they reported large differences in the extruded microstructure between ZKQX6000 and ZQX600. With the presence of a small amount of Zr (<0.1at.%), the extruded microstructure became more than one order of magnitude smaller than that of the Zr-free alloy. This substantial change in recrystallized microstructure was attributed to the dispersion of Zr-containing β'_1 particles that precipitate dynamically during extrusion at the elevated temperature, 350 °C (Oh-ishi et al., 2009b). This work triggered questions on the role of Zr in ZK60 compared to binary Z6 alloy. As mentioned earlier, a small amount of Zr is microallyed to Mg–Zn alloy as a grain size refiner in order to refine the grain size of cast ingots. So the roles of Zr on the refinement of the recrystallized microstructure and also its influence on the phase equilibrium in Mg–Zn system have long been overlooked. Bhattacharjee et al. (2012) recently revisited

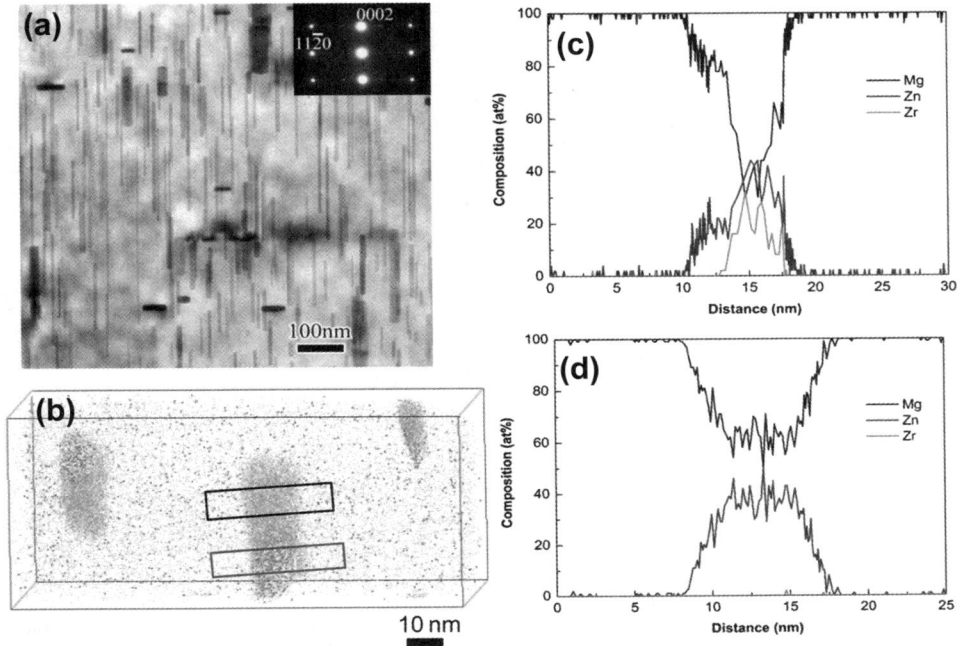

Figure 74 (a) TEM image of the low-temperature aging heat-treated ZK60 alloy at 160 °C for 120 h from the ⟨10 0⟩ direction and its corresponding 3DAP map (b); (c and d) show the composition profile of the Zr and Zn-rich regions of a selected β_1' precipitate indicated in (b) by a blue and black rectangle respectively. Ref: Scr. Mater. 2012: 67: 967. (For color version of this figure, the reader is referred to the online version of this book.)

this issue and found that Zr stabilized the β_1' precipitates even at the solution treatment temperature for Mg–Zn binary alloy, i.e. 350 °C. Zr substitute Zn of MgZn β_1' making the composition to Mg(Zn,Zr) but the structure is identical to the β_1' phase in the binary alloy. These β_1' precipitates that are insoluble even at the solution treatment temperature are spherical, and they act as heterogeneous nucleation sites for β_1' rods as shown in **Figure 74**. Thus, Zr atoms are detected only at one ends of the β_1' rods.

The Mg–Sn system is also a precipitation-hardening system that the relatively large solid solubility at high temperature becomes almost negligible at room temperature. The equilibrium Mg_2Sn precipitate forms from the solid solution with a coarse microstructure so the precipitation hardening is very moderate and sluggish. However, the Mg_2Sn phase has a high melting point of ~ 770 °C, which is higher than most of MgxRey compounds. This makes the Mg–Sn-based alloys potentially attractive for high-temperature applications. The moderate age-hardening response can be enhanced by the Zn addition (Sasaki et al., 2006b). Mendis et al. reported that the trace addition of Na to Mg–Sn–Zn alloy further enhanced the age-hardening response including the precipitation kinetics (Mendis et al., 2006). The age-hardening response was further improved by optimizing the composition of Mg–Sn–Al–Zn alloys (Elsayed et al., 2013a). Elsayed et al. added a trace amount of Na to the optimized Mg–Sn–Al–Zn (TAZ1030) alloy and enhance not only the peak hardness but also the precipitation kinetics (Elsayed et al., 2013b). **Figure 75** shows the 3DAP map of the peak-aged TAZ10330 + Na alloy. Na clusters can be seen with red sphere. In direct contact with these clusters, Mg_2Sn precipitates can be observed, suggesting Sn clusters that form in the early stage of

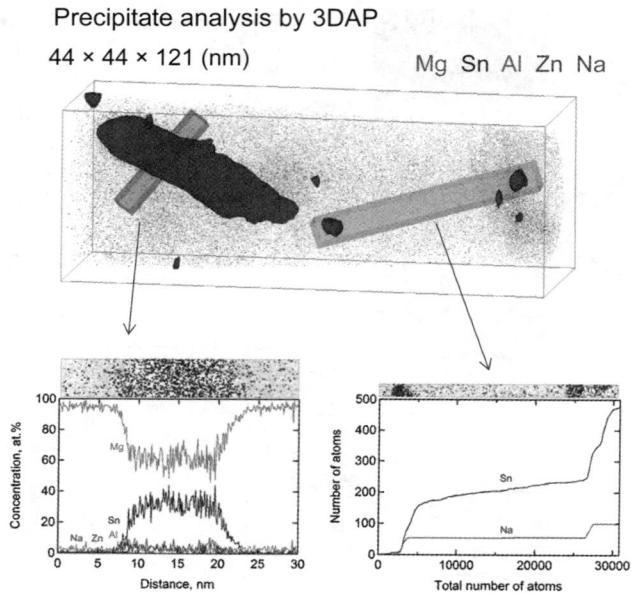

Precipitate analysis by 3DAP

44 × 44 × 121 (nm)

Mg Sn Al Zn Na

Figure 75 Three-dimensional AP mapping of peak-aged Mg–10Sn–3Al–1Zn–0.1Na at 160 °C: (a) Sn-rich Mg_2Sn zones, with different alloying element distributions; (b) magnified map showing the composition of a Sn-rich region; (c) compositional profile of the main alloying elements at the matrix/precipitate interface; (d) corresponding ladder diagram for Na and Sn atoms across a selected volume. From Elsayed et al. (2013b). (For color version of this figure, the reader is referred to the online version of this book.)

precipitation provided the heterogeneous nucleation sites for the Mg_2Sn precipitates. The noteworthy finding was that both the kinetics and the hardening increment are significantly enhanced by the trace addition of Na.

15.4.5.2 Quantitative Analysis of Precipitates in Mg Alloys

Rare earth elements (RE) are common additives to magnesium alloys to improve creep properties. All Mg–RE systems show large temperature-dependence in solubility in the α-Mg region, so they are precipitation hardening. However, the age-hardening response of Mg–RE systems are very small because of sluggish diffusion of RE and low number density of precipitates. Most of Mg_xRE_y compounds have large misfit with the matrix, so the nucleation barrier is very large, thereby the number density of precipitates become very small. The addition of Zn to Mg–RE alloys, e.g. Mg–Gd and Mg–Y, results in significant enhancement in the age-hardening response (Nie et al., 2005; Yamasaki et al., 2005; Honma et al., 2007). The enhancement has been attributed to the fine dispersion of metastable plate-like precipitates or long period stacking-ordered (LPSO) phase. The LPSO phase was first reported in rapidly solidified Mg–2Y–1Zn (at.%) alloy (Honma et al., 2007; Kawamura et al., 2001), and now it is known to be rather common stable phase observed in cast and aged alloys (Yamasaki et al., 2007). The first 3DAP investigation of magnesium alloy was performed for the Mg–2Y–1Zn alloy in which a large fraction of LPSO phase existed in nanocrystalline Mg grains (Ping et al., 2002). **Figure 76** shows TEM and 3DAP analysis result of the nanocrystalline Mg containing a large fraction of LPSO phase. The grain size in about ~70 nm and contrast arising from stacking faults on basal planes is observed in the bright

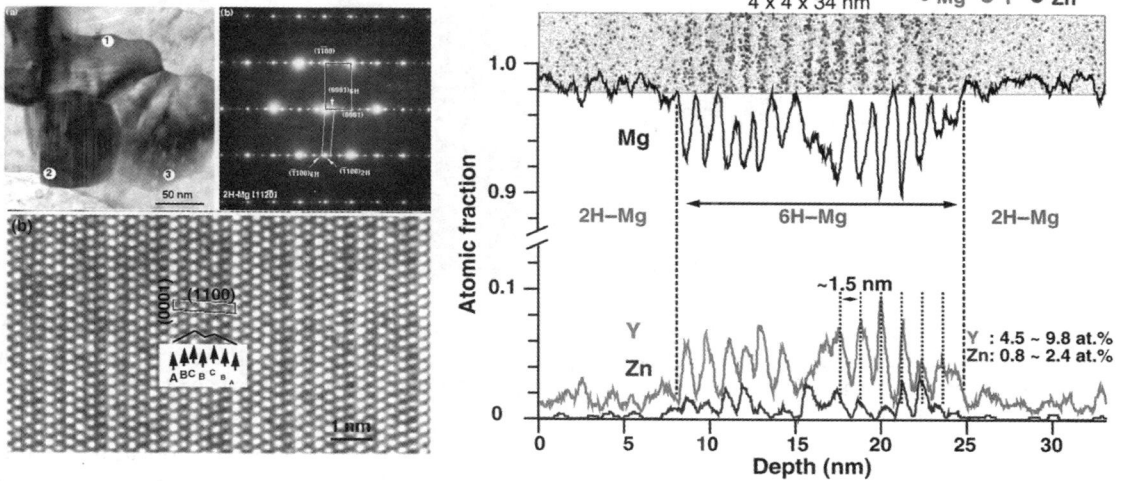

Figure 76 (a) Bright-field TEM image, revealing that a high density of stacking faults exists within the nanosized grains. (b) Selected-area electron diffraction pattern taken from grain 2. (c) HREM image of the 6H–Mg phase, and (d) 3DAP map of Zn and Mg atoms. Ref: D. H. Ping, K. Hono, Y. Kawamura, and A. Inoue, Philos. Mag. Lett. 2002: 82: 543. (For color version of this figure, the reader is referred to the online version of this book.)

field image in **Figure 76(a)**. The diffraction pattern shows the stacking period is six times larger than the unit cell of Mg. The HREM image shows periodic change in contrast in the basal planes. The 3DAP map in **Figure 76(d)** shows enrichment of Y and Zn on the same planes and the concentrations of Y and Zn in these planes are quantitatively characterized. This 3DAP result was consistent with the HAADF image in which the planes rich in Y and Zn were observed with two atomic layers' thickness (Abe et al., 2002). At that time, there were speculations as to the chemical compositions of the LPSO phase and this work has clearly shown the periodic stacking of Zn and Y-rich layers on the basal plane of hcp-Mg alloy.

When RE/Zn ratio is lower than 0.5 and the RE composition is no more than 2 at.%, plate-like precipitate with similar internal ordered structure with the LPSO phase precipitate. Ping et al. characterized the aged Mg–RE–Zn–Zr alloy using 3DAP and reported plate-like GP zones of nearly monoatomic layer from in the peak-hardness condition (Ping et al., 2003). Using selected area diffraction pattern, he concluded that there is an internal chemical order within the planes. This type of structure was later found to be rather common in Mg–Gd–Zn (Nie et al., 2008), Mg–Ca–Zn (Oh et al., 2005a; Oh-ishi et al., 2009a) and Mg–Ca–Al (Jayaraj et al., 2010; Homma et al., 2011) alloys. Both RE and Ca are oversized with respect to Mg. The precipitation of equilibrium phase from the supersaturated solid solution is difficult due to the structural difference of the Mg_xRE_y and Mg_2Ca phases with the matrix phase. As both RE and Ca have positive enthalpy of mixing with limited solubility, the supersaturated RE and Ca tend to form clusters prior to the precipitation of the equilibrium phases. If Zn is added in the alloys, Zn segregation next to RE and Ca atoms reduces the strain field, so they end up with the formation of internal order within the basal plane. In order to reduce the interfacial and strain energies, the RE–Zn and Ca–Zn precipitates tend to lie on the basal planes coherent with the matrix. Therefore, the single-layer plate of Mg–RE–Zn, Mg–Ca–Zn and Mg–Ca–Al alloys can be referred as (internally) ordered GP zones. **Figure 77** shows 3DAP map of Mg, Ca and Zn in an aged Mg–0.3Ca–0.6Zn (at.%) alloy. The basal planes can be resolved in the Mg map and a number of fine plate-like precipitates

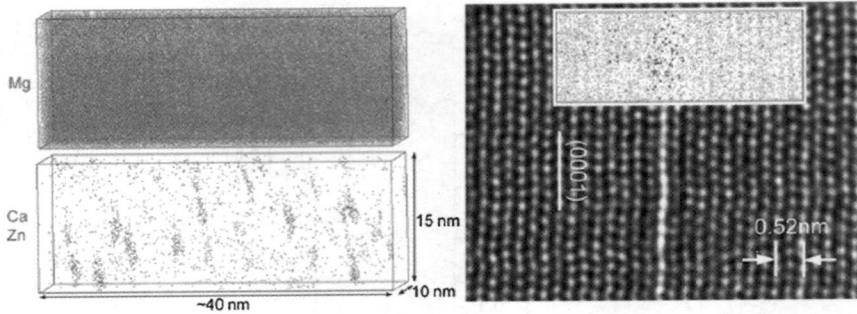

Figure 77   (a) 3DAP map sand (b) HAADF image of Mg–0.3Ca 0.6Zn alloy aged at 200 °C for 1.4 h (peak-aging condition). The atom map for Ca and Zn taken from the region near the precipitate is superimposed on the HAADF image. From Homma et al. (2011). (For color version of this figure, the reader is referred to the online version of this book.)

enriched with Ca and Zn can be seen in the Ca and Zn map. The HAADF image clearly shows that the precipitate is a monoatomic layer while the 3DAP map shows a diffuse nature. This is because of the difference in the evaporation field of Zn and Mg. Therefore, accurate atomic layer resolution is lost near the precipitate. This example also shows the importance of the complementary use of 3DAP and TEM/STEM. Modern STEM with aberration corrector has atomic layer resolution, so the structural information should be obtained using TEM/STEM. On the other hand, the quantitative analysis of local compositions can be better done with 3DAP with a subnanometer resolution.

In order to strengthen hcp-Mg with precipitation hardening, it is considered to be effective to dispers plate-like precipitates on prismatic planes (Nie, 2001). In fact, the precipitation of prismatic plates gives rise to significant precipitation hardening in WE54 (Mg–5wt%Y–4wt%RE) alloy (Nie and Muddle, 2000). However, the most of metastable precipitates in Mg–RE and Mg–Ca-based alloy precipitate on basal planes. Mendis et al. found the In addition to Mg–Ca alloy change the habit plane of the plate-like precipitate to the prismatic plate as shown in **Figure 78** (Mendis et al., 2011). This change in the habit plane is considered to be due to the difference in the strain. The atomic radius of In is nearly the same as that of Mg, so the strain due to the formation of Ca–In clusters will be totally different from those for Ca–Zn and Ca–Al. This would have caused the selection of a different habit plane. By the control of the habit planes of the plate-like precipitate, designing of highly age-hardenable alloy may be possible. The microalloying effect in Mg–Ca is recently reviewed by Mendis et al. (Mendis et al., 2012).

15.4.6   Amorphous and Nanocrystalline Alloys

Physical properties of metallic materials are sensitive to their microstructures; thus controlling microstructures by heat treatment, thermomechanical treatment and plastic deformation has been a popular method for improving mechanical, magnetic and electrical properties. The scale of the microstructures sometimes reach a nanometer scale, e.g. GP zones in high-strength aluminum alloys or heavily deformed lamellar structure in pearlitic steel wires. It has been known for a long time that the strength of materials becomes high when the grain size is refined. The relationship between the strength and the crystal grain size is known as Hall–Petch relations, and this has been used as a guiding principle for developing high-strength metallic materials in the past. In the conventional structural materials, the limit of the grain size refinement was above a 1 µm range using industrially viable processing methods. Due to the advent of new laboratory level nonequilibrium processing, such as mechanical alloying,

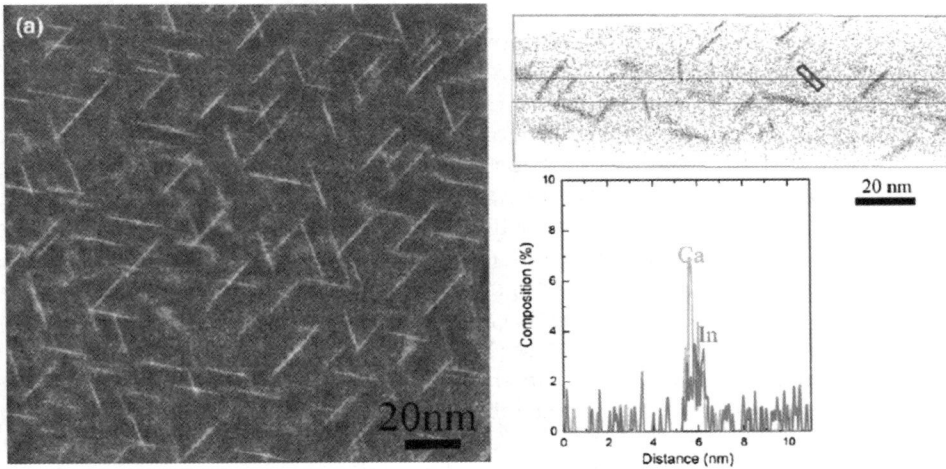

Figure 78 HAADF–STEM images of Mg–0.3 Ca–1.0 In alloy aged to peak hardness with an electron beam parallel to (a) [0001] and (3 DAP tomography results for the Mg–0.3 Ca–1.0 In alloy) aged to peak hardness (2 h at 473 K at 200 °C) (a) from the volume investigated, and (b) composition profile from the particle boxed in (a). From Mendis et al. (2011). (For color version of this figure, the reader is referred to the online version of this book.)

severe plastic deformation, and vapor deposition, it became possible to make the grain size much finer than that was obtained in an industrial scale. These laboratory scale nanomaterials have demonstrated unprecedented level of strength and unique magnetic properties as described in Chapter 19, this volume. The ultimate case of nanocrystalline alloy is amorphous in which there are no long-range order or grain boundaries. By using crystallization of amorphous alloy, nanocrystalline microstructure is often formed, and some of these materials find industrial applications such as nanocrystalline soft magnetic materials and nanocrystalline hard magnetic materials. In this section, we review the role of AP techniques that made substantial contributions in understanding the evolution of nanocrystalline materials.

15.4.6.1 *Amorphous Alloys and Metallic Glasses*

Amorphous alloys can be obtained either by rapid solidification or intense straining of the alloys composed of two or more constituent elements with negative heat of mixing. Near the eutectic composition between a primary solid solution and an intermetallic compound, the liquid phase becomes metastable within the composition range between the two instability lines of the solid/liquid phases. So if B element is forced to dissolve in the primary solid solution exceeding the instability line, the metastable liquid phase (amorphous solid) may appear by the solid-state transformation. This amorphization can be seen in mechanical alloying of various binary or multicomponent alloy systems. The negative heat of mixing between major constituent elements is essential for the formation a short-range order in the amorphous solid, many alloys that exhibit extraordinary glass-forming ability have icosahedral short-range ordering. The alloys that exhibit distinct glass transition during heating are called "metallic glass" analogous to oxide glass. Amorphous alloys and metallic glasses show common features like high strength at an elastic limit of about 2%, no plastic deformation occurs by dislocation motion so they are inherently brittle in tension, although some plasticity appears in some bulk metallic glasses by shear band deformation. Another common feature of amorphous alloys is an excellent corrosion resistance because of the uniformity of the composition, and soft magnetic properties because of the zero magnetocrystalline anisotropy. Some of these features find applications and the extraordinary glass-forming ability that can form bulk metallic glass has been the center of intensive research in the past two decades as described in Chapter 4, Volume I.

In order to understand the origin of the extraordinary glass-forming ability of particular compositions of alloys, extensive investigations have been done to analyze the structure of these glasses. However, the structures of glasses are defined by the short-range ordering of different atomic species, which can be better studied by the scattering methods in view of statistical accuracy. In order to obtain the information on short-range ordering using AP tomography, we need to assume that the positions of atoms are rigorously reconstructed, but this is not the case because of unavoidable trajectory aberrations and errors in reconstruction. In addition, the detection efficiency of 3DAP is limited by the open area ratio (see **Figure 14**) of the multichannel plate (MCP) detector, $\sim 60\%$. This causes another unreliability in getting information of short-range ordering using AP tomography. When amorphous solids are observed by APT, the distribution of atoms looks uniform as shown in **Figure 79**. However, some amorphous alloys show heterogeneous distribution of atoms in nanoscale dimension because of the presence of metastable miscibility gaps caused by the large difference in the pairs of negative enthalpy of mixing or by the addition of elements with positive enthalpy of mixing (Abe et al., 2006). These phase separations are sometimes attributed to the origin of the large plastic strain in the compression deformation, as share bands can be effectively refined. **Figure 80** shows typical example of the inhomogeneous distribution of Ag atoms that form clusters in the Cu–Zr metallic glass (Oh et al., 2005a,b). As seen from this example, APT has been used to detect the local chemical inhomogeneity in glassy alloys that is below the detection limit using TEM.

APT played critical roles in understanding the crystallization of amorphous alloys. There are three types of crystallization modes in amorphous alloys: polymorphous, eutectic and primary crystallizations. The crystallization mode is related to the compositions of alloys, at a given temperature (see **Figure 81**). The primary crystallization occurs without partitioning of solute elements when the free energy for the liquid phase is higher than that of the solid phase. The polymorphous crystallization occurs by changing the site of atoms in a few atomic distances at the amorphous/crystal interface without solute partitioning or long diffusion of solute atoms; hence, this type of crystallization is similar to the massive transformation in solid–solid reaction. The growth of the crystalline product is interfacial-controlled and thus the growth rate is fast in the polymorphous crystallization mode, resulting in relatively gross final microstructure. If the composition of the amorphous alloy is close to the eutectic where the driving force for the precipitations of a and b phases are nearly equal, eutectic crystallization occurs, by which two phases appear with a lamellar structure with a discontinuous reaction. This crystallization mode is most frequently observed in bulk metallic glasses, because good glass-forming alloys tend to have the compositions close to the eutectic composition. The amorphous alloy reaches equilibrium after crystallizing with the eutectic crystallization mode, thus this crystallization reaction occurs in one stage.

The amorphous alloy with the compositions between the polymorphous and eutectic regions, i.e. between b and c, and d and e in **Figure 81**, crystallizes with the primary crystallization mode. The composition of the primary crystal is determined so that the chemical driving force for nucleation is the maximum. Since the concentration of the nuclei is quite different from that of the matrix amorphous phase, diffusion and partitioning of solute atoms must be involved in this process. This crystallization process is called *primary crystallization*, and the microstructure obtained by this process is in general very fine because its grain growth is diffusion-controlled. The concentration profile of solute during the growth of the primary α particles is schematically shown in **Figure 82**. Since the size of crystal formed by

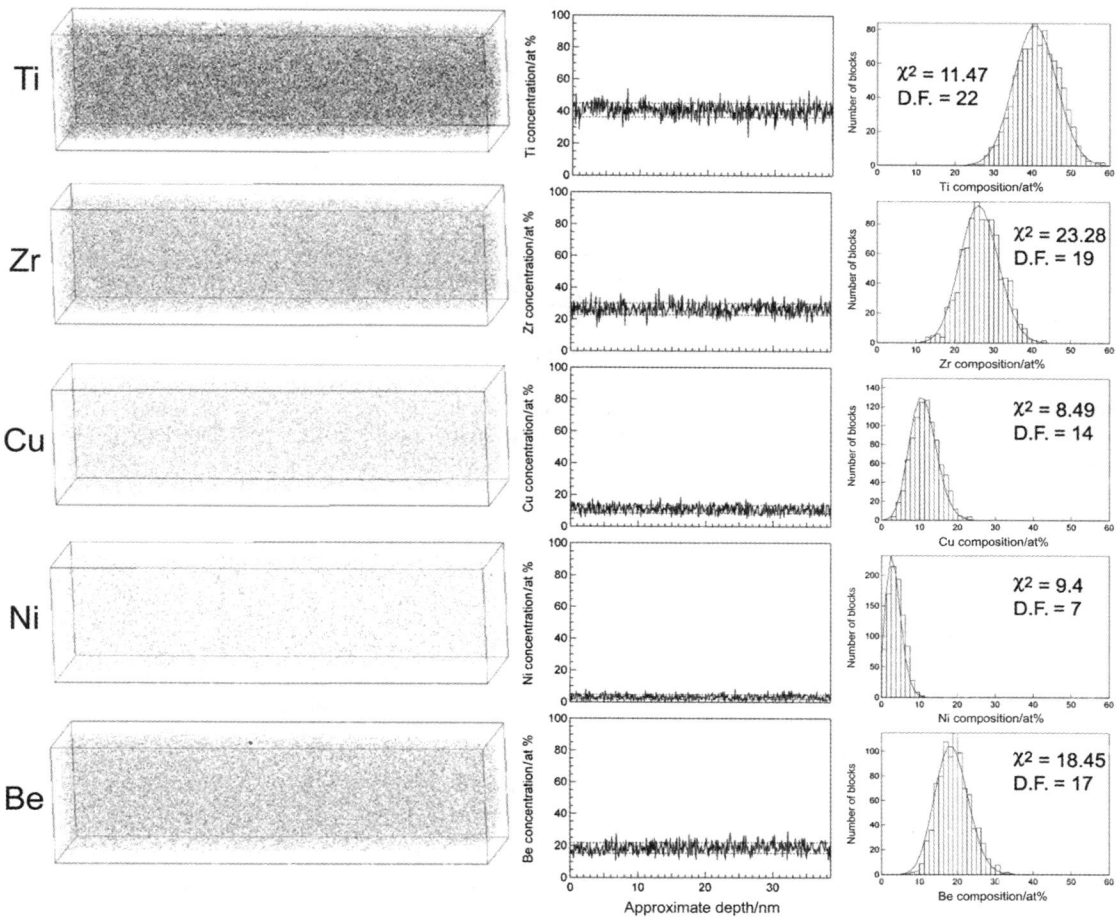

Figure 79 (a) Three-dimensional atom maps obtained by 3DAP analysis of the sample cast at 1273 K; (b) composition profiles and (c) frequency distributions calculated from whole region with the binomial distribution. (Ref: Ohkubo, T; Nagahama, D; Mukai, T; Hono, K. J. Mater. Res. 2007: 22: 1406). (For color version of this figure, the reader is referred to the online version of this book.)

the primary crystallization is often in nanoscale dimension, the experimental measurement of the local equilibrium at the interface is very difficult, but there is a such an experimental observation in Al–Ni–Ce amorphous/nanocrystalline alloy as shown in **Figure 82**. Initially, solute atom will build up at the α/amorphous interface. As the volume fraction of the α phase increases, solute partitioning into the amorphous phase progresses because they are rejected from the α phase. When solute concentration in the amorphous phase becomes uniform and reaches x_B, the volume fraction of the α phase becomes constant and the system reach the metastable equilibrium between α and the remaining amorphous phase.

In addition to these three crystallization modes, if thermodynamic conditions are met, phase decomposition may occur in the glassy state prior to the crystallization reaction. However, the phase decomposition in glassy state is possible only when the free-energy composition curve of the liquid

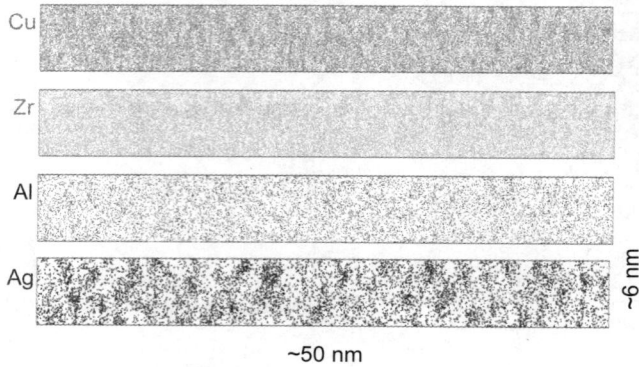

Figure 80 3DAP elemental maps for the as-cast $Cu_{43}Zr_{43}Al_7Ag_7$ alloy in a probed volume of 6 nm $\times$ 6 nm $\times$ 50 nm. The chemical fluctuations are shown clearly in the Ag elemental map. From Oh et al. (2005a). (For color version of this figure, the reader is referred to the online version of this book.)

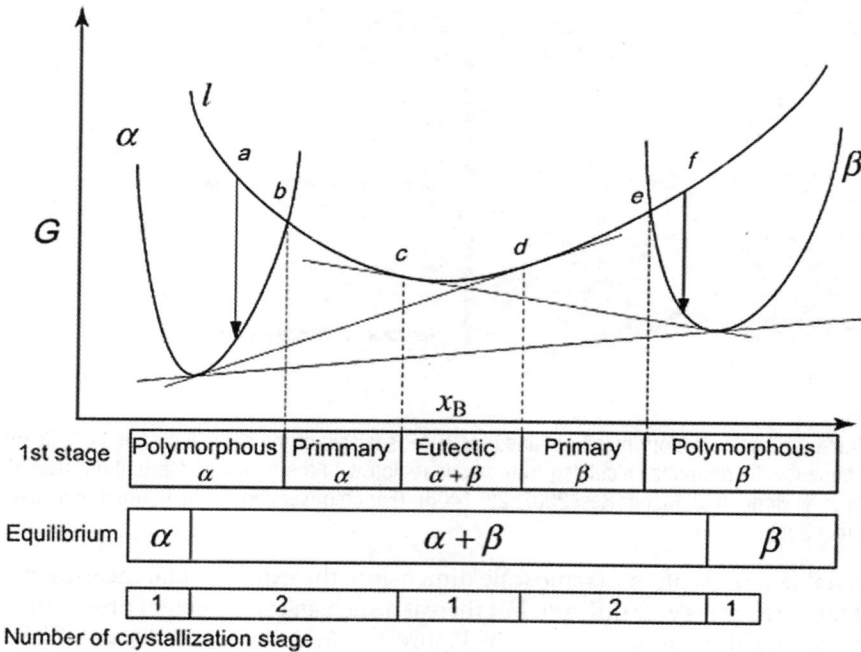

Figure 81 Hypothetical free-energy composition diagrams to explain possible modes of crystallization from an amorphous alloy. Hono (2002).

phase is curved upward due to large positive enthalpy of mixing of alloying elements in glassy state as shown in **Figure 81**. As the alloys with excellent glass-forming abilities (bulk metallic glasses) are mostly multicomponent alloys with the combination of the elements with negative entropy of mixing with strong short-range ordering tendency, miscibility gap due to positive mixing enthalpy hardly

(a)

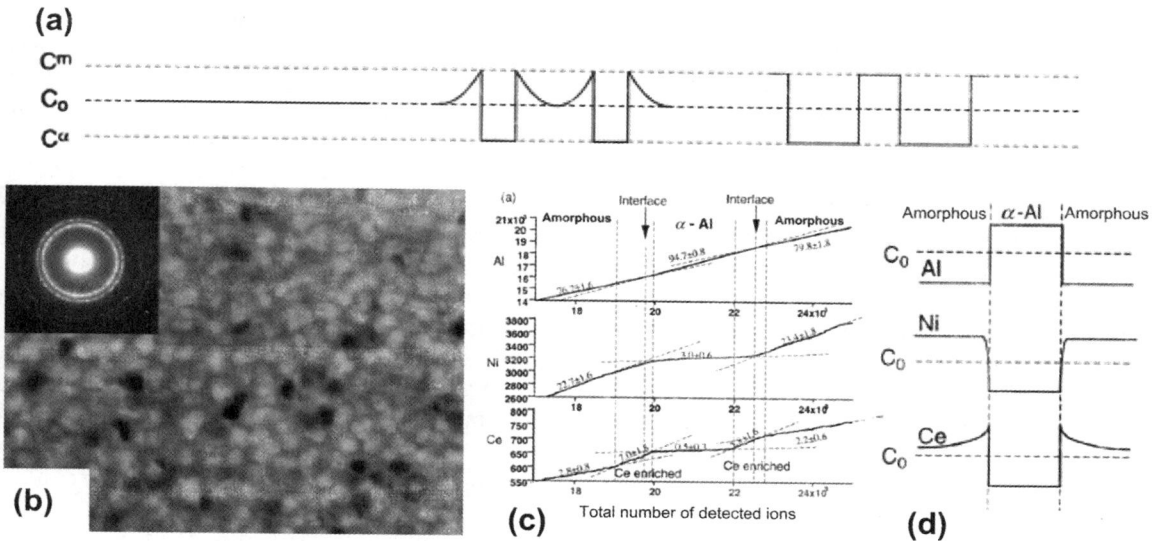

(b)

(c)   Total number of detected ions

(d)

Figure 82   (a) The evolution of concentration depth profile of a primary crystal from an amorphous phase to the two-phase equilibrium. (b) Nanocrystalline microstructure observed by the primary crystallization of an Al–10Ni–3Ce melt-spun ribbon, (c) AP integral depth profile (ladder diagram) through an α-Al nanocrystal in an Al–10Ni–3Ce alloy. (d) Schematic concentration depth profile deduced from the ladder diagram. From Hono (2002).

occurs. However, if the differences in the pair-mixing enthalpies are large in multicomponent alloys, miscibility gap can appear even in the alloys composed of atom pairs with all negative enthalpy of mixing (Abe et al., 2006). Although there are several reports on phase decomposition in glassy state of amorphous alloys, to the author's knowledge, there is little convincing evidence that the phase decomposition in fact occurs in the glassy state prior to the nucleation event. For example, the modulated microstructure used as evidence for phase decomposition in Ti–Zr–Be (Tanner and Ray, 1990) and Fe–Zr–B (Kim et al., 1995) are both too coarse to explain them as a result of the spinodal decomposition in a glassy state. In fact, 3DAP investigation combined with XRD, TEM and SAXS confirmed that the Ti–Zr–Be alloy does not undergo phase-decomposition in the supercooled liquid state, but they crystallize by the eutectic crystallization mode (Nagahama et al., 2003; Kumar et al., 2006). Relatively large-scale microstructure heterogeneity in Zr–Cu–Ni–Al metallic glass was also concluded to be the artifact during TEM specimen preparation (Chen et al., 2009c). A model work using Ag–Cu–Zr ternary alloy showed that glass-forming range and miscibility gap do not overlap (Kundig et al., 2006). The phase decomposition proposed in Zr-based amorphous (Schneider et al., 1996) alloy is very difficult to differentiate from nanocrystallization. To differentiate glass phase separation and primary crystallization, AP analysis for local compositional measurements and TEM for determining the onset of crystallization must be used complementarily (Martin et al., 2004; Kundig et al., 2005).

15.4.6.2 Nanoquasicrystalline Alloys

Multicomponent metallic glasses with excellent glass-forming ability and high thermal stability have attracted much attention in recent years due to the feasibility of obtaining bulk amorphous alloys using conventional solidification techniques at cooling rates as low as 100 K/s (Inoue, 1999; Johnson, 1999).

Among various bulk metallic glasses, Zr-based alloys exhibit a large supercooled liquid region, $\Delta T_x = T_x - T_g$, where T_x and T_g are the crystallization and glass transition temperatures, respectively (Zhang et al., 1991; Inoue et al., 1994; Pecker and Johnson, 1993). In general, there is correlation between the glass-forming ability and ΔT_x. Since ΔT_x is the difference between the crystallization and glass transition temperatures, the alloy with good glass-forming ability should have some retardation force against crystallization. This suggests that the nucleation barrier for crystallization tends to be high for the metallic glasses with a good thermal stability. On the other hand, nanocrystallization should occur only when the nucleation barrier is small, by which high nucleation rate can be achieved. However, there are several reports that nanocrystallization occurs even from the alloys with good glass-forming ability (Inoue et al., 1993; Fan and Inoue, 1997; Schneider et al., 1996). Dispersion of nanocrystalline particles in bulky alloy may lead to development of high-strength alloys; thus, nanocrystallization of bulk amorphous alloy has been studied with great scientific interest. A peculiar crystallization reaction where nanoquasicrystalline particle precipitate out from amorphous phase has been reported recently. This reaction is called *nano-quasicrystallization*. For example, the nanocrystallizaiton of some Zr-based bulk metallic glasses are now explained with nanoquasicrystallization. In the melt, the quasicrystalline local structure is believed to be a most stable structural unit. Thus the amorphous alloys with good glass-forming ability tend to contain icosahedral local structure (Takagi et al., 2001; Shimono and Onodera, 2007). Then, the quasicrysalli-zaiton can spontaneously occur by the growth of such quasicrystalline local structure. For the analysis of the quasicrystalline nanostructure and the influence of oxyge on it has widely been studied using 3DAP (Takagishi et al., 2010; Sakuraba et al., 2012; Li et al., 2013).

15.4.6.3 *Nanocrystalline Soft Magnetic Materials*

If an alloying element with a large positive enthalpy of mixing with the major alloying element is alloyed, phase separation in the glass may occur. Based on this, Kunding et al. showed convincing evidence for two-phase metallic glass (Kundig et al., 2004). However, these alloys do not show high glass-forming ability, so only ribbon samples can be prepared using melt-spinning. So glass-phase separation can be easily found in marginally glass form ring alloys, e.g. in Cu-containing Fe–Si–B–Nb and Fe–Zr–B alloys (Hono et al., 1999), in which Cu has large negative enthalpy of mixing with Fe. In these alloy systems, Cu atoms form cluster in the amorphous state, which is a sort of phase decomposition in the glassy state. Phase separation in an amorphous precursor is particulaly useful for controling the nanocrystalline microstructure of indusrial useful materials.

Nanocrystallization phenomena from amorphous precursor had been known for a long time, it did not find any applications until Yoshizawa et al. (1988) reported excellent soft magnetic properties from nanocrystalline $Fe_{77.5}Si_{13.5}B_9Nb_3Cu_1$ alloy. A typical electron micrograph and FIM image of the $Fe_{73.5}Si_{13.5}B_9Nb_3Cu_1$ nanocrystalline soft-magnetic alloy produced by crystallizing a melt-spun amorphous ribbon is shown in **Figure 83** (Hono et al., 1992g). This consists of the nanocrystalline $D0_3$ ordered Fe_3Si phase with an average grain size of approximately 10 nm. The grains are randomly oriented without any preferred orientation. In addition to the typical grains of ~ 10 nm, smaller grains can be recognized as indicated by the arrowheads. These are the FCC-Cu particles that precipitate in the early stage of crys-tallization (Hono et al., 1993b). Yoshizawa et al. (Yoshizawa and Yamauchi, 1990) reported that a combined addition of Nb and Cu is required for producing such nanocrystalline microstructure; thus the role of Nb and Cu for nanocrystallization has been of great interest. In order to investigate the role of these additives in nanocrystallization, the microstructural evolution by the crystallization of an amorphous $Fe_{73.5}Si_{13.5}B_9Nb_3Cu_1$ alloy has been investigated by 1DAP (Hono et al., 1992g) and more recently by 3DAP (Hono et al., 1999; Ohnuma et al., 2000). Three phases are present in the optimally heat-treated nanocrystalline microstructure as shown in **Figure 83(d)** (Hono et al., 1992g; Hono et al., 1999). The $D0_3$

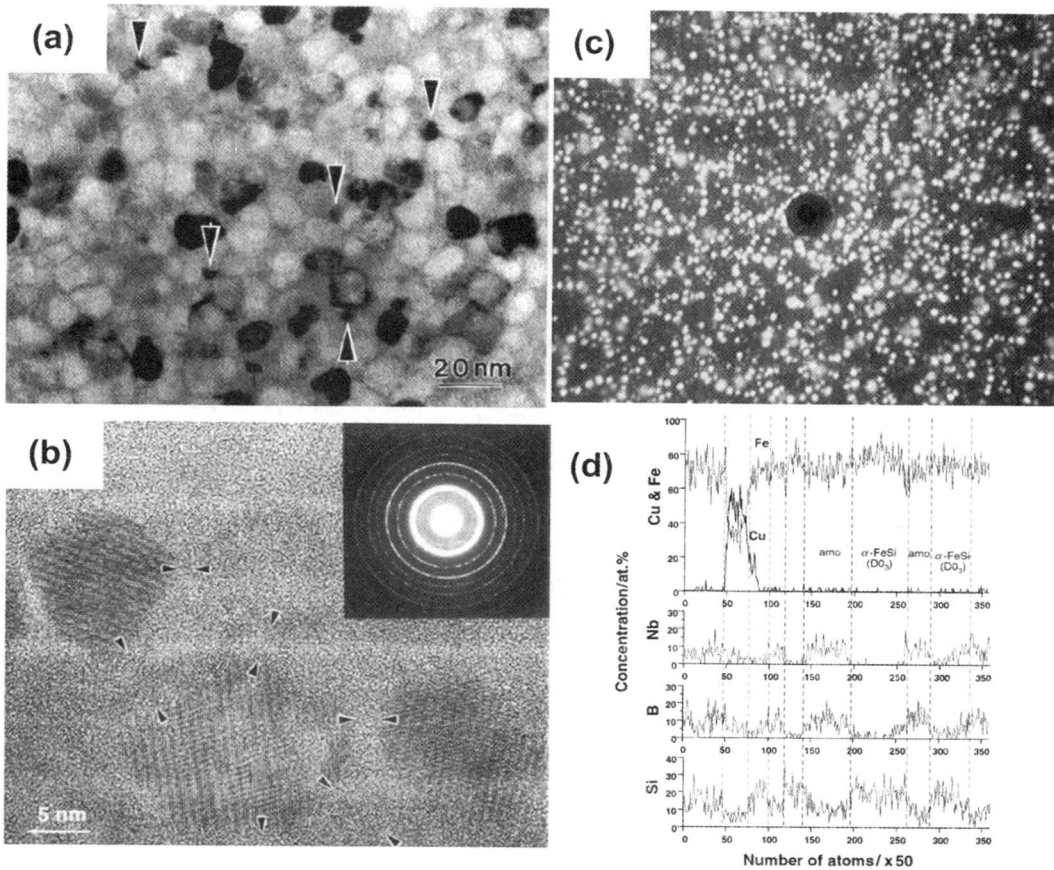

Figure 83 (a) Bright-field TEM image of nanocrystalline $Fe_{73.5}Si_{13.5}B_9Nb_3Cu_1$ alloy processed by crystallizing the amorphous precursor. The grains indicated by arrows are FCC-Cu. (b) HREM image and selected area diffraction pattern. The arrowheads indicate the residual amorphous phase surrounding α-Fe nanocrystals. (c) FIM images show clear contrast for nanocrystalline α-Fe (dark) and residual amorphous (bright). (d) 1DAP concentration profile showing the local compositions of each phase. From Hono et al. (1999). (For color version of this figure, the reader is referred to the online version of this book.)

phase contains ~ 20–25 at.%Si with little Nb and Cu. A few atomic percent of boron remains in this phase, and this appears to be a common feature of the α-Fe particles crystallized from boron-containing Fe-based amorphous alloys (Zhang et al., 1996). The remaining amorphous phase is enriched in B and Nb containing a small amount of Si. In addition to these two major phases, a Cu-enriched particle is observed. This phase is strongly enriched in Cu ($\sim 60\%$ or higher) but still contains appreciable amounts of the other elements. A separate nanobeam electron diffraction study revealed that the Cu-enriched particles were FCC-Cu in this stage (Hono et al., 1993b). This phase appears as a grain having a diameter of approximately 5 nm as indicated by the arrowheads in **Figure 83(a)**.

It was confirmed that the as-melt-spun alloy was a uniform amorphous phase, and neither structural nor chemical ordering was detected within the limitation of the spatial resolution of APFIM and TEM techniques (Hono et al., 1999). Here, it should be noted that AP data do not include any information

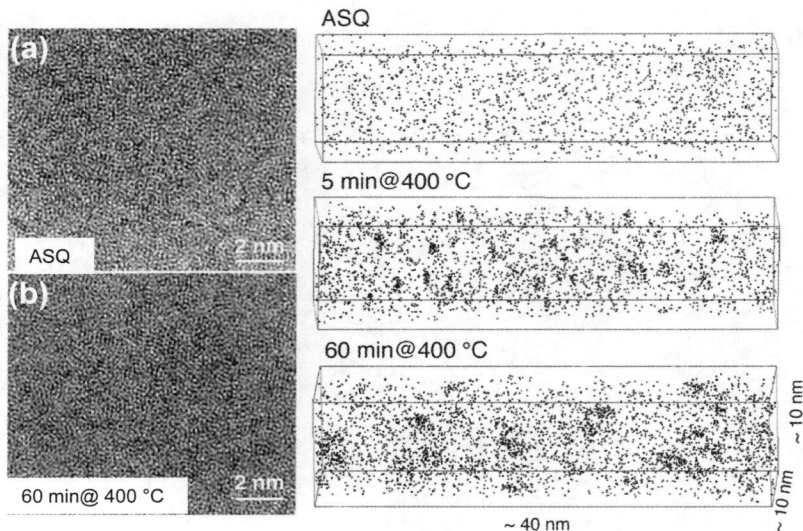

Figure 84 3DAP elemental maps of Cu of Fe73.5Si13.5B9Nb3Cu1 alloy as-quenched and annealed for 5 and 60 min at 400 °C. HREM images of the same alloy as-quenched and annealed for 60 min at 400 °C are also shown. From Hono et al. (1999). (For color version of this figure, the reader is referred to the online version of this book.)

on the chemical short-range order (CSRO), because APFIM data do not retain the information on the nearest neighbor atoms. Since AP is sensitive in local compositional change in a nanometer scale, it can detect solute clusters, which are formed even prior to the crystallization reaction. 3DAP elemental maps of Cu within analyzed volumes of $10 \times 10 \times 40$ nm in an as-melt-spun sample and the samples annealed for 5 and 60 min at 400 °C are shown in **Figure 84**, together with their HREM images (Hono et al., 1999). In the as-melt-spun specimen, Cu distribution is uniform, confirming that the as-melt-spun specimen is a chemically homogeneous solid solution. In the specimen annealed for 5 min at 400 °C, heterogeneous distribution of Cu atoms is apparent, indicating that clustering of Cu atoms occurs. After a 60 min annealing at 400 °C, clustering of Cu atoms is observed more clearly. The number density of the clusters decreases, while the composition of Cu in the clusters increases. Separate TEM observation results confirmed that no crystallization occurs up to 60 min at 400 °C, and thus it was concluded that the clustering observed in this analysis occurs in the amorphous phase. The number of atoms in each cluster was in the range of 50–100 (taking the detection efficiency of the MCP detector into consideration), and the size of the clusters is approximately 3 nm. The number density of the Cu clusters estimated from the analyzed volume is in the order of 10^{24} m^{-3}. The concentration of the Cu cluster has been estimated to be approximately 12 at.% Cu initially, but it increases as the clusters grow in size. The structure of these clusters was not identified by HREM, because the HREM image did not give any fringe contrast corresponding to any crystalline structure at these stages. Based on the extended X-ray absorption fine structure (EXAFS) measurement results, Sakurai et al. (1994) and Ayers et al. (1998) reported that Cu atoms form clusters having near-fcc symmetry from the very early stage of the heat treatment. Thus, the Cu clusters observed in the 3DAP data are believed to have an fcc-like local structure, but they do not appear to be distinct fcc-Cu in the initial stage. Evidence for Cu clustering prior to the onset of the crystallization reaction in this alloy was also obtained by more recent differential scanning calorimetry (DSC) and small angle neutron scattering (SANS) results (Ohnuma et al., 1999).

Earlier studies by Hono et al. employing 1DAP (Hono et al., 1992g) failed to prove that these Cu clusters serve as heterogeneous nucleation sites for the Fe–Si primary crystals. Thus, they concluded that the Fe–Si primary crystals nucleate from the Cu-depleted (Fe-enriched) amorphous matrix as originally proposed by Yoshizawa and Yamauchi (1990). On the other hand, Ayers et al. (1998) proposed that the Cu clusters that form in the amorphous matrix should serve as heterogeneous nucleation sites for the Fe–Si primary crystals. Although the earlier APFIM and EXAFS investigations convincingly showed the presence of Cu clusters prior to the onset of the crystallization reaction, the way Cu clusters stimulate the nucleation of the α-Fe had been a mere speculation until the alloy was studied by 3DAP (Hono et al., 1999). **Figure 85** shows 3DAP elemental maps near a Cu precipitate in the $Fe_{73.5}Si_{13.5}B_9Nb_3Cu_1$ alloy annealed at 550 °C for 10 min. The Cu precipitate is in direct contact with the Fe–Si particle, which suggests that the Cu cluster directly serves as heterogeneous nucleation sites for the Fe–Si primary crystals. Similar results were obtained by Warren et al. using 3DAP (Warren et al., 1999). The clustering of Cu atoms in the glassy state is reasonably understood because the free-energy curve of the liquid phase in Fe–Cu binary system has a miscibility gap. Based on these AP studies, the nanocrystalline microstructure evolution process in FINEMET-type alloy has been proposed. The initial amorphous phase is a chemically uniform amorphous solid solution. By annealing, Cu clusters appear in the fully amorphous matrix, which are thought to have the FCC-like local structure according to EXAFS results (Ayers et al., 1998). The concentration of Cu in the clusters appears to be much lower than 100 at.% at the beginning, but it gradually increases and the clusters eventually evolve to FCC-Cu. These Cu particles act as heterogeneous nucleation sites for α-Fe primary crystals, from which Nb and B atoms are rejected. The rejected solutes are enriched in the remaining amorphous phase, resulting in

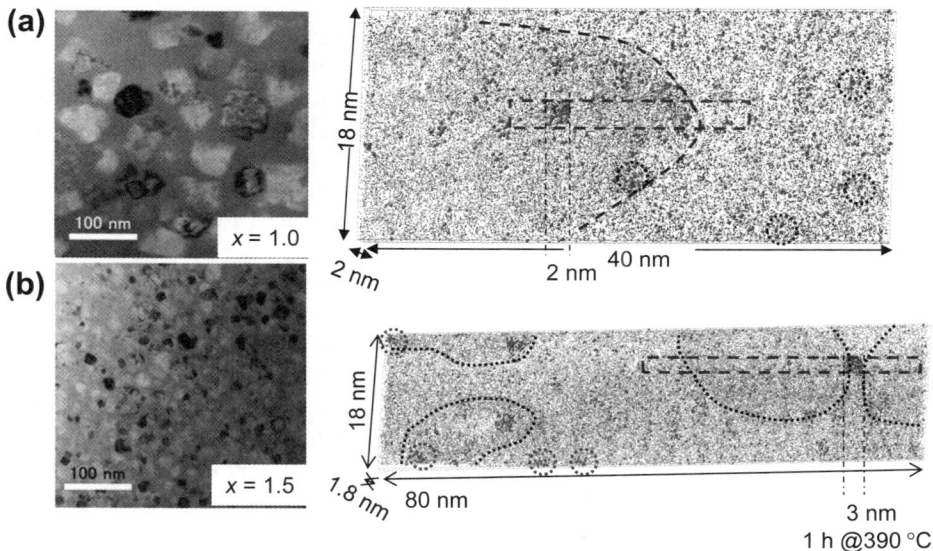

Figure 85 TEM bright field image and 3DAP elemental maps Fe, B, Cu atoms in melt-spun and annealed (Fe0.85B0.15) 100-xCux alloys (a) $x = 1.0$, and (b) $x = 1.5$. For $x = 1.0$, Cu clusters can be seen both in amorphous and α-Fe phase, indicating Cu precipitation occurred after the crystallization of a-Fe, while for $x = 1.5$, Cu clusters acted as heterogeneous nucleation sites for α-Fe, so nanocrystalline structure formed. From Warren et al. (1999). (For color version of this figure, the reader is referred to the online version of this book.)

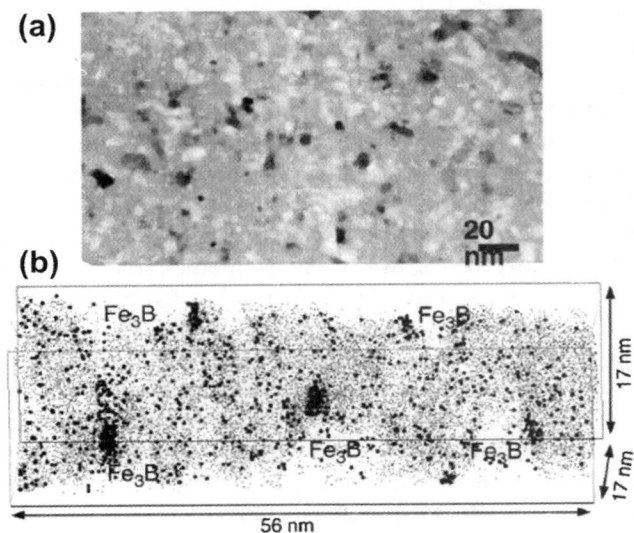

Figure 86 (a) TEM and (b) 3DAP elemental maps of Cu in $Fe_{73.5}Si_{13.5}B_9Nb_3Cu_1$ alloy annealed at 530 °C for 30 min. From Warren et al. (1999).

stabilization of the remaining amorphous phase. This retards further growth of the α-Fe nanocrystals. Si partitions in the α-Fe phase as it grows, ending up with Fe–20Si with the DO_3 structure.

A number of attempts have been made to improve the soft magnetic properties of the $Fe_{73.5}Si_{13.5-}B_9Nb_3Cu_1$ alloy by modifying the alloy composition (see **Figure 86**) and substituting some of the alloying elements. In these investigations, controlling the number density of Cu was crucial to refine the nanocrystalline structure. The interplay between the Cu clustering and the final microstructure has been investigated by combined use of 3DAP and TEM, and these investigations made substantial contributions on the development of nanocrystalline soft magnetic materials with improved properties (Ohnuma et al., 2000, 2003). With the similar concept of obtaining soft magnetic properties from nanocrystalline microstructure, various Fe-based and FeCo-based nanocrystalline magnetic materials were developed. To attain the amorphous precursor, the alloy must contain glass-forming elements, such as Zr and B. Then, these alloys should crystallize by the primary crystallization mode as described above. In order to introduce heterogeneous nucleation size, most of the Fe- and FeCo-based soft magnetic materials contain small amounts of Cu. Such exaples are Fe–Zr–B (Yoshizawa and Yamauchi, 1990), FeCo–Zr–B (Willard et al., 1998), Fe–B–Cu (Ohta and Yoshizawa, 2007), Fe–Si–B–P–Cu (Makino et al., 2009) alloys. The complementary uses of 3DAP and TEM made a substantial roles in understanding the microstructure evolutions and the roles of alloying elements in Fe–Zr–B(–Cu) (Sakurai et al., 1994; Warren et al., 1999; Ohnuma et al., 2003), FeCo–Zr–B(–Cu) (Makino et al., 2009; Ping et al., 2001; Ohodnicki et al., 2009). Among various derivatives of nanocrystalline soft magnetic alloys, only Fe–Si–B–Nb–Cu alloys are used commercially, as appropriate glass-forming ability in air is essential for industrial viability.

15.4.6.4 *Nanocrystalline Permanent Magnets*

Coehoorn et al. (1988) reported relatively high coercivity and high remnance can be obtained from the nanocrystalline alloy produced from crystallizing $Nd_{4.5}Fe_{77}B_{18.5}$ amorphous alloy, which is neodymium-poor and boron-rich compared to the stoichiometry of the $Nd_2Fe_{14}B$ phase, $Nd_{11.8}Fe_{82.3}B_{5.9}$. The microstructure was composed of nanoscale grains of magnetically soft Fe_3B and magnetically hard $Nd_2Fe_{14}B$ phases, and unexpectedly high remnance for isotropic material, as much as $0.8M_s$, was obtained due to the remnance enhancement effect of exchange-coupled magnetic grains. This type of material was considered to be suitable for a magnet whose performance and price is intermediate between low-cost low-performance ferrites and high-cost high-performance Nd–Fe–B anisotropic magnets (Archambault and Pere, 1999). Various microalloying elements have been added to enhance the coercivity while optimizing remnance and 3DAP combined with TEM have been used to understand the roles of these alloying elements, e.g. Co and Ga (Ping et al., 1998), Cu and Nb (Ping et al., 1999a), Cu and Zr (Kajiwara et al., 2001) and Co and Cr (You et al., 2006). Among these additives, of particular interest was the effect of Cu and Nb. Ping et al. (1999a, 1999b) reported the combined addition of Cu and Nb substantially refines the $Fe_3B/Nd_2Fe_{14}B$ nanocomposite structure, thereby improving hard magnetic properties of $Nd_{4.5}Fe_{77}B_{18.5}$-base alloy. They chose these alloying elements to be analogous with the effect of Cu and Nb in the above description of Fe–Si–B–Nb–Cu nanocrystalline soft magnets. For example, **Figure 86(a)** shows a TEM bright field micrograph of the Fe–Si–B–Nb–Cu alloy annealed at 530 °C for 30 min (Ping et al., 1999a). The crystalline particles observed in this micrograph are all Fe_3B (bct structure) embedded in the amorphous matrix, and the second stage crystallization of the $Nd_2Fe_{14}B$ phase does not occur yet in this stage. **Figure 86(b)** shows a 3DAP elemental map of Cu and Nd of the same specimen, where small and large black spots correspond to Nd and Cu atoms, respectively. Fe_3B particles can easily be recognized based on the atomic distribution of Nd because the Nd atoms are rejected from the Fe_3B phase. Each Cu cluster is located at the interface between the Fe_3B crystals and the amorphous matrix. This result suggests that Cu-enriched clusters serve as heterogeneous nucleation sites for the Fe_3B primary crystals. From more detailed 3DAP analysis results, Nd atoms were also found to be enriched in the Cu cluster region. This is because Cu and Nd have strong affinity each other. The total concentration of Cu and Nd in the cluster was found to be less than 15%; thus, it is unlikely that they have their own distinct structure, and they are presumed to be still amorphous. For the formation of Fe_3B crystals, Fe atoms do not have to diffuse too much because the matrix concentration of Fe (76.8 at.%) is very close to that for the Fe_3B phase. For the formation of the Fe_3B phase, B atoms have to diffuse in by rejecting the Nd atoms. Since Nd has a strong affinity with Cu (large negative enthalpy of mixing), when Cu atoms aggregate, Nd atoms are attracted to the Cu atoms, resulting in a cosegregation of Cu and Nd. As a result, the enrichment of B and the depletion of Nd occur at the Cu/amorphous interface, which makes a chemical environment for the Fe_3B phase formation adjacent to the Cu clusters. Unlike Fe–Si–B–Nb–Cu and Fe–Zr–B–Cu nanocrystalline soft magnetic alloys, the Cu clusters in Nd–Fe–B–Cu alloys do not develop to FCC-Cu, because the Cu-enriched clusters dissolve into the $Nd_2Fe_{14}B$ phase when the remaining amorphous phase is crystallized. Since the size of the secondary crystal phase, $Nd_2Fe_{14}B$, is controlled by the initial distribution of Fe_3B particles, a high nucleation density of the Fe_3B primary crystals induced by the Cu clusters contributes to the refinement of the grain size in the final $Fe_3B/Nd_2Fe_{14}B$ nanocomposite microstructure.

One of the shortcomings of $Fe_3B/Nd_2Fe_{14}B$ nanocomposites is their relatively low coercivity as mentioned in the previous section. The α-Fe/$Nd_2Fe_{14}B$ nanocomposites that can be produced from the alloys with compositions ranging from 8 to 9 at.%Nd and 5 to 6 at.%B tend to have higher coercivities than those for $Fe_3B/Nd_2Fe_{14}B$ nanocomposites, and thus the α-Fe/$Nd_2Fe_{14}B$ system has been more extensively studied. However, the glass-forming ability of the alloys with this composition range is not good; thus the melt-spun ribbons are partially crystallized during rapid solidification with a typical wheel surface velocity of 10–20 m/s. For optimization of the magnetic properties, partially crystallized melt-spun specimens must be used as starting material, which is subsequently annealed for short period of time (<10 min) at around 1000 K. Hence, the magnetic properties of α-Fe/$Nd_2Fe_{14}B$ nano-composites are very sensitive to both cooling rate and post-annealing temperature. From this example, it can be easily imagined that optimization of the magnetic properties of α-Fe/$Nd_2Fe_{14}B$ nano-composite magnets is not easy in an industrial scale. Effect of various microalloying elements has also been investigated intensively. Co is commonly added to increase the Curie temperature of the $Nd_2Fe_{14}B$ phase. In order to improve the coercivities and energy products of α-Fe/$Nd_2Fe_{14}B$ nanocomposites, additions of Ga (ref), Nb (Hadjipanayis et al., 1995), Cu and Nb (ref) and Si (Wecker et al., 1995) are reported to be effective. It is interesting to note that most of these elements are also effective in improving the hard magnetic properties of Fe_3B/$Nd_2Fe_{14}B$ nanocomposites. For these investigations, 3DAP played crucial roles. Although Fe_3B/Nd_2Fe14B-type nanocomposite microalloyed with Ti and C were commercialized with unique properties like superior corrosion resistance, easy magnetization medium performance with relatively low cost (http://www.hitachi-metals.co.jp/products/auto/el/pdf/hg-a22-b.pdf), it did not find major applications so far.

15.4.7   Permanent Magnets

15.4.7.1   Nd–Fe–B Magnets

In the previous section, we showed examples of applications of 3DAP on understanding the nano-crystallization mechanism from amorphous precursors. Although the nanocrystalline Nd–Fe–B pro-cessed by rapid solidification find applications as raw materials for bonded magnets, high-performance permanent magnets are processed by a powder metallurgy route, as the orientation of magnetic easy axis must be strongly aligned so that high remanence can be achieved (anistropic magnets). Nd–Fe–B-based anisotropic sintered magnets are currently used as the highest performance permanet magnets for motors and actuators where the highest energy density is reqired. The interest in the coercivity mechanism of such Nd–Fe–B sintered magnets has recently been revived due to the increasing demand for developing Dy-free high coercive Nd–Fe–B anisotropy magnets for the traction motors of (hybrid) electric vehicles (Hono and Sepehri-Amin, 2012). The coercivity of sintered magnets has been attributed to the formation of a thin continuous nonferromagnetic Nd-rich layer surrounding the $Nd_2Fe_{14}B$ grains (Kronmüller and Schrefl, 1994). Since the coercivity of Nd–Fe–B permanent magnets is extremely sensitive to their microstructures, especially the GB structure and chemistry, understanding the microstructure–coercivity relationships of Nd–Fe–B permanent magnets is essential to develop a higher coercivity magnet. Recent investigations on the GB phase of both commercial and experimental Nd–Fe–B magnets using 3DAP have shed light on the coercivity mechanism of Nd–Fe–B magnets (Li et al., 2009; Sepehri-Amin et al., 2012).

The typical microstructure of sintered magnets is shown in **Figure 87**. The crystal grains of the major $Nd_2Fe_{14}B$ phase are observed with a uniform gray contrast. The thin GBs show bright contrast indi-cating the enrichment of Nd atoms. In addition, brightly imaging grains are observed at the triple junctions of the GBs, which are either metallic FCC-Nd or neodymium oxides (FCC-NdO_x or Nd_2O_3). The chemistry of the GBs and the interfaces between the $Nd_2Fe_{14}B$ crystal and Nd-rich phases are considered to influence the coercivity substantially. They have been characterized by TEM in the past (Makita and Yamashita, 1999; Fidler and Bernardi, 1991; Vial et al., 2002), but AP characterization of sintered magnets has not been successful until the site-specific specimen preparation method using the FIB technique was developed recently. Li et al. (2009) reported limited AP analysis results of the GB phase using voltage-pulse AP. Although the specimens containing a GB were successfully prepared, the low yield of successful analyses using voltage-pulse limited the application of the AP technique to

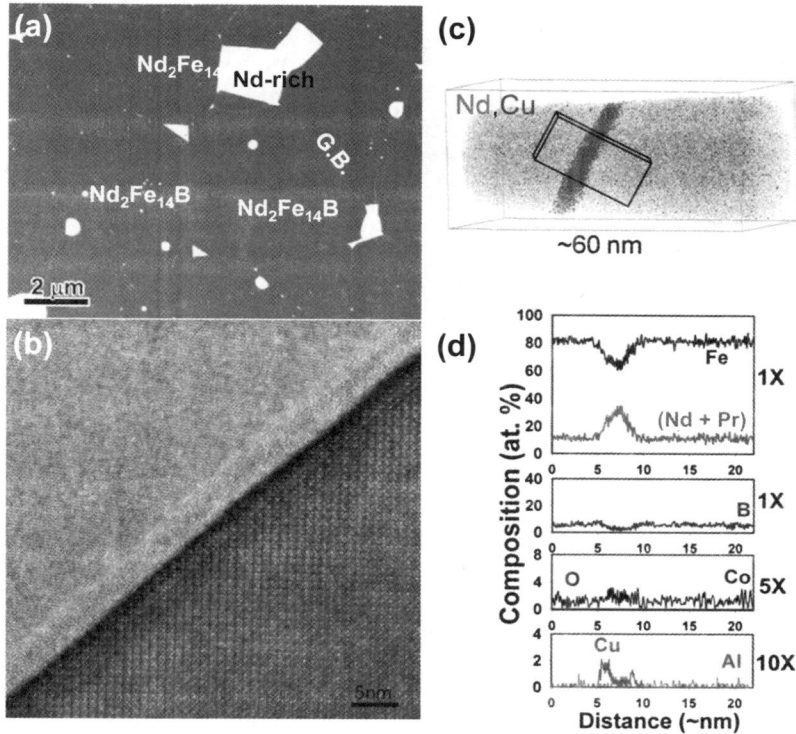

Figure 87 (a) Typical SEM back-scattered electron image of a commercial sintered magnet and (b) HREM image of a GB, (c) 3DAP map of Nd and Cu atoms including a GB, and (d) quantitative concentration profile calculated from the inset of (c). From Sepehri-Amin et al. (2012). (For color version of this figure, the reader is referred to the online version of this book.)

commercial sintered magnets. However, recent implementation of ultraviolet (UV) laser to 3DAP substantially improved the yield of successful AP analyses, thereby contributed the advances in understandings of the microstructures of Nd–Fe–B sintered magnets (Sepehri-Amin et al., 2011a). **Figure 87(b)** shows HREM image of a GB observed in optimally heat-treated Nd–Fe–B-based sintered magnets (Li et al., 2009; Sepehri-Amin et al., 2012). A layer of about 3 nm in thickness with the amorphous structure can be seen along the GB. In order to understand the magnetism of the GB phase, quantitative chemical analysis of the thin layer is necessary. **Figure 87(c)** shows 3DAP map of Nd and Cu and (c) shows concentration depth profile crossing the GP phase in perpendicular direction as shown in the inset figure. In commercial Nd–Fe–B sintered magnets, about 0.1 at.%Cu is microalloyed. The 3DAP maps of the GB phase indicate that the Cu is segregated at the interfaces between the amorphous GB phase and the $Nd_2Fe_{14}B$ phases. The quantitative concentration profile indicate that as much as 65 at.% of Fe was contained in the thin amorphous phase. Base on this observation, the authors pointed out the possibility of ferromagnetism of the GB phase. In earlier investigations, this amorphous GB phase was assumed to be non-ferromagnetic, but the AP investigations strongly suggest that it could be ferromagnetic. This triggered active debate on the mechanism of coercivity of sintered magnets (Gutfleisch and Franco, 2012). 3DAP was employed to characterize a newly developed fine grain-sized sintered magnet recently, reporting two types of GB phase, one is thick layer with high Nd

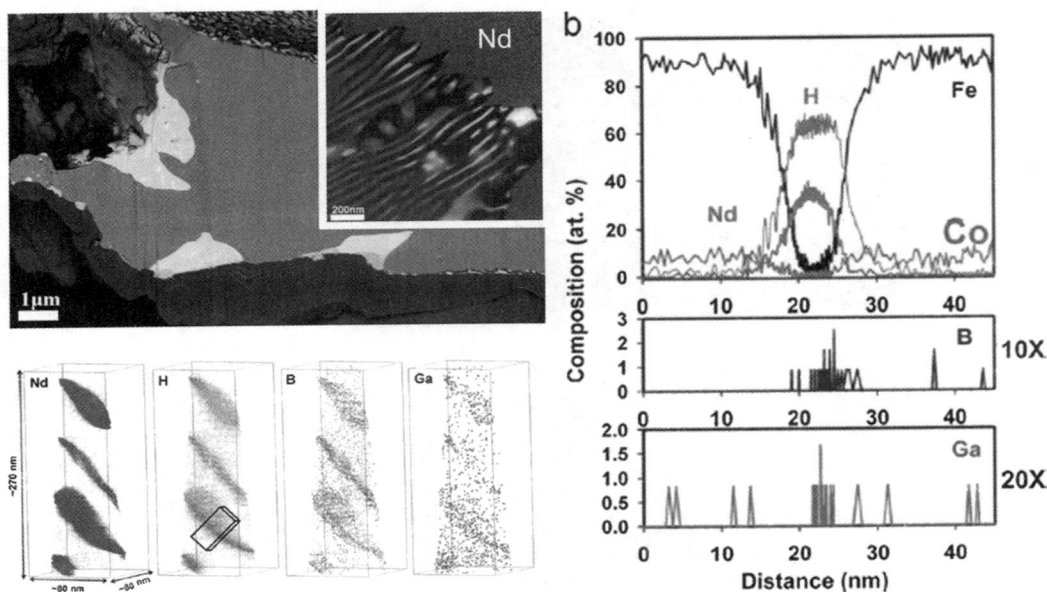

Figure 88 Analyses of microalloying element distribution during the hydrogenation desorption process. From Sepehri-Amin et al. (2010a). (For color version of this figure, the reader is referred to the online version of this book.)

content and another is thin layer similar to that observed in the sintered magnets (Sepehri-Amin et al., 2011a). 3DAP was also applied to the detailed analysis of the grain boundaries in Dy diffusion-processed sintered magnets (Author).

In spite of recent effort on refining the grain size of sintered magnets, reduction of the grain size below 1 µm is extremely difficult. Hydrogen disproportionation desorption recombination (HDDR) process is an industrially used technique to process ultrafine grained anisotropic Nd–Fe–B powder (Takeshita and Morimoto, 1996; Gutfleisch et al., 2003). However, typical coercivity values reported for Nd–Fe–B-based HDDR powders were much too low and have long been a mystery for the HDDR magnets. Based on 3DAP analysis results of the GBs of HDDR magnets, Li concluded that the $Nd_2Fe_{14}B$ grains are ferromagnetically coupled (Li et al., 2008). Based on the magnetic domain wall observation using Lorentz TEM, they concluded that the coercivity of the HDDR magnet is controlled by the pinning of magnetic domains at the GBs and the formation of the Nd-rich GB phase causes a stronger pinning force. The pinning force arises from the difference in the magnetocrystalline anisotropy between the hard and soft phases. If the rare earth content in the GB layer can be increased to decouple exchange coupling, the coercivity would increase substantially. Using 3DAP, they also investigated the effect of Ga microalloying on the coercivity of HDDR magnets (Sepehri-Amin et al., 2010a). 3DAP is particularly unique in its analytical capability of hydrogen, so it can play a critical role in analyzing microalloying elements during the hydrogenation desorption process as shown in **Figure 88**. The SEM image shows NdH_2 and Fe lamellar regions that develop from the surface of an $Nd_2Fe_{14}B$ particle. Using the site-specific specimen preparation, it is possible to prepare needle-like specimen containing the NdH_2/Fe lamella. The 3DAP elemental maps indicate both B and Ga are partitioned in the NdH_2 phase. The concentration profile calculated from the inset box show the hydrogen concentration of NdH_2 can be quantitatively characterized by 3DAP.

To increase the coercivity of HDDR powder, Sepehri-Amin et al. (2010b) proposed the modification of GB chemistry by diffusing the Nd–Cu eutectic alloy melt from the surface of the powders. On annealing the HDDR powders mixed with the eutectic $Nd_{70}Cu_{30}$ alloy powder at $\sim 700\ °C$, $Nd_{70}Cu_{30}$ melts and diffuses into the powders through the GBs, leading to the formation of an Nd-rich layer along the GBs, resulting in the enhancement of the coercivity. The microstructure change before and after the Nd-diffusion process was characterized using TEM and 3DAP. In spite of a large increase in coercivity, the Dy concentration in the GB phase was only 20 at.%, containing a large amount of ferromagnetic elements. Based on these results, they suggest further coercivity enhancement should be possible by modifying the $Nd_{70}Cu_{30}$ eutectic diffusion process. In fact, they extended this to hot-deformed anisotropic magnets. Based on the 3DAP analysis results of the intergranular phases in a series of hot-deformed magnets with different Nd contents in the overall alloy compositions, Liu et al. found direct correlation between the coercivity and the Nd concentration in the intergranular layer using 3DAP (Liu et al., 2013). By applying the above-mentioned $Nd_{70}Cu_{30}$ eutectic diffusion process, they modified the chemical compositions and thickness of the intergranular phases and enhanced the coercivity substantially.

Figure 89 shows TEM and 3DAP analysis results of the hot-deformed Nd–Fe–B anistropic magnet that was diffusion processed using the $Nd_{70}Cu_{30}$ eutectic alloy (Sepehri-Amin et al., 2013). (a) TEM bright field image of the hot-deformed anisotropic Nd–Fe–B magnet that was diffusion processed using an $Nd_{70}Cu_{30}$ eutectic alloy. Large expansion of the GB phase along c-axis is observed, which is

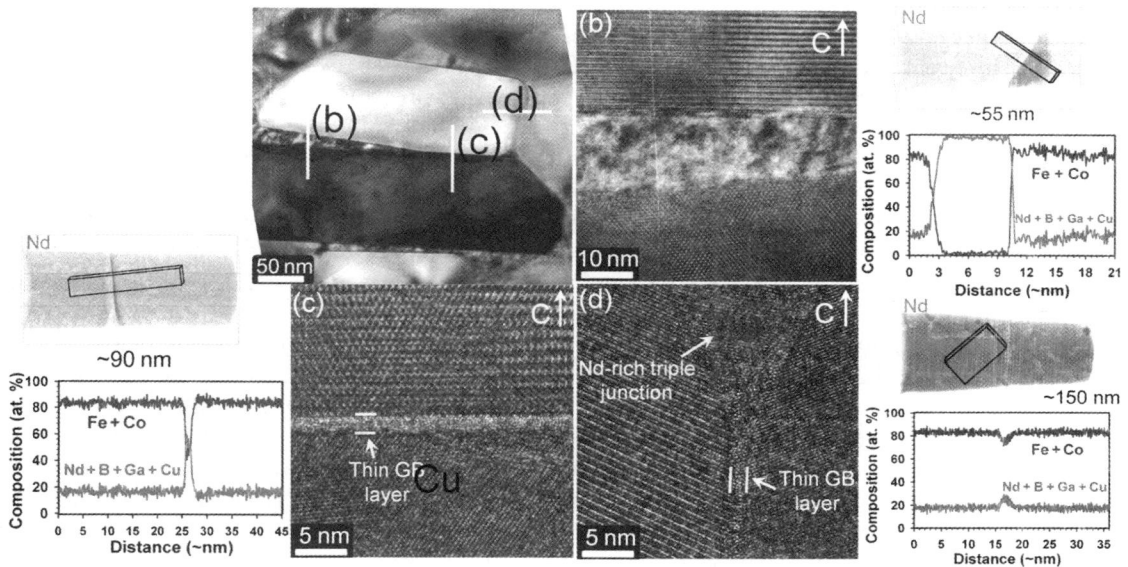

Figure 89 (a) TEM bright field image of the hot-deformed anisotropic Nd-Fe-B magnet that was GB diffusion processed using an Nd70Cu30 eutectic alloy. Large expansion of the GB phase along c-axis is observed, which is attributed to the thickened N-rich phase as shown in (b). The AP analyses indicate that this thick phase is essentially pure Nd. The thin intergranular phase (c) also contains about 60 at.% Nd. The intergranular phases (b) and (c) are believed to be non-ferromagnetic. However, the intergranular phase along c-axis contains only a small amount of Nd ($\sim 30\%$) with $\sim 70\%$ of Fe + Co, indicating the grains are exchange-coupled in the lateral direction. From Depehri-Amin et al. (2013). (For color version of this figure, the reader is referred to the online version of this book.)

attributed to the thickened Nd-rich phase as shown in (b). The AP analyses indicate that this thick phase is essentially pure Nd. The thin intergranular phase (c) also contains about 60 at.% Nd. The intergranular phases (b) and (c) are believed to be nonferromagnetic. However, the intergranular phase along c-axis contains only a small amount of Nd ($\sim$30%) with $\sim$70% of Fe + Co, indicating the grains are exchange coupled in the lateral direction.

15.4.8   Device Analysis

One of the key variables that control the performance of silicon-based semiconductor is the distribution of the dopant concentration in a given device after the standard manufacturing process. It has been always dream of electronic device manufacturers to characterize with good spatial and depth resolution. As mentioned earlier, early application of AP was limited to conductive materials, at least at the temperature of analyses (20 to 70 K). With the introduction of wide-angle-pulsed AP and FIB -thinning techniques, the adoption of this technique for 3D semiconductor structures (Koelling et al., 2011; Lauhon et al., 2009; Hono et al., 2011b) has expanded rapidly. Lauhon et al. (2009) provide an overview of other applications and made a case for 3DAP by comparing the capabilities of SIMS, TEM and EELS in terms of probe size, resolution and fidelity of measuring very low concentrations. In addition, they also discussed the flexibility of using FIB and bottoms-up approach to make these devices amenable for 3DAP analyses. With these developments, it is quite possible to interrogate the transistors, nanowires (Agrawal et al., 2011), isotopic modulated (see Section 15.3) multilayers (Moutanabbir et al., 2011), and field-effect transistors. One of the recently published examples is discussed below (Takamizawa et al., 2011) to demonstrate the application of 3DAP to semiconductor devices.

With ever-increasing need for reducing the size of electronic devices, some of the recent work shows variability of the performance. For example, Takamizawa et al. (2011) investigated the reason for variability in n- and p-type metal-oxide-semiconductor field-effect transistors (MOSFETs) devices. One of the hypotheses, they considered, is the local changes in the dopant concentration in these devices. In the past, these variations were detected by secondary ion mass spectroscopy (SIMS), but the resolution required in the lateral direction is not good. In this work, these researchers used 3DAP to understand these lateral variations by comparing the results with the SIMS measurements in the depth direction (see **Figure 90**). The 3DAP analysis was consistent with the SIMS data in the depth direction. However, the boron concentration profiles appear to larger scatter. In order to evaluate these changes quantitatively, a planar area at a fixed depth (30 nm) from the surface was taken for both p- and n-channels. The variations of boron and arsenic atoms were analyzed further. The measured frequency distribution of concentration was compared with a binomial distribution expected for random solid solution. These analyses showed that boron distribution is indeed consistent with random solid solution. This result correlated well with the performance of n-channel as expected by random dopant fluctuations. In contrast, the concentration profile analyses suggested clustering of arsenic atoms in the p-channel at similar depth levels. Although, mechanism for these clustering was not identified, the above example shows the power of 3DAP for the new generation of semiconductor devices.

The second example is related to GB segregation of dopants in fin field-effect transistors (Takamizawa et al., 2012). The performance of these transistors is very sensitive to the processing route. A typical 3DAP data form from a device, manufactured by low-energy-FIB direct-deposition technique, is shown in **Figure 91**. In the first step, the fins were created in a silicon substrate. Then, boron was introduced on the surface using self-regulatory plasma doping (SRPD) technique. In the next step, low-energy-FIB -deposition technique was used to deposit an amorphous silicon layer on fin structures. This low-energy deposition technique allows one to control the type of Si isotope deposited on the surface.

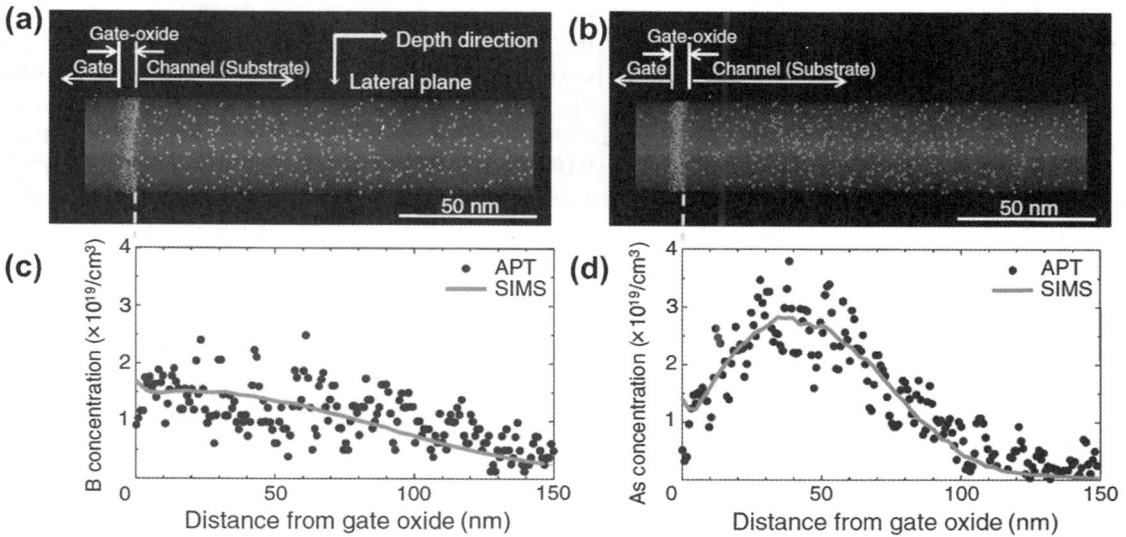

Figure 90 2D Projection of the 3D atom maps of boron (green), arsenic (purple) and oxygen (blue) in (a) n-MOSFET and (b) p-MOSFET device. Concentration profile (in the units of atoms cm^{-2}) as a function of distance from surface for (c) B in n-MOSFET and (d) As in p-MOSFET. From Takamizawa et al. (2011). (For color version of this figure, the reader is referred to the online version of this book.)

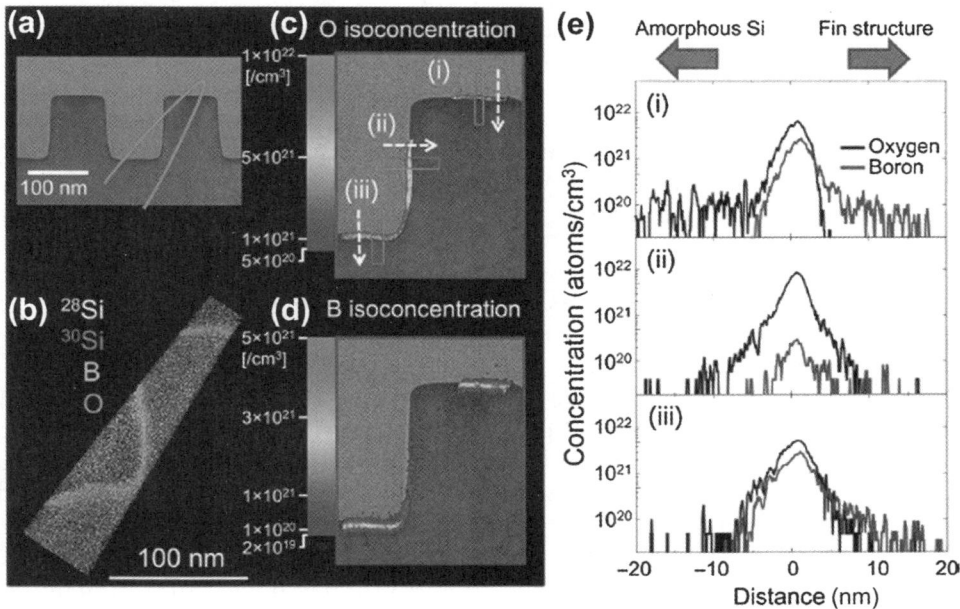

Figure 91 Characterization of interfaces between finned Si-substrate and amorphous Si deposits. (a) Region of interest from the device showing the position of the AP tip; (b) atom map of two forms of Si (^{28}Si and ^{30}SI), B and O; (c) the orientation of the composition profile shown in (e); (d) concentration of oxygen and boron along the fin boundary. Takamizawa et al. Appl. Phys. Lett. 100, 093502 (2012). (For color version of this figure, the reader is referred to the online version of this book.)

As a result, the concentration of ^{28}Si ion was enhanced in the deposit region. With this approach, the authors delineated the interfaces between the fin (^{30}Si) and the deposit (^{28}Si) and interrogated the effectiveness of boron deposition at the sidewalls. After the fabrication of the above device, the authors made an AP specimen using the FIB technique. The orientation of the tip with reference to the device geometry was designed to intersect top and bottom flat surfaces as well as the sidewall surfaces of the fin as shown in **Figure 91(a)**. Atom maps in **Figure 91(b)** were constructed by separating the ^{28}Si and ^{30}Si ions to image the substrate and the deposit distinctly. The atom maps also show preferential segregation of B and O at these interfaces. In order to quantify, composition profiles were calculated across the top (noted as i), sidewall (noted as ii) and bottom (noted as iii) interfaces. The analyses showed that the boron and oxygen segregations were consistent for both bottom and top interfaces. However, the sidewall surfaces show one order of magnitude lower concentration of boron than top and bottom interfaces. Although, the levels are acceptable from the transistor's functional point of view, the authors believe that higher performance can be achieved if they can increase the boron activation at the sidewalls. The above example demonstrates the enormous potential of designing and processing conditions based on 3DAP analyses to control the interface chemistry for targeted performance of semiconductor devices.

Increasing the areal magnetic recording density in hard-disk drive requires a continuous evolution of read sensors with a lower resistance, a higher sensor output, and a higher bit resolution. In the history of hard-disk drives, development of high output magnetoresistive sensors led to the growth of recording density starting from anisotropic magnetoresistance (AMR), giant magnetoresistance (GMR) and tunnerling magnetoresistance (TMR). AMR is known to be the intrinsic properties of ferromagnetic materials, but both GMR and TMR appears from artificially fabricated multilayer structure composed of ferromagnetic (FM) and nonmagnetic (NM) layers. The GMR devices used as read sensors have the exchange-biased spin valve structure, which is the stack of PtMn/CoFe/Ru/CoFe/Cu/CoFe. The IrMn is an antiferromagentic (AF) phase, which is exchange coupled with CoFe. Through Ru, the magnetization of bottom CoFe layer is pinned through AF exchange coupling with the pinned IrMn/CoFe. This is the basic structure of in-plane GMR sensor, where current is applied parallel to the planer direction of the films. Although this is the most important structure for a read sensor, there are little AP investigations of exchange-biased spin valve structure. Earlier AP work analyzed FM/NM/FM trilayer structure deposited on Si post (Larson et al., 2000b). These investigations demonstrated that atomic layer resolution could be obtained, so that the intermixing at interfaces can be accurately analyzed using 3DAP. Since it was difficult to prepare AP specimen from actual exchange-biased spin valves, the thermal stability of PtMn/CoFe interfaces was studied using a model PtMn/CoFe multilayer films (Larson et al., 2004b). Since lift-out and microsampling technique had not been developed in these days, the AP analysis of actual devices was not possible.

In 2005, Djayaprawira et al. reported large room temperature TMR of 230% in polycrystalline spin-valve-type magnetic tunnel junctions (MTJs) using MgO as tunnel barrier layer and FeCoB as FM layer (Djayaprawira et al., 2005). The technical merit of FeCoB-based MTJs is that they can be sputter deposited on amorphous substrates because the FeCoB layers are intially amorphous and then crystallized by post-annealing heat treatment so that CoFe make template growth from MgO (001)-textured polycrystalline layer (Ikeda et al., 2008). Pinitsoontorn et al. (2008) characterized Si sub/Ta (3 nm)/PtMn (15.4 nm)/CoFe(2 nm)/Ru (0.8 nm)/CoFeB (2.3 nm)/MgO (1.2 nm)/CoFeB (2.6 nm)/NiFeCr (60 nm) multilayer and reported lateral variations in the MgO composition of the order of tens of nanometers. They also reported that B is not uniformly distributed within the CoFeB layers, but has tended to segregate to the interfaces. Larson et al. investigated the concentration change aross the SiO2/Ta(10)/IrMn(40)/CoFeB(0.35)/CoFe(5)/Mg(0.8)/MgO(2.7)/CoFe(2)/CoFeB(30) exchange-biased spin valves and reported the Mn segregation at the CoFe/MgO interface after annealing (Larson et al., 2011b) in accordance with the TEM results reported by Ikeda et al. (2008).

With ever increasing need for high areal density in magnetic recording requires a continuous evolution of read sensors with a lower resistance, a higher sensor output and a higher bit resolution. Low-resistance magnetoresistive (MR) devices are required of the read sensors for impedance matching with preamplifiers, for lower electric noises, and for high-frequency data transfer (Takagishi et al., 2010). A resistance-area product (RA) below 0.1 $\Omega\mu m^2$ is required for read sensors for the recording densities exceeding 2 Tbit/in, which is a big challenge for currently used MTJs. On the other hand, current-perpendicular-to-plane giant magnetoresistance (CPP-GMR) spin-valves (SVs) composed of all metallic layers can easily achieve low RA values of typically below 0.05 $\Omega\mu m^2$. However, the MR output from the CPP-GMR SVs with conventional FM materials such as CoFe alloy is only $\sim 1\ \mu\Omega m^2$, which is too small for sensor applications. Recently, relatively large MR outputs were reported from the spin-valves with thin polycrystalline Heusler alloy FM layers (nm) with a modest annealing temperature (Carey et al., 2011), and from epitaxial pseudo SVs (Sakuraba et al., 2012; Li et al., 2013). The half-metallic transport properties of the Co-based Heusler alloys are very sensitive to the chemical disordering of the Heusler alloy layer, interface structure and interdiffusion of alloying elements, and thus rigorous structure/chemistry analysis of CPP-GMR using Heusler alloy is essential to develop the devices with higher MR output. Takahashi et al. characterized $Co_2Mn(Ge,Ga)$-based CPP-GMR and reported that no structural change and intermixing can be attributed to the MR degradation by high-temperature annealing (Takahashi et al., 2013). **Figure 92 (a)** shows HAADF/STEM image and (b) AP tomography of MgO(sub)/Cr(20 nm)/Ag(10 nm)/Co_2MnSi(9 nm)/Ag(5 nm)/Co_2MnSi (9 nm)/Au(2 nm)/Cu(cap) current-perpendicular-to-plane giant magnetoresistive device that shows large MR output ($\sim 58\%$) (Kodzuka, 2012). In the AP tomography, the full multilayer stack is reproduced. The zigzagged surface profile for the tip apex is due to the difference in the evaporation field of each material.

The large difference in the evaporation field often cause specimen rupture due to the local concentration of field stress; thus the multilayer stuck containing high melting temperature layers such as Ru and Ta often cause poor yield of successful analysis. However, the deformed reconstruction can be corrected by selecting small volume of interest as shown in **Figure 92(c)**. This enlarged atom map from the upper Co2MnSi/Ag spacer interface show the interface is atomically sharp and no intermixing occurs at the interface.

15.4.9 Ceramics Materials

Ceramics materials can be crystalline or amorphous and its constitutions can be carbide, oxide and nitrides or their combinations. They can be used for structural and functional applications in a monolithic form or as coatings. Ceramics thin films are often used as dielectric and insulator materials in various electronic devices. Their inherent properties are also modified by doping or alloying and by changing the spatial arrangement of constituent phases. The phase transformations in ceramics materials are also driven by reconstructive or displacive mechanisms. As the microstructure sensitively change structural and functional properties, the ability to interrogate structure and elemental distribution in three dimensions is useful for developing tailored ceramics for a given application (Kingery et al., 1976). Early APFIM work on oxides can be traced back to the studies of high-T_c superconductors (Andren, 1996) by voltage pulsing. This was possible, since these oxides become conductive at the analyses temperature. Despite the recent remarkable advances in the laser-assisted AP, little work has been carried out on the analyses of insulating materials. A few successful examples were all on thin oxide films, such as MgO and Al_2O_3 barriers in magnetic tunneling junctions (Chiaramonti et al., 2008;

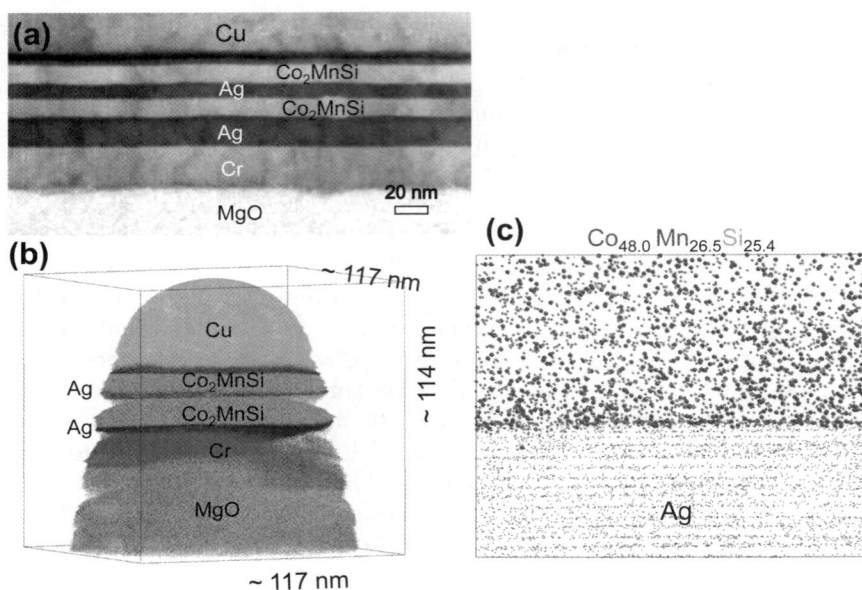

Figure 92   (a) HAADF/STEM image and (b) AP tomography of MgO(sub)/Cr(20 nm)/Ag(10 nm)/Co₂MnSi(9 nm)/Ag(5 nm)/Co₂MnSi (9 nm)/Au(2 nm)/Cu(cap) current-perpendicular-to-plane giant magnetoresistive device and (b) enlarged atom map from the upper Co₂MnSi/Ag spacer interface show the interface is atomically sharp and no intermixing occurs at the interface. Kodzuka et al. Ph.D. thesis, Univ. Tsukuba (2012). (For color version of this figure, the reader is referred to the online version of this book.)

Kuduz et al., 2004), HfO_2 high dielectric (high-k) ultrathin films on Si (Pinitsoontorn et al., 2008) and oxide films that are formed on tip surfaces (Pinitsoontorn et al., 2008; Yoon et al., 2008). Although Larson et al. (2008) reported the feasibility of insulator analyses using green laser, the work did not demonstrate reconstruction of 3D tomography of the nanostructure within the specimen. Using UV laser-assisted AP, Chen et al. (2009b) first demonstrated AP tomography that reconstructed 3D nanocrystalline structure of the specimen prepared from sintered bulk ceramics. In order to demonstrate the atomic resolution 3D tomography, they selected nanocrystalline Y_2O_3 stabilized ZrO_2–$MgAl_2O_4$ bulk polycrystalline ceramics as the standard sample. A piece of microsample was cut out from a bulk sintered sample and lifted out on the top of a tungsten needle. Then it was sharpened to a needle specimen using the annular Ga^+ ion milling method. The length of the ZrO_2–$MgAl_2O_4$ tip was about 3 μm, so this is entirely different from the prior 3DAP analysis of thin film oxides. The presence of nanocrystalline Mg and Al-rich grains ($MgAl_2O_4$) is clearly observed in the AP tomography in **Figure 93**, and its nanocrystalline feature is consistent with that observed by TEM. Note that the segregation of Mg, Al and Y along the ZrO_2/ZrO_2 grain boundaries can be clearly discerned in the AP tomography as well as in the concentration profile (b). The gradually increasing curvature of the tip is reconstructed in the AP tomography, indicates that the electric field was radially applied from the tip apex to the detector, resulting in accurate reconstruction of 3D atom tomography.

As shown in the previous sections, 3DAP has been shown to be particularly useful when we study early stage of phase separation in metallic materials. Similarly, 3DAP has been demonstrated to be useful in detecting phase separation that was not clearly characterized with TEM. **Figure 94** shows AP

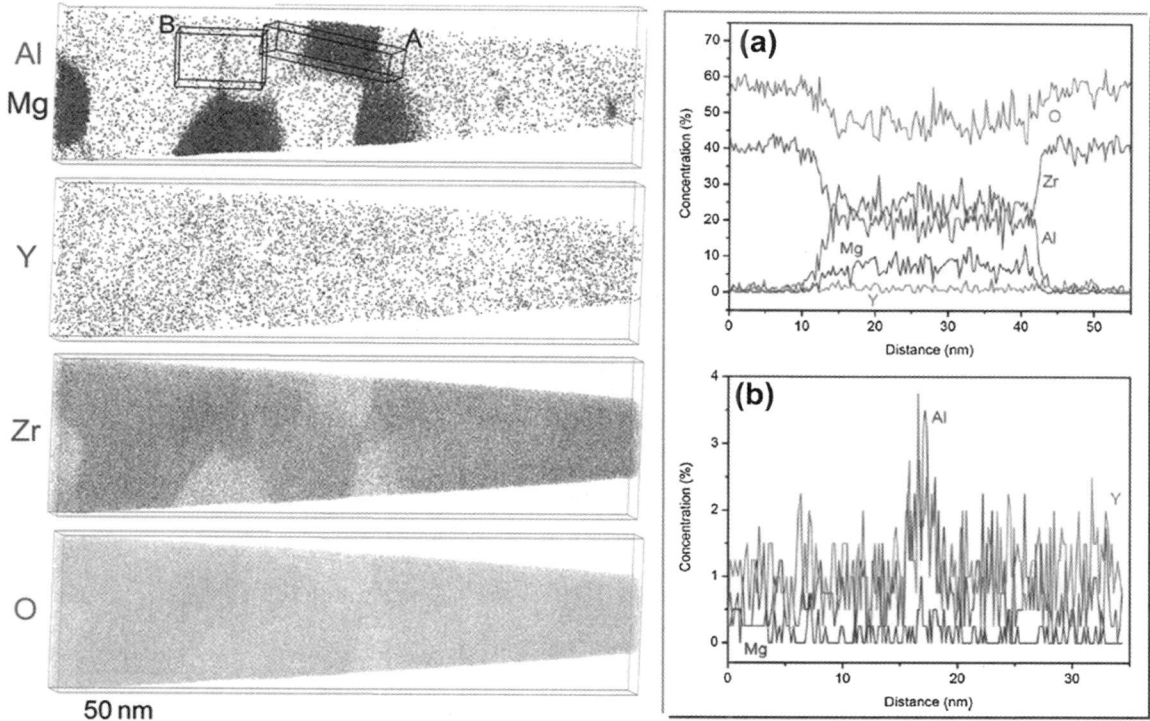

Figure 93 (a) AP tomography of 2.4 mol.% Y_2O_3–79.2 mol.% ZrO_2–18.4 mol.% $MgAl_2O_4Mg$ (green), Al (blue) and Y atoms with a volume of $15 \times 65 \times 240$ nm³. (a) Concentration depth profiles calculated from the selected volumes A and B in (a). Ref: Y. M. Chen et al. Scr. Mater. 61, 693 (2009). (For color version of this figure, the reader is referred to the online version of this book.)

tomography of Y in yttria (Y_2O_3) stabilized zirconia (ZrO_2) in the as-sintered and long-term aged sample and the concentration profiles calculated from the selected regions (Hono et al., 2011b). Yttria-stabilized zirconia is an insulator at room temperature, but it exhibits high ion conductivity at high temperature and is considered as a promising candidate for the ion conductor of a solid-oxide fuel cell (SOFC). However, the degradation of the ion conductivity during a long-time use at high temperature around 1000 °C is a problem in its application. One proposed mechanism for the ion conductivity degradation is the phase separation in the Yttria-stabilized zirconia at the service temperature. The AP tomography taken from an as-sintered 8 mol%Y_2O_3–ZrO_2 sample indicates (see **Figure 94(a)**) homogeneous distribution of Y. However, after annealing for 2000 h at 1000 °C, the AP tomography indicates that the distribution of Y is not homogeneous and that the sample is in the early stage of phase separation (see **Figure 94(b)**). As shown here, AP tomography of insulating ceramics is achievable with a high spatial resolution using the UV femtosecond laser. Using UV laser, rather quantitative analysis of Gd and Y doped Ceria was performed (Li et al., 2010, 2011c).

Marquis et al. (2010) reported 3DAP analysis result of alumina showing carbon segregation along grain boundaries using green laser and they stressed that the 3DAP analysis of insulators can be possible even with green laser. However, the merit of using short-wave length laser for the analysis of wide band gap oxide is apparent as demonstrated using MgO whose band gap is 7.8 eV, much higher than the photon

As sintered

2 × 20 × 85 nm

1000 °C × 2000 h

2 × 20 × 85 nm

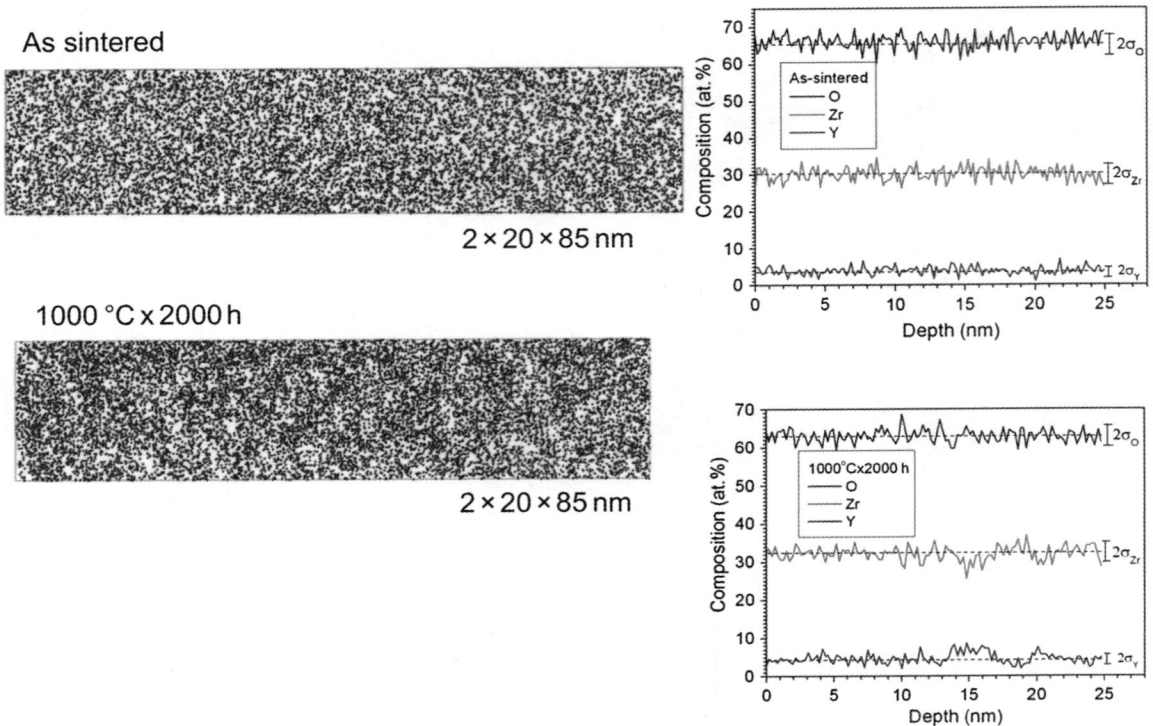

Figure 94 Atom probe tomography of Y in 8 mol% Y_2O_3–ZrO_2: (a) as-sintered sample and (b) sample annealed for 2000 h at 1000 °C. From Hono et al. (2011b). (For color version of this figure, the reader is referred to the online version of this book.)

energy of UV laser 3.5 eV (Chen et al., 2011b; Mazumder et al., 2011). Chen et al. compared the filed evaporation rate using UV and green laser at the same standing voltage and reported that the laser-assisted field evaporation is orders of magnitude faster with UV laser. **Figure 95(a)** shows AP mass spectrum of pure MgO single crystal obtained by 343 nm laser at 32 K with laser power of 0.3 µJ/pulse. All the mass peaks are well defined and three isotopes of Mg, i.e. ^{24}Mg, ^{25}Mg, and ^{26}Mg, are all clearly resolved. The mass resolution of Mg^{2+} peaks at FWHM was about 300. The AP tomography obtained from the MgO specimen is shown in **Figure 95 (b)**, suggesting that the Mg and O are uniformly detected. The reconstruction of the specimen taper means that the ions that were field evaporated from the oxide tip apex were projected to the detector due to the presence of the projection field. To explain the laser-assisted field evaporation of MgO with the UV laser, whose photon energy is lower than the band gap, Tsukada et al. (2011b) proposed the laser-induced holes accumulation as the prerequisite for field evaporation. According to their calculations, the hole on the sample surface could substantially decrease the minimum required energy for a field evaporation. In other words, the band gap at the surface with holes is much lower than that of bulk MgO, so the band gap at tip apex could be substantially lower than the band gap of bulk sample. While thermal activation by laser-assisted field evaporation is making contribution for the laser-assisted field evaporation of oxides, Tsukada et al. (2011b) and Tamura et al., (2012) proposed that hole-accumulations by direct electron excitation make significant role in UV laser-assisted field evaporation. Otherwise, it is hard to

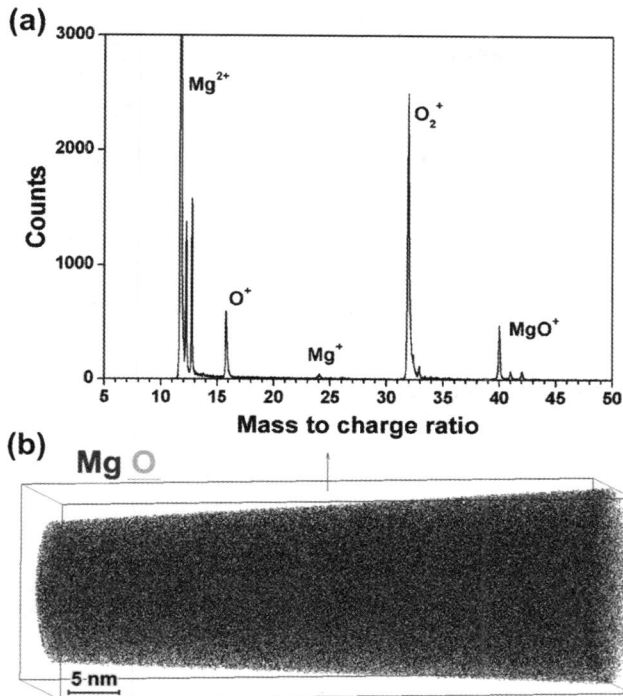

Figure 95 (a) Atom probe mass spectrum of pure MgO single crystal obtained by 343 nm laser at 32 K with laser power of 0.3 μJ/pulse. The FWHW of Mg^{2+} is 300. (b) Atom probe tomography obtained from MgO, showing uniform distribution of Mg and O.Y. M. Chen et al. Ultramicroscopy 111, 562 (2011). (For color version of this figure, the reader is referred to the online version of this book.)

explain the mass resolution exceeding 1000. Silaeva et al. (2013) developed a numerical model of laser-assisted field evaporation and accounted for the effects of field penetration, hole and electron movement, and laser absorption to explain the fs laser-assisted field evaporation.

Thermal barrier coating of heat resistant alloys is a very important technology for improving the service temperature of gas turbines and jet engines. Since the oxide analysis has been demonstrated to be feasible using laser-assisted 3DAP, the characterization of thermally grown oxides (TGOs) is developing as new application area of AP (Chen et al., 2012; Baik et al., 2013). Although concentration of oxygen does not match with the expected stoichiometry of oxides, the concentration profiles show overall growth mechanism of TGOs.

15.5   Future Directions

Thanks to the recent advances in laser-assisted 3DAP, the application area of the AP technique has expanded to a wide variety of materials including metals, ceramics, semiconductors and their devices. Although the analysis volume of AP has expanded significantly compared to those for 1DAP and early 3DAP, it is still limited to a nanometric volume ($\sim 50 \times 50 \times 100$ nm^3). The purpose of using 3DAP is to characterize atomic distributions with the near-atomic resolution in three-dimensional space to

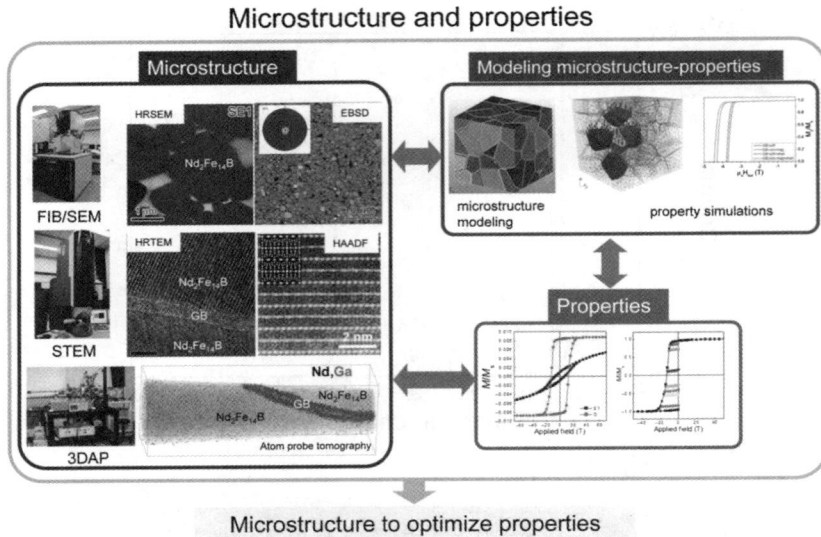

Figure 96 Multiscale characterization and understanding structure–property relationships to develop higher performance materials. (For color version of this figure, the reader is referred to the online version of this book.)

explain the origin of nano/microstructure developments and their correlations with mechanical and functional properties. However, properties of materials are often governed by the microstructure rather than atomic scale features, so it is of utmost importance to use the AP technique together with other microscopy techniques, such as TEM, SEM and sometime even with OEM. AP characterization often fails to see the forest for the tree. In this sense, the importance of "multiscale characterization" cannot be too much emphasized.

Figure 96 shows how multiscale characterization can be performed using SEM, TEM and 3DAP. Modern SEM with various imaging and spectroscopic capabilities can show microstructure details with subnanometer resolution including orientation mapping using EBSD. Modern SEM is equipped with FIB, so TEM specimen can be prepared from any specific area of the sample. Using TEM, bright field, dark field, electron diffraction, EDS, EELS spectroscopy can be performed In particular, modern STEM shows atomic column resolution with direct data interpretation. So the projection images of atomic columns can be obtained. However, the quantitative chemical information is hard to obtain with near-atomic scale resolution using TEM/STEM. On the other hand, 3DAP can give near-atomic resolution 3D atom tomography, from which rather quantitative analysis of local composition can be obtained. From this information, we can thoroughly characterize the microstructure feature with near-atomic resolution to correlate this to properties. Sometimes, it is difficult to directly correlate the microstructure and properties. In such a case, modeling of microstructure and simulation of properties expected from this microstructure is quite useful. Modern analytical facilities require large funding, and the investment to these facilities can be justified only if users can demonstrate that the analytical results can be used for the development of new materials with higher performance. To fully utilize the AP technique, it should be complemented with TEM and SEM/FIB. Stand-alone use of 3DAP technique gives little information on the forest for the tree.

References

Al-Kassab, T., Macht, M.-P., Naundorf, V., Wollenberger, H., Chambreland, S., Danoix, F., Blavette, D., 1996. Appl. Surf. Sci. 94/95, 306.

Aaronson, H.I., Reynolds, W.T., Purdy, G.R., 2004. Coupled-solute drag effects on ferrite formation in Fe–C–X systems. Metall. Mater. Trans. A 35A, 1187–1210.

Abe, E., Kawamura, Y., Hayashi, K., Inoue, A., 2002. Acta Mater. 50, 3845.

Abe, T., Shimono, M., Hashimoto, K., Hono, K., Onodera, H., 2006. Scr. Mater. 55, 421.

Agrawal, R., et al., 2011. Characterzing atomic composition and dopant distribution in wide band gap semiconductor nanowires using laser assisted atom probe tomography. J. Phys. Chem. C 115, 17688–17694.

Almazouzi, A., Numakura, H., Koiwa, M., Hono, K., Sakurai, T., 1997. Site occupation preference of Fe in Ni$_3$Al: an atom probe study. Intermetallics 5, 37–43.

Andersson, J.O., et al., 2002. Thermo-Calc and DICTRA computational tools for materials science. CALPHAD 26, 273–312.

Andren, H.O., 1996. Atom probe field ion microscopy of High-Tc superconductors. J. Phys. IV France 5. C5-C197 to C5-C204.

Archambault, V., Pere, D., 1999. Mat Res. Soc. Symp. Proc. 577, 153.

Auger, P., Pareige, P., Akamatsu, M., Blavette, D., 1995. APFIM investigation of clustering in neutroin-irradiated Fe–Cu alloys and pressure vessel steels. J. Nucl. Mater. 225, 225–230.

Ayers, J.D., Harris, V.G., Sprague, J.A., Elam, W.T., Jones, H.N., 1998. Acta Mater 46, 1861.

Baabu, S.S., et al., 2005. Fine scale precipitation in Haynes 214 alloy. In: Howe, J.M., Laughlin, D.E., Lee, J.K., Dahmen, U., Soffa, W.A. (Eds.), Proc. of International Conference on Solid—Solid Phase Transformations, vol. 2, pp. 523–528.

Babu, S.S., 2009. Thermodynamic and kinetics models for describing microstructure evolution during joining of advanced materials. Int. Mater. Rev. 54, 333–367.

Babu, S.S., et al., 1996a. Atom probe field ion microscopy investigation of CMSX-4 Ni-base superalloy laser beam welds. J. Phys. IV 6. C5-253-C5-258.

Babu, S.S., et al., 1996b. Microstructural development in PWA1480 electron beam welds—an atom probe field ion microscopy study. Appl. Surf. Sci. 94/95, 280–287.

Babu, S.S., David, S.A., Vitek, J.M., Miller, M.K., 1996c. Phase stability and atom probe field ion microscopy of type 308 stainless steel weld metal. Metall. Mater. Trans. A 27A, 763–774.

Babu, S.S., et al., 1997. Welding of nickel-base superalloys single-crystals. Sci. Technol. Welding Joining 2, 79–88.

Babu, S.S., et al., 2001. Characterization of the microstructure evolution in a nickel base superalloy during continuous cooling. Acta Mater. 49, 4149–4160.

Babu, S.S., et al., 2005. In-situ observations of lattice parameter fluctuations in austenite and transformation to bainite. Metal. Mater. Trans. A 36A, 3281–3289.

Babu, S.S., Hono, K., Sakurai, T., 1994. Atom probe field ion microscopy study of the partitioning of substitutional elements during tempering of a low-alloy steel martensite. Metall. Mater. Trans. A 25, 499–508.

Babu, S.S., David, S.A., Vitek, J.M., Miller, M.K., 1995. Phase stability and atom probe field ion microscopy of type 308 stainless steel weld metal. Appl. Surf. Sci. 87–88, 207–215.

Baik, S.I., Yin, X., Seidman, D.N., 2013. Scripta Mater. 68, 909.

Bault, B., et al., 2011. Ultramicroscopy 111, 683.

Bettles, C.J., Gibson, M.A., Venkatesan, K., 2004. Scr. Mater. 51, 193.

Bhadeshia, H.K.D.H., 1985. Diffusional formation of ferrite in iron and its alloys. Prog. Mater. Sci. 29, 321–386.

Bhadeshia, H.K.D.H., 1989. Theoretical analysis of changes in cementite composition during tempering of bainite. Mater. Sci. Technol. 5, 131–137.

Bhadeshia, H.K.D.H., 1993. High-strength (5 GPA) steel wire—an atom probe study. Appl. Surf. Sci. 67, 328–333.

Bhadeshia, H.K.D.H., 2001. Bainite in Steels, second ed. Institute of Materials, London.

Bhattacharjee, T., Mendis, C.L., Sasaki, T.T., Ohkubo, T., Hono, K., 2012. Scr. Mater. 67, 967.

Bhattacharjee, T., Mendis, C.L., Oh-ishi, K., Ohkubo, T., Hono, K., 2013. Mater. Sci. Eng. A 575, 231.

Bigot, A., Danoix, F., Auger, P., Blavette, D., Menand, A., 1996a. Appl. Surf Sci. 94/95, 261.

Bigot, A., Danoix, F., Auger, P., Blavette, D., Reeves, A., 1996b. Mater. Sci. Forum 217–222, 695.

Bigot, A., Auger, P., Chambreland, S., Blavette, D., Reeves, A., 1997. Microsc. Microanal. Microstruct. 8, 103.

Biswas, et al., 2011. Acta Mater. 59, 6187.

Blavette, D., 1992. The role of atom-probe techniques in the investigation of some selected metallurgical problems. Surf. Sci. 266, 299–309.

Blavette, D., Auger, P., 1990. Fine scale investigation of some phenomena in metallic alloys by field-ion microscopy and atom probe microanalysis. Microsc. Microanal. Microstruct., 481–492 (appeared in Fench—Swedish conference on atomic level microstructure of inorganic solids).

Blavette, D., Deconihout, B., Bostel, A., Sarrau, J.M., Bouet, M., Menand, A., 1993. Rev. Sci. Instrum. 64, 2911.

Blavette, D., Letellier, L., Deconihout, B., 1993. Microscopy and microanalysis of interfaces on a sub-nanometric scale with atom probe. Ann. Chimie Sci. Materiaux vol. 18, 303–310 (appeared in 3rd Japan/France Materials Science Seminar On Nanostructure And Properties Of Materials (Jfmss-3)).

Blavette, D., Deconihout, B., Chambreland, S., Bostel, A., 1998. Ultramicroscopy 70, 115.

Blavette, D., Cadel, E., Fraczkiewicz, A., Menand, A., 1999. Three-dimensional atomic scale imaging of impurity segregation to line defects. Science 286, 2317–2319.

Boll, T., et al., 2007. Investigation of the site occupation of atoms in pure and doped TiAl/Ti$_3$Al intermetallic. Ultramicroscopy 107, 769–801.

Booth-Morrison, C., et al., 2008. Effects of solute concentration on kinetic pathways in Ni–Al–Cr alloys. Acta Mater. 56, 3422–3438.

Boyd, J.D., Nicholson, R.B., 1971. Acta Metall. 19, 1101.

Boyer, R., 1998 "Titanium and Titanum Alloys, " Metals Handbook, Desk Edition, second ed., ASM International, pp. 574–584.

Brenner, S.S., Kowalik, J., Hua, M.J., 1991a. Surf Sci. 246, 1991.

Brenner, S.S., Kowalik, K., Hua, M.J., 1991b. Surf Sci. 246, 210.

Burke, M.G., et al., 2006. Quantitative characterization of nano-precipitates in irradiated low alloy steels: advances in the application of FEG-STEM quantitative microanalysis to real materials. J. Mater. Sci. 41, 4512–4522.

Caballero, F.G., et al., 2008. Redistribution of alloying elements during tempering of a nanocrystalline steel. Acta Mater. 56, 188–199.

Caballero, F.G., et al., 2007. Atomic scale observations of bainite transformation in a high carbon high silicon steel. Acta Mater. 55 (1), 381–390.

Caballero, F.G., et al., 2009. New experimental evidence on the incomplete transformation phenomenon in steel. Acta Mater. 57, 8–17.

Caballero, F.G., Miller, M.K., Garcia-Mateo, C., 2010a. Tracking solute atoms during bainitic reaction in nanocrystalline steel. Mater. Sci. Technol. 26, 889–898.

Caballero, F.G., Miller, M.K., Garcia-Mateo, C., 2010b. Carbon supersaturation of ferrite in a nanocrystalline bainitic steel. Acta Mater. 58, 2338–2343.

Caballero, F.G., et al., 2010c. Examination of carbon partitioning into austenite during tempering of bainite. Scr. Mater. 63, 442–445.

Cadel, E., Fraczkiewicz, A., Blavette, D., 2004. Sukuki effect on {001} stacking faults in boron doped FeAl intermetallics. Scr. Mater. 51, 437–441.

CAMECA, May 2011. IVAS 3.6, User Guide.

Cao, L., et al., 2013. Clustering behavior in an Al-Mg-Si-Cu alloy during natural ageing and subsequent under-ageing. Mater. Sci. Engg. A. 559, 257–261.

Capdevila, C., Miller, M.K., Russell, K.F., Chaoa, J., González-Carrasco, J.L., 2008. Phase separation in PM 2000™ Fe-base ODS alloy. Mater. Sci. Eng. A490, 277–288.

Capdevilla, C., et al., 2008. Phase separation in PM 2000™ Fe-base ODS alloy: experimental study at the atomic level. Mater. Sci. Eng. A A490, 277–288.

Carey, M.J., Maat, S., Chandrashekariaih, S., Katine, J.A., Chen, W., York, B., Childress, J.R., 2011. Co MnGe-based current-perpendicular-to-the-plane giant-magnetoresistance spin-valve sensors for recording head applications. J. Appl. Phys. 109, 093912.

Ceresara, S., Dirusso, E., Fiorini, P., Giarda, A., 1969/70. Mater. Sci. Eng. 5, 220.

Cerezo, A., Godfrey, T.J., Smith, G.D.W., 1988. Rev. Sci. Instrum. 59, 862.

Cerezo, A., Godfrey, T.J., Grevenor, C.R.M., Hetherington, H.G., Hoyle, R.M., Jakubovics, J.P., Liddle, J.A., Smith, G.D.W., Worrall, G.M., 1989. J. Microsc. 154, 215.

Cerezo, A., Godfrey, T.J., Hyde, J.M., Sijbrandij, S.J., Smith, G.D.W., 1994. Appl. Surf. Sci. 76/77, 374.

Cerezo, A., Hirosawa, S., Rozdilsky, I., Smith, G.D.W., 2003. Combined atomic-scale modeling and experimental studies of nucleation in the solid state. Philos. Trans. A Math. Phys. Eng. Sci. 361, 463–477.

Cerezo, A., Smith, G.D.W., Clifton, P.H., 2006. Appl. Phys. Lett. 88, 154103.

Chakrabarti, D.J., Cheong, B.K., Laughlin, D.E., 1998. In: Das, K. (Ed.), Automotive Alloys II. TMS, p. 27.

Chance, J., Ridley, N., 1981. Chromium partitioning during isothermal transformation of a eutectoid steel. Metall. Trans. A. 12, 1205–1213.

Chatterjee, D.K., Entwistle, K.M., 1973. J. Inst. Metals 101, 53.

Chellali, M.R., et al., 2012. Triple junction transport and the impact of grain boundary width in nanoscrystalline Cu. Nano Lett. 12, 3448–3454.

Chen, Y.M., Ohkubo, T., Kodzuka, M., Morita, K., Hono, K., 2009a. Scr. Mater. 61, 693–696.

Chen, Y.M., Ohkubo, T., Kodzuka, M., Morita, K., Hono, K., 2009b. Scr. Mater. 61, 693.

Chen, L.Y., Zeng, Y.W., Cao, Q.P., Park, B.J., Chen, Y.M., Hono, K., Vainio, U., Zhang, Z.L., Kaiser, U., Wang, X.D., Jiang, J.Z., 2009c. J. Mater. Res. 24, 3116.

Chen, Y.M., Ohkubo, T., Hono, K., 2011a. Ultramicroscopy 111, 562.

Chen, Y.M., Ohkubo, T., Hono, K., 2011b. Ultramicroscopy 111, 562–566.

Chen, Y., Reed, R.C., Marquis, E.A., 2012. Scr. Mater. 67, 779.

Chiaramonti, A.N., Schreiber, D.K., Egelhoff, W.F., Seidman, D.N., Petford-Long, A.K., 2008. Appl. Phys. Lett. 93, 103113.

Clarke, A.J., et al., 2008. Carbon partitioning to austenite from martensite or bainite during the quench and partition (Q&P) process: a critical assessment. Acta Mater. 56, 16–22.

Clarke, A.J., et al., 2009. Low temperature age hardening in U-13 at.% Nb: an assement of chemical redistrobution mechanisms. J. Nucl. Mater. 393, 282–291.

Coehoorn, R., de Mooij, D.B., Duchateau, J.P.W.B., Buschow, K.J.H., 1988. J. Phys. C8, 669.

Collins, D.L.W., 1957/58. J. Inst. Metals 86, 325.

Da Costa, et al., 2005. Rev. Sci. Instrum. 76, 013304.

Cottrell, A.H., Bilby, B.A., 1949. Dislocation theory of yielding and strain ageing of iron. Proc. Phys. Soc. Lond. A 62, 49–62.

Cozar, R., Pineau, A., 1973. Morphology of γ' and γ'' precipitates and thermal stability of Inconel 718 type alloys. Metall. Trans. 4, 47–59.

Danoix, F., Auger, P., 2000. Atom probe studies of the Fe-Cr system and stainless steels aged at intermediate temperature: a review. Mater. Character. 44, 177–201.

Danoix, F., Grancher, G., Bostel, A., Blavette, D., 2007a. Standard deviations of composition measurements in atom probe analyses. Part 1 conventtional 1D atom probe. Ultramicroscopy 107, 734–738.

Danoix, F., Grancher, G., Bostel, A., Blavette, D., 2007b. Standard deviations of composition measurements in atom probe analyses. Part II: 3D atom probe. Ultramicroscopy 107, 739–743.

Danoix, F., Lefebvre, W., Holmestad, R., 2009. Composition of β'' precipitates in Al–Mg–Si alloys by atom probe tomography and first principles methods. J. Appl. Phys. 106, 123527.

de Castro, V., et al., 2011. Stability of nanoscale secondary phases in an oxide dispersion strengthened Fe–12Cr alloy. Acta Mater. 59, 3927–3936.

Deconihout, D., Bostel, A., Bas, P., Chambreland, S., Letellier, L., Danoix, F., Blavette, D., 1994. Appl. Surf. Sci. 76/77, 145.

Deconihout, B., Vurpillot, F., Gault, B., Da Costa, G., Bouet, M., Bostel, A., Blavette, D., Hideur, A., Martel, G., Brunel, M., 2007a. Surf. Inter. Anal. 39, 278.

Deconihout, B., Vurpillot, F., Gault, B., Da Costa, G., Bouet, M., Bostel, A., Blavette, D., Hideur, A., Martel, G., Brunel, M., 2007b. Proc. 49th International Field Emission Symposium, 2005. Surf. Inter. Anal. 29, 278.

De Geuser, F., Lefebvre, W., Blavette, D., 2006. 3D atom probe study of solute atoms clustering during natural ageing and pre-ageing of an Al-Mg-Si alloy. Philos. Mag. Lett. 86 (04), 227–234.

Detor, A.J., Miller, M.K., Schuh, C.A., 2007. Measuring grain-boundary segregation in nanocrystalline alloys: direct validation of statistical techniques using atom probe tomography. Philos. Mag. Lett. 87 (8), 581–587.

Devaraj, A., et al., 2009. Three-dimensional morphology and composition of omega precipitates in a binary Ti-Mo alloy. Scr. Mater. 61, 701–704.

Djayaprawira, D.D., Tsunekawa, K., Nagai, M., Maehara, H., Yamagata, S., Watanabe, N., Yuasa, S., Suzuki, Y., Ando, K., 2005. Appl. Phys. Lett. 86, 092502.

Dmitrieva, O., et al., 2011. Chemical gradients across phase boundaries between martensite and austenite steel studied by atom probe tomography and simulation. Acta Mater. 59, 364-274.

Drachsel, W., von Alvensleben, L., Melmed, A.J., 1989. J. Phys. 50, C8–C541.

Duguay, S., et al., 2010. Direct imaging of boron segregation to extended defects in silicon. Appl. Phys. Lett. 97 (paper no: 242104).

Dutta, I., Allen, S.M., 1991. J. Mater. Sci. Lett. 10, 323.

Eaton, H.C., Bayuzick, R.J., 1978. Surf. Sci. 70, 408.

Edwards, G.A., Stiller, K., Dunlop, G.L., 1994. Appl. Surf. Sci. 76/77, 219.

Edwards, G.A., Stiller, K., Dunlop, G.L., Couper, M.J., 1998. Acta Mater. 46, 3893.

Elsayed, F.R., Sasaki, T.T., Mendis, C.L., Ohkubo, T., Hono, K., 2013a. Mater. Sci. Eng. A 566, 22.

Elsayed, F.R., Sasaki, T.T., Mendis, C.L., Ohkubo, T., Hono, K., 2013b. Scr. Mater. 68, 797.

Fan, C., Inoue, A., 1997. Mater. Trans. JIM 38, 1040.

Fasth, J.E., Loberg, B., Norden, H., 1967. J. Sci. Instrum. 44, 1044.

Faulkner, R.G., 1996. Segregation to boundaries and interfaces in solids. Int. Mater. Rev. 41, 198–208.

Fidler, J., Bernardi, J., 1991. J. Appl. Phys. 70, 6456.

Fischer, R., Frommeyer, G., Schneider, A., 2001. Atom probe field ion microscopy investigations on antiphase boundaries and super disclocations in NiAl alloyed with chromum. Phys. Stat. Solidi 186, 115–121.

Frommeyer, G., Fischer, R., Deges, J., Rablbauer, R., Schneider, A., 2004. APFIM Investigations on site occupancies of the ternary alloying elements Cr, Fe and Re in Ni Al. Ultramicroscopy 101, 139–148.

Galt, B., Moody, M.P., Cairney, J.M., Ringer, S.P., 2012. Atom Probe Microscopy. Springer.

Gao, X., Zhu, S.M., Muddle, B.C., Nie, J.F., 2005. Scr. Mater. 53, 1321.

Garcia, C.I., et al., 2003. Application of grain boundary engineering in lead-free Green Steel. ISIJ International, 43, 2023–2027.

Garg, A., Howe, J.M., 1991. Acta Metall. Mater. 39, 1939.

Gault, B., et al., 2009. Advances in the calibration of atom probe tomographic reconstruction. J. Appl. Phys. 105 (paper no: 034913).

Gault, B., et al., 2012. Impact of directional walk on atom probe microanalysis. Ultramicroscopy 113, 182–191.

Gault, B., Vurpillot, F., Bostel, A., Menand, A., Deconihout, B., 2005. Appl. Phys. Lett. 86, 094101.

Gault, B., Vurpillot, F., Vella, A., Gilbert, M., Menand, A., Blavette, D., Decohihout, B., 2006. Rev. Sci. Instrum. 77, 043705.

Gault, B., Geuser, F., Stephenson, L.T., Moody, M.P., Muddle, B.C., Ringer, S.P., 2008. Estimation of the reconstruction parameters for atom probe tomography. Microsc. Microanal. 14, 296–305.

Gavriljuk, V.G., 2003. Decomposition of cementite in pearlitic steel due to plastic deformation. Mater. Sci. Eng. A345, 81–89.

Gemma, R., Al-Kassab, T., Kirchheim, R., Pundt, A., 2009. Ultramicroscopy 109, 631–636.

Geng, W.T., Ping, D.H., Gu, Y.F., Gui, C.Y., Harada, H., 2007. Stability of nanoscale co-precipitates in superalloy: a combined first-principles and atom probe tomography study. Phys. Rev. 76 (224102).

Gerold, V., 1954. Z. Metallkd. 45, 599.

Ghosh, G., Olson, G.B., 2002. Precipitation of paraequilibrium cementite: experiments, and thermodynamic and kinetic modeling. Acta Mater. 50, 2099–2119.

Gladman, T., 2002. The Physical Metallurgy of Microalloyed Steels. Institute of Materials, London.

Guo, H., Purdy, G.R., 2008. Scanning transmission electron microscopy study of interfacial segregation of Mn during the formation of a partitioned grain boundary ferrite in a Fe-C-Mn-Si alloy. Metall. Mater. Trans. A 39A, 950–953.

Gupta, A.K., Gaunt, P., Chaturvedi, M.C., 1987. Philos. Mag. A55, 375.

Gupta, A.K., Marois, P.H., Lloyd, D.J., 1996. Mater. Sci. Forum 217-222, 801.

Gutfleisch, O., Franco, V., 2012. Scr. Mater. 67, 521.

Gutfleisch, O., Khlopkov, K.a, Teresiak, A.a, Müller, K.-H.a, Drazic, G.b, Mishima, C.c, Honkura, Y., 2003. IEEE Trans. Magn. 39, 2926.

Hadjipanayis, G.C., et al., 1995. IEEE Trans. Magn. 31, 3596.

Haley, D., et al., 2011. Atom probe trajectory mappong using experiemntal tip shape measurements. J. Microsc. 244, 170–180.

Hall, E.O., 1968. J. Inst. Metals 96, 21.

Han, K., van Genderen, M.J., Bottger, A., Zandbergen, H.W., Mittemeijer, E.J., 2001. Initial stages of Fe–C martensite decomposition. Philos. Mag. A—Phys. Condens. Matter Struct. Defects Mech. Prop. 81, 741–757.

Harada, H., Murakami, H., 1999. Design of Ni-base superalloys. In: Saito, T. (Ed.), Computational Materials Design, Springer Series in Materials Science, vol. 34, pp. 39–69.

Hardy, H.K., 1950-1951. J. Inst. Metals 78, 169.

Hardy, H.K., 1954-1955. J. Inst. Metals 83, 17.

Hasegawa, N., Hono, K., Okano, R., Fujimori, H., Sakurai, T., 1993. Appl. Surf. Sci. 67, 507.

Hattestrand, M., Andren, H.O., 1999. Boron distribution in 9-12% chromium steels. Mater. Sci. Eng. A 270, 33–57.

Hättestrand, M., Andren, H.-O., 1999. Boron distribution in 9-12% chromium steels. Mater. Sci. Eng. A270, 33–37.

Hättestrand, M., Andren, H.-O., 2001a. Microstructural development during aging of an 11% chromium steel alloyed with copper. Mater. Sci. Eng. A318, 94–101.

Hättestrand, M., Andren, H.-O., 2001b. Influence of strain on the precipitation reactions during creep of an advanced 9% chromium steel. Acta Mater. 49, 2123–2128.

Hinchliffe, C.E., Smith, G.D.W., 2001. Strain aging of pearlitic steel wire during post-drawing heat-treatments. Mater. Sci. Technol. 17, 148–154.

Hofer, P., Miller, M.K., Babu, S.S., David, S.A., Cerjack, H., 2000. Atom probe field ion microscopy investigation of boron containing martensitic 9% chromium steel. Metall. Mater. Trans. A 31A, 975–984.

Hofer, P., Miller, M.K., Babu, S.S., David, S.A., Cerjak, H., 2002. Investigation of boron distribution in martensitic 9% Cr Creep Resistant steel. ISIJ Int. 42, S62–66.

Homma, T., Nakawaki, S., Oh-ishi, K., Hono, K., Kamado, S., 2011. Acta Mater. 59, 7662.

Hondros, E.P., Seah, M.P., 1983. Interfacial and surface microchemistry. In: Chan, R.W., Haasen, P. (Eds.), Physical Metallurgy, third ed. Elsevier (may be need update with the new version).

Honeycombe, R.W.K., Bhadeshia, H.K.D.H., 1995. Steels: Microstructure and Properties, second ed. Edward Arnold, London.

Honma, T., et al., 2004a. Coincidence Doppler broadening and 3DAP study of the pre-precipitation stage of an Al–Li–Cu–Mg–Al alloy. Acta Mater. 52, 1997–2003.

Honma, et al., 2004b. Acta Mater. 52, 1997.

Honma, et al., July 2012. Metall. Mater. Trans. A 43A (rai T).

Honma, T., Ohkubo, T., Kamado, S., Hono, K., 2007. Acta Mater. 55, 4137.

Hono, K., 2002. Nanoscale microstructural analysis of metallic materials by atom probe field ion microscopy. Prog. Mater. Sci. 47, 621–729.

Hono, K., et al., 1992a. Early stage phase decomposition of Al–7.8 at.% Li alloy studied by APFIM. Acta Metall. Mater. 40, 3027–3034.

Hono, K., et al., 1992b. Determination of site occupation probability of Cu in Ni_3Al by atom probe field ion microscopy. Acta Metall. Mater. 40, 419–425.

Hono, K., Sepehri-Amin, H., 2012. Scr. Mater. 67, 503.

Hono, K., Hashizume, T., Hasegawa, Y., Hirano, K., Sakurai, T., 1986. Scr. Metall. 20, 487.

Hono, K., Sakurai, T., Pickering, H.W., 1989. Metall. Trans. 20A, 1585.

Hono, K., Maeda, Y., Li, J.-L., Sakurai, T., 1992c. J. Magn. Magn. Mater. 110, L254.

Hono, K., Numakrua, H., Szabo, I.A., Chiba, A., Sakurai, T., 1992d. Determination of site preference of Cu and Ge in Ni₃Al. Surf. Sci. 266, 358–363.

Hono, K., Babu, S.S., Hiraga, K., Okano, R., Sakurai, T., 1992e. Acta Metall. 40, 3027.

Hono, K., Sano, N., Sakurai, T., 1992f. Surf Sci. 266, 350.

Hono, K., Hiraga, K., Wang, Q., Inoue, A., Sakurai, T., 1992g. Acta Metall. Mater. 40, 2137.

Hono, K., Sano, N., Babu, S.S., Okano, R., Sakurai, T., 1993a. Acta Metall 41, 829.

Hono, K., Li, J.-L., Ueki, Y., Inoue, A., Sakurai, T., 1993b. Appl. Surf Sci. 67, 398.

Hono, K., Sakurai, T., Polmear, I.J., 1994. Scr. Metall. Mater. 30, 695.

Hono, K., Ping, D.H., Ohnuma, M., Onodera, H., 1999. Acta Mater. 47, 997.

Hono, K., Ohnuma, M., Murayama, M., Nishida, S., Yoshie, A., Takahashi, T., 2001. Cementite decomposition in heavily drawn pearlite steel wire. Scr. Mater. 44, 977–983.

Hono, K., Mendis, C.L., Sasaki, T.T., Oh-ishi, K., 2010. Scr. Mater. 63, 710.

Hono, K., Ohkubo, T., Chen, Y.M., Kodzuka, M., Oh-ishi, K., Sepehri-Amin, H., Li, F., Kinno, T., Tomiya, S., Kanitani, Y., 2011a. Ultramicroscopy 111, 576.

Hono, K., et al., 2011b. Broadening the applications of the atom probe technique by ultraviolet femotosecond laser. Ultramicroscopy 111, 576–583.

Horita, Z., Matsumura, S., Baba, T., 1995. Gernal formulation for ALCHEMI. Ultramicroscopy 58, 327–335.

Houard, J., et al., 2010. Optical near-field absorption at a metal tip far from plasmonic resonance. Phy. Rev. B. 81, (no. 125411).

Humphreys, E.S., et al., 2004a. Molybdenum accumulation at ferrite: austenite interfaces during isothermal transformation of an Fe–0.24 Pct C–0.93 Pct Mo alloy. Metall. Mater. Trans. A 35A, 1223–1235.

Hutchinson, C.R., Fuchsmann, A., Brechet, Y., 2004. The Diffusional formation of ferrite from austenite in Fe–C–Ni alloys. Metall. Mater. Trans. A 35A, 1211–1221.

Hwang, J.Y., et al., 2009. Compositional variations between different generations of γ' precipitates forming during continuous cooling of a commercial nickel-base superalloy. Metall. Mater. Trans. A 40A, 3059–3068.

Hyde, J.M., et al., 1995a. Spinodal decomposition in Fe–Cr alloys—experimental study at the atomic level and comparison with computer models—2: development of domain size and composition amplitude. Acta Metall. Mater. 43, 3403–3413.

Hyde, J.M., et al., 1995b. Spinodal decomposition in Fe–Cr alloys—experimental study at the atomic level and comparison with computer models—3: development of morphology. Acta Metall. Mater. 43, 3415–3426.

Ikeda, S., Hayakawa, J., Ashizawa, Y., Lee, Y.M., Miura, K., Hasegawa, H., Tsunoda, M., Matsukura, F., Ohno, H., 2008. Appl. Phys. Lett. 93, 082508.

Inoue, A., 1999. Bulk amorphous alloys. Mater. Sci. Found. vols. 4 and 6 (Trans Tech Zülich).

Inoue, A., Chen, S., Oikawa, E., Zhang, T., Masumoto, T., 1993. Mater. Lett. 16, 108.

Inoue, A., Kawase, D., Tsai, A.P., Zhang, T., Masumoto, T., 1994. Mater. Sci. Eng. A178, 255.

Isheim, D., et al., 2008. Copper-precipitation hardening in a non-ferromagnetic face-centered cubic austenitic steel. Scr. Mater. 59, 1235–1238.

Isheim, D., Hellman, O.C., Seidman, D.N., Danoix, F., Bostel, A., Blavette, D., 2001. Atomic-scale study of a transition phases precpitate and its interfacial chemistry in an Fe-15 at.% Mo – 5 at.% V alloy. Microsc. Microanal. 7, 424–434.

Islamgaliev, R.K., Chmelik, F., Kuzel, R., 1997. Thermal stability of submicron grained copper and nickel. Mater. Sci. Eng. A A237, 43–51.

Jaramillo, R.A., et al., 2005a. Effect of 30 tesla magnetic field on phase transformations in a bainitic high-strength steel. Scr. Mater. 52, 461–466.

Jaramillo, R.A., et al., 2005b. Investigation of austenite decomposition in high-carbon high-strength Fe–C–Si–Mn steel under 30 Tesla magnetic field. In: Howe, J.M., Laughlin, D.E., Lee, J.K., Dahmen, U., Soffa, W.A. (Eds.), Proc. of International Conference on Solid – Solid Phase Transformations, vol. 1, pp. 873–878.

Jayaraj, J., Mendis, C.L., Ohkubo, T., Oh-ishi, K., Hono, K., 2010. Scr. Mater. 63, 831.

Jayaram, R., Miller, M.K., 1998. Microstructural characterization of 5 to 9 pct Cr-2 pct W–V–Ta Martensitic steels. Metall. Mater. Trans. A. 29A, 1551–1558.

Jayaram, R., Russell, K.F., Miller, M.K., 1993. An atom probe study of the substitutional behavior of beryllium in NiAl. Appl. Surf. Sci. 67, 316–320.

Johnson, W.L., 1999. MRS Bull. 24, 42.

JP2774884, USP5270552, United States Patent: "method for separating specimen and method for analyzing the specimen separated by the specimen separating method," US Patent Number 5,270,552, 1993.

Kajiwara, K., Hono, K., Hirosawa, S., 2001. Mater. Trans. 42, 1858.

Källqvist, J., Andren, H.-O., 1999. Microanalysis of a stabilized austenitic steel after long term aging. Mater. Sci. Eng. A270, 27–32.

Kanno, M., Suzuki, H., Kanoh, O., 1980. J. Jpn. Inst. Metals Mater. 44, 1139.

Karlsson, L., 1988. Non-equilibrium grain boundary segregation of boron in austenitic stainless steel –III. Computer simulations. Acta Metall. 36, 25–34.

Karlsson, L., Norden, H., 1988a. Non-equilibrium grain boundary segregation of boron in austenitic stainless steel –II. Fine scale segregation behavior. Acta Metall. 36, 13–24.

Karlsson, L., Norden, H., 1988b. Non-equilibrium grain boundary segregation of boron in austenitic stainless steel –IV. Precipitation behavior and distribution of elements at grain boundaries. Acta Metall. 36, 35–48.

Karlsson, L., Andren, H.O., Norden, H., 1982. Grain boundary segregation in an austenitic stainless steel containing boron—an atom probe study. Scr. Metall. 16, 297–302.

Karlsson, L., Norden, H., Odelius, H., 1988. Non-equilibrium grain boundary segregation of boron in austenitic stainless steel –I. Large scale segregation behavior. Acta Metall. 36, 1–12.

Karnesky, R.A., Subrack, C.K., Seidman, D.N., 2007. Best-fit ellipsoids of atom-probe tomographic data to study coalescence of gamma prime (L12) precipitates in Ni–Al–Cr. Scr. Mater. 57, 353–356.

Kawamura, Y., Hayashi, K., Inoue, A., Masumoto, T., 2001. Mater. Trans. 42, 1172.

Kellogg, G.L., 1981. Determining the field emitter temperature during laser irraadiation in the pulsed laser atom probe. J. Appl. Phys. 52, 5320–5328.

Kellogg, G.L., Tsong, T.T., 1980. J. Appl. Phys. 51, 1184.

Kelly, T.F., Larson, D.J., Miller, M.K., Flinn, J.E., 1999. Three dimensional atom probe investigation of vanadium nitride precipitates and the role of oxygen and boron in rapidly solidified 316 stainless steel. Mater. Sci. Eng. A270, 19–26.

Kelly, T.F., Larson, D.J., Thompson, K., Alvis, R.L., Bunton, J.H., Olson, J.D., Gorman, B.P., 2007. Annu. Rev. Mater. Res. 37, 681.

Khachaturyan, A.G., Lindsey, T.F., Morris, J.W., 1988. Theoretical investigation of the precipitation in δ' in Al–Li. Metall. Trans. A 19A, 249–258.

Khalid, F.A., Edmonds, D.V., 1992. An atom probe FIM study of interphase precipitation in model alloy steel. Surf. Sci. 266, 424–432.

Khalod, F.A., Edmonds, D.V., 1993. Interphase precipitation in microalloyed engineering steels and model alloy. Mater. Sci. Technol. 9, 384–396.

Kim, K.Y., Noh, T.H., Kang, I.K., Kang, T., 1995. Mater. Sci. Eng. A179/180, 552.

Kimura, H., Hashiguti, R., 1961. Acta Metall. 1076, 9.

Kingery, W.D., Bowen, H.K., Uhlmann, D.R., 1976. Introduction to Ceramics, second ed. Wiley-Interscience.

Kobayashi, Y., Takahashi, J., Kawakami, K., 2011. Anomalous distribution in atom map of solute carbon in steel. Ultramicroscopy 111, 600–603.

Kodzuka, M., 2012. Ph.D. thesis, Graduate School of Pure and Applied Sciences, University of Tsukuba.

Kodzuka, M., Ohkubo, T., Hono, K., Matsukura, F., Ohno, H., 2009. Ultramicroscopy 109, 644.

Kodzuka, M., Ohkubo, T., Hono, K., 2011. Ultramicroscopy 111, 557.

Koelling, S., et al., 2011. Characteristics of cross-sectional atom probe analyses on semiconductor structures. Ultramicroscopy 111, 540–545.

Kou, S., 2003. Welding Metallurgy, second ed. Wiley-Interscience.

Krakauer, B.W., Seidman, D.N., 1998. Subnanometer scale study of segregation at grain boundaries in an Fe(Si) alloy. Acta Mater. 46, 6145–6616.

Kronmüller, H., Schrefl, T., 1994. J. Magn. Magn. Mater. 129, 66.

Kuduz, M., Schmitz, G., Kirchheim, R., 2004. Ultramicroscopy 101, 197.

Kumar, G., Nagahama, D., Ohnuma, M., Ohkubo, T., Hono, K., 2006. Scr. Mater. 54, 801.

Kundig, A.A., Ohnuma, M., Ping, D.H., Ohkubo, T., Hono, K., 2004. Acta Mater. 52, 2441.

Kundig, A.A., Ohnuma, M., Ohkubo, T., Hono, K., 2005. Acta Mater. 53, 2091.

Kundig, A.A., Ohnuma, M., Ohkubo, T., Abe, T., Hono, K., 2006. Scr. Mater. 55, 449.

Larde, R., et al., 2009. Structural analyses of a (Pt/Co)3/IrMn Multilayer: Investigation of subnanometric layers by tomographic atom probe. J. Appl. Phys. 105 (paper no: 084307).

Larson, D.J., Miller, M.K., 1998. Precipitation and segregation in $\alpha_2 + \gamma$ titanium aluminides. Mater. Sci. Eng. A 250, 65–71.

Larson, D.J., Liu, C.T., Miller, M.K., 1997a. Boron solubility and boride compositions in $\alpha_2 + \gamma$ titanium aluminide. Intermetallics 5, 411–414.

Larson, D.J., Liu, C.T., Miller, M.K., 1997b. Microstructural characterization of segregation and precipitation in $\alpha_2 + \gamma$ titanium aluminide. Mater. Sci. Eng. A 240, 220–228.

Larson, D.J., Liu, C.T., Miller, M.K., 1997c. Tungsten segregation in $\alpha_2 + \gamma$ titanium aluminide. Intermetallics 5, 497–500.

Larson, D.J., Foord, D.T., Petford-Long, A.K., Anthony, T.C., Rozdilsky, I.M., Cerezo, A., Smith, G.D.W., 1998. Ultramicroscopy 75, 147.

Larson, D.J., Foord, D.T., Petford-Long, A.K., Cerezo, A., Smith, G.D.W., 1999a. Nanotechnology 10, 45.

Larson, D.J., Foord, D.T., Petford-Long, A.K., Liew, H., Glamire, M.G., Cerezo, A., Smith, G.D.W., 1999b. Ultramicroscopy 79, 287.

Larson, D.J., Petford-Long, A.K., Cerezo, A., Smith, G.D.W., 1999c. Acta Mater. 47, 4019.

Larson, D.J., Liu, C.T., Miller, M.K., 1999d. The alloying effects of tantalum on the microstructure of an $\alpha_2 + \gamma$ titanium aluminide. Mater. Sci. Eng. A 270, 1–8.

Larson, D.J., et al., 1999e. Atom probe field ion microscopy of Zr-doped poly-synthetically twinned titanium aluminides. In: Kim, Y.W., Dimiduk, D.M., Loretto, M.H. (Eds.), Gamma Titanium Aluminides. TMS Annual Meeting, pp. 133–139.

Larson, D.J., Martens, R.L., Kelly, T.F., Miller, M.K., Tabat, N., 2000a. J. Appl. Phys. 87, 5989.

Larson, D.J., Clifton, P.H., Tabata, N., Cerezo, A., Petford-Long, A.K., Martens, R.L., Kelly, T.F., 2000b. Appl. Phys. Lett. 77, 726.

Larson, D.J., Maziasz, P.J., Kim, I.-S., Miyahara, K., 2001. Three-dimensional atom probe investigation of nanoscale titanium oxygen clustering in an oxide-dispersion strengthened Fe-12Cr-3W-0.4Ti+Y_2O_3 ferritic alloy. Scr. Mater. 44, 359--364.

Larson, D.J., Petford-Long, A.K., Ma, Y.Q., Cerezo, A., 2004a. Acta Mater. 52, 2847.

Larson, D.J., Ladwig, P.F., Chang, A., Ulfig, R.M., Kelly, T.F., 2004b. Microsc. Microanal. 10 (Suppl. 2), 518CD.

Larson, D.J., Alvis, R.L., Lawrence, D.F., Prosa, T.J., Ulfig, R.M., Reinhard, D.A., Clifton, P.H., Gerstl, S.S.A., Bunton, J.H., Lenz, D.R., Kelly, T.F., Stiller, K., 2008. Microsc. Microanal. 14 (Suppl. 2), 1254.

Larson, D.J., Geiser, B.P., Prosa, T.J., Grestl, S.S.A., Reinhard, D.A., Kelly, T.F., 2011a. Improvements in planar feature reconstructions in atom probe tomography. J. Microsc. 243, 15–30.

Larson, D.J., Marquis, E.A., Rice, P.M., Prosa, T.J., Geiser, B.P., Yang, S.-H., Parkin, S.S.P., 2011b. Scr. Mater. 64, 673.

Lauhon, L.J., et al., 2009. Atom probe tomography of semi-conductor materials and device structures. MRS Bull. 34, 738–743.

Lefebvre, W., Loiseau, A., Menand, A., 2002. Field evaporation behavior in the g phase in Ti-Al during analysis in the tomographic atom probe. Ultramicoscopy 92, 77–87.

Li, W.F., Ohkubo, T., Hono, K., Nishiuchi, T., Hirosawa, S., 2008. Appl. Phys. Lett. 93, 052505.

Li, W.F., Ohkubo, T., Hono, K., 2009. Acta Mater. 57, 1337.

Li, F., Ohkubo, T., Chen, Y.M., Kodzuka, M., Ye, F., Ou, D.R., Mori, T., Hono, K., 2010. Scr. Mater. 63, 332.

Li, T., et al., 2011a. Characterization of oxidation and reduction of a platinum-rhodium alloy by atom probe tomography. Catal. Today 175, 552–557.

Li, F., Ohkubo, T., Chen, Y.M., Kodzuka, M., Hono, K., 2011b. Ultramicroscopy 111, 589–594.

Li, F., Ohkubo, T., Chen, Y.M., Kodzuka, M., Hono, K., 2011c. Ultramicroscopy 111, 589.

Li, S., Takahashi, Y.K., Furubayashi, T., Hono, K., 2013. Appl. Phys. Lett. 103, 042405.

Liu, C.T., Sikka, V.K., 1986. Nickel aluminides for structural use. J. Metals 38, 19–21.

Liu, Z.G., Frommeyer, G., Kreuss, M., 1991a. Investigation on APBs and dissociation of super dislocations in TiAl by APFIM. Philos. Mag. Lett. 64, 117–124.

Liu, Z.G., Alkassab, T., Haasen, P., 1991b. Phase separation in a Ni-Be alloy, as studied by atom probe field ion microscopy. Surf. Sci. 246, 329–335.

Liu, J., Sepehri-Amin, H., Ohkuboa, T., Hioki, K., Hattori, A., Schrefl, T., Hono, K., 2013. Acta Mater. 61, 5387.

Lorimer, G.W., 1978. In: Russel, K.C., Aaronson, H.I. (Eds.), Precipitation Processes in Solid. The Metallurgical Society AIME, p. 87.

Lorimer, G.W., Nicholson, R.B., 1966. Acta Metall. 14, 1009.

Lundin, L., 1995. High Resolution Microanalysis of Creep Resistant 9–12% Chromium Steels, Ph. D. thesis, Department of Physics, Chalmers University, Göteborg, Sweden.

Lundin, L., Andren, H.O., 1996. Observation of molybdenum-nitrogen clustering in highly alloyed martensite. Appl. Surf. Sci. 94–95, 320–325.

Lundin, L., Richarz, B., 1995. Atom–probe study of phosphorous segregation to the carbide/matrix interface in an aged 9% chromium steel. Appl. Surf. Sci. 87/88, 194–199.

Lütjering, G., Williams, J.C., 2010. Titanium. Springer.

Makino, A., Men, H., Kubota, T., Yubuta, K., Inoue, A., 2009. J. Appl. Phys. 105 (07A308).

Makita, K., Yamashita, O., 1999. Appl. Phys. Lett. 74, 2056.

Maloney, S.K., Hono, K., Polmear, I.J., Ringer, S.P., 1999. Scr. Mater. 41, 1031.

Marceau, R.K.W., et al., 2010. Solute Clustering in Al-Cu-Ag alloys during the early stages of elevated temperature ageing. Acta Mater. 58, 4923–4939.

Marquis, E.A., 2008. Core/shell structures of oxygen-rich nanofeatures in oxide-dispersion strengthened Fe-Cr Alloys. Appl. Phys. Lett. 93, 181904.

Marquis, E.A., et al., 2010. Probing the improbable: imaging C atoms in alumina. Mater. Today 13, 34–36.

Marquis, E.A., Gault, B., 2008. Determination of the tip temperature in laser assisted atom-probe tomography using charge state distributions. J. Appl. Phys. 104 (084914).

Martin, I., Ohkubo, T., Ohnuma, M., Deconihout, B., Hono, K., 2004. Acta Mater. 52, 4427.

Maruyama, N., Smith, G.D.W., 2002. Effect of nitrogen and carbon on the early stage of austenite recrystallization in iron-niobium alloys. Mater. Sci. Eng. A327, 34–39.

Maruyama, N., Uemori, R., Terada, Y., Tamehiro, H., 1996. Form of Nb at an early srage of recovery and recrystallization in austenite of hot-deformed steel. J. Jpn. Inst. Metals 60, 1051–1057.

Mazumder, B., Vella, A., Deconihout, B., Al-Kassab, T., 2011. Ultramicroscopy 111, 571.

Melmed, A.J., Carroll, J.J., 1984. J. Vac. Sci. Technol. 2A, 1388.

Menand, A., 1999. Atom probe investigations of fine-scale features in Ti–Al alloys. In: Kim, Y.W., Dimiduk, D.M., Loretto, M.H. (Eds.), Gamma Titanium Aluminides, TMS Annual Meeting, pp. 111–124.

Menand, A., Zapolsky-Tatarenko, H., Nerac-Partaix, A., 1998. Atom probe investigations of TiAl alloys. Mater. Sci. Eng. A250, 55–64.

Mendis, C.L., Bettles, C.J., Gibson, M.A., Hutchinson, C.R., 2006. Mater. Sci. Eng. A 435, 163.

Mendis, C.L., Oh-ishi, K., Hono, K., 2008. Scr. Mater. 57, 485.

Mendis, C.L., Oh-ishi, K., Kawamura, Y., Honma, Y., Kamado, S., 2009. Acta Mater. 57, 749.

Mendis, C.L., Oh-ishi, K., Ohkubo, T., Hono, K., 2011. Scr. Mater. 64, 137.

Mendis, C.L., Oh-ishi, K., Hono, K., 2012. Metall. Mater. Trans. 43A, 3978.

Meslin, E., et al., 2010. Kinetic of solute clustering in neutron irradiated ferritic model alloys and a frech pressure vessel steel unvestigated by atom probe tomography. J. Nucl. Mater. 399, 137–145.

Miao, W.F., Laughlin, D.E., 2000. Metall. Mater. Trans. 31A, 361.

Miller, M.K., 1999a. Atom Probe Tomography: Analysis at the Atomic Level. Kluwer Academic/Plenum Publisheers.

Miller, M.K., 1999b. Atom probe studies of phase transformations in nickel-based superalloys. In: Koiwa, M., Ohtsuka, K., Miyazaki, T. (Eds.), Proceedings of Solid–Solid Phase Transformations, pp. 73–76.

Miller, M.K., 2000. Atom Probe Tomography: Analysis at the Atomic Level. Kluwer Academic.

Miller, M.K., Burke, M.G., 1992. An atom probe field ion microscopy study of neutron irradiated pressure vessel steels. J. Nucl. Mater. 195, 68–82.

Miller, M.K., Burke, M.G., 1993. An APFIM/AEM characterization of Alloy-X750. Appl. Surf. Sci. 67, 292–298.

Miller, M.K., et al., 1978. Atom probe microanalytical studies of some commercially important steels. Surf. Sci. 70, 470–484.

Miller, M.K., et al., 1979. Quantitative determination of alloy element partitioning in pearlitic steels by atom probe analysis. Ultramicroscopy 4, 368-368.

Miller, M.K., et al., 1995. Spinodal decomposition in Fe–Cr alloys—experimental study at the atomic level and comparison with computer models—I: Introduction and Methodology. Acta Metall. Mater. 43, 3385–3401.

Miller, M.K., et al., 2001. Atom probe tomography of 14Kh2MFA Cr-Mo-V steel surveillance specimens. Micron 32, 749–755.

Miller, M.K., et al., 2002. Effect of stress relief temperature and cooling rate on pressure vessel steel welds. Mater. Sci. Eng. A327, 76–79.

Miller, M.K., et al., 2003. Mater. Sci. Eng. A A353, 140–145.

Miller, M.K., et al., 2009. Evolution of the nanostructure of VVER-1000 RPV materials under neutron and post irradiation annealing. J. Nucl. Mater. 385, 615–622.

Miller, M.K., Jayaram, R., 1992. An APFIM TEM study of crept model Ni–Mo–Ta–Al superalloys. Surf. Sci. 366, 316–321.

Miller, M.K., Parish, C.M., 2011. Role of alloying elements in nanostructured ferritic steels. Mater. Sci. Technol. 27, 729–734.

Miller, M.K., Smith, G.D.W., 1989. Atom Probe Microanalysis: Principles and Applications to Materials Problems. Materials Research Society, Pittsburgh.

Miller, M.K., Smith, G.D.W., 1989. Atom Probe Microanalysis: Principles and Applications to Materials Problems. MRS Publication.

Miller, M.K., Beaven, P.A., Smith, G.D.W., 1981. A study of the early stages of tempering of iron-carbon martensites by atom probe field ion microscopy. Metallurgical Mater. Trans. A 12, 1197–1204.

Miller, M.K., Beaven, P.A., Brenner, S.S., Smith, G.D.W., 1983. An atom probe study of the aging of iron-nickel-carbon martensite. Metallurgical Mater. Trans. A 14, 1021–1024.

Miller, M.K., Jayaram, R., Lin, L.S., Cetel, A.D., 1994. APFIM characterization of single-crystal PWA 1480 nickel-base superalloy. Appl. Surf. Sci. 76/77, 172–176.

Miller, M.K., Cerezo, A., Hetherington, M.G., Smith, G.D.W., 1996. Atom Probe Field Ion Microscopy. Oxford University Press, Oxford.

Miller, M.K., Babu, S.S., Burke, M.G., 1999. Intragranular precipitation in Alloy 718. Mater. Sci. Eng. A270, 14–18.

Miller, M.K., Babu, S.S., Burke, M.G., 2002. Comparison of the phase compositions in Alloy 718 measured by atom probe tomography and predicted by thermodynamic calculations. Mater. Sci. Eng. A327, 84–88.

Miller, M.K., Kenik, E.A., Russell, K.F., Heatherly, L., Hoelzer, D.T., Maziasz, P.J., 2003. Atom probe tomography of nanoscale particles in ODS ferritic alloys. Mater. Sci. Eng. A353, 140–145.

Miller, M.K., Russell, K.F., Thompson, G.B., 2005. Ultramicroscopy 102, 287.

Miller, M.K., Longstreth-Spoor, L., Kelton, K.F., 2011. Detecting density variations and nanovoids. Ultramicroscopy 111, 469–472.

Miller, M.K., Kelly, T.F., Rajan, K., Ringer, S.P., April 2012. The future of atom probe tomography—review Article. Mater. Today.

Moody, M., et al., 2009. Qualification of tomographic reconstruction in atom probe by advanced spatial distribution map techniques. Ultramicroscopy 109, 815–824.

Moons, T., Ratchev, P., De Smet, P., Verlinden, B., Van Houtte, P., 1996. Scr. Mater. 35, 939.

Mottura, A., et al., 2010. Aom probe tomography analysis of the distribution of rhenium in nickel alloys. Acta Mater. 58, 931–942.

Moutanabbir, O., et al., 2011. Ultravilet-laser atom-probe tomographic three-dimensional atom-by-atom mapping of isotopically modulated Si nanoscopic layers. Appl. Phys. Lett. 98 (Paper No: 013111).

Müller, E.W., Krishnaswamy, S.V., 1968. Rev. Sci. Instrum. 39, 83.

Müller, E.W., Krishnaswamy, S.V., 1974. Rev. Sci. Instrum. 45, 1053.

Müller, E.W., Panitz, J.A., McLane, S.B., 1968. Rev. Scient Instrum 39, 83.

Murakami, H., Harada, H., Bhadeshia, H.K.D.H., 1994. The location of atoms in Re-containing and V-containing multicomponent nickel-base single-crystal superalloys. Appl. Surf. Sci. 76, 177–183.

Murayama, M., Hono, K., 1999. Acta Mater. 47, 1537.

Murayama, M., Hono, K., Saga, M., Kikuchi, M., 1998. Mater. Sci. Eng. A250, 127.

Murayama, M., Hono, K., 2001. Role of Ag and Mg on the precipitation of T-1 phase in an Al-Cu-Li-Mg-Ag alloy. Scr. Mater. 44, 701–706.

Murayama, M., Katayama, Y., Hono, K., 1999. Microstructure evolution in a 17-4 PH stainless steel after aging at 400°C. Metall. Mater. Trans. A. 30A, 345–353.

Murayama, M., et al., 2001. The effect of Cu additions on the precipitation kinetics in an Al-Mg-Si alloy with excess Si,. Metall. Mater. Trans. A. 32, 239–246.

Nag, S., et al., 2011. Novel mixed-mode phase transition involving a composition-dependent displacive component. Phys. Rev. Lett. 106, 245701.

Nag, S., Banerjee, R., Hwang, J.Y., Harper, M., Fraser, H.L., 2009a. Elemental partitioning between alpha and beta phases in the Ti–5Al–5-Mo–3Cr–0.5Fe (Ti-5553) alloy. Philos. Mag. 89, 535–552.

Nag, S., Banerjee, R., Fraser, H.L., 2009b. Intra-granular alpha precipitation in Ti–Nb–Zr–Ta biomedical alloys. J. Mater. Sci. 44, 808–815.

Nag, S., et al., 2009c. ω-assisted nucleation and growth of α precipitates in the Ti-5Al-5Mo-5V-3Cr-0.5Fe β titanium alloy. Acta Mater. 57, 2136–2147.

Nagahama, D., Ohkubo, T., Hono, K., 2003. Scr. Mater. 49, 729.

Nagai, Y., et al., 2001. Role of vacancy-solute complex in the initial age hardening in an Al-Cu-Mg alloy. Acta Mater. 49, 913–920.

Nagai, et al., 2001. Acta Mater. 49, 913.

Nagai, et al., 2002. Philos. Mag. 82, 1559.

Ng, Y.S., Tsong, T.T., McLane, S.B., 1979. Atom-probe investigation of surface segregation in Ni–Cu, stainless steel 410 and Pt–Cu alloys. Surf. Sci. 84, 31–53.

Nie, J.F., 2001. Scr. Mater. 48, 1009.

Nie, J.F., Muddle, B.C., 1997. Scr. Mater. 37, 1475.

Nie, J.F., Muddle, B.C., 2000. Acta Mater. 48, 1691.

Nie, J.F., Gao, X., Zhu, S.M., 2005. Scr. Mater. 53, 1049.

Nie, J.F., Gao, X., Zhu, S.M., 2005. Scr. Mater. 53, 1049.

Nie, J.F., Oh-ishi, K., Gao, X., Hono, K., 2008. Acta Mater. 56, 6061.

Niewieczerzai, D., Oleksy, C., Szczepkowicz, A., 2012. Multi-scale similation of field ion microscopy images—image compression with and without the tip shank. Ultramicroscopy 112, 1–9.

Nishikawa, O., Kimoto, M., 1994. Appl. Surf. Sci. 76/77, 424.

Oh, J.C., Ohkubo, T., Mukai, T., Hono, K., 2005a. Scr. Mater. 53, 675.

Oh, J.C., Ohkubo, T., Kim, Y.C., Fleury, E., Hono, K., 2005b. Scr. Mater. 53, 165.

Oh-ishi, K., Hono, K., Shin, K.S., 2008. Mater. Sci. Eng. A 496, 425.

Oh-ishi, K., Watanabe, R., Mendis, C.L., Hono, K., 2009a. Mater. Sci. Eng. A 526, 177.

Oh-ishi, K., Mendis, C.L., Honma, T., Kamado, S., Ohkubo, T., Hono, K., 2009b. Acta Mater. 57, 5593.

Oh-ishi, K., Mendis, C.L., Ohkubo, T., Hono, K., 2011. Ultramicroscopy 111, 715.

Ohkubo, T., Hono, K., 2010. Unpublished Research, National Institute for Materials Science, Tsukuba, Japan.

Ohmori, Y., Tamura, I., 1992. An interpretation of the carbon redistribution during aging of high-carbon martensite. Metallurgical Trans. A 23, 2147–2158.

Ohmori, Y., Ogo, T., Nakai, K., Kobayashi, S., 2001. Effects of ω-phase precipitation on β-α, α'' transformations in a metastable β titanium alloy. Mater. Sci. Eng. A 312, 182–188.

Ohnuma, M., Ping, D.H., Abe, T., Onodera, H., Hono, K., Yoshizawa, Y., 2003. J. Appl. Phys. 93, 9187.

Ohnuma, M., Hono, K., Onodera, H., Pedersen, Linderoth, S., 1999. Nanostruct. Mater. 12, 693.

Ohnuma, M., Hono, K., Linderoth, S., Pedersen, J.S., Yoshizawa, Y., Onodera, H., 2000. Acta Mater. 48, 4783.

Ohodnicki Jr., P.R., Qin, Y.L., Laughlin, D.E., McHenry, M.E., Kodzuka, M., Ohkubo, T., Hono, K., Willard, M.A., 2009. Acta Mater. 57, 87.

Ohsaki, S., Yamazaki, K., Hono, K., 2005a. Scr. Mater. 48, 1569.

Ohsaki, S., Hono, K., Hidaka, H., Takaki, S., Hono, K., 2005b. Scr. Mater. 52, 271.

Ohta, M., Yoshizawa, Y., 2007. Jpn. J. Appl. Phys. 46, L477.

Okuda, H., et al., 1998. Nature of an endothermic peak of as-quenched Al-11.8 mol% Li alloys. Mater. Trans. JIM 39, 62–68.

Olson, G.B., Bhadeshia, H.K.D.H., Cohen, M., 1990. Coupled diffusional/displacive transformations; part II. Solute trapping. Metall. Trans. A. 21A, 805–809.

Ortner, S.R., Grovenor, C.R.M., 1988. Scr. Metall. 22, 843.

Ortner, S.R., Grovenor, C.R.M., Shollock, B.A., 1988. Scr. Metall. 22, 843.

Osamura, K., Nakamura, T., Kobayashi, A., Hashizume, T., Sakurai, T., 1986. Acta Metall. 34, 1563.

Panitz, J.A., 1978. Prog. Surf Sci. 8, 219.

Pareige, P., Van Duysen, J.C., Auger, P., 1993. An APFIM study of the microstructure of a ferritic alloy after high fluence neutron irradiation. Appl. Surf. Sci. 67, 342–347.

Pareige, P., Welzel, S., Auger, P., 1996. Clustering effects under irradiation in Fe–0.1%Cu alloy: an atomic scale investigation with tomographic atom probe. J. Phys. IV 6, 229–234.

Park, S.C., Lim, J.D., Eliezer, D., Shin, K.S., 2003. Mater. Sci. Forum 419, 159.

Pashley, D.W., Rhodes, J.W., Sendorek, A., 1966. J. Inst. Metals 94, 41.

Pashley, D.W., Jacobs, M.H., Vietz, J.T., 1967. Philos. Mag. 16, 51.

Pecker, A., Johnson, W.L., 1993. Appl. Phys. Lett. 63, 2342.

Peet, M., et al., 2004. Three-dimensional atom probe analysis of carbon distribution in low-temperature bainite. Scr. Mater. 50, 1277–1281.

Pereloma, E.V., et al., 2006. Fine-scale microstructural investigations of warm rolled low-carbon steels with and without warm rolled low-carbon steels with and without Cr, P and B addtions. Acta Mater. 54, 4539–4551.

Pickering, F.B., 1975. High-strength, Low-alloy Steels-A Decade of Progress. In: Proceedings of Microalloying, vol. 75. Union Carbide Corporation.

Ping, D.H., Hono, K., Hirosawa, S., 1998. J. Appl. Phys. 83, 7769.

Ping, D.H., Hono, K., Kanekiyo, H., Hirosawa, S., 1999a. Acta Mater. 47, 4641.

Ping, D.H., Hono, K., Kanekiyo, H., Hirosawa, S., 1999b. J. Appl. Phys. 85, 2448.

Ping, D.H., Wu, Y.Q., Hono, K., Willard, M.A., McHenry, M.E., Laughlin, D.E., 2001. Scr. Mater. 45, 781.

Ping, D.H., Hono, K., Kawamura, Y., Inoue, A., 2002. Phil. Mag. Lett. 82, 543.

Ping, D.H., Hono, K., Nie, J.F., 2003. Scr. Mater. 48, 1017.

Pinitsoontorn, S., Cerezo, A., Petford-Long, A.K., Mauri, D., Folks, L., Carey, M.J., 2008. Appl. Phys. Lett. 93, 071901.

Pogatscher, S., et al., 2011. Acta Mater. 59, 3352.

Pogatscher, S., et al., 2012. Influence of interrupted quenching on artificial ageing of Al-Mg-Si alloys. Acta Mater. 60, 4496–4505.

Polmear, I.J., 1964. Trans. Metall. Soc. AIME 230, 1331.

Raabe, D., Ohsaki, S., Hono, K., 2009. Acta Mater. 57, 5254.

Radmilovic, V., Thomas, G., Shiflet, G.M., Starke, E.A., 1989. Scr. Mater. 23, 1141.

Rajan, K., 2011. Atom probe tomography—a high throughout screening tools for the atomic scale chemistry. Comb. Chem. High Throughput Screeing 14, 198–205.

Ratchev, P., Verlinden, B., DeSmet, P., VanHoutte, P., 1998. Acta Mater. 46, 3523.

Read, H.G., 1997. Tempering of cold-drawn dual phase a + a' steel wire. Scr. Mater. 37, 151–157.

Read, H.G., Reynolds Jr., W.T., Hono, K., Tarui, T., 1997. APFIM and TEM studies of drawn pearlitic wire. Scripta Materialia 37, 1221–1230.

Reed, R.C., 2006. The Superalloys: Fundamentals and Applications. Cambridge University Press.

Reich, L., Murayama, M., Hono, K., 1998. Acta Mater. 46, 6053.

Reich, L., Ringer, S.P., Hono, K., 1999. Philos. Mag. Lett. 79, 639.

Ringer, S.P., Hono, K., 2000. Mater. Character. 44, 101.

Ringer, S.P., Hono, K., Sakurai, T., 1995. Metall. Mater. Trans. 26A, 2207.

Ringer, S.P., Sakurai, T., Polmear, I.J., 1997. Acta Mater. 45, 3731.

Rissig, L., et al., 2005. Comparison of oxide measurement techniques applied to Ti6Al4V. Mater. Character. 55, 153–159.

Ruban, A.V., Skriver, H.L., 1997. Calculation of substitution in ternary γ'-Ni3Al; Temperature and Composition effects. Phys. Rev. B 55, 856–874.

Saito, Y., et al., 1999. Novel ultra-high straining process for bulk materials – development of the accumlative roll-bonding (ARB) process. Acta Mater. 47, 579–583.

Sakuraba, Y., Ueda, M., Miura, Y., Sato, K., Bosu, S., Saito, K., Shirai, M., Konno, T.J., Takanashi, K., 2012. Appl. Phys. Lett. 101, 252408.

Sakurai, T., Hashizume, T., Jimbo, A., 1984. High-Performance, focusing-type time-of-flight atom probe with a channeltron as a signal detector. Appl. Phys. Lett. 44, 38–40.

Sakurai, M., Matsuura, M., Kim, S.H., Yoshizawa, Y., Yamauchi, K., Suzuki, K., 1994. Mater. Sci. Eng. A179, 469.

Sankaran, R., Laird, C., 1974. Mater. Sci. Eng. 14, 217.

Sasaki, T.T., Oh-ishi, K., Ohkubo, T., Hono, K., 2006a. Scr. Mater. 53, 251.

Sauvage, X., Lefebvre, W., Genevois, C., Ohsaki, S., Hono, K., 2009. Scr. Mater. 60, 1056–1061.

Schleiwies, J., Schmitz, G., 2002. Mater. Sci. Eng. A 327, 94–100.

Schlesiger, R., Oberdorfer, C., Würz, R., Greiwe, G., Stender, P., Artmeier, M., Pelka, P., Spaleck, F., Schmitz, G., 2010. Design of a laser-assisted tomographic atom probe at Münster University. Rev. Sci. Instrum. 81, 043703.

Schneider, S., Thiyagarajan, P., Johnson, W.L., 1996. Appl. Phys. Lett. 68, 493.

Sepehri-Amin, H., Li, W.F., Ohkubo, T., Nishiuchi, T., Hirosawa, S., Hono, K., 2010a. Acta Mater. 58, 1309.

Sepehri-Amin, H., Ohkubo, T., Nishiuchi, T., Hirosawa, S., Hono, K., 2010b. Scr. Mater. 63, 1124.

Sepehri-Amin, H., Ohkubo, T., Nishiuchi, T., Hirosawa, S., Hono, K., 2011a. Ultramicroscopy 111, 615.

Sepehri-Amin, H., Ohkubo, T., Shima, T., Hono, K., 2012. Acta Mater. 60, 819.

Sepehri-Amin, H., Ohkubo, T., Nagashima, S., Yano, M., Shoji, T., Kato, A., Schrefl, T., Hono, K., 2013. Acta Mater. http://dx.doi.org/10.1016/j.actamat.2013.07.049.

Seto, K., Larson, D.J., Warren, P.J., Smith, G.D.W., 1999. Grain boundary segregation in boron added interstitial free steels studied by 3-dimebsioal atom probe. Scr. Mater. 40, 1029–1034.

Shekhter, A., Aaronson, H.I., Miller, M.K., Ringer, S.P., Pereloma, E.V., 2004. Effect of aging and deformation on the microstructure and properties of Fe-Ni-Ti maraging steel. Metall. Mater. Trans. A 35A, 973–983.

Sherman, D.H., et al., 2007. Characterization of the carbon and retained austenite distributions in martensitic medium carbon, high silicon steel. Metall. Mater. Trans. A 38A, 1698–1711.

Shimono, M., Onodera, H., 2007. Mat. Sci. Eng. 449-451, 717.

Sieloff, D.D., Brenner, S.S., Burke, M.G., 1986. Atom probe field-ion microscopy of conventionally cast and melt-spun nickel aluminide alloys containing boron and hafnium. J. Phys. Colloques 47. C7-C289 -; C7–293.

Silaeva, E.P., Shcheblanov, N.S., Itina, T.E., Vella, A., Houard, J., Sévelin-Radiguet, N., Vurpillot, F., Deconihout, B., 2013. Appl. Phys. A 110, 703.

Silcock, J.M., 1960-1961. J. Inst. Metals 89, 203.

Silcock, J.M., Heal, T.J., Hardy, H.K., 1955. J. Inst. Metals 84, 23.

Smith, G.D.W., Cerezo, A., Godfrey, T.J., Wilde, J., Venker, F.M., 1998. Atom probe studies of the nano chemistry of steels. Microsc. Microanal. 4 (Suppl. 2), 104–105.

Soffa, W.A., Laughlin, D.E., 1989. Decomposition and ordering processes involving thermodynamically first-order order-disorder transformations. Acta Metall. 37, 3019–3029.

Srinivasarao, B., Oh-ishi, K., Ohkubo, T., Hono, K., 2009. Acta Mater. 57, 3277.

Stender, P., Oberdorfer, C., Artmeier, M., Pelka, P., Spaleck, F., Schmitz, G., 2007. Ultramicroscopy 107, 726.

Stephenson, L.T., et al., 2007. Microsc. Microanal. 13, 448.

Stiller, K., Danoix, F., Bostel, A., 1996. Investigation if precipitation in a new maraging stainless steel. Appl. Surf. Sci. 94/95, 326–333.

Stiller, K., Hättestrand, M., Danoix, F., 1998. Precipitation in 9Ni–12Cr–2Cu maraging steels. Acta Mater. 46, 6063–6073.

Stiller, K., Warren, P.J., Hansen, V., Angenete, J., Gjonnes, J., 1999. Mater. Sci. Eng. A270, 55.

Stone, H., et al., 2008. Synchrotron X-ray studies of Austenite and Bainitic Ferrite. Proc. Royal Soc. A 464, 1009–1027.

Sudbrack, C.K., 2004. The influence of tungsten on the chemical composition of a temporally evolving nanostructure of a model Ni–Al–Cr superalloy. Microsc. Microanal. 10, 355–365.

Sudbrack, C.K., et al., 2006. Temporal evolution of the nanostructure and phase compositions in a model Ni-Al-Cr alloy. Acta Mater. 54, 3199–3210.

Sudbrack, C.K., et al., 2007. Compositional pathways and capillary effects during isothermal precipitation in a nondilute Ni-Al-Cr alloy. Acta Mater. 55, 119–130.

Sudbrack, C.K., et al., 2008. Effects of a tungsten addition on the morphological evolution, spatial correlations and temporal evolution of a model Ni-Al-Cr superalloy. Acta Mater. 56, 448–463.

Suh, I.S., Park, J.K., 1995. Scr. Metall. Mater. 33, 205.

Takagi, T., Ohkubo, T., Hirotsu, Y., Murty, B.S., Hono, K., Shindo, D., 2001. Appl. Phys. Lett. 79, 485.

Takagishi, M., Yamada, K., Iwasaki, H., Fuke, H.N., Hashimoto, S., 2010. Magnetoresistance ratio and resistance area design of CPP-GMR film for 2–5 Tb/in read sensors. IEEE Trans. Magn. 46, 2086–2089.

Takahashi, Y.K., Hase, N., Kodzuka, M., Itoh, A., Koganezawa, T., Furubayashi, T., Li, S., Varaprasad, B.S. D.Ch. S., Ohkubo, T., Hono, K., 2013. J. Appl. Phys. 113, 223901.

Takamizawa, H., et al., 2011. Channel dopant distribution in Metal-Ozide-Semiconducor Field-Effect transducers analyzed by laser-assisted atom probe tomography. Appl. Phys. Express 4 (paper no: 036601).

Takamizawa, H., Shimizu, Y., Nozawa, Y., Toyama, T., Morita, H., Yabuuchi, Y., Ogura, M., Nagai, Y., 2012. Appl. Phys. Lett. 100, 093502.

Takeshita, T., Morimoto, K., 1996. J. Appl. Phys. 79, 5040.

Tamura, H., Tsukada, M., McKenna, K.P., Shluger, A.L., Ohkubo, T., Hono, K., 2012. Phys. Rev. B 86, 195430.

Tanner, L., Ray, R., 1990. Scr. Metall. 14, 657.

Taylor, J.A., Parker, B.A., Polmear, I.J., 1978. Metal Sci. 12, 478.

Taylor, K.A., Chang, L., Olson, G.B., Smith, G.D.W., Cohen, M., Vandersande, J.B., 1989. Spinodal decomposition during aging of Fe–Ni–C martensites. Metall. Trans. A 20, 2717–2737.

Terasaki, H., Komizo, Y., 2011. In-situ study of microstructure development during phase transformation in welding process. Solid State Phenomenon 172–174, 1261–1266 (in the conference proceedings of "Solid-Solid Phase Transformations in Inorganic Materials, " Edited by Yves Bréchet, Emmanuel Clouet, Alexis Deschamps, Alphonse Finel and Frédéric Soisson, 2011).

Thomas, G.J., 1961. Inst. Metals 94/95, 261.

Thompson, G.B., Miller, M.K., Fraser, H.L., 2004. Ultramicrosocpy 100, 25.

Thompson, K., Lawrence, D.J., Larson, D.J., Olson, J.D., Kelly, T.F., Gorman, B.P., 2007. Ultramicroscopy 107, 131.

Thomson, R.C., 2000. Characterization of carbides in steels using atom probe filed ion microscopy. Mater. Character. 44, 219–233.

Thuilier, O., et al., 2006. Atom probe tomography of the austenite-ferrite interphase boundary composition in a model alloy Fe–C–Mn. Scr. Mater. 55, 1071–1074.

Thuvander, M., Andren, H.-O., 2000. APFIM studies of grain and phase boundaries. Mater. Characterization 44, 87–100.

Thuvander, M., Stiller, K., 2000. Microstructure of a boron containing high purity nickel-based alloy 640. Mat. Sci. Engg. A. 281, 96–103.

Thuvander, M., Stiller, K., Olsson, E., Influence of heat treatment on the grain boundary microstructure in a Ni-16Cr-10Fe-0.022C model material, Mater. Sci. Technol. 15, 237–245.

Tomokhina, I.B., et al., 2011. Three-dimensional atomic scale analysis of microstructures formed in high strength steel. Mater. Sci. Technol. 27, 739–741.

Torster, M., et al., 2011. Scr. Mater. 64, 817.

Tsong, T.T., 1990. Atom-probe Field Ion Microscopy. Cambridge University Press, Cambridge.

Tsong, T.T., Ng, Y.S., Krishnaswamy, S.V., 1978. Quantification of atom-probe FIM data and an application to investigation of surface segregation of alloys. Appl. Phys. Lett. 32, 778–780.

Tsong, T.T., McLane, S.B., Kinkus, T.J., 1982. Rev. Sci. Instrum 53, 1442.

Tsukada, M., Tamura, H., McKenna, K.P., Shluger, A.L., Chen, Y.M., Ohkubo, T., Hono, K., 2011a. Ultramicroscopy 111, 567–570.

Tsukada, M., Tamura, H., McKenna, K.P., Shluger, A.L., Chen, Y.M., Ohkubo, T., Hono, K., 2011b. Ultramicroscopy 111, 567.

Valiev, R.Z., 2001. Developing SPD methods for processing bulk nanostructured materials with enhanced properties. Metals Mater. Int. 7, 413–420.

Vanbakel, G.P.E.M., Hariharan, K., Seidman, D.N., 1995. On the structure and chemistry of Ni3Al on an atomic scale via atom probe field ion microscopy. Appl. Surf. Sci. 90, 95–105.

vanGenderen, M.J., Sijbrandij, S.J., Bottger, A., Mittemeijer, E.J., Smith, G.D.W., 1997. Atom probe analysis of initial decomposition of Fe-N martensite. Mater. Sci. Technol. 13, 806–812.

Venker, F.M., Smith, G.D.W., Cerezo, A., Sijbrandij, S.J., De Hosson, J.Th. M., 1998. Precipitation processes in stainless maraging steel Sandvik 1RK91. In: Proceedings of the European Conference on Electron Microscopy, Dublin, 1996. Edited and Published by the Committee of European Societies of Microscopy, Brussels, Belgium, pp. I192–I193.

Vial, F., Joly, F., Nevalainen, E., Sagawa, M., Hiraga, K., Park, K.T., 2002. J. Magn. Magn. Mater. 242–245, 1329.

Vietz, J.T., Polmear, I.J., 1966. J. Inst. Metals 94, 410.

Viswanathan, G.B., et al., 2011. Precipitation of ordered phases in metallic solid solutions: a synergistic clustering and ordering process. Scr. Mater. 65, 485–488.

Vitek, J.M., et al., 1991. Low temperature aging behavior of type 308 stainless steel weld metal. Acta Metall. Mater. 39, 503–516.

Wagner, R., 1982. Field-ion Microscopy in Materials Science. Springer-Verlag, Berlin.

Warren, P.J., Grovenor, C.R.M., Crompton, J.S., 1992. Surf Sci. 266, 342.

Warren, P.J., Todd, I., Davies, H.A., Cerezo, A., Gibbs, M.R.J., Kendall, D., Major, R.V., 1999. Partitioning behaviour of Al in a nanocrystalline Fe71.5Si13.5B9Nb3Cu1Al2 alloy. Scr. Mater. 41 (11), 1223–1227.

Warren, P.J., Cerezo, A., Smith, G.D.W., 2000. J. Jpn. Inst. Light Metals 50, 255.

Waugh, A.R., Boyes, E.D., Southon, M.J., 1976. Surf Sci. 61, 109.

Wecker, J., et al., 1995. Appl. Phys. Lett. 67, 563.

Wilde, J., Cerezo, A., Smith, G.D.W., 2000. Three-dimensional atomic-scale mapping of a cottrell atmosphere around a dislocation in iron. Scr. Mater. 43, 39–48.

Willard, M.A., Laughlin, D.E., Mchenry, M.E., Thoma, D., Sickafus, K., Cross, J.O., Harris, V.G.J., 1998. Appl. Phys. 84, 6773.

Williams, C.A., et al., 2010. Nanoscale characterization of ODS-Eurofer 97 steel: an atom probe tomography study. J. Nucl. Mater. 400, 37–45. Wire drawing reference.

Wolverton, C., 1999. Philos. Mag. Lett. 79, 683.

Yamasaki, M., Anan, T., Yoshimoto, S., Kawamura, Y., 2005. Scr. Mater. 53, 799.

Yamasaki, M., Nishijima, M., Hiraga, K., Kawamura, Y., 2007. Acta Mater. 55, 6798.

Yoon, K.E., Seidman, D.N., Antonie, C., Bauer, P., 2008. Appl. Phys. Lett. 93, 132502.

Yoshizawa, Y., Yamauchi, K., 1990. Mater. Trans. JIM 31, 307.

Yoshizawa, Y., Oguma, S., Yamauchi, K., 1988. J. Appl. Phys. 64, 6044.

You, C.Y., Ping, D.H., Hono, K., 2006. J. Mag. Mag. Mater. 299, 136.

Yu, M.-S., Chen, H., 1992. In: Chen, H., Vasudevan, V. (Eds.), Kinetics of Ordering Transformation in Solids, p. 307.

Yu, X., et al., 2010. Characterization of microstructural strengthening of the heat-affected-zone of a blast resistant Naval steel. Acta Mater. 58, 5596–5609.

Yu, X., et al., 2011. Strength recovery in a high strength steel during multiple weld thermal simulations. Accepted for Publication in Metall. Mater. Trans. A 42, 3669–3679.

Zhang, T., Inoue, A., Masumoto, T., 1991. Mater. Trans. JIM 32, 1005.

Zhang, Y., Hono, K., Inoue, A., Makino, A., Sakurai, T., 1996. Acta Mater. 44, 1497.

Zhang, H.W., Ohsaki, S., Mitao, S., Hono, K., 2006. Mat. Sci. Eng. A 421, 191–199.

Zurob, H.S., et al., 2009. Kinetic transition during non-partitioned ferrite growth in Fe–C–X alloys. Acta Mater. 57, 2781–2792.

Biography

Kazuhiro Hono is NIMS Fellow and Director of Magnetic Materials Unit at National Institute for Materials Science and Professor of Materials Science at the University of Tsukuba. He obtained his B.S. from Tohoku University in 1982, and Ph.D. in Materials Science and Engineering from the Pennsylvania State University in 1988. His research interests are mirostructure-property relationships of metallic materials, in particular of magnetic materials.

Dr. Babu obtained his bachelors degree in metallurgical engineering from PSG College of Technology, Coimbatore, India and his master's degree in industrial welding metallurgy-materials joining from Indian Institute of Technology, Madras. He obtained his PhD in materials science and metallurgy from University of Cambridge, UK in 1992. He also worked as a research associate in the prestigious Institute for Materials Research, Sendai, Japan before joining ORNL in 1993. From 1993 to 1997, he held joint researcher position with ORNL, University of Tennessee and The Penn State University. From 1997 to 2005, he worked as an R&D staff at ORNL. From 2005 to 2007, Suresh held a senior level technology leader position in the area of engineering and materials at Edison Welding Institute, Columbus, Ohio. From 2007 to 2013, Suresh served as Professor of Materials Science and Engineering and Director of NSF I/UCRC Center for Materials Joining Science for Energy Applications, at The Ohio State University. In 2013, Suresh was appointed as UT/ORNL chair of advanced manufacturing at the University of Tennessee, Knoxville, TN. In this role he acts as a bridge to the ORNL's expertise and infrastructure including manufacturing demonstration facility to develop a collaborative research and education ecosystem locally and deploy engineering solutions to manufacturing industries.

Dr. Babu has 21 years of experience in the area of advanced manufacturing, additive manufacturing, physical metallurgy, as well as, computational materials modeling. His work relates to welding metallurgy, solid-state joining, ultrasonic additive manufacturing, laser/electron beam assisted additive manufacturing, phase transformation issues related to low-alloy steels, inclusion formation, nonequilibrium solidification, and application of computational thermodynamics and kinetics to corrosion issues. He is also involved in the application of state-of-the-art characterization tools including atom probe tomography; synchrotron diffraction and neutron diffraction for understanding interaction between weld thermal cycles, phase stability and diffusion in complex alloys, as well as, energy storage materials. Dr. Babu has published more than 140 journal papers and numerous conference proceedings.

16   Dislocations

David Rodney, Institut Lumière Matière, Université Lyon 1, CNRS UMR 5306, France
Joël Bonneville, Département de Physique et Mécanique des Matériaux, Institut P', Université de Poitiers, CNRS UPR 3346, France

16.1   Introduction

The response of crystalline materials to an external force can be roughly classified as elastic, that is reversible, or plastic when a permanent deformation remains after the removal of the external force. While the elastic behavior of crystalline solids has been characterized and understood since the pioneering work of Hooke in 1678, that is more than three centuries ago, it is only since 1934 and the theoretical works of Orowan, Polyani and Taylor that the plastic behavior of crystalline solids has been successfully linked to the existence of linear defects within the crystals. These defects, called dislocations as we will see hereafter in more details, inherently exist in the crystals, usually resulting from the solidification process. They have been directly observed for the first time in a transmission electron microscope only in 1956 in the independent works of Hirsch and Bollmann. The density of dislocations, expressed as the dislocation length per unit volume, varies considerably, from almost zero in the so-called dislocation-free silicon ($\rho \approx 10^4$ m/m^3), to nearly 10^{17} m/m^3 in heavily deformed face-centered cubic (FCC) polycrystals. Interestingly, a density of 10^{15} m/m^3 represents a dislocation length equal to approximately three times the distance of the earth to the moon in a cm^3 sample!

A dislocation line consists of a local atomic rearrangement of the crystalline lattice, which is therefore described at the nanometer scale, while dislocation motion can be over distances that are at the micrometer scale. The movement of dislocations constitutes the fundamental basis for understanding the plastic behavior of crystalline materials. The early description of dislocations was performed in the framework of continuum linear elasticity, which proved very fruitful and successful in understanding the individual and collective behaviors of dislocations, related to macroscopic phenomena, such as the yield stress and strain hardening. This approach however does not account for the processes that occur directly inside the dislocation core region. Treating this region requires an atomistic description of the crystal. The first model of a dislocation core dates back to the work of Peierls in 1940. Nowadays, dislocation cores are treated with numerical computer simulations that account with different levels of approximation for electronic effects in the crystal cohesion.

In this Chapter, we will cover the main aspects of dislocations and their cores, starting with the elasticity treatment (Sections 16.2 and 16.3), continuing with analytical models of the dislocation core

(Section 16.4) and finishing with dislocations in two particular crystallographic structures, FCC (Section 16.5) and body-centered cubic (BCC) (Section 16.6).

16.2   Plastic Deformation and Dislocations

16.2.1   Phenomenology of Plastic Deformation

16.2.1.1   *Stress and Strain Tensors—Definitions*

Let us consider a solid body submitted to an arbitrary force system from which we extract an elementary tetrahedron, as shown schematically in **Figure 1**. Once the tetrahedron was extracted, it deforms due to the removal of the stress state inside the solid body. To recover its initial size and geometry, it is necessary to apply forces on each of its sides. Each elementary force $d\vec{F}$ acting on a lateral surface of area $d\vec{S} = dS_k \cdot \vec{x}_k$ (with outward normal) can be decomposed into a component $dF_k = d\vec{F} \cdot \vec{x}_k$ normal to the surface and a tangential force $dF_t = d\vec{F} \cdot \vec{t}$, $\vec{t}$ being a unit vector perpendicular to $\vec{x}_k$. $\vec{t}$ can be further decomposed along the $\vec{x}_i$- and $\vec{x}_j$-axes leading to two tangential forces along the i- and j-directions, respectively. The associated stresses are defined by

$$\sigma_{ij} = \frac{dF_i}{dS_j} \left(\text{N·m}^{-2} = \text{Pa in SI units} \right),$$ (1)

where the subscript i indicates the direction of the force component and the index j designates the surface normal upon which the force acts, with $i = 1, 2, 3$ and $j = 3$ in **Figure 2**. Since the tetrahedron is under static equilibrium, all applied forces on the four tetrahedron faces must sum up to zero and the components dF_i of the force $d\vec{F}$ acting on the base surface ABC are therefore given by

$$dF_i = \sigma_{ij} dS_j,$$ (2)

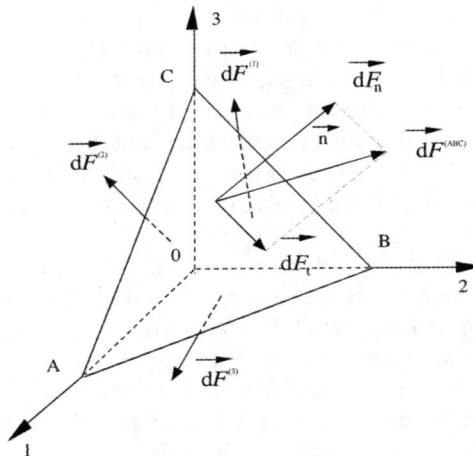

Figure 1   Infinitely small tetrahedron used to define the local stress state. The superscript in parentheses for the force $d\vec{F}$ indicates the face normal on which the force is applied.

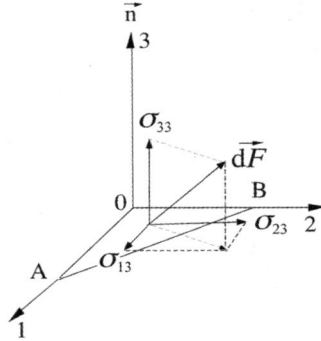

Figure 2   Illustrative decomposition of an arbitrary applied force $d\vec{F}$ on a surface dS_j and associated σ_{ij} stress components. In the proposed example, $d\vec{F}$ is applied on the face of normal 3.

where Einstein summation convention is implied for j and the equation is true for each value of $i = 1, 2$ and 3. Using tensor notation, $d\vec{F}$ can be written as $d\vec{F} = \overline{\overline{\sigma}} \cdot d\vec{S}$, where $\overline{\overline{\sigma}}$ is the so-called 'stress tensor'. $\overline{\overline{\sigma}}$ is a second-rank tensor with nine components, but only six of them are independent since the equilibrium condition implies no net moment, i.e. $\sigma_{ij} = \sigma_{ji}$. Therefore, since the tetrahedron can be considered as infinitesimal and the force system is entirely arbitrary, the most general stress state in a solid body is determined by the knowledge of three normal stresses and three shear stresses or equivalently, by calculating the eigenvalues and eigenvectors of $\overline{\overline{\sigma}}$, resulting in three normal stresses and three associated principal axes.

Let us assume that the above stress tensor generates for a material point in the tetrahedron initially located at $\overrightarrow{OM} = \vec{x}(x_1, x_2, x_3)$ a displacement $\vec{u}(u_1, u_2, u_3)$, while the material point initially located at $\overrightarrow{OM'} = \vec{x}'(x_1', x_2', x_3')$, with $\vec{x}' = \vec{x} + d\vec{x}$, undergoes a displacement $\vec{u}'(u_1', u_2', u_3')$, with $\vec{u}' = \vec{u} + d\vec{u}$, where $d\vec{u}$ is the relative displacement of M with respect to M'. The distortions due to applied stresses are defined as

$$d\vec{u} = \overline{\overline{\beta}} \cdot d\vec{x} \quad \text{with} \quad \beta_{ij} = \frac{\partial u_i}{\partial x_j}, \tag{3}$$

where, similarly to the stress, index i indicates the direction of displacement while index j indicates the normal to the local displaced surface. Like $\overline{\overline{\sigma}}$, $\overline{\overline{\beta}}$ is a second-rank tensor. However, contrary to $\overline{\overline{\sigma}}$, $\overline{\overline{\beta}}$ is not necessarily symmetric because distortions do not include deformations only, but may also include rotations; in the latter case $\beta_{ij} = -\beta_{ji}$ (see **Figure 3**).

As seen in **Figure 3**, the strain components correspond to the symmetric part of $\overline{\overline{\beta}}$ and the strain tensor $\overline{\overline{\varepsilon}}$ is defined for small distortions as

$$\overline{\overline{\varepsilon}} = \frac{1}{2}\left(\overline{\overline{\beta}} + \overline{\overline{\beta}}^T\right) \tag{4}$$

where $\overline{\overline{\beta}}^T$ is the transpose of $\overline{\overline{\beta}}$. $\overline{\overline{\varepsilon}}$ is symmetric of rank two and the ε_{ij} strain component writes

$$\varepsilon_{ij} = \frac{1}{2}\left(\beta_{ij} + \beta_{ji}\right) = \frac{1}{2}\left(\frac{\partial u_i}{\partial x_j} + \frac{\partial u_j}{\partial x_i}\right) = \frac{1}{2}\left(u_{i,j} + u_{j,i}\right), \tag{5}$$

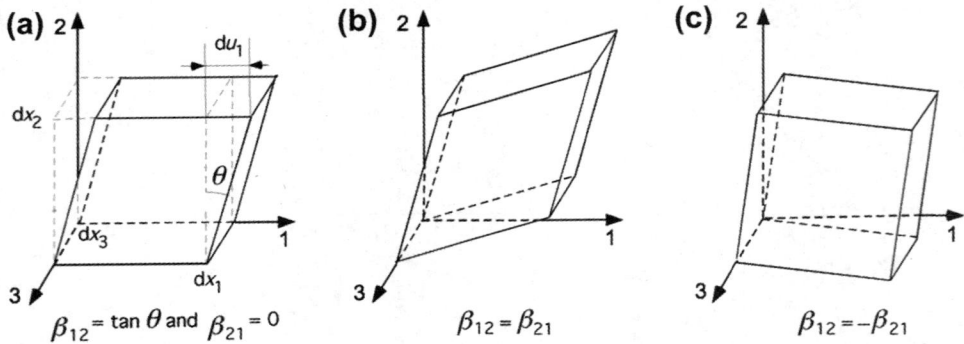

Figure 3   Schematic representations of possible distortions. (a) Simple shear, (b) Pure shear, (c) Pure rotation.

using the notation $\partial u_i / \partial x_j = u_{i,j}$. The antisymmetric part of $\bar{\bar{\beta}}$,

$$\bar{\bar{\omega}} = \frac{1}{2}\left(\bar{\bar{\beta}} - \bar{\bar{\beta}}^T\right) \tag{6}$$

describes rotations, which may result from the applied forces. As for $\bar{\bar{\sigma}}$, $\bar{\bar{\varepsilon}}$ can be diagonalized by calculating eigenvalues and eigenvectors, the latter vectors being the same as those of $\bar{\bar{\sigma}}$ in the case of linear elasticity.

The stress and strain tensors were introduced independently. In the framework of linear elasticity for infinitesimal strains, stress components are proportionally connected to strain components by the elastic coefficients (Hooke's law).

$$\sigma_{ij} = c_{ijkl}\varepsilon_{kl}, \tag{7}$$

where Einstein summation notation is again used. Since $\bar{\bar{\sigma}}$ and $\bar{\bar{\varepsilon}}$ are second-rank tensors, $\bar{\bar{\bar{c}}}$ is a fourth-rank tensor with 81 components. In the framework of isotropic linear elasticity, however, the number of elastic coefficients drastically reduces down to two independent elastic coefficients only. A common form of Hooke's law for isotropic materials is

$$\sigma_{ij} = 2\mu\varepsilon_{ij} + \lambda\varepsilon_{kk}\delta_{ij} \tag{8}$$

where δ_{ij} is Kronecker delta function and the proportionality coefficients μ and λ are called Lamé constants; μ is also often called the "shear modulus". Inverse useful equations, that give the strain components as a function of stress components, are

$$\varepsilon_{ij} = \frac{1}{E}\left[(1 + v)\sigma_{ij} - v\sigma_{kk}\delta_{ij}\right] = \frac{1}{2\mu}\sigma_{ij} - \frac{v}{E}\sigma_{kk}\delta_{ij}. \tag{9}$$

where E is Young's modulus and v Poisson's ratio. We may further proceed by calculating the elastic energy density stored in an isotropic body subjected to an external stress-field (see Appendix 1)

$$w = \frac{1}{2}\sigma_{ij}\varepsilon_{ij} \tag{10}$$

where σ_{ij} are the stresses corresponding to the applied forces that produce the ε_{ij} elastic strains.

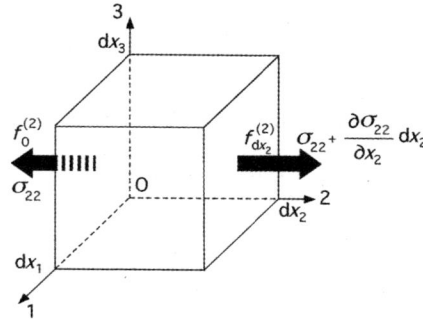

Figure 4 Infinitesimal prism of sides dx_1, dx_2 and dx_3 used to establish the equilibrium stress conditions. The prism is considered at equilibrium. The equilibrium condition is illustrated only for a stress applied of a face of normal 2 in direction 2; i.e. σ_{22}.

At rest, stresses and strains cannot be arbitrary and must meet some specific conditions, the so-called equilibrium conditions of classic elasticity. A simple way to establish these relationships consists in considering an infinitesimal rectangular prism of sides dx_1, dx_2 and dx_3 as in **Figure 4**

The force component of the σ_{22} stress acting at the origin on the parallelepiped is: $f_o^{(2)} = \sigma_{22}dx_1dx_3$, where (2) is a reminder of the stress direction, while at distance dx_2 it amounts to:

$$f_{dx_2}^{(2)} = \left(\sigma_{22} + \frac{\partial \sigma_{22}}{\partial x_2} dx_2 \right) dx_1 dx_3, \tag{11}$$

that is, when no body force is acting, the resulting force component in direction 2 due to σ_{22} is

$$f_{dx_2}^{(2)} - f_o^{(2)} = \frac{\partial \sigma_{22}}{\partial x_2} dV \quad \text{with} \quad dV = dx_1 dx_2 dx_3. \tag{12}$$

Summing all other stress contributions in direction 2 and setting the equilibrium conditions at rest yield

$$\left(\frac{\partial \sigma_{21}}{\partial x_1} + \frac{\partial \sigma_{22}}{\partial x_2} + \frac{\partial \sigma_{23}}{\partial x_3} \right) dV = 0, \tag{13}$$

which, with the derivative notation introduced in Eqn (5), can be written in the following form

$$\sigma_{21,1} + \sigma_{22,2} + \sigma_{23,3} = 0. \tag{14}$$

By analogous reasoning, a similar result is obtained in directions 1 and 3 arriving finally at the set of three equations:

$$\sigma_{ij,j} = 0 \quad \text{for} \quad i = 1, 2 \text{ and } 3 \text{ with summation convention for index } j. \tag{15}$$

Equation (15) can be expressed in the more contracted manner as div $\overline{\overline{\sigma}} = 0$, which in this form reads as a set of three equations. Equation (15) leads to the equilibrium equations for both the strain and displacement fields. Introducing Hooke's law, (Eqn (8)) into Eqn (15) gives an equilibrium equation for strains:

$$2\mu \varepsilon_{ij,j} + \lambda \varepsilon_{kk,j} \delta_{ij} = 0, \tag{16}$$

from which with Eqn (5) it directly follows for the displacements:

$$2\mu\left(u_{i,jj} + u_{j,ij}\right) + \lambda u_{k,ki} = 0. \tag{17}$$

Using the gradient operator ∇, these equations can be rewritten as

$$\mu\nabla^2 u_i + \left(\lambda + \mu\right)\nabla_i\left(\nabla \cdot \overrightarrow{u}\right) = 0 \quad \text{for} \quad i = 1, 2 \text{ and } 3, \tag{18}$$

or, in vector notation,

$$\mu\nabla^2 \overrightarrow{u} + \left(\lambda + \mu\right)\nabla\left(\nabla \cdot \overrightarrow{u}\right) = 0. \tag{19}$$

16.2.1.2 *Plastic Deformation—Experimental Background*

Any solid body submitted to an external force system undergoes deformation that, depending on the force level and distribution, can be of different nature. Let us take the simple example of a sample deformed by an axial tensile force applied at constant displacement rate. The first response is called elastic (see **Figure 5**) and corresponds to an unequivocal relationship between stress and strain. As previously seen for crystalline materials, in the limit of small strains, this relation is linear according to Hooke's law. The elastic stage corresponds to a distortion of the atomic bonds and is entirely reversible. This stage is followed by a deformation stage, called "anelastic", for which no permanent strain exists after unloading. The stress–strain curves associated with loading–unloading cycles show closed hysteresis loops indicative of energy dissipation. The relation between stress and strain is no longer linear and the hysteresis arises from a phase lag between strain and stress. This deformation stage may begin at strains lower than 10^{-5} and is currently studied through the so-called "internal friction" or "mechanical spectroscopy" techniques (Zener, 1948; Gibbons, 1964; Nowick and Berry, 1972).

With further increasing load, i.e. stress, a permanent deformation develops after sample unloading and the related stage is called plastic. The slope of the stress–strain curve then usually drastically decreases in comparison with the elastic slope and the departure from the linear behavior is now clearly visible on the curve. A subtle distinction may also exist when the onset of plastic strain is controlled by a mechanism that is different from the mechanism actually controlling the ductile behavior. Then, a microplastic stage is distinguished at incipient plasticity before the actual macroplastic stage. It must

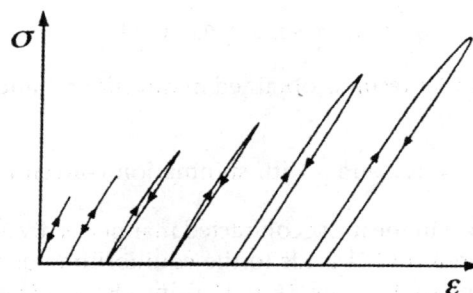

Figure 5 Stress–strain curve showing the different deformation stages of a crystalline material deformed in constant strain-rate experiment. Note that, in order to be discernible, the elastic and anelastic stages are greatly exaggerated in stress and strain, respectively. The arrows indicate the deformation path.

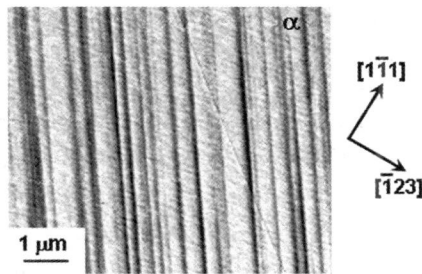

Figure 6 Traces observed by atomic force microscopy at the surface of a Ni$_3$Al single crystal sample deformed at room temperature up to about 2% plastic strain. Crystallographic orientations are reported and the [$\bar{1}$23] orientation is the compression axis. Traces coincide with (111) planes (α indicates the end of a slip trace). Courtesy of C. Coupeau (2012). (For color version of this figure, the reader is referred to the online version of this book.)

be noticed that the transition stresses from elastic to anelastic and anelastic to plastic stages are dependent on the deformation history of the sample, that is, these stresses vary with sample straining and successive deformation cycles may not necessarily yield the same critical stresses. In addition, the transition between elastic, anelastic and plastic stages is generally very gradual, so that the critical stress for macroplasticity is often conventionally defined at a given plastic strain, typically 2×10^{-3}.

The purpose of the present chapter is not to give an exhaustive review of the historical lines of thought that have originated the concept of dislocations in crystalline materials (for a concise review, see Hirth (1985)). For this, we shall use as an illustrative example, the examination of the surface of a plastically deformed sample, either by optical or atomic force microscopy. An example is given in **Figure 6** for a single crystalline sample of Ni$_3$Al deformed at constant strain-rate in compression at 77 K. Prior to deformation, the sample was carefully polished without any detectable surface defect. It is clearly observed that plastic deformation is accompanied by the appearance of lines. These lines, together with the surface relief, suggest that plastic deformation occurs by the shearing of the sample along planes, with large regions that have slipped past each other. An illustrative representation of **Figure 6** is given in **Figure 7**. In many cases, the interface between two translated blocks is planar and the corresponding surface is referred to as the glide plane. The observed lines, called slip traces or slip markings, therefore result from a step at the intersection of the glide plane and the sample surface.

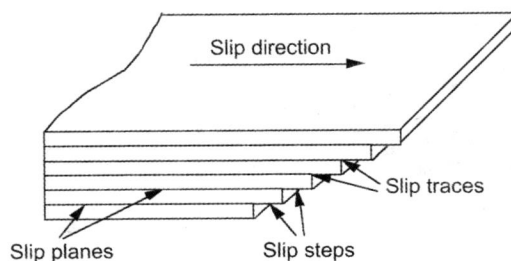

Figure 7 Schematic representation of the surface of the deformed sample shown in **Figure 6**. The surface observed in **Figure 6** corresponds to that on which steps are formed.

The preceding observations also require three general remarks:

- glide is not the unique plastic deformation mode,
- plastic deformation proceeds by shear and
- slip traces are indicative of the heterogeneity of plastic deformation, which are predominantly localized on a few glide planes separated by undeformed regions.

16.2.1.3 Theoretical Elastic Limit

A first interpretation of the above experimental observation is to consider that the different portions of the sample have moved in a rigid manner with respect to each other. This implies that the displacement of all atoms located above the glide plane with respect to all atoms located below the glide plane occurs simultaneously. Taking into account the periodic structure of crystalline materials and assuming, as schematized in **Figure 8**, that the upper part of the sample is shifted by a lattice-translation vector b, the applied shear stress τ requires:

- to be periodic with b,
- to be symmetric with respect to x, the shift or disregistery between the upper and lower half-crystals and
- to vanish at stable, $x = 0$, and unstable, $x = b/2$, equilibrium positions.

A simple solution was given by Frenkel (1926):

$$\tau = \tau_{th} \sin\left(2\pi \frac{x}{b}\right), \tag{20}$$

where τ is the shear stress and τ_{th} a constant that corresponds to the maximum sustainable shear stress. In the framework of linear elasticity, at small elastic strains, the shear strain τ is related to the shear angle θ by:

$$\tau = \mu\, \theta, \tag{21}$$

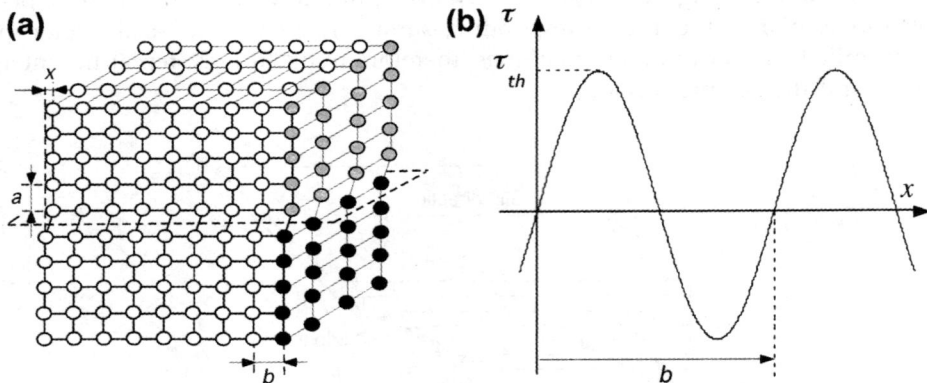

Figure 8 (a) Schematic representation of a three-dimensional cubic lattice. The upper part of the lattice is partially shifted by a distance x with respect to the lower part to illustrate the shear process. Deformation takes place on a plane represented by dashed lines. (b) Periodic function used to model the stress τ that produces the plane-on-plane displacement.

where μ is the shear modulus and $\theta = x/h$, h being the interplanar spacing. In the limit of small strain, a first-order Taylor expansion of Eqn (20) yields

$$\tau \cong \tau_{th} \, 2\pi \frac{x}{b} \equiv \mu \frac{x}{h}. \tag{22}$$

Therefore, the maximum theoretical shear stress τ_{th}, which is obtained at a shift $x = b/4$ in Eqn (20) is related to the shear modulus by the expression

$$\tau_{th} = \frac{\mu}{2\pi} \frac{b}{h}. \tag{23}$$

τ_{th} is also called the "theoretical shear stress" or "Frenkel stress". For the FCC structure, taking $b = a/\sqrt{6}$ and $h = a/\sqrt{3}$ where a is the lattice parameter (see Section 16.5 for details about the FCC structure), we have

$$\tau_{th} \approx \frac{\mu}{10}. \tag{24}$$

Experimental values of the stress at which plastic deformation starts strongly disagree with the prediction of Eqn (23). For instance in Cu, with $\mu = 45$ GPa, $\tau_{th} \approx 5$ GPa while values of the order of a few MPa are obtained experimentally. Other approaches using different interatomic-force laws have been used to estimate τ_{th} (for more detail see for instance Hirth and Lothe (1982)), but, as a rule, τ_{th} stays rather high: $\mu/4 < \tau_{th} < \mu/30$, that is, several orders of magnitude higher than experimental values. Therefore, the hypothesis that glide takes place homogeneously at the same time along the entire glide plane cannot explain the plastic deformation.

16.2.2 Concept of Dislocation

16.2.2.1 Concept of Localized Shear-Introduction to Mathematical Dislocations

Another possible approach to interpret **Figure 6** would consist in dividing the sample in elementary cubes of sides dx_j (**Figure 9(a)**). All cubes are further cut into slices of thickness ∂x_j, for instance parallel to

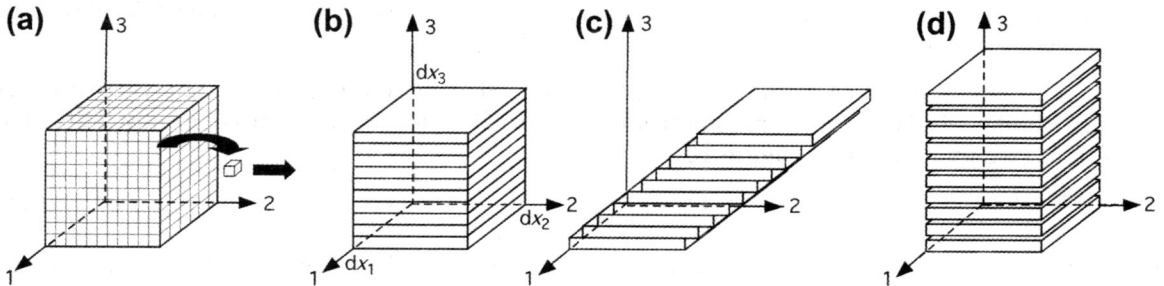

Figure 9 (a) Cutting of the deformed sample into elementary cubes to elaborate an understanding for the traces observed in **Figure 6**. (b) Cutting of the elementary cubes into slices of thickness ∂x_i ($i = 3$ in the present case). (c) Relative displacement of the slices with respect to each other in the shear direction ($j = 2$ in the figure). (d) Particular case when adding (represented case) or removing matter is necessary, that is when both the slice normal and slice displacement have similar index. Case in (d) may illustrate a diffusion process.

the slip plane (see **Figure 9(b)**). Then, a relative displacement ∂u_i is applied at the cut surfaces Γ between successive layers ∂x_j of all cubes in the shear direction i (see **Figure 9(c)**) and we further proceed to the limit $\partial x_j \rightarrow 0$, while maintaining the ratio $\partial u_i / \partial x_j$ constant. In doing this, it is given to each cube the local plastic strain corresponding to its position in the deformed sample without applying any stress.

Let us now define by du_i^p the displacement in the i direction of the cube face noted j. It follows that

$$du_i^p = \int_0^{dx_j} \frac{\partial u_i}{\partial x_j} \partial x_j = \left(\frac{\partial u_i}{\partial x_j} \right) dx_j, \tag{25}$$

where dx_j is the distance in the initial state between the sheared faces. Setting

$$\beta_{ij}^p = \frac{\partial u_i}{\partial x_j}, \tag{26}$$

one has

$$du_i^p = \beta_{ij}^p dx_j. \tag{27}$$

With this notation, as for the elastic distortions, the first subscript i indicates the shear direction while the second subscript j denotes the normal of the displaced faces. It is easy to see that a generalization of Eqn (27) in three dimensions, which takes into account all possible plastic distortions, leads to

$$d\overrightarrow{u}^p = \overline{\overline{\beta}}_p \cdot d\overrightarrow{x} \tag{28}$$

where $\overline{\overline{\beta}}_p$ is the plastic displacement tensor. It must be noticed that, in the general case where β_{ii} components exist, some matter has to be added or removed (see **Figure 9(d)**). We have now to rebuild the whole sample by adjoining all deformed cubes. Two cases may happen:

● all cubes are directly adjustable to each other without creating any crack, that is, all cubes were submitted to a similar plastic distortion and no elastic stress subsists in the plastically deformed sample;
● certain cubes do not match with each other and it is necessary to apply stresses to deform elastically these cubes to recover a compact sample. Once the compact sample is rebuilt, all applied stresses are removed and the final shape of the sample is similar to that before cutting. However, local stresses now exist in the sample.

In the latter case, rebuilding a compact sample requires that the joining faces of consecutive cubes undergo similar displacements. This implies that the total distortion of each cube must fulfill the conditions (see for an example **Figure 10**)

$$\varepsilon_{ijk}\beta_{km,i} = 0, \tag{29}$$

where $\beta_{km,i} = \partial \beta_{km}/\partial x_i$ and β_{km} is here the total distortion tensor, which now includes both elastic and plastic distortions due to the elastic stresses that were applied to rebuilt the compact sample. In Eqn (29), ε_{ijk} is the permutation operator such that:

● $\varepsilon_{ijk} = 1$, if ijk is an even permutation of $(1, 2, 3)$,
● $\varepsilon_{ijk} = -1$, if ijk is an odd permutation of $(1, 2, 3)$ and
● $\varepsilon_{ijk} = 0$, otherwise.

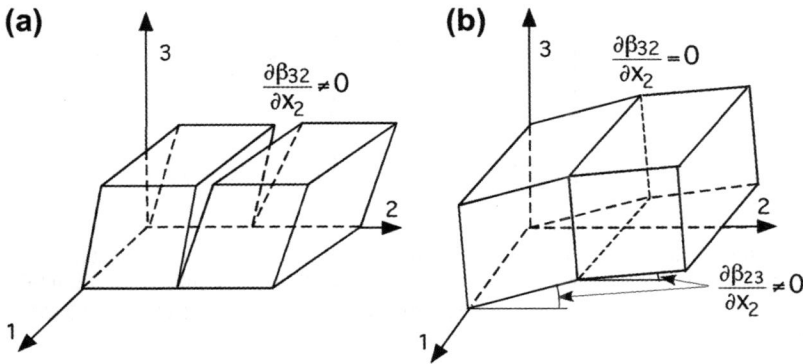

Figure 10 Three-dimensional illustration of the necessary conditions for rebuilding a compact sample after the successive operations of **Figure 9**. The conditions are indicated in the figure.

Equation (29) can be rewritten in a more general form for a deformation without crack as

$$\operatorname{curl}\overline{\overline{\beta}}^{T} = 0, \quad \text{with} \quad \overline{\overline{\beta}}^{T} = \overline{\overline{\beta}}_{e}^{T} + \overline{\overline{\beta}}_{p}^{T}, \tag{30}$$

where $\overline{\overline{\beta}}^{e}$ is the elastic distortion tensor corresponding to local elastic stresses. We therefore have

$$\operatorname{curl}\overline{\overline{\beta}}_{p}^{T} = -\operatorname{curl}\overline{\overline{\beta}}_{e}^{T}. \tag{31}$$

In general, the distribution of plastic displacements in a deformed sample does not obey the relation $\operatorname{curl}\overline{\overline{\beta}}_{p}^{T} = 0$, so that elastic distortions also exist in the sample. It is worth noting that $\operatorname{curl}\overline{\overline{\beta}}_{p}^{T} = 0$ corresponds to the first mentioned case (all cubes directly adjustable). Elastic distortions induce elastic stresses, which exist in the deformed sample even without external applied stress. The corresponding elastic stresses are defined by

$$\overline{\overline{\alpha}} = \operatorname{curl}\overline{\overline{\beta}}_{e}^{T} \tag{32}$$

and using Stoke's theorem, one has for the flux of $\overline{\overline{\alpha}}$ across a surface S

$$\int_{S}\overline{\overline{\alpha}}\,\mathrm{d}\overrightarrow{S} = \int_{S}\operatorname{curl}\overline{\overline{\beta}}_{e}^{T}\,\mathrm{d}\overrightarrow{S} = \oint_{C}\overline{\overline{\beta}}_{e}^{T}\mathrm{d}\overrightarrow{x} = \oint_{C}\mathrm{d}\overrightarrow{u}^{e}. \tag{33}$$

Let us call $\overrightarrow{b}$ the flux of $\overline{\overline{\alpha}}$ across the surface S, then we can write according to Eqns (30) and (23)

$$\overrightarrow{b} = \int_{S}\overline{\overline{\alpha}}\,\mathrm{d}\overrightarrow{S} = \oint_{C}\mathrm{d}\overrightarrow{u}^{e} = -\oint_{C}\mathrm{d}\overrightarrow{u}^{p}. \tag{34}$$

Assuming that the circuit C intersects the Γ cut-surfaces that were previously shifted by a displacement vector $\mathrm{d}\overrightarrow{u}_{p}$, as schematized in **Figure 11**, integration of Eqn (34) yields:

- $\overrightarrow{b} = 0$, for Γ cut-surfaces that completely cut the C circuit;
- $\overrightarrow{b} = -\int_{C}\mathrm{d}\overrightarrow{u}^{p}$ for a Γ cut-surface that ends along a line D inside the C circuit.

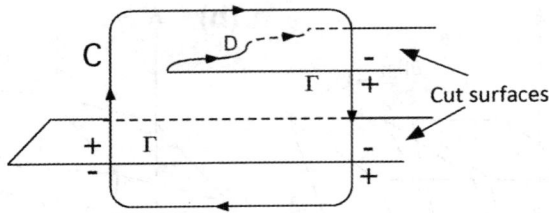

Figure 11   Schematic representation of Eqn (34). d $\vec{S}$ is the surface enclosed by the circuit (C) across which the flux of elastic displacement is evaluated. The Γ surfaces represent cut surfaces that either cross entirely the circuit C or stop inside (corresponding for instance to the slip trace labeled α in **Figure 6**). The Γ surface ending within the C circuit gives rise to a line, labeled D, that crosses the surface d $\vec{S}$.

In the latter case, $\vec{b}$ represents the relative displacement of the negative face of the Γ surface with respect to the positive face. The line D bounds the cut surface over which a shear displacement has occurred and is called a dislocation (Love, 1920), the displacement vector $\vec{b}$ its Burgers vector and C the Burgers circuit. The C circuit is oriented and, consequently, both the Γ cut-surface and the line D are also oriented. The line D is oriented by assigning a unit vector $\vec{\xi}$ tangent to the line, which points into the paper. It must be noticed that if $\vec{\xi}$ is reversed, the sense of $\vec{b}$ is also reversed since the sense of the circuit C is also reversed. On the basis of the above definitions, $\overline{\overline{\alpha}}$ is often referred to as the tensor of dislocation density and

$$\mathrm{div}\overline{\overline{\alpha}} = \mathrm{div}\left(\mathrm{curl}\,\overline{\overline{\beta}}_e^T\right) = 0, \tag{35}$$

which implies that $\vec{b}$ is constant along a dislocation line, that is, a dislocation line cannot end within a crystal. Another way to see this property is that, by construction, a dislocation bounds a surface and thus cannot end inside the crystal. Consequently, dislocation lines can only exist as closed loops or end at free surfaces or reconnect with other dislocation lines. In the latter case, it follows that if several dislocations meet at a node, one must have

$$\sum_i \vec{b}_i = 0 \tag{36}$$

when all $\vec{\xi}_i$ line dislocation vectors are oriented toward the node. This equation is often called the principle of Burgers vector conservation. It constitutes a strong link between dislocation networks and electric circuits.

To summarize, the concept developed above suggests that plastic deformation results from localized shears. Plastic deformation is therefore not homogeneous as previously assumed. Glide starts in some region of the sample and progressively extends from there. It is possible at every moment to localize the boundary between the sheared region that has moved on one side of the glide plane with respect to the other side and the corresponding boundary is the dislocation line. Unlike the previous scenario where an entire plane was sheared simultaneously, here only local movements of a few atoms along the dislocation line are now involved, significantly reducing the critical stress for plastic deformation.

16.2.2.2   *Transposition to a Crystalline Structure*
Orowan (1934), Polanyi (1934) and Taylor (1934) proposed in 1934 a transposition of the mathematical dislocation (**Figure 11**) to a dislocation in a real crystal (**Figure 12**). The transposition is given here for an edge dislocation in a simple cubic crystalline structure.

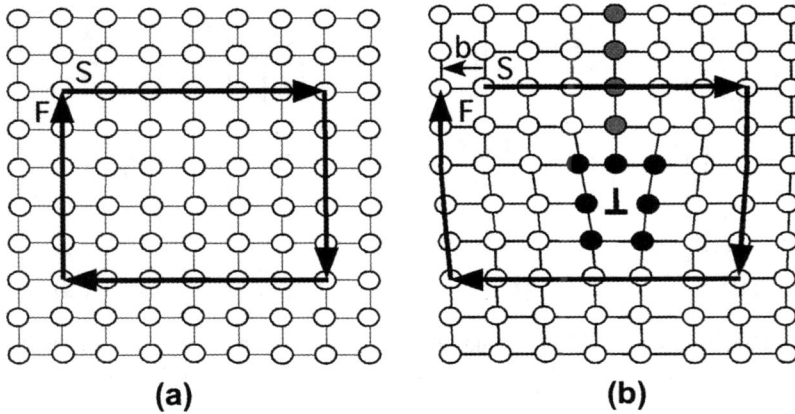

Figure 12 Transposition of **Figure 11** in a cubic crystalline material. The C circuit of **Figure 11**, called now the Burgers circuit, is represented and oriented according to the SF/RH convention. (a) Perfect reference crystal. (b) With an ending Γ line inside the C circuit. With the present convention, the dislocation line points into the paper and the resulting Burgers vector is the "closure failure" of the circuit. The extra half-plane is colored in gray, while atoms enclosing the dislocation core are in black.

Using the conventions of Section 16.2.2.1, the Burgers vector $\vec{b}$ is given by the line integral taken in the right-handed sense relative to the unit vector $\vec{\xi}$ of the dislocation line, pointing into the paper in **Figure 11**:

$$\vec{b} = \oint_C \frac{\partial \vec{u}^e}{\partial \ell} d\ell \tag{37}$$

where $\vec{u}^e$ is the elastic displacement around the dislocation. As we shall see hereafter, because of energy considerations, the Burgers vector $\vec{b}$ usually corresponds to a lattice translation of the Bravais lattice. With our conventions, the extra half-plane shown in **Figure 12** is located in the direction of the normal $\vec{n}$ of the glide plane defined by $\vec{n} = \vec{\xi} \times \vec{b}$. An implicit and convenient rule is the so-called start-finish (SF)/right-hand (RH) convention where the Burgers circuit is first drawn in the perfect reference crystal and subsequently reported in the dislocated crystal around the dislocation line (**Figure 12**). The Burgers vector $\vec{b}$ is then defined as the "closure failure" vector of the Burgers circuit.

The dislocation of **Figure 12** can be introduced in a crystal by the following virtual steps. Consider the cylinder in **Figure 13**. First, perform a cut of the cylinder parallel to its axis as shown in **Figure 13(a)**, creating in this way an arbitrary cut-surface Γ with two surfaces S₋ and S₊ bounded by a line D. Then, move rigidly the S₋ surface relative to the S₊ surface by a shear displacement vector $\vec{b}$ perpendicular to cylinder axis, glue the two surfaces back together and remove the external stress applied during the operation (**Figure 13(b)**). The line defect that bounds the Γ cut-surface is an edge dislocation, i.e. $\vec{b}$ is perpendicular to the dislocation line. A similar result is obtained by separating the two surfaces S₋ and S₊ by a vector $\vec{b}$ perpendicular to the cut surface and by inserting an extra plane of matter (**Figure 13(c)**). As a result, once the two sides of the cut-surface are glued back together and provided that $\vec{b}$ corresponds to a perfect lattice vector, the cut-surface is no longer discernible. The as-created dislocation is called a Volterra dislocation after the work of Volterra (1907). For the screw dislocation, $\vec{b}$ is parallel to the dislocation line vector $\vec{\xi}$ and, according to our convention, it is right handed if $\vec{b}$ points in the same direction as $\vec{\xi}$ (**Figure 13(d)**). In general, the dislocation line is arbitrarily curved

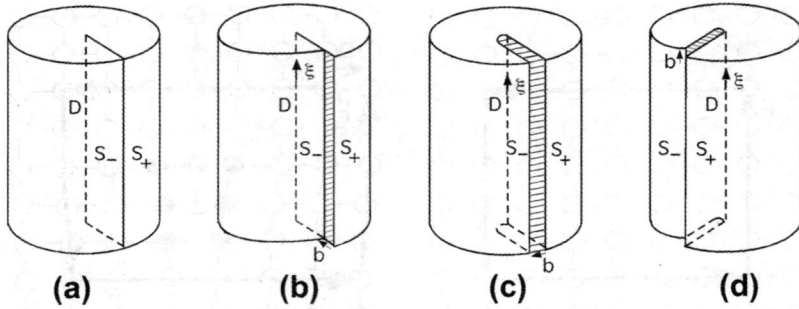

Figure 13   General method to introduce a Volterra dislocation in a solid body. In the present case, the translation is in the cut plane and parallel to the cylinder axis. For a Volterra dislocation, the translation vector between the two cut surfaces is constant.

and the dislocation is called "mixed", the angle between $\vec{\xi}$ and $\vec{b}$ being neither 0 (screw) nor $\pi/2$ (edge). The glide plane of both the edge and the mixed dislocations is unambiguously determined by $\vec{\xi} \times \vec{b}$, i.e. all motions out of this plane is nonconservative, while all planes are potential glide planes for a perfect screw dislocation. For screw dislocations, the plane of motion is determined by further considerations such as core effects, dissociations or kink height. The combination of a Burgers vector and a glide plane is called a glide system.

Note that other types of dislocation can be imagined with, for instance, nonconstant displacement vector along the dislocation line (Somigliana, 1914, 1915) by deforming the edge of the S_ surface when displacing it.

16.2.2.3   Plastic Strain Associated with Dislocation Motion

Dislocations are line defects whose motion produces plastic deformation. The plastic shear strain resulting from the motion of one dislocation can be calculated by considering the schematic of **Figure 14**.

Once the dislocation has been created at one edge of the crystal (**Figure 14(a)**) and has entirely crossed the crystal by traveling a distance ℓ, the corresponding plastic shear is $\gamma = b/h$, where h is the

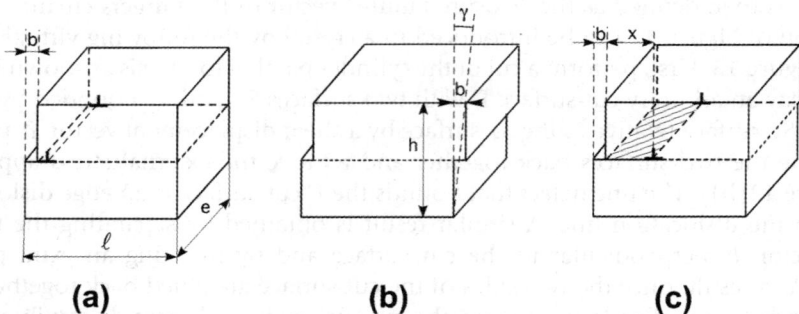

Figure 14   Plastic shear strain associated with dislocation motion. (a) First step: introduction of the dislocation in the crystal. (b) Final stage: the dislocation has entirely crossed the crystal. (c) Intermediate stage: the dislocation has moved over a distance x in the crystal.

crystal height (**Figure 14(b)**). If now, the dislocation is moved by a distance x only, the plastic shear is given by (**Figure 14(c)**)

$$\gamma = \frac{b}{h} \times \frac{x}{\ell},$$ (38)

which can be rewritten

$$\gamma = \frac{bxe}{h\ell e} = \frac{bA}{V}$$ (39)

where e is the dislocation length, A the area swept by the dislocation and V the volume of the crystal. If we assume that N dislocations are simultaneously mobile and travel approximately the same mean distance $\bar{x}$, in other words have similar mean free paths, then the total resulting plastic shear is

$$\gamma = N\frac{bA}{V} = \rho_m b\,\bar{x}, \text{with } \rho_m \text{ the density of mobile dislocations: } \rho_m = \frac{Ne}{V}.$$ (40)

Equation (40) expresses the plastic shear strain produced at the microscale by the motion of dislocations. Assuming for the sake of simplicity the geometry in **Figure 15**, γ can be related to the macroscopic plastic strain $\varepsilon = dh/h$ measured experimentally.

The sample elongation dh along the tensile axis due to N dislocations of Burgers vector $\vec{b}$ is $dh = N\,b\,\cos\lambda$, where λ is the angle between the tensile axis and the direction of the slip vector $\vec{b}$, so that

$$\varepsilon = Nb\cos\lambda/h.$$ (41)

The area swept by the dislocations is $A/\cos\phi$, which with Eqn (40) yields

$$\gamma = N\frac{b}{V} \times \frac{A}{\cos\phi} \text{ with now } V = hA.$$ (42)

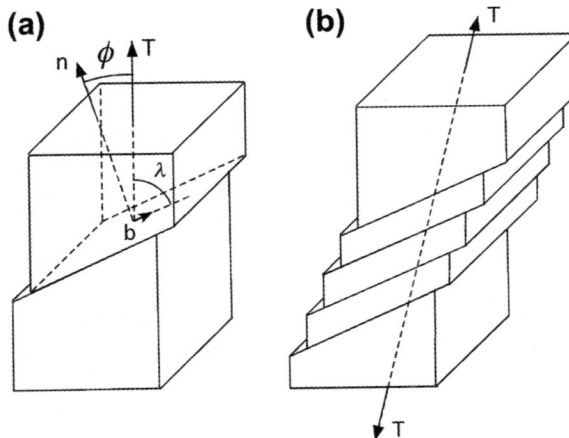

Figure 15 (a) Geometrical representation of the plastic shear strain produced by dislocation motion and related macroscopic crystal strain for tensile test condition. (b) Lattice rotations associated with plastic shear during a tensile test.

A combination of Eqns (41) and (42) gives the relation

$$\varepsilon = \gamma \cos \lambda \cos \phi \qquad (43)$$

where $\cos \lambda \cos \phi = m$ is called Schmid factor. A similar relation also holds for the stress: $\tau = m\,\sigma$, where τ is the resolved shear stress (RSS) in the glide plane of the dislocations, which is also expressed as $\tau = (\vec{b} \cdot \overline{\overline{\sigma}}) \cdot \vec{n}/b$, where $\overline{\overline{\sigma}}$ is the stress tensor, $\vec{b}$ the dislocation Burgers vector and $\vec{n}$ the normal to the glide plane.

More complex "orientation" factors have to be considered in the general case of polycrystalline materials, that not only depend of the crystallographic structure but also on grain deformation hypotheses (Taylor, 1938; Sachs, 1928). It can be seen in **Figure 15(b)** that, for single glide, the load axis rotates with plastic strain. The rotation occurs for compression and tensile tests, but the direction of rotation is different. During a compression test, the compression axis rotates toward the glide plane normal, while, during a tensile test, the axis rotates in the direction of the Burgers vector. Crystallographic rotations have consequences regarding the plastic deformation of polycrystals. Polycrystals are composed of single crystalline grains that are usually of various crystallographic orientations, leading with plastic deformation to constraints between the grains. Because a single grain must accommodate the deformations of all its neighbors, it should in general be able to satisfy all possible ε_{ij} strain components of the strain tensor. In addition, plastic strain only corresponds to a change in sample geometry and takes place at constant volume: $\sum_i \varepsilon_{ii} = 0$, so that only five components of the strain tensor are independent. It implies that, if only dislocation glide is considered, at least five independent glide systems are required to plastically deform a polycrystalline material. This condition is called the Von Mises criterion.

A combination of Eqns (40) and (43) gives

$$\varepsilon = m\,\rho_m b\bar{x}, \qquad (44)$$

where the mean distance $\bar{x}$ traveled by the dislocations can be expressed as $\bar{v}\,t$, where $\bar{v}$ and t are the mean dislocation velocity and time spent for a dislocation to cover the distance $\bar{x}$, respectively. Substituting $\bar{x}$ in Eqn (44), and taking its derivative with respect to time yields Orowan equation (Orowan, 1940)

$$\dot{\varepsilon} = m\,\rho_m b\bar{v}. \qquad (45)$$

This simple equation is very useful and often used in the framework of thermal activation theories of the plastic strain rate, with $\bar{v} = \bar{v}(T, \tau)$ where T is the absolute temperature (Schoeck, 1980). It is however a very rough approximation in consideration of the complexities of dislocation networks and atomic processes (Krausz and Faucher, 1982).

As a remark, it can also be roughly estimated that the necessary stress to produce a plastic strain equivalent to that of **Figure 8** by moving a dislocation (**Figure 16**) is well below the theoretical shear strength. Indeed, neglecting thermal activation, the work done by the applied stress should be approximately the same in the two cases for achieving the same plastic strain: $W_{\sigma\text{-}th} = W_{\sigma\text{-}dis} = Fx$, x being the working distance of F. As seen in Section 16.2.1.2, the theoretical shear stress works on a displacement distance $x = b/4$, which corresponds to a work $W_{\sigma\text{-}th} = F_{th}b/4$, while the work required to achieve the same plastic strain by moving the dislocation is $W_{\sigma\text{-}dis} = F_{dis}x$ where x is the dislocation displacement, usually of the order of several thousands of b. While this demonstration is not rigorous, it however simply suggests than one can expect $F_{dis} \ll F_{th}$.

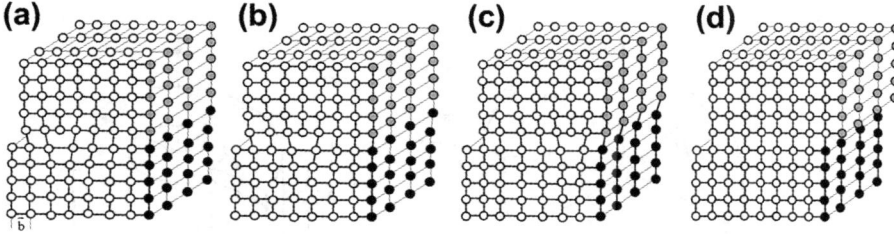

Figure 16   Atomic displacements associated with dislocation motion to produce a plastic shear equivalent that of **Figure 8(a)**.

16.3   Elasticity Associated with Dislocations

16.3.1   Stress Field and Energy Associated with Straight Dislocations

In this paragraph, we give, in the framework of isotropic linear elasticity, the strain and stress fields associated with straight dislocations. The reader may find more general treatments for straight dislocations of arbitrary Burgers vector in an arbitrary anisotropic media (Eshelby et al., 1953; Bacon et al., 1978). The crystal is considered as a continuum medium with two independent elastic coefficients, μ the shear modulus and ν Poisson's ratio. The use of linear elasticity implicitly implies that stresses and strains are additive. We shall therefore consider hereafter only screw and edge straight dislocations, since as will be shown below the treatment of a mixed dislocation consists simply of adding the contributions of both screw and edge components.

16.3.1.1   Displacements, Stress and Strain Fields
16.3.1.1.1   Screw Dislocation
The displacement field about a screw dislocation can be derived from the geometry presented in **Figure 17**. Both the dislocation line and the Burgers vector are oriented along the positive direction of the z-axis. The displacements can be expressed either in Cartesian or cylindrical coordinates, the latter being more convenient.

Since the displacement is produced by shear in the z-direction, there will be no displacement in the x- and y-directions and

$$u_r = u_\theta = 0, \quad u_z = \frac{b\theta}{2\pi}. \tag{46}$$

It can be easily checked that Eqn (46) satisfy the displacement equilibrium condition (Eqn (19)) established for linear elasticity. The corresponding strains are obtained by differentiation

$$\varepsilon_{\theta z} = \frac{1}{2}\left(\frac{1}{r}\frac{\partial u_z}{\partial \theta} + \frac{\partial u_\theta}{\partial z}\right) = \frac{b}{4\pi r}, \tag{47}$$

the other components being zero except $\varepsilon_{z\theta}$ ($=\varepsilon_{\theta z}$), due to the symmetry of the strain tensor $\bar{\bar{\varepsilon}}$.

For an infinite cylinder, the associated stress components are obtained from Hooke's law

$$\sigma_{\theta z} = 2\mu\varepsilon_{\theta z} = \frac{\mu b}{2\pi r} = \sigma_{z\theta}. \tag{48}$$

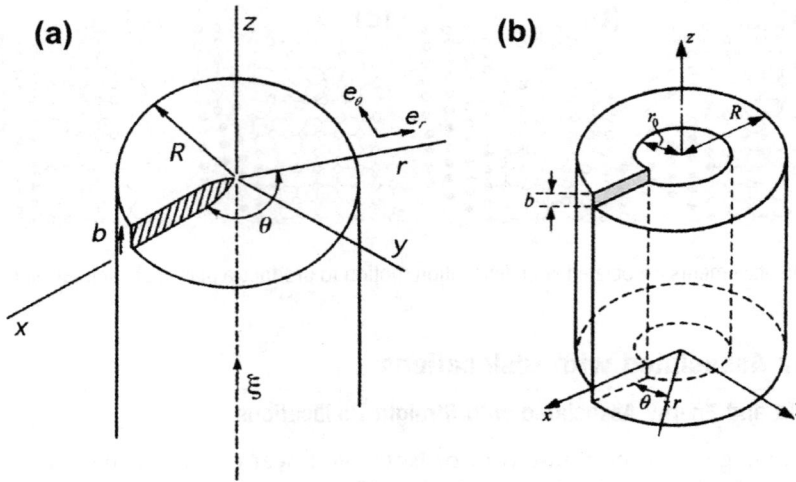

Figure 17 Screw dislocation along the z-axis of a cylinder of radius R. (a) In a perfect cylinder. (b) Bounded by two cylinders of inner and outer radii r_o and R, respectively.

It must be noticed that both the strain and stress vary like $1/r$ and diverge when $r \to 0$, that is close to the dislocation line where elastic theory is no longer valid. As a consequence, the core of the dislocation line is often excluded by considering a hollow cylinder with inner and outer radii r_o and R, respectively (**Figure 17(b)**). For models of the dislocation core, see Section 16.4.

In the case of a finite isotropic cylinder in the z-direction, the $\sigma_{z\theta}$ stress component produces a torque term at the end surfaces of the cylinder, so that the boundary condition at the free surface

$$\sigma_{ij}dS_j = 0 \tag{49}$$

is not fulfilled. Consequently, the stress components $\sigma_{z\theta}$ and $\sigma_{\theta z}$ must be corrected by a term that ensures the free-stress state of the end surfaces of the cylinder. On the upper surface of the cylinder, for $\ell \gg R$, the torque term has the form

$$M_z = \int_0^{2\pi} \int_{r_o}^{R} (r\sigma_{z\theta})\, d\theta r dr = \frac{\mu b}{2}\left(R^2 - r_o^2\right), \tag{50}$$

where inner and outer radii r_o and R are considered to avoid divergences (**Figure 17(b)**). It should be noted that for the sake of simplicity, the dependence of $\sigma_{z\theta}$ on r_o and R is neglected in the above equation, as usually done in the literature (see Hirth and Lothe, 1982). The torque M_z must be counterbalanced by an opposite torque term $M_z' = -M_z$. An opposite situation exits for the lower surface of the cylinder. Consequently, we have to superimpose to the displacement of the infinite solution a displacement u_θ' that produces the torque M_z' at $z = \pm \ell/2$ on the base surfaces of the cylinder of finite length ℓ. This field is given in Timoshenko and Goodier (1951):

$$u_z' = u_z' = 0, \quad u_\theta' = -Arz. \tag{51}$$

where A is a geometric factor. This new displacement leads to the additional strain and stress fields

$$\varepsilon'_{z\theta} = \varepsilon'_{\theta z} = -Ar/2 \quad \text{and} \quad \sigma'_{z\theta} = \sigma'_{\theta z} = -\mu Ar, \tag{52}$$

that induce the torque

$$M'_z = \int\limits_0^{2\pi} \int\limits_{r_o}^R r\sigma'_{z\theta} \, d\theta dr = -\frac{\mu A}{2}\pi\left(R^4 - r_o^4\right). \tag{53}$$

By setting that the condition $M_z + M'_z = 0$ must be fulfilled, we obtain for A.

$$A = \frac{b\left(R^2 - r_o^2\right)}{\pi\left(R^4 - r_o^4\right)}, \text{ which at the limit of } r_o \ll R \text{ gives } A \cong \frac{b}{\pi R^2}. \tag{54}$$

We can now rewrite the strain and stress fields for a screw dislocation in a finite elastic continuum medium by superposing the two contributions

$$\varepsilon_{\theta z} = \varepsilon_{z\theta} = \frac{b}{4\pi r}\left(1 - \frac{2r^2}{R^2}\right) \quad \text{and} \quad \sigma_{\theta z} = \sigma_{z\theta} = \frac{\mu b}{2\pi r}\left(1 - \frac{2r^2}{R^2}\right). \tag{55}$$

The correction is called the Eshelby twist (Eshelby and Stroh, 1951), which has been evidenced by transmission electron microscopy observation in the high-resolution mode (Baluc, 1990; Mills et al., 2006). In this experiment, a screw dislocation is seen edge-on in a thin foil of a few nanometers thick. The dislocation is parallel to the electron beam and its two extremities interact with the thin foil surface. It is observed on the experimental images that the atomic columns have a diffuse contrast on one side of the dislocation, which can only be reproduced on the simulated images by considering the Eshelby twist.

16.3.1.1.2 Edge Dislocation

The geometry of an edge dislocation is presented in **Figure 18**. The treatment of the displacement field of the edge dislocation is less straightforward than that of the screw dislocation, because the

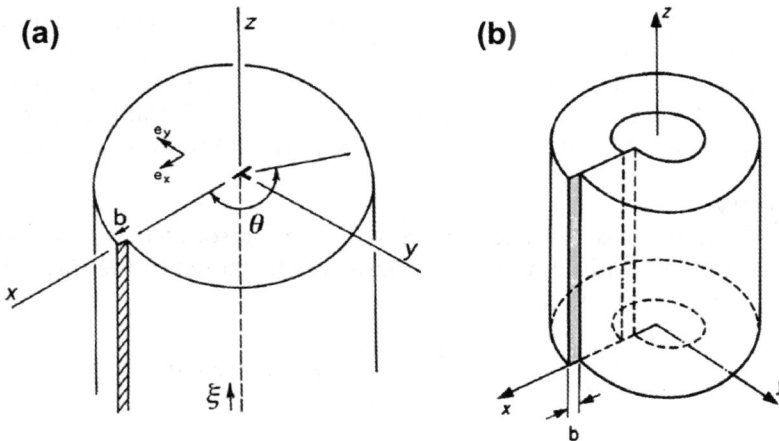

Figure 18 Edge dislocation along the z-axis of a cylinder of radius R. (a) In a perfect cylinder. (b) Bounded by two cylinders of inner and outer radius r_o and R, respectively.

displacement has now two components u_r and u_θ, while $\partial/\partial z = 0$ and $u_z = 0$. Displacements and strains in the z-direction are thus zero and the problem can be treated in "plane strain", as for the screw dislocation. The derivation of the displacement fields is rather tedious and will not be reproduced here.

A complete demonstration can be found, for instance, in Teodosiu (1982) and an elegant demonstration in Eshelby (1966). It comes out for an infinite isotropic medium (see also Hirth and Lothe (1982))*

$$u_r = -\frac{b}{2\pi}\left(\frac{1-2v}{2(1-v)}\sin\theta\ln r - \frac{\sin\theta}{4(1-v)} - \theta\cos\theta\right)$$

and

$$u_\theta = -\frac{b}{2\pi}\left(\frac{1-2v}{2(1-v)}\cos\theta\ln r + \frac{\cos\theta}{4(1-v)} + \theta\sin\theta\right)$$

(56)

from which, by making use of Eqn (5), we obtain the components of the strain tensor in cylindrical coordinates

$$\varepsilon_{rr} = \varepsilon_{\theta\theta} = -\frac{b(1-2v)}{4\pi(1-v)}\frac{\sin\theta}{r}$$

$$\varepsilon_{r\theta} = \frac{b}{4\pi(1-v)}\frac{\cos\theta}{r}, \; \varepsilon_{zz} = \varepsilon_{rz} = \varepsilon_{\theta z} = 0$$

(57)

The stresses directly follow from Hooke's law (Eqn (8))

$$\sigma_{rr} = \sigma_{\theta\theta} = -D\frac{\sin\theta}{r}, \sigma_{zz} = -2vD\frac{\sin\theta}{r},$$

$$\sigma_{r\theta} = D\frac{\cos\theta}{r}, \sigma_{rz} = \sigma_{\theta z} = 0,$$

(58)

with $D = \mu b/2\pi(1-v)$. As for the screw dislocation, the stress field associated with an edge dislocation varies as $1/r$ and diverges when approaching the dislocation core. Consequently, the dislocation core requires special theoretical treatments that will be presented hereafter (see Section 16.4). It is also worth noting that, as previously mentioned, the stress and strain tensors of the screw and the edge dislocations do not have common terms; i.e. both tensors are orthogonal. Therefore, in the framework of isotropic linear elasticity for which the principle of superposition applies, the stress and strain fields of a mixed dislocation are obtained by calculating independently the stress and strain fields associated with the screw and edge components of the Burgers vector, respectively, and summing both contributions.

16.3.1.2 Elastic Energy

With the stress and strain fields in hand, the elastic energy associated with a screw and an edge dislocation can be calculated with Eqn. A.9. The elastic energy stored in the elastic field of a straight screw dislocation in an isotropic medium is

$$E = \iiint w\mathrm{d}V = \frac{1}{2}\iiint(\sigma_{\theta z}\varepsilon_{\theta z} + \sigma_{z\theta}\varepsilon_{z\theta})\mathrm{d}V.$$

(59)

---

* This problem is usually solved by starting from the determination in polar coordinates of the stress-field characteristics of an edge dislocation using Airy stress function for plane strain condition. The displacements are further derived from Eqn (9) and from Eqn (5) by integration.

Because the stress and strain fields diverge at the limit $r \to 0$, the above integral requires that the elastic energy is only calculated from a radius r_0 of the dislocation line where linear elasticity is valid. The matter inside the cylinder of radius r_0 will be referred hereafter to as the core of the dislocation, such as in **Figure 17(b)**. Thus, integration of Eqn (59) will only provide the elastic energy outside the cylinder of radius r_0 and the total stored elastic energy will be obtained by further adding a compensatory term corresponding to the core energy inside the cylinder of radius r_0. The elastic energy per unit length may be obtained by considering the screw dislocation either infinite or of finite length. In the former case, Eqns (48) and (59) give

$$E_{\ell(\text{inf})} = \int_0^1 \mathrm{d}z \int_0^{2\pi} \mathrm{d}\theta \int_{r_0}^R \frac{\mu b^2}{4\pi r} \mathrm{d}r = \frac{\mu b^2}{4\pi} \ln \frac{R}{r_0}, \tag{60}$$

while for the latter case, with Eqns (55) and (59), we have

$$E_{\ell(\text{fin})} = \int_0^1 \mathrm{d}z \int_0^{2\pi} \mathrm{d}\theta \int_{r_0}^R \frac{\mu b^2}{4\pi r} \left(1 - \frac{2r^2}{R^2}\right)^2 \mathrm{d}r. \tag{61}$$

Integration of the above equation yields in the limit of $R \gg r_0$

$$E_{\ell(\text{fin})} = \frac{\mu b^2}{4\pi} \left(\ln \frac{R}{r_0} - 1\right). \tag{62}$$

It is seen that $E_{\ell(\text{fin})} < E_{\ell(\text{inf})}$, which indicates that the free-surface terms influence the elastic energy. This result is expected since a surface relaxation in a finite medium should decrease the elastic energy.

A similar calculation can be carried out for the edge dislocation. By restricting ourselves to the case of an infinite edge dislocation in an isotropic medium, it is found with Eqns (A.9), (48) and (59) and by applying the similar integral procedure as for the screw dislocation

$$E_{\ell} = \frac{\mu b^2}{4\pi(1-\nu)} \ln \frac{R}{r_0}. \tag{63}$$

A comparison between Eqns (60) and (63) indicates that the elastic energy per unit length of the screw and edge dislocations differs by the term $1/(1-\nu)$. For isotropic material, ν is usually of the order of $1/3$, so that the elastic energy associated with an edge dislocation is about 1.5 larger than that of a screw dislocation.

It was shown that no interaction exists between straight parallel screw and edge dislocations. Consequently, the elastic energy of the mixed dislocation is obtained by the sum of the elastic energy of screw and edge components with Burgers vectors $b_s = b\cos\psi$ and $b_e = b\sin\psi$, respectively, where b is the magnitude of the total Burgers vector $\vec{b}$ of the dislocation and ψ the angle between the dislocation line $\vec{\xi}$ and $\vec{b}$. From Eqns (60) and (63) and assuming a similar core radius r_0 for screw and edge dislocations, we have for the elastic energy per unit length associated with the stress field of a mixed dislocation

$$E_{\ell} = \frac{K b^2}{4\pi} \ln \frac{R}{r_0}. \tag{64}$$

with $K = \mu\left(\cos^2\psi + \dfrac{\sin^2\psi}{1-\nu}\right)$. It is seen that all E_ℓ energy terms are proportional to b^2. Two dislocations of Burgers vector b have therefore a lower energy than a single dislocation of Burgers vector $2b$. As a result, stable dislocations are generally those with Burgers vectors corresponding to the shortest translations of the Bravais lattice of the crystalline structures. Dislocations with large Burgers vectors will tend to dissociate into shorter and more stable components. Also, all E_ℓ diverge as $R \to \infty$. It is clear, however, that R is always bounded at least by the crystal size. Because crystals usually contain a high density of dislocations, a reasonable approximation consists in taking for R half the average distance between dislocations. E_ℓ is not very sensitive to R since its dependence is logarithmic, but it must be emphasized that E_ℓ has generally a high value, of the order of a few eV (electron-volts) per atomic plane along the dislocation line. As a consequence, dislocations will not be spontaneously nucleated in a crystal by thermal fluctuations. Another consequence discussed below in the context of line tension (LT) is that dislocations will always tend to minimize their length in order to minimize their energy.

16.3.1.3 Notion of Dislocation Core

As repeatedly mentioned, the divergence of the stress and strain fields, when approaching the dislocation line, requires the introduction of an inner cutoff radius, r_o, for calculating the elastic energy. As emphasized by Saada and Douin (1991), the lack of a robust model for describing dislocation cores prevents a genuine estimate of the actual inner cutoff radius. The knowledge of r_o is of prime importance in the description of several elementary plasticity mechanisms and will be hereafter the subject of a dedicated Section (Section 16.4). We shall present here only phenomenological approaches, which will be useful for further theoretical developments in this Chapter.

The first arising question concerns the possible existence of hollow core dislocations within a crystal. For this, let us consider the case of an edge dislocation, whose elastic energy per unit length outside the hollow core is given by Eqn (63). The surface energy due to the hollow core of radius r would be per unit length: $2\pi r\gamma$, where γ is the crystal surface energy. The total energy of the configuration per unit length is then

$$E_\ell = \frac{\mu b^2}{4\pi(1-\nu)}\ln\frac{R}{r} + 2\pi r\gamma, \tag{65}$$

where it is implicitly assumed that $r > r_0$. E_ℓ is minimum for

$$\left(\frac{\partial E_\ell}{\partial r}\right)_{r=r_o} = 0 \quad \text{that is for} \quad r_o = \frac{\mu b^2}{8\pi^2(1-\nu)\gamma}, \tag{66}$$

where r_o is the equilibrium radius of the hollow core, referred to as Frank's radius (Frank, 1951a). An estimate of γ can be obtained from Eqn (23): $\gamma \sim \mu a/10$, since the theoretical shear strength τ_{th} may be roughly approximated by the necessary stress for cleavage leading to the production of two free surfaces. Inserting γ into Eqn (66) and taking $\nu = 1/3$, one has

$$r_o \approx 3b^2/16a. \tag{67}$$

Since a prerequisite to have a hollow core in a crystal is that $r_o \geq a$, one finally obtains the following condition for the Burgers vector of such dislocation

$$b \geq \frac{4a}{\sqrt{3}}. \tag{68}$$

Consequently, dislocations with hollow core may exist in crystals, but will have in general large Burgers vectors, which are only observed in very specific crystallographic structures (Klemenz, 1998). The cutoff radius r_o is usually chosen so that the elastic energy outside r_o equals the total energy of the dislocation, that is, the elastic energy plus the core energy. It is therefore necessary to have an estimate of the energy E_c corresponding to the core of the dislocation. In a first approximation, one can calculate a lower limit of E_c by assuming that for $r \leq r_o$ the strain and stress fields do not diverge close to the dislocation core, but remain constant at the values corresponding to $r = r_o$ This would lead, for instance in the case of the screw dislocation, to a core energy E_c per unit length

$$E_c = \frac{1}{2}(2\sigma_{z\theta}\varepsilon_{z\theta})\,\pi\,r_o^2 = \frac{\mu b^2}{8\pi}. \tag{69}$$

It should be noted that r_o is still small, since linear elasticity theory can be considered valid up to elastic deformation corresponding the theoretical shear strength $\tau_{th} \sim \mu/10$ (Eqn (24)), which combined with Eqns (8) and (48) gives

$$\varepsilon_{z\theta} = \frac{\sigma_{z\theta}}{2\mu} = \frac{b}{2\pi r_o} \approx 0.1 \quad \text{and} \quad r_o \sim 1.5b. \tag{70}$$

An upper limit of E_c can be estimated by assuming a melting core, in which case the heat energy of melting is $E_c \sim \mu b^2/5$ (Bragg, 1948). Both limit values of E_c are of course not exact and, for the sake of simplicity, in consideration of the energy term in Eqn (62), one may take for E_c a mean value of $\mu b^2/4\pi$. Using this latter value and Eqns (62) and (70), the total energy per unit length of a screw dislocation in a finite isotropic medium is given by

$$E_{\ell-tot} = E_{\ell-fin} + E_c = \frac{\mu b^2}{4\pi}\ln\frac{\alpha R}{b} \quad \text{with} \quad \alpha \sim 1.5. \tag{71}$$

Finally, the energy per unit length of a dislocation of mixed character is usually given by the general expression:

$$E = \frac{K(\psi)b^2}{4\pi}\ln\frac{\alpha R}{b} \tag{72}$$

where K is defined in Eqn (64).

16.3.2 Force on Dislocations

For nucleating a dislocation loop of Burgers vector $\vec{b}$ in a crystal, work must be done not only against the self-stress of the dislocation, which corresponds to the self-energy calculated above, but also against the other stress-fields present in the crystal. Let us consider that this stress field arises from an applied external stress characterized by a stress tensor $\bar{\bar{\sigma}}$. Creation of the dislocation against the stress field $\bar{\bar{\sigma}}$ requires a stress onto the cut surface that does work

$$E = \int_S \vec{b}\cdot(\bar{\bar{\sigma}}\cdot\vec{dS}), \tag{73}$$

where S is the area swept by the dislocation loop. If we further move an element $\mathrm{d}\vec{L}$ of the dislocation loop over a distance $\mathrm{d}\vec{\ell}$, the extra work that must done is

$$\delta E = \vec{b} \cdot (\overline{\overline{\sigma}} \cdot \mathrm{d}\vec{S}) \tag{74}$$

where $\mathrm{d}\vec{S}$ is the area swept by the dislocation increase: $\mathrm{d}\vec{S} = \mathrm{d}\vec{\ell} \times \mathrm{d}\vec{L}$, which may be regarded as an increment of the cut-surface (see **Figure 19**). Using the commutativity in Eqn (74) and inserting $\mathrm{d}\vec{S}$, we obtain:

$$\delta E = (\vec{b} \cdot \overline{\overline{\sigma}}) \cdot (\mathrm{d}\vec{\ell} \times \mathrm{d}\vec{L}), \tag{75}$$

which, with properties of the mixed product, can be rewritten in the form

$$\delta E = -(\vec{b} \cdot \overline{\overline{\sigma}}) \cdot (\mathrm{d}\vec{L} \times \mathrm{d}\vec{\ell}) = -[(\vec{b} \cdot \overline{\overline{\sigma}}) \times \mathrm{d}\vec{L}] \cdot \mathrm{d}\vec{\ell}. \tag{76}$$

As the variation δE of E depends only on the displacement $\mathrm{d}\vec{\ell}$ of the segment $\mathrm{d}\vec{L}$, which is arbitrary, the elementary force acting on each element of $\mathrm{d}\vec{L}$ the dislocation is thus

$$\mathrm{d}\vec{F} = -\nabla_{\mathrm{d}\vec{\ell}}\delta E = \left(\vec{b} \cdot \overline{\overline{\sigma}}\right) \times \mathrm{d}\vec{L}. \tag{77}$$

This equation, which is not limited to linear elasticity, was first obtained by Peach and Koehler (1950). Then, the force $\vec{F}$ resulting from an external force on a dislocation line of length L is

$$\vec{F} = \left(\vec{b} \cdot \overline{\overline{\sigma}}\right) \times \vec{L}, \tag{78}$$

which shows that it is perpendicular to the dislocation line. This result is obvious if one considers that moving the dislocation along its line does not change the configuration. The force per unit length $\vec{F}/L$ acting on a dislocation defined by $(\vec{\xi}, \vec{b}, \vec{n})$ (see Section 16.2.2.2) can be decomposed into a component F_g/L in the glide plane

$$F_g/L = [(\vec{b} \cdot \overline{\overline{\sigma}}) \times \vec{\xi}] \cdot \frac{\vec{b}}{b} = (\vec{b} \cdot \overline{\overline{\sigma}}) \cdot \vec{n} \tag{79}$$

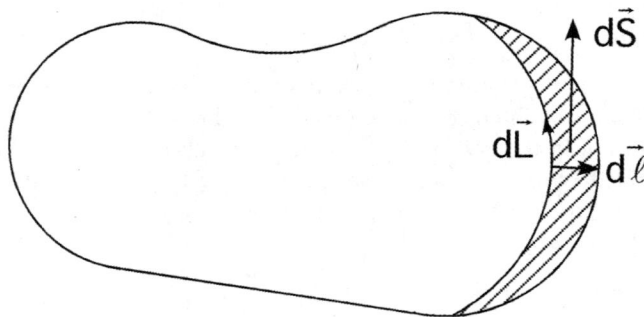

Figure 19   Increase in cut area $\mathrm{d}\vec{S}$ by moving an element $\mathrm{d}\vec{L}$ of a dislocation line over a distance $\mathrm{d}\vec{\ell}$.

that moves the dislocation within the glide plane. This force per unit length is simply the RSS introduced in Section 16.2.2.3. Equation (79) can also be written as

$$F_g/L = \sigma_{ij}b_i n_j \tag{80}$$

which indicates that only shear stresses do work for moving a dislocation onto its glide plane. The second component F_m/L is perpendicular to the glide plane

$$F_m/L = \left[(\vec{b} \cdot \bar{\bar{\sigma}}) \times \vec{\xi} \right] \cdot \vec{n}, \tag{81}$$

which corresponds to a climb force.

16.3.3 Elastic Interaction

16.3.3.1 Between Dislocations

We previously gave the expression of the work done for the nucleation of a dislocation loop in presence of a stress field resulting from an external applied force. A similar expression can be derived when introducing a dislocation in a crystalline solid in the presence of an existing dislocation. Let us call $\bar{\bar{\sigma}}_1$ the stress tensor originating from the existing dislocation of Burgers vector $\vec{b}_1$, labeled 1 hereafter, on the cut-surface S_2 of the dislocation being nucleated with Burgers vector $\vec{b}_2$, labeled 2 hereafter. Analogous to Eqn (73), the work done to balance the self-stress field of both the dislocation 2, $\bar{\bar{\sigma}}_2$, and the dislocation 1, $\bar{\bar{\sigma}}_1$, can be written in the general form

$$E = \int_0^{b_2} \int_{S_2} (\bar{\bar{\sigma}}_2 \cdot \vec{dS}_2) \cdot d\vec{u}_2 + \int \int_{S_2} (\bar{\bar{\sigma}}_1 \cdot \vec{dS}_2) \cdot \vec{b}_2, \tag{82}$$

where the first term of the right-hand side is identified as the self-energy of dislocation 2 and the second term represents the interaction between dislocations 2 and 1. Interaction energies can be developed from Eqn (82). However, developments for arbitrary curved dislocations are rather tedious and, except for special cases, the integrals can only be solved numerically. Even when analytical solutions exist, the results are lengthy (Nabarro, 1967; Hirth and Lothe, 1982), beyond the scope of the present work. Of particular interest however for this chapter in what follows is the interaction energy between straight and parallel dislocations. In isotropic elasticity, by considering two dislocations of Burgers vectors $\vec{b}_1$ and $\vec{b}_2$ separated by a distance r, we have from Eqn (82) an energy per unit length (Nabarro, 1952)

$$\begin{aligned} \frac{E}{L} = &-\frac{\mu}{2\pi} \left[(\vec{b}_1 \cdot \vec{\xi})(\vec{b}_2 \cdot \vec{\xi}) + \frac{(\vec{b}_1 \times \vec{\xi}) \cdot (\vec{b}_2 \times \vec{\xi})}{(1-\nu)} \right] \ln \frac{r}{R_o} \\ &- \frac{\mu}{2\pi(1-\nu)r^2} \left[(\vec{b}_1 \times \vec{\xi}) \cdot \vec{r} \right] \left[(\vec{b}_2 \times \vec{\xi}) \cdot \vec{r} \right], \end{aligned} \tag{83}$$

where R_o is a reference position introduced to make dimensionless the argument of the logarithmic term and which can be chosen equal to r_o. In polar coordinates, the resulting radial interaction force per unit length between the two dislocations is:

$$f_r = -\frac{\partial(E/L)}{\partial r} = \frac{\mu}{2\pi r}(\vec{b}_1 \cdot \vec{\xi})(\vec{b}_2 \cdot \vec{\xi}) + \frac{\mu}{2\pi(1-\nu)r}(\vec{b}_1 \times \vec{\xi}) \cdot (\vec{b}_2 \times \vec{\xi}). \tag{84}$$

Note that the first term of the right-hand side in Eqn (84) is the radial interaction force due to the dislocation screw components, while the second term is the radial interaction force due to the dislocation edge components. The angular interaction force per unit length is:

$$f_\theta = -\frac{1}{r}\frac{\partial(E/L)}{\partial\theta} = \frac{\mu}{2\pi(1-\nu)r^3}\{(\vec{b}_1\cdot\vec{r})[(\vec{b}_2\times\vec{r})\cdot\vec{\xi}] + (\vec{b}_2\cdot\vec{r})[(\vec{b}_1\times\vec{r})\cdot\vec{\xi}]\}, \tag{85}$$

which gives rise to a torque between the two dislocations. In Eqn (85), the cross products $\vec{b}_i\times\vec{r}$ yield vectors perpendicular to $\vec{b}_i$, so that the scalar products with $\vec{\xi}$ will have zero values for screw dislocations. The interaction force between two parallel straight screw dislocations of Burgers vectors $\vec{b}_1$ and $\vec{b}_2$ consists only of the radial term per unit length (**Figure 20(a)**):

$$f_r = \frac{\mu b_1 b_2}{2\pi r}, \tag{86}$$

and taking $|\vec{b}_1| = |\vec{b}_2| = |\vec{b}|$, we have

$$f_r = \pm\frac{\mu b^2}{2\pi r}. \tag{87}$$

Note that a similar result would be obtained from Eqns (48) and (80). Hence, in isotropic elasticity, the interaction force between two screw dislocations is radial only. It has however been shown that in anisotropic elasticity, a torque term exists between two screw dislocations (Teutonico, 1964). This missing interaction force term in isotropic theory may sometimes lead to erroneous results and it has been shown in a number of situations that this term can be of prime importance (Yoo, 1987). Equation (87) also indicates that, as for electrostatic charges, dislocations with the same Burgers vector repeal each other, while dislocations of opposite Burgers vectors attract each other. Because perfect screw dislocations can freely glide in the crystals since they do not have a specific glide plane, they will either annihilate or repel each other to infinity.

It is also of interest to consider the interaction between two parallel edge dislocations. If $\vec{b}_1$ and $\vec{b}_2$ are parallel to the x-axis and the dislocation lines are parallel to the z-axis, such as schematized in **Figure 20(b and c)**, one finds with Eqns (84) and (85) that the dislocation at the origin exerts on a dislocation of polar coordinates (r,θ) forces given by:

$$f_r = \frac{\mu b_1 b_2}{2\pi(1-\nu)r} \quad\text{and}\quad f_\theta = \frac{\mu b_1 b_2}{2\pi(1-\nu)r}\sin2\theta. \tag{88}$$

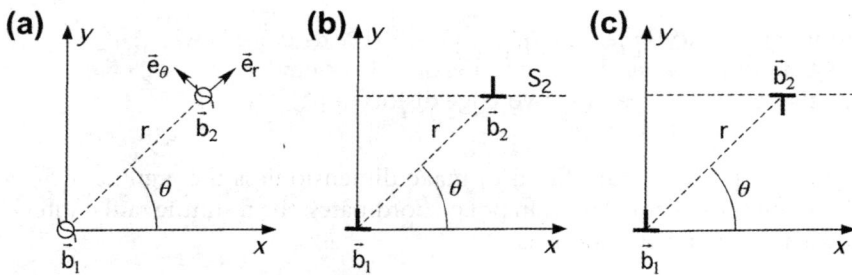

Figure 20 Schematic representations of dislocation pairs. (a) Screw dislocations parallel to the z-axis. (b) and (c) edge dislocations along the z-axis on parallel glide planes with $\vec{b}_1\cdot\vec{b}_2 > 0$ and $\vec{b}_1\cdot\vec{b}_2 < 0$, respectively.

An edge dislocation can move by glide only in the plane containing its line and its Burgers vector and the resulting force component in the glide plane is, by considering the projections of f_r and f_θ,

$$f_g = f_r \cos\theta - f_\theta \sin\theta = \frac{\mu b_1 b_2}{2\pi(1-\nu)r} \cos\theta \cos 2\theta \tag{89}$$

Note that, as for the screw dislocation, a similar equation can be found from Eqn (80) and the stress component σ_{xy} of an edge dislocation in Cartesian coordinates. Thus, for two dislocations of the same sign, that is if $\vec{b}_1 \cdot \vec{b}_2 > 0$, f_g is positive for all values of $\theta < \pi/4$ and the dislocations will repeal each other while they attract each other if $\theta > \pi/4$. The two values $\theta = 0$ and $\theta = \pi/2$ are therefore two positions of stable equilibrium, while $\theta = \pi/4$ corresponds to an unstable equilibrium. The results are reversed when $\vec{b}_1 \cdot \vec{b}_2 < 0$. It follows that edge dislocations of same sign will preferentially organize vertically one above the other to form low-angle tilt boundaries. Edge dislocations of opposite sign will form dislocation dipoles $\theta < \pi/4$. Their interaction energy can be calculated by integrating Eqn (89). A dislocation of Burgers vector $\vec{b}_2$ is nucleated at the surface of the crystal at a distance R and is brought to a distance r from another dislocation of Burgers vector $\vec{b}_1$ already present (**Figure 20(c)**). By neglecting surface effects and assuming that the moving dislocation is confined to glide in a y-plane, the interaction energy per unit length between the moving and the fixed dislocation is

$$E_{int} = \frac{\mu b_1 b_2}{2\pi(1-\nu)} \int_{x=R}^{x=r\cos\theta} \frac{\cos\theta \cos 2\theta}{r} dx, \tag{90}$$

which yields after integration the interaction energy per unit length

$$E_{int} = \frac{\mu b^2}{2\pi(1-\nu)} \left(\ln\frac{r}{R} - \frac{1}{2}\cos 2\theta + \frac{1}{2} \right). \tag{91}$$

Therefore, the total strain energy per unit length of the edge dislocation dipole at an angle of $\pi/4$ is given by

$$E_{dip} = 2E_\ell + E_{int} = \frac{\mu b^2}{2\pi(1-\nu)} \left(\ln\frac{r}{r_0} - \frac{1}{2} \right) \tag{92}$$

which depends only on the distance r between the two dislocations. Thus, each dislocation can be considered as contained in a cylinder of radius r equal to the distance between the two dislocations and relaxes the stress field of the other dislocation beyond this cylinder.

A climb force also exists between the two edge dislocations

$$f_m = f_r \sin\theta + f_\theta \cos\theta = \frac{\mu b_1 b_2 \sin\theta}{2\pi(1-\nu)r} (2 + \cos 2\theta), \tag{93}$$

which will be efficient at high temperature, typically at $T > T_m/2$ (T_m is melting temperature), when diffusion processes dominate. For a dislocation dipole $\vec{b}_1 \cdot \vec{b}_2 < 0$ and f_m contributes to the dipole annihilation.

16.3.3.2 Frank Criterion for Dislocation Junctions

We have already seen that a consequence of the dislocation self-energies being proportional to b^2 is that the Burgers vectors usually correspond to short translation vectors of the crystalline structure. Another consequence concerns dislocation interactions when two dislocations react to form a third dislocation segment. From the conservation of Burgers vectors Eqn (36), the Burgers vector of the new dislocation is just the sum of the Burgers vectors of the impeding dislocations: $\vec{b}_3 = \vec{b}_1 + \vec{b}_2$. Energetic considerations also imply that the formation of the junction will be energetically favorable if the so-called Frank criterion is fulfilled:

$$|\vec{b}_3|^2 < |\vec{b}_1|^2 + |\vec{b}_2|^2, \tag{94}$$

This criterion is consistent with the observation made above that two dislocations attract each other, and thus may form a junction, if $\vec{b}_1 \cdot \vec{b}_2 < 0$, since in this case $|\vec{b}_3|^2 = |\vec{b}_1 + \vec{b}_2|^2 = |\vec{b}_1|^2 + |\vec{b}_2|^2 + 2\vec{b}_1 \cdot \vec{b}_2 < |\vec{b}_1|^2 + |\vec{b}_2|^2$. More local criteria, however, must be considered for dissociated dislocations (see Section 16.5) where interactions between partial dislocations can become predominant.

16.3.3.3 Interactions with Free Surfaces, Image Dislocations

For the sake of simplicity, we have so far mainly considered that the dislocations were in an infinite continuous medium and neglected surface effects. The question that can be addressed is: "what is the interaction between a dislocation and a free surface in a medium of finite size." For this, let us consider the simple case of an infinite screw dislocation parallel to a surface S and located at a distance d from S (**Figure 21**). As seen above, the self-energy of the dislocation per unit length is given by Eqn (60)

$$E = \frac{\mu b^2}{4\pi} \ln \frac{d}{r_o}, \tag{95}$$

which decreases when d decreases. Consequently, S attracts the dislocation and the attractive force can be calculated by taking into account that the free surface must be stress free. To solve this problem,

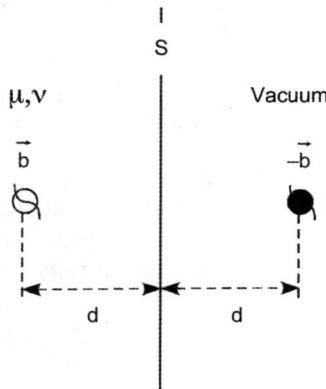

Figure 21 Screw dislocation of Burgers vector $\vec{b}$ parallel to a planar free surface S (the dislocation points into the paper). On the right, true dislocation in a crystal of elastic coefficients μ and v and on the left, "image" dislocation of Burgers vector $-\vec{b}$ in vacuum.

which is equivalent to that of the method of image charges, we shall consider a dipole of screw dislocations separated by a distance $2d$ in an infinite solid medium of elastic coefficients μ and ν.

According to Eqn (48) and the superposition principle, the stress field in the middle plane of two screw dislocations will be zero, which fulfills the zero stress plane condition. Therefore, we can theoretically cut along this plane, i.e. the free surface, and the force can be calculated assuming the existence of a negative hypothetical dislocation on the other side of the surface. It follows that the force, called an image force, experienced by the dislocation near the free-surface is (Eqn (87) with $r = 2d$)

$$f = -\frac{\mu b^2}{4\pi d}. \tag{96}$$

Calculations are slightly more complicated for an edge dislocation, since it is necessary to introduce appropriate dislocation distributions at the interface to account for the angular forces that produce shear stresses. Nevertheless, a similar result holds for an edge dislocation and one has

$$f = -\frac{\mu b^2}{4\pi(1 - \nu)d}. \tag{97}$$

Image forces become especially important when they exceed the Peierls stress (see Section 16.4) since dislocations can move spontaneously and annihilate at surfaces even in the absence of external applied stresses. This can lead to a depletion of surface dislocations (Ungar et al., 1982) and play a role in low-dimensional samples (Veyssière, 1989; Couret et al., 2000).

Another practical case is to consider two adjacent elastic media of elastic coefficients μ_1 and μ_2. As previously, we shall consider the simple case of an infinite screw dislocation of Burgers vector $\vec{b}_1$ parallel to the separation surface S and located at a distance d from the surface (**Figure 22**).

Likewise, an image screw dislocation is symmetrically positioned with respect to S in medium 2, but to account for the difference in shear moduli $\mu_2 \neq \mu_1$, the image dislocation has a Burgers vector $\vec{b}_2$. Let us consider the points M_1 and M_2 infinitely close to S in medium 1 and 2, respectively (see **Figure 22(b) and (c)**). The stress field in M_1 is due to the actually existing and virtual image dislocations (**Figure 22(b)**), that is

$$\sigma_{\theta z} = \frac{\mu_1(b_1 + b_2)}{4\pi d} \quad \text{and the associated displacement is } u_z = \frac{b_1 \theta}{2\pi} + \frac{b_2(\pi - \theta)}{2\pi}. \tag{98}$$

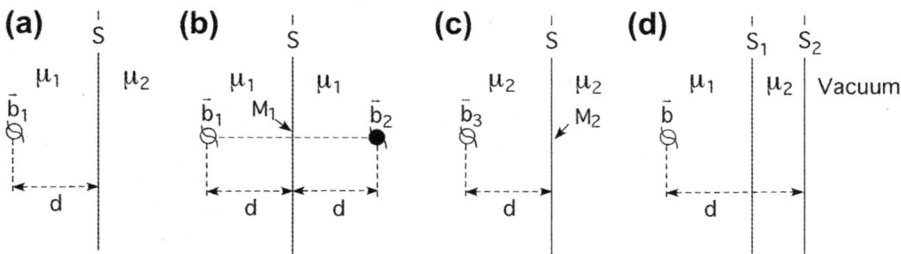

Figure 22 Screw dislocation of Burgers vector $\vec{b}_1$ parallel to a planar surface separating two media of different elastic coefficients (μ_1, ν_1) and (μ_2, ν_2). The dislocation points into the paper. (a) Actual configuration. (b) Introduction of an "image" dislocation of Burgers vector $\vec{b}_2$ considering that both materials have the same μ_1 elastic coefficient. (c) Introduction of a "virtual" Burgers vector $\vec{b}_3$ considering that both materials have the same μ_2 elastic coefficient. (d) Case of an oxide layer.

At M_2, the image dislocation does not exist and, because the existing dislocation is considered as lying in the medium of elastic coefficient μ_2 (**Figure 22(c)**), a virtual Burgers vector $\overrightarrow{b_3}$ must be introduced such that

$$\sigma_{\theta z} = \frac{\mu_2 b_3}{4\pi d} \text{ and the associated displacement is } u_z = \frac{b_3 \theta}{2\pi}. \tag{99}$$

Applying the continuity condition at S for both the stresses and displacements yields

$$b_2 = b_1 \frac{\mu_2 - \mu_1}{\mu_1 + \mu_2} \quad \text{and} \quad b_3 = b_1 \frac{2\mu_1}{\mu_1 + \mu_2}. \tag{100}$$

Therefore, if $\mu_1 > \mu_2$, $b_2 < 0$ and the dislocation is attracted by the surface between the two media and the opposite is true for $\mu_1 < \mu_2$. This result can also be intuitively established by considering the dislocation self-energy, which is proportional to μ, i.e. the dislocation energy would be lower in the medium of smaller μ.

The situation can be further complicated by considering the case of an oxide layer as presented in **Figure 22(d)**. The mathematical development is far more complex and will not be treated here (Head, 1953). Qualitatively, it can be said that the dislocation decreases its self-energy by getting closer to the metal/oxide interface, which in turn leads to an increase of the energy of the oxide layer. Therefore, it exists for the dislocation an equilibrium position at a distance that depends of the ratio μ_2/μ_1, roughly of the order of the thickness of the oxide layer.

16.3.3.4 *Concept of Line Tension*

16.3.3.4.1 *Definition*

The nonvanishing self-energy of a dislocation implies that it will tend to decrease its energy by reducing its length to adopt a shape of minimum energy length. To characterize this property one attributes to a dislocation, by analogy with the surface tension of liquids, an LT T defined by Hirth et al. (1966)

$$T = \lim_{\partial \ell \to 0} \frac{\partial E}{\partial \ell} \tag{101}$$

where ∂E is the energy variation of the dislocation due to a change $\partial \ell$ of its length. The LT can be interpreted as the tangential force that must be applied to a point of the dislocation line for keeping its shape after cutting the dislocation at that point. Let us consider a small dislocation segment of length ℓ of a dislocation loop of Burgers vector $\overrightarrow{b}$. The dislocation loop is assumed at equilibrium under the action of the RSS τ in its glide plane, which is due to an externally applied force f_e. According to the Peach and Koehler relationship (Eqn (78)), the force acting on the dislocation segment ℓ is $f_p = \tau b \ell$. Because the dislocation loop is at rest, the dislocation segment ℓ is also in equilibrium and there exists a restoring force f_r, such as $f_r = -f_p$. This restoring force can be replaced by two tangential forces f_t acting at the extremities of the dislocation segment ℓ as schematically shown in **Figure 23**.

Assuming that the dislocation segment ℓ can be assimilated to an elastic string, the work done to increase its length by an increment $d\ell$ is $\partial E = f_t \, d\ell$, that is $f_t = \partial E/d\ell$. By analogy with Eqn (101), the LT T can be identified with f_t, i.e. $T \equiv f_t$.

We have seen that, even in isotropic elasticity, the self-energy E per unit length of a dislocation line varies with its orientation and continuously changes from edge to screw character. The exact determination of T under the influence of an external stress is rather tedious, but an approximate analytical result for an orientation-dependent LT can be obtained using suitable simplifying assumptions. We

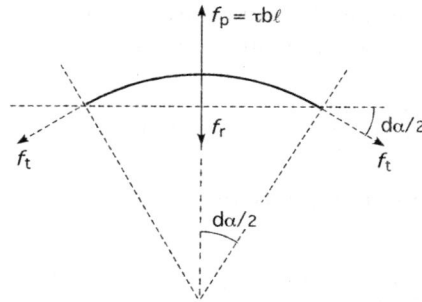

Figure 23 Illustration of the LT as equivalent to the tangential forces applied at the extremities of a dislocation segment.

shall assume that a single energy value per unit length, which depends only on the local dislocation character, can be assigned to each dislocation segment along the dislocation line. In doing so, we neglect the interactions between dislocation segments since a rigorous determination of T should take into account the exact shape of the dislocation line.

A relation between the LT and energy of the dislocation can be established by considering a small curvature of an initially straight and short dislocation segment of length L (**Figure 24**). The energy variation ΔE due to the small bow-out is

$$\Delta E = \int_0^L E(\theta)\,ds - \int_0^L E(0)\,dx \tag{102}$$

where $E(0)$ and $E(\theta)$ are the line energy of the straight and curved segments, respectively. For a small bow-out, $E(\theta)$ can be expanded in a Taylor series in θ:

$$E(\theta) = E(0) + \theta\left(\frac{dE}{d\theta}\right)_{\theta=0} + \frac{\theta^2}{2}\left(\frac{d^2 E}{d\theta^2}\right)_{\theta=0} + O(\theta^3) \tag{103}$$

Because at equilibrium the second term of the right-hand side is zero in Eqn (103) and we neglect higher-order terms, Eqn (102) can be rewritten

$$\Delta E = \int_0^L (E(\theta) - E(0))\,ds + \int_0^L E(0)\,(ds - dx) = \left(\frac{d^2 E}{d\theta^2}\right)_{\theta=0}\int_0^L \frac{\theta^2}{2}\,ds + E(0)\int_0^L dL \tag{104}$$

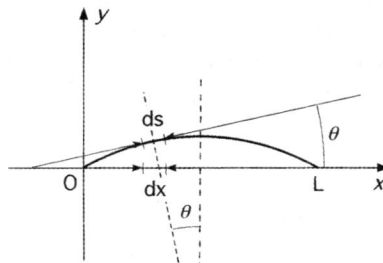

Figure 24 Geometrical representation of a dislocation segment to calculate the influence of dislocation curvature on the LT.

with $dL = ds - dx$. The elementary segment ds along the curved line can be expressed as a function of the angle θ. Considering that for small curvature $\tan \theta \sim \theta$, one has

$$ds = \sqrt{dx^2 + dy^2} = \left(1 + \frac{\theta^2}{2}\right)dx, \quad \text{which gives} \quad \frac{\theta^2}{2}ds = (ds - dx)\frac{ds}{dx} \approx dL. \tag{105}$$

Reporting this latter expression in Eqn (104), one obtains

$$\Delta E = \left(E + \frac{d^2E}{d\theta^2}\right)_{\theta=0} \int_0^L dL = \left(E + \frac{d^2E}{d\theta^2}\right)_{\theta=0} \Delta L \tag{106}$$

where ΔL is the increase in length of the dislocation segment due to its curvature. Comparing this expression with Eqn (101), the relation between the orientation-dependent LT $T(\theta)$ and the energy per unit length E of a dislocation can be written under the general form (deWit and Koehler, 1959)

$$T(\theta) = E + \frac{d^2E}{d\theta^2}. \tag{107}$$

The self-energy per unit length of a dislocation line E depends on orientation through the angle ψ between the line and the Burgers vector. For straight dislocations, E depends on ψ according to Eqn (72) so that one obtains for T by substituting θ to ψ in Eqn (107)

$$T(\psi) = \frac{\mu b^2}{8\pi(1-\nu)}\ln\frac{R}{r_o}[2 - \nu + 3\nu\cos 2\psi]. \tag{108}$$

As remarked by Friedel (1964), in many cases the second term on the right-hand side of Eqn (107) alters the LT only by a small numerical factor and can be neglected in rough estimates. Thus, in this case, the LT is equal to the energy of the dislocation per unit length and for a dislocation loop of diameter λ, T is given by

$$T = \frac{Kb^2}{4\pi}\ln\frac{\lambda}{r_o} \tag{109}$$

where $K = \mu/(1-\nu)$ for an edge dislocation and $K = \mu$ for a screw dislocation, as defined in Eqn (64). Similar result is obtained for dislocations in a crystal that are separated by a mean spacing distance λ (Friedel, 1964). For an arbitrary three-dimensional distribution of dislocations, the so-called Frank network, the dislocation density ρ and the mean spacing distance λ are related by $\lambda^2\rho = 1$. The LT is little sensitive to the dislocation density and using values of $\rho \sim 10^{10}$ m/m^3 (well-annealed fcc single crystal) and of $\rho \sim 10^{16}$ m/m^3 (strongly work-harden crystal) does not change T by more than a factor 2. Current values used in the literature in the framework of linear isotropic elasticity are therefore in the range μb^2 to $\mu b^2/2$.

16.3.3.4.2 Dislocation Nucleation

We have seen that the energy per atomic plane threaded by the dislocation line is high (usually a few eV). A rough estimate of the critical stress necessary to nucleate a dislocation loop can be performed using the LT concept. The work done W_σ by the applied shear stress τ for nucleating the dislocation loop must compensate the energy E_L required to create the dislocation line. For a dislocation loop of radius R, this work is

$$W_\sigma = \tau b \pi R^2.$$ (110)

The energy of a dislocation loop can be estimated according to Eqn (106) as

$$E_L = \int\limits_0^{2\pi} T(\psi) R d\psi$$ (111)

by replacing $T(\psi)$ with Eqn (108), one obtains after integration

$$E_L = \frac{\mu b^2 R (2-\nu)}{4(1-\nu)} \ln \frac{2R}{r_0}$$ (112)

and by taking $\nu \sim 1/3$, one finally has

$$E_L \approx \frac{\mu b^2 R}{2} \ln \frac{2R}{r_0}.$$ (113)

Without thermal activation, the dislocation loop can be nucleated only if $\Delta E = E_L - W_\sigma \leq 0$, that is with $r_0 \sim b$

$$\frac{2R}{b} \ln \frac{2R}{b} - \frac{\tau}{\mu} \pi \left(\frac{2R}{b}\right)^2 \leq 0.$$ (114)

The critical stress to nucleate a dislocation loop therefore corresponds to the limit case when $\Delta E = 0$. By numerically solving Eqn (114), one obtains the minimum required stress

$$\tau = \frac{\mu}{e\pi},$$ (115)

where e is the base of natural logarithm, for a critical radius $R \sim be/2$. A comparison of Eqns (23) and (115) shows that the two stresses are of the same order of magnitude, which similarly indicates that the experimental stresses reached for plastically deformed crystalline materials are in general not sufficient to create dislocations, except in very special cases, such as under the tip of a nanoindenter (Tromas et al., 1999, 2006; Montagne et al., 2009).

16.3.3.4.3 Dislocation Multiplication—Frank–Read Source (Frank and Read (1950))
There is a direct experimental evidence that dislocations multiply during plastic deformation, but as noted above, dislocations can hardly be produced by the applied stresses that are experimentally measured. A mechanism for dislocation multiplication has been proposed based on existing dislocations in the crystals before any deformation. This multiplication mechanism, called "Frank–Read source" (Frank and Read, 1950), considers a free dislocation segment of length $d\ell$ of a dislocation line whose ends are efficiently pinned at dislocation nodes, precipitates or sites where the dislocation leaves the glide plane. Under the action of an increasing RSS τ in its glide plane, the dislocation segment $d\ell$ bows out and its radius of curvature R decreases (**Figure 25**).

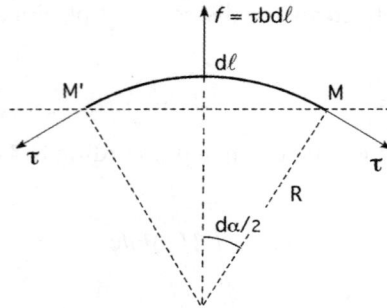

Figure 25 Geometrical representation of the curvature of a dislocation segment submitted to an applied stress σ.

According to the Peach and Koehler relation, the force acting on the dislocation segment due to the shear stress τ is $f = \tau b d\ell$ and is balanced by the restoring line-tension forces f_t. These forces are related by

$$2T \sin (d\alpha/2) = \tau b d\ell = \tau b R d\alpha, \tag{116}$$

where $d\alpha$ is the angle at which the segment $d\ell$ is seen from the center of curvature C (**Figure 25**). For small dislocation segment $\sin(d\alpha/2) \approx d\alpha/2$ and Eqn (116) rewrites

$$R = \frac{T}{\tau b}. \tag{117}$$

This equation is just the one-dimensional equivalent of Laplace–Young equation (Lamb, 1928), which relates the radius of curvature of a bubble to its surface tension and internal pressure. If one assumes in addition that T is orientation independent, then the radius of curvature of each segment $d\ell$ of the dislocation is the same and the whole dislocation line forms an arc of a circle. The forces f and f_t concomitantly increase with τ until R reaches a maximum value, corresponding to half the distance ℓ between anchor points. In other words, equilibrium between the applied and line-tension forces exists only when

$$R = \frac{T}{\tau b} \geq \frac{\ell}{2} \quad \text{or} \quad \text{equivalently } \tau \leq \frac{2T}{b\ell}. \tag{118}$$

Above $2T/b\ell$, the Peach–Koehler force becomes larger than the line-tension force and the dislocation line evolves as schematically shown in **Figure 26**. A dislocation loop is produced and the initial dislocation line is recovered, which allows the production of another dislocation loop. This process can continue until the applied stress is sufficient to counterbalance the LT force and the back stress resulting from the emitted dislocations. The necessary stress to multiply dislocations by this process is usually small. For a common dislocation density $\rho \sim 10^{10}$ m/m^3, the average length of free dislocation line is of the order of 10 μm, which with $T = \mu b^2$ yields $\tau \approx 6.10^{-5}$ μ, that is about 2.4 mPa for Cu.

Dislocation multiplication may also occur by a similar mechanism when edge dislocations are constrained to move by climb. In this mechanism, edge components are postulated to climb due to the sum of the applied and osmotic forces. The latter forces arise from a nonequilibrium concentration of vacancies in the vicinity of the dislocations that emit or absorb vacancies for reestablishing equilibrium conditions. In the presence of strong pinning, the dislocation segments bow out normal to their glide

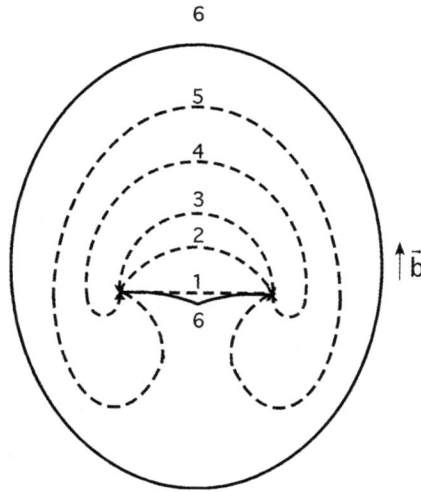

Figure 26   Diagrammatic representation of successive dislocation positions in a Frank–Read source.

planes to form circular loops until a critical configuration is reached. As for Frank–Read sources, the critical configuration occurs when the LT T of the dislocations balances the climb forces. These climb sources were first proposed by Bardeen and Herring (1952).

16.4   Models of Dislocation Cores

So far, we have considered dislocations in continuous, homogeneous, elastic media. This very fruitful approach has however one remarkable shortcoming already met several times in Sections 16.2 and 16.3: the elasticity theory predicts an unphysical divergence of the stress and strain fields near the dislocation line, forcing the definition of a dislocation core region where elasticity cannot be applied. In this region, the atomic nature of the crystal cannot be ignored to treat dislocations realistically. The aim of the present section is to outline some theoretical models, which incorporate a dislocation core.

As first recognized by Peierls (1940), crystals, simply because they are periodic and made of atoms, impose a resistance to the glide of their dislocations, called lattice resistance or lattice friction. The latter is materialized by the so-called Peierls potential, the energy of a dislocation as it moves in the crystal. As cartooned in **Figure 27(a)**, the Peierls potential has the periodicity of the crystal in the glide direction and contains minima, which correspond to stable equilibrium positions of the dislocation. Minima are surrounded by regions of stability called Peierls valleys, separated by energy maxima, or Peierls barriers, which are the cause for the lattice friction. Application of a RSS τ_A produces a plastic work τ_A by per unit length of dislocation, where y is the dislocation position. As illustrated in **Figure 27(b)**, the Peierls potential is then tilted, which decreases the Peierls barrier until it vanishes at an applied stress, called the Peierls stress. At that point, the lattice resistance is overcome and the dislocation starts to glide spontaneously in the crystal, even in absence of thermal fluctuations. If crystals were continuous and not made of atoms, the Peierls potential would be flat and the Peierls stress would be zero. The Peierls stress measured experimentally is the clearest signature of an interaction between the crystal lattice and a dislocation core.

(a)   **(b)**

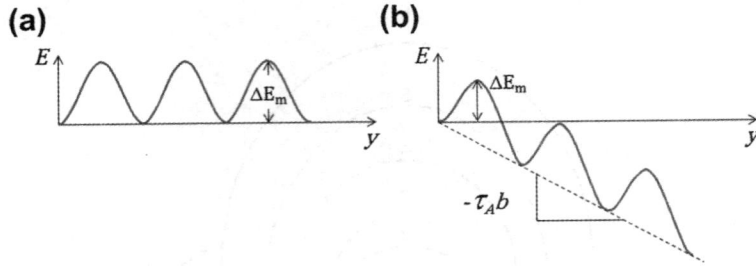

Figure 27 Schematic representation of a Peierls potential (a) in an unstressed crystal and (b) under an applied RSS τ_A. (For color version of this figure, the reader is referred to the online version of this book.)

The most straightforward way to account for dislocation cores is probably to perform atomistic simulations. Atoms are then considered individually and their mutual interactions are modeled using one of the various theories that treat the electronic contribution to the crystal cohesion with different levels of approximation (Finnis, 2003; Kaxiras, 2003). The models range from simple and mostly generic interatomic potentials, such as the Lennard–Jones potential, to sophisticated and highly computer demanding first-principle theories such as the density functional theory (DFT). Dislocation cores are obtained by finding the atomic configurations that minimize the potential energy of a crystal containing a dislocation. Stress may also be applied to measure the Peierls stress, or alternatively, the Peierls potential can be computed from the minimum energy path of a dislocation between two Peierls valleys. For reviews, see Bulatov and Cai (2006) and Bacon et al. (2009). The advantage of atomistic simulations is to recover the full anisotropic elasticity in regions of small deformations and to account for nonlinear interactions inside dislocation cores. Their drawback is that they are mostly numerical and usually do not allow for analytical developments.

In the present section, we restrict ourselves to models that can be solved analytically at least to some degree. The first of such models is the Peierls–Nabarro (PN) model, named after its first two contributors, Peierls (1940) and Nabarro (1947). This model has its own limitations but includes the main physical aspects of dislocations in crystals, i.e. a finite core width and a lattice friction. We also address kinks, small dislocation segments that allow dislocations to move from one Peierls valley to the next, using the LT model introduced in Section 16.3.3.4. Atomistic simulations will be mainly addressed in later sections when we consider specific lattice structures, FCC in Section 16.5 and BCC in Section 16.6.

16.4.1 The PN Model

16.4.1.1 *Misfit or Generalized Stacking Fault Energy*

When two half-crystals are shifted with respect to one another along a given plane, the crystal energy is a periodic function with minima each time the perfect crystal is recovered. In modern terms, this misfit energy is called a generalized stacking fault (GSF) energy, or γ-energy, after the work of Vitek (1968).

More precisely, the γ-energy is defined for a plane, for example the {111} plane of the FCC lattice. The crystal is cut in two halves along this plane and the top part is displaced blockwise with respect to the bottom part by a misfit vector $\vec{f}$, also called a disregistery vector. The two parts are then joined back together and the atoms are let to relax in the direction perpendicular to the shear plane. This procedure is analog to Frenkel's homogeneous shearing process considered in Section 16.2.1.3. Varying $\vec{f}$ over a periodic unit cell of the plane leads to an energy surface called a γ-surface. Varying $\vec{f}$ along a given direction only produces a γ-line. For example the $\langle 110 \rangle \{111\}$ γ-line is the misfit energy when shearing a crystal between two successive $\{111\}$ planes along a $\langle 110 \rangle$ direction. Examples of γ-surfaces and γ-lines are shown in **Figure 28**.

GSF energies are widely used to characterize the energetics of crystals in shearing conditions. Global minima correspond to the perfect crystal, while local minima indicate stable stacking faults, i.e. the possibility of altering the stacking sequence of a crystal while recovering a stable configuration. The most famous example is probably the $a/6\langle 112 \rangle\{111\}$ stacking fault in FCC crystals (see Section 16.5 and **Figure 28**).

GSF energies have been obtained for a number of crystalline structures with various cohesive models. Being periodic, their first-order shape is a sinusoidal function and their derivative corresponds to the stress needed to induce a homogeneous shear between two planes. It therefore corresponds to Frenkel stress expression given in Eqn (20). We may therefore write for a misfit f in a given direction along a given plane

$$\gamma(f) = \frac{\tau_{th}b}{\pi} \sin^2\left(\pi\frac{f}{b}\right) \tag{119}$$

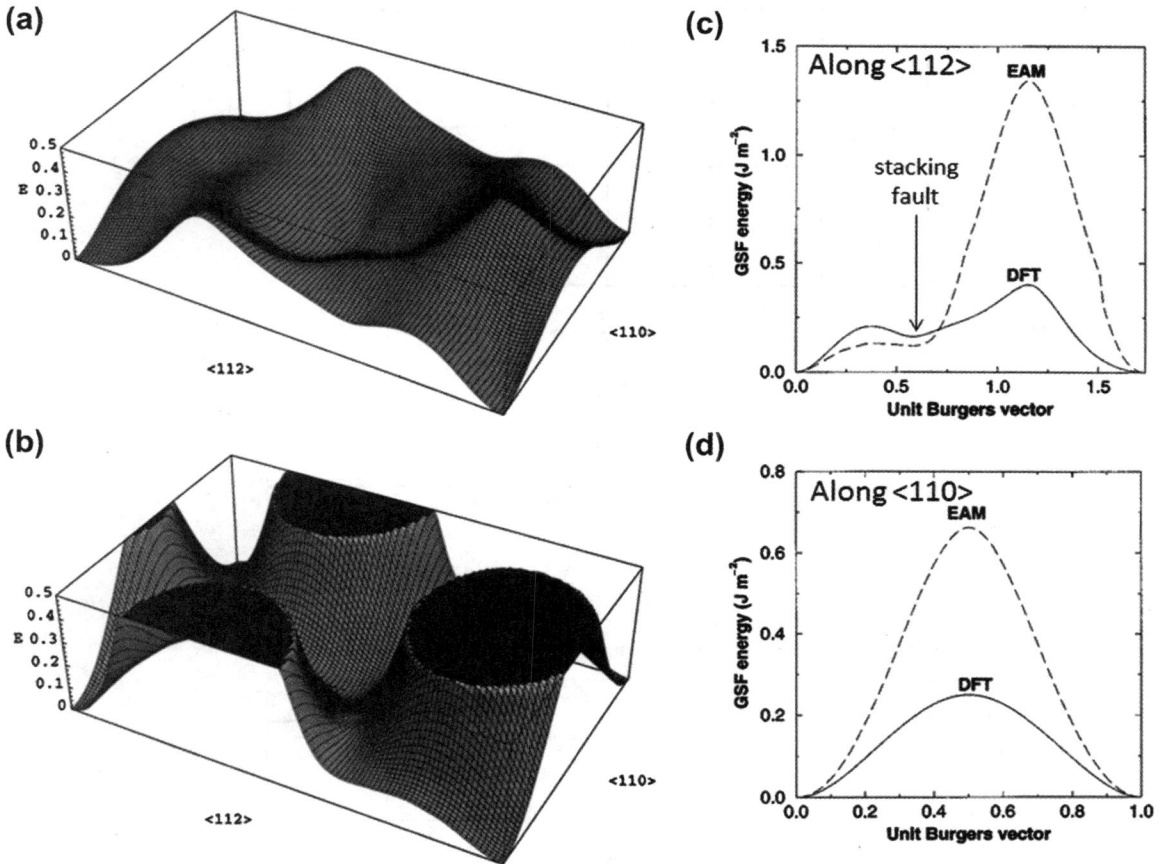

Figure 28 Examples of (a) γ-surfaces and (b) γ-lines parallel to a $\{111\}$ plane in FCC aluminum. Energies are computed using either an interatomic potential (denoted EAM for Embedded Atom Method) or first-principle electronic calculations (denoted DFT for Density Functional Theory). Reproduced with permission from Lu et al. (2000a).

and

$$\tau = \frac{\partial \gamma}{\partial f} = \tau_{th} \sin\left(2\pi \frac{f}{b}\right) \equiv T(f).$$ (120)

16.4.1.2 The PN Equation

The PN model was first proposed by Peierls (1940) and Nabarro (1947) and improved and refor-mulated by many authors including Eshelby (1949), Hirth and Lothe (1982) and Joós and Duesbery (1997a, b). See also Nabarro (1997) and Schoeck (2005) for reviews.

The model can be seen as an extension of Frenkel's misfit procedure where the half-crystals are not sheared rigidly, but inhomogeneously as a result of a distribution of infinitesimal Burgers vectors in the shear plane. More precisely, as illustrated in **Figure 29(a)**, we consider two isotropic linear elastic

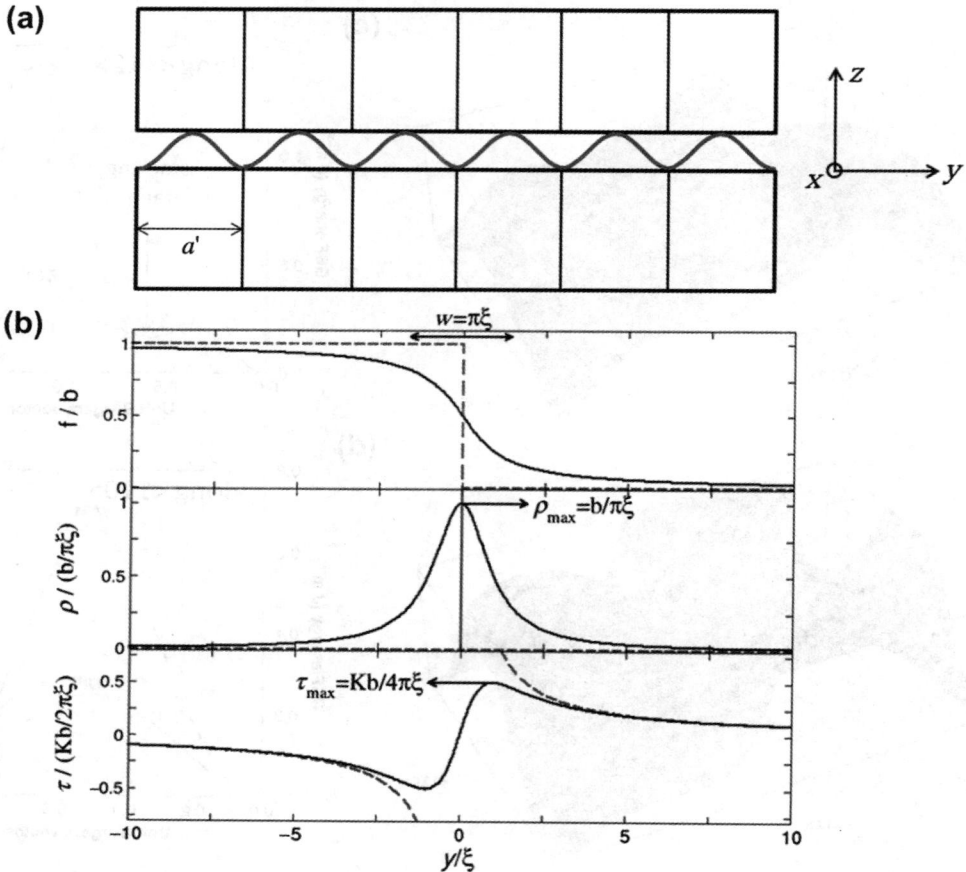

Figure 29 PN model: (a) representation of an elastoplastic crystal made of two elastic half-spaces joined by a misfit ener-gy, (b) Equilibrium misfit f, Burgers vector density ρ and internal RSS τ in case of a sinusoidal misfit energy. The dashed curves correspond to a Volterra dislocation. (For color version of this figure, the reader is referred to the online version of this book.)

half-crystals joined together by a γ-energy along a plane, the glide plane. Any difference of displacement f between the two planes has an energy cost $\gamma(f)$ given by Eqn (119). A dislocation with Burgers vector b is positioned inside the glide plane with a line along the x-axis and a cut plane in the $y < 0$ part of the xy-plane shown in **Figure 29(a)**. In the PN model, the dislocation core is assumed to spread in the glide plane only.

If we set $r_0 \to 0$, a Volterra dislocation as discussed in Sections 16.2 and 16.3 has a punctual core with a disregistery $f = b$ for $y < 0$ and $f = 0$ for $y > 0$, i.e. $f(y) = b\,H(y)$, with H the Heaviside step-function. The Burgers vector is then a delta function, $b\delta(y)$. Both disregistery and Burgers vector of a Volterra dislocation are shown in **Figure 29(b)**. The internal stress produced by this dislocation in the glide plane is proportional to $1/y$ and diverges at $y = 0$, a consequence of applying linear elasticity down to the dislocation line, as discussed in Section 16.3.

In contrast, within the PN model, the dislocation core may spread in the glide plane. The Burgers vector is then replaced by a distribution of infinitesimal Burgers vectors $\rho(y)$, such that $\rho(y)dy$ is an infinitesimal Burgers vector positioned at y. To maintain the full Burgers vector b, we have the normalization condition

$$\int_{-\infty}^{+\infty} \rho(y')\mathrm{d}y' = b. \tag{121}$$

By definition (see Section 16.2), a Burgers vector $\rho(y)dy$ induces in the crystal a disregistery between the upper and lower half-crystals $\mathrm{d}f(y') = \rho(y)\mathrm{d}yH(y' - y)$ and

$$\rho(y') = -\frac{\mathrm{d}f}{\mathrm{d}y'}. \tag{122}$$

Each infinitesimal Burgers vector produces a stress field in the crystal. In the framework of linear elasticity, we can use the superposition principle and Eqns (48) and (58). Considering only the displacements along the Burgers vector, we find that the RSS produced by the distribution of Burgers vectors in the glide plane is the sum of the contributions from all infinitesimal dislocations

$$\tau(y) = \int_{-\infty}^{+\infty} \frac{K}{2\pi} \frac{1}{y - y'} \rho(y')\mathrm{d}y' \tag{123}$$

with $K = \mu[\sin^2\Psi/(1 - v) + \cos^2\Psi]$ for isotropic crystals, where Ψ is the angle between the Burgers vector and the dislocation line as introduced in Eqn (64): $K = \mu/(1 - v)$ for an edge dislocation, i.e. if the disregistery is in the y-direction perpendicular to the dislocation line and $K = \mu$ for a screw dislocation, i.e. if the disregistery is in the x-direction parallel to the dislocation line.

At equilibrium, this local stress must be balanced by the restoring stress between the two half-crystals, $T[f(y)]$. Using Eqns (122) and (123), we arrive at the famous integrodifferential equation known as the PN equation

$$\tau(y) = \int_{-\infty}^{+\infty} \frac{K}{2\pi} \frac{1}{y - y'} \frac{\mathrm{d}f}{\mathrm{d}y'}\mathrm{d}y' = T[f(y)]. \tag{124}$$

This equation cannot be solved in its general form. However, if we assume the sinusoidal shape of Eqn (120), Eqn (124) has a well-known soliton like solution

$$f(y) = -\frac{b}{\pi}\tan^{-1}\left(\frac{y}{\xi}\right) + \frac{b}{2} \text{ with } \xi = \frac{Kb}{4\pi\tau_{th}}, \tag{125}$$

for the boundary conditions $f(-\infty) = b$, $f(+\infty) = 0$ and by symmetry $f(0) = b/2$. The disregistery field is shown in **Figure 29(b)**. The change of disregistery across the dislocation line is not abrupt as with a Volterra dislocation, but spread in the glide plane. The core width may be defined by the intersections of the tangent at $y = 0$ with the axis $f = 0$ and $f = b$ (Frank and van der Merwe, 1949). We obtain $w = \pi\xi$. The parameter ξ thus represents the core width of the dislocation. The core is wider for edge dislocations than for screw dislocations because of the larger value of K, and the core width is a decreasing function of the ideal shear strength, i.e. stronger crystals have narrower dislocations. The corresponding Burgers vector density is shown in the central panel of **Figure 29(b)**, where the Dirac function of the Volterra dislocation is replaced by a Lorentzian distribution, $\rho(x) = b/\pi\xi\, \xi^2/(y^2 + \xi^2)$. Inserting Eqn (125) into Eqn (123), the stress produced by the dislocation in its glide plane is expressed as

$$\tau(x) = \frac{Kb}{2\pi}\frac{y}{y^2 + \xi^2}. \tag{126}$$

with K defined in Eqns (64) and (123). As shown in the lower panel of **Figure 29(b)**, the Volterra stress field is recovered far away from the dislocation core where the stress in Eqn (126) scales as $1/y$. At short distance, on the other hand, the spreading of the dislocation core avoids the divergence of the stress field and removes the singularity at $y = 0$. The stress is now maximum at $y = \pm\xi$ and decreases near the dislocation line to be zero at $y = 0$. Note that if $\xi \to 0$, all the results of the Volterra dislocation are recovered. To contrast with the Voltera dislocation, the dislocation modeled within the PN model is sometimes called a Peierls dislocation (Schoeck, 2005).

16.4.1.3 *Peierls Potential and Peierls Stress*

One limitation of the above approach is that the equilibrium disregistery has translational invariance along the y-axis: if $f(y)$ is solution of Eqn (124), any function $f(y - y')$ is also solution, whatever the value of y'. The reason is that we assumed in Eqns (121) and (122) a continuous distribution of Burgers vectors and therefore a continuous disregistery. The consequence of the translational invariance is that the dislocation energy, its Peierls potential, is a constant, independent of the dislocation position and the Peierls stress, related to the derivative of the Peierls potential, is zero. However, we should note that a misfit can physically be left only at the location of atomic planes perpendicular to the glide plane. In-between two planes, there is only vacuum and therefore, no way to sense the disregistery.

To include discreteness in the dislocation energy, Peierls (1940) and Nabarro (1947) first, and later Christian and Vitek (1970), Hirth and Lothe (1982), Joós et al. (1994), Joós and Duesbery (1997a, b), Joós and Zhou (2001) proposed to compute the dislocation energy not as an integral, but as a discrete sum of energy costs due to the misfits at each atomic plane across the glide plane. If we assume a crystal with atomic planes perpendicular to the glide plane, as in **Figure 29(a)**, with planes positioned at $y = (m + 1/2)a'$ (a' is the lattice parameter in the shear direction) and a dislocation centered at y', the plane misfits are $f((m + 1/2)a' - y')$, with a corresponding energy cost $\gamma(f((m + 1/2)a' - y'))$ (for a generalization to planes nonperpendicular to the glide plane and considerations on accounting for discreteness, see Lu et al. (2000b)). The dislocation energy also includes a part coming from the elastic distortion of the two half-spaces, which is independent of the dislocation position because the two half-spaces are continuous. For isotropic crystals, the elastic energy noted here E_{elas} is given by Eqn (64). The energy per unit length of the dislocation as a function of its position, i.e. its Peierls potential, is given thus by

$$E(y) = E_{elas} + \sum_{m=-\infty}^{+\infty} \gamma \left(f \left(\left(m + \frac{1}{2} \right) a' - y \right) \right) a'. \tag{127}$$

Here, it is assumed that the equilibrium disregistery is unchanged during translation of the dislocation, i.e. there is no atomic relaxation during dislocation motion, an obvious approximation whose consequences are discussed in Schoeck (1999) and Lu et al. (2000b). As expected from Peierls' argument, the above function is periodic with period a', the distance between atomic planes across the glide plane, and has a nonzero Peierls stress. If γ is further assumed sinusoidal, Joós and Duesbery (1997a) showed that the Peierls potential can be expressed analytically

$$E(y) = E_{elas} + \frac{Kb^2}{4\pi} \frac{\sin h(2\pi\xi/a')}{\cos h(2\pi\xi/a') - \cos(2\pi y/a')}. \tag{128}$$

The Peierls potential is minimum for $y = ma'$, that is the equilibrium position for the dislocation is at mid-distance between atomic planes, and the Peierls barrier is when the dislocation center is aligned with an atomic plane. The minimum dislocation energy in Eqn (128) can be written $E_0 = E_{elas} + E_c$, with the first term being the elastic self-energy of the dislocation and the second the core energy expressed as

$$E_c = \frac{Kb^2}{4\pi} \frac{\sin h(2\pi\xi/a')}{\cos h(2\pi\xi/a') + 1}. \tag{129}$$

This term varies from 0 in very narrow cores to $Kb^2/4\pi$ in very wide cores. It may therefore be an order of magnitude (or more) smaller than the elastic term, which dominates the dislocation line energy. In order to highlight the energy contribution, which varies with the dislocation position, the Peierls potential can be rewritten as

$$E(y) = E_0 + \Delta E(y) \text{ with } \Delta E(y) = E_c \frac{1 - \cos(2\pi y/a')}{\cos h(2\pi\xi/a') + \cos(2\pi y/a')}. \tag{130}$$

The amplitude of ΔE is the Peierls barrier

$$\Delta E_m = \frac{Kb^2}{2\pi} \frac{1}{\sin h(2\pi\xi/a')}. \tag{131}$$

The Peierls barrier diverges for narrow cores and decreases exponentially when the core width increases. The Peierls barrier may therefore be smaller, of the same order or larger than the dislocation line energy. In the next section, we will take $\Delta E_m/E_0 = 1$ as a reference. Examples of Peierls potentials,

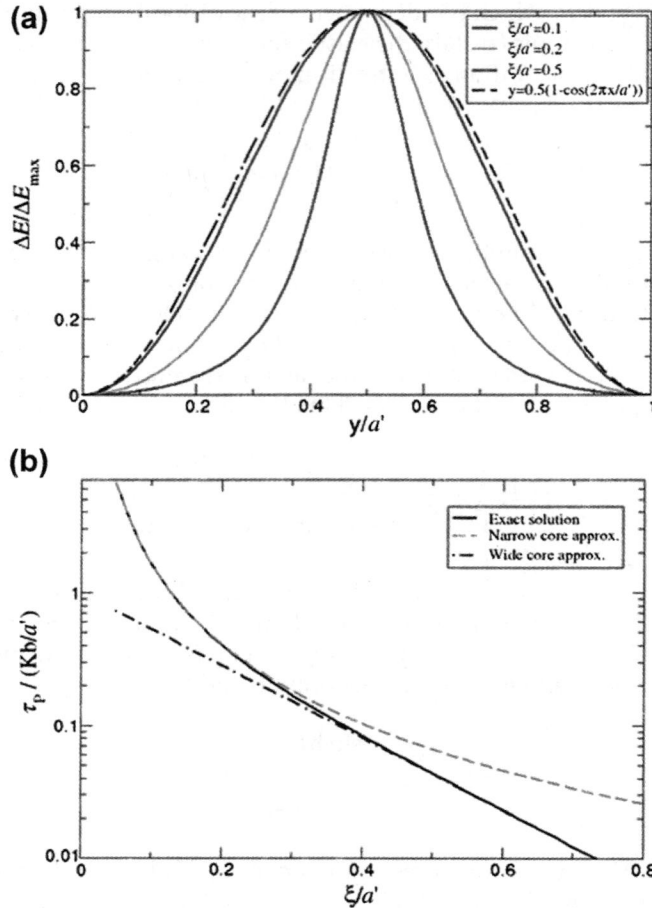

Figure 30 (a) Peierls potential and (b) Peierls stress predicted by the PN model for a sinusoidal misfit energy. Different values of the core width parameter ξ are considered. (For color version of this figure, the reader is referred to the online version of this book.)

scaled by the Peierls barrier, are shown in **Figure 30(a)**. As ξ/a' increases, i.e. when the dislocation core widens, the Peierls potential evolves from a Dirac to a sinusoidal function. As seen in **Figure 30(a)**, for $\xi/a' > 1$, which may be taken as the limit of wide cores, the Peierls potential is well approximated by the sinusoidal function.

$$E_{SG}(\gamma) = \left(E_{elas} + \frac{Kb^2}{4\pi} \right) + \frac{Kb^2}{2\pi} e^{-\left(2\pi\xi/a'\right)} \left(1 - \cos\left(2\pi\gamma/a'\right)\right). \tag{132}$$

This function is denoted as E_{SG} in reference with the sine Gordon model, discussed below. If a RSS is applied to the dislocation, the energy per unit length of dislocation becomes $E_{tot}(\gamma) = E(\gamma) - \tau_A b\gamma$. The new equilibrium position of the dislocation is given by $E'(\gamma_{eq}) = \tau_A b$. This equation has a solution as long as $\tau_A b$ is less than the maximum derivative of the Peierls potential. Above this threshold, no equilibrium position exists, or equivalently, the Peierls barrier has disappeared and the dislocation can glide without limit within the crystal. The threshold stress is therefore the Peierls stress, τ_P. There is no general analytic expression for the Peierls stress with a sinusoidal potential. It is given by Joós and Duesbery (1997a)

$$\tau_P = \tau(\gamma_M) \tag{133}$$

with

$$\tau(\gamma) = \frac{1}{b}\frac{dE}{d\gamma} = \frac{Kb}{2a'}\frac{\sin \mathrm{h}(2\pi\xi/a')\sin(2\pi\gamma/a')}{\left[\cos \mathrm{h}(2\pi\xi/a') - \cos(2\pi\gamma/a')\right]^2} \tag{134}$$

and

$$\gamma_M = \frac{a'}{2\pi}\cos^{-1}\left(\frac{1}{2}\left\{-\cos \mathrm{h}(2\pi\xi/a') + \left[8 + \cos \mathrm{h}^2(2\pi\xi/a')\right]^{1/2}\right\}\right). \tag{135}$$

The Peierls stress is shown in **Figure 30(b)** as a function of ξ, with two approximations.

$$\left.\begin{array}{l}\text{for narrow cores } \left(\dfrac{\xi}{a'} < 0.25\right),\ \ \tau_P \approx \dfrac{3\sqrt{3}}{8}\dfrac{Kb}{a'}\left(\dfrac{a'}{2\pi\xi}\right)^2 \\[4mm] \text{for wide cores } \left(\dfrac{\xi}{a'} > 0.4\right),\ \ \tau_P \approx \dfrac{Kb}{a'}e^{-2\pi\frac{\xi}{a'}}\end{array}\right\}. \tag{136}$$

The Peierls stress decreases exponentially with the core width when the latter is wide. The lattice friction is therefore important only for narrow dislocation cores, as is indeed observed by direct atomistic simulations (see Sections 16.5 and 16.6).

In this section, we have shown that accounting for the discrete atomic nature of matter inside the dislocation core removes the singularity of the stress field near the dislocation line while retaining the long-range linear elastic stress field. We have also seen that the PN model predicts a Peierls stress which decreases exponentially with the core width. We have restricted ourselves to sinusoidal γ-surfaces in order to obtain analytical results. However, the PN model can also be used with more complex γ-surfaces, in particular those obtained by atomistic calculations (see Kaxiras and Duesbery (1993) for an early example). This approach has been shown to predict dislocation properties in good agreement with values obtained from direct atomistic calculations. For examples in Si, see Joós et al. (1994), Ren et al. (1995) and Hansen et al. (1995); in Al, see Lu et al. (2000a), in NiAl, see Schoeck (2001), in Cu, Schoeck and Krystian (2005) and in Ni, Schoeck (2006).

The main limitation of the PN model is probably that the dislocation core can spread in a single plane only and is therefore necessarily planar. This restriction is not an issue for edge dislocations because their Burgers vector is perpendicular to their line implies planar spreading, but the situation is different for screw dislocations that do not have a definite glide plane. As will be seen in Section 16.6, screw dislocations in BCC crystals for instance, spread in several nonparallel planes. Extensions of the PN model to nonplanar cores have been proposed but require simplifying assumptions (Zhang et al., 1989; Ngan,

1995, 1997). Alternatively, one can use numerical approaches, particularly the so-called Peierls–Nabarro Galerkin model (Denoual, 2007). But nonplanar and narrow cores are more reliably treated using direct atomistic simulations, as discussed in Section 16.6. Other extensions of the PN model have been proposed, to account for elastic anisotropy (Schoeck, 1994), for displacements both parallel and perpendicular to the Burgers vector (Schoeck, 2005), to treat more reliably narrow dislocation cores (Bulatov and Kaxiras, 1997) and atomic relaxations during dislocation motion (Lu et al., 2000b).

16.4.2 Kinks

In materials with significant Peierls stress, dislocations do not move by rigid translations, but rather by the nucleation and propagation of kink pairs. As illustrated in **Figure 31(a)**, a kink pair is created when part of the dislocation is activated to a neighboring Peierls valley. A kink is therefore a very short dislocation segment, of length a', the distance between Peierls valleys, connecting the dislocation between two valleys. Expansion of the kink pair along the dislocation transfers the dislocation in the next Peierls valley, allowing the defect to glide in the crystal. At high temperature, thermal activation is strong enough that the dislocation does not feel the Peierls potential and is in an athermal regime, but at lower temperature, dislocation motion becomes thermally activated. For most stresses, the rate-controlling mechanism is then the kink-pair nucleation because the kinks, having usually a very small lattice friction, move very fast along the dislocation. In this regime, the dislocation is predominantly straight at the bottom of its Peierls valley and forms infrequently a kink pair, which expands rapidly. Only at very low stresses is a kink motion along the dislocation line the limiting process. The kinks may then reach an equilibrium density along the dislocation line, as discussed by Lothe and Hirth (1959). In the following however, we will consider only the regime where dislocation motion is controlled by a kink-pair nucleation.

Viewed as a dislocation segment, a kink is defined by a line direction and a Burgers vector. If perpendicular to the dislocation line, a kink on a screw dislocation is of edge character and vice versa for an edge dislocation. Two kinks in a kink pair have opposite signs and are expected to attract each other from elasticity theory as discussed in Section 16.3.3.1. If a dislocation changes Peierls valleys several times, a succession of same-sign kinks that repel each other is created (see **Figure 31(b)**).

In the present section, we discuss kink models, and in particular the elastic theory of kinks and the LT model. Both models allow the prediction of the activation energy for the formation of a kink pair on a dislocation line, which is the energy barrier that controls the glide of high-Peierls stress dislocations at low temperatures (Caillard and Martin, 2003).

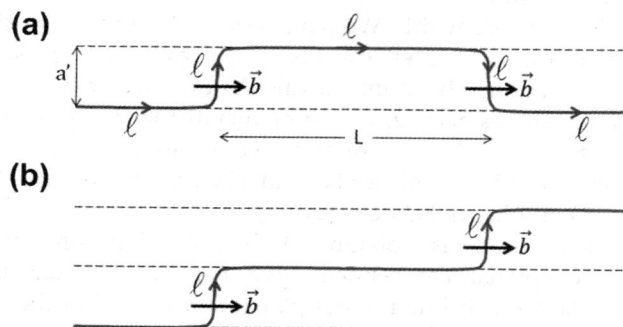

Figure 31 Schematic representation of (a) a kink pair on a screw dislocation, (b) two successive kinks. (For color version of this figure, the reader is referred to the online version of this book.)

16.4.2.1 Elastic Energy of Kinks

The calculation of the elastic self-energy of kinks and of their interaction was first developed by Eshelby (1962) and Seeger and Schiller (1962). We consider here abrupt kinks of length a', the distance between Peierls valleys, perpendicular to the dislocation line. The dislocation is of either screw or edge character. The excess elastic energy due to a kink pair of separation L on an otherwise straight and infinite dislocation is due to the self-energy of the kinks and to the interaction energy between parallel segments (the interaction between perpendicular segments vanishes on screw and edge dislocations as shown in Section 16.3.1.1). Calculations are detailed in Hirth and Lothe (1982) and Koizumi et al. (1993). The excess energy of the kink pair, developed to the first power of a'/L, may be written as

$$E_{KK} = 2E_K + W_{KK}. \tag{137}$$

The single kink self-energy, E_K, is identified as the part of E_{KK}, which does not depend in the kink separation. Its expression is

$$\left.\begin{array}{l} E_K = \dfrac{\mu b^2 a'}{4\pi(1-\nu)}\left[\ln\dfrac{a'}{e\rho} - (1-\nu)\right] \text{ for a (edge) kink on a screw dislocation} \\[4mm] E_K = \dfrac{\mu b^2 a'}{4\pi}\left[\ln\dfrac{a'}{e\rho} - \dfrac{1}{(1-\nu)}\right] \text{ for a (screw) kink on an edge dislocation} \end{array}\right\}. \tag{138}$$

The interaction energy between the kinks, W_{KK}, is

$$\left.\begin{array}{l} W_{KK} = -\dfrac{\mu b^2 a'^2}{8\pi L}\dfrac{1+\nu}{1-\nu} \text{ for two opposite kinks on a screw dislocation} \\[4mm] W_{KK} = -\dfrac{\mu b^2 a'^2}{8\pi L}\dfrac{1-2\nu}{1-\nu} \text{ for two opposite kinks on an edge dislocation} \end{array}\right\}. \tag{139}$$

This latter energy is negative and increases with increasing L. Kinks of opposite signs thus attract each other, as expected. For kinks of same sign, the result is the same but with the sign reversed, i.e. kinks of same sign repel each other. As discussed in Hirth and Lothe (1982), the kink self-energy is not well defined because it depends on the choice of the cutoff length ρ since a' is small and of the same order as ρ. Also, this energy should include a core contribution as discussed below and the simple fact of considering a segment as short as a' as a dislocation segment may be doubtful. On the other hand, the kink–kink interaction energy is well defined and does not depend on the details of the kinks as long as a'/L is small. In particular, the interaction energy does not depend on whether the kinks are abrupt (perpendicular to the dislocation line) or oblique.

The interaction force between kinks is given by $F_K = -dW_{KK}/dL$:

$$\left.\begin{array}{l} F_{KK} = -\dfrac{\mu b^2 a'^2}{8\pi L^2}\dfrac{1+\nu}{1-\nu} \text{ for two opposite kinks on a screw dislocation} \\[4mm] F_{KK} = -\dfrac{\mu b^2 a'^2}{8\pi L^2}\dfrac{1-2\nu}{1-\nu} \text{ for two opposite kinks on an edge dislocation} \end{array}\right\}. \tag{140}$$

This force scales as $1/L^2$ and has the same dependence on the kink–kink separation as the Coulomb force between charged particles. It decreases more rapidly than the interaction between infinite dislocations, proportional to $1/L$ (see Section 16.3.3). The interaction between kinks is therefore more short-ranged than between infinite dislocations.

To represent more faithfully the energy of a kink pair, one should also include a contribution from the dislocation core. The latter reflects the fact that a kink extends across a Peierls valley and therefore has a penalty due to the Peierls potential. This contribution, noted ΔP below, can be modeled by an integral over the Peierls potential (Koizumi et al., 1993). For an abrupt kink, we have

$$\Delta P = \int_0^{a'} \Delta E(y) \mathrm{d}y. \tag{141}$$

where ΔE, introduced in Eqn (130), is the part of the Peierls potential, which depends on the dislocation position. Under stress, the energy of the kink pair has an extra energy term which corresponds to the work of the applied stress, $\tau_A b a' L$, where τ_A is the RSS and $a'L$ the area swept by the kink pair. The kink pair energy, or rather its enthalpy, is thus

$$H_{KK}(L) = 2(E_K + \Delta P) - \eta \frac{\mu b^2 a'^2}{8\pi L} - \tau_A b a' L. \tag{142}$$

with $\eta = (1 + v)/(1 - v)$ for a kink pair on a screw dislocation and $\eta = (1 - 2v)/(1 - v)$ for a kink pair on an edge dislocation. The first two terms on the right-hand side are the twice the formation energy of a single isolated kink, including the elastic and core contributions. Only the last two terms depend on the kink-pair separation. The kink enthalpy H_{KK} increases at small L, has a maximum at $L^* = \sqrt{\eta \mu b a'/8\pi \tau_A}$ and decreases at large L because of the work of the applied stress. The formation and separation of a kink pair is thus an unstable process and the maximum enthalpy, which is the activation energy that must be supplied by thermal fluctuations in order to create the kink pair, is $H^*_{KP} = H_{KK}(L^*)$, expressed as

$$H^*_{KP} = 2(E_K + \Delta P) - b a' \sqrt{\eta \frac{\mu b a'}{8\pi} \tau_A}. \tag{143}$$

At zero applied stress, the activation enthalpy is equal to the formation energy of two isolated kinks because $L^* \rightarrow \infty$ in this limit. As the applied stress increases, L^* decreases as $1/\sqrt{\tau_A}$ and H^*_{KP} decreases as $-\sqrt{\tau_A}$. At large applied stresses, the kinks become too close in the activated state to be treated individually and the present model breaks down. In this limit, the two kinks must be considered as a single entity, as discussed in next section.

16.4.2.2 *LT Model*

16.4.2.2.1 *LT Model*

In the LT model, the dislocation is materialized by an elastic line moving on its γ-surface. The dislocation is no longer restricted to be straight and its profile is described by a continuous function, $y(x)$, defined in the glide plane as illustrated in **Figure 32(a)**. The energy of an infinitesimal segment of dislocation, such as the AB segment in **Figure 32(a)**, has two contributions. The first comes from the LT on the dislocation, introduced in Section 16.3.3.4. Here, we will neglect the dependence of the dislocation character and will consider a constant LT equal to the dislocation line energy per unit length, noted E_0 in Eqn (130). The second contribution to the energy comes from the Peierls potential

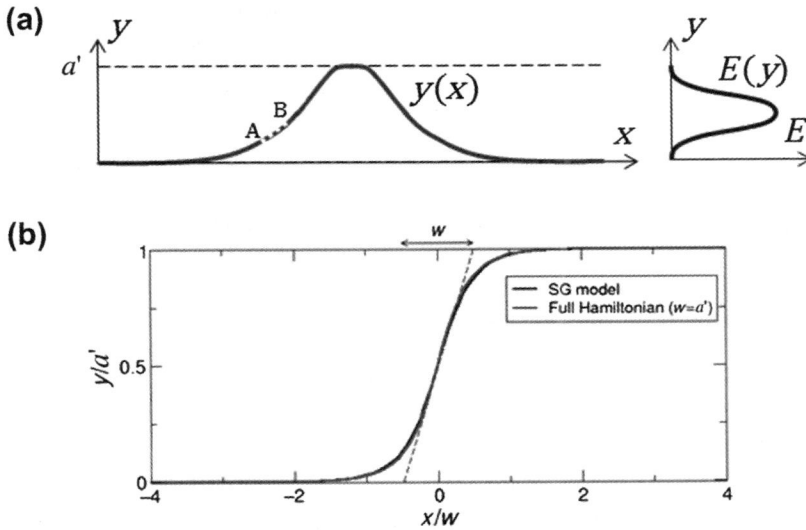

Figure 32   (a) LT model of a dislocation of shape y(x) and (b) equilibrium profiles for two different approximations (see text for details). (For color version of this figure, the reader is referred to the online version of this book.)

and accounts for the energy penalty of having dislocation segments away from the bottom of the Peierls valley. The energy per unit length of the AB segment can thus be written

$$E(y(x)) = E_0 + \Delta E(y(x)), \tag{144}$$

with ΔE, the part of the Peierls potential, which depends on the dislocation position, as in Eqn (130).

The excess energy with respect to a straight dislocation lying at the bottom of the Peierls valley is given by integrating Eqn (144) along the length of the dislocation (Dorn and Rajnak, 1964)

$$E_{\text{excess}} = \int_{-\infty}^{+\infty} \left\{ (E_0 + \Delta E(y(x))) \sqrt{1 + \left(\frac{dy}{dx}\right)^2} - E_0 \right\} dx, \tag{145}$$

The equilibrium profile of the dislocation minimizes the excess energy. It is thus an extremum of the above functional integral and must satisfy Euler–Lagrange equation

$$\frac{d}{dx}\frac{\partial L}{\partial y'} - \frac{\partial L}{\partial y} = 0, \tag{146}$$

with $L(y, y') = (E_0 + \Delta E(y(x)))\sqrt{(1 + (dy/dx)^2)} - E_0$. This condition can be integrated, yielding a first-order differential equation

$$1 + \left(\frac{dy}{dx}\right)^2 = \left(1 + \frac{\Delta E(y)}{E_0}\right)^2. \tag{147}$$

Using this relation, the equilibrium dislocation energy can be rewritten in different ways

$$E = \frac{1}{E_0} \int\limits_{-\infty}^{+\infty} \left[E(\gamma)^2 - E_0^2\right] dx = E_0 \int\limits_{-\infty}^{+\infty} \left(\frac{d\gamma}{dx}\right)^2 dx. \tag{148}$$

We consider first the case of a single kink on a dislocation. Solving Eqn (147) with the boundary conditions $\gamma(-\infty) = 0$, $\gamma(+\infty) = a'$ and by symmetry $\gamma(0) = a'/2$ yields the equilibrium kink shape. Using Eqn (148), we then obtain the kink energy. This energy is the LT equivalent of Eqn (138). As discussed below, the advantage of the LT model is to treat more accurately the excess energy of close kinks but its disadvantage is to ignore long-range elastic interaction between the kinks. Dorn and Rajnak (1964) showed that if the Peierls potential is sinusoidal, as in Eqn (132), the kink shape and energy can be obtained analytically. Other simple potentials, for instance piece-wise parabolic (Eshelby, 1962), can also be treated analytically, as reviewed by Guyot and Dorn (1967). Using the full Peierls potential obtained from the PN model in Eqn (130), Joós and Zhou (2001) obtained semianalytical results.

Here, we consider only a sinusoidal Peierls potential. Before treating the full nonlinear differential equation in Eqn (147), we will address another well-known approximation, valid for wide kinks. In this limit, $d\gamma/dx \ll 1$ and if we further assume that $\Delta E/E_0 \ll 1$, the excess energy can be simplified to

$$E_{SG} = \int\limits_{-\infty}^{+\infty} \left\{ \frac{E_0}{2} \frac{d^2\gamma}{dx^2} + \Delta E(\gamma(x)) \right\} dx. \tag{149}$$

We call this energy E_{SG} in reference with the sine-Gordon (SG) equation, the well-known differential equation obtained from the Euler–Lagrange condition applied to the above excess energy. The present approximation can also be obtained from the LT model by neglecting the change of dislocation length in the Peierls contribution to the excess energy. The SG approximation thus does not take fully into account the change of dislocation length when the dislocation bows out. We will see however that, provided $\Delta E/E_0 \ll 1$, the SG model yields results almost identical to the full energy in Eqn (145).

In the present approach, since the dislocation line is continuous and the energy is computed as an integral, the energy of a kink has translational invariance and there is no resistance to the motion of a kink along the dislocation line. In real crystals, the discrete nature of the atoms will again induce a resistance to the kink motion, and a so-called secondary Peierls stress, which may control the expansion of a kink pair at low applied stress, as discussed by Lothe and Hirth (1959) and Joós and Duesbery (1997b). In order to introduce a lattice resistance to the kink motion, the kink energy must be computed as a discrete sum instead of an integral, as in the PN model. However, in metals, the secondary Peierls stress is very small and can be neglected in all practical conditions. We will not address this aspect here, and the reader is referred to Joós and Duesbery (1997b) for more information on that topic.

16.4.2.2.2   SG Equation

Consistent with Eqn (132), the Peierls potential is approximated by a sinusoidal law

$$\Delta E(\gamma) = \frac{\Delta E_m}{2} \left(1 - \cos\left(2\pi\gamma/a'\right)\right). \tag{150}$$

The Euler–Lagrange condition yields a second-order differential equation

$$E_0 \frac{d^2 \gamma}{dx^2} = \frac{d\Delta E}{d\gamma} = \frac{\Delta E_m \pi}{2a'} \sin(2\pi\gamma/_{a'}).$$ (151)

In scaled form, this equation can be rewritten as $d^2\Phi/d\theta^2 = \sin\Phi$, the time-independent part of the SG equation (Peyrard and Dauxois, 2004). If the above equation is discretized, i.e. if the second derivative is replaced by a finite difference, this equation becomes another very famous equation, called the Frenkel–Kontorova equation after the work of Frenkel and Kontorova (1938) (see Braun and Kivshar (2004) for a review). The sine Gordon equation has a soliton solution for a single kink

$$\gamma(x) = a'\left(1 - \frac{1}{\pi} \cos^{-1}\tanh\left(\sqrt{\frac{2\Delta E_m}{E_0}} \pi \frac{x}{a'}\right)\right) = \frac{2a'}{\pi}\tan^{-1}\exp\left(\sqrt{\frac{2\Delta E_m}{E_0}}\pi\frac{x}{a'}\right).$$ (152)

The kink width can be defined from the maximum slope of the profile, in the same way as for the dislocation core width in the PN model. We obtain

$$w = \frac{a'}{(d\gamma/dx)_{x=0}} = a'\sqrt{\frac{E_0}{2\Delta E_m}}.$$ (153)

Inserting the dislocation profile into Eqn (147), we obtain the kink formation energy

$$E_K = \frac{2a'}{\pi}\sqrt{2\Delta E_m E_0}.$$ (154)

The profile from the sine Gordon equation is shown in **Figure 32(b)**. It is similar to that obtained from the PN model but steeper, with a kink width a factor π smaller than the dislocation core (compare Eqn (153) with Eqn (125)). Also, we see that dislocations with a higher Peierls barrier ΔE_m have narrower kinks and higher kink energy. The reason is the increased energy penalty for placing dislocation segments near the top of the Peierls barrier.

16.4.2.2.3 Full Sinusoidal Model
Equivalent expressions are obtained with the full nonlinear first-order equation in Eqn (147) with a sinusoidal Peierls potential (Dorn and Rajnak, 1964; Joós and Zhou, 2001). The equilibrium kink profile is

$$\left. \begin{array}{l} \gamma(x) = \dfrac{a'}{\pi}\sin^{-1}\dfrac{1}{\sqrt{\cos h^2\left(\dfrac{\pi x}{c}\right) + \dfrac{a'^2}{4c^2}\sin h^2\left(\dfrac{\pi x}{c}\right)}} \quad \text{if} \quad x < 0 \\[3em] \gamma(x) = a'\left(1 - \dfrac{1}{\pi}\sin^{-1}\dfrac{1}{\sqrt{\cos h^2\left(\dfrac{\pi x}{c}\right) + \dfrac{a'^2}{4c^2}\sin h^2\left(\dfrac{\pi x}{c}\right)}}\right) \quad \text{if} \quad x > 0 \end{array} \right\} \quad \text{with} \quad c = a'\sqrt{\dfrac{E_0}{2\Delta E_m}}.$$ (155)

The resulting kink width is

$$w = \frac{a'}{(dy/dx)_{x=0}} = \frac{a'\sqrt{\dfrac{E_0}{2\Delta E_m}}}{\sqrt{1 + \dfrac{\Delta E_m}{2E_0}}}. \tag{156}$$

and the kink energy is

$$E_K = \frac{E_0 a'}{\pi}\left[\sqrt{\frac{2\Delta E_m}{E_0}} + \left(2 + \frac{\Delta E_m}{E_0}\tan^{-1}\sqrt{\frac{\Delta E_m}{2E_0}}\right)\right]. \tag{157}$$

The kink profile is shown in **Figure 32(b)** and is undistinguishable from the SG result when $w = a'$. We can see by comparing Eqns (153) and (156) that the kink width is slightly smaller with the full sinusoidal model, but the difference is sensible only if ΔE_m is large compared to E_0.

16.4.2.3   *Kink Pair Formation Energy*

Under an applied stress, the dislocation glides by the thermally activated nucleation and propagation of kink pairs. As discussed above, the kinks are well-separated in the activated state when the applied stress is low and elasticity theory can be applied. However, at higher applied stresses, the critical distance between kinks in the activated state decreases and the kinks cannot be treated individually. They merely form an excursion, or bulge, on the dislocation, as illustrated in **Figure 33**. Formation of this critical bulge is then the controlling process. Beyond this configuration, the bulge develops to become fully formed kinks that move apart from each other on the dislocation line in the same way as at lower stresses. As shown below, the critical high-stress bulge configuration can be modeled using the LT model.

Under an applied stress τ_A less than the Peierls τ_P, the straight dislocation has a stable equilibrium position shifted from $y = 0$ to $y = y_0$, with y_0 solution of

$$\Delta E'(y_0) = \tau_A b. \tag{158}$$

The excess energy has an extra term, which corresponds to the mechanical work done when the dislocation bulges. The excess energy becomes

$$E_{excess} = \int_{-\infty}^{+\infty}\left\{(E_0 + \Delta E(y(x)))\sqrt{1 + \left(\frac{dy}{dx}\right)^2} - (E_0 + \Delta E(y_0)) - \tau_A b(y(x) - y_0)\right\}dx. \tag{159}$$

The equilibrium profile of the dislocation remains an extremum of the excess energy and satisfies Euler–Lagrange condition, which is integrated as (Dorn and Rajnak, 1964)

$$1 + \left(\frac{dy}{dx}\right)^2 = \left(\frac{E_0 + \Delta E(y)}{E_0 + \Delta E(y_0) + \tau_A b(y - y_0)}\right)^2. \tag{160}$$

The above equation has a trivial solution $y = y_0$, the stable straight equilibrium dislocation. The unstable solution, which describes the bulge, cannot be obtained analytically except if the Peierls potential is piecewise parabolic (Seeger and Schiller, 1962; Guyot and Dorn, 1967). When the Peierls

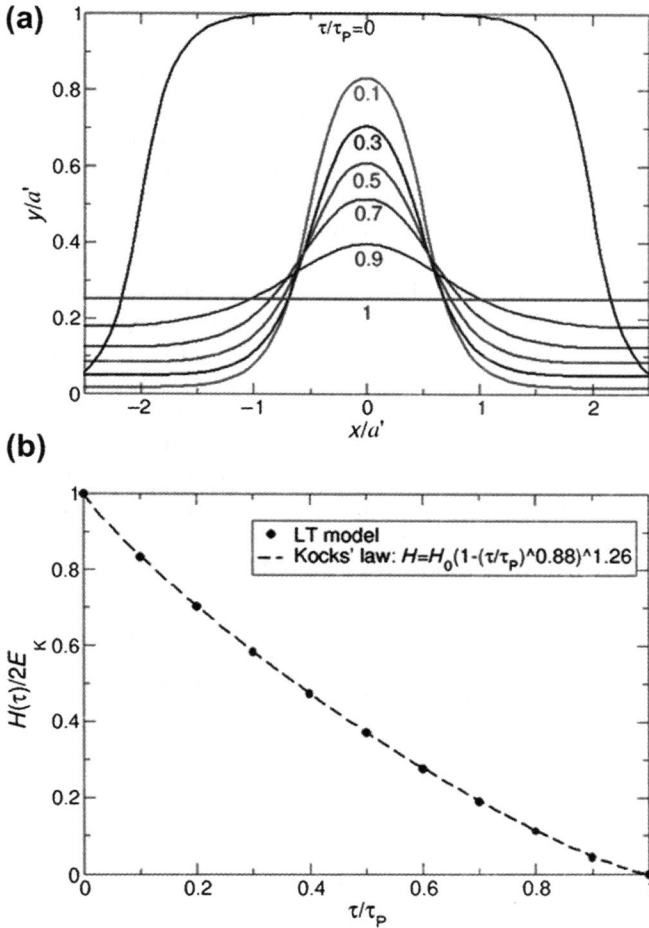

Figure 33 Kink-pair nucleation in the LT model: (a) unstable bulge at different applied stresses and (b) activated enthalpy as a function of applied stress. The Peierls potential is sinusoidal as in Eqn (150) and $E_0 = E_m$. Energies are scaled by twice the kink-pair formation energy (Eqn (157)), distances by a' (the separation between Peierls valleys) and stresses by the Peierls stress $\tau_P = \pi E_m / ba'$. (For color version of this figure, the reader is referred to the online version of this book.)

potential is sinusoidal, approximations have to be used (for example, see Büttiker and Landauer (1979) for a high-stress approximation). Equation (160) can also be integrated numerically to obtain the unstable bulge profile, which, inserted in Eqn (158), provides the activation enthalpy for the nucleation of a kink pair. Examples of critical bulges are shown in **Figure 33(a)**. **Figure 33(b)** shows the activation enthalpy as a function of applied stress. Dorn and Rajnak (1964) and Guyot and Dorn (1967) have shown that when scaled by the kink-pair formation energy as in **Figure 33(b)**, the activation enthalpy is somewhat insensitive to the height and shape of Peierls potential, at least as long as the Peierls potential does not depend on the applied stress. Rodney and Proville (2009) have shown by atomistic calculations that the Peierls potential may depend on the applied stress, which results in a more nonlinear variation of the activation enthalpy with applied stress. The enthalpy then tends to decrease rapidly at small stresses and more gradually at higher stresses, to become zero at the Peierls stress as in **Figure 33(a)**.

The LT model neglects the elastic interaction between kinks. As a result, the force between well-separated kinks decays exponentially with their separation, instead of as $1/L^2$ when elasticity is accounted for. The elastic model of Section 16.4.2.1 and the present LT model have therefore distinct regions of applicability, the former at low stresses and the latter at high stresses. Seeger (1981) proposed that the transition between models could explain nonmonotonous variations of the flow stress with temperature, the so-called hump, observed experimentally in particular in BCC metals (Caillard and Martin, 2003). In order to bridge both regimes, Koizumi et al. (1994) and Suzuki et al. (1995) considered oblique kinks and applied the elasticity theory to obtain the excess energy of a kink pair as a function of kink angle, kink height and kink separation. Minimizing the excess energy with respect to the first two variables and maximizing the energy with respect to the third variable, Koizumi et al. (1994) and Suzuki et al. (1995) found the activation enthalpy as a function of applied stress. The main advantage of this model is to recover the elastic model of Section 16.4.2.1 in the limit of low stresses and to recover the LT model at high stresses, if the LT is fitted appropriately. The enthalpy then varies monotonically with a sinusoidal Peierls potential. On the other hand, the authors observed a non-monotonous behavior when the Peierls potential, instead of having a single maximum between Peierls valleys, has a camel-hump shape with two maxima separated by a metastable state. Such Peierls potentials have been considered in LT models (Dorn and Rajnak, 1964; Guyot and Dorn, 1967) using parametric potentials that can be varied from single- to double-hump shapes. Camel-hump potentials have also been obtained by direct atomistic calculations (see Takeuchi and Kuramoto (1975) and Takeuchi (1979) for early calculations) but not with more recent and more accurate electronic structure calculations (Ventelon and Willaime, 2007). This origin of the discontinuity is therefore still largely unclear (Caillard and Martin, 2003).

A final word in this section is to mention that, for fitting purposes, the variation of the activation enthalpy is often approximated by Kocks law (Kocks et al., 1975)

$$H_{KK}^* = 2E_K \left(1 - \left(\frac{\tau_A}{\tau_P} \right)^p \right)^q \tag{161}$$

where p and q are fitting parameters. Comparing with Eqn (143), we see that the elastic model predicts $p = 1/2$ and $q = 1$. Fitting the LT enthalpy in **Figure 33**, we find $p = 0.88$ and $q = 1.26$, close to values obtained experimentally (Tang et al., 1998).

16.5   Planar Dislocation Cores: Case of FCC Dislocations

16.5.1   FCC Stacking and Slip Systems

If the Volterra procedure detailed in Section 16.2.1.2 to create a dislocation is applied in a crystal, we see that the dislocation Burgers vector must be a vector of the host lattice in order to recover a perfect crystal along the cut plane behind the dislocation. This is a necessary condition to concentrate the dislocation energy along its line and not along the entire cut plane, which would lead to prohibitively high energies. In order to further minimize the dislocation line energy, proportional to b^2 (see Section 16.3.1.2), the Burgers vector should be as small as possible. In practice, Burgers vectors observed in crystals are most often the shortest lattice vectors of the host crystal. Also, we have seen in Section 16.4

that the lattice friction is smaller in planes of larger interplanar distance. We therefore expect dislocations to glide in the most widely spaced planes, which are the densest planes of the lattice.

FCC crystals follow both of the above rules. Conventional FCC dislocations have $a/2\langle 110 \rangle$ Burgers vectors (a, the lattice parameter), the shortest vector of the FCC lattice, as illustrated in the inset of **Figure 34(a)** and are called perfect dislocations. They glide in $\{111\}$ planes, the dense planes of the lattice, with a negligibly small lattice friction. For that reason, they are also called glissile dislocations.

An FCC lattice can be constructed in a $\langle 111 \rangle$ direction by stacking hexagonal planes of atoms. A hexagonal plane, such as plane "a" in **Figure 34(a)**, is made of two families of regular triangles, pointing either upward (Δ) or downward (∇). To create a hexagonal "b" plane above plane "a", we simply place an atom above the center of gravity of, say, all upward triangles at a height $a\sqrt{2}/3$. Each upward triangle with its new atom then forms a regular tetrahedron. The third plane, called plane "c", is created by placing an atom above the downward triangles of plane "a" at a height $2a\sqrt{2}/3$, or equivalently, at height $a\sqrt{2}/3$ above the upward triangles of plane "b". If we repeat this sequence, we produce a stacking of hexagonal planes …abcabc…, which forms an FCC lattice.

In this procedure, we have produced the densest possible packing of spheres, as first stated in Johannes Kepler conjecture (1611), prompted by the problem of how to best stack cannonballs on navy ships (Szpiro, 2003). The other well-known close-packed lattice, the hexagonal close-packed (HCP) lattice, is obtained by a binary sequence of hexagonal planes …ababab…, instead of a ternary sequence.

The FCC lattice can also be seen as a compact arrangement of regular tetrahedra that may have two orientations, illustrated in **Figure 34(a)**. Looking at the lattice down a $\langle 111 \rangle$ direction, the tetrahedra may point toward us, their base being an upward triangle and their summit in the $\{111\}$ plane above the base. We will say that the tetrahedron is seen "from above". The other tetrahedra have a downward triangle as a base and a summit in the plane below the base. They are seen "from below". We can check that all four faces of the tetrahedra are parallel to one of the four $\{111\}$ planes of the FCC lattice and that the six edges are the six $a/2\langle 110 \rangle$ vectors of the lattice. All slip systems of glissile dislocations, composed of an $a/2\langle 110 \rangle$ Burgers vector and a $\{111\}$ slip plane, may therefore be visualized within a tetrahedron,

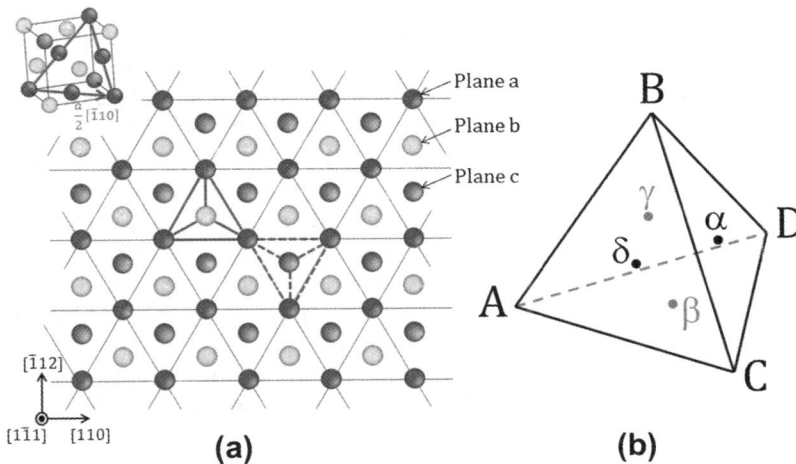

Figure 34 (a) FCC stacking and (b) Thompson tetrahedron. In (a), a ternary stacking of compact hexagonal planes forms an FCC lattice. In (b), Thompson tetrahedron shows the FCC crystallographic orientation and dislocation slip systems. (For color version of this figure, the reader is referred to the online version of this book.)

called Thompson tetrahedron after the work of Thompson (1953). Conventionally, the tetrahedron vertices are denoted by roman letters, ABCD, while faces are referred to by their center of gravity, denoted by Greek letters, $\alpha\beta\chi\delta$, which correspond to the opposite vertex, as shown in **Figure 34(b)**. We may attach a crystallographic frame of reference to the crystal and assign Miller indices to all vectors and planes. On the other hand, dislocation reactions can be unequivocally described once the orientation of the Thompson tetrahedron is given. We will thus refer here mostly to Thompson tetrahedron notations. For tetrahedra with full crystallographic indices, the reader is referred to Hirth and Lothe (1982).

If we consider a tetrahedron that points upward as in **Figure 34(b)**, the three base vertices, ACD, belong to a hexagonal plane, say a plane of type "a". Then, vertex B belongs to the plane "b" above the base. Useful for the following, we should note that if we translate all atoms of an "a"-plane by a vector of type $\overrightarrow{A\beta}$, the atoms ends up directly below atoms of a "b"-plane; or said in other words, we have transformed the "a"-plane into a "b"-plane. The same operation transforms a "b"-plane into a "c"-plane and a "c"-plane into an "a"-plane. Translating by $\overrightarrow{A\beta}$, or equivalently by $\overrightarrow{C\beta}$ or $\overrightarrow{D\beta}$, thus transforms a plane into the plane just above in the stacking. These vectors are of the type $a/6\langle112\rangle$. They are not vectors of the FCC lattice and are called Shockley partials (Heidenreich and Shockley, 1948). We can check that inversely, a plane translated by either $\overrightarrow{\beta A}$, $\overrightarrow{\beta C}$ or $\overrightarrow{\beta D}$ is transformed into the plane just below in the stacking, i.e. "c" into "b", "b" into "a" and "c" into "a". Naturally, the same reasoning also applies to Shockley partials in other {111} planes of the lattice.

16.5.2   Dissociation of Perfect Dislocations

16.5.2.1   *Stacking Fault*
The γ-surfaces in **Figure 28(a)** and **(b)** correspond to FCC dislocations since they show the energetics of shear in {111} planes of the FCC lattice. The γ-lines in **Figure 28(d)** are along a $\langle112\rangle$ direction, which is the direction of a Shockley partial of the $\overrightarrow{A\beta}$ type. We can see in this figure that the γ-line has an intermediate local minimum, which corresponds precisely to a shear of $a/6\langle112\rangle$, i.e. a shear by a Shockley partial. This local minimum means that a crystal sheared by a Shockley partial is in a state of metastable equilibrium. The crystal then contains a fault in its stacking sequence. To see this, we can start with a perfect stacking sequence ...abcabc|abcabc... and shear rigidly the half-crystal above the vertical bar "|" by a vector of the $\overrightarrow{A\beta}$-type, as in the construction of the γ-line. From our above discussion, we know that this shear transforms all planes in the upper half-crystal into the plane just above in the stacking sequence. We thus obtain the sequence ...abcabc|bcabca... This sequence would be the same if we removed an "a"-plane from the stacking and is called an intrinsic stacking fault. In the new sequence, first-neighbor distances are unchanged compared to the perfect crystal, resulting in a relatively low energy cost. Locally around the fault, there is a binary sequence of hexagonal planes bcbc, which corresponds to the HCP stacking. The stacking fault may thus be seen in different ways, either as a hexagonal plane missing in the stacking or as a local HCP region. The energy of a stacking fault is defined per unit area, usually denoted γ, and varies from metal to metal: it is relatively high in aluminum (~165 mJ m^{-2}) and low in Silver (~16 mJ m^{-2}).

Shifting the two half-crystals by the vector opposite, $\overrightarrow{\beta A}$ for instance, transforms the planes into the planes just below in the stacking, resulting in the sequence ...abcabc|cabcab... There are then two planes of same type on top of each other, with atoms closer than in the perfect crystal. This configuration is of high energy and is unstable, as can be seen in **Figure 28(d)**, where this shear results in an energy maximum (the negative shear is seen on the right-hand side of the γ-line).

Other faults can be introduced in the FCC stacking, namely extrinsic stacking faults (Frank, 1951b) which correspond to the insertion of an extra plane in the stacking, for example ...abcbabc. There are also twin planes, which correspond to a mirror symmetry in the crystal, for example …abcabcbacbac… (see Hirth and Lothe (1982) for a more detailed description of stacking faults in the FCC lattice).

16.5.2.2 Dissociation Reaction
16.5.2.2.1 Dissociation Rule
Glissile dislocations as introduced in Section 16.5.1 make use of the intrinsic stacking fault to reduce their line energy. The reaction is called dissociation and corresponds to a decomposition of the perfect Burgers vector into two Shockley partial Burgers vectors. We have seen in Section 16.2 the conservation of Burgers vectors. A vector such as $\overrightarrow{AC}$ in **Figure 34(c)** may thus decompose into other vectors such that their sum is $\overrightarrow{AC}$. The dissociation reaction involves the Shockley partials mentioned above and is of the type

$$\overrightarrow{AC} = \overrightarrow{\beta C} + \overrightarrow{A\beta} \tag{162}$$

or with crystallographic indices

$$\frac{a}{2}[110] = \frac{a}{6}\left[12\overline{1}\right] + \frac{a}{6}[211] \tag{163}$$

As illustrated in **Figure 35**, the perfect $\overrightarrow{AC}$-dislocation splits into two dislocations with Burgers vectors $\overrightarrow{A\beta}$ and $\overrightarrow{\beta C}$. If the cut plane is on the left of the dislocation, the part of the crystal on this side is sheared by $\overrightarrow{AC}$, or equivalently, by the two partial Burgers vectors, and the perfect stacking is recovered. In-between the two partials however, the crystal is sheared by the $\overrightarrow{A\beta}$ front partial only, resulting locally in a stacking fault.

The order of the partials matters. In the configuration shown in **Figure 35**, the crystal between the partials is sheared by $\overrightarrow{A\beta}$, resulting in a metastable-stacking fault. If the front partial was $\overrightarrow{\beta C}$, the crystal between the partials would be sheared by this vector, resulting in the unstable high-energy configuration of two same planes on top of each other. The front partial is thus the one which creates an intrinsic stacking fault. There are different ways to express this dissociation rule which identifies the front partial, but the underlying principle is that the front partial is the one which transforms the glide plane into the plane just above in the stacking, the direction of "above" and "below" being defined according to the normal to the cut plane, as in Section 16.2. The two basic cases are considered in **Figure 36**. The first, in

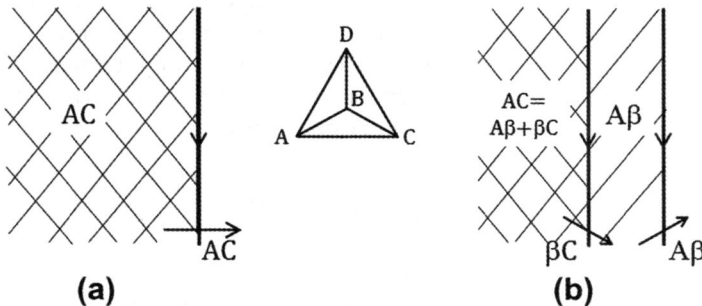

Figure 35 Dissociation of a perfect edge dislocation: (a) before dissociation, (b) after dissociation. The shears along the cut plane of the dislocation are shown as dashed regions. The Thompson tetrahedron indicates the crystallographic orientation of the lattice.

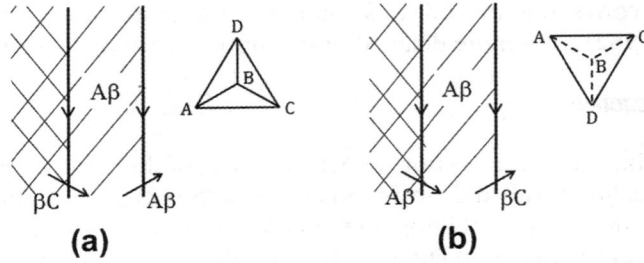

Figure 36 Order of the Shockley partials in a dissociated dislocation. Two cases are considered, depending on whether the Thompson tetrahedron is seen (a) from above or (b) from below.

Figure 36(a), is the same as above. The Thompson tetrahedron is seen from above and vertex B is above the ACD glide plane. The vector $\overrightarrow{A\beta}$ then transforms the glide plane into the plane just above, i.e. places the A vertex directly below the B vertex, and $\overrightarrow{A\beta}$ is the front partial. In the second case, the Thompson tetrahedron is seen from below and vertex B is now below the ACD glide plane. The bottom plane contains B and the top plane contains A, C and D. The front partial has thus to place B directly below one of these vertices and is $\overrightarrow{\beta C}$.

Another way to state the dissociation rule, adapted from Hull and Bacon (2001) is the following. Looking down the glide plane, if the Thompson tetrahedron is seen from above and the dislocation has a positive line sense pointing down (as in **Figure 36(a)**), the partials are such that the Greek letters are on the outside, that is, the left partial is $\overrightarrow{\beta C}$ (β is on the left of the left partial) and the right partial is $\overrightarrow{A\beta}$ (β is on the right of the right partial). Inversely, if the tetrahedron is seen from below, greek letters are on the inside, i.e. $\overrightarrow{A\beta}$ on the left and $\overrightarrow{\beta C}$ on the right. Applying this rule is sometimes delicate and reverting to the more physical concept of transforming a plane into the plane just above may be easier.

16.5.2.2.2 Dissociation Distance

The partials inside a dissociated perfect dislocation are submitted to two types of forces. First, they interact elastically and according to Frank's criterion (see Section 16.3.3.2), since the product of their Burgers vectors is always positive, this interaction is expected to be repulsive. On the other hand, if the partials move apart from each other, they increase the surface of the stacking fault and increase the stacking fault energy, which amounts to $E = \gamma d$ per unit length of dislocation if the partials are separated by d. The restoring force $-\partial E/\partial d = -\gamma$ is attractive. From the interaction stress between dislocations in isotropic elastic media (see Eqns (86) and (88)), the elastic interaction force between partial dislocations is known, and equilibrating this repulsive force with the attractive stacking fault restoring force, we arrive at a stable equilibrium distance between partials

$$d = \frac{\mu b^2}{24\pi\gamma}\frac{2-\nu}{1-\nu}\left(1 - \frac{2\nu\cos2\Psi}{2-\nu}\right) \tag{164}$$

where Ψ is the angle between the perfect Burgers vector and the dislocation line. We can check that the dissociation distance for an edge dislocation ($\Psi = \pi/2$) is larger than for a screw dislocation ($\Psi = 0$). The main term in Eqn (164) is $\mu b/\gamma$, which measures the relative strength of the elastic repulsion (μb) compared to the stacking fault energy (γ). This ratio, rather than just the stacking fault energy, controls the dissociation distance.

16.5.2.3   Dislocation Modeling

Dissociated dislocations are a very good application of the PN model introduced in Section 16.4 because of their planar core, one of the main assumptions of the model. It is however necessary to consider a two-dimensional model with disregisteries both parallel and perpendicular to the dislocation line to account for the Shockley partials. Dislocations of different characters have been modeled using parameterized (Schoeck and Krystian, 2005) and computed (Carrez et al., 2007) γ-surfaces. The Burgers vector distribution shows two peaks corresponding to the Shockley partials separated by a region of constant shear, which corresponds to the stacking fault.

Dissociated dislocations are also simulated using direct atomistic simulations. The first simulations of a dislocation core in an FCC crystal (copper) was performed by Cotterill and Doyama (1966), followed by Perrin et al. (1972). These authors showed that the simulations reproduced the dissociation of an edge dislocation in two Shockley partials separated by a stacking fault. Since then, FCC dislocations have been simulated in a number of situations. One has to bear in mind when using an interatomic potential that if the stacking fault energy was not explicitly adjusted during the fitting of the potential, it will be much smaller than the experimental value. However, all recent interatomic potentials reproduce fairly well the stacking energy.

Most calculations are based on interatomic potentials because a relatively large number of atoms, exceeding the current limit of electronic structure calculations, is needed to model a glissile dislocation due to its dissociation. One exception is aluminum, where dislocations are hardly dissociated (Woodward et al., 2008). **Figure 37** shows an example of dissociated edge dislocation in copper obtained with the interatomic potential developed by Mishin et al. (2001). **Figure 37(a)** shows the crystal along the $\langle 112 \rangle$ direction of the dislocation line. Along the $\langle 110 \rangle$ direction of the Burgers vector, the crystal is made of a binary stacking of planes and the Burgers vector encompasses two such planes, such that the undissociated dislocation shown in **Figure 37(a)** appears to have two extra half-planes (but the two planes are not equivalent). After dissociation (**Figure 37(b)**), the two extra half-planes are separated and directly above the two partial dislocations, whose edge component is half the perfect Burgers vector. Seen from above, when only three planes near the glide plane are shown as in **Figure 37(c)**, we see how the ternary "abc" sequence of hexagonal planes ahead of the dislocation progressively becomes a binary sequence inside the dislocation core and returns to a ternary sequence behind the dislocation.

16.5.3   Partial Dislocations

16.5.3.1   Stair-Rod Partials

Shockley partials are not the only partial Burgers vectors of the FCC crystals (for a review article, see Amelynckx (1979)). The most common partials are summarized in **Table 1**. Among them are stair-rod dislocations called Lomer–Cottrell and Hirth partials. They form when a dissociated dislocation changes glide plane. This situation occurs in several instances: for example, when two dislocations meet at the intersection of their glide planes to form junctions (Bulatov et al., 1997; Zhou et al., 1998; Rodney and Phillips, 1999; Shenoy et al., 2000; Madec et al., 2003) or when a dissociated dislocation locally climbs and acquires an extended jog (Hirsch, 1962; Amelynckx, 1979; Rodney and Martin, 2000).

Let us consider the configuration shown in **Figure 38**, where an edge dislocation with an $\overrightarrow{AC}$ Burgers vector has two arms in different glide planes, ACD and ABC. This configuration is geometrically possible because the $\overrightarrow{AC}$ vector is contained in both glide planes. The Burgers vectors of the various dislocation segments can be determined from the dissociation rule given in Section 16.5.2.2.1. The result is that the Burgers vector changes on each partial when the dislocation changes glide plane. Because of the conservation of the Burgers vector, there must be a segment of dislocation along the line

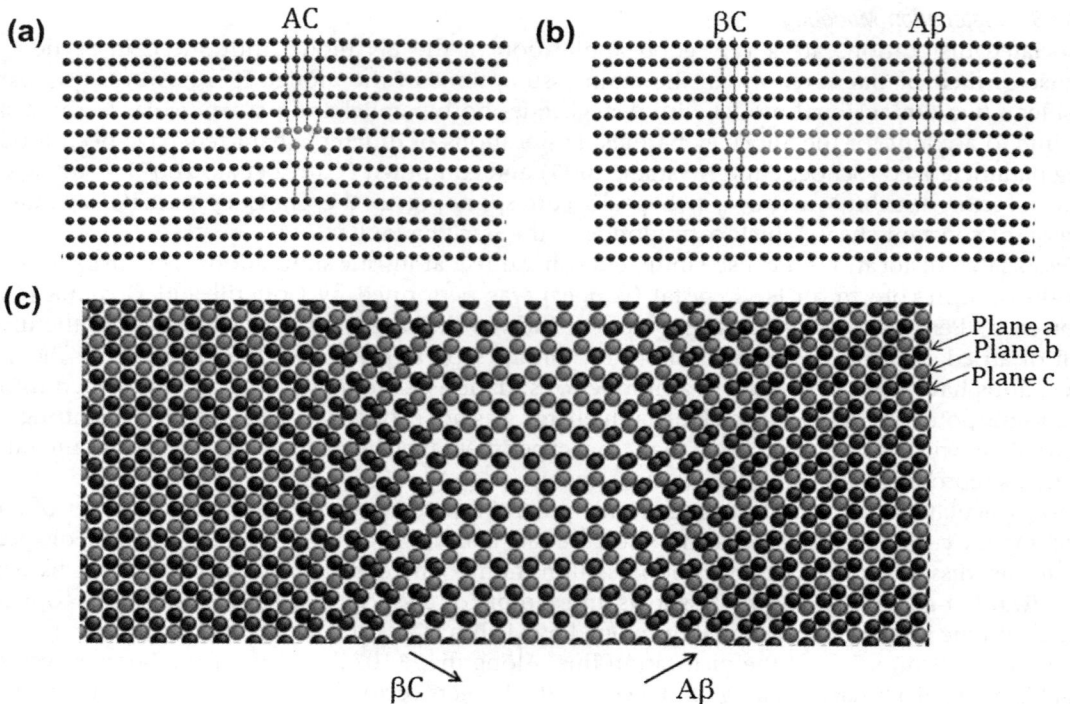

(a) AC

(b) βC   Aβ

(c)

Plane a
Plane b
Plane c

βC   Aβ

Figure 37 Atomistic simulation of a dissociated edge dislocation in copper. The dislocation is shown (a) before and (b) after dissociation in projection along the dislocation line. The crystal is seen from above the glide plane in (c). Only three {111} planes around the glide plane are shown with different colors, highlighting how the stacking evolves from ternary outside the dislocation core to binary inside the core. (For color version of this figure, the reader is referred to the online version of this book.)

Table 1   Main partial Burgers vectors of the FCC lattice

Name	Vector	Miller index	Norm b^2
Shockley	$A\beta$	$a/6\langle 121\rangle$	$a^2/6$
Lomer-Cottrell	$\delta\beta$	$a/6\langle 110\rangle$	$a^2/18$
Hirth	$\delta\beta/AC$	$a/3\langle 001\rangle$	$a^2/9$
Frank	βB	$a/3\langle 111\rangle$	$a^2/3$

of intersection between glide planes, which connects the partials and balances the Burgers vectors. In **Figure 38(a)**, the connecting vector is: $\overrightarrow{\beta C} - \overrightarrow{\delta C} = \overrightarrow{A\delta} - \overrightarrow{A\beta} = \overrightarrow{\beta\delta}$. This vector is of the type a/$6\langle 110\rangle$, equal to a third of a perfect Burgers vector. It joins two adjacent face centers in the Thompson tetrahedron and is called a Lomer–Cottrell partial.

The nature of the stair-rod dislocation depends on the orientation of the dislocation and on whether the angle between the two arms is acute, as in **Figure 38(a)**, or obtuse, as in **Figure 38(b)**. In latter case, the Thompson tetrahedron is seen from below in the ABC plane and the order of the partials is reversed compared to **Figure 38(a)**. The connecting partial becomes $\overrightarrow{A\delta} - \overrightarrow{\beta C} = \overrightarrow{A\beta} - \overrightarrow{\delta C}$, noted $\overrightarrow{\beta\delta}/\overrightarrow{AC}$

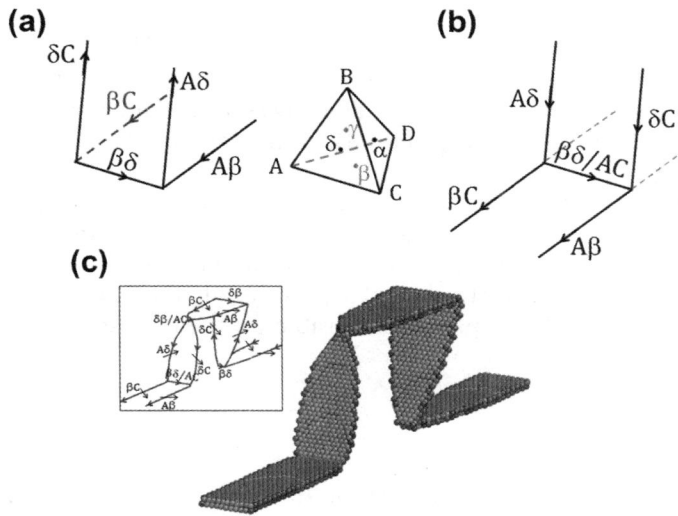

Figure 38 Stair-rod partial dislocations: structure of a dissociated edge dislocation changing glide plane with an (a) acute or (b) obtuse angle between planes. (c) Atomistic structure of an extended jog on an edge dislocation in copper with details of the Burgers vectors in the inset. Only atoms with a non-FCC local environment are shown. (For color version of this figure, the reader is referred to the online version of this book.)

(Hirsch, 1962; Hirth and Lothe, 1982). This vector of the type $a/3\langle 001\rangle$ is called a Hirth partial. It is longer than the Lomer–Cottrell vector and joins the middle of two edges in the Thompson tetrahedron, AC and BD in present case.

Stair-rod dislocations form when an edge dislocation locally climbs and acquires a pair of superjogs. An example of atomistic structure, obtained with an interatomic potential (Mishin et al., 2001) in copper, is shown in **Figure 38(c)**. The inset details the Burgers vectors of all partials. Hirth partials form on one superjog and Lomer–Cottrell partials on the other. With the Hirth vector being longer than the Lomer–Cottrell vector, we expect the Hirth segments to be shorter because of their higher line energy. As seen in **Figure 38(c)**, this is not the case. Both Hirth and Lomer–Cottrell segments are almost closed on one side of the jog and extended on the other side. Such configurations are referred as dissociated and undissociated nodes, or opened and closed nodes. The reason is that, as seen in Section 16.3.1.2, the line energy of a screw dislocation is smaller than that of an edge. As a result, dislocations try, when they can, to reorient their line along their screw orientation to minimize their line energy. This effect is responsible for instance for the fact that Frank–Read sources are not circular but elliptical as shown in **Figure 26**, with a longer axis along the Burgers vector direction, which favors the screw character. Reorientations also occur in dissociated dislocations when their line is interrupted, for example at a free surface, or on a junction, a grain boundary or when the dislocation changes glide plane as in **Figure 38(c)**. As seen in the inset, the Shockley partials of a dissociated dislocation point toward each other. Therefore, if the partials rotate to orient their line along their Burgers vector, they will move toward each other on one side and away from each other on the other side. The result is oblique V-shaped partials that have been observed in dissociated dislocations ending on free surfaces both in transmission electron microscopy (TEM) (Hazzledine et al., 1975; Georges and Champier, 1980) and atomistic simulations (Rasmussen et al., 1997a, b). In **Figure 38(c)**, we can check from the orientation of the Shockley partials that this LT effect tends to close the nodes on one side of the superjog and open them up on the other side, as observed in the atomic structure.

Dissociated nodes also form in dislocation junctions, when two dislocations join to form a third dislocation lying at the intersection of their glide planes. Dissociated junction structures can be found in the works of Bulatov et al. (1997), Zhou et al. (1998), Rodney and Phillips (1999), Shenoy et al. (2000) and Madec et al. (2003).

Finally, we wish to mention the Lomer dislocation, a perfect edge dislocation with for instance, a $\overrightarrow{DB}$ Burgers vector and an $\overrightarrow{AC}$ line. This dislocation forms in Lomer–Cottrell junctions (Rodney and Phillips, 1999) and after absorption of crystalline defects (Rodney and Martin, 2000). It has a compact pentagonal core and a noncompact {001} glide plane. As a result, the Peierls stress of this dislocation is high and the resulting low mobility is partly responsible for the high strength of the Lomer–Cottrell dislocation.

16.5.3.2 Frank Partial
16.5.3.2.1 Frank Loops

Another important Burgers vector is the Frank vector, of the type $a/3\langle 111 \rangle$. In the Thompson tetrahedron, this vector is of the type $\overrightarrow{\beta B}$, joining the center of gravity of a face to the opposite vertex. Such Burgers vectors form when vacancies cluster along a given {111} plane. As illustrated in **Figure 39(a)**, the vacancies locally remove part of the {111} plane. If the crystal collapses, an intrinsic stacking fault forms inside the platelet. This fault is the same as that inside a dissociated dislocation, but the Burgers vector bounding the fault is different. If we draw a Burgers circuit around the edge of the platelet, we find a discontinuity equal to the collapsing vector, i.e. the $a/3\langle 111 \rangle$ Frank vector perpendicular to the platelet. The defect is called a Frank loop. This defect is sessile, i.e. in absence of diffusion or shuffling,

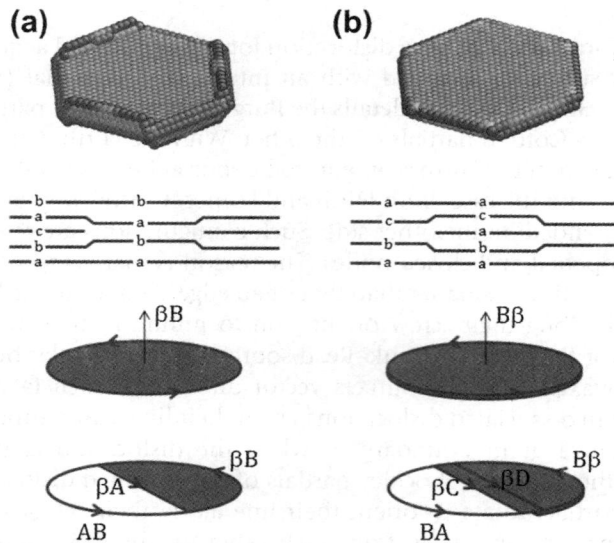

Figure 39 (a) Vacancy and (b) interstitial Frank loops. The cartoons show the FCC stacking inside the two types of loops, their Burgers vectors and unfaulting mechanisms. (For color version of this figure, the reader is referred to the online version of this book.)

the loop cannot glide in the crystal because that would require a motion of the stacking fault perpendicular to itself.

Frank loops are known to unfault, i.e. to transform into perfect dislocation loops. The reaction occurs by dissociation of the Frank vector into a Shockley partial and a perfect Burgers vector, of the type

$$\overrightarrow{\beta B} = \overrightarrow{AB} + \overrightarrow{\beta A}. \tag{165}$$

The $\overrightarrow{\beta A}$ Shockley partial is contained in the plane of the loop and can thus glide across the loop surface. As illustrated in the lower panel of **Figure 39(a)**, the partial progressively transforms the Frank vector into the perfect $\overrightarrow{AB}$ Burgers vector and removes the stacking fault. After the reaction, a perfect loop, with an $\overrightarrow{AB}$ Burgers vector remains. The driving force for this reaction is that, for large loops, the energy of the perfect loop is lower than that of the Frank loop (Hull and Bacon, 2001).

When interstitial atoms cluster in a {111} plane, as shown in **Figure 39(b)**, an extrinsic stacking fault forms. The Burgers vector on the edge of the platelet remains a Frank vector, but with a direction reversed compared to a vacancy loop. Interstitial Frank loops may also unfaulty. Extrinsic stacking faults can be viewed as two intrinsic stacking faults on adjacent {111} planes. The dissociation reaction in Eqn (165) remains valid, but the Shockley partial $\overrightarrow{A\beta}$ is now made of two Shockley partials on adjacent {111} plane according to

$$\overrightarrow{A\beta} = \overrightarrow{\beta C} + \overrightarrow{\beta D}. \tag{166}$$

The cores of the two Shockley partials may be separated and they may sweep the surface of the Frank loop successively, but the two cores may also join to form a single core, extended over 2 {111} planes, called a D-Shockley partial (for "double Shockley partial") (Weertman and Weertman, 1967). Atomistic simulations (Rodney, 2004) showed that interstitial Frank loops may unfault upon interaction with a glissile dislocation by both mechanisms, either successive sweeping by two separated Shockley partial and single sweep by a D-Shockley partial, depending on the interaction configuration.

16.5.3.2.2 Stacking Fault Tetrahedron

Triangular vacancy Frank loops may transform into a particular defect, called a stacking fault tetrahedron (SFT), observed in low stacking-fault energy metal and alloys, for instance gold (Silcox and Hirsch, 1959) and copper (Singh et al., 2001), after treatments such as rapid quenching or irradiation that produces vacancy supersaturations. An SFT consists of a regular tetrahedron of stacking faults on the 4 {111} planes of the FCC lattice, with Lomer–Cottrell partial dislocations running along the 6 $\langle 110 \rangle$ edges. One mechanism to produce an SFT was proposed by Silcox and Hirsch (1959) and involves the transformation of a triangular Frank loop. An example from an atomistic simulation in copper is shown in **Figure 40**. The triangular Frank loop is unstable. Its $\overrightarrow{\beta B}$ Burgers vector dissociates along the three edges of the loop into a Lomer–Cottrell and a Shockley partial. The reactions are of the type of Eqn (165) but the Shockley partial is in the inclined {111} plane which contains the edge, rather than the plane of the loop. Burgers vectors are shown in the inset of **Figure 40**. The Shockley partials glide along the inclined planes and meet two by two along $\langle 110 \rangle$ directions where they form Lomer–Cottrell partials according to reactions of the type $\overrightarrow{\delta B} + \overrightarrow{B\alpha} = \overrightarrow{\delta\alpha}$. The final configuration is a regular tetrahedron with four stacking faults by Lomer–Cottrell stair rods on all sides. Also, interestingly, during the formation of the SFT, the Shockley partials have a convex shape because of the restoring force due to the stacking fault, instead of a concave shape, often depicted (Hirth and Lothe, 1982; Hull and Bacon, 2001).

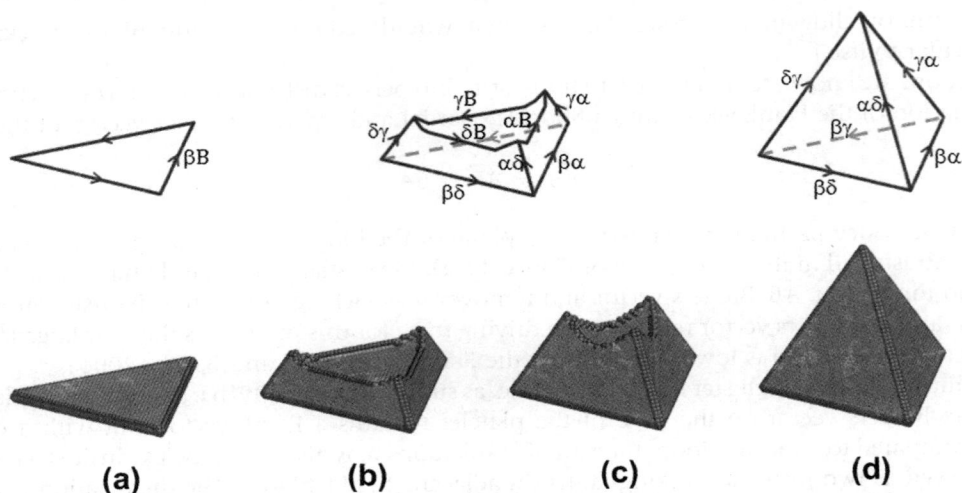

Figure 40 Formation of a SFT via the Silcox and Hirsch mechanism simulated in copper. The upper insets detail the Burgers vector of all partials. (For color version of this figure, the reader is referred to the online version of this book.)

16.5.4 Cross Slip

A direct consequence of dislocation splitting is the confinement of screw dislocations in particular planes. The dissociations may be planar with well-defined partials, such as in the FCC structure when the stacking fault energy is not too high (Section 16.5.2), or may exist as narrow core extensions onto several planes like in the BCC structure (Section 16.6). In particular, screw dislocations in the FCC structure, which do not have a specific glide plane when they are perfect, are confined to {111} dense planes when they dissociate. This may lead with increasing plastic strain to high dislocation storage when no recovery process is active. Mixed and edge dislocations can escape from their glide planes through a climb process that requires diffusion, but screw dislocations cannot. A dissociated screw dislocation can however escape from its initial {111} glide plane to continue to glide on another {111} glide plane by a mechanism called "cross slip". This mechanism cannot be avoided in the plasticity of crystalline materials and is very often invoked to explain dynamic recovery at intermediate temperature. Dynamic recovery is reflected by a decreasing strain-hardening rate with increasing plastic deformation during constant strain rate experiments or in steady-state condition during creep tests. A typical example is the stage III of parabolic hardening of FCC single crystals, which follows stage II of high linear hardening (Honeycomb, 1968). Cross slip may also be responsible for strain hardening by developing additional obstacles to moving dislocations (Jackson, 1983) or by exhausting mobile dislocations (Veyssière, 2000). Cross slip only concerns screw dislocations since for such a process to occur both the line and Burgers vector of the dislocation must be parallel to the intersection line between the two glide planes. Two classes of cross slip are basically distinguished depending on the further ability of the cross-slipping dislocation to redissociate or not in the cross-slip plane, that is whether cross slip takes place between compact or noncompact planes. Intermediate situations have been proposed where meta-stable spreading of dislocation cores is introduced in the cross-slip plane to explain in situ TEM observations (Couret and Caillard, 1991).

In this section, we shall concentrate on cross-slip models that have been proposed to model this elementary dislocation mechanism in the FCC structure. In this crystallographic structure, a stacking fault cannot exist concomitantly onto two planes (Mott, 1952) without a stair-rod dislocation bounding the stacking faults in each plane along the intersection between the two planes, as discussed in Section 16.5.3.1. Consequently, the cross-slip process requires either the existence of such a stair-rod dislocation or the recombination of the dissociated dislocation. It follows that cross slip involves the overcoming of an energy barrier, which has been theoretically modeled by various authors. Phenomenological descriptions also exist and are reviewed first.

16.5.4.1 Fleischer Model (Fleischer, 1959)

This model is based on the dissociation of the leading Shockley partial in the primary plane into a stair-rod dislocation and a Shockley partial in the cross-slip plane (**Figure 41**). Consequently, this cross-slip process does not require a complete recombination of the stacking fault ribbon, which is continuously transmitted from the primary to the cross-slip plane. For this, we shall consider hereafter the cross slip of a $\frac{a}{2}[101]$ screw dislocation initially lying in the $(\bar{1}\bar{1}1)$ primary plane onto the $(\bar{1}11)$ cross-slip plane, so that the dissociation schemes are:

- in the primary plane: $\frac{a}{2}[101] \to \frac{a}{6}[2\bar{1}1] + \frac{a}{6}[112]$, in Thomson notation: $\overrightarrow{DA} \to \overrightarrow{\gamma A} + \overrightarrow{D\gamma}$,
- in the cross-slip plane: $\frac{a}{2}[101] \to \frac{a}{6}[1\bar{1}2] + \frac{a}{6}[211]$, in Thomson notation: $\overrightarrow{DA} \to \overrightarrow{\beta A} + \overrightarrow{D\beta}$.

Cross slip may take place either at acute or at obtuse dihedron and the leading Shockley partial in the primary plane must dissociate into a stair-rod dislocation that remains immobile at the intersection between the primary and cross-slip planes, and a Shockley partial that is free to glide onto the cross-slip plane, according to,

$$\frac{a}{6}[112] \to \frac{a}{6}[\bar{1}01] + \frac{a}{6}[211], \quad \text{in Thomson notation}: \quad \overrightarrow{D\gamma} \to \overrightarrow{\beta\gamma} + \overrightarrow{D\beta},$$

or

$$\frac{a}{6}[112] \to \frac{a}{3}[010] + \frac{a}{6}[1\bar{1}2], \quad \text{in Thomson notation}: \quad \overrightarrow{D\gamma} \to \overrightarrow{AD}/\overrightarrow{\beta\gamma} + \overrightarrow{\beta A},$$

respectively. It is seen that for both cases the dissociation of the leading Shockley partial in the primary plane is not energetically favorable according to Frank criterion (Eqn (94)). The two resulting partial

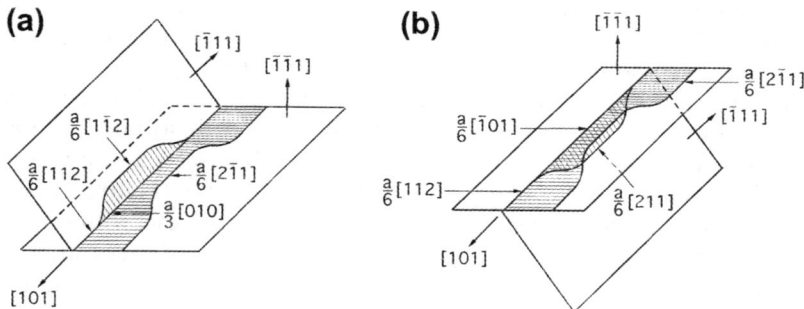

Figure 41 Schematic illustration of the cross slip of an extended dislocation in the FCC crystalline structure via a stair-rod mode. (a) Obtuse angle. (b) Acute angle.

dislocations, i.e. the stair-rod and the Shockley partials, attract each other and thus, thermal activation is required for the dissociation to occur. After dissociation of the leading Shockley partial in the primary plane, the stair-rod and the remaining Shockley partials in the primary plane are attractive and their reaction gives the second Shockley partial needed in the cross-slip plane to complete the process. It was considered that, according to Frank criterion (Eqn (94)), this cross-slip process was more favorable than the recombination of the Shockley partials in the primary plane. With the same consideration, the process at an acute angle should be preferred over the process at an obtuse angle. However, as soon as the tree-fold configuration is formed, the two Shockley partials are repulsive, which may continuously help in the cross-slip process at an obtuse dihedron, since the two partials are moving in the same direction. The same is not true at an acute dihedron because the two Shockley partials must first get closer before moving away from each other, leading to an increase in the configuration energy.

Both cases were theoretically treated by considering infinite parallel partial dislocations (Marcinkowski et al., 1974). It was demonstrated that the critical stress for spontaneous cross slip is always smaller for the acute case than for the obtuse case, which is indeed related to the smaller value of the Burgers vector associated with the stair-rod dislocation. The coalescence case was also examined and it was shown that it has always higher energy than the acute case. If this is valid for infinite parallel dislocations, it does not hold when local partial recombination is considered (Stroh, 1954). This model has nevertheless recently received some support from atomistic simulations that show that the saddle point configuration for cross slip occurs before the dislocation is entirely recombined in the primary plane, while a stacking fault extension does exit in the cross-slip plane (Duesbery, 1998). Cross slip of Shockley partials by the stair-rod mode has been experimentally confirmed in Cu-8at.%Si (Clarebrough and Forwood, 1975) and in Ni$_3$V (Vanderschaeve and Escaig, 1983). In the latter study, a theoretical calculation performed for the specific case of the ordered Ni$_3$V microstructure shows that the activation energy is very large, $E_{cs} \sim 7$ eV at $\tau/\mu = 5.10^{-3}$, but can be significantly reduced by the internal stress field of neighboring partial dislocations.

16.5.4.2 *Washburn Model*

The theoretical arrangements where a moving dislocation intersects a fixed dislocation to form a dislocation interaction that lies in the cross-slip plane of either the intersecting or the intersected dislocations were examined by Whelan (1958). The purpose was to understand the formation of hexagonal dislocation networks observed by TEM in deformed stainless steel. Whelan noticed that, in the case where the intersecting dislocation can move onto the cross-slip plane of the intersected dislocation, the cross-slip process is natural. The necessary constrictions are automatically formed at the junction between dislocations and no activation energy is therefore required. A constriction is a point on a dissociated dislocation where the partial dislocations are brought together.

This description has been subsequently developed by Washburn (1965) to propose a cross-slip process that involves the motion of the constricted node onto the cross-slip plane (see **Figure 42**). It is assumed that a moving dislocation (M) with Burgers vector $\overrightarrow{DA}$ dissociated in the primary β glide plane according to the dissociation reaction: $\overrightarrow{DA} = \overrightarrow{\beta A} + \overrightarrow{D\beta}$, intersects a forest dislocation (F) with Burgers vector $\overrightarrow{BD}$ dissociated in the γ plane such as $\overrightarrow{BD} = \overrightarrow{\gamma D} + \overrightarrow{B\gamma}$ (**Figure 42(a)**). When the two dislocations meet, the partial dislocations $\overrightarrow{D\beta}$ and $\overrightarrow{\beta A}$ spontaneously react with the partial dislocation $\overrightarrow{\gamma D}$ following the reaction $\overrightarrow{\gamma D} + \overrightarrow{D\beta} + \overrightarrow{\beta A} \rightarrow \overrightarrow{\gamma A}$. Thanks to the preexisting constriction in O, only one constriction needs to be formed to move the mobile dislocation onto the cross-slip plane. The driving force is supplied by the LT of the intersected dislocation, whose shortening is expected to compensate for the energy necessary to produce the constriction. This mechanism is therefore expected to operate predominantly at low temperature, when thermal activation is small. Of course the subsequent motion

(a)

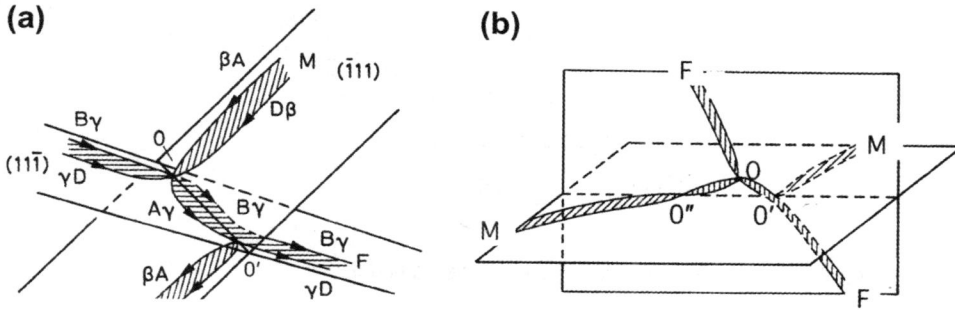

(b)

Figure 42 Cross-slip scheme according to Washburn's description. (a) Detail of reaction junction before cross slip. (b) After cross slip.

of the moving dislocation segments is determinant for a complete cross-slip process. This mechanism has been identified during high-voltage TEM in situ straining experiment performed at room temperature in a Cu/Al$_2$O$_3$ thin foil (Caillard and Martin, 1976).

16.5.4.3 Schoeck–Seeger and Wolf Models

In the original model proposed by Schoeck and Seeger (1955), the activation energy for cross slip is taken as the energy of a screw dislocation where the two partials are recombined over a critical length $2\ell_o$ (**Figure 43**). The cross-slip activation energy E_{cs} writes

$$E_{cs} = E_c + 2\ell_o E_o, \tag{167}$$

where E_o is the energy decrease per unit length of a screw dislocation dissociated into partials (Seeger and Schöck, 1953) and E_c the energy contribution due to the constrictions at both ends of the recombined portion of the dislocation line. Modeling is here based on linear-elastic continuum theory and the problem of cutoff radii is treated in the frame of the PN model (see Section 16.4).

The critical length $2\ell_o$ of recombined segment must be long enough to become unstable onto the cross-slip plane under the RSS τ_o acting in this plane, which implies a high activation energy. For instance, E_{cs} is nearly 8 eV for Cu at a stress of about 10^{-4} μ and a dissociation width of 4.5b. In E_{cs}, the energy constriction E_c is only 0.84 eV and the main contribution arises from the required recombination of the dissociated dislocation in the primary plane over a length $2\ell_o$ of 32 b.

Wolf (1960) improved the above model by taking into account, firstly, the possible dissociation in the cross-slip plane of the recombined screw dislocation segment and, secondly, the existence of nonrelaxed dislocation pileups to achieve the critical stress necessary for dislocation recombination

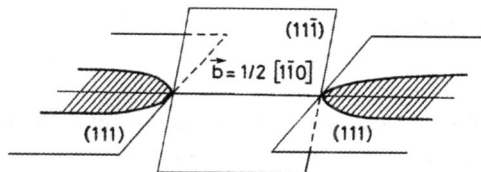

Figure 43 Schematic representation of a cross-slipping dislocation according to Schoeck–Seeger model.

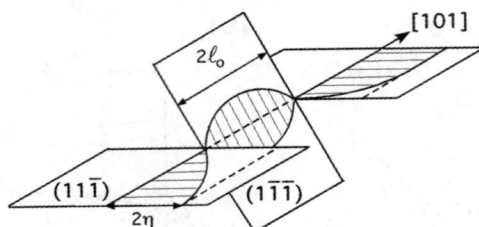

Figure 44 Asymmetrical constrictions assuming blocking of the leading partial.

along the length $2\ell_o$. The first partial is now pressed by the pileup against a strong linear obstacle leading to an asymmetrical shape of the configuration in the primary plane (**Figure 44**). Applying the model to the stress τ_{III} at the onset of stage III leads to an expression of activation energy having the form (Haasen, 1967)

$$E_{cs} = -A \ln \frac{\tau_{III}}{\tau_{III}^o} \tag{168}$$

where τ_{III}^o is the critical stress at 0 K and A is a factor that strongly depends on the stacking fault energy γ and to a lesser extent on the number n of dislocations in the pileup. Two parameters are therefore adjustable for fitting the experimental results. As a rule, it is necessary to introduce high n values, from 23 to 108 (Seeger et al., 1959), while γ is also found rather high, but may significantly vary from one study to another without clear justification.

It must be noticed that, to our knowledge, such dislocation pileups have never been observed in pure FCC single crystals. Using dedicated straining experiments combined with neutron irradiations to pin dislocation microstructures under load, Mughrabi (1968) reported TEM observations of dislocation pileups that never contain more than 20 dislocations. In addition, the dislocations of the pileups are stacked against Lomer–Cottrell barriers, which are not in the screw orientation. One of the successes of the model is that assuming thermal activation, that is

$$\dot{\gamma} = \dot{\gamma}_o \exp\left(-\frac{E_{cs}}{kT}\right), \tag{169}$$

the model correctly predicts the linear dependence of $\ln \tau_{III}$ with T or with $\ln \dot{\gamma}$, which are experimentally obtained. As a final remark, the associated cross-slip activation volume is given by

$$V = -\left(\frac{\partial E_{cs}}{\partial \tau}\right)_T = \frac{A}{\tau} \tag{170}$$

and diverges as the stress tends to zero.

16.5.4.4 *Friedel–Escaig Model*

Friedel (1956) suggested that the successive stages for initiating cross slip in the FCC structure differ from the above scenario. The basic idea is that once the screw dislocation recombines at the intersection of the two {111} planes, forming a constriction which therefore belongs to the two glide planes, the dislocation splits again in the new glide plane (**Figure 45**). The two partial dislocations should therefore

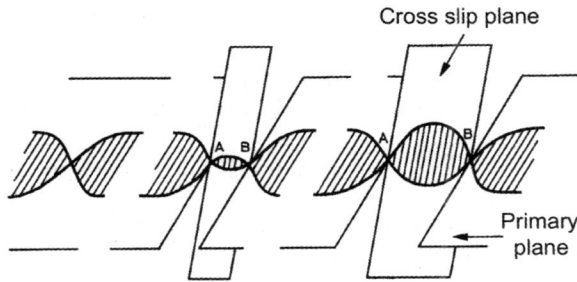

Figure 45 Successive stages of cross slip in the Friedel–Escaig model.

recombine only locally on a very short length, thus avoiding the high recombination energy over the length $2\ell_o$. Preexisting constrictions resulting from dislocation interactions may also be privileged sites for initiating cross-slip events. Here, cross slip develops once the cross-slipping dislocation segment reaches a sufficient length to bend and become unstable under the action of the shear stress in the new glide plane. As pointed out by Friedel (1964) at the saddle-point configuration, the two constrictions should no longer interact and the activation energy for this cross-slip process is at most $2E_c$.

The cross-slip mechanism described by Friedel, today referred to as the Friedel–Escaig cross-slip mechanism, has been modeled by Escaig (1968a, b, 1974) in the frame of the LT formalism. The constriction energies in the primary and in the cross-slip planes are calculated similarly to the treatment given by Stroh (1954). Escaig demonstrated that for such a cross-slip process the activation energy never exceeds twice the constriction energy and may occur under an applied stress only, without dislocation pileups. Another major result of Escaig's modeling is that, in the range of stress of experimental interest, the contribution of the bending stress in the cross-slip plane is negligible, the cross-slipping segment being only slightly curved at the saddle-point position. The shear stresses that govern cross slip are those affecting the splitting widths in both the primary and the cross-slip planes, called the "Escaig stresses". This leads to particular stress orientation effects that have been semiquantitatively verified experimentally by Bonneville and Escaig (1979). The "Escaig stresses" have been successfully used to explain the complex dependences in orientation and sign of the critical resolved shear stress (CRSS) of some $L1_2$ intermetallic compounds exhibiting a flow stress anomaly (Umakoshi et al., 1984; Paidar et al., 1984). Detailed theoretical developments can be found in Escaig (1968b) or in more concise form in Escaig (1974). An asymptotic development of the activation energy at low stresses ($\tau b / \gamma \ll 1$) gives (Escaig, 1968b)

$$E_{cs,i} = 2iAd_o^2\gamma\left[\left(1 - \frac{b_o}{d_o}\right)^2 - \alpha_i\frac{\tau b}{\gamma}\right] \tag{171}$$

where the index $i = 1$ or 2, depending on whether cross slip takes place starting from an existing constriction, $i = 1$, or if two constrictions are needed, $i = 2$. In Eqn (171), A is a logarithmic function of the effective dissociation width in the cross-slip plane, $A \approx 1$ for Cu and 0.5 for Al, d_o is the stress-free dissociation width, b_o the distance at which the two partials may be considered as recombined. b_o is of the order of the magnitude of Burgers vector b for Cu (Bonneville et al., 1988) and $b/2$ for Al (Carrard and Martin, 1988). The α_i coefficients include the orientation factors, which can be tediously calculated using spherical trigonometry or by a more elegant and simple manner in terms of crystallographic indexes (Kubin et al., 2009). These coefficients are also weakly dependent of d_o/b_o. An important underlying assumption to the model is that the cross slip is considered as a heterogeneous process, which takes place at some particular spots in the crystal where the leading partial dislocation is held against an obstacle in the primary plane (Kubin, 2013).

In this model, at the limit of low stresses, the activation volume is given by

$$V_{cs,i} = 2iA\alpha_i d_o^2 b, \tag{172}$$

which is of the order of a few times $d_o^2 b$. Therefore, in contrast with the Schoeck–Seeger–Wolf model, the activation volume achieves a finite value even in the zero stress limit.

Using a dedicated technique where an experimental situation is created such that bursts of cross slips control yielding, activation volumes of $280 \pm 65\ b^3$ were measured at room temperature for [214]-oriented Cu single crystals deformed in compression (Bonneville et al., 1988). In the frame of the Friedel–Escaig model, this value corresponds to dissociation widths $d_o = 6.5 \pm 0.7\ b$ for $i = 1$ and $d_o = 6.1 \pm 0.6\ b$ for $i = 2$, in good agreement with those measured by transmission electron microscopy, $d_o = 7 \pm 2.3\ b$, using weak beam techniques (Stobbs and Sworn, 1971). The associated experimental activation energy is $E_{cs} = 1.15 \pm 0.37$ eV, which is intermediate between the two activation energies predicted by the model, $E_{cs,1} = 0.86 \pm 0.13$ eV and $E_{cs,2} = 1.58 \pm 0.2$ eV, making use of the preceding d_o values. More recent measurements (Couteau et al., 2011) using the same technique in Cu single crystals, but different deformation conditions, found lower activation volumes and activation energies, suggesting that Eqns (171) and (172) are only valid in a very limited range of low stresses.

16.5.4.5 Other Calculations based on the Friedel–Escaig Scenario

Saada (1991) revisited the constriction energy in the LT formalism using anisotropic elasticity. He also considered that the leading partial may be blocked against a hard linear obstacle, leading to an asymmetrical constriction in the primary plane (**Figure 44**). In the extreme case where the leading Shockley partial remains entirely straight at the intersection between glide planes and the trailing partial is pushed against for local recombination, the constriction energy is raised by a factor $\sqrt{2}$ with respect to the symmetrical case. Saada also pointed out the importance of the core radius, showing that the core width largely influences the final results.

The most elaborated theoretical calculations of cross slip using the Friedel–Escaig scenario were undoubtedly performed by Püschl (2002). The calculations are based on linear-elasticity continuum theory. Essential improvements regarding previous cross-slip calculations are (1) the determination of proper cutoff radii and approach distances for recombination of partial dislocations by treating the core configuration using the PN model (see Section 16.4), (2) the use of fully anisotropic treatment for elastic interactions, (3) the choice of realistic, not too restrictive, dislocation shapes.

Analytical solutions do not exist and the reader is strongly advised to refer to the excellent review paper of Püschl (2002) for a detailed presentation of the results. In short, Püschl's computations confirmed that the Escaig stress gives rise to a much stronger effect for cross slip than the bow-out stress and that Escaig values for stress-free constriction energy are in reasonable agreement with those obtained using the more elaborated theoretical treatment, except for small values of d_o/b (**Figure 46**).

16.5.4.6 Atomistic Studies

Atomistic simulations have been widely used in the recent years to study the elementary mechanisms and interactions of dislocations. Atomistic simulations are a powerful tool for predicting dislocation-core structures on an atomic scale, but can suffer from serious artifacts, in particular depending on the determination of the interatomic potentials to which they are very sensitive. It is also often unrealistic

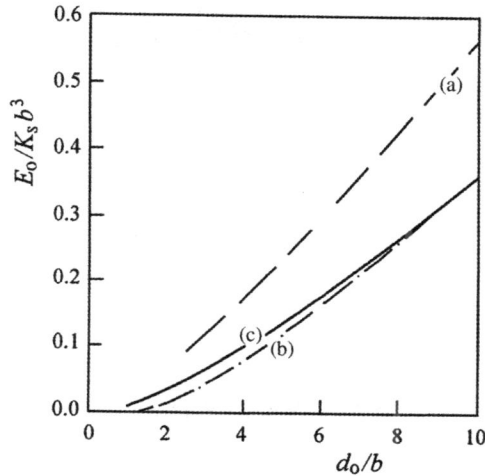

Figure 46 Normalized stress-free constriction energies as a function of normalized splitting width d_0/b. (a) Duesbery et al. (1992). (b) Escaig (Eqn (171)). (c) Püschl (2002).

to simulate thermally activated processes by standard molecular dynamics, which allow simulations on the 10^{-9} s time scale only. To avoid the latter issue, Rasmussen et al. (1997a, b) used a sophisticated method based on a configuration space-patch technique, called the nudged elastic band (NEB) method (Jónsson et al., 1998), which allows the determination of the minimum energy transition path of a dissociated screw dislocation from one glide plane to another. The clear advantage of this method is that there is no need to speculate on the cross-slip mechanism, which is an outcome of the simulation. Only the initial and final states of the dissociated screw dislocation need to be given. Selected configurations obtained with the same approach as Rasmussen et al. (1997a) are shown in **Figure 47**.

The observed scenario unambiguously corresponds to the cross-slip mechanism suggested by Friedel and Escaig. It was also evidenced that the subsequent redissociation in the cross-slip plane creates two nonequivalent twisted constrictions. The constrictions are referred to as "edge-like" and "screw-like" (**Figure 48**). This result was already mentioned by Duesbery et al. (1992). The formation energies associated with the two types of constriction were calculated for Cu. It was found that the edge-like constriction has a formation energy of about 4 eV, whereas surprisingly the screw-like constriction has a negative formation energy of nearly -1 eV (note that the total energy change for cross-slip remains positive at a large value of 3 eV). Hence, the barrier energy for cross slip with two independent constrictions was estimated of the order of 3 eV. It was shown that the barrier energy can be noticeably reduced to ≈ 1 eV if a preexisting constriction at an elementary jog is considered (Vegge et al., 2001), in better agreement with experimental results (Bonneville et al., 1988).

Atomistic simulations were also performed by Rao et al. (1999) in Ni. These simulations also confirmed the Friedel–Escaig mechanism. Moreover, it was shown that constrictions exhibit diffuse core structures, that is, no complete constriction of the stacking fault occurs and the core is spread in both planes. The authors emphasized that two types of constriction with privileged edge and screw type character must be distinguished. The estimated activation energy for Cu is, as currently obtained in atomistic simulations, very large, but is drastically reduced by assuming a preexisting constriction. In this case, the estimated values are $E_{cs} = 1.18 \pm 0.1$ eV at stress level $\tau = 10^{-3}$ μ for a splitting

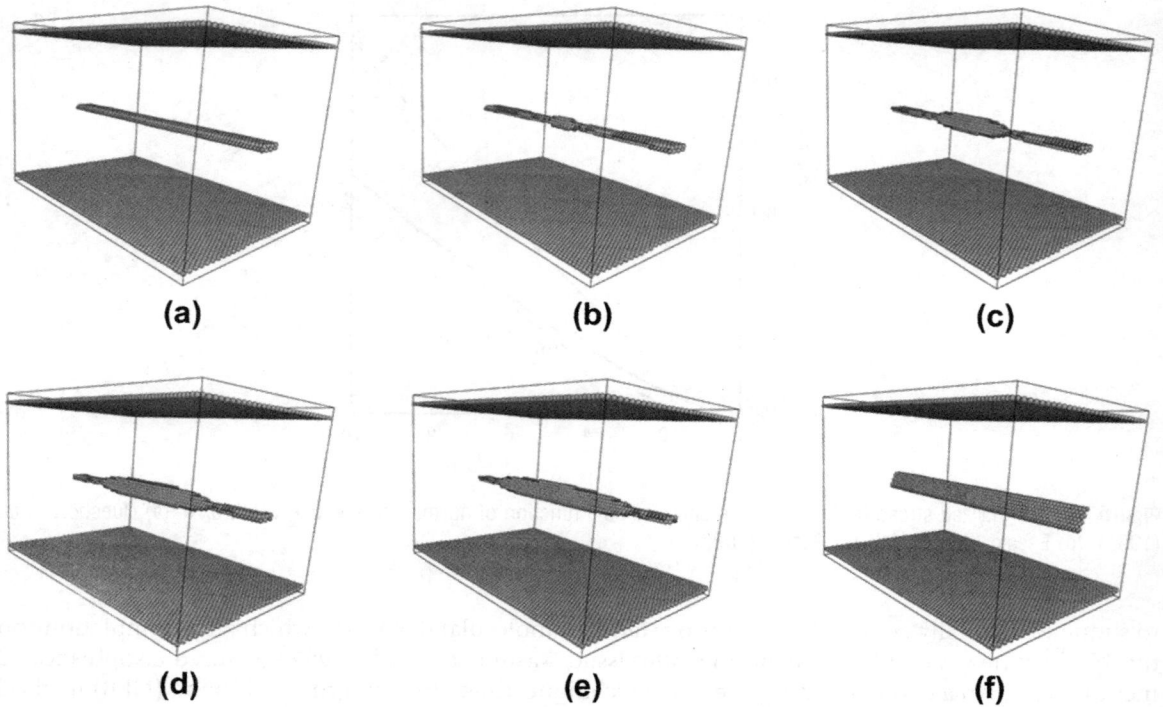

Figure 47 Configurations of minimum energy path for cross slip according to atomistic calculations performed using the NEB method in copper. Only atoms with a non-FCC local environment are shown. (For color version of this figure, the reader is referred to the online version of this book.)

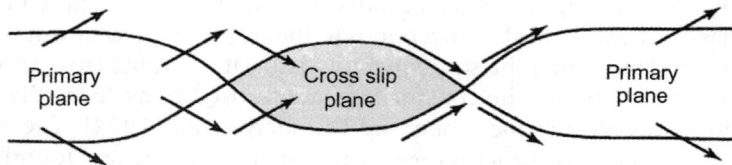

Figure 48 Edge-like and screw-like constrictions in a cross-slipping dislocation.

width $d/b = 6.5 \pm 1.3$. The associated activation volume has however a very small value of about 20 b^3. Rao et al. (2009) also recently revisited the model proposed by Washburn (1965) of cross-slip nucleation near junctions. While the new proposed cross-slip mechanism at dislocation–dislocation interactions has significantly lower values of activation energy compared to the Friedel–Escaig process, E_{cs} is found to follow a scaling law of the Stroh type

$$E_{cs} \propto \frac{d}{b} \left(\ln \frac{d}{b} \right)^{1/2}. \tag{173}$$

We would also like to mention the two-dimensional atomistic simulations performed by Duesbery (1998) using Finnis–Sinclair-type interatomic potentials for Cu. A striking feature obtained in these

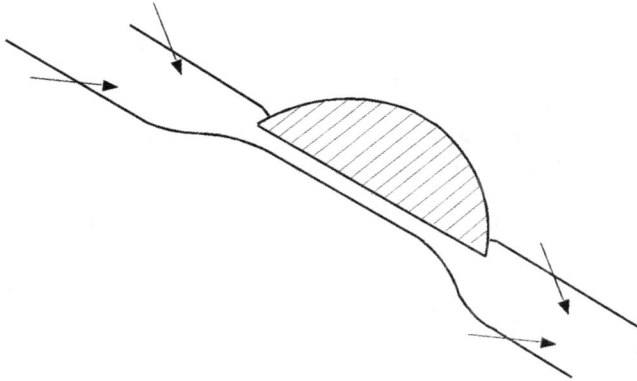

Figure 49 Saddle point configuration from Duesbery's atomistic simulations using finnis-sinclair interatomic potential (Duesbery, 1998). Note that, akin to Fleischer model, full constriction is not required.

simulations is that, contrary to presumptions based on the continuum theory of dislocations, cross slip can occur regardless of the extension of the dislocation in the primary plane, provided the stress is sufficiently large. The calculations show that there is a lower limit of 1.55 b for the distance between the two Shockley partial dislocations of a screw dislocation in Cu. Therefore, akin to Fleischer model at an obtuse dihedron, a full constriction is not required for cross slip (see **Figure 49**).

Similar results were obtained by Lu et al. (2004) for Ag, where increasing negative Escaig stress, i.e. narrowing stress, cannot yield separation width smaller than 1.7 b for straight dislocations. Finally, it should be noted that a Fleischer-type cross-slip process was reported recently in atomistic simulations by Bitzek et al. (2008) in nanocrystalline FCC metals.

16.6   High-Peierls Stress Dislocations: Case of BCC Screw Dislocations

16.6.1   Core of the Screw Dislocation

Plasticity in BCC crystals is remarkably different from that in FCC and other close-packed crystals (Kubin, 1982; Christian, 1983; Duesbery, 1989; Taylor, 1992). The yield stress in BCC crystals depends strongly on the temperature and strain rate at low temperatures, indicating that plasticity is controlled by thermally activated processes in this temperature regime (Caillard and Martin, 2003). Also, there are well-known deviations from Schmid law. The latter, valid in close-packed metals, states that plastic slip begins in a glide system when the RSS reaches a critical value, known as the CRSS, independently of the sign of the stress and of all other components of the stress tensor (Schmid and Boas, 1950). However, as first observed in BCC α-Fe and β-brass by Taylor (1928), the yield stress in BCC crystals is not symmetric in tension and compression and depends on the sense of the applied stress; an effect related to the twinning (T)/antitwinning (AT) asymmetry (see Section 16.6.2.1). Also, the yield stress depends not only on the RSS but also on nonglide stresses, i.e. stress components that produce no Peach–Koehler force on the dislocation, such as shear stresses perpendicular to the dislocation Burgers vector.

We will show in the present section that all above effects may be explained by the fact that the core of screw dislocations in BCC crystals is not planar but spread in several nonparallel planes (Hirsch, 1960). This effect is the first historical example of macroscopic plastic properties directly related to atomic-scale core processes.

16.6.1.1   *The* ⟨111⟩*-Zone*

The most common Burgers vector in BCC crystals is $b = a/2\langle 111 \rangle$ (a, the cubic lattice parameter), the smallest lattice vector of the BCC lattice, in agreement with the elastic energy criterion seen in Section 16.3.1.2. As illustrated on **Figure 50**, the BCC stacking along a ⟨111⟩ direction is made of a ternary sequence of hexagonal planes. The period is b, with the three planes positioned at reference altitudes 0, $b/3$ and $2b/3$. The ⟨111⟩ direction is an axis of three-fold symmetry, at the intersection of three {110} and three {112} planes, forming the so-called ⟨111⟩-zone. As shown in **Figure 50(b)**, {110} and {112} planes alternate around the ⟨111⟩ direction with an angle of 30° between them.

Along the ⟨111⟩ direction, the BCC crystal can be seen as a hexagonal lattice of ⟨111⟩ columns, with a lattice parameter $a\sqrt{2/3}$. Atoms in first-neighbor columns are not at the same altitude and appear in different colors in **Figure 50(b)**. As illustrated in **Figure 50(c)**, atoms in three first-neighbor columns forming an upward triangle (Δ) are at reference altitudes 0, $b/3$ and $2b/3$ when turning clockwise in the triangle, while it is the opposite in downward triangles (∇). The lattice thus forms helices inside the triangles with a different chirality depending on the orientation of the triangle. This property, symbolized by circular arrows in **Figure 50(b)**, will be important when discussing dislocation cores in the next section.

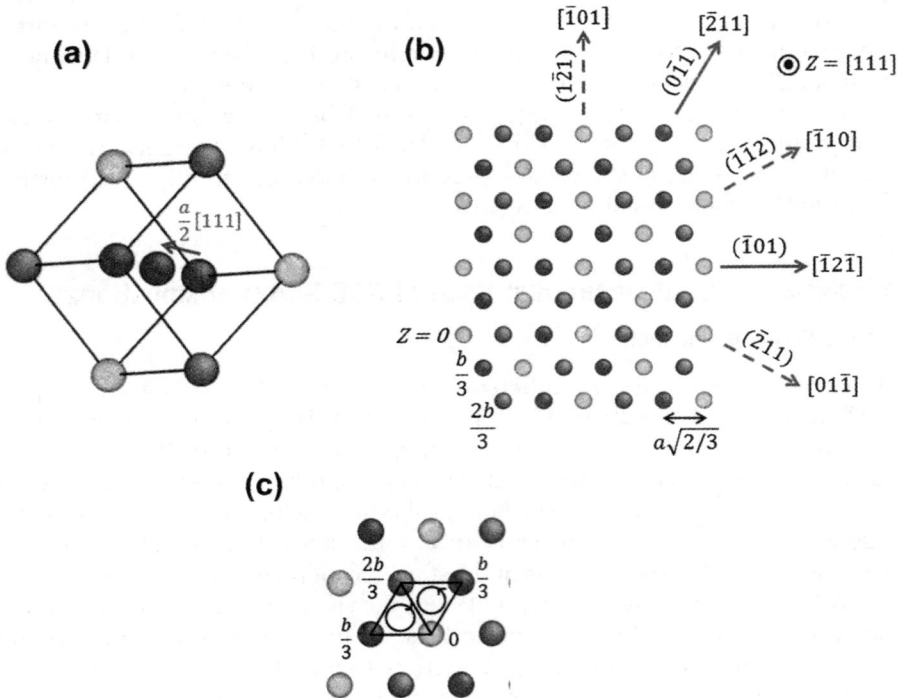

Figure 50   The ⟨111⟩-zone. (a) BCC unit cell, (b) BCC stacking along the [111] direction. Atoms at different heights appear in different colors. (c) Detail of the stacking showing that neighboring [111] columns form helices with different chirality depending on the orientation of the triangles. In (b), ⟨112⟩ and ⟨110⟩ directions are shown with solid and dashed vectors respectively. The corresponding plane indexes are given above the arrows in the direction of the plane normal. (For color version of this figure, the reader is referred to the online version of this book.)

16.6.1.2   *Easy and Hard Screw Dislocation Cores*

The screw dislocation with Burgers vector $b = a/2\langle 111 \rangle$ is parallel to $\langle 111 \rangle$ columns. Symmetry arguments and atomistic simulations showed that the dislocation core is centered on a triangle of first-neighbor $\langle 111 \rangle$ columns, with an orientation (Δ or ∇) depending on the sign of the Burgers vector ($+b$ or $-b$). To see this, consider the displacement field of an infinite screw dislocation (Eqn (46))

$$u_z = \frac{b\theta}{2\pi}. \tag{174}$$

with the z-axis along the $\langle 111 \rangle$ direction and θ, the azimuth with respect to the $[\bar{1}2\bar{1}]$ x-direction as shown in **Figure 51(a)**. If a dislocation with a $+b$ Burgers vector is centered on a downward triangle (see **Figure 51(a)**), the three columns of the central triangle have heights and azimuths $(0, -\pi/2)$, $(b/3, \pi/6)$ and $(2b/3, 5\pi/6)$. According to Eqn (174), they are displaced along the $\langle 111 \rangle$ direction by $-b/4$, $b/12$ and $5b/12$, respectively. Their altitudes in the deformed crystal are thus $-b/4$, $5b/12$ and $13b/12$. The last term is equivalent to $b/12$ because of the periodicity along $\langle 111 \rangle$ columns and if we add $b/4$ to all three heights to shift them between 0 and b, we obtain finally 0, $2b/3$ and $b/3$. These heights are the same as in the undeformed crystal but in reversed order, since now, as indicated by the circular arrow in **Figure 51(a)**, the altitude increases when turning clockwise in the triangle. Distances between atoms in the core triangle are thus unchanged with respect to the undeformed crystal. By way of contrast, if the

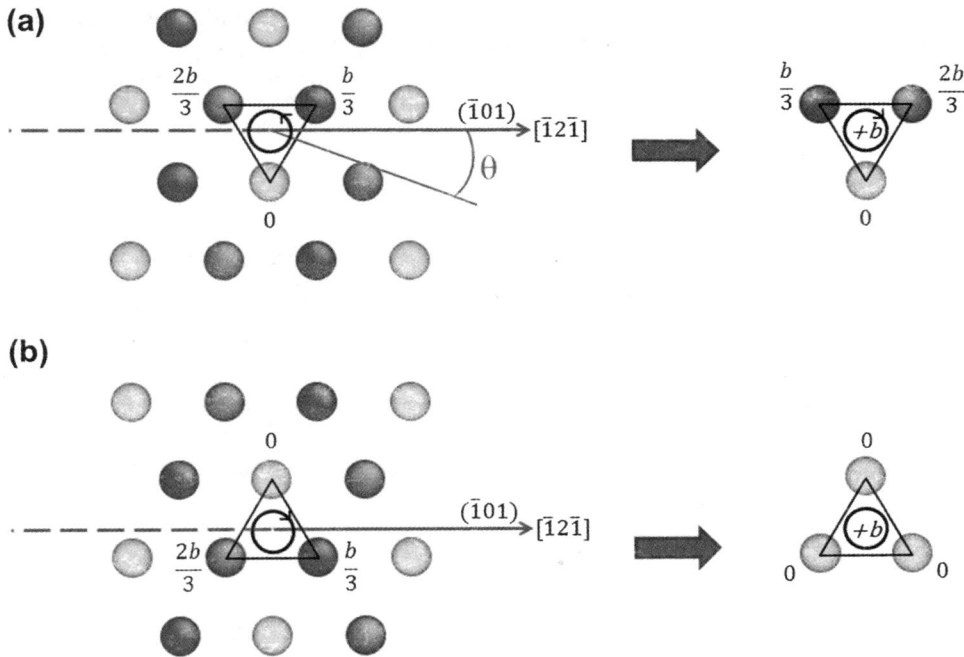

Figure 51   Easy and hard dislocation cores. (a) Dislocation centered on a downward triangle, with a cut surface on the right shown as a dashed line, (b) same dislocation when centered on an upward triangle. Left-hand side: atomic columns in the undeformed crystal; RH side: in the deformed crystal. Atomic colors scale with the height. Circular arrows show the direction of increasing height. Courtesy of Ventelon (2008). (For color version of this figure, the reader is referred to the online version of this book.)

$+b$ dislocation is centered on an upward triangle, the heights and azimuths of the central columns are $(2b/3, -5\pi/6)$, $(b/3, -\pi/6)$ and $(0, \pi/2)$, with displacements $-5b/12$, $-b/12$ and $b/4$, resulting in heights in the deformed crystal all equal to $b/4$ (or equivalently, equal to 0 as noted in **Figure 51(b)**). Atoms at the same altitude are much closer than in the undeformed crystal, resulting in a high-energy configuration. The same reasoning shows that a screw dislocation with negative Burgers vector $(-b)$ inverts the sense of rotation in an upward triangle and places the three columns at the same height in downward triangles, i.e. opposite to the $+b$ dislocation.

Among the two possibilities, one ($+b$ dislocation in downward triangle or $-b$ dislocation in upward triangle) is of much lower energy than the other and is called easy core while the high-energy configuration is called hard core. In atomistic simulations, depending on the model used to represent atomic interactions, the hard core is either saddle or a maximum-energy configuration. In the following, we will only consider stable easy cores. Also, when a screw dislocation moves in a crystal, it does so between stable positions, thus between two triangles of same orientation, with a step length $a' = a\sqrt{2/3}$, the distance Peierls valleys.

16.6.1.3 Symmetrical and Asymmetrical Screw Dislocation Cores

The screw dislocation core configuration was elucidated mainly by atomistic simulations (see Ito and Vitek (2001) and Vitek (2004) for recent reviews). Experimentally, the most adapted technique, High-Resolution Transmission Electron Microscopy, allows to visualize only displacements perpendicular to the dislocation line (edge component), which are very small in BCC screw dislocations. The best way to visualize the dislocation core from atomistic simulations is to use differential displacement maps (see the reviews by Duesbery (1989), Vitek (1992) and Duesbery and Vitek (1998)). The arrows in **Figure 52** linking two first-neighbor $\langle 111 \rangle$ columns have lengths proportional to the difference in displacement

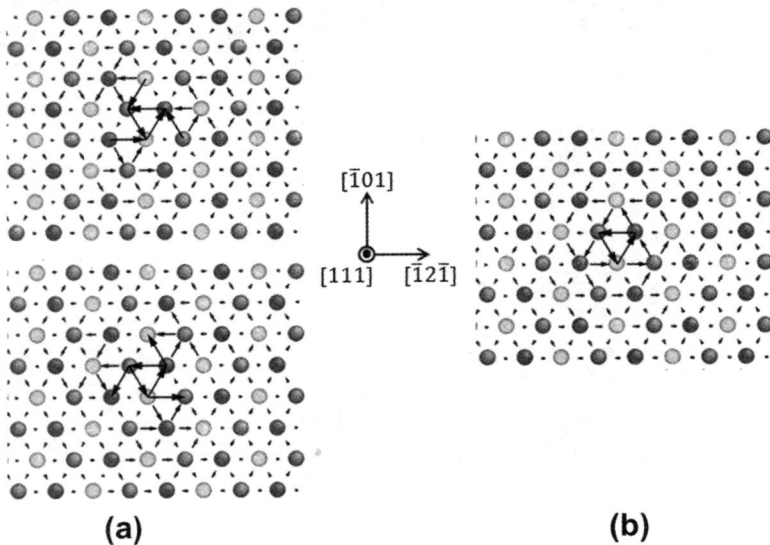

(a)   **(b)**

Figure 52   (a) Asymmetrical and (b) symmetrical screw dislocation cores, showed by differential displacement maps. The structures were obtained by atomistic calculations in α-Fe, using many-body interatomic potentials developed by Simonelli et al. (1993) in (a) and Mendelev et al. (2003) in (b). (For color version of this figure, the reader is referred to the online version of this book.)

induced by the dislocation in the $\langle 111 \rangle$-direction between the two columns. Differential maps therefore show only the screw component of the displacement field. Vector lengths are normalized such that a difference of $b/3$, which occurs between the columns of the core triangle, corresponds to an arrow of length equal to the intercolumn spacing $(a\sqrt{2/3})$. Arrows are oriented from the column of smaller displacement to the column of larger displacement. Also, periodicity along the $\langle 111 \rangle$ columns is used to show displacements only between 0 and b. Interestingly, the algebraic sum of differential displacements along a circuit that encompasses the dislocation core is equivalent to a Burgers circuit (Section 16.2.2.2) and is necessarily equal to b.

The easy core of the BCC screw dislocation can adopt two configurations depending on the material and the energy model. Examples of both cores are shown in **Figure 52**. In both cases, differential displacements are equal to $b/3$ in the core triangle, resulting in the chirality inversion explained above. In the first possible configuration (upper configuration in **Figure 52(a)**), the arrow directly to the right of the core triangle in the central $(\overline{1}01)$ plane corresponds to a displacement close to $b/3$, while on the left-hand side, the differential displacement is almost zero. The same asymmetry is found in the two other $\{110\}$ planes that leave the core triangle. The core is thus asymmetrically extended in the three $\{110\}$ planes of the [111] zone, and is called an asymmetrical core. The wording is important here: this core is "extended" (or "spread") and not "dissociated" as in FCC metals because there is no stable stacking fault in $\{110\}$ planes in the BCC lattice (Vitek, 1992; Duesbery, 1998). As a result, the shear in the extensions is not constant but decreases continuously and the extension is limited to one or two lattice spacings only, as seen in **Figure 52(a)**. Also in contrast with FCC crystals, this core is nonplanar due to its extension in the three nonparallel $\{110\}$ planes of the $\langle 111 \rangle$ zone.

By way of contrast, the second configuration (**Figure 52(b)**) is called a symmetrical core because differential displacements are symmetrical on both sides of the core triangle. This core has also been called nonplanar, or sixfold extended in contrast with the asymmetrical core, which is threefold extended. These denominations may however be misleading. The displacement field associated with the symmetrical dislocation is very close to the linear anisotropic elastic solution of the Volterra dislocation (Eqn (174)). The present core is therefore neither nonplanar, nor extended, but only highly compact. The impression of extension in $\{110\}$ planes given in **Figure 52(b)** comes from the fact that $\{110\}$ planes are more dense than $\{112\}$ planes, they have therefore a larger interspacing, resulting in larger differential displacements.

Both cores are invariant by the threefold symmetry of the $\langle 111 \rangle$ axis, but the asymmetrical core is not invariant with respect to the $[\overline{1}01]$ diad, a symmetry operation of the BCC lattice, whereas the symmetrical core is invariant by this symmetry operation. As a result, the asymmetrical core has two variants, related by the diad symmetry but otherwise identical (see the upper and lower configurations in **Figure 52(a)**) and is called degenerate or sometimes polarized (Seeger and Wüthrich, 1976; Seeger, 1995; Yang et al., 2001; Cai et al., 2004). On the other hand, the symmetrical core has only one variant and is called nondegenerate. The degree of asymmetry of a core can be quantified using a polarization parameter (Wang et al., 2003; Ventelon and Willaime, 2010).

Which core configuration is adopted in a given crystal depends on the material and for the purpose of atomistic simulations, depends also on the energetic model used to describe interatomic interactions. Duesbery and Vitek (1998) proposed a criterion based on γ-surfaces, or rather based on the $\langle 111 \rangle \{110\}$ γ-line, the crystal energy when displacing two halves of the crystal in opposite $\langle 111 \rangle$ directions parallel to a $\{110\}$ plane (see Section 16.4 for more details on γ-surfaces). Examples of γ-lines are shown in **Figure 53**. Duesbery and Vitek (1998) proposed to view a degenerate core as split into three fractional dislocations of Burgers vector $b/3$. This is an idealization because, the stacking fault being unstable, no true partial dislocations form in the dislocation core and the shear decreases

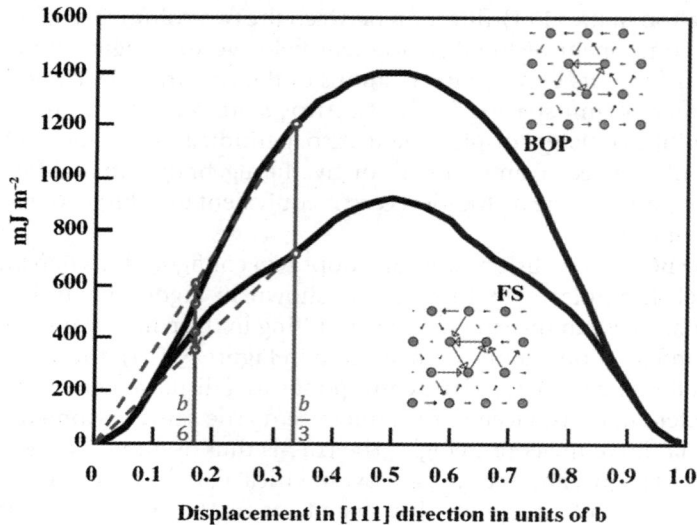

Figure 53 γ-lines for shear along the [111] direction and parallel to the ($\overline{1}$01) plane, obtained with different energy models predicting different screw dislocation core structures. Adapted and reproduced with permission from Vitek (2004). (For color version of this figure, the reader is referred to the online version of this book.)

continuously away from the core. In this model, the energy cost of the splitting in the three fractional dislocation is $3\gamma(b/3)$. The nondegenerate core then consists of six fractional dislocations of Burgers vector $b/6$, with an energy cost of $6\gamma(b/6)$. This is another strong assumption since we have seen that nondegenerate cores are not extended. Comparing the two energies, we conclude that the degenerate core is favored when $\gamma(b/3) < 2\gamma(b/6)$ and the degenerate core otherwise. **Figure 53** shows examples of γ-lines obtained with two different energy models and illustrates how to apply the criterion. We see that the criterion predicts the correct core structure in both cases. The criterion was shown to apply to many systems, modeled with interatomic potentials (Duesbery and Vitek, 1998; Ventelon and Willaime, 2010) and ab initio calculations (Frederiksen and Jacobsen, 2003; Ventelon and Willaime, 2010), with only one exception to date (Ventelon and Willaime, 2010). We should however insist on the fact that the criterion gives only a trend since the cores are not dissociated.

Early simulations, based on generic pair potentials (Vitek, 1974) as well as more recent calculations using many-body potentials predict a degenerate core. For examples in Fe, see Wen and Ngan (2000), Marian et al. (2004) and Chaussidon et al. (2006); in Ta, see Yang et al. (2001); in Mo, see Ito and Vitek (2001). The nondegenerate core is obtained much less frequently with interatomic potentials, but examples exist: in Ta, see Ito and Vitek (2001), in Fe, see Domain and Monnet (2005) and Chaussidon et al. (2006). However, the more recent calculations based on electronic structure all predict nondegenerate core structures in pure transition metals. For DFT calculations in Mo and Ta, see Ismail-Beigi and Arias (2000), Woodward and Rao (2002); in Fe, see Frederiksen and Jacobsen (2003) and Domain and Monnet (2005); in Mo with bond order potentials, see Mrovec et al. (2004) and in Fe, see Mrovec et al. (2011). There is therefore now a large consensus that the core of the screw dislocation is symmetrical and nondegenerate in pure transition metals. In alloys, the situation is different. It was for example shown that alloying W with Re changes the core from symmetric to asymmetric (Romaner et al., 2010). Also the core structure depends on the pressure (Yang et al., 2001) and temperature (Chaussidon et al., 2006).

16.6.2 Screw Dislocations under Stress

So far, we have discussed the core of screw dislocations in BCC crystals in absence of applied stress. The question now is as to how does this dislocation glide. From the discussion in Section 16.4, we expect the dislocation to have a large Peierls stress because the core is compact with a narrow core width. For a long time, the high Peierls stress of the BCC screw dislocation was in fact attributed to the threefold extension of the asymmetrical core, which requires to constrict the nonplanar extensions before the dislocation can glide. However, the symmetrical core, which is not extended, also has a high Peierls stress. The main origin of the high Peierls stress is therefore probably the compactness of the core. Also, having a high Peierls stress, screw dislocations are expected to glide by a kink pair mechanism at low temperatures (see Section 16.4.2), consistent with the strong temperature effect mentioned in introduction of the present section. Finally, we would like to understand the origin of the asymmetry between tensile and compressive elastic strengths reported experimentally.

The above points are addressed below, starting with the T/AT asymmetry, followed by the Peierls stress, Peierls potential and ending up with the kink pair mechanism.

16.6.2.1 T/AT Asymmetry

In the BCC lattice, {110} planes are mirror planes while {112} planes are not. As shown in **Figure 50(b)**, ⟨111⟩ columns on either side of a {110} plane are at the same altitude and the crystal is symmetrical with respect to the {110} plane. As a result, shearing a BCC crystal parallel to a {110} plane along a ⟨111⟩ direction is equivalent whether you shear in the positive or negative directions. By contrast, the BCC lattice is not symmetrical with respect to {112} planes. As seen in **Figure 50(b)** and 54(b), the ⟨111⟩ columns on either side of a {112} plane are not at the same altitude and the {112} plane is not a mirror plane for the BCC lattice. Moreover, when viewing the crystal along a ⟨110⟩ direction, for instance along $[\bar{1}10]$ as in **Figure 54(a)**, we see that dense $[\bar{1}\bar{1}1]$ columns are inclined with respect to the $(\bar{1}\bar{1}2)$ plane. Consequently, shearing the crystal in the positive ⟨111⟩ direction as indicated in **Figure 54(a)** straightens up the $[\bar{1}\bar{1}1]$ columns and brings the atoms inside the columns closer, while shearing in the negative direction flattens out the $[\bar{1}\bar{1}1]$ columns and separates the atoms. Interatomic interactions being asymmetrical and steeper in compression than in tension, nonlinear effects appear when shearing a BCC crystal parallel to a {112} plane and one direction, called T direction, is easier than the other, called AT direction. Asymmetry is also visible on the γ-lines when shear is operated parallel to {112} planes (Frederiksen and Jacobsen, 2003; Ventelon and Willaime, 2010).

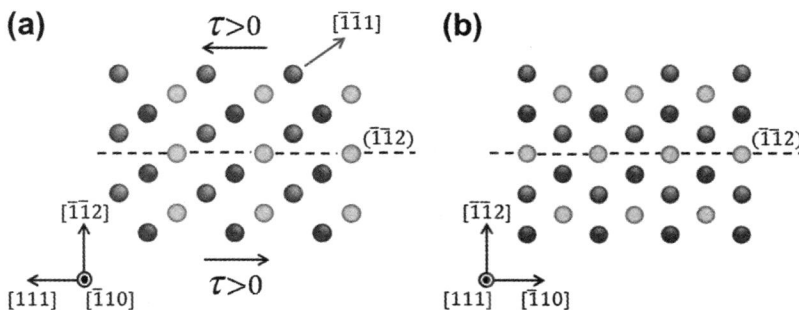

Figure 54 T/AT asymmetry. (a) A BCC crystal is projected along the $[\bar{1}01]$ direction and (b) along the [111] direction, with $(\bar{1}\bar{1}2)$ horizontal planes in both cases. Atoms in different {111} planes appear in different colors. (For color version of this figure, the reader is referred to the online version of this book.)

If a BCC crystal is sheared along a $\langle 111 \rangle$ direction parallel to a given plane, called Maximum Resolved Shear Stress (MRSS) plane, making an angle χ with respect to a $(\bar{1}01)$ reference plane, the stress tensor in the $x = [\bar{1}2\bar{1}]$, $y = [\bar{1}01]$ and $z = [111]$ reference frame of **Figure 55(a)** is

$$\overline{\overline{\sigma}} = \begin{bmatrix} 0 & 0 & -\tau \sin\chi \\ 0 & 0 & \tau \cos\chi \\ -\tau \sin\chi & \tau \cos\chi & 0 \end{bmatrix}. \tag{175}$$

When χ varies around the [111] zone, the crystal is sheared alternatively in T and AT directions, as shown in **Figure 55(a)**. The effect is strongest when the MRSS plane is a $\{112\}$ plane, but persists for all shear planes except $\{110\}$ planes where symmetry is recovered. As shown in **Figure 55(a)**, the [111] zone can thus be decomposed in alternating T and AT regions. **Figure 55(a)** corresponds to a positive shear along [111]. If the sign of the stress is reversed, the T and AT regions are simply interchanged.

Consequences of the T/AT asymmetry are seen up to the macroscopic scale, where different yield stresses are measured in tension and compression. Consequences are also seen on the slip geometry. For instance in Nb–Mo alloys deformed at 77 K (Christian, 1983), when shear is applied in the T direction ($\chi < 0$), the slip plane is systematically $\{110\}$, while in the AT direction ($\chi > 0$), slip becomes

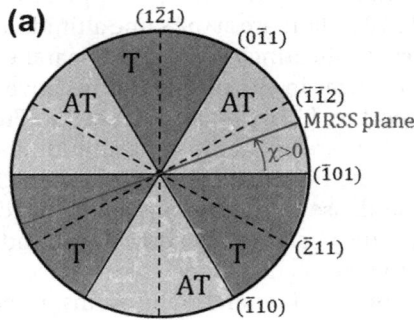

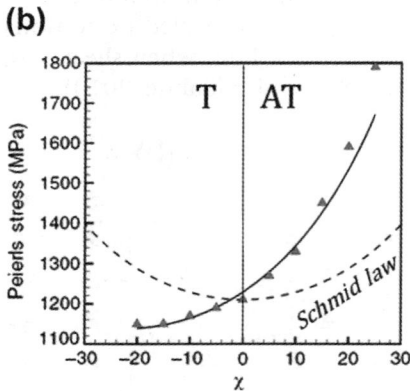

Figure 55 Non-Schmid effects related to the T/AT asymmetry. (a) Twinning (T) and antitwinning (AT) regions as a function of the orientation of the MRSS plane in the [111] zone for a positive shear stress along the [111] direction. (b) Atomistic calculation of Peierls stress in a model crystal of α-Fe as a function of χ and comparison with Schmid law. Reproduced with permission from Chaussidon et al. (2006). (For color version of this figure, the reader is referred to the online version of this book.)

noncrystallographic, or also called pencil glide: the average slip plane does not follow a dense crystallographic plane of the lattice but is parallel to the MRSS.

16.6.2.2   *Peierls Stress and Deviations from Schmid Law*

As introduced in Section 16.4, the Peierls stress is the minimum resolved shear stress to apply to force a dislocation to glide in an otherwise perfect crystal without the help of thermal fluctuations. Experimentally, the Peierls stress is obtained as the zero-temperature limit of the yield stress of well-annealed crystals (Kuramoto et al., 1979; Brunner and Diehl, 1992). The Peierls stress may also be measured in atomistic simulations by subjecting a simulation cell containing a dislocation to increasing applied stresses until the dislocation starts to glide spontaneously. **Figure 55(b)** shows an example of Peierls stress dependence on χ angle from the work of Chaussidon et al. (2006). Similar curves can be found for Mo and Ta in Ito and Vitek (2001), Yang et al. (2001), Woodward and Rao (2002), Vitek (2004), Vitek et al. (2004), Mrovec et al. (2004). Owing to the symmetry of the BCC lattice, only χ angles between $-30°$ and $+30°$ need to be considered. When $\chi = -30°$, shear is applied in the T direction parallel to the $(\bar{2}11)$ plane, while when $\chi = +30°$, shear is parallel to the $(\bar{1}\bar{1}2)$ plane in the AT direction. It should be noted that, usually, degenerate cores do not have a well-defined Peierls stress because when the stress increases, there is first a lower critical stress where the core restructures itself but remains sessile and an upper critical stress where the dislocation really starts to glide (Chaussidon et al., 2006).

A first observation is that the Peierls stresses reported in **Figure 55(b)** are much larger than those extrapolated from experiments (in α-Fe, the experimental Peierls stress is 400 mPa). The origin of this discrepancy, which is beyond the scope of the present Chapter, is the object of much speculation.

In **Figure 55(b)**, the dislocation glides in the reference $(\bar{1}01)$ plane and has a Burgers vector $[00b]$. A stress tensor as given in Eqn (175) has an RSS $(\vec{b} \cdot \bar{\bar{\sigma}}) \cdot \vec{n} = \tau \cos \chi$. If Schmid law applies, slip would start when the RSS reaches a critical value independent of χ. We would thus have $\tau_P(\chi) \cos \chi = \text{Cste}$ or

$$\tau_P = \frac{\tau_P(\chi = 0)}{\cos \chi}. \tag{176}$$

The Peierls stress would therefore be symmetrical with respect to $\chi = 0$. But, as seen in **Figure 55(b)**, this is clearly not the case here. The Peierls stress is lower in the T region ($\chi < 0$) than in the AT region ($\chi > 0$). This deviation from Schmid law is a direct consequence of the T/AT asymmetry. Vitek et al. (2004) suggested that this asymmetry can be accounted for at least phenomenologically by considering the CRSS as a linear combination of the shear stresses resolved in all {110} planes of the [111] zone. Based on their simulations, they found that only the shear stress resolved in the $(\bar{1}01)$ slip plane and in the $(\bar{1}10)$ plane on the border of the T region influence the Peierls stress, resulting in a yield surface

$$\tau_P = \frac{\tau_P(\chi = 0)}{\cos(\chi) + A \cos(\chi + 60°)}. \tag{177}$$

with A, a fitting parameter. The solid curve in **Figure 55(b)** shows the resulting fit (with $A = 0.61$), which indeed closely follows the atomistic calculations. We should note however that the above relation is only phenomenological since the physical origin of the asymmetry is the T/AT asymmetry of the BCC lattice and not an influence of the stress resolved in a {110} plane different from the glide plane, although this point has not been checked explicitly to the best of our knowledge.

Deviations from Schmid law can have another origin coming from stresses perpendicular to the Burgers vector, i.e. such that $\overline{\overline{\sigma}} \cdot \overrightarrow{b} = 0$. Such stresses are called nonglide stresses because they produce no Peach Koehler force on the dislocation. Duesbery (1984a, b), Duesbery and Vitek (1998) and Ito and Vitek (2001) showed that shear stresses perpendicular to the Burgers vector have a much stronger influence than tensions and compressions perpendicular to the slip plane, although the latter can still affect the Peierls stress (see Chaussidon et al., 2006). In order to decouple the effect of shear stresses parallel and perpendicular to the Burgers vector, Ito and Vitek (2001) computed the Peierls in cells subjected to a background pure shear in the (111) plane. The applied stress is then

$$\overline{\overline{\sigma}} = \begin{bmatrix} -\lambda \cos 2\chi & -\lambda \sin 2\chi & -\tau \sin \chi \\ -\lambda \sin 2\chi & \lambda \cos 2\chi & \tau \cos \chi \\ -\tau \sin \chi & \tau \cos \chi & 0 \end{bmatrix}. \tag{178}$$

The shear stress τ parallel to the Burgers vector was increased at constant λ until the dislocation started to glide. Ito and Vitek (2001) considered two BCC crystals (Mo and Ta). They observed a dependence of the Peierls stress on nonglide stresses in both materials, but with very different trends. For instance, the Peierls stress for an maximum resolved shear stress plane (MRSSP) parallel to a {110} plane decreases with λ in Mo but increases in Ta. This is in contrast with the effect of parallel shear stresses, which are very similar in different materials. The influence of nonglide stresses can be traced back to the edge component in the dislocation core (Duesbery, 1989; Ito and Vitek, 2001), i.e. displacements perpendicular to the $\langle 111 \rangle$ atomic columns. Such components are very small (of the order to a few hundredth of Angstroms), but they are sensitive to nonglide stresses and differ largely in cores modeled with different energy models.

Above the Peierls stress, the slip plane followed by the dislocation is correlated with the core structure. Most degenerate cores have a zig–zag motion with an average plane corresponding to the {112} plane in the T region (the $(\overline{2}11)$ plane in **Figure 55(a)**), resulting from alternate elementary steps in the two {110} planes on each side of the {112} plane as illustrated in **Figure 56(a)**. During each elementary step, the dislocation transforms from one variant to the other. Only {112} average planes are observed because elementary steps in the {110} plane in the AT region have never been observed. As a result, the average plane cannot be a {110} plane, even if the latter is the MRSS plane. By contrast, nondegenerate cores can glide on {110} planes because they have a single core configuration, as was observed in the works of Domain and Monnet (2005) and Chaussidon et al. (2006). Experimentally, most pure BCC metals have {110} glide planes, confirming that the most probable core structure is nondegenerate. In alloys, the change of core structure reported in W(Re) alloys is correlated to a change of glide plane, from {110} to {112} (Romaner et al., 2010; Li et al., 2012).

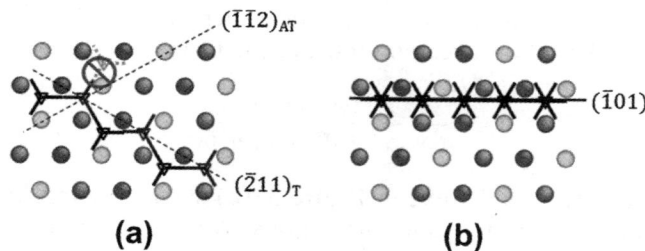

Figure 56 Glide mechanisms of screw dislocations with (a) degenerate and (b) nondegenerate cores. (For color version of this figure, the reader is referred to the online version of this book.)

16.6.2.3 Peierls Potential

Another way to approach the Peierls stress is to consider the Peierls potential of the dislocation. In BCC crystals, stable positions are at the center of triangles of given orientation. The periodicity of the Peierls potential is thus the distance between such triangles, $a\sqrt{2}/3$ (see **Figure 50(b)**). **Figure 57** shows examples of Peierls potentials obtained with different energy models. There are important differences between energy models, even in a given material. As seen in **Figure 57**, the interatomic potential predicts an energy barrier much lower than electronic structure based calculations with a camel-hump shape discussed in Section 16.4. The metastable intermediate configuration, which does not correspond to the hard core configuration (it is not centered on a triangle), is referred to as a split configuration (Suzuki, 1968; Takeuchi, 1979). Using generic potentials, Takeuchi (1979) showed that such config-urations are systematically predicted by interatomic potentials that predict a nondegenerate core structure. By contrast, potentials with a degenerate core structure predict Peierls stresses with a single energy barrier and no metastable split configuration. The existence of such split configuration is however most likely an artifact of interatomic potentials since electronic structure-based calculations predict in pure BCC metals nondegenerate core structures with an unstable split configuration.

Peierls potentials were also computed under stress by Rodney and Proville (2009). Contrary to the simple picture of a fixed Peierls potential only tilted by an applied stress as shown in **Figure 27**, this work showed that the Peierls stress is not constant but depends on the applied stress, reflecting an evolution of the core structure under stress. This dependence invalidates the classical relation used between the maximum derivative of the Peierls potential and the Peierls stress and as a result, the Peierls stress should not be predicted from the stress-free Peierls potential. Moreover, the path followed by the

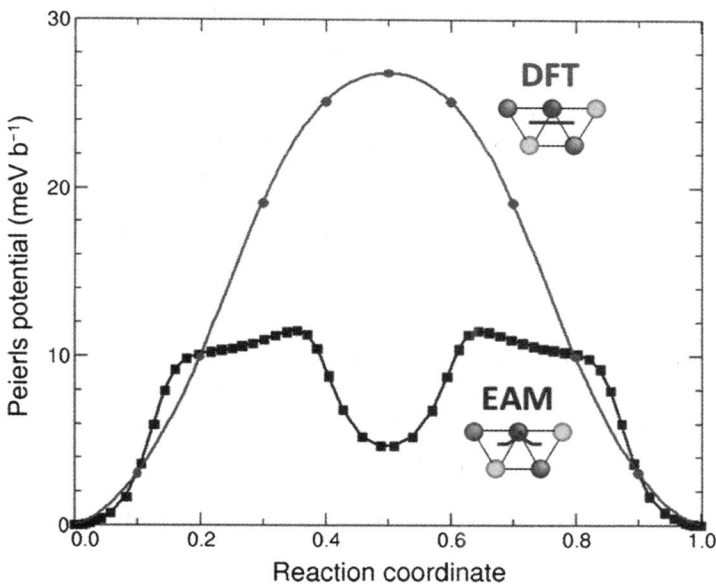

Figure 57 Peierls potentials for a screw dislocation in α-Fe with different energy models: circles were obtained using ab initio DFT calculations, squares are from the many-body potential of Mendelev et al. (2003) referred to as EAM for Embedded Atom Method. The diagrams show the trajectories of the dislocation core along the paths. Courtesy of Ventelon (2008). (For color version of this figure, the reader is referred to the online version of this book.)

core of the screw dislocation may not be planar and it is not clear how to properly take the derivative of the Peierls potential with respect to the dislocation position.

16.6.2.4 Kink-Pair Mechanism

BCC screw dislocations having a high Peierls stress are expected to glide by the kink-pair mechanism at low temperatures. As explained in Section 16.4.2, a kink pair on a screw dislocation is formed of two nonsymmetrical kinks, with a tensile and a compressive kink. Moreover, if the dislocation core is degenerated, several types of kinks must be considered because a kink may join two sections of the dislocation with the same or different variants. Duesbery (1983a, b), Wen and Ngan (2000) and Ngan and Wen (2001) showed by atomistic calculations that kinks of lower energy join dislocation segments with different polarizations. This result was expected from the change of polarization undergone by the dislocation when changing Peierls valley, as mentioned in Section 16.6.2.2.

For non-degenerate cores, the dislocation has only one variant, which limits the number of kink configurations to be considered. There are only two types of kinks, either compressive or tensile. Single kink formation energies have been computed in α-Fe using an interatomic potential (Ventelon et al., 2009) and bond-order potentials (Mrovec et al., 2011). Both calculations show that the tensile kink has a formation energy higher than the compressive kink, but the values differ with both energy models (0.08 and 0.57 eV for the compressive and tensile kinks respectively with the interatomic potential and 0.53 and 0.62 eV with the bond order potential). These data should be compared with the experimental estimate for the kink pair formation energy (sum of the formation energy of the two kinks) of 0.6 eV obtained by Brunner and Diehl (1991). The interatomic potential is closer but neglects magnetism, which strongly limits its realism for α-Fe. The reason why the more accurate bond order potential strongly overestimates the kink pair formation energy is not clear at this point. Another difference between the two calculations is that with the interatomic potential, both kinks have a large width, of the order of $20b$, while with the bond order potential, the kinks are much more compact and extend over 2 to 3 Burgers vectors only. This is consistent with the LT model presented in Section 16.4.2.2 and with the Peierls potentials predicted by both methods. The Peierls potential of the interatomic potential is shown in **Figure 57** while the bond-order potential predicts a Peierls potential close to the DFT calculation and therefore much higher than the interatomic potential: the bond-order potential having a higher Peierls barrier predicts narrower kinks because the energy penalty for placing dislocation segments near the Peierls barrier is higher.

Finally, the activation enthalpy for kink pair formation as a function of applied RSS has been computed by atomistic simulations using several interatomic potentials (Rodney, 2007; Rodney and Proville, 2009; Gordon et al., 2010). The general shape is the same as shown in **Figure 55**, although usually more convex, which can be related to the stress-dependence of the Peierls potential. The LT model has also been generalized to two dimensions by Edagawa et al. (1997) in order to treat the nonplanar core and motion of BCC screw dislocations. The dislocation line is then parameterized as $(y(x),z(x))$ and the Peierls potential is defined not just along the y-axis as in Section 16.4.2.2 but inside the entire yz-plane as a function $E(y,z)$. Edagawa et al. (1997) using simple sinusoidal functions to represent the Peierls potential were able to reproduce the kink-pair mechanism including the dependence of the Peierls stress on the angle between the MRSS plane and the glide plane.

Appendix 1

The elastic energy associated with $\bar{\bar{\sigma}}$ and $\bar{\bar{\varepsilon}}$ can be simply computed, without loss of generality, by considering the elastic deformation of a cube of side dx_0. Because in linear elasticity the superposition principle is assumed to hold true, we first calculate the reversible work done by progressively applying a force f_1, from 0 to F_1, on the surface of normal $\overrightarrow{x}_1$ of the cube. During this isothermal process, the cube is only elastically deformed by a quantity δu_1 that varies with f_1 from 0 to du_1 in the $\overrightarrow{x}_1$ direction. The corresponding elemental work writes

$$\delta W_1 = f_1 \delta u_1. \tag{A1}$$

Using Eqns (2), (3) and (9), we have $f_1 = \sigma'_{11} dx_o^2$ and $\delta u_1 = \delta \varepsilon_{11} dx_o$ with $\delta \varepsilon_{11} = \delta \sigma_{11}/E$ so that, Eqns (A1)–(1) can be rewritten as

$$\delta W_1 = \sigma'_{11} \partial u_1 dx_o^2 = \sigma'_{11} \frac{d\sigma'_{11}}{E} dx_o^3, \tag{A2}$$

where the prime symbol accounts for the progressive application of f_1. By integrating Eqn (A2), the elastic energy stored per unit volume is expressed as

$$w_1 = \int_0^{\sigma_{11}} \frac{\sigma'_{11}}{E} d\sigma'_{11} = \frac{1}{2E} \sigma_{11}^2 \tag{A3}$$

We now progressively apply a force f_2 on the surface of normal $\overrightarrow{x}_2$, while maintaining F_1 constant. The work done per unit volume by the displacement δu_2 due to f_2 in presence of F_1 is

$$w_2 = \int_0^{\sigma_{22}} \frac{\sigma'_{22}}{E} d\sigma'_{22} + \int_0^{\sigma_{22}} \sigma_{11} du'_1, \tag{A4}$$

where the first term in the RH side of Eqn (A4) is equivalent to that of Eqn (A4) while the second term takes into account the fact that f_2 induces a displacement du'_1 of F_1. We now have $du'_1 = -\nu d\sigma'_{22}/E$ and integration of Eqn (A4) yields

$$w_2 = \frac{\sigma_{22}^2}{2E} - \frac{\nu \sigma_{11} \sigma_{22}}{E}. \tag{A5}$$

We finally apply a force f_3 along the third axis. Proceeding similarly, the work done w_3 by unit volume due to f_3, while maintaining F_1 and F_2 constant, is

$$w_3 = \frac{\sigma_{33}^2}{2E} - \frac{\nu(\sigma_{11}\sigma_{33} + \sigma_{11}\sigma_{33})}{E}. \tag{A6}$$

The elastic energy stored per unit volume by successively applying the forces F_1, F_2 and F_3 is equivalent to the work $w = w_1 + w_2 + w_3$ produced by these forces to elastically deform the cube

$$w = \frac{1}{2E}\left(\sigma_{ii}^2 - 2\nu \sum_{j>i} \sigma_{ii}\sigma_{jj}\right), \text{ with summation over } i \text{ and } j \text{ understood.} \tag{A7}$$

Introducing the associated strains in Eqn (A7) and using Eqn (8), leads to

$$w = \frac{1}{2} \sigma_{ii} \varepsilon_{ii}, \tag{A8}$$

which represents the stored elastic energy density calculated in the principal axes of both the stress and the strain. However, w being a scalar, it is invariant by transformation of coordinate axes, i.e. w is independent of the coordinate system in which it is calculated. It follows that, in the framework of isotropic linear elasticity, the elastic energy density can be expressed by the general formula

$$w = \frac{1}{2}\sigma_{ij}\varepsilon_{ij}. \tag{A9}$$

References

Amelynckx, S., 1979. Dislocations in particular structures. In: Nabarro, F.R.N. (Ed.), 1979. Dislocations in Solids, vol. 7. Elsevier, Amsterdam, pp. 43–111.

Bacon, D.J., Barnett, D.M., Scattergood, R.O., 1978. Anisotropic continuum theory of lattice defects. Prog. Mater. Sci. 33, 51–262.

Bacon, D.J., Osetsky, Y.N., Rodney, D., 2009. Dislocation-obstacle interactions at the atomic level. In: Hirth, J.P., Kubin, L. (Eds.), 2009. Dislocations in Solids, vol. 15. Elsevier, Amsterdam, pp. 1–90.

Baluc, N.L., 1990. Contribution à l'étude des défauts et de la plasticité d'un composé intermétallique ordonné: Ni3Al. PhD Ecole Polytechnique Fédérale de Lausanne.

Bardeen, J., Herring, C., 1952. In: Shockley, W. (Ed.), Imperfections in Nearly Perfect Crystals. John Wiley & Sons, Inc., New York, pp. 261–289.

Bitzek, E., Brandl, C., Derlet, P.M., Van Swygenhoven, H., 2008. Dislocation cross-slip in nanocrystalline fcc metals. Phys. Rev. Lett. 100, 235501.

Bonneville, J., Escaig, B., 1979. Cross-slipping process and the stress-orientation dependence in pure copper. Acta Metall. 27, 1477–1486.

Bonneville, J., Escaig, B., Martin, J.L., 1988. A study of cross-slip activation parameters in pure copper. Acta Metall. 36, 1989–2002.

Bragg, W.L., 1948. Symposium on Internal Stress. Institute of Metals, London. pp. 221.

Braun, O.M., Kivshar, Yu.S., 2004. The Frenkel–Kontorova Model: concepts, Methods, and Applications. Springer, Berlin Heidelberg New York.

Brunner, D., Diehl, J., 1991. Temperature and strain-rate dependence of the tensile flow stress of high-purity α-iron below 250 K. I, stress/temperature regime III. Phys. Status Solidi A 124, 455–464.

Brunner, D., Diehl, J., 1992. Extension of measurements for the tensile flow stress of high-purity alpha-iron single crystals to very low temperatures. Z. Metallkd. 83, 828–834.

Bulatov, V.V., Cai, W., 2006. Computer Simulations of Dislocations. Oxford University Press, Oxford.

Bulatov, V.V., Kaxiras, E., 1997. Semidiscrete variational Peierls framework for dislocation core properties. Phys. Rev. Lett. 78, 4221–4224.

Bulatov, V., Abraham, F.F., Kubin, L., Devincre, B., Yip, S., 1997. Connecting atomistic and mesoscale simulations of crystal plasticity. Nature 391, 669–672.

Büttiker, M., Landauer, R., 1979. Nucleation theory of overdamped soliton motion. Phys. Rev. Lett. 43, 1453–1456.

Cai, W., Bulatov, V.V., Chang, J., Li, J., Yip, S., 2004. Dislocation core effects on mobility. In: Nabarro, F.R.N., Hirth, J.P. (Eds.), 2004. Dislocations in Solids, vol. 12. Elsevier, Amsterdam, pp. 1–80.

Caillard, D., Martin, J.L., 1976. In: Brandon, D.G. (Ed.), 1976. Electron Microscopy, vol. 1. TAL International Publishing Company, Jerusalem, pp. 587–593.

Caillard, D., Martin, J.L., 2003. Thermally-activated Mechanisms in Crystal Plasticity. Elsevier, Amsterdam.

Carrard, M., Martin, J.L., 1988. A study of (001) glide in [112] aluminium single crystals II. Microscopic mechanism. Philos. Mag. A58, 491–505.

Carrez, P., Ferré, D., Cordier, P., 2007. Implications for plastic flow in the deep mantle from modeling dislocations in MgSiO3 minerals. Nature 446, 68–70.

Chaussidon, J., Fivel, M., Rodney, D., 2006. The glide of screw dislocations in bcc Fe: atomistic static and dynamic simulations. Acta Mater. 54, 3407–3416.

Christian, J.W., 1983. Some surprising features of the plastic deformation of body-centered cubic metals and alloys. Metall. Trans. A 14, 1237–1256.

Christian, J.W., Vitek, V., 1970. Dislocations and stacking faults. Rep. Prog. Phys. 33, 307–411.

Clarebrough, L.M., Forwood, C.T., 1975. Direct observation of cross slip of Shockley partial dislocations. Phys. Status Solidi A32, K15–K16.

Cotterill, R.M.J., Doyama, M., 1966. Energy and atomic configuration of complete and dissociated dislocations. Phys. Rev. 145, 465–478.

Couret, A., Caillard, D., 1991. Dissociations and friction forces in metals and alloys. J. Phys. III 1, 885–907.

Couret, A., Crestou, J., Caillard, D., Clément, N., Coujou, A., Molénat, G., 2000. In situ straining experiments in TEM: application to the study of high performance materials. In: Lépinoux, D., Mazière, D., Pontikis, V., Saada, G. (Eds.), Multiscale Phenomena in Plasticty: from Experiments to Phenomenology, Modelling and Materials Engineering. Appl. Sci. E, 367, pp. 185–194.

Couteau, O., Kruml, T., Martin, J.L., 2011. About the activation volume for cross-slip in Cu at high stresses. Acta Mater. 59, 4207–4215.

Denoual, C., 2007. Modeling dislocation by coupling Peierls–Nabarro and element-free Galerkin methods. Comput. Methods Appl. Mech. Eng. 196, 1915–1923.

deWit, G., Koehler, J.S., 1959. Interaction of dislocations with an applied stress in anisotropic crystals. Phys. Rev. 116, 1113–1120.

Domain, C., Monnet, G., 2005. Simulation of screw dislocation motion in iron by molecular dynamics simulations. Phys. Rev. Lett. 95, 215506.

Dorn, J.E., Rajnak, S., 1964. Nucleation of kink pairs and the Peierls' mechanism of plastic deformation. Trans. Metall. Soc. AIME 230, 1052–1064.

Duesbery, M., 1969. The influence of core structure on dislocation mobility. Philos. Mag. 19, 501–526.

Duesbery, M.S., 1983a. On kinked screw dislocations in the b.c.c. lattice—I. The structure and Peierls stress of isolated kinks. Acta Metall. 31, 1747–1758.

Duesbery, M.S., 1983b. On kinked screw dislocations in the b.c.c. lattice—II. Kink energies and double kinks. Acta Metall. 31, 1759–1770.

Duesbery, M.S., 1984a. On non-glide stresses and their influence on the screw dislocation core in body-centred cubic metals. I. The Peierls stress. Proc. R. Soc. A 392, 145–173.

Duesbery, M.S., 1984b. On non-glide stresses and their influence on the screw dislocation core in body-centred cubic metals. II. The dislocation core. Proc. R. Soc. A 392, 175–197.

Duesbery, M.S., 1989. The dislocation core and plasticity. In: Nabarro, F.R.N., Hirth, J.P. (Eds.), 1989. Dislocations in Solids, vol. 8. Elsevier, Amsterdam, pp. 67–174.

Duesbery, M.S., Louat, N.P., Sadananda, K., 1992. The mechanics and energetics of cross-slip. Acta Metall. Mater. 40, 149–158.

Duesbery, M.S., 1998. Dislocation motion, constriction and cross-slip in FCC metals. Modell. Simul. Mater. Sci. Eng. 6, 35–49.

Duesbery, M.S., Vitek, V., Bowen, D.K., 1973. The effect of shear stress on the screw dislocation core structure in body-centred cubic lattices. Proc. R. Soc. A 332, 85–111.

Duesbery, M.S., Vitek, V., 1998. Plastic anisotropy in b.c.c. Trans. Met. Acta Mater. 46, 1481–1492.

Edagawa, K., Suzuki, T., Takeuchi, S., 1997. Motion of a screw dislocation in a two-dimensional Peierls potential. Phys. Rev., B 55, 6180–6187.

Escaig, B., 1968a. Dislocation Dynamics. McGraw-Hill, New York pp 655.

Escaig, B., 1968b. Sur le glissement dévié des dislocations dans la structure cubique à faces centrées. J. Phys. 29, 225–239.

Escaig, B., 1974. Dissociation and mechanical properties. Dislocation splitting and the plastic glide process in crystals. J. Phys. 35 (C7). C7–C151-C7–166.

Eshelby, J.D., 1949. Edge dislocations in anisotropic materials. Philos. Mag. 40, 903–912.

Eshelby, J.D., Stroh, A.H., 1951. Dislocations in thin plates. Philos. Mag. 42, 1401–1405.

Eshelby, J.D., 1962. The interaction of kinks and elastic waves. Proc. R. Soc. A 266, 222–246.

Eshelby, J.D., 1966. A simple derivation of the elastic field of an edge dislocation. Br. J. Appl. Phys. 17, 1131–1135.

Eshelby, J.D., Read, W.T., Shockley, W., 1953. Anisotropic elasticity with applications to dislocation theory. Acta Metall. 1, 251–259.

Finnis, M., 2003. Interatomic Forces in Condensed Matter. Oxford University Press, Oxford.

Fleischer, R.L., 1959. Cross slip of extended dislocations. Acta Metall. 7, 134–135.

Frank, F.C., 1951a. Capillarity equilibria of dislocated crystals. Acta Crystallogr. 4, 497–501.

Frank, F.C., 1951b. Crystal dislocations – elementary concepts and definitions. Philos. Mag. 42, 809–819.

Frank, F.C., Read, W.T., 1950. Multiplication processes for slow moving dislocations. Phys. Rev. 79, 722–723.

Frank, F.C., van der Merwe, J.H., 1949. One-dimensional dislocations. I. Static theory. Proc. R. Soc. A 198, 205–216.

Frederiksen, S.L., Jacobsen, K.W., 2003. Density functional theory studies of screw dislocation core structures in bcc metals. Philos. Mag. 83, 365–375.

Frenkel, J., 1926. Zur theorie der Elastizitätsgrenze und der Festigheit kristal linischer Körper. Z. Phys. 37, 572–609.

Frenkel, J., Kontorova, T., 1938. The model of dislocation in solid body. J. Exp. Theor. Phys. 8, 1340.

Friedel, J., 1956. Les Dislocations. Gauthier-Villars, Paris.

Friedel, J., 1964. Dislocations. Pergamon Press, New York.

Georges, A., Champier, G., 1980. On the cross-slip of isolated dislocations at the surface of silicon crystals. Scr. Metall. 14, 399–403.

Gibbons, D.F., 1964. In: Vogel (Ed.), Resonance and Relaxation in Metals. Plenum Press, New York.

Gordon, P.A., Neeraj, T., Li, Y., Li, J., 2010. Screw dislocation mobility in BCC metals: the role of the compact core on double-kink nucleation. Modell. Simul. Mater. Sci. Eng. 18, 085008.

Guyot, P., Dorn, J., 1967. A critical review of the Peierls mechanism. Can. J. Phys. 45, 983–1015.

Haasen, P., 1967. The stacking fault energies of the noble metals. In: Hasiguti, N.Y. (Ed.), Lattice Defects and Their Interactions, pp. 437–454.

Hansen, L.B., Stokbro, K., Lundqvist, B.I., Jacobsen, K.W., Deaven, D.M., 1995. Nature of dislocations in silicon. Phys. Rev. Lett. 75, 4444.

Hazzledine, P., Karnthaler, H., Wintmer, E., 1975. Non-parallel dissociation of dislocations in thin foils. Philos. Mag. 32, 81–97.

Head, A.K., 1953. The interaction of dislocations with boundaries. Philos. Mag. 44, 92–94.

Heidenreich, R.D., Shockley, W., 1948. Report of a Conference on Strength of Solids. Physical Society, London. p. 57.

Hirsch, P.B., 1960. Comment. Proceedings of the Fifth International Conference on Crystallography. Cambridge University Press. p. 139.

Hirsch, P.B., 1962. Extended jogs in dislocations in face-centred cubic metals. Philos. Mag. 7, 67–93.

Hirth, J.P., Jøssang, T., Lothe, J., 1966. Dislocation energies and the concept of line tension. J. Appl. Phys. 37, 110–116.

Hirth, J.P., Lothe, J., 1982. Theory of Dislocations, second ed. John Wiley and Sons, New York.

Hirth, J.P., 1985. A brief history of dislocation theory. Metall. Trans. A 16A, 2085–2090.

Honeycomb, R.W.K., 1968. The Plastic Deformation of Metals. St. Martin's Press, New York.

Hull, D., Bacon, D.J., 2001. Introduction to Dislocations. Butterworth-Heinemann, Oxford.

Ismail-Beigi, S., Arias, T.A., 2000. Ab initio study of screw dislocations in Mo and Ta: a new picture of plasticity in bcc transition metals. Phys. Rev. Lett. 84, 1499–1502.

Ito, K., Vitek, V., 2001. Atomistic study of non-Schmid effects in the plastic yielding of bcc metals. Philos. Mag. 81, 1387–1407.

Jackson, P.J., 1983. The role of cross-slip in the plastic deformation of crystals. Mater. Sci. Eng. 57, 39–47.

Joós, B., Duesbery, M.S., 1997a. The peierls stress of dislocations: an analytic formula. Phys. Rev. Lett. 78, 266–269.

Joós, B., Duesbery, M.S., 1997b. Dislocation kink migration energies and the Frenkel–Kontorova model. Phys. Rev., B 55, 11161.

Joós, B., Ren, Q., Duesbery, M.S., 1994. Peierls–Nabarro model of dislocations in silicon with generalized stacking-fault restoring forces. Phys. Rev., B 50, 5890–5998.

Joós, B., Zhou, J., 2001. The Peierls–Nabarro model and the mobility of the dislocation line. Philos. Mag. A 81, 1329–1340.

Jónsson, H., Mills, G., Jacobsen, K.W., 1998. Nudged elastic band method for finding minimum energy paths of transitions. In: Berne, B.J., Cicotti, G., Coker, D.F. (Eds.), Classical and Quantum Dynamics in Condensed Phase Simulations. World Scientific, Singapore. (Chapter 16).

Kaxiras, E., 2003. Atomic and Electronic Structure of Solids. Cambridge University Press, Cambridge.

Kaxiras, Duesbery, 1993. Free energies of generalized stacking faults in Si and implications for the brittle-ductile transition. Phys. Rev. Lett. 70, 3752–3755.

Klemenz, C., 1998. Hollow cores and step-bunching effects on (001) YBCO surfaces grown by liquid-phase epitaxy. J. Cryst. Growth 187, 221–227.

Kocks, U.F., Argon, A.S., Ashby, M.F., 1975. Thermodynamics and kinetics of slip. In: Chalmers, B., Christian, J.W., Massalski, T.B. (Eds.), 1975. Progress in Material Science, vol. 19. Pergamon Press, Oxford.

Koizumi, H., Kirchner, H.O.K., Suzuki, T., 1993. Kink pair nucleation and critical shear stress. Acta Metall. Mater. 41, 3483–3493.

Koizumi, H., Kirchner, H.O.K., Suzuki, T., 1994. Nucleation of trapezoidal kink pairs on a Peierls potential. Philos. Mag. 69, 805–820.

Krausz, A.S., Faucher, B., 1982. Energy-barrier systems in thermally activated plastic flow. Rev. Deform. Behav. Mater.s 4, 105–180.

Kubin, L.P., 1982. The low-temperature mechanical properties of B.C.C. Metals and their alloys. Rev. Deform. Behav. Mater. 4, 181–275.

Kubin, L., Hoc, T., Devincre, B., 2009. Dynamic recovery and its orientation dependence in face-centered cubic crystals. Acta Mater. 57, 2567–2575.

Kubin, L., 2013. Dislocations, Mesoscale Simulations and Plasticity. Oxford University Press, Oxford.

Kuramoto, E., Aono, Y., Kitajima, K., Maeda, K., Takeuchi, S., 1979. Thermally activated slip deformation between $0 \cdot 7$ and 77 K in high-purity iron single crystals. Philos. Mag. A 39, 717–724.

Lamb, H., 1928. Statics, Including Hydrostatics and the Elements of the Theory of Elasticity, third ed. Cambridge University Press, Cambridge, England.

Li, H., Wurster, S., Motz, C., Romaner, L., Ambrosch-Draxl, C., Pippan, R., 2012. Dislocation-core symmetry and slip planes in tungsten alloys: ab initio calculations and microcantilever bending experiments. Acta Mater. 60, 748–758.

Lothe, J., Hirth, J.P., 1959. Dislocation dynamics at low temperatures. Phys. Rev. 115, 543–550.

Love, A.E.H., 1920. A Treatise on the Mathematical Theory of Elasticity, third ed. Cambridge University Press, London.

Lu, G., Kioussis, N., Bulatov, V.V., Kaxiras, E., 2000a. Generalized-stacking-fault energy surface and dislocation properties of aluminum. Phys. Rev., B 62, 3099–3108.

Lu, G., Kioussis, N., Bulatov, V.V., Kaxiras, E., 2000b. The Peierls-Nabarro model revisited. Philos. Mag. Lett. 80, 675–682.

Lu, G., Bulatov, V.V., Kioussis, N., 2004. On stress assisted dislocation constriction and cross-slip. Int. J. Plast. 20, 447–458.

Madec, R., Devincre, B., Kubin, L., Hoc, T., Rodney, D., 2003. The role of the collinear interaction in dislocation-induced hardening. Science 301, 1879–1882.

Marcinkowski, M.J., Sadananda, K., Olson, N.J., 1974. Cross slip of extended dislocations in disordered face-centered-cubic alloys. Cryst. Lattice Defects 5, 187–198.

Marian, J., Cai, W., Bulatov, V.V., 2004. Dynamic transitions from smooth to rough to twinning in dislocation motion. Nat. Mater. 3, 158–163.

Mendelev, M.I., Han, S., Srolovitz, D.J., Ackland, G.J., Sun, D.Y., Asta, M., 2003. Development of new interatomic potentials appropriate for crystalline and liquid iron. Philos. Mag. 83, 3977–3994.

Mills, M.J., Baluc, N.L., Sarosi, P.M., 2006. HRTEM of dislocations cores and thin-foil effects in metals and intermetallic compounds. Microsc. Res. Tech. 69, 317–329.

Mishin, Y., Mehl, M.J., Papaconstantopoulos, D.A., Voter, A.F., Kress, J.D., 2001. Structural stability and lattice defects in copper: ab initio, tight-binding, and embedded-atom calculations. Phys. Rev., B 63, 224106.

Montagne, A., Tromas, C., Audurier, V., Woirgard, J., 2009. A new insight on reversible deformation and incipient plasticity during nanoindentation test in MgO. J. Mater. Res. 24, 883–889.

Mott, N.F., 1952. A theory of work-hardening of metal crystals. Philos. Mag. 43, 1151–1178.

Mrovec, M., Nguyen-Manh, D., Pettifor, D.G., Vitek, V., 2004. Bond-order potential for molybdenum: application to dislocation behavior. Phys. Rev., B 69, 094115.

Mrovec, M., Nguyen-Manh, D., Elsässer, C., Gumbsch, P., 2011. Magnetic bond-order potential for iron. Phys. Rev. Lett. 106, 246402.

Mughrabi, H., 1968. Electron microscope observations on the dislocation arrangement in deformed copper single crystals in the stress-applied state. Philos. Mag. 18, 1211–1217.

Nabarro, F.R.N., 1947. Dislocations in a simple cubic lattice. Proc. Phys. Soc., London 59, 256–272.

Nabarro, F.R.N., 1952. The mathematical theory of stationary dislocations. Adv. Phys. 1, 269–394.

Nabarro, F.R.N., 1967. Theory of Crystal Dislocations. Clarendon Press, Oxford.

Nabarro, F.R.N., 1997. Fifty-year study of the Peierls–Nabarro stress. Mater. Sci. Eng., A 234–236, 67–76.

Ngan, A.H.W., 1995. On generalizing the Peierls–Nabarro model for screw dislocations with non-planar cores. Philos. Mag. Lett. 72, 207–213.

Ngan, A.H.W., 1997. A generalized Peierls–Nabarro model for nonplanar screw dislocation cores. J. Mech. Phys. Solids 45, 903–921.

Ngan, A.H.W., Wen, M., 2001. Dislocation kink-pair energetics and pencil glide in body-centered-cubic crystals. Phys. Rev. Lett. 87, 075505.

Nowick, A.S., Berry, B.S., 1972. Anelastic Relaxation in Crystalline Solids. Academic Press, New York.

Orowan, E., 1934. Zur Kristallplastizität. Z. Phys. 89, 605–659.

Orowan, E., 1940. Problems of plastic gliding. Proc. Phys. Soc., London 52, 8–22.

Paidar, V., Pope, D.P., Vitek, V., 1984. A theory of the anomalous yield behavior in L1$_2$ ordered alloys. Acta Metall. 32, 435–448.

Peach, M., Koehler, J.S., 1950. The forces exerted on dislocations and the stress fields produced by them. Phys. Rev. 80, 436–439.

Peierls, R., 1940. The size of a dislocation. Proc. Phys. Soc., London 52, 34–37.

Perrin, R.C., Englert, A., Bullough, R., 1972. Extended defects in copper and their interactions with point defects. In: Gehlen, P.C., Beeler, J.R., Jaffee, R.I. (Eds.), Interatomic Potentials and Simulation of Lattice Defects. Plenum, New York, pp. 509–524.

Peyrard, M., Dauxois, T., 2004. Physique des Solitons. CNRS Editions, Paris.

Polanyi, M., 1934. Über eine Art Gitterstörung, die einen Kristall plastisch machen könnte. Z. Phys. 89, 660–664.

Püschl, W., 2002. Models for dislocation cross-slip in close-packed crystal structures: a critical review. Prog. Mater. Sci. 47, 415–461.

Rao, S., Parthasarathy, T.A., Woodward, C., 1999. Atomistic simulation of cross-slip processes in model fcc structures. Philos. Mag. 79, 1167–1192.

Rao, S., Dimiduk, D.M., El-Awady, J.A., Parthasarathy, T.A., Uchic, M.D., Woodward, C., 2009. Atomistic simulations of cross-slip nucleation at screw dislocation intersections in face-centered cubic nickel. Philos. Mag. 89, 3351–3369.

Rasmussen, T., Jacobsen, K.W., Leffers, T., Pederson, O.B., Srinivasan, S.G., Jónsson, H., 1997a. Atomistic determination of cross slip passway and energetics. Phys. Rev. Lett. 79, 3676.

Rasmussen, T., Jacobsen, K.W., Leffers, T., Pedersen, O.B., 1997b. Simulations of the atomic structure, energetics, and cross-slip of screw dislocations in copper. Phys. Rev., B 56, 2977–2990.

Ren, Q., Joós, B., Duesbery, M.S., 1995. Test of the Peierls–Nabarro model for dislocations in silicon. Phys. Rev., B 52, 13223–13228.

Rodney, D., 2004. Molecular dynamics simulation of screw dislocations interacting with interstitial frank loops in a model FCC crystal. Acta Mater. 52, 607–614.

Rodney, D., 2007. Activation enthalpy for kink-pair nucleation on dislocations: comparison between static and dynamic atomic-scale simulations. Phys. Rev., B 76, 144108.

Rodney, D., Martin, G., 2000. Dislocation pinning by glissile interstitial loops in a nickel crystal: a molecular-dynamics study. Phys. Rev., B 61, 8714–8725.

Rodney, D., Phillips, R., 1999. Structure and strength of dislocation junctions: an atomic level analysis. Phys. Rev. Lett. 82, 1704–1707.

Rodney, D., Proville, L., 2009. Stress-dependent Peierls potential: influence on kink-pair activation. Phys. Rev., B 79, 094108.

Romaner, L., Ambrosch-Draxl, C., Pippan, R., 2010. Effect of Rhenium on the dislocation core structure in Tungsten. Phys. Rev. Lett. 104, 195503.

Saada, G., 1991. Cross-slip and work hardening of f.c.c. crystals. Mater. Sci. Eng. A137, 177–183.

Saada, G., Douin, J., 1991. On the stability of dissociated dislocations. Philos. Mag. Lett. 64, 67–70.

Sachs, G., 1928. Zur Ableitung einer Fliessbedingung. Z. Ver. Dtsch. Ing. 72, 734–736.

Schmid, E., Boas, W., 1950. Plasticity of Crystals. Hughes, London.

Schoeck, G., Seeger, A., 1955. Activation energy problems associated with extended dislocations. The Physical Society – Bristol Conference on Defects in Crystalline Solids, 340–346.

Schoeck, G., 1980. Thermodynamics and thermal activation of dislocations. In: Nabarro, F.R.N. (Ed.), Dislocations in Solids. North-Holland Publishing Company, pp. 65–163.

Schoeck, G., 1994. The generalized Peierls–Nabarro model. Philos. Mag. A 69, 1085–1095.

Schoeck, G., 1999. The Peierls energy revisited. Philos. Mag. A 79, 2629–2636.

Schoeck, G., 2001. The core structure of dissociated dislocations in NiAl. Acta Mater. 49, 1179–1187.

Schoeck, G., 2005. The Peierls model: progress and limitations. Mater. Sci. Eng., A 400-401, 7–17.

Schoeck, G., 2006. The core structure of dislocations: Peierls model vs. atomic simulation. Acta Mater. 54, 4865–4870.

Schoeck, G., Krystian, M., 2005. The Peierls energy and kink energy in FCC metals. Philos. Mag. 84, 949–966.

Seeger, A., Schöck, G., 1953. Die aufspaltung von versetzungen in metallen dichtester kugelpackung. Acta Metall. 1, 519–530.

Seeger, A., Berner, R., Wolf, H., 1959. Die experimentelle Bestimmung von Stapelfehlerenergien kubisch-flächenzentrierter Metalle. Z. Phys. 155, 247–262.

Seeger, A., Wüthrich, C., 1976. Dislocation relaxation processes in body-centred cubic metals. Il Nuovo Cimento B 33, 38–75.

Seeger, A., 1981. The temperature and strain-rate dependence of the flow stress of body-centered cubic metals – a theory based on kink–kink interactions. Z. Metallkd. 72, 369–380.

Seeger, A., 1995. The flow stress of high-purity refractory body-centred cubic metals and its modification by atomic defects. J. Phys. IV France 5. C7–C45-C7–65.

Seeger, A., Schiller, P., 1962. The formation and diffusion of kinks as the fundamental process of dislocation movement in internal friction measurements. Acta Metall. 10, 348–357.

Silcox, J., Hirsch, J.B., 1959. Direct observations of defects in quenched gold. Philos. Mag. 4, 72–89.

Simonelli, G., Pasianot, R., Savino, E., 1993. Embedded-atom-method interatomic potentials for bcc-iron. Mater. Res. Soc. Symp. Proc. 291, 567–572.

Singh, B.N., Edwards, D.J., Toft, P., 2001. Effect of neutron irradiation and post-irradiation annealing on microstructure and mechanical properties of OFHC-copper. J. Nucl. Mater. 229, 205–218.

Somigliana, C., 1914. Sulla teoria delle distorsioni elastiche. Atti Accad. Naz. Lincei 23, 463–472.

Somigliana, C., 1915. Sulla teoria delle distorsioni elastiche. Atti Accad. Naz. Lincei 24, 655–666.

Szpiro, G.G., 2003. Kepler's Conjecture. John Wiley & Sons, New York.

Shenoy, V.B., Kukta, R.V., Phillips, R., 2000. Mesoscopic analysis of structure and strength of dislocation junctions in FCC metals. Phys. Rev. Lett. 84, 1491–1494.

Stobbs, W.M., Sworn, C.H., 1971. The weak beam technique as applied to the determination of the stacking-fault energy of copper. Philos. Mag. 24, 1365–1381.

Stroh, A.N., 1954. Constrictions and jogs in extended dislocations. Proc. Phys. Soc., London B67, 427–436.

Suzuki, H., 1968. Dislocation Dynamics. McGraw-Hill, New York.

Suzuki, T., Koizumi, H., Kirchner, H.O.K., 1995. Plastic flow stress of b.c.c. transition metals and the Peierls potential. Acta Metall. Mater. 43, 2177–2187.

Takeuchi, S., 1979. Core structure of a screw dislocation in the b.c.c. lattice and its relation to slip behaviour of α-iron. Philos. Mag. A 39, 661–671.

Takeuchi, S., Kuramoto, E., 1975. Thermally activated motion of a screw dislocation in a model b.c.c. crystal. J. Phys. Soc. Jpn. 38, 480–487.

Tang, M., Kubin, L.P., Canova, G.R., 1998. Dislocation mobility and the mechanical response of b.c.c. single crystals: a mesoscopic approach. Acta Mater. 46, 3221–3235.

Taylor, G.I., 1928. The deformation of crystals of β-brass. Proc. R. Soc. London A 118, 1–24.

Taylor, G.I., 1934. The mechanisms of plastic deformation of crystals. Proc. R. Soc. London A 145, 362–387.

Taylor, G.I., 1938. Plastic strain in metals. J. Inst. Met. 62, 307–324.

Taylor, G., 1992. Thermally-activated deformation of s and alloys. Prog. Mater. Sci. 36, 29–61.

Teodosiu, C., 1982. Elastic Models of Crystal Defects. Springer-Verlag, Berlin. p. 112.

Teutonico, L.J., 1964. The interaction of dislocations in anisotropic face-centred cubic crystals. Philos. Mag. 10, 401–421.

Thompson, N., 1953. Dislocation nodes in face-centred cubic lattices. Proc. Phys. Soc. B 66, 481–492.

Timoshenko, S., Goodier, J.N., 1951. Theory of Elasticity, second ed. McGraw-Hill, New York.

Tromas, C., Colin, J., Coupeau, C., Girard, J.C., Woirgard, J., Grilhe, J., 1999. Pop-in phenomenon during nanoindentation in MgO. Eur. Phys. J.-Appl. Phys. 8, 123–128.

Tromas, C., Gaillard, Y., Woirgard, J., 2006. Nucleation of dislocations during nanoindentation in MgO. Philos. Mag. 86, 5595–5606.

Umakoshi, Y., Pope, D.P., Vitek, V., 1984. The asymmetry of the flox stress in $Ni_3(Al, Ta)$ single crystals. Acta Metall. 32, 449–456.

Ungar, T., Mughrabi, M., Wilkens, M., 1982. An X-Ray line-broadening study of dislocations near the surface and in the bulk of deformed copper single crystals. Acta Metall. 30, 1861–1867.

Vanderschaeve, G., Escaig, B., 1983. Cross-slip of twinning dislocations on crossing domain boundaries in ordered Ni_3V. Philos. Mag. A48, 265–277.

Vegge, T., Rasmussen, T., Leffers, T., Pedersen, O.B., Jacobsen, K.W., 2001. Atomistic simulations of cross-slip of jogged screw dislocations in copper. Philos. Mag. 81, 137–144.

Ventelon, L., Willaime, F., 2007. Core structure and Peierls potential of screw dislocations in α-Fe from first principles: cluster versus dipole approaches. J. Comput.-Aided Mater. Des. 14, 85–94.

Ventelon, L., 2008. Simulation ab initio des coeurs de dislocation vis dans le Fer, Thèse de doctorat, Université Lyon 1.

Ventelon, L., Willaime, F., Leyronnas, P., 2009. Atomistic simulation of single kinks of screw dislocations in a-Fe. J. Nucl. Mater. 386–388, 26–29.

Ventelon, L., Willaime, F., 2010. Generalized stacking-faults and screw dislocation core-structure in bcc iron: a comparison between ab initio calculations and empirical potentials. Philos. Mag. 90, 1063–1074.

Veyssière, P., 2000. Yield stress anomalies in ordered alloys: a review of microstructural findings and related hypotheses. Mater. Sci. Eng. A309, 44–48.

Veyssière, P., 1989. Transmission electron microscope observation of dislocations in ordered intermetallic alloys and the flow stress anomaly. In: Liu, C.T., Taub, A. I., Stoloff, N.S., and Koch, C.C. (eds). Mrs Symposium Proceedings on High Temperature Ordered Intermetallic Alloys 133, 175–181.

Vitek, V., 1974. Theory of core structures of dislocations in bcc metals. Cryst. Lattice Defects 5, 1–34.

Vitek, V., 1968. Intrinsic stacking faults in body-centred cubic crystals. Philos. Mag. A 18, 773–786.

Vitek, V., 1992. Structure of dislocation cores in metallic materials and its impact on their plastic behavior. Prog. Mater. Sci. 36, 1–27.

Vitek, V., 2004. Core structure of screw dislocations in body-centred cubic metals: relation to symmetry and interatomic bonding. Philos. Mag. 84, 415–428.

Vitek, V., Mrovec, M., Bassani, J.L., 2004. Influence of non-glide stresses on plastic flow: from atomistic to continuum modeling. Mater. Sci. Eng., A 365, 31–37.

Volterra, V., 1907. Sur l'équilibre des corps élastiques multiplement connexes. Ann. Sci. Ec. Norm. Supér. 24, 401–517.

Wang, G., Strachan, A., Cagin, T., Goddard, W.A., 2003. Atomistic simulations of kinks in 1/2a $\langle 111 \rangle$ screw dislocations in bcc tantalum. Phys. Rev., B 68, 224101.

Washburn, J., 1965. Intersection cross slip. Appl. Phys. Lett. 7, 183–185.

Wen, M., Ngan, A., 2000. Atomistic simulation of kink-pairs of screw dislocations in body-centred cubic iron. Acta Mater. 48, 4255–4265.

Weertman, J., Weertman, J.R., 1967. Elementary Dislocation Theory. Macmillan, London. p. 99.

Whelan, M.J., 1958. Dislocation interactions in face-centered cubic metals, with particular reference to stainless steel. Proc. Phys. Soc., A 249, 114–137.

Wolf, H., 1960. Die Aktivierungsenergie für dis Quergleitung aufgespaltener Schraubenversetzungen. Z. Phys. 15A, 180–193.

Woodward, C., Rao, S.I., 2002. Flexible *ab initio* boundary conditions: simulating isolated dislocations in bcc Mo and Ta. Phys. Rev. Lett. 88, 216402.

Woodward, C., Trinkle, D.R., Hector, L.G., Olmsted, D.L., 2008. Prediction of dislocation cores in aluminum from density functional theory. Phys. Rev. Lett. 100, 045507.

Yang, L.H., Soderlind, P., Moriarty, J.A., 2001. Accurate atomistic simulation of (a/2) $\langle 111 \rangle$ screw dislocations and other defects in bcc tantalum. Philos. Mag. A 81, 1355–1385.

Yoo, M.H., 1987. Stability of superdislocations and shear faults in $L1_2$ ordered alloys. Acta Metall. 35, 1559–1569.

Zener, C., 1948. Elasticity and Anelasticity of Metals. Univ. of Chicago Press.

Zhang, H.Z., Mann, E., Seeger, A., 1989. Approximate solution of the Peierls equation for screw dislocations with threefold symmetry. Phys. Status Solidi B 153, 465–477.

Zhou, S.J., Preston, D.L., Lomdahl, P.S., Beazley, D.M., 1998. Large-scale molecular dynamics simulations of dislocation intersection in copper. Science 279, 1525–1527.

Biography

David Rodney performed his Ph.D. between the CEA Saclay, France, and Brown University, USA. After a post-doc at ONERA, France, he joined the Grenoble Institute of Technology in 2001 as assistant Professor. In 2008, he spent a sabbatical at M.I.T. and became Professor at the University of Lyon in 2013. His research is focused on the multiscale modelling of plasticity in crystalline metals, amorphous glasses and fibrous entangled materials. He is associate editor for *Acta Materialia*, Elsevier, and member of editorial board of *Scientific Reports*, Nature Publishing group.

Joël Bonneville passed his Ph.D. thesis in 1978 at the University of Lille I in France. After two years as a sales engineer (CVC company, Rochester, USA), he joined in 1980 the laboratory of Physics Metallurgy at the Ecole Polytechnique Fédérale de Lausanne (Switzerland), where he obtained in 1985 a Doctorate in Materials Science. Since 1999, he is Professor at the University of Poitiers, France. His research is focused on the understanding of materials mechanical properties in terms of their microstructures. His relatively new research topics include the characterisation and modelling of deformation behaviour of quasicrystalline materials and related metal matrix composites.

17 Plastic Deformation of Metals and Alloys

N. Hansen, Danish–Chinese Center for Nanometals, Section for Materials Science and Advanced Characterization, Technical University of Denmark, Roskilde, Denmark
C.Y. Barlow, Institute for Manufacturing, University of Cambridge, Cambridge, UK

Glossary

Bamboo incidental dislocation boundary An incidental dislocation boundary observed at large strain that is connected to lamellar boundaries on two edges.

Bamboo structure Two roughly parallel GNBs linked by IDBs.

Cell A roughly equiaxed volume in which the dislocation density is well below the average and which is rigidly rotated from its neighboring volumes.

Cell block (CB) A contiguous group of cells in which the same set of glide systems operate.

Cell boundaries Low-angle dislocation boundaries that surround the cells and are classified as incidental dislocation boundaries.

Climb Nonconservative dislocation motion involving interaction with vacancies to allow a dislocation to move out of its glide plane.

Cross-slip Transfer of screw dislocations from one glide plane to another which contains the same slip direction.

Crystallite A crystal volume with different orientation from its neighbors.

Deformation band A wide region subdivided by cell blocks rotated relative to neighboring regions from which it is separated by parallel transition bands.

Dense dislocation wall (DDW) A single, nearly planar deformation induced boundary with a high dislocation density, classified as a geometrically necessary boundary, enclosing a cell block at small to intermediate strains.

Domain A coarse region subdivided by cell blocks rotated relative to neighboring regions from which it is separated by diffuse boundaries.

Extrinsic dislocations in boundary Segments of matrix dislocations lying partly in the plane of a grain boundary.

Geometrically necessary boundary (GNB) A boundary whose angular misorientation is controlled by the difference in glide-induced lattice rotations in the adjoining volumes.

Grain A crystal within a polycrystalline sample.

High-angle boundary (HAB) A boundary with misorientation above 15°.

Incidental dislocation boundary (IDB) A dislocation boundary formed by the mutual and statistical trapping of glide dislocations.

Intrinsic dislocations in boundary Dislocations which essentially constitute a boundary.

Lamellar boundary (LB) A single, nearly planar boundary, classified as a GNB, enclosing a narrow cell block at large strain.

Microband (MB) Plate-like regime formed by two closely spaced DDWs and defining the edge of a cell block.

Microband (MB), **first generation** MBs which are developing through the splitting of dense dislocation walls (DDWs) over part or all of their length.

Microband (MB), **second generation** MBs which are narrow plate like zones passing through one or more pre-existing dislocation cells along one of the crystallographic slip planes. Shear offsets are formed where these MBs intersect other walls

Misorientation Change in crystallographic orientation across a boundary.

Misorientation angle The orientation difference across a boundary is characterized by the misorientation angle and the misorientation axis. Taking into account the symmetry of the lattice (e.g. 24 alternative descriptions for cubic symmetry) the smallest of all misorientation angles is selected. This angle is called the misorientation or disorientation angle, and is used with the appropriate corresponding misorientation axis.

S-band A band of localized glide that intersects parallel groups of DDWs, MBs or LBs creating a string of S-shaped perturbations in those boundaries. The length of an S-band is generally shorter than a grain diameter.

Subgrain (SG) A nearly dislocation-free volume surrounded by higher-angle boundaries.

Shear band A region of intense local shear that typically spans several grains.

Taylor lattice (TL) Relatively uniform distribution of mainly edge dislocations composed of one or more sets of dislocations of alternating sign.

Texture The distribution of crystallographic orientations within a sample.

Transition band A narrow region separating two deformation bands, subdivided by small pancake-shaped cells and carrying a cumulative orientation change.

17.1   Introduction

Plastic deformation of metals involves macroscopic changes to the geometrical shape of samples and structures. The way in which the material responds to imposed stresses and strains is determined by its microstructure and texture. This chapter focuses on relationships between the microstructure of plastic deformation and mechanical properties, with particular regard to flow stress. The fundamental mechanisms of plastic deformation are first outlined, leading to discussion of the way in which deformation structures evolve from the interactions between single dislocations. The characterization and analysis of the complex structures, which are characteristic of polycrystalline deformation, and are also observed in single crystals, are then described. All these factors are then drawn together to make correlations between a wide range of microstructural features and mechanical properties.

The discussion is limited to deformation of fcc and bcc alloys of medium to high stacking-fault energies and containing no more than a low volume fraction of second-phase particles. Only cold deformation is considered, and at strain-rates where diffusional processes can be regarded as negligible. We consider strain introduced by a variety of deformation processes, including monotonic processes such as tension, torsion, rolling and extrusion, together with less conventional processes such as high-pressure torsion (HPT) and friction for higher strains.

Plastic deformation takes place predominantly by the movement of dislocations. Accumulation of dislocations within the material is accompanied by progressive refinement of the microstructure, which is seen as reduction in the size of subgrains and grains. This is a consequence of rotation of the material in a heterogeneous way. Analysis of the deformation structures therefore requires data not only on the spacing between microstructural features but also on the crystallographic orientations within the structure.

The required level of analysis has been enabled by the development of a range of sophisticated techniques, which are described briefly. Central to this has been Transmission Electron Microscopy (TEM) with its associated range of analytical techniques.

The quantification of structural parameters forms the basis for an analysis of strengthening mechanisms and parameters especially related to the presence of dislocations and boundaries, which range from dislocation boundaries to grain boundaries. The effect of structural parameters on the flow stress is then derived and related to the observed work-hardening stages.

17.2 Plastic Deformation Processes

Plastic deformation is an intrinsic part of the processing of most metals. The aim is to achieve shape change by means of externally applied stress, whilst at the same time causing a controlled alteration in material mechanical properties. In this section we will briefly review the main classes of metalworking processes, concentrating on their relationship with development of microstructure.

17.2.1 Classification and Characterization

Plastic forming processes are traditionally divided broadly into hot and cold deformation, and into specific individual processes such as rolling, extrusion, wire-drawing, as summarized in **Table 1**. These metal-working processes may have different objectives; for example, breaking down a cast structure by hot-rolling, or shaping a component by stretch-forming. During hot-deformation or during the subsequent cooling the metal may recrystallize, leaving a material that is almost dislocation-free on the microstructural scale. However, worked metal or alloy more typically contains crystallites and dislocation structures that are characteristic of the process parameters such as stress, strain, strain-rate and temperature. The term crystallite is a structural parameter defined as a crystal volume with different orientation from its neighbors. Crystallites may be cells, subgrains or grains, and be surrounded by low, medium or high-angle boundaries, or a mixture. Besides these structures, a crystallographic texture characterizes the deformed metal. The microstructure and the texture may be homogeneous throughout the worked sample, but more typically gradients will build up giving rise to heterogeneity of both microstructure and texture. The characteristic length-scales of the heterogeneous structures may vary,

Table 1   Metal-forming processes

Process type	Macro-scale and microscale effects
First-stage shaping processes: hot deformation.	
Forging; rolling; extrusion	Large-scale shape change
	Removes porosity
	Dynamic recrystallization
	Solutionize to give chemically homogeneous material
	Distribute second-phase particles parallel to working direction
	Generates uniform equiaxed crystallite structure of required grain size
	Crystallographic texture generated
Second-stage shaping processes: cold deformation.	
Forging; rolling; wire-drawing; sheet-forming processes including deep drawing, stretch forming, bending, shearing	Finer-scale shape change
	Work-hardening: Formation of deformation microstructures including single dislocations, dislocation walls, twins
	Generally produces crystallite structures elongated parallel to straining direction
	Generates texture

and can for example be dependent on the sample size, the grain size, or a scale dependent on the distribution of dislocations and second-phase particles. In all cases, the sample is then found to contain residual elastic stress. The deformation structures are influenced by the presence of second-phase particles, which are present in the material. These particles may take a wide range of sizes, distributions and species, and detailed consideration of all possible contributions to deformation behavior is outside the scope of this chapter. Some mention will be made of the way in which low concentrations of small particles alter the evolution of deformation microstructures, using examples of dispersion-strengthening.

In recent years, a number of nontraditional processes have been added, all based on shear or plane strain compression. These have been developed specifically in order to produce very fine-grain structures in materials by applying very large plastic strains (5–10, or higher), with the aim of increasing strength. These processes are characterized by the desire to introduce high internal strains whilst substantially retaining the original sample dimensions. The aim is therefore purely to introduce large amounts of redundant work into the materials. The characteristics of the main processes are outlined in **Table 2**.

17.2.2   Process Parameters

The broad distinction between hot and cold working relates to the way in which the structure develops and the mechanical properties change. In hot working the material does not work-harden because there is a high rate of recovery involving a range of processes: the material reaches steady state with a low or zero rate of damage accumulation. Hot working is therefore favored when large amounts of deformation are required. The absence of work-hardening means that hot-working processes predominantly involve compression and shear and not tension, since work-hardening is required to stabilize tensile deformation. With cold metal-working processes involving significant tension (such as wire-drawing)

Table 2   Principal nontraditional deformation processes

Process	Process outline
Accumulative roll bonding (Saito et al., 1999; Tsuji et al., 2003, 2004)	Sheet is rolled, stacked, rolled repeatedly. Wire-brushing assists bonding between stacked sheets
Cyclic extrusion compression (Tsuji et al., 2003, 2004)	Material extruded back and forth through die
Equal channel angular extrusion (Tsuji et al., 2003, 2004; Valiev et al., 2000)	Material repeatedly extruded around bend (90° or other angle)
High-pressure torsion (Tsuji et al., 2003, 2004)	Sample rotated in die, introducing high shear strain without altering sample dimensions
Multidirectional deformation (Tsuji et al., 2003, 2004; Armstrong et al., 1982)	Typically compression, performed sequentially along three orthogonal axes.
Repetitive corrugation and straightening (Tsuji et al., 2003, 2004; Huang et al., 2001)	Rolling process in which deformation is introduced using shaped rolls and removed on subsequent pass on flat rolls.

the amount of reduction that can be supplied in a single stage is limited by work-hardening behavior. Cold-deformation processes involving high strain, such as those listed in **Table 2**, therefore apply compression and shear rather than tension.

Strain rate effects are linked with temperature effects using the Zener–Holloman relationship:

$$Z = \dot{\varepsilon} \exp(Q/RT), \tag{1}$$

where Z is the Zener–Holloman parameter, $\dot{\varepsilon}$ is the strain rate, Q the activation energy, R the gas constant and T the absolute temperature (Kalpakjian and Schmid, 2003).

The parameter is applicable particularly to the characterization of metal-working processes where it can help with predictive modeling of process parameters However, it should be noted that altering the strain rate may change the nature of the deformation; this is discussed further in Section 17.3.4.

17.2.3 Process Efficiency

The mechanical work done on the sample undergoing shape change is given by the volume integral over the strain range:

$$W = \int_V \sigma \, d\varepsilon. \tag{2}$$

This is the minimum energy, which is needed to deform the material. In practice the energy required is always higher, because deformation is accompanied by *redundant work*, which results in energy dissipation and storage.

The deformation needed to introduce a required shape change in a sample of material can be analyzed using the first law of thermodynamics:

$$\Delta U = W + Q, \tag{3}$$

where ΔU is the change in internal energy of the body, W is the mechanical work done on the sample and Q is the heat effect associated with the deformation.

The plastic deformation is typically accompanied by a temperature rise, so Q has a negative value in Eqn (3). In most cases, this term dominates the energy requirement for deformation, so that most of the mechanical energy supplied is dissipated as heat.

The internal energy change ΔU relates to energy stored in the sample, predominantly in the form of dislocations and walls resulting in a variety of microstructural features. The amount of redundant work is dependent on the material, temperature, and also by the characteristics of the metalworking process.

Very few metalworking processes produce completely homogeneous deformation of the material. When the levels of plastic deformation in cold-worked material do vary spatially, the material is always found to contain residual elastic stress. The residual stresses balance through the material: regions of tensile stress are matched by regions of compressive stress elsewhere. The stresses are frequently of the order of the material yield stress. The energy associated with such strain fields is small, and is included in the term ΔU.

The above analysis includes only the energy balance for the system narrowly defined as the sample under deformation. In practice, extra energy is also expended in overcoming external friction with any surfaces (e.g. dies, rolls) with which the material is in contact.

17.3   Role of Dislocations in Plastic Deformation

17.3.1   Strain Using Dislocations

Plastic deformation of metals takes place predominantly by shearing: lattice planes in the material slide over each other, allowing macroscopic shape change without appreciably affecting the ordering and arrangements of atoms within the structure. The stress to cause plastic deformation can be reduced by a factor of 1000 if, rather than moving complete lattice planes simultaneously, deformation can be localized by the movement of line defects, which are dislocations (e.g. Hull and Bacon, 2011). For this reason, plastic deformation of metals depends on the generation and subsequent movement of dislocations. Metals, even in the annealed state, contain a statistical density of dislocations (which can be determined using thermodynamic principles) (Honeycombe, 1975), which is sufficient to allow plastic deformation to take place by this mechanism.

A characteristic intrinsic parameter of a dislocation is its Burgers Vector **b**, which is the amount of shear that it can produce by moving through the material. Dislocations can take a range of geometries (Hull and Bacon, 2011), and are classified into edge-type (movement resulting in deformation normal to the line defect) and screw-type (deformation parallel to the line defect). Mixed dislocations have intermediate character.

17.3.2   Dislocation Energy

Dislocations have intrinsic elastic strain energy, which may be estimated by integrating the contributions resulting from the displacements of atoms from their equilibrium lattice positions. This displacement decreases with distance r from the dislocation core, and is modeled as being proportional to $1/r$ (Hull and Bacon, 2011). The governing equation for the energy U_D per unit length of dislocation is:

$$U_D = \frac{G\mathbf{b}^2}{4\pi} f(v) \ln \frac{R}{r_0}, \tag{4}$$

where G is the shear modulus and **b** the Burgers vector, and $f(v)$ is a constant involving Poisson's ratio, v. This takes the value 1 for screw dislocations, and $\{1/(1-v)\}$ for edge dislocations. For a typical value of

$v = 0.35$, this leads to the energy of an edge dislocation being greater than that of a screw dislocation by a factor of about 1.5 (Hull and Bacon, 2011). R and r_0 are distances from the dislocation core, which have to be determined. The smaller of these values, r_0, is defined as the core radius, which is of the order of the atomic spacing, equivalent also to the Burgers vector **b**. The appropriate value to be used for the upper cutoff radius R is less clearly defined, since for a single dislocation embedded within a single crystal the disruption to atomic positions extends to the edge of the crystal. However, in a deformed structure that typically contains a large number of dislocations in low-energy configurations (Bever et al., 1973), the influence of one dislocation can be approximated as extending only as far as the next dislocation. The average dislocation spacing λ for a uniformly distributed three-dimensional array of dislocations of density ρ is approximated by

$$\lambda \approx \frac{1}{\sqrt{\rho}}. \tag{5}$$

Dislocations more commonly cluster into low-energy arrays (Section 17.5.3), and the appropriate value to choose for the dislocation spacing is then the spacing of these features. Fortunately the equation for dislocation energy U_D is relatively insensitive to the exact value of R, and for most purposes an average value is used, which gives an empirically appropriate solution. The appropriate value of R is usually of the order of 1000**b**, corresponding to a few tenths of µm.

In conclusion, the line energy per unit length for a dislocation, often termed the *line tension*, may be written as

$$U_D \approx 0.5 \ f(v) \ Gb^2. \tag{6}$$

One consequence of this function is that dislocations tend to have the minimum available value of **b** in the structure, which is the shortest separation between atoms in a close-packed direction.

A further consequence relates to deformation microstructures. Because systems tend to adopt low-energy configurations, they will minimize their total dislocation line length. This is achieved by dislocations being straight rather than curved, and by dislocations interacting and annihilating whenever possible. These factors contribute the development of low-energy dislocation structures (LEDS) which are discussed in Section 17.5.3.

17.3.3 Crystallography of Slip

In order to generate plastic strain, dislocations glide on close-packed planes. These planes are the most widely separated in a structure, so by a simple geometric argument it can be seen that the stress required for shear deformation is lowest for sliding these planes over each other. Dislocations therefore generally move in close-packed directions, and on close-packed planes. Conservative motion of dislocations takes place when dislocations are able to glide in the required direction on a slip plane that contains both their line vector and the Burgers vector; the passage of the dislocation leaves perfect material behind.

Crystals contain a number of possible slip systems (crystallographically equivalent combinations of slip direction and slip plane containing that slip direction). The applied stress direction determines which systems are activated: they will normally be the direction and plane in which the resolved component of the shear stress reaches its highest value (Honeycombe, 1975). Crystals may deform by single slip (one slip system dominating), or multiple slip (two or more slip systems operating simultaneously). The two regimes are distinct: with single slip, dislocations pass through the material without significantly hindering each other; with multiple slip there is immediately conflict as dislocations interact with each other. These give rise to different work-hardening rates. Dislocations cannot pass through each other without interacting and leaving segments of themselves on the other dislocation, *jogs*; these are in general incompatible with the slip plane of the host dislocation, and will leave trails of point defects if they are dragged through the lattice (Hull and Bacon, 2011). Dislocation movement of this type is referred to as nonconservative. Dislocations trapped within the material and intersecting a slip plane may be termed forest dislocations.

Slip systems in face-centered cubic, fcc, materials take the form $\langle 110 \rangle \{1\bar{1}1\}$; there are 12 distinguishable, equivalent systems. This large number of fully close-packed slip systems allows fcc materials to exhibit high ductility at all temperatures and under all loading conditions.

Body-centered cubic, bcc, materials contain $\langle 111 \rangle$ close-packed directions, but there are no fully close-packed planes, so slip systems in these materials are more variable. The most commonly observed plane is $\{110\}$, but other planes that have been identified are $\{112\}$ (12 slip systems) and $\{123\}$ (24 slip systems) (Honeycombe, 1975; Hull and Bacon, 2011).

The behavior of hexagonal close packed, hcp, materials is heavily influenced by the geometry of the unit cell, in the form of the ratio between c and a, the principal crystallographic unit cell parameters. Slip is predominantly in close-packed directions $\langle 11\bar{2}0 \rangle$, but the slip plane is not always the (0001) basal close-packed plane. A wide range of slip systems has been observed, using the pyramidal planes $\{10\bar{1}1\}, \{10\bar{1}2\}, \{11\bar{2}1\}, \{11\bar{2}2\}$ (Honeycombe, 1975; Hull and Bacon, 2011). Hcp materials will not be considered further in this chapter but it has recently been observed (Yang et al., 2010) that graded microstructures evolving during cold rolling of Ti can be interpreted within the same universal framework for structural evolution presented below for fcc and bcc metals.

Plastic deformation begins when the stress on dislocations reaches a critical level, the critical resolved shear stress τ_{CRSS}. The macroscopic yield stress σ_y required to cause plastic strain is naturally related to this critical stress experienced by the dislocation on its slip plane. However, estimating the relationship between the two requires knowledge of the range of crystal orientations within the material, i.e. the texture.

17.3.4 Stress for Dislocation Movement

The stress τ_P to move a dislocation through an otherwise perfect lattice is given by the Peierls–Nabarro equation:

$$\tau_P = 3G \exp - (2\pi w/\mathbf{b}), \tag{7}$$

where G is the lattice shear modulus and $\mathbf{b}$ the dislocation Burgers vector; w is the dislocation width, typically up to about 10 atomic spacings in metals (Hull and Bacon, 2011).

The stress to move a dislocation is increased by any lattice imperfections, either through interaction between elastic strain fields of the dislocation and the imperfections, or through physical obstacles to dislocation movement, such as incoherent precipitates and dispersed particles, high-angle grain boundaries and twin boundaries (Kubin, 2013).

A feature of impedance to dislocation movement arising from elastic interactions is that given sufficient applied stress, the dislocations are still able to move, although other deformation or stress-release mechanisms may intervene, such as twinning or fracture. By contrast, dislocations cannot penetrate the physical obstacles. At grain boundaries, this leads to pileups and in some cases the

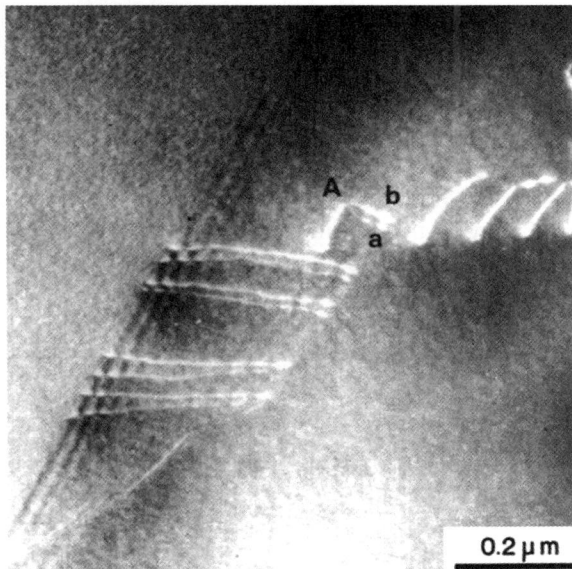

Figure 1 Grain boundary dislocations in lightly deformed steel. Dissociation of matrix dislocation A into two extrinsic dislocations a and b. Hansen (1985); Figure 7.

dislocations may enter the interface and become incorporated into the structure (e.g. extrinsic dislocations in grain boundaries, **Figure 1**) (Ralph et al., 1981). Dislocations can form pileups and arrays around second-phase particles, but they can bypass small particles by cross-slipping.

Interdislocation interaction to form jogs is a special case. Jogs can move by nonconservative processes involving the generation of point defects, but if alternative mechanisms are available (such as dislocation multiplication by bowing) then they may be immobile.

17.3.5 Temperature and Strain-Rate Effects

The governing equation for the interrelationship between temperature and strain-rate has been introduced as the Zener–Holloman parameter in 17.2.2. However, it should be noted that when materials have the possibility of using a range of crystallographically different slip systems as described in 17.3.3 then they can demonstrate more complex temperature and strain-rate dependence. A familiar example is that of pure or low-carbon iron, ferrite, which has a bcc structure, and which can switch between ductile and brittle behavior in the vicinity of room temperature. The effect occurs in bcc materials because operation of even the lowest-stress slip systems requires some thermal activation, and this can become quenched-out at low temperatures.

17.3.6 Dislocation Multiplication and Formation of Boundaries

17.3.6.1 Frank-Read Sources

As plastic strain increases, so also does the stress required to cause plastic deformation; this is the phenomenon of work-hardening. A proportion of dislocations do not disappear from the structure after moving through it and causing plastic deformation; they become trapped inside it. Plastic deformation is therefore accompanied by an increase in dislocation density in the material. Annealed material typically contains a dislocation density of 10^8–10^{10} m/m^3; this will typically rise to about 10^{13}–10^{14} on straining to about 0.1, and continues to increase to above 10^{15} as the strain increases further.

Dislocations are nucleated at Frank-Read sources, which involve a section of dislocation length L which is pinned at two points (Hull and Bacon, 2011). Sources may be at boundaries (low-angle or high-angle boundaries; interphase boundaries), between incoherent particles, or at pinned dislocation sections within grains, where L may be the spacing of forest dislocations, as defined in Section 17.3.3. Sources operate when the applied stress is high enough to bend the dislocation section into a semi-circular arc. For a dislocation link of length L, where L is the separation between pinning points, the required stress to cause this supercritical bowing is given by Hull and Bacon (2011).

$$\tau_{FR} = \frac{U_D}{\mathbf{b}L/2}. \tag{8}$$

Together with Eqn (6) this gives an approximate value for operation of a Frank-Read source of

$$\tau_{FR} \approx \frac{G\mathbf{b}}{L}. \tag{9}$$

17.3.6.2 Formation of Boundaries

Dislocations within the structure can reduce their elastic strain energy by forming arrays or boundaries. The boundaries are associated with changes in the lattice orientation, and may form cell, subgrain or grain boundaries within the material. For small-angle boundaries, the misorientation angle across a dislocation boundary is approximated by $\theta = \mathbf{b}/h$ where h is the spacing of dislocations in the wall. Boundaries may be tilt or twist in character, depending on whether the dislocations are edge or screw respectively. With increasing plastic strain, the misorientation between neighboring crystallites increases and at large strain, deformation-induced high-angle boundaries supplement original grain boundaries. For an ideal dislocation boundary free of long-range stresses its misorientation angle (θ), axis (R) and its Burgers vector content are related according to Frank's formula (Frank, 1950):

$$B = (r \times R)2 \sin \theta/2, \tag{10}$$

where the vector, r, represents an arbitrary straight line lying in the plane of a boundary which contains the dislocation network and B is the sum of the Burgers vectors of all the dislocations intersected by r.

17.3.7 Influence of Stacking-Fault Energy

The evolution of the dislocation structure depends on the nature of deformation and on the material, in particular whether it has a low or high stacking-fault energy, and how many slip systems are operating. The stacking-fault energy (SFE) determines how easy it is for dislocations to cross-slip, that is, to move from one glide plane to another glide plane, which also contains the dislocation's Burgers vector. High SFE materials are characterized by easy cross-slip. This is because dislocations in low stacking-fault energy materials are to a greater or lesser extent dissociated (separated into two or more partial dislocations, which individually do not cause perfect lattice displacement) (Honeycombe, 1975; Hull and Bacon, 2011) and cross-slip is much less easy. The number of slip systems is relevant

Figure 2 Dark field TEM micrograph of 99.98% aluminum strained to 0.16. Well-defined GNBs labeled A and B are separated by a cell-block structure containing cell walls and loose dislocations. Barlow et al. (1985); Figure 7.

here because there must be additional highly-stressed slip systems available on to which dislocations can move.

The cross-slip process is important as it affects both the evolution of the deformation structure and the flow stress. In a metal where cross-slip is easy the dislocations can have a high mobility in three dimensions (3D), which facilitates the formation of 3D structures such as dislocation cells; an example is shown in **Figure 2** (Barlow et al., 1985). However, if cross-slip is difficult then the deformation structures tend to be planar, forming a so-called *Taylor lattice*, as shown in **Figure 3** (Hughes, 1993). Cross-slip may also reduce strain-hardening on individual slip planes, because dislocations can move onto neighboring slip planes. Dislocation interaction and annihilation can then take place by a process that has been termed *dynamic recovery*, which is accompanied by a decrease in strain-hardening rate. Cross-slip is implicated as a cause for the transition from linear hardening in Stage II to parabolic hardening in Stage III.

17.3.8 Solute Effects

The presence of atoms of different chemical species in solution in the host lattice can influence deformation behavior in a number of ways. These arise not only from the physical size and shear modulus difference between the atoms but also in some cases from alterations to the electronic structure, which affect such properties as the SFE and bond energy (Haasen, 1996).

Most solid solutions involve atoms of roughly the same diameter, which substitute for host atoms in the lattice. The maximum size difference that can be tolerated in a solid solution is about 15% (Massalski, 1996). The resulting elastic strain fields increase the stress required for dislocation movement by increasing lattice friction according to the relationship:

$$\tau_{SS} = G\delta^{3/2}C^{1/2}, \tag{11}$$

where G is the shear modulus, δ is the atomic size difference defined as $\Delta r/r$ where r is the solute atomic size and Δr the size difference between solute and solvent, and C is the solute concentration.

Substitutional solute atoms may alter the evolution of strain microstructure in a number of different ways. A change in the slip pattern from *distributed glide* to *planar glide* has been observed when adding

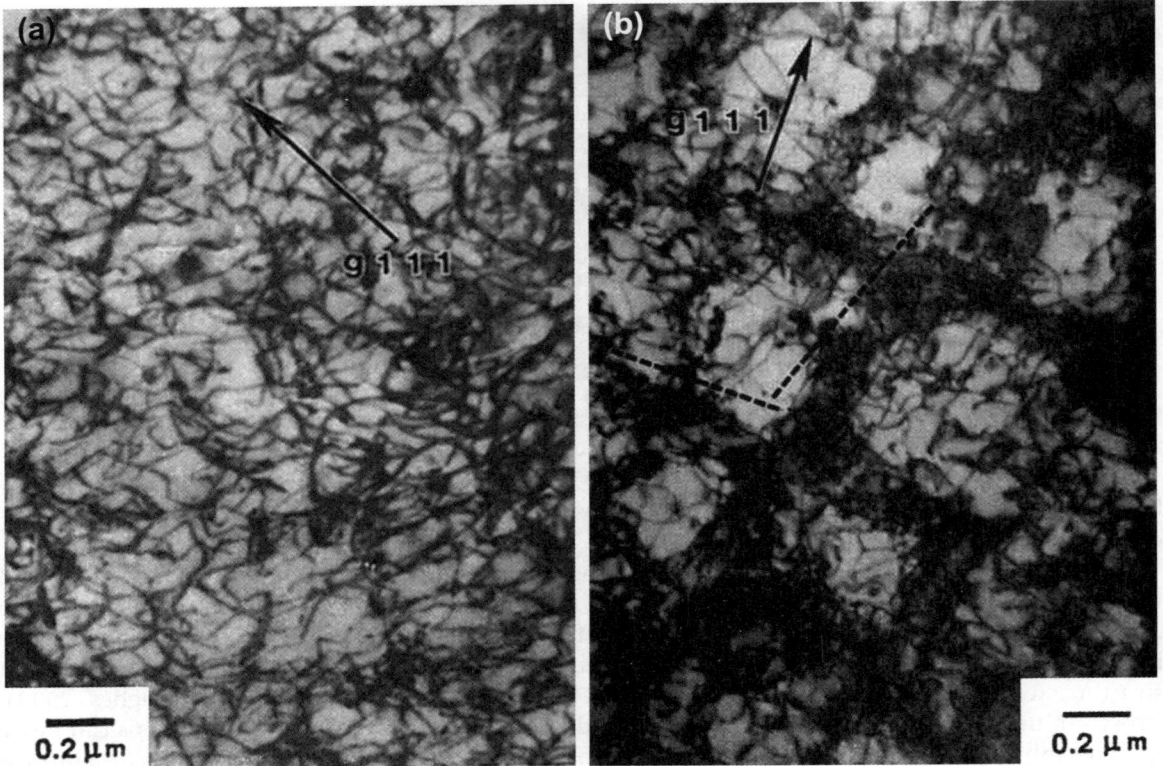

Figure 3   Multi-Burgers vector Taylor lattice in Al–5%Mg following 10% strain. (a) The reflecting vector suggests random dislocations. (b) Organization along {111} planes whose traces are denoted by dashed lines. Hughes (1993); Figures 2 and 3.

magnesium to aluminum alloys (Hughes and Hansen, 2004). An important effect of solutes is to increase the dislocation density in a metal at a given strain, thereby refining the structure and increasing the strength. An example of this has been observed in nickel deformed by HPT (Zhang et al., 2011). The effect of the solute atoms here is to inhibit dynamic recovery, and may result mainly from a reduction in the ability to cross-slip (Argon, 2008; Hong and Laird, 1990).

   There are relatively few examples of alloys where the solute atoms are sufficiently small that they can fit into interstitial sites in the lattice, but they have enormous significance. Steels fall into this category: the interstitial carbon, nitrogen and hydrogen atoms all have strong influences on dislocation movement. By contrast with substitutional atoms, the interstitials are relatively mobile and can diffuse even at room temperature. Interstitial atoms introduce elastic strain fields, and will move to regions where these strain fields can be reduced by interaction with strain fields of opposite sign within the lattice. They can migrate to dislocation cores and accumulate at high concentrations, forming Cottrell atmospheres (Hirth, 1996). They can thus have a very profound influence on mechanical properties, since even a low atomic fraction of interstitial element can pin dislocations and raise the shear yield stress. Once dislocations have broken free from these local concentrations of interstitial atoms then the effect on lattice fraction is negligible, so there is a yield drop (Haasen, 1996). Interstitial solid-solution hardening affects deformation mechanisms and is responsible for Luders band formation in steels (Cottrell, 1963).

17.3.9   Twinning

Shear strain can be achieved by the formation of mechanical twins, which is a very different mechanism from dislocation movement (Hull and Bacon, 2011; Cottrell, 1963; Hosford, 2005). Twinning is a localized shear process, involving the cooperative shifting of lattice planes into crystallographically different configurations. It occurs rapidly and abruptly, and is often accompanied by audible clicks. An example of a structure showing mechanical twins is shown in **Figure 4** (Zhang et al., 2008a).

Mechanical twinning can take place in addition to or in preference to dislocation glide under certain conditions. It most commonly occurs when dislocation glide is difficult for whatever reason, and is a very process-sensitive deformation mechanism. Low dislocation mobility is typically found at low temperatures and high strain-rates. Crystal structure is a major factor: hcp materials frequently demonstrate twinning, and it may be the dominant mechanism. It is common also in bcc materials, particularly at low temperatures. It is less common in fcc metals, although twinning has been observed for deformation at very low temperatures (e.g. Cu at 4 K) or high strain-rates. Other systems in which dislocation mobility is impeded include solid solutions, and twinning is found for deformation of some fcc alloys, for example Ag–Au at room temperature (Cottrell, 1963). A somewhat distinct type of twinning on a nano- or microscale can be found supplementing the dislocation structure in ultrahigh strained materials (Huang et al., 2003).

17.4   Experimental Techniques for Characterization

Deformation of a metal causes changes in the microstructure and in the crystallographic texture. As illustrated in **Figure 5**, length scales associated with dislocation structures range from grain-size features (typically 0.1–10 mm) down to perhaps tens of atoms (1–10 nm). Because of this large range, characterization normally requires a combination of techniques.

Dislocation structures are composed of characteristic features that relate to the way dislocations are stored. Typically, the structure is subdivided by dislocation boundaries that are associated with different amounts of crystallographic misorientation. Low-, medium- and high-angle boundaries are usually defined as providing misorientations of less than $5°$, $5–15°$ and above $15°$ respectively. Loose dislocations are normally present between the dislocation boundaries, and their configurations may be affected by the presence of precipitates or other second-phase particles. The microstructural evolution may be very sensitive to the nature and distribution of such particles, which should therefore be included as part of the full microstructural characterization of the material. In parallel with the evolution of microstructure with increasing strain, a crystallographic texture develops that affects the microstructural evolution, and vice versa. Structural characterization of the deformed state should therefore include both microstructure and texture.

17.4.1   Classification of Techniques

The complete analysis and characterization of deformation microstructures require a range of techniques, spanning morphological aspects, crystallographic orientation and chemical information. The key microstructural features and the principal techniques for their analysis are listed in **Table 3**.

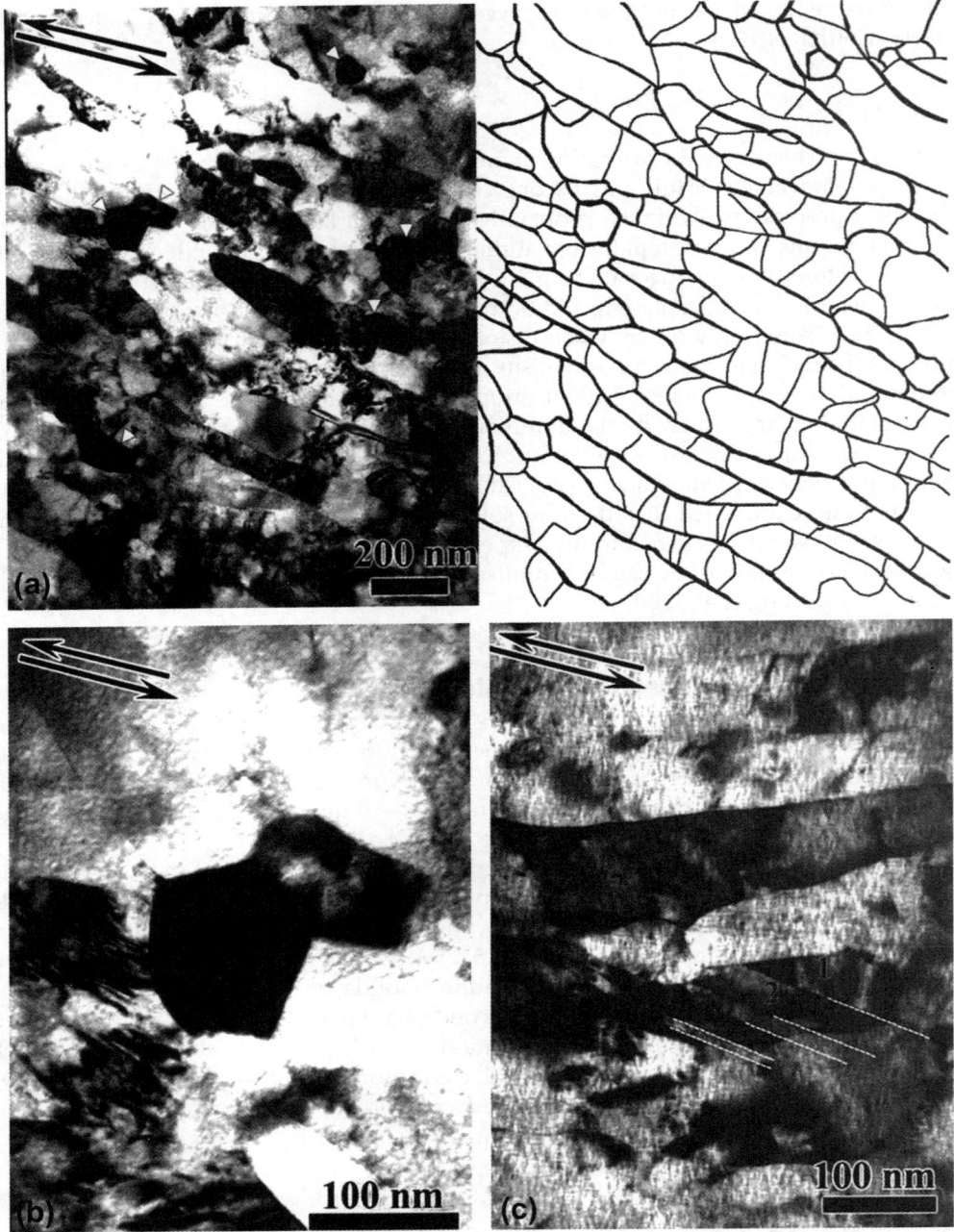

Figure 4   Twin boundaries in nickel deformed by high pressure torsion to a strain of 12. Twin boundaries are marked with a dashed line. Zhang et al. (2008a); Figure 3(c).

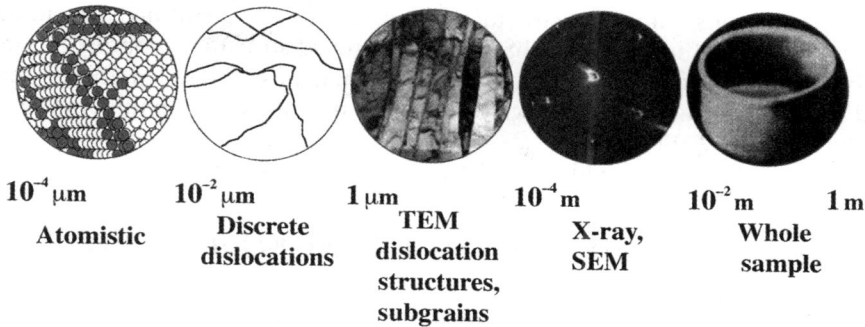

$10^{-4}\,\mu m$ $10^{-2}\,\mu m$ $1\,\mu m$ $10^{-4}\,m$ $10^{-2}\,m$ $1\,m$

Atomistic   **Discrete dislocations**   **TEM dislocation structures, subgrains**   **X-ray, SEM**   **Whole sample**

Figure 5   Length scales relevant to characterization of deformation structures. Courtesy D.A. Hughes.

Table 3   Key microstructural features and techniques for their analysis

Microstructural feature	Principal techniques available	Typical size of feature
Grain, subgrain and cell sizes and aspect ratios	LM, SEM, TEM (light, scanning electron, transmission electron microscopy) X-ray	Grain sizes: Typically 10 μm up to mm Subgrain and cell sizes: 0.1–10 μm
Nature and distribution of second-phase particles	LM SEM TEM	Sizes 0.5–10 μm Sizes 0.1–10 μm Sizes 0.01–3 μm
Macrotexture	Neutron, X-ray,	Whole specimen
Microtexture (local)	SEM, TEM, X-ray	Subgrain size 0.1–5 μm
Macroscopic grain and subgrain orientation (with respect to external stress)	SEM, TEM X-ray	Grain sizes: Typically 10 μm up to mm Subgrain sizes: 0.1–10 μm
Orientation across grain and subgrain boundaries	SEM, TEM	Spatial resolution required to be at least that of grain or subgrain size.
Dislocations	TEM, X-ray	5–20 nm
Structure of dislocation boundaries (width; tilt or twist misorientation)	TEM	2–100 nm
Macroscopic stress fields	Neutron, X-ray	Up to sample dimensions: 10 mm or more
Local stress fields	TEM, SEM, X-ray	2–20 nm

In addition to these, chemical microstructural analysis can be used to assist in phase identification, and in assessing the presence of chemical segregation. A range of techniques is available, often based on emission of characteristic X-rays from atoms excited by electrons during imaging in TEM or Scanning Electron Microscopy (SEM). This aspect will not be further considered here.

TEM and SEM form the basis of a large number of associated techniques (Loretto, 1993), which have increased in scope and sophistication over the years. There has been a trend toward increased automation of analytical techniques, with useful advances in resolution and ease of use. A consequence of these advances has been the possibility of providing a unified description of structural evolution under different loading conditions. Important structural parameters can be quantified, providing the basis for analytical modeling and simulation. This has led to improved structural property models. In parallel with these developments, increase in computer processing power has led to the introduction of modeling and simulation down to atomic scale of microstructural evolution under mechanical loading (Hansen and Juul Jensen, 2011).

Some of the these techniques are illustrated in the following sections.

17.4.2   Examples of Use of the Principal Experimental Techniques for Key Parameters

17.4.2.1   Dislocation Density

Dislocation number density (measured as number of lines per unit area, or length per unit volume) in thin foils can be measured directly using TEM. Measurements are experimentally challenging (problems include sample preparation, and finding imaging conditions for all dislocations), and thin films may not be representative of bulk material (dislocations may move to foil surfaces and be lost). Indirect methods include the use of etching to produce pits where dislocations meet the specimen surface (Honeycombe, 1975; Hull and Bacon, 2011); such methods can be much more straightforward than TEM but often of lower accuracy. X-ray line profile analysis is a versatile technique that can be used for determining dislocation densities in bulk samples. When employed as high-angular resolution three-dimensional X-ray diffraction it can provide spatial dislocation density information on the level of individual grains, or even subgrains (Ungar and Pantleon, 2009).

Estimates of total dislocation densities in bulk samples can be made by densitometry, making use of the fact that there is a small volume expansion associated with the presence of dislocations. The magnitude of this is approximately equivalent to one atomic volume per close-packed atomic plane cut by a dislocation (Hirth, 1996).

17.4.2.2   Grain and Subgrain Sizes

Grains and subgrains are imaged in SEM and TEM by virtue of the orientation changes between adjacent subgrains (Forwood and Clarebrough, 1991). The development of orientation-based diffraction techniques in TEM and SEM has allowed large-area mapping of subgrain sizes and crystallographic orientations to be performed rapidly. The use of Electron Backscatter Diffraction in conjunction with SEM is particularly suitable for this. The technique combines high spatial resolution (20 nm grains may be analyzed (Field, 2005)) with an angular resolution that is typically 1.5°, and large areas may be analyzed using automated methods.

Individual subgrains within a bulk polycrystalline sample may be identified using high angular resolution three-dimensional X-ray diffraction, and their evolution under in situ loading may be followed dynamically (Ungar and Pantleon, 2009).

17.4.2.3   Texture

Macroscopic texture measurements can be made using neutron scattering or X-ray diffraction, both of which can be used to generate orientation data for the whole sample (Randle and Engler, 2000). The particularly high penetration of neutrons allows bulk measurements to be made, and opens the possibility for in situ measurements to be obtained (Juul Jensen et al., 1984). Microscopic or local-scale texture analysis is performed using electron diffraction techniques in conjunction with TEM or SEM to obtain crystallographic orientations of individual grains (Randle, 1992). Highly penetrating hard X-rays from a synchrotron can be used for three dimensional X-ray diffraction, allowing in situ observation of individual grains during straining (Ungar and Pantleon, 2009).

17.4.2.4　Second-Phase Particles

The mechanism for revealing particles and distinguishing them from the matrix varies between different techniques, and can include surface relief in LM and SEM (difference in hardness; chemical etching); crystallographic analysis (e.g. dark field imaging in TEM); atomic number contrast in SEM and TEM. In all cases, stereological analysis must be used to convert observed particle distributions to a true 3D distribution (Kurzydlowski and Ralph, 1995).

17.4.2.5　Local Strain (Stress) Fields

Local stress fields cannot be revealed directly, but can be analyzed using the elastic strain fields, which they cause. They can be imaged directly in TEM as a result of their effect on local orientation, and also by local changes to the lattice parameter (de Graef, 2003). Although this allows high spatial resolution together with good sensitivity to strain, the requirement for preparation of thin-film samples for TEM is a limitation. It has clearly been demonstrated that stresses in thin films differ from those in bulk samples because proximity to free surfaces allows them to relax (Clement et al., 2004). EBSD can be used to provide data from SEM specimens (Wilkinson et al., 2010). Bulk samples can be analyzed using neutron diffraction or three-dimensional high-resolution X-ray diffraction (Poulsen, 2004).

17.4.2.6　Boundary Characterization

A complete description of a boundary requires (Forwood and Clarebrough, 1991; Randle, 1992):

(1) Knowledge of the orientation of the grains on either side of the boundary
(2) Knowledge of the orientation of the boundary plane with respect to both grains
(3) Description of the structure of the boundary

(1) and (2) are quantitative parameters; (3) is qualitative and is of secondary importance in its effect on mechanical properties.

(1) Grain orientations can be measured with high accuracy ($0.1°$) in TEM using semiautomated analysis of Kikuchi diffraction lines (Liu, 1994), or in SEM by EBSD with accuracy of up to about $1.5°$ (Randle, 1992) in a fully automated way. The misorientation between two crystals can be defined either by a rotation matrix, or by specifying an *axis-angle pair*, being a rotation axis and a clockwise rotation angle about that axis. It should be noted that because of the high degree of symmetry of metals, the description is not unique; in cubic crystals, for example, there are 24 equivalent descriptions of the same crystal pair relationship (e.g. Bhadeshia, 1987).
(2) Knowledge of the boundary plane can be combined with the misorientation data to allow the atomic-scale structure of the boundary to be identified. This can be helpful in understanding some aspects of dislocation interaction with boundaries (Sutton and Balluffi, 1995).
(3) Low-angle boundaries evolve from loose tangles of dislocations, and the misorientation angle increases as dislocations are absorbed; how far this has proceeded is indicated by the *width* of the boundary (Sutton and Balluffi, 1995).

17.4.2.7　Stored Energy

The stored energy in a deformed metal or alloy is typically small but it can be determined with high precision by differential scanning calorimetry (DSC). However, making use of the assumption that the deformation microstructures are stress-screened so that long-range stresses are negligible, the stored energy can also be determined by quantitative structural analysis. This analysis is based on the characteristics of the deformation microstructure composed of dislocations, which can be randomly distributed or present in boundaries with a wide range of misorientation angles. The stored energy is also present in the form of high-angle boundaries that can be the original grain boundaries and deformation-induced boundaries. The distinction between dislocation boundaries and high-angle boundaries is not well defined and typically an angle of 15° is chosen to arbitrarily separate the contribution of the two types of boundaries.

The energy stored in dislocations can be estimated as the energy per unit length of dislocation line multiplied by the dislocation density. This means that a measurement of the dislocation density can give an estimate of the stored energy.

The energy stored in boundaries (γ) is expressed by the Read–Shockley equation:

$$\begin{cases} \gamma = \gamma_m \left(\dfrac{\theta}{\theta_m} \right) \left[1 - \ln \left(\dfrac{\theta}{\theta_m} \right) \right] & \theta < \theta_m \\ \gamma = \gamma_m & \theta > \theta_m \end{cases}, \tag{12}$$

where θ is the misorientation angle across the boundary, γ_m is the grain boundary energy per unit boundary area and θ_m is taken to 15°. The energy γ is typically a minimum value as some dislocations are redundant and have a net Burgers vector which is near zero. Such dislocations do not contribute to the misorientation, and so are excluded from calculations correlating the dislocation density and the boundary angle.

The two techniques, calorimetry and a microstructural analysis, must be considered as complementary. The former is most suitable when analysis is performed on the scale of the sample, whereas the latter must be used to determine local differences in stored energy on the grain scale. These aspects are discussed further in Section 17.5.3.3.

17.4.2.8 *Chemical Analysis*

A range of techniques can be used to determine chemical compositions in conjunction with microscopic imaging. TEM-based techniques include Electron Energy-Loss Spectroscopy, EELS, and Parallel Electron Energy-Loss Spectroscopy, PEELS (Robertson et al., 2011). Atomic scale analysis of materials can be performed using atom probe techniques (Robertson et al., 2011). Specimens for this are prepared in the form of very sharp needles and are subjected to an electric field strong enough to ionize individual atoms and remove them from the material. Time-of-flight mass spectroscopy is used to analyze these individual species, and when used in conjunction with a position-sensitive detector the original locations of the atoms on the sample can be deduced (Robertson et al., 2011).

17.5   Development of Deformation Microstructures

17.5.1   Classification and Terminology

During plastic deformation a small fraction of the mechanical energy (a few percent or less) is stored in the metal primarily in the form of dislocations, which form characteristic 2D and 3D configurations, and which have a profound influence on mechanical properties (Bever et al., 1973; Godfrey et al., 2005). The formation of 3D structures requires dislocation motion in directions other than those contained in the primary glide plane. This may be achieved by intersecting glide planes, or by dislocations *climbing* out of their primary glide plane. The key factors that affect dislocation mobility in single-phase materials are:

Stacking fault energy;
Presence of solute atoms;
Temperature and strain rate;
Deformation mode and grain orientation; and
Strain path changes.

In multiphase materials, additional factors include:

Size, number, spacing and distribution of second-phase particles;
Nature of the interface between the second-phase particle and the host lattice; and
Mechanical properties of the second-phase particles.

A classification of deformation microstructures in accordance with the mobility of dislocations is described in Hughes and Hansen (2004). In this classification metals and alloys are separated into two groups: (1) Wavy glide materials in which 3D structures evolve, and (2) Planar glide materials in which dislocations are stored in 2D arrays. These two types of structures have been termed Cell Block (CB) structures and Taylor Lattice structures, respectively, and are illustrated in **Figures 2 and 3**. In parallel with this general classification, terminology for the individual microstructural features is provided in the glossary preceding this article.

A second classification is based on a length scale, which may be either the scale of the whole sample, or the scale of individual grains. On the scale of the whole sample, stress and strain gradients may lead to local changes in microstructure and crystallography e.g. through-thickness variations in rolled samples and center-to-surface variations in torsion samples. On the grain scale, relevant features are that grains can subdivide macroscopically and microscopically, and grain-to-grain interaction can lead to changes in microstructure and crystallography at or near grain boundaries.

A third classification is based on crystallographic orientation. The crystallographic orientations of grains and crystals affect the evolution of microstructure and texture in addition to local texture effects.

17.5.2   Assemblies of Dislocations

It is a general observation that the dislocations generated during deformation assemble into characteristic configurations, which typically take the form of dislocation boundaries. These boundaries may form a cell structure, or may form characteristic extended planar boundaries. Observations indicate that there is an absence of significant stress/strain fields associated with these dislocation boundaries. It has therefore been suggested that they represent LEDS. Their characteristics have been analyzed on the basis of the LEDS hypothesis (Kuhlmann-Wilsdorf, 1989), which has proved to be a powerful tool. According to this hypothesis, the configurations of glide dislocations will alter in order to approach the lowest possible energy per unit length of dislocation line. The ability of dislocations to reach their lowest-energy configurations is constrained by a number of factors, mainly relating to dislocation mobility. The factors include the number of available slip systems, dislocation mobility (for both glide and climb), and the frictional stress.

17.5.3   Development of Dislocation Structures

17.5.3.1   Dislocation Boundaries

During plastic deformation, dislocations multiply and are stored inside the material. The dislocation strain energy is reduced by forming largely dislocation-free regions separated by dislocation-rich walls.

The accumulation of dislocations in dislocation boundaries makes these boundaries the key feature in the characterization of deformation microstructure. Dislocation cells and cell boundaries have been studied since the introduction of TEM but detailed structural characterization in the 1980 and 1990s (Barlow et al., 1985; Liu et al., 2002; Bay et al., 1992; Hansen, 1990; Hughes, 1993; Bay et al., 1989; Hansen and Hughes, 1995; Kuhlmann-Wilsdorf, 1995; Leffers, 1995) showed that cell boundaries coexist with dislocation boundaries, which have been termed Dense Dislocation Walls (Barlow et al., 1985). These boundaries have a completely different morphology as can be seen in **Figure 6** (CB structure) (Liu and Hansen, 1995a). In contrast to the short cell boundaries with an almost random orientation in the structure the DDWs are extended, nearly planar boundaries. These extended boundaries have varying characteristics but a common feature is that they delineate regions that are further subdivided by cell boundaries. Both types of boundaries are rotation dislocation boundaries.

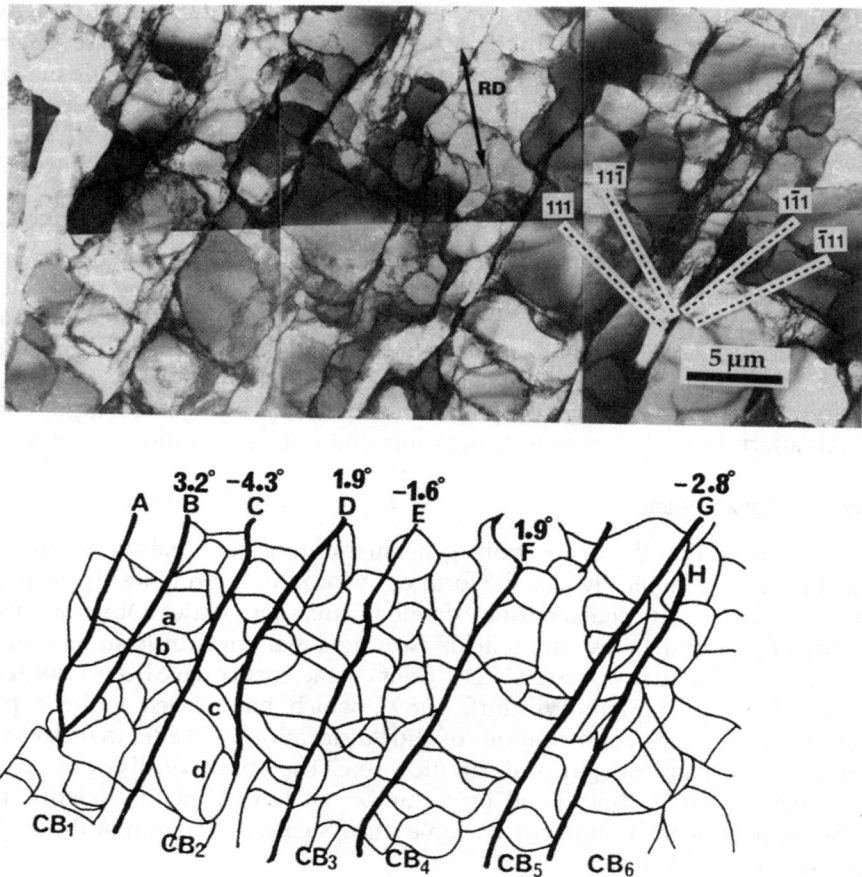

Figure 6 TEM image and a sketch of a microstructure in a grain of a 10% cold-rolled specimen of pure aluminum (99.996%) in longitudinal plane view. One set of extended noncrystallographic dislocation boundaries is marked A, B, C, etc., and their misorientations are shown. These boundaries form CBs marked CB1, CB2, etc., which are subdivided by cell boundaries marked a, b, c, etc. The rolling direction is marked RD and the dashed lines are traces of {111} planes. Liu and Hansen (1995a).

The two types of boundaries may be formed by different mechanisms. The cell boundaries may form by mutual trapping of glide dislocations and, as such, have been termed incidental dislocation boundaries (IDBs) (Kuhlmann-Wilsdorf and Hansen, 1991). Extended boundaries may have their origin in a different range of active slip systems in neighboring regions called CBs (see **Figure 6**). Each CB has been assumed (Bay et al., 1992) to deform by four or fewer active slip systems, i.e. falling short of the five required for homologous deformation according to the Taylor model (Taylor, 1938). However, a group of CBs may collectively fulfill the Taylor criterion. This slip pattern will cause a strain in the CB that is different from the macroscopic strain. Strain differences are accommodated by the formation of cell-block boundaries which, therefore, have been termed geometrically necessary boundaries (GNBs) (Kuhlmann-Wilsdorf and Hansen, 1991). It has been suggested that the number of simultaneously acting slip systems is a result of two competing effects (Bay et al., 1992). The flow stress is lowered with a reduction in slip systems because the number of intersecting dislocations and, thus, the number of jogs, decreases. On the other hand, stress screening will become more effective as the number of different Burgers vectors increases (Kuhlmann-Wilsdorf, 1989).

The slip pattern discussed above is not the only one that can cause grain subdivision and formation of GNBs. The slip pattern may also follow the Taylor model, which requires that the strain in the different regions in a grain equates to the macroscopic strain, but GNBs can still form if different sets of slip systems operate in neighboring regions or if the total shear amplitude is partitioned differently among a common set of systems in neighboring regions (Wert et al., 1995). Finally, GNBs may form due to a nonuniform strain in the interior of the grains (see, for example, Becker and Panchanadeesvaran, 1995), or near grain boundaries and triple junctions (Barlow et al., 1985; Randle et al., 1996; Zisman and Rybin, 1998).

This classification into IDBs and GNBs adds to the classical distinction between dislocations as statistically stored (redundant) and geometrically necessary (nonredundant) dislocations (Nye, 1953; Ashby, 1970; Cottrell, 1964). The statistically stored dislocations are assumed to be present in the form of tangles, dipoles and multipoles, which will not give rise to a significant lattice rotation because their net Burgers vector is practically zero. By contrast, the geometrically necessary dislocations are assumed to assemble in configurations characterized by a lattice rotation and a net Burgers vector. The storage of these dislocations can be random or the dislocations can be assembled in boundaries in which the lattice rotation is related to the misorientation angle across the boundary. Distinguishing between statistically stored and geometrically necessary dislocations is not possible through a microscopic examination of the microstructure. However, differentiation has been attempted using X-ray micro-diffraction of deformed samples, where streaking of Laue spots is interpreted as due to an elastic strain gradient or to a distribution of geometrically necessary dislocations. By contrast, splitting of the Laue pattern indicates the presence of angular misorientations between neighboring volumes, so implying the presence of GNBs (Barabash et al., 2009).

The structural evolution with increasing strain is most characteristic in wavy (distributed) glide materials, such as fcc metals with medium to high SFE, e.g. Cu, Ni and Al (Malin and Hatherly, 1979; Bay et al., 1992; Hughes and Hansen, 1993; Huang, 1998), and bcc metals, e.g. iron (Jago and Hansen, 1986). In these metals, a wavy slip band structure indicates a three-dimensional mobility of dislocations by cross-slip. However, a characteristic structural evolution (although more difficult to observe) also takes place in planar glide materials such as low-SFE metals, e.g. austenitic stainless steel (Hughes, 1998) and some alloys, e.g. Al–Mg (Hughes, 1993). A typical microstructure for a wavy glide material is seen in **Figure 6**, where the structure is subdivided into CBs and cells. **Figure 7** shows grain subdivision in a planar glide material (warm rolled Al–5%Mg), where the extended dislocation walls now form blocks, instead of cells, and contain a more uniform distribution of dislocations called a Taylor lattice.

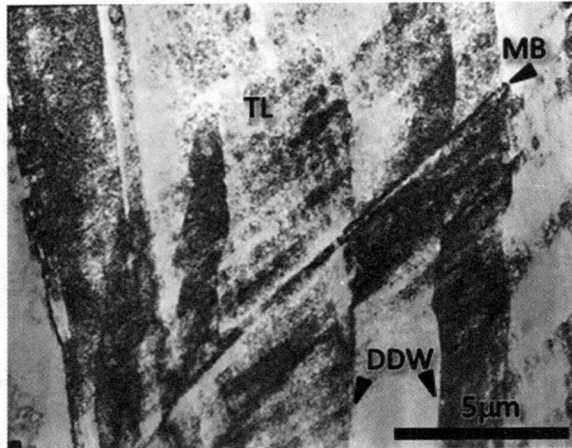

Figure 7 Dislocation microstructure in Al–5%Mg warm rolled at 350 °C to a 30% reduction in longitudinal plane view. Extended boundaries are seen that form a Taylor lattice (TL) block structure. The extended boundaries are marked DDW and MB. Hughes and Godfrey (1998).

The microstructures shown in **Figures 6 and 7** are typical deformation structures. However, a change in process parameters can affect the slip pattern and the deformation microstructure. For example, in wavy glide metals, the tendency for planar glide increases with decreasing temperature (and increasing strain rate), whereas a planar glide metal after hot deformation shows microstructures typical of wavy glide (Liu et al., 1998; Theyssier et al., 1995; Duly et al., 1996; Hughes and Godfrey, 1998).

In the analyses of different types of dislocation configurations the low-energy dislocation structure hypothesis gives principles and guide lines, but it does not allow prediction of specific dislocation arrangements because of the innumerable dislocation configurations, which can be constructed from any given dislocation population. This fact together with observations in both single crystals and polycrystals shows that there is a clear correlation between the characteristics of the deformation microstructure and the crystallographic orientation of the volume in which it evolves initially. The suggestion (hypothesis) that different underlying mechanisms determine the evolution of IDBs and GNBs, respectively, has been supplemented by many experimental observations and physically based models which will be discussed in the following sections. This discussion will focus on grain subdivision in medium to high SFE metals as Al, Ni and Cu deformed at room temperature, i.e. materials that form a CB structure.

17.5.3.2   Grain Subdivision

The subdivision of grains and crystals will be described in the following, omitting discussion of structural features on the sample scale. The analysis focuses on the grain scale, where the macroscopic scale subdivision is into deformation bands and transition bands, and subdivision on a smaller, microstructural scale, is into CBs and cells.

17.5.3.2.1   Macroscopic Subdivision

The macroscopic subdivision of crystals and grains has been investigated since the discovery by Barrett and Levenson (1940) of deformation bands in cold-compressed single crystals and polycrystals of aluminum of certain crystal orientations. Many different techniques have been applied, and the sub-divided structure is generally described as being composed of well-defined wide deformation bands (or matrix bands) separated by more narrow regions, termed transition bands. During deformation, the matrix bands rotate in different directions. Orientation differences between neighboring bands are accommodated in transition bands characterized by a relatively high cumulative misorientation across them (Dillamore et al., 1972; Hsun, 1963). For reviews in this area, see, for example, Gil-Sevillano et al. (1981), Liu and Hansen (1998), Kuhlmann-Wilsdorf (1999), Chin (1973) and Wert et al. (1999).

The many experimental conditions, and an inconsistent nomenclature, make it difficult to synthesize the many observations of the macroscopic breakup of crystals and grains. However, a number of single-crystal experiments have been analyzed in detail and have shown a consistent picture of the macro-scopic breakup, which is illustrated in **Figure 8** (Liu and Hansen, 1998).

The macroscopic breakup of single crystals into regions that extend over several hundreds of micrometers reflects rotation changes on a macroscopic scale. Such changes can be caused by the imposed deformation conditions (e.g. geometry and friction), which may also cause macroscopic subdivision in polycrystals with a large grain size. However, as the grain size becomes smaller,

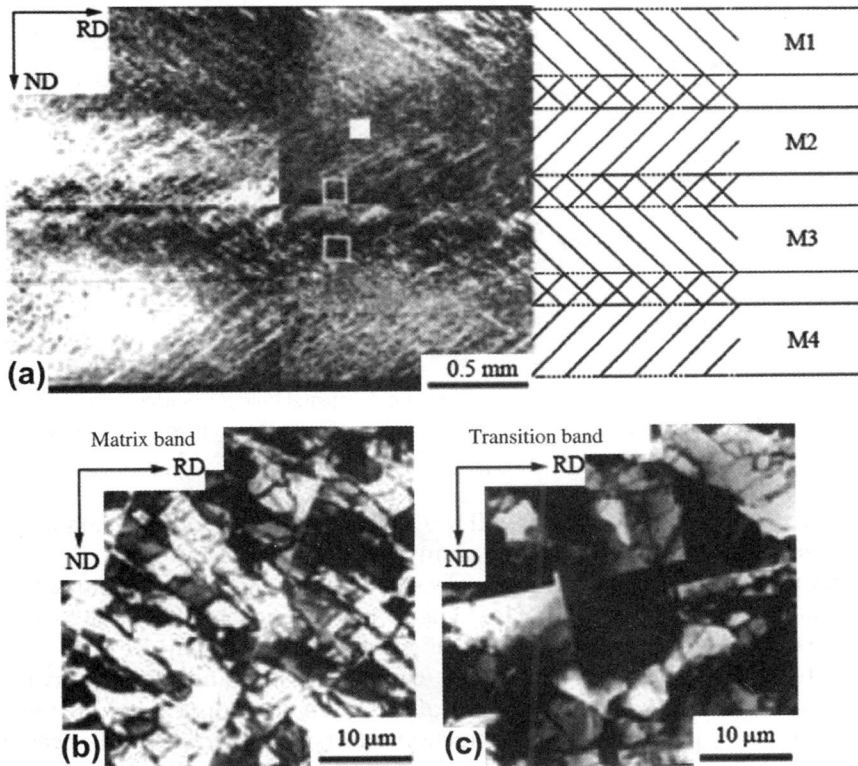

Figure 8 Subdivision of a cube-oriented aluminum crystal cold rolled 30%. (a) SEM channeling contrast image showing the subdivision of the crystal into four matrix bands marked M1, M2, etc., and three transition bands. (b) and (c) TEM micrographs revealing the microstructure in a matrix and a transition band are marked by squares in (a). The rolling direction and the normal direction are marked RD and ND, respectively. Liu and Hansen (1998).

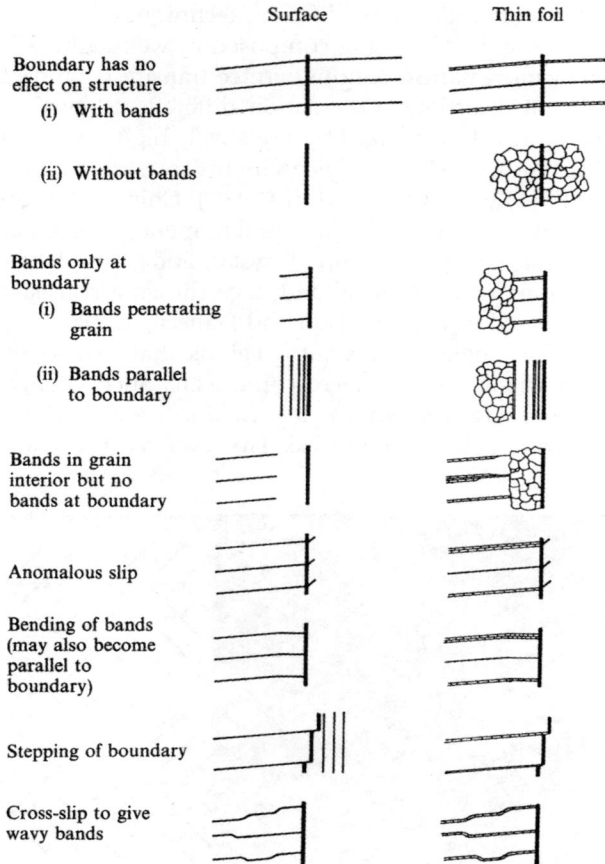

Figure 9   Classification of structures at grain boundaries in cold-deformed aluminum. Surface structures seen by SEM or optical microscopy in the left column correspond to structures seen in TEM in the right column. Barlow et al. (1985).

conditions for macroscopic rotation changes will be difficult to obtain. For such smaller grain sizes it is possible that interactions between neighboring grains may start to play a role.

Grain interaction effects may manifest themselves by local changes in microstructure and crystallography in the grain boundary region. Such changes have been observed by SEM of surface relief structures and by TEM of microstructures at or near grain boundaries (Barlow et al., 1985). **Figure 9** presents a schematic characterization of structures, comparing SEM and TEM observations. In most cases, the deformation structures that are typical of the grain interiors extend all the way to grain boundaries. In some cases, however, grain interaction effects are reflected by changes in the microstructure and local crystallography close to boundaries and triple junctions. Such regions are sometimes referred to as rims. TEM observations of grain boundary regions in tensile-deformed Al and Cu are shown in **Figure 10** (Huang and Winther, 2007) and **Figure 11** respectively. Grain interaction has also been studied by EBSD parallel and perpendicular to grain boundaries where grain interaction results in perturbations that reflect local changes in crystallography (Randle et al., 1996). For specimens

Figure 10 TEM image showing that a uniform dislocation structure extends from the grain interior all the way to the grain boundary in a tensile-deformed Al. The grain boundary is marked by a dotted line. The dashed line on the micrograph indicates the trace of the primary slip plane. Huang and Winther (2007).

deformed to 5% reduction in thickness by cold rolling, significant perturbations are observed only at triple junctions, and not in grain boundary regions. However, when the reduction is increased to 30%, perturbations are observed at many grain boundaries (**Figure 12**). These perturbations may extend a few micrometers into the grain from the boundary, so forming a narrow rim region compared with a grain size of about 300 μm. In order to quantify such regions and their dependence on grain size and strain, more extensive studies are needed, which are now made possible by the use of advanced EBSD techniques.

The studies of grain interaction have been coupled with an analysis of the structural evolution in grains in high-purity copper deformed in tension to a strain of 25% (Thorning et al., 2005). A relatively coarse grain size has been chosen, which allows grain breakup and lattice rotations to be tracked as a function of strain by EBSD analysis. As shown in **Figure 13**, a few small domains of special orientations provide evidence of grain interaction. However, the crystal rotation for the large domains covering most of the grains is consistent with the rotation direction predicted by the Taylor model in some, but not all, cases (Thorning et al., 2005).

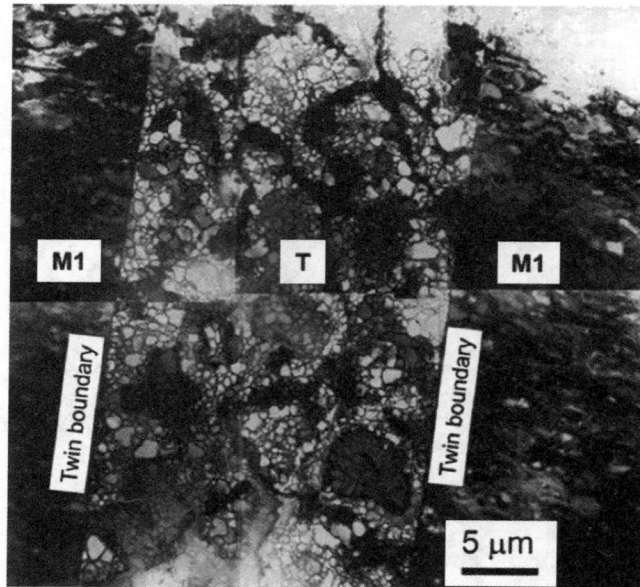

Figure 11 TEM image taken from the cross section of a tensile deformed Cu. A uniform cell structure is formed in the twin (T), which is in contrast to the structure in the matrix parts (M1). There is no clear effect of twin boundaries on the structural evolution in the twin and matrix parts. Figure 11 from Hansen et al. (2011).

17.5.3.2.2 Microscopic Subdivision

The microstructural studies have shown that the CBs are typical building blocks that form the deformation microstructure both in single crystals and polycrystals (Bay et al., 1992). At low and medium strains, the CBs contain many cells but as the strain increases the number of cells per CB decreases. The rate of decrease depends on the crystallographic orientation of the crystal or the grain (Liu and Hansen, 1995a). The boundary spacing decreases with strain and can typically reach 50–300 nm depending on the material and the deformation process. The spacings can be reduced further to the order to 10–20 nm in metals deformed to extreme strain, for example by friction and surface mechanical attrition. In such samples, structures typical of dislocation glide may be supplemented with deformation twins (Lu and Hansen, 2009).

In parallel with the structural refinement which accompanies increased strain, more and more dislocations are stored in boundaries, leading to an increase in the average misorientation angle across boundaries and an increase in the number of high-angle boundaries. The structural morphology also changes, and this follows the shape change of the bulk material, for example forming a lamellar structure in rolling and a fibrous structure in wire drawing. The structural refinement is material-dependent. For example in aluminum with a high SFE and easy cross-slip the smallest crystallite size is about 200–300 nm after rolling to ultrahigh strain, whereas in copper the minimum size may be about 100 nm. Addition of elements in solid solution can, however, increase dislocation pinning, and hence reduce the amount of annihilation during deformation. Using this mechanism it has been possible to reduce the crystallite size in aluminum to about 100 nm and in copper to about 5–10 nm, by cryomilling (Han and Lavernia, 2005) and friction wear (Hughes and Hansen, 2001), respectively.

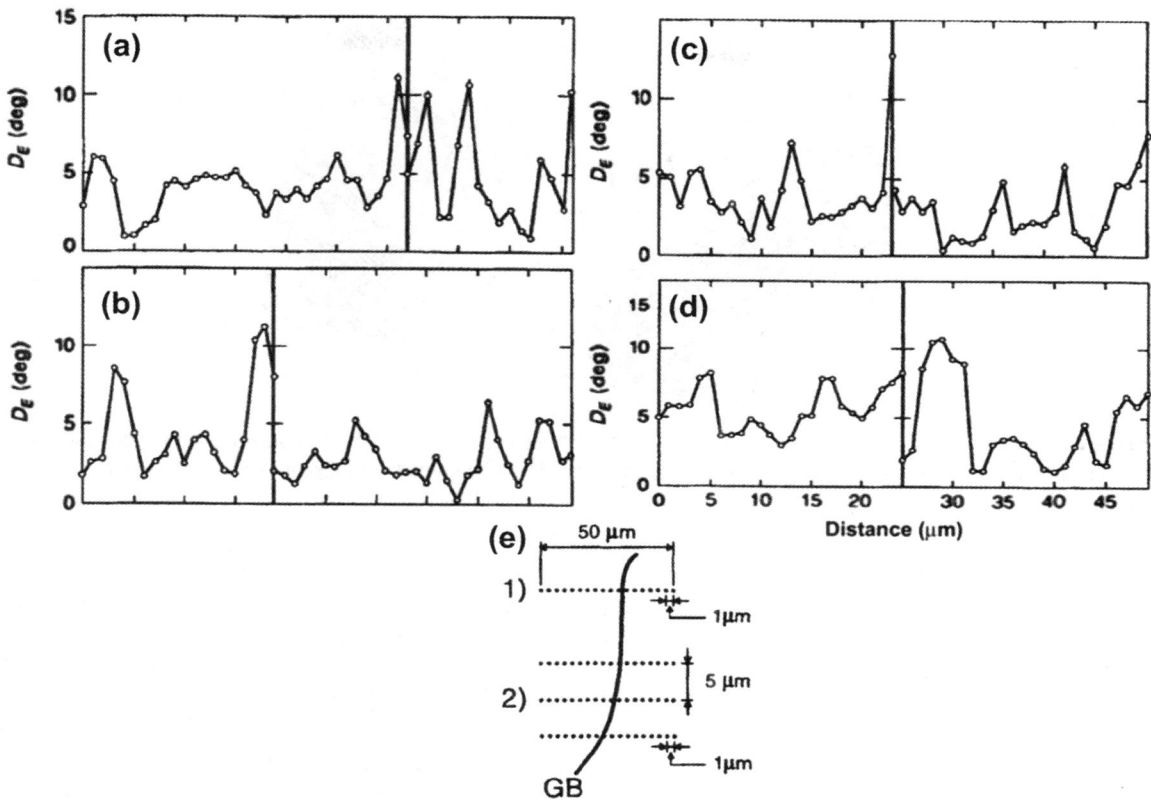

Figure 12 Examples of orientation scans perpendicular to a grain boundary, sampled according to the schedule (a, b, c, d and e) shown in the figure in 30% deformed, coarse-grained sample. Randle et al. (1996); Figure 7.

Figure 14 (Hughes and Hansen, 2001) illustrates a graded nanostructure structure found in copper following friction deformation.

In parallel with the decrease in boundary spacing, the misorientation angle across IDBs and GNBs increases. A main finding when moving to large strain is that the misorientation across deformation-induced boundaries can reach and exceed values of about 15°, i.e. the deformation-induced boundaries supplement the original high-angle boundaries. The average misorientation angle of boundaries approaches saturation at values of ∼40° and the frequency of high-angle boundaries can be as high as 70–80% (Pantleon, 2011; Mishin et al., 2003).

An important part of the microstructural evolution at increasing strain is the occurrence of microstructural transition. Examples of this are the transition from a two-dimensional carpet structure in Stage II (Steeds, 1966) to a CB structure in Stage III, and the transition during rolling from a structure of bands (Dense Dislocation Walls and Microbands) oriented about 40° to the rolling plane to a lamellar structure at large strain oriented almost parallel to the rolling plane. As an illustration, the former has been related to the occurrence of cross-slip, whereas the latter has been related to the occurrence of localized glide (Hughes and Hansen, 1993).

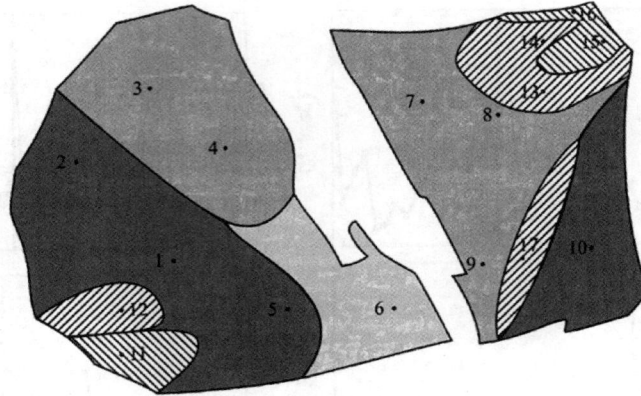

Figure 13 Subdivision of a coarse grain in copper deformed in tension at $\varepsilon_{vm} = 0.08$. The gray-shaded regions are coarse domains, and the hashed regions are grain interaction domains. The white domain represents an annealing twin. Thorning et al. (2005).

17.5.3.3 Stored Energy

The stored energy is an important parameter in a deformed structure. It can be measured precisely by DSC and estimated indirectly based on a quantitative microstructural analysis. Experimental comparison of the two values can be used to evaluate the assumptions on which the analysis is based.

The stored energy of a deformation microstructure subdivided by IDBs and GNBs can be calculated based on the Read–Shockley equation (see Section 17.4.2.7) to obtain average boundary energy values for IDBs (γ^{IDB}) and GNBs (γ^{GNB}). As their respective surface areas are S_V^{IDB} and S_V^{GNB}, the total stored energy in boundaries E_b is (Godfrey et al., 2007):

$$E_b = S_V^{IDB}\gamma^{IDB} + S_V^{GNB}\gamma^{GNB}. \tag{13}$$

In addition to E_b energy is also associated with dislocations present between boundaries, which can be calculated in accordance with Eqn (13). This contribution is typically small in medium to high stacking-fault energy materials.

The evolution in stored energy with increasing strain measured by calorimetry (DSC) and by microscopy (TEM) is shown in **Figure 15** (Godfrey and Liu, 2009) for polycrystalline pure nickel. The samples have been deformed by cold rolling (CR) to $\varepsilon_{vm} = 4.5$ and to a somewhat higher strain by accumulative roll bonding (ARB) arising from the additional redundant shear characteristic of the ARB process. **Figure 15** illustrates that accumulation of stored energy takes place at a decreasing rate as the strain increases, demonstrating that a smaller and smaller part of the energy expended on deformation is stored in the material. **Figure 15** also shows that the TEM data are lower than the DSC data by a factor of two or more. Such a discrepancy has also been observed in an earlier experiment (Ungar et al., 1984), in which the stored energy was calculated in a deformed (001) oriented single crystal of copper. The microstructural analysis used dislocation density measured by TEM and X-ray line broadening, with an estimated value for the dislocation line energy per unit length. This discrepancy cannot be accounted for by line energy alone and other causes have been explored. One is an unaccounted contribution from unscreened long-range stresses that (as predicted by the LEDS hypothesis) may reach the level of the yield stress. This is the highest stress, which can be present in the material, since any higher stresses will

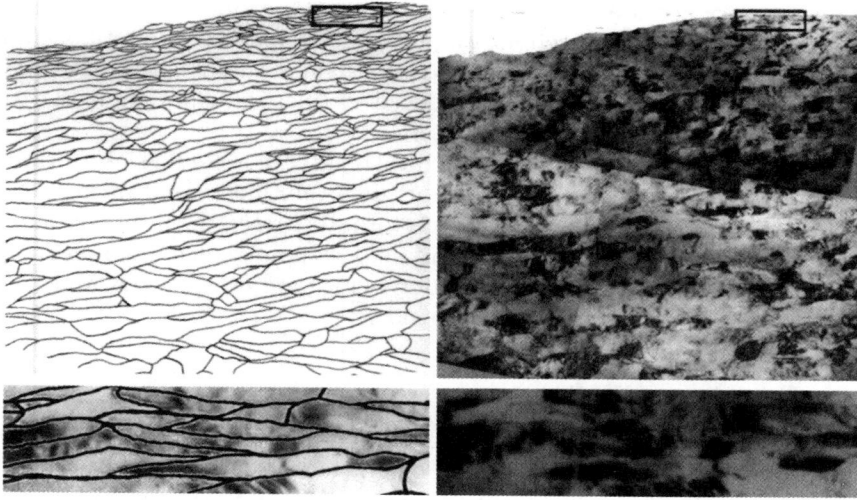

Figure 14 Graded nanostructures in Cu produced by friction deformation, with schematic. Hughes and Hansen (2001); Figure 12.16 from Metalworking.

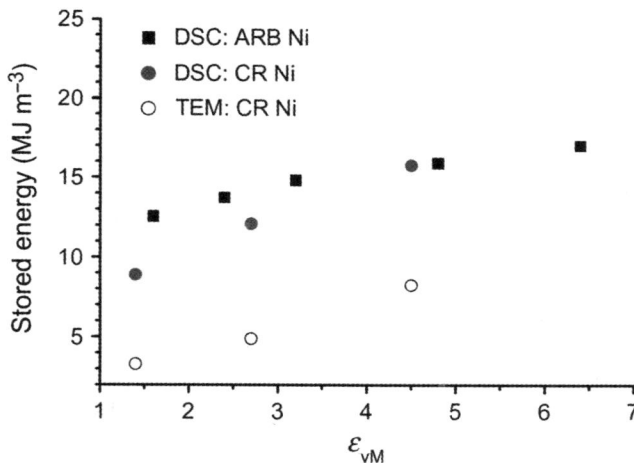

Figure 15 Stored energy as function of strain in Nickel determined by DSC. Godfrey and Liu 2009; Figure 1. (For color version of this figure, the reader is referred to the online version of this book.)

be relaxed by yielding. A theoretical analysis of such a contribution (Hansen and Kuhlmann-Wilsdorf, 1986) has shown it to reach one fifth of the total energy stored in dislocations. However, the magnitude of the internal stresses can also be measured experimentally. This has been demonstrated in recent studies (Levine et al., 2011; Jakobsen et al., 2007) by three-dimensional X-ray microscopy with sub-micrometer spatial resolution in a synchrotron. The sample examined (Levine et al., 2011) was a single (001)-oriented crystal of 99.999% Cu, which was deformed in compression to a true flow stress of 210 MPa and a true strain of 28%. The deformed structure was subdivided by dislocation boundaries forming cells with a diameter of about 1 μm. The magnitude of the long-range stress, taken as the stress difference between cell walls and cell interiors, was about 40 MPa, which is about one fifth of the flow stress. This is in agreement with the earlier analysis (Hansen and Kuhlmann-Wilsdorf, 1986) and also with the LEDS hypothesis that a structure introduced by plastic deformation is a stress-screened structure with a fairly low level of long-range stresses.

In other studies (Knudsen et al., 2008; Godfrey and Liu, 2009), a number of structural features are discussed that can reduce the observed difference between calorimetry and microstructural observations. However, the quantification of such features is for future research, and at present the two techniques must be considered as complementary. Calorimetry is suitable on the sample scale, whereas the microstructural analysis must be used for an estimation of the local stored energy for example at or near grain boundaries, at triple junctions and in shear bands.

17.5.4   Dislocation Structures and Models

An important part of the structural analysis is to quantify microstructural parameters and their relationship with the applied strain and stress. These parameters are typically averaged over the whole sample volume, but can also be restricted to the scale of a grain. Making use of a separation of boundaries into IDBs and GNBs, typical parameters are:

(1) boundary spacing and width;
(2) boundary misorientation angle and the plane on which the boundary lies; and
(3) crystal orientation with respect to the deformation axes.

In the following, the focus will be on the evolution of boundary spacing and misorientation angles with increasing strain. The effect of crystal orientation on dislocation structures will be discussed in Section 17.5.5.

The way in which boundary spacing and misorientation angle change as a function of plastic strain is illustrated in **Figure 16** for cold-rolled nickel; with increasing strain there is a monotonic decrease in spacing and increase in misorientation. This evolutionary trend can continue to large strains without saturation. The distribution of misorientation angles is illustrated in histograms in **Figure 17** (Hughes and Hansen, 2000). The distribution can include both low- and high-angle boundaries, and the number of very high-angle boundaries increases with increasing strain, creating a bimodal distribution. HPT can be used to introduce a larger strain than in rolling, and the evolution in GNB spacing and angle for this is shown together with data for nickel deformed by rolling and torsion in **Figure 18(a–c)** (Zhang et al., 2008a). Note that saturation is approached in both spacing and angle at very large strain, respectively 60 nm for the spacing and about 40° for the angle. As the strain increases, the structural subdivision by IDBs and GNBs is observed to continue up to the largest strain. There is a continuous increase in the fraction of high-angle boundaries, which at large strain can approach saturation at about 60–80%, see **Figure 19** (Mishin et al., 2003). The evolution in boundary spacing and angle is also revealed in **Figure 20** (Hughes and Hansen, 2004) in which a power-law relationship is found between strain and the boundary area of GNBs per unit volume (S_v). The evolution of the misorientation angle is illustrated in **Figure 21** (Hughes and Hansen, 2004), yielding a power-law relationship for both IDBs and GNBs, although with a significantly different exponent for each.

The rate of evolution of spacing and angle of boundaries decreases with increasing stress and strain (see **Figure 16**). This evolution illustrates that a continuous storage of glide dislocations in IDBs and GNBs is counterbalanced by dynamic recovery mechanisms by which both dislocation and boundaries are removed. These mechanisms may be related to an increased dislocation mobility due to cross-slip

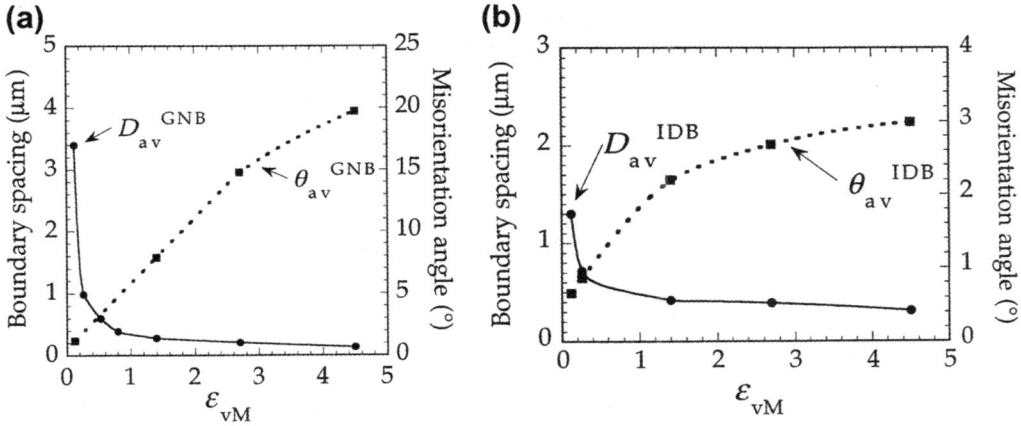

Figure 16   Boundary spacing and misorientation angle as a function of strain for cold-rolled nickel for (a) GNBs and (b) IDBs. Hughes and Hansen (2004).

and to an enhanced frequency of dislocations interacting with low- and high-angle boundaries. The annihilation rate is expected to increase with increasing stress and strain, since this increases the dislocation density and reduces the spacing between boundaries. An increase in misorientation angle may also enhance dynamic recovery, since there is evidence that high-angle boundaries may act as especially efficient sinks for dislocations. Dynamic recovery as a result of interactions between individual dislocations is an inherent part of work-hardening models and is well characterized. On the other hand, annihilation of dislocations by interactions between dislocations and dislocation boundaries can be explored theoretically through molecular dynamics (MD) simulations, but there are few experimental observations. However, new techniques are under development, which allow in situ observation of such interactions when small specimens containing selected types of boundaries are loaded in a transmission electron microscope (Zhang et al., 2012).

Dynamic recovery by removal of dislocations induced boundaries is also to be explored. This phenomenon is illustrated in **Figure 22** (Hughes and Hansen, 2000) for GNBs in cold-rolled nickel. These boundaries are of medium to high angle, and in **Figure 22(a)**, the number of GNBs across a sample is plotted as a function of strain. The slope of this plot gives a net rate of creating and removing GNBs. Initially there is a very rapid increase in the number of GNBs but after a strain of 0.8, there is a net decrease in the number. Such a transition naturally alters the rate of change of boundary spacing with respect to the strain, as shown in **Figure 22(b)**.

The mechanism by which GNBs can be removed during deformation is at present unknown. However, it is observed in **Figure 22(c)** that a sharp rolling texture evolves during deformation. This macroscopic texture consists of a number of components that are in the form of thin lamellae separated by GNBs of medium to high angle. This means that removal of GNBs by a coalescence mechanism appears not to be feasible. It remains a possibility, however, that GNBs can be removed by motion of triple junctions, which are present in large numbers in a highly strained structure. A possible mechanism for such a process has recently been discovered during annealing of a lamellar structure in cold-rolled aluminum (Yu et al., 2011). This structure contains numerous so-called Y-junctions formed by three lamellar boundaries, and it has been observed that such boundaries are mobile at relatively low temperatures. However, the extent to which such junctions can move during cold deformation is still under investigation.

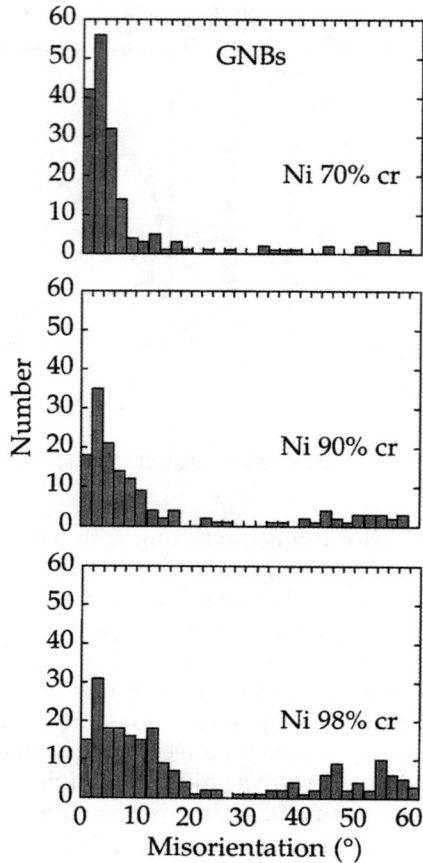

Figure 17 Distribution of misorientation angles. Hughes and Hansen (2000); Figure 5.

The classification of dislocation boundaries has led to a theoretical analysis of the evolution of misorientations in the lattice during plastic deformation (Pantleon, 1998). The buildup of such misorientations requires an excess of dislocations of one sign over dislocations of the opposite sign such an excess can be created by stochastic or deterministic processes acting separately or in combination. The stochastic evolution of misorientations is related to the existence of a bias in the dislocation fluxes caused by statistical fluctuations (Nabarro, 1994; Pantleon, 1998). Excess dislocations are trapped in dislocation boundaries during their passage and assuming that the boundary spacing does not change significantly during boundary formation, the rate of change in the misorientation angle across the boundaries is $pb\Delta j$ (Pantleon, 1998, 1999), where **b** is the Burgers vector, Δj is the bias in the dislocation flux and p is the capture probability. The probability is taken to be $p = d/\lambda$, where d is the boundary spacing and λ is the mean free path of moving dislocations (Pantleon, 1999). The deterministic evolution of misorientations is in addition to the stochastic evolution (Pantleon, 1998, 1999). Two different deterministic mechanisms have been proposed (Pantleon, 1999). One is related to the assumption of different slip activities on each side of the boundary that lead to a certain imbalance in the plastic strain between neighboring sides. This again leads to a geometrical bias in the dislocation

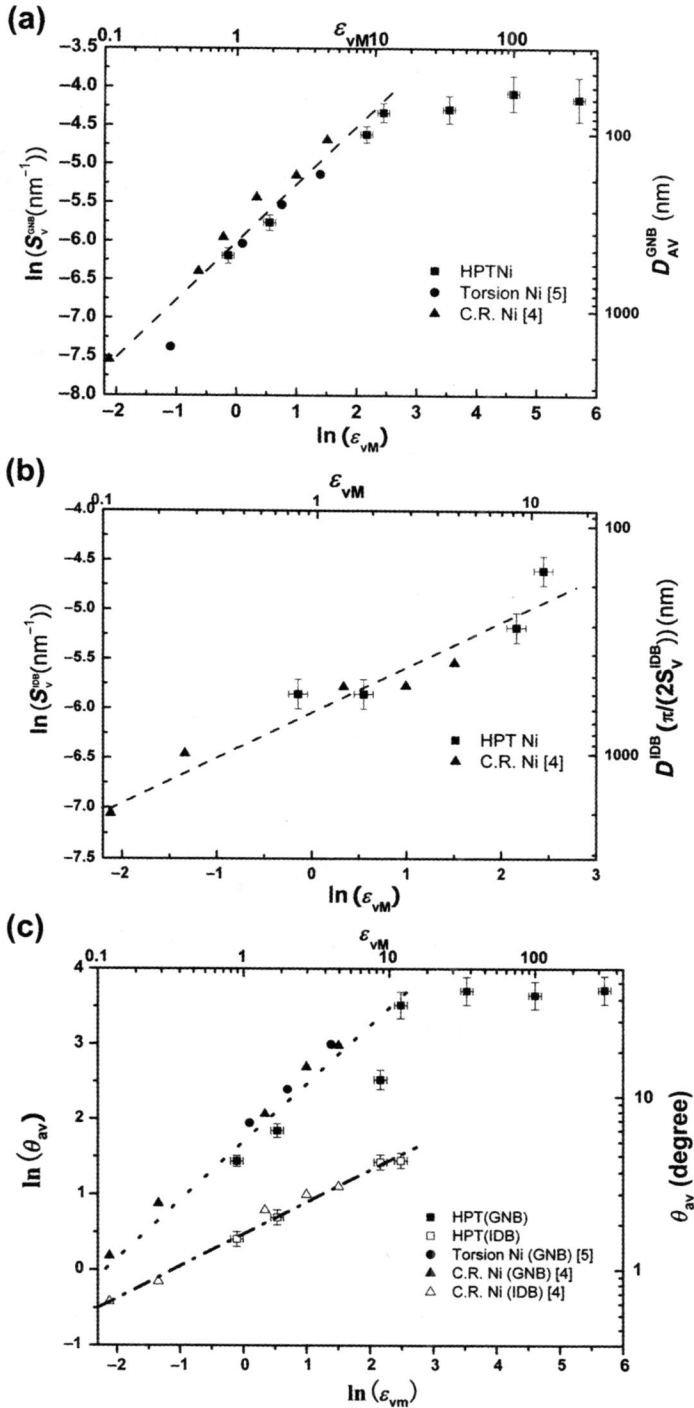

Figure 18 GNB spacing in nickel as a function of strain for (a) high pressure torsion and (b) rolling and torsion. (c) shows misorientation of IDBs and GNBs in nickel as a function of strain for high-pressure torsion, torsion and rolling.

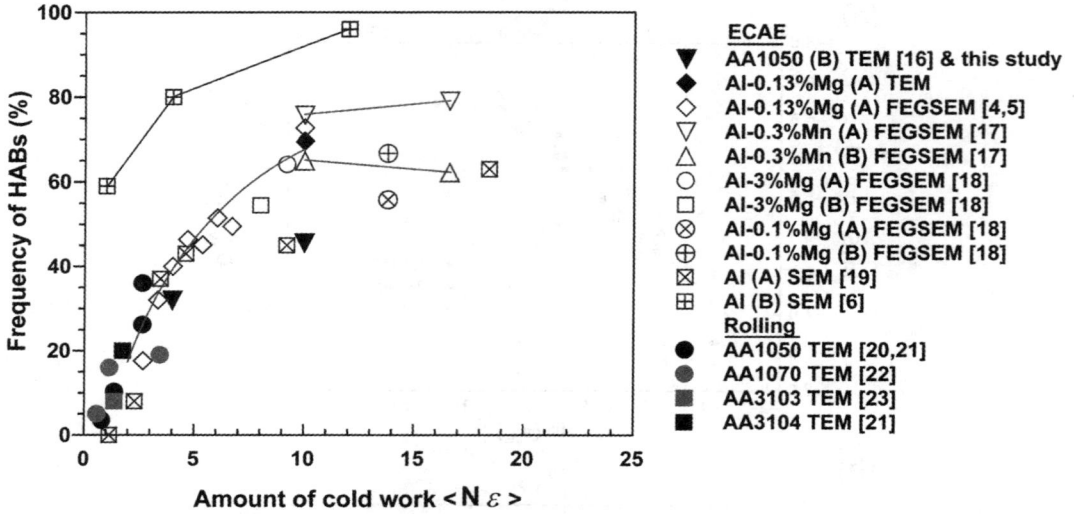

Figure 19 Fraction of HAGBs in Al alloys deformed by ECAE and rolling. Mishin et al. (2003); Figure 8.

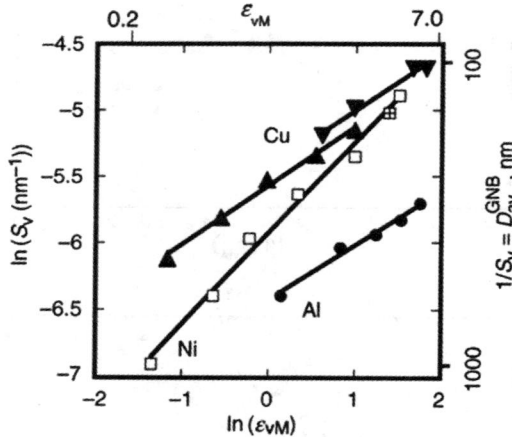

Figure 20 Power-law relationship between strain and the boundary area of GNBs per unit volume (S_v). Hughes and Hansen (2004); Figure 16.

flux Γj between the activated slip systems and to a rate of change in misorientation angle, which equals $b\Gamma j$. In addition, there is a deterministic contribution due to the formation of extrinsic dislocations. These dislocations form as a result of dislocations interacting with the boundary as they pass through it (Pantleon, 1999).

Bringing together the processes discussed here and the classification of dislocation boundaries, it is suggested that cell boundaries or IDBs are formed by statistical fluctuations, whereas GNBs form by

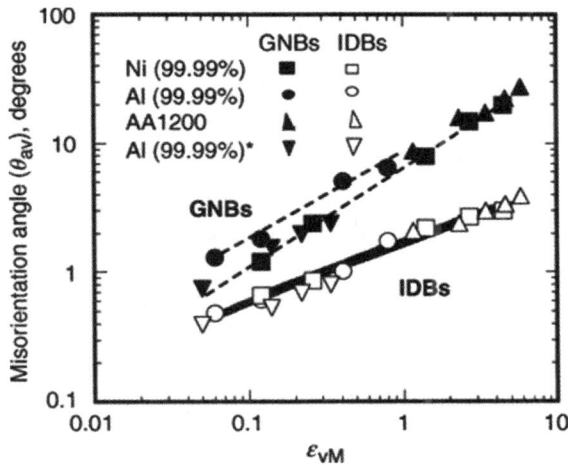

Figure 21 Power-law relationships between strain and misorientation angle for IDBs and GNBs. Hughes and Hansen (2004); Figure 17.

a joint operation of statistical fluctuations, activation imbalance, and formation of misfit dislocations. **Figure 23** shows a sketch of the two types of boundaries (Pantleon and Hansen, 2001a). Looking at the change in misorientation angle with increasing strain, calculations predict a much more rapid increase in misorientation angle when the formation is deterministic compared to the stochastic formation (Pantleon, 1999). This trend agrees with the experimental measurement of misorientation angles as a function of strain for IDBs and GNBs (**Figure 24**; Pantleon and Hansen, 2001b). In this figure, the calculated curves are fitted to the experimental data using a single parameter, which is the average boundary misorientation angle measured at a strain of one. **Figure 24** shows a rather good agreement between theory and experiment for IDBs. The average predicted angle follows a square root dependence on plastic strain ε. This dependency is followed up to large strains indicating that the development of IDBs is purely statistical and not affected by deterministic contributions. This is in contrast to the misorientation angle across GNBs where square root dependence is found at small strains turning into a linear relationship at large strain. This change relates to the empirical power low dependence where the exponent is about 0.7 (Hansen et al., 2006).

Good agreement is found between experiments and models developed using dislocation mechanisms to explain the evolution in boundary angles with strain, but there are some exceptions. First, the distribution of misorientation angles at high strain can be bimodal, so violating the scaling hypothesis. Second, the prediction of an unlimited increase in misorientation angle with strain does not match the observation of a saturation angle of about 40° in nickel deformed by HPT (Zhang et al., 2008b).

The change in a boundary angle distribution from low to high strain has been analyzed using the assumption that the high-angle boundaries observed at large strain may have their origins in two different mechanisms. One is a microstructural mechanism that relates to the formation of dislocation boundaries in which the misorientation angles steadily increase with strain. The other is a textural mechanism that relates to the texture evolution in different parts of a subdivided grain, which means that the presence of high-angle boundaries may relate directly to the orientation of different texture components and their neighbor–neighbor relationships. Such relationships have been analyzed for 10

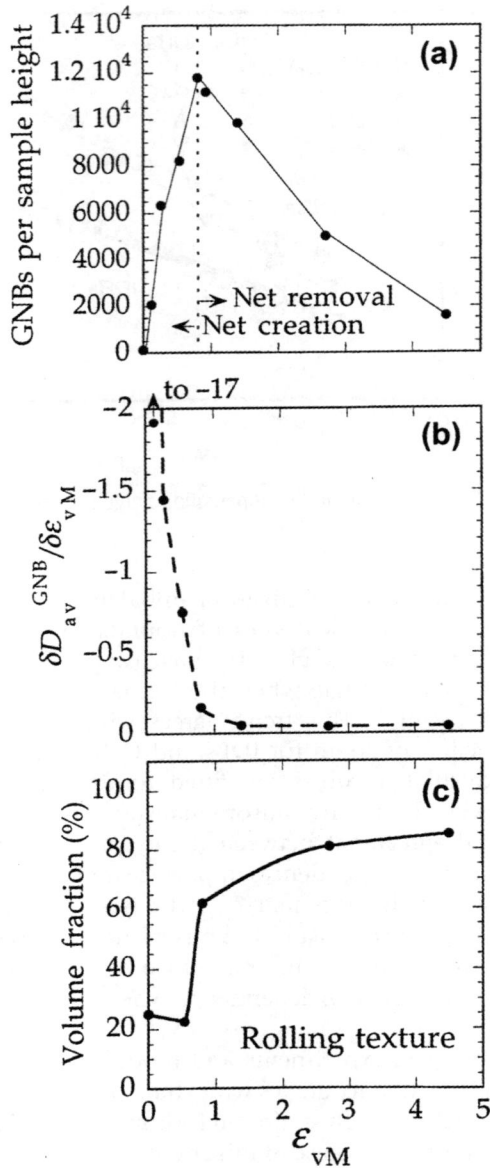

Figure 22 GNBs in cold-rolled nickel. (a) Number of GNBs as a function of strain. (b) Rate of change of number of GNBs. (c) Development of texture during deformation. Hughes and Hansen (2000); Figure 17.

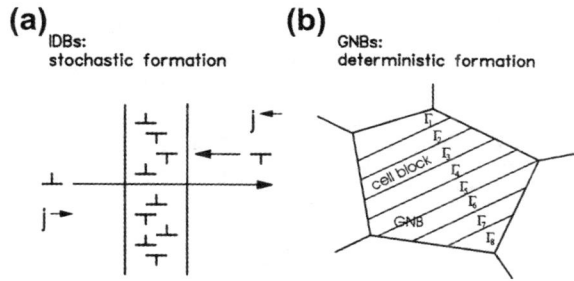

Figure 23 Stochastic and deterministic evolution of IDB and GNB. (a) Dislocation wall (IDB) with penetrating dislocation fluxes of a single slip system. (b) Cell block boundaries (GNBs) with different slip system activities Γ in the individual cell blocks. Pantleon and Hansen (2001a).

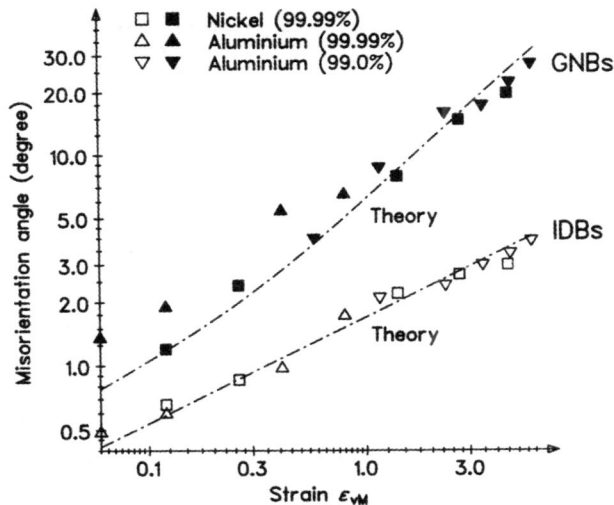

Figure 24 Experimental measurements of misorientation angles as a function of strain for IDBs and GNBs. Pantleon and Hansen (2001b).

components and variants of the typical rolling texture of fcc metals (Hughes and Hansen, 1997). These ideal orientations can give up to 44 different neighbor–neighbor relationships, each characterized by an angle/axis pair. Based on these 44 permutations, a distribution of misorientation angles has been calculated, see **Figure 25** (Hughes and Hansen, 1997), showing an angular range from 40 to 60°. Predictions of probability distributions of misorientation angles formed by microstructural mechanisms at medium and high strains are also plotted in this figure. A comparison of **Figure 25(a) and (b)** shows that very different distributions and ranges of angles are produced by the two mechanisms, which provide qualitative support of the experimental data. However, a combination of the two distributions into a total distribution with the correct amount of each population is required for a quantitative prediction.

(a)

(b)

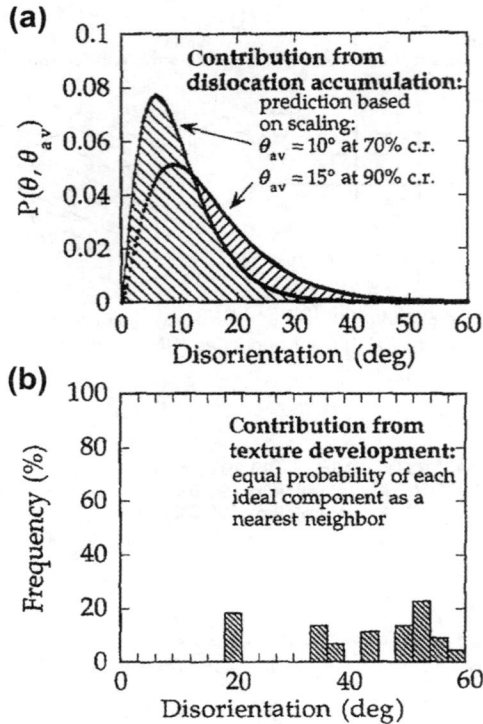

Figure 25 Misorientation distributions predicted as a function of strain for two different mechanisms: microstructural and textural. Hughes and Hansen (1997); Figure 3(a) and (b).

The observation that the average angle of GNBs approaches saturation at about 40° at large strain has been analyzed (Pantleon, 2011), with the conclusion that it can be a consequence of the restriction of experimentally measured misorientation angles to angles below a certain maximum value imposed by crystalline symmetry. For cubic symmetry, this results in a limiting value for the average GNB misorientation angle of 40.74°. However, as it is observed that dynamic recovery leads to saturation in both misorientation angle and boundary spacing at comparable strains (Zhang et al., 2008a), the underlying physically based mechanisms for the saturation must also be analyzed.

The above analysis, using average values of boundary spacing and angle as structural parameters, has been extended by applying a scaling hypothesis. This hypothesis is based on the assumption that similar underlying mechanisms control the formation of deformation microstructures. This is supported by the experimental observations, which show, for example, that the deformation takes place predominantly by dislocation glide and that dislocations are stored in LEDS.

The scaling hypothesis provides a general tool for analysis of structural parameters as demonstrated in analyses of the evolution in boundary spacing and angles with increasing strain (Hughes and Hansen, 1997). The distributions of these parameters at a given state depend on strain, but the scaling hypothesis demonstrates that the distributions may be represented by a strain-independent distribution using as a scaling parameter either the average boundary spacing or the average misorientation angle at each strain. **Figure 26** (Hughes and Hansen, 2004) shows that the scaled distribution of

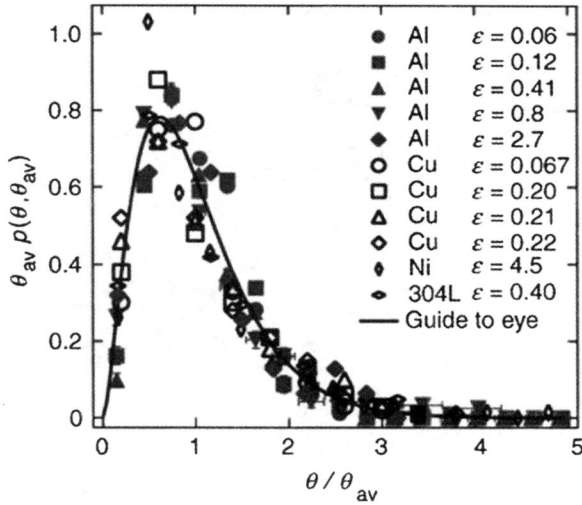

Figure 26 Scaled distribution of misorientation angles for different materials, plastic strain, temperature and deformation condition. Hughes and Hansen (2004); Figure 18.

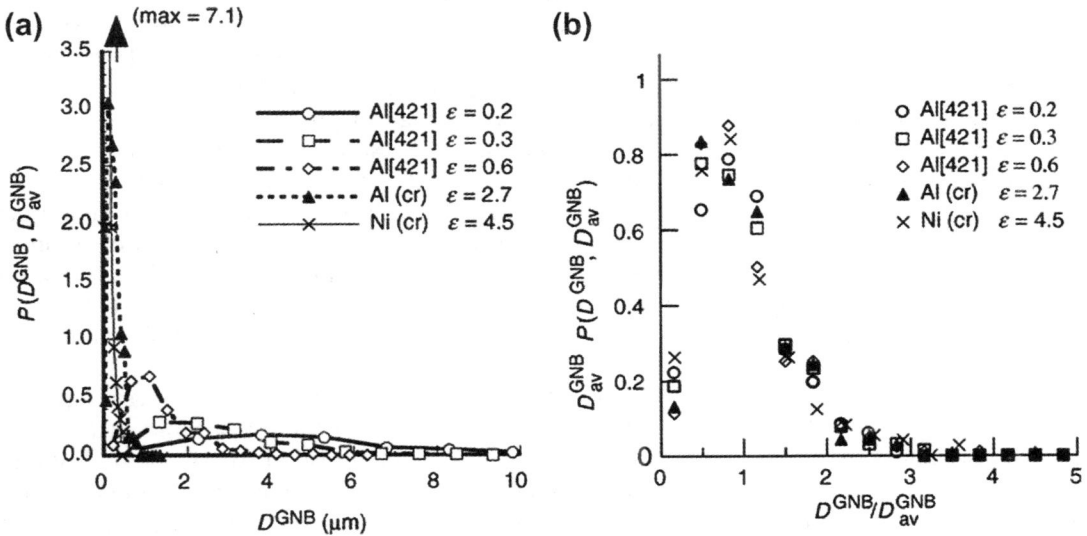

Figure 27 Probability distribution of GNB spacings (a and b) scaled by average spacings. Hughes and Hansen (2004); Figure 20.

misorientation angles of IDBs is independent of the material type, plastic strain temperature and deformation conditions. The example given in **Figure 27** (Hughes and Hansen, 2004) shows that the probability distributions of the GNB spacings collapse into a single function when scaled by their average spacings for a wide variety of conditions. In this figure, the minimum average GNB spacing at

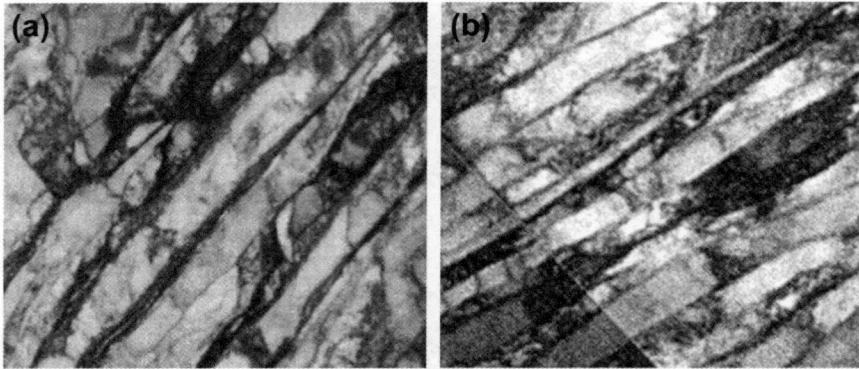

Figure 28   TEM showing structural refinement in cold-rolled Ni following a scaling relationship (a) Strain 0.8, average GNB spacing 0.5 µm. (b) Strain 4.5, average GNB spacing 0.13 µm. The structure is refined by a factor of 4 but is otherwise very similar. Barlow and Hansen (2007); Figure 12.11.

large strain is about 100 nm. However, the scaling hypothesis has also found application in the analysis of finer structures of the order of 60 nm in nickel deformed by HPT and 5–10 nm in copper deformed in friction (Hughes and Hansen, 2001; Godfrey and Hughes, 2004). In **Figure 28** an example is given of scaling of structures in nickel deformed to low and high strain.

The distribution functions used when applying the scaling hypothesis have been analyzed theoretically in the case of misorientation-angle distributions. The angular distribution can be described by a Gauss, a Raleigh, or a Maxwell distribution depending on whether interaction takes place between one, two, or three sets of dislocations, respectively (Pantleon and Hansen, 2001b). These cases show scaling behavior. The preceding analysis is based on statistical assumptions and geometric arguments on the premise that only small misorientations are present. For larger angles, misorientation distributions have been obtained from orientation distributions proposed using statistical assumptions about orientations (Pantleon and Hansen, 2001b). It is found, however, that the misorientation-angle distribution in such a case can also be described by a Raleigh or a Maxwell distribution. In **Figure 29** (Pantleon and Hansen, 2001b) experimental distributions of misorientation angles are compared with theoretically derived distributions, showing for both IDBs and GNBs that the Raleigh distribution gives a good fit.

The finding that the probability density distributions for boundary spacing exhibit a scaling behavior has been combined with the general experimental observation that there is a rapid increase in the number of boundaries in the low strain regime followed by a net removal or coalescence of boundaries at medium and high strain (Hughes and Hansen, 2000). A geometric model has been developed that relates both to formation and removal of GNBs, and it has been found that this model can account for the observation of scaling if various assumptions are met. For example in the low-strain regime, where net boundary creation dominates, the experimentally observed shape for the distribution of spacings can be maintained if it assumed that new boundaries form between existing GNBs with a probability proportional to the spacing between the GNBs, or that new boundaries are only formed between existing boundaries with a spacing greater than the average value.

A general approach to the microstructural analysis is to apply the principle of similitude (Kuhlmann-Wilsdorf, 1962), which suggests that any dislocation arrangement will be under equilibrium under a stress $n\,\tau$ if shrunk in size by a factor $1/n$. This principle has also been expressed as an

(a)

(b)

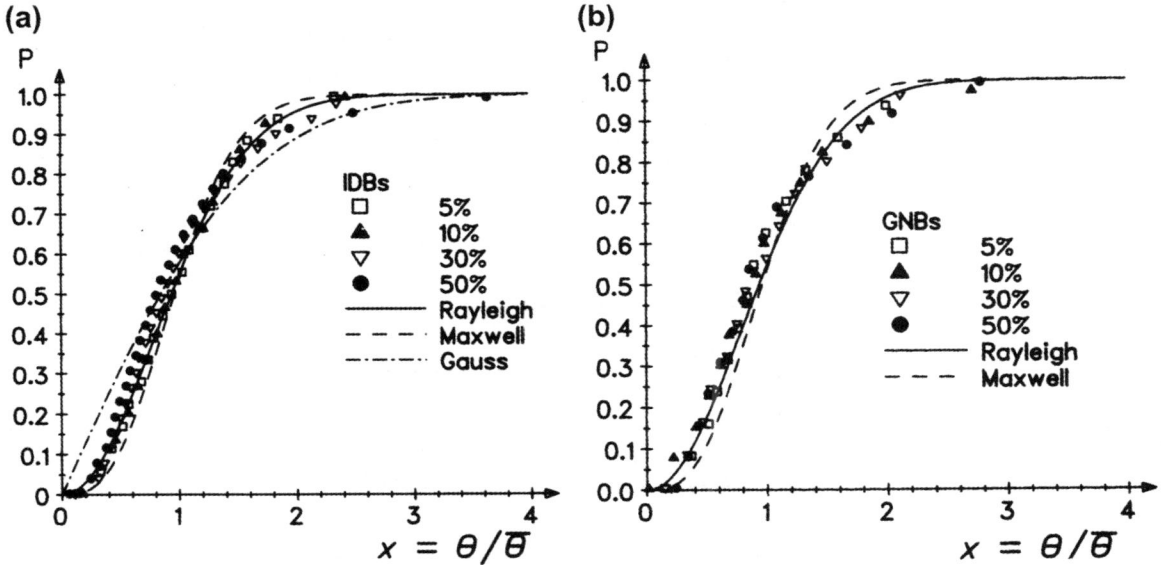

Figure 29 Accumulated frequencies of normalized misorientation angles for (a) IDBs and (b) GNBs for different rolling reductions of 99.99% Al. The lines show the probabilities P that a normalized misorientation angle is less than theta/theta bar. Pantleon and Hansen (2001b); Figures 5 and 6.

inverse relationship between the flow stress τ and the average dimension of the dislocation patterns d (Kubin, 2013):

$$\tau = KGb \, \frac{1}{d}, \tag{14}$$

where G is the shear modulus, $\mathbf{b}$ the Burgers vector and K is a constant. Similar equations have been proposed for the relationship between the flow stress and the reciprocal cell or subgrain size (Holt, 1970; Staker and Holt, 1972; Hansen, 1992). The value of K has been determined to be 10 for a wide range of materials (Staker and Holt, 1972), and was found to be about 8 for copper and aluminum (Hansen and Kuhlmann-Wilsdorf, 1986). For a large data compilation of materials containing subgrains, K has been determined to be 23 (Raj and Pharr, 1986). In this analysis the flow stress is the axial stress, which converted into a shear stress is 7.5, taking the Taylor factor as 3.06. The principle of similitude has also been expressed (Hansen and Hughes, 1995) by the relationship:

$$D \, \theta/\mathbf{b} = C, \tag{15}$$

where D is the spacing between a group of boundaries, θ is their average angle of misorientation, $\mathbf{b}$ is the magnitude of the Burgers vector and C is a constant. Different types of microstructures formed by tension and rolling, respectively, have been classified according to the value of C, which has been found to be almost constant in the range of 50–80 rad for a cell structure, and to be independent of strain (Hansen and Hughes, 1995). This constancy of C shows that the principle of similitude is a relatively good assumption for the cell structure in the strain range considered.

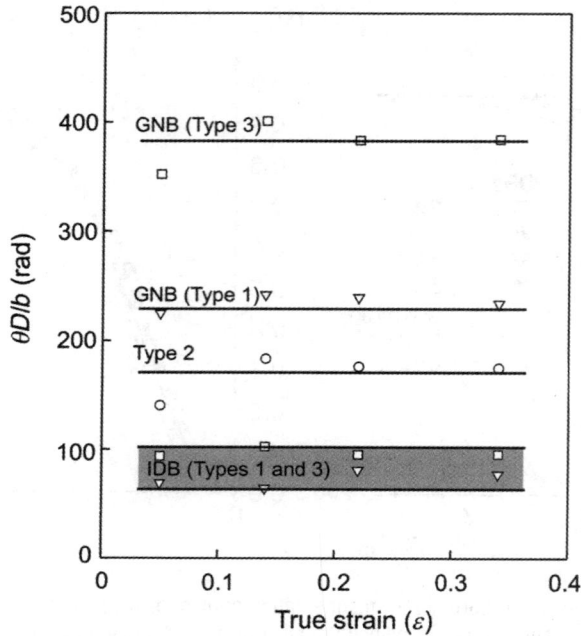

Figure 30 Similitude, expressed by the relationship $\theta D/b$, for GNBs and IDBs in Type 1 and Type 3 grains, and for all boundaries in Type 2 grains. Hansen et al. (2006); Figure 3.

The principle of similitude has been applied in an analysis of boundary structures characterizing Type I, II and III structures (described in Section 17.5.5.1) in aluminum strained in tension, see **Figure 30** (Hansen et al., 2006). It is seen that IDBs in Type I and Type III structures have very similar behaviors whereas the C values for the remaining boundaries are higher, showing significant variations but being fairly constant for the individual boundaries. Note, however, the difference between the cell structures in IDBs and the Type II structure, which is classified as a cell structure. The latter structure is subdivided on a larger scale by higher angle boundaries (GNBs) and a finer scale by low-angle boundaries (IDBs), whereas the cell boundaries in the IDB structure are all low angle (see also Section 17.8). The principle of similitude has also been applied to IDBs in Ni cold rolled over a large strain range. **Figure 31** shows an almost constant value of C for the IDBs, in good agreement with **Figure 30**.

17.5.5   Crystallography and Microstructure

17.5.5.1   *Grain/Crystal Orientation*

Microstructural subdivision as a result of plastic strain may be significantly affected by the crystallographic orientation of the crystal or grain. This has been demonstrated in many single crystal studies (Wróbel et al., 1988; Driver et al., 1994; Godfrey et al., 1998a, 1998b; Liu and Hansen, 1995b). Such an orientation effect has also been established for polycrystals (Al, Ni, Cu) deformed in tension, compression or in cold rolling (Le et al., 2012).

An example of tensile strained Cu (grain size about 100 μm) is given in **Figure 32** (Huang and Winther, 2007), where three different structural types have been identified (Huang et al., 2001; Huang

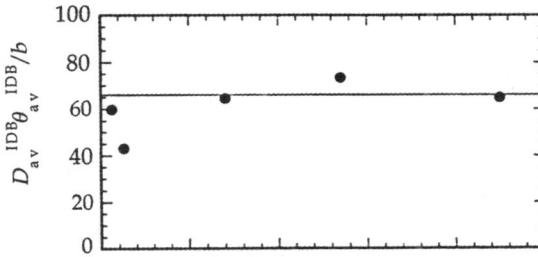

Figure 31 The similitude relationship $\theta D/b = $ constant for cold-rolled Ni. Hughes and Hansen (2000).

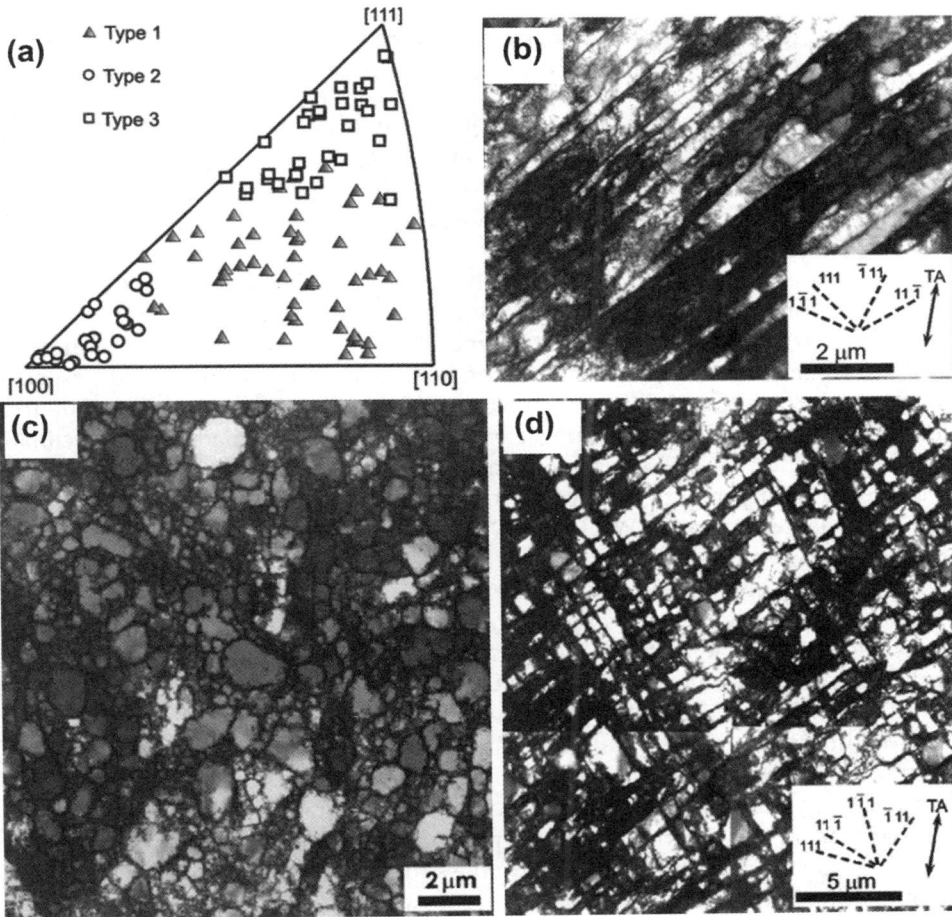

Figure 32 Grain-orientation-dependent structure types in a tensile-strained 99.99 pct Cu (grain size = 90 μm). (a) TA orientations of 120 grains, which have been examined. Three symbols, filled triangle, open circle, and open square, are used to separate the grains developing type 1, type 2, and type 3 structures, respectively. (b) through (d) TEM images to illustrate the characteristic features of (b) type 1, (c) type 2, and (d) type 3 structures. Among these images, the ones shown in (b) and (d) were taken from the section//TA, and that shown in (c) was taken from the section ⊥ TA. Huang and Winther (2007).

and Winther, 2007). The type 1 structure (**Figure 32(b)**) is a CB structure with CB boundaries (which are GNBs) aligned approximately with the slip planes (within 10°). The grains forming the type 1 structure have their tensile axes in the middle part of the stereographic triangle (**Figure 32(a)**). The GNBs are straight and parallel, and thus have a well-defined macroscopic orientation with respect to the tensile axis (TA). The slip plane to which the GNBs are related is in most cases the primary slip plane defined by the largest Schmid factor value and, in some cases, the conjugate slip plane for grains near the [100]–[111] boundary (where the Schmid factor of the conjugate slip system is similar to that of the primary one). The Schmid factor provides an indication of the slip system (slip plane and slip direction) on which the resolved shear stress τ is greatest, and is given as

$$\tau = \sigma \cos \phi \, \cos \lambda, \tag{16}$$

where σ is the applied tensile or compressive stress, ϕ is the angle between the slip direction and the TA and λ is the angle between the normal to the slip plane and the TA.

The type 2 structure (**Figure 32(c)**) is a cell structure without GNBs. The grains showing this type of structure have their tensile axes in a small area near the [100] corner (**Figure 32(a)**). The cell boundaries are randomly oriented in the cross section, as seen in **Figure 32(c)**, and are extended along the TA in the longitudinal section (for examples of micrographs for tensile-deformed Cu of different grain sizes, see Huang, 1998; Huang et al., 2001). Therefore, the type 2 structure in Cu can be described as a 3-D cylindrical cell structure.

The type 3 structure (**Figure 32(d)**) is also a CB structure similar to type 1, but the GNBs deviate substantially from the slip planes (>10°). The grains developing the type 3 structure have tensile axes in the upper part of the triangle toward the [111] corner. Each GNB is made up of a number of long or short segments. The majority of the segments are oriented in similar directions with respect to the TA, but segments also exist that have different macroscopic orientation and form angles of up to about 20° with the dominating segments. These features give rise to a fairly clear macroscopic inclination of the GNBs over a long distance, but present a somewhat curved appearance on a local scale.

Minor morphological differences were found in the type 2 structure between Cu, Ni, and Al. Recent observations of tensile-deformed Ni (Wei et al., 2011) showed a similar 3-D cylindrical cell structure to what is seen in Cu. However, the type 2 structure in tensile-deformed high-purity Al is characterized by a 3-D equiaxed cell structure (Huang and Winther, 2007).

The three structure types and their grain orientation dependence have also been identified in cold-rolled samples. **Figure 33** (Hansen et al., 2011) shows the results for cold-rolled copper. Clusters of orientation showing a type 1 structure are seen within 15–20° of the ideal Goss, brass and rotated cube orientations. Grains within 15° of the cube orientation develop a type 2 structure, which shows elongated cell boundaries parallel to the rolling direction (RD) in the longitudinal (RD–ND) section. Clusters of orientation showing a type 3 structure are seen at greater than 15° from the cube orientation, and extending between S and Cu orientations. A clear demonstration of the crystallographic dependence in the alignment of the CB boundaries in type 1 structures formed during rolling of aluminum is seen in the study by Lin et al. (2009). By looking at the structures formed in different symmetry variants of the same grain orientation it was found that in every case symmetry variants of the same type showed CB alignment in the same direction.

17.5.5.2 Local Texture

The analysis of a dependency between the deformation microstructures and the crystallographic orientation of the volumes in which they develop has been extended from an effect of grain or

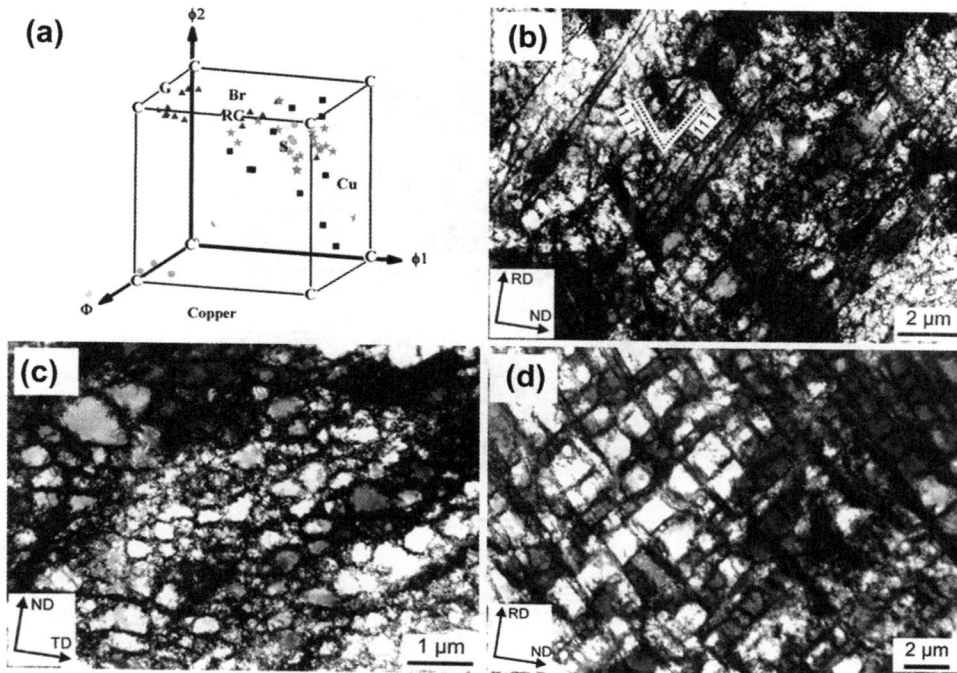

Figure 33 Grain-orientation-dependent structure types in cold-rolled 99.99% copper. (a) 3-D Euler space showing grain orientations: triangles are Type 1; circles are Type 2; squares Type 3. (b), (c), (d) show TEM images of Types 1, 2 and 3 respectively. Hansen et al. (2011); Figure 7.

crystal orientation at low and medium strain (Stage III) to an effect of local texture in samples that have been deformed to large strain (Stage IV). An example of such structure is shown in **Figure 34** (Hansen, 2001) of aluminum rolled to a true strain of 5.0, where subdivision into thin lamellae is shown together with the evolution in misorientation angles and local texture. The local texture components are the typical rolling texture components (copper, S, brass and cube). The structure also demonstrates that it is not an entire grain that rotates into a new position as the deformation texture evolves. Individual parts of a grain can rotate toward different orientations, and the mutually misorientated lamellae comprise the macroscopic texture components. To explore the relationship between the local texture and the characteristics of the deformation microstructure a sample was rolled to a lower true thickness strain of 2 and the microstructure in individual lamellae characterized. It was found (Xing et al., 2006) that there was no striking morphological difference that can be related to the local crystallographic orientation. However, when analyzing the crystallographic parameters, there was a significant difference between bands of different textured components.

A general observation from all these studies is that both the average GNB and IDB misorientation angles are smaller in the bands of rolling texture components than in bands of other texture components, in which there is a higher stored energy. There are also distinct differences between different rolling texture components, for example in the spread of misorientation angles.

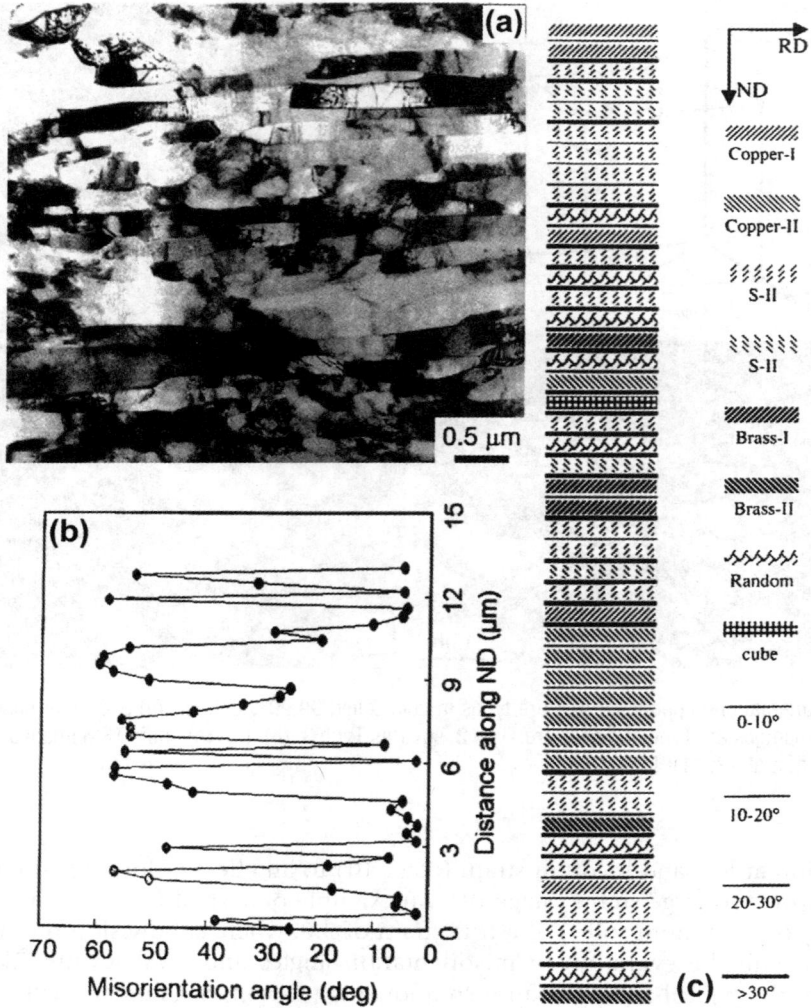

Figure 34 Commercial aluminum rolled to strain of 5. (a) The microstructure is subdivided by lamellar boundaries almost parallel to the trace of the rolling plane. (b) The misorientation angle across the lamellar boundaries measured along the normal direction to the rolling plane. (c) Rolling texture components in the lamellar structure. Hansen (2001); Figure 5.

17.6   Slip System Dependence

The orientation dependence of the main dislocation structure types can be explained in terms of an underlying dependence on the active slip systems, which suggests that combinations of slip systems leading to structure types 1, 2, and 3 can be identified (see Section 17.5.5.1). This further implies that if similar slip system combinations are activated under different circumstances, e.g. different combinations of crystal/grain orientation and deformation mode, the structure type should be the

Table 4   Relationship between slip classes and structure types

Slip class	Crystallographic GNB plane	Structure type
Single slip	{111}	Type 1
Coplanar slip	{111}	Type 1
Two sets of codirectional slip	no GNBs, only cells	Type 2
One set of codirectional slip symmetric	{101} or {010}	Type 3
Dependent coplanar and directional slip	{315}, {441}, or {115}	Type 3

same. Analysis of the active slip systems in crystals or grains with the different types of structures has been conducted for tension and rolling (Winther and Huang, 2007), focusing on cases where the active slip systems can be predicted fairly unambiguously by the Taylor model and Schmid factors. The result is the identification of five classes of slip system combinations, which lead to specific structure types. **Table 4** (Hansen et al., 2011) summarizes the relationships between slip classes and structure types.

17.6.1   Slip Classes and Structure Types

For the type 1 structure, it is found that the GNBs always align with a highly active slip plane (Huang and Winther, 2007). When conducting an even more detailed study of the type 1 structure, it is found that it falls into two subcategories, both having GNBs aligned with {111} planes but with different systematic small deviations from the ideal slip plane. These deviations are rotations of a few degrees (up to $10°$) around a specific crystallographic axis, which may be either a $\langle 112 \rangle$ or a $\langle 110 \rangle$ axis (Winther and Huang, 2007). The existence of these two subcategories implies an underlying difference in slip systems, which is the basis for the definition of two different slip classes. Detailed analysis of a number of cases leads to the conclusion that the difference lies in the number of active slip systems on the slip plane so that deviations around $\langle 112 \rangle$ axes result from one active system while the $\langle 110 \rangle$ axes originate from activation of two systems in the same slip plane (often referred to as coplanar systems (Winther and Huang, 2007)).

For the type 2 structure, the common feature of the slip systems is the activation of codirectional pairs of slip systems, involving two systems on two different slip planes but with the same slip direction. When comparing different cases, it is concluded that at least two sets of codirectional slip systems (a total of four systems) must be activated to produce a type 2 structure. The sign of the dislocations gliding on the slip systems also plays a role (Winther and Huang, 2007).

For the type 3 structure, the number of subcategories is the highest and the difference between subcategories is much larger than for type 1. **Table 4** lists the different GNB planes observed and divides them into two subcategories, each of which is associated with activation of a slip class. Note that in all cases the GNB planes lie far from a slip plane, in agreement with the definition of the type 3 structure. The two slip classes consist of the following combinations of slip systems:

(1) One set of codirectional slip systems. The GNBs align with a plane containing the common slip direction and bisecting the angle between the two slip planes. The GNB lies closer to the more active slip plane and the signs of the slip systems control whether the GNB bisects the acute or obtuse angle between the slip planes.

(2) So-called dependent coplanar and codirectional slip, which is a combination of three systems on two slip planes where one system is codirectional and coplanar to the other two. Several such slip system combinations may be activated in the same grain, possibly sharing some of the slip systems, which gives rise to a number of characteristic GNB planes. For more details, see Winther and Huang (2007).

The relationships between slip classes and structure types are of universal nature in the sense that they are not restricted to any deformation mode or grain orientation. The identification of these relationships establishes a new framework for the analysis and interpretation of deformation microstructures.

By means of the slip class concept, some of the fluctuations observed between grains of similar orientation or within a grain can be better understood. The occurrence of both type 1 and 3 boundaries in the rolling texture (Huang and Winther, 2007) is, for example, due to activation of both a coplanar set of slip systems and a codirectional set, leading to type 1 and 3 boundaries, respectively. It is often observed that either the type 1 or 3 boundary is more clearly developed. In most cases, however, the second set can also be detected. These variations are attributed to local variations in the relative activities of the slip systems, possibly caused by minor strain variations or ambiguities. Analogously, one or two sets of type 1 GNBs are found in grains of Goss or Brass (Huang and Winther, 2007) orientation, depending on the relative activities on the two slip planes with which the GNBs align. It is, however, emphasized that these variations have their origin in fluctuations in the relative activities of a fixed set of slip systems, which can be predicted from the grain orientation, rather than activation of new systems, which can only be predicted by analyzing detailed interactions with neighboring grains.

17.6.2   Similarity between Single Crystals and Polycrystals

The classification into three main types of structures, of which types 1 and 3 can be subdivided further according to the exact plane of the GNBs, has been presented and analyzed above for grains in polycrystals. The classification applies equally well to single crystals, and furthermore, the structure type follows the crystal or grain orientation in almost the same way. Based on observation and on analysis as above, it is found that single crystals and grains of similar orientations will in general develop the same structure type. This is evidence of activation of similar slip systems. It is particularly interesting that type 1 structures are found in both single crystals and grains in the middle of the triangle. It is generally accepted that these single crystals deform in single glide, while the Taylor model often applied to polycrystals predicts up to eight active systems. The structure observations suggest that the grains have one dominant slip system, which may, however, be accompanied by minor activity on other systems. It should be noted that the lattice rotations of individual grains of these orientations have also been observed to be in good agreement with the operation of just one dominant system (Winther et al., 2004; Winther, 2008). However, some minor discrepancies are observed when considering the exact GNB plane in tension around the $\langle 111 \rangle$ orientation. Furthermore, the expectation is that single crystals fairly close to $\langle 111 \rangle$ and $\langle 100 \rangle$ inside the triangle will also deform in single glide, whilst grains in polycrystals of these orientations develop type 3 and 2 structures. The origin of this discrepancy is likely to be activation of a larger number of slip systems in grains in polycrystals compared to single crystals. This is due to the restrictions imposed by the neighboring grains combined with the availability of a number of slip systems with almost equal Schmid factors in these orientation ranges. Similar discrepancies between single crystals and polycrystals have not been reported for rolling, probably because of the inherent tendency of rolling to induce multislip in all crystals or grains.

17.6.3 Structure of Dislocation Boundaries

The internal structure of dislocation boundaries can be investigated using standard TEM techniques to identify the Burgers vectors present in the boundaries. These Burgers vectors can then be related to the active slip systems in the volumes adjacent to the boundary. An example is shown in **Figure 35** of almost planar dislocation boundaries parallel to a slip plane that formed during tensile straining of a copper single crystal in Stage II hardening (Hughes et al., 2001). These extended boundaries are identical to the carpet structures described in earlier studies (Steeds, 1966). Note in this figure that dislocation cells have started to form between the GNBs. Analysis of the GNBs shows that they contain several Burgers vectors including the primary $[\bar{1}01]$ as well as secondary $[\bar{1}10]$ and $[011]$. This observation of the variety of Burgers vectors is similar to the observation by Steeds (1966). A second example (McCabe et al., 2004) is a study of GNBs present in pure polycrystalline copper, which has been cold-rolled to a reduction in thickness of 7.5%. The structure contains two sets of GNBs, one of which is aligned with the slip plane, and the other which is not. Analysis of the former reveals three sets of dislocations belonging to the slip systems with the highest Schmid factors. In addition, dislocation interactions forming Lomer dislocations have been observed (McCabe et al., 2004).

The close correlation between the dislocation content of boundaries and the active slip systems has also been observed by characterization of Burgers vectors, dislocation lines and relative densities in cell boundaries in near-cube grains in aluminum rolled to 10% strain (Wei et al., 2011). Practically all the dislocations have screw character with Burgers vectors corresponding to the slip systems, which are active. Such a correlation has also been investigated in an extensive study of the dislocation content in slip plane aligned GNBs (Type I boundary) in grains of 45° ND-rotated cube orientation in heavily cold-rolled pure aluminum (Hong et al., 2013). The investigation by TEM shows that dislocations with all six Burgers vectors of $1/2\langle 110 \rangle$ type expected for fcc crystals are observed but the dislocations from the four slip systems expected to be active dominate in the boundary. The dislocations not corresponding to one of the active slip systems are primarily believed to be the result of dislocation relations in the boundary, supplementing Lomer locks, which are also identified. Redundant dislocations have not been observed in the GNBs analyzed (Hong et al., 2013), but storage of such dislocations in the IDBs must be expected.

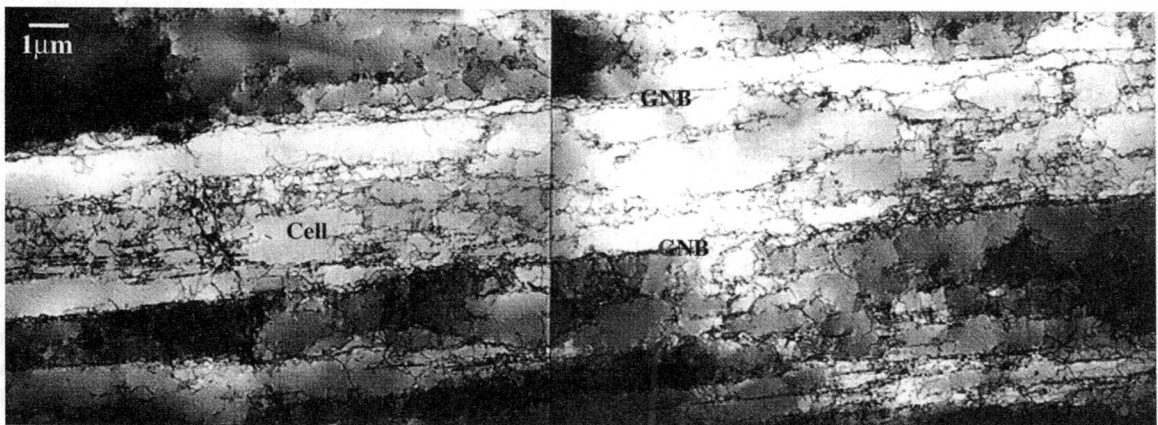

Figure 35 Extended planar GNBs in Cu single crystal oriented for single slip. Cells and loose dislocations are seen between the GNBs. Hughes et al. (2001).

An alternative method for analyzing boundaries is to use the Frank formula (see Section 17.3.6.2) which relates boundary misorientation to dislocation content in order to make estimates about the dislocation density within a boundary. The approach requires experimental analysis of the axis/angle pair of the boundary (R/θ). In general, a minimum of three independent Burgers vectors are required to accommodate the misorientations across the boundary. For a Burgers vector $\mathbf{b} = 1/2\langle 110 \rangle$ in the fcc lattice, 16 combinations of three Burgers vectors are possible, where each combination defines a boundary. Different criteria have been proposed to select between the combinations, such as:

> the boundary shall be of low energy, i.e. a high area density of dislocations; or the dislocations in the boundary shall relate directly to the slip systems and the slip activity in the adjoining volumes.

In applying the latter criterion the slip activity in a given volume has been calculated (Wert et al., 1995) using its crystallographic orientation together with the stress and strain conditions imposed by using either a conventional polycrystal plasticity model such as the full constraint Bishop–Hill model (Bishop and Hill, 1951) or a Schmid factor analysis. Such analysis has been applied to boundaries in pure aluminum rolled to a reduction of 30% ($\varepsilon = 0.41$) in the two cases where the boundaries are and are not aligned with slip planes. It was found (Wert et al., 1995) that a substantial fraction of the dislocations in the boundaries relate to the primary and the conjugate slip systems. In addition to the dislocations that contribute to the boundary misorientation angle, the boundaries also contain a significant fraction of redundant dislocations. This density has been estimated to reach up to a third of the total dislocation density in the boundary. These estimates have been made based on a comparison between the calculated dislocation number density and the total number of dislocations derived from densitometry measurements (see Section 17.4.2.1) on aluminum samples rolled to a comparable reduction in thickness (Wert et al., 1995).

With increasing strain the misorientation angle across dislocation boundaries increases and their width decreases. This sharpening of the structure may lead to a reduction in the density of redundant dislocations. An example of a boundary with no redundant dislocations is illustrated in **Figure 36(a)** for copper deformed to a very high strain. For such boundaries, the distance between individual dislocations, D, can be determined, corresponding to an area fraction of dislocations of 1/2. For the boundary in **Figure 36(a)**, this area density is equal to $\theta/\mathbf{b}$, where θ is obtained experimentally. This agreement demonstrates that a minimum dislocation density can be accounted for on the basis of the boundary misorientation.

Extra dislocations may, however, be present close to the boundary, as can be seen in **Figure 36(b)**. The misorientation angle of this boundary is about $14°$ and $D = 1.05$, which agrees with the analysis for **Figure 36(a)**.

The boundaries in **Figure 36** are tilt walls i.e. a low energy configuration, however, with some end stresses being present. Extra dislocations may, however, be located close to the boundary as observed in **Figure 36(b)**. These dislocations contribute to the lattice bending and can therefore not be classified as redundant dislocations. Rather, they must be glide dislocations, which have been stopped by the boundary acting as an obstacle to slip. In other cases, such glide dislocations may interact with boundary dislocations and become incorporated in the structure, e.g. as extrinsic dislocations in grain boundaries (Ralph et al., 1981) as illustrated in **Figure 1**. The characterization of deformation-induced boundaries has been discussed in detail elsewhere in this volume (Wilde, 2013). The deformation-induced high-angle boundaries are generally classified as high energy boundaries—a more useful description than the term nonequilibrium boundaries commonly found in the literature. The energies

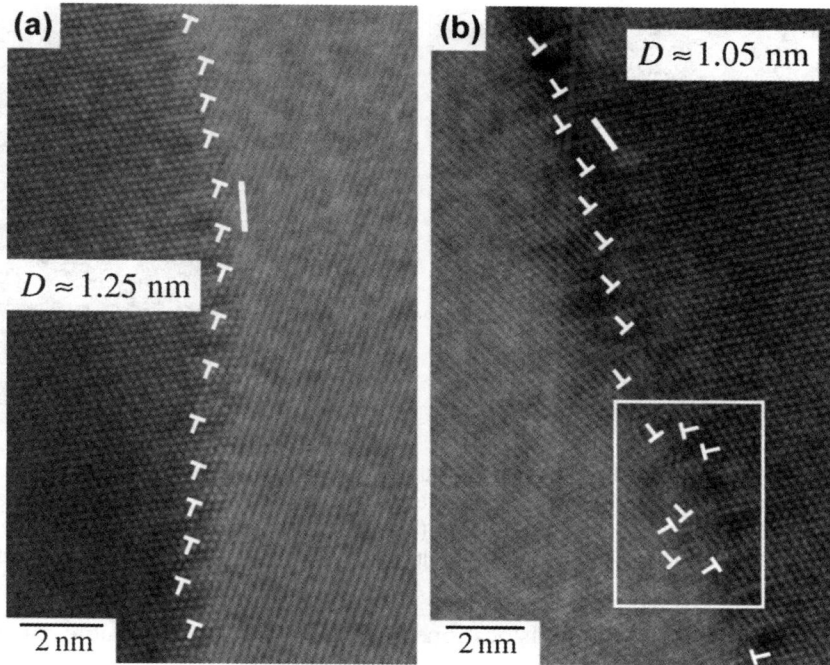

Figure 36 (a) Boundary containing no redundant dislocations in highly strained Cu. (b) Extra dislocations can be seen close to this boundary. Ikeda et al. (2004); Figure 2.

in such boundaries are higher than in relaxed grain boundaries formed by recrystallization. The increase in boundary energy can have its origin in dislocations at or near the grain boundary and due to the existence of long-range- and short-range internal stresses. The magnitudes of such stresses are limited by the flow stress of the matrix and they may be determined by X-ray diffraction in a synchrotron (Levine et al., 2011). Another method is geometric phase analysis (GPA) in an HREM (Wilde et al., 2011), where strain fields at grain boundaries have been found (e.g. in Ag–Pd alloys) in the range of 1.5–2 nm, which is 2–3 times the value observed for a relaxed boundary. Another method to characterize deformation induced high-angle boundaries is to measure boundary diffusivities (Wilde, 2013) with a radiotracer technique. This has led to a classification of boundaries into high-energy boundaries, however of different types representing a spectrum of diffusivities and structures. However, it is also found that boundaries fall into the category of general high-angle grain boundaries, with diffusivity and probably structure being similar to those of relaxed grain boundaries. In conclusion the behavior of deformation-induced high-angle boundaries may as a first approximation be similar to relaxed boundaries; however, local strain fields may in some cases add to the boundary energy per unit volume. Such an effect may be taken into account for example when calculating stored energy of deformed structures based on structural parameters (see Section 17.4.2.7).

Analysis of the structural evolution by applying scaling of the boundary spacing at different strains has demonstrated that the same scaling law can apply over a structural scale from the micrometer dimension to the nanometer scale (Godfrey and Hughes, 2004). This indicates that the basic mechanisms for grain subdivision by dislocation multiplication and storage apply, independent of the

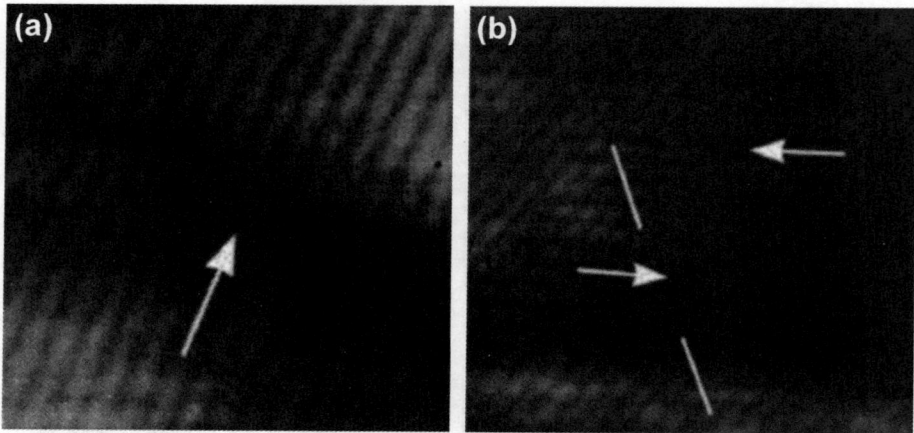

Figure 37 HREM cross section of 5–20 nm layers produced by sliding of copper showing the projected edge component of mixed 60° lattice dislocations. The lattice fringes are marked by straight lines and extra half planes of lattice fringes are marked with arrows. In (b) two separate extra half planes of (111) fringes are marked. No extra half plane is observed on the (11-1) fringes indicated by straight lines. **Figure 12** from Hughes and Hansen (a and b only). Hughes and Hansen (2003); Figure 12 (part).

structural scale. This means that basic dislocation theory may also be applicable for nanostructured metals processed by plastic deformation.

The role of dislocations when the structural scale is reduced to the nanometer dimension has been explored in fine structures produced in copper by sliding friction (Hughes and Hansen, 2003). Observations of dislocations in such fine structures require more specialized TEM techniques compared to submicrometer structures. The difference is caused by the high degree of overlapping diffraction contrasts that obscure the images of the dislocations. Sources of this contrast include Moiré fringes from overlapping crystallites and inclined boundaries, as well as the size of the dislocation strain field compared to the crystallite size.

Figure 37 shows HREM (high-resolution electron microscopy) micrographs of several layers divided by nearly parallel GNBs in a friction sample of copper. In total 41 dislocations were characterized within these layers. Of these, 34 could be clearly identified as 60° lattice glide dislocations lying on either the (111) or $(\overline{11}1)$ glide planes. The total dislocation density observed in **Figure 37** is about 10^{16} m^{-2}, which is one or two orders of magnitude larger than in conventional cold-worked metals. Additional dislocations may also be present that are invisible under these viewing conditions.

In summary, the detailed analysis of dislocation structures in the preceding section has demonstrated the dominating role of dislocations in the development of deformation microstructures on multiple length scales down to about 10 nm.

17.7   Microstructure and Local Texture

The grain breakup during plastic deformation into small crystallites that have a large spread in orientations does not lead to new major texture components, suggesting that Taylor-like kinematics are valid on average. Instead, subdivision creates some minor components and a spread in the texture (Hughes and Hansen, 1997).

Local crystallographic measurements show that the volume fraction of the individual texture components is composed of micrometer and submicrometer-sized volumes distributed throughout the deformed microstructure. Examples of such patterns of individual crystallites are given in **Figure 38** (Hughes et al., 1998). In this figure the color shading regions represent association with specific ideal texture components. A local orientation is classified as an ideal component if it has a misorientation of less than 15° related to the ideal component. An orientation not within 15° of any ideal component is taken to be a random orientation.

Figure 38 shows that in general the high-angle boundaries are distributed widely throughout the distances measured. At the same time there are quiet regions with only low to medium-angle boundaries and regions with clusters of high-angle boundaries. For the case of fcc rolling these quiet regions were associated with average grain orientations including Goss and brass components, whereas for bcc rolling these regions were primarily associated with rotated cube orientations. In contrast to the quiet areas, the clusters of high-angle boundaries in fcc rolling are associated with alternations, which include finely distributed regions with the variants of S, copper and brass plus occasionally a Goss orientation.

The different deformation textures that develop reflect the different crystal structures fcc and bcc as expected. However, both the fcc and bcc metals show similar trends with regard to the number of high-angle boundaries and texture components.

17.8 Hot Deformation

The structural evolution described in previous sections characterizes cold and warm deformations. A high-temperature microstructural modification may be observed but many microstructural features are common to both cold and hot deformations, e.g. cells, dense dislocation walls, CBs, microbands and subgrains (Duly et al., 1996; McQueen et al., 1989; Doherty et al., 1997).

Characteristic hot-deformation conditions are temperatures higher than 0.6 Tm (where Tm is the absolute melting temperature) and strain rates in the range 10^{-2}–10^2 s^{-1} and it is observed that both temperature, strain rate and strain affect characteristic structural parameters. This behavior is for example expressed in an inverse relationship between the boundary spacing and the Zener–Hollomon parameter (Theyssier et al., 1995) (see Section 17.2.2). In contrast the misorientation angle is much less affected by temperature and strain rate but it is significantly affected by the strain as both the misorientation angle and the misorientation angle distribution are observed to increase with increasing strain (Juul Jensen et al., 1998).

The structural evolution during hot deformation will in the following be described by the way of an example, the alloy AA 3104 (Al one pct. Mg, one pct. Mn) deformed by plane strain compression at 510 °C ($\sim$0.8 Tm) and a strain rate of 5 s^{-1} (Liu, 1998). The low strain structure is illustrated in **Figure 39 (a)** (Hansen, 2001), showing a CB structure with extended dense dislocation walls and microband forming an angle of about 35° with the RD in this longitudinal section. Between the dislocation boundaries a loose dislocation structure is present. However, such a clear similarity between low- and high-temperature deformation structures at low strains changes significantly as the strain is increased and the typical lamellar structure evolving during cold rolling is replaced by an almost equiaxed structure as shown in **Figure 39(b) and (c)**. This morphological change may have its origin in different factors: (1) an increased concentration of vacancies; (2) an increased mobility or dislocation boundaries as their misorientation angles increase with increasing strain; and (3) an energy gain as the boundary energy per unit volume is reduced in parallel with the evolution of the equiaxed structure.

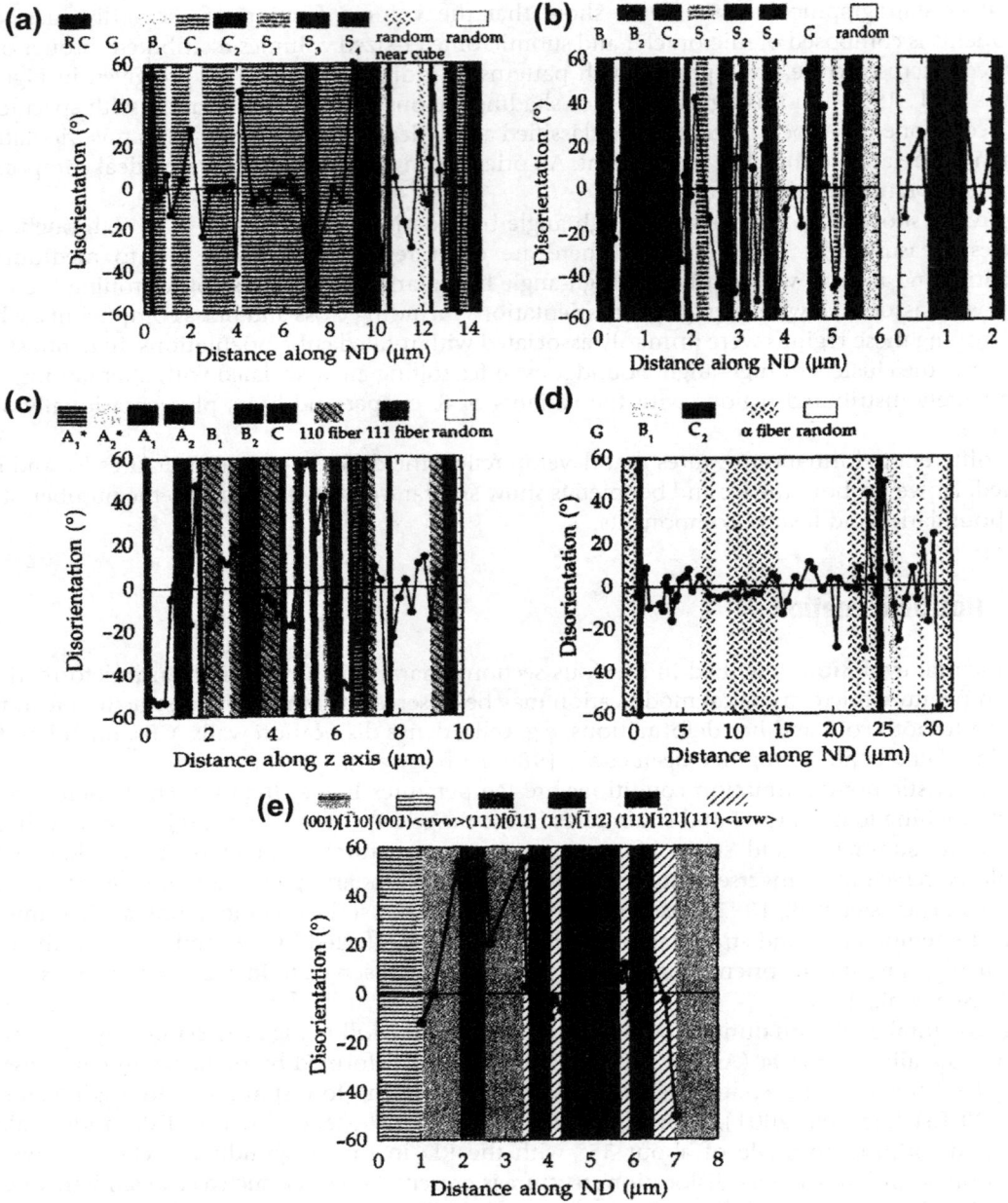

Figure 38 Individual crystallites related to texture components. (a) Nickel 70% cold rolled; (b) nickel 98% cold rolled; (c) nickel deformer by torsion; (d) aluminum 90% cold rolled; (e) tantalum 90% cross-rolled. Hughes and Hansen (1997); Figure 6.

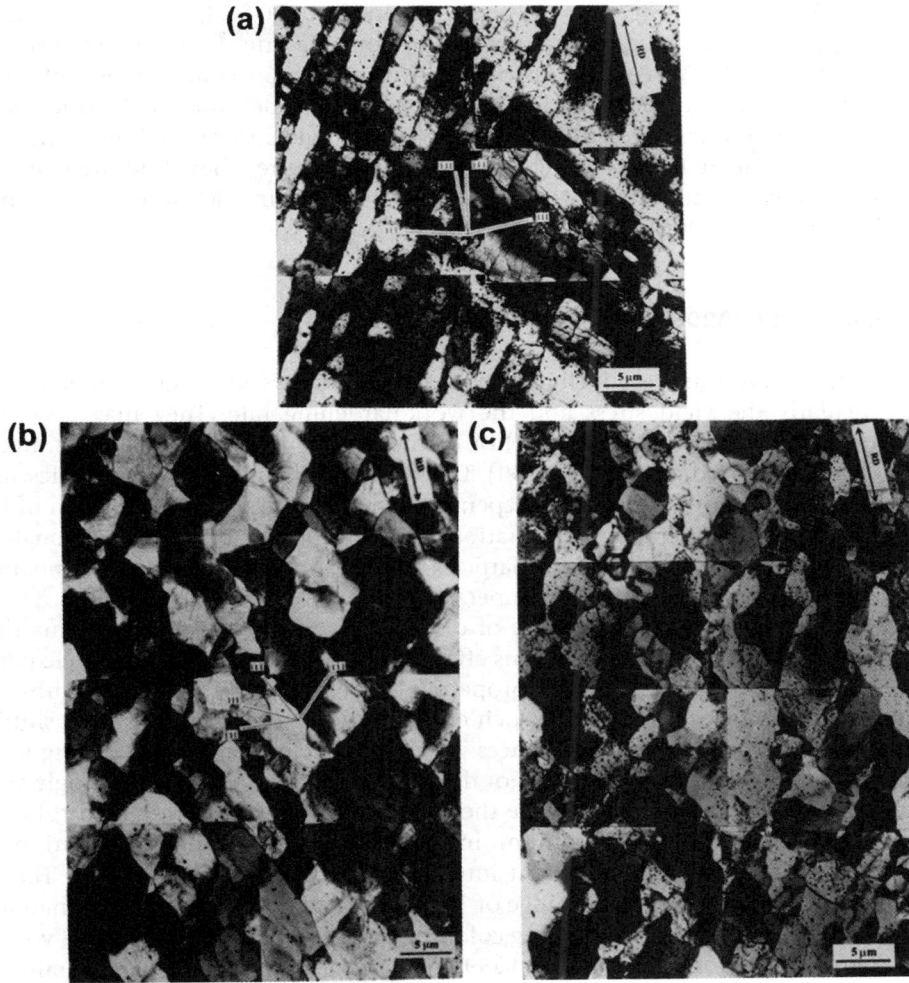

Figure 39 TEM micrographs of Al–1%Mg–1%Mn (AA3104) plane strain compressed at 510 °C. (a) Strain 0.15. The structure is divided by two sets of boundaries neither of which is parallel to a slip plane trace. (b) Strain 0.6. The structure has evolved into a parallelogram-shaped configuration. (c) Strain 1.7. The structure is divided into almost equiaxed subgrains. Hansen (2001); Figure 8.

A characteristic feature of cold deformation is the clear relationship between the characteristic of the deformation microstructure and the crystallographic orientation of the volume in which it evolves. Such a relationship can also be found in hot-deformed structures. In a study of IF steel cold-rolled to 10 and 30% reductions, grains of various orientations have been studied by TEM showing CBs delineated by extended GNBs with characteristics that depend on the grain orientation. An example of macroscopic orientations in ferrite using a rolling temperature of 1073 K (0.59 Tm) shows strong correlation between planes of the extended boundaries and the active slip planes for a given grain orientation (Haldar et al., 2004). Other examples are (1) equiaxed CBs formed within cube-oriented grains in an aluminum alloy AA 5005 (0.60 Mg, 0.14 Si) tensile strained at 250 °C (Cizek, 2010) and (2) micro-bands along the slip planes with high Schmid factors in noncube-oriented grains and cell structures in cube-oriented grains in a Ni-30 Fe alloy deformed by plane strain compression between 700 and 900 °C (Taylor et al., 2012). In both studies it is observed that deformed cube-oriented grains are subdivided on two length scales. On the grain scale by larger-angle dislocation boundaries (DDWs and GNBs) and on a finer scale by IDBs forming a uniform cell structure. These findings confirm previous findings in cube-oriented grains in coarse-grained pure aluminum deformed in tension at room temperature (Hansen et al., 2006).

17.9   Influence of Second-Phase Particles on Structure

Second-phase particles are routinely included in metals and alloys in order to enhance mechanical properties, particularly the yield stress and the work-hardening rate. They may arise from phase transformation (e.g. carbide particles in steels) or by mechanical alloying, sometimes involving inert particles (e.g. Thoria-Dispersed nickel, TD nickel). The influence on mechanical properties, as a result of impedance to dislocation movement, varies depending on the nature and distribution of the second-phase particles. For room-temperature deformation, particles that are distributed through grain interiors are the most significant. By contrast, particles that are mainly on grain boundaries become increasingly important for enhancing high-temperature behavior and creep properties.

Second-phase particles disturb the passage of dislocations through the metallic matrix by introducing additional barriers and strain fields. This effect depends on a number of particle parameters as for example size, volume fraction, shape and properties of the interphase interface (Ashby, 1972). It is not the intention in this chapter to analyze such effects in detail only to describe in simple systems how a fine dispersion of fine particles influences the microstructural evolution during plastic deformation up to high strains and how this behavior deviates from that observed in a single-phase metal. The effects are illustrated using as an example the structural evolution during cold rolling and cold drawing of dispersion-strengthened aluminum, in which the dispersed phase is hard, plate-shaped aluminum oxide particles with an incoherent interface with the aluminum matrix. The equivalent spherical diameter of the particle is in the range of 30–50 nm and their volume fraction in the range of 0.2–3.8 vol. %. The structural evolution during cold rolling of a sample containing 3.8 vol. of Al_2O_3 is illustrated in **Figure 40** (Barlow and Hansen, 1989) as schematic figures covering a strain range up to 90% reduction in thickness ($\varepsilon = 2.3$).

At low strain ($\varepsilon = 0.04$) the structure is characterized by dislocation boundaries and tangles around the particles as shown in **Figure 41** (Barlow and Hansen, 1989). In areas where the particle density is low, dislocations boundaries have already formed, whereas in areas of high-particle density many particles are centers of dislocation activity. These observations indicate the importance of particles, which even in small concentrations may accelerate the formation of dislocation boundaries while in higher concentrations they can reduce the tendency of dislocations to arrange in a low-energy boundary structure.

At medium strains (0.16, 0.69) dislocation boundaries tend to replace the dislocation tangles in the high particle-density areas and the mean spacing between boundaries decreases rapidly with increasing strain as shown in **Figure 42** (Barlow and Hansen, 1989).

At high strains (1.2, 2.3) the structure begins to show signs that dynamic recovery affects the structural characteristics. The positions of the dislocation walls are much less influenced by the alumina particles. Some dislocation boundaries have disintegrated and dislocation tangles and nets of

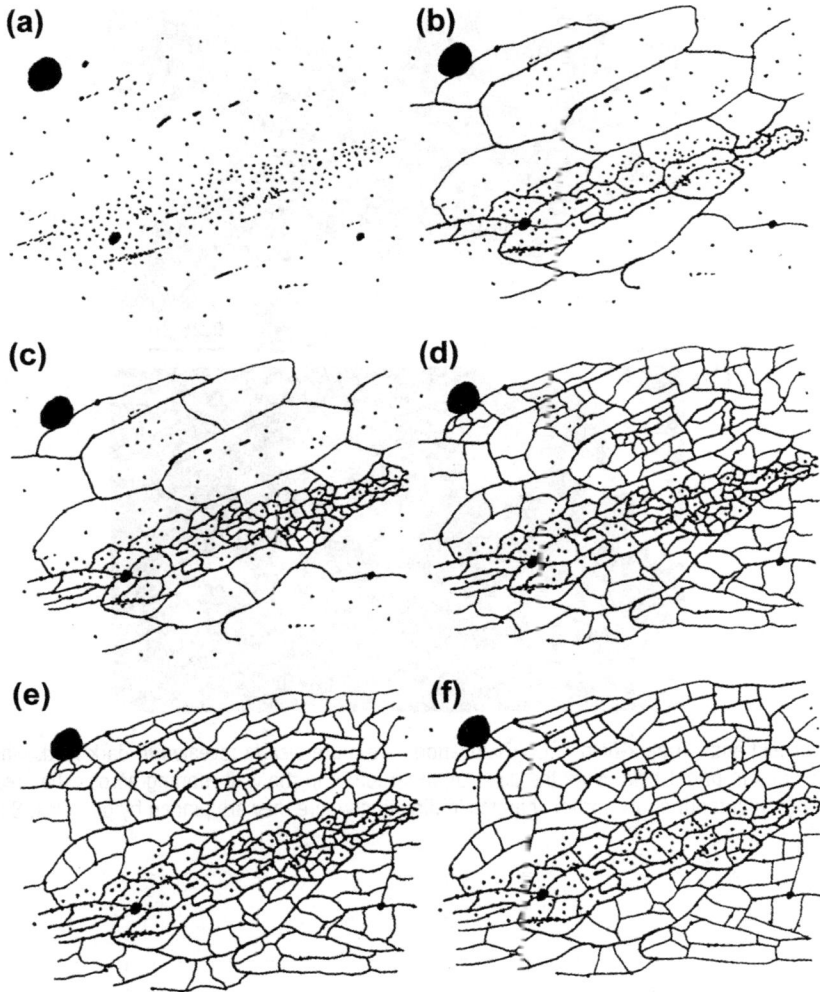

Figure 40 Schematic pictures of the evolution of the deformation structure with plastic deformation. (a) Undeformed, (b) 0.04 strain, (c) 0.16 strain, (d) 0.69 strain, (e) 1.2 strain, (f) 2.3 strain. Barlow and Hansen (1989); Figure 7.

a qualitatively different nature have formed. The overall dislocation density appears to fall and a resultant reduction in lattice strain shows up in an increase in the incidence of Kikuchi lines (Barlow and Hansen, 1989).

The microstructural evolution at medium strain in the aluminum containing oxide particles has some similarity to that observed in aluminum without particles. In both cases the dislocations are stored in low-energy dislocation boundaries and a well-defined boundary structure develops where the spacing between the boundaries decreases with increasing strain. However, the spacing decreases much more rapidly at low to medium strains when particles are present, whereas both materials reach a limiting spacing of about 0.3 µm, see **Figure 42**. The boundary spacing and subgrain size are

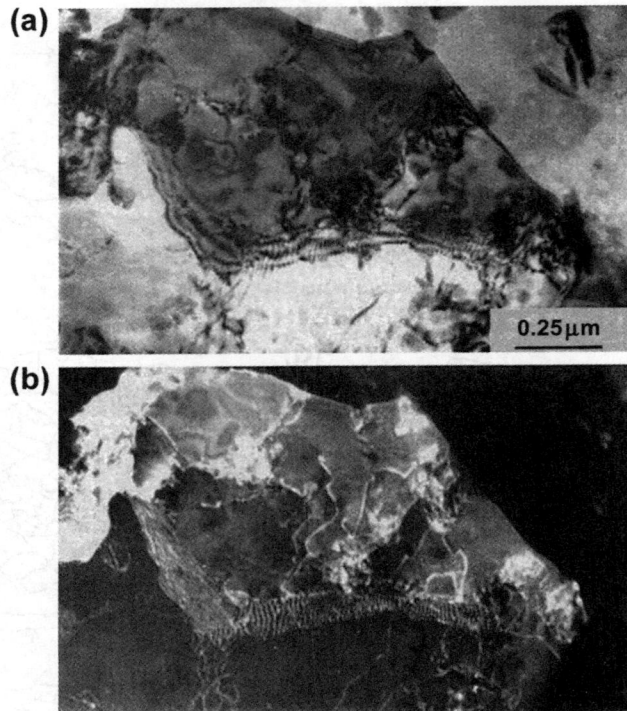

Figure 41 (a) Bright-field and (b) dark-field TEM. Dislocations are more clearly revealed in dark field, whilst alumina particles are more obvious in bright field. Note the dislocation structure in the wall running across the area. The single dislocations in the subgrain interior show abrupt changes in direction where they are pinned by particles. Barlow and Hansen (1989); Figure 2.

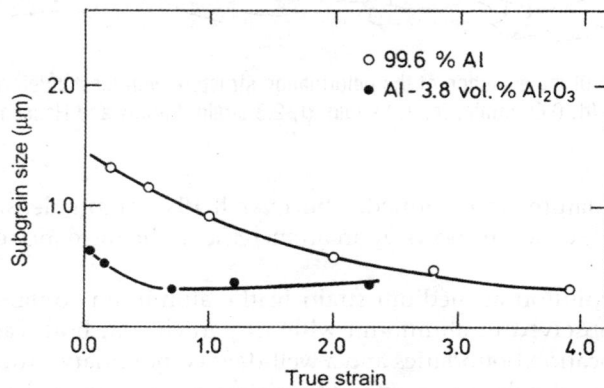

Figure 42 Comparison of subgrain sizes as a function of true plastic strain for cold-rolled 99.6% Al and Al containing 3.8% alumina particles. Barlow and Hansen (1989); Figure 9.

synonymous in this figure. Note also that there is a tendency for the boundary spacing in the particle-containing material to increase with strain in the high-strain range. A supplementary parameter to the boundary spacing is the misorientation angle of boundaries, which in the particle-containing metal reaches a limiting value of about 30° at a strain of 2.7 (Barlow and Liu, 1999; Barlow et al., 2002). The distribution of angles does not change at high strains, implying that some steady state is likely to have been reached.

The structural evolution from low to high strain observed when deforming the oxide-containing metal by cold-rolling also characterizes materials deformed by cold-drawing. The boundary spacing for a number of materials containing from 0.2 to 3.8 vol. %. Al_2O_3 decreases with a drawing strain to about 0.2–0.3 μm (Hansen, 1969). The enhancement in the rate of development of the deformation microstructure when small hard particles are present may have a number of different causes. At small strain, geometrically necessary dislocations are stored, but with increasing strain these dislocations become low in number compared to the density of statistically stored dislocations (Ashby, 1972). Therefore, the increase in the dislocation density when particles are present may have its cause in an enhancement of the frequency of cross-slip and multislip. This is reflected in a significant refinement in the slip line pattern on the surface (Hansen and Juul Jensen, 1987) and the absence of any indication of localized slips e.g. in S-bands (Barlow and Hansen, 1989). As a result, when a fine dispersion of particles is introduced, cells and subgrains replace the evolution of a CB structure. In addition, the activation of many slip systems and introduction of many Burgers vectors may lead to LEDS. Looking at the observation of an increase in subgrain size and decrease in hardness at high strain (**Figure 42**) a cause may be dynamic recovery by dislocation annihilation and removal of boundaries by triple junction motion. However, as the work-hardening rate is very low at high strain, flow instability during processing and testing may also cause structural changes and softening. It is also noteworthy that the amount of structural refinement at large strain is not affected by the presence of oxide, with subgrain sizes reaching a minimum value of about 0.3 μm independent of the oxide content. This is in strong contrast with what is observed with the presence of elements in solid solution (Zhang et al., 2011), which significantly reduce the boundary spacing over the whole strain range.

17.10   Structure/Mechanical Property Relationships, and Modeling

17.10.1   Work-Hardening Stages

The plastic deformation of metals and alloys is typically characterized by a number of distinct work-hardening stages. The deformation begins with Stage I planar glide where glide occurs on one system with a very low work-hardening rate. In single crystals oriented for multiple glide and in polycrystals, Stage I is typically insignificant.

Stage II is characterized by a high and almost constant work-hardening rate. Stage II can extend to a fairly large strain in planar glide metals, whereas in wavy glide metals like aluminum with a high SFE, Stage II is absent at ambient temperatures.

Stage III is reflected in a parabolic shape of the stress–strain curve showing a rapid decrease in the work-hardening rate as the stress is increased. Different mechanisms for example cross-slip may cause the transition from Stage II to Stage III.

Stage IV is characterized by a low and approximately constant but decreasing work-hardening rate and it can extend to large strains. At the limit of high strain, the hardening rate can decrease to almost zero, and this is characteristic of Stage V. The preceding structural analysis (Section 17.5.4) covers the work-hardening Stages III, IV and V—all showing a decrease in the hardening rate.

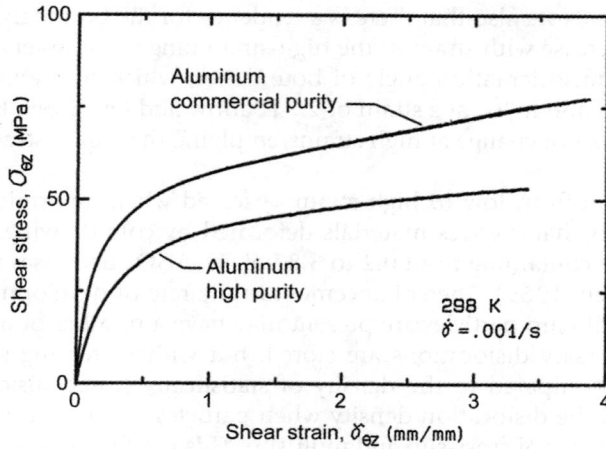

Figure 43 Stress–strain curves of high purity and commercial purity aluminum deformed in torsion at 296 K and 10^{-3} s^{-1}. Hughes (1986).

The work-hardening stages are illustrated in the stress–strain curves for (1) aluminum deformed to medium strain by torsion (**Figure 43**, Hughes, 1986), (2) low-carbon iron deformed by wire drawing and tested in tension at room temperature (**Figure 44**, Langford and Cohen, 1969) and (3) pure Ni deformed by HPT, torsion and rolling at room temperature (**Figure 45**, Zhang et al., 2008a). These curves illustrate rapid decreasing work-hardening rate in Stage III (**Figure 43**) and almost constant hardening rate in Stage IV (**Figure 44**), which approach saturation in Stage V at very large strains (**Figure 45**). The transition from Stage III to Stage IV hardening in aluminum is shown in **Figure 46** (Hughes, 1986), which is based on **Figure 43**. Such stress–strain relationships are typically observed (Gil-Sevillano et al., 1981; Lloyd and Kenny, 1982; Hughes and Nix, 1988; Rollett, 1988; Liu and Bassim, 1993; Zehetbauer and Seumer, 1993). The transition between these two regimes takes place

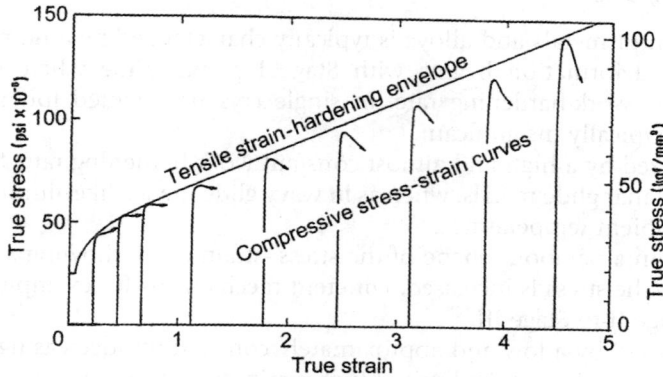

Figure 44 Stress–strain curve for low-carbon iron deformed by wire drawing and tested in tension at room temperature. Gil-Sevillano et al. (1981); Figure 2.29.

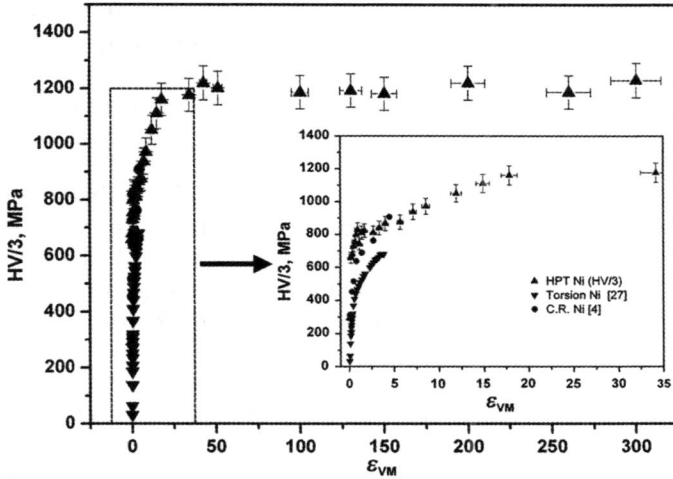

Figure 45 Stress–strain relationships for HPT Ni, cold-rolled Ni (Hughes and Hansen, 2000) and torsion-deformed Ni (Hughes and Nix, 1988). Zhang et al. (2008); Figure 11.

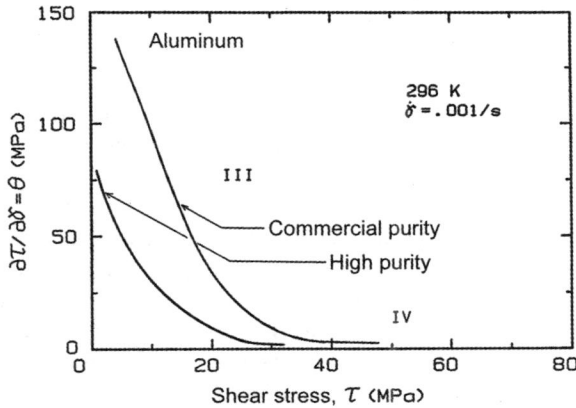

Figure 46 Work-hardening rate as a function of resolved shear stress for aluminum. A Taylor factor of 1.65 was used. Hughes (1986).

gradually over a strain from 0.8 to 1.4. The hardening in Stage IV persists beyond strains of 5 and 6 (Hughes and Nix, 1988; Zimmer et al., 1983). The work-hardening behavior can be described by phenomenological relationships (Hecker and Stout, 1984), which predict continued parabolic hardening (Hollomon) or saturation at a stress where the hardening curve approaches zero hardening (Voce hardening).

The different work-hardening stages relate to the observed microstructural evolution with increasing stress and strain from a carpet structure in Stage II to a cell and CB structure in Stage III. There is a continuous decrease in the cell and the CB size in Stage III, for example toward a lamellar structure in rolling and a fibrous structure in drawing in Stage IV. Finally, depending on loading conditions, there may be an evolution toward saturation where equiaxed subgrains and grains may replace the directional structure (Zhang et al., 2008a). Dislocation-based mechanisms may dominate throughout the different stages, but other mechanisms may under certain conditions also be operative, including localized shear, twinning and grain boundary sliding (Lu and Hansen, 2009; Wilde, 2013).

The work-hardening stages correlate with the structural evolution (see Section 17.5.2), which suggests that relationships can be formulated between the flow stress and selected structural parameters. In the following, such relationships will be derived and discussed focusing on yield and flow stress at ambient temperature, covering a range from low to very high strain. Key structural parameters are the dislocation density, together with some boundary parameters, in particular the spacing between and misorientation angle across boundaries. The boundaries considered include strain-induced dislocation boundaries, classified as IDBs or cell boundaries, and GNBs, together with preexisting high-angle grain boundaries.

The present analysis is based on hardening contributions of both IDBs and GNBs in a CB structure at medium strains (Stage III) or in a lamellar or fibrous structure at large strain (Stage IV and above). This is somewhat in contrast to previous analyses of the hardening stages: e.g. (1) focusing on the dislocation structure in an analysis of Stage III (Gil-Sevillano, van Houtte and Aernoudt, 1981; Kocks and Mecking, 2003; Argon, 2008); (2) relating the buildup of long-range stresses associated with an increase in the misorientation angle across cell boundaries in an analysis of Stage IV (Argon and Haasen, 1993) and (3) incorporating dislocations related to the evolution of the misorientation (disorientation) angle across IDBs and GNBs when interpreting the lower hardening rate in Stage IV (Pantleon, 2005).

17.10.2 Strengthening Effect of Boundaries

Because of their importance to the mechanical behavior of polycrystalline samples, we will examine in more detail the ways in which boundaries and dislocations interact. The behavior may depend on the angle of misorientation across the boundary.

For low-angle boundaries, dislocations may be able to pass through the boundary. The stress required is inversely proportional to the spacing of the dislocations in the boundary (Kuhlmann-Wilsdorf, 1989). Analytically, this is equivalent to dislocations passing through an array of forest dislocations. The manner in which dislocations effectively travel through the structure can take place by bowing between the pinning points, which are jogs. With high-angle grain boundaries, the spacing between intrinsic dislocations is so small (if indeed it still has physical reality) that dislocations cannot pass through by a bowing mechanism. However, high-angle boundaries commonly do act as dislocation sources, though the stresses required for this are high enough that grain boundaries are regarded as a polycrystal strengthening mechanism.

For high-angle boundaries such as grain boundaries, the boundary will normally act as a barrier to slip causing a pileup of dislocations. The stress concentration when a slip plane meets the boundary can initiate slip in a neighboring grain and the glide can then continue. This barrier effect is generally expressed as a proportionality between the yield stress and the inverse square root of the grain size (Hall, 1951; Petch, 1953).

In special cases, there is evidence that dislocations may enter a boundary from one grain and trigger release of dislocations from the boundary into the other grain. For particular orientation relationships, dislocations may effectively pass through the boundary using such a mechanism. In both cases, dislocation fragments are left as extrinsic dislocations in the boundary (Sutton and Balluffi, 1995). These particular instances require a high degree of correspondence between the crystallographic orientation of both the slip plane and the slip direction in the crystals on either side of the boundary.

17.10.3 Relating Flow Stress with Microstructural Parameters

In cell-forming metals, experimental evidence has demonstrated (e.g. Gil-Sevillano, van Houtte and Aernoudt, 1981; Raj and Pharr, 1986; Hansen, 1994) that an empirical relationship exists between the true flow stress σ and the average diameter D of cells and subgrains:

$$\sigma - \sigma_0 = K_1 Gb/D^n, \tag{17}$$

where σ_0 is the lattice friction stress, G is the shear modulus and $\mathbf{b}$ is the Burgers vector. The exponent n normally lies in the range between 0.5 and 1.0, with values at the upper end of the range characteristic of dislocation cell structures and values at the lower end characteristic of subgrain structures. Only semiquantitative theories for the relationship expressed in Eqn (17) have been developed. K_1 is regarded as a constant, although in fact a large range of values has been observed (see Eqn (14)). For example, studies on single crystals (Neuhaus and Schwink, 1992) have shown that K_1 can be affected by the thickness of the cell walls, which implies that it is not independent of the material system. For structures where similitude can be applied, D can be replaced by the boundary misorientation angle, so that for a cell structure and taking $n = 1$, Eqn (17) can also be written

$$\sigma - \sigma_0 = K_1 G\theta/C. \tag{18}$$

In cell-forming metals the flow stress can also be related to the total dislocation density ρ_t by the relationship:

$$\sigma - \sigma_0 = M\alpha\mathbf{b}G\sqrt{\rho_t}, \tag{19}$$

where M is the Taylor factor which takes the crystallographic texture into account, and the number α is of order 0.2–0.4 for tensile stresses. Apart from the variation in α, which decreases slowly with increasing flow stress, the relationship in Eqn (19) is rather insensitive to changes in dislocation structure, grain size, purity, prestrain, strain rate and temperature (see for example Mecking and Kocks, 1981).

Equations (17) and (19) represent single-parameter relationships and can give a satisfactory description of the increase in flow stress at low and medium strains (Gil-Sevillano et al., 1981). However, as a CB structure forms and evolves with increasing strain, the single-parameter relationships are replaced by models, which include separate representations of the strengthening effect of cell boundaries and GNBs.

The first approach is based on the proposal (Kuhlmann-Wilsdorf and Hansen, 1991) that GNBs, which are associated with a relatively high boundary misorientation angle, may have a strengthening effect σ_{GNB}, which is equivalent to that provided by ordinary grain boundaries. For these, the flow stress σ at a particular strain is related to the grain size D by a Hall–Petch relationship (Hall, 1951; Petch, 1953):

$$\sigma - \sigma_0 = K_{HP}\left(\frac{1}{D}\right)^{1/2}, \tag{20}$$

where K_{HP} is constant at a given strain. Eqn (20) has also been applied to describe the effect of grain size on other measures of mechanical properties, especially the hardness (which is given approximately by 3σ).

σ_{GNB} may therefore be represented by the relationship:

$$\sigma_{GNB} - \sigma_0 = K_2 G\mathbf{b}\left(\frac{1}{D_{GNB}}\right)^{1/2},$$

(21)

where D_{GNB} is the CB size and $K_2 G\mathbf{b}$ is equal to K_{HP} in Eqn (20).

Based on the assumption that σ_{GNB} is linearly additive to the strength contribution σ_C from ordinary dislocation cells of size D_C, and using a value of $n = 1$ for the exponent in Eqn (17), the flow stress can now be written as:

$$\sigma - \sigma_0 = \frac{K_1 G\mathbf{b}}{D_C} + K_2 \left(\frac{G\mathbf{b}}{D_{GNB}}\right)^{1/2}.$$

(22)

The second approach is based on Eqn (19), but uses a single combined value for dislocation density for all the microstructural features, which have their origin in dislocations. This value ρ_t therefore includes not only the density of dislocations distributed in the material between the dislocation boundaries, ρ_i but also estimates for the dislocation content in both the cell boundaries, ρ_C and the GNBs, ρ_{GNB}. The model treats these boundary dislocations as if they were uniformly distributed through the material in line with previous analysis of dislocation strengthening in Stage III. The total dislocation density can be written as:

$$\rho_t = \rho_i + \rho_C + \rho_{GNB}.$$

(23)

For 2D boundaries, both ρ_C and ρ_{GNB} are equal to $S_V \rho_A$ where S_V is the boundary area per unit volume and ρ_A is the dislocation density per unit area of boundary. For 3D cells and CBs of uniform size D, S_V is by a simple geometric argument given by K/D where K is a number typically equal to 3. For GNBs with planar spatial orientations, $S_V = 1/D^*_{GNB}$ where D^*_{GNB} is the true (perpendicular) distance between GNBs. For both cell boundaries and GNBs, $\rho_A = 1/h$ where h is the spacing between dislocations in a simple wall (Read and Shockley, 1950). If the angle of misorientation θ across such a wall is small then h can be approximated as $\mathbf{b}/\theta$ and ρ_C and ρ_{GNB} can be expressed in terms of D and θ. For a well-developed structure consisting of 3D cells and GNBs in a planar orientation, then the dislocation density ρ_i within cell interiors is close to zero and the total dislocation density can be given by:

$$\rho_t \approx \frac{K}{D_C}\frac{\theta_C}{\mathbf{b}} + \frac{1}{D^*_{GNB}}\frac{\theta_{GNB}}{\mathbf{b}},$$

(24)

where θ_C and θ_{GNB} are the angles of misorientation across cell boundaries and GNBs respectively. As has already been mentioned in Section 17.5.3, it should be noted that these dislocation densities represent lower limits. This is because in addition to the dislocations required by the Frank formula, dislocation walls contain redundant dislocations that do not contribute directly to the misorientation. In addition, the assumption that h can be approximated as $\mathbf{b}/\theta$ needs verification, especially for larger values of θ.

Bearing in mind these limitations, Eqns (19) and (24) may now be combined to give the deformation-induced strengthening contribution as:

$$\sigma - \sigma_0 = M\alpha\sqrt{\mathbf{b}}\sqrt{\frac{K\theta_C}{D_C} + \frac{\theta_{GNB}}{D^*_{GNB}}}.$$

(25)

The two approaches expressed by Eqns (22) and (25) have been applied in a study (Hansen, 1994) where the flow stress and the microstructural parameters have been determined for high purity aluminum (99.996%), which has been cold-rolled to reductions of 5, 10 and 30%. The two approaches both gave values in good agreement with the measured flow stress representing Stage III hardening at low to medium strain.

For medium to large strain deformation (Stage III and IV), analysis of boundary parameters shows that the structural dimensions shrink with increasing strain and that misorientations across boundaries encompass the whole range of values from low to high angles. As predicted by a Hall–Petch relationship the flow stress will increase as the boundary spacing decreases, but the effect of the variation in misorientation angle is not well established. For diffuse IDBs, in which the boundary width contributes substantially to the boundary volume fraction, it is a standard practice to account for their contribution to strength via their contribution ρ_b to the total dislocation density, by a related analysis to that described above in Eqn (23):

$$\rho_t = \rho_i + \rho_b, \tag{26}$$

ρ_b is again equal to $S_V\rho_A$.

For a mixed tilt and twist wall, ρ_A is given by $1.5\theta/\mathbf{b}$ so $\rho_b = 1.5S_V\theta/\mathbf{b}$.

The flow stress contribution from IDBs can therefore be written as:

$$\sigma_{IDB} = M\alpha G\mathbf{b}\sqrt{1.5S_V\theta_{IDB}/\mathbf{b} + \rho_i}, \tag{27}$$

where θ_{IDB} is the average misorientation associated with IDBs.

This treatment assumes that the contributions of the two boundary types to the flow stress can be expressed by two equations. Both boundary types trap dislocations, so their misorientation rises as the strain increases. The GNBs also separate regions with differing slip, and their misorientation angles increase so that they become analogous with grain boundaries. This transition from a penetrable to an impenetrable barrier to dislocation glide imply the existence of a critical angle at which slip may progress for example by generation of dislocations from the boundary (Li, 1963). This critical angle has been estimated to 3–5° (Kamikawa et al., 2009) based on a comparison of experimental and calculated values for the flow stress of aluminum in the deformed and annealed state. Therefore, when modeling the flow stress of medium and large strain, the strength contribution of the GNBs may be written as:

$$\sigma_{GNB} = K_{HP}/\sqrt{D_{GNB}}, \tag{28}$$

in which K_{HP} is the slope of the straight line relating the flow stress of a polycrystalline metal to the inverse square root of the grain size.

The flow stress can now be expressed as the sum of the strength contributions for the IDBs and the GNBs by combining Eqns (27) and (28) and including the lattice friction stress σ_0 to give:

$$\sigma = \sigma_0 + \sigma_{IDB} + \sigma_{GNB} = \sigma_0 + M\alpha G\mathbf{b}\sqrt{1.5S_V\theta_{IDB}/\mathbf{b} + \rho_i} + K_{HP}/\sqrt{D_{GNB}}. \tag{29}$$

For large strain deformation, the deformation microstructures resulting from monotonic deformation are lamellar structures. TEM investigation allows distinction to be made between IDBs and GNBs. However, at very high strains or after strain path changes (e.g. in cyclic deformation) the structure may become more equiaxed, and is subdivided by a mixture of low- and high-angle boundaries. The Hall–Petch equation for the flow stress may now be written:

$$\sigma_f = \sigma_0 + M\alpha G\sqrt{1.5\mathbf{b}\,S_V\theta_{LAB}(1-f)} + K_{HP}\sqrt{f\left(\frac{S_V}{2}\right)},\qquad(30)$$

where S_V is the total boundary area per unit volume, θ_{LAB} is the average misorientation angle of low-angle boundaries and f is the proportion of high-angle boundaries. This analysis illustrates the importance of both dislocation strengthening and high-angle boundary strengthening. However, frequently the flow stress structural analysis of deformed metals is based on the use of a single structural parameter, the boundary spacing D_B. Taking S_V equal to $2/D_B$ (again by a geometric argument, taking into account the spatial distribution of the boundaries), Eqn (30) may be written:

$$\sigma_f = \sigma_0 + \left[M\alpha G\sqrt{3\mathbf{b}\theta_{LAB}(1-f)} + K_{HP}\sqrt{f}\right]D^{-1/2},\qquad(31)$$

which can be simplified to

$$\sigma_f = \sigma_0 + k_2 D_B^{-1/2}.\qquad(32)$$

The contribution of GNBs is generally expressed by a Hall–Petch relationship derived on the basis of a pile-up mechanism. This raises the question of the minimum volume required for the formation of a pileup in front of the boundary. However, calculations (Nieh and Wadsworth, 1991) have shown that a lower bound for the boundary spacing may be of the order of 10–15 nm (Li and Chou, 1970; Li and Liu, 1967). Flow stress–structure relationships based on contributions from dislocation and boundary strengthening may therefore be valid even for extremely fine structural scales.

17.10.4   Relating Flow Stress Models with Experiments

17.10.4.1   Flow Stress of Tensile-Deformed Aluminum in Stage III ($\varepsilon \leq 0.4$)
The significant effect of grain orientation on the evolution of the deformation microstructure suggests that the hardening rate for a grain in a polycrystalline sample depends on its crystallographic orientation, an effect which is also found for single crystals. This implies that the grain interaction is small, a suggestion that is supported by the observation that the strengthening parameters (misorientation angle and boundary spacing) evolve at significantly different rates for the three structural types described above. For each of the three orientations in a polycrystalline aluminum sample strained in the range $\varepsilon = 0.05$–0.34 the strengthening parameters have been determined and used in a calculation of the dislocation density (Hansen and Huang, 1998). Then for each type of grain the shear stress–strain relationship has been obtained by taking the shear stress proportional to the square root of the dislocation density:

$$\tau_i = \alpha GB\sqrt{\rho_i}.\qquad(33)$$

Experiments show a decrease in the value of the number α with strain (Hansen and Huang, 1998) in agreement with a theoretical production (Kuhlmann-Wilsdorf, 1989). For the present strain range an average value $\alpha = 0.24$ has been used. This will lead to underprediction at low strain and an overprediction at large strain. Each of the three types of grain shows characteristic shear stress–strain

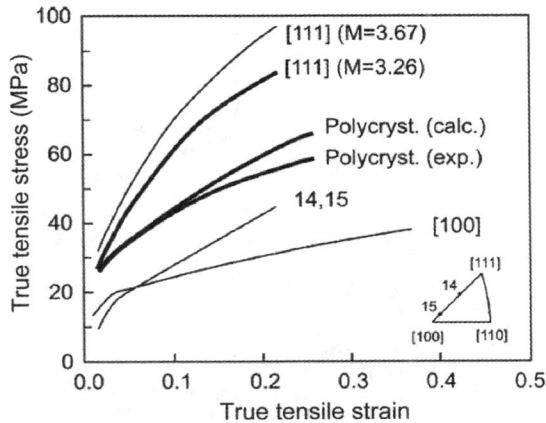

Figure 47 Tensile flow-stress curves for single crystals with orientation corresponding to the three structural types. Based on these curves the stress–strain curve for a polycrystalline sample has been calculated. Hughes (1986).

relationships similar to those obtained by the testing of single crystals of representative orientations (Hansen and Huang, 1998). This finding illustrates the dominating effect of grain orientation. An estimate of the stress–strain curve of the polycrystal can then be constructed, based on a combination of single crystal data, which are weighted based on a quantitative texture analysis of the polycrystal. As shown in **Figure 47** good agreement between calculation and experiment has been found.

17.10.4.2 Flow Stress of Cold-Rolled Ni in Stages III and IV ($\varepsilon \leq 5$)

The stress-strain response of cold-rolled polycrystalline nickel consists of an initially high-hardening rate that decreases smoothly with increasing strain, characteristic of Stage III, to a low almost constant hardening rate at large strains, characteristic of Stage IV, but a saturation stress (Stage V) is not observed (**Figure 50**).

The microstructural parameters describing the CBs were identified as the spacing and misorientation angles of IDBs and GNBs (see **Figure 48**). Strength-determining parameters are chosen based on the structural information together with a detailed description of the morphology. An observed scaling behavior across the hardening ranges strongly suggests that these same parameters affect the flow stress in all the stages. Eqn (29) has therefore been used to calculate the flow stress–strain relationship.

The individual strength contribution of IDBs and GNBs based on Eqn (29) is shown in **Figure 49**. Note the different contributions to the strain-hardening behavior from the IDBs and GNBs. The calculated flow stress is given in **Figure 50**, together with Vickers hardness measurements and experimental data (involving rolling followed by tensile deformation) from the work of Nuttall and Nutting (1978) and Zimmer et al. (1983). In the latter work the low purity of the commercial nickel may explain the higher observed flow stress. Very good agreement is obtained for both the shape and magnitude of the calculated stress–strain curve. **Figure 50** also shows that a saturation flow stress is not approached.

17.10.4.3 Hardness of Ni Deformed by HPT in Stages III, IV and V ($\varepsilon \leq 300$)

Deformation microstructures produced in Ni (99.5%) by HPT demonstrate structural length scales greater than 50 nm. At medium strains the structure is a typical lamellar structure and its evolution

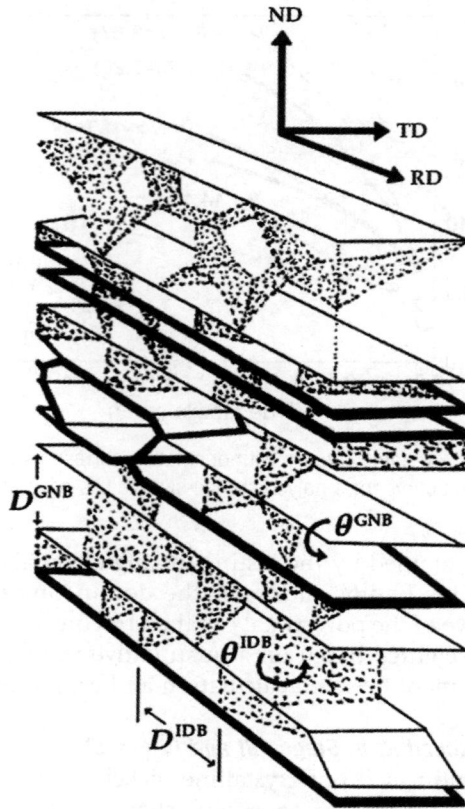

Figure 48 Schematic 3D drawing of the strain microstructure showing the structural parameters related to flow stress. Sheets of extended GNBs are seen with stippled low-angle (bamboo) IDBs bridging between them. High-angle GNBs are represented by heavy line weight and medium angle by medium line weight. Hughes and Hansen (2000).

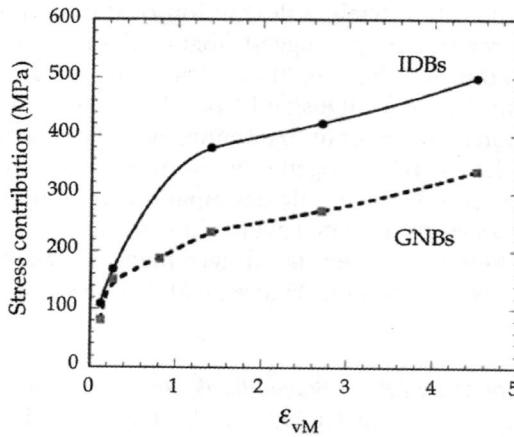

Figure 49 Individual flow stress contributions for IDBs and GNBs. Hughes and Hansen (2000).

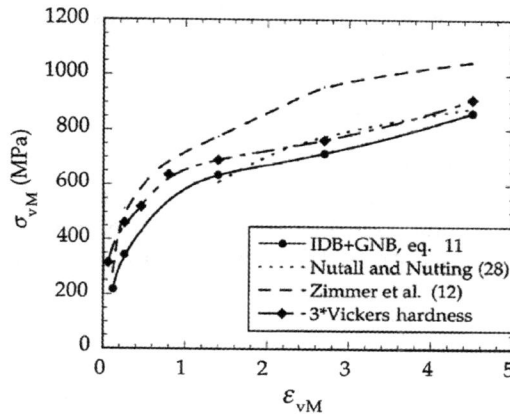

Figure 50 Flow stress calculations compared with Vickers hardness data and literature data. The trend in the flow stress from Stages III–IV, calculated by the microstructural contributions, matches the trends in hardness and measured strength. Hughes and Hansen (2000).

follows the universal pattern of structural subdivision. However at a very large strain there is a clear trend that the lamellar structure is replaced by an equiaxed structure while maintaining some of the directional appearance characteristic of structures at lower strains. As such high strains ($\varepsilon > 34$) (Zhang et al., 2008b) both the average boundary spacing and the average misorientation angle approach saturation at about 50 nm and about 45°, respectively.

The flow stress (estimated as one third of the Vickers hardness) has been shown in **Figure 45** (Zhang et al., 2008a) as a function of strain. It is seen that significant hardening take place at low and medium strains; however, for strains larger than about 12, the stress–strain relationship levels off toward saturation. Also plotted in **Figure 45** are data for cold-rolled Ni (Hughes and Hansen, 2000) and torsion-deformed Ni (Hughes and Nix, 1988) showing that the stress–strain curves for torsion-deformed Ni and cold-rolled Ni are comparable in the strain range 2–5, where the strain-hardening rate is near constant. Finally, **Figure 45** shows that for strains in the small to medium range, typical Stage III and IV hardening takes place, whereas at large strains the hardening rate is very small, illustrating a transition from Stage IV to Stage V hardening. For torsion-deformed Ni the rate of work-hardening in Stage IV (strains between 1.7 and 12) is about $4 \times 10^{-4}\,G$ (G is the shear modulus, 79 GPa), in agreement with the average value of approximately $2 \times 10^{-4}\,G$ for many cubic materials (Rollett, 1988). Flow stress calculations in accordance with Eqn (33) show good agreement with the data in **Figure 52**.

17.10.4.4 Flow Stress of Cold-Drawn Steel Wire

Cold-drawn steel wires can have microstructures with a structural scale in the range from about 10 to 100 nm. These dimensions represent the spacing between hard cementite lamella separating volumes of ferrite containing solutes, especially carbon atoms. During deformation, dislocations are stored in the ferrite and their density can increase to the order of $10^{16}\,\mathrm{m}^{-2}$ or even higher (Zhang et al., 2011b). The flow stress at a strain of about 5 is approximately 5 GPa and the following strengthening mechanisms may be operative: (1) dislocation strengthening, (2) boundary strengthening (expressed by a Hall–Petch relationship) and (3) solid-solution hardening, σ_S. On the assumption of additive strengthening, the strength contributions may be written as the sum of these three contributions (Zhang et al., 2011b):

$$\sigma - \sigma_0 = M\alpha bG\sqrt{\rho_t} + K(2d)^{-0.5} + \sigma_S, \tag{34}$$

where d is the width of the ferrite lamellae.

Good agreement has been found between the estimated and the measured flow stress, which illustrates the robustness of the relationships between flow stress and the microstructural parameters in deformed metals and alloys as expressed by the strengthening equations presented in Section 17.10.3.

17.10.5   Flow Stress Anisotropy

The crystallographic texture is an important source of anisotropy, but in addition, it has been shown that the evolving GNBs also contribute substantially by being a set of closely spaced parallel planar boundaries with a strong directionality. This has been exemplified by the flow stress anisotropy in rolled aluminum sheets, where it has been systematically found that the flow stress in the rolling plane increases with increasing angle between the rolling and the transverse direction for aluminum (Juul Jensen and Hansen, 1990; Eardley et al., 2002; Li et al., 2006). This is illustrated in **Figure 51** (Rollett et al., 1992), where the flow stress has been normalized by the Taylor factor to separate a microstructure and a textural effect on the flow stress anisotropy. The orientation distribution of the GNBs (**Figure 51(b)**) shows a strong directionality, which was assumed to be the cause of the microstructural anisotropy. In support of this assumption it was found that cross-rolling neither introduced a directional boundary orientation (**Figure 51(c)**) nor anisotropy (**Figure 51(a)**). Averaging over the whole sample, the GNBs tend to be spread around the sample planes that carry most macroscopic stress, although the distribution of planes in the sample coordinate system can be quite wide. For rolling, the most stressed planes are inclined at about 45° to the RD in the longitudinal section and perpendicular to the RD in the rolling plane. While the GNBs on average align with the most highly stressed planes, the exact GNB plane depends on the grain orientation and the slip systems, as discussed in Section 17.5. The effect of GNBs on the flow stress anisotropy therefore can be modeled both on the sample average scale and on the grain average scale, in both cases assuming that the contribution from the GNBs follows a Hall–Petch relationship. When modeling on the sample average scale, the contributions from texture and GNBs are calculated separately for the entire sample and added linearly (Hansen and Juul Jensen, 1992). When modeling on the grain average scale, it becomes possible to consider the coupling between the texture (grain orientation) and the GNB alignment. This has been done by introducing anisotropic critical resolved shear stresses (given by a Hall–Petch type relationship) into traditional polycrystal plasticity models (Rollett et al., 1992; Winther et al., 1997; Peeters et al., 2001; Wilson and Bate, 1996). The distance parameter in the Hall–Petch relation is the spacing between GNBs from the point of view of each slip system; i.e. the distance is much larger for slip approximately parallel to the GNB than for slip perpendicular to the GNB. The anisotropic critical resolved shear stresses are then used instead of the conventional isotropic ones in the Taylor model to calculate the flow stress in different testing directions (Winther, 2005).

17.10.5.1   Flow Stress Anisotropy in Cold-Rolled Aluminum

The experimental macroscopic tensile flow stress σ along different directions in the sheets (99.5% Al) rolled to prestrains in the range 0.05–0.2 is shown in **Figure 52**. The degree of anisotropy reflects both

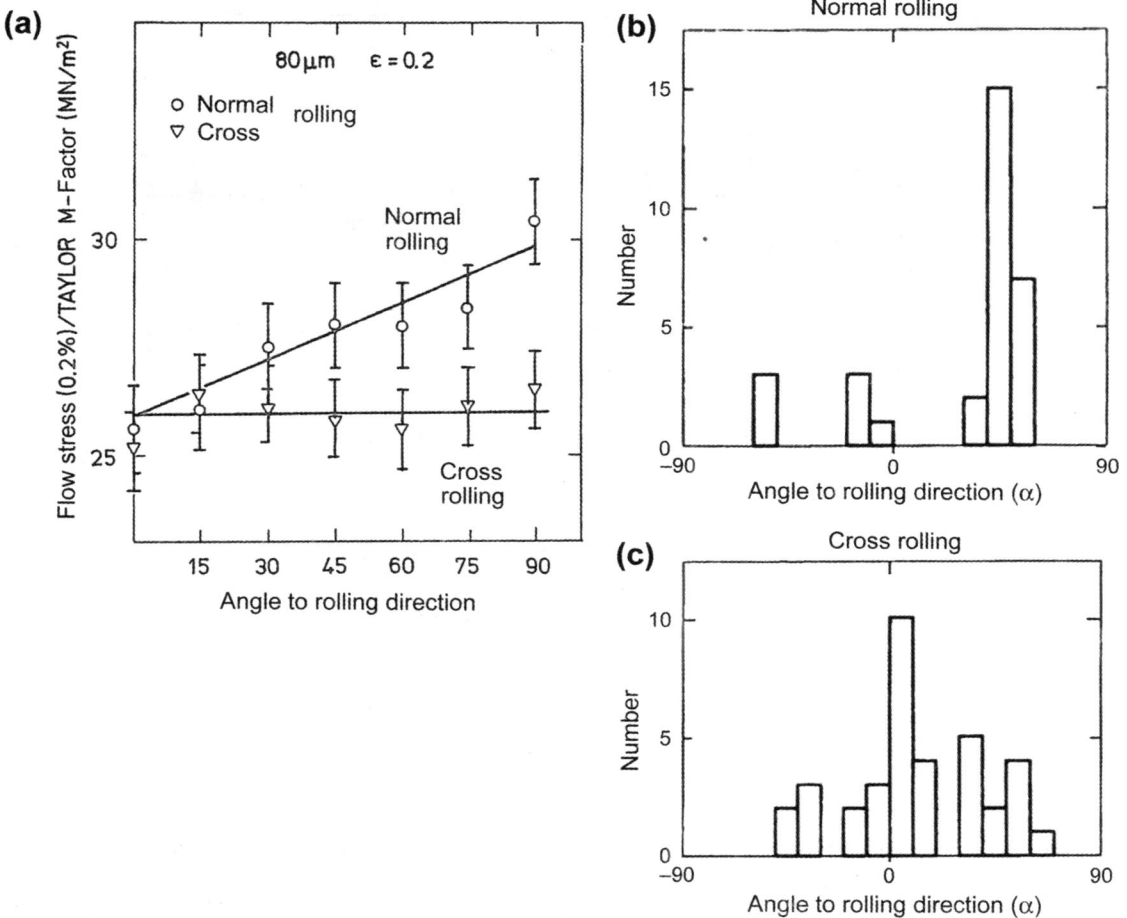

Figure 51 (a) Plot of flow stress versus angle: the flow stress has been normalized by the Taylor factor. (b) Plot of orientation distribution of microbands in normal rolled aluminum at strain 0.2 in the RD–ND plane. (c) Plot of orientation distribution of microbands (as in (b)) in cross-rolled aluminum at strain 0.2. Rollett et al. (1992); Figure 2.

the texture and the microstructure. Previous work (Hansen and Juul Jensen, 1987; Juul Jensen and Hansen, 1990) has, however, established that the anisotropy expected arising solely from the relatively weak deformation-induced textures in the materials considered here is negligible compared with the experimentally observed anisotropy.

Figure 52 (Li et al., 2006) also shows the model predictions based on the microstructural parameters and the measured textures. It is seen that the agreement is good for all prestrains. Furthermore, investigation of the degree of anisotropy in combination with the absolute flow stress level as a function of rolling prestrain constitutes a critical test of the model as discussed in the following.

First of all, the basic assumption for this model, that the mechanical properties are determined by both the GNBs and the IDBs in an additive manner, is confirmed. The degree of anisotropy is controlled

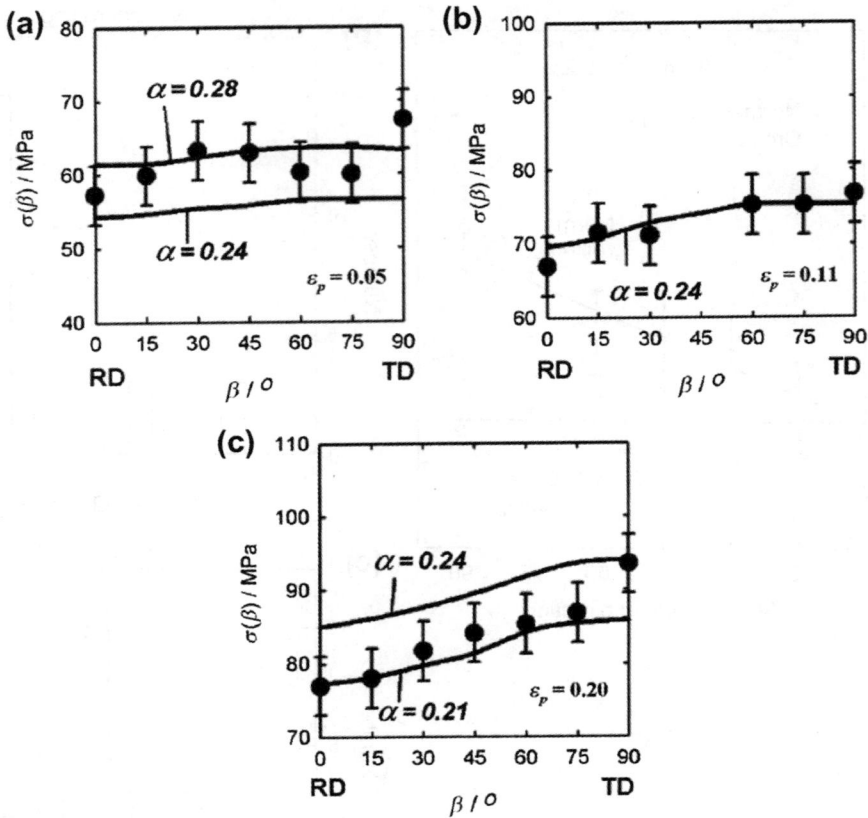

Figure 52 Model calculations of tensile flow stress of aluminum (99.5% purity) as function of angle to rolling direction β in the rolling plane compared with experimental data (0.2% offset) for three different rolling prestrains (a, b and c). In model calculations, boundary spacing for individual slip system is calculated along the slip direction. Hansen et al. (2005).

by the GNBs while the absolute stress level is controlled by a combination of GNBs and IDBs. The good agreement of both absolute stress level and the magnitude of the observed increase in flow stress with the angle to the RD for all prestrains indicates that the two contributions are both predicted correctly. With respect to the method for calculation of the boundary spacing experienced by the individual slip system, it is seen that the curves representing the spacing along the slip direction agree well with the experimental data, taking into account the uncertainties in the data.

The IDB contribution has been calculated based on a general dislocation theory, involving the square root of the dislocation density. It is also important that the parameter α, which takes into account the spatial distribution and mutual screening of the dislocations, is varied. As the dislocation density goes up this parameter decreases. As mentioned above, the values employed in **Figure 52** have been determined independently from monotonic tensile tests of aluminum (Hansen and Huang, 1998). This figure illustrates the importance of the variation in α (from 0.28 to 0.21) by comparing the curves from **Figure 52** with the spacing taken along the slip direction with similar curves using a constant α value (0.24), representing the average (Hansen, 2005).

The anisotropic strength contribution from GNBs was originally calculated according to the Hall–Petch relationship taking the boundary strength, represented by the factor k_2 in Eqn (32), to be constant. However, the resistance of GNBs may in fact increase with increasing strain as their angles and thereby their strength as obstacles to dislocation movement increase.

17.10.6 Mechanical Behavior in Nanoscale Microstructures

The mechanical behavior of ultrafine grained and nanocrystalline materials has been reviewed in the present volume (Wilde, 2013). In the preceding sections the strength–structure relationships at medium to high strain have been analyzed more specifically in terms of dislocations and boundary strengthening in materials with a structural scale reaching from the submicrometer to the nanometer scale and good agreement has been found between estimates and experiments, although not in all cases. For example, it has been found in nanostructured aluminum cold-rolled to a high strain by ARB that the measured flow stress is 259 MPa compared to an estimate of 190 MPa (Huang, 2009). This difference points to the operation of additional strengthening mechanisms in the nanoscale structure. One such mechanism has been identified by repeated low-temperature annealing and slight cold-rolling of ARB aluminum in the as-processed condition. The stress–strain curves of samples in the different conditions are shown in **Figure 53** (Huang, 2009). Comparing the stress–strain curves of the as-processed and the annealed sample (curves 1 and 2) it is seen that after low-temperature annealing the yield stress has increased slightly while the total elongation to failure has decreased significantly. This is in contrast to the expected behavior after annealing, which is a decrease in strength and an increase in elongation to failure. Also unexpectedly, a strength decrease and an increase in elongation to failure followed a subsequent deformation by cold-rolling (curve 3). When such annealing and deformation treatments were repeated, this behavior was also repeated. This phenomenon has been explained by testing samples of different purity (Huang et al., 2006; Hung et al., 2005; Bowen et al., 2004) and produced by different techniques e.g. ARB (Huang et al., 2006) and ECAE (Hung et al., 2005; Bowen et al., 2004) suggesting that causes must be related to the characteristics of the fine scale

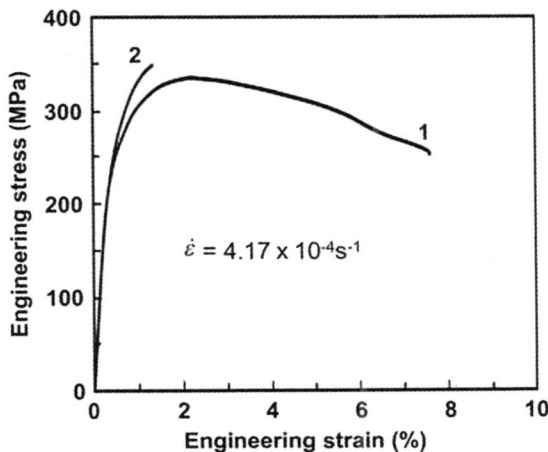

Figure 53 Stress–strain curves for 99.2% ARB aluminum. Curve 1: Six ARB cycles giving equivalent strain of 4.8. Curve 2: Same material as 1 plus annealing at 150 °C for 30 min. Huang et al. (2009); Figure 2.

microstructure. Further investigation by TEM of annealed and deformed samples have shown that the major changes in microstructural parameters are a decrease in the interior dislocation density during annealing and a reintroduction of dislocations during the cold-rolling step. This has led to the suggestion (Huang et al., 2006) that dislocations that can easily glide are annealed out, which means that a higher yield stress is required to activate new dislocation sources during loading. Consequently, when dislocations are reintroduced, the yield stress is decreased as new dislocation sources are now available. A correlation between the density of dislocation sources and the strength is typically observed in nanoscale metal, both experimentally (Greer et al., 2005; Shan et al., 2008) and by atomic scale simulations (Schøitz, 2004).

The operation of a source-hardening mechanism has also been observed in samples of aluminum and IF steel processed by ARB, which has been slightly deformed by cold-rolling (Huang et al., 2008) i.e. without a preceding material's annealing step. In both cases, a decrease in yield stress is found, indicating that there is a deficiency of dislocation sources in the as-processed conditions. The dislocation source hardening observed in nanoscale structures can also be found in large-scale structures, for example in aluminum deformed and annealed to a grain size of a few micrometers (Kamikawa et al., 2009). An additional finding is the occurrence of localized shear deformation, causing a yield drop and formation of Lüders bands. These different and new observations suggest that the mechanical behavior of nanoscale and submicrometer microstructure combining deformation and annealing treatments may be a fruitful future research area.

17.10.7 Second-Phase Particles

The structural evolution during plastic deformation of aluminum containing a fine dispersion of oxide particles has been analyzed in Section 17.9. In the following the stress–strain behavior will be characterized for samples deformed at low to medium strain in tension and from low to high strain by drawing and rolling. Examples of stress–strain curves for samples containing from 0.2 to 3.8 vol. % of aluminum oxide with an equivalent spherical diameter of 30–50 nm are given in **Figure 54** for tension (Hansen, 1970) and in **Figure 55** for drawing (Hansen, 1969). Both figures show a high work-hardening rate at low and medium strain, whereas the work-hardening rate at medium to large strain is unaffected by the presence of oxide particles. The strengthening parameters are proposed as the dislocation density and the spacing between boundaries (cell boundaries and subgrain boundaries). The dislocations stored in the deformed structure are geometrically necessary dislocations with a density ρ_g in addition to statistically stored dislocations with a density ρ_m. The total dislocation density is then

$$\rho = \rho_g + \rho_m. \tag{35}$$

ρ_g for dislocations in the form of loops can be approximated (Ashby, 1972).

$$\rho_g = \frac{8f}{d_a \mathbf{b}} \varepsilon, \tag{36}$$

where f is the volume fraction of particles, d_a is the particle diameter in the slip plane and ε is the tensile strain. ρ_m can be approximated by

$$\rho_m \approx \frac{2\varepsilon}{\mathbf{b}c}, \tag{37}$$

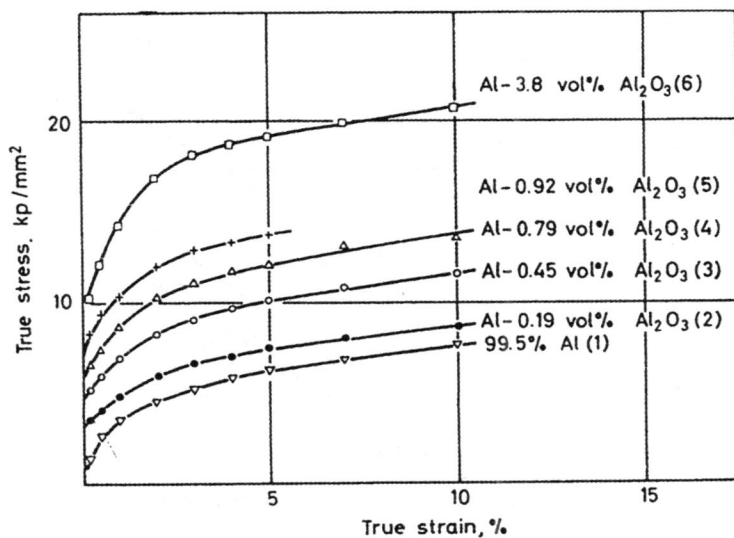

Figure 54   Stress–strain curves at room temperature for Al and Al-alumina. Hansen (1970); Figure 2.

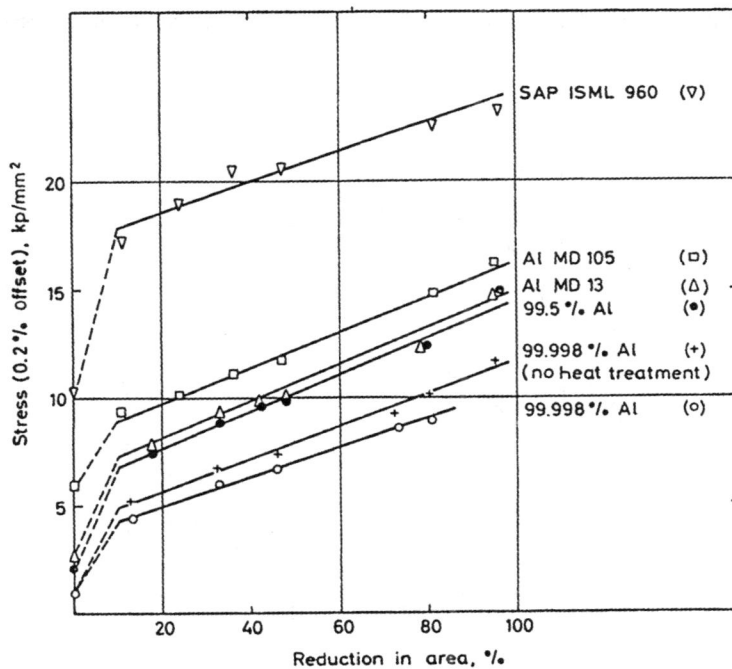

Figure 55   Flow stress as a function of drawing strain for Al and Al-alumina. Hansen (1969); Figure 13.

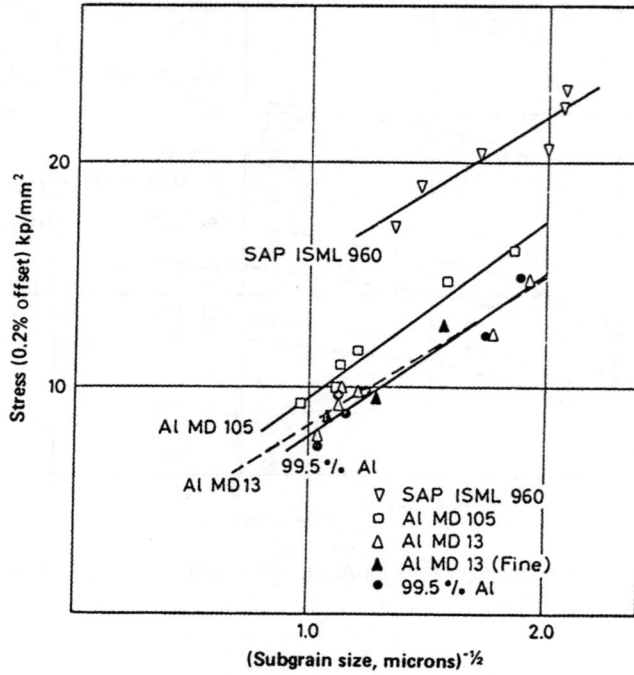

Figure 56   Flow stress as a function of reciprocal subgrain size for 99.5% Al and for Al-alumina. Hansen (1969); Figure 15.

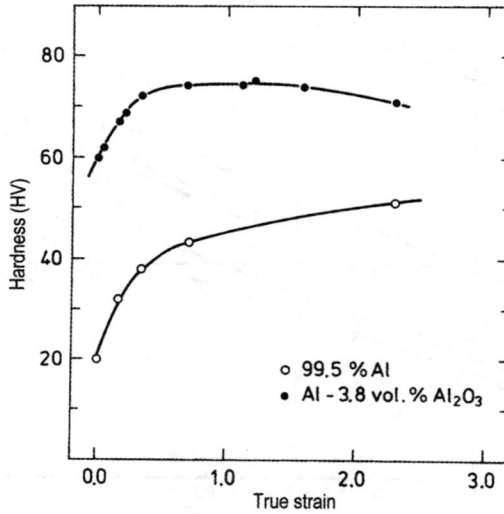

Figure 57   Hardness as a function of cold rolling strain for Al–3.8% alumina and 99.5% Al. Barlow and Hansen (1989); Figure 11.

where c is the slip distance. Based on Eqn 19 the flow stress is

$$\tau - \tau_0 \approx ML\mathbf{b}G\sqrt{\frac{8f}{d_a\mathbf{b}} + \frac{2}{\mathbf{b}L}}\sqrt{\varepsilon}. \tag{38}$$

For deformed structures where the strengthening parameter is the spacing between all boundaries or subgrain boundaries, the flow stress is assumed to be proportional to the inverse square root of the spacing in accordance with Eqn (32).

The stress–strain relationship in **Figure 54** has been analyzed in accordance with Eqn (38) on the assumption that the rapid work hardening in the low-strain region relates to the generation of geometrically necessary dislocations. In accordance with Eqn (38) the hardening rate increases rapidly with increasing volume fraction of oxide particles and an estimate of the rate based on Eqn (38) is in good agreement with experiment (Hansen, 1970). At higher strain, the work-hardening rate is practically independent of the presence of oxide particles in accordance with the prediction (Ashby, 1970) that a significant contribution to the flow stress from geometrically necessary dislocation is limited to the low-strain regime.

The stress–strain relationship in **Figure 55** (Hansen, 1969) has been analyzed in accordance with Eqn (32) on the assumption that the dominating strengthening mechanism is boundary strengthening. For aluminum and the oxide containing metals the flow stress is plotted in (**Figure 56**) (Hansen, 1969) against the reciprocal square root of the subgrain size showing no significant effect of an addition of oxide particles to aluminum.

An unexpected result in the structural analysis (Section 17.9) is a significant effect of dynamic recovery at large strain where there is a tendency to structural coarsening and a decrease in hardness as shown in **Figure 57** (Barlow and Hansen, 1989). These observations point to the existence of additional dynamic recovery processes at high strain, for example by a combination of dislocation annihilation and removal of boundaries by triple junction motion. However, as the work-hardening rate is negligible, flow instability during processing and testing may also lead to structural changes and some softening. The effect of small particles is also of interest here, with the observation that although the development of the deformation structure is accelerated at low strain, the subgrain size at large strain approaches a limiting value that is insensitive to the presence of fine particles (Barlow and Hansen, 1989). This behavior differs from the effect of elements in solid solution, see Section 17.9.

17.11   Conclusion and Outlook

TEM studies and scaling analysis of deformation microstructures in medium- and high-stacking fault fcc and bcc metals and alloys have demonstrated the dominant role of dislocations in the development of deformation microstructures down to the nanoscale.

Quantification and analysis of structural parameters has shown for a variety of metals and processes that the microstructural evolution follows a universal and hierarchical pattern of grain subdivision on multiple length scales by the formation of dislocation boundaries and high-angle boundaries.

Energetic and crystallographic guidelines have been applied in the analysis of deformation microstructures. The underlying assumption is that they are LEDS free of long-range stresses and that dislocations accumulate in configurations with a grain-orientation dependency related to the underlying slip systems, which also control the development of the deformation texture.

The development of the deformation microstructure has been related to the observation of work-hardening stages and to the flow stress up to ultrahigh strains. Based on a number of measurable structural parameters and assuming additional strength contributions from dislocation and boundary strengthening, flow stress predictions are in good agreement with the experimentally observed stress values and hardening rates. The analysis has revealed a clear interrelationship between microstructure, texture, strain hardening and flow stress. This present analysis must be extended to cover hardening and softening mechanisms (e.g. dynamic recovery) controlling the mechanical responses of the material to different loading conditions. This analysis must include experiments and models for the evolution of structural parameters down to the nanometer dimension in combination with mechanical testing of micron and submicron size specimens clarifying deformation mechanisms in such structures. There is also a need for further analyses and modeling of the structural and hardening response to the addition of solute elements and fine particles and to investigate transition regions between work-hardening stages extending to high and ultrahigh strains.

Acknowledgments

Niels Hansen gratefully acknowledges the support from the Danish National research foundation (Grant No. DNRF86-5) and the National Natural Science Foundation of China (Grant No. 51261130091) to the Danish-Chinese Center for Nanometals, within which part of this work has been performed.

Furthermore the authors are most grateful to Helle Hemmingsen and Tianbo Yu for assistance with preparing this manuscript.

References

Argon, A.S., 2008. Strengthening Mechanisms in Crystal Plasticity. Oxford University Press, Oxford.

Argon, A.S., Haasen, P., 1993. A new mechanism of work-hardening in the late stages of large-strain plastic-flow in fcc and diamond cubic-crystals. Acta Metall. 41, 3289–3306.

Armstrong, P.E., Hockett, J.E., Sherby, O.D., 1982. Large strain multidirectional deformation of 1100 aluminum at 300K. J. Mech. Phys. Solids 30, 37–58.

Ashby, M.F., 1970. Deformation of plastically non-homogeneous materials. Philos. Mag. 21, 399–424.

Ashby, M., 1972. First report on deformation-mechanism maps. Acta Metall. 20, 887–897.

Barabash, R.I., Ice, G.E., Liu, W., Barabash, O.M., 2009. Polychromatic microdiffraction characterization of defect gradients in severely deformed materials. Micron 40, 28–36.

Barlow, C.Y., Hansen, N., 1989. Deformation structures in aluminum containing small particles. Acta Metall. 37, 1313–1320.

Barlow, C.Y., Hansen, N., Liu, Y.L., 2002. Fine scale structures from deformation of aluminium containing small alumina particles. Acta Mater. 50, 171–182.

Barlow, C.Y., Hansen, N., 2007. Metalworking (Chapter 12). In: Groza, J.R., Shackelford, J.F. (Eds.), Materials Processing Handbook. CRC Press, Taylor and Francis.

Barlow, C.Y., Liu, Y.L., 1999. High-strain deformation of aluminium containing small alumina particles. In: Bilde-Sørensen, J.B., Carstensen, J.V., Hansen, N., et al. (Eds.), Proceedings of the 20th Risø International Symposium on Materials Science. Risø National Laboratory, Roskilde, Denmark, pp. 261–268.

Barlow, C.Y., Bay, B., Hansen, N., 1985. A comparative investigation of surface relief structures and dislocation microstructures in cold-rolled aluminum. Philos. Mag. A 51, 253–275.

Barrett, C., Levenson, L., 1940. The structure of aluminum after compression. Trans. Am. Inst. Min. Metall. Eng. 137, 112–126.

Bay, B., Hansen, N., Hughes, D.A., Kuhlmann-Wilsdorf, D., 1992. Overview No-96-Evolution of Fcc deformation structures in polyslip. Acta Metall. Mater. 40, 205–219.

Bay, B., Hansen, N., Kuhlmann-Wilsdorf, D., 1989. Deformation structures in lightly rolled pure aluminium. Mater. Sci. Eng. A113, 385–397.

Becker, R., Panchanadeeswaran, S., 1995. Effects of grain interactions on deformation and local texture in polycrystals. Acta Metall. Mater. 43, 2701–2719.

Bever, M.B., Holt, D.L., Titchener, A.L., 1973. Stored Energy in Cold Rolled Work. Progress in Materials Science, vol. 17. Pergamon Press, Oxford.

Bhadeshia, H.K.D.H., 1987. Worked Examples in the Geometry of Crystals. Institute of Materials, London.

Bishop, J., Hill, R., 1951. A theoretical derivation of the plastic properties of a polycrystalline face-centred metal. Philos. Mag. 42, 1298–1307.

Bowen, J., Prangnell, P., Juul Jensen, D., Hansen, N., 2004. Microstructural parameters and flow stress in Al-0.13% Mg deformed by ECAE processing. Mater. Sci. Eng. A Struct. Mater. Prop. Microstruct. Process. 387, 235–239.

Chin, G.Y., 1973. The role of preferred orientation in plastic deformation. In: Reed-Hill, R.E. (Ed.), The Inhomogeneity of Plastic Deformation. American Society for Metals, Metals Park, Ohio, pp. 83–112.

Cizek, P., 2010. Dislocation boundaries and disclinations formed within the cube-oriented grains during tensile deformation of aluminium. Acta Mater. 58 (17), 5820–5833.

Clement, L., Pantel, R., Kwakman, L.F.T., Rouviere, J.L., 2004. Strain measurements by convergent-beam electron diffraction: the importance of stress relaxation in lamella preparations. Appl. Phys. Lett. 85, 651–653.

Cottrell, A.H., 1963. Dislocations and Plastic Flow in Crystals, second ed. Oxford University Press.

Cottrell, A.H., 1964. The Mechanical Properties of Matter. Wiley, New York.

de Graef, M., 2003. Introduction to Conventional Transmission Electron Microscopy. Cambridge University Press, Cambridge.

Dillamore, I.L., Morris, P., Smith, C.J.E., Hutchins, W.B., 1972. Transition bands and recrystallization in metals. Proc. Roy. Soc. Lond. A Math. Phys. Sci. 329, 405–420.

Doherty, R.D., Hughes, D.A., Humphreys, F.J., et al., 1997. Current issues in recrystallization: a review. Mater. Sci. Eng. A Struct. Mater. Prop. Microstruct. Process. 238, 219–274.

Driver, J.H., Juul Jensen, D., Hansen, N., 1994. Large-strain deformation structures in aluminum crystals with rolling texture orientations. Acta Metall. Mater. 42, 3105–3114.

Duly, D., Baxter, G., Shercliff, H., Whiteman, J., Sellars, C., Ashby, M., 1996. Microstructure and local crystallographic evolution in an Al-1 wt% Mg alloy deformed at intermediate temperature and high strain-rate. Acta Mater. 44, 2947–2962.

Eardley, E.S., Humphreys, F.J., Court, S.A., Bate, P.S., 2002. Microstructure and plastic anisotropy in rolled AA1200. Mater. Sci. Forum 396-4, 1085–1090.

Field, D.P., 2005. Improving the spatial resolution of EBSD. Microsc. Microanal. 11 (S02), 52–53.

Forwood, C.T., Clarebrough, L.M., 1991. Electron Microscopy of Interfaces in Metals and Alloys. IOP Publishing Ltd., Bristol.

Frank, F.C., 1950. Report of the Symposium on the Plastic Deformation of Crystalline Solids. Carnegie Institute of Technology and Office of Naval Research, Pittsburgh, pp. 150–152.

Gil-Sevillano, J., van Houtte, P., Aernoudt, E., 1981. Large strain work-hardening and textures. Prog. Mater. Sci. 25, 69–134.

Godfrey, A., Juul Jensen, D., Hansen, N., 1998a. Slip pattern, microstructure and local crystallography in an aluminium single crystal of brass orientation {110}<112>. Acta Mater. 46, 823–833.

Godfrey, A., Juul Jensen, D., Hansen, N., 1998b. Slip pattern, microstructure and local crystallography in an aluminium single crystal of copper orientation {112}<111>. Acta Mater. 46, 835–848.

Godfrey, A., Hughes, D.A., 2004. Physical parameters linking deformation microstructures over a wide range of length scales. Scr. Mater. 51, 831–836.

Godfrey, A., Hansen, N., Juul Jensen, D., 2007. Microstructural-based measurement of local stored energy variations in deformed metals. Metall. Mater. Trans. A Phys. Metall. Mater. Sci. 38A, 2329–2339.

Godfrey, A., Liu, Q., 2009. Stored energy and structure in top-down processed nanostructured metals. Scr. Mater. 60, 1050–1055.

Godfrey, A., Cao, W.Q., Hansen, N., Liu, Q., 2005. Stored energy, microstructure, and flow stress of deformed metals. Metall. Trans. A 36A, 2371–2378.

Greer, J., Oliver, W., Nix, W., 2005. Size dependence of mechanical properties of gold at the micron scale in the absence of strain gradients. Acta Mater. 53, 1821–1830.

Haasen, P., 1996. Mechanical properties of solid solutions (Chapter 23). In: Cahn, R.W., Haasen, P. (Eds.), Physical Metallurgy, fourth ed. North Holland.

Haldar, A., Huang, X., Leffers, T., Hansen, N., Ray, R.K., 2004. Grain orientation dependence of microstructures in warm rolled IF steel. Acta Mater. 52, 5405–5418.

Hall, E., 1951. The deformation and ageing of mild steel .3. discussion of results. Proc. Phys. Soc. Lond. B 64, 747–753.

Han, B.Q., Lavernia, E.J., 2005. Deformation mechanisms of nanostructured alloys. Adv. Eng. Mater. 7, 457–465.

Hansen, N., 1969. Microstructure and flow stress of aluminium and dispersion strengthened aluminium-aluminium oxide products drawn at room temperature. Trans. TMS AIME 246, 2061–2068.

Hansen, N., 1970. Dispersion strengthening of aluminium—aluminium oxide products. Acta Metall. 18, 137–145.

Hansen, N., 1985. Polycrystalline strengthening. Metall. Trans. A 16A, 2167–2190.

Hansen, N., 1990. Cold deformation microstructures. Mater. Sci. Technol. 6, 1039–1047.

Hansen, N., 1992. Viewpoint Set No. 20. Microstructure and flow-stress of cell forming metals. Scr. Metall. Mater. 27, 947–950.

Hansen, N., 1994. In: Andersen, S.I., et al. (Eds.), Flow Stress and Microstructural Parameters, Numerical Predictions of Deformation Processes and the Behaviour of Real Materials. Risø National Laboratory, Roskilde, Denmark, pp. 325–334.

Hansen, N., 2001. New discoveries in deformed metals. Metall. Trans. A 32, 2917–2935.

Hansen, N., 2005. Boundary strengthening in undeformed and deformed polycrystals. Mater. Sci. Eng. A409, 39–45.

Hansen, N., Huang, X., 1998. Microstructure and flow stress of polycrystals and single crystals. Acta Mater. 46, 1827–1836.

Hansen, N., Hughes, D.A., 1995. Analysis of large dislocation populations in deformed metals. Phys. Stat. Solidi (A) 149, 155–172.

Hansen, N., Juul Jensen, D., 1987. Effect of metallurgical parameters on texture and microstructure. In: Sachdev, A.K., Embury, J.D. (Eds.), Formability and Metallurgical Structure. TMS, Warrendale, PA, pp. 119–136.

Hansen, N., Juul Jensen, D., 1992. Flow-stress anisotropy caused by geometrically necessary boundaries. Acta Metall. Mater. 40, 3265–3275.

Hansen, N., Juul Jensen, D., 2011. Deformed metals—structure, recrystallisation and strength. Mater. Sci. Technol. 27, 1229–1240.

Hansen, N., Kuhlmann-Wilsdorf, D., 1986. Low-energy dislocation-structures due to unidirectional deformation at low temperatures. Mater. Sci. Eng. 81, 141–161.

Hansen, N., Huang, X., Pantleon, W., Winther, G., 2006. Grain orientation and dislocation patterns. Philos. Mag. 86, 3981–3994.

Hansen, N., Huang, X., Winther, G., 2011. Effect of grain boundaries and grain orientation on structure and properties. Metall. Mater. Trans. A Phys. Metall. Mater. Sci. 42A, 613–625.

Hecker, S.S., Stout, M.G., 1984. Strain hardening of heavily cold rolled metals. In: Krauss, G. (Ed.), Deformation, Processing, and Structure. American Society for Metals, Metals Park Ohio, pp. 1–13.

Hirth, J.P., 1996. Dislocations (Chapter 20). In: Cahn, R.W., Haasen, P. (Eds.), Physical Metallurgy, fourth ed. North Holland.

Holt, D.L., 1970. Dislocation cell formation in metals. J. Appl. Phys. 41, 3197–3201.

Honeycombe, R.W.K., 1975. The Plastic Deformation of Metals. Edward Arnold, UK.

Hong, S., Laird, C., 1990. Mechanisms of slip mode modification in fcc solid-solutions. Acta Metall. Mater. 38, 1581–1594.

Hong, C.S., Huang, X., Winther, G., 2013. Dislocation content of geometrically necessary boundaries aligned with slip planes in rolled aluminium. Philos. Mag.. http://dx.doi.org/10.1080/14786435.2013.805270.

Hosford, W.F., 2005. Physical Metallurgy. CRC Press.

Hsun, H., 1963. Annealing of silicon-iron single crystals. In: Himmel (Ed.), Recovery and Recrystallization of Metals. Interscience Publishers, New York, pp. 311–378.

Huang, J.Y., Zhu, Y.T., Jiang, H., Lowe, T.C., 2001. Microstructures and dislocation configurations in nanostructured Cu processed by repetitive corrugation and straightening. Acta Mater. 49, 1497–1505.

Huang, X., 2009. Tailoring dislocation structures and mechanical properties of nanostructured metals produced by plastic deformation. Scr. Mater. 60, 1078–1082.

Huang, X., Winther, G., 2007. Dislocation structures. Part I. Grain orientation dependence. Philos. Mag. 87, 5189–5214.

Huang, X., Winther, G., Hansen, N., Hebesberger, A., Vorhaueer, A., Pippan, R., Zehetbauer, M., 2003. Microstructures of nickel deformed by high pressure torsion to high strains. Mater. Sci. Forum Thermec'2003 426–432, 2819–2824.

Huang, X., 1998. Grain orientation effect on microstructure in tensile strained copper. Scr. Mater. 38, 1697–1703.

Huang, X., Hansen, N., Tsuji, N., 2006. Hardening by annealing and softening by deformation in nanostructured metals. Science 312, 249–251.

Huang, X., Kamikawa, N., Hansen, N., 2008. Increasing the ductility of nanostructured al and Fe by deformation. Mater. Sci. Eng. A Struct. Mater. Prop. Microstruct. Process. 493, 184–189.

Huang, X., Tsuji, N., Pantleon, W., 2009. Strengthening mechanisms, instabilities and property optimization in nanostructured metals. In: Grivel, J.-C., Hansen, N., Huang, X., et al. (Eds.), Proceedings of the 30th Risø International Symposium on Materials Science. Risø National Laboratory, Roskilde, Denmark, pp. 59–80.

Hughes, D.A., 1986. PhD dissertation: Strain Hardening of FCC Metals and Alloys at Large Strains Stanford University.

Hughes, D.A., 1993. Microstructural evolution in a non-cell forming metal: Al–Mg. Acta Metall. Mater. 41, 1421–1430.

Hughes D.A., 1998. Personal Communication.

Hughes, D.A., Godfrey, A., 1998. Dislocation structures formed during hot and cold working. In: Bieler, T.R., et al. (Eds.), Hot Deformation of Aluminum Alloys II. TMS, Warrendale PA, pp. 23–36.

Hughes, D.A., Chrzan, D.C., Liu, Q., Hansen, N., 1998. Scaling of misorientation angle distributions. Phys. Rev. Lett. 81, 4664–4667.

Hughes, D.A., Hansen, N., 1993. Microstructural evolution in nickel during rolling from intermediate to large strains. Metall. Trans. A Phys. Metall. Mater. Sci. 24, 2021–2037.

Hughes, D.A., Hansen, N., 1997. High angle boundaries formed by grain subdivision mechanisms. Acta Mater. 45, 3871–3886.

Hughes, D.A., Hansen, N., 2000. Microstructure and strength of nickel at large strains. Acta Mater. 48, 2985–3004.

Hughes, D.A., Hansen, N., 2001. Graded nanostructures produced by sliding and exhibiting universal behavior. Phys. Rev. Lett. 87, 135503.

Hughes, D.A., Hansen, N., 2003. Deformation structures developing on fine scales. Philos. Mag. 83, 3871–3893.

Hughes, D.A., Hansen, N., 2004. ASM Handbook: Plastic Deformation Structures. In: Metallography and Microstructures, vol. 9. ASM International, Materials Park, Ohio, USA, p. 292.

Hughes, D.A., Khan, S., Godfrey, A., Zbib, H., 2001. Internal structures of deformation induced planar dislocation boundaries. Mater. Sci. Eng. A Struct. Mater. Prop. Microstruct. Process. 309, 220–226.

Hughes, D.A., Nix, W., 1988. The absence of steady-state flow during large strain plastic-deformation of some fcc metals at low and intermediate temperatures. Metall. Trans. A Phys. Metall. Mater. Sci. 19, 3013–3024.

Hull, D., Bacon, D.J., 2011. Introduction to Dislocations, fifth ed. Butterworth-Heinemann, Oxford.

Hung, P., Sun, P., Yu, C., Kao, P., Chang, C., 2005. Inhomogeneous tensile deformation in ultrafine-grained aluminum. Scr. Mater. 53, 647–652.

Ikeda, K., Yamada, K., Takata, N., et al., 2004. Atomic structure of grain boundaries in ARB processed copper. In: Gundlach, C., et al. (Eds.), Proceedings of the 25th Risø International Symposium on Materials Science: Evolution of Deformation Microstructures in 3D, pp. 357–362 (Roskilde, Denmark).

Jago, R.A., Hansen, N., 1986. Grain-size effects in the deformation of polycrystalline iron. Acta Metall. 34, 1711–1720.

Jakobsen, B., Poulsen, H.F., Lienert, U., Pantleon, W., 2007. Direct determination of elastic strains and dislocation densities in individual subgrains in deformation structures. Acta Mater. 55, 3421–3430.

Juul Jensen, D., Hansen, N., 1990. Flow-stress anisotropy in aluminum. Acta Metall. Mater. 38, 1369–1380.

Juul Jensen, D., Lyttle, M.T., Hansen, N., 1998. Hot and cold deformed aluminium: deformation microstructure and recrystallization behaviour. In: Bieler, T.R., Lalli, L.A., MacEwen, S.R. (Eds.), Hot Deformation of Aluminum Alloys 2. The Minerals, Metals and Materials Society, Warrendale, PA, pp. 9–21.

Juul Jensen, D., Hansen, N., Kjems, J.K., Leffers, T., 1984. In-situ texture measurements by neutron diffraction used in a study of recrystallization kinetics. In: Hessel Andersen, N., Eldrup, M., Hansen, N., et al. (Eds.), Proceedings of the 5th Risø International Symposium on Metallurgy and Materials Science, pp. 325–332 (Roskilde, Denmark).

Kalpakjian, S., Schmid, S.R., 2003. Manufacturing Processes for Engineering Materials, fourth ed. Prentice Hall, Upper Saddle River, New Jersey (Chapter 2).

Kamikawa, N., Huang, X., Tsuji, N., Hansen, N., 2009. Strengthening mechanisms in nanostructured high-purity aluminium deformed to high strain and annealed. Acta Mater. 57, 4198–4208.

Knudsen, T., Cao, W.Q., Godfrey, A., Liu, Q., Hansen, N., 2008. Stored energy in nickel cold rolled to large strains, measured by calorimetry and evaluated from the microstructure. Metall. Mater. Trans. A Phys. Metall. Mater. Sci. 39A, 430–440.

Kocks, U.F., Mecking, H., 2003. Physics and phenomenology of strain hardening: the FCC case. Prog. Mater. Sci. 48, 171–273.

Kubin, L.P., 2013. Dislocations, Mesoscale Simulations and Plastic Flow. Oxford University Press, Great Clarendon Street, Oxford, OX2 6DP, UK.

Kuhlmann-Wilsdorf, D., Hansen, N., 1991. Geometrically necessary, incidental and subgrain boundaries formed during cold deformation. Scr. Metall. Mater. 25, 1557–1562.

Kuhlmann-Wilsdorf, D., 1989. Theory of plastic-deformation—properties of low energy dislocation-structures. Mater. Sci. Eng. A 113, 1–41.

Kuhlmann-Wilsdorf, D., 1962. A new theory of workhardening. Trans. Met. Soc. AIME 224, 1047–1061.

Kuhlmann-Wilsdorf, D., 1995. Technological high strain deformation of 'wavy glide' metals and LEDS. Phys. Stat. Solidi (A) 149, 225–241.

Kuhlmann-Wilsdorf, D., 1999. Overview No. 131 "Regular" deformation bands (DBs) and the LEDS hypothesis. Acta Mater. 47, 1697–1712.

Kurzydlowski, K.J., Ralph, B., 1995. The Quantitative Description of the Microstructure of Materials. CRC Press.

Langford, G., Cohen, M., 1969. Microstructure of Armco-iron subjected to severe plastic drawing. Trans. ASM 82, 623.

Le, G.M., Godfrey, A., Hong, C.S., Huang, X., Winther, G., 2012. Orientation dependence of the deformation microstructure in compressed aluminum. Scr. Mater. 66, 359–362.

Leffers, T., 1995. Long-range stresses associated with boundaries in deformed metals. Phys. Stat. Solidi (A) 149, 69–84.

Levine, L.E., Geantil, P., Larson, et al., 2011. Disordered long-range internal stresses in deformed copper and the mechanisms underlying plastic deformation. Acta Mater. 59, 5803–5811.

Li, J.C.M., 1963. Petch relation and grain-boundary sources. Trans. TMS AIME 227, 239.

Li, J.C.M., Chou, Y., 1970. Role of dislocations in flow stress grain size relationships. Metall. Trans. 1, 1145–1159.

Li, J.C.M., Liu, G., 1967. Circular dislocation pile-ups I. Strength of ultra-fine polycrystalline aggregates. Philos. Mag. 15, 1059–1063.

Li, Z.J., Winther, G., Hansen, N., 2006. Anisotropy in rolled metals induced by dislocation structure. Acta Mater. 54, 401–410.

Lin, F.X., Godfrey, A., Winther, G., 2009. Grain orientation dependence of extended planar dislocation boundaries in rolled aluminium. Scr. Mater. 61, 237–240.

Liu, C., Bassim, M., 1993. Dislocation substructure evolution in torsion of pure copper. Metall. Trans. A Phys. Metall. Mater. Sci. 24, 361–367.

Liu, Q., Hansen, N., 1995a. Geometrically necessary boundaries and incidental dislocation boundaries formed during cold deformation. Scr. Metall. Mater. 32, 1289–1295.

Liu, Q., Hansen, N., 1995b. Deformation microstructure and orientation in FCC crystals. Phys. Stat. Solidi (A) 149, 187–199.

Liu, Q., Hansen, N., 1998. Macroscopic and microscopic subdivision of a cold-rolled aluminium single crystal of cubic orientation. Proc. Roy. Soc. Lond. A Math. Phys. Eng. Sci. 454, 2555–2591.

Liu, Q., 1994. A simple method for determining orientation and misorientation of the cubic-crystal specimen. J. Appl. Cryst. 27, 755–761.

Liu, Q., Huang, X., Lloyd, D.J., Hansen, N., 2002. Microstructure and strength of commercial purity aluminium (AA 1200) cold-rolled to large strains. Acta Mater. 50, 3789–3802.

Liu, Q., Jensen, D.J., Hansen, N., 1998. Effect of grain orientation on deformation structure in cold-rolled polycrystalline aluminium. Acta Mater. 46, 5819–5838.

Liu, W., 1998. Report No. Risø I-1362 (EN), Risø National Laboratory.

Lloyd, D., Kenny, D., 1982. The large strain deformation of some aluminum-alloys. Metall. Trans. A Phys. Metall. Mater. Sci. 13, 1445–1452.

Loretto, M.H., 1993. Electron Beam Analysis of Materials. Springer Verlag, Heidelberg.

Lu, K., Hansen, N., 2009. Structural refinement and deformation mechanisms in nanostructured metals. Scr. Mater. 60, 1033–1038.

Malin, A., Hatherly, M., 1979. Deformation of copper and low stacking-fault energy, copper-base alloys. Metals Technol. 6, 308–319.

Massalski, T.R., 1996. Structure and stability of alloys (Chapter 3). In: Cahn, R.W., Haasen, P. (Eds.), Physical Metallurgy, fourth ed. North Holland.

McCabe, R., Misra, A., Mitchell, T., 2004. Experimentally determined content of a geometrically necessary dislocation boundary in copper. Acta Mater. 52, 705–714.

McQueen, H.J., Solberg, J.K., Ryum, N., Nes, E., 1989. Evolution of flow stress in Al during ultra-high straining at elevated temperature. Philos. Mag. 60, 473–485.

Mecking, H., Kocks, U., 1981. Kinetics of flow and strain-hardening. Acta Metall. 29, 1865–1875.

Mishin, O.V., Juul Jensen, D., Hansen, N., 2003. Microstructures and boundary populations in materials produced by equal channel angular extrusion. Mater. Sci. Eng. A Struct. Mater. Prop. Microstruct. Process. 342, 320–328.

Nabarro, F., 1994. The coefficient of work-hardening in stage-iv. Scr. Metall. Mater. 30, 1085–1087.

Neuhaus, R., Schwink, C., 1992. On the flow stress of [100]- and [111]-oriented Cu–Mn single crystals: a transmission electron microscopy study. Philos. Mag. A 65, 1463–1484.

Nieh, T., Wadsworth, J., 1991. Hall–Petch relation in nanocrystalline solids. Scr. Metall. Mater. 25, 955–958.

Nuttall, J., Nutting, J., 1978. Structure and properties of heavily cold-worked fcc metals and alloys. Metal Sci. 12, 430–438.

Nye, J., 1953. Some geometrical relations in dislocated crystals. Acta Metall. 1, 153–162.

Pantleon, W., 1998. On the statistical origin of disorientations in dislocation structures. Acta Mater. 46, 451–456.

Pantleon, W., 1999. Dislocation boundaries: formation, orientation and implications. In: Bilde-Sørensen, J.B., Carstensen, J.V., Hansen, N., et al. (Eds.), Proceedings of 20th Risø International Symposium on Materials Science, pp. 123–146 (Roskilde, Denmark).

Pantleon, W., Hansen, N., 2001a. Disorientations in dislocation boundaries: formation and spatial correlation. Mater. Sci. Eng. A 309–310, 246–250.

Pantleon, W., Hansen, N., 2001b. Dislocation boundaries—the distribution function of disorientation angles. Acta Mater. 49, 1479–1493.

Pantleon, W., 2005. Disorientations in dislocation structure. Mater. Sci. Eng. A Struct. Mater. Prop. Microstruct. Process. 400, 118–124.

Pantleon, W., 2011. Disorientations after severe plastic deformation and their effect on work-hardening. Mater. Sci. Forum 667–669, 205–210.

Petch, N., 1953. The cleavage strength of polycrystals. J. Iron Steel Inst. 174, 25–28.

Peeters, B., Seefeld, M., Teodosiu, et al., 2001. Work-hardening/softening behaviour of b.c.c. polycrystals during changing strain paths: I. An integrated model based on substructure and texture evolution, and its prediction of the stress–strain behaviour of an IF steel during two-stage strain paths. Acta Mater. 49, 1607–1619.

Poulsen, H.F., 2004. Three-dimensional X-ray Diffraction Microscopy. Mapping Polycrystals and Their Dynamics. Springer, Berlin.

Raj, S., Pharr, G., 1986. A compilation and analysis of data for the stress dependence of the subgrain size. Mater. Sci. Eng. 81, 217–237.

Ralph, B., Ecob, R.C., Porter, A.J., Barlow, C.Y., Ecob, N.R., 1981. The structure of grain boundaries and their affect on mechanical properties. In: Hansen, N., Horsewell, A., Leffers, T., Lilholt, H. (Eds.), Proceedings of the 2nd Risø International Symposium, Denmark, pp. 111–124 (Roskilde, Denmark).

Randle, V., Engler, O., 2000. Introduction to Texture Analysis. Taylor and Francis.

Randle, V., Hansen, N., Juul Jensen, D., 1996. The deformation behaviour of grain boundary regions polycrystalline aluminium. Philos. Mag. A Phys. Condens. Matter Struct. Defects Mech. Prop. 73, 265–282.

Randle, V., 1992. Microtexture Determination and Its Applications. Institute of Materials, London.

Read, W., Shockley, W., 1950. Dislocation models of crystal grain boundaries. Phys. Rev. 78, 275–289.

Robertson, I.M., Schuh, C.A., Vetrano, J.S., et al., 2011. Towards an integrated materials characterization toolbox. J. Mater. Res. 26, 1341–1383.

Rollett, A.D., 1988. Strain Hardening at Large Strain in Aluminum Alloys. Ph.D. thesis. Philadelphia: Drexel University.

Rollett, A.D., Juul Jensen, D., Stout, M.G., 1992. Modelling the effect of microstructure on yield anisotropy. In: Andersen, S.I., Bilde-Sørensen, J.B., Hansen, N., et al. (Eds.), Proceedings of the 13th Risø International Symposium on Materials Science, pp. 93–110 (Roskilde, Denmark).

Saito, Y., Utsunomiya, H., Tsuji, N., Sakai, T., 1999. Novel ultra-high straining process for bulk materials—development of the accumulative roll-bonding (ARB) process. Acta Mater. 47, 579–583.

Schiøtz, J., 2004. Atomic-scale modeling of plastic deformation of nanocrystalline copper. Scr. Mater. 51, 837–841.

Shan, Z.W., Mishra, R.K., Asif, S.A.S., Warren, O.L., Minor, A.M., 2008. Mechanical annealing and source-limited deformation in submicrometre-diameter Ni crystals. Nat. Mater. 7, 115–119.

Staker, M.R., Holt, D.L., 1972. The dislocation cell size and dislocation density in copper deformed at temperatures between 25 and 700 °C. Acta Metall. 20, 569–579.

Steeds, J.W., 1966. Dislocation arrangement in copper single crystals as a function of strain. Proc. Roy. Soc. Lond. A Math. Phys. Sci. 292, 343–372.

Sutton, A.P., Balluffi, R.W., 1995. Interfaces in Crystalline Materials: Sections 12.5, 12.6. Oxford University Press, Oxford.

Taylor, A.S., Cizek, P., Hodgson, P.D., 2012. Orientation dependence of the substructure characteristics in a Ni–30Fe austenitic model alloy deformed in hot plane strain compression. Acta Mater. 60, 1548–1569.

Taylor, G., 1938. Plastic strain in metals. J. Inst. Metals 62, 307–324.

Theyssier, M., Chenal, B., Driver, J., Hansen, N., 1995. Mosaic dislocation-structures in aluminum crystals deformed in multiple slip at 0.5 to 0.8t(m). Phys. Stat. Solidi A Appl. Res. 149, 367–378.

Thorning, C., Somers, M.A.J., Wert, J.A., 2005. Grain interaction effects in polycrystalline Cu. Mater. Sci. Eng. A 397, 215–228.

Tsuji, N., Saito, Y., Lee, S.-H., Minamino, Y., 2003. ARB (accumulative roll-bonding) and other new techniques to produce bulk ultrafine grained materials. Adv. Eng. Mater. 5, 338–344.

Tsuji, N., Huang, X., Nakashima, H., 2004. Microstructure of metals and alloys deformed to ultrahigh strains. In: Gundlach, C., Haldrup, K., Hansen, N., et al. (Eds.), Proceedings of the 25th Risø International Symposium on Materials Science, pp. 147–170 (Roskilde, Denmark).

Ungar, T., Pantleon, W., 2009. Identifying boundary structures by X-ray diffraction. In: Grivel, J.-C., Hansen, N., Huang, X., et al. (Eds.), Proceedings of the 30th Risø International Symposium on Materials Science, pp. 157–170 (Roskilde, Denmark).

Ungar, T., Mughrabi, H., Ronnpagel, D., Wilkens, M., 1984. X-ray line-broadening study of the dislocation cell structure in deformed [001]-orientated copper single-crystals. Acta Metall. 32, 333–342.

Valiev, R.Z., Islamgaliev, R.K., Alexandrov, I.V., 2000. Bulk nanostructured materials from severe plastic deformation. Prog. Mater. Sci. 45, 103–189.

Wei, Y.L., Godfrey, A., Liu, W., Liu, Q., Huang, X., Hansen, N., Winther, G., 2011. Dislocations, boundaries and slip systems in cube grains of rolled aluminium. Scr. Mater. 65, 355–358.

Wert, J.A., Liu, Q., Hansen, N., 1995. Dislocation boundaries and active slip systems. Acta Metall. Mater. 43, 4153–4163.

Wert, J.A., Huang, X., Inoko, F., Okada, T., Hansen, N., 1999. Observation and analysis of deformation microstructures in a [110] Al single crystal deformed in tension. In: Bilde-Sørensen, J.B., Carstensen, J.V., Hansen, et al. (Eds.), Proceedings of the 20th Risø International Symposium on Materials Science, pp. 529–534 (Roskilde, Denmark).

Wilde, G., 2013. Physical metallurgy of nanocrystalline metal (Chapter 28). In: Physical Metallurgy, fifth ed. Elsevier.

Wilkinson, A.J., Karamched, P.S., Britton, T.B., 2010. High resolution EBSD—3D strain tensors, and geometrically necessary dislocation distributions. In: Hansen, N., Juul Jensen, D., Nielsen, S.F., et al. (Eds.), Proceedings of the 31th Risø International Symposium on Materials Science, pp. 187–200 (Roskilde, Denmark).

Wilson, D.V., Bate, P.S., 1996. Internal elastic strains in an IF steel following changes in strain path. Acta Mater. 44 (8), 3371–3383.

Winther, G., 2005. Effect of grain orientation dependent microstructures on flow stress anisotropy modelling. Scr. Mater. 52, 995–1000.

Winther, G., 2008. Slip systems extracted from lattice rotations and dislocation structures. Acta Mater. 56, 1919–1932.

Winther, G., Huang, X., 2007. Dislocation structures. part II. slip system dependence. Philos. Mag. 87, 5215–5235.

Winther, G., Juul Jensen, D., Hansen, N., 1997. Modelling flow stress anisotropy caused by deformation induced dislocation boundaries. Acta Mater. 45, 2455–2465.

Winther, G., Margulies, L., Schmidt, S., Poulsen, H., 2004. Lattice rotations of individual bulk grains part II: correlation with initial orientation and model comparison. Acta Mater. 52, 2863–2872.

Wrobel, M., Dymek, S., Blicharski, M., Gorczyca, S., 1988. Microstructure and texture of rolled (110)[001] copper single-crystals. Textures Microstruct. 10, 9–19.

Xing, Q., Huang, X., Hansen, N., 2006. Recovery of heavily cold-rolled aluminum: effect of local texture. Metall. Mater. Trans. A Phys. Metall. Mater. Sci. 37A, 1311–1322.

Yang, D.K., Cizek, P., Hodgson, P.D., Wen, C.E., 2010. Microstructure evolution and nanograin formation during shear localization in cold-rolled titanium. Acta Mater. 58, 4536–4548.

Yu, T.B., Hansen, N., Huang, X., 2011. Recovery by triple junction motion in aluminium deformed to ultrahigh strains. Proc. Roy. Soc. A Math. Phys. Eng. Sci. 467, 3039–3065.

Zehetbauer, M., Seumer, V., 1993. Cold work-hardening in stage-IV and stage-V of fcc metals .1. experiments and interpretation. Acta Metall. Mater. 41, 577–588.

Zhang, H.W., Huang, X., Hansen, N., 2008a. Evolution of microstructural parameters and flow stresses toward limits in nickel deformed to ultrahigh strains. Acta Mater. 56, 5451–5465.

Zhang, H.W., Huang, X., Hansen, N., Pippan, R., Zehetbauer, M., 2008b. Strengthening of nickel deformed by high pressure torsion. Nanomater. Sev. Plast. Deform. IV 584–586 (Pts 1 and 2), 417–421.

Zhang, H.W., Lu, K., Pippan, R., Huang, X., Hansen, N., 2011. Enhancement of strength and stability of nanostructured Ni by small amounts of solutes. Scr. Mater. 65, 481–484.

Zhang, X.D., Godfrey, A., Huang, X., Hansen, N., Liu, Q., 2011. Microstructure and strengthening mechanisms in cold-drawn pearlitic steel wire. Acta Mater. 59, 3422–3430.

Zhang, X.D., Godfrey, A., Winther, G., Hansen, N., Huang, X., 2012. Plasticity of submicron-sized crystals studied by in-situ Kikuchi diffraction and dislocation imaging. Mater. Charact. 70, 21–27.

Zimmer, W., Hecker, S., Rohr, D., Murr, L., 1983. Large strain plastic-deformation of commercially pure nickel. Metal Sci. 17, 198–206.

Zisman, A., Rybin, V., 1998. Mesoscopic stress field arising from the grain interaction in plastically deformed polycrystals. Acta Mater. 46, 457–464.

Biography

Dr. Techn. Niels Hansen is senior scientist in the Danish-Chinese Center for Nanometals and he is former head of the Materials Research Department at Risø National Laboratory (now DTU), Denmark. His field is materials science and engineering with special emphasis on the evolution of microstructures and mechanical properties during plastic deformation of metals and alloys from low to ultra-high strain. He is presently involved with industrial problems, for example friction and wear of mechanical components in large wind turbines.

Dr Claire Barlow specializes in materials engineering with particular interest in the relationships between microstructure and mechanical properties in a wide range of materials. As Senior Lecturer in the Department of Engineering, Cambridge University she teaches materials topics to undergraduate and postgraduate students. Research work on metals has centred on transmission electron microscopy of dislocation structures in cold-deformed aluminium alloys. She is increasingly involved with industrial projects on the sustainable use of materials resources, looking at how design, processing and choice of material affect their lifecycle impact.

18 Fatigue of Metals

Pedro Peralta, School for Engineering of Matter, Transport and Energy, Arizona State University, Tempe, AZ, USA
Campbell Laird, Department of Materials Science and Engineering, University of Pennsylvania, Philadelphia, PA, USA

18.1 Introduction: History, Fatigue Approaches and Nomenclature

The classes of loads to which engineering materials and structures are commonly subjected in service are static or steady loads, repeated loads, impact loads and combinations of these three; for example, it is possible and quite common to have repeated impact loads or repeated loads in combination with a steady load. Documented efforts to understand and control the resistance of materials to fracture by repeated or cyclic loads have been on going for more than 180 years with ever-increasing intensity. In 1983, Battelle Laboratory under contract to United States (US) Government agencies performed detailed economic analyses of the costs of fractures in materials such as metals, wood and glass, and efforts to prevent it. These costs to US industry amounted to $119 billion /year, that is, a meaningful fraction of the Gross National Product at the time. It is unlikely that the percentage is much different today since the largest fraction of fractures in metals is associated with cyclic loads. The vast literature on fatigue, which has been driven by the severity of the problem, has been documented by Mann (1970, 1978) up to 1960 and by Laird (1996) up to 1996.

Service-related cyclic loads may produce stresses that vary from zero to a positive extremum, from a positive or negative extremum to another positive or negative extremum, or from a negative extremum to a positive extremum. The nature of the stress range is conventionally indicated by the ratio of the minimum to maximum stress. This is typically called the load ratio and denoted R. Given this definition, when the negative and positive loads are numerically equal, an alternating load would be designated $R = -1$. Fatigue failure from repeated nominally negative stresses, that is, compression, is known to occur although it is unlikely in such circumstances that the material actually sees only negative stresses. A large fraction of the experimental effort to understand mechanisms of failure under cyclic stress has employed alternating stress with $R = -1$.

The term "fatigue" has been applied to the phenomenon of fracture under repeated stresses, dating from the latter half of the nineteenth Century. While it is admittedly not a proper descriptive term, it has become so thoroughly established in the literature that no worker in the field has had the temerity or the energy to discard it. The term "progressive failure" is more precisely descriptive of the action of cyclic stresses on a component or material. The stresses needed to produce fatigue failure lie well below the ultimate tensile strength and for many engineering materials, below the yield stress as well. In pure metals and alloys, cyclic loading can cause hardening and the stresses required to cause failure in numbers of cycles that are easily accessible experimentally are usually greater than the yield strength of the annealed metal. However, recent advances on testing techniques have shown that failure can occur at a very high number of cycles, that is, 10^8–10^{10}, under stress levels much lower than previously expected.

The earliest reported tests on fatigue appear to be those of Albert (1838), which were performed in Germany with welded chains used in mine hoists. The British were very active in the early days of the field in association with the development of railroads and bridges, being represented by such well-known personalities as Rankine, Hodgkinson, and Fairbairn. Wöhler justly receives credit for the concept of the *S–N curve* and the *fatigue or endurance limit*, by which is meant that the cyclic stress that may be applied to a given material for an indefinitely large number of cycles without producing rupture. The endurance limits measured using some of the traditional definitions, for example, cyclic stresses for which lives are longer than 10^6–10^7 cycles have been under scrutiny recently, given the consistent experimental observation of failure at an extremely high number of cycles and stresses lower than the established fatigue limit for a wide variety of materials. Wöhler may have been slow in publishing; however, like the other early workers, he was ingenious in designing machines for loading his specimens. He was forced to run his tests at slow speeds and the highest speed available to him appears to have been 72 rpm in his rotating bending machine. Thus, it took many years to accumulate measurements at long lives (e.g. a test of 10^8 cycles would take about 3 years).

It is interesting that Moore and Kommers (1927), who wrote one of the first books on fatigue, dedicated it to... "the many distinguished British investigators, who have been foremost in forwarding our knowledge of the fatigue phenomena of metals..." but, with American evenhandedness, in the frontispiece, showed a portrait of Wöhler. In more recent decades, it is apparent that now, as then, the nationality of those interested in fatigue has followed the center of gravity of the global economy, just as to be expected from a critically fatal phenomenon associated with structural systems and mechanisms.

Bauschinger, who is remembered best for reversing the stress once, also carried out experiments in which he reversed the stress many times (Bauschinger, 1886). He essentially discovered *cyclic strain hardening* and *cyclic softening*, where the latter term implies the softening produced by cyclic stresses of hardening previously introduced into a metal by monotonic straining. Ewing and Humphrey (1902) initiated a different approach to fatigue by microscopically studying slip band (SB) behavior when specimens were subjected to reversed stress. They found that an increase in the number of stress cycles produced additional slip lines that had not been visible before, and that the first ones to have formed showed a tendency to broaden. Ultimately cracks were observed to form in the broadened bands. They thus discovered the phenomenon of *persistent slip bands* (PSBs) although these SBs were not thus termed until decades later when Thompson and Wadsworth (1958) discovered that they reappeared in the same location after the test had been interrupted, the surface repolished, and the test restarted.

The stress-based approach to fatigue design rooted in the S–N curve dominated the technology of fatigue up to the time of World War II and later. However, in the 1950s, Manson (1953), Coffin (1956), Coffin and Read (1956), and Coffin and Tavernelli (1958) reported test results in which strain rather than stress was used as the control mode of the test. These investigations not only gave rise to the well-known strain-life correlation named after the investigators (the *Coffin–Manson Law*), but also permitted *cyclic stress–strain response* to be explored. If a test is conducted at constant strain amplitudes, the stress necessary to enforce that strain repeatedly can be measured. Usually, but not always depending on material, the specimen ultimately attains a constant stress amplitude, termed the *saturation stress*. A plot of the saturation stress amplitude as a function of the applied strain amplitude (total or plastic strain) defines the *cyclic stress–strain curve* (CSSC), analogous to the monotonic stress–strain curve. This analogy gave rise to the definition of fatigue properties similar to monotonic mechanical properties and by the 1970s, industries were formulating specifications in terms of these properties. Up to then, published fatigue properties were expressed primarily in terms of the *fatigue or endurance limit* and no other. An elaborate "strain-based" approach to fatigue design has been developed and is now widely used (Fatigue Design Handbook, Society of Automotive Engineers). The study of cyclic deformation also provided a different route to the understanding of fatigue mechanisms.

Although it had long been recognized that fatigue life was dominated by the initiation and propagation of cracks, there were essentially no measurements of crack growth kinetics up to the early 1960s. At that time Paris and coworkers (Paris et al., 1961; Paris & Erdogan, 1963) suggested that the rate of fatigue crack growth should be characterized in terms of *linear elastic fracture mechanics* and based on the range of the *stress intensity factor*. This approach circumvented the difficult problem of defining the elastic–plastic behavior at the crack tip and permitted the correlation of measurements obtained in different specimens and loading systems. It produced an explosion of effort on the measurement of crack growth in a wide variety of metals and alloys, which has dominated fatigue research in the past 40 years and has defined yet more fatigue properties, such as the *threshold stress intensity range* for fatigue crack propagation, and the exponent m of the dominant region of the so-called *Paris law curve*. This curve is a plot of crack growth rate versus the stress intensity range, which tends to show a well-defined region where it appears linear when plotted in log–log coordinates, indicating a power-law behavior. These findings gave rise to the interesting problem of obtaining quantitative links between the elastic–plastic behavior at and around the crack tip, which would control the local mechanisms of crack propagation, and the macroscopic linear elastic fields used to characterize crack growth kinetics. Emphasis has been placed on predicting the value of m in terms of basic material properties.

The emphasis on fatigue crack propagation produced yet another approach to fatigue design based on the tolerance of the structure to an existing crack or flaw. This approach, also known as *damage tolerance*, is connected to the *"fail-safe" philosophy* of design, meaning that no single member of the structure is "critical" for its survival, and that there is sufficient redundancy in the structure so that the failure of one part can be sustained by the others until the partial failure is detected during inspection. The crack propagation approach, which is economically applicable to expensive and safety-critical structures like aircraft, calls for costly methods of nondestructive testing that have to be applied during periodic inspections to detect fatigue cracks, often when they are quite short, for example, 0.15 mm deep for engine components or less than 0.80 mm for airframes. The high cost of these inspections drives efforts to improve the predictability of fatigue crack propagation and the residual strength of a cracked structure. The older stress-based and strain-based fatigue methodologies, the *endurance*-based approaches, and fracture mechanics are both required at the present day in fatigue design, the former for predicting the durability of a structure or component, the latter for controlling its safety. This has become even more critical because of infrastructure aging in the United States and many other countries, as well as aging of military and commercial aircraft fleets.

Reliable prediction of fatigue failure can be obtained only by a thorough understanding of the physical mechanisms involved. For engineering materials, with their complicated microstructures and mechanical behavior, it is a formidable challenge to formulate a quantitative theory that relates elementary processes to the observable life under fatigue loading, although so much experimental data exist for crack propagation kinetics that reasonable estimates of residual useful life can be obtained if the geometry of the preexisting flaw is known. Nonetheless, characterization and modeling advances in the past two decades, particularly at the microscopic scale, have shown great promise toward the formulation of such a quantitative theory and great strides have been made to achieve it (Dunne et al., 2007; Anahid et al., 2011). Understanding of cracking mechanisms, rooted in the early work of Gough and coworkers (Gough, 1933) who performed research on single crystals, has benefited from the modern approach of using single crystals in conjunction with tools, such as Transmission Electron Microscopy (TEM) and Scanning Electron Microscopy (SEM), for research into fracture mechanisms and dislocation behavior. These key techniques have been complemented with additional capabilities to study cyclic deformation and fatigue fracture in polycrystalline materials, such as in situ straining stages to observe cyclic deformation as it develops in real time at the microscale, as well as characterization approaches such as electron backscattering diffraction (EBSD) and electron channeling contrast (ECC), which supplement SEM to allow mapping of grains and dislocation structures using local crystallographic information. Micromachining techniques such as focused ion beam (FIB), combined with SEM in dual-beam configurations, have also become a staple of characterization techniques and have allowed for in situ studies of crack initiation from microscale flaws with controlled geometry (Marx et al., 2006).

If the structure or specimen is "smooth," meaning that it does not contain a flaw that can immediately develop into a growing fatigue crack under the action of the applied cyclic loads, then the phenomena of fatigue are complex. At the start of life, the material undergoes *cyclic hardening*, which conditions the material to form cracks. Such conditioning usually causes the strain to localize, with considerable local enhancement of the plastic strain that stimulates both crack initiation and growth. Usually the localization will be emphasized by a stress concentrator with a stress concentration factor $K_t > 1$, either in the structural component, for example, a weld or reentrant corner, or in the material itself (nonmetallic inclusion or other microstructural feature). However, localization can also occur by dislocation mechanisms in smooth specimens of pure metals. No clear position has yet emerged to distinguish fatigue-produced surface roughness from a defined growing crack. However, a crack is usually regarded as "proper," that is, engaged in propagation, if it is ~ 3–5 µm in depth. The slow development of such small cracks is described by the study of "short" cracks, which refers to cracks that are either: (1) physically small, that is, their length is short compared with the size of the component where they reside, (2) mechanically small, that is, their length is short in relation to the size of the localized plastic zone that has favored their development, and (3) microstructurally small, when their length is small in relation to a significant microstructural length scale, such as the average grain size. After the crack has grown to a length of many tens of micrometers, the behavior of the plastic zone at the crack tip controls the crack growth rate, and the size of the plastic zone becomes small in relation to the size of the crack (long crack growth). Ultimate failure then occurs by unstable crack growth when the applied maximum stress intensity reaches the value of the fracture toughness of the material, K_{Ic}. The overall progression of fatigue damage spans several length and timescales, and is influenced by a variety of both intrinsic and extrinsic factors, as illustrated in **Figure 1** (Vasudevan et al., 2001).

The effects of microstructural parameters, such as porosity and inclusions, make the fatigue problem even more complex in engineering materials. The present chapter does not deal with engineering aspects of fatigue except for the brief summary given above and the discussion of performance indicators, but rather focuses on the basic phenomena of fatigue and their mechanisms in metals and alloys. Nonetheless, examples of research performed using engineering alloys will be described as needed to provide insight into phenomena that are of fundamental nature within the field. Experimental work is key to elucidate the physical mechanisms behind the various fatigue phenomena mentioned above. Hence, a brief description of fatigue testing techniques will be provided next.

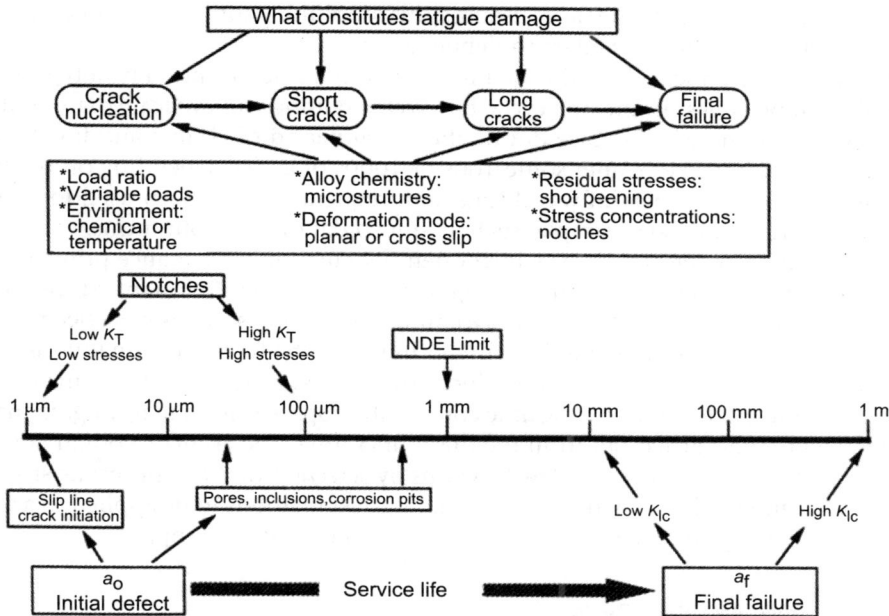

Figure 1 Stages of fatigue damage along with their corresponding length scales and the factors that affect them. Source: From Vasudevan et al., 2001, courtesy of Elsevier.

18.2 Fatigue Testing

To obtain repeatable results in tension testing, especially for elongation to monotonic fracture, the tension test is standardized with respect to the shape of the specimen. Fatigue behavior and properties are so complex that no generally acceptable testing standards apply universally, except that fracture mechanics specimens are frequently used for measuring fatigue crack propagation. In that regard, standards for cyclic deformation and fatigue crack propagation testing have been formulated by professional societies, such as the American Society for Testing and Materials (ASTM), for example, ASTM E606 for strain-controlled fatigue testing and E647 for fatigue crack growth testing. These standards, although widely used, do not cover all the needs of the fatigue research community; therefore, fatigue tests have been conducted in many different ways, and minor differences in testing can produce significant differences in behavior. Hence, scatter of the results is endemic to fatigue as much from testing reasons as from variability among specimens. The most popular testing methods are as follows:

18.2.1 Constant Amplitude Stress Tests

In engineering materials where both yield and ultimate tensile strengths are high, fatigue can occur at long life when applied stresses are low relative to these basic mechanical properties. Consequently, both the elastic and plastic strains are small. Such materials can be subjected to the full stress amplitude at the first cycle, and the test can be run truly at constant stress amplitude until the specimen fails. Complete fracture of the specimen is normally taken to define failure.

If the material is in the annealed condition, the strain amplitude produced by a typical value of the stress amplitude will be largest in the first cycle, but such straining will cyclically harden the specimen and the strain will continuously decrease during the initial period of cycling and finally reach a saturation value, or even become zero, after some 100s or 1000s of cycles depending on the material and the stress amplitude. A wide variety of waveforms and frequencies have been used for fatigue testing, but those preferred are usually kept simple, such as sinusoidal or sawtooth forms.

In work conducted on older machines, which often operated on a resonance principle, a specimen might undergo a complex series of work-up cyclings before the desired "constant" stress amplitude was attained. However, since the mid-1960s, servo-hydraulic closed-loop systems have been employed. The specimen is loaded by means of a hydraulic cylinder in which the high-pressure fluid is controlled by a servo-valve actuated by an electrical input produced by a closed-loop control system with a computer interface. Such a testing mode offers very flexible control and sophistication in testing, so that a constant amplitude test can be truly carried out from the onset of cycling. The ability to obtain a wide range of loads from several metric tons down to a few Newtons by selection of the appropriate size of hydraulic cylinder, servo-valve and load cell enables servo-hydraulic test methods to be applied to a wide range of specimen sizes and configurations, from macrostructures to miniature samples.

18.2.2   Increasing Stress Amplitude Tests

In softer engineering metals and alloys, or in model materials used in research, such as metallic single crystals, the stress needed to cause fatigue fracture at long life may be high enough to strain the material to an unacceptably high level if applied to the full level at the first cycle. The specimen might become destructively deformed, for example. To avoid this very large initial strain, the stress amplitude must be worked up gradually, say in 30–100 cycles, to the desired final stress amplitude, which may then be kept constant for the remainder of the experiment. These relatively high initial strain amplitudes may well affect the fatigue behavior of the material.

To avoid such a history effect on the life behavior, or for other reasons, an experimenter may choose to ramp up the alternating stress over a much longer interval, say 20,000 cycles, to the desired final stress amplitude, at a controlled constant rate of increase for the stress amplitude. Tests where the stress amplitude is the primary control variable, either by keeping it constant as described in Section 18.2.1 or by varying it as described in this section, are typically called "stress" or "load" control tests. The two terms will be used interchangeably in this chapter, although it is important to keep in mind that the two terms are equivalent only when changes in cross-section area of the specimen are small. This is typically the case during fatigue testing because of the small strains that are usually involved, particularly for metallic materials.

18.2.3   Constant Plastic Strain Amplitude Tests

Although most structures are subject to load cycling, suggesting that constant stress amplitude tests would be most representative of service fatigue, in practice fatigue cracks tend to occur in regions of stress concentration. In such regions the straining behavior is constrained by the rest of the structure and the stress–strain response is more complicated than under simple stress control, which thus represents an extreme of behavior. As Landgraf (Landgraf, 1977) among others has pointed out, the normally used test conditions represent extremes of completely unconstrained (or load-cycling conditions) and completely constrained (or strain-cycling conditions). In actual engineering structures, stress–strain gradients do exist, and there is usually a certain degree of structural constraint of the material at critical locations. Such a condition tends to be similar to strain control. Therefore, this type of test has become increasingly favored in the past 40 years, for both low strain and high strain fatigue testing. Furthermore, tests at high stress under stress control are difficult to conduct because the specimen creeps uncontrollably, even at room temperature, because of "cyclic creep". *Cyclic creep* or *Ratchetting* is the deformation produced by an alternating stress superimposed on a mean stress. For reviews, see Kennedy (1962) and Lorenzo (1983). An overview of modeling approaches can be found in Chaboche (1991), Bari and Hassan (2001), and Abdel-Karim (2010).

Accordingly, for well-behaved tests conducted at high strains, tests in strain control are necessary. It is also necessary to choose between control in *total* strain amplitude, or *plastic* strain amplitude. In tests under total strain control, cyclic hardening causes the elastic component of the strain to increase at the expense of the plastic strain. However, since hardening usually saturates rapidly, tests in total strain control do end up amounting to tests in plastic strain control. With the advent of sophisticated electronic control of servo-hydraulic testing equipment, it is now more readily possible to keep the plastic strain amplitude constant from the onset of cycling, a control mode that is widely practiced.

It is generally accepted that plastic strains are necessary for the development of fatigue damage despite the fact that, in long-life fatigue, the plastic strain amplitude is a small fraction of the total strain amplitude, and may be so small as to be difficult to measure. Because of the fundamental importance of plastic strain in fatigue, some workers choose to use plastic strain-controlled tests even at relatively long lives (plastic strain amplitude 5×10^{-5}–5×10^{-4}), but at the cost of much greater effort because of the great resolution needed to measure the strain. Tests conducted at strain amplitudes in the range 5×10^{-4}–10^{-2} would be considered high strain fatigue tests, in which the lives would typically be less than 100,000 cycles to failure. Reports of fatigue tests at strain amplitudes $> 10^{-2}$ are uncommon.

The relative sizes of the elastic and plastic components of the cyclic strain are used to define formally the distinction between low and high strain fatigue. The dividing point is that total strain amplitude for which the elastic and plastic components of the strain are equal. The life associated with this point is termed the *transition life*, N_t. If the plastic strain is greater than the elastic strain, the test is defined as *high strain fatigue*, and *low strain fatigue* for vice versa.

18.2.4 Variable Amplitude Tests

To simulate more realistically the loading history of the components of actual structures, tests are also performed with highly complicated loading sequences. In such tests, it becomes difficult to define a stress cycle, and reversals of the stress tend to be counted instead. Special techniques have been developed to characterize such loading histories (Wetzel, 1971). The physics of load interaction in fatigue under variable amplitudes is not fully understood yet and extremely complicated; it will not be treated in detail here, and only a few of the experimental and modeling efforts undertaken to understanding it better will be briefly mentioned. An overview of experimental results in copper single crystals tested under variable strain amplitudes can be found in Ma and Laird (1989e).

Test frequency is an important variable both for research purposes and in both scientific and practical matters. For example, frequency effects can be pronounced in materials with strain rate sensitivity or because of environmental concerns. From a practical point of view, long tests are more economical if they can be run at high frequencies. For high strain fatigue tests, frequencies in the range from 0.1 Hz to a few Hz are convenient.

One of the limitations of conventional servo-hydraulic equipment is the difficulty of controlling tests at high frequencies, and such tests are generally conducted at frequencies less than 50 Hz, but it is now possible to perform load-controlled tests with frequencies up to 1 kHz, although the corresponding strains have to be small. This has made long-life testing practical in servo-hydraulic equipment. Using ultrasonic resonance techniques with horn-shaped specimens or appropriately gripped cylindrical specimens, fatigue experiments can also be performed with frequencies in the range of about 20 kHz (Tschegg and Stanzl, 1981). Recent progress using these techniques include the ability to perform 3-point bending as well as torsion fatigue experiments at similar frequencies (Bathias, 2006). Such testing is valuable for obtaining results at long life, but specimen heating because of the high rate of heat generation is a problem and requires special techniques for avoiding it, for example, using air cooling to lower the sample's temperature. This has proven viable for the low stress amplitudes that lead to extremely long fatigue lives in metallic samples.

18.2.5 Multiaxial Testing

The ability to perform testing with loads applied independently on two or more axes is another important development on fatigue testing, since this allows to approximate better the states of stress that can be expected in components with complex geometries, complex loads or both. This has been made possible by the development of electronic/computer control of servo-hydraulic equipment that accounts for several independent variables and their potential "cross-talk", as well as new structural designs for the load frames themselves. There are two common configurations for these load frames: axial-torsion and biaxial.

In the axial-torsion frames, there is an axial actuator very similar to that used in conventional uniaxial frames; however, in tandem with this actuator there is a torsion actuator that allows superimposing a twisting motion to the axial one, and a load cell able to measure force and torque is used to obtain the applied loads. Strains can be controlled with available extensometers that measure normal and shear strains, so either load or strain control modes are possible.

In the biaxial load frames, axial actuators are mounted at 90° from one another in the plane of the frame, which is typically shaped as a hexagon to make it as stiff as possible for all directions within that plane. Most of these load frames have two actuators in each axis that can be controlled to move in a variety of ways. Typically, the motion of the two actuators in each axis is such that the center of the specimen does not move, which minimizes interactions between the two load axes and also facilitates sample monitoring. The loads on each axis can be controlled independently from one another and strain control along each axis is also possible using biaxial extensometers.

The typical specimen for axial torsion testing is a thin-walled tubular sample, so that the shear stresses produced by torsion can be considered approximately constant over the cross-section of the specimen. In the case of biaxial testing, the sample has a cruciform geometry. Examples are shown in **Figure 2**.

Each configuration has advantages and disadvantages. The biaxial specimen has a planar gage area, which is convenient for in situ monitoring of defects and to correlate microstructure to damage evolution, since the gage area is easy to characterize using techniques that require flat surfaces, for example, EBSD or ECC. However, in this loading configuration the principal axes of stress are fixed and only the effect of having load amplitudes in both axes be in phase (IP) or out of phase (OP) can be studied. Note that IP loads correspond to proportional loading, whereas OP loads lead to non-proportional loading. The axial-torsion configuration, on the other hand, is less convenient to relate local microstructure to damage evolution, since the curved surface makes it harder to use many characterization techniques like the ones mentioned above, which are becoming more typical. Nonetheless,

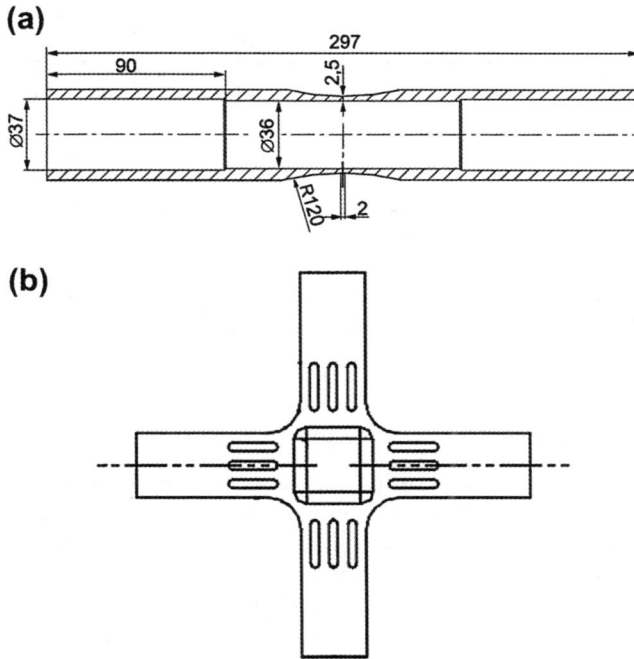

Figure 2 (a) Cross-section of a typical thin-walled tubular sample used for axial torsion testing. Dimensions in mm. From Döring et al. (2006). Courtesy of Elsevier. (b) Cruciform specimen used for biaxial testing, from unpublished work by P. Peralta.

it allows to study changes on material behavior as the loads go from pure torsion to pure axial load (although pure shear can be induced under biaxial loads by having the loads in the two axes be equal in magnitude but opposite in signs at all times) and also the effect of variable principal axes of stress, since the application of OP axial force and torsion will lead to principal axes that vary in direction as a function of time. Examples of load paths that have been used to study effects from axial-torsion loading conditions are shown in **Figure 3**.

The testing procedures described above can provide a wealth of data that must be interpreted and condensed to extract the information required to evaluate the parameters that control fatigue behavior of materials. These parameters are discussed in the next section.

18.3 Performance Parameters of Fatigue

The monotonic tensile properties of materials are described by such well-known parameters as the yield or flow stress (often evaluated at 0.2% strain), the ultimate tensile strength, the percent reduction in area, the true fracture stress (defined as the final load divided by the final area of the neck), and the true fracture ductility (given by $\ln(A_o/A_f)$, where A_o is the initial cross-sectional area of the specimen, and A_f the final area at the point of fracture). Properties parallel to these have been defined for the fatigue behavior of materials, dealing with both stress–strain response and fracture behavior. For more information on this subject, see Mitchell (1979), Lampman (1996), and Suresh (1998).

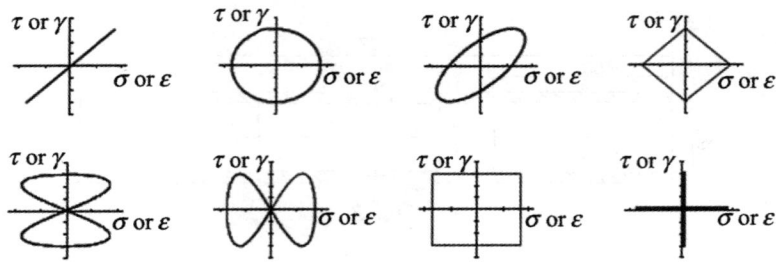

Figure 3 Examples of stress (or strain paths) that can be induced using axial-torsion testing. The case in the upper left corner corresponds to IP loads, where the directions of the principal axes are fixed. Source: Redrawn from a figure originally from Döring et al., 2006. Courtesy of Elsevier.

18.3.1 Cyclic Stress–Strain Behavior

Since the dislocation substructure of crystalline metals is not stable under the application of cyclic loads, their *stress–strain response* can be drastically altered when subjected to cyclic strains. Depending on the initial state of the metal (whether it has been hardened by some metallurgical process, or softened by annealing) and its test condition, the specimen may: (a) *cyclically harden*, (b) *cyclically soften*, (c) maintain its flow stress, when it is said to be *cyclically stable*, or (d) show mixed behavior (soften or harden, depending on strain amplitude, or both harden and soften, as life progresses). Examples of some of these types of behavior are shown in **Figure 4**, for tests under strain control: if the stress amplitude required to enforce the strain amplitude increases as life unfolds, the metal undergoes hardening. This behavior is typical of annealed metals and alloys, and the softer steels and aluminum alloys, (b) if the stress required to enforce the strain decreases with successive cycles, the phenomenon is called cyclic softening. Cold-worked metals and steels that have inherited dense dislocation structures from their processing history tend to behave this way. The stress–time plots shown in **Figure 4(a) and (b)** are *cyclic strain hardening* and *softening curves*, respectively. Some metals (in rare instances) are cyclically stable, in which case their monotonic stress–strain behavior adequately describes their cyclic response.

Whatever the nature of the material, cycled under control of the strain amplitude, it will eventually attain (almost always) a steady-state stress response. To construct a CSSC, one connects the tips of the stabilized hysteresis loops from similar specimen tests at several controlled strain amplitudes (see **Figure 5**).

The CSSC can be compared directly with the monotonic stress–strain curve to quantitatively assess cyclically induced changes in mechanical behavior. Typical examples taken from Mitchell (1979) are shown in **Figure 6**; materials that cyclically harden and others that soften are shown. Note that, for extreme examples, meaning materials containing metallurgical structures well removed from equilibrium, softening can be considerable. Thus, the "cyclic yield stress", defined at some elapsed strain, just as for monotonic deformation, can be a mere 50% of the monotonic yield strength.

If the CSSC is represented by a power-law function, such as

$$\sigma_a = K(\varepsilon_p)^n, \tag{1}$$

where σ_a is the *steady-state* (saturation) stress amplitude, and ε_p is the plastic strain amplitude. The parameters K and n can then be used to describe the cyclic stress–strain response. K is termed the *cyclic*

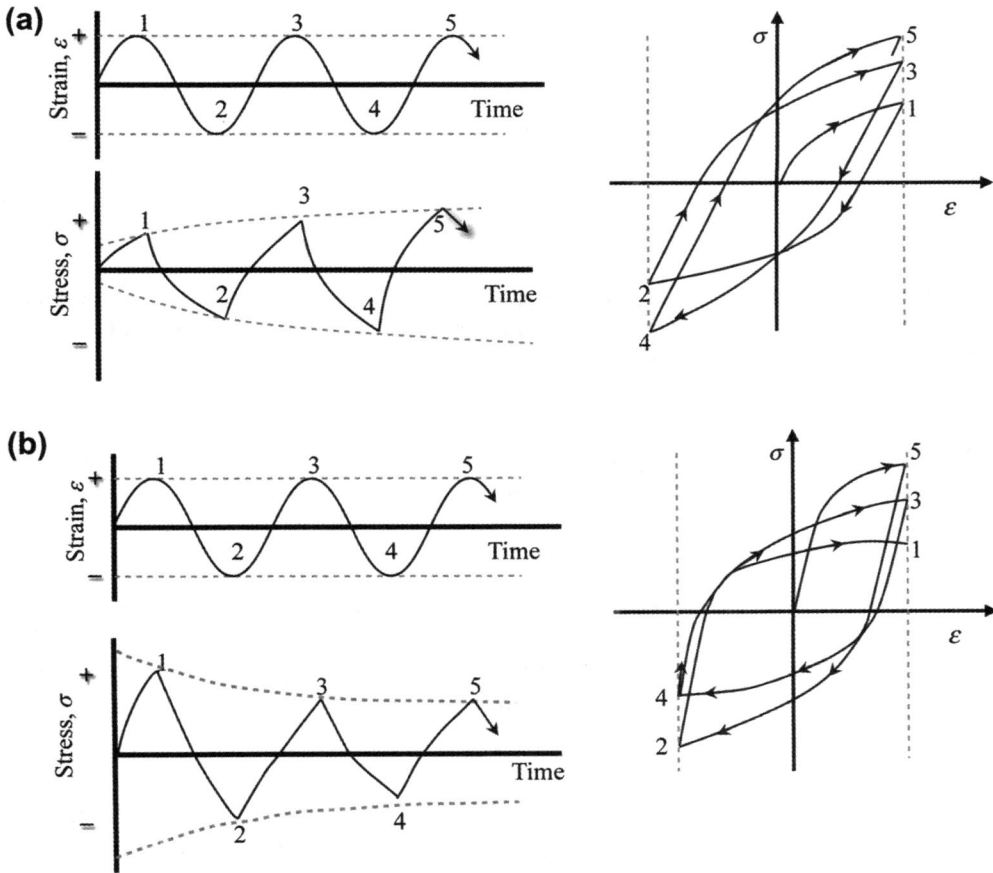

Figure 4 Cyclic response under controlled-strain-amplitude cycling showing the control condition, the stress response and hysteresis loops associated with the increasing cycles for (a) cyclic hardening and (b) cyclic softening. Source: Redrawn from **Figure 1** in Laird (1996) from an original figure by Mitchell (Mitchell, 1979). Printed with permission from ASM International.

strength coefficient and *n* the *cyclic strain hardening exponent.* In cyclically softening materials, where completion of softening may be doubtful, σ_a is conventionally defined at 50% of life. Some materials may show such large stress variations during life that the CSSC may be undefinable. The value of *n* usually varies between 0.10 and 0.20.

18.3.2 Fatigue Life Behavior

Because of the old-fashioned emphasis on the S_a–N_f curve, where S_a is the stress amplitude and N_f is cycles to failure, fatigue life data are generally most available in this form. For a compilation of such data, see Forrest's book (Forrest, 1962). Newer data sets can be found in the Metallic Material Properties Development and Standardization Handbook (MMPDS-06) (FAA, 2011).

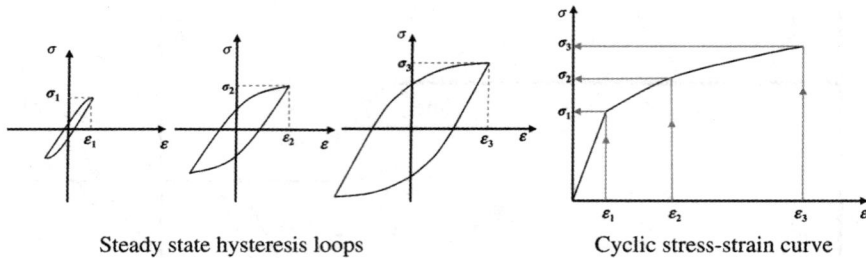

Steady state hysteresis loops   Cyclic stress-strain curve

Figure 5 Construction of the CSSC by joining tips of stabilized hysteresis loops. In the CSSC, the total strain has been used for the abscissa. If the loop half-width at zero stress had been used, the CSSC would be plotted for the plastic strain. Source: Redrawn from Figure 2 in Laird (1996), after a figure by Mitchell (1979). Printed with permission from ASM International.

Around the turn of the century, Basquin showed that the $S_a–N_f$ plot could be linearized with full log coordinates, and thereby established the exponential description of fatigue life:

$$S_a = \sigma_f(N_f)^b, \tag{2}$$

where σ_f is the *fatigue strength coefficient* and b is the *fatigue strength exponent*, often called after Basquin; these parameters are fatigue properties of the metal.

Recalling the Coffin–Manson finding that plastic strain life data may also be linearized with log–log coordinates, we have the well-known so-called *Coffin–Manson law*:

$$\varepsilon_p = \varepsilon_f(N_f)^c, \tag{3}$$

where ε_p is the plastic strain amplitude, ε_f is the *fatigue ductility coefficient*, and c is the *fatigue ductility exponent*. The parameter ε_f correlates quite well with the true strain to fracture in a monotonic test, and c varies between approximately -0.5 and -0.7 for most metals. The fatigue ductility exponent can approach -1 or -2 if the strain localization shown by the material is especially marked.

Fatigue life behavior is now usually displayed in a log–log plot of strain versus cycles (or reversals $= 2N_f$) to failure, and such a plot is shown in **Figure 7**, along with the fatigue strength and ductility properties from the four fatigue parameters that have been introduced: fatigue strength coefficient σ_f and exponent b, and fatigue ductility coefficient ε_f and exponent c.

Before the discovery that "conventional" fatigue limits, that is, those measured to lives of $10^6–10^7$ cycles, did not hold for cycling at very long lives $(10^9–10^{10})$ (see Mughrabi, 2006 for a review), the conventional fatigue limit was (and still is) added to the four parameters defined above as another defining index of performance in materials that are expected to possess such a limit. It would be more accurate to use the term "fatigue strength" at a given life, which is the terminology used for materials where no conventional fatigue limit has been measured, as has been the case for many face-centered cubic (fcc) metals.

Regarding the fatigue limit itself, it has been postulated that all materials that require plasticity for their fatigue failure mechanism are likely to possess a fatigue limit. In materials that localize strain via PSBs, this threshold was proposed to be the stress for PSB formation (Laird, 1976) or, more generally, for strain localization. The evidence from very long fatigue cycling experiments indicates that the applied stress threshold for strain localization is lower than the one measured using conventional

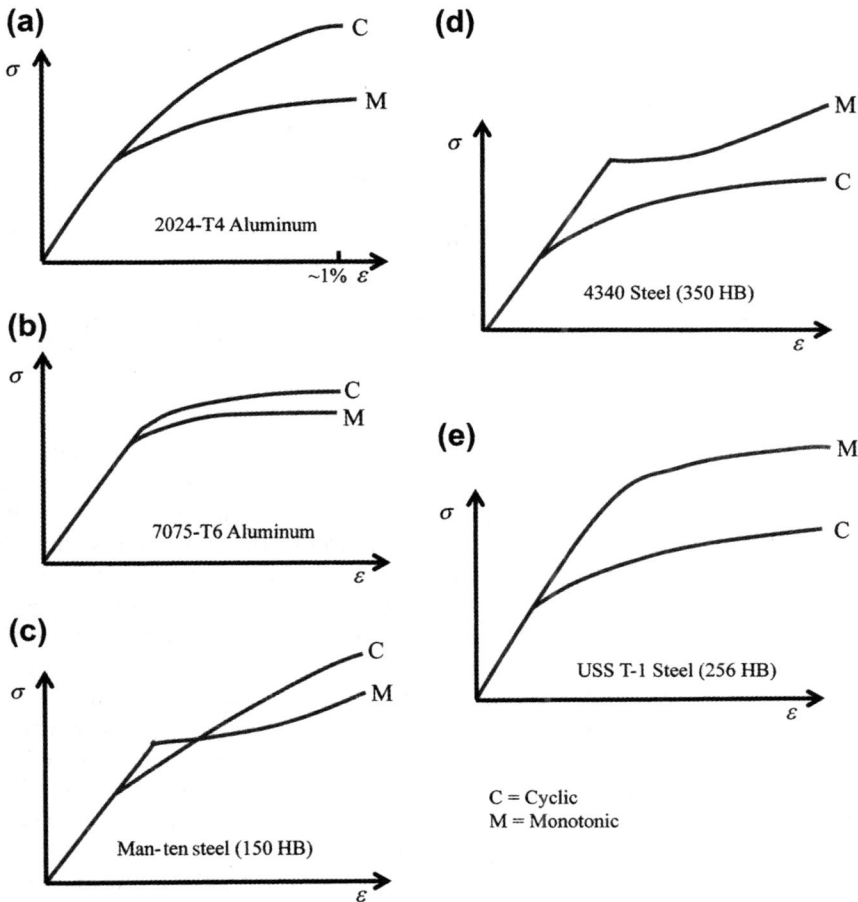

Figure 6 Cyclic stress–strain response, curve C, compared with the monotonic stress–strain curve M, for typical engineering metals. If the C curve lies above the M curve, cyclic hardening has occurred; cyclic softening has occurred for vice versa. The Brinell hardness of the steels are indicated. Soft steels generally harden and quenched and tempered steels generally soften. Source: Plotted with data from **Figure 3** in Laird (1996) from an original figure by Mitchell (1979). Printed with permission from ASM International.

strain- and stress-controlled tests (Mughrabi, 2006; Weidner et al., 2010). However, it is still quite reasonable to expect that metals may last indefinitely if strain localization does not occur, and the current evidence suggests that this would require reducing the amount of irreversible deformation to zero (Mughrabi, 2006; Weidner et al., 2010). Yet another parameter defining fatigue behavior, the *transition life*, N_t, is shown on **Figure 7**, and, as already pointed out, it marks the formal distinction between high and low strain fatigue, for a particular material. In the (hypothetical) example shown, N_t is low but not unusually so, and would be typical of a brittle metal. The transition life is nearer 100,000 cycles for ductile metals.

The use of the performance parameters described above, which have been obtained using uniaxial testing for the most part, and their corresponding empirical relationships to describe multiaxial fatigue

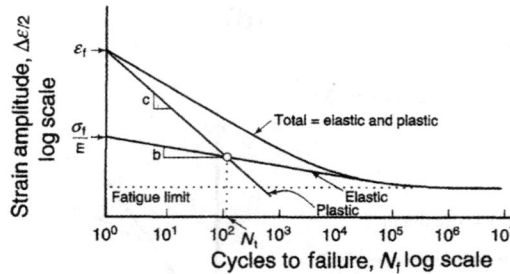

Figure 7 Plot of plastic strain, elastic strain and total strain versus cycles to failure, showing the "fatigue ductility and strength" parameters, as well as the "conventional" fatigue limit and the transition life, N_t. Source: After Laird, 1996. Courtesy of Elsevier.

behavior are not straightforward (Fatemi and Shamsaei, 2011). The use of equivalent stress and strain measures, for example, von Mises stress or strain, to "convert" a multiaxial state into a uniaxial one with "equivalent" behavior runs into difficulties, particularly under nonproportional loading, since the material behavior is different, even when the maximum amplitudes of each stress component are kept the same as in an equivalent proportional loading experiment. One of these effects is the fact that some materials experience hardening under nonproportional loads beyond what is observed for an experiment under proportional loads with the same individual load amplitudes along each axis (Döring et al., 2006; Fatemi and Shamsaei, 2011). This is compounded by the fact that the effect is material dependent. This is attributed to the variable directions of the principal axes under nonproportional loading and the way these variable load axes lead to interactions among different slip systems at the grain level (Fatemi and Shamsaei, 2011). Softening of the additional hardening has been reported once loads are switched back to be IP (Döring et al., 2006). The other effect is that fatigue life tends to be lower under nonproportional loads as compared with results under proportional loading with the same equivalent stress or strain. Examples of these effects are shown in **Figure 8**.

Note in **Figure 8(a)** the significant additional hardening in SS with OP loading as compared with the IP experiment, whereas the Ti response was mostly unaffected by going to the nonproportional loading condition. This is no doubt a reflection on the difference in slip behavior at the local level produced by the different crystal structures of these two materials. Regarding **Figure 8(b)**, load paths that lead to sustained values of one of the loads seem to cause the most damage, and hence the shorter lives (diamond, double loop and square paths). Fatigue performance parameters are harder to obtain under these conditions, and other approaches are needed. In particular, critical plane methodologies, whereby the physical plane that experiences the most damage as loads are applied, have found some success and they have the advantage of providing information on the orientation of nucleating cracks (Suresh, 1998). A measure of fatigue damage that is commonly used for this purpose, among many others, is the Fatemi-Socie parameter, given by

$$\frac{\Delta\gamma_{max}}{2}\left(1+k\frac{\sigma_{nn,max}}{\sigma_y}\right), \tag{4}$$

where $\Delta\gamma_{max}/2$ is the maximum shear stress amplitude, σ_y is the monotonic yield strength and $\sigma_{nn,max}$ is the maximum normal stress acting on the plane with maximum shear strain over a loading cycle and the parameter k mediates the effect of the normal stress (Fatemi and Shamsaei, 2011).

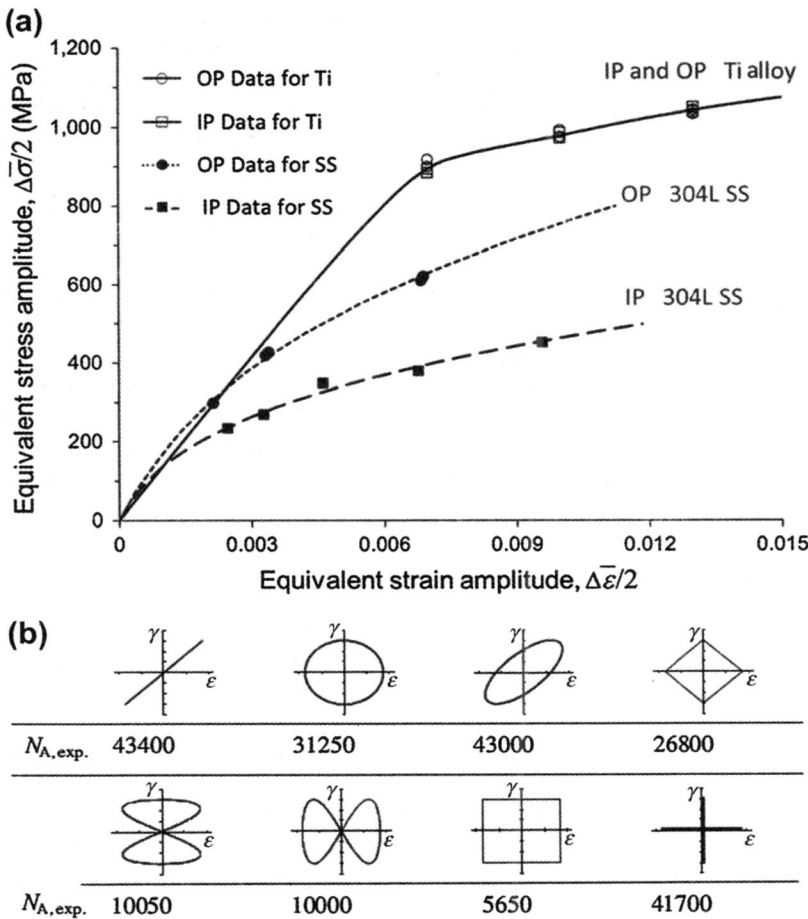

Figure 8 (a) Comparison of IP and 90° OP equivalent stress–strain data for Ti-6.5Al-3.4Mo (Ti) and 304L stainless steel (SS). From Fatemi and Shamsaei (2011). (b) Experimentally measured fatigue lives for samples of S460N steel tested with an axial strain amplitude of 1.73×10^{-3} and a shear strain amplitude of 3×10^{-3}. Source: Taken from Döring et al., 2006. Both figures were courtesy of Elsevier. (For color version of this figure, the reader is referred to the online version of this book.)

The brief overview provided above clearly indicates that fatigue performance of materials is strongly dependent on applied loads through the effects those loads have on cyclic deformation, which "conditions" the material for damage accumulation and eventual crack nucleation and growth. Therefore, understanding of cyclic deformation is key to elucidate fatigue performance, so this topic is discussed in the next section.

18.4 Cyclic Deformation

Cyclic deformation is usually studied in fatigue by tests conducted in total or plastic strain control. Since fatigue strains are generally low in relation to those employed in monotonic deformation, the stress is also low at the start of cycling if the material is soft or annealed. However, the specimen typically hardens quickly giving rise to the phenomenon known as *rapid hardening*. With continued cycling the hardening rate declines and eventually falls to zero, at which point the specimen is termed as being in *saturation*. The dislocation evolution mechanisms during rapid hardening and the dislocation behavior by which a specimen carries plastic strain in saturation without hardening further are usually different. These aspects of cyclic deformation are, therefore, treated separately. Often, as mentioned in the Introduction, *strain localization* occurs during cyclic deformation, and it is this phenomenon that is particularly destructive in promoting fatigue cracking. It has received much attention in fatigue studies and is accordingly emphasized here.

The fundamental aspects of cyclic deformation and fatigue cracking have repeatedly been reviewed (Laird, 1977, 1979, 1981, 1983; Mughrabi, 1980; Mughrabi et al., 1983; Laird et al., 1986; Gerold and Meier, 1987; Basinski and Basinski, 1992; Laird, 1996; Peralta, 2005). However, fatigue research continues actively and there have been new results on the cyclic deformation and fatigue of both wavy and planar slip alloys, which have provided additional insights into the mechanisms of fatigue. For reviews of early work, the reader is referred to Thompson and Wadsworth (1958), Grosskreutz (1971), and Grosskreutz and Mughrabi (1975); for engineering aspects of cyclic deformation, see Mitchell (1979) and Lampman (1996).

18.4.1 Phenomenological Behavior and Dislocation Structures

Rapid hardening ends when the *saturation stress* is reached. This saturation stress, when plotted against the applied plastic (or total) strain amplitude, defines the CSSC as identified in Section 18.3. The CSSC for copper single crystals deforming in single slip tested under total or plastic strain control is shown in **Figure 9**.

This curve, which is typical of fcc pure metals and most of their substitutional alloys, is seen to consist of a three-stage curve centered around a prominent *plateau* (which lies at a stress amplitude of 28 MPa for Cu). This curve differs in its form in minor ways for the different fcc metals, although the *plateau stress* normalized by the shear modulus has an approximately constant value of $\sim 6.6 \times 10^{-4}$ for several fcc metals (Mughrabi, 1980; Li et al., 2011). Since the fatigue behavior of copper has been studied extensively, many of the examples cited here will be taken from copper studies, but the understanding claimed for this material can be broadly applied to a wide range of fcc metals. Materials with other crystal structures will be briefly reviewed as well.

The saturated dislocation structures, which are shown as insets in **Figure 9**, differ for the various regions of the CSSC: A, below the plateau, where *loop patches* are present; B, in the plateau, where there is a combination of loop patches and PSBs; C, above the plateau, where a structure made of 100% PSBs starts transitioning to a *cell* structure. It is to be anticipated, therefore, that the rapid hardening behavior will reflect these differences. One method of showing these differences is by plotting and comparing the "cyclic hardening curves", that is, those obtained by plotting the peak stress per cycle conducted in constant strain amplitude against the number of cycles or *cumulative plastic strain*. This frequently used quantity, which is specific for fatigue, is defined as the plastic strain that occurred in all previous cycles summed without regard to sign. For constant strain cycling, the cumulative plastic strain is the product of four times the plastic strain amplitude and the number of cycles. The behavior that corresponds to regions A and B and applies to single slip orientations appears to be understood best. Under these conditions, the dislocation structures produced during rapid hardening consist of dense and irregular loop patches, as shown in **Figure 9**, strung out in the orientations of edge dislocations. Loop patches are made up of edge dislocation dipoles in which the positive and negative dislocations of the pair typically

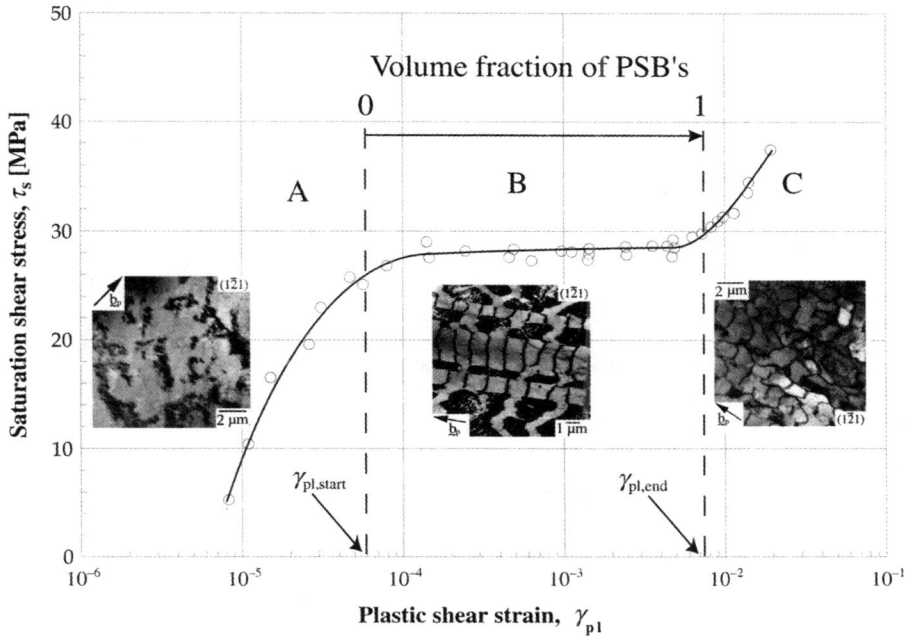

Figure 9 Cyclic stress–strain curve for a copper single crystal oriented for single slip, showing the three stages of the curve and the prominent plateau. Typical dislocation structures for each regime are shown. Source: Numerical data for the CSSC courtesy of H. Mughrabi. After Peralta, 2005. Courtesy of Elsevier.

end up being separated by about 10 nm. These "loop patches" are separated by regions of low dislocation density called "channels". The combination of loop patches and channels is often called *vein structure* (Llanes and Laird, 1993). They are formed most likely by a variety of mechanisms, shown in **Figure 10**.

These mechanisms include mutual trapping of positive and negative edge dislocations traveling on different atomic slip planes in different directions (**Figure 10(a)**), screw dislocation pairs gliding on different atomic planes and drawing out an edge dislocation dipole between them (**Figure 10(b)**) and/ or by a double cross-slip mechanism (**Figure 10(c)**).

The trapped edge dislocations position themselves at their 45° equilibrium position (**Figure 10(a)**) and can flip-flop past one another as the stress is cycled. They act as barriers to other dislocations and can be swept up into clumps: the loop patches. Because the stress is cycled in fatigue many times, positive and negative dislocations have many opportunities of encountering and trapping each other, and loop patches rarely show excess dislocations of one sign. Continued trapping events gradually refine the dipoles to smaller and smaller separations. The typical three-dimensional (3-D) appearance of these structures is shown in **Figure 11**.

At the low magnification used to obtain **Figure 11**, the individual loops cannot be seen, just their patches. Under normal viewing in TEM, close to the Bragg condition, the dislocation images can be as wide as the separation of the positive and negative edge dislocations in the dipole and it may be difficult to resolve an isolated dipole. They can, of course, be resolved separately using the weak-beam technique (Antonopoulos et al., 1976), even when the dipoles are clumped in a loop patch. Nonetheless, the

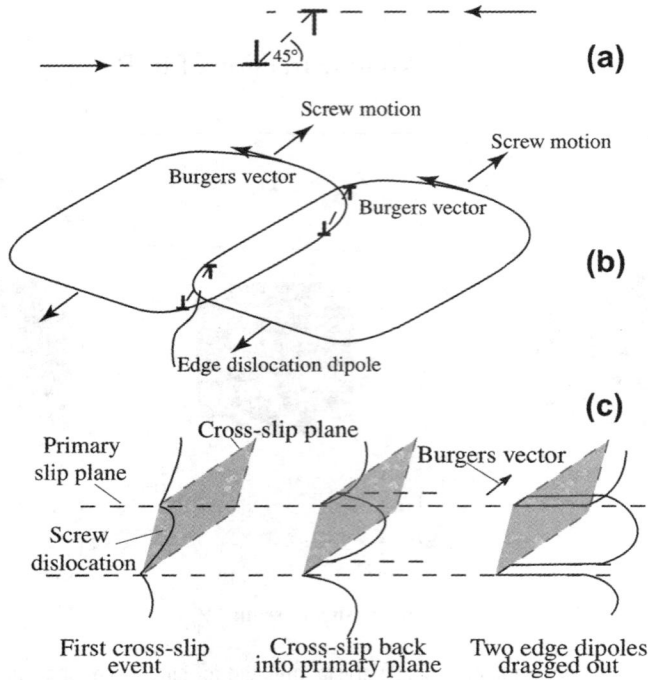

Figure 10 Mechanisms for loop patch formation: (a) random trapping of edge dislocations gliding on different planes, (b) trapping of the edge components of dislocation loops gliding on different planes and (c) formation of edge dipoles by double cross-slip of a screw dislocation. Source: From Peralta, 2005 after a figure from Laird, 1996. Courtesy of Elsevier.

Figure 11 Three-dimensional view of loop patches corresponding to region A of the CSSC for a copper single crystal. $\gamma_{pl} = 2.6 \times 10^{-5}$, $\tau_s = 19.8$ MPa. Source: After Laird et al., 1986. Courtesy of Elsevier.

uniform contrast across loop patches in TEM pictures reveals their dipolar nature, since this implies that they have no net Burgers vector. The 3-D view shown in **Figure 11** also suggests the vein-like structure by which the loop patches and channels are often called. In addition, note how loop patches and channels extend along a direction perpendicular to the primary Burgers vector $\underline{\mathbf{b}}_p$, showing the edge nature of the loops.

Typical cyclic hardening curves for low and moderate strain amplitudes are shown in **Figure 12**.

It is interesting that the saturation stress and thus the shape of the CSSC appear to be quite insensitive to crystal orientation (Cheng and Laird, 1981d), for load axes within the standard stereographic triangle. The effect of orientation within this range is indicated by the parameter Q, defined as the ratio of the Schmid factor* for the second most highly stressed slip system to that of the primary system. If $Q < 90\%$, the slip tends to be dominantly single; for $Q > 90\%$, secondary slip becomes more evident, but the deformation remains dominated by primary slip (Cheng and Laird, 1981d). The rapid hardening rate is not affected by orientation for a wide range of crystal orientations provided the amplitude is low (**Figure 12(a)**). However, if the amplitude increases, the rate of hardening increases as Q increases (**Figure 12(b)**). Under these circumstances, the loop patches can acquire significant populations of secondary dislocation dipoles.

For the majority of crystal orientations in which the cyclic deformation is dominated by single slip, the results of TEM show that rapid hardening at low strain amplitudes is primarily caused by the accumulation of dislocations in the primary slip system (for reviews of early work, see Laird, 1977, 1979, 1983, Mughrabi et al., 1983). The most recent work has been reviewed by Li et al. (Li et al., 2011). With increasing strain amplitude, and particularly for amplitudes greater than 2×10^{-3}, the increasing hardening rate is associated with increasing amounts of secondary slip. There is much evidence for this behavior; the dependence of rapid hardening rate on orientation shown in **Figure 12** is one form of it. Mughrabi reviewed bulk measurements of magnetic properties in fatigued nickel single crystals (Mughrabi, 1980). Deviations in the magnetic measurements from the symmetry characteristics of crystals containing primary dislocations occur with increasing cycles at strain amplitudes corresponding to the upper half of the plateau, and these reflect the growing contribution of secondary dislocations. Additionally, Ackermann et al (1984) showed that the strain amplitude of 2×10^{-3}, which lies near the low strain end of the plateau in the CSSC (6×10^{-5}–7.5×10^{-3}), represents a threshold. Below this amplitude, TEM showed the loop patches to be dominated by primary dislocations. Above this amplitude, the density of secondary dislocations increased significantly as the amplitude increased. The role of secondary dislocations, in bundling with the primary ones and to some extent controlling the vein morphology, can be seen in the early work of Hancock and Grosskreutz (1969). The amplitude chosen for their study lay at the top end of the plateau.

Thus, early in rapid hardening, and especially at low amplitudes, the dislocation structures consist almost entirely of primary dislocation dipoles. The dipoles collect initially as "*unit loop patches*", that is, juvenile loop patches that form at low stresses when dislocation densities are low (Buchinger et al., 1986). The unit loop patches are well matched in the sense of dipoles, but are rather ragged and have a low volume fraction. At higher stresses the volume fraction of loop patches increases, apparently by dipoles occupying the space between closely situated unit loops (Buchinger et al., 1984) and the structure gradually becomes similar to that seen in **Figure 11**, which applies to ~ 20 MPa. By the time the plateau stress is reached, the volume fraction of loop patches is about 50% and the channels between the loop patches become more sinuous while remaining interconnected.

---

* The Schmid factor is defined as the ratio of the shear stress, resolved on the plane and in the direction of the relevant slip system, to the uniaxial applied stress.

(a)

(b)

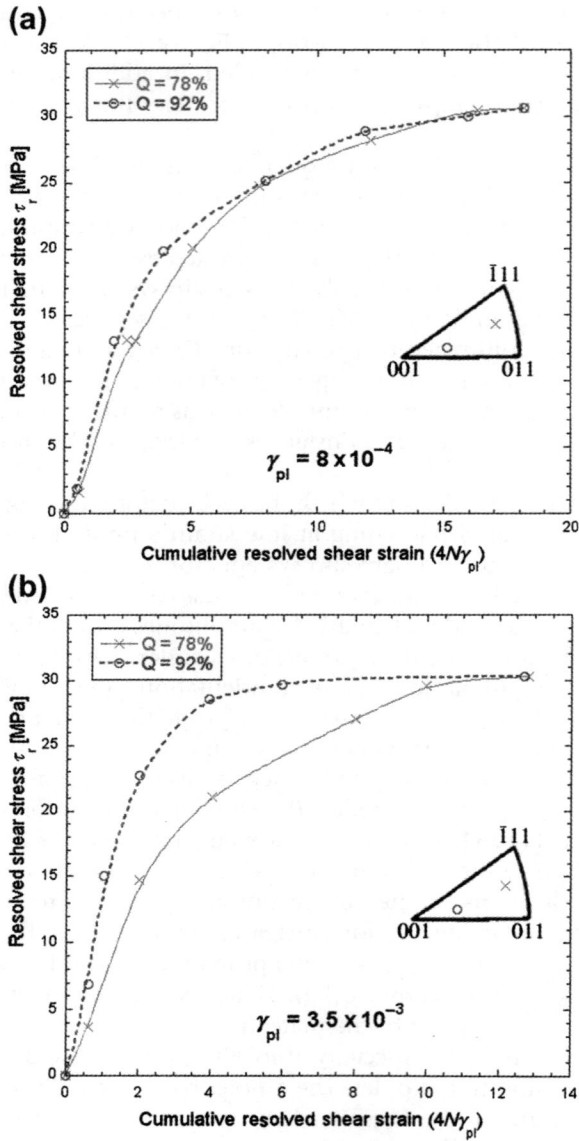

Figure 12 Cyclic hardening curves for single crystals of different orientations specified by the orientation parameter Q: (a) at low strain amplitude; (b) at higher strain amplitude. Note the much higher hardening rate for the more multislip condition (nearer [001]) when the strain amplitude is high. Source: Plotted with data from Cheng and Laird, 1981d. (For color version of this figure, the reader is referred to the online version of this book.)

(a)

Before   After

(b)

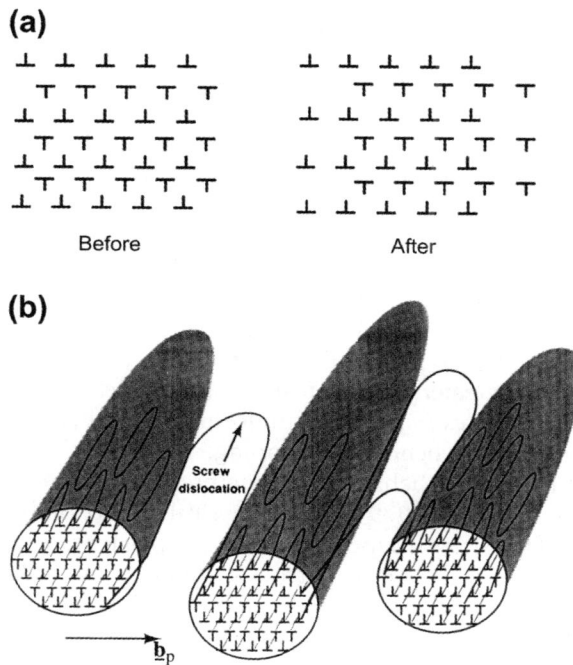

Figure 13 Models of plastic deformation in loop patches. (a) "Flip-flop" motion of edge dislocations inside the patches. (b) Screw dislocations gliding in the channels. Note the Taylor lattice morphology of the loop patches. Source: After Peralta, 2005 from a figure in Laird, 1996. Courtesy of Elsevier.

During the accumulation of loop patches, the slip lines observed on the surface of the specimen are long, straight and uniformly distributed, indicating homogeneous shear. These observations indicate that loop patches and channels are deforming cooperatively, by a "flip-flop" motion of edge dislocations in the loop patches and by screw dislocations gliding in the channels (Laird, 1983; Kuhlmann-Wilsdorf and Laird, 1977, 1980). This widely accepted model (Kuhlmann-Wilsdorf and Laird, 1977; Mughrabi and Wang, 1982) for the dislocation behavior is schematically illustrated in **Figure 13**.

18.4.2 Models of Rapid Hardening Behavior: Loop Patches, Channels and PSBs

Interpretation of the dislocation behavior during rapid hardening has been of interest for many years. Early models used to explain the behavior have previously been reviewed (Laird, 1983). In those models, interpretation has been made via analysis of the friction and back stresses acting on the dislocations (Kuhlmann-Wilsdorf and Laird, 1977; Kuhlmann-Wilsdorf, 1979a,b; Kuhlmann-Wilsdorf and Laird, 1979). The friction stress acting on the dislocations, being dependent on point defect hardening, Peierls forces and jog dragging, is independent of straining direction, but the back stress changes sign during each half cycle. The back stress reaches its maximum value at maximum applied strain, thus acting to lower the yield stress in the reversed direction. As it is of elastic nature because of dislocation interactions, it soon decreases on straining in the direction favored by it, and then reverses again so as to oppose the imposed strains. Measurements of the friction and back stresses have been made by analyzing hysteresis loops using the *Cottrell method* (Kuhlmann-Wilsdorf and Laird, 1979). The results show that the friction and back stresses (the sum of which equals the applied stress) increase during rapid hardening in parallel, the friction stress leading by 1–2 MPa. From this analysis the friction stress is separated into two parts, one equal in magnitude to the back stress (and presumed to have the same physical cause) and the other, smaller, part due primarily to jog dragging by the screw dislocations shuttling in the channels between the loop patches (Kuhlmann-Wilsdorf and Laird, 1979).

Further analysis of the behavior of these stresses provides additional evidence that the loop patches fully participate in the cyclic deformation via the coordinated loop flipping illustrated in **Figure 13a** (Kuhlmann-Wilsdorf, 1979a,b). That is, the loop patches are idealized as *Taylor lattices*, in which the dislocations are regarded as being infinitely long rather than consisting of dipolar loops as they actually do (see **Figure 13(b)**), and they are conceived to deform by slipping rows of oppositely signed dislocations past each other (compare **Figure 13(a) and (b)**). In practice, the Taylor lattices are not ideal, the density of loops being greater on the outsides of the patches than on the insides, the loops being of finite length, and the lattices may not be uniform enough everywhere to prevent individual loop flipping, in coordination with synchronized flipping.

A detailed theory of the Taylor lattice behavior and interpretation of the friction and back stresses has been proposed by Kuhlmann-Wilsdorf (1979a,b). The physical essence of the theory is that, at the strain limits of the cycles, the bulk of the stress is supported by the loop patches, so that the back stress roughly equals one half of the stress at which the loop patches would deform in the absence of friction since they occupy only half the volume. The part of the friction stress that is equal to the back stress comes about because the dislocation motions within the loop patches are irreversible (in the mechanical sense) for every dislocation as soon as it flips with its neighbors of opposite sign. Thus this component of the friction stress, along with the back stress, is given by the flipping stress of the Taylor lattice (Kuhlmann-Wilsdorf, 1979a,b). Many aspects of rapid hardening have been worked out, including anelastic effects and the shapes of hysteresis loops (Kuhlmann-Wilsdorf, 1979a,b), such as the apparent reduction of the shear modulus when the applied stresses are lower than the flipping stress. See Laird (1996) and Wilkens et al. (1980) for more details.

It is necessary to set this model in the proper context of rapid hardening. As mentioned above, hardening takes place by multiplication of primary dislocations, of which many become trapped as edge dislocation dipoles. Patches of dipoles then act as obstacles to the motion of the primary glide dislocations. At this early stage of hardening, the screw dislocations glide over long distances in the channels between loop patches and the Taylor sublattices flip relative to each other several times the interloop spacing. In this slip, the passing stress controls the hardening (Kuhlmann-Wilsdorf and Laird, 1977; Kuhlmann-Wilsdorf, 1979a). Later in rapid hardening, when the volume fraction of loop patches has built up considerably, the channel screw dislocations glide over shorter distances and the loops may not accomplish a single flip, but merely oscillate about their equilibrium positions. If the specimen is called upon to strain more than the channels and loop patches can accommodate at this point, PSBs occur by strain localization. A model for the mechanism of this transition, which is stimulated by secondary dislocations, has been provided by Kuhlmann-Wilsdorf and Laird (1980).

Figure 14 shows schematically the model for the transition from loop patches to PSBs, which is considered to take place by clearing the dislocations from the interiors of the loop patches, and allowing the outer skin to become the PSB walls.

In this model, the loop patches are converted into a series of quite uniformly separated, densely packed dipolar dislocation walls, which, observed in elevation (e.g. by a $(1\bar{2}1)$ TEM section perpendicular to the primary slip plane), look quite like a ladder structure. Secondary slip, stimulated by the self-stresses of the primary dislocations (Kuhlmann-Wilsdorf and Laird, 1980), occurring along the

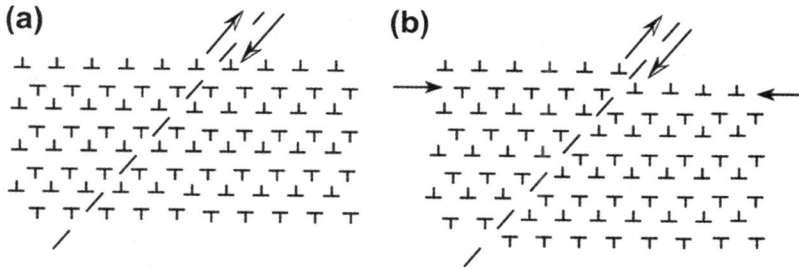

Figure 14 Annihilation of dislocations in the interior of loop patches via secondary slip, indicated by a dotted line. (a) Before secondary slip; (b) positive and negative Taylor sublattices aligned after secondary slip. Source: After Peralta, 2005. Courtesy of Elsevier.

dotted trace shown in **Figure 14**, acts to translate the sublattices of the oppositely signed dislocations into positions favorable for them to annihilate by glide. Residual loops and debris are then swept up into the walls to form the well-known ladder structure of the PSBs. In this type of model, then, the loop patches decompose from the inside, a process that might take a number of cycles. By another point of view, PSBs have been considered to form from outside the loop patches, by a process creating a shear band penetrating many patches. This type of process may well occur if the crystal is suddenly called upon to bear a heavier strain than its current PSBs are experiencing (by a sudden increase in amplitude). The "cords" reported by Mecke (1974) and others (Wang and Laird, 1989) may well be incipient PSBs created by external penetration of the loop patches. In either case, the formation of PSBs during rapid hardening is preceded by an overshooting of the stress followed by softening before reaching saturation, as shown in **Figure 15**.

The maximum shear stress, about 32 MPa for copper, is the nucleation stress for PSBs. This stress decreases at higher strain amplitudes within the plateau (Mughrabi, 1981; Mecke et al., 1982; Blochwitz and Kahle, 1980) or in single crystals oriented for double slip (Gong et al., 1996) because of the higher density of secondary dislocations present, which eases PSB formation.

The effect of secondary dislocations on the extent of hardening in crystals oriented mainly for single slip before PSBs nucleate does not appear to be large. This can be gaged from results of ramp-loading tests. Neumann (Neumann, 1968) reported early on the occurrence of strain bursts if copper single crystals are fatigued under ramp-loading conditions where the controlled load amplitude is gradually increased from 0 to a maximum of 32 MPa (the stress at which PSBs initiate). The nature of the strain bursts is shown in **Figure 16**, that is, there are periodic episodes of high strain lasting for a number of cycles with intervening periods of low strain.

It is known that the dislocation structures produced by ramp loading consist of regular loop patches (Yan et al., 1986); however, they contain a much higher, and more uniform, density of secondary dislocations than the loop patches produced by regular tests under strain control. The strain bursts indicate instabilities in the developing structure of the loop patches, which change discontinuously, burst by burst, to more stable arrangements, depending on the level of the applied stress. Almost certainly the change consists in a refinement of the loop patches, involving a decrease in the dipole width so that the passing stress for the edge dislocations increases, and brings the structure into stability with the load increase.

After ramp loading to 32 MPa, PSBs, the agent by which saturation is obtained, normally develop at a rate that depends on the frequency of cycling, a low frequency, for example, 2 Hz, encourages PSB

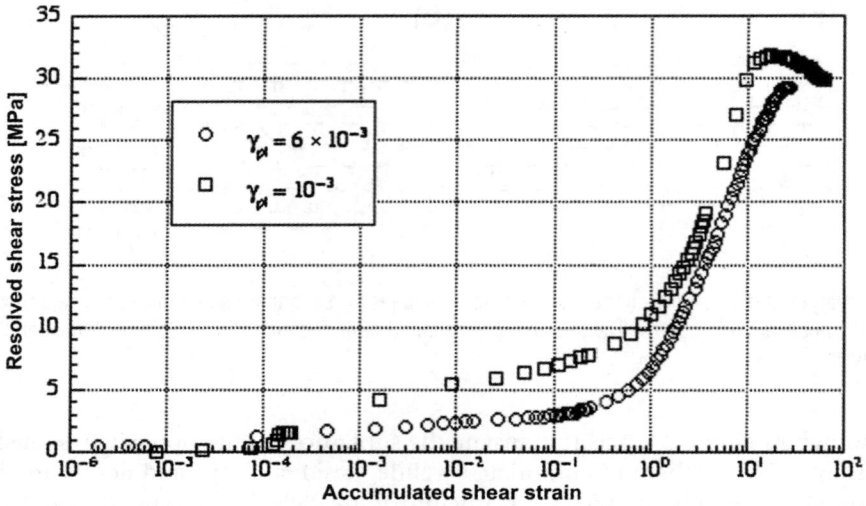

Figure 15 Cyclic hardening for copper single crystals oriented for single slip at plastic shear strain amplitudes right before (6×10^{-5}) and within the plateau (10^{-3}). Note the stress overshooting for the sample tested at $\gamma_{pl} = 10^{-3}$. Source: After Peralta, 2005. Courtesy of Elsevier.

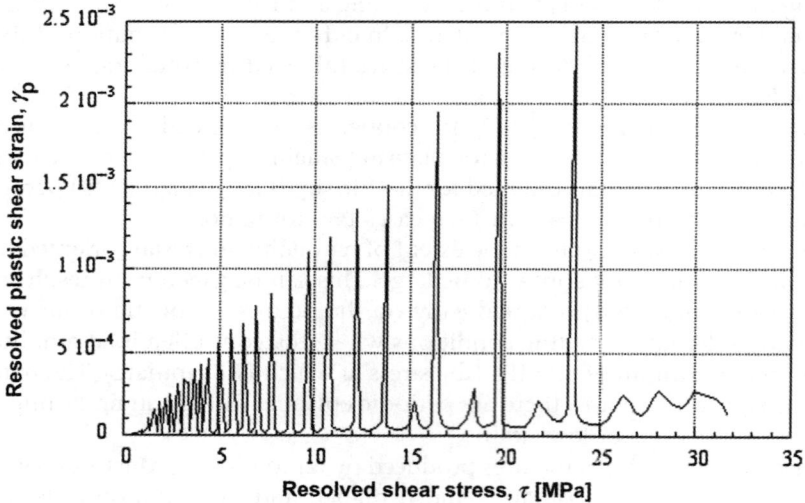

Figure 16 Strain bursts observed in a copper single crystal ramp loaded to a shear stress of 32 MPa in 23 170 cycles. Source: Plotted with data taken from Neumann, 1968. After Peralta, 2005. Courtesy of Elsevier.

formation (Yan et al., 1986). This compares with a stress of 28 MPa, which is required to produce PSBs in a regular test in strain control. The ~10% increment in flow stress produced by ramp loading is significant and can be attributed to the fact that loop patches produced by ramp loading are unable to transform so easily, in spite of their secondary dislocation content, because they are harder and more uniform.

Note that under total or plastic strain control copper samples show tension–compression asymmetry in the stress, that is, a mean stress is present. Under symmetric stress control, a mean strain is then present and cyclic creep (ratchetting) occurs, that is, the mean strain increases monotonically with the number of cycles. The resulting plastic strain usually produces cell structures (Lorenzo and Laird, 1984). When a small mean stress is used to counteract cyclic creep, saturation is observed for stresses below the plateau, suggesting that the dislocation structure in this case is related to loop patches (Jameel, 1997). On the other hand, PSBs seem to be unstable in single crystals under stress control (Jameel, 1997). Nucleation shear stresses as low as 19 MPa have been reported under this load condition (Melisova et al., 1997) and have been attributed to the presence of higher levels of secondary slip. Saturation is not reached in a test conducted at the plateau stress: the plastic strain increases continuously until failure, although this also seems to depend on how the stresses are applied, for example, the use of ramp loading over 1000 cycles to increase the stress from 0 up to a resolved shear stress of 28 MPa has been reported to stabilize the cyclic behavior under stress control (Melisova et al., 1997). Nonetheless, the behavior observed under regular stress control suggests that PSBs nucleate continuously during the life of the specimen (Jameel, 1997). Therefore, Masing behavior, that is, the self-similarity of the loading portions of the hysteresis loops (Mughrabi, 2001), is not observed because the microstructure is in constant flux (Jameel et al., 2001).

Additional evidence showing that secondary slip stimulates PSB formation can be found, for example, in polycrystals, where PSBs are frequently found to nucleate adjacent to twins, often parallel to them when the grain adjoining the twin has its primary system parallel to the twin boundary (TB) (Llanes and Laird, 1992). In such a situation the stress concentration of the twin, because of the need to accommodate elastic and plastic strains across the TB, acts to raise the stresses of the secondary system (Peralta et al., 1994a; Sangid et al., 2011). An example of a PSB lying adjacent to a TB is shown in **Figure 17**.

Dipolar walls have also been observed emanating from TB steps (Llanes and Laird, 1992), which is also likely the result of stress concentrations associated with strain mismatches as well as the corner-like

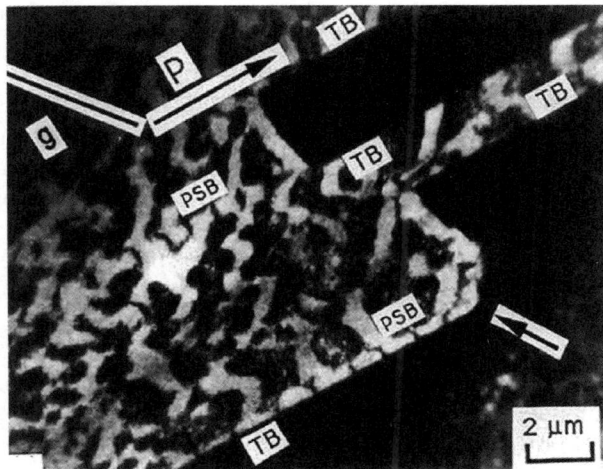

Figure 17 Dipolar wall formation stimulated by stress concentrations from TBs in copper. Note the PSBs parallel and adjacent to TBs. The reflection $\mathbf{g} = (11\bar{1})$ was excited on the matrix, while the twinned regions were near the Bragg condition and so appear dark, but contain loop patches. Source: Taken from Llanes and Laird, 1992. Courtesy of Elsevier.

geometry of these locations. In addition, the fact that these steps are also related to noncoherent sections of the twin is also likely to play a role, given that the noncoherent sections will also have higher energy (Sutton and Ballufi, 1995; Sangid et al., 2011).

The cumulative strain (transition point) at which loops in the patches cease to slip during rapid hardening has been treated semiquantitatively by Cheng and Laird (1981d); this treatment amounts to an adjustment of Kuhlmann-Wilsdorf's theory and describes the variation of the transition point with strain amplitude, and also the dependence of the friction and back stresses on strain amplitude and crystal orientation. The initial model of Kuhlmann-Wilsdorf assumed constancy of the friction and back stresses throughout the plateau of the CSSC, and this assumption is invalid.

Neumann has modeled loop patch geometry by computer techniques (Neumann, 1986) and has confirmed the Taylor lattice morphology. His results agree in general terms with those of Kuhlmann-Wilsdorf, but again there are some minor differences, for example, Neumann computes a different network configuration for the Taylor lattice. In a regular Taylor lattice, the dislocation cores (viewed in the x–y plane, the z axis corresponding to the dislocation lines) form a two-dimensional centered rectangular lattice. As shown in **Figure 13(a)**, the lattice has an aspect ratio of about 2, which is the preferred aspect ratio and slightly different from the aspect ratio selected by Kuhlmann-Wilsdorf (1979b) in her treatment of loop patch structure. In addition, Neumann discovered that finitely sized loop patches are not always stable (Taylor lattices are formally considered to be infinite), and there are some shapes of the Taylor lattice that decompose spontaneously at zero applied stress. Note that the shapes of the loop patches in **Figure 11** can be regarded as assemblies of the diamond shape, one of the stable configurations.

It is of interest to explore whether the shapes predicted by the computer simulations can be observed in simple form. The outlines of loop patches can best be observed in $(1\bar{2}1)$ sections, that is, normal to both the primary Burgers vector, $[\bar{1}01]$, and the slip plane normal, (111). A typical example is shown in **Figure 18**.

The shapes turn out to be irregular, but have facets that roughly conform to the quadrupole stability requirement. It will be seen also that the loop patches are quite uniformly separated and clear channels exist between them at angles of about 20 and $35°$ with respect to the primary slip plane. These channels suggest a possible role of secondary dislocations in loop patch morphology. The structure shown in **Figure 18** was obtained after long cyclic exposures obtained with a high frequency machine, when secondary dislocations were customarily observed by TEM (Buchinger et al., 1984). Also, even for low frequencies (Ma and Laird, 1988) have found secondary dislocations to appear in loop patches at amplitudes equal to, or lower than, the threshold reported by Ackermann et al. (1984) if cycling is extended long enough. The presence of these secondary dipoles exerts a stabilizing influence on the loop patches and causes deviations from the simple morphologies.

Since secondary dislocations are well known to exist in loop patches, models of their behavior that deal only with primary dislocations must be simplistic. Dickson et al. (Dickson et al., 1986, 1988; Turenne et al., 1988), taking secondary dislocations into account, developed a model to explain the specific crystallographic directions adopted by the borders of loop patches and dislocation dipolar walls. If more than one slip system is involved in developing a dislocation structure (and two systems seem to operate in many fatigue conditions even at fairly low amplitudes), the structure consists of the appropriate stacking of the dipoles from the contributing systems. The modes in which this can be done are illustrated schematically in **Figure 19**.

If the loops with the Burgers vector $\mathbf{b}_1$ are extended in the direction n_1, and those with Burgers vector $\mathbf{b}_2$ in the direction n_2, then the dislocation dipoles can be stacked in walls (or loop patch borders) in the three directions: $a = n_1 + n_2$, $c = n_1 - n_2$, and $b = a \times c$, that is, the walls or borders adopt a direction that

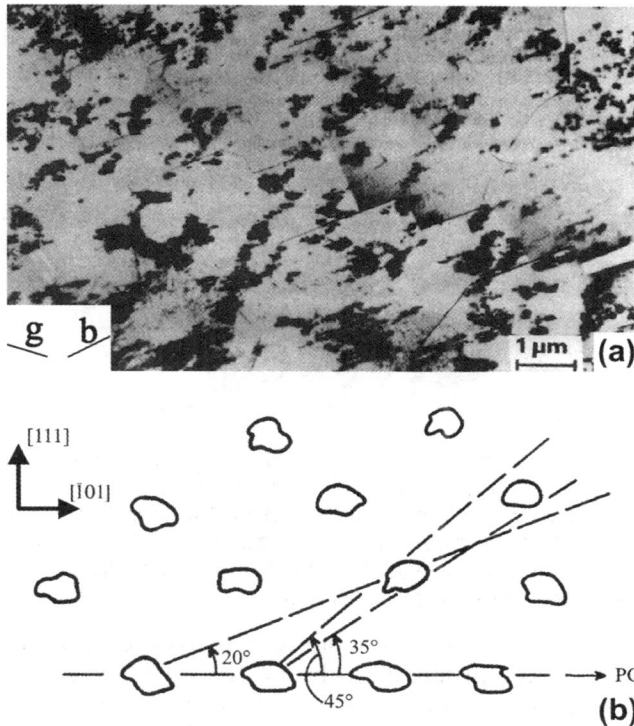

Figure 18 The outlines of loop patches and the structure of the channels separating them in monocrystalline copper viewed along the [1$\bar{2}$1] direction: (a) actual structure and (b) loop patch alignment indicated schematically. Source: After Laird, 1996. Courtesy of Elsevier.

"averages" the dipolar configurations. Dickson et al. (1986) and Dickson et al. (1988) have had considerable success in explaining irrational crystallographic orientations taken up by walls and veins. Very pronounced orientations can be produced in loop patches if a crystal is first cycled at high amplitudes to introduce a cell structure based on several Burgers vectors, and the crystal is subsequently cycled at low amplitudes to convert the cells into well-faceted loop patches, which inherit the secondary dislocations (Ma and Laird, 1988).

Mughrabi (1981) modeled cyclic plasticity in loop patches by a composite model, in which he regards the loop patches and channels as hard and soft phases, respectively. He then considers the compatible shear deformation in the direction of the primary Burgers vector under the condition of uniform total strain γ_{tot}. Mughrabi points out that this is the basic idea underlying Masing's model of polycrystal plasticity, where the different grains are considered as a large number of components that have different flow stresses. When the two-phase structure produced by rapid hardening is exposed to stress, the softer channel elements (phase #2) will yield plastically when $\gamma_{tot} = \tau_2/G$, where τ_2 is the yield stress of the channels and G the shear modulus. After yielding of the channel elements, the applied stress will increase further with an apparent shear modulus $f_1 G$ (where f_1 is the volume fraction of the loop patches) until the harder loop patches (phase #1) flow plastically at $\gamma_{tot} = \tau_1/G$ (where τ_1 is the yield stress of the loop patches). At this point the channels will already have undergone a plastic shear

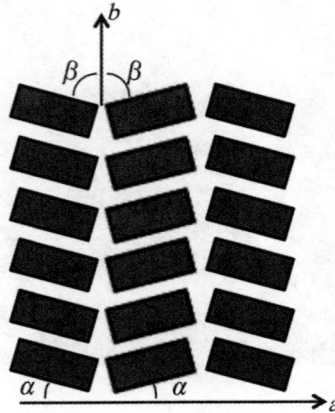

Figure 19 Schematic 2-D drawing of possible 3-D stacking arrangements that can be constructed from dipolar loops having two different Burgers vectors. The long segments of the loops correspond to the two systems, as seen in the plane containing both orientations of long segments. There are other ways to stack these loops, in addition to that shown, which has its stacking direction normal to the paper; the other two stacking directions are parallel to the directions indicated by *a* and *b*. Source: Adapted from Laird, 1996 after Turenne et al., 1988, courtesy of Elsevier. (For color version of this figure, the reader is referred to the online version of this book.)

strain $\gamma_{tot} = (\tau_1 - \tau_2)/G$. Subsequently, the shear stress, τ, remains constant and is now equal to the macroscopic flow stress of the composite, given by $\tau = f_1\tau_1 + f_2\tau_2$, where $f_2 = 1 - f_1$ is the volume fraction of the channels. To apply this model, Mughrabi (1981) employed dislocation mechanisms for calculating τ_1 and τ_2. These mechanisms were (1) screw dislocation bowing in the clear channels between loop patches and (2) loop flipping in the loop patches. To judge from the many previous opinions expressed about these mechanisms (Basinski et al., 1969; Kuhlmann-Wilsdorf and Laird, 1977, 1979; Brown, 1980), there seems to be a strong consensus about dislocation behavior in the loop patches produced by rapid hardening.

18.4.3 Saturation Behavior and Strain Localization

The CSSC for copper single crystals, shown in **Figure 9**, has been described in terms of three regions A, B and C. Region A corresponds to a deformation regime in which the slip is uniformly dispersed, the slip lines long and even, and the dislocation microstructures consist of loop patches. In this region, the slip is carried uniformly by the loop patches and channels, via the mechanisms discussed above, that is, Taylor lattice flipping and by screw dislocation motion, respectively. Since the strain is small, the dislocation microstructures are stable and the deformation is mechanically reversible. The stress reaches saturation.

Within region B, the plateau of the CSSC, PSBs form and carry the strain at the expense of the loop patch matrix structure. A threshold plastic shear strain of 6×10^{-5} is required to initiate PSB formation (Mughrabi, 1978; Li et al., 2011) and at this low strain a single PSB alone is needed to carry the strain. As the strain amplitude is increased within the plateau region, the volume fraction of PSBs increases. Since a constant stress is required to strain the PSBs, the stress is constant and defines the plateau stress. At the high strain end of the plateau, 7.5×10^{-3} plastic shear strain, the whole gauge section of the specimen becomes filled with PSBs (inset in the C region of **Figure 9**). At intermediate strain amplitudes, the volume fraction of PSBs varies linearly with the applied strain (Winter, 1974; Finney and Laird, 1975). The value of the strain at the high end of the plateau, about 1%, is commonly viewed as representing the *average* local strain in the PSBs, while the value of the plastic strain at the low end of the plateau is taken to be that of the matrix. Within the plateau, then, the applied strain is related to the PSB and matrix strain, and their respective volume fractions, by the rule of mixtures (Winter, 1974; Mughrabi, 1980). The properties of PSBs have been studied by many workers, mainly in copper (for reviews, see Laird, 1977; Mughrabi, 1980; Laird, 1983; Basinski and Basinski, 1992; Li et al. 2011).

Since, within the plateau regime, nearly all the applied strain is carried by the PSBs, they completely traverse the specimen as lamellae parallel to the primary glide plane, and thus they determine bulk deformation behavior. Although it can be inferred from the CSSC and the plateau that there is a constant local strain in the PSBs, the deformation in the PSBs is not constant and varies widely, in several different ways in both space and time. The evidence for such behavior comes from measurements using interferometry techniques ((Finney and Laird, 1975; Cheng and Laird, 1981d; Laird et al., 1981); and (Witmer et al., 1987; Ma and Laird, 1989a,b,c,d,e)) and also from atomic force microscopy (AFM) measurements (see Holste, 2004, Mughrabi, 2009 and references therein). After cycling a single crystal to saturation, the former workers would interrupt a test, repolish the surfaces of the crystal on which the PSBs were evident, subject the crystal to a quarter cycle of strain and then study the slip offsets of the PSBs by two-beam interferometry. Since the dislocation structure of the PSBs is known by etch-pitting studies (Winter, 1974; Basinski et al., 1980) and by ECC (Li et al., 2011) to extend virtually unchanged through the bulk of the crystal, the slip offsets observed must reflect the bulk behavior. Typical "well-behaved" PSBs, observed by interferometry, are shown in **Figure 20(a)**. These bands show a reasonably uniform concentration of strain ($\sim 1\%$) within micro-PSBs of which five are visible in the field of view. Note that PSBs often occur as macro-bands, just visible to the naked eye, but they are composed of smaller bands, often separated by matrix structure, termed micro-PSBs, sometimes very narrow and appearing as a single step of coarse slip, but mostly in wider packets such as those shown in **Figure 20(a)**. The PSBs shown in **Figure 20(a) and (b)** were polished off so as to allow the observation of slip during a single reversal (or a few), since they tend to have complex morphologies after thousands of cycles because of the accumulated effect of much localized slip and they usually protrude from the surface, which makes interferometry measurements difficult to carry out on an unpolished surface.

Examples of micro-PSBs where large steps formed in very narrow bands are shown in **Figure 20(b)**. The crystal shown in this figure had been cycled 2763 3/4 cycles (i.e. interrupted from compression), repolished and then subjected to 1 1/4 cycles, thus being returned to the state of zero stress and zero strain, but with a net 1/4 tensile strain from the point of repolishing. Note that the steps do not all have the same sign and thus some of the steps were formed in the previous compression stroke; they were thus wholly or partially reversed. In light of the height of the offset and the narrowness of the bands, the investigators (Laird et al., 1981) opined that the plastic strain concentration factor for these micro-PSBs would be about 1000. These values of strain localization are much greater than the widely accepted average value of 1%.

Figure 20(a) shows another interesting feature of PSB behavior: the strain within the PSB is not uniform but tends to be further localized near the matrix-PSB boundary. Note that the slip offsets are highest in such regions. This observation explains the tendency of PSBs to initiate cracks at the PSB–matrix interface, because the damage mechanism responds to the magnitude of the local strain.

Cheng and Laird, 1981b,a,c found not only that the plastic strain is not uniform, even at the upper end of the plateau, where the gauge length of the crystal is essentially a single large PSB, but also that the

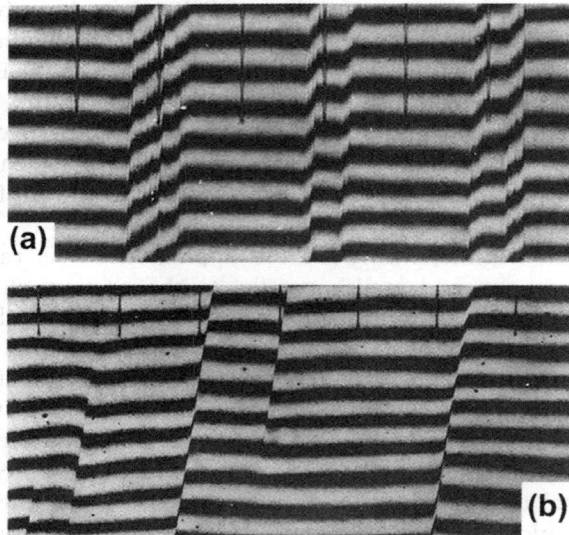

Figure 20 Interferograms of copper single crystals after repolishing and straining for a quarter cycle, showing strain concentrations at micro-PSBs: (a) plastic shear strain amplitude of 1.25×10^{-3} for 30,000 cycles, repolished, strained in tension and (b) plastic shear strain of 2.5×10^{-3}. History given in text. Note that the shear strain is not uniform along the individual lines. Fiducial marks distant 100 μm apart. Source: Taken from Laird et al., 1981. Courtesy of Elsevier.

magnitude of the localized strain in the PSB that produced the fatal crack increases with an increase of the applied strain. Therefore, the fatigue lives of the crystals decreased as the strain increased, so that normal Coffin–Manson life behavior was observed with a fairly steep slope, ~ -0.78 (Cheng and Laird, 1981b). It is important to note that Cheng and Laird cycled in constant plastic strain control. The life behavior, therefore, was different from that observed by Hunsche and Neumann (1986), who employed the gradual ramping approach of testing their specimens. This caused their crystals to be conditioned with structures similar to those that apply at the low strain end of the plateau. Therefore, when Hunsche and Neumann subsequently converted their test control to plastic strain, the PSBs behaved more uniformly and the crystal lives, although showing large scatter, tended to be constant. History effects are well known to be important in controlling fatigue lives.

Testing under inert environments performed by Witmer et al. (1987), which led to extended fatigue life, followed by interferometry and surface observations of PSB behavior revealed clearly that PSBs evolve with extended cycling, which confirmed some of the earlier observations by Wang and Mughrabi (1984), but showed that old PSBs do not become inactive because of secondary hardening, as postulated by the latter authors, but rather continue localizing strain until a final value of 3×10^{-3} is achieved and the surface of the old PSBs show intrusions and extrusions that are fine and evenly spaced, and they carry strain in a more reversible manner on a macroscopic scale as compared with the "new" micro-PSBs that localize much higher levels of strain (Witmer et al., 1987). Measurements of PSB activity on sample surfaces using AFM (Holste, 2004; Mughrabi, 2009) have confirmed these findings and shown that this behavior extends to other fcc materials such as nickel, while also revealing new phenomena. The mechanisms that lead to strain localization in PSBs via dislocation motion have been studied extensively and will be briefly reviewed in the next section.

18.4.4   Models of Dislocation Behavior in Persistent Slip Bands

Our understanding of how the localized deformation is accomplished in PSBs by dislocation mechanisms is based mainly on observations of dislocation structures by TEM, although recent work using ECC has shed light into behavior that occurs at larger length scales (see Li et al., 2011 for a review of this work). Seen from a direction normal to the Burgers vector of the primary slip system on which the PSB has formed, the appearance of typical narrow PSBs is shown in the inset for region B in **Figure 9**. The PSBs are the ladder like structures embedded in the matrix of loop patches. The rungs of the ladder, the PSB walls, consist of primary edge dislocation dipoles that gather together to form some type of multipole viewed, perhaps rather idealistically, as dipolar walls. The dipolar nature can be seen because the background contrast of the electron micrograph does not vary from side to side of the walls. The average separation of the walls is about 1.4 μm and their spacing is not precisely uniform.

The relation of the PSBs to the matrix structure of loop patches is such that at low strain amplitudes, the PSBs occupy a relatively small volume fraction of the crystal. With increase of strain amplitude, the rungs of the PSBs extend until the whole structure comprises dislocation dipolar walls (see insets for regions B and C in **Figure 9**).

How the dislocations glide in the PSBs to carry the localized strain has been the object of much study. There appear to be two different, but overlapping, viewpoints. According to one viewpoint (Finney and Laird, 1975; Mughrabi et al., 1979) dislocations are considered to bow out from free links in the dipolar PSB walls and glide across the channel to the adjacent walls, where they become trapped and partially annihilated by interacting with dislocations of opposite sign. The screw dislocations at either side of these bowing loops are then liberated to glide down the channels until they encounter oppositely signed screw dislocations, which can mutually annihilate by cross-slip. Thus an equilibrium between dislocation multiplication and annihilation is considered to exist in the PSBs, and the wall structure is kept stable.

The bulk of the PSB strain is carried by screw dislocations, and the value of the strain is limited by the maximum density of screw dislocations arranged in groups of the same sign. There is a problem with this link-bowing mechanism, however: if the channel screw dislocations are all continually annihilated, then the walls would build up by acquisition of edge dislocations from the bowing. Mughrabi and coworkers overcome this objection by requiring the dipoles of smallest spacing (the near-miss entrapments of oppositely signed dislocations) to spontaneously dissolve by point defect emission, thereby also explaining the high point defect content of the PSBs. This self-annihilation by climb maintains the stable structure of the walls.

According to the other viewpoint, the role of the screw dislocations in the PSB channels is emphasized at the expense of the edge-bowing mechanism (Kuhlmann-Wilsdorf and Laird, 1977; Laird et al., 1981). The dipolar structure of the walls accounts for the prominence of screw dislocation motion in the channels by a theory of *low energy dislocation structure* (LEDS). These screw dislocations, gliding in groups of the same sign, move in coordination between PSB walls, so as to lay down simultaneously edge dislocations of opposite sign in dipolar configuration at the walls. The detailed manner in which this could be done is shown in **Figure 21**, a schematic representation of three adjacent dipolar walls. Only the left-hand wall is shown with its dipolar structure; the other two walls are left blank to show the coordinated behavior of the screw dislocations.

The screw dislocations A, B, C in channel 2 are laying down edge dislocations at the walls, and the coordinated motions of A′, B′, C′ are doing likewise in channel 1, but their atomistic glide planes are stepped with respect to those in channel 2. Thus considering the center wall, the negative edge

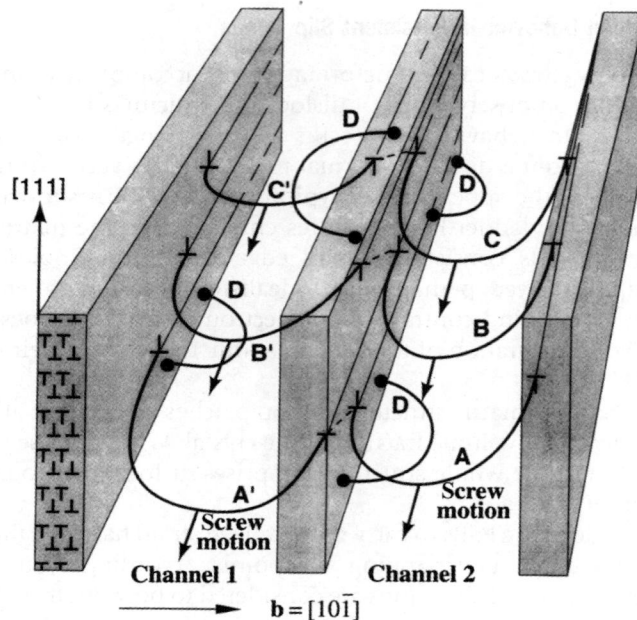

Figure 21 Models of dislocation behavior in PSBs, showing both the link-bowing mechanism, D, and the coordinated motions of screw dislocations A, B, and C and A′, B′ and C′ across the walls. It is understood that A′, B′ and C′ are positioned lower than A, B and C so that their edge dislocations match as dipoles across the center wall (connected by dotted line). The dipolar wall structure shown in the left wall applies to the others as well. Source: After Laird, 1996, courtesy of Elsevier.

dislocation deposited by C′ is being matched as a dipole by the positive edge dislocation of C. The matching pairs in the dipoles are joined by dotted lines to show their "connection" across the wall thickness. Under the action of the applied strain, the screw dislocations are gliding toward the reader, with their bowing likewise directed. As the strain is cycled, the screw dislocations reverse their directions of travel, picking up edge dislocations or laying them down, in coordination. The particular possibility of the structure shown is considered a likely one, based on the LEDS principle, because these coordinated screw dislocations maintain the dipolar balance of the wall structure and thus the lowest energy configuration of the dislocations. No annihilation by point defect action is necessary, since the dislocation structure is maintained purely by a glide process.

The most reasonable position seems to be that both the coordinated screw dislocation mechanism and the dislocation link-bowing mechanism occur, but to different degrees (Laird et al., 1981). Therefore, **Figure 21** shows both mechanisms, and the experimental evidence seems to indicate both. In this connection, it is difficult to use regular TEM for assessing this point because the screw dislocations escape from the foil during specimen preparation. However, Mughrabi and his coworkers (Mughrabi et al., 1979; Mughrabi, 1981) used neutrons to irradiate specimens retained under load so as to keep the screw dislocations in place during TEM specimen preparation by pinning them with irradiation-induced defects. Representative photographs taken parallel to the slip plane are shown in **Figure 22**. **Figure 22(a)** shows a region in which link bowing is clearly dominant, and **Figure 22(b)** shows regions where screw dislocations, bowed under stress, are

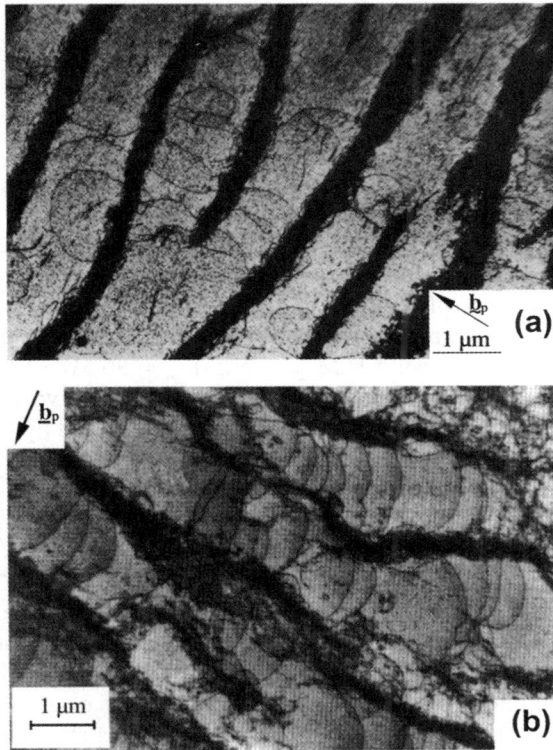

Figure 22 Dislocation arrangements in a PSB from a fatigued copper single crystal (neutron irradiated in the stress-applied state: section parallel to the primary glide plane). (a) Screw dislocation bowing and (b) coordinated screw dislocation glide. Source: Taken from Laird et al., 1986, originating in work by Mughrabi and coworkers. Courtesy of Elsevier.

grouped with the same sign as can be seen by the similarity of their bowing, and they are coordinated across the channel walls.

Models have been offered to explain the flow stress of the plateau, that is, PSB deformation, on the basis of the dislocation models described above (Kuhlmann-Wilsdorf and Laird, 1977; Mughrabi, 1981). These models have a great deal in common; however, the composite approach employed by Mughrabi has the virtue of simplicity and is described as follows:

Consider a PSB at saturation with the applied stress/strain at zero. As the stress is applied, the crystal first deforms elastically and then begins to yield, initially in the channels because of their lower dislocation density and their large Frank-Read bowing distance, that is, the width of the channel. As the stress builds up, the walls will begin to yield, but by this time the channels will already have experienced a plastic strain given by the difference of the flow stresses of the walls and channels divided by the shear modulus (see equations for the composite model used to describe the deformation of the vein structure at the end of Section 18.4.2).

The large local strains, perceived as easily measurable slip offsets at the surface, imply that the plastic strain is ultimately continuous through the wall/channel structure of the PSBs, and this result supports the assumption that the *total* strain is the same in the walls and channels. This means that the

edge dislocations laid down at the walls infiltrate them, dipoles flip within the walls and new dislocations bow out of them. On the assumption of continuity of total strain, Mughrabi expresses the applied
stress τ_a as

$$\tau_a = f_w \tau_w + f_c \tau_c, \tag{5}$$

where f_w and f_c are the volume fractions of walls and channels, respectively, and τ_w and τ_c are the flow
stresses of the walls and channels. That is, the rule of mixtures applies.

There is a slight difference in the *plastic* strain of the walls and channels because the flow stress in the
walls is higher, so that their elastic strain is greater. Kuhlmann-Wilsdorf and Laird (1977) expressed
the flow stress of the walls in terms of the local stresses adjacent to dipolar walls. Mughrabi measured
the flow stresses in the walls and channels by examining the curvatures of dislocations pinned in each
of the regions by neutron irradiation using TEM. He found $\tau_w = 90$ MPa and the average $\tau_c \approx 15$ MPa,
that is, in close approximation to the Frank-Read stress. Since the volume fractions of walls and
channels are typically 20 and 80%, respectively, in the section analyzed by him (e.g. **Figure 22(b)**), the
rule of mixtures correctly predicts the plateau stress as ~ 28 MPa. The volume fractions observed in
$[1\bar{2}1]$ sections taken perpendicular to the walls (the "ladder" structures) generally show somewhat
lower values, for example, 10 and 90%, but the volumes affected by elastic stresses extend beyond the
volumes actually occupied by dislocations.

Mughrabi and coworkers (see Schwarz and Mughrabi, 2006; Mughrabi, 2009 and references therein)
as well as Brown (2006) have attempted to refine these models by carrying out analytical and numerical
analyses of the stresses required to move dislocations in the channels produced by the dipolar walls in
PSBs, accounting for the stress required to bow the dislocations (the Orowan stress) as well as the
stresses resulting from interactions between dislocations passing one another in the channel. The
numerical calculations presented in Schwarz and Mughrabi (2006) showed good agreement between
the passing stress and the experimentally measured flow stress for PSBs; however, these authors
regarded this as a disagreement, given that the flow stresses inside the channels are known to be reduced
as compared with the overall flow stress because of internal back stresses and the contribution of the
wall to the overall flow stress is needed (see the composite model in the previous paragraph). They
argued statistical variations associated with effects of dislocation annihilation and variability of separation distances between interacting dislocations could account for the discrepancy.

It was noted above that spontaneous dissolution of edge dipoles by point defect emission allowed
PSB walls to remain stable and also produced high densities of point defects. In addition, since much of
cyclic deformation is produced by screw dislocations, which will be dragging jogs and emitting point
defects, point defect concentrations can be expected to be high in fatigue (Kuhlmann-Wilsdorf and
Laird, 1979). This high production of point defects has been verified by measurements of the electrical
resistivity (Polák, 1970; Charsley, 1981) and is also reflected in a large temperature dependence of
fatigue hardening (Broom and Ham, 1959).

There are other implications of heavy point defect production. For example, the fact that PSBs
usually protrude from the surface is taken to be connected with enhanced point defect production in
PSBs. This point has been discussed and developed into a detailed model by Essmann et al. (1981) and
Essmann (1982), who argue that vacancy production dominates the point defect formation, thereby
causing the protrusion. See Mughrabi (2009) for a recent review of this model and its latest
developments.

However, there is other evidence that interstitials can produce significant effects in other situations.
For example, Lee et al. (1981) cycled Al-15% Ag alloy aged so as to contain Guinier-Preston (GP) zones

at 77 K and found that γ' precipitates had nucleated and grown during a fatigue life (at high strain) lasting only the order of 10 min. The only mechanism that could explain these surprising precipitate–growth kinetics was point defect enhancement of the interdiffusivity (the precipitates that formed were silver rich); short circuit diffusion mechanisms were not adequate without enhanced volume diffusivity. Since it is unlikely that vacancies would be mobile at such low temperature, interstitials seem the likely cause for the behavior.

The dipolar wall structures of the PSBs follow from the dominance of primary dislocations in them. At the upper end of the plateau, where multislip begins to occur, and especially in polycrystals, more complex wall structures are observed. However, fatigue cells represent the typical dislocation structure at the end of the plateau (region C) for single crystals oriented for single slip and will be discussed next.

18.4.5 Fatigue Cells

Multiple slip in regime C leads to the decomposition of the wall structure of PSBs into cellular bands and then to cells (**Figure 9**). The cell structure in a copper single crystal is shown in 3-D in **Figure 23**.

Dipolar walls with high secondary dislocation content as well as cells can be seen on the primary slip plane (111), whereas the more common cell structure dominates the view on the $(1\bar{2}1)$ plane. The cell structure forms early during rapid hardening and the cells themselves are very stable: their size can increase when the strain is decreased and vice versa (Laird et al., 1986). Cells are composed of walls with several Burgers vectors, they can be either elongated or equiaxed and they surround a region of low dislocation density (Feltner and Laird, 1967a). Note from **Figure 23** and also from **Figure 9** that misorientations can be observed among fatigue cells as differences in contrast from cell to cell in the TEM images. These misorientations have been explained as the result of lattice rotations normal to the primary slip plane that alternate their senses from one group of cells to another (Kuhlmann-Wilsdorf, 1975). Nonetheless, the strain produced by the cell structure can be considered homogeneous, since

Figure 23 Cell structures observed in 3-D in a copper single crystal saturated at a plastic shear strain amplitude of 1.45×10^{-2} (Region C, dislocation structures neutron pinned in the unloaded state). Source: Taken from Laird et al., 1986, courtesy of Elsevier.

slip traces on the surface of the samples are uniform (Laird et al., 1986). Deformation is produced by dislocations that bow out from the walls and glide inside the cell until they are eliminated by cross-slip through an encounter with a dislocation with opposite sign. This deformation favors the nucleation and propagation of intergranular cracks (Ma et al., 1990; Peralta et al., 1999a). Besides cells, which form at high strains, other multiple slip structures can form at intermediate strains in polycrystals and in single crystals oriented for double or multiple slip. The effect of load axis orientation for these cases is significant and it is quite relevant to the understanding of cyclic deformation behavior of polycrystalline materials and the effect of crystallographic texture; therefore, it will be addressed in the next section.

18.4.6   Effect of Load Axis Orientation

As discussed above (see Section 18.4.1), all single slip orientations present similar cyclic behavior, with minor differences related to the content of secondary dislocations (Cheng and Laird, 1981d). On the other hand, significant differences exist for double and multiple slip orientations, that is, at the edges and corners of the standard stereographic triangle, respectively. Some of the early work included studies on the following orientations: <111> (Lepistö et al., 1986; Villechaise et al., 1991), <100> (Jin and Winter, 1984a; Wang et al., 1997), <110> (Jin, 1987; Li et al., 1998) as well as a wide variety of double slip orientations (Jin and Winter, 1984b; Li et al. 1999c, b, a). See the recent review by Li et al. (2011) for more details and references. **Figure 24** shows examples of dislocation structures for some of these orientations. The cyclic behavior of double and multislip orientations also leads to differences in their CSSCs, particularly in regards to the presence of a plateau. Several double slip orientations and crystals loaded along <110> have shown plateaux in their CSSCs (Li et al., 2011), and the corresponding behavior for all orientations within the standard triangle is summarized in **Figure 25**.

Crystals oriented along <100> have high hardening rates and reach saturation stresses than increase monotonically with the plastic strain amplitude (Gong et al., 1997); therefore, no plateau is present in the CSSC, as pointed out in **Figure 25**.

The usual dislocation structure observed for loading along <001> has dipolar walls at 90° from each other and at least two Burgers vectors are present. This arrangement is known as the "maze" structure (**Figure 24**) and Charsley and Kuhlmann-Wilsdorf (1981) proposed that its morphology was controlled by the equilibrium of dislocation dipoles in the walls and at their intersections. This is a biwall variant of PSB walls (Kuhlmann-Wilsdorf and Laird, 1977). By the well-known LEDS principle, groups of similarly signed edge dislocations can reduce their energy if they arrange themselves as a tilt wall. Such a wall can further reduce the energy of the system by pairing with a similar wall, but of oppositely signed dislocations (i.e. a dipolar wall). But even dipolar walls have remnant stresses, which may be reduced by other dislocations, where they end. Thus where a dipolar wall ends, it will attract another dipolar wall oriented at 90° to its plane, as schematically illustrated in **Figure 26**. Possible L and T wall intersections and the dislocation configurations within them are shown in this figure.

Note that the dislocations are pairwise in equilibrium not only within each wall, but from wall to wall along the diagonal lines indicated. Note also in **Figure 24** that the intersecting walls have free ends, L-bends and T intersections, and no other kinds of intersections. In the T-intersection, there is repulsion between the opposed horizontal walls (**Figure 26**) of the T, thus leading to a skew configuration. On the basis of such considerations, Kuhlmann-Wilsdorf and Charsley are able to understand the frequency of occurrence of the various intersections and other properties of the dislocation structure. The agreement of these predictions with the observations of dislocation morphology lends support to the model.

An alternative explanation has been proposed by Li et al. (2010b). These authors postulated that the formation of the maze structure was intimately linked to the activation of the critical slip system and the

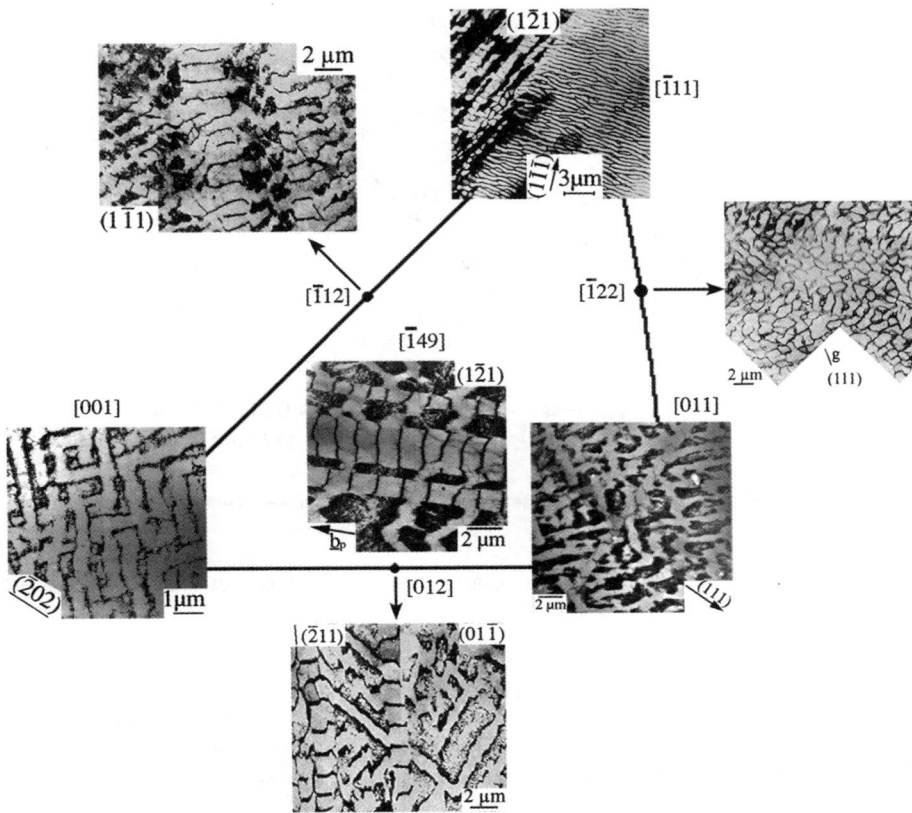

Figure 24 Dislocation structures at intermediate strains for orientations in the standard stereographic triangle. The [100] structure is from Wang et al. (1997). The [Ī11] structure is from Lepistö et al. (1986). Structures for double slip axes [012], [Ī22] and [Ī12] are from Jin and Winter (1984b). Images for [Ī49] and [011] are from Peralta (2005). Source: Courtesy of Elsevier.

formation of PSBs on both the primary and critical slip systems (PSBs have been observed to coexist with the maze structure by Jin and Winter, 1984a). Then the dislocations in these bands ([Ī01] and [101]) would start reacting with the PSBs intercept to form dislocations with [001] and [100] Burgers vectors that lie on (001) and (100) planes, as shown schematically in **Figure 27**, and these would form the walls. These authors, however, did not offer an explanation for the different configurations observed in the maze structure that have been explained by the LEDS model described above.

Crystals oriented along <110>, on the other hand, have a hardening behavior similar to crystals oriented for single slip and a plateau is present in the CSSC (Jin, 1987; Li et al., 2010b, 2011), as shown in **Figure 25**. However, dislocation structures develop where at least two Burgers vectors are present. An example is shown in **Figure 24**, where a PSB can be seen along with incipient maze structure. An interesting feature of crystals loaded along <110> is that they present deformation bands (DBs) that extend over long portions of the sample that are reminiscent of kink bands predicted by Rice (1987) in his analysis of crack tip fields in ideally plastic single crystals, in the sense

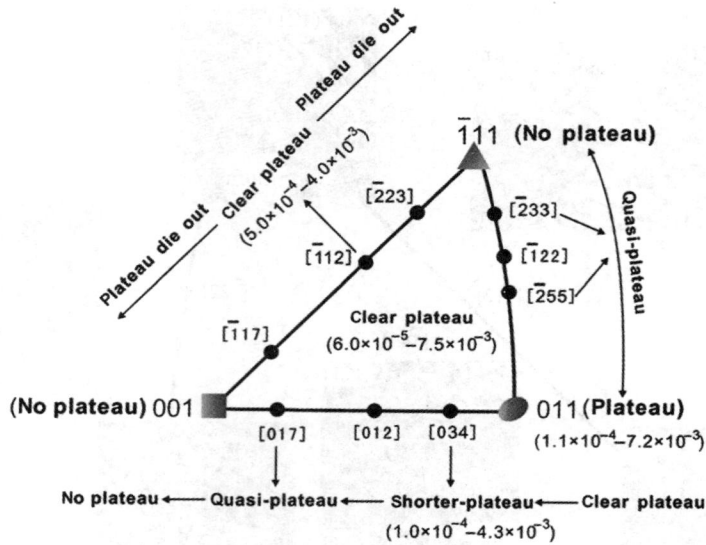

Figure 25   Effect of loading axis orientation on plateau behavior for the CSSC of copper single crystals. Source: Figure reproduced from Li et al., 2011, courtesy of Elsevier.

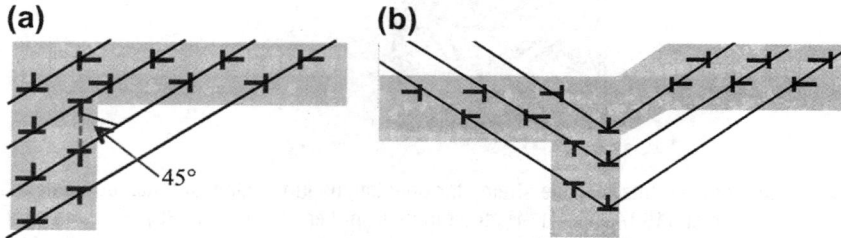

Figure 26   Possible configurations of intersecting, perpendicular dipolar walls and the arrangement of dislocations within them to permit pairwise equilibrium: (a) the preferred L shape and (b) the popular, skewed T. Source: Redrawn from Laird, 1996 with data from Kuhlmann-Wilsdorf and Charsley, 1981. Printed with permission from Taylor and Francis.

that shear seems to be carried along <111>directions slipping on {$\bar{1}10$} planes (see Li et al., 2010b, 2011 and references therein). These DBs start as stacked PSBs and eventually become dipolar walls. According to Li et al. (2010b, 2011) the need to carry additional deformation on certain regions where secondary slip cannot be easily activated leads to formation of additional PSBs locally, which, at some point, reach 100% local volume fraction. At that point, the shear stress along the <111> {$\bar{1}10$}-habit system of the DB will drive the PSB walls to align so that more strain can be carried, as shown schematically in **Figure 28**.

Crystals loaded along <111> show a hard cyclic response, since the Schmid factor of the most stressed systems is low. The cyclic hardening has a steep rate and a maximum can be followed by softening, probably because of the growth of fatigue cracks activated by the high stress reached during hardening (Lepistö et al., 1986), and the CSSC does not show a plateau (**Figure 25**) The dislocation

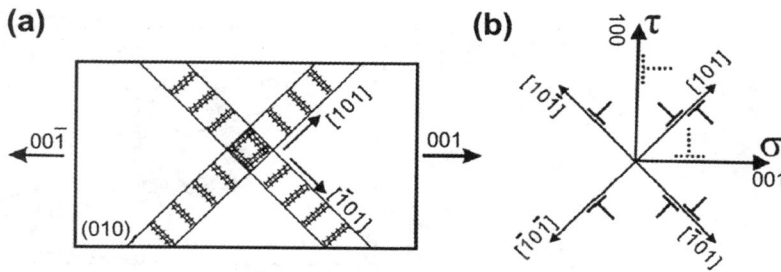

Figure 27 Mechanism for the formation of the maze structure, as proposed by Li et al. (2010b). (a) Intersection of PSBs on two different slip systems. (b) Dislocation reactions. Source: Figure from Li et al., 2010b, courtesy of Elsevier.

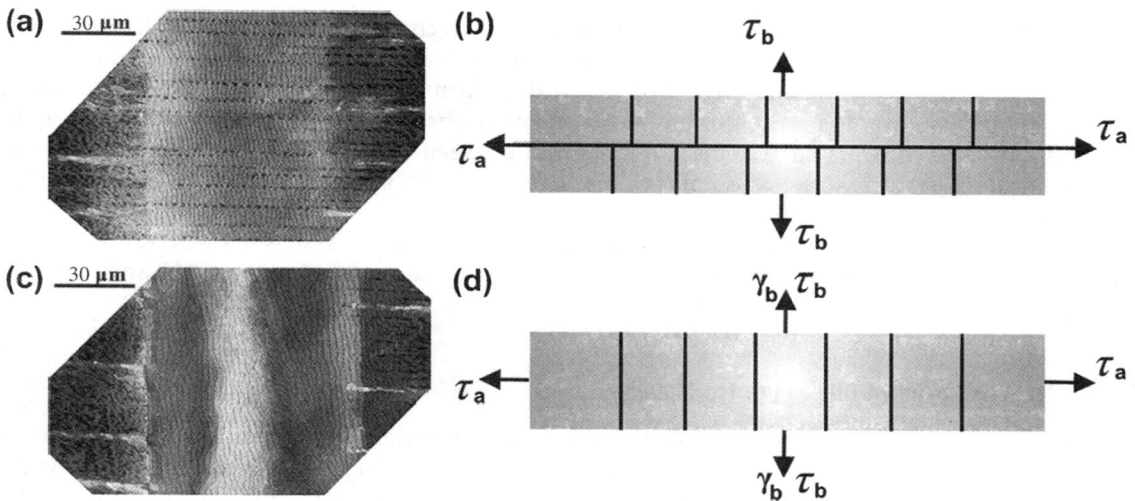

Figure 28 Mechanism for the formation of a DB in a <011> Cu single crystal. (a) Early DB. Note the stack of PSBs. (b) Misaligned dipolar walls inside the PSBs in the early DB. (c) Well-developed DB. Note how the dipolar walls extend along the length of the DB. (d) Aligned dipolar walls in the well-developed DB. Source: From Li et al., 2010b, courtesy of Elsevier.

structures for this orientation favor multislip structures, with walls normal to the loading axis and cellular shear bands at 45° from it (Lepistö et al., 1986, see **Figure 24**). Mechanisms for the formation of the cell and wall structure for this orientation have been discussed by Li et al. (2010b), who indicated that the activation of a primary and two secondary coplanar slip systems results in the three Burgers vectors needed to produce cells with the geometry shown in **Figure 29**.

Single crystals oriented for double slip can have different hardening responses depending on the nature of the interaction between the two active slip systems (Li et al., 1999c,b,a and Li et al. 2011). In particular, Li et al. (2011) pointed out that the conjugated system does not affect the plateau behavior of their CSSCs. These authors summarized the effect of double slip orientations on the plateau behavior, as outlined in **Figure 25**, as follows: on the <001>-<$\bar{1}$11> edge and taking <$\bar{1}$12> as the center, the plateau behavior of the CSSCs gradually disappears as the load axis moves toward either

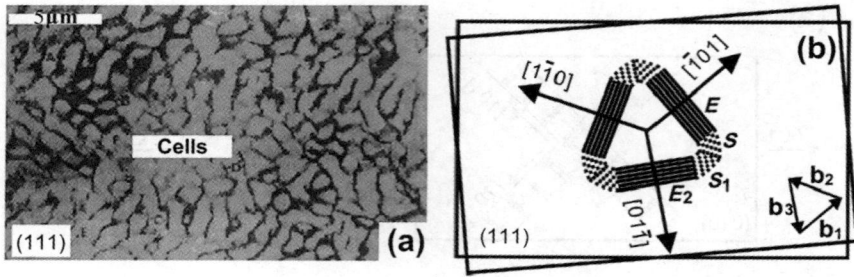

Figure 29 (a) Cells observed in a Cu single crystal loaded along [111]. (b) Proposed cell wall geometry and their relationship to active Burgers vectors. Source: After Li et al., 2010b. Courtesy of Elsevier.

corner of the triangle. On the <001>-<011> edge, the CSSCs change as follows when the orientation goes from [011] to [001]: clear plateau, shorter plateau, quasi-plateau and no plateau. On the <011>-<$\bar{1}$11> edge, the behavior ranges from the wide plateau for the [011] load axis to the complete absence of a plateau for [$\bar{1}$11]. The quasi-plateau behavior observed in between for orientations that display double slip in coplanar systems is a transition between the two extremes. In general, combinations of structures are seen: cells, maze, and sometimes PSBs in one or two slip planes (see **Figure 24**).

The behavior described so far has been discussed for copper single crystals. However, the behavior translates fairly well to other pure fcc metals, including Ni, Ag and Au (see Li et al., 2011 and references therein). This will be briefly discussed next.

18.4.7   Cyclic Behavior of Other Pure fcc Metals

The prominent plateau behavior has been observed in these materials, in addition to copper, and it can be attributed to PSBs, which show the typical ladder-like structure for all of them. Differences include the range of plastic strain amplitudes where the plateau is present, as well as the plateau stress. Typical values of the plateau stress for Cu, Ni, Ag and Au are given in **Table 1**, along with Stacking Fault Energy (SFE) and the shear modulus G.

Note that copper and silver have very similar behavior in terms of the strain range for the plateau and their plateau stresses normalized by their corresponding shear moduli. Nickel has a similar normalized plateau stress, but a slightly higher lower plastic strain for the plateau, perhaps because of its ferromagnetic nature (Holste, 2004) and the presence of a reverse magnetostriction effect (Zhou et al., 2005,

Table 1   Parameters related to plateau behavior in different fcc single crystals

Material	SFE [mJ m^{-2}]	G[GPa]	γ_{start}	γ_{end}	$\frac{\gamma_{end}}{\gamma_{start}}$	$\tau_{plateau}$ [MPa]	$\frac{\tau_{plateau}}{G}$
Nickel	150	76	$\approx 10^{-4}$	$\approx 7.5 \times 10^{-3}$	75	50–52	6.7×10^{-4}
Copper	42	44	6×10^{-5}	7.5×10^{-3}	125	28–30	6.6×10^{-4}
Silver	20	30	6×10^{-5}	7.5×10^{-3}	125	18–20	6.3×10^{-4}
Gold	50	27	$\leq 1.1 \times 10^{-3}$	$\geq 2.2 \times 10^{-3}$	N/A	23–24	8.7×10^{-4}

Source: Data taken from Peralta, 2005; Mughrabi, 1981 and Li et al., 2010c.

2006). The latter can provide a significant contribution of inelastic strain at low strain amplitudes (Zhou et al., 2005), so that higher strains might be needed to increase the plastic strain carried by dislocations before PSBs can nucleate. Regarding gold, not enough data were available to determine the full extent of the plateau regime and the normalized plateau stress is higher than for the other three fcc metals. However, the dislocation structures in gold within the plateau regime are essentially the same as in the other materials and the reason for the difference on the normalized plateau stress is yet to be discerned (Li et al., 2010c).

The behavior of PSBs have also been examined in Ni single crystals using an approach similar to that reviewed in Section 18.4.3, except that AFM was used to measure slip offsets rather than interferometry (Holste, 2004). Spatially resolved measurements and statistics on the distribution of strain in PSBs and SBs were obtained from these measurements and produced interesting results. In particular, the author reports that that only a fraction of the cumulative PSB volume is active during one half cycle and that this fraction is about one-third, independent of the number of cycles, after the cumulative slip traces are polished away, and independent of the temperature. The whole volume occupied by PSBs eventually becomes included in the slip process because all the cumulative slip traces appear again after a large enough number of cycles. An important conclusion from this study is that PSBs have slip activity and inactivity regions and time periods, so their contributions vary as function of time and space. In addition, the study reports that the average strain amplitude of SBs after one half cycle is three times the upper plateau strain limit and independent of the imposed strain amplitude. Furthermore, the ratio between the average SB strain and the upper limit of the plateau was found to be independent of temperature.

The orientation dependence of the hardening behavior and the dislocation structures in Ni and Ag are also essentially identical to those of Cu as discussed in Section 18.4.6. More details can be found in the recent review by Li et al. (2011) and the comprehensive review on Ni behavior by Holste (2004). The absence of Al in **Table 1** is notorious, and the reason it cannot be put together in the same group as the rest of the fcc metals in that table is because its cyclic behavior and corresponding dislocation structures are different. In particular, recent work has confirmed earlier reports that the CSSC does not show a plateau (see Li et al., 2010a and references therein). In addition, the dislocation structures are different from the point of view that cells tend to form readily, because of the very high value of SFE for this material. Strain localization occurs in cellular bands (Li et al., 2010a) and the ladder-like structure of PSBs is not commonly observed at room temperature, but Zhai et al. (1996) reported convincing evidence that they can be present. However, the cellular band structure is much more common (Zhai et al., 1996; Li et al., 2010a).

This completes the discussion of cyclic behavior in pure fcc metallic single crystals and provides a framework for comparison with other materials. In that regard, the behavior of materials with other crystal structures will be discussed in the next section.

18.4.8   Cyclic Hardening in Metals Other than fcc

The cyclic deformation of metals with crystal structures other than fcc has not received nearly as much attention. However, the basic studies have been conducted and reviewed from time to time (Laird, 1977, 1996; Mughrabi et al., 1979; Mughrabi, 1980, 2009). Most investigations pertain to bcc metals. Workers have also explored hcp materials such as α-Ti (Tan et al., 1994, 1998; Zhang et al. 1998) and Mg (Yu et al. 2011b, a), because of the interest in engineering alloys based on these metals.

The factors of temperature, strain rate and small amounts of impurities, especially of interstitials, play large roles in the cyclic deformation of bcc metals, but are not so important for fcc metals. For

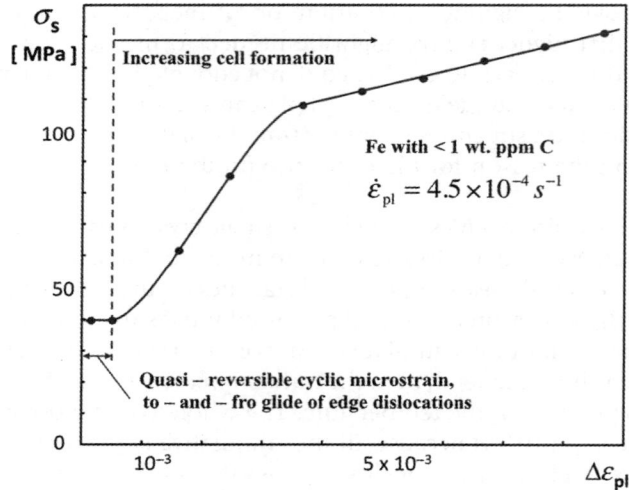

Figure 30 CSSC of pure α-iron single crystals at 295 K. Source: Replotted with data from Mughrabi and coworkers Mughrabi et al., 1979. Printed with permission from ASTM.

example, the cyclic response of copper containing several percent of substitutional alloying element might be quite close from that of pure copper (Woods, 1973), but this depends on the solute and its concentration and the effect on key variables such as SFE. Under slow strain rates, bcc metals can behave similarly to fcc metals, but under other conditions their behavior is quite different from that of fcc metals. The details for bcc metals, many of which have been elucidated by Mughrabi and his coworkers (Mughrabi et al., 1976; Mughrabi and Wothrich, 1976; Mughrabi et al., 1979), are as follows:

For *pure* single crystals of bcc metals cycled at low temperatures and at a strain rate larger than 10^{-4} s^{-1}, the CSSC is typical of that of iron shown in **Figure 30**.

Unlike copper, a plateau produced by PSBs does not occur. Instead, for the cyclic plastic strain range $\Delta\varepsilon_{pl} \leq 5 \times 10^{-4}$, cyclic hardening and microstructural changes are negligible (Mughrabi et al., 1979). The dislocation arrangement is dominated by screw dislocations in low density and has been explained by assuming that only edge dislocations move to-and-fro as the strain is reversed in a nonhardening mechanically reversible manner (Mughrabi et al., 1976, 1979; Mughrabi and Wothrich, 1976). Recall that the thermally activated glide of the screw dislocations in bcc metals is strongly impeded at ambient temperature and below because of the nonplanarity of their cores, giving rise to a behavior reflecting typical features of the low temperature mode of unidirectional deformation of bcc metals. At higher strain amplitudes, cyclic hardening is pronounced and is associated with the formation of equiaxed cells. The slip traces are generally diffuse and in single crystals oriented for single slip do not follow the trace of the primary plane. Mughrabi and his coworkers (Mughrabi et al., 1979) claim that, at high strain rates, the effective stress component characteristic of thermally activated dislocation glide dominates the contribution of the athermal component to the total flow stress, i.e., that due to the elastic interaction of the dislocations.

One of the most interesting observations is that cyclically deformed bcc crystals undergo shape changes (Mughrabi et al., 1976, 1979; Mughrabi and Wothrich, 1976; Mughrabi, 2009; Nine 1970, 1972; Neumann 1975). These changes occur as the contribution of screw dislocations to the cyclic straining increases as the plastic strain amplitudes grow larger and the strain transitions from micro- to macro-plasticity under stresses high enough to force screw dislocations to move (Mughrabi, 2009). Then, the slip plane asymmetry of screw dislocations with respect to tension and compression because of their nonplanar cores leads to these shape changes in single crystals, as has been observed in α-iron and niobium (Mughrabi, 2009). The shape changes that occur are impressive, being easily visible to the naked eye. The cross-section of a cylindrical single crystal cycled in push–pull will change from circular to elliptical and because of the constraints of the grips, additional couples are applied to the specimen that can give it an "S" shape. More complicated shape changes have been reported on α-iron crystals of orientations other than that of single slip (Ikeda, 1980). Mughrabi (Mughrabi, 2009) points out that it is possible to analyze these shape changes quantitatively in terms of the fraction f of plastic strain accommodated by slip because of screw dislocations. This author provides an example for α-iron single crystals under cyclic load at room temperature where the fraction f could be determined as $f = 0.03$, 0.06, and 0.2 for axial strain ranges $\Delta\varepsilon_{pl}$ of 5×10^{-4}, 1.5×10^{-3}, and 6×10^{-3}, respectively. Note that in this case the slip plane asymmetry is the leading cause of irreversibility of cyclic plasticity.

Slip plane asymmetry can be overcome by cycling at higher temperatures, at lower strain rates ($\sim 10^{-5} \, s^{-1}$), and with interstitial impurities present. Investigations conducted under these conditions have been performed (Mughrabi et al., 1979). The CSSC for an α-iron crystal containing 30 wt. ppm carbon tested at a strain rate of $2.5 \times 10^{-5} \, s^{-1}$ is shown in **Figure 31**.

Unlike specimens tested at higher strain rates, considerable cyclic hardening was observed down to a plastic strain range of 10^{-4}. Shape changes were much less pronounced and the contribution of the athermal component of the flow stress, i.e., that because of the elastic interaction of the dislocations, dominated the effective stress component at all strain amplitudes (Mughrabi et al., 1979). Mughrabi and his coworkers observed slip lines parallel to the trace of the primary slip plane, and sometimes these were intense enough to be labeled PSBs. Moreover, their density increased with increasing strain amplitude. Congruent with their appearance, the slope of the CSSC flattened to a gently sloping plateau. Mughrabi et al. were not convinced they had evidence of localized deformation in the PSBs (Mughrabi et al., 1979), but it is almost certain that strain localization does occur because of the

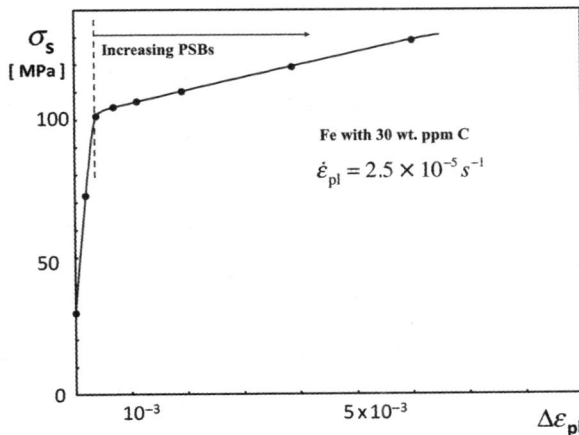

Figure 31 CSSC of impure α-iron single crystals tested stepwise with respect to amplitude and at slow strain rate. Note the difference from that of **Figure 30**. Source: Replotted with data from Mughrabi and coworkers (Mughrabi et al., 1979). Printed with permission from ASTM.

widespread observation of PSBs in polycrystalline bcc metals in early work (see Mughrabi, 2009 and references therein). In low carbon steels these PSBs do not have the typical ladder-like structure, but rather an "open" cell structure in channels with low dislocation density, particularly at moderately elevated temperatures, since the slip behavior becomes more like that in fcc metals (Mughrabi, 2009). Note that the difference in behavior with respect to the pure metals is certainly driven in part by dynamic strain aging induced by the interstitial carbon atoms. In particular, there is preferential retardation of the motion of edge dislocations because of the carbon atoms and dislocation mobilities along different directions become comparable, as is the case for fcc metals (Mughrabi, 2009). Thus the overall behavior is seen to be similar in some respects to that of cyclically deformed fcc crystals.

Dislocations in single crystals of iron-based solid solutions have been studied as well. In that regard, single crystals of a ferritic stainless steel with 13 wt.% Cr have been studied by Sesták et al. (1987), crystals with 26 wt.% Cr by Magnin and Driver (1979), Magnin et al. (1981), with 30 wt.% Cr by Kaneko et al. (1999) and with 35 wt.% Cr as well as fine precipitates by Li and Umakoshi (2003). Iron–Silicon single crystals with 0.5–3 wt.% Si were studied by Sesták et al. (1988) and crystals with 3 wt.% Si by Mori et al. (1979). The reader is referred to these articles for further details on the cyclic behavior of these bcc solid solutions.

A few investigations of the cyclic deformation of hcp single crystals have been made (Stevenson and Sande, 1974; Kwadjo and Brown, 1978). Kwadjo and Brown working with magnesium have shown that PSBs form at suitable strain amplitudes and the properties of the PSBs are quite similar to those of fcc metals. However, at high strains, twinning is extensive and significantly affects the hardening. This manifests itself in kinks that can be observed, for example, in the compression half of the hysteresis loops, because of the tension–compression asymmetry for twinning in hcp metals (Stevenson and Sande, 1974). After a few cycles the loops studied by Stevenson and Sande (1974) returned "back to normal" without "tails" indicating higher hardening rates in compression because of detwinning. However, a recent study by Yu et al. (2011a) on cyclic deformation of Mg single crystals loaded in tension compression along [0001] provided direct evidence of the asymmetric shape of the hysteresis loops (low slope in tension and high slope in compression) and in situ observations revealed that a twinning–detwinning mechanism was responsible for it.

Regarding Ti single crystals, some of the more systematic work dates to the mid-1990s and late 1990s (Gu et al., 1994; Tan et al., 1994, 1998), which is surprising given the technological interest in this material. The cyclic behavior has been studied as a function of orientations by the authors mentioned above, and both the CSSC and the dislocation structures have been studied under uniaxial load as a function of orientation of the load axis. In that regard, CSSCs for single crystals of different orientation are shown in **Figure 32**.

Note the wide range of responses and the strong effect of the orientation of load axis in the cyclic response. A short plateau seems to be present for orientation 3 and a clear inflection point (or a fairly short plateau) can be seen in curve 4. In addition, a plateau was reported by Tan et al. (1998) for a load axis within the triangle and close to $11\bar{2}0$, which is an indication that strain localization also occurs in this material. Gu et al. (1994) reported that orientation 3 presented dislocation lines connected to form branches and loose groups of dislocation dipoles, but no twinning, whereas orientation 4 showed dense dislocation clusters, fine structured twins and stacking faults. The lack in twinning in orientation 3 is probably a contributing factor for its low hardening compared with many of the other orientations. In general, these authors reported a wide variety of dislocation structures in crystals with different orientation including planar dipolar arrays, dislocation loops, dislocation networks and cells. This variability is no doubt linked to the various slip systems that can be activated for different orientations

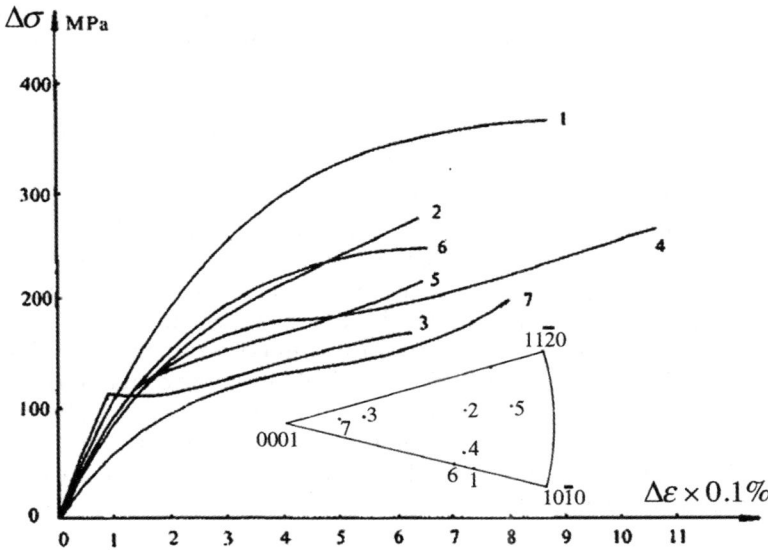

Figure 32 Cyclic stress-strain curves of α-Ti single crystals as a function of orientation of the load axis, given by the standard stereographic triangle in the inset. Source: Image from Gu et al., 1994, courtesy of Elsevier.

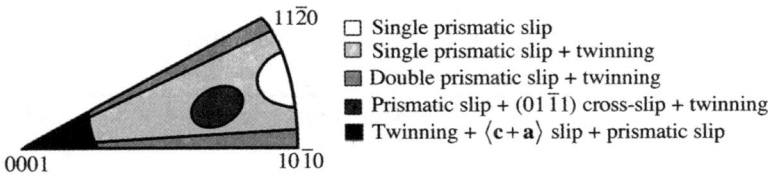

Figure 33 Cyclic deformation mechanisms as a function of orientation in monocrystalline α-Ti. Source: Plotted with data from Tan et al., 1998. Printed with permission from Springer. (For color version of this figure, the reader is referred to the online version of this book.)

and their interaction with twinning, as shown in **Figure 33**. This clearly indicates that behavior of hcp metals can differ significantly from that of the fcc materials, and it deserves some more detailed study.

A clear understanding of the cyclic behavior of single crystals is fundamental to elucidate the cyclic deformation and eventual failure mechanisms of pure polycrystalline materials, since they represent a first approximation to engineering metallic alloys. This important topic will be addressed in the next section.

18.4.9 Cyclic Deformation of Polycrystalline Metals

The understanding of cyclic deformation in polycrystalline metals has lagged that of single crystals because it is much more complicated and, for a given type of behavior, the plastic strains are lower in polycrystals than in single crystals. Therefore, experimental demands are more severe. Hence, early work (e.g. Feltner and Laird, 1967a,b) tended to focus on high strain fatigue behavior (for a review see Laird, 1983). Lukáš and Klesnil (1973) showed the way in greatly improving strain resolution and increasing the fatigue lives for which studies of cyclic deformation could be made. Their results encouraged other workers to run tests in plastic strain control at very small amplitudes, 10^{-5}–10^{-3}. Strain resolution in cyclic deformation has usually lagged that in monotonic deformation because the gauge length must be kept small in fatigue to avoid buckling during compression reversals. Servo-hydraulic test methods brought significant improvements in extensometers and the electronics needed to excite and record from them. In addition, studies of polycrystalline deformation received impetus from results on single crystals and there now exists a large body of results on polycrystals and greatly improved understanding (Lukáš and Kunz, 1994; Holste, 2004; Peralta, 2005; Zhang and Jiang, 2006).

The suggestion that the connection between the cyclic response of mono- and polycrystalline material might be much simpler than that in monotonic deformation was first made when hardening mechanisms in stages I and II/III were identified with low and high strain fatigue respectively (Feltner and Laird, 1967a,b). Bhat and Laird (1978) extended the connection when they claimed, on the basis of Lukáš and Klesnil (1973) data, that the CSSC of polycrystalline copper contained a plateau at low strain amplitudes and that the plateau levels in polycrystals and single crystals could be related by the Taylor factor. These views were initially difficult to sustain because, as new results were gathered at low strain amplitudes, the CSSCs of polycrystalline copper showed great variation. An overview of results gathered in the early 1980s is shown in **Figure 34**.

It is apparent that, for this copper, in strain control, there is no plateau, but a tendency to one is shown over the range of strains to be expected from the CSSC of copper single crystals. Similar behavior has been found by others (see Mughrabi and Wang, 1988 and references therein). Mughrabi and Wang (1988) also noted that the first PSB to be observed at the specimen surface occurred at the plateau stress divided by the optimum Schmid factor, indicating that, at low levels of strain, individual grains were behaving much like isolated crystals. Blochwitz et al. (1996) carried out similar work by studying SB traces in polycrystalline Ni samples using EBSD. They collected statistics on the slip traces that could be

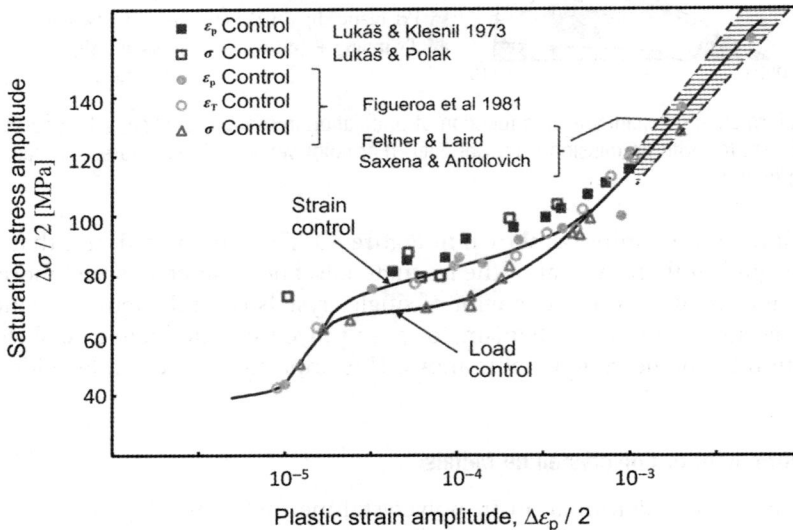

Figure 34 The CSSC of polycrystalline copper at low strains for tests under various control forms, compared with results of Lukáš and Klesnil (1973) at low strain and of other workers at high strain. Source: Replotted with data from Figueroa et al., 1981; Figueroa and Laird, 1981. (For color version of this figure, the reader is referred to the online version of this book.)

explained by the highest resolved shear stress produced by uniaxial stress, and found that this assumption could be used to explain slip in 90% of the grains they surveyed. Despite these indications that grains in conventional polycrystals (grain size a few tens of micrometers) deform similarly to single crystals, reports of the presence of a plateau in the CSSC of polycrystalline fcc metals remained sporadic.

In the early 1980s Rasmussen and Pedersen (1980) claimed a plateau for their polycrystals. As shown in **Figure 34**, a plateau occurs in load control in circumstances where the strains early in life are permitted to be large. Such large strains create cell structures within which there are ample mobile dislocations to carry the strain when saturation is reached. Otherwise, in strain control, the dislocation structures are rather similar to those observed in single crystals except perhaps for a somewhat greater density of secondary dislocations. That is: loop patches are observed at low strains, a mixture of loop patches and dipolar walls are found at intermediate strains and cell structures occur at higher strains, $>5 \times 10^{-4}$ (Figueroa et al., 1981; Figueroa and Laird, 1981). The ladder structure of real PSBs has repeatedly been observed in fcc polycrystals, both at the surface in fcc materials (Lukáš et al., 1966) and in the bulk (Winter et al., 1981; Wang and Laird, 1988; Llanes and Laird, 1992; Morrison and Chopra, 1994), in impure bcc polycrystals (Pohl et al., 1980) both at the surface and in the bulk and in ferritic stainless steels (Petrenec et al., 2006). Note that Mughrabi (2009) has pointed out that the ladder-like structure in bcc materials with low impurities is rare (see Section 18.4.8) and that reasons for its presence are yet to be understood.

With the passage of time and further investigations, the cyclic response of polycrystals seems to be even more complicated. A convincing plateau in the CSSC of polycrystalline nickel has been reported by Morrison and Chopra (1994) and by Buque et al. (1999). Wang and Laird (1988) were able to induce a plateau in the CSSC of polycrystalline copper when they ramp-treated their specimens first, thereby producing a more uniform distribution of loop patches from grain to grain, thereby enhancing PSB formation when the stress/strain were high enough to require their production. However, Llanes et al. (1993) were unable to reproduce this CSSC with another batch of copper although it was supposedly treated in similar fashion with regard to annealing and grain size. A prominent plateau was found in the CSSC of polycrystalline Al-4% Cu alloy containing shearable precipitates, which correlated well with that for single crystals containing the same kind of precipitate, but in this alloy the localized strains are extremely high (Lee and Laird, 1983).

The explanation for the wide range of polycrystalline response observed by the different workers in different metals and even in the same metal originated in systematic efforts to study grain size effects (Llanes et al., 1993). Llanes and coworkers found the CSSC curve of large-grained copper to lie at *higher* stresses (by 20 MPa) than that for smaller grained copper. They attributed this behavior to the occurrence of a strong {111}-{200} texture induced in the copper by a combination of the mechanical pretreatment of the raw material, for example, circular bar, and the high annealing temperatures needed to obtain the large grain size. These multislip orientations caused enhanced cyclic hardening in the large grain specimens as compared with that in the smaller grain specimens in which the texture was not so pronounced. This interpretation was supported by critical experiments in which the same texture was obtained in a slab of copper large enough to permit specimens to be cut in different directions with respect to the axes of the main texture components, so as to vary the dominant crystallographic orientation of the grains parallel to the load without otherwise varying the metal (Peralta et al., 1995b). For choices of specimen directions in which single slip orientations parallel to the load axis were favored (Peralta et al., 1995b, 1999b), the stresses fell considerably and the normally expected hierarchy with respect to the flow stress of differently grain-sized specimens was reestablished. Typical results are shown in **Figure 35**. Note in **Figure 35(a)** the pronounced texture for single slip orientations obtained as a results of the process used to manufacture the samples (cutting specimens at 15° from the

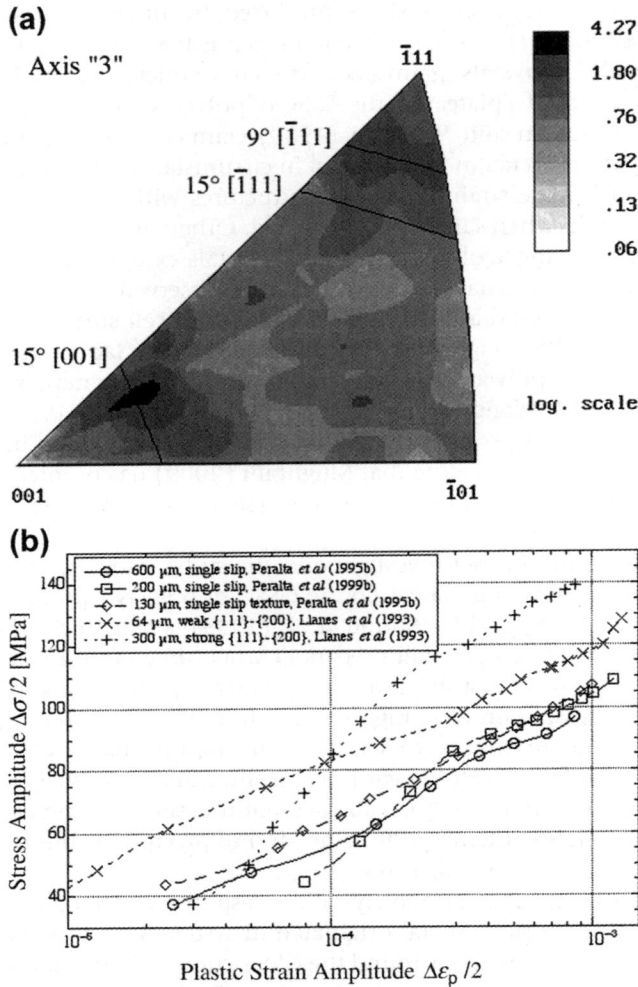

Figure 35 (a) Inverse pole figure for the loading axis of Cu samples heat-treated to have a 200 µm grain size from Peralta et al. (1999b), courtesy of Elsevier. (b) CSSCs (obtained by step-testing) for large, medium and small grained copper, one set plotted with data from Llanes et al. (1993) having variable {111}-{200} texture and the other set plotted with data from Peralta et al. (1995b), Peralta et al. (1999b) having a more "single slip" texture, as shown in (a).

axis of the original rod). Regarding the cyclic response on **Figure 35(b)**, note that all samples with the "single slip" texture had a cyclic response that was much lower than that of the samples with the {111}-{200} texture. In addition, the response of the former samples tends to be almost independent of grain size for strain amplitudes larger that 2×10^{-4}.

Rather similar conclusions can be found in Morrison's work on polycrystalline nickel (Morrison, 1994), in which the texture details vary in a different fashion with respect to grain size. Note that Mughrabi and Wang (1988) had pointed out that grain size does not affect the CSSC of polycrystalline copper significantly in samples with random texture. This was also observed by Zhang and Jiang

(2006), who extended the study of grain size and texture effects on cyclic behavior of polycrystalline copper to multiaxial loading conditions, where variations on the cyclic response with changes on the load axis correlated well with anisotropy expected from the presence of texture.

The presence of a plateau in the CSSC of polycrystals would require a significant volume fraction of PSBs in the samples. Now, given that TBs seem to stimulate PSB formation and secondary slip (**Figure 17**), the study of the stress interaction at TBs as a function of crystallographic orientation appeared interesting. Peralta et al. (1994a) developed a simple model to account for the increase of the stresses adjacent to a TB because of elastic strain compatibility requirements in copper. They showed that grain orientation could have a pronounced effect on the compatibility stresses next to a TB and found that the stress-concentrating effects of a twin reach a maximum when the load axis is parallel to a $\langle 111 \rangle$ direction. That is, the effect of the twin depends on the crystallographic orientation of the grain in which the twin is embedded and, therefore, is a function of texture. Thus the $\{111\}$ textures that produce enhanced cyclic hardening in polycrystals are reinforced in their effect by the presence of twins in metals with medium and low SFE. These effects had been studied by many authors before Peralta et al. (1994a) and have been updated to include other effects. See Peralta et al. (1994a), Sangid et al. (2011) and references therein for more details. A summary is shown in **Figure 36**.

Peralta et al. (1995a) have shown using the elastic strain incompatibility model that this conclusion can also be extrapolated to other types of boundaries. In an example taken from Peralta et al. (1993), they cite a bicrystal containing a twist boundary about the $[\bar{1}49]$ direction, a boundary that could be expected (say) in an fcc polycrystal containing a texture related to single slip orientations, and another bicrystal with a $[110]$ asymmetrical tilt boundary, consonant with a multislip type of texture. The crystallography of these boundaries is shown in **Figure 37**.

Their results showed that the compatibility stresses at the latter can be about four times higher than those in the former case and a significant fraction of the applied stress. The additional stresses found at grain boundaries can in part be responsible for: (1) the presence of more "advanced" dislocation structures adjacent to them, meaning that cells and mazes have been observed close to GBs in grains with veins and PSBs in the bulk, as can be seen in **Figure 38** (Figueroa and Laird, 1981; Llanes, 1992; Llanes and Laird, 1993); and (2) the observation of enhanced cross-slip in polycrystalline metals with low SFE (Thompson, 1972; Li and Laird, 1994b).

Given that local stresses can be very important in fatigue, mostly at low values of the plastic strain (long fatigue life), the correlation between texture, grain size, and the type of grain boundary/misorientation in polycrystals (studied by several authors from a "geometrical" point of view (Pan and Adams, 1994; Garbacz et al. 1995)) must be fully taken into account if a reasonable model of poly-crystalline behavior is to be developed.

Another factor that could play a significant role in polycrystalline deformation is the elastic anisotropy of the metal. Kitagawa et al. (1986) showed, using finite element methods, that the compatibility stresses that appear at grain boundaries for a given misorientation depend on the anisotropy. Peralta and his coworkers (Peralta et al., 1993, 1994b, 1995a, 1994a, 1995b) attempted to estimate (Peralta et al., 1995b) the stresses around the boundaries of bicrystals for several fcc metals as a function of their *anisotropy factor* (AF), defined as follows:

$$AF = \frac{S_{44}}{2(S_{11} - S_{12})}, \tag{6}$$

where S_{ij} are the components of the compliance tensor for a cubic crystal expressed in Voigt (contracted) notation. The values of AF for typical fcc metals and alloys are shown in **Table 2**.

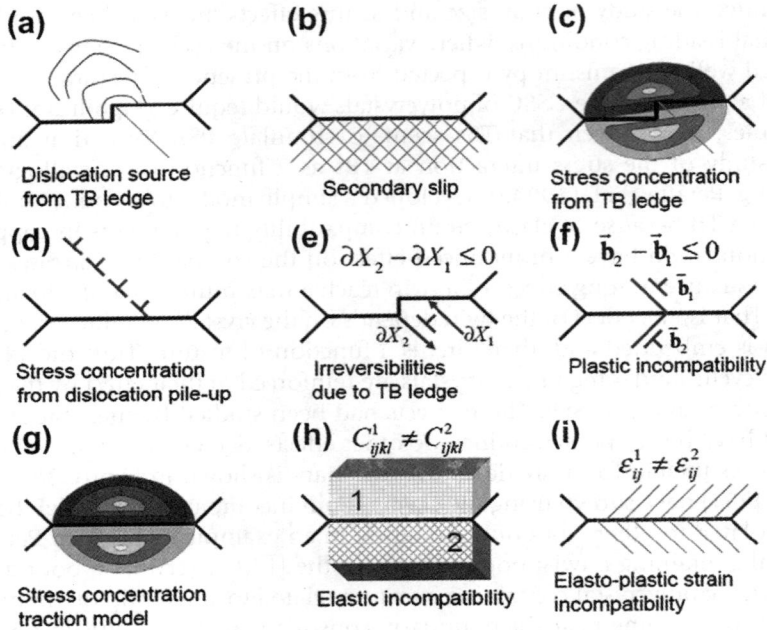

Figure 36 Mechanisms of stress concentration and secondary slip initiation at TBs. (a) Noncoherent ledges and steps act as effective sources for dislocation nucleation (Boettner et al., 1964). (b) Secondary slip is activated because of the stress concentration at the TB (Boettner et al., 1964), which is more prevalent in low SFE materials (Thompson, 1972). (c) Stress concentration because of ledges and steps (Boettner et al., 1964). (d) Dislocation pile-ups at the TBs lead to stress concentration, which is more dominant in larger grains (Thompson, 1972). (e) Ledges and steps at the TB result in irreversible slip (Kim and Laird, 1978a). (f) Plastic incompatibility at the microscale as a result of interaction of incoming and outgoing glissile dislocations with the TB (Lim and Raj, 1985). (g) Stress concentration because of traction at the surface and TBs (Heinz and Neumann, 1990). (h) Stress concentration because of elastic strain incompatibility at the TB (Peralta et al., 1994a). (i) Extension of (h) to include plastic strain incompatibility (Blochwitz and Tirschler, 2005). Source: From Sangid et al., 2011, courtesy of Elsevier. (For color version of this figure, the reader is referred to the online version of this book.)

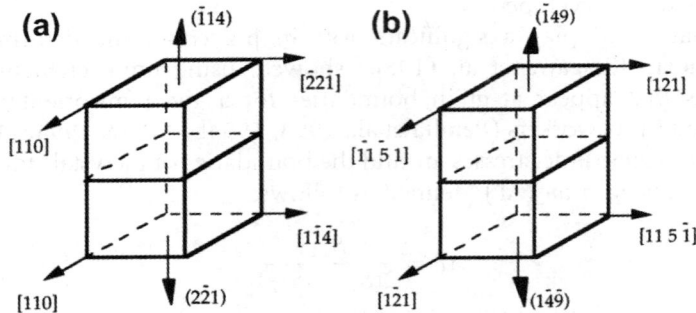

Figure 37 Bicrystal misorientations used in the elastic calculations of Peralta et al. (1993): (a) [110] asymmetrical tilt boundary, and (b) [$\bar{1}$49] twist boundary.

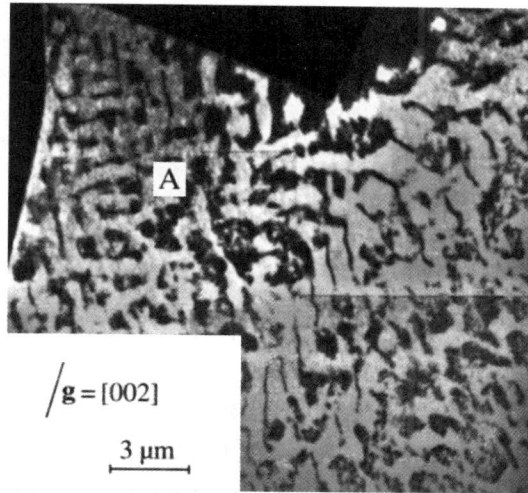

Figure 38 Maze structure (marked A) close to the GB of a copper polycrystal tested under strain control at $\varepsilon_{pl} = 2.7 \times 10^{-4}$, $\mathbf{g} = (002)$. Source: After Llanes and Laird, 1993. Courtesy of Elsevier.

Table 2 Anisotropy factors and stacking fault energies (SFE) for representative fcc metals and copper alloys of wavy and planar slip mode, respectively

Material	AF	SFE [mJ m^{-2}]
Al	1.22	250.0
Ni	2.41	150.0
Au	2.85	50.00
Ag	3.02	20.00
Cu	3.27	42.00
Cu-5 at% Al	3.39	24.00
Cu-10 at% Al	3.51	12.00
Cu-23 at% Zn	3.79	15.00

Source: From Peralta et al., 1995a.

Note from these values and the associated values of SFE that the anisotropy of the metals and the slip mode are somewhat coupled, because there is a tendency for metals with a high AF to have a low SFE. This is interesting since, as noted above, cross-slip and secondary slip in general have been observed to be enhanced near grain boundaries of polycrystalline metals with low SFE (Thompson, 1972; Li and Laird, 1994b).

To investigate the role of anisotropy in polycrystalline cyclic stress–strain response, Peralta et al. (1995b) chose the maximum principal stress in the plane parallel to the boundary as the parameter to characterize the strength of the elastic interactions. The results expressed as a function of the AF and for the pair of archetypal tilt and twist boundaries shown in **Figure 37** are given in **Figure 39**.

The maximum principal stress at the two interfaces increases with the AF, as expected, but the behavior is different for the two types of boundaries. The stress for the twist boundary is low and

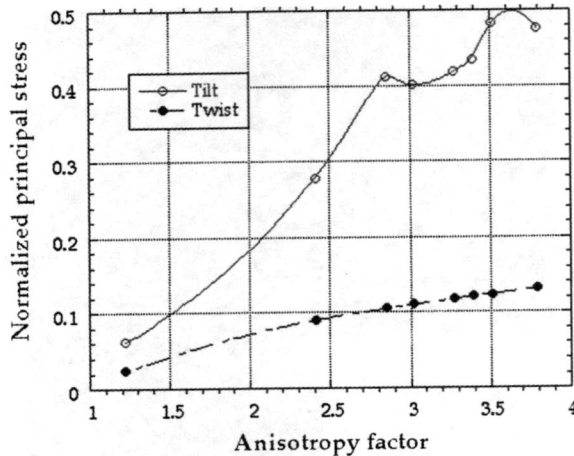

Figure 39 Maximum "in-plane" principal stress for the two boundaries shown in **Figure 37** for the various fcc metals and solid solution alloys listed in **Table 2** and represented here through the specific values of their AF. Source: Replotted with data from Peralta et al., 1995a. (For color version of this figure, the reader is referred to the online version of this book.)

increases monotonically with the AF, whereas the stress for the tilt boundary is larger and shows oscillations. The twist boundary is associated with high index directions, and thus the property contributions of the principal directions become averaged. The tilt boundary is associated with a low index axis and this can make the stresses more sensitive to changes in stiffness with respect to orientation. These results show that the interactions between adjacent grains not only depend on the structure and crystallography of the boundary but also on the basic elastic properties of the material. The results from the archetypal boundaries have been compared with those obtained from actual grain boundary misorientations measured in copper polycrystals by Obergfell et al. (2001) and it was found that the results from the archetypal boundaries compare reasonably well to those from actual grain boundaries in copper.

It was noted above that many convincing plateaux had been observed in the CSSCs of polycrystals for aluminum (alloy) and nickel (Morrison and Chopra, 1994; Buque et al., 1999), the latter in samples with random texture. The discussion provided above makes it easy to speculate that the lower elastic interactions in the polycrystalline structures of these metals (0.06 for Al and 0.28 for Ni vs 0.42 in Cu—see the normalized in-plane principal stress shown in **Figure 39**) favor the presence of plateaus in their CSSCs, whereas the high anisotropy of copper works against localization of deformation. Note that Mughrabi (2001) has pointed out that, although PSBs are observed in polycrystals, a plateau is not generally observed because of the constraints in the bulk of the sample. Nevertheless, these constraints can be lessened under certain conditions. In nickel, the low elastic anisotropy, as discussed above, as well as the high SFE could homogenize the dislocation structures and produce enough PSBs to obtain a plateau. Copper polycrystals, on the other hand, with higher elastic anisotropy and lower SFE, do not show a plateau under conventional testing. Measurements at $\varepsilon_{pl} = 1 \times 10^{-3}$ of a 22% volume fraction of PSBs in polycrystalline nickel versus 11% in a copper polycrystal support this idea (Buque et al. 1999). Obergfell et al. (2001) used ramp loading to promote PSBs in copper samples where the fiber texture had been deviated from the loading axis (see **Figure 35**). Narrow but well-defined plateaux were observed, as shown in **Figure 40**.

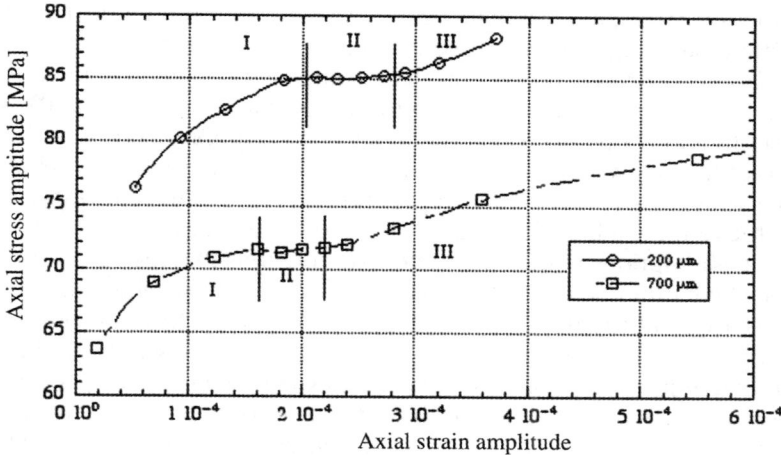

Figure 40 Cyclic stress-strain curves of polycrystalline copper samples after ramp loading showing changing plateau behavior as a function of grain size. Source: After Obergfell et al., 2001, courtesy of Elsevier.

The plateau stress changed as a function of grain size and small variations were reported: 85–92 MPa for 200 μm grains and 71–73 MPa for 700 μm grains and the plateaux were fairly narrow (a linear scale for strain amplitude is needed to see it clearly, as is the case for the plateau found in nickel by Buque et al., 1999). No plateau was detected for a grain size below 150 μm. Reasons for this behavior need to be investigated. Observations using TEM showed a higher density of PSBs in the bulk of copper grains during the narrow plateau regime. Given all these factors, it seems that a reliable model for the cyclic response of polycrystals that is able to account for strain localization at the macroscopic level is still out of reach. Some models used for monotonic deformation have been applied to fatigue (for a comparison between cyclic and monotonic deformation, see Laird, 1996). However, their use presents some difficulties, as discussed next.

Given all the complex interactions that influence the cyclic response of polycrystals, it is not surprising that our understanding of the behavior is rather modest; in particular, a good model to extrapolate the cyclic behavior of polycrystals from that of a single crystal seems to be still out of reach. Some of the models used for monotonic behavior have been applied to fatigue with a moderate degree of success by Lukáš and Kunz (1994), mostly on the conversion of S–N data of fcc polycrystals and the CSSC of pure iron. The fundamental premise of these models is to find a conversion factor between the resolved shear stress τ and the resolved shear strain γ, of individual grains and the stress σ and the axial plastic strain ε of the polycrystal, according to the following formulas:

$$\tau = \frac{\sigma}{\overline{M}} \quad \text{and} \quad \gamma = \varepsilon \overline{M} \tag{7}$$

The conversion factor $\overline{M}$ lies between 2.24 and 3.06 for polycrystals with randomly oriented grains (Kocks, 1970). The lower value of $\overline{M}$ corresponds to a calculation performed assuming that all the grains have the same stress, which is in turn equal to the applied stress. This is known as the Sachs model. The higher value of $\overline{M}$ is obtained by assuming that the strain in all grains is the same and equal to the macroscopic strain. This is known as the Taylor model.

On the Taylor model for rate independent plastic behavior, as it can be safely assumed for copper and nickel in the strain rate range typically used to study cyclic deformation, each grain is assumed to deform by the activation of the five slip systems that minimize the product of the resolved shear stress on each system and its corresponding shear strain (local plastic work) (Kocks, 1970). As a consequence of the elastic and plastic anisotropy of the grains, the isostrain assumption (Taylor) violates equilibrium and the estimations obtained by using it represent an upper bound of the actual behavior. The isostress assumption (Sachs), on the other hand, leads to conditions where strain compatibility among grains is violated, and thus the result is a lower bound of polycrystalline behavior.

Given the implicit assumptions in the two models, it is obvious that the Taylor model should apply better to situations where multiple slip is involved, and Sachs to those where single slip is dominant. High-cycle fatigue is a typical situation where the plastic strains are low and polycrystals deform by single slip processes for a significant range of plastic deformations (see discussion offered above); therefore, Sachs should apply in this case. Nevertheless, experimental evidence seems to indicate that the Taylor model offers a better approximation to the behavior of polycrystals (Wang and Laird, 1988; Lukáš and Kunz, 1994; Buque et al., 1999; Peralta et al., 1999b; Obergfell et al., 2001). Reasons for this are yet to be elucidated. See Peralta (2005) for further comments on the use and limitations of these models as applied to fatigue.

The cyclic behavior of polycrystals of pure magnesium (Yu et al., 2011b) and commercial purity α-Ti (Zhang et al., 1998) has been studied as well, and twinning is found to play a significant role in both materials. Space limitations preclude further discussion here and the reader is referred to these articles (and references therein) for more details.

It is clear from the brief review on polycrystalline behavior offered above that understanding of both single crystal behavior, which describes what happens inside the grains, and interactions between these crystals at the grain boundaries, are necessary to understand cyclic deformation of polycrystals. In that regard, studies in samples with an individual boundary (bicrystals) are of interest and will be discussed below.

18.4.10   Cyclic Deformation of Bicrystals

The introduction of a grain boundary through the use of bicrystal specimens is an appropriate first step to build up a fundamental understanding of the transition in cyclic behavior from single crystals to polycrystals. This approach has been indeed undertaken by many investigators, for example, see Peralta and Laird (1997) and references therein for some of the early work and Zhang and Wang (2008) for a review of more recent work. Most of the bicrystal experiments described in these references were carried out in such a way that the tensile axis is either contained in the boundary plane or is perpendicular to it, with some work also done in grain boundaries inclined with respect to the load axis (Zhang and Wang, 2008). The processes of cyclic hardening and saturation which precedes the nucleation and growth of the fatigue crack itself can be affected by all types of boundary and these processes usually determine the final outcome of failure.

Peralta and Laird (1997) studied bicrystals with the grain boundary perpendicular to the applied load using two crystallographic configurations: isoaxial bicrystals (where the crystallographic axis parallel to the load is the same for both grains) and nonisoaxial bicrystals. These authors had 90° and 180° twist boundaries in the isoaxial bicrystals, that is, where the orientation of one grain can be obtained from the other by a rotation of either 90° or 180° about the common crystallographic axis perpendicular to the grain boundary, in this case a $[\bar{1}49]$. This choice of load axis leads to a Schmid factor ≈ 0.5 for both grains, which implies strong single slip conditions and allowed a direct

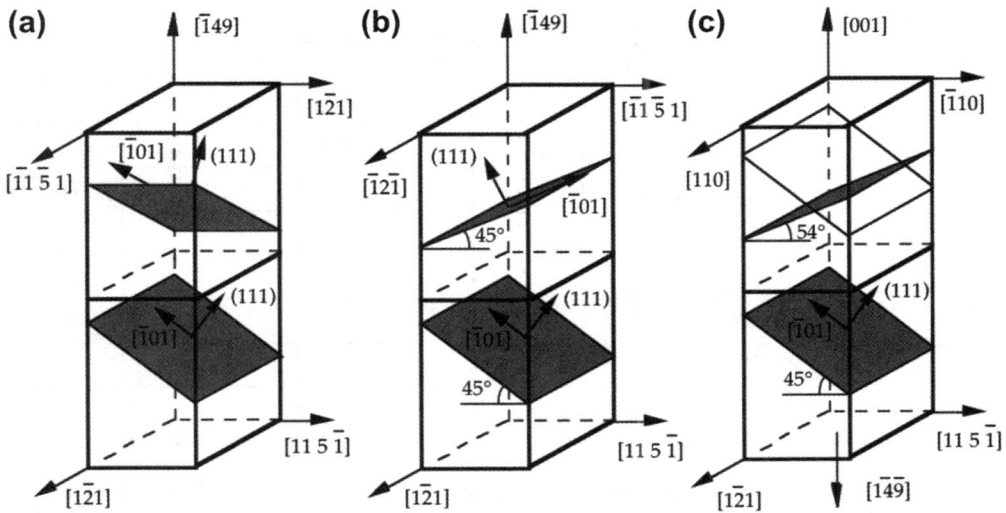

Figure 41 Crystallography and primary slip geometry of bicrystals used by Peralta and Laird (1997). (a) 90° twist boundary; (b) 180° twist boundary; (c) $[\bar{1}49]/[001]$ (HS) misorientation. Source: Courtesy of Elsevier.

comparison with results on copper single crystals oriented for single slip, since they tested their samples under strain control. The also used a "hard–soft" (HS) bicrystal configuration, where one grain had a load axis parallel to [001] ("hard" cyclic response because of multiplicity of slip) and the other was parallel to $[\bar{1}49]$ ("soft" cyclic response, because of single slip), leading to a nonisoaxial sample. The crystallography of these bicrystals is shown in **Figure 41**.

The design of these bicrystals allowed Peralta and Laird (1997) to study the effect of the presence of the boundary on the cyclic behavior, in terms of the compatibility of both elastic and plastic deformation as well as the way plastic strain is distributed between grains with different cyclic hardening rates, as was the case of the HS configuration. Examples of hardening curves for all three configurations for the same plastic strain amplitude within the plateau and the saturation states as compared with copper single crystal data are shown in **Figure 42**.

The results in **Figure 42(a)** indicate that the hardening behavior of the 180° twist and the HS bicrystals is very similar to that of single crystals: rapid hardening, followed by overshooting to nucleate PSBs and finally a softening to the saturation stress. The "overshooting" of the stress and the softening that follows are because of the breakdown of loop patches to form PSBs (see **Figure 15**). The fact that this happened on the sample with the 180° twist boundary is not surprising, since it is elastically and plastically compatible at the macroscopic level. The behavior of the HS sample indicates that the grain with the single slip orientation dominated. The sample with the 90° twist boundary was the hardest, and reached a saturation stress that was higher than that of a single crystal with the same load axis.

The results in **Figure 42(b)** indicate that the samples with 180° twist boundary behaved almost like single crystals, although the saturation stress was slightly higher for all three values of plastic strain amplitude used. In the plateau regime, the 90° twist sample was harder than the HS sample, which in turn was harder than the 180° twist.

All samples showed some degree of secondary slip at the grain boundary region for strain amplitudes within the plateau, forming a so-called grain boundary affected zone (GBAZ), and this correlates well

(a)

(b)

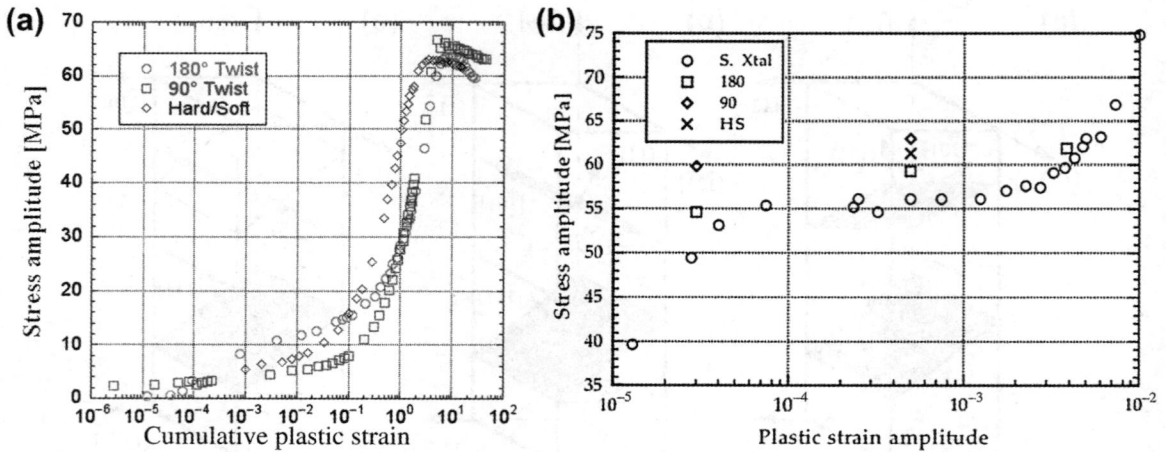

Figure 42 (a) Cyclic hardening curve for the bicrystals configurations tested by Peralta and Laird (1997) for $\Delta\varepsilon_p/2 = 5 \times 10^{-4}$. Plotted with their data. (b) Saturation states of the bicrystals as a function of strain amplitude compared with the CSSC of a copper single crystal (note that axial stress amplitudes are used). Source: From Peralta and Laird, 1997, courtesy of Elsevier. (For color version of this figure, the reader is referred to the online version of this book.)

with the fact that all of them had slightly higher saturation stresses than the single crystals. In particular, samples with the 180° twist misorientation developed secondary slip at the intersection of PSBs with the boundary. This was likely the result of microscopic, rather than macroscopic, incompatibilities (Peralta and Laird, 1997), i.e., slip on the primary slip plane could not be transmitted across the boundary, since the primary slip planes in the two grains were not parallel (see **Figure 41**b). The HS sample showed secondary slip as well and a GBAZ. The slip behavior of these samples was somewhat similar to that observed by Vehoff et al. (1987) in a "HS" (<111>-<123>) nickel bicrystal. They found thick PSBs in the lower (soft) grain, and only a few SBs in the upper (hard) grain. They also reported additional slip close to the grain boundary for that sample.

The specimen with the 90° twist boundary had the largest GBAZ, which also explains the fact that it had the hardest response. The GBAZ for this sample is shown in **Figure 43**.

Note in **Figure 43** that at least three slip traces are visible in the lower grain and two are visible in the upper one and the GBAZ represents at least 16% of the gauge length of the sample. The extra slip results in additional hardening at the GBAZ, which can lead to lower levels of plastic strain at the GBAZ for a given stress level. Peralta and Laird (1997) argued that plastic strain was then transferred to the bulk of the grains, leading to higher overall stresses. These results show that there can be a significant difference in the behavior of isoaxial bicrystals with the boundary perpendicular to the tensile axis as the misorientation between the grains is changed.

Regarding the HS bicrystal, one of the interesting aspects in the behavior of this sample is the fact that the two grains have quite different tensile axes, and as individual single crystals they can show very different cyclic hardening behaviors. This raises the question of how they behave when they are put together to form a bicrystal. Peralta and Laird (1997) placed strain gages to measure the evolution of plastic strain in each and compared it with the overall behavior as measured by an extensometer that included both grains.

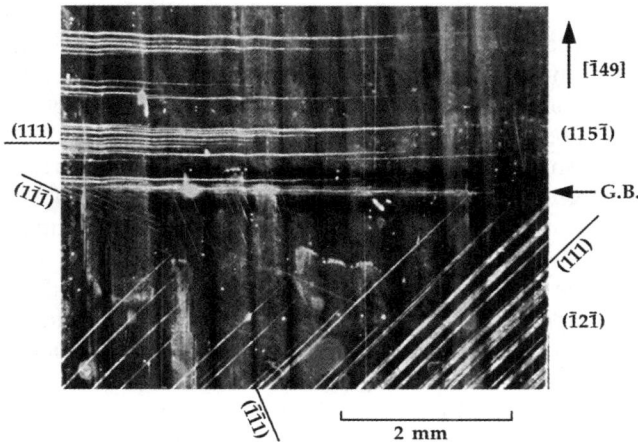

Figure 43 Slip patterns produced at $\Delta\varepsilon_p/2 = 5 \times 10^{-4}$ at the boundary of a bicrystal with a 90° twist boundary. The primary slip plane, that is, the one with the highest Schmid factor, is (111) for both grains. Source: After Peralta and Laird, 1997, courtesy of Elsevier. Load axis is vertical.

The evolution of plastic strain for each grain and the overall bicrystal versus the number of cycles is shown in **Figure 44**.

The soft grain dominated the deformation at first, as expected, until its plastic strain reached a maximum value of approximately 9×10^{-4}, at about 140 cycles, when the overall plastic strain reached the nominal value for the test (5×10^{-4}). Note that the hard grain had started to deform plastically as well, but the plastic strain carried by it went to a minimum when the soft grain was at its maximum. From then on, the plastic strain in the soft grain started to decrease, only to be compensated

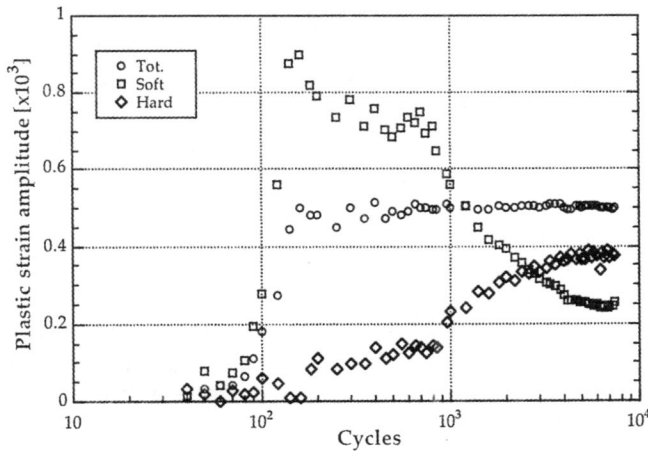

Figure 44 Plastic strain amplitude for sample HS and its component single crystals as a function of the number of cycles. Source: After Peralta and Laird, 1997, courtesy of Elsevier.

by an increase in the plastic strain carried by the hard grain. It is likely that the high value of the plastic strain in the soft grain induced a level of hardening high enough to compete with the enhanced hardening produced by the multiple slip orientation in the hard grain, and then the plastic strain could start to be "redistributed" according to the individual cyclic hardening levels in each grain. Peralta and Laird (1997) indicated that the process of strain localization likely affected the measurement after the soft grain reached a plastic strain amplitude of 7×10^{-4}, and that the results beyond that are not completely reliable. Nonetheless, note that both grains had reached saturation and it is likely that this is representative of the steady state, where the soft grain carried about five times the plastic strain of the hard grain.

The results discussed above clearly indicate that the issue of compatibility, at both the elastic and plastic levels, plays a very important role in the cyclic behavior of bicrystals. In addition, differences in cyclic hardening rates for adjacent crystals can also affect the overall behavior significantly. These issues need to be accounted for when trying to understand the cyclic behavior of polycrystalline materials and they have been studied over a wide variety of bicrystals configurations (boundaries parallel, inclined and perpendicular to the load) and several orientations of the individual grains by Zhang and Wang (2008), who looked into how to predict the CSSC of bicrystals given knowledge of the single crystal behavior and have reviewed the models to account for the effects of the GBAZ on the CSSC. The reader is referred to that extensive review for more information on this topic. Copper tricrystals have also been studied (Jia et al., 1998, 1999). Those readers interested in the cyclic behavior of bicrystals of hcp metals are referred to the work by Tan et al. (1995), Xiaoli et al. (1995) on α-Ti bicrystals.

18.4.11   Cyclic Deformation in Alloys

As noted from the sections above, a wide range of phenomena in the fatigue of fcc metals is now understood, at least qualitatively. Studies of substitutional alloy behavior that involve single crystals have not been as common, but reports of sustained attempts to understand such alloys are available. For substitutional alloys of low concentration of solute (Woods, 1973; Wilhem and Everwin, 1980), the behavior has been found rather similar to that of pure metals, even to such details as the value of the plateau stress in the CSSC, and the nature of the dislocation structures.

When the solute concentration is significantly greater, so that the alloy behaves with a planar slip mode (for discrimination of wavy and planar slip modes, see Hong and Laird,1990e), the dislocation structure of the alloy is considerably different from that of a wavy metal (Lukáš and Klesnil, 1970, 1971; Youssef, 1970; Buchinger et al., 1986; Li et al., 2011). The typical ladder structure and loop patches that are produced in the cyclic deformation of pure metals have not been observed in planar slip metals at room temperature.

One of the controversies concerning the cyclic deformation behavior of planar slip alloys was whether or not strain localization exists. Yan et al. (1986), employing the same interferometric method as previously used on copper, observed strain localization in Cu-16 at% Al, contrary to previous reports (Lukáš and Klesnil, 1970, 1971; Youssef, 1970); they also observed the existence of a plateau in the cyclic stress–strain response. However, the alloy they used hardens sluggishly, exhibits burst phenomena, and shows strain-aging behavior to a marked degree. It is possible that saturation was not attained in their experiments. Interferometric methods involve tests interruptions and are very time-consuming because the specimen preparation technique is demanding. The strain-aging phenomena that occurred during such lengthy delays could and probably did affect the subsequent straining behavior when attempts were made to measure localized strain. The interactions of all these complex phenomena have since been understood through more extensive investigations (Abel et al., 1979; Hong and Laird 1990d,c,a, 1991b,c; Hong et al., 1992; Gong et al., 1999; Wang et al., 1999; Wu et al., 2001; Li et al., 2011) and a fairly clear picture of the behavior has emerged:

In a typical planar-slip alloy, such as Cu-16 at% Al, studied by Hong and his coworkers, as well as (Wu et al., 2001), the friction stress acting on the dislocations by reason of the alloy content is very high and the application of a stress of about 19 MPa is needed simply to produce yielding (Hong and Laird, 1990a). The slip is confined to bands and persists in them for a while, as the metal gradually hardens. This hardening causes the localized strain to decrease gradually and eventually disappear, and new SBs appear elsewhere, generally accompanied by a strain burst. These bursts can readily be seen in the shapes of the hysteresis loops and in load-cycle recordings for which the specimen is cycled in strain control. The typical cyclic stress–strain response during hardening for a Cu-7 at% Al, while burst behavior is occurring, is shown in **Figure 45**.

At low amplitudes the hardening rate is extremely small and bursts are few and far between. Slip is localized in the active bands for a long time. As the amplitude increases, the bursts become more numerous and the hardening rate increases. At the highest amplitude, the bursts are so numerous that they are no longer distinguishable. Thus the slip gradually percolates throughout the gauge length until bands occupy the whole crystal. This temporary persistence of the slip gave rise to the term *"persistent Lüders bands"* (PLBs) to describe the regions of currently active slip (Buchinger et al., 1986). Bands that were previously active and had become quiescent can be reactivated if the other volumes of the specimen have undergone slipping, temporary persistence, gradual hardening, and subsequent quiescence.

The shapes of stress–strain hysteresis loops during hardening fluctuate between round-topped and sharply pointed, the latter occurring while the PLBs are settled, the former after a burst has occurred. This behavior is connected with effects of solute and its segregation. In freshly deformed areas,

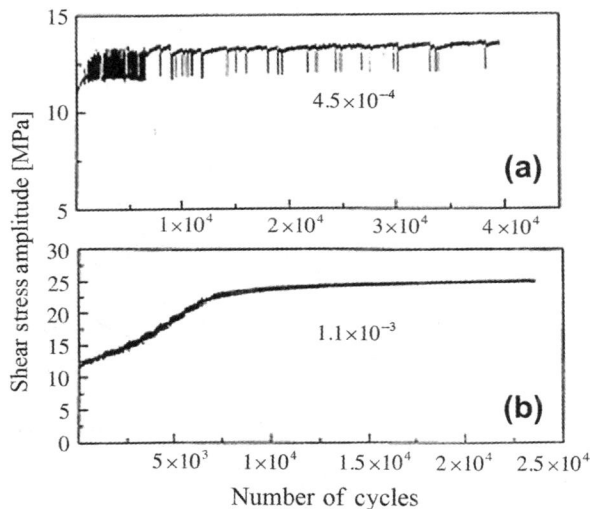

Figure 45 Strain bursts in a [$\bar{1}$12] Cu-7 at% Al crystal under various plastic shear strain amplitudes. (a) 4.5×10^{-4} and (b) 1.1×10^{-3}. Source: After Wu et al., 2001, courtesy of Elsevier. Note that the yield strength (11–12 MPa) is lower than in Cu-16 at% Al (19 MPa), as expected.

dislocations can be pinned during the quiet moments of a cycle. Once they break free, their reduced friction stress allows them to run with limited resistance by back stresses until the limit of the strain cycle causes them to stop. After hardening and dislocation accumulation have developed in an active band, the back stresses reassert their values and the pointed loop shape returns.

It is interesting to find that, in single crystals of Cu-16 at% Al, saturation behavior is not observed and earlier claims to a CSSC containing a plateau must be attributed to overoptimism on the part of the investigators that hardening had been exhausted. The hardening behavior is therefore presented in the form of cyclic hardening curves—see **Figure 46**.

At low amplitudes, the tests were interrupted without reaching saturation or failure. At higher amplitudes, hardening also never seemed to stop but cracks formed and grew instead (the final "drop-offs" in the curves were caused by crack propagation). Similar results were obtained by Wu et al. (2001) on the same alloy. Note in **Figure 46** that the initial flow stress, representing the solute friction contribution, has been subtracted off so that actually the stresses are much higher (except at low amplitudes) than those observed in copper (see Hong and Laird, 1990b for a treatment of the friction and back stress behavior in Cu-16 at% Al alloy). Consequently cracks grow very rapidly as soon as they form. Studies on other Cu solid solutions, for example, Cu-30 wt% Zn by Gong et al. (1999); Wang et al. (1999) have confirmed that saturation does not occur in single crystals of these alloys oriented for single slip for plastic shear strain amplitudes above 6×10^{-4} (Hong and Laird, 1990d; Wang et al., 1999). However, saturation has been reported for Cu-30 wt% Zn for plastic shear strains below 3×10^{-4} (Wang et al., 1999) and for Cu-7 at% Al for all plastic strain amplitudes (Wu et al., 2001, see **Figure 45**). The behavior can be explained in terms of the dislocation structures, which are quite different from those observed in wavy slip materials. At low strains, $\Delta\gamma_{pl}/2 < 3 \times 10^{-4}$, single crystals of Cu-30 wt% Zn have structures that resemble those observed during monotonic deformation of the same material. They consist of various kinds of dislocation multipoles, that is, two rows of dislocations with opposite signs. It is also interesting that secondary slip is also active. Examples of the different dislocation multipoles as well as secondary SBs are shown in **Figure 47**. Note the 2D dislocation arrangement.

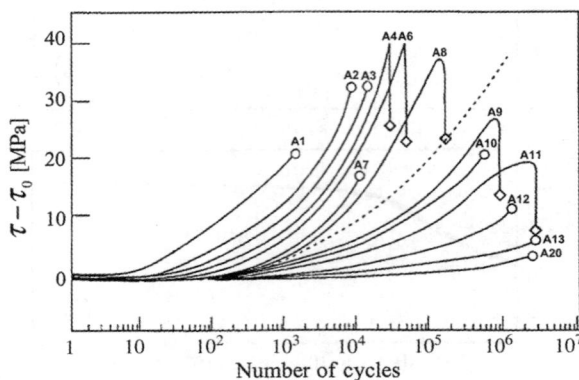

Figure 46 Cyclic hardening behavior of Cu-16 at% Al single crystals, oriented for single slip, cycled under strain control at various strain amplitudes. Here, τ is the shear stress and $\tau_0 = 19$ MPa is the critical resolved shear stress. Plots are smoothened from the effects of strain bursts to represent average hardening behavior. Source: Taken from Hong and Laird, 1990b, courtesy of Elsevier.

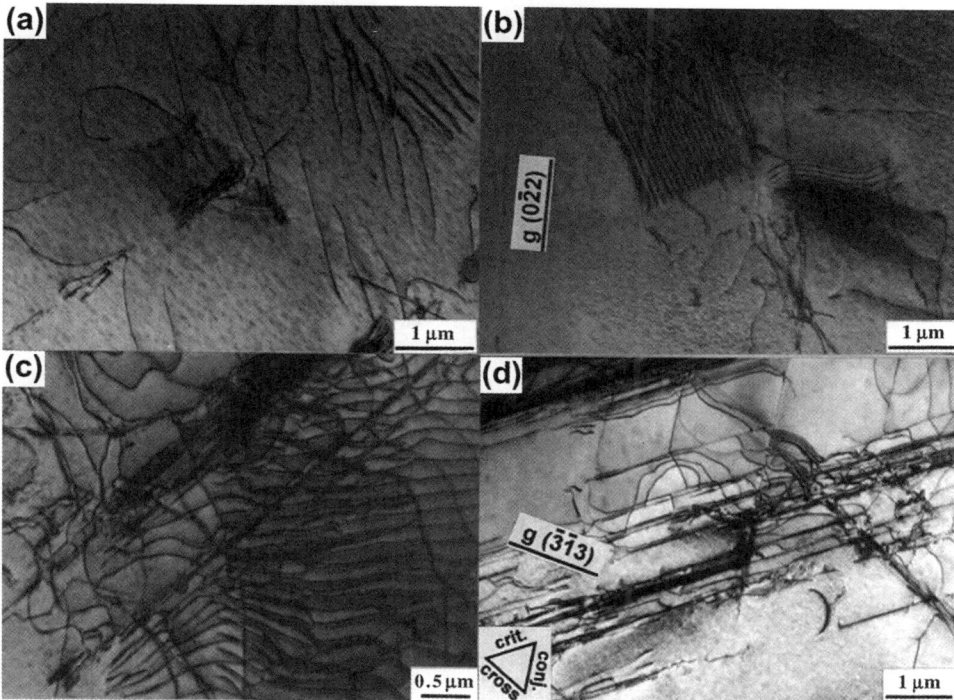

Figure 47 Dislocation structure in the (111) slip plane of a Cu-30 wt% Zn single crystal tested at $\Delta\gamma_{pl}/2 = 1.5 \times 10^{-4}$. (a) Closely spaced multipoles; (b) regular multipoles; (c) cross-gridded multipoles; (d) secondary SBs. Source: After Gong et al., 1999. Courtesy of Elsevier.

The presence of secondary slip at the low stress reported, 18 MPa (Wang et al., 1999), can be attributed to the low stress needed to generate secondary dislocations via Frank-Read sources, as compared with the cyclic flow stress, and a weak latent hardening effect. However, the presence of solute leads to a high friction stress that restricts the motion of these dislocations; therefore, they do not play a significant role at low stresses (Gong et al., 1999). Saturation behavior has been attributed to mechanically reversible strain produced by the flipping of dislocation multipoles. Gong et al. (1999) introduced a model based on this mechanism and found that multipoles become unstable if their internal stresses are higher than the friction stress. They calculated that the maximum stress at which saturation could be attained is 24 MPa, about twice the friction stress for Cu-30 wt% Zn. This agrees well with the stress amplitude at the last strain level where saturation was observed. At higher strains, where there is no saturation, dislocation arrangements are constantly evolving, and they need to be related to both the plastic strain and the stress amplitudes at which the test is stopped.

Dislocation arrangements for three levels of stress in single crystals of Cu-16 at% Al tested at $\Delta\gamma_{pl}/2 > 9 \times 10^{-4}$ are shown in **Figure 48** (Inui et al., 1990). Rafts of dipoles and multipoles are formed both from perfect and partial dislocations (see Buchinger et al., 1986, for examples). These rafts are organized as well-defined 2D bands at low stresses (**Figure 48(a)**) with well-paired dislocations. As hardening proceeds, the stress increases and so does the dislocation density (**Figures 48(b) and 48(c)**), with primary dislocations as the main component.

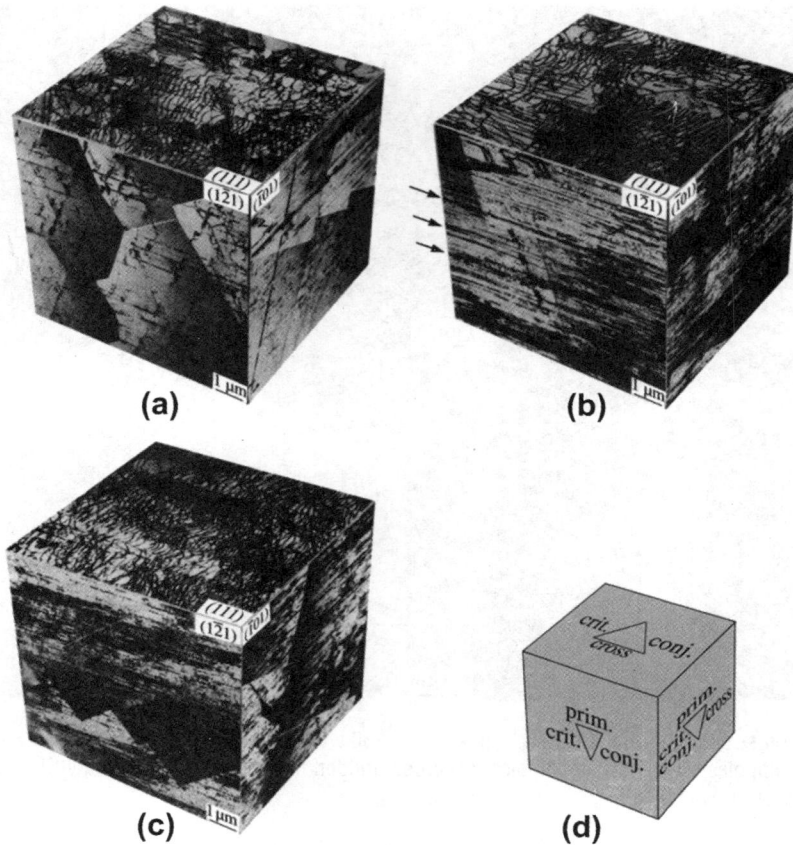

Figure 48 Three-dimensional views of dislocation structures in a planar slip, Cu-16 at% Al single crystal cyclically hardened to: (a) 28 MPa, (b) 34 MPa and (c) 41 MPa. Slip traces for each plane of observation are given in (d). Arrows in (b) show PLBs. Source: Taken from Inui et al., 1990. Courtesy of Elsevier.

Secondary slip is active from early stages of cycling and becomes progressively more so as hardening proceeds. In the later stages of hardening, the back stress begins to increase abruptly, presumably because formation of Lomer-Cottrell locks accelerates and there is increasing evidence of the formation of small circular prismatic loops probably associated with point defects produced by intersecting slip. The structures formed in these planar slip alloys of low SFE amount to large, irregular Taylor lattices, in which there are large variations in the extent of localized strain, both spatially and temporally.

Because of the planar slip nature of the alloy and the difficulty of cross-slip, the dislocation agglomeration that occurs in wavy metals cannot occur. The Taylor lattice is therefore extended in three dimensions and since this is the main dislocation structure, the flow stress can be explained by the passing stress of the oppositely signed edge dislocations in the multipoles:

$$\tau_p = \frac{Gb}{8\pi(1-\nu)d},$$

(8)

(a)   **(b)**

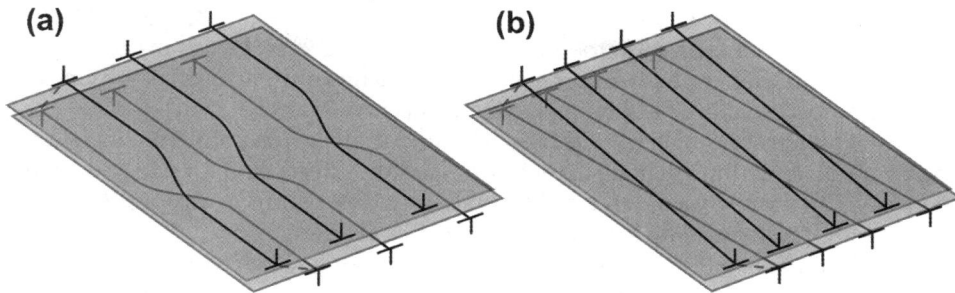

Figure 49 Schematic representations of multipole structures on the primary slip plane of planar slip alloys: (a) kinked dipoles and (b) cross-grid multipoles. Source: As presented by Peralta, 2005 from a figure redrawn from Hong and Laird (1990b).

where G is the shear modulus, b is the length of the Burgers vector, v is Poisson's ratio and d is the separation of the slip planes. Hardening is accomplished by the gradual reduction in d as shown in **Figure 48**. In wavy-slip metals the friction and back stresses are coupled via the behavior of the loop patches (Kuhlmann-Wilsdorf, 1979a,b). In Cu-16 at% Al, the friction stress, dominated by solute effects, remains fairly constant and hardening is mainly accomplished by an increase in the back stress, which is to be expected in multipolar structures (Hong and Laird, 1990a). The question as to why the friction stress does not increase with cycling in spite of the multipoles is answered by assuming that the geometry of dipoles in the multipole can be modeled as shown in **Figure 49(a)**. That is, the dipoles are kinked, in the sense that the relative orientation of the dislocations is changed at some intermultipole point.

The slipping stress (and thus the friction stress) would be greatly reduced for a dipole with the kinked configuration, and the zipping–unzipping of the dipole seems a more appropriate term in describing the motion of the dipole. Also, zipping would be much easier than flipping of the whole dipole. With application of a stress that is lower than that needed to separate the dipole, the cross-over point will move to-and-fro depending on the sign of the applied stress.

Copper cannot behave in such a fashion. If the dipolar partners in a loop patch of pure copper were kinked as in **Figure 49**, cross-slip would easily occur at the cross-over point because of the high SFE and the small spacing between the partners. Thus the dipole would become subdivided and the kinks would not be stable. In copper, therefore, the equality between the back and the friction stresses holds for the deformation of the Taylor lattice.

Another common type of structure observed in planar slip alloy is the cross-grid type of multipole shown in **Figure 49(b)**. Although this type of multipole is less stable than the normal (unkinked) multipole, it has the advantage of easily accommodating large cyclic strain, and they are more frequently observed in the heavily deformed areas (Hong and Laird, 1990b). In less heavily deformed areas, such a multipole could be decomposed with the application of a stress large enough to overcome the sum of the friction stress because of the solute and the friction stress for zipping–unzipping. As such it would not contribute much to the hardening, unless it was constrained by obstacles such as Lomer-Cottrell locks or other dislocation structures. Under such circumstances they could contribute to the hardening by their elastic interaction (the back stress), which is unaffected by the cross-grid geometry. For a more detailed discussion of the hardening in such alloys, see Hong and Laird (1990b), Gong et al. (1999), Wu et al. (2001).

Single crystals of American Iron and Steel Institute (AISI) 316L stainless steel, also fcc in structure, and also planar in its slip mode in spite of having a higher SFE than Cu-16 at% Al alloy (20 mJ m^{-2}

vs < 5 mJ m^{-2}), behave rather similarly in their cyclic response (Li and Laird, 1994a) except that stainless steel manages to saturate its hardening and, on prolonged cycling, transforms its dislocation structure gradually from planar arrays into ragged loop patches quite like those observed in copper. Such transformation occurs more rapidly in polycrystals (Li and Laird, 1994b).

Just as studies of planar slip single crystals lagged behind those of wavy slip pure metals, much less is also known about polycrystalline planar slip behavior, and the effects of grain size and texture. There is a strong tendency for monocrystals and polycrystals to show common dislocation structures and mechanisms, and the results may be briefly summarized as follows:

In early work, Thompson and Backofen (1971) studied the fatigue behavior of polycrystalline 70-30 brass as a function of grain size, presenting their results in terms of fatigue life, which limited the information about cyclic hardening and saturation behavior that could be extracted from them. They found that the fatigue life of brass depended on grain size within a certain range, below which the cycles to failure were insensitive to grain size. This was interpreted as a result of the planarity of slip and the effect that the longer slip length of large grains has on the stress concentrations at grain boundaries. Such an interpretation involves "pile-ups" that are not favored in cyclic deformation where dislocation pairing limits long range stress fields. Moreover, Li and Laird (1994b) have found wavy-type dislocation structures in near-boundary regions of stainless steel. Laird et al. (1986) and Feltner and Laird (1967a,b) studied polycrystalline Cu-16 at% Al, and Lukáš and Klesnil (1973) worked on Cu–Zn. Li and Laird (1994b) and Polák et al. (1994) studied polycrystalline 316L stainless steel, but all these studies, with the exception of the last one, were limited to one grain size and even though texture was sometimes mentioned in their analyses, it was not properly characterized as to derive proper conclusions about its effect in this kind of material. The works of Li and Laird (1994b) and Polák et al. (1994) show yet another contradiction in the results related to strain localization, since the former workers did not find a plateau in their polycrystalline samples, even though they used ramp loading, whereas the latter group found a plateau for the two grain sizes they used, without any mechanical pretreatment. Furthermore, the strain range of the plateau they found seemed to depend on grain size. A texture effect could be responsible for the absence of the plateau in the CSSCs reported by Li and Laird and there might be differences of interstitial content as well.

These results indicate, however, that saturation does occur in polycrystals, that hardening occurs more rapidly than in single crystals (although unevenly as well, because of strain bursts (Laird et al., 1986; Yan et al., 1986)), probably because of multislip effects stimulated by adjacent grains, and that strain localization also occurs. Thus PLBs do eventually transform into PSBs. More research to determine the effects and interactions of texture and grain size in planar slip alloys is needed. It is likely that a similar synergism between the variables will be present in these materials, and perhaps to a stronger degree than in wavy slip metals.

Aged alloys containing small precipitates that are so dense and closely spaced that they must be sheared by mobile dislocations are very susceptible to localized deformation in PSBs, although the PSBs are structurally different from those in pure metals. Rapid hardening is considerable as tests are initiated, the flow stress reaches a peak, and softening then sets in. Since saturation does not occur, there is no generally accepted definition of flow stress for the CSSC. Lee and Laird (1982b,a) chose to define the CSSC in terms of the peak stress, and it is this curve that is shown in **Figure 50**. Since the peak marks the onset of strain localization and the first appearance of PSBs, it is not surprising that a plateau is observed in the CSSC. However, for this system, the onset of the plateau occurs at 1.8×10^{-5}, and the high strain end occurs at $\sim 2 \times 10^{-3}$. Both of these strains are much lower than those of pure metals, because the localized strains in the PSBs are greater.

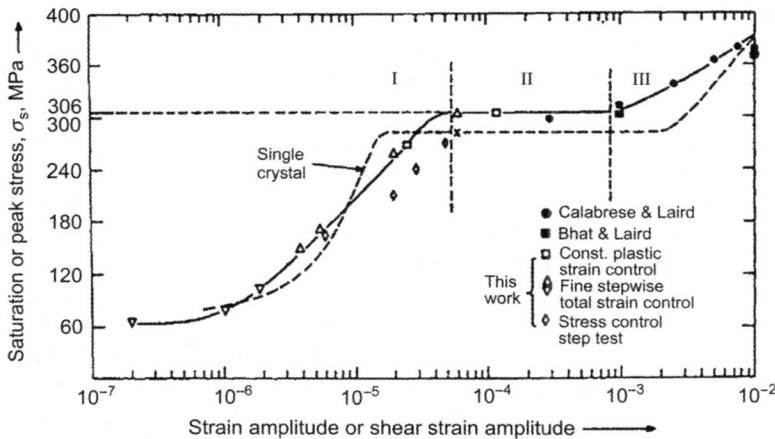

Figure 50 The cyclic stress-strain curve for polycrystalline Al-4% Cu alloy containing shearable precipitates (θ'') compared with that for the same alloy in monocrystalline form. Note that the ordinate scale should be divided by ~3 to apply to the shear stress of the monocrystal; the abscissa applies to both forms of material. Source: Taken from Laird, 1996 from an original figure by Lee, 1980. Courtesy of Lee.

During rapid hardening, the slip was observed to be fine and homogeneous (Lee and Laird, 1983). For strains below those of the plateau, saturation behavior occurred and softening was not observed; rather the slip remained homogeneous for hundreds and thousands of cycles and a fatigue limit was defined by the lower end of the plateau. As usual, PSBs occurred for strains corresponding to the plateau. Interferometric techniques were used to measure the heights of slip offsets at the PSBs and thus to measure the localized strain (Lee and Laird, 1983). It was found to be very large, varying from 0.3 to 0.6. As softening developed, the strain became more localized. Initially the most numerous PSB offsets were about 30 nm high with a few groups of micro-PSBs being closely clumped and showing a combined height of ~300 nm. With continued cycling the extremes of the PSB offsets did not change, but the distribution of offsets became more uniform (Lee, 1980) and progressed toward higher values. Observations using TEM indicated that the PSBs are quite narrow, typically 0.25 μm thick, and densely occupied by dislocations and precipitates. The softening occurred only after the PSBs formed and was attributed to degradation of the precipitates by disordering. This "disordering" can occur by a number of mechanisms. Since the shearable θ'' precipitates in Al–Cu alloy are ordered, the scrambling of the crystals by repeated cutting events could eliminate the component of the order hardening, consistent with the kinetics of softening (Lee et al., 1981). Also, the roughening of the precipitate–matrix interface could reduce the chemical hardening contribution to the overall flow stress. It is also likely that some of the precipitates were chopped up to the point of dissolution and, locally, the material reverted to the solutionized condition.

It is interesting that the volume fractions of PSBs are always very low in the Al–Cu alloy within the plateau. At strains greater than those of the plateau, deformation continues to be localized in PSBs but they develop on more than one slip system (Lee and Laird, 1982b,a). Intense kink bands were also observed. The PSBs really constitute micro-PSBs, and like those of pure metals, they are limited in length—typically 400 μm. Nevertheless the PSBs pass right through the crystal as a string of overlapping short segments (Lee, 1980).

The dislocations in the PSBs have roughly equal densities of edge and screw dislocations (Lee and Laird, 1982b,a). In Al–Cu they tend to be somewhat denser at the interface between the PSB and matrix because the PSBs are slightly misoriented with respect to the matrix. There is a slight tendency for dislocations to clump, but more often they are uniformly arranged. Apparently the highly localized strains can be accommodated by motion of many of the dislocations and by a balance between their multiplication and annihilation.

Although the broad aspects of cyclic deformation are similar in different alloys containing shearable precipitates, the details are often quite dissimilar. For example, Gerold, Wilhelm and their coworkers (Wilhem et al., 1979; Vogel et al., 1980, 1982; Wilhem and Everwin, 1980; Wilhem, 1981) have carried out an elegant study of Al–Zn–Mg alloy containing shearable precipitates, η', and find many aspects of mechanical response similar to those described above. However, in the PSBs of Al–Zn–Mg, the precipitates dissolve and leave the cleared channel supersaturated in alloy elements, explaining the softening that occurs on extended cycling. Even within a single alloy system, a wide variety of behavior can be observed. For example, in Al–Ag of high solute concentration and activity, Al-15% Ag, a structure initially containing GP zones, which one would expect to be cut up and dissolved, did not behave in such a manner. Rather the alloy underwent cyclic strain-induced formation of more stable precipitates, γ' and γ, which cause enhanced strengthening rather than softening (Laird et al., 1978, 1981). Thus, tests conducted at lower strains (and which therefore lasted longer, and provided more opportunity for precipitate nucleation and growth) produced more hardening than at higher strains and a reduction in strain localization. This behavior caused the CSSC to have a negative slope over part of its range (Laird et al., 1978). On the other hand, Gerold and Meier (1987), who studied an Al-5% Ag alloy of much lower solute activity, found that PSBs formed and the GP zones were dissolved in them in association with cyclic softening. For precipitate behavior in a wider range of alloys hardened with shearable precipitates, see the work of Gerold and his coworkers (Gerold and Steiner, 1982; Steiner et al. 1983; Lerch and Gerold, 1985).

When the precipitates are large, the cyclic hardening behavior is entirely different. A typical example of such an alloy is Al-4% Cu containing θ' precipitates. Such precipitates consist of plates on {001} habit planes, roughly 1 μm in diameter and the three families of plates serve to divide up the material into small cubical volumes separated by reasonably dislocation-impermeable particles. In such a case, the cyclic deformation is homogeneous, slip lines are hardly visible, hardening is rapid, saturation is extremely stable and softening does not occur (Bhat and Laird, 1979a,b; Horibe and Laird 1983, 1985a,b). Since PSBs are not possible, the CSSC does not contain a plateau.

The cyclic hardening and saturation behaviors of metals containing (somewhat) impenetrable precipitates have been interpreted in a variety of ways, in early work in terms of *geometrically necessary dislocations* (Calabrese and Laird, 1974). The idea here is that the dislocations are confined to the matrix between the particles that resist deformation and cannot be cut by dislocations (actually the situation is more complex; at high strains the particles do become cut, as can be seen by anti-phase domain boundaries (APDBs) left behind in the ordered θ' particles Horibe and Laird, 1983). Accordingly, the dislocations in the matrix are arrayed at the interfaces of the particles to accom-modate the curvatures introduced by the incompatible deformation of the two phases. The increase in the flow stress is associated with the increase of the dislocation density. In crystals oriented for single slip, the dislocations will be dominated by primary dislocations, and they can provide strong back stresses through tilt wall arrangements at the θ' interfaces. A principle of LEDS theory is that multiple slip permits more complete pairing of oppositely signed dislocations as well as energy reductions between dislocation arrangements of different Burgers vectors. This approach can explain the reduced flow stress of the multislip crystals as compared with that of single-slip crystals. Horibe and Laird (1983) report seeing the equivalent of "Labyrinth" structure at the θ' interfaces. In addition, Li et al. (2011) have pointed out that understanding of the slip mode requires the consideration of not only for the SFE, but also the short range order, as given by the APDB energy. They proposed that the difference between the SFE and the APDB energy is the appropriate parameter to use to parameterize slip mode effects on cyclic dislocation structures: in pure metals γ_{SFE} is high and γ_{APDB} is zero, so the difference $\Delta\gamma = \gamma_{SFE} - \gamma_{APDB}$ is high (wavy slip). As alloying elements are added γ_{SFE} goes down and γ_{APDB} goes up; hence, $\Delta\gamma$ becomes small and even negative, indicating a transition from wavy to planar slip.

An alternative approach to understanding dislocation behavior is that of *nonlinear dynamics*, an approach to dealing with dislocation populations as a whole in conditions away from equilibrium, which corresponds to those of cyclic straining. Glazov and Laird (1995) have shown that in the relatively small volumes of matrix accessible to dislocations between impenetrable precipitates, the self-organization of the dislocations acts to arrange them at the extremities of their glide space, that is, at the particle interfaces, provided the particles are approximately 1 μm apart. If the cutting-resistant particles are arranged more densely (and therefore with decreased separations), so as to limit further the glide volume (or distance—Glazov and Laird, 1995 reported a one-dimensional problem), then the available dislocation populations can carry the applied strain without a tendency to rearrange or to clump together. This behavior applies, for example, to thoria-dispersed (T-D) nickel in which a dislocation population is inherited from the processing and the thoria particles are both dense and resistant to cutting. This explains the observations of Bhat and Laird (1979b), who found the starting dislocation structures in T-D nickel extremely stable to cyclic deformation at both room and elevated temperatures.

Engineering alloys based on the simple systems treated above, and other fcc systems, show many differences in detail, depending on the specifics of their microstructures. For example, engineering aluminum alloys typically contain three levels of precipitates: constituent particles, the largest; dispersoids, of intermediate size; and G-P zones, the smallest and most responsible for the superior strength of the alloys. The fatigue behavior is determined by a compromise between the effects of the different microstructural components: the constituent particles and dispersoids act to homogenize the strain during cyclic deformation, while the G-P zones encourage PSB formation, slip localization and zone dissolution. The complexities of the behavior still warrant further research (Laird, 1977; Starke and Lutjering, 1979), but the larger particles seem to be effective in limiting cyclic softening.

18.4.12 Dislocation Patterning in Cyclic Deformation

Reference to the treatment of dislocation structures by nonlinear dynamics in the previous section calls for comments on the application of such techniques to fatigue. There is an excellent review by Kubin (1993) on the nonlinear dynamics approach as well as a recent review on scaling of dislocation structures for monotonic and cyclic deformation (Sauzay and Kubin, 2011). The former review covers classical models of dislocation patterning such as those of Holt (1970) and Kocks (1985), the reaction-plus-diffusion approach and the approach by numerical simulations. The latter review covers relationships between characteristic length scales of dislocation structures and their relationship to flow stresses and hardening.

The reaction-plus-diffusion approach to dislocation patterning in cyclic deformation borrows heavily from physical chemistry, where self-organized spatial arrangements have been observed in nonlinear physicochemical systems under strongly nonequilibrium conditions (Glazov et al., 1995). This has been pioneered by Walgraef and Aifantis (1985), Aifantis (1986), and has been complemented by numerical simulations of the solutions of the model equations (Schiller and Walgraef, 1988). Dislocations are classified as either mobile or immobile and their densities define two reacting species, whereas the interactions between them, in terms of trapping, releasing, annihilation, and so on, define quasi-chemical reactions. Dislocation mobility is modeled by diffusion-like terms and systems of differential equations are then set up to follow the evolution of the density of each species. Stress-like terms force the response and the coupling between the equations is given by nonlinear terms. Solutions to these differential equations have been able to provide matches to the dislocation structures seen in PSBs, dynamic aging effects and strain bursts (Glazov et al., 1995).

The approach is promising, since dynamic processes play an important role in fatigue. However, the meaning of the adjustable parameters used, the nature of the interaction terms and the use of a diffusion-like model to describe dislocation mobility in general cases need better physical justification. In that regard, the recent review by Sauzay and Kubin (2011) does not include many results using this method, but rather those of discrete dislocation dynamics (DDD). This is an approach where individual dislocations are tracked and their interactions are governed by well-known equations describing elastic stresses around screw and edge dislocations as well as effects of jogs and kinks, and so on. The foundation of the DDD method is well grounded on the physical mechanisms that are known to govern dislocation interactions and can be used to carry modeling of dynamic processes. The basic issues with this technique as applied to cyclic deformation is the high dislocation density induced by fatigue and the need to use many load cycles to let the structures evolve, both of which lead to a high computational expense (the computational issues of DDD are similar to those of other multibody techniques such as molecular dynamics). In that regard, the reaction-plus-diffusion approach has the advantage of using dislocation densities as variables, which means that the computational cost is lower; both techniques could find a place in a well-formulated multiscale approach to dislocation patterning in fatigue.

This brief overview of the modeling of dislocation patterning naturally leads to the question of how the fatigue process has been modeled at larger length scales. This will be addressed briefly in the next section.

18.4.13 Mesoscale Modeling of Cyclic Plasticity

As mentioned in the introduction, the last two decades have brought significant advances on modeling and simulation of cyclic plasticity and its effects on fatigue damage accumulation and crack nucleation. These advances are rooted on the formulation of constitutive models that account for the Bauschinger effect through the use of a back stress term in the yield function or the flow rule that describes how the plastic strain increment is calculated. The main effect of this back stress term is to move the center of the flow surface as deformation evolves, leading to the tension compression asymmetry of the flow stress induced by the Bauschinger effect. This has led to the term *kinematic hardening* to describe this type of model. Some of these early models, particularly the ones with nonlinear formulations, were reviewed by Chaboche (1991) as applied to simulations of metallic materials at the macroscale. Recent work has focused on improving these early nonlinear models to perform accurate prediction of the cyclic response of metallic materials under multiaxial loading, particularly under nonproportional loading paths. Evaluations of the many models proposed to achieve these goals have been reviewed from time to time, for example, Bari and Hassan (2001), Abdel-Karim (2010).

In the case of single crystals, modeling of the cyclic plasticity has taken advantage of mature formulations to model single crystal plasticity under monotonic loading, from kinematic assumptions, for example, the multiplicative decomposition of the deformation gradient **F** and expressions for the plastic part of the velocity gradient **L** in terms of the crystallography of the active slip systems, to the hardening formulations for both rate dependent and independent behavior. The reader is referred to the vast literature on these topics for more details, for example, Asaro (1983), Bassani (1994), Simo and Hughes (2000). Emphasis will be placed here on the kinematic hardening details that pertain to the modeling of cyclic behavior. In that regard, cyclic plasticity of fcc single crystals that display strain localization is PSBs is challenging to model, as is any plastic phenomenon that localizes deformation, since the low hardening rates can create problems with the local tangent stiffness that is used in numerical solutions via finite elements. Hence, modeling the plateau in the CSSC of copper and other fcc metals has proven a difficult task, despite the fact that the mechanisms leading to the presence of the plateau in terms of dislocation structures have been well known for some time. A brief overview of work to develop kinematic hardening rules for single crystals is offered by Xu and Jiang (2004), as well as earlier efforts to use crystal plasticity for fatigue modeling.

A model that provided a fairly good approximation to the rapid hardening behavior as well as the CSSC of copper single crystals was proposed by Xu and Jiang (2004). These authors offered a rate independent model where the yield criterion is based on the Schmid law modified by a back stress, that is,

$$f^{(\alpha)} = \left| \tau^{(\alpha)} - x^{(\alpha)} \right| - k^{(\alpha)} \leq 0, \tag{9}$$

where $\tau^{(\alpha)}$ is the resolved shear stress on slip system α, $k^{(\alpha)}$ is the shear flow stress of the slip system α and $x^{(\alpha)}$ is the back stress on that slip system. These authors used a modified Armstrong-Frederick kinematic hardening rule such that

$$x^{(\alpha)} = \sum_{i=1}^{I} x_i^{(\alpha)} \quad \text{and} \quad \dot{x}_i^{(\alpha)} = c_i r_i \sum_{\beta} H_{\alpha\beta} \dot{\gamma}^{(\beta)} - c_i x_i^{(\alpha)} \sum_{\beta} \left| \dot{\gamma}^{(\beta)} \right| \quad (i = 1, 2, ..., I), \tag{10}$$

where c_i and r_i are material and history dependent parameters and $\dot{\gamma}^{(\beta)}$ is the shear strain rate in slip system β. The hardening matrix $H_{\alpha\beta}$ follows the model by Bassani and Wu (1991), that is, it is taken to be diagonal, where the magnitude of the diagonal terms depends on the shear strain on all slip systems. Xu and Jiang (2004) use five components of the kinematic hardening rule ($I = 5$). Their formulation seemed to be based on pure kinematic hardening so there is no evolution of the shear flow stress $k^{(\alpha)}$ with plastic strain. An important component of this model is that the coefficients c_i and r_i are functions of the strain evolution, so that the strain dependent hardening behavior can be included in the simulations. This is key to obtain the plateau in the CSSC.

Examples of some of their results are shown in **Figure 51**.

Note from **Figure 51(a)** that the model matches the result at the higher strain fairly well, but the agreement is not as good as the plastic strain decreases. Nonetheless, the overall behavior is captured to the point that the CSSC can be reproduced. In that regard, the results shown in **Figure 51(b)** indicate that the model captures multislip behavior well and the plateau in the CSSC is reproduced fairly well, although the simulated plateau starts at a strain somewhat higher than that observed in experiments.

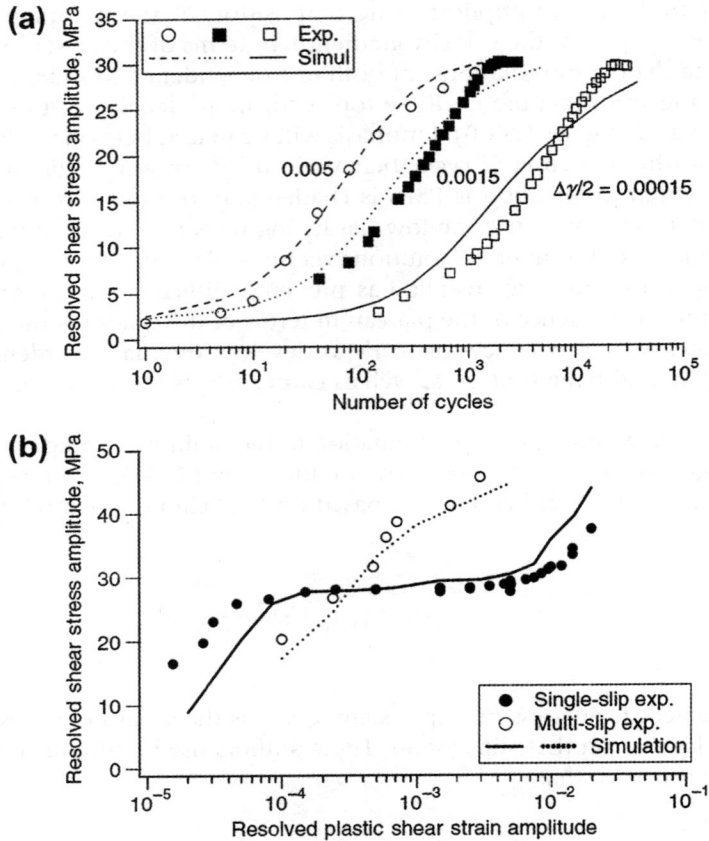

Figure 51 Comparison of modeling and experimental results for simulations of cyclic plasticity in single crystals. (a) Rapid hardening curves for copper single crystals oriented for single slip, tested under strain control within the plateau and (b) CSSC for both single slip and multiple slip orientations. Source: After Xu and Jiang, 2004, courtesy of Elsevier.

The authors tested their model by comparing it to experimental hysteresis loops as well as high/low load sequences and obtained good agreement. The choice of a rate independent crystal plasticity framework is quite appropriate, despite the fact that is computationally more involved. In the experience of one of the present authors, the use of viscoplastic regularization techniques, that is, a rate dependent framework that tends to rate independency in the limit when the rate sensitivity exponent goes to zero (power-law flow rules are quite common for this) could lead to numerical problems, particularly in stiff constitutive models such as conventional crystal plasticity (see Amirkhizi and Nemat-Nasser, 2007 and references therein) and, in our experience, could produce lingering, and artificial, rate dependencies in some cases.

Note that approximately 20 material parameters, some of which were functions of a surface memory parameter, were needed to accomplish the simulations described above, which illustrates the difficulty of modeling the pronounced strain localization in copper single crystals. In addition, the approach is partially phenomenological as far as strain localization is concerned, from the point of view that some of the equations are chosen so that observed behavior can be reproduced. Ideally, equations should come from the mechanisms controlling the behavior of the dislocation structures responsible for the strain localization or from fits to models that reproduce such behavior in a multiscale framework. Further research is needed to accomplish that goal.

Simulations of cyclic plasticity in single crystals open the door to modeling of plasticity in polycrystalline materials starting at the microscale, that is, simulate a polycrystal as an ensemble of monocrystalline grains where the microstructure is fully resolved. This, in turn, provides excellent opportunities to study variability in fatigue behavior associated with microstructural heterogeneity, which is key to understand damage accumulation and crack nucleation under cyclic loading. Significant progress has been achieved in this regard, thanks not only to improvements in computational power, but also to experimental techniques that allow characterizing the geometry and crystallography of microstructure in 2D, for example, conventional EBSD, as well as in 3-D, for example, serial sectioning combined with EBSD in an FIB or X-ray tomography at bright X-ray sources. In addition, these characterization techniques have led to the development of tools to create "artificial" microstructures that share the same statistical characteristics of the actual materials, for example, grain size and shape distribution, crystallographic texture and grain boundary misorientation distributions, among others. See Nygårds and Gudmundson (2002), Saylor et al. (2004), Zhang et al. (2004), Brahme et al. (2006) for examples of a vast literature on this topic.

Several authors (see Dunne et al., 2007; Anahid et al., 2011 and references therein) have used constitutive models and characterization tools similar to those described above to study fatigue damage at the microscale in polycrystalline materials, with emphasis on locations where strain localizes. However, it is important to keep in mind the potential effects of the modeling choices made to simulate the behavior of a material of interest. In that regard, there is an interesting study performed by Dingreville et al. (2010) where the effects of microstructure representation and choice of constitutive model on cyclic plasticity are examined. The authors used the general approach shown in **Figure 52**.

The process outlined in **Figure 52** is fairly descriptive of this type of simulation, that is, the microstructure is characterized and digitized, and their characteristics are used to make either faithful representations of it or simplified ones, using Voronoi tessellations, polygons or ellipses. At the same time, mechanical behavior is measured and used to calibrate and validate the model. Finally, the model is used to study the effects of microstructure representation and hardening model on the simulated stress–strain response at the macroscale as well as the strain localization at the microscale. These authors found the details of the microstructure representation did not affect significantly the predicted stress–strain and ratcheting curves as long as the crystallographic orientation distribution "fed" to the artificial microstructure is reasonable. However, the prediction of the detailed distribution of plastic strain at the microstructural level was quite sensitive to the specific microstructural representation and to the choice of hardening formulation (see **Figure 53** for an example of the latter).

The results of **Figure 53** show how for different cycles (the first eight), the evolution of the sites with the highest levels of plastic strain depends on the hardening model chosen. These results have important implications for damage predictions based on plastic strain accumulation. The level of detail necessary to adequately simulate the microplastic ratcheting response of a microstructure was found by Dingreville et al. (2010) to be dependent on the scale and phenomenon of interest. A more realistic description of the microstructure is essential to describe localized mechanical behavior at the microscopic scale that can lead to crack nucleation. This important issue will be discussed in the next section.

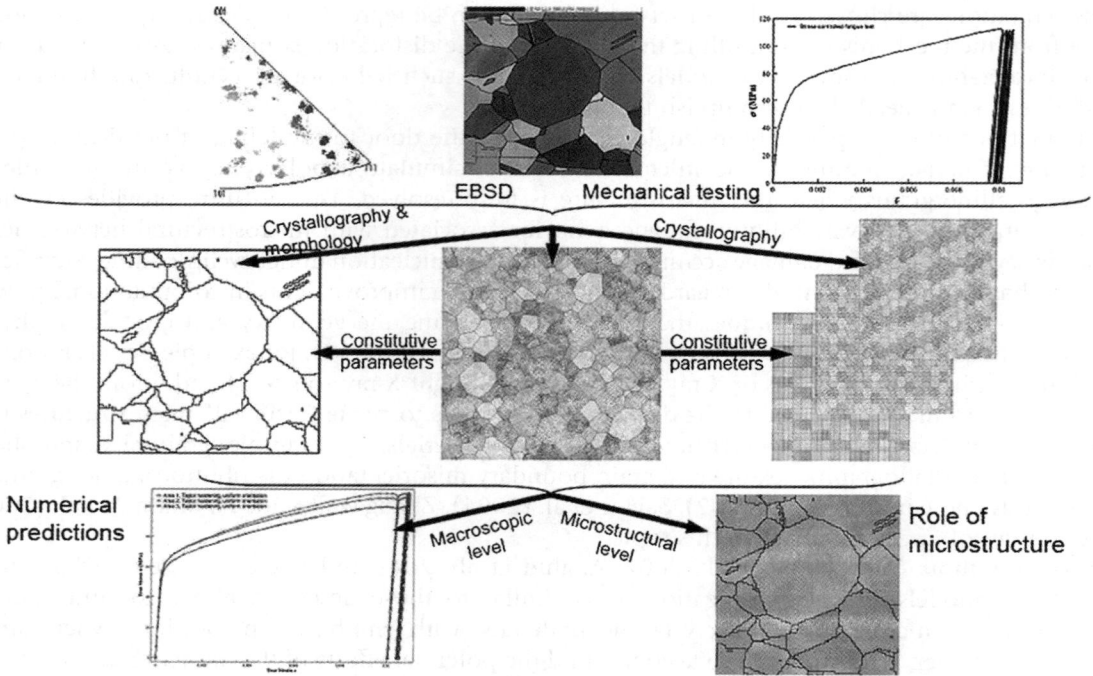

Figure 52 Framework for coupling microstructure characterization, mechanical testing, and numerical simulations to study cyclic plasticity at the microscale. Source: After Dingreville et al., 2010, courtesy of Elsevier.

18.5   Fatigue Crack Initiation in Ductile Metals

The initiation of fatigue cracks is defined, to a large extent, by the determination of the investigator, and the precision and resolution of the techniques used to find them. A mechanical engineer may either regard them as an inevitable development of the manufacturing process consisting of flaws that simply grow during subsequent service, or view them at the scale of ~ 1 mm on the basis of performance in nondestructive evaluation. A materials scientist would define them in terms of a failure mechanism associated with a microstructural feature, such as a PSB, a grain boundary, nonmetallic inclusion or other kind of stress concentrator or strain localizer. Even in smooth specimens, it is widely accepted that fatigue cracks initiate early in the fatigue life; however, the earliest point of crack detection seems to depend strongly on the skill of the investigator.

As noted in the introduction, where design philosophies were outlined, the emphasis on fatigue crack initiation varies widely. Our understanding, in quantitative terms, of fatigue crack initiation has improved considerably in the past two decades, but more experimental and modeling work is certainly needed.

The present section deals with models for fatigue crack initiation and their dependence on dislocation structure (and thus strain amplitude) and microstructure, on the assumption that the specimen contains no gross strain concentrator, and the initiation mechanisms arise purely from the

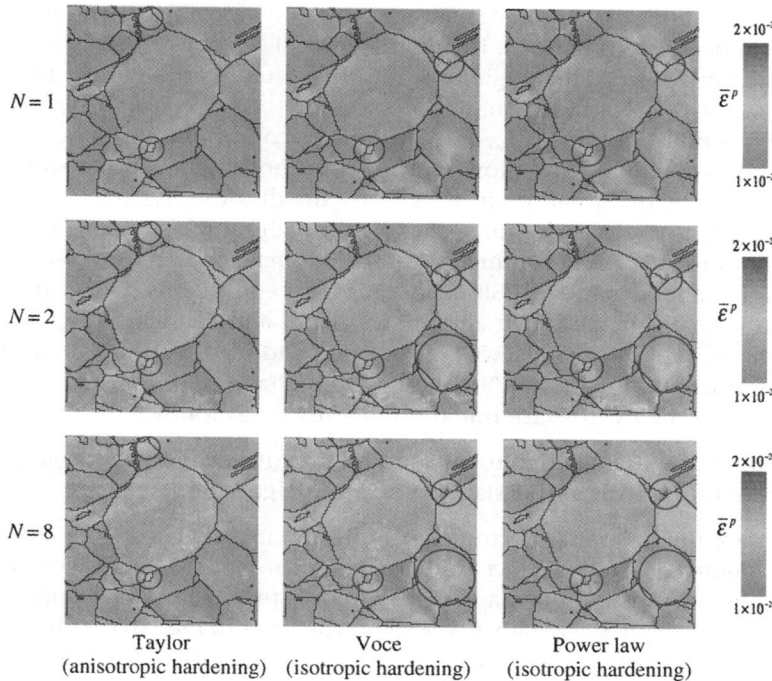

Figure 53 Effect of hardening model on plastic strain localization. The circles correspond to the highest equivalent local plastic strain. Source: After Dingreville et al., 2010, courtesy of Elsevier. (For color version of this figure, the reader is referred to the online version of this book.)

deformation processes of the metal. In actual structures, the hope is to keep fatigue stresses at low levels generally, and this usually causes fatigue to occur (if it does) at stress concentrators. Crack initiation at these locations, however, can be affected by stress and strain gradients, although the behavior of a smooth specimen cycled at a stress equivalent to that in the stress concentration can be used as a first approximation. Fatigue crack nucleation under high cycle fatigue (HCF) conditions will be reviewed first, and then the mechanisms under very high cycle fatigue (VHCF) conditions will be discussed.

18.5.1 Fatigue Crack Initiation and Surface Roughness: The Phenomena

In order for fatigue cracks to nucleate at all, the applied stress or strain in a fatigued specimen must be above the fatigue limit, or at some time in its life have been above it. However, as mentioned in the introduction, the stress level that represents a fatigue limit is certainly much lower than previously expected, as revealed by experiments under VHCF conditions. When HCF conditions are satisfied, PSBs, or other manifestations of strain localization, form and intensify. Cracks ultimately form at the surface both in the PSBs and at other regions, such as twin or grain boundaries, on which the PSBs impinge. Under VHCF conditions, cracks can nucleate inside the specimen, usually at a local stress concentrator such as an inclusion or constituent particle for engineering materials or a grain boundary for a pure metal. Usually, the cracks are associated with localized strain and its interaction with the microstructure. In the absence of a subsurface flaw that can provide a site for crack initiation, the free surface is the preferred site for crack initiation under HCF conditions. The discussion will center on HCF from this point on.

Cracks will initiate in PSBs or other regions of localized strain (say a narrow twin that has a more favorable Schmid factor than its host grain, and essentially becomes a PSB in itself) as long as the straining satisfies the conditions for localized strain, namely, corresponding to the plateau region of the CSSC for single crystals, or the deformation equivalent for polycrystals. If the stress or strain is high enough to homogenize the deformation, in which case the dislocation structure will consist of cells, either locally near grain boundaries or more generally throughout the structure, then cracks initiate at the grain boundaries instead. These comments particularly refer to metals of wavy slip character. In single phase alloys of planar slip mode, PSBs (or their planar equivalents, PLBs), are the preferred sites of crack nucleation both at low and high amplitudes (Laird and Feltner, 1967). The mechanisms of crack initiation have been repeatedly reviewed (Thompson and Wadsworth, 1958; Laird and Duquette, 1972; Brown, 1977; Fine and Ritchie, 1979; Suresh, 1998; Zhang et al., 2011).

The phenomena associated with crack initiation in PSBs are as follows:

1. PSBs initiate, localize the strain within them and start to protrude. This happens early in life, and the typical appearance of the PSBs at this stage is shown in **Figure 54**.

A *protrusion* is a surface uplift, several micrometers in height, usually occupying the width of a micro-PSB, like those in **Figure 54**, and distinct from an *extrusion*, which is narrower, more pointed and deserves more detailed treatment. Often, a wide PSB and its protrusion may contain several *intrusions* and extrusions, but these features generally occur somewhat later in life. An example of a macro-PSB in this condition is shown in **Figure 55**. Note that the crystal had been given 120,000 cycles when the SEM photograph was recorded, a large fraction of life.

2. After 5–10% of life has been expended, numerous small cracks (fissures) can be found within the PSBs, and the cracks can be considered to have initiated. Typical evidence for this behavior is shown in **Figure 56**, comprising an SEM photograph of a single crystal prepared by the "sharp corner" technique. This technique is employed on specimens of square cross-section, in which the cracks tend to initiate preferentially along the corners of the gauge section. After cycling, the surface appearance is as shown in **Figure 55**.

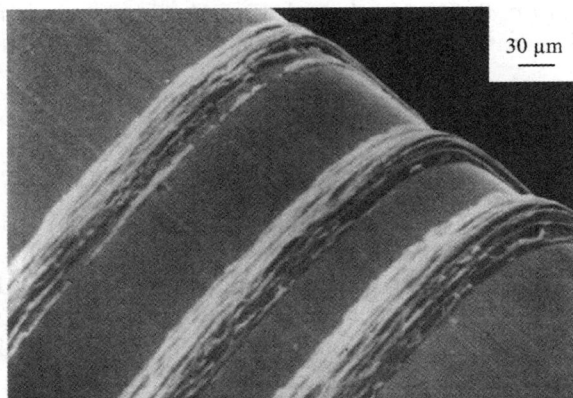

30 µm

Figure 54   Actual appearance of PSBs after much accumulated strain (30,000 cycles at 2×10^{-3} strain) showing protrusion and complex PSB morphology. Source: Taken from Ma and Laird, 1986, courtesy of Elsevier.

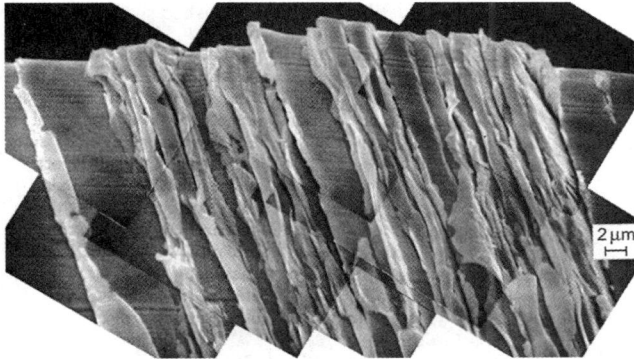

Figure 55 A macro-PSB in a copper single crystal, oriented for single slip, fatigued at room temperature for 120,000 cycles at a plastic shear strain amplitude of 2×10^{-3}. The macro-PSB consists of many micro-PSBs, separated by regions of matrix structure (loop patches). There are numerous extrusions and intrusions superimposed on an overall protrusion. Source: Taken from Ma and Laird, 1989b,a,d. Courtesy of Elsevier.

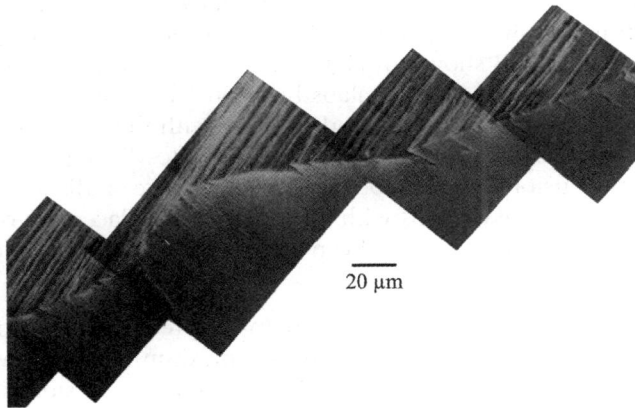

Figure 56 Numerous Stage I cracks visible in a "sharp corner" section. Copper single crystal cycled at a shear strain amplitude of 2×10^{-3} for 30,000 cycles. Source: Courtesy of Ma, 1987.

Since the extrusions fold over the surface, they hide the cracks and it is not possible to observe how deep the cracks are. The cracks can be revealed and their depths measured by carefully removing a thin layer from the side surface of the specimen and polishing that surface without affecting the cracks, a difficult technique that reveals the cracks in a 90° section. In early work, a *taper section* might be employed to "magnify" the cracks geometrically, but given the high resolution of the SEM, taper sections are not needed, except perhaps for special purposes. The sharp corner technique was developed by Basinski and Basinski (1984).

3. The morphology of extrusions is shown in **Figure 57**. Some workers, including the present writers, view them as thin tongues of metal, schematically indicated in **Figure 57**.

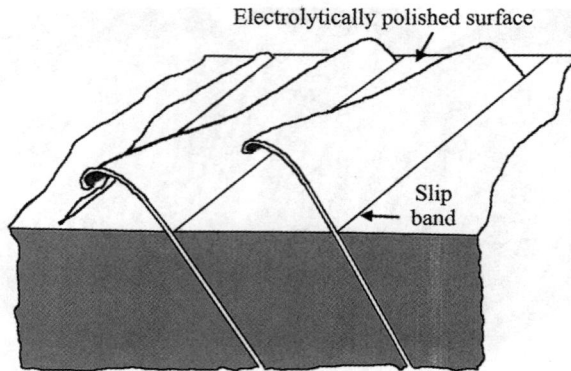

Figure 57   Schematic representation of a slip-band, or micro-PSB, extrusion. Source: Redrawn from Laird, 1996 after an image originally by Forsyth, 1957. Printed with permission from Royal Society Publishing.

Other workers, such as Suresh (Suresh, 1998) report them as having a triangular cross-section (base width ∼1–2 μm, height 2–3 μm); no doubt, a great variety of shapes and sizes has been observed (see Laird and Duquette, 1972). Extrusions grow at high rates, 1–10 nm/cycle, whereas the growth rate of the kind of protrusions shown in **Figure 55** is an order of magnitude slower. The overall protrusion visible in the macro-PSB shown in **Figure 56** is connected, at least in part, with shearing of micro-PSBs and the "sliding" of the matrix slabs between. The kinetics of such behavior is rather "stochastic", and the overall protrusion height tends to scale with the width of the PSB. It is necessary to have a different nomenclature for this type of protrusion, to distinguish it from that shown in **Figure 54**, for which "protrusion" should be retained to make a distinction. The writers offer *"bulging"*. Such bulging would be associated with a net inward displacement on the opposite side of the single crystal. The term *"encroachment"* has been employed by Ma and Laird (1989b,a,d) to label such a phenomenon.

4. The strain localized within groups of PSBs is macroscopically reversible, in that the local strain at the tensile maximum is identical to that at the compressive maximum, at least within the time interval of several cycles or even hundreds of cycles. Clearly it cannot be reversible all the time, or bulging and encroachments would not occur. Moreover, the local strain within a micro-PSB appears to increase linearly with the overall strain (Finney and Laird, 1975) during a typical straining excursion, and reverses with the cycling of the strain, again for most of the time. The evidence shown in **Figure 20(b)** indicates that micro-PSBs do not deform wholly reversibly all of the time. There is an interplay and redivision of strain between the micro-PSBs. At the finest distribution of slip within a PSB, the deformation is definitely not strictly reversible. Evidence for this has been presented by Finney and Laird (1975) using interferometry techniques in single crystal copper, and further details have been explored using AFM techniques in nickel (Holste, 2004). These studies showed roughening of the PSB. Irreversibility of slip is an important parameter in fatigue that can be used to unify behavior across several regimes (HCF and VHCF), as discussed by Mughrabi (2009).

This roughening evidence indicates that lamellae defined by the atomistic slip within the PSBs undergo constrained slip in the direction of the primary Burgers vector, and produce notch–peak topology by the micro-irreversibility of the slip. This roughness will be superimposed on the protrusion and the initial stages of it can be seen in **Figure 54**.

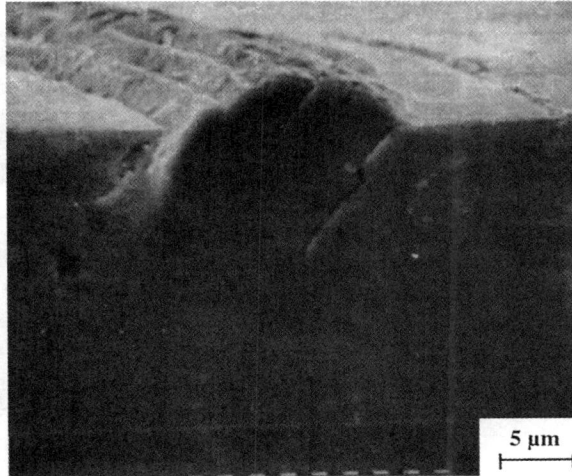

Figure 58 Fatigue crack initiation in a PSB for a copper single crystal fatigued for 60k cycles at a $\Delta\gamma_{pl}/2 = 2 \times 10^{-3}$ at room temperature, showing preferred initiation at the PSB matrix interfaces. Source: Taken from Ma and Laird, 1989b,a,d. Courtesy of Elsevier.

5. As noted in the previous section, optical interferometry indicates that a greater than average local strain develops at (or near) the PSB–matrix interface. This "double" localization of slip increases the kinetics of crack initiation at the PSB–matrix interface. Counts of crack nuclei reported by Neumann and Tonnessen (1987) indicate that cracks located at the PSB–matrix interfaces outnumber all the others by about six to one. Bear in mind that this result applies to copper crystals in which the ramp-loading method of starting the test had been used, and the results could be skewed by the test method. However, even in crystals cycled in constant amplitude, cracks occur frequently at the PSB–matrix interfaces. A typical example of such behavior can be seen in the "sharp corner" section shown in **Figure 58**.

Note the protrusion containing minor cracks, a leading crack on the acute angle side of the PSB and a lesser crack on the obtuse angle side. These cracks are parallel to the primary slip plane and as such belong to a class of short cracks termed *Stage I cracks* because they propagate at 45° to the tensile axis and are strongly influenced by the shear stress, as found decades ago by Gough (1933).

6. The fatigue crack initiation behavior in planar slip metals, which form PLBs rather than PSBs, is different from that in wavy slip metals. Since the localized strain does not persist, but moves around the gauge section, the fatigue life increases, at least in Cu-16 at% Al (Hong and Laird, 1991b), as compared with that in copper. In association with the uniform slip distribution, a very uniform hill and valley surface morphology develops in Cu-Al crystals with continued cycling. This morphology, which is shown in **Figure 59**, is related to strain burst behavior (Hong and Laird, 1990b), because it develops when bursts are active.

The hill and valley morphology is seen in **Figure 59** side-on in a square sectioned specimen so that the profile can be observed directly. The crystal was oriented with the primary Burgers vector parallel to the side facing the photographer in **Figure 59**, and the morphology forms on the face at which the slip steps emerge. The wavelength of the morphology increases with strain amplitude because the strain bursts are

(a)

(b)

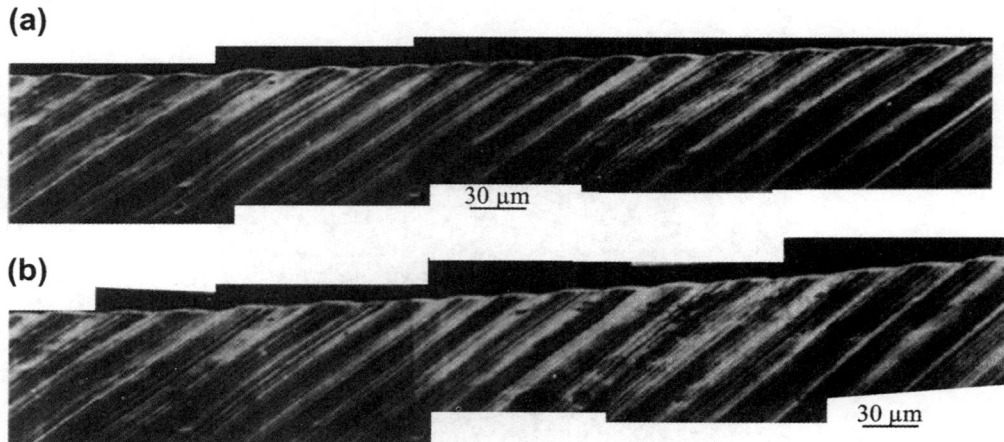

Figure 59 The uniform hill and valley surface morphology observed on Cu-Al single crystals oriented for single slip, fatigued at room temperature. Sample cycled at an (average) plastic shear strain amplitude of 4.9×10^{-3} for (a) 16 500 cycles; (b) 37 200 cycles. No significant changes occurred between (a) and (b). Source: Taken from Hong and Laird, 1990b. Courtesy of Elsevier.

larger in the early stages of life (Hong and Laird, 1990b). When the bursts die out, as they do later in life, the morphology stabilizes (compare (a) and (b) in **Figure 59**). The development of the protrusions observed in copper does not occur in Cu-16 at% Al single crystals but extrusions are seen in polycrystals. In the closing stages of life, numerous cracks initiate in the valleys of the surface morphology and propagate rapidly because the stress amplitude is high, on account of the long-drawn-out and pronounced cyclic hardening (Hong and Laird, 1991b).

18.5.2   Fatigue Crack Initiation in Persistent Slip Bands – Mechanisms

The origin of fatigue cracks in metals and alloys of high purity is often rationalized by mechanisms of the type championed by Wood (1958). The basic premise of the mechanism is that repeated cycling of the material leads to different amounts of net slip on different atomistic glide planes. The irreversibility of shear displacements along the SBs then results in roughening of the surface of the material and the gradual development of the roughening into notch-peak morphology. The valleys in the morphology function as micronotches and the effect of stress concentration at the root of the valleys promotes additional slip. This step is likely to be more intense in tension than in compression because the micronotches can close in compression and defocus the stress concentration, further enhancing slip irreversibility. Thus fatigue cracks initiate. In 1996, the bulk of the evidence favored this mechanism for HCF (Laird, 1996) and, in the opinion of the present writers, it does so today.

A quantitative statistical model for random slip leading to the formation of hills and valleys on the surfaces of fatigued metals was published by May (1960a,b), where the model to explain the Coffin–Manson law was initially offered, with the implicit assumption that the model applied to high strain fatigue, which it does not, because cracks form by a different mechanism at grain boundaries. Rosenbloom and Laird (1993), who studied slip irreversibility in single crystals very early in life with emphasis on crack initiation, have applied May's model to PSB behavior and find it to give reasonable predictions of the cycles required to initiate cracks in PSBs of copper. The observations of these authors also cast doubt on some of the details of the surface roughening model of Essmann et al. (1981), who base their model on dislocation irreversibility judged from intrabulk behavior of PSBs, not of PSBs where they intersect the surface (Mughrabi et al., 1983). Nonetheless, this model has been refined and is used to explain phenomena associated with fatigue damage (Mughrabi, 2009).

The mechanisms of crack initiation by slip irreversibility, as they relate to the variations in localized strain from one PSB to another, have been explored by Cheng and Laird (1981a,c). By repolishing and reloading, followed by interferometric observations on one side of a fatigued single crystal, coupled with observations of cracking in the same PSBs on the opposite (unpolished) side of the crystal, these authors have documented the properties of the PSBs in which the crack nucleates. As noted previously, for copper single crystals subject to an applied strain within the plateau of the CSSC, there will be a certain distribution of strains localized in the PSBs, and not a constant strain equivalent to that of the upper end of the plateau. The crack nucleates in the "fatal band" that contains the highest localized strain (Cheng and Laird, 1981a,c). It is found that the length of the slip offset in a fatal band (nb, i.e. n Burgers vectors) and the applied plastic shear strain amplitude, γ_p, are related as $\gamma_p = C(nb)^{0.78}$, where C is a constant (Cheng and Laird, 1981a,c). The orientation of the crystal is found to affect only the step height, the volume fraction of the PSBs only slightly, and the slip offsets not at all within some scatter (see also the data reported by Holste, 2004). It is the magnitude of the slip offset that controls the crack nucleation behavior if the cyclic stress is uniaxial, and thus there is no effect of orientation on the cycles to nucleate a crack, assuming of course the wide range of orientation within the standard triangle over which the slip remains single. The situation for multislip orientations will be different if PSBs are not present.

There are many competing models for crack initiation. When the thin type of extrusion was first reported in the 1950s (Forsyth, 1953) it caused great excitement in the context of initiation. It was soon discovered to exist in all sorts of ductile materials including silver chloride (Forsyth, 1957) and in all kinds of metals and microstructural features, including TBs and grain boundaries (see Laird and Duquette, 1972, for details), which should have given a useful hint for the mechanism of extrusion formation. Most significantly, the formation of an intrusion–extrusion pair (meaning an extrusion was observed in conjunction with a small crack) was identified by Forsyth and Stubbington (1955) and by Cottrell and Hull (1957). Many distinguished investigators have offered ingenious explanations for the formation of extrusions and of extrusion–intrusion pairs in terms of dislocation behavior. But it turns out that the thin extrusion has been somewhat misleading. Dickson et al. (1993) have presented the most convincing argument that extrusions come after the fact of initiation rather than before.

Consider that a specimen contains a small stage I crack. During the compression portion of the cycle, the extrusion is produced by a rubbing/burring process on the crack face that subtends an obtuse angle with the specimen surface. The fracture surface rubbing also tends to push the extrusions away from the crack. During the tensile part of the cycle (or possibly the load-increasing part), an extrusion can form on the other face, if there is sufficient local closure to produce the required fracture surface rubbing. The net effect is that more numerous and larger extrusions are formed on the crack faces subtending an obtuse angle with the surface. This mechanism appears to explain all the facts of extrusions known to the present writers, for example, the rapidity of their formation, their ability to form in conjunction with grain boundary cracks, their formation at soldered joints oriented at 45° to the stress axis, their presence in fatigue of ultrafine grained materials (Agnew and Weertman, 1998), and their formation under a wide variety of testing conditions.

Another group of mechanisms consists of vacancy dipole models. There is no evidence known to the writers in which point defects directly contribute to crack initiation at low temperatures, including room temperature. It does appear reasonable, however, to suppose that vacancy clusters produced by cyclic deformation are responsible for the swelling of the material that produces protrusions, and which can therefore contribute to roughness (see Suresh, 1998 for more details on vacancy models). Repetto and Ortiz (1997) proposed a model of cracking at PSBs that leads to their elongation because of pair annihilation, and vacancy generation. The model also includes vacancy diffusion that accounts for surface motion resulting from the outward flux of vacancies. Their numerical simulations show that this flux causes the surface to recede, which contributes to the formation of grooves at the PSB/matrix interface.

Micromechanical models have been proposed to describe intrusion and extrusion formation because of glide on parallel planes, but with more systematic selection of slip in tension and compression than applies to random slip models (Liu and Ito, 1969; Liu and Lin, 1979). In bcc systems, where slip in tension and compression can occur on different types of slip plane, it is possible that this type of model, suitably modified, may operate. Irreversibility of slip in these materials is certainly due in large part because of this asymmetry (Mughrabi, 2009). Other models that have been proposed, for example, that by Venkataraman et al. (1990) and more recently by Sangid et al. (2011), depend on some kind of storage of elastic energy to initiate cracks. In the latter, the energy storage is justified in terms of PSBs interacting with defects and leading to pile-up of dislocations at micro-structural barriers. Other models based on damage accumulation via continuous damage mechanics have been proposed for crack nucleation in single crystals, for example, the work by Kalnaus and Jiang (2006).

In engineering metals, many of the considerations described above apply to the crack initiation behavior. For example, in aluminum alloys containing shearable precipitates, PSBs may well occur and behave similarly in principle to PSBs in pure copper. If the microstructure is such that the strain is homogenized, for example, when the precipitates are large and impenetrable, then most likely grain boundary initiation will occur. In hard metals containing nonmetallic inclusions or constituent particles, cracks initiate at the inclusions either from a stress concentration effect or from low mechanical properties of the inclusion itself or its interface with the matrix, either one of which may crack. See Starke and Lutjering (1979) and also Bozek et al. (2008) for more recent work.

18.5.3   Grain-Boundary Crack Initiation

At strain amplitudes greater than those needed to produce PSBs, dislocation dipolar walls or more complex cell structures are general throughout a specimen and the deformation is broadly homoge-neous. Only in planar slip materials is the deformation sufficiently confined to bands to give rise to SB cracking, as noted above. More usually, notches develop at grain boundaries and cracks form in them. The mechanisms of this cracking were explored by interferometric observations of cracking morphology at grain boundaries in conjunction with efforts to study the nature of the boundaries that acted as nucleation sites (Kim and Laird, 1978b,a). At the beginning of life a small step is formed in tension, but it is canceled by the compression stroke. With continued cycling, the amplitude of the boundary step increases and resists complete cancellation in compression. Eventually a step 1.5 μm high, having a sharp root radius, develops and the crack grows along the grain boundary into the material from this step (Kim and Laird, 1978a,c). A step can have nearly as high a stress concentration as a crack. However, a crack would need to start growing from the step to be considered initiated. Only a relatively small fraction of grain boundaries were observed to be susceptible to this mechanism of failure. The "vulnerable" boundaries were identified as those separating highly misoriented grains, the dominant slip systems of which were directed over large distances at the intersection of the boundary with the surface, as schematically shown in **Figure 60**.

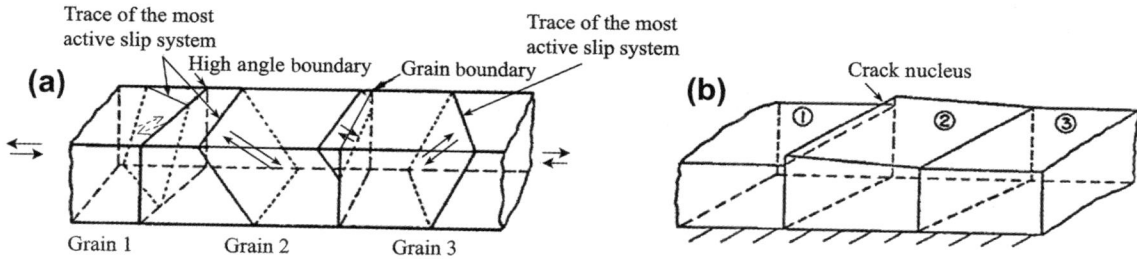

Figure 60 Schematic representation of the crack nucleation process in high strain fatigue. (a) The active slip systems in either grains 1 or 2, or in both, are directed at their boundary. (b) After cycling, a step forms at the boundary between grains 1 and 2, but not 2 and 3. Source: From Kim and Laird, 1978a, courtesy of Elsevier.

The ratcheting mechanism by which the step develops was attributed to cross-slip resulting from the presence of a boundary-induced stress gradient.

The phenomenon of grain boundary cracking has been studied in more detail by the use of bicrystal specimens. In particular, Peralta et al. (1998) showed using grain boundary notches for the bicrystals shown in **Figure 41** that the longest slip length of the local slip systems active at the nucleation site is what counts, whether or not this slip system is the most active in the bulk of the grain. An example is shown in **Figure 61**.

The results in **Figure 61(a)** show that the corner where the crack nucleated had significant slip length along the primary slip plane, as expected given the choice of notch location. The crack nucleation shown in **Figure 61(b)** is for an identical sample, except that the notch was located such that the slip length of the primary slip plane was minimum toward that side of the sample. Note the significant amount of secondary slip activated, particularly along the $(1\bar{1}\bar{1})$ plane. The thickness of the SBs suggests that they were PSBs hitting the boundary at the corner of the side and the notch surfaces. In either case, the corner where the crack nucleated had significant slip length for the slip systems that were active around the notch.

Further experimental work on bicrystals, including characterization of dislocation structures around GBs using ECC, as reviewed by Zhang and Wang (2008), has confirmed the results shown above, whereby intergranular cracks are nucleated at high-angle grain boundaries at the intersections of SBs and the surface. These authors also pointed out that as the misorientation angle of the boundary decreases, PSBs can eventually cross the boundary, leading to enhanced strain localization within them because of an increase in slip length and crack nucleation at the PSB. Details can be found in Zhang and Wang (2008) and references therein.

The grain boundary step mechanism of crack nucleation and the associated crystallographic requirements are not specific to pure metals, and thus the mechanism may be a general one, subject of course to modifications in complex engineering materials (Hodgson, 1968; Coffin and McMahon, 1979; Starke and Lutjering, 1979). The asymmetric slip of screw dislocations in tension and compression can lead to severe shape changes of bcc single crystals in fatigue as described above. Such changes will act to aggravate the formation of boundary steps and cracking. It has been shown on fatigued polycrystalline α-iron specimens (Mughrabi 1975, 1977) that near surface grains do suffer shape changes similar to those in single crystals and cause surface roughness leading to crack nucleation. Effects of slip geometry and twinning in hcp polycrystalline materials can be also pronounced, as is the case for polycrystalline Mg, where cracking because of SB intersections with grain boundaries as well as from twin tips has been observed (Yu et al., 2011b).

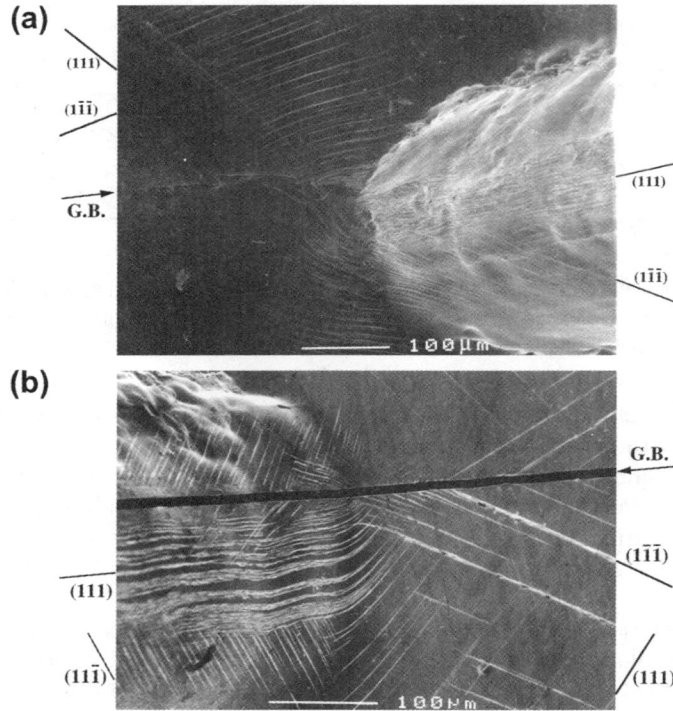

Figure 61 Crack nucleation sites in copper bicrystals with 180° twist boundaries about [$\bar{1}$49]. (a) Samples with a notch located such that the primary slip plane had the maximum slip length toward the notch. (b) Sample with a notch 180° from the one in (a), so that the primary slip plane had the minimum slip length toward the notch. Specimens tested at an axial plastic strain amplitude of 5×10^{-4}. Source: After Peralta et al., 1998, courtesy of Elsevier.

Modeling of crack initiation in polycrystals must include, by necessity, issues associated with the effect of grain boundaries and their potential effects as strain concentration sites. However, as pointed out by McDowell and Dunne (2010), mesoscale simulations based on crystal plasticity cannot yet account for dislocation based mechanisms, for example, PSB–boundary interactions, like the ones discussed above. Instead, the simulation results are used to obtain fatigue indicator parameters (FIPs) that reflect the driving force for nucleation and short crack growth at the local level. These authors point out the maximum value of the plastic shear strain range is a good indicator for nucleation of cracks in surface grains, and also the Fatemi-Socie parameter (see Eqn (4)). One particularly interesting aspect of the work presented by McDowell and Dunne (2010) (see also references therein), is that these FIPs need to be calculated in a nonlocal manner, that is, instead of using the value at the point where the simulation predicts a maximum, the volume average of the parameter is calculated over a chosen length scale, typically given by microstructural considerations. This avoids numerical problems, as the numerical noise produced by the numerical discretization is smoothed out (regularization) and it also targets length scales associated with crack embryos (McDowell and Dunne, 2010). An example of this work is shown in **Figure 62**.

Predictions of this type are computationally expensive because of the need to reproduce the microstructure in enough detail to make reliable calculations of the extrema of variables of interest,

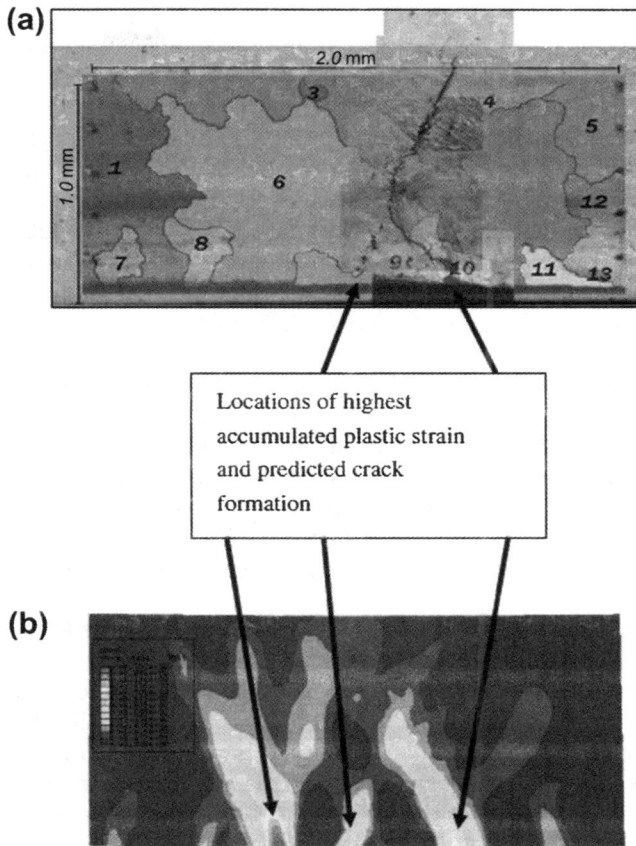

Figure 62   (a) Experimentally observed crack nucleation and initial growth under plane stress conditions. (b) Predicted site of fatigue crack nucleation and crack orientation based on the accumulated plastic strain. Note that some of the sites are close to grain boundaries. Source: After McDowell and Dunne, 2010. Courtesy of Elsevier. (For color version of this figure, the reader is referred to the online version of this book.)

including the FIPs discussed above (see also the discussion in Section 18.4.13). This, in turn, makes it difficult to perform simulations for the hundreds of thousands (or millions) of cycles needed to model HCF directly. Methods to "accelerate" simulations in a manner that still lead to reliable results is a matter of active current research, and it is a problem that needs to be solved if behavior is to be modeled at the extremely long lives that result from VHCF conditions. The fatigue behavior under these conditions will be discussed next.

18.5.4   Crack Nucleation under Very High Cycle Fatigue Conditions

By the late 1990s and early 2000s there was convincing evidence that the conventional fatigue limit observed in many metallic materials at lives of 10^6–10^7 cycles did not mean "infinite" life, but rather a fairly pronounced inflexion in the S–N curves of these materials (Bathias, 1999) and that failure was possible for stresses below the fatigue limit at lives $>10^7$ cycles. In the next two decades after that, a fair amount of work was carried out to try to elucidate the mechanisms behind these observations. The results of this work clearly pointed out that Wöhler-type S–N curves or Coffin–Manson fatigue life diagrams (see **Figure 7**) can exhibit a lower fatigue limit in the VHCF range and they take a step-wise form, and need to be considered as multi-stage fatigue life diagrams (Mughrabi, 2006; Pyttel et al., 2011). It was also found that as the stress was reduced below the conventional fatigue limit and lives went into the VHCF regime, the origin of crack initiation changed from the surface to the inside of the sample, producing the well-known "fisheye" fracture, particularly at nonmetallic inclusions (Mughrabi, 2006; Pyttel et al., 2011). However, crack nucleation sites that cannot be attributed to inclusions have also been observed (Chai, 2006; Pyttel et al., 2011). It is important to point out that, under VHCF conditions, crack nucleation can account for up to 99% of the fatigue life, which makes it the life controlling mechanism.

The behavior under VHCF conditions has led to the classification of materials into two types: type I refers to ductile materials where failure is driven by pure plasticity rather than because of the effect of inclusions. Pure metals certainly fall into this category. Type II refers to materials with nonmetallic inclusions and processing defects that can become sites for crack nucleation, particularly in the inside of the samples. Many engineering alloys fall into this category (Mughrabi, 2006; Pyttel et al., 2011). Schematic fatigue life diagrams for the two types of material are shown in **Figure 63** in the form of Coffin–Manson plots (plastic strain amplitude $\Delta\varepsilon_{pl}/2$ versus N_f), from which S–N curves could be obtained by using the CSSC.

The diagram shown in **Figure 63(a)** has four ranges, given by (I) the low cycle fatigue (LCF) Coffin–Manson range; (II) the PSB threshold related to the HCF plastic strain fatigue limit, that is, the strain below which PSBs do not appear; (III) the transition from the HCF limit to the VHCF range and (IV) the threshold corresponding to the VHCF limit, where there is no irreversibility associated with plastic strain. The four stages in **Figure 63b** are similar, that is, (I) is also the Coffin–Manson range (or Basquin range, depending on the strength of the material), (II) is the conventional fatigue limit, (III) is the transition from HCF to VHCF and (IV) is the very high cycle (and true) fatigue limit. Mughrabi (2006) points out that there is reason to believe such limit does exist. The transitions between the different ranges, of course, will be gradual with some scatter and overlap.

Note the transition in failure mechanism from the surface to the inside of the sample as life increases for type II materials. The mechanisms for stages (III) and (IV) may or may not be similar to those present in type I materials. A discussion of the mechanisms for type II materials, as important as they are since they apply to many engineering materials, is beyond the scope of this chapter and the reader is referred to the reviews by Mughrabi (2006), Pyttel et al. (2011) and references therein for more information.

Following Mughrabi (2006), copper will be used for the discussion of the VHCF behavior of type I materials given the vast literature on fatigue behavior for that material. Results for pure nickel can be found in Stöcker et al. (2011). In addition, see the review by Pyttel et al. (2011) for information on other materials.

Now, given that the LCF and HCF behaviors for copper are well known, the focus of any inquiry must be the extension of range II and the mechanisms controlling ranges III and IV.

Regarding the extension of range II, note that at stress levels below the PSB threshold, cyclic strain localization in PSBs will, in principle, not occur, when cyclic saturation is achieved. Therefore, loop patches and channels (the vein structure, see Section 18.4.1) will be present and the deformation produced by cyclic slip will be homogeneous. However, this cyclic slip, which is assumed to occur randomly at very low strain amplitudes, still produces a nonnegligible amount of irreversible slip,

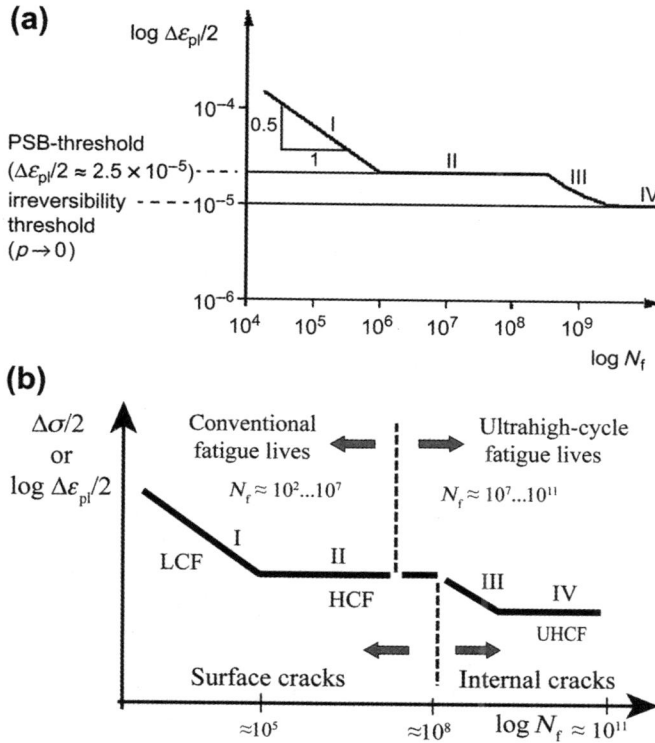

Figure 63 (a) Schematic Coffin–Manson type fatigue life diagram for ductile type I materials. (b) Schematic fatigue life diagram of type II materials containing inclusions. Source: After Mughrabi, 2006, courtesy of Elsevier.

because of dislocation glide that is not reversed as the strain changes sign (Mughrabi, 2006). This will leave slip steps at the surface and a gradual roughening of the surface will occur. Mughrabi (2006) estimated the extent of range II for copper using a relationship between the root mean square (quadratic mean) of the surface roughness and the local plastic shear strain, via the following equation:

$$\sqrt{\langle x^2 \rangle} = \sqrt{4N\gamma_{pl,loc}pbh},$$ (11)

where N is the number of cycles, $\gamma_{pl,loc}$ is the resolved shear strain amplitude associated with the local axial plastic strain, p is the fraction of plastic strain that is irreversible, b is the magnitude of the Burgers vector and h is the interplanar spacing. If the roughness is measured and $\gamma_{pl,loc}$ can be estimated, and assuming that a critical value of $4N\gamma_{pl,loc}\,p$ is needed for crack nucleation that is the same above and below the PSB threshold, Mughrabi (2006) estimated that range II should extend from 10^7 to 10^9 cycles. He considered this range to be a lower bound, which implies that finite lives on type I materials can extend well into the VHCF regime.

Regarding regime III, one mechanism proposed by Mughrabi (2006) requires reaching a critical state of surface roughness. At that point, some valleys in the rough surface profile will act as strong enough

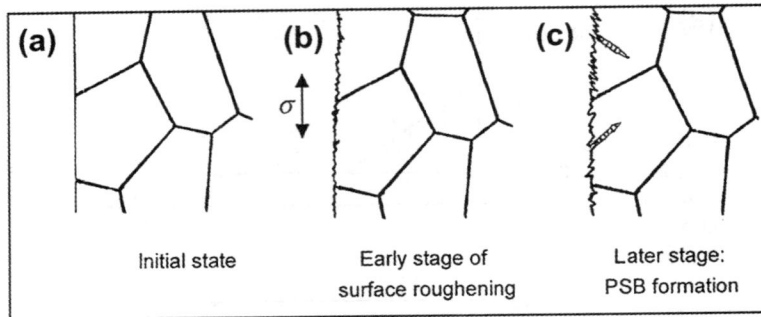

Figure 64 Schematic crack nucleation in polycrystalline copper under VHCF conditions. (a) Initial state, (b) early stage of surface roughening and (c) PSB formation at a later stage, which will be followed by stage I crack initiation. Source: After Mughrabi, 2006, courtesy of Elsevier.

stress concentrations to increase the local stress to a level equal to or higher than the PSB threshold. This would then lead to the formation of an "embryonic" PSB, as shown in **Figure 64**.

Despite the fact that the embryonic PSBs will only penetrate to a shallow depth into the material, they will accelerate the surface roughening as strain is localized. This increase in surface roughening will eventually result in stage I crack nucleation, even though the macroscopic stress level is below the threshold for PSB formation (Mughrabi, 2006).

These mechanisms were studied in more detail experimentally by Stanzl-Tschegg and Schönbauer (2010) and Weidner et al. (2010) using ultrasonic testing, as well as FIB, ECC and TEM characterization techniques. Stage I cracks and localization of slip were found below the known PSB threshold of 63 MPa, not only on surface grains, but also in inside grains as well, as shown in **Figure 65**.

These results indicate that PSB structures can indeed be found at stresses lower than the expected threshold. In this regard, note that Stanzl-Tschegg and Schönbauer (2010) found that the plastic strain for PSB nucleation was actually lower than expected, since they found that PSB could appear at plastic strain amplitudes as low as 3.6×10^{-6} (see **Figure 63(a)**, where the expected value is given). The conclusion is that PSB formation depends not only on the plastic strain level but also on the number of cycles as irreversible plastic deformation is accumulated. In addition, cracks can initiate subsurface as well, just like in type II materials, but driven by pure plasticity.

This offers some insight into observations of interior crack nucleation sites in type II materials that cannot be linked to a nonmetallic inclusion. In that regard, nucleation in some of these cases can be linked to the presence of a "supergrain" (see Huang et al., 2010 and references therein), which is a cluster of grains with small angle grain boundaries among them where strain can localize to a level much higher than in other sites of the microstructure. Note that Stanzl-Tschegg and Schönbauer (2010), Weidner et al. (2010) also report the presence of grain boundary cracking, most likely because of the interaction of SBs (or local PSBs) interacting with high-angle grain boundaries. Finally, Stanzl-Tschegg and Schönbauer (2010) report that the stage I cracks might not necessarily lead to ultimate failure and that a stress level higher than that needed to nucleate them is required to make these short cracks transition into long ones. The issues related to fatigue fracture, which are an extremely important aspect of the overall fatigue phenomenon, will be discussed in the next section.

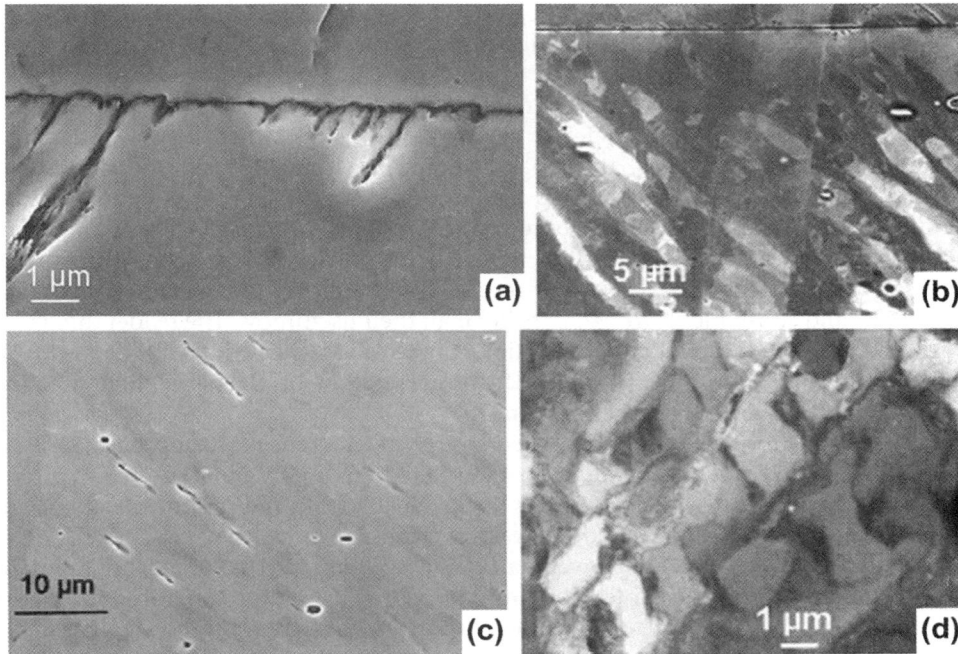

Figure 65 Characterization of fatigue cracking and dislocation structures in polycrystalline copper samples tested below the PSB threshold. (a) Stage I cracks at the surface from an FIB cross-section at a location where the local stress amplitude was about 57 MPa, that is, about 6 MPa below the PSB threshold. (b) Region of localized cyclic shear deformation in slip lamellae from a surface grain, about 15 MPa below the PSB threshold (ECC image). (c) Stage I "microcracks" in an interior grain, about 1.5 MPa below the PSB threshold (FIB cross-section). (d) Dislocation structure in an interior grain, with ladder-like slip lamella, about 9 MPa below the PSB threshold. Source: From Weidner et al., 2010. Courtesy of Elsevier.

18.6 Fatigue Crack Propagation

18.6.1 Macroscopic Behavior of Fatigue Crack Propagation

The great bulk of research into crack propagation in the past 30 years (dating from the application of *linear elastic fracture mechanics*, LEFM, to crack propagation, which provided a rational basis for growth measurements) has been empirical with the aim of defining crack propagation rate as a function of the stress intensity, that is, measuring crack growth kinetics. Propagation for long cracks, under variable loading, after overloads, in different environments, and for different modes of stressing has been explored in great detail. Studies of long crack growth, that is, when the crack length is much greater than the microstructural length scale or the plastic zone of the stress raiser (if any) that nucleated the crack, often permit a direct comparison between the behaviors of large structures and small laboratory specimens via the elastic stress intensity factor range, $\Delta K = Y(a, \text{load}, \text{geometry})\Delta\sigma\sqrt{\pi a}$, where a is the crack length, $\Delta\sigma$ is the stress range and Y is a "shape factor" that depends on crack length, specimen geometry and loading configuration. Comparisons between the behavior of cracks in structures (or samples) of different sizes are typically made when the values of ΔK are the same for both length scales.

This similitude only applies when the plastic region at the crack tip is small in relation to the length of the crack and the size of the structure, that is, under *small-scale yielding* conditions. Generally, the small extent of the plasticity does not significantly perturb the elastic crack tip stress field in the laboratory specimen, thereby allowing direct use of laboratory results for predicting the behavior of large engineering structures, containing bigger defects, but where the stress intensity is similar because the operating stresses are lower. There are handbooks available providing LEFM solutions for a wide variety of structure and specimen geometries and loading modes, for example, the well-known Stress Intensity Factor Handbook. It is not the purpose here to review in detail the theoretical basis of this approach (see Suresh, 1998), or the great body of work in which it has been applied, nor the theories of crack propagation that have been spawned by its results (for example, the sustained investigations of Weertman (1979), and subsequent works), in spite of all their usefulness. The reader interested in such matters is referred to the many excellent publications of the ASTM, the review of Miller (1987), and the book by Suresh (1998), to cite a few. Rather the focus here will be on the basic phenomena and those aspects that give insight directly into the mechanism.

Fatigue crack propagation could not be covered entirely by the long crack approach, however. Crack growth rates have been recorded as being faster than LEFM analyses would predict, as attempts were made to measure the growth kinetics of small cracks. Such cracks are typical of fatigue in smooth specimens and high strength materials under low stresses and their growth behavior is life determining. The reasons for cracks growing faster or slower than LEFM would predict are many, but (chemical effects aside) generally are connected with a loss of similitude (Davidson and Lankford, 1984) that occurs when stress levels are too high and general yielding occurs or when cracks are so small as to be affected strongly by small-scale plasticity effects, such as PSBs, or by microstructural features. In addition, Vasudevan and coworkers (Vasudevan et al. 2001; Sadananda and Vasudevan, 2003; Sadananda et al., 2003, 2007) have postulated that, in addition to some of the effects mentioned above, the fact that many short cracks are growing on a residual stress field also leads to their anomalous behavior, since in their model of fatigue crack growth both K_{max} and ΔK play key roles on controlling the kinetics of crack growth. Consequently it is necessary to consider the behavior of short cracks.

A common method of presenting kinetic data on the crack propagation of long cracks and short cracks is to plot the fatigue crack propagation rate, da/dN, versus the stress intensity range, ΔK, on a log–log scale, as shown in **Figure 66**.

Such a curve has become as familiar as the S–N curve. Three regimes of da/dN are widely recognized: (I) The threshold region where the propagation rate is of the order of an atomic spacing per cycle or less. At low enough stress intensities (the "threshold" ΔK_{th}) the crack can be considered immobile. This applies in general to long cracks, but not to short cracks, as shown in **Figure 66**. (II) A mid-region where the Paris equation $da/dN = C(\Delta K)^m$ is considered to hold. (III) A high rate region in which the mechanisms of failure are of the quasi-static type and the maximum stress intensity approaches the critical value for static failure, K_{Ic}. This region is one in which general yielding may occur, and the residual strength of a cracked member is of concern. Since it involves static failure mechanisms, it will not be treated here.

In regime I, where the stress intensity is low, a crack may typically develop along an active slip plane, like a stage I crack in a copper single crystal, and may involve PSB formation ahead of the crack. In this case the PSB is short compared with the length of the crack. Since the PSBs are formed on the most highly stressed slip plane (having the most favorable Schmid factor), the orientation of this crack is usually near 45° to the stress axis. To obtain a Stage I crack, the stress intensity must be low and such propagation is associated with regime I described above. However, it is noteworthy that many materials, including steels, do not propagate in the Stage I, 45° manner, and even cracks in

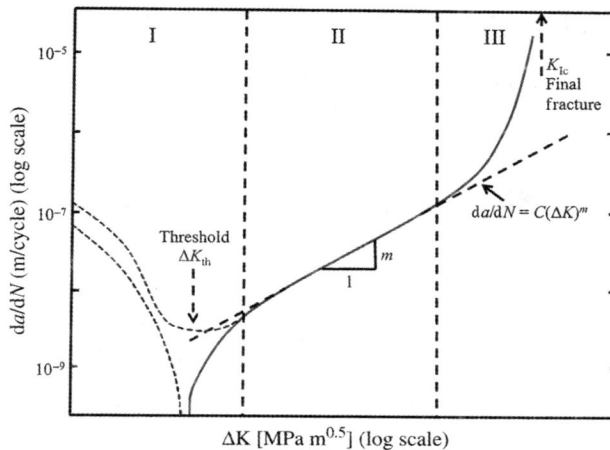

Figure 66 Schematic plot of fatigue crack propagation rate da/dN versus stress intensity range, ΔK, depicting three regimes of crack propagation, I, II and III. Curves for two types of short cracks able to propagate below the threshold are also shown (dotted lines): one crack slows down as ΔK increases until it arrests and the other one slows down at first, but then accelerates and starts behaving like a long crack.

copper, which normally propagate by Stage I in air, can be induced to propagate normal to the direction of stressing by testing in vacuum (Neumann et al., 1977). Note that the crack may stop growing altogether if the stress intensity is too low, and thus the threshold stress intensity range, ΔK_{th}, shown in **Figure 66**, is defined.

When a Stage I crack is propagating in a copper single crystal, the PSB hosting it may well go right through the crystal. Under these conditions LEFM will not apply and the plasticity of the specimen, for purposes of analyzing growth, must be taken into account by *elasto-plastic fracture mechanics* (*EPFM*). Such a crack may be considered a short crack, especially since the great part of the fatigue life of the crystal is taken up with the crack growing physically as a short crack (Ma and Laird, 1989b,d,c).

In regime II, cracks usually propagate by a mechanism of plastic deformation that leaves characteristic fatigue striations on the fracture surface, provided the metal is a well-behaved ductile deformer and does not introduce another mechanism of failure such as intergranular fracture or a progressive brittle fracture. Fracture surface striations can, in some circumstances, be suppressed by testing in vacuum (Laird, 1979), but at high stresses, a ductile metal will form regular striations even in a good vacuum (Laird and Smith, 1963).

Kitagawa and Takahashi (1976) compared high stress and low stress fatigue fracture behavior on a diagram similar to **Figure 67(a)**.

The line given by ΔK_{th} represents the low stress threshold condition below which a crack should not grow if LEFM assumptions are valid. They will be invalid of course, if the plastic zone at the crack tip is no longer small in relation to the size of the crack, and this occurs to an increasing extent when the term $\Delta\sigma$ increases above about two-thirds of the cyclic yield stress, or exceeds the threshold for PSB formation in a polycrystal, in a push–pull test, if that is the factor that determines the fatigue limit (it may not, the fatigue limit may be higher for some reason). The reader must be aware that conventional fatigue limits lie generally below the monotonic yield stress of engineering materials, but in a pure copper polycrystal, the conventional fatigue limit may be several times the monotonic 0.2%

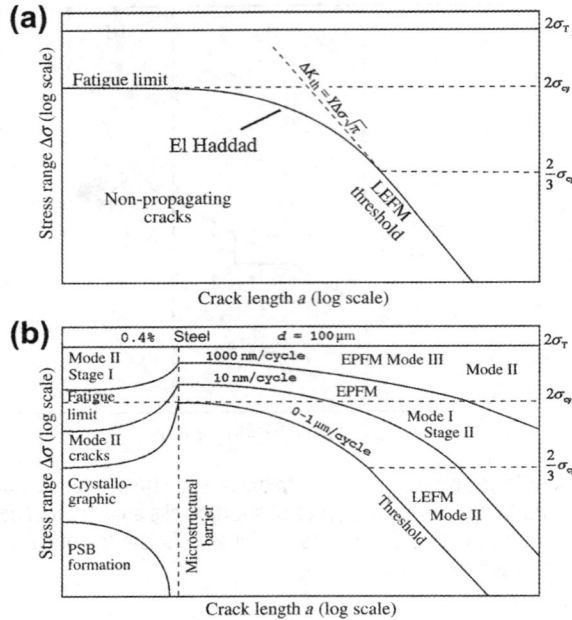

Figure 67 Fatigue failure boundary maps: (a) the Kitagawa–Takahashi diagram (Kitagawa and Takahashi, 1976) and (b) the Brown (Brown, 1986) diagram with crack speed contours. Both plot the stress range versus the crack length. Source: Redrawn and modified from Laird, 1996 based on an original figure from Miller, 1987. Printed with permission of John Wiley and Sons.

flow stress of a well-annealed copper. This, of course, changes for VHCF conditions as discussed in Section 18.5.4. A second line of the Kitagawa–Takahashi plot is the conventional fatigue limit itself, approximated to the cyclic yield stress range (Miller, 1987). The lines for the fatigue limit, and the threshold stress intensity are then joined up by using the El Haddad's relationship, that is,

$$\Delta\sigma_{EH} = \frac{\Delta K_{th}}{\sqrt{\pi(a + a_0)}} \quad \text{and} \quad a_0 = \frac{1}{\pi}\left(\frac{\Delta K_{th}}{\Delta\sigma_0}\right), \tag{12}$$

where $\Delta\sigma_0$ is the stress range corresponding to the fatigue limit (Ciavarella and Monno, 2006). This relationship matches the asymptotic behavior of the fatigue limit regime (where a crack might be absent) to the threshold regime governed by the a_0 constant, which is in principle a material property, at least for long cracks. The curve defined by this relationship defines a threshold for crack propagation or self-arrest that divides the Kitawaga–Takahashi map into a lower area with nonpropagating cracks leading to "safe life," and an upper area where crack propagation is possible, resulting in finite life (Ciavarella and Monno, 2006). Generalizations to the Kitawaga–Takahashi diagram that include alternatives to El Haddad's expression to transition between the fatigue limit and the crack threshold regimes have been studied by Ciavarella and Monno (2006).

Figure 67(a) was subsequently reanalyzed and extended by Brown (1986) to the form shown in **Figure 67(b)**, to include contours of crack speed and fatigue fracture modes (Stages I and II). The *Modes*

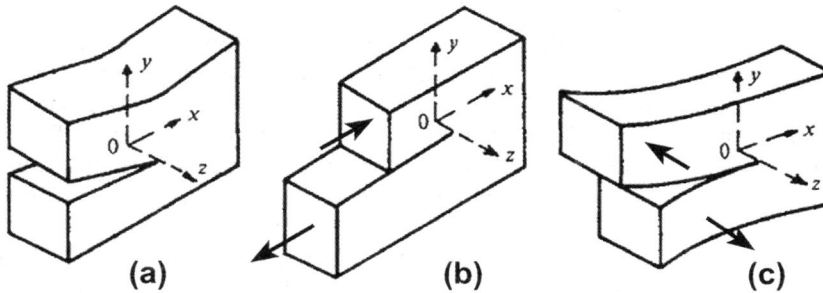

Figure 68   The three basic modes of stressing a crack: (a) Tensile opening (Mode I); (b) in-plane sliding (Mode II) and (c) anti-plane shear (Mode III).

I to III shown on the diagram correspond not to the fatigue fracture stages but to the fracture stressing modes shown in **Figure 68**.

In many situations these modes may well operate in combination, that is, mix mode fracture. For example, a thumb-nail Stage I crack growing in a PSB under push–pull stress will be subjected to Mode I by the uniaxial stress, Mode II in the direction of the primary Burgers vector at the front of the crack and by Mode III at the sides of the "nail". It is customary, in complex situations, to describe the mode according to the dominant one of the three. For example, the Stage I crack growing in a PSB would be described under Mode II.

Brown associated in **Figure 67(b)** the fatigue limit and the ΔK_{th} lines of **Figure 67(a)** with the contour of ∼zero crack growth rate. Note that, on the left of the diagram, short cracks can initiate and grow at stresses greater than two-thirds of the cyclic yield stress. However, if a microstructural feature, such as a grain boundary or a patch of a second phase, provides a large barrier to growth, then the crack may stop growing, giving rise to the phenomenon of *nonpropagating cracks*. Even at stresses above the fatigue limit, a short crack growing in a single grain can be retarded by a microstructural barrier, but it will accelerate again once the barrier is overcome. On the other hand, it is evident from the Brown map, as Miller (1987) points out, that cracks growing under conditions of LEFM only increase in speed as the crack length increases, assuming of course that the applied stress range is constant and the expression for ΔK given in **Figure 67(a)** applies. Variations in crack growth can come about for geometrical reasons, for example, a crack that has nucleated and grown in a stress concentration may grow out of it and, entering a region where it is no longer affected by the higher stresses, becomes a nonpropagating crack.

The above overview serves to establish the context of scientific interest in Stage I and II crack propagation. The phenomena and mechanisms of these fatigue fracture modes are described as follows:

18.6.2   Short Crack Growth – Stage I

Interest in research on short crack growth has grown enormously in the past 2 decades and there have been many investigations of growth behavior in both engineering and pure metals (see Miller and De Los Ríos, 1986; Ritchie and Lankford, 1986; Ravichandran et al., 1999 and the proceedings of the Triannual Fatigue conferences). Space does not allow full treatment of the many interesting interactions that can occur between small cracks and microstructural features in complex materials. The focus here will be on pure metals in which the behavior is well-defined and only a few examples of studies in engineering materials are mentioned.

By delicate techniques of specimen preparation, TEM studies have been conducted on short Stage I cracks in preexisting PSBs in copper (Katagari et al., 1977). These studies show that cracks can penetrate to depths many times the inter-wall spacing of the PSBs without altering their dislocation structure significantly. Observations of this kind are not confined to pure metals: cracks in age-hardened alloys have been shown to propagate this way (Wilhem et al., 1979; Vogel et al., 1980). Thus, a propagating Stage I crack conditions the metal in the PSB ahead of it and follows its path. This has also been observed to occur in copper single crystals tested under load control at the plateau stress (Jameel, 1997; Jameel et al., 2003).

The implication of results of this type is that the PSB deformation processes, which act first to form notch-peak topography containing the crack embryo, now can lead to Stage I propagation. As noted in Section 18.4, screw dislocations are the main agent for carrying strain in the PSBs of copper and other fcc metals with wavy slip and they are distributed with roughly equal probability on all the slip planes within the PSB. A notch, that is, a crack embryo, serves to concentrate slip in tension along a narrow group of slip planes at the base of the notch. That is, a stress concentration attracts the screw dislocations into the planes at the tip of the notch, the dislocations cross-slipping from their regular planes to meet the notch. In compression, the notch closes, the stress concentration is not sensed, and the PSB screw dislocations return to the other end of their channel more equally distributed on the planes in the PSB. Repetition causes crack intrusion, Stage I growth; the mechanism is illustrated schematically in **Figure 69**.

The mechanism shown in **Figure 69** was offered (Laird, 1979) before it was understood that (and why) narrow extrusions develop after the crack has formed, but this mechanism shows that extrusion formation accompanies the process of crack growth. It represents, therefore, a current mechanism of extrusion formation in terms of dislocation behavior.

Mughrabi (1980) has offered helpful suggestions for explaining the absence of (or reduced tendency to) Stage I cracking in steels, and perhaps in other bcc metals. He cites the work on α-iron of Katagari et al. (1979) who have observed dislocation cell structures at the tips of young cracks formed in a 45° mode, and points out that, in typical fatigue conditions, where the strain rates are quite fast, bcc transition metals do not usually deform by PSBs, since they deform by the low temperature mode of deformation. Clearly, a capacity for conventional Stage I growth is tied very closely with PSB formation.

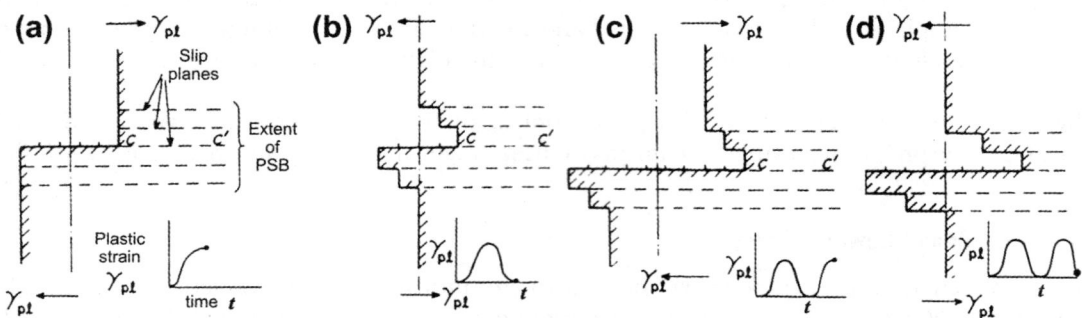

Figure 69 Schematic representation of Stage I propagation in a PSB, via the deformation produced in a surface normal to the primary Burgers vector. A notch, omitted for clarity from (a), serves to concentrate the strain during tension in or near slip plane C–C'. During compression, (b) and (d), the strain is more evenly distributed on the atomistic slip planes, shown as dotted lines. The inserted γ_{pl} versus time plot indicates the loading sequence corresponding to the slip behavior. Source: Taken from Laird, 1979. Courtesy of ASM International.

Also, the role of the environment, in limiting the reversibility of slip at the crack tip by chemical interactions with "fresh" surfaces produced by slip steps, will have important effects on crack growth kinetics.

Because of the environmental problem, few workers have been bold enough to advance theories for Stage I growth. Of course, insofar as the mechanisms of Stage I and Stage II are common, any theory for Stage II growth that emphasizes plasticity mechanisms for growth should frequently apply to Stage I. Purushothaman and Tien (1978) and Tien and Purushothaman (1978) have treated Stage I growth by an interesting discontinuous growth mechanism analogous to ledge growth in phase transformations. There is experimental support for such a mechanism: Stanzl (1980) investigated near-threshold growth for variously oriented copper single crystals and observed large variations in growth kinetics. She also observed that the fracture surfaces were not planar on a macroscopic scale and that the crack propagation direction changed frequently. Furthermore, SEM examination showed that the crack propagated along at least two planes. These effects in combination with step changes in growth kinetics (often observed as "scatter") are the hallmarks of a ledge mechanism. The interaction of Stage I cracks with multislip systems is still an unsolved problem in fatigue, even for wavy slip metals. The morphologies of Stage I faceted fracture surfaces for Cu-16 at% Al single crystals have been reported by Hong and Laird (1991a). Even in this planar slip metal, there are interesting effects of multislip on the fracture surface morphology.

The phenomena of Stage I growth in copper single crystals are now much better understood through application of the sharp corner technique (Basinski and Basinski, 1984; Ma and Laird, 1989b,d,c) or its equivalent, precision sectioning through crystals containing Stage I cracks (Hunsche and Neumann, 1986; Neumann and Tonnessen, 1987). The sharp corner technique has the unique advantage of permitting the population of Stage I cracks, or any one of them, to be studied during the whole fatigue life, whereas precision sectioning, say by using FIB techniques, sacrifices the crystal, and growth behavior can then be treated only statistically. However, as shown in **Figure 65**, these sectioning techniques can be very helpful to understand Stage I cracking in polycrystalline materials.

The growth behavior goes as follows: the cracks initiate fairly early in life, and a sizable population of them is established by 10,000 cycles, but they are still small at this stage. Ma and Laird (1986) have documented the distributions of small crack sizes in fatigued copper single crystals and treated the statistical aspects of their competitive growth. Ma and Laird (1989b,d,c) collected statistical information on crack growth kinetics of single crystals tested under constant strain amplitudes within the plateau regime. The growth kinetics of these cracks can be expected to vary with specimen history—for example, by ramp loading at the start in the Neumann manner. However, for the constant amplitude testing method used and for the results shown here, the average growth rate for *much* of Stage I life does not differ very much for crystals cycled at different amplitudes. Such behavior reflects the average localized strain of the PSBs. However, the maximum growth rate for each specimen, reflecting the highest level of the local strain, increases as the strain increases. The crack growth rate was found to diminish slightly as cycling proceeds. The most likely explanation at present is that the population of PSBs is not actually constant during life and it takes a good fraction of life to become established (secondary hardening). In addition, this behavior is also consistent with the spatiotemporal variability of PSB activity reported by Holste (2004). Later arriving PSBs act to deprive older PSBs, which contain the cracks, of some of the localized strain they carried at earlier stages of life. This deprivation operates in an average sense, not by nearest neighbor interactions.

After slow Stage I crack growth has occurred, with an average rate of about 0.1 nm/cycle, crack acceleration takes place, the higher the amplitude the earlier, giving rise to shorter lives at higher amplitudes. According to regular LEFM such an acceleration would be surprising because, under strain control, the stress intensity does not vary with crack length. The reason for the acceleration is connected with the localized strain of the PSBs and how it interacts with the growing cracks. It turns out (Ma and Laird, 1989b,d,c) that there is little effect of crack growth on the local strain behavior until the cracks become so numerous that their stress shadows fall on each other. There is a critical spacing of about 10 μm within which the effect becomes strong. Since the PSB volume fraction increases with applied strain amplitude, the crack population also increases proportionately. Therefore, the higher the strain amplitude, the sooner the cracks start to compete with each other for localized strain, and the quicker failure ensues. Naturally the longest cracks survive this process and rob their lesser neighbors of the strain they formerly carried.

Evidence for such behavior is shown in **Figure 70** in the form of both sharp corner observations (a) and PSB slip behavior on that surface of the specimen for which the primary Burgers vector is most normally oriented (b).

Figure 70(a) shows a series of Stage I cracks and the longer ones have robbed the smaller crack of strain. In this experiment, a sharp corner section was prepared for observation and then the crystal was returned to the machine for further cycling. Reactivation of the PSBs produced slip markings once again on the polished face of the "sharp corner". Note how the slip markings connected with the PSB of the shortest crack have stopped short of the crack, quenching its growth. The cracks are now close enough for the acceleration phase to begin. **Figure 70(b)** gives a different view of localized strain deprivation. Here the top face was repolished (slightly) but without eliminating the cracks. On restarting the test, elliptical shadows free of slip markings are cast around the larger members of the crack population, indicating the extent of the shadowed volume. On average, the larger cracks grow a few 100 μm in depth before the acceleration process takes over, and complete failure develops rapidly. As the Stage I crack propagates, multislip conditions develop in the remainder of the cross-section, and the crack turns normal to the tensile stress axis.

Naturally, if a Stage I crack is propagating under LEFM conditions, meaning that the PSBs at the crack tip are small in relation to the crack size then the detailed kinetics will be different from those described above, which apply to single crystals. An attempt at analyzing monocrystalline Stage I growth by EPFM

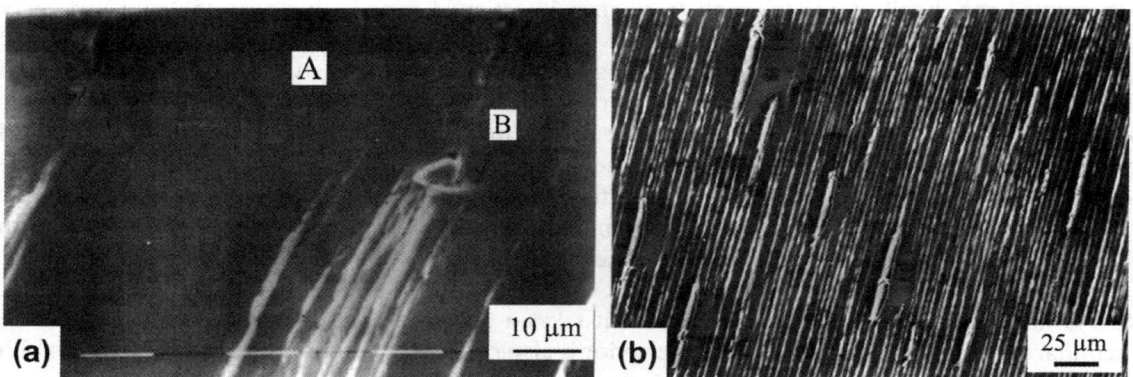

Figure 70 Competition for plastic strain between growing Stage I cracks in a copper single crystal and adjacent PSBs: (a) sharp corner technique. Crack A finally fell within the strain relaxation region caused by crack B, and stopped growing. The PSB immediately ahead of its tip became inactive. (b) Stress shadows causing PSB inactivity around more significant Stage I cracks. Crystal originally cycled at 2×10^{-3} and then at 8×10^{-3}, next repolished, and finally cycled at 24 MPa for 341 kilocycles. Source: Taken from Laird, 1996 after an image originally from Ma, 1987. Courtesy of Ma.

has been made by Ma and Laird (1989b,d,c). The behavior of interacting Stage I cracks have also been studied in copper single crystals tested under load control (Jameel et al., 2003). These authors also showed that larger crack populations are present under load control than under plastic strain control for the same level of accumulated plastic strain because of the presence of a larger number of micro-PSBs, as they nucleate continuously during the life of the specimen under constant stress amplitudes.

In polycrystals, even those of pure metals, Stage I propagation becomes complicated because of the issues related to individual crystallography of grains and the presence of grain boundaries. As a short Stage I crack grows in mix mode along a slip plane it can approach a grain boundary, when it will typically slow down and one of three things can happen: the crack can arrest, it can propagate into the next grain, typically following another slip plane in the neighboring grain (Zhai et al., 2000), or it can follow the grain boundary (see Peralta et al., 2005 for examples of some of these cases in pure Ni).

Studying the mechanisms behind this behavior in polycrystals is complicated because of the intrinsic 3-D nature of the slip geometry of individual grains, as well as their elastic and plastic anisotropy, all of which are affected by the individual grain orientations. This is compounded significantly by the difficulty to characterize the local behavior within individual grains in samples with grain sizes that are typical of engineering materials, for example, from a few to a few tens of micrometers. One approach to simplify this problem is the use of "multicrystalline" samples, that is, specimens with grains large enough so that the small length scale is no longer a problem, which allows using more conventional metallographic techniques for characterization (Peralta et al., 2005). Although useful insight can be obtained from these experiments, they have the disadvantage of limiting the number of "cases" that can be studied, since only a few grains and grain boundaries are present in each sample, which limits the amount of data, and understanding, that can be obtained from these experiments. This is particularly so given the fact that the typical scatter displayed by short cracks requires statistical information on microstructural effects to be analyzed thoroughly.

Studies in actual polycrystals are, therefore, the best way to obtain a full picture. In that regard, Zhai et al. (2000) performed experiments in an Al–Li alloy with elongated grains to relate the change in slip geometry across a grain boundary to the deflection experienced by a Stage I crack as it crosses the boundary. These authors defined two angles between slip planes in the adjacent grains that control crack growth across the boundary, as shown in **Figure 71**.

In **Figure 71**, the angle β corresponds to the angle between the traces of the slip planes in each grain on the surface of the sample (the plane of observation) and can be thought of as a tilt angle. On the other hand, the angle α corresponds to the angle between the traces of the slip planes on the plane of the boundary and can be interpreted as a twist angle. Zhai et al. (2000) pointed out that the angle α is particularly important, since it represents the major resistance to propagation of a crack across a grain boundary. These authors explain that a large value of α will lead to a large wedge-shaped area (a–c in **Figure 71**) on the boundary plane that represents a "ligament" that has to be fractured for the crack to cross to grain 2. The larger this area, the more resistance there should be for the crack to cross the boundary, as additional energy has to be spent breaking the ligament. These authors also collected evidence that Stage I cracking across grain boundaries will occur preferentially on slip systems that minimize the angle of twist, which provides a simple model to predict crack path and local retardation. The angle of tilt also contributes to crack retardation by reducing the driving force for crack growth, from a fracture mechanics point of view.

Wen and Zhai (2011) have recently studied these effects in more detail and so have Schaef et al. (2011). The latter authors used FIB tomography in a Ni-based superalloy CMSX-4 to visualize Stage I cracks in 3-D and also used the FIB to machine micronotches to nucleate and propagate short cracks

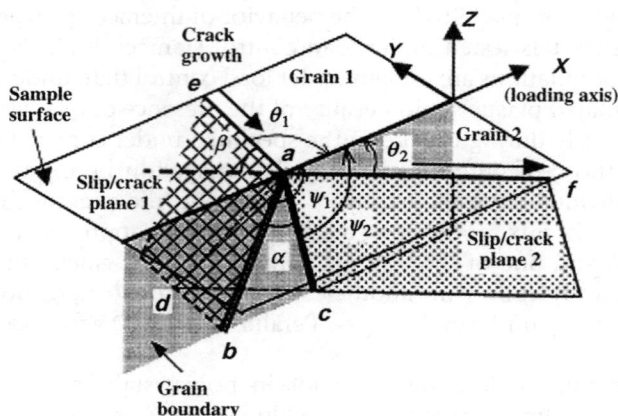

Figure 71   Schematic slip geometry a Stage I crack growing along slip plane 1 in grain 1 as it transfers onto plane 2 in grain 2. The crack growth is controlled by angles α and β. Source: After Zhai et al., 2000, courtesy of Elsevier.

such that they were able to exert some control on the slip planes where the crack propagated and also the distance to the grain boundary. Therefore, they were able to keep constant most of the crack parameters for different boundaries and they demonstrated the reproducibility of their techniques. These authors chose to do 3-D tomography on boundaries that were selected to minimize and maximize crack growth resistance across the boundary as per the geometrical model by Zhai et al. (2000) discussed in the previous paragraph. Therefore, one boundary was such that the angle of twist was small for the potentially preferred slip planes that can carry the crack (those with high Schmid factors given the applied load) and another one where that was not the case. The 3-D views of these cracks are shown in **Figure 72**.

Schaef et al. (2011) reported that the crack with the low value of twist angle slowed down as it approached the boundary, but eventually continued, whereas the other one actually stopped for 2000 cycles before propagating through the boundary. The results in **Figure 72(a)** indicate that for the small angle of twist the boundary propagated onto another (111) slip plane on the adjacent grain and it can be seen clearly that the traces of the crack planes are almost parallel at the boundary. In the case of the crack with the larger angle of twist, **Figure 72(b)**, the crack went almost straight through onto a plane that is not parallel to a slip plane. These authors provided evidence that the crack actually grew in zigzag using small steps on two slip planes. This reduced the amount of material that had to be broken in the large ligament resulting from the large twist angle and represents a mechanism by which Stage I cracks can propagate across boundaries without having to minimize the angle of twist.

There are a variety of other phenomena of interest on short crack growth; for example modeling of Stage I crack growth in polycrystals is certainly important (see, e.g. Kunkler et al., 2008 and references therein). However, because of space limitations they will not be addressed here. Nonetheless, one noteworthy development on characterization techniques derives from the use of in situ testing inside powerful X-ray sources, capable of resolving 3-D details of short cracks and microstructure as loading is applied while simultaneously providing time resolved data. Examples of the data that can be collected with such techniques are shown in **Figure 73**.

Significant advances on the understanding of crack nucleation and short propagation are likely to results from data sets such as the one shown in **Figure 73**, as full 3-D views of cracks interacting with

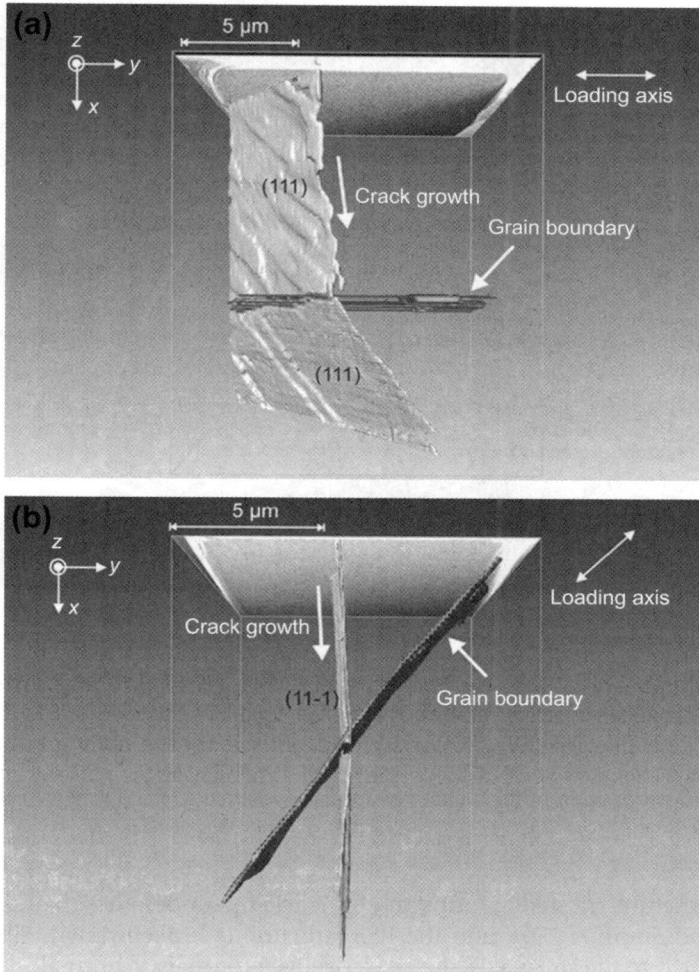

Figure 72 Three-dimensional FIB tomography of short cracks traversing grain boundaries in a Ni-based superalloy. (a) Small twist angle. (b) Large twist angle. Source: After Schaef et al., 2011. Courtesy of Elsevier.

microstructure as a function of applied cycles can be obtained. See Rannou et al. (2010) for another example.

As the length of Stage I cracks increases, they eventually kink to grow perpendicular to the applied load, and become long cracks, which will be discussed below.

18.6.3 Long Crack Growth – Stage II

How deep a Stage I crack grows in a smooth specimen depends on the applied stress. In low strain fatigue, a Stage I crack can penetrate to the depth of several grain diameters before the increase of the stress intensity with the crack length promotes slip on systems other than the primary one in any grain.

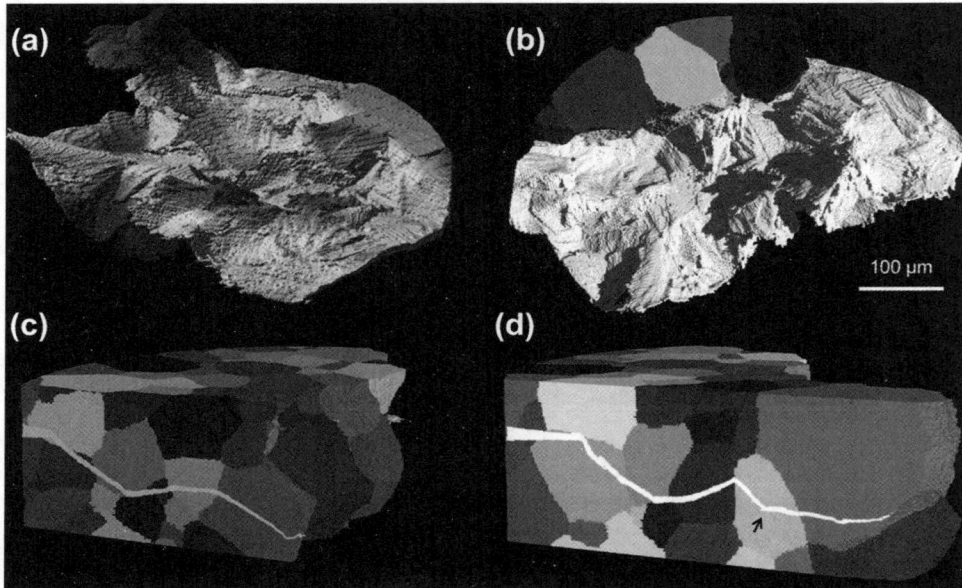

Figure 73 Three-dimensional rendition of a fatigue crack in a titanium alloy after 75.5 kilocycles (load axis is vertical). (a) The colors represent the vertical crack position: blue = 0 μm, red = 180 μm. (b) Three-dimensional rendering of the grains intersecting the fracture surface: some grains have been made transparent to illustrate the crack interactions with the local microstructure. The different colors correspond to different grain orientations. (c and d) 2-D views of the crack path on different planes cut in the bulk of the sample. Changes in the direction of the crack can be correlated with the presence of grain boundaries (c), but can also be observed within a grain (arrow in d). Source: After Herbig et al., 2011, courtesy of Elsevier. (For color version of this figure, the reader is referred to the online version of this book.)

A dislocation cell structure normally forms at the crack tip under these conditions as frequently observed (Wilkins and Smith, 1970), and the PSB structure is broken down. Since slip is no longer confined to planes at 45° to the stress axis, the crack begins to propagate normal to the stress axis, and is then in Stage II growth by definition. In high strain fatigue the stress intensity is so large at a crack nucleus that the crack will almost immediately propagate by the Stage II process (Laird and Smith, 1962, 1963; Kim and Laird, 1978a).

The transition from Stage I to II has been of concern for many years. Cheng and Laird (1983) explored this transition in copper single crystals, both theoretically and experimentally. They have taken as the criterion for the transition the situation in which the plastic zone at the crack tip in the PSB (the crack has grown quite long and therefore now does interfere with the regular PSB structure) raises the shear stress on the second most highly stressed, nonprimary plane, to that of the plateau in the CSSC. They chose the plateau stress because they had no information concerning latent hardening in fatigue. Latent hardening was later studied in sufficient detail by Wang and Laird (1989) to show that the model needs only minor modification for the actual flow stress on slip planes intersecting the primary. When the transition criterion is satisfied, a SB can emanate from the crack tip at a large angle to the primary slip plane, and the duplex slip at the crack tip changes the plane of the crack from 45° to the stress axis to one approximately normal to it (Cheng and Laird, 1983).

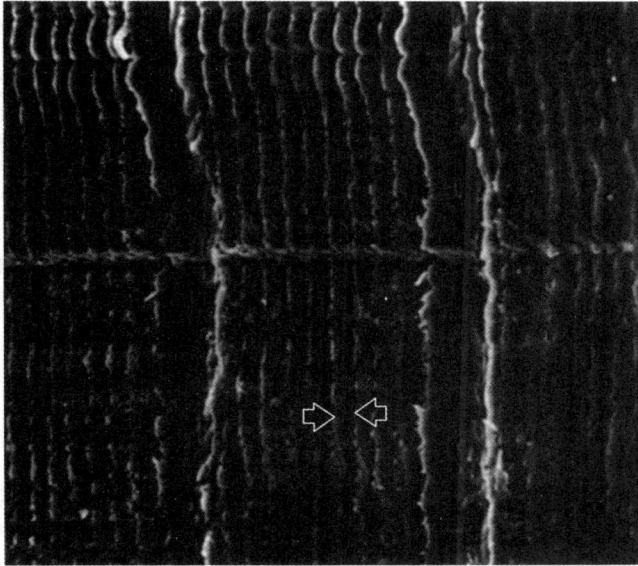

Figure 74 Typical ductile fatigue striations. Note that each striation, indicated by a double-headed arrow, consists of a flat region and trench. These striations were produced by a push–pull load sequence of 10 cycles at ±0.0025 and 1 cycle at ±0.005 plastic strain. Direction of crack advance from left to right. Source: Taken from Laird and De La Veaux, 1977. Courtesy of Springer.

Stage II fracture surfaces are recognized by the observation of fatigue striations on them provided the propagation rate is large enough for the striations to be resolved. Striations occur in two morphological types, "ductile" striations, by far the most common, and "brittle" striations, which occur in strongly hardened metals, or metals prone to brittle fracture mechanisms. Typical striations of the ductile variety are shown in **Figure 74**.

These striations were produced under conditions of variable loading and therefore vary in width (Laird and De La Veaux, 1977). Observations of this type were first used to prove that each striation was associated with one load cycle (Forsyth and Ryder, 1961). The morphology of the striation can vary considerably in any material and some materials have such a distorted morphology as to be unrecognizable. The variations in striation morphology and their dependence on the details of the slip processes have been repeatedly discussed (Laird 1967, 1979). It has also long been understood that cycling in vacuo at low or intermediate stress intensities suppresses the appearance of the striations while lowering the crack propagation rate by an order of magnitude or more. Welding at the crack tip during the compression strokes of the cycle and/or an influence of the absence of the environment on slip localization at the crack tip may provide the explanation for such behavior (see also Bowles and Schijve, 1983).

The regular ductile striation has a profile consisting of a more or less flat region (actually curved when the striation spacing is small) bounded by a trench (**Figure 74**). Both the morphology of the striation and its one-to-one connection with each load cycle were major clues to unraveling the mechanism of Stage II growth. Furthermore, the trench is formed on the side of the striation in the direction of crack propagation (**Figure 74**). By making sections through cracks in specimens

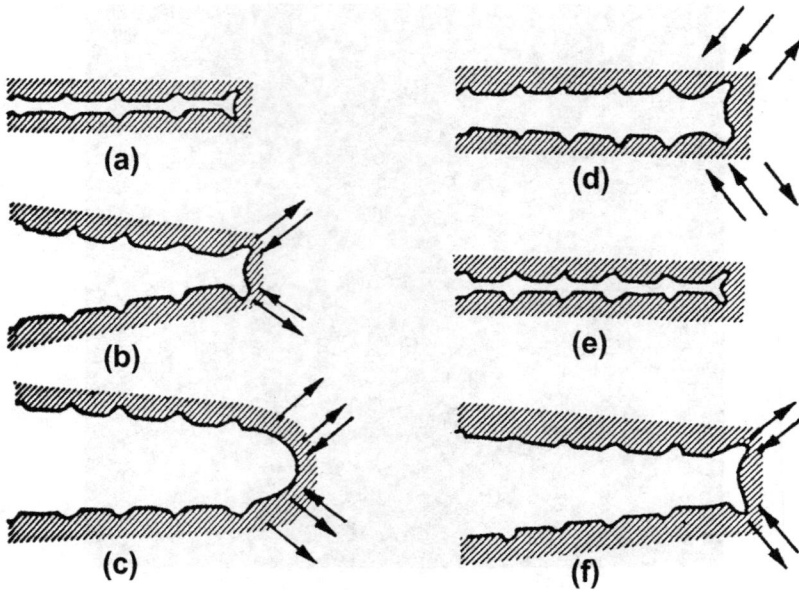

Figure 75 Schematic representation of the plastic blunting process of fatigue crack propagation in Stage II (a) zero load; (b) small tensile load; (c) maximum tensile load of the cycle; (d) closure; (e) maximum compression load of the cycle and (f) small tensile load in the succeeding cycle. The double arrowheads in (c) and (d) signify the widening of SBs at the crack in these stages of the process. Source: After Laird, 1967. Courtesy of ASTM.

unloaded from different parts of the straining cycle, Laird and Smith (1962) provided direct evidence for the purely ductile model of crack propagation. This evidence was refined and led to the model for crack growth shown in **Figure 75**, termed the "plastic blunting process" (Laird, 1967).

Although this model was proposed some time ago, the authors believe it is still an accurate representation of the ductile fatigue crack growth mechanism, as follows:

The initially unloaded crack tip is shown in **Figure 75(a)**. As tensile loading is applied, slip is concentrated in sharp bands at the double notch of the tip. Depending on the detailed morphology of the striation, the crack tip can also be singly pointed (Laird 1967, 1979). With continued tensile deformation, the crack progresses a distance proportional to the sliding off. This model has been directly supported by in situ observations in single crystals (Neumann et al., 1977). On reversing the loading direction, the slip in the zones is reversed, the crack faces close, and the new crack surface just created is folded by buckling of material at the crack tip, producing the double notch (i.e. a trench on each fracture face) and unit of crack advance shown in **Figure 75**. Depending on the planarity and coarseness of slip, the details of this process can vary considerably. For example, it is frequently observed that a trench on one face of the fracture matches with a ridge on the other (Laird, 1967).

The basic similarity between the mechanisms of Stage I and Stage II growth will be noted, in that both involve crack advance by a "slipping-off" process. The difference between them is that only one slip system needs to be involved in Stage I, but at least duplex slip occurs in the Stage II crack plastic zones.

Several authors have published alternative representations of the plastic blunting process in single crystals (Shijve, 1967; Pelloux, 1969; Neumann, 1974a,b; Kuo and Liu, 1976; Neumann et al., 1977).

They all involve the essential "slipping-off" process of the plastic blunting mechanism. Differences may be perceived because alternating slip on specific intersecting slip planes has been proposed. In particular, for high symmetry orientations of the crystal with respect to the stress axis, the detailed crystallography of the slip may influence the striation morphology.

In addition, work to characterize and model crack growth via plastic blunting has also been performed for intergranular cracking, via bicrystals experiments, where effects of slip geometry on crack advance and crack front shape have been explored (Peralta and Laird, 1998; Peralta et al., 1999a). The contrast of this case with respect to other situations is that the slip geometry at either side of the grain boundary can change significantly depending on the misorientation between the grains. Peralta and Laird (1998) offered a kinematic model based on a modification to the model proposed by Neumann (1974b) and, by making different assumptions regarding the interaction of SBs with the crack tip, deduced that for crack growing in Mode I, the crack advance is proportional to $\cos \phi \sin^2 \phi$, where ϕ is the angle of inclination of the slip plane with respect to the crack plane. This parameter reaches a maximum for $\phi = 54.74°$, suggesting that there is an optimum slip geometry for crack propagation. This angle coincides with the inclination of slip planes with respect to cracks growing on the (001) plane of a copper single crystal, which Peralta and Laird (1998) pointed out has been observed experimentally to be optimum for fatigue crack growth. This led them to use that parameter as a measure of efficiency of crack growth via plastic blunting in bicrystals given their slip geometry, with some success. They also pointed out that the model could be applied to single crystals as well.

One purpose of a crack propagation theory is to describe accurately the kinetics of crack propagation of the type shown in **Figure 66**. The kinematic models briefly summarized in Peralta and Laird (1998) and the one offered by the authors themselves, as described above, suffer from the drawback that they cannot be used to predict crack growth kinetics. In that sense, there have been many theoretical studies resulting in equations of the form of the Paris relation and its modifications to account for the effect of the load ratio.

Alternatives to the classical approach to explain and account for the load ratio effect by using closure to calculate an effective value of ΔK to use as a driving force have been offered by Vasudevan and coworkers (see Sadananda and Vasudevan, 2004 and references therein). These authors have postulated, as has been mentioned elsewhere in this chapter, that fatigue crack propagation is essentially a two-parameter phenomenon and that both ΔK and K_{max} are needed to propagate a fatigue crack. In this framework, fatigue crack growth has thresholds for both ΔK and K_{max} and the overall behavior needs to be plotted in a map like the one in **Figure 76**.

The reader is referred to the reference given above (and citations therein) for more details on the implication of this model vis-à-vis the closure concept, particularly regarding plasticity induced closure (see Solanki, 2004 for a review on this topic), since there is still debate within the community as to what the actual effect of this phenomenon is in the engineering treatment of fatigue crack propagation (see contrasting opinions in Solanki, 2004 and Vasudevan et al., 2001). In the context of the scope given to this chapter, that is, basic processes for pure metals and their solid solutions, the two-parameter model is fairly compatible with the physics of the basic plastic blunting mechanism that controls the behavior in most pure metals and ductile alloys, from the point of view that K_{max} would control the blunting of the tip, whereas ΔK would be responsible for resharpening it (see **Figure 75**).

Another development from the two parameter approach is that driving forces to quantify crack growth kinetics have been modified from the basic Paris law, and its various modifications to include the load ratio, to depend explicitly on ΔK and K_{max} in expressions such as $da/dN \propto \Delta K^p K_{max}^q$, where p and q can be obtained by performing fits on crack growth data that includes the load ratio as a parameter. See Kujawski (2001), Noroozi et al. (2005) for examples.

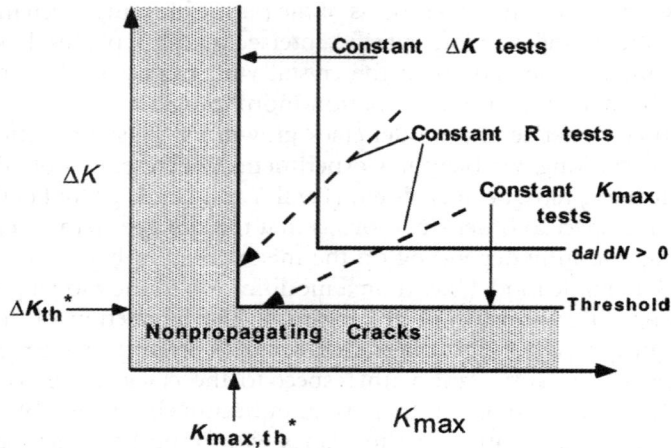

Figure 76 Two-dimensional representation of fatigue crack growth rate data in terms of ΔK and K_{max}. Note the individual thresholds for each of these variables, with the K_{max} threshold being greater than the ΔK threshold for many materials. Source: After Sadananda and Vasudevan, 2004, courtesy of Elsevier.

Now, returning to the mechanisms controlling crack growth kinetics via the basic Paris law, there are two classes of problem to handle: (1) the crack propagation rate is proportional to the square of the stress intensity range, which corresponds to ductile Stage II growth, and (2) the rate is proportional to the stress intensity range to a power greater than two, of which four is a popular value, but still higher values have been observed.

It is clear that there is a close relationship between stage II crack advance and the plastic strain around the tip, for example, Neumann (1974a), Peralta and Laird (1998), Suresh (1998). This has led to models of fatigue crack growth based on either ductility exhaustion or plastic blunting. In ductility exhaustion, fatigue cracks grow by damage accumulation ahead of a stationary tip until the material breaks. This is likely to play an important role in materials with planar slip character, where saturation is not achieved and stress increases continuously with cycling (Hong and Laird, 1990d; Li and Laird, 1994b; Wang et al., 1999). In plastic blunting, fatigue crack growth takes place via the deformation process at the tip, as shown in **Figure 75**. This should dominate in materials with wavy slip behavior, since they present saturation of cyclic strains, which implies that crack tip deformation can be sustained in a mechanically reversible fashion. In single crystals with high symmetry orientations this is dominated by SBs at the crack tip, acting either in a simultaneous (Laird, 1967) or alternating fashion (Neumann, 1974a), but the corresponding mechanisms in polycrystals, where the crack tip may be contained in a grain with arbitrary orientation, are not as straightforward.

Plastic blunting is accepted as one of the main mechanisms of fatigue crack growth in one-phase metallic materials (Suresh, 1998). Furthermore, finite element simulations by Tvergaard (2004) and Levkovitch et al. (2005) have shown that cracks can indeed grow via this mechanism and have been able to reproduce the geometry changes in the crack tip observed by Laird (1967). Therefore, one would expect that a relationship exists between crack advance, plastic blunting and applied loads such that the kinetics of stage II crack growth, that is, Paris' law, can be deduced from a mesoscopic framework based on appropriate kinematics, that is, the relationship between strains ahead of the tip and the crack advance. As mentioned above, kinematic models cannot predict crack growth kinetics, while the finite element work by Tvergaard (2004) did not address the kinetics of crack growth and did not account for strain localization, which excludes modeling plastic blunting via shear bands.

Connections between plastic blunting and applied loads have been provided at the expense of simplifying the kinematics. For example, Kanninen et al. (1977) and Davidson (1984) offered models based on superdislocations emitted from a crack tip, where the crack advance was proportional to the effective Burger's vector. This ignores large strains and finite SB widths, which are included in other models (Neumann, 1974b; Peralta and Laird, 1998), and does not predict a length scale based on crack tip fields.

The length scale associated with crack advance should arise naturally from the strain field ahead of the crack tip if plastic blunting is the main mechanism. The Crack Tip Opening Displacement (CTOD) has typically been used for this, and it has been related to fatigue crack growth either via dimensional arguments like those used by McClintock (1967) or experimentally, as proposed by Tanaka et al. (1984). The kinematical connection among crack tip strain, CTOD and crack advance has been explored in models proposed by Neumann (1974b), Peralta and Laird (1998) and Tvergaard (2004), among many others. Neumann's model suggested $da/dN \approx (2/3-1)$CTOD, Peralta and Laird's resulted in $da/dN \approx (0.35-0.5)$CTOD and Tvergaard obtained $da/dN \approx 0.33$CTOD. The fact that plastic blunting is predicted to give a Paris exponent $m = 2$ is linked to this correlation, because the appropriate length scale is assumed to be either the CTOD or the plastic zone radius, both of which are proportional to ΔK^2 according to LEFM (Kanninen and Popelar, 1985). However, plasticity at the tip should preclude LEFM and neither the CTOD nor the plastic zone size is related *directly* to the displacements ahead of the tip; hence, a length scale linking these displacements to crack advance is not readily available.

Obtaining $m > 2$ via plastic blunting is a difficult problem. Ductility exhaustion has been used to address this issue, for example, the model by Jiang and Feng (2004), as well as empirical corrections to CTOD approaches (Mercer et al., 2003). Discrete dislocation models have resulted in m between 4 and 8 (Deshpande et al., 2002). Other work, for example, Fine (1980), Kwun and Fine (1980), Nicholls (1994), has been based on the plastic work needed to advance the crack, which bypasses the kinematics and leads to $m = 4$. Weertman (2007) proposed that the shielding produced by dislocations in the plastic zone ahead of the crack tip can lead to higher values of m, but the use of a Taylor expansion in his approach to account for this effect leads to only integer values of m. Note, however, that m typically ranges between 2 and 4 and most experimental values are not integers. The need to resolve this discrepancy has led to experimental studies of strains ahead of fatigue cracks, as reflected in the extensive literature in the subject, for example, Tomkins and Biggs (1969), Chan et al. (1986), Davidson (1986), Gerberich et al. (1990), McClung and Davidson (1991), Zhang and Edwards (1992), Peralta et al. (2007).

The approach used by Ding et al. (2007), which is based in a modification of ductility exhaustion, is interesting, since the fatigue damage parameter these authors used was predicted solely from the cyclic deformation behavior of the material at a point ahead of the crack tip, pointing out, once again, that there is a close connection between fatigue crack growth and cyclic plasticity. However, these authors did not address the problem of predicting the Paris exponent, despite the fact that they obtained good agreement between measured and predicted crack growth rates.

Peralta et al. (2007) performed experiments using in situ loading, SEM and digital image correlation to study fatigue crack tips in pure nickel polycrystals to explore the kinematics of fatigue crack growth and its relationship to the Paris exponent, as a key parameter for kinetics. They found that deformation around the tips of the fatigue cracks in their samples was produced by localized strain along SBs that carried mostly pure shear. They also found instances of a single SB or two present ahead of the tip. The double SB cases, which can be linked to the double forward slip typically observed ahead of fatigue cracks in single crystals, could be either simultaneous (Laird, 1967) or alternating (Neumann, 1974a), indicating that fatigue crack propagation in polycrystals can occur by either of these mechanisms. Examples are given in **Figure 77**.

Note in **Figure 77(a)** there is clear localization along a DB, but the band does not extend all the way to $y = 0$, where the crack tip is. This indicates that a second DB is developing at the crack tip, "cutting off" a preexisting band on the process. Part of the shear strain of the new band can be seen in **Figure 77(b)**. Therefore, the crack tip was developing a dual SB configuration, in a process similar to the alternating slip model proposed by Neumann (1974b). The sample corresponding to **Figure 77(c) and (d)** also has a dual shear band configuration and the two bands have approximately the same origin, which coincides with the crack tip location, and they do not interfere with each other. Hence, they must act simultaneously, forming a dual shear band configuration as predicted by the model by Laird (1967).

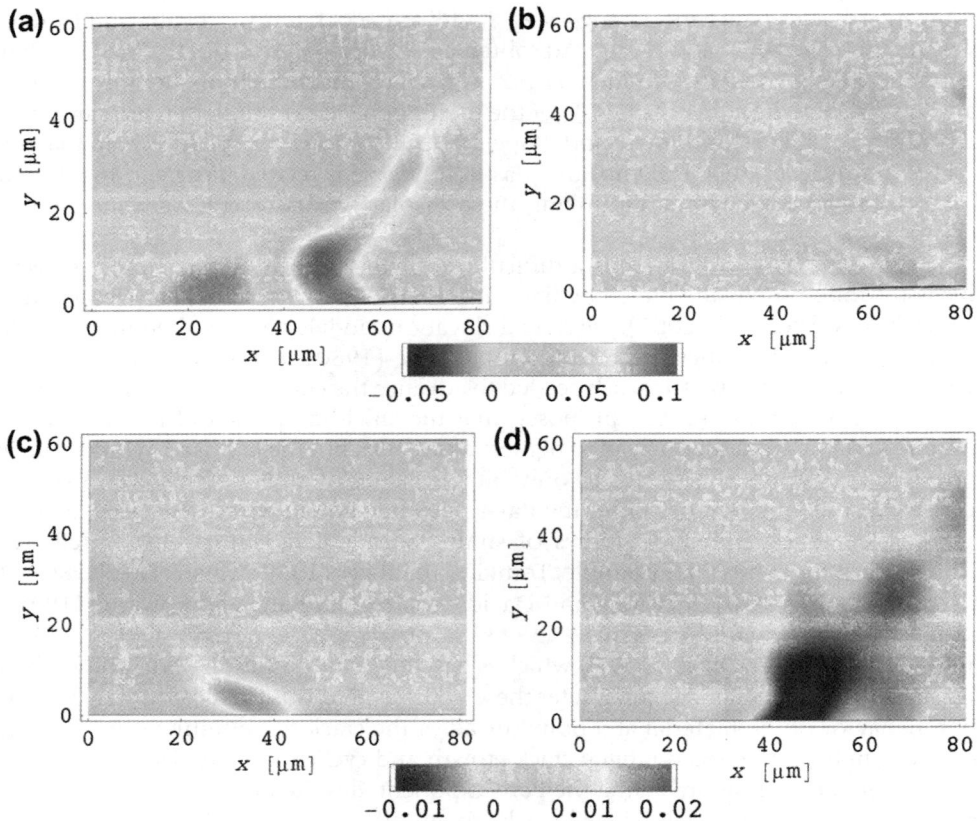

Figure 77   Strain field ahead of fatigue crack tips measured in pure polycrystalline nickel for a coordinate system parallel (x') and perpendicular (y') to the SB with the most strain. (a and b) Sample tested at $\Delta K = 17.3$ and $K_{max} = 19.3$ MPa.m$^{1/2}$ (a) $-\partial u'/\partial y'$; (b) $-\partial v'/\partial x'$. (c and d) Sample tested at $\Delta K = 10.7$ and $K_{max} = 11.9$ MPa.m$^{1/2}$ (c) $\partial u'/\partial y'$; (d) $-\partial v'/\partial x'$. The components u' and v' refer to displacements parallel to x' and y', respectively. Source: After Peralta et al., 2007, courtesy of Elsevier. (For color version of this figure, the reader is referred to the online version of this book.)

Peralta et al. (2007) characterized the deformation ahead of the crack tips using a nonlocal parameter given by the area integral of the opening strain (normal strain parallel to the applied load for their Mode I cracks ε_{op}), that is,

$$\varepsilon_{int} = \int_A \varepsilon_{op} dA. \tag{13}$$

They called this quantity the "integrated strain" and it was found to vary with ΔK following a power law with an exponent approximately equal to the Paris exponent measured experimentally. This implies that the crack growth rate in these samples is approximately proportional to the integrated strain. Therefore, the integrated strain must contain a length scale that is proportional to crack advance per cycle. By performing the integration with some simplifying assumptions, Peralta et al. (2007) found that the integrated strain was proportional to the accumulated net displacement u' along the length of the SBs, and, therefore, so is the crack growth rate. This is consistent with the kinematical models by Neumann (1974b), Peralta and Laird (1998).

Given that strain was localized, the domain of integration for Eqn (13) becomes the area of the bands, taken by Peralta et al. (2007) as approximate rectangles of width h and length l. The latter was assumed to be proportional to the plastic zone radius, that is, $l = \alpha r_p$. Using the mean value theorem, these authors found that

$$\varepsilon_{int} = \int_0^{\alpha r_p} \int_0^h \left(\frac{\partial u'}{\partial y'}\right) f(\theta) dy' dx' = f(\theta) h \alpha r_p \bar{\gamma}, \tag{14}$$

where $f(\theta)$ is a function of the angle of inclination of the band and $\bar{\gamma}$ is the average shear strain in the band. The width of the band was found to have a fairly weak dependence on ΔK, so the crack advance per cycle has to be proportional to $r_p \bar{\gamma}$, which is in turn proportional to the accumulated displacement in the band. Analysis of the spatial dependence of the strain fields led Peralta et al. (2007) to assume that the Hutchison-Rice-Rosengren (HRR) fields (Kanninen and Popelar, 1985) could be used to obtain the ΔK dependence of $\bar{\gamma}$. That assumption, along with small-scale yielding, led them to

$$\frac{da}{dN} \propto r_p \bar{\gamma} \propto \Delta K^2 \Delta K^{\frac{2}{1+n'}} = \Delta K^{2(1+\frac{1}{1+n'})}, \tag{15}$$

where n' ($0 \leq n' \leq 1$) is the cyclic hardening exponent. Therefore, in this framework the Paris exponent is given, to a first approximation, by $m = 2(1 + \frac{1}{1+n'})$, which results in $3 \leq m \leq 4$ for $0 \leq n' \leq 1$. This produces a variable value of the Paris exponent that is well within the range of experimentally observed values, where the product $r_p \bar{\gamma}$ is the main contributor to the crack growth rate and represents a characteristic length that is consistent with dimensional analysis, but resolves the paradox associated with the LEFM predictions of $m = 2$. These authors analyzed Paris exponents for a wide variety of materials and found that the probability of finding a value of m between 2 and 4 was about 77% and the probability to find m between 3 and 4 was about 50%, with just a 27% probability of finding a value of m between 2 and 3. This indicates that the simple expression given by Eqn (15) accounts for a large fraction of experimentally observed values of the Paris exponent. The formulation, however, has drawbacks, related to load ratio effects and quantitative predictions of the crack growth rate, among others. These issues are discussed in detail by Peralta et al. (2007).

Paris exponents higher than 4 are not covered by the formulation given above and are typically associated with the effect of quasi-static damage mechanisms that can be present in engineering alloys. However, Peralta et al. (2012) used a framework similar to the one given above to analyze long fatigue cracks in β-annealed Ti-6Al-4V and deduced that the integrated strain obtained over the cyclic plastic zone size is proportional to the crack rate and that

$$\frac{\mathrm{d}a}{\mathrm{d}N} \propto \varepsilon_{\mathrm{int}}^C = \int_{A_{\mathrm{cyclic}}} \varepsilon_{\mathrm{op}} \mathrm{d}A \propto \Delta K^4 \Delta K^{\frac{2}{1+n'}} = \Delta K^{2(2+\frac{1}{1+n'})}. \tag{16}$$

This implies that for $0 \leq n' \leq 1$, $5 \leq m \leq 6$, and they found good agreement with experimental measurements of m for the alloy (close to 6). The main difference with the results for nickel was the lack of shear bands within the area of integration, which led to different assumptions for the area integral. They also found that the domain of integration made a difference and that the cyclic plastic zone radius was the appropriate length scale to use, which indicates, as expected, that the processes involved on fatigue crack growth are contained within that length scale.

Note that these results for pure Ni and Ti-6Al-4V, although interpreted by Peralta et al. (2007) and Peralta et al. (2012) in the spirit of plastic blunting mechanisms, are not incompatible with ductility exhaustion approaches. In any case, the use of a nonlocal parameter is consistent with modeling strategies for fatigue damage accumulation using integrals of FIPs over relevant length scales (see McDowell and Dunne, 2010). In the case of fatigue crack growth, the integrated strain described above has a clear physical connection to net displacements ahead of a crack tip.

Regarding fatigue crack growth under VHCF conditions, note that the crack nucleation stage strongly dominates the life, and the behavior of short cracks is key to determine the fate of long cracks. One can envision short cracks that grow and then arrest, as the local driving force becomes lower than the long crack threshold (see the example in **Figure 66**). This seems to be the case for the Stage I cracks reported by Stanzl-Tschegg and Schönbauer (2010) in polycrystalline copper, which needed an applied stress to propagate and transform into long cracks that was higher than that needed to nucleate them. The mechanisms for long crack growth as well as estimations of life from crack growth kinetics under VHCF for type II materials have been addressed by Paris and coworkers (Marines-Garcia et al., 2007, 2008; Huang et al., 2010), among others.

18.7   Additional Topics

This chapter has been written with a desire to balance well-established results with significant recent developments. This is difficult to achieve, particularly given the large volume of high quality work that has been carried out since the previous edition of this chapter; hence, some topics well worth detailed consideration could not be addressed because of space limitations. In that regard, temperature effects on cyclic deformation have not been described, but the reader can use some of the references cited here to obtain more information, particularly the reviews by Holste (2004), Li et al. (2011) and references therein, see also the review on scaling behavior of dislocation structures by Sauzay and Kubin (2011).

Another important topic is the effect of microstructural length scales on cyclic behavior, particularly as grains become smaller than 1 μm. In his review, Holste (2004) points out that given the intrinsic length scales of the dislocation structures observed in Ni and other fcc metals, there should be a change of the cyclic deformation mechanisms for grains smaller than 0.5 μm. Materials with ultrafine grain

(UFG) size certainly qualify for this, as well as those with nanocrystalline grains (size < 100 nm). The cyclic behavior of these materials is extremely interesting and has been shown to depend strongly on processing route, intrinsic material properties such as SFE and microstructure stability (some materials experience grain growth under cyclic loads, see Canadinc et al. (2011)), and they have a tendency for cyclic softening (Malekjani et al., 2011). The reader is referred to the review by Mughrabi and Höppel (2010) and references therein for more information and the recent paper by Lukáš et al. (2011) for the behavior of UFG copper under VHCF conditions.

Another related issue is the interaction of physical and microstructural length scales, for example, fatigue behavior of thin films, where the volume to surface ratio is low, and the synergy between the intrinsic physical length scale (film thickness) and, say, grain size. See the paper by Zhang et al. (2006), where it was found that when either of these length scales goes below about 1 μm in copper thin films, the typical dislocation wall and cell structures found in coarse-grained copper samples no longer develop and are replaced by individual dislocations. In addition, they found that the typical surface damage features observed in fatigued bulk metallic samples, such as extrusions and cracks near extrusions, are gradually suppressed and replaced by damage localized at interfaces, such as cracks and voids along grain and TBs, in agreement with Holste (2004).

As the microstructure length scale goes even smaller, further deviations in behavior with respect to bulk materials should be observed. From that point of view, the fatigue of bulk metallic glasses, with their amorphous structure, has been studied as well. These materials are quite interesting, because of the lack of crystalline structure and the facts that their plastic response is controlled by strain localization in shear bands and that they show little or no hardening after yielding, all of which have important implications for fatigue behavior. The reader can find more information in the following references Menzel and Dauskardt (2006), Gary Harlow et al. (2008), Menzel and Dauskardt (2008), Wang et al. (2008), Trexler and Thadhani (2010), Wang et al. (2010).

The behavior of engineering alloys also offers a wealth of interesting phenomena and approaches, including probabilistic issues (Jha et al., 2009), multiscale and multistage modeling of fatigue damage (Xue et al. 2007), and, of course, the extremely important issue of environmental effects. The interested reader can see the previous edition of this chapter for a brief overview and (Gupta and Agnew, 2008; Ro et al., 2008) for some recent work on the effect of environment on crack growth in Al alloys, for example.

Acknowledgments

During preparation of this chapter, P.P. received support from the Air Force Office of Scientific Research, under the 2006 Structural Health and Prognosis MURI, and the Department of Energy, for which he gives warm thanks. He is also very grateful to his wife J. Ortiz for her editorial help with the text and references, as well as her patience for the long hours he spent away from domestic duties writing this chapter.

Both authors would also like to express their utmost gratitude to numerous generations of graduate students who have provided unremitting stimulation.

References

Abdel-Karim, M., 2010. Int. J. Plast. 26 (5), 711.
Abel, A., Wilhem, M., Gerold, V., 1979. Mater. Sci. Eng. 37, 187.
Ackermann, F., Kubin, L.P., Lepinoux, J., Mughrabi, H., 1984. Acta Metall. 32 (5), 715.

Agnew, S.R., Weertman, J.R., 1998. Mater. Sci. Eng. A244, 145.

Aifantis, E.C., 1986. Mater. Sci. Eng. 81, 563.

Albert, W.A.J., 1838. Archiv für Mineralogie, Geognosie, Bergbau und Hüttenkunde, vol. 10, 215.

Amirkhizi, A.V., Nemat-Nasser, S., 2007. Int. J. Plast. 23, 1918.

Anahid, M., Samal, M.K., Ghosh, S., 2011. J. Mech. Phys. Solids 59 (10), 2157.

Antonopoulos, J.G., Brown, U., Winter, A.T., 1976. Philos. Mag. 34, 549.

Asaro, R.J., 1983. Adv. Appl. Mech. 23, 1.

Bari, S., Hassan, T., 2001. Int. J. Plast. 17, 885.

Basinski, S.J., Basinski, Z.S., Howie, A., 1969. Philos. Mag. 19, 899.

Basinski, Z.S., Korbel, A.S., Basinski, S.J., 1980. Acta Metall. 28, 191.

Basinski, Z.S., Basinski, S.J., 1984. Scr. Mater. 18, 851.

Basinski, Z.S., Basinski, S.J., 1992. Prog. Mater. Sci. 36, 89.

Bassani, J.L., Wu, T.-Y., 1991. Proc. R. Soc. London, Ser. A 435, 21.

Bassani, J.L., 1994. Adv. Appl. Mech. 30, 191.

Bathias, C., 1999. Fatigue Fract. Eng. Mater. Struct. 22, 559.

Bathias, C., 2006. Int. J. Fatigue 28 (11), 1438.

Bauschinger, J., 1886. *Mitt. Mech. Tech. Lab.* München.

Bhat, S.P., Laird, C., 1978. Scripta Metall. 12, 687.

Bhat, S.P., Laird, C., 1979a. Acta Metall. 27, 1861.

Bhat, S.P., Laird, C., 1979b. Fatigue Eng. Mater. Struct. 1, 59.

Blochwitz, C., Kahle, E., 1980. Cryst. Res. Technol. 15, 977.

Blochwitz, C., Brechbühl, J., Tirschler, W., 1996. Mater. Sci. Eng. A 210 (1–2), 42.

Blochwitz, C., Tirschler, W., 2005. Cryst. Res. Technol. 40, 32.

Boettner, R.C., McEvily, J.A.J., Liu, Y.C., 1964. Philos. Mag. 10, 95.

Bowles, C.Q., Schijve, J., 1983. Fatigue Mechanisms. Am. Soc. Test. Mater.

Bozek, J.E., Hochhalter, J.D., Veilleux, M.G., Liu, M., Heber, G., Sintay, S.D., Rollett, A.D., Littlewood, D.J., Maniatty, A.M., Weiland, H., Christ, R.J., Payne, J., Welsh, G., Harlow, D.G., Wawrzynek, P.A., Ingraffea, A.R., 2008. Modell. Simul. Mater. Sci. Eng. 16 (6), 065007.

Brahme, A., Alvi, M.H., Saylor, D., Fridy, J., Rollett, A.D., 2006. Scr. Mater. 55, 75.

Broom, T., Ham, R.K., 1959. Proc. Roy. Soc. A251, 186.

Brown, L.M., 1977. Met. Sci. 11, 315.

Brown, L. M., 1980. Int. Conf. on Dislocation Modelling in Physical Systems, Gainesville, Florida.

Brown, L.M., 2006. Philos. Mag. 86 (25–26), 4055.

Brown, M.W., 1986. The Behavior of Short Fatigue Cracks. Inst. of Mech. Engrs, London.

Buchinger, L., Stanzl, S., Laird, C., 1984. Philos. Mag. A. 50, 275.

Buchinger, L., Cheng, A.S., Stanzl, S., Laird, C., 1986. Mater. Sci. Eng. 80, 155.

Buque, C., Holste, C., Schwab, A., 1999. 20th Risø International Symposium on Materials Science: Deformation Induced Microstructures - Analysis and Relation to Properties. Risø National Laboratory, Roskilde, Denmark (Risø National Laboratory).

Calabrese, C., Laird, C., 1974. Mater. Sci. Eng. 13, 141.

Canadinc, D., Niendorf, T., Maier, H.J., 2011. Mater. Sci. Eng. A 528 (21), 6345.

Chaboche, J.L., 1991. Int. J. Plast. 7, 661.

Chai, G., 2006. Int. J. Fatigue 28 (11), 1533.

Chan, K.S., Lankford, J., Davidson, D.L., 1986. J. Eng. Mater. Technol. 108 (3), 201.

Charsley, P., 1981. Mater. Sci. Eng. 47, 181.

Charsley, P., Kuhlmann-Wilsdorf, D., 1981. Philos. Mag. A 44 (6), 1351.

Cheng, A.S., Laird, C., 1981a. Fatigue Eng. Mater. Struct. 4, 331.

Cheng, A.S., Laird, C., 1981b. Mater. Sci. Eng. 51, 55.

Cheng, A.S., Laird, C., 1981c. Fatigue Eng. Mater. Struct. 4, 343.

Cheng, A.S., Laird, C., 1981d. Mater. Sci. Eng. 51, 111.

Cheng, A.S., Laird, C., 1983. Mater. Sci. Eng. 60, 177.

Ciavarella, M., Monno, F., 2006. Int. J. Fatigue 28 (12), 1826.

Coffin, L.E., 1956. Trans. Am. Soc. Mech. Eng. 78, 527.

Coffin, L.E., Read, J.H., 1956. International Conference on Fatigue of Metals. Institute of Mechanical Engineers, London.

Coffin, L. E., Tavernelli, J., 1958. General Electric Company. Research Lab.

Coffin, L.E., McMahon, C.J., 1979. Metall. Trans. 1, 3443.

Cottrell, A.N., Hull, D., 1957. Proc. R. Soc: A 242, 211.

Davidson, D.L., 1984. Acta Metall. 32 (5), 707.

Davidson, D.L., Lankford, J., 1984. IUTAM Eshelby Memorial Symposium. Cambridge University Press.

Davidson, D.L., 1986. Eng. Fract. Mech. 25 (1), 123.

Deshpande, V.S., Needleman, A., Van der Giessen, E., 2002. Acta Mater. 50, 831.

Dickson, J.I., Handfield, L., L'Espérance, G., 1986. Mater. Sci. Eng. 81, 477.

Dickson, J.I., Bande, H., L'Espérance, G., 1988. Acta Metall. A101, 75.

Dickson, J. I., Bailon, J. E., Xia, J., Bureau, M., 1993. Fatigue 93, EMAS, W. Midlands, Warley.

Ding, F., Zhao, T., Jiang, Y., 2007. Eng. Fract. Mech. 74 (13), 2014.

Dingreville, R., Battaile, C.C., Brewer, L.N., Holm, E.A., Boyce, B.L., 2010. Int. J. Plast. 26 (5), 617.

Döring, R., Hoffmeyer, J., Seeger, T., Vormwald, M., 2006. Int. J. Fatigue 28 (9), 972.

Dunne, F.P.E., Wilkinson, A.J., Allen, R., 2007. Int. J. Plast. 23 (2), 273.

Essmann, U., Gösele, U., Mughrabi, H., 1981. Philos. Mag. A 44, 405.

Essmann, U., 1982. Philos. Mag. 45, 171.

Ewing, J.A., Humphrey, J.C.W., 1902. Philos. Trans. R. Soc., A 200, 241.

FAA, 2011. Metallic Materials Properties Development & Standardization (MMPDS-06). Batelle, Columbus, OH.

Fatemi, A., Shamsaei, N., 2011. Int. J. Fatigue 33 (8), 948.

Feltner, C.E., Laird, C., 1967a. Acta Metall. 15, 1633.

Feltner, C.E., Laird, C., 1967b. Acta Metall. 15, 1621.

Figueroa, J.C., Bhat, S.P., De La Veaux, R., Murzenski, S., Laird, C., 1981. Acta Metall. 29, 1667.

Figueroa, J.C., Laird, C., 1981. Acta Metall. 29, 1679.

Fine, M.E., Ritchie, R.O., 1979. Fatigue and Microstructure. Am. Soc. Metals.

Fine, M.E., 1980. Metall. Trans. A 11 (3), 365.

Finney, J.M., Laird, C., 1975. Philos. Mag. 31 (2), 339.

Forrest, P.G., 1962. Fatigue of Metals. Pergamon, Oxford.

Forsyth, P.J.E., 1953. Nature 171, 172.

Forsyth, P.J.E., Stubbington, A., 1955. J. Inst. Met. 84, 175.

Forsyth, P.J.E., 1957. Proc. Roy. Soc. A242, 198.

Forsyth, P.J.E., Ryder, D.A., 1961. Metallurgia 51, 117.

Garbacz, A., Ralph, B., Kurzydlowski, K.J., 1995. Acta Metall. Mater. 43, 1541.

Gary Harlow, D., Liaw, P.K., Peter, W.H., Wang, G., Buchanan, R.A., 2008. Acta Mater. 56 (13), 3306.

Gerberich, W.W., Davidson, D.L., Kaczorowski, M., 1990. J. Mech. Phys. Solids 38 (1), 87.

Gerold, V., Steiner, D., 1982. Scr. Mater. 16, 405.

Gerold, V., Meier, B., 1987. Fatigue 87, West Midlands.

Glazov, M., Llanes, L.M., Laird, C., 1995. Phys. Status Solidi A 149, 297.

Glazov, M.V., Laird, C., 1995. Acta Metall. Mater. 43 (7), 2849.

Gong, B., Wang, Z.R., Wang, Z.-G., Zhang, Y.W., 1996. Mater. Sci. Eng. 210 (1–2), 94.

Gong, B., Wang, Z., Wang, Z.-G., 1997. Acta Mater. 45 (4), 1365.

Gong, B., Wang, Z., Wang, Z.-G., 1999. Acta Mater. 47 (1), 317.

Gough, H. J., 1933. Proceedings of The American Society Testing Materials. vol. 33, p. 3.

Grosskreutz, J.C., 1971. Phys. Status Solidi 47, 11.

Grosskreutz, J.C., Mughrabi, H., 1975. Constitutive Equations in Plasticity. MIT Press, Cambridge, Mass.

Gu, H., Guo, H., Chang, S., Laird, C., 1994. Mater. Sci. Eng. A 188 (1–2), 23.

Gupta, V., Agnew, S., 2008. Mater. Sci. Eng. A 494 (1–2), 36.

Hancock, J.R., Grosskreutz, J.C., 1969. Acta Metall. 17, 77.

Heinz, A., Neumann, P., 1990. Acta Metall. Mater. 38, 613.

Herbig, M., King, A., Reischig, P., Proudhon, H., Lauridsen, E.M., Marrow, J., Buffière, J.-Y., Ludwig, W., 2011. Acta Mater. 59 (2), 590.

Hodgson, B., 1968. Met. Sci. J 2, 235.

Holste, C., 2004. Philos. Mag. 84 (3–5), 299.

Holt, D.L., 1970. J. Appl. Phys. 41, 3197.

Hong, S.I., Laird, C., 1990a. Mater. Sci. Eng. A124, 183.

Hong, S.I., Laird, C., 1990b. Mater. Sci. Eng. A128, 155.

Hong, S.I., Laird, C., 1990c. Acta Metall. Mater. 38 (11), 2085.

Hong, S.I., Laird, C., 1990d. Mater. Sci. Eng. A128, 55.

Hong, S.I., Laird, C., 1990e. Acta Metall. Mater. 38 (8), 1581.

Hong, S.I., Laird, C., 1991a. Met. Trans. 22A, 415.

Hong, S.I., Laird, C., 1991b. Fatigue Fract. Eng. Mater. Struct. 14, 143.
Hong, S.I., Laird, C., 1991c. Mater. Sci. Eng. 142, 1.
Hong, S.I., Inui, H., Laird, C., 1992. Acta Metall. Mater. 40, 394.
Horibe, S., Laird, C., 1983. Acta Metall. 31 (10), 1567.
Horibe, S., Laird, C., 1985a. Acta Metall. 33 (5), 819.
Horibe, S., Laird, C., 1985b. Mater. Sci. Eng. 72 (2), 149.
Huang, Z., Wagner, D., Bathias, C., Paris, P.C., 2010. Acta Mater. 58 (18), 6046.
Hunsche, A., Neumann, P., 1986. Acta Metall. 34, 207.
Ikeda, S., 1980. 5th Int. Congress on Strength of Metals and Alloys. Pergamon Press, Oxford.
Inui, H., Hong, S.I., Laird, C., 1990. Acta Metall. Mater. 38 (11), 2261.
Jameel, M. A., 1997. Ph. D. Thesis, Philadelphia, University of Pennsylvania.
Jameel, M.A., Peralta, P., Laird, C., 2001. Mater. Sci. Eng. A297, 48.
Jameel, M.A., Peralta, P., Laird, C., 2003. Mat. Sci. Eng. A342, 279.
Jha, S.K., Larsen, J.M., Rosenberger, A.H., 2009. Eng. Fract. Mech. 76 (5), 681.
Jia, W.P., Li, S.X., Wang, Z.G., Li, G.Y., 1998. J. Mater. Sci. Lett. 17 (23), 2009.
Jia, W.P., Li, S.X., Wang, Z.-G., Li, X.W., Li, G.Y., 1999. Acta Mater. 47 (7), 2165.
Jiang, Y., Feng, M., 2004. J. Eng. Mater. Technol. 126, 77.
Jin, N.Y., Winter, A.T., 1984a. Acta Metall. 32 (8), 1173.
Jin, N.Y., Winter, A.T., 1984b. Acta Metall. 32 (7), 989.
Jin, N.Y., 1987. Philos. Mag. Lett. 56 (1), 23.
Kalnaus, S., Jiang, Y., 2006. Eng. Fract. Mech. 73 (6), 684.
Kaneko, Y., Mimaki, T., Hashimoto, S., 1999. Acta Mater. 47, 165.
Kanninen, M.F., Atkinson, C., Feddersen, C.E., 1977. Cyclic Stress–strain and Plastic Deformation Aspects of Fatigue Crack Growth. ASTM STP 637. ASTM, Philadelphia. American Society of Testing and Materials: 122.
Kanninen, M.F., Popelar, C.H., 1985. Advanced Fracture Mechanics. Oxford University Press, New York.
Katagari, K., Omura, A., Koyanagi, K., Awatani, J., Shiraishi, T., Kaneshiro, H., 1977. Metall. Trans. A 8, 1769.
Katagari, K., Awatani, J., Omura, A., Koyanagi, K., Shiraishi, T., 1979. Fatigue Mechanisms. Am. Soc. Test. Mater.
Kennedy, A.J., 1962. Processes of Creep and Fatigue in Metals. J. Wiley and Sons, London.
Kim, W.H., Laird, C., 1978a. Acta Metall. 26, 789.
Kim, W.H., Laird, C., 1978b. Mater. Sci. Eng. 33, 225.
Kim, W.H., Laird, C., 1978c. Acta Metall. 26, 777.
Kitagawa, H., Takahashi, S., 1976. International Conference on Mechanical Behavior of Materials (ICM2). Am. Soc. Metals.
Kitagawa, K., Asada, H., Monzen, R., Kikuchi, M., 1986. Suppl. Trans. Jap. Inst. Metals 27, 827.
Kocks, U.E., 1970. Metall. Trans. 1, 1121.
Kocks, U.E., 1985. Dislocations and Properties of Real Materials. The Inst. of Physics, London, 125.
Kubin, L., 1993. In: Mughrabi, H. (Ed.), Materials Science and Technology, vol. 6. VCH, Weinheim, p. 137.
Kuhlmann-Wilsdorf, D., 1975. In: Thompson, A.W. (Ed.), Work Hardening in Tension and Fatigue. The Metallurgical Society, AIME, New York, 1.
Kuhlmann-Wilsdorf, D., Laird, C., 1977. Mater. Sci. Eng. 27, 137.
Kuhlmann-Wilsdorf, D., 1979a. Mater. Sci. Eng. 39, 127.
Kuhlmann-Wilsdorf, D., 1979b. Mater. Sci. Eng. 39 (2), 231.
Kuhlmann-Wilsdorf, D., Laird, C., 1979. Mater. Sci. Eng. 37, 111.
Kuhlmann-Wilsdorf, D., Laird, C., 1980. Mater. Sci. Eng. 46, 209.
Kuhlmann-Wilsdorf, D., Charsley, P., 1981. Philos. Mag. 44, 1351.
Kujawski, D., 2001. Int. J. Fatigue 23, 733.
Kunkler, B., Duber, O., Koster, P., Krupp, U., Fritzen, C., Christ, H., 2008. Eng. Fract. Mech. 75 (3–4), 715.
Kuo, A.S., Liu, H.W., 1976. Scripta Metall. 10 (8), 723.
Kwadjo, R., Brown, L.M., 1978. Acta Metall. 26 (7), 1117.
Kwun, S.I., Fine, M.E., 1980. Scripta Metall. 14 (1), 155.
Laird, C., Smith, G.C., 1962. Philos. Mag. 7, 847.
Laird, C., Smith, G.C., 1963. Philos. Mag. 8, 1945.
Laird, C., 1967. Fatigue Crack Propagation. ASTM STP 415. ASTM, Philadelphia. American Society of Testing and Materials: 131.
Laird, C., Feltner, C.E., 1967. Trans. Am. Inst. Met. Eng. 239, 1074.
Laird, C., Duquette, D. J., 1972. NACE-2.
Laird, C., 1976. Mater. Sci. Eng. 22, 231.
Laird, C., 1977. Work Hardening in Tension and Fatigue. The Metallurgical Society AIME.

Laird, C., De La Veaux, R., 1977. Met. Trans. 8A, 657.

Laird, C., Langelo, V.J., Hollrah, M., Yang, N.C., De La Veaux, R., 1978. Mater. Sci. Eng. 32, 137.

Laird, C., 1979. Mechanisms and Theories of Fatigue. American Society of Metals.

Laird, C., 1981. Metallurgical Treatises. The Metallurgical Society AIME.

Laird, C., Finney, J.M., Kuhlmann-Wilsdorf, D., 1981. Mater. Sci. Eng. 50, 127.

Laird, C., 1983. Dislocations. North-Holland Company, Amsterdam.

Laird, C., Charsley, P., Mughrabi, H., 1986. Mater. Sci. Eng. 81, 433.

Laird, C., 1996. In: Cahn, R.W., Haasen, P. (Eds.), Physical Metallurgy. Elsevier Science, Amsterdam, p. 2294.

Lampman, S.R. (Ed.), 1996. ASM Handbook: Fatigue and Fracture. ASM International, Materials Park, OH.

Landgraf, R.W., 1977. Proceedings of Fatigue Seminars: Fundamental and Applied Aspects. Saabgarden, Remforsa, Sweden.

Lee, J. K., 1980. Ph.D. Thesis. University of Pennsylvania.

Lee, J.K., Bhat, S.P., De La Veaux, R., Laird, C., 1981. Int. J. Fract. 17, 121.

Lee, J.K., Laird, C., 1982a. Mater. Sci. Eng. 54 (1), 53.

Lee, J.K., Laird, C., 1982b. Mater. Sci. Eng. 54 (1), 39.

Lee, J.K., Laird, C., 1983. Philos. Mag. A 47 (4), 579.

Lepistö, T.K., Kuokkala, V.-T., Kettunen, P.O., 1986. Mater. Sci. Eng. 81, 457.

Lerch, B., Gerold, V., 1985. Acta Metall. 33, 1709.

Levkovitch, V., Sievert, R., Svendsen, B., 2005. Int. J. Fract. 136, 207.

Li, P., Li, S., Wang, Z.G., Zhang, Z.F., 2010a. Metall. Mater. Trans. A 41 (10), 2532.

Li, P., Li, S.X., Wang, Z.G., Zhang, Z.F., 2010b. Acta Mater. 58 (9), 3281.

Li, P., Li, S.X., Wang, Z.G., Zhang, Z.F., 2010c. Mater. Sci. Eng. A 527 (23), 6244.

Li, P., Li, S.X., Wang, Z.G., Zhang, Z.F., 2011. Prog. Mater. Sci. 56 (3), 328.

Li, X.W., Wang, Z.-G., Li, G.Y., Wu, S.D., Li, S.X., 1998. Acta Mater. 46 (13), 4497.

Li, X.W., Wang, Z.-G., Li, S.X., 1999a. Mater. Sci. Eng. A269 (1–2), 166.

Li, X.W., Wang, Z.-G., Li, S.X., 1999b. Mater. Sci. Eng. A265 (1–2), 18.

Li, X.W., Wang, Z.-G., Li, S.X., 1999c. Mater. Sci. Eng. A260 (1–2), 132.

Li, X.W., Umakoshi, Y., 2003. Scr. Mater. 48, 545.

Li, Y.-F., Laird, C., 1994a. Mater. Sci. Eng. A186, 65.

Li, Y.-F., Laird, C., 1994b. Mater. Sci. Eng. A186, 87.

Lim, L.C., Raj, R., 1985. Acta Metall. 33, 1577.

Liu, T.H., Ito, Y.M., 1969. J. Mech. Phys. Solids 17, 511.

Liu, T.H., Lin, S.R., 1979. Fatigue Mechanisms. Am. Soc. Test Mater, Philadelphia.

Llanes, L., 1992. Ph.D. Thesis. University of Pennsylvania.

Llanes, L., Laird, C., 1992. Mater. Sci. Eng. A157, 21.

Llanes, L., Laird, C., 1993. Mater. Sci. Eng. A161, 1.

Llanes, L., Rollett, A.D., Laird, C., Bassani, J.L., 1993. Acta Metall. Mater. 41 (9), 2667.

Lorenzo, F. (1983). Ph.D. Thesis. University of Pennsylvania.

Lorenzo, F., Laird, C., 1984. Acta Metall. 32 (5), 671.

Lukáš, P., Klesnil, M., Krejci, J., Rys, P., 1966. Phys. Status Solidi 15, 71.

Lukáš, P., Klesnil, M., 1970. Phys. Status Solidi 37, 833.

Lukáš, P., Klesnil, M., 1971. Phys. Status Solidi A5, 247.

Lukáš, P., Klesnil, M., 1973. Mater. Sci. Eng. 11, 345.

Lukáš, P., Kunz, L., 1994. Mater. Sci. Eng. A189, 1.

Lukáš, P., Kunz, L., Navrátilová, L., Bokůvka, O., 2011. Mater. Sci. Eng. A 528 (22–23), 7036.

Ma, B.-T., Laird, C., 1989a. Acta Metall. 37 (2), 349.

Ma, B.-T., Laird, C., 1989b. Acta Metall. 37 (2), 325.

Ma, B.-T., Laird, C., Radin, A.L., 1990. Mater. Sci. Eng. A123, 159.

Ma, B.T., Laird, C., 1986. Small Fatigue Cracks. The Metallurgical Soc., Warrendale, PA.

Ma, B. T., 1987. Ph.D. Thesis. University of Pennsylvania.

Ma, B.T., Laird, C., 1988. Mater. Sci. Eng. A102, 247.

Ma, B.T., Laird, C., 1989c. Acta Metall. 37, 369.

Ma, B.T., Laird, C., 1989d. Acta Metall. 37, 337.

Ma, B.T., Laird, C., 1989e. Acta Metall. 37 (2), 357.

Magnin, T., Driver, J.H., 1979. Mater. Sci. Eng. 39, 175.

Magnin, T., Fourdex, A., Driver, J.H., 1981. Phys. Status Solidi A 65, 301.

Malekjani, S., Hodgson, P.D., Cizek, P., Hilditch, T.B., 2011. Acta Mater. 59 (13), 5358.
Mann, Y.F., 1970. Bibliography on the Fatigue of Materials, Components and Structures. Pergamon, Oxford.
Mann, Y.F., 1978. Bibliography on the Fatigue of Materials, Components and Structures. Pergamon, Oxford.
Manson, S. S., 1953. Tech. Note 2933, N.A.C.A.
Marines-Garcia, I., Paris, P., Tada, H., Bathias, C., 2007. Mater. Sci. Eng. A 468-470, 120.
Marines-Garcia, I., Paris, P.C., Tada, H., Bathias, C., Lados, D., 2008. Eng. Fract. Mech. 75 (6), 1657.
Marx, M., Schaf, W., Vehoff, H., Holzapfel, C., 2006. Mater. Sci. Eng. A 435-436, 595.
May, A.N., 1960a. Nature 186, 573.
May, A.N., 1960b. Nature 185, 303.
McClintock, F.A., 1967. Fatigue Crack Propagation. ASTM STP 415. ASTM, Philadelphia. American Society of Testing and Materials: 170.
McClung, R.C., Davidson, D.L., 1991. Eng. Fract. Mech. 39 (1), 113.
McDowell, D.L., Dunne, F.P.E., 2010. Int. J. Fatigue 32 (9), 1521.
Mecke, K., 1974. Phys. Status Solidi A 25A (2), K93.
Mecke, K., Blochwitz, C., Kremling, U., 1982. Cryst. Res. Technol. 17, 1557.
Melisova, D., Weiss, B., Stickler, R., 1997. Scr. Mater. 36 (9), 1061.
Menzel, B.C., Dauskardt, R.H., 2006. Acta Mater. 54 (4), 935.
Menzel, B.C., Dauskardt, R.H., 2008. Acta Mater. 56 (13), 2955.
Mercer, C., Shademan, S., Soboyejo, W.O., 2003. J. Mater. Sci. 38, 291.
Miller, K.J., De Los Ríos, E.R., 1986. The Behavior of Short Cracks. Inst. of Mech. Engrs, London.
Miller, K.J., 1987. Fatigue Fract. Eng. Mater. Struct. 10, 75.
Mitchell, M., 1979. In: Meshii, M. (Ed.), Mechanisms and Theories of Fatigue. American Society of Metals, p. 385.
Moore, H.F., Kommers, J.B., 1927. The Fatigue of Metals. McGraw-Hill, New York.
Mori, H., Tokuwame, M., Miyazaki, T., 1979. Philos. Mag. A 40, 409.
Morrison, D.J., 1994. Mater. Sci. Eng. A187, 11.
Morrison, D.J., Chopra, V., 1994. Mater. Sci. Eng. A177, 29.
Mughrabi, H., 1975. Z. Metallkd. 66, 719.
Mughrabi, H., Herz, K., Stark, X., 1976. Acta Metall. 24, 659.
Mughrabi, H., Wothrich, C., 1976. Philos. Mag. 33, 963.
Mughrabi, H., 1977. Surface Effects in Crystal Plasticity Leyden, Noordholt.
Mughrabi, H., 1978. Mater. Sci. Eng. 33, 207–223.
Mughrabi, H., Ackermann, F., Herz, K., 1979. In: Fong, J.T. (Ed.), Fatigue Mechanisms. ASTM, Philadelphia, p. 69.
Mughrabi, H., 1980. 5th International Conference on Strength of Metals and Alloys. Pergamon Press.
Mughrabi, H., 1981. In: Brulin, O., Hsieh, R.K.T. (Eds.), Continuum Models of Discrete Systems 4. North Holland Publishing Co., Amsterdam, p. 241.
Mughrabi, H., Wang, R., 1982. Symposium on "Defects and Fracture", Tuczno, Poland.
Mughrabi, H., Wang, R., Differt, K., Essmann, U., 1983. Fatigue Mechanisms.
Mughrabi, H., Wang, R., 1988. International Colloquium on Basic Mechanisms in Fatigue of Metals. Czechoslovak Academy of Science, Brno, Czech Republic.
Mughrabi, H., 2001. In: Buschow, K.H.J., Cahn, R.W., Flemings, M.C., Ilschner, B., Kramer, E.J., Mahajan, S. (Eds.), Encyclopedia of Materials: Science and Technology. Elsevier, Amsterdam.
Mughrabi, H., 2006. Int. J. Fatigue 28 (11), 1501.
Mughrabi, H., 2009. Metall. Mater. Trans. B 40 (4), 431.
Mughrabi, H., Höppel, H.W., 2010. Int. J. Fatigue 32 (9), 1413.
Neumann, P., 1968. Z. Metallkd. 59 (2), 927.
Neumann, P., 1974a. Acta Metall. 22, 1155.
Neumann, P., 1974b. Acta Metall. 22, 1167.
Neumann, P., Vehoff, H., Fuhlrott, H., 1977. ICF 4. Univ. of Waterloo Press.
Neumann, P., 1986. Mater. Sci. Eng. 81 (1–2), 465.
Neumann, P., Tonnessen, A., 1987. Fatigue 87. EMAS Warley, W. Midlands.
Neumann, R., 1975. Zeit. Metallkunde 66, 26.
Nicholls, D.J., 1994. Eng. Fract. Mech. 48 (1), 9.
Nine, H., 1970. Scr. Mater. 4, 887.
Nine, H., 1972. Philos. Mag. 26, 1409.
Noroozi, A., Glinka, G., Lambert, S., 2005. Int. J. Fatigue 27 (10–12), 1277.
Nygårds, M., Gudmundson, P., 2002. Comput. Mater. Sci. 24, 513.

Obergfell, K., Peralta, P., Martinez, R., Michael, J.R., Llanes, L., Laird, C., 2001. Int. J. Fatigue 23, S207.

Pan, Y., Adams, B.L., 1994. Scripta Metall. Mater. 30, 1055.

Paris, P.C., Gomez, M.P., Anderson, W.P., 1961. Trend Eng. 13, 9.

Paris, P.C., Erdogan, F., 1963. Basic Eng. 85, 528.

Pelloux, R. M. N. (1969). ICF 2, Brighton.

Peralta, P., Schober, A., Laird, C., 1993. Mater. Sci. Eng. A169, 43.

Peralta, P., Llanes, L., Bassani, J., Laird, C., 1994a. Philos. Mag. A 70 (1), 219.

Peralta, P., Llanes, L., Czapka, A. and Laird, C., 1994b. MRS Fall Meeting.

Peralta, P., Li, Y., Llanes, L., Laird, C., 1995a. In: Chu, S.N.G., Liaw, P.K., Arsenault, R.J., Sadananda, K., Chan, K.S., Gerberich, W.W., Chau, C.C., Kung, T.M. (Eds.), Micromechanics of Advanced Materials. TMS, Warrendale, PA, p. 157.

Peralta, P., Llanes, L., Czapka, A., Laird, C., 1995b. Scr. Metall. Mater. 32 (11), 1877.

Peralta, P., Laird, C., 1997. Acta Mater. 45 (7), 3029.

Peralta, P., Laird, C., 1998. Acta Mater. 46 (6), 2001.

Peralta, P., Laird, C., Ramamurty, U., Suresh, S., Campbell, G., King, W.E., Mitchell, T.E., 1998. Small Fatigue Cracks: Mechanics, Mechanisms and Appliations. Elsevier, Oahu, HI, USA.

Peralta, P., Laird, C., Mitchell, T.E., 1999a. Mater. Sci. Eng. A264, 215.

Peralta, P., Obergfell, K., Llanes, L., Laird, C., Mitchell, T.E., 1999b. Int. J. Fatigue 21, S247.

Peralta, P., Dickerson, R., Dellan, N., Komandur, K., Jameel, M.A., 2005. J. Eng. Mater. Technol. 127, 23.

Peralta, P., Choi, S.-H., Gee, J., 2007. Int. J. Plast. 23, 1763.

Peralta, P., Villarreal, T., Atodaria, I., Chattopadhyay, A., 2012. Scr. Mater. 66, 13.

Peralta, P.D., 2005. In: Robert, W.C., Merton, C.F., Bernard, I., Edward, J.K., Subhash, M., Patrick, V. (Eds.), Encyclopedia of Materials: Science and Technology, Second ed. K. H. J. B. Editors-in-Chief. Elsevier, Oxford, p. 1.

Petrenec, M., Polak, J., Obrtlik, K., Man, J., 2006. Acta Mater. 54 (13), 3429.

Pohl, K., Mayr, P., Macherauch, E., 1980. Scr. Mater. 14, 1167.

Polák, J., 1970. Scr. Mater. 4, 761.

Polák, J., Obrtlík, K., Hájek, M., 1994. Fatigue Fract. Eng. Mater. Struct. 17 (7), 773.

Purushothaman, S., Tien, J., 1978. Mater. Sci. Eng. 34, 241.

Pyttel, B., Schwerdt, D., Berger, C., 2011. Int. J. Fatigue 33 (1), 49.

Rannou, J., Limodin, N., Réthoré, J., Gravouil, A., Ludwig, W., Baïetto-Dubourg, M.-C., Buffière, J.-Y., Combescure, A., Hild, F., Roux, S., 2010. Comput. Methods Appl. Mech. Eng. 199 (21–22), 1307.

Rasmussen, K.V., Pedersen, O.B., 1980. Acta Metall. 28, 1467.

Ravichandran, K.S., Ritchie, R.O., Murakami, Y. (Eds.), 1999. Small Fatigue Cracks: Mechanics, Mechanisms and Applications. Kidlington. Elsevier, Oxford, UK.

Repetto, E.A., Ortiz, M., 1997. Acta Mater. 45 (6), 2577.

Rice, J.R., 1987. Mech. Mater. 6, 317.

Ritchie, R.O., Lankford, J., 1986. Small Fatigue Cracks. The Metallurgical Soc. Inc.

Ro, Y., Agnew, S.R., Gangloff, R.P., 2008. Metall. Mater. Trans. A 39 (6), 1449.

Rosenbloom, S.N., Laird, C., 1993. Acta Metall. Mater. 41, 3473.

Sadananda, K., Vasudevan, A., 2003. Fatigue Fract. Eng. Mater. Struct. 26, 835.

Sadananda, K., Vasudevan, A.K., Kang, I.W., 2003. Acta Mater. 51, 3399.

Sadananda, K., Vasudevan, A.K., 2004. Acta Mater. 52 (14), 4239.

Sadananda, K., Vasudevan, A., Phan, N., 2007. Int. J. Fatigue 29 (9–11), 2060.

Sangid, M.D., Maier, H.J., Sehitoglu, H., 2011. Int. J. Plast. 27 (5), 801.

Sauzay, M., Kubin, L.P., 2011. Prog. Mater. Sci. 56 (6), 725.

Saylor, D.M., Fridy, J., El-Dasher, B.S., Jung, K.-Y., Rollett, A.D., 2004. Metall. Mater. Trans. A 35A, 1969.

Schaef, W., Marx, M., Vehoff, H., Heckl, A., Randelzhofer, P., 2011. Acta Mater. 59 (5), 1849.

Schiller, C., Walgraef, O., 1988. Acta Metall. 36, 563.

Schwarz, K.W., Mughrabi, H., 2006. Philos. Mag. Lett. 86 (12), 773.

Sesták, B., Vicherková, Z., Novák, V., Libovicky, S., Brádler, J., 1987. Phys. Status Solidi A 104, 79.

Sesták, B., Novák, V., Libovicky, S., 1988. Philos. Mag. A 57, 353.

Shijve, J., 1967. Fatigue Crack Propagation.

Simo, J.C., Hughes, T.R.J., 2000. Computational Inelasticity. Springer-Verlag, New York.

Solanki, K., 2004. Eng. Fract. Mech. 71 (2), 149.

Stanzl, S.E., 1980. Scr. Metall. 14, 749.

Stanzl-Tschegg, S.E., Schönbauer, B., 2010. Int. J. Fatigue 32 (6), 886.

Starke, J.E.A., Lutjering, G., 1979. Fatigue and Microstructure. Am. Soc. Metals.

Steiner, D., Beddoe, R., Gerold, V., Kostorz, G., Schmekzer, R., 1983. Scr. Metall. 17, 733.

Stevenson, R., Sande, J.B.V., 1974. Acta Metall. 22 (9), 1079.

Stöcker, C., Zimmermann, M., Christ, H.J., 2011. Int. J. Fatigue 33 (1), 2.

Suresh, S., 1998. Fatigue of Materials. Cambridge University Press, Cambridge.

Sutton, A.P., Ballufi, R.W., 1995. Interfaces in Crystalline Materials. Oxford University Press, New York.

Tan, X., Gu, H., Zhang, S., Laird, C., 1994. Mater. Sci. Eng. A189, 77.

Tan, X., Gu, H., Wang, Z., 1995. Mater. Sci. Eng. A 196 (1–2), 45.

Tan, X., Gu, H., Laird, C., Munroe, N.D.H., 1998. Metall. Mater. Trans. A 29A, 507.

Tanaka, K., Hoshide, T., Sakai, N., 1984. Eng. Fract. Mech. 19 (5), 805.

Thompson, A.W., Backofen, W.A., 1971. Acta Metall. 19, 597.

Thompson, A.W., 1972. Acta Metall. 20, 1085.

Thompson, N., Wadsworth, N.J., 1958. Adv. Phys. 7 (25), 72.

Tien, J., Purushothaman, S., 1978. Mater. Sci. Eng. 34, 247.

Tomkins, B., Biggs, W.D., 1969. J. Mater. Sci. 4, 544.

Trexler, M.M., Thadhani, N.N., 2010. Prog. Mater. Sci. 55 (8), 759.

Tschegg, E., Stanzl, S., 1981. Acta Metall. 29, 33.

Turenne, S., L'Espérance, G., Dickson, J.I., 1988. Acta Metall. 36, 459.

Tvergaard, V., 2004. J. Mech. Phys. Solids 52, 2149.

Vasudevan, A.K., Sadananda, K., Glinka, G., 2001. Int. J. Fatigue 23, S39.

Vehoff, H., Laird, C., Duquette, D.J., 1987. Acta Metall. 35, 2877.

Venkataraman, G., Chung, Y.W., Nakasone, Y., Mura, T., 1990. Acta Metall. Mater. 38, 31.

Villechaise, P., Mendez, J., Violan, P., 1991. Acta Metall. Mater. 39 (7), 1683.

Vogel, W., Wilhem, M. and Gerold, V., 1980. Int. Conf. on Strength of Metals and Alloys, Aachen, Pergamon.

Vogel, W., Wilhem, M., Gerold, V., 1982. Acta Metall. 30, 21.

Walgraef, D., Aifantis, E.C., 1985. Int. J. Eng. Sci. 23 (12), 1351.

Wang, G.Y., Liaw, P.K., Yokoyama, Y., Inoue, A., Liu, C.T., 2008. Mater. Sci. Eng. A 494 (1–2), 314.

Wang, G.Y., Liaw, P.K., Yokoyama, Y., Freels, M., Inoue, A., 2010. Int. J. Fatigue 32 (3), 599.

Wang, R., Mughrabi, H., 1984. Mater. Sci. Eng. 63, 147.

Wang, Z., Laird, C., 1988. Mater. Sci. Eng. 100, 57.

Wang, Z., Laird, C., 1989. Metall. Trans. 20A, 2033.

Wang, Z., Gong, B., Wang, Z.-G., 1997. Acta Mater. 45 (4), 1379.

Wang, Z., Gong, B., Wang, Z.-G., 1999. Acta Mater. 47 (1), 307.

Weertman, J., 1979. Fatigue and Microstructure. Am. Soc. Metals.

Weertman, J., 2007. Mater. Sci. Eng. A 468-470, 59.

Weidner, A., Amberger, D., Pyczak, F., Schönbauer, B., Stanzl-Tschegg, S., Mughrabi, H., 2010. Int. J. Fatigue 32 (6), 872.

Wen, W., Zhai, T., 2011. Philos. Mag. 91 (27), 3557.

Wetzel, R. M., 1971. Ph.D. Thesis. University of Waterloo.

Wilhem, M., Nageswararao, M., Meyer, R., 1979. Am. Soc. Test. Mater. Spec. Tech. Publ. 675, 214.

Wilhem, M., Everwin, P., 1980. Int. Conf. on Strength of Metals & Alloys, Aachen, Pergamon.

Wilhem, M., 1981. Mater. Sci. Eng. 48, 91.

Wilkens, M., Herz, K., Mughrabi, H., 1980. Z. Metallkd. 71, 376.

Wilkins, M.A., Smith, G.C., 1970. Acta Metall. 18, 1035.

Winter, A.T., 1974. Philos. Mag. 30, 719.

Winter, A.T., Pedersen, O.B., Rasmussen, K.V., 1981. Acta Metall. 29, 735.

Witmer, D.E., Farrington, G.C., Laird, C., 1987. Acta Metall. 35 (7), 1895.

Wood, W.A., 1958. Philos. Mag. 3, 692.

Woods, P.J., 1973. Philos. Mag. 28 (1), 155.

Wu, X.M., Wang, Z.G., Li, G.Y., 2001. Mater. Sci. Eng. A314, 39.

Xiaoli, T., Haicheng, G., Wang, Z.-G., 1995. Mater. Sci. Eng. A196, 45.

Xu, B., Jiang, Y., 2004. Int. J. Plast. 20 (12), 2161.

Xue, Y., McDowell, D., Horstemeyer, M., Dale, M., Jordon, J., 2007. Eng. Fract. Mech. 74 (17), 2810.

Yan, B.D., Hunsche, A., Neumann, P., Laird, C., 1986. Mater. Sci. Eng. 79 (2), 9.

Youssef, 1970. Phys. Status Solidi A3, 801.

Yu, Q., Zhang, J., Jiang, Y., 2011a. Philos. Mag. Lett. 91 (12), 757.

Yu, Q., Zhang, J., Jiang, Y., 2011b. Mater. Sci. Eng. A 528 (25–26), 7816.
Zhai, T., Martin, J.W., Briggs, G.A.D., 1996. Acta Mater. 44 (5), 1729.
Zhai, T., Wilkinson, A.J., Martin, J.W., 2000. Acta Mater. 48, 4197.
Zhang, C., Suzuki, A., Ishimaru, T., Enomoto, E., 2004. Metall. Mater. Trans. A 35A, 1927.
Zhang, G.P., Volkert, C.A., Schwaiger, R., Wellner, P., Arzt, E., Kraft, O., 2006. Acta Mater. 54 (11), 3127.
Zhang, J., Jiang, Y., 2006. Int. J. Plast. 22 (3), 536.
Zhang, P., Qu, S., Duan, Q.Q., Wu, S.D., Li, S.X., Wang, Z.G., Zhang, Z.F., 2011. Philos. Mag. 91 (2), 229.
Zhang, Y.H., Edwards, L., 1992. Mater. Charact. 29 (3), 313.
Zhang, Z.F., Gu, H.C., Tan, X.L., 1998. Mater. Sci. Eng. A252, 85.
Zhang, Z.F., Wang, Z.G., 2008. Prog. Mater. Sci. 53 (7), 1025.
Zhou, D., Moosbrugger, J.C., Jia, Y., Morrison, D.J., 2005. Int. J. Plast. 21 (12), 2344.
Zhou, D., Moosbrugger, J.C., Morrison, D.J., 2006. Int. J. Plast. 22 (7), 1336.

Further Reading

Encyclopedia of Materials, 2010. Elsevier Science.
Stephens, R.I., Fatemi, A., Stephens, R.R., Fuchs, H.O., 2001. Metal Fatigue in Engineering, Second ed. Wiley and Sons.
Frost, N.E., Marsh, K.J., Pook, L.P., 1974. Metal Fatigue. Clarendon Press, Oxford.
Lemaitre, J., Desmorat, R., 2005. Engineering Damage Mechanics: Ductile, Creep, Fatigue and Brittle Failures. Springer-Verlag.

Biography

Pedro Peralta received his B.S. degree in Mechanical Engineering from Universidad Simón Bolívar (Caracas, Venezuela) in 1989, his M.S. degree in Mechanical Engineering and Applied Mechanics in 1994 and his Ph.D. in Materials Science and Engineering in 1996, both from the University of Pennsylvania. From 1996 to 1998 he was a Director Funded Postdoctoral Fellow at the Center for Materials Science at Los Alamos National Laboratory (LANL). In 1998 he joined the faculty in the Mechanical and Aerospace Engineering Department at Arizona State University (ASU). He is currently a Professor of Mechanical, Aerospace and Materials Engineering at ASU and a visiting scientist in the Materials Science and Technology Division at LANL. His research focuses on experiments and modeling of microstructure effects on shock loading, fatigue and creep of metals and ceramics. Prof. Peralta received an NSF CAREER award in 2000 for research on fatigue crack propagation and the Orr award from ASME in 2005 for best paper on fatigue and fracture in the Journal of Engineering Materials and Technology. He is a member of ASME, TMS and USACM.

Campbell Laird, currently Emeritus Professor at the University of Pennsylvania (Penn), began his engineering career with a commission in the Royal Engineers. He received his B.A. (1959), M.A. (1962) and Ph.D. (1963) degrees from Cambridge University in the field of Materials Science and Engineering (MSE). Having obtained a junior faculty position at Christ's College, Cambridge, he took a Leave of Absence at the Ford Motor Co Scientific Laboratory (1963-1968) and returned to an academic career at Penn via a Visiting Professorship at Ohio Sate University. He has enjoyed the study of Fatigue and Fracture (amongst other topics) for his entire career, and the largest majority of his over 300 publications deal with cyclic deformation and fatigue failure mechanisms. He has twice served as head of the MSE Department at Penn, and has served on many professional committees. He was awarded the Darwin Prize at Cambridge for his Dissertation, two Citations in the MITRE report for his research, one in the top 20, the Outstanding Paper Award, Acta Metallurgica, and was made Plenary Lecturer at the conference FATIGUE 2000. He was elected Fellow of AIMMM(TMS), of ASM, and of the Institute of Diagnostic Engineers.

19 Magnetic Properties of Metals and Alloys

Michael E. McHenry and David E. Laughlin, Department of Materials Science and Engineering, Carnegie Mellon University, Pittsburgh, PA, USA

19.1 Magnetic Field Quantities and Properties Survey

19.1.1 Introduction

We begin this chapter by reviewing the magnetic properties which are relevant to soft and hard magnetic materials. We develop the subject by considering

(1) a summary of technical magnetic properties of importance for soft and hard magnetic materials in the context of a few illustrative examples;
(2) definitions of important magnetic phenomena which are relevant to the determination of these properties;
(3) a survey of alloys within an historical context as to their development;
(4) an extrapolation to the future to highlight some emerging areas of magnetic metals and alloys relevant to evolving technological needs with an emphasis on materials important to energy applications.

The magnetization, M, or magnetic induction, B, is an extensive material property and the field, H, is an intensive variable. These ($M(H)$ or $B(H)$) form a set of *conjugate variables* which define the *magnetic work* in a thermodynamic analysis. It is a goal to describe the response function $B(H)$ for a material system. This response function can be single valued as shown in **Figure 1(a)** which would be the case for reversible equilibrium thermodynamics or two valued (hysteretic) as shown in **Figure 1(b)** which is the case for irreversible nonequilibrium thermodynamics. The equilibrium response is generally nonlinear and depends on the process by which atomic dipoles rotate into the direction of the field. The *permeability* is the slope of the $B(H)$ curve. The easier it is to magnetize a material for a given applied field, the higher the permeability. The hysteresis curve is the starting point for discussing technical magnetic properties of soft and hard magnetic materials and distinguishing between hard and soft magnets.

Two thermodynamic states exist when all the dipoles are aligned in the direction of a positive field or negative field. The value of the magnetization, M, when all of the dipoles are aligned is called the *saturation magnetization, M_s*. A *hysteresis loop* (**Figure 1(b)**) is a plot of the magnetization of a material as it is cycled from positive fields to negative and back. Important states on a hysteresis loop include the *remanent magnetization, M_r*[1] which is the magnetization remaining when the driving field is reduced to

---

[1] $B_r = M_r$ (cgs units); $= \mu_o M_r$ (mksa units).

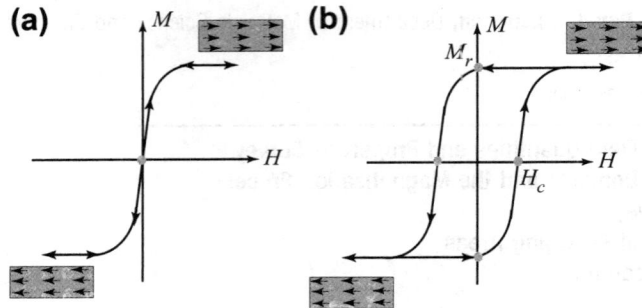

Figure 1   (a) Magnetic response, *M(H)*, for a reversible ferromagnetic system of dipoles. (b) Hysteretic magnetic response, *M(H)*, for an irreversible ferromagnetic system of dipoles. (For color version of this figure, the reader is referred to the online version of this book.)

zero. The *coercive field*, H_c, is the reverse field necessary to drive the magnetization to zero after being saturated.

The value of the coercive field is used to distinguish between hard magnets and soft magnets. Hard or *permanent magnets* require very large fields to switch and are therefore useful for a variety of applications taking advantage of the remanent induction. Soft magnetic materials are characterized by a low coercivity and therefore a narrow loop. Since they are easy to reverse, they are ideal for high-frequency operation.

19.1.1.1   Soft Magnet Introduction
Box 1 summarizes applications and technical properties of soft magnetic materials. **Michael Faraday** (1791–1867) observed that an applied voltage to one copper coil resulted in a voltage across a second coil wound on the same iron core. Faraday's law of induction (1831) defines the induced voltage

$$V = N \frac{d\Phi}{dt} \tag{1}$$

in terms of the number of turns, N, in a primary exciting coil and the rate of change in the magnetic flux, Φ, caused by an alternating current (AC). This physical law is the basis for many inductive components including power transformers which is a prototypical application of soft magnetic materials. The 1884 demonstrations of an open (Gaulard and Gibbs) and closed (Deri, Blathy, and Ziperman) core *transformer* was followed by use of closed core transformers to supply power to Edison lamps which illuminated the 1885 Budapest exhibition. Transformers now (1) are key energy conversion devices in power distribution systems, (2) provide the basis for operation of many electromechanical devices, and (3) are components of modern integrated circuits.

19.1.1.2   Hard Magnet Introduction
Permanent magnets provide large fields in a confined space. The strength of a magnet is characterized by the amount of energy it stores. Unlike a battery, this energy is always available as the permanent magnet does no net work. Initially, a field is applied to align domains to obtain a net magnetization. For a permanent magnet, a wide *magnetic hysteresis loop* reflects a large *coercive field* which stabilizes the moment with respect to field fluctuations. In most applications, a large magnetization is also desirable. Figures of

---

Box 1   Applications and technical properties of soft magnetic materials

Important applications of soft magnetic materials include the following:

(1) Inductors and inductive components, low- and high-frequency transformers
(2) AC machines, motors and generators
(3) Magnetic lenses for particle beams and magnetic amplifiers
(4) High-frequency inductors and absorbers
(5) Magnetocaloric materials
(6) Magnetic sensors.

   The desired technical properties of interest for soft magnetic materials include the following:

(1) High permeability: Permeability $\mu = \frac{B}{H} = (1 + \chi)$ is the materials parameter describing the slope of the flux density, B, as a function of the applied field, H. High-permeability materials can produce very large flux changes at small fields.
(2) Low hysteresis loss: Hysteresis loss is the energy consumed in cycling a material between $\pm H$. The energy consumed per cycle is the area in the hysteresis loop. The power loss of an AC device includes a term equal to the frequency multiplied by the hysteretic loss per cycle. At high frequencies, eddy current losses are intimately related to the material's resistivity, ρ.
(3) Large saturation and remnant magnetizations: A large saturation induction, B_s, is desirable because it represents the ultimate response of soft magnetic materials.
(4) High Curie temperature: The ability to use soft magnetic materials at elevated temperatures depends on the Curie temperature of the material.

---

merit include the *coercive field*, *remnant magnetization* and *energy product*, $(BH)_{max}$. A high T_c is desired for a greater operational temperature range. Applications of *permanent magnets* include small motors, loud speakers, electronic tubes, focusing magnets for charged particle beams and mechanical work devices.

   The energy product, $(BH)_{max}$, is defined as the maximum value of the product of B and H taken in the second quadrant of a hysteresis curve. This has units of energy per unit volume of the material. For a material with a square hysteresis loop, $(BH)_{max}$, is the product of the remanence and coercivity.

   Figure 2 shows (a) the schematic demagnetization curves for important permanent magnet materials and (b) the evolution of $(BH)_{max}$ over time. The first permanent magnet dates back to 600 BC

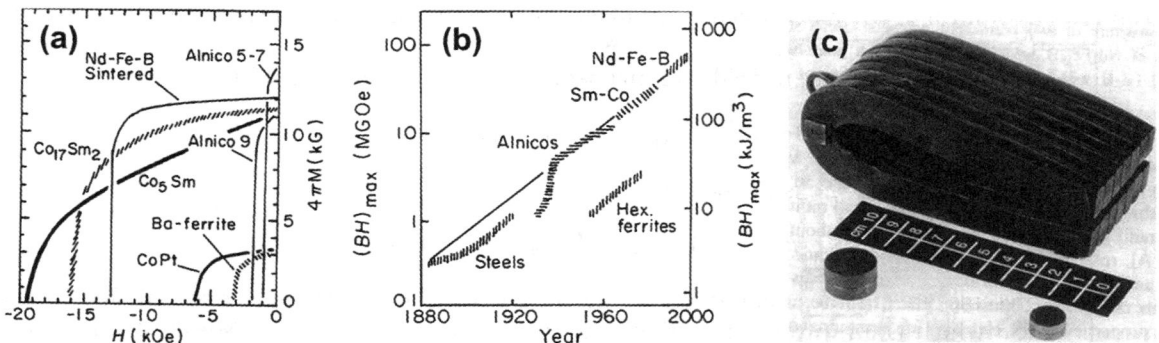

Figure 2   (a) Schematic demagnetization curves for some permanent magnet materials. (b) The evolution of the energy product, $(BH)_{max}$ over time. (c) size reduction of magnets. (O'Handley, 2000; Coey, 2010).

when lodestone, Fe_3O_4, was used as the first compass needle. This was followed by iron. C steel was reported in 1600 AD in the work "De Magnete" of Gilbert (1958).

Improvements in $(BH)_{max}$ accompanied development of alloy, oxide and fine-particle ferromagnets. Recently, rare-earth transition metal (RETM) materials represent a major development. Energy products were ~ 5 kJ/m^3 for steels available circa 1900 and now approach ~ 1000 kJ/m^3 in state-of-the-art *rare earth permanent magnet* (REPM) systems. REPM materials have large magnetocrystalline anisotropies which are at the root of large coercivity. Mixtures of rare-earth and transition metals species result in large remanent and saturation magnetizations.

19.1.2 Dipoles and Magnetization

19.1.2.1 Definitions of Field Quantities

We begin discussing magnetic properties of materials by defining macroscopic field quantities.[2] Two fields, the *magnetic induction*, $\vec{B}$, and the *magnetic field*, $\vec{H}$, are vectors. In many cases, the induction and the field will be collinear (parallel) and we can treat them as scalar quantities, B and H.[3]

In a vacuum, the magnetic induction, $\vec{B}$, is related to the magnetic field, $\vec{H}$:

$$\vec{B} = \mu_0\vec{H} \quad | \quad \vec{B} = \vec{H} \tag{2}$$

where the *permeability of the vacuum*, μ_0, is $4\pi \times 10^{-7}$ H/m in SI (mksa) units. This quantity is taken as one in cgs units. In cgs units, the induction and field are the same. In SI (mksa) units, we assign a permeability to the vacuum, so the two are proportional.

In a magnetic material, the magnetic induction can be enhanced or reduced by the material's *magnetization*, $\vec{M}$ (defined as net dipole moment per unit volume), so that

$$\vec{B} = \mu_0(\vec{H} + \vec{M}) \quad | \quad \vec{B} = \vec{H} + 4\pi\vec{M} \tag{3}$$

where the magnetization, $\vec{M}$, is expressed in linear response theory as

$$\vec{M} = \chi_m\vec{H}. \tag{4}$$

The constant of proportionality is called the *magnetic susceptibility*, χ_m. The magnetic susceptibility which relates two axial vector quantities is a second-rank polar tensor. For most discussions (whenever B and H are collinear or when interested in the magnetization component in the field direction), we can view the susceptibility as a scalar.

Considering a scalar induction, field and magnetization, we express $B = \mu_r H$ as

$$B = \mu_0(1 + \chi_m)H \quad | \quad B = (1 + 4\pi\chi_m)H \tag{5}$$

and the *relative permeability*, μ_r, is defined as

$$\mu_r = \mu_0(1 + \chi_m) \quad | \quad \mu_r = (1 + 4\pi\chi_m) \tag{6}$$

μ_r thus represents an enhancement factor of the flux density in a magnetic material due to the magnetization which is an intrinsic materials property. If we have $\chi_m < 0$, we speak of *diamagnetic*

---

[2] Selected formulas are introduced in SI (mksa) units followed by cgs units.
[3] For many discussions, it will be sufficient to treat field quantities as scalars, when this is not the case, vector symbols will be explicitly used.

response and for $\chi_m > 0$ (and no *collective magnetism*), we speak of *paramagnetic response*. A *superconductor* is a material which acts as a *perfect diamagnet* so that $\chi_m = -1$ or $\chi_m = -1/4\pi$.

19.1.2.2 *Magnetic Dipole Moments—Definitions*

A *magnetic dipole moment* originates from a circulating charge (**Figure 3**). Concepts relating circulating charge, angular momentum and dipole moments are as follows:

(1) A dipole moment for a circulating charge is defined formally as

$$\vec{\mu} = IA\vec{u}_{\vec{r}\times\vec{J}} = \int_V \vec{r} \times \vec{J} dV \tag{7}$$

where $\vec{r}$ is the position vector of the charged particle about the origin for the rotation, $\vec{J}$ is the current density of the orbiting charge, I is the current due to the circulating charge, $A = \pi r^2$ is the area swept out by the circulating charge, V is the volume, and $\vec{u}_{\vec{r}\times\vec{J}}$ is the unit vector normal to the area, A.

(2) The magnetic dipole moment is proportional to the axial *angular momentum* vector. Let $\vec{\Pi}$ be a general angular momentum vector. The fundamental relationship between magnetic dipole moment and the angular momentum vector is

$$\mu = g\frac{e}{2m}\Pi \quad | \quad \mu = g\frac{e}{2mc}\Pi \tag{8}$$

where g is called the *Lande g-factor*. For an orbiting electron, the constant $g = 1$. The dipole moment associated with *spin angular momentum* has $g = 2$.

(3) In quantum mechanics, every electron has a dipole moment associated with its spinning charge density (spin) and its orbit about the nucleus (orbit). Angular momentum is quantized in units of **Planck's constant** divided by $2\pi (\hbar = \frac{h}{2\pi} = 1.05 \times 10^{-34} J - s = 1.05 \times 10^{-27} erg - s)$. A fundamental unit of magnetic dipole moment, the *Bohr magneton*, is defined as follows:

$$\mu_B = \frac{e}{2m}\hbar \quad | \quad \mu = \mu_B = \frac{e}{2mc}\hbar \tag{9}$$

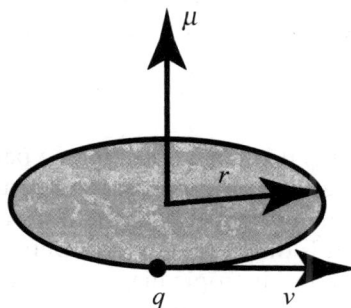

Figure 3 Geometry of a charged particle orbiting at a distance *r*, with a linear velocity, *v*. The particle orbit sweeps out an area, *A*, and gives rise to a dipole moment, $\vec{\mu}$. (For color version of this figure, the reader is referred to the online version of this book.)

The Bohr magneton has the following value:

$$\mu_B = 9.27 \times 10^{-24} Am^2 \left(\frac{J}{T}\right) \quad | \quad \mu = \mu_B = 9.27 \times 10^{-21}\frac{erg}{G} \tag{10}$$

(4) An atomic dipole moment, μ_{atom}, is calculated by summing all the electron dipole moments for an atom in accordance with Hund's rules.

(5) For a collection of identical atoms, the magnetization, M, is

$$M = N_a\mu_{atom} \tag{11}$$

where N_a is the number of dipole moments per unit volume.

(6) The potential energy of a dipole moment in the presence of a field is

$$E_p = -\vec{\mu}\cdot\vec{B} = -\mu \cdot B \cos \theta \tag{12}$$

where θ is the angle between the dipole moment and $\vec{B}$. The magnetization (or other field quantity) multiplied by another field has units of energy per unit volume. In quantum mechanical systems, the component of the dipole moment vector projected along the field direction is quantized and only particular values of the angle θ are allowed.

19.1.2.3 Magnetization and Dipolar Interactions

We now turn to the definition of the magnetization, M.

Magnetization, M, is the net dipole moment per unit volume.

It is expressed as

$$M = \frac{\Sigma_{atoms} \times \mu_{atom}}{V} \tag{13}$$

where V is the volume of the material.[4] Magnetization is an extrinsic materials property that depends on the constituent atoms in a system, the individual dipole moments, and how the dipole moments add vectorially. Collinear dipoles can add or subtract depending on whether they are parallel or antiparallel. This can give rise to many interesting types of collective magnetism.

A *paramagnet* is a material where permanent local atomic dipole moments are aligned randomly.

In the absence of an applied field, H_a, the magnetization of a paramagnet is precisely zero since the sum of randomly oriented vectors is zero. This emphasizes the importance of the word "net" in the definition of magnetization. A permanent nonzero magnetization does not necessarily follow from having permanent dipole moments. It is only through a coupling mechanism which acts to align the dipoles in the absence of a field that a macroscopic magnetization is possible.

Two dipole moments interact through dipolar interactions that are described by an interaction force analogous to the Coulomb interaction between charges. If we consider two collinear dipoles pointing

---

[4] Magnetization can also be reported as specific magnetization which is net dipole moment per unit weight.

in a direction perpendicular to $\vec{r}_{12}$, the potential energy between the dipoles can be expressed as follows:

$$E_p = \frac{\pm\mu_1\mu_2}{4\pi\mu_0 r_{12}^3} \quad | \quad E_p = \frac{\pm\mu_1\mu_2}{r_{12}^3} \tag{14}$$

which is the familiar Coulomb's law.[5]

Free poles collect at surfaces to form the North and South poles of a magnet (**Figure 4(b)**). Magnetic flux lines travel from the N to the S pole of a permanent magnet causing a self-field, the *demagnetization field*, H_d, outside the magnet. H_d can act to demagnetize a material for which the magnetization is not strongly tied to an *easy magnetization direction (EMD)*.

Most hard magnets have a large *magnetocrystalline anisotropy* which acts to fix the magnetization vector along particular crystal axes. As a result, they are difficult to demagnetize. A soft magnet can be used as permanent magnet if its shape is engineered to control the path of the demagnetization field to not interact with the magnetization vector. Fe is an example of a soft magnetic material with a large magnetization. To use Fe as a permanent magnet, it is often shaped into a *horseshoe magnet* (**Figure 4(c)**) where the return path for flux lines is spatially far from the material's magnetization.

19.1.2.4 Magnetic States

In systems (ionic compounds and rare earths) with localized atomic orbitals responsible for the atomic magnetic dipole moments, discrete magnetic states can be calculated using quantum mechanical rules called *Hund's rules* (Hund, 1927). Systems with delocalized electron states are treated within the band theory of solids.

For localized electrons assigned to a particular atom, discrete magnetic states are calculated using quantum mechanical rules. The general angular momentum vector, Π, has contributions from *orbital angular momentum*, $\vec{L}$, and *spin angular momentum*, $\vec{S}$, both quantized in units of $\hbar$. Orbital and spin angular momentum are summed using *Hund's rules* (Hund, 1927) as shown in **Figure 5** and tabulated in **Table 1**.

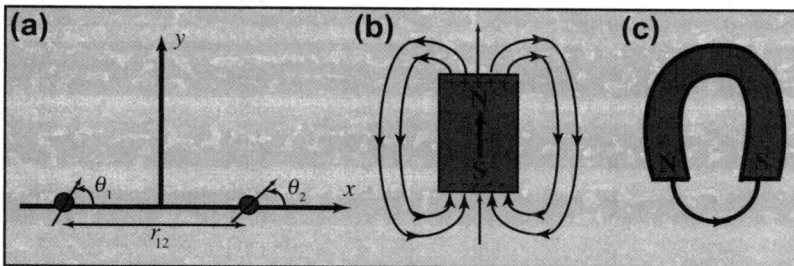

Figure 4 Geometry of (a) two coplanar dipole moments used to define dipolar interactions, (b) net surface dipole moments at the poles of a permanent magnet with a single domain and (c) flux return path for magnetic dipoles in a horseshoe magnet. (For color version of this figure, the reader is referred to the online version of this book.)

[5] Some authors (Cullity, e.g.) use p to denote dipole moment.

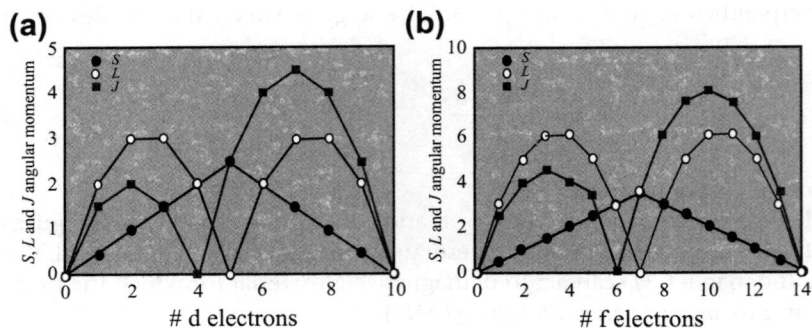

Figure 5 J, L, and S quantum numbers for (a) TM^{2+} and (b) RE^{3+} ions. (For color version of this figure, the reader is referred to the online version of this book.)

Table 1   Ground state multiplets of TM and RE ions from van Vleck (*The Theory of Electric and Magnetic Susceptibilities*, Oxford Univ. Press, 1932, 243)

	Ion	S	L	J	Term	n_{eff} $g[J(J+1)]^{1/2}$	Obs.	n_{eff} $g[S(S+1)]^{1/2}$
d-shell electrons								
1	Ti^{3+}, V^{4+}	1/2	2	3/2	$^2D_{3/2}$	1.55	1.70	1.73
2	V^{3+}	1	3	2	$^3F_2$	1.63	2.61	2.83
3	V^{2+}, Cr^{3+}	1/3	3	3/2	$^4F_{3/2}$	0.77	3.85	3.87
4	Cr^{2+}, Mn^{3+}	2	2	0	$^5D_0$	0	4.82	4.90
5	Mn^{2+}, Fe^{3+}	3/2	0	5/2	$^5S_{5/2}$	5.92	5.82	5.92
6	Fe^{2+}	2	2	4	$^5D_4$	6.7	5.36	4.90
7	Co^{2+}	3/2	3	9/2	$^4F_{9/2}$	6.63	4.90	3.87
8	Ni^{2+}	1	3	4	$^3F_4$	5.59	3.12	2.83
9	Cu^{2+}	1/2	2	5/2	$^2D_{5/2}$	3.55	1.83	1.73
10	Cu^+, Zn^{2+}	0	0	0	$^1S_0$	0	0	0
f-shell electrons								
1	Ce^{3+}	1/2	3	5/2	$^2F_{5/2}$	2.54	2.51	
2	Pr^{3+}	1	5	4	$^5H_4$	3.58	3.56	
3	Nd^{3+}	3/2	6	9/2	$^4I_{9/2}$	3.62	3.3	
4	Pm^{3+}	2	6	4	$^5I_4$	2.68	–	
5	Sm^{3+}	5/2	5	5/2	$^6H_{5/2}$	0.85 (1.6)*	1.74	
6	Eu^{3+}	3	3	0	$^7F_0$	0 (3.4)*	3.4	
7	Gd^{3+}, Eu^{3+}	5/2	0	5/2	$^8S_{7/2}$	7.94	7.98	
8	Tb^{3+}	3	3	6	$^7F_6$	9.72	9.77	
9	Dy^{3+}	5/2	5	15/2	$^6H_{15/2}$	10.63	10.63	
10	Ho^{3+}	2	6	8	$^5I_8$	10.60	10.4	
11	Er^{3+}	3/2	6	15/2	$^4I_{15/2}$	9.59	9.5	
12	Tm^{3+}	1	5	6	$^3H_6$	7.57	7.61	
13	Yb^{3+}	1/2	3	7/2	$^2F_{7/2}$	4.53	4.5	
14	Lu^3+, Yb^3+	0	0	0	$^1S_0$	0	–	

The *ground state multiplet* including the m_l and m_s eigenstates allows us to calculate the components of the orbital, L, spin, S, and total angular momentum, J. The magnitudes of orbital and spin angular momenta are summed over a multielectron shell as follows:

$$L = \sum_{i=1}^{n} (m_l)_i \hbar \quad S = \sum_{i=1}^{n} 2(m_s)_i \hbar \tag{15}$$

The projection of the total angular momentum vector, $\vec{J} = \vec{L} + \vec{S}$, along an applied field direction is subject to quantization conditions that require $(J = L + S)$, where J is $J = |L - S|$ for less than half-filled shells and $J = |L + S|$ for greater than half-filled shells. To determine the occupation of eigenstates of S, L, and J, we use Hund's rules:

(1) We fill m_l states (which are $2l + 1$-fold degenerate) in such a way as to first maximize total spin.
(2) We fill m_l states first in such a way as to first maximize total spin.

We consider the ions of transition metal series, TM^{2+}, i.e. ions which have given up 2s electrons to yield a $3d^n$ outer shell configuration in **Figure 5(a)**. The ground state J, L and S quantum numbers for rare-earth, RE^{3+}, ions are shown in **Figure 5(b)**.

Defining L, S and J specifies the ground-state multiplet written compactly in the spectroscopic term symbol:

$$^{2S+1}L_J \tag{16}$$

where L is the symbol for orbital angular momentum ($L = 0 = S$, $L = 1 = P$, $L = 2 = D$, $L = 3 = F$, etc.) and $2S + 1$ and J are the numerical. Cr^{3+} with $L = 3$, $S = 3/2$ and $J = 3/2$ would be assigned the term symbol, $^4L_2^3$. We can relate the permanent local atomic moment vector with the total angular momentum vector, $\vec{J}$:

$$\vec{\mu} = \gamma \hbar \quad \vec{J} = -g(J, L, S)\mu_B \vec{J} \tag{17a}$$

where γ is the *gyromagnetic factor* and $g = g(J, L, S)$ is the *Lande g-factor*:

$$g(J, L, S) = \frac{3}{2} + \frac{1}{2}\left[\frac{S(S+1) - L(L+1)}{J(J+1)}\right] \tag{17b}$$

Table 1 tabulates the ground-state multiplets for transition-metal and rare-earth cation species that are prevalent in interesting ionic systems.

The Lande g-factor accounts for precession of angular momentum and quantum mechanical rules for projection onto the field axis (Russell and Saunders, 1925). For identical ions with angular momentum J, we define an effective magnetic moment in units of μ_B:

$$P_{eff} = g(J, L, S)[J(J+1)]^{1/2} \tag{18}$$

In many systems, the orbital angular momentum is quenched. The *quenched-orbital angular momentum* refers to the fact that the orbital angular moment vector is strongly tied to a crystalline EMD. For this reason to a good approximation, we can take $L = 0$ and $J = S$. In this case, $g = 2$ and $p_{eff} = 2[S(S+1)]^{1/2}$. This is true for transition metals and their simple oxides.

Systems with delocalized electron states are treated within the *band theory* of solids. This is important for transition metals, rare earths, and their alloys. In such systems, energy levels[6] form a continuum of

---

[6] We use ε to denote the energy per electron.

states over a range of energies called an *energy band*. The *density of states* (per unit volume), $g(\varepsilon)$, is defined so that $g(\varepsilon)d\varepsilon$ represents the number of electronic states (per unit volume) in the energy range from ε to $\varepsilon + d\varepsilon$:

$$g(\varepsilon)d\varepsilon = \frac{1}{V}\frac{dN_e}{d\varepsilon}d\varepsilon \qquad (19)$$

Figure 6(a) shows the density of states for free electrons with a characteristic $\varepsilon^{1/2}$ energy dependence. Many other forms for $g(\varepsilon)$ are possible with different potentials.

Figure 6(b) shows the density of states for free electrons where the spin degeneracy is broken by a *Zeeman energy* due to an applied or internal (exchange) field. We divide the density of states, $g(\varepsilon)d\varepsilon$, by two, placing half the electrons in *spin-up states* and the other half in *spin-down states*. Spin-up electrons have potential energy lowered by $-\mu_B H$ where H is an applied, H_a, or internal exchange, H_{ex}, field. Spin-down electrons have their potential energy increased by $\mu_B H$. We integrate each density of states separately to yield a different number of electrons per unit volume in spin-up and spin-down bands, respectively:

$$n_\uparrow = \frac{N_\uparrow}{V}\int_0^{\varepsilon_F} g_\uparrow(\varepsilon)d\varepsilon \quad n_\downarrow = \frac{N_\downarrow}{V}\int_0^{\varepsilon_F} g_\downarrow(\varepsilon)d\varepsilon \qquad (20)$$

The magnetization, net dipole moment per unit volume is then

$$M = (n_\uparrow - n_\downarrow)\mu_B \qquad (21)$$

The magnetization of elemental ferromagnets can be calculated using atomic dipole moments from energy band theory and atomic density from crystallography. Band theory gives the number of spin-up and spin-down electrons per unit volume or per atom. **Figure 7** shows the schematic densities of states for (a) Fe and (b) Ni (from O'Handley, 2000). In each case, we have a separate spin-up and spin-down density of states for the s- and d-electrons, respectively. The s-electrons are most important for conduction and have little splitting in energy between their spin-up and spin-down density of states. Thus, they do not contribute much to the net dipole moment. The d-electrons have very substantial splitting in energy between their spin-up and spin-down density of states. Thus, they have the largest contribution to the net dipole moment (**Table 2**).

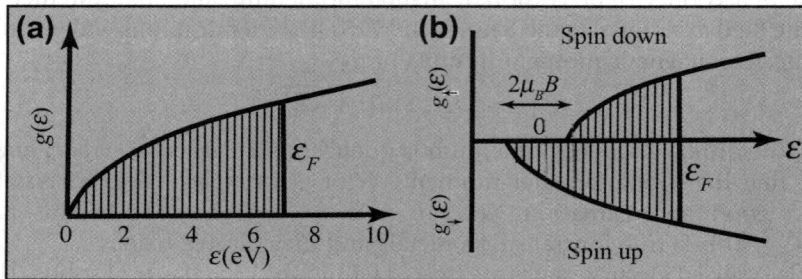

Figure 6 (a) Free electron density of states and (b) free electron density of states where the spin degeneracy is broken by a Zeeman energy due to an applied or internal (exchange) field. (For color version of this figure, the reader is referred to the online version of this book.)

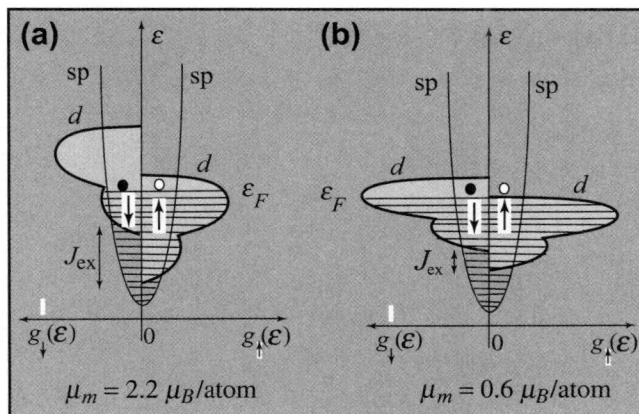

Figure 7 Schematic densities of states for (a) Fe and (b) Ni used for the illustration of how to calculate the atomic dipole moment for each. Adapted from O'Handley (2000). (For color version of this figure, the reader is referred to the online version of this book.)

Table 2   Transition metal ion spin and dipole moments ($L = 0$)

d-electrons	1	2	3	4	5
Cations	Ti^{3+}, V^{4+}	V^{3+}	V^{2+}, Cr^{3+}	Cr^{2+}, Mn^{3+}	Mn^{2+}, Fe^{3+}
S	1/2	1	3/2	2	5/2
μ (μ_B)	1	2	3	4	5
d-electrons	6	7	8	9	10
Cations	Fe^{2+}	Co^{2+}	Ni^{2+}	Cu^{2+}	Cu^+, Zn^{2+}
S	2	3/2	1	1/2	0
μ (μ_B)	4	3	2	1	0

We can integrate the densities of states for (a) Fe and (b) Ni (**Figure 7**) to determine the net dipole moment per atom and determine the magnetization for each. This is done in **Box 2**. Notice one other feature of the densities of states for Fe and Ni. Ni is an example of a *strong ferromagnet* in that the Fermi level only passes through one of the d-electron spin bands. Fe is an example of a *weak ferromagnet* in that the Fermi level passes through both d-electron spin bands.

In dilute alloy solutions (having a valency difference $\Delta Z \leq 1$), a rigid band model can be employed to explain alloying effects on magnetic moment. Rigid band theory assumes that d-bands do not change much in alloys but just get filled or emptied to a greater or lesser extent depending on composition. The magnetic moment of the solvent matrix remains independent of concentration. A moment reduction of $\Delta Z \mu_B$ is predicted at the solute site. The resulting average magnetic moment per solvent atom is the concentration weighted average of that of the matrix and that of the solute:

$$\mu = \mu_{\text{matrix}} - \Delta Z C \mu_B \qquad (22)$$

where C is the solute concentration and ΔZ is the valency difference between solute and solvent atoms. This relationship is the basis for explaining the Slater–Pauling curve (**Figure 8**) (Slater, 1937; Pauling,

Box 2 Dipole moments and magnetizations in Fe and Ni

For Fe, we calculate the number of spin-up and spin-down electrons to be

$$N_\uparrow \int_0^{\varepsilon_F} g_d^\downarrow \, d\varepsilon = 4.62 \quad N_\downarrow \int_0^{\varepsilon_F} g_d^\uparrow \, d\varepsilon = 2.42$$

and its net atomic dipole moment is then

$$\mu_{atom} = (N_\uparrow - N_\downarrow)\mu_B - (4.62 - 2.42)\mu_B = 2.2\mu_B$$

Given that Fe is bcc with $a_0 = 0.28664$ nm, the atomic density for bcc Fe is

$$N_a = \frac{2 \text{ atoms/cell}}{(0.288664 \times 10^{-9})^3 \text{ m}^3} = 8.49 \times 10^{28} \text{ atoms/m}^3$$

and the magnetization is therefore calculated to be

$$M = N_a\mu_{atom} = 8.49 \times 10^{28} \text{ atoms/m}^3 \times 2.2 \times 9.27 \times 10^{-24} \text{ A} - \text{m}^2 = 1.73 \times 10^6 \text{A/m}$$

and $\mu_0 M = 2.17$ T.
Repeating for Ni, we calculate the number of spin-up and spin-down electrons to be

$$N_\uparrow = \int_0^{\varepsilon_F} g_d^\uparrow \, d\varepsilon = 5 \quad N_\downarrow = \int_0^{\varepsilon_F} g_d^\uparrow \, d\varepsilon = 4.4$$

and its net atomic dipole moment is then

$$\mu_{atom} = (N_\uparrow - N_\downarrow)\mu_B = (5 - 4.4)\mu_B = 0.6\mu_B$$

Given that Ni is fcc with $a_0 = 0.3524$ nm, the atomic density for bcc Fe is

$$N_a = \frac{4 \text{ atoms/cell}}{(0.3524 \times 10^{-9})^3 \text{ m}^3} = 1.69 \times 10^{29} \text{ atoms/m}^3$$

and the magnetization is therefore calculated to be

$$M = N_a\mu_{atom} = 9.14 \times 10^{28} \text{atoms/m}^3 \times 0.6 \times 9.27 \times 10^{-24} \text{ A} - \text{m}^2 = 0.508 \times 10^6 \text{ A/m}$$

and $\mu_0 M = 0.64$ T.

1938). **Figure 8(b)** shows the band structure determination of the Slater–Pauling curve for Fe–Co alloys that is an alloy discussed below. **Figure 8(b)** shows the band theory prediction of the average (spin only) dipole moment in Fe–Co as a function of composition is in good quantitative agreement with the experimentally derived Slater–Pauling curve.

(a)

(b)

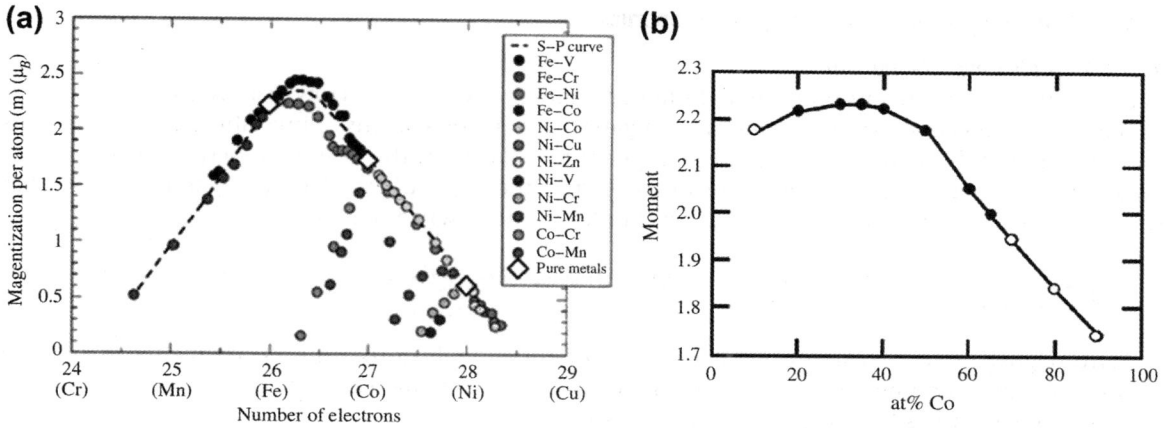

Figure 8 (a) Slater–Pauling curve for Fe alloys and (b) spin-only Slater–Pauling curve for an ordered Fe–Co alloy as determined from LKKR band structure calculations (MacLaren et al., 1998). (For color version of this figure, the reader is referred to the online version of this book.)

For transition metal impurities that introduce a strongly perturbing, highly localized potential, Friedel proposed a *virtual bound state* (VBS) (Friedel, 1958) model to explain departure from the simple relationship for the compositional dependence of dipole moment above. Friedel suggested that when the perturbation due to alloying is strong enough, a bound state will be subtracted from the full d-band majority spin and moved to higher energies. When the VBS lies below the Fermi energy, the moment of the alloy will change slowly with concentration. As the VBS moves above the Fermi level, however, a dramatic change in the magnetic moment is predicted to take place as electrons in the VBS empty into previously empty minority spin states. At this point for every additional solute atom added, a fivefold degenerate majority spin 3d VBS will empty into five unoccupied minority spin 3d states. The change in average magnetic moment is predicted to be

$$\mu = \mu_{\text{matrix}} - (\Delta Z + 10)C\mu_B \tag{23a}$$

Magnetic moment suppression is given by the VBS as follows:

$$\frac{\mathrm{d}\mu}{\mathrm{d}C} = -(\Delta Z + 10)\mu_B \tag{23b}$$

The VBS model predicts a moment reduction of $-6\mu_B$ for V or Nb additions to Co, e.g. the Friedel model was extended by Malozemoff et al. (1984) regarding the effect of the valence of the solute species on the magnetization in late transition metal metalloid/early transition metal systems. VBS effects have been looked at in amorphous magnets (Corb and O'Handley, 1985; Ghemawat et al., 1989). In particular, when early transition metal glass formers are used, these reduce the induction of the amorphous phase. Current efforts discussed below seek to reduce early transition metal and metalloid glass former concentrations to optimize the inductions in amorphous and nanocomposite systems (Miller et al., 2010c; Kernion et al., 2011).

19.1.3 Temperature-Dependent Magnetic Properties

19.1.3.1 Collective Magnetism

Dipolar interactions are important in defining demagnetization effects. However, they are too weak to explain the existence of a spontaneous magnetization in a material at any appreciable temperature. This is because thermal energy at relatively low temperature will destroy the alignment of dipoles. To explain a spontaneous magnetization, it is necessary to describe the origin of an internal magnetic field or other strong magnetic interaction that act to align dipoles in the absence of a field.

> A *ferromagnet* is a material for which an *internal field* or an equivalent *exchange interaction* acts to align atomic dipole moments parallel to one another in the absence of an applied field ($H = 0$).

Ferromagnetism is a *collective phenomenon* since individual atomic dipole moments interact to promote parallel alignment with one another. The interaction giving rise to the collective phenomenon of ferromagnetism has been explained by two models:

(1) *Mean Field Theory* considers the existence of a nonlocal internal magnetic field, called the *Weiss field*, which acts to align magnetic dipole moments even in the absence of an applied field, H_a.

(2) *Heisenberg Exchange Theory* considers a local (nearest neighbor) interaction between atomic moments (spins) mediated by direct or indirect overlap of the atomic orbitals responsible for the dipole moments. This acts to align adjacent moments in the absence of an applied field, H_a.

Both these theories help us to explain the T-dependence of the magnetization.

The Heisenberg theory lends itself to convenient representations of other collective magnetic phenomena, such as *antiferromagnetism, ferrimagnetism, helimagnetism*, etc., illustrated in **Figure 9**.

> An *antiferromagnet* is a material for which dipoles of equal magnitude on adjacent nearest-neighbor atomic sites (or planes) are arranged in an antiparallel fashion in the absence of an applied field.

Antiferromagnets have zero magnetization in the absence of an applied field because of the vector cancellation of adjacent moments. They exhibit temperature-dependent collective magnetism, though, because the arrangement of the dipole moments is not random but precisely ordered.

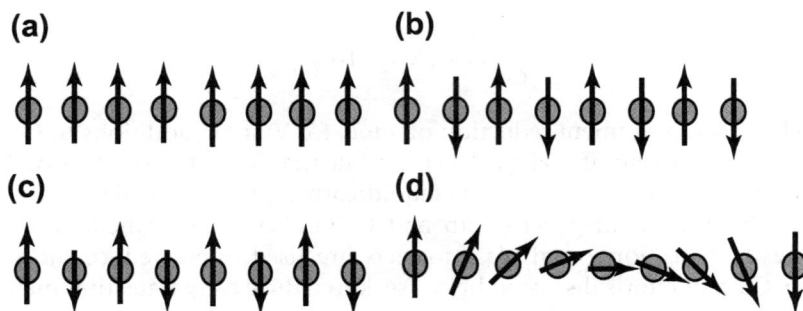

Figure 9 Atomic dipole moment configurations in a variety of magnetic ground states: (a) ferromagnet, (b) antiferromagnet, (c) ferrimagnet, and (d) noncollinear spins in a helimagnet. (For color version of this figure, the reader is referred to the online version of this book.)

A *ferrimagnet* is a material having two (or more) sublattices, for which the magnetic dipole moments of *unequal* magnitude on adjacent nearest-neighbor atomic sites (or planes) are also arranged in an antiparallel fashion.

Ferrimagnets have nonzero magnetization in the absence of an applied field because their adjacent dipole moments do not cancel.[7] All the collective magnets described thus far are *collinear magnets,* meaning that their dipole moments are either parallel or antiparallel. It is possible to have ordered magnets for which the dipole moments are not randomly arranged, but are not parallel or antiparallel. The *helimagnet* of **Figure 9(d)** is an example of a noncollinear ordered magnet. Other examples of noncollinear ordered magnetic states include the triangular spin arrangements in some ferrites (Yafet and Kittel, 1952).

The temperature dependence of the magnetization in a system with collective magnetic response is calculated within the context of statistical mechanics. We consider the ordering effect of an applied and internal field and the disordering effect of thermal fluctuations. We illustrate this for ferromagnetism in terms of *mean field theory* as introduced by Pierre Weiss in 1907 (Weiss, 1907). Weiss postulated an *internal magnetic field* (the Weiss field), $\vec{H}_{\text{int}}$, which acts to align atomic dipole moments even in the absence of an external applied field, H_a. The premise of mean field theory is that the internal field is directly proportional to the magnetization of the sample:

$$\vec{H}_{\text{int}} = \lambda \vec{M} \tag{24a}$$

The constant of proportionality, λ, is called the *Weiss molecular field constant.*

A statistical mechanical treatment (the canonical ensemble) of the quantum theory of paramagnetism describes the magnetization as a function of H/T in terms of a *Brillouin function:*

$$M = N g J \mu_B B_J(x) \quad x = \frac{g J \mu_B B}{k_B T} \tag{24b}$$

with

$$B_J(x) = \frac{2J+1}{2J} \coth\left[\frac{(2J+1)x}{2J}\right] - \frac{1}{2J}\coth\left[\frac{x}{2J}\right] \tag{24c}$$

For spin only angular momentum, $J = S$ this reduces to

$$M = N\mu \frac{\exp x - \exp(-x)}{\exp x + \exp(-x)} = N\mu \tanh x \quad x = \frac{\mu_B(\mu_0 H)}{k_B T}$$

and in the classical limit, $J = \infty$ reduces to

$$M = M_s\left(\coth(x) - \frac{1}{x}\right) = M_s L(x) x = \frac{\mu_{\text{atom}}\mu_0 H}{k_B T}$$

---

[7] Ferrimagnets are named after a class of magnetic oxides called ferrites.

where $L(x)$ is the *Langevin function*. To consider a ferromagnet, we treat the problem in analogy to a paramagnet but now consider the superposition of applied and internal magnetic fields. We conclude

$$M = M_s\left(\coth(x') - \frac{1}{x'}\right) = M_s L(x') \quad x' = \frac{\mu_0 \mu_{atom}}{k_B T}[H_a + \lambda M] \tag{24d}$$

for a collection of classical dipole moments. Similarly, $M = N_m\langle\vec{\mu}_{atom}\rangle$ and

$$\frac{M}{N_m \mu_{atom}} = \frac{M}{M_s} = L\left(\frac{\mu_0 \mu_{atom}}{k_B T}[H_a + \lambda M]\right) \tag{24e}$$

This transcendental equation has solutions with a nonzero magnetization (*spontaneous magnetization*) in the absence of an applied field. We show this graphically considering $M(H = 0)$ the variables

$$b = \frac{\mu_0 \mu_{atom}}{k_B T}[\lambda M] \quad T_c = \frac{N_m \mu_0 (\mu_{atom})^2 \lambda}{3 k_B} \tag{24f}$$

Notice that T_c has units of temperature. Notice also that

$$\frac{b}{T_c} = \frac{3}{T}\left[\frac{M(0)}{M_s}\right] \quad \frac{M(0)}{M_s} = \frac{bT}{3T_c} = L(b) \tag{24g}$$

The two equations for the reduced magnetization (i.e. $\frac{M(0)}{M_s} = \frac{bT}{3T_c}$ and $\frac{M(0)}{M_s} = L(b)$) can be solved graphically, for any choice of T by considering the intersection of the two functions $\frac{b}{3}\left(\frac{T}{T_c}\right)$ and $L(b)$. For $T > T_c$, the only solution is for $M = 0$, i.e. paramagnetic response.

For $T < T_c$, we obtain solutions with a nonzero, spontaneous magnetization, the defining feature of a ferromagnet. For $T = 0$ to $T = T_c$, we can determine the spontaneous magnetization graphically as the intersection of our two functions $\frac{b}{3}\left(\frac{T}{T_c}\right)$ and $L(b)$. This allows us to determine the zero field magnetization, $M(0)$ as a fraction of the spontaneous magnetization as a function of temperature. **Figure 10(c)** shows $\frac{M(0,T)}{M_s}$ to decrease monotonically from 1 at 0 K to 0 at $T = T_c$, where T_c is called the *ferromagnetic Curie temperature*. At $T = T_c$, we have a higher order phase transformation from a ferromagnetic phase to paramagnetic phase. In summary, mean field theory for ferromagnets predicts the following:

(1) For $T < T_c$, collective magnetic response gives rise to a spontaneous magnetization even in the absence of a field. This spontaneous magnetization is the defining feature of a ferromagnet.
(2) For $T > T_c$, the misaligning effects of temperature serve to completely randomize the direction of the atomic moments in the absence of a field. The loss of the spontaneous magnetization defines the return to paramagnetic response.
(3) In zero field, the ferromagnetic to paramagnetic phase transition is higher order (first order in a field).

19.1.3.2 *Landau Theory of Magnetic Phase Transitions*
The *Landau theory* also describes collective magnetism in systems that undergo paramagnetic to ferromagnetic *phase transitions* in terms of an *order parameter*.

The order parameter, η, quantifies a new physical property of the system that arises from a phase transformation.

(a) **(b)** **(c)**

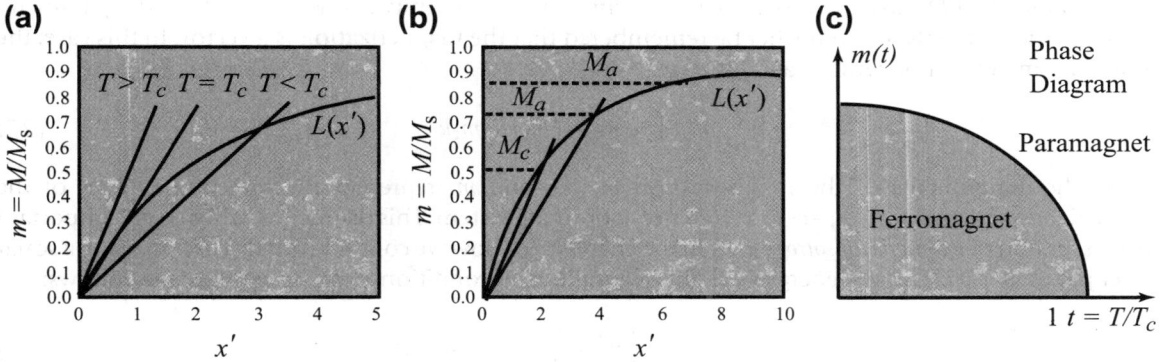

Figure 10 (a) Intersection between the curves, $\frac{b}{3}\left(\frac{T}{T_c}\right)$, and $L(b)$ for $T < T_c$ gives a nonzero, stable ferromagnetic state and (b) the locus of $M(T)$ determined by intersections at temperatures $T < T_c$, (c) reduced magnetization, m, vs reduced temperature, $t = \frac{T}{T_c}$, as derived from (b). (For color version of this figure, the reader is referred to the online version of this book.)

The order parameter is defined such that it is zero on one side of the transformation (disordered) and finite and positive on the other side of the transformation. It approaches unity when the transformation nears completion. The order parameter is determined from the condition that the free energy is a minimum. In 1937, Landau (1937) was the first to expand the *free energy* in a Taylor's series about the *Curie temperature* T_C in a ferromagnet:

$$G = G_0 + \left(\frac{\partial G}{\partial \eta}\right)_{\eta=0} \eta + \frac{1}{2}\left(\frac{\partial^2 G}{\partial \eta^2}\right)_{\eta=0} \eta^2 + \frac{1}{6}\left(\frac{\partial^3 G}{\partial \eta^2}\right)_{\eta=0} \eta^3 + \dots \tag{25a}$$

This can also be written as

$$G = G_0 + a\eta + b\eta^2 + c\eta^3 + \dots \tag{25b}$$

where it may be assumed that the expansion is valid away from the transition temperature. To express equilibrium, it is required that

$$\left(\frac{\partial G}{\partial \eta}\right) \quad \text{and} \quad \left(\frac{\partial^2 G}{\partial \eta^2}\right) > 0 \tag{25c}$$

and therefore $a = 0$. For magnetic materials, the order parameter is related to the magnetization, M. Above the Curie temperature, this value goes to zero and below the Curie temperature, M rises until it reaches its maximum value at 0 K, which can be written as $M(0)$. $M(0)$ is also called the *spontaneous magnetization*. In order to define an order parameter which varies from 0 to 1, we can define the *reduced magnetization* as follows:

$$m(T) = \frac{M(T)}{M(0)} \tag{26}$$

The value of $m(T)$ thus is 0 above the Curie temperature and unity at 0 K. This is the order parameter for magnetic materials. It can further be remembered that the magnetization is a vector. In this case, the order parameter can be written as

$$m(T) = m_x\vec{i} + m_y\vec{j} + m_z\vec{k} \tag{27}$$

where the components of the magnetization m_x, m_y and m_z represent the direction cosines of the magnetization vector with respect to a reference coordinate system. This distinction will become important in a discussion of *magnetic anisotropy* later in the chapter. For now we continue our discussion using a scalar order parameter, m. The free energy (in light of equilibrium conditions) can be expressed as follows:

$$G = G_0 + \frac{a}{2}m^2 + \frac{b}{4}m^4 + \ldots b > 0 \tag{28a}$$

There are no odd-order terms in this expansion because the free energy must be invariant with respect to *time reversal symmetry*. Time reversal symmetry refers to the fact that changing the direction (sign) of the magnetization must yield the same free energy. We now consider the conditions for thermodynamic equilibrium:

$$\left(\frac{\partial G}{\partial m}\right) = am + bm^3 = 0 \quad \left(\frac{\partial^2 G}{\partial m^2}\right) = a + 3bm^2 \tag{28b}$$

which can be true for $m = 0$, the paramagnetic state ($a > 0$) and for $m^2 = -\frac{a}{b}$, the ordered state for which a spontaneous magnetization exists and $a < 0$. Examination of the second derivative allows us to infer that for $a < 0$ m is nonzero (ferromagnetic); for $a > 0$, m is zero (paramagnetic) and $a = 0$ at T_c. **Figure 11(a)** shows the Gibb's free energy functional dependence on m for a paramagnet, ferromagnet and at T_c and **Figure 11(b)** shows the temperature dependence of the spontaneous magnetization. To first order, we can reflect the sign change occurring at T_c by writing

$$a = a_\theta(T - T_c) \quad \text{and} \quad m^2 = \frac{a_\theta(T_c - T)}{b} \quad m = \left(\frac{a_\theta}{b}\right)^{1/2}(T_c - T)^{1/2} \tag{28c}$$

The magnetic phase transition (ferromagnetic to paramagnetic) that occurs at the Curie temperature is an example of a *higher-order phase transition*.

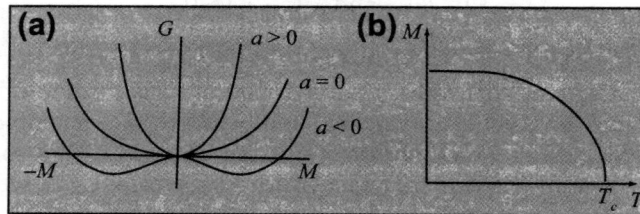

Figure 11   (a) Gibb's free energy functional dependence on m for a paramagnet, ferromagnet and at T_c. (b) Resulting temperature dependence of the spontaneous magnetization within the Landau theory. (For color version of this figure, the reader is referred to the online version of this book.)

The Landau expansion can be expanded to include the conjugate thermodynamic parameters (including applied fields) of the order parameter. When the order parameter is the magnetization, we can write

$$G = G_0 + \frac{a}{2}m^2 + \frac{b}{4}m^4 + \cdots - \mu_0 M H \qquad (29)$$

where in a scalar formalism we take M and H as parallel. This is an example of a free energy functional that gives rise to a *first-order phase transition* in field.

It is often of interest to determine the Curie temperature of a magnetic material from magnetization vs temperature data taken in a constant applied field. There are a variety of fitting procedures for this including taking the temperature at which an inflection point in such a curve occurs as an estimation of T_c. A more satisfying method for determining the Curie temperature is in terms of the *Arrott plot* (Arrott, 1957; Belov, 1956) which is rooted in a consideration of the first-order phase transition in the presence of an applied field, H. Taking the derivative of the function above with respect to m and setting it equal to zero yields

$$a(T)m + b(T)m^3 = \mu_0 H = a_\theta(T - T_c)m + b(T)m^3 \qquad (30)$$

At T_c, a plot of m^3 vs H is therefore predicted to be linear.

The influence of an applied field on the magnetic transition can be rationalized as follows. **Figure 12(a)** shows the previous Gibb's free energy functional dependence on m. **Figure 12(b)** shows the same with the inclusion of a field (Zeeman) energy term. In zero applied field, both domains ($\pm M$) have equal free energies, thus either or both domains may form. In the presence of an applied field, one domain (that with its magnetization parallel to the applied field) is favored over the other.

The Landau theory can be extended in several ways to consider other types of phase transitions. These include

(1) explicitly considering the magnetization as a vector-order parameter (magnetic anisotropy);
(2) considering spatial derivatives of the magnetization in the expansion (micromagnetics);
(3) including additional work terms in the free energy with or without coupling terms to the magnetization (magnetostriction for example).

These are the subject of several of the discussions below.

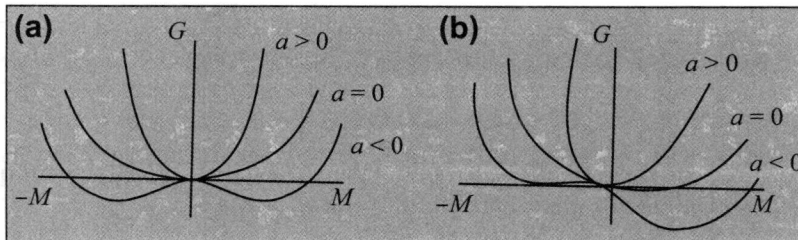

Figure 12 (a) Gibb's free energy functional dependence on *m*. (b) The same with the inclusion of a field (Zeeman) energy term. (For color version of this figure, the reader is referred to the online version of this book.)

19.1.4 Angularly Dependent Magnetic Properties: Magnetic Anisotropy

Anisotropy refers to the phenomenon that certain properties of single crystal materials depend on the direction in which they are measured. *Magnetic anisotropy* includes (i) *magnetocrystalline anisotropy*; (ii) *shape anisotropy* (magnetostatic); (iii) *induced anisotropy*: (a) stress; (b) annealing; (c) deformation; (d) irradiation.

Only magnetocrystalline anisotropy is intrinsic, depending only on the symmetry of the magnetic material. Most ferromagnetic materials derive from cubic or uniaxial materials. The magnetic free energy is expressed in a tensor formalism:

$$E_a = k_{mn}M_m M_n + k'_{mnop}M_m M_n M_0 M_p + k'_{mnopqr}M_m M_n M_0 M_p M_q M_r + \ldots \tag{31}$$

where the ks are second-, fourth-, sixth-, etc. rank tensors. M's are axial vectors.

Only even powers of M occur in the expansion since E_a is invariant to the *time reversal operation, R*. We express the magnetization in direction cosines:

$$M_i = M_S \cos \alpha_i = M_S \left(\cos \alpha_1 \vec{i} + \cos \alpha_2 \vec{j} + \cos \alpha_3 \vec{k} \right) \tag{32a}$$

and we can construct the *dyadic product* of M_i with itself to be

$$M_m M_n = M_S^2 \begin{pmatrix} \cos^2 \alpha_1 & \cos \alpha_1 \alpha_2 & \cos \alpha_1 \alpha_3 \\ \cos \alpha_2 \alpha_1 & \cos^2 \alpha_2 & \cos \alpha_2 \alpha_3 \\ \cos \alpha_3 \alpha_1 & \cos \alpha_3 \alpha_2 & \cos^2 \alpha_3 \end{pmatrix} \tag{32b}$$

If the paramagnetic phase is cubic, the second-rank tensor is a scalar as shown:

$$k'_{mn}M_m M_n = k_1 M_S^2 \left(\cos^2 \alpha_1 + \cos^2 \alpha_2 + \cos^2 \alpha_3 \right) = k_1 m_S^2 = k'_{mn}\delta_{mn} \tag{32c}$$

where δ_{mn} is the *Kronecker delta*. For cubic materials, the first term in the expansion is constant and we must go to the fourth order to obtain information on anisotropy. For hexagonal symmetry, the second-rank anisotropy tensor has the following form:

$$k'_{mn} \begin{pmatrix} k_1 & 0 & 0 \\ 0 & k_1 & 0 \\ 0 & 0 & k_3 \end{pmatrix} \tag{33a}$$

$$k'_{mn}M_m M_n = M_s^2 k_1 + M_s^2 (k_3 - k_1)\cos^2 \alpha_3 \tag{33b}$$

Thus in hexagonal, or any *uniaxial materials*, we need to include second-order terms in the expansion. Since $\cos^2 \alpha_3 = 1 - \sin^2 \alpha_3$, the last expression can be written as follows:

$$E_\alpha = K_0 + K_1 \sin^2 \theta \tag{33c}$$

where $\theta = \alpha_3$ is the more common symbol used in the literature to denote the one unique angle. When fourth-order terms are included in the expansion, the anisotropy is

$$E_a = K_0 + K_1 \sin^2 \theta + K_2 \sin^4 \theta \tag{33d}$$

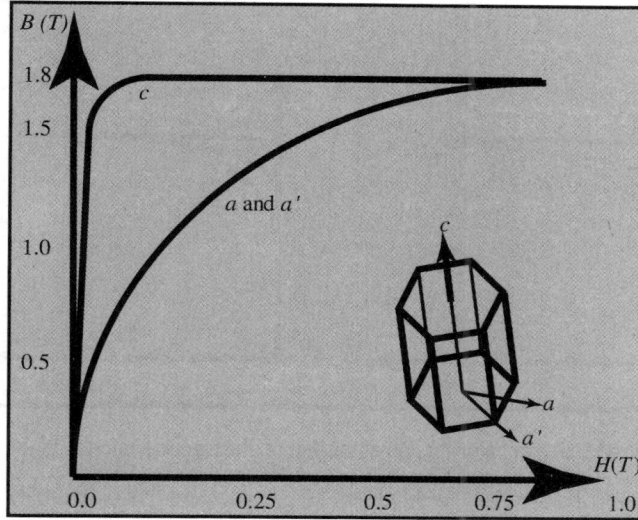

Figure 13 *M–H* curves for a single crystal of *hcp* Co along different directions. (For color version of this figure, the reader is referred to the online version of this book.)

19.1.4.1 Uniaxial Symmetry

The work performed when a field, *H*, is applied to a magnetic material is *HdM* with d*M* is the change in magnetization. Figure 13 shows *M* vs *H* plot for hexagonal Co. With *H* along the [0001] direction, *M* changes more easily than when perpendicular to this direction. Thus the [0001] is the *EMD*. The area between two *H*(*M*) curves (the work difference to saturate) is the *magnetocrystalline anisotropy energy* with units of energy per unit volume. Since a large amount of work must be performed to align *M* along basal plane directions (⊥ to the EMD), these are *hard axes*. Materials with a single easy axis have uniaxial symmetry.

The magnetocrystalline anisotropy energy of a hexagonal material is written to first order in Eqn (19.33c), where θ is the angle from the [0001] axis. For positive K_1, the minimum energy is for $\theta = 0°$ and the maximum for $\theta = 90°$. If K_1 is negative, the minimum energy is in the basal plane. The expression for the energy denoted in Eqn (19.33c) is a first-order expansion in terms of $\sin^2 \theta$. If we add the second term, we obtain Eqn (19.33d), where the second-order term, K_2, allows more complicated variation of energy with angle. The three extrema for this equation are at $\theta = 0°$, $\theta = 90°$ and at an intermediate value depending on K_1 and K_2. The third solution allows for easy directions to lie on a cone at an angle from the vertical [0001] direction.

The field associated with magnetic anisotropy is called the *anisotropy field*, H_K:

$$H_K = \frac{2K_1}{M_s} \tag{34}$$

If $K_2 = 0$, H_K is the field necessary to saturate the material along the hard direction and serves as an upper bound for the coercivity.

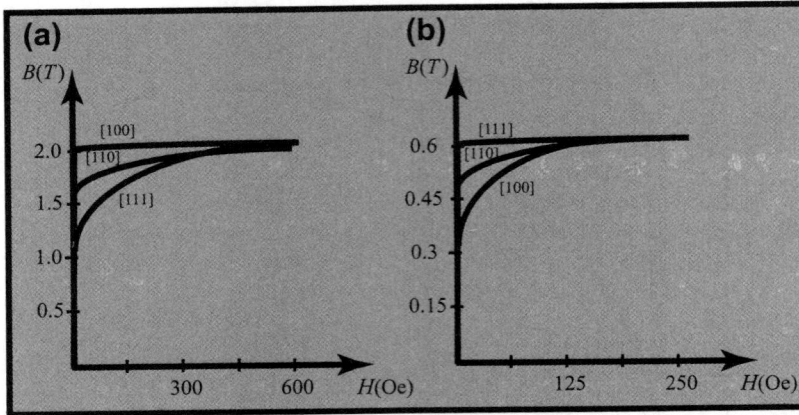

Figure 14   *H* vs *M* plots for Fe and Ni. (For color version of this figure, the reader is referred to the online version of this book.)

19.1.4.2   Cubic Symmetry

H vs *M* plots for Fe and Ni are shown in **Figure 14**. For Fe, the $\langle 100 \rangle$ directions are the easy axis directions, whereas for Ni, the easy axes are the $\langle 111 \rangle$ directions. Thus, Fe has six easy axes ($\pm \langle 100 \rangle$) and Ni has eight easy axes ($\pm \langle 111 \rangle$).

The cubic anisotropy energy is written to sixth-order terms in direction cosines as

$$E_a = K_1 \left(\alpha_1^2 \alpha_2^2 + \alpha_2^2 \alpha_3^2 + \alpha_3^2 \alpha_1^2 \right) + K_2 \left(\alpha_1^2 \alpha_2^2 \alpha_3^2 \right) \tag{35}$$

where αs are direction cosines of angles between *M* and the $\langle 100 \rangle$ axes. For the case where $K_2 = 0$, it can be shown that the minima are along the $\langle 100 \rangle$ if $K_1 > 0$ and along the $\langle 111 \rangle$ if $K_1 < 0$. This is the case of Fe and Ni, respectively. In these cases, the intermediate axes are along $\langle 110 \rangle$ and the maxima (hard axes) are $\langle 111 \rangle$ if $K_1 > 0$ and along the $\langle 100 \rangle$ if $K_1 < 0$ (i.e. opposite the easy axes).

Consider Fe with saturation induction, $B_s = 2.158$ T and cubic magnetocrystalline anisotropy with $K_1 = 4.8 \times 10^5$ erg/cm^3. If we consider the magnetization oriented along [111], then $E_a = 1.28 \times 10^5$ erg/cm^3. We can calculate the anisotropy field, H_K, as follows:

$$H_K = \frac{2K_1}{M_s} = \frac{9.6 \times 10^5 \text{ erg/cm}^3}{\frac{21580}{4\pi}} = 559 Oe$$

This is the field necessary to saturate when oriented along the hard axis. Recognizing $B = \mu_r H$, the relative permeability is

$$\mu_r = \frac{4\pi M_s}{H_K} = \frac{M_s^2}{2K_1} = \frac{21580}{2 \times 559} = 19.3$$

For $M \parallel [100]$, μ_r is unbounded if there are not impediments to domain wall motion (see also 19.2).

19.1.4.3   Coupling of Magnetic Properties to Stress Fields: Magnetostriction

The coupling of magnetic and mechanical properties is termed *magnetostriction*. Magnetoelastic interactions result from coupling magnetic anisotropy and elastic response. Magnetostriction can cause dissipative energy loss in magnetic materials. Coupling of residual stress in alloys with nonzero

(a)

(b)

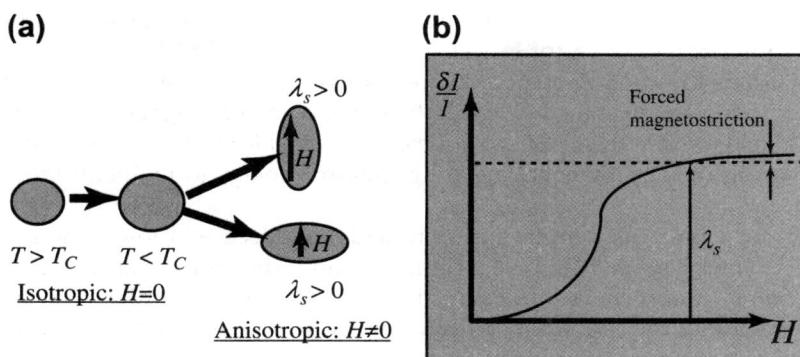

Figure 15 Schematic of (a) isotropic and anisotropic magnetostriction and (b) magnetostrictive strain as a function of the magnitude of the applied field, *H*. (For color version of this figure, the reader is referred to the online version of this book.)

magnetostriction coefficient λ (units of strain) results in stress-induced anisotropy that can reduce the ease of magnetization, lower μ and increase H_c.[8] In invar alloys, magnetostrictive volume changes cancel the thermal expansion coefficient making material dimensions temperature independent (Gignoux, 1995).

A change in dimension during magnetization by an applied field is *anisotropic magnetostriction*. Along the direction of the applied magnetic field, the sample elongates or shrinks and perpendicular to the field, the sample shrinks or elongates (**Figure 15(a)**). An *isotropic magnetostriction* or *exchange magnetostriction* arises from coupling of the exchange energy to the interatomic spacing. This occurs during cooling through T_c without an applied magnetic field. For anisotropic magnetostriction, there are both longitudinal and transverse effects that add to zero volume change. A tensor property, magnetostriction is tied to the crystallography.

Magnetostrictive strains, $\frac{\delta l}{l}$ (where δl is the magnetostrictive displacement), for typical transition metal ferromagnets are $\sim 10^{-5}$–10^{-6} at saturation. The magnetostriction coefficient in the saturated state is the *saturation magnetostriction*, λ_s. Beyond saturation, there is a small increasing *forced magnetostriction* typically measured on a logarithmic scale with field, H. Co has the largest magnetostriction of the elements with a value of $\frac{dl}{l} \sim 60 \times 10^{-6}$ at room temperature. For interesting alloys, $\frac{dl}{l}$ can be large; $\frac{dl}{l} \sim 10^{-3}$ for Terfenol, $Tb_{1-x}Dy_xFe_2$.

19.1.5   Microscopic Magnetization and Domains

A *magnetic domain* is the macroscopic volume over which atomic magnetic moments are aligned.

For a ferromagnet, when $H_a = 0$, the existence of a spontaneous magnetization requires the existence of domains. It is perhaps surprising that ferromagnetic materials can exist in a "virgin state" for which the magnetization is zero in the absence of an applied field. This is understood by ferromagnetic domain theory. In a typical magnetic material, a macroscopic volume contains many domains. Each domain has a spontaneous magnetization of magnitude, M_s. In the absence of an applied field, the magnetization vectors are randomly aligned from domain to domain (just like in a paramagnet, atomic

---

[8] Strain is defined as the change in length divided by the original length, a dimensionless quantity.

dipoles were random). Taking a vector, sum of the magnetization over many domains may yield zero sample magnetization because of vector cancellation.

The energetic reasons for domain formation are discussed below. We can qualitatively understand this by recognizing that magnetic flux line leaves the north pole of a magnet and enters the south pole. This gives rise to a field outside the magnet, the *demagnetization field*, H_d, which would like to misalign the dipole moments in the ferromagnet. It requires internal energy to maintain the alignment of the dipoles.

A configuration for which the demagnetization field is reduced will lower the total energy of the system. For two domains (**Figure 16**), we significantly reduce the return path which is necessary to be taken by fringing fields. By applying successively more domains, we can further reduce the *magnetostatic self-energy* to nearly zero. In the case where we have two long domains and two closure domains,[9] the magnetization makes a nearly circuitous path reducing the demagnetization field to nearly zero. There is no free lunch, though. Each boundary between domains requires that we pay an energy associated with a *domain wall*. The configuration of domains and walls ultimately depends on the balancing of these two energies.

Magnetic domains were first postulated to exist by Weiss (1907). In his epoch paper, he postulated their existence in order to explain how a ferromagnet could exist in a demagnetized state below its Curie temperature. A random array of regions with differing directions of magnetization could be imagined to exist and the regions would align themselves with application of a magnetic field.

Magnetic domains exist because of the lowering of the symmetry of a material when it passes from the paramagnetic state to the ferromagnetic state. However, their specific arrangement in the material is dependent on the various energies associated with the ferromagnetic state. These include the following:

(1) Exchange
(2) Magnetostatic
(3) Applied fields
(4) Magnetocrystalline anisotropy
(5) Magnetostriction.

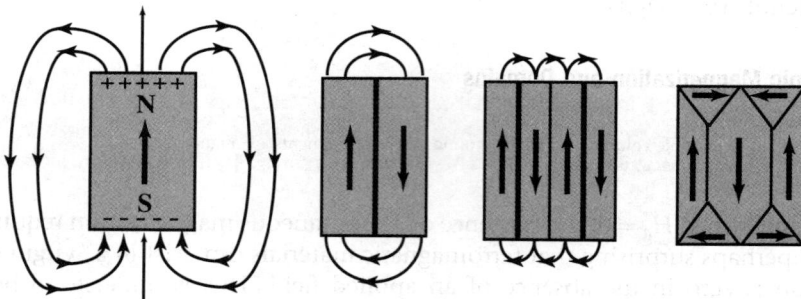

Figure 16 Reduction in the demagnetization field as a result of the introduction of magnetic domains into a ferromagnetic sample. (For color version of this figure, the reader is referred to the online version of this book.)

---

[9] This often happens in cubic materials with $\langle 100 \rangle$ easy axes.

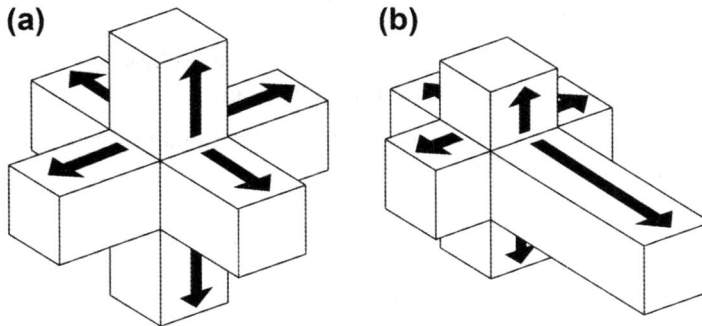

Figure 17 Random domains in a ferromagnetic material, with the [100]-type directions as easy direction of magnetization (a) a case where the net magnetization is zero and (b) where the net magnetization is nonzero.

When cooling a cubic material with $\langle 100 \rangle$ easy axes of magnetization from above the Curie temperature, if all domains have equal probability of forming their arrangement may be thought to look like **Figure 17a**.

Figure 17(a) illustrates the configurations for which the total magnetization would be zero. If however the arrangement of domains were as shown in **Figure 17(b)**, the magnetization would be nonzero, with the largest component along the y-axis. The material could magnetize this way if there were a field applied during the cooling process. Such field processing can therefore be important in determining the magnetic remnant state and subsequent magnetization processes.

Breakup into domains is undesirable in *permanent magnet* materials because it reduces the useful *remnant induction*. For this reason, permanent magnet materials are engineered to have large magnetic anisotropies and consequently large coercive fields. Magnetic anisotropy can also be developed by judiciously chosen shapes. The shape of a *horseshoe magnet* was developed to prevent the magnetic stray fields from demagnetizing a permanent magnet made from a relatively soft magnetic material.

19.1.6 Energies Determining the Magnetic Domain Size

It can be seen from the schematic that the stray field in the first configuration of **Figure 16** is much larger than the second. The second arrangement of domains lowers this energy. However, a new feature appears, namely the wall between the two domains. This *domain wall* has an energy of its own, so if the material in is to break up into two domains, the decrease in stray field energy must be greater than the increase in energy due to the domain wall.

Further subdivision into more domains is possible as seen in **Figure 16**. The ultimate size of the domains is determined as calculated below based on the balance between savings of magnetostatic energy and the cost in wall energy. The magnetostatic energy can be further reduced by ensuring *flux closure* through the introduction of *closure domains* at the surface of the sample (Landau and Lifschitz, 1935) as illustrated in the last frame of **Figure 16**.

In more complicated domain structures, the principles of minimizing the total energy associated with free poles and walls applies. These can be complicated by sample geometries, internal interfaces, surface morphological features and nonmagnetic inclusions.

19.1.7 Domain Wall Configurations

There are a variety of configurations for domain walls (Schafer and Hubert, 1998) and their geometries depend on a variety of considerations including the following:

(1) *The exchange stiffness, A*, which is a measure of the energy penalty for rotating a spin against the exchange energy in a material. It is defined as

$$A_{ex} = A \frac{nJS^2}{a} \tag{36a}$$

where n is the number of atoms (dipoles) per unit cell, A is a proportionality constant and a is the lattice constant. This derives from the Heisenberg nearest-neighbor exchange energy:

$$E_{ex} = -2JS^2 \cos \phi_{ij} = -2A \cos \left(\frac{d\phi}{dx} \right) \tag{36b}$$

where the first expression is for a discrete lattice with ϕ_{ij}, the angle between adjacent spins. The second expression is the continuum expression.

(2) The crystal symmetry and the magnetocrystalline anisotropy energy density.
(3) The material geometry. In particular, domains in bulk samples and thin films are fundamentally different because of demagnetization effects. The magnetostatic energy is often a primary determinant of the wall configurations, especially in thin samples.
(4) Volume and surface pinning effects.

Figure 18 illustrates a few examples of domain wall configurations in magnetic materials. **Figure 18(a)** illustrates a 180° domain wall which accommodates a complete magnetization reversal in a material. **Figure 18(b)** illustrates 90° domain walls which provide a closed path for magnetic flux within a sample, minimizing free pole magnetostatic effects. **Figure 18(c)** shows curved walls and **Figure 18(d)**, the spike domains. Each of these offers a means of mixing the free pole density on a finer scale which can be of energetic advantage in thin samples. The geometry of the reversal in a wall can also differ from bulk to thin-film samples as discussed in the context of *Bloch walls* and *Neel walls*.

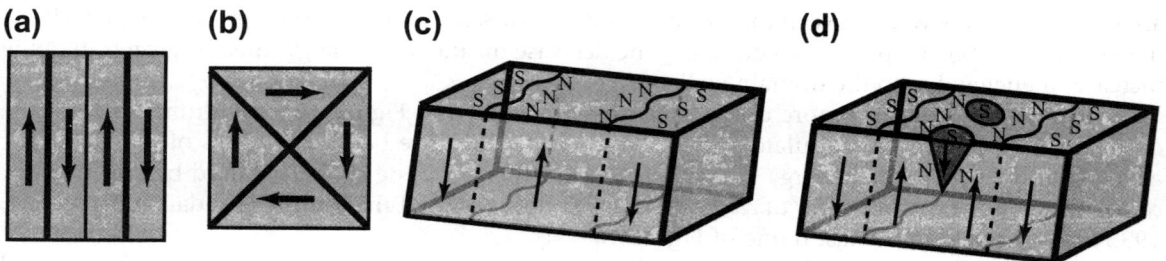

Figure 18 Examples of domain wall configurations in magnetic materials (a) 180°, (b) 90°, (c) curved domain walls and (d) surface spike domains. (For color version of this figure, the reader is referred to the online version of this book.)

(a) **(b)**

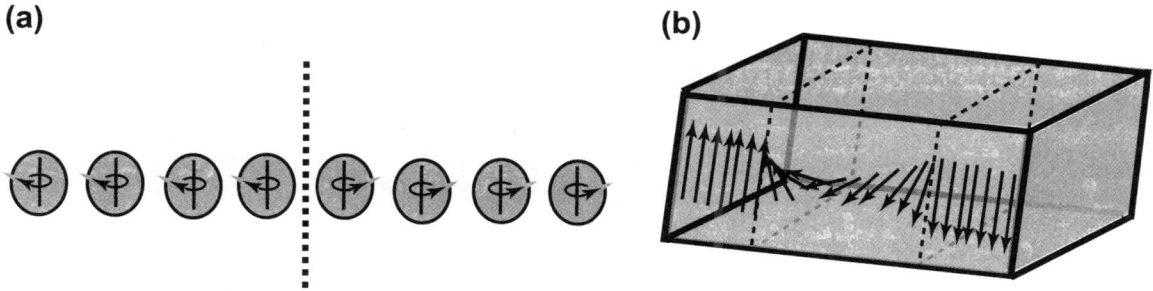

Figure 19 Magnetic domains separated (a) by an atomically sharp domain wall and (b) by a wall with a gradual 180° rotation of the dipole moments in the wall. (For color version of this figure, the reader is referred to the online version of this book.)

19.1.8 Energetics of a 180° Domain Wall

We now discuss the relative values of energies in determining the geometry of the domain wall. A possible wall structure is one with an abrupt change (**Figure 19(a)**). This abrupt change has high energy because of the large value of the *exchange energy* across the wall. This energy is estimated using

$$E_{ex} = -2J_{ex}\vec{S}_i \cdot \vec{S}_j \tag{37a}$$

and calculating the exchange energy for parallel and antiparallel spins:

$$E_{ex}(\uparrow\uparrow) = -2J_{ex}S^2 \cos(0) = -2J_{ex}S^2 \quad E_{ex}(\uparrow\downarrow) = -2J_{ex}S^2 \cos(180) = 2J_{ex}S^2$$

The excess energy of the abrupt change wall is proportional to $4J_{ex}S^2$. By spreading the spin reversal over n equal steps of angle $\frac{\pi}{n} = \frac{180}{n}$, a lower wall energy is obtained (**Figure 19(b)**). The potential energy between adjacent dipole moments is

$$E_{ex}(\text{small angle}) = -2J_{ex}S^2 \cos\left(\frac{\pi}{n}\right) \tag{37b}$$

If the angle θ between nearest-neighbor dipole moments is small, then we can use the small angle Taylor series expansion for $\cos\theta$:

$$\cos\theta \approx \left(1 - \frac{\theta^2}{2}\right) = \left(1 - \frac{\pi^2}{2n^2}\right)$$

and therefore,

$$E_{ex} = -2J_{ex}S^2\left(1 - \frac{\pi^2}{2n^2}\right)(\Delta E_{ex}) = 2J_{ex}S^2\left(\frac{\pi^2}{2n^2}\right) = J_{ex}S^2\left(\frac{\pi}{n}\right)^2 \tag{37c}$$

and to obtain the total exchange energy over the $n+1$ spins in the wall,

$$U_{ex} = (n+1)(\Delta E_{ex}) = J_{ex}S^2\left(\frac{\pi}{n}\right)^2(n+1) \approx \frac{J_{ex}S^2\pi^2}{n} \tag{37d}$$

As $n \to \infty$, U_{ex} goes to zero. Thus, if this were the only energy term, the wall would be infinitely wide. However, all the spins, in the domain wall, are not aligned in easy axes directions, so there will be a magnetocrystalline anisotropy energy term which must be taken into account.

Since we are ultimately interested in normalizing energies per unit area of a domain wall, we can consider the contribution of the exchange energy to the wall energy. The exchange energy per unit area is found by dividing the above energy by a^2, where a is the lattice parameter of the material since a^2 represents the area per spin:

$$\frac{U_{ex}}{a^2} = \frac{J_{ex}S^2\pi^2}{na^2} = \gamma_{ex} \tag{37e}$$

Hence the exchange energy is inversely related to the wall thickness. A small exchange energy gives rise to a larger wall width.

The magnetocrystalline anisotropy energy within the wall can be approximated by multiplying the width of the wall times the anisotropy constant, K_1:

$$\gamma_a = naK_1 \tag{37f}$$

The total wall energy is then the sum of the contributions from exchange and anisotropy:

$$\gamma_{tot} = \gamma_{ex} + \gamma_a = \frac{J_{ex}S^2\pi^2}{na^2} + naK_1 \tag{37g}$$

We can now determine the value of n (and therefore the width of the wall) which minimizes this total energy:

$$\frac{d\gamma_{tot}}{dn} = 0 = -\frac{J_{ex}S^2\pi^2}{n^2a^2} + aK_1 \quad n_{eq} = \left(\frac{J_{ex}S^2\pi^2}{a^3K_1}\right)^{1/2} \tag{37h}$$

and we can see that increasing JS^2 increases n and increasing K_1 decreases the value of n. The thickness of a domain wall is given as follows:

$$\delta_\omega = n_{eq}a = \pi\left(\frac{J_{ex}S^2}{aK_1}\right)^{1/2} = \pi\sqrt{\frac{A_{ex}}{K_1}} \tag{37i}$$

The total energy of the wall is given as follows:

$$\gamma_{tot} = \gamma_{ex} + \gamma_a = 2\left(\frac{J_{ex}S^2\pi^2K_1}{a}\right)^{1/2} = 2\pi(A_{ex}K_1)^{1/2} \tag{37j}$$

Box 3 shows that the wall energy is equally divided between exchange and magnetocrystalline energy and calculates the domain wall width for Fe.

19.1.9 Energetics of 180° Domain Walls in Thin Films

In a thin film, there are two ways of rotating the magnetization within the wall, as shown in **Figure 20**. The *Bloch wall* (**Figure 20(a)**) has high demagnetization energy since the magnetization

Box 3 Calculation of the domain wall thickness for crystalline Fe.

The domain wall thickness may be written as

$$\delta_\omega = n_{eq}a = \pi\left(\frac{J_{ex}S^2}{aK_1}\right)^{1/2}$$

where n_{eq} is the value of n which minimizes the total wall energy. For Fe,

$$J_{ex}S^2 = 2.16 \times 10^{-21}\ \text{J}, \quad K_1 = 4.2 \times 10^4\ \text{J/m}^3 \quad a = 2.86 \times 10^{-10}\ \text{m}$$

and we can calculate δ_ω^{Fe} to be

$$\delta_\omega^{Fe} = \pi\left(\frac{J_{ex}S^2}{aK_1}\right)^{1/2} = 42\ \text{nm}$$

This value is ≈ 150 lattice spacings.

vector points out of the thin film. For a Bloch wall, the rotation of the magnetization is parallel to the wall surfaces. An alternative configuration is shown in **Figure 20(b)**. Here, the magnetization vector rotates in the film, thereby minimizing the demagnetization energy. This type of wall is called a *Neel wall*. For a Neel wall, the rotation of the magnetization is normal to the wall surfaces. Typically the Neel wall is present in films less than ~ 10 nm thick. For thicker films, the Bloch wall may be present.

The demagnetization effects in domain walls have important implications for the stability of the domain wall as a function of the thickness of magnetic thin-film samples. We consider these stability arguments next. The energies and widths of these walls have been reported by Middelhoek (1963). For

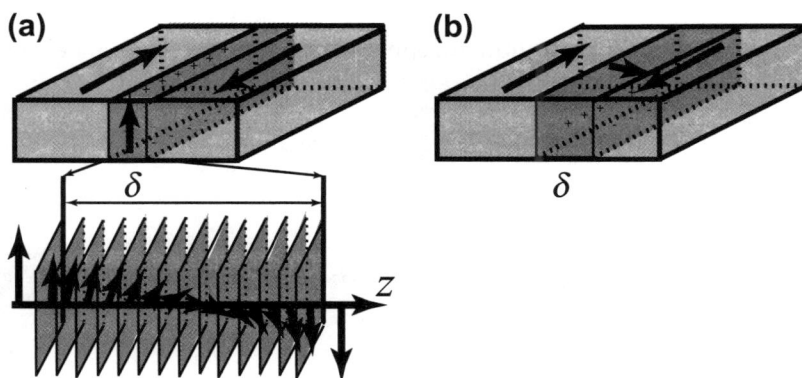

Figure 20 Geometries of 180° domain walls: (a) a Bloch wall and (b) a Neel wall. (For color version of this figure, the reader is referred to the online version of this book.)

a 180° Bloch domain wall parallel to the easy direction, as is illustrated in **Figure 20(a)**, the direction of magnetization within wall can be expressed as

$$\theta(x) = \left(\frac{\pi}{\delta_\omega}\right)x \quad -\frac{\delta_\omega}{2} < x < \frac{\delta_\omega}{2} \tag{38}$$

where δ_ω is the thickness of domain wall and θ is the angle between the magnetization and a direction in the plane of the wall and perpendicular to the plane of the film.

When the thickness of specimens is comparable to the domain wall width, it is not correct to neglect demagnetization effects at intersections of the domain wall and the specimen surface. The importance of this for thin films was first noted by Neel. The lowest energy configuration is the Neel domain wall, where the domain wall magnetization rotates from one domain direction to the next without leaving the plane of the film. A 180° Bloch domain wall parallel to the easy direction is illustrated in **Figure 20(b)**. In the Neel domain wall, the magnetization rotates about the axis perpendicular to the plane of the film. **Box 4** considers the energetics and the domain wall thickness for a Bloch and Neel wall.

Figure 21 shows the results of calculations of domain wall energies and widths for an $Fe_{40}Co_{40}Nb_4B_{13}Cu$ nanocomposite alloy. For these calculations, the magnetocrystalline anisotropy energy density is taken as 1 J/m^3, the exchange stiffness is 2×10^{11} J/m and the experimentally determined magnetization is 1.088×10^6 A/m. **Figure 21(b)** shows that when the film thickness is <36 nm, the wall energy of Bloch wall is higher than that of the Neel wall. Therefore, the Neel-type domain wall is energetically favored in the thinnest films. The wall width is roughly 28–40 nm in Bloch walls when the film thickness is larger than $\sim$36 nm (Long et al., 2008).

19.1.10 Magnetostatic Energy

To evaluate the magnetostatic energy (stray field energy), it is easier to use spheres, as their demagnetization factor is well known and does not vary with size. **Figure 22** shows how the stray field differs between a particle with one domain and one with two domains. The magnetostatic energy in the single-domain spherical particle is given as

$$^1E_{MS} = \frac{2}{9}\mu_0 M_s^2 \pi R^3 \tag{39}$$

The magnetostatic energy for the two-domain sphere is estimated to be one half that of the single-domain particle; hence its total energy is

$$^2E_{MS} + \gamma\pi R^2 = \frac{1}{9}\pi_0 M_s^2 \pi R^3 + \gamma\pi R^2 \tag{40}$$

The *single-domain particle size* is the size of the largest particle that can remain single domain, and also known as the *monodomain size*.

The single-domain particle size is seen to be

$$R_{SD} = \frac{9\gamma}{\mu_0 M_s^2} = \frac{36\mu_0\sqrt{(AK_1)}}{\mu_0^2 M_s^2} \tag{41}$$

Box 4 Calculation of the domain wall thickness for Bloch and Neel walls

The exchange energy can be expressed as

$$E_{ex} = A\left(\frac{d\theta}{dx}\right) = A\left(\frac{\pi}{\delta_\omega}\right)^2$$

where A is the exchange stiffness. The mean anisotropy energy in the wall is written as

$$E_K = \frac{1}{\delta_\omega}\int_{-\delta_\omega/2}^{\delta_\omega/2}\delta_\omega K\cos^2\theta dx = \frac{K}{2}\delta_\omega$$

where K is the anisotropy constant. The demagnetization energy is calculated assuming a wall volume approximated by a cylinder with an elliptical cross-section, and expressed as

$$E_d = \frac{1}{2}\mu_0 H_d M_{wall}\quad H_d = -N_d M_{wall}\quad N_d = \frac{\delta_\omega}{\delta_\omega + D}$$

where H_d is the demagnetization field, M_{wall} is the average magnetization in the cylinder and D is the film thickness:

$$E_d = \frac{\mu_0}{2}\frac{\delta_\omega}{\delta_\omega + D}M_{wall}^2 = \frac{\mu_0}{4}\frac{\delta_\omega}{\delta_\omega + D}M_s^2$$

where M_s is the saturation magnetization. The total wall energy density sums the exchange, anisotropy and demagnetization energy densities. Multiplying by the wall width,

$$\gamma = \left(A\left(\frac{\pi}{\delta_\omega}\right)^2 + \frac{K}{2} + \frac{\pi_0\delta_\omega}{4(\delta_\omega + D)}M_s^2\right)\delta_\omega = A\frac{\pi^2}{\delta_\omega} + \frac{K}{2}\delta_\omega + \frac{\mu_0\delta_\omega}{4(\delta_\omega + D)}M_s^2$$

as before energy is minimized with respect to the domain wall thickness to yield

$$\frac{\partial\gamma}{\partial\delta_\omega} = 0 = -A\left(\frac{\pi^2}{\delta_\omega^2}\right) + \frac{K}{2} + \frac{\mu_0(\delta_\omega^2 + 2\delta_\omega)D}{4(\delta_\omega + D)^2}M_s^2$$

for a thick film, the third term is neglected and the wall energy and wall width are the following:

$$\gamma_B = 0 = \sqrt{2\pi}\sqrt{AK}\quad \delta_\omega = \sqrt{2}\pi\sqrt{\frac{A}{K}}$$

The variation of magnetization angle for a Neel wall differs from a Bloch wall in the rotation direction. The total wall energy density is calculated without ignoring the demagnetization term and the expression and solved numerically for the Neel wall width.

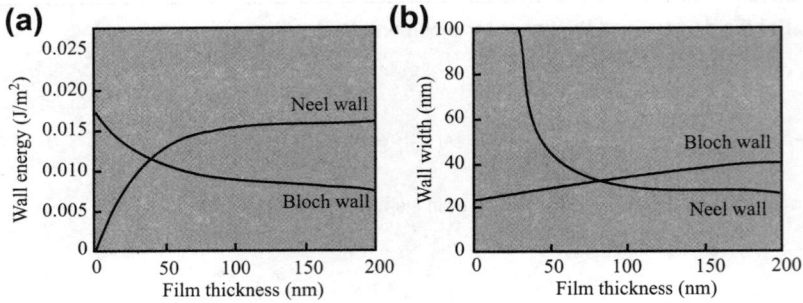

Figure 21 Domain wall calculations: (a) wall energy of a Bloch and Neel wall as a function thickness and (b) domain wall width for an $Fe_{40}Co_{40}Nb_4B_{13}Cu$ nanocomposite alloy. (For color version of this figure, the reader is referred to the online version of this book.)

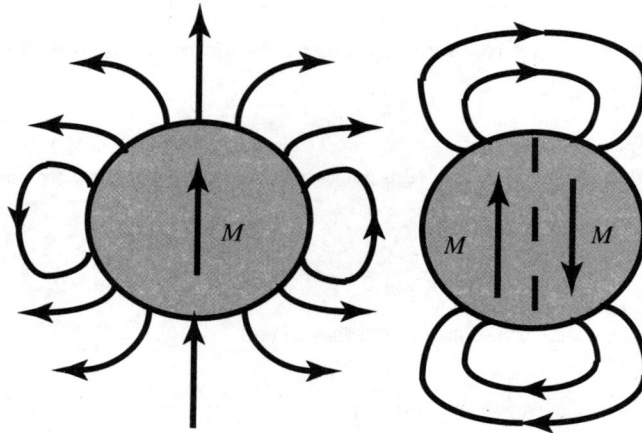

Figure 22 Monodomain spherical particle (a) and break up into two domains to reduce the magnetostatic energy. (For color version of this figure, the reader is referred to the online version of this book.)

Using typical values for these parameters, this size is 6 nm for Fe and 16 nm for Ni. **Box 5** calculates the single-domain size for crystalline Fe.

19.1.11   Other Length Scales in Ferromagnets

19.1.11.1   *Equilibrium Size of Domains*

The equilibrium size of domains in a sample depends on among other variables the size of the sample. The relation is given as

$$D_{eq} = \frac{\sqrt{(\gamma_\omega L)}}{M_s} \tag{42}$$

Box 5 Calculation of the single-domain size for crystalline Fe

The single-domain size for crystalline Fe is calculated as follows:

$$R_{SD} = \frac{9\gamma}{\mu_0 M_s^2} = \frac{36\mu_0\sqrt{AK_1}}{\mu_0^2 M_s^2} = \frac{36 \times (4\pi \times 10^{-7})\sqrt{((8.3 \times 10^{-12}) \times (5 \times 10^4))}}{(2.158T)^2} \approx 6 \text{ nm}$$

---

where L is the size of the sample. As L decreases, the equilibrium domain size decreases but not as rapidly. Eventually a size is reached where there should only be one domain in the particle. Thus, the single domain particle size is R_{SD}.

For Fe, the magnetic domain size in a sample of $L = 10^{-2}$ m is of the order of 5×10^{-6} J/m^3 or about 5 µm. Since the wall energy depends on K_1, it can be seen that decreasing the magnetocrystalline anisotropy decreases the equilibrium size of a domain.

19.1.11.2 Ferromagnetic Exchange Length
The ferromagnetic exchange length is the distance over which a perturbation in the spins of a material can be detected due to the changing (flipping) of a single spin. This value is written as follows:

$$\Gamma_{ex} = \left(\frac{A_{ex}}{\mu_0 M_s^2}\right)^{1/2} \tag{43}$$

19.1.11.3 Magnetic Hardness Parameter
The quotient of the ferromagnetic exchange length to the magnetic domain wall width divided by π is a parameter that is called the *magnetic hardness parameter*, κ. It is dimensionless and is given as

$$\kappa = \left(\frac{K_1}{\mu_0 M_s^2}\right)^{1/2} \tag{44}$$

where κ is a measure of the relative strengths of the magnetocrystalline anisotropy and magnetostatic energies. For $\kappa > 1$, the material is termed a *permanent magnet*. If $\kappa < 1$, the material is a *soft magnet*.

19.1.11.4 Superparamagnetic Particle Size
Below a certain size, the thermal energy is greater than the anisotropy energy of a ferromagnetic particle, which causes the magnetization to be unstable in the particle with respect to thermally activated switching. This means that it can take on different orientations with time. The probability per unit time that M switches is given as

$$P = f_0 \exp\left(\frac{\Delta f V}{\kappa_B T}\right)$$

where f_0 is an attempt frequency on the order of 10^9 Hz. The value of Δf depends on whether there is strong shape or strong magnetocrystalline anisotropy. For the former, $\Delta f = \Delta N \mu_0 M_S^2$ and for the latter, it equals the anisotropy constant, K_1.

It can be shown that the size of a particle with K_1 of 10^5 J/m^3 that will prevent switching within 1 s is about 6 nm, whereas the size of a particle with the same K_1 that will prevent switching for 1 year is about 7.5 nm.

These size particles are easily made, and current magnetic recording media are pushing the limits of superparamagnetism. It can be shown that if the anisotropy constant is raised to 3×10^6 J/m^3, the size of a particle that will prevent switching for 1 year is about 2.4 nm. The material FePt with the $L1_0$ structure (Space Group P4/mmm, Pearson Symbol tP2) has approximately this value of a magnetocrystalline anisotropy constant.

19.2 Magnetic Domains and the Magnetization Process

There are two ways to magnetize a polydomain ferromagnetic material, namely, *domain wall displacement* and *rotation of the magnetization vector*. Consider the process as shown schematically in **Figure 23**.

Considering **Figure 23**, if the applied field, H, is directed at an arbitrary angle with respect to the magnetization, as shown in (b), for small H, magnetic wall displacement occurs. As H increases, the fraction of the domains with M pointing up increases until in (c) only a single domain exists. Going from (d) to (e) occurs by M rotating toward H until saturation at (f). The application of the field H does not change the direction of M anymore but it does increase the magnitude of M until it reaches the value of its saturation at 0 K. This process in which the entropy due to finite temperature is overcome by the magnetic field is called the *paraprocess*.

Now consider what happens if H is applied perpendicular to the easy axis. There is no force available to move the magnetic domain wall since it is perpendicular to the field. Thus rotation of the moments occurs from the start and the response as shown in **Figure 23(g)**. The magnetocrystalline anisotropy tends to keep M_s aligned in the c-direction but interaction with H tends to turn M toward H. The total energy is

$$E = K_1 \sin^2 \theta - HM \sin \theta \tag{46}$$

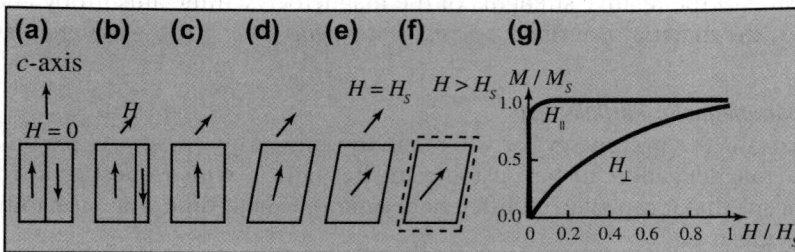

Figure 23 Steps in the magnetization process (a) virgin state, (b) wall motion, (c) monodomain, (d) rotation, (e) saturation and application of a field exceeding that required to saturate. (For color version of this figure, the reader is referred to the online version of this book.)

where θ is measured from the c-axis. To determine the equilibrium configuration, we take the first derivative and obtain

$$\frac{\partial E}{\partial \theta} = 2K_1 \sin \theta \cos \theta - HM \cos \theta = 0 \quad \sin \theta = \frac{HM}{2K_1} \tag{47}$$

The susceptibility in the direction H perpendicular to c equals the component of M along H divided by the magnitude of H, that is:

$$\chi_\perp = \frac{M_s \sin \theta}{H} = \frac{M_s^2}{2K_1} \tag{48}$$

At saturation, $\theta = 90°$ and $\sin \theta = 1$. This occurs when

$$H = H_K = \frac{2K_1}{M_s} \quad H = H_K + \frac{4K_2}{M_s} \tag{49}$$

where H_K is called the anisotropy field of the material, the first expression is first order (when $K_2 = 0$) and the second includes a nonzero K_2 (see Section 19.1.4.2).

19.2.1 Frequency-Dependent Switching of Magnetic Materials

19.2.1.1 Bulk Materials

When a magnetic material is cycled, the area of the hysteresis loop transversed corresponds to the energy loss and consequent heat generated during this cycling process. These energy losses are referred to as *core losses* and are generated due to irreversible magnetization processes occurring in the material. The core losses tend to increase significantly with increased cycling frequency and maximum saturation induction according to the Steinmetz (1984) equation:

$$P_{\text{tot}} = Cf^\alpha B_m^\beta \tag{50}$$

where P_c is the total core losses per unit volume or mass, C is the constant, α and β are called *Steinmetz coefficients*, f is the cycling frequency, and B_m is the maximum induction during magnetic cycling. While these coefficients are material and geometry dependent, typical values are $\alpha \sim 2$ and $\beta \sim 1-2$.

Since soft magnetic materials are used in AC circuits, the dynamic response as a function of frequency is very important. This response is dictated by magnetic loss mechanism. The total power loss, P_{tot}, can be partitioned as

$$P_{\text{tot}} = P_h + P_{\text{ec}} + P_{\text{an}} \tag{51}$$

which is the sum of hysteresis, P_h, classical *eddy current*, P_{ec}, and excess or anomalous eddy current loss, P_{ec}, components. Loss separation is a useful way to analyze each loss term as if they were independent of each other. The static part of loss represents that determined by the quasi-static hysteresis loop at a given maximum induction and varies linearly with frequency. Classical eddy current losses depend on the electrical resistivity, sample geometry and dimensions compared to the *skin depth* for EM radiation. These losses increase with the square of frequency. For clockwise eddy currents in a monodomain cylinder uniformly magnetized along the $\pm z$-axis:

$$P_{\text{ec}} \sim \frac{(\mu_0 M_0)^2 R^2}{8\rho} f^2 \tag{52}$$

where M_0 is the magnetization, R is the radius, ρ is the resistivity and f is the frequency of switching. It can then be seen that eddy currents become increasingly important at higher frequencies and can be mitigated by control of magnet dimension (R) and the material's resistivity. The magnetization is typically required to be large.

Anomalous eddy currents associated with domain walls give rise to losses which increase at a power of frequency that is >2. Ultimately, in a material in which all sources of losses have been minimized, losses will occur at a frequency of a radiofrequency (RF) magnetic field that corresponding to the precession of atomic dipole moments about a static magnetic field, the *ferromagnetic resonance* (FMR) frequency.

19.2.1.2 Relaxation Processes in Magnetic Fluids

Ferrofluids are colloidal suspensions of magnetic nanoparticles (MNPs) in liquids that have properties important to applications discussed below. Particles in a ferrofluid have a magnetization aligned along the EMD. Dipole moments align in direction of applied field, either by the rotation of the magnetization vector out of the EMD or by the rotation of the particle with the magnetization remaining fixed along the EMD.

The magnetization decay with time after field removal occurs by either a *Neel process* associated with magnetization rotation or a *Brownian process* associated with particle rotation. **Figure 24** illustrates these processes for a hypothetical two-dimensional nanoparticle. Both processes have particle size-dependent relaxation times: τ_N and τ_B.

For many applications, Brownian relaxation controls heat transfer from monodomain MNPs, excited by a time-varying magnetic field. This heat is transferred to the liquid by rotational processes. Of interest is the power dissipation in a ferrofluid. The response to an AC field can be expressed as

$$B = B_0 \exp(i[\omega t - \delta]) \quad M = M_0 \exp(i[\omega t + \delta]) \tag{53a}$$

where field quantities are collinear scalar quantities and the response functions B and M lag the stimulus H by a phase angle, δ, due to dissipative losses in the material. The magnetic permeability and susceptibility in SI units are as follows:

$$\mu = \frac{B}{H} = \frac{B_0}{H_0}\exp(i\delta) = \mu_{DC}\exp(i\delta) \quad \chi = \frac{M}{H} = \frac{M_0}{H_0}\exp(i\delta) = \chi_o\exp(i\delta) \tag{53b}$$

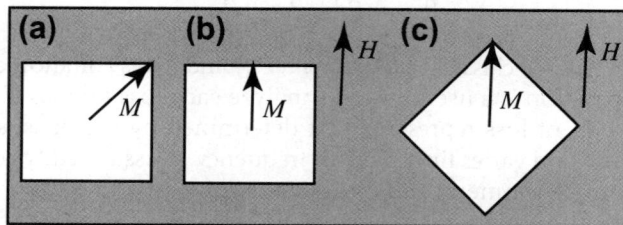

Figure 24   Two-dimensional square nanoparticle with a [11] EMD (a) $H = 0$ equilibrium configuration, (b) Neel rotation of $\vec{M}$ against anisotropy and (c) Brownian rotation with $\vec{M}$ along the EMD. (For color version of this figure, the reader is referred to the online version of this book.)

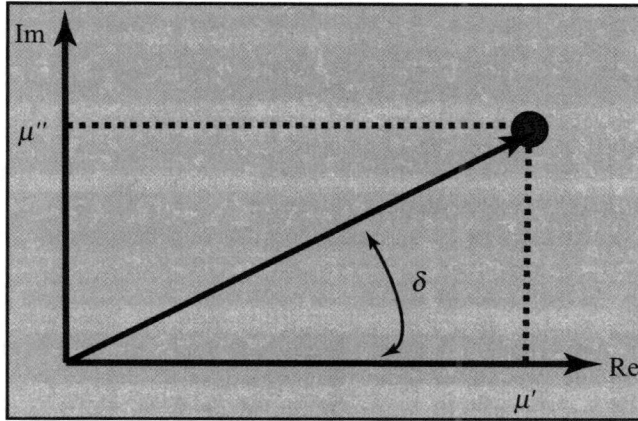

Figure 25 Argand diagram of the complex permeability. (For color version of this figure, the reader is referred to the online version of this book.)

The complex permeability or susceptibility is depicted in an *Argand diagram*. **Figure 25** illustrates the complex permeability in an Argand diagram. The complex permeability is

$$\mu = \mu' + i\mu'' \quad \mu' = \mu_{DC}\cos(\delta) \quad \mu'' = \mu_{DC}\sin(\delta) \tag{53c}$$

Similarly, the susceptibility is expressed as follows:

$$\chi = \chi' + \chi'' \quad \chi' = \chi_0\cos(\delta) \quad \chi'' = \chi_0\sin(\delta) \tag{53d}$$

We define a *loss tangent*, $\tan\delta$ and *quality factor*, Q as follows:

$$\tan\delta = \frac{\mu''}{\mu'} = \frac{\chi''}{\chi'} \quad Q = \frac{1}{\tan\delta} \tag{53e}$$

When the magnetization is switched, it relaxes on a timescale called the *magnetic relaxation time*, τ. This is determined by dissipative processes that occur in switching. For an equilibrium DC magnetization, $M_0 = \chi_0 H_0$ and time-dependent magnetization $M(t) = M_0\exp(i[\omega t - \delta])$, with an equilibrium susceptibility, χ_0 the relaxation of the magnetization the first-order relaxation of the magnetization is

$$\frac{\partial Mt}{\partial t} = \frac{1}{\tau}(M_0(t) - M(t)) \tag{54a}$$

Substituting M_0 and $M(t)$ into this partial differential equation yields

$$i\omega M_0(i[\omega t + \delta]) = \frac{1}{\tau}(M_0 - M_0\exp(i[\omega t - \delta])) \tag{54b}$$

from which we can arrive at

$$\chi = \frac{\chi_0}{1 + i\omega\tau} = \frac{\chi_0(1 - i\omega\tau)}{1 + (\omega\tau)^2} \tag{54c}$$

where

$$\chi' = R_e[\chi] = \frac{\chi_0}{1+(\omega\tau)^2} \quad \chi'' = \text{Im}[\chi] = -\chi_0\frac{\omega\tau}{1+(\omega\tau)^2} \tag{54d}$$

The preceding relationships are *Kramers–Kronig relationships* (Kramers, 1927). At $\omega = 0$, the real component of the susceptibility reduces to the DC value χ and the imaginary component is zero. The imaginary component reflects frequency-dependent losses.

The differential form of the first law of thermodynamic (conservation of energy) is

$$dU' = \delta q' - \delta w' \tag{55a}$$

where dU' is the differential internal energy, $\delta q'$ = heat absorbed by the system and $\delta w'$ = work done by the system. For an adiabatic process ($\delta q' = 0$) and for work done $\delta w' = -\vec{H}\cdot d\vec{M} = MdH$, where the last scalar expression derives from the assumption of collinear fields H and B. In SI units, ignoring work done on the vacuum,

$$\Delta U' = -\mu_0 \oint MdH \tag{55b}$$

and for excitation by a time varying magnetic field, the power dissipation is given by

$$P = f\Delta U' = -\mu_0 f \oint MdH \tag{55c}$$

Consider an alternating magnetic field of the form $H = H_0 \exp(iwt)$. We are concerned with the real part of this complex function which is

$$H(t) = \text{Re}[H_0 \exp(i\omega t)] = H_0 \cos(\omega t)] \tag{56a}$$

where H_0 is the field amplitude and $f = \frac{\omega}{2\pi}$ is the frequency of the field. Since the magnetization is $M = \chi H$, the time variation of the magnetization is

$$M(t) = \text{Re}[\chi H_0 \exp(i\omega t)] = H_0(\chi' \cos(\omega t) + \chi'' \sin(\omega t)) \tag{56b}$$

where χ' and χ'' are the in- and out-of-phase components of the susceptibility. The adiabatic change in internal energy for one cycle of the magnetization is

$$\Delta U = -\mu_0 \oint MdH = 2\mu_0 H_0^2 \chi'' \int\limits_0^{2\pi/\omega} \sin^2(\omega t)dt = \mu_0\pi\chi'' H_0^2 \tag{56c}$$

and the power loss is

$$P = \mu_0\pi\chi''fH_0^2 \tag{56d}$$

In general, relaxation processes have time constants which can be expressed as

$$\tau = \tau_0 \exp\left(\frac{-E_A}{\kappa_B T}\right) \tag{57a}$$

where E_A is the activation energy barrier appropriate for the physical process equal to the product of an energy density and material volume. The denominator expresses the thermal energy. For rotation of MNPs in the fluid, the *Brownian time constant* is

$$\tau_B = \frac{3\eta V_H}{k_B T} \tag{57b}$$

where η is the fluid viscosity containing the MNPs and the associated volume, V_H, is

$$V_H = \left(1 + \left(\frac{\delta}{R}\right)^3\right) V_M \quad V_M = \frac{4}{3}\pi R^3 \tag{57c}$$

with R, the MNP radius, V_M, the volume, and δ, an incremental surfactant radius.

The Neel time constant energy barrier is the *magnetic anisotropy energy density, K*, which determines the EMD. We define a constant, Γ, as the ratio of KV_M to the thermal energy:

$$\Gamma = \frac{KV_M}{k_B T} \tag{58a}$$

For rotation of the magnetization out of the EMD, the *Neel time constant* is

$$\tau_N = \frac{\sqrt{\pi}}{2}\tau_0 \frac{\exp\Gamma}{\Gamma^{1/2}} = \frac{\sqrt{\pi}}{2}\tau_D \frac{\exp\Gamma}{\Gamma^{3/2}} \quad \tau_D = \Gamma\tau_0 \tag{58b}$$

Material properties that tune the relaxation and therefore dissipation in the ferrofluid are fluid viscosity for the Brownian relaxation and magnetic anisotropy for the Neel relaxation. In both cases, a particle volume is a parameter. Parallel Brownian and Neel relaxation processes have an effective relaxation time:

$$\frac{1}{\tau} = \frac{1}{\tau_B} + \frac{1}{\tau_N} \tag{58c}$$

If the ferrofluid can be made of monodisperse particles, then the size can be chosen so that relaxation is dominated by the dissipative Brownian motion. The power dissipation for this Brownian motion is given by

$$P = \pi\mu_0 x_0 H_0^2 f \frac{2\pi f\tau}{1 + (2\pi f\tau)^2} \tag{59a}$$

H_0 and χ_0 are experimentally chosen parameters as is the RF of the exciting field. To quantify the power loss, we need a reasonable estimate of initial susceptibility χ_0. For monodomain, super-paramagnetic MNPs, it obeys the *Langevin relationship*:

$$x_0 = x_i \frac{3}{\xi}\left(\coth\xi - \frac{1}{\xi}\right) \xi = \frac{\mu_0 M_d H V_M}{k_B T} \quad M_d = \frac{M_s}{\phi} \tag{59b}$$

M_s is the saturation magnetization in the MNP and ϕ is the volume fraction of the magnetic material in the ferrofluid. The initial susceptibility χ_i is

$$x_i = \frac{\mu_0 \phi M_d^2 V_M}{3k_B T} \tag{59c}$$

Temperature rise is calculated from a volume average heat capacity of MNPs and fluid, c_p,

$$\Delta T = \frac{P\Delta t}{c_p} \tag{59d}$$

19.2.2   The Hierarchy of Length Scales in a Magnetic Material

Describing magnetization curves for a ferromagnet involves discussing equilibrium and nonequilibrium (reversible and irreversible) magnetization processes and several contributions to the magnetic energy density. **Figure 26** considers magnetic dipoles on several different length scales. Collective magnetic phenomena consider the magnitude of the magnetic moments and their coupling with one another (i.e. the direction of the moments). This description is on the atomic scale of the atomic orbital and spin angular momentum and exchange interactions.

In ferromagnets, atomic dipole moments are parallel at 0 K. The next length scale of importance in describing a ferromagnetic material is that of magnetic domains. A magnetic domain is a macroscopic volume which might contain $\sim 10^{15}$ atoms with magnetic moments aligned. Each domain is a tiny magnet with a dipole moment:

$$\mu = N\mu_{atom} \tag{60}$$

where N is the number of atoms in the domain. In a typical polycrystalline material with $\sim 100-1000$ μm grains, individual grains will contain $\sim 10^5$ magnetic domains whose moments are randomly oriented in the virgin state. Between two adjacent domains, individual atomic dipoles are viewed rotating from the orientation of the dipole moment in one domain to that of the other. The region over which this rotation occurs is called the *domain wall*. In our hierarchy of length scales, we

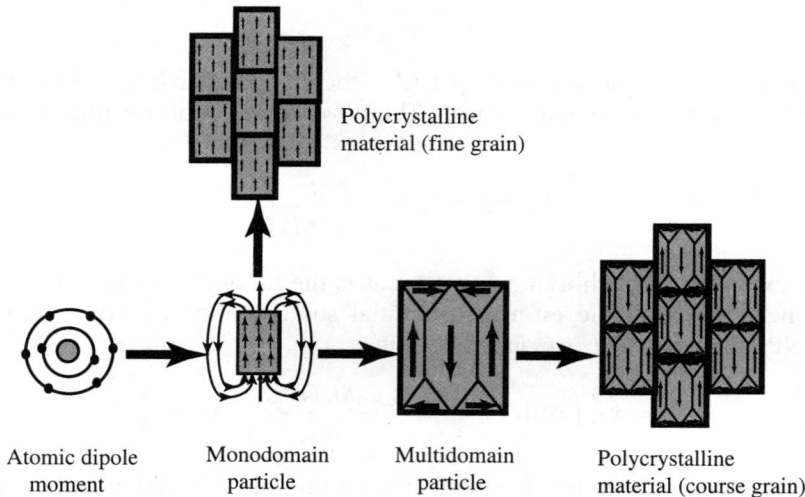

Polycrystalline
material (fine grain)

Atomic dipole   Monodomain   Multidomain   Polycrystalline
moment   particle   particle   material (course grain)

Figure 26   Hierarchy of magnetic length scales. (For color version of this figure, the reader is referred to the online version of this book.)

might take 1000 or more individual crystalline grains to form a polycrystalline aggregate in 1 cm^3 of material. The sizes chosen here are subject to variation with a variety of material parameters but give a feel for the order of magnitude of the size of various entities.

19.3 Alloy Survey

We next survey important alloys systems with interesting soft and hard magnetic properties. We discuss (a) attractive intrinsic magnetic properties of the alloy system; (b) phase relations in the system; (c) synthesis and processing techniques to develop microstructures of interest for particular technical magnetic properties; and (d) performance issues in some prototypical applications. **Table 3** summarizes structures, room temperature and 0 K saturation magnetizations and Curie temperatures for elemental ferromagnets (O'Handley, 2000).

Fe, Co, Ni and Gd are the only elemental ferromagnets at room temperature. Elemental Fe and Ni are cubic materials used as soft ferromagnets because of their high symmetry, relatively low values of *cubic magnetocrystalline anisotropy* and low magnetostriction coefficients. These properties can be improved upon by alloying as discussed below. Steels (Fe–C alloys) can serve as reasonably good hard magnetic materials as well if second-phase impurities and grain size are controlled to increase *domain wall pinning*. Co has two allotropic forms, *fcc* α-Co and hexagonal close-packed (*hcp*) ε-Co. α-Co is a relatively soft magnetic material. ε-Co has *uniaxial magnetocrystalline anisotropy* with the *c*-axis *EMD* making it a hard magnetic material. Both hard and soft Fe were developed early as magnetic materials because of the natural abundance and consequent low cost of Fe and the importance of the mechanical properties of steels in many applications. Fe has ⟨100⟩ EMDs and Ni has ⟨111⟩ EMDs.

Gd is also a hexagonal material with uniaxial magnetocrystalline anisotropy. It has an [0001] EMD. It is the only rare earth which is ferromagnetic at room temperature. Because its Curie temperature is close to room temperature, its alloys are widely investigated for magnetocaloric applications as discussed below. Dy is another example of an *hcp* ferromagnet but with a Curie temperature well below room temperature.

19.3.1 Fe and Steels Structure, Properties, and Applications

19.3.1.1 Phase Diagram and Physical Properties Survey for Fe

Understanding structure–properties relationships in Fe for magnetic applications requires knowledge of both thermal and physical properties of Fe. We also need to know the phase transformation

Table 3   Structures, room temperature and 0 K saturation magnetizations and Curie temperatures for elemental ferromagnets (O'Handley, 1987)

Element	Structure	M_S (290 K) (emu/cm^3)	M_S (0 K) (emu/cm^3)	n_B (μB)	T_c (K)
Fe	bcc	1707	1740	2.22	1043
Co	hcp, fcc	1440	1446	1.72	1388
Ni	fcc	485	510	1.72	627
Gd	hcp	–	2060	7.63	292
Dy	hcp	–	2920	10.2	88

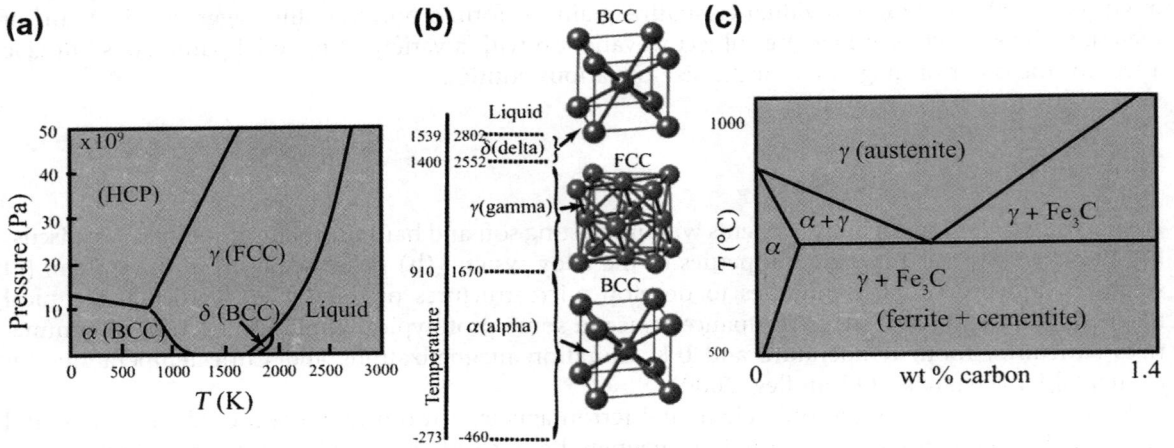

Figure 27 (a) P–T phase diagram for Fe (Dinsdale, 1991) and (b) crystal structures at atmospheric pressure (Guy and Wren, 1974), (c) schematic of the Fe–C eutectoid as a portion of the Fe–C phase diagram. (For color version of this figure, the reader is referred to the online version of this book.)

temperatures and details of the binary phase diagrams of elemental Fe and other alloying additions, e.g. the FeC-phase diagram (**Figure 27**).

Figure 27(a) shows the unary phase diagram for elemental Fe to have several important allotropes (or polymorphs). Allotropes are named with successive letters in the Greek alphabet, e.g. α, β, γ, δ… Notice that there is no β-iron on the diagram. Before about 1922, the paramagnetic *bcc* iron that formed above the Curie temperature of ferromagnetic *bcc* iron was identified as the β-phase. However, since it was believed that there was no change in the crystal structure after the ferromagnetic magnetic iron transformed to paramagnetic iron, it was concluded that the α to β transformation was not a phase change. Implicit was a definition of phase that only included structure and composition. But paramagnetic *bcc* Fe is a different phase than ferromagnetic α-Fe since it has different properties and symmetry. See below and Massalski and Laughlin (2009) from which the following discussion is derived.

Figure 27(b) summarizes crystal structures of elemental Fe at atmospheric pressure (this figure ignores the magnetic symmetry). The γ-phase of iron and steels is called *austenite*. Austenite is a high-temperature phase and has an *fcc* [close packed] structure. The α-phase is called *ferrite*. Ferrite is a common constituent in steels and has a *bcc* structure (less densely packed than *fcc*). **Table 4** gives

Table 4 Selected thermal and physical properties of bcc Fe including phase transformation temperatures (Bozorth, 1993)

Property	Value
Density, ρ(g/cm^3) at 293 K	7.874
Thermal expansion coefficient $\times$ 10^6 at 293 K	11.7
Lattice constant, a_o (nm)	0.2861
Melting point, T_m, (°C)	1539
$\alpha \to \gamma$ transition temperature, $T_\alpha \to \gamma$, (A$_3$) on heating (°C)	910
$\gamma \to \delta$ transition temperature, $T_\alpha \to \gamma$, (A$_4$) on cooling (°C)	1400

a summary of thermal and physical properties of Fe including phase-transformation temperatures (Bozorth, 1993).

Allotropic transformations in iron are examples of additional contributions to structure stability from magnetic transitions that can bring about phase changes in a unary system. The temperature-/pressure-phase diagram for iron indicates that the stable phase, at ambient temperatures and pressures, is the α iron. It has the *bcc* structure and remains stable up to about 910 °C (1183 K). On heating above 910 °C, α iron transforms to the *fcc* γ iron, but it reverts again to *bcc* (δ) at 1140 °C. The δ-phase is thus a continuation of the α-phase in the sequence $\alpha \rightarrow \gamma \rightarrow \delta \rightarrow liquid$. Application of pressure changes the temperatures of these transitions (which can be calculated using the Clapeyron equation). A substantial pressure also makes possible another structure modification, to the hcp ε-form of Fe, above approximately 15.2 MPa ($\sim$110 K bar).

The trends with temperature at ambient pressure of the closest interatomic distances (d) and volumes per atom (Ω) are shown in **Figure 28(a)** (Barrett and Massalski, 1980). The d values in the close-packed *fcc* structure are larger than in the *bcc*, but comparable Ω values are slightly smaller. Hence, the *bcc* is more open and allows for more vibrational choices, making the vibrational entropy larger in the *bcc* than in the *fcc* phase. The fact that volume per atom differences between the two structures are quite small indicates that the total energies are similar, and confirms that the bonding between atoms in the two different crystal structures of iron remains essentially metallic.

Several interesting questions about the allotropic changes in Fe include the following:

(1) Why is the *bcc* form stable at low temperatures? Usually, a close-packed allotrope (e.g. *fcc* or *hcp*) is stable at low temperatures because enthalpy is generally lowered by close packing.
(2) Why does the *bcc* phase transform to the close-packed *fcc* phase on heating? This seems backward because usually on heating, a more open, and therefore higher entropy phase, is observed to form at elevated temperatures.
(3) When the *fcc* γ reverts to the *bcc* δ at even higher temperatures, what is the cause of this? Is the large vibrational entropy of the *bcc* δ the only factor?
(4) Why is there no β-phase in the sequence?

In answering the last question, we note that currently both the paramagnetic and ferromagnetic states of *bcc* iron are most commonly designated by the same Greek letter α. However, strictly speaking, it can be

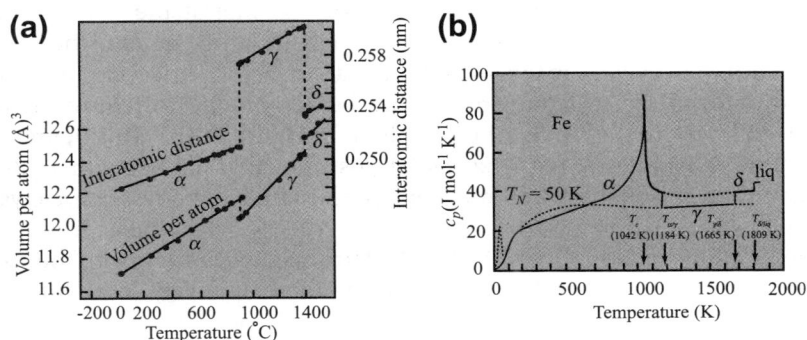

Figure 28   (a) Atomic volumes and lattice spacings and (b) heat capacity as a function of temperature for elemental solid Fe (Barrett and Massalski, 1980). (For color version of this figure, the reader is referred to the online version of this book.)

(a)   **(b)**

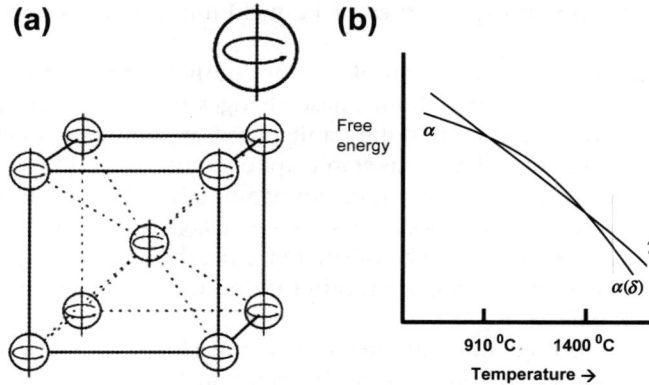

Figure 29   (a) A *bcc* arrangement of atoms with [001] oriented magnetization, showing that its symmetry is I4/mmm (Laughlin et al., 2000); (b) *G(T)* curves for α and γ Fe.

argued that the ferromagnetic to paramagnetic transition ($\alpha \rightarrow \beta$) is a phase change because a symmetry change accompanies the magnetic change. This is so, because the ferromagnetic form is not truly cubic because of its magnetic symmetry (**Figure 29(a)**). Considerations of magnetism also explain the free energy vs temperature curves of **Figure 29(b)** discussed below.

The loss of ferromagnetism on heating takes place by what is called a higher order transition, which occurs over a range of temperatures from 600 K to nearly 1042 K, as can be seen from the large peak in the specific heat curve for α-iron corresponding to the magnetic disordering in **Figure 28(b)**. The temperature associated with the highest point of the peak corresponds to the largest number of the magnetic moments becoming random (or becoming aligned on cooling) and is designated as the Curie temperature, T_{Curie} (~ 1042 K ($769\,^\circ$C)). This magnetic transition and the accompanying increase in entropy contribute to the reappearance of the *bcc* form of Fe at higher temperatures (δ).

The *fcc* γ-phase undergoes a magnetic change below about 50 K, namely an antiferromagnetic transition. The temperature of this transition is called the *Neel temperature*, T_N. In antiferromagnetically (AF) ordered materials, the atomic dipoles become aligned in opposite directions with respect to neighboring spins. This has been demonstrated in micron-size metastable γ-iron particles embedded in Cu which remain *fcc* on cooling. In an antiferromagnetic arrangement, the opposite magnetic moments cancel with the overall magnetization equal to zero. However, this antiferromagnetic arrangement contributes to the structure energy, and heat is needed to remove it. The transition is indicated by the small peak in the curve for the γ-phase in **Figure 28(b)**. This figure shows the trends for both the *fcc* and *bcc* Fe phases, with extrapolations into metastable regions (shown by dashed lines), which seem quite reasonable. Both magnetic states stabilize the respective crystal structures at low temperatures.

Question 1 can be addressed as follows. At 0 K, the ferromagnetic ordering in the *bcc* α-phase causes the internal energy (and enthalpy) to be lower than the internal energy (and enthalpy) of the anti-ferromagnetic (and paramagnetic) fcc phase. This is so because the large exchange energy of the aligned magnetic moments of the ferromagnetic *bcc* iron greatly reduces its internal energy. Thus, even though the ferromagnetic *bcc* iron is less close-packed than *fcc* γ, it is the equilibrium phase at low temperatures.

The answers to questions 2 and 3 are more complex and involve the influence of entropy of each of the phases, as well as their internal energies. Here, we need to consider the behavior of the specific heats with temperature for the α- and γ-phases in more detail (**Figure 28(b)**). Both curves show a peak corresponding to a magnetic transition. We consider two aspects of the entropy of the α- and γ-phases, namely, their vibrational entropy and their magnetic spin entropy. The vibrational entropy usually is larger for *bcc* structures than for *fcc* structures because the atoms vibrate with a higher frequency in the more close-packed structures. This is the reason that *bcc* structures are usually more stable than the *fcc* ones at the high temperatures. In the case of Fe, we must also take into account the entropy due to the disordering of the spins.

The low-temperature antiferromagnetic to paramagnetic transition in γ-Fe increases its configurational spin entropy to the extent that the overall entropy for the close-packed γ-phase becomes greater than that of the more loosely packed ferromagnetic α-phase. This excess entropy (due to the disordering of the magnetic spins at the Neel temperature) is the cause of the appearance of the γ-phase at higher temperatures, as can be seen considering the following for the γ-phase:

$$G^\gamma = H_0^\gamma + \int_0^T C_p dT - T \int_0^T \frac{C_p}{T} dt \tag{61}$$

From Eqn (61) and **Figure 28(b)**, we conclude that the $-TS$ term eventually causes the γ-phase to have a lower Gibbs Energy at higher temperatures, since this term continues to decrease as the temperature increases. Hence the *fcc* phase replaces the *bcc* one at about 910°, that is the $\alpha \to \gamma$ structural transformation occurs (or better $\beta \to \gamma$). This *bcc* to *fcc* transformation on heating is opposite to the more usual *fcc* to *bcc* one, because, in the case of iron, it is the entropy due to the spin disordering that determines the equilibrium phase, not the vibrational entropy term.

Figure 29(b) shows free energy versus temperature curves of these transformations. At 910 °C, the G_γ curve crosses below the G_α curve, producing a phase change. However, the excess specific heat of the α-phase in the vicinity of T_c makes the α-phase TS term important and, thus, the G_α curve recrosses the G_γ at 1400 °C. The large entropy due to the randomizing of the spins in the α-phase along with its larger vibrational entropy allows the *bcc* phase to reappear, this time labeled as the δ-phase.

If the ferromagnetic to paramagnetic transition of the α-phase were slightly lower in temperature, its large negative TS term would stabilize the α-relative to the γ-phase and *fcc* iron might never form at elevated temperatures. Also, although the exact form of the c_p trends at very low temperatures has not been determined experimentally, it can be argued that if γ-Fe were not AF at the very low temperatures, it might never be stable enough to form as the higher temperature phase because ferromagnetic *bcc* Fe has low enthalpy and paramagnetic *bcc* Fe has higher entropy. This demonstrates the importance of magnetic transformations in iron in determining the equilibrium phase stability. If the antiferromagnetic Neel transition temperature in γ iron were higher, or if the ferromagnetic to paramagnetic T_c in α iron were a little lower, *fcc* iron would never be a stable phase! In terms of **Figure 29**, the G_α curve may shift to the left, making the *bcc* structure stable at all temperatures.

19.3.1.2   *Fe Properties and Performance*

The magnetic softness of Fe is related to the purity of Fe. For very pure $\langle 100 \rangle$ oriented single crystals of Fe, magnetic permeabilities of 10^6 have been attained in the laboratory. Typical commercial alloys have permeabilities less than 10^4, though this can be increased to 10^5 by annealing in vacuum or hydrogen. Such annealing serves to remove most of the C, S, N and O impurities in Fe. These elements are

---

Box 6 Pinning of domain walls by grain boundaries

A domain wall can be viewed as a planar defect with an energy per unit area, γ_{wall}, and a thickness, δw. It will be ideally pinned by a grain boundary of similar thickness, δg. If $\delta g > \delta \omega$, then the pinning force is $f_p \sim \frac{\delta w}{\delta g}$. The coercivity H_c of the material will be proportional to the summed pinning force in a unit volume. This will depend linearly on grain boundary area per unit volume. For grains of diameter, D_g,

$$H_c \sim f_p \sim \frac{\delta w}{\delta g} \frac{4\pi \frac{D_g^2}{2}}{\frac{4}{3}\pi \frac{D_g^3}{2}} = \frac{\delta w}{\delta g} \frac{6}{D_g}$$

---

relatively insoluble in solid Fe but form second-phase carbides, sulfides, nitrides and oxides, respectively, which can act as impediments to the motion of magnetic domain walls.

Prototypical soft Fe has a saturation induction, $B_s = 2.22$ T, a cubic magnetocrystalline anisotropy energy density, $K_1 = 4.8 \times 10^4$ J/m and a magnetostriction coefficient, $\lambda_{111} \sim -20 \times 10^{-6}$ (20 ppm). Coercivities of ~ 1 Oe are achievable in commercially rolled Fe. A typical commercial alloy is Armco iron which is Fe sheet rolled from ingot with typical impurity levels of 0.014 wt% C, 0.007 wt% N, 0.15 wt% O, 0.003 wt% Si, 0.005 wt% P, 0.025 wt% S, 0.03 wt% Mn, and 0.003 wt% Al. The biggest disadvantages of soft Fe for many applications are that it is also mechanically soft and has low electrical resistivity. The latter makes it unattractive for high-frequency applications for which eddy current losses are prohibitive. Mechanical strength can be improved by reducing the grain size but at a cost of coercivity because of consequent grain boundary pinning of domain walls.

Understanding domain wall pinning is fundamental to understanding coercivity in soft Fe. The *Herzer curve*, which will be discussed in the context of amorphous and nanocomposite materials below, shows the important relationship between the coercivity and the grain size of crystalline magnetic material. Notable for understanding alloy development in soft magnetic Fe is the $\frac{1}{D_g}$ dependence of the coercivity on D_g at large grain sizes. **Box 6** develops a relationship between pinning and grain diameter for large grained materials.

19.3.1.3 Materials Application: Soft Fe for Magnetic Lenses

A magnetic lens consists of a coil of Cu or superconducting wires within Fe-pole pieces. Current through the coils creates an exciting field in the bore of the pole pieces. Electrons (or other charge particles) are deflected by a *Lorentz force*:

$$\vec{F} = -e(\vec{v} \times \vec{B}) \tag{62}$$

where e is the electron charge and $\vec{v}$ is the electron velocity. Pole piece design is aimed at providing a rotationally symmetric magnetic field that is inhomogeneous in such a way that it is weak in the center of a gap and stronger close to the bore. This causes electrons close to the central axis of the lens to be less strongly deflected than those passing far from the axis, allowing a beam of parallel electrons to be focused into a spot. This requires that large magnetic fields and their distributions are confined to narrow regions comparable to the lens focal length. Electron lenses quality suffers from aberrations resulting from geometrical lens precision and inhomogeneous magnetic properties of pole materials.

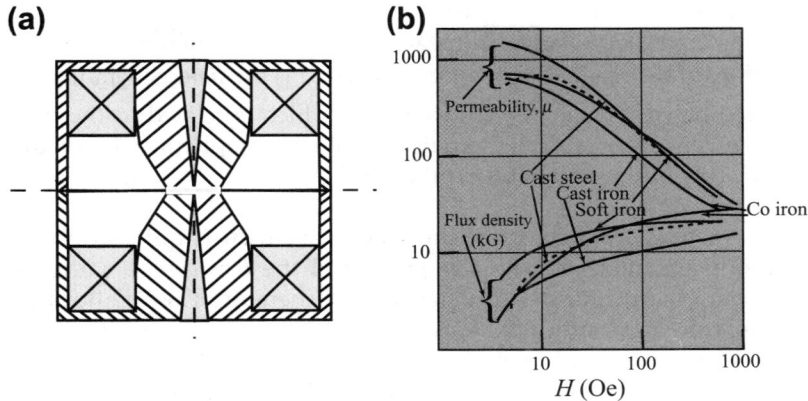

Figure 30 (a) Schematic of a symmetrical objective lens of an electron microscope. (b) Magnetic properties of commercial materials used for magnetic lens circuits (Reicke, 1982). (For color version of this figure, the reader is referred to the online version of this book.)

Aberrations limit the spatial resolution of an image in the image plane of a transmission electron microscope, for example.

Magnetic lens design for electron microscopy depends on the operating mode of the lens. The fundamental operating modes in electron microscopy require *objective lens, projector lens* and *condenser lens*. **Figure 30(a)** shows the schematic of a symmetrical objective lens of an electron microscope. Rotationally symmetric magnetic fields can be used to focus electrons in a manner similar to how glass lenses focus light (Reicke, 1982). The first design of objective lenses for electron microscopy (Busch, 1926; Reicke and Ruska, 1966) used high-permeability pole pieces which were tapered to concentrate the magnetic flux in the working area of the lens. Fe-shrouded lenses are the commonly adopted magnetic lens.

Although details of the lens design can differ for different operating modes, the design invariably considers flux concentration at the lens working point, geometrical optical aberrations (which result from inhomogeneity in the flux distribution) and general concerns as to flux paths in the magnetic circuit as a whole. The components of a lens include the pole pieces systems with casing, the lens coil, stigmators, aperture alignment systems and exchange systems for control of vibrations and stray fields.

Electron optical design of a magnetic lens has the following considerations (Reicke, 1982):

(1) Specifying the field which differs for objective, projector and condenser lenses.
(2) Specifying the focal length and acceptable values for geometrical aberrations.
(3) Choice of *gap width, s,* and *bore diameter, D* and the number of *ampere-turns*, NI for the exciting field.
(4) Design of the shape of the magnetic circuit to prevent saturation of the pole pieces and achieve tolerable stray fields.
(5) Provide alignment apertures, stigmators and beam deflection systems.
(6) Design to avoid environmental disturbance including shielding AC stray fields and mechanical decoupling to avoid vibrations.

The issues of intended field, gap geometry, and shape of the magnetic circuit pose the most restrictions on the magnetic materials.

Ferromagnetic materials for lens components are chosen so that none of the materials in the magnetic circuit saturate and their relative permeabilities can be maintained at ~ 100. A characteristic parameter for the scaling of the magnetic circuit is $\frac{I}{\sqrt{V}}$, where I is the field generating current and V is the relativistically corrected beam voltage, the dispersion of which is related to the flux in the magnetic circuit. For an objective lens, geometrical aberration results from spherical, chromatic and diffraction aberrations. The latter two of these depend on the fluctuations in V and therefore the uniformity of the field.

The choice of ferromagnetic materials for the magnetic circuit is quite important and distinctly different considerations are made for the *pole pieces* and the *lens core*. For the pole pieces, high-saturation inductions are the primary consideration and therefore Co–Fe alloys such as *Permendur* are chosen as they provide maximum flux densities. For the lens core, where lower magnetizations and higher permeabilities are desirable, soft Fe is chosen. **Figure 30(b)** shows the field dependence of the magnetic properties of some commercial materials used for magnetic lens circuits (Reicke, 1982). Specification of the field dependence of the induction and permeability for fields less than saturating fields are required for magnetic circuit design. Predictable magnetic response is of paramount importance.

19.3.1.4 *Steels*

19.3.1.5 *Phase Diagram of the Binary Fe–C System*
The Fe–C phase diagram is shown in **Figure 27(c)** for compositions in the eutectoid portion of the phase diagram. If extended to 6.7 wt% C (corresponding to 25 at.%), the intermetallic phase Fe₃C called **cementite** is observed. Steels are restricted to C concentrations of up to 1.4 wt% and the portion of the Fe–C phase diagram pertaining to steels and technical or pure Fe is the simpler one illustrated in **Figure 27(c)**. Fe₃C is called *cementite* and has an orthorhombic crystal structure. Cementite is a magnetically hard phase. Carbides are magnetic hardening agents in technical Fe.

Figure 31(a) shows the portion of the Fe–C phase diagram relevant to processing low-C and ultra-low-C steels (Chin and Wernick, 1986). The *maximum solubility* 0.0218 wt% of C in α-Fe occurs at the *eutectoid transition temperature*, 727 °C. Typically low-C and ultra-low-C steels are kept to concentrations near or below this solubility limit so as to limit the amounts of second-phase carbides in the steel. In the

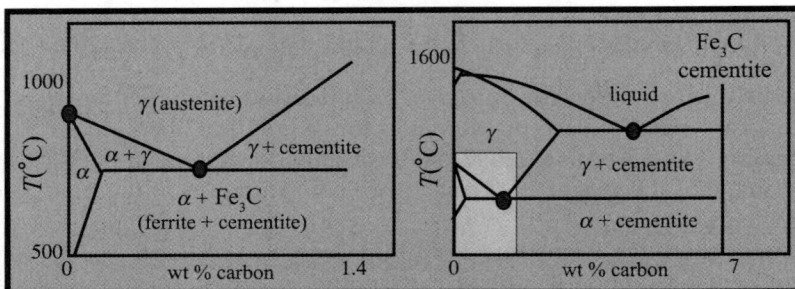

Figure 31 (a) Portion of the Fe–C phase diagram relevant to processing low-C and ultralow-C steels (Chin and Wernick, 1986). (b) Schematic *I–T* diagram illustrating the onset of various types of ferrite precipitation. (For color version of this figure, the reader is referred to the online version of this book.)

discussion of steel microstructures, it is important to define several other important metastable phases and phase mixtures which are observed in steel. The first, *pearlite*, is a two-phase, lamellar structure having alternating layers of ferrite (88 wt%) and cementite (12%). Pearlite forms by a eutectoid reaction:

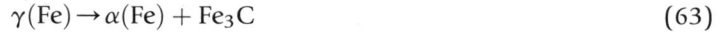

$$\gamma(\text{Fe}) \rightarrow \alpha(\text{Fe}) + \text{Fe}_3\text{C} \tag{63}$$

when austenite is slowly cooled below 727 °C. In this reaction, austenite of *eutectoid composition* 0.77 wt% transforms to the two-phase equilibrium mixture where the alternating layers (lamellae) of α-ferrite and cementite grow cooperatively.

19.3.1.6 Phase Diagram and Physical Properties of Low-C Steels

Low-carbon steels are FeC alloys for which the C concentration is kept to less than ~ 0.1 wt% so that the maximum C content lies close to or within the single-phase α-ferrite (ferromagnetic, bcc) phase field at the eutectoid temperature (**Figure 27(c)**). ASTM A848 is an example of an ultra-low-carbon steel (C < ~ 0.02 wt%), where C content allows the eutectoid transformation to be completely avoided. Thus, the characteristic pearlite and bainite obtained in medium- and high-carbon steels are not typically observed in low-carbon steels used for magnetic applications. This choice minimizes the second phase, which is important from the standpoint of both induction- and structure-sensitive properties such as permeability and coercive force.

Alloying and impurity elements in low-carbon steels have a variety of effects on the overall performance of the material. From the standpoint of technical magnetic properties, the nonmetallic elements which are present interstitially such as C, O, S, and N have the largest effect. Removal of such impurity atoms can result in higher permeabilities and reduced coercive force and several strategies for purifying the iron exist (Chin and Wernick, 1986). The allowable degree of purity is dictated primarily by the material requirements of the intended application. Compositional additives that can be used for controlling the magnetic aging behavior can be found in the discussion on magnetic aging below. Although impurity elements typically degrade the technical magnetic properties of freshly prepared (before magnetic aging) magnetic iron alloys, such impurities may be desirable in order to tailor the features of the microstructure for other design purposes. In particular, impurities or alloying elements such as Cr, Mo, and Mn can help to achieve a refined grain size due to a solute drag effect. The presence of P can result in improved machinability.

19.3.2 AlNiCo Permanent Magnets: Precipitation and Spinodal Decomposition

Fe and subsequently C steels were used as permanent magnets (Gilbert, 1958). Hardening mechanisms in C steels are largely associated with pinning of domain walls by carbide precipitates. W and Cr steels were developed in mid- to late-1800s. The 1917 development of Co steels (Honda and Saito, 1920) permanent magnets was followed by the discovery of AlNiCo permanent magnets (Mishima, 1931). This major breakthrough realized the potential for refining microstructures through *spinodal decomposition*. Alloys near the composition Fe_2NiAl and alloys with Co substituted for Fe are called *Alnico magnets*. Alnico magnets still have significant markets today. Fe_2NiAl has a *pseudobinary phase diagram* described in terms of a single-composition variable as illustrated in **Figure 32**. Alnico alloys typically contain 8–12% Al, 15–26% Ni, 5–24% Co, <6% Cu, <1% Ti, with the remainder Fe. The low cost of raw materials makes these particularly economical permanent magnet materials. Alnico magnets can have remanent inductions as large as 1 T, coercivities as large as 2000 Oe and energy products as large as ~ 10 MGOe.

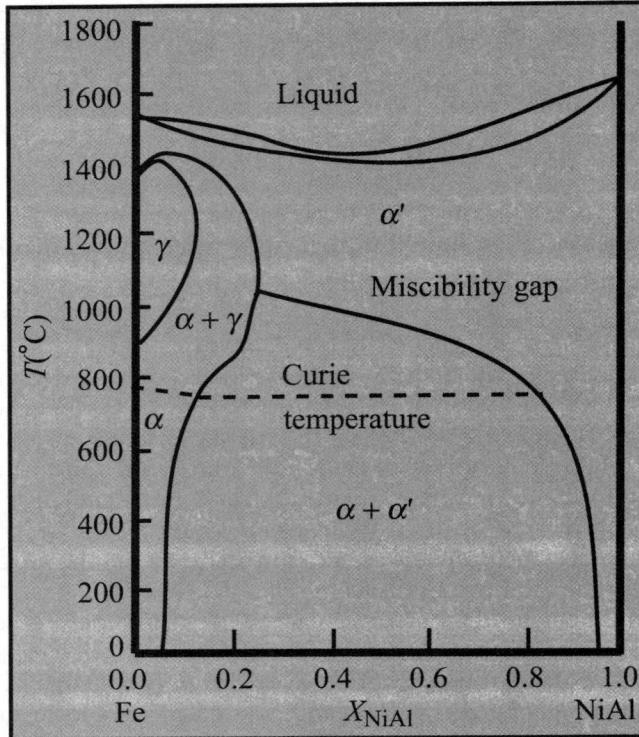

Figure 32 Pseudobinary phase diagram for Fe$_2$NiAl (McCurrie, 1982). (For color version of this figure, the reader is referred to the online version of this book.)

The pseudobinary PD for Alnico exhibits a miscibility gap between the Fe and NiAl components. NiAl is a near stoichiometric compound due to strong Ni–Al bonds. Alloys below the maximum in the gap will decompose into an Fe-rich α- and NiAl-rich α-phase. The α'-phase has some Fe solubility to low temperature, therefore it designates the low-temperature NiAl-rich phase and the high-temperature solution. This decomposition reaction can be written as $\alpha' = \alpha' + \alpha$ where it is implied that the new α' is of a different composition. Spinodals (not shown) from regular solution theory exist in a system that exhibits tendency to cluster (in this case into Fe- and NiAl-rich regions). The Ni- and Al-rich regions subsequently order to the B2 phase. This phase forms coherently with *bcc* Fe. The fine microstructures which can be developed by spinodal decomposition give rise to development of regions of mono-domain Fe separated by regions providing excellent pinning by the weakly magnetic NiAl phase.

Permanent magnets require large *magnetic anisotropy* to develop significant H_c. In Alnico magnets, much of the anisotropy is derived from *shape anisotropy*, by controlling the free pole distribution of the high-moment particles in the alloy. Alnico magnets are further drawn so that the high-moment particles deform to take acicular shapes from which significant shape anisotropy can be derived. **Box 7** calculates the typical values of the shape anisotropy (O'Handley, 2000) for Fe in Alnico magnets. Spinodal decomposition and thermomechanical processing then allow Alnico magnets to take advantage of shape anisotropy effects originally used in Fe horseshoe magnets to retain hard magnetic properties. In this case, it is the nanostructures which retain single domains.

Box 7 Calculation of Shape Anisotropy in Alnico Magnets

Shape anisotropy in Alnico magnets derives from acicular particles of the high-moment phase and the difference in demagnetization factors along long and short particle dimensions. The magnetostatic anisotropy energy density, E_{ms}, can be calculated as follows:

$$E_{ms} = \frac{\mu_0 \Delta N \langle M \rangle \Delta M_s}{2}$$

for typical Alnico magnets $\mu_0 \langle M \rangle \sim 2.1$ T for Fe particles. $\Delta N \sim 0.5$ and $\mu_0 \Delta M_s \sim 2.1$ T, considering the typical aspect ratios for the particles and differences in the saturation magnetizations of the particles and the Ni–Al matrix. Using these parameters, value of the magnetic anisotropy can be calculated:

$$K = E_{ms} = 1.1 \times 10^5 \text{ J/m}^3 = 1.1 \times 10^6 \text{ ergs/cm}^3$$

It is instructive to estimate an upper bound for this anisotropy if the even larger magnetization differences and particle aspect ratios were possible. For this, we consider the case of FeCo particles in a nonmagnetic matrix for which needle-like particles with $\delta N = 1.0$ could be developed. For such materials, $\mu_0 \langle M \rangle \sim 2.5$ T for FeCo particles and $\mu_0 \Delta M_s \sim 2.5$ T could be achieved in a nonmagnetic matrix and

$$K = E_{ms} = 2.5 \times 10^5 \text{ J/m}^3 = 2.5 \times 10^6 \text{ ergs/cm}^3$$

19.3.3 Fe–Si Alloys

19.3.3.1 Si Steels

Silicon (Si) steels are alloys of Fe and Si having important electrical applications (motors, transformers, etc.). With low-C steels, *nonoriented (NO) steels* and *grain-oriented (GO) Si steels* are referred to as *electrical steels*. NO Si steels have applications in rotating machinery; GO Si steels are used in devices (e.g., transformers) where an EMD parallel to a roll direction is desired. GO Si steels are further subdivided into *regular grain-oriented (RGO)* and *high-permeability grain-oriented (HGO)* materials (Boll, 1994). The drive to develop Si steels was motivated by the large increase in the resistivity of Fe with Si additions (**Figure 33(a)**). **Figure 33(b)** shows the variation of properties with Si content in Fe–Si steel.

Studies of Fe–Si date back to 1885 (Hopkinson, 1885). Fe–Si alloys with 2–4.5 wt% Si were the most important soft magnetic material in both volume and market value by 1934 (Chen, 1986). Before 1950, electrical sheet for transformers was produced by melting, ingot casting and hot rolling. By 1951 (Bozorth, 1993), Si–Fe transformer sheet was produced by cold rolling. Polycrystalline NO Fe–Si materials were followed by the GO materials first with (110)/[001] (Goss texture) and later (100)/[001] texture. Efforts to reduce losses in transformer-grade Si steels have focused on (1) improving Goss texture in GO steel and (2) reducing eddy current losses by decreasing thickness and surface treatment.

Si steels are commonly rated on the basis of core loss (the combined power lost due to (a) magnetic hysteresis, (b) eddy currents and (c) anomalous losses). Limiting hysteresis losses requires a magnetically soft material. Magnetic softness is rooted in (1) low-magnetocrystalline anisotropy, (2) low-magnetostrictive coefficients, and (3) maximizing the magnetic domain wall mobility. Eddy current losses are limited, along with (3) by having large electrical resistivities. Si additions to Fe have beneficial effects that include (1) improved magnetic softness and (2) increase in electrical resistivity (Chen, 1986). This is coupled with the disadvantageous (1) decrease in Curie temperature and saturation magnetization and (2) embrittlement in alloys with >2% Si. Embrittlement increases approaching the Fe_3Si intermetallic with the DO_3 structure (**Figure 34**). Thermomechanical processing is used to

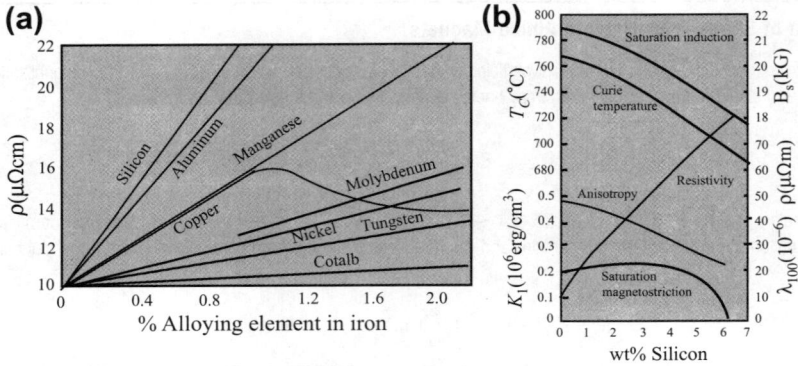

Figure 33 (a) Effect of alloying additions on the resistivity of Fe and (b) the variation of magnetic properties with Si concentration in Si steels. (For color version of this figure, the reader is referred to the online version of this book.)

develop a texture that aligns magnetic easy axes. Impurity removal (e.g. C, N, S, O and B) reduces impediments to magnetic domain wall motion by second-phase impurities. Surface modification of Si–Fe sheet decreases domain size, and reduces anomalous eddy current losses.

In Fe–Si alloys, magnetic moment reduction with Si concentration (from 2.2 µB per atom) can be predicted from dilutional arguments. T_c decreases gradually up to 2 wt% and at an increasing rate for

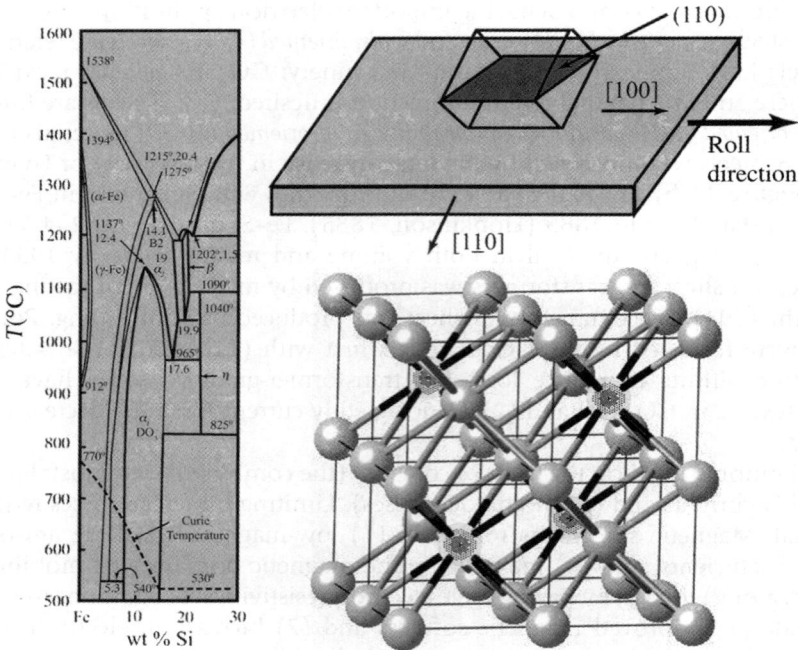

Figure 34 (a) Fe–Si phase diagram. (b) Goss texture schematic. (c) DO_3 structure of Fe_3Si. (For color version of this figure, the reader is referred to the online version of this book.)

>2 wt% Si. The first cubic magnetocrystalline anisotropy constant, K_1, decreases linearly for <5 wt% Si concentration. For Fe–Si, with 3.2 wt% Si, at RT, [100] is the EMD. For Fe–Si with 3.5 wt% Si, $\lambda_{100} = 24 \times 10^{-6}$ and $\lambda_{111} = -2.3 \times 10^{-6}$. Both λ_{100} and λ_{111} decrease in magnitude with Si concentration with a zero crossing in λ_{111} occurring at ~5 wt% Si.

Processing developments for GO Si steel include promoting polycrystalline Goss (110)/[001] recrystallization texture. This is not ideal, but is practically and economically viable. A figure of merit is the average deviation angle of the [100] axis from the roll direction of cold reduced materials. Evolution of GO Si steels have focused on efficient primary crystallization grain growth inhibitors, e.g. MnS (McCurrie, 1982). Improved GO laminates have MnS and AlN grain growth inhibitors and achieve 7–9 mm grain diameters and 30, the average deviation angles (as compared to 2–4 mm and 70).

GO Si steel has a glass film with the main component being forsterite (Mg_2SiO_4) produced prior to secondary recrystallization. This is an electrical insulator and induces a tensile stress in the Fe–Si core. The latter reduces magnetostrictive losses (and transformer hum), decreases domain wall spacing, and promotes favorable domain orientations with respect to the roll axis (limiting anomalous losses). GO Si steel processing to reduce eddy current and anomalous losses includes (1) a reduction in sheet thickness from the original 0.35–0.23 mm now standard, and scratching or laser scribing to form 180° magnetic domain walls parallel to the roll direction.

Sendust has composition $Fe_{1-x-y}Si_x Al_y$, with the optimum composition occurring near the zero-anisotropy zero-magnetostrictive composition with $x = 0.1$ and $y = 0.05$. **Figure 35** shows how these compositions were discovered by determining constant magnetic anisotropy (a) and (b) magnetostriction contours in the Fe–Si–Al ternary system. Because of the appearance of the brittle DO_3 structure in these high Si containing alloys, Sendust is very brittle and is used in powder or dust form in applications such as magnetic recording heads.

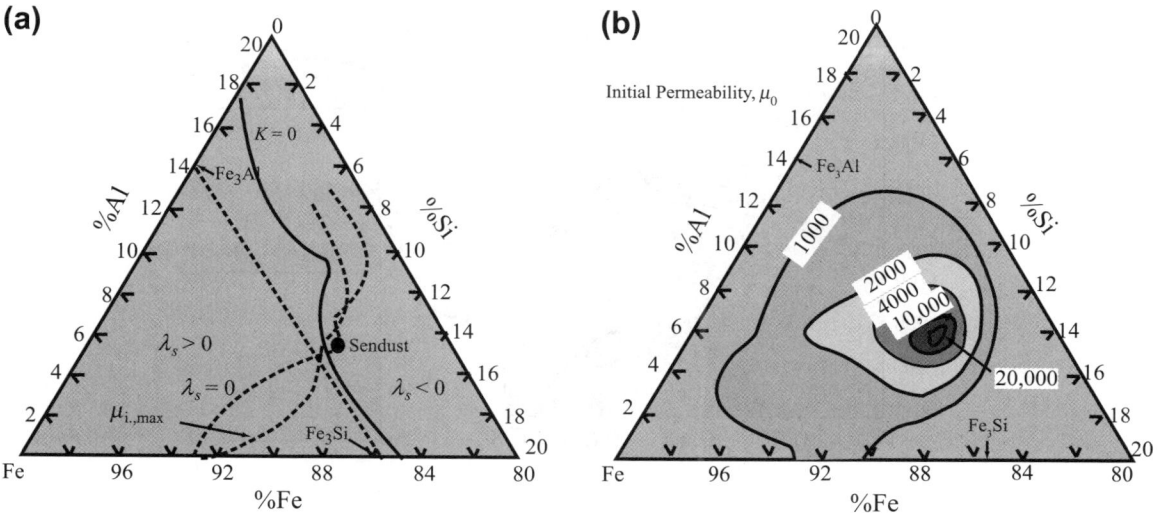

Figure 35 (a) Magnetic anisotropy and (b) magnetostriction in an Fe–Si–Al ternary (O'Handley, 1987). (For color version of this figure, the reader is referred to the online version of this book.)

19.3.4 Fe–Co Alloys

FeCo-based alloys are technologically important materials, for soft magnetic applications that require high-saturation inductions and high-temperature operation. The largest RT saturation induction ($B_s \sim 2.5$ T) is documented for bulk FeCo-based binary alloys. This was illustrated earlier within the context of energy band theory and the Slater–Pauling curve. Energy band theory predicts well the moment variation in the Fe–Co alloy system. This alloy system is one for which the peak in the Slater–Pauling occurs, making alloys in this system, those with the highest inductions of transition metal systems. **Figure 36(a)** illustrates (courtesy Paul Ohodnicki) within the framework of the Bragg–Williams model and the shift in phase boundaries in the presence of a large static magnetic field.

Processing magnetic materials in fields can lead to dramatic changes in properties and is used a processing tool in many important magnetic materials. With the ability to generate large magnetic fields, it is now possible to contribute enough magnetic work to make significant changes in phase boundaries in magnetic materials systems. **Figure 36(a)** shows the calculated shifts in the Fe–Co-phase boundaries in experimentally accessible 50 T fields. The shift in phase boundaries can also be realized by including surface energy effects in free energy calculations. Such surface energy terms become noticeable in magnetic nanostructures. Suppression of phase transformations in *metastable nanostructures* can produce materials with properties that are not obtainable in equilibrium structures. Recent examples are found in the suppression of nucleation of the stable γ-phase in Co–Fe-based nanocomposites at compositions where the binary Fe–Co-phase diagram would predict that the α- and γ-phases should coexist (Ohodnicki et al., 2008c, 2008d, 2009b). Fe–Co alloys will be discussed below in the context of nanocrystals and nanocomposite systems.

The Curie temperature, T_c, of the low-temperature disordered bcc α-phase is such that ferromagnetic behavior persists to temperatures approaching 1273 K. At equiatomic composition, the extrapolated T_c of the α-phase is significantly higher than the temperature above which ferromagnetic behavior is no

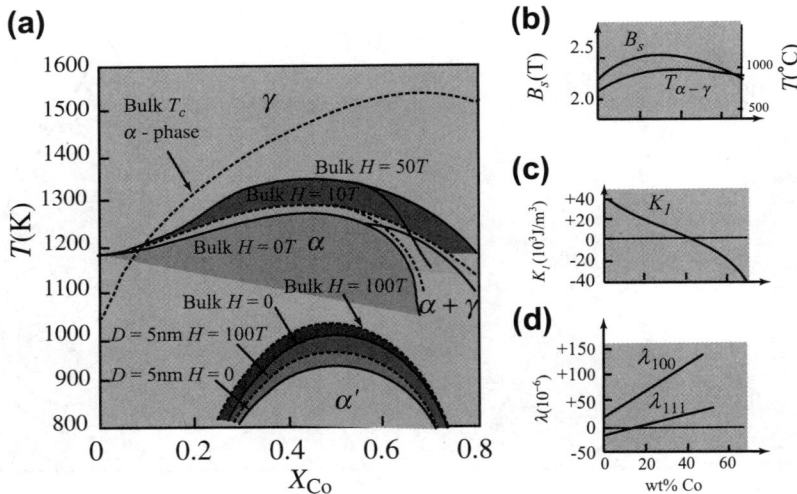

Figure 36 (a) FeCo phase diagram and shifts in phase boundaries in a magnetic field, (b) saturation magnetization and Curie temperature, (c) magnetocrystalline anisotropy, K_1 and (d) magnetostriction, λ, in the FeCo system. Reproduced from Pfeiffer and Radeloff (1980). (For color version of this figure, the reader is referred to the online version of this book.)

longer observed. The abrupt loss of ferromagnetism results from the structural phase transformation from chemically disordered ferromagnetic *bcc* α-Fe–Co to the high-temperature paramagnetic *fcc* γ-phase. If the low-temperature *bcc* phase could be stabilized relative to the γ-phase, ferromagnetism would persist to higher temperatures. In α' FeCo, the chemical order–disorder phase transformation is a higher order transition where the equilibrium value of the chemical order parameter falls continuously to zero with increasing temperature. The phase transition is from an ordered CsCl structure to a disordered *bcc* phase.

Figure 36(b) illustrates the variation of saturation magnetization, T_c, (c) magnetocrystalline anisotropy, K_1, and (d) magnetostriction, λ, in FeCo alloys. In choosing a binary alloy composition in the binary $Fe_{1-x}Co_x$ system, it is important to consider the maximum induction which occurs near $x = 0.3$, the minimum magnetocrystalline anisotropy occurring near $x = 0.5$, and finally compositions which minimize magnetostrictive coefficients. In the FeCo alloys magnetostriction coefficients, λ_{100} and λ_{111} are both substantial near the equiatomic composition where the magnetocrystalline anisotropy vanishes. Across this alloy system, T_c's are larger than Fe or Co.

Consideration of alloy resistivity (important for determining eddy current losses) and alloy additions which influence mechanical properties is used to design alloys for rotor applications in electric aircraft engines. Magnetostriction considerations are also known to be quite important in the design of soft magnetic materials for inductive devices, transformers and a host of other applications. Notable commercial alloys include *Permendur*™, an $Fe_{1-x}Co_x$ alloy with $x = 0.5$, near the minimum magnetocrystalline anisotropy, *Supermendur*™ and *Hiperco-50*™ a similar $Fe_{1-x}Co_x$ alloy with $x = 0.5$, but with a 2% V addition to increase strength and electrical resistivity. Small Nb additions to Hiperco-50™ have led to grain size refinement in the alloy *Hiperco-50 HS*™ which improves the mechanical properties of these materials as well as their degradation in long-term aging at elevated temperatures (Fingers et al., 2002).

19.3.5 Ni–Fe Alloys

Fe–Ni alloys are among the most important soft magnetic alloy systems. Ni-rich alloys are called Permalloys. There are three important permalloy compositions. The first is the 78% Ni Permalloy. This alloy has a zero-magnetostriction coefficient. Variations of this alloy are sold under the names, Supermalloy, Mu metal and Hi-mu 80. The second is the 65% Ni Permalloy. This alloy has $K_1 \sim 0$, and exhibits a strong response to field annealing. The third is the 50% Ni Permalloy. This alloy has higher B_s, strong response to field annealing and produces square-loop materials.

Figure 37(a) shows the saturation magnetization and Curie temperature, magnetocrystalline anisotropy, K_1 and magnetostriction, λ, in the FeNi system for *fcc* or *fcc*-derivative alloys. Details of the magnetic properties and response to field annealing to induce anisotropies in permalloys depend on whether the material has a disordered *fcc* or ordered $L1_2$ Ni_3Fe intermetallic phase structure. These structures are illustrated in **Figure 37(b)**. Crystallographic texture in permalloys is often developed by suitable *thermomechanical processing* to take advantage of the strong variation in properties with direction. These *fcc*-derivative alloys have [111] EMDs and strong anisotropy in magnetostriction coefficients. Rolling can also benefit the high-frequency magnetic properties which are limited by eddy current losses. **Figure 37(c)** shows the frequency dependence of the effective permeability for permalloy as a function of thickness.

In Fe–Ni-based nanocomposite systems, similar nanostructure phenomena are exhibited as noted for Fe–Co. This is observed in Fe-rich alloys (Greer and Whitaker, 2002) where the nucleation of the equilibrium α-phase is suppressed in favor of the metastable γ-phase. This can also have profound effects on technical magnetic properties because on the Fe-rich side of the Fe–Ni-phase diagram, there is a strong compositional dependence of the Curie temperature, T_c, on composition in the γ-phase

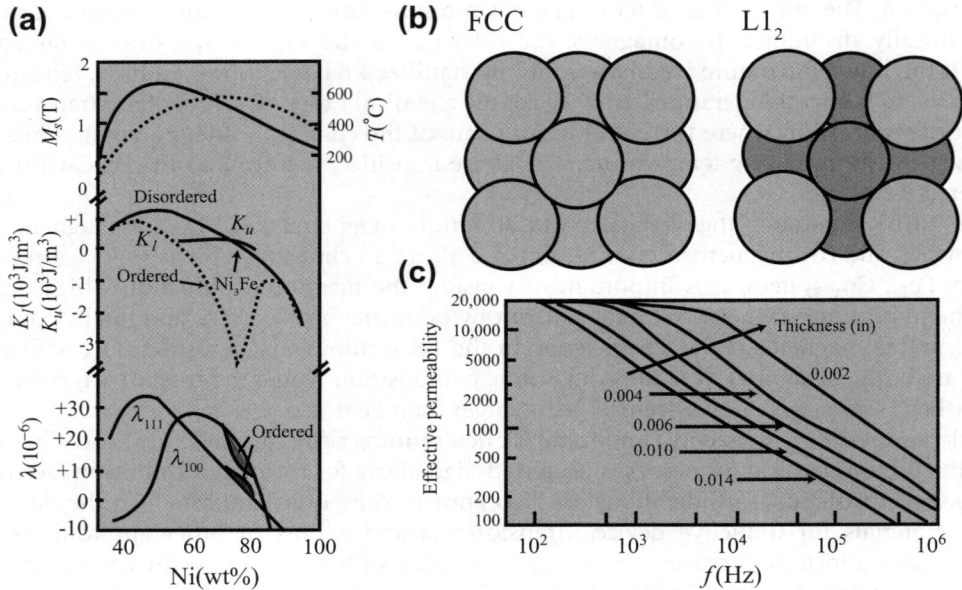

Figure 37   (a) Saturation magnetization and Curie temperature, magnetocrystalline anisotropy, K_1 and magnetostriction, λ, in the FeNi system (O'Handley, 1987); (b) the disordered *fcc* and ordered superlattice structure of the L1$_2$ Ni$_3$Fe intermetallic phase and (c) frequency dependence of the effective permeability for permalloy as a function of thickness. (For color version of this figure, the reader is referred to the online version of this book.)

(Reuter et al., 1989). **Figure 38(a)** illustrates the Fe–Ni binary phase diagram (Massalski, 1990) with **Figure 38(b)** showing the information on the compositional dependence of the Curie temperature, T_c (X_{Ni}). This T_c (X_{Ni}) behavior for the γ-phase can be extrapolated to metastable regions of the Fe–Ni-phase diagram where desirable T_cs near 100 °C for magnetocaloric (Miller et al., 2010b) and biomedical applications (McNerny et al., 2010) are predicted to occur near the 27% Ni composition.

It can be seen that the γ-phase of Fe–Ni has appreciable magnetization on the Fe-rich side of the phase diagram. Because of the variations in K_1 and λ with composition, interesting variations in properties can be observed as described for the Permalloy alloys. Another important set of alloys in the Ni$_x$Fe$_{1-x}$ family is the *invar alloys*, which are based on the 36% Ni alloys. The *invar effect* or *invar anomaly* occurs because of the magnetoelastic effects near its Curie temperature. Magnetoelastic effects (magnetostriction) give rise to spontaneous changes in the lattice parameters as a function of temperature. Like the *magnetocaloric effect* (MCE), these are also largest where the temperature dependence of the magnetization is the largest, i.e. near the Curie temperature. In invar alloys, the magnetostrictive volume change can be tuned to precisely cancel the thermal expansion coefficient, near room temperature, making the material dimensions temperature independent (Gignoux, 1995).

19.3.6   Giant Magnetostrictive Materials

19.3.6.1   *Terfenol and Galfenol*

In soft magnetic materials, magnetostriction can be a deleterious property for certain applications (i.e. it is the source of loss in transformers which is accompanied by "transformer hum"). In other

(a)

(b)

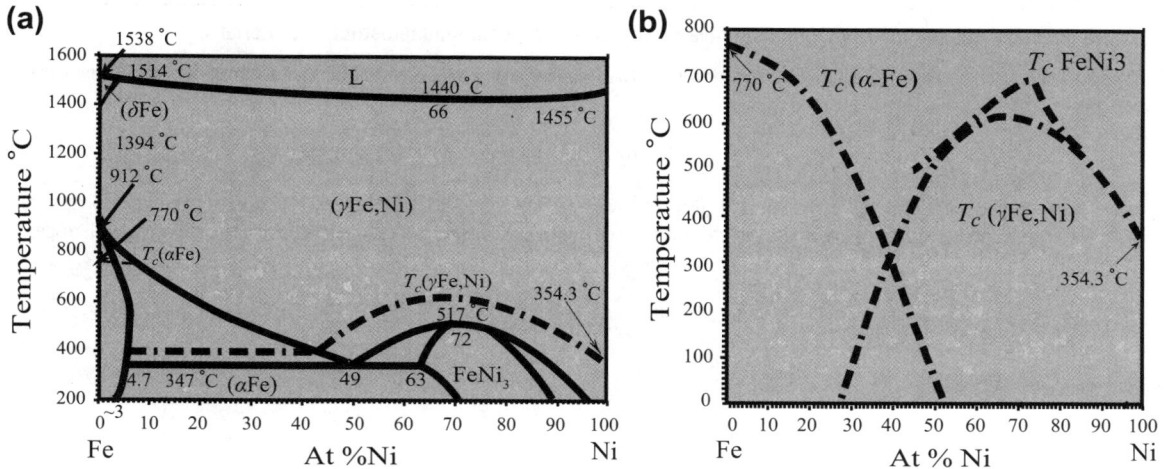

Figure 38 Fe–Ni phase diagram with T_c composition curves (Massalski, 1990) and (b) higher resolution depiction of T_c composition curves. (For color version of this figure, the reader is referred to the online version of this book.)

applications, however, a large value of the magnetostrictive coefficient is critical to the operation of devices. For example, the generation of magnetostatic waves and consequent magnetoacoustic emission is of paramount importance in sonar applications. Materials with large values of magnetostrictive coefficients are often found by looking for large magnetoelastic coupling coefficients. However, as illustrated in **Box 8**, the magnetostrictive coefficient is also predicted to be large in materials compositions where the mechanical properties are soft and the elastic compliance $C \sim 0$.

Large $\frac{dl}{l}$'s are observed in interesting rare-earth/transition metal alloys. For the cubic *Laves phase* material *Terfenol*, $Tb_{1-x}Dy_xFe_2$. *Terfenol-D* $\frac{dl}{l} \sim 10^{-3}$, is an alloy with the composition $Tb_{0.3}Dy_{0.7}Fe_{1.9}$ is a state-of-the-art magnetostrictive material. Terfenol was first developed in the 1970s by the US Naval Ordnance Laboratory and further developed in the 1980s at Ames Laboratory. Terfenol is named after its components terbium (ter), iron (fe), and Naval Ordnance Laboratory (nol). The D in Terfenol-D reflects the dysprosium additions which increase the magnetostrictive coefficients. Terfenol-D has a saturation magnetostriction as high as 1000 ppm.

The reciprocal effect to magnetostriction is the *Villari effect* or *inverse magnetostriction*. It is the change in a material's magnetization when subjected to a stress, σ. For a material with a nonzero magnetostriction coefficient, a mechanical stress can alter the magnetic domain structure of a material. This in turn influences the effective magnetic anisotropy, the permeability and the coercivity of a magnetic material.

Magnetostrictive strain is observed to saturate as a function the applied field, H, in Terfenol (Clark, 1980). Because the effective anisotropy constant is strongly dependent on stress, the approach to saturation of the magnetostriction follows the magnetization of Terfenol-D at room temperature for compressive preloads from 4 to 39 MPa (Wun-Fogle et al., 1999). A theory of the mechanism of magnetostriction in Terfenol-D has been presented by (James and Kinderlehrer, 1994).

Changes in magnetization with stress can be predicted based on *Le Chatelier's principle*. A material with a positive magnetostriction coefficient will elongate along the axis of magnetization and one with a negative magnetostriction coefficient will contract. Therefore, a material with a positive magnetostriction coefficient, under a tensile stress along the direction of magnetization, will exhibit an

---

Box 8 Example calculation of magnetostrictive strain for a tetragonal magnetostrictive material

Consider a tetragonal material having a [001] EMD. We can construct a Landau free energy consisting of elastic, $C\varepsilon^2$, and magnetoelastic αBem^2 terms. We take

$$G = G_0 + C\varepsilon^2 + \alpha Bem^2$$

where α is the geometric constant; $\alpha = 1$ for $m \parallel$ [001] and $\alpha = -0.5$ for $m \perp$ [001], $C = \frac{3V_0}{4}(C_{11} - C_{12})$ is the elastic compliance constant, B is the magnetoelastic coupling constant, m is the reduced magnetization and ε is the strain. Minimizing the free energy with respect to ε yields

$$\frac{\partial G}{\partial \varepsilon} = 0 = 2C\varepsilon + \alpha B\varepsilon$$

and therefore,

$$\lambda = \varepsilon_{eq} = \frac{B}{2C}$$

for $m \parallel$ [001] and

$$G^{[100]} - G^{[001]} = -\frac{3B\varepsilon}{2}$$

and therefore, the effective uniaxial anisotropy energy density is

$$K_u^{eff} = K_u - \frac{3B\varepsilon}{2}$$

---

increased magnetization and under a compressive stress, a decreased magnetization in the direction of a field at a fixed nonsaturating value of the field. These conclusions are true for all but the virgin remnant state where the magnetization is zero.

Terfenol has a saturation magnetization that is approximately 1.0 T at room temperature. Many of Terfenol's material constants vary widely depending on the initial and final magnetic states. Terfenol's initial applications were in sonar systems but have since been applied in magnetomechanical sensors, actuators, and acoustic and ultrasonic transducers. It has also been considered for use in fuel injectors for diesel engines because of the high stresses that can be produced. It is also employed in devices for which passive energy absorption is desired.

Another alloy of recent interest is the shape memory alloy (SMA) *galfenol* with compositions near Fe_3Ga (Clark et al., 2003). Again, galfenol is named after its components gallium (gal), iron (fe), and Naval Ordnance Laboratory (nol). Galfenol is a more recently invented magnetostrictive and is currently under active development. Galfenol has magnetostrictions only 25–33% that of Terfenol-D, but because it doesn't have rare-earth components, it is much less susceptible to corrosion. It can therefore more robust to be used in more challenging environments. Galfenol is also of interest for sonar applications. This interesting magnetostrictive behavior in galfenol results from a maximum in the magnetoelastic coupling constant of Fe with increasing Ga concentration, combined with a temperature-dependent elastic shear modulus that becomes small near 27% Ga (Clark et al., 2003).

19.3.6.2 *Ferromagnetic Shape Memory Effects in Heusler Alloys*

The *shape memory effect* has been explored for a variety of applications, including notable applications in dentistry and biomedicine (Van Moorleghem et al., 1998). A *shape memory alloy* (SMA) is the one that can undergo large plastic deformations and then return to its original shape upon heating. SMAs can exhibit both superelasticity and the shape memory effect. Superelasticity refers to materials for which strains exceeding 10% can be recovered elastically. Superelasticity is a strain-induced transformation. Shape memory effects were first noted in AuCd alloys in 1932 (Olander, 1932) and then again in CuZn alloys in 1956 (Hornbogen, 1956). Of widespread current interest is the NiTi alloy Nitinol. This alloy was also discovered at the Naval Ordnance Laboratory in 1962, by William Buehler and Frederick Wang.

Shape memory effects derive from *martensite* to *austenite* phase transformations. These terms derive from phases of Fe, which are used generically to describe a low-temperature, low-symmetry deformed phase and a high-temperature, high-symmetry undeformed phase. This phase transformation is diffusionless. Since it doesn't require long-range atomic movement, it can be made potentially reversible with temperature. This only happens when the structure of the two phases are closely related by symmetry. In practice, there is *thermal hysteresis* in these transformations and four temperatures are used to define this response as illustrated in **Figure 39(a)**. $f(T)$ is the volume fraction of martensite in the material. Starting with pure martensite, $f = 1$, at low temperature on heating to the temperature A_s the transformation to austenite starts and finishes at A_f. On cooling pure austenite, $f = 0$; at high temperature, the transformation to martensite begins at the temperature M_s and finishes at M_f.

The shape memory effect is described with reference to **Figure 39(b)**. In the elastic region, the $\sigma(\varepsilon)$ response is linear until the elastic limit (A) after which nonlinear plastic deformation occurs until (B). At point (B), the stress is released and the material returns to a zero-stress state with the permanent plastic deformation (C) caused by martensitic distortions. This plastic deformation can then be removed through thermal treatment to arrive back to (0,0).

There has been much recent interest in the development of actuator materials capable of large strains, appreciable thrust and rapid response time. Some representatives of the *Heusler alloy* family exhibit a cubic to tetragonal martensitic phase transformation (MPT). Since this transformation can occur between two ferromagnetic states, and involves large strains, these materials belong to a class called *ferromagnetic shape memory alloys* (FSMA). Unlike magnetostrictive materials (e.g. Terfenol-D), which

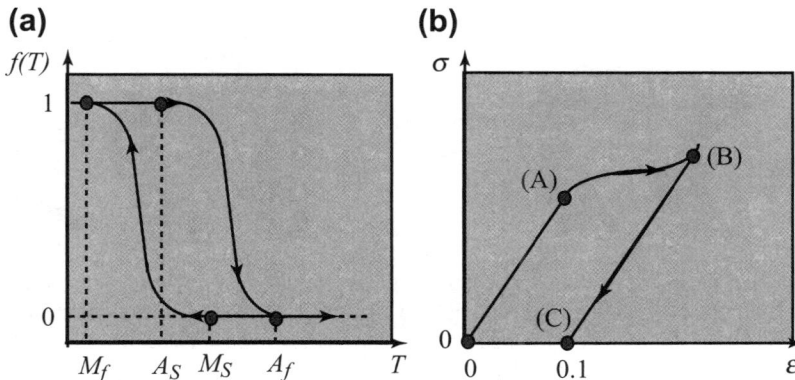

Figure 39   (a) Thermal hysteresis and temperatures in a martensitic phase transformation and (b) shape memory effects in a stress–strain curve for an SMA. (For color version of this figure, the reader is referred to the online version of this book.)

derive their magnetostrictive strains from large magnetoelastic coupling coefficients, FSMAs can have vanishing elastic compliance upon approaching the MPT temperature and correspondingly large strains.

Magnetic field-induced strain of several percent in Ni_2MnGa Heusler alloys (Ullakko et al., 1996; James et al., 1999) are attributed to the motion of martensitic twin boundaries. The mechanism for the large strains in FMSA is explained in terms of the response of martensitic twin boundaries to a magnetic field pressure as illustrated schematically in **Figure 40** (O'Handley, 2000). At high temperatures, the materials exist in a high-symmetry *austenite* phase and undergo a *displacive phase transformation* (DPT) at the MPT temperature, T_M. At low temperatures, the material has transformed into a low-symmetry *martensite phase*. This DPT is accompanied by a large crystallographic distortion. **Figure 40(a)** shows the schematic crystal structure in the austenite. **Figure 40(b)** shows the configuration of distorted cells in the martensite phase.

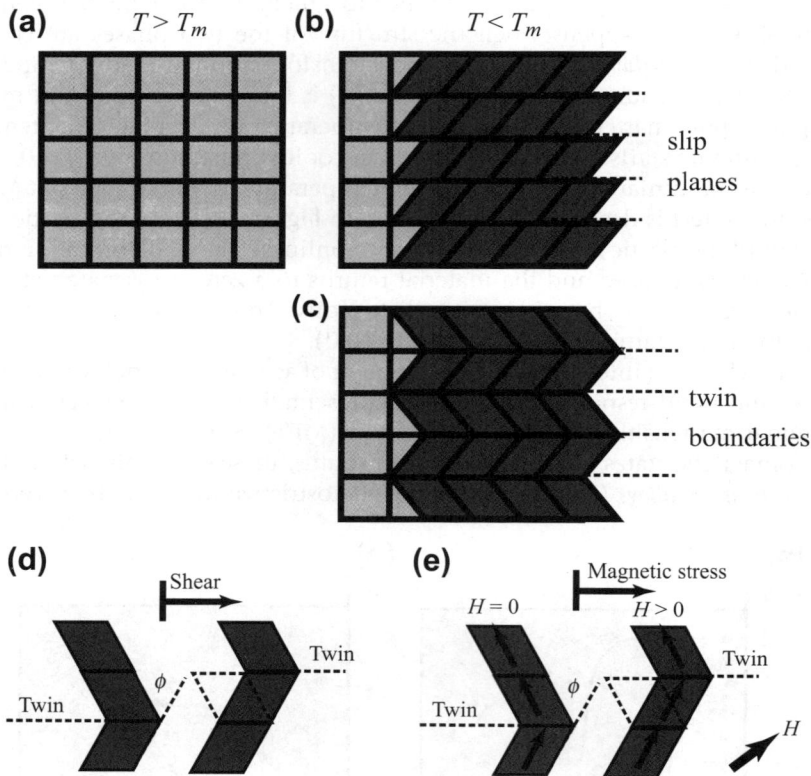

Figure 40 Since a martensitic phase is lower symmetry, there will typically be a uniaxial magnetocrystalline anisotropy determining the direction of the magnetization in the martensite (a) unstrained crystal, (b) strained crystal with slip planes, (c) strained crystal with twin planes, (d) motion of twinned planes under a shear. and For fields, $\mathbf{H} < \mathbf{H_K}$, the normal component of the field will exert a magnetic pressure on the twin boundary in a manner similar to the magnetic pressure on a domain wall. If the boundaries are mobile then this pressure will cause one of the twin variants to grow at the expense of others (e). This in turn can result in large field induced deformation of the material. After O'Handley, 2000. (For color version of this figure, the reader is referred to the online version of this book.)

The crystallographic distortion can involve large strains and large strain energies. The strain energies can be accommodated by slip (**Figure 40(b)**) or twinning (**Figure 40(c)**). In a material where the strain of the MPT is accommodated in twins, mechanical response to a stress can take place by twin boundary motion. **Figure 40(d)** shows such a deformation in a material subjected to mechanical shear. Since the martensitic phase has lower symmetry, there will typically be a uniaxial magnetocrystalline anisotropy determining the direction of the magnetization in the martensite as illustrated in **Figure 40(e)**. For fields, $H < H_K$, the normal component of the field will exert a magnetic pressure on the twin boundary in a manner similar to the magnetic pressure on a domain wall. If the boundaries are mobile, then this pressure will cause one of the twin variants to grow at the expense of others (**Figure 40(e)**). This can result in large field-induced deformation of the material.

The temperatures of ferromagnetic (T_c) and structural martensitic (T_M) phase transitions of the FSMAs are important to determining the temperature-dependent magnetoelastic properties of the materials. The magnetic and structural phase transformations can be characterized by discontinuities in the temperature and/or field dependence of its properties. Such properties include strain, magnetization, resistivity, enthalpy and magnetocrystalline anisotropy energy. When ferromagnetic materials undergo a symmetry lowering structural phase transformation, the crystallographic orientation of the magnetization vectors within the domains displays a corresponding change. Thus the magnetic anisotropy and other properties of the magnetic material can be used as a probe of the change in state in an FMSA (Chu et al., 2000, 2001).

The temperature dependence of the magnetization of the as-grown and postannealed single crystals with a nearly stoichiometric Ni$_2$MnGa composition is shown in **Figure 41(a)** (Chu et al., 2001). An experimental protocol was employed where (1) the sample was first cooled from 385 to 5 K in zero field; (2) then its DC magnetic moment was measured as a function of increasing temperature as the sample was heated above the ferromagnetic T_c in an applied field of 0.08 T; (3) finally, the crystal was cooled to 5 K under the same applied field. In comparison with the as-grown sample, the onset temperatures of the austenite to martensite, T_M, and martensite to austenite, T_A, transitions in the postannealed sample were significantly enhanced from 177.5 to 221.5 K and from 202.5 to 224.5 K, respectively.

The postannealing process reduces the temperature hysteresis for the structural phase transition from 25 to <4 K. This is a signature of the mobility of the twin boundaries in the as-grown and annealed

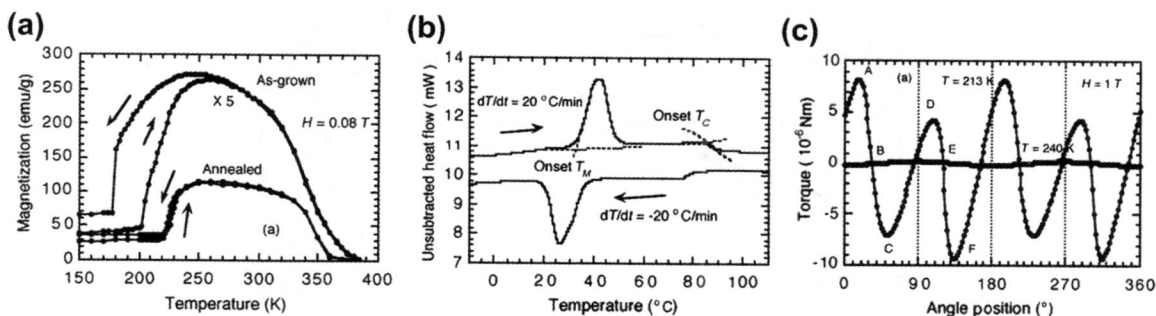

Figure 41   (a) Temperature dependence of the magnetization of the as-grown and postannealed single crystals with a nearly stoichiometric Ni$_2$Mn Ga composition; (b) differential scanning calorimetry (DSC) curve for a crystal of composition Ni$_{52.7}$Mn$_{22.6}$Ga$_{24.7}$ and (c) magnetic torque curves for a stoichiometric crystal in the austenite and martensite phases (Chu et al., 2001).

states. Both the hysteresis in the magnetomechanical response and the field induced strain rate reflect this mobility. The postannealing effect can be interpreted in terms of a reduction of compositional fluctuations and/or second phase in the as-grown sample. High-temperature annealing made the sample more homogenous in composition and microstructure. A sharp step in a magnetization curve is a signature of a cooperative-phase transition. The elimination of a possible contamination phase may result in a decrease of the magnetic coercivity because an impurity phase in a ferromagnetic material can serve as a pinning phase and increase the coercivity. A magnetic hysteresis measurement in the post-annealed sample shows its coercivity to be 50 Oe, which is much smaller than the 250 Oe coercivity of the as-grown sample.

Figure 41(b) (Chu et al., 2001) shows the temperature hysteresis as well as the heat evolved during the structural transition as measured by differential scanning calorimetry (DSC). Again the calorimetric results probe the hysteresis in the MPT. Evidence for a spin reorientation in a stoichiometric sample was also obtained by magnetic torque measurements (**Figure 41(c)**). Following a zero-field cooling, the torque is shown as a function of angle between the applied field (1 T) and a reference orientation. A large value from peak to peak on the torque curve implies that in comparison with the cubic austenite structure (at 340 K), the tetragonal martensite structure (at 213 K) exhibits a strong magnetic anisotropy.

Vasilev et al. (1999) have constructed a free energy functional that describes the energy behavior of a magnetoelastic material that undergoes a cubic to tetragonal martensitic transformation. This functional uses two order parameters, the elastic strain tensor and the magnetization vector, and has been used to predict a magnetic phase diagram for $Ni_{2-x}Mn_xGa$. This phase diagram is illustrated in **Figure 42** and has five phase fields. These are a tetragonal paramagnet, an axial tetragonal ferromagnet, a conical tetragonal ferromagnet, a cubic paramagnet and a cubic ferromagnet.

For stoichiometric Ni_2MnGa, the ferromagnetic cubic phase with the $L2_1$ Heusler structure has $\langle 111 \rangle$ easy axes. The low-temperature tetragonal phase has $\langle 001 \rangle$ easy axes with three possible orientations of the tetragonal phase with respect to the cubic parent phase. There is a close correlation between crystallographic twins and magnetic domain walls in the martensitic state.

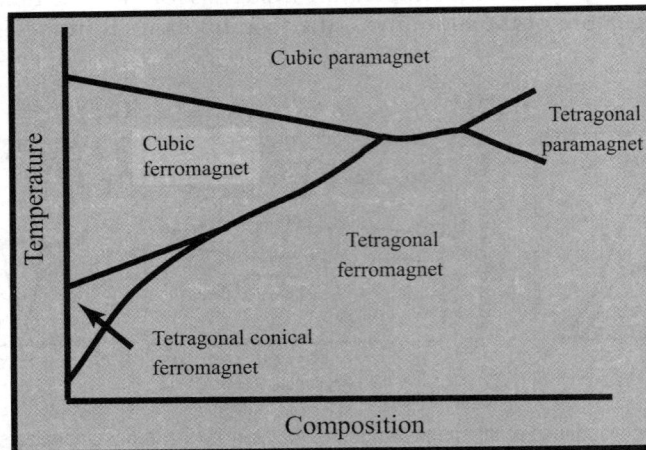

Figure 42   Magnetic phase diagram for $Ni_{2-x}Mn_xGa$ (Vasilev et al., 1999). (For color version of this figure, the reader is referred to the online version of this book.)

Magnetic domain configurations in the Ni_2MnGa FMSA have been observed by means of Lorentz microscopy and noninterferometric phase reconstruction methods (DeGraef et al., 2001). So-called *cross-tie walls* were observed in the thinnest sections of the cubic austenite phase. In thicker regions for the cubic austenite phase, 180° domain walls are most prevalent, along ∼300 nm spaced regular arrays of 71° domain walls. The observations indicate that there is a critical thickness below which the nature of the magnetic domains changes consistent with the observations of *Bloch wall* and *Neel wall* transitions as a function of thickness. In the low-temperature tetragonal martensite phase, the domain configurations correlated with the EMDs for this uniaxial material.

19.3.7 $L1_0$ Permanent Magnets

Binary Fe–Pt and Co–Pt alloys are studied as magnetic materials for both bulk and thin-film magnetic recording applications. Alloys with low-symmetry crystal structures possess large K_us that can be exploited to yield H_cs in appropriate microstructures. Equiatomic Fe–Pt and Co–Pt possess large K_us (for CoPt : $K_u = 4.9 \times 10^7$ ergs/cm^3 and for FePt : $K_u = 6.6 \times 10^7$ ergs/cm^3) due to their tetragonal $L1_0$ crystal structures. Metastable hexagonal derivative structures in Co_3Pt thin films also possess large uniaxial magnetic anisotropies in alloys with less of the expensive Pt (Harp et al., 1993).

FePt, CoPt and other alloys crystallizing in the $L1_0$ structure have hard magnetic properties deriving from a tetragonal crystal structure. FePt adopting the $L1_0$ prototype structure has the $P4/mmm$ space group with Fe in the $1a[(0,0,0)]$ and $1c\left[\left(\frac{1}{2},\frac{1}{2},0\right)\right]$ and Pt in the $2c\left[\left(0,\frac{1}{2},\frac{1}{2}\right)\right]$ special positions in a four-atom supercell. **Figure 43** illustrates the relationship between the ordered tetragonal $L1_0$ and *fcc* phases. Anisotropy is derived from atomic ordering and variation of the tetragonal c/a ratio.

This $L1_0$ structure has alternate stacking of (001) planes of Fe(Co) and Pt atoms. Elemental Pt exhibits Stoner enhance paramagnetic susceptibility but in structures, where Pt is in contact with a ferromagnetic species, Pt can also have an appreciable moment (McHenry and MacLaren, 1991; McHenry et al., 1991). The dipole moment for Pt atoms is induced by polarization effects mediated both by conduction electrons through direct exchange interaction with Co or Fe atoms. Further large spin–orbit interactions ascribed to Pt s outer shell electrons gives rise to the high-magnetocrystalline anisotropy of such materials.

The FePt and CoPt systems were first studied by Jellinghaus (1936) who observed the highest energy products known at the time (Bozorth, 1993). Engineering interest in FePt and CoPt thin films arise from applications in extremely high-density recording and permanent magnets. The magnetic properties of FePt and CoPt alloys strongly depend on annealing temperature, time and composition due to the

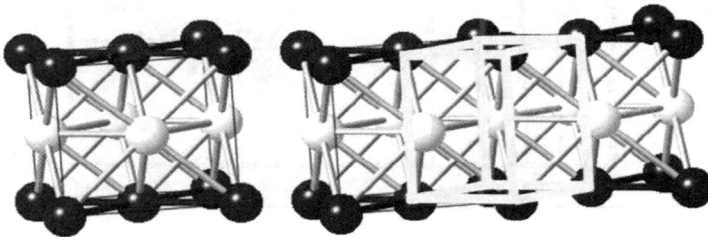

Figure 43 $L1_0$ prototype structure (DeGraef and McHenry, 2007).

influence of each on the atomic ordering fraction of the anisotropic tetragonal phase. Recent efforts have concentrated on developing optimum microstructures to harness the large magnetic anisotropies in FePt and CoPt alloys to develop coercivities. A focus is on nanometer-sized, exchange-decoupled grains approaching monodomain sizes.

19.3.8 Rare Earth Transition Metal Permanent Magnets

Important *RETM alloy* permanent magnets have properties that depend on the T to R ratio. Alloys where the ratio is 2 often adopt a cubic Laves phase structure which does not promote permanent magnetic properties. For alloys with ratios above 2, low-symmetry crystal structures can result in large uniaxial magnetocrystalline anisotropy impacting the coercivity, H_c, of permanent magnets. The magnitude of the coercivity depends further on the microstructure of the magnet. The *magnetocrystalline anisotropy* must be understood with respect to crystal structure.

R metals (Dy, Nd, Pr, Sm, etc.) have metallic radii of ~ 0.175–0.185 nm, $\sim 30\%$ larger than early and ~ 50–60% larger than late-transition metals. Crystal structures exist in RT systems with stoichiometries of RT_2, RT_3, R_6T_{23}, RT_5, R_2T_{17}, R_3T_{29}, RT_{12}, etc. RT phase diagrams often contain many line compounds (with limited solubilities). **Figure 44** shows a schematic RT phase diagram (approximating the Sm–Co system.). Near-line compounds occur at the stoichiometries: R_2T_{17}, RT_5, R_2T_7, RT_3, RT_2, R_9T_4, and R_3T. In state-of-the-art Sm–Co permanent magnets, a two-phase microstructure relies on the large magnetic anisotropy of Sm Co_5 and the high moment of Sm_2Co_{17}. Alloys of composition $SmCo_{7.7}$ have achieved the largest magnetic energy products (highest stored magnetic energy) in this system.

Properties of *RT* permanent magnets derive from the magnetic ground states of the rare-earth species. Other properties derive from the low-symmetry crystal structures and the separation of the T atom planes. It is important to understand how the coupling between the dipoles on the R and T sites are influenced by the choice of R and T species and constituents. Just as in strengthening in steels *interstitial modification* can be used to great benefit in optimizing properties in hard magnets.

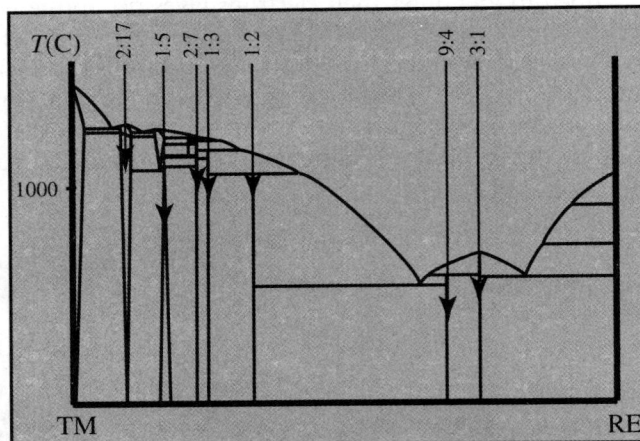

Figure 44 Schematic RT phase diagram (approximating the Sm–Co system). (For color version of this figure, the reader is referred to the online version of this book.)

19.3.8.1 Magnetic Dipole Moments and Coupling

The magnetic dipole moments in RT permanent magnets depend on several significant variables related to the two species—the atomic spacings and the coupling between the dipoles. These aspects are summarized as follows:

(1) R atom moments derive from the localized 4f electrons and thus R dipole moments are well described by the Hund's rule ground state, which do not differ from the R^{3+} species tabulated above.
(2) T dipole moments derive from T d-bands. Since rare-earth planes push T planes further apart, this narrows the T bands, decreasing band widths, W, in comparison with respect to their exchange splitting, J_{ex}. This serves to increase the T dipole moments in cases where weak ferromagnetic bands are transformed into strong bands.
(3) The large R species require larger volumes. Since magnetization is net dipole moment per unit volume, this larger volume reduces the magnetization.
(4) R and T dipole coupling changes from ferromagnetic for *light rare earths* (<7 f electrons) to antiferromagnetic for *heavy rare earths* (>7 f electrons).

The coupling between the dipole moments can be transmitted by direct or RKKY exchange interactions. RKKY exchange is mediated through the conduction electron gas associated, for example, with sp conduction electrons of the magnetic atoms in rare earths. This indirect exchange is transmitted by polarization of the free electron gas and influences the coupling to the local f-electron dipoles.

For atoms with open shells, however, the occupation of angular momentum states in accordance with Hund's rules can lead to anisotropic charge distributions. This is especially pronounced in rare-earth species. Depending on the details of the filling of the 4f states, the resulting charge density can be oblate, prolate or nearly spherical (i.e. for Gd^{3+}). For a material to exhibit magnetic anisotropy, both the orbital angular momentum (L_z) and the crystalline electric field need to have less than spherical symmetry. The crystal field causes the orbital angular momentum to be strongly tied to particular crystallographic directions. The coupling of spin angular momentum to the orbital angular momentum comes about through the f *spin–orbit interaction*. Spin–orbit interactions give rise to an interaction energy, the *spin–orbit energy*:

$$E_{so} = -\lambda \vec{L} \cdot \vec{S} \tag{64}$$

where λ is the phenomenological parameter called the *spin–orbit coupling constant*. For $\lambda > 0$, the lowest spin–orbit energy occurs for spin's aligning parallel to the orbits and for $\lambda < 0$ for spin's aligning antiparallel to the orbits. The spin–orbit interactions can lead to an energy lowering for spin dipole moments aligned with the orbitals and consequently with specific crystallographic directions! This lower energy for dipoles in particular crystallographic directions is the origin of *magnetocrystalline anisotropy*.

19.3.8.2 Curie Temperatures

The Curie temperatures for RT permanent magnets depend on the composition and the sign of the coupling between the R and T species dipoles. Wohlfarth (1979) considered the variation of T_c with R species in R–Co compounds using a Landau theory (discussed above). The model was modified to account for the large R dipole moments and the coupling between the R and Co species by replacing the first Landau coefficient, $a(T)$ with a coefficient $a'(T)$ defined as

$$a'(T) = a(T) - \frac{\alpha}{T} \quad \alpha = \frac{N\mu_B^2 J_{RCo}^2 [g-1]^2 J(J+1)}{3k_B} \tag{65}$$

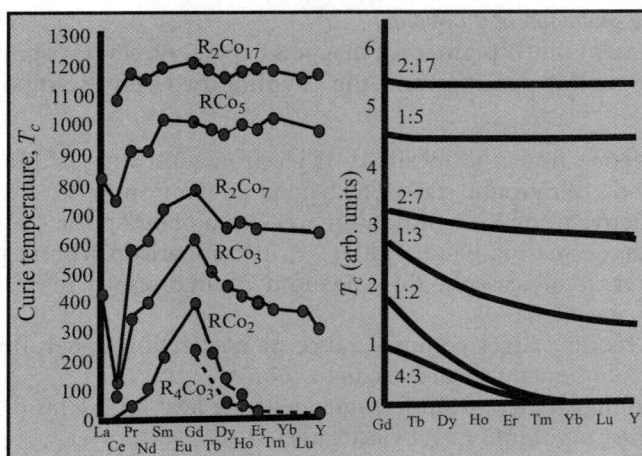

Figure 45 Variation of the Curie temperature in *R*–Co and *R*–Y compounds and those predicted from Wohlfarth's (Wohlfarth, 1979) Landau theory formalism. (For color version of this figure, the reader is referred to the online version of this book.)

where $N\,R$ atoms have moments $gJ\mu_B$ and J_{RCo} is the R–Co exchange constant which changes sign on moving from light to heavy rare-earth species. In general, J_{RT}, the R–T exchange constant to changes sign on moving from light to heavy rare earths due to the change in spin–orbit interactions between oblate and prolate 4f electron distributions. The parameter α can equivalently be expressed more compactly in terms of the *DeGennes factor*, D:

$$\alpha = \frac{N\mu_B^2 J_{RCo}^2 D}{3k_B} \quad D = [g-1]^2 J(J+1) \tag{66}$$

The scale for the T_cs was set by considering Y–Co compounds for which $D = 0$ since Y has an empty f-shell. **Figure 45** shows the variation of T_c in R–Co and R–Y compounds and those predicted from Wohlfarth's (1979) Landau theory. T_cs are largest in the compounds with the largest Co content. Gd gives the strongest coupling to Co. Unfortunately Gd is not a good choice for permanent magnet systems because its half-filled f-shell does not allow for magnetocrystalline anisotropy because there f-electron energy cannot be decreased by a symmetry lowering distortion.

19.3.8.3 *Magnetocrystalline Anisotropy*

There are also several considerations necessary to understand the origin of magnetocrystalline anisotropy in RT permanent magnets. For the large magnetic anisotropies required to develop high coercivities in permanent magnets, these include the following:

(1) The compound's crystal system should be uniaxial rather than cubic.
(2) The anisotropy should give rise to an easy axis ($K_u > 0$) rather than an easy plane ($K_u < 0$), since for an easy plane, the magnetization can rotate without rotating through a hard direction! It is therefore easier to demagnetize.
(3) The anisotropy associated with the R species depends on the 4f-electron charge distribution. RE^{3+} ions have oblate or prolate ellipsoidal charge distributions depending on whether the f-shell is more or less than half-filled (Coehorn, 1991).

(4) The anisotropy associated with the R species depends on the details of the *crystalline electric field* seen by the R species in their special positions.

(5) The anisotropy associated with the T species can be greatly enhanced if the structure has planes for which are 2- as opposed to 3d. T planes can be isolated by the incorporation of the large R species in the structure or through interstitial modifications which can serve to expand the lattice anisotropically.

In a quantum mechanical framework, the energy of electronic states for atoms experiencing a crystal field can be summarized in terms of a *crystal field Hamiltonian*:

$$H = H_0 - eV_{el} + \lambda \vec{L} \cdot \vec{S} \tag{67}$$

where H_0 is the Hamiltonian giving rise to electronic energy levels in the absence of the crystal field. V_{el} is the electrostatic Coulomb potential due to ions (viewed as point charges) surrounding the ion for which crystal field energy levels are to be calculated and $\lambda \vec{L} \cdot \vec{S}$ is the aforementioned *spin–orbit coupling*. The electrostatic potential at an ion of interest and the resulting total electrostatic energy is written as follows:

$$V(r, \theta, \phi) = \sum_j \frac{q_j}{|\vec{R}_j - \vec{r}|} \quad V(r, \theta, \phi) = \sum_i \sum_j \frac{q_i q_j}{|\vec{R}_j - \vec{r}|} \tag{68a}$$

where $\vec{r}$ is the position vector to the ion of interest, $\vec{R}_j$ is the position of the jth ion surrounding the ion of interest and q_j is the charge of the jth surrounding ion. The total energy involves summing over all ions in the structure $(\sum_i)$ where q_i is the charge of the ith reference ion. The solution to crystal field problems is aided by expanding in terms of powers of r and *spherical harmonics* or *tesseral harmonics*, since the potential must be a solution to *Laplace's equation*. An expansion in spherical harmonics is

$$V(r, \theta, \phi) = \sum_{l,m} C r^l Y_l^m(\theta, \phi) \tag{68b}$$

where C is a constant and $Y_l^m(\theta, \phi)$ are the spherical harmonics.

19.3.8.4 *Extrinsic Magnetic Properties*

For applications, it is the extrinsic properties, coercivity, remnant induction and the derivative energy product that are most important as figures of merit. These are determined by the intrinsic properties and the magnetization reversal mechanism. The reversal mechanism is intimately related to the microstructure. Processes involved in the reversal of the magnetization in the second quadrant of the hysteresis loop include nucleation of a reverse domain(s) at a defect, reversible growth of a reverse domain, and irreversible motion of domain walls passed pinning centers.

In real materials, reversal takes place by nucleation of reverse domains on microstructural imperfections or inhomogeneities such as defects, surfaces, interfaces, and so on, and coercivity is less than that predicted by a rotational mechanism alone.

Brown's paradox states that only a fraction of the anisotropy field is practically realizable in coercivity.

Values of the coercivity do not typically exceed 50% of the anisotropy field in the best permanent magnet microstructures.

Microstructural development is very important to determining the ultimate properties of a permanent magnet. Microstructural development can be thought of as attempting to control two distinct aspects of the magnetic microstructure. These are the following:

(1) Engineering the free pole distribution (demagnetization factors) in the microstructure. This was seen in the discussion of Alnico magnets above.
(2) Engineering the spatial variation of magnetic anisotropy by the distribution and texture of low and high K phases. Domain wall pinning is maximized in regions of low K and isolated particles can be forced to reverse by rotation. Texture is important for achieving optimal alignment of easy and hard axes.

As a practical limitation, these are very difficult microstructural features to control and require clever applications of phase relations and phase diagrams coupled with state-of-the-art processing techniques.

Kronmuller's formalism parameterizes the contributions to the coercivity as follows:

$$H_c = \alpha \frac{2K_u}{\mu_0 M_s} - N_d M_s \qquad (69)$$

where K_u is the uniaxial anisotropy and N_d is a spatially averaged effective demagnetization factor associated with the switching entities (i.e. isolated particles). α is the microstructural parameter that can further be expressed as $\alpha = \alpha_K \alpha_\psi$ where α_K describes the spatial variations in the anisotropy and α_ψ describes the texture variations.

19.3.8.5 *Selected Low Symmetry RT Systems*
Table 5 summarizes the structural information for some noncubic R–T phases.

19.3.8.6 *SmCo₅*
$SmCo_5$ is a premiere permanent magnet. This hexagonal material (**Figure 46**), with the largest value of magnetocrystalline anisotropy, was first reported in 1967, by Strnat et al. (1967). Nonmagnetic RT_5 compounds were first synthesized by Wallace et al. (Wallace, 1960) of the University of Pittsburgh and subsequently Carnegie Mellon University. In many applications, Sm_2Co_{17} with larger T fractions and larger inductions are more attractive. However, the Sm_2Co_{17} phases do not achieve magnetocrystalline anisotropy comparable to $SmCo_5$.

In state-of-the-art Sm–Co permanent magnets, two-phase microstructures rely on the large magnetic anisotropy of the $SmCo_5$ and the high moment of the Sm_2Co_{17} materials. Alloys of composition $SmCo_{7.7}$ have achieved the largest magnetic energy products (highest stored magnetic energy) in this

Table 5   Summary of structural information for some noncubic RT phases

Type	System	Space group	Z	a, c (nm)
$SmCo_5$	Hex	P6/mmm	1	0.51, 0.41
β-Sm_2Co_{17}	Hex	P6₃/mmm	2	0.84, 0.82
α-Sm_2Co_{17}	Rhom	R $\bar{3}$m	3	0.843, 1.2222
$Nd_2Fe_{14}B$	Tetra	P$\frac{4_2}{m}$nm	4	0.88, 1.22
Nd_3Fe_{29}	Mono	A2/m	2	1.05, 0.97 (b = 0.85)
α-$Sm_2Fe_{17}N_3$	Rhom	R $\bar{3}$m	3	0.843, 1.222
$Sm(Fe,Ti)_{12}N$	Tetra	I 4/mmm	2	0.893, 0.522

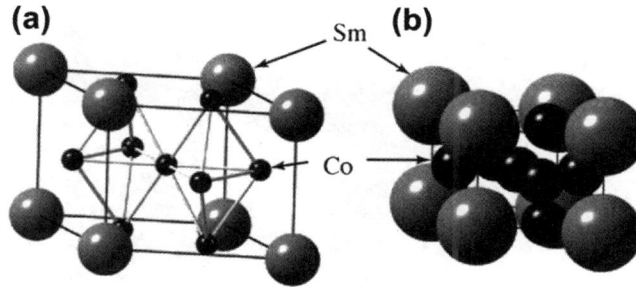

Figure 46 Hexagonal unit cell of CaCu$_5$ (a) ball and stick and (b) space-filling depictions. Reproduced from (DeGraef and McHenry (2007).

system. In these nonstoichiometric two-phase materials, some Co is substituted for by Fe, Cu and/or Zr (Buschow, 1988).

SmCo$_5$ is a template for many important permanent magnets. CaCu$_5$, with the hexagonal P6/mmm (#191) space group, is the prototype for SmCo$_5$ (Wernick and Geller, 1959). **Figure 46** shows the formula unit of SmCo$_5$ in the unit cell (DeGraef and McHenry, 2007), to have eight Ca atoms in each cell, each shared with eight other cells for a total of one Ca. There are $1 + 8(1/2) = 5$ Cu atoms per unit cell. The tetrahedral interstices in SmCo$_5$ are empty. Rare-earth intermetallics are important for H storage in these tetrahedral interstices.

19.3.8.7   Dumbell Substitutions: α-Sm$_2$Co$_{17}$ and β-Sm$_2$Co$_{17}$

Other important REPMs are related to the SmCo$_5$ structure through *dumbell substitutions* replacing a R atom with a pair of T atoms. The stoichiometry is changed to enrich the T content at the expense of the R species. Stadelmeier (1984) proposed the formula, $R_{m-n}T_{5m+2n}$, where m and n are integers, to describe RT compounds formed dumbell transformations. When $m = 1$ and $n = 0$, we describe the parent $RT5$ structure. For other structures, m represents the number of RT_5 formula units and n, the number of dumbell substitutions within the m units. A stoichiometry of R_2T_{17} is obtained for $m = 3$ and $n = 1$. If $m = 2$ and $n = 1$, an RT_{12} is the compound obtained. If $m = 5$ and $n = 3$, an R_3T_{29} compound is obtained. The transformation by which one-third of the R atoms in the CaCu$_5$ structure of SmCo$_5$ are replaced with pairs of transition metal atom *dumbells* is represented as follows:

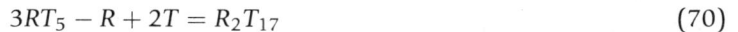

$$3RT_5 - R + 2T = R_2T_{17} \qquad (70)$$

where a single rare earth (Sm) is removed from three units of SmCo$_5$ and replaced by a transition metal (Co) dumbell, $2T$, to yield the Sm$_2$Co$_{17}$ compound. Phases with this stoichiometry exist in both hexagonal and rhombohedral variants. Pairs of T atoms are arranged along the c-axis at dumbell sites (**Figure 47**).

The Th$_2$Ni$_{17}$ structure (P6$_3$/mmc #194) is shown in **Figure 47**. It is the structure of the β-Sm$_2$Co$_{17}$ phase. β-Sm$_2$Co$_{17}$ has lattice constants $a = 0.84$ nm and $c = 0.81$ nm, respectively. **Figure 47** shows the structure of the β-Sm$_2$Co$_{17}$, a single unit cell depicted in (a) space filling and (b) ball and stick formats.

Two to seventeen phase magnet microstructures are multiphase microstructures with each phase having an important role in developing hard magnetic properties. The SmCo$_5$ phase develops on the six

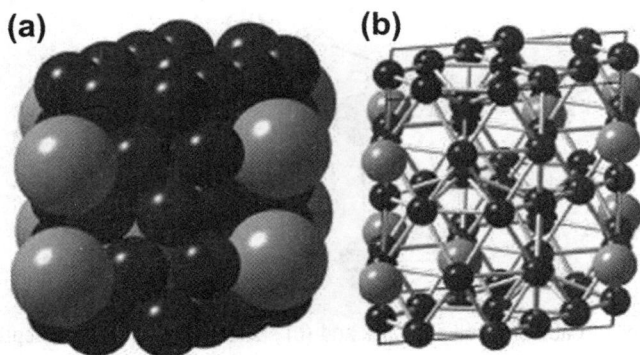

Figure 47 β-Sm_2Co_{17} phase unit cell, depicted in (a) space-filling and (b) ball and stick formats. The hexagonal prismatic representation with three cells is shown projected along the [001] direction. Reproduced from DeGraef and McHenry (2007).

equivalent pyramid planes of the 2:17R phase. A small amount of Zr stabilizes the 2:17 phase with Fe substitutions (partial substitution of Fe for Co increases the magnetization of the material, but a corresponding Sm_2Fe_{17} does not possess a large T_c). Sm_2Co_{17} adopts the 2:17R structure (**Figure 48**) at room temperature, but the 2:17H structure can be retained by rapid quenching. The presence of Fe is also thought to promote the formation of a cellular structure illustrated in **Figure 49**. $SmCo_5$ is metastable at room temperature, but can generally be retained.

In magnetic systems, it is often interesting to look mixed Fe and Co as the transition metal species. This is motivated by the binary Fe–Co systems, where a magnetization larger than pure Fe or pure Co is attainable. Also, since Fe is cubic and Co has an *hcp* structure, their alloys can often be engineered to show preferences for low-symmetry phases, influencing the resulting magnetocrystalline anisotropy. Studies have been performed in phases with the Sm_2Co_{17} structure. Deportes et al. (1976) studied $Y_2(Co_{1-x}Fe_x)_{17}$ alloys and Herbst et al. (1982) studied T site selection in $Nd_2(Co_{1-x}Fe_x)17$ alloys.

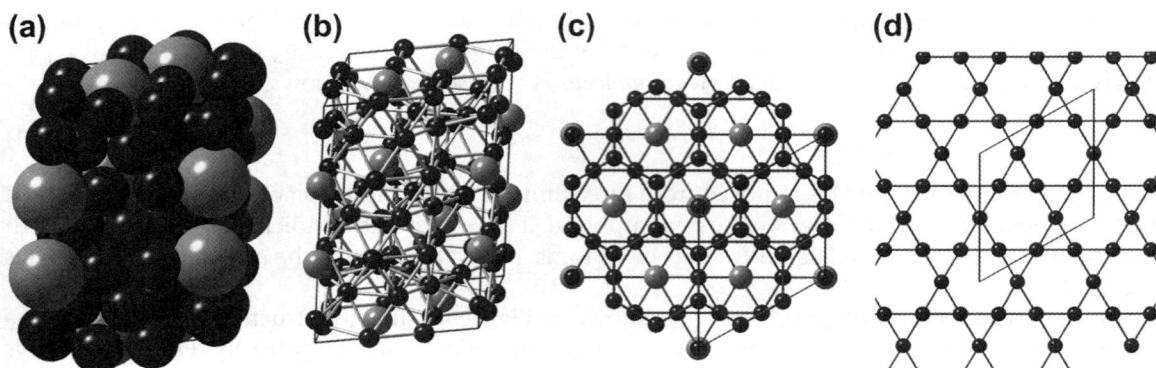

Figure 48 The structure of the α-Sm_2Co_{17} phase, a single rhombohedral unit cell depicted in (a) space-filling and (b) ball and stick formats. (c) The hexagonal prismatic representation projected along [001]. (d) The Co atom Kagome net in the $z = 1/6$ plane. Reproduced from DeGraef and McHenry (2007).

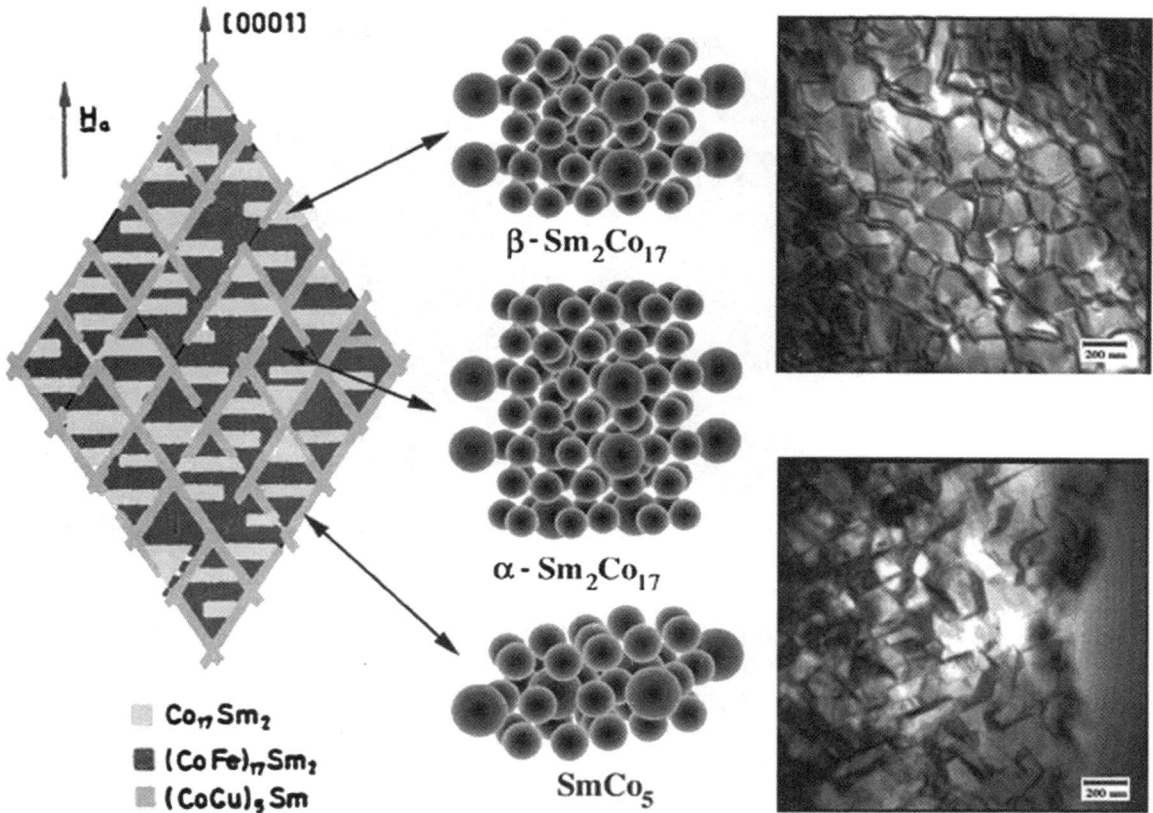

Figure 49 Cartoon and typical microstructures of Sm_2Co_{17} magnetic materials with a cellular structure along with an actual TEM micrograph showing the cell structure. Reproduced from DeGraef and McHenry (2007).

19.3.8.8 The Tetragonal Nd₂Fe₁₄B Phase

The $Nd_2Fe_{14}B$ (2:14:1) phase is the most important tetragonal permanent magnet material because of its large magnetocrystalline anisotropy and large magnetic induction (Herbst et al., 1984; Givord et al., 1984). It is also less costly than the SmCo-based magnets. $Nd_2Fe_{14}B$ has a $P\frac{4_2}{m}nm$ (D_{4h}^{14}, #136) space group with $a = 0.88$ nm and $c = 1.22$ nm. Since it has four formula units per unit cell, the $Nd_2Fe_{14}B$ cell contains 68 atoms. Herbst et al. (1984) solved the magnetic structure by *neutron diffraction*. **Figure 50** shows the $Nd_2Fe_{14}B$ single tetragonal unit cell. For a review of properties, the reader is referred to Herbst (1991).

19.3.8.9 Monoclinic R₃(Fe, Co)₂₉ Phases

The monoclinic $R_3(Fe, Co)_{29}$ phase is a lower symmetry RT magnets for which the hard axis is not orthogonal to a basal plane because the crystal system has all nonorthogonal basis vectors (Cadogan et al., 1994). Their structures result from the dumbell substitution:

$$5RT_5 - 2R + 4T = R_3T_{29} \tag{71}$$

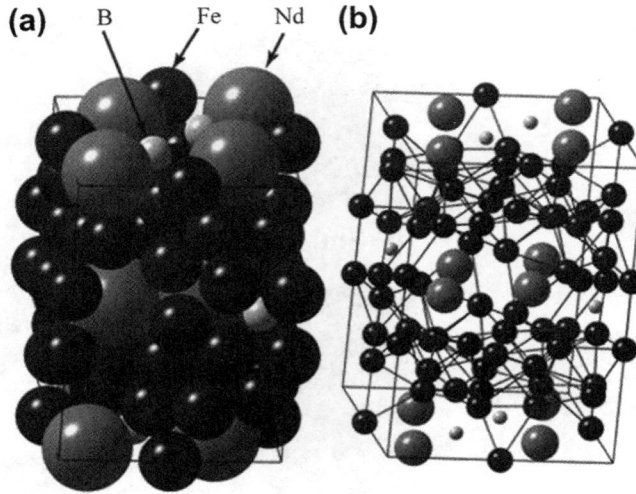

Figure 50 $Nd_2Fe_{14}B$ unit cell depicted in (a) space-filling and (b) ball and stick formats. Reproduced from DeGraef and McHenry (2007).

The R_3T_{29} prototype has an A2/m (#12) space group with two inequivalent R sites and 11 inequivalent T sites, respectively. Relationships between the A, B and C lattice constants in the monoclinic unit cell and a, b and c lattice constants in the unit cell of the similar 1:5 derivative structure exist:

$$B = 3^{1/2}a; \quad A = (4a_2 + c_2)^{1/2}; \quad C = (a_2 + 4c_2)^{1/2} \tag{72}$$

A single cell of the R_3T_{29} phase has two formula units (64 atoms). This structure is formed by alternative stacking of 1:12 and 2:17 type segments. The R_3T_{29}s are the example of a phase that is stabilized by ternary additions. The composition of the phases are reported as $R_3(Fe,M)_{29}$ with M being a larger early transition metal. Substitution of other magnetic transition metals for Fe (notably Co) is also possible, but do not impact the stability of the phase.

For the example of $Nd_3Fe_{27.5}Ti_{1.5}$, illustrated in **Figure 51**, the phase has lattice constants $a = 1.06382$ nm, $b = 0.85892$ nm, and $c = 0.97456$ nm, respectively. A monoclinic tilting angle, $\beta = \arctan\left(\frac{2a}{c}\right)$, for the structure illustrated is 96.93°. **Figure 51** illustrates the structure of the $Nd_3Fe_{27.5}Ti_{1.5}$ with a monoclinic unit cell.

The $R_3(Fe,TE)_{29}$ (3:29) (*TE* (early transition metal) = Ti, etc.) compounds have been reported as potential high-temperature permanent magnet applications. Synthesis structure and properties of the 3:29 phase magnets (Shah et al., 1998, 1999) were examined in the $(Pr_3(Fe_{1-x}Co_x)_{27.5}Ti_{1.5}$ ($x = 0$, 0.1, 0.2, 0.3, 0.4 and 0.5) system with up to 50% Co substitution for Fe. Co substitutions in alloys of composition $(Pr_3(Fe_{1-x}Co_x)_{27.5}Ti_{1.5}$ have been shown to increase T_c, induction and anisotropy field in these magnets. With larger T content, the magnetic exchange and consequent T_cs of these magnets can be increased. The attractiveness of a high T:R ratio in this phase is mitigated by the fact that 1.5 of 29 T atoms are replaced by Ti to stabilize the metastable phase.

Site selection with substitution of Ti for Fe in $(Pr_3(Fe_{1-x}Co_x)_{27.5}Ti_{1.5}$ ($x = 0$, 0.1, 0.2, 0.3, 0.4) magnets was studied (Harris et al., 1999) using EXAFS and neutron diffraction showing that Ti substitutes in the 4g and 4i special positions, consistent with the observations (Hu et al., 1996) in

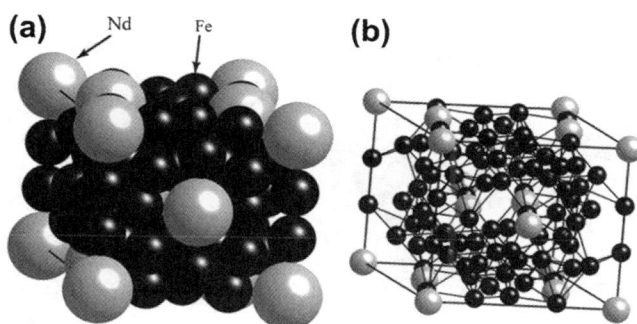

Figure 51 The structure of $Nd_3(Fe,Ti)_{29}$, a single unit cell depicted in (a) space-filling and (b) ball and stick formats. Reproduced from DeGraef and McHenry (2007).

$(Pr_3Fe_{27.5}Ti_{1.5})$ materials. The distribution of Ti between the 4g and two 4i sites agreed with observations (Yelon and Hu, 1996) on $(Pr_3Fe_{27.5}Ti_{1.5})$ materials.

19.3.8.10 Interstitial Modifications

Interstitial modification can improve the properties of REPM materials. Interstitials should have covalent radii less than ~ 0.1 nm (Skomski, 1996) to occupy interstitial sites (often octahedral) in the REPM lattice. B has a covalent radius of 0.088 nm, but, it has a strong preference for trigonal prismatic coordination which can dictate the structure as in 2:14:1 magnets. C and N with covalent radii of 0.077 and 0.070 nm, respectively, often occupy interstitial sites in an REPM structure. These promote volume expansions as large as 8%. Anisotropic volume expansion by interstitial modification can increase K_u.

Dramatic effects on the magnetization and of the Curie temperatures of interstitially modified REPM materials are observed especially for those containing Fe as the T species. Larger spacing between the Fe atoms causes narrowing and closing of the majority spin Fe d-band, increasing Fe's dipole moment. The increased separation of Fe atoms also favorably influences exchange and increases T_c.

For N interstitial modification (Coey and Sun, 1990) of the Sm_2Fe_{17} phase, gas-phase reaction with fine particles is a typical synthesis route. A nitrogenation reaction is

$$2Sm_2Fe_{17} + (3 - \delta)N_2 = 2Sm_2Fe_{17}N_{3-\delta} \tag{73}$$

The nitride disproportionates, above 720 K, by the reaction

$$2Sm_2Fe_{17}N_3 = 2SmN + Fe_4N + 13Fe \tag{74}$$

N occupies large octahedral interstices in the nitrides. Structural features of the modification in the rhombohedral Sm_2Fe_{17} phase are explained considering a volume expansion of the parent α-Sm_2Co_{17} phase lattice with occupation of octahedral interstitial sites. **Figure 52(a)** shows three formula units in the single unit cell of the rhombohedral Sm_2Co_{17} phase. Here, nine N atoms (highlighted) are incorporated into octahedral interstices to yield the compound $Sm_2Co_{17}N_3$.

The N interstitials sit at the 9e special positions in the $R\bar{3}m$ space group with the sites of the R and T atoms the same as in the parent phase. **Figure 52(b)** shows a polyhedral setting, illustrating the N octahedral coordination polyhedra. Notice the vertex sharing polyhedra connected along [100] and

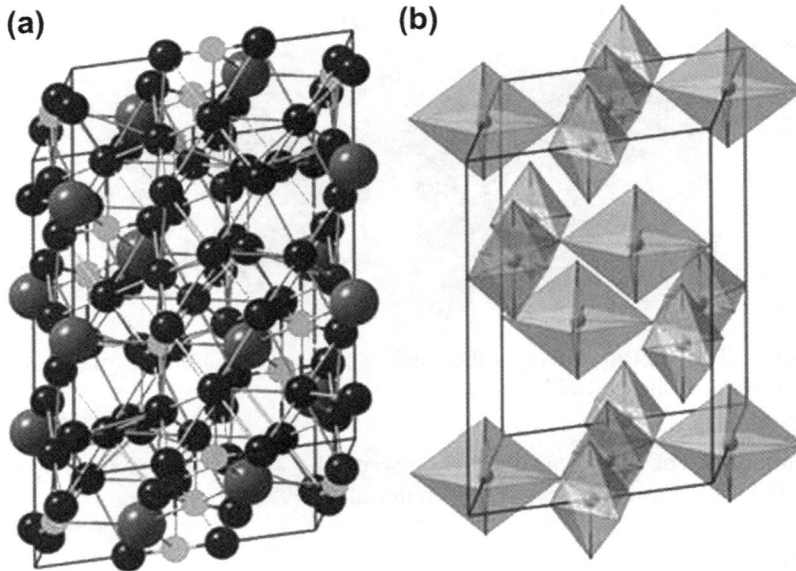

Figure 52 $Sm_2Co_{17}N_3$ structure (a) single unit cell, with three formula units of the rhombohedral Sm_2Co_{17} and 9 N atoms (highlighted) in octahedral interstices; (b) polyhedral setting, illustrating N octahedral coordination polyhedra. Reproduced from DeGraef and McHenry (2007).

[010] directions in this structure. Interstitial modification with C to form $Sm_2Fe_{17}C_x$ has been demonstrated. Solid-state diffusion is used for interstitial modification (Skomski and Coey, 1993).

Interstitial modification of the $Sm(Fe,Ti)_{12}$ phase structures has been investigated by Yang et al. (1993). Here N occupies large octahedral interstices. The $Sm(Fe,Ti)_{12}$ phase requires one of the 12 T sites be occupied by Ti in order to stabilize the Fe containing compound. The structural features of the modification in the tetragonal $Sm(Fe,Ti)_{12}$ phase can also be explained by considering a volume expansion of the parent phase with occupation of octahedral interstitials.

Interstitial modification of the $Sm_3(Fe,Ti)_{29}$ phase structures (Cadogan et al., 1994) includes hydrogen, nitrogen and carbon interstitial modifications. $Sm_3(Fe,Ti)_{29}N_5$ alloys have been synthesized with notable increase in the anisotropy fields, saturation magnetization and T_cs.

19.3.9 Amorphous and Nanocomposite Materials

Amorphous and more recently nanocomposite materials have been investigated as magnetically soft materials for many applications. The term *nanocomposite* will be used for alloys that have a majority of grain diameters in the range from ~1 to 50 nm embedded in an amorphous matrix. Interest in nanocomposite soft magnetics derives from synergy in the properties of the amorphous and crystalline phases from which it is composed. Interesting properties in amorphous and nanocomposite materials derive from chemical and structural variations on a nanoscale which are important for developing unique magnetic properties. We review structure → properties relationships in amorphous and nanocomposite material. Nanocomposites are discussed further below.

Amorphous alloys are topologically disordered. Crystalline alloys have long-range (periodic) order in their atomic and (usually) spin positions. The magnitudes of the spin dipole moments are typically uniform and exchange interactions are discrete as a result of the periodicity. Amorphous alloys have short-range atomic order but lack long-range order and have disordered spin positions, spin magnitudes and exchange interactions. Attributes of amorphous alloys which derive from disorder are the following:

(1) More possible spin configurations, including noncollinear arrangements.
(2) More free volume in the amorphous state than in the crystalline state influences the size of dipoles and net magnetization.
(3) The metastable thermodynamic state is promoted by alloying with glass formers which reduce net dipole moments by dilution and/or bonding.
(4) Magnetic species can have different coordination polyhedra and different nearest-neighbor symmetries than in their crystalline counterparts.
(5) Strong positional disorder cause large fluctuations in exchange. There can be competing ferromagnetic and antiferromagnetic exchange interactions.
(6) Amorphous alloys possess random magnetic anisotropy.

These contribute to rich magnetic phase diagrams.

19.3.9.1   The Structure of Amorphous Materials

Table 6 classifies amorphous and nanocrystalline alloys by *short-range order*, *long-range order*, and the ordering length scales (O'Handley, 1987). Crystalline alloys are designated macrocrystalline, microcrystalline, or nanocrystalline. Amorphous alloys with local order similar to crystalline counterparts are known as *amorphous I alloys*, whereas amorphous alloys with noncrystalline local order belong to the amorphous II type.

In *amorphous solids*, atomic positions lack crystalline (periodic) or quasi-crystalline order but have short-range order. Amorphous metals are usually structurally and chemically homogeneous, which gives them isotropic properties attractive for many applications. Chemical and structural homogeneity can lead to isotropic magnetic properties that are important in materials for many inductive components.

In amorphous metals, atomic correlations extend only to a few coordination shells (0.1–0.5 nm out from the central atom), resulting in significant broadening of peaks and fewer features in X-ray diffraction patterns. In *nanocrystalline alloys*, finite size effects give rise to *Scherrer broadening* of the XRD peaks. For a particle size of 1 nm (i.e. about four unit cells), the peaks have broadened so much that they overlap and the high-angle peaks are no longer resolved. The peak broadening is a signature of the *nanocrystalline structure*.

Atomic distances in an amorphous solid can be described by the *pair correlation function*, $g(r)$. The pair correlation function is defined as the probability that a pair of atoms are separated by a distance, r.

Table 6   Classification of materials by range and type of atomic order

SRO	Range of SRO	LRO	Range of LRO	Material classification
Crystalline	$\geq$10 μm	Crystalline	$\geq$10 μm	Macrocrystalline
Crystalline	100 nm–10 μm	Crystalline	$\geq$100 nm	Microcrystalline
Crystalline	$\leq$100 nm	Crystalline	$\leq$100 nm	Nanocrystalline
Crystalline	~1 nm	No LRO		Amorphous I
Noncryst	~1 nm	No LRO		Amorphous II
Noncryst		Quasi-periodic	~1 μm–0.1 m	Quasi-crystalline

We consider N atoms in a volume Ω; let $\mathbf{r}_1, \mathbf{r}_2, ..., \mathbf{r}_N$ represent the positions of these atoms with respect to an arbitrary origin. The distance $r = |\mathbf{r}_i - \mathbf{r}_j|$ is the length of the vector connecting atoms i and j. Related is the spatially dependent atomic density, $\rho_{atom}(r)$, defined as

$$\rho_{atom}(r) = \frac{N}{\Omega} g(r), \tag{75a}$$

and the *radial distribution function* ($RDF(r)$), defined in terms of the atomic density:

$$RDF(r) = 4\pi r^2 \rho_{atom}(r) dr, \tag{75b}$$

The $RDF(r)$ represents the number of atoms between the distances r and $r + dr$.

The functions $\rho_{atom}(r)$ and $g(r)$ are determined from scattering experiments using wavelengths on the order of atomic distances. A completely structurally disordered material (e.g. a gas) has a uniform probability of finding neighboring atoms at all possible distances (larger than twice the atomic radius), leading to a featureless $g(r)$. In a crystalline solid, $g(r)$ is represented by a set of *delta* functions related to the discrete distances between pairs of atoms (i.e. a diffraction pattern). In amorphous alloys, broad peaks in $g(r)$ reflect the presence of short-range order.

19.3.9.2 *Amorphous Metal Synthesis*

Figure 53(a) illustrates that it is possible to cool alloys fast enough to avoid the nose of the liquid to solid phase transformation in a *TTT diagram*. This requires solidification fast enough to preclude nucleation of the stable crystalline phase; in this case, a phase with an amorphous structure forms. Such a material is called a *metallic glass*. Eutectic alloys where the liquid phase is already stable to low temperatures are ideal. Also of interest is the return to thermodynamically stable phases (or intermediate metastable phases) in the process of crystallization (or nanocrystallization) in **Figure 53(b)**.

The pioneering work of Duwez et al. (1960) was followed by discovery of many metallic glass systems produced by rapid solidification. Amorphous alloy synthesis typically requires cooling rates $>10^4$ K/s (McHenry et al., 1999). Rapid solidification techniques include *splat quenching, melt spinning*, etc. In melt spinning, alloys at temperatures typically >1300 K are cooled to room temperature in 1 ms, at cooling rates of $\sim 10^6$ K/s. **Figure 54** illustrates the melt-spinning process, where an alloy charge is placed in a crucible with a small hole at one end. The alloy is typically induction melted. Surface tension keeps the melt in the crucible until an inert gas overpressure pushes the melt through the hole onto a rotating Cu wheel. The stream rapidly solidifies into $\sim 20\,\mu m$ thick amorphous ribbons. Reviews of melt spinning include those by Liebermann (1983), Davies (1985), and Boettinger and Perepezko (1985). Many other techniques have been used in amorphous metal synthesis including the following:

- *Rapid Solidification Processing*: Amorphous alloys can be produced by rapid solidification processing routes, typically requiring cooling rates $\geq 10^4$ K/s. Examples of these techniques include splat quenching, melt spinning, etc.
- *Solidification of Bulk Amorphous Alloys*: Bulk amorphous alloys are formed by conventional solidification with slow cooling rates. A *large glass forming ability* (GFA) allows producing amorphous materials with much larger dimensions (up to centimeter).
- *Powder Synthesis*: Amorphous metals may be synthesized as powders or compacted to form bulk alloys with an amorphous structure. Techniques include *plasma torch synthesis* (Turgut, 1999), *gas atomization*, and *mechanical milling*. Rapid solidification in the gas phase (e.g. ultrasonic gas

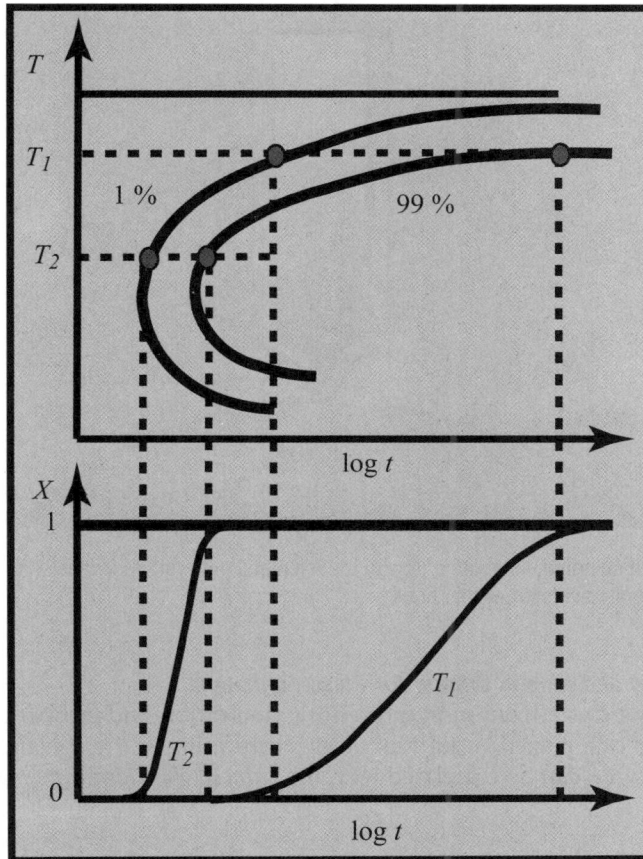

Figure 53 TTT diagram for a hypothetical eutectic alloy and Volume fraction transformed versus time for the crystallization reaction (Mazaleyrat and Barrue, 2001). (For color version of this figure, the reader is referred to the online version of this book.)

atomization) or splatting on a substrate can lead to the formation of nanometer-sized glassy droplets, *nanoglasses* (Gleiter, 1989).

- *Solid-State Mechanical Processing*: In *mechanical alloying*, the energy of the milling process and constituent thermodynamic properties determine whether amorphization will occur. Mechanical alloying is also a means of synthesizing amorphous alloys by solid-state reaction of two crystalline elemental metals in multilayer systems with a fine interlayer thickness (Johnson et al., 1985).
- *Amorphization by Irradiation*: Crystalline alloys can be made amorphous by irradiation by energetic particles beam (Matteson and Nicolet, 1982). Amorphization is an effect that is often observed at high particle fluencies in radiation damaged materials (Sutton, 1994).
- *Thin Film Processing*: *Thin film deposition* techniques were shown as early as 1963 (Mader et al., 1963) to produce amorphous alloys under certain deposition conditions.

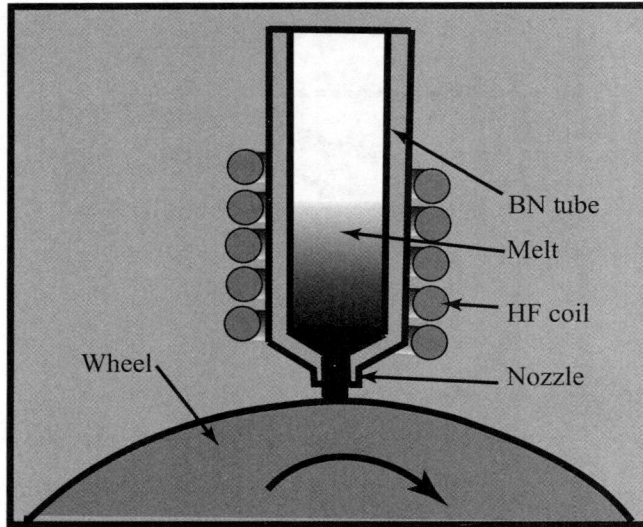

Figure 54 Schematic of melt spinning, illustrating flow of molten metal onto rotating wheel. (For color version of this figure, the reader is referred to the online version of this book.)

19.3.9.3 Thermodynamic and Kinetic Criteria for Glass Formation

GFA involves suppressing crystallization by preventing nucleation and growth of the stable crystalline phase. The solidification of a eutectic liquid involves partitioning of the constituents so as to form the stable crystalline phase. GFA can be correlated with the *reduced glass forming temperature*, T_{rg}, defined as

$$T_{rg} = \frac{T_g}{T_L} \tag{76}$$

where T_L and T_g are the liquidus and glass transition temperatures, respectively. Below the glass transition temperature, T_g, the atomic mobility is too small for diffusional partitioning of alloy constituents.

The thermodynamic condition for glass formation is described by the T_0 construction illustrated in **Figure 55**, for an A–B eutectic system. For compositions between the T_0 curves, the liquid phase can lower its free energy only through the partitioning of the chemical components, nucleating an A-rich or a B-rich region that expels the other (B or A) constituent as it grows. This nucleation and growth process requires long-range diffusion to continue. If an alloy can be quenched below its glass transition temperature, T_g, in the region of compositions between two T_0 curves, then the atomic motion necessary for this partitioning will not be possible and the material will retain the configuration of the liquid. The GFA is increased for materials where the reduced glass forming temperature, T_{rg}, is large.

Massalski (1981) presented thermodynamic and kinetic considerations for the synthesis of amorphous metals. The criteria suggested for *partionless freezing* (no composition change) of a liquid to form a metallic glass are the following:

(1) *Quenching to below the T_0 Curve:* The T_0 curve is the temperature below which there is a thermodynamic driving force for partitionless freezing and since this requires no long atomic diffusion,

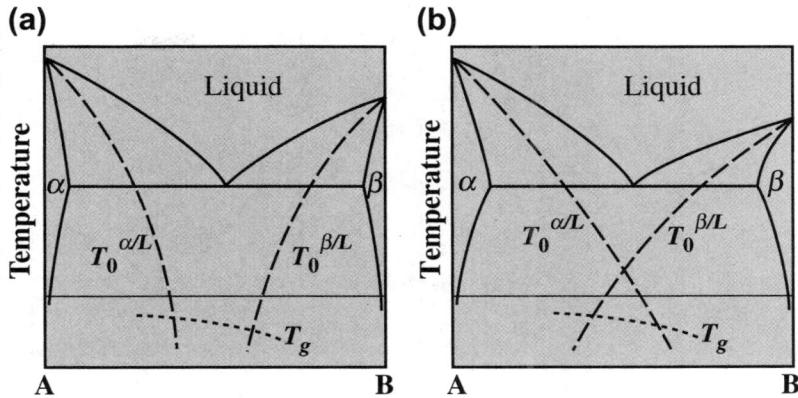

Figure 55 (a) T_0 construction for an *AB* binary alloy with a deep eutectic. (b) In an alloy for which the T_0 curves intersect above T_g, partitionless solidification is not possible (Willard, 2000). (For color version of this figure, the reader is referred to the online version of this book.)

the liquid freezes into a solid of the same composition. The T_0 curve is the locus of *T*-composition points where the liquid- and solid-phase free energies are equal.

(2) *Morphological Stability:* It depends on the comparison of imposed heat flow and the velocity of the interface between the amorphous and liquid phases.

(3) *Heat Flow:* To prevent segregation, liquid supercooling must exceed L/C, where *L* is the latent heat of solidification and *C* is the specific heat of the liquid.

(4) *Kinetic Criteria:* A critical cooling rate, R_c, for quenching of the liquid is empirically known to depend on the *reduced glass forming temperature*, T_{rg}.

(5) *Structural:* Atomic size ratios with difference exceeding $\sim 13\%$ (consistent with *Hume–Rothery rules*) retard the diffusion necessary for partitioning.

Criterion (1) is a condition on the supercooling of the liquid. Criterion (3) requires that heat must be transported quickly enough from the moving solidification front. This is determined by heat transfer between the amorphous solid phase and the wheel and depends on the wheel conductivity, speed and the degree of wetting by the liquid. Criterion (4) defines a critical cooling rate required to prevent the *nucleation and growth* of the crystalline phase. Deep eutectics occur in systems with large positive heats of mixing and consequent atomic size differences motivating the criterion (5).

Figure 56 illustrates examples of two alloy systems that exhibit relatively deep eutectics and can be rapidly solidified to form a metallic glass. The Fe–Zr system is an example of a eutectic in a late transition metal/early transition metal system. The Fe–B system is an example of a eutectic in a (transition) metal/metalloid system.

19.3.9.4 *Magnetic Dipole Moments and Their Ordering*

The most important determinants in the magnitude of magnetic dipole moments in amorphous alloys is alloy chemistry followed by positional disorder. The issue of the influence of alloy chemistry on magnetic dipole moments and magnetization, previously illustrated in the Slater–Pauling curve, can be addressed in a more quantitative manner in a variety of empirical models. Weak solutes, as defined by a valence difference $\Delta Z \leq 1$, are explained by the previously discussed *rigid band model* while a *VBS*

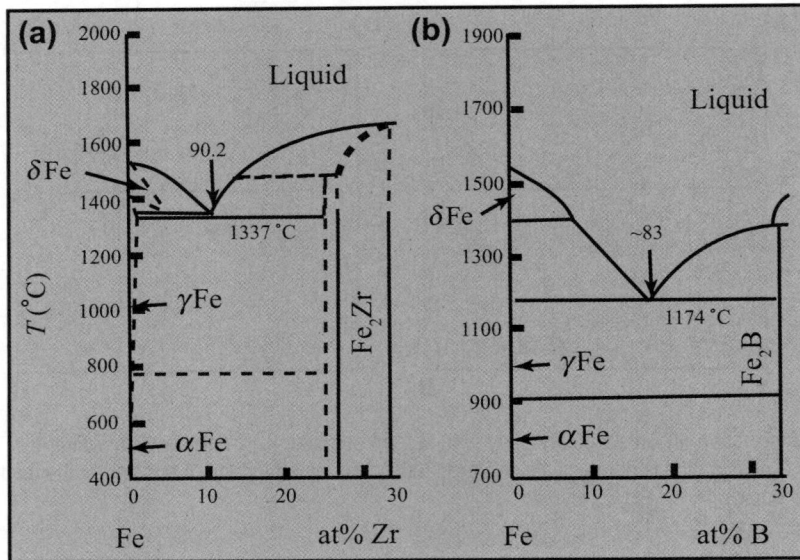

Figure 56 Eutectic phase diagrams for the (a) Fe–Zr and (b) Fe–B systems. In both cases, the alloys can be rapidly solidified to form a metallic glass (Willard, 2000). (For color version of this figure, the reader is referred to the online version of this book.)

model is employed when the solute perturbing potential is strong, $\Delta Z \geq 2$. TM moment reduction due to sp-d bonding with metalloid glass formers is in addition to dilution effects! Therefore, amorphous magnet inductions are significantly reduced as compared with elemental counterparts. Moment reduction due to coordination bonding has been discussed by Corb and O'Handley (1985) and magnetic valence ideas have been developed by Williams (Williams et al., 1983).

The absence of crystalline periodicity in amorphous materials leads to distributed local exchange interactions and sometimes competing exchange interactions. This leads to richness in the possible magnetic structures in amorphous materials. The collinear magnetism found in crystals, i.e. ferromagnetism, antiferromagnetism and ferrimagnetism are all observed in amorphous materials. In addition, noncollinear magnetism can also be observed in the form of *speromagnetism, asperomagnetism* and *sperimagnetism*. Competing exchange interactions can also lead to frustration and spin–glass behavior (Coey, 1978). These configurations are depicted in **Figure 57**.

19.3.9.5   *Random Exchange*

As compared with bulk crystalline materials, amorphous alloys typically have reduced Curie temperatures, due to alloying with glass forming elements (O'Handley, 1987). Amorphous alloys also have large distributions of interatomic spacings resulting in distributed exchange interactions which alter the mean field description of $M(T)$ (Chien, 1978).

Distributed exchange is predicted on the basis of a distribution of nearest-neighbor distances and *Bethe–Slater curve*. Observations of distributed hyperfine fields in *Mossbauer spectroscopy* are consistent with this. Handrich and Kobe incorporated distributed exchange into mean field theory to predict $M(T)$ for amorphous magnets. The mean field theory for the temperature dependence of the magnetization in amorphous alloys was proposed by Handrich and Kobe (Kobe, 1969; Handrich, 1969).

(a)   **(b)**   **(c)**   **(d)**

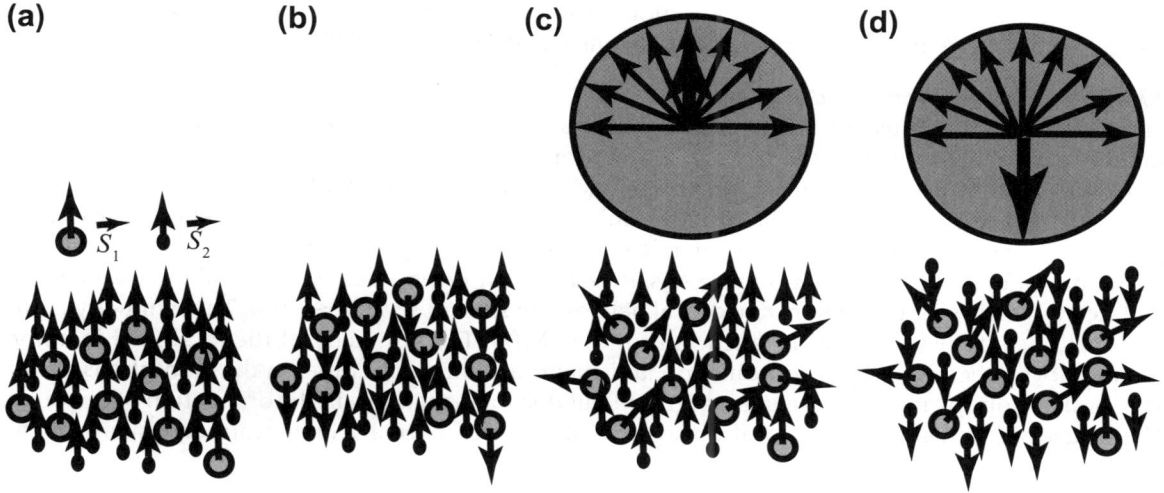

Figure 57 Spin configurations possible in amorphous magnets, (a) ferromagnet, (b) antiferromagnet or ferrimagnet, (c) speromagnet and (d) sperimagnet. (For color version of this figure, the reader is referred to the online version of this book.)

To consider the effects of positional and chemical disorder in amorphous alloys, we are guided by simple Taylor series expansions of the exchange energy:

$$J = J_0 + \left(\frac{\partial J}{\partial r}\right)\Delta r \quad J = J_0 + \left(\frac{\partial J}{\partial C}\right)\Delta C \tag{77}$$

where the first expression reflects positional (R) and the second chemical disorder (C). The starting point for each of these is a description of $J(C)$ or $J(R)$ in similar crystalline alloys. For $J(C)$, we can be guided by the Slater–Pauling curve and for $J(R)$, we can be guided by the Bethe–Slater curve. In the case of positional disorder, the first-order effects can be predicted by recognizing that for Fe, $\left(\frac{\partial J}{\partial r}\right) > 0$, for C_o, $\left(\frac{\partial J}{\partial r}\right) \sim 0$ and for Ni, $\left(\frac{\partial J}{\partial r}\right) < 0$ in their pure crystals according to the Bethe–Slater curve.

Considering positional disorder, the mean field theory description of the Curie temperature can be modified to yield the following:

$$T_c = \frac{NS(S+1)}{3k_B} \int J(r)g(r)\mathrm{d}r \tag{78}$$

where $g(r)$ represents the previously defined pair correlation function. Another approximate method for arriving at the positional disorder (Δr) is to use the Scherrer peak broadening in an amorphous XRD pattern.

The temperature dependence of the magnetization for amorphous alloys has been fit using a two-parameter exchange fluctuation mean field theory (Gallagher et al., 1999). This is shown to give better fit $m(T)$ data than a single-parameter model for amorphous alloys. The modification made to the Handrich–Kobe (Handrich, 1969; Kobe, 1969) model defined two δ-parameters, δ_+ and δ_- to yield the mean field equation:

$$m(T) = \frac{1}{2}\left(B_s[(1 + \delta_+)x] + (B_s[(1 + \delta_-)x])\right) \tag{79}$$

where δ_+ and δ_- can differ due to the position on the Bethe–Slater curve. The *Gallagher* model gives a better quantitative fit to the temperature dependence of the magnetization.

19.3.9.6 *Random Magnetic Anisotropy*

The discussion of an effective magnetocrystalline anisotropy in both amorphous and nanocrystalline materials is deeply rooted in the notion of a random local anisotropy as presented by Alben et al. (1978b, 1978a). The original basis for this model is in the problem of averaging over randomly oriented local anisotropy axes. The random anisotropy model leads to a small effective magnetic anisotropy due to the statistical averaging of the magnetocrystalline anisotropy.

In amorphous alloys, the notion of a *crystal field* (which determined K in crystalline materials) is replaced by the concept of a short-range local field on the scale of several ~ 1 nm. The symmetry of the local field determines the local magnetic anisotropy, $K_1(r)$. The symmetry of the local field in amorphous alloys depends on local coordination and chemical short-range order in the amorphous alloys. The local anisotropy may be large but is averaged out due to fluctuations in the orientation of easy axes. The statistical averaging of the local anisotropy takes place over a length scale equivalent to the ferromagnetic exchange length, L_{ex}.

As in the macroscopic picture of a domain wall, the local anisotropy correlation length is determined by a balance of exchange and anisotropy energy densities. Considering these two terms in the *magnetic Helmholtz free energy*, F_V (i.e. ignoring field and demagnetization terms) can be expressed as

$$F_v = \int_V A(r)|\nabla\vec{m}|^2 - K_1(r)[\vec{m}\cdot\vec{n}]^2 \tag{80}$$

where $A = A(r)$ is the local exchange stiffness, $\vec{m} = \frac{\vec{M}(r)}{M_s}$ is the reduced magnetization, $\vec{n}$ is the unit vector in the local direction of an easy axis, and $K_1 = K_1(r)$ is the leading term in the expansion of the local magnetic anisotropy. If we denote the length scale of chemical or structural fluctuations in the amorphous material as l (this will be on the order of 1 nm for a typical amorphous material). For this case, $L_{ex} \gg l$ (or D_g) in soft materials and a statistical averaging of $K_1(r)$ over a volume of $\sim L_{ex}^3$ is warranted. Considering a random walk through a volume L_{ex}^3, which samples all local anisotropies, leads to a scaling of $K_1(r)$ by $\left(\frac{L_{ex}}{l}\right)^{3/2}$ to arrive at an effective anisotropy K_{eff}. This scaling is responsible for the substantial reduction of the magnetic anisotropy in amorphous alloys. In TM-based amorphous alloys where $K_1(r)$ is small to begin with and $L_{ex} \gg l$, then K_{eff} is very small in good agreement with experimental observations for amorphous transition metal alloys. On the other hand for amorphous rare earth-based materials where $K_1(r)$ is large and L_{ex}^3 may sample only a few fluctuations in n, then K_{eff} can remain quite large (Alben et al., 1978a, 1978b). Amorphous rare-earth alloys with substantial coercivities are often observed.

Consideration of the benefits of nanocrystalline alloys for soft magnetic applications include the coercivity and the permeability. Reduction of coercivity and the related increase in permeability are both desirable properties that can be found in select amorphous and nanocrystalline alloys. The extension of the random anisotropy model by Herzer (1997) to nanocrystalline alloys has also been used as the premise for explaining the soft magnetic properties of these materials (**Figure 58**). The Herzer argument for effective anisotropies in nanocrystalline materials builds on the arguments of the random anisotropy model for amorphous alloys presented above.

Herzer considers a characteristic volume whose linear dimension is the magnetic exchange length, $L_{ex} \sim \left(\frac{A}{K}\right)^{1/2}$. The Herzer argument considers N grains, with random easy axes, within a volume of Lex3

(a)

(b)

Grain size, D

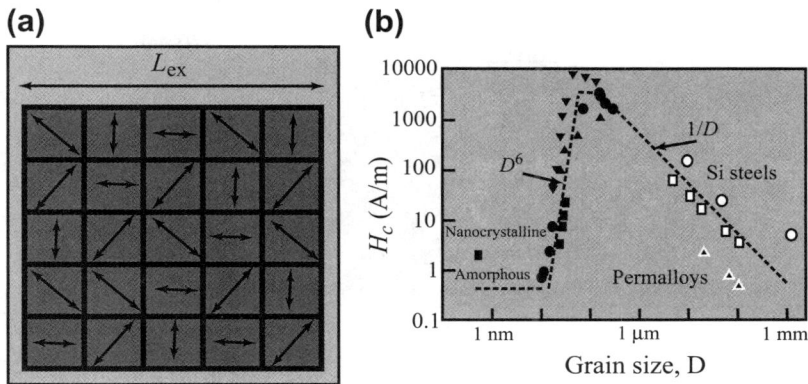

Figure 58 (a) Cartoon illustrating N nanocrystalline grains of dimension D, in a volume L_{ex}^3 and (b) the dependence of coercivity on grain size in the Herzer model. (For color version of this figure, the reader is referred to the online version of this book.)

to be exchange coupled. Since the easy axes are randomly oriented, a random walk over all N grains will yield an effective anisotropy which is reduced by a factor of $\left(\frac{1}{N}\right)^{1/2}$ from the value K for any one grain, thus $K_{eff} = \frac{K}{\sqrt{N}}$. The number of grains in this exchange-coupled volume is $N = \left(\frac{L_{ex}}{D}\right)^3$, where D is the average diameter of individual grains. Treating the anisotropy self-consistently, then

$$K_{eff} \sim KD^{3/2} \sim \left[\frac{K_{eff}}{A}\right]^{3/2} \sim \left[\frac{K^4 D^6}{A^3}\right] \tag{81}$$

Since the coercivity can be taken as proportional to the effective anisotropy, this analysis leads to Herzer's prediction that the effective anisotropy and therefore the coercivity should grow as the sixth power of the grain size: $H_c \sim H_K \sim D^6$. For such a reduction in the coercivity to be realized, Herzer noted that the nanocrystalline grains must be exchange coupled. This is to be contrasted with uncoupled particles that have an exchange length comparable to the particle diameter and are susceptible to superparamagnetic response.

19.3.9.7 Pair Order Anisotropy

Induced anisotropy is also very important in tailoring the properties of amorphous and nanocomposite magnets. Many ferromagnetic materials exhibit a uniaxial anisotropy when they are processed in a magnetic field. An *induced anisotropy* parallel (or perpendicular) to the applied magnetic field direction can be obtained. Controlling the magnetic anisotropy in amorphous soft magnetic materials is important to technical applications of the materials. Induced anisotropy includes (1) roll anisotropy, (2) stress annealing anisotropy, (3) field-induced anisotropy, etc.

Field-induced anisotropy is obtained by annealing in a magnetic field. The mechanism requires the presence of different atomic species that are thought to form ordered atomic pairs aligned with, or perpendicular to the field as illustrated schematically in **Figure 59**. **Figure 59(a)** shows the hypothetical atomic arrangements in a random equiatomic A–B alloy. Here the atoms are distributed among sites randomly. **Figure 59(b)** shows an ordered equiatomic A–B alloy for comparison. **Figure 59(c)** shows

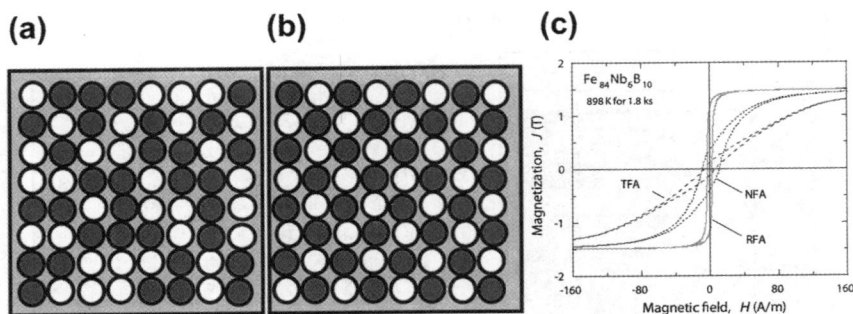

Figure 59 (a) Random atoms in an A–B alloy system, (b) an ordered equiatomic A–B alloy, (c) directional pair ordering. From Suzuki and Herzer (2012). (For color version of this figure, the reader is referred to the online version of this book.)

the same alloy with directional pair ordering. In directional pair ordering, more atomic pairs A–B pairs point in the direction (or perpendicular to) the field direction.

Field annealing yields induced uniaxial anisotropy in the direction of the applied field. This can be used to tailor the hysteresis loop to influence the permeability. This can result in the annealing field in a direction transverse to the magnetic path will shear the loop and result in a nearly linear $B–H$ response. The magnetization process is dominated by the rotation of moment vectors within domains, and domain wall motion is minimized. This is important in minimizing the power loss due to irreversible wall motion past defects.

Important technical magnetic properties of amorphous magnetics also derive from *random magnetic anisotropy*. The magnetic anisotropy in amorphous materials is typically low. Interesting properties also derive from zero magnetostriction in Co-based alloys.

19.3.9.8 Examples of Amorphous Metal Alloy Systems

Common glass forming systems include Group IIa-transition metal, actinide-transition metal, and early transition metal (TE)-Group IIa alloy systems (Ausleos and Elliot, 1983). There are several classes of alloy systems that have been shown to have commercial potential. The first of these, *metal–metalloid amorphous systems* include simple metal, early transition metal and late transition metal–metalloid systems. The second class is the *RETM amorphous systems,* and the third class is the *late transition metal (TL)–early transition metal (TE) systems*. In this section, we discuss each of these classes as well as a few interesting multicomponent systems.

19.3.9.8.1 Metal–Metalloid Systems

Metal–metalloid systems include early (TE) or late (TL) transition metals along with *metalloids* (M = C, B, P, Si, etc.). Eutectic compositions are found near 20–30at% M in typical TL-M systems. The eutectic alloy composition is often bracketed by a solid solution and an intermetallic alloy with composition richer in M. There may also be other M-rich high-temperature phases and/or metastable intermetallic phases.

TL-M systems are among the important amorphous systems for magnetic applications. In these alloys, considerations of GFAs and the desire to maintain high early transition metal (typically Fe or Co and sometimes Ni) concentrations are paramount. A series of eutectics that form near the Fe- and Co-rich edges of TL-M binary phase diagrams are summarized in **Table 7**.

Table 7   Glass forming ability parameters (eutectic composition and temperature, and solubility) in binary TL-M systems (McHenry et al., 1999)

Binary alloy	x_e (at%)	x_e (wt%)	T_e (° C)	Solubility of X at 600 (° C) (at%)	Terminal phases
Fe–B	17	3.8	1174	0	Fe, Fe_2B
Co–B	18.5	4.0	1110	0	Co, Co_2B, (Co_3B)
Cr–C	14.0	3.6	1530	0	Co, C_6Cr_{23}
Fe–P	17	10.2	1048	1	Fe, Fe_3P
Co–P	19.9	11.5	1023	0	Co, Co_2P
Fe–Si	33	20	1200	10	Fe, β-Fe_2Si, (Fe_3Si)
Co–Si	23.1	12.8	1204	8	Co, α-Co_2Si, (Co_3Si)

19.3.9.8.2 Rare-Earth Transition Metal Systems

Amorphous *RETM systems* have been studied widely, in part because of their importance as *magneto-optic materials*. $Co_{80}Gd_{20}$ amorphous alloys were the first materials considered for magneto-optic recording. Chaudhari et al. (1973) discovered the phenomenon of perpendicular magnetic anisotropy (PMA) in amorphous GdCo films. In a presumably isotropic amorphous material, this anisotropy was puzzling. Atomic structure anisotropy (ASA) was proposed as a source of the large PMA. This anisotropy results from preferential ordering of atomic pairs in the amorphous materials. Another model (Gambino and Cuomo, 1978) proposed selective resputtering to explain the ASA. In 1992, Harris and Sachidanandam (1985) employed the polarization properties of EXAFS to measure and describe the anisotropic atomic structure and relate it to the amplitude of the PMA in amorphous TbFe. They reported a direct measure of the ASA in a series of amorphous TbFe films and correlated it with the growth conditions and the magnetic anisotropy energy (Harris and Pokhil, 2001).

19.3.9.8.3 Early Transition Metal–Late Transition Metal Systems

Early transition metal–late transition metal binary alloy systems can have eutectics on both the TE- and TL-rich sides of the phase diagram. Of technological importance for magnetic applications are the TL-rich eutectics, since Fe, Co and Ni, the ferromagnetic transition metals, are TL species. The TL-rich eutectics are of interest in that they typically occur at 8–20 at% of the TE species. These alloys do not have as deep a eutectic, making them harder to synthesize, but they do have larger T_{x1} temperatures, making the resulting amorphous alloys more stable.

Eutectics forming near the Fe- and Co-rich edges of TL–TE binary phase diagrams are summarized in **Table 8**. Terminal alloy compounds and other phases in proximity to the eutectic in these systems include Fe_2Zr, Fe_2Hf, Co_2Zr and Co_2Hf phases, which all have the cubic $MgCu_2$ Laves phase structures. The Fe_2Ta, Co_2Ta, Fe_2Nb and λ-Fe_7Hf_3 phases have the hexagonal $Mg\,Zn_2$ Laves phase structures. Other compounds include Fe_3Zr, δ-Co_4Zr and $Co_{23}Hf_6$ (with the cubic Th_6Mn_{23} structure).

19.4   Current and Emerging Areas

19.4.1   Nanocomposite Magnets

Magnetic properties stemming from chemical and structural variations on a nanoscale are among those most significantly impacted by amorphous metals. Recent research has focused on *nanocomposite* and *bulk amorphous alloys* for many applications. A *metal/amorphous nanocomposite* is produced by

Table 8   Glass forming ability parameters (eutectic composition and temperature, and solubility) in binary TL-TE systems (McHenry et al., 1999)

Binary alloy	x_e (at%)	x_e (wt%)	T_e (°C)	Solubility of X at 600 (°C) (at%)	Terminal phases
Fe–Zr	9.8	15.1	1337	0	Fe, Fe_3Zr
Co–Zr	9.5	14	1232	0	Co, γ-Co_5Zr, (δ-Co_4Zr)
Fe–Hf	7.9	21.9	1390	0	Fe, λ-Fe_7Hf_3
Co–Hf	11	27.2	1230	0.5	Co, Co_7Hf_2, $Co_{23}Hf_6$
Fe–Nb	12.1	18.6	1373	0	Fe, Fe_2Nb
Co–Nb	13.9	20.3	1237	0.5	Co, Co_3Nb
Fe–Ta	7.9	21.7	1442	0	Fe, Fe_2Ta
Co–Ta	13.5	32.4	1276	3	Co, Co_2Ta

nanocrystallization of a multicomponent amorphous precursor, to yield a nanocrystalline phase embedded in an amorphous matrix. Important nanocomposites have nanocrystalline grains of a (*bcc*, **D0$_3$** or CsCl) Fe(Co)X phase, consuming 20–90% by volume in an amorphous matrix. A typical nanocomposite microstructure (Willard, 2000) is illustrated in **Figure 60**. **Table 9** lists the magnetic nanocomposite and bulk amorphous systems. Magnetic applications also exist for *bulk amorphous alloys*, which can often be synthesized at cooling rates as low as 1 K/s.

Understanding *primary nanocrystallization* is important in developing nanocomposites (McHenry et al., 2003). Fe-based metallic glass crystallization is the most widely studied process (Luborsky, 1977). Typical commercial Fe-based metallic glass alloys are *hypoeutectic* (Fe-rich), and are observed to crystallize in a two-step process. A primary crystallization reaction (Am $\rightarrow$ Am$'$ + α-Fe) is followed by secondary crystallization of the glass former enriched amorphous phase, Am$'$ (Koster and Herold, 1981). Luborsky and Lieberman (1978) studied the crystallization kinetics of Fe_xB_{1-x} alloys using *DSC*. Activation energies for crystallization were determined to be the largest for the eutectic compositions.

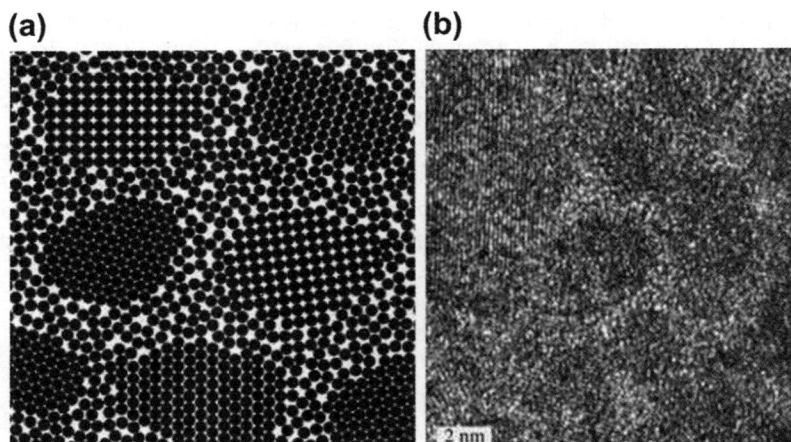

Figure 60   (a) Cartoon of a nanocomposite with nanocrystals embedded in an amorphous matrix. (b) High-resolution TEM image of a nanocrystallized HITPERM material (Willard et al., 1998).

Table 9   Examples of nanocrystalline and bulk amorphous alloy systems

Alloy composition	Citation	Alloy composition	Citation
Fe–Si–B–Nb–Cu	(Yoshizawa et al., 1988)	Fe–Si–B–Nb–Au	(Kataoka et al., 1989)
Fe–Si–B–V–Cu	(Sawa and Takahashi, 1990)	Fe–(Zr,Hf)–B	(Suzuki et al., 1990)
Fe–(Ti, Zr, Hf, Nb, Ta)–B–Cu	(Suzuki et al., 1991c)	Fe–Si–B–(Nb, Ta, Mo, W, Cr, V)–Cu	(Yoshizawa and Yamauchi, 1991)
Fe–P–C–(Mo, Ge)–Cu	(Fujii et al., 1991)	Fe–Ge–B–Nb–Cu	(Yoshizawa et al., 1992)
Fe–Si–B–(Al, P, Ga, Ge)–Nb–Cu	(Yoshizawa et al., 1992)	Fe–Zr–B–Ag	(Kim et al., 1993)
Fe–Al–Si–Nb–B	(Watanabe et al., 1993)	Fe–Al–Si–Ni–Zr–B	(Chou et al., 1993)
Fe–Si–B–Nb–Ga	(Tomida, 1994)	Fe–Si–B–U–Cu	(Sovák et al., 1995)
Fe–Si–B–Nd–Cu	(Muller et al., 1996)	Fe–Si–P–C–Mo–Cu	(Liu et al., 1996)
Fe–Zr–B–(Al, Si)	(Inoue and Gook, 1996)	Fe–Ni–Zr–B	(Kim et al., 1996)
Fe–Co–Nb–B	(Kraus et al., 1997)	Fe–Ni–Co–Zr–B	(Inoue et al., 1997)
Fe–Co–Zr–B–Cu	(Willard et al., 1998)		

Amorphous alloy precursors to nanocomposites are typically based on ternary (or higher order) systems. These are often variants of TL/TE/M systems. In many cases, a small amount of a fourth element (making the alloys *quaternary*) such as Cu, Ag, or Au can be added to promote nucleation of the nanocrystalline phase. Five- and six-component systems are also commonplace if more than one metalloid and/or early transition metal species are used as glass formers. A matrix of typical elements in many amorphous phases is given by (Willard, 2000)

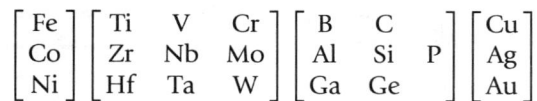

$$\begin{bmatrix} Fe \\ Co \\ Ni \end{bmatrix} \begin{bmatrix} Ti & V & Cr \\ Zr & Nb & Mo \\ Hf & Ta & W \end{bmatrix} \begin{bmatrix} B & C \\ Al & Si & P \\ Ga & Ge \end{bmatrix} \begin{bmatrix} Cu \\ Ag \\ Au \end{bmatrix}$$

Magnetic metal/amorphous nanocomposites (McHenry et al., 1999) have excellent soft magnetic properties as measured by the figures of merit of combined magnetic induction and permeability, high-frequency magnetic response, and retention of magnetic softness at elevated temperatures. Applications have been identified for the patented Fe–Si–B–Nb–Cu alloys (tradename FINEMET) (Yoshizawa et al., 1988) and Fe MBCu alloys (tradename NANOPERM) (Suzuki et al., 1990). Another nanocomposite $(Fe_{1-x}Co_x)_{88}M_7B_4Cu$ (M = Nb, Zr, Hf) soft magnetic material is known as HITPERM (Willard et al., 1998); HITPERM has a superior high-temperature magnetic induction.

Two routes are possible for inducing anisotropy in nanocrystalline alloys. The first is *field annealing*, where a previously crystallized alloy is annealed in an applied field at a temperature below the secondary crystallization temperature. In the second, called *field crystallization*, the amorphous precursor is field annealed and the nanocrystalline grains are precipitated in the presence of the field. In addition to pair ordering, the field crystallization technique holds potential for altering the crystalline texture of the precipitated grains. This would allow magnetocrystalline anisotropy as well as induced anisotropy to be controlled.

19.4.1.1 *Crystallization of Metallic Glasses*

We categorize *crystallization reactions* from a parent amorphous phase. In analogy with nondiffusional transformations in solids, the discussion of crystallization in amorphous alloys and *micromechanisms of crystallization* follows the work of Koster and Herold (1981). We illustrate transition metal–metalloid metallic glasses. We can, however, use this to discuss crystallization of other amorphous alloys. We use the Fe–B-phase diagram (**Figure 61**) as an example.

Consider a chemically homogeneous amorphous solid. Phase transformations can be diffusional or nondiffusional. Nondiffusional phase transformations can occur where the parent phase and final phase are both amorphous, i.e.

$$\beta(\text{amorphous}) \rightarrow \alpha(\text{amorphous}) \tag{82a}$$

An example is a pseudo-*MPTs* in metallic glasses (Corb and O'Handley, 1985) where change in local short-range order from *fcc*- to *hcp*-like is inferred from thermal hysteresis in magnetic anisotropy.

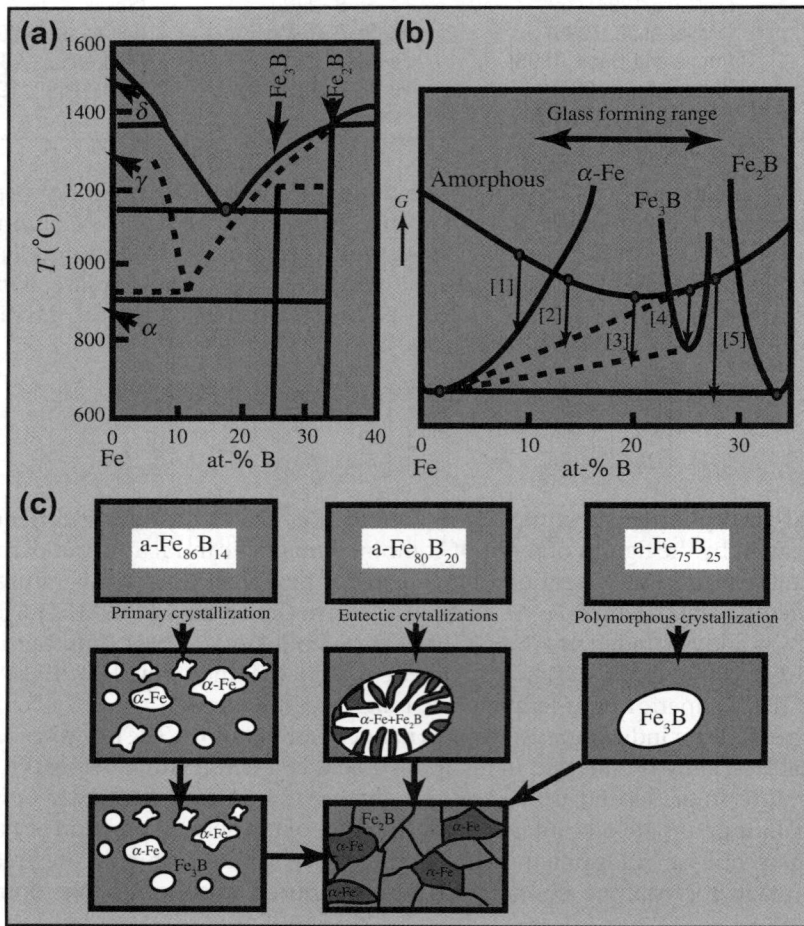

Figure 61   (a) Fe–B phase diagram, (b) $G(X)$ curves and crystallization reactions and (c) microstructure development during crystallization or other decomposition reactions. (For color version of this figure, the reader is referred to the online version of this book.)

Diffusional decomposition reactions are reported where the parent phase and final phase are both amorphous, i.e.

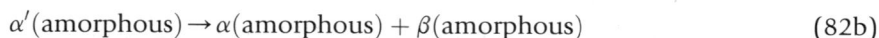

$$\alpha'(\text{amorphous}) \rightarrow \alpha(\text{amorphous}) + \beta(\text{amorphous}) \tag{82b}$$

This can occur by *nucleation and growth* or other mechanisms. An example is amorphous phase *spinodal decomposition* that is observed in Zr–Ti–Be metallic glasses.

Crystallization reactions require the parent phase to be amorphous and one or more of the final phases crystalline. The *polymorphic crystallization* is a nondiffusional phase transformation where an amorphous parent phase crystallizes to a final phase of the same composition according to the reaction:

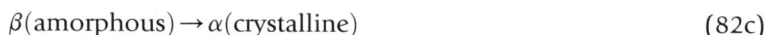

$$\beta(\text{amorphous}) \rightarrow \alpha(\text{crystalline}) \tag{82c}$$

The stable crystalline phase could be a supersaturated alloy or a metastable or stable crystalline compound. Polymorphic crystallization is not a common alloy primary crystallization route as they requires compositions that are close to that of either the pure element or compounds. Further decomposition of the crystalline phases, e.g. precipitation form the supersaturated solid, is the possible route to equilibrium structures.

Crystallization reactions that are diffusional can be classified as *primary crystallization* or *eutectoid crystallization*. Primary crystallization is a diffusional phase transformation where the parent phase is amorphous and the final phases are a crystalline phase and an amorphous phase depleted of the primary component of the crystalline phase. Examples in a simple A–B eutectic system would be:

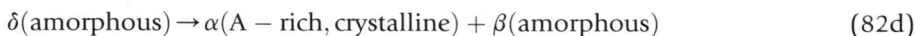

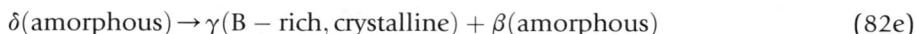

$$\delta(\text{amorphous}) \rightarrow \alpha(\text{A} - \text{rich, crystalline}) + \beta(\text{amorphous}) \tag{82d}$$

$$\delta(\text{amorphous}) \rightarrow \gamma(\text{B} - \text{rich, crystalline}) + \beta(\text{amorphous}) \tag{82e}$$

Equilibrium is achieved through subsequent secondary crystallization of the new amorphous phase.

A diffusional phase transformation where the parent phase is amorphous and the final phases are both crystalline phases and an amorphous phase depleted of the primary component of the crystalline phase is called *eutectoid crystallization*. A typical eutectoid reaction is given by

$$\gamma(\text{amorphous}) \rightarrow \alpha(\text{A} - \text{rich, crystalline}) + \beta(\text{B} - \text{rich, crystalline}) \tag{82f}$$

Eutectoid crystallization is analogous to eutectic crystallization, in that it requires mutual diffusion between the A-rich α- and B-rich β-phases. **Figure 61(a)** illustrates a binary Fe–B phase diagram on the Fe-rich end of the diagram. **Figure 61(b)** illustrates hypothetical free energy curves for crystalline and amorphous phases in the Fe–B alloy system. Here equilibrium crystalline phases on the iron-rich side of the phase diagram include α-iron, and the compounds Fe_3B and Fe_2B (analysis can be made more complicated by considering the possibility of metastable γ-iron, and Fe_2B).

For amorphous alloys of various compositions, [1] to [5] *crystallization reactions* can be described. Reaction [1] is an example of *polymorphic crystallization* of α-iron. Reaction [4] is an example of *polymorphic crystallization* of Fe_3B. Reaction [2] represents the *primary crystallization* of α-iron. Reaction [3] represents the simultaneous *eutectoid crystallization* of α-iron and Fe_3B. Reaction [5] represents the simultaneous *eutectoid crystallization* of α-iron and Fe_2B. These last two reactions are examples of *discontinuous reactions*. As can be seen in **Figure 61(b)**, these reactions have the largest driving force and are therefore thermodynamically favored. On the other hand, they require two components to separate into two new phases and therefore should take more time than *polymorphic reactions*, where no diffusion is required, or even *primary crystallization* where solute is expelled to an existing amorphous phase which becomes enriched in glass formers.

The replacement of B by Si increases the activation energy barriers in Fe-based metallic glasses (Ramanan and Fish, 1982). The primary crystallization temperature, T_{x1}, has been studied as a function of transition metal substitution, X, in $M_{78-x}X_xSi_{10}B_{12}$ compounds (Donald and Davies, 1978). Variations in T_{x1} are explained using the *Hume–Rothery rules*, correlating T_{x1} with the cohesive energies of the pure X species and the atomic size. Two important observations were that (1) Cu additions, which promote the nucleation of the primary nanocrystals by clustering, resulted in significant reductions in T_{x1}, with additions as small as 0.5–1.0 at%; and (2) early transition metal (e.g., Zr, Hf, and Mo) additions impede growth and result in the largest primary crystallization temperatures.

Primary crystallization is important in developing nanocomposite microstructures, as microstructures can be stabilized with nanocrystals embedded in an amorphous matrix that is more stable than the parent phase. Increases in magnetization often accompany primary crystallization which can be probed in thermomagnetic measurements [Johnson et al., 2001]. Inhomogeneity in the composition of the new amorphous phase after primary crystallization results from the build-up of early transition metal species near the nanocrystal amorphous phase interface which can be *probed by atomic probe field ion microscopy* (Ping et al., 2001; Ohodnicki et al., 2009b). Nanocrystallization has been analyzed using Johnson–Mehl–Avrami nucleation and growth kinetics with morphology indices identifying the micromechanisms (Hsiao et al., 1999; Hsiao et al., 2002).

It is only through secondary or higher order crystallization reactions that the amorphous phase is completely consumed. Interesting primary and secondary crystallization reactions often yield metastable crystalline phases in pathways toward eventual equilibration. Secondary crystallization can often have sluggish kinetics because of the complicated crystal structures of the product phase. Of particular recent interest has been in phases with stoichiometries $Fe_{23}B_6$ (Zhang et al., 2003; Long et al., 2007a) and $Fe_{23}Zr_6$ and their stabilities (Ohodnicki et al., 2008a).

19.4.1.2 *Figures of Merit for Nanocomposite Magnets*

Figure 62 shows the figures of merit for soft magnets (McHenry et al., 1999), the saturation induction and permeability. Permeability is inversely proportional to coercivity, H_c. Improving soft magnetic properties requires tailoring chemistry microstructural optimization, recognizing that H_c decreases with grain size (D_g) for $D_g > \sim 0.1$–1 μm, where D_g exceeds the domain wall width (δw). Grain boundaries impede wall motion so Hiperco, etc. are widely used in inductive applications due to high flux density, high permeability and low core losses.

Coercivity mechanisms (McHenry et al., 1999) indicate that for $D_g > \sim 100$ nm, H_c decreases rapidly with decreasing D_g (**Figure 58**). It is understood that when the domain wall thickness is large, fluctuations in magnetic anisotropy on the scale of D_g are irrelevant to wall pinning. FeCo-based nanocomposites (HITPERM) (McHenry et al., 1999; Willard et al., 1998) exhibit excellent soft magnetic properties through exploiting this Herzer model (Herzer, 1997). Alloys near equiatomic Fe–Co ratios are standard bearers for high induction with grains much smaller than domains (Long et al., 2007b). The potential for nanocomposites alloys to push the envelope for high-f magnetic components, increasing power densities and allowing smaller volume components is pushing recent soft magnet development.

The *bcc*-derivative nanocrystals embedded in an amorphous matrix have high saturation induction and low core losses unobtainable in amorphous or large grained crystalline alloys. Increased glass former concentration in the intergranular amorphous phase impedes diffusion, prevents grain

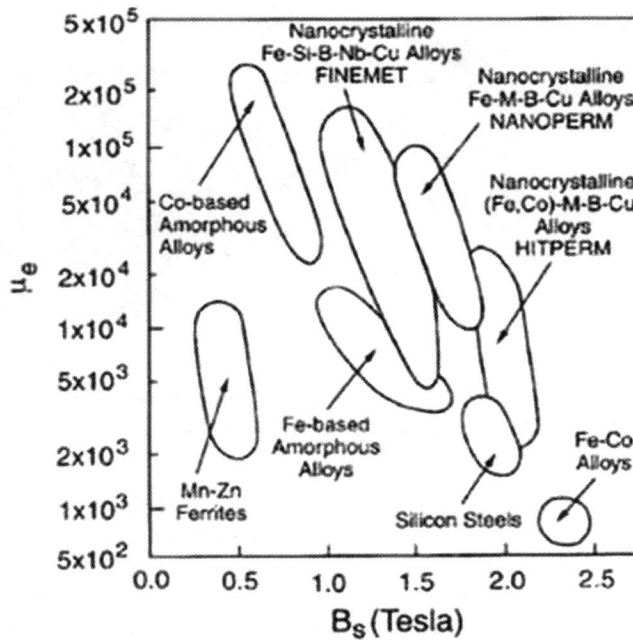

Figure 62 Figures of merit for soft magnets, saturation induction and permeability (McHenry et al., 1999).

coarsening and allows nanostructure retention (McHenry et al., 1999). HITPerm alloys are of particular interest for high-temperature applications.

19.4.1.3 Applications of Nanocomposite Magnets

There is an increasing demand for miniaturization and higher power density in electronics operating at high frequency and high temperature. It is desirable that magnetic materials have improved properties such as low core loss, high-saturation magnetic flux density, high Curie temperature, T_c, and linear magnetization as a function of field at high frequencies. Previously mentioned soft magnets do not satisfy these new requirements because of high core losses at high frequencies, relatively low-saturation induction, and/or relatively low-temperature stability of magnetic properties. Amorphous alloys have properties desirable in applications such as pulse transformers and high-frequency inductors but are generally not stable at elevated temperatures.

Materials for power electronic applications must compete with Si steels and their high-saturation flux density ($B_s \sim 2$ T). Si steels, prevalent in 60 Hz transformers, have high losses at frequencies >1 kHz. Applications requiring operation at higher frequencies have turned to low loss materials such as FINEMET. FINEMET, however, has a relatively low $B_s \sim 1.24$ T. Nanocomposites can provide high flux density approaching that of Si steels, tunable permeabilities, and low losses (Yoshizawa et al., 1988; Willard et al., 1998; Iwanabe et al., 1999) comparable to ferrites due to unique nanostructures with higher B_s. Fe-based nanocomposites can have large inductions but relatively low T_cs (Suzuki et al., 1990; Suzuki et al., 1991a, 1991b, 1991c). Co additions to Fe-based nanocomposites increase flux density and improve magnetic properties at elevated temperatures.

(a)   **(b)**

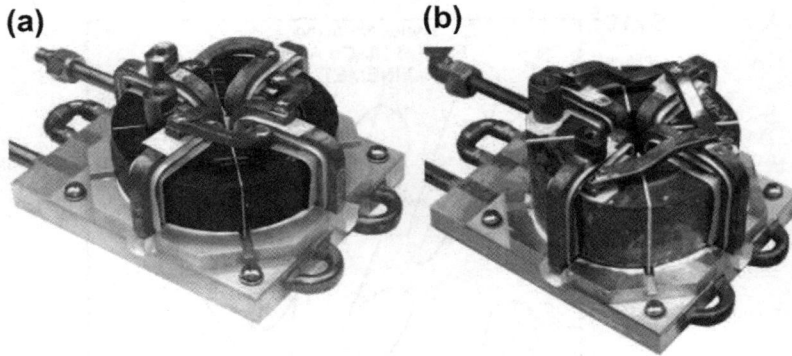

Figure 63   Size comparison of a (4.23 in. OD) converter constructed from (a) HITPERM and (b) FINEMET (Long et al., 2008). (For color version of this figure, the reader is referred to the online version of this book.)

A promising HITPERM alloy has been reported that exhibited low losses and applicability in a dc–dc converter (Long, 2008). Recent developments have shown that the early transition (TE) metal content could be reduced without significantly increasing the core loss to achieve higher flux B_s (Miller et al., 2010b). Other alloys, with P additions, have completely eliminated TE elements in order to compete with Si steels (Makino et al., 2009) but have been targeted for lower frequency applications.

New materials with high B_s and low losses at higher frequencies can reduce inductor size in high-frequency applications increasing efficiency through weight reduction for dc–dc converters in electric vehicles (Long et al., 2008). A HITPERM nanocrystalline alloy was developed for high-power inductors for Army electric vehicle applications (Long et al., 2008). A core wound (**Figure 63**) was used to construct a 25 μH inductor for use in a 25 kW DC–DC converter with a reduction in the overall size (by ~30%) compared with FINEMET. The nanocomposite's ability to operate at 20 kHz as compared with 1 kHz offers 10-fold or more size reduction compared to Si steel.

Devices using high-induction nanocomposite cores at 20–100 kHz with significant core size reductions can translate directly to cost savings (Nikolov and Valchev, 2009; Long et al., 2008; Reass, 2008). In addition to power converters, this can impact inverters and power transformers. A high-frequency, high-power transformer, using an Fe–Si-based (FINEMET) nanocomposite, constructed demonstrated dramatic reductions in weight (300×) and electrical losses (1000×) as compared to Si-steel transformers (Reass, 2008).

State-of-the-art Fe-rich alloys have losses of 500 kW/m^3 (100 kHz, 0.2 T) (Kernion et al., 2011) comparing favorably to Mn–Zn ferrites (500 kW/m^3, 100 kHz, 0.2 T). Benefits of cold- and hot rolling to reduce ribbon cross-sections may result in increases in operating frequency. An issue for long-term high-temperature material stability is polymorphic secondary crystallization (Kernion et al., 2011) at reduced glass former concentrations.

Co-rich nanocomposite materials (Ohodnicki et al., 2008c, 2008d, 2009b) also have significant promise for high-frequency applications (Dweyedi et al., 2011). They have small magnetostrictions, strong field annealing response (Ohodnicki et al., 2008b, 2009a) (**Figure 64(a)**) and superior mechanical properties. They also have promise for sensing applications using the Giant Magneto-impedance (GMI) effect (Chaturvedi et al., 2011; Laurita et al., 2011) (**Figure 64(b)**).

GMI is a large variation of the impedance of a magnetic conductor in a field. GMI is observed in soft magnets with large electrical conductivities and is the largest in materials where the thickness is

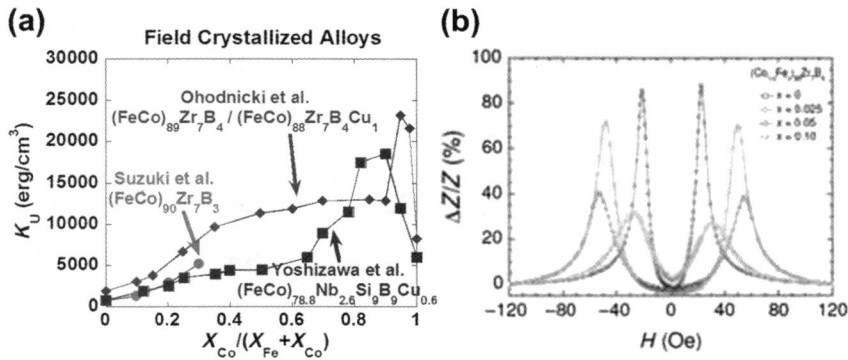

Figure 64 (a) Field-induced anisotropy as a function of Fe, Co-ratio for HITPERM nanocomposites (Ohodnicki et al., 2008b) and (b) GMI as a function of field for Co-rich alloys (Chaturvedi et al., 2011). (For color version of this figure, the reader is referred to the online version of this book.)

comparable to the skin depth. Most attention has been paid to improving GMI effect by tuning alloy compositions or thermal annealing in an applied magnetic field. GMI can also be changed by sample shape.

19.4.2 Magnetic Nanoparticles and Core Shell Structures

Nanomaterials research impacts many diverse applications (McHenry and Laughlin, 2000). Magnetic properties are distinctly different at the nanoscale because many magnetic length scales are on the order of 10–100 nm. Engineering magnetic nanostructures allows for tailoring magnetic properties. MNPs have therefore been studied for a variety of applications over the last decades. Many new applications of MNPs require high-frequency switching and power absorption of materials that have oxidation stability. This is achieved in some interesting MNP systems having metallic magnet cores with thin adherent protective oxide shells (Turgut et al., 1999b). Not only the thin shells passivate the particles but also the oxide shells are more easily functionalized. MNPs amenable to functionalization for synthesis of aqueous magnetic fluids (ferrofluids) make them attractive for biomedical applications and incorporation into polymer nanocomposites.

Synthesis of MNPs is performed in the solid, liquid or gaseous states. Solid-state synthesis begins with bulk solids (often micron-sized powder precursors) and transforming them to nanostructures by mechanical milling and thermal processing. Other approaches include plasma torch synthesis, chemical vapor deposition, molecular beam epitaxy, focused ion beam milling, and lithography. These methods allow synthesis of large quantities of material but often with polydisperse size distributions.

Plasma torch synthesis has been used to produce large quantities of MNPs. Plasma torch synthesis is a high-throughput method for MNP synthesis. Thermal plasma processing (Turgut, 1999) has been used to produce a variety of MNP chemistries:

(1) Alloy nanopowders (Fe–Co and Fe–CoV) (Turgut et al., 1997, 1999b; Turgut, 1999).
(2) Carbon-coated nanocrystalline powders (Fe–Co and Fe–Co–V) nanopowders using acetylene as an auxiliary gas source (Turgut, 1999).
(3) Nitrogen martensite (γ-FeN$_x$) nanoparticles using nitrogen/ammonia as the auxiliary gas source (Turgut et al., 1999a).

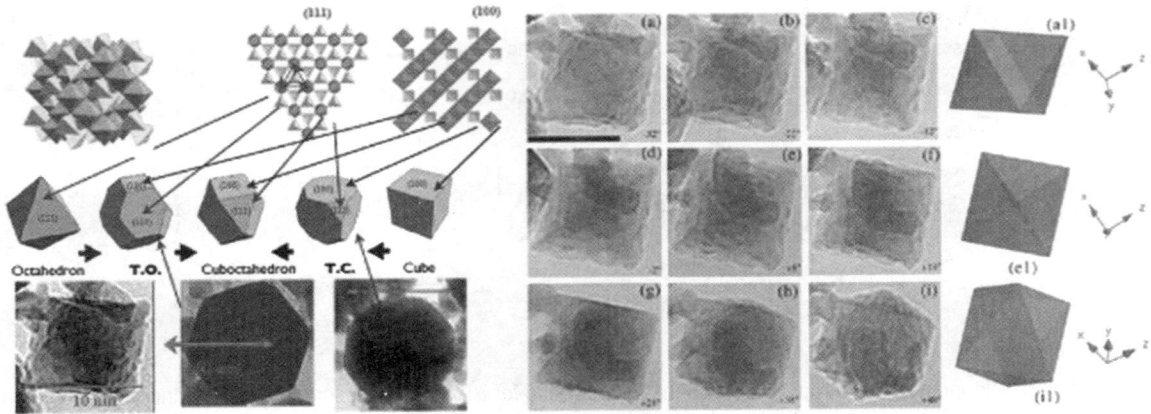

Figure 65   Left: (top) Coordination polyhedral in spinel (Blue = B-site cation, Yellow = A-site cation); (middle) cuboctahedral shapes and (bottom) TEM observations of nanoparticles (Swaminathan, 2005). Right: 2D projected morphologies [(a)–(i)] at different tilt angles for small octahedral nanoparticle. (a1), (e1) and (i1) are projections of an octahedron. Scale bar on (a) is 20 nm. (For color version of this figure, the reader is referred to the online version of this book.)

(4) Oxide-coated (CoFe$_2$O$_4$–FeCo) core shell nanoparticles (Turgut et al., 1999b).
(5) Mixed ferrites (Fe$_3$O$_4$, MnFe$_2$O4, (MnZn)Fe$_2$O$_4$, NiFe$_{2a}$nd (NiZn)Fe$_2$O$_4$ for high-frequency magnetic applications) (Son et al., 2002, 2003; Swaminathan et al., 2005).

The crystallography of terminating faces determines the symmetry of atomic environments. In oxides, as the polyhedral environments at the surface differ from the bulk, so do the cationic crystal fields that determine magnetocrystalline anisotropy and ferrimagnetic superexchange interactions. These changes increased relative importance in MNPs and can influence the magnetic properties and surface activity. **Figure 65** illustrates the surface chemistry and facets of some ferrite nanoparticles.

Chemical synthetic approaches involve assembling atoms, molecules and particles to produce materials at a nano- or macroscopic scale (McNerny, 1999). These methods are preferred for monodisperse particles to exploit size-dependent properties. Monodisperse particles are uniform in size, shape and internal structure and have a *coefficient of variation* (standard deviation/mean size) less than about 10%. Chemical synthesis in gas and liquid states benefits from diffusion coefficients which are orders of magnitude larger than in the solid state, making chemical homogenization much quicker. Liquid-phase chemical synthesis has been conducted in a range of aqueous or nonaqueous solvents. Dissolved metal precursors form thermodynamically unstable supersaturated solutions, from which homogeneous (in solution) or heterogeneous (on vessel walls or impurities) nucleation of MNPs takes place. The LaMer process (**Figure 66**) is a common technique for chemical synthesis of monodisperse MNPs (McNerny et al., 2010; McNerny, 1999).

19.4.2.1   *Properties of Magnetic Nanoparticles*

Faceted MNPs with core shell nanostructures (Collier et al., 2009) are important for several applications discussed below. For biomedical applications developing chemistries to functionalize with antibodies to promote attachment of the MNPs to body tissues is also desirable. This requires knowledge of (1) synthesis of core–shell nanoparticles; (2) crystallographic facets of MNPs; (3) orientation relationships

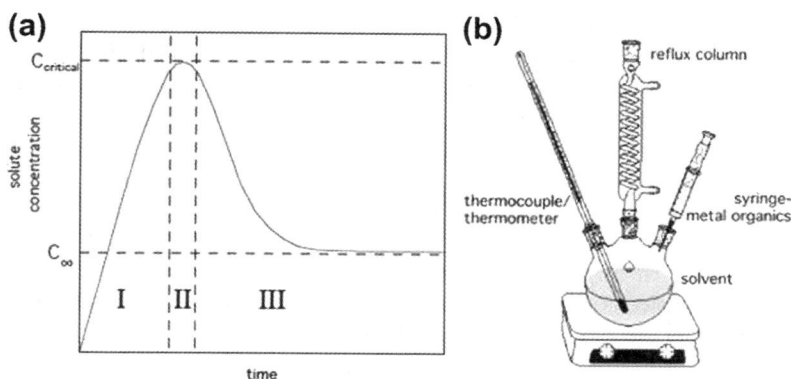

Figure 66 (a) Schematic of the La Mer process (a) (I) prenucleation, (II) nucleation, and (III) growth stages of chemical synthesis and (b) chemical synthesis apparatus (McNerny, 1999).

oxide shells and metallic cores; (4) properties of the nanocomposites and (5) performance in RF heating and magnetoelastic applications. Applications include the efficient point source heating of metallic nanoparticles (Habib et al., 2008; Ondeck et al., 2009; Sawyer et al., 2009) for thermoablative cancer therapies, curing polymers and tagging implanted tissue scaffolds to track degradation and changing the shape for regenerative medicine (Miller et al., 2009).

A magnetic property important in many applications is a large saturation magnetization. In particular in biomedical applications, concerns over toxicity suggest accomplishing functions with lower concentrations and with particles with passivated surface. The Slater–Pauling curve shows Fe–Co alloys to have the largest M_s (270 emu/g) of transition metal alloys, thus reducing the ferric content required to produce a comparable magnetic response to that of magnetite. Magnetite is also a commonly used magnetic material for applications with a reasonably high M_s (93 emu/g for bulk materials) and polar surfaces that are advantageous for functionalization (McNerny, 1999).

Important magnetic properties that impact biomedical applications include saturation magnetization, magnetic anisotropy and the exchange interactions. The magnetization of Fe–Co core shell MNPs is determined by the chemistry and the core/shell volume ratios. Prior work using C-arc methods produced MNPs demonstrating the compositional dependence of the magnetization in Fe_xCo_{1-x} C-coated MNPs (Gallagher et al., 1996). The largest moments were observed at nearly equiatomic compositions with deviation from Slater–Pauling predictions due to carbide formation. Recent work (Jones et al., 2010) has shown the decrease in magnetization of MNPs as a function of controlled consumption of the core by the oxide shell. FeCo exchange interactions are also large so that the variation of magnetization with temperature is weak near room temperature.

In ferrites, the strength and number of oxygen-mediated J_{AB} and J_{BB} superexchange bonds determines the temperature dependence of the magnetization and the collinear or noncollinear (Yafet and Kittel, 1952) alignment of cation dipoles. The symmetry of surface polyhedra also determines surface magnetic anisotropy (Jones et al., 2010). In ferrite MNPs and ferrite shells on FeCo, surface and interface anisotropy can be an important means of tuning overall MNP magnetic anisotropies. The overall magnetic anisotropy of FeCo/ferrite MNPs can be tuned by choosing the core composition.

Thus, the core/shell radius ratio can be used as a means of controlling the total magnetic anisotropy.

19.4.2.2 *Applications of Nanoparticles in Biomedicine*

An increasing interest in using MNPs for biological and medical applications has developed (Pankhurst et al., 2003). Researcher faces new challenges and opportunities for using MNPs in biomedical applications. A highlight of current research is the challenges to functionalizing the materials to achieve biocompatibility for a host of applications. Ferromagnetic as well as superparamagnetic particles coated or encapsulated with polymers or liposomes are used for magnetic labeling.

Magnetite, Fe_3O_4, and hematite, Fe_2O_3, have been used most commonly because of their ionic surfaces which aid functionalization and the fact that they are chemically inert contributing to biocompatibility. Materials optimization for specific applications has demanded further efforts and consideration of a wider variety of materials where metals and alloys can play a role. Emerging techniques, therapies, and tools are made possible by developments in nanotechnology. Topics of current interest include the following:

(1) Magnetic resonance imaging (MRI) enhancement
    (a) Principles of nuclear magnetic resonance (NMR)
    (b) Materials for MRI enhancement
(2) Thermoablative cancer therapies—RF and microwave materials
(3) Magnetic beads for bioseparation, cell sorting and drug delivery
    (a) MNPs for bone marrow and blood detoxification
    (b) Magnetic bead tracking of blood flow and stem cells
    (c) Targeted drug delivery
(4) Magnetoelastic actuators, sensors and MNP cell tagging
(5) Nanocapsid templating of MNPs
(6) Magnetic microorganisms.

In this section, we review a few selected applications where metallic nanoparticles are playing a significant role. These include MRI enhancement, RF heating applications, cell tagging and magnetoelastic applications. Of interest to several applications of MNPs is the *Neel relaxation* of superparamagnetic nanoparticles in an AC exciting field that allows MNPs to act as point heat sources. The RF power losses can cause local heating useful in *magnetic hyperthermia*. When nanoparticles can be placed near target cells, they can be heated from a distance by RF fields generated in a coil assembly. The targeting of cancer cells need not be particularly selective because cancer cells typically die at temperatures $\sim 2\,°C$ lower than those of normal cells. Current issues in developing this concept further is the need for highly efficient magnetic nanocrystals, stable under in-vivo conditions for noninvasive cancer therapies. Thermoablative cancer therapy is rooted in the RF heating of ferrofluids discussed below.

19.4.2.2.1 *MRI Enhancement*

Neel relaxation of superparamagnetic MNPs in AC fields is used in *MRI*. Magnetostatic stray fields of MNPs affect proton relaxation providing enhancement of MRI contrast. Resonance phenomena include *electron paramagnetic resonance (EPR)*, *electron spin resonance (ESR)*, *FMR* and *NMR*. In all absorption of electromagnetic radiation accompanies excitation between quantum states. For EPR, the quantum states are the $2J$ + onefold degenerate ground state of a quantum paramagnet. For ESR, the quantum states are the $2S$ + onefold degenerate states of an atom for which $L = 0$. In FMR, exchange interactions influence the resonance phenomena. NMR is a resonance phenomena associated with nuclear dipole moments. Just as electrons have spin dipole moments, protons also have dipole moments. Excitation between states of the dipole moments of protons in the nucleus gives rise to NMR.

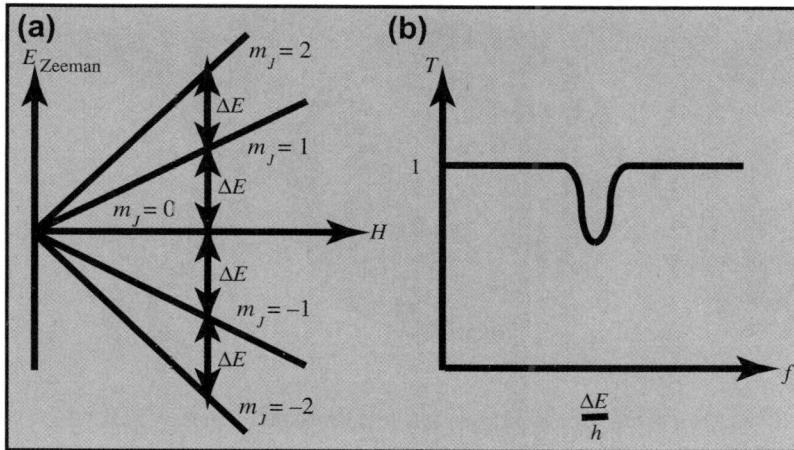

Figure 67 (a) Zeeman splitting of the $2J+1$-fold degenerate states vs applied field, H (b) resonant absorption seen in reduced transmission of EM radiation of the resonant frequency. (For color version of this figure, the reader is referred to the online version of this book.)

The *Zeeman splitting* lifts the degeneracy of magnetic states and increases linearly with applied field (**Figure 67(a)**). Atoms of interest exhibit a transmission coefficient for EM radiation that use a monotonic function of frequency, f. At a resonant frequency $f = \frac{\Delta E}{h} \left(\omega = \frac{\Delta E}{\hbar}\right)$, EM radiation is strongly absorbed and the transmission coefficient diminished. Spatially resolved transmission in NMR is used as an imaging tool.

MRI is improved by using contrast agents that increase sensitivity or specificity of MRI processes. A difference in proton density and their relaxation in biological environments provide the mechanism for contrast. Contrast agents include superparamagnetic macromolecular compounds, superparamagnetic iron oxide and rare-earth metal ion (Gd) complexes (Weinmann et al., 1984). Paramagnetic metal ions reduce the T_1 relaxation of protons in water. This enhances the intensity of the NMR signal making images brighter. The use of *superparamagnetic iron nanoparticles* have been shown to be more effective than Gd contrast agents.

19.4.2.2.2 Magnetic Beads for Bioseparation, Cell Sorting and Drug Delivery

Bioseparation is the isolation of biological substances including molecules, cells and cell components, and organisms. Magnetic bioseparation may be accomplished by two routes. First, the species to be separated may have a large enough magnetic dipole moment (e.g. red blood cells) so as to be directly separated by a magnetic field. Second, nonmagnetic cells or biomolecules are modified by attaching MNPs so that they can be manipulated by an external field. In *magnetic bioseparation*, a substance that is desired to be isolated is bound to an MNP through suitable surface chemistry. The material responds to a magnetic field used to move the substance.

Magnetic particles of micron-size dimensions, called *magnetic beads*, can be embedded in other bioactive materials. They serve as a bioactive agent and can be localized or retrieved with a magnet. This can shorten purification steps. Using magnetic bioseparation decreases times needed for target substance recovery. Magnetic separation technologies have also been used for environmental remediation in removing oil spills from waterways. In many bioseparation applications, ideal particles are

(a)   **(b)**   **(c)**

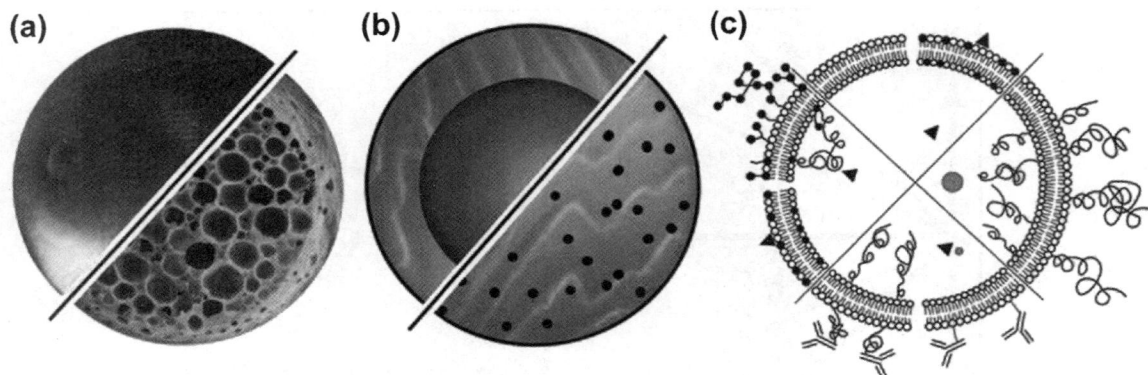

Figure 68   (a) Magnetic microspheres, (b) microcapsules and (c) liposomes. Courtesy: U. Hafeli.

superparamagnetic. Superparamagnetic particles are not hysteretic and do not have a remnant magnetization, so they return to zero magnetization after field removal. Typically, superparamagnetic particles are nanosized and micron-sized beads are composites.

Magnetic particles for biomagnetic applications typically fall into three classes. These are (1) *magnetic microspheres* and *magnetic nanospheres*, (2) *magnetic microcapsules* and (3) *magnetic liposomes* illustrated schematically in **Figure 68**. Magnetic bioseparation-based technologies have applications in mRNA extraction, separation of DNA from solutions or gels, and isolation of plasmid DNA. MNPs have also been used for bone marrow and blood detoxification; tracking of blood flow and stem cells and drug delivery. *Dynabeads*™ are monodisperse polymer beads with a uniform dispersion of super-paramagnetic γ-Fe$_2$O$_3$ or Fe$_3$O$_4$ coated with a thin polymer. Dynabeads have well-defined surfaces for the adsorption or coupling of bioreactive molecules and can be added to a suspension and bind to cells, nucleic acids, proteins or other biomolecules. The resulting target–bead complex is removed using a magnet.

19.4.2.2.3 Magnetoelastic Actuators, Sensors and MNP Cell Tagging

A *smart material* is one which is capable of two functions, e.g. a material capable of both actuation and sensing is an example of a smart material. Exciting new biomedical applications of smart materials include magnet–polymer composites which are being used in such applications as artificial muscle (Ramanujan and Lao, 2006). Attempts to synthesize artificial muscles have ranged from a robot-like metallic actuators to soft actuators. Hydrogels are water swollen cross-linked polymer networks. These are a promising materials class for producing soft actuators.

Since most gels are homogenous and shrink or swell uniformly without a dramatic change in shape, efforts have focused on means to promote larger anisotropic shape changes. Szabo et al. (1998) have reported magnetic field-sensitive gels where colloidal magnetic particles are dispersed in the gels. These are called *ferrogels* and constitute composites which combine the magnetic properties of the fillers and the elastic properties of hydrogels. In these materials, an instantaneous shape distortion is observed which disappears quickly when the external field is removed (**Figure 69**).

When ferrogels are placed into a spatially inhomogeneous magnetic field, forces act on the magnetic particles. Due to the strong mechanical coupling between the particles and polymer chains, the

(a)

(b)

(c)

Ferrogel

Electromagnet

(d)

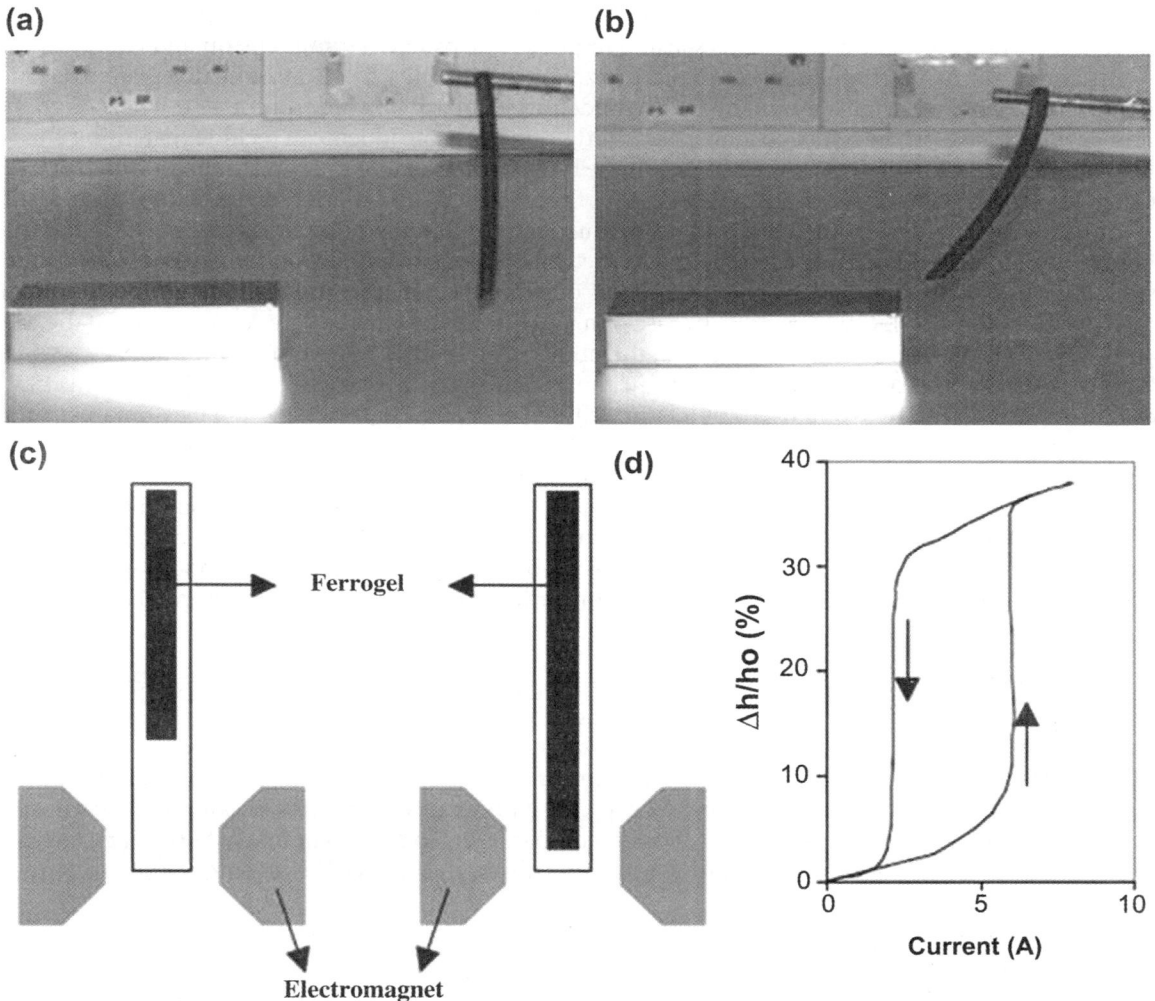

Figure 69   (a) Experimental manifestation of the response of a ferrogel (a) in no field and (b) in a field provided by a permanent magnet (Ramanujan and Lao, 2006). (c) Elongation of ferrogel in no magnetic field and in maximal magnetic field; and (d) elongation versus current density (hysteresis). (For color version of this figure, the reader is referred to the online version of this book.)

composite responds as a single entity. The coupling of the hydrogel and magnetic particles has applications in soft actuators.

A *superelastic polymer* is an elastomer, silicone, or gel embedded with micro- or nanomagnetic particles. A composite superelastic polymer is capable of responding to a magnetic field with a smooth muscle-like motion. The composite can also be classified as a *smart material* as it is capable of both actuation and sensing it own motion. For every concentration of magnetic particles in the composite, there is a threshold field strength at which the deflection increases rapidly. When the material is elongated, its resistivity is changed remarkably, so that the strain in the actuator can be sensed by the current passed at constant voltage. This sensing response is sensitive in both tension and compression.

An important tissue engineering problem is proving the viability of a scaffold delivery system. This requires determining details about scaffold remodeling such as degradation time and where degradation products travel in the body. The fate of scaffold degradation products can include remaining within local tissue or traveling to extreme locations in the body. It is desirable to monitor the in vivo behavior of scaffold remodeling through suitable tagging of the scaffold cells to monitor their fate over time.

Imaging with the use of MNPs has many advantages. MNP tracking also has the added benefit of being bioinert as well as simple to measure. Magnetic fields do not pose deleterious effects in biological systems. This bioinert property of magnetic flux is one of the chief advantages of MNPs tracking. MNP tracking has additional advantages due to ease of measurement. An immediate consequence of the ease with which magnetic information is read is the ability to miniaturize such detection devices. Typical hard disk drives employ materials that exhibit the so-called giant magnetoresistive (GMR) effect and can distinguish domains as small as $0.0065\ \mu m^2$ (Millen et al., 2005). In order to image domains of this size, the materials composing the sensor must also be on the same order of magnitude in size as well. This miniaturization is the foundation for an ultimate method of imaging.

Once a scaffold has been implanted for regeneration, an array of GMR sensors could be implanted near the scaffold. It has been shown theoretically that the use of a GMR sensor for detection of a MNP is feasible and is predicted to distinguish particles sizes of down to 10 nm in size (Tondra et al., 2000). A GMR array would allow adequate resolution for tracking MNPs as the scaffold degraded.

19.4.2.2.4 MNPs for Thermoablative Cancer Therapies

An important biomedical application of the RF heating of ferrofluids is in magnetic hyperthermia. The choice of the frequency of the alternating magnetic field is a design consideration. The power dissipation scales linearly (or as a higher power) of frequency. In biomedical applications, the frequency is often chosen to be noninvasive. This means that the EM radiation does not interact strongly with body fluids or tissues. RFs of 100–300 kHz have been chosen for cancer therapies because they are noninvasive but interact strongly with suitably chosen nanoparticles. To the extent that MNPs can be targeted to specific cells or if specific cells die at lower temperatures, this is a viable technique for cancer therapies, for example.

The expression for power dissipation has a H_0^2·dependence and therefore, the choice of field amplitude is also an important consideration for promoting dissipation. Typical fields are on the order of ~ 100 Oe. High-amplitude (~ 1000 Oe) AC fields have been used in recent therapies. **Figure 70** shows an example of an RF coil designed for a high-amplitude RF radiation source for inductive heating.

Most work on ferrofluids for thermoablative cancer therapies have focused on iron oxides, magnetite, Fe_3O_4, and maghemite, γ-Fe_2O_3. Calculations of Maenosono and Saita (2006) argue for the efficiency of Fe–Pt nanoparticles showing that the comparative heating rates (at noninvasive frequencies, ~ 300 kHz, AC field amplitudes of 50 mT and nanoparticle volume fractions $\phi = 0.1$) that exceed those of magnetite and maghemite. FeCo nanoparticles have significantly larger heating rates but at sizes that are argued to be too large to allow colloidal suspension. However, calculations considered only FeCo alloys with small magnetocrystalline anisotropy ($K = 1.5 \times 10^3$ J/m³). **Figure 71(a)** shows comparative heating rates as a function of particle size for the materials considered in Maenosono and Saita (2006). Important observations are (1) the material magnetization sets the scale for heating rate and (2) magnetic anisotropy (entering the expression for Neel relaxation) determines the particle diameter, D, where a maximum heating rate occurs.

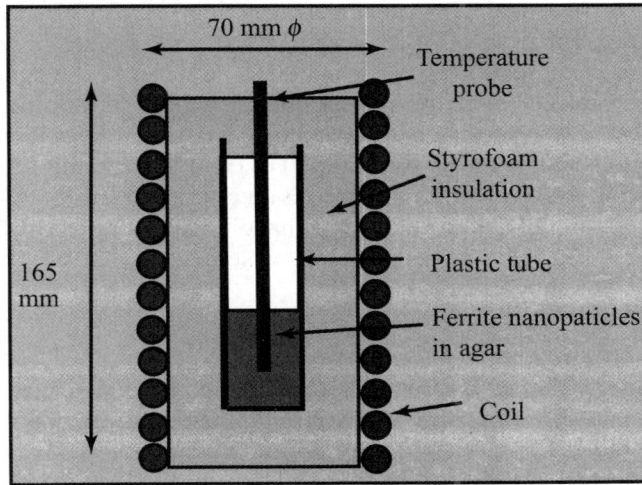

Figure 70 Schematic of an RF coil designed for inductive heating. (For color version of this figure, the reader is referred to the online version of this book.)

Figure 71(b) shows calculations for FeCo alloys with much larger Ks, achievable by varying composition (Habib et al., 2008). In $Fe_{1-x}Co_x$ ($x = 0.250.5$) alloys, K varies between 0 and 4×10^5 J/m^3, while the induction is relatively constant (2.2–2.5 T). These heating rates are twice those of the oxides or Fe–Pt and are peaked at smaller particle diameters. The materials are thus predicted to have superior heating rates at sizes appropriate for colloidal suspension. Particles that are too large settle due to gravitational forces.

Magnetic shape anisotropy in acicular Fe–Co particles offers another possible mechanism for stabilizing colloidal Fe–Co-based ferrofluids. Fe–Co-based materials that exploit shape anisotropy were discussed earlier in the context of Alnico magnets. Because of their large magnetization, shape

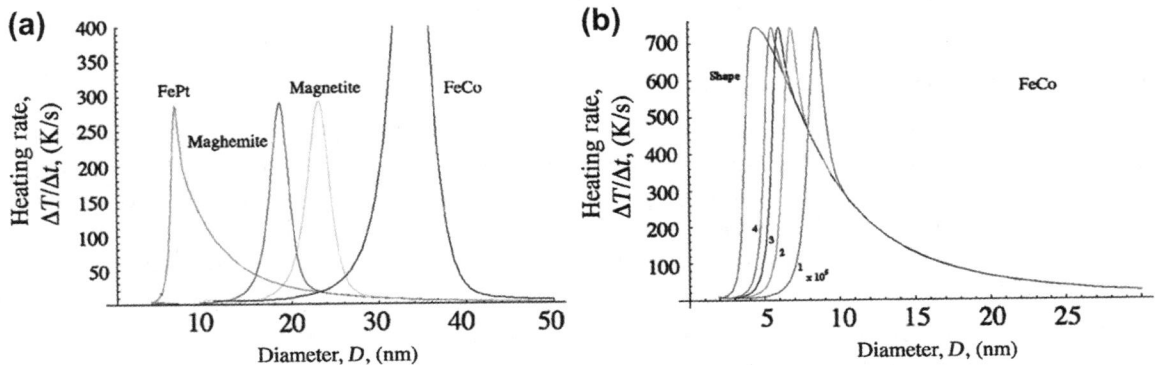

Figure 71 (a) Comparative heating rates as a function of particle size for nanoparticle materials considered in Maenosono and Saita (2006) and (b) for FeCo nanoparticles of different compositions and acicular equiatomic FeCo nanoparticles with shape anisotropy (Habib et al., 2008).

anisotropy on the order of $K_s = 1 \times 10^6$ J/m^3 are achievable in Fe–Co nanoparticles with moderate aspect ratios. Fe–Co nanoparticles have other advantages in forming stable protective cobalt-ferrite shells. Cobalt ferrite has been shown to be suitable for water-based magnetic fluids, suggesting Fe–Co/CoFe$_2$O$_4$ acicular core shell structures for use in water-based ferrofluids.

Recently, the Rosensweig (2002) formalism has been generalized to consider T-dependent material properties in a self-regulated heating theory [OHS$^+$09]. For Fe–Co MNPs, this can be ignored because of their high Curie temperatures, T_cs. However, T_c tuning is an important area of recent study (Miller et al., 2010a, 2010b; McNerny et al., 2010). γ-FeNi MNPs have tunable Curie temperatures which can be employed for self-regulated heating in hyperthermia (McNerny et al., 2010). This allow for control of heating at a therapeutic target temperature.

19.4.2.3 *Applications of Nanoparticles in Microelectronics*

The reflow of small solder volumes is critical to high interconnect densities needed for electronic packages. Efforts to eliminate Pb from consumer electronics due to health concerns of Pb toxicity have led adoption of Pb-free solders for packaging applications. An example is SAC305 (3 wt%Ag; 0.5 wt% Cu, remaining Sn) which has a higher reflow temperature and longer reflow times, due to higher melting points, and poor wetting compared to Sn–Pb eutectics. This increases risk of printed circuit board delamination and blistering.

Localized heating, promoted by MNPs to cause solder reflow, can aid efficient processing of Pb-free solders (Habib et al., 2010). MNPs subjected to AC magnetic fields dissipate by the relaxation processes discussed above. Unlike eddy current losses, these result in significant power loss in submicron- and nano-sized magnetic materials (Miller et al., 2009). Nanocomposites incorporating MNPs in solder paste have been shown to enhance induction losses (Habib et al., 2010).

Figure 72(a) shows the thermal images of the spatial T-profile in solder balls at various time intervals in an AC magnetic field of 500 Oe at 280 kHz. **Figure 72(b)** shows the temporal T-profiles derived from this data. When MNPs are incorporated into solder paste, the resulting composite shows a significant temperature rise in AC magnetic fields sufficient to cause solder reflow. Reflow times vary systematically with MNP concentration in the composite. Reflow of solder–MNP composite paste ($\sim$5 g) at modest MNP loadings on Si, glass and Ni/Au substrates has been demonstrated. However, for processing of smaller solder volumes ($\sim$0.05 g) higher MNP concentration and field parameters are required due to the increased fraction of heat lost to the surrounding at smaller volumes.

19.4.3 Magnetocaloric Materials

Applications of thermodynamics to heat engines, refrigerators, heat pumps, etc. rely on efficient use of thermodynamic cycles. Most of these cycles, e.g. *Carnot cycles*, are based on use of PV work in the expansion/compression of a fluid (liquid and/or gas) system. Here we consider magnetic work and the consequent MCE in magnetic refrigeration cycles. The MCE was first described by Warburg (1881). MCE provides a ways to realize refrigeration from ultralow temperatures to room temperature and above. We discuss the following:

(1) Attaining low temperatures by adiabatic demagnetization.
(2) MCEs with paramagnetic salts.
(3) MCEs with superparamagnetic particles.
(4) MCEs using ferromagnetic to paramagnetic phase transitions.
(5) Magnetic refrigeration cycles.

(a)

(b)

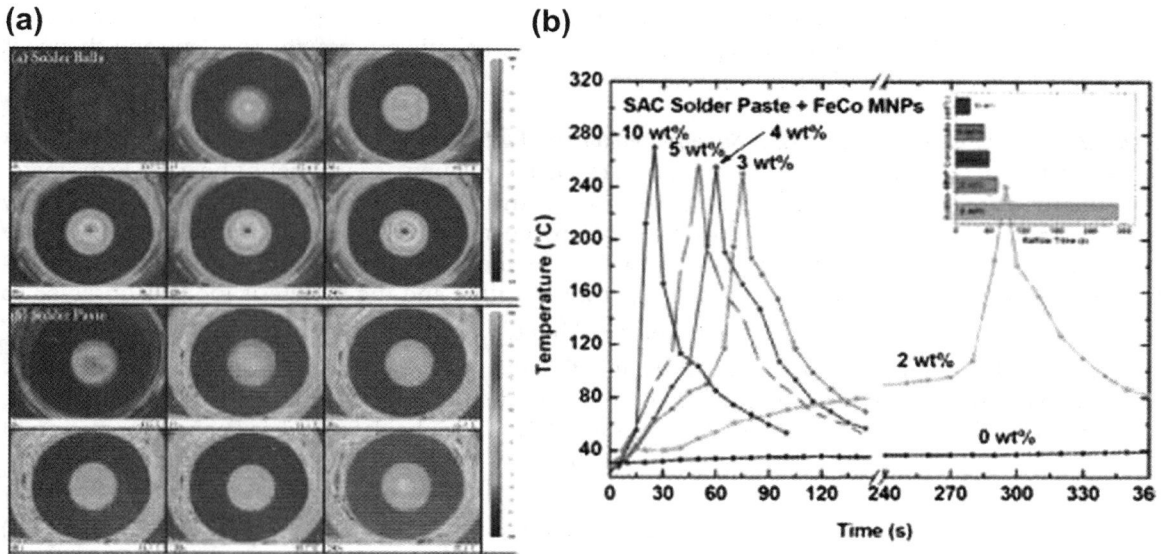

Figure 72   (a) Thermal images showing T-profile in solder balls and (b) $T(t)$ for different wt% solder-MNP composites in AC fields of 500 Oe at 280 kHz, inset shows time to reflow for different wt% samples. From Habib (Ph.D. Thesis, 2011). (For color version of this figure, the reader is referred to the online version of this book.)

19.4.3.1 *Attaining Low Temperatures by Adiabatic Demagnetization*

The use of paramagnetic salts in adiabatic demagnetization processes dates back to the 1920s and the work of Giauque (1927) and Debye (1926). Recall that paramagnetism results from the existence of permanent magnetic moments on atoms. In a paramagnetic material in the absence of a field, the local atomic moments are uncoupled and randomly oriented, these dipole moments will align themselves with a magnetic field. The magnetic susceptibility of a paramagnet obeys the *Curie Law*:

$$\chi = \frac{M}{H} = \left(\frac{N_m \mu_0 (\mu_{\text{atom}})^2}{3k_B T} \right) = \frac{C}{T} \tag{83}$$

with N_m, the dipole density and μ_{atom}, the atomic dipole moment. In a paramagnetic system, the spin entropy, S, is $S = R \ln(n + 1)$, where n is the number of unpaired electrons per formula unit. Adiabatic demagnetization has the following steps:

(1) Cool a paramagnetic salt to ~ 1 K using liquid Helium (LHe) using $\textbf{He}(g)$ as a heat transfer medium.
(2) Isothermally magnetize the salt in fields of ~ 1 T.
(3) Adiabatically isolate the salt by evacuating, i.e. removing the $\textbf{He}(g)$.
(4) Turn of the field (adiabatic demagnetization).

The spin entropy change of a paramagnetic salt during adiabatic demagnetization is

$$\Delta S = -\left(\frac{C}{T^2} \right) \int_H^0 H dH = \frac{C}{2} \left(\frac{H}{T} \right)^2 \tag{84}$$

where since $C > 0$ the temperature of the paramagnetic salt decreases. Spin entropy is decreased on magnetization and increased on demagnetization. In an adiabatic process, the spin entropy change is opposed by the thermal(lattice) entropy change.

19.4.3.2 Magnetocaloric Effects with Superparamagnetic Particles

Superparamagnetic response refers to thermally activated switching of a magnetic fine particle moment. This thermally activated switching is described by an Arrhenius law for which the activation energy barrier is $K_u\langle V\rangle$, where K_u is the magnetic anisotropy energy density and $\langle V\rangle$ is the average particle volume. The switching frequency becomes larger for smaller particle size. Above a blocking temperature, T_B, the switching time is less than the experimental time and the hysteresis loops is observed to collapse. For experimental times of ~ 1 h, the blocking temperature roughly satisfies $T_B = \frac{K_u\langle V\rangle}{30k_B}$.

Above T_B, the magnetization scales with H and T in the same way as does a classical paramagnetic material, with the exception that the inferred dipole moment is a particle moment and not an atomic moment but a cluster moment. The magnetization, $M(H, T)$ can be described as Langevin equation of state:

$$\frac{M}{M_s} = L(a) = \coth(a) - \frac{1}{a} \tag{85}$$

where $\alpha = \frac{\mu H}{k_B T}$ and moment μ is given by the product $M_s\langle V\rangle$, where M_s is the spontaneous magnetization and $\langle V\rangle$ is the average particle volume. A limiting form for the equation of state is the standard Curie law. The associated MCE for superparamagnetic particles is therefore analogous to that of a paramagnetic salt except for two important differences:

(1) We consider a particle rather than atomic dipole moment and the density is a particle rather than atomic density. The Curie constant scales accordingly.
(2) The largest effect occurs near the blocking temperature, T_B, rather than on approaching 0 K.

The large moment allows for potentially more entropy change associated with the adiabatic demagnetization of the superparamagnetic particles. By tuning the particle size and therefore blocking the temperature, one can influence the temperature (range) where the MCE is significant and raise it above the very low (LHe) temperatures typically accessible with paramagnetic salts.

19.4.3.3 Magnetocaloric Effects in Ferromagnetic to Paramagnetic Phase Transitions

Many modern day refrigerants take advantage of the entropy change in a phase transition that occurs at or near the temperature at which the refrigerator is to operate. Often this is a liquid- to gas-phase transition and sometimes a solid-/solid-phase transition. Another significant phase transition would be the ferromagnetic- to paramagnetic-phase transition that occurs at the ferromagnetic Curie temperature which will be discussed further in terms of the Landau Theory below. Of the elemental ferromagnets, the MCE Gd is of most interest since its Curie temperature is close to room temperature!

A figure of merit for refrigerants is the *refrigeration capacity, RC*, which is a measure of the effective cooling capacity of a refrigerator, expressed in Btu/h or in tons of refrigeration (1 ton of refrigeration = 13 898 kJ/h = 3.861 kW). Another measure of the RC for magnetic materials takes the product of the entropy change, dS and multiplies in by the incremental temperature change, dT. The dS is determined from integrating $M(H)$ isotherms and dT would represent the increment between the hot and cold sides of the refrigerator.

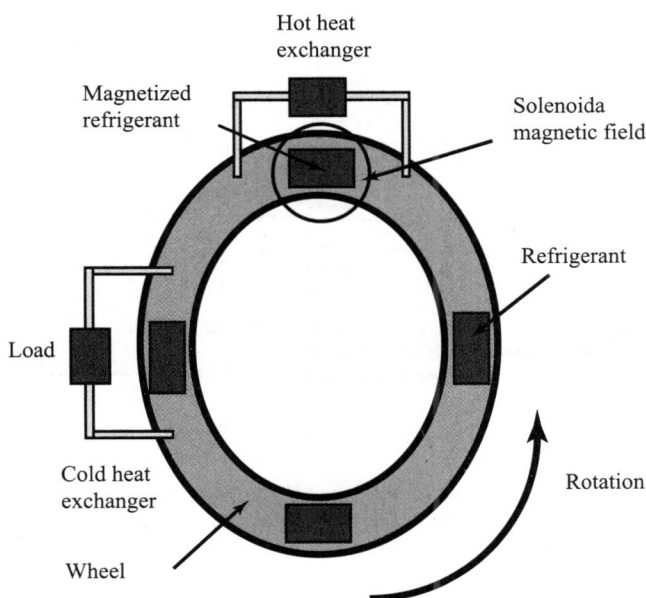

Figure 73 Magnetocaloric device example: rotary active magnetic liquefier. (For color version of this figure, the reader is referred to the online version of this book.)

19.4.3.4 *Magnetic Refrigeration Cycles*

An example of a thermodynamic cycle that takes advantage of the MCE is the *Joule–Brayton cycle*. **Figure 73** shows an example of a rotary active magnetic liquefier taking advantage of adiabatic demagnetization in a Joule–Brayton cycle. The steps in the Joule–Brayton cycle are the following:

(1) Magnetize and thereby warm the magnetic solid.
(2) Remove heat with a cooling fluid.
(3) Demagnetize and thereby cool the magnetic solid.
(4) Absorb heat from a cooling load (adiabatic demagnetization).

19.4.3.5 *Selected Recent Research on Magnetocaloric Effects*

The MCE explains the fact that magnetic materials heat when placed in a magnetic field and cool upon removal from the field. This effect was first observed by Warburg (1881) in iron. The magnitude of the effect in elemental ferromagnets, like Fe is 0.5–2 C per T. Recently, Gd–Ge–Si alloys have been shown to have much larger effects of $\sim$3–4 °C per T. The search for new MCE materials is a very active area of current research.

A large MCE is the figure of merit for the efficacy of a magnetic refrigerant. With an increasing field, the magnetic entropy decreases and heat is transferred from the magnetic system to the environment in an isothermal process. With a decreasing field, the magnetic entropy increases and heat is absorbed from the lattice system to the magnetic system in an adiabatic process. Both a large, isothermal entropy change and adiabatic temperature change characterize the prominent MCE (Bruck, 2005; Gschneidner

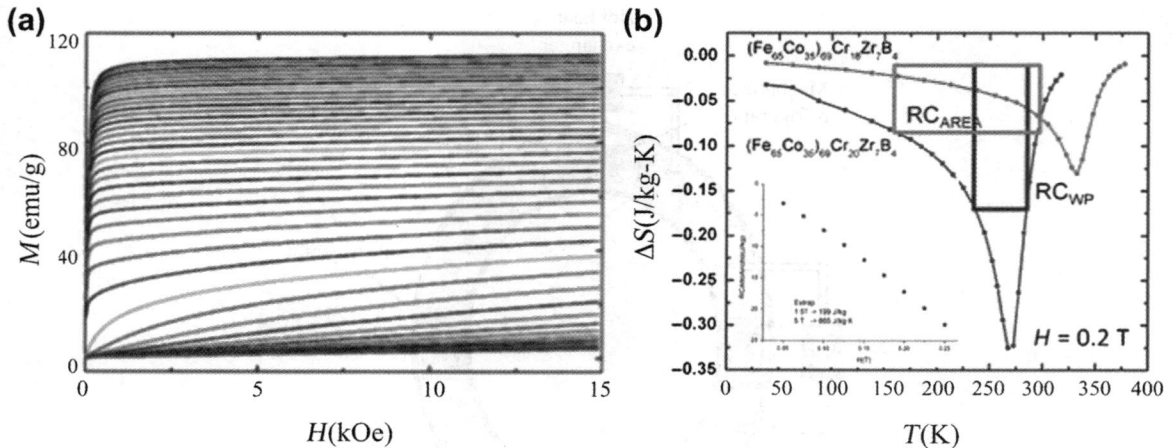

Figure 74 MCE data for an FeCo-based amorphous alloy. (a) M(H) isotherms from 40 to 400 K, (b) integrated magnetic entropy changes for the curves of (a) with method for determining RC shown and (inset) RC as a function of applied field. Courtesy: A. Colletti and K. Miller. (For color version of this figure, the reader is referred to the online version of this book.)

et al., 2005). The importance of the lattice system has led much recent literature in MCE to consider MCEs in materials in the vicinity of simultaneously occurring magnetic and structural phase transitions.

Nanocomposite magnetic materials have been among the systems of interest for MCE applications (Johnson and Shull, 2006; Franco et al., 2007). The ability to tune the Curie temperature by alloying the amorphous phase in these materials makes them of interest for a variety of refrigerant applications at different temperatures. **Figure 74** shows the unpublished MCE data for an FeCo-based amorphous alloy. **Figure 74(a)** shows the $M(H)$ isotherms from 40 to 400 K; (b) shows the integrated magnetic entropy changes for the curves of (a) with the method for determining the *refrigeration coefficient* (RC) shown and the inset shows the RC as a function of applied field (courtesy A. Colletti and K. Miller). The magnetic entropy is determined by integrating the Maxwell relation described above (Franco et al., 2007). The RC is defined as the specific (maximum) energy that can be absorbed over a range of temperatures in J/kg. The RC is a figure of merit for a magnetocaloric material. Values approaching at 1 kJ/kg at $H = 5$ T are observed for this particular material. The largest area under the dS curve from T(hot) to T(cold) is the RC. There are two suggested methods of getting the RC from $\Delta S(T)$ data: (1) full width at half max of the peak as a function of temperature and (2) finding the largest rectangle which fits under the curve (Wood and Potter, 1985).

Giant MCE has been observed in materials exhibiting first-order phase transitions, including $Gd_5Si_2Ge_2$, $LaFe_{13-x}M_x$ (M = Si, Al), MnFePAs, $MnAs_{1-x}Sb_x$, $Mn_{1-x}Fe_xAs$, and Ni Mn X (X = Ga, Sn). Large MCEs also are observed in some rare-earth metals, alloys, compounds and the manganites (Phan and Yu, 2007). Second-order phase transitions are also useful in limiting the hysteresis in the MCE. Over the past few years, the MCE and magnetic refrigeration materials have been investigated extensively and several kinds of magnetic-refrigerant prototype instruments have been demonstrated. This remains a very active area of current research in magnetic materials. Of particular interest is the suspension of MNPs in fluid systems capable of transporting absorbed heat away.

Suppression of phase transformations in metastable nanostructures can be used to produce materials with properties not obtainable in equilibrium. In Fe-rich, Fe–Ni-based nanocomposite systems [11], the nucleation of the equilibrium *bcc* phase is suppressed in favor of the metastable γ-phase. This has profound effects on technical magnetic properties because on the Fe-rich side of the Fe–Ni phase diagram, there is a strong compositional dependence of the Curie temperature, T_c on composition in the γ-phase [12]. For certain Naval applications, the T_c of the particles should lie between ambient, ocean water temperature and the desired upper limit operating temperature of a device. A magneto-caloric thermodynamic cycle can be used for such an exploitation of the ferromagnetic to paramagnetic phase transformation in γ-Fe–Ni. New research has focused on the stabilization of γ-phase nano-structures in magnetic nanopowders and nanostructures produced by primary crystallization of amorphous precursors, powdering materials and producing aqueous ferrofluids suitable for magne-tocaloric cooling applications near room temperature.

19.4.4 Electromagnetic Interference Absorbers

Materials designed to absorb electromagnetic noise in specific frequency ranges is another active area of new research in *electromagnetic interference* (*EMI*) absorption. This research requires understanding ways to tune magnetic and dielectric properties of EM absorbers for specific applications. This requires choosing appropriate intrinsic magnetic and dielectric materials properties and the influence of morphology and microstructure on their frequency-dependent absorption spectrum.

EMI is electrical noise transmitted by electric motors, machines, generators, circuits and other sources that interferes with the function of another electronic device such as a circuit, computer or transmission line. Electromagnetic radiation is classified by its energy, E, frequency, f, and wavelength, λ, related through *Einstein's formula* relating the photon energy to frequency $v = 2\pi f$:

$$E = hv = \frac{hc}{\lambda} \tag{86}$$

where h is the *Planck's constant* ($h = 6.626 \times 10^{-34}$ J $- s = 6.626 \times 10^{-27}$ erg $- s$) and for EM radiation, $v = \frac{c}{\lambda}$ and c is the velocity of light, $c = 3 \times 10^{10}$ cm/sec $= 3 \times 10^8$ m/.

High-frequency electronic devices exacerbate the need for EMI absorbers. Examples of these applications include the following:

(1) Telecommunications
(2) Consumer applications
(3) Biomedical applications
(4) Military and civilian security applications.

Mechanisms for EMI shielding are *reflection EMI shielding and absorption EMI shielding*. In reflection shielding, the mobile charge carriers in metals or conductive polymers interact with the EM field and the radiation is reflected. In absorption shielding, the electric or magnetic dipoles in dielectric or magnetic materials interact with the EM waves and the EM waves are transformed into heat in the absorbing material (Lian, 2007).

Reflection and attenuation properties are determined by the complex permeability ($\mu' - i\mu''$) and permittivity ($\varepsilon' - i\varepsilon''$). In absorption, the loss is proportional to the product, $\sigma\mu$, of conductivity, σ, and permeability, μ. For reflection, the loss is a function of the ratio $\frac{\sigma}{\mu}$; absorption loss increases with frequency and reflection loss decreases. The sum of all the losses gives the total loss. Good conductors such as Ag and Cu have excellent reflection characteristics, whereas materials like Superpermalloy and Mumetal have excellent absorption characteristics due to their high permeabilities.

19.4.4.1 Principles for Designing EMI Absorbers with Magnetic Materials

The effectiveness of EMI shielding is quantified by the reflection loss, Γ given in dB:

$$\Gamma(\text{dB}) = -20 \log_{10} \left[\frac{Z_A - Z_0}{Z_A + Z_0} \right] \quad Z_A = Z_0 \sqrt{\frac{\mu_e}{\varepsilon_e}} \tanh \left(\frac{2\pi f}{cd \frac{\mu_e}{\varepsilon_e}} \right) \tag{87}$$

where Z_0 is the impedance of free space (377 Ω), Z_A is the absorber impedance at the free space–material interface, μ_e and ε_e are the effective permeability ($\mu' - i\mu''$) and permittivity ($\varepsilon' - i\varepsilon''$) of the absorber medium and d is the absorber thickness. When the absorber impedance is equal to the impedance of air, there is no reflection. A reflection loss of -20 dB corresponds to 99% wave absorption and is considered an impedance matching situation (Liu, 2004) atypical target performance of EMI absorbers. For gigahertz-range absorbers, it is favorable to have matching permeability and permittivity. Increasing the permeability or thickness will increase the loss. Changing permeability can change properties without adding additional size and weight. The permeability of the materials depends on composition and microstructural features such as particle size and porosity.

Magnetic materials for EMI shielding include *spinel ferrite* and *hexagonal ferrites*, metallic (and amorphous metallic) magnetic materials, and combinations of both. Desired properties include high *permeability*, a high *Curie temperature*, high *saturation magnetization*, high induction, and a high *resistivity* to limit eddy current losses. High permeability requires considering a variety of other materials and morphological properties including *magnetocrystalline anisotropy*, *shape anisotropy*, *induced anisotropy* and *magnetostriction*. The consideration and matching of magnetic and *dielectric properties* increasingly lead to the consideration of composite and *nanocomposite* structures to exploit attractive properties of two or more engineering materials.

Ferrite materials have many properties which make them useful for EMI wave absorbing applications. Ferrites are ceramics having high permeability, temperature stability, and low cost which provides many design advantages. However, they have low inductions as compared to metals and cubic ferrites do not have sufficient magnetic anisotropy to function at high frequencies. Their high resistivity does allow them to operate without the frequency limitations of eddy current losses. To have superior EMI absorbers at the highest frequencies, it would be desirable to have the large inductions of metals combined with the large resistivities of the ferrites along with the engineering of large magnetic anisotropies. This has spurred the studies of ferromagnetic nanocomposites and *metamaterials* often exploiting high-induction metal–high-resistivity oxide core shell structures often with morphologies designed to exploit shape anisotropy.

The complex impedance and an EMI absorber depend on the *dielectric permittivity* and the *magnetic permeability*. The complex permeability depends on the chemical composition of the alloy, as well as microstructure, particle size, and porosity of the compact (Nakamura et al., 2000). The permeability can be described by contributions from domain wall motion and spin rotation processes. Domain wall motion dominates in the initial permeability and is sensitive to the microstructure, particle size, and particle loading in the compact. Spin rotation dominates at frequencies above 100 MHz and depends on the anisotropy field and particle loading. Thus optimizing the microstructure, particle size, anisotropy field, and loading can lead to a large permeability.

To design state-of-the-art absorbers, a variety of fundamental considerations must be taken into account. In the following sections, we develop the physical principles necessary to tailor the properties of an EMI absorber. These are summarized as follows:

(1) *DC Magnetic Shielding*: DC shielding of a soft magnetic material is determined by its permeability. If the fields to be shielded are sufficiently large, saturation induction is an issue. At higher frequencies, considerations of eddy current and anomalous eddy current losses and resonant absorption are important considerations in assessing the frequency-dependent magnetic response.

(2) *EMI Absorption and Skin Depth*: In general, we wish to limit the amount of the material to that required to perform the engineering function. In the case of EMI absorbers, the important length scale of note is the *EM skin depth*. The skin depth is the depth to which losses occur in a material and to which the material responds to an oscillating field. The skin depth is a function of the magnetic permeability, the conductivity and the frequency of the exciting EM radiation. Solutions of *Maxwell's equations* for fields penetrating into a permeable medium give rise to an exponential decay of the applied field over the *classical skin depth*, δ, which is given by the relationship:

$$\delta = 5030\sqrt{\frac{\rho}{\mu f}} \tag{88}$$

where ρ is the resistivity in Ω – cm, μ is the relative permeability and f is the frequency of the EM radiation in Hz. Since most of the losses are confined to the skin depth, it is desirable to design absorbers with dimensions approximating the skin depth at the frequency to be absorbed. This illustrates the importance of high resistivity for high-frequency absorbers. For very high frequencies, e.g. 1 GHz and reasonable permeabilities of 100–1000, e.g. a good insulator will have nanoscale dimensions to match its skin depth. High-frequency absorbers require tailoring microstructures at the nanoscale.

(3) *Magnetic Anisotropies*: To tune the permeability, it is important to engineer magnetic anisotropy. In many state-of-the-art EMI absorption materials, the important types of magnetic anisotropy include (a) magnetocrystalline anisotropy, (b) shape anisotropy, (c) magnetoelastic anisotropy and (d) induced anisotropy.

(4) *FMR*: The FMR, f_r, is a spin resonance at microwave frequencies where transverse EM field energy is absorbed at the precessional frequency of atomic dipoles about a static field. In the absence of magnetic anisotropy, this resonance frequency depends on the magnetization of the material as described by *Snoek's Law*:

$$\mu' f_r \sim M_s \tag{89}$$

where μ' is the real magnetic permeability and M_s is the saturation magnetization. In the absence of other loss mechanisms, the FMR frequency represents a material's ultimate EMI absorption frequency. EMI absorber design recognizing the role of the FMR frequency then requires materials with large magnetizations and restricts the permeability. It is desirable to fix the internal magnetization direction with a suitable magnetic anisotropy to ensure that switching takes place by nonlossy rotational mechanisms and the real permeability is

$$\mu' = \frac{M_s}{H_K} \quad H_K = \frac{2K}{M_s} \tag{90}$$

with H_k, the anisotropy field and K, the magnetic anisotropy energy density.

(5) *Magnetic Losses*: Power losses in a ferromagnetic material are described by

$$P_{\text{Total}} = P_{\text{hys}} + P_{\text{ec}} + P_{\text{anomalous}} \tag{91}$$

where the first term, the *hysteretic power loss*, has a linear dependence on frequency; the second term, the *classical eddy current power loss*, has a square dependence on frequency; and the third term, the *anomalous power loss*, has a power >1 dependence on frequency. It is apparent that the eddy current losses dominate at high frequencies. Approaches to limiting these losses include (1) choice of strongly exchange-coupled soft magnetic materials to limit hysteretic losses; (2) choice of high-resistivity materials or composite structures with insulating coatings; and (3) limiting the dimensionality of the material to limit the spatial extent over which eddy currents flow.

(6) *Morphology and Microstructure*: Often, composites can give better response than a single material. The morphology of the magnetic species and the microstructure of the composite must be carefully engineered to optimize the properties. Morphological and microstructural considerations include (a) skin depth matching and insulation, and (b) monodomain particles, engineering shape anisotropy and/or magnetoelastic anisotropy. To control losses, the eddy current must be confined to run in a dimension comparable to the skin depth. This requires insulating coatings in three-dimensional structures and insulation parallel to the induction in two-dimensional or one-dimensional structures.

19.4.5 Magnetic Sensors

Magnetic metals and alloys are important in sensor applications. Existing magnetic field sensors use a variety of physical phenomena including inductive pick-up coils, Hall probes, magnetoresistive elements, magneto-optic devices, flux-gates and superconducting quantum interference devices (SQUIDs) (Ripka, 2001). Magnetic sensors are classified by their field sensing ranges.

Low-field sensors are typically larger and can require careful and costly sample preparation to correct for demagnetization effects and for the Earth's magnetic field. A SQUID is the most sensitive low-field sensor. SQUID magnetometers can detect fields as low as femtotesla. They require cryogenic cooling, making testing expensive and sizes cumbersome. Other sensitive low-field sensors include alternating field gradient magnetometers (AGM) and vibrating sample magnetometers (VSM).

In many applications, the sensor size is of paramount importance. In space applications, where the price per unit mass of devices launched is a primary concern, sensor size is a critical design issue (Diaz-Michelena, 2009). This is the subject of considerable efforts in miniaturization (Acuna, 2002). The size of current AGMs and VSMs is an obstacle for applications in orbit measurements, planetary characterizations and space exploration. Similarly, many biomedical applications are limited by sensor size. *In vivo* sensing applications, for example, have significant physical size constraints. Miniaturization of magnetic sensing devices has fueled the rapid increase in data storage capacity.

Figure 75 summarizes the current magnetic sensor technologies in terms of minimal detectable field and dynamic range. Applications require measuring fields anywhere from 10^{-15} to 10^4 T, a range of 19 orders of magnitude! An *SQUID* magnetometer, with sensitivities of 1 fT, can be used in such sensitive applications as mapping currents in the brain in the field *of magnetoencephalography* (*MEG*). MEG signals result from ionic currents flowing in neuron dendrites during synaptic transmission. These measurements are now used in research and clinical settings. MER tomography gives sensitivities of 0.1–1 microGauss (μG). A typical magnetic noise floor in a city environment is 10^8 fT. As

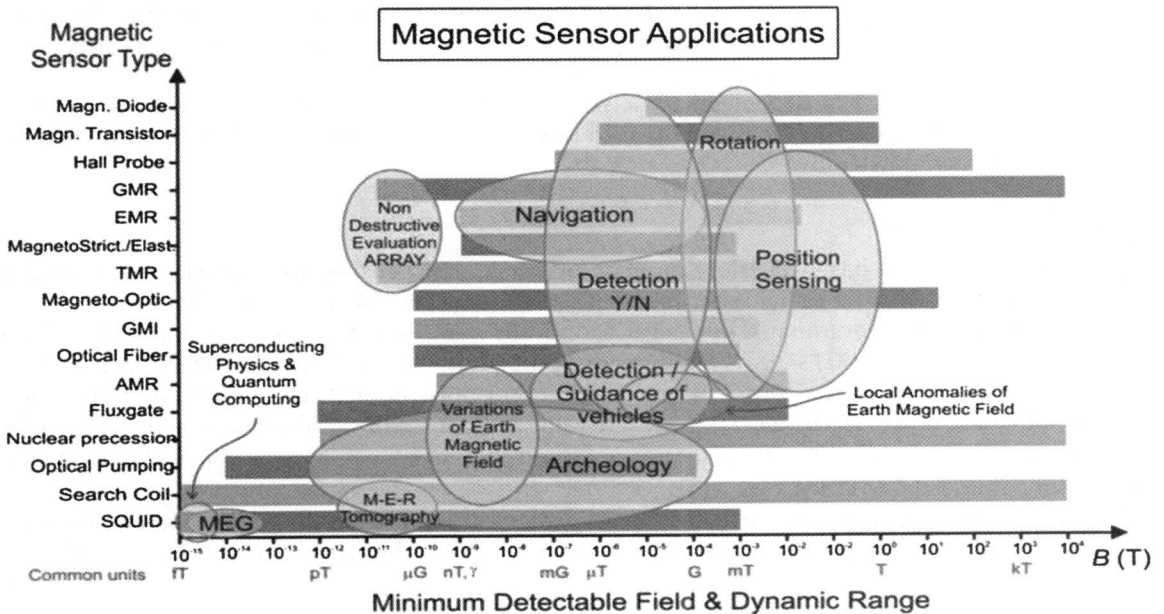

Figure 75 Magnetic sensors technologies and applications categorized by sensor type and magnetic properties (minimum detectable field and dynamic ranges (Diaz-Michelena, 2009)). (For color version of this figure, the reader is referred to the online version of this book.)

a result, measurements of the smallest fields require sophisticated magnetic shielding and shielded rooms.

Measurement of the earth's magnetic field for mapping and as a navigational tool uses fluxgate magnetometers. Fluxgates measure fields from mT to 10s of pT in DC to 10s of kHz frequencies. Fluxgates for geomagnetic field mapping have dynamical ranges of 64000 nT, and resolutions in the order of tens of nT to pT and *low-noise density* (several $pT/\sqrt{Hz}$ at 1 Hz). A typical fluxgate mass is 0.5 kg and 2 W of power consumption (Diaz-Michelena, 2009).

Magnetometer design and development has sought smaller size, lower power consumption, and lower cost for similar performance (Lenz and Edelstein, 2006). Recent innovations aimed at improving size, power, and cost have included the use of piezoresistive cantilevers and magnetometers based on electron-tunneling effects, *magnetic tunnel junctions* (*MTJs*). Magnetic fields are vectors with both magnitude and direction. Sensors differ in their ability to measure the magnitude and direction of the field. A scalar sensor measures the field's total magnitude but not its direction, while an omnidirectional sensor measures the magnitude of the component of magnetization that lies along its sensitive axis. Bidirectional sensors include direction in its measurements and vector magnetic sensor incorporates two or three bidirectional detectors.

Detection and measurement of magnetic fields can be classified by distinguishing whether the measurement technique is direct or examines a spatial derivative of the field in question:

(1) A *magnetometer* is a device that detects magnetic fields directly. Typically this uses a simple induction loop.

(2) An *axial gradiometer* is a device in which two magnetometers placed in series axially. The gradiometer then measures the difference in magnetic flux between those points in space. It is therefore a measure of the first spatial derivative of the field.

(3) A *planar gradiometer* has two magnetometers placed next to each other in the same plane. It measures the "difference in the differences" in flux between two points. It is therefore a measure of the second spatial derivative.

19.4.5.1 Noise Levels

A goal of recent effort has been development of small, inexpensive, low-power, low-frequency, magnetic sensors to sensitive to fields between 1 nT and 1 pT (Egelhoff et al., 2009). This range of field measurement is currently dominated by fluxgates, optically pumped magnetometers, and SQUIDS. The development of picoTesla MTJ sensors requires controlling noise levels. Noise characteristics have been recently modeled based on the contributions to the sensor noise floor. These include amplifier noise, Johnson and shot noise, electronic $1/f$ noise, thermal magnetic white noise and magnetic $1/f$ noise (Egelhoff et al., 2009). One $pT/\sqrt{Hz}$ at 1 Hz is a current goal which would be a factor of ~ 100 below the current commercial MTJ sensors.

Electrical noise is the random fluctuation in a signal. Noise can be produced by a variety of effects. Thermal noise and shot noise are inherent to all devices, while other types depend on other physical phenomena. We explore a few of these here:

(1) *Johnson–Nyquist noise* is synonymous with *thermal noise* and is generated by the random thermal motion of charge carriers in a conductor. The power spectral density of thermal noise, nearly constant throughout the frequency spectrum, is called *white noise*. Thermal noise can be modeled with Gaussian distribution of amplitudes. The root mean square (RMS) voltage due to thermal noise V_n, generated in a resistance R (ohms) over a bandwidth, Δf (Hz), is

$$V_n = \sqrt{4k_B TR\Delta f} \tag{92}$$

where k_B is the Boltzmann's constant (J/K) and T is the resistor's absolute temperature (K).

(2) *Shot noise* is the random fluctuation of current which results from the fact that current is carried by discrete charges.

(3) $1/f$ *noise* is also called flicker noise. It is the portion of the noise spectrum which falls off at high frequencies. In magnetic sensors, this noise results from fluctuations of magnetic properties on a nanoscale. One example of a magnetic $1/f$ noise source is the random telegraph noise which is generated from the motion of magnetic domain walls through pinning sites. This noise is called *Barkhausen noise*.

The noise level can be quantified (1) as an electrical power in watts or dBm, (2) as an RMS voltage equal to the noise standard deviation in volts, or (3) characterized by a probability distribution and noise spectral density $N_0(f)$ in watts per hertz. The frequency dependence of the noise is expressed in units of $\left(\frac{\text{power}}{\text{frequency}}\right)^{1/2}$. In a resistor, the power is proportional to V^2 and noise units are V/Hz. In an inductive component, power is proportional to B^2 and noise units are $T/\sqrt{Hz}$.

19.4.5.2 Measurement of Fields

Common field measurement techniques include a Hall sensors, fluxgate magnetometers, and magnetoresistive sensors. Several sensors use the *Lorentz force*, or *Hall effect* which describes the force $\vec{F}_L$

experienced by a particle with charge q moving with velocity v in a magnetic field, $\vec{B}$: $\vec{F}_L = q\vec{v} \times \vec{B}$. The measurable transverse voltage in a conductor carrying a longitudinal current in a field results from the *Hall effect* (Hall, 1879). For a geometry where a current is induced by an electric field, E_x, in the x-direction, a magnetic field is applied in the z-direction and the Hall voltage (and electric field) are along the y-direction, the *Hall coefficient*, R_{Hall}, is

$$R_{\text{Hall}} = \frac{E_y}{j_x B_z} \tag{93}$$

Measuring a Hall voltage, and determining longitudinal current density from *Ohm's Law*, $j_x = \sigma_0 E_x$, the Hall coefficient can be determined for a material. This depends on the charged conducting particles (electron and hole) densities in a material. The Hall voltage is proportional to the applied field. A material with a known Hall coefficient can be used in a *Hall sensor* to accurately measure the magnitude of a normal applied field. A *Hall effect sensor* is a *transducer* with an output voltage that varies with field and used in switches, position, speed, and current detection applications.

Magnetoresistive sensors have historically been important sensors for field and magnetization measurements. The *magnetoresistance* is the change in the electrical resistance of a material in response to an applied field. The *magnetoresitive* (MR) ratio is the ratio of the change in resistance in a field to the resistance in zero field (Spaldin, 2003):

$$\text{MR} = \frac{R_H - R_0}{R_0} = \frac{\Delta R}{R_0} \tag{94}$$

Positive magnetoresistance (MR > 0) is a larger and negative (MR < 0) a smaller resistance in a field. Normal magnetoresistance, from the Hall effect, is used to sense fields. Ferromagnetic *anisotropic magnetoresistance (AMR)* is discussed below.

A *fluxgate magnetometer* is a sensor with good field sensitivity related to the magnetic core material (Gordon and Brown, 1972; Lenz, 1924). A *single-axis fluxgate magnetometer* is of a toroidal magnetic core wound with a magnetizing drive coil. Another sensing coil is wound around an axis in which the magnetic field component is measured. A fluxgate compares the current required to saturate a material in one direction with that in the normal direction. The difference in the two currents is proportional to the field. An ideal core will have linear response to saturation. A sense coil detects harmonics of a fundamental drive coil frequency which are stronger at the abrupt permeability change at saturation. Largest sensitivity occurs for a high-permeability material. Linear $M(H)$ response is achieved by controlling a uniaxial anisotropy. A uniaxial anisotropy induced by field annealing an amorphous or nanocomposite material serves this purpose.

Fluxgates were developed during World War II to detect submarines from low-flying aircraft. Modern applications include electronic compasses for navigation of aircraft and oil-well borehole measurements to navigate while digging deep oil wells. Because of their sensitivity (typically ~ 0.5–1.0 nT), fluxgates can be used to follow gradients in the earth's magnetic field. The development of two-axis and three-axis fluxgates has led to their use as navigational tools and in magnetic surveying. Variations in the earth's magnetic field result from the geometry of the earth's magnetic poles and the magnetic properties of the materials in the Earth's core and crust. NASA has employed fluxgates to measure magnetospheric and solar system magnetic fields.

19.4.5.3 Measurement of Magnetization

Commercial magnetometers (Foner, 1996) can be broadly classified as measuring the magnetization by (a) direct techniques, including measuring the force on a sample placed in a nonuniform field, e.g.

Faraday, Guoy, Kahn balances and (b) indirect techniques, including measurements based on magnetic induction due to relative motion between the sample and the detector, e.g. VSMs and SQUIDs (Long, 2008). Magnetometers are characterized by their sensitivity and a range of other features such as vector or scalar operation, bandwidth, heading error, size, weight, power, cost and reliability. These characteristics determine the suitable applications for the magnetometer.

Lorentz force magnetometers are simple, small, lightweight, low-cost, and low-power-consumption devices for measuring vector magnetic fields. *Xylophone magnetometers*, based on a classical xylophone resonator, are linear sensors with a wide dynamic range from nanotesla to tesla. A *search-coil magnetometer* exploits Faraday's law of induction. The sensitivity of a search coil depends on the permeability of the core and the area and number of turns of the coil. Solenoidal search coils sense time-varying magnetic fields. This is the principle of operation of a *pick-up coil*. A time-varying magnetization $\frac{dM}{dt}$ is sensed by the induced voltage. Other low-field sensor technologies include nuclear precession, optically pumped, and fiber-optic magnetometers.

An example of the use of a pick-up coil sensor is in the *VSM* or Foner magnetometer (Foner, 1959). The principle of operation of a VSM is based on placing a magnetic sample in a uniform magnetic field. The sample dipole moment is made to undergo a periodic sinusoidal motion at a fixed frequency (typically 60 Hz, i.e. a loudspeaker frequency), f, using a transducer drive head to vibrate a sample rod. The vibrating magnetic dipole moment (through Faradays law of induction) induces a voltage, V, in a sensitive set of pick-up coils placed between the pole pieces of the electromagnet. This signal, proportional to the magnetization, is amplified and monitored. The VSM is calibrated using a standard sample of known moment such as an Ni (ferromagnetic) or a Pt (diamagnetic) sphere.

New applications for atomic magnetometers include detection of biomagnetic signals (Bison et al., 2003; Xia et al., 2006), nuclear magnetization (Lee et al., 2006; Savukov and Romalis, 2005) and magnetic particles (Xu et al., 2006). The combination of sensitivity with ease of use in atomic magnetometers makes them a promising alternative for detecting weak magnetic fields. One of the first chip-scale atomic magnetometer constructed at NIST (Schwindt et al., 2004) was based on magnetically sensitive *coherent population trapping* (CPT) resonances at atomic hyperfine frequencies. Since the CPT resonance is optically driven, there are no magnetic fields applied to the instrument that might interfere with other sensors.

Commercially available magnetoresistive sensors can exploit *AMR* of ferromagnets. In AMR, the change in resistance is different when the field is applied parallel to the current than when normal to the current direction. AMR sensors can measure both linear and angular position and displacement in the Earth's field. Typical devices are Ni–Fe thin films deposited on an Si wafer using 2–3% resistance changes, in a field, in patterned resistive strip (Ciureanu and Middelhoek, 1992).

These types of sensors have found much recent importance after the discovery of *GMR*. **Figure 76(a)** shows the schematic of FeCr (001) superlattices for which GMR was first observed; (b) shows the typical resistivity vs applied field behavior for a GMR material; and (c) shows the magnetic state of the material in terms of its field-dependent magnetization, $M(H)$. The term giant refers to values of MR $\geq \sim 25\%$ to several 100%. MR values 10–100 times larger have been observed in oxides and other materials exhibiting *colossal magnetoresistance* (CMR). CMR materials typically require very large bias fields limiting their applicability.

AGMs are another sensitive technique which is used to measure magnetization. A thin-film AGM has been designed for use in detecting extraterrestrial minerals (Sanz et al., 2011; Wise et al., 2011) and biomedical applications (Jones et al., 2011) of tagging tissue scaffold. The sensor system is shown in **Figure 77(a)**. The sensor head consists of a permanent FePt or CoCrPt magnet deposited using

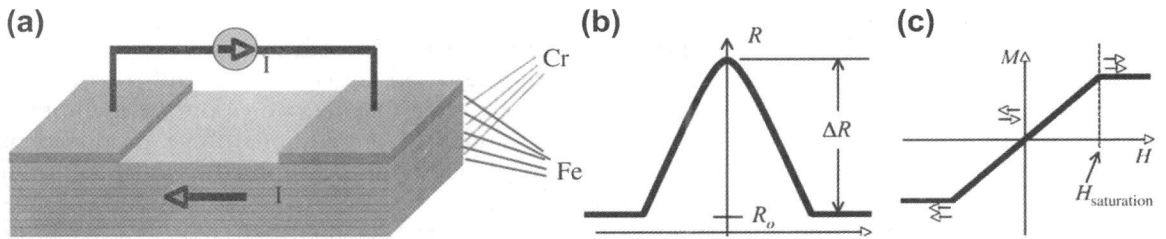

Figure 76   (a) Schematic FeCr (001) superlattices (Baibich et al., 1988); (b) typical resistivity vs applied field behavior and (c) $M(H)$ response for a GMR sensor.

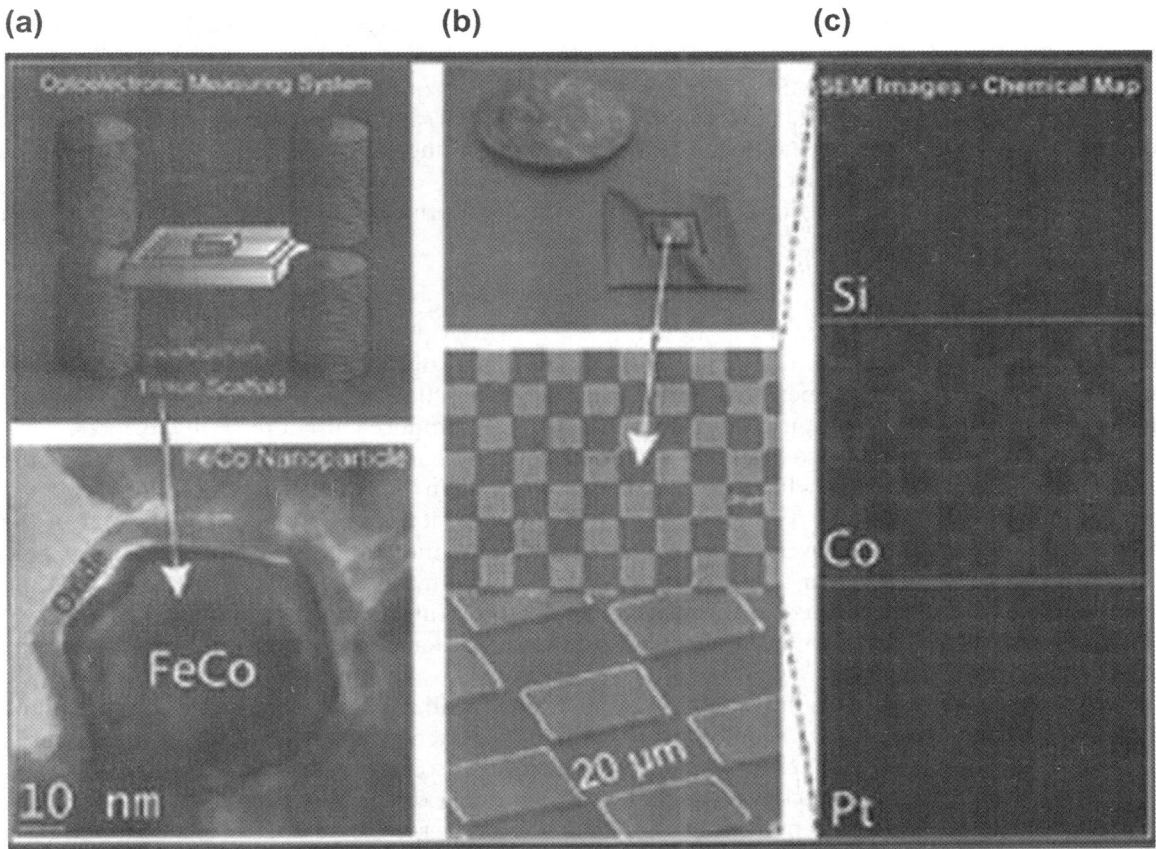

Figure 77   (a) AGM thin-film resonator sensor geometry; (b) membrane, the checkerboard and (c) the SEM determined chemistry of the films (April 1, 2011 cover of the *J. Appl. Phys.*). (For color version of this figure, the reader is referred to the online version of this book.)

a thin-film sputtering to create a checkerboard pattern onto an Si substrate. **Figure 77(b)** shows the membrane and the checkerboard and **Figure 77(c)**, the SEM determined chemistry.

The sensor shown in **Figure 77(a)** has coils creating a driving field gradient. The correct geometries and spatial arrangement of the coils will generate a magnetic force on the membrane in the z-direction, causing it to vibrate in mechanical resonance. An additional force, proportional to the detected material's magnetic moment and the field gradient is determined from shifts in the mechanical resonance frequency. When excited at the resonance frequency, the membrane deformation is measured using a high-resolution optical detection method. Light emitted from a light emitting diode is guided through an optical fiber and is reflected on the surface of the magnet. Reflected light is detected by a bundle of Polymethyl methacrylate (PMMA) fibers and transmitted to a photodiode. The resulting photocurrent is amplified and a lock-in amplifier is used to measure the detection signal.

Strains can also be the basis for sensing exploiting giant magnetostriction in important rare-earth/Fe alloys like Terfenol-D [$(Dy_{0.7}Tb_{0.3})Fe_2$]. The shape of the magnetostriction-applied field curve yields little magnetostriction at low fields making such materials unsuitable for low-field magnetometry. However, it has been shown that in tension, a significant slope is observed at low fields (Schatz et al., 1994). Novel sensors have been developed using surface micromachining of Si with micro-electromechanical system (MEMS) fabrication techniques (Benson et al., 1995). In a magnetostrictive magnetometer, a thin-Terfenol-D film is deposited on an MEMS cantilever. The magnetostriction of the Terfenol-D causes the cantilever to deflect. The deflection is a function of the field strength and can be measured by optical beam deflection or capacitance changes.

GMI as discussed above is observed in materials with large permeabilities including amorphous and nanocomposite materials.

19.4.6 Magnetic Recording

The rapid development of information technology has been greatly enabled by the rapid growth in the areal recording density of magnetic hard drives (Plumer et al., 2001). This has been made possible by advances in the materials and processing techniques used to produce current recording devices.

Magnetic recording devices consist of two main magnetic systems: the heads which write or read the information and the media on which the information is stored. In today's media, information is written in the perpendicular mode by a single pole write head when it applies a magnetostatic field in the perpendicular direction and reverses the magnetization in the medium. In **Figure 78(a)**, the magnetization is stored perpendicular to the disk plane because the magnetic easy axis of materials in the recording layer has been aligned in this direction. When the information is read back, a GMR (see above) read sensor is moved to where the information is stored and it reads the magnetization in the recording layer.

The information is stored on the media in bits. **Figure 78(b)** illustrates two magnetic domains in the opposite direction (Park, 2010). One bit consists of 50–75 magnetic grains and each of the grains is surrounded by a nonmagnetic oxide film which magnetically isolates them from each other. The bits are aligned along a track and tracks are configured concentrically on the hard disk. In **Figure 78(b)**, the transition region between written bits is delineated and it can be seen to be directly related to the grain size of the film. In turn, media noise is directly related to the transition width as well as the grain size distribution (Park, 2010).

The early generations of computers had the information stored on tapes or hard disks that had been coated with a slurry that contained elongated particles of γ-Fe_2O_3 (maghemite). The particles coercivity

(a)

(b)

a: transition parameter

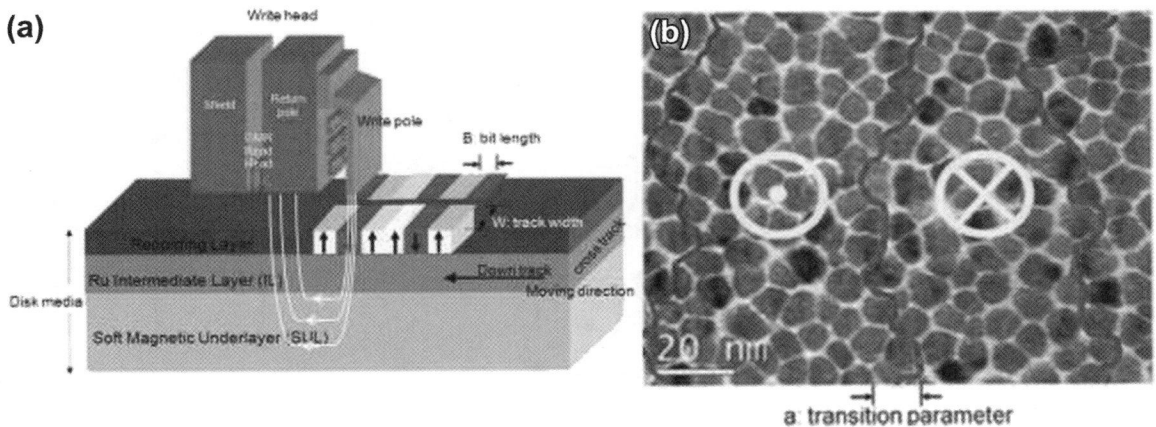

Figure 78 (a) Schematic of perpendicular recording scheme and (b) schematic of a bit drawn on micrograph of an actual thin film media. Courtesy: Dr. S. Park (Park, 2010). (For color version of this figure, the reader is referred to the online version of this book.)

was determined by shape anisotropy. Later, thin films of Co alloys were introduced and as of the time writing of this chapter, thin-Co-alloy films are still utilized.

The development of media for magnetic storage is a good example of the use of the basic principles of materials science. The media must consist of small magnetic grains that are strongly oriented (textured) with their easy axes in the plane of the disk for longitudinal recording or normal to the plane of the disk for perpendicular recording. Furthermore, the grains should be of uniform size and should be magnetically isolated from each other.

As mentioned above, particulate media and the first of the thin-film hard drives were longitudinal media. In the particulate media, the particles had their long axes in the plane of the film. In the first commercialized thin-film media, the films were produced in such a way that the c-axis of the Co-alloy grains were in the plane of the disk. Since the magnetic easy axis of Co alloys is the c-axis, a method had to be devised to grow the Co alloys with the c-axis in (or nearly in) the plane of the disk. This was done by using Cr underlayers. Cr (bcc) naturally sputters with a (110) texture and this gives rise by epitaxy to a $(10\bar{1}1)$ texture of the Co alloy. In this case, the c-axis was about 30° out of the plane of the disk. Later it was found that by depositing the Cr films at elevated temperatures (270–300 °C), they would obtain the (200) texture, which in turn produced (by epitaxy) the $(11\bar{2}0)$ texture in the Co alloy. This texture allowed for the c-axis to be in the plane of the film (**Figure 79**) (Laughlin and Wong, 1991). Later the underlayers were grown with the (112) texture (with the help of NiAl seedlayers) (Laughlin et al., 1995) and the Co alloys deposited with the $(10\bar{1}0)$ texture (Laughlin et al., 2009).

Since the turn of the century, information in hard disk media has been stored in the so-called perpendicular mode, in which the easy axes of the magnetic grains in the films are perpendicular to the plane of the disk. Co alloys have the hcp structure and their c-axis is usually the magnetically soft axis. Since the anisotropy of a thin film tends to lie in the plane of the film (because of shape effects), the magnetization of the Co alloy has to be lowered by the addition of Cr. This decreases the magnetization and hence lowers the magnetostatic energy of the film.

A schematic cross-section of perpendicular media is shown in **Figure 80**. Here there are several layers of films, each with their specific purpose. The layer directly under the magnetic layer is Ru (or an alloy of

(a)   **(b)**   **(c)**

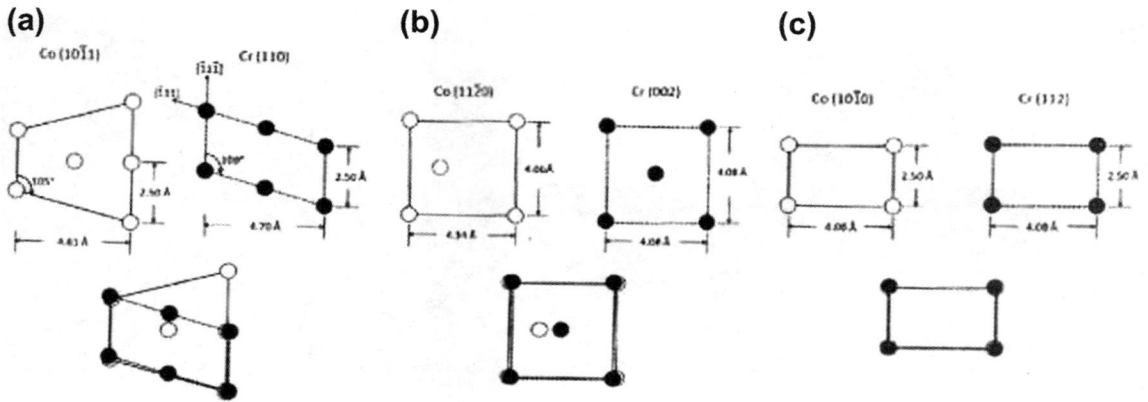

Figure 79 Schematics of the atomic matching of (a) the (10$\bar{1}$1) planes of *hcp* Co and (110) planes of Cr; (b) (1$\bar{1}$20) planes of *hcp* Co and (200) planes of Cr and (c) (10$\bar{1}$0) planes of *hcp* Co and (112) planes of Cr (Laughlin and Wong, 1991).

Ru) which has the *hcp* structure and a slightly larger lattice parameter than the Co alloys. When deposited in the perpendicular manner, the larger Ru lattice tends to stretch the Co alloy lattice in the basal plane, since this is the plane of epitaxy in perpendicular recording. Since Co alloys have a negative magnetostrictive constant, a perpendicular anisotropy is induced in the Co alloy over and above its already strong magnetocrystalline anisotropy in the same direction.

Another function of the Ru layer is to control the grain size and the positioning of the oxide phase in the recording layer. The deposited Ru films have domes on their surface which controls where the oxide

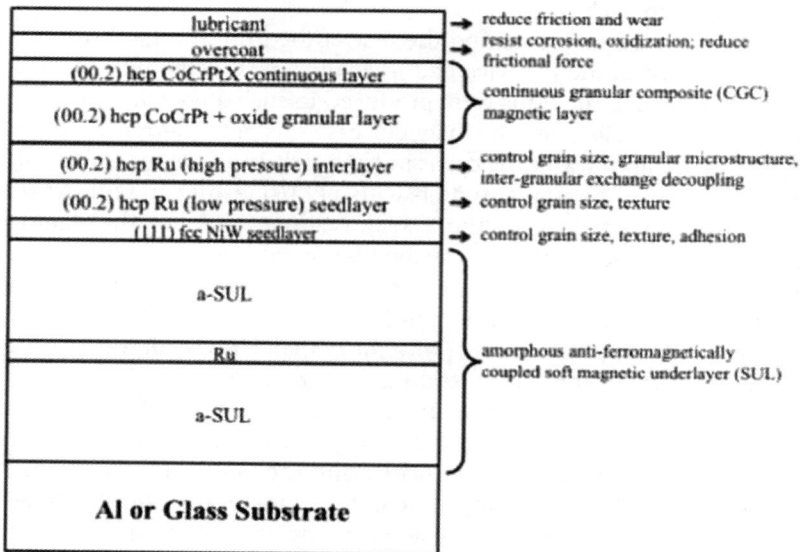

Figure 80 A schematic of the structure of the thin-film layers of a typical perpendicular magnetic recording media. Courtesy of Dr. H. Yuan (Yuan, 2009).

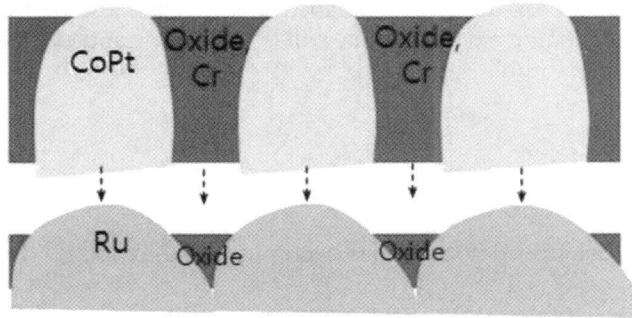

Figure 81 A schematic showing the role of the Ru domes in controlling the microstructure of the Co recording layer. Courtesy of Dr. S. Park (Park, 2010). (For color version of this figure, the reader is referred to the online version of this book.)

in the recording layer forms (**Figure 81**). The domes form best in films that are sputtered under a higher Ar pressure. In the figure, the Ru domes can be seen to allow for the oxide to form in the valleys during the deposition of the Co-alloy/oxide layer. The oxides then grow around the Co-alloy grains magnetically isolating them from other Co-alloy grains.

The grain size of the storage layer is controlled by a combination of the percentage of oxide and the grain size of the Ru underlayer. In turn, this is controlled by processing variables (sputtering power, Ar back-pressure, etc.) and sometimes by an additional underlayer.

Beneath the Ru layer, there is often a layer such as an Ni–W alloy which deposits with a very strong (111) *fcc* texture and because of its similar atomic arrangement with the (0001) planes of HCP materials, it increases the subsequent Ru- and Co-alloy films texture by means of epitaxy. The texture of the Co-alloy film is crucial to good recording properties and it is measured by the full width half maximum of an X-ray rocking curve.

In most perpendicular media, there exists a *soft magnetic underlayer* (SUL) which acts to close the magnetic flux with the head. Amorphous soft magnetic materials with high permeability, such as CoZrTa, are commonly used as the SUL. The thin Ru layer in the middle of the SUL couples the two SULs in an antiferromagnetic fashion and acts to pin any magnetic domains that exist in that layer, thereby minimizing its contribution to the noise (**Figure 80**).

The type of media described above is often termed *conventional perpendicular magnetic media*. It has been used in hard drives for computers over a decade now. A limit in the lower bound of the Co-alloy grain size is approaching since the grains will soon be small enough to become superparamagnetic. This will require the introduction of a new magnetic media: one with larger magnetocrystalline anisotropy (Wood, 2000). A candidate for this is FePt alloys, which will be able to sustain magnetization down to about 3 nm grains. There are many challenges with FePt. First, since they will need to have very high magnetocrystalline anisotropy, they will be very difficult to write on. Schemes such a heat-assisted magnetic recording (Rottmayer et al., 2006) may be needed if the new media is used. Another challenge is producing the high anisotropy: to do this, the FePt alloy must be highly ordered atomically. This entails high-temperature processes, which may not be compatible with the rest of the magnetic recording device. Research on these alloys continues.

Closing Remarks

This chapter has introduced metals and alloys used for soft and hard magnets and surveyed some current applications. This field has continued to innovate a progress over many decades. The choices of materials and applications are by no means exhaustive and reflect the interest and expertise of the authors. They are convinced that in a decade, there will be more new and exciting additions to this field and hope to contribute to its future.

19.5   Further Reading

19.5.1   Magnetic Materials

(1) **Ferromagnetism** by Richard M. Bozorth has been published by the IEEE Press as a classic reissue (Bozorth, 1993).
(2) Chen discusses the development of soft magnetic metal alloys (Chen, 1986).
(3) O'Handley (1987) offers a more modern materials perspective.
(4) Cullity and Graham (2009) is the update of the classic text on magnetic materials.
(5) Coey (2010) is a modern text on magnetic materials.
(6) Coey (1996) discusses permanent magnet materials.
(7) The book edited by Plumer et al. (2001) discussed data storage technologies.

19.5.2   Structure

(1) DeGraef and McHenry (2007) discuss the structure of materials with many magnetic materials examples.
(2) Cullity (1978) is the classic text on X-ray diffraction by the author of the famous book on magnetic materials.
(3) Barrett and Massalski (1980) is a classic text on structure of materials.

19.5.3   Phase Equilibria

(1) Prince (1966) is a classic text discussing phase diagrams.
(2) Porter and Easterling (2000) discusses phase equilibria and the kinetics of phase transformations.
(3) Massalski (1990) is the handbook of Binary alloy phase diagrams

19.5.4   Current Research and Applications

There are many good journals covering magnetism and magnetic materials. In particular, the *Journal of Applied Physics* annually publishes the peer-reviewed articles from the Magnetism and Magnetic Materials (MMM) Conference, the largest international scientific conference on the subject. The *IEEE Transaction on Magnetics* similarly publishes peer-reviewed articles from the Intermag conferences which are more focused on applications. The two are held simultaneously every three years as the MMM/Intermag Conference. Many materials societies also have topical groups and symposia dedicated to magnetic materials. The Materials Society (TMS) is one such example.

References

Acuna, M.H., 2002. Space-based magnetometers. Rev. Sci. Instrum. 73, 3717–3736.
Alben, R., Becker, J.J., Chi, M., 1978a. J. Appl. Phys. 49, 1653.
Alben, R., Budnick, J., Cargill, G.S., 1978b. Metallic Glasses. ASM, Metals Park, OH, 304.

Arrott, A.S., 1957. Criterion for ferromagnetism from observations of magnetic isotherms. Phys. Rev. 108, 1394–1396.

Ausleos, M., Elliot, R.I. (Eds.), 1983. Magnetic Phase Transitions, Springer Series in Solid State Sciences, vol. 48. Springer-Verlag, Berlin, Heidelberg, New York, Tokyo.

Baibich, M.N., Broto, J.M., Fert, A., Nguyen Van Dau, F., Petroff, F., Eitenne, P., Creuzet, G., 1988. Giant magnetoresistance of (001)Fe/(001)Cr magnetic superlattices. Phys. Rev. Lett. 61, 2472.

Barrett, C., Massalski, T.B., 1980. Structure of Metals, Crystallographic Methods, Principles and Data. In: International Series on Materials Science and Technology, third revised ed., vol. 35. Pergamon, Oxford.

Belov, K.P., 1956. Fiz. Met. Metalloved. 27, 261.

Benson, R.C., Murphy, J.C., Charles Jr., H.K., 1995. Miniature sensors based on microelectromechanical systems. Johns Hopkins APL Tech. Dig. 16 (3), 311–318.

Bison, G., Wynands, R., Weis, A., 2003. Dynamical mapping of the human cardiomagnetic field with a room temperature, laser-optical sensor. Opt. Express 11, 904–909.

Boettinger, W.J., Perepezko, J.H., 1985. Fundamentals of rapid solidification. In: Rapidly Solidified Crystalline Alloys. Proceedings of a TMS-AIME North East Regional Meeting, May 1–3.

Boll, R., 1994. Chapter 14: soft magnetic metals and alloys. In: Buschow, K.H.J. (Ed.), Materials Science and Technology, a Comprehensive Treatment, vol. 3B. VCH, Weinheim, p. 399.

Bozorth, Richard M., 1993. Ferromagnetism. IEEE Press, Piscataway, N. J.

Bruck, E., 2005. Developments in magnetocaloric refrigeration. J. Phys. D: Appl. Phys. 38, R381–R391.

Busch, H., 1926. Berechnung der Bahn von Kathodenstrahlen im Axialsymmetrischen Elektromagnetischen Felde. Ann. Phys. (Lpz.) 18, 974–993.

Buschow, K.H.J., 1988. In: Wohlfarth, E.P., Buschow, K.H.J. (Eds.), Ferromagnetic Materials. Elsevier Science Publishers, Amsterdam.

Cadogan, M., Li, H.S., Margarian, A., Dunlop, J.B., Ryan, D.H., Collocott, S.J., Davis, R.L., 1994. New rare-earth intermetallic phases $R_3(Fe, M)_{29}X_n$ – (R Ce, Pr, Nd, Sm, Gd, M Ti, V, Cr, Mn, and X H, N, C). J. Appl. Phys. 76, 6138.

Chaturvedi, A., Laurita, N., Phan, M.H., McHenry, M.E., Srikanth, H., 2011. Giant Magnetoimpendance and field sensitivity in amorphous and nanocrystalline $(Co_{1-x}Fe_x)_{89}Zr_7B_4$ (x = 0,0.025,0.05,01) ribbons. J. Appl. Phys. 109, 07B508.

Chaudhari, P., Cuomo, J.J., Gambino, R.J., 1973. Amorphous films for magnetic bubble and magneto-optic applications. Appl. Phys. Lett. 22, 337–339.

Chen, Chih-Wen, 1986. Magnetism and Metallurgy of Soft Magnetic Materials. Dover, New York.

Chien, C.L., 1978. Mossbauer study of a binary amorphous ferromagnet: $Fe_{80}B_{20}$. Phys. Rev. B 18, 1003–1015.

Chin, G.Y., Wernick, J.H., 1986. In: Wohlfarth (Ed.), Soft Magnetic Metallic Materials in Ferromagnetic Materials. North-Holland, New York.

Chou, T., Igarashi, M., Narumiya, Y., 1993. Soft magnetic properties of microcrystalline Fe–Al–Si–Ni–Zr–B alloys. J. Magn. Soc. Jpn. 17, 197–200.

Chu, S.Y., Cramb, A., De Graef, M., Laughlin, D.E., McHenry, M.E., 2000. The effect of field cooling and field orientation on the martensitic phase transformation in a Ni_2MnGa single crystal. J. Appl. Phys. 87, 5777–5779.

Chu, S.Y., Gallagher, R., Graef, De, McHenry, M.E., 2001. Structural and magnetic phase transitions in Ni_2MnGa ferromagnetic shape-memory alloy crystals. IEEE Trans. Magn. 37, 2666–2668.

Ciureanu, P., Middelhoek, S., 1992. Thin Film Resistive Sensors. Institute of Physics Publishing, New York.

Clark, A.E., 1980. Chapter 7: magnetostrictive rare earth-Fe2 compounds. In: Wohlfarth, E.P. (Ed.), Ferromagnetic Materials, vol. 1. North-Holland, Amsterdam, p. 531.

Clark, A.E., Hathaway, K.B., Wun-Fogle, M., Restorff, J.B., Lograsso, T.A., Keppens, V.M., Petculescu, G., Taylor, R.A., 2003. Extraordinary magnetoelasticity and lattice softening in bcc Fe–Ga alloys. J. Appl. Phys. 93, 8621.

Coehorn, R.L., 1991. In: Long, G.J., Grandjean, F. (Eds.), Supermagnets: Hard Magnetic Materials. Kluwer (chapter 8).

Coey, J.M.D., 1978. Amorphous magnetic order. J. Appl. Phys. 49, 1646–1652.

Coey, J.M.D. (Ed.), 1996. Rare Earth Permanent Magnets. Oxford Science Publications, Clarendon Press, Oxford.

Coey, J.M.D., 2010. Magnetism and Magnetic Materials. Cambridge University Press, Cambridge, UK.

Coey, J.M.D., Sun, H., 1990. Improved magnetic properties by treatment of iron-based rare earth intermetallic compounds in ammonia. J. Magn. Magn. Mater. 87, L251–L254.

Collier, K.N., Jones, N.J., Miller, K.J., Qin, Y.L., Laughlin, D.E., McHenry, M.E., 2009. Controlled oxidation of FeCo magnetic nanoparticles (MNPs) to produce faceted FeCo/ferrite nanocomposites for RF heating applications. J. Appl. Phys. 105, 07A328–330. http://dx.doi.org/10.1063/1.3054376.

Corb, B.W., O'Handley, R.C., 1985. Magnetic properties and short-range order in Co–Nb–B alloys. Phys. Rev. B 31, 7213–7219.

Cullity, B.D., 1978. Elements of X-ray Diffraction. Addison-Wesley Publishing Company, Inc., Reading, MA.

Cullity, B.D., Graham, C.D., 2009. Introduction to Magnetic Materials, second ed. IEEE Press, John Wiley and Sons, Hoboken, NJ.

Davies, H.A., 1985. In: Steeb, S., Warlimont, H. (Eds.), Rapidly Quenched Metals, vol. 1. North Holland, Amsterdam.

Debye, P., 1926. Ann. Phys. (Lpz.) 81, 1154.

DeGraef, M.E., McHenry, M.E., 2007. The Structure of Materials. Cambridge University Press, Cambridge.

DeGraef, M.E., Willard, M.A., McHenry, M.E., Zhu, Y., 2001. In-situ Lorentz TEM cooling study of magnetic domain configurations in Ni_2MnGa. IEEE Trans. Magn. 37, 2663–2665.

Deportes, J., Givord, D., Lemaire, R., Nagai, H., Yang, Y.J., 1976. Influence of substitutional pairs of cobalt atoms on magnetocrystalline anisotropy of cobalt-rich rare-earth compounds. J. Less Common Met. 44, 273–279.

Diaz-Michelena, M., 2009. Small magnetic sensors for space applications. Sensors 9, 2271–2288.

Dinsdale, T., 1991. SGTE data for pure elements. CALPHAD 15 (4), 317–425.

Donald, I.W., Davies, H.A., 1978. Prediction of glass forming ability for metallic systems. J. Non Cryst. Solids 30, 77–85.

Duwez, P., Willens, R.H., Klement, W., 1960. Continuous series of metastable solid solutions in silver–copper alloys. J. Appl. Phys. 31, 1136–1137.

Dweyedi, S., Markandeyulu, G., Ohodnicki, P.R., Leary, A., McHenry, M.E., 2011. Stress-MI and domain studies in Co-based nanocrystalline ribbons. J. Magn. Magn. Mater. 323, 1929–1933.

Egelhoff, W.F., Pong, P.W.T., Unguris, J., McMichael, R.D., Nowak, E.R., Edelstein, A.S., Burnette, J.E., Fischer, G.A., 2009. Sens. Actuators A 155, 217225.

Fingers, R.T., Carr, R.P., Turgut, Z., 2002. Effect of aging on magnetic properties of Hiperco^{TM} 27, Hiperco^{TM} 50, and Hiperco^{TM} 50 HS^{TM} alloys. J. Appl. Phys. 91, 7848.

Foner, S., 1959. Versatile and sensitive vibrating-sample magnetometer. Rev. Sci. Instrum. 30, 548–557.

Foner, S., 1996. The vibrating sample magnetometer: experiences of a volunteer (invited). J. Appl. Phys. 79, 4740.

Franco, V., Conde, C.F., Blazquez, J.S., Conde, A., 2007. Magnetocaloric effect in Mn-containing Hitperm-type alloys. J. Appl. Phys. 102 (013908).

Friedel, J., 1958. Theory of magnetism in transition metals. Nuevo Cim. Suppl. 7, 287.

Fujii, Y., Fujita, H., Seki, A., Tomida, T., 1991. Magnetic properties of fine crystalline Fe–P–C–Cu–X alloys. J. Appl. Phys. 70 (10), 6241–6243.

Gambino, R.J., Cuomo, J.J., 1978. Selective resputtering induced anisotropy in thin films. J. Vac. Sci. Tech. 15 (2), 296–301.

Gallagher, K.A., Johnson, F.A., Kirkpatrick, E., Scott, J.H., Majetich, S., McHenry, M.E., 1996. Synthesis, structure, and magnetic properties of Fe–Co alloy nanocrystals. IEEE Trans. Magn. 32, 4842–4844.

Gallagher, K.A., Willard, M.A., Laughlin, D.E., McHenry, M.E., 1999. Distributed magnetic exchange interactions and mean field theory description of temperature dependent magnetization in amorphous $Fe_{88}Zr_7B_4Cu_1$ alloys. J. Appl. Phys. 85, 5130–5132.

Ghemawat, A.M., McHenry, M.E., O'Handley, R.C., 1989. Magnetic moment suppression in rapidly solidified Co–TE–B alloys. J. Appl. Phys. 63, 3388–3390.

Giauque, W.F., 1927. Paramagnetism and the third law of thermodynamics. Interpretation of the low-temperature magnetic susceptibility of gadolinium sulfate. J. Am. Chem. Soc. 49, 1870–1877.

Gignoux, D., 1995. Magnetic properties of metallic systems. In: Buschow, K.H.J. (Ed.), Chapter 5 of Materials Science and Technology, A Comprehensive Treatment. Electronic and Magnetic Properties of Metals and Ceramics, vol. 3. VCH, Weinheim, pp. 367–453.

Gilbert, W., 1958. De Magnete (P. Mottelay, Trans.). Dover Publications, Inc., New York.

Givord, D., Li, H.S., Moreau, J.M., 1984. Magnetic properties and crystal structure of $Nd_2Fe_{14}B$. Solid State Commun. 50, 497–499.

Gleiter, H., 1989. Nanocrystalline materials. Prog. Mat. Sci. 33, 223–315.

Gordon, D.I., Brown, R.E., 1972. Recent advances in fluxgate magnetometry. IEEE Trans. Magn. MAG-8, 76–82.

Greer, A.L., Whitaker, I.T., 2002. Transformations in primary crystallites in (Fe, Ni)-based metallic glasses. Mater. Sci. Forum 386–388, 77–88.

Gschneidner, K.A., Pecharsky, V.K., Tsokol, A.O., 2005. Recent developments in magnetocaloric materials. Rep. Prog. Phys. 68, 1479–1539.

Guy, A.G., Wren, J.J., 1974. Elements of Physical Metallurgy. Addison-Wesley.

Habib, A.H., 2011. Magnetic nanoparticle based solder composites for electronic packaging applications. Ph.D. Thesis. Carnegie Mellon University.

Habib, A., Ondeck, C.L., Chaudhary, P., Bockstaller, M.R., McHenry, M.E., 2008. Evaluation of iron–cobalt/ferrite core-shell nanoparticles for cancer thermotherapy. J. Appl. Phys. 103, 07A307.

Habib, A., Ondeck, M.G., Miller, K.J., Swaminathan, R., McHenry, M.E., 2010. Novel solder-magnetic particle composites, their reflow using AC magnetic fields. IEEE Trans. Magn. 46, 2187–2190.

Hall, E., 1879. On a new action of the magnet on electric currents. Am. J. Math. 2, 28792.

Handrich, K., 1969. A simple model for amorphous and liquid ferromagnets. Phys. Status Solidi 32, K55.

Harp, G.R., Weller, D., Rabedeau, T.A., Farrow, R.F.C., Toney, M.F., 1993. Magnetooptical Kerr spectroscopy of a new chemically ordered alloy—Co_3Pt. Phys. Rev. Lett. 71 (15), 2493.

Harris, V.G., Pokhil, T., 2001. Selective-resputtering-induced perpendicular magnetic anisotropy in amorphous TbFe films. Phys. Rev. Lett. 87 (36), 067207.

Harris, A.B., Sachidanandam, R., 1985. Orientational ordering of icosahedra in solid C_{60}. Phys. Rev. B 46 (8), 4944–4957.

Harris, V.G., Huang, Q., Shah, V.R., Markandeyulu, G., Rama Rao, K.V.S., Huang, M.Q., Sirisha, K., McHenry, M.E., 1999. Neutron diffraction and extended X-ray absorption fine structure studies of $Pr_3(Fe_{1-x}Co_x)_{27.5}Ti_{1.5}$ permanent magnet compounds. IEEE Trans. Magn. 35, 3286–3288.

Herbst, J.F., 1991. $R_2Fe_{14}B$ materials: intrinsic properties and technological applications. Rev. Mod. Phys. 63, 819–898.

Herbst, J.F., Croat, J.J., Lee, R.W., 1982. Neutron diffraction studies of $Nd_2(Co_xFe_{1-x})_{17}$ alloys: preferential site occupation and magnetic structure. J. Appl. Phys. 53, 250–256.

Herbst, J.F., Croat, J.J., Pinkerton, F.E., 1984. Relationships between crystal structures and magnetic properties in Nd_2Fei_4B. Phys. Rev. B 29, 4176–4178.

Herzer, G., 1997. In: Buschow, K.H.J. (Ed.), Chapter 3 of Handbook of Magnetic Materials, vol. 10. Elsevier Science, Amsterdam, p. 415.

Honda, K., Saito, S., 1920. Sci. Rep. Tohoku Imp. Univ. 9, 417.

Hopkinson, J., 1885. Magnetization of iron. Trans. Roy. Soc. (Lond.) A A 176, 455.

Hornbogen, E., 1956. Z. Metallkd. 47, 47.

Hsiao, A., McHenry, M.E., Laughlin, D.E., Kramer, M.J., Ashe, C., Okubo, T., 2002. The thermal, magnetic and structural characterization of the crystallization kinetics of amorphous and soft magnetic materials. IEEE Trans. Magn. 38, 2946–2948.

Hsiao, A., Turgut, Z., Willard, M.A., Selinger, E., Lee, M., Laughlin, D.E., McHenry, M.E., Hasegawa, R., 1999. Crystallization and nanocrystallization kinetics of Fe- and Fe(Co)-based amorphous alloys. MRS Res. Symp. Proc. 577, 551.

Hu, Z., Yelon, W.B., Kaligirou, O., Psycharis, V., 1996. Site occupancy and lattice changes on nitrogenation in $Nd_3Fe_{29-x}Ti_xN_y$. J. Appl. Phys. 80, 2955–2959.

Hund, F., 1927. Linienspektren und periodische system der elemente. Springer, Berlin.

Inoue, A., Gook, J.S., 1996. Effect of additional elements on the thermal stability of supercooled liquid in $Fe_{72-x}Al5Ga_2P_{11}C_6B_4M_x$ glassy alloys. Mat. Trans. JIM 37 (1), 32–38.

Inoue, A., Zhang, T., Itoi, T., Takeuchi, A., 1997. New Fe–Co–Ni–Zr–B amorphous alloys with wide supercooled liquid regions and good soft magnetic properties. JIM 38 (4), 359–362.

Iwanabe, H., Lu, B., McHenry, M.E., Laughlin, D.E., 1999. Thermal stability of the nanocrystalline Fe–Co–Hf–B–Cu alloy. J. Appl. Phys. 85, 4424–4426.

James, R.D., Kinderlehrer, M., 1994. Theory of magnetostriction with application to terfenol-D. J. Appl. Phys. 76, 7012–7014.

James, R.D., Tickles, R., Wuttig, M., 1999. Large field-induced strains in ferromagnetic shape memory materials. Mat. Sci. Eng. A A273–A275, 320–325.

Jellinghaus, W., 1936. Zeit. Krist. 133, 33.

Johnson, W.L., Atzmon, M., Van Rossum, M., Dolgin, B.P., Yeh, X.L., 1985. In: Steeb, S., Warlimont, H. (Eds.), Rapidly Quenched Metals, vol. 1. Elsevier, North Holland, New York.

Johnson, F., Hughes, P., Gallagher, R., McHenry, M.E., Laughlin, D.E., 2001. Structure and thermomagnetic properties of new FeCo-base nanocrystalline ferromagnets. IEEE Trans. Magn. 37, 08K909.

Johnson, F., Shull, R.D., 2006. Amorphous-FeCoCrZrB ferromagnets for use as high temperature magnetic refrigerants. J. Appl. Phys. 99, 8.

Jones, N.J., McNerny, K.L., Wise, A.T., Sorescu, M., McHenry, M.E., Laughlin, D.E., 2010. Observations of oxidation of oxidation mechanisms and kinetics in facetted FeCo nanoparticles. J. Appl. Phys. 107, 09A304.

Jones, N.J., McNerny, K.L., Sokalski, V., Diaz-Michelena, M., Laughlin, D.E., McHenry, M.E., 2011. Fabrication of thin films for a small alternating gradient field magnetometer for biomedical magnetic sensing applications. J. Appl. Phys. 109, 07E512.

Kataoka, N., Matsunaga, T., Inoue, A., Masumoto, T., 1989. Soft Magnetic properties of BCC Fe–Au–X–Si–B (X = early transition metal) alloys with fine grain structure. Mat. Trans. JIM 30, 947–950.

Kernion, S.J., Miller, K.J., Shen, S., Keylin, V., Huth, J., McHenry, M.E., 2011. High induction low loss FeCo-based nanocomposite alloys with reduced metalloid content. IEEE Trans. Magn. 73 (10) to appear.

Kim, K.S., Driouch, L., Strom, V., Jonsson, B.J., Rao, K.V., Yu, S.C., 1996. Magnetic properties of glassy $Fe_{91-x}Zr_7B_2\,Ni_x$. IEEE 32 (5), 5148–5150.

Kim, K.-S., Yu, S.-C., Kim, K.-Y., Noh, T.-H., Kang, I.-K., 1993. Low temperature magnetization in nanocrystalline $Fe_{88}Zr_7B_4Cu_1$ alloy. IEEE Trans. Magn. 29, 2679–2681.

Kobe, S., 1969. Phys. Status Solidi 41, K13.

Koster, U., Herold, U., 1981. Glassy Metals I. In: Topics in Physics, vol. 46. Springer-Verlag, Berlin.

Kramers, R. de L., 1927. La diffusion de la lumiere par les atomes. Atti Cong. Intern. Fisica, (Transactions of Volta Centenary Congress) Como 2, 545557.

Kraus, I., Haslar, V., Duhaj, P., Svec, P., Studnicka, V., 1997. The structure and magnetic properties of nanocrystalline $Co_{21}\,Fe_{64-x}\,Nb_xB_{15}$ alloys. Mat. Sci. Eng. A-Struct. 226, 626–630.

Landau, L.D., 1937. theory of phase transformations. i. Zh. Eksp. Teor. Fiz. 7, 19.

Landau, L.D., Lifschitz, E., 1935. on the theory of the dispersion of magnetic permeability in ferromagnetic bodies. Phys. Z. Sowjetunion 8, 153–169.

Laughlin, D.E., Wong, B.Y., 1991. The crystallography and texture of Co-based thin film deposited on Cr underlayers. IEEE Trans. Magn. 27 (6), 4713–4717.

Laughlin, D.E., Cheong, B., Feng, Y.C., Lambeth, D.N., Lee, L.L., Wong, B.Y., 1995. The control of microstructural features of thin films for magnetic recording. Scr. Metall. Mater. 33, 1525–1536.

Laughlin, D.E., Willard, M.A., McHenry, M.E., 2000. Phase transformations and evolution in materials (chapter). In: Turchi, P., Gonis, A. (Eds.), Magnetic Ordering: Some Structural Aspects. The Minerals, Metals and Materials Society, pp. 121–127.

Laughlin, D.E., Yuan, H., Yang, E., Wang, C., 2009. Application of texture analysis edited by A.D. Rollett. Ceramic Trans. 201.

Laurita, N., Chaturvedi, A., Leary, A., Bauer, C., Miller, C., Phan, M.H., McHenry, M.E., Srikanth, H., 2011. Magnetoimpedance effect in soft ferromagnetic amorphous ribbons coated with magnetic metals. J. Appl. Phys. 109.

Lee, S.K., Sauer, K.L., Seltzer, S.J., Alem, O., Romalis, M.V., 2006. Subfemtotesla radio-frequency atomic magnetometer for detection of nuclear quadrupole resonance. Appl. Phys. Lett. 89, 214106.

Lenz, J.E., Edelstein, A.S., 2006. Magnetic sensors and their applications. IEEE Sensors J. 6.

Lenz, J.E., 1924. A review of magnetic sensors. Proc. IEEE 78 (6), 973.

Lian, L., 2007. Effect of Nd content on natural resonance frequency and microwave permeability of $Nd_2Fe_{14}B$/a-Fe nanocomposites in 26.5–40 GHz frequency range. J. Alloys Compd. 441, 301–304.

Liebermann, H.H., 1983. In: Luborsky, F.E. (Ed.), Chapter 2 in Amorphous Metallic Alloys. Butterworths, London.

Liu, J., 2004. A GHz range electromagnetic wave absorber with wide bandgap made of $FeCo/Y_2O_3$ nanocomposites. J. Magn. Magn. Mater. 271, L147–L152.

Liu, T., Gao, Y.F., Xu, Z.X., Zhao, Z.T., Ma, R.Z., 1996. Compositional evolution and magnetic properties of nanocrystalline $Fe_{81.5}Cu_{0.5}Mo_{0.5}P_{12}C_3Si_{2.5}$. JAP 80 (7), 3972–3976.

Long, J., 2008. FeCoB and FeZrSi-based Nanocomposite Soft Magnetic Alloys and Application (Ph.D. thesis), Carnegie Mellon Univ., Pittsburgh, PA.

Long, J., Ohodnicki, P.R., Laughlin, D.E., McHenry, M.E., 2007a. Structural studies of secondary crystallization products of the $Fe_{23}B_6$-type in a nanocrystalline FeCoB-based alloy. J. Appl. Phys. 101, 09N114.

Long, J., Qin, Y., Nuhfer, N.T., De Graef, M., Laughlin, D.E., McHenry, M.E., 2007b. Magnetic domain observations in Fe Co -base nanocrystalline alloys by Lorentz microscopy. J. Appl. Phys. 101, 09N115.

Long, J., McHenry, M.E., Urciuoli, D., Keylin, V., Huth, J., Salem, T., 2008. Nanocrystalline material development for high-power inductors. J. Appl. Phys. 103, 07E705.

Luborsky, F.E., 1977. Amorphous magnetism II. In: Chapter "Perspective on Application of Amorphous Alloys in Magnetic Devices". Plenum Press, New York, pp. 345–368.

Luborsky, F.E., Lieberman, H., 1978. Crystallization kinetics of Fe–B amorphous alloys. Appl. Phys. Lett. 33, 233.

Maenosono, S., Saita, S., 2006. Theoretical assessment of FePt nanoparticles as heating elements for magnetic hyperthermia. IEEE Trans. Magn. 42, 1638–1642.

MacLaren, J.M., Schultess, T.C., Butler, W.H., Sutton, R.A., McHenry, M.E., 1998. Magnetic and electronic properties of Fe/Au multilayers and interfaces. J. Appl. Phys.

Mader, S., Widmer, H., dHuerle, F.M., Nowick, A.S., 1963. Metastable alloys of Cu–Co and Cu–Ag thin films deposited in vacuum. Appl. Phys. Lett. 3 (11), 201–203.

Makino, A., Men, H., Kubota, T., Yubuta, K., Inoue, A., 2009. New excellent soft magnetic FeSiBPCu nanocrystallized alloys with high B_s of 1.9 T From Nanohetero-amorphous phase. IEEE Trans. Magn. 45, 4302–4305.

Malozemoff, A.P., Williams, A.R., Moruzzi, V.L., 1984. Band-gap theory of strong ferromagnetism: application to concentrated crystalline and amorphous Fe- and Co-metalloid alloys. Phys. Rev. B 29, 1620–1632.

Massalski, T.B., 1981. Relationships between metallic glass formation diagrams and phase diagrams. In: Masumoto, T., Suzuki, K. (Eds.), Proc. 4th Int. Conf. on Rapidly Quenched Metals. The Japan Institute of Metals, pp. 203–208.

Massalski, T.B., 1990. Binary Alloy Phase Diagrams, second ed. ASM International. pp. 1735–1738.

Massalski, T.B., Laughlin, D.E., 2009. The surprising role of magnetism on the phase stability of Fe (Ferro). CALPHAD: Comput. Coupling Phase Diagrams Thermochem. 33, 3–7.

Matteson, S., Nicolet, M.A., 1982. In: Picraux, S.T., Choyke, W.J. (Eds.), Metastable Materials Formation by Ion Implantation. Elsevier, New York.

Mazaleyrat, F., Barrue, F.R., 2001. Soft amorphous and nanocrystalline magnetic materials. Handbook of Advanced Electronic and Photonic Materials and Devices. In: Nalwa, H.S. (Ed.), Nanostructured Materials. Academic Press.

McCurrie, R.A., 1982. Ferromagnetic Materials Structure and Properties. Academic Press, San Diego.

McHenry, M.E., MacLaren, J.M., 1991. Iron and chromium monolayer magnetism in noble-metal hosts: systematics of local moment variation with structure. Phys. Rev. B 43, 10611.

McHenry, M.E., MacLaren, J.M., Clougherty, D.P., 1991. Monolayer magnetism of 3d transition metals in Ag, Au, Pd, and Pt hosts: systematics of local moment variation. J. Appl. Phys. 70, 10611.

McHenry, M.E., Laughlin, D.E., 2000. Nano-scale materials development for future magnetic applications. Acta Mater. 48, 223.

McHenry, M.E., Willard, M.A., Laughlin, D.E., 1999. Amorphous and nanocrystalline materials for applications as soft magnets. Prog. Mat. Sci. 44, 291.

McHenry, M.E., Johnson, F., Okumura, H., Ohkubo, T., Ramanan, V.R.V., Laughlin, D.E., 2003. The kinetics of nanocrystallization and microstructural observations in FINEMET, NANOPERM and HITPERM nanocomposite magnetic materials. Scr. Mater. 48, 881–887.

McNerny, K.L., 1999. Chemical Synthesis of α-FeCo and Metastable γ-FeNi Magnetic Nanoparticles with Tunable Magnetic Properties for Study of RF Heating and Magnetomechanical Responses in Polymeric Systems (Ph.D. thesis), Carnegie Mellon Univ., Pittsburgh, PA.

McNerny, K.L., Kim, Y., Laughon, D.E., Mchenry, M.E., 2010. Synthesis of monodisperse γ-Fe–Ni magnetic nanoparticles with tunable Curie temperatures for self-regulated hyperthermia. J. Appl. Phys. 107, 09A312.

Middelhoek, S., 1963. Domain walls in thin Ni–Fe films. J. Appl. Phys. 34, 1054.

Millen, R., Kawaguchi, T., Granger, M., Porter, M., 2005. Giant magnetoresistive sensors and superparamagnetic nanoparticles: a chip-scale detection strategy for immunosorbent assays. Anal. Chem. 77.

Miller, K.J., Soll-Morris, H.B., Collier, K.N., Swaminathan, R., McHenry, M.E., 2009. Induction heating of FeCo nanoparticles for rapid RF curing of epoxy composites. J. Appl. Phys. 105, 07E714.

Miller, K.J., Colleti, A., Papi, P.J., McHenry, M.E., 2010a. Fe–Co–Cr nanocomposites for applications in self-regulated rf heating. J. Appl. Phys. 107, 09A313.

Miller, K.J., Sofman, M., McNerny, K., McHenry, M.E., 2010b. Metastable γ-FeNi nanostructures with tunable Curie temperatures. J. Appl. Phys. 107, 09A305.

Miller, K.J., Wise, A., Leary, A., Laughlin, D.E., McHenry, M.E., Keylin, V., Huth, J., 2010c. Increased induction in nanocomposite materials with reduced glass-formers. J. Appl. Phys. 107, 09A316.

Mishima, T., 1931. Stahl u. Eisen 53, 79.

Muller, M., Mattern, N., Kuhn, U., 1996. Correlation between magnetic and structural properties of nanocrystalline soft magnetic alloys. J. Magn. Magn. Mater. 157/158, 209–210.

Nakamura, T., et al., 2000. J. Appl. Phys. 88 (1), 348–353.

Nikolov, G.T., Valchev, V.C., 2009. Nanocrystalline magnetic materials vs. ferrites in power electronics. Proc. Earth Planetary Sci. 1, 1357–1361.

O'Handley, R.C., 1987. Physics of ferromagnetic amorphous alloys. J. Appl. Phys. 62 (10), R15–R49.

O'Handley, R.C., 2000. Modern Magnetic Materials, Principles and Applications. John-Wiley and Sons, New York.

Ondeck, C.L., Habib, A.H., Sawyer, C.A., Ohodnicki, P., Miller, K.J., Chaudhary, P., McHenry, M.E., 2009. Theory of magnetic fluid heating with an alternating magnetic field with temperature dependent materials properties for self-regulated heating. J. Appl. Phys. 105, 07B324.

Olander, A., 1932. An electrochemical investigation of solid cadmium–gold alloys. J. Am. Chem. Soc. 54, 3819–3833.

Ohodnicki, P.R., Cates, N.C., Laughlin, D.E., McHenry, M.E., Widom, M., 2008a. Ab initio theoretical study of magnetization and phase stability of (Fe, Co, Ni)$_{23}$B$_6$ and (Fe, Co, Ni)$_{23}$ Zr structures of Cr$_{23}$ C$_6$ and Mn$_{23}$Th$_6$ prototypes. Phys. Rev. B 78, 144414.

Ohodnicki, P.R., Long, J., Laughlin, D.E., McHenry, M.E., Keylin, V., 2008b. Composition dependence of field induced anisotropy in ferromagnetic (Co, Fe)$_{89}$Zr$_7$B$_4$ and (Co, Fe)$_{88}$Zr$_7$B$_4$Cu amorphous and nanocrystalline ribbons. J. Appl. Phys. 104.

Ohodnicki, P.R., McWilliams, H., Laughlin, D.E., McHenry, M.E., Keylin, V., 2008c. Phase evolution of three-phase fcc, hcp, and amorphous soft magnetic nanocomposite alloy Co$_{89}$Zr$_7$B$_4$. J. Appl. Phys. 103, 07E740.

Ohodnicki, P.R., Park, S.Y., Laughlin, D.E., McHenry, M.E., Keylin, V., Willard, M.A., 2008d. Crystallization and thermal-magnetic treatment of Co-rich HiTPerm-type alloys. J. Appl. Phys. 103, 07E729.

Ohodnicki, P.R., McHenry, M.E., Laughlin, D.E., Keylin, V., Huth, J., 2009a. Temperature stability of field induced anisotropy in soft ferromagnetic Fe, Co-based amorphous and nanocomposite ribbons. Acta Mater. (accepted).

Ohodnicki, P.R., Qin, Y.L., Laughlin, D.E., McHenry, M.E., Kodzuka, M., Ohkubo, T., Hono, K., Willard, M.A., 2009b. Composition and non-equilibrium crystallization in partially devitrified Co-rich soft magnetic nanocomposite alloys. Acta Mater. 57, 87–96.

Pankhurst, Q.A., Connolly, J., Jones, S.K., Dobson, J., 2003. J. Phys. D: Appl. Phys. 36, R167.

Park, S., 2010. Extending the Recording Density in Perpendicular Recording Media (Ph.D. Thesis), Carnegie Mellon Univ., Pittsburgh, PA.

Pauling, L., 1938. The nature of the interatomic forces in metals. Phys. Rev. 54, 899.

Pfeiffer, F., Radeloff, C., 1980. Soft magnetic Ni–Fe and Co–Fe alloys-some physical and metallurgical aspects. J. Magn. Magn. Mater. 19, 190–207.

Phan, M.H., Yu, S.C., 2007. Review of the magnetocaloric effect in manganite materials. J. Magn. Magn. Mater. 308, 325–340.

Ping, D.H., Wu, Y.Q., Hono, K., Willard, M.A., Laughlin, D.E., McHenry, M.E., 2001. Microstructural characterization of Fe$_{44}$Co$_{44}$Zr$_7$B$_4$Cu$_1$ nanocrystalline alloys. Scr. Mater. 45, 781–786.

Plumer, M., Van Ek, J., Weller, D., Ertl, G. (Eds.), 2001. The Physics of Ultrahigh-density Magnetic Recording. Springer Series in Surface Sciences, vol. 41. Springer, Pittsburgh, PA.

Porter, D.A., Easterling, K., 2000. Phase Transformations in Metals and Alloys, second ed. Nelson Thornes Ltd.

Prince, A., 1966. Alloy Phase Equilibria. Elsevier Publishing Company, Amsterdam.

Ramanan, V.R.V., Fish, G., 1982. Crystallization kinetics in Fe–B–Si metallic glasses. J. Appl. Phys. 53, 2273–2275.

Ramanujan, R.V., Lao, L.L., 2006. The mechanical behavior of smart magnet-hydrogel composites. Smart Mater. Struct. 15, 952–956.

Reass, W.A., 2008. Components and technologies for high frequency and high average power converters. In: High Megawatt Power Converter Technology R&D Roadmap Workshop. National Institute of Standards and Technology.

Reicke, W.D., 1982. Practical lens design. In: Hawkes, P.W. (Ed.), Topics in Current Physics: Magnetic Electron Lenses, vol. 18. Springer-Verlag, Berlin (chapter 4).

Reicke, W.D., Ruska, E., 1966. A 100-kV transmission electron microscope with single-field condenser objective. In: Proc. 6th Int. Cong. Elec. Mic., Kyoto, vol. 1, pp. 19–20.

Reuter, K.B., William, D.B., Goldstein, J.I., 1989. Determination of the Fe–Ni phase diagram below 400 °c. Metall. Trans. A 20, 719–725.

Ripka, P., 2001. Magnetic Sensors and Magnetometers. Artech House, Inc., Norwood, MA.

Rosensweig, R.E., 2002. Heating magnetic fluid with alternating magnetic field. J. Magn. Magn. Mater. 252, 370–374.

Rottmayer, R.E., Batra, S., Buechel, D., Challener, W.A., Hohlfeld, J., Kubota, Y., Lei, L., Lu, B., Mihalcea, C., Mountfield, K., Pelhos, K., Peng, C., Rausch, T., Seigler, M.A., Weller, D., Yang, X., 2006. Heat-assisted magnetic recording. IEEE Trans. Magn. 42, 2417–2421.

Russell, H.N., Saunders, F.A., 1925. New regularities in the spectra of the alkaline earths. Astrophys. J. 61, 38.

Sanz, R., Cerdan, M., Wise, A., McHenry, M.E., Daz-Michelena, M., 2011. Temperature dependent magnetization and remanent magnetization in pseudo-binary $(Fe_2TiO_4)_x$—$(Fe_3O_4)_{1-x}$ $(0.30_ix_i\ 1.00)$ titanomagnetites. IEEE Trans. Magn. (page to appear).

Savukov, I.M., Romalis, M.V., 2005. NMR detection with an atomic magnetometer. Phys. Rev. Lett. 94, 123001.

Sawa, T., Takahashi, Y., 1990. Magnetic properties of Fe–Cu–(3d transition-metals)–Si–B alloys with fine-grain structure. J. Appl. Phys. 67, 5565–5567.

Sawyer, C.A., Habib, A.H., Collier, K.N., Miller, K.J., Ondeck, C.L., McHenry, M.E., 2009. Modeling of temperature profile during magnetic thermotherapy for cancer treatment. J. Appl. Phys. Issue Virtual J. Biol. Phys. Res. 105, 07B320.

Schafer, R., Hubert, A., 1998. Magnetic Domains. Springer, Heidelberg.

Schatz, F., Hirscher, M., Schnell, M., Flik, G., Kronmuller, H., 1994. Magnetic anisotropy and giant magnetostriction of amorphous TbDyFe films. J. Appl. Phys. 76, 5380–5382.

Schwindt, P.D.D., Knappe, S., Shah, V., Hollberg, L., Kitching, J., Liew, L.A., Moreland, J., 2004. Chip-scale atomic magnetometer. Appl. Phys. Lett. 85, 6409–6411.

Shah, V.R., Markandeyulu, G., Rama Rao, K.V.S., Huang, M.Q., Sirisha, K., McHenry, M.E., 1998. Structural and magnetic properties of $Pr_3(Fe_{1-x}Co_x)_{27.5}Ti_{1.5}$ $(x = 0.0, 0.1, 0.2, 0.3)$. J. Magn. Magn. Mater. 190, 233.

Shah, V.R., Markandeyulu, G., Rama Rao, K.V.S., Huang, M.Q., Sirisha, K., McHenry, M.E., 1999. Effects of Co substitution on magnetic properties of $Pr_3(Fe_{1-x}Co_x)_{27.5}Ti_{1.5}$ $(x = 0-0.3)$. J. Appl. Phys. 85, 4678.

Skomski, R., 1996. Interstitial Modification, Chapter 4 in Rare Earth Permanent Magnets. Oxford Science Publications, Clarendon Press, Oxford.

Skomski, R., Coey, J.M.D., 1993. Nitrogen diffusion in Sm_2Fe_{17} and local elastic and magnetic-properties. J. Appl. Phys. 73 (11), 7602–7611.

Slater, John C., 1937. Electronic structure of alloys. J. Appl. Phys. 8, 385.

Son, S., Swaminathan, R., McHenry, M.E., 2003. Structure and magnetic properties of RF thermally plasma synthesized Mn and Mn–Zn ferrite nanoparticles. J. Appl. Phys. 93, 7495–7497.

Son, S., Taheri, M., Carpenter, V., Harris, V.G., McHenry, M.E., 2002. Synthesis of ferrite and nickel ferrite nanoparticles using radio-frequency thermal plasma torch. J. Appl. Phys. 91, 7589–7591.

Sovák, P., Petrovič, P., Kollár, P., Zatroch, M., Knoć, M., 1995. Structure and magnetic properties of an $Fe_{73.5}Cu_1U_3Si_{13.5}B_9$ nanocrystalline alloy. J. Magn. Magn. Mater. 140–144, 427–428.

Spaldin, N., 2003. Magnetic Materials Fundamentals and Device Applications. Cambridge University Press, Cambridge.

Stadelmeier, H.H., 1984. Structural classification of transition metal rare earth boride permanent compounds between T and T_5R. Z. Metallkd. 75, 227–230.

Steinmetz, C.P., 1984. On the law of hysteresis. IEEE Proc. 72, 196–221.

Strnat, K., Hoffer, G., Olson, J.C., Ostertag, W., Becker, J.J., 1967. A family of new cobalt-base permanent magnet materials. J. Appl. Phys. 38, 1001.

Sutton, R.A., 1994. Irradiation Defects in Oxide Superconductors: Their Role in Flux Pinning (Ph.D. thesis), Carnegie Mellon Univ., Pittsburgh, PA.

Suzuki, K., Herzer, J., 2012. Magnetic-field induced anisotropies and exchange softening in fe-rich nanocrystalline soft magnetic allopys. Scripta Met. 67, 548–553.

Suzuki, K., Kataoka, N., Inoue, A., Makino, A., Masumoto, T., 1990. High saturation magnetization and soft magnetic properties of bcc Fe–Zr–B alloys with ultrafine grain structure. Mat. Trans. JIM 31, 743–746.

Suzuki, K., Kikuchi, M., Makino, A., Inoue, A., Masumoto, T., 1991a. Changes in microstructure and soft magnetic properties of an $Fe_{86}Zr_7B_6Cu_1$ amorphous alloy upon crystallization. Mat. Trans. JIM 32, 961–968.

Suzuki, K., Makino, A., Inoue, A., Masumoto, T., 1991b. Soft magnetic properties of nanocrystalline bcc Fe–Zr–B and Fe–M–B–Cu (M = transition metal) alloys with high saturation magnetization (Invited). J. Appl. Phys. 70, 6232–6237.

Suzuki, K., Makino, A., Kataoka, N., Inoue, A., Masumoto, T., 1991c. High saturation magnetization and soft magnetic properties of bcc Fe–Zr–B and Fe–Zr–B–M (M = transition metal) alloys with nanoscale grain size. Mat. Trans. JIM 32, 93–102.

Swaminathan, R., 2005. Influence of Surface Structure on the Magnetic Properties of RF Plasma Synthesized NiZn Ferrite Nanoparticles (Ph.D. thesis), Carnegie Mellon Univ., Pittsburgh, PA.

Swaminathan, R., McHenry, M.E., Calvin, S., Diamandescu, L., 2005. Surface structure model of cuboctahedrally truncated ferrite nanoparticles. J. Am. Cer. Soc.

Szabo, D., Szeghy, G., Zrinyi, M., 1998. Shape transition of magnetic field sensitive polymer gels. Macromolecules 31, 6541–6548.

Tomida, T., 1994. Crystallization of an Fe–Si–B–Ga–Nb amorphous alloy. Mat. Sci. Eng. A179/180, 521–525.

Tondra, M., Porter, M., Lipert, R., 2000. Model for detection of immobilized superparamagnetic nanosphere assay labels using giant magneto-resistive sensors. J. Vacuum Sci. Technol. 18, 1125–1129.

Turgut, Z., 1999. Thermal Plasma Synthesis of Coated FeCo–FeCoV Nanoparticles as Precursors for Compacted Nanocrystalline Bulk Magnets. (Ph.D. thesis), Carnegie Mellon Univ., Pittsburgh, PA.

Turgut, Z., Huang, M.Q., Gallagher, K., Majetich, S.A., McHenry, M.E., 1997. Magnetic evidence for structural phase transformations in Fe–Co alloy nanocrystals produced by a carbon arc. J. Appl. Phys. 81, 4039–4041.

Turgut, Z., Ferguson, D.E., Huang, M.Q., Wallace, W.E., McHenry, M.E., 1999a. Thermal plasma synthesis of γ-FeN$_x$ nanoparticles as precursors for Fe$_{16}$N$_2$ synthesis by annealing. MRS Res. Symp. Proc. 577, 399–404.

Turgut, Z., Nuhfer, N.T., Piehler, H.R., McHenry, M.E., 1999b. Magnetic properties and microstructural observations of oxide coated FeCo nanocrystals before and after compaction. J. Appl. Phys. 85, 4406–4408.

Ullakko, K., Huang, J.K., Kantner, C., O'Handley, R.C., Kokorin, V.V., 1996. Large magnetic-field-induced strains in Ni$_2$Mn Ga single crystals. J. Appl. Phys. 69, 1966–1968.

Van Moorleghem, W., Chandrasekaran, M., Reynaerts, D., Peirs, J., Van Brussel, H., Schirber, J.E., Venturini, E.L., Kwak, J.F., 1998. Shape memory and superelastic alloys: the new medical materials with growing demand. Biomed. Mat. Eng. 8, 55–60.

Vasilev, A.N., Bozhko, A.D., Khovailo, V.V., et al., 1999. Phys. Rev. 59, 1113.

Wallace, M.G., 1960. Intermetallic compounds between lanthonons and transition metals of the first long period 1. Preparation, existence and structural studies. J. Phys. Chem. Sol. 16 (1–2), 123–130.

Warburg, E., 1881. Magnetische Untersuchungen. Ann. Phys. (Lpz.) 13, 14164.

Watanabe, H., Saito, H., Takahashi, M., 1993. Soft magnetic properties and structures of nanocrystalline Fe–Al–Si–Nb–B alloy ribbons. Trans. Magn. Soc. Jpn. 8, 888–894.

Weinmann, H.J., Brasch, R.C., Press, W.R., et al., 1984. Am. J. Roentgenol. 142, 619–624.

Weiss, Pierre, 1907. L'hypothese du Champ Moleculaire et la Propriete Ferromagnetique. J. Phys. 6, 661.

Wernick, J.H., Geller, S., 1959. Transition element rare earth compounds with the CaCu$_5$ structure. Acta Cryst. 12 (9), 662–665.

Willard, M.A., 2000. Structural and magnetic characterization of HITPERM soft magnetic materials for high temperature applications. (Ph.D. Thesis), Carnegie Mellon Univ., Pittsburgh, PA.

Willard, M.A., Laughlin, D.E., McHenry, M.E., Sickafus, K., Cross, J.O., Harris, V.G., Thoma, D., 1998. Structure and magnetic properties of (Fe$_{0.5}$Co$_{0.5}$)$_{88}$Zr$_7$B$_4$Cu$_1$ nanocrystalline alloys. J. Appl. Phys. 84, 6773–6777.

Williams, A.R., Moruzzi, V.L., Malozemoff, A.P., Terakura, K., 1983. Generalized Slater-Pauling curve for transition metal magnets. IEEE Trans. Magn.

Wise, A., Saenko, M., Velazquez, M., Laughlin, D.E., Daz-Michelena, M., McHenry, M.E., 2011. Phase evolution in the Fe$_3$O$_4$—Fe$_2$Ti O$_4$ pseudo-binary system and its implications for remanent magnetization in Martian minerals. IEEE Trans. Magn. page to appear.

Wohlfarth, E.P., 1979. First and second order transitions in some metallic ferromagnets. J. Appl. Phys. 50, 7542–7544.

Wood, R., 2000. The feasibility of magnetic recording at 1 terabit per square inch. IEEE Trans. Magn. 36, 36.

Wood, M.E., Potter, W.H., 1985. Cryogenics 25, 667.

Wun-Fogle, M., Restorff, J.B., Leung, K., Cullen, J.R., 1999. Magnetostriction of terfenol-d heat treated under compressive stress. IEEE Trans. Magn. 35, 3817–3819.

Xia, H., Baranga, A.B.A., Hoffman, D., Romalis, M.V., 2006. Magnetoencephalography with an atomic magnetometer. Appl. Phys. Lett. 89, 211104.

Xu, S.J., Donaldson, M.H., Pines, A., Rochester, S., Budker, D., Yashchuk, V.V., 2006. Application of atomic magnetometry in magnetic particle detection. Appl. Phys. Lett. 89, 224105.

Yafet, Y., Kittel, C., 1952. Antiferromagnetic arrangements in ferrites. Phys. Rev. 87 (2), 290–294.

Yang, Y.C., Pan, Q., Zhang, X.D., Zhang, C.L., Li, Y., Ge, 1993. Structural and magnetic properties of RMo$_{1.5}$Fe$_{10.5}$N$_x$. J. Appl. Phys. 74, 4066–4071.

Yelon, W.B., Hu, Z., 1996. Neutron diffraction study of lattice changes in Nd$_2$Fe$_{17-x}$Si$_x$(C$_y$). J. Appl. Phys. 78, 7196–7201.

Yoshizawa, Y., Bizen, Y., Yamauchi, K., Sugihara, H., 1992. Improvement of magnetic properties in Fe-based nanocrystalline alloys by addition of Si, Ge, C, Ga, P, Al elements and their applications. Trans. IEE Jpn. 112A, 553–558.

Yoshizawa, Y., Oguma, S., Yamauchi, K., 1988. New Fe-based soft magnetic alloys composed of ultrafine grain structure. J. Appl. Phys. 64, 6044–6046.

Yoshizawa, Y., Yamauchi, K., 1991. Magnetic properties of Fe–Cu–Cr–Si–B, Fe–Cu–V–Si–B, and Fe–Cu–Mo–Si–B alloys. Mat. Sci. Eng. A 133, 176–179.

Yuan, H., 2009. Study of Composite Thin Films for Applications in High Density Data Storage (Ph.D. thesis), Carnegie Mellon Univ., Pittsburgh, PA.

Zhang, Y., Blazquez, J.S., Conde, A., Warren, P.J., Cerezo, A., 2003. Magnetic properties of FeCuCrSiB, FeCuVSiB, FeCuMoSiB, alloys. Mat. Sci. Eng. A 353, 176–179.

Biography

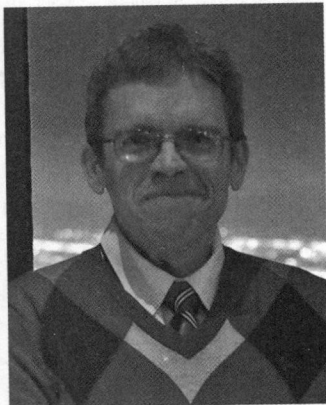

Michael E. McHenry is Professor of Materials Science and Eng. (MSE) at Carnegie Mellon University. He has a B.S. from Case Western Reserve in 1980. He was a 1980–1983 Process Engineer at U.S. Steel Lorain Works. In 1988 he earned a Ph.D from MIT. He was a 1988–1989 Director's Funded Post-doctoral Fellow at Los Alamos Lab. His research involves rapid solidification processing, plasma and solution synthesis of nanoparticles, magnetic field of processing materials, structural and magnetic properties characterization as a function of field, temperature and frequency. He directed a Multidis-ciplinary University Research Initiative (MURI) on magnetic materials for aircraft power applications and currently leads an ARPA-E program in magnetic materials for power electronics. He has served as proceeding Editor, Publication Chair and a member of the Program Committee for the Magnetism and Magnetic Materials (MMM) and Intermag Conferences. He was a 2013 IEEE Distinguished Lecturer and currently serves as 2014 Intermag Conference co-General Chair. He is a 2014 TMS Distinguished Scientist. He has published over 260 papers and owns two patents in the field. He has co-authoredthe textbook "Structure of Materials", Cambridge University Press, 2007, second edition 2012.

David E. Laughlin is the ALCOA Professor of Physical Metallurgy in the Department of Materials Science and Engineering at Carnegie Mellon University, Pittsburgh, PA. He obtained his B.S. in Metallurgical Engineering from Drexel University in 1969 and his Ph.D. in Metallurgy and Materials Science from MIT in 1973. He has taught at CMU since 1974. He is Principal Editor of *Metallurgical and Materials Transactions* and has co-edited eight books. His research has centered on the structure of materials as observed by electron microscopy, phase transformations and magnetic materials. He has published more than 450 peer reviewed research papers and is co-inventor on eleven US patents. Laughlin is a Fellow of TMS and ASM International.